AN

ANGLO-SAXON DICTIONARY

BASED ON THE MANUSCRIPT COLLECTIONS

OF THE LATE

JOSEPH BOSWORTH, D.D., F.R.S.

RAWLINSONIAN PROFESSOR OF ANGLO-SAXON
IN THE UNIVERSITY OF OXFORD.

EDITED AND ENLARGED

BY

T. NORTHCOTE TOLLER, M.A.

LATE FELLOW OF CHRIST'S COLLEGE, CAMBRIDGE;
AND SMITH PROFESSOR OF ENGLISH IN THE OWENS COLLEGE, MANCHESTER.

OXFORD UNIVERSITY PRESS

Oxford University Press, Ely House, London W.1

GLASGOW NEW YORK TORONTO MELBOURNE WELLINGTON
CAPE TOWN IBADAN NAIROBI DAR ES SALAAM LUSAKA ADDIS ABABA
DELHI BOMBAY CALCUTTA MADRAS KARACHI LAHORE DACCA
KUALA LUMPUR SINGAPORE HONG KONG TOKYO

ISBN 0 19 863101 4

FIRST EDITION 1898
REPRINTED 1929, 1954, 1972
1973

REPRODUCED AND PRINTED BY PHOTOLITHOGRAPHY AND BOUND IN
GREAT BRITAIN AT THE PITMAN PRESS, BATH

PREFACE.

WITH the issue of the last part of this work comes the necessity for some additions to the Preliminary Notice that accompanied Parts I and II. In that Notice it was mentioned that Dr. Bosworth's MS. for so much of the Dictionary as was contained in Part II was incomplete, and a similar remark applies with more force to the succeeding parts: little, indeed, was added in the MS. to what was already contained in the previous edition. If with corresponding parts of this previous edition the later part of the present one be compared, it will be seen that much had to be done in order to get together the additional material that finds its place in the new work. As the editor could not devote his time exclusively to the Dictionary, the length of the interval between the date of appearance of Part II and that of Part IV may seem not inexcusably great. It has, however, been so great that in some respects alterations have occurred in matters with which the Dictionary is concerned. Fresh material has been brought to light, or old material has been brought forth in more accessible form; the views on many points connected with the language that are now held, are not those of fifteen years ago, and there will be certainly some points in work done fifteen years ago that now will need revision. There will also be other points that need revision, but which cannot plead this excuse: mistakes and omissions, to some extent, are almost inevitable. Revision required under one or other head will be attempted in a supplement, which will be prepared as soon as possible.

In the course of the work some alterations have been made in the plan adopted by Dr. Bosworth. One of the difficulties connected with the cataloguing of English words preserved in works written before 1100 is due to the variety of forms which a word may take according to the time at which, or the locality in which, the MS. where it occurs was written. The Old-English specimens are scattered over centuries, and belong to different parts of England; naturally the form of a word is not always the same in the earlier and in the later specimen of the same locality, or in the contemporary specimens of different localities. In the earlier part of the Dictionary the different forms of a word are given separately, in the later part they are collected under a single form; e.g. in the former case words having the mutation of *eá* may appear under each of the forms which the varieties of that mutation (*ê*, *iê*, *î*, *ŷ*) admit of, in the latter one form alone (*î*) is given. Slight alterations, too, will be found noted in the list of references.

With regard to the marks used to distinguish difference in the vowels it may be noticed that *eá*, *eó* are employed in all cases where the short *ea*, *eo* are not meant, e.g. *sceóp*, Goth. *skóp*, has the same symbols as *leóf*, Goth. *liubs*, etc.

My thanks are due to Professor Skeat for the readiness which he has always shown to answer an appeal for help in a difficulty; to Professor Kluge and to Professor Heyne for very helpful criticism of the earlier parts of the Dictionary. To the former I am indebted not only for pointing out omissions, but for the assistance he has given in remedying them. He very kindly sent me a copy of the glosses cited under the abbreviation Germ., and further gave the Delegates of the University Press the opportunity, which they accepted of acquiring

a

a collection of Anglo-Saxon words that he had made. These words were drawn for the most part from sources already utilized for the Dictionary, but it was an advantage to have even the same material noted by another. As an example of this it may be remarked that between thirty and forty of the passages cited under S were taken from Professor Kluge's notes, and the number would have been larger had not, as already stated, Professor Kluge's criticism called attention to omissions in the earlier part of the work. To the late Dr. Grein my obligations are very great. He has done so much to remove the difficulties of one of the most difficult parts of the vocabulary—the poetical—that he has earned the gratitude of every one who attempts to work in the same field as the author of the Sprachschatz der Angelsächsischen Dichter.

In conclusion, it may not be out of place to refer to some of the difficulties which are met with in an attempt to compile an Anglo-Saxon Dictionary. The Anglo-Saxon remains are varied in respect to the subjects of which they treat, and the technical terms peculiar to some of these subjects, e.g. law, require the knowledge of a specialist. The poetical vocabulary, again, as a part of the language almost lost in later times presents many difficulties. Even where at first sight it might seem that the solution of difficulties would be most certainly furnished—in the case of glosses to Latin words—the expectation is not always realized, and at times the gloss is the only authority for both the English and the Latin word. And throughout there is the difficulty of realizing the condition of those who used the language and thus of appreciating the significance of the language they used. It is hoped, however, that the numerous citations given under many words, by shewing the actual use of those words, may help to the appreciation of their significance, and so supplement the often necessarily imperfect explanations afforded by the Modern English words that are used as the nearest equivalents to the old forms. Further, English philology has become so extensive a study that to keep pace with its developments is a task that might occupy so much time as to leave comparatively little for other work. To compile an Anglo-Saxon Dictionary calls for so much in the compiler that some leniency towards shortcomings may perhaps be looked for by any one who attempts the labour.

EXPLANATION OF REFERENCES.

———◦⊰⊱◦———

In the following list a want of uniformity may be noticed in the case of some of the contractions used. This is due partly to modifications of Dr. Bosworth's forms, which it seemed convenient to make; partly to different conditions in respect to texts cited, which have been brought about while the work was in progress: some texts, that existed in MS. only, have been printed; of others, that were already printed, new editions have appeared, which were more convenient to refer to than were the old. Cross references are given below in these cases. Double references are given to passages cited from the poetry, to English editions and to Grein's Bibliothek der Angelsächsischen Poesie; in the later the contractions used are those to be found in Grein's Lexicon, and they are given together at the end of this list.

Where a reference to any citation consists of more than one part (e. g. Bt. —; Fox —), the several parts are separated by a semi-colon: where after a citation several references are given, these are separated by a colon.

When consecutive citations are taken from the same work the full reference is given only with the first (e. g. Bt. is not repeated where consecutive citations are taken from Boethius; or if the reference be of one part, e. g. Nar. —, the Nar. is not repeated).

A. D. Altenglische Dichtungen der MS. Harl. 2253, herausgegeben von K. Böddeker, Berlin, 1878.

A. P. v. Allit. Pms.

A. R. The Ancren Riwle, edited for the Camden Society (No. lvii.) by J. Morton, 1853. Quoted by page and line.

Abus. Codex Junii 23, fol. 60, in the Bodleian Library. See Wanley's Catalogue, p. 37, and Engl. Stud. viii. 62.

Ælf. Ep. 1st = L. Ælfc. P. (q.v.).

Ælf. Test. Ælfric on the Old Testament in Sweet's Anglo-Saxon Reader (1st ed.).

Ælfc. Gen. Thw. The preface to Genesis in Thwaites' edition of the Heptateuch. v. Gen.

Ælfc. Gl. Codex Junii 71, in the Bodleian Library. See Wanley's Catalogue, p. 96. Printed by Somner (Som.) at the end of his Dictionary, and again by Wright in A Volume of Vocabularies (Wrt. Voc.). In the early part of the Dictionary the page of the MS., and the page and number of the word in Somner and in Wright are given, but later the reference is to Wright only (Wrt. Voc. i.).

Ælfc. Gl.; Zup. Ælfric's Grammatik und Glossar, herausgegeben von Julius Zupitza, Berlin, 1880. Quoted by page and line.

Ælfc. Gr. Ælfric's Grammar, referred to at first in the edition by Somner, printed with his Dictionary (Som.), later in that of Zupitza (Zup. v. preceding explanation). Quoted by section of the Grammar, and by page and line of the editions.

Ælfc. pref. Gen. = Ælfc. Gen. Thw.

Ælfc. T. or **Ælfc. T. Lisle.** A Saxon treatise concerning the Old and New Testament. . . . Now first published in print with English of our times by William L'Isle, London, 1623. Quoted by page and line.

Ælfc. T. Grn. The same text, in vol. i. of Bibliothek der Angelsächsischen Prosa, herausgegeben von Chr. Grein. 1872. See also Ælf. Test.

Æqu. Vern. This contraction (used, but not explained, by Lye) seems to refer to the Anglo-Saxon abridgement of Bede's *De Natura Rerum* in MS. Cotton. Tiberius, B. V. (see Wanley's Catalogue, p. 216). It is printed in Popular Treatises on Science, edited for the Historical Society of Science by Wright, London, 1841 (Wrt. popl. science); and again in the 3rd vol. of Cockayne's Leechdoms (Lchdm. III). The later quotations are taken from the latter edition. v. Equin. vern.

Al. The Life of St. Alexius, edited by F. J. Furnivall, E.E.T.S., No. 69, 1878. Quoted by line.

Alb. resp. Albini responsa ad Sigewulfi interrogationes. For a text and MSS. see Anglia, vol. vii. pp. 1 sqq.

Ald. Sancti Aldhelmi Opera, edited by J. A. Giles, Oxford, 1844. Quoted by page.

Alex. The Alliterative Romance of Alexander, edited by J. Stevenson, Roxburghe Club, 1849. Quoted by line.

Alex. (Skt.). The same, edited by W. W. Skeat, E.E.T.S., No. lxvii., 1866. Quoted by line.

Alis. King Alisaunder, in Weber's Metrical Romances, vol. i., Edinburgh, 1810. Quoted by line.

Allit. Pms. Early English Alliterative Poems, edited by R. Morris, E.E.T.S., No. 1, 1864. Quoted by page and line.

Am. and Amil. Amis and Amiloun, in Weber's Metrical Romances, vol. ii.

An. Lit. Anecdota Literaria, edited by Thomas Wright, London, 1844. Quoted by page and line.

An. (or **Anal.**) **Th.** or **Th. An.** (**Anal., Anlct.**). Analecta Anglo-Saxonica, by Benjamin Thorpe, London, 1846. Quoted by page and line.

And. = St. And. (q.v.).

Andr. Grm. See Grm[m]. A. u. E.

Andr. Kmbl. The Poetry of the Codex Vercellensis, edited by J. M. Kemble for the Ælfric Society. Part 1. The Legend of St. Andrew, London, 1844.

Andr. Recd. The same poem edited for the Record Commission by Benjamin Thorpe, but not published. See Glos. Epnl. Recd.

Andrews' Old English Manor. The Old English Manor, a study in English Economic History, by Charles McLean Andrews, Baltimore, 1892.

Anglia. Anglia, Zeitschrift für Englische Philologie. Halle, 1878–

Anlct. v. An. Th.

Ap. (Apol.) Th. *or* **Th. Ap. (Apol.).** The Anglo-Saxon Version of the Story of Apollonius of Tyre, from a MS. in the Library of C. C. C., Cambridge (v. Wanley's Catalogue, p. 146), edited by Benjamin Thorpe, London, 1834. Quoted by page and line.

App. (Lib.) Scint. v. Scint.

Apstls. Crd. An interlinear version of the Apostles' Creed on folio 199 a of the MS. referred to as Ps. Lamb. (q.v.).

Apstls. Kmbl. The Fates of the Twelve Apostles in The Poetry of the Codex Vercellensis. Part II. v. Andr. Kmbl.

Apstls. Recd. The same poem edited for the Record Commission. v. Andr. Recd.

Arth. and Merl. Arthour and Merlin, a Metrical Romance edited by W. D. Turnbull, Abbotsford Club, 1838. Quoted by line.

Ass. B. Assumpcio Beate Marie, edited by J. R. Lumby, E.E.T.S., No. 14, 1866. Quoted by line.

Ath. Crd. *or* **Athan.** An interlinear version of the Athanasian Creed, folios 200 a–202 b of the MS. referred to as Ps. Lamb. (q.v.). Quoted by paragraph.

Ayenb. Dan Michel's Ayenbite of Inwyt, in the Kentish Dialect, 1340, edited by R. Morris, E.E.T.S., No. 23, 1866.

Bailey. An Universal Etymological English Dictionary, by N. Bailey, 10th edition, London, 1742.

Basil admn.; Norm. The Anglo-Saxon Remains of St. Basil's Admonitio ad filium spiritualem, edited by the Rev. Henry W. Norman, 2nd edition, London, 1849. Quoted by chapter, and by page and line.

Bd. de nat. rm. (rerum). *See under* Æqu. Vern.

Bd.; M. The Old English Version of Bede's Ecclesiastical History of the English People, edited by Thomas Miller, E.E.T.S., Nos. 95, 96, 1890–1891. Quoted by book and chapter, and by page and line.

Bd.; S. Baedae Historia Ecclesiastica a gloriosissimo veterum Anglo-Saxonum rege Aluredo Saxonice reddita, cura et studio Johannis Smith, Cantabrigiae, 1722. Quoted as in previous work.

Bd.; Whel. (Whelc.). Bedae Venerabilis Historia Ecclesiastica Anglorum, Anglo-Saxonice ex versione Ælfredi Magni Gentis et Latine, cura Abrahami Wheloci, Cantabrigiae, 1644.

Ben. Vocabularium Anglo-Saxonicum, opera Th. Benson, Oxoniae, 1701.

Beo. Kmbl. The Anglo-Saxon poems of Beowulf, the Traveller's Song and the Battle of Finnesburh, edited by John M. Kemble, 2nd edition, London, 1835.

Beo. Th. The Anglo-Saxon Poem of Beowulf, edited by Benjamin Thorpe, Oxford, 1855.

Beves. Sir Beves of Hamtune, edited by E. Kölbing, E.E.T.S., Nos. xlvi., xlviii., 1885–1886. Quoted by line.

Blickl. Gl. (Gloss.). Glosses taken from a copy of the Roman Psalter in the library at Blickling Hall. Printed at the end of the Blickling Homilies. See next paragraph.

Blickl. Homl. *or* **Homl. Blick.** The Blickling Homilies, edited by R. Morris, E.E.T.S., Nos. 58, 63, 1874–1876. Quoted by page and line.

Boutr. (Btwk.) Scrd. Screadunga. Anglo-Saxonica maxi-

mam partem inedita publicavit C. G. Bouterwek, Elberfeld, 1858. Quoted by page and line.

Brand. Popular Antiquities of Great Britain, edited, from the materials collected by John Brand, by W. C. Hazlitt. Three vols. London, 1870.

Bridf[r]. Bridferth's Enchiridion contained in MS. No. 328 in the Ashmolean Library (see Wanley's Catalogue, p. 103). Quoted by folio. This MS. is printed in Anglia viii. 298–337, and later references are to this edition by page and line.

Bt.; Fox. King Alfred's Anglo-Saxon version of Boethius de Consolatione Philosophiae, edited by the Rev. S. Fox. Bohn's Antiquarian Library, London, 1864. Quoted by chapter and paragraph, and by page and line.

Bt. Met. Fox *and* **Bt. Tupr.** The Anglo-Saxon metrical version of the metrical portions of Boethius, with a verse translation by M. Tupper. At the end of the previous work. Quoted by number of metre and line.

Bt.; Rawl. Boethii Consolationis Philosophiae libri v Anglo-Saxonice redditi ab Ælfredo ; ad Apographum Junianum expressos edidit Christophorus Rawlinson, Oxoniae, 1698. Quoted by chapter and paragraph, and by page and line.

Btwk. Cædmon's Biblische Dichtungen, herausgegeben von K. W. Bouterwek. Erster Theil, Gütersloh, 1854. The references are to the Anglo-Saxon piece 'De officiis diurnalium et nocturnalium horarum,' preface, pp. cxciv-ccxxii. Quoted by page and line.

Btwk. Scrd. v. Boutr. Scrd.

Byrht. Th. The poem on the battle of Maldon in Thorpe's Analecta Anglo-Saxonica. Quoted by page and line.

C. L. Castel off Love, edited by R. F. Weymouth, Philol. Soc., 1864. Quoted by line.

C. M. Cursor Mundi, edited by R. Morris, E.E.T.S. Quoted by line.

C. R. Ben. An Anglo-Saxon version of the Benedictine Rule contained in a MS. in the library of Corpus Christi College, Cambridge. See Wanley's Catalogue, p. 122. Quoted by chapter. In the latter part of the Dictionary references are given to the work noticed under R. Ben., in which this MS. is used.

Cambr. MS. Ps. = Ps. Spl. C. (q.v.).

Canon. Hrs. Appendix to Hickes' Letters to a Popish Priest. Quoted by page and line. The piece is printed in Select Monuments of the Doctrine and Worship of the Catholic Church in England before the Norman Conquest, by E. Thompson, London, 1875 (2nd edition).

Cant. Ab. (Abac., Habac., Abac. Lamb.). A gloss of Habakkuk, 3, 2–19, contained in the same MS. as Ps. Lamb. (q.v.) on folios 189–191. Quoted by verse.

Cant. Abac. Surt. A gloss of the same material as the preceding, printed in An Anglo-Saxon and Early English Psalter, edited by J. Stevenson, Surtees Soc., No. 19. Quoted by page and line.

Cant. (Cantic.) An. A gloss of the song of Hannah (I. Sam. 2, 1–10), contained in the same MS. as Ps. Lamb. (q.v.) on folios 185 b–186 b. Quoted by verse.

Cant. Es. A gloss of Isaiah 12, 1–6, contained in the same MS. as the preceding, on folio 184. Quoted by verse.

Cant. Ez. (Cant. Ezech. Lamb.). A gloss of Isaiah 38, 10–20, contained in the same MS. as the preceding, on folios 184 b–185 b. Quoted by verse (in some instances the folio of the MS. is also given).

Cant. M. (Moys., Moys. Lamb.). A gloss of Exodus 15, 1–19, contained in the same MS. as the preceding, on folios 186 b–189. Quoted as in the preceding.

Cant. M. ad fil. (Moys. Isrl. Lamb.). A gloss of Deuteronomy 32, 1–43, contained in the same MS. as the preceding, on folios 191–195. Quoted as in the preceding.

Cant. Mar. A gloss of Luke 1, 46–55, contained in the same

MS. as the preceding, on folios 198–198 b. Quoted by verse.

Cant. Moys. Ex. (Cantic. Moys.); **Thw.: Cant. Moys. Thw.** A gloss of Exodus 15, 1–19, at the end of Thwaites' Heptateuch.

Cant. Zach. A gloss of Luke 1, 68–79, contained in the same MS. as Ps. Lamb. (q. v.) on folios 197–198. Quoted by verse.

Cart. Eadgif. R. A charter of Queen Eadgifu, v. Chart. Th. 201.

Cath. Ang. (Angl.). Catholicon Anglicum, edited by S. J. Herrtage, E.E.T.S., No. 75, 1881. Quoted by page.

Cd.; Th. (*later* **Cd. Th.**). Cædmon's Metrical Paraphrase of parts of the Holy Scripture, in Anglo-Saxon, by Benjamin Thorpe, London, 1832. Quoted at first by folio, and by page and line, later by page and line.

Chart. Erl. A Handbook to the Land Charters and other Saxonic Documents, by John Earle, M.A., Oxford, 1888. Quoted by page and line.

Chart. (Ch.) Th. Diplomatarium Anglicum Aevi Saxonici, by Benjamin Thorpe, London, 1865. Quoted by page and line.

Chauc. The abbreviations used in connexion with Chaucer are not given as not requiring explanation.

Chr.; Erl. Two of the Saxon Chronicles parallel with supplementary extracts from the others, edited by John Earle, M.A., Oxford, 1865. Quoted by year, and by page and line.

Chr.; Gib. Chronicon Saxonicum, Latine et Anglo-Saxonice, cum notis Edmundi Gibson, Oxon., 1692.

Chr.; Ing. The Saxon Chronicle, with an English translation and notes, by the Rev. James Ingram, 1823.

Chr.; Th. The Anglo-Saxon Chronicle, according to the several original authorities. Edited, with a translation, by Benjamin Thorpe, Master of the Rolls Series, 1861. Quoted by year, and by page, line, and column.

Chron. Abing. Chronicon Monasterii de Abingdon. Edited by Rev. J. Stevenson, Master of the Rolls Series, 1858.

Chron. Vilodun. Chronicon Vilodunense, sive de vita et miraculis sanctae Edithae, cur. W. H. Black. Quoted by line.

Cl. and Vig. Dict. An Icelandic-English Dictionary, based on the MS. collections of the late Richard Cleasby, enlarged and completed by Gudbrand Vigfusson, Oxford, 1874.

Cod. Dip. B. Cartularium Saxonicum: a collection of Charters relating to Anglo-Saxon History, by Walter de Gray Birch, London, 1883–1893. Quoted by volume, page and line.

Cod. Dip. (Dipl.) Kmbl. Codex Diplomaticus Aevi Saxonici, opera Johannis M. Kemble. Publications of the English Historical Society, 1839–1848. Quoted by volume, page and line.

Cod. Exon. v. Exon. Th.

Coll. Monast. Th. *or* **Wrt.** Colloquium ad pueros linguae Latinae locutione exercendos ab Ælfrico compilatum. Printed in Thorpe's Analecta (v. An. Th.), or in Wright's Vocabularies (v. Wrt. Voc. i.). Quoted by page and line.

Confess. Pecc. (Peccat.). A gloss of a 'Confessio pro peccatis ad Deum,' contained in the same MS. as Ps. Lamb. (q. v.) on folios 182 b–183 b.

Corp. Gl. (ed.) Hessels. An eighth-century Latin-Anglo-Saxon Glossary preserved in the Library of Corpus Christi College, Cambridge, edited by J. H. Hessels, Cambridge, 1890. Quoted by page and number of word.

Cot. In the earlier part of the Dictionary several glossaries found among the Cotton MSS. are referred to by this abbreviation. These glossaries are printed in Wrt. Voc. i., ii., to which works later references are given; in

a supplement to the Dictionary similar references will be found to replace the abbreviation in question.

D. Arth. Morte Arthure; or the Death of Arthur, edited by Edm. Brock, E.E.T.S., No. 8, 1871. Quoted by line.

Dep. Rich. Richard the Redeles, an Alliterative Poem on the Deposition of Richard II, edited by W. W. Skeat, E.E.T.S., No. 54, 1873. Quoted by passus and line.

Destr. Tr. The Gest Historiale of the Destruction of Troy, edited by G. A. Panton and D. Donaldson, E.E.T.S., Nos. 39, 56. Quoted by line.

Deut. The Anglo-Saxon version of Deuteronomy in Thw. Hept. (q. v.) or in Bibliothek der Angelsächsischen Prosa, herausgegeben von Chr. Wilh. Mich. Grein, erster Band, 1872. Quoted by chapter and verse.

Dial. v. Gr. Dial.

Dief. Vergleichendes Wörterbuch der Gothischen Sprache, von Dr. Lorenz Diefenbach, 1851.

Dietr. Dietrich's Commentatio de Kynewulfi poetae aetate, Marburg, 1859–1860.

Dóm. L. Be Dómes Dæge, an Old English version of the Latin poem ascribed to Bede. Edited with other short poems from the MS. in the Library of Corpus Christi College, Cambridge, by J. R. Lumby, E.E.T.S., No. 65, 1876. Quoted by page and line.

E. D. S. (Publ.). The publications of the English Dialect Society.

E. E. T. S. The publications of the Early English Text Society.

E. G. English Gilds, edited by Miss L. Toulmin Smith, E.E.T.S., No. 40, 1870. Quoted by page and line.

Earle A.S. Lit. Anglo-Saxon Literature. By John Earle. London: Society for Promoting Christian Knowledge, 1884.

Ecclus. The book of Ecclesiasticus.

Elen. Grm. v. Grmm. A. u. E.

Elen. Kmbl. The Poetry of the Codex Vercellensis, edited for the Ælfric Society by J. M. Kemble. Part II. Elene and Minor Poems, London, 1856.

Engl. Stud. Englische Studien. Organ für englische Philologie. Herausgegeben von Dr. Eugen Kölbing.

Ep. Gl. (Gloss. Ep.). The Epinal Glossary, Latin and Old-English of the eighth century. Edited by Henry Sweet. Printed for the Philological and Early English Text Societies, 1883. Quoted by page, column and line.

Equin. vern. An Anglo-Saxon summary of Bede's De Temporibus, referred to in Wanley's Catalogue under the heading De equinoctio vernali. It is printed in Lchdm. iii. pp. 232–280, and the quotations from the work are, except in the earlier part of the Dictionary, from this printed form. v. Æqu. Vern.

Erf. Gl. A Latin-Anglo-Saxon Glossary contained in a MS. preserved in the Amplonian library at Erfurt. Printed in the oldest English Texts, edited by Henry Sweet, E.E.T.S., No. 83, 1885.

Ettm. Lexicon Anglosaxonicum, edidit Ludovicus Ettmüllerus. Quedlinburgii et Lipsiae, 1851.

Ettm. Poet. Anglosaxonum poetae atque scriptores prosaici. Edidit Ludovicus Ettmüllerus. Quedlinburgii et Lipsiae, 1850.

Ex. The Anglo-Saxon version of Exodus. v. Deut.

Exod. Thw. v. preceding.

Exon.; Th. (*later* **Exon. Th.**). Codex Exoniensis. A Collection of Anglo-Saxon poetry, from a MS. in the library of the Dean and Chapter of Exeter, by Benjamin Thorpe, London, 1842. Quoted at first by folio, and by page and line, later by page and line.

Fer. Sir Ferumbras, edited by S. J. Herrtage, E.E.T.S., No. xxiv., 1879. Quoted by line.

a 3

Fins. Th. The Anglo-Saxon poem of the Fight at Finnesburg, edited by Benjamin Thorpe. In the same volume with Beo. Th. (q. v.).

Fl. a. Bl. Floriz and Blaunchefur, edited by J. R. Lumby, E.E.T.S., No. 14, 1866. Quoted by line.

Frag. Kmbl. A Fragment, Moral and Religious, contained in the Poetry of the Codex Vercellensis, edited by J. M. Kemble (v. Elen. Kmbl.).

Frag. Phlps. Fragment of Ælfric's Grammar, Ælfric's Glossary, and a Poem on the Soul and the Body, in the orthography of the twelfth century, edited by Sir T. Phillipps, London, 1838.

Frag. Recd. The same poem as Frag. Kmbl., printed with Andr. Recd. (q. v.).

Fulg. S. Fulgentii Regulae Monachorum, an Anglo-Saxon gloss of the Latin work contained in MS. Cott. Tib. A. 3 (see Wanley's Catalogue, p. 91).

Gam. The Tale of Gamelin, edited by W. W. Skeat, Oxford, 1884. Quoted by line.

Gaw. Sir Gawayne and the Green Knight, edited by R. Morris, E.E.T.S., No. 4, 1864. Quoted by line.

Gen. The Anglo-Saxon version of Genesis. v. Deut.

Gen. and Ex. The Story of Genesis and Exodus, edited by R. Morris, E.E.T.S., No. 7, 1865. Quoted by line.

Gen. pref. Thw. The Anglo-Saxon preface to Genesis in Thw. Hept. Quoted by page and line.

Germ. Die Bouloneser Angelsächsischen Glossen zu Prudentius. Herausgegeben von Dr. Alfred Holder. In vol. xi. (new series) of Germania. Quoted by page and number preceding the gloss. v. Gl. Prud., Gl. Prud. H., Glos. Prudent. Recd.

Gl. Amplon. Glossae Amplonianae, ed. Oehler in Jahn's Jahrb. 13, 1847.

Gl. E. A Latin-Anglo-Saxon Glossary contained in MS. Cott. Cleopatra A III. (v. Wanley's Catalogue, p. 238). Printed in Wrt. Voc. ii. pp. 70 sqq., whence, except at the beginning, quotations are taken.

Gl. M. An Anglo-Saxon Gloss of Aldhelm's De laude virginitatis, published in Mone's Quellen und Forschungen, Leipzig, 1830. Quoted by page. See Hpt. Gl., where the same gloss is referred to.

Gl. Mett. Glossae Mettenses in Mone Anzeiger, 1839.

Gl. Prud. (1). Glosses to Prudentius in Mone Anzeiger, 1839. Quoted by number of gloss. From the same MS. as that given under Germ.

Gl. Prud. (2). The same abbreviation as the preceding has also sometimes been used for another work, which elsewhere is referred to as Glos. Prud. (q. v.) or simply Prud. The quotations, however, in this case are by paragraph.

Gl. Prud. H. This is the gloss given under Germ. (q. v.). The quotations are by folio instead of by page.

Gl. Wülck. v. Wülck.

Glos. Brux. Recd. An Anglo-Saxon Vocabulary taken from a MS. in the Royal Library at Brussels. It is printed in Wrt. Voc. i. pp. 62 sqq., and to this edition alone, except in the earlier part of the Dictionary, references are given.

Glos. Epnl. Recd. The Epinal Glossary printed (but not published) in Appendix B of An Account of the most important Public Records of Great Britain (Publications of the Record Commissioners), London, 1836.

Glos. Prud. or Prud. Englische Übersetzungen der lateinischen Erklärungen von Bildern zur Psychomachie des Prudentius entlehnt : (A) einer Hs. im Britischen Museum, Cotton. Cleop. C. viii, (B) einer Cambridger Hs., Corpus Christi College 23, published by J. Zupitza in Zeitschrift für deutsches Alterthum, vol. 8 (new series), 1876. Quoted by paragraph and MS.

Glos. Prudent. Recd. The glosses given under Germ., printed in the same work as the Glos. Epnl. Recd.

Gloss. Ep. v. Ep. Gl.

Glostr. Frag. Legends of Saint Swithun and Sancta Maria Ægyptiaca, published by John Earle, M.A., London, 1861.

Gospel of Nicodemus. Quoted from The Apocryphal New Testament. Printed for William Hone, 1820. Tenth edition, London, 1872.

Goth. Gothic; the text referred to has been Die Gothischen Sprachdenkmäler, herausgegeben von H. F. Massmann. v. Dief.

Gow. Confessio Amantis of John Gower, edited by R. Pauli, London, 1857. Quoted by volume, page and line.

Greg. Die englische Gregorlegende, herausgegeben von F. Schulz, Königsberg, 1876. Quoted by line.

Gr. (Greg.) Dial. The Anglo-Saxon version of Gregory's Dialogues. Quoted from Lye. v. Wanley's Catalogue, p. 71.

Grff. Althochdeutscher Sprachschatz von Dr. E. G. Graff. Berlin, 1834–1842.

Grm. (Grmm. Gr.). Deutsche Grammatik von Dr. Jacob Grimm. 2. Ausgabe.

Grm[m]. A. u. E. (And. u. El.). Andreas und Elene. Herausgegeben von Jacob Grimm. Cassel, 1840.

Grm[m]. D. M. Deutsche Mythologie, von Jacob Grimm. Zweite Ausgabe, Göttingen, 1844.

Grm[m]. Gesch. D. S. (Gsch.). Geschichte der deutschen Sprache, von Jacob Grimm. 3. (2.) Ausgabe, Leipzig, 1868.

Grm. Mythol. The first edition of Grmm. D. M.

Grm[m]. R. A. Deutsche Rechtsalterthümer, von Jacob Grimm. 2. Ausgabe, Göttingen, 1854.

Guthl. (Gu.); Gdwin. The Anglo-Saxon version of the Life of St. Guthlac, Hermit of Crowland, edited by C. W. Goodwin, London, 1848. Quoted by chapter (Guthl.) and by page and line (Gdwin.).

H. (K.) de visione Isaiae. The reference is to Wanley's Catalogue, p. 27, l. 9; the passage will be found Wulfst. 44, 23.

H. M. Hali Meidenhad, edited by O. Cockayne, E.E.T.S., No. 18, 1866. Quoted by page and line.

H. R. Legends of the Holy Rood, edited by R. Morris, E.E.T.S., No. 46, 1871. Quoted by page and line.

H. S. Robert of Brunne's Handling Sinne, edited by F. J. Furnivall, Roxburghe Club, 1862. Quoted by line.

H. Z. (Hpt., Hpt. Zeit[sch].). Zeitschrift für deutsches Alterthum, herausgegeben von Moritz Haupt.

Hall. (Halliw., Halwl.) Dict. A Dictionary of Archaic and Provincial Words, by J. O. Halliwell. Seventh edition, London, 1872.

Handl. Synne. v. H. S.

Harl. Gl. 978. This glossary is printed at p. 139 of Wrt. Voc. i.

Havel. The Lay of Havelok the Dane, edited by W. W. Skeat, E.E.T.S., No. iv., 1868. Quoted by line.

Hél. Héliand. Herausgegeben von Moritz Heyne. Paderborn, 1866.

Heli. Schmel. Heliand. Poema Saxonicum seculi noni. Edidit J. A. Schmeller, 1830.

Hem. (Heming.). Hemingi Chartularium Eccl. Wigorniensis, edidit T. Hearne, Oxon., 1723. Tom. ii.

Herb.; Lchdm. i. An Anglo-Saxon Herbarium printed in Lchdm. i. Quoted by section and paragraph (Herb.), and by page and line. See Lchdm.

Hexam. (Hex.); Norm. The Anglo-Saxon version of the Hexameron of St. Basil, edited by H. W. Norman.

2nd edition, London, 1849. Quoted by chapter (Hexam.), and by page and line (Norm.).

Hick. Thes. Linguarum veterum septentrionalium thesaurus, auctore G. Hickesio, Oxoniae, 1705.

Hick. Diss. Ep. (Hickes' Diss.). G. Hickesii de antiquae litteraturae septentrionalis utilitate dissertatio epistolaris, Oxoniae, 1703. Contained in vol. i. of the preceding.

Hom. = O. E. Homl.

Hom. de Comp. Cord. Cited by Dr. Bosworth from Lye.

Hom. 8 Cal. Jan. This homily is printed in Homl. Th. i. 28. [v. ge-þryle, the reference to which = Homl. Th. i. 34, 34.]

Homl. As[s]. Angelsächsische Homilien und Heiligenleben, herausgegeben von Bruno Assman, Kassel, 1889. [Bibliothek der Angelsächsischen Prosa, begründet von C. W. M. Grein, 3. Band.] Quoted by page and line of section.

Homl. Blick. v. Blickl. Homl.

Homl. in nat. Innoc. This homily is printed in Homl. Th. i. 76. [v. ærst, the reference to which = Homl. Th. i. 78, 18.]

Homl. Pasc. Daye. A Sermon of the Paschall Lambe to be spoken unto the people at Easter. Imprinted (with other works of Ælfric) at London by John Daye, 1567.

Homl. Pasc. Lisl. The same homily as the preceding, published in 1623 by Lisle. The homily is printed in Homl. Th. ii. 262.

Homl. Skt. Ælfric's Metrical Lives of Saints, edited by W. W. Skeat, E.E.T.S., Nos. 76, 82, 94, 1881-85-90. Quoted by volume, homily and line.

Homl. Th. The Homilies of Ælfric, edited by B. Thorpe for the Ælfric Society, London, 1844-1846. Quoted by volume, page and line.

Horn (K[ing] Horn). King Horn, edited by J. R. Lumby, E.E.T.S., No. 14, 1866. Quoted by line.

Hpt. v. H. Z.

Hpt. Gl. Die Angelsächsischen Glossen in dem Brüsseler Codex von Aldhelms Schrift De Virginitate, published in vol. ix. of Haupt's Zeitschrift, by K. Bouterwek. Quoted by page and line.

Hpt. Zeit[sch]. v. H. Z.

Hymn. ad Mat. Hymnus ad Matutinos Dies Dominicos, contained in fols. 195-196 of Ps. Lamb. Quoted by verse.

Hymn. in Dedic. Eccles. (Hymn.). The piece referred to will be found printed in Homl. Th. ii. 576 sqq.

Hymn. L. = Hymn. ad Mat.

Hymn. Lye = Hymnarium in Cott. MS. Jul. A. 6.

Hymn. Surt. Anglo-Saxon Hymnarium, edited by Rev. J. Stevenson, Surtees Society, vol. xxiii., 1851. Quoted by page and line.

Hymn. T. P. An Anglo-Saxon gloss of Dan. 3, 57-88, contained in the same MS. as Ps. Lamb. on folios 196-197. Quoted by verse.

Icel. Icelandic; the forms are taken from Cleasby and Vigfusson's Dictionary.

Invent. Crs. Recd. The poem in the Codex Vercellensis on the finding of the Cross (v. Elen. Kmbl.), edited for the Record Commission by Benjamin Thorpe, but not published. See Andr. Recd.

Jamieson. Jamieson's Dictionary of the Scottish Language, abridged by J. Johnstone. A new edition by J. Longmuir, Edinburgh, 1877.

Japx. Gysbert Japicx, a Friesian poet, who wrote about 1650.

Jellinghaus. Die Westfälischen Ortsnamen nach ihren Grundwörtern, von H. Jellinghaus. Kiel und Leipzig, 1896.

Jn. The Gospel of St. John. v. Mt.

Job Thw. A portion of Ælfric's homily on Job (v. Homl. Th. ii. 446) printed in Thw. Hept. Quoted by page and line.

Jos. (1). The Anglo-Saxon version of the book of Joshua. v. Deut.

Jos. (2). (Jos. of Arith.). Joseph of Arimathie, edited by W. W. Skeat, E.E.T.S., No. 44, 1871.

Josc. For the passage under *sliten* cited from Joscelin by Lye, see Lk. Spt. p. 2, 11. For Joscelin's Dictionary see Wanl. Cat. p. 101.

Jud. (1). The Anglo-Saxon version of the book of Judges. v. Deut.

Jud. (2) (Jud. Thw.). Where the quotation is by page and line the reference is to the matter printed in Thw. Hept. at the end of the book of Judges.

Jud. (3). See under the contractions used in Grein's Dictionary.

Jud. Civ. Lund. Judicia Civitatis Lundoniae. L. Ath. v.; Th. i. 228.

Judth.; Thw. (*later* Judth. Thw.). The poem of Judith printed at the end of Thw. Hept. Quoted by section (Judth.), and by page and line (Thw.).

Jul. (Juliana). The Liflade of St. Juliana, edited by O. Cockayne and T. Brock, E.E.T.S., No. 51, 1872. Quoted by page and line.

K. Alis. v. Alis.

Kath. The Life of Saint Katherine; in the earlier part of the Dictionary reference is to the edition of Rev. J. Morton, later to that of Dr. E. Einenkel, E.E.T.S., No. 80, 1884. Quoted by line. The correspondence of lines in the two editions is marked in the later.

Kent. Gl. Kentische Glossen des neunten Jahrhunderts, published in Zeitschrift für deutsches Alterthum, vol. ix., new series, by J. Zupitza. These glosses, from MS. Cott. Vesp. D 6, are on the book of Proverbs, and in the earlier part of the Dictionary the abbreviation used is Prov.; in this case the quotation is by chapter, in the other by the number of the gloss.

Ker. Kero, the name assumed to be that of the author of a glossary, and of a gloss of the Benedictine Rule, in the Alemannic dialect.

Kil. Etymologicum Teutonicae linguae, sive dictionarium Teutonico-Latinum, studio et opera Corn. Kiliani Dufflaei, Antverpiae, 1599.

King Horn. v. Horn.

Kmbl. Cod. Dipl. v. Cod. Dipl. Kmbl.

Kmbl. Sal. and Sat. v. Salm. Kmbl.

L.; Th. The following contractions refer to the matter contained in Ancient Laws and Institutes of England, edited by Benjamin Thorpe, and printed under the direction of the Commissioners on the Public Records of the Kingdom, 1840. Quoted by (section and) paragraph (L. —), and by volume, page and line (Th.) :—

L. A. G. Alfred and Guthrum's Peace.

L. Ælfc. C. Canons of Ælfric.

L. Ælfc. E. Ælfric's Epistle, 'Quando dividis Chrisma.'

L. Ælfc. P. Ælfric's Pastoral Epistle.

L. Ædelb. = L. Ethb.

L. Ædelst. = L. Ath.

L. Alf. Extracts from Exodus, prefixed to Alfred's Laws.

L. Alf. pol. Laws of King Alfred.

L. Ath. i-v. Laws of King Athelstan.

L. C. E. Ecclesiastical Laws of King Cnut.

L. C. F. Constitutiones de Foresta of King Cnut.

L. C. S. Secular Laws of King Cnut.

L. de Cf. De Confessione (Canons enacted under King Edgar).

L. E. B. Ecclesiastical Compensations (Bôt).

L. Ecg. C. Ecgberti Confessionale.

L. Ecg. E. Excerptiones Ecgberti.

L. Ecg. P. i-iv. Ecgberti Poenitentiale (libri iv.).

L. Ecg. P. addit. Additamenta to the preceding.

L. E. G. Laws of Edward and Guthrum.

L. E. I. Ecclesiastical Institutes.

L. Ed. Laws of King Edward.

L. Ed. C. Laws of King Edward the Confessor.

L. Edg. i, ii. Laws of King Edgar, (i) ecclesiastical, (ii) secular.

L. Edg. C. Canons enacted under King Edgar.

L. Edg. H. Laws of King Edgar (How the Hundred shall be held).

L. Edg. S. Supplement to King Edgar's Laws.

L. Edm. B. Laws of King Edmund (of betrothing a woman).

L. Edm. C. ,, ,, ,, (Concilium Culintonense).

L. Edm. E. ,, ,, ,, (Ecclesiastical).

L. Edm. S. ,, ,, ,, (Secular).

L. Eth. i-ix. Laws of King Ethelred.

L. Ethb. Laws of King Æthelbirht of Kent.

L. Ff. Of Forfang.

L. H. Laws of King Henry I.

L. H. E. Laws of Hlothhære and Eadric.

L. I. P. Institutes of Polity.

L. In. Laws of King Ine.

L. M. I. P. Modus Imponendi Poenitentiam.

L. M. L. Mercian Law.

L. N. P. L. Law of the Northumbrian Priests.

L. O. Oaths.

L. O. D. Ordinance respecting the Dúnsætas.

L. P. M. Of Powerful Men.

L. Pen. Of Penitents.

L. R. Ranks.

L. R. S. Rectitudines Singularum Personarum.

L. Th. C. Theodori Capitula et Fragmenta.

L. Th. P. Theodori Liber Poenitentialis.

L. Wg. Wergilds.

L. Wih. Laws of King Wihtræd.

L. Wil. i-iv. Laws of William the Conqueror.

L. Const. W. Wilkins' (v. Wilk.) edition of the text cited as L. I. P. in Thorpe's Laws.

L. Eádg., L. Eádg. Suppl., L. Eccles., L. Ecg. P. A.= L. Edg., L. Edg. S., L. E. I., L. Ecg. P. addit.

L. Edw. Conf. Schmid. The Laws of King Edward the Confessor in Schmid's A. S. Gesetz. (q.v.).

L. H. R.= H. R.

L. Lund.= L. Ath. v.

L. M. 1, 2, 3. Three books on medicine, contained in Lchdm. ii. Quoted by book and section; in the latter part of the Dictionary the references are to Lchdm. only.

L. Med. ex Quadr.= Med. ex Quadr.

L. N. F. Altenglische Legenden, neue Folge, herausgegeben von C. Horstmann, Heilbronn, 1881. Quoted by page and line.

L. S. Lives of Saints, edited by C. Hortsmann, E.E.T.S., No. 87, 1887. Quoted by page and line of poem.

L. Th. Thorpe's edition of the Laws given under L.; Th.

Lambd. Lambard's edition of the Laws printed in 1568.

Laym. Laȝamon's Brut, edited by F. Madden, Society of Antiquaries, London, 1847. Quoted by line.

Lchdm. Leechdoms, Wortcunning, and Starcraft of early England, edited by O. Cockayne, Master of the Rolls Series, 3 vols. London, 1864–1866. Quoted by volume, page and line.

Leo A. S. Names. A treatise on the local nomenclature of the Anglo-Saxons, translated from the German of Prof. H. Leo, London, 1852.

Leo A. Sax. Gl. Angelsächsisches Glossar von H. Leo, Halle, 1877.

Lev. The Anglo-Saxon version of the book of Leviticus. v. Deut.

Lk. The Gospel of St. Luke. v. Mt.

LL. Th.= L. Th.

Lupi Serm. v. Wulfst.

Lye. Dictionarium Saxonico- et Gothico-Latinum. Auctore Eduardo Lye. Edidit Owen Manning, London, 1772.

M. H. The MS. so quoted has now been printed. v. Homl. Skt.

Man. ed. Furn. (F). Robert Manning's History of England, edited by F. J. Furnivall, Rolls Series, London, 1887. Quoted by line.

Mand. The Voiage and Travaile of Sir John Maundeville, edited by J. O. Halliwell, London, 1883.

Manip. Vocab. Levins' Manipulus Vocabulorum, a riming Dictionary, 1570, edited by H. B. Wheatley, E.E.T.S., No. 27, 1867.

Mann. Manning's edition of Lye's A. S. Dict., particularly the Supplement.

Mapes. The Latin Poems commonly attributed to Walter Map, edited by T. Wright, Camden Soc., No. xvi., 1841. Quoted by page and line.

March. A comparative grammar of the Anglo-Saxon language, by F. A. March, New York, 1873.

Marg. Seinte Margarete, edited by O. Cockayne (in the same volume as the next). Quoted by line.

Marh. Seinte Marherete, þe meiden ant martyr, edited by O. Cockayne, E.E.T.S., No. 13, 1866. Quoted by page and line.

Martyr. (Martyrol.). Martyrologium in Bibl. C. C. C. Cant. D. 5. v. Wanl. Catal. p. 106. Alterum exemplar, mutilum licet, multa tamen continens quae in superiori desiderantur, occurrit in Bibl. Cott. Jul. A. 10, v. Wanl. Catal. p. 185. The MSS. thus referred to by Lye are used by Cockayne in Shrn. pp. 44–156, and from this edition most passages are taken in the Dictionary. The quotation by month and day of Martyr. makes reference to Shrn. easy.

Med. ex Quadr. An Anglo-Saxon version of the Medicina de Quadrupedis of Sextus Placitus, printed in Lchdm. i. Quoted by section and paragraph.

Med. Pec. For the passage given under *ágotenes* with this abbreviation see L. Ecg. C. 2; Th. ii. 136, 20.

Menol. Fox. Menologium seu Calendarium Poeticum, ex Hickesiano Thesauro, edited by S. Fox, London, 1830. Quoted by line.

Met[r]. Homl. English Metrical Homilies from MSS. of the 14th century, edited by J. Small, Edinburgh, 1862. Quoted by page and line.

Mid. York. Gl. A glossary of words pertaining to the dialect of Mid-Yorkshire, by C. C. Robinson, E.D.S., 1876.

Migne. Lexicon Manuale ad Scriptores mediae et infimae Latinitatis, par M. L'Abbé Migne, Paris, 1866.

Min. The Poems of Laurence Minot, edited by J. Hall, Oxford, 1887. Quoted by number of poem (or of page) and line.

Mirc. Instructions for Parish Priests by John Myrc, edited by E. Peacock, E.E.T.S., No. 31, 1868. Quoted by line.

Misc. An Old English Miscellany, edited by R. Morris, E.E.T.S., No. 49, 1872. Quoted by page and line.

Mk. The Gospel of St. Mark. v. Mt.

Mobr. Venerabilis Baedae Historia Ecclesiae Gentis Anglorum, cura G. H. Moberly, Oxon., 1869.

Mod. Confit. Confessio et oratio ad Deum, MS. Cott. Tib. A. 3, fol. 44, v. Wanl. Cat. p. 195. See an edition of this piece, Anglia xi. 112–115.

Mod. Lang. Notes. Modern Language Notes, Baltimore.

Mone. Mone's Quellen und Forschungen zur Geschichte der teutschen Literatur und Sprache, Leipzig, 1830.

Mone A. A copy of the same glossary as Glos. Brux. Recd. printed in Mone.

Mone B. A copy of the same glossary as Hpt. Gl. printed in Mone.

Morris Spec. i. Specimens of Early English, edited by R. Morris, Part I. Oxford, 1882. Quoted by page and line of section.

Mort A. Morte Arthure, edited by E. Brock, E.E.T.S., No. 8, 1865. Quoted by line.

Morte Arthure (Halliwell). From a MS. quoted in Halliwell's Dictionary.

Mt. The Gospel of St. Matthew. Several editions of various versions of the Gospels are referred to, for a detailed notice of which see Prof. Skeat's preface to his edition of St. Mark's Gospel in the series noted below under Kmbl.

Bos. The Gothic and Anglo-Saxon Gospels with the versions of Wycliffe and Tyndale, edited by J. Bosworth, London, 1865.

Foxe. The Gospels of the fower Euangelistes, translated in the olde Saxon tyme out of Latin into the Vulgare toung of the Saxons, London, printed by John Daye, 1571. This work was published by Fox, the Martyrologist.

Hat. The Hatton MS. in the Bodleian Library, at Oxford, marked 38. See Wanl. Cat. p. 76.

Jun. Quatuor D. N. Jesu Christi Euangeliorum versiones perantiquae duae, Gothica scil. et Anglo-Saxonica ; illam ex Codice Argenteo depromsit Franciscus Junius, hanc curavit Thomas Mareschallus, Dordrechti, 1665.

Kmbl. The Gospel according to Saint Matthew in Anglo-Saxon and Northumbrian Versions. Cambridge, 1858. The work was begun by J. M. Kemble and completed by Mr. Hardwick. The other Gospels were edited by Prof. Skeat, who in 1887 edited this Gospel also.

Lind. MS. Cott. Nero D. 4. The Latin Text was written in the island of Lindisfarne. See Wanl. Cat. p. 250.

Rl. MS. Bibl. Reg. I. A. xiv. See Wanl. Cat. p. 181.

Rush. MS. Auct. D. ii. 19, in the Bodleian Library at Oxford. The MS. was at one time in the possession of John Rushworth, deputy-clerk to the House of Commons during the Long Parliament, and was by him presented to the Bodleian Library. See Wanl. Cat. p. 31.

Skt. v. Kmbl.

Stv. An edition of the Lindisfarne and Rushworth Gospels was published by the Surtees Society (Nos. 28, 39, 43, 48, 1854–1865), the first volume being edited by Rev. J. Stevenson, the last three by G. Waring.

Th. The Anglo-Saxon version of the Holy Gospels, edited by B. Thorpe, London and Oxford, 1842.

War. v. Stv.

N. Dictionarium Saxonico-Anglicum Laurentii Noelli, in the Bodleian Library. See Wanl. Cat. p. 102.

N. P. Nugae Poeticae. Select pieces of Old English popular poetry, edited by J. O. Halliwell, London, 1844. Quoted by page.

Nar. Narratiunculae Anglice conscriptae, edited by O. Cockayne, London, 1861. Quoted by page and line.

Nat. S. Greg. Els. An English-Saxon Homily on the Birthday of St. Gregory, translated into Modern English by Elizabeth Elstob, London, 1709 (cf. Homl. Th. ii. 116). Quoted by page and line.

Nath. (Nathan). Nathanis Judaei legatio ad Tiberium Caesarem. It is contained in a MS. preserved in the University Library at Cambridge, described in Wanl. Cat. p. 152, and has been edited in Publications of the Cambridge Antiquarian Society by C. W. Goodwin, Cambridge, 1851. v. St. And., under which abbreviation references by page and line are given except in the earlier part of the Dictionary.

Nicod. (Nic.) ; Thw. (Nicod. Thw.). An Anglo-Saxon version of the Gospel of Nicodemus, printed in Thw. Hept.

Num. The Anglo-Saxon version of the book of Numbers. v. Deut.

O. and N. An Old English poem of the Owl and the Nightingale, edited by F. H. Stratmann, Krefeld, 1868. Quoted by line.

O. E[ngl.] Homl. Old English Homilies, edited by R. Morris, E.E.T.S., first series, Nos. 29, 32 ; second series, No. 53, 1867–1868, 1873. Quoted by series, page and line.

O. E. Misc. = Misc.

O. Frs. Old Frisian ; the forms are taken from Altfriesisches Wörterbuch von Karl von Richthofen, Göttingen, 1840.

O. H. Ger. Old High German. v. Grff.

O. L. Ger. Old Low German ; the references are mostly to Kleinere altniederdeutsche Denkmäler, herausgegeben von M. Heyne, Paderborn, 1877.

O. Nrs. v. Icel.

O. Sax. v. Hél.

Obs. Lun. De obseruatione lune, printed from MS. Cott. Tib. A. iii. fol. 30 b in Lchdm. iii. 184.

Octo Vit. cap[it]. A homily De octo vitiis et de xii. abusivis, in a MS. of the Bodleian, Cod. Jun. 24, p. 329 (Wanl. Cat. p. 42). It is printed in O. E. Homl. i. 296–304. Cf. also Homl. Skt. i. 16, 246–384.

Octov. Octovian Imperator in Weber's Metrical Romances, vol. iii., 1810. Quoted by line.

Off. episco[op.]. The reference seems to be to the matter printed in Thorpe's Laws from Cod. Jun. 121, and referred to as L. I. P. (e. g. *â-wildian* will be found, Th. ii. 322, 15).

Off. reg[um]. The same MS. as the preceding seems sometimes to be referred to, e. g. *efen-wel* occurs Th. ii. 324, 2 : but *bæc-slitol* I have noted only Wulfst. 72, 16, where the MS. is Cod. Jun. 99 (Wanl. Cat. p. 27).

Orm. The Ormulum, edited by R. M. White, Oxford, 1852. Quoted by line.

Ors. ; Bos. King Alfred's Anglo-Saxon version of the compendious history of the world by Orosius, edited by J. Bosworth, London, 1859. Quoted by book and chapter (Ors.), and by page and line (Bos.).

Ors. Hav. The edition of Orosius by Havercamp, Leyden, 1738.

Ors. ; Swt. King Alfred's Orosius, edited by H. Sweet, E.E.T.S., No. 79, 1883. Quoted by book and chapter (Ors.), and by page and line (Swt.).

Ottf. Otfrid's Krist, edited by E. G. Graff, Königsberg, 1831.

P. B. Beiträge zur Geschichte der deutschen Sprache und Literatur, herausgegeben von H. Paul und W. Braune.

P. L. S. Early English Poems and Lives of Saints, edited by F. J. Furnivall, Philol. Soc., 1862. Quoted by number of piece and line (or stanza).

P. R. L. P. Political, Religious, and Love Poems, edited by F. J. Furnivall, E.E.T.S., No. 15, 1866. Quoted by page and line.

P. S. The Political Songs of England, from the reign of John to that of Edward II, edited by T. Wright, Camden Soc., No. vi., 1839. Quoted by page and line.

Palgrv. Eng. Com. Palgrave's Rise and Progress of the English Commonwealth, London, 1834.

Pall. Palladius on Husbondrie, edited by B. Lodge and S. T. Herrtage, E.E.T.S., Nos. 52 and 72, 1872 and 1879. Quoted by page and line of book.

Parten. The Romans of Partenay, edited by W. W. Skeat, E.E.T.S., No. 22, 1866. Quoted by line.

Past.; Hat. An Anglo-Saxon version of Gregory's Pastoral Care, contained in a MS. (Hatton 20) preserved in the Bodleian Library. Quoted by chapter and paragraph of an edition of the Cura Pastoralis by J. Stephen, London, 1629 (Past.), and by folio and line of MS. (Hat.).

Past.; Swt. The Anglo-Saxon version of Gregory's Pastoral Care from the Hatton MS. and the Cotton MSS., edited by H. Sweet, E.E.T.S., Nos. 45 and 50, 1871–1872. Quoted by chapter (Past.), and by page and line (Swt.).

Peccat[orum] Medic[ina] = L. Pen. (e. g. *âspîwan* may be found L. Pen. 5; Th. ii. 278, 22).

Pegge's Kenticisms. An Alphabet of Kenticisms by Samuel Pegge, 1735. E. D. S., 1876.

Piers [P.]. The Vision of William concerning Piers the Plowman (Text B), E. E. T. S., No. 38, 1869. [Texts A and C are Nos. 28 and 54.] Quoted by passus and line.

Piers P. Crede. Pierce the Ploughman's Crede, edited by W. W. Skeat, E.E.T.S., No. 30, 1867. Quoted by line.

Pl. Cr. = Piers P. Crede.

Pol. Songs Wrt. = P. S.

Pr. C. The Pricke of Conscience, by R. Rolle de Hampole, edited by R. Morris, Philol. Soc., 1863. Quoted by line.

Pref. [Ælfc.]. Thw. Ælfric's preface to Genesis in Thw. Hept. Quoted by page and line.

Pref. (Procem.) R. Conc[ord]. Prohemium regularis concordie Anglicae nationis monachorum (MS. Cott. Tib. A. 3, v. Wanl. Cat., p. 193). This is edited in Anglia, vol. xiii. p. 365, and in the later part of the Dictionary this edition is referred to.

Prehn's Rätsel des Exeterbuches. Komposition und Quellen der Rätsel des Exeterbuches, von Dr. August Prehn, Paderborn, 1883.

Proclam. H. III. The only English Proclamation of Henry III, edited by Alex. J. Ellis, Philol. Soc., 1868.

Prompt. [Parv.]. Promptorium Parvulorum, sive Clericorum, dictionarius Anglo-Latinus princeps, auctore fratre Galfrido, recensuit Albertus Way, Camden Soc., Nos. xxv., liv., lxxxix., 1843–1865. Quoted by page.

Prov. Glosses on the book of Proverbs, which are printed as noticed under Kent. Gl. Quoted by chapter (and verse).

Prov. Kmbl. Anglo-Saxon Apothegms in Salm. Kmbl. (q. v.) Part III. pp. 258–268. Quoted by number.

Prud. v. Gl. Prud. (2).

Ps. An Early English Psalter, edited by J. Stevenson, Surtees Soc., Nos. 16, 19, 1843–1847. Quoted by psalm and verse.

Ps. Grn. The edition of the metrical version of Psalms 51–150 in Grein's Bibliothek der Angelsächsischen Poesie. 2. Band. Göttingen, 1858.

Ps. Lamb. An interlinear version of the Psalms in a MS. preserved in the library of Lambeth Palace. It is thus described by Wanley: Psalterium D. Hieronymi Gallicum, Astericis et obolis, punctisque Musicis subjectis notatum, una cum interlineata Versione Saxonica, Catalogue, p. 268.

Ps. Spl. Psalterium Davidis Latino-Saxonicum vetus. A Johanne Spelmanno D. Hen. fil. editum. E vetustissimo exemplari MS. in Bibliotheca ipsius Henrici, et cum tribus aliis non multo minus vetustis collatum, Londini, 1640. The MS. used by Spelman subsequently was in the library at Stowe, and has been described by Dr. O'Conor in his account of that library. Afterwards it passed into the possession of Lord Ashburnham. Of the three collated MSS., which Spelman refers to under the letters C, T, M, the first is in the University Library at Cambridge, see Wanl. Cat. p. 152; the second is in the library of Trin. Coll. Camb., and has been edited by F. Harsley, E.E.T.S., No. 92, 1889 (Eadwine's Canterbury Psalter); the third is Arundel MS. No. 60 in the British Museum. The printed edition, as regards C and T, was collated with those MSS. for Dr. Bosworth by Dr. Aldis Wright, and many corrections were made.

Ps. Stev. or Surt. An Anglo-Saxon Psalter (printed from MS. Cott. Vesp. A. 1), edited by J. Stevenson, Surtees Soc., Nos. 16, 19, 1843–1847.

Ps. Th. Libri Psalmorum versio antiqua Latina; cum paraphrasi Anglo-Saxonica, partim soluta oratione, partim metrice composita. E Cod. MS. in Bibl. Regia Parisiensi adservato descripsit et edidit B. Thorpe, Oxonii, 1835.

Ps. Trin. Camb. = Ps. Spl. T.

Ps. Vos[sii]. An interlinear version of the Psalms in a MS. given by Isaac Vossius to Francis Junius (MS. Bodl. Junius 27). See Wanl. Cat. p. 76.

R., Lye. Ælfric's Vocabulary, transcribed by or for Junius from a MS. in the possession of Reubens the painter, v. Wanl. Cat. p. 96. It was printed by Somner at the end of his Dictionary, and will be found in Wrt. Voc. i. 15.

R. Ben. Die Angelsächsische Prosabearbeitung der Benedictinerregel, herausgegeben von A. Schröer, Kassel, 1885. Quoted (at first by chapter, later) by page and line.

R. Ben. Interl. The Rule of S. Benet. Latin and Anglo-Saxon interlinear version. Edited by H. Logeman, E.E.T.S., No. 90, 1888. Quoted (at first by chapter, later) by page and line.

R. Brun[ne]. Peter Langtoft's Chronicle (as illustrated and improved by Robert of Brunne), published by Thomas Hearne, Oxford, 1725. Quoted by page and line.

R[eg.] Conc[ord]. v. Pref. R. Conc.

R. Glouc. Robert of Gloucester's Chronicle, published by Thomas Hearne, Oxford, 1724. Quoted by page and line.

R. R. The Romaunt of the Rose, formerly attributed to Chaucer. Quoted by line.

R. S. Religious Songs, edited by Thomas Wright, Percy Soc., vol. xi., 1843. Quoted by number of piece and line.

Rask Hald. Björn Halldórsson's Icelandic-Latin Dictionary, edited by Rask, 1814.

Recd.; Wrt. Voc. v. Glos. Brux. Recd.

Rel[iq.] Ant[iq.]. Reliquiae Antiquae. Scraps from Ancient Manuscripts, edited by T. Wright and J. O. Halliwell. 2 vols., London, 1845. Quoted by volume, page and line.

Rich. Richard Coer de Lion, in Weber's Metrical Romances, vol. ii. 3–278. Quoted by line.

Rol. H. Richard Rolle of Hampole and his followers, edited by C. Horstmann. 2 vols., London, 1895. Quoted by volume, page and line.

Rood Kmbl. The Holy Rood, a poem in the Vercelli MS., published with Elen. Kmbl. (q. v.).

Rood Recd. The same poem as the preceding, printed as Andr. Recd. (q. v.).

Rtl. Rituale Ecclesiae Dunelmensis (Latin and interlinear Anglo-Saxon versions), Surtees Soc., No. 10, 1839. Quoted by page and line.

Runic Inscrip. Kmbl. On Anglo-Saxon Runes. By J. M. Kemble. Archaeologia, published by the Society of Antiquaries, vol. xxviii., London, 1840. Quoted by page and line.

Runic pm. Kmbl. A poem printed in the above paper. Quoted by page and line.

S. de Fide Cathol. This homily is printed Homl. Th. i. 274.

Salm. Kmbl. Anglo-Saxon Dialogues of Salomon and Saturn, by J. M. Kemble. Printed for the Ælfric Society, London, 1845–1848. The poetical part is quoted by line, the prose by page and line.

Sax. Engl. The Saxons in England. A History of the English Commonwealth till the period of the Norman Conquest, by J. M. Kemble. 2 vols., London, 1876.

Schmid [A. S. Gesetz.]. Die Gesetze der Angelsachsen. Herausgegeben von Dr. R. Schmid, Leipzig, 1858.

Scint. Defensoris Liber Scintillarum, with an interlinear Anglo-Saxon version, edited by E. W. Rhodes, E.E.T.S., No. 93, 1889. Quoted (at first by chapter, later) by page and line of the interlinear version. App. [Lib.] Scint. refers to the matter in pp. 223–236 of this edition.

Scint. de Praedest. = Scint., pp. 226–228.

Scóp. Th. The Scóp or Gleeman's Tale printed in Beo. Th.

Scot. Scottish. v. Jamieson.

Seebohm Vill. Comm. The English Village Community examined in its relations to the Manorial and Tribal Systems, by F. Seebohm, London, 1890.

Serm. Creat. = Homl. Th. i. 8–28. (v. ge-dæman, *where read* ge-clæman.)

Shor[eham]. The Religious Poems of William de Shoreham, edited by T. Wright, Percy Soc. vol. xxviii., 1849. Quoted by page.

Shrn. The Shrine. A Collection of occasional papers on dry subjects, by O. Cockayne, London, 1864–1870. Quoted by page and line.

Skt. Dict. An Etymological Dictionary of the English Language, by W. W. Skeat, Oxford, 1879–1882.

Solil. Soliloquia Augustini Selecta et Saxonice reddita ab Ælfredo Rege, MS. Cott. Vitell. A. 15, fol. 1. (Printed in Shrn. pp. 163–204.) v. Wanl. Cat. p. 218.

Som. Dictionarium Saxonico-Latino-Anglicum, by E. Somner, Oxon., 1659.

Somn. De somniorum diuersitate (MS. Cott. Tib. A. iii. fol. 25 b) *and* De somniorum eventu (v. Wanl. Cat. p. 40). The two pieces are printed Lchdm. iii. pp. 198–214, 168–176. Quoted by number of paragraph in the two combined. (In the later part of the Dictionary the references are to Lchdm. iii.)

Soul Kmbl. The departed soul's address to the body, a poem in the Vercelli MS. published with Elen. Kmbl. (q. v.).

Soul Recd. The same poem as the preceding, printed with Andr. Recd. (q. v.).

Spec. Specimens of Lyric Poetry composed in England in the reign of Edward I, edited by T. Wright, Percy Soc., vol. iv., 1842. Quoted by page and line.

St. And. Anglo-Saxon Legends of St. Andrew and St. Veronica, Cambridge Antiquarian Society, Cambridge, 1851.

Swt. A. S. Prim. An Anglo-Saxon Primer by H. Sweet, Oxford, 1882.

Swt. [A. S.] Rdr. An Anglo-Saxon Reader, in prose and verse, by H. Sweet, Oxford, 1876.

Techm. Internationale Zeitschrift für allgemeine Sprachwissenschaft, begründet und herausgegeben von F. Techmer, Leipzig. Quoted by volume, page and line.

Te Dm. Lamb. (Te Deum; Lamb.). An interlinear version of the Te Deum in the same MS. as Ps. Lamb. It is also cited as Hymn ad Mat.

Te Dm. Lye. v. Wanl. Cat. p. 222.

Te Dm. Thomson. A version of the same in Thomson's Select Monuments of the Doctrine and Worship of the Catholic Church in England before the Norman Conquest, 1849.

Text. Rof. Textus de Ecclesia Roffensi. v. Wanl. Cat. p. 273.

Th. An[al.] (Anlct.). v. An. Th.

Th. Ap[ol.]. v. Ap. Th.

Th. Ch[art.] (Diplm.). v. Chart. Th.

Th. Lapbg. A History of England under the Anglo-Saxon Kings, translated from the German of Dr. J. M. Lappenberg by B. Thorpe, London, 1845.

Th. Ll. v. L.; Th.

Thw. Hept. Heptateuchus, Liber Job, et Evangelium Nicodemi; Anglo-Saxonice. Historiae Judith Fragmentum; Dano-Saxonice. Edidit Edwardus Thwaites, Oxoniae, 1698. v. Wanl. Cat. pp. 67–68, 152.

Torrent of Portugal. An English Metrical Romance, edited by J. O. Halliwell, London, 1842.

Tr. and Cr. Chaucer's Troilus and Creseyde. Quoted by book and line.

Tract. de Spir. Septif. A homily De Septiformi Spiritu. See Wulfst. 50–56.

Trev. Polychronicon Ranulphi Higden, with the English translation of John Trevisa. Rolls Series, 1865–1886. Quoted by volume, page and line.

Trist. Die Nordische und die Englische Version der Tristan-Saga, herausgegeben von E. Kölbing, Heilbronn, 1882–1883. Quoted by line.

Txts. The Oldest English Texts, edited by H. Sweet, E.E.T.S., No. 83, 1885. Quoted by page and number of gloss (or by line).

Tynd. Tyndal's version of the New Testament.

V. Ps. = Ps. Vos.

Vit. Swith. See either Glostr. Frag. or Homl. Skt. i. 21.

W. Cat. = Wanl. Cat.

W. F. (Wells Frag.). MS. of the A.S. version of the Benedictine Rule in the possession of the Chapter at Wells, printed in R. Ben.

W. S. West-Saxon.

Wald. Two leaves of King Waldere's Lay, published by George Stevens, Copenhagen. Quoted by line.

Wanl. Cat[al.]. Wanley's Catalogue of Anglo-Saxon MSS., forming the third volume of Hickes' Thesaurus, Oxoniae, 1705.

Wht. Dict. White and Riddle's Latin-English Dictionary.

Wick. v. Wyc.

Wicklif Select Wrks. Select English Works of John Wyclif, edited by T. Arnold, Oxford, 1869–1871. Quoted by volume and page.

Wilk. Leges Anglo-Saxonicae Ecclesiasticae et Civiles, edited by D. Wilkins, London, 1721. Quoted by page and line.

Will. The Romance of William of Palerne, edited by W. W. Skeat, E.E.T.S., No. i., 1867. Quoted by line.

Wrt. Biog. Brit. A. Sax. Biographia Britannica Literaria; or Biography of Literary Characters of Great Britain and Ireland. Anglo-Saxon Period. By Thomas Wright, London, 1842.

Wrt. Popl. Science. Popular Treatises on Science written during the Middle Ages, edited by Thomas Wright, London, 1841. Quoted by page and line.

Wrt. Provncl. Dictionary of Obsolete and Provincial English, compiled by Thomas Wright, London, 1837.

Wrt. Spec. v. Spec.

Wrt. Voc. [i.]. A volume of Vocabularies, edited by Thomas Wright. Privately printed, 1857. Quoted by page and number of gloss.

Wrt. Voc. ii. A second volume of Vocabularies, edited by Thomas Wright. Privately printed, 1873. Quoted by page and line.

Wülck. [Gl.]. Anglo-Saxon and Old English Vocabularies, by Thomas Wright. Second edition, edited by R. P. Wülcker, London, 1884. Quoted by column and line.

Wulfst. Wulfstan. Sammlung der ihm zugeschriebenen Homilien, herausgegeben von A. Napier, Berlin, 1883. Quoted by page and line.

Wyc. The Holy Bible in the earliest English versions made from the Latin Vulgate by John Wycliffe, edited by Forshall and Madden, Oxford, 1850.

York. Gl. A Glossary of words pertaining to the Dialect of Mid-Yorkshire. E. D. S. Pub., 1876.

Zacher. Das Gothische Alphabet Ulfilas und das Runenalphabet. Eine sprachwissenschaftliche Untersuchung von Julius Zacher, Leipzig, 1855.

EXPLANATION OF THE PRINCIPAL CONTRACTIONS.

[Contractions used in Grein's Lexicon Poeticum are given separately on the next page.]

Ælfc. Gl; Som; Wrt. Voc. Ælfric's Glossary given at the end of Somner's Dictionary, and in A Volume of Vocabularies, edited by Thomas Wright (First Series, 1857).

Ælfc. Gr; Som. Ælfric's Grammar, at the end of Somner's Dictionary.

Ælfc. T. Grn. Ælfric de veteri et de novo testamento, in Grein's edition of the Heptateuch.

Alis. King Alisaunder, in Weber's Metrical Romances.

Andr. Kmbl. The Legend of St. Andrew, edited by Kemble for the Ælfric Society.

An. Lit. Anecdota Literaria, by T. Wright.

Apstls. Kmbl. The Fates of the Twelve Apostles; a fragment, in Part II. of The Poetry of the Codex Vercellensis, edited by Kemble for the Ælfric Society.

A. R. The Ancren Riwle, edited for the Camden Society by J. Morton.

Ayenb. Dan Michel's Ayenbite of Inwyt, edited for the Early English Text Society by R. Morris.

Basil admn. The Anglo-Saxon remains of St. Basil's Admonitio ad filium spiritualem, edited by H. W. Norman.

Bd; S. The Anglo-Saxon version of Bede's Ecclesiastical History, edited by Smith.

Beo. Th. The Anglo-Saxon poem of Beowulf, edited by Thorpe.

Blickl. Gl. Blickling Glosses, at the end of the Blickling Homilies.

Blickl. Homl. The Blickling Homilies, edited for the Early English Text Society by R. Morris.

Boutr. Scrd. Screadunga, edited by C. G. Bouterwek.

Bt; Fox. King Alfred's Anglo-Saxon version of Boethius De Consolatione Philosophiæ, edited by Fox (in Bohn's Antiquarian Library).

Bt. Met. Fox. The Anglo-Saxon version of the Metres of Boethius, at the end of the previous work.

Byrht. Th. The Battle of Maldon, in Thorpe's Analecta Anglo-Saxonica.

Cant. Abac. Canticum Abacuc Prophetæ, in Ps. Lamb., q. v.

Cant. Moys. Canticum Moysis, at the end of Thwaites' Heptateuch.

Cd; Th. Cædmon's Metrical Paraphrases of parts of the Holy Scriptures, edited by Thorpe.

Chart. Th. Diplomatarium Anglicum Ævi Saxonici, edited by Thorpe.

Chauc. Chaucer.

Chr; Erl. Two of the Saxon Chronicles, edited by Earle.

Cod. Dipl. Kmbl. Codex Diplomaticus Ævi Saxonici, edited by Kemble.

Coll. Monast. Th. Ælfric's Colloquy, in Thorpe's Analecta Anglo-Saxonica.

Confess. Peccat. Confessio Peccatorum, in Ps. Lamb., q. v.

Cot. Lye. A MS. of the Cotton Library quoted by Lye in his Dictionary.

Deut. Deuteronomy, in Thwaites' Heptateuch.

E. D. S. English Dialect Society.

Elen. Kmbl. Elene, or the Recovery of the Cross, edited by Kemble for the Ælfric Society.

Ex. Exodus, in Thwaites' Heptateuch.

Exon. Th. Codex Exoniensis, edited by Thorpe.

Fins. Th. The Fight at Finnesburg, at the end of Thorpe's Beowulf.

Frag. Kmbl. A Fragment, Moral and Religious, in Part II. of The Poetry of the Codex Vercellensis, edited by Kemble.

Gen. Genesis, in Thwaites' Heptateuch.

Grff. Althochdeutscher Sprachschatz von Graff.

Grmm. A. u. E. Andreas und Elene, herausgegeben von Jacob Grimm.

Grmm. D. M. Deutsche Mythologie von Jacob Grimm. Zweite ausgabe.

Grmm. Gesch. D. S. Geschichte der Deutschen Sprache von Jacob Grimm. Dritte auflage.

Grn. R. A. Deutsche Rechtsalterthümer von Jacob Grimm. Zweite ausgabe.

Guthl; Gdwin. The Anglo-Saxon version of the Life of St. Guthlac, edited by C. W. Goodwin.

Halliw. Dict. Halliwell's Dictionary of archaic and provincial words.

Herb. Herbarium in Vol. I. of Saxon Leechdoms.

Hexam. Norm. The Anglo-Saxon version of the Hexameron of St. Basil, edited by H. W. Norman.

Homl. Skt. Ælfric's Lives of Saints, edited for the Early English Text Society by W. W. Skeat.

Homl. Th. The Homilies of Ælfric, edited for the Ælfric Society by B. Thorpe.

Hpt. Gl. Angelsächsische Glossen, von Bouterwek mitgetheilt in Haupts Zeitschrift ix. (quoted from Leo's Angelsächsische Glossar).

H. R. Legends of the Holy Rood, edited for the Early English Text Society by R. Morris.

Hymn. Surt. The Latin Hymns of the Anglo-Saxon Church, edited for the Surtees Society by J. Stevenson.

Icel. Icelandic; the references being to Cleasby and Vigfusson's Icelandic Dictionary.

Jn. Skt. The Gospel of St. John, edited by Skeat. v. Mk. Skt.

Jos. Joshua, in Thwaites' Heptateuch.

Jud. Judges, in Thwaites' Heptateuch.

Judth; Thw. The poem of Judith, at the end of Thwaites' Heptateuch.

Jul. The Liflade of St. Juliana, edited for the Early English Text Society by Cockayne.

L. Alf; Th. The Laws of King Alfred, in Thorpe's Ancient Laws and Institutes. The other contractions, being the same as those used by Thorpe, are not given here.

Laym. The Brut of Laȝamon, edited by Sir F. Madden.

Lchdm. i. ii. iii. Leechdoms, Wortcunning and Starcraft of Early England, edited by Cockayne (Master of the Rolls' series, 3 vols.).

Lev. Leviticus, in Thwaites' Heptateuch.

Lk. Skt. The Gospel of St. Luke, edited by Skeat. v. Mk. Skt.

L. M; Lchdm. v. Lchdm. (L. M. = Liber Medicinalis.)

L. Med. ex Quadr; Lchdm. v. Lchdm.

Menol. Fox. Menologium or Poetical Calendar of the Anglo-Saxons, edited by Fox.

Met. Homl. English Metrical Homilies from MSS. of 14th century, edited by J. Small.

Mk. Skt. Lind. or Rush. The Gospel of St. Mark in Anglo-Saxon and Northumbrian versions, edited for the Syndics of the University Press by W. W. Skeat. (Lind. = Lindisfarne MS. Rush. = Rushworth MS.)

Mt. Kmbl. Lind. or Rush. The Gospel of St. Matthew, in Anglo-Saxon and Northumbrian versions, edited by Kemble. v. preceding.

Nar. Narratiunculæ Anglice Conscriptæ, edited by Cockayne.

Nicod; Thw. The Anglo-Saxon version of the Gospel of Nicodemus, at the end of Thwaites' Heptateuch.

Num. Numbers, in Thwaites' Heptateuch.

O. and N. An Old English poem of the Owl and the Nightingale, edited by Stratmann.

O. E. Homl. Old English Homilies, edited for the Early English Text Society by R. Morris.

O. Frs. refers to Altfriesisches Wörterbuch von Dr. Karl Freiherrn von Richthofen.

O. H. Ger. v. Grff.

Orm. The Ormulum, edited by Dr. White.

Ors; Swt. or Bos. The Anglo-Saxon version of Orosius, edited by Sweet or by Bosworth.

O. Sax. The Old Saxon poem of the Heliand.

Past; Swt. King Alfred's version of Gregory's Pastoral Care, edited for the Early English Text Society by Sweet.

Piers P. The Vision concerning Piers the Plowman, B-text, edited for the Early English Text Society by W. W. Skeat.

P. L. S. Early English Poems and Lives of Saints, edited by F. J. Furnivall.

Prompt. Parv. Promptorium parvulorum sive clericorum, lexicon Anglo-latinum princeps, edited for the Camden Society by Way.

EXPLANATION OF THE PRINCIPAL CONTRACTIONS.

Prov. Kmbl. Anglo-Saxon Apothegms given by Kemble in Anglo-Saxon Dialogues of Salomon and Saturn, Part III. (Ælfric Society's publications).

Ps. Lamb. Lambeth Psalter. The references are taken from the copy made under Dr. Bosworth's direction.

Ps. Spl. Psalterium Davidis Latino-Saxonicum vetus, a Johanne Spelmanno editum. (Dr. Bosworth's copy has been collated with the original MSS. and has thus been corrected in many places.)

Ps. Surt. Anglo-Saxon and Early English Psalter, edited for the Surtees Society by Stevenson.

Ps. Th. Libri Psalmorum versio antiqua Latina ; cum paraphrasi Anglo-Saxonica, edidit Benjamin Thorpe.

R. Ben. Anglo-Saxon version of the Benedictine Rule (quoted from Lye).

R. Brun. Peter Langtoft's Chronicle translated and continued by Robert Manning of Brunne, edited by Hearne.

Rel. Ant. Reliquiæ Antiquæ, edited by Wright and Halliwell.

R. Glouc. Robert of Gloucester's Chronicle, edited by Hearne.

Rood Kmbl. The Holy Rood; a Dream, in Part II. of The Poetry of the Codex Vercellensis, edited for the Ælfric Society by Kemble.

Rtl. Rituale Ecclesiæ Dunelmensis, edited for the Surtees Society by Stevenson. (Compare Skeat's collation in the Philological Society's Transactions.)

Runic pm. Kmbl. Runic poem printed by Kemble in Archæologia, vol. 28.

Salm. Kmbl. Anglo-Saxon Dialogues of Salomon and Saturn, edited for the Ælfric Society by Kemble.

Schmid. A. S. Ges. Die gesetze der Angelsachsen, herausgegeben von Reinh. Schmid.

Scot. Scottish; the references being to Jamieson's Dictionary.

Shrn. The Shrine; a collection of occasional papers on dry subjects, by Cockayne.

Soul Kmbl. The Departed Soul's Address to the Body, in Part II. of The Poetry of the Codex Vercellensis, edited for the Ælfric Society by Kemble.

St. And. Anglo-Saxon Legends of St. Andrew and St. Veronica (Publications of the Cambridge Antiquarian Society).

Swt. A. S. Rdr. An Anglo-Saxon Reader in prose and verse, by Henry Sweet.

Th. An. Analecta Anglo-Saxonica, by Benjamin Thorpe.

Th. Ap. The Anglo-Saxon version of the story of Apollonius of Tyre, edited by Thorpe.

Trev. Polychronicon Ranulphi Higden, with the English translation of John Trevisa (Master of the Rolls' series).

Will. The romance of William of Palerne, edited by W. W. Skeat (Early English Text Society).

Wrt. popl. Science. Popular Treatises on Science written during the Middle Ages, edited by Wright.

Wrt. Voc. A Volume of Vocabularies, edited by Wright. (First Series, Liverpool, 1857.)

CONTRACTIONS USED BY GREIN.

Ælf. Tod. Poem on the death of Alfred, son of Ethelred, given in the Chronicle under the year 1036.

Æðelst. Poem on the victory of Athelstan, taken from the Chronicle.

Alm. Almosen, from the Codex Exoniensis, p. 467.

An. The legend of St. Andrew.

Ap. The fates of the Apostles, from the Codex Vercellensis.

Az. Azarias, from the Codex Exoniensis, p. 185.

B. Beowulf.

Bo. Botschaft des Gemahls, from the Codex Exoniensis, p. 472.

By. The death of Byrhtnoth.

Crä. Manna cræftas, from the Codex Exoniensis, p. 292.

Cri. Cynewulfs Crist, from the Codex Exoniensis, p. 1.

Dan. Daniel, in Thorpe's Cædmon, p. 216.

Deór. Deors Klage, from Codex Exoniensis, p. 377.

Dôm. Dômes dæg, from Codex Exoniensis, p. 445.

Edg. Eádgár; poems from the Chronicle, under the years 973, 975.

Edm. Eádmund, from the Chronicle, under the year 942.

Edw. Eádweard, from the Chronicle, under the year 1065.

El. Elene, from the Codex Vercellensis.

Exod. Exodus, in Thorpe's Cædmon, p. 177.

Fä. Fæder lārcwidas, in Codex Exoniensis, p. 300.

Fin. The fight at Finnsburg.

Gen. Genesis, in Thorpes' Cædmon, p. 1.

Gn. C. Versus gnomici (Cotton MS.).

Gn. Ex. Versus gnomici, from Codex Exoniensis, p. 333.

Gú. Legend of St. Guthlac, from Codex Exoniensis, p. 104

Hö. Höllenfahrt Christi, from Codex Exoniensis, p. 459.

Hy. Hymnen und Gebete.

Jud. The poem of Judith.

Jul. The legend of St. Juliana, in Codex Exoniensis, p. 242.

Kl. Klage der Frau, in Codex Exoniensis, p. 442.

Kr. Das heilige Kreuz, from the Codex Vercellensis.

Leás. Bî manna leáse, from the Codex Vercellensis.

Men. Menologium.

Met. The metres of Alfred.

Môd. Manna môd, in the Codex Exoniensis, p. 313.

Pa. Panther, in the Codex Exoniensis, p. 355.

Ph. Phönix, in the Codex Exoniensis, p. 197.

Phar. Pharao, in the Codex Exoniensis, p. 468.

Ps. Psalms, from Thorpe's edition.

Ps. C. The 50th psalm, from one of the Cotton MSS.

Rä. Riddles from the Codex Exoniensis.

Reb. Rebhuhn, from the Codex Exoniensis, p. 365.

Reim. Reimlied, from the Codex Exoniensis, p. 352.

Ruin. Ruine, from the Codex Exoniensis, p. 476.

Rûn. Runenlied, in Archæologia, vol. 28.

Sal. Salomo und Saturn ; see above Salm. Kmbl.

Sat. Crist und Satan, in Thorpe's Cædmon, p. 265.

Seef. Seefahrer, in the Codex Exoniensis, p. 306.

Seel. Reden der Seelen, in the Codex Exoniensis, p. 367 ; see also above, Soul Kmbl.

Sch. Wunder der Schöpfung, in the Codex Exoniensis, p. 346.

Vîd. Vidsîð, in Codex Exoniensis, p. 318.

Vy. Manna wyrde, in Codex Exoniensis, p. 327.

Wal. Walfisch, in Codex Exoniensis, p. 360.

Wand. Wanderer, in Codex Exoniensis, p. 286.

A.

A. It is not necessary to speak of the form of what are often called Anglo-Saxon letters, as all Teutonic, Celtic, and Latin manuscripts of the same age are written in letters of the same form. There is one exception: the Anglo-Saxons had, with great propriety, two different letters for the two distinct sounds of our *th*: the hard þ in *thin* and soo*th*, and the soft ð in *thine* and soo*the*, vide Þ, þ. 2. The indigenous Pagan alphabet of our Anglo-Saxon forefathers, called Runes, it must be particularly observed, not only represents our letters, but the names of the letters are significant. The Runes are chiefly formed by straight lines to be easily carved on wood or stone. For instance, the Rune ᚫ âc is not only found in inscriptions on wood and stone, but in Anglo-Saxon MSS. and printed books. In manuscripts and in books, it sometimes denotes the letter a; and, at other times, *the oak*, from its Anglo-Saxon name, âc *the oak*. v. ÂC, and RÚN.

B. The *short* or unaccented Anglo-Saxon a is contained in the following words, which are represented by modern English terms of the same import, having the sound of *a* in *man;* as Can, man, span, hand, land, sand, camp, dranc, *etc.* 2. The short a is often found in the final syllables of inflections, -a, -an, -as, -aþ, *etc.* It generally appears in the radix before a doubled consonant, as swamm *a fungus*, wann *wan*; or two different consonants, as mp, mb, nt, nc, ng, *etc.*—Camp, lamb, plante, dranc, lang, *etc.* 3. The radical short a can only stand before a single consonant and *st, sc*, when this single consonant and these double letters are again followed, in the inflections or formative syllables, by *a, o, u* in nouns; and by *a, o, u, e* in adjectives; and *a, o, u,* and *ia* in verbs; as Dagas, daga from dæg, hwalas from hwæl, fatu from fæt, gastas from gæst, ascas from æsc; *adj.* Smales, smale, smalost, smalu, from smæl *small*: Lates, latu, latost, from læt *late*: Stapan, faran, starian, wafian. Grimm's Deut. Gram. vol. i. p. 223, 2nd edit. 1822. In other cases, the short or unaccented æ is used instead of a. See ÆE in its alphabetical order. 4. The remarks in 3. are of great importance in declining words, for monosyllables, ending in a single consonant, in *st* or *sc*, change the æ into a, whenever the consonant or consonants are followed by *a, o, u* in nouns, and *a, o, u, e* in adjectives, vide ÆE. 5. It must be remembered then, that a short a cannot stand in a word (1) when it ends in a single consonant, that is, when no inflections of *a, o, u* in nouns follow; as in Stæf, fræt: (2) when in nouns a single consonant is followed by *e*; as Stæfes, stæfe, wæter: (3) when the word has any other double consonants besides *st, sc*, though followed by *a, o, u*; as Cræft, cræfta, ægru *n. pl. of* æg: (4) in contracted words, when æ is not in the last syllable; as Æcer, *pl.* æceras, æcerum, contracted æcras, æcrum; wæpen, *pl.* wæpenu; mægen, *pl.* mægenu, contracted wæpnu, and mægnu. 6. Though I have given in C. 3. the reasons, which Grimm assigns for making the prefixed a-, long, I believe it is generally short in *A. Sax.* as in *Eng.* a-bide = *A. Sax.* a-bídan = bídan, so a-cende = cende:—Ic todæg cende [cende Surt; acende Spl. T; Th.] ðé *ego hodie genui te*, Ps. Spl. 2, 7. A-beran = beran *to bear*:—Hefige byrðyna aberan ne mæg *a man is not able to bear heavy burdens*, Mt. Bos. 23, 4. Ne bere ge sacc *nolite portare sacculum*, Lk. Bos. 10, 4. A-biddan = biddan *to ask, pray*:—Abiddaþ [biddaþ Cott.] hine *pray to him*, Bt. 42; Fox 258, 21. Ic bidde ðé, Drihten *I pray to thee*, Lord, Gen. 19, 18. It is evident by these examples that words have the same meaning with and without the prefixed a-: this a- was not prominent or long, and therefore this prefix is left unaccented in this Dictionary. 7. a- prefixed, sometimes denotes *Negation, deterioration*, or *opposition*, as *From, out, away*; thus awendan *to turn from, subvert*, from wendan *to turn*; amód *out of* or *without mind, mad*; adón *to do away, banish*, composed of a *from*, dón *to do*, vide ÆE. The prefixed a- does not always appear to alter the signification: in this case it is generally omitted in modern English words derived immediately from Saxon,—thus, Aberan *to bear*; abrecan *to break*; abítan *to bite*. The prefixed a-, in such cases, seems to add some little force or intensity to the original signification of the word to which it is joined,—thus, fǽran *to make afraid*; terrere: a-fǽran *to terrify, dismay, astound*; exterrere, perterrere, consternare, stupefacere.

C. The *long* Anglo-Saxon â is accented, and words containing this long or accented â are now represented by English terms, with the vowel sounded like o in *no* and *bone*. The following words have either the same or an analogous meaning, both in English and Anglo-Saxon: Hâm *home*, ân *one*, bân *bone*, hân *hone*, stân *stone*, sâr *sore*, râp *rope*, lâr *lore*, gâst *ghost*, wrât *wrote*. Sometimes the accented or long â is represented in English by *oa*; as Âc *an oak*, gâd *a goad*, lâd *load*, râd *road*, brâd *broad*, fâm *foam*, lâm *loam*, sâpe *soap*, âr *oar*, bâr *boar*, hâr *hoar*, bât *boat*, gât *goat*, âta *oat*, âþ *oath*, lâþ *loath*. Occasionally â becomes *oe* in English; as Dâ *a doe*, fâ *a foe*, tâ *toe*, wâ *woe*; but the *oe*, in these words, has the sound of o in *no*. The same may be said of *oa* in *oak, goad*. Hence it appears that the Anglo-Saxon â is represented by the modern English o, oa, and oe, which have the sound of o in *no* and *bone*; as Râd *rode* (*p.* of ride), râd *a road*, and dâ, *a doe*. Deut. Gram. von Jacob Grimm, vol. i. pp. 358, 397, 398, 3rd edit. 1840. 2. The long â is often changed into ǽ; as Lâr *lore*, lǽran *to teach*, ân *one*, ǽnig *any*. 3. The following is a precise summary from Grimm of the prefixed â-, long or accented. The prefixed â- is long because it is a contraction and represents the preposition æf *of, off, from, away, out of*, or the preposition on *on, in, upon, into*, or as the *Lat.* in and *Eng.* un; as â-dúne for æf-dúne, â-wendian for æf-wendian, â-drǽdan for on-drǽdan, â-gean for on-gean, â-týnan *to unshut, open*, Ps. Spl. 38, 13, for on-týnan, un-týnan *to open*. Â, as an inseparable particle, is long because it represents the inseparable prefixed particles ar, ur, ir, in *O. H. Ger.* and *O. Sax.* commonly expressing the meaning of the Latin prepositions ab, ex, ad, etc: *A. Sax.* â-hebban, *O. H. Ger.* ur-hefan *elevare*; *A. Sax.* â-fyllan, *O. H. Ger.* ar-fullan *implere*; *A. Sax.* â-beran, *O. H. Ger.* ar-peran *ferre, efferre*; *A. Sax.* â-scínan, *O. H. Ger.* ir-scínan *clarescere*. The peculiar force which this particle imparts to different verbs may correspond (1) to the Latin ex *out*, as â-gangan *to go out*; exire: (2) to the English *up*, as â-hleápan *to leap up*; exsilire: â-fyllan *to fill up*; implere: (3) it expresses the idea of an origin, *becoming, growing*, â-blacian *to blacken, to become black*; â-heardian *to grow hard*: (4) it corresponds to the Latin *re,* as â-geban *reddere*, â-lôsian *redimere*, â-sêcan *requirere*: (5) it is often used merely to render a verb transitive, or to impart a greater force to the transitive meaning of the simple verb,—â-beódan *offerre*, â-ceapian *emere*, â-lecgan *ponere*, â-sleán *occidere*: (6) it is used with intransitive verbs, where it has hardly any meaning, unless it suggests the *commencement* or *beginning* of the action, as â-hleahan *ridere*, â-sweltan *mori*: (7) it expresses the *end, aim*, or *purpose* of an action, as â-dômian *condemnare*, â-biddan *deprecari*, â-wirþan *perire*. But, after all, it must be borne in mind, that the various shades of its meaning are innumerable, and that, even in one and the same compound, it often assumes different meanings. For further illustration we must therefore refer to the compounds in which it occurs, Grm. ii. 818–832. I have, in justice to Grimm, given his motives for marking the prefixed â- long: I believe, however, it is short. See B. 6.

-a, affixed to words, denotes *A person, an agent*, or *actor*, hence, *All nouns ending in* -a *are masculine, and make the gen. in* an; as from Cum *come* [*thou*], cuma *a person who comes*, or *a guest*: Swíc *deceive* [*thou*], swíca *a traitor*: Worht *wrought*, wyrhta *a workman, wright*: Fôregeng *foregoing*, fôregenga *a foregoer*: Beád or gebêd *a supplication, praying*, beáda *a person who supplicates* or *prays*: Bytl *a beetle* or *hammer*, bytla *a hammerer, builder*. Some *abstract nouns*, and words denoting *inanimate things*, end in -a; and these words, having the same declension as those which signify *Persons* or *actors*, are masculine; as Hlísa, an; *m. fame*: Tíma, an; *m. time*: Líchama, an; *m. a body*: Steorra, an; *m. a star*: Gewuna, an; *m. a custom, habit*.

a; *prep. acc.* To, for; in:—A worlda world *to* or *in an age of ages*; in seculorum seculum, Ps. Th. 18, 8, = on worlda world, Ps. Lamb. 20, 5, = on worulda world, Ps. Th. 103, 6.

Â, aa, aaa; *adv.* Always, ever, for ever; hence the *O. Eng.* AYE, *ever*; semper, unquam, usque:—Ac â sceal ðæt wiðerwearde gemetgian

but ever must the contrary moderate, Bt. 21; Fox 74, 19. Ân God â on ēcnysse *one God to all eternity* [lit. *one God ever, in eternity*], Homl. Th. ii. 22, 32. Â on ēcnisse *usque in æternum*, Jos. 4, 7. Ic â ne geseah '*I not ever saw*' = *I never saw*, Cd. 19; Th. 24, 10; Gen. 375. Â = ǽfre: Nû, sceal beón â on Iî abbod *now, there shall always [ever] be an abbot in Iona*, Chr. 565; Th. 33, 2, col. 2. Nû, sceal beón ǽfre on Iî abbod *now, there shall ever [always] be an abbot in Iona*, Chr. 565; Th. 32, 11; 33, 4, col.1. He biþ aa [áá MS.] ymbe þæt ân *he is for ever about that one [thing]*, L. Th. ii. 310, 25. Aa on worulda woruld *semper in seculorum seculum*, Ps. Th. 105, 37. Nû and aaa [ááá MS.], to worulde bûton ǽghwilcum ende *now and ever, to a world without any end*, Bt. 42; Fox 260, 15. Â world *for ever*, Ex. 21, 6. Â forþ *ever forth, from thence*, Bt. Tupr. 303, 31. [The original signification seems to be a flowing, referring to time, which every moment flows on, hence *ever, always*, also to ǽ, eá *flowing water, a river*. In Johnston's Index Geog. there are nineteen rivers in Europe with the name of Aa = Â.]

â, *indecl; f. A law*; lex:—Dryhtnes â *the Lord's law*, Andr. Recd. 2387; An. 1196. vide Ǽ.

aac, e; *f. An oak* :—Aac-tûn *Acton Beauchamp, Worcestershire*, Cod. Dipl. 75; A. D. 727; Kmbl. i. 90, 19. v. Âc-tûn.

aad *a pile* :—He mycelne aad gesomnode *he gathered a great pile*, Bd. 3, 16; S. 542, 22. v. âd.

âǽðan *to lay waste*; vastare, Gen. 1280 : â ēðan, Cd. 64; Th. 77, 24. v. ēðan.

aam, es; *m. A reed of a weaver's loom*, Exon. 109 a; Th. 417, 22; Rä. 36, 8; Cod. Lugd. Grn. v. âm.

aar *honour* :—In aar naman *in honore nominis*, Bd. 2, 6; S. 508, note 43: 5, 11; S. 626, note 36. v. ÂR; f.

aaþ *an oath* :—He ðone aaþ gesæh *he saw the oath*, Th. Dipl. A.D. 825; p. 71, 12. v. Âþ.

a-bacan, ic -bace, ðû -bæcest, -bæcst, he'-bæceþ, -bæcþ, *pl*. -bacaþ; *p.* -bóc, *pl.* -bócon; *pp.* -bacen *To bake*; pinsere, coquere :—Se hlâf þurh fýres hætan abacen *the bread baked by the heat of the fire*, Homl. Pasc. Daye, A.D. 1567, p. 30, 8; Lisl. 4to, 1623, p. 4, 16; Homl. Th. ii. p. 268, 9.

a-bâd *expected, waited* :—And abâd swâ ðeáh seofon dagas *expectavitque nihilominus septem alios dies*, Gen. 8, 12. v. abîdan.

a-bǽd, abǽdon *asked*; *p.* of abiddan.

a-bǽdan; *p.* -bǽdde; *pp.* -bǽded *To restrain, repel, compel*; avertere, repellere, cogere, exigere :—Is fira ǽnig, ðe deáþ abǽde *is there any man, who can restrain death?* Salm. Kmbl. 957; Sal. 478. Ðæt oft wǽpen abǽd his mondryhtne *which often repels the weapon for its lord*, Exon. 114 a; Th. 437, 24; Rä. 56, 12. v. bǽdan.

a-bǽligan; *p.* ode; *pp.* od *To offend, to make angry*; irritare, offendere :—Sceal gehycgan hæleða ǽghwylc ðæt he ne abǽlige bearn waldendes *every man must be mindful that he offend not the son of the powerful*, Cd. 217; Th. 276, 27; Sat. 195. v. a-belgan, a-bylgan.

a-bǽr *bore* or *took away*; sustulit, Ps. Spl. 77, 76; *p.* of a-beran.

ABAL, afol, es; *n. Power of body, strength*; vigor, vires, robur corporis :—Dîn abal and cræft *thy strength and power*, Cd. 25; Th. 32, 9; Gen. 500. [*Orm.* afell : *O. H. Ger.* aval, n : *O. Nrs.* afl, n. *robur, vis* : *Goth.* abrs *strong* : *Grk.* ὄβρῐμος.]

a-bannan; *p.* -beónn, *pl.* -beónnon; *pp.* -bannen. **I.** *to command, order, summon*; mandare, jubere :—Abannan to beadwe *to summon to battle*, Elen. Grm. 34. **II.** *to publish, proclaim; with* ût *to order out, call forth, call together, congregate, assemble*; edicere, avocare, citare :—Aban ðû ða beornas ût of ofne *command thou the men out of the oven*, Cd. 193; Th. 242, 32; Dan. 428. Ðâ hēt se cyng abannan ût ealne þeódscipe *then the king commanded to order out [to assemble] all the population*, Chr. 1006; Erl. 140, 8. v. bannan.

a-barian; .*p.* ede; *pp.* ed [a, barian *to make bare*; bær, se bara; *adj.* bare] *To make bare, to manifest, discover, disclose*; denudare, prodere, in medium proferre :—Gif ðû abarast ûre sprǽce *si sermonem nostrum profers in medium*, Jos. 2, 20: R. Ben. Interl. 46: Cot. 80.

a-bât *bit, ate* :—He abât *he ate*, MS. Cott. Jul. E. vii. 237; Salm. Kmbl. 121, 15; *p.* of a-bîtan.

abbad, abbod, abbud, abbot, es; *m* : abboda, an; *m*. **I.** *an abbot*; abbās,—the title of the male superior of certain religious establishments, thence called abbeys. The word *abbot* appears to have been, at first, applied to any member of the clerical order, just as the French Père and English Father. In the earliest age of monastic institutions the monks were not even priests: they were merely religious persons, who retired from the world to live in common, and the abbot was one of their number, whom they elected to preside over the association. In regard to general ecclesiastical discipline, all these communities were at this early time subject to the bishop of the diocese, and even to the pastor of the parochial district within the bounds of which they were established. At length it began to be usual for the abbot to be in orders; and since the sixth century monks generally have been priests. In point of dignity an abbot is generally next to a bishop. A minute account of the different descriptions of abbots may be found in Du

Cange's Glossary, and in Carpentier's supplement to that work :—Se ârwurða abbad Albînus *the reverend abbot Albinus*, Bd. pref. Riht is ðæt abbodas fæste on mynstrum wunian *it is right that abbots dwell closely in their minsters*, L. I. P. 13; Th. ii. 320, 30. Hēr Forþrēd abbud forþfērde *in this year abbot Forthred died*, Chr. 803; Erl. 60, 13. Se abbot Saxulf *the abbot Saxulf*, Chr. 675; Ing. 50, 15. Swâ gebireþ abbodan *as becometh abbots*, L. Const. W. p. 150, 27; L. I. P. 13; Th. ii. 320, 35. **II.** *bishops were sometimes subject to an abbot, as they were to the abbots of Iona* :—Nû, sceal beón æfre on Iî abbod, and nâ biscop; and ðan sculon beón underþeódde ealle Scotta biscopas, forðan ðe Columba [MS. Columban] was abbod, nâ biscop *now, in Iî [Iona], there must ever be an abbot, not a bishop; and to him must all bishops of the Scots be subject, because Columba was an abbot, not a bishop*, Chr. 565; Th. 32, 10–16, col. 1. [*Laym.* abbed : *O. Frs.* abbete : *N. Ger.* abt : *O. H. Ger.* abbat : *Lat.* abbās; *gen.* abbātis *an abbot* : *Goth.* abba : *Syr.* אַבָּא *father, from Heb.* אָב ab *father, pl.* אָבוֹת abot *fathers*.] DER. abbad-dôm, -hâd, -isse, -rîce : abboda.

abbad-dôm *an abbacy*. v. abbud-dôm.

abbad-hâd *the state* or *dignity of an abbot*. v. abbud-hâd.

abbadisse, abbodisse, abbatisse, abbudisse, abedisse, an; *f.* [abbad *an abbot, isse a female termination, q. v.*] *An abbess*; abbatissa :—Riht is ðæt abbadissan fæste on mynstrum wunian *it is right that abbesses dwell closely in their nunneries*, L. I. P. 13; Th. ii. 320, 30; L. Const. W. 150, 21: Bd. 3, 8; S. 531, 14: Guthl. 2; Gdwin. 16, 22: Bd. 3, 11; S. 536, 38.

abbad-rîce *an abbacy*. v. abbod-rîce.

Abban dûn, e; *f. Abingdon, in Berkshire*, Chr. 985; Ing. 167, 5. v. Æbban dûn.

abbod *an abbot*, L. I. P. 13; Th. ii. 320, 30. v. abbad.

abboda; *m. An abbot*; abbas :—Swâ gebireþ abbodan *as becometh abbots*, L. I. P. 13; Th. ii. 320, 35. v. abbad.

abbod-rîce, abbot-rîce, es; *n. The rule of an abbot, an abbacy*; abbatia :—On his tíme wæx ðæt abbodrîce swîðe rîce *in his time the abbacy waxed very rich*, Chr. 656; Ing. 41, 1. On ðis abbotrîce *in this abbacy*, Chr. 675; Ing. 51, 12.

abbodysse *an abbess*, Guthl. 2; Gdwin. 16, 22. v. abbadisse.

abbot *an abbot*, Chr. 675; Ing. 50, 15. v. abbad.

abbud *an abbot*, Chr. 803; Erl. 60, 13: Bd. 5, 23; S. 645, 14. v. abbad.

abbud-dôm, es; *m.* [= abbod-rîce, q. v.] *An abbacy, the rule* or *authority of an abbot*; abbātia, abbātis jus *vel* auctoritas :—Abbuddômes, *gen.* Bd. 5, 1; S. 613, 18. Abbuddôme, *dat.* 5, 21; S. 642, 37.

abbud-hâd, es; *m. The state* or *dignity of an abbot*; abbatis dignitas :—Munuchâd and abbudhâd ne syndon getealde to ðysum getele *monkhood and abbothood are not reckoned in this number*, L. Ælf. C. 18; Th. ii. 348, 31.

abbudisse, an; *m. An abbess* :—Ðâ sealde seó abbudisse him sumne dǽl ðære moldan *tunc dedit ei abbatissa portiunculam de pulvere illo*, Bd. 3, 11; S. 536, 38. v. abbadisse.

a-beág *bowed down*, Beo. Th. 1555; B. 775; *p.* of a-bûgan.

a-bealh *angered*, Cd. 222; Th. 290, 4; Sat. 410. v. a-belgan.

a-beátan; *p.* -beót; *pp.* -beáten *To beat, strike*; tundere, percellere :—Stormum abeátne *beaten by storms*, Exon. 21 b; Th. 58, 26; Cri. 941. v. beátan.

a-beden *asked*, Nicod. 12; Thw. 6, 15: Bd. 4, 10; S. 578, 31; *pp.* of a-biddan.

abedisse, an; *f. An abbess*; abbatissa :—Ðære abedissan betæhton *committed to the abbess*, Chr. 1048; Erl. 181, 28. v. abbadisse.

a-bêgan; *p.* de; *pp.* ed; *v. trans. To bend, bend down, bow, reduce, subdue*; incurvare, redigere, subigere :—Weorþe heora bæc swylce abêged eác dorsum illorum semper incurva, Ps. Th. 68, 24: Chr. 1073; Erl. 212, 1: 1087; Th. 356, 10. v. bêgan.

a-bêgendlîc; *adj. Bending*; flexibilis, Som. v. a-bêgan.

a-behôfian; *p.* ode *To behove, concern*; decere :—Mid mâran unrǽde ðone him abehôfode *with more animosity than it behoved him*, Chr. 1093; Th. 360, 4. v. be-hôfian.

a-belgan, ic -belge, ðû -bilgst, -bilhst, he -bylgþ, -bilhþ, *pl.* -belgaþ; *p.* -bealg, -bealh, *pl.* -bulgon; *pp.* -bolgen, *v. trans.* [a, belgan *to irritate*] *To cause any one to swell with anger, to anger, irritate, vex, incense*; ira aliquem tumefacere, irritare, exasperare, incendere :—Ne sceal ic ðē abelgan *I would not anger thee*, Salm. Kmbl. 657; Sal. 328. Oft ic wífe abelge *oft I irritate a woman*, Exon. 105 b; Th. 402, 20; Rä. 21, 32. He abilhþ Gode *he will incense God*, Th. Dipl. 856; 117, 20. Ic ðē abealh *I angered thee*, Cd. 222; Th. 290, 4; Sat. 410: Beo. Th. 4550; B. 2280. God abulgan *Deum exacerbaverunt*, Ps. Th. 77, 41: Ex. 32, 29. Nû hig me ablugon habbaþ *irascatur furor meus contra eos*, Ex. 32, 10. He him abolgen wurþeþ *he will be incensed against them*, Cd. 22; Th. 28, 4; Gen. 430. Wæs swýðe abolgen *erat graviter offensus*, Bd. 3, 7; S. 530, 8.

a-beódan; *p.* -beád; *pp.* -boden; *v. a.* [a, beódan *to order*] *To announce, relate, declare, offer, command*; referre, nuntiare, annuntiare,

edicere, offerre, jubere:—Ðæt he wolde ðæt ǽrende abeódan *that he would declare the errand*; Ors. 4, 6; Bos. 86, 20: Cd. 91; Th. 115, 14; Gen. 1919: 200; Th. 248, 9; Dan. 510.

a-beófian *To be moved* or *shaken, to tremble*; moveri, contremere:—Ealle abeofedan eorþan staðelas *movebuntur omnia fundamenta terræ*, Ps. Th. 81, 5. v. beofian.

a-beornan; *p.* -bearn, -barn, *pl.* -burnon; *pp.* -bornen, *v. intrans.* *To burn*; exardere:—Fýr abarn *exarsit ignis*, Ps. Th. 105, 16. v. beornan.

a-beran; *p.* -bær; *pp.* -boren. I. *to bear, carry, suffer*; portare, ferre:—Ðe man aberan ne mæg *which they are not able to bear*, Mt. Bos. 23, 4. Hí ne mágon nán earfoða aberan *they cannot bear any troubles*, Bt. 39, 10; Fox 228, 3: Andr. Kmbl. 1912; An. 958: Ps. Th. 54, 11. II. *to take* or *carry away*; tollere, auferre:—Abær hine of eowdum sceápa *sustulit eum de gregibus ovium*, Ps. Spl. 77, 76: Ps. Grn. 50, 12. v. beran.

a-berd, -bered; *adj.* *Sagacious, crafty, cunning*; callidus, Wrt. Voc. 47, 36: Lchdm. iii. 192, 10: 188, 26: 186, 17.

a-berend-líc; *adj.* [berende *bearing*] *Bearable, tolerable, that may be borne*; tolerabilis:—Aberendlíc broc *bearable affliction*, Bt. 39, 10; Fox 228, 4, note 5.

a-berstan; *p.* -bearst, *pl.* -burston; *pp.* -borsten [a, berstan] *To burst, break, to be broken*; perfringi. v. for-berstan.

a-bet; *adv. Better*; melius:—Hwæðer ðé se ende abet lícian wille *whether the end will better please thee*, Bt. 35, 5; Fox 166, 23. v. bet.

a-beþecian; *subj.* ðú abeþecige; *p.* ode; *pp.* od [be, þeccan *to cover*] *To uncover, detect, find hidden, to discover, disclose*; detegere:—Búton ðú hit forstele oððe abeþecige *unless thou steal it, or find (it) hid*, Bt. 32, 1; Fox 114, 9.

a-bicgan; *p.* -bohte; *pp.* boht; *v. a.* [a, bycgan *to buy*] *To buy, pay for, recompense*; emere, redimere:—Gif fríman wið fríes mannes wíf geligeþ, his wérgelde abicge *if a freeman lie with a freeman's wife, let him buy her with his wergeld*, i. e. *price*, L. Ethb. 31; Th. i. 10, 7. v. a-bycgan.

a-bídan, ic -bíde, ðú -bídest, -bítst, -bíst, he -bídeþ, -bít, *pl.* -bídaþ; *p.* -bád, *pl.* -bidon; *pp.* -biden; *v. intrans.* *To* ABIDE, *remain, wait, wait for, await*; manere, sustinere, expectare:—Hý abídan sceolon in sin-nihte *they must abide in everlasting night*, Exon. 21 b; Th. 99, 28; Cri. 1631. Hér sculon abídan bán *here the bones shall remain*, 99 a; Th. 370, 18; Seel. 61. Abád swá ðeáh seofon dagas *expectavit nihilominus septem alios dies*, Gen. 8, 12. We óðres sceolon abídan *alium expectamus*? Mt. Bos. 11, 3. Ic abád [anbídode Spl.] hælu ðíne *expectabam salutare tuum*, Ps. Surt. 118, 166. Sáwla úre abídyþ Driht *anima nostra sustinet Dominum*, Ps. Spl. C. 32, 20. Windes abídon *ventum expectabant*, Bd. 5, 9; S. 623, 19. Ðær abídan sceal maga miclan dómes *there the being [Grendel] shall await the great doom*, Beo. Th. 1959; B. 977: Exon. 115 b; Th. 444, 27; Kl. 53. [Laym. abiden; *p.* abad, abed, abeod, abod, abaod, abide, *pl.* abiden.] v. bídan.

a-biddan, ic -bidde, ðú -bidest, -bitst, he -bit, -byt, -bitt, *pl.* -biddaþ; *p.* -bæd, *pl.* -bædon; *pp.* -beden *To ask, pray, pray to, pray for, obtain by asking* or *praying*; petere, precari, postulare, exorare, impetrare:—Wilt tú wit unc abiddan *drincan vis petamus bibere*? Bd. 5, 3; S. 616, 30. Abiddaþ [Cott. biddaþ] hine eáþmódlíce *pray to him humbly*, Bt. 42; Fox 258, 21. Se ðe hwæt to lǽne abit *qui quidquam mutuo postulaverit*, Ex. 22, 14. Ne mihte ic lýfnesse abiddan *nequaquam impetrare potui*, Bd. 5, 6; S. 619, 8. Ða sendon hý tuá heora ǽrendracan to Rómánum æfter friðe; and hit abiddan ne mihtan *then they sent their ambassadors twice to Rome for peace; and could not obtain it*, Ors. 4, 7; Bos. 87, 39. He abiddan mæg ic ðé lǽte duguða brúcan *he may obtain by prayer that I will let thee enjoy prosperity*, Cd. 126; Th. 161, 5; Gen. 2660. v. biddan.

a-bifian, -bifigan; *p.* ode, ede; *pp.* od, ed *To be moved* or *shaken, to tremble*; moveri, contremere:—For ansýne écan Dryhtnes ðeós eorþe sceal eall abifigan *a facie Domini mota est terra*, Ps. Th. 113, 7. v. bifian.

a-bilgþ, a-bilhþ *anger, an offence*. v. a-bylgþ.

a-biran *to bear, carry*; portare, Bd. 1, 27; S. 491, 31. v. a-beran.

a-bísegien *should prepossess*, Bt. 35, 1; Fox 154, 32. v. abýsgian.

a-bit *prays*, Ex. 22, 14; *pres.* of a-biddan.

a-bítan, ic -bíte, ðú -bítest, -bítst, he -bíteþ, -bít, *pl.* -bítaþ; *p.* -bát, *pl.* -biton; *pp.* -biten; *v. a.* *To bite, eat, consume, devour*; mordere, arrodere, mordendo necare, comedere, devorare:—Gif hit wildeór abítaþ, bere forþ ðæt abitene and ne agife *si comestum a bestia, deferat ad eum quod occisum est, et non restituet*, Ex. 22, 13. He abát his suna *he ate his children*, Salm. Kmbl. p. 121, 15. Ðæt se wód-freca were-wulf tó fela ne abíte of godcundre heorde *that the ferocious man-wolf devour not too many of the spiritual flock*, L. I. P. 6; Th. ii. 310, 31. Míne scép sind abitene *my sheep are devoured*, Homl. Th. i. 242, 10. Ðú his ne abítst *non comedas ex eo*, Deut. 28, 31. v. bítan.

a-biterian, -bitrian; *p.* ode; *pp.* od *To make sour* or *bitter*; exacerbare. v. biterian, biter *bitter*.

a-bi-tweónum; *prep. dat. Between*; inter:—Ic wiht geseah horna abitweónum [hornum bitweónum, Grn; Th.] húðe lǽdan *I saw a creature bringing spoil between its horns*, Exon. 107 b; Th. 411, 19; Rä. 30, 2. [Sansk. abhi: Zend aibi.] v. bi-tweónum.

a-blácian, -blácigan; *p.* ode; *pp.* od *To be* or *look pale, grow pale*; pallere, obrigescere:—Ablácodon *obriguerunt*, Ex. 22, 16? Lye. Ic blácige *palleo*, Ælfc. Gr. 26, 2; Som. 28, 42. Blácian from blícan, *p.* blác *to shine*: blǽcan *to bleach, whiten, fade*. Observe the difference between blác, blǽc *pallid, bleak, pale*, and blæc, blaces, se blaca *black, swarthy*. DER. blácian *pallere*.

a-blǽcan; *p.* -blǽhte; *pp.* -blǽht [a, blǽcan *to bleach*] *To bleach, whiten*; dealbare, Ps. Vos. 50, 8: 67, 15.

a-blǽcnes, -ness, e; *f. A paleness, gloom*; pallor, Herb. 164; Lchdm. i. 294, 3, note 6. v. æ-blǽcnys.

a-blǽndan *to blind, deaden, benumb*. v. ablendan.

a-blann *rested*; *p.* of a-blinnan *to leave off*.

a-bláwan; *p.* -bleów; *pp.* -bláwen *To blow, breathe*; flare, efflare:—On ableów *inspiravit*, Gen. 2, 7. Út abláwan *to breathe forth*, Hexam. 4; Norm. 8, 20. Nǽfre mon ðæs hlúde býman abláweþ *never does a man blow the trumpet so loudly*, Exon. 117 b; Th. 451, 27; Dóm. 110. God ðá geworhte mannan and ableów on his ansýne líflícne blǽd *God then made man and blew into his face the breath of life*, Hexam. 11; Norm. 18, 25.

a-bláwung, e; *f. A blowing*. v. bláwung.

a-blend, se a-blenda; *adj. Blinded*; cæcatus:—Wénaþ ða ablendan mód *the blinded minds think*, Bt. 38, 5; Fox 206, 6. v. *pp.* of a-blendan.

a-blendan; *p.* -blende, *pl.* -blendon; *pp.* -blended, -blend; *v. a.* *To blind, make blind, darken, stupify*; cæcare:—Ða gyldenan stánas ablendaþ ðæs módes eágan *the golden stones blind the mind's eyes*, Bt. 34, 8; Fox 144, 34. Swá bióþ ablend *so are blinded*, 38, 5; Fox 206, 1. Ic sýne ablende bealo-þoncum *I blinded their sight by baleful thoughts*, Exon. 72 b; Th. 270, 22; Jul. 469. He ablende hyra eágan *excæcavit oculos eorum*, Jn. Bos. 12, 40. Ablended in burgum *blinded as I am in these dwellings*, Andr. Kmbl. 155; An. 78. Wæs ablend *was blinded*, Mk. Bos. 6, 52: Num. 14, 44. v. blendan.

a-bleóton *sacrificed*; *p. pl.* of a-blótan.

a-bleów *blew*; *p.* of a-bláwan.

a-blícan; *p.* -blác, *pl.* -blicon; *pp.* -blicen; *v. n. To shine, shine forth, to appear, glitter, to be white, to astonish, amaze*; dealbari, micare:—Sóþlíce on rihtwísnysse ic ablíce *ego autem in justitia apparebo [micabo]*, Ps. Spl. T. 16, 17. Ofer snáw ic beó ablicen *super nivem dealbabor*, Ps. Spl. 50, 8.

a-blícgan; *p.* ede; *pp.* ed *To shine, to be white, to astonish*; consternare:—Ic eom ablícged *consternor*, Ælfc. Gr. 37; Som. 39, 42.

a-blignys, -nyss, e; *f. An offence*. v. a-bylgnes.

a-blindan *to blind*, Abus. 1, Lye. v. a-blendan.

a-blinnan; *p.* -blann, *pl.* -blunnon; *pp.* -blunnen *To cease, desist*; cessare, desistere, Ps. Spl. 36, 8: Bd. 4, 1; S. 563, 16.

a-blisian; *p.* ode; *pp.* od *To blush*; erubescere:—Óþ eówre lyþre mód ablísige *donec erubescat incircumcisa mens eorum*, Lev. 26, 41.

a-blótan; *p.* -bleót, *pl.* -bleóton; *pp.* -blóten *To sacrifice*; immolare. v. blótan.

a-blýsgung, -blýsung, e; *f. The redness of confusion, shame*; pudor, R. Ben. 73.

a-boden *told*; *pp.* of a-beódan *to bid, tell*.

a-bogen *bowed*; *pp.* of a-búgan, -beógan *to bow, bend*.

a-boht *bought*; *pp.* of a-bicgan *to buy*.

a-bolgen *angered*, Ex. 32, 10; *pp.* of a-belgan *to offend, anger*.

a-boren *carried*; *pp.* of a-beran *to bear*.

a-borgian; *p.* ode; *pp.* od *To be surety, to undertake for, to assign, appoint*; fidejubere:—Gif he nite hwá hine aborgie, hæfton hine *if he know not who will be his borh, let them imprison* [lit. *have, detain*] *him*, L. Ath. i. 20; Th. j 210, 8.

a-bracian; *p.* ode; *pp.* od *To engrave, emboss*; cælare:—Abracod cælatum, Cot. 33.

a-bradwian *To overthrow, slay, kill*; prosternare, occidere, Beo. Th. 5232; B. 2619. v. a-bredwian.

a-bræc *broke*; *p.* of a-brecan *to break*.

a-bræd, -brægd *drew*, Mt. Bos. 26, 51; *p.* of a-bredan, a-bregdan *to move, drag, draw*.

a-breát; *p.* -breót, *pl.* -breóton *To break, kill*; frangere, concidere, necare:—Abreót brim-wísan, brýd aheorde *slew the sea-leader, set free his bride*, Beo. Th. 5852; B. 2930. v. a-breótan.

a-brecan; *p.* -brece, ðú -bricst, he -bricþ; *p.* -bræc, *pl.* -brǽcon; *pp.* -brocen *To break, vanquish, to take by storm, to assault, destroy*; frangere, effringere, expugnare:—Abrecan ne meahton reced *they might not break the house*, Cd. 115; Th. 150, 14; Gen. 2491. He Babilone abrecan wolde *he would destroy Babylon*, Cd. 209; Th. 259, 10; Dan. 689. Hú ǽnig man mihte swylce burh abrecan *how any man could take such a town*, Ors. 2, 4; Bos. 44, 16. DER. brecan.

a-bredan, he -brit -brideþ, -bret -bredeþ; *p.* -bræd, *pl.* -brudon; *pp.* -broden; *v. a. To move quickly, remove, draw, withdraw*; vibrare, destringere, eximere, retrahere:—Abræd hys swurd, *exemit gladium suum*, Mt. Bos. 26, 51. Gif God abrit *if God remove*, Bt. 39, 3; Fox

216, 5. Of môde abrit ðæt micle dysig *he removes from his mind that great ignorance*, Bt. Met. Fox 28, 155; Met. 28, 78. Hond up abræd *he raised his hand*, Beo. Th. 5144; B. 2575. Lâr Godes is abroden of breóstum *the knowledge of God is withdrawn from your breasts*, Cd. 156; Th. 194, 31; Exod. 269. v. bredan.

a-bredwian; *p.* ade; *pp.* ad *To overthrow, slay? kill?* prosternare? occidere?—Ðeáh ðe he his bróðor bearn abredwade [abradwade Th.] *although he had overthrown [exiled? killed?] his brother's child*, B. 2619.

a-brêgan; *p.* de; *pp.* ed *To alarm, frighten*; terrere :—Mec mæg grîma abrêgan *a phantom may frighten me*, Exon. 110 b; Th. 423, 7; Rä. 41, 17. Abrêgde, *p.* Bd. 3, 16; S. 543, 12; Ps. Spl. T. 79, 14.

a-bregdan; *p.* -brægd, *pl.* -brugdon; *pp.* -brogden *To move quickly, vibrate, remove, draw from, withdraw*; vibrare, destringere, eximere, retrahere :—Ðê abregdan sceal deáþ sâwle ðîne *death shall draw from thee thy soul*, Cd. 125; Th. 159, 22; Gen. 2638. Hwonne of heortan hunger oððe wulf sâwle and sorge abregde *when from my heart hunger or wolf shall have torn both soul and sorrow*, 104; Th. 137, 22; Gen. 2277. Hine of gromra clommum abrugdon *they drew him from the clutches of the furious*, 114; Th. 150, 4; Gen. 2486. v. bregdan.

â-brêmende *ever-celebrating*, Exon. 13 a; Th. 24, 20; Cri. 387. v. brêman.

a-breótan; *p.* -breát, *pl.* -bruton; *pp.* -broten *To bruise, break, destroy, kill*; frangere, confringere, concidere, necare :—Billum abreótan *to destroy with bills*, Cd. 153; Th. 190, 14; Exod. 199. Yldo beám abreóteþ *age breaks the tree*, Salm. Kmbl. 591; Sal. 295. Hine seó brimwylf abroten hæfde *the sea-wolf had destroyed him*, Beo. Th. 3203; B. 1599. Stânum abreótan *lapidare*, Elen. Kmbl. 1017; El. 510.

a-breóðan; *p.* -breáþ, *pl.* -bruðon; *pp.* -broðen *To unsettle, ruin, frustrate, degenerate, deteriorate*; perdere, degenerare :—Hæleþ oft hyre hleór abreóðeþ *a man often unsettles her cheek*, Exon. 90 a; Th. 337, note 18; Gn. Ex. 66. Abreóðe his angin *he frustrated his enterprise*, Byrht. Th. 138, 59; By. 242. Hî abruðon ða ðe he toþohte *they frustrated that which he had thought of*, Chr. 1004; Ing. 178, 1. Eálâ ðû abroðene folc *degener O populus*, Ælfc. Gr. 8; Som. 8, 10. *Hic et hæc et hoc nugas* ðæt is abroðen on Englisc, Ælfc. Gr. 9, 25; Som. 11, 2.

abret, abrit *takes away*, Bt. 39, 3; Fox 216, 5. v. abredan.

a-brocen *broken.* v. a-brecan.

a-broden, a-brogden *opened, freed, taken away.* v. abredan, abregdan.

abrotanum = ἀβρότονον *southernwood*, Herb. 135; Lchdm. i. 250, 16. v. sûðerne-wudu.

a-broten? *crafty, silly, sluggish*; vafer, fatuus, socors :—Abroten vel dwæs *vafer*, Ælfc. Gl. 9; Som. 56, 114. Abroten? *for* abroðen.

a-broðen *degeneratus*; *pp.* of a-breóðan.

a-broðennes, -ness, e; *f.* *Dulness, cowardice, a defect, backsliding*; ignavia, pusillanimitas. DER. a-broðen.

a-brugdon *withdrew*, Cd. 114; Th. 150, 4; Gen. 2486; *p. pl.* of a-bregdan.

a-bruðon *frustrated*, Chr. 1004; Ing. 178, 1; *p. pl.* of a-breóðan.

a-bryrdan; *p.* -bryrde; *pp.* -bryrded, -bryrd, *v. trans. To prick, sting, to prick in the heart, grieve*; pungere, compungere :—Nâ ic ne beó abryrd, God mîn *non compungar, Deus meus*, Ps. Spl. 29, 14. v. bryrdan.

a-bryrdnes, -ness, e; *f.* *Compunction, contrition*; compunctio, contritio. v. bryrdnys, a-byrdan.

a-brytan; *p.* -brytte; *pp.* -brytt *To destroy*; exterminare, Ps. Spl. C. 36, 9. v. brytan.

a-bûfan; *adv.* [a + be + ufan] ABOVE; supra :—Swâ wæ ǽr abûfan sǽdan *as we have before above said*, Chr. 1090; Th. 358, 15. DER. bûfan.

a-bûgan; *p.* -beág, -beáh, *pl.* -bugon; *pp.* -bogen *To bow, bend, incline, withdraw, retire*; se vertere, declinare, inclinare, averti :—Abûgaþ eádmôdlîce *inclinate suppliciter*, Coll. Monast. Th. 36, 3. Ac ðê firina gehwylc feor abûgeþ *but from thee each sin shall far retire*, Exon. 8 b; Th. 4, 22; Cri. 56. Ðǽr fram sylle abeág medu-benc monig *there many a mead-bench inclined from its sill*, Beo. Th. 1555; B. 775. v. bûgan.

a-bulgan = abulgon *angered*, Ps. Th. 77, 41; *p.* of a-belgan.

a-bunden *ready*; expeditus, Cot. 72; *pp.* of a-bîndan. v. bîndan.

a-bûtan, -bûton; *prep. acc.* [a + be + ûtan] ABOUT, *around, round about*; circa :—Ðû tæcst Israhela folce abûtan ðone mûnt *thou shalt take the people of Israel around the mountain*, Ex. 19, 12. Abûton hî *circa eos*, Mk. Bos. 9, 14. Abûton stân *about a stone*, L. N. P. L. 54; Th. ii. 298, 16.

a-bûtan, -bûton; *adv.* ABOUT; circa :—Besæt ðone castel abûtan *beset the castle about*, Chr. 1088; Th. i. 357, 29. Besǽton ðone castel abûton *they beset the castle about*, Chr. 1090; Th. i. 358, 25.

a-bycgan, -bicgan; *p.* -bohte, *pl.* -bohton; *pp.* -boht [a, bycgan *to buy, procure*]. I. *to buy, pay for*; emere, redimere, L. Ethb. 31; Th. i. 10, 7. II. *to perform, execute*; præstare :—Aþ abycgan *jusjurandum præstare*, L. Wih. 19; Th. i. 40, 18.

a-byffan; *p.* ode; *pp.* od *To mutter*; mutire, Cot. 134. v. byffan.

a-bŷgan, *v. trans. To bow, bend*; incurvare, Grm. ii. 826. v. a-bêgan.

a-bŷgendlíc; *adj. Bending, flexible*; flexibilis. DER. un-abŷgendlíc.

a-bylgan, -byligan, -bylgean; *p.* de; *pp.* ed *To offend, anger, vex*; offendere, irritare, exacerbare :—Hî hine oft abylgdon [MS. -dan] *ipsi sæpe*

---

exacerbaverunt eum, Ps. Th. 105, 32. Ða môd abylgeaⁿ ûra ðara nŷhstena *animos proximorum offendere*, Bd. 3, 19; S. 548, 17; Hy. 6, 22. v. a-belgan.

a-bylg-nes, æ-bylig-nes, æ-bylig-nys, -ness, e; *f.* [abylgan *to offend*] *An offence, scandal, anger, wrath, indignation*; offensa, ira, indignatio :—He him abylgnesse oft gefremede *he had oft perpetrated offence against him*, Exon. 84 a; Th. 317, 25; Môd. 71.

a-bylgþ, -bilgþ, -bilhþ, e; *f.* *An offence, wrong, anger*; offensa, injuria, ira :—He sceal Cristes abilgþe wrecan *he ought to avenge offence to Christ*, L. Eth. 9, 2; Th. i. 340, 13; L. Pen. 16; Th. ii. 284, 6. v. æ-bylgþ.

a-byligd, e; *f.* *Anger*; indignatio, Ps. Th. 77, 49. v. a-bylgþ.

a-byrgan, -byrgean, -byrian *To taste*; gustare :—We cŷðaþ eów ðæt God ælmihtig cwæþ his âgenum mûðe, ðæt nân man he môt abyrgean nânes cynes blôdes. Ælc ðæra ðe abyrgþ blôdes ofer Godes bebod sceal forwurþan on ëcenysse *we tell you that God Almighty said by his own mouth, that no man may taste any kind of blood. Every one who tastes blood against God's command shall perish for ever*, Homl. intitul. Hêr is hælwendlíc lâr, Bibl. Bodl. MSS. Junii 99, fol. 68. Se wulf for Gode ne dorste ðæs hæfdes abyrian *the wolf durst not, for God, taste the head*, Homl. Brit. Mus. MSS. Cot. Julius, E. 7, fol. 203, Bibl. Bodl. MSS. Bodley 343. v. byrgan.

a-bŷsgian, -bŷsgan, -bŷsean, -bîsegian; *p.* ode, ade; *pp.* od, ad [a, bŷsgian *to busy*] *To occupy, preoccupy, prepossess*; occupare :—Ðeáh uuþeáwas oft abîsegiaþ ðæt môd *though imperfections oft prepossess the mind*, Bt. 35, 1; Fox 154, 32. Biþ hyra seó swîþre symble abŷsgod ðæt hî unrihtes tiligeaþ *dextera eorum dextera iniquitatis*, Ps. Th. 143, 9. Biþ hyra seó swîþre symble abŷsgad *dextera iniquitatis*, 143, 13.

a-bŷsgung, -bîsgung, e; *f.* *Necessary business, employment*; occupatio, Past. 18, 1; Hat. MS. 25 a, 27, 29, 30.

a-bŷwan; *p.* de; *pp.* ed; *v. trans. To adorn, purify, clarify*; exornare, purgare :—Beóþ monna gǽstas beorhte abŷwde þurh bryne fŷres *the souls of men are brightly adorned [clarified] through the fire's heat*, Exon. 63 b; Th. 234, 24; Ph. 545. v. bŷwan.

AC, ach, ah, oc; *conj.* I. *but*; sed :—Ne com ic nâ towurpan, ac gefyllan *non veni solvere, sed adimplere*, Mt. Bos. 5, 17. Brytwalas fultumes bǽdon wið Peohtas, ac hî næfdon nænne *the Brito-Welsh begged assistance against the Picts, but they had none*, Chr. 443; Erl. 11, 34. II. *for, because*; nam, enim, quia :—Ne se aglǽca yldan þôhte, ac he gefêng hraðe slǽpendne rinc *nor did the wretch mean to delay, for he quickly seized a sleeping warrior*, Beo. Th. 1484; B. 740. Ðû ne þearft onsittan wîge, ac nê-fuglas [wig, eácne MS.] blôdig sittaþ þicce gefylled *thou needest not oppress with war, because carrion birds sit bloody quite satiated* (lit. *thickly filled*), Cd. 98; Th. 130, 12; Gen. 2158. III. *but also, but yet*; sed etiam, sed et, sed tamen :—Nâ læs weoruld men, ac eác swylce ðæt Drihtnes eowde *not only men of the world, but also* [sed etiam Bd.] *the Lord's flock*, Bd. 1, 14; S. 482, 25. Ða cwican nô genihtsumedon ðæt hî ða deádan bebyrigdon, ac hwæðere ða ðe lifigende wæron nôht dôn woldon *the living were not sufficient to bury the dead, but yet those who were living would do nothing*, Bd. 1, 14; S. 482, 32 : 2, 7; S. 509, 13. Ac swylce tunge mîn ælce dæge smeáþ rightwîsnysse ðîne *sed et lingua mea tota die meditabitur justitiam tuam*, Ps. Spl. 70, 26. [R. Glouc. Orm. ac : Laym. ac, æc, ah : Scot. ac : O. Sax. ak : O. H. Ger. oh : Goth. ak.]

ac; *adv. interrogative, Why, whether*; nonne, numquid :—Ðâ ðû gehogodest sæcce sêcean, ac ðû gebettest mǽrum þeódne *when thou resolvedst to seek warfare, hadst thou compensated the great prince?* Beo. Kmbl. 3976; B. 1990. Ac [ah MS.] ætfileþ ðê seld unrihtwîsnesse *numquid adhæret tibi sedes iniquitatis?* Ps. Surt. 93, 20. Ac hwâ dêmeþ *who shall judge?* Salm. Kmbl. 669; Sal. 334. Ac forhwon fealleþ se snâw *why falleth the snow?* 603; Sal. 301.

ac-, v. ag-, ag-lǽca, ah-, ah-lǽca.

ÁC, æc; *g.* e; *f.* I. *an* OAK; quercus, robur :—Ðeós âc *hæc quercus*, Ælfc. Gr. 8; Som. 7, 46. Sume âc astâh *got up into an oak*, Homl. Th. ii. 150, 31. *acc.* Âc *an oaken ship*, Runic pm. 25; Kmbl. 344, 21. Geongre âce *of a young oak*, L. M. I, 38; Lchdm. ii. 98, 9. Of ðære âc [*for* âce], Kmbl. Cod. Dipl. iii. 121, 22. II. âc; *g.* âces; *m. The Anglo-Saxon Rune* ᚪ=a, the name of which letter, in Anglo-Saxon, is âc *an oak*, hence, this Rune not only stands for the letter a, but, for âc *an oak*, as ᚪ byþ on eorþan elda bearnum flǽsces fôdor *the oak is on earth food of the flesh to the sons of men*, Hick. Thes. vol. i. p. 135; Runic pm. 25; Kmbl. 344, 15. Âcas twegen *two A's*, Exon. 112 a; Th. 429, 26; Rä. 43, 10. [R. Glouc. ôk : Chauc. ôk, âke, oak : O. Frs. êk : Dut. eek, eik : North Frs. ik : L. Ger. eke : N. Ger. eiche : M. Ger. eich : O. Ger. eih : Dan. eg : Swed. ek : O. Nrs. eik. Grn. starting from Goth. ayuk in âiw-dup, i. e. âiw-k-dup εἰς τὸν αἰῶνα, supposes a form ayuks, contracted to âiks, the equivalent of which would be âc, which would, therefore, indicate a tree of long durability.]

a-cǽgan *to name.* v. a-cîgan.

a-cænned = a-cenned *brought forth*; *pp.* of acennan.

a-cænnednys, -cænnys *nativity.* v. a-cennednes.

a-cærran *to avert;* acærred *averted.* v. a-cerran.

a-calan; *p.* -cōl, *pl.* -cōlon *To become cold;* algere, frigescere :—Nô acōl for ðȳ egesan *he never became cold for the terror,* Andr. Grm. 1267. v. calan.

ACAN; ic ace, ðū æcest, æcst, he æceþ, æcþ, *pl.* acaþ; *p.* ôc, *pl.* ôcon; *subj.* ic, ðū, he ace; *pp.* acen; *v. n.* To AKE, *pain;* dolere :— Gif mannes midrif [MS. midrife] ace *if a man's midriff ake,* Herb. 3, 6; Lchdm. i. 88, 11: Herb. Cont. 3, 6; Lchdm. i. 6; 3, 6. Acaþ mîne eágan *my eyes ake,* Ælfc. Gr. 36, MS. D; [mistiaþ=acaþ, Som. 38, 48]; dolent mei oculi, Mann. [*Laym. p.* oc: *R. Glouc. p.* ok: *Chauc.* ake: *N. L. Ger.* aken, æken.]

Âcan-tûn, es; *m.* [âcan=âcum, *pl. d.* of âc *an oak,* tûn *a town*] *Acton, Suffolk* :—Ðæt hit cymþ to Âcantûne; fram Âcantûne [MS. Âcyntûne] ðæt hit cymþ to Rigindûne *till it comes to Acton; from Acton till it comes to Rigdon,* Th. Diplm. A.D. 972; 525, 22–24. v. Âc-tûn, *and* ðæt *adv.*

acas, e; *f.* : acase, axe, an; *f. An axe;* securis :—Acas, Mt. Lind. Stv. 3, 10. Acase, Lk. Rush. War. 3, 9 [id. Lind. Acasa, *a Northumbrian form*]. Axe, Mt. Rush. Stv. 3, 10. v. æx.

âc-beám, es; *m. An oak-tree;* quercus, Ettm. p. 51.

âc-cærn, âc-corn *an acorn.* v. æcern.

accutian? *to prove;* probare :—Accuta me *proba me,* Ps. Spl. M. 138, 22.

âc-cyn, -cynn, es; *n.* [âc *oak,* cyn *kind*] *A species of oak;* ilex, Mann.

âc-drenc, -drinc, es; *m. Oak-drink, a kind of drink made of acorns;* potus ex quercus glandibus factus. v. âc, drenc.

ace *ake, pain.* DER. acan *to ake.* v. ece.

a-cealdian; *p.* ode; *v. intrans. To be* or *become cold;* algere, frigescere, Past. 58, 9. v. a-cōlian, calan.

a-ceápian; *p.* ode; *pp.* od *To buy.* v. ceápian.

a-cearfan *to cut off* :—Acearf *abscindet,* Ps. Spl. C. 76, 8. v. a-ceorfan.

a-cēlan; *p.* de; *v. intrans. To be* or *become cold;* algere, frigescere :—Ðæs þearfan ne biþ þurst acēled *the thirst of this desire is not become cold,* Bt. Met. Fox 7, 34; Met. 7, 17. v. cēlan, calan.

Acemannes burh, burg; *g.* burge; *d.* byrig, beri; *f.* : ceaster, cester; *g.* ceastre; *f.* [æce *ake,* mannes *man's,* ceaster *or* burh *city or fortress*] *Bath, Somersetshire* :—Hēr Eádgâr to rîce fēng at Acemannes byrig, ðæt is at Baðan *here,* A.D. 972, *Edgar took the kingdom at Akeman's burgh, that is at Bath,* Chr. 972; Th. 225, 18, col. 3. On ðære ealdan byrig, Acemannes ceastre; ac beornas Baðan nemnaþ *in the old burgh, Akeman's Chester; but men call it Bath,* Chr. 973; Ing. 158, 26. At Acemannes beri *at Akeman's bury,* Ing. 158, note g. v. Baðan.

acen *pained.* v. acan.

âcen *oaken.* v. æcen.

a-cennan, ðū -censt, he -cenþ; *p.* -cende; *pp.* -cenned; *v. a.* To bring forth, produce, beget, renew; parere, gignere, renovare, renasci :—Swâ wîf acenþ bearn *as a woman brings forth a child,* Bt. 31, 1; Fox 112, 2. On sârnysse ðū acenst cild *in dolore paries filios,* Gen. 3, 16. Ðâ se Hælend acenned wæs *cum natus esset Jesus,* Mt. Bos. 2, 1. Crist wæs acenned [MS. acennyd] on midne winter *Christ was born in mid-winter,* Menol. Fox 1; Men. 1. Gregorius wæs of æðelborenre mægþe acenned *Gregory was born of a noble family,* Homl. Th. ii. 118, 7. Eal edniwe, eft acenned, synnum asundrad *all renewed, born again, sundered from sins,* Exon. 59 b; Th. 214, 19; Ph. 241. Ðonne se mōna biþ acenned [geniwod, v. geniwian] *when the moon is changed [born anew],* Lchdm. iii. 180, 19, 22, 28. v. cennan.

a-cenned-lîc; *adj. Native;* nativus, Cot. 138.

a-cennednes, -cennes, -cennys, -cænnednys, -cænnys, -ness, e; *f. Nativity, birth, generation;* nativitas, ortus :—Manega on his acennednysse gefagniaþ *multi in nativitate ejus gaudebunt,* Lk. Bos. 1, 14: Ps. Spl. 106, 37.

a-ceócian? *p.* ode; *pp.* od *To choke;* suffocare. v. a-þrysman.

a-ceócung, e; *f. A consideration;* ruminatio, Wrt. Voc. 54, 62. v. a-ceósung.

a-ceorfan; *p.* -cearf, *pl.* -cūrfon; *pp.* -corfen *To cut off;* abscidere, succidere, concidere :—Of his ansȳne ealle ic aceorfe, ða ðe him feóndas syndon *concidam inimicos ejus a facie ipsius,* Ps. Th. 88, 20.

a-ceósan; *p.* -ceás, *pl.* -curon; *pp.* -coren *To choose, select;* eligere. DER. ceósan.

a-ceósung [MS. aceócung], e; *f. A consideration;* ruminatio, Wrt. Voc. 54, 62.

acer *a field,* Rtl. 145, 18. v. æcer.

a-cerran; *p.* -cerde; *pp.* -cerred *To turn, return;* vertere, reverti :—Úton acerran ðider ðær he sylfa sit, sigora waldend *let us turn thither where he himself sitteth, the triumphant ruler,* Cd. 218; Th. 278, 6; Sat. 217.

a-cerrednes, -ness, e; *f. An aversion.* v. a-cerran.

ach *but;* sed :—Ach ðæs weorodes eác *but of the host also,* Andr. Recd. 3182; An. 1594. v. ac; conj.

âc-hâl; *adj. Oak-whole* or *sound, entire;* roboreus, integer, Andr. Grm. 1700.

a-cigan; *p.* de; *pp.* ed *To call;* vocare, evocare :—Acîgde of corþre cyninges þegnas *he called the thanes of the king from the band,* Beo. Th. 6233; B. 3121. Sundor acîgde *called him alone, in private,* Elen. Kmbl. 1203; El. 603. Hine acîgde ût *evocavit eum,* Bd. 2, 12; S. 513, 19.

ac-lēc-cræft, es; *m.* [ac-lēc=ag-lēc *miseria,* cræft *ars*] *An evil art;* ars mala vel perniciosa :—Ðū ðē, Andreas, aclēccræftum lange feredes *thou, Andrew, hast long betaken thyself to evil arts,* Andr. Kmbl. 2724; An. 1364.

a-clēnsian; *p.* ode; *pp.* od *To cleanse, purify;* mundare :—Hyra nân næs aclēnsod, būton Naaman se Sirisca *nemo eorum mundatus est, nisi Naaman Syrus,* Lk. Bos. 4, 27.

Âc-leá=Âc-leáh; *g.* -leáge; *f.* [âc *an oak,* leáh *a lea, ley, meadow; acc.* leá=leáh, *q. v.*] *The name of a place, as Oakley* :—Sinoþ wæs gegaderod æt Âcleá *a synod was assembled at Acley* or *Oakley,* Chr. 789; Ing. 79, 14. Âcleá, Chr. 782; Erl. 57, 6: 851; Erl. 67, 26; 68, 3.

âc-leáf, es; *n. An oak-leaf;* quercus folium :—Âcleáf, Lchdm. iii. 311: L.M. 3, 8; Lchdm. ii. 312, 19.

a-cleopian; *p.* ode; *pp.* od *To call, call out;* clamare, exclamare. DER. cleopian, clypian.

aclian; *p.* ode; *pp.* od [acol, acl *excited by fear*] *To frighten, excite;* terrere, terrore percellere. DER. ge-aclian.

âc-melu, *g.* -meluwes; *n. Acorn-meal;* querna farina, L.M. 1, 54; Lchdm. ii. 126, 7.

âc-mistel, e; *f. Oak mistletoe;* quercus viscum :—Genîm âcmistel *take mistletoe of the oak,* L.M. 1, 36; Lchdm. ii. 88, 4.

a-cnyssan; *p.* ede; *pp.* ed *To expel, drive out;* expellere. v. cnyssan.

a-cofrian; *p.* ode; *pp.* od *To recover;* e morbo consurgere, convalescere :—Acofraþ *will recover,* Lchdm. iii. 184, 15.

acol, acul, acl; *adj. Excited, excited by fear, frightened, terrified, trembling;* agitatus, perterritus, pavidus :—Wearþ he on ðam egesan acol worden *he had through that horror become chilled, trembling,* Cd. 178; Th. 223, 24; Dan. 124. Forht on mōde, acul for ðam egesan *fearful in mood, trembling with dread,* 210; Th. 261, 14; Dan. 726. Acol for ðam egsan *trembling with terror,* Exon. 42 b; Th. 143, 20; Gū. 664. Forht and acol *afraid and trembling,* Cd. 92; Th. 117, 18; Gen. 1955. Wurdon hie ðâ acle *they then became terrified,* Andr. Kmbl. 2678; An. 1341. Fyrd-leóþ galan aclum stefnum *they sung a martial song with loud excited voices,* Cd. 171; Th. 215, 4; Exod. 578.

a-cōlian; *p.* ade, ode; *pp.* ad, od *To become cool, cold, chilled;* frigescere :—Ræst wæs acōlad *his resting-place was chilled,* Exon. 119 b; Th. 459, 28; Hö. 6. Ðonne biþ ðæt werge lîc acōlad *then shall be the accursed carcase cooled,* Exon. 100 a; Th. 374, 12; Seel. 125. v. cōlian.

acolitus=ἀκόλουθος *A light-bearer;* lucifer :—Acolitus is se ðe leóht berþ æt Godes þēnungum *acolite is he who bears the light at God's services,* L. Ælf. P. 34; Th. ii. 378, 7: L. Ælf. C. 14; Th. ii. 348, 4. v. hâd II. *state, condition;* ordo, gradus, etc.

acol-mōd; *adj. Of a fearful mind, timid;* pavidus animo :—Eorl acolmōd *a chief in trembling mood, fearful mind,* Exon. 55 b; Th. 195, 36; Az. 166. Þegnas wurdon acolmōde *the thanes were chilled with terror,* Andr. Kmbl. 753; An. 377.

acordan; *p.* ede; *pp.* ed *To* ACCORD, *agree, reconcile;* reconciliare, Chr. 1119; Ing. 339, 30.

a-coren *chosen;* *pp.* of a-ceósan. v. ceósan, gecoren.

a-corenlîc; *adj. Likely to be chosen;* eligibilis :—Biþ swîðe acorenlîc *is very estimable,* Past. 52, 8; Swt. 409, 36.

a-corfen *carved;* *pp.* of a-ceorfan.

a-costnod *tried;* *pp.* of a-costnian. v. costnian.

a-cræftan; *p.* de; *pp.* ed *To devise, plan, contrive as a craftsman;* excogitare :—Úton ðeáh hwæðere acræftan hū we heora, an ðyssa nihta, mâgan mæst beswîcan *let us however plan how we can, in this night, most weaken them,* Ors. 2, 5; Bos. 47, 19.

a-crammian; *p.* ode; *pp.* od *To cram, fill;* farcire. v. crammian.

a-creópian; *p.* ede; *pp.* ed *To creep;* serpere, scatere :—Ðâ læfdon hîg hit [Manhu] sume, oþ hit morgen wæs, and hit wearþ wyrmum acreóped *dimiserunt quidam ex eis usque mane, et scatere cœpit vermibus,* Ex. 16, 20.

a-crimman; *p.* -cramm, *pl.* -crummon; *pp.* -crummen *To crumble;* friare :—Acrummen *in micas fractus,* Cot. 88: 179: 193.

âc-rind, e; *f. Oak-rind* or *bark;* querna cortex :—Nîm âcrinde *take oak-bark,* Lchdm. iii. 14, 1.

acs *an axe.* v. æx, acas.

Acsa, Axa, an; *m? The river Axe.* v. Acsan mynster.

Acsan mynster, Ascan mynster, Axan minster, es; *n.* [Acsa, an; *m? the river Axe;* mynster *a monastery: Flor.* Axanminster: *Hunt.* Acseminster] AXMINSTER *in Devonshire;* hodie *Axminster,* in agro Devoniensi; ita dictum quod situm est ad ripam fluminis *Axi* :—Se Cynewulf rîcsode xxxi wintra, and his lîc liþ æt Wintan ceastre, and ðæs æðelinges æt Ascan [Acsan, Gib. 59, 3; Ing. 71, 28] mynster *Cynewulf reigned thirty-one years, and his body lies at Winchester, and the prince's at Axminster,* Chr. 755; Erl. 50, 32: Th. 86, 13, col. 1.

acse *ashes,* Cot. 40. v. asce.

acsian, acsigan ; *p.* ode ; *pp.* od *To ask, ask for, demand* ; rogare, expostulare, exigere :—Môt ic acsian, Bd. 4, 3 ; S. 568, 26. Cômon corþrum miclum cuman acsian *they came in great multitudes to demand the strangers,* Cd. 112 ; Th. 148, 8 ; Gen. 2453 : Lk. Bos. 20, 40. Híg hine acsodon ðæt bigspell *they asked him the parable,* Mk. Th. 4, 10. Hû mæg ænig man acsigan *how can any man inquire ?* Bt. 35, 1 ; Fox 156, 6. v. ascian.

acsung, e ; *f. An asking, a question, an inquiry, inquisition, interrogation, that which is inquired about, information* ; interrogatio :—Uneáþe ic mæg forstandan ðíne acsunga *I can scarcely understand thy questions,* Bt. 5, 3 ; Fox 12, 16. v. ascung.

ác-treó, -treów, es ; *n. An oak-tree* ; quercus :—Under áctreó *under the oak-tree,* Exon. 115 a ; Th. 443, 10 ; Kl. 28.

Ác-tûn, es ; *m.* [ác *oak,* tûn *a town*] ACTON, *Staffordshire ?*—Æt Áctûne at Acton, Th. Diplm. A.D. 1002 ; 546, 27. v. aac.

a-cucian *to revive* [cuc = cwic, Cd. 65 ; Th. 78, 23 = Ors. 2, 1 ; Bos. 38, 8]. v. a-cwician.

acul *frightened,* Cd. 210 ; Th. 261, 14 ; Dan. 726. v. acol.

á-cuma OAKUM ; putaman :—Ácuman *putamina,* Mone p. 398 ; B. 3231. v. ácumba.

a-cuman ; *p.* -cám, -com, *pl.* -cámon, -cômon ; *pp.* -cumen, -cymen *To come, bear* ; venire, ferre, sustinere :—Wæs of fere acumen *he had come from the vessel,* Cd. 75 ; Th. 93, 12 ; Gen. 1544. Ðæt land híg ne mihte acuman *non sustinebat eos terra,* Gen. 36, 7. Ge hyt ne mágon nú acuman *non potestis portare modo,* Jn. Bos. 16, 12.

á-cumba, an ; *m:* æ-cumbe, an ; *n?* [cemban *to comb*]. I. *oakum, that which is combed, the coarse part of hemp,—Hards, flax, tow* ; stuppa = στύππη, στύπη [v. heordas *stuppæ,* R. 68] :—Afyl ða wunde, and mid ácumban beswede *fill the wound, and swathe up with tow,* L.M. 1, 1 ; Lchdm. ii. 22, 21. Æcumbe *stuppa,* Ælfc. Gl. 64 ; Som. 69, 2 ; Wrt. Voc. 40, 36. II. *the thing pruned or trimmed, properly of trees, and figuratively of other things, hence,—Prunings, clippings, trimmings* ; putamen, hinc,—putamina non solum arborum sunt, verum omnium rerum purgamenta. Nam quicquid ex quacumque re projicitur, putamen appellatur :—Ácumba *putamen,* Mone B. 3702. Ácumban *putamina,* 3703, p. 407. III. *reduced to ashes, it was used as a substitute for* σπόδιον = σποδός, *Wood ashes* ; spodium Græcorum nihil aliud est, quam radix Alcannæ combusta, officinæ ustum ebur ejus loco substituunt :—To sealfe, nim. ácumban, cneówholen *for a salve, take the ashes of oakum, butcher's broom,* L.M. 1, 33 ; Lchdm. ii. 80, 11. Ácumba *ashes of oakum,* 1, 47 ; Lchdm. ii. 120, 14.

a-cumend-líc ; *adj. Tolerable, beárable* ; tolerabilis :—Acumendlícre byþ Sodoma lande and Gomorra on dômes dæg, ðonne ðære ceastre *tolerabilius erit terræ Sodomorum et Gomorrhæorum in die judicii quàm illi civitati,* Mt. Bos. 10, 15.

a-cumendlícness, e ; *f. The possibility to bring anything to pass* ; possibilitas. v. cumende ; *part. of* cuman.

a-cunnian ; *p.* ode ; *pp.* od *To prove* ; probare :—Ðú acunnodest [MS. acunnudyst] us God *probasti nos Deus,* Ps. Spl. C. 65, 9. v. cunnian.

a-curon *chose* ; *p. pl. of* a-ceósan.

a-cwædon *said,* Ps. Th. 72, 6 ; *p. of* a-cwedan.

a-cwælon *died,* Chr. 918 ; Erl. 104, 13 ; *p. pl. of* a-cwelan.

a-cwæþ *spoke,* Cd. 30 ; Th. 40, 14 ; Gen. 639. *p. of* a-cweðan.

a-cwalde *killed,* Ps. Vos. 104, 27 ; 134, 11, = a-cwealde ; *p. of* a-cwellan.

a-cwân *melted, decayed,* Bd. 2, 7 ; S. 509, 29 ; *p. of* a-cwînan.

a-cwanc *quenched,* Chr. 1110 ; Ing. 331, 30 ; *p. of* a-cwincan.

a-cwealde *killed,* Cd. 69 ; Th. 84, 25 ; Gen. 1403 ; *p. of* a-cwellan.

a-cweccan ; *p.* -cwehte ; *pp.* -cweht *To move quickly, to shake, vibrate* ; movere, quatere, vibrare :—Æsc acwehte *he shook the ash,* i. e. *the lance,* Byrht. Th. 140, 59 ; By. 310.

a-cwelan, he -cwilþ, *pl.* -cwelaþ ; *p.* -cwæl, *pl.* -cwælon ; *pp.* -cwolen, -cwelen, *v. n. To die, perish* ; mori :—Ða fixas acwelaþ *pisces morientur,* Ex. 7, 18. Ofercumen biþ he ær he acwele *he will be overcome ere he dies,* Exon. 90 b ; Th. 340, 10 ; Gn. Ex. 114. Monige men hungre acwælon *many men died of hunger,* Chr. 918 ; Erl. 104, 13.

a-cwellan ; *p.* -cwealde ; *pp.* -cweald *To kill, destroy* ; interficere, necare :—Freá wolde on ðære to-weardan tíde acwellan *the Lord would destroy them in the coming time,* Cd. 64 ; Th. 77, 31 ; Gen. 1283. Ic wille mid flóde folc acwellan *I will destroy the folk with a flood,* 64 ; Th. 78, 21 ; Gen. 1296. Acwelleþ ða wyrmas *killeth the worms,* Herb. 137 ; Lchdm. i. 254, 22. Ða ðe égor-here eorþan tuddor eall acwealde *when the water-host destroyed all the progeny of earth,* Cd. 69 ; Th. 84, 25 ; Gen. 1403. Wíges heard wyrm acwealde *the bold one in battle slew the worm, the dragon,* Beo. Th. 1777 ; B. 886. Steóp-cilda feala stundum acwealdon *pupillos occiderunt,* Ps. Th. 93, 6.

a-cwelledness, e ; *f. A quelling, killing* ; occisio. DER. cwellan.

a-cwencan ; *p.* de, te, *pl.* don, ton ; *pp.* ed, d, t *To quench, extinguish, put out* ; extinguere :—Bæd ðæt hí ðæt leóht acwencton *prayed that they would put out the light,* Bd. 4, 8 ; S. 575, 40, note, MS. B. Úre leóhtfatu synt acwencte *lampades nostræ extinguuntur,* Mt. Bos. 25, 8. Fyr ne byþ acwenced *ignis non extinguitur,* Mk. Bos. 9, 44.

a-cweorran ; *p.* -cwear, *pl.* -cwurron ; *pp.* -cworren *To eat* or *drink immoderately, to glut, guzzle* ; ingurgitare :—Swá swá mihti acworren fram wíne *tanquam potens crapulatus a vino,* Ps. Spl. T. 77, 71.

ác-wern, es ; *n. The name of an animal, a squirrel* ; scirra, sciurus, Ælfc. Gl. 19 ; Som. 59, 9.

a-cwerren, -cworren *drunk* ; *pp. of* a-cweorran.

a-cweðan, he -cwyþ ; *p.* -cwæþ, *pl.* -cwædon ; *pp.* -cweden *To say, tell, answer* ; dicere, eloqui, respondere :—Ðæt word acwyþ *that word says,* Beo. Th. 4099 ; B. 2046. Word acwæþ, wuldres aldor *he spake the word, the chief of glory,* Cd. 30 ; Th. 40, 14 ; Gen. 639. Ðæt me acweden syndon *quæ dicta sunt mihi,* Ps. Th. 121, 1. v. cweðan.

a-cwician ; *p.* ode ; *pp.* od *To quicken, revive, to come to life* ; vivificare, reviviscere :—On ðínre mild-heortnesse me scealt acwician *in misericordia tua vivifica me,* Ps. Th. 118, 159. Ða acwicode ic hwon *then I revived a little,* Bd. 5, 6 ; S. 619, 29.

a-cwilþ *perishes* :—Nea-cwilþ *perishes not,* Bt. 13 ; Fox 38, 29. v. a-cwelan.

a-cwînan ; *p.* -cwân, *pl.* -cwinon ; *pp.* -cwinen *To waste* or *dwindle away, decline, become extinct* ; tabescere :—Ðæt fýr acwân and adwæsced wæs *the fire declined and was extinguished,* Bd. 2, 7 ; S. 509, 29.

a-cwincan ; *p.* -cwanc, *pl.* -cwuncon ; *pp.* -cwuncen *To vanish, become extinguished, quenched* ; extingui, evanescere :—Se môna acwanc *the moon was extinguished,* i. e. *eclipsed,* Chr. 1110 ; Ing. 331, 30.

a-cwinen *quenched.* v. a-cwînan.

a-cwolen *died,* Chr. 918 ; Gib. 105, 37, note a. v. a-cwelan.

a-cworren *drunk,* Ps. Spl. T. 77, 71 ; *pp. of* a-cweorran.

a-cwucian *to quicken.* v. a-cwician.

a-cwylan *to die,* L. H. E. 6 ; Th. i. 30, 3. v. a-cwelan.

acxan *ashes,* Ors. 1, 3 ; Bos. 27, 32. v. axe, asce.

a-cýd *said, confirmed,* R. Ben. 27. v. a-cýðan.

a-cyrran ; *p.* -cyrde ; *pp.* -cyrred, -cyrd *To avert* ; avertere :—Ne ðú næfre gedést, ðæt ðú mec acyrre from Cristes lofe *thou shalt never do so, that thou avert me from the love of Christ,* Exon. 67 b ; Th. 251, 2 ; Jul. 139. Acyrred from Cristes æ *turned from Christ's law,* 71 b ; Th. 267, 6 ; Jul. 411.

a-cyrrednes, -cerrednes, -ness, e ; *f. A turning, aversion, a turning from, apostacy, revolting* ; aversio. DER. a-cyrred. v. a-cyrran.

a-cýðan ; *p.* -cýðde ; *pp.* -cýðed, -cýd *To show, announce, confirm* ; manifestare, annuntiare, confirmare :—Yrre acýðan *iram manifestare, irasci,* Ps. Th. 88, 39. Ær he hine acýðan môte *ere he can show himself,* Exon. 89 b ; Th. 336, 15 ; Gn. Ex. 49. Torn acýðan *to make known* or *show one's affliction,* Exon. 78 a ; Th. 293, 8 ; Wand. 113. Ðær me wæs yrre ðín on acýðed *in me confirmata est ira tua,* Ps. Th. 87, 7.

ÁD, aad, es ; *m. A funeral pile, pile, heap* ; rogus, congeries :—Ða onbærnde he ðone ád *then kindled he the pile,* Bd. 3, 16 ; S. 542, 25. Ád stód onæled *the pile was* [*stood*] *kindled,* Cd. 141 ; Th. 176, 35 ; Gen. 2922. Hêt ád onælan *he commanded to kindle the funeral pile,* Exon. 74 a ; Th. 277, 13 ; Jul. 580. Mycelne aad [ád MS. B. T.] gesomnode on beámum *advexit plurimam congeriem trabium,* Bd. 3, 16 ; S. 542, 22. [*Kath.* ad: O. Ger. eit *ignis, rogus.* v. *Lat.* æs-tus: *Grk.* αἶθος: *Sansk.* edh-as *wood for fuel,* from the *Sansk.* root indh *to light, kindle.*] DER. ád-fær, -fýr, -lêg, -loma.

a-dælan ; *p.* ede ; *pp.* ed, *To part, divide, separate* ; partiri, dividere, separare :—He sceal wesan of eorþan feor adæled *he shall be far parted from the earth,* Cd. 106 ; Th. 140, 4 ; Gen. 2322. Ða wæron adælede ealle of ánum *these were parted all from one,* 12 ; Th. 14, 13 ; Gen. 218 : Ps. Th. 54, 20. v. dælan.

a-deádan, -deádian ; *p.* ode ; *pp.* od *To fail, decay, die, mortify, lay waste, destroy* ; fatiscere, Herb. 35, Lye : Cot. 90.

a-deáf ; *adj. Deáf* ; surdus, Ben. v. deáf.

a-deáfian ; *p.* ode, ede ; *pp.* od, ed *To become* or *wax deaf* ; surdescere, obsurdescere :—Adeáfede *obsurduit,* Ælfc. Gl. 100 ; Som. 77, 13 ; Wrt. Voc. 55, 17.

a-deáfung eárena *A deafening of the ears* ; surditas. v. a-deáf.

ádel *a disease,* Exon. 48 b ; Th. 167, 23 ; Gú. 1064. v. ádl.

adela, an ; *m. Filth* ; cœnum :—Ðæt hér yfle adelan stinceþ *that here ill smells of filth,* Exon. 110 b ; Th. 424, 1 ; Rä. 41, 32. [addle-pool *a pool near a dunghill: Scot.* adill, addle *foul and putrid water: N. Ger.* adel, m. cœnum: *Holst.* addeln *lotium pecudum.*] DER. adeliht, adel-seáþ.

a-delfan ; *p.* -dealf, -dylf, *pl.* -dulfon ; *pp.* -dolfen *To dig, delve:* fodere, effodere :—Cleopatra hêt adelfan hyre byrigenne *Cleopatra ordered her burying place to be dug,* Ors. 5, 13 ; Bos. 113, 22. Seáþ adealf *lacum effodit,* Ps. Spl. 7, 16 : Bd. 3, 2 ; S. 524, 16. Oþ ðæt biþ seáþ adolfen *donec fodiatur fovea,* Ps. Th. 93, 12 : Bd. 3, 9 ; S. 533, 23.

adeliht ; *adj. Dirty, filthy* ; cœnosus, Cot. 48.

adel-seáþ, es ; *m. A sewer, gutter, sink* ; cloaca. v. adul-seáþ.

adelyng *a prince,* Joh. Brompt. ad ann. 907. v. æðeling.

a-dêman ; *p.* de ; *pp.* ed *To judge, adjudge, doom, deem, try, abjudicate, deprive* ; examinare, abjudicare, judicio facto relegare :—Lícode Gode hire ða hálgan sáule eác swylce mid longre hire líchoman untrymnesse adêmde and asodene beón *it pleased God that her holy soul should also be tried and seethed with long sickness of her body,* Bd. 4, 23 ;

S. 595, 15. Ðú adémest me fram duguðe *thou deprivest me of good*, Cd. 49; Th. 63, 14; Gen. 1032. v. déman.

a-deorcian; *p.* ode, ade; *pp.* od, ad *To obscure, dim, darken, hide;* obscurare :—Adeorcad *obscuratus*, Som. v. deorcian.

adesa, eadesa, an; *m. An addice* or *adze, a cooper's instrument;* ascia, Bd. 4, 3; S. 567, 26: Wrt. Voc. p. 84, 62.

ád-fær, *nom. acc; g.* -færes; *pl. nom.* -faru; *n. The pile-way, the way to the funeral pile;* iter rogi :—Ðæt we hine gebringen on ádfære *that we may bring him on the way to the pile,* Beo. Th. 6012; B. 3010.

ád-fýr, es; *n. A pile-fire;* ignis rogi :—Abraham ádfýr onbran *Abraham kindled a pile-fire,* Cd. 162; Th. 203, 4; Exod. 398.

a-dihtan; *p.* -dihte, -dihtode; *pp.* -dihtod, -diht *To compose, edit, write;* facere, componere. v. dihtan.

a-dilegian, -dilgian, -dylegian; *p.* ode; *pp.* od; *v. a.* [a, dilgian *to destroy*] *To abolish, blot out, destroy, do away;* abolere, delere :—His sáwul biþ adilegod of his folce *delebitur anima illa de populo suo,* Gen. 17, 14. Ic adilgige hí *delebo eos,* Ps. Lamb. 17, 43. Adilga me of ðínre béc *dele me de libro tuo,* Ex. 32, 32 : Ps. Th. 68, 29 : 108, 13, 14. Adilgode, Ps. Th. 17, 40.

a-dimmian; *p.* ode; *pp.* od, ad *To dim, darken, obscure, make dull;* obscurare :—Ðeáh heora mód sie adimmad *though their mind be obscured,* Bt. 24, 4; Fox 84, 28: Ps. Th. 68, 24.

ÁDL, ádel; *g.* ádle, *f.:* ádle, an; *f. A disease, pain, a languishing sickness, consumption;* morbus, languor :—Wæs seó ádl þearl, hát and heorogrim *the disease was sharp, hot and very fierce,* Exon. 47 a; Th. 160, 30; Gú. 951. Seó mycle ádl *the great disease, leprosy;* elephantiasis, Som. Ne hine dréfeþ ádl *disease does not afflict him,* Beo. Th. 3476; B. 1736. Ðé to heortan hearde grípeþ ádl unlíðe *fell disease gripes thee hard at heart,* Cd. 43; Th. 57, 32; Gen. 937. Ðé untrymnes ádle gongum býsgade *infirmity has afflicted thee through attacks of disease,* Exon. 47 b; Th. 163, 8; Gú. 990. He ðíne ádle ealle gehælde *sanavit omnes languores tuos,* Ps. Th. 102, 3. Ðæt ádla hí gehæl don *ut languores curarent,* Lk. Bos. 9, 1. Hú manega ádla *how many diseases?* Bt. 31, 1; Fox 110, 29: Bd. 3, 12; S. 537, 6. Laman legeres ádl *the palsy.* v. leger. [*Orm.* adl *disease.* Probably akin to the *Sansk.* root indh *to burn.*] DER. feorh-ádl, fót-, horn-, in-, lungen-, mónaþ-: ádl-ian, -íc, -ig, -þracu, -wérig.

ádle, an; *f. A disease;* morbus :—Ne yldo ne ádle *neither age nor disease,* Exon. 112 a; Th. 430, 7; Rä. 44, 4. v. ádl.

ád-lég, es; *m. The flame of the funeral pile;* flamma rogi :—Ádlég æleþ flæsc and bán *the flame of the pile burns flesh and bones,* Exon. 59 a; Th. 213, 9; Ph. 222.

ádlian, -igan; *p.* ode; *pp.* od *To ail, to be sick, to languish;* ægrotare, languere :—Ðæt se ylca biscop án ádliende mæden gebiddende gehælde *ut idem episcopus puellam languentem orando sanaverit,* Bd. 5, 3; S. 615, 35. Ic ádlige *langueo,* Ælfc. Gr. 26, 2; Som. 28, 46.

ádlic, ádlig; *adj.* [ádl *disease,* líc *like*] *Sick, ill, diseased, corrupted, putrid;* morbidus, ægrotus, tabidus, vitiatus, putidus. Hence ADDLE *egg;* putidum ovum :—Ádlige men *languentes homines,* Bd. 3, 2; S. 524, 32. Ádlig *æger vel ægrotus,* Wrt. Voc. 45, 59.

ád-loma, -lama? an; *m. One crippled by the flame?* cui flamma claudicationem attulit? —Earme ádloman *poor wretches, i. e.* diaboli, Exon. 46 a; Th. 156, 33; Gú. 884.

ádl-þracu, -þræce; *f. The force or virulence of disease;* morbi impetus :—Seó ádlþracu *the force of disease,* Exon. 46 b; Th. 159, 31; Gú. 935. v. þræc.

ádl-wérig; *adj. Weary with sickness;* morbo fatigatus :—Fonde his mon-dryhten ádlwérigne *he found his master weary with sickness,* Exon. 47 b; Th. 162, 25; Gú. 981.

a-dolfen *dug,* Ps. Th. 93, 12; *pp.* of a-delfan.

a-dón; *p.* -dyde; *impert.* -dó; *v. a. To take away, remove, banish;* tollere, ejicere :—Ne mágon ðé nú heonan adón hyrste ða reádan *the red ornaments may not now take thee hence,* Exon. 99 a; Th. 370, 14; Seel. 57. Ðæt hý God ðanon adó to heora ágnum lande *that God will bring them thence to their own land,* Ors. 3, 5; Bos. 56, 37. Adó ða buteran *remove the butter,* L. M. 1, 36; Lchdm. ii. 86, 22. Adó of ða buteran *take off the butter,* 86, 19. Flód adyde mancinn *a flood destroyed mankind,* Ælfc. T. 5, 25: Gen. 7, 23; 9, 11. Adó ðas wylne *ejice ancillam hanc,* Gen. 21, 10: Bt. 16, 1; Fox 50, 10: Ps. Th. 68, 14.

a-drædan; *p.* -dréd; *pp.* -dræden *To fear;* timere :—He adréd ðæt folc *timuit populum,* Mt. Bos. 14, 5.

a-dræfan, -dréfan; *p.* de; *pp.* ed *To drive away;* expellere :—Ða wearþ adræfed deórmód hæleþ *then was driven away the beloved hero,* Chr. 975; Th. i. 228, 22; Edg. 44. He adréfed wæs *ejectus est,* Gen. 3, 24. Osræd wæs of ríce adréfed *Osred was banished from his kingdom,* Chr. 790; Th. 99, 20, col. 2.

a-dreág, -dreáh *bore,* Exon. 25 b; Th. 74, 6; Cri. 1202; *p.* of a-dreógan.

a-dréd *feared,* Mt. Bos. 14, 5; *p.* of a-drædan.

a-dréfed *driven,* Chr. 790; Th. 99, 20, col. 2, = a-dráfed; *pp.* of a-dræfan.

adreminte, an; *f. The herb feverfew;* parthenium = παρθένιον, Prior 78.

a-drencan; *p.* -drencte; *pp.* -drenced; *v. a. To plunge under, to im-*

---

merse, drown; immergere :—Wolde hine adrencan on ðære eá *would drown him in the river,* Bt. 16, 2; Fox 52, 36. Caines ofspring eall wearþ adrenced on ðam deópan flód, ðe adyde mancinn *Cain's offspring were all drowned in the deep flood, which destroyed mankind,* Ælfc. T. 5, 24. Heora feóndas flód adrencte, Ps. Th. 105, 10: Ex. 14, 28.

a-dreógan, -driógan; *ic* -dreóge, ðú -dreógest, -drýhst, he -dreógeþ, -drýhþ; *p.* -dreág, -dreáh, *pl.* -drugon; *pp.* -drogen. I. *to act, perform, practise;* agere, perficere :—He adreág unrihte þing *gessit iniqua,* Hymn. Bibl. Cott. Jul. A. 6. Ðe his iufan adreógeþ *who practises his love,* Exon. 33 b; Th. 107, 24; Gú. 63. Líf adreógan *agere vitam,* Hexam. 3; Norm. 4, 29. II. *to bear, suffer, endure;* pati, sustinere :—Hí adreógan mágan *they may bear,* Bt. 40, 3; Fox 238, 27. Ic ne mæg adreógan ðíne seófunga *I cannot tolerate thy lamentations,* Bt. 11, 1; Fox 30, 20. Ðæt hie ðe eáþ mihton drohtaþ adreógan *that they might the easier endure their way of life,* Andr. Kmbl. 737; An. 369. Earfeðu ðe he adreág *the pains that he endured,* Exon. 25 b; Th. 74, 6; Cri. 1202. Earfeðo ðe he adreáh *the pains that he endured,* Andr. Kmbl. 2971; An. 1488. v. dreógan.

a-dreógendlic; *adj. Bearable;* tolerabilis; *part.* of a-dreógan, -líc.

a-dreópan; *ic* -dreópe, ðú -drýpst, he -drýpþ; *p.* -dreáp, *pl.* -drupon; *pp.* -dropen *To shed drop by drop;* guttatim effundere :—Nú is mín swát adropen *now is my blood sprinkled,* An. 1427, note. v. a-þráwan.

a-dreósan; *ic* -dreóse, ðú -drýst, he -dreóseþ, -drýst; *p.* -dreás, *pl.* -druron; *pp.* -droren *To fall, decline;* labi, deficere :—Ne biþ se hlísa adroren *fame will not decline,* non erit fama tædio affecta, Exon. 95 a; Th. 355, 19; Reim. 79.

a-drífan, æ-drífan; *ic* -drífe, ðú -drífest, -drífst, he -drífeþ, -drífþ, -dríft, *pl.* -drífaþ; *p.* -dráf, *pl.* -drifon; *pp.* -drifen *To drive, stake, expel, pursue, follow up;* agere, pellere, expellere, repellere, sequi, prosequi :—Ða Walas adrifon sumre eá ford ealne mid scearpum pílum greátum innan ðam wætere *the Welsh staked all the ford of a certain river with great sharp piles within the water,* Chr. Introd; Th. 5, 35. Rihtwísnyssa his ic ne adráf fram me *justitias ejus non repuli a me,* Ps. Spl. 17, 24. Adrífe ðæt spor út of his scíre *let him pursue the track out of his shire,* L. Ath. v. § 8, 4; Th. i. 236, 23. Adrifene fatu *graven or embossed vessels,* Ælfc. Gl. 67; Som. 69, 99. v. drífan.

a-drigan, -drygan, -drygean, -drugian, -druwian; *p.* de, ode; *pp.* ed, od *To dry, dry up, rub dry, wither;* abstergere, siccare, exsiccare :—Hlúde streámas on Æthane ealle ðú adrigdest *tu exsiccasti fluvios Ethan,* Ps. Th. 73, 15.

a-drincan; *p.* -dranc, *pl.* -druncon; *pp.* -druncen *To be immersed, extinguished, quenched by water, to be drowned;* immergi, exstingui, aquis suffocari :—Lígfýr adranc *the fire-flame was quenched,* Cd. 146; Th. 182, 18; Exod. 77. Mycele má moncynnes adranc on ðam wætere *many more of mankind were drowned in the water,* Bd. 3, 24; S. 556, 36.

a-driógan, -dríóhan *to bear,* Bt. 40, 3; Fox 238, 22; MS. Cott. The Bodl. MS. has a-dríóhan. v. a-dreógan.

a-drogen *done, finished;* transactus, peractus; *pp.* of a-dreógan.

a-dronc, -droncen, *for* a-dranc, -druncen; *p.* and *pp.* of a-drincan.

a-drugian; *p.* ode; *pp.* od *To dry;* siccari :—Ða sóna adrugode se stream *alveus siccatus est,* Bd. 1, 7; S. 478, 13. v. a-drigan.

a-druncen *drowned;* *pp.* of a-drincan.

a-druwian; *p.* ode; *pp.* od *To dry up;* siccari :—Ðæt ða wætera wæron adruwode ofer eorþan *quod aquæ cessassent super terram,* Gen. 8, 11. Eorþan brádnis wæs adruwod *exsiccata erat superficies terræ,* 8, 13. v. a-drigan.

a-drygan, -dryggean *to dry,* Past. 13, 1; Hat. MS. 16 b, 6. v. adrigan.

adul-seáþ *a sewer, sink;* cloaca, Wrt. Voc. 36, 42. v. adelseáþ.

a-dumbian; *p.* ode, ede; *pp.* od, ed; *v. n. To hold one's peace, to keep silence, to become mute* or *dumb;* obmutescere :—Adumba and gá of ðisum men *obmutesce et exi de homine,* Mk. Bos. 1, 25. Adumbiaþ ða fácnfullan weoloras *muta efficiantur labia dolosa,* Ps. Th. 30, 20. Ic adumbede *obmutui,* Ps. Spl. 38, 3. Ic adumbode, Ps. Lamb. 38, 10.

a-dún, -dúne; *adv. Down, adown, downward;* deorsum :—Adún of ðam wealle *down from the wall,* Bd. 1, 12; S. 481, 21. Ða óðre ða dura brǽcon adúne *the others broke the doors down,* Chr. 1083; Th. 352, 19. Adúne asetton (*they*) *put down, deposed,* Bd. 4, 6; S. 573, 35. He adúne astáh *descendit,* Ps. Spl. 71, 6: 87, 4.

a-dún-weard; *adv. Downward;* deorsum :—Scotedon adúnweard mid arewan *they shot their arrows downward,* Chr. 1083; Th. i. 352, 14.

a-dwǽscan; *p.* ede, te; *pp.* ed, t; *v. a.* [a, dwǽscan *to quench*] *To quench, put out, staunch, appease;* extinguere :—Smeócende flex he ne adwǽscþ *linum fumigans non extinguet,* Mt. Bos. 12, 20. Ðæt fýr adwǽsced wæs *flammæ extinctæ sunt,* Bd. 2, 7; S. 509, 29. Adwǽscton *extinguerent,* 4, 8; S. 575, 41. Adwǽscton ðínum feóndum *extinctis tuis hostibus,* 2, 12; S. 514, 7. Efne swá he mid wætre ðone weallendan lég adwǽsce *even as he with water the raging flame quenches,* Exon. 122 a; Th. 467, 23; Alm. 6. Eall mín unriht adwǽsc *omnes iniquitates meas dele,* Ps. Ben. 50, 10.

a-dwelian; *p.* -dwelede, -dwealde; *pp.* -dweled, -dweald [a, dwelian *to err*] *To seduce, lead into error;* seducere :—Woldon adwelian

mancyn fram heora Drihtene *they would seduce mankind from their Lord,* L. Ælf. P. 29; Th. ii. 374, 31.

a-dwínan; ic -dwíne, -dwínest, -dwínst, he -dwíneþ, -dwínþ, *pl.* -dwínaþ; *p.* -dwán, *pl.* -dwinon; *pp.* -dwinen *To dwindle* or *vanish away;* vanescere. v. dwínan.

a-dýdan, -dýddan; *p.* -dýdde; *pp.* -dýded, -dýd; *v. a.* [a, dýdan *to die*] *To put to death, to destroy, kill, mortify;* perdere, occidere:—Wolde híg adýddan *would destroy them,* Ælfc. T. 22, 19. Ðæt ic náteshwon nelle heonon forþ eall flǽsc adýdan mid flódes wæterum *that I will not, by any means, henceforth destroy all flesh with the waters of a flood,* Gen. 9, 11. Ælc þing ðe líf hæfde þearþ adýd *everything which had life was destroyed,* Gen. 7, 23.

a-dydest, *hast banished;* expulisti, Ps. Lamb. 59, 12; *p.* of a-dón.

a-dylegian; *pres.* ic -dylegige; *p.* ode; *pp.* od *To destroy;* delere:—Ic adylegige deleo; ic adylegode [adeligode Som.] *delevi;* adylegod *deletum,* of ðam is gecweden letum [=lethum *death;* Grk. λήθη *oblivio*] deáþ, ðe adylegaþ líf *I destroy; I destroyed; destroyed,* deletum, *from which is derived [called]* letum *death, which destroyeth life,* Ælfc. Gr. 26; Som. 28, 32, 33. v. a-dilegian, dilgian.

a-dylf *effodit,* Ps. Th. 7, 15, = a-dealf; *p.* of a-delfan, *q. v.*

Æ. The short or unaccented Anglo-Saxon **æ** has a sound like *ai* in main and fairy, as appears from these cognate words:—Wæl *wail,* brædan *to braid,* nægel *a nail,* dæg, spær, læt, snæce, mæst, æsp, bær, *etc.*   **2.** The short or unaccented **æ** stands only (1) before a single consonant; as Stæf, hwæl, dæg: (2) a single consonant followed by *e* in nouns; Stæfes, stæfe, hwæles, dæges, wæter, fæder, æcer: (3) or before *st, sc, fn, ft;* Gæst, æsc, hræfn, cræft: (4) before *pp, bb, tt, cc, ss;* Æppel, cræbba, hæbben, fætte, fættes; wræcca, næsse: (5) before double consonants, arising from the inflection of monosyllabic adjectives:—Lætne, lætre, lætra, from læt *late;* hwætne, hwætre, hwætra from hwæt *quick.*   **3.** In the declension of monosyllabic nouns and adjectives, **e** is rejected from the short or unaccented **æ**, and becomes **a,** when a single consonant, or *st, sc,* is followed by *a, o, u* in nouns, and by *a, o, u, e* in adjectives; as Stæf, *pl.* stafas, *g.* stafa, *d.* stafum; hwæl, *pl.* hwalas; dæg, *pl.* dagas. *adj.* Læt *late; g. m. n.* lates; *d.* latum; se lata *the late;* latost, latemest, *latest:* Smæl *small; g. m. n.* smales; *d.* smalum; se smala *the small,* etc. See short *a* in B. 3, p. 1, col. 1.   **4. æ-,** prefixed to words, like **a-,** often denotes *A negative, deteriorating* or *opposite signification,* as *From, away, out, without,* etc. Like a, **ge,** etc. **æ** is sometimes prefixed to *perfect tenses and perfect participles* and other words without any perceptible alteration in the sense; as Céled, æ-céled *cooled.*   **5.** The Anglo-Saxon Rune for **æ** is ᚨ, which is also put for æsc *an ash-tree,* the name of the letter. v. æsc.

B. The long or accented **ǽ** has the sound of *ea* in *meat, sea.* The **ǽ** is found in the following words, which are represented by English terms of the same signification, having *ea* sounded as in *deal, fear;* Dǽl, fǽr, drǽd, lǽdan, brǽdo, hǽto, hwǽte, hǽþ, hǽðen, clǽne, lǽne, sǽ, ǽr, hǽlan, lǽran, tǽcan, tǽsan, tǽsel, wǽpen, *etc.*   **2.** The **ǽ** is known to be long, and therefore accented, when in monosyllables, assuming another syllable in declining, **ǽ** is found before a single consonant or *st, sc,* and followed in nouns by *a, o, u,* and in adjectives by *a, o, u,* or *e;* as Blǽda *fruits;* blǽdum *Dwæs dull; g. m.* dwǽses. The **ǽ** is often changed into **á;** as Stǽnen *stony,* stán *a stone;* lǽr, lár *lore.*

Æ; *indecl. f. Law, statute, custom, rite, marriage;* lex, statutum, ceremoniæ, ritus, matrimonium :—God him sette **ǽ** ðæt ys open lagu *God gave them a statute that is a plain law,* Ælfc. T. 10, 20. Ǽ Drihtnes *the law of the Lord,* Ps. Spl. 18, 8; Mt. Bos. 26, 28. God is wísdóm and ǽ woruldbúendra *God is the wisdom and law of the inhabitants of the world,* Bt. Met. Fox 29, 165; Met. 29, 83. Cristes **ǽ** *the Gospel.* Bútan ǽ oððe útlaga *an outlaw,* Ælfc. Gr. 47; Som. 48, 44. Seó æftere **ǽ** *Deuteronomy,* Bd. 1, 27. Húslfatu hálegu ða ǽr Israela in ǽ hæfdon *the holy vessels which the Israelites formerly used in their rites,* Cd. 212; Th. 262, 29; Dan. 751. Wircaþ his bebodu and his **ǽ** and his dómas *observa præcepta ejus et ceremonias atque judicia,* Deut. 11, 1. Stýrde unryhtre **ǽ** *he reproved the unlawful marriage,* Exon. 70 a; Th. 260, 14; Jul. 297. [O. Sax. ẹo, m: O. Frs. â, ê, ewe, ewa, f: Ger. ehe, f. *matrimonium:* M. H. Ger. ewe, ê, f: O. H. Ger. ewa, êha, êa, f: Sansk. eva, m. *course, manner.*]

ǽ; *indecl. f. Life;* vita :—Ðæt hí ne meahtan acwellan cnyhta ǽ *that they might not destroy the young men's lives,* Exon. 55 a; Th. 195, 32; Az. 164.

ǽ; *indecl. f. A river, stream;* rivus, torrens :—On ðære ǽ ðú hý drencst *thou shalt give them to drink of the stream;* torrente potabis eos, Ps. Th. 35, 8. v. eú.

ǽ *alas!* Æ, Hy. 1, 1, = eá, Lamb. MS. fol. 183 b, line 11. v. ǽlá, æálá, eálá.

æálá; *interj. O! alas!* O, eheu :—Æálá ðú Scippend *O! thou Creator,* Bt. Met. Fox 4, 1; Met. 4, 1. v. eúlá, ǽlá.

a-eargian; *p.* ode, ade; *pp.* od [a, eargian *torpescere*] *To become slothful;* segnis fieri :—Hý ondrǽdan, gif hí hwílum ne wunnon, ðæt hý tó raðe a-eargadon *they dreaded, if they did not sometimes wage war, that they should too soon become slothful.* Ors. 4, 13; Bos. 100, 20.

ǽ-bǽr *notorious,* L. Eth. vi. 36; Th. i. 324, 11. v. ǽ-ber.

Æbban dún, Abban dún, e; *f.* [Æbba, an; *m:* or Æbbe, an; *f:* dún *a down* or *hill;* Æbba's or Æbbe's *down* or *hill*] ABINGDON; Abindoniæ oppidum in agro Bercceriensi :—His líc líþ on ðam mynstre æt Abban dúne *his body lies in the monastery at Abingdon,* Chr. 981; Th. 234, 34, col. 1.

a-ebbian; *p.* a-ebbode; *pp.* a-ebbad, ge-ebbod; *v. intrans. To ebb away, recede;* recedere :—Ðæt wæter wæs a-ebbad [a-ebbod MS. C. T; ge-ebbod Cant.] feala furlanga from ðám scipum *the water had ebbed many furlongs from the ships,* Chr. 897; Ing. 123, 19. v. ebbian.

æbbung, e; *f. An* EBBING; recessus aquarum :—Sǽ-æbbung *a bay;* sinus, Wrt. Voc. 41, 63. v. ebba.

ǽ-bebod, es; *n. Law, injunction of the law, command;* lex, legis mandatum :—Ðú me ǽbebod ǽrest settest *tu legem posuisti mihi,* Ps. Th. 118, 102.

ǽ-béc *law books, books of the law;* juris codices, Cot. 126.

ǽ-ber, ǽ-bǽr; *adj. Clear and evident by proof, manifest, apparent, notorious;* apricus, manifestus :—Se æbera þeóf *the notorious thief,* L. Edg. ii. 7; Th. i. 268, 22. Æbǽre manslagan *notorious homicides,* L. Eth. vi. 36; Th. i. 324, 11.

æbesen, æbesn *pasturage;* pasnagium, L. In. 49; Th. i. 132, 18, note 46. v. æfesen.

æ-bilgan, æ-bilian *to make angry;* exasperare, Ps. Spl. 67, 7. v. a-belgan.

æ-bilignes, -ness, e; *f. Indignation, anger;* indignatio, Apol. Th. v. æ-bylignes.

æ-blǽcnys, -nes, -ness, e; *f. A paleness;* pallor :—Wið æblǽcnysse ðæs líchaman *for paleness of the body,* Herb. 164, 2; Lchdm. ii. 294, 3.

ǽ-blǽc; *adj. Pale, wan, whitish, bleak;* pallidus. v. blǽc, blác.

æ-blǽcing, æ-blécnys *paleness.* v. æ-blǽcnys, blácung.

ǽ-bod, es; *m. A business;* negotium :—Æbodas *pragmatica negotia,* Ælfc. Gl. 12; Som. 57, 94.

ǽ-boda, an; *m. A messenger of the law;* legis nuntius :—Ðá wæs frófre gǽst onsended eádgum ǽbodan *then the spirit of comfort was sent to the blessed messenger of the law, i. e. the preacher of the gospel,* Exon. 46 b; Th. 158, 15; Gú. 909.

ǽ-brec [eá *water,* bræc] *A catarrh, rheum;* rheuma. v. brecan.

æbs, e; *f? A fir-tree;* abies, Ælfc. Gr. 5; Som. 4, 45; 9, 26; Som. 11, 18.

æ-bylg, es; *n. Anger;* ira, indignatio, Exon. 50 b; Th. 176, 17; Gú. 1211. v. æ-bylgþ.

æ-bylgan, -bylgian *To make angry;* exasperare, Ps. Spl. 65, 6. v. a-belgan.

æ-bylgþ, -bylþ, -bylygþ, e; *f:* es; *n?* [bylgþ, v. belgan] *An offence, a fault, scandal, wrong, anger, wrath, indignation;* offensa, injuria, ira, indignatio :—To æbylgþe *for offence,* Ors. 4, 1; Bos. 76, 27. He sende on hí graman æbylygþe hys *misit in eos iram indignationis suæ,* Ps. Spl. 77, 54. Cristenum cyningce gebyraþ swýðe rihte ðæt he Godes æbylþe wrece *Christiano regi jure pertinet ut injurias Deo factas vindicet,* L. C. S. 40; Th. i. 400, 10. v. a-bylgþ, a-byligd.

æ-bylignes, -ness -nys, -nyss, e; *f. Indignation, wrath;* indignatio :—Æbylignes yrres ðínes *indignatio iræ tuæ,* Ps. Th. 68, 25. He sende on hí graman æbylignysse hys *misit in eos iram indignationis suæ,* Ps. Spl. 77, 54. v. a-bylgnes.

ǽc *also,* Th. Dipl. A. D. 804–829; 460, 9: 461, 18, 33. v. eác.

ǽc, e; *f. An oak;* quercus :—Of ðære æce [MS. ǽc] andlang heges to ðæm wege *from the oak and along the hedge to the road,* Kmbl. Cod. Dipl. iii. p. 78, 7. v. ác.

ǽcan *to eke,* Solil. 11. v. écan.

æcced, es; *n. Vinegar;* acetum, Jn. Lind. War. 19, 30. v. eced.

æce, ace, es; *m. An ake, pain;* dolor :—Eal ðæt sár and se æce onwæg alǽded wæs *all the sore and ake were (led) taken away,* Bd. 5, 3; S. 616, 35: 5, 4; S. 617, 22. DER. acan *to ake.* v. ece.

ǽce; *adj. Eternal;* æternus :—Ðæt we ge-earnian ǽce dreámas *that we may obtain eternal delights,* Ps. C. 156. v. éce.

æced, es; *n. Vinegar* :—Onféng ðe Hǽlend ðæt æced *the Saviour received the vinegar,* Jn. Rush. War. 19, 30. v. eced.

æced-fæt, es; *n. An acid-vat, a vinegar-vessel;* acetabulum, Wrt. Voc. 25, 21. v. eced-fæt.

æced-wín, es; *n.* ACID-WINE; murratum vinum, Mk. Lind. War. 15, 23.

æ-céled *cooled; pp.* of æ-célan = a-célan. DER. célan.

æcelma, an; *m. A chilblain;* mula, L. M. 1, 30; Lchdm. ii. 70, 16.

ǽcen = ácen; *adj. Oaken, made of oak;* quernus, Cot. 165.

ǽcen, eácen; *pp.* of eácan *to increase.* v. eácan.

ÆCER, æcyr, es; *m.* I. *a field, land, what is sown, sown land;* ager, seges :—For ðam is se æcer gehǽten Acheldemah *propter hoc vocatus est ager ille Haceldama,* Mt. Bos. 27, 8. Hér ys seó bót, hú ðú meaht ðíne æceras betan *here is the remedy, how thou mayest improve thy fields,* Lchdm. i. 398, 1. Of ðære æcere *from the field,* Bt. Met. Fox 12, 3; Met. 12, 2. Æcera þúsend *a thousand fields,* 14, 10; Met. 14, 5.   II. a definite quantity of land which, in *A. Sax.* times, a yoke of oxen could plough in a day, an ACRE, *that is* 4840 *square yards;* jugeri spatium, jugerum, a jugo quod tantum fere spatii uno jugo boum arari posset: *also ager* — Ger. acker *an acre* :—Ælce dæg ic sceal erian fulne æcer oððe máre *omni die debeo arare integrum jugerum* [MS. *agrum*]

aut plus, Coll. Monast. Th. 19, 21. Ðæt is se teóða æcer, eal swâ seó sulh hit gegâ *that is the tenth acre, all as the plough goes over it*, L. C. E. 8; Th. i. 366, 6. Æceras *jugera*, Cot. 109. [*O. Sax.* akkar: *O. Frs.* ekker: *O. Ger.* ahhar: *N. Ger.* acker *a field, an acre: Goth.* akrs: *O. Nrs.* akr: *Lat.* ager: *Grk.* ἀγρός: *Sansk.* ajra *a plain.*]

æcer-ceorl, es; *m. A field-churl, a farmer, ploughman; agricola.* DER. æcer *a field*, ceorl *a free husbandman.*

æcer-man, æcer-mon; *g.* æcer-mannes; *m. A field-man, farmer; agricola,* Ælfc. Gl. 5.

æcern, æcirn, es; *n.* [æc = âc *oak*, corn *corn*] *The corn* or *fruit of an oak, an* ACORN, *a nut; glans:—*Æcern *glans,* Ælfc. Gl. 46; Som. 65, 7. Æcirnu, *pl. nom.* Gen. 43, 11. [*Spenser, Grafton,* acornes, *pl.: N. Dut.* aker *in* aker-boom: *N. L. Ger.* ecker, *m. n.: N. Ger.* ecker, *pl.* eckern, *m. n. glans* quernea or *fagea: Goth.* akran, *n. fructus: Dan.* agern, *n: Norw.* aakorn: *O. Nrs.* akarn, *n. glans silvestris.*]

æcer-spranca, æcer-spranga, an; *m.* [æcer, spranca, an; *m. a* shoot, sprout] *Young shoots springing up from acorns, saplings, the holm oak, scarlet oak; ilex:—*Æcer-spranca *ilex,* Ælfc. Gr. 9, 61; Som. 13, 48.

æcest = æcst *akest, 2nd pers. sing. pres. of* acan.

æceþ = æcþ *aketh, 3rd pers. sing. pres. of* acan.

æchir *an ear of corn,* Mt. Rush. Stv 12, 1. v. ear.

æ-ciorfan *to cut to pieces,* Ps. Spl 128, 4. v. a-ceorfan.

æcirnu *nuts,* Gen. 43, 11. v. æcern.

æc-læca, an; *m.* [æc = ag, *q. v.*] *A wretch, miscreant, monster; miser, perditus, monstrum,* Elen. Grm. 901; El. 902. v. ag-læca.

æ-cræft, es; *m. Law-craft and its result; legis peritia et vires inde oriundæ:—*Æcræft eorla *law-craft of men,* Elen. Kmbl. 869; El. 435: Cd. 173; Th. 217, 7; Dan. 19.

æ-cræftig; *adj. Law-crafty, one skilled in law, a lawyer, scribe; legis peritus:—*Him æcræftig andswarode *to them the skilled in law answered,* Cd. 212; Th. 262, 10; Dan. 742.

æcse *an axe,* Bd. 4, 3; S. 567, 26. v. æx.

æcst *akest, 2nd pers. sing. pres. of* acan.

æcþ *aketh, 3rd pers. sing. pres. of* acan.

æcumbe *oakum; stuppa,* Wrt. Voc. 40, 36. v. âcumba.

æcyr *a field:—*Blôdes æcyr *sanguinis ager,* Mt. Foxe 27, 8. v. æcer.

æcyrf, e; *f. That which is cut off, a fragment, piece; recisura, fragmentum:—*Ðara treówa æcyrf and lâfe forbærnde wæron *the offcuttings and leavings of the wood were burnt,* Bd. 3, 22; S. 552, 13. v. cyrf, ceorfan.

æd-, *prefixed to words, denotes Anew, again, as the Latin re:—*Ædsceaft *re-generation.* v. ed-.

æddran *kidneys; renes,* Ps. Spl. C. 7, 10. v. ædre.

æder-seax, ædre-seax, es; *n. A vein-knife, a lancet; lancetta,* Cot. 92.

æd-fæst [eád *substance,* fæst *fast, fixed*] *Goods, property; bona:—*Ædfæst tæht to healdenne *property taken to hold, a pledge,* Ælfc. Gl. 14; Som. 58, 8.

æd-leán *a reward,* Th. Diplm. A. D. 804–829; 459, 11. v. ed-leán.

ædr *vein, artery,* Ps. Th. 72, 17. v. ædre, ædre.

ædre; *adv. Quickly, promptly, at once, forthwith; illico, confestim, statim, protinus:—*Him ðá ædre God andswarede *God answered him forthwith,* Cd. 42; Th. 54, 4; Gen. 872. Wille ðé ða andsware ædre gecýðan *I will quickly let you know the answer,* Beo. Th. 714; B. 354. Nú ðú ædre const siþ-fæt mínne *now thou comprehendest at once my journey,* Exon. 52 b; Th. 184, 29; Gû. 1351. [*O. H. Ger.* atar: *O. Sax.* adro: *O. Frs.* edre *velociter.*] v. ædre.

ædre, æddre, édre, an; *f.* ædr, e; *f.* I. a channel for liquids, *An artery, a vein, fountain, river; arteria, vena, fons, rivus; v.* wæter-ædre:—Feorh alêton þurh ædra wylm *they let life forth through the fountain of their veins,* Exon. 72 b; Th. 271, 6; Jul. 478. Blêdaþ ædran *the veins shall bleed,* Salm. Kmbl. 290; Sal. 144. Swât ædrum sprong *blood sprang from the veins,* Beo. Th. 5925; B. 2966. II. *a nerve, sinew, kidney; nervus,* ren:—Wæron míne ædra ealle tolýsde *renes mei resoluti sunt,* Ps. Th. 72, 17. Ðú canst míne ædre ealle *tu possedisti omnes renes meos,* 138, 11. Ðá for ðam cýle him gescuncan ealle ædra *then all his sinews shrank because of the cold,* Ors. 3, 9; Bos. 64, 39. [*Plat.* ader: *O. Frs.* eddere, eddre: *O. Dut.* adere: *Ger.* ader: *M. H. Ger.* âder: *O. H. Ger.* âdara: *Dan.* aare: *Swed.* åder: *Norw.* aader: *O. Nrs.* æd, *f.*] DER. wæter-ædre.

ædre-seax *a vein-knife, lancet.* v. æder-seax.

ædre-weg, es; *m. A drain way, a vein, an artery; arteria, vena.* v. ædre, weg *a way.*

æ-drífan *to expel,* Ps. Spl. T. 42, 2: 43, 26. v. a-drífan.

æd-sceaft, e; *f. A regeneration, new creation; regeneratio:—*Hí ælce geáre weorþaþ to ædsceafte *they become every year a new creation,* Bt. 34, 10; Fox 150, 16. v. edsceaft.

Ædwines clif, EDWIN'S CLIFF, Chr. 761; Ing. 73, 15.

æd-wist *substance; substantia, essentia.* v. æt-wist.

æd-wít, es; *n. A reproach; opprobrium:—*Æd-wít manna *opprobrium hominum,* Ps. Spl. C. T. 21, 5. v. ed-wít.

æd-wítan *To reproach; exprobare:—*Æd-wioton him *improperabant ei,* Mt. Lind. Stv. 27, 44. v. ed-wítan.

æf, af, of; *prep. Of, from; ab, de.* v. compound æf-lást and in of-.

æf-æst, es; *n. Envy; invidia:—*Bútan æfæste *sine invidia,* Bd. 5, 22; S. 644, 13. v. æf-æst.

æ-fæst, -fest; *adj.* [æ *law,* fæst *fast, fixed*] *Firm in observing the law, religious, pious; tenax observandi legem, religiosus, pius, justus:—*Æfæst hæleþ *a pious man,* Cd. 59; Th. 72, 6; Gen. 1182. Æfæste men *pious men,* 86; Th. 108, 7; Gen. 1802. We æfæstra dæde dêman *we consider the deeds of the pious,* Exon. 40 a; Th. 133, 30; Gû. 497. Wæs he æfæst and ârfæst *was he devout and good?* Bd. 3, 14; S. 539, 33. v. æw-fæst.

æ-fæsten, es; *n. A legal fast; legitimum jejunium:—*III æfæstenu fæste he *tribus legitimis jejuniis jejunet,* L. Ecg. C. 4; Th. ii. 138, 1.

æ-fæstnes, -festnes, -nys, -ness, e; *f. Firmness in the law, religion; religio:—*He wæs mycelre æfæstnesse wer *he was a man of much religion,* Bd. 4, 31; S. 610, 7: 2, 9; S. 510, 30, 32.

æf-dæl; *g.* -dæles; *pl. nom.* -dalu; *n.* [æf, dæl *a vale*] *A descent; descensus:—*To æfdæle *ad descensum,* Lk. Lind. War. 19, 37. v. of-dæl.

æfdon *performed, executed,* Exon. 27 b; Th. 83, 16; Cri. 1357, = æfndon, *p. pl. of* æfnan.

æ-felle, a-felle; *adj.* [æ, fell *a skin*] *Barked, peeled, skinned; decorticatum,* Ælfc. Gl. 115; Som. 80, 34; Wrt. Voc. 61, 14.

ÆFEN, æfyn, æfen, es; *m. The* EVEN, *evening, eventide; vesper, vespera:—*Syððan æfen cwom *after evening came,* Beo. Th. 2475; B. 1235. Æfen ærest *vesperum primum,* Cd. 8; Th. 9, 7; Gen. 138. Æfena gehwâm *in each of evenings,* 148; Th. 184, 16; Exod. 108. Æt æfenne, on æfenne, *or* to æfenne, *at even, in the evening,* Ps. Spl. 29, 6. [*Laym.* aefen: *Orm.* efen: *Gow. Chauc.* even: *N. Dut.* avond: *M. Dut.* avont, *m.: Plat.* abend, *m.: O. Sax.* âband, *m.: O. Frs.* âvend, *m.: Ger.* abend, *m.: M. H. Ger.* âbent, *m.: O. H. Ger.* âpand, âbant, âbunt, *m.: Dan.* aften, *m.: Swed.* afton, *m.: Icel.* aptan, aftan, *n.: confr. Grk.* ὀψέ.]

æfen-dreám, es; *m. Even-song; vespertinus cantus.* v. æfen.

æfen-fela *as many; totidem,* Deut. 9, 11. v. efen-feola.

æfen-gebéd, es; *n. An evening prayer, evening service:—*Æfen-gebéd *vespertinum officium,* Ælfc. Gl. 34; Som. 62, 50.

æfen-gereord, e; *f. An evening meal, a supper; cœna,* Ælfc. Gl. 58; Som. 67, 87; Wrt. Voc. 38, 13.

æfen-gereordian; *p.* ode; *pp.* od *To sup* or *take supper; cœnare.* v. gereordian *to take food.*

æfen-gifl, -giefl, es; *n. Evening food, supper; cœna:—*Hí sécaþ ðæt hie fyrmest hlynigen æt æfengieflum [-giflum MS. C.] *quærunt primos in cœnis recubitus,* Past. 1, 2; MS. Hat. 6 b, 20: 44, 3; MS. Hat. 61 b, 22.

æfen-glóm, es; *m. The evening gloom* or *twilight; crepusculum:—*From æfenglôme ôþ ðæt eástan cwom dægrêdwôma *from evening twilight there came the rush of dawn from the east,* Exon. 51 b; Th. 179, 21; Gû. 1265.

æfen-grom; *adj. Fierce in the evening; vespere ferox:—*Grendel cwom eatol, æfengrom *Grendel came terrible, fierce at eve,* Beo. Th. 4154; B. 2074.

æfen-hlytta, an; *m. A fellow, consort, companion* or *mate; consors,* Ælfc. Gr. 9, 44; Som. 13, 6.

æfen-hrepsung, e; *f. The evening close; vesper.* v. hrepsung *closing.*

æfen-lác, es; *n. An evening sacrifice; vespertinum sacrificium:—*Swylce ahafenes handa mínra, ðonne ic æfenlác secge *elevatio manuum mearum sacrificium vespertinum,* Ps. Th. 140, 3.

æfen-lécan *to match; imitari.* v. efen-lécan.

æfen-lécan; *p.* -lǽhte; *pp.* -lǽht *To grow towards evening; advesperascere:—*Hit æfenlǽcþ *advesperascit,* Lk. Bos. 24, 29.

æfen-lécend *an imitator.* v. efen-lécend.

æfen-leóht, es; *n. Evening light; vespertina lux:—*Siððan æfen-leóht under heofenes hâdor beholen weorþeþ *after the evening light is concealed under heaven's serenity,* Beo. Th. 831; B. 413.

æfen-leóþ, es; *n. An evening song; vespertinus cantus:—*Atol æfenleóþ *a dreadful evening song,* Cd. 153; Th. 190, 18; Exod. 201.

æfen-líc; *adj. Vespertine, of the evening; vespertinus,* Ps. Spl. 140, 2.

æfen-mete; *m. Evening meat, supper; cœna,* Cot. 42.

æfen-rest, e; *f. Evening rest; vespertina requies:—*Sum sâre ongeald æfenreste *one paid dearly for his evening rest,* Beo. Th. 2508; B. 1252.

æfen-rima, an; *m.* [æfen *vesper,* rima *margo, labrum*] *Twilight; crepusculum.* v. rima *a rim, margin.*

æfen-sang, es; *m.* EVEN-SONG, *vespers; vespertinus cantus,* L. Ælf. C. 19; Th. ii. 350, 7.

æfen-sceóp, -scóp, es; *m. An evening bard; vespertinus cantor:—*Eald æfensceóp ic bringe *I bring an old evening bard,* Exon. 103 a; Th. 390, 21; Rä. 9, 5.

æfen-scíma, an; *m. Evening splendour; vespertinus splendor,* Cd. 112; Th. 147, 31; Gen. 2448.

æfen-spræc, e; *f. Evening speech; vespertina loquela:—*Gemunde æfenspræce he *remembered his evening speech,* Beo. Th. 1522; B. 759.

æfen-steorra, an; *m. The evening star; Hesperus; the Grk.* Ἕσπερος [*Lat.* vesper]. *the evening star, is called by Hesiod a son of Astræus*

and Eos, and was regarded by the ancients the same as the morning star, whence both Homer and Hesiod call him the bringer of light, ἐωσ-φόρος, Il. xxii. 318 : xxiii. 226. The Romans designated him by the names Lucifer and Hesperus, to characterise him as the morning or evening star :—Se steorra ðe we hâtaþ æfensteorra, ðonne he biþ west gesewen, ðonne tácnaþ he æfen. Færþ he ðonne æfter ðære sunnan on ðære eorþan sceade, oþ he ofirnþ ða sunnan hindan, and cymþ wið fóran ða sunnan up, ðonne hâten we hine morgensteorra (q. v.) forðam he cymþ eástan up, bodaþ ðære sunnan cyme *the star which we call the evening star, when it is seen westwardly, then it betokens the evening. It then goes after the sun into the earth's shade, till it runs off behind the sun, and comes up before the sun, then we call it the morning star, because it comes up in the east, and announces the sun's approach*, Bt. 39, 13 ; Fox 232, 34. Se môna, mid his blâcan leóhte, dunniaþ ðone beorhtan steorran, ðe we hâtaþ morgensteorra : ðone ilcan we hâtaþ ôðre naman, æfensteorran *the moon, with his pale light, obscures the bright star, which we call the morning star : the same we call by another name, the evening star*, 4 ; Fox 8, 3.

æfen-þénung, e ; *f. An evening service* or *duty, evening repast, supper* ; cœna, R. Concord 8. v. þegnung.

æfen-þeówdôm, es ; *m. An evening service* or *office* ; vespertinum officium, Ælfc. Gl. 34 ; Som. 62, 50.

æfen-tíd, e ; *f. The eventide, evening* ; vespertina hora :—Seó æfen-tíd ðæs dæges *the eventide of the day*, Dial. 1, 10. On æfen-tíd *at eventide*, Cd. 111 ; Th. 146, 19 ; Gen. 2424.

æfen-tíma, an ; *m. Evening time, eventide* ; vespertinum tempus :—Ðá æfentíma wæs, he férde to Bethanîam *cum jam vespera esset hora, exiit in Bethaniam*, Mk. Bos. 11, 11.

æfen-tungel, es ; *m. n. The evening star* ; hesperus. v. tungel.

ÆFER ; *adv.* EVER, *always* ; unquam, semper :—Æfer ge fliton ongên God *semper contentiose egistis contra Deum*, Deut. 31, 27. v. æfre.

æfesen, æfesn, æbesen, æbesn, e ; *f. Pasturage, the charge for pigs going into the wood to fatten on acorns* ; pasnagium, pretium propter porcos in quercetum admissos :—Gif mon nîme æfesne on swînum *if [a man] any one take pasturage on swine*, L. In. 49 ; Th. i. 132, 18.

æf-ést, æf-æst, æfst, es ; *n.* [æf, of = ab, ést *gratia*] *Without favour or good-will, hence, Envy, spite, enmity, zeal, rivalry, emulation* ; livor, invidia, odium, zelus, æmulatio :—Æfst and oferhygd *envy and pride*, Cd. 1 ; Th. 3, 1 ; Gen. 29. Eald-feóndes æfest *the old fiend's envy*, Exon. 61 b ; Th. 225, 5 ; Pł. 401. Æféstes *livoris*, Mone B. 2699, p. 386. Heora æfstu ealle sceamien *they all shall be ashamed of their enmities*, Ps. Th. 69, 4. Fore æfstum *from envy*, Exon. 43 a ; Th. 144, 27 ; Gú. 684. Æféstum onæled *inflamed with envy*, Exon. 84 a ; Th. 316, 3 ; Môd. 43. [*O. Sax.* ab-unst, *f. invidia : O. Frs.* ev-est *invidia : Ger.* ab-gunst, *f. invidia : O. H. Ger.* ap-anst, ap-unst, *m. invidia, livor, zelus, rancor.*]

æ-fest ; *adj.* [æ *law*, fæst *fast, fixed*] *Fast* or *firm in the law, religious, devout* ; religiosus :—Wæs se mon swýðe æfest *erat vir multum religiosus*, Bd. 4, 24 ; S. 598, 20. Ongunnon æfeste leóþ wyrcean *religiosa poemata facere tentabant*, id ; S. 596, 38. v. æ-fæst, æw-fæst.

æf-ést-ful ; *adj. Full of envy* ; invidia plenus, invidiosus :—He is swîðe æféstful for ðînum gôde *he is very full of envy at thy prosperity*, Th. Apol. 14, 24. v. æf-ést.

æf-éstian, -éstigan ; *p.* ode ; *pp.* od *To envy, be envious of* or *at* ; invidere :—Ðes iunga man ne æféstigaþ on nánum þingum, ðe he hér gesihþ *this young man is envious at nothing, which he here seeth*, Th. Apol. 14, 25 : Cot. 119. v. æf-ést.

æf-éstig, æfstig ; *adj. Envious, emulous, jealous* ; invidus, æmulus :—Sum eald and sum æféstig ealdorman *an old and an envious nobleman*, Th. Apol. 14, 19. v. æf-ést.

æ-festlíce ; *adv. Religiously* ; religiose. v. fæstlíce.

æf-éstnes, -ness, -nys, -nyss, e ; *f. Envy, spite* ; invidia, malignitas. DER. æf-ést.

æ-festnes, -ness, e ; *f. Religion, devotion* ; religio :—Ða ðe to æfestnesse belumpon *quæ ad religionem pertinebant*, Bd. 4, 24 ; S. 597, 1. v. æ-fæstnes.

Æffric ; *def. m.* Æffrica ; *adj. African* ; Afer :—Severus Câsere se wæs Æffrica cynnes *Severus Cæsar genere Afer*, Bd. 1, 5 ; S. 476, 5. v. Affric.

æf-íst *envy* ; invidia, Mt. Lind. Stv. 27, 18. v. æf-ést.

æf-lást, es ; *m.* [æf = af *from*, lást *a course*] *A wandering away ?* aberratio, Cd. 166 ; Th. 207, 27 ; Exod. 473.

æfnan ; *p.* de ; *pp.* ed *To perform, execute, labour, show* ; patrare, facere, laborare, præstare :—His dômas æfnaþ *they fulfil his judgments*, Exon. 32 b ; Th. 102, 29 ; Cri. 1680. Gif hý woldun his bebodu æfnan *if they would execute his judgments*, 54 a ; Th. 152, 29 ; Gú. 816. Wile corlscipe æfnan *he wishes to show his dignity*, 87 a ; Th. 327, 3 ; Wîd. 141. Æfdon unsofte *for æfndon ?* 27 b ; Th. 83, 16 ; Cri. 1357. DER. ge-æfnan. v. efnan.

æfne ; *interj. Behold* ; ecce :—Æfne sôþlíce sôþfæstnysse ðú lufudest *ecce enim veritatem dilexisti*, Ps. Spl. 50, 7. v. efne ; *interj.*

æfnian ; *p.* ode ; *pp.* od *To grow towards evening* ; vesperascere, Dial. 1, 10.

æfnung, e ; *f. Evening* ; vespera :—Heó com ða on æfnunge eft to Nôe *illa venit ad eum [Noe] ad vesperam*, Gen. 8. 11 : Homl. Th. ii. 266, 5, 6.

æfre, æfer ; *adv. Ever, always* ; unquam, semper :—Nolde æfre *nolebat unquam*, Cd. 72 ; Th. 89, 14 ; Gen. 1480. Ne sceal æfre gehéran *nor shall I ever hear*, 216 ; Th. 275, 14 ; Sat. 171. Nú ic eóm orwêna ðæt· unc seó ædyl-stæf æfre weorþe gifede ætgædere *now I am hopeless that the staff of our family will ever be given to us two together*, 101 ; Th. 134, 12 ; Gen. 2223. Ðú æfre wære *tu semper fuisti*, Exon. 9 b ; Th. 8, 2 ; Cri. 111. Æfre forþ *sempiternum*, Cd. 220 ; Th. 282, 35 ; Sat. 297. Æfre to aldre *in æternum*, 38 ; Th. 51, 1 ; Gen. 820. æfre = â, q. v.

æ-fremmende ; *part. Fulfilling the law, religious* ; legis præcepta conficiens, religiosus :—Ic læran wille æfremmende ðæt ge eówer hús gefæstnige *I will teach that you, the laws fulfilling, should make firm your house*, Exon. 75 a ; Th. 281, 18 ; Jul. 648.

æfst *envy*, Past. 13, 2 ; Hat. MS. 17 a, 12 : Cd. 1 ; Th. 3, 1 ; Gen. 29. v. æf-ést.

æfstian ; *p.* ode ; *pp.* od *To hasten* ; festinare, accelerare. v. éfstan.

æfstig ; *adj. Envious, emulous* ; æmulus :—Æfstig wið ôðra manna yflu *æmulus contra aliena vitia*, Past. 13, 2 ; MS. Hat. 17 a, 11. v. æf-éstig.

æft ; *adv.* AFT, *behind*, as go aft = go astern, *Afterwards, again* ; postea, iterum :—Moises cwæþ æft• to Israela folce *Moses said afterwards to the people of Israel*, Deut. 28, 15. Æft uferan dôgum *afterwards in later days*, Beo. Th. 4406 note ; B. 2200. Ðæt hí æft to him cômen *that they would come to him again*, Bt. Met. Fox 1, 130 ; Met. 1, 65. v. eft.

æftan ; *adv. Behind* ; post, pone :—Earn æftan hwît *the eagle white behind*, Chr. 937 ; Th. i. 206, 29 ; Æðelst. 63, col. 1. DER. be-æftan.

æftan-weard ; *adj. Coming after, following* ; posterior :—Rinc biþ on ôfeste, se mec onþýþ æftanweardne *the man is in haste, who urges me following*, Exon. 125 a ; Th. 480, 3 ; Rä. 63, 5. v. weard II ; *adj.*

æft-beteht *re-assigned*, R. Ben. 4. v. eft-betæht.

æftemest, -myst, -most ; *adj. superlative of* æfter,—*After-most, last* ; postremus, novissimus :—Deós bôc is æftemyst on ðære bibliobêcan *this is the last book of the Bible*, Ælfc. T. 31, 22 ; Grn. Ælfc. T. 16, 3. Ðonne he sylf mid ðam fyrmestan dæle wið ðæs æftemestan flúge *when he himself with the first part should flee towards the hindermost*, Ors. 4, 6 ; Bos. 85, 20 : Mk. Bos. 12, 22 : Jn. Bos. 7, 37.

æften-tíd, e ; *f.* [æftan *after*] *Evening, eventide* ; vespertinum tempus, vesper :—Æt morgenes gancg wið æftentíd ealle ða dêman Drihten healdeþ *exitus matutini et vespere delectaberis*, Ps. Th. 64, 9.

æfter ; *prep.* [æft, q. v ; er, q. v.] *dat* ; *rarely acc.* I. local and temporal *dat.*—AFTER ; post :—Ne far ðú æfter fremdum godum *go not thou after strange gods*, Deut. 6, 14. Æfter þrím monþum *after three months*, Gen. 38, 24. Æfter dagum *after those days*, Lk. Bos. 1, 24. Cumaþ æfter me *venite post me*, Mt. Bos. 4, 19. Æfter þrým dagum [MS. dagon] *he aríse post tres dies resurgam*, Mt. Bos. 27, 63. Ða eóde ðæt wíf æfter him *then the wife went after him*, Bt. 35, 6 ; Fox 170, 13. Hâm staðeledon, ân æfter ôðrum *they established a home, one after another*, Cd. 213 ; Th. 266, 22 ; Sat. 26. Æfter ðâm wordum werod eall arâs *after those words all the host rose*, Cd. 158 ; Th. 196, 29 ; Exod. 299 : Exon. 28 b ; Th. 86, 24 ; Cri. 1413. Wunder æfter wundre *wonder after wonder*, Beo. Th. 1866 ; B. 931 : Cd. 8 ; Th. 9, 19 ; Gen. 144 : Cd. 46 ; Th. 59, 15 ; Gen. 964 : Cd. 143 ; Th. 178, 1 ; Exod. 5 : Cd. 148 ; Th. 184, 18 ; Exod. 109 : Cd. 227 ; Th. 304, 14 ; Sat. 630 : Exon. 16 a ; Th. 36, 8 ; Cri. 573 : Exon. 18 a ; Th. 44, 31 ; Cri. 711 : Exon. 117 a ; Th. 449, 32 ; Dôm. 80 : Exon. 117 a ; Th. 450, 3 ; Dôm. 82 : Exon. 124 a ; Th. 476, 20 ; Ruin. 10 : Beo. Th. 170 ; B. 85 : Beo. Th. 238 ; B. 119 : Apstls. Kmbl. 163 ; Ap. 82 : Andr. Kmbl. 175 ; An. 88 : Andr. Kmbl. 265 ; An. 133 : Exon. 39 b ; Th. 130, 22 ; Gú. 442 : Exon. 40 b ; Th. 134, 5 ; Gú. 503 : Elen. Kmbl. 859 ; El. 430 : Elen. Kmbl. 977 ; El. 490 : Exon. 118 a ; Th. 454, 10 ; Hy. 4, 30. 2. extension over space or time,—*Along, through, during* ; κατά, per :—Sæton æfter beorgum *they sat along the hills*, Cd. 154 ; Th. 191, 9 ; Exod. 212. His wundra geweorc, wíde and síde, brême æfter burgum *his works of wonder, far and wide, famed through towns*, Exon. 45 b ; Th. 155, 4 ; Gú. 855. Deáh ic fela for him æfter woruldstundum wundra gefremede *though I performed many miracles for them during my time in this world*, Elen. Kmbl. 725 ; El. 363 : Exon. 55 b ; Th. 196, 18 ; Az. 176 : Judth. 10 ; Thw. 21, 17 ; Jud. 18 : Salm. Kmbl. 233 ; Sal. 116 : Exon. 108 a ; Th. 412, 25 ; Rä. 31, 5. 3. mode or manner,—*According to, by means of* ; secundum, propter :—Æfter dôme ðínum geliffæsta me *secundum judicium tuum vivifica me*, Ps. Lamb. 118, 149. He hæfþ mon geworhtne æfter his onlícnesse *he has created man after [secundum] his own image*, Cd. 21 ; Th. 25, 19 ; Gen. 396. Ðæt sweord ongan æfter heaðoswâte wanian *the sword began to fade away by the warsweat [in consequence of the hot blood]*, Beo. Th. 3216 ; B. 1606 : Exon. 19 b ; Th. 50, 20 ; Cri. 803 : Andr. Kmbl. 156 ; An. 78 : Exon. 45 b ; Th. 154, 27 ; Gú. 849 : Bt. Met. Fox 20, 93 ; Met. 20, 47 : Exon. 110 a ; Th. 421, 8 ; Rä. 40, 15 : Beo. Th. 5499 ; B. 2753 : Cd. 28 ; Th. 37, 19 ; Gen. 59. 4. object,—*after, about* ; propter, ob, de :—Hæleþ frægn æfter æðelum *a chief asked after the heroes*, Beo. Th. 670 ; B. 332. Him æfter deórum men dyrne langaþ *he longs secretly after the dear man*, Beo. Th. 3762 ; B. 1879. Gróf æfter golde *he dug after gold*, Bt. Met. Fox 8, 113 ; Met. 8, 57 : Elen. Kmbl. 1346 ; El. 675 : Beo. Th. 2648 ; B. 1322 : Beo. Th. 2688 ; B. 1342 : Cd. 15 :

Th. 18, 33; Gen. 282: Cd. 15; Th. 19, 14; Gen. 291: Cd. 92; Th. 117, 20; Gen. 1956: Cd. 98; Th. 130, 3; Gen. 2154: Cd. 203; Th. 251, 30; Dan. 571: Elen. Kmbl. 1653; El. 828: Andr. Kmbl. 74; An. 37: Beo. Th. 4913; B. 2461: Beo. Th. 4917; B. 2463: Beo. Th. 4528; B. 2268.   **II.** *acc;* cum accusativo, *After, above, according to; post, super, secundum:*—Æfter ðâs dagas *post hos dies,* Lk. Lind. War. 1, 25. He eorþan æfter wæter ærest sette *qui fundavit terram super aquas,* Ps. Th. 135, 6. Stefne mîne gehêr æfter mildheortnesse dîne, Drihten *vocem meam audi secundum misericordiam tuam, Domine,* Ps. Lamb. 118, 149. [*O. Sax.* aftar, after: *O. Frs.* efter, after: *O. Dut. N. Dut.* achter: *Ger.* after, *only in compnd: M. H. Ger.* after: *O. H. Ger.* aftar: *Goth.* aftra *backward, again: Dan.* efter: *Swed.* efter: *O. Nrs.* eptir, eftir, *prep;* aptr, aftr, *adv. back, again: Sansk.* apara.]

æfter; *adv. After, then, afterwards; post, postea, exinde:*—Æfter siddan *ever afterwards, from thenceforth,* Cd. 26; Th. 35, 6; Gen. 550. Æfter to aldre *for ever after,* Cd. 22; Th. 28, 15; Gen. 436. Dæm eafera wæs æfter cenned *a son was afterwards born to him,* Beo. Th. 24; B. 12. Word æfter cwæþ *then he spake these words,* Beo. Th. 636; B. 315. Ær oððe æfter *sooner or later,* Exon. 32 b; Th. 103, 22; Cri. 1692. Ic wât æfter nû hwâ mec serede ofer flôdas *now afterwards I know who conveyed me over the floods,* Andr. Kmbl. 1808; An. 906. Dær sceal ylda cwealm æfter wyrþan *then must slaughter of men take place afterwards,* 364; An. 182. Swâ ðas foldan fædme bewîndeþ ðes eástrodor and æfter west *quantum ortus distat ab occasu,* Ps. Th. 102, 12.

æftera, æftra; *adj. compar. of* æfter, —*Hinder, next, second; posterior, sequens, alter, secundus:*—Dŷ æfteran dæge *sequenti die,* Lk. Bos. 13, 33. Dæs æfteran monþes *mensis secundi,* Ex. 16, 1. On ðam forman dæge ðæs æftran monþes *primo die mensis secundi,* Num. 1, 18. Seó æftre, *i. e.* eá, Ethiopia land belîgeþ ûton *the next river encompasses the country of Ethiopia,* Cd. 12; Th. 15, 4; Gen. 228. Siddan ic ongon on ðone æftran ânseld bûgan *after I had begun to live in this second hermitage,* Exon. 50 b; Th. 176, 22; Gû. 1214.

æfter-boren [= æftergenga, *q. v.*] *part. Born after the father's death; posthumus,* Ælfc. Gr. 47; Som. 48, 32.

æfter-cweðan; *p.* -cwæþ; *pp.* -cweden *To speak after, repeat, to answer, revoke, renounce, abjure; repetere, revocare:*—Bebeád he ðæt him mon lengran cwidas beforan cwæde, and he symle gedêfelîce æftercwæþ *he ordered longer sayings to be spoken before him, and he always repeated them properly,* Bd. 5, 2; S. 615, 15. His brôðer griþ eall æftercwæþ *his brother renounced all peace,* Chr. 1094; Th. 360, 23. Æftercweðendra lof *the praise of the after-speaking* [*post mortem laudantium*], Exon. 82 b; Th. 310, 10; Seef. 72.

æfter-eala, an; *m. After-ale, small beer:*—Æfter-eala *sapa,* Ælfc. Gl. 33; Som. 62, 22; Wrt. Voc. 28, 5.

æfter-fæce; *adv.* [æfter *after,* and the *dat.* of fæc *a space*] *Afterwards, after that;* postmodum. v. fæc.

æfter-folgere, es; *m. A follower;* successor, Ors. 3, 11; Bos. 74, 36.

æfter-folgian; *p.* ode; *pp.* od *To follow after, pursue;* subsequi, persequi:—Him æfterfolgiende wæron *they were pursuing him,* Ors. 1, 10; Bos. 32, 25.

æfter-fylging, e; *f. A following after, a sequence;* sectatio, successio. v. fylging.

æfter-fylian, -filian; *p.* de; *pp.* ed *To follow or come after, to succeed;* sequi, prosequi, subsequi:—Dæs sæs smyltnys æfterfyligeþ *serenitas maris prosequetur,* Bd. 3, 15; S. 541, 35. Dæs æfterfiliendan tâcnes *signi sequentis,* Ex. 4, 8.

æfter-fyligend, -fylgend, es; *m. One who follows or succeeds, a follower;* successor:—Ac Oswald his æfterfyligend hî ge-endade swâ we ær beforan sædon *sed successor ejus Oswaldus perfecit ut supra docuimus,* Bd. 2, 20; S. 521, 36: Bd. 5, 23; S. 646, 2.

æfter-fylignes, -ness, e; *f. A following after, a succession, succeeding;* successio. v. fylignes.

æfter-gân [gân *to go*] *To follow after;* subsequi, Past. 15, 2?

æfter-gencnys, -nyss, e; *f.* [gengnys *a going*] *Extremity;* extremitas, R. Ben. Interl. 7.

æfter-genga, an; *m.* [genga *goer*] *One who goes or follows after, a follower;* successor, posthumus:—Æftergenga *posthumus,* æfter boren, se ðe biþ geboren æfter bebyrgedum fæder *one who is born after the father has been buried,* Ælfc. Gr. 47; Som. 48, 32. Dû me ne derige, ne mînum æftergengum *ne noceas mihi et posteris meis,* Gen. 21, 23.

æfter-gengnys, -nyss, e; *f. Succession;* posteritas. v. æfter-gencnys.

æfter-gild, -gyld, es; *n. An after-payment, a paying again or in addition;* secunda *vel* iterata compensatio, L. C. S. 24; Th. i. 390, 7.

æfter-hætu, e; *f.* [æfter, hætu *heat*] *After-heat;* insequens calor:—Mid ungemetlîcum hærfest-wætan and æfterhæte *from heavy harvest-rains and after-heat,* Ors. 3, 3; Bos. 55, 23.

æfter-hŷrigean; *p.* de; *pp.* ed *To follow another's example, to imitate, resemble;* imitari:—He wilnode æfterhŷrigean *he wished to imitate,* Bd. 3, 18; S. 545, 44.

æfter-leán, es; *n. An after-loan, reward, recompense, retribution;* præmium, merces:—Þearl æfterleán *hard retribution,* Cd. 4; Th. 5, 24; Gen. 76.

æfter-lîc; *adj. After, second;* secundus, Cot. 191.

æfterra *second;* secundus:—Se æfterra deáþ *the second death,* Bt. 19; Fox 70, 18. Sende he eft æfterran sîðe ærenddracan *he sent messengers again a second time,* Bd. 2, 12; S. 513, 10. v. æftera.

æfter-râp, es; *m. An AFTER-ROPE, a crupper;* postilena, Ælfc. Gl. 20; Som. 59, 54.

æfter-rîdan; *p.* -râd, *pl.* -ridon; *pp.* -riden *To ride after;* equo insequi:—Hîg ða sôna æfterridon ídelum færelde *secuti sunt eos per viam,* Jos. 2, 7.

æfter-ryne, es; *m. An encountering, meeting, running against one;* occursus:—Æfterryne his ôþ to heáhnesse his *occursus ejus usque ad summum ejus,* Ps. Spl. 18, 7.

æfter-sang, es; *m. The after-song;* posterior cantus:—Mid ðam æftersange *with the after-song,* L. Ælf. P. 31; Th. ii. 376, 6.

æfter-singend, es; *m. An after-singer;* succentor, Wrt. Voc. 28, 21.

æfter-spræc, e; *f. After-speech or claim;* repostulatio, L. O. 7; Th. i. 180, 23.

æfter-sprecan; *p.* -spræc, *pl.* -spræcon; *pp.* -sprecen [sprecan *to speak*] *To claim;* petere, repetere:—Âgnung biþ nêr ðam ðe hæfþ, ðonne ðam ðe æftersprecþ *possession is always nearer to him who has, than to him who claims,* L. Eth. ii. 9; Th. i. 290, 21.

æfter-spyrian, -spyrgean; *p.* ede; *pp.* ed *To inquire after, examine;* examinare:—Gif ge hit willaþ æfterspyrian *if ye will examine it,* Bt. 16, 2; Fox 52, 8. v. spyrian.

æfter-weard *After,* AFTERWARD, *following;* posterior, secundus:—Gif he me æfterweard weorþeþ *if he shall be after* [*afterward*] *me,* Exon. 104 b; Th. 397, 3; Râ. 16, 14. v. æfte-weard, weard; *adj.*

æfter-weardnes, -ness, e; *f. Posterity;* posteritas, Cot. 149.

æfter-wearþ beón *To be away, absent,* Bd. 3, 15; S. 542, note 6. v. æfweard.

æfter-yldo, -yld, e; *f.*   **I.** *after-age, old age;* ætas provecta:—Ne mâgon ða æfteryld in ðam ærestan blæde geberan *they may not produce* [*show*] *old age in their first strength* [*youth*], Exon. 39 b; Th. 132, 3; Gû. 467.   **II.** *an after-age, after-time;* posterius ævum:—Swâ nænig æfteryldo syððan gemunan mæg *so no after-age since can remember,* Bd. 1, 14; S. 482, 22.

æfte-weard; *adj.* [= æfter] *After, back, late, latter, full;* posterior:—Æfteweard lencten *full spring,* Wrt. Voc. 53, 27. Æfteweard heáfod *the back of the head,* 42, 43. Drihten ðê gesett nâ on æfteweard *the Lord will not set thee in the after-part,* Deut. 28, 13.

æfte-wearde; *adv.* [æfter, wearde, weardes] *Afterward, after, behind;* post, pone:—Dû gesihst me æftewearde *thou shalt see me behind,* Ex. 33, 23.

æf-þanc, es; *m:* æf-þanca, -þonca, -þunca, an; *m. Offence, insult, grudge, displeasure, envy, zeal;* simultas, offensa, odium, zelus:—Swindan me dyde æfþanca mîn *tabescere me fecit zelus meus,* Ps. Spl. M. 118, 139. Æfþonca gefylled *full of grudges,* Exon. 83 b; Th. 315, 4; Mod. 26. Eald æfþoncan edniwedan *they have renewed old grudges,* 72 b; Th. 271, 20; Jul. 485. Æfþancum herian *to vex with insults,* Cd. 102; Th. 135, 3; Gen. 2237.

æftyr *after, according to;* secundum, Mt. Bos. 9, 29. v. æfter **I. 3.**

æf-weard, æf-ward; *adj. Absent, distant;* absens:—Lîcumlîce æfward *corporaliter absens,* Bd. 3, 15; S. 542, 6.

æf-weardnes, -ness, e; *f. Absence, removal, posterity;* absentia:—For ðînre æfweardnesse *because of thy absence,* Bt. 10; Fox 28, 28.

æf-werdelsa, an; *m. Damage, detriment, loss;* detrimentum, damnum, L. Alf. 27; Th. i. 50, 28. v. æf-werdla.

æf-werdla, æf-wyrdla, æ-wyrdla, a-wyrdla, an; *m.* [æf *of,* wyrdan *to corrupt*] *Damage, injury, loss, the amercement for it;* detrimentum, jactura, damnum:—Þolie ðone æfwerdlan [æfwyrdlan MS. H.] *let him bear the damage,* L. In. 40; Th. i. 126, 16: R. Ben. 2: Cot. 104.

æ-fyllende; *adj.* [æ = law, fyllende *part. of* fyllan *to fill, fulfil*] *Following the law, faithful;* legem exsequens:—Seó circe æfyllendra *the church of the faithful,* Exon. 18 a; Th. 44, 17; Cri. 704.

æfyn, es; *m. The evening:*—On æfyn *at evening,* Cd. 17; Th. 20, 22; Gen. 313. v. æfen.

æ-fyrmþa; *pl. f.* [æ, fyrmþ, e; *f. washing*] *Ablutions, the sweepings of a house, the refuse of things or things of no value;* ablutiones, quisquiliæ:—Æfyrmþa [MS. æfyrmþe] *quisquiliæ,* Ælfc. Gr. 13; Som. 16, 22.

ÆG, æig; *g.* æges; *pl. nom. acc.* ægru; *g.* ægra; *d.* ægrum, ægerum; *n. An* EGG; *ovum*:—Gif hit [cild] æges bitt *if he ask for an egg,* Homl. Th. i. 250, 9. Ðæt æg [æig MS.] getácnaþ ðone hâlgan hiht *the egg betokens the holy hope,* i. 250, 11. Gif he bit æg si *petierit ovum,* Lk. Bos. 11, 12. Genim hænne æges geolocan *take the yolk of a hen's egg,* L. M. 1, 2; Lchdm. ii. 38, 6. Sceáwa nû on ânum æge, hû ðæt hwîte ne biþ gemenged to ðam geolcan, and biþ hwæðere ân æg *look now on an egg, how the white is not mingled with the yolk, and yet it is one egg,* Homl. Th. i. 40, 27, 28. On æge biþ gioleca on middan *in an egg the yolk is in the middle,* Bt. Met. Fox 20, 338; Met. 20, 169. Of ægerum *from eggs,* Exon. 59 a; Th. 214, 2; Ph. 233. Ægru lecgan *to lay eggs,* Som.

121. **Æges hwîte** *white of an egg.* Æmettan ægru genîm *take emmet's eggs*, L. M. 1, 87; Lchdm. ii. 156, 6. [*Ger.* ei, *n* : *M. H. Ger.* ei, *g.* eies, eiges, *pl.* eiger, *n* : *O. H. Ger.* ei, *g.* eics, eiges, *pl.* eigir, *n* : *Dan.* äg, *n* : *Swed.* ägg, *n* : *O. Nrs.* egg, *n*.]

æg, e ; *f. water, water land, an island.* v. æge, îgg.

æg- used in composition,—*water, sea* ; aqua, mare. DER. æg-flota, æg-weard. v. îg-.

æg- *Ever, always* ; semper : either a contraction of the prefixes â, æ, with a *g* added, as æg, or derived from aa = â, âwa, æw. It is used in compound pronouns and adverbs, as,—æg-hwâ, æg-hwær, æg-hwilc, *etc* ; but, in its place, we also find the prefix â-, as,—â-hwær, â-hwilc, *etc.* Both æg- and â- impart to their compounds a sense of universality.

ægan *to own*, Ps. Spl. T. 78, 12 : 138, 12. v. âgan.

æge *fear* ; timor, terror, Chr. 1006, Th. 257, 41. v. ege.

æge *the island* ; insulam :—Æt eðelinga æge *at the island of nobles* ; apud nobilium insulam, Sim. Dunelm. an. 888. v. Æðelinga îgg.

Ægeles birg *Aylesbury*, Chr. 571 ; Th. 32, 29. v. Ægles burg.

Ægeles ford, Egeles ford, es ; *m. Ailsford*, Chr. 1016 ; Th. 279, 16, col. 2 : 1016 ; Th. 282, 10, col. 2.

Ægeles þrep *Aylesthorpe*, Chr. 455 ; Th. 21, 32. v. Ægles þrep.

ægen ; *adj. Own* ; proprius, Bt. 14, 2 ; Fox 44, 23. v. âgen.

æger-felma, an ; *f. Film of an egg* ; membrana vitellum complectens :—Genîm ðonne ægerfelman *then take film of egg*, L. M. 1, 11 ; Lchdm. ii. 54, 21.

ægerum *from eggs*, Exon. 59 a ; Th. 214, 2 ; Ph. 233. v. æg.

æ-gewrítere, es ; *m.* [*æ law*, gewrítere *a writer*] *A writer* or *composer of laws* ; legum conditor, Prov. 8.

æg-flota, an ; *m. A floater on the sea, sailor, ship* ; nauta, navis, Andr. Kmbl. 515 ; An. 258. v. flota.

æg-hwâ ; *m. f : neut.* æg-hwæt ; *gen.* æg-hwæs [â + ge + hwâ] *Every one, everything* ; quisque, quicunque :—Æghwâ secge *let every one say*, Exon. 88 b ; Th. 333, 5 ; Vy. 97 : 125 a ; Th. 482, 4 ; Rä. 66, 2. Æghwæt heó gefôn mæg *whatever she may seize*, Bt. 25 ; Fox 88, 14. God æghwæs wealt *God governs everything*, Bt. 35, 4 ; Fox 160, 14. Þearfum æghwæs oftugon *ye denied the poor everything*, Exon. 30 a ; Th. 92, 8 ; Cri. 1505. Se fugol is on hiwe æghwæs ænlîc *the bird is in aspect every way unique*, 60 a ; Th. 219, 24 ; Ph. 312. Æghwæs orwîgne *wholly defenceless*, 72 a ; Th. 268, 18 ; Jul. 434.

æg-hwær, â-hwær ; *adv.* [â + ge + hwær]. I. *everywhere* ; ubique :—God æghwær is eall, and nâhwâr todæled *God is everywhere all, and nowhere divided*, Homl. Th. i. 286, 27. Hî ðâ farende æghwær bodedon *illi profecti prædicaverunt ubique*, Mk. Bos. 16, 20. Æghwær sindon hiora gelîcan *they are everywhere like them*, Bt. Met. Fox 10, 116 ; Met. 10, 58. II. *in every respect, in every way* ; omnino :—Eofore eom æghwær cênra *I am in every respect bolder than a wild boar*, Exon. 110 b ; Th. 423, 9 ; Rä. 41, 18 : Ps. Th. 102, 14.

æg-hwæt *whatever* ; quodcunque. v. æg-hwâ.

æg-hwæðer ; *pron.* [â + ge + hwæðer]. I. *of two, either, each, both* ; uterque :—Æghwæðer ôðerne earme beþehte *they embraced each other*, Andr. Kmbl. 2029 ; An. 1017. Beámas twegen ðara æghwæðer efngedælde heáhþegnunga hâliges gâstes *two pillars, each of which shared alike the high services of the holy spirit*, Cd. 146 ; Th. 183, 21 ; Exod. 94. II. *of many, every one, each* ; unusquisque :—Heora æghwæðrum *to each, to every one of them*, Beo. Th. 3277 ; B. 1636. Æghwæðer ge lengre fæc ðysses lîfes ðê forgîfan ge ðê eác ðæs êcan lîfes inganges wyrþne gedôn *et hujus vitæ longiora spatia concedere et ingressu te vitæ perennis dignum reddere*, Bd. 3, 13 ; S. 539, 2. Æghwæðer ge—ge *et—et*, 2, 16 ; S. 519, 34.

æg-hwanan, -hwanon, -hwonon, -hwanone, -hwonene ; *adv. Everywhere, every way, on all sides* ; undique :—Æghwanan mid wæterum ymbseald *undique aquis circumdata*, Bd. 4, 19 ; S. 588, 28. Hî æghwanon to him cômon *conveniebant ad eum undique*, Mk. Bos. 1, 45. Æghwonan ymb-boren mid brondum *on every side surrounded with brands*, Exon. 74 a ; Th. 277, 14 ; Jul. 580. Æghwanon, Ælfc. Gr. 45 ; Som. 46, 57. Hine æghwonan ælmihtig God [MS. Good] gehealdeþ *Almighty God keeps him everywhere*, Bt. Met. Fox 7, 89 ; Met. 7, 45. Æghwonon *everywhere*, Bd. 4, 13 ; S. 582, 44. Æghwanone, 3, 6 ; S. 528, 18. Æghwonene, 3, 15 ; S. 541, 42.

æg-hwâr, æg-hwêr *everywhere*, Ors. 4, 1 ; Bos. 76, 38. v. æg-hwær.

æg-hwider, -hwyder ; *adv. On every side, every way* ; quaquaversum :—Æghwider ymb swâ swâ Edwines rîce wære *quaquaversum imperium regis Æduini pervênerat*, Bd. 2, 16 ; S. 519, 38. Æghwider wolde wîde toscrîðan *it would everywhere widely wander*, Bt. Met. Fox 20, 184 ; Met. 20, 92.

æg-hwilc, -hwelc, -hwylc ; *adj.* [â + ge + hwý + lîc] *Every, all, whosoever, whatsoever, every one* ; quicunque, unusquisque, omnis :—Æghwylc dæg *every day*, Mt. Bos. 6, 34. Æghwylce geáre *every year*, Bd. 2, 16 ; S. 519, 23. Hêr is æghwylc eorl ôðrum getrýwe *here is every man true to the other*, Beo. Th. 2460 ; B. 1228. Æghwylcum mâððum gescealde *he gave a present to every one*, Beo. Th. 2104 ; B. 1050. Æghwylcne ellpeódigra *unumquemque alienorum*, Andr. Kmbl. 51 ; An. 26. Wrediaþ fæste æghwilc ôðer *each supports the other firmly*, Bt. Met. Fox 11, 69 ;

Met. 11, 35. Æghwelce dæg *on every day*, Bt. Met. Fox 14, 9 ; Met. 14, 5. Æghwylc wille lîfes tiligan *every one wishes to cultivate life*, Exon. 27 a ; Th. 81, 4 ; Cri. 1318. Ðû æghwylces canst *thou art knowing in every matter*, Andr. Kmbl. 1016 ; An. 508.

æg-hwonene ; *adv. On every side* ; ubique :—Ða ýða æghwonene ðæt scyp fyldon *the waves filled the ship on every side*, Bd. 3, 15 ; S. 541, 42. v. æg-hwanan.

æg-hwyder *every way.* v. æghwider.

æg-gift, e ; *f. A legal gift, restitution* ; legalis dos, restitutio, Cart. Eadgir R.

æ-gilde, æ-gylde, a-gilde, a-gylde ; *adv.* [*æ without*, gild *payment*] *Without compensation* ; sine compensatione :—Gif he gewyrce ðæt hine man afylle, licge ægilde *if he so do that any man fell him down, let him be without compensation*, L. Eth. vi. 38 ; Th. i. 324, 24 : L. E. G. 6 ; Th. i. 170, 13 : L. C. S. 49 ; Th. i. 404, 14 : L. Eth. v. 31 ; Th. i. 312, 12.

Ægiptisc *Egyptian.* v. Ægypte, Egiptisc.

æg-lâc, es ; *n. Misery, trouble, torment* ; miseria, tribulatio, cruciatus, Elen. Grm. 1188. v. ag-lâc.

æg-lêca, an ; *m. A miserable being, wretch, monster* ; miser, perditus, monstrum :—Atol æglæca *the fell wretch*, Beo. Th. 1188 ; B. 592 : Cd. 216 ; Th. 274, 28 : Sat. 161 : Andr. Kmbl. 2717 ; An. 1361. v. ag-lêca.

æ-gleáw ; *adj. Skilled in the law, learned, wise* ; legis peritus, sagacissimus, sapientissimus :—Ðæt andswarode him sum ægleáw *respondit quidam ex legis peritis*, Lk. Boş. 11, 45. Ealde ægleáwe *elders skilled in laws*, Menol. Fox 37 ; Men. 19. Ðæt scell ægleáwra findan *that a more learned man must find out*, Andr. Kmbl. 2965 ; An. 1485.

æg-lêca, an ; *m. A wretch, miscreant*, Cd. 214 ; Th. 269, 14 ; Sat. 73. v. ag-lêca.

Ægles burg, Ægeles burg, [burh] ; *g.* burge ; *f* : Ægles byrig, e ; *f.* AYLESBURY, *in Buckinghamshire* :—Cûþwulf genom Ægeles burg *Cuthwulf took Aylesbury*, Chr. 571 ; Erl. 18, 13. Genam Ægles burh *id.* Th. 32, 29, col. 2. Genam Ægles byrig *id.* Th. 33, 27, col. 1. Betweóx Byrnewuda and Ægles byrig *betwixt Bernwood and Aylesbury*, 921 ; Th. 194, 19.

Ægles ford, es ; *m.* AYLESFORD *on the Medway near Maidstone, Kent*, Chr. 455 ; Ing. 15, 15. v. Ægeles ford.

Ægles þrep, es ; *n.* [þorp *a village*] AYLESTHORPE, *a village near Aylesford, Kent*, Chr. 455 ; Ing. p. 15, note *h* ; Th. 20, 39.

Ægles wurþ, es ; *m. The village of* EYLESWORTH, *Northamptonshire*, Chr. 963 ; Ing. 155, 9.

æg-lîm, es ; *m.* [æg *an egg*, lîm *lime, glue*] EGG-LIME, *the sticky part* or *white of an egg* ; ovi viscum :—Æglîm *glara*, Ælfc. Gl. 81 ; Som. 72, 119.

æg-moran ; *pl. f. Eye-roots* ; nervi quibus oculus cum cerebro connectitur :—Ðe beóþ on ðan ægmoran sâra *which are sores in the eye-roots*, Lchdm. iii. 98, 5. v. more.

ægnes þonces *of his own accord* ; sponte, ultro. v. âgen.

ægnian ; *p.* ede ; *pp.* ed ? *To frighten, vex* ; terrere, tribulare :—Ægnian mid yrmþum *to frighten with misery*, Cd. 156 ; Th. 194, 23 ; Exod. 265.

ægru *eggs*, L. M. 1, 87 ; Lchdm. ii. 156, 6. v. æg.

ægsa, an ; *m. Fear* ; timor, Mt. Rush. Stv. 14, 26. v. egsa.

æg-ðer [= æg-hwæðer] ; *pron. Either, each, both* ; uterque, ambo :—Ægðer byþ gehealden *ambo conservantur*, Mt. Bos. 9, 17. Ægðer ðara eorla *each of the men*, Andr. Kmbl. 2103 ; An. 1053. Heora ægðer *either* or *both of them*, each, Gen. 21, 31. On ægðre hand, on ægðere healfe *on either hand* or *half, on both sides*, Ors. 1, 11 ; Bos. 34, 40 : 1, 14 ; Bos. 37, 33. On ægðre healfe weard *towards both sides*, Ælfc. Gr. Ægðer ge—and, ge, both—and, as well—as :—Ægðer ge hâdes, ge êðeles þolige *let him forfeit both degree and country*, L. C. S. 41 ; Th. i. 400, 14. Ægðer ge heonan ge ðanan *both here and there*. Hî hatedon ægðer ge me ge mînne fæder *they hated both me and my father*, Jn. Bos. 15, 24.

æg-weard, e ; *f. Sea-ward, sea-guard* or *guardianship* ; maris [litoris] custodia :—Ic ægwearde heóld *I hold guard*, Beo. Th. 488, note ; B. 241. v. weard.

æg-wyrt, e ; *f. Egg-wort, dandelion* ; leontodon taraxacum, Lacn. 40 ; Lchdm. iii. 28, 26.

æ-gylde ; *adv. Without compensation*, L. E. G. 6 ; Th. i. 170, 13. v. æ-gilde.

æ-gylt, -gilt, es ; *m.* [æ, gylt *guilt, fault*] *A breach* or *violation of the law, a trespass, fault* ; delictum :—Ægiltas iúguþ-hâdes mînes ne gemun ðû *delicta juventutis meæ ne memineris*, Ps. Spl. T. 24, 7.

æ-gype, -gipe ; *adj. Trifling, worthless* ; nugalis :—Forðon hî dydan Drihtnes spræce æghwæs ægype *quia exacerbaverunt eloquium Domini*, Ps. Th. 106, 10.

Ægypte *Egypt*, Bd. 4, 24 ; S. 598, 11. v. Egypte.

æ-hiwnes, -ness, e ; *f. Paleness, gloom* ; pallor, deficientia coloris :—Wið æblæcnysse and æhiwnesse ðæs lîchoman *for paleness and discoloration of the body*, Herb. 164 ; Lchdm. i. 294, 3.

æ-hlýp, -hlîp, es ; *m.* [æ *law*, hlýp *a leap*] *A transgression, breach of the law, an assault* ; legis transgressio, aggressus :—Se ðe æ-hlîp gewyrce

whoever commits an assault, L. Ath. v. § 1, 5; Th. i. 230, 10. þurh æ-hlýp *by a violation of the law,* L. Eth. v. 31; Th. i. 312, 11. v. æt-hlýp.

æht, e; *f. Valuation, estimation, deliberation, council;* æstimatio, deliberatio, consilium:—Fira bearn æht besittaþ *the sons of men sit in council,* Andr. Kmbl. 820; An. 410. Biscopas and bóceras and ealdormen æht besæton *bishops and scribes and princes sat in council,* Andr. Kmbl. 1216; An. 608. v. eaht *deliberation, council.*

æht, e; *f.* [ēhtan *to persecute*] *Persecution, hostility;* persecutio, hostilitas:—Ðá wæs æht boden Sweóna leódum *then was persecution announced to the people of the Swedes,* Beo. Th. 5907; B. 2957. [*Ger.* acht, *f. proscriptio: M.H.Ger.* āhte, æhte: *O.H.Ger.* āhta, *f. persecutio.*]

æht, e; *f.* [æhte = áhte *had;* p. of *ágan to own, possess.*] **I.** *possessions, property, lands, goods, riches, cattle;* opes, substantia, possessio, greges:—He hæfde mycele æhta *erat habens multas possessiones,* Mk. Bos. 10, 22. Esau nam ealle his æhta, and eall ðæt he æhte *Esau took all his goods, and all that he possessed,* Gen. 36, 6. Grúndleás gítsung gilpes and æhta *bottomless avarice of glory and possessions,* Bt. Met. Fox 7, 30; Met. 7, 15. Israéla æhta *the Israelites' possessions,* Cd. 174; Th. 218, 23; Dan. 43. Genam on eallum dǽl æhtum sínum *he took a part of all his possessions,* 74; Th. 90, 23; Gen. 1499. Ealle his æhta *omnem substantiam ejus,* Ps. Th. 108, 11. **II.** *possession, power;* possessio, potestas:—His miht and his æht ofer middangeard gebledsod *his might and power is blessed throughout the earth,* Andr. Kmbl. 3432; An. 1720. Ágan us ðis wuldres leóht eall to æhte *let us get all this light of glory into our possession,* Cd. 219; Th. 280, 11; Sat. 254. On ágene æht syllan *in possessionem dare,* Ps. Th. 104, 10, 39; 110, 4. [*Scot.* aucht: *O.H.Ger.* ēht, *f:* Goth. aíhts, *f: O.Nrs.* ætt, átt *family.*] DER. gold-, máðum-, staðol-, wan-, won-.

æhta *eight,* Chr. 1070; Th. 345, 32. v. eahta.

æhte *had, owned, possessed.* v. áhte; *p. of* ágan.

æhte land, es; *n.* [æht *property*] *Landed property;* terra possessionis:—Forðon ðe Peohtas heora æhte land ðætte Angle ǽr hæfdon eft onféngon *nam Picti terram possessionis suæ quam tenuerunt Angli receperunt,* Bd. 4, 26; S. 602, 29.

æhte man, mann, es; *pl.* men; *m. A husbandman, a farmer, ploughman;* colonus:—Laboratores sind yrplingas and æhte men *labourers are ploughmen and husbandmen,* Ælfc. T. 40, 20.

æhtere, es; *m. An estimator, a valuer;* æstimator, Ælfc. Gl. 114; Som. 80, 25.

æhte swán, es; *m.* [æht *property,* swán *swain or herdsman: O.H.Ger.* sweino *a herdsman*] *A cowherd, swineherd, who belongs to the property of his lord;* bubulcus, porcarius qui in peculio domini est, L. R. S. 7; Th. i. 436, 22.

æht-gesteald, es; *n. Possession;* possessio:—He ða brýdlufan sceal to óðerre æhtgestealdum idese sēcan *he must seek conjugal love in the possession of another woman,* Exon. 67 b; Th. 249, 22; Jul. 115.

æht-gestreón, es; *n. Possessions, riches;* possessio, divitiæ:—Ðonne líg eal þigeþ eorþan æhtgestreón *when the flame devours all the possessions of the earth,* Exon. 63 a; Th. 232, 13; Ph. 506.

æht-geweald, es; *m. n. Possession, power, the power of the possessor;* potestas possessoria:—Cwæþ he his sylfes sunu syllan wolde on æhtgeweald *he said that he would give his own son into their power,* Andr. Kmbl. 2221; An. 1112. Ðú usic bewrǽce in æhtgewealda *tu nos tradidisti in potestatem,* Exon. 53 a; Th. 186, 28; Az. 26.

æhtian [æht *persecution*] *to persecute;* persequi. v. ēhtan.

æht-spēdig; *adj. Wealthy, rich;* locuples, opulentus:—Se is betra ðonne ðú, æhtspēdigra feoh-gestreóna *he is better than thou, richer in money-treasures,* Exon. 67 a; Th. 248, 26; Jul. 101.

æhtung, e; *f. Estimation, valuing;* æstimatio, Ælfc. Gl. 114; Som. 80, 26. v. eahtung.

æht-wela, an; *m. Wealth, riches;* opes, divitiæ:—Gelufian eorþan æhtwelan *to love earth's riches,* Exon. 38 a; Th. 125, 24; Gú. 359: Apstls. Kmbl. 167; Ap. 84.

æht-welig; *adj. Rich, wealthy;* locuples:—Sum wæs æhtwelig geréfa *there was a wealthy count,* Exon. 66 a; Th. 243, 29; Jul. 18.

æ-hwǽr; *adv. Everywhere;* ubique, Ps. Th. 88, 31. v. á-hwǽr.

æ-hwyrfan *To turn from, avert;* avertere, Ps. Spl. T. 53, 5. v. a-hwerfan, hwyrfan, hweorfan.

æig, es; *n. An egg;* ovum:—Ðæt æig getácnaþ hiht: ǽrest hit biþ æig, and seó mōdor siððan mid hihte bret ðæt æig to bridde *the egg betokens hope: first it is an egg, and the mother then with hope cherishes the egg to a young bird,* Homl. Th. i. 250, 22–24. v. æg.

æl-; *prefix.* **I.** = eal *all;* totus, omnis, as æl-beorht, æl-ceald, *etc.* **II.** æl- = el-, ele-, *foreign;* peregrinus, as æl-fylce, æl-wihta, *etc.*

æl, e; *f. An awl;* subula:—Hwanon sceó-wyrhtan æl *unde sutori subula,* Coll. Monast. Th. 30, 33: L. Alf. 11; Th. i. 46, 10. Æl *subula,* Ælfc. Gl. 1; Som. 55, 27; Wrt. Voc. 16, 2. v. al.

æl, es; *m. Oil;* oleum:—Ða sceolon beón æle bracene *they must be beaten up with oil,* Lev. 6, 21. v. ele.

ÆL, es; *m. An EEL;* anguilla:—Hwilce fixas gefēhst ðú? ǽlas and hacodas *what fishes catchest thou? eels and haddocks,* Coll. Monast. Th. 23, 33. Ac seó þeód ðone cræft ne cūðe ðæs fiscnōðes nymþe to ǽlum ánum *sed piscandi peritia genti nulla nisi ad anguillas tantum inerat,* Bd. 4, 13; S. 582, 43. Smæl ǽl *a small eel,* Cot. 161. [*Plat. Dut. Ger.* aal, *m: M.H.Ger.* O.H.Ger. āl, *m:* Swed. āl, *m: Dan.* aal, *m: O.Nrs.* āll, *m.*] DER. ǽl-net, ǽle-puta.

æ-lá *O!*—Ælá Drihten *O Lord,* Hy. 1, 1. v. eálá, æálá.

æ-lædend, es; *m.* [æ *lex,* lædend *lator,* from lǽdan *ferre,* to move or propose a law] *A lawgiver;* legislator, Ps. Spl. 9, 21.

æ-lǽrende; *part. Teaching the law;* legem docens:—Siððan him nǽnig wæs ǽlǽrendra ōðer betera *since there was none other of those teaching the law better than he,* Elen. Kmbl. 1009; El. 506.

æ-lǽten *divorced,* L. C. E. 7; Th. i. 364, 23; = a-lǽten; *pp.* of a-lǽtan.

ælan; *p.* de; *pp.* ed; *v. a. To kindle, set on fire, burn, bake;* accendere, urere, comburere, coquere:—Ne ælaþ hyra leóhtfæt *neque accendunt lucernam,* Mt. Bos. 5, 15. Úton wircean us tigelan and ælan hīg on fȳre *faciamus lateres et coquamus eos igni,* Gen. 11, 3. Fȳr æleþ uncyste *the fire burns the vices,* Exon. 63 b; Th. 233, 17; Ph. 526. Flǽsc and bán ádlēg æleþ *the fire of the pile burns flesh and bones,* Exon. 59 a; Th. 213, 9; Ph. 222. Brond biþ ontyhte, æleþ ealdgestreón *let the brand be kindled, consume the old treasure,* 19 b; Th. 51, 8. DER. in-ælan, on-.

æl-beorht *All-bright, all-shining:*—Engel ælbeorht *an all-bright angel,* Cd. 190; Th. 237, 13; Dan. 337: Exon. 15 a; Th. 32, 1; Cri. 506: 21 b; Th. 58, 2; Cri. 929: 53 b; Th. 188, 27; Az. 52. Hwīlum cerreþ eft on up rōdor ælbeorhta lēg *the all-bright flame returns sometimes again up to the sky,* Bt. Met. Fox 29, 104; Met. 29, 51. v. eall-beorht.

ælc; *adj.* [ā + ge + líc] *Each, any, every, all;* quisque, quivis, unusquisque, omnis:—Ælc gōd treów byrþ gōde wæstmas *omnis arbor bona fructus bonos facit,* Mt. Bos. 7, 17. Ælc wæs on twegra sestra gemete *capientes singulæ metretas binas,* Jn. Bos. 2, 6. Ælc hine selfa begrindeþ gástes dugeðum *each deprives himself of his soul's happiness,* Cd. 75; Th. 91, 32; Gen. 1521. Ælc flǽsc *omnis caro,* Ps. Th. 64, 2. Ælces monnes *of every man,* Bt. Met. Fox 26, 236; Met. 26, 118. Ælcum cuique, Andr. Kmbl. 3067; An. 1536. On ælcere tīde *omni tempore,* Lk. Bos. 21, 36. In ælce tīd *in æternum,* Exon. 13 b; Th. 25, 26; Cri. 406. Ælce dæg *each day,* Bt. Met. Fox 27, 15; Met. 27, 8. [*Plat. Dut.* elk *each, every one.*]

æl-ceald; *adj.* [æl = eal] *All cold, most cold;* usquequaque frigidus:—Meahtest weorþan æt ðǽm ælcealdan steorran ðone Saturnus hātaþ *you might be at that all-cold star which they call Saturn,* Bt. Met. Fox 24, 37; Met. 24, 19.

ælcor; *adv. Elsewhere, besides, otherwise;* alias, præter, nisi, aliter:—Forðon ðam bisceope ne wæs alýfed ælcor būtan on myran rīdan *non enim licuerat pontificem sacrorum præter in equa equitare,* Bd. 2, 13; S. 517, 7. Ælcor alias, Ælfc. Gr. 38; Som. 41, 67. v. elcor.

ælcra, *adv. Otherwise;* aliter, R. Ben. 62. v. ælcor.

æl-cræftig; *adj. All-powerful, all-mighty;* omnipotens:—Nán þing nis ðín gelíca, ne hūru ǽnig ælcræftigre *nothing is like unto thee, nor is any one more all-powerful,* Bt. Met. Fox 20, 76; Met. 20, 38.

æld *fire,* Exon. 22 a; Th. 59, 30; Cri. 960. v. æled.

æld *age,* Exon. 45 a; Th. 152, 11; Gú. 807. v. ældu.

ældan *To delay, forbear, postpone, conceal:*—Ældyst, Ps. Spl. C. 88, 37. Ælde, Ps. Surt. 77, 21: Mt. Rush. Stv. 25, 5: Bd. 1, 27; S. 491, 31; MS. B. v. yldan.

ælde men;—Ælda bearnum *for the sons of men,* Exon. 21 b; Th. 58, 18; Cri. 937. Ænig ælda cynnes *any one of the race of men,* 19 a; Th. 49, 4; Cri. 780: 44 b; Th. 151, 16; Gú. 796. Mid ældum *with men,* 13 b; Th. 25, 25; Cri. 406. v. ylde.

ælding *delay,* Mt. Rush. Stv. 24, 48. v. ylding.

ældo, aldu *the elders;* seniores, Mt. Lind. Stv. 21, 23. v. ældu.

ældran; *pl. Parents;* parentes:—Míne ældran, Ps. C. 65; Ps. Grn. ii. 278, 65. v. yldra.

ældru, aldro, aldro *parents,* Mk. Rush. War. 13, 12: Lk. Rush. War. 2, 27, 41, 43. v. ældran.

ældu, æld, e; *f.* **I.** *age, old age;* sæculum, senectus:—In ða ǽrestan ældo *in his first age,* Exon. 34 a; Th. 108, 30; Gú. 80. On ælde *in senectute,* Ps. C. 142: Ps. Surt. 91, 15: 70, 18. **II.** *an age, century;* ævum, centuria:—Þurh ælda tíd *per sæcula sæculorum,* Exon. 45 a; Th. 152, 11; Gú. 807. Wið ælda *against the age,* 81 a; Th. 305, 16; Fä. 89. v. yldu.

ælecung, e; *f. An allurement, a blandishment;* blandimentum, C. R. Ben. 2.

æled, g. ældes; *m.* [*pp.* of ælan] *Fire, conflagration;* ignis, incendium:—Æled wæs micel *the fire was great,* Cd. 186; Th. 231, 6; Dan. 243. Hát biþ monegum egeslíc æled *the dreadful fire shall be hot to many,* Exon. 63 a; Th. 233, 9; Ph. 522. Æled weccan *to light a fire,* Cd. 140; Th. 175, 26; Gen. 2901. Ældes fulle *full of fire,* Exon. 22 a; Th. 59, 30; Cri. 960. [*O.Sax.* eld, *m. ignis: O.Nrs.* eldr, *m. ignis.*]

æled-fȳr, es; *n. Flame of fire;* incendii flamma, Exon. 61 a; Th. 223, 27; Ph. 366.

æled-leóma, an; *m. A gleaming fire, fire-brand;* ignis micans, Beo. Th. 6241; B. 3125.

ælednys, -nyss, e; *f. A burning;* incendium. v. æled *a fire.*

æ-leng; *adj. Long, protracted, lengthy, troublesome;* longus, molestus:—Me þincþ ðæt ðé þincen tó ælenge ðás langan spell *methinks that these long discourses appear to thee too lengthy,* Bt. 39, 4; Fox 218, 6.

éle-puta, an; *m. An* EEL-POUT; capito:—Hwilce fixas geféhst ðú? mynas and æleputan *what fishes catchest thou?* minnows *and* eel-pouts, Coll. Monast. Th. 23, 33. [*Plat.* aalput *or* putte: *Dut.* aalpuit *or* puit aal, *m. a young eel,* eel-pout.] v. myne.

æ-léten, æ-lǽten, a-lǽtən; *part.* [from a-lǽtan *to let go*] *One let go, divorced;* repudiata uxor:—Ne on ælǽten ǽnig cristen mann ǽfre ne gewífige *nor with one divorced let any christian man ever marry,* L.C.E. 7; Th. i. 364, 23.

ÆLF, es; *An* ELF; genius, incubus:—Wið ælfe gníd myrran on wín *against an elf rub myrrh in wine,* L.M. 2, 65; Lchdm. ii. 296, 9. Ylfe, *pl. nom. m.* Beo. Th. 224; B. 112. v. ylfe. [*Plat.* elf: *O. Dut.* alf: *Ger.* elf, *m;* elbe, *f;* alp, *m. nightmare,* Grm. Wörterbch. iii. 400; i. 200, 245; Grm. Mythol. 249: *M. H. Ger.* alp, alf, *m. pl;* elbe, *f: O. H. Ger.* alp, *m: Dan.* elv: *Swed.* elf: *O. Nrs.* álfr, *m.*] DER. ælf-ádl, -cyn, -nóþ, -réd = rǽd, -sciéne, -scínu, -scýne, -siden, -sogoða, -þone: ylfe: ælfen, elfen, dún-, feld-, múnt-, sǽ-, wudu-, wylde-.

ælf-ádl, e; *f. Elf-disease;* ephialtæ morbus:—Wið ælfádle *against elf-disease,* L.M. 3, 62; Lchdm. ii. 344, 20.

ǽl-fǽle *All-fell, very baleful;* omnino perniciosus:—Áttor ælfǽle *very baleful poison,* Andr. Kmbl. 1539; An. 771. v. eal-felo.

ælf-cynn, es; *n. The elf-kind, the race of elves, elfin race;* ephialtum genus, Som. Lye:—Wyrc sealfe wið ælfcynne *work a salve against the elfin race,* L.M. 3, 61; Lchdm. ii. 344, 7.

-ælfen, -elfen, e; *f. A fairy, nymph;* nympha. *It is found only in compound words, as* Múnt-ælfen *a mountain nymph;* oreas = ὀρειάς, ἄδος:—Wudu-elfen *a wood nymph;* dryas, *etc,* Wrt. Voc. 60, 14–19. v. -en.

ǽl-fer, es; *n.* [=-fær, *n.*] *The whole army;* totus exercitus:—Ymbwícigean mid ǽl-fere Æthanes byrig *to surround with the whole army the town of Etham,* Cd. 146; Th. 181, 24; Exod. 66.

Ælf-nóþ, es; *m.* [ælf, nóþ *boldness, courage*] *Ælfnoth, elf courage;* nomen viri præclari in audacia, Byrht. Th. 137, 8; By. 183.

Ælfred, Alfriþ, Aldfriþ, Ealdfriþ, es; *m.* [æl *all;* ald, eald *old:* fred = friþ *peace:* v. Ælfréd] *Alfred the wise, king of Northumbria for twenty years,* A.D. 685–705. *He was educated in Ireland for the Church, and was the first literary king of the Anglo-Saxons;* Lat. Ælfrédus, Alfrid, Alfrídus, Bd. 4, 26; S. 175, 4: Aldfrídus, Bd. 5, 2; S. 183, 6: Aldfrithus, Chr. 685; Gib. 45, 24:—Féng Ælfréd [MS. Ealdfriþ] æfter Ecgfriþe tó ríce, se mon wæs se gelǽredesta on gewrítum, se wæs sæd ðæt his bróðor wǽre Oswies sunu ðæs cyninges *Ecgfrith was succeeded in the kingdom by Alfred, who was said to be his brother, and a son of king Oswy, and was a man most learned in scripture;* successit Ecgfrido in regnum Alfrid, vir in scripturis doctissimus, qui frater ejus et filius Osuiu regis esse dicebatur, Bd. 4, 26; S. 603, 6–8. A.D. 685, Hér man ofslóh Ecgferþ, and Ælfréd [MS. Aldfriþ *Aldfrithus*] his bróðor féng æfter him to ríce *here,* A.D. 685, *they slew Ecgferth, and Alfred his brother succeeded* [took] *to the kingdom after him,* Chr. 685; Erl. 41, 29. On Ælfredes [MS. Aldfriþes *Aldfrithi*] tídum ðæs cyninges *in temporibus Aldfridi regis,* Bd. 5, 1; S. 614, 20. Hér Ælfred [MS. Aldfriþ] Norþanhymbra cining forþférde *here,* A.D. 705, *Alfred, king of the Northumbrians, died,* Chr. 705; Erl. 43, 3?.

Ælfréd, es; *m.* [ælf *an elf;* réd = rǽd *counsel, wise in counsel:* v. Ælfred] *Alfred;* Alfrédus. **I.** Alfred the Great, born A.D. 849, grandson of Egbert, and fourth son of king Ethelwulf, reigned thirty years, A.D. 871–901:—Ðá, A.D. 871, féng Ælfréd, Æðelwulfing, to West Seaxna ríce ... And ðes geáres wurdon ix folcgefeoht gefohten wið ðone here on ðam cineríce be súþan Temese; bútan ðam ðe hí Ælfréd, ... and ealdormen, and ciningas þægnas, oft ráda on riden, ðe man náne rímde *then,* A.D. 871, *Alfred, son of Ethelwulf, succeeded to the kingdom of the West Saxons ... And this year nine great battles were fought against the army in the kingdom south of the Thames; besides which, Alfred ... and aldormen, and king's thanes, often rode raids on them, which were not reckoned,* Chr. 871; Erl. 77, 3–10. A.D. 897, Ðá hét Ælfréd cyning timbrian lange scipu ongeán ðas æscas [MS. æsceas] ða wǽron fulneáh twá swá lange swá ða óðre; ... ða wǽron ægðer ge swiftran ge untealran, ge eác heárran [MS. heárra] ðonne ða óðru; næron hí náwðær ne on Frysisc gesceapen ne on Denisc; bútan swá him sylfum þúhte ðæt hí nytwyrðe beón meahton *then,* A.D. 897, *king Alfred commanded long ships to be built against the Danish ships* [æscas] *which were full nigh twice as long as the others; ... they were both swifter and steadier, and also higher than the others; they were shapen neither as the Frisian nor as the Danish, but as it seemed to himself that they might be most useful,* 897; Th. 175, 37, col. 2—177, 5, col. 2. Ðæs ilcan geáres, hét se cyning [Ælfréd] faran to Wiht ... Ðá geféngon hý ðara scipa twá, and ða men [MS. mæn] ofslógon ... Ða ylcan sumere, forwearþ ná læs ðonne xx scipa mid mannum mid ealle be ðam súþ riman *in the same*

year [A.D. 897], *the king* [Alfred] *commanded his men to go to Wight ... They then took two of the ships, and slew the men ... In the same summer, no less than twenty ships, with men and everything* [*of the Danes*], *perished on the south coast,* Chr. 897; Th. 177, 5, col. 2—179, 3, col. 1. A.D. 901, Hér gefór Ælfréd cyning vii Kł Nouembris ... and ðá féng Eádweard, his sunu to ríce *here died king Alfred, on the twenty-sixth of October ... and then Edward* [*the Elder*], *his son, succeeded to the kingdom,* Chr. 901; Th. 179, 14–18, col. 2. **II.** Though the talents and energy of Alfred were chiefly occupied in subduing the Danes, and in confirming his kingdom, he availed himself of the short intervals of peace to read and write much. He selected the books best adapted for his people, and translated them from Latin into Anglo-Saxon. In translating he often added so much of his own, that the Latin text frequently afforded only the subject, on which he wrote most interesting essays, as may be seen in his first work, Boethius de Consolatione Philosophiæ. 1. *Boethius was probably finished about* A.D. 888. In his preface, he thus speaks of his book and of his other occupations:—Ælfréd, Cyning [MS. Kuning] wæs wealhstód ðisse béc, and hie of béc Lédene on Englisc wende ... swá swá he hit ða sweotolost and andgitfullícost gereccan mihte, for ðæm mistlícum and manigfealdum weoruld bísgum, ðe hine oft ægðer ge on móde ge on líchoman bísgodan. Ða bísgu us sint swíðe earfoþ ríme, ðe on his dagum on ða rícu becómon, ðe he underfangen hæfde; and ðeáh, ða he ðas bóc hæfde geleornode, and of Lædene to Engliscum spelle gewende, and geworhte hí eft to leóðe, swá swá heó nú gedón is *king Alfred was translator of this book, and turned it from book Latin into English ... as he the most plainly and most clearly could explain it, for the various and manifold worldly occupations, which often busied him both in mind and in body. The occupations are to us very difficult to be numbered, which in his days came upon the kingdoms which he had undertaken; and yet, when he had learned this book, and turned it from Latin into the English language, he afterwards put it into verse, as it is now done,* Bt. prooem; Fox viii. 1–10. 2. *Alfred,* having supplied his people with a work on morality in Boethius, *next translates for them the Historia Anglorum of his learned countryman Bede, about* A.D. 890. This was the King's work, for the Church says in Ælfric's Homilies, about A.D. 990:—'Historia Anglorum' ða ðe Ælfréd cyning of Lédene on Englisc awende *Historia Anglorum, which king Alfred turned from Latin into English,* Homl. Th. ii. 116, 30–118, 1. 3. *The third book which Alfred translated, about* A.D. 893, was the Compendious History of the World, written in Latin by the Spanish monk Orosius in A.D. 416. There is the best evidence, that the voyages of Ohthere and Wulfstan were written by the king, for we read that,—Ohthere sæde Alfréde cyninge, ðæt he ealra Norþmanna norþmest búde *Ohthere told king Alfred that he dwelt northmost of all Northmen,* Ors. 1, 1; Bos. 19, 25. Wulfstan also uses the language of personal narrative,—Burgenda land wæs on us bæcbord *we had* [lit. *there was to us;* erat nobis] *the land of the Burgundians on our left,* Ors. 1, 1; Bos. 21, 44. This is the longest and most important specimen of Alfred's own composition. 4. We have undoubted evidence of the date of *Alfred's Anglo-Saxon translation of Gregory's Pastoral Care,* for the king thus speaks of archbishop Plegmund,—Ic hie geliornode æt Plegmunde mínum ærcebiscepe *I learnt it from Plegmund my archbishop,* Introduction to Gregory's Pastoral, Oxford MS. Hatton 20, fol. 2. Plegmund was raised to the archbishopric in 890: Alfred was engaged with the invasion of Hastings till he was conquered in 897; Alfred, therefore, had only leisure to translate the Pastoral *between the expulsion of Hastings in 897, and his own death in 901.* It was certainly translated by Alfred, for he distinctly states,—Ðá ongan ic, ongemang óðrum mislícum and manigfealdum bísgum ðisses kyneríces, ða bóc wendon on Englisc, ðe is genemned on Læden Pastoralis, and on Englisc Hierde bóc, hwílum word be worde, hwílum andgit of andgite *then began I, among other different and manifold affairs of this kingdom, to turn into English the book, which is called in Latin Pastoralis, and in English Herdman's book, sometimes word for word, and sometimes meaning for meaning,* Oxford MS. Hatton 20, fol. 2.

æl-fremd, æl-fremed; *adj. Strange, foreign;* alienus, alienigena:—Bearn ælfremde, Ps. Spl. 17, 47: 18, 13: 107, 10: 82, 6: Lk. Bos. 17, 18.

Ælfríc, es; *m.* [ælf, ríc] *Ælfric;* Ælfricus. 1. Ælfric of Canterbury, the grammarian, was of noble birth, supposed to be the son of the earl of Kent. He was a scholar of Athelwold, at Abingdon, about 960. When Athelwold was made bishop of Winchester, he took Ælfric with him and made him a priest of his cathedral. Ælfric left Winchester about 988 for Cerne in Dorsetshire, where an abbey was established by Æthelmær. Ic Ælfríc, munuc 'and mæssepreóst ... wearþ asend, on Æðelrédes dæge cyninges, fram Ælfeáge biscope, Aðelwoldes æftergengan, to sumum mynstre, ðe is Cernel gehaten, þurh Æðelmæres ðæs þegenes *I Ælfric, monk and mass-priest ... was sent, in king Æthelred's day, from bishop Ælfeah, Æthelwold's successor, to a minster, which is called Cerne, at the prayer of Æthelmær the thane,* Homl. Th. i. 2, 1–5. He is said to have been bishop of Wilton, and he was elected archbishop of Canterbury. A.D. 995, Hér Siric arcebisceop forþférde, and Ælfríc,

Wiltunscîre bisceop wearþ gecoren on Easterdæi on Ambresbyri, fram Æðelrēde cinge, and fram eallan his witan *in this year*, A. D. 995, *archbishop Sigeric died, and Ælfric, bishop of Wiltshire, was chosen on Easter-day at Amesbury, by king Æthelred, and all his witan*, Chr. 995; Th. 243, 36, col. 2—245, 3, col. 2. *This Ælfric was a very wise man, so that there was no more sagacious man in England. Then went Ælfric to his archiepiscopal see, and when he came thither, he was received by those men in orders, who of all were most distasteful to him, that was, by clerks*, Chr. 995; Th. ii. 106, 20–24. Ælfric speaks strongly against the transubstantiation in the Eucharist, which gave his Homilies so great an importance in the eyes of the English reformers: v. hûsel. He died A. D. 1006, Hēr forþfērde Ælfríc arcebisceop *in this year, archbishop Ælfric died*, Chr. 1006; Th. 255, 35, col. 2. The preceding is the most probable biography of Ælfric, archbishop of Canterbury. Others have been written in Pref. to Homl. Th. i. pp. v–x: Lchdm. iii. pref. pp. xiv–xxix, *etc.* A list of his numerous books is given in Wright's Biographia Britannia Literaria, A. Sax. Period, pp. 485–494, and in Homl. Th. i. pp. vii–ix. 2. Ælfric Bata was the pupil of the preceding Ælfric, the grammarian. In the title of the MS. in St. John's College, Oxford, we read,—' Hanc sententiam Latini sermonis olim Ælfricus abbas composuit, qui meus fuit magister, sed tamen ego Ælfric Bata multas postea huic addidi appendices,' Wanl. Catal. p. 105, 4–7. It appears that in the time of Lanfranc, when the newest Romish doctrines relating to transubstantiation *etc.* were imposed upon the English Church by the Norman prelates, Ælfric Bata was regarded as an opponent of that doctrine, Wrt. Biog. Brit. A. Sax. p. 497.

ælf-sciéne, -sciéno; *adj. Beautiful, like an elf* or *nymph, of elfin beauty;* formosus ut genius vel nympha :—Mæg ælfsciéno = ides ælfsciéno *O woman of elfin beauty!* Cd. 86; Th. 109, 23; Gen. 1827: Cd. 130; Th. 165, 11; Gen. 2730.

ælf-scînu; *adj. Shining like an elf* or *fairy, elfin-bright, of elfin beauty;* splendidus ut genius vel nympha :—Iudiþ ides ælf-scînu *Judith, the woman of elfin beauty*, Judth. 9; Thw. 21, 11; Jud. 14.

ælf-siden, -sidenn, e; *f. The influence of elves* or *of evil spirits, the nightmare;* impetus castalidum, diaboli incubus :—Ðis is se hâlga drænc wið ælfsidene and wið eallum feóndes costungum *this is the holy drink against elfin influence and all temptations of a fiend*, Lacn. 11; Lchdm. iii. 10, 23. Wið ælfsidenne, L. M. 1, 64; Lchdm. ii. 138, 23.

ælf-sogoða, an; *m.* [sogeða *juice*] *A disease ascribed to fairy influence, chiefly by the influence of the* castalides, dûnelfen, *which were considered to possess those who were suffering under the disease, a case identical with being possessed by the devil, as will appear from the forms of prayers appointed for the cure of the disease,*—Deus omnipotens expelle a famulo tuo omnem impetum castalidum; *and further on,*—Expelle diabolum a famulo tuo, L. M. 3, 62; Lchdm. ii. 348, 11. v. ælf, sogeða, sogoða.

ælf-þone, an; *f? Enchanter's nightshade;* circæa lutetiana :—Wið ælfádle nîm ælfþonan nioðowearde *against elf disease take the lower part of enchanter's nightshade*, L. M. 3, 62; Lchdm. ii. 344, 21.

æl-fylc, es; *n.* [æl, folc]. I. *a foreign land;* aliena provincia :—Ðæt hie on ælfylce on Danubie stæðe wîcedon *till they encamped in the foreign land on the banks of the Danube*, Elen. Kmbl. 72; El. 36. II. *foreigners, a foreign army, an enemy;* peregrinus exercitus, hostes :—Ðæt he wið ælfylcum ēðelstōlas healdan cûðe *that he could keep his paternal seats against foreigners*, Beo. Th. 4731; B. 2371. [*Icel.* fylki, *n.*]

æl-grēne *all-green*, Cd. 10; Th. 13, 3; Gen. 197: Cd. 74; Th. 91, 24; Gen. 1517: Bt. Met. Fox 20, 155; Met. 20, 78. v. eal-grēne.

æl-gylden *all-golden.* v. eal-gylden.

é-lîc; *adj. Belonging to law, lawful;* legalis, legitimus, Bd. 1, 27, resp. 8; S. 495, 29. Tyn ǽlîcan word *the ten commandments*, Som.

æling, e; *f. Burning, burning of the mind, ardour;* ardor, flagrantia animi :—Ðŷ læs ælinge ûtadrîfe selflícne secg *lest burning desires should excite the self-complacent man*, Bt. Met. Fox Introd. 11; Met. Eink. 6.

æling *weariness;* tædium, Bt. pref. Cot; Rawl. viii. notes, line 10.

æll-beorht *all-bright*, Exon. 26 b; Th. 78, 20; Cri. 1277. v. eall-beorht.

æll-mihtig *all-mighty*, Cd. 17; Th. 20, 19; Gen. 311. v. eall-meahtig.

æll-reord *foreign speaking, barbarous*, Bd. 1, 13; S. 481, 44. v. el-reord.

æll-þeódignes, -nys, -ness, e; *f. A going* or *living abroad, a pilgrimage*, Bd. 1, 23; S. 485, 38. v. æl-þeódignes.

ællyfta *the eleventh;* undecimus, Bd. 1, 34; S. 499, 35. v. endlefta.

æl-mǽst *adv. Almost;* fere, Chr. 1091; Th. 359, 12. v. ealmǽst.

Ǽl-meahtig *Almighty* :—Habbaþ we Fæder æl-meahtigne *we have the Almighty Father*, Exon. 19 a; Th. 47, 22; Cri. 759: Ps. C. 50, 85; Ps. Grn. ii. 278, 85: 50, 97; Ps. Grn. ii. 279, 97. v. eall-mihtig.

Ǽl-mehtig *Almighty*, Hy. 8, 14. v. eall-mihtig.

ælmes-feoh, g. -feós; *n. Alms, alms' money;* pecunia eleemosynæ, L. R. S. 2; Th. i. 452, 13.

ælmes-georn; *adj. Diligent in giving alms, benevolent;* beneficus, liberalis :—Sum biþ ǽr-fæst and ælmesgeorn *one is honest and diligent in*

giving alms, Exon. 79 a; Th. 297, 13; Crä. 67. Sum man Tobias gehāten, swîðe ælmesgeorn *a man, whose name was Tobias, very diligent in giving alms*, Ælfc. T. 21, 24.

ælmes-lond *land given in frankalmoigne.* v. almes-lond.

ÆLMESSE, ælmysse, an; *f.* ALMS, *almsgiving;* eleemosyna :—Ðæt ofer sî and to lâfe sellaþ ælmessan *quod superest date eleemosynam*, Bd. 1, 27; S. 489, 30. Hwæt is us to sprecanne hû hî heora ælmessan dǽle *de faciendis portionibus et adimplenda misericordia nobis quid erit loquendum*, 1, 27; S. 489, 25. Ðæt ðîn ælmesse sŷ on dîglum *ut sit eleemosyna tua in abscondito*, Mt. Bos. 6, 4. Sōþlíce ælmessan dô *sic facias eleemosynam*, 6, 3. Ðonne he ælmessan dǽleþ *when he deals alms*, Exon. 62 a; Th. 229, 10; Ph. 453. Syle ælmyssan *give alms*, Cd. 203; Th. 252, 31; Dan. 587. Ælmessan dǽlan *or* syllan *or* dôn *to give or distribute alms;* eleemosynam dare, facere, Mt. Bos. 6, 2, 3. [*Scot.* almous : *O. Sax.* alamôsna, *f: O. Frs.* ielmisse : *Ger.* almosen, *n: M. H. Ger.* almuosen, *n: O. H. Ger.* alamuosan, *n: Dan.* almisse : *Swed.* almosa : *O. Nrs.* almusa, ölmusa, *f: from the Grk.* ἐλεημοσύνη.]

Ælm-hâm, es; *m. Elmham, Norfolk*, Kmbl. Cod. Dipl. 759; 59, 17.

Ǽl-miht; *adj. Almighty;* omnipotens :—Wiston Drihten ælmihtne *they knew the Almighty Lord*, Cd. 182; Th. 228, 1, note a; Dan. 195.

Ǽl-mihteg *Almighty;* omnipotens :—Ic hæbbe me geleáfan to ðam ælmihtegan Gode *I have confidence in the Almighty God*, Cd. 26; Th. 34, 27; Gen. 544.

Ǽl-mihtig, -mihti *Almighty* :—Se Ælmihtiga *the Almighty*, Beo. Th. 184; B. 92: Andr. Kmbl. 497; An. 249: Elen. Grm. 1146: Exon. 9 b; Th. 8, 22; Cri. 121: Cd. 191; Th. 239, 10; Dan. 368: Hy. 10, 1: Bt. Met. Fox 9, 97; Met. 9, 49: Menol. Fox 187; Men. 95: Salm. Kmbl. 68; Sal. 34: Ps. Th. 69, 6: Bd. 3, 15; S. 541, 19: Gen. 17, 1: 35, 11: 48, 3: Ex. 6, 3: Job Thw. 167, 27. Ælmihti, Bt. Met. Fox 13, 144; Met. 13, 72: Th. Dipl. 125, 20. Se ælmihtiga God is unasecgendlîc and unbefangenlîc, se ðe ǽghwær is eall, and nâhwar todǽled *the Almighty God is unspeakable and incomprehensible, who is everywhere all, and nowhere divided*, Homl. Th. i. 286, 26. v. eall-mihtig.

æl-myrca, an; *m. All sallow, a black man, an Ethiopian;* omnino fuscus, Æthiops :—On ælmyrcan ēðel-rîce *in the realm of the Ethiopian*, Andr. Kmbl. 863; An. 432.

ælmysse, an; *f. Alms*, Cd. 203; Th. 252, 31; Dan. 587. v. ælmesse.

æl-net, es; *n. An eel net;* rete anguillare :—Gesomnedon ða ælnet ǽghwonon ðe hî mihton *retibus anguillaribus undique collectis*, Bd. 4, 13; S. 582, 44.

ǽlpig; *adj.* [= ân-lípig, ân-lêpig, *from* ân *one*, hleáp *a leap*] *Each, single;* unicus :—Ðæt næs ân ǽlpig hîde, ne ân gyrde landes *that there was not one single hide, nor one yard of land*, Chr. 1085; Th. i. 353, 12. [*Laym.* alpi, ǽlpi *single, only:* Relq. Ant. W. on alpi word *one single word*, ii. 275, 3.]

ælr *an alder-tree;* alnus. v. alr, alor.

æl-reord, æl-reordig *of foreign speech, barbarous;* exterus, barbarus. v. el-reord, el-reordig.

æl-tǽw, -tǽwe, -teow; *comp.* re; *sup.* est; *adj. All good, excellent, entire, sound, healthful, perfect, honest;* omnino bonus, sanus :—Fîndest ðû æltæwe hǽlo *thou shalt find perfect healing*, Herb. 1, 29; Lchdm. i. 80, 7; MS. B. Næfþ nô æltæwne ende *has no good end*, Bt. 5, 2; Fox 10, 29. Full æltæwe geboren *born quite* [*full*] *sound or healthy*, 38, 5; Fox 206, 22. Oððe ǽnig þing ǽr wǽre oððe æltæwre *if anything were before or more excellent*, Bt. 34, 2; Fox 136, 8. Ealle ða æltæwestan ofslôgen *they slew all the best men*, Ors. 4, 4; Bos. 81, 16. v. eal-teaw.

æl-tæwlíce; *adv. Well, perfectly;* bene. v. æl-tæw, -lîce.

æl-teaw, -teow *All good, sound, perfect;* omnino bonus, sanus :—Fîndest ðû ælteowe [æltæwe MS. B.] hǽlo *thou shalt find perfect healing*, Herb. 1, 29; Lchdm. i. 80, 7: Hy. 2, 13. v. æl-tæw.

æl-þeód, -þiód, e; *f. A foreign nation, foreign people, foreigners* :—Ðonne ða rîcan beóþ oððe on ælþeóde oððe on hiora âgenre gecŷððe *when the rich are among foreigners or in their own country*, Bt. 27, 3; Fox 98, 34. v. el-þeód.

æl-þeódelíce; *adv. Among foreigners, abroad;* peregre :—Swâ se man ðe ælþeódelíce fērde *sicut homo qui peregre profectus*, Mk. Jun. 13, 34.

æl-þeódig, æl-þiódig; *adj. Strange, foreign;* exterus, peregrinus, barbarus :—On ælþeódig folc *to a foreign people*, Bt. 27, 3; Fox 98, 22. Ælþeódigra manna gisthûs *foreign men's guest house, an inn*, Wrt. Voc. 58, 51. Ælþeódige men acwealdon *advenam interfecerunt*, Ps. Th. 93, 6. Ne geunret ðû ælþeódige, ge wǽron ælþeódie on Egipta lande *advenam non contristabis, advenæ enim et ipsi in terra Ægypti*, Ex. 22, 21. Ðâm ælþeódegan *to the foreigners*, Bt. 27, 3; Fox 100, 2. v. el-þeódig.

æl-þeódiglíce; *adv. In foreign parts, among foreigners;* peregre, Ælfc. Gr. 38; Som. 41, 26–28.

æl-þeódignes, -ness, -nyss, e; *f. A being* or *living abroad, a pilgrimage* :—On stôwe ælþeódignysse mînra *in loco peregrinationis meæ*, Ps. Spl. 118, 54: Gen. 12, 10: Bd. 4, 23; S. 593, 11.

æl-þeódine *foreign, a proselyte*, Mt. Bos. 23, 15; *for* æl-þeódigne, *acc. s. of* æl-þeódig.

æl-þeódung, e; _f. A being_ or _living abroad_; peregrinatio, Bd. 4, 23; S. 593, 15.

æl-þióðig _foreign_, Bt. 39, 2; Fox 212, 17. v. æl-þeóðig.

æl-walda _the all-powerful_, Cd. Jun. 6, 10. v. eal-wealda.

æl-wihta; _pl._   **I.** _strange creatures, monsters_; alieni generis entia, monstra:—Ðæt ðǽr gumena sum ælwihta eard ufan cunnode _that a man from above explored there the dwelling of strange creatures_, Beo. Th. 3004; B. 1500.   **II.** _all created things_; omnia creata:—Helm ælwihta, engla scippend _the protector of all created things, the creator of angels_, Andr. Kmbl. 236; An. 118. v. eall-wihta.

æ-melle; _adj. Unsavoury, without taste_; insipidus, Cot. 116.

æmelnys, æmylnys, -nyss, e; _f. Loathsomeness, weariness, disdain, falsehood, unfaithfulness, false dealing, treason_; fastidium, tædium:—Hneppade sáwle mín for þrece oððe for æmelnysse _dormitavit anima mea præ tædio_, Ps. Lamb. 118, 28.

æ-men; _adj._ [æ _without_, man _man_] _Unmanned, depopulated, desolate_; hominibus nudus, non habitata—Stód seó dýgle stów ídel and æmen _the secret spot stood void and desolate_, Exon. 35 a; Th. 115, 9; Gú. 187.

æmete, æmette, æmetta, an; _f. An_ EMMET, _ant_; formica:—Æmete _formica_, Wrt. Voc. 23, 78. Æmettan ægru gením _take emmet's eggs_, L. M. 1, 87; Lchdm. ii. 156, 6. Æmytte _formica_, Somn. 108. Nime æmettan _take emmets_, L. M. 3, 34; Lchdm. ii. 328, 7. [æ = a _from, off, away_; mete _meat, food_; Grm. (Gr. ii. 88) thinks it is connected with O. H. Ger. emizîc _assiduus_; ameiza _formica_: O. Nrs. ami _labour_: A. Sax. æmetigian _otiosus_; æmtegian _vacare_.]

æmet-hwíl, e; _f._ [æmetta _leisure_, hwíl _while, time_] _Leisure, sparetime, respite_; otium, Ælfc. Gr. 8; Som. 8, 1.

æmet-hyll, æmett-hyll, es; _m. An_ EMMET-HILL, _ant-hill_; formicetum, Past. 28, 3; Hat. MS. 37 a, 3.

æmetig; _adj. Vacant, empty, barren_; vacuus:—Hit æmetig læg _it lay barren_, Ors. 1, 10; Bos. 34, 16. v. æmtig.

æmetta _rest_, Bt. proœm; Fox viii. 13. v. æmta.

æmettig _idle_, Solil. 13. v. æmtig.

æmnitta, an; _m. A balance_; statera. v. emnettan, emnian _to make equal_.

æ-mód; _adj._ [æ _without_, mód _mind_] _Out of mind, mad, dismayed, discouraged_; amens:—Forðam Rómáne wǽron swá æmóde, ðæt hý ne wéndon ðæt hí ða burh bewérian mihton _because the Romans were so out of heart, they thought that they could not guard the city_, Ors. 3, 4; Bos. 56, 12.

æmta, emta, æmetta, an; _m. Quiet, leisure, rest_; quies:—Ic ne æmtan nabbe _I have no leisure_, Bt. 39, 4; Fox 218, 9. Be his æmettan _by his leisure_, Bt. proœm; Fox viii. 13.

æmtegian _to be at leisure_, Past. 18, 4; Hat. MS. 26 b, 16. v. æmtian.

æmtian, æmtegian, æmtigean; _p._ ode; _pp._ od _To be at leisure, to be vacant_; otiosum esse:—Æmtiagaþ and geseóþ forðan ðe ic eom God _vacate et videte quoniam ego sum Deus_, Ps. Spl. C. 45, 10; Ælfc. Gr. 33; Som. 37, 14.

æmtig, æmteg, emtig, æmetig, emetig, æmettig; _adj. Vacant_, EMPTY, _free, idle_; vacuus, inanis:—Seó eorþe wæs æmtig _terra erat vacua_, Gen. 1, 2. Gefylde sáwle æmtige _satiavit animam inanem_, Ps. Spl. 106, 9: Mt. Bos. 12, 44: Bd. 4, 3; S. 567, 5. Híg synd emtige _they are idle_, Ex. 5, 8. Æmtege wífmen _unmarried women_, Past. 21, 8, Lye. cf. æmete.

æmtigean _to be at leisure_, Ælfc. Gr. 33; Som. 37, 14. v. æmtian.

æ-múða [æ _without_, múða _a mouth_] _cæcum intestinum_, Wrt. Voc. 44, 64.

æmyce, æmyrce; _adj. Excellent, singular_; egregius, Cot. 74.

æmylnys, -nyss, e; _f. Weariness_; tædium, Pref. R. Conc. v. æmelnys.

æmytte _an 'emmet'_; formica, Somn. 108. v. æmete.

æn _one_; unus:—Wyrc ðé nú ænne arc _now make thee an ark_, Gen. 6, 14: Mt. Bos. 5, 36. v. án.

ænde _and_, L. Wih. 8; Th. i. 38, 16. v. and.

ændemes, ændemest _likewise, equally_; pariter, Bt. 41, 1; Fox 244, 12. v. endemes.

ændian; _p._ ode; _pp._ od _To end_; finire, Solil. 12. v. endian.

ændlefen _eleven_; undecim:—He ætýwde ændlefene _he appeared to the eleven_, Mk. Bos. 16, 14. v. endleofan.

ændlyfta _eleventh_, Bd. 2, 14; S. 517, 23. v. endlyfta.

æne; _adv. Once, alone_; semel, solum:—Nú ic æne begann to sprecanne to mínum Drihtne _quia semel cœpi, loquar ad Dominum meum_, Gen. 18, 31. Oft, nalles æne _often, not once_, Beo. Th. 6030; B. 3019. Æne on dæge _once in the day_, Bt. Met. Fox 8, 35; Met. 8, 18. Ic ðé æne abealh, éce Drihten _I alone angered thee, eternal Lord_, Cd. 222; Th. 290, 4; Sat. 410. v. áne _once_.

æneg, ænegu _any_:—Ænegu gesceaft _any creature_, Bt. 35, 4; Fox 160, 26: Cd. 26; Th. 34, 17; Gen. 539. v. ænig.

æn-ette _solitude_; solitudo, Dial. 2, 3. v. án-ád, án-æd.

ænforléten; _part. Clothed?_ amictus? Ps. Spl. T. 103, 2; _amissus?_ and not _amictus_. v. ánforlǽten; _pp._ of án-forlǽtan.

ænga _Single, sole_; unicus:—Fram ðam ængan hláforde _from the sole lord_, Salm. Kmbl. 766; Sal. 382. v. ánga.

ænge; _def._ se ænga; _adj. Narrow, troubled, anxious_; angustus, anxius:—Ðes ænga stede _this narrow place_, Cd. 18; Th. 23, 9; Gen.

---

356. Is me ænge [MS. ænige] gást innan hreðres _anxiatus est in me spiritus meus_, Ps. Th. 142, 4. v. ange, enge.

ænge; _adv. Narrowly, sadly_; anguste, anxie, triste, Ps. Th. 136, 8.

ængel _an angel_, Ps. Spl. 8, 6: 34, 7. v. engel.

Ænglisc _English_; Anglicus:—Hér synd on ðam íglande fíf geþeódu, Ænglisc, Brytwylsc, Scottysc, Pihttisc, and Bóclǽden _here are in the island five languages, English, Brito-Welsh, Scottish, Pictish, and Book-Latin_, Chr. Th. 3, 5, col. 1. v. Englisc.

ængum, Beo. Th. 952; B. 474, = ænigum _to any_; _dat._ of ænig.

ænig, æneg, áni; _adj._ [æn = án _one_, -ig _adj. termination_; ánig, g = y, _Eng._ any] ANY, _any one_; ullus, quisquam, aliquis:—Ðæt ænig man ænig fæt þurh ðæt templ bǽre _that any man should bear any vessel through the temple_, Mk. Bos. 11, 16. Mæg ænig þing gódes beón of Nazareth _a Nazareth potest aliquid boni esse?_ Jn. Bos. 1, 46. Æniges sceates _of any treasure_, Cd. 25; Th. 32, 15; Gen. 503. Monnes ænges _of any man_, Exon. 10 b; Th. 13, 9; Cri. 200. Næs ðǽr ænigum gewin _there was no toil for any one_, Andr. Kmbl. 1776; An. 890. Ængum ne mæg se cræft losian _the skill may not desert any one_, Bt. Met. Fox 10, 71; Met. 10, 36. DER. nǽnig _none_.

æn-íge, æn-ígge _one-eyed_:—Gif he hí gedó ænígge _if he make them one-eyed_, L. Alf. 20; Wilk. 30, 11: Cot. 179. v. án-eáge.

æniht [æn = án _one_, -iht _adj. termination_] _Anything_; quicquam:—Æniht _quicquam_, Jn. Lind. War. 11, 49. In mec ne hæfeþ æniht _in me non habet quicquam_, Jn. Rush. War. 14, 30. v. stániht, -ig, -ihtig.

æninga; _adv. Of necessity, by all means_, Bd. 4, 16; S. 584, 32: 5, 19; S. 640, 16: Andr. Kmbl. 439; An. 220. v. áninga.

æn-lépnes, ness, e; _f. Solitude, privacy_; solitudo. v. án-lépnes.

æn-líc; _adj._ [án _one_, líc _like_] ONLY, _singular, incomparable, excellent, beautiful, elegant_; unicus, egregius, elegans, pulcher:—He hæfde án swíðe ænlíc wíf _he had a very excellent wife_, Ælfc. T. 33, 15; Fox 166, 30. Ænlíces hiwes _of an excellent shape_, Ælfc. T. 33, 15. Ðeáh hió ænlícu sý _though she be beautiful_, Beo. Th. 3887; B. 1941. Eal wæs ænlícra ðon mæge stefn areccan _all was more excellent than voice can tell_, Exon. 52 a; Th. 181, 17; Gú. 1294. Cynn Fabiane forðan hit ealra Rómána ænlícost wæs _because the Fabian family was the highest in rank of all the Romans_, Ors. 2, 4; Bos. 43, 28. v. án-líc.

æn-líce; _adv._ ONLY, _singularly, elegantly_; eleganter, Coll. Monast. Th. 35, 37.

æn-lípie = æn-lípige _singulos_, Ps. Lamb. 7, 12. v. æn-lípig.

æn-lípig, -lýpig, -lépig; _adj._ [án _one_, hlíp, hlýp] _Each, every, singular, solitary, private_; singuli, solus:—Þurh ænlípige dagas _per singulos dies_, Ps. Spl. 41, 15. Be ænlípigum mannum _per singulos viros_, Jos. Grn. 7, 14: C. R. Ben. 22. v. án-lípig.

ænne _one_; unum:—Ðú ne miht ænne locc gedón hwítne _non potes unum capillum album facere_, Mt. Bos. 5, 36; _acc._ of æn = án, _q. v._

æ-not; _adj._ [æ _without_, not _use_] _Useless, of no use, unprofitable_; inutilis:—Ðæt hit ænote weorþe _that it be useless_, L. Eth. vi. 34; Th. i. 324, 7.

a-eóde _happened_; evenit:—Swá hit sóþlíce aeóde _so it truly happened_, H. de visione Isaiæ. _p._ of a-gán.

æpel-sceal, -scel, e; _f. An apple-shale or film about the kernels or pips_; pomi scheda, Cot. 43.

æpel-tre _an apple-tree_; malus, Wrt. Voc. 79, 79. v. æppel-treów.

æplian; _p._ ede; _pp._ ed _To make into the form of apples_, Elen. Kmbl. 2517; El. 1260. v. æpplian.

ÆPPEL, æpl, appel, apl, eapl, es; _m: nom. acc. pl. m._ æpplas; _nom. acc. pl. n._ æppla.   **I.** _an_ APPLE, _fruit generally_, Ors. Eng. 1. 3; Bos. 63, note 1; malum, pomum:—Æples gelícnes _likeness of an apple_, Exon. 59 a; Th. 213, 26; Ph. 232. Æppel unsælga, deáþ-beámes ofet _the unblest apple, fruit of the tree of death_, Cd. 30; Th. 40, 10; Gen. 637. Ða reádan appla _the red apples_; mala Punica, Past. 15, 5; MS. Hat. 19 b, 28. Nǽnig móste heora hrórra hrím æpla gedígean _none of their hardy fruits could withstand the frost_; occidit moros in pruina, Ps. Th. 77, 47. Gením brembel-æppel _take a bramble-fruit_, i. e. _a blackberry_, L. M. 1, 64; Lchdm. ii. 138, 27.   **II.** _what is round as an apple, the apple of the eye, a ball, bolus, pill_; quidvis globosum, pupilla, globus, bolus, pilula:—On ðæs siwenígean eágum beóþ ða æpplas hále, ac ða brǽwas greátiaþ _in lippi oculis pupillæ sanæ sunt, sed palpebræ grossescunt_, Past. 11, 4; MS. Hat. 15 a, 18. Hí scilde swá geornlíce swá swá man déþ ðone æpl on his eágan _he protected them as carefully as a man does the apple of his eye_, Bt. 39, 10; Fox 228, 13. Írenum aplum _with iron balls_, Salm. Kmbl. 56; Sal. 28. [_Orm._ appell: _R. Glouc._ appel: _Gow._ apple: _O. Frs._ appel, _m. malum, pomum_: _N. Dut. L. Ger._ appel, _m_: _Ger. M. H. Ger._ apfel, _m_: _O. H. Ger._ aphul, aphol, _m_: _Dan._ æble, _n_: _Swed._ æple, _n_: _O. Nrs._ epli, _n_: _Wel._ aval: _Ir._ abhall, ubhall: _Gael._ abhal, ubhal: _Manx_ ooyl: _Corn. Arm._ aval: _Lith._ obolys: _O. Slav._ jabluko.] DER. æppel-bǽre, -bearo, -cyrnel, -fealu, -hús, -leáf, -sceal, -screáda, -þorn, -treów, -tún, -wín: brembel-æppel, eág-, eorþ-, fíc-, finger-, palm-, wudu-.

æppel-bǽre; _adj. Apple-bearing, fruit-bearing_; pomifer:—Æppel-bǽre treów _lignum pomiferum_, Gen. 1, 11; Hexam. 6; Norm. 12, 5.

æppel-bearo, -bearu; _g._ -bearwes; _d._ -bearwe; _acc._ -bearo; _pl. nom. acc._ -was; _g._ -wa; _d._ -wum; _m. An orchard_; pomarium, Ps. Th. 78, 2.

æppel-cyrnel, es; *n. A pomegranate*; malogranatum, malum Punicum, Cot. 128.

æppelder, æppeldor *an apple-tree*. v. apulder.

æppel-fealu; *g. m. n.* -fealuwes; *adj. Apple-fallow, apple* or *reddish yellow*; flavus ut pomum :—Mearas æppelfealuwe *bay steeds*, lit. *apple-coloured steeds*, Beo. Th. 4336; B. 2165. DER. fealo, fealu, wes; *n.*

æppel-hús, es; *n. An apple-house, a place for fruit generally*; pomarium, Wrt. Voc. 58, 55.

æppel-leáf, es; *n. An apple-leaf*. v. appel-leáf.

æppel-sceal, e; *f. A film about the kernels of an apple*. v. æpel-sceal.

æppel-screáda *Apple-shreds, apple-parings*; pomi præsegmina, quisquiliæ, Wrt. Voc. 22, 13; *nom. pl.* of æppel-screáð. v. screáð.

æppel-þorn *an apple-thorn, a crab-tree*. v. appel-þorn.

æppel-treów, es; *n. An apple-tree*; malus. v. æppel-tre.

æppel-tún, es; *m. An apple-garden, orchard*; pomarium, Ælfc. Gl. 24? Somn. 299.

æppel-win, es; *n. Apple-wine, cider*; pomaceum, Cot. 117.

æppled, æpled; *part.* APPLED, *made into the form of apples, made into balls or bosses*; in pomorum formam redactus :—Æpplede gold *appled gold*, Exon. 63 a; Th. 232, 14; Ph. 506; 75 b; Th. 283, 30; Jul. 688. Æplede gold, Elen. Kmbl. 2517; El. 1260. v. *pp.* of æpplian.

æpplian, æplian; *p.* ede; *pp.* ed [æppel *an apple*] *To make into the form of apples, to make into balls or bosses*; in pomorum formam redigere, globosum facere, Exon. 63 a; Th. 232, 14; Ph. 506; 75 b; Th. 283, 30; Jul. 688: Elen. Kmbl. 2517; El. 1260.

æppuldre, æpuldre, an; *f. An apple-tree*; malus. v. apuldre.

æppuldre-tún, es; *m. An apple-tree inclosure, apple-orchard*; pomarium. v. apulder-tún.

æppyl *an apple*, Ælfc. Gr. 6; Som. 5, 57; MS. C. v. æppel.

æps, æsp, e; *f.* æpse, æspe, an; *f. An asp* or *aspen-tree, a species of poplar*; populus tremula :—Æps sicomorus, vel celsa, Wrt. Voc. 33, 27: Cot. 165. Nim æps-rinde *take asp-rind*, L. M. 3, 39; Lchdm. ii. 332, 7. Genim æpsan *take asp-tree*, I, 36; Lchdm. ii. 86, 6. [*Chauc.* aspe: *Prompt. parv.* aspe: *O. Frs.* aspe, *f.*: *Ger.* espe, *f.* populus tremula: *M. H. Ger.* aspe, *f.*: *O. H. Ger.* aspa, *f.*: *O. Nrs.* espi, *n.*]

æpsenys, -nyss, e; *f. Disgrace, dishonour, shame*; dedecus, Scint. 56.

æps-rind; *f. Asp-rind*; populi tremulæ cortex, L. M. 3, 39; Lchdm. ii. 332, 7. DER. æps.

ær, es; *m.* [ær = ear, *q. v.*] *Ocean*; *pl. The waves of the ocean* :—Ofer æra gebland *over the mingling of the waves*, Chr. 937; Th. i. 202, 38, col. I. v. ear, ear-gebland.

ær, es; *n. Brass*; æs :—Siððan folca bearn æres [MS. ærest] cúðon and isernes *since then the sons of men have known brass and iron*, Cd. 52; Th. 66, 22; Gen. 1088: Wrt. Voc. 8, 27. v. ár.

ær; *comp. m.* æra, ærra; *f. n.* ære, ærre; *sup.* ærest; *adj. Early, former, preceding, ancient*; prior, præcedens, antiquus :—On ærne mergen *in early morning*; primo mane, Mt. Bos. 20, I: Mk. Bos. 16, 9: Jn. Bos. 21, 4: Ps. Spl. 5, 3, 4. Fram ærne mergen óþ æfen *from early morning till evening*, Bd. 2, 14; S. 518, 8. Swá he wæs gyrstan dæg and æran dæg *sicut erat heri et nudius tertius*, Gen. 31, 5. Ðæs æran tácnes *prioris signi*, Ex. 4, 8. Forlýst he his ærran gód *he loses his former good*, Bt. 35, 6; Fox 170, 22. Of deáþe woruld awehte in ðæt ærre líf *awoke the world from death into the former life*, Elen. Kmbl. 609; El. 305: Exon. 113 b; Th. 436, 11; Rä. 54, 12. On ðysse ærran béc *præcedente libro*, Bd. 4, 1; S. 563, 18. Ærran dæges *dies antiqui*, Ps. Th. 142, 5: Beo. Th. 1819; B. 907. Weorpe ærest stán *primus lapidem mittat*, Jn. Bos. 8, 7. Se hér-búendra hearpan ærest hlyn awehte *who first of dwellers here awoke the sound of the harp*, Cd. 52; Th. 66, 5; Gen. 1079. Se æreste wæs Enos háten *the first was called Enos*, 50; Th. 64, 24; Gen. 1055. Wæs seó æreste costung ofercumen *the first temptation was overcome*, Exon. 39 a; Th. 128, 22; Gú. 408. In ða ærestan ældu *in the first age*, 34 a; Th. 108, 29; Gú. 80. Ða ærestan ælda cynnes *the first of the race of men*, 47 a; Th. 160, 23; Gú. 948. Ðú ealt oncneówe, ða ærestan eác ða néhstan *tu cognovisti omnia, antiqua et novissima*, Ps. Th. 138, 3. Æt ærestan *at the first*; primo, L. Alf. pol. 1; Th. i. 60, 2: Exon. 10 a; Th. 49, 15; Cri. 786. DER. ær-bel, -cwide, -dæd, -dæg, -deáþ, -fæder, -gestreón, -geweorc, -gewinn, -gewyrht, -ing, -morgen, -mergen, -sceaft, -wéla, -woruld.

ær, eár, ér; *sup.* ærost, ærest, ærst; *adv.* ERE, *before, sooner, earlier, formerly, already, some time ago, lately, just now, till, until*; antea, prius, mane, mature, dudum :—Gang ær *vade prius*, Mt. Bos. 5, 24. He wæs ær ðonne ic *ille erat prius quam ego*, Jn. Bos. I, 15, 30. Ær on morgen *early in the morning*, Cd. 224; Th. 297, 10; Sat. 515: Ps. Th. 18, 5: Ex. 12, 22. Nóht micle ær *non multo ante*, Bd. 4, 23; S. 593, 21. Hwéne ær *scarcely before, just before*, Bt. 23; Fox 78, 25. Swýðe ær *very early*; valde mane, Mk. Bos. 16, 2: 1, 35. Tó ær *too soon*, Exon. 45 a; Th. 152, 30; Gú. 816. Hwonne ær *how soon? when? quando?* Ps. Th. 40, 5. Ærost *first*, Gen. 19, 33. Swá hit engel gecwæþ ærest on Ebresc *as the angel said it first in Hebrew*, Exon. 9 b; Th. 9, 11; Cri. 133: 88 b; Th. 333, 15; Gn. Ex. 4. Him cenned wearþ Caïnan ærest *to him was born Cainan first*, Cd. 57; Th. 70, 7;

Gen. 1149. Mon wæs to Godes anlícnesse ærest gesceapen *man was at first shapen to God's image*, 75; Th. 92, 16; Gen. 1529. Ðá ic hér ærest com *when I first came here*, 129; Th. 164, 8; Gen. 2711: Beo. Th. 1236; B. 616. [*Laym.* ær, ære, ear: *Orm.* ær: *R. Glouc.* er: *Wyc. Chauc. Piers* er: *T. More* ere: *O. Sax.* ér prius, antea: *O. Frs.* ér: *Ger.* eher prius, antea: *O. H. Ger.* ér antea, dudum, prius, quondam: *Goth.* air diluculo, mane: *O. Nrs.* ár olim, mane.]. DER. ær-boren, -gedón, -genemned, -gód, -gystran-dæg, -líce, -wacol.

ér; *conj.* ERE, *before that*; antequam, priusquam :—Ær heó wordum cwæþ *ere she said in words*, Cd. 222; Th. 290, 3; Sat. 409. Ær hie to setle gong *ere she went to her seat*, Beo. Th. 4043; B. 2019. Ær ge furður féran *ere that ye further proceed*, 510; B. 252. Ær hie on tú hweorfon *before they departed from one another*, Andr. Kmbl. 2102; An. 1052. [*O. Sax.* ér priusquam: *M. H. Ger. O. H. Ger.* ér priusquam.]

ér; *prep. d. Before*; ante :—Ær his swylt-dæge *before his death-day*, Cd. 62; Th. 74, 12; Gen. 1221. Ær dægréde *before dawn*, 223; Th. 294, 4; Sat. 466. Ær sunnan his nama sóþfæst standeþ, byþ his setl ær swylce ðonne móna *ante solem permanebit nomen ejus, et ante lunam sedes ejus*, Ps. Th. 71, 17. Ær ðam flóde *ante diluvium*, Mt. Bos. 24, 38. Ær ðé *before him*, Bt. 41, 3; Fox 246, 26. Ær ðam *before that, before*; antequam, Mt. Bos. 6, 8: Exon. 61 a; Th. 224, 22; Ph. 379. Ær ðam ðe *before that which, till*; priusquam, Ps. Spl. 38, 18: Mt. Bos. 12, 20. [*O. Sax.* ér *ante*: *M. H. Ger. O. H. Ger.* ér *ante.*]

æra; *adj. Earlier, former*; prior, præcedens :—Ðæs æran tácnes *prioris signi*, Ex. 4, 8: Gen. 31, 5. v. ær; *adj.*

ér-ádl; *f. Early-disease*; præmaturus morbus :—Ðá ærádl nímeþ *when early disease takes them*, Exon. 89 a; Th. 335, 10; Gn. Ex. 31.

æra gebland [ær = ear *sea*] *The agitation of the sea*, Chr. 937; Th. 202, 38, col. I; ear in col. 2, and p. 203, 38, col. I; eár in col. 2. v. ear-gebland.

ér-boren; *p. part. First-born*; primogenitus, Cd. 47; Th. 59, 33; Gen. 973.

æerce-biscop, ærce-bisceop, es; *m. An archbishop*, Bd. 2, 3; S. 504, 35. v. arce-bisceop.

æerce-diácon, es; *m. An archdeacon*. v. arce-diácon.

ér-cwide, es; *m. Prophecy*; prophetia? nuntii vel doctoris loquela ?—He ærcwide onwreáh [MS. onwearh] *he revealed the prophecy*, Exon. 83 a; Th. 313, 23; Mód. 4.

ér-dæd, e; *f. Former conduct, a past deed*; ante-actum :—Wyt witodlíce be uncer ér-dædum onfóþ nos *duo quidem juste, nam digna factis recipimus*, Lk. Bos. 23, 41: Bd. I, 6; S. 476, 24, note.

ér-dæg, es; *m.* I. *early day, early morn*; matutinum, mane, prima lux :—Mid ærdæge *at early day*, Andr. Kmbl. 440; An. 220: 3048; An. 1527: Cd. 121; Th. 155, 19; Gen. 2575. On uhtan mid ærdæge *in the morning at early day*, Beo. Th. 253; B. 126. To ðam ærdæge *on that morn*, Cd. 153; Th. 100, 12; Exod. 198. II. *in pl. Early days, former days*; dies prisci :—On ærdagum *in former days*, Cd. 119; Th. 153, 23; Gen. 2543: Exon. 92 a; Th. 6, 4; Cri. 79. [*O. Sax.* an érdagun *priscis diebus*: *O. Nrs.* í árdaga *primis temporibus, olim.*]

ér-deáþ, es; *m. Early death*; mors immatura :—Regnþeófas dælaþ yldo, oððe ér-deáþ *the great thieves find age, or early death*, Cd. 169; Th. 212, 14; Exod. 539.

ærdjan, ærdyan *to inhabit* [ærd = eard *earth, dwelling*] :—Ærdydon habitabant, Bd. 2, 9; S. 510, 15. v. eardian.

ærdon = ærudon? *from* ærnan; *p. de To run, run away*; currere :—He gehleóp and his bróðru mid him begen ærdon *he fled and both his brothers ran away with him*, Byrht. Th. 137, 25; By. 191.

ærdung, e; *f.* [eard *a dwelling*] *A tabernacle*, Ps. Spl. T. 18, 5. v. eardung.

æ-réfnan *to bear*, Ps. Spl. T. 24, 5. v. a-réfnan.

éren, éryn, ærn; *adj. Made of brass, brazen*; æneus :—Wirc ane érenan næddran *fac serpentum æneum*, Num. 21, 8. Ærnum bemum *with brazen trumps*, Cd. 154; Th. 191, 18; Exod. 216: Ors. 2, 8; Bos. 52, 16: Ælfc. Gr. 9; Som. 4, 60.

éren-byt, -bytt, e; *f.* [byt *a butt, vessel*] *A brass pan* or *vessel*; lenticula, Wrt. Voc. 25, 17.

érend, érende, érynd, es; *n: pl. nom. acc.* érendu, érendo *An* ERRAND, *a message, an embassy, news, tidings, an answer, business, care*; nuntium, mandatum, negotium, cura :—Ne mæg ðæs érendes ylding wyrþan *there may not be a delay of this errand*, Andr. Kmbl. 429; An. 215. He his hláfordes érende secgan sceolde *he should tell his lord's message*, Bd. 2, 9; S. 511, 19. Hí hæfdon nyt érend *they had a profitable errand*, 5, 10; S. 624, 21: 3, 6; S. 528, 17: L. C. S. 76; Th. i. 418, 5. He sent on his érenda *he sends on his errands*, Bt. 39, 13; Fox 234, 25. Híg lægdon érende *they imposed an errand*, Chr. 1065; Th. 332, 25, col. 2. He aboden hæfde Godes érendu *he had announced God's messages*, Exon. 43 a; Th. 145, 17; Gú. 696: 51 b; Th. 179, 31; Gú. 1270. Hí lufedon Godes érendo *they loved God's errands*, 34 b; Th. 111, 27; Gú. 133. [*Laym.* arend, erend, *as in* arend-rake, erend-mon:

C

Orm. ernde: *R. Glouc.* ernde, erinde: *O. Sax.* árundi, *n. message:* *M. H. Ger.* árant, érende, *m. message: O. H. Ger.* áranti, áronti, árunti, *m. nuntius; f. verbum, mandatum: Dan.* ærinde, ærend: *Swed.* ærende: *O. Nrs.* örundi, erendi, *n. negotium: Sansk.* ír *ire, to go.*] v. ár *a messenger.*

æren-dæg, es; *m.* [*contracted for* on ærran dæg *on a former day*] *The day before, yesterday;* pridie, Ælfc. Gl. 96; Wrt. Voc. 53, 31. v. dæg.

ærend-bóc, e; *f. A letter, message;* epistola, litteræ:—Hí ne mihton arǽdan engles ǽrendbéc *they might not interpret the angels' messages,* Cd. 212; Th. 261, 32; Dan. 735. v. ǽrend-gewrit.

ærend-gást, es; *m. A spiritual messenger, an angel;* nuntius spiritus, angelus:—Godes ǽrendgást *God's spiritual messenger,* Cd. 104; Th. 138, 23; Gen. 2296.

ærend-gewrit, ǽrend-writ, es; *n. A message or report in writing, a letter, an epistle, letters mandatory, a brief writing, short notes, a summary;* epistola:—Hí sendon ǽrendgewrit *mittunt epistolam,* Bd. 1, 13; S. 481, 41. On, forþgeonge ðæs ǽrendgewrites *in processu epistolæ,* 1, 13; S. 481, 43: Bt. Met. Fox 1, 125; Met. 1, 63. Ærend-gewrit *epistola* vel *pictacium,* Wrt. Voc. 46, 64: 61, 21. Þurh his ǽrendgewritu *by his letters,* Bd. pref; S. 472, 22.

ærendian; *p.* ede; *pp.* ed *To go on an errand, to carry news, tidings, or a message, to intercede, to treat for anything, to plead the cause;* nuntium ferre, mandatum deferre, intercedere, annuntiare:—He mæg unc ǽrendian *he may bear our messages,* Cd. 32; Th. 41, 31; Gen. 665. Ða ǽrendracan, ðe his cwale ǽrndedon [Whel. ǽrenddedon] *the messengers, who had treated for his death,* Bd. 2, 12; S. 515, 4.

ærend-raca, ǽrend-wreca, an; *m.* [ǽrend *an errand;* raca, wreca *from* reccan *to tell,* wrecan *to utter*] *A messenger, ambassador, an apostle, angel;* nuntius, apostolus, angelus:—Se ǽrendraca nys mǽrra ðonne se ðe hine sende *non est apostolus major eo qui misit eum,* Jn. Bos. 13, 16. Sende he ǽrendracan *misit legatarios,* Bd. 5, 21; S. 642, 34. Gabriél Godes ǽrendraca *Gabriel God's angel,* Hy. 10, 12. Ærendraca, Bd. 2, 9; S. 510, 27: 2, 12; S. 513, 8; 515, 3: 1, 12; S. 480, 25. Ærendraca *an apostle,* Wrt. Voc. 42, 1. Ærendraca *unnytnesse a tale-bearer,* Cot. 139. Gesibbe ǽrendracan *messengers of peace;* caduceatores vel pacifici, Wrt. Voc. 36, 6.

ærendran *messengers;* nuntii:—Ædele ǽrendran andswarodon [*Grn.* ǽrendracan] *the noble messengers answered,* Cd. 111; Th. 147, 4; Gen. 2434.

ærend-secg, es; *m. An errand-deliverer, a messenger;* legatus, nuntius:—Ic, on his gearwan, geseó ðæt he is ǽrend-secg uncres Hearran I, *by his habit, see that he is the messenger of our Lord,* Cd. 30; Th. 41, 17; Gen. 658.

ærend-secgan *to deliver a message;* nuntium deferre. v. secgan.

ærend-spræc, e; *f. A verbal message;* nuntiatio:—Ærendspræce abeódan *to announce a verbal message,* Exon. 123 a; Th. 472, 13; Rä. 61, 15.

ærendung, e; *f. A command;* mandatum, C. R. Ben. 38.

ærend-wreca, an; *m. A messenger, ambassador;* nuntius, legatus:—Hí onsendon ǽrendwrecan *miserunt nuntios,* Bd. 1, 12; S. 480, 25. He sende ǽrendwrecan in Gallia ríce *he sent ambassadors into the kingdom of the Gauls,* 2, 6; S. 508, 33. v. ǽrendraca.

ærend-writ, es; *n. A letter;* epistola, Bd. 5, 21; S. 642, 34, note. v. ǽrend-gewrit.

æren-geát, *for* earn-gǽt *a goat-eagle;* harpe = ἅρπη, Ælfc. Gl. 17; Wrt. Voc. 21, 62. v. earn-geát.

ærer; *adv. Before:*—Ærer hit gewyrþe *before it comes to pass,* Bt. 41, 2; Fox 244, note 8. v. ǽror.

æ-rest, es; *m.: e; f. The resurrection:*—On lífes ǽreste *in resurrectionem vitæ,* Jn. Bos. 5, 29; Andr. Grm. 780: Exon. 37 b; Th. 122, 29; Gú. 313. v. æ-rist.

ærest; *adj. First,* ERST; primus:—Weorpe ǽrest stán *primus lapidem mittat,* Jn. Bos. 8, 7: Cd. 52; Th. 66, 5; Gen. 1079. v. ǽr; *adj.*

ærest; *adv. First, at first;* primum, primo:—Him cenned wearþ Cainan ǽrest *to him was born Cainan first,* Cd. 57; Th. 70, 7; Gen. 1149: 75; Th. 92, 16; Gen. 1529. v. ǽr.

ær-fæder; *indecl. in sing. but sometimes gen.* -fæderes *and dat.* -fædere *are found; pl. nom. acc.* -fæderas; *gen.* a; *dat.* um; *m. A forefather, father;* propator, pater, Beo. Th. 5238; B. 2622.

ær-fæst; *adj. Honourable, good, gracious, merciful,* Judth. 11; Thw. 24, 15; Jud. 190. v. ár-fæst.

ær-fæstnys, -nyss, e; *f. Honesty, goodness, piety;* pietas:—Aidanus wæs mycelre ǽrfæstnysse and gemetfæstnysse mon *Aidan was a man of much piety and moderation,* Bd. 3, 3; S. 525, 31. v. ár-fæstnes.

ærfe *an inheritance,* Heming, pp. 104, 105. v. yrfe.

ær-geára; *adv. Heretofore, of old;* olim, Salm. Kmbl. 860; Sal. 429; Bt. Met. Fox 20, 104; Met. 20, 52. v. geára.

ær-geblond *the sea agitation.* v. ǽra gebland, ear-gebland.

ær-gedón; *adj. Done before;* anteactus, prior:—Wæs seó ǽhtnysse unmǽtre and singalre eallum ðám ǽrgedónum *quæ persecutio omnibus*

fere anteactis diuturnior atque immanior fuit, Bd. 1, 6; S. 476, 24: 1, 12; S. 481, 25.

ær-genemned; *pp. Before-named;* prænominatus. v. ge-nemnan.

ær-gescod; *pp. Brass-shod, shod with brass;* ære calceatus:—Bill ǽrgescod *a brass-shod bill,* Beo. Th. 5548; B. 2777.

ær-gestreón; es; *n. Ancient treasure;* thesaurus antiquitus repositus:—Ðǽr wæs fela in ðam eorþ [-scræfe] ǽrgestreóna *there were many ancient treasures in that earth-cave,* Beo. Th. 4457; B. 2232: 3518; B. 1757: Exon. 22 b; Th. 62, 5; Cri. 997: Cd. 98; Th. 129, 22; Gen. 2147.

ær-geweorc, es; *n. An ancient work;* antiquum opus:—Enta ǽrgeweorc *the ancient work of giants,* Beo. Th. 3362; B. 1679: Andr. Kmbl. 2471; An. 1237.

ær-gewinn, es; *n. An ancient struggle, former agony;* antiquum certamen, pristina agonia:—Earmra ǽrgewinn *the former agony of the wretched ones,* Rood Kmbl. 37; Kr. 19.

ær-gewyrht, es; *n. A former work, a deed of old;* opus pristinum, facinus olim commissum:—Ða byre siððan grimme onguldon gafulrædenne þurh ǽrgewyrht *the children since have bitterly paid the tax through the deed of old,* Exon. 47 a; Th. 161, 17; Gú. 960: Elen. Kmbl. 2599; El. 1301. *Nom. pl.* ærgewyrhtu, Exon. 26 a; Th. 76, 18; Cri. 1241.

ær-glæd; *adj. Brass-bright, gleaming with brazen arms;* armis æneis coruscans, Cd. 158; Th. 196, 17; Exod. 293.

ær-gód; *adj. Good before others, of prime goodness;* præ ceteris bonus:—Æðeling ǽrgód *a prince good before others,* Beo. Th. 260; B. 130: 2662; B. 1329. Íren ǽrgód *iron of prime goodness,* 1982; B. 989.

ær-gystran-dæg *ere-yesterday, the day before yesterday;* nudius tertius. v. gysternlíc dæg. gyrstan-dæg.

ærian *to plough:*—Hwilc man aþohte ǽrust myd sul to ærienne [MS. æriende] *what man thought first of ploughing with a plough?* Anlct. 113, 27. v. erian.

æ-riht, es; *n.* [æ *law,* riht *right*] *Law-right, law;* jus legum, jus:—Ða ðe fyrngewritu sélest cunnen, ǽriht *who the old writings best know, your own law,* Elen. Kmbl. 749; El. 375: 1176; El. 590.

æring, e; *f. The early dawn, day-break;* diluculum:—In ǽringe, æfter leóhtes cyme *at early dawn, after light's coming,* Exon. 68 a; Th. 252, 9; Jul. 160: Mk. Lind. War. 1, 35. v. ǽr; *adv.*

æ-risc, e; *f.* [eá *running water,* risc *a rush*] *A water-rush, bulrush;* scirpus, Ælfc. Gl. 42; Wrt. Voc. 31, 31. v. eá-risc.

æ-rist, æ-ryst, æ-rest, es; *m:* e; *f. A rising up, the resurrection;* resurrectio: — Drihtnes ærist *the resurrection of the Lord,* Menol. Fox 116; Men. 58. Æfter ǽriste *after resurrection,* Exon. 64 a; Th. 235, 18; Ph. 559. Dú mín setl swylce oncneówe and mínne ærist æfter gecýþdest *tu cognovisti sessionem meam et resurrectionem meam,* Ps. Th. 138, 1: Hy. 10, 55. Ærist gefremede *accomplished his resurrection,* Exon. 48 b; Th. 168, 6; Gú. 1073. Ðonne ǽriste ealle gefremmaþ *when all shall accomplish their resurrection,* 63 a; Th. 231, 26; Ph. 495. [*Goth.* urrists. *f.*]

ærist = ǽrest; *adv. First:*—Mec sewong ǽrist cende *the field first brought me forth,* Exon. 109 a; Th. 417, 10; Rä. 36, 2: *sup. of* ǽr; *adv.*

ær-lést, e; *f. Dishonour, impiety, cruelty, a disgraceful deed:*—Hwelce ǽrléste Nerón worhte *what disgraceful deeds Nero wrought,* Bt. Met. Fox 9, 2; Met. 9, 1. v. ár-leást.

ær-líce, ár-líce; *adv.* [ǽr *ere, before,* líce] EARLY *in the morning;* diluculo, mane, Jn. Lind. War. 8, 2.

ærm; *adj. Poor;* pauper:—On dǽre ǽrman byrig *in that poor city,* Chr. 1011; Th. i. 269, 1, col. 1: 1014; Th. i. 272, note 1, 3. v. earm.

ær-margen, es; *m. The early morning, the day-break,* Ps. Surt. 56, 9; 107, 3: 118, 148. v. ǽr-morgen.

ær-morgen, -mergen, es; *m. The early morning, day-break;* primum mane, matutinum, diluculum:—On ǽrmorgen *in the early morning,* Bt. Met. Fox 28, 72; Met. 28, 36. Ærmorgenes gancg wið æstentíd *exitus matutini et vespere,* Ps. Th. 64, 9. On ærmergen *diluculo,* 107, 2: 56, 10: Bd. 1, 34; S. 499, 27. Ærmyrgen *mane,* Ælfc. Gl. 94; Wrt. Voc. 53, 2. [*O. Nrs.* ár-morgin.]

ærn, ern, es; *n. A place, secret place, closet, an habitation, a house, cottage;* locus, locus secretior, domus, casa:—Bireþ into his ærne beareth into his habitation, L. In. 57; Th. i. 138, 16. [*O. Nrs.* rann, *n.*] DER. bere-ærn [-ern] *a barley place, barn,* blác-, blæc-, bléc-, breáw-, carc-, cweart-, cwert-, dóm-, eást-, eorþ-, fold-, gæst-, gest-, gyst-, heal-, hédd-, holm-, hord-, mædel-, medo-, meðel-, mold-, norþ-, slǽp-, súþ-, þrýþ-, west-, wín-.

-ærn, -ern, es; *n.* [ærn *a place*] is generally used as a termination, and denotes *a place:* thus, Eorþ-ærn, es; *n. An earth-place* or *house, the grave:*—Open wæs ðæt eorþ-ærn *the grave was open,* Exon. 120 a; Th. 460, 18; Hö. 19: 119 b; Th. 459, 22; Hö. 3; Th. 460, 4; Hö. 12. Dóm-ern *a judgment-place, judgment-hall, court of justice,* Mt. Bos. 27, 27. Hédd-ern *a heeded-place, store-house, cellar,* Lk. Bos. 12, 24.

-ærn; *adj. termination def.* se -ǽrna, *m;* -ǽrne, *f. n.* v. -ern.

ǽrn *brazen* :—Ǽrnum bēmum *with brazen trumpets*, Cd. 154; Th. 191, 18; Exod. 216. v. ǽren.

ærnan; *p.* de; *pp.* ed; *v. intrans. To run*; currere :—Ǽrnan *to run*, Bd. 5, 6; S. 618, 42: S. 619, 12. Ǽrnaþ hȳ *they run*, Ors. 1, 1; Bos. 22, 36. DER. ge-ærnan. v. yrnan.

ǽrnddedon = ǽrendedon; *p.* of ǽrendian *To go on an errand*; nuntium ferre, Bd. 2, 12; S. 515, 4.

ǽrne *Early* :—On ǽrne mergen *primo mane*, Mt. Bos. 20, 1; *acc. sing. m.* of ǽr, *adj.*

ærne-weg, es; *m.* [ærnan *to run*, weg *a way*] *A running-way, a way fit for running on, a broad road*; via cursui apta, platea :—Æt sumes ærneweges ende *at the end of some course*, Bt. 37, 2; Fox 188, 9. Gescroepe ærneweg *via apta cursui equorum*, Bd. 5, 6; S. 618, 41.

ǽrnian *to earn*. v. ge-ærnian.

ǽrning, e; *f. A running, riding*; cursus, equitatio :—Ða ðe hiora ærninge trēwaþ *those who trust in their running*, Bt. 37, 2; Fox 188, 10: Bd. 5, 6; S. 619, 15.

ǽrnung, e; *f. An* EARNING, *stipend, hire, wages*; merces. v. earnung.

ǽron; *adv. Before*; antea :—Ic hyt ǽron nyste *I knew it not before*, Nicod. 12; Thw. 6, 22. v. ǽr; *adv.*

ǽror, ǽrror; *prep. dat. Before*; ante, priusquam :— Næs ǽror ðe [MS. aworþe] ǽnegu gesceaft *there was not before thee any creature*, Bt. Met. Fox 20, 81; Met. 20, 41.

ǽror, ǽrror, ǽrur, ǽrer; *adv. Before, formerly*; antea, prius :—Weras on wonge wibed setton, neáh ðam ðe Abraham ǽror rǽrde *the men placed an altar in the plain, near that which Abraham had reared before*, Cd. 90; Th. 113, 7; Gen. 1883. Se ðe fela ǽror fyrena gefremede *he who before had committed many crimes*, Beo. Th. 1623; B. 809. Nemne we ǽror mǽgen fāne gefyllan *unless we before may fell the foe*, 5302; B. 2654. Ðæt hió eft cume, ðǽr hió ǽror wæs *that it again comes where it was before*, Bt. Met. Fox 13, 152; Met. 13, 76. Ǽror, *on his lífdagum before, in the days of his life*, 26, 174; Met. 26, 87: Exon. 35 b; Th. 114, 32; Gú. 181: Ps. Th. 77, 3; 91, 8: 134, 11: 135, 21: 145, 4: Menol. Fox 330; Men. 166. v. ǽr; *adv.*

ǽrost; *adv. First*, Byrht. Th. 135, 27; By. 124: Gen. 19, 33. v. ǽr.

ǽrra, ǽrre; *adj. Former, earlier*, Exon. 113 b; Th. 436, 11; Rä. 54, 12: Menol. Fox 213; Men. 108: Elen. Kmbl. 609; El. 305. v. ǽra.

ǽrra geóla *the ere* or *former Yule month, December*, Menol. Fox 439; Men. 221. v. geóla.

ǽrra líða *the ere* or *former Litha, June*, Menol. Fox 213; Men. 108. v. líða.

ǽrror; *adv. Before, formerly* :—We iú in heofonum hæfdon ǽrror wlite and weorþmynt *we once in heaven had formerly beauty and dignity*, Cd. 216; Th. 274, 9; Sat. 151: 220; Th. 283, 4; Sat. 299. v. ǽror.

ǽrror; *prep. dat. Before*; ante :—Cymeþ eástan up ǽror [MS. æst ror] sunnan, and eft æfter sunnan on setl glídeþ *comes up from the east before the sun, and again after the sun glides to his seat*, Bt. Met. Fox 29, 52; Met. 29, 26. v. ǽror.

ǽrs *The buttocks, the hind part*; anus, podex :—Open-ærs *a medlar*, Wrt. Voc. 32, 50; Som. 64, 116. v. ears.

ǽr-sceaft, e; *f. An old creation, an ancient work*; pristina creatio, priscum opus, Exon. 124 a; Th. 477, 1; Ruin. 16.

ǽrsc-hen *a quail*, Ælfc. Gl. 38; Wrt. Voc. 29, 42. v. ersc-hen.

ǽrst *first*; primo, Homl. in nat. Innoc. p. 36, = ǽrost. v. ǽr; *adv.*

ǽr-ðam, ǽr-ðon *before that*, Mt. Bos. 6, 8: Exon. 61 a; Th. 224, 22; Ph. 379. v. ǽr; *prep.*

ǽr-ðam-ðe *before that which, till*, Mt. Bos. 12, 20. v. ǽr; *prep.*

ǽrur; *adv. Before*; antea :—Swā he him ǽrur, hér on ðyssum lífe, ge-earnaþ *as he for himself before, here in this life, earneth*, Rood Kmbl. 214; Kr. 108: Ps. Th. 115, 3. v. ǽror.

ǽr-wacol; *adj. Early awake*; diluculo vigil :—For hwí eart ðú ðus ǽrwacol *why art thou thus early awake?* Apol. Th. 19, 5.

ǽr-wéla, an; *m.* [ǽr ere, before, wéla *wealth*] *Ancient wealth*; divitiæ antiquitus accumulatæ, Beo. Th. 5488; B. 2747.

ǽr-woruld, e; *f. The former world*; pristinus mundus :—Ðonne weorþeþ sunne sweart gewended, on blódes hiw, seó ðe beorhte scān ofer ǽrworuld *then the sun shall be turned swart, to hue of blood, which shone brightly over the former world*, Exon. 21 b; Th. 58, 17; Cri. 937.

ǽryn *brazen*; æreus :—Ðú gesettest swā swā bogan bræsenne [ǽrynne, Spl. C.] earmas míne *posuisti ut arcum æreum brachia mea*, Ps. Lamb. 17, 35: Ps. Spl. C. 106, 16. v. ǽren.

ǽrynde, es; *m. An interpreter*; interpres :—Ðæra byrla ealdor forgeat Iosepes ǽrynde *prepositus pincernarum oblitus est Josephi interpretis sui*, Gen. 40, 23.

ǽrynd-writ *a letter*, Lye. v. ǽrend-gewrit.

ǽryr; *adv. Before*; prius, C. Jn. 1, 30, Lye. v. ǽror.

æ-ryst, es; *m*: e; *f. The resurrection* :—Ða secgeaþ ðæt nán æryst ne sȳ *qui dicunt non esse resurrectionem*, Mt. Bos. 22, 23; 27, 53. v. á-rist.

ǽryst; *adv. First*; primum, primo, Ps. Th. 104, 15. v. ǽrest.

ǼS, es; *n. Food, meat, carrion, a dead carcase*; esca, cibus, pabulum, cadaver :—Earn ǽses georn *the eagle eager for food*, Byrht. Th. 134, 60; By. 107. Léton him behíndan ðone earn ǽses brúcan *they left behind*

them *the eagle to eat of the carrion*, Chr. 938; Th. i. 207, 30, col. 2; Æðelst. 63. Ǽse wlanc *exulting in carrion*, Beo. Th. 2668; B. 1332: Ps. Th. 146, 10. [Dut. aas, *n. esca, cadaver*: Ger. aas, *n. esca, cadaver*: M. H. Ger. ās, *n*: O. H. Ger. ās, *n. esca*: Dan. aas, *n*: Swed. as, *n.*]

ǼSC; *g.* æsces; *pl. nom. acc.* æscas, ascas; *g.* æsca, asca; *d.* æscum, ascum; *m.* I. *an ash-tree*; fraxinus excelsior :—On ðone æsc *to the ash-tree*, Cod. Dipl. Apndx. 461; A. D. 956; Kmbl. iii. 450, 3. Ǽsc *fraxinus*, Ælfc. Gl. 45; Som. 64, 98. II. *the Anglo-Saxon Rune* ᚫ = æ, the name of which letter in Anglo-Saxon is æsc *an ash-tree*, hence this Rune not only stands for the letter æ, but for æsc *an ash-tree*, as,—ᚫ byþ oferheáh, eldum dȳre, stíþ staðule *the ash-tree is over-high, dear to men, firm in its place*, Hick. Thes. vol. i. p. 135; Runic pm. 26; Kmbl. 344, 23. Se torhta æsc *the remarkable Rune* æsc, Exon. 112 a; Th. 429, 24; Rä. 43, 9. III. *an ash-spear, a spear, lance*; hasta fraxina, hasta :—Byrhtnoþ wānd wācne æsc *Byrhtnoth brandished his slender ashen spear*, Byrht. Th. 132, 68; By. 43: 140, 59; By. 310. Ðe ðé æsca tír æt gúðe forgeaf *who to thee gave glory of spears in battle*, Cd. 97; Th. 127, 10; Gen. 2108. Asca, *g. pl.* Exon. 78 a; Th. 292, 15; Wand. 99. Ǽscum *with spears*, Beo. Th. 3548; B. 1772: Andr. Kmbl. 2195; An. 1099. IV. *because boats were made of ash,—a small ship, a skiff, a light vessel to sail or row in*; navis, navigium, dromo :—Hét Ælfréd cyng timbrian langscipu ongén ða æscas *king Alfred commanded to build long ships against those ships*, Chr. 897; Th. i. 174, 41. Ǽsc *dromo*, Wrt. Voc. 63, 34: 56, 24. [O. H. Ger. asc, *m*: O. Nrs. askr, *m. arbor, fraxinus, vas ligneum, navis, gladius*, Egils.] DER. daroþ-æsc, ceaster-: æsc-rind.

æ-scære; *adj.* [æ = a, scær, *p.* of sceran *to shear, cut*] *Without tonsure, uncut, untrimmed, neglected*; intonsus, incultus, neglectus :—Deóplíc dǽdbót biþ, ðæt lǽwede man swā æscære beó, ðæt íren ne cume on háre, ne on nægle *it is a deep penitence, that a layman be so untrimmed that scissors [iron] come not on hair, nor on nail*, L. Pen. 10; Th. ii. 280, 20. v. a-scære.

æsc-berend, es; *m.* [æsc *a spear*, berende *bearing*, *part.* from beran *to bear*] *A spear* or *lance-bearer, a soldier*; hastifer :—Eorre æscberend *the fierce spear-bearer*, Andr. Kmbl. 93; An. 47: 2153; An. 1078. Ealde æscberend *the old spear-bearer*, 3072; An. 1539.

æsc-berende; *part. Spear-bearing*; hastam gerens :—Wígena æscberendra *of warriors bearing spears*, Cd. 94; Th. 123, 7; Gen. 2041.

æsce; *g.* æscean; *f. Ashes* :—Forðon ic anlíc ætt æscean hláfe *quia cinerem sicut panem manducabam*, Ps. Th. 101, 7: 147, 5. v. asce.

æsce; *f. Search, inquisition, examination, inquiry, trial* of *asking after any matter* or *thing*; interrogatio, investigatio, disquisitio :—Hæfdon ealle ða æscean *all should have the search*, L. Ath. 5; Th. i. 230, 18.

æsceda, an; *m. A farrago, mixture, perfume*; migma, Wrt. Voc. 38, 53.

æscen *A vessel made of ash-wood, such as a bottle, bucket, pail*, etc; lagena :—Ǽscen ðe is óðre namon hrygilebuc gecleopad *an ascen, its other name is called Rigelbuc*, q. back-bucket, Heming, p. 393.

æscen; *adj. Ashen, ash, made of ash*; fraxineus. v. æsc, -en.

Ǽsces dún, e; *f.* [æsc *ash-tree*, dún *a hill*] ASHDOWN, *the hill of the ash-tree, on the Ridgeway in Berkshire, where Alfred and his elder brother, king Ethelred, first routed the Danes*; 'dicitur Latine mons fraxini,' Asser.—Hér gefeaht Æðeréd cyning and Ælfréd, his bróður, wið ealne ðone here, on Ǽsces dúne A. D. 871, *here fought king Ethelred and Alfred, his brother, with all the army [of the Danes], on Ashdown*, Chr. 871; Th. 139, 5, col. 1.

æsc-here, es; *m. A spear-band, company armed with spears, a ship* or *naval-band*; exercitus hastifer, exercitus navalis, Byrht. Th. 133, 53; By. 69.

æsc-holt, es; *nom. pl.* -holt; *n. Ash-wood, an ash-wood spear*; lignum fraxineum, hasta fraxinea :—Ǽscholt asceóc *shook his ashen spear*, Byrht. 138, 35; By. 230: Beo. Th. 665; B. 330.

æscian *to ask*; interrogare, Jud. Civ. Lund. v. acsian.

æsc-man, -mann, es; *m. A ship-man, sailor*, and hence *a pirate*; nauta, pirata :—Ægþer ge æscmanna ge óðerra *both of the ship-men and of the others*, Chr. 921; Th. 195, 15: Cot. 155.

æsc-plega, an; *m.* [plega *play*] *The play of spears, war*; hastarum ludus, prœlium :—Æt ðam æscplegan, Judth. 11; Thw. 24, 31; Jud. 217.

æsc-rind, e; *f. Ash-bark*; fraxini cortex :—Nim æscrinde *take ash-bark*, Lchdm. iii. 14, 1. Wel æscrinde *boil ash-bark*, ii. 78, 5.

æsc-róf; *adj. Spear-famed, distinguished in battle, illustrious, noble*; hasta clarus, in prœlio strenuus, illustris, nobilis :—Eorlas æscrófe *illustrious nobles*, Judth. 12; Thw. 26, 20; Jud. 337: Elen. Grm. 276: 202.

æsc-stéde, es; *m. The ash-spear place, place of battle*; hastæ locus, pugnæ locus :—Hí witan fundian æscstéde *they strive to know the battle place*, Exon. 83 b; Th. 314, 20; Mód. 17.

æsc-þræc; *g.* -þræce; *pl. nom. g. acc.* -þraca; *f. Spear-strength, brunt of spears, a battle*; hastæ vis, hastarum impetus, prœlium :—Æt æscþræce, Cd. 98; Th. 130, 2; Gen. 2153.

æsc-þrote, an; *f*: -þrotu; *e*; *f.* [æsc *ash*, þrote *a throat*] ASH-THROAT, *vervain*; verbenaca, verbena officinalis, Prior, p. 242: vocabularies give the *Lat.* ferula *the fennel-giant*, but verbenaca *vervain* seems more probable from the following quotations,—Herba uermenaca [= uerbenaca, Herb. 4, = verbenaca: *Lat.* = berbena, 67, = verbena, *Lat.*] ðæt is æscþrotu

the herb verbena, that is ash-throat [=*vervain*], Herb. cont. 4, 1; Lchdm. i. 8, 1. Niðeweardre æscþrotan *of the netherward* [*part of*] *vervain*, L. M. 3, 72; Lchdm. ii. 358, 16. Ním æscþrotan *take vervain*, 1, 88; Lchdm. ii. 156, 22. Æscþrotan, 1, 43; Lchdm. ii. 108, 6. Æscþrote, nom. Herb. 4, 1; Lchdm. i. 90, 1. Æscþrotu, L. M. 1, 47; Lchdm. ii. 120, 9: 2, 53; Lchdm. ii. 274, 9. Man æscþrote nemneþ *one nameth it vervain*, Herb. 4, 1; Lchdm. i. 90, 3. Gením æscþrote *take vervain*, 101, 3; Lchdm. i. 216, 11: L. M. 3, 61; Lchdm. ii. 344, 9: Lchdm. iii. 28, 14.

æsc-tír, es; m. *Spear-glory, glory in war;* hastæ gloria, belli gloria, Cd. 95; Th. 124, 27; Gen. 2069.

æsc-wert, e; *f. Ash-wort, vervain;* verbena, Mone C. 3; p. 442, 24.

æsc-wíga, an; *m. A spear-warrior;* bellator hastifer :—Eald æscwíga *an old spear-warrior*, Beo. Th. 4090; B. 2042. Æscwígan, nom. pl. Elen. Grm. 260.

æsc-wlanc; *adj. Spear-proud;* hasta superbus, Leo 104.

ÆSP, e; f. æspe, an; *f. An* ASP or *aspen-tree;* populus tremula :—Æspan rind *the rind of the asp-tree*, L. M. 1, 47; Lchdm. ii. 116, 1. v. æps.

æspen; *adj.* ASPEN, *belonging to the asp-tree;* populeus. DER. æsp.

æ-spring, æ-springe, æ-sprynge, es; *n.* [æ *water*, spring *a spring*] *A water-spring, fountain;* aquæ fons, fons :—Se æðela fugel æt ðam æspringe wunaþ *the noble fowl remains at the fountain*, Exon. 57 a; Th. 204, 28; Ph. 104. Æspringe útawealleþ *of clife a fountain springs out of a cliff*, Bt. Met. Fox 5, 23; Met. 5, 12. Ealle æsprynge *all springs*, Exon. 55 a; Th. 194, 5; Az. 134: 93 b; Th. 351, 8; Sch. 77. v. eá-spring.

æ-springnes, -ness, e; *f.* [aspringan *to fail*] *A failing, fainting;* defectio, Ps. Spl. T. 118, 53. v. a-sprungennes.

æstel, es; *m. A tablet, a table for notes, a waxed tablet;* indicatorium, astula, pugillaris. Du Cange says astula = *tabula sectilis*, referring to pugillares, under which he gives the following quotation from Cassander in Liturgicis, p. 53,—'Inter instrumenta sacra numerantur pugillares aurei sive argentei. . . . Propriè pugillares sunt tabulæ, in quibus scribi consuevit, quæ Græcè πινακίδια dicuntur.' In St. Luke i. 63, αἴτησας πινακίδιον, *postulans pugillarem*, is in the A. Sax. Gospels, gebedenum wex-brede *a waxed tablet being asked for.* William of Malmsbury may have alluded to one of these waxed tablets in Gesta Reg. ii. § 123,—'Cum pugillari aureo in quo est manca auri.' It is most probable then that Alfred's æstel consisted of two waxed tablets, joined together by a hinge, and framed or covered with gold to the value of fifty mancuses. When these waxed tablets were closed, being framed or covered with gold, they would have a splendid and costly appearance, worthy the gift of a king :—Æstel *indicatorium*, Ælfc. Gr. 8; Som. 7, 63 : Cot. 214 : Ælfc. Gl. 19? Lye. Ða ongan ic [Ælfrēd cyning] ða bóc wendan on Englisc, ðe is genemned on Læden *Pastoralis*, and on Englisc *Hierde-bóc*, hwílum word be worde, hwílum andgit of andgite, swá swá ic hie geliornode æt Plegmunde mínum Ærcebiscepe, and æt Assere mínum Biscepe, and æt Grimbolde mínum Mæsse-Prióste, and æt Iohanne mínum Mæsse-Preóste. Siððan ic hie ða geliornod hæfde, swá swá ic hie forstód, and swá ic hie andgitfullícost areccean meahte, ic hie on Englisc awende, and to ǽlcum Biscep-stóle on mínum Ríce wille áne onsendan, and on ǽlcre biþ án Æstel, se biþ on fíftegum Mancessan. Ond ic bebióde, on Godes naman, ðæt nán mon ðone Æstel from ðære béc ne dó, ne ða bóc from ðæm Mynstre *then I* [*Alfred king*] *began to translate into English the book, which is called in Latin Pastoralis, and in English Herdsman's book, sometimes word by word, sometimes meaning for meaning, as I learned it from Plegmund my archbishop, and from Asser my bishop, and from Grimbold my presbyter, and from John my presbyter. After I had then learned it, so that I understood it as well as my understanding would allow me, I translated it into English, and I will send one copy to each bishop's see in my kingdom; and on each one there shall be one tablet, which shall be worth fifty mancuses. And in God's name, I command that no man take the tablet from the book, nor the book from the minster*, Past. Hat. MS. Pref.

æsul, es; *m. An ass;* asinus, Mt. Rush. Kmbl. 21, 2. v. esol.

æ-swáp, es; *n. pl.* æswápa *Sweepings, dust;* peripsema, purgamentum. v. a-swáp.

æ-swíc, æ-swýc, e-swíc, es; *m.* [æ *law*, swíc *an offence*] *An offence, a scandal, stumbling-block, sedition, deceit;* scandalum :— Ne biþ him æswíc *non est illis scandalum*, Ps. Th. 118, 165 : Ps. Spl. 118, 165 : 48, 13 : 49, 21, C. To æswýce *in scandalum*, Ps. Th. 105, 26.

æ-swíca, an; *m.* æ-swícend, es; *m. An offender of the law, a deceiver, hypocrite, apostate;* hypocrita, apostata. v. swíca.

æ-swícian; *p.* ode; *pp.* od *To offend, to depart from the law, to dissemble;* scandalizare, deficere ab aliquo :—Gyf ðín swýðre eáge ðē æswície *si oculus tuus dexter scandalizat te*, Mt. Bos. 5, 29. v. a-swícian?

æ-swícung, e; *f. An offence;* scandalum :—Ðú settest æswícunge *ponebas scandalum*, Ps. Spl. 49, 21. v. æ-swíc.

æ-swind; *adj. Idle;* iners, Cot. 108. v. a-swind.

æ-swutol; *adj.* [æ *law*, sweotol *manifest, clear, open*] *One who makes the law clear, a lawyer;* legisperitus. v. sweotol.

æ-swýc, es; *m. An offence;* scandalum, Ps. Th. 105, 26. v. æ-swíc.

æ-syllend, es; *m.* [æ *law*, syllende *giving*] *A lawgiver;* legislator, Ps. Spl. 83, 7.

ÆT; *prep.*   **I.** *with the dative;* cum dativo AT, *to, before, next, with, in, for, against;* apud, juxta, prope, ante, ad, in, contra :—Sittende æt tollsceamule *sitting at the seat of custom*, Mt. Bos. 9, 9. Æt fruman worulde *at the beginning of the world*, Exon. 47 a; Th. 161, 7; Gú. 955. Wæs seó treów lufu hát æt heortan *the true love was hot at heart*, 15 b; Th. 34, 8; Cri. 539. Ge ne cómon æt me *ye came not to me*, Mt. Bos. 25, 43. Æt selde *before the throne*, Cd. 228; Th. 306, 12; Sat. 663. Ic áre æt him fínde *I may find honour with them*, Exon. 67 a; Th. 247, 19; Jul. 81. Ic nú æt feáwum wordum secge *I now say in few words*, Bd. 3, 17; S. 545, 14. Is seó bót gelong eal æt ðe ánum *the expiation is all ready with thee alone*, Exon. 10 a; Th. 10, 16; Cri. 153. Ne mihton hí áwiht æt me ǽfre gewyrcean *they might not ever do anything against me*, Ps. Th. 128, 1. Ðe him æt blisse beornas habbaþ *which men have for their merriment*, Exon. 108 b; Th. 414, 4; Rä. 32, 15.   **2.** *because you approach a person or thing when you wish to take something away,* as they say in Lancashire, Nottinghamshire, *etc. Take this at me*, i. e. *from me*, hence,—*Of, from;* a, ab, de :—Anýmaþ ðæt pûnd æt hym *tollite ab eo talentum*, Mt. Bos. 25, 28. Leorniaþ æt me *learn by coming near me, learn at, of,* or *from me;* discite a me, Mt. Bos. 11, 29. Æt his sylfes múþe at or *from his own mouth*, Bd. 3, 27; S. 558, 40. Æt ðam wífe *from the woman*, Cd. 33; Th. 44, 31; Gen. 717. Ic gebád grynna æt Grendle *I endured snares from Grendel*, Beo. Th. 1864; B. 930: Ps. Th. 27, 18.   **3.** *the names of places are often put in the dat. pl. governed by* æt, *the preposition is then, as in Icelandic, not translated, and the noun is read as singular* :—Ðe mon hæt æt Hæðum *which they call Haddeby;* quem vocant Hæthe, Ors. i. 1, § 19; Bos. Eng. 47, note 57. In monasterio, quod situm est in civitate æt Baðum [MS. Bathun], Kmbl. Cod. Dipl. cxciii; vol. i. 237, 1.   **II.** *very rarely used with the accusative;* cum accusativo To, *unto, as far as;* ad, usque ad :—Æt sæstreámas *ad mare*, Ps. Th. 79, 11. Æt Ác-leá *at Oakley*, Chr. 789; Ing. 79, 14. v. Ác-leá.   **III.** *sometimes* æt *is separated from its case* :—Ðonne wile Dryhten sylf dǽda gehýran æt ealra monna gehwám *then will the Lord himself hear of the deeds from all sorts of men* [ab omnium hominum quocunque], Exon. 99 b; Th. 372, 15; Seel. 93. [*O. Sax.* at : *O. Frs.* et, it : *O. H. Ger.* az : *Goth.* at : *O. Nrs.* at.]

æt *ate;* comedit :—He æt *he ate*, Gen. 3, 6; *p. of* etan *to eat.*

æt-, prefixed to words, like the *prep.* æt, denotes *at, to,* and *from;* ad-, ab-. v. æt; *prep.* I. 2.

ǽt, es; *m.* ǽt, e; *f.* [æt, *p.of* etan *to eat*].   **I.** *food;* cibus, esca :—Ætes on wénan *in hope of food*, Cd. 151; Th. 188, 9; Exod. 165. He us ǽt giefeþ *he gives us food*, Exon. 16 b; Th. 38, 9; Cri. 604. Oft he him ǽte heóld *he often gave them food*, Exon. 43 a; Th. 146, 12; Gú. 708: Cd. 200; Th. 247, 32; Dan. 506.   **II.** *eating;* esus, manducatio :—*After ǽte after eating*, Exon. 61 b; Th. 226, 13; Ph. 405. Hí to ǽte útgewítaþ *ipsi dispergentur ad manducandum*, Ps. Th. 58, 15 : Andr. Kmbl. 2148; An. 1075. [*Orm.* aet : *O. Sax.* ât, n : *O. Frs.* ét, n : *O. H. Ger.* âz, n : *O. Nrs.* át, n. *esus*.] v. etan.

ǽta, an; *m. An eater;* edax. DER. self-ǽta, q. v.

æt-arn *ran away*, Gen. 39, 12; *p. of* æt-irnan.

æt-bær *bore, produced*, Cd. 202; Th. 249, 31; Dan. 538; *p. of* æt-beran.

æt-befón, ic -befó; *subj.* ic, he -befó [æt, be, fón] *To take to, attach;* deprehendere, capere, invenire :—Gif hwá befó ðæt him losod wæs, cenne se ðe he hit ætbefó hwanon hit him cóme *if any one attach that which he had lost, let him with whom he attaches it declare whence it came to him*, L. Eth. ii. 8; Th. i. 288, 15: L. C. S. 23; Th. i. 388, 22. v. be-fón, æt-fón.

æt-beón *To be at* or *present;* adesse :—Ætbeón ðe we biddaþ *adesse te deposcimus*, Hymn Surt. 14, 26.

æt-beran; *p.* -bær, *pl.* -bǽron *To bear* or *carry to, bring forward, produce, bear away* or *forth;* afferre, proferre, efferre :—Hió Beówulfe medo-ful ætbær *she to Beowulf the mead-cup bore*, Beo. Th. 1253; B. 624. He wundor manig fór men ætbær *he many a wonder produced before men*, Cd. 202; Th. 249, 31; Dan. 538. Hí hyne ætbǽron to brimes faroðe *they bore him away to the sea-shore*, Beo. Th. 55; B. 28: 4261; B. 2127: 5222; B. 2614. Ðæt [wǽpen] to beadu-lǽce ætberan meahte *might bear forth that* [*weapon*] *to the game of war*, 3127; B. 1561.

æt-berstan, ic -berste, he -birsteþ, -byrst; *p.* -bærst, *pl.* -burston; *pp.* -borsten *To break out* or *loose, to escape, get away;* erumpere, evadere :—Ða ætbærst him sum man *evasit homo quidam*, Gen. 14, 13. Ða fíf cyningas ætburston *fugerunt enim quinque reges*, Jos. 10, 16. Ðæt he ðanon ætberste *that he escape thence*, L. C. E. 2; Th. i. 358, 25.

æt-bredan, he ætbryt; *p.* -brǽd, *pl.* -brudon; *pp.* -broden, -breden; *v. a. To take away, withdraw, set at liberty, enlarge, release, rescue;* tollere, eripere :—Se deófol ætbryt ðæt word *diabolus tollit verbum*, Lk. Bos. 8, 12. Ge ætbrudon ðæs ingehýdes cǽge *tulistis clavem scientiæ*, 11, 52. Ðæt ðe he hæfþ him biþ ætbroden *quod habet auferetur ab eo*, Mt. Bos. 13, 12 : 21, 43: Ex. 22, 10. Ðe hys wealas him ætbrudon *quem abstulerant servi ejus*, Gen. 21, 25. DER. bredan.

æt-bredendlíc; *adj.* [æt-bredende, *part. of* ætbredan *to take away*]

Taking away; ablativus :—Ætbredendlíc is *ablativus :* mid ðam casu biþ geswutelod swá hwæt swá we ætbredaþ óðrum, oððe swá hwæt swá wé underfóþ æt óðrum, oððe hwanon we faraþ,—Fram ðisum menn ic underféng feóh *ab hoc homine pecuniam accepi.* Fram ðisum láreówe ic gehýrde wísdóm *ab hoc magistro audivi sapientiam.* Fram ðære byrig ic rád *ab illa civitate equitavi.* Fram cyninge [MS. kynincge] ic com *a rege veni,—ablative is* ablativus : *with this case is shewn whatsoever we take away from others, or whatsoever we receive from others, or whence we proceed :—From this man I received money. From this teacher I heard wisdom. I rode from that city. I came from the king,* Ælfc. Gr. 7; Som. 6, 27-32.

æt-broden *Taken away;* ablatus :—Him biþ ætbroden *shall be taken away from him,* Mt. Bos. 13, 12 : 21, 43; *pp. of* æt-bredan.

æt-bryidan; *p.* ede; *pp.* ed *To take away;* auferre :—Ðæs óðres áþ ðe mon his orf æt-bryideþ *the oath of the other from whom the cattle is taken away,* L. O. 3; Th. i. 178, 16, = æt-bredan. DER. bryidan.

æt-byrst *he will escape;* evadet, Basil. 7; Norm. 5, 12; *fut. of* æt-berstan.

æt-clifian; *p.* ode; *pp.* od; *v. intrans. To cleave to, adhere;* adhærere, Ps. Vos. 101, 6.

æt-dón, ic æt-dó; *p.* -dide; *subj.* ic, ðú, he -dó; *pp.* -dón, -dén *To take away, deprive;* eripere :—Ðæt nán preósta óðrum ne ætdó ǽnig ðara þinga *that no priest deprive another of any of those things,* L. Edg. C. 9; Th. ii. 246, 10.

éte, an; *f. pl.* ǽtan; *g.* ǽtena *Oats;* avena sativa, L. M. 1, 35; Lchdm. ii. 84, 5. v. áte.

æt-écan, -ýcan; *p.* -écte; *v. trans.* [æt to, at, eácan to eke] *To add to; increase;* addere, adjicere :—He ætécte *addidit,* Bd. 3, 27; S. 559, 33 : Mt. Rush. Stv. 6, 27.

æt-eglan; *p.* ede; *pp.* ed; *v. intrans. To inflict pain, torment, trouble, grieve;* molestum quid injicere :—Ne mæg him ǽnig fácen feónd æteglan *any deceitful fiend may not inflict grief upon him,* Ps. Th. 88, 19.

æt-eom, -eart, -is, -ys [æt at, eom am] *I am present;* adsum :—Ðæt ríp æt-is [æt-ys, Jun.] *adest messis,* Mk. Bos. 4, 29. v. wesan *to be.*

æ-teorian; *p.* ode; *pp.* od *To fail, be wanting;* deficere :—Æteorode se heofonlíca mete *the heavenly food* [manna] *failed,* Jos. 5, 12.

æt-eówedniss, e; *f. A revelation;* revelatio :—To æteówednisse cynna *ad revelationem gentium,* Lk. Rush. War. 2, 32.

æt-eówian, -eówigan; *p.* de, ede; *impert.* -eów; *pp.* ed. **I.** *v. trans. To shew, display, manifest, declare;* ostendere, manifestare :—Æteów ðínne andwlitan *ostende faciem tuam,* Ps. Th. 79, 4, 7; 84, 6. God æteówde me *Deus ostendit mihi,* Ps. Spl. 58, 11 : Mt. Bos. 13, 26. He geseah dríge stówe æteówde *he saw the dry places displayed,* Cd. 8; Th. 10, 31; Gen. 165. **II.** *v. intrans. To appear;* apparere, manifestari :—Æteówige drígnis *appareat arida,* Gen. 1, 9. Æteów fór Effr im *appare coram Effrem,* Ps. Th. 79, 2. v. eáwan.

æt-eówigendlíce; *adv. Evidently, demonstratively;* demonstrative,— æt-eówigende; *part. of* æt-eówian, -eówigan.

ætern; *Venomous, poisonous ;* venenosus :—Wið ǽlcum ǽternum swile *for every venomous swelling,* L. M. 1, 45; Lchdm. ii. 112, 24. v. ǽtren.

ǽternes, -ness, e; *f. Venomousness, full of poison;* venenositas. v. ǽtern.

æt-éwung, e; *f. A shewing, manifesting, epiphany;* manifestatio, Wrt. Voc. 16, 49.

æt-fæstan; *p.* -fæste; *pp.* -fæsted; *v. trans.* [æt, fæstan *to fasten*] *To fix, fasten, drive into, afflict with, inflict on;* impingere, infigere :—Hí míne sáwle synne ætfæsten *they inflict sin on my soul,* Ps. Th. 142, 12. He him ætfæste éce edwít *opprobrium sempiternum dedit illis,* 77, 66. Bitere ætfæsted *bitterly afflicted,* 136, 8. Ne mágon we him láþ ætfæstan *we cannot afflict him with pain,* Andr. Kmbl. 2694; An. 1349.

æt-fealh adhæsit, Ps. Th. 118, 25; *p. of* æt-felgan.

æt-feallan; *p.* -feól, *pl.* -feóllon; *pp.* -feallen *To fall away;* cadere :— Healf wer ðǽr æt-fealþ *one half of the wer there falls away,* L. O. D. 5; Th. i. 354, 21.

æt-fecgan; *p.* -feah; *v. trans. To seize;* apprehendere :—Me ætfeah fyrhtu helle *fear of hell seized me,* Ps. Th. 114, 3.

æt-fele *Adhesion;* adhæsio ? :—Mín is ætfele mihtigum Drihtne *mihi autem adhærere Deo,* Ps. Th. 72, 23. v. æt-feolan.

æt-felgan; *p.* -fealh, *pl.* -fulgon; *pp.* -folgen; *v. intrans. To cleave on, adhere, stick to;* adhærere :—Mín sáwul fióre ætfealh *adhæsit pavimento anima mea,* Ps. Th. 118, 25 : 118, 31: Beo. Th. 1941; B. 968 : Ps. Spl. C. 62, 8.

æt-feng, es; *m. Attaching;* comprehensio :—Be yrfes ætfenge *of attaching cattle,* L. Ath. i. 9; Th. i. 204, 9. DER. æt-fón.

æt-feohtan; *p.* -feaht, *pl.* -fuhton. **I.** *to fight against, contend;* oppugnare :—Ætfeohtan mid frumgárum *to fight against the patriarchs,* Cd. 97; Th. 127, 25; Gen. 2116. **II.** *to feel earnestly, grope;* contendere, tentare circum :—Folmum ætfeohtan *with his hands to contend* or *grope,* Exon. 87 b; Th. 328, 15; Vy. 18.

æt-feolan, -fiolan; *p.* -fæl, *pl.* -fǽlon, -félon; *pp.* -folen, -feolen *To adhere, cleave* or *hang on, insist upon, stick to, continue;* insistere, ad-

hærere :—Ætfeole mín tunge fæste gómum *adhæreat lingua mea faucibus meis,* Ps. Th. 136, 5. Is ætfeolen eác mín bán flǽsce mínum *adhæserunt ossa mea carni meæ,* Ps. Th. 101, 4. Ætfélon [MS. ætfelun] *vel* ætclofodon [MS. -fodun] *adhæserunt,* Ps. Surt. 101, 6. Me sóþlíce ætfeolan Gode gód is *mihi autem adhærere Deo bonum est,* 72, 28. Ætfeolan wæccum and gebédum *to continue in watchings and prayers,* Bd. 4, 25; S. 601, 2. DER. felan, feolan.

æt-ferian; *p.* ede; *pp.* ed; *v. trans. To carry out, take away, bear away;* auferre :—Ic ðæt hilt feóndum ætferede *I bore the hilt away from the foes,* Beo. Th. 3342; B. 1669.

æt-fiolan *to stick to, continue;* adhærere. v. æt-feolan.

æt-fleón; *p.* -fleáh, *pl.* -flugon; *pp.* -flogen [æt, fleón *to flee*] *To flee away, escape by flight, eschew;* aufugere :— Ic ána ætfleáh *I alone escaped,* Job Thw. 165, 30. Nán þing ætfleón ne mihte *nothing might remain,* Jos. 10, 35 : L. C. S. 78; Th. i. 420, 7.

æt-flówan; *p.* -fleów, *pl.* -fleówon; *pp.* -flówen; *v. intrans. To flow to* or *together, to increase;* affluere :—Gyf wélan ætflówon *si divitiæ affluant,* Ps. Spl. 61, 10.

æt-fón [æt to, fón *to seize*] *To claim, lay claim, attach;* deprehendere, capere :—Gif se ágend hit eft ætfó *if the owner afterwards lay claim to it,* L. H. E. 7; Th. i. 30, 8 : 16; Th. i. 34, 6 : L. Ed. 1; Th. i. 160, 8.

æt-foran; *prep. dat.* [æt at, foran *fore*] *Close before, close by, before, at;* ante, pro, coram :—Ætforan eágan ðíne *ante oculos tuos,* Ps. Spl. 5, 5 : 13, 7 : Byrht. Th. 132, 14; By. 16. Sæt ætforan ðam dómsetle *sedit pro tribunali,* Jn. Bos. 19, 13.

æt-foran-weall, es; *m. The outer wall, out-works, a bulwark before a castle;* antemurale. v. weall; *m.*

æt-fylgan; *p.* de; *pp.* ed *To adhere to, stick to;* adhærere :—Ne ætfyligeþ ðé áhwǽr fácn ne unriht *numquid adhæret tibi sedes iniquitatis,* Ps. Th. 93, 19.

æt-gædere; *adv.* [æt, gædrian = gadrian *to gather*] *Together;* una, simul :—Twá beóþ ætgædere gríndende, Lk. Bos. 17, 35; *tweye* [wymmen] *schulen be gryndinge to gidere,* Wyc. His mǽgþe biþ ætgædere *his kindred is together,* Bt. Met. Fox 20, 320; Met. 20, 160. Gáras stódon samod ætgædere *the javelins stood altogether,* Beo. Th. 664; B. 329. Blód and wæter bú tú ætgædre *blood and water both together,* Exon. 70 a; Th. 260, 5; Jul. 292. Bismærede ungket [= uncit] men, bá ætgædre *they* [men] *reviled us two, both together,* Runic Inscr. Kmbl. 354, 30. DER. gædere.

æt-gár, es; *m.* [æt, gár *a spear*] *A short spear* or *javelin, a kind of dart* or *other weapon to cast at the enemy;* framea, Cot. 188 : 86. [O. Frs. etger : M. H. Ger. aziger : O. H. Ger. azkér : O. Nrs. atgeirr.]

æt-gebicgan; *p.* -bohte; *pp.* -boht [æt, gebycgan *to buy*] *To buy for himself;* emere :—He hí æft æt ðam ágende sínne willan æt-gebicge *let him afterwards buy her at her owner's will,* L. Ethb. 82; Th. i. 24, 4.

æt-gebrengan; *p.* -gebrohte; *pp.* -gebroht; *v. trans. To bring* or *lead to;* adducere :—He ætgebrenge, ðe him sealde *let him bring the person who sold it him,* L. H. E. 7; Th. i. 30, 8.

æt-geníman; *p.* -genam, *pl.* -genámon; *pp.* -genumen *To take away by force, to pluck out, withdraw, deliver, rescue;* eripere, Cot. 77.

æt-giefa, -geofa, an; *m.* [æt *food,* gifa *a giver*] *A food-giver, feeder;* cibi dator :—Oþ ðæt se fugel his ætgiefan eáþmód weorþeþ *till that the bird becomes obedient to his feeder,* Exon. 88 b; Th. 332, 26; Vy. 91 : 90 b; Th. 339, 22; Gn. Ex. 98.

æt-gifan; *p.* -geaf, -gaf, *pl.* -geáfon, géfon; *pp.* -gifen [æt to, gifan] *To give to, render, afford;* tribuere, afferre :—Ic him líf-wraðe lytle meahte ætgifan æt gúðe *I could render to him little life-protection in the conflict,* Beo. Th. 5748; B. 2878.

æt-gongan [æt at, gangan *to go*] *To go to, approach;* accedere :—Hét hie of ðam líge neár ætgongan *he bade them from the flame to approach nearer,* Exon. 55 b; Th. 197, 1; Az. 183.

æt-grǽpe; *adj. Grasping at, seizing;* prehendens :—Ðǽr him aglǽca ætgrǽpe wearþ *where the miserable being seized him,* Beo. Th. 2542; B. 1269.

æt-habban; *p.* -hæfde; *pp.* -hæfed *To retain, detain, withhold;* retinere, detinere, Scint. 10. DER. habban.

ǽðan *To overflow, deluge, lay waste* :—Cwæþ ðæt he wolde eall á ǽðan ðæt on eorþan wæs *said that he would for ever lay waste all that was on the earth,* Cd. 64; Th. 77, 24; Gen. 1280. v. ǽðan.

ǽðel- *noble;* nobilis :—v. the compounds æðel-boren, -borennes, -cund, etc. *from* æðele *noble.*

ǽðel, es; *m. A native country, country, land;* patria, terra :—In ðeossum ǽðele *in this country,* Cd. 215; Th. 271, 21; Sat. 108. On ǽðelum, *d. pl.* Menol. Fox 236; Men. 119. v. éðel.

ÆÐELBALD, es; *m.* [æðele, bald *bold, brave*] *Æthelbald;* Æthelbaldus; *the eldest son of Æthelwulf. Æthelbald, the eldest brother of Alfred, was king of Wessex for five years, from A. D.* 855-860 :—A. D. 855, ðá féngon Æðelwulfes ii suna to ríce; Æðelbald to Westseaxna ríce, and Æðelbryht to Cantwara ríce *then, A. D.* 855, *Æthelwulf's two sons succeeded to the kingdom; Æthelbald to the kingdom of the West Saxons, and Ethelbert to the kingdom of Kent,* Chr. 855; Th. 129, 16-19, col. 1.

A. D. 860, hēr, Æðelbald cyning forþfērde *here*, A.D. 860, *king Æthelbald died*, Chr. 860; Erl. 71, 3.

æðel-boren; *part. Noble-born, free-born, noble*; natu nobilis, nobili genere natus, nobilis:—Sum æðelboren man *homo quidam nobilis*, Lk. Bos. 19, 12. Æðelborene cild *vel* freóbearn *liberi*, Ælfc. Gl. 91; Wrt. Voc. 51, 67: Apol. Th. 19, 21. v. beran.

æðel-borennes, -ness, e; *f. Nobleness of birth*; nobilitas:—Ic ðīne æðelborennesse geseó *I see the nobleness of thy birth*, Apol. Th. 15, 18.

Æðelbryht, -berht, -briht, es; *m.* [æðele, bryht *bright, excellent*. v. beorht]. 1. *Ethelbert king of Kent, for fifty-six years, from* A.D. 560–616. Ethelbert was converted to Christianity by the preaching of St. Augustine: v. Augustinus:—A. D. 560 [MS. 565], hēr, fēng Æðelbryht [MS. Æðelbriht] to Cantwara rīce *here*, A. D. 560, *Ethelbert succeeded to the kingdom of Kent*, Chr. 565; Erl. 17, 18. Đā wæs ymb syx hund wintra and ·syxtyne winter fram Drihtnes mennyscnesse, ðæt wæs ymb ān and twentig wintra wæs ðe Agustinus, mid his geferum, to lǣranne on Angel þeóde sended wæs, ðæt Æðelbryht Cantwara cyning æfter ðam hwīlendlīcan rīce ðæt he six and fiftig wintra wundorlīce hæfde, and ðā to ðam heofonlīcan rīce mid gefeán astāh *anno·ab incarnatione Dominica sexcentesimo decimo sexto, qui est annus vicesimus primus, ex quo Augustinus cum ſociis ad prædicandum genti Anglorum missus est, Æthelbryhtus [Æthelberht] rex Cantuariorum, post regnum temporale, quod quinquaginta et sex annis gloriosissime tenuerat, æterna cœlestis regni gaudia subiit*, Bd. 2, 5; S. 506, 5–9. Hēr forþfērde Æðelbryht [MS. Æðelberht] Cantware cining, se rīxade lvi wintra *here*, A.D. 616, *Ethelbert king of the Kentish people died, who reigned fifty-six years*, Chr. 616; Erl. 21, 37. 2. Æðelbryht, es; *m. Ethelbert the second; Æthelbryhtus, the second son of Æthelwulf. This Ethelbert, after the lapse of 239 years from the death of Ethelbert the first in 616, became king of Kent, Essex, Surrey, and Sussex, for five years, from 855 to 860; he succeeded to Wessex on his brother's death, in 860, and reigned five years more over these five counties, from 860 to 865; he was therefore king for ten years, from A. D. 855–865*:—A.D. 855, ðā fēngon Æðelwulfes ii suna to rīce; Æðelbald to Westseaxna rīce; and Æðelbryht to Cantwara rīce, and to Eástseaxena rīce, and to Sūþrigean, and to Sūþseaxena rīce *then*, A.D. 855, *Æthelwulf's two sons succeeded to the kingdom; Æthelbald to the kingdom of the West Saxons, and Ethelbert to the kingdom of Kent, and to the kingdom of the East Saxons, and to Surrey, and to the kingdom of the South Saxons*, Chr. 855; Th. 129, 16–22, col. 1. · A.D. 860, hēr, Æðelbald cyning forþfērde, and fēng Æðelbryht to eallum ðam rīce his brōðor, and se Æðelbryht [MS. Æðelbriht] rīcsode v geár. *here*, A.D. 860, *king Æthelbald died, and Ethelbert succeeded to all the kingdom [Wessex] of his brother, and Ethelbert reigned five years*, Chr. 860; Erl. 71, 3–10.

æðel-cund; *adj. Of noble kind* or *origin, noble*; nobilis originis:—Æðelcunde mægþ *the noble woman*, Exon. 119 b; Th. 459, 18; Hö. 1.

æðel-cundnes, -ness, e; *f. Nobleness, nobility*; nobilitas:—Mid micelre æðelcundnesse *with great nobleness*, Bt. 19; Fox 68, 31.

æðel-cyning, es; *m. The noble king*, used for *Christ*; rex nobilis, Christus:—Cristes onsȳn, æðelcyninges wlite *Christ's countenance, the noble king's aspect*, Exon. 21 a; Th. 56, 27; Cri. 907. Æðelcyninges rōd *the cross of the noble king*, Elen. Kmbl. 437; El. 219; Andr. Kmbl. 3354; An. 1681.

æðel-duguþ, e; *f. A noble attendance*; comitatus nobilis:—Hine ymbūtan æðelduguþ, eádig engla gedryht *around him a noble attendance, a blessed train of angels*, Exon. 22 b; Th. 62, 36; Cri. 1012.

æðele, eðele; *comp.* -ra; *sup.* -ast, -est, -ust; *adj.* I. *noble, eminent, not only in blood or by descent, but in mind, excellent, famous, singular*; nobilis, generosus, præstabilis, egregius, excellens:—Se eorl wæs æðele *the earl was noble*, Cd. 59; Th. 72, 5; Gen. 1182. He sægde Habrahame, æðeles geþingu *he told to Abraham the promises of the noble*, Andr. Kmbl. 1512; An. 757. Æðelan cynnes *of noble race*, Cd. 154; Th. 192, 6; Exod. 227. Æðelre gebyrde *of noble birth*, Bd. 2, 15; S. 518, 37. Æðelan cempan *to the noble champion*, Andr. Kmbl. 460; An. 230. Đære æðelan [cwēne] *to the noble lady*, Elen. Kmbl. 1085; El. 545. Wuldriaþ æðelne ordfruman *they glorify the noble origin*, Exon. 13 b; Th. 25, 17; Cri. 402. Æðelum stencum *with sweet odours*, 64 a; Th. 237, 7; Ph. 586: Cd. 75; Th. 92, 24; Gen. 1533. Đone æðelan Albanum *Albanum egregium*, Bd. 1, 7; S. 476, 34. He wæs on his mōde æðelra ðonne on woruld gebyrdum *he was in his mind more noble than in worldly birth*, Bd. 3, 19; S. 547, 26. Of ðam æðelestan cynne *of the most noble·race*, 3, 19; S. 547, 25. Æðelast tungla *the noblest of stars*, Exon. 57 a; Th. 204, 6; Ph. 93: Ps. Th. 84, 10. Æðelust bearna *the noblest of heroes*, Elen. Kmbl. 950; El. 476. II. *noble, vigorous, young*; nobilis, novellus:—Đīne bearn swā elebeámas æðele weaxen *thy children grow like young olive-trees*; sicut novellæ olivarum, Ps. Th. 127, 4: 143, 14. Swā swā æðele plantunga *sicut novellæ plantationes*, Ps. Spl. 143, 14. [*O. Sax.* edili: *O. Frs.* ethel, edel: *Dut. Ger.* edel: *M. H. Ger.* edele: *O. H. Ger.* edili: *Dan. Swed.* ädel: *O. Nrs.* aðal, *n. natura, ingenium*]. DER. emn-æðele, ge-, on-, un-.

··**Æðelflæd**, e; *f.* [æðele, flæd] *Æthelfled; Æthelfleda.* The eldest and most intellectual daughter of king Alfred the Great, and sister of king Edward the Elder. She married Æthelred, a Mercian nobleman, who was made viceroy of Mercia by king Alfred. He died in A.D. 912, Chr. Erl. 100, 30, and his widow Æðelflæd governed Mercia most efficiently for about ten years:—Hēr com Æðelflæd, Myrcna hlæfdige, on ðone hālgan æfen Inuentione Sanctæ Crucis, to Scergeate, and ðǣr ða burh getimbrede; and, ðæs ilcan geáres, ða æt Bricge *here*, A.D. 912, *Æthelfled, the lady of the Mercians, came to Scergeat [Sarrat?] on the holy eve of the Inventio Sanctæ Crucis [May third], and there built the burgh; and in the same year, that at Bridgenorth*, Chr. 912; Th. 187, 5–10, col. 1: Chr. 913; Th. 186, 11–37, col. 2: Chr. 917; Th. 190, 37, col. 2–192, 1, col. 2: Chr. 918; Th. 192, 7, col. 2: Th. Diplm. A.D. 886–899, 138, 5–11: 138, 29–32. Æthelfled died at Tamworth in A.D. 922. Đā on ðæm setle Eádweard cyng ðǣr sæt [æt Steanforde], ðā gefór Æðelflæd his swystar æt Tameworþige, xii nihtum ǣr middum sumera. Đā gerád he ða burg æt Tameworþige; and him cierde to eall se þeódscype on Myrcna lande, ðe Æðelflæde ǣr underþeóded wæs *then, while king Edward was tarrying there [at Stamford], Æthelfled his sister died at Tamworth, twelve nights before midsummer. Then rode he to the borough of Tamworth; and all the population in Mercia turned to him, which before was subject to Æthelfled*, Chr. 922; Erl. 108, 22–26.

æðelian; *p.* ode; *pp.* od; *v. trans. To ennoble, improve*; nobilitare. DER. ge-æðelian, un-.

æðel-īc; *adj.* [æðele *noble*, līc *like*] *Noble, excellent*; egregius:—Æðelīc onginn *a noble beginning*, Andr. Kmbl. 1775; An. 890. Stenc æðelīcra eallum eorþan frætwum [MS. frætwa] *a nobler odour than all earth's ornaments*, Exon. 96 a; Th. 358, 19; Pa. 48.

éðe-līc; *adj.* [éðe = eáðe *easy*; *adj.* līc *like*] *Easy*; facilis:—Gif ðū ne wilt us geþafian in swā æðelīcum þinge *si non vis assentire nobis in tam facili causa*, Bd. 2, 5; S. 507, 26. v. eáðelīc.

æðel-īce; *adv. Nobly, elegantly*; nobiliter, insigniter, Cot. 77. v. æðel-līce.

æðeling, es; *m.* [æðele, -ing *son of, originating from*]. I. *the son of a king, one of royal blood, a nobleman, used also in poetry for the king, God, and Christ*; regia suboles, vir nobilis:—Se iunga æðeling *regius juvenis*, Bd. 2, 12; S. 514, 27: 3, 21; S. 550, 40: 2, 14; S. 517, 22. Æðelinges bearn *the prince's child*, Beo. Th. 1780; B. 888. Be sumum Rômāniscum æðeling *by a certain Roman nobleman*, Bt. 16, 2; Fox 52, 19. Crist Nergende! wuldres Æðeling! *Saviour Christ! Prince of Glory!* Exon. 10 a; Th. 10, 26; Cri. 158. Đā se Æðeling cwom in Betlem *when the Prince came in Bethlehem*, 14 a; Th. 28, 18; Cri. 448. Æðelstān cyning and his brōðor eác, Eádmund æðeling *king Æthelstan and his brother also, Edmund the noble*, Chr. 938; Th. 200, 33; Æðelst. 3. Éce is se æðeling *the creator [atheling] is eternal*, Exon. 60 b; Th. 220, 12; Ph. 319: 119 b; Th. 459, 21; Hö. 3. Stōd æfter mandrihtne eard and éðel, æfter ðam æðelinge [*his*] *land and dwelling-place stood after [waiting for] the man-lord, the chieftain*, 207; Th. 256, 10; Dan. 638. II. *man generally, in pl. men, people, used in a good and noble sense, as a derivative of æðele noble*; homo, homines:—Đæs æðelinges ellen dohte *the man's courage was good*, Cd. 56; Th. 78, 4; Gen. 1288. Đa nū æðelingas, ealle eorþ-bűend, Ebréi hātaþ *which people now, all dwellers upon earth, call Hebrews*, 79; Th. 99, 17; Gen. 1647. Hēht him ceósan æðelingas *he commanded him to choose men*, 90; Th. 112, 9; Gen. 1868: 58; Th. 70, 31; Gen. 1161. DER. sib-.

Æðelinga īgg, éig, e; *f. The island of nobles, Athelney*; nobilium insula:—Æt Æðelinga īgge *apud nobilium insulam*, Chr. 878; Th. 146, 42, col. 2. Wid..., Th. 148, 31, col. 2: Chr. 879; Th. 148, 30, col. 3.

æðel-līc; *adj. Noble*; nobilis, Andr. Kmbl. 1775; An. 890. v. æðel-īc, æðele.

æðel-līce, æðel-īce; *adv. Nobly*; nobiliter:—Wæs se wer on hālgum gewritum æðellīce gelǣred *vir erat sacris litteris nobiliter instructus*, Bd. 5, 23; S. 646, 17: 4, 26; S. 603, 9: 2, 1; S. 501, 8.

æðel-nes, -nys, -nyss, e; *f. Nobility*; nobilitas, Bd. 2, 20; S. 522, 7: Ps. Th. 118, 142, [MS. æðeles.]

æðelo; *indecl. in sing; pl. nom. acc.* æðelu, æðelo; *gen.* æðela; *dat.* æðelum; *n. Nobility, pre-eminence, origin, family, race, nature, talents, genius*; nobilitas, principatus, origo, natales, prosapia, natura, indoles, ingenium:—Ic lǣre ðæt ðū fægenige óðerra manna gódes and heora æðelo *I advise that thou rejoice in other men's good and their nobility*, Bt. 30, 1; Fox 108, 31. His æðelo bióþ on ðam mōde *his nobility is in the mind*, 30, 1; Fox 110, 1. Ryht æðelo biþ on ðam mōde, næs on ðam flǣsce *true nobility is in the mind, not in the flesh*, Bt. 30, 2; Fox 110, 19. Him frumbearnes riht freóbrōðor óþþah, eád and æðelo *his own brother had withdrawn from him his wealth and pre-eminence*, Cd. 160; Th. 199, 15; Exod. 339. Ealdaþ eorþan blǣd æðela gehwylcre *earth's produce of every nature grows old*, Exon. 33 a; Th. 104, 26; Gū. 14. Hwæt his æðelu sīen *which his origin is*, 69 b; Th. 259, 23; Jul. 286. Sindon him æðelum óðere twegen beornas geborene brōðorsibbum *to him in his family are two other men born in brotherly relationship*, Andr. Kmbl. 1377; An. 689. Þurh ðīne wordlǣde æðelum ēcne *through thy discourse great with talents*, 1271; An. 636. He eówer æðelu can *he*

knows your nobility, Beo. Th. 790; B. 392: 3745; B. 1870. DER. fæder-ædelo, riht-.

Æðelréd, Æðelréd, Æðeréd, es; *m.* [æðele *noble,* réd *counsel*] *Æthelred, a Mercian nobleman, the viceroy* or *governor of the Mercians;* Æthelred, Æthelrēdus. He married Æthelfléd, the eldest and most intellectual daughter of king Alfred the Great. He styles himself *sub-regulus* in subscribing his name to a charter of king Alfred, A. D. 889,—Ego Æthelréd, subregulus et patricius Merciorum, hanc donationem signo crucis subscripsi, Th. Diplm. 136, 21. His wife simply writes,—Ego Æthelfléd consensi, Th. Diplm. 136, 23. Ríxiendum ussum Dryhtene ðæm Hǽlendan Crist. Æfter ðon ðe agán wæs ehta hund wintra and syx and hund nigontig efter his acennednesse, and ðý feówerteóðan gebonn-gēre [v. geban **II**], ðá ðý gēre gebeón [*p. of* gebannan] Æðelréd ealder-man alle Mercna weotan tosomne to Gleaweceastre, biscopas, and aldermen, and alle his duguþe; and ðæt dyde be Ælfrēdes cyninges gewitnesse and leúfe *under the rule of our Lord Jesus Christ. When 896 winters were passed after his birth, and in the fourth indiction year, then in that year Æthelred alderman assembled all the witan of the Mercians together at Gloucester, bishops, and aldermen, and all his nobility; and did that with the knowledge and leave of king Alfred,* Th. Diplm. A. D. 896; 139, 4-16. Æthelred died in A. D. 912. Hér gefór Æðelréd, ealdorman on Myrcum *here,* A. D. 912, *died Æthelred, alderman of the Mercians,* Chr. 912; Erl. 101, 46. *His widow, Æthelfléd, governed Mercia about ten years, with great vigour and success, under her brother, king Edward the Elder,* Chr. 922; Erl. 108, 22-26. v. **Æðelfléd.**

Æðelréd, Æðeréd, es; *m.* [æðele, réd = réd *counsel*]. **1.** *Æthelred, third son of Æthelwulf, and brother of Alfred the Great. Æthelred was king of Wessex for five years,* A. D. 866-871; Æthelred, Æthelrēdus:—Hér fēng Æðelréd to West Seaxna ríce *here,* A. D. 866, *Æthelred succeeded to the kingdom of the West Saxons,* Chr. 866; Erl. 73, 1. Æfter Eástron gefór Æðelréd [MS. Æðeréd] cining; and he rícsode [MS. ríxade] v geár *after Easter* [A. D. 871] *king Æthelred died; and he reigned five years,* 871; Erl. 77, 1. **2. Æðelréd** *Æthelred Atheling, the second son of Edgar. Æthelred was king of Wessex, Mercia, and Northumbria, for thirty-eight years,* A. D. 978-1016:—Hér, Æðelréd æðeling fēng to ðam ríce *here* [A. D. 978] *Æthelred Atheling succeeded to the kingdom,* Chr. 978; Th. 232, 3, col. 1. A. D. 1016, Ðá gelamp hit ðæt se cyning Æðelréd forþférde *then, A. D. 1016, it happened that king Æthelred died,* 1016; Erl. 155, 15. **3. Æðelréd, Æðeréd** *Æthelred, a Mercian nobleman,* Th. Diplm. A. D. 896; 139, 11: Chr. 912; Erl. 101, 46. v. **Æðelréd.**

Æðelstán, es; *m.* [æðele, stán *stone*] *Athelstan, the eldest son of Edward the Elder. Athelstan, who gained a complete victory over the Anglo-Danes in the battle of Brunanburh, in A. D. 937, was king of Wessex fourteen years and ten weeks, from A. D. 925-940:—*A. D. 925, hér, Eádweard cyning [MS. cing] forþférde and Æðelstán his sunu fēng to ríce *here,* A. D. 925, *king Edward died, and Athelstan his son succeeded to the kingdom,* Chr. 925; Erl. 110, 19. A. D. 940, hér, Æðelstán cyning forþférde, and Eádmund Æðeling fēng to ríce, and Æðelstán cyning rícsode xiv geár, and teon wucan *here,* A. D. 940, *king Athelstan died, and Edmund Atheling succeeded to the kingdom, and king Athelstan reigned fourteen years and ten weeks,* Chr. 940; Th. 209, 13-23, col. 1.

æðel-stenc, es; *m. A noble odour;* odor nobilis, Exon. 58 b; Th. 211, 10; Ph. 195.

æðel-tungol, es; *m. A noble star;* sidus nobile, Exon. 60 a; Th. 218, 5; Ph. 290: 52 a; Th. 181, 4; Gú. 1288.

Æðel-wulf, es; *m.* [æðele *noble,* wulf *a wolf*] *Æthelwulf;* Æthelwulfus; *eldest son of Egbert and father of Alfred the Great. Æthelwulf was king of Wessex, from A. D. 837* (v. Ecg-bryht) *-855:—*A. D. 837 [MS. 836], hér, Ecgbryht cyning forþférde, and fēng Æðelwulf his sunu to Westseaxna ríce A. D. 837, *king Ecgbryht died, and Æthelwulf his son succeeded to the kingdom of the West Saxons,* Chr. 836; Th. 117, 34, col. 1. A. D. 855, hér, Æðelwulf cyning gefór *here,* A. D. 855, *king Æthelwulf died,* Chr. 855; Erl. 68, 24.

Æðeréd, es; *m. The name of a king and a Mercian nobleman,* Chr. 867; Th. 130, 22, cols. 1, 2, 3; Th. 131, 22, cols. 1, 3: Chr. 912; Erl. 100, 30. v. **Æðelréd 1, Æðelréd.**

æt-híde, æt-hýde *Put out of the hide, skinned, bowelled;* excoriatus, Cot. 42.

æt-hindan; *adv. At the back, behind, after;* a tergo, pone, post :—Se cyning férde him æthindan *the king went after them,* Chr. 1016; Th. i. 282, 17.

æt-hleápan; *p.* -hleóp, *pl.* -hleópon; *pp.* -hleápen; *v. intrans. To leap out, to flee, escape, get away;* aufugere, evadere :—Ðēh þréla hwylc hláforde æthleápe *a domino suo servus si quis aufugerit,* Lupi Serm. 1, 13; Hick. Thes. ii. 103, 4.

æt-hlýp, es; *m.* [æt *to,* hlýp *a leap*] *An assault;* aggressus, assultus :—For ðam æthlýpe *for the assault,* L. Ath. i. 6; Th. i. 202, 22. v. æ-hlýp.

ÆÐM, éðm, es; *m. A vapour, breath, a hole to breathe through, a smell;* halitus, spiritus, vapor :—Hreðer éðme weóll *his breast heaved with*

breathing, Beo. Th. 5180; B. 2593. Hú síd se swarta éðm seó *how vast the black vapour may be,* Cd. 228; Th. 309, 4; Sat. 704. [*Plat.* ádem, ám, *m* : *O. Sax.* áðom, *m* : *O. Frs.* ethma, ádema, ôm, *m* : *Dut.* ádem, *m* : *Ger.* athem, odem, *m* : *M. H. Ger.* ātem, *m* : *O. H. Ger.* ātam, ātum, *m. spiritus,* ἀτμή *vapor* : *Sansk.* ātman *breath, soul.*] v. bréþ.

éðmian; *p.* ode; *pp.* od [éðm *vapour*] *To raise vapour, boil, to be heated, to be greatly moved;* exæstuare, Scint. 30.

æt-hredan; *p.* -hredde; *v. eripere* :—Ic æthrede oððe ahredde *eripio,* Ælfc. Gr. 28, 3; Som. 30, 63.

æt-hreppian, Ettm. æt-hræppian, Som; *p.* ode; *pp.* od *To rap at, to knock, dash about;* impingere. v. hrepian.

æt-hrínan; *p.* -hrán, *pl.* -hrinon; *pp.* -hrinen *To touch, take, move;* tangere, apprehendere, movere :—Ðæt ic æt-hríne ðín *ut tangam te,* Gen. 27, 21. He æt-hrán hyre hand *tetigit manum ejus,* Mt. Bos. 8, 15. Se uncléna gást hine æt-hrínþ *spiritus apprehendit eum,* Lk. Bos. 9, 39. Nellaþ híg ðá mid heora fingre æt-hrínan *digito autem suo nolunt ea movere,* Mt. Bos. 23, 4.

æ-þrýt; *adj. Troublesome, tedious;* molestus, Equin. vern. 38.

æ-þrýtnes, -ness, e; *f. Trouble;* molestia, Lye. v. a-þrotennes.

æt-hwá; *pron. Each;* quisque :—Se is æt-hwám freónd *which is to each a-friend,* Exon. 95 b; Th. 356, 22; Pa. 15.

æt-hwega, æt-hwega *Somewhat, about, in some measure, a little;* aliquantum, aliquantulum, aliquatenus, R. Ben. interl. 73. Scíres wínes drince æt-hwæga *let him drink somewhat of pure wine,* L. M. 2, 59; Lchdm. ii. 284, 5. Æt-hwega yfel wǽte biþ gegoten on ðæt lim *whatever evil humour is secreted on the limb,* L. M. 2, 59; Lchdm. ii. 284, 28. v. hwæt-hwæga in hwæt, hwega.

æt-hweorfan; *p.* -hwearf, *pl.* -hwurfon; *pp.* -hworfen [æt, hweorfan *to turn*] *To turn, return;* accedere, reverti :—Hwílum on beorh æt-hwearf *sometimes he turned to the mount,* Beo. Th. 4587; B. 2299.

æt-hwón; *adv. Almost;* pæne, fere. v. hwón.

æt-hýde *Put out of the hide, skinned;* excoriatus. v. æt-híde.

æt-irnan; *p.* -arn, *pl.* -urnon; *pp.* -urnen; *v. intrans. To run away;* egredi :—Ðá ætarn he út *et egressus est foras,* Gen. 39, 12. v. yrnan.

æt-is *is present;* adest, Mk. Bos. 4, 29; *3rd pres. of* æt-eom.

æt-íwednes, e; *f. A shewing, manifestation;* ostensio :—Wæs on wéstenum ôþ ðone dæg hys ætíwednessum on Israhel *erat in desertis usque in diem ostensionis suæ ad Israel,* Lk. Bos. 1, 80. v. æt-ýwnys.

æt-lǽdan; *p.* de; *pp.* ed *To lead out, drive away;* abigere :—Ðæt ðú ætlæddest me míne dôhtra *ut clam me abigeres filias meas,* Gen. 31, 26.

æt-lǽtnes, e; *f. Desolation, destruction;* desolatio, Somn. 323.

æt-licgan; *p.* -læg, *pl.* -lǽgon; *pp.* -legen *To lie still* or *idle;* inutilem jacere :—Ðæt Godes feoh ne ætlicge *ne Dei pecunia jaceat,* Ælfc. Gr. pref; Som. 1, 27.

æt-lútian [lútan *to lurk*] *To lie hid;* latere, Jud. 4, 18.

Ætne, es; *m. Etna,* Bt. 15; Fox 48, 20: 16, 1; Fox 50, 5. v. Etna.

æt-níman; *p.* -nam, *pl.* -námon; *pp.* -numen *To take from, to take away;* demere, adimere :—Ne wolde him beorht fæder bearn ætníman *the glorious father would not take the child away from him,* Cd. 162; Th. 204, 5; Exod. 414.

æt-nýhstan; *adv. At last;* tandem, Bd. 2, 2; S. 502, 26. v. nýhst.

ætol, ætol-man, ætul-man *A glutton;* edax. v. etol.

ǽton *ate,* Mt. Bos. 13, 4; *p. of* etan.

ǽtor *Poison;* venenum. v. átor-cyn, átor.

ǽtor-cyn, -cynn, es; *n. The poison-kind;* veneni genus :—Ætorcyn gewurdon onwæcned *the poison-kinds arose,* Salm. Kmbl. 437; Sal. 219. v. átor, *etc.*

ǽtren, ǽttren, ǽtern, ǽttern; *adj. Poisonous;* venenosus :—Ǽttren wæs ellorgǽst *the strange guest was poisonous,* Beo. Th. 3238; B. 1617. Me of bôsme fareþ ǽtren onga *from my bosom comes a poisonous sting,* Exon. 106 b; Th. 405, 18; Rä. 24, 4: Ps. Th. 139, 3. Him æt heortan stód ætterne ord [*sc.* gáres] *the poisonous point* [*of the spear*] *stood in his heart,* Byrht. Th. 136, 4; By. 146: Frag. Kmbl. 37; Leás. 20: L. M. 1, 45; Lchdm. ii. 112, 24.

ǽtren-môd; *adj. Venom-minded;* malitiosus :—Ǽtrenmôd mon *a venom-minded man,* Exon. 91 b; Th. 343, 26; Gn. Ex. 163.

ǽtrian, ǽttrian; *p.* ede; *pp.* ed; *v. trans.* [ǽtor = átor *poison*] *To poison, envenom;* venenare :—For ǽtredum gescotum *from poisoned arrows,* Ors. 3, 9; Bos. 68, 38; MS. C.

æt-rihte; *adv.* [æt *at,* rihte *rightly, justly, well*] *Rightly* or *justly at, near, at hand, almost;* pæne, haud multum abest quin :—Ætrihte wæs gúþ getwǽfed, nympe mec God scylde *the contest had almost been finished, had not God shielded me,* Beo. Th. 3319; B. 1657. Wæs him ende-dôgor ætryhte *his final day was near,* Exon. 49 b; Th. 171, 12; Gú. 1125: 47 a; Th. 162, 4; Gú. 970.

æt-rihtes; *adv. By and by, presently;* mox. v. æt-rihte; *adv.*

æt-ryhte *Nearly, almost;* pæne, Exon. 47 a; Th. 162, 4; Gú. 970: Exon. 49 b; Th. 171, 12; Gú. 1125. v. æt-rihte.

æt-sacan; *p.* -sóc, *pl.* -sócon; *pp.* -sacen; *v. a. n. To deny, disown, abjure;* negare, detestari, abjurare :—Ða ætsacaþ ðæs ærýstes *qui negant esse resurrectionem,* Lk. Bos. 20, 27; L. Ath. i. 4;

Th. i. 202, 2: i. 6; Th. i. 202, 12, 13. Đâ ætsôc he *at ille negavit*, Mk. Bos. 14, 68: Lk. Bos. 22, 57. Đâ ætsôc he and swerede *tunc cœpit detestari et jurare*, Mt. Bos. 26, 74. Đâ ongan he ætsacan and swerian *ille autem cœpit anathematizare et jurare*, Mk. Bos. 14, 71. v. sacan.

æt-sæcst *shalt deny; fut. of* æt-sacan :—Þriwa đû me ætsæcst *ter me negabis*, Mk. Bós. 14, 72: Lk. Bos. 22, 34, 61. v. sacan.

æt-samne ; *adv. In a sum, together* :—Begen æt-samne *both together*, Chr. 937 ; Th. 218, col. 1 ; Æđelst. 58. Ealle ætsamne *all together*, Ps. Th. 148, 12. v. æt-somne.

æt-sceófan *To shove away*; removere, Leo 239. v. scúfan.

æt-sittan ; *p.* -sæton, *pl.* -sæton ; *pp.* -seten ; *v. intrans. To sit by, to remain, stay, wait* ; adsidere :—Đâ ætsæton đa Centiscan đǽr beæftan *then the Kentish men remained there behind*, Chr. 905 ; Th. 180, 31, col. 1.

æt-slídan ; *p.* -slád, *pl.* -slidon ; *pp.* -sliden [*æt from, away*; v. æt I. 2: slídan *labi*] *To slip* or *slide away*; labi, elabi :—Ic ætslíde *labor*, Ælfc. Gr. 29 ; Som. 33, 43 : 35 ; Som. 38, 10. Đæt hira fôt ætslíde *ut labatur pes eorum*, Deut. 32, 35.

æt-somne, æt-samne ; *adv. In a sum, at once, together* ; una, simul, pariter :—Eardiaþ ætsomne *habitant simul*, Deut. 25, 5. Ic gongan gefregn gingran ætsomne *I have understood that the disciples went together*, Cd. 224 ; Th. 298, 2 ; Sat. 526. Wǽr is ætsomne Godes and monna *a covenant is together of God and men*, Exon. 16 a ; Th. 36, 29 ; Cri. 583. Blôd and wæter bû tû ætsomne ût bicwôman *blood and water both together came out*, 24 a ; Th. 68, 34 ; Cri. 1113. Tyne ætsomne *ten together*, Beo. Th. 5687 ; B. 2847. Ealle ætsomne *omnes pariter*, Bd. 2, 13 ; S. 515, 38 : Ps. Th. 87, 17. v. somne.

æt-speornan, -spornan, đû -spyrnst, he -spyrnþ ; *p.* -spearn, *pl.* -spurnon ; *pp.* -spornen ; *v. trans. To stumble, spurn at, dash* or *trip against, mistake* ; cæspitare, offendere ad aliquid, impingere :—He ætspyrnþ *he stumbleth*; offendit, Jn. Bos. 11, 9, 10. Đe-læs đe đîn fôt æt stáne ætsporne *ne forte offendas ad lapidem pedem tuum*, Mt. Bos. 4, 6. Đe-læs đû ætspurne [Lamb. ætsporne] æt stáne fôt đînne *ne forte offendas ad lapidem pedem tuum*, Ps. Spl. 90, 12. Ætspornen [MS. ætspurnan] ic wæs *offensus fui*, Ps. Lamb. 95, 10.

æt-springan, -sprincan ; *p.* -sprang, -spranc, *pl.* -sprungon ; *pp.* -sprungen ; *v. intrans. To spring out* ; prosilire :—Blôd ætspranc *the blood sprang out*, Beo. Th. 2247 ; B. 1121.

æt-springnes, -ness, e ; *f. A springing out, falling off, despondency* ; defectio, defectio animi, Ps. Spl. T. 118, 53.

æt-spurne *offendas*, Ps. Spl. 90, 12 ; *subj. p. of* æt-speornan, q. v.

æt-spyrning *An offence, a stumbling, stumbling-block*; offensio, scandalum. DER. speornan.

ætst *shalt eat*; comedes :—Đu ætst *thou shalt eat*, Gen. 3, 17 ; *for* ytst Gen. 3, 18. DER. etan *to eat*.

æt-standan ; ic -stande, đû -standest, -stentst, he -standeþ, -stént, -stynt, *pl.* -standaþ ; *p.* -stôd, *pl.* -stôdon ; *pp.* -standen. I. *v. intrans. To stand, stand still, stop, stand near, rest, stay, stand up* ; stare, adstare, restare, requiescere :—Íren on wealle ætstôd *the iron stood in the wall*, Beo. Th. 1787 ; B. 891. Đâ ætstôd se Hǽlend *then Jesus stood still*, Mk. Bos. 10, 49. Ætstôd đæs blôdes ryne *stetit fluxus sanguinis*, Bós. 8, 44. Đâ ætstôd se arc *requievit arca*, Gen. 8, 4 : Ps. Th. 106, 24 : Lk. Bos. 7, 14. Ætstôdon cyningas [Ps. Th. 2, 2, arîsaþ] *kings stood up*; adstiterunt reges, Ps. Spl. 2, 5, 4. Ic ætstande *resto*, Ælfc. Gr. 24 ; Som. 25, 62 : Ælfc. T. 37, 6 : L. Eth. ii. 9 ; Th. i. 290, 3. II. *v. trans. To stop*; obturere, claudere :—Gif se mîcgđa ætstanden sý *if the water be stopped*, Herb. 7, 3 ; Lchdm. i. 98, 5. Hî habbaþ ætstandene ǽdran *they have stopped veins*, 4, 4 ; Lchdm. i. 90, 11.

æt-stapan ; *p.* -stôp, *pl.* -stôpon ; *pp.* -stapen *To step forth, approach*; accedere :—He forþ ætstôp *he stepped forth*, Beo. Th. 1495 ; B. 745.

æt-steal, -steall, -stæl, es ; *m.* : *pl. nom. acc.* -stalas [*at a place, a fixed place*] *Station, camp station*; sedes, statio :—Æt đam ætstealle *at the camp station*, Wald. 37 ; Vald. 1, 21. Æt-stælle *at the place*, Exon. 35 a ; Th. 112, 26 ; Gú. 150. v. stæl.

æt-stent *shall stand* ; consistet :—Seó eá ætstent on hire ryne *the river shall stand in its course*, Jos. 3, 13 ; *fut. of* æt-standan, q. v.

æt-stillan ; *p.* ede ; *pp.* ed *To still* ; compônere :—Sió cwacung sôna biþ ætstilled *the quaking will soon be stilled*, L. M. 1, 26 ; Lchdm. ii. 68, 11.

æt-swerian ; *p.* -swôr ; *pp.* -sworen *To forswear, deny with an oath* ; abjurare, L. In. 35 ; Th. i. 124, 11, note.

æt-swimman ; *p.* -swamm, *pl.* -swummon ; *pp.* -swummen *To swim out, swim* ; enatare, Chr. 918 ; Ing. 132, 17, note *m.* v. æt ; *prep.* 2.

ætten *should eat*, L. In. 42 ; Lambd. 8, 5 ; Wilk. 21, 24 ; *for* ǽten. v. etan *to eat*.

ætter, ættor, es ; *n. Poison*; venenum. v. átor.

ætter-berende ; *part. Poison-bearing, poisonous, venomous*. v. átter-berende.

ætter-loppe, an ; *f.* [átor *poison*, loppe *a silk worm, spinner of a web*] *A spider*; aranea :—And a-ýdlian ođđe aswarcan ođđe acwînan ođđe aswindan đû dydest swá swá ætterloppan ođđe ryngan sáwle his *et*

tabescere fecisti sicut araneam animam ejus, Ps. Lamb. 38, 12 ; and thou madist his lijf to faile as an yreyne [*Lat.* aranea *a spider*], Wyc. v. átor-loppe.

æt-þringan *To take away, deprive of* ; eripere :—Đa đê feorh ætþringan *who may deprive thee of life*, Andr. Kmbl. 2742 ; An. 1373.

ættren, ættern ; *adj. Poisonous* ; venenosus, Beo. Th. 3238 ; B. 1617 : Byrht. Th. 136, 4 ; By. 146 : Frag. Kmbl. 37 ; Leás. 20. v. ætren.

ættrian ; *p.* ede ; *pp.* ed ; *v. trans. To poison, envenom* ; venenare, Pref. R. Conc. v. ætrian.

ættryn ; *adj. Poisonous* ; venenosus :—Ættrynne ord *the poisonous point*, Byrht. Th. 133, 8 ; By. 47. v. ætren.

æt-wæg *took away*, Beo. Th. 2401 ; B. 1198 ; *p. of* æt-wegan.

æt-wæsend, -wesend, -weosend [*æt at*, wesende *being* ; *part. of* wesan *to be*] *At hand, approaching, hard by* ; imminens, Cot. 107.

æt-wegan ; *p.* -wæg, *pl.* -wǽgon ; *pp.* -wegen *To take away* ; auferre :—Hama ætwæg sigle *Hama took away the jewel*, Beo. Th. 2401 ; B. 1198. v. wegan.

æt-wéla, an ; *m. Abundance of food, a feast* ; copia cibi, Exon. 100 a ; Th. 374, 8 ; Seel. 123.

æt-wenian ; *p.* ede ; *pp.* ed [*æt from*, wenian *to wean*] *To deliver from, wean* ; dissuescere, seducere, ablactare :—Đe híg deófum ætweneþ *who weaneth them from devils*, L. C. S. 85 ; Th. i. 424, 13.

æt-wesan ; *p.* ic, he -wæs, *pl.* -wǽron [*æt at*, wesan *to be*] *To be present* ; adesse :—Wilferþ ætwæs, eác swylce ætwǽron úre brôđru *Wilfrid adfuit, adfuerunt et fratres nostri*, Bd. 4, 5 ; S. 572, 12. [*Goth.* at-wisan.]

æt-wîndan ; *p.* -wánd, *pl.* -wúndon ; *pp.* -wúnden *To wind off, turn away, escape, flee away* ; aufugere :—Ic âna ætwánd *effugi ego solus*, Job Thw. 165, 27 ; Grn. Iob 1, 16 : Beo. Th. 289 ; B. 143. Ic lǽte híg ætwîndan to wuda *dimitto eos avolare ad silvam*, Coll. Monast. Th. 26, 3.

æt-wist, æd-wist, ed-wist, e ; *f.* [æt, wist *substantia, cibus*] *Substance, existence, being, presence* ; substantia, præsentia :—God heora ǽhta and ætwist on-genîmeþ *God takes their wealth and substance away*, Cd. 60 ; Th. 73, 21 ; Gen. 1208. Se gǽst lufaþ onsýn and ætwist yldran hádes *the spirit loves the aspect and substance of elder state*, Exon. 40 a ; Th. 132, 11 ; Gú. 471. Him đæt Crist forgeaf đæt hý môtan his ætwiste brûcan *Christ gave them that they might enjoy his presence*, 13 b ; Th. 24, 29 ; Cri. 392 : Gen. 7, 4.

æt-wîtan ; *p.* -wát, *pl.* -witon ; *pp.* -witen *To reproach, blame, upbraid* ; imputare, improperare, exprobrare :—Ne sceolon me on đǽre þeóde þegenas ætwîtan *the thanes of this people shall not reproach me*, Byrht. Th. 138, 15 ; By. 220. Siđđan Gúþláf and Ósláf ætwiton weána dǽl *since Guthlaf and Oslaf reproached him for a part of their woes*, Beo. Th. 2304 ; B. 1150: Ps. Th. 88, 44 : 73, 17 : Ps. Spl. 31, 2. v. edwîtan.

æt-ýcan ; *p.* -ýcte ; *pp.* -ýced, -ýct [æt, ýcan, ecan *to eke*] *To add to, augment, increase* ; adjicere :—Se gesîþ ætýcte eác swylce his bênum, đæt he his teáras geát *the earl also added to his intreaties, that he shed tears*, Bd. 5, 5 ; S. 617, 40 : 4, 5 ; S. 573, 13.

æt-ýcnys, -ýcnys, -nyss, e ; *f. An increase, addition* ; augmentum :—Mid ætýcnysse *cum augmento*, Bd. 1, 27 ; S. 490, 24 : 3, 22 ; S. 553, 14.

æ-týnan ; *p.* de ; *pp.* ed ; *v. a.* [æ=a=on, un un ; týnan *to shut*] *To open* ; aperire :—Dura heofones he ætýnde *januas cœli aperuit*, Ps. Spl. 77, 27. v. a-týnan.

æt-ys *is present* ; adest, Mk. Jun. 4, 29. v. æt-eom.

æt-ýwan ; *p.* de ; *pp.* ed. I. *v. trans. To shew, reveal, manifest* ; ostendere, manifestare :—Đû me ætýwdest earfođes feala *ostendisti mihi tribulationes multas*, Ps. Th. 70, 19 : Exon. 121 b ; Th. 465, 34 ; Hö. 114 : Judth. 11 ; Thw. 24, 6 ; Jud. 174. Đá him wearþ on slæpe swefen ætýwed *then was a dream revealed to him in sleep*, Cd. 199 ; Th. 247, 13 ; Dan. 496 : Exon. 31 a ; Th. 96, 19 ; Cri. 1576. II. *v. intrans. To appear* ; apparere, manifestari :—Ealle ætýwaþ *omnes apparuerint*, Ps. Th. 91, 6. Deóful ætýwde *the devil appeared*, Andr. Kmbl. 2338 ; An. 1170. Nolde ǽfre siđđan ætýwan *would not ever afterwards appear*, Cd. 73 ; Th. 89, 16 ; Gen. 1481. v. æteówian.

æt-ýwnys, -nyss, æt-ýwedness, æt-eówedness, æt-íwedness, e ; *f. A shewing, manifestation, laying open, a declaration* ; ostensio :—Seó ætýwnys heofonlîces wundres *miraculi cœlestis ostensio*, Bd. 3, 11 ; S. 535, 23. Mid monigra heofonlîcra wundra ætýwnysse *miraculorum multorum ostensione*, Bd. 1, 26 ; S. 488, 10. Oþ ætýwednessum, Lk. Foxe 1, 80.

æw, ǽwe, es ; *n.* [ǽ *law*]. I. *law*, what is established by law, hence *wedlock, marriage, a marriage vow* ; lex, matrimonium :—Đætte ryht ǽw gefæstnod wǽre *that just law might be settled*, L. In. pref ; Th. i. 102, 9 : 1 ; Th. i. 102, 16. Rihtum ǽwe *legitimo matrimonio*, Bd. 4, 6 ; S. 573, 21, note. Se man đæt ǽwe brycþ *homo qui adulterium committit*, L. M. I. P. 15 ; Th. ii. 268, 28. II. a female bound by the law of marriage, *a wife* ; conjux legitima, uxor justa :—Se đe hæfþ ǽwe *he who has a wife* ; qui legitimam uxorem habet, L. M. I. P. 17 ; Th. ii. 270, 6. Gif ceorl wiđ ôđres riht ǽwe hæmþ *si maritus cum alterius legitima uxore adulteraverit*, 18 ; Th. ii. 270, 10. Se man, đe his riht

ǽwe forlǽt, and ṓðer wíf nímþ, he biþ ǽwbreca *the man who forsakes his lawful wife* [suam legitimam uxorem], *and takes another woman* [aliam mulierem], *he is an adulterer,* L. Ecg. P. ii. 8; Th. ii. 184, 21. Gif hwylc man wið ṓðres riht ǽwe hǽmþ, oððe wíf wið ṓðres gemǽcan, fǽste vii geár *if any man commit adultery with the lawful wife* [cum legitima uxore] *of another, or a woman* [mulier] *with the husband of another, let the fast be seven years,* ii. 10; Th. ii. 186, 6. vide ǽ.

ǽw; *adj. Lawful, legitimate, related by the law of marriage, married;* legitimus, nuptus, germanus :—Mid his ǽwum wífe *with his lawful wife,* L. Alf. pol. 42; Th. i. 90, 26, 29. Ǽwe gebrṓðru *brothers of the same marriage, own brothers;* germani fratres, Bd. 1, 27; S. 490, 28.

ǽwan, ðú ǽwest *To despise, contemn, scorn;* spernere, aversari :—Ða ðú ǽfre ne ǽwest *ea tu nunquam spernis,* P. C. 129.

ǽw-breca, -brica, ǽw-bryca, -an; *m.* [ǽw *marriage,* breca *a breaker*] *A breaker of the marriage vow, an adulterer;* adulter :—Se ðe his ǽwe forlǽt, and nímþ ṓðer wíf, he biþ ǽwbryca [Wilk. ǽwbrica] *he who leaves his wife, and taketh another woman, he is an adulterer,* L. M. I. P. 16; Th. i. 268, 30.

ǽw-bryce, es; *m. A breaking of the marriage vow, adultery;* adulterium :—Wið ǽghwylcne ǽwbryce *against all kind of adultery,* L. C. E. 24; Th. i. 374, 10: L. C. S. 51; Th. i. 404, 20: L. Edm. S; Th. i. 246, 8.

ǽwda, an; *m. A witness, one who affirms the truth by oath;* fidejussor, consacramentalis :— Hæbbe him in áþe ṓðerne ǽwdan gṓdne *let him have with him in the oath another good witness,* L. Wih. 23; Th. i. 42, 8. Mid gṓdum ǽwdum *by good witnesses,* L. H. E. 2; Th. i. 28, 2.

ǽwda-man, -mann, es; *m. A witness;* fidejussor, consacramentalis :— Rím ǽwdamanna *a number of witnesses,* L. H. E. 5; Th. i. 28, 12. v. ǽwda.

ǽwe, es; *n. Law;* lex, L. M. I. P. 15; Th. ii. 268, 28. v. ǽw.

ǽ-welm, -wellm, -wylm, -wylme, -wielme, es; *m.* [eá *water,* wælm *a welling or boiling up*] *A welling up of water, spring, fountain, source, head of a river, beginning;* aquæ fons :—Swá sum mical ǽwelm and díóp *as some great and deep spring,* Bt. 34, 1; Fox 134, 10. Seó eá cymþ eft to ðam ǽwelme *the river comes again to the source,* Fox 134, 17. Ðe mæg geseón ðone hluttran ǽwellm *who can behold the clear fountain,* 35, 6; Fox 166, 25. Gif he gesión mǽge ǽðelne ǽwelm ǽlces gṓdes [MS. goodes] *if he may see the noble fountain of all good,* 23, 7; Met. 23, 4: 20, 517; Met. 20, 259. Andlang Lígan ṓþ hire ǽwylm *along the Lea unto its source,* L. A. G. 1; Th. i. 152, 9. Ðǽre ǽwylme [MS. L. ǽwielme] is neáh ðǽre eá Rínes *whose spring is near the river Rhine,* Ors. 1, 1; Bos. 18, 25. God is ǽwelm and fruma eallra gesceafta *God is the beginning and origin of all creatures,* Bt. Met. Fox 29, 161; Met. 29, 81. v. eá-wylm.

**ǽ-wén; *adj.* [*ǽ without,* wén *hope*] *Doubtful, uncertain;* dubius :—And eów biþ eówre líf ǽwéne *and your life will be doubtful to you,* Deut. 28, 66.

ǽwen-brṓðor *a brother of the same marriage, an own brother;* germanus, Cot. 97. v. ǽw; *adj.*

**ǽ-werd; *adj.* [*ǽ law,* werd *from* werdan *to corrupt*] *Perverse, froward, averse;* perversus. v. wyrdan *to corrupt.*

ǽ-werdla, an; *m. Damage, injury,* L. In. 42; Th. i. 128, 10. v. ǽ-wyrdla.

**ǽw-fæst; *adj.* Firm in observing the law, religious, bound by the law, married;* religiosus, vinculo nuptiarum constrictus :—Ǽwfæst religiosus, Scint. 28. Ǽwfæst man *a married man,* L. C. S. 51; Th. i. 404, 21. v. ǽ-fæst.

ǽw-fæsten, es; *n.* [ǽw *law,* fæsten *a fast*] *A fixed or legal fast;* legitimum jejunium :—To ǽwfæstene *for the legal fast,* Rubc. Lk. Bos. 3, 1 a, notes, p. 578.

ǽw-fæst-man *a man bound by law, a married man;* vinculo nuptiarum constrictus, L. C. S. 51; Th. i. 404, 21.

ǽw-festnys, -nyss, e; *f. Religion, piety;* religio, pietas. v. ǽfestnes.

**ǽ-wintre; *adj.* [ǽ = ǽn = án *one*] *Of one winter or year, continuing for a year.* v. án-wintre.

ǽ-wintre-cyning, es; *m. A king or ruler for one winter or year, a consul;* consul. v. winter; g. wintres.

ǽwisc, e; *f. A dishonour, disgrace, offence;* dedecus, scandalum :— Cwæþ ðæt him to micel ǽwisce wǽre *said that it would be much disgrace to them,* Ors. 4, 6; Bos. 86, 26. On ǽwisce *in scandalum,* Ps. Th. 68, 23. [*Goth.* aiwisks, *n. dedecus.*]

**ǽwisc; *adj.* Disgraced, ashamed, abashed;* dedecoratus. v. ǽwisc-mṓd.

**ǽwisc-berende; *part. Bearing disgrace, unchaste, lewd, unclean, shameless, impudent;* impudicus. v. ǽwisc, berende *bearing.*

**ǽwisc-mṓd; *adj. Disgraced in mind, ashamed, abashed;* dedecoratus animo, pudore suffusus :—Ides, ǽwiscmṓd, andswarode *the woman, disgraced in mind, answered,* Cd. 42; Th. 55, 18; Gen. 896. Ðæt he ǽwiscmṓd eft síðade, heán, hyhta leás *that he abashed returned, depressed, void of hopes,* Exon. 46 a; Th. 157, 23; Gú. 896: 80 b; Th. 302, 16; Fä. 37. Gewiton hym ða Norþmen Dyflin sécan ǽwiscmṓde *then the Northmen departed, abashed in mind, to seek Dublin,* Chr. 938; Th. 207, 16, col. 1; Æðelst. 56.

ǽwisc-nys, -ness, e; *f. Disgrace, obscenity, filthiness, a blushing for shame, reverence;* dedecus, obscenitas, pudore suffusio, reverentia :— Ǽwiscnys *reverentia,* Ps. Spl. C. 34, 30. On ǽwiscnesse *openly, as not being ashamed to be seen;* in propatulo, Cot. 110, 202.

ǽ-wita, an; *m.* [ǽ *lex,* wita *gnarus homo, sapiens*] *One skilled in the law, a counsellor;* legis peritus, consiliarius :—Ealdum ǽwitan ageaf andsware *gave answer to the old counsellor,* Elen. Kmbl. 907; El. 455.

**ǽw-líc; *adj. Lawful;* legitimus. v. ǽ-líc.

**ǽwnian; *p.* ode; *pp.* od [ǽw *marriage*] *To marry, wed;* connubio jungere, Leo 104. DER. be-ǽwnian.

ǽ-writere, es; *m. A writer, composer or framer of laws;* legum conditor, Prov. 8.

ǽwul *A wicker-basket with a narrow neck for catching fish, a* WEEL; nassa, Ælfc. Gl. 102; Som. 77, 85; Wrt. Voc. 56, 9.

**ǽwum-boren; *part. Lawfully born, born in wedlock;* legitimo matrimonio natus :—Ǽt his dehter ǽwum-borene *with his lawfully-born daughter,* L. Alf. pol. 42; Th. i. 90, 28. v. ǽw.

**ǽwunge; *adv. Openly, publicly;* manifeste :—On ǽwunge *openly, abroad, in the sight of all;* in propatulo. v. eáwunga, eáwunge.

ǽ-wylm, es; *m. A spring, fountain, source* :—Andlang Lígan ṓþ hire ǽwylm *along the Lea unto its source,* L. A. G. 1; Th. i. 152, 9. v. ǽ-welm.

ǽ-wyrdla, -werdla, an; *m. Damage, detriment, injury;* detrimentum :— He sóna mycle wonunge and ǽwyrdlan wæs wyrcende ðǽre mǽrwan cyrican weaxnesse *magno tenellis ibi adhuc ecclesiæ crementis detrimento fuit,* Bd. 2, 5; S. 506, 37: 1, 3; S. 475, 21: Herb. 141; Lchdm. i. 262, 11. v. ǽf-wyrdla.

ǽ-wyrp, es; *m.* [ǽ = a *from,* wyrp *a cast,* from wyrpan *or* weorpan *to cast*] *A cast-away, throwing away;* abjectus, abjectio :—Ǽwyrp folces *abjectio populi,* R. Ben. 7.

ÆX—ÆCS, æsc, acas, e; *f:* acase, axe, an; *f.* what is brought to an edge, *An* AXE, *a hatchet, pickaxe;* securis, ascia :—Eallunga ys seó æx to ðæra treówa wurtrumum asett *jam enim securis ad radicem arborum posita est,* Mt. Bos. 3, 10. Mid æxum *with axes,* Ps. Th. 73, 6. On æxe *in securi,* Ps. Spl. 73, 7. Forðon seó æx [MS. H. sió æsc; seó eax B.] biþ melda, nalles þeóf *because the axe is an informer, not a thief;* quia securis acclamatrix potius est, non fur, L. In. 43; Th. i. 128, 23. [*O. Sax.* acus, *f: N. Dut.* akse, *f: Ger.* axt, *f: M. H. Ger.* ackes, *f: O. H. Ger.* achus, *f: Goth.* aqizi, *f: Dan.* ökse: *Swed.* yxa: *O. Nrs.* öx, *f: Lat.* ascia, *f: Grk.* ἀξίνη.]

æx, e; *f. An axis;* axis, Ælfc. Gr. 9, 28; Som. 11, 45. v. eax.

æxe, an; *f. Ashes,* Ps. Spl. T. 101, 10. v. axe, asce.

**æxian; *p.* ode *To ask;* rogare :—Ǽxodon *asked;* interrogaverunt, Ps. Spl. T. 136, 3. v. acsian.

**af- = æf- = of- *of, from, away from;* de, ex, ab. v. æf-, of-: af-god *an idol.*

**a-féded; *part.* [*for* a-féded; *pp.* of a-fédan *to feed, nourish*] *Fed, nourished, brought up, educated;* nutritus, Bd. 1, 27; S. 489, 37.

**a-fæged, -fægd; *part. Depicted, drawn;* depictus :—Bǽron ankicnysse Drihtnes Hǽlendes on brede afægde and awritene *ferebant imaginem Domini Salvatoris in tabula depictam,* Bd. 1, 25; S. 487, 4. v. a-fægrian.

**a-fægniende *rejoicing,* = fægniende; *part.* of fægnian.

**a-fægrian; *p.* ode; *pp.* od *To make fair or beautiful, to adorn, embroider;* depingere, ornare :—Mid missendlícum blóstmum wyrta afægrod *variis herbarum floribus depictus,* Bd. 1, 7; S. 478, 22.

**a-fǽlan, -fǽllan; *p.* de; *pp.* ed *To overturn, overthrow, cast out, drive out, cause to stumble, offend;* evertere, prosternere, ejicere, scandalizare, Mt. Rush. Stv. 21, 12; Mk. Rush. War. 3, 23: Mt. Rush. Stv. 18, 6. v. ge-fælan.

**a-fǽman; *p.* de; *pp.* ed *To foam out, breathe out;* exspumare, exhalare :—Múþ ic ontýnde mínne wíde, ðæt me mín oreþ út afǽmde *os meum aperui, et exhalavi spiritum,* Ps. Th. 118, 131.

**a-féran; *p.* de; *pp.* ed [a, féran *to terrify*] *To make greatly afraid, to affright, terrify, dismay, astound;* exterrere, perterrere, consternare, stupefacere :—Ðæt heó afǽre fleógan on nette *that she may terrify flies into her net,* Ps. Th. 89, 10. Folc wæs afǽred *the folk was affrighted,* Cd. 166; Th. 206, 3; Exod. 446: Exon. 63 b; Th. 233, 15; Ph. 525: Mk. Bos. 9, 6, 15: Lk. Bos. 24, 4. Híg wurdon ealle afǽrede *erant omnes exterriti,* Gen. 42, 35: Ex. 20, 18.

**a-fœrþ *he shall lead out,* Ps. Spl. 51, 5. v. afaran II.

**a-fæstan; *p.* -fæste; *pp.* -fæsted *To fast;* jejunare :—He afæste to ǽfenes *he fasted till evening,* Bd. 3, 23; S. 554, 32: 3, 27; S. 559, 13.

**afǽstla; *interj. O certainly! O assuredly! O certe* :—Afǽstla, and hi lá hi, and wella well, and þyllíce ṓðre syndon Englisc interjections *O certainly, and alas, and well well, and such other are English interjections,* Ælfc. Gr. 48; Som. 49, 28.

**a-fæstnian; *p.* ode; *pp.* od *To fix, fasten or make firm, to strengthen, fortify, confirm, betroth, espouse, inscribe;* munire, firmare, consignare libris, infigere :—Ðæt we hí móton afæstnian on ðé *that we may fix them* [our eyes] *on thee,* Bt. 33, 4; Fox 132, 31: Bt. Met. Fox 20, 525; Met. 20, 263. Hú afæstnod wæs feld-húsa mǽst *how that greatest of*

field-houses was fastened, Cd. 146; Th. 183, 2; Exod. 85: 173; Th. 218, 17; Dan. 40. Ðe he on fíf bócum afæstnode *which he inscribed in five books*, Hexam. 1; Norm. 2, 18: Deut. 32, 23. Afæstnod ic eom *infixus sum*, Ps. Spl. 68, 2.

a-fandelíc *probable*. v. a-fandigendlíc.

a-fandian, -fandigean; *p.* ode, ude, ade; *pp.* od, ud, ad; *v. a.* To prove, try, to make a trial, to discover by trying, to experience; probare, tentare, experiri :—Ðú afandodest heorte míne *probasti cor meum*, Ps. Spl. 16, 4. Lá líceteras, cunne ge afandian heofones ansýne and eorþan, húmeta ná afandige ge ðas tíde? *hypocritæ, faciem cæli et terræ nostis probare, hoc autem tempus quomodo non probatis?* Lk. Bos. 12, 56. Ðú hit hæfst afandad be ðé selfum *thou hast experienced it of thyself*, Bt. 31, 1; Fox 112, 19. Seolfor afandod eorþan *argentum probatum terræ*, Ps. Spl. 11, 7: 80, 7. Afandud, Gen. 43, 23. Afanda hwæðer Freá wille *make a trial whether the Lord will*, Cd. 101; Th. 134, 23; Gen. 2229.

a-fandigendlíc, -fandelíc, -fandodlíc; *adj.* What may be tried, proved, probable; probabilis, Scint. de prædest.

a-fandung, e; *f.* A trying; probatio, experientia, Scint. v. fandung.

a-fangen *taken, received*; assumptus, Mk. Bos. 16, 19. v. a-fón.

afara *a son*, Chr. 937; Th. 200, 41, col. 1; Æðelst. 7. v. eafora.

a-faran, he -færþ; *p.* -fór, *pl.* afóron; *pp.* -faren. I. *v. n.* To depart, march, to go out of or from a place; exire, egredi :—Hie of Egyptum út afóron *they marched out from Egypt*, Cd. 173; Th. 217, 14; Dan. 6. II. *v. act.* To remove, lead out; emigrare :—Afærþ ðé *emigrabit te*, Ps. Spl. 51, 5.

a-feallan; *p.* -feól, -feóll, *pl.* -feóllon; *pp.* -feallen To fall down; cadere :—Ðæt hús afeóll *domus cecidit*, Lk. Bos. 6, 49: Cd. 202; Th. 251, 1; Dan. 557: Jud. 16, 30. Wearþ afeallen Æðelrædes eorl *Ethelred's earl fell* [*in the battle*], Byrht. Th. 137, 46; By. 202.

a-feccan To receive; accipere :—He afecþ [MSS. C. T. onféhþ] me *acceperit me*, Ps. Spl. 48, 16.

a-fédan; *p.* -fédde; *pp.* -féded, -féd To feed, nourish, rear, bring up; nutrire, cibare, alere, pascere :—Heó bearn afédeþ *she nourishes her child*, Salm. Kmbl. 746: Sal. 372: Ps. Th. 135, 26: 83, 3. Ðæt ðú hí afédde mid ðý Godes worde *that thou didst feed them with the word of God*, Bd. 3, 5; S. 527, 34: Ors. 1, 6; Bos. 29, 10: Ps. Th. 94, 7: 99, 3: Andr. Kmbl. 1177; An. 589. He wæs aféded *he was brought up*, 1367; An. 684. He wæs aféded and gelǽred *he was reared and taught*; nutritus atque eruditus est, Bd. 5, 20; S. 642, 16. Wearþ lafeðe geóguþ aféded *to Japhet was youth brought up*, Cd. 78; Th. 96, 34; Gen. 1604: 82; Th. 102, 29; Gen. 1707. Ic eom aféd *pascor*, Ælfc. Gr. 33; Som. 36, 44. Ðá híg afédde wǽron *quibus adultis*, Gen. 25, 27.

a-féhþ *receives*; suscipit, Ps. Spl. 47, 3. DER. a-féhan. v. féhan, fón.

a-fellan; *p.* de; *pp.* ed To fell; cædere, prosternere, L. In. 43; Th. i. 128, 23. v. a-fyllan.

a-felle *barked*; decorticatum, R. 115. v. æ-felle.

Afen, Afn, e; *f.* Afene, an; *f.* I. AVON, *the name of a river in Somersetshire* :—Eást óþ Afene múþan *east at the Avon's mouth*, Chr. 918; Th. 190, 4. II. *also of other rivers in different parts of England* :—Into Afenan múþan *into Avon's mouth*, Chr. 1067; Th. 342, 5.

aféng, aféngon *took*, Ps. Spl. 47, 8: 118, 16; *p.* of a-fón.

a-feohtan; *p.* -feaht, *pl.* -fuhton; *pp.* -fohten. I. *to fight against, attack, assail*; impugnare, expugnare :—Bryttas Ongel þeóde afuhton *the Britons fought against the English nation*, Bd. 5, 23; S. 647, 1: 4, 26; S. 602, 25. Hí afuhton me *expugnaverunt me*, Ps. Th. 108, 2: Ps. Grn. 34, 1. II. *to tear* or *pluck out*; evellere :—Ǽr hit afohten foldan losige *priusquam evellatur*, Ps. Th. 128, 4. v. feohtan.

a-feóll *fell*; cecidit, Lk. Bos. 6, 49; *p.* of afeallan.

a-feormian, -igan; *p.* ode; *pp.* od; *v. trans.* [a intensive, feormian *to cleanse*] *To cleanse, clean thoroughly, purge, wash away*; mundare, emundare, permundare, diluere :—Mid besmum afeormod *scopis mundatus*, Lk. Bos. 11, 25. He afeormaþ his þyrscelflóre *permundabit aream suam*, Mt. Bos. 3, 12. Hyt ðone magan ealne afeormaþ *it purges the whole stomach*, Herb. 60, 3; Lchdm. i. 162, 19. Ic afeormige *diluo*, Ælfc. Gr. 28, 3; Som. 30, 49. Hit afeormaþ ðæ nebcorn *it will cleanse away all the face pimples*, Herb. 22, 3; Lchdm. i. 118, 24.

a-feormung, e; *f.* A cleansing, purging; purgatio, Scint. 2.

a-feorran, -ferran, -firran, -fyrran; *p.* de, ode; *pp.* ed, od To remove, take away, expel; removere, elongare, amovere, auferre :—Ðæs líchoman fæger and his streón mágon beón afeorred *the fairness of the body and its strength may be taken away*, Bt. 32, 2; Fox 116, 31. Ðú afeorrodyst fram me freónd and nýhstan *elongasti a me amicum et proximum*, Ps. Spl. C. 87, 19: Cd. 219; Th. 282, 9; Sat. 284.

a-feorsian, -fersian, -firsian, -fyrsian; *p.* ode; *pp.* od. I. *v. trans.* To remove, take away, expel; removere, elongare, expellere :—Ðe afeorsiaþ hine fram ðé *qui elongant se a te*, Ps. Spl. 72, 26: L. C. E. 4; Th. i. 360, 29. II. *v. intrans.* To go away, depart; emigrare :—Ic ná afeorsie *non emigrabo*, Ps. Spl. 61, 6.

afera *a son*, Cd. 95; Th. 123, 31; Gen. 2054. v. eafora.

a-féran; *p.* de; *pp.* ed To affright, terrify; perterrere, Chr. 1083; Th. 352, 9. . a-fǽran.

a-ferian, -igan; *p.* ede; *pp.* ed To take away, remove, withdraw; auferre, amovere, subducere, cum averiis vel curru vehere, averiare :—Ðæt ðú ðe aferige of ðisse folcsceare *that thou withdraw thyself from this people*, Cd. 114; Th. 149, 19; Gen. 2477. He aferede *he bore away*, Andr. Kmbl. 2355; An. 1179: Ps. Th. 135, 25: Menol. Fox 47; Men. 23. Gif he aferaþ *if he remove*; si averiat, L. R. S. 4; Th. i. 434, 8. He sceal aferian [MS. auerian = averian = aferian] *he shall remove*; debet averiare, 432, 10. v. a-feorran.

a-ferran; *p.* de; *pp.* ed To remove, take away; elongare, removere :—Gást háligne fram me aferredne *the holy spirit taken from me* [*acc. absol.*], Ps. C. 97: Bt. 39, 11; Fox 230, 19. v. a-feorran.

a-ferscean [a, fersc *fresh*] *To freshen, to become fresh*; salsuginem deponere :—Swá swá of ðære sǽ cymþ ðæt wæter innon ða eorþan and ðǽr afersceaþ *thus from the sea the water enters into the earth and then becomes fresh*, Bt. 34, 6; Fox 140, 18.

a-fersian *to take away*; removere. v. a-feorsian.

a-festnian *to fix, fasten*; munire, firmare. v. a-fæstnian.

a-fétigan *to beat with the feet, to praise, applaud*; plaudere :— Ic afétige *plaudo*, Ælfc. Gr. 28, 4; Som. 31, 28.

Affric; *def. m.* Affrica; *adj.* AFRICAN; Afer, Africanus :—Severus se Cásere Affrica *Severus Cæsar Afer*, Bd. 1, 5; S. 476, 5, note. Fóron Rómane on Affrice, *acc. pl.* the Romans went against [*upon*] *the African people*, Ors. 4, 6; Bos. 84. 24: 5, 4; Bos. 105, 2: 5, 7; Bos. 106, 22. On Africum *among the African people*, 6, 1; Bos. 115, 31.

Affrica; *indecl*: but Lat. Affrica, *gen.* æ; *acc.* am; *f.* Africa :—Asia and Affrica togæðere licgaþ *Asia and Africa lie together*, Ors. 1, 1; Bos. 15, 14. Ðære Affrica norþ-west gemǽre *the north-west boundary of Africa*, id; Bos. 16, 4. Nú wille we ymbe Affrica *now will we* [*speak*] *about Africa*, id; Bos. 24, 26. Hý ða þrý dǽlas on þreó tonemdon—Asiam, and Európam, and Affricam *they named the three parts by three names—Asia, and Europe, and Africa*, id; Bos. 15, 5: 5, 11; Bos. 109, 23: 6, 30; Bos. 126, 32.

Affrican, es; *m.* An African; Africanus :—Regulus feaht wið Affricanas *Regulus fought against Africans*, Bt. 16, 2; Rawl. 33, 19. v. African.

af-god, es; *n.* [af = of = æf a, ab; god, n. a heathen god] An idol, an image; idolum. [Platt. Dut. afgod, m: O.H. Ger. apcot, n: M.H.Ger. abgot, n. m: Ger. abgott, m: Goth. afguþs impius: Dan. Swed. afgud, m: O. Nrs. afguð, m.] v. god; n.

af-godnes, -ness, e; *f.* Idolatry, the worshipping of images; idololatria. v. af, god, es; n. a heathen god; -nes, -ness.

a-findan; *p.* -fánd, *pl.* -fúndon; *pp.* -fúnden To find, detect, feel, experience; invenire, deprehendere, experiri, sentire :—Ðe he Godes eorre afinde *though he felt God's anger*, Ps. C. 25. Ic afinde *experior*, Ælfc. Gr. 31; Som. 35, 55. Ðis wíf wæs afunden on unrihton hǽmede *hæc mulier deprehensa est in adulterio*, Jn. Bos. 8, 4: Bt. 35, 5; Fox 162, 31.

a-firhtan *to affright*; exterrere :—Hí flugon afirhte to múntum *they fled affrighted to the mountains*, Gen. 14, 10. v. a-fyrhtan.

a-firran; *p.* de; *pp.* ed To remove, take away, put away, expel; elongare, amovere, auferre :—Ðæt he him afirre frécne geþohtas *that he put away from him wicked thoughts*, Cd. 219; Th. 282, 9; Sat. 284. Crist heó afirde *Christ expelled them*, 214; Th. 269, 3; Sat. 67: Ps. Spl. T. 87, 19. v. a-feorran.

a-firsian; *p.* ode; *pp.* od To take away, remove; longefacere, removere :—He afirsode fram us unrihtwísnysse *longefecit a nobis iniquitates*, Ps. Spl. M. 102, 12. v. a-feorsian.

a-fleón, he -flíhþ; *p.* -fleáh, *pl.* -flugon; *pp.* -flogen. I. *v. intrans.* To flee away; effugere :—Gást afliþþ *the spirit fleeth away*, Exon. 40 a; Th. 132, 20; Gú. 475: 58 a; Th. 208, 13; Ph. 155. II. *v. trans.* To drive away, put to flight; fugare :—Hí aflogene wǽron *they were put to flight*, Jud. 6, 14. DER. fleón.

a-fleótan *to float off, scum, clarify, purify liquor by scumming*; despumare. DER. fleótan.

a-fleów *overflowed*, Ors. 5, 4; Bos. 105, 9; *p.* of aflówan.

a-flian *to put to flight*; fugare, Herb. 96, 2; Lchdm. i. 208, 20. v. a-fligan.

a-flíeman; *p.* de; *pp.* ed To cause to flee, to banish :—Síe he aflíemed *let him be* [*as one*] *banished*, L. Alf. pol. 2; Th. i. 60, 17. v. a-flýman, ge-fléman.

a-fligan; *p.* de; *pp.* ed [a, fligan] To drive away, put to flight; fugare, arcere :—Sóna hit ðone fefer afligeþ *it will soon put the fever to flight*, Herb. 37, 2; Lchdm i. 138, 5. Aflian [MS. B. afiigan] *to put to flight*, 96, 2; Lchdm. i. 208, 20. Ic aflige míne fýnd *arcesso inimicos meos*, Ælfc. Gr. 28, 2; Som. 30, 43. Afliged beón *to be driven away*, R. Ben. cap. 48. Afliged mon *an apostate*, Prov. 6.

a-fliung, e; *f.* A fleeing; rejectio :—Mete-afliung *a rejecting of meat*; atrophia, Ælfc. Gl. 10; Som. 57, 41; Wrt. Voc. 19, 44.

a-flogen *driven away*, Jud. 6, 14; *pp.* of a-fleón.

a-flówan; *p.* -fleów, *pl.* -fleówon; *pp.* -flówen To flow from, flow over; effluere :—Etna fýr afleów up *the fire of Etna flowed over*, Ors. 5, 4; Bos. 105, 9.

a-flyge, es; m. [a, flyge a flight] A flying, flight; volatus. [Ger. flug, Grm. Wörterbuch; fuga?]

a-flŷman; p. de; pp. ed; v. trans. [a, flŷman] To cause to flee, put to flight, drive away, banish, scatter, disperse; fugare, in fugam vertere, ejicere, pellere, dispergere:—He swâ manigne man aflŷmde he caused so many men to flee, Byrht. Th. 138, 61; By. 243. Đû me aflŷmst tu me ejicis, Gen. 4, 14. Wurdon twegen æðelingas aflŷmde of Sciððian two noblemen were driven from Scythia, Ors. 1, 10; Bos. 32, 34. Sŷ he aflŷmed let him be [as one] banished, L. Alf. pol. 2; Th. i. 60, 17, note. And eall his weored oððe ofslægen wæs oððe aflŷmed ejusque totus vel interemptus vel dispersus est exercitus, Bd. 2, 20; S. 521, 13.

afol, es; n. Power; vires, robur:—Eallum his afole with all his power, L. I. P. 2; Th. ii. 304, 22. v. abal.

a-fôn; p. -fêng, pl. -fêngon; pp. -fangen, -fongen To receive, take, take up, hold up, support, seize, lay hold of; suscipere, assumere, corripere, occupare, tradere:—We afêngon mildheortnysse ðîne on midle temple suscepimus misericordiam tuam in medio templi, Ps. Spl. 47, 8: 118, 116. Afônde suscipiens, 146, 6. He wæs on heofonum afangen assumptus est in cælum, Mk. Bos. 16, 19. Hyre se aglæca ageaf andsware, forht afongen to her the wretch gave answer, seized with fear, Exon. 70 a; Th. 261, 24; Jul. 320: 25 a; Th. 73, 3; Cri. 1184. Đæt Johannes wæs afongen quod Johannes traditus esset, Mt. Rush. Stv. 4, 12.

a-fônde taking up, raising up; suscipiens, Ps. Spl. 146, 6; part. of a-fôn.

afor, es; adj. Vehement, dire, hateful, rough, austere; vehemens, atrox, odiosus, asper, austerus, acerbus:—Iudiþ, egesfull and afor Judith, dreadful and vehement, Judth. 12; Thw. 25, 13; Jud. 257. Afrum onfengum with their dire attempts, Exon. 40 a; Th. 133, 15; Gû. 490. Đæt [sǽd] byþ þreóhyrne, and hyt byþ afor and sweart the seed is three-cornered, and it is rough and swarthy, Herb. 181, 1; Lchdm. i. 316, 11. [Goth. abrs strong: O. Nrs. æfr sævus, vehemens, ferox.] v. nefre.

a-fôr, -fôron departed, Ors. 2, 4; Bos. 45, 14: Cd. 173; Th. 216, 14; Dan. 6; p. of a-faran.

afora a son, Chr. 937; Th. 200, 41, col. 3; Æðelst. 7. v. eafora.

afor-feorsian; p. ode; pp. od To defer, delay, prolong; prolongare:—Eardbiggengnes [MS. eardbiggendes] mîn aforfeorsode is incolatus meus prolongatus est, Ps. Spl. 119, 5; Lambeth has, Eardbegengnes oððe elþeódignys mîn afeorrad oððe gelængd is, Ps. 119, 5; my pilgrimaging is drawen along, Wyc. v. feorsian.

a-forhtian; p. ode; pp. od [a intensive, forhtian to fear] To be very much afraid, to tremble with fear, to be affrighted, amazed; expavescere:—Đâ aforhtode Isaac micelre forhtnisse expavit Isaac stupore vehementi, Gen. 27, 33.

â-forþ; adv. [â always, forþ forth] Always, continually, daily, still; indies, Cot. 115.

aforud exalted; exaltatus. v. ofer-ge-aforud.

a-frêfran; p. ede; pp. ed To comfort, console; consolari:—God eáðe mæg afrêfran feásceaftne God can easily comfort the distressed, Exon. 10 b; Th. 11, 23; Cri. 175: 13 a; Th. 23, 13; Cri. 368. He mec þurh engel oft afrêfreþ he through his angel oft comforteth me, 37 a; Th. 121, 10; Gû. 286. We weorþaþ afrêfrede facti sumus sicut consolati, Ps. Th. 125, 1: 118, 52: Andr. Kmbl. 1275; An. 638.

a-frêfrian; p. ode; pp. od To comfort, console; consolari:—Forwyrnde beón afrêfrod sâwle mîn renuit consolari anima mea, Ps. Spl. 76, 3.

a-freoðan; p. ede; pp. ed To froth; spumare:—Læt afreoðan yt froth, L. M. 1, 47; Lchdm. ii. 118, 27. [O. Nrs. froða, frauð froth; spuma.]

Africa = Affrica Africa; Africa:—Affrica onginþ Africa begins, Ors. 1, 1; Bos. 24, 35. v. Affrica.

African, Affrican, es; m. An African; Africanus:—Đâ he feaht wið Africanas, he hæfde sige ofer ða Africanas when he fought against Africans, he gained a victory over the Africans, Bt. 16, 2; Fox 52, 39: 54, 1.

Africanisc, Afrisc; adj. Belonging to Africa, African; Africanus:—Africanisc æppel [MS. -isca, -ple] a pomegranate; malum Punicum, Cot. 133.

Afrisc; adj. African; Africanus:—Afrisc meówle an African maid, Cd. 171; Th. 215, 7; Exod. 579.

a-froefred comforted; consolatus, Mt. Rush. Stv. 5, 4, = a-frêfred; pp. of a-frêfran.

a-fûl, es; n. A fault; culpa. v. fûl.

a-fûlian; p. ode; pp. od; v. n. To become foul, to putrefy, be defiled; putrescere, putrefieri, inquinari, Scint. 66: 17. v. fûlian.

a-fúnden found, discovered; Jn. Bos. 8, 4: Bt. 35, 5; Fox 162, 31; pp. of a-findan.

a-fúndennis, -niss, e; f. An experiment, an invention, a discovery; experimentum, R. Ben. interl. 59.

a-fŷlan; p. ede; pp. ed; v. a. [a, fûl foul, unclean] To foul, defile, pollute, to make filthy, to corrupt; inquinare, contaminare, fœdare:—Yfel biþ ðæt man mid flǽsc-mete hine sylfne afŷle it is sinful that any one defile himself with flesh-meat, L. C. S. 47; Th. i. 402, 24; Past. 54, 1. Afŷled fœdatus, Procem. Greg. Dial. v. ge-fŷlan, a-fûlian.

a-fyllan; p. de; pp. ed [a, fyllan to fill] To fill up or full, replenish, satisfy; replere, implere:—Afyllaþ ða eorþan replete terram, Gen. 9, 1. He ne mæg ða gîtsunga afyllan he cannot satisfy the desires, Bt. 16, 3;

Fox 56, 16. Fŷres afylled with fire filled, Exon. 30 b; Th. 95, 26; Cri. 1563: Cd. 215; Th. 271, 4; Sat. 100: Beo. Th. 2040; B. 1018: Ps. Th. 128, 5.

a-fyllan = a-fellan; p. de; pp. ed; v. a. [a, fyllan, fellan to fell] To fell, to strike or beat down, to overturn, subvert, lay low, abolish, slay; cædere, occidere, prosternere, dejicere, demoliri, comprimere, abrogare:—Gif mon afelle [MS. B. afylle] on wuda wel monega treówa if any one fell in a wood a good many trees, L. In. 43; Th. i. 128, 19. Drihten afylþ ðîne fŷnd the Lord will strike down thine enemies, Deut. 28, 7. Hî to eorþan afyllaþ ðê ad terram prosternent te, Lk. Bos. 19, 44: Salm. Kmbl. 595; Sal. 297. Afylde hine he felled him, Salm. Kmbl. 917; Sal. 458. Wæs Waldendes lof afylled the supreme ruler's praise was suppressed, Chr. 975; Th. 228, 10; Edg. 38. Hû man mæg unlage afyllan how one may abolish unjust laws, L. C. S. 11; Th. i. 382, 8. Gif hwâ ôðres ryht afylle if any one suppress another's right, L. Ath. i. 17; Th. i. 208, 16: L. Eth. vi. 8; Th. i. 316, 26. Đæt hine man afylle that any one slay him, 38; Th. i. 324, 23: v. 31; Th. i. 312, 12. v. be-fyllan, ge-.

a-fyran; p. ede; pp. ed To remove, take away, expel; amovere, elongare, Exon. 43 b; Th. 147, 1; Gû. 720. v. a-fyrran.

a-fŷran; p. de; pp. ed, yd [a, fŷran castrare] To castrate; castrare:—Afŷred olfend a dromedary, a kind of swift camel; dromeda MS. Twegen afŷryde men duo eunuchi, Gen. 40, 1.

a-fŷrd, es; m. A eunuch; spado, Cot. 189. v. a-fŷrida.

a-fyrhtan; p. -fyrhte; pp. -fyrhted, -fyrht To affright, terrify; terrere, exterrere, perterrere, timore afficere:—He afyrhted wearþ he was affrighted, Exon. 52 a; Th. 181, 29; Gû. 1300: Andr. Kmbl. 3057; An. 1531. Wæran mid egsan ealle afyrhte with dread were all affrighted, Cd. 222; Th. 288, 22; Sat. 385. Đa weardas wǽron afyrhte custodes exterriti sunt, Mt. Bos. 28, 4: Bd. 3, 16; S. 543, 12, MS. T. Afirhte, Gen. 14, 10. v. a-forhtian.

afŷrida, afŷryda, an; m. [a-fŷred; pp. of a-fŷran] A eunuch, a castrated animal, servant, courtier; eunuchus, servus:—Se afŷrida the servant, courtier [eunuch], Gen. 39, 1. Hî sealdon Iosep Putifare ðam afŷrydan Faraones vendiderunt Joseph Putiphari eunucho Pharaonis, 37, 36.

a-fyrran, -fyran; p. ede, de; pp. ed [a from, fyrr far] To remove, take away, expel, deliver; amovere, avertere, elongare, auferre, eripere:—Nǽddran hî afyrraþ serpentes tollent, Mk. Bos. 16, 18. Beóþ afyrrede are taken away, Ps. Spl. 57, 8. Đû afyrdest of Jacobe ða graman hæftnêd avertisti captivitatem Jacob, Ps. Th. 84, 1. Đû me afyrdest frŷnd ða nýhstan elongasti a me amicum et proximum, 87, 18: 88, 36: Bd. 2, 20; S. 522, 23: 4, 11; S. 579, 34. Afyrrinde gefeoht oððe oþ ende eorþan auferens bella usque ad finem terræ, Ps. Spl. C. T. 45, 9. Afyrr me feóndum mînum eripe me de inimicis meis, Ps. Th. 142, 10. Afyr, 118, 22: 53, 5. Ic ðê wolde cwealm afyrran I would remove death from thee, Exon. 28 b; Th. 87, 17; Cri. 1426. Dreám wæs afyrred joy was removed, 42 a; Th. 142, 9; Gû. 641. He hæfde feóndas afyrde he had the fiends expelled, 43 b; Th. 147, 1; Gû. 720. v. a-feorran.

a-fyrsian; p. ode; pp. od; v. a. [a, fyrsian to remove] To remove farthest away, drive away, dispel; pellere, propellere, auferre:—He afyrseþ gâst ealdormanna aufert spiritum principum, Ps. Spl. 75, 12: 45, 9. Đe deófla afyrseþ which drives devils away, L. C. E. 4; Th. i. 360, 29. v. a-feorsian, a-fyrran.

a-fŷryda a eunuch; eunuchus:—Đam afŷrydan Faraones eunucho Pharaonis, Gen. 37, 36. v. afŷrida.

a-fŷsan; p. de; pp. ed. I. to hasten; festinare, tendere:—Feor afŷsan and forþ gangan to hasten away and to go forward, Byrht. Th. 131, 4; By. 3. II. to hasten away, impel, accelerate, incite, excite, make ready; incitare, accelerare, paratum vel promptum reddere:—Đonne he afŷsed biþ when he hastened away, Exon. 65 a; Th. 241, 11; Ph. 654. To heofonum biþ môd afŷsed to heaven is the spirit impelled, 65 b; Th. 241, 17; Ph. 657: 59 b; Th. 217, 3; Ph. 474: Rood Kmbl. 247; Kr. 125: Exon. 119 a; Th. 457, 22; Hy. 4. 27. Swâ ǽr wæter flcówan, flôdas afŷsde as the waters flowed before, the excited floods, 22 b; Th. 61, 17; Cri. 986.

ag, es; n? Wickedness; nequitia:—Hî þohton and hî sprǽcon ag cogitaverunt et locuti sunt nequitiam, Ps. Spl. T. 72, 8. [Goth. aglo, f. trouble: O. Nrs. agi, m. terror: Grm. ii. 503, 20.] DER. ag-lâc, ag-lǽc, -lǽca, -lâc-hâd, -lǽc-cræft, -lǽc-wîf.

âga; m. A possessor, an owner; possessor. v. un-âga.

a-gæf returned; reddidit, Cd. 196; Th. 244, 24; Dan. 453; p. of a-gifan.

a-gǽlan; p. de; pp. ed. I. v. trans. To hinder, occupy, detain, delay, neglect; impedire, retardare, morari, negligere:—Đæt he ne agǽle gǽstes þearfe that he delay not his spirit's welfare, Exon. 19 b; Th. 51, 16; Cri. 817. Me ðiós siccetung hafaþ agǽled this sighing has hindered me, Bt. Met. Fox 2, 9; Met. 2, 5. Ic mîne tîd-sangas oft agǽlde I have often neglected my canonical hours, L. De Cf. 9; Th. ii. 264, 11. Astrecceaþ agǽledan honda remissas manus erigite, Past. 11, 1; Cot. MS. And swâ eall ðæt folc wearþ mid him ânum agǽled and all the people were so occupied with him alone, Ors. 3, 9; Bos. 68, 24. II. v. intrans. To hesitate, be careless; cunctari, indiligens esse:—He wihte ne agǽlde ðæs ðe þearf wæs þeódcyninges he

was not careless about anything that was needful for the king, Chr. 1066; Th. 335, 15, col. 1; Edv. 33.

a-gælende; *part. enchanting*; incantans, Ps. Vos. 57, 5. v. a-galan.

a-gælwed *astonished*; consternatus, Bt. 34, 5; Fox 140, 9; MS. Cot. v. a-gelwan.

a-gǽn *gone, past*; præteritus, Cart. Uuerfriþ in app. ad Bædam, S. 772, 1, 4. v. a-gán.

a-gǽþ *happens* :—Hit agǽþ eall swá *it happens so as* [*also*], Deut. 13, 2. v. agán, gán, hit gǽþ.

a-galan; he -gælþ; *p.* -gól, *pl.* -gólon; *pp.* -gaien [a, galan *to sing*] *To sing, chant*; canere, cantare :—He fúsleóþ agól *he sang the death-song*, Exon. 52 b; Th. 183, 1; Gú. 1320. Fyrdleóþ agól wulf on walde *a war-song sung the wolf in the wood*, Elen. Kmbl. 54; El. 27: Beo. Th. 3047; B. 1521.

a-gālan *To loose, dissolve*; remittere, Past. 11, 1; Hat. MS. 14 b, 24. v. agǽlan.

a-gan *began*; cœpit, Mk. Bos. 6, 7; *p.* of a-ginnan.

a-gán; *p.* -eóde; *pp.* -gán [a *from, away*, gán *to go*].   I. *to come to pass, happen*; præterire, transire :—Ǽr his tíd agá [tíde ge MS.] *before his time come to pass*, Exon. 82 a; Th. 310, 3; Seef. 69; [Grn. Gloss.] Ðá sæternes dæg wæs agán *cum transivisset sabbatum*, Mk. Bos. 16, 1. Ǽfen-fela nihta agáne wǽron *totidem noctes transierunt*, Deut. 9, 11: Andr. Kmbl. 293; An. 147: Elen. Kmbl. 2452; El. 1227. Swá hit sóþlíce a-eóde *so it truly happened*, K. de visione Isaiæ.   II. *to come forth; provenire* :—Him upp agá horn on heafde *a horn comes forth on his head*, Ps. Th. 68, 32.   III. *to approach to any one to solicit him*; procedere ad aliquem sollicitandi causa :—Ne meahton heora brego-weardas agán *might not approach their lords*, Cd. 131; Th. 166, 14; Gen. 2747.

ÁGAN, to áganne; *pres. part.* ágende; *pres. indic.* ic, he áh, ðú áhst, *pl.* ágon, ágan, águn; *p.* ic, he áhte, ðú áhtest, *pl.* áhton; *subj.* ic, ðú, he áge, *pl.* ágen; *p.* ic áhte, *pl.* áhten; *pp.* ágen.   I. *to own, possess, have, obtain*; possidere, habere, percipere :—Ðe micel ágan willaþ *who desire* [*will*] *to possess much*, Bt. 14, 2; Fox 44, 13. Nú ic áh mǽste þearfe *now I have the utmost need*, Byrht. Th. 136, 60; By. 175. Gesyle eall ðæt ðú áge *vende quæcumque habes*, Mk. Bos. 10, 21. Ðú ðe áhst dóma geweald *thou that hast power of dignities*, Elen. Kmbl. 1448; El. 726. Áh him lífes geweald *he hath power over life*, Andr. Kmbl. 1036; An. 518: Cd. 103; Th. 137, 8; Gen. 2270. Wuna ðǽm ðe ágon *dwell with those who own thee*, Cd. 104; Th. 138, 18; Gen. 2293: 221; Th. 287, 3; Sat. 361. Ðæt híe heofonríce ágan *that they shall possess heaven's kingdom*, 22; Th. 27, 33; Gen. 427. Hí águn *they possess*, Exon. 33 b; Th. 106, 33; Gú. 50. Ðæt ic' éce líf áge *ut vitam æternam percipiam*, Mk. Bos. 10, 17. He sealde eall ðæt he áhte *vendidit omnia quæ habuit*, Mt. Bos. 13, 46; Ps. Th. 147, 3: Beo. Th. 5210; B. 2608. Hí gewyrhto áhton *they possessed merits*, Cd. 196; Th. 244, 7; Dan. 444. Áhton, Ps. Th. 118, 79. Ðæt hí sige áhten *that they had the victory*, Bd. 3, 2; S. 524, 28. Dóm ágende *possessing power*, Andr. Kmbl. 1139; An. 570: Exon. 68 a; Th. 253, 26; Jul. 186. Ðeáh he feoh-gestreón áhte *although he possessed riches*, Exon. 66 b; Th. 245, 13; Jul. 44.   II. *to make another to own or possess, hence,—to give, deliver, restore*; dare in possessionem, reddere, rependere :—Éðelstówe ðé ic ágan sceal *I shall give thee a dwelling-place*, Cd. 130; Th. 164, 34; Gen. 2724. On hand ágan *to deliver in hand*, Ors. 3, 11? Ágan út *to have or find out*. Lett ágan út, hú fela *permit to find out, how many*, Chr. 1085; Th. 353, 5. [Ágan is the first of the following twelve Anglo-Saxon verbs,—ágan, cunnan, dugan, durran, magan, mótan, munan, nugan, sculan, þurfan, unnan, witan, which are called *præterito-præsentia*, because they take their *new infinitives and their present tenses* from the perfects of strong verbs *with their inflections*. These new infinitives form their *p.* tenses regularly in accordance with the weak conjugations. Thus, the new infinitive ágan has *pres.* ic, he áh = ág, *pl.* ágon; *p.* áhte = ágde, *pl.* áhton = ágdon. The *inf.* ágan and the *pres.* áh, *pl.* ágon [*for* ígon], retaining preterite inflections, are taken from the *p.* of a strong verb, ascertained from áh [*Goth.* áih], which shews the á of the *p. singular* in the sixth class of Grimm's division of strong verbs [Grm. i. p. 837; Koch i. p. 253], and requires by analogy, with other verbs of the same class, the *inf.* ígan, the *p. pl.* igon, and the *pp.* igen. Thus we find the original verb ígan; *p.* áh, *pl.* igon; *pp.* igen. But in ágan the á of the singular *indef.* is kept in the *pl. inf.* and *pp.* The weak *p.* áhte = ágde, *pl.* áhton = ágdon are formed regularly from the weak *infin.* ágan. The same *præterito-præsens* may be generally observed in the following cognate words :—

	inf.	*pres.*	*pl.*	*p.*
Engl.	owe, *possidere*,			ought.
Laym.	agen,	ah,	agen,	ahte.
O.Sax.	ēgan,	[ēh],	ēgun,	ēhta.
O.Frs.	āga, hāga,	āch,	āgon,	āchte.
O.H.Ger.	eigan,		eigumēs.	
Goth.	áigan,	áih,	áigum,	áihta.
O.Nrs.	eiga,	ā,	eigum,	átta.]

DER. ágen, -frigea, -nama, -nyss, -slaga : ágend, -freá, -líce : áhni-an, ágni-an, -end, -endlíc : ge-ágnian, ge-ágnigendlíc : ágenung : ǽht, e; *f.* ǽhteland, -man, -swán : ǽhtige.

ágan, Cd. 216; Th. 274, 1; Sat. 147; *g. d. acc. etc.* of áge, an; *f. property*.

a-gangan; *pp.* -gangen, -gongen *To go or pass by or over, to happen, befal*; præterire, evenire :—Ðá wæs agangen, geára hwyrftum, tú hund and þreó *there were passed, in the circuits of years, two hundred and three*, Elen. Kmbl. 1; El. 1: Chr. 974; Th. 224, 33; Edg. 10. Swá hit agangen wearþ *how it had befallen*, Beo. Th. 2473; B. 1234. Wæs ðæs mǽles mearc agongen *the limit of the time was passed*, Cd. 83; Th. 103, 17; Gen. 1719: Exon. 39 b; Th. 130, 20; Gú. 441.

áge, an; *f. Property*; possessio, proprium :—Ðe he to ágan nyle *which he will not have for his property*, Cd. 216; Th. 274, 1; Sat. 147. Ðe ðé gedafenode ágan to habbanne *quem te conveniebat proprium habere*, Bd. 3, 14; S. 540, 26.

áge, Mk. Bos. 10, 17; *subj. s.* of ágan *to own*.

a-geaf *gave up*, Jn. Bos. 19, 30; *p.* of agifan.

a-geald *rewarded*, Beo. Th. 3335; B. 1665; *p.* of agildan.

a-geán; *prep. Towards*; adversus, Chr. 1052; Th. 314, 23. v. on-geán.

ageán-féran; *p.* ed *To go again, return*; reverti, Chr. 1070; Th. 344, 31. v. ongeán-faran.

ageán-hwyrfan *To turn again, to return*; redire, Mk. Jun. 6, 31. v. agén-hwyrfan.

a-geara, -gearwa *prepared*; paratus.   v. gearwa *in* gearo; *adj.*

a-gearwian *To prepare*; parare.   v. gearwian.

a-geat *understood*, Ps. Spl. 118, 95; *p.* of a-gitan.

a-geát *poured out*, Cd. 47; Th. 60, 20; Gen. 984.   v. a-geótan.

a-géfan; *3rd pl. perf.* of a-gifan, *for* a-géfon, Menol. Fox 160.

a-geldan; *p.* -geald, *pl.* -guldon; *pp.* -golden *To pay, render*; reddere :—Scilling agelde *let him pay a shilling*, L. H. E. 11, 12; Th. i. 32, 5, 9. v. a-gildan.

a-geldan; *pp.* -geald [Grn.] *To punish*; punire :—Wurdon teónlíce tóðas idge [MS. to þas idge] ageald *the greedy teeth were harmfully punished*, Exon. 61 b; Th. 226, 19; Ph. 408.

a-gelwan; *p.* ede; *pp.* ed *To stupefy, astonish*; stupefacere, consternare :—Ðá wearþ ic agelwed *then I was astonished*, Bt. 34, 5; Fox 140, 9.

a-gén; *prep. acc. Against*; adversum, contra :—Se ðe nis agén eów, se is for eów *qui non est adversum vos, pro vobis est*, Mk. Bos. 9, 40. Ðín bróðor hæfþ ǽnig þing agén ðé *frater tuus habet aliquid adversum te*, Mt. Bos. 5, 23. v. on-geán; *prep.*

a-gén; *adv.* AGAIN, *anew, also*; iterum, denuo, et :—Ðe ðé slihþ on ðín gewenge, wend óðer agén *qui te percutit in maxillam, præbe et alteram*, Lk. Bos. 6, 29. Ðá wende he on scype agén *then he went into the ship again*, 8, 37, 40. Wæs forworht agén *was punished anew*, Cd. 214; Th. 269, 21; Sat. 76. v. on-geán; *adv.*

ágen; *adj.* [*originally the pp.* of ágan *to own, possess*].   I. OWN, *proper, peculiar*; proprius :—Sécþ his ágen wuldor *gloriam propriam quærit*, Jn. Bos. 7, 18. Godes ágen bearn *God's own child*, Cd. 213; Th. 265, 20; Sat. 10: 109; Th. 144, 27; Gen. 2396: Bd. 3, 14; S. 539, 19. Hire ágenes húses *of her own house*, Bt. Met. Fox 13, 60; Met. 13, 30. Binnan heora ágenre hýde *within their own skin*, Bt. 14, 2; Fox 44, 23. On eówerne ágenne dóm *in your own decision*, Andr. Kmbl. 677; An. 339. On his ágenum dagum *in diebus ejus*, Ps. Th. 71, 7. His ágnum willan *on his own accord*, Ors. 4, 11; Bos. 98, 6. Ágna gesceafta *thy own creatures*, Bt. Met. Fox 20, 28; Met. 20, 14: Bt. 14, 2; Fox 44, 36. Dínes ágenes þonces *of thine own choice*, Bt. 8; Fox 26, 12.   II. used substantively *The property owned, or one's own property*; proprium :—Agife man ðam ágen-frigean his ágen *let his own be rendered to the proprietor*, L. C. S. 24; Th. i. 390, 7: L. Eth. ii. 10; Wilk. 106, 38. [*Chauc.* owen : *Laym.* agan : *Plat.* egen : *O. Sax.* ēgan : *O. Frs.* ein, ain, eigen, egen : *Ger. M. H. Ger.* eigen : *O. H. Ger.* eikan, eigan : *Goth.* áigin, *n.* and áihts, *f.* oùσía : *O. Nrs.* eigin.] v. ágan.

agén-arn *met*; occurrit, Mk. Bos. 5, 2; *p.* of agén-yrnan.

agén-bewendan; *p.* de; *pp.* ed *To turn again, return*; reverti :—And ðá he hine eft agén-bewende *and then he turned himself again*, Mk. Bos. 14, 40.

agén-cuman; *p.* -com, *pl.* -cómon; *pp.* -cumen *To come again*; redire :—Ðá se Hǽlend agén-com *cum rediisset Iesus*, Lk. Bos. 8, 40.

ágend, es; *m.* [*part.* of ágan *to own*] *An owner, a possessor, the Lord*; possessor, proprietarius, Dominus :—Þreóm hundum scillinga gylde se ágend *with three hundred shillings let the owner pay*, L. H. E. 1; Th. i. 26, 9: 3; Th. i. 28, 5. Ágendes ēst *the owner's favour*, Beo. Th. 6142; B. 3075. Wuldres Ágend *the Lord of glory*, Exon. 25 b; Th. 73, 32; Cri. 1198: 14 b; Th. 29, 32; Cri. 471. Se Ágend *the Lord*; Dominus, Cd. 158; Th. 196, 21; Exod. 295.

ágend-freá, an; *m. The owning lord, possessor*; dominus, possessor :—He heofona is and disse eorþan ágend-freá *he is the owning Lord of heaven and of this earth*, Cd. 98; Th. 129, 10; Gen. 2141: Beo. Th. 3770; B. 1883.

ágend-freán; *acc. f. A mistress*; dominam :—Heó [Agar] ongan

sea, heaven and earth, marked out with his own hands, 1499; An. 751: R. Concord. 2.

amel, es; *m. A vessel for holy water;* amula, vas lustrale, Cot. 2.

a-meldian; *p.* ode; *pp.* od *To betray, make known;* prodere, indicare:—Ic ameldige *prodo,* Ælfc. Gr. 28, 8; Som. 33, 4. He hine ameldode *prodidit eum,* Bd. 3, 14; S. 539, 46. Đá wǽron hí đǽr ameldode *proditi sunt,* 4, 16; S. 584, 26: Jos. 9, 17. v. meldian.

ameos =ἄμμεως *of ammi* or *bishop-wort; gen. of* ammi.

a-merian, -myrian; *p.* ode, ede; *pp.* od, ed *To examine, purify [generally said of melted metal];* examinare, purgare, merum reddere:—Óđer dǽl sceal beón amered on đam fýre, swá hér biþ sylfor *the other part shall be proved in the fire, as silver here is,* Bt. 38, 4; Fox 204, 1. Đæt seolfor đe biþ seofon síđum amered *argentum examinatum septuplum,* Ps. Th. 11, 7: Exon. 63 b; Th. 234, 22; Ph. 544: 65 a; Th. 240, 3; Ph. 633: Elen. Kmbl. 2621; El. 1312: Ps. Spl. 11, 7: 16, 4. Gepin ánne cuculere fulne ameredes huniges *take a spoon-full of purified honey,* Herb. 106; Lchdm. i. 220, 12. Fýre đú us amyrdest swá swá amyred biþ seolfor *igne nos examinasti sicut examinatur argentum,* Ps. Spl. 65, 9. Amerodest *examinasti,* Ps. Lamb. 65, 9.

a-merran *to hinder, trouble, disturb,* Bt. Met. Fox 8, 87; Met. 8, 44. v. a-myrran.

a-metan; *p.* -mæt, *pl.* -mǽton; *pp.* -meten; *v. trans.* [a, metan *to measure*]. I. *to mete, measure, measure out;* metiri, emetiri:—His micelnesse ne mæg nán monn ametan *his greatness no man can measure,* Bt. 42; Fox 258, 13. Mid hondum amet *measure with [thy] hands,* Cd. 228; Th. 308, 30; Sat. 700. Đæt súsl amǽte *that he should measure his torment,* 229; Th. 310, 13; Sat. 725. Đæt đú hús ameten hæbbe *that thou hast measured the house,* 228; Th. 309, 16; Sat. 710: Bd. 4, 23; S. 596, 26. II. *to measure out to any one, to allot, assign, bestow;* aliquid alicui emetiri, ex mensura dare, largiri:—Ametan wolde wrece be gewyrhtum wóh-fremmendum *would mete out punishment according to their deeds to the doers of wickedness,* Bt. Met. Fox 9, 70; Met. 9, 35. Ǽr me gife unscynde mægen-cyning amæt *before the powerful king measured out to me a blameless grace,* Elen. Kmbl. 2493; El. 1248. III. *to measure out, plan, form, make;* emetiri, formare, confingere:—Đú amǽte mundum đínum ealne ymbhwyrft and upráđor *thou measuredst with thine hands the whole circumference and the firmament above,* Elen. Kmbl. 1456; El. 730.

a-metan; *p.* -mette; *pp.* -mett; *v. trans.* [a, metan *to paint*] *To paint, depict, adorn;* pingere, depingere, ornare:—Swelce he hit amete and atiefre on his heortan *quasi in corde depingitur,* Past. 21, 3; Hat. MS. 30 b, 26. Firmamentum [fæstnes] mid manegum steorrum amett *the firmament adorned with many stars,* Bd. de nat. rm; Wrt. popl. scienc. 10, 12; Lchdm. iii. 254, 9.

amet-hwíl, e; *f. Leisure;* otium, Ælfc. Gr. 8; Som. 8, 1, MS. D. v. æmet-hwíl.

a-middan; *adv.* [a=on *in, into;* mid *middle*] *In the middle, into the midst;* in medium:—Arís, and stand hér amiddan *surge, et sta in medium,* Lk. Bos. 6, 8.

ammi, ami; *g.* ameos; *n. Ammi, an African umbelliferous plant, millet, bishopwort;* ammi Copticum [ἄμμι; *g.* ἄμμεως]:—Đeós wyrt đe man ami, and óđrum naman milium, nemneþ *this wort which is named ammi, and by another name millet,* Herb. 164, 1; Lchdm. i. 292, 20. Óđer swilc ameos *as much more of ammi,* L. M. 2, 14; Lchdm. ii. 192, 7.

a-molsnian; *p.* ode, ade; *pp.* od, ad *To corrupt, putrefy;* putrefacere, Som. v. molsnian.

amore, an; *f. A kind of bird;* avis quædam, scorellus, Cot. 160.

Amorreas, *pl.* *g.* a *The Amorites;* Amorrhæi:—Seon cyning Amorrea Sehon regem Amorrhæorum, Ps. Th. 135, 20.

ampella, ampolla, ampulla, an; *m. A vial, bottle, flask, flagon;* ampulla, lecythus, lenticula:—Ampella *vel* ele-fæt *an oil-flask,* lecythus = ληκυθος [MS. legithum], Cot. 119. Ampella *vel* crog *lenticula,* 124. [Ger. ampel, *f.:* O. H. Ger. ampulla, ampla, *f.:* O. Nrs. ampli, hömpull, *m.*]

ampre, an; *f. Sorrel* or *dock;* rumex, Lchdm. iii. 12, 25. v. ompre.

a-munan; ic, he -man, đú -manst, *pl.* -munon; *p.* -munde, *pl.* -mundon; *pp.* -munen *To think of, mind, consider, be mindful of, have a care for;* cogitare, reputare, memor esse, providere:—Hwæt is se mann, đe đú swá miclum amanst? *quid est homo, quod memor es ejus?* Ps. Th. 8, 5. Cwǽdon hí, đæt hie đæs ne amundon đe má đe eówre geferan *they said, that they no more minded it than did your companions,* Chr. 755; Th. 84, 36, col. 3. v. munan.

a-mundian; *p.* ode; *pp.* od *To protect, defend;* tueri, tutari, Æthelfl. Test; Th. Diplm. A.D. 972; 522, 28. v. mundian.

a-mundon *thought of, minded,* Chr. 755; Th. 84, 36, col. 3; *p. of* a-munan.

a-myrdrian; *p.* ede; *pp.* ed *To murder, kill;* occidere, interficere, trucidare:—Đæt man sý amyrdred *that a man be murdered,* L. C. S. 57; Th. i. 406, 25. v. myrđrian.

a-myrgan; *p.* de; *pp.* ed; *v. trans.* [a, myrgan *to be merry*] *To make merry, to gladden, cheer;* exhilarare, lætificare:—Béc syndon breme: hí

amyrgaþ módsefan manna gehwylces of þreánýdlan đisses lífes *books are famous: they cheer the mind of every one from the necessary affliction of this life,* Salm. Kmbl. 479; Sal. 240.

a-myrian; *p.* ede, ode; *pp.* ed, od *To examine;* examinare, Ps. Spl. 65, 9. v. a-merian.

a-myrran, -merran; *p.* de; *pp.* ed [a, myrran *impedire*]. I. *to hinder, impede, obstruct, check, disturb;* impedire, turbare, obstruere:—Đæs wéla amerþ and lǽt đa men *this wealth obstructs and hinders those men,* Bt. 32, 1; Fox 114, 3. He ofslóh fætta heora, and gecorene Israhéla he amyrde *occidit pingues eorum, et electos Israhel impedivit,* Ps. Spl. C. 77, 35. Me habbaþ hringa gespong síđes amyrred *the binding of these rings hath impeded me in my course,* Cd. 19; Th. 24, 18; Gen. 378. He đæs eorles earm amyrde *he checked the earl's arm,* Byrht. Th. 136, 43; By. 165. II. *to dissipate, spend, distract, defile, mar, corrupt, spoil, destroy;* dissipare, perdere, consummare, corrumpere, devorare, distrahere:—Đá he hæfde ealle amyrrede *postquam omnia consummasset,* Lk. Bos. 15, 14, 30. Ne amyrþ he hys méde *non perdet mercedem suam,* Mt. Bos. 10, 42. Đeós gitsung hafaþ gumena gehwelces mód amerred *this covetousness has corrupted the mind of every man,* Bt. Met. Fox 8, 87; Met. 8, 44: 22, 8; Met. 22, 4. Eorþe wæs amyrred *corrupta est terra,* Ex. 8, 24: Ors. 3, 10; Bos. 69, 39. Ic amyrre *distraho,* Ælfc. Gr. 28, 5; Som. 32, 10.

an; *prep. In, among, into, to; in,* ad; followed by *dat.* or *acc:—*An ferþe *in the spirit,* Ps. C. 50, 110; Ps. Grn. ii. 279, 110: 50, 157; Ps. Grn. ii. 280, 157. Hió biþ eallunga an hire selfre *she is altogether in herself,* Bt. Met. Fox 20, 440; Met. 20, 220. An folcum *among the people,* Ps. C. 50, 5; Ps. Grn. ii. 276, 5. Dó glêda an glêdfæt *put embers into a chafing dish,* L. M. 3, 62; Lchdm. ii. 346, 3. Đæt ic an forþgesceaft féran môte *that I may come to a future state,* Ps. C. 50, 52; Ps. Grn. ii. 278, 52. v. on.

an *I give,* Alfd. Will 14, 4; *he gives,* Cd. 141; Th. 176, 22; Gen. 2915. v. unnan.

an- is used in composition. I. for A. Sax. and *against, in return;* contra, re-; as an-sacan *to strive against, to contradict;* repugnare, contradicere: an-swarian *to answer;* respondere. II. for un-, denoting *privation;* as an-bindan *to unbind;* absolvere. III. for on, in *in, to;* as an-wadan *to invade;* invadere: an-fón *to take to one's self;* accipere. Sometimes an- appears scarcely to alter the meaning of the word before which it is placed.

-an, -anne, v. -anne, in alphabetical order, and TO; *prep.* IV. The termination of most Anglo-Saxon verbs is in -an; but -án is found, which seems to be contracted from aa, agan, ahan, as,—gán *to go,* from gaan: smeán *to consider,* from smeagan: sleán *to slay,* from sleahan, *etc.* The termination of verbs in -ôn, appears to be a contraction from ahan, ohan, as,—fón *to take,* from fahan: gefeón *to rejoice,* from gefeohan: teón *to draw,* from teohan, *etc.* Mrch. § 247*.

ÁN, I. *m. f. n.* ONE; unus, una, unum: *gen. m. n.* ánes; *f.* ánre *of one;* unius: *dat. m. n.* ánum; *f.* ánre *to one;* uni: *acc. m.* áne, ǽnne; *f.* áne, *n.* án *one;* unum, unam, unum: *instr. m. n.* áne; *f.* ánre *with one;* uno, unâ, uno: *pl. nom. acc. m. f. n.* áne *each, every one, all;* unusquisque, una-quæque, unum-quodque; singuli, æ, a: *gen. m. f. n.* ánra *of every one, all;* singulorum, arum, orum: *dat. m. f. n.* ánum *to every one, all;* singulis: *instr.* ánum *with all: def.* se ána; seó, đæt áne *the one: gen.* đæs, đære, đæs ánan *of the one: dat.* đam, đære, đam ánan *to the one: acc.* đone, đa ánan, đæt án *the one: instr.* m. n. đý ánan; *f.* đære ánan *with the one; adj.:—*Án *of* đám *unus ex illis,* Mt. Bos. 10, 29. Án wæs on Ispania *one was in Spain,* Ors. 4, 9; Bos. 92, 19. God geworhte áne mannan, Adam, of láme *God created one man, Adam, of earth,* Homl. Th. i. 12, 28. He is án God *Deus unus est,* Mk. Bos. 12, 29. Đis is án đara gerǽdnessa *this is one of the ordinances,* L. Eth. ix. 1; Th. i. 340, 2. II. *alone, only, sole, another;* solus, alius: with these meanings it is used *definitely,* and *generally written* ána, *m. and sometimes* aina, ánna, ánga, *q. v:—*Án God ys gód *God alone is good;* solus [unus] est bonus, Deus, Mt. Bos. 19, 17. Đæt ge forlǽton me ǽnne, and ic ne eom ána *ut me solum relinquatis, et non sum solus,* Jn. Bos. 16, 32. God ána wát hú his gecynde biþ, wífhádes đe weres *God alone knows how its sex is, [the sex of] female or male,* Exon. 61 a; Th. 223, 6; Ph. 355. Đæt ge aina [ge á má, Grn.] gebróđra hæfdon *quod alium haberetis vos fratrem,* Gen. 43, 6. 2. *sole, alone of its kind, singular, unique, without an equal;* unicus, eximius:—Án sunu, mǽre meotudes bearn *the only Son, illustrious child of the Creator,* Exon. 128 a; Th. 492, 7; Rä. 81, 10: Hy. 8, 14; Hy. Grn. ii. 290, 14: Bt. Met. Fox 21, 19, 25; Met. 21, 10, 13, 16. Đæt wæs án foran eald-gestreóna *that was before a singular old treasure,* Beo. Th. 2920; B. 1458. Đæt wæs án cyning, æghwæs orleáhtre *that was a singular king, faultless in everything,* 3775; B. 1885. III. *a certain one, one;* quidam; v. sum:—Án man hæfde twegen suna *homo quidam habebat duos filios,* Mt. Bos. 21, 28. *In this sense it is used as* sum *in the parallel passage.—*Sum man hæfde twegen suna *homo quidam habuit duos filios,* Lk. Bos. 15, 11. 2. *sometimes, though rarely,* án *may be used as the English article* a, an. It does not, however, appear to be generally used as an indefinite article,

but more like the *Moes.* ain, or the *Lat.* unus.—When a noun was used indefinitely by the Saxons, it was without an article prefixed; as,—Þeódríc wæs Cristen *Theoderic was a Christian,* Bt. 1; Fox 2, 7. **3.** *in the following examples it seems to be used for the indefinite article a, an* :— Ân engel bodade ðám hyrdum ðæs heofonlícan cyninges acennednysse *an angel announced to the shepherds the birth of the heavenly king,* Homl. Th. i. 38, 3. Ðár beó ân mann stande *there shall be a man standing,* Chr. 1031; Ing. 206, 5; Erl. 162, 7. Ðá stód ðar ân Iudeisc wer, ðæs nama wæs Nichodēmus *then stood there a Jewish man, whose name was Nicodemus,* Nicod. 11; Thw. 5, 38. On ânum reste-dæge *on a rest-day* or *sabbath,* Lk. Bos. 24, 1; Jn. Bos. 20, 1. Sceollon ænne tíman gebídan *must wait* [*abide*] *a time,* L. C. E. 18; Th. i. 370, 18: Ors. 3, 7; Bos. 61, 36. Wirc ðé nú ænne arc *now make for thee an ark,* Gen. 6, 14. Âne lytle hwíle *a little while,* Bt. 7, 1; Fox 16, 4. Cynríc ofslógon ænne Bryttiscne cyning *Cynric slew a British king,* Chr. 508; Ing. 21, 6. **IV.** *each, every one, all;* unus-quisque, una-quæque, unum-quodque; singuli, -æ, -a. It is in this sense that it admits of a plural form*: nom. acc. pl. m. f. n.* âne; *gen. m. f. n.* ânra; *dat. m. f. n.* ânum :— Ânra gehwá, ânra gehwylc *every one,* or, literally, *every one of all.* Swelte ânra gehwilc for his ágenum gilte *unusquisque pro peccato suo morietur,* Deut. 24, 16. Ânes hwæt, Bt. 18, 3; Fox 64, 30, denotes *anything,* literally ' *anything of all,*' and is used adverbially for *at all, in any degree.* ¶ *One, other,* &c. ân æfter ânum *one after another,* Jn. Bos. 8, 9: Salm. Kmbl. 771; Sal. 385. To ânum to ânum *from one to the other, only;* duntaxat. Ðæt ân, *or for ân this one thing, for one thing, only;* tantummodo, Mk. Bos. 5, 36. Hý forbærndon ânne finger, and ânne *they burnt off one finger, and then another,* Ors. 2, 3; Bos. 42, 15. Ete ænne and ænne *let him eat one and another, one after another,* Herb. 1, 20; Lchdm. i. 76, 24. On ân *in one, continually, ever,* Gen. 7, 12: Cd. 140; Th. 175, 9; Gen. 2892. DER. nân [=ne + ân *n + one*] *none, no one;* nullus [ne-ullus].

ân; *adv. Only;* tantum :—Cweþ ðín ân word *speak thy word only;* tantum dic verbo, Mt. Bos. 8, 8. v. ÂN II.

âna; *m. One, sole, single, solitary;* unus, unicus, solus, solitarius: *nom. f. n.* âne *one, etc;* una, unum : *gen. m. f. n.* ânan *of one;* unius = unici, unicæ, unici : *dat.* ânan *to one;* uni = unico, unicæ, unico : *acc. m. f.* ânan *one;* unum, unam; *def. numeral adj.* ðæt (treów, *n.*] se âna is ealra beáma beorhtast geblówen *that is the one of all the trees most brightly flourishing,* Exon. 58 b; Th. 209, 27; Ph. 177. God âna on ēcnysse ríxaþ *one God ruleth to eternity,* Homl. Th. i. 28, 23. v. ÂN II.

ân-âd, ân-æd, es; *n.* [ân *unus,* âd = eád, eáþ *desertus, vastus,* Ett: *Goth.* áuþs ἔρημος *desertus* : v. DER. eáðe; *adj.*] *Solitude, a desert;* solitudo, desertum :—On ðam ânáde *in the desert,* Exon. 37 a; Th. 122, 12; Gú. 304: 37 b; Th. 123, 24; Gú. 327. On ânæðe *in a desert,* 122 b; Th. 471, 22; Rä. 61, 5. [*O. Sax.* ênôdi, einôdi, *f. n. solitudo :* Ger. einöde, *f. desertum, solitudo: M. H. Ger.* einoede, *f;* einoete, einôte, *n: O. H. Ger.* einôdi, *f;* einoti, *n. solitudo, desertum.*]

an-ælan; *p.* -ælde; *pp.* -æled, -æld [an, ælan *to light*] *To kindle, inflame, enlighten;* accendere, incendere, inflammare, illuminare :—Mid andan ðære rihtwísnesse anæld *kindled with a zeal of righteousness,* Chr. 694; Th. 66, note 2 : R. Concord. 5. v. on-ælan, in-ælan.

an-æðelian; *p.* ode, ade; *pp.* od, ad; *v. trans.* [an = un *not,* æðelian *to ennoble*] *To dishonour, degrade;* ignobilem reddere :—And ðonan wyrþ anæðelad óþ-ðæt he wyrþ unæðele *and thence becomes degraded till he is unnoble,* Bt. 30, 2; Fox 110, 22: Bt. Met. Fox 17, 53; Met. 17, 27. v. un-æðelian.

ânan, ânum *by this alone, only; dat. of* ân *one.*

anan-beám, es; *m. The spindle-tree, prick-wood, prick-timber;* euonymus Eúropæus, L. M. 1, 32; Lchdm. ii. 78, 13.

ana-wyrm, es; *m.* [ana = an, in *in,* as in *Goth.* anahneiwan *inclinare;* wyrm *a worm*] *An intestinal worm;* lumbricus :—Gif anawyrm on men weaxe *if an intestinal worm grow in a man,* L. M. 1, 46; Lchdm. ii. 114, 13, 18, 23.

an-bærnys, on-bærnys, -nyss, e; *f.* [v. on-bærning, in-bærnis] *Incense, frankincense;* incensum, thus :—Sý gereht gebéd mín swá swá anbærnys *dirigatur oratio mea sicut incensum,* Ps. Spl. 140, 2.

an-be-lédan; *p.* -lédde; *pp.* -léded, -léd *To lead* or *bring in;* inducere. DER. belédan, lédan.

an-bestingan; *p.* -bestang, *pl.* -bestungon; *pp.* -bestungen *To thrust in;* immittere, intromittere :—Ða anbestungne [Cot. MS. anbestungnan] saglas *intromissi* [*scil. circulis*] *vectes,* Past. 22, 1; Hat. MS. 33 a, 22.

an-bíd, es; *n. Awaiting, expectation;* expectatio, mora :—Ðær wæron ærendracan on anbíde *there ambassadors were in waiting,* Ors. 3, 9; Bos. 68, 44. Næs ic on náuht [ne, áht, áuht] ídlum anbíde, ðeáh hit me lang anbíd þúhte, ðá ðá ic anbídode Godes fultumes *expectans, expectavi Dominum,* Ps. Th. 39, 1. Earmra anbíd *the expectation of the miserable,* Cd. 169; Th. 212, 2; Exod. 533: Elen. Kmbl. 1767; El. 885. v. on-bíd.

an-bídian; *p.* ode, ude; *pp.* od *To abide, wait, wait for, expect;* morari, commorari, expectare :—Wolde ðær on ælþeódignisse anbídian *ut peregrinaretur ibi,* Gen. 12, 10. Me anbídiaþ rihtwíse óþ-ðæt ðú afyldest

me *me expectant justi donec retribuas mihi,* Ps. Spl. 141, 10. Ic anbídude hine *expectabam eum,* 54, 8.

an-bídung, es; *m. An abiding, tarrying, awaiting, expectation;* commoratio, expectatio :—Wícode þreó niht on anbídunge *moratus est tres dies,* Jos. 3, 1. Hwylc is anbídung mín *quæ est expectatio mea?* Ps. Spl. 38, 11.

an-bindan; ic -binde, ðú bindst, he -bint, *pl.* -bindaþ; *p.* -band, ðú -bunde, *pl.* -bundon; *pp.* -bunden; *v. a.* [an = un un-, bindan *to bind*] *To* UNBIND, *untie;* solvere, absolvere, religare :—Seó wiðerwearde wyrd anbint and gefreóþ ælc ðara ðe hió togeþiþ *adverse fortune unbinds and frees every one of those whom she adheres to,* Bt. 20; Fox 72, 2. v. on-bindan, in-bindan.

an-biscopod; *part. Unbishoped, unconfirmed;* non confirmatus ab episcopo, L. Edg. C. 15; Wilk. 83, 40. v. un-biscopod.

ân-boren; *part. Only-born, only-begotten;* unigenitus :—Ðæt in Bethlême cyning ânboren cenned wære *that in Bethlehem the only-begotten king was born,* Elen. Kmbl. 783; El. 392. Exon. 16 b; Th. 39, 6.

an-bróce, an; *f. Material, wood, timber;* materies, tignum :—Æðele anbróce *noble material,* Elen. Grm. 1029, note, p. 161.

an-bryrdan; *p.* -bryrde; *pp.* -bryrded, -bryrd; *v. a. To prick, goad, vex;* compungere, stimulare :—He hêhtende wæs menn wanspendinne, and anbryrdne heortan *persecutus est hominem inopem, et compunctum corde,* Ps. Spl. 108, 15. v. on-bryrdan, in-bryrdan.

an-bryrdnes, -ness, e; *f. Compunction, remorse;* compunctio, C. R. Ben. 70. v. on-bryrdnes.

ân-búende; *part. Dwelling alone;* anachoreticam vitam agens :—Eáhteþ ânbúendra *persecutes those dwelling alone,* Exon. 33 b; Th. 107, 15; Gú. 59.

an-búgan; *p.* -beáh, -beág, *pl.* -bugon; *pp.* -bogen; *v. intrans. To bend* or *bow one's self in, submit to any one;* se inflectere, se submittere alicui :—To ðon ðæt hí him anbugon *that they might submit to him,* Ors. 1, 12; Bos. 36, 25. v. on-búgan.

anbyht-scealc, ombiht-scealc, onbyht-scealc, es; *m.* [ambeht *an office,* scealc *a servant*] *An official servant, a servant;* minister, servus :—Hraðe fremedon anbyhtscealcas swá him heora ealdor bebeád *the official servants quickly did as their lord bade them,* Judth. 10; Thw. 21, 27; Jud. 38. v. ombiht-scealc, onbyht-scealc.

an-byrdnys, nyss, e; *f.* [an *contra,* byrdnys *status*] *Resistance;* repugnantia :—Gif ænig man anbyrdnysse* beginþ *if any man begin resistance,* L. Edg. S. 14; Th. i. 276, 31. v. geán-byrdan.

an-byrignys, -nyss, e; *f. A tasting, taste;* gustus, Ælfc. Gl. 70; Som. 70, 51; Wrt. Voc. 42, 59. v. byrignes.

ân-cænned; *def.* se ân-cænneda; *part. Only-begotten;* unigenitus :— To árwurþianne [MS. tarwurþianne, v. weorþianne = wurþianne in weorþian I] ðínne, ðone sóðan and ðone âncænnedan, Sunu *to honour thy, the true and only-begotten, Son,* Te Dm. Thomson 35, 12. v. ân-cenned.

ân-cenda = ân-cenneda *only-begotten,* Exon. 99 a; Th. 370, 2; Seel. 51. v. ân-cenned.

ân-cenned; *def.* se ân-cenneda; *part.* [ân *unus,* cennan *gignere*] *Only-begotten;* uni-genitus :—Âncenned Sunu *only-begotten Son,* Exon. 14 b; Th. 29, 18; Cri. 464. Se âncenneda Sunu *the only-begotten Son,* Jn. Bos. 1, 18: 3, 16.

ancer; *g.* ancres; *m. An anchor;* ancora, Wrt. Voc. 73, 84. v. ancor.

âncer, es; *m. An anchoret, hermit;* anachoreta :—Mid ðý he leornode be ðám âncerum *when he learnt concerning the anchorets,* Guthl. 2; Gdwin. 18, 22. v. âncor.

âncer-líc; *adj. Anchoretic, like a hermit;* anachoreticus, Som. v. âncor-líc.

ancer-líf, es; *n. An anchoret's* or *hermit's life;* anachoretica vita, Bd. 4, 28; S. 605, 6. v. âncor-líf.

ancer-man, -mann, es; *m. An anchor-man, the man in charge of the anchor;* proreta, Ælfc. Gl. 104; Som. 77, 126. v. ancor-man.

âncer-setl, -settl, es; *m. An anchoret's cell, hermitage;* anachoretæ sedes :—Twegen hálige menn, on âncersettle wuniende, wæron forbearnde *two holy men, dwelling in a hermitage, were burned,* Chr. 1087; Th. 354, 23: Guthl. 4; Gdwin. 26, 10.

ancer-streng, es; *m. An anchor-string, a cable;* ancorarius funis, Solil. 4.

ancleow, es; *m. The* ANCLE; talus :—Ancleow *talus,* Ælfc. Gl. 75; Wrt. Voc. 44, 74. Lytel ancleow *taxillus,* 75; Wrt. Voc. 45, 1. [*Dut.* anklauuw, enklauuw, enkel: *Ger. M. H. Ger.* enkel, *m: O. H. Ger.* anchal, *m;* anchala, *f: Dan. Swed.* ankel: *O. Nrs.* ökul, ökli, *m.*]

an-cnáwan *To recognise;* agnoscere, Ælfc. Gr. 28, 1; Som. 30, 31. v. on-cnáwan.

ancor, ancer, oncer; *g.* ancres; *m.* [ancôra = ἄγκυρα: uncus = ὄγκος *a hook,* v. DER.] *An anchor;* ancora :—Ðín ancor is git on eorþan fæst *thine anchor is yet fast in the earth,* Bt. 10; Fox 30, 5. On ancre fæst *fast at anchor,* Beo. Th. 611; B. 303. On ancre rád *rode at anchor,* 3771; B. 1883. Ða ancras *the anchors,* Bt. 10; Fox 30, 10, 13: Bd. 3, 15; S. 541, 40. Ýþmearas ancrum fæste *ships* [*wave-horses*] *fast with anchors,* Exon. 20 b; Th. 54, 6; Cri. 864. [*Chauc.* ancre: Plat. Dut. Ger. M. H. Ger.* anker, *m: O. H. Ger.* anchar, *m: Dan.* anker, *m: Swed.*

ankare, m: O. Nrs. akkéri, m: Lat. ancora: Grk. ἄγκυρα: Lith. inko-
ras; from the Sansk. anka a hook.]

âncor, âncer; g. âncres; m. An anchoret, hermit; anachoreta:—Sléfleás âncra scrúd hermits' sleeveless garment, Ælfc. Gl. 63; Som. 68, 111.
[O. Sax. ênkoro, m: O. H. Ger. einchoranar, m: Grk. ἀναχωρητής.]

ancor-bend, es; m. An anchor-band or cord or rope. v. oncer-bend.

âncor-líc; adj. Anchoretic, like a hermit; anachoreticus. DER. v. âncor a hermit, líc like.

âncor-líf, âncer-líf, es; n. An anchoret's or hermit's life, a solitary life;
anachoretica vita, Bd. 4, 28; S. 605, 11.

ancor-man, ancer-man, -mann, es; m. An anchor-man, the man in
charge of the anchor; ancorarius, proreta, Ælfc. Gl. 83; Som. 73, 66:
104; Som. 77, 126.

ancor-ráp, es; m. An anchor-rope, a cable. v. oncyr-ráp.

ancor-setl, es; n. An anchor-seat, the fore-castle of a ship, the prow;
prora, Ælfc. Gl. 104; Som. 78, 11.

âncor-stôw, e; f. An anchoret's or hermit's cell, a solitary place;
anachoretæ mansio, solus locus, Bd. 5, 12; S. 627, 26.

ancra, an; m. An anchor, ballast; ancora vel saburra, Ælfc. Gl. 83;
Wrt. Voc. 48, 21. v. ancor.

âncra, an; m. An anchoret, hermit; anachoreta, solitarius, Ælfc. Gl.
69; Som. 70, 20.

ancre, an; f. [antre?] Radish; raphănus = ῥάφανος:—Ancre, ðæt is
rædic raphanus, Mone A. 493. v. ontre.

anc-sum, anc-sum-líc troublesome. v. ang-sum, ang-sum-líc.

an-cuman; p. -com, pl. -cômon; pp. -cumen, -cymen To come, arrive;
advenire:—Ðá he west ancom [westan com, MS.] when he came to the
west, Cd. 90; Th. 113, 9; Gen. 1884. DER. cuman.

ân-cummum; adv. [ân one, cummum the dat. of cuma a comer] One
by one, singly; singulatim, Jn. Lind. War. 21, 25.

ân-cyn; g. m. n. -cynnes; f. -cynre; adj. [ân one, only; cyn proprius]
Only; unicus:—Ðé seó [MS. se] hálige andett geladung,—ðínne sóðan
and âncynne sunu te sancta confitetur ecclesia,—tuum verum et unicum [=
proprium] filium, Te Dm. Lye. v. ân-líc.

and; prep. dat. acc.   I. with the dative; cum dativo With;
cum:—Emb eahta niht and feówerum after eight nights with four [twelve
nights], Menol. Fox 419; Men. 211. Ymb twentig and fíf nihtum after
twenty with five nights, i. e. after twenty-five nights, 373; Men. 188.   II.
with the accusative; cum accusativo Against, before, on, into; contra,
apud, in; κατά:—Hæfdon dreám and heora ordfruman had joy before
their creator [apud creatorem], Cd. 1; Th., 2, 2; Gen. 13. Ðæt is cræft
eágorstreámes, wætres and eorþan, and on wolcnum eác that is the
power of the sea, of water on earth, and also in the clouds, Bt. Met. Fox
20, 245; Met. 20, 123. Ýþ up færeþ, ôfstum wyrceþ wæter and wealfæsten the wave goes up [and] rapidly makes [worketh] the water into a
wall [wall-fastness], Cd. 157; Th. 195, 27; Exod. 283. [O. Sax. ant
usque ad: O. Frs. anda, and in, on: Goth. and against: O. H. Ger. ant:
O. Nrs. and contra: Lat. ante: Grk. ἀντί, ἄντα: Lith. ant on, upon:
Sansk. anti opposite, against, before. Thus and seems to be connected
with Goth. andi end, A. Sax. ende frontier, boundary, and Sansk. anta
end, boundary, limit, border, which is probably derived from the Sansk.
root ant, and to bind; hence near or with, and that which is with or near,
may be against.]

and; conj. AND; et, atque, ac:—Gesceóp God heofenan and eorþan
creavit Deus cœlum et terram, Gen. 1, 1. Cum and geseóh veni et vide,
Jn. Bos. 1, 46. And swá forþ and so forth; et cætera, Ælfc. Gr. 25;
Som. 26, 59.

and- [Goth. anda-: Icel. and-, önd-: Grk. ἀντι-] in composition denotes
opposition,—Against, without; contra:—And-bita, and-beorma without
barm, what was unleavened; azymos = ἄ-ζυμος, Cot. 17. And-saca an
adversary, apostate, Cd. 23; Th. 28, 27; Gen. 442. And-swaru an
answer, Beo. Th. 5713; B. 2860.

anda, onda, an; m. emotion of mind,—Malice, envy, hatred, anger, zeal,
annoyance, vexation; animi emotio,—rancor, invidia, indignatio, ira, zelus,
molestia:—Anda rancor, Ælfc. Gl. 89; Som. 74, 93. Næfst ðú nânne
andan to nânum þinge thou hast not any envy to anything, Bt. 33, 4;
Fox 128, 18. Hyne for andan sealdon per invidiam tradidissent eum,
Mt. Bos. 27, 18. Nyste nænne andan know not any hatred, Bt. 35, 6;
Fox 168, 10. For hwilcum líþrum andan ex prava aliqua invidia,
L. M. I. P. 12; Th. ii. 268, 11: Bt. Met. Fox 20, 72; Met. 20, 36.
Habbaþ andan betweóh him have enmity between them, 28, 104; Met. 28,
52. On andan in hatred, Beo. Th. 1421; B. 708: Cd. 191; Th. 237,
28; Dan. 344. Manigum on andan for vexation to many, Elen. Grm.
969. For ðæm andan his rihtwísnes [-nesse MS. Cot.] per zelum justitiæ,
Past. 17, 1; Hat. MS. 21 b, 28. [O. Sax. ando, m. indignatio, ira, zelus:
O. H. Ger. anado, ando, m. zelus: O. Nrs. andi, m. halitus oris, spiritus,
animus.] DER. andian: andig.

ân-dæge; adj. [ân one, dæg a day] For one day, lasting a day; diurnus, unius diei:—Næs ðæt ândæge níþ that was no one-day evil, Exon.
92 a; Th. 345, 25; Gn. Ex. 195. Sǽ-weall astâh, uplang gestód ân-
dægne fyrst the sea-wall arose, [and] stood erect one day's space, Cd. 158;

Th. 197, 9; Exod. 304. Ðe hire ândæges eágum starede who daily
gazed on her with his eyes, Beo. Th. 3874; B. 1935.

andættan to confess, Th. Anlct. v. andettan.

ân-daga, an; m. [dæg a day = daga, q. v.] A fixed day, a time appointed, a day or term appointed for hearing a cause; dies dictus, dies
constitutus:—Gesette me ânne ândagan constitue mihi tempus, Ex. 8, 9:
9, 5: Gen. 18, 14. Ðæt gehwilc spræc hæbbe ândagan hwænne heó
gelǽst sý that every suit have a term when it shall be brought forward,
L. Ed. prôœm; Th. i. 158, 6: 11; Th. i. 164, 21: L. Edg. H. 7; Th. i.
260, 13: L. C. S. 19; Th. i. 386, 14. [O. Sax. ên-dago, m. dies statutus,
fatalis,—terminus vitæ: O. Nrs. ein-dagi dies oculatus, tempus præscrip-
tum, a verbo eindaga certum tempus definire.]

ân-dagian; p. ode; pp. od; v. a. To appoint a day or term, to cite;
diem dicere, L. Edg. H. 7; Th. i. 260, 12. DER. ge-ân-dagian. v. ân-
daga.

and-beorma, an; m. That which is without barm, unleavened, unleavened bread, the feast of unleavened bread; azyma:—Andbita vel
[and-]beorma azyma, Cot. 17. v. beorma, and-bita.

and-bídian; p. ode; pp. od To expect; expectare:—Ðe andbídiaþ ðé
qui expectant te, Ps. Spl. 68, 8. Andbídiaþ wíldeór on þurste heora expectabunt onagri in siti sua, 103, 12. v. an-bídian.

and-bídung; e; f. Expectation; expectatio:—Nâ ðú gescend me
fram andbídunge míne non confundas me ab expectatione mea, Ps. Spl.
118, 116. v. an-bídung.

and-bita, an; m. That which is unleavened, unleavened bread, the feast
of unleavened bread; azyma:—Andbita vel and-beorma azyma, Cot. 17.
[Goth. unbeistei, f. ἄζυμον.]

and-cwis, -cwiss, e; f. An answer; responsum:—Andcwis ageaf gave
answer, Exon. 47 b; Th. 163, 26; Gú. 999.

anddetan To confess; confiteri:—Hyra synna anddetende confitentes
peccata sua, Mk. Bos. 1, 5. v. andetan.

and-efn; adj. [and against, eáw = ǽw lawful, legitimate] Arrogant,
presumptuous, proud; arrogans, Scint. 46.

Andefera, an; m. ANDOVER, a market town in the north west of Hamp-
shire built on the east bank of the river Ande or Anton; oppidum in agro
Hamtunensi:—Hí ðá lǽddon Ânláf to Andeferan they then led Anlaf to
Andover, Chr. 994; Th. 242, 27, col. 1; Th. 243, 26, col. 1, 12, col. 2.
To Andefron, Th. 242, 26, col. 2. [Dun. Andeafara: Kni. Andever.]
About the year 1164 Simeon Durham writes it Andeafara = Ande-eá-fara
a farer over the river Ande, on the bank of which Andover is built, v. fara
a traveller, faran to go, travel, sail. From the A. Sax. of the MS. Cott.
Tiber. B. IV. to Andefron, of Knighton Andever, about 1395, and from
the present name Andover = Ande + ôfer, another derivation may be supposed,—Ande the river Ande, and ôfer; g. ôfres; d. ôfre; m. a margin,
bank, that is a town on the bank of the river Ande.

and-efn, es; n. [and, efen even] An equality, a proportion, measure,
an amount; proportio:—Be hire andefne by its proportion, Bt. 32, 2;
Fox 116, 14.

andet, andett, e; f. Confession, praise, honour, glory; confessio. v.
comp. wlite-andet, andetnes.

andetan To confess, acknowledge, give thanks or praise; confiteri:—Ic
ðé on folcum andete confitebor tibi in populis, Ps. Th. 56, 11: 98, 3:
104, 1: 135, 27. v. andettan.

andetla, an; m. A confession; confessio, L. Alf. pol. 22; Th. i. 76, 4.

andetnes, -ness; andetnys, -nyss, e; f. A confession, acknowledgment,
profession, giving of thanks or praise, praise, honour, glory; confessio:—
In andetnesse in confessione, Bd. 4, 25; S. 599, 42. Seó andetnes ðe
we Gode andettaþ the confession that we confess to God, L. E. I. 30; Th.
ii. 426, 33. Ðe his naman neóde sealdon him andetnes ǽghwǽr habban
ad confitendum nomini tuo, Ps. Th. 121, 4. Is upp-ahafen his andetnes,
heáh ofer myclum heofone and eorþan confessio ejus super cœlum et
terram, 148, 13: 95, 6. Andetnysse and wlite ðú scrýddest confessionem
et decorem induisti, Ps. Spl. 103, 2.

andetta, an; m. One who confesses, a confessor, an acknowledger; confessor:—Se ðæs sleges andetta síe who is a confessor of the slaying, L. Alf.
pol. 29; Th. i. 80, 7.

andettan, andetan, ondettan, ondetan; p. and-ette [and = Lat. re,
contra; Grk. ἀντί; hâtan to command, promise] To confess, acknowledge,
give thanks or praise; fateri, confiteri:—Gif he wille and cunne his dǽda
andettan if he will and can confess his deeds, L. De. Cf. 2; Th. ii. 260,
18, 16. Ic andette Ælmihtigum Gode I confess to Almighty God, 6;
Th. ii. 262, 20. Seó andetnes ðe we Gode ânum andettaþ, dèþ hió us
ðæt to gôde the confession that we confess to God alone, it doth this for
our good, L. E. I. 30; Th. ii. 426, 33. Drihtne andette confitebatur
Domino, Lk. Bos. 2, 38. Folc ðé andetten confiteantur tibi populi,
Ps. Th. 66, 5. Ealra godena Gode andettaþ confitemini Domino omnium
dominorum, 135, 28. [O. Sax. and-hêtan, ant-hêtan præcipere, vovere:
O. H. Ger. ant-heizan proponere, spondere, polliceri, vovere.] DER. anddetan: ge-andettan, -ondettan: andet, -an, -la, -nes, -ta, -tere, -ting.

andettean to confess; confiteri, Bd. 1, 1; S. 474, 3. v. andettan.

andettere, es; m. A confessor; confessor:—Ðæt Albanus hæfde ðone

Cristes andettere mid him *confessorem Christi penes Albanum latere*, Bd. 1, 7; S. 477, 7.

andetting, es; *m. A confession, profession;* confessio, professio. v. andettan.

and-feng, an-, on-, es; *m. A taking to one's self, taking up, a receiving, defence, defender;* assumptio, susceptio, susceptor, Lk. Bos. 9, 51: Ps. Spl. 90, 2: Cd. 218; Th. 279, 28; Sat. 245: Ps. Spl. 88, 18. v. an-feng, on-feng.

and-fenga, -fengea, -fencgea, [ond-], an; *m. A receiver, undertaker, defender;* susceptor :—Is andfenga Drihten sâwle mînre *Dominus susceptor est animæ meæ*, Ps. Th. 53, 4: 118, 114. Ðû me, God, eart andfengea *tu, Deus, susceptor meus es*, 58, 18: 143, 2. Andfencgea, 58, 9.

and-fenge, -fencge; *adj. That which can be received, acceptable, approved, fit;* acceptabilis, acceptus, aptus :—Asette his hand ofer ðære offrunge heáfod, ðonne biþ heó andfenge *ponet manum super caput hostiæ, et acceptabilis erit*, Lev. 1, 4. Bodian Drihtnes andfenge gêr *prædicare annum Domini acceptum*, Lk. Bos. 4, 19: 4, 24. Nys andfenge Godes rîce *non est aptus regno Dei*, 9, 62. Andfencge *acceptus:* andfengra *acceptior*, Ælfc. Gr. 43; Som. 44, 47.

and-fengend, es; *m. A receiver, undertaker, defender;* susceptor :— Ûre andfengend is Iacobes God *susceptor noster Deus Jacob*, Ps. Th. 45, 6.

and-fengnes, -ness, on-, e; *f. A receiving, reception, a place for receiving, a receptacle;* receptaculum, Bd. 2, 9; S. 510, 12: Cot. 190. v. on-fangennes.

and-findende; *part. Finding, getting;* nanciscens, Cot. 138.

and-gelôman, and-lôman; *pl. m. Implements, tools, utensils;* instrumenta, Cot. 104. v. ge-lôma.

and-get, es; *n. The understanding, intellect;* intellectus, Bt. 39, 4; Fox 216, 28. v. and-git.

andgete; *adj. Manifest;* manifestus, Exon. 26 a; Th. 76, 22; Cri. 1243; [*perhaps we should read* or-gete: v. l. 1238.]

andget-full, andgit-full; *adj. Sensible, discerning, knowing;* intelligentiæ plenus, intelligens, intelligibilis :—Ðæt ænig mon sîe swa andgetfull [andgitfull, MS. Cot.] *that any man is so discerning*, Bt. 39, 9; Fox 226, 1: R. Ben. 7: 63.

and-giet, es; *n. understanding, intellect, knowledge;* intellectus :—Ic ðec, mon, ærest geworhte, and ðê andgiet sealde *I first wrought thee, O man, and gave thee understanding*, Exon. 28 a; Th. 84, 30; Cri. 1381: 117 a; Th. 449, 16; Dôm. 72. v. and-git.

andgiet-tâcen, es; *n. A sensible token;* intelligibile signum :—Ge on wolcnum ðæs andgiettâcen mâgon sceáwigan *ye may behold a sensible token of this in the clouds*, Cd. 75; Th. 93, 3; Gen. 1539.

and-git, -giet, -gyt, -get, [ond-, on-], es; *n.* [and, git=get, *p. of* gitan *to get*]. I. *the understanding, the intellect;* intellectus :— Þurh ðæt andgit, man understent ealle ða þing, ðe he gehŷrþ oððe gesihþ *by the understanding, man comprehends [understands] all the things, which he hears or sees*, Homl. Th. i. 288, 21. Þurh ðæt andgit, seó sâwul understent *through the understanding, the soul comprehends [understands]*, 288, 28. Ðær ðæt gemynd biþ, ðær biþ ðæt andgit and se willa *where the memory is, there is the understanding and the will*, 288, 26. Ðæs andgites mæþ *the measure of the understanding*, Bt. 41, 4; Fox 250, 23. Andgit *intellectus*, Ælfc. Gl. 69; Som. 70, 28: Exon. 28 a; Th. 84, 30; Cri. 1381: Ps. Th. 31, 10. II. *understanding, knowledge, cognizance;* intellectus, cognitio, agnitio :—Ic ðê sylle andgit *intellectum dabo tibi*, Ps. Th. 31, 9: 91, 5. Forðan biþ andgit æghwær sêlest *therefore is understanding everywhere best*, Beo. Th. 2122; B. 1059. Nolde ic hiora andgit ænig habban *non agnoscebam eos*, Ps. Th. 100, 4. III. *sense, meaning, one of the senses;* sensus :—Hwîlum [he sette] andgit of andgite *sometimes [he put] meaning for meaning*, Bt. procm; Fox viii. 3. Ða fif andgitu ûre lîchaman, ðæt is, gesihþ and hlyst, swæcc and stenc and hrepung *the five senses of our body, that is, sight and hearing, taste and smell and touch*, Homl. Th. ii. 550, 10.

andgitan; *p.* -geat; *pp.* -giten *To perceive, understand;* animadvertere, Cot. 3. v. on-gitan.

and-gite, -giete, an; *f. The intellect, understanding, knowledge;* intellectus, cognitio. v. ond-giete.

andgit-fullîc; *adj. Fully* or *clearly understood, intelligible;* omnino intellectus, intelligibilis :—Ælc stemn is oððe andgitfullîc oððe gemenged. Andgitfullîc stemn is ðe mid andgite biþ geclypod, swâ swâ is, Ic hêrige ða wæpnu, and ðone wer *arma virumque cano,—every voice is either intelligible or confused. Intelligible voice is what is spoken with understanding, as, Arms and the man I sing*, Ælfc. Gr. 1; Som. 2, 32–34.

andgit-fullîce; *comp.* or; *sup.* ost; *adv. Sensibly, clearly, plainly, distinctly, intelligibly;* intelligenter :—Swâ swâ he hit andgitfullîcost gereccan mihte *as he most clearly might explain it*, Bt. procm; Fox viii. 4.

andgit-leás; *adj. Foolish, senseless, doltish;* stolidus, insipiens :—Geonge men and andgitleáse man sceal swingan *young men and foolish must be beaten [one shall beat]*, L. M. I. P. 14; Th. ii. 268, 26.

andgit-lîc; *adj. Sensible, intelligible;* intelligibilis, Solil. 11.

andgit-lîce; *adv. Clearly;* liquido, Cot. 123. v. andgit-fullîce.

andgitol; *adj. understanding;* intelligibilis. v. andgyttol.

andgit-tâcen, es; *n. a sensible token.* v. andgiet-tâcen.

and-gyt, es; *n. the intellect, understanding, knowledge;* intellectus, cognitio :—Ðâm nis andgyt *quibus non est intellectus*, Ps. Spl. 31, 11: 118, 73. Ne mâgon andgyt habban? *nonne cognoscent?* Ps. Th. 52, 5: 66, 2. v. and-git.

andgyttol, andgytol; *adj. understanding, intelligent, sensible;* intelligens, intelligibilis, R. Ben. 7: 63. v. andget-full.

and-hétan; *p.* -hêtte *to confess;* confiteri :—He his gyltas Gode andhêtte *he confessed his offences to God*, Ps. C. 50, 29; Ps. Grn. ii. 277, 29. v. andettan.

andian, -igan; *part.* -igende; ic andie, andige, ðû andast, he andaþ, andgaþ, *pl.* andiaþ; *p.* ode; *pp.* od [anda *envy*] *To envy;* invidere :—Ic andige on ðê *invideo tibi*, Ælfc. Gr. 41; Som. 43, 58: 26; Som. 29, 3. Andgaþ *invidet*, Prov. 28.

andig; *adj. Envious;* invidus, Scint. 15.

andigende; *part. envying*, R. Ben. interl. 55. v. andian.

and-lang, -long, [ond-]; *adj. All-along, throughout, continuous, extended;* per totum, continuus, in longum porrectus :—Wæs andlangne dæg swungen *was beaten all day long*, Andr. Kmbl. 2550; An. 1276: Chr. 937; Th. 202, 27, col. 2; Ædelst. 21: Beo. Th. 4237; B. 2115.

and-lang, ond-long, on-long; *prep. only gen. On length*, ALONG, *by the side of;* in longum, per :—Læte yrnan ðæt blôd nyðer andlang ðæs weofudes *decurrere faciet sanguinem super crepidinem altaris;* he will let the blood run down along the altar, Lev. 1, 15. Andlang ðæs [MS. ðas] wêstenes *along the desert*, Jos. 8, 16. Andlang ðara nægla *along the nails*, Bd. 3, 17; S. 544, 30. Ðæt wæter wyrþ to eá, ðonne andlang eá to sæ *the water runs to the river, then along the river to the sea*, Bt. 34, 6; Fox 140, 20. Andlang Mæse *along the Mase*, Chr. 882; Th. 150, 22, col. 2, 3. Andlang dîces *along the dike*, Cod. Dipl. Apndx. 442; A. D. 956; Kmbl. iii. 438, 18.

and-leán, ond-leán, es; *n. Retribution, retaliation;* retributio, talio :—Hî sculon onfôn wrâþlîc andleán *they shall receive dire retribution*, Exon. 20 a; Th. 52, 12; Cri. 832. DER. leán.

and-leofen, -lifen, -lyfen, es; *n.* I. *living, food, sustenance, nourishment, pottage;* victus, alimenta, pulmentum :—Mon to andleofne eorþan wæstmas hâm gelædeþ *man for sustenance brings home earth's fruits*, Exon. 59 a; Th. 214, 22; Ph. 243. Ðû winnan scealt and dîne andlifne selfa gerêcan *thou shalt labour and thyself get thy sustenance*, Cd. 43; Th. 57, 25; Gen. 933. Sealde him andlyfene *dedit eis alimenta*, Gen. 47, 17: Bd. 1, 27, resp. 8; S. 494, 16. Sealde ealle hyre andlyfene *misit totum victum suum*, Mk. Bos. 12, 44. II. *that by which food is procured, money, wages, alms;* stipendium, stips :—Ðæt he mihte dæghwâmlîce andleofene onfôn *ut quotidianam ab eis stipem acciperet*, Bd. 5, 2; S. 615, 3. Beóþ êþhylde on eówrum andlyfenum *estote contenti stipendiis vestris*, Lk. Bos. 3, 14.

and-lîcnis, -niss, e; *f. A likeness, similitude;* imago :—God gesceóp man to his andlîcnisse *creavit Deus hominem ad imaginem suam*, Gen. 1, 27. v. an-lîcnes.

and-lôman, and-lûman; *pl. m. Utensils, vessels;* utensilia, vasa, Ælfc. Gl. 22: R. Ben. interl. 31. v. and-gelôman.

and-long; *adj. All-along, throughout;* per totum :—Andlonge niht *all night long*, Exon. 51 b; Th. 179, 14; Gû. 1261: Beo. Th. 5383; B. 2695. v. and-lang.

and-mitta, an; *m.* [and, mitta *a measure*] *A weight, a standard weight;* exagium. v. an-mitta.

an-drædan; *part.* an-drædende *To fear*, Cd. 156; Th. 194, 25; Exod. 266. v. on-drædan.

Andreas; *m. indecl. but* Andreæ *and* Andrea *are found in dat. as in Lat. and Grk.* Andrew; Andreas. [*Lat.* Andreas; *g. dat.* Andreæ; *m.*= Ἀνδρέας; *g. ov; dat.* ᾳ; *m. from* ἀνδρεία; *g. as manliness, manly strength* or *courage, from* ἀνήρ; *g.* ἀνδρὸς *a man*] :—Andreas, Simónes brôðer Petres *Andreas, frater Simonis Petri*, Ἀνδρέας, ὁ ἀδελφὸς Σίμωνος Πέτρου, Jn. Bos. 1, 40. Hî cômon on Andreas hûs *venerunt in domum Andreæ*, ἦλθον εἰς τὴν οἰκίαν Ἀνδρέου, Mk. Bos. 1, 29. Fram Bethsaida, Andreas ceastre and Petres *a Bethsaida, civitate Andreæ et Petri*, Jn. Bos. 1, 44. Philippus sæde hit Andreæ *Philippus dicit Andreæ*, Φίλιππος λέγει τῷ Ἀνδρέᾳ, 12, 22. Ðâ ðæt Andrea earmlîce þûhte *then that seemed pitiful to Andrew*, Andr. Kmbl. 2271; An. 1137. Ðær Andrea ongete wearþ wîgendra þrym *there the glory of the warriors became known to Andrew*, 3136; An. 1571. Ðis Gôdspel sceal on Andreas mæsse-dæg *this Gospel must be on St. Andrew's day*, Rubc. Mt. Bos. 4, 18–22, Notes, p. 574.

and-reccan; *p.* -reahte; *pp.* -reaht *To relate;* referre :—Ic mæg andreccan spræce *I can relate a tale*, Bt. Met. Fox 26, 3; Met. 26, 2. v. reccan.

an-drece-fæt, es; *n.* [drecan *vexare*, fæt *vas*] *A pressing-vat, a wine* or *oil vat;* emistis? *vel* trapetum, *scil.* torcular ad uvas *vel* olivas premendas, Mann; Ælfc. Gl. 26; Wrt. Voc. 25, 22.

Andred, es; *m. The name of a large wood in Kent, also the city of* ANDRED *or* Andrida: Andredes ceaster, e; *f. the Roman station* or *city of Andred, Pevensey* or *Pemsey Castle*, Sussex: Andredes leág, e; *f.*

ANDREDSLEY: Andredes weald, es; *m.* ANDRED'S WEALD, *a large wood in Kent, extending into Sussex* [v. Sandys Gavel. Ind. p. 340] :—Hine ðá Cynewulf on Andred adræfde *then Cynewulf drove him into Andred,* Chr. 755; Th. 82, 9, col. 2. Hér Ælle and Cissa ymbsæton Andredes ceaster *in this year Ælle and Cissa besieged Andredescester,* 491; Th. 24, 19, col. 2. On ðone wudu ðe is genemned Andredes leáge *into the wood which is called Andredsley,* 477; Th. 22, 40, col. 1. Se múþa [Limene] is on eásteweardre Cent, on ðæs ilcan wuda eást ende ðe we Andred hátaþ. Se wudu is westlang and eástlang cxx míla lang oððe lengra, and xxx míla brád. Seó eá, ðe we ǽr embe sprǽcon, líð út of ðam wealde *the mouth [of the Limen] is in the east of Kent, at the east end of the same wood which we call Andred. The wood is, along the east and along the west,* 120 *miles long, or longer, and thirty miles broad. The river, of which we before spoke, flows out from the weald,* Chr. 893; Th. 162, 29, col. 3.

Andredes ceaster, leág, weald. v. Andred, es; *m.*

an-drysen-líc, -drysn-líc, [on-]; *adj. Terrible;* terribilis :—Swýðe heáh God and swýðe andrysnlíc ofer ealle godas *Dominus summus, terribilis super omnes deos,* Ps. Th. 46, 2; Past. 15, 2; Hat. MS. 19 a, 26. v. dryslíc.

an-drysne, on-drysne; *adj.* I. *terrible, fearful, dreadful;* terribilis, horrendus :—Wearþ ðæt andwyrde swíðe andrysne *that answer was very fearful,* Ors. 5, 3; Bos. 104, 3. II. *as causing fear, venerable, venerated, respectable;* verendus, reverendus :—Ne biþ he náuðer ne weorþ, ne andrysne *he is neither honourable, nor respectable,* Bt. 27, 1; Fox 94, 22: Ors. 5, 12; Bos. 112, 13.

an-drysno, *dat. pl.* an-drysnum; *f. Fear, awe, reverence;* timor, metus, reverentia :—For andrysnum *from reverence,* Beo. Th. 3596; B. 1796. v. on-drysno.

and-saca, ond-, an; *m. A denier, renouncer, an apostate, opposer, enemy;* negator, renunciator, adversarius :—Ofer eorþan andsaca ne wæs *there was not an opposer on the earth,* Cd. 208; Th. 258, 2; Dan. 669. Godes andsaca *an opposer* or *a forsaker of God,* 23; Th. 28, 27; Gen. 442: Beo. Th. 3369; B. 1682. Godes andsacan *God's enemies,* Cd. 219; Th. 281, 10; Sat. 269: Exon. 31 a; Th. 97, 22; Cri. 1594. Mid ðám andsacum *with the apostates,* Cd. 17; Th. 21, 6; Gen. 320. v. saca.

and-sacian, -sacigan, -sacigan; *p.* ode; *pp.* od *To strive against, to deny, refuse, gainsay, forsake, abjure;* impugnare, negare, recusare, abjurare :—Ne mæg ic andsacigan *I cannot deny,* Bt. 10; Fox 26, 24. v. sacian.

and-sæc, es; *m?* [and-; sacu, sæc *strife, contention*] *Contention, resistance, denial, refusal;* contentio, repugnantia, contradictio, negatio :—Borges andsæc *inficiatio* vel *abjuratio,* Ælfc. Gl. 14; Som. 58, 16. Be borges andsæce *concerning a refusing of a pledge,* L. In. 41; Th. i. 128, 1, note 1. Ðe ðæs upstíges andsæc fremedon *who made denial of the Ascension,* Exon. 17 b; Th. 41, 14; Cri. 655: Elen. Grm. 472.

and-sǽte; *adj.* [and *against,* sǽtan *to lie in wait*] *Odious, hateful, abominable;* exosus, perosus, Ælfc. Gr. 33; Som. 36, 60: Ælfc. Gl. 84; Som. 73, 101; Wrt. Voc. 49, 9.

and-speornan *to stumble,* Mt. Kmbl. Rush. 4, 6. v. on-speornan.

and-spyrnes, -ness, e; *f. An offence;* scandalum, Mt. Rush. Stv. 16, 23.

and-standan [and, standan *to stand*] *To sustain, abide, stand by, bear;* sustinere :—Andstandende ongeán *contending against,* R. Ben. 1.

and-swarian, an-, ond-, on-; *p.* ede, ode, ude; *pp.* ed, od; *v. a. n. To give an answer, to* ANSWER, *respond;* respondere :—Ðá ne mihton híg him nán word andswarian *non poterant ei respondere verbum,* Mt. Bos. 22, 46. Andswarode ic *I answered,* Bt. 26, 2; Fox 92, 18. Him se yldesta andswarode *the chiefest answered him,* Beo. Th. 522; B. 258: Andr. Kmbl. 519; An. 260: Cd. 38; Th. 51, 16; Gen. 827. Him englas andswaredon *the angels answered him,* 117; Th. 152, 25; Gen. 2525. Andswarodon, 111; Th. 147, 5; Gen. 2434. DER. swarian, ond-, geand-: swerian.

and-swaru, ond-, e; *f.* [and, swaru *a speaking*] *An* ANSWER; responsum :—Andswaru líðe *a soft answer,* Scint. 77. Grim andswaru *a fierce answer,* Beo. Th. 5713; B. 2860. Hí aféngon andsware *illi acceperunt responsum,* Mt. Bos. 2, 12. Andsware bídan wolde *would await an answer,* Beo. Th. 2991; B. 1493: Exon. 10 b; Th. 12, 11; Cri. 184: Bt. Met. Fox 22, 86; Met. 22, 43. Nú sceal he sylf faran to incre andsware *now he must come himself for your answer,* Cd. 27; Th. 35, 19; Gen. 557.

and-swerian; *p.* ade, ede, ode; *pp.* ed, od *to answer* :—Ðá him andsweradan gástas *then the ghosts answered him,* Cd. 214; Th. 268, 6; Sat. 51. Andsweredon, Elen. Grm. 397. v. and-swarian.

and-sýn, e; *f. A face;* facies :—Woldon hí ðæt hí mihton geholene beón fram andsýne ðæs cyninges *they wished that they might be hidden from the face of the king,* Bd. 4, 16; S. 584, 25. v. an-sýn.

and-þwǽre; *adj. Perverse, froward, athwart, cross;* perversus. v. and against, þwǽre *quiet.*

and-timber, an-, on-, es; *n. Matter, materials, substance, a theme;* materies, materia, thema :—Lengran feóndscipes andtimber *longioris inimicitiæ materies,* Bd. 4, 21; S. 590, 19. Antymber [MSS. C. and D.

antimber] *materies, materia,* Ælfc. Gr. 12; Som. 15, 54. Antimber *thema,* 9, 1; Som. 8, 21. v. timber.

and-warde; *adj. Present;* præsens :—Ðis andwarde líf manna on eorþan *vita hominum præsens in terris,* Bd. 2, 13; S. 516, 14. v. and-weard.

and-wardnys, -nyss, e; *f. Presence;* præsentia :—Bútan óðra bisceopa andwardnysse *sine aliorum episcoporum præsentia,* Bd. 1, 27; S. 491, 40. v. and-weardnes.

and-wealcan *to roll;* volvere, Th. Anlct. v. on-wealcan.

and-weald, es; *m. Power, right* or *title to anything* :—Ðæt he wolde habban andweald ongeán God *that he would have power against God,* Homl. Th. i. 10, 25: Ps. Spl. 19, 7: 113, 2: Ælfc. Gl. 13; Som. 57, 121. v. án-weald, onweald.

and-weard, -werd, -warde; *adj. Present;* præsens :—Ðér is Dryhten andweard *where the Lord is present,* Exon. 48 b; Th. 167, 7; Gú. 1056. Andweard Gode *present with God,* 30 b; Th. 95, 29; Cri. 1564. Fór ðé andweardne *before thee present,* Cd. 40; Th. 54, 2; Gen. 871: Andr. Kmbl. 2449; An. 1226. Oþ ðisne andweardan dæg *usque in hunc præsentem diem,* Mt. Bos. 28, 15. On ðis andweardan lífe *in this present life,* Bt. 10; Fox 26, 30. Ða scearpþanclan wítan ðone twydǽledan wísdóm hlutorlíce tocnáwaþ, ðæt is, andweardra þinga and gástlícra wísdóm *the sharp-minded wise men knew clearly the twofold wisdom, that is, the wisdom of things temporal [present] and spiritual,* MS. Cot. Faust, A. x. 150 b; Lchdm. iii. 440, 30. [*O. Sax.* aud-ward *præsens: O. H. Ger.* ant-wart: *Goth.* and-wairþs.] DER. and-warde, and-wardnys, and-weardlíce, and-weardnes.

and-weard-líce; *adv. Presentially, in the presence of, present;* præsentialiter :—Ðe hine andweardlíce gesáwon *who saw him present,* Bd. 4, 17; S. 585, 30: Elen. Grm. 1141.

and-weardnes, -ness, and-weardnys, and-wardnys, -nyss, e; *f. Presentness, presence, present time;* præsentia, præsens tempus, præsens :—Wæs ic swýðe for his andweardnesse afyrhted *ejus præsentia eram exterritus,* Bd. 4, 25; S. 600, 42. On andweardnysse *in præsenti,* 1, 1; S. 474, 1.

and-wendan; *p.* -wende; *pp.* -wended *to change;* mutare. DER. wendan. v. on-wendan.

and-wendednys, a-wændednys, -nyss, e; *f.* [and, wended, *pp. of* wendan *to turn,* nes] *A changing, change;* mutatio, Ps. Spl. 76, 10. v. on-wendednes.

and-weorc, ond-weorc, an-weorc, es; *n. Matter, substance, material, metal, a cause of anything;* materia, cæmentum, metallum, causa :—He ðæt andweorc of Adames líce aleoðode *he dismembered the substance from Adam's body,* Cd. 9; Th. 11, 16; Gen. 176. Ðæt leád is hefigre ðonne ǽnig óðer andweorc *plumbum cæteris metallis est gravius,* Past. 37, 3; Hat. MS. 50 a, 16. Búton andweorce *without cause,* Bt. 10; Fox 30, 2: Bt. Met. Fox 17, 32; Met. 17, 16.

and-werd; *adj. Present;* præsens :—On ðisum andwerdan dæge *on this present day,* Homl. Th. ii. 284, 5. v. and-weard.

and-werdan, and-wirdan, and-wyrdan, ond-wyrdan; *p.* de; *pp.* od [and, word *a word: Goth.* and-waúrdyan *to answer,* waúrd *a word: Ger.* antwort *an answer*] *To answer;* respondere :—Abram hire andwerde *Abram ei respondit,* Gen. 16, 6.

and-wirdan; *p.* de; *pp.* od *to answer;* respondere :—Ðæt wíf andwirde *the woman answered,* Gen. 3, 2. v. and-werdan.

and-wís; *adj. Expert, skilful;* gnarus, expertus :—Yfeles andwís *expert in evil,* Exon. 69 a; Th. 257, 8; Jul. 244. DER. wís.

and-wísnes, -ness, e; *f. Experience, skilfulness;* experientia. DER. and, wísnes. v. wís wise.

and-wlata, an; *m. The face, forehead,* Herb. 75, 6; Lchdm. i. 178, 16: 101, 2; Lchdm. i. 216, 9. v. and-wlita.

and-wlita, an-wlita, an; *m:* and-wlite, es; *n. The face, countenance, personal appearance, forehead, form, surface;* facies, vultus, aspectus, frons, forma, superficies :—Hleór bolster onféng, eorles andwlitan *the bolster received his cheek, the hero's face,* Beo. Th. 1382; B. 689: Exon. 24 a; Th. 69, 20; Cri. 1123: Bt. Met. Fox 31, 33; Met. 31, 17. Leóht andwlitan ðínes *lumen vultus tui,* Ps. Spl. 4, 7: Ps. Th. 89, 8. Ealle gesceafta onfóþ æt Gode andwlitan *all creatures receive form from God,* Bt. 39, 5; Fox 218, 15. On andwlitan wídre eorþan *on the face of the wide earth,* Cd. 67; Th. 81, 21; Gen. 1348. He hæfde blácne andwlitan *he had a pale countenance,* Bd. 2, 16; S. 519, 34. [*Plat.* antlaat, *n: N. H. Ger.* antlitz, *n: M. H. Ger.* antlütze, antlitze: *O. H. Ger.* antluzi: *O. Nrs.* andlit, *n.*]

and-wlítan; *p.* -wlát, *pl.* -wliton; *pp.* -wliten *To look upon;* intueri :—Nó ðæt hí mósten in ðone Écan andwlítan *that they might not look on the Eternal,* Cd. 221; Th. 288, 10; Sat. 378. DER. wlítan.

and-wlite, es; *n. The countenance, face;* vultus, facies :—Efennysse geseah andwlite his *æquitatem vidit vultus ejus,* Ps. Spl. T. 10, 8. v. and-wlita.

and-wráþ; *adj. Hostile;* infensus :—Ðam dracan he andwráþ leofaþ *he lives hostile to the serpent,* Exon. 95 b; Th. 356, 26; Pa. 17. DER. wráþ.

and-wyrdan, ond-wyrdan *to answer*, Ps. Th. 101, 21: 118, 42: Ors. 1, 10; Bos. 32, 20. v. and-werdan.

and-wyrde, es; *n. An answer;* responsum :—Hêtan him ꝺæt andwyrde secgan *they commanded them to deliver this answer*, Ors. 1, 10; Bos. 32, 23: Cd. 27; Th. 36, 17; Gen. 573: Elen. Grm. 544: 618. v. and-swaru.

and-wyrding, e; *f. A consent, an agreement, a conspiring, conspiracy;* conspiratio, Cot. 46.

and-yttan *To confess, praise, thank;* confiteri :—Ic andytte ꝺê *ego confiteor tibi*, Mt. Bos. 11, 25. v. andettan.

âne, æne; *adv.* [ân *one, with the adverbial* -e] *Once, once for all, only, alone;* semel, solum, tantum :—Is ꝺys âne mâ *this is once more*, Andr. Kmbl. 984; An. 492. Ic bydde ꝺê, ꝺæt ꝺû lǽte me sprecan âne feáwa worda *I pray thee, that thou let me speak only* [*once for all*] *few words*, Nicod. 11; Thw. 5, 40. Ic ꝺê æne abealh, êce Drihten *I alone angered thee, eternal Lord*, Cd. 222; Th. 290, 4; Sat. 410.

ân-eáge, ân-êge, ân-îge, ân-îgge; *adj.* [ân *one*, eage *an eye*] *One-eyed, blind of one eye;* monoculus, luscus :—Gif he hî gedô âneáge *if he make them one-eyed*, L. Alf. 20; Th. i. 48, 25, note. Gif hîg ânêge gedô *si luscos eos fecerit*, Ex. 21, 26.

ân-ecge; *adj. One-edged, having one edge;* unam habens aciem :—Ânecge sweord *a one-edged sword;* machæra, Ælfc. Gl. 52; Som. 66, 48; Wrt. Voc. 35, 36.

ân-êge; *adj. One-eyed* :—Gif hîg ânêge gedô *si luscos eos fecerit*, Ex. 21, 26. v. ân-eáge.

ân-êged; *part. One-eyed, blinded of one eye;* monoculus, monophthalmus, luscus :—Gif he hî gedô ânêgede *if he make them one-eyed*, L. Alf. 20; Th. i. 48, 25, note: Ælfc. Gl. 71; Som. 70, 76; Wrt. Voc. 43, 9.

a-neglod; *part. Nailed, fastened with nails, crucified;* clavis fixus, crucifixus, Som. v. næg-lian.

a-nêhst *at last, in the last place;* ad ultimum, ultimo. v. a-nîhst.

a-nemnan; *p.* de; *pp.* ed *To declare;* pronuntiare :—Godes spel-bodan eal anemdon *God's messengers declared all*, Exon. 33 a; Th. 104, 25; Gû. 13. v. nemnan.

ânes, âness, e; *f. A oneness, an agreement;* unitas :—Gewearþ him and ꝺam folce on Lindesige ânes *there was an agreement between him and the people in Lindsey*, Chr. 1014; Th. 274, 13. v. ân-nes.

ânes *of one, g. m. n. of* ân :—Ânes bleós *of one colour;* unicolor. Ânes geáres *of one year.* Ânes hiwes *of the same hue or shape.* Ânes wana *wanting of one, as* ânes wana twentig *twenty wanting one, nineteen.*

a-nescian, -hnescian; *p.* ode; *pp.* od *To make nesh, to weaken;* emollire :—He sceolde ꝺa ânrêdnesse anescian *poterat constantiam ejus emollire*, Bd. 1, 7; S. 477, 44. v. hnescian.

an-færeld *a journey;* iter, Nathan. 2. v. on-færeld.

ân-fáh; *adj. Of one colour;* unicolor. v. fâg.

an-fangen *received;* *pp.* of an-fôn.

an-fangennes, -ness, e; *f. A receiving, receptacle;* acceptio, susceptio, receptaculum, R. Ben. 2. v. on-fangenes.

ân-feald; *adj.* [ân *one*, feald *fold*] ONE FOLD, *simple, single, one alone, singular, peculiar, matchless;* simplex :—Swâ mid þrŷfealdre swâ mid ânfealdre lâde *either with a threefold or with a simple exculpation*, L. C. E. 5; Th. i. 364, 2: 5; Th. i. 362, 10. Ânfeald âþ *a simple oath*, L. C. S. 22; Th. i. 388, 11. Ânfeald getel *the singular number*, Ælfc. Gr. 13; Som. 16, 25. Ân-feald gcwin *single combat*, R. Ben. interl. 1. Ꝺa ânfealdan stræcan *those who are uniformly strict*, Past. 42, 1; Hat. MS. 57 b, 25.

ânfeald âþ *a simple oath*, L. C. S. 22; Th. i. 388, 11, note b. v. âþ, III.

ânfeald-lîce; *adv. Singly, simply, without intermission;* simpliciter, R. Ben. 52.

ânfeald-nes, -ness, e; *f. Oneness, unity, simplicity, singleness;* simplicitas :—Ymbe ꝺa ânfealdnesse ꝺare godcundnesse *concerning the oneness of the divine nature*, Bt. 35, 5; Fox 164, 18: 39, 5; Fox 218, 19. Ꝺâ hwîle ꝺe hî heora ânrêdnesse geheóldan him betwênan and ânfealdnysse *while they had agreement and simplicity amongst themselves*, Ors. 5, 3; Bos. 104, 1.

an-feng, es; *m. A taking to one's self, a receiving, defence, defender;* assumptio, susceptio, susceptor :—Drihtnes anfeng ûre *Domini assumptio nostra*, Ps. Spl. 88, 18. He anfeng mîn *ipse susceptor meus*, 61, 2: Runic pm. 3; Hick. Thes. i. 135; Kmbl. 340, 1. v. and-feng.

an-fenga, an; *m. A receiver, an undertaker;* susceptor. v. and-fenga.

an-fenge; *adj. Acceptable, fit.* v. and-fenge.

an-fênge *shouldest have taken*, Cd. 42; Th. 54, 10; *p. subj.* of an-fôn.

an-fengednes, -ness, e; *f. A receiving;* acceptio. v. on-fangenes.

ân-fête; *adj. One-footed, with one foot;* monopodius, Exon. 114 b; Th. 439, 9; Rä. 59, 1.

an-fêde *in walking*, Bt. 36, 5; Fox 180, 20. v. fêde.

an-filt, on-filt *An* ANVIL; incus, Ælfc. Gr. 28, 6; Som. 32, 34: Ælfc. Gl. 50; Som. 65, 128; Wrt. Voc. 34, 56. [*Plat.* ambolt, ambult, *m: Dut.* aanbeeld, aenbeld, *n: O. H. Ger.* anafalz.]

an-findan *to discover, find;* deprehendere, Cot. 61. v. on-findan.

ân-floga, an; *m. Lonely flying;* solitarie volans, solivagus, Exon. 82 a; Th. 309, 25; Seef. 62.

an-fôn; *p.* -fêng; *pp.* -fangen *To take, take to one's self, receive, perceive, comprehend;* accipere, suscipere, sumere, percipere, recipere :—Ꝺû sceonde æt me anfênge *thou shouldest have taken to thyself shame from me*, Cd. 42; Th. 54, 10; Gen. 875: Exon. 112 a; Th. 429, 12; Rä. 43, 3: Ps. C. 50, 135; Ps. Grn. ii. 280, 135. To anfônne *to receive*, Bd. 3, 6; S. 528, 4. v. on-fôn.

an-forht; *adj. Fearful, timid;* timidus :—Ne þearf ꝺonne ænig anforht [MS. unforht] wesan *no one then need be fearful*, Rood Kmbl. 232; Kr. 117. DER. forht.

ân-for-lǽtan; ic -lǽte, ꝺû -lǽtest, -lætst, he -lǽteþ, -lêteþ, *pl.* -lǽtaþ; *p.* -lêt, -leórt, -leót, *pl.* -léton; *pp.* -lǽten *To leave alone, lose, relinquish, forsake;* amittere :—Ꝺû nû ân-forlête *thou hast now lost*, Bt. 7, 3; Fox 20, 12: Bd. 1, 27, resp. 3; S. 490, 25: 4, 10; S. 578, 34. v. ân; *adv.* and forlâtan.

an-funden *found, taken;* *pp.* of an-findan.

ang-, *a prefix*, as in ang-breóst, ang-môd, ang-môdnes, ang-sum, *etc.* from ange *narrow, vexed.*

ânga, ænga, ênga, *m;* ânge, *f. n; def. adj.* I. *one and no more, only, sole, single, singular;* unicus, ullus, quisquam :—Se ânga hyht *the sole hope*, Exon. 62 a; Th. 227, 14; Ph. 423: 96 b; Th. 360, 1; Pa. 73. Ꝺû eart dôhtor mîn ânge for eorþan *thou art my only daughter on earth*, 67 a; Th. 248, 13; Jul. 95. Abraham wolde gesyllan his swǽsne sunu, ângan ofer eorþan yrfelâfe *Abraham would give his dear son, his sole hereditary remnant on earth*, Cd. 162; Th. 203, 13; Exod. 403. Cain gewearþ to ecgbanan ângan brêꝺer *Cain was the murderer of his only brother*, Beo. Th. 2529; B. 1262. II. *any, every one, all;* quisque. In this sense it admits of a plural :—Secge me nû, hwæꝺer ꝺû æfre gehŷrdest, ꝺæt wîsdôm ângum ꝺara eallunga þurhwunode *tell me now, whether thou hast ever heard, that wisdom always remained to any of them*, Bt. 29, 1; Fox 102, 9. v. ân, II, IV.

an-gan *began*, Cd. 23; Th. 28, 26; Gen. 442. v. an-ginnan.

ang-breóst, es; *n.* [ange *narrow, contracted, troubled;* breóst *a breast*] *An asthma, a difficulty of breathing, breast-anguish;* asthma :—Wiꝺ angbreóste *against breast-anguish*, L. M. 1, 15; Lchdm. ii. 58, 15.

ange, ænge, enge, onge; *adj. Narrow, straitened, vexed, troubled, sorrowful;* angustus, anxius, vexatus, tristis :—Ꝺes ænga stede *this narrow place*, Cd. 18; Th. 23, 9; Gen. 356. Ufan hit is enge *it is narrow above*, Exon. 116 a; Th. 446, 14; Dôm. 22. Ꝺâ wæs ꝺam cynge swîꝺe ange on his môde *then the king was greatly troubled in his mind*, Ors. 2, 5; Bos. 48, 14. [*N. Ger. M. H. Ger.* enge *angustus: O. H. Ger.* angi: *Goth.* aggwus: *O. Nrs.* öngr: *Lat.* angustus: *Grk.* ἐγγύς: *Sansk.* aṇhu *narrow.*]

angeán; *prep. Against;* contra :— Hý him brohtan angeán ehta hund M fêꝺena *they brought against him eight hundred thousand foot*, Ors. 3, 9; Bos. 68, 9. v. on-geán; *prep.*

angel; *g.* angles; *m. A hook, a fishing-hook;* hamus :—Wurp ꝺînne angel tô *mitte hamum*, Mt. Bos. 17, 27. Swâ swâ mid angle fisc gefangen biþ *as a fish is caught by a hook*, Bt. 20; Fox 72, 11. [*Plat. Dut. Ger. M. H. Ger.* angel, *m: O. H. Ger.* angul, *m: O. Nrs.* öngull, *m.*]

Angel; *gen. dat. acc.* Angle; *f. Anglen in Denmark, the country between Flensburg and the Schley from which thc Angles came into Britain;* Angulus, nomen terræ quam Angli ante transitum in Britanniam coluerunt :—Of Angle cômon Eást-Engle *from Anglen came the East-Angles*, Chr. 449; Ing. 15, 1. Ꝺæt land, ꝺe man Angle hæt *the land, which is called Anglen*, Ors. 1, 1; Bos. 18, 37. Hî ꝺâ sendon to Angle *they then sent to Anglen*, Chr. 449; Th. 20, 12. v. Engel.

angel *an angel;* angelus, Ps. Spl. 33, 7. v. engel.

Angel-, *English;* Anglicanus,—*as in the following compounds:*—Angel-cyning, -cynn, -þeód.

Angel-cyning, es; *m. An Angle or English king*, Bd. 3, 8; S. 531, 8: 3, 9; S. 533, 8. v. Engle.

Angel-cynn, es; *n. The Angle or English race;* Anglorum gens, Bd. pref; S. 471, 23: 4, 16; S. 584, 13. v. Engle.

ân-geld, es; *n. A single payment or compensation*, L. In. 56; Th. i. 138, 9: L. Edg. ii. 7; Th. i. 268, 19, MS. G. v. ân-gild.

an-gelic; *adj. Like, similar;* similis :—Ꝺonne ne finst ꝺû ꝺǽr nâuht angelîces *then thou wilt not find there anything of like*, Bt. 18, 3; Fox 66, 11. v. ge-lîc.

Angel-þeód, e; *f. The English people;* Anglorum gens, Bd. 5, 24; S. 646, 34. 37. v. Engle.

angel-twicce, an; *f. A red worm used for a bait in angling or fishing;* lumbricus :—Rên-wyrm *vel* angel-twicce *lumbricus*, Ælfc. Gl. 24; Som. 60, 30; Wrt. Voc. 24, 31. [twachel *the dew-worm*, Halwl. Dict.]

ân-genga, -gengea, an; *m.* [ân *unus, solus;* gengan *ire*] *A lone-goer, a solitary;* solivagus, solitarius :—Blôdig wæl eteþ ângenga *the lone-goer will eat my bloody corpse*, Beo. Th. 902; B. 449. Fela fyrena atol ângengea oft gefremede *many crimes the foul solitary oft perpetrated*, 332; B. 165.

ân-ge-trum, es; *n.* [ân *unicus, eximius;* ge-trum *cohors, caterva*] *A singular company;* unica cohors, eximia caterva :—Micel ângetrum *a great* [*and*] *singular company*, Cd. 160; Th. 199, 6; Exod. 334.

án-geweald, es; *m. Power, empire, dominion;* potestas, imperium, dominatio:—Hyne ðære helle sealde on ángeweald *gave him into the power of hell,* Nicod. 29; Thw. 17, 1. v. án-weald, ge-weald.

angil *a hook,* Coll. Monast. Th. 23, 11. v. angel.

án-gild, -geld, -gyld, es; *n.* [án *one,* gild *a payment, compensation*]. I. *a single payment or compensation, the single value of property claimed or in dispute,—a rate fixed by law, at which certain injuries, either to person or property, were to be paid for;* simplex compensatio:—Forgylde ðæt ángylde *let him pay for it with a single compensation,* L. Alf. pol. 6; Th. i. 66, 3: 22; Th. i. 76, 7: L. In. 22; Th. i. 116, 12. Forgylde ðæt yrfe ángylde *let him pay for the property with a single recompense,* L. Ath. v. § 8, 4; Th. i. 236, 24: L. Edg. H. 6; Th. i. 260, 7: L. Edg. ii. 7; Th. i. 268, 19: L. Eth. iii. 4; Th. i. 294, 17: L. O. D. 4; Th. i. 354, 15: Th. Diplm. A. D. 883; 130, 18–131, 5. II. *the fixed price or rate at which cattle and other goods were received as currency;* æstimatio, pretium:—Gif we ðæt ceáp-gild arǽraþ be fullan ángylde *if we raise the market-price [of cattle] to the full fixed price,* L. Ath. v. § 6, 4; Th. i. 234, 17.

an-gildan; *p.* -geald, *pl.* -guldon; *pp.* -golden *To pay for, repay, atone for;* rependere, pœnas dare:—Sum sáre angeald ǽfen-reste *one sorely paid for his evening rest,* Beo. Th. 2507; B. 1251: Ors. 6, 23; Bos. 124, 13. v. on-gildan.

an-gin, -ginn, -gyn, on-gin, es; *n. A beginning, attempt, resolve, purpose, design, undertaking, opportunity;* initium, principium, conatus, inceptum, cœptum, occasio:—Ælc angin *every beginning,* Bt. 5, 3; Fox 12, 18. Ðis synd sára angin *initium dolorum hæc,* Mk. Bos. 13, 8. Se ána Scyppend næfþ nán anginn, ac he sylf is anginn ealra þinga *the Creator alone hath not any beginning, but he is himself the beginning of all things,* Hexam. 13; Norm. 22, 3. On anginne *in principio,* 1; Norm. 2, 26. Bútan anginne *without beginning,* Exon. 9b; Th. 8, 1; Cri. 111. Synt ðæra sára anginnu *sunt dolorum initia,* Mt. Bos. 24, 8. Gif ðú ðæt angin fremest *if thou perfect that attempt,* Cd. 27; Th. 36, 27; Gen. 578. Ðá geseah Iohannes sumne cniht swíðe glæd on móde and on anginne cáf *there John saw a certain youth cheerful in mind and quick in design,* Ælfc. T. 33, 17. Abreóðe his angin *may his design perish,* Byrht. Th. 138, 59; By. 242: Cd. 178; Th. 223, 26; Dan. 125: R. Ben. 69. [*O. Sax.* angin *initium.*]

an-ginnan; *p.* -gan, *pl.* -gunnon; *pp.* -gunnen *To begin, undertake;* incipere:—Angan hine gyrwan *began to prepare himself,* Cd. 23; Th. 28, 26; Gen. 442: Bt. Met. Fox 1, 118; Met. 1, 59. v. on-ginnan.

an-gitan; *p.* -geat; *pp.* -giten *To get, lay hold of, seize;* assequi, corripere, invadere:—Hine se bróga angeat *terror seized him,* Beo. Th. 2587; B. 1291. v. on-gitan.

Angle; *g.* a; *dat.* um; *pl. m. The* ANGLES, who came from Anglen [v. Angel = Engel *Anglen*] in Denmark, and occupied the greater part of England, from Suffolk to the Frith of Forth, including Mercia. Bede says,—Ðæt mynster Æbbercurníg, ðæt is geseted on Engla lande *the minster, Abercorn, that is seated in the land of the Angles,* or Engla land = England, Bd. 4, 26; S. 602, 35. Abercorn is on the south coast of the Frith of Forth, and at the mouth of the river Carron, where the Roman wall of Severus began, and extended to the Frith of Clyde. Bede wrote his history about A. D. 731, at which time Abercorn was within the bounds of Engla land = England:—Ðæt land, ðætte Angle ǽr hæfdon *the land, that the Angles formerly had,* Bd. 4, 26; S. 602, 30. To Anglum *to the Angles,* Chr. 443; Th. 18, 33, col. 1; 19, 30, col. 1. Ðá cómon ða menn of þrým mægþum Germanie,—of Eald-Seaxum, of Anglum, of Iotum *then came the men from three tribes of Germany,—from Old-Saxons, from Angles, from Jutes,* Chr. 449; Th. 20, 18–21, col. 1.

Angle; *g. d. acc.* of Angel *Anglen:*—Ðæt land, ðe man Angle hǽt *the land, which they call Anglen,* Ors. 1, 1; Bos. 18, 37. v. Engel, Ongel.

Angles ég, e; *f.* [íg *an island*] ANGLESEY, so called after it was conquered by the English: it was anciently called Mona:—Hugo eorl wearþ ofslagen innan Angles ége *earl Hugo was slain in Anglesey,* Chr. 1098; Ing. 317, 31.

ang-mód, ancg-mód; *adj.* [ange *vexed,* mód *mind*] *Vexed in mind, anxious, sad, sorrowful;* anxius, sollicitus, tristis, R. Ben. 64.

ang-módnes, -ness, e; *f. Sadness, sorrowfulness;* tristitia. v. ange *vexed,* módnes, módignes *pride.*

ang-nægl, es; *m. An* AGNAIL *or* ANGNAIL, *a whitlow, a sore under the nail;* paronychia = παρωνῠχία, dolor ad ungulam. [*Frs.* ongneil: *O. H. Ger.* ungnagal.] v. ange *vexed,* nægel a nail.

angnes, -ness, angnis, -niss, angnys, -nyss, e; *f.* [ange *angustus, anxius;* -nes] *Narrowness, anxiety, distress, sorrow, trouble, anguish;* angustiæ, anxietas, tristitia, ærumna:—Angnes módes *anxietas animi,* Somn. 354. On angnisse mín *in ærumna mea,* Ps. Spl. T. 31, 4. Geswinc and angnys gemétton me *tribulatio et angustiæ invenerunt me,* Ps. Spl. 118, 143. v. angsumnes.

an-golden *repaid, requited; pp.* of an-gildan. v. gildan.

Angol-þeód, e; *f. The English nation;* gens Anglorum, Bd. 5, 21; S. 642, 31. v. Angel-þeód.

angol-twæcce; *g.* -twæccean; *f. An earth-worm:*—Genim angoltwæc-cean *take an earth-worm,* L. M. 1, 39; Lchdm. ii. 100, 8. v. angel-twicce.

an-gríslíc, -grýslíc, on-gríslíc; *adj. Grisly, horrible, dreadful, horrid;* horridus, terribilis, horrendus:—Micel and angríslíc *magnus et terribilis,* Ps. Spl. 88, 8: Ps. Th. 104, 33. DER. gríslíc.

an-grysen-líce; *adv. Terribly;* terribiliter, Nicód. 26; Thw. 14, 22. v. an-gríslíc.

ang-set, es; *m?* ang-seta, an; *m? A disease with eruptions, a carbuncle, pimple, pustule, an eruption, St. Anthony's fire;* carbunculus:—Angset *vel* spring *carbunculus,* Ælfc. Gl. 9; Som. 57, 9; Wrt. Voc. 19, 19. Angseta *furunculus vel anthrax,* Ælfc. Gl. 12; Som. 57, 69; Wrt. Voc. 20, 12: Ælfc. Gl. 64; Som. 69, 19; Wrt. Voc. 40, 51.

ang-sum, anc-sum; *adj. Narrow, strait, troublesome, hard, difficult;* angustus, difficilis:—Eálá hú neara and hú angsum is ðæt geat, and se weg ðe to life gelædt; and swýðe feáwa synt ðe ðone weg findon *quam angusta porta, et arcta via est, quæ ducit ad vitam; et pauci sunt qui inveniunt eam,* Mt. Bos. 7, 14.

ang-sumian; *p.* ode; *pp.* od *To vex, afflict, to be solicitous;* vexare, angere, sollicitus esse. DER. angsum.

ang-sum-líc *troublesome, anxious;* tristis, sollicitus. v. ang-sum.

ang-sum-líce; *adv. sorrowfully;* triste. v. angsumlíc.

ang-sumnes, -ness, -nys, ang-sumnis, -niss, -nys, -nyss, e; *f. Troublesomeness, sorrow, anxiety, anguish;* angustiæ, ærumna:—Geswinc and ang-sumnes gemétton me *tribulatio et angustiæ invenerunt me,* Ps. Spl. M. 118, 143. We gesáwon hys angsumnisse *nos vidimus angustiam animæ illius,* Gen. 42, 21: Jos. 7, 7. v. angnes.

ángum *to any,* Bt. 29, 1; Fox 102, 9. v. ánga.

án-gyld, es; *n. A single payment or compensation,* L. Alf. pol. 6; Th. i. 66, 3: 22; Th. i. 76, 7: L. In. 22; Th. i. 116, 12. v. án-gild.

an-gyn *a beginning,* Mk. Bos. 1, 1. v. an-gin.

an-gytan [an, gytan *to get*] *To find, discover, understand, know;* invenire, intelligere, R. Ben. 2. v. on-gitan.

an-hafen *lifted up, exalted,* Bd. 3, 6; S. 528, 9. v. an-hebban.

án-haga, -hoga, an; *m. One dwelling alone, a recluse;* solitarius, solitarie habitans *vel* degens:—Ðǽr se ánhaga eard bihealdeþ *ibi solitarius natalem locum tenet,* Exon. 57a; Th. 203, 20; Ph. 87. Ic eom ánhaga *I am a recluse,* 102b; Th. 388, 1; Rä. 6, 1: Beo. Th. 4725; B. 2368. To ðam ánhagan *against the solitary,* Andr. Kmbl. 2701; An. 1353.

an-hagian; *p.* ode; *pp.* od *To be at leisure,* R. Ben. 58. v. on-hagian.

an-healdan; *p.* -heóld, *pl.* -heóldon; *pp.* -healden *To hold, keep;* tenere, servare, præstare:—Gesceaft fæste sibbe anhealdaþ *creatures keep firm peace,* Bt. Met. Fox 11, 84; Met. 11, 42.

an-hebban, -hæbban; *p.* -hóf, *pl.* -hófon; *pp.* -hafen *To heave up, lift up, exalt, raise up, take away, remove;* elevare, erigere, exaltare, sublimare, attollere, auferre:—Ðæt ðú ðé ne anhebbe on ofermetto *that thou lift not up thyself with arrogance,* Bt. 6; Fox 14, 34. Mid ða heánnesse ðæs eorþlícan ríces anhafen *regni culmine sublimatus,* Bd. 3, 6; S. 528, 9. v. on-hebban.

an-hefednes, -ness, e; *f. Exaltation;* exaltatio, C. R. Ben. 7.

án-hende; *adj. One-handed, lame, imperfect, weak;* unimanus, Ælfc. Gl. 77; Som. 72, 25; Wrt. Voc. 45, 58.

án-hoga, an; *m.* [án-wuniende] *A lone dweller, recluse:*—Geworden ic eom swá swá spearwa ánhoga oððe ánwuniende on efese oððe on þecene *factus sum sicut passer solitarius in tecto,* Ps. Lamb. 101, 8. Se ánhoga *the recluse,* Exon. 60b; Th. 222, 10; Ph. 346: 47a; Th. 162, 3; Gú. 970. v. án-haga.

an-hón *to hang;* suspendere. v. on-hón.

án-horn, es; *m:* án-horna, an; *m. A unicorn;* unicornis, monoceros = μονόκερως:—Ánhornes *unicornis,* Ps. Surt. 91, 11. Ðonne ánhorna *sicut unicornis,* Ps. Th. 91, 9: [MS. ónhornan], 77, 68.

an-hrædlíce *unanimously,* Ps. Spl. 82, 5. v. án-rǽdlíce.

an-hreósan *to rush upon;* irruere. v. on-hreósan.

án-hydig; *adj. One or single minded, steadfast, firm, constant, stubborn, self-willed;* firmus, constans, pervicax:—Elnes ánhydig *steadfast in courage,* Exon. 45b; Th. 156, 3; Gú. 869: Elen. Grm. 828. Ánhydig eorl *the stubborn chieftain,* Exon. 55b; Th. 196, 28; Az. 181: 100a; Th. 377, 11; Deór. 2. Wearþ ðá ánhydig *then he became self-willed,* Cd. 205; Th. 254, 1; Dan. 605.

an-hyldan *to incline;* inclinare, R. Ben. in procem. v. on-hyldan.

an-hyrian *To emulate;* æmulari:—Ne anhyre ðú *noli æmulari,* Ps. Spl. T. 36, 8. v. onhyrian.

án-hyrne; *adj. One-horned, having one horn;* unicornis:—Ánhyrne deór *unicornis, vel monoceros, vel rhinoceros,* Ælfc. Gl. 18; Som. 58, 129; Wrt. Voc. 22, 43.

án-hyrned; *p. part. One-horned, having one horn;* unicornis:—Biþ upahafen swá swá ánhyrnedes deóres mín horn *exaltabitur sicut unicornis cornu meum,* Ps. Lamb. 91, 10: 77, 69.

án-hyrnende; *pres. part. Having one horn;* unicornis:—Fram hornum ánhyrnendra *a cornibus unicornium,* Ps. Spl. 21, 20: 77, 75: 91, 10: Ps. Lamb. 21, 22.

áni *any,* Bt. 38, 3; Fox 200, 27 [MS. Bod.] v. ǽnig.

a-nídan; *p.* -nídde; *pp.* -níded, *pl.* ·nídde = nídede *To force,* Chr. 823; Th. 110, 33, col. 1. v. a-nýdan.

án-íge, -ígge; *adj. One-eyed :*—Ánige *luscus,* Cot. 122. Gif he hí gedó anígge *if he make them one-eyed,* L. Alf. 20; Th. i. 48, 25. v. án-eáge.

a-níhst; *adv.* [a = on *in, ad;* níhst *ultimus*] *At last, in the last place;* ad ultimum, ultimo :—Ne wǽron ðæt gesíða ða sǽmestan, ðeáh ðe ic hý aníhst nemnan sceolde *they were not the worst of comrades, though I should name them last,* Exon. 86 b; Th. 326, 9; Wíd. 126.

a-niman, -nyman; *p.* -nam, *pl.* -námon; *pp.* -numen [a *from,* niman *to take*] *To take away, remove;* tollere, capere :—Animaþ ðæt pûnd æt hym *take the talent from him,* Mt. Foxe 25, 28. Animan wolde *would take,* Fins. Th. 43; Fin. 21.

áninga, æninga, ánunga; *adv.* [án *one,* inga] *One by one, singly, at once, clearly, plainly, entirely, altogether, necessarily, by all means, at all events;* per singula, singulatim, plane, prorsus, omnino, necessario, ad omnem eventum :—Woldon áninga ellenrófes mód gemiltan *they would entirely subdue the bold man's mind,* Andr. Kmbl. 2785; An. 1394. Gif ða cnihtas áninga ofslagene beón sceoldan *si necesse esset pueros interfici,* Bd. 4, 16; S. 584, 32: Beo. Th. 1272; B. 634: Judth. 12; Thw. 25, 9; Jud. 250: Jn. Lind. War. 21, 25: Bt. Met. Fox 18, 11; Met. 18, 6.

a-niðerian; *p.* ode; *pp.* od [a *intensive,* niðerian *to thrust down*] *To put down, condemn, damn;* deorsum trudere :—Ðá wurþe he aniðrod mid Iudas *then let him be cast down with Judas,* Chr. 675; Ing. 52, 12.

an-læc *A respect, regard, consideration;* respectus, Ælfc. Gr. 28, 5; Som. 31, 67.

an-lǽdan; *p.* de *To lead on* or *to;* adducere :—Ðær eorp-werod an-lǽddon *there led on the swarthy host,* Cd. 151; Th. 190, 5; Exod. 194. v. on-lǽdan.

án-lǽtan [án *alone,* lǽtan *to let*] *To let alone, forbear, relinquish;* relinquere, Cd. 30; Th. 40, 24; Gen. 644.

Án-láf, es; *m. Olaf, king of Dublin, defeated at Brunanburh,* Chr. 937; Th. 201, 29, col. 3: 202, 37; Æðelst. 26.

án-laga; *adj. Alone, solitary, without company;* solitarius, Cot. 198.

anlang cempa, an; *m. A regular soldier;* miles ordinarius, gregarius, Cot. 136.

án-lápe; *adj. Going alone, one by one;* singuli :—Ánlápum oððe syndrigum hond gesette *singulis manus imposuit,* Lk. Lind. War. 4, 40. Ða síe awritten ánlápum *quæ scribantur per singula,* Jn. Lind. War. 21, 25. v. án-lépe.

án-lápum; *adv. One by one;* per singula, singulatim, Jn. Lind. War. 21, 25. v. án-lápe, án-lépe.

an-lec *a respect,* Ælfc. Gr. 28, 5; Som. 31, 67, MS. D. v. anlæc.

án-leger; *adj.* [án *one,* leger *jacens*] *Lying with one person;* unicubus :—Ánlegere wífman *a woman with one husband;* unicuba, R. 8.

an-leofa, an; *m. I. food, nourishment;* victus, cibus :—Beón beraþ árlícne anleofan *bees carry delicious food,* Frag. Kmbl. 36; Leás. 20. **II.** *a gift, alms, wages;* stips, Ælfc. Gl. 4; Som. 55, 105.

án-lépe, -lépig, -lípig, -lýpig, [æn-] ; *adj.* [án *one;* hleáp, hlýp *a running, leap*] *Going alone, solitary, private, alone, singular, one, each one;* solivagus, solitarius, privatus, solus, singularis, unus, singulus :—Nis nán ðe eallunga wel dó, nó forðon ánlépe *non est qui faciat bonum, non est usque ad unum,* Ps. Th. 13, 2. Ánlépra *ælc each one,* Bt. Met. Fox 25, 111; Met. 25, 56. [*Ger.* einläufig, einläuftig *solivagus, singularis.*]

án-lépig; *adj. Solitary, private, alone.* v. án-lípig.

án-lépnes, -ness, e; *f. Solitude, loneliness;* solitudo :—Ne tala ðú me, ðæt ic ne cunne ða ánlépnesse ðínes útsetles *think not thou, that I know not the loneliness of thy outsitting,* Bd. 2, 12; S. 513, 41.

an-líc, on-líc; *adj. Like, similar, equal;* similis, æqualis :—Forðam ys heofena ríce anlíc ðam cyninge *ideo assimilatum est regnum cœlorum homini regi,* Mt. Bos. 18, 23. Ðæt he bióþ swíðe anlíc *that he is very like,* Bt. 37, 1; Fox 186, 11. Nis under wolcnum Drihtne ænig anlíc? *quis in nubibus æquabitur Domino?* Ps. Th. 88, 5: 57, 4: 72, 18: 112, 5. [*Ger.* æhnlich *similis :* M. H. Ger. anelîch: O. H. Ger. anagalîh: Goth.* analeiks: O. Nrs.* álíkr.]

án-líc, æn-líc; *adj.* [án *one,* líc *like*] ONLY, *singular, incomparable, excellent, elegant, beautiful;* unicus, eximius, egregius, elegans, pulcher :—He is mín anlíca sunu *unica est mihi filius,* Lk. Bos. 9, 38. Andett seó geladung ðínne sóðan and ánlícan sunu *confitetur ecclesia tuum verum et unicum filium,* Ps. Lamb. fol. 195 a, 12: Te Dm. Thomson 37, 12. Ic spearuwan swá some gelíce gewearþ, ánlícum fugele *factus sum sicut passer unicus,* Ps. Th. 101, 5: Exon. 56 a; Th. 198, 12; Ph. 9: Beo. Th. 507; B. 251. Gesete fram deóflum oððe fram leónum ánlícan oððe ánnysse míne *restitue a leonibus unicam meam,* Ps. Lamb. 34, 17; restore thou myn oon lijf aloone [darling] fro líouns, Wyc.

an-lícast *most like,* Ps. Th. 78, 2: 89, 4, 10: 91, 11; *sup.* of an-líc.

an-líce, on-líce; *adv. In like manner, similarly;* similiter :—Anlíce swá swá *sicut,* Ps. Th. 123, 6. Ðæm anlícost, ðe ... *in a manner most like to his, that ...,* Bt. Met. Fox 20, 337; Met. 20, 169.

án-líce ONLY. v. æn-líce.

an-lícnes, on-lícnes, and-lícnis, -lícness, -lícnýss, e; *f.* **I.** *a like-*

---

ness, image, similitude, resemblance;* imago, similitudo :—Mon wæs tó Godes anlícnesse ǽrest gesceapen *man was to God's image first shapen,* Cd. 75; Th. 92, 15; Gen. 1529. Hwæs anlícnys ys ðis? *cujus est imago hæc?* Mt. Bos. 22, 20. God gesceóp man to his andlícnisse *creavit Deus hominem ad imaginem suam,* Gen. 1, 27. On ðæs mannes sáwle is Godes anlícnyss *in the soul of the man is God's image,* Hexam. 11; Norm. 18, 21. Uton gewyrcan mannan to úre anlícnysse and to úre gelícnysse *faciamus hominem ad imaginem nostram et similitudinem nostram,* 11; Norm. 18, 14, 20, 21, 25. God worhte Adam to his anlícnysse. On hwilcum dǽle hæfþ se man Godes anlícnysse on him? On ðære sáwle, ná on ðam líchaman. Ðæs mannes sáwl hæfþ on hire gecynde ðære Hálgan Þrynnysse anlícnysse; forðan ðe heó hæfþ on hire þreó þing, ðæt is gemynd, and andgit and willa *God made Adam in his own likeness. In which part has man the likeness of God in him? In the soul, not in the body. The soul of man has in its nature a likeness to the Holy Trinity; for it has in it three things, these are memory, and understanding, and will,* Homl. Th. i. 288, 14–19. **II.** *a parable;* parabola :—Ic on anlícnessum ontýne mínes sylfes mûþ *aperiam in parabolis os meum,* Ps. Th. 77, 2. v. big-spell, gelícnes, II. **III.** *an image, statue, idol, stature, height;* statua, simulacrum, statura :—He wundoragræfene anlícnesse geseh *he beheld a wondrously-carved image,* Andr. Kmbl. 1425; An. 713. Tobrec hira anlícnyssa *confringes statuas eorum,* Ex. 23, 24: Cd. 119; Th. 154, 33; Gen. 2565. Anlícnes *agalma, vel iconisma, vel idea,* Ælfc. Gl. 81; Som. 72, 123. Hwylc mæg ícan ane elne to his anlícnesse? *quis potest adjicere ad staturam suam cubitum unum?* Lk. Bos. 12, 25.

án-lípie = án-lípige *solitary, private,* Bd. 1, 15; S. 483, 45. v. án-lípig.

án-lípig, -lýpig; *adj.* [án *one;* hlíp, hlýp] *Going alone, solitary, private, singular, alone;* solitarius, privatus, singularis, solus, tantus :—Se ðá ánlýpig [MS. ánlýpi] awunode on syndrige stówe fram ðære cyricean *qui tum in remotiore ab ecclesia loco solitarius manebat,* Bd. 4, 30; S. 609, 1. Cynelíco getimbro and ánlípige [MS. ánlípie] *publica ædificia et privata,* 1, 15; S. 483, 45. He nánwiht on hand nyman wolde bútan his ágene gyrde ánlípige *nonnisi virgam tantum habere in manu voluit,* 3, 18; S. 546, 32. v. án-lépe.

an-lútan; *p.* -leát, *pl.* -luton; *pp.* -loten *To bend down, to incline;* se inclinare, R. Ben. 53. v. on-lútan.

án-lýpig, -lýpi; *adj. Solitary, private,* Bd. 4, 30; S. 609, 1. v. án-lípig.

an-medla, on-medla, on-mædla, an; *m. Pride, pomp, arrogance, presumption;* superbia, fastidium, arrogantia, præsumptio :—For ðam anmedlan ðe hie ǽr drugon *for the arrogance which they before had practised,* Cd. 214; Th. 269, 16; Sat. 74. Ðú for anmedlan in ðím bære [MS. bère] húsl-fatu hálegu on hand werum *thou, in thy presumption, barest for a possession the holy sacrificial vessels into the hands of men,* Cd. 212; Th. 262, 22; Dan. 748.

an-mitta, an; *m. A measure, bushel;* mensura, modius :—Habbaþ rihte anmittan *habete justam mensuram,* Lev. 19, 35. Hæbbe ælc man rihte anmittan, and rihte wǽgan, and rihte gemetu on ælcum þingum *pondus habebis justum et verum, et modius æqualis et verus erit tibi,* Deut. 25, 15. v. mitta.

an-mód, on-mód; *adj.* [*Ger.* anmüt *gratus,* Grimm] *Steadfast, eager, bold, courageous, daring, fierce;* constans, alacer, animosus :—For wæs anmód, rófe rincas *the folk were steadfast, renowned men,* Cd. 80; Th. 99, 23; Gen. 1650: 80; Th. 100, 10; Gen. 1662. Feónd wæs anmód *the foe was courageous,* 153; Th. 190, 23; Exod. 203. Ðá wearþ yrre an-mód cyning *then the daring king was wroth,* 141; Th. 229, 29; Dan. 224. Úr byþ anmód *a bull is fierce,* Runic pm. 2; Hick. Thes. i. 135; Kmbl. 339, 7.

án-mód; *adj.* [án *one;* mód *mood, mind*] *Of one mind, unanimous;* unanimis :—Ðú sóþlíce man ánmód *tu vero homo unanimis,* Ps. Spl. 54, 14: 67, 6. Ealle ánmóde *all with one mind,* Andr. Kmbl. 3128; An. 1567. Hie ðá ánmóde ealle cwǽdon *then they all with one mind said,* 3200; An. 1603: 3274; An. 1640: Elen. Grm. 397: 1118. [*Ger.* einmütig *unanimis :* M. H. Ger. einmuot: O. H. Ger. einmuoti *unanimis, constans.*]

án-módlíce; *adv. Unanimously, with one accord;* unanimiter :—Hí ánmódlíce cómon *they came with one accord,* Jos. 11, 4: Exon. 12 b; Th. 21, 25; Cri. 340. Gesamnodon hí ealle ánmódlíce [MS. ánmódlíc] *congregati sunt pariter,* Jos. 9, 2.

án-módnes, -módness, e; *f. Unity, unanimity;* unitas, unanimitas, Som.

ann *he gives :*—Ðe he ann *he gives thee,* Ps. Th. 74, 7 = an; *pres.* of unnan.

-anne, -enne, -ende *the termination of the declinable infinitive in the dat. governed by* to, *as,*—Ondréd to faranne *timuit ire,* Mt. Jun. and Th. 2, 22, but the B. MS. of A. D. 995 has farende, also Foxe, Bos. and the Rl. MS. about A. D. 1145. The Lind., about A. D. 957, has farenne [MS. færenne]. Alýfe me to farenne *permitte me ire,* Mt. Bos. 8, 21, and B. MS. about A. D. 995. Sometimes -ende is found, because -enne = ende, as in the preceding example farende about A. D. 995. The

most usual form is -anne, from the infin. -an; *g.* -annes; *dat.* -anne. v. TO; *prep.* **IV. 2**: also -enne and -ende, and Grm. iv. III.

ân-ne *alone;* solum :—Ðæt ge forlæton me ânne *that ʒe leeue me aloone,* Wyc; ut me solum relinquatis, Jn. Bos. 16, 32. v. ân, **II.**

ân-nes, ân-nys, ânes, -ness, e; *f.* **I. ONENESS,** *unity;* unitas :— Geleáfa sôþlíce se geleáffulla ðes is; ðæt ânne God on Þrynnesse and Þrynnesse on Annesse we ârwurþian *fides autem catholica hæc est; ut unum Deum in Trinitate et Trinitatem in Unitate veneremur,* Ps. Lamb. fol. 200 a, 13. On ða ânnysse ðære hâlgan cyrican *in unitate sanctæ ecclesiæ,* Bd. 2, 4; S. 505, 7; 4, 5; S. 572, 1. We andettaþ Þrynnesse in Ânnesse efenspêdiglíce, and Ânnesse on ðære Þrynnesse *confitemur Trinitatem in Unitate consubstantialem, et Unitatem in Trinitate,* 4, 17; S. 585, 37: Exon. 76 a; Th. 286, 5; Jul. 727: Hy. 8, 41; Hy. Grn. ii. 291, 41. Gesete fram deôflum oððe fram leónum ânlícan oððe ânnysse mîne *restitue a leonibus unicam meam,* Ps. Lamb. 34, 17; restore thou myn oon lijf aloone [darling] fro liouns, Wyc. **II.** *a covenant, an agreement;* conventio :—Gewearþ him and ðam folce on Lindesige ânes *there was an agreement between him and the people in Lindsey,* Chr. 1014; Th. 274, 13, col. 1. **III.** *loneliness, solitude;* solitudo :—Ânnys ðæs wîdgillan wêstenes *the solitude of the wide desert,* Guþl. 3; Gdwin. 20, 20.

ân-nyss, e; *f. Oneness, unity, agreement, solitude;* unitas, conventio, solitudo, Bd. 2, 4; S. 505, 7. v. ân-nes.

anoða? *fear, amazement;* formido. v. onoða.

ân-pæþ, es; *nom. pl.* -paðas; *m. A single path, a pass, lonely way;* solitaria via :—Enge ânpaðas, uncûþ gelâd *narrow passes, an unknown way,* Beo. Th. 2824; B. 1410: Cd. 145; Th. 181, 8; Exod. 58.

ânra *of every one; g. pl. of* ân *one,* q. v.

ân-ræd; *adj.* [ân *one,* ræd *counsel*] *One-minded, unanimous, agreed, persevering, resolute, prompt, vehement;* unanimus, firmus consilii, confidens, audax, vehemens :—And ðonne beón hîg ânræde *and when they be unanimous,* L. Ath. iv. 7; Th. i. 226, 19. Ðis swefen ys ânræde *somnium unum est,* Gen. 41, 25. Ealle ânræde to gemænra þearfe *all unanimous for the common need,* L. Edg. C. 1; Th. ii. 244, 4. Wæs seó mæg ânræd and unforht *the maid was resolute and fearless,* Exon. 74 b; Th. 278, 21; Jul. 601. Eft wæs ânræd mæg Hygeláces *Hygelac's kinsman was resolute again,* Beo. Th. 3062; B. 1529: Byrht. Th. 133, 2; By. 44.

ân-rædlíce, -rêdlíce; *adv.* [ân, ræd *opinion, advice,* líce] *Unanimously, resolutely, constantly;* unanimiter, constanter :—Hî þohton ânrædlíce [MS. ânhrædlíce] *cogitaverunt unanimiter,* Ps. Spl. 82, 5. Ðe ânrædlíce wile his sinna geswîcan *who resolutely desires to abstain from his sins,* L. Pen. 17; Th. ii. 284, 17. Ânrædlíce wrêgende *constanter accusantes,* Lk. Bos. 23, 10.

ân-rêdnes, -rêdnes, -nys, -ness, -nyss, e; *f.* [ân *one,* rædnes *opinion*] *Unanimity, concord, agreement, constancy, steadfastness, diligence, earnestness;* concordia, constantia :—Hî heora ânrædnesse geheóldan him betwênan *they had agreement among themselves,* Ors. 5, 3; Bos. 103, 44. Brôðerlíc ânrædnys *brotherly unanimity,* Scint. 11. Ânrædnys gôdes weorces *constancy of good works,* Oct. vit. cap. Scint. 7; Job Thw. 167, 33. *Opposed to* twŷrædnes, un-geræednes *dissention,* q. v.

ânra-gehwâ, ânra-gehwilc *every one;* unusquisque, Deut. 24, 16. v. ân, **IV.**

ân-reces; *adv. Continually, forthwith,* Chr. 1010; Th. 262, 34. v. ân-streces.

ân-rêdlíce *unanimously,* Jud. Thw. 161, 27. v. ân-rædlíce.

ân-rêdnes *unanimity, constancy,* Bd. 1, 7; S. 477, 43. v. ân-rædnes.

an-rine, es; *m.* [an *in,* ryne *a course*] *An inroad, incursion, assault;* incursio :—Fram anrine *ab incursu,* Ps. Spl. 90, 6.

an-sacan; *p.* -sôc, *pl.* -sôcon; *pp.* -sacen *To strive against, resist, deny;* impugnare, repugnare, negare :—Se ðe lyhþ, oððe ðæs sôðes ansaceþ *he that lieth, or the truth resisteth,* Salm. Kmbl. 365; Sal. 182: L. In. 46; Th. i. 130, 14, 15. v. on-sacan.

an-sæc, es; *m?* *Contention, resistance;* contentio, repugnantia :—Bûtan ansæce *without resistance,* Chr. 796; Ing. 83, 5. v. and-sæc.

an-sægdnes, an-segdnes, -ness, e; *f.* [ansægd *affirmed; pp. of* an-secgan] *A thing which is vowed,* or *devoted, an oblation, a sacrifice;* sacrificium, Bd. 1, 7; S. 477, 39. v. onsægdnes.

an-sæte *odious, hateful;* exosus, perosus, Ælfc. Gl. 84; Som. 73, 101; Wrt. Voc. 49, 9. v. and-sæte.

an-sceát, -sceót, es; *m?* *The bowels;* exentera = ἔντερα, *pl. n,* Cot. 73.

an-scôd *unshod;* discalceatus. v. un-scôd.

an-scûnian *to shun;* evitare, Bt. 18, 1; Fox 60, 20. v. onscûnian.

an-scûniend-líc, an-scûnigend-líc *abominable;* abominabilis. v. on-scûniendlíc.

an-secgan; *p.* -sægde, -sæde; *pp.* -sægd, -sæd *To charge against, affirm,* L. Edg. ii. 4; Wilk. 78, 12. v. on-secgan.

ân-seld, es; *m.* [ân *only,* seld *dwelling*] *A solitary dwelling, an hermitage;* habitatio solitaria :—Ic ongon on ðone ânseld bûgan *I began to dwell in this hermitage,* Exon. 50 b; Th. 176, 23; Gû. 1214.

an-sendan; *p.* -sende *To send forth, send;* emittere, mittere :—Ne mægen hî leóhtne leóman ansendan *they cannot send forth a clear light,*

Bt. Met. Fox 5, 10; Met. 5, 5: Ps. C. 50, 16; Ps. Grn. ii. 277, 16. v. on-sendan.

an-settan *to impose,* Bt. 39, 10; Fox 228, 4. v. on-settan.

an-sién, e; *f. aspect, figure* :—Idesa ansién *the aspect of the females,* Cd. 64; Th. 76, 22; Gen. 1261. Ansién ðyses middan-geardes *the figure of this world,* Past. 51, 2. v. an-sŷn, **II.**

an-sín, e; *f. a view, sight, figure* :—Ðîn môd wæs âbîsgod mid ðære ansíne ðissa leásena gesælþa *thy mind was occupied with the view of these false goods,* Bt. 22, 2; Fox 78, 10: Bd. 5, 13; S. 633, 5. Gûþlác wæs on ansíne mycel *Guthlac was tall in figure,* Guthl. 2; Gdwin. 18, 1. v. an-sión, **II.**

an-sión, e; *f. a sight* :—Ne aweorp ðû me fram ansióne ealra ðînra miltsa *cast me not away from the sight of all thy mercies,* Ps. C. 50, 95; Ps. Grn. ii. 279, 95. v. an-sŷn, **III.**

an-speca, on-speca, an; *m.* [spæc *a speech*] *A speaker against, an accuser, a persecutor;* persecutor. v. an = and *Against,* spæca *a speaker.*

an-spel, -spell, es; *n.* [an, spel *a speech*] *A conjecture;* conjectura, Cot. 56.

an-spilde; *adj.* [an = and *against,* spild *destruction*] *Anti-destructive, salutary;* salutaris :—Ðæt biþ anspilde lyb wið eágena dimnesse *that is a salutary medicine for dimness of eyes,* L. M. 1, 2; Lchdm. ii. 30, 14.

an-spræce; *adj. One speaking, speaking as one,* Ps. Th. 40, 7. v. -spræce.

an-standan; *p.* an-stôd, *pl.* an-stôdon; *pp.* an-standen. **I.** *to stand against, resist, withstand, to be firm* or *steadfast;* adversari. **II.** *to stand upon, inhabit, dwell;* insistere, habitare. v. on-standan.

an-standende; *part. One standing alone* :—Ânstandende, ân-stonde, oððe munuc *one standing alone,* or *a monk,* Ælfc. Gl. 3?

ân-stapa, an; *m. A lone wanderer;* solivagus, Exon. 95 b; Th. 356, 21; Pa. 15.

ân-stealled *one-stalked* :—Nim bête, ðe biþ ânstealled *take beet, which is one-stalked,* Lchdm. iii. 70, 2. v. ân-steled.

ân-steled, an-stealled *One-stalked, having one handle* or *stalk;* unicaulis, L. M. 1, 1; Lchdm. ii. 20, 15: Lchdm. iii. 70, 2.

an-stellan; *p.* -stealde, -stalde; *pp.* -steald *To cause, establish, appoint;* instituere, constituere :—Ic ðæs orleges ôr anstelle *I cause the beginning of that strife,* Exon. 102 a; Th. 386, 10; Rä. 4, 59. v. on-stellan.

ân-stonde *one standing alone, a monk.* v. ân-standende.

ân-stræc; *adj.* [ân *one;* strec *stretch, from* streccan *to stretch?*] *Of one stretch, constant, resolute, determined;* pertinax :—Ða ânstræcan sint to monianne *admonendi sunt pertinaces,* Past. 42, 2; Hat. MS. 58 a, 24.

ân-streces; *adv.* [ân *one;* streces, *gen. of* strec *a stretch*] *At one stretch, with one effort, continually;* sine intermissione :—And fôron on ânstreces dæges and nihtes *and went at one stretch day and night,* Chr. 894; Th. 170, 25.

ân-sûnd, on-sûnd; *adj.* [ân *sole, entire, wholly;* sûnd *sound*] *Sound, entire, unhurt;* sanus, integer, incolumis :—Hrôf âna genæs ealles ânsûnd *the roof alone was saved wholly sound,* Beo. Th. 2004; B. 1000. Gehwâ ânsûndan and ungewemmedne [geleáfan] healde *quisque integram inviolatamque* [*fidem*] *servaverit,* Ps. Lamb. fol. 200 a, 7. Beóþ ðá gebrosnodan bân mid ðam flæsce ealle ânsûnde eft geworden *then the corrupted bones together with the flesh will all again be made sound,* Hy. 7, 89; Hy. Grn. ii. 289, 89. Seó heofon is sinewealt and ânsûnd *heaven is circular and entire,* Bd. de nat. rer; Wrt. popl. scienc. 1, 17. v. on-sûnd.

ân-sûndnes, -ness, e; *f.* [ân, sûnd, nes] *Wholeness, soundness, integrity;* integritas :—Ânsûndnesse lufigend *a lover of integrity,* Wanl. Catal. 292, 34.

an-swarian; *p.* ode; *pp.* od *To answer;* respondere :—Ic answarige *ego respondebo,* Ps. Spl. 118, 42. v. and-swarian.

ân-swêge; *adj.* [ân *one,* swêg *a sound*] *Of the same sound, agreeing in sound, consonant;* consonus :—Ânswêge sang *symphonia,* Ælfc. Gl. 34; Wrt. Voc. 28, 40.

an-sŷn, -sîn, -sién, -sión; on-, e; *f.* [an, sŷn *sight, vision*]. **I.** *a face, countenance;* facies, vultus :—His ansŷn sceán swâ swâ sunne *facies ejus resplenduit sicut sol,* Mt. Bos. 17, 2. Beforan ðîne ansŷne *ante faciem tuam,* Lk. Bos. 7, 27. Gûþlác wæs wlitig on ansŷne *Guthlac was handsome in countenance,* Guthl. 2; Gdwin. 18, 3. God ableów on his ansŷne líflícne blæd *God blew into his face the breath of life,* Hexam. 11; Norm. 18, 25. Fleóþ his ansŷne *fugiant a facie ejus,* Ps. Th. 67, 1. Gedô ðæt hiora ansŷn âwa sceamige *imple facies eorum ignominia,* 82, 12. Ansŷn ðîn *vultus tuus,* 88, 14. Ic bidde ðînre ansŷne *deprecatus sum faciem tuam,* 118, 58. Ansŷn ŷwde *shewed his countenance,* Beo. Th. 5660; B. 2834. **II.** *a view, aspect, sight, form, figure;* aspectus, conspectus, visus, visio, species, forma, figura :—Fæger ansŷne *fair in aspect,* Runic pm. 11; Hick. Thes. i. 135; Kmbl. 341, 19. Ðîn môd wæs âbîsgod mid ðære ansíne ðissa leásena gesælþa *thy mind was occupied with the view of these false goods,* Bt. 22, 2; Fox 78, 10. For ðînre ansŷne *in conspectu tuo,* Ps. Th. 68, 20: 108, 14. Se Hâlega Gâst astâh lîchamlícre ansŷne, swâ ân culfre *descendit Spiritus Sanctus corporali specie, sicut columba,* Lk. Bos. 3, 22: Cot. 74. Ansién ðyses middan-geardes *figura hujus mundi,* Past. 51, 2. **III.**

a thing to be looked upon, *a sight*; spectaculum :—Ðisse ansýne Alwealdan þanc gelimpe *for this sight may thanks to the Almighty take place*, Beo. Th. 1860; B. 928. Seó ansín wearþ mycel wundor Rômānum *the sight was a great wonder to the Romans*, Ors. 6, 7; Bos. 120, 3. IV. a view or sight producing desire or longing, and hence,—*a desire of anything, want* or *lack of anything*; desiderium, defectus :—Swā eorþan biþ ansýn wæteres *sicut terra sine aquâ*, Ps. Th. 142, 6. [*O. Sax.* ansiun, *f. aspectus : Plat.* anseen, *n : Dut.* aanzien, *n : Ger.* ansehen, *n. aspectus, forma : M. H. Ger.* ansiune, *n : O. H. Ger.* anasiuni, *n.*]

an-tállíc, an-tālíc; *adj.* [an = un *not*, tāllíc *blamable*] *Unblamable, undefiled*; irreprehensibilis, immaculatus :—Æ Drihtnes antālíc *lex Domini immaculata*, Ps. Spl. 18, 8.

Antecrist, es; *m. Antichrist*; Antichristus :—Ðonne cymþ se Antecrist, se biþ mennisc mann and sóþ deófol *then Antichrist shall come, who is human being* [*man*] *and true devil*, Homl. Th. i. 4, 14. Ðes deófol, ðe is gehāten Antecrist, ðæt is gereht þwyrlíc Crist, is ord ælcere leásunge and yfelnysse *this devil, who is called Antichrist, which is interpreted opposed Christ, is the origin of all leasing and evil*, Homl. Th. i. 4, 21. Togeánes Antecriste *against Antichrist*, Ælfc. T. 6, 22 : Job Thw. 166, 8.

antefn = antefen, e; *f?* es; *n?* [ἀντί *opposite*, φωνή *a voice*] *An antiphon, anthem, a hymn sung in alternate parts*; antiphona, cantus Ecclesiasticus alternus :—Is ðæt sægd, ðæt hí ðysne letanían and antefn geleóþre stæfne sungan *fertur, quia hanc litaniam consona voce modularentur*, Bd. 1, 25; S. 487, 24.

ant-fenge; *adj. Acceptable*; acceptabilis, R. Ben. 5. v. and-fenge.

an-þracian *to fear, to be afraid, to dread*; revereri, horrere :—Ic onginne to anþracigenne *I begin to dread*; horresco, Ælfc. Gr. 35; Som. 38, 4 : Ps. Spl. 69, 2. v. on-þracian.

an-prǽclíc? *adj. Horrible, terrible, fearful*; horridus, horribilis, terribilis, Hymn?

ân-tíd, e; *f. The first hour*; hora prima :—Ymb ân-tíd ôðres dôgores *about the first hour of the second day*, Beo. Th. 443; B. 219.

an-timber; *g. -timbres; n. Matter, materials, substance, a theme*; materies, materia :—Ungehiwod antimber *rudis atque informis materia*, Alb. resp. 15, 22. v. and-timber.

antre, an; *f. Radish?* raphanus, raphanis sativa :—Dô ðonne betonican and antran *add then betony and ontre* [*radish?*], L. M. 2, 51; Lchdm. ii. 266, 3. Ancre [antre?], ðæt is rædic *raphanus*, Mone A. 493. v. ontre.

an-trumnys *infirmity*; infirmitas. v. un-trumnes.

an-tymber *matter*, Ælfc. Gr. 12; Som. 15, 54. v. an-timber.

an-týnan; *p.* de; *pp.* ed [an = un *un-*, týnan *to inclose*] *To unclose, open*; recludere, aperire :—Ic antýne on bigspellum mûþ mínne *aperiam in parabolis os meum*, Ps. Spl. 77, 2. v. un-týnan, on-týnan.

a-numen *taken away*; *pp.* of a-niman.

anunga *zeal, an earnest desire, jealousy*; zelus, Jn. Rush. War. 2, 17.

ânunga; *adv. Entirely, necessarily, by all means*; plane, prorsus, omnino, Beo. Th. 1272; B. 634. v. âninga.

an-wadan; *p.* -wôd *To invade, enter into*; invadere :—Hie wlenco anwôd *pride invaded them*, Cd. 173; Th. 217, 3 ; Dan. 17. v. on-wadan.

ân-wald, es; *m. Sole power, jurisdiction, rule* :—Ðæt se Cāsere eft ânwald ofer hí âgan môste *that the Cæsar might again obtain power over them*, Bt. Met. Fox 1, 123; Met. 1, 62. Se ânwald Godes Ælmihtiges *the power of Almighty God*, 9, 95; Met. 9, 48 : Exon. 63 a; Th. 232, 23; Ph. 511: Lk. Bos. 23, 7: Bd. 4, 32; S. 611, 15: Ors. 2, 1; Bos. 38, 11. v. ân-weald.

ân-walda, an; *m. A sole ruler, the sole ruler of the universe* :—Him to Ânwaldan âre gelýfde *in him as sole ruler reverently trusted*, Beo. Th. 2548; B. 1272. Ealra Ânwalda, eorþan and heofones *ruler of all, of earth and heaven*, Exon. 110 a; Th. 422, 10; Rä. 41, 4 : Cd. 227; Th. 305, 5; Sat. 642. v. ân-wealda.

ân-waldan *to have sole power over, to exercise absolute rule*; solam potestatem habere, dominari :—He ðone ânwaldeþ *he rules it*, Bt. Met. Fox 29, 154. v. wealdan.

ân-waldeg? *adj. Having sole power, powerful*; solus potens :—Ðæt se síe ânwaldegost *that he is most powerful*, Bt. 36, 5 ; Fox 180, 16.

an-walg, -wealg; *adj. Entire, whole, sound*; integer, Past. 52, 2. v. on-walg.

an-wann *fought against*; *p.* of an-winnan.

ân-weald, ân-wald, es; *m. Single, sole, monarchical*, or *royal power, empire, dominion, jurisdiction, rule, government, bidding*; solius dominatus, unius imperium, monarchia, potestas, imperium, ditio, dominatio, jus, arbitrium, nutus :—Me is geseald ælc ânweald *data est mihi omnis potestas*, Mt. Bos. 28, 18. Ânweald Godes is *potestas Dei est*, Ps. Spl. 61, 11. Ðín ânweald *dominatio tua*, Ps. Th. 144, 13 : 135, 20 : 118, 91: Ors. 2, 1; Bos. 38, 15 : Bd. 1, 3; S. 475, 12. Cyning biþ ânwealdes georn *a king is desirous of power*, Exon. 89 b; Th. 337, 4; Gn. Ex. 59. Mid ðínum âgenum ânwealde *by thine own power*, Bt. 33, 4 ; Fox 128, 13. Hí synd heora sylfes ânwealdes *illi sunt sui juris*, Bd. 5, 23; S. 647, 4. On his ânwealde *ad ejus nutum*, Gen. 42, 6. [*O. Nrs.* einwald, *n. singularis potestas, monarchia.*] DER. wealdan.

ân-wealda, ân-walda, an; *m.* [ân *one, sole*; wealda, walda *a ruler*]

The one or *sole ruler of a province* or *of the universe, a sovereign, governor, magistrate, a power*; qui solus dominatur, monarcha, dominus, gubernator, magistratus, potestas :—Se Ânwealda hæfþ ealle his gesceafta befangene and getogene *the governor has caught hold of, and restrained all his creatures*, Bt. 21; Fox 74, 5. Ânwealda Ælmihtig *Almighty Ruler*, Rood Kmbl. 303; Kr. 153. Ðonne híg lǽdaþ eów to ânwealdum *cum inducent vos ad potestates*, Lk. Bos. 12, 11. [*O. Nrs.* einwaldi, *m. solus dominus.*]

an-wealg *whole*. v. an-walg.

an-wealglíce; *adv. Wholly, soundly*; integre, Past. 33, 5 ; Hat. MS. 42 a, 33.

an-wealgnes, -ness, e; *f. Wholeness, soundness, entireness*; integritas. v. on-walhnes.

an-weg *away*; inde, exinde. v. on-weg.

an-weorc, es; *n. Material, cause*; materia, causa :—Bûton anweorce *without cause*, Bt. 30, 2 ; Fox 110, 16. v. and-weorc.

ân-wíg, es; *n? m?* [ân *one*, wíg *a contest*] *A single combat, a duel*; certamen singulare :—Ðǽr gefeaht Mallius ânwíg wið ænne Galliscne mann *there Mallius fought a single combat with a man of Gaul*, Ors. 3, 4; Bos. 56, 15 : 3, 6; Bos. 57, 42. Hí gefuhton ânwíg *they fought a duel*, Ors. 3, 9; Bos. 67, 32.

ân-wíg-gearo, -gearu; *g. m. n.* -wes, -owes *f.* -re, -rwe; *adj.* [gearo *prepared*] *Prepared for single combat*; ad singulare certamen paratus :—Wæs þeáw hyra, ðæt hie oft wǽron ânwíggearwe *it was their custom, that they oft were for single combat prepared*, Beo. Th. 2499; B. 1247. v. gearo; *adj.*

ân-wíglíce; *adv. In single combat*; singularis certaminis modo :—Ân-wíglíce feohtende *fighting in single combat*, Cot. 186.

ân-wille, *def.* se ân-willa; *adj.* [ân *one*, willa *a will*] *Having one will, following one's own will, self-willed, obstinate, stubborn*; pertinax, obstinatus, contumax :—Ânwilla *obstinatus, pertinax*, Ælfc. Gl. 90; Wrt. Voc. 51, 29. Sint to maníanne ða ânwillan *admonendi pertinaces*, Past. 42, 1; Hat. MS. 57 b, 23.

ân-willíce; *adv. Obstinately, stubbornly, pertinaciously*; pertinaciter :—Ic tô ânwillíce winne wið ða wyrd *I too pertinaciously attack fortune*, Bt. 20; Fox 70, 20 : Past. 7, 2; Hat. MS. 12 a, 15.

ân-wilnes, -ness, e; *f. Obstinacy, self-will, contumacy*; pertinacia, protervia, Past. 32, 1; Hat. MS. 40 a, 16, 25.

an-winnan; *p.* -wann *To fight against, to attack*; impugnare :—Him onwann [MS. L. anwann] *fought against them*, Ors. 3, 7; Bos. 61, 7.

ân-wintre, æ-wintre; *adj.* [ân *one*, winter *a winter*] *Of one year, one year old, continuing for a year*; hornus = horinùs = ὥρινος from ὥρα, hornotínus, anniculus :—Ðæt lamb sceal beón ânwintre *erit agnus anniculus*, Ex. 12, 5.

ân-wíte, es; *n. A simple* or *single fine, a mulct* or *amercement*; simplex mulcta :—Ealle forgielden ânwíte *let them all pay a single fine*, L. Alf. pol. 31; Th. i. 80, 17.

an-wlǽta, -wlāta, an; *m. A livid bruise*; sugillatio, livor :—Wið wundspringum and anwlātan *ad livores et sugillationes*, Med. ex quadr. 7; Lchdm. i. 356, 20. v. wlǽtan.

an-wlita, an; *m. The countenance, face*; vultus, facies, Ælfc. Gl. 70; Som. 70, 44. v. and-wlita.

an-wlíte, es; *n.* [an = un *un-*, wlite *decus*] *Disgrace*; dedecus :—Sconde oððe anwlite *dedecus*, Cot. 66, Lye.

an-wlitegian; *p.* ode; *pp.* od [an = un *un-*, wlitigian *to form*] *To unform, change the form of anything*; deformare :—Ða he þwaraþ and gewlitegaþ; hwílum eft unwlitegaþ [MS. Cot. anwlitegaþ] *these it tempers and forms; sometimes again it unforms*, Bt. 39, 8 ; Fox 224, 9.

an-wlô, an-wlôh; *adj.* [an = un *without*, wlôh *a fringe, ornament*] *Untrimmed, neglected, without a good grace, deformed, ill-favoured*; inornatus, deformis :—Ðín ríce restende biþ an-wlôh *thy kingdom shall remain neglected*, Cd. 203; Th. 252, 27; Dan. 585.

an-wôd *invaded*, Cd. 173; Th. 217, 3 ; Dan. 17; *p.* of an-wadan.

an-wreón; *pl.* -wreáh, *pl.* -wrugon; *p.* -wrogen [an = un *un-*, wreón *to cover*] *To uncover, reveal*; revelare, R. Ben. 3. v. un-wreón, on-wreón.

an-wrigenys, -nyss, e; *f.* [an = un, wrigen, nys] *A revealing, disclosing, an opening, a sermon, homily*; explicátio, expositio. v. wrigen; *pp.* of an-wreón *to cover*.

ân-wunian; *part.* -wuniende; *p.* ode; *pp.* od *To dwell* or *be alone*; esse solitarius, Ps. Lamb. 101, 8.

ân-wuniende; *part. Dwelling alone, being alone*; solitarius :—Geworden ic eom swā swā spearwa ânhoga oððe ânwuniende on efese oððe on þecene *factus sum sicut passer solitarius in tecto*, Ps. Lamb. 101, 8.

an-wunigende; *part. Dwelling in, inhabiting;* inhabitans, Bt. Met. Fox 7, 93; Met. 7, 47; *part. pres.* of an-wunigan = ou-wunian, *q. v.*

anxsumnes, -ness, e; *f. Anxiety*, Somn. 87; 133. v. angsumnes.

a-nýdan; *p.* -nýdde; *pp.* -nýded, *pl.* -nýdede = -nýdde [a *from*, nýdan *to compel*]. I. *to repel, thrust* or *beat back, keep from, restrain, constrain, force*; repellere, extorquere :—Hí fram his mâgum ǽr mid unrihte anýdde wǽron *they had formerly been unjustly forced from his kinsmen*, Chr. 823; Th. 111, 34. II. *with* ût *to expel, to drive*

out; expellere, depellere, exigere :—Ic anýde hîg ût on fremde folc *I will drive them out among a strange people,* Deut. 32, 21.

a-nyman; *impert.* a-nymaþ ge *To take away;* tollere :—Anymaþ ðæt pûnd æt hym *take away that pound from him,* Mt. Bos. 25, 28 : Hick. Thes. i. 192, 16, col. 2. v. a-niman.

an-ȳwan; *p.* de; *pp.* ed *To shew, demonstrate;* ostendere, demonstrare, R. Ben. 7, 11. v. eáwan.

apa, an; *m. An* APE; simia :—Wið apan bîte *against bite of an ape,* Med. ex quadr. 11, 7; Lchdm. i. 366, 24 : Ælfc. Gl. 19; Som. 59, 18; Wrt. Voc. 22, 59.

a-pǽcan; *p.* -pǽhte; *pp.* -pǽht *To seduce, mislead;* seducere :—Gif hwǽ ðîres mannes folgere fram him apǽce *si quis alius hominis pedisequam ab eo seducat,* L. M. I. P. 23; Th. ii. 270, 31.

a-pǽran *to pervert, turn from;* evertere, pervertere. v. for-pǽran.

a-parian; *p.* ode; *pp.* od *To apprehend, take;* deprehendere :—Seó wæs aparod on unriht-hǽmede *deprehensa est in adulterio,* Jn. Bos. 8, 3.

apelder-tûn, es; *m. An apple-tree garden.* v. apulder, apulder-tûn.

ap-flôd, es; *m. The low tide;* ledo, æstus maris, Martyr. 20, Mar. v. nêp-flôd.

a-pinsian; *p.* ode; *pp.* od, ud *To ponder, weigh, estimate;* ponderare, pensare :—Ðá ðá he ðæra Judea misdǽda ealle apinsode *when he estimated all the misdeeds of the Jews;* cum Judeæ singula delicta pensarentur, Past. 53, 3. DER. pinsian.

apl, es; *m;* nom. acc. pl. aplas, *m;* nom. acc. pl. apla, *n. An apple, a ball* :—Ða reádan appla [MS. C. apla] *mala Punica,* Past. 15, 5; Hat. MS. 19 b, 28 : Salm. Kmbl. 55; Sal. 28. v. appel.

a-plantian; *p.* ode; *pp.* od *To plant, transplant;* plantare, transplantare :—God ðá aplantode wynsumnisse orcerd *plantaverat autem Dominus Deus paradisum voluptatis,* Gen. 2, 8. Ge sǽdon ðissum treówe, Sý ðû awyrtwalod, and aplantod on sǽ *dicetis huic arbori, Eradicare, et transplantare in mare,* Lk. Bos. 17, 6.

Apollinus; *gen.* Apollines; *m. Apollo;* Apollo, Ínis; *m.* [=’Απόλλων, ωνος; *m.*] :—Wæs se Apollinus ǽdeles cynnes, Iôbes eafora *this Apollo was of noble race, the son of Jove,* Bt. Met. Fox 26, 67; Met. 26, 34. Apollines dôhtor *Apollo's daughter,* 26, 64; Met. 26, 32 : Bt. 38, 1; Fox 194, 12, 19.

apostata, an; *m. An apostate;* apostata :—Hér syndon apostatan *here are apostates,* Lupi Serm. i. 19; Hick. Thes. ii. 105, 1.

apostol, es; *m: also like the Lat.* Apostolus; *g.* -i; *m. One sent, an apostle;* apostolus [=ἀπόστολος, ἀπό *from,* στέλλω *to send*] :—Se eádiga apostol Simon *the blessed apostle Simon,* Homl. Th. ii. 492, 7. He apostolas geceás, ðæt sind ǽrendracan *he chose apostles, that are messengers,* Ælfc. T. 26, 17. Ðá gesáwon ða apostolas Drihten *then the apostles saw the Lord,* Homl. Th. ii. 494, 28. Ða apostoli becômon to ðære byrig *the apostles came to the city,* 494, 14 : 482, 18, 25, 27. Æt ðæra apostola fôtum *at the apostles' feet,* 488, 4. Ðá fleáh ðæt folc eal to ðám apostolum *the folk then all fled to the apostles,* 492, 12. Se ealdorman ðá ða apostolas mid him to ðam cyninge Xerxes gelǽdde *the general then led the apostles with him to the king Xerxes,* 486, 3. Ðæra twelf apostola naman *duodecim apostolorum nomina,* Mt. Bos. 10, 2 : Cd. 226; Th. 300, 27; Sat. 571 : Menol. Fox 242; Men. 122. DER. ealdor-apostol.

apostol-hâd, es; *m. The apostolic office;* apostolatus :—Se apostolhâd *the apostolic office,* Apstls. Kmbl. 28; Ap. 14. Gesette bisceop ðám leódum and gehálgode þurh apostolhâd *set a bishop over the people and hallowed him through the apostolic office,* Andr. Kmbl. 3300; An. 1653.

apostolic; *def. m.* -a, *f. n.* -e; *adj. Apostolic;* apostolicus :—Ðá ongunnon hí ðæt apostolíce líf ðære frymþelícan cyricean onhýrigean *cœperunt apostolicam primitivæ ecclesiæ vitam imitari,* Bd. 1, 26; S. 487, 31. Se papa ðe on ðam tíman ðæt apostolíce setl gesæt *the pope who at that time occupied the apostolic seat,* Homl. Th. ii. 120, 10.

appel, es; *m;* nom. acc. pl. applas, *m;* nom. acc. pl. appla; *n. An apple* :—Ða reádan appla *the red apples;* mala Punica, Past. 15, 5; Hat. MS. 19 b, 28. v. æppel.

appel-leáf, es; *n.* [lit. *apple-leaf*] *A violet;* viola, viola odorata, Harl. Gl. 978. v. æppel-leáf.

appel-screáda APPLE-SHREDS, *apple-parings.* v. æppel-screáda.

appel-þorn, es; *m. An* APPLE-THORN, *a crab-tree;* pirus malus, Cod. Dipl. Apndx. 460; A. D. 956; Kmbl. iii. 448, 20.

appel-treów *an apple-tree.* v. apple-treów.

appel-tûn *an apple-garden, orchard.* v. apple-tûn.

apple-treów, es; *n. An apple-tree;* pomus, malus, Ælfc. Gr. 5? v. æppel-treów.

apple-tûn, es; *m. An orchard;* pomarium, Cot. 146. v. æppel-tûn.

Aprêlis; *m. April;* Aprîlis mensis :—Aprêlis mônaþ *the month April,* Menol. Fox 112; Men. 56.

aprotane, an; *m. The herb southernwood, wormwood;* abrotonum =ἀβρότονον [artemisia, Lin.] :—Genim aprotanan *take wormwood,* L. M. 1, 16; Lchdm. ii. 60, 1.

apulder, apuldor; es, *n? An apple-tree;* malus, Wrt. Voc. 32, 47 : L. M. 1, 23; Lchdm. ii. 66, 1 : 1, 36; Lchdm. ii. 86, 6. Sûr-melsc

[MS. -melst] apulder *malus matiana* [MS. *matranus*],—*pyrus malus,* Lin, *a sour-sweet apple-tree, a souring apple-tree,* Wrt. Voc. 32, 48. Swête [MS. swîte] apulder *a sweet apple-tree;* malomellus, 32, 49.

Apulder, es; *m.* [*in paludibus*] APPLEDORE, *a village in Kent, near Tenterden* :—Æt Apuldre *at Appledore,* Chr. 893; Th. 164, 10 : 894; Th. 166, 41, col. 1. Æt Apoldre *at Appledore,* Th. Diplm. A. D. 1032; 328, 23. [O. Dut. polder, *m. palus marina pratum litorale : ager, qui est fluvio aut mari eductus, aggeribus obsepitur,* Kil.]

Apulder-comb, es; *m.* [*in paludibus vallis*] APPLEDORE COMBE, *Isle of Wight;* nomen loci in insula *Vecti,* Mann.

apulder-tûn, es; *m. An apple-tree inclosure, an apple-orchard;* malorum hortus, arborum pomiferarum hortus, Cot. 146.

apuldor-rind, apuldre-rind, e; *f. Apple-tree rind;* mali cortex :—Nim apuldorrinde *take apple-tree rind,* L. M. 1, 38; Lchdm. ii. 98, 7 : 3, 47; Lchdm. ii. 338, 12 : Med. ex quadr. 8; Lchdm. i. 358, 14.

apuldre, an; *f. An apple-tree;* malus :—Deós apuldre *hæc malus,* Ælfc. Gr. 6, 9; Som. 5, 57. v. apulder.

apuldur *an apple-tree.* v. apulder.

a-pullian; *p.* ode; *pp.* od *To pull;* vellere. v. pullian.

Aquilegia; *indecl.* [Aquileia=’Ακυληΐα] *Aquileia in Gallia Transpadana, north of the Adriatic* :—Maximus abád æt Aquilegia ðære byrig *Maximus encamped at the town Aquileia,* Ors. 6, 36; Bos. 131, 21.

ÂR, es; *n.* ORE, *brass, copper;* æs, g. ǽris; *n.* v. ǽres :—Brǽs oððe ár æs, Ælfc. Gr. 5; Som. 4, 59. Israhéla folc is geworden nû me to áre on mínum ofne *versa est mihi domus Israel in æs in medio fornacis,* Past. 37, 3; Hat. MS. 50 a, 6. Grêne ár *green copper, brass;* orichalcum, Cot. 14. [O. Sax. êrin, *adj. æneus :* Ger. *n. metallum,* æs : M. H. Ger. O. H. Ger. êr, *n.* æs : Goth. aiz, *n.* æs : Dan. erts : Swed. ör *a copper coin :* O. Nrs. eir, *n.* æs : Sansk. ayas *ferrum.*] DER. ár-fæt, -geótere, -gescôd, -gesweorf, -geweorc, -glæd, -sápe, -smiþ : ǽren : óra.

ÂR, e; *f.* I. *honour, glory, rank, dignity, magnificence, respect, reverence;* honor, dignitas, gloria, magnificentia, honestas, reverentia :— Sý him ár and onwald *be to him honour and power,* Exon. 65 b; Th. 241, 28; Ph. 663. Ne wolde he ǽnige áre wîtan *nor would he ascribe any honour,* Bd. 2, 20; S. 521, 29. He sundor líf wæs fôreberende eallum ðám árum *he was preferring a private life to all honours,* Bd. 4, 11; S. 579, 8. Nyton náne áre on nánum men *they know no respect for any man,* Bt. 35, 6; Fox 168, 25. Be ðære cirican áre *according to the rank of the church,* L. Alf. pol. 42; Th. i. 90, 10. He on his ágenum fæder áre ne wolde gesceáwian *he would not look with reverence on his own father,* Cd. 76; Th. 95, 18; Gen. 1580. II. *kindness, favour, mercy, pity, benefit, use, help;* gratia, favor, misericordia, beneficium, auxilium :—He gemunde ðá ða áre ðe he him ǽr forgeaf, wîc-stede wê-ligne *he remembered then the favour which he before had conferred upon him, the wealthy dwelling-place,* Beo. Th. 5205; B. 2606. Ne mihte earmsceapen áre findan *nor might the poor wretch find pity,* Andr. Kmbl. 2260; An. 1131. Him wæs ára þearf *to him was need of favours,* Cd. 97; Th. 128, 12; Gen. 2125. To gôdre áre *to good use,* Herb. 2, 9; Lchdm. i. 82, 21 : Bd. 3, 5; S. 527, 14. Eallum to áre ylda bearnum *for the benefit of all the sons of men,* Jul. A. 2. (Vid. Price's Walton, ci. note 34.) Leáf and gærs grôweþ eldum to áre *leaves and grass grow for the benefit of men,* Bt. Met. Fox 20, 199; Met. 20, 100. Ðǽr is ár gelang fira gehwylcum *there is help ready to every man,* Andr. Kmbl. 1958; An. 981. III. *property, possessions, an estate, land, ecclesiastical living, benefice;* bona, possessio, fundus, beneficium :—He plihte to him sylfum and ealre his áre *he acts at peril of himself and all his property,* L. Eth. ix. 42; Th. i. 350, 3 : Ors. 1, 1; Bos. 20, 32. Hwîlum be áre, hwîlum be ǽhte *sometimes in estate, sometimes in goods,* L. Eth. vi. 51; Th. i. 328, 11 : L. C. S. 50; Th. i. 404, 18. Se ðe sitte on his áre on lîfe *he who lives on his property during life,* L. Eth. iii. 14; Th. i. 298, 9 : L. Eth. vi. 4; Th. i. 316, 1, 3. Ðæt hí him andlyfne and áre forgeáfen for heora gewinne *that they should give them food and possessions for their labour,* Bd. 1, 15; S. 483, 19. [Laym. ære, are : Orm. are : O. Sax. êra : O. Frs. êre : Dut. eer : Ger. ehre, *f :* M. H. Ger. êre : O. H. Ger. êra : Dan. äre : Swed. ära : O. Nrs. æra.]

ÂR, es; *m. A messenger, legate, herald, apostle, angel, minister, servant, man, soldier;* nuntius, legatus, præco, apostolus, angelus, minister, vir :—Ðes ár sægeþ *this messenger sayeth,* Cd. 32; Th. 42, 34; Gen. 682 : Beo. Th. 5559; B. 2783. Stîðlíce clypode Wicinga ár *the herald of the Vicings firmly proclaimed,* Byrht. Th. 132, 34; By. 26. Ædelcyninges ár *the noble King's messenger* [*Christ's apostle*], Andr. Kmbl. 3354; An. 1681. Hie hêton lǽdan ût hálige áras *they commanded him to lead out the holy messengers* [*angels*], Cd. 112; Th. 148, 14; Gen. 2456: Exon. 15 a; Th. 31, 29; Cri. 503. Fæder ælmeahtig his áras hider onsendeþ *the almighty Father will send his angels hither,* Exon. 19 a; Th. 47, 23; Cri. 759. Ðá afyrhted wearþ ár [Gûþláces] *then* [*Guthlac's*] *servant was affrighted,* 52 a; Th. 181, 30; Gû. 1301. Lǽt gebîdan beornas dîne, áras *let thy warriors, thy men, await,* Andr. Kmbl. 799; An. 400. [O. Sax. êru, *m :* Goth. áirus, *m :* O. Nrs. árr, *m.* from the Sansk. rjot îr *to go.*] v. ǽrend.

ÂR, e; *f. An* OAR; remus :—Drȳgaþ his ár on borde *his oar becomes*

dry on board, Exon. 92 a ; Th. 345, 15 ; Gn. Ex. 188. Sume hæfdon lx ára *some had sixty oars*, Chr. 897 ; Th. 174, 43, col. 1. Særófe árum bregdaþ ýþbord [MS. yþborde] neáh *brave seamen draw the vessel near with oars*, Exon. 79 a ; Th. 296, 26 ; Crä. 57. [*Havl.* ár : *Chauc.* oore : *Dan.* aare : *Swed.* are : *O. Nrs.* ár, *f.*] DER. ár-blæd, -gebland, -wéla, -widde, -ýþ.

ár *before* :—Ǽrist odde ár *primo*, Mt. Kmbl. Lind. 20, 1. v. ǽr.

ára = geára? *adv. Formerly* ; quondam :—Ðú me ára, God, ǽrest lǽrdest of geóguþháde *Deus, docuisti me a juventute mea*, Ps. Th. 70, 16.

a-rád *rode* :—He út arád *he rode out*, Ors. 3, 7 ; Bos. 62, 22 ; *p.* of a-rídan.

a-rǽcan ; *p.* -rǽhte, -rǽcte ; *pp.* -rǽht. I. *to reach, get at* ; prehendere, attingere :—Ðæt man arǽcan mihte *that one could reach*, Chr. 1014 ; Ing. 193, 19. II. *to hold forth, reach out, hand* ; porrigere :—Arǽce me ða bóc *porrige mihi librum*, Ælfc. Gr. 28, 5 ; Som. 31, 47. v. rǽcan.

a-rǽd, -réd, es ; *m.* [a intensive, rǽd *counsel*] *Counsel, welfare, safety* ; consilium, commodum, salus :—Smeágende ymbe heora sáwla arǽd [aréd, MS. B ; rǽd, MS. D] *considering about their souls' welfare*, L. Edm. E. pref. Th. i. 244, 6.

a-rǽd ; *def.* se a-rǽda ; *adj. Counselling, consulting, wise, prudent* ; sagax, prudens :—Hwǽr is nú se fóremǽra and se arǽda Rómwara heretoga *where is now the illustrious and prudent consul of the Romans?* Bt. 19 ; Fox 70, 6.

a-rǽd *uttered*, Bt. 23 ; Fox 78, 20, note 8, = a-rǽded, *pp.* of a-rǽdan.

a-rǽdan, -rédan ; *p.* -rǽdde, -rédde, -réde ; *pp.* -rǽded, -rǽd, -réd [réd *counsel*]. I. *to take counsel, care for, appoint, determine* ; consilium capere, consulere alicui, decernere, definire :—Sende gewrit, on ðám he gesette and arǽdde *misit literas, in quibus decrevit*, Bd. 2, 18 ; S. 520, 33. Gif hit eallinga ðus arǽded sí *si omnimodis ita definitum est*, 4, 9 ; S. 577, 29. Ða dómas ða ðe fram fæderum arǽdde and gesette wǽron *quæque definierunt canones patrum*, 4, 5 ; S. 572, 18. Hwæðere ðis betwyh heom arǽddon *his tamen conditionibus interpositis*, 4, 1 ; S. 564, 15. He symble þearfum arǽde *semper pauperibus consulebat*, 3, 9 ; S. 533, 25. II. *to conjecture, guess, prophesy, interpret, utter* ; conjectare, divinare, prophetizare, interpretari, eloqui :—Ne mihton arǽdan men engles ǽrend-béc *men might not interpret the angel's messages*, Cd. 212 ; Th. 261, 30 ; Dan. 734. And him to cwǽdon, Arǽd *et dixerunt ei, Prophetiza*, Mk. Bos. 14, 65. Ða se wísdom ðis spell arǽd hæfde *when wisdom had uttered this speech*, Bt. 23 ; Fox 78, 20, note 8 : Exon. 76 b ; Th. 286, 24 ; Wand. 5. v. rǽdan, *p.* rǽdde.

a-rǽdnis *a condition*, Bd. 4, 4 ; S. 571, 11. v. a-rédnes.

a-rǽfnan, -réfnan ; *p.* ede, ed ; *pp.* ed *To endure, bear, suffer* ; sustinere, tolerare, perferre :—Ðæt he ðæt sár mihte geþyldelíce mid smylte móde aberan and arǽfnan *ut patienter dolorem ac placida mente sustineret*, Bd. 4, 31 ; S. 610, 27. Ðonne hí ðæt mægen ðære unmǽtan hǽto arǽfnan ne mihton *cum vim fervoris immensi tolerare non possent*, 5, 12 ; S. 627, 41. Ic þrówade and arǽfnde *pertuli*, 2, 6 ; S. 508, 21 : Andr. Kmbl. 1632 ; An. 817. Sáwl mín symble arǽfnede *sustinuit anima mea*, Ps. Th. 129, 5 : 68, 21 : 64, 7. v. rǽfnan.

a-rǽfnian ; *p.* ade ; *pp.* ad. I. *to endure, bear, suffer, support* ; sustinere, pati, supportare :—Ic arǽfnige *sustineo*, Ps. Th. 129, 4. Ðæ fordon ic edwít for ðé oft arǽfnade *quoniam propter te supportavi improperium*, 68, 8. II. *to ponder in mind* or *heart* ; animo versare, ponderare :—Maria sóþlíce heóld ealle ðás word, arǽfniende on hire heortan *but Mary kept all these words, pondering them in her heart*, Homl. Th. i. 30, 35. v. a-rǽfnan.

a-rǽfniende, -rǽfnigende ; *part. Bearing in mind, considering, pondering*, Homl. Th. i. 42, 17, 30. v. a-rǽfnian.

a-rǽfniendlíc ; *adj. Possible, tolerable* ; possibilis, tolerabilis. DER. *part.* arǽfniende, líc.

a-rǽman ; *p.* de ; *pp.* ed. I. *v. trans. To raise, lift up, elevate* ; excitare, erigere, elevare :—Ða ge mihton rǽdan, and eów arǽman on ðám *which ye may read, and elevate yourselves in them*, Ælfc. T. 31, 15. II. *v. intrans. To raise* or *lift up one's self, to arise* ; se erigere, se elevare, surgere :—Dæges þriddan ord arǽmde *the beginning of the third day arose*, Cd. 139 ; Th. 174, 10 ; Gen. 2876 : 162 ; Th. 203, 29 ; Exod. 411. [*O. H. Ger.* ráma *sustentaculum, columen*.] DER. up-a-rǽman, rǽman.

a-rǽran ; *p.* de ; *pp.* ed ; *v. trans.* [a, rǽran *to rear, raise*] *To rear up, raise up, lift up, exalt, set up, build up, create, establish* ; erigere, excitare, resuscitare, extollere, ædificare, creare :—Ðone stán arǽrde to mearce *lapidem erexit in titulum*, Gen. 28, 18, 22. Arǽrende þearfan *lifting up the poor* ; erigens pauperem, Ps. Spl. 112, 6. Gyld of golde arǽrde *reared up an idol of gold*, Cd. 180 ; Th. 226, 23 ; Dan. 175. Arǽrde Cristes róde *reared up Christ's rood*, Exon. 35 a ; Th. 112, 27 ; Gú. 150. Ic arǽre ðis tempel binnan þrím dagum *excitabo hoc templum in tribus diebus*, Jn. Bos. 2, 19, 20. Ic hine arǽre on ðám ýtemestan dæge *ego resuscitabo eum in novissimo die*, 6, 44, 54. Weá wæs arǽred *woe was raised up*, Cd. 47 ; Th. 60, 26 ; Gen. 987. Se ðe fóre duguðe wile dóm arǽran *who desires before his nobles to exalt his dignity*, Exon. 87 a ; Th. 327, 2 ; Wid. 140 : Beo. Th. 3411 ; B. 1703. Ða wæs æ

Godes riht arǽred *then was God's right law set up*, Andr. Kmbl. 3288 ; An. 1647. Weofod arǽrde *ædificavit altare*, Gen. 22, 9. Eardas rúme Meotud arǽrde for mon-cynne *the Creator established spacious lands for mankind*, Exon. 89 a ; Th. 334, 14 ; Gn. Ex. 16.

a-rǽrnes, -ness, e ; *f. A raising, an exaltation* ; exaltatio :—Heora hrýre wearþ Athénum to arǽrnesse *their fall was the raising of the Athenians*, Ors. 3, 1 ; Bos. 53, 42.

a-rǽsan *to rush* ; irruere, Anlct.

a-ráfian *To unrove, uniravel, unwind* ; dissolvere :—Aráfaþ ðæt cliwen ðære twífaldan heortan *unwinds the clew of the double heart* ; dissolvit corda duplicitatis involuta, Past. 35, 5 ; Hat. MS. 46 b, 1.

a-rás *arose* ; surrexit, Gen. 19, 1. v. a-rísan.

áras *messengers*, Exon. 15 a ; Th. 31, 10 ; Cri. 493. v. ár.

a-rásade = résade *suspicabatur*, Bd. 4, 1 ; S. 564, 48, note.

a-rásian ; *p.* ode, ade ; *pp.* od, ad ; *v. trans.* [a, rásian *to raise, uncover*] *To lay open, discover, explore, detect, reprove, correct, seize* ; detegere, invenire, explorare, corripere, reprehendere, intercipere :—God hæfþ arásod úre unrihtwísnissa *Deus invenit nostras iniquitates*, Gen. 44, 16. Arásian *explorare*, Gr. Dial. 2, 14. Ðǽr hý arásade, reótaþ and beofiaþ, fóre freán forhte *there they detected, shall wail and tremble, afraid before the Lord*, Exon. 25 b ; Th. 75, 31 ; Cri. 1230. Hæleþ wurdon acle arásad for ðý rǽse *the men were seized with fear on account of its force*, 74 a ; Th. 277, 27 ; Jul. 587. Se ðe wilnaþ hiera unþeáwas arásian *qui eorum culpas corripere studet*, Past. 35, 3 ; Hat. MS. 45 b, 6 : 35, 5 ; Hat. MS. 46 a, 20. Beón arásod *reprehendi*, Fulg. 5. Arásad wæs *interceptus est*, Cot. 109. Arásod beón on hefygtímum gyltum *gravioris culpa noxæ teneri*, R. Ben. 25 : 34.

ár-blæd, es ; *n. The oar-blade* ; palmula remi, Ælfc. Gl. 103 ; Wrt. Voc. 56, 38.

arc, es ; *m.* : earc, erc, e ; *f.* : earce, an ; *f. A vessel to swim on water, the* ARK, *a coffer, small chest* or *box* ; arca, cista, cistella, cibotium = κιβώτιον :—Ðá ætstód se arc *tunc requievit arca*, Gen. 8, 4. Wirc ðé nú senne arc *fac tibi arcam*, 6, 14. Þreó hund fǽðma biþ se arc on lenge, and fíftig fǽðma on brǽde, and þrittig on heáhnisse *trecentorum cubitorum erit longitudo arcæ, quinquaginta cubitorum latitudo, et triginta cubitorum altitudo illius*, 6, 15. Se arc wæs geférud ofer ða wæteru *arca ferebatur super aquas*, 7, 18. [*Laym.* archen, arche, *dat* : *Dut.* ark, *f* : *Ger. M. H. Ger.* arche, *f* : *O. H. Ger.* archa : *Goth.* arka : *Dan.* ark : *O. Nrs.* örk, *f.*] v. earc.

arce- *chief* = ἀρχι = ἀρχός, a prefix ; v. arce-bisceop :—Hér Ælfríc arcebisceop férde to Róme æfter his arce[-pallium] *this year archbishop Ælfric went to Rome after his arch-pallium*, Chr. 997 ; Th. 247, 2, col. 2. = Wið ðan ðe he scolde gifan heom ðone arce [MS. erce] *on condition that he should give them the arch-pallium*, 996 ; Th. 244, 42, note. = Forðí ðæt he scolde heom ðone pallium gifan *on condition that he should give them the pallium*, 996 ; Th. 245, 11, note.

arce-bisceop, arce-bysceop, arce-biscop, ærce-bisceop, erce-biscop, es ; *m. The chief bishop*, ARCHBISHOP ; archiepiscopus [= ἀρχι-επίσκοπος from ἀρχι = ἀρχός *a leader, chief* ; ἐπίσκοπος. v. bisceop] :—Honorius se arcebisceop gehálgode Thoman his diácon, to bisceope *archbishop Honorius consecrated Thomas his deacon, as bishop*, Bd. 3, 20 ; S. 550, 21 : 4, 1 ; S. 563, 6, 8, 12, 29.

arce-bisceop-ríce, arce-biscop-ríce, es ; *n. An* ARCHBISHOPRIC ; archiepiscopatus :—To ðam arcebisceopríce *to the archbishopric*, Chr. 994 ; Th. 242, 38. Ðæt arcebiscopríce on Cantwara byrig *the archbishopric of Canterbury*, 1114 ; Th. 370, 15.

arce-diácon, archi-diácon, ærce-diácon, es ; *m. An* ARCHDEACON, *a bishop's vicegerent* ; archidiáconus [= ἀρχι-διάκονος, from ἀρχός *a chief, leader*, and διάκονος *a deacon*] :—Becom Benedictus to freóndscipe ðæs hálgan weres and ðæs gelǽredestan, Bonefacii archidiácones *Benedictus pervenit ad amicitiam viri doctissimi ac sanctissimi, Bonifacii videlicet archidiaconi*, Bd. 5, 19 ; S. 638, 14. Arcediácon *archidiaconus*, Ælfc. Gl. 69 ; Wrt. Voc. 42, 27.

arce-stól, es ; *m.* [arce *chief*, stól *a stool*] *An archiepiscopal see* or *seat* ; sedes archiepiscopalis :—Æt his arcestóle on Cantwara byrig *at his archiepiscopal see in Canterbury*, Chr. 1115 ; Th. 371, 5 : 1119 ; Th. 372, 32.

ár-cræftig ; *adj.* [ár *respect*, cræftig *crafty*] *Skilful* or *quick in shewing respect, respectful, polite* ; morigerus, obsequens :—Árcræftig ár *a respectful messenger, a prophet*, Cd. 202 ; Th. 250, 23 ; Dan. 551.

arctos ; *acc.* arcton ; *f.* [ἄρκτος, ov, *m. f.* a bear ; ἄρκτος, *f. the constellation* Ursa Major, called also ἄμαξα, carles wǽn *the churl's wain* : the bright star in Boötes is denominated by ancient astronomers and poets Ἀρκτοῦρος, *the bear-ward*. *The constellation* Ursa Major ; arct-os, -us, i ; *f.* = ἄρκτος, *f* :—Arcton hátte án tungol on norþ dǽle, se hæfþ seofon steorran, and is for ðí óðrum naman geháten, *septemtrio*, ðone hátaþ lǽwede menn carles wǽn. Se ne gǽþ nǽfre adúne under ðyssere eorþan, swá swá óðre tunglan dóþ, ac he went abútan, hwílon adúne and hwílon up, ofer dæg and ofer niht *one constellation is called arctos in the north part, which has seven stars, and for that is called by another name, septemtrio, which untaught men call the churl's wain. It never goes down under this earth, as the other constellations do, but one*

while it turns down and another while up, over day and over night, Bd. de nat. rerum; Wrt. popl. science 16, 3-7; Lchdm. iii. 270, 9-15.

árde; *dat.* [=arce MS?] *A mark of honour, badge of office, the pallium*, Cht. 997; Ing. 172, 7. v. árod.

ardlíce; *adv.* [arod *quick*, líce] *Quickly, immediately*; prompte, cito:—Éfstaþ nū ardlíce *persequimini cito*, Jos. 2, 5: Gen. 14, 14: 22, 11.

are, es; *m. A court-yard*; area, Alb. resp. 48.

áre, an; *f. Honour, honesty, favour, benefit, pity, mercy*; honor, honestas, gratia, beneficium, misericordia:—Áre [MS. aare] cyninges dóm æghwær lufade *honor regis judicium diligit*, Ps. Th. 98, 3. Mid áran *with honours*, Cd. 155; Th. 193, 12; Exod. 245. Árna ne gýmden *they had no regard of honour*, 113; Th. 148, 20; Gen. 2459. Us is ðínra árna þearf *to us is need of thy mercies*, Exon. 11 b; Th. 16, 19; Cri. 255. Árna gemyndig *mindful of benefits*, Cd. 98; Th. 130, 22; Gen. 2163: Beo. Th. 2379; B. 1187. We ðec árena biddaþ *we pray thee for thy mercies*, Exon. 53 a; Th. 186, 6; Az. 15. v. ár *honour*.

a-reáfian; *p.* ode; *pp.* od [a *from*, reáfian *to tear*] *To tear from, tear asunder, separate*; diripere:—Brim [MS. bring] is areáfod *the sea is separated*, Cd. 158; Th. 196, 12; Exod. 290.

a-reaht, -reht *put forth, spoken, explained*, Exon. 24 a; Th. 69, 23; Cri. 1125: Bt. 36, 2; Fox 174, 3; *pp.* of a-reccan.

a-recan *to recount*:—Hit nis nánum men aléfed, ðæt he mæge arecan ðæt ðæt God geworht hæfþ *it is not permitted to any man, that he may recount that which God has wrought*, Bt. 39, 12; Fox 232, 10. v. a-reccan.

a-reccan, -recan, -reccean; ic -recce, ðū -reccest, -recest, he -receþ, -recþ; *p.* -reahte, -rehte; *impert.* -rece; *pp.* -reaht, -reht; *v. trans.* I. *to put forth, stretch out, strain, raise up*; extendere, expandere, erigere:—Hondum slógun, folmum areahtum and fýstum eác *they struck with their hands, with outstretched palms and fists also*, Exon. 24 a; Th. 69, 23; Cri. 1125. Areahtum eágum *attonitis oculis*, Prov. 16, Lye. He mæg of woruf-torde ðone þearfendan areccan *de stercore erigens pauperem*, Ps. Th. 112, 6: 144, 15. II. *to put forth, relate, recount, speak out, express, explain, interpret, translate*; proponere, exponere, enarrare, eloqui, exprimere, disserere, interpretari, reddere:—Ðara sume we areccan wyllaþ *some of which we will relate*, Bd. 5, 12; S. 627, 7: Menol. Fox 138; Men. 69. Ðá se Wisdóm ðá ðis spell areht [MS. Cot. areaht] hæfde *when Wisdom then had spoken this speech*, Bt. 36, 2; Fox 174, 3: 39, 3; Fox 214, 14: Bt. Met. Fox 8, 3; Met. 8, 2. Wordum gereccan [MS. Cot. areccan] *to express in words*, Bt. 20; Fox 70, 28. Arece us ðæt bigspell *edissere nobis parabolam*, Mt. Bos. 13, 36: 15, 15. Arece us ðæt gerýne *explain to us the mystery*, Exon. 9 a; Th. 5, 24; Cri. 74: 49 a; Th. 169, 16; Gú. 1095: Cd. 202; Th. 250, 5; Dan. 542. Án ærendgewrit of Lǽdene on Englisc areccean *to translate an epistle from Latin into English*, Past. pref. Hat. MS. III. *to set in order, adorn, deck?* expedire, expolire, comere?—Areaht síe *expoliatur*, Cot. 77, Lye: Exon. 94 a; Th. 353, 9; Reim. 10.

a-reccean; *p.* -reahte, -rehte; *pp.* -reaht, -reht; *v. trans. To tell out, relate, recount, express, translate*; enarrare, eloqui, exprimere, reddere:—Hwá is ðæt ðe eall ða yfel, ðe hí dónde wǽron, mæge areccean *who is there that can relate all the evils which they did?* Ors. 1, 8; Bos. 31, 24: Hy. 3, 17; Hy. Grn. ii. 281, 17. Án ærendgewrit of Lǽdene on Englisc areccean *to translate an epistle from Latin into English*, Past. pref. v. a-reccan.

a-reccende; *part. Explaining*; exponens, Bd. 1, 27, resp. 8; S. 494, 35. v. a-reccan.

a-receþ, -recþ *raises up*; erigit, Ps. Th. 144, 15: Ps. Spl. 145, 7. v. a-reccan.

a-réd *counsel*, L. Edm. E. pref; Th. i. 244, 6, MS. B. v. a-rǽd.

a-rédad *discovered*, R. Ben. 61; *pp.* of a-rédian.

a-reddan *to liberate*. v. a-hreddan.

a-réde *cared for*, Bd. 3, 9; S. 533, 25, =a-redde=a-rǽdde; *p.* of a-rǽdan, *q. v.*

a-rédian; *p.* ode; *pp.* od, ad *To make ready, provide, furnish, execute, find, to find the way to any place, reach*; parare, præparare, exsequi, invenire, pervenire aliquo:—Us is þearf ðæt we arédian ðæt úre hláford wille *it behoves us that we provide that which our lord wants*, L. Ath. v. § 8, 9; Th. i. 238, 25. Smeáge man hū man mæge rǽd arédian þeóde to þearfe *let it be considered how advantage may be provided for the behoof of the nation*, L. Eth. vi. 40; Th. i. 324, 28: L. C. S. 11; Th. i. 382, 6. Arédod *furnished*, Som. Woruld-gerihta mon arédian mæge Gode to gecwémnysse *secular rights may be executed to the pleasure of God*, L. Edg. S. 2; Th. i. 272, 24. Hí arédian ne mágon, ðæt hí aslépen *they cannot find out that they may slip*, Bt. Met. Fox 13, 16; Met. 13, 8. Arédad beón *inveniri*, R. Ben. 61. Ðæt ðú ne mæge ðíne wegas arédian *ut non dirigas vias tuas*, Deut. 28, 29. Ðú ne mihtest gyt fulrihtne weg arédian *thou hast not yet been able to find the most direct way*, Bt. 22, 2; Fox 78, 8: 40, 5; Fox 240, 22: Bt. Met. Fox 23, 19; Met. 23, 10. Oferdruncen man ne mæg to his húse arédian *a drunken man is not able to find the way to his house*, Bt. 24, 4; Fox 84, 31. Ic ne mæg út arédian *I cannot find the way out*, 35, 5; Fox 164, 14. Ðú eart cumen

innon ða ceastre, ðe ðū ǽr ne mihtest arédian *thou art come into the city, which thou couldest not reach before*, 35, 3; Fox 158, 11.

a-rédnes, -rǽdnis, -ness, e; *f. A degree, condition, covenant*; consultum, conditio:—Ðá geþafedon hí ðære arédnesse *ea conditione consenserunt*, Bd. 1, 1; S. 474, 20. Ðæt wíf he onféng ðære arédnesse *uxorem ea conditione acceperat*, 1, 25; S. 486, 33.

a-rédod *furnished*, Som. v. a-rédian.

a-réfnan *to endure*:—Ic aréfnde *sustinui*, Ps. Spl. C. 68, 25. v. a-ráefnan.

a-reht *spoken*, Bt. 36, 2; Fox 174, 3; *pp.* of a-reccan.

árena of *mercies*, Exon. 53 a; Th. 186, 6; Az. 15, =árna; *gen. pl.* of áre, *q. v.*

a-reódian; *p.* ode; *pp.* od [a, reódian *to redden*] *To become red, to redden, blush*; erubescere:—His andwlita eal areódode *all his countenance became red*, Apol. Th. 21, 26.

a-reósan; *p.* -reás, *pl.* -ruron; *pp.* -roren *To fall down, perish*; decidere, corruere:—Ic areóse [MS. areófe] be gewyrhtum fram feóndum mínum on ídel *decidam merito ab inimicis meis inanis*, Ps. Spl. 7, 4. v. a-hreósan.

a-rétan; ic -réte, he -réteþ, -rét; *p.* -rétte; *pp.* -réted, -rét; *v. trans.* [a, rétan *to comfort*] *To exhilarate, comfort, delight, restore, refresh, set right*; exhilarare, lætificare, reficere:—Ic monigra mód aréte *I exhilarate the mind of many*, Exon. 102 b; Th. 389, 12; Rä. 7, 6. Seó hwætnes ðæs líchoman geblissaþ ðone mon and arét *the vigour of the body rejoices and delights the man*, Bt. 24, 3; Fox 84, 8. Ðæt ge bróðor míne wel arétten *that ye should well cherish my brethren*, Exon. 30 a; Th. 91, 33; Cri. 1501. Æghwylcum wearþ mód aréted *every one's mind was delighted*, Judth. 11; Thw. 24, 2; Jud. 167. Hí hæfdon ðæt mód arét *they had restored or refreshed the mind*, Bt. titl. xxii; Fox xiv, 5. Ðú me hæfst arétne on ðam tweóne *thou hast set me right in the doubt*, Bt. 41, 2; Fox 246, 12: 22, 1; Fox 76, 12, MS. Cot.

arewe, an; *f. An arrow*; sagitta:—Sume scotedon adúnweard mid arewan *some shot downward with arrows*, Chr. 1083; Erl. 217, 19.

Arewe, Arwe, an; *f.* [arewe *arrow*] ARROW, the name of a river in several counties, called so either from its *swiftness* or *straightness*, also *the Orwell*; fluvii nomen:—Se here gewende ða fram Lundene, mid hyra scypum, into Arewan [MS. Laud. Arwan] *the army [of the Danes] went then from London, with their ships, into the river Orwell [in Suffolk]*, Chr. 1016; Erl. 157, 14. Gibson says of Orwell,—Hunc suspicor antiquitus fuisse pronunciatum *Arwel*, tum quod Saxonicum *A* sequentibus sæculis transiit in *O*, tum etiam quod oppidum est ad ejus ripam situm, *Arwerton* dictum; accedit quod *Harewich* ad oram hujus fluminis, olim *Arwic*, non ut conjectat Camd. *Herewic*, dici posset, Gib. Chr. Explicatio 13, col. 1.

áre-weorþ *honourable, venerable*; honore dignus, honorabilis, venerabilis, Lye. v. ár-weorþ.

ár-fæst, ǽr-fæst; *adj.* [ár *honour*, fæst *fast*] *Honourable, honest, upright, virtuous, good, pious, dutiful, gracious, kind, merciful*; honestus, probus, bonus, pius, propitius, clemens, misericors:—Árfæste rincas *honourable chieftains*, Cd. 90; Th. 113, 29; Gen. 1894: 136; Th. 171, 9; Gen. 2825. Wæs he se mon ǽfæst and árfæst *he was the religious and pious man*; vir pietatis et religionis, Bd. 3, 14; S. 539, 33. Wes ðú ðínum yldrum árfæst simle *be thou always dutiful to thy parents*, Exon. 80 a; Th. 300, 25; Fä. 11. Ongan ðá ródera wealdend árfæst wið Abraham sprecan *then began the gracious Ruler of the skies to speak with Abraham*, 109; Th. 145, 13; Gen. 2405. Drihten biþ árfæst his folces lande *Dominus propitius erit terræ populi sui*, Deut. 32, 43: Exon. 11 b; Th. 15, 32; Cri. 245. Ðæt Drihten him árfæst and milde wǽre *that the Lord might be to him merciful and mild*, Bd. 4, 31; S. 610, 31.

ár-fæstlíce; *adv. Honestly, piously*; honeste, pie. DER. árfæst, líce.

ár-fæstnes, ár-fæstnys, ǽr-fæstnys, -ness, e; *f. Honourableness, honesty, goodness, piety, clemency, mercifulness*; honestas, probitas, pietas, clementia, misericordia:—Ðæt he wæs mycelre árfæstnesse and ǽfæstnesse wer *quod vir esset multæ pietatis ac religionis*, Bd. 4, 31; S. 610, 7. Seó godcunde árfæstnys *pietas divina*, 5, S. 512, 24: 3, 13; S. 539, 1. Mid ða upplícan árfæstnesse *apud supernam clementiam*, 5, 23; S. 649, 8: Jos. 6, 17. For ðínre árfæstnesse *of thy clemency*, Hy. 8, 24; Hy. Grn. ii. 290, 24.

ár-fæt, es; *n. A brazen vessel*; æramentum, labrum:—Fyrmþa árfata *baptismata æramentorum*, Mk. Bos. 7, 4. Hálgode ðæt árfæt *labrum sanctificavit*, Lev. 8, 11.

ar-faran *To go away, depart*; abire:—Ar-faraþ, Bt. Met. Fox 20, 25: Met. 20, 13 *suggests* an-faraþ, *taking an as an adv. away*, without referring to any authority.

ár-fæst *merciful*, Ps. Spl. 102, 3. v. ár-fæst.

ár-ful, ár-full; *adj. Venerable, respectful, favourable, merciful, mild*; honorabilis, venerabilis, propitius, reverens:—Ic Æðelbald wæs beden from ðæm árfullan bisceope Milrede *I Æthelbald have been solicited by the venerable bishop Milred*, Th. Diplm. A. D. 743-745; 28, 22. Se ðe árful biþ eallum unrihtwísum ðínum *qui propitiatur omnibus iniquitatibus tuis*, Ps. Spl. M. 102, 3. Cristenra manna gehwilc beó árful fæder

E

and mēder *Christianorum quivis reverenter habeat patrem et matrem*, Wulfst. parǽn. 7.

árful-líce; *adv. Mildly, gently*; clementer :—Iosep híg oncneów árfullíce *Joseph clementer resalutavit eos*, Gen. 43, 27.

arg; *adj. Wicked, depraved, bad*; malus, pravus. ☞ An impure word only found in the Lindisfarne Gospels or the Durham Book :—Cneórisse yflo and arg *an evil and wicked generation*; generatio mala et adultera, *i. e.* prava, pigra, etc. Mt. Kmbl. Lind. 12, 39. Arg *peccatrix*, Mk. Skt. Lind. 8, 38. [*Plat. Dut. Ger. Franc. Dan. Swed.* arg: *Grk.* ἀργός *idle*: *Icel.* argr *effeminatus, pavidus, ignavus, malus, detestabilis.*] v. earg.

ár-gebland, es; *m. The mingling of the oars, the sea disturbed by the oars, the oar-disturbed sea*; remorum commixtio, mare remis turbatum, Andr. Kmbl. 765; An. 383. v. ár.

ár-geótere, es; *m.* [*ár brass*, geótere *a pourer*] *A caster* or *pourer of brass, melter of brass, brass-founder*; ærarius :—Đá wæs sum árgeótere, se mihte dón anlícnessa *there was a certain brass-founder, who could make images*, Ors. 1, 12; Bos. 36, 26.

ár-gesweorf, es; *m. Brass filings*; limatura æris, L. M. 1, 34; Lchdm. ii. 80, 22. v. gesweorf, sweorfan.

ár-geweorc, es; *n. Brass-work*; æramentum, Cot. 79.

ár-gifa, an; *m. A benefit-giver*; beneficiorum dator, Exon. 78 b; Th. 294, 6; Crä. 11.

ár-glæd *bright with brass.* v. ǽr-glæd.

arhlíce *disgracefully, basely* :—Eádwine eorl wearþ ofslagen arhlíce fram his ágenum mannum *earl Eadwine was basely slain by his own men*, Chr. 1071; Erl. 210, 14; Th. 347, 12. v. earhlíce from earg, earh **II.** *evil, vile.*

ár-hwæt; *g. m. n.* -hwates; *f.* -hwætre; *adj.* [*ár honour*, hwæt *eager, brisk*] *Eager* or *desirous of honour, bold, valiant*; honoris cupidus, fortis :—Wealas ofercómon eorlas árhwate *the men eager for glory overcame the Welsh*, Chr. 937; Erl. 115, 22; Th. 208, 9, col. 2; Æðelst. 73.

árian; *to árianne*; *part.* ende, gende; *p.* ede, ode; *pp.* ed, od; *v. a.* [*ár honour*]. **I.** *to give honour, to honour, reverence, have in admiration*; honorare, honorificare, venerari :—Is to árianne *is to be honoured*, Bt. 32, 2; Fox 116, 14. Onsægednys lófes áreþ me *sacrificium laudis honorificabit me*, Ps. Spl. T. 49, 24. He áraþ đa gódan *he honoureth the good*, Bt. 41, 2; Fox 246, 19. Ic árode đē ofer ealle gesceafta *I honoured thee over all creatures*, Exon. 28 a; Th. 84, 33; Cri. 1383. Se ríca Rómāna wita and se ároda *the rich and honoured senator of the Romans*, Bt. Met. Fox 10, 89; Met. 10, 45. **II.** *to regard, care for, spare, have mercy, pity, pardon, forgive*; consulere, propitium esse, misereri, parcere :—He þearfum árede *he cared for the poor*, Bd. 3, 9; S. 533, 25. Ac árodon heora lífe *but they spared their lives*, Jos. 9, 21; Beo. Th. 1201; B. 598. Búton him se cyning árian wille *unless the king will pardon him*, L. In. 36; Wilk. 20, 39; Th. i. 124, 19. Ára ambehtum [MS. onbehtum] *pity thy servants*, Exon. 13 a; Th. 23, 17; Cri. 370. DER. ge-árian.

Arianisc, Arrianisc; *adj.* ARIAN, *belonging to Arius, an Alexandrian, who lived in the fourth century* :—Se Arrianisca gedweolda arás *the Arian heresy arose*, Bd. 1, 8; S. 479, 27, 18, 33. On đam Arianiscan gedwolan *in the Arian heresy*, Ors. 6, 31; Bos. 127, 43.

a-rídan; *p.* -rád, *pl.* -ridon; *pp.* -riden *To ride*; equitare :—He út of đam mann-werode arád *he rode out from the crowd*, Ors. 3, 7; Bos. 62, 22. v. rídan.

a-riddan, đú -riddest [a-, riddan] *To rid, deliver*; liberare, repellere :—For hwy me đú ædrífe odđe ariddest *quare me reppulisti?* Ps. Spl. T. 42, 2. v. a-hreddan.

áriende, árigende *sparing*; parcens. v. árian.

a-riht; *adv.* ARIGHT, *right, well, correctly*; probe, recte :—Gif man hit ariht asmeáþ *if one considereth it right*, L. Edg. C. 13; Th. ii. 246, 21. v. riht.

a-ríman; *p.* de; *pp.* ed *To number, count, enumerate*; numerare, enumerare, dinumerare, recensere :—He aríman mæg regnas scúran dropena gehwelcne *he can count every drop of the rain-shower*, Cd. 213; Th. 265, 21; Sat. 11: Ps. Th. 89, 13: 146, 5. Hí arímdon ealle bán míne *dinumeraverunt omnia ossa mea*, Ps. Spl. C. 21, 16; Past. 16, 1; Hat. MS. 20 b, 4.

ár-ing, árung, e; *f. Honour, respect*; honoratio :—Búton áringe *without honour*, Ors. 5, 10; Bos. 108, 41.

a-rinnan; *p.* -ran, *pl.* -runnon; *pp.* -runnen *To run out, pass by, to disappear*; effluere, præterire :—Đæt sý [MS. sie] cwide arunnen *that the word be run out*, Salm. Kmbl. 960; Sal. 479. v. rinnan, yrnan, a-yrnan.

a-rísan; *part.* arísende; *p.* arás, *pl.* arison; *pp.* arisen; *v. n. To ARISE, rise, rise up, rise again, to come forth, originate*; surgere, exsurgere, resurgere, provenire, oriri :—Ic aríse *surgo*, Ælfc. Gr. 28, 5; Som. 31, 49. Micel aríseþ dryht-folc to dóme *a great multitude shall arise to judgment*, Exon. 23 a; Th. 64, 22; Cri. 1041. Đý þryddan dæge arísan *tertia die resurgere*, Mt. Bos. 16, 21: Exon. 23 a; Th. 64, 2; Cri. 1031. Ýdel is eów ǽr leóhte arísan *vanum est vobis ante lucem surgere*, Ps. Spl. 126, 3. He arás sóna *surrexit*, Gen. 19, 1. Đá arison đa þrí weras *surrexerunt tres viri*, Gen. 18, 16. Weorod eall arás *the band all arose*, Beo. Th. 6053; B. 3030. Storm upp arás *the storm rose up*, Andr. Kmbl. 2474; An. 1238. Sindon costinga monge arisene *many temptations are arisen*, Th. 104, 20; Gú. 10. Arisen wæs sunne *exortus est sol*, Mk. Lind. War. 4, 6.

a-ríseþ *it behoveth*; oportet :—Đætte aríseþ sunu monnes *for it bihoueth mannis sone*, Wyc: Lk. Lind. Rush. War. 9, 22; *quia oportet filium hominis*, Vulg. v. gerísan.

Arius [=Ἄρειος], Arrius; *g.* ii; *acc.* um; *m. A presbyter of Alexandria, founder of the Arians, born in Cyrenaica, Africa, and died in A. D. 336* :—Đá cwæþ Arrius đæt Crist, Godes Sunu, ne mihte nā beón his Fæder gelíc, ne swá mihtig swá he; and cwæþ, đæt se Fæder wǽre ǽr se Sunu, and nam býsne be mannum, hú ǽlc sunu biþ gingra đonne se fæder on đisum lífe.... He wolde dón Crist læssan đonne he is, and his Godcundnysse wurþmynt wanian *then Arius said that Christ, the Son of God, could not be equal to his Father, nor so mighty as he; and said, that the Father was before the Son, and took example from men, how every son is younger than his father in this life.... He would make Christ less than he is, and diminish the dignity of his Godhead*, Homl. Th. i. 290, 3–8, 22, 23. Hý amánsumodon đǽr [on đære ceastre Nicea A. D. 325] đone mæsse-preóst Arrium, forđan đe he nolde gelýfan đæt đæs lífigendan Godes Sunu wǽre ealswá mihtig swá se mǽra Fæder is *they there [in the city of Nice A. D. 325] excommunicated the mass-priest Arius, because he would not believe that the Son of the living God was as mighty as the great Father is*, L. Ælf. C. 3; Th. ii. 344, 2–4.

ariwe *an arrow*; sagitta. v. arewe.

ár-leás; *def.* se ár-leása; *adj.* [*ár, leás*]. **I.** *void of honour, honourless, disgraceful, infamous, wicked, impious*; inhonestus, impius, infamis :—Him árleáse cyn andswarode *the honourless race answered him*, Cd. 114; Th. 149, 15; Gen. 2475: 91; Th. 116, 10; Gen. 1934. Hleór geþolade árleásra spátl *my face endured the spittle of the impious*, Exon. 29 a; Th. 88, 7; Cri. 1436: Elen. Kmbl. 1668; El. 836. Đa árleásan *the impious men*, Andr. Kmbl. 1117; An. 559. Wið đam árleásestan eretice *against the most wicked heretic*, Bd. 4, 17; S. 585, 43. Forweorþaþ se árleása *the wicked perisheth*, Ps. Spl. 9, 5: Ps. Lamb. 1, 4, 5. Đú scealt hweorfan árleás of earde đínum *thou shalt depart infamous from thy dwelling*, Cd. 48; Th. 62, 24; Gen. 1019: Exon. 28 b; Th. 87, 25; Cri. 1430. **II.** *pitiless, merciless, cruel*; crudelis :—Maximianus, árleás cyning, cwealde cristne men *Maximian, the cruel king, slew Christian men*, Exon. 65 b; Th. 243, 1; Jul. 4.

árleáslíce; *adv.* [árleás, líce] *Wickedly, impiously*; impie :—Ic ne dyde árleáslíce *nec impie gessi*, Ps. Th. 17, 21: Ps. Spl. 17, 23: Exon. 40 b; Th. 136, 7; Gú. 537.

árleás-nes, -ness, e; *f.* [árleás *honourless, wicked*, -nes, -ness] *Wickedness, acts of wickedness, impiety*; iniquitas :—Æfter mænigo árleásnyssa heora *secundum multitudinem impietatum eorum*, Ps. Spl. 5, 12: 64, 3. Seó wíldeórlíce árleásnes Bretta cyninges *feralis impietas regis Brittonum*, Bd. 3, 9; S. 533, 7: 3, 19; S. 548, 18.

ár-leást, ǽr-lēst, e; *f.* [*ár honor, honestas, gratia*, -leást] *Dishonour, impiety, cruelty, a disgraceful deed*; inhonestas, impietas, crudelitas, flagitium :—Árleásta fela *many disgraceful deeds*, Bt. Met. Fox 9, 12; Met. 9, 6.

ár-líc; *adj.* [*ár honour*, líc *like*]. **I.** *honest, honourable, noble, becoming, proper*; honestus, decorus, honorabilis, nobilis :—Árlíc bisceopsetl *an honourable bishop-seat*, Bd. 3, 7; S. 530, 1: Ors. 2, 8; Bos. 51, 11. Is nú árlíc đæt we ǽfestra dǽde dēmen *it is now becoming that we consider the deeds of the pious*, Exon. 40 a; Th. 133, 29; Gú. 497. **II.** *applied to food of a high quality, Delicious*; delicatus, suavis :—Đa beón beraþ árlícne anleofan,—hafaþ hunig on múþe, wynsume wist *the bees produce delicious food,—have honey in the mouth, a pleasant food*, Frag. Kmbl. 36; Leás. 20: Ps. Th. 95, 8. DER. un-árlíc.

árlíce; *adv. Honourably, honestly, properly, mercifully*; honorifice, honeste, decenter, misericordi vel propitio animo :—He hine árlíce bebyride *honorifice eum sepelivit*, Bd. 4, 22; S. 591, 20: Bt. 16, 2; Fox 52, 31: Cd. 127; Th. 162, 23; Gen. 2685. Waldend usser gemunde Abraham árlíce *our Lord remembered Abraham mercifully*, 121; Th. 156, 9; Gen. 2586.

ár-líce; *adv.* [= ǽr *early*] *Early*; diluculo, mane, Mk. Lind. War. 16, 2: Lk. Lind. War. 24, 1: Jn. Rush. War. 8, 2. v. ǽr-líce.

arm; *adj. Miserable*; miser :—Arm leód *miserable people*, Chr. 1104; Th. 367, 15. v. earm.

armēlu *Field* or *wild rue, which is called* Mōly [=μῶλυ] *in Cappadocia and Galatia, and by some* Harmāla; *hence the botanical name* =pēgānum harmāla, Lin. vol. ii. p. 327, =πήγανον ἄγριον *wild rue* :—Armēlu wyl on buteran to sealfe *boil wild rue in butter to a salve*, L. M. 1, 64; Lchdm. ii. 140, 4.

ár-morgen *early dawn*, Jn. Lind. War. 18, 28: 20, 1. v. ǽr-morgen.

arn ran, Mk. Bos. 5, 6; *p.* of yrnan.

árna *of honours, of mercies*, Exon. 11 b; Th. 16, 19; Cri. 255; *gen. pl.* of áre, *q. v.*

arod, es; *n? A species of herb, probably* arum =ἄρον; herbæ genus,

arum :—Nim lybcornes leáf, oððe arod *take a leaf of saffron, or arod,*
L. M. 3, 42; Lchdm. ii. 336, 10. Gehwæde arodes wôses *a little of the
ooze of arum,* Lchdm. iii. 2, 23.

arod; *adj. Quick, swift, ready, prepared;* celer, velox, promptus, paratus :—Ðá wearþ sum to ðam arod, ðæt he in ðæt bûrgeteld nêþde *then
one became ready for this, that he ventured into the bower-tent,* Judth. 12;
Thw. 25, 24; Jud. 275. [*O. Nrs.* ördugr, örðigr *arduus, difficilis, acer,
vehemens.*] v. earu.

árod *honoured,* Bt. Met. Fox 10, 89; Met. 10, 45; *pp.* of árian, *q. v.*

árod, es; *m?* [árian *to honour*] *A mark of honour, badge of office,
the pallium given by the pope to a bishop or archbishop;* honoris vel
muneris signum :—Hêr Ælfríc arcebisceop fêrde to Rôme æfter his árde
[? arce, MS. *q. v.*] *this year archbishop Ælfric went to Rome after his pallium,* Chr. 997; Ing. 172, 7. v. arce-.

arodlíce, arudlíce, ardlíce; *adv. Quickly, immediately;* cito, sine
mora :—Hí hebbaþ swíðe arodlíce ða earce up *arcam sine mora elevant,*
Past. 22, 2; Hat. MS. 33 b, 9.

arodscipe, es; *m. Quickness, swiftness, readiness, dexterity;* velocitas,
dexteritas, promptitudo :—Oft mon biþ swíðe rempende and ræsþ swíðe
dollíce on ælc weorc and hrædlíce, and ðeáh wênaþ men ðæt hit síe for
arodscipe and for hwætscipe *sæpe præcipitata actio velocitatis efficacia
putatur,* Past. 20, 1; Hat. MS. 29 b, 5. DER. un-arodscipe.

aron *estis,* Mt. Kmbl. Lind. 5, 11, = earon.

árra *of favours, mercies, grace,* Cd. 131; Th. 166, 20; Gen. 2750;
gen. pl. of ár.

Arrian, es; *m. Arius; Arrianus* :—Arrianes gedwola *the heresy of
Arius,* Bt. Met. Fox 1, 80; Met. 1, 40. v. Aríus.

Arrianisc *Arian,* Bd. 1, 8; S. 479, 18, 27, 33. v. Arianisc.

Arrius, ii; *m. Arius,* L. Ælf. C. 3; Th. ii. 344, 3. v. Aríus.

ár-sápe, an; *f.* [ár *ore, brass;* sápe = sáp, *p.* of sípan *stillare*] *Verdigris;* ærugo :—Nim ársápan *take verdigris,* Lchdm. iii. 14, 31.

ár-sceamu, e; *f. Verecundia* :—Árscame, *acc.* Ps. Th. 68, 19.

Ár-scyldingas, a; *pl. m. The honoured Skyldings, Danes,* Beo. Th. 933;
B. 464 : 3425; B. 1710.

ars-gang, es; *m.* [ears *anus,* gang *a passage*] *Ani foramen,* anus.
v. ears-gang.

ár-smiþ, es; *m.* [ár *brass,* smiþ *a smith*] *A copper-smith, a brazier, a
worker in brass;* faber ærarius, Coll. Monast. Th. 30, 1.

ár-stæf, *gen.* -stæfes; *pl. nom. acc.* -stafas; *m. Favour, kindness,
benefit, help;* gratia, beneficium, auxilii latio :—Fæder alwalda mid
árstafum eówic gehealde síða gesunde *may the all-ruling Father hold
you with kindness safe on your ways,* Beo. Th. 639; B. 317. For árstafum ðú usic sóhtest *thou hast sought us for help,* 920; B. 458 : Exon.
107 a; Th. 409, 5; Rä. 27, 24. v. ár, stæf.

art *:—*Art vel arþ es, Jn. Lind. War. 1, 19. v. eom.

arþ *art,* Mk. Lind. Rush. War. 14, 70 : Jn. Lind. Rush. War. 1, 19.
v. eom.

ár-þegn, ár-þeng, es; *m.* [ár *honour,* þegen *a servant*] *A servant* or
minister by his place or *employment;* servus, minister honorabilis :—Cumena árþegn *the servant of guests,* Bd. 4, 31; Whel. 361, 14.

arudlíce *quickly.* v. arodlíce, ardlíce.

árung, e; *f.* I. *an honouring, a reverence;* honoratio. II.
a regarding, sparing, pardoning; remissio. v. ár *honour,* árian.

Arwan :—Into Arwan *into the river Orwell,* Chr. 1016; Laud. MS;
Erl. 157, 1. v. Arewe.

arwe *an arrow.* v. arewe.

ár-wéla, an; *m.* [ár *an oar,* wéla] *The wealth of oars, the sea;* divitiæ remorum, mare, Andr. Kmbl. 1705; An. 855.

ár-weorþ; *adj.* [ár *honour,* weorþ *worth, worthy*] *Honour-worth,
honourable, venerable;* honorabilis, venerabilis, venerandus. v. ár-wurþ,
ár-wyrþ.

ár-weorþe; *adv. Honourably;* honorifice, Bd. 2, 20; S. 522, 1,
MS. B. v. ár-wurþlíce.

ár-weorþian, -wurþian, -wyrþian; *p.* -ode; *pp.* -od [ár *honour,*
weorþian *to hold worthy*] *To hold worthy of honour, to give honour to,
to honour, reverence, worship;* honorare, honorificare, honorem referre,
venerari :—He ongan árweorþian ða þrówunge háligra martyra *incepit
honorem referre cædi sanctorum,* Bd. 1, 7; S. 479, 1. Ðæt mynster
seó cwén swýðe lufode and árwyrþode *regina monasterium multum
diligebat et venerabatur,* 3, 11; S. 535, 15; Jn. Bos. 5, 23 : Deut.
5, 16.

ár-weorþig; *adj. Venerable, reverend;* reverendus. v. árwurþig.

ár-weorþlíc; *adj. Venerable;* venerabilis. v. ár-wurþlíc.

ár-weorþlíce; *adv. Honourably, reverently, solemnly, kindly;* honorifice, reverenter, solemniter, clementer, R. Ben. 58, Lye : Bd. 3, 19;
S. 547, 8 : 1, 27, resp. 8; S. 495, 17 : Gen. 45, 4. v. ár-weorþe,
-wurþlíce, -wyrþlíce.

ár-weorþnes, ár-wyrþnes, -ness, e; *f.* [ár *honour,* weorþnes *worthiness*]
Honour-worthiness, honour, dignity; honor, dignitas, reverentia :—Æfter
árwyrþnesse swá micles biscopes *juxta venerationem tanto pontifice dignam,* Bd. 3, 17; S. 544, 3, col. 2. Gif ðú nú gemunan wilt eallra ðara

árwyrþnessa *if thou now wilt be mindful of all the honours,* Bt. 8; Fox
24, 20. Mid árweorþnesse *with honour, honourably,* R. Ben. 6, 61.

ár-weorþung, e; *f. Honour, reverence;* honor, reverentia :—On ár-
weorþunge *in honore,* Ps. Lamb. 48, 21. v. ár-wurþung.

ár-widðe, an; *f?* [ár *an oar,* widðe *withe*] *An oar-withe, a willow
band to tie oars with;* struppus :—Árwidðe *vel* strop *struppus,* Ælfc. Gl.
103; Som. 77, 117; Wrt. Voc. 56, 37.

arwunga, arwunge; *adv. Gratuitously;* gratis :—Arwunga ge on-
fêngun, arwunge ge sellaþ *gratis accepistis, gratis date,* Mt. Kmbl. Rush.
10, 8. v. earwunga.

ár-wurþ, -wyrþ; *def.* se árwurþa; seó, ðæt árwurþe; *adj.* [ár *honour,*
weorþ *worth*] *Honour-worth, honourable, venerable, reverend;* honorabilis, honorandus, venerabilis, venerandus :—Se árwurþa wer *vir venerabilis,* Bd. 4, 18; S. 586, 22 : 5, 1; S. 613, 11. Se gôda biþ simle
árwyrþe *the good is always honourable,* Bt. 39, 2; Fox 212, 23. Árwurþe wudewe [MS. wurdewe] *or* nunne *nonna,* Ælfc. Gl. 69; Som.
70, 21; Wrt. Voc. 42, 30. Se árwurþesta Godes andettere *reverentissimus Dei confessor,* Bd. 1, 7; S. 478, 20. Ða árwurþan bán *honoranda
ossa,* 3, 11; S. 535, 16. Ðæt árwurþe bæþ *lavacrum venerabile,* 3, 11;
S. 535, 34.

ár-wurþian, -wurþigean; *p.* ode; *pp.* od; *v. a. To give honour to, to
honour, reverence, worship;* honorare, honorificare, venerari :—Onsægednys lôfes árwurþaþ me *sacrificium laudis honorificabit me,* Ps. Spl.
49, 24. Ðæt ealle árwurþion [árwurþigeon, Jun.] ðone Sunu, swá swá
hig árwurþiaþ [árwurþigeaþ, Jun.] ðone Fæder; se ðe ne árwurþaþ ðone
Sunu, ne árwurþaþ he ðone Fæder *ut omnes honorificent Filium, sicut
honorificant Patrem; qui non honorificat Filium, non honorificat Patrem,*
Jn. Bos. 5, 23 : Bd. 5, 19; S. 637, 6. To árwurþianne [MS. tarwurþi-
enne, v. weorþianne = wurþianne, in weorþian I] ðinne, ðone sôðan and
ðone áncænnedan, Sunu *to honour thy, the true and only begotten, Son,*
Te Dm. Thomson 35, 12. Geleáfa sôþlíce se geleáffulla ðes is; ðæt
ánne God on Þrýnnesse and Þrýnnesse on Ánnesse we árwurþian *fides
autem catholica hæc est; ut unum Deum in Trinitate et Trinitatem in
Unitate veneremur,* Ps. Lamb. fol. 200 a, 15. Árwurþa ðinne fæder and ðíne
môdur *honora patrem tuum et matrem,* Deut. 5, 16. v. ár-weorþian.

ár-wurþig *reverend.* v. ár-weorþig, ár-weorþ.

ár-wurþigean *to honour, reverence;* honorificare, Jn. Jun. 5, 23.
v. árwurþian.

ár-wurþlíc; *adj. Venerable;* venerabilis :—Árwurþlíc on to seónne
venerabilis aspectu, Bd. 2, 16; S. 519, 35. v. ár-weorþ, -wurþ.

ár-wurþlíce; *adv. Honourably, reverently, kindly, solemnly, mildly;*
honorifice, solemniter, reverenter, clementer :—Hí swíðe árwurþlíce onfangene wæron *they were very honourably received,* Bd. 2, 20; S. 522, 1 :
3, 19; S. 547, 8 : 5, 19; S. 637, 33. Fram cyricean ingonge árwurþlíce ahabban *ab ingressu ecclesiæ reverenter abstinere,* Bd. 1, 27, resp. 8;
S. 495, 17. Ða grêtte hig árwurþlíce *quos ille clementer allocutus est,*
Gen. 45, 4. v. ár-weorþe, -weorþlíce.

ár-wurþung, e; *f. Honour, reverence;* honor, reverentia :—Bryngaþ
Drihtne árwurþunge *afferte Domino honorem,* Ps. Spl. T. 28, 2 : Ps. Spl.
48, 12. v. ár-weorþung.

ár-wyrþ; *adj. Honourable, venerable;* honorabilis, venerandus, Bt.
39, 2; Fox 212, 23 : Elen. Kmbl. 2256; El. 1129. v. ár-weorþ.

ár-wyrþian; *p.* ode; *pp.* od *To honour, reverence,* Bd. 3, 11; S. 535,
15. v. ár-weorþian.

ár-wyrþlíce; *adv. Honourably, reverently, solemnly, kindly,* R. Ben.
58. v. ár-wurþlíce.

ár-wyrþnes, -ness, e; *f. Dignity,* Bd. 3, 17; S. 544, 3, col. 2. v. ár-
weorþnes.

a-rýpan; *p.* de, te; *pp.* ed, d, t *To tear off, to rip;* evellere, abscindere :—He me of hýd arýpeþ *he tears off my hide from me,* Exon. 127 a;
Th. 488, 15; Rä. 76, 7. v. be-rýpan.

ár-ýþ, e; *f. An oar-wave;* unda remis pulsata :—Hærn eft onwand,
árýða geblond *the tide turned back, the commotion of the oar-waves,* Andr.
Kmbl. 1063; An. 532.

a-sæcgan; *p.* -sægde, -sæde; *pp.* -sægd, -sæd *To speak out, relate, tell,
say, express, explain, announce, proclaim;* edicere, effari, exprimere,
referre, enarrare, annunciare :—Ne mæge we næfre asæcgan, hú ðú æðele
eart, êce Drihten *we may never express, how excellent thou art, everlasting Lord,* Hy. 3, 13; Hy. Grn. ii. 281, 13. v. a-secgan.

a-sǽd *said out, related, told,* Bd. 4, 22; S. 590, 32; *pp.* of a-secgan, *q. v.*

a-sǽdon *said out, related, told,* Ors. 4, 6; Bos. 86, 33; *p.* of a-secgan.

a-sǽlan; *p.* -sælde; *pp.* -sæled [a, sælan *to bind*] *To bind fast, bind;*
astringere, ligare :—Synnum asæled *bound fast by sins,* Elen. Kmbl. 2485;
El. 1244; Cd. 100; Th. 132, 18; Gen. 2195 : 166; Th. 207, 21; Exod.
470.

a-sændan; *p.* -sænde; *pp.* -sænd *To send forth, to send,* Apol. Th.
6, 16 : 13, 5. v. a-sendan.

a-sáh *set, sank,* Chr. 1012; Th. 268, 30, col. 1; 269, 28, col. 1; 26,
col. 2; *p.* of a-sígan.

asal, asald *an ass,* Mt. Lind. Stv. 18, 6 : 21, 2. v. esol.

a-sánian; *p.* ode; *pp.* od *To languish, grow weak, diminish;* langues-

F 2

cere, laxari :—Næfre ic lufan sibbe forlǽte asánian *never will I permit the love of my kin to languish,* Exon. 50 a ; Th. 172, 23 ; Gú. 1148.

asaru *Asarabacca, folefoot, hazelwort ;* asárum Europæum = ἄσαρον, L. M. 2, 14 ; Lchdm. ii. 192, 7.

a-sáwan ; *p.* -seów, -siów, *pl.* -seówon ; *pp.* -sáwen *To sow ;* seminare, obserere, Bt. Met. Fox 20, 499 ; Met. 20, 250. v. sáwan.

asca *dust ;* pulvis, Mk. Lind. Rush. War. 6, 11. v. asce.

asca, ascas, ascum :—Asca *of ash spears,* Exon. 78 a ; Th. 292, 15 ; Wand. 99. v. æsc.

a-scacan *to shake off, to shake, brandish ;* excutere, Ps. Th. 67, 10. v. asceacan.

a-scádan *to separate,* L. Wih. 3 ; Th. i. 36, 19. v. asceádan.

a-sceacan *to shake,* Exon. 58 a ; Th. 207, 20 ; Ph. 144 : Ps. Spl. 7, 13. v. a-sceacan.

a-scǽre ; *adj.* [a, scær ; *p. of* sceran *to cut, shear*] *Without tonsure, untrimmed ;* intonsus, incultus, Peccatorum Medicina 8. v. æ-scære.

a-scafan ; *p.* -scóf, *pl.* -scófon ; *pp.* -scafen, -scæfen *To shave ;* abradere, obradere :—Ascæfen *obrasus,* Cot. 148. v. scafan.

a-scamian ; *p.* ode ; *pp.* od *To be ashamed, to make ashamed or abashed ;* erubescere, pudore confundere :—Ná ascamien on me *non erubescant in me,* Ps. Spl. 68, 8. Hí ascamode swíciaþ on swíman *they wander abashed in giddiness,* Exon. 26 b ; Th. 79, 31 ; Cri. 1299. v. scamian.

Ascan mynster *Axminster,* Chr. 755 ; Th. 86, 13, col. 1. v. Acsan mynster, Axan mynster.

ASCE, æsce [g. æscean], acse, ahse, axe, axse, æxe, an ; *f.* ASH, *ashes ;* cinis :—On dære ascan *in the ashes,* Exon. 59 a ; Th. 213, 27 ; Ph. 231 : 60 a ; Th. 217, 24 ; Ph. 285. Gebreadad weorþeþ eft of ascan *it becomes formed again from [its] ashes,* 61 a ; Th. 224, 9 ; Ph. 373. Ascan and ýslan *ashes and embers,* 64 a ; Th. 236, 18 ; Ph. 576 : 65 a ; Th. 240, 33 ; Ph. 648. [O. H. Ger. asca, f. cinis : Goth. azgo, f : O. Nrs. aska, f.]

a-sceacan, -scacan, -scæcan ; he -sceaceþ, -sceacþ, -scæceþ, -scaceþ ; *p.* -sceóc, -scóc, *pl.* -sceócon, -scacon, -sceacen, -scacen. **I.** *to shake off, remove ;* excutere :—Asceacaþ þæt dust of eówrum fótum *excutite pulverem de pedibus vestris,* Mk. Bos. 6, 11. **II.** *to be removed, forsake, desert, flee ;* excuti, fugere, aufugere, deserere :—Asceacen [Lamb. ofascacen] ic eom *excussus sum,* Ps. Spl. C. 108, 22. Ðæt Iacob wæs asceacen *quod fugeret Jacob,* Gen. 31, 22. He asceacen wæs fram Æðelréde *he had deserted from Æthelred,* Chr. 1001 ; Ing. 174, 15. **III.** *to shake, brandish, to be shaken ;* vibrare, quatere, concuti, labefieri, infirmari :—His swurd he acwecþ oððe asceaceþ *gladium suum vibrabit,* Ps. Lamb. 7, 13. He ascæceþ feðre *it shakes its plumage,* Exon. 58 a ; Th. 207, 20 ; Ph. 144 : Ps. Spl. 7, 13. Offa æscholt asceóc *Offa shook his ashen spear,* Byrht. Th. 138, 35 ; By. 230. Wilsumne regn wolcen bringceþ, and donne ascaceþ God sundoryrfe *pluviam voluntariam segregabis, Deus, hereditati tuæ, etenim infirmata est,* Ps. Th. 67. 10.

a-sceádan, -scádan ; *p.* -scéd, *pl.* -scédon ; *pp.* -sceáden, -scáden ; *v. a.* [a *from,* sceádan *to divide*] *To separate, disjoin, exclude, distinguish ;* separare, segregare :—Ic mec ascéd ðara scylda *I separated myself from the guilt,* Elen. Kmbl. 937 ; El. 470 : 2623 ; El. 1313. And he hine from nýtenum ascéd *and he distinguished him from beasts,* L. E. I. 23 ; Th. ii. 420, 8. Hí of ciricean gemánan ascádene síen *they from the church communion shall be excluded,* L. Wih. 3 ; Th. i. 36, 19. Ðæt eálond is feor asceáden fram Hibernia *insula ab Hibernia procul secreta est,* Bd. 4, 4 ; S. 570, 40.

a-sceáf *expelled,* Cd. 55 ; Th. 68, 11 ; Gen. 1115 ; *p. of* a-scúfan.

a-scealian ; *p.* ode ; *pp.* od [a *from,* scealu *a scale*] *To pull off the scales or bark, to scale, bark ;* decorticare, Cot. 79.

a-sceamian *to be ashamed.* v. a-scamian.

a-scearpen *to sharpen,* Ps. Surt. 63, 4. v. a-scirpan.

a-scéd *separated,* Elen. Kmbl. 937 ; El. 470 ; *p. of* a-sceádan.

a-sceofen *expelled,* = a-scofen, Bd. 4, 12 ; S. 581, 17 ; *pp. of* a-scúfan.

a-sceónung, e ; *f. Detestation, abomination ;* abominatio, Mk. Bos. 13, 14. v. a-scúnung.

a-sceóp *gave,* Cd. 161 ; Th. 201, 32 ; Exod. 381. v. a-sceppan.

a-sceortian, -scortian ; *p.* ode ; *pp.* od *To be short, to grow short, shorten, elapse, diminish, fail ;* breviare, effluere :—Ðæt wæter asceortode *the water failed,* Gen. 21, 15. Ten þúsend geára ascortaþ *ten thousand years will elapse,* Bt. 18, 3 ; Fox 66, 12.

a-sceótan ; he -scýt, -scýtt ; *p.* -sceát, *pl.* -scuton ; *pp.* -scoten [a, sceótan *to shoot*] *To shoot forth, shoot, shoot out, fall ;* jaculari, cum impetu erumpere :—Hie ne mehton from him nænne flán 'asceótan *they could not shoot an arrow from them,* Ors. 6, 36 ; Bos. 132, 8. Ne ascýtt Sennacherib flán into ðære byrig Hierusalem *Sennacherib shall not shoot arrows into the city of Jerusalem,* Homl. Th. i. 568, 31. Ða eágan of his heáfde ascuton, and on eorþan feóllan *the eyes shot out of his head, and fell on the earth,* Bd. 1, 7 ; S. 478, 38.

a-sceppan ; *p.* -sceóp, -scóp, *pl.* -sceópon, -scópon ; *pp.* -sceapen, -scapen *To create, appoint, give ;* creare, designare :—Him God naman

niwan asceóp *God gave him a new name,* Cd. 161 ; Th. 201, 32 ; Exod. 381.

a-scerian *to cut from, separate.* v. a-scirian.

a-scerpan *to sharpen.* v. a-scirpan.

ASCIAN, acsian, ahsiàn, axian ; *p.* ode ; *pp.* od. **I.** *to* ASK, *to ask for, to demand, inquire, to call, summon before one ;* interrogare, postulare, exigere :—Ðe ðú me æfter ascast *which thou askest about,* Bt. 39, 4 ; Fox 216, 26, 29. Ne ascige ic nú ówiht bi ðam bitran deáþe mínum *I demand now nothing for my bitter death,* Exon. 29 b ; Th. 90, 16 ; Cri. 1475. He ongan hine ahsian *he began to call him,* Cd. 40 ; Th. 53, 18 ; Gen. 863. **II.** *to obtain, experience ;* nancisci, experiri :—He weán ahsode *he obtained woe,* Beo. Th. 2417 ; B. 1206 : 851 ; B. 423. [*Orm.* asskenn : *Laym.* axien : *O. Sax.* escón : *O. Frs.* askia, aschia : *Dut.* eischen : *Ger.* heischen : *M. H. Ger.* eischen : *O. H. Ger.* eiscón : *Dan.* äske : *Swed.* äska : *O. Nrs.* æskja *optare* : *Sansk.* ish *to wish, desire.*]

a-scilian ; *p.* ede ; *pp.* ed [a *from,* scel *a shell*] *To take off the shell, to shell ;* enucleare, Cot. 171.

a-scínan ; *p.* -scán, *pl.* -scinon ; *pp.* -scinen *To shine forth, to be clear, evident ;* clarescere, elucere :—Hwylc wǽre his líf cúþlícor ascíneþ *vita qualis fuerit certius clarescat,* Bd. 5, 1 ; S. 613, 14. Ða ðær ascán beáma beorhtast *then there shone the brightest of beams,* Exon. 52 a ; Th. 180, 20 ; Gú. 1282.

a-scirian, -scyrian ; *p.* ede ; *pp.* ed, ud ; *v. a.* [a, scirian *to share*] *To cut from, separate, divide, part, sever ;* separare, sejungere, excommunicare, destinare :—He ascirede Adames bearn *he separated Adam's sons,* Deut. 32, 8. Ascyrud beón fram mannum *moveri ab hominibus,* Somn. 280. Ascyred and asceáden scylda gehwylcre *sundered and set apart from every sin,* Elen. Kmbl. 2623 ; El. 1313 : Exon. 31 b ; Th. 98, 16 ; Cri. 1608. Ðæt he scyle from his Scippende ascyred weorþan to deáþe niðer *that he shall be separated from his Creator by death beneath,* Exon. 31 b ; Th. 99, 2 ; Cri. 1618.

a-scirigendlíc *disjoining, disjunctive.* v. a-scyrigendlíc.

a-scirpan, a-scyrpan, a-scerpan, a-scearpan ; *p.* te, tun ; *pp.* ed *To sharpen ;* exacuere :—Swíðor ablendaþ ðæs módes eágan ðonne hí hí ascirpan *they rather blind the eyes of the mind than sharpen them,* Bt. 34, 8 ; Fox 144, 34. v. scerpan.

ascirred = ascired *separated from, saved,* Bt. 20 ; Fox 72, 6 ; *pp. of* a-scirian.

a-scofen *banished,* R. Ben. 63. v. a-scúfan.

a-scóp *gave,* Ors. 1, 8 ; Bos. 31, 16. v. a-sceppan.

a-scortian *to shorten,* Bt. 18, 3 ; Fox 66, 12. v. a-sceortian.

a-scræp *he scraped ;* radebat, Job 2, 8 ; Thw. 166, 33 ; *p. of* a-screopan.

a-screádian ; *p.* ode ; *pp.* od *To prune, lop ;* præsecare, Anlct. Gl. DER. screádian.

a-screncan ; *p.* -screncte ; *pp.* -screnct [a, screncan *to supplant*] *To supplant* :—Ne eft sió þræsþing ðæs líchoman ðæt mód ne ascrence mid upahæfenesse *ne aut istos afflicta caro ex elatione supplantet,* Past. 43, 9 ; Hat. MS. 60 b, 3.

a-screopan ; *p.* -scræp, *pl.* -scræpon ; *pp.* -screpen *To scrape off, scrape ;* radere :—Ascræp ðone wyrms of his líce *testa saniem radebat,* Job 2, 8 ; Thw. 166, 33. v. screopan.

a-screpan, -scrypan ; *pp.* en *To bear, cast or vomit out ;* egerere, Cot. 71. v. a-screopan.

a-scrincan ; *p.* -scranc, *pl.* -scruncon ; *pp.* -scruncen *To shrink ;* arescere. v. scrincan.

a-scrypan *to cast out.* v. a-screpan.

asc-þrotu *fennel-giant.* v. æsc-þrote, an ; *f.*

a-scúfan, -sceófan ; *p.* -sceáf, *pl.* -scufon ; *pp.* -scofen, -sceofen [a *from,* scúfan *to shove*] *To drive away, expel, banish, repel, shove away ;* expellere, pellere, abigere, extrudere, emittere :—Forþ ascúfan *to drive forward,* Exon. 129 b ; Th. 498, 1 ; Rä. 87, 6. Me cearsorge of móde asceáf þeóden usser *our Lord has driven anxious sorrow from my mind,* Cd. 55 ; Th. 68, 11 ; Gen. 1115. He wæs asceofen and adrifen of his biscop-setle *pulsus est a sede sui episcopatus,* Bd. 4, 12 ; S. 581, 17.

ascung, e ; *f. An asking, a question, an interrogation, inquiry, inquisition ;* interrogatio, inquisitio :—Ðæs sædes corn biþ simle aweaht mid ascunga *the grain of this seed is always excited by inquiry,* Bt. Met. Fox 22, 81 ; Met. 22, 41 : Bt. Met. Fox 5, 3 ; Fox 12, 16. v. acsung.

a-scúnian ; *p.* ode ; *pp.* od ; *v. a.* [a *away,* scúnian *to shun*]. **I.** *to avoid, shun, fly from ;* evitare, reprobare :—He mót þyllic ascúnian *he must shun the like,* L. C. S. 7 ; Th. i. 380, 9 : L. Ed. 4 ; Th. i. 162, 6. **II.** *to hate, detest ;* odisse, detestari :—Esau ascúnode Iacob *oderat Esau Jacob,* Gen. 27, 41. Ða ascúnodon híg hine *oderant eum,* Gen. 37, 4. **III.** *to accuse, reprove, convict ;* arguere :—Hwylc eówer ascúnaþ me for synne *quis ex vobis arguet me de peccato?* Jn. Bos. 8, 46.

a-scúniendlíc ; *adj. Detestable, abominable ;* detestabilis :—Beforan Gode ys ascúniendlíc *abominatio est ante Deum,* Lk. Bos. 16, 15.

a-scúnung, a-sceónung, e ; *f. An execration, abomination, a detesta-**

tion; execratio, abominatio:—Ge geseóþ ðære toworpennysse asceó-nunge [ascūnunge, Jun.] *videritis abominationem desolationis,* Mk. Bos. 13, 14 : Ps. Spl. 58, 14.

a-scuton *shot out,* Bd. 1, 7 ; S. 478, 38 ; *p. pl. of* a-sceótan.

a-scyled *taken out of the shell, shelled;* enucleatus, Cot. 75 ; *pp. of* a-scilian.

a-scyndan [a *from,* scyndan *to hasten*] *To separate, remove, take away;* tollere, elongare :—Ðú ascyndest fram me freónd *elongasti a me amicum,* Ps. Spl. M. 87, 19.

a-scyrian *to separate,* Elen. Kmbl. 2623 ; El. 1313. v. a-scirian.

a-scyrigendlíc; *adj.* [ascirigende *disjoining,* from ascirian] *Disjoining, disjunctive;* disjunctivus, Ælfc. Gr. 44 ; Som. 45, 43.

a-scyrigendlíce; *adv. Disjunctively, severally;* disjunctive, Ælfc. Gr. 44 ? Lye.

a-scyrpan *to sharpen,* Ps. Th. 126, 5 : Ps. Spl. C. 63, 3. v. a-scirpan.

a-sealcan; *pp.* asolcen *To languish, to be* or *become weak, idle, slothful, remiss;* languescere, remittere, desidiosum fieri :—Ne lǽt ðú ðe ðín mód asealcan wǽrfæst willan nínes *let not thou thy mind languish [to be] observant of my will,* Cd. 99 ; Th. 130, 30; Gen. 2167. Asolcen fram gódre drohtnunge *slothful for good living,* Homl. Th. i. 306, 11 : 340, 35. Asolcen *accidiosus?* vel *tediosus,* Ælfc. Gl. 114 ; Som. 80, 18 ; Wrt. Voc. 60, 52. Asolcen *dissolutus, desidiosus,* R. Ben. 48. Asolcen *deses,* Ælfc. Gr. 9, 26 ; Som. 11, 10. Asolcen *iners,* Cot. 108. Asolcen *remissus, ignavus,* Scint. 16.

a-seárian; *p.* ode; *pp.* od *To become dry, to sear, dry up;* arescere, Lchdm. iii. 355, 24.

a-seáþ *seethed.* p. of a-seóðan.

a-sécan, -sécean; *p.* -sóhte; *pp.* -sóht [a, sécan *to seek*]. I. *to search* or *seek out, to seek for, to require, demand;* eligere, requirere, petere aliquid ab aliquo :—Aséan ða sélestan *to seek out the best,* Elen. Kmbl. 2035 : El. 1019 : 813 ; El. 407. Mid swá mycle fóreseónysse wæs ðæs líchoman clǽnnesse asóht *tanta provisione est munditia corporis requisita,* Bd. 1, 27, resp. 8 ; S. 496, 8. Wyllaþ me lífes aséean *they will demand my life,* Ps. Th. 118, 95. II. *to seek, go to, explore;* adire, explorare :—Ðæt fýr georne aséceþ innan and útan eorþan sceátas *the fire shall eagerly seek the tracts of earth within and without,* Exon. 22 b ; Th. 62, 20 ; Cri. 1004.

a-secgan, -sæcgan; *p.* -sægde, -sǽde; *pp.* -sægd, -sǽd [a *out,* secgan *to say*] *To speak out, declare, express, tell, say, relate, explain, announce, proclaim;* edicere, effari, exprimere, referre, enarrare, annunciare :—Ic him mín ǽrende asecgan wille *I will relate to him my errand,* Beo. Th. 693 ; B. 344. Heofonas asecgaþ wuldor Godes *cœli enarrant gloriam Dei,* Ps. Spl. C. 18, 1. Wundor asecgan *miraculum enarrare,* Bd. 3, 2 ; S. 524, 39. Gif seó gemyndelíc wíse asǽd biþ *if that memorable thing be told,* 4, 22 ; S. 590, 32 : Bt. 34, 8 ; Fox 144, 22 : 35, 1 ; Fox 154, 18. Him engel Godes eall asægde *God's angel told him all,* Cd. 179 ; Th. 225, 19 ; Dan. 156. Ðá asǽdon his geféran *then said his companions,* Ors. 4, 6 ; Bos. 86, 33. Oþ ðæt ic asecge *donec annunciem,* Ps. Th. 70, 17.

a-secgendlíc; *adj,* *That which may be spoken, expressible;* effabilis, Som.

a-sellan; *p.* -sealde; *pp.* -seald *To expel, banish, deliver;* expellere, relegare, tradere, Cd. 215 ; Th. 270, 14 ; Sat. 90. v. sellan.

a-sendan, ic -sende, ðú -sendest, -sendst, -senst, he -seṇt, -sendeþ, *pl.* -sendaþ; *p.* -sende ; *pp.* -sended, -send *To send forth, send out, send;* emittere, mittere :—Asend gást ðínne and biþ gescapen *emitte spiritum tuum et creabuntur,* Ps. Spl. 103, 31. Ðonne ðú of líce aldor asendest *when thou sendest forth life from thy body,* Cd. 134 ; Th. 168, 29 ; Gen. 2790. Drihten asent hungor on eów and þurst and nǽcede *the Lord shall send forth on you hunger and thirst and nakedness,* Deut. 28, 48. Ðæt he wolde asendan his ǽncennedan Sunu *that he would send his only-begotten Son,* Homl. Th. ii. 22, 3 : Ps. Spl. 105, 15. Ic eom asend *ego missus sum,* Lk. Bos. 1, 19. DER. sendan.

a-séngan *for* a-sénian [a, sénian *to see*] *To shew, discover, manifest;* manifestare, perspicuum facere :—Ðe ic aséngan ne mæg *which I may not discover,* Exon. 70 a ; Th. 261, 11 ; Jul. 313.

a-seón, ic -seó, ðú -síhest, -síhst, he -síheþ, -síhþ, *pl.* -seóþ; *p.* -sáh, *pl.* -sigon, -sihon; *impert.* -seóh; *pp.* -sigen, -sihen [a *from, out;* seón, síhan *to strain*] *To strain out;* percolare :—Aseóh ðone drenc, and dó ðonne mele fulne buteran *strain out the drink, and then add* [do] *a basin full of butter,* L. M. 1, 36 ; Lchdm. ii. 86, 16.

a-seóðan; *p.* -seáþ, *pl.* -sudon; *pp.* -soden *To boil, seethe, scorch, to purify by seething;* coquere :—Swá man seolfor aseóðeþ mid fýre *as one seethes silver by fire,* Ps. Th. 65, 9. Ðé ic geceás on ðam ofne ðe ðú on wǽre asoden, ðæt wæs on ðínum iermþum *elegi te in camino paupertatis,* Past. 26, 1 ; Hat. MS. 35 a, 6. Ðæt heó mid longre hire líchoman un-trumnesse asodene beón *that she should be purified by the long suffering of her body,* Bd. 4, 23 ; S. 595, 15. Ealle we lǽtaþ to viii healf-marcum asodenes goldes *we estimate all at eight half-marks of pure gold,* L. A. G. 2 ; Th. i. 154, 2.

a-seów, -siów *sowed,* Bt. 33, 4 ; Fox 132, 26 ; *p. of* a-sáwan.

a-setan *to appoint, design;* destinare, R. Conc. pref.

a-seted, -sett *set, placed, stored, built,* Beo. Th. 1338; B. 667: Mt. Bos. 3, 10 ; *pp. of* a-settan.

a-sēðan; *p.* -sēðde; *pp.* -sēðed *To affirm, confirm;* affirmare, confirmare :—Sume [adverbia] syndon ad vel confirmativa, mid ðám we asēðaþ úre sprǽce *some adverbs are affirmative or confirmative, with which we affirm our speech,* Ælfc. Gr. 38 ; Som. 40, 16.

a-sēðan *to boil.* v. seóðan.

a-setnys, -nyss, e ; *f. What is set or fixed, a statute, law;* constitutio, statum :—Eádmundes cyninges asetnysse *king Edmund's institutes,* L. Edm. E. 1 ; Th. i. 244, 1.

a-settan; *p.* -sette; *pp.* -seted, -sett. I. *to set, put, place, appoint, lay, set up, erect, build, to set or take, to plant;* ponere, statuere, constituere, instituere, collocare, deponere, desumere, plantare :—He asette his swíðran hand under Abrahames þeóh *posuit manum sub femore Abraham,* Gen. 24, 9. He hæfde Grendle togeánes seleweard aseted *he had set a hall-ward against Grendel,* Beo. Th. 1338 ; B. 667. Eallunga ys seó æx to ðære treówa wurtrumum asett *jam enim securis ad radicem arborum posita est,* Mt. Bos. 3, 10. Hēht ðá asettan líc on eorþan *he then commanded to place the body upon the earth,* Elen. Kmbl. 1750 ; El. 877. Ac heó hire ðǽr wíc asette *ibique sibi mansionem instituit,* Bd. 4, 23 ; S. 593, 26 : Exon. 108 a ; Th. 411, 27 ; Rä. 30, 6. Hēt ǽnne weall asettan *he ordered a wall to be built,* Ors. 6, 15 ; Bos. 122, 34. Hēt hí eft asettan *he bade her again be taken,* Exon. 69 a ; Th. 256, 14 ; Jul. 231. Ic on neorxna wonge niwe asette treów mid telgum *I planted in paradise a new tree with branches,* Cd. 223 ; Th. 295, 5 ; Sat. 481. II. *síþ* asettan *to make a journey;* iter facere :—He in helle ceasl síþ asette *he made his journey into the jaws of hell,* Andr. Kmbl. 3404 : An. 1706 : Exon. 103 a ; Th. 391, 26 ; Rä. 10, 11.

a-sette *set, placed, built,* Bd. 4, 23 ; S. 593, 26 ; *p. of* a-settan.

asicyd; *part.* [a *from,* súcan *to suck*] *Taken from suck, weaned;* ablactatus :—Swá swá asicyd ofer módor *sicut ablactatus super matre,* Ps. Spl. M. C. 130, 4.

a-siftan; *p.* -sifte; *pp.* -sift *To sift;* cribrare :—Asift þurh cláþ *sift through a cloth,* L. M. 1, 2, 21 ; Lchdm. ii. 36, 7. v. siftan.

a-sigan; *p.* -sáh, *pl.* -sigon; *pp.* -sigen *To decline, go down, fall down;* delabi, occidere :—Ðæt, mid ðam dynte, he nyðer asáh *that, with the blow, he fell down,* Chr. 1012 ; Th. 268, 30, col. 1 ; 269, 28, col. 1 ; 269, 26, col. 2. Lǽt ðínne sefan healdan freán dómas, ða ðe hér men forlǽtaþ asígan *let thy mind observe the Lord's decrees, which here men permit to decline,* Exon. 81 a ; Th. 304, 24 ; Fä. 75.

a-sigen *fallen;* *pp. of* a-sígan.

a-sindrian; *p.* ode; *pp.* od *To sunder, separate.* v. a-syndran.

a-singan; *p.* -sang, *pl.* -sungon; *pp.* -sungen [a, singan] *To sing;* canere :—Ðæt man asinge *that a man sing,* Ps. Th. 91, 1 : Beo. Th. 2323 ; B. 1159 : Bd. 3, 27 ; S. 559, 12.

Asirige *The Assyrians;* Assyrii :—Ðæt synd Asirige and Rómáne *these are the Assyrians and the Romans,* Ors. 2, 5 ; Bos. 49, 14. v. Assyrias.

a-sittan ; *p.* -sæt, *pl.* -sǽton ; *pp.* -seten *To dwell together;* considere :—Secgas, mid sigecwēn, aseten hæfdon, on Crēca land *the men had a dwelling together with the victorious queen, in the land of the Greeks,* Elen. Kmbl. 1993 ; El. 998. v. sittan II.

a-slacian; -slæcian; *p.* ode, ade, ude; *pp.* od, ad, ud *To slacken, loosen, untie, remit, dissolve, enervate;* laxare, remittere, solvere, dissolvere, dimittere, hebetare, enervare, Cot. 103 : 169 : Prov. 19 : 10. v. slacian.

a-slacigendlíc; *adj. Remissive;* remissivus :—Sume [adverbia] syndon remissiva, ðæt synd aslacigendlíce [lytlum *paulatim,* softe *suaviter, etc.*] *some* [adverbs] *are remissiva, that is remissives, etc.* Ælfc. Gr. 38 ; Som. 40, 29.

a-slacigendlíce; *adv. Slackly, remissly;* remisse, Ælfc. Gr. 38 ? Lye.

aslád *slipped away.* v. aslídan.

a-slæccan; *p.* -slæcte ; *pp.* -slæced, -slæct *To slacken, loosen, remit;* laxare, remittere. v. slæccan, slacian.

a-slæcian; *p.* ude; *pp.* ud *To dissolve;* dimittere, Cot. 62. v. a-slacian.

a-slægen *struck,* Lye. v. a-sleán.

a-slápan; *p.* -slép, *pl.* -slépon ; *pp.* -slápen [a, slápan = slǽpan *to sleep*] *To be sleepy, begin to sleep, fall asleep;* dormitare :—Mín sáwl aslép *dormitavit anima mea,* Ps. Th. 118, 28.

a-sláwian; *p.* ode; *pp.* od *To be heavy, dull, sluggish;* torpescere, Ors. 4, 13 ; Bos. 100, 20.

a-sleán; *p.* -slóh, *pl.* -slógon; *pp.* -slegen, -slagen, -slægen *To strike, beat, hammer, to fix, erect;* ferire, icere, cædere, figere, ponere :—On býman aslegenum [Lamb. onaslagenum], Ps. Spl. 97, 6 ; *in tubis ductilibus,* Vulg; in trumpis beten out, Wyc. Hí aslógan án geteld *tetenderunt tentorium,* Bd. 3, 17 ; S. 543, 33, col. 1 : 5, 6 ; S. 619, 26. Ðe of his líchoman aslegen wæs *that was struck off his body,* Bd. 3, 12 ; S. 537, 34. v. sleán. DER. on-asleán; *pp.* on-aslagen.

a-slépen = a-sleópen *slip away,* Bt. Met. Fox 13, 18 ; Met. 13, 9. v. a-slúpan.

a-slídan; ic -slíde, ðú -slídest, -slíst, he -slídeþ, -slít, *pl.* -slídaþ; *p.* -slád,

pl. -slidon; *pp.* -sliden *To slide* or *slip away*; labare :—Ne aslît his fôt *non supplantabuntur gressus ejus*, Ps. Th. 36, 31. Ðæt mîn fôt asliden wǽre *motus est pes meus*, 93, 17. Asliden beón *labi*, Scint. 13, 24, 78.

a-slítan, -slýtan; *p.* -slát, *pl.* -sliton; *pp.* -slyten, -sliten; *v. a.* [*a from*, slítan *to slit*] *To cleave, rive, destroy, cut off*; discindere, diruere, abscindere :—Aslât ða tûnas ealle *destroyed all the villages*, Bd. 3, 16; S. 542, 20. Mildheortnysse his aslýteþ of cneórysse on cynrine *misericordiam suam abscindet a generatione in generationem*, Ps. Spl. 76, 8.

a-slôh, -slôgon *struck, fixed*, Bd. 3, 17; S. 543, 33, col. 1; *p.* of a-sleán.

a-slúpan, -sleáp, *pl.* -slupon; *p.* -slopen *To slip away*; elabi :—Lǽt ðê aslúpan sorge of breóstum *let sorrow slip away from thy breast*, Cd. 134; Th. 169, 7; Gen. 2796. Ðæt hî ǽfre him of aslêpen [=a-sleópen] *that they may ever slip from them*, Bt. Met. Fox 13, 18; Met. 13, 9.

a-slýtan; *p.* -slát; *pp.* -slyten; *v. trans. To cut off* :—Aslýteþ *abscindet*, Ps. Spl. 76, 8. *v.* a-slítan.

a-smeágan, -smeán; *p.* -smeáde; *pp.* -smeád *To look closely into, examine, trace out, elicit, meditate upon, consider, contemplate, ponder, judge, deem, be of opinion, think*; perscrutari, investigare, indagare, elicere, contemplari, pensare, censere :—Nû ne mǽge we asmeágan hû God of ðam lâme flǽsc worhte and blôd, bân and fell, fex and nǽglas *now we cannot trace out how of the loam God made flesh and blood, bones and skin, hair and nails*, Homl. Th. i. 236, 15. Stîge mîne ðû asmeádest *semitam meam investigasti*, Ps. Spl. 138, 2: R. Ben. 55. Asmeágende *indagantes*, Cot. 104. Asmeáde *elicuit*, Cot. 77. Gif man hit ariht asmeáþ *if one rightly considers it*, L. Edg. C. 13; Th. ii. 246, 21. Ic dême oððe ic asmeáge *censeo*, Ælfc. Gr. 26, 2; Som. 28, 51.

a-smeágung, e; *f. Investigation, meditation*; scrutinium, investigatio, meditatio :—Þurh asmeágunge bôclîcre snotornesse *through investigation of book-like wisdom*, Apol. Th. 3, 16.

a-smiðian; *p.* ode; *pp.* od; *v. trans. To forge, make, work as a smith*; fabricare :—Asmiðod *fabricatus*, Cot. 82.

a-smorian; *p.* ede, ode; *pp.* ed, od; *v. trans. To smother, choke, strangle, suffocate;* suffocare :—Asmoraþ ðæt word *suffocat verbum*, Mt. Rush. Stv. 13, 22. Hî hine on his bedde asmoredan and aþrysemodan *they smothered and stifled him on his bed*, Ors. 5, 4; Bos. 105, 5. Ðæt ge ne blôd ne þicgen, ne asmored [MS. H. asmorod] *that ye taste not blood, nor [what is] strangled*, L. Alf. 49; Th. i. 56, 26.

a-snǽsan, -snásan; *p.* de; *pp.* ed; *v. trans.* I. *to hit* or *strike against, to stake oneself upon anything;* impingere :—Gif befóran eágum asnáse [MS. H. asnǽse] *if he stake himself before his eyes*, L. Alf. pol. 36; Th. i. 84, 14. II. *to wrest anything from another?* extorquere, L. Noel, Lye. DER. on-snǽsan, ona-.

a-snîðan; *p.* -snáþ, *pl.* -snidon; *pp.* -sniden; *v. trans. To cut off;* amputare. *v.* snîðan *to cut.*

a-soden *sodden, boiled, tried by seething*, Bd. 4, 23; S. 595, 15; *pp.* of a-seóðan.

a-sogen *sucked*, Cot. 193; *pp.* of a-súgan.

a-sôht *sought out, searched*, Bd. 1, 27, resp. 8; S. 496, 8; *pp.* of a-sêcan.

a-solcen, a-swolcen; *part. Idle, lazy, dissolute, slow, slothful;* remissus, desidiosus, Homl. Th. i. 306, 11. *v.* a-sealcan.

a-solcennys, -nyss, e; *f. Idleness, sloth, slothfulness, sluggishness, laziness;* ignavia, desidia, pigritia :—Heora liðnys is asólcennys and nýtennys *their mildness is sloth and ignorance*, Homl. Th. ii. 46, 11: 220, 21. Se sixta heáfodleáhter is asolcennys *the sixth chief sin is slothfulness*, 218, 22. Þurh ûre asolcennysse *through our sluggishness*, Th. Diplm. A. D. 970; 240, 12: Homl. Th. i. 602, 8.

a-spanan; *p.* -spôn, -speón, *pl.* -spônon, -speónon; *pp.* -spanen, -sponen; *v. trans. To allure from, entice, induce, urge, persuade, introduce secretly;* allicere, illicere, impellere, persuadere, attrahere, subintroducere :—Gif he ða cwêne gespannan [MS. B. aspanan] and gelǽran mihte, ðæt heó brúcan wolde his gesynscipes *si reginæ posset persuadere ejus uti connubio*, Bd. 4, 19; Whel. 304, 42, note. Hér aspôn Æðelwald ðone here to unfriþe *in this year Æthelwald allured the army to a violation of the peace*, Chr. 905; Th. 180, 18, col. 1. Hine Hannibal aspôn, ðæt he ðæt gewinn leng ongan *Hannibal induced him to carry on the war longer*, Ors. 4, 11; Bos. 97, 15. He aspeón him fram ealle *he enticed all from him*, 1, 12; Bos. 35, 19: 2, 2; Bos. 41, 8: 5, 2; Bos. 102, 21. Aspeón ôðerne bisceop *subintroduxit alium episcopum*, Bd. 3, 7; S. 530, 4.

a-spáw *vomited out; p.* of a-spíwan.

a-spêdan; *p.* -spêdde; *pp.* -spêded, -spêdd *To speed, prosper;* prosperare :—Wîtum aspêdde *made prosperous by their sufferings*, Andr. Kmbl. 3261; An. 1633.

a-spelian; *part.* a-speliende; *p.* ode, ade; *pp.* od, ad *To supply another's room, to be deputy* or *proxy for another, represent another;* vicario munere fungi, vicem *vel* locum alicujus supplere :—He môste his hláford aspelian *he might represent his lord*, L. R. 3; Th. i. 192, 3: R. Ben. 58. Aspelad beón *to have one's place supplied by another;* excusari, R. Ben. 35.

a-spendan; *p.* de; *pp.* ed [a, spendan *to spend*] *To spend entirely,*

consume, squander, to spend, expend, lay out, bestow, distribute; consumere, dissipare, expendere, sumptum facere, erogare, impertiri :—Ðonne hys gestreón beóþ ðus eall aspended *when his property is thus all entirely spent*, Ors. 1, 1; Bos. 22, 43. Ic aspende yfele *distraho*, Ælfc. Gr. 47; Som. 48, 52. Ic aspende [asende MS.] oððe gife *impertior*, 37; Som. 39, 13. Aspendan þearfum *to spend on the poor;* erogare pauperibus, R. Ben. interl. 58: Scint. 1.

a-speón *enticed, secretly introduced*, Ors. 1, 12; Bos. 35, 19: Bd. 3, 7; S. 530, 4. *v.* a-spanan.

a-sperian *to track, trace, investigate;* investigare, Prov. 20. *v.* a-spyrian.

aspide, es; *m. An asp, viper, serpent;* aspis, ïdis; *f.* = ἀσπίς, ίδος; *f. a sort of serpent remarkable for rolling itself up in a spiral form: a negative, and σπίζω to extend*, Scapulæ Lexicon :—Aspidas *aspides*, Ps. Th. 139, 3. Anlîc nǽdran, ða aspide ylde nemnaþ *like a serpent, which men call an asp*, Ps. Th. 57, 4. Spl. Lamb. in Ps. 57, 4 *have* nǽdran *instead of* aspide. Ðû ofer aspide miht gangan *thou mayest go over an asp* [super aspidem], Ps. Th. 90, 13; Lamb. *has* ofer nǽdran, 90, 13.

a-spirian, -spirigan; *p.* ede; *pp.* ed *To search, trace* :—Aspirige hit ût *let him trace it out*, L. Ath. iv. 2; Th. i. 222, 14. *v.* a-spyrian.

a-spíwan, -spáw, *pl.* -spiwon; *pp.* -spiwen *To spew out, vomit forth;* evomere, vomere :—Aspau=aspáw *evomuit*, Cot. 78: Peccat. Medic. 5.

a-spôn *allured, induced*, Chr. 905; Th. 180, 18, col. 1: Ors. 4, 11; Bos. 97, 15. *v.* a-spanan.

a-spreádan; *p.* de; *pp.* ed [=a-sprǽdan] *To spread forth, extend;* prætendere :—Aspreád mildheortnysse ðîne *prætende misericordiam tuam*, Ps. Spl. T. 35, 11. *v.* sprǽdan.

a-sprecan; *p.* -spræc, *pl.* -sprǽcon; *pp.* -sprecen [a, sprecan] *To speak out, speak;* eloqui, loqui :—Hwylc mæg ǽfre mihta Drihtnes asprecan and aspyrian *quis loquetur potentias Domini?* Ps. Th. 105, 2. Ðû asprǽce *locutus es*, 59, 5: 58, 12: 73, 21.

a-spreótan; *p.* -spreát, *pl.* -spruton; *pp.* -sproten; *v. intrans.* [a, spreótan] *To sprout forth, break forth;* progerminare, erumpi, eructare :—Swá unefne is eorþe þicce, syndon ðas môras myclum asprotene *sicut crassitudo terræ erupta est super terram*, Ps. Th. 140, 9.

a-sprettan *to sprout out;* germinare, pullulare, Solil. 9. *v.* a-sprýtan.

a-sprian; *v. a. To lay before, shew?* prætendere, Bd. 4, 19.

a-sprincan; *p.* -spranc, *pl.* -spruncon; *pp.* -spruncen *To spring up, arise;* oriri, exoriri :—Aspruncen is on þýstrum leóht *exortum est in tenebris lumen*, Ps. Spl. 111, 4: C. R. Ben. 7. *v.* a-springan.

a-sprindlad; *part.* [=a-springlad? from springan *to spread*, or sprengan *to burst open*] *Torn asunder, ripped up;* diruptus, L. M. 2, 24; Lchdm. ii. 216, 7.

á-spring *a water-spring, fountain;* scaturigo, Hom. de Comp. Cordis, Lye. *v.* ǽ-spring.

a-springan, -spryngan, -sprincan; *p.* -sprang, *pl.* -sprungon; *pp.* -sprungen; *v. intrans.* I. *to spring up, arise, originate, break forth;* surgere, assurgere, oriri, exoriri, rumpi, prorumpi :—Aspryngþ rihtwîsnys *orietur justitia*, Ps. Spl. 71, 7: R. Ben. 69. Asprang *ortum traxit*, Lupi Serm. 3, 7. Ðâ asprungon ealle wyllspringas ðære micelan niwelnisse *rupti sunt omnes fontes abyssi magnæ*, Gen. 7, 11. II. *to spring out, lack, fail, cease, fall away;* deficere, desinere :—Asprang gâst mîn *deficit spiritus meus*, Ps. Spl. C. 76, 3. Asprong hâlig *deficit sanctus*, 11, 1: 72, 19. Ne ðâm fore yrmþum ðe ðǽr inwuniaþ lîf aspringeþ *nor, through sorrows, shall life fail to them that dwell therein*, Exon. 32 b; Th. 103, 8; Cri. 1685: 30 b; Th. 94, 11; Cri. 1538. Wróht wæs asprungen *strife had ceased*, Cd. 5; Th. 6, 4; Gen. 83: Ps. Th. 54, 10. Ðæt hî ne asprungan fram heora geleáfan *ne a fide deficerent*, Bd. 2, 9; S. 511, 6.

a-sprît *shall sprout out*, Gen. 3, 18. *v.* a-sprýtan.

a-spruncen *arisen;* *v.* a-sprincan.

a-sprungennes, -sprungennýs, -ness, e; *f.* [asprungen *failed, ceased; pp.* of a-springan] *An eclipse, deficiency, failing, fainting, exhaustion;* eclipsis, defectio :—Wæs geworden sunnan asprungennys *facta erat eclipsis solis*, Bd. 3, 27; S. 558, 10. Asprungynnes nam me *defectio tenuit me*, Ps. Spl. C. 118, 53.

a-spryngan *to spring up, arise*, Ps. Spl. 71, 7. *v.* a-springan.

a-sprýtan, -sprítan; *p.* -sprýtte, -sprîtte; *pp.* -sprýted *To sprout out, cause to sprout out;* germinare :—Þornas and bremelas heó asprît ðê *spinas et tribulos germinabit tibi*, Gen. 3, 18. *v.* sprýtan, spryttan.

a-spýlian, -spýligan; *p.* ode; *pp.* od *To cleanse, wash, purify;* abluere :—Swîn nyllaþ aspýlian [aspýlian MS. Cot.] on hluttrum wæterum *swine will not wash in pure waters*, Bt. 37, 4; Fox 192, 27. [*Plat.* afspölen: *Dut.* afspoelen: *Ger.* abspülen.]

a-spyrgan *to search, explore, investigate*, Exon. 92 b; Th. 348, 16; Sch. 23. *v.* a-spyrian.

a-spyrgeng, e; *f. An inventing, invention;* adinventio, Cot. 186.

a-spyrian, -spyrigan, -spyrigean; *p.* ede; *pp.* ed *To search, explore, trace, discover, explain;* investigare, indagare, explorare, enucleare :—Se ðe nele, be his andgites mǽðe, ða bôclîcan gewritu aspyrian, hû hî to

Criste belimpaþ *he who will not, according to the measure of his understanding, search the book-writings, how they refer to Christ*, Homl. Th. ii. 284, 30. Aspyrige hit ût *let him trace it out*, L. Ath. iv. 2; Th. i. 222, 14, note 33. Ðæt mihte ðæra twegra tweón aspyrian *that might discover the difference of the two*, Salm. Kmbl. 870; Sal. 434: Elen. Kmbl. 932; El. 467. Ic aspyrige *enucleo*, Ælfc. Gr. 26, 6; Som. 29, 18; Ps. Th. 105, 2.

assa, an; *m*: **asse**, es; *m. A male ass; asinus*:—Se assa geseah ðone engel *asinus cernebat angelum*, Num. 22, 23, 25. Beót ðone assan *verberabat asinum*, 22, 23, 25. Gif ðû gemête ðînes feóndes assan, læd hine to him *si occurreris inimici tui asino erranti, reduc ad eum*, Ex. 23, 4: 23, 5. Wîlde assan *wild asses*; onagri, Ps. Spl. C. 103, 12. Ðâ feóll se assa adûne *tum concidit asinus*, Num. 22, 27. He hæfde on olfendum and on assum micele æhta *he had great possessions in camels and in asses*, Gen. 12, 16: 22, 5. [O. Nrs. asni, *m. asinus*.] v. asse, esol.

Assan dûn, e; *f*. [assan, dûn *a hill*: 'Assendun S. Hovd. *i.e.* vertente Florent. *mons asini*,' Gib.] *Assingdon or Ashingdon, in Essex*:—Se cyning offêrde hî innon Eást-Seaxan, æt ðære dûne ðe man hæt Assandûn *the king overtook them in Essex, at the hill which is called Assingdon*, Chr. 1016; Th. 282, 19, col. 2: 1020; Th. 286, 16, 19, col. 1.

asse, an; *f*: **assen**, e; *f. A she-ass; asina*:—Uppan assan folan sittende *sedens super pullum asinæ*, Jn. Bos. 12, 15. Finde gyt âne assene ye [*two*] *shall find a she-ass*, Mt. Bos. 21, 2. Rît uppan tamre assene *rides on a tame she-ass*, 21, 5. Læddon ða assene to him *adduxerunt asinam*, 21, 7.

Asse-dun; *adj*. [asse *asina*; *or* asce *ash*, cinis; dun *dun or grey*, fuscus] ASS-DUN *or* ASH-DUN, *of a dun or dark colour*; dosinus, cinereus:—Assedun *dosinus vel cinereus*, Ælfc. Gl. 79; Wrt. Voc. 46, 39. 'Glossæ Isidori: *Dosius vel dosinus, equus asinini pili*,' Du Cange.

ass-myre, an; *f. A mare ass, she-ass*; asina:—And xx assmyrena *and twenty of mare asses*, Gen. 32, 15.

Assyria, æ; *f. Assyria*, Cd. 12; Th. 15, 13; Gen. 232.

Assyrias; *gen.* Assyria, Assiria; *dat.* Assyrium; *pl. m. The Assyrians*; Assyrii:—Assyria ealdorduguþ *the people of the Assyrians*, Judth. 12; Thw. 26, 4; Jud. 310.

Assyrige; *gen.* a; *dat.* um; *pl. m. The Assyrians*; Assyrii:—Ðæt synd Assyrige and Rômâne *these are the Assyrians and the Romans*, Ors. 2, 5; Bar. 77, 31. v. Assyrias.

ast *a kiln*; siccatorium:—Cyln oððe ast *siccatorium*, Ælfc. Gl. 109; Som. 78, 132. v. cyln.

a-stælan [a, stælan *to steal*] *To steal out, to seduce*; obrepere:—Ðæt me næfre deófol on astælan ne mæge *that the devil may never secretly creep on me* [*seduce me*], L. De. Cf. 9; Wilk. 88, 49. v. stelan.

a-stænan; *p.* de; *pp.* ed *To adorn with stones or gems*; lapidibus *vel* gemmis ornare:—Gimmum astæned *adorned with gems*, Salm. Kmbl. 128; Sal. 63. Mid deórwyrþum gimmum astæned *de lapide pretioso ornata*, Ps. Th. 20, 3. Astæned gyrdel *a girdle set with stones*, Cot. 201.

a-stâh *ascended*, Chr. 1012; Th. 268, 29, col. 2; *p.* of a-stîgan.

a-standan; *p.* -stôd, *pl.* -stôdon; *pp.* -standen. **I.** *to stand up, get up, rise up, rise*; exsurgere, resurgere, surgere:—Ðâ astôd he semninga *exsurrexit repente*, Bd. 2, 9; S. 511, 20. He up astandeþ of slæpe *he rises up from sleep*, Exon. 96 a; Th. 358, 4; Pa, 40. Eft lifgende up astôdon *they stood up living again*, 24 b; Th. 71, 18; Cri. 1157. **II.** *to insist, persist, continue*; persistere, instare:—Ðæt hî on ðam geleáfan sôþfæstnysse symle fæstlíce astôdon and awunedon *ut in fide veritatis persisterent semper ac proficerent*, Bd. 2, 17; S. 520, 21, note: 4, 25; S. 599, 31. Híg astôdon *illi instabant*, Lk. Bos. 23, 23.

a-steápan, -steópan, -stêpan; *p.* -steápde, -steápte; *pp.* -steáped, -steapt *To deprive, bereave, as children of their parents*; orbare, orphanum reddere:—Sîen bearn his asteápte *fiant filii ejus orphani*, Ps. Surt. 108, 9. [O. H. Ger. stiufan *orbare*, arstiufan *viduare*: Swed. stufwa, stubba *to cut off*: O. Nrs. stýfa *abrumpere, abscindere*.]

a-stellan; *p.* -stealde, -stalde; *pp.* -steald; *v. a. To set forth, to set, place, afford, supply, appoint, establish, ordain, undertake, undergo, begin*; statuere, collocare, instituere, præbere, stabilire, fundare, suscipere, inire:—Bîsene astellan *exemplum præbere*, Past. 3, 1; Hat. MS. 8 b, 5. Asteald to býsne *set for an example*, Ors. 2, 4; Bos. 44, 33. Crist hit astealde and tæhte *Christ established and taught it*, Homl. Th. ii. 582, 29. Heofonas, and môna, and steorran, ða ðû astealdest *cœlos, lunam et stellas, quæ tu fundasti*, Ps. Th. 8, 4. Astealde ðæt gewin *undertook the war*, Ors. 2, 5; Bos. 46, 26. Stephanus ðône martyrdôm astealde *Stephen suffered* [*underwent*] *martyrdom*, Homl. Th. i. 50, 2. Ðone fleám ærest astealde Þurcytel *Thurkytel first began the flight*, Chr. 1010; Th. 262, 43. DER. up-a-stellan. v. stellan.

a-stemnian; *p.* nede; *pp.* ned [a *from*, stemnian *to build*] *To proceed from a foundation, to found, build, erect*; condere:—Ðe hî sylf astemnedon *which they themselves built*, Bd. Pref; S. 472, 17.

a-steópan *to bereave*. v. a-steápan.

a-steorfan; *p.* -stearf, *pl.* -sturfon; *pp.* -storfen *To die*; mori:—

Færunge astorfen *sideratus vel ictuatus*, Ælfc. Gl. 114; Som. 80, 29; Wrt. Voc. 61, 9: Wanl. Catal. 43, 17.

a-stêpan; *p.* -stêpte; *pp.* -stêped, -stêpt *to bereave, as children of their parents*, Gr. Dial. 1, 2: Ps. Vos. 108, 8. v. a-steápan.

a-stêpnes, -ness, e; *f. A privation*; orbatio, Cot. 187.

a-stêpte *bereaved, orphans*, Ps. Vos. 108, 8. v. a-stêpan, a-steápan.

astered *disturbed, stirred, moved*; *pp.* of a-sterian.

a-sterfan; *p.* de; *pp.* ed *To cause death, kill, destroy*; necare, eradicare, Mt. Rush. Stv. 15, 13. v. a-styrfan.

a-sterian; *p.* ede; *pp.* ed *To agitate, stir, move*; commovere, movere:—He astereþ ðone rôdor ænd ða tungla *it moves the sky and the stars*, Bt. 39, 8; Fox 224, 6, note. v. a-styrian.

asterion, es; *n.* [= ἀστέριον] *The herb pellitory, so called from its star-like form*; astericum, Herb. 61; Lchdm. i. 164, 1, 10.

a-stîfian; *p.* ede, ode; *pp.* ed *To stiffen, grow or wax stiff*; obrigere, Cot. 146. His sine astîfode *his sinew stiffened*, Gen. 32, 32.

a-stîfician, -stîficigan; *p.* ode; *pp.* od; *v. a. To eradicate, extirpate, destroy, exterminate*; eradicare:—Ðæt he astîficige unþeáwas *that he exterminate vices*, Bt. 27, 1; Fox 94, 23.

a-stîgan, ic -stîge, ðû -stîgest, -stîhst, he -stîgeþ, -stîhþ, *pl.* -stîgaþ; *p.* -stâg, -stâh, *pl.* -stigon; *impert.* -stîh; *pp.* -stigen [a, stîgan *to go*]. **I.** *to go, come, step, proceed, climb*; ire, venire, gradi, procedere, scandere:—Hwider sceal ðæs monnes môd astîgan *whither shall the mind of man go*, Exon. 32 b; Th. 103, 21; Cri. 1691. Egsa astîgeþ *dread shall come*, 102 a; Th. 385, 24; Rä. 4, 49. Word-hleóðor astâg *the sound of words came*, Andr. Kmbl. 1416; An. 708: Bd. 4, 3; S. 568, 2. Se Hâlega Gâst astâh *líchamlîcre* ansýne *the Holy Spirit came in bodily form*, Lk. Bos. 3, 22. Se môt wuldres dreám astîgan *he may climb the delight of glory*, Exon. 84 b; Th. 317, 30; Môd. 73: Ps. Th. 79, 10. Ic astîge *scando*, Ælfc. Gr. 28, 6; Som. 32, 30. **II.** *to go in any direction*: **1.** *generally indicated by a preposition* or *adverb, hence to rise, ascend, descend, etc*; surgere, ascendere, descendere:—Ðe þurh oferhyd up astîgeþ *who comes up through pride*, Cd. 198; Th. 247, 11; Dan. 495. He from helle astâg *he came from hell*, Exon. 48 b; Th. 168, 14; Gû. 1077. Ðæt he mid ðam dynte nyðær astâh *that he came down with the blow*, Chr. 1012; Th. 268, 29, col. 2. Astîgaþ [Spl. C. upastîgaþ] mûntas, and niðer astîgaþ feldas on stôwe *the mountains ascend, and the fields go down into their place*; ascendunt montes et descendunt campi in locum, Ps. Lamb. 103, 8. Moises âna astîhþ to Drihtne *Moses alone goes to the Lord*; solus Moyses ascendit ad Dominum, Ex. 24, 2. Astîh on Fasgan mûntes cnæpp *go to the top of mount Pisgah*; ascende cacumen Phasgæ montis, Deut. 3, 27. He astâh on scyp *he went into a ship*; ascendit in naviculam, Mt. Bos. 8, 23: 9, 1. He nyðer astîhþ swâ swâ rên on flýs, and swâ swâ niðer astîhþ droppetung, droppende ofer eorþan *he shall come down as rain on a fleece, and as falling* [*rain*] *comes down, dropping over the earth*; descendet sicut pluvia in vellus, et sicut stillicidium stillantium [MS. stillicidia stillantia] super terram, Ps. Lamb. 71, 6. **2.** *but sometimes the direction is indicated in the sentence without a preposition*:—Hire môd astâh *her mind rose*, Cd. 101; Th. 134, 35; Gen. 2235: 205; Th. 253, 18; Dan. 597. He astîgeþ swâ se rên fealleþ on flýs *he shall come as the rain falleth on a fleece*; descendet sicut pluvia in vellus, Ps. Th. 71, 6.

a-stîgend, es; *m. A rider*; ascensor:—Hors and astîgend [MS. astîgende] aweorpeþ on sîe *equum et ascensorem dejecit in mare*, Cant. Moys. Ex. 15, 1; Thw. 29, 6. v. stîgan.

a-stîgnes, -ness, e; *f. An ascent, ascending*; ascensus, Ps. Spl. T. 103, 4.

a-stîh *go, ascend*, Deut. 3, 27; *impert.* of a-stîgan.

a-stîhst, a-stîhþ *ascendest, ascends*, Jn. Bos. 3, 13; *2nd and 3rd pres.* of a-stîgan.

a-stihtan; *p.* -stihte; *pp.* -stiht [a, stihtan *to dispose*] *To determine on*; decernere:—Fleám wearþ astiht *flight was determined on*, Chr. 998; Th. 246, 22. v. stihtan.

a-stintan; *p.* -stint, -stunton; *pp.* -stunten = -stinted, Som. Lye. = -stint = -stynt *To make dull, to blunt, stint, assuage*; hěbětare, obtundere, Scint. 12: Cot. 101. v. a-stynt, stintan.

a-stirian *to move, remove, agitate, stir up, raise*, Lk. Bos. 6, 48. v. a-styrian.

astîdian; *p.* ode, ude; *pp.* od, ud [a *intensive*, stîdian *to become hard*] *To become hard, dry, dry up, wither*; indurare, arescere:—Astîdude swâ swâ tigle miht mîn *my strength dried up as a tile*, Ps. Spl. 21, 14. Hit astîdaþ and drugaþ *induret et arescat*, 89, 6.

a-stôd *stood up, insisted*, Bd. 2, 9; S. 511, 20: Lk. Bos. 23, 23; *p.* of a-standan.

a-stondnes, -ness, e; *f. An existence, a subsistence*; subsistentia:—Âna God on þrým astondnessum *one God in three subsistences*; unum Deum in tribus subsistentiis, Bd. 4, 17; S. 585, 38.

a-storfen; *part. Starved, like a dead body*; cadaverosus, Wanl. Catal. 43, 17. v. a-steorfan.

a-streahte, -streaht *stretched out*; *p. and pp.* of a-streccan.

a-streccan; ic -strecce, ðú -strecest, he -strecþ; *p.* -streahte, -strehte; *impert.* -strece; *pp.* -streaht, -streht; *v. a.* To stretch out, to extend, prostrate, or lay low, to prostrate oneself, bow down; extendere, expandere, prosternere, se prosternere, adorare:—Ðe leas he astrecce his hand ne forte mittat manum suam, Gen. 3, 22: 22, 12. He neowol astreaht feól on ða flóre he fell stretched prostrate on the floor, Bt. Met. Fox I, 159; Met. I, 80. Ðá feóll Abram astreht to eorþan cecidit Abram pronus in faciem, Gen. 17, 3. Astrehte hine to eorþan adoravit in terram, Gen. 18, 2: Mt. Bos. 18, 26, 29: Mk. Bos. 3, 11.

a-stregdan; *p.* -stregde; *pp.* -stregd [a, stregdan to sprinkle] To sprinkle, scatter, strew; aspergere:—Ðú astregdest me mid hysopon asperges me hyssopo, Ps. Spl. T. 50, 8.

astreht, astrehte prostrated; *pp. and p.* of a-streccan.

astrengd Malleable; ductilis, Ælfc. Gl. 115; Som. 80, 46; Wrt. Voc. 61, 24.

a-strícan; *p.* -strác, *pl.* -stricon; *pp.* -stricen To strike; percutere. v. strícan.

a-striénan, -strýnan; *p.* -strýnde; *v. a.* To engender, procreate, beget; gignere:—Hie ðá ongunnon bearn astriénan they began then to beget children, Cd. 46; Th. 59, 19; Gen. 966. He bearn astrýnde he begat children, 57; Th. 70, 5; Gen. 1148. v. streónan, strýnan.

astrihilthet [astre a house, hold a master, þeowet a fine? Mann.] A fine levied on a householder; compensatio facta a domino mansionis, L. Ed. C. 26; Th. i. 454, 2, MS. L.

a-stundian To ASTOUND, grieve, suffer grief, to bear; dolere, R. Ben. 36, Mann.

a-stýfecigan to exterminate, Bt. 27, 1; Fox 94, 23, note 9. v. a-stífician.

a-styltan to astonish; stupescere. v. styltan.

a-stynt made dull; hĕbĕtātus, Cot. 101. v. a-stintan.

a-styrfan; *p.* de; *pp.* ed To cause death, kill, slay; necare:—Stánum astyrfed slain with stones, Exon. 10 b; Th. 12, 27; Cri. 192. v. a-sterfan.

a-styrian, -stirian; *p.* ode, ede; *pp.* od, ed To remove, move, agitate, stir violently, stir up, raise; amovere, removere, movere, commovere:—Astyre fram me wítu ðíne amove a me plagas tuas, Ps. Spl. 38, 13: 118, 29: Rood Recd. 59; Kr. 30. Drihten astyrede ða wéstan stówe commovit Dominus desertum, Ps. Th. 28, 6: 17, 7. Simle ðonne ðér án tweó ofadón biþ, ðonne biþ ðér unrím astyred always when there is one doubt removed, then is there an innumerable multitude raised, Bt. 39, 4; Fox 216, 19.

a-styrred starred; stellatus, Scint. 58.

a-styrung, e; *f.* A motion; motus, Lye. v. stirung.

a-suand = a-swand weakened. v. a-swindan.

a-súcan, -súgan; *p.* -seác, -seág, *pl.* -sucon, -sugon; *pp.* -socen, -sogen To suck; sugere:—Asogen wǽre sugeretur, Cot. 193. Sina beóþ asocene [Exon. asogene] the sinews shall be sucked, Soul Kmbl. 217; Exon. 99 b; Th. 373, 19; Seel. 111. v. súcan.

a-sudon seethed; *p. pl.* of a-seóðan.

a-súgan to suck, Exon. 99 b; Th. 373, 19; Seel. 111. v. a-súcan.

asundran, asundron; *adv.* ASUNDER, apart, alone, privately; seorsum:—Eall he hys leorning-cnihtum asundron rehte seorsum discipulis suis disserebat omnia, Mk. Bos. 4, 34. v. sunder.

a-sundrian, -syndrian; *p.* ode, ade; *pp.* od, ad [a from, sundrian to sunder] To put asunder, to sunder, separate, disjoin, sever; separare:—Se deáþ asundraþ líc and sáwle death separates body and soul, Exon. 98 a; Th. 367, 7; Seel. 4: 50 a; Th. 172, 27; Gú. 1150. Asundrod fram synnum separated from sins, Elen. Kmbl. 2615; El. 1309. Asundrad, Exon. 59 a; Th. 214, 20; Ph. 242.

a-sungen sung, Beo. Th. 2323; B. 1159; *pp.* of a-singan.

a-sund = a-swand languished, Cot. 101; *p.* of a-swindan.

a-súrian; *p.* ode; *pp.* od To be or become sour, tart, bitter; acescere, Cot. 10: 177. v. súrian.

a-swéman to wander about; vagari, Exon. 52 b; Th. 183, 12; Gú. 1326. [vide H. Z. x. 315.]

a-swǽpþ sweeps away, Past. 36, 8; Hat. MS. 48 b, 16; *pres.* of a-swápan.

a-swǽrnung, -swárnung, e; *f.* Bashfulness, confusion; verecundia:—Aswǽrnung [aswǽrnunga MS. aswárnung Ps. Lamb.] mín ongeán me is verecundia mea contra me est, Ps. Spl. 43, 17. v. sceamu.

a-swáf wandered away; exorbitavi, exorbitavit; *p.* of a-swífan.

a-swámian; *p.* ode; *pp.* od To languish, fail, cease; tabescere, deficere [H. Z. x. 315], Cd. 19; Th. 24, 12; Gen. 376.

a-swand languished away, Ps. Lamb. 106, 26; *p.* of a-swindan.

a-swáp, es; *n; pl.* -swápa Sweepings, dust; peripsema, = περίψημα, purgamentum. v. a-swápan.

a-swápan; he -swápþ, -swǽpþ; *p.* -sweóp, *pl.* -sweópon; *pp.* -swápen To sweep off, clean; verrere, mundare:—Hit aswǽpþ aweg ðæt yfel abstergat mala, Past. 36, 8; Hat. MS. 48 b, 16; Exon. 106 b; Th. 405, 21; Rä. 24, 5. Aswópen clǽne mundatus, Mt. Rush. Stv. 12, 44. v. swápan.

a-swarcan To languish, consume; tabescere:—A-ýdlian oððe aswar- ~an oððe acwínan oððe aswindan ðú dydest swá swá ætterloppan oððe

ryngan sáwle his tabescere fecisti sicut araneam animam ejus, Ps. Lamb. 38, 12.

a-swarcian; *p.* ode; *pp.* od To confound, dismay, abash, fear; confundere, revereri:—Ðon gescynde and aswarcode [MS. aswarcod] beóþ cum confusi et reveriti fuerint, Ps. Spl. 70, 26.

a-swárnian; *p.* ode; *pp.* od To be confounded; confundi:—Ðæt hí aswárnian that they be confounded, Ps. Spl. 85, 16. v. a-swarcian.

a-swárnung, e; *f.* Bashfulness, Ps. Lamb. 43, 16. v. a-swǽrnung.

a-swearc languished, failed, Jos. 2, 11; *p.* of a-sweorcan.

a-sweartian; *p.* ode; *pp.* od To blacken, darken, to be made SWARTHY or black, obscured, darkened; denigrari:—Ðæt gold biþ asweartod aurum obscuratur, Past. 18, 4; Hat. MS. 26 b, 8.

a-swebban; *p.* -swefede, -swefde; *pp.* -swefed; *v. a.* [a intensive, swebban to put to sleep] To sooth, appease, set at rest, put to death, destroy; sopire, sedare, necare, dolere:—He ðone storm aswefede and gestilde tempestatem sopivit, Bd. 3, 15; S. 542, 5: Exon. 58 b; Th. 210, 15; Ph. 186. Sweordum aswebban to put to death with swords, Andr. Kmbl. 143; An. 72. He his ealdordóm synnum aswefede his eldership he had destroyed by sins, Cd. 160; Th. 199, 9; Exod. 336.

a-swefecian; *p.* ade; *pp.* ad To eradicate; eradicare:—Aswefecad eradicata, Cot. 75: 199.

a-swefed, -swefede, -swefedon; *pp. and p.* of a-swebban.

a-swellan; *p.* -sweall, *pl.* -swullon; *pp.* -swollen To swell; tumere:—Se earm wæs swíðe aswollen the arm was much swollen, Bd. 5, 3; S. 616, 7. v. swellan.

a-sweltan; *p.* -swealt, *pl.* -swulton; *pp.* -swolten To die; mori, Cot. 147: 62. v. sweltan.

a-swengan; *p.* -swengde; *pp.* -swenged To shake out or off, to cast forth; excutere:—He aswengde Pharaon in ðæm reádan sǽ excussit Pharaonem in Mari Rubro, Ps. Surt. 135, 15.

a-sweorcan; *p.* -swearc, *pl.* -swurcon; *pp.* -sworcen [a, sweorcan to dim, darken] To languish, fail; caligare, elanguere:—Aswearc úre mód elanguit cor nostrum, Jos. 2, 11.

a-sweorfan; *p.* -swearf, *pl.* -swurfon; *pp.* -sworfen To rub off, to file off, polish; expolire:—To asworfenum ðran, to gesworfenum ðran sub expolita, Glos. Prudent. Recd. 142, 19. v. sweorfan.

a-sweotole; *adv.* Clearly; manifeste, Bt. 34, 4; Fox 138, 16. v. sweotol.

a-swerian; *p.* -swór, *pl.* -swóron; *pp.* -sworen; *v. a.* To swear; jurare:—Ðæs deópne áþ Drihten aswór juravit Dominus veritatem, Ps. Th. 131, 11. Ðæt he hine for hóle ǽr ne aswóre non frustrabitur eam, 131, 11. DER. swerian.

a-swícan; *p.* -swác, *pl.* -swicon; *pp.* -swicen; *v. a.* [a from, swícan to go] To go away from any one, to desert any one, to deceive, betray, offend; desciscere, deficere ab aliquo, prodere, scandalizare:—Ne aswíc sundorwíne do not desert a particular friend, Exon. 80 b; Th. 301, 34; Fä. 29. Eádríc aswác his cynehláforde Eadric betrayed his royal lord, Chr. 1016; Erl. 158, 5. Gif ðín swíðre hand ðé aswíce si dextra manus tua scandalizat te, Mt. Bos. 5, 30.

a-swícian; *p.* ode; *pp.* od To offend; scandalizare:—Gyf ðín swíðre eáge ðé aswíce [aswikie, Hat. MS.] si oculus tuus dexter scandalizat te, Mt. Kmbl. Rl. 5, 29.

a-swífan; *p.* -swáf, *pl.* -swifon; *pp.* -swifen To wander out of the way, to wander about; exorbitare, Cot. 76: 188. v. swífan.

a-swind, -ǽswind; *adj.* Slothful, sluggish, idle; iners, Cot. 108.

a-swindan; *p.* -swand, *pl.* -swundon; *pp.* -swunden [a away, swindan to languish] To languish away, to enervate, pine, consume away, to decay, perish, dissolve; tabescere, torpescere, consumi:—Hwý ge swá aswundene sión why are ye so enervated? Bt. 40, 4; Fox 238, 31. Ðýlæs ealle gesceafta aswindaþ lest all creatures perish, Bt. 33, 4; Fox 130, 34. Aswindan me dyde anda mín tabescere me fecit zelus meus, Ps. Spl. C. 118, 139: 111, 9: 106, 26. Aswunden reses, Ælfc. Gr. 9, 26; Som. 11, 11. A-ýdlian oððe aswarcan oððe acwínan oððe aswindan ðú dydest swá swá ætterloppan oððe ryngan sáwle his tabescere fecisti sicut araneam animam ejus, Ps. Lamb. 38, 12.

a-swindung, e; *f.* Idleness, sloth; desidia. DER. aswind.

a-swógan; *p.* -swég, *pl.* -swégon; *pp.* -swógen [a, swógan to rush] To rush into, invade, overrun, choke; irruere, invadere, occupare, suffocare:—We witon ðæt we lufiaþ ðone æcer ðe ǽr wæs mid þornum aswógen, and æfter ðæm ðe ða þornas beóþ aheáwene and se æcer biþ onered, bringþ gódne wæstm we know that we love the land which before was overrun with thorns, and after that the thorns are dug out and the land is ploughed up, brings good fruit, Past. 52, 9; Hat. MS. 81 b, 23.

a-swolcen idle; iners, Cot. 108. v. solcen.

a-swollen swollen, Bd. 5, 3; S. 616, 7. v. a-swellan.

a-swond = a-swand he weakened, enervated; enervavit, Cot. 71; *p.* of a-swindan.

a-swondennes, -ness, e; *f.* Slothfulness; inertia. v. a-swundennes.

a-swópen swept, cleaned:—Aswópen clǽne mundatus, Mt. Rush. Stv. 12, 44. v. a-swápan.

a-sworetan; *p.* te; *pp.* ed To sigh, draw a deep breath; suspi-

rare :—He hefiglíce asworette *graviter suspiravit*, Bd. 3, 11 ; S. 536, 33. v. sworetan.

a-sworfen *polished*, Glos. Prudent. Recd. 142, 19 ; *pp.* of a-sweorfan.

a-swunan ; *p.* -swan, *pl.* -swônon ; *pp.* -swunen *To swoon*; deficere animo. v. a-swâmian.

a-swunden *weakened, slothful*, Ælfc. Gr. 9, 26 ; Som. 11, 11 ; *pp.* of æ-swindan.

a-swunden-líce ; *adv. Slothfully*; segniter. v. a-swunden.

a-swundennes, -ness, -nys, -nyss, e ; *f. Slothfulness, idleness*; inertia :—His líf toscægde fram ussa tíde aswundennysse *vita illius a nostri temporis segnitia distabat*, Bd. 3, 5 ; S. 526, 35. v. a-swindan.

a-swýðerian, -swýðrian ; *p.* ade ; *pp.* ad *To make heavy* or *grievous, aggravate, increase, make stronger*; gravare, aggravare, ingravare, augere. v. swiðrian.

a-syndran, -syndrian ; ic asyndrige ; *p.* ede, ode ; *pp.* ed, od [a *from*, syndrian *to sunder, part*] *To put* ASUNDER, *to separate, disjoin, sever*; separare :—Ic com mann asyndrian ongên his fæder *veni separare hominem adversus patrem suum*, Mt. Bos. 10, 35 : Ps. Spl. 67, 10. Se deáþ asyndreþ líc and sáwle *death sunders body and soul*, Soul Kmbl. 7 ; Seel. 4. v. a-sundrian.

a-syndrung, e ; *f. A division, separation, divorce*; divortium, Cot. 68.

at- *at*; apud, ad ; *used in composition for* æt-, as in at-ýwan, *p.* -ýwde ; at-âwian, *p.* -âwode *ostendere*, Ps. Spl. T. 77, 14. v. at-âwian.

a-tæfran, -tiefran, -tifran ; *p.* ede ; *pp.* ed *To depict, paint*; depingere :—Ic hæbbe atæfred *I have depicted*, Past. 65 ; Hat. MS.

at-âwian ; *p.* ode ; *pp.* od *To shew*; ostendere :—He atâwode him *ostendit eis*, Ps. Spl. T. 77, 14. v. æt-eówian, æt-ýwan.

at-berstan ; *p.* -bærst, *pl.* -burston ; *pp.* -borsten *To break out, escape*; erumpere, Chr. 607 ; Ing. 30, 9. v. æt-berstan.

ÂTE, æte ; *gen.* âtan ; *pl.* âtan ; *gen.* âtena ; *f.* OATS, *tares, darnel, cockle*; avena fatua, Lin. lolium :—Nim âtena grátan *take groats of oats*, Lchdm. iii. 292, 24. Genim mela ætena *take meal of oats*, L. M. 1, 35 ; Lchdm. ii. 84, 5 : Chr. 1124 ; Th. 376, 6. Âte *lolium*, Cot. 126. Âtan or lasor *tares*; zizania, Cot. 204. [*Frs.* ôat : *O. Nrs.* ât *food*.]

a-teáh *drew out* or *away, went, came*, Exon. 29 b ; Th. 91, 19 ; Cri. 1494 : Beo. Th. 1537 ; B. 766 ; *p.* of a-teón.

a-tefred *painted*, Solil. 4. v. a-tæfran.

ate-gâr, es ; *m. A javelin*; framea. v. æt-gâr.

atel *dire, terrible* :—Se atela gǽst *the dire spirit*, Exon. 34 a ; Th. 109, 9 ; Gú. 87. v. atol, *adj.*

a-telan *to reckon*, Bt. 8 ; Fox 24, 21 ; *for* a-tellan.

atelíc ; *adj.* [=atol, líc] *Dire, terrible, horrid, foul, loathsome*; dirus, terribilis, horridus, deformis, fœdus :—Norþ-Denum stôd atelíc egesa *over the North-Danes stood dire terror*, Beo. Th. 1572 ; B. 784. Unwlitig swile and atelíc *tumor deformis*, Bd. 4, 32 ; S. 611, 17. v. atol.

a-tellan ; *p.* -tealde, *pl.* -tealdon ; *pp.* -teald ; *v. trans.* [a, tellan] *To tell out, enumerate, reckon, explain, interpret*; dinumerare, numerare, interpretari :—Hwylc wât ânweald yrres dínes, and for ege dínum graman dínum atellan *quis novit potestatem iræ tuæ, et pro timore tuo iram tuam dinumerare?* Ps. Spl. C. 89, 13. Gif ðú nú atellan wilt ealle ða blíþnessa wið ðám unrótnessum *if thou wilt now reckon all the enjoyments against the sorrows*, Bt. 8 ; Fox 24, 21, note 6. Wit gesáwon swefen, ac wyt nyton hwá hyt unc atelle *nos duo somnium vidimus, et non est qui interpretetur nobis duobus*, Gen. 40, 8.

atelucost, R. Ben. 1 ; *for* atelícost ; *sup.* of atelíc *foul*.

a-temian ; *p.* ede ; *pp.* ed [a *intensive*, temian *to tame*] *To tame thoroughly, make very tame* or *gentle, to subdue, tame*; edomare :—Atemiaþ hira líchoman *edomant carnem*, Past. 46, 2 ; Hat. MS. 66 a, 10. Sum sceal wildne fugel atemian *one shall tame the wild bird*, Exon. 88 b ; Th. 332, 15 ; Vy. 85 : 89 b ; Th. 336, 11 ; Gn. Ex. 46 : Bt. Met. Fox 13, 38 ; Met. 13, 19 : 13, 71 ; Met. 13, 36. DER. un-atemed.

a-tendan ; *p.* de ; *pp.* ed ; *v. trans.* [a *intensive*, tendan *to tind, set on fire*] *To set on fire, kindle, inflame*; accendere, incendere, inflammare :—Hí atendon hiora herebeácen *they kindled their war-beacons*, Chr. 1006 ; Th. 256, 24, col. 1. Hí mid fýre atendan woldan *they wished to set it on fire*, Chr. 994 ; Th. 241, 32, col. 2.

a-tendend, es ; *m. An incendiary, inflamer, inciter*; incensor, inflammator, Scint. 78.

a-tendinc =atending, e ; *f. A fire-brand, an incentive, a provoking*; incentivum, Scint. 81.

a-teón ; ic -teó, ðú -týhst, he -týhþ, -tíhþ, -tíþ, *pl.* -teóþ ; *p.* -teáh, *pl.* -tugon ; *pp.* -togen [a *from, out*; teón *to tow, draw*]. **I.** *v. trans.* generally with a preposition : *to draw out* or *away, pull out, lead out, pluck, draw*; abstrahere, extrahere, ejicere, educere, trahere, draw :—For ðam ðe he wolde ateón ðé fram Drihtne *quia voluit te abstrahere a Domino*, Deut. 13, 10. Ðonne he atíþ hine, Ps. Surt. 9, 30. Ðonne he fram atíhþ [atýgþ MS. C.] hine *dum abstrahet eum*, Ps. Spl. second 9, 11. Seó mæg ateón ǽlces cynnes áttor út of men *which can draw poison of every kind out of man*, Ors. 5, 13 ; Bos. 113, 33. Mid atogenum swurde *evaginato gladio*, Num. 22, 22. He ateáh rib of sídan *he extracted a rib from his side*, Cd. 9 ; Th. 11, 19 ; Gen. 177. Lǽt, ðæt

ic ateó ða egle of dínum eágan *sine ejiciam festucam de oculo tuo*, Lk. Bos. 6, 42. Gif ðú up atýhst and awyrtwalast of gewitlocan leása gesælþa *if thou pluckest up and rootest out of thy mind false felicities*, Bt. Met. Fox 12, 49 ; Met. 12, 25. Ða ic ðec from helle ateáh *when I drew thee from hell*, Exon. 29 b ; Th. 91, 19 ; Cri. 1494 : 124 b ; Th. 479, 4 ; Rä. 62, 2. Múþ mín ic ontýnde, and ic ateáh to [to geteáh MS. C.] gást *os meum aperui, et attraxi spiritum*, Ps. Spl. 118, 131. Hig ne mihton hit ateón *non valebant illud trahere*, Jn. Bos. 21, 6. **II.** *to treat, use, dispose of, employ*; tractare, uti, adhibere :—Ðú ðín âgen môst mennen ateón swâ ðín môd freóþ *thou mayest treat thine own maidservant as thy mind inclines* (*liketh*), Cd. 103 ; Th. 136, 14 ; Gen. 2258. Ða his fýnd hine ne meahton ateón swâ hý woldon *when his enemies might not treat him as they would*, Ps. Th. arg. 9. Ateóh hyne swylce bróðer *tracta eum sicut fratrem*, Scint. 60 : Nicod. 14 ; Thw. 7, 7. Hú híg sceoldon ðæs Hǽlendes wurþ ateón *how they should dispose of the Saviour's price*, Mt. Bos. 27, 7. **III.** *intrans.* or with a cognate noun : *to draw to any place, betake oneself anywhere, go, come, make a journey* or *expedition*; se recipere, meare, proficisci, ire, venire, iter facere :—Siððæt se hearmscaða to Heorute ateáh *after the injurious scather came to Heorot*, Beo. Th. 1537 ; B. 766. Wíg-síþ ateáh *went on a warlike expedition*, Cd. 96 ; Th. 126, 13 ; Gen. 2094 : 167 ; Th. 208, 28 ; Exod. 490 : 208 ; Th. 256, 34 ; Dan. 650 : Exon. 37 a ; Th. 120, 15 ; Gú. 272.

a-teorian, -teorigan ; *p.* ede, ode ; *pp.* ed, od ; *v. intrans. To fail, become weary, cease, leave off*; deficere, fatiscere, cessare, desistere :—Geteorigende ateoraþ *deficientes deficient*, Ps. Spl. 36, 21. Ateorode hálig *defecit sanctus*, Ps. Spl. 11, 1. Ateorode on sáre líf mín *defecit in dolore vita mea*, 30, 12. Híg ateoredon smeágende mid smeáunge defecerunt scrutantes scrutinio, Ps. Lamb. 63, 7. Ateorodun *defecerunt*, 9, 7 : Cot. 69 : Greg. Dial. 1, 1 : R. Ben. interl. 53.

a-teorigendlíc ; *adj.* [a-teorigende *part.* of a-teorigan *to fail*, líc] *Failing, fleeting, perishable*; caducus, fugax :—Seó yld is geteald to ǽfnunge ðises ateorigendlícan middaneardes *that age is considered as the evening of this fleeting world*, Homl. Th. ii. 266, 6.

a-teorung, e ; *f. A failing, fainting, weariness*; defectio, fatigatio. v. ge-teorung.

at-eówad, -eówed ; *part. Shewn, made known*; ostensus. v. æt-eówian.

áter *poison*; venenum. v. âtor.

áter-drinca, an ; *m. A poisonous potion* or *drink, poison*; potio venenata, venenum, Cot. 24. v. âtor, *etc.*

a-terian ; *p.* ede ; *pp.* ed *To fail, become weary*; deficere, fatigare :—Atered *fatigatus*, Ælfc. Gl. 87 ; Wrt. Voc. 50, 20 : R. Ben. interl. 53. v. a-teorian.

áter-láðe, an ; *f. The plant cock's leg*; panicum crus galli. Betonica ? Cot. 24. v. átter-láðe.

áter-líc ; *adj. Poison-like*; veneno similis :—Áterlíc *vel biter gorgoneus*, Cot. 98, = âtor-líc.

áter-tán, es ; *m. A poisonous rod, twig*; vimen venenosum :—Ecg wæs íren, átertánum fáh *the edge was iron, tainted with poisonous twigs*, Beo. Th. 2923 ; B. 1459.

ÂÞ, es ; *m.* **I.** *an* OATH, *a swearing*; juramentum :—Ðú agyltst ðine áþas *reddes juramenta tua*, Mt. Bos. 5, 33. Ðá behêt he mid áþe *pollicitus est cum juramento*, 14, 7, 9. He áþ swereþ þurh his selfes líf *he sweareth an oath by his own life*, Cd. 163 ; Th. 205, 5 ; Exod. 431 : Ps. Th. 131, 11. Hí sealdon unwillum hálige áþas *they gave unwillingly holy oaths*, Bt. Met. Fox 1, 49 ; Met. 1, 25. Gif ðæt geswutelod wǽre, oððe him áþ burste, oððe ofercýðed wǽre *if that were made evident, or an oath failed to them, or were out-proven*, L. Ed. 3 ; Th. i. 160, 20. Nú on worulde hér monnum ne deriaþ máne áþas *now here in the world wicked [false] oaths do not inflict injury on men*, Bt. Met. Fox 4, 96 ; Met. 4, 48. Mid unforedan áþe *with an unbroken oath*; pleno juramento, L. Wil. ii. 3 ; Th. i. 489, 25. Ðæt he ðonne áþ funde gif he mǽhte ungecorenne *that he bring forward the oath of persons unchosen if he could*, L. Ed. 1 ; Th. i. 158, 18. **II.** *every accusation must be verified by oath : the accused and his witness then replied also upon oath; thus,* **1.** Ðæs áþ ðe his ǽhte bryideþ, ðæt he ne áðþ ne for hete ne for hôle :—' On ðone Drihten, ne teó ic N. ne for hete ne for hôle ne for unrihtre feohgyrnesse ; ne ic nán sôþre nát ; búte swá mín secga me sǽde, and ic sylf to sôþe talige, ðæt he mínes orfes þeóf wǽre ' *The oath of him, who takes his [own] property, that he does it neither for hatred nor for envy :—* ' *By the Lord, I accuse not N. neither for hatred nor for envy, nor for unlawful lust of gain ; nor know I anything soother ; but as my informant to me said, and I myself in sooth think, that he was the thief of my property.*' **2.** Ðæs óðres áþ ðe he is unscyldig :—' On ðone Drihten, ic eom unscyldig, ǽgþer ge dǽde ge dihtes æt ðære tíhtlan ðe N. me tíhþ ' *The other's oath that he is guiltless :—* ' *By the Lord, I am guiltless, both in deed and purpose, of the accusation of which N. accuses me.*' **3.** His geféran áþ ðe him mid standaþ :—' On ðone Drihten, se áþ is clǽne and unmǽne ðe N. swór ' *His companion's oath who stands with him :—* ' *By the Lord, the oath is clean and unperjured which N. has sworn,*'

L. O. 4-6; Th. i. 180, 8-19.   III. Ânfeald áþ [lâd] *a simple oath* [*exculpation*]; simplex juramentum [purgatio] hoc est, accipiat duos, et sit ipse tertius, et sic jurando conquirat simplicem purgationem. Þrýfeald áþ *a threefold oath*; triplex juramentum, hoc est, accipiat quinque, et ipse sit sextus, L. C. S. 22; Th. i. 388, 11, 12, and note b. [*Plat*. ēd : *O. Sax*. ēđ : *O. Frs*. eth, ed : *Dut*. eed : *Ger*. eid : *M. H. Ger*. eit ; *gen*. eides : *O. H. Ger*. eid : *Goth*. aiþs : *Dan*. eed : *Swed*. ed : *O. Nrs*. eiđr, *m*.] v. ânfeald áþ.

áþ-brice, es ; *m. A breaking of an oath, perjury;* perjurium, Wulf. 8.

á đe, á đý *Ever the;* unquam eo :—Á đe, á đý deórwyrþran *ever the more precious*, Bt. 14, 2 ; Fox 44, 2. Á đý mâ *ever the more*, Bt. 40, 2 ; Fox 236, 30. Á đý betera *ever the better*, Bt. 13 ; Fox 38, 9. v. đý.

a-þecgan ; *p*. -þegde ; *pp*. -þeged, -þegd *To receive;* recipere, excipere, Exon. 100 b ; Th. 380, 3, 12 ; Rä. 1, 2, 7.

áþe-gehât *an oath*, Ælfc. Gl. 13 ; Som. 57, 119 ; Wrt. Voc. 20, 56. v. áþ-gehât.

a-þegen ; *part*. [a, þegen ; *pp. of* þecgan *sumere*] *Full, stuffed out;* distentus, Cot. 63.

a-þencan, -þencean ; *p*. -þohte ; *pp*. -þoht.   I. *to think out, devise, invent;* excogitare :—Gif we hit mægen wihte aþencan *if we may devise it in any way*, Cd. 21 ; Th. 26, 2 ; Gen. 400 : 179 ; Th. 224, 35 ; Dan. 146 : Ors. 1, 10 ; Bos. 33, 28.   II. *to think, intend;* cogitare, intendere, velle :—He đis ellenweorc âna aþohte *to gefremmanne he thought this bold work to perform alone*, Beo. Th. 5280 ; B. 2643.

a-þenian ; *p*. ede, ode ; *pp*. ed, od ; *v. a.* [a *out*, þenian *to stretch*].   I. *to stretch out, extend, distend, expand, stretch;* tendere, extendere, expandere :—Aþene đîne hand, and he hî aþenede *extende manum tuam, et extendit*, Mt. Bos. 12, 13 : Ps. Th. 59, 7 : 103, 3. Gif se maga aþened sîe *if the stomach be distended*, L. M. cont. 2, 2 ; Lchdm. ii. 158, 4. Bogan his he aþenede *arcum suum tetendit*, Ps. Spl. 7, 13.   II. *to prostrate;* prosternere :—Hî aþenedon hî *they prostrated themselves*, Mt. Bos. 2, 11.   III. *to stretch, apply;* intendere :—He đa geornlîce his môd aþenode on đa þing, đe he gehýrde *ille sollicitus in ea, quæ audiebat, animum intendit*, Bd. 4, 3 ; S. 567, 45.

a-þenung, e ; *f. An extending, extension;* extensio. v. a-þenian.

a-þeódan ; *p*. -þeódde ; *pp*. -þeóded [a *from*, þeódan *to join*] *To disjoin, separate;* disjungere :—Aþeódde from Gode *disjuncti a Deo*, Gr. Dial. 2, 16.

a-þeóstrian ; *p*. ode, ade, ede ; *pp*. od *To overcloud, to be eclipsed;* obumbrare, obscurare :—Aþeóstrade *obscuravit*, Ps. Surt. 104, 28 : Chr. 538 ; Th. 28, 6, col. 2, Cott. Tiber. A. vi ; col. 3, Cott. Tiber. B. 1. v. a-þýstrian.

a-þeótan ; he -þýteþ ; *p*. -þeát, *pl*. -þuton ; *pp*. -þoten *To wind, sound, blow;* inflare, canere :—Næfre mon đæs hlúde horn aþýteþ, ne býman ablâweþ *never so loudly one sounds a horn, nor blows a trumpet*, Exon. 117 b ; Th. 451, 26 ; Dóm. 109. v. þeótan.

áđer *either;* alter, Ors. 3, 9 ; Bos. 68, 11. v. áđor.

a-þéstrian ; *p*. ode ; *pp*. od *To be eclipsed;* obscurari :—Seó sunne aþéstrode *the sun was eclipsed*, Chr. 538 ; Th. 29, 4, col. 1 ; Bodl. Laud. 636. v. a-þýstrian.

áđexe, an ; *f. A lizard, newt;* lacerta, Som. [*O. Sax.* egithassa : *Dut.* hagedisse : *Ger.* eidechse : *M. H. Ger.* egedehse : *O. H. Ger.* egidehsa.] v. efete.

áþ-fultum, es ; *m.* [áþ *an oath*, fultum *a help, support*] *The support to an oath*, i. e. *the supporters of an oath, those who support one's oath, who will swear for another as witnesses;* sacramentales :—Freónd-leás weofod-þên, đe áþfultum næbbe *a friendless servant of the altar, who has no support to his oath*, L. C. E. 5 ; Th. i. 362, 19 : L. Eth. ix. 22 ; Th. i. 344, 23.

áþ-gehât, áþe-gehât, es ; *n.* [áþ *an oath*, gehât *a promise*] *A promise on oath, sacred pledge, an oath;* sacramentum :—Áþ-wed *vel* áþe-gehât *sacramentum*, Ælfc. Gl. 13 ; Som. 57, 119 ; Wrt. Voc. 20, 56. v. áþ-wed.

a-þierran ; *p*. de ; *pp*. ed *To wash off* or *away, rinse, make clean, purge, clear;* diluere :—Hit is þearf, đæt sió hond sîe ær geclænsad, đe wille đæt fenn of óđerre aþierran *necesse est ut esse munda studeat manus, quæ diluere sordes curat*, Past. 13, 1 ; Hat. MS. 16 b, 8.

a-þiéstrian ; *p*. ode ; *pp*. od *To overcloud, to be eclipsed;* obscurari :—Seó sunne aþiéstrode *the sun was eclipsed*, Chr. 538 ; Th. 28, 6, 11, col. 1. v. a-þýstrian.

a-þindan ; *p*. -þand, *pl*. -þundon ; *pp*. -þunden *To puff up, swell, inflate;* intumescere :—He đâ đone aþundenan sæ gesmylte *tumida æquora placavit*, Bd. 5, 1 ; S. 614, 8. Gif he aþunden sý *if he be swollen*, Herb. 1, 21 ; Lchdm. i. 76, 27. Aþindaþ *occurs in* Ps. Th. 106, 25 *as a translation of* tabescebat ; *the translator confounded* tabescere *with* tumescere. v. þindan.

a-þindung, e ; *f. A swelling* or *puffing up;* tumor, Som. v. a-þindan.

a-þístrian ; *p*. ode ; *pp*. od *To overcloud, to be eclipsed;* obscurari :—Seó sunne aþístrode *the sun was eclipsed*, Chr. 540 ; Ing. 22, 22 : Bt. Met. Fox 6, 8 ; Met. 6, 4. v. a-þýstrian.

Athlans ; *m.* [῎Ατλας, αντος, *m.*] *Mount Atlas, in West Africa;* Atlas mons :—Hyre west-ende is æt đæm beorge, đe man Athlans nemneþ *his west end is at the mountain, which is called Atlas*, Ors. 1, 1 ; Bos. 16, 6.

áþ-loga, an ; *m. A perjurer;* perjurus, Exon. 31 b ; Th. 98, 10 ; Cri. 1605.

a-þoht, es ; *m.* [a *out*, þoht *a thought*] *A thinking out, an excogitation, a device, an invention;* commentum, Cot. 35.

a-þohte, -þoht *thought out, thought*, Beo. Th. 5280 ; B. 2643 ; *p. and pp. of* a-þencan.

a-þolian ; *p*. ode, ude ; *pp*. od *To sustain, endure, suffer;* sustinere, perdurare, pati :—Hwylc aþolaþ *quis sustinebit?* Ps. Spl. 129, 3 : Exon. 27 a ; Th. 81, 8 ; Cri. 1320 : Solil. 4. Đæt him frêcne on feorh aþolude *that their soul in them suffered violently;* anima eorum in ipsis defecit, Ps. Th. 106, 4.

a-đol-ware ; *gen*. -wara ; *dat*. -warum ; *pl. m. Citizens;* cives, Exon. 92 a ; Th. 346, 6 ; Gn. Ex. 201.

áđor ; *pron. Either the one or the other, both;* alter, alteruter, uterque :—And se đe áđor fulbrece *and he who violates either*, L. C. E. 2 ; Th. i. 358, 20 : L. Ed. 2 ; Th. i. 160, 11 : Hy. 10, 42 ; Hy. Grn. ii. 293, 42. On áđrum *on both*, Cot. 214. On áđre hand *on either hand*, Ors. 1, 14 ; Bos. 37, 32. v. áwđer.

a-þracian ; *p*. ode ; *pp*. od *To fear;* conturbari, horrescere, Ps. Spl. 6, 10 : 34, 4. v. þracian.

a-þræstan ; *p*. -þræste ; *pp*. -þræst *To wrest out;* extorquere, Cot. 73. v. þræstan.

a-þræt *irksomeness;* tædium. v. a-þreát.

a-þráwan ; *p*. -þreów, *pl*. -þreówon ; *pp*. -þráwen [a, þráwan *to throw*].   I. *to throw forth, to spill;* effundere :—Is mîn swât aþráwen [MS. aþrowen] *my blood is spilt*, Andr. Kmbl. 2850 ; An. 1427.   II. *to twist, wreath, twine;* contorquere :—Aþráwenan gold-þræddas *twisted gold-threads*. Aþráwenum þrædum *with twisted threads*, Cot. 50.

a-þreát, -þræt, es ; *m. Irksomeness, disgust;* tædium :—Eów wæs lungre aþreát *you had soon disgust [at this]*, Elen. Kmbl. 736 ; El. 368. v. a-þreótan.

a-þreótan ; *indef.* hit aþrýt ; *p*. -þreát, *pl*. -þruton ; *pp*. -þroten.   I. *impers. To weary, irk, displease, be loathsome, irksome to any one;* tædere, pigere :—Me aþrýt *it wearies me, I am weary*, Ælfc. Gr. 33 ; Som. 37, 19. Hwî ne læte ge eów đonne aþreótan *why then let ye [it] not to be loathsome to you?* Bt. 32, 2 ; Fox 116, 8. Ne sceal đæs aþreótan þegn môdigne, đæt he wîslîce woruld fulgonge *it must not irk therefore an energetic man, that he wisely passes his life*, Exon. 92 b ; Th. 347, 31 ; Sch. 21. Hý tô ær aþreát, đæt hý waldendes willan læsten *it too soon displeased them, that they should execute their sovereign's will*, 45 a ; Th. 152, 30 ; Gú. 816 : Bt. Met. Fox 29, 82 ; Met. 29, 40.   II. *pers. To loathe, dislike, be weary of anything;* pertæsum esse :—Se cyning wæs aþroten his ællreorde gespræce *rex pertæsus erat barbaræ loquelæ*, Bd. 3, 7 ; S. 530, 4.

a-þrescan ; *p*. -þræsc, *pl*. -þruscon ; *pp*. -þroscen, -þroxen [a, þerscan *to thresh, beat*] *To rob, spoil;* spoliare, expilare :—Aþroxen *spoliatus*.

a-þriéttan ; *p*. -þriétte ; *pp*. -þriétted *To weary, loathe any one;* tædio afficere aliquem :—Ic đê hæbbe aþriét mid đis langan spelle *I have wearied thee with this long discourse*, Bt. 39, 12 ; Fox 232, 19.

a-þringan ; *p*. -þrang, -þrong, *pl*. -þrungon ; *pp*. -þrungen [a *out*, þringan *to throng*].   I. *to throng* or *press out* or *forth, to urge out, to urge, to throng* or *press away* or *out of sight, to conceal;* extrudere, celare :—Ne mihte ic of đære heortan heardne aþringan stýlenne stân *I could not press out from his heart the hard and steely stone*, Salm. Kmbl. 1008 ; Sal. 505. Aþrungen, ût-aþrungen *celatum*, Cot. 33.   II. *to rush forth, to rush;* prorumpere :—Ic of enge up aþringe *I rush up from the narrow place*, Exon. 101 b ; Th. 383, 18 ; Rä. 4, 12.

a-þrintan ; *p*. -þrant, *pl*. ... -þrunton ; *pp*. -þrunten [a *out*, þrintan *to swell*] *To swell up;* tumere :—Ic đa wiht geseah, womb wæs aþrunten *I saw the creature, its belly was swollen up*, Exon. 109 b ; Th. 419, 7 ; Rä. 38, 2.

a-þroten *loathed*, Bd. 3, 7 ; S. 530, 4 ; *pp. of* a-þreótan.

a-þrotennes, -þrotenes, -ness, e ; *f. Tediousness, loathsomeness, wearisomeness;* tædium, Cot. 91.

a-þrotsum ; *adj.* [a-þroten *pp. of* a-þreótan *to trouble*, -sum] *Troublesome, irksome, wearisome;* tædiosus, pertæsus :—Aþrotsum is *pertæsum est*, Cot. 188.

a-þrowen = a-þráwen *thrown forth, spilt*, Andr. Kmbl. 2850 ; An. 1427 ; *pp. of* a-þráwan.

a-þrówian ; *p*. ode ; *pp*. od *To suffer;* pati. v. þrówian.

a-þroxen *spoiled, robbed;* spoliatus ; *pp. of* a-þrescan.

a-þrungen ; *part. Concealed;* celatum, Cot. 33 ; *pp. of* a-þringan.

aþrunten *swollen up*, Exon. 109 b ; Th. 419, 7 ; Rä. 38, 2 ; *pp. of* a-þrintan.

aþryd ; *part. Robbed, pilled;* expressus, expilatus, Cot. 73 ; *pp. of* a-þrýþan.

a-þrysman, -þrysemian ; *p*. ede, ode ; *pp*. ed, od *To suffocate with smoke* or *vapour, to suffocate, stifle;* fumo suffocare :—Hî hine on his bedde asmoredan and aþrysmodon *they smothered and stifled him in his bed*, Ors. 5, 4 ; Bos. 105, 6. Sunne wearþ adwæsced, þreám aþrysmed *the sun was darkened, stifled by sufferings*, Exon. 24 b ; Th. 70, 5 ; Cri. 1134. v. þrysman.

a-þrýt *wearies*, Ælfc. Gr. 33; Som. 37, 19. v. a-þreótan.

a-þrýþian; *p.* -þrýþede; *pp.* -þrýþed, -þryd [a *away*, þryþian from þrýþ *force*] *To force from, rob, pillage*; exprimere, expilare :—Aþryd *expressus, expilatus*, Cot. 73 : 74.

áþ-stæf, es; *m.* [áþ *oath*, stæf] *An oath*; juramentum, Ps. Spl. C. 104, 8.

áþ-swaring, -swerung, e; *f. An oath-swearing*; juramentum :— Gemindig wæs áþswaringe his *memor fuit juramenti sui*, Ps. Spl. 104, 8. Mid áþswerunge *with oath-swearing*, Chr. 1070; Th. 344, 27.

áþ-swaru, e; *f. An oath-swearing, a solemn oath, an oath*; juramentum :—For heora áþsware *because of their oath*, Jos. 9, 18. Ðæt he lange gehét mid áþsware *what he long had promised on oath*, Cd. 170; Th. 213, 26; Exod. 558 : Ps. Th. 88, 3. Áþsware pytt *the well of the oath, Beersheba*, Gen. 46, 1.

áþ-sweord, es; *n.* [áþ *an oath*, sweord *sword*] *A sword-oath, a warrior's oath, an oath*; jusjurandum :—Ðonne bióþ brocene áþsweord eorla *then will be broken the oaths of the warriors*, Beo. Th. 4134; B. 2064.

áþ-swerung *an oath*, Chr. 1070; Th. 344, 27. v. áþ-swaring.

áþ-swyrd, es; *n. An oath*; juramentum :—Gemyndig wæs áþswyrdes [MS. áþswyrde] his *memor fuit juramenti sui*, Ps. Surt. 104, 9. v. áþ-sweord.

ÁÐUM, es; *m. A son-in-law, a daughter's husband, a brother-in-law, a sister's husband*; gener; sororis, ut et patris, sororis maritus :—Áðum *gener*, Ælfc. Gr. 8; Som. 7, 18. Hæfst ðú suna oððe dóhtra oððe áðum *habes filios aut filias aut generum*, Gen. 19, 12. Cwæþ to his twám áðumum *locutus est ad generos suos*, 19, 14 : Exon. 66 b; Th. 246, 22; Jul. 65. Fór to ðam cynge his áðume *went to the king his sister's husband*, Chr. 1091; Th. 359, 6. [Ger. eidam *a daughter's husband*: M. H. Ger. eidem, *id*: O. H. Ger. eidum, eidam, eidem, *id*.]

a-þunden *swollen*, Bd. 5, 1; S. 614, 8; *pp.* of a-þindan.

a-þundenes, -ness, e; *f. A tumour, swelling, puffing up*; tumor :— Wið lifre swyle and aþundenesse *for swelling and puffing up of the liver*, L. M. cont. 2, 18; Lchdm. ii. 160, 18. Wið aþundenesse magan windigre *for windy swelling of the stomach*, 2, 11; Lchdm. ii. 158, 23. DER. aþindan, þindan; *pp.* þunden *swollen*.

a-þwǽgen *washed*, Bd. 4, 19; S. 588, 9; *pp.* of a-þweán.

a-þwǽnan; *p.* de; *pp.* ed [a *away*, þwǽnan *to soften, diminish*] *To soften, diminish, lessen, abate, take away*; diminuere, demere :—Seó sealf wile ðone swile aþwǽnan *the salve will diminish the swelling*, L. M. 3, 39; Lchdm. ii. 332, 25.

a-þwát *disappointed*, Ps. Spl. 131, 11. v. a-þwítan.

a-þweán; ic -þweá, -þweah, ðú -þweahst, -þwyhst, -þwehst, he -þwýhþ, -þwehþ, *pl.* -þweáþ; *p.* -þwóh, *pl.* -þwógon; *pp.* -þwegen [a *from, out*; þweán = þweahan *to wash*] *To wash out, to wash, cleanse, baptize, anoint*; abluere, luere, lavare, baptizare, unguere :—Gif ðú aþweán wylt *if thou wilt wash out*, Guthl. 5; Gdwin. 32, 8. Aþweah me *lava me*, Ps. Spl. 50, 3. Ðú aþweahst me *lavabis me*, 50, 8. He þegnas mid ða hálgan wyllan fulluht-bædes aþwóh *milites sacrosancto fonte abluebat*, Bd. 4, 13; S. 582, 13 : 3, 7; S. 529, 14 : 1, 7; S. 478, 41. Wǽtere aþwegen and bebaðod *lotus aqua*, 1, 27; S. 496, 17 : 4, 19; S. 588, 9. Ðæt híg aþwegene wǽren *ut baptizarentur*, Lk. Bos. 3, 12. Aþwóg *unxit*, Jn. Lind. War. 12, 3.

áþ-wed, -wedd, es; *n.* [áþ *an oath*, wed *a pledge*] *A pledge on oath, a solemn pledge*; sacramentum :—Áþ-wed *vel* áþe-gehát *sacramentum*, Ælfc. Gl. 13; Som. 57, 119; Wrt. Voc. 20, 56. v. áþ-gehát.

a-þwegen *washed*, Bd. 1, 7; S. 478, 41; *pp.* of a-þweán.

a-þweran; *p.* -þwær, *pl.* -þwǽron; *pp.* -þworen *To shake or stir together with a churn-staff* [A. Sax. þwiril], *to churn*; bacillo agitare :— Aþweran buteran *butyrum agitare*, Som. Aþwer buteran *churn butter*, L. M. 1, 45; Lchdm. ii. 112, 25. v. þweran.

a-þwitan; *p.* -þwát, *pl.* -þwiton; *pp.* -þwiten [a, þwítan *to cut off*] *To disappoint*; frustrari :—Ná aþwát [bewægde C.] him *non frustrabitur eum*, Ps. Spl. 131, 11.

a-þwóh, -þwógon *washed*, Bd. 4, 13; S. 582, 13; *p.* of a-þweán.

áþ-wyrþe; *adj. Worthy of an oath, worthy of credit*; dignus qui juret :—Gif he áþwyrþe biþ *if he be oath-worthy*, L. In. 46; Th. i. 130, 14 : L. Ed. 3; Th. i. 160, 21.

á ðý *ever the*; unquam eo, Bt. 13; Fox 38, 9. v. ðý.

a-þýan; *p.* de; *pp.* ed *To press*; premere :—Wel on aþýdum sceapes smeruwe *boil in pressed sheep's grease*, L. M. 1, 8; Lchdm. ii. 54, 1. v. þýan.

aþýdum *pressed*, L. M. 1, 8; Lchdm. ii. 54, 1; *dat.* of aþýed = aþýd. v. aþýan.

a-þylgian; *p.* ode; *pp.* od *To sustain, bear, be patient, wait patiently*; sustinere :—For ǽ ðínre ic aþylgode ðé *propter legem tuam sustinui te*, Ps. Spl. 129, 4. Aþylgode sáwle mín on worde his *sustinuit anima mea in verbum ejus*, 129, 5. v. þyldigean.

a-þynnian, -þinnian; *p.* ade; *pp.* ad *To thin*; tenuare. DER. þynnian, þyn.

a-þýstrian, -þístrian, -þeóstrian, -þiéstrian, -þéstrian, -þystrian; *p.* ode, ade; *pp.* od *To overcloud, to be obscured or eclipsed*; obnubilare, obscurari :— Sýn aþýstrode eágan heora *obscurentur oculi eorum*, Ps. Spl. 68, 28. Seó sunne aþýstrade *the sun was eclipsed*, Ors. 6, 2; Bos. 117, 14.

Aþýstrade *obnubilavit*, Bd. 5, 13; S. 633, 34. Ðonne aþeóstriaþ ealle steorran *then all the stars are darkened*, Bt. 9; Fox 26, 15. Byþ sunne aþeóstrod, Mk. Bos. 13, 24. Hér sunne aþýstrode *here the sun was eclipsed*, Chr. 538; Ing. 22, 18 : 540; Ing. 22, 22. DER. þýstrian.

a-þýteþ *sounds*, Exon. 117 b; Th. 451, 26; Dóm. 109. v. a-þeótan.

a-þýwan; *p.* de; *pp.* ed [a *from*, þýwan *to drive*] *To lead or drive from, to discard*; ejicere :—He hý raðe aweg aþýwde *he soon drove them away*, Ors. 6, 36; Bos. 131, 28.

a-tiarian *to fail*; deficere, Prov. 3. v. a-teorian.

a-tiefran, -tifran; *p.* ede; *pp.* ed *To paint, describe by painting*; depingere :—Ealle ða hearga Israhéla folces wǽron atiefrede [MS. C. atifred : MS. Oth. atiefred] on ðæm wage *universa idola domus Israel depicta erant in pariete*, Past. 21, 3; Hat. MS. 30 a, 23. He atiefreþ [MS. C. atifreþ] ðæs þinges onlícnesse on his móde ðe he ðonne ymbsmeáþ *in corde depingitur quidquid fictis imaginibus deliberando cogitatur*, Past. 21, 3; Hat. MS. 30 b, 27 : 30 b, 26.

a-tihtan; *p.* -tihte; *pp.* -tihted, -tiht *To attract, incite*, Bt. 32, 1; Fox 114, 3. v. a-tyhtan.

a-tíhþ, a-tíþ *draws away, draws to*; abstrahit, attrahit, Ps. Spl. second 9, 11 : Ps. Surt. 9, 30. v. a-teón.

a-tihting *intention, an aim*; intentio, Scint. 6, 7. v. a-tyhtan.

a-tillan; *p.* de; *pp.* ed *To touch*; tangere, R. Ben. interl. 7. v. tillan.

a-timbrian, -timbran; *p.* ode, ede; *pp.* od, ed *To erect, build*; ædificare :—Hét ða burh atimbrian *ordered to build the city*, Ors. 3, 9; Bos. 65, 21 : 66, 40; 67, 39 : 6, 30; Bos. 127, 34. Búr atimbran *to build a bower*, Exon. 108 a; Th. 411, 26; Rä. 30, 5.

a-tión; *p.* -teáh, *pl.* -tugon; *pp.* -togen *To draw out, pull out*; abstrahere, extrahere :—Atió of ðæm æcere fearn and þornas *let him pull out from the field fern and thorns*, Bt. Met. Fox 12, 3; Met. 12, 2 : 22, 53; Met. 22, 27.

at-íwan; *p.* ede; *pp.* ed *To appear*; apparere :—Atíwede cométa *a comet appeared*, Chr. 1066; Th. 330, 38. v. æt-ýwan.

a-togen *drawn out*, Num. 22, 22; *pp.* of a-teón.

atol, es; *n. Terribleness, terror, horror, wretchedness*; diritas, terror, horror, niiseria :—Sceal atol þrówian *must suffer terror*, Cd. 222; Th. 289, 10; Sat. 395. Is ðes windiga sele atole gefylled *this windy hall is filled with horror*, 216; Th. 273, 16; Sat. 137 : Exon. 26 a; Th. 77, 33; Cri. 1266.

ATOL, atul, atel, eatol; *adj. Dire, terrific, terrible, horrid, foul, loathsome*; dirus, atrox, terribilis, horridus, fœdus, teter :—Atol ǽglǽca *the dire miscreant*, Beo. Th. 1188; B. 592: Andr. Kmbl. 2625; An. 1314. Atol is ðín onseón *horrid is thine aspect*, Cd. 214; Th. 268, 26; Sat. 61. Atol mid ǽgum *terrific with his eyes*, 229; Th. 310, 18; Sat. 728. Atol ýða gewealc *the terrible rolling of the waves*, 166; Th. 206, 21; Exod. 455 : Beo. Th. 1700; B. 848: Exon. 81 b; Th. 306, 11; Seef. 6. Se atola *the horrid one* [*the devil*], Cd. 222; Th. 290, 10; Sat. 413. In ðeossum atolan ǽðele *in this horrid country*, 215; Th. 271, 20; Sat. 108. Atole gástas *horrid ghosts*, 214; Th. 268, 7; Sat. 51. Gúþrinc geféng atolan clommum *the warrior seized in her horrid clutches*, Beo. Th. 3008; B. 1502. [*Orm.* atell *foul, corrupt: O. Nrs.* atall, ötul *fierce*; atrox.] DER. atelíc.

atolíc; adj. [atol, líc] *Dire, horrid, loathsome*; dirus, horridus, deformis, Bd. 4, 23; S. 611, note 17. v. atelíc.

átor, áttor, áter, átter, ǽtor, ǽtter, ǽttor; *gen.* átres, áttres; *n. Poison, venom*; venenum :—Átres drync *the drink of poison*, Andr. Kmbl. 105; An. 53. Áttre gelícost *most like to poison*, Cd. 216; Th. 274, 32; Sat. 162. Flór áttre weól *the floor boiled with venom*, 220; Th. 284, 8; Sat. 318. Áttru *venena*, Scint. 28. Wið áttrum *against poisons*, Ps. Th. 57, 4: Bd. 1, 1; S. 474, 39: Bd. 4, 23; S. 595, 1. Wið fleógendum átre *for flying venom*, L. M. 1, 45; Lchdm. ii. 112, 24. [*Orm.* atterr; *Laym.* atter; *Piers* attre: *Plat.* etter, eiter, *m. n: O. Sax.* ētar, ettar, *m: O. Dut. Dut.* etter, *m: Ger.* eiter, *n. m: M. H. Ger.* eiter, *n: O. H. Ger.* eitar, *n: Dan.* edder, *n: Swed.* etter, *n: Norw. O. Nrs.* eitr, *n.* Cf. *M. H. Ger.* eiten *to burn: Sansk.* i-n-dh and the *A. Sax.* ád *a funeral pile: O. H. Ger.* eit *ignis*, áttor then would seem to mean *a cause of burning, a pricking pain*.]

átor-berende; *part. Venom-bearing*; venenifer, L. M. 2, 1; Lchdm. ii. 176, 5. v. átter-berende.

átor-coppe, an; *f. A spider*; aranea. v. áttor-coppe.

átor-cræft, es; *m. Poison-craft, the art of poisoning, sorcery*; veneficium, Lye.

átor-cyn, es; *n. The poison-kind*; veneni genus, Salm. Kmbl. 437; Sal. 219. v. ǽtor-cyn.

átor-drinc, es; *m. Poisonous drink, poison*; potio venenata, venenum. v. áttor-drinca.

átor-drinca *poisonous drink, poison*. v. áttor-drinca.

a-torfian; *p.* ode; *pp.* od *To throw forth, to throw*; jactare, Mt. Hat. 12, 24, Lye. v. torfian.

átor-láðe, an; *f. The cock's spur grass*; panicum crus galli, v. áttor-láðe, L. M. 45; Lchdm. ii. 110, 8; 114, 11.

átor-líc *poison-like*; veneno similis. v. áter-líc.

ător-loppe, an; *f.* [ātor, loppe *a silkworm, spinner of a web*] *A spider, spider's web;* aranea. v. ǣtter-loppe.

ător-sceaða *a venomous destroyer.* v. āttor-sceaða.

ător-spere, es; *n. A poisoned spear;* telum venenatum. v. āttor-spere, Exon. 105 a; Th. 399, 10; Rä. 18, 9.

ător-tân, es; *m. A poisonous rod;* ramus venenosus. v. āter-tân, Beo. Th. 2923; B. 1459.

a-tredan, *p.* -træd, *pl.* -trædon; *pp.* -treden *To tread, twist from* or *out, extort;* extorquere :—Atred him ða giltas ût *extort his sins from him,* L. De Cf. 3; Th. ii. 260, 21.

a-treddan, *p.* de; *pp.* ed *To investigate, search, examine* or *explore carefully;* scrutari, investigare :—Ðæt ic ðîn bebod beorht atredde *scrutabor mandata tua,* Ps. Th. 118, 69 : 138, 2. v. treddan.

a-trendlian; *p.* ode; *pp.* od *To trundle, roll;* volutare, provolvere, Bt. Met. Fox 5, 33; Met. 5, 17.

ātren-môd *venom-minded;* malitiosus. v. ǣtren-môd.

âtrian *to poison, envenom;* venenare. v. ǣtrian.

ātter; *gen.* āttres; *n. Poison, venom;* venenum :—Ðæt ātter wæs sôna ofernumen *the poison was soon detected,* Bd. 1, 1; S. 474, 39. v. ātor.

ātter-berende; *part. Venom-bearing;* venenifer :—Wǣtan ātterberendum *by venom-bearing humours,* L. M. 2, 1; Lchdm. ii. 176, 5. v. ātor, *etc.*

ātter-coppe, an; *f.* [ātor *poison,* copp *a head*] *A spider;* aranea :— Swindan ðû dydest swâ swâ āttercoppan sâwle his *tabescere fecisti sicut araneam animam ejus,* Ps. Spl. T. 38, 15. v. āttor-coppe.

ātter-lâðe, an; *f. The cock's spur grass;* panicum crus galli :—Ātterlâðe *venenifuga* [venom-loather], Wrt. Voc. 30, 38. v. ātor, *etc.*

āttor; *gen.* āttres; *n. Poison, venom;* venenum, Beo. Th. 5423; B. 2715 : Ps. Spl. 13, 5. v. ātor.

āttor-coppe, an; *f. A spider;* aranea :—Loppe, fleónde næddre, *vel* āttorcoppe *a spider,* Wrt. Voc. 24, 1. Āttorcoppe—wið āttorcoppan bîte *a spider—for spider's bite,* Herb. 4, 9; Lchdm. i. 92, 5, 6 : Med. ex Quadr. 4, 10; Lchdm. i. 344, 15. v. ātor, *etc.*

āttor-drinca, an; *m. A poisonous drink, poison;* potio venenata, venenum, Martyrol. ad 11 Junii.

āttor-, ātter-lâðe, an; *f. The cock's spur grass, atterlothe* [venom-loather]; panicum crus galli :—Wið āttre, betonican and ða smalan āttorlâðan dô on hâlig wæter *against poison, put betony and the small atterlothe into holy water,* L. M. 1, 45; Lchdm. ii. 110, 8; 114, 11 : Herb. 45, 1; Lchdm. i. 148, 4 : L. M. 1, 1; Lchdm. ii. 22, 15. Ātterlâðe *venenifuga,* Ælfc. Gl. 40; Som. 63, 88; Wrt. Voc. 30, 38. v. ātor, *etc.*

āttor-sceaða, an; *m. A poisonous destroyer, a venomous dragon, serpent;* hostis venenosus, draco venenosus, serpens :—Bûtan ðam āttorsceaðan *save to the venomous destroyer,* Exon. 96 a; Th. 357, 24; Pa. 33 : Beo. Th. 5670; B. 2839. v. ātor, *etc.*

āttor-spere, es; *n. A poisoned spear;* telum venenatum :—Eglum āttorsperum *with dire poisoned spears,* Exon. 105 a; Th. 399, 10; Rä. 18, 9. v. ātor, *etc.*

a-tuge *might draw away;* abstraheret, Bd. 4, 24; S. 598, 19; *p. subj. of* a-teón.

atul; *adj. Dire, terrible, horrid* :—In ðæt atule hûs *into that dire house,* Exon. 40 b; Th. 136, 1; Gû. 534 : Andr. Kmbl. 106; An. 53 : Ps. Th. 118, 123. v. atol.

a-tydran; *p.* ede; *pp.* ed *To procreate, create;* procreare, gignere, Elen. Kmbl. 2555; El. 1279. v. tydran.

a-tŷhst *drawest out,* Bt. Met. Fox 12, 49; Met. 12, 25. v. a-teón.

a-tyhtan, -tihtan; *p.* -tyhte, -tihte; *pp.* -tyhted, -tyht, -tiht. I. *to persuade, solicit, incite, attract, allure;* persuadere, allicere, incitare :— Ðâ wæs ofer Mûntgiop monig atyhted Gota, gylpes full *then was allured over the Alps many a Goth, full of arrogance,* Bt. Met. Fox 1, 16; Met. 1, 8. Ðe beóþ atihte to ðâm sôðum gesǣlþum *who are intent upon* [attracted to] *the true felicities,* Bt. 32, 1; Fox 114, 3. II. *to produce, procreate;* procreare, gignere :—Wîga is of dumbum twâm atyhted *a warrior is produced from two dumb ones,* Exon. 113 a; Th. 433, 27; Rä. 51, 3. v. tyhtan.

a-tŷhþ *draws away;* abstrahit, *3rd sing. pres. of* a-teón.

a-tymbrian, -tymbran; *p.* ode, ede; *pp.* od, ed *To erect, build;* ædificare :—Se Cênwalh hêt atymbran [atymbrian MS. Laud.] ða ealdan cyrican on Wintanceastre *Cenwalh ordered to build the old church at Winchester,* Chr. 643; Ing. 38, 1 : 919; Ing. 133, 17. v. a-timbrian.

a-tŷnan; *p.* -tŷnde; *pp.* -tŷned, -tŷnd; *v. a.* I. [a *away, out;* tŷnan *to inclose, shut*] *to shut out, exclude;* excludere :—Ne beóþ ût fram ðê atŷnde *ut non excludantur,* Ps. Th. 67, 27. II. [a = on, un *un,* tŷnan] *to un-shut, open;* aperire :—Nâ ic atŷnde mûþ mînne *non aperui os meum,* Ps. Spl. 38, 13. Atŷn us *aperi nobis,* Lk. Bos. 13, 25. v. on-tŷnan, un-tŷnan.

a-tyrian *to fail;* deficere. v. a-teorian.

at-ŷwan; *p.* de; *pp.* ed *To shew;* ostendere :—He atŷwde him *ostendit eis,* Ps. Spl. C. 77, 14. v. æt-ŷwan, ŷwan.

Augustînus, i; *m; Lat.* [Augustînus is correct in the quotations from the titles of the two following chapters of Bede, but in the A. Sax. text

it is Agustînus] *St. Augustine, the missionary sent by Pope Gregory to England,* A. D. 597, *and died May* 26, 605; Augustînus :—Ðæt se hâlga Papa Gregorius Augustînum sende Angel-þeóde to bodiganne Godes word *ut sanctus Papa Gregorius Augustînum ad prædicandum genti Anglorum verbum Dei miserit,* Bd. 1, 23, titl; S. 485, 14. Augustînus cumende on Breotone *Augustînus veniens Brittaniam,* 1, 25, titl; S. 486, 10. Hêr com Augustînus and his gefêran to Engla lande *here,* A. D. 597, *Augustine and his companions came to England,* Chr. 597; Th. 35, 41, col. 2 : 596; Th. 34, 37, col. 1; 35, 36, cols. 1, 2.

Augustus, i; *m; Lat.* I. *the first Roman Emperor.* v. Agustus. II. *the month of August;* mensis Augustus :—On ðam monþe ðe man Augustum nemneþ *in the month which is named August,* Herb. 7, 1; Lchdm. i. 96, 23. v. Agustus.

â-uht, es; *n. Aught, anything;* aliquid :—Eálâ, ðæt on eorþan âuht fæstlîces weorces ne wunaþ ǣfre *alas, that on earth aught of permanent work does not ever remain,* Bt. Met. Fox 6, 32; Met. 6, 16. Ðe âuht oððe nâuht âuðer worhte *which could either make aught or naught,* 20, 83; Met. 20, 42. Hwý biþ his ânwald ðŷ mâra *why will his power be by aught the greater?* 16, 40; Met. 16, 20 : Bt. 35, 5; Fox 164, 6, 10.

â-uht; *adv. At all, by any means;* omnino, ullo modo :—Âuht ne gebêtaþ hiora scearpnesse *nor by any means improve their sharpness,* Bt. Met. Fox 21, 46; Met. 21, 23 : 6, 12; Met. 6, 6. v. â-wuht, â-wiht.

a-urnen *run out, passed,* Cd. 79; Th. 98, 6; Gen. 1626. v. a-yrnan.

âuðer *either, each,* Bt. Met. Fox 29, 19; Met. 29, 10. v. âwðer.

ÂWA, âwo; *adv. Always, ever, for ever;* semper, unquam, usque :— Âwa *always,* Ps. Th. 143, 13. Âwa *usque,* 70, 16 : 138, 15 : Elen. Kmbl. 1899; El. 951. Ne wile heó âwa ðæs sîþes geswîcan *nor will it ever desist from its course,* Salm. Kmbl. 646; Sal. 322. Âwa *to feore in seculum,* Ps. Th. 51, 8 : 65, 6. On ǣcnesse, âwa *in æternum,* 118, 89. Âwa *to worlde in seculum seculi,* 71, 19 : 144, 1. Âwa *to worulde usque in seculum,* 130, 5 : 132, 4. Âwa *to ealdre for evermore,* Exon. 93 a; Th. 348, 22; Sch. 32 : Beo. Th. 1914; B. 955. [O. Sax. êo *unquam, semper* : O. H. Ger. êo, io *unquam, semper* : Goth. aiw *semper* : Lat. ævum *an age* : Grk. αἰεί, ἀεί *always* : αἰών *an age.*] vide â.

a-wacan; *p.* -wôc, *pl.* -wôcon; *pp.* -wacen; *v. intrans.* I. *to AWAKE;* expergisci, expergefieri, evigilare :—Awôc of ðam slǣpe *awoke from sleep,* Gen. 9, 24. Awôc Pharao *expergefactus est Pharao,* 41, 4, 7. II. *to wake into being, to arise, be born;* oriri, provenire, nasci :—Twâ þeóda awôcon *two nations arose,* Cd. 124; Th. 158, 11; Gen. 2615. v. wacan.

a-wacian; *p.* ode; *pp.* od *To awake;* expergisci, expergefieri, evigilare :—Of hefegum slǣpe awacode *e gravi somno expergefactus est,* Gen. 45, 26. v. wacian.

a-wâcian, -wâcigan; *p.* ode; *pp.* od; *v. intrans. To grow weak* or *effeminate, to languish, decline, fail, fall away, relax, to be indolent;* infirmari, deficere, recedere :—Awâcode mid langre ealdunge *weakened with old age,* Gr. Dial. 2, 15. Awâciaþ on ðære costnunge tîman *in tempore tentationis recedunt,* Lk. Bos. 8, 13. Ðæt ne awâcodon wereda Drihtne *that they might not fall away from the Lord of hosts,* Cd. 183; Th. 229, 20; Dan. 220. Gif he nâ ne awâcaþ *if he never relax,* L. Pen. 12; Th. ii. 280, 29. v. ge-wâcian, on-.

a-wacnian, -wæcnian; *p.* cnede, cenede; *pp.* cned, cened; *v. intrans.* I. *to AWAKEN, come to life again, revive;* evigilare, expergefieri, reviviscere :—On dagunge he eft acwicode [awacenede MSS. Ca. O.] *diluculo revixit,* Bd. 5, 12; S. 627. 13. II. *to arise, spring, have one's origin;* suscitari, oriri, nasci :—Of ðam frumgârum folc awæcniaþ *from these patriarchs shall spring a people,* Cd. 104; Th. 138, 14; Gen. 2291. Eall heora gewinn awacnedon ǣrest fram Alexandres epistole *all their wars first arose from Alexander's letter,* Ors. 3, 11; Bos. 72, 19. v. wæcnan, on-wæcnan, on-wæcnian.

a-wǣcan; *p.* -wæcte, -wæhte; *pp.* -wæced, -wæct, -wæht *To weaken, fatigue;* debilitare, fatigare :—Awæht *defessus,* Hymn. Awæht *porrectus,* Cot. 157.

a-wæccan *To awake;* suscitare, Mt. Rush. Stv. 3, 9. v. a-weccan.

a-wæcnan; *p.* ede; *pp.* ed; *v. intrans. To awake, rise up, be born;* evigilare, suscitari, nasci :—Nû is ðæt bearn cymen, awæcned *now is that child come, risen up,* Exon. 8 b; Th. 5, 9; Cri. 67.

a-wæcnian; *p.* ode; *pp.* od *To awaken, arise, spring;* evigilare, oriri :—Awæcniaþ, Cd. 104; Th. 138, 14; Gen. 2291. v. a-wacnian.

a-wǣgan; *p.* de; *pp.* ed; *v. trans. To deceive, delude, frustrate, disappoint, cause to fail;* eludere, frustrari, irritum facere :—Ðæt is sôþ ðæt ðû ǣr awǣgdest *that is true which thou before didst frustrate,* Homl. Th. ii. 418, 18. Ǣr awǣged sîe worda ǣnig *ere any word be made to fail,* Andr. Kmbl. 2876; An. 1441. Awǣged ne dô ðû wedd *irritum ne facias fœdus,* Hymn, Lye. v. wǣgan, ge-wǣgan.

a-wǣh *weighed out, weighed to;* appendit, Gen. 23, 16. v. a-wegan.

a-wǣht *weakened, wearied; pp. of* a-wǣcan.

a-wǣhte *aroused;* suscitavit, Bd. 4, 23; S. 596, 14. v. a-weccan.

a-wǣlan; *p.* ede, de, te; *pp.* ed. I. *v. trans. To roll away, roll back, roll to;* revolvere, advolvere :—Awǣlede ðone stân *revolvit lapidem,* Mt. Rush. Stv. 28, 2. Awǣlte ðone stân *advolvit lapidem,*

Mk. Rush. War. 15, 46.   **II.** *to move violently, vex, afflict;* vexare :—Awǽled *vexatus,* Mk. Rush. War. 5, 18.

a-wǽndan; *p.* de; *pp.* ed *To turn from* or *away, to translate;* avertere, transferre :—Ðonne awǽnt Driht hæftnunge folces his *cum averterit Dominus captivitatem plebis suæ,* Ps. Spl. 13, 11. v. a-wendan, wǽndan.

a-wǽndednys, -nyss, e; *f. A change;* mutatio, Ps. Lamb. 76, 11. v. awendednys.

a-wǽnian; *p.* ede; *pp.* ed *To wean from;* ablactare :—Swá swá awǽned cild *sicut ablactatus,* Ps. Lamb. 130, 2.

a-wǽrged, -wǽrgd; *p.; def. m.* -wǽrgda *Accursed;* maledictus :— Wit ðæs awǽrgdan wordum gelýfdon *we two believed the words of the accursed one,* Cd. 222; Th. 290, 16; Sat. 416. v. a-wyrged.

a-wǽscen *washed;* lotus; *pp.* v. wascan.   DER. un-a-wǽscen.

a-wéstan; *p.* -wéste; *pp.* -wésted; *v. trans. To waste, lay waste, eat up;* vastare, carpere :—Swá swá oxa gewunaþ to awǽstenne gærs *quo modo solet bos herbas carpere,* Num. 22, 4. v. a-wéstan.

a-wanian; *p.* ode; *pp.* od *To diminish;* diminuere. v. wanian.

a-wannian; *p.* ode; *pp.* od *To wax wan* or *pale;* pallescere :—Awannod *pallidus factus,* Greg. Dial. 1, 2.

â-wâr; *adv.* [=â-wǽr= â-hwǽr] *Anywhere;* alicubi :—Swilce he âwâr wǽre, ǽrðan þe he geboren wǽre *as if he were anywhere, before he was born,* Homl. Th. ii. 244, 19.

a-wariged; *part. Accursed;* maledictus. v. a-werged; *pp. of* a-wergian: awyrged; *pp.* of a-wyrgian.

a-wǽrnian; *p.* ode; *pp.* od *To be confounded;* confundi, Ps. Spl. M. 85, 16. v. a-swǽrnian.

a-warpen; *pp. cast out;* ejectus, Ps. Spl. 108, 9. v. a-worpen; *pp. of* a-weorpan.

a-weaht, a-weahte *awaked, excited, raised up,* Ps. Th. 77, 65 : Bd. 3, 5; S. 526, 34; *pp. and p.* of a-weccan.

a-weallan; ic -wealle, ðú -weallest. -wylsſt, he -wealleþ, -wealþ, -wylþ, *pl.* -weallaþ; *p.* -weól, -weóll, *pl.* -weóllon; *pp.* -weallen; *v. intrans. To boil* or *bubble up, break forth, stream* or *gush forth, well out, flow forth, issue;* ebullire, erumpere, emanare :—Swá ǽspringe út awealleþ of clife hǽrum *so a water-spring wells out of a hoary cliff,* Bt. Met. Fox 5, 24; Met. 5, 12 : Ps. Th. 103, 10 : Ex. 8, 3 : Andr. Kmbl. 3045; An. 1525. Ða fruman aweallaþ Deorwentan streámes *Deruentionis fluvii primordia erumpunt,* Bd. 4, 29; S. 607, 11. Is ðæt eác sǽd, ðæt wylle aweólle *fertur autem, quia fons ebullierit,* Bd. 5, 10; S. 625, 23 : Exon. 17a; Th. 39, 20; Cri. 625.   DER. weallan.

a-weardian; *p.* ode, ede; *pp.* od, ed; *v. trans. To ward off, defend, protect;* defendere, protegere :—Hí hí sylf aweardedon *they defended themselves,* Ors. 5, 3; Bar. 182, 19.   DER. weardian.

a-wearpan= a-weorpan *to cast away;* projicere :—Dust ðæt awearpþ wind *pulvis quem projicit ventus,* Ps. Spl. 1, 5.

a-weaxan; *p.* -weóx, -wôx; *pp.* -weaxen; *v. intrans. To wax, grow, arise, come forth;* crescere, oriri, provenire :—Him aweaxeþ wynsum gefeá *to them shall grow winsome delight,* Exon. 26a; Th. 77, 7; Cri. 1253 : Ps. Th. 128, 4 : Exon. 103a; Th. 391, 24; Rä. 10, 10 : 103b; Th. 392, 6; Rä. 11, 3 : Elen. Kmbl. 2450; El. 1226.

a-web, es; *n. The cross threads in weaving, called the woof* or *weft;* subtegmen, Cot. 161.

a-weccan, -weccean; ic -wecce, ðú -wecest, -wecst, he -wecceþ, -weceþ, -wecþ, *pl.* -weccaþ, -wecceaþ; *p.* -weahte, -wehte, *pl.* -weahton, -wehton; *impert.* -wec, -wece, *pl.* -wecceaþ; *pp.* -weaht, -weht; *v. trans.*   **I.** *to awake, arouse from sleep, awake from death;* e somno excitare, resuscitare :—Hí awehton hine *excitaverunt eum,* Mk. Bos. 4, 38. Ðá wearþ aweaht Drihten swá he slǽpende *excitatus est tamquam dormiens Dominus,* Ps. Th. 77, 65. Ic hine awecce *resuscitabo eum,* Jn. Bos. 6, 40. Se Fæder awecþ ða deádan *Pater suscitat mortuos,* 5, 21. He manige men of deáþe awehte *he awoke many men from death,* Andr. Kmbl. 1167; An. 584. Awecceaþ deáde *suscitate mortuos,* Mt. Bos. 10, 8.   **II.** *to excite, rouse, stir up, call forth, raise up, raise up children;* excitare, concitare, suscitare, resuscitare :—To ælmessan and to gódra dǽda fylignessum he hí aweahte ge mid wordum ge mid dǽdum *ad eleemosynas operumque bonorum executionem et verbis excitabat et factis,* Bd. 3, 5; S. 526, 34. Awehte wǽlnîþ Babilónes brego *deadly hatred excited the prince of Babylon,* Cd. 174; Th. 218, 28; Dan. 46. Ðæs sǽdes corn biþ simle aweaht mid ascunga, eác siððan mid gódre lâre, gif hit grówan sceal *the grain of this seed is always excited by inquiry, and moreover by good instruction, if it shall grow,* Bt. Met. Fox 22, 80; Met. 22, 40. Awehte ða windas of heofenum *excitavit ventos de cælo,* Ps. Th. 77, 26. Awece ðíne mihte *excita potentiam tuam,* 79, 3. Hí his yrre aweahtan *in ira concitaverunt eum,* 77, 58, 40 : Cd. 52; Th. 66, 7; Gen. 1080. Aweccaþ wôpdropan *calls forth tears,* Salm. Kmbl. 567; Sal. 283. He aweahte gewitnesse on Iacobe *suscitavit testimonium in Jacob,* Ps. Th. 77, 6. Ic awecce wið ðé óðerne cyning *I will raise up against thee another king,* Elen. Kmbl. 1851; El. 927. Aweccende fram eorþan wǽdlan *suscitans a terra inopem,* Ps. Spl. 112, 6. Awece me *resuscita me,* 40, 11. He mæg bearn aweccan [aweccean Mt. Bos. 3, 9] *potens est suscitare filios,* Lk. Bos. 3, 8. Hys bróðor sǽd awecce *suscitet semen fratri suo,* 20, 28.

a-wece *arouse, raise up,* Ps. Spl. C. T. 40, 11; *impert.* of a-weccan.

a-wecgan, -wegan; *p.* -wegde, -wegede; *pp.* -weged; *v. trans. To move, remove, shake;* movere, amovere, commovere, agitare :—Ne mihton awecgan Iob of his módes ânrǽdnysse *might not move Job from his constancy of mind,* Job Thw. 167, 33; Andr. Kmbl. 1005; An. 503. Hí ne mihton hine awecgan *they could not move it,* Homl. Th. ii. 164, 31. Môd biþ aweged of his stede *the mind is removed from its place,* Bt. 12; Fox 36, 18: Bt. Met. Fox 7, 48; Met. 7, 24. Winde aweged [MS. awegyd] hreód *arundinem vento agitatam,* Mt. Bos. 11, 7. v. wecgan.

a-wecþ *awakes, raises up,* Jn. Bos. 5, 21; *3rd pers. pres. of* a-weccan.

a-wédan; *p.* -wédde; *pp.* -wéd; *v. n. To be mad, to rage, to be angry, to go* or *wax mad, revolt, apostatize;* in furorem agi :—Awéddon ða nýtena *the cattle became mad,* Ors. 5, 10; Bos. 108, 31. Se ðe for sleápe awéd *phreneticus* = φρενιτικός, Ælfc. Gl. 78; Som. 72, 40; Wrt. Voc. 45, 72. v. wédan.

a-wefan; *p.* -wæf, *pl.* -wǽfon; *pp.* -wefen *To weave;* texere :—Wyrmas ne awǽfon *worms did not weave,* Exon. 109a; Th. 417, 23; Rä. 36, 9 : Jn. Bos. 19, 23.

a-weg; *adv.* AWAY, *out;* (this is its meaning both in and out of composition); auferendi vim habet :—Ðá eóde he aweg *autem abiit,* Mt. Bos. 19, 22. Ge drehnigeaþ ðone gnæt aweg *ye strain the gnat out;* excolantes [ex *out,* colare *to filter, strain*] culicem, Mt. Bos. 23, 24. He hí raðe aweg aþýwde *he quickly drove them away,* Ors. 6, 36; Bos. 131, 28 : Ps. Th. 77, 57. v. on-weg.

aweg-adrífan *to drive* or *chase away;* expellere, Ps. Spl. C. 35, 13. v. a-drífan.

aweg-áferian *to carry away, to cart away;* evehere, Cot. 205.

aweg-alúcan [aweg *away,* alúcan *to lock out, separate*] *To shut* or *lock out, to separate;* discludere, Cot. 67.

a-wegan; *p.* -wæg, -wæh, *pl.* -wǽgon; *pp.* -wegen; *v. trans.*   **I.** *to lift up, take* or *carry away;* levare, auferre :—Hí â sibbe gelǽraþ, ða ǽr wonsǽlge awegen habbaþ *they shall ever advise peace, which the unblest have before taken away,* Exon. 89a; Th. 334, 25; Gn. Ex. 21 : Homl. Th. i. 308, 17.   **II.** *to weigh out, weigh to any one;* appendere :—Abraham ðá awæh feówer hund scillinga seolfres *Abraham appendit quadringentos siclos argenti,* Gen. 23, 16. Eálá gif míne synna and mín yrmþ wǽron awegene on ânre wǽgan *utinam appenderentur peccata mea et calamitas in statera,* Job 6, 2; Thw. 167, 18.

a-wegan; *p.* -wegede, -wegede *To move, shake :*—Aweged, Bt. 12; Fox 36, 18: Bt. Met. Fox 7, 48; Met. 7, 24 : Mt. Bos. 11, 7. v. a-wecgan.

aweg-animan *to take away;* sufferre, Jn. Bos. 20, 1. v. a-niman.

aweg-awyltan *to roll away;* revolvere, Mk. Bos. 16, 4. v. a-wyltan.

aweg-beran *to bear, carry* or *convey away;* asportare, Ælfc. Gr. 47; Som. 48, 37. v. beran.

aweg-cuman *to go away, to leave, escape;* dimittere :—Sume aweg-cómon *some escaped,* Ors. 3, 3; Bos. 55, 26. v. cuman.

a-weged *moved,* Bt. 12; Fox 36, 18; *pp. of* a-wegan *to move.*

a-wegen *taken away, weighed as in a balance,* Job 6, 2; Thw. 167, 18. v. a-wegan *to weigh.*

aweg-gân *to go away;* abire :—Ongan aweg-gân *began to go away,* Bd. 4, 22; S. 591, 1. v. gân.

aweg-geniman *to take away;* auferre. v. geniman.

aweg-gewítan; *p.* -gewât, *pl.* -gewiton; *pp.* -gewiten *To go away, depart;* discedere :—Ic eom aweg-gewiten *I am gone away,* Ors. 2, 4; Bos. 44, 36. v. ge-wítan.

aweg-gewítenes, -ness, e; *f. A going away, departure;* abscessio :— Æfter þrím geárum Willfreþes aweg-gewitenesse *post tres abscessionis Vilfridi annos,* Bd. 4, 12; S. 581, 30. v. gewítan.

aweg-lǽtan *to let* [go] *away, let escape;* abire permittere, L. C. S. 29; Th. i. 392, 14. v. lǽtan.

aweg-onwendan *to turn* or *move away;* amovere, Ps. Spl. C. 65, 19. v. on-wendan.

aweg-weorpan *to cast* or *throw away;* abjicere. v. aweg, weorpan.

a-wegyd *shaken,* Mt. Bos. 11, 7. v. a-wecgan.

a-weht *awaked, aroused; pp. of* a-weccan.

a-wehte *awaked, excited,* Andr. Kmbl. 1167; An. 584: Ps. Th. 77, 26; *p. of* a-weccan.

a-wehtnes, -ness, e; *f. An awaking, a stirring up, excitation, quickening, encouraging;* excitatio :—To awehtnesse lífiendra monna of sáule deáþe *ad excitationem viventium de morte animæ,* Bd. 5, 12; S. 627, 5.

awel *an awl;* subula, fuscinula, harpago = ἁρπάγη, Cot. 84: 13. v. al.

a-wellan; *p.* de; *pp.* ed *To cause to bubble, to well;* facere ut aliquid ferveat vel ebulliat :—Hreðor innan wæs wynnum awelled *the breast within was welled with joy,* Andr. Kmbl. 2037; An. 1021. v. a-weallan.

a-wend *turned, translated; pp. of* a-wendan :—Seó bôc is on Englisc awend *the book is turned* [translated] *into English,* Homl. Th. ii. 358, 30.

a-wendan; ic -wende, ðú -wendest, -wenst, he -wendeþ, -went, *pl.* -wendaþ; *p.* -wende; *pp.* -wended, -wend, -went.   **I.** *v. trans. To turn away* or *off, avert, remove, to turn upside down, turn, change, translate, pervert;* avertere, vertere, mutare, transferre, subvertere :—

Ansýne ðýn awendst ðú *faciem tuam avertis*, Ps. Spl. 43, 27 : Ps. Th. 73, 11 : 103, 27 : 101, 2 : 77, 38. Heó awent hyre hús and sécþ geornlíce ðþ heó hine fint *sche turneth vpsodoun the hous and sekith diligently til sche fynde it*, Wyc; Lk. Bos. 15, 8. He wæter awende to wínlícum drence *he turned water into winelike drink*, Ælfc. T. 27, 7 : Ps. Spl. 101, 28 : Gen. 19, 26 : Cd. 14; Th. 17, 13; Gen. 259 : Jn. Bos. 10, 35. 'Historia Anglorum' ða ðe Ælfréd cyning of Lédene on Englisc awende [*Bede's*] *Historia Anglorum, which king Alfred translated from Latin into English*, Homl. Th. ii. 116, 30–118, 1. Ðeáh ðe seó bóc on Englisc awend sý *though the book be translated into English*, 118, 5. Ne nim ðú lác, ða awendaþ rihtwísra word *nec accipies munera, quæ subvertunt verba justorum*, Ex. 23, 8.   **II. v. intrans.** To turn or direct oneself, to turn from, go, depart; se vertere, ire :—Ðæt hý, mid sume searawrence, from Xerse awendan [awende MS.] *that they would by some stratagem turn from Xerxes*, Ors. 2, 5; Bos. 47, 41. Hí awendon aweg *they turned away*, Ps. Th. 77, 57. v. wendan.

a-wended-líc, -wende-líc, -wendend-líc; *adj.* [awended *changed, pp.* of awendan, líc] *Movable, changeable, alterable, mutable;* mobilis, Alb. resp. 42.

a-wendednys, a-wændednys, -nyss, e; *f. A change, alteration;* commutatio :—Ná sóþ is him awendednys *non enim est illis commutatio*, Ps. Spl. 54, 22 : 88, 50.

a-wendelíc-nes, -ness, e; *f. Mutableness, mutability, changeableness, inconstancy;* mutabilitas, Som. [a-wendedlíc *changeable,* -ness].

a-wendincg, e; *f. An overthrowing, a change, ruin;* subversio, Scint. 61.

a-wenian; *p.* ede; *pp.* ed *To wean;* ablactare :—Ær ðone, ðæt acennede bearn, awened sí *quoadusque, qui gignitur, ablactatur*, Bd. 1, 27, resp. 8; S. 493, 33. v. wenian.

a-went, -wenþ, -wendeþ *turns*, Lk. Bos. 15, 8. v. a-wendan.

a-weódian, -weódigan; *v. a. To weed, root or rake up, to destroy;* sarculare :—Ðæt man aweódige unriht *that one should root up injustice*, L. C. S. 1; Th. i. 376, 7.

a-weól *flowed forth*, Cot. 72. v. a-weallan.

a-weorpan, -wurpan, -wyrpan; ðú -wyrpst, he -wyrpþ; *p.* ic, he -wearp, ðú -wurpe, *pl.* -wurpon; *impert.* -weorp, -wurp, -wyrp ðú; *pp.* -worpen; *v. a.* [a *from,* weorpan *to throw*] *To throw or cast from or down, to cast away or off, cast out, to degrade, reject, divorce;* abjicere, dejicere, projicere, ejicere, propellere, repellere, reprobare, repudiare :—Ðæt he ðec aweorpe of woruldríce *that he shall cast thee from thy worldly kingdom*, Cd. 203; Th. 253, 1; Dan. 589. Ðú awurpe hí, ða hí wæron upahafen *dejecisti eos, dum allevarentur*, Ps. Spl. 72, 18 : 79, 9 : Ps. Th. 72, 14. Is wærgðu [wærgða MS.] aworpen *the curse is cast off*, Exon. 9 a; Th. 7, 8; Cri. 98 : Bt. Met. Fox 23, 12; Met. 23, 6 : Bd. 3, 24; S. 557, 44 : Mt. Bos. 12, 28. Ða woldon senatus hine aweorpan *then would the senate degrade him*, Ors. 3, 10; Bos. 70, 36 : Bt. 37, 4; Fox 192, 10. Ne aweorp ðú me *ne projicias me*, Ps. Spl. 70, 10. Mannes sunu gebyreþ beón aworpen *oportet filium hominis reprobari*, Mk. Bos. 8, 31. Aworpen wíf *a divorced wife*, L. Ælf. C. 7; Th. ii. 346, 6. Aworpen man biþ á unnyt *homo apostata, vir inutilis*, Past. 47, 1; Hat. MS. 68 a, 23. Used also with the prepositions on *into*, as awurpan on *to cast into*, Mt. Foxe 13, 50. Fram *from*, Mt. Bos. 5, 29, 30. Út *out*, Mt. Bos. 13, 48. Under *below*, Bt. 37, 4; Fox 192, 10.

a-weorpnis, -niss, e; *f. A casting off, putting away, divorce;* repudium, Mt. Rush. Stv. 19, 7. v. a-worpenes. DER. weorpan.

a-weorþan, a-wurþan, ic -weorþe, -wurþe, ðú -wyrst, he -weorþeþ, -wyrþeþ, -wurþeþ, -wyrþ, *pl.* -weorþaþ, -wurþaþ; *p.* -wearþ, *pl.* -wurdon; *pp.* -worden; *v. intrans.* [a *from, away,* weorþan *to become*] *To cease to be, become insipid or worthless;* evanescere :—Gyf ðæt sealt awyrþ *if the salt become insipid*, Mt. Bos. 5, 13 : Lk. Bos. 14, 34. Ðú awordena raca, Mt. Bos. 5, 22.

a-weosung, e; *f. The being, essence,* or *subsistence of a thing;* subsistentia, essentia, Cot. 170. v. wesan.

a-weóx *waxed, increased*, Ors. 1, 3; Bos. 27, 25. v. a-weaxan.

á-wér *anywhere, in any wise*, Bt. Met. Fox 8, 28; Met. 8, 14 : Bt. 3; Fox 20, 14. v. á-hwǽr.

a-werd, es; *m. A spoiled or worthless fellow;* vappa, Ælfc. Gl. 9; Som. 56, 113; Wrt. Voc. 18, 61, = a-wered = a-werded; *pp.* of a-werdan.

a-werdan; *p.* de; *pp.* ed; *v. trans. To injure, corrupt, violate, destroy;* lædere, corrumpere, vitiare, violare. v. a-wyrdan.

a-wered *protected, worn; pp.* of a-werian I and III.

a-wergian, -wirgean, -wyrgian; *p.* de; *pp.* ed [a, wergian *to curse*] *To accurse, curse, condemn, malign;* maledicere, condemnare, malignari :—Helle dióful, awerged in wítum *hell's devil, accursed to torments*, Andr. Kmbl. 2599; An. 1301 : Gen. 8, 21 : Ps. Spl. 73, 4.

a-werian, -wergan, -wergean; *p.* ede; *pp.* ed; *v. trans.*   **I.** *to ward off, defend, restrain, protect, cover;* defendere, prohibere, protegere :—Ðæt he hine eáþ awerian mæge *that he may easily defend him*, L. C. S. 20; Th. i. 388, 2. He hine awerede *he defended himself*, Ors. 3, 9; Bos. 68, 23, 29 : 5, 3; Bos. 103, 25 : Ps. Th. 105, 24. Ðú mín heáfod scealt on gefeohtdæge feóndum awergean *obumbrasti caput meum in die belli*, 139, 7. Ðú me oft aweredest wyrigra gemótes *protexisti*

me a conventu malignantium, 63, 2 : 55, 11. Ðeáh hit mon awerge wírum útan *though it be covered with wires without*, Exon. 111 a; Th. 424, 30; Rä. 41, 47.   **II.** *to ward off from oneself, spurn from oneself;* aspernari :—Aweredon ða óðre *aspernabantur ceteros*, Lk. Rush. War. 18, 9.   **III.** *to wear, wear out;* terere, deterere :—Awered *tritus*, R. Ben. 55. v. werian.

a-werpan *to cast away;* projicere :—Awerp from ðé *projice abs te*, Mt. Rush. Stv. 5, 29. v. a-weorpan.

a-wersian *to make worse;* deterius facere, Cart. Edwardi R. v. wyrsian.

áwesc-nis, -niss, e; *f. Disgrace, blushing for shame, reverence*, Ps. Surt. 34, 26. v. ǽwisc-nys.

a-wést; *part. Wasted, laid waste, waste, desert;* vastatus, desertus :—Awést wearþ *was laid waste*, Ors. 3, 9; Bos. 66, 17, 19, 21 : Ps. Spl. T. 68, 30. v. a-wéstan.

a-wéstan; *p.* -wéste; *pp.* -wésted [-wéstd], -wést [a *intensive,* wéstan *to waste*] *To waste, lay waste, depopulate, ravage, destroy;* vastare, devastare, desertum facere, desolare :—Hí awéste *eam vastavit*, Jn. 10, 39. Hí ealle Ægypta awéston *they laid waste all Egypt*, Ors. 1, 10; Bos. 32, 26. Troia awésted wæs *Troy was laid waste*, 2, 2; Bos. 40, 28. Eall seó þeód awést wearþ *all the nation was laid waste*, 3, 9; Bos. 66, 17, 19, 21. Sý wunung heora awést *fiat habitatio eorum deserta*, Ps. Spl. T. 68, 30. Widútan awést híg sweord *swerd with outforth schal waaste* [*destroy*] *hem*, Wyc; foris vastabit eos gladius, Cant. Moys. Isrl. Lamb. 193 a, 25. His stede oððe stówe híg awéston *locum ejus desolaverunt*, Ps. Lamb. 78, 7.

a-wéstendnes, -ness, e; *f. A wasting, a laying waste;* vastatio, Som. v. a-wéstan, a-wéstende, *part*; ness.

a-wéstnis, -niss, e; *f.* [a-wést *wasted,* ness] *Desolation;* desolatio, Lk. Rush. War. 21, 20.

áwian; *p.* ode; *pp.* od [= eówan, ýwan] *To shew;* ostendere. v. atáwian, Ps. Spl. T. 77, 14.

a-wierdan *to corrupt;* corrumpere :—He awiert ðæt mód *corrumpit animum*, Past. 53, 5. v. a-wyrdan.

a-wierged; *def. m.* -wiergeda, -wiergda; *pp. Accursed, wicked;* maledictus, malignus, Past. 65, 4 ? v. a-wyrged.

â-wiht, â-wyht, â-wuht, â-uht, âht, es; *n.* [â *semper,* wiht *creatura, animal, aliquid*] AUGHT, *anything;* aliquid :—Unc gemǽne ne sceal elles áwiht to us *two shall not be aught else common*, Cd. 91; Th. 114, 16; Gen. 1905 : Ps. Th. 55, 9. Handa hí habbaþ, ne hió hwǽdere mágon gegrápian gódes áwiht *they have hands, and yet they may not touch anything of good*, Ps. Th. 113, 15 : 58, 3 : 65, 16 : Bt. Met. Fox 9, 124; Met. 9, 62. Nafast ðú for áwiht ealle þeóda *pro nihil habebis omnes gentes*, Ps. Th. 58, 8. Ðæt hí geseón ne mágon áwiht *ne illi videant aliquid*, 68, 24. v. ná-wiht, náht.

â-wiht, â-wyht, â-wuht, â-uht, âht; *adv. At all, by any means;* omnino, ullo modo :—Ne lata ðú áwiht *do not thou tarry at all*, Ps. Th. 69, 7 : 77, 10, 12 : 134, 19. Me ðæt riht ne þinceþ, ðæt ic óleccan áwiht þurfe Gode æfter góde ænegum *to me it seems not right, that I at all need cringe to God for any good*, Cd. 15; Th. 19, 13; Gen. 290.

a-wildian; *p.* ode; *pp.* od; *v. intrans. To become wild* or *fierce;* silvescere, efferari, Off. Episcop. 7.

a-willan; *p.* de; *pp.* ed *To cause to bubble, to boil;* facere ut aliquid ferveat vel ebulliat, coquere, decoquere :—Awilled meolc *boiled milk, pottage;* juta [jura ?], Cot. 168. Awilled wín *vel* cyren *new wine, just pressed from the grape,* or *new wine boiled till half evaporated;* dulcisapa, Cot. 62, 168. v. a-wyllan, cyren.

a-windan; ic -winde, ðú -wintst, -winst, he -wint, *pl.* -windaþ; *p.* -wand, *pl.* -wundon; *pp.* -wunden [a, windan *to wind*].   **I. v. trans.** *To wind, bend;* plectere, torquere :—Hí him onsetton þyrnenne helm awundenne *imponunt ei plectentes spineam coronam*, Mk. Bos. 15, 17.   **II. v. trans.** *To strip off;* detrahere :—Gif him mon ðonne awint of ða cláþas *if any man should strip off the clothes from him*, Bt. 37, 1; Fox 186, 10 : Mt. Rush. War 25, 44; Met. 25, 22.   **III.** *v. intrans. To whirl* or *slip off;* labi :—Gif sió æcs ðonne awient [awint, Cot.] of ðæm hielfe *if the axe then slip from the handle*, Past. 21, 7; Hat. MS. 32 b, 6.

a-windwian, -wyndwian *to winnow, blow away;* ventilare, Ps. Spl. 43, 7. v. windwian.

a-winnan; *p.* -wan, *pl.* -wunnon; *pp.* -wunnen *To labour, contend, gain, overcome;* laborare, contendere, acquirere, nancisci, superare :—Ælc wís mon scyle awinnan ægðer ge wið ða réðan wyrde ge wið ða wínsuman *every wise man ought to contend both against the severe fortune and against the pleasant*, Bt. 40, 3; Fox 238, 16. Ealles ðú ðæs wíte awunne *for all this thou hast gained suffering*, Exon. 39 b; Th. 130, 18; Gú. 440. Súsl wæs awunnen *the pain was overcome*, Cd. 208; Th. 257, 8; Dan. 654. DER. winnan.

a-wint *strips off, slips off.* v. a-windan.

a-wirdan *to destroy*, Leo 254. v. a-wyrdan.

a-wirgan; *p.* de; *pp.* ed *To strangle;* strangulare :—Gelícost ðam ðe he hine sylfne hæfde unwitende awirged *as if he had voluntarily strangled himself*, Ors. 6, 36; Bos. 131, 38. v. a-wyrgan.

a-wirgean; *p.* de; *pp.* ed *To accurse, curse;* maledicere :—Nelle ic awirgean ða eorþan *nolo maledicere terræ,* Gen. 8, 21. Awirgede woruldsorga *ye execrable worldly cares,* Bt. 3, 1; Fox 4, 25. v. a-wergian, a-wyrgian.

a-wirgnis, -niss, e; *f. A curse, cursing;* maledictio :—Sette ge awirgnisse uppan Hebal dūne *ponite maledictionem super montem Hebal,* Deut. 11, 29. v. a-wyrgednes.

āwisc-ferinend, es; *m.* [āwisc = æwisc *disgrace,* ferinian = firenian *to sin*] *One who sins disgracefully, a publican;* qui turpiter pēccat, publicanus, Cot. 204.

a-wisnian; *p.* ade; *pp.* ad *To be dry, to become dry,* wizen; arescere :—Awisnade *vel* oferdrugade *aruit,* Lk. Lind. War. 8, 6. v. wisnian.

a-wlǽtan; *p.* -wlǽtte; *pp.* -wlǽted *To defile,* foedare, Hymn: Mod. Confit. 1.

a-wlancian; *p.* ode; *pp.* od *To come in youthful strength, to exult, to be proud;* exultare, Leo 262. v. wlancian.

āwo; *adv. Always, ever;* semper, unquam :—Âwo *ever,* Exon. 26 b; Th. 78, 9; Cri. 1271: 32 a; Th. 101, 25; Cri. 1664. Siððan âwo *ever after,* 48 a; Th. 164, 24; Gū. 1016. Âwo to ealdre *for evermore,* 14 b; Th. 30, 13, note; Cri. 479. v. âwa.

a-wōc *awoke, arose,* Gen. 9, 24; *p.* of a-wacan.

a-wōdian *to root up.* v. awēodian.

a-woffian; *p.* ode; *pp.* od *To rave, be delirious, frantic;* delirare :—Awoffod *phreneticus,* Leo 266. v. woffian.

awōh; *adv.* [a, wōh *crooked*] AWRY, *unjustly, wrongfully, badly; the same as* mid wōge *with injustice, or unjustly;* tortè, obliquè, malè :—Gif mon ðæt trod awôh drîfe *if one wrongfully pursue the footstep* [*tread*], L. O. D. 1; Th. i. 352, 10. Ðæt man ǽr awôh tosomne gedydon *which they before unjustly joined together,* L. Edm. B. 9; Th. i. 256, 11.

a-worden; *pp.* of a-weorþan; *def. m.* awordena *become worthless* :—Ðū aworðena *raca,* Mt. Bos. 5, 22.

a-worpen *cast off, away,* Exon. 9 a; Th. 7, 8; Cri. 98; *pp. of* a-weorpan.

a-worpenes, -worpennys, -worpnes, -ness, -nyss, e; *f. A rejection, casting away, reprobation, reproving;* abjectio :—Ic eom aworpennys folces *ego sum abjectio plebis,* Ps. Spl. 21, 5. v. a-weorpnis; forwyrpnes.

a-worpen-lîc; *adj. Damnable;* damnabilis, Past. 52, 8.

a-wōx *waxed, grew, rose,* Exon. 103 b; Th. 392, 6; Rä. 11, 3; *p.* of a-weaxan.

a-wrǽc, -wrǽcon *related,* Exon. 17 a; Th. 40, 3; Cri. 633; *p.* of a-wrecan.

a-wrǽstan, -wrēstan; *p.* -wrǽste; *pp.* -wrǽst *To wrest from, to extort;* extorquere, Cot. 78. v. wrǽstan.

a-wrāt *wrote,* Bd. 5, 23; S. 648, 27; *p.* of a-wrîtan.

a-wrāþ *bound up,* Bd. 4, 22; S. 590, 36; *p.* of a-wrîðan.

a-wreāh *discovered,* Ps. Spl. 97, 3; *p.* of a-wreóhan. v. a-wreón.

a-wrecan; *p.* -wrǽc, *pl.* -wrǽcon; *pp.* -wrecen. I. *to drive away;* pellere, expellere :—Ðara ðe he of lîfe hēt awrecan *of those whom he bade to drive from life,* Exon. 130 a; Th. 498, 11; Rä. 87, 11. II. *to hit, strike;* icere, percutere :—Awrecen wǽlpîlum *hit with darts of death,* Exon. 49 b; Th. 171, 15; Gū. 1127: 51 b; Th. 179, 11; Gū. 1260. III. *to relate, recite,* sing; narrare, enarrare, canere :—Bi ðon Iob giedd awrǽc *of whom Job related his lay,* Exon. 17 a; Th. 40, 3; Cri. 633: 84 a; Th. 316, 20; Môd. 51: Beo. Th. 3452; B. 1724: 4223; B. 2108. IV. *to avenge, revenge;* ulcisci :—Gif hine hwā awrecan wille *if any one will avenge him,* L. Ath. i. 20; Th. i. 210, 10, note 20. v. wrecan.

a-wrecen *banished, driven away;* extorris, Cot. 212: 5; *pp.* of a-wrecan.

a-wregennes *a discovery.* v. a-wrigenes.

a-wrehte, a-wreht *aroused, awoke;* suscitavit, suscitatus, Jn. Bos. 12, 1; *p.* and *pp.* of a-wreccan.

a-wreón, -wreóhan, -wrióhan, -wrión; *p.* -wreáh, *pl.* -wrugon; *pp.* -wrogen; *v. a.* [a *not,* wreón *to cover*] *To uncover, discover, disclose, open, reveal;* revelare :—Se Sunu hit awreón wyle *the Son will reveal it,* Lk. Bos. 10, 22. Ðū ðâs þing lytlingum awruge *revelasti ea parvulis,* 10, 21. Drihten awreáh rihtwîsnysse hys *Dominus revelavit justitiam suam,* Ps. Spl. 97, 3. Awreóh Drihtne weg ðînne *revela Domino viam tuam,* Ps. Lamb. 36, 5. DER. wreóhan, wreón.

a-wrēstan *to wrest from, extort;* extorquere. v. a-wrǽstan.

a-wreðian; *p.* ede; *pp.* ed; *v. a.* [a, wreðian *to support*] *To support, underprop, sustain;* sustentare :—Agustînus fram Gode awreðed wæs *Augustin was sustained by God,* Bd. 2, 3; S. 505, 1. He, mid his crycce hine awrediende, hām becom *he, with his crutch supporting himself, came home,* Bd. 4, 31; S. 610, 18; Past. 17, 11; Hat. MS. 25 a, 20: Exon. 37 a; Th. 121, 27; Gū. 295.

a-wrigen *revealed,* Lk. Bos. 2, 35; *pp.* of a-wrîhan.

a-wrigenes, -wregennes, -ness, e; *f. A discovery, revelation;* revelatio :—To þeóda awrigenesse *ad revelationem gentium,* Lk. Bos. 2, 32.

a-wrîhan; *p.* -wrāh, *pl.* -wrigon; *pp.* -wrigen [a *not, un-;* wrîhan *to cover*] *To uncover, reveal;* revelare :—Stefn Drihtnes awrîhþ þiccetu *vox Domini revelabit condensa,* Ps. Spl. 28, 8. Awrigene synd grūndweallas [grundfeallas MS.] ymbhwyrftes eorþan *revelata sunt fundamenta orbis terrarum,* 17, 17: Lk. Bos. 2, 35.

a-wringan; *p.* -wrang, *pl.* -wrungon; *pp.* -wrungen *To wring out, to squeeze out, express;* exprimere, Cot. 196. v. wringan.

a-wriðhan, -wrión *to uncover, reveal;* revelare :—Awrióh Drihtne weg ðînne *revela Domino viam tuam,* Ps. Spl. T. 36, 5. v. a-wreón, wreón.

a-wrîtan; *p.* -wrāt, *pl.* -writon; *pp.* -writen; *v. a.* [a, wrîtan *to engrave, write*]. I. *to write out* or *down, to transcribe, describe, compose;* transcribere, describere, conscribere, contexere :—Ðæs hâlgan fæder and biscopes Sancti Cuþberhtes lîf ǽrest eroico metro and æfter fǽce gerǽde worde ic awrât *I wrote out the life of the holy father and bishop, St. Cuthbert, first in heroic metre, and after a space in prose,* Bd. 5, 23; S. 648, 27. Eall þurh endebyrdnesse ic awrât *cuncta per ordinem transcriberē curavi,* 5, 23; S. 648, 11. Nū hæbbe we awriten ðære sûþ *now have we described the south,* Ors. 1, 1; Bos. 17, 42. Leviticus ys genemned *Ministerialis* on Lŷden, ðæt ys þēnungbôc, for ðam ðara sacerda þēnunga sind ðâr âwritene *Leviticus is called in Latin Ministerialis, that is servicebook, because the services of the priests are described therein,* Lev. pref. Ðâm ðæt hâlige gewrit awriten *in quibus scriptura sancta contexta est,* Bd. 5, 23; S. 648, 43. Wēndest ðū ðæt awriten nǽre *thoughtest thou that it was not written,* Cd. 228; Th. 307, 8; Sat. 676: Ps. Th. 138, 14. Sum biþ list-hendig to awrîtanne word-gerŷnu *one is cunning to write down word-mysteries,* Exon. 79 b; Th. 299, 2; Crä. 96. Ðara abbuda stǽr and spell ðysses mynstres on twâm bôcum ic awrât *I wrote a history and narrative of the abbots of this monastery in two books,* Bd. 5, 23; S. 648, 30: 5, 23; S. 649, 11. II. *to inscribe;* inscribere, inscriptione ornare :—Wæs se beám bôcstafum awriten *the beam was inscribed with letters,* Elen. Kmbl. 182; El. 91. III. *to carve, delineate, draw;* sculpere, delineare :—Sindon awritene [MS. awriten] on wealle wuldres þegnas *upon the wall are carved the thanes of glory,* Andr. Kmbl. 1451; An. 726. Hî bǽron anlîcnysse Hǽlendes on brede afǽgde and awritene *they bore the Saviour's likeness figured and drawn on a board;* ferebant imaginem Domini Salvatoris in tabula depictam, Bd. 1, 25; S. 487, 4.

a-wrîðan; *p.* -wrāþ, *pl.* -wriðon; *pp.* -wriðen [a, wrîðan *to wreathe, bind*]. I. *to bind up, bind, wreathe;* alligare, torquere :—Hî me gyrene awriðon [MS. awriðan] *posuerunt mihi laqueos,* Ps. Th. 118, 110. Sylfa his wûnda awrâþ *he bound up his wounds;* sua vulnera ipse alligavit, Bd. 4, 22; S. 590, 36. II. *to unbind, loosen;* solvere :—Ðæt he awrîðe bearn fordôndra *ut solveret filios interemptorum,* Ps. Spl. 101, 21.

a-wruge *revealedst,* Lk. Bos. 10, 21; *p.* of a-wreóhan. v. a-wreón.

a-wrungen *wrung;* *pp.* of a-wringan.

a-wrygen = a-wrigen *discovered;* *pp.* of a-wrîhan.

a-wrygenes = a-wrigenes *a discovery, revealing.* v. a-wrigenes.

āwðer = â-hwæðer; *adj. pron. Either, each, one or other;* alter, alteruter :—Ne uncer âwðer *not either of us;* neuter [*ne-uter*] nostrum, Exon. 129 b; Th. 496, 29; Rä. 85, 22. Ða tungl âwðer [MS. auðer] ôðres rene â ne gehrîneþ, ǽr ðam ðæt ôðer of gewîteþ *the stars never touch each other's course, before the other goes away,* Bt. Met. Fox 29, 19; Met. 29, 10: 20, 84; Met. 20, 42: Bt. 6; Fox 16, 3.

â-wuht [= â-wiht] *Aught, anything; at all, by any means;* aliquid; omnino, ullo modo :—Ne meahte on ðære eorþan âwuht libban *nor might aught live on the earth,* Bt. Met. Fox 20, 214; Met. 20, 107: 11, 18; Met. 11, 9: 18, 14; Met. 18, 7: Cd. 25; Th. 32, 1; Gen. 496. v. â-wiht, nâ-wuht.

awul *an awl;* fuscinula *vel* tridens, Ælfc. Gl. 31; Som. 61, 78; Wrt. Voc. 17, 8. v. al.

a-wunden *bent,* Mk. Bos. 15, 17; *pp.* of a-windan.

a-wundrian; *p.* ade; *pp.* ad *To make a wonder of;* vertere quasi miraculi ad modum :—Eów sceal ðæt leás awundrad weorþan *the falsehood shall be made a wonder of for you,* Invent. Crs. Recd. 1161.

a-wunian; *p.* ode, ade; *pp.* od, ad [a, wunian *to dwell*] *To abide, remain, continue, insist;* manere, permanere, insistere :—Ðeós sibb awunade on Cristes cyrican *hæc pax mansit in ecclesia Christi,* Bd. 1, 8; S. 479, 26. He lēt hit on his bôsme awunian *he let it remain in his bosom,* 3, 2; S. 525, 14. He on hâlgum gebêdum astôd and awunode *he insisted and continued in holy prayers,* 4, 25; S. 599, 31. Hreówe awunian *pœnitentiæ insistere,* 4, 25; S. 600, 11.

a-wunnen *overcome,* Cd. 208; Th. 257, 8; Dan. 654; *pp.* of a-winnan.

a-wurpan *to cast away;* projicere :—Awurp hî fram ðē *projice eam abs te,* Mt. Bos. 5, 30. v. a-weorpan.

a-wurpon *cast off,* Bd. 3, 24; S. 557, 44; *p. pl.* of a-weorpan.

a-wurþan, ic -wurþe, he -wurþeþ, pl. -wurþaþ; p. -wearþ, pl. -wurdon; pp. -worden To cease to be, become insipid or worthless; evanescere:—Ðæt ge awurþaþ [wurþaþ MS.] that ye perish [cease to be], Deut. 4, 26. v. a-weorþan.

a-wurtwarian; p. ude; pp. ud To root up; exterminare:—Awurtwarude hine exterminavit eam, Ps. Spl. M. 79, 14. v. a-wyrt-walian.

a-wygedne, Exon. 74 b; Th. 279, 21, note; Jul. 617; for awyrgedne accursed; pp. of a-wyrgian.

â-wyht [= â-wiht] Aught, anything; at all:—Ne hî for âwyht eorþan cyste ða sêlestan geseón woldan pro nihilo habuerunt terram desiderabilem, Ps. Th. 105, 20: 103, 9: 113, 14.

a-wyllan, -willan, -wellan; p. de; pp. ed; v. trans. To cause to bubble, to boil; facere ut aliquid ferveat vel ebulliat, coquere, decoquere:—Genim awylled hunig take boiled honey, Herb. 1, 20; Lchdm. i. 76, 23. Awylled wîn defrutum, Lye. v. wyllan.

a-wyltan; p. -wyltede, -wylte; pp. -wylted = -wyltd = -wylt; v. a. To roll, roll away, revolve; devolvere, volutare:—Ðæt hîg awylton ðone stân ut devolverent lapidem, Gen. 29, 3. Awylt rolled away, Lk. Bos. 24, 2.

a-wylþ shall bubble up; ebulliet, Ex. 8, 3. v. a-weallan.

a-wyltne rolled away, Lk. Bos. 24, 2; acc. s. m. of a-wylt; pp. of a-wyltan.

a-wyndwian to blow away; ventilare:—We awyndwiaþ [windwiaþ, Lamb.] fŷnd ûre ventilabimus inimicos nostros, Ps. Spl. 43, 7. v. a-windwian.

a-wyrcan; p. -wyrhte; pp. -wyrht To do, effect; facere, agere:—Riht awyrce let him do right, L. H. E. 8; Th. i. 30, 13. Ðæt ðû me gewissige bet ðonne ic awyrhte tó ðê that thou wouldest direct me better than I have done towards thee, Bt. 42; Fox 260, 6. DER. wyrcan.

a-wyrdan, -werdan; p. -wyrde; pp. -wyrded, -wyrd; v. trans. To injure, corrupt, destroy; lædere, corrumpere, vitiare, violare:—Ðe he sylf awyrde whom he himself had injured, Homl. Th. i. 4, 24. Æðeling manig wundum awyrded many a noble injured with wounds, Beo. Th. 2230; B. 1113. Gif spræc awyrd weorþ if speech be injured, L. Ethb. 52; Th. i. 16, 5. Ðŷlæs hî [scil. wæstmas] rênes scûr awyrde lest the shower of rain should destroy them [i. e. the fruits], Exon. 215, 2; Ph. 247. [O. H. Ger. ar-wartian violare, vitiare, fœdare, adulterare, corrumpere, depravare.] DER. wyrdan.

a-wyrdla, an; m. Damage; detrimentum. v. æ-wyrdla, æf-werdla.

a-wyrdnys, -nyss, e; f. Hurt, injury, damage, ruin, destruction; læsio, labes, damnum:—Crist mihte, bûtan awyrdnysse his lima, nyðerasceótan Christ could, without injury of his limbs, cast himself down, Homl. Th. i. 170, 22. Awyrdnyss labes, Ælfc. Gr. 9, 27; Som. 11, 25: 13; Som. 16, 5.

a-wyrgan, -wirgan; p. de; pp. ed To strangle, suffocate, corrupt, injure, violate; strangulare, suffocare, corrumpere, lædere, violare:—He hine sylfne hæfde awirged he had strangled himself, Ors. 6, 36; Bos. 131, 38. Wommum awyrged corrupted with sins, Cd. 169; Th. 211, 26; Exod. 532: Exon. 30 b; Th. 95, 24; Cri. 1562: 105 b; Th. 401, 25; Rä. 21, 17. [Ger. erwürgen strangulare: O. H. Ger. arwurgian id.]

a-wyrgda, an; m. [the def. pp. of a-wyrgian to curse] The cursed, the devil; diabolus, Cd. 220; Th. 284, 3; Sat. 316.

a-wyrged cursed; malignus, maledictus, Mt. Bos. 25, 41. v. a-wyrgian.

a-wyrgedlíc; adj. Wicked, evil; malignus:—Awyrgedlíc geþanc a wicked thought, Nicod. 20; Thw. 10, 11.

a-wyrgednes, a-wyrgednys, a-wirgnis, -niss, e; f. A cursedness, wickedness, a curse, reviling; malignitas, maledictio:—Ðæs mid awyrgednesse [of awyrgednysse, Ps. Spl. C.] mûþ full is cujus maledictione os plenum est, Ps. Lamb. second 9, 7: 13, 3: Deut. 11, 29; Th. Diplm. A. D. 970; 243, 16. DER. wyrgednes.

a-wyrgendlíc; adj. Detestable, abominable; detestabilis, Nathan. 7.

a-wyrgian; p. -wyrgede; pp. -wyrged, -wyrgd To curse, execrate, malign; execrari, maledicere, malignari:—Ðû awyrgedest his cynegyrdum maledixisti sceptris ejus, Cant. Abac. Lamb. 3, 14; Ps. Spl. 73, 4. Nelle ic awirgean ða eorþan nolo maledicere terræ, Gen. 8, 21. The perfect participle signifies execrable, wicked, detestable; execrabilis, maledictus, malignans:—Gewîtaþ nû, awirgede woruldsorga depart now, execrable worldly cares, Bt. 3; Fox 4, 25. Gewîtaþ ge awyrgede fram me on ðæt êce fŷr discedite a me maledicti in ignem æternum, Mt. Bos. 25, 41: Exon. 30 a; Th. 93, 2; Cri. 1520. Of ðam awyrgedan wrâðan sweorde de gladio maligno, Ps. Th. 143, 11. Seó gegaderung ðara awyrgdra consilium malignantium, 21, 14. The devil is called Se awyrgda the accursed, Cd. 220; Th. 284, 3; Sat. 316. Se awyrgeda gâst the accursed spirit, Guthl. 7; Gdwin. 44, 12. Se awyrgda wulf the accursed wolf, Exon. 11 b; Th. 16, 20; Cri. 256. v. a-wergian.

âwyrn; adv. Before? antea, olim? Fox; Manning says,—perhaps for âhwær, anywhere, in any place; alicubi:—Ne hŷrde ic guman âwyrn [gumena fyrn, Grn.] ænigne ær æfre bringan sêlran lâre I have not heard before any other man ever bring better lore, Menol. Fox 200.

a-wyrpan; p. -wearp, pl. -wurpon; pp. -worpen To cast away, cast out, reject, take away; projicere, repellere, auferre:—Tó awyrpanne ût auferant, Ps. Th. 39, 16. Ahola hit ût, and awyrp hit fram ðê erue eum [oculum], et projice abs te, Mt. Jun. 5, 29; Ps. Th. 50, 12; Ps. Grn. ii. 149, 50, 12. v. a-weorþan.

a-wyrþ loses its strength, becomes insipid, Mt. Bos. 5, 13. v. a-weorþan.

a-wyrþian? [a intensive, wyrþian to glorify] To give honour to, to glorify; glorificare, Cant. Moys. Lye. v. weorþian.

a-wyrt-walian; p. ode; pp. od; v. a. [a out, wyrtwalian to root, to fix roots] To root up, eradicate, extirpate, exterminate; eradicare, supplantare:—Ælc plantung byþ awyrtwalod omnis plantatio eradicabitur, Mt. Jun. 15, 13. Ðelæs ge ðone hwæte awyrtwalion ne forte eradicetis triticum, 13, 29: Lk. Bos. 17, 6: Bt. Met. Fox 12, 51; Met. 12, 26: Ps. Th. 36, 9. Awyrtwala hine supplanta eum, Ps. Spl. 16, 14.

a-wystelan, a-wystlan to hiss, lisp, whistle; sibilare. v. hwistlan.

Axa-mûþa, an; m. Exmouth, Chr. 1049; Th. 307, 37. v. Exan mûþa.

axan = oxan oxen; boves:—Sceáp and axan oves et boves, Ps. Spl. 8, 7. v. oxa.

axan ashes, Lev. 1, 16. v. axe.

Axan minster Axminster, Devon, Lye. v. Acsan mynster.

ax-baken; part. Baked in ashes; subcinericius, Gr. Dial. 1, 11.

axe an axe, Mt. Rush. Stv. 3, 10. v. acas, acase.

axe, an; f. Ash, ashes; cinis:—Swâ swâ dust oððe axe as dust or ashes, Bt. 33, 4; Fox 130, 9: Bt. Met. Fox 20, 211; Met. 20, 106. On ðære stôwe ðe man ða axan gît in loco in quo cineres effundi solent, Lev. 1, 16. Bearwas wurdon tó axan and tó ŷslan the groves became ashes and embers, Cd. 119; Th. 154, 9; Gen. 2553. v. asce.

axian, axigan, axigean; p. ode; pp. od To ask; interrogare:—He axode he asked, Ors. 2, 5; Bos. 46, 43. Ic axige me rædes consulo, Ælfc. Gr. 28, 3; Som. 31, 2. Ic axige percunctor [= percontor], 25; Som. 27, 6: Mt. Foxe 22, 46. v. acsian, ascian.

axiendlíc, axigendlíc; adj. Interrogative, inquiring, inquisitive; interrogativus:—Gif ic cwede, hwâ dyde ðis? quis hoc fecit? ðon biþ se [hwâ quis] interrogativum, ðæt is axigendlíc, Ælfc. Gr. 18; Som. 21, 27.

axigean to ask; interrogare:—Ne nân ne dorste hyne axigean neque ausus fuit quisquam eum interrogare, Mt. Foxe 22, 46. v. axian.

axode asked, Ors. 2, 5; Bos. 46, 43; p. of axian.

axse, an; f. Ashes; cinis:—On axsan gehwyrfeþ in cinerem convertit, Bd. 4, 25; S. 600, 34. v. asce.

axung inquiry, Scint. 16. v. acsung.

a-ŷdlian; p. ode; pp. od To make useless, Ps. Lamb. 38, 12. v. a-îdlian.

a-ŷdlig; adj. Void, empty, idle, vain; vacuus, irritus, vanus. v. îdel.

a-yrnan, he -yrnþ; p. -arn, pl. -urnon; pp. -urnen [a out, yrnan to run] To run over, to pass or go over, pass, go; præterire, decurrere:—To nâhte hîg becumaþ swâ swâ a-yrnende wæter ad nihilum deveniunt tamquam aqua decurrens, Ps. Lamb. 57, 8. Swâ neáh wæs þûsend wintra a-urnen so near was a thousand winters gone, Chr. 973; Th. 226, 5, col. 1; Edg. 16: Cd. 79; Th. 98, 6; Gen. 1626. A-urnenre tîde in or at a declining time, the time being far spent or gone. A-urnen biþ is run out, passed, Som.

a-ŷtan; p. -ŷtte; pp. -ŷted [a from, ŷtan = ûtian to out] To expel, drive out; expellere:—He ðâ a-ŷtte ða Swegen ût he then drove Sweyn out, Chr. 1047; Th. 304, 4, col. 2. DER. ŷtan, ûtian.

azima, orum; pl. n. Lat. Unleavened; infermentata, azŷma' [= τὰ ἄζυμα, ἀ without, ζύμη fermentation]:—Freólsdæg azîmorum, se is gecweden eástre dies festus azŷmorum, qui dicitur pascha; ἡ ἑορτὴ τῶν ἀζύμων, ἡ λεγομένη πάσχα, Lk. Bos. 22, 1. Se dæg azîmorum dies azŷmorum; ἡ ἡμέρα τῶν ἀζύμων, Lk. Bos. 22, 7.

B

The sound of b is produced by the lips; hence it is called a labial consonant, and has the same sound in Anglo-Saxon as in English. In all languages, and especially in the dialects of cognate languages, the letters employing the same organs of utterance are continually interchanged. In Anglo-Saxon, therefore, we find that b interchanges with the other labials, f and p:—Ic hæbbe I have, he hæfþ he hath. When words are transferred into modern English, b is sometimes represented by f or v:—Beber or befor a beaver; Ober, ofer, over. 2. In comparing the Anglo-Saxon aspirated labial f with the corresponding letter in Old Saxon, the sister dialect, we find that the Old Saxons used a softer aspirated labial ð = bh. This softer aspirated ð generally occurs as a medial letter between two vowels; as,—

O. Sax.		A. Sax.		Eng.
graƀan	=	grafan	=	engrave
klioƀan	=	cleófan	=	cleave
geƀan	=	gifan	=	give

3. The Runic letter ᛒ not only stands for the letter B, b, but also for the name of the letter in Anglo-Saxon beorc the birch-tree. v. beorc.

bâ, bû both; nom. f. n. acc. m. f. n. of begen:—Ða idesa bâ both the women, Judth. 11; Thw. 23, 22; Jud. 133. Wæter and eorþe, sint on gecynde cealda bâ twâ water and earth, both the two are by nature cold,

Fox 20, 152; Met. 20, 76. Bysmæredon uncit [*Inscription* Bismærede ungket] men, bâ ætgædere *they* [*men*] *reviled us two, both together*, Runic Inscrip. Kmbl. 354, 30.

baan, es; *n. A bone*:—Ne tobræcan ða baan *they broke not the bones*, Homl. Daye 55, 17; Th. *has*, Ne tobræcon ða bân, Homl. ii. 280, 9. v. bân.

Babilôn, e; *f*: Babilônie, Babilônige, an; *f*: Babilôn, Babylôn, es; *f*. [v. wim-man, es; *f*.] *Babylon*; Babylôn, ōnis; *f*. This celebrated city of antiquity, in Mesopotamia, was built on both banks of the Euphrates. Its foundation by Nimrod is mentioned immediately after the Deluge, Gen. 10, 9, 10: 11, 9:—Nimrod [MS. Membrað], se ent, ongan ǽrest timbrian Babilônia; and Ninus, se cyning æfter him, and Sameramis, his cwên, hî ge-endade æfter him, on middeweardum hire rîce. Seó burh wæs getimbred on fildum lande, and on swîðe emnum. And heó wæs swîðe fæger on to lôcianne, and heó is swîðe rihte feówerscŷte. And ðæs wealles mycelnyss and fæstnyss, is ungelŷfedlîc to secgenne: ðæt he is l elna brâd, and ii hund elna heáh, and his ymbgang is hund seofantig mîla, and seofeþan dǽl ânre mîle . . . Seó ylce burh Babylônia, seó ðe mǽst wæs, and ǽrest ealra burga, seó is nû læst and wēstast *Nimrod, the giant, first began to build Babylon; and, after him, king Ninus, and then Semiramis, his queen, finished it in the middle of her reign. The city was built on open and very level land. It was very fair to look upon, and it is quite a true square. The greatness and firmness of the wall, when stated, is hardly to be believed. It is fifty ells broad, and two hundred ells high, and its circumference is seventy miles, and the seventh part of a mile* . . . *This very city of the Babylonians, which was the greatest and first of all cities, is now the least and most desolate*, Ors. 2, 4; Bos. 44, 17–31. Babilôn wæs mǽrost burga *Babylon was the greatest of cities*, Cd. 209; Th. 259, 19; Dan. 694. Babilône weard *the guardian of Babylon*, 177; Th. 222, 14; Dan. 104: 178; Th. 223, 9; Dan. 117. Þurh Babilônian burh *through the city of Babylon*, Ors. 2, 4; Bos. 44, 11. Babilônes brego *the ruler of Babylon*, Cd. 174; Th. 218, 30; Dan. 47. Se wæs Babylônes brego *he was the ruler of Babylon*, 79; Th. 98, 20; Gen. 1633. Ofer flôdas Babilônes *super flumina Babylonis*, Ps. Surt. 136, 1: Ps. Spl. 136, 1. Dôhtor Babylônes earm *filia Babylonis misera*, Ps. Surt. 136, 8: Ps. Spl. 136, 11. In Babilône *in Babylon*, Cd. 82; Th. 102, 28; Gen. 1707. On ðære þeóde, ðe swâ hâtte bresne Babilônige *in the country, that was so called powerful Babylon*, 180; Th. 226, 18; Dan. 173. [*Heb.* בָּבֶל *bâbĕl the city of Belus: Grk.* Βαβυλών, ῶνος; *f: Lat.* Babýlôn, ōnis; *f*.]

Babilônia *Babylon, acc. Grk*, Ors. 2, 4; Bos. 44, 17. v. Babilôn.

Babilônie, an; *f. Babylon*, Ors. 2, 4; Bos. 44, 11. v. Babilôn.

Babilônige *Babylon*, Cd. 180; Th. 226, 18; Dan. 173. v. Babilôn.

Babilônis *of Babylon, gen. Lat.* Ps. Th. 86, 2. v. Babilôn.

Babilônisc; *def.* se Babilônisca, seó, ðæt Babilônisce; *adj. Babylonish*; Babylônÿcus:—Dôhtor se Babilônisce wrǽcce [MS. babilonisca wrǽcca] *filia Babilonis misera*, Ps. Lamb. 136, 8.

Babilônisca, an; *m. Babylon*; Babýlôn, ōnis; *f*:—Ofer flôd Babilôniscan *super flumina Babilonis*, Ps. Lamb. 136, 1. DER. Babilônisc.

Babylôn *Babylon*, Cd. 79; Th. 98, 20; Gen. 1633. v. Babilôn.

baca *of backs*; *gen. pl. of* bæc.

BACAN; ic bace, ðû bacest, bæcest, bæcst, becest, becst, he baceþ, bæceþ, beceþ; *pl.* bacaþ; *p.* ic bôc, ðû bôce, ðû bôcon; *pp.* bôcon; *v. a. To* BAKE; torrere, pinsere, coquere:—Fîf bacaþ on ânum ofene *quinque in uno clibano coquant*, Lev. 26, 26. Hî bôcon melu *coxerunt farinam*, Ex. 12, 39. [*Orm.* bakenn: *Chauc.* bake: *Wyc.* bake; *p.* boke; *pp.* bakun: *Scot.* baike *to bake*; *pp.* baiken; bakster *a baker*: *O. Sax.* bakan: *N. Frs.* backe: *Dut.* bakken: *Ger.* backen: *M. Ger.* bachen: *O. H. Ger.* pachan; *p.* puoch; *pp.* pachanēr: *Dan.* bage: *Swed. O. Nrs.* baka *to roast*: *Sansk.* bhak-tas *cooked, from* bhaj *to cook*.] DER. bæcere, bæcestre: bacen, niw-, ofen-.

bacen *baked*; *pp. of* bacan.

bac-slitol, es; *m. A backbiter*; detractor, Off. reg. 15. v. bæc-slitol.

bacu *backs*; *nom. acc. pl. of* bæc:—Hî me towendon heora bacu *they turned their backs on me*, Bt. Met. Fox 2, 29; Met. 2, 15.

bâd, e; *f.* [from bǽdan *compellere*] *A pledge, stake, a thing distrained*; pignus:—Gif bâd genumen sŷ, ðonne begyte ða bâde hâm *if a pledge be taken, then shall he obtain the pledge home again, or back*, L. O. D. 3; Th. i. 354, 6, 7. DER. bâdian; nêd-bâd, nŷd. v. wed, wedd.

bâd *expected, waited*, Cd. 132; Th. 167, 32; Gen. 2774; *p. of* bîdan.

Baddan-burh, -burge; *d.* -byrig; *f.* BADBURY, *Dorsetshire, formerly Baddanburgum*; Baddanburgus in quo castra metatus est Eadweardus Ælfredi fil, An. 901; haud longe a Winburna, in agro Dorsetensi:—He gewîcode æt Baddanbyrig wið Winburnan *he encamped at Badbury near Winburn*, Chr. 901; Th. 178, 26.

Badecan wylle, an; *f.* [*Badec's well: Flor.* A. D. 1114, Badecanwella] BAKEWELL, *Derbyshire*:—Fôr on Peac-lond to Badecan wyllan [MS. wiellon] *went into the Peak to Bakewell*, Chr. 924; Erl. 110, 12.

bâdian; *p.* ode; *pp.* od; *v. a. To pledge, seize, take by way of a pledge*; pignerare, pignus auferre:—Of ǽgðran stæde on ôðer man môt

bâdian, bûte man elles riht begytan mǽge *from one shore to the other one may take a pledge, unless he can get justice in another way*, L. O. D. 2; Th. i. 354, 3.

Bæbba-burh *Bamborough*, Chr. 1093; Th. 360, 6: 1095; Th. 362, 12. v. Bæbban burh.

Bæbban burh, Chr. 993; Th. 241, 17, col. 1. v. Bebban burh.

BÆC; *g.* bæces; *pl. nom. acc.* bacu, bæc; *g.* baca; *d.* bacum; *n. A* BACK, dorsum, tergum [dorsum *is opposed to* venter, *especially in animals, and* tergum *to* frons, v. hricg]:—Mînra feónda bæc ðû onwendest to me *inimicorum meorum dedisti mihi dorsum*, Ps. Th. 17, 38. Fŷnd mîne ðû sealdest me on bæc *vel* hricc *inimicos meos dedisti mihi dorsum*, Ps. Spl. 17, 42; myn enemys thou ʒeue to me bac, Wyc. 17, 41. Ða wendon hî me heora bæc *to then turned they their backs to me*, Bt. 2; Fox 4, 13. Hî me towendon heora bacu *they turned their backs on me*, Bt. Met. Fox 2, 29; Met. 2, 15. Ǽr hî bacum tobreden *before they turn their backs to each other*, Exon. 92 a; Th. 345, 20; Gn. Ex. 192. ¶ On bæc *retro*, Jn. Bos. 6, 66: *and* under bæc *retrorsum*, Ps. Spl. 43, 12: *at his back, behind, backward*, v. under-bæc. Clǽne bæc hæbban *to have a clean back, to be free from deceit*, L. A. G. 5; Th. i. 156, 6. Gang on bæc, Mt. Bos. 4, 10. Gâ on bæc *go behind* or *away*; vade retro, Mk. Bos. 8, 33. [*Orm.* bac, bacch: *Chauc.* back: *O. Sax.* bak, *n*: *N. Frs.* beck, *n*: *O. Frs.* bek, *n*: *O. Ger.* pacho, bacho, *m*: *O. Nrs.* bak, *n*: *Scot.* back *a body of followers*. Is it allied to the root in bîgan *to bow*, as the *N. Ger.* buckel *dorsum* is to biegen?] DER. ofer-bæc, on-, under-.

bæc-bord, es; *m. The* larboard *or left-hand side of a ship, when looking towards the prow or head*; navigii sinistra pars:—Burgenda land wæs us on bæcbord *the land of the Burgundians was on our larboard* or *left*, Ors. 1, 1; Bos. 21, 44. [*Plat. Dut.* bakboord *the larboard*.]

bêce *a beech-tree*, Som. Lye. v. bêce.

bæcere, es; *m. A* BAKER; pistor, Ælfc. Gl. 50; Som. 65, 109; Wrt. Voc. 34, 38. [*Plat. Dut.* bakker: *Ger.* bäcker: *Dan. Swed.* bagere: *O. Nrs.* bakari.] v. bacan.

bæce-ring, es; *m. A grate formed as a ring used for baking, a gridiron*; craticula, Cot. 99.

bæc-ern, es; *n.* [bæc from bacan *to bake*, ern *a place*] *A baking-place, a bakehouse*; pistrinum, Ælfc. Gl. 50; Som. 65, 110; Wrt. Voc. 34, 39.

bæcest *bakest*, = bacest, *2nd sing. pres. of* bacan.

bæcestre, bæcistre, bæcystre, an; *f? m.* [bacan *to bake*, heó bæc-eþ; estre, v. -isse] *A woman who bakes*; pistrix: but because afŷrde men performed that work which was originally done by females, this occupation is here denoted by a feminine termination; hence, *a baker*; pistor:—Ða gelamp hit ðæt twegen afŷryde men agylton wið heora hlâford, Egypta cynges byrle and his bæcistre *ecce accidit ut peccarent duo eunuchi, pincerna regis Ægyptorum, et pistor, domino suo*, Gen. 40, 1. Ðara ôðer bewiste his byrlas, ôðer his bæcestran *illorum alter pincernis præerat, alter pistoribus*, 40, 2. Bæcistra ealdor *pistorum magister*, 40, 16, 20. Bæcestre *a baker*; pistor, Ælfc. Gr. 28, 1; Som. 30, 36.

bæceþ *baketh*, = baceþ, *3rd sing. pres. of* bacan.

bæc-hûs, es; *n. A* BAKEHOUSE; pistrinum, Ælfc. Gl. 22? v. bæc-ern.

bæcling; *adv.* Only used with on, *On the back, backwards, behind*; retrorsum:—On bæcling *retrorsum*, Ps. Th. 113, 5. On bæclincg, 43, 12, 19. Cer ðe on bæcling *turn thee behind me*, Cd. 228; Th. 308, 26; Sat. 698. v. ears-ling, hinder-ling.

bæc-slitol, es; *m.* [bæc *a back*; slitol *a biter*, from sliten, *pp. of* slîtan *to slit, bite*] *A backbiter*; detractor, Off. reg. 15.

bæcst *bakest*; bæcþ *bakes*. v. bacan.

bæc-þearm, es; *m. The entrails*; anus, longanon:—Wrt. Voc. 283, 60. Bæcþearmas ðe bowels; extales, Ælfc. Gr. 13; Som. 16, 23. Bæcþearm *vel* snædel *extales*, Ælfc. Gl. 74; Som. 71, 66; Wrt. Voc. 44, 48. Bæcþearmes ûtgang *morbus, fortasse, ani procidentia*, Som. v. snædel.

bæcystre *a baker*; pistor:—Bæcystra ealdor *pistorum magister*, Gen. 41, 10. v. bæcestre.

bæd, *pl.* bǽdon *asked, besought*, Cd. 94; Th. 122, 12; Gen. 2025: 37; Th. 48, 24; Gen. 780; *p. of* biddan.

Bæda-ford-scír *Bedfordshire*, Chr. 1011; Th. 267, 4, col. 2. v. Bedan ford-scír.

bǽdan; *p.* de; *pp.* ed *To constrain, compel, require, solicit*; cogere, compellere, exigere, postulare, flagitare:—Ðæs his lufu bǽdeþ *whom his love constrains*, Exon. 90 b; Th. 339, 27; Gn. Ex. 100. Mǽru cwên bǽdde byras geonge *the illustrious queen solicited her young sons*, Beo. Th. 4040; B. 2018. [*O. Sax.* bêdian *cogere aliquem ad aliquid*: *O. H. Ger.* ga-peitian: *Goth.* báidjan: *O. Nrs.* beiða *petere, postulare*.] DER. a-bǽdan, ge-.

bædd *a bed*, Vit. Swith. v. bed.

bêdde, an; *f? A thing required, tribute*; exactum, Cot. 73.

bêdde *solicited*, Beo. Th. 4040; B. 2018; *p. of* bǽdan.

bæddel, es; *m. A hermaphrodite*; hermaphroditus:—Wǽpen-wîfestre *vel* scritta *vel* bæddel *hermaphroditus*, Ælfc. Gl. 76; Som. 71, 125; Wrt. Voc. 45, 28. v. wǽpen-wîfestre, scritta.

bædd-ryda, an; *m. One bedridden*; clinicus, Vit. Swith. v. bedreda.

F

bǽdel *a beadle*, Som. Lye. v. býdel.

bǽdend, es; *m. A vehement* or *earnest persuader, a solicitor, stirrer;* impulsor, Cot. 115.

bǽde-wēg, -wīg, es; *n. A cup;* poculum :—Heó scencte bittor bǽdewēg *she poured out the bitter cup*, Exon. 47 a; Th. 161, 13; Gū. 958.

bǽdling, es; *m.* [bedd *a bed*] *A delicate fellow, tenderling, one who lies much in bed;* homo delicatus :—Bǽdlingas *effeminate men;* μαλακοί, Cot. 71: 1 Cor. 6, 9.

bǽdling, es; *m.* [from bǽdan *to compel, solicit*] *A carrier of letters* or *orders;* tabellarius, Som.

bǽd-þearm, es; *m. Mentera, entera?* = ἔντερα, *pl. n. exentera?* Bǽdþearm *seems to be an error of the copyist for* bæcþearm, Ælfc. Gl. 76; Som. 71, 122; Wrt. Voc. 45, 27.

bǽdzere, bæzere, es; *m:* bezera, an; *m. A baptist, baptizer;* baptista :—Hie cwǽdun, sume Iohannes se bǽdzere *illi dixerunt, alii Ioannem Baptistam*, Mt. Rush. Stv. 16, 14: 3, 1. v. fulluhtere.

bǽfta, an; *m. The after part, the back;* tergum :—Ic geseah ðone bæftan *I saw the back*, Gen. 16, 13.

bǽfta; *adv. Behind;* post, Gen. 32, 24. v. bæftan; *adv.*

bæftan, beftan; *prep. dat.* [be-æftan, *q. v.*] I. *after, behind;* post, pone :—Gang bæftan me *vade post me*, Mt. Bos. 16, 23. II. *behind, without;* sine :—Bæftan ðam hláforde *without the master*, Ex. 22, 14.

bæftan, bæfta; *adv.* [be-æftan, *q. v.*] *After, behind, hereafter, afterwards;* postea :—Git synd fíf hungor gēr bæftan *adhuc quinque anni residui sunt famis*, Gen. 45, 11. He āna belāf ðǽr bæfta *he alone was left there behind*, Gen. 32, 24. Mycel ðæs heres ðe mid hyre bæftan wæs *much of the army that was behind with her*, Ors. 1, 10; Bos. 33, 23.

bæftan-sittende; *part. Idle;* reses, Ælfc. Gr. 9, 26; Som. 11, 11.

bēg *a collar* :—Wearm lim gebundenne bēg hwílum bersteþ *the warm limb sometimes escapes from the bound collar*, Exon. 102 b; Th. 387, 20; Rä. 5, 8. v. beáh.

bǽga *of both*, Th. Diplm. A. D. 804–829; 462, 17. v. begen.

Bǽgere, Bǽgware; *gen.* a; *dat.* um; *pl. m. The Bavarians;* Bavarii, the Boiari, *or* Bajuvarii, *whose country was called* Boiaria, *its German name is* Baiern, *now called the kingdom of Bavaria* :—Mid Bǽgerum *with the Bavarians*, Chr. 891; Th. 160, 24. Hī Maroaro habbaþ, be westan him, Þyringas, and Behemas, and Bǽgware healfe *they, the Moravians, have, on their west, the Thuringians, Bohemians, and part of the Bavarians*, Ors. 1, 1; Bos. 18, 42.

bēh *a crown*, Ælfǽdæ Test. v. beáh.

BǼL, es; *n.* I. *fire, flame;* ignis, flamma :—Hǽfde landwara līge befangen, bǽle and bronde *he had enveloped the inhabitants of the land with flame, with fire and brand*, Beo. Th. 4633; B. 2322: 4606; B. 2308. Bǽles cwealm in helle *the torment of the fire in hell*, Andr. Kmbl. 2374; An. 1188. II. *the fire of a funeral pile, in which dead bodies were burned, a funeral pile;* rogus, pyra :—Ǽr he bǽl cure ere he chose the pile *[the fire of the pile]*, Beo. Th. 5629; B. 2818. Bǽl biþ onǽled *the pile is kindled*, Exon. 59 a; Th. 212, 26; Ph. 216. [*Piers.* bal : *O. Nrs.* bāl, *n. a fire, funeral pile.*]

bǽl-blǽse, an; *f. Blaze of a flame;* flammæ candor *vel* ardor, Exon. 42 b; Th. 142, 22; Gū. 648.

bǽl-blys, e; *f. Blaze of a fire;* flammæ ardor, Cd. 184; Th. 230, 12; Dan. 232: 162; Th. 203, 9; Exod. 401.

bǽlc, es; *m.* I. *a* BELCH; eructatio, Mann. II. *the stomach, pride, arrogance;* stomachus, superbia, arrogantia :—He him bǽlc forbígde *he bent their pride*, Cd. 4; Th. 4, 15; Gen. 54: Judth. 12; Thw. 25, 18; Jud. 267.

BǼLC, es; *m. A covering;* tegmen, peristroma, tabulatum :—He bǽlce oferbrǽdde byrnendne heofon *he overspread with a covering the burning heaven*, Cd. 146; Th. 182, 9; Exod. 73. [*N. Ger.* gebälk, es; *n. the beams* or *timber of a house* : *Icel.* bálkr.]

bǽlcan *to cry out;* vociferari :—He bǽlceþ *he cries out*, Exon. 83 b; Th. 315, 8; Mód. 28. [*Plat.* bölken : *N. Frs.* balckien : *N. Dut.* balken : *Ger.* bolken.]

bǽldan *to animate, encourage;* animare, instigare :—Ðū þeóde bǽldest to beadowe *thou encouragest the people to strife*, Andr. Kmbl. 2373; An. 1188. v. byldan.

bǽldu, e; *f. Confidence;* fiducia, Mt. Rush. Stv. 14, 27.

bǽl-egsa, an; *m. Terror of flame?* flammæ terror?—Bǽlegsan [bell egsan MS.] hweóp *he threatened with terror of flame*, Cd. 148; Th. 185, 12; Exod. 121.

bǽl-fýr, es; *n. A funeral fire;* rogi ignis :—Bǽlfýra mǽst *greatest of funeral fires*, Beo. Th. 6278; B. 3143 : Exon. 74 a; Th. 277, 12; Jul. 579.

bǽlg, bǽlig, es; *m. A bulge, bag;* bulga, Cot. 27. v. belg.

bǽlig-nis, -niss, e; *f.* [from belgan *to be angry, to make angry*] *An injury;* injuria, Mt. Lind. Stv. 20, 13.

bǽl-stede; es; *m. A funeral pile place;* rogi locus, Beo. Th. 6185; B. 3097.

bǽl-þræc; *g.* -þræce; *pl. nom. g. acc.* -þraca; *f. Force of fire;* flammæ impetus :—Æfter bǽlþræce *after the fire's force*, Exon. 59 b; Th. 216, 19; Ph. 270.

bǽl-wudu, es; *m. Wood of the funeral pile;* rogi lignum, Beo. Th. 6216; B. 3112.

bǽl-wylm, es; *m. Fire's heat;* flammæ æstuatio, Exon. 70 b; Th. 262, 22; Jul. 336.

bǽm *for* bām; *dat. of* begen *both*, Bt. 38, 5, MS. Cott; Fox 206, 15 : Th. Diplm. A. D. 804–829; 463, 3. v. begen.

bǽnc *a bench*, Som. Lye. v. benc.

bǽnd, es; *m. A band;* vitta :—Healfne bǽnd gyldenne [*dederunt*] dimidiam vittam auream, Text. Rof. 111, 3; Th. Diplm. A. D. 950; 501, 35 : Text. Rof. 110, 23; Th. Diplm. A. D. 950; 501, 20. v. bend.

Bǽnesing-tūn *Bensington*, Chr. 571; Th. 32, 29, col. 1. v. Bensingtūn.

BǼR; *g. m. n.* bares; *f.* bǽrre : *d.* barum : *acc.* bǽrne : *pl. nom.* baru : *acc.* bare; *dat.* barum; *def.* se bara; seó, ðæt bare; *adj.* BARE, *naked, open;* nudus :—On bǽr líc *on the bare body*, Exon. 125 a; Th. 482, 7; Rä. 66, 4. On barum sondum *on bare sands*, Bt. 34, 10; Fox 148, 24. Wit hēr baru standaþ unwered wǽdo *we stand here naked, unprotected by garments*, Cd. 38; Th. 50, 20; Gen. 811. [*Plat. Dut. Ger.* baar *nudus, promptus, merus, manifestus* : *M. H. Ger.* bar *nudus* : *O. H. Ger.* par, bar : the *Goth.* form is not found, but would be basis or basus : *Dan. Swed.* bar : *O. Nrs.* berr : *Slav.* bos : *Lith.* bosus; then the radical consonants would be b–s, not b–r; therefore the word is not connected with beran *ferre.* v. Grm. Wrtbch. i. 1055.] v. berie.

bǽr, *pl.* bǽron *bore*, Cd. 24; Th. 31, 2; Gen. 479 : 178; Th. 223, 18; Dan. 121; *p. of* beran.

bǽr, e; *f.* I. *a* BIER; feretrum :—Sîe seó bǽr gearo *let the bier be ready*, Beo. Th. 6202; B. 3105. Gefǽrenne man brohton on bǽre *they brought a dead man on a bier*, Elen. Kmbl. 1742; El. 873. II. *a couch, pallet, litter;* grabatus :—On his þegna handum on bǽre boren wæs *manibus ministrorum portabatur in grabato*, Bd. 5, 19; S. 640, 22. [*Chauc. Wyc.* bere : *Plat.* baar, *f.* : *O. Sax.* bāra, *f.* : *O. Frs.* bēre, *f.* : *Dut.* baar, *f.* : *Ger.* bahre, *f.* : *M. H. Ger.* bâre, *f.* : *O. H. Ger.* bāra, *f.* : *Dan.* baar, *f.* : *Icel.* bör, *f.* : *Lind. Rush.* bēr, beer, Lind. Rush. DER. ge-bǽran.

bǽran; *p.* de; *pp.* ed *To bear, bear oneself;* ferre, transferre :—He ne geþafode, ðæt ǽnig man ǽnig fæt þurh ðæt templ bǽre, Mk. Bos. 11, 16; *he suffride not, that ony man schulde bere a vessel thurʒ the temple*, Wyc. DER. ge-bǽran.

bǽr-beáh; *g.* -beáges; *m. A bearing-ring, ring;* anulus, Exon. 108 b; Th. 414, 18; Rä. 32, 22.

bǽr-disc, es; *m.* [bǽr, disc *a dish*] *A dish bier* or *tray, a frame on which several dishes were brought to table at once, a course, service;* ferculum, Wrt. Voc. 26, 64.

bǽre *a bier;* feretrum, Wrt. Voc. 49, 26. v. bǽr.

-bǽre *an adjective termination* signifying *Producing, bearing*, from beran *to bear, produce;* as, wæstm-bǽre *fruit-bearing, fruitful;* frugifer : æppel-bǽre *apple-bearing;* pomifer : horn-bǽre *horn-bearing;* corniger : leóht-bǽre *light-bearing.* [*Plat. Dut.* -baar : *Ger.* -bar : *M. H. Ger.* -bǽre : *O. H. Ger.* -pâri.] v. bora.

bǽre-flór, es; *m. A barley-floor, barn-floor, threshing-floor;* hordei area, area :—Þurh-clǽnsaþ his bǽreflór *permundabit aream suam*, Mt. Kmbl. Rush. 3, 12. v. bere.

bǽrende *bearing;* *part.* of bǽran. v. berende.

bǽr-fót; *adj.* BAREFOOT or *that goeth barefooted;* nudipes, Peccat. Med. 8. [*Ger.* barfusz.]

bǽrlíc, es; *m? Barley;* hordeum :—Man sǽlde ðæt æcer-sǽd bǽrlíc to six scillingas *one sold the acre-seed of barley for six shillings*, Chr. 1124; Th. 376, 5. v. bere.

bǽr-líce; *adv. Openly, nakedly*, BARELY; palam, Jn. Lind. War. 6, 29.

bǽrm *a bosom, lap;* sinus, Som. Lye. v. bearm.

bǽr-man, -mann, es; *nom. pl.* bǽrmenn; *d.* bǽrmannum; *m. A man who bears, a bearer, carrier, porter;* bajulus :—Ða bǽrmenn gesetton heora fótlǽst *the porters set their footstep*, Jos. 3, 15.

bǽrn *a barn*, Wrt. Voc. 84, 55. v. bern.

bǽrnan; *p.* bǽrnde; *pp.* bǽrned; *v. a. To kindle, light, set on fire, to* BURN, *burn up;* accendere, urere, comburere, exurere :—Bǽrnaþ nū eówer blǽcern *light now your lamp*, Bd. 4, 8; S. 576, 5. Hī bǽrndon gecorene *they burned the chosen*, Exon. 66 a; Th. 243, 26; Jul. 16. [*Plat.* brennen; *p.* brende *ardere, urere* : *Dut.* branden; *p.* brande *id* : *O. Dut.* bernen; *p.* bernde; branden; *p.* brande *id* : *Ger.* brennen; *p.* brannte; *but* brinnan; *p.* brann *ardere* : *M. H. Ger.* brennen; *p.* brante *urere* : *O. H. Ger.* brennan; *p.* branta; prennan; *p.* pranta *id* : *O. Sax.* brinnan, brennan : *Goth.* brannjan; *p.* brannida : *Dan.* brände *ardere, urere* : *Swed.* bränna *urere* : *O. Nrs.* brenna; *p.* brendi *id.*] DER. forbǽrnan, ge-, on-. v. byrnan, beornan.

bǽrnes, bǽrnis, -ness, e; *f. A burning;* incendium, Bd. 1, 6; S. 476, 25. DER. an-bǽrnis, -bǽrnys, in-, on-.

bǽrnet, bǽrnyt, bernet, es; *n.* I. *a combustion, burning up;* combustio :—He wudu gelogode to his sunu bǽrnytte *he laid in order the wood for the burning of his son*, Gen. 22, 9. II. *arson;* incendium :—Hūsbryce and bǽrnet . . . is bótleás *bootless is . . . house-breaking and arson*, L. C. S. 65; Th. i. 410, 5. DER. wudu-bǽrnet.

bærning, berning, e; *f. A* BURNING; adustio :—Sylle bærninge wið bærninge *reddat adustionem pro adustione,* Ex. 21, 25.

bærnyt *a combustion, burning,* Gen. 22, 9. v. bærnet.

-bǽro, -bǽru *a bearing.* v. forþ-, ge-, on-.

bærs, bears, es; *m. A perch;* perca, lupus :—Bærs *lupus vel scardo,* Ælfc. Gl. 101; Som. 77, 58; Wrt. Voc. 55, 63. [*Dut.* baars, *m:* *Ger.* bars, barsch, *m.*]

bærst *burst,* Byrht. Th. 140, 6; By. 284; *p. of* berstan.

bærstlian; *p.* ode; *pp.* od *To break, burst;* crepare :—Bærstlaþ *crepuerit,* Cot. 39. v. brastlian.

bær-synnig, -sinnig, -suinnih, -sunig; *adj.* [bær *bare, open;* synnig *sinful, wicked*] *Openly-wicked;* used substantively, *an open or public sinner, a publican;* apertus *vel* publicus peccator, publicanus :—Sîe ðê swǽ bærsynnig *sit tibi sicut publicanus,* Mt. Lind. Stv. 18, 17: 21, 32: Mk. Lind. War. 2, 16: Lk. Lind. War. 15, 1: Mt. Lind. Stv. 9, 10. [*O. Nrs.* ber-syndugr.]

bǽrwe *a grove,* Som; *dat. of* bearo.

BÆST, es; *m? n? The inner bark of a tree, of which ropes were made;* tilia :—Bǽst *vel* lind *tilia,* Lye. [*Plat. Dut.* bast, *m. bark: O. Dut.* bast, *m. signifies the bark of a tree and also a rope; because the inner part of the linden or lime-tree was mostly used for making ropes: Ger. M. H. Ger.* bast, *m. bark: O. H. Ger.* past, *m: Dan.* bast, *m: Swed.* bast, *n: O. Nrs.* bast, *n.* The word is probably to be derived from bindan *to bind,* v. Grm. Wrtbch. i. 1148.]

bǽsten; *adj. Made of bast,* BAST; tiliaceus :—Híg ðá hine gebundon mid twǽm bǽstenum rǎpum *then they bound him with two bast ropes,* Jud. Grn. 15, 13.

bǽstere *a baptizer;* baptista :—Bǽstere *baptista,* Mt. Lind. Stv. 3, 1. v. bǽdzere.

bǽswi [= basu *purple*] *A scarlet robe;* coccinum, Cot. 208.

bǽtan; *p.* bǽtte; *pp.* bǽted; *v. a. To bridle, rein in, restrain, curb, bit;* frenum equo vel asino injicere, frenare, cohibere :—Esolas bǽtan to *bridle asses,* Cd. 138; Th. 173, 25; Gen. 2866. Gif he ǽr þweores windes bǽtte *if he first restrained the perverse wind,* Bt. 41, 3; Fox 250, 16. [*O. H. Ger.* beizian *mordere facere, infrenare: O. Nrs.* beita.] DER. ge-bǽtan, ymbe-.

bǽte, es; *n. A* BIT *of a bridle, a bridle, trappings, harness;* lupatum, frenum. v. gebǽte, gebǽtel.

BÆÞ, es; *pl. nom. acc.* baðu; *g.* baða; *d.* baðum, baðan, baðon; *n.* I. *a* BATH; balneum, balneatio :—Bæþ hǽte weól *the bath boiled* [*welled*] *with heat,* Exon. 74 a; Th. 277, 16; Jul. 581. On hǎtum baðum *in hot baths,* Bd. 4, 19; S. 588, 6. II. *a font;* fons lustralis :—Hú hí hine bǽdan fullwihtes bǽðes *how they had asked him for a font of baptism,* Ors. 6, 34; Bos. 130, 30. [*Plat.* bad, *n: O. Sax.* bath, *n: Dut. Ger.* bad, *n: M. H. Ger.* bat *n: gen.* bades, *n: O. H. Ger.* bad, *n: Dan. Swed.* bad: *O. Nrs.* bað, *n.*] DER. fýr-bæþ, seolh-: Baðan *Bath.*

bǽdere, es; *m. A baptist;* baptista, Grm. i. 253, 38. v. bædzere.

bæþ-hûs, es; *n. A* BATH-HOUSE; thermarum domus :—Bæþhûs *balnearium vel thermarium,* Ælfc. Gl. 109; Som. 79, 13; Wrt. Voc. 58, 54. Bæþhûs *vel* bæþstôw *thermæ,* Ælfc. Gl. 107; Som. 78, 75; Wrt. Voc. 57, 53. v. bæþ-stôw.

bǽðian; *p.* ode; *pp.* od *To bathe,* Som. Lye. v. baðian.

bæþ-stede, es; *m. A place of baths;* thermarum locus :—Bæþstede *thermæ vel gymnasium,* Ælfc. Gl. 55; Som. 67, 7; Wrt. Voc. 37, 5.

bæþ-stôw, e; *f. A bathing-place;* thermarum locus :—Bæþhûs *vel* bæþstôw *thermæ,* Ælfc. Gl. 107; Som. 78, 75; Wrt. Voc. 57, 53. v. bæþ-hûs.

bæþ-weg, es; *m. A bath-way, the sea;* via balnei, mare :—Brecan ofer bæþweg *to break over the bath-way,* Andr. Kmbl. 445; An. 223. Bæþweges blǽst *a blast or wind of the sea, a sea breeze, the south wind.* Sûþwind *is so called,* Cd. 158; Th. 196, 11; Exod. 290.

bǽting, bêting, e; *f. A cable, a rope, anything that holds or restrains;* funis, retinaculum :—Lǽtan ða bêtinge [Cot. bætinge] *to slip the cable,* Bt. 41, 3; Fox 250, 15.

bǽtte *restrained,* Bt. 41, 3; Fox 250, 16; *p. of* bǽtan.

bǽzera, bǽzere *a baptizer,* Mt. Rush. Stv. 11, 11, 12. v. bǽdzere.

bala-níþ, es; *m. Baleful malice, evil,* Ps. C. 50, 151; Ps. Grn. ii. 280, 151. v. bealo-níþ.

balca, an; *m. A* BALK, *beam, bank, a ridge;* trabs, porca, terra inter duos sulcos congesta :— On balcan lecgan *to lay in ridges,* Bt. 16, 2; Fox 54, 2. [*Piers P. Chauc.* balke *trabs: Plat.* balk, *m. id: O. Sax.* balko, *m: Dut.* balke, *m: Ger. M. H. Ger.* balke, *m: O. H. Ger.* baicho, balko, *m: Dan.* bjälke: *Swed.* bjelke: *O. Nrs.* bálkr, *m; but cf. also Gaelic* balc *a ridge of earth between two furrows,* Grm. Wrtbch. i. 1089.]

balcettan *to belch,* Som. Lye. v. bealcettan.

bald; *adj.* BOLD, *audacious, adventurous, confident;* audax, confidens :—Bald breóst-toga *a bold chief,* Salm. Kmbl. 369; Sal. 184. Hilde calla bald bord upahôf *the bold war-herald raised his shield,* Cd. 156; Th. 193, 27; Exod. 253. Wǽron hí ðe baldran gewordene *confidentiores facti,* Bd. 1, 12; S. 481, 17. v. beald.

-bald, -bold; as the incipient or terminating syllable of proper names denotes *Bold, courageous, honourable;* audax, virtuosus :—Baldwin *from* bald, *and* win *a contest, battle.* Cûþbold, Cûþbald *from* cûþ *known,* bald *bold.* Eádbald *happily bold,* from eád *or* eádig *and* bald.

balde; *adv. Boldly, freely, confidently, instantly;* audacter, libere, fidenter, instanter, prone, statim, sine mora :—Hie balde gecwǽdon *they said boldly,* Cd. 182; Th. 228, 11; Dan. 200. v. bealde.

bald-lîce *boldly;* fortiter :—He baldlîce beornas lǽrde *he boldly exhorted the warriors,* Byrht. Th. 140, 60; By. 311. v. beald-lîce.

bald-lîcost; *sup. Most bravely;* fortissime :—Ðe baldlîcost on ða bricge stôp *who stept on the bridge most bravely,* Byrht. Th. 134, 2; By. 78. v. beald-lîce.

baldor, es; *m.* [*the comp. of* bald *is* baldor *more bold, courageous, honourable, hence*] *A prince, ruler;* princeps, dominus :—*thus,* Gumena baldor *a ruler of men,* Cd. 128; Th. 163, 4; Gen. 2693: Judth. 9; Thw. 21, 8; Jud. 9. Rinca baldor, 12; Thw. 26, 21; Jud. 339. Wîgena baldor *a prince of warriors,* 10; Thw. 22, 5; Jud. 49. v. bealdor.

baldra *bolder,* Bd. 1, 12; S. 481, 17. v. bald, beald.

baldsamum, i; *n. Balsam, balm;* balsamum :—Swá swá mon hêddern ontŷnde ða baldsami *quasi opobalsami cellaria esse viderentur aperta,* Bd. 3, 8; S. 532, 19. v. balsam.

balewa, an; *m. The baleful or wicked one, Satan;* Satanas, Diabolus :—Swá inc se balewa hêt *as the baleful one desired you,* Cd. 224; Th. 295, 11; Sat. 484.

balewe *wicked* :—Se inc forgeaf balewe geþohtas *he inspired you with wicked thoughts,* Cd. 224; Th. 295, 19; Sat. 488. v. bealo.

ballîce *boldly* :—Ballîce *audacter,* Mk. Lind. War. 15, 43. v. bald-lîce, beald-lîce.

balo *bale, evil,* Lye. DER. balo-cræft. v. bealo.

balo-cræft, es; *m. A pernicious, wicked, or magic art;* ars perniciosa *vel* magica, Bt. Met. Fox 26, 150; Met. 26, 75. v. bealo-cræft.

balsam, es; *n.* [balsamum, balsamum, i; *n.*] *Balsam, balm;* balsamum :—Balsames blǽd *the balsam's fruit;* carpo balsami, Ælfc. Gl. 48; Som. 65, 54; Wrt. Voc. 33, 50. Balsames teár *the tear or juice of the balsam-tree;* opobalsamum, Ælfc. Gl. 48; Som. 65, 55; Wrt. Voc. 33, 51. Hêddern ða balsamum on wǽre *a store-house in which was balm,* Bd. 3, 8; S. 532, 19, note.

bals-minte, an; *f.* BALSAM-MINT, *spear-mint, water-mint;* sisymbrium: *q.* mentha aquatica, Lin. Ælfc. Gl. 43; Som. 64, 52; Wrt. Voc. 31, 62.

balw; *g. m. n.* es; *f.* re *Miserable, wicked;* malus, Beo. Th. 1958; B. 977. v. bealo.

balzam *balsam* :—Se sceal on balzame beón *it shall be of balsam,* L. M. 2, 64; Lchdm. ii. 288, 23. v. balsam.

bǽm *with both,* Hexam. 2; Norm. 4, 22: Cd. 6; Th. 8, 23; Gen. 128; *dat. of* begen.

ban, bann, es; *n. A command, edict, interdict;* mandatum, edictum, interdictum, Grm. 3rd edit. i. 359, 8. v. ge-ban.

BÁN, baan, es; *pl.* bán; *n. A* BONE; os :—Ðis ys nû bán of mínum bánum *hoc nunc os ex ossibus meis,* Gen. 2, 23. Moises nam Iosepes bán mid him *tulit Moyses ossa Ioseph secum,* Ex. 13, 19: Cd. 9; Th. 12, 9; Gen. 182. Híg synt innan fulle deádra bána *intus plena sunt ossibus mortuorum,* Mt. Bos. 23, 27. Bán mîne *my bones,* Ps. Spl. 6, 2: Exon. 110 a; Th. 421, 14; Rä. 40, 18: 125 b; Rä. 68, 3: Beo. Th. 5149; B. 2578. [*Plat.* been, *n. os, crus: O. Sax.* O. Frs. bên, *n: Dut.* been, *n: Ger. M. H. Ger.* bein, *n: O. H. Ger.* pein, *n: Dan.* been: *Swed.* ben: *O. Nrs.* bein, *n.* In *Goth.* the word is preserved only in baina-bagms *a bone-tree, cornel-tree,* for σνκαμυνος. Thus, all the *Teut.* languages have the same word, the chief and oldest signification of which is os *a bone.* This is the only meaning it has in *A. Sax.* where scanca is used for *crus;* also in *O. Nrs.* the meaning *crus* is very rare, the more common word being leggr *a leg.* The *Sansk. Lat. Grk.* and the *Slav.* languages use a totally different root.—*Sansk.* asthi os: *Lat.* os: *Grk.* ὀστέον: the *Slav.* branch kost, *Boh.* kost, *Pol.* kosc, all with an initial k. Grimm, Wrtbch. i. 1381, suggests, if crus could be proved to be the original meaning of bán, it might be related to βαίνειν, in the same way as *Sansk.* asthi to στῆναι.] DER. breóst-bán, cin-, elpen-, brycg-, wîdo-, ylpen-.

BANA, bona, an; *m. A killer, murderer, manslayer,* also applied to *the devil;* interfector, occisor, homicida, diabolus :—Ðam wearþ Weohstán bana *to whom Weohstan became a murderer,* Beo. Th. 5220; B. 2613: Cd. 144; Th. 180, 3; Exod. 39. Banena byre *the son of the murderers,* Beo. Th. 4112; B. 2053. Hie nǽfre his banan folgian noldon *they never would follow his murderer,* Chr. 755; Th. 84, 33, col. 1: L. Ethb. 23; Th. i. 8, 7: L. H. E. 2, 3, 4; Th. i. 28, 1, 5, 7. On banan fæðme *in the embrace of the murderer,* i. e. *the devil,* Andr. Kmbl. 1232; An. 616. [*O. Sax.* bano: *O. Frs.* bona: *O. H. Ger.* bano: *O. Nrs.* bani.] DER. aldor-bana [-bona], brôðor-, dǽd-, ecg-, feorh-, ferhþ-, fugel-, gást-, hand-, mûþ-, ord-, rǽd-, sûsl-.

bán-beorgas; *pl. m. Bone defences, greaves;* ossium præsidia, ocreæ, Cot. 17: 145.

bán-brice, -bryce, es; *m. A* BONE-BREAKING or *fracture of a bone;*

ossis fractura :—Wið bānbryce genim ðysse ylcan wyrte wyrttruman *for fracture of a bone take roots of this same plant*, Herb. 15, 3; Lchdm. i. 108, 9.

BANC, e; *f. A bench*, BANK, *hillock*; tumulus, Som. v. benc.

bān-cōfa, an; *m. A bone-dwelling, the body*; ossium cubile, corpus :—Wæs se bāncōfa ādle onǣled *the body was inflamed with disease*, Exon. 46 b; Th. 159, 16; Gū. 927.

Bancorena burh, Bancorna burh; *g.* burge; *d.* byrig; *Bangor, in Wales*; civitas Bangor :—Swȳðest of Bancorena [Bancorna, B.] byrig *most chiefly from the city of Bangor*, Bd. 2, 2; S. 502, 39, note.

ban-cōða, an; *m.*: -cōþ, -cōðu, e; *f.*: -cōðe, an; *f.* [ban, bana *a killer*, cōða *a disease*] *A baneful disease, a fatal* or *deadly malady, erysipelas*; lethalis morbus, ignis sacer :—Wæs him inbogen bittor bancōða *a bitter malady was fixed in him*, Exon. 47 b; Th. 163, 23; Gū. 998. Wið bancōðe, ðæt is ōman, nim eolonan *for the baneful disease, that is erysipelas, take elecampane*, L. M. 1, 39; Lchdm. ii. 102, 16.

band *bound*, Cd. 143; Th. 178, 22; Exod. 15; *p.* of bindan.

banda, an; *m. A householder, husband*, Som. Lye. v. bonda.

bān-fæt; *g.* -fætes; *pl.* nom. acc. -fatu; *n. The bone vessel, the body*; ossium vas, corpus, Exon. 59 a; Th. 213, 23; Ph. 229.

ban-fāh, -fāg; *adj.* [ban, bana *a killer*, fāg *stained*] *Death* or *murder stained*; homicidio pollutus, lethifer, Beo. Th. 1564; B. 780.

bān-gebrec, es; *n. A bone-breaking*; ossium fractio, Andr. Kmbl. 2882; An. 1444.

bān-helm, es; *m. A bone-helm, shield*; ossium galea, clipeus, Fins. Th. 60; Fin. 30.

bān-hring, es; *m. A bone-ring, a neck-bone*; ossium artus, vertebra :—Ðæt hire wið halse heard grāpode, bānhringas bræc *against her neck it griped her hard, broke the bone-rings*, Beo. Th. 3138; B. 1567.

bān-hūs, es; *n. The bone-house, the chest, body*; ossea domus, pectus, corpus :—He ðæt bānhūs gebrocen hæfde *he had broken the bone-house, the breast*, or *body*, Beo. Th. 6285; B. 3147. *Hence* bānhūses weard *the body's guard, the mind*, Cd. 169; Th. 211, 9; Exod. 523.

Baningas; *pl. m. The Banings, people mentioned in the Gleeman's tale* :—Becca weóld Baningum *Becca ruled the Banings*, Scōp Th. 39; Wīd. 19.

bān-leás; *adj. Bone-less, without bones*; ossibus carens, Exon. 112 b; Th. 431, 19; Rä. 46, 3.

bān-loca, an; *m. A bone inclosure, the skin, body*; ossium clausura, caro :—Ðȳ-læs se ord ingebuge under bānlocan *lest the point enter in under the skin*, Exon. 19 a; Th. 48, 10; Cri. 769.

BANNAN, bonnan; ic banne, ðū bannest, banst, benst, he banneþ, banþ, benþ, *pl.* bannaþ; *p.* bēn, bēnn, beón, beónn, *pl.* beónnon; *pp.* bannen *To summon*; jubere, citare, convocare :—Leóde tosomne bannan *to summon the people together*, Andr. Kmbl. 2189; An. 1096. Elen. Grm. 45. [O. Frs. banna, bonna : Ger. M. H. Ger. bannen *edicere, interdicere, prohibere, expellere* : O. H. Ger. pannan : Goth. bandwjan *significare, innuere* : O. Nrs. banna *prohibere, interdicere*.] DER. a-bannan, ge-: ge-ban.

bannuc-camb, es; *m.* [camb *a comb*] *A wool-comb*; pecten textorium :—Bannuccamb *pecten*, Ælfc. Gl. 111; Som. 79, 77. DER. cimban.

bān-rift, bān-ryft; *pl. n. Bone coverings, greaves*; tibialia, ossium velamen, ocreæ, Cot. 174. v. bān-beorgas.

ban-segn, es; *m. A banner, an ensign*; vexillum, Cot. 23. v. treuteru.

bān-sele, es; *m. A bone-house* or *dwelling, the body*; ossium aula, corpus :—Gǣst and bānsele *soul and body*, Exon. 117 b; Th. 451, 12; Dōm. 102.

banst, he banþ *summonest, summoneth*; 2nd *and* 3rd *pers. pres. of* bannan.

bān-wærc, es; *n. Grief, pain*, or *ache in the bones*; ossium dolor. v. bān *a bone*, wærc *pain*.

bān-wyrt, e; *f. Bone-wort, a violet, perhaps the small knapweed*; viola, centaurea minor :—Bānwyrt hæbbe croppan *bone-wort hath bunches of flowers*, L. M. 2, 51; Lchdm. ii. 266, 5. Bānwyrt *centaurea minor*, Ælfc. Gl. 44; Som. 64, 85; Wrt. Voc. 32, 21. Sió greáte bānwyrt *the great bone-wort*, L. M. 3, 8; Lchdm. ii. 312, 19 : 1, 1; Lchdm. ii. 22, 15 : 1, 25; Lchdm. ii. 66, 17, 20 : 1, 31; Lchdm. ii. 74, 24 : 1, 36; Lchdm. ii. 86, 21 : 1, 59; Lchdm. 130, 11 : 1, 63; Herb. 165, 1; Lchdm. i. 294; 7 : 152, 1; Lchdm. i. 276, 24 : Lchdm. iii. 16, 6.

baorm *bosom* :—On baorm *in sinu*, Jn. Rush. War. 13, 23. v. bearm.

bar, es; *m. A bear*; ursus. v. bera.

BĀR, es; *m. A* BOAR; aper :—Cyng Willelm forbeád sleán ða heortas swylce eác ða bāras *king William forbade men to kill the stags, and also the boars*, Chr. 1087; Ing. 296, 12. Ic gefeó heortas, and bāras, and rann, and rægan, and hwīlon haran *capio cervos, et apros, et damas, et capreas, et aliquando lepores*, Coll. Monast. Th. 21, 31 : Ælfc. Gr. 8; Som. 7, 14 : Ps. Lamb. 79, 14. [*Dut.* beer: *M. H. Ger.* bēr: *O. H. Ger.* pēr.]

barda, an; *m. A beaked ship, a ship pointed with iron*; rostrata navis, Mone A. 131.

bare *bare, naked*, Cd. 37; Th. 48, 30; Gen. 783; *acc. pl.* of bær, *adj.*

barenian; *p.* ode; *pp.* od *To make bare*; denudare :—Sand barenodon *made bare the sand*, Cd. 166; Th. 207, 22; Exod. 470, note.

barian; *p.* ede; *pp.* ed *To make bare, discover, disclose*; denudare, prodere, in medium proferre. DER. a-barian.

barm *a bosom* :—On barme *in sinu*, Jn. Rush. War. 1, 18. v. bearm.

barn *a child*, Th. Diplm. A. D. 830; 465, 30. v. bearn.

barn *burned*, Ex. 3, 2; *p.* of beornan.

Baroc-scír, e; *f. The bare oak shire* or BERKSHIRE, *so called from a polled oak in Windsor forest, where public meetings were held*, Brompt. p. 801. It was most commonly written by the Anglo-Saxons—Barruc, Bearruc, and Bearwucscīre, Chr. 860; Th. 130, 3.

bār-spere, es; *n. A* BOAR SPEAR; venabulum :—Bārspere *vel* huntigspere *venabulum*, Ælfc. Gl. 51; Som. 66, 22.

bār-spreót, es; *m. A boar spear*; venabulum. v. bār-spere.

barþ, es; *m. A kind of ship, a light vessel to sail* or *row in*; dromo :—Æsc *vel* barþ *dromo*, Ælfc. Gl. 103; Som. 77, 102; Wrt. Voc. 56, 24. v. æsc.

Barton *Barton, a corn village*; frumentaria villa. v. bere-tūn.

basilisca, an; *m. A basilisk*; basiliscus :—Ðū ofer aspide miht eáðe gangan and bealde nū basiliscan tredan *super aspidem et basiliscum ambulabis*, Ps. Th. 90, 13.

Basilius; *g.* Basilies; *m. Basil, bishop of Cæsarea* = Καισάρεια :—Basilius se eádiga wæs swīðe hālig bisceop, on Cessarean byrig, on Grēciscre þeóde, manegra munuca fæder, munuchādes him sylf. He wæs swȳðe gelǣred and swȳðe mihtig lareów, and he munuc regol gesette mid swȳðlīcre drohtnunge. He wæs ǣr Benedictus, ðe us bōc awrāt on Lēdenre sprǣce leóhtre be dǣle ðonne Basilius, ac he tymde swāðeáh to Basilies tǣcinge for his trumnysse. Basilius awrāt āne wundorlīce bōc, be eallum Godes weorcum, ðe he geworhte on six dagum, 'Exameron' gehāten, swīðe deópum andgite. And he awrāt ða lāre ðe we nū willaþ on Englisceum gereorde secgean *Basil the blessed* [*born* A. D. 328, *died* 379] *was a very holy bishop in the city of Cæsarea, a province belonging to Greece, the father of many monks, himself of the monkhood. He was a very learned and a very mighty teacher, and he appointed monastic canons with strict conduct. He was before Benedict* [*born* A. D. 480, *died* 540], *who wrote us a book in the Latin language more clear in part than Basil, but yet he appealed to the teaching of Basil for his confirmation. Basil wrote a certain wonderful book concerning all the works of God which he wrought in six days, called the 'Hexameron,' with a very deep understanding. And he wrote the advice which we now wish to tell in the English language*, Basil prm; Norm. 32, 1–14. Sancti Basilii Exameron [= ἑξάμερον], ðæt is, be Godes six daga weorcum *the Hexameron of holy Basil, that is, concerning the six days' works of God*, Hexam. 1; Norm. 1, 1–3.

basing, es; *m. A short cloak, a cloak*; chlamys = χλαμύς, pallium :—Ic geseah wurm-reádne basing *I saw a purple* [*worm* or *shell-fish reddened*] *cloak*; vidi pallium coccineum, Jos. 7, 21.

Basing, es; *m. The name of a place, Basing, old Basing, near Basingstoke, Hampshire*; nomen oppidi ita hodie vocatum in agro Hantoniensi :—Wið ðone here æt Basingum *with the army at Basing*, Chr. 871; Th. 138, 28, col. 2; 139, 27, col. 1, 2.

bāsnian, bāsnan; *p.* ode; *pp.* od *To expect, await*; exspectare :—Gestōd ðæt folc bāsnende *stabat populus exspectans*, Lk. Lind. War. 23, 35. Bāsnode hwæt him gifeðe wurde *he awaited what should befall him*, Andr. Kmbl. 2131; An. 1067. DER. ge-bāsnian.

bāsnung, e; *f. Expectation*; exspectatio, Lk. Lind. War. 21, 26.

baso, basu, e; *f. Purple*; purpura, Cot. 85. DER. brūn-baso, wealh-. v. basu.

baso, basu *a berry*; bacca, Grm. i. 244, 36.

baso-popig, es; *n?* [*astula regia*, Glos. Brux. Recd. 40, 57; Mone A. 354; Wrt. Voc. 66, 65] *Corn* or *red poppy*; papaver rhœas, L. Prior, p. 279.

Basterne *The people of Sarmatia in Europe* or *upper Hungary*; Bastarnæ. Lye.

basu: *g. m. n.* -wes; *f.* -re: *pl. nom. m. f. n.* -we: *def. m.* se baswa; *adj. Purple, crimson*; purpureus, phœniceus, coccineus :—Sum brūn, sum basu *part brown, part purple*, Exon. 60 a; Th. 218, 17; Ph. 296. Baswe bōcstafas *crimson characters*, Cd. 210; Th. 261, 10; Dan. 724. Basu hǣwen *of purple colour* or *hue, of scarlet* or *crimson colour*, Cot. 117. [Grimm, Wrtbch. i. 1243, connects the word with *Goth.* basi *a berry*: *Ger.* beere: *A. Sax.* berie.]

basu, e; *f. A scarlet robe*; coccinum, Grm. i. 254, 2. v. baso.

basuian; *p.* ode; *pp.* od *To be clad in purple*; purpura vestiri. v. basu.

baswa stān, es; *m.* [basu *purple*, stān *stone*] *A topaz, a precious stone varying from a yellow to a violet colour*; topazium :—Ofer gold and ðone baswon stān [=baswan stān] *super aurum et topazion*, Ps. Spl. 118, 127.

baswe *crimson* :—Baswe bōcstafas *crimson letters*, Cd. 210; Th. 261, 10; Dan. 724; *pl.* of basu, *adj.*

bat, e; *f.* I. *contention, strife*; contentio, R. Ben. 21. II.

a bat, club, staff, stick; fustis, Som. [*O. Nrs.* beit, *f;* lamina explanata *a thin board, plank.*]

BÂT, e; *f:* es; *m. A* BOAT, *ship, vessel*; linter, scapha, navicula:—Ðeós bât glîdeþ on geofene *this boat glideth over ocean,* Andr. Kmbl. 992; An. 496. He bât gestâg *he ascended a boat,* Exon. 52 a; Th. 181, 33; Gû. 1302. [*Plat.* boot, *n: Dut.* boot, *f: Ger.* boot, *n: Dan.* baad, *c: Swed.* bât, *m: Icel.* bátr, *m.* cymba, *navicula.*] DER. mere-bât, sæ̂-, wudu-.

bât *bit;* momordit, Beo. Th. 1488; B. 742; *p. of* bîtan.

bât, e; *f.* What can be bitten,—*Food;* esca, Ettm. 305. [*Icel.* beit, *f. pascuum;* beita, *f. esca:* bât; *p. of* bîtan *to bite.*]

baða *of baths,* Exon. 57 b; Th. 205, 10; Ph. 110; *gen. pl. of* bæþ.

Baðan [*dat. pl. of* bæþ *a bath, q.v.*], Baðan-ceaster; *g.* -ceastre; *acc.* -ceastre, -ceaster; *f. The city of Bath, Somersetshire,* so called from its baths; Bathoniæ urbs a balneis dicta, in agro Somersetensi:—Baðan, Baðon, Baðun, *for* Baðum, æt Baðum, Cod. Dipl. 170; Kmbl. i. 207, 5, *at the Baths,* or, as we now say, *at Bath* or *Bath* [v. æt, *prep.* I. 3, before names of places]; apud balneas, *vel* apud Bathoniam, *vel* apud urbem Bathoniæ. Æt Baðan, Chr. 1106; Erl. 241, 1. On Baðan, Th. Diplm. A. D. 1060; 379, 14: 436, 8. Æt Baðun, Cod. Dipl. 354; A. D. 931; Kmbl. ii. 177, 7. In monasterio, quod situm est in civitate æt Baðun, Cod. Dipl. 193; A. D. 808; Kmbl. i. 237, 1. In illa famosa urbe, quæ nominatur calidum balneum, id est æt ðæm hâtum baðum, Cod. Dipl. 290; A.D.864; Kmbl. ii.80,8. Eádgâr wæs to cyninge gehâlgod on ðære ealdan byrig, Acemannes ceastre; eác, óðre worde, beornas Baðan nemnaþ *Edgar was consecrated king in the old town, Akemansceaster; also, by another word, men name Bath,* Chr. 973; Th. 224, 22, col. 1; Edg. 5. Genâmon þreó ceastra,—Gleawan-ceaster and Ciren-ceaster and Baðan-ceaster *they took three cities,—Gloucester, Cirencester, and Bath,* Chr. 577; Erl. 18, 20. v. Ace-mannes burh.

baðian, beðian, beðigean, ic -ige, -yge; *p.* ode, ede; *pp.* od. I. *v. trans. To wash, foment, cherish;* lavare, fovere:—Hî baðedon ðone lîchoman *they washed the body,* Bd. 4, 19; S. 589, 38. Wit unc in ðære burnan baðodan *we two washed ourselves in that brook,* Exon. 121 b; Th. 467, 2; Hö. 132. II. *v. intrans. To* BATHE; lavari, balneare, aquis se immergere:—Seldon heó baðian wolde *she would seldom bathe,* Bd. 4, 19; S. 588, 6. Gesihþ baðian brimfuglas *he sees sea-fowls bathing,* Exon. 77 a; Th. 289, 12; Wand. 47. Baðiendra manna hûs ðær hî hî unscrêdaþ inne *apodyterium, domus, qua vestimenta balneantium ponuntur,* Ælfc. Gl. 55; Som. 67, 9. DER. bi-baðian. v. bæþ.

baðo *baths,* Bd. 1, 1; S. 473, 22; *acc. pl. of* bæþ.

bâtian; *p.* ode; *pp.* od *To* BAIT *or lay a bait for a fish, to bait a hook;* inescare, Som.

bât-swân, es; *m. A* BOATSWAIN; scaphiarius, proreta. v. bât *a boat;* swân *a swain, servant.*

bâtwâ, bûtâ, bûtû, bûtwû; *adj.* [bâ *both,* twâ *two*] BOTH THE TWO, *both :*—Bâtwâ Adam and Eue *both Adam and Eve,* Cd. 37; Th. 47, 24; Gen. 765: Gen. 26, 35. v. begen.

bât-weard, es; *m.* [bât *boat,* weard *keeper*] *Keeper* or *commander of a ship;* navis custos:—He ðæm bâtwearde swurd gesealde *he gave a sword to the keeper of the ship,* Beo. Th. 3804; B. 1900.

BE [*abbreviated from* big = bî, *q.v.*]; *prep. dat. and instr.* 1. BY, *near to, to, at, in, on, upon, about, with;* juxta, prope, ad, secus, in, cum :—Be wege *by the way,* Mk. Bos. 8, 3. Wunode be Iordane *he dwelt by Jordan,* Cd. 91; Th. 116, 6; Gen. 1932. Be grûnde wôd *went on the ground,* Exon. 106 a; Th. 404, 29; Râ. 23, 15. Be ŷþláfe *along the leaving of the waves,* Beo. Th. 1136; B. 566. Ic be grûnde græfe *I dig along the ground,* Exon. 106 a; Th. 403, 3; Râ. 22, 4. Be fullan *in full;* abundanter, Ps. Th. 30, 27. Be eallum *with all, altogether,* L. Ath. v. § 8, 2; Th. i. 236, 12. Ne mæg he be ðŷ wedre wesan *he may not be in the open air,* Exon. 90 b; Th. 340, 18; Gn. Ex. 113. Be ðam strande *upon the strand* or *shore,* Mt. Bos. 13, 48. Ne leofaþ se man be hláfe ânum, ac be ælcon worde, ðe of Godes mûþe gæþ *non in solo pane vivit homo, sed in omni verbo, quod procedit de ore Dei,* Mt. Bos. 4, 4. Byrgan be deáðum *to bury with the dead,* Exon. 82 b; Th. 311, 27; Seef. 98. 2. *of, from, about, touching, concerning;* de, quoad :—Be ðam cilde *of* or *concerning the child,* Mt. Bos. 2, 8. Be hlîsan *of* or *about fame,* Bt. titl. xviii. xix; Fox xiv. 1. Gramlîce be Gode spræcan *male locuti sunt de Deo,* Ps. Th. 77, 20. Be his horse Bucefal *about his horse Bucephal,* Ors. 3, 9; Bos. 67, 39. Ahsiaþ be ealdum dagum *interrogate de diebus antiquis,* Deut. 4, 32. Mæg ic be me sylfum sôþ gied wrecan *of myself I can relate a true tale,* Exon. 81 b; Th. 306, 1; Seef. 1. Ic ðis gid be ðê awræc *I recited this strain of thee,* Beo. Th. 3451; B. 1723. Nysse ic be ðære [rôde] riht *I did not know the right about the cross,* Elen. Kmbl. 2479; El. 1241. 3. *for, because of, after, by, through, according to;* pro, propter, per, secundum :—He sette word be worde *he set word for word,* Bt. proœm; Fox viii. 3. Be hyra weorcum *for their works,* Exon. 26 b; Th. 79, 13; Cri. 1290. Ðû scealt sunu âgan, bearn be brŷde ðînre *thou shalt have a son, a child, by thy bride,* Cd. 106; Th. 140, 11; Gen. 2326. Forlæded be ðam lygenum *misled by the lies,* 28; Th. 37, 31; Gen. 598. Ðæt ic meahte ongitan be ðam gealdre Godes bearn *that I might comprehend,*

through that lore, God's child, Exon. 83 a; Th. 313, 26; Môd. 6. Hie, be wæstmum, wîg curon *they, according to his strength, choose each warrior,* Cd. 155; Th. 193, 8; Exod. 243. Nâ ðû be gewyrhtum ûrum woldest us dôn *thou wouldst not do to us according to our sins* [secundum peccata nostra], Ps. Th. 102, 10. 4. *beside, out of;* e, ex :—Ic ðê læde be ðam [bi ðæm MS. Cott.] wege *I should lead thee out of the way,* Bt. 40, 5; Fox 240, 23. Genam hine æt eowde ûte be sceápum *tulit eum de gregibus ovium,* Ps. Th. 77, 69. 5. *sometimes* be *is separated from its case :*—Be dæges leóhte *at the light of day* or *at daylight,* Exon. 107 b; Th. 410, 17; Râ. 28, 17. Be fæder lâre *through the father's counsel,* Beo. Th. 3905; B. 1950. Ûre bân syndon toworpene be helwarena hæfte neódum *dissipata sunt ossa nostra secus infernum,* Ps. Th. 140, 9. Mîn bibod ðû bræce be ðînes bonan worde *thou didst break my command through the word of thy destroyer* [*the devil*], Exon. 28 a; Th. 85, 21; Cri. 1394. ¶ Be ânfealdum *single.* Be twîfealdum *twofold,* Ex. 22, 4. Be ðam mæstan *at the most.* Be ðam ðe as, Gen. 3, 6. [*Orm. Lœym. R. Glouc.* Piers P. bi: *Chauc. Wyc.* by: *Plat.* bî: *O. Sax.* bi, be: *O. Frs.* bî, be: *Dut.* by: *Ger.* bei: *M. H. Ger.* bî: *O. H. G.* bî, pî: *Goth.* bi: *Sansk.* abhi?]

be-, bi-, big-, and bî- are often used as prefixes. I. when prefixed to verbs, be- and bi- either give an intensive signification to a transitive verb, or change an intransitive into a transitive verb, as,—Sprengan *to sprinkle,* be-sprengan *to be-sprinkle;* lecgan *ponere,* be-lecgan *im-ponere;* settan *to set, put,* be-settan *to be-set, surround;* fôn *to seize,* be-fôn *to surround;* gangan *to go,* be-gangan *to exercise;* reótan *plorare,* be-reótan *de-plorare.* 2. *they have a privative sense, as,—*Be-niman *to deprive,* be-reáfian *to bereave,* be heáfdian *to behead.* 3. sometimes they do not indicate any perceptible variation in the sense, as,—Be-cuman *to come,* be-sencan *to sink.* 4. be-, bi-, big- have the same effect when prefixed to *substantives, adjectives, and adverbs.* II. the accented bî- and big-, as prefixes, generally have the original sense of the preposition *by,* as,—Bî-cwide, big-cwide *a by-saying, proverb;* bî-spell, big-spell *a by-story, parable;* bî-wærlan *to pass-by;* big-standan *to stand-by.* vide I. 2.

BEÁCEN, bêcen, bêcn, bêcun; *g.* beácnes; *n. A* BEACON, *sign, token, standard;* signum, significatio, typus, vexillum, portentum, miraculum; in specie de sancta cruce et de sole :—Leóht eástan com beorht beácen *light came from the east a bright beacon,* Beo. Th. 1144; B. 570. He beácen onget *he perceived the sign,* Cd. 198; Th. 246, 33; Dan. 488. Wæs beácen boden *the token was announced,* Andr. Kmbl. 2403; An. 1203. Beácnes cyme *the beacon's* [*the sun's*] *coming,* Exon. 57 b; Th. 205, 4; Ph. 107. Segn genom beácna beorhtost *he took an ensign brightest of standards,* Beo. Th. 5547; B. 2777. [*O. Sax.* bôkan: *O. Frs.* bâken: *O. H. Ger.* pouchan.] DER. fore-beácen, freoðo-, heofon-, here-, sige-, sigor-, wundor-: beácn, -ian, -ung: bêcn-an, -ian: bîcn-ian: bŷcn-an, -endlîc, -iend, -iendlîc.

beácen-stân, es; *m. A stone whereon the beacon fire was made, a stone* or *tower whereon to set the beacon fire;* specula, pharus, Cot. 88.

beácne *to a sign,* Cd. 80; Th. 100, 19; Gen. 1666; *dat. of* beácen.

beácneng *a beckoning* or *nodding, a speaking by tropes* or *figures;* nutus, Cot. 139: tropologia, Cot. 201. v. beácnung.

beácnian, bŷcnian, bîcnian; *p.* ode; *pp.* od. I. *to* BECKON, *nod;* innuere:—He wæs bîcniende him *erat innuens illis,* Lk. Bos. 1, 22, 62: 5, 7. II. *to shew, indicate;* indicare, typice significare :—Swâ fenix beácnaþ *as the phœnix shews,* Exon. 65 a; Th. 240, 30; Ph. 646. Ðisses fugles gecynd beácnaþ hû hî beorhtne gefeán healdaþ *this bird's nature indicates how they possess bright joy,* Exon. 61 b; Th. 225, 14; Ph. 389. DER. ge-beácnian, -bêcnan.

beáceniend-lîc, bŷcniend-lîc, bŷcnend-lîc; *adj. Allegorical;* allegoricus :—Ic sette âne bôc beácniendlîcre race be Cristes cyricean *unum librum explanationis allegoricæ de Christo et ecclesia composui,* Bd. 5, 23; S. 648, 5.

beácnung, bŷcnung, beácneng, e; *f.* I. *a* BECKONING or *nodding;* nutus, Cot. 139. II. *a speaking by tropes* or *figures;* tropologia, Cot. 201.

beád *a prayer;* oratio. v. gebêd, beáda.

beád, es; *m. A table;* mensa :—Of beád *de mensa,* Lk. Lind. War. 16, 21. Beádas, Mt. Kmbl. Lind. 21, 12. v. beód.

beád *commanded,* Cot. 111; Th. 147, 1; Gen. 2432; *p. of* beódan.

beáda, an; *m. A counsellor, persuader, an exhorter* or *intreater;* suasor. v. beád.

Beada-ford-scîr, e; *f. Bedfordshire :*—Cnut wende him ût þurh Buccingahâmscîre into Beadafordscîre *Canute went out through Buckinghamshire into Bedfordshire,* Chr. 1016; Th. 279, 16, col. 1. v. Bedan ford-scîr.

BEADO, beadu; *g. d.* beadowe, beadwe, beaduwe; *f. Battle, war, slaughter, cruelty;* pugna, strages :—Gûþ-Geáta leód, beadwe heard *the War-Goths' prince, brave in battle,* Beo. Th. 3082; B. 1539. Wit ðære beadwe begen ne onþungan *we both prospered not in the war,* Exon. 129 b; Th. 497, 2; Râ. 85, 23. Beorn beadwe heard *a man brave in battle,* Andr. Kmbl. 1963; An. 584. Ðû þeóde bealdest to

beadowe *thou encouragest the people to slaughter,* Andr. Kmbl. 2373; An. 1188. [*O. H. Ger.* badu-, pato-: *O. Nrs.* böð, *f. a battle: Sansk.* badh *to kill.*]

beado-cræftig; *adj. War-crafty, skilful in war, warlike;* bellicosus :— Beadocræftig beorn *a chief skilful in war,* Exon. 78 b; Th. 295, 28; Crä. 40. v. beadu-cræftig.

beado-gríma, -grímma, an; *m. A war-mask, helmet;* bellica larva, cassis :—Ða ðe beadogrímman býwan sceoldon *those who should prepare the war-helmet,* Beo. Th. 4506; B. 2257. v. beadu-gríma.

beado-hrægl, es; *n. A war-garment, coat of mail;* bellica vestis, lorica :—Beadohrægl on breóstum læg *the coat of mail lay on my breast,* Beo. Th. 1108; B. 552. v. beadu-hrægl.

beado-leóma, an; *m. A war-gleam, sword;* stragis flamma, ensis :— Ðæt se beadoleóma bítan nolde *that the war-gleam would not bite,* Beo. Th. 3050; B. 1523. v. beadu-leóma.

beado-méce, es; *m. A battle-sword, sword of slaughter;* pugnæ ensis :—Ðæt hine·nó beadomécas bítan ne meahton *that no battle-sword might bite it,* Beo. Th. 2912; B. 1454. v. beadu-méce.

beado-rinc, es; *m. A soldier;* bellicosus vir :—Betst· beadorinca *the best of soldiers,* Beo. Th. 2222; B. 1109: Judth. 12; Thw. 25, 24; Jud. 276. v. beadu-rinc.

beado-róf; *adj. War-renowned, bold in war;* in pugna strenuus :— Beornas beadorófe *war-renowned warriors,* Apstls. Kmbl. 155; Ap. 78. v. beadu-róf.

'beado-searo; *gen.* -searewes, -searwes; *n. A war-train, an engine* or *weapon of war;* bellicus apparatus :—Þurh ða heora beadosearo wǽgon *through which their war-train had moved,* Cd. 1/0; Th. 214, 21; Exod. 572. v. beadu-searo.

beado-wǽpen; *gen.* -wǽpnes; *dat.* -wǽpne; *n. A war-weapon;* bellica arma :—Ic beadowǽpen bere *I bear a war-weapon,* Exon. 104 b; Th. 396, 11; Rä. 16, 3. Ic swelgan onginne beadowǽpnum *I begin to swell with war-weapons,* 105 a; Th. 399, 8; Rä. 18, 8. v. beadu-wǽpen.

beado-wég, -wége, es; *n. A war-cup, contest, discussion;* poculum certaminis, certamen :—Him betwih beadowég [MS. beadowíg] scencton ðæs heofonlícan lífes *dum sese alterutrum cælestis vitæ poculis ebriarent* [MS. *debriarent*], Bd. 4, 29; S. 607, 17. v. beadu-wég, bǽde-wég.

beado-weorc, es; *n. A war-work, warlike operation;* bellicum opus :— Ic eom beadoweorca sæd *I am tired of war-works,* Exon. 102 b; Th. 388, 4; Rä. 6, 2: Chr. 937; Th. 205, 40, col. 1, 2; Æðelst. 47. v. beadu-weorc.

Beado-wulf, es; *m. Beowulf,* Th. Anlct. v. Beówulf.

beadu; *gen.* beaduwe; *f. Battle, war, etc.* Andr. Kmbl. 1963; An. 984. v. beado and the following compounds.

beadu-cáf; *adj. Battle-prompt, ready for battle;* ad pugnam expeditus, Exon. 100 b; Th. 380, 20; Rä. 1, 11.

beadu-cræft, es; *m. War-craft, strength in war;* bellica vis :—Ðé gúþgewinn þurh hǽðenra hilde wóman, beorna beaducræft, geboden wyrþeþ *a war-contest will be offered to thee through the heathens' battle rush, the war-craft of heroes,* Andr. Kmbl. 437; An. 219.

beadu-cræftig, beado-cræftig; *adj. War-crafty, warlike;* bellicosus :— Fugel beaducræftig *the warlike bird,* Exon. 60 a; Th. 217, 26; Ph. 286. Beaducræftig beorn Bartholameus *a warlike chief, Bartholomeus,* Apstls. Kmbl. 87; Ap. 44.

beadu-cwealm, es; *m. A war-death, violent death;* nex :—Ðǽr he sáwulgedál beaducwealm gebád *there he awaited the separation of the soul, a war-death,* Andr. Kmbl. 3400; An. 1704.

beadu-folm, e; *f. A war* or *bloody hand;* bellica manus :—Nán íren blóðge beadufolme onberan wolde *no iron would impair his bloody war-hand,* Beo. Th. 1984; B. 990.

beadu-grim; *adj. War-grim, war-furious;* in pugna atrox, Leo 114.

beadu-gríma, an; *m. A war-mask, helmet.* v. beado-gríma.

beadu-hrægl, es; *n. A war-garment;* bellica vestis, lorica. v. beado-hrægl.

beadu-lác, es; *n. Play of battle, battle, war;* stragis actio, pugna :— Ænig mon to beadulâce ætberan meahte *any man might bear forth to the play of battle,* Beo. Th. 3126; B. 1561. To ðam beadulâce *to the battle-play,* Andr. Kmbl. 2238; An. 1120.

beadu-leóma, an; *m. A war-gleam, sword;* stragis flamma, ensis. v. beado-leóma.

beadu-mægen; *gen.* -mægnes; *n. Battle-strength, military power;* militaris vis, exercitus stragem faciens :—Beadumægnes rǽs, grím-helma gegrind *the rush of battle-strength, the crash of grim helmets,* Cd. 160; Th. 198, 28; Exod. 329.

beadu-méce, es; *m. A battle-sword, sword of slaughter;* pugnæ ensis. v. beado-méce.

beadu-rǽs, es; *m. A battle-rush, onset;* pugnæ impetus :—Biter wæs se beadurǽs *the onset was bitter,* Byrht. Th. 134, 68; By. 111.

beadu-rinc, es; *m. A soldier;* bellicosus vir, miles :—Beadurincum wæs Rôm gerýmed *Rome was laid open by the soldiers,* Bt. Met. Fox 1, 36; Met. 1, 18. v. beado-rinc.

beadu-róf; *adj. War-renowned, bold in war;* in pugna strenuus :—

Beadurôfes beácn *a beacon of the war-renowned,* Beo. Th. 6301; B. 3161. He hǽlo and frôfre beadurôfum abeád *he offered safety and comfort to the bold in war,* Andr. Kmbl. 191; An. 96. v. beado-róf.

beadu-rún, e; *f. A war-secret, quarrel;* jurgiosum arcanum, rixa :— Hûnferþ onband beadurûne *Hunferth unbound the war-secret,* Beo. Th. 1006; B. 501.

beadu-scearp; *adj. Battle-sharp, sharp in fight, applied to a sword;* ad pugnam acutus :—Cyning wælseaxe gebrǽd biter and beaduscearp *the king drew his deadly knife bitter and battle-sharp,* Beo. Th. 5401; B. 2704.

beadu-scrúd, es; *n.* [scrúd *clothes*] *Warlike apparel, warlike garmen a coat of mail;* bellicum vestimentum, lorica :—Beaduscrúda betst míne breóst wereþ *the best of warlike garments defends my breast,* Beo. Th. 910; B. 453.

beadu-searo; *gen.* -searewes, -searwes; *n. A war-train, an engine* or *weapon of war;* bellicus apparatus. v. beado-searo.

beadu-serce, an; *f. A war-shirt, coat of mail;* bellica tunica, lorica :— Ic gefrægn sunu Wihstânes beran beadusercean *I heard that Wihstan's son bore the coat of mail,* Beo. Th. 5503; B. 2755.

beadu-þreát, es; *m. A war-host, an army;* exercitus, Elen. Kmbl. 62; El. 31.

beadu-wǽpen; *gen.* -wǽpnes; *dat.* -wǽpne; *n. A war-weapon;* bellica arma. v. beado-wǽpen.

beadu-wang, es; *m. A battle-plain;* pugnæ campus :—On beadu-wange *on the battle-plain,* Andr. Kmbl. 825; An. 413.

beadu-wég *a war-cup, contest, discussion.* v. beado-wég.

beadu-weorc, es; *n. A war-work, warlike operation;* bellicum opus. v. beado-weorc.

beadu-weorca, an; *m. A war-worker, soldier;* miles, Grm. ii. 449, 34. **Beadu-wulf** *Beowulf.* v. Beado-wulf.

be-æftan; *prep.*   I. *after, behind;* post, pone :—Be-æftan *contracted to* bæftan, *q. v.*   II. *without;* sine :—Beæftan ðære menego *sine turba,* Lk. Bos. 22, 6.

be-æftan; *adv. Behind, after, hereafter;* post, pone, postea :—Ðǽr beæftan forlêt eall *left there all behind,* Ors. 2, 4; Bos, 45, 14. Ðæt ic wille hêr beæftan sweotolor gereccan *that I will hereafter more clearly shew,* Bt. 11, 1; Fox 30, 29.

beærn *a son,* Ps. Spl. T. 28, 1. v. bearn.

be-ǽwnian; *p.* ode; *pp.* od *To join in marriage, marry, wed;* legitime despondere :—Bewedded and beǽwnod *wedded and married,* Chr. 1052; Th. 314, 38. v. ǽwnian.

beaf *a gad-fly;* œstrus = οἴστρος, Leo 118.

beaftan, beaftian; *p.* beaftode, beafte, *pl.* beaftodon, beafton; *pp.* beaftod *To lament;* lamentare :—We mid hondum beafton *lamentavimus,* Mt. Lind. Stv. 11, 17. v. beofian.

beág *a ring, crown;* anulus, corona, Exon. 91 a; Th. 341, 24; Gn. Ex. 131. v. beáh.

beág *gave way,* Exon. 124 a; Th. 477, 2; Ruin. 17; *p. of* búgan.

beágian, biágian; *p.* ode; *pp.* od *To crown, to set a garland on;* coronare :—Of wuldre and weorþmynt ðú beágodest hine *gloria et honore coronasti eum,* Ps. Spl. 8, 6.

beáh, beág, bǽh, bêg, bêh; *gen.* beáges; *dat.* beáge; *pl.* beágas; *m.* [beáh, beág; *p. of* búgan *to bend*] *Metal made into circular ornaments, as A ring, bracelet, collar, garland, crown;* anulus, armilla, diadema, corona. Bracelets were worn about the arms and wrists; rings on the fingers, round the ankles, the neck, and about the head. See Guide to Northern Archæology, by the Earl of Ellesmere, 8vo. 1848, p. 54; also Weinhold, Altnordisches Leben, 8vo. Berlin, 1856, p. 185. These being valuable were probably used in early times as means of exchange or as money; hence the origin of ring-money. v. Sir Wm. Betham's Essay in the Trans. of Rl. Ir. Acd. and Gent's. Mag. April 1837, pp. 372, 373, and May, p. 499 :—Ic nyme ðînne hring and ðînne beáh and ðînne stæf, ðe ðú on handa hæfst *capiam anulum tuum et armillam et baculum, quem manu tenes,* Gen. 38, 18, 25. Gehwearf in Francna fæðm cyninges se beáh *the collar of the king went into the grasp of the Franks,* Beo. Th. 2427; B. 1211. Sceal bryde beág *a ring shall be for a bride,* Exon. 91 a; Th. 341, 24; Gn. Ex. 131. He beágas dǽlde *he distributed bracelets,* Beo. Th. 161; B. 80. Ic frinan wille beága bryttan *I will ask the distributor of bracelets,* Beo. Th. 709; B. 352. Brúc ðisses beáges *make use of this collar,* Beo. Th. 2436; B. 1216. Se beorhta beág hlífaþ ofcr heáfde *the bright garland rises over the head,* Exon. 64 b; Th. 238, 10; Ph. 602. Under gyldnum beáge *under a golden crown,* Beo. Th. 2330; B. 1163. To ðam beáge *to the crown,* Bt. 37, 2; Fox 188, 11. Se beáh gôdes [Cot. MS. beág goodes] *the crown of good,* 37, 2; Fox 188, 21. [*O.Sax.* bôg, *m: M. H. Ger.* bouc, *m: O. H. Ger.* pouc, *m: O. Nrs.* baugr, *m.*] DER. earm-beáh, -beág, heals-, rand-, scanc-, wuldor-.

beáh *submitted,* Chr. 1015; Th. 276, 22; *p. of* búgan.

beáh-gifa, beág-gifa, -gyfa, an; *m. A ring-giver, a giver of ring* or *bracelet money;* anulorum *vel* armillarum largitor :—Se geonga gewât Eádgâr of lífe, beorna beáhgifa *the young Edgar, ring-giver of men,*

departed from life, Chr. 975; Th. 226, 36, col. 2: Byrht. Th. 140, 19; By. 290: Elen. Grm. 100: 1199: Beo. Th. 2208; B. 1102.

beáh-gifu, e; *pl. nom. acc.* a; *gen.* a, ena; *f. A ring-gift, distribution of rings* or *bracelets;* armillarum largitio:—Geongne ædeling sceolan gōde gesíðas byldan tp beáhgife *good companions should exhort a young prince to a distribution of bracelets*, Menol. Fox 490; Gn. C. 15.

beáh-hord, es; *n. A ring-hoard*, Beo. Th. 1792; B. 894.

beáh-hroden [hroden; *pp. of* hreóðan] *Crown-adorned, adorned with bracelets;* armillis *vel* diademate ornatus:—Beáh-hroden [MS. beághroden] cwēn *a queen adorned with bracelets*, Beo. Th. 1251; B. 623.

beáh-sel, es; *n. Hall of bracelets;* domus *vel* aula in qua armillas dominus largitur, Andr. Kmbl. 3312; An. 1659.

beáh-sele, es; *m. Idem*, Beo. Th. 2358; B. 1177.

beáh-þegu, e; *f. A ring-receiving;* armillarum acceptio:—Æfter beáhþege *after the receiving of rings*, Beo. Th. 4358; B. 2176.

beáh-wriða, an; *m. A ringed wreath, armlet, bracelet;* armilla = armilla, quæ brachialis vocatur, *Cic* :—Oft hió beáhwriðan secge sealde *oft she gave a ringed wreath to the warrior*, Beo. Th. 4041; B. 2018.

beal *bellowed, roared; p. of* bellan.

beala-níþ, es; *m. Baleful malice, evil, wickedness*, Ps. C. 50, 111; Ps. Grn. ii. 279, 111. v. bealo-níþ.

bealcan *to emit, utter, pour out;* eructare :—Dæg ðam dæge bealceþ word *dies diei eructat verbum*, Ps. Spl. 18, 2. v. bealcettan.

bealcettan, belcettan, bealcan; *p.* te; *pp.* ted *To belch, utter, send forth, emit;* eructare, dicere, emittere :—Swēte to bealcetenne *pleasant to belch*, Bt. 22, 1; Fox 76, 32. Bealcetteþ heorte mín word gōd *eructat cor meum verbum bonum*, Ps. Spl. 44, 1. Bealcettaþ weleras mīne lofsang *eructabunt labia mea hymnum*, Ps. Spl. 118, 171.

BEALD, bald; *adj.* BOLD, *brave, confident, of good courage;* validus, strenuus, fortis, constans, audax, fidens, bono animo, liber :—He beald in gebēde bídsteal gifeþ *he confident in prayer maketh a stand*, Exon. 71 a; Th. 265, 28; Jul. 388. Beald reordade, eádig on elne *brave he spake, happy in courage*, Exon. 47 b; Th. 163, 24; Gū. 998. He healdeþ Meotudes æ beald in breóstum *bold in his breast he holds the law of the Creator*, Exon. 62 b; Th. 229, 20; Ph. 458. Hí beóþ bealde, ða ðe beorhtne wlite Meotude bringaþ *they will be of good courage, who bring a bright aspect to the Creator*, Exon. 23 b; Th. 66, 25; Cri. 1077. [*Goth.* balþs: *O. Sax.* bald: *O. Frs.* balde, *adv. quickly: O. H. Ger.* bald: *O. Nrs.* ballr.] DER. cyning-beald, cyre-, un-.

bealde, balde; *adv. Boldly, freely, instantly;* audacter, libere, fiducialiter, fidenter, instanter, prone, statim, sine mora :—Of Basan cwæþ bealde Drihten *dixit Dominus ex Basan*, Ps. Th. 67, 22. Bletsige mīne sáwle bealde Dryhten *benedic anima mea Dominum*, Ps. Th. 102, 2 : 65, 18 : 66, 4 : 67, 24 : 72, 16 : 118, 130. Balde, Cd. 182; Th. 228, 11; Dan. 200 : Ps. Th. 113, 25 : 133, 3 : 149, 8.

bealdian; *p.* ode; *pp.* od *To be brave, bear oneself bravely;* strenue *vel* fortiter se gerere :—Swā bealdode bearn Ecgþeówes *thus the son of Ecgtheow bore himself bravely*, Beo. Th. 4360; B. 2177.

beald-líce, bald-líce, bal-líce; *adv.* BOLDLY, *instantly, earnestly, saucily;* audenter, statim :—Ic bealdlíce mínum hondum slóg *I boldly slew with my hands*, Exon. 73 a; Th. 272, 1; Jul. 492. Aoth bleów bealdlíce his horn *Aod statim insonuit buccina*, Jud. 3, 27 : 3, 21.

bealdor, baldor, es; *m. A hero, prince;* princeps :—Wedera bealdor *prince of the Weders*, Beo. Th. 5127; B. 2567. Is hláford mín beorna bealdor *my lord is the prince of men*, Exon. 52 b; Th. 183, 24; Gū. 1332. v. baldor.

bealg *was angry*, Exon. 68 a; Th. 253, 25; Jul. 185; *p. of* belgan.

bealh *was angry, irritated; p. of* belgan.

beallucas *testiculi*, Wrt. Voc. 283, 57.

BEALO, bealu, balu; *gen.* bealowes, bealwes, bealuwes, baluwes; *dat.* bealuwe, bealwe, baluwe, bealo; *acc.* bealu, balu, bealo; *instr.* bealwe, bealuwe; *pl. gen.* bealwa, bealuwa, baluwa; *dat. instr.* balawum; balawun; *n.* I. BALE, *woe, harm, evil, mischief;* malum, calamitas, pernicies, damnum, noxa, tribulatio :—Hæfdon bealo *they had woe*, Cd. 214; Th. 269, 10; Sat. 71. Bealowes gāst *spirit of evil* [diabolus], Cd. 228; Th. 307, 19; Sat. 682. Oft heó to bealwe bearn afēdeþ *often she nourisheth her child to woe*, Salm. Kmbl. 745; Sal. 372. Him to bealwe *to their own harm*, Exon. 24 a; Th. 68, 19; Cri. 1106. Bealwe gebǣded *by calamity compelled*, Beo. Th. 5644; B. 2826. Ne ondrǣde ic ðínra wíta bealo *I dread not the evil of thy torments*, Exon. 68 b; Th. 255, 9; Jul. 211. II. *wickedness, depravity;* malities, nequitia :—Me wið blóðhreówes weres gehǣle *preserve me against the wickedness of the blood-thirsty man*, Ps. Th. 58, 2. [*O. Sax.* balu : *O. Frs.* balu : *O. H. Ger.* balo : *Goth.* balweins *punishment, pain: O. Nrs.* böl : *Slav.* böl *pain*.] DER. aldor-bealo [-bealu], ealdor-, feorh-, firen-, folc-, helle-, hreðer-, leód-, mán-, morþ-, morþor-, niht-, sweord-, þeód-, un-, wíg-.

bealo-ben, -benn, e; *f. A baleful wound.* v. bealu-ben.

bealo-blonden; *pp. Mixed with bale, pernicious;* pernicie mixtus, perniciosus :—Bealoblonden níþ *pernicious hate*, Exon. 92 a; Th. 345, 30; Gn. Ex. 198.

bealo-clom, -clomm, es; *m :* e; *f. A dire chain.* v. bealu-clom.

bealo-cræft, balo-cræft, es; *m. A wicked, pernicious,* or *magic art;* perniciosa *vel* magica ars, Bt. Met. Fox 26, 150; Met. 26, 75.

bealo-cwealm, es; *m. A pernicious* or *violent death;* perniciosa *vel* violenta mors, Beo. Th. 4523; B. 2265.

bealo-dǣd, bealu-dǣd, e; *f. A wicked, evil,* or *sinful deed;* peccatum :—Ðæt hý bealodǣde gescomeden *that they felt shame for a sinful deed*, Exon. 27 a; Th. 80, 4; Cri. 1302.

bealo-ful, -full; *def.* se bealo-fulla; *adj.* BALEFUL, *dire, cursed, wicked;* pestiferus, facinorosus, scelestus, malitiosus :—Bealofull *baleful*, Judth. 10; Thw. 22, 15; Jud. 63. Se bealofulla hýneþ heardlíce *the baleful one hardly oppresseth*, Exon. 11 b; Th. 16, 27; Cri. 259. Heó ðone bealofullan alēde mannan *she laid down the odious man*, Judth. 10; Thw. 23, 2; Jud. 100. Biter bealofullum *bitter to the baleful*, Exon. 21 a; Th. 56, 31; Cri. 909.

bealo-fús; *adj. Inclined to sin;* peccandi pronus, Exon. 94 b; Th. 354, 23; Reim. 50.

bealo-hycgende; *part. Intending evil;* perniciem moliens :—Æghwæðrum wæs bealo-hycgendra brōga fram ōðrum *to either of them, intending evil, was a fear of the other*, Beo. Th. 5123; B. 2565.

bealo-hydig; *adj. Intending evil, baleful-minded;* perniciem moliens, Beo. Th. 1450; B. 723.

bealo-inwit, es; *n. Guile, deceit.* v. bealu-inwit.

bealo-leás; *adj. Void of evil, innocent;* innocens, Exon. 89 b; Th. 335, 27; Gn. Ex. 39.

bealo-níþ, beala-níþ, bala-níþ, es; *m. Baleful malice, evil, wickedness;* pravum *vel* perniciosum studium, pernicies, calamitas :—Him on breóstum bealoníþ weóll *baleful malice boiled in his breast*, Beo. Th. 5422; B. 2714. Bebeorh ðē ðone bealoníþ *keep from thee that baleful evil*, Beo. Th. 3520; B. 1758.

bealo-ráp, es; *m. A pernicious cord;* dirus laqueus, Exon. 13 a; Th. 23, 7; Cri. 365.

bealo-searu; *g.* -searwes; *n. A wicked machination* or *snare;* malitiosa machinatio, Exon. 72 b; Th. 270, 30; Jul. 473.

bealo-síþ, bealu-síþ, es; *m.* I. *an evil fortune, misfortune, calamity;* calamitas, adversa fortuna :—Bealosíþa hwōn *few* [of] *misfortunes*, Exon. 81 b; Th. 307, 24; Seef. 28. II. *a destructive* or *deadly path, death;* fatale iter, mors, Cd. 143; Th. 178, 1; Exod. 5.

bealo-sorg, e; *f. Baleful sorrow;* dirus ægritudo *vel* mæror, Exon. 61 b; Th. 226, 21; Ph. 409.

bealo-spell, es; *n. A baleful message* or *tale;* perniciei nuntius, Cd. 169; Th. 210, 5; Exod. 510.

bealo-þanc, -þonc, es; *m. A baleful* or *wicked thought;* prava *vel* malitiosa cogitatio, Exon. 72 b; Th. 270, 22; Jul. 469.

bealo-ware; *gen.* -wara, *pl. m. Baleful inhabitants, criminals;* scelesti. v. bealu-ware.

bealu, balu; *adj. Baleful, pernicious, wicked, malicious;* dirus, perniciosus, pravus, malus, malitiosus :—Awrítaþ hie on his wæpne bealwe bōcstafas *they cut baleful letters upon his weapon*, Salm. Kmbl. 325; Sal. 162. v. bealo.

bealu-ben, -benn, e; *f. A baleful wound;* lethale vulnus, Cd. 154; Th. 192, 27; Exod. 238.

bealu-clom, -clomm, es; *m :* e; *f. A dire chain;* dirum vinculum :—Under bealuclommum *under dire chains*, Exon. 120 b; Th. 463, 5; Hö. 65.

bealu-dǣd, e; *f. An evil deed*, Elen. Kmbl. 1027; El. 515. v. bealo-dǣd.

bealu-inwit, es; *n. Guile, deceit;* dolus, Ps. Th. 54, 24.

bealu-síþ, es; *m. A destructive* or *deadly path, death;* fatale iter, mors, Cd. 143; Th. 178, 1; Exod. 5. v. bealo-síþ.

bealu-ware; *gen.* -wara; *pl. m. Baleful inhabitants, criminals;* scelesti :—Ðæt ic bealuwara weorc gebiden hæbbe *that I have endured the work of criminals*, Rood Kmbl. 155; Kr. 79.

BEÁM, es; *m.* I. *a tree;* arbor :—Se beám bude wyrda geþingu *the tree boded the councils of the fates*, Cd. 202; Th. 250, 11; Dan. 545 : 23; Th. 30, 18; Gen. 468 : 24; Th. 31, 1; Gen. 478. On ðæs beámes blēdum *on the branches of the tree*, 200; Th. 248, 4; Dan. 508 : Exon. 114 a; Th. 437, 14; Rä. 56, 7. On ðam beáme *on the tree*, Cd. 24; Th. 31, 11; Gen. 483 : Exon. 57 b; Th. 206, 6; Ph. 122. Forlǣtaþ ðone ænne beám *abstain from the one tree*, Cd. 13; Th. 15, 19; Gen. 235 : 25; Th. 31, 28; Gen. 492. Twegen beámas stōdon ofætes gehlædene *two trees stood laden with fruit*, 23; Th. 30, 2; Gen. 460 : Exon. 56 a; Th. 200, 4; Ph. 35. Ic beámas fylle *I fell the trees*, 101 a; Th. 381, 11; Rä. 2, 9. II. *the tree, cross;* patibulum, crux :—Wæs se beám bōcstafum awriten *the cross was inscribed with letters*, Elen. Kmbl. 181; El. 91: Exon. 24 a; Th. 67, 17; Cri. 1090. Se ðe deáþes wolde biteres onbyrigan on ðam beáme *who would taste of bitter death on the cross*, Rood Kmbl. 226; Kr. 114: Cd. 224; Th. 296, 30; Sat. 510. He on ðone hálgan beám ahongen wæs *he was hung on the holy cross*, Exon. 24 a; Th. 67, 25; Cri. 1094 : 29 a; Th. 88, 29; Cri. 1447. III. *a column, pillar;* columna :—Hæfde wuldres beám

werud geléded *the pillar of glory had led the host*, Cd. 170; Th. 214, 10; Exod. 566: 148; Th. 184, 22; Exod. 111. God hét him fýrenne beám befóran wísian *God commanded a pillar of fire to point out the way before them*, Ps. Th. 104, 34. Him befóran fóron beámas twegen *two pillars went before him*, Cd. 146; Th. 183, 20; Exod. 94.   **IV.** *wood, a ship*; lignum, navis :—Ic of fædmum cwom brimes and beámes *I came from the clutches of sea and ship*, Exon. 103 b; Th. 392, 13; Rä. 11, 7.   **V.** *a* BEAM, *splint, post, a stock of a tree*; trabs, stipes :—Se beám biþ on ðínum ágenum eágan *trabs est in oculo tuo*, Mt. Bos. 7, 4. Bunden under beáme *bound under a beam*, Exon. 126 a; Th. 485, 9; Rä. 71, 11. Ðú ne gesyhst ðone beám on ðínum ágenum eágan *trabem in oculo tuo non vides*, Mt. Bos. 7, 3, 5. Heora ærenan beámas ne mihton fram Galliscum fýre forbærnede weorþan *their brazen beams could not be destroyed by the fire of the Gauls*, Ors. 2, 8; Bos. 52, 16. Of beáme *de stipite*, Cot. 63.   **VI.** *in composition,* anything proceeding in a right line, hence,—*A ray of light, a* sun-BEAM; radius :—Cométa, se steorra, scán swilce sunne-beám *a comet, the star, shone like a sun-beam*, Chr. 678; Erl. 41, 5.   **VII.** in the Northumbrian Gospels beám is put for býme *a trumpet*; tuba :—Mið beám *cum tuba*, Mt. Kmbl. Lind. 24, 31. [*Tynd.* beame: *Chauc. Wyc.* beme: *R. Glouc.* beam, bem: *Laym.* beam, bem: *O. Sax.* bôm, *m*: *N. Frs.* baem, beamme, bjemme: *O. Frs.* bâm, *m*: *Dut.* boom, *m*: *Ger.* baum, *m*: *M. H. Ger.* boum, *m*: *O. H. Ger.* poum, *m*: *Goth.* bagms, *m*: *Icel.* baðmr, *m*.] DER. beg-beám, ceder-, deáþ-, ele-, fíc-, firgen-, gár-, gleó-, sige-, wer-, wudu-, wyn-.

Beám-dún, Beán-dún, e; *f.* BAMPTON, *Devonshire*; oppidum situm esse arbitror in agro Devoniensi, qua Somersætensibus adjacet, et vocari hodie *Bampton*, Gibson Chr. Explicatio, p. 14, col. 1 :—Hér Cynegils and Cwichelm gefuhton on Beámdúne *in this year Cynegils and Cwichelm fought at Bampton*, Chr. 614; Th. 38, 38, cols. 2, 3. [beám *a tree*; dún *a hill, down*; collis stipitibus seu trabibus refertus, Gibson.]

Beám-fleót, es; *m.* The name of places now called *Beamfleet* [*Beamfled*, Hunt.] *Bamfleet, Benfleet, Essex*; æstuarii nomen in agro Essexiensi, hodie *Benfleet* :—Hie fóron eást to Beámfleóte *they marched east to Benfleet*, Chr. 894; Erl. 91, 15.

beámian; *p.* ede; *pp.* ed *To shine, to cast forth rays* or *beams like the sun*; radiare, Som.

beám-sceadu, e; *f. A tree-shade, the shade of a tree*; arborum umbra :—Gewitan him ðá gangan under beámsceade *then they retired under the tree-shade*, Cd. 40; Th. 53, 10; Gen. 859. Hí slépon under beámsceade *they slept under the tree-shade*, Bt. Met. Fox 8, 55; Met. 8, 28.

beám-telg, es; *m. Dye of a tree* [*ink*]; tinctura arborea [atramentum scriptorium] :—Fugles wyn beámtelge swealg *the bird's joy* [i. e. *the pen*] *swallowed dye of a tree*, Exon. 107 a; Th. 408, 9; Rä. 27, 9.

BEÁN, bién, e; *f. A* BEAN, *all sorts of pulse*; faba, legumen :—Beán pisan *a vetch*, Cot. 34: 122. [*Plat. Dut.* boon, *f*: *Ger.* bohne, *f*: *M. H. Ger.* bône, *f*: *O. H. Ger.* pôna, *f*: *Dan.* bönne: *Swed.* böna: *O. Nrs.* baun, *f*: *Lat.* faba, *f*.]

beán-belgas, beán-coddas; *pl. m.* [beán *a bean*, belg or codd *a bag*] *Bean-pods, husks, cods* or *shells*; fabarum sacculi, siliquæ :—Of ðám beáncoddum *de siliquis*, Lk. Bos. 15, 16: Cot. 200.

beand, es; *m. A band, bond*; vinculum :—On beandon *in bonds* or *captivity*; in vinculis, Chr. 1069; Erl. 207, 15. v. bend.

Beán-dún, e; *f. Bampton, Devonshire*, Chr. 614; Th. 38, 38, col. 1; 39, 37, col. 1; Erl. 20, 36; 21, 35. v. Beám-dún.

beánen; *adj. Beany, belonging to beans*; fabarius :—Beánene melewe BEAN-MEAL, Herb. 155, 3; Lchdm. i. 282, 9.

beán-scealas BEAN-SHELLS; siliquæ, quisquiliæ, Cot. 200.

Bearan burh; *gen.* burge; *dat.* byrig; *f. Banbury, Oxfordshire.* v. Beran burh.

BEARD, es; *m.*   **I.** *a* BEARD; barba :—Ne beard ne sciron nec radetis barbam, Lev. 19, 27; nether ge schulen schave the beerd, *Wyc.* Smyringc niðerfeól on bearde, bearde Aarones *unguentum descendit in barbam, barbam Aaronis*, Ps. Lamb. 132, 2.   **II.** *the Anglo-Saxons were proud of their beards, and to shave a layman by force was a legal offence* :—Gif mon ðone beard ofascire, mid xx scillinga gebéte. Gif he hine gebinde, and ðonne to preoste bescire, mid LX scillinga gebéte *if a man shave off the beard, let him make amends* [*boot*] *with* XX *shillings. If he bind him, and then shave him like a priest, let him make amends* [*boot*] *with* LX *shillings*, L. Alf. pol. 35; Th. i. 84, 8. [*Laym.* baerd: *Plat. Dut.* baard, *m*: *Frs.* berd, bird, *m*: *Ger.* bart, *m*: *Icel.* bart, *n*.]

beard-leás; *adj.* BEARDLESS; imberbis. Used as a noun, it denotes those *without a beard*, as *a youth, stripling*, also *a hawk* or *buzzard*; ephebus, buteo :—Beardleás *ephebus*, vel *buteo*, Ælfc. Gl. 87; Som. 74, 51; Wrt. Voc. 50, 33.

BEARG, bearh, es; *m. A castrated boar, a barrow pig*; majalis :—Amæsted swín, bearg bellende on bôc-wuda *a fattened swine, a barrow pig* [*castrated boar*] *grunting in beech woods*, Exon. 111 b; Th. 428, 10; Rä. 41, 106. Bearh *majalis*, Ælfc. Gl. 20; Som. 59, 31; Wrt. Voc. 22, 72. [*Plat.* borg, *m. a castrated boar pig*: *Dut.* barg, *m*: *Frs.* baerg, *m*: *Ger.* borg-schwein: *O. H. Ger.* barc, barg, *m. porcus castratus*.]

bearg, bearh *saved, secured*, Exon. 55 a; Th. 195, 21; Az. 159; *p. of* beorgan.

bearh *saved*, Cd. 124; Th. 158, 29; Gen. 2624; *p. of* beorgan.

bearht *bright*, Ps. Spl. 22, 7. v. beorht.

bearhtm, es; *m. A noise, tumult, clamour, sound, cry*; fragor, strepitus, tumultus, clamor :—Ic on ðisse byrig bearhtm gehýre *I hear a tumult in this city*, Cd. 109; Th. 145, 16; Gen. 2406. v. breahtm, brecan *to break*.

bearhtm, es; *m. Brightness, glittering, scintillation, twinkling, glance*; claritas, splendor, nitor, scintillatio, acies :—Eágena bearhtm forsiteþ and forsworceþ *the brightness of the eyes vanishes and darkens*, Beo. Th. 3537; B. 1766. Ðæt biþ an eágan bearhtm [MS. bryhtm] *that is in the twinkling of the eye, in a moment*, Bd. 2, 13; S. 516, 20. DER. bearht, beorht *bright*.

bearhtm-hwíl, byrhtm-hwíl, e; *f. A twinkling while, a moment*; oculi nictus tempus, momentum :—On ánre byrhtmhwíle *in momento temporis*, Lk. Bos. 4, 5.

bearhtnes *brightness.* v. beorhtnes.

bearm, es; *m. The bosom, lap*; sinus, gremium :—On eówerne bearm *in sinum vestrum*, Lk. Bos. 6, 38. Iosep hí nam of ðæs fæder bearme *Ioseph eos tulit de gremio patris*, Gen. 48, 12: Cd. 216; Th. 274, 12; Sat. 153. Ðá wæs fæger foldan bearm *then was earth's bosom fair*, Beo. Th. 2278; B. 1137. Alédon leófne þeóden on bearm scipes *they laid the beloved chief in the ship's bosom*, Beo. Th. 70; B. 35: Exon. 101 b; Th. 382, 28; Rä. 4, 3. [*Chauc.* barme *the bosom*: *O. Sax.* barm, *m.* sinus, gremium: *O. Frs.* barm-bracco *a lap-dog*: *O. H. Ger.* barm, *m.* barms, *m*: *Icel.* barmr, *m.* I. *the brim of anything*; ora, margo; II. *the bosom*; gremium: from beran, beoran *to bear, to carry in folded arms*, or *on the bosom*.]

bearm-cláþ, es; *n. A* BARME-CLOTH [*Chauc. The Milleres Tale*, 3237], *a bosom-cloth, an apron*; sinui imposita mappula :—Bearmcláþ *mappula*, Wrt. Voc. 26, 68.

bearm-rægl, es; *m. A bosom-garment*; sinui imposita vestis *vel* mappula, Wrt. Voc. 26, 28.

bearn, es; *n. A* BEARN, *child, son, issue, offspring, progeny*; natus, infans, puer, filius, soboles, proles :—Bearn Godes *Son of God*, Elen. Kmbl. 1624; El. 814. Nú is ðæt bearn cymen *now is that child come*, Exon. 8 b; Th. 5, 8; Cri. 66. Híg næfdon nán bearn *non erat illis filius*, Lk. Bos. 1, 7. Þurh bearnes gebyrd *through the birth of a child*, Exon. 8 b; Th. 3, 18; Cri. 38. Beón mid bearne *gravidam esse*, Somn. 370. Bearn soboles vel *proles*, Ælfc. Gl. 91; Som. 75, 19. Geáta bearn *the sons of the Goths*, Beo. Th. 4374; B. 2184. He Noe gebletsade and his bearn *he blessed Noah and his sons*, Cd. 74; Th. 91, 1; Gen. 1505. Ðys cynd Israéla bearna naman *hæc sunt nomina filiorum Israel*, Ex. 1, 1. Geseah his bearna bearn *vidit filios filiorum suorum*, Job Thw. 168, 35. Ge Godes bearn, bringaþ Gode ramma bearn *filii Dei, afferte Domino filios arietum*, Ps. Th. 28, 1. [*Piers* barn *a child*: *Scot. and Northumb.* bairn: *O. Sax.* barn, *n*: *O. Frs.* bern, *n*: *O. H. Ger.* barn, *n*: *Goth.* barn, *n*: *Dan. Swed. Icel.* barn, *n. a child*: *what is borne, from* beran *to bear*.] DER. cyne-bearn, dryht-, folc-, freó-, frum-, god-, hælu-, húsel-, sige-, þryþ-, woruld-. v. beran.

bearn, es; *n. A barley-place, a* BARN; horreum :—He gadereþ hys hwǽte on his bearn *congregabit triticum suum in horreum*, Mt. Kmbl. Hat. 3, 12. v. bere-ærn.

be-arn *occurred*, Wanl. Catal. 154, 5; *p. of* be-yrnan.

bearn *burned, consumed*; *p. of* beornan.

bearn-cennung, e; *f. Child-birth*; puerperium. v. cenning, *from* cennan *parere*.

bearn-eácen [bearn *a child*, eacen *increased*] *Increased, pregnant*; auctus, gravidus :—Bearneácen wíf þrówaþ micel earfoðu *a pregnant woman suffers much trouble*, Bt. 31, 1; Fox 112, 2, note 2, Cott: L. Alf. pol. 9; Th. i. 66, 23. DER. eácan.

bearn-eácnung, e; *f. Generation, conception, pregnancy*; genitura, conceptio, prægnatio. v. eácnung.

bearnende *burning*; ardens, Jn. Lind. War. 5, 35. v. bernende; *part. of* byrnan.

bearn-gebyrdo; *indecl. f. Child-bearing*; partus :—Hyre eald Metod éste wǽre bearngebyrdo *to her the ancient Creator was gracious in her child-bearing*, Beo. Th. 1896; B. 946.

bearn-gestreón, es; *n. Child-procreation*; liberorum procreatio :—Ðæt ic þolian sceal bearngestreóna : ic wið brýde ne mót hǽmed habban *that I shall lack child-procreation : with a bride I may not have intercourse*, Exon. 105 b; Th. 402, 9; Rä. 21, 27.

bearn-leás; *adj. Childless*; absque liberis :—Bearnleásne ge habbaþ me gedónne *absque liberis me esse fecistis*, Gen. 42, 36: Ex. 21, 22.

bearn-lést; *f. Childlessness, want of children*; liberorum defectus *vel* orbitas, eorum conditio qui liberis carent :—For bearnléste *for want of children*, Bt. 11, 1; Fox 32, 6.

bearn-lufe, an; *f. Child-love, love of one's own* or *of an adopted child*; liberorum amor, filii sui *vel* adoptivi amor :—Hine on bearnlufan habban wolde *eum loco adoptivi haberet*, Bd. 5, 19; S. 638, 4.

bearn-myrþra, an; *m. A child-murderer, an infanticide;* liberorum interfector, Lupi Serm. i. 19; Hick. Thes. ii. 105, 5.

bearn-teám, es; *m. A succession of children, issue, posterity;* liberorum ordo *vel* successio, soboles :—Ðæt hí to raðe woldon fultumleáse beón æt hiora bearnteámum *that they should very soon be without help from posterity*, Ors. 1, 14; Bos. 37, 19. [*Scot.* barn-teme, bairn-time *a brood of children, all the children of one mother.*]

BEARO, bearu; *gen.* bearwes; *dat.* bearwe, bearowe, bearuwe; *acc.* bearo; *pl. nom. acc.* bearwas; *gen.* -wa; *dat.* -wum; *m. A grove, wood;* nemus *vel* lucus, silva, virgultum :—Se hálga bearo sette *the holy man planted a grove*, Cd. 137; Th. 172, 7; Gen. 2840. Wæter wynsumu bearo ealne geondfarað *pleasant waters pervade all the grove*, Exon. 56 b; Th. 202, 10; Ph. 67. Bearu *nemus* vel *lucus*, Wrt. Voc. 32, 38. Se fugel of ðæs bearwes beáme gewíteþ *the fowl departs from the tree of the grove*, Exon. 57 b; Th. 206, 5; Ph. 122 : 58 a; Th. 207, 27; Ph. 148. Wíc mid bearuwe ymbsealde *mansions surrounded with a grove*, Bd. 5, 2; S. 614, 31. In bearwe, on bearwe *or* on bearowe *in a wood*, Cot. 109. Heó begeát gréne bearwas *she gained the green groves*, Cd. 72; Th. 89, 13; Gen. 1480. [Heyne says a *bearing* or a *fruit-bearing tree, hence trees in general, a wood: O. Nrs.* börr, *m. arbor.*] DER. æppel-bearo, sun-, wudu-.

Bearocscýre, Bearucscýre, Bearwucscíre *Berkshire.* v. Barocscír.

bearo-næs, -næss, es; *m. A woody shore* or *promontory;* litus nemorosum :—Trædaþ bearonæssas *they tread the woody promontories*, Exon. 114 b; Th. 439, 5; Rä. 58, 5.

bearowe *in a wood*, Menol. Fox 496; Gn. C. 18. v. bearo.

bears *a perch;* lupus. v. bærs.

bear-swinig; *adj. openly wicked, a publican,* Lk. Rush. War. 3, 12 : 15, 1. v. bær-synnig.

bearu *a grove*, Wrt. Voc. 32, 38. v. bearo.

bearug *a barrow-pig.* v. bearg.

bearuwe *with a grove*, Bd. 5, 2; S. 614, 31. v. bearo.

bearwas, bearwe, bearwes, Exon. 57 b; Th. 206, 5; Ph. 122. v. bearo.

BEÁTAN; *part.* beátende; ic beáte, ðú beátest, býtst, he beáteþ, být, *pl.* beátaþ; *p.* beót, *pl.* beóton; *pp.* beáten. I. *to BEAT, strike, lash, dash, hurt;* percutere, tundere, verberare, cædere, pulsare, quatere, lædere :—Agynþ beátan hys efenþeówas *cœperit percutere conservos*, Mt. Bos. 24, 49. Hwî beátst ðú me *quid me cædis?* Jn. Bos. 18, 23. Ðá Balaam beót ðone assan *cum Balaam verberaret asinam*, Num. 22, 23. Streámas staðu beátaþ *streams beat the shores*, Exon. 101 a; Th. 382, 4; Rä. 3, 6. Sǽ on staðu beáteþ *the sea lashes against the shore*, Bt. Met. Fox 6, 30; Met. 6, 15. Beóton brimstreámas *the sea-streams dashed*, Andr. Kmbl. 477; An. 239 : 3084; An. 1545. Ne se bryne beót mæcgum *nor did the burning hurt the youths*, Cd. 187; Th. 232, 24; Dan. 265. II. to beat with the feet,—*to tread, trample, tramp;* calcare, proculcare :—Se mearh burhstede beáteþ *the steed tramps the castle-place*, Beo. Th. 4522; B. 2265. [*Ger.* boszen *to beat: M. H. Ger.* bôzen *id: O. H. Ger.* pôzan *id: O. Nrs.* bauta *id.*] DER. a-beátan, ge-, of-, ofa-, to-.

beátere, es; *m. A* BEATER, *fighter, champion;* pugil, Ælfc. Gr. 9, 8.

beáw-hyrnet = beó-hyrnet, -hyrnett, e; *f. A bee-hornet, gad-fly, horse-fly;* œstrus = οἶστρος :—Beáw-hyrnet *œstrus* [MS. beáw-hyrnette œstrum, *acc?*], Ælfc. Gl. 22; Som. 59, 108; Wrt. Voc. 23, 64. v. beó, hyrnet.

be-baðian, bi-baðian; *p.* ode; *pp.* od *To bathe, wash;* luere, abluere, lavare :—Wætere aþwegen and bebaðod *lotus aqua*, Bd. 1, 27; S. 496, 17.

Bebba-burh *Bamborough*, Chr. 1095; Th. 361, 39, 40 : 362, 1. v. Bebban burh.

Bebban burh, Chr. 547; Th. 28, 25; 29, 24 : 641; Th. 49, 3 : 993; Th. 240, 17; 241, 16, col. 2 : Bæbba-burh, Chr. 1093; Th. 360, 6 : Bebba-burh, Chr. 1095; Th. 361, 39, 40 : *gen.* -burge; *dat.* -byrig; *acc.* -burg, -burh; *f.* BAMBOROUGH, *in Northumberland:*—Bebba oppidum in provincia Northanhymbrorum :—Hér Ida féng to ríce, ðonon Norþanhymbra cyne-cyn onwóc, and ríxode twelf geár. He timbrode Bebban burh, seó wæs ǽrost mid hegge betýned, and ðǽr æfter mid wealle *here* [A. D. 547] *Ida began to reign, from whom arose the royal race of the Northumbrians, and reigned twelve years. He built Bamborough, which was at first inclosed by a hedge, and afterwards by a wall*, Chr. 547; Erl. 16, 7-10. From Bebban byrig *from Bamborough*, Chr. 926; Th. 199, 31. Ðá becom Penda, Myrcna cyning, to ðære cynelícan byrig, seó is nemned Bebban burh *then came Penda, king of the Mercians, to the royal city, which is named Bamborough*, Bd. 3, 16; S. 542, 18 : 3, 6; S. 528, 28. Hér wæs Bæbban burg tobrocon, and mycel herehúðe ðǽr genumen *here* [A. D. 993] *Bamborough was destroyed, and much spoil was there taken*, Chr. 993; Erl. 133, 1. [Bebba, æ; *f. Lat:* Bebbe, an; *f. Bebba, the name of a queen:* burh *a borough, corporate town;* hence Bebban burh *Bebba's burgh* or *city;* Bebbæ urbs. Bede calls it,—'Urbs regia, quæ a Regina quondam vocabulo Bebba cognominatur,' Bd. 3, 6; S. 109, 22. We thus see that the town had its name from queen Bebba. It is probable that king Ida, who built the town, did not give it this name; but his grandson, Ædilfrid, as Nennius says,—'Eadfered [= Ædilfrid] dedit uxori suæ [urbem], quæ

vocatur Bebbab, et de nomine suæ uxoris suscepit nomen, id est Bebbanburch,' Nenn. 63, ed. Stevens; Bd. Gidl. 187, note 1. Bebban burh was written in succeeding ages,—Bebbanburc, Flor. A. D. 1117 : Bebanburgh, Bebamburgh, Babanburch, Hunt. A. D. 1148 : Babbanburch, Bebbanburc, Dun. A. D. 1164 : Babanburch, Ric. A. D. 1184 : Bebbamburg, Hovd. A. D. 1204 : Bamburgh, Kni. A. D. 1395 : now, in 1873, Bamborough.]

bebeád *commanded*, Elen. Kmbl. 1417; El. 710; *p. of* be-beódan.

be-beódan, bi-beódan; *part.* be-beódende, he be-být; *p.* be-beád, *pl.* be-budon; *impert.* be-beód; *pp.* be-boden. I. *to give a by-command or a gentle command, but generally to command, order;* jubere, præcipere, mandare :—He hys englum bebýt *angelis suis mandavit*, Lk. Bos. 4, 10. Bebeód Iosue *præcipe Iosue*, Deut. 3, 28 : Ps. Th. 67, 26 : Ex. 16, 16. Swá him God bebeád *as God commanded him*, Frag. Kmbl. 75; Leás. 39. Hí bebudon him *præceperunt illi*, Bd. 4, 24; S. 597, 35. Ðǽm laudbúendum is beboden, ðæt ealles ðæs ðe him on heora ceápe geweaxe, híg Gode ðone teóðan dǽl agyfen *to farmers it is commanded, that of all which increases to them of their cattle, they give the tenth part to God*, L. E. I. 35; Th. ii. 432, 27. II. *to offer, give up, commend;* offerre, commendare, mandare :—Dú scealt leófes líc forbærnan and me lác bebeódan *thou shalt burn the beloved's body and offer it me as a sacrifice*, Cd. 138; Th. 173, 9; Gen. 2858. On handa ðíne ic bebeóde gást mínne *in manus tuas commendo spiritum meum*, Ps. Spl. 30, 6 : Hy. 4, 5; Hy. Grn. ii. 283, 5 : Ps. Th. 132, 4. III. *to announce;* nuntiare, pronuntiare :—He bebeád wyrd gewordene *he announced the event that had passed*, Cd. 197; Th. 245, 29; Dan. 470. v. beódan.

be-beódend, es; *m. One who commands, a master;* præceptor, Lk. Bos. 5, 5 : 9, 33.

be-beódendlíc gemet, beódendlíc gemet, es; *n. The imperative mood;* modus imperativus :—Ðæt óðer modus is imperativus, ðæt is bebeódendlíc; mid ðam gemete we hátaþ óðre menn dón sum þingc, oððe sum þingc þrówian,—Rǽd ðú *lege*, rǽde he *legat*, beswing ðis cild *flagella istum puerum*, sí he beswungen *flagelletur*. Ðis gemet sprecþ forþwerd, and næfþ nænne *præteritum*, forðande nán mann ne hǽt dón ðæt ðe gedón biþ *the other mood is the imperative, that is the commanding; with this mood we order other people to do something, or to suffer something,—Read thou, let him read, beat this child, let him be beaten. This mood speaketh directly* [*forthward* or *to those present*], *and has no preterite, because no man commands to do what is done*, Ælfc. Gr. 21; Som. 23, 20-24.

be-beorgan; *p.* -bearg, *pl.* -burgon; *pp.* -borgen *To defend oneself, to take care;* cavere ab aliqua re :—He him bebeorgan ne con wóm *he cannot defend himself against the evil*, Beo. Th. 3497; B. 1746 : 3520; B. 1758.

beber *a beaver*, Som. Lye. v. befor.

be-beran; he -byreþ; *p.* -bær *To bear* or *carry to, provide, supply;* afferre, instruere :—Gif man mannan wǽpnum bebyreþ *if one supply a man with weapons*, L. Ethb. 18; Th. i. 6, 19. v. beran.

be-biddan *to command.* v. biddan.

be-bindan; *p.* -band, -bond, *pl.* -bundon; *pp.* -bunden [be, bindan, *q. v.*] *To bind in* or *about;* inligare, Bd. 3, 11; S. 536, note 9.

be-birgan, -birigan; *p.* de; *pp.* ed *To bury;* sepelire :—Mín fæder me byd ðæt ic hine bebirgde *pater meus adjuravit me, ut eum sepelirem*, Gen. 50, 5 : 50, 6. He hine bebirigde *he buried him*, Ælfc. T. Grn. 6, 2. Hine bebirgdon *sepelierunt eum*, Gen. 50, 13. Bebirged *sepultus*, 50, 14. Ðǽr wæs Isaac bebirged, and ðǽr líþ eác Lia bebirged *ibi sepultus est Isaac, ibi et Lia condita jacet*, 49, 31. v. be-byrgan.

be-birigan; *p.* de; *pp.* ed *To bury*, Gen. 49, 29. v. be-byrigan.

be-blonden; *pp. infected, dyed;* infectus, tinctus. v. blandan.

be-bod, bi-bod, es; *pl. nom. acc.* u, o; *gen.* a; *dat.* um; *n. A command, mandate, decree, order;* mandatum, jussum :—Hwilc ðære geógoþe gleáwost wǽre bôca bebodes *which of the youth was most skilful in the precepts of books*, Cd. 176; Th. 221, 2; Dan. 82. Eall ðín bebodu *omnia mandata tua*, Ps. Th. 118, 172. Ealra beboda mǽst *primum omnium mandatum*, Mk. Bos. 12, 28. Hí brǽcon bebodo *they broke the commandments*, Cd. 188; Th. 234, 28; Dan. 299.

be-bodan *to command*, Ps. Spl. 67, 31. v. be-beódan.

be-boden *commanded, commended;* *pp. of* be-beódan.

be-bohte *sold*, Cd. 226; Th. 301, 5; Sat. 577; *p. of* be-bycgean.

be-bond *bound*, Bd. 3, 11; S. 536, note 9; *p. of* be-bindan.

be-boren-inniht *born within a country, free of a country, native;* municipalis, Cot. 136. v. beran.

be-brecan, he, heó -briceþ, -bricþ; *p.* -bræc, *pl.* -brǽcon; *pp.* -brocen *To break off, deprive by breaking, to break to pieces, consume;* carpendo spoliare, confringere, consumere :—Beám heó abreóteþ and bebriceþ telgum *it crusheth the tree and deprives it of its twigs*, Salm. Kmbl. 592; Sal. 295. Bebrocene wǽron ealle hyra hláfas *consumpti erant omnes eorum panes*, Gr. Dial. 2, 21.

be-bregdan; *p.* -brægd, *pl.* -brugdon; *pp.* -brogden *To pretend;* simulare, Lk. Lind. War. 20, 20. v. bregdan.

be-briceþ, -bricþ *breaks off, deprives by breaking*, Salm. Kmbl. 592; Sal. 295. v. be-brecan.

be-brocen *broken, consumed,* Gr. Dial. 2, 21; *pp. of* be-brecan.

be-brugdon *they pretended,* Lk. Lind. War. 20, 20; *p. of* be-bregdan.

be-búgan, bi-búgan; *p.* -beág, *pl.* -bugon; *pp.* -bogen.   **I.** *to avoid;* avertere, evitare :—Ne meahte he ða gehðu bebúgan *he could not avoid the sorrow,* Elen. Kmbl. 1215; El. 609: Ps. Th. 138, 17.   **II.** *to surround, encircle, encompass ;* circumire, circumcingere :—Swá wæter bibúgeþ ðisne beorhtan bósm *so far as the water encircles this bright expanse,* Exon. 95 b; Th. 356, 4; Pa. 6: Cd. 190; Th. 236, 16; Dan. 322.   **III.** *to reach, extend ;* pertinere :—Swá bebúgeþ gebod geond Brytenrícu Sexna cyninges [MS. kyninges] *so far as the command of the king of the Saxons extendeth through Britain,* Menol. Fox 457; Men. 230: Beo. Th. 2451; B. 1223.

be-byogean, -bycgan; *part.* -bycgende; *p.* -bohte; *pp.* -boht *To sell, to set or put to sale ;* vendere :—On gold bebycgean *to sell for gold,* Bd. 2, 12; S. 514, 39. Iudas bebohte bearn wealdendes on seolfres sinc *Judas sold the child of the Almighty for a heap of silver,* Cd. 226; Th. 301, 5; Sat. 577: Ps. Th. 43, 14: 104, 15: Beo. Th. 5591; B. 2799.

be-byrd *garnished with nails, set with spikes ;* clavatus, Cot. 49, Som. Lye.

be-byreþ *supplies,* L. Ethb. 18; Th. i. 6, 19; *pres. of* be-beran.

be-byrgan, be-birgan; *p.* de; *pp.* ed *To bury ;* sepelire :—Bebyrgeþ bán and ýslan *buries bones and embers,* Exon. 60 a; Th. 217, 26; Ph. 286: Gen. 23, 19. To bebyrgenne *sepelire,* Mt. Bos. 27, 7: Jn. Bos. 19, 40. v. byrgan.

be-byrian ; *p.* ede, ide; *pp.* ed *To bury ;* sepelire :—Ðæt hí móston ða deádan bebyrian *that they might bury the dead,* Ors. 3, 1; Bos. 54, 29. Hine árlíce bebyride *eum honorifice sepelivit,* Bd. 4, 22; S. 591, 20. v. byrian.

be-byrigan, be-birigan; *p.* ede; *pp.* ed *To cover with a mound, to bury ;* tumulare, sepelire :—Bebirigaþ me *sepelite me,* Gen. 49, 29. Ða bán ðe ðǽr bebyrigede wǽron *ossa quæ ibidem fuerant tumulata,* Bd. 4, 10; S. 578, 10: 2, 1; S. 500, 15. v. byrgan.

be-byrigean *to bury,* Mt. Bos. 8, 21, 22: Bd. 4, 11; S. 580; 3. v. byrgan, byrigan.

be-byrigedness, -ness, e; *f.* *A burying ;* sepultura :—Æfter monigum geárum his bebyrigednesse *post multos ejus sepulturæ annos,* Bd. 4, 32; S. 611, 27. v. be-byrignys.

be-byrignys, -nyss; be-byrigednes, -ness, e; *f.* *A burying ;* sepultura :—Ne wæs ǽnig se ðe bebyrignysse sealde ðám ðe acwealde wǽron *nec erat qui interemptos sepulturæ traderet,* Bd. 1, 15; S. 484, 3.

be-být *commands,* Lk. Bos. 4, 10; *3rd pres. of* be-beódan.

beo, becc, es; *m.* *A brook ;* BECK *or small rapid stream ;* rivulus :—Of ðan bece [MS. bæce] *from the beck,* Kmbl. Cod. Dipl. iii. 121, 16.

Bec *an abbey in Normandy :*—Teodbald, ðe was abbot in ðe Bec *Theobald, who was abbot of Bec,* Chr. 1140; Th. 383, 40.

béc *books,* Hy. 7, 20; Hy. Grn. ii. 287, 20. v. bóc.

be-cæfian, be-cefian; *p.* ede; *pp.* ed *To embroider, ornament, decorate ;* phalerare :—Becæfed *phaleratus,* Cot. 84. v. cæfian.

be-carcan *to take care of ;* accurare, Som. Lye. v. carc *care.*

becc *a beck, brook.* v. bec.

-beco, -bec, -beck, *used for the name of places, or as a termination to the names of places, denotes the situation to be near a brook or river.*

becca, an; *m.* *A* BECK, *pick-axe, mattock ;* ligo, marra, Ælfc. Gl. 2; Som. 55, 42.

béce, bæce, beóce, an; *f.* *A beech-tree, a tree bearing mast ;* fagus, æsculus :—Béce *fagus,* Wrt. Voc. 285, 21. v. bócce, bóc.

be-ceápian ; *p.* ode; *pp.* od *To sell ;* vendere :—He sceolde ealle his wélan beceápian *he should sell all his wealth,* Homl. Th. i. 62, 3. Se ðe sóþfæstnysse beceápaþ wið feó *he who sells truth for money,* ii. 244, 24. Hí beceápodon heora ǽhta *they sold their possessions,* i. 316, 4, 11, 31. Beceápa ealle ðíne ǽhta *sell all thy possessions,* ii. 400, 12. v. be-cýpan, ceápian.

be-ceásan ; *p.* -ceós, *pl.* -ceóson; *pp.* -ceásen *To attack, fight, combat ;* oppugnare, contendere, Leo 131. v. ceásan, ceás *strife.*

be-cefian ; *p.* ede; *pp.* ed *To ornament, embroider,* Lye. v. be-cæfian.

bécen *a beacon,* Mk. Skt. Lind. 13, 22. v. beácen.

bécen ; *adj.* BEECHEN, *made of beech ;* fagineus :—Bécen *fagineus,* Ælfc. Gl. 45; Som. 64, 101; Wrt. Voc. 32, 36.

be-ceorfan ; *p.* -cearf, *pl.* -curfon; *pp.* -corfen *To* BECARVE, *cut off, to cut or pare away ;* amputare, præcidere :—Ðá hét he hine heáfde beceorfan *then he ordered to cut off his head,* Bd. 1, 7; S. 478, 3.

be-ceorian ; *p.* ode; *pp.* od *To complain ;* obmurmurare, R. Ben. 5. v. ceorian.

be-ceówan, bi-ceówan; *p.* -ceáw, *pl.* -cuwon; *pp.* -cowen *To chew, gnaw ;* corrodere :—Biþ swyra becowen [bicowen, Exon.] *the neck is gnawed,* Soul Kmbl. 218; Seel. 111.

be-cerran, -cyrran; *p.* de; *pp.* ed *To turn, turn round ;* vertere, convertere, Bt. Met. Fox 13, 156; Met. 13, 78. v. be-cyrran, cyrran.

becest *bakest* = bacest; *2nd pers. pres. of* bacan.

beceþ *baketh* = baceþ; *3rd pers. pres. of* bacan.

be-clǽmed ; *part. p.* BECLAMMED, *glued to or together, emplastered, plastered over ;* glutinatus, Som. v. be-clemman.

be-clǽnsian ; *p.* ode; *pp.* od *To cleanse ;* purgare, Lye. v. clǽnsian.

be-clemman ; *p.* de; *pp.* ed *To fetter, bind, tie, inclose, glue together,* BECLAM ; vincire, includere, glutinare :—Ðeáh he hie mid fíftigum clúsum beclemme *though he inclose it with fifty bonds,* Salm. Kmbl. 143; Sal. 71. Beclæmed *glutinatus,* Lye.

be-clingan ; *p.* -clang, *pl.* -clungon; *pp.* -clungen [clingan, I. *to wither,* II. *to adhere*] *To* BECLING, *surround, inclose ;* circumcludere, includere :—Clommum beclungen *inclosed in bands,* Elen. Kmbl. 1388; El. 696.

be-clísan ; *p.* de; *pp.* ed *To inclose ;* includere, Leo 126. v. be-clýsan.

be-clísing, e; *f.* *An inclosed place, a cell ;* cella, Leo 126. v. be-clýsing, be-clýsan.

be-clýpian, be-cleopian, be-clepian ; *p.* ede, ode, ade; *pp.* ed, od, ad *To accuse, summon, sue at law ;* accusare, in judicium vocare, judicio compellere :—Ær he clǽne sý ǽlcere spæce, ðe he ǽr beclyped wæs *before he be clear of every suit, in which he had been previously accused,* L. C. S. 28; Th. i. 392, 12: 31; Th. i. 394, 29: 73; Th. i. 414, 23.

be-clyppan, bi-clyppan ; *p.* -clypte; *pp.* -clypt *To clip, embrace ;* amplecti, Ps. Th. 118, 61: Mk. Bos. 9, 36. v. clyppan.

be-clýsan ; *p.* de; *pp.* ed *To close in, to shut in, to inclose, to shut ;* includere, concludere, claudere :—He beclýsde Iohannem on cwearterne *inclusit Johannem in carcere,* Lk. Bos. 3, 20: Ps. Spl. 30, 10: Jos. 10, 18. Híg hyra eágan beclýsdon *oculos suos clauserunt,* Mt. Bos. 13, 15: Exon. 12 b; Th. 20, 26; Cri. 323.

be-clýsing, e; *f.* *A cell.* v. be-clísing.

bécn, es; *n.* *A sign, beacon ;* signum :—Mín gebéd nú gyt bécnum standeþ ðæt him on wísum is wel lýcendlíce *adhuc est oratio mea in beneplacitis eorum,* Ps. Th. 140, 8: Beo. Kmbl. 6314;· B. 3161. v. beácen.

bécnan ; *p.* ede; *pp.* ed *To indicate, denote, signify ;* indicare, significare :—Ðe we mid ðæm bridle bécnan tiliaþ *which we will denote by the bridle,* Bt. Met. Fox 11, 158; Met. 11, 79: Exon. 110 a; Th. 421, 31; Rä. 40, 26: 106 b; Th. 407, 5; Rä. 25, 10. v. beácnian.

be-cnáwan ; *p.* -cneów, *pl.* -cneówon; *pp.* -cnáwen *To know ;* cognoscere, C. R. Ben. 25. v. on-cnáwan.

bécniendlíce ; *adv. Allegorically or by parable ;* allegorice, Som. v. bécnan.

bécnunog, e; *f.* *A sign, token ;* significatio :—Ðú bécnuncge sealdest ðám ðe ege ðínne elne healdaþ *dedisti metuentibus te significationem,* Ps. Th. 59, 4.

bécnydlíc ; *adj. Allegorical ;* allegoricus :—Bécnydlícre gerecednesse *explanationis allegoricæ,* Bd. 5, 23; S. 648, 5, note. v. bécnan.

be-cnyttan ; *v. a. To knit, bind or tie, inclose ;* ligare :—Ðe seó molde on becnit wæs *in which the mould was inclosed,* Bd. 3, 10; S. 534, 29, note. v. cnyttan, cnittan.

be-com *came, was come,* Beo. Th. 231; B. 115; *p. of* be-cuman.

be-corfen ; *part. p. Cut off, beheaded ;* truncatus :—Becorfen wæs heáfde capite truncatus est, Bd. 1, 27; S. 491, 19. v. be-ceorfan.

be-crafian ; *p.* ode, ede; *pp.* od, ed *To crave.* · v. crafian.

be-creópan ; *p.* -creáp, *pl.* -crupon; *pp.* -cropen *To bring secretly, to creep ;* irrepere :—Ðæt he síe becropen on carcern *that he should be secretly led to prison,* Bt. Met. Fox 25, 71; Met. 25, 36.

becst *bakest* = bacest; *2nd pers. pres. of* bacan.

be-cuman, he -cymþ; *p.* -com, -cwom, *pl.* -cómon, -cwómon; *pp.* -cumen; *v. intrans.*   **I.** *to* BECOME, *happen, befall, meet with, fall in with ;* contingere, evenire, supervenire, incidere :—Syððan niht becom *after it had become night, or night had come,* Beo. Th. 231; B. 115. Oft becymþ se ánwéald ðisse worulde to swíðe gódum monnum *often cometh the power of this world to very good men,* Bt. 39, 11; Fox 228, 18. Ðǽm gódum becymþ ánfeald ýfel *to the good happens unmixed evil,* Bt. 39, 9; Fox 224, 29. Him ðæs grim leán becom *this grim retribution happened to them,* Cd. 2; Th. 3, 36; Gen. 46. Him becómon fela yrmþa *much misery befell them,* Ælfc. T. 41, 21. Becom *evenit,* Ælfc. Gr. 33; Som. 37, 18. He becom on ða sceaðan *he fell among thieves,* Lk. Bos. 10, 30: R. Ben. 65.   **II.** *to come, enter, come or attain to, come together ;* venire, ingredi, pervenire, attingere, concurrere :—In ða ceastre becuman meahte *thou mightest come into the city,* Andr. Kmbl. 1858; An. 931. Hannibal to ðam lande becom *Hannibal came to that land,* Ors. 4, 8; Bos. 90, 14. Gehlýde mín to ðe becume *clamor meus ad te perveniat,* Ps. Th. 101, 1. Ic eft up becom éce dreámas *I again on high attained to eternal joys,* Cd. 224; Th. 297, 4; Sat. 512. Becumen sí *concurratur,* R. Ben. 43. Becumendum to Segor *venientibus in Segor,* Gen. 13, 10.

bécun *a beacon,* Mk. Skt. Rush. 13, 22. v. beácen.

be-cunnian ; *p.* ode; *pp.* od *To assay, prove, try ;* experiri. v. cunnian.

be-cweðan ; ðú -cwíst, he -cwiþ ; *p.* -cwæþ, *pl.* -cwǽdon; *pp.* -cweden, -cweðen.   **I.** *to say, assert ;* dicere :—Swá ðú worde becwíst *as thou sayest by word,* Andr. Kmbl. 386; An. 193: 419; An. 210.   **II.** *to reproach ;* exprobrare :—Hí becwǽdon *exprobraverunt,* Ps. Th. 88, 44.   **III.** *to* BEQUEATH, *to give by will ;* legare :—Ealle ða mynstra and ða cyrican wǽron givene and becweðene Gode *all the minsters and churches were given and bequeathed to God,* Chr. 694; Th. 66, 6, note 2: Th. Diplm. A. D. 830; 465, 16.

be-cwom, *pl.* -cwómon *came, fell,* Cd. 160; Th. 199, 26; Exod. 344; *p. of* be-cuman.

be-cwyddod; *part. p.* [be, cwiddian *to speak*] *Bespoken, deposited;* depositum, Ælfc. Gl. 14; Som. 58, 9.

be-cyme, es; *m.* A BY-COMING, *an event* or *coming suddenly;* eventus:—Ðæs gehátes and ðæs wítedómes sóþ se æfterfyligenda becyme ðara wísena geséðde and getrymde *cujus promissi et prophetiæ veritatem sequens rerum astruxit eventus,* Bd. 4, 29; S. 607, 35.

be-cymþ *happens,* Bt. 39, 9; Fox 224, 29. v. be-cuman.

be-cýpan; ic -cýpe, ðú -cýpest, -cýpst, he -cýpeþ, -cýpþ, *pl.* -cýpaþ; *p.* ic, he -cýpte, ðú -cýptest, *pl.* -cýpton; *pp.* -cýped, -cýpt *To sell;* vendere:—Ðú becýptest folc ðín *vendidisti populum tuum,* Ps. Spl. 43, 14. Gif hwá becýpþ his dóhtor *si quis vendiderit filiam suam,* Ex. 21, 7. Iosep becýped wæs *venundatus est Ioseph,* Ps. Spl. 104, 16: Mt. Bos. 10, 29. v. cýpan.

be-cyrran; *p.* -cyrde; *pp.* -cyrred, -cyred, -cyrd; *v. trans.* To turn to, *to give up, deliver, betray;* vertere, transferre ad:—Ælfmær hí becyrde *Ælfmær betrayed it,* Chr. 1011; Th. 266, 23. v. be-cerran.

BED, bedd, es; *n.* I. *a* BED, *couch, pallet;* stratum, lectus:—Hí ðá inasendon ðæt bed, ðe se lama on læg, Mk. Bos. 2, 4; *thei senten doun the bedd, in whiche the sike man lay,* Wyc. To ðínum bedde *to thy bed,* Gen. 16, 2. II. *a bed in a garden;* pulvillus *vel* areola in hortis: *used in compounds,* as Wyrt-bedd *a wort bed,* Herb. 7, 1; Lchdm. i. 96, 22: Hreód-bedd *a reed bed,* 8, 1; Lchdm. i. 98, 13. [*Plat. O. Sax. Dut.* bed, *n:* *Ger.* bett, bette, *n:* *M. H. Ger.* bette, *n:* *O. H. Ger.* petti, *n: Goth.* badi, *n: Dan.* bed: *Swed.* bädd, *n: O. Nrs.* beðr, *m.* According to Grm. Wrtbch. i. 1722 connected with *A. Sax.* biddan: *Goth.* bidjan? for which he suggests the original meaning *to lie on the ground;* humi jacere.] DER. bed, bedd, -bolster, -clýfa, -cófa, -felt, -ian, -ing, -ling, -reáf, -reda [-rida], -rest, -stede, -þen, -tíd: gebed, -clýfa, -scipe.

bed *asked:*—Ic bed *petii,* Ps. Spl. 26, 7, = bæd; *p.* of biddan.

BÉD, es; *nom. acc. pl.* bédu, bédo; *n.* A *prayer, supplication, religious worship;* oratio, supplicatio, Dei cultus:—Ðæt he sceolde ða bédu [*MS. B.* byldo *constancy*] anescian *that he should diminish* [*weaken*] *the prayers,* Bd. 1, 7; S. 477, 43. Béd is chiefly found in composition, as in,—Béd-hús *a place for prayer,* béd-dagas *prayer-days, Rogation-days.* The original word béd *a prayer* was superseded by ge-béd *a prayer, q. v.* [*Orm.* bede *a prayer;* *acc. pl.* bedess: *Laym. acc. s.* bede, bode *a prayer;* *dat. s.* ibede; *nom. pl.* beden: *R. Glouc. acc. pl.* bedes *prayers;* *Piers acc. pl.* bedes *prayers,*—' if I bidde any bedes:' *Piers and Chauc. also* bedes,—' a peire of bedes,'—*a set of beads* or *small balls of glass etc. on a string, for counting prayers: O. Sax.* beda; *gen. s.* bede; *dat. s.* bedu: *O. Frs.* bede: *M. H. Ger.* bete: *O. H. Ger.* beta.] DER. béd-dagas, -hús, -ríp: gebéd, -dagas, -hús, -man, -ræden, -stów. v. biddan.

Beda, an; *m. Venerable Bede, born at Monkton by Jarrow, near the mouth of the Tyne, in* A. D. 674. He wrote his *Historia Ecclesiastica gentis Anglorum about* A. D. 731, *and died May 26, at the age of* 61, *in* 735.—He *gives the following account of himself, according to king Alfred's Anglo-Saxon version, made about* 890:—Ic Beda, Cristes þeów, and Mæsse-Preóst ðæs Mynstres ðara eádigra Apostola Petrus and Paulus, ðæt is æt Wira-múþan [*Wearmouth*] and on Gyrwum [*Jarrow*], wæs acenned on sundor-lande ðæs ylcan Mynstres.—Mid ðý ic wæs seofon wintre, ða wæs ic mid gýmenne mínra maga seald to fédanne and to læranne ðam árwurþan Abbude Benedicte, and Ceolfriþe æfter ðon and syððan ealle tíd mínes lífes on ðæs ylcan Mynstres eardunge, ic wæs dónde, and ealle geornnesse ic sealde to leornianne and to smeágianne hálige gewríto and betwyh gehald regollíces þeódscipes and ða dæghwámlícan gýmenne to singanne on cyricean me symble swéte and wynsum wæs ðæt ic oððe [leornode oððe] lærde oððe write.—And ða ðý nigonteoðan geáre mínes lífes ðæt ic Deáconháde onféng; and ðý þrittigoðan geáre Mæsse-Preóst-háde. And æghwæðerne þurh þenunge ðæs árwurþan biscopes Johannes þurh hæse and bebod Ceolferþes ðæs Abbudes.—Of ðære tíde ðæs ðe ic Mæssepreósthåde onféng óþ nigon and fiftig wintra mínre yldo, ic ðás béc for mínre nýdþearfe and mínra freónda of geweorcum árwurþra Fædera wrát and sette ge eác swylce to mægwlite andgytes and gástlícra gerecenessa ic to ætýcte [*Ego Bæda, famulus Christi, et Presbyter Monasterii beatorum Apostolorum Petri et Pauli, quod est ad Viuræmuda et Ingyruum, natus sum in territorio ejusdem Monasterii.* —*Cum essem annorum septem* [A. D. 674 + 7 = 681] *cura propinquorum datus sum educandus reverentissimo Abbati Benedicto, ac deinde Ceolfrido cunctumque ex eo tempus vitæ in ejusdem Monasterii habitatione peragens, omnem meditandis Scripturis operam dedi atque inter observantiam disciplinæ regularis et quotidianam cantandi in ecclesia curam semper aut discere aut docere aut scribere dulce habui.*—*Nonodecimo autem vitæ meæ anno* [A. D. 674 + 19 = 693] *Diaconatum, tricesimo gradum Presbyteratus* [A. D. 674 + 30 = 704]. *Utrumque per ministerium reverentissimi Episcopi Johannis jubente Ceolfrido Abbate suscepi.*—*Ex quo tempore accepti Presbyteratus usque ad annum ætatis meæ quinquagesimum nonum* [A. D. 674 + 59 = 733], *hæc in Scripturam sanctam meæ meorumque necessitati ex opusculis venerabilium Patrum breviter adnotare sive etiam ad formam sensus et interpretationis eorum superadjicere curavi,* Bd. 5, 23; S. 647, 18-35. Hér forþférde Beda *here,* A. D. 735 [*MS.* 734],

Bede died, Chr. 734; Th. 77, 20, col. 1, 2, 3. *Anno* 735, *Bæda Presbyter obiit,* Bd. S. 224, 5. Sanctes Bedan bán restaþ on Gyrwa-wíc *saint Bede's bones rest in Jarrow,* L. Ælf. C. 6; Th. ii. 344, note 4, 3.

be-délan, -délan, bi-délan; *p.* -dælde, -délde; *pp.* -déled, -déled *To deprive, bereave of anything, to deliver, release, free from anything;* privare, orbare, sejungere, liberare, expertem reddere:—Wuldres bedæled *deprived of honour,* Salm. Kmbl. 760; Sal. 379. Nele hí God æfre góde bedælan *Dominus non privabit eos bonis,* Ps. Th. 83, 13. Be ðære lyfte bedæled *aere privatus,* Bd. de nat. rerum; Wrt. popl. scienc. 17, 11. Hí bióþ ælces cræftes bedælde *they are destitute of all ability,* Bt. 36, 6; Fox 180, 28. Hwí sceal ic beón bedæled ægþer mínra sunena *cur utroque orbabor filio?* Gen. 27, 45. Gesælige sáwle sorgum bedælde *happy souls released from cares,* Cd. 220; Th. 282, 34; Sat. 296.

Beda-ford *Bedford,* Chr. 915; Th. 191, 26, col. 1. v. Bedan ford.

bédan *to offer,* Chr. 1011; Th. 267, 12, col. 1. v. beódan III.

Bedan ford, Beda-ford, Bedcan ford, Bede-ford, Bedican ford, Biedcan ford, es; *m: dat.* -forde, -forda [*Hunt.* A. D. 1148 Bedeford: *West.* 1377 Bedford: *Kni.* 1395 Bedforde, Bedeforde: bedan = bedum *lectis,* ford *vadum:* lectos et diversoria ad vadum sonans, *Camd.*] BEDFORD; oppidi nomen:—Ða yldestan men to Bedan forda hyrdon *the first men belonged to Bedford,* Chr. 918; Ing. 133, 2. Eádward cyning fór to Bedan forda *king Edward went to Bedford,* 919; Ing. 133, 13. Hie gedydon æt Bedan forda *pervenirent ad Bedanfordam,* Chr. 921; Gib. 107, 40.

Bedan ford-scír, Bæda-ford-scír, Beada-ford-scír, Bede-ford-scír, e; *f.* BEDFORDSHIRE; comitatus nomen:—Hí hæfdon ofergán Bedan fordscíre *they had subjugated Bedfordshire,* Chr. 1011; Th. 266, 5, col. 2. Wende him út into Bedan fordscíre *egressus est in Bedanfordsciram,* 1016; Th. 278, 16, col. 1.

Bedan heáfod, es; *m. Beda's head, Bedwin? in Wiltshire,* Chr. 675; Erl. 37, 6. v. Biedan heáfod.

bed-bolster; *gen.* -bolstres; *m.* A *pillow, bolster;* plumacium:—Bed-bolster *plumacium,* Ælfc. Gl. 27; Som. 60, 103; Wrt. Voc. 25, 43.

Bedcan ford *Bedford,* Chr. 571; Th. 32, 27, col. 1. v. Bedan ford.

bed-clýfa, bedd-clýfa, bed-cleófa, bed-cófa, an; *m.* A *bed-chamber, closet;* cubile hominis, cubiculum:—Gang into ðínum bedclýfan *intra in cubiculum tuum,* Mt. Bos. 6, 6.

bed-cófa, an; *m.* A *bed-place;* cubiculum:—Bed-cófa *vel* búr *cubiculum,* Ælfc. Gl. 27; Som. 60, 99: Lk. Bos. 12, 3. v. bed-clýfa.

bedd *a bed;* stratum, lectus, Cd. 101; Th. 134, 33; Gen. 2234. v. bed.

bedd *bid, command,* Lev. 6, 20, = bid, bidd; *impert.* of biddan.

béd-dagas; *pl. nom. m. Prayer-days, Rogation-days;* orandi dies, Rogationis dies, Wanl. Catal. 20, 12.

bedd-clýfa *a bed-chamber;* cubiculum, Gen. 43, 30. v. bed-clýfa.

beddian, beddigan; *p.* ode; *pp.* od *To prepare* or *make a bed;* sternere:—Ic strewige oððe beddige *I make* or *prepare a bed,* Ælfc. Gr. 28, 1; Som. 30, 34. Féde þearfan, and beddige him *feed the needy, and make a bed for them,* L. Pen. 14; Th. ii. 282, 16.

bedding, beding, e; *f.* I. BEDDING, *covering of a bed;* stramentum, stratum, Ælfc. Gl. 111; Som. 79, 60:—Mid mínum teárum míne bedding ic beþweá *lacrimis meis stratum meum rigabo,* Ps. Lamb. 6, 7. II. *a bed;* lectus:—Gyf ic astíge on bedinge stræte mínre *si ascendero in lectum strati mei,* Ps. Spl. 131, 3.

bedd-reáf *bed-clothes.* v. bed-reáf.

bedd-redda, bedd-rida, an; *m. One bed-ridden;* clinicus, Ælfc. Gl. 77; Som. 72, 28. v. bed-reda.

bedd-rest, bed-rest, e; *f.* A *bed-rest, a bed;* lectus:—Me Sarran beddreste gestáh *Sarah ascended my bed,* Cd. 129; Th. 164, 16; Gen. 2715: 102; Th. 135, 25; Gen. 2248.

-béde *exorable.* DER. eáþ-béde, *q. v.*

be-deáglian, bi-deáglian; *p.* ode; *pp.* od *To hide, cover, conceal, keep close* or *secret;* occultare, abscondere:—Me ne meahte monna ænig bideáglian hwæt he hogde *nobody could conceal from me what he meditated,* Exon. 51 a; Th. 177, 12; Gú. 1226. v. be-díglian.

be-deaht = be-þeaht *covered,* Judth. 11; Thw. 24, 29; Jud. 213; *pp.* of be-þeccan.

Bede-ford *Bedford,* Chr. 1010; Th. 264, 12, col. 1. v. Bedan ford.

Bede-ford-scír *Bedfordshire,* Chr. 1011; Th. 266, 5, col. 1. v. Bedan ford-scír.

be-déglad, bi-déglad *hidden, obscured,* Exon. 57 a; Th. 204, 15; Ph. 98; *pp.* of be-díglian.

be-délan; *p.* -délde; *pp.* -déled *To deprive;* privare:—Duguðum bedéled *deprived of dignity,* Cd. 215; Th. 272, 19; Sat. 122. v. be-dælan.

be-delfan; *p.* -dealf, *pl.* -dulfon; *pp.* -dolfen *To dig in* or *around, to bury, inter;* circumfodere, sepelire:—Óþ ic hine bedelfe *usque dum fodiam circa illam,* Lk. Bos. 13, 8. Bedealf hyt on eorþan *he buried it in the earth,* Mt. Bos. 25, 18. Bedolfen, Elen. Kmbl. 2159; El. 1081.

be-delfing, e; *f.* A *digging about;* ablaqueatio:—Niðerwart treówes bedelfing *a digging about the lower part of a tree,* Ælfc. Gl. 60; Som. 68, 16; Wrt. Voc. 39, 2.

beden *prayed,* Bd. 3, 5; S. 527, 28: Th. Diplm. A. D. 743-745; 28, 22; *pp.* of biddan.

Bederices weorþ, es; *m.* [Bederices *Bederic's*, weorþ *worth, town,* or *residence*] *Bederic's worth* or *town, so called because the manor formerly belonged to Bederic, who bequeathed it to Edmund the king and martyr, hence it was subsequently called* Eádmundes burh, *St. Edmund's bury:*—On Bedericeswyrþe *at Bedericesworth,* Will 23; Th. Diplm. A.D. 970; 517, 26. *At an earlier date, in* A.D. 958, *Ælfgar records,*—Ic an ðat lond into Beodricheswrþe to Seynt Eádmundes stówe *I give the land at Bedericsworth to St. Edmund's place,* Th. Diplm. 506, 12. v. Eádmundes·burh.

Bedewinda, an; *m.* BEDWIN, *Wilts:*—Ic, Ælfréd, West-Seaxena cining [MS. cingc], an Eádweade, mínum yldran suna, ðæs landes æt Bedewindan I, *Alfred, king of the West-Saxons, give the land at Bedwin to Edward, my elder son* [lit. *made a grant of the land at Bedwin*], Alfd. Will 14, 10.

bed-felt, es; *m? A bed-covering;* lecti pannus, lodix, R. Ben. 55.

bēd-hús, es; *n.* [bēd *a prayer,* hús *a house*] *A chapel, an oratory, a place for prayer;* oratorium, Fulg. 43.

Bedican ford, es; *n.* Bedford, Chr. 571; Ing. 26, 12. v. Bedan ford.

be-dícian; *p.* ode; *pp.* od; *v. a.* To BEDIKE, *to mound, to fortify with a mound;* aggere munire:—Bedícodon ða buruh útan *they embanked the city without,* Chr. 1016; Th. 280, 8, col. 1.

be-didrian; *p.* ode; *pp.* od *To deceive;* decipere:—Wéndon ge, ðæt ge mihton bedidrian mínne gelícan *think ye, that ye could deceive one like me?* Gen. 44, 15. DER. be-dyderian, dyderian.

be-dielf *dug,* Mt. Foxe 25, 18, *for* be-dealf; *p. of* be-delfan.

be-díglian, -díhlian, -deáglian; ic -díglige; *p.* -díglode; *pp.* -díglod, -díhlod; *v. a.* To hide, cover, conceal, keep close or secret; occultare, abscondere:—Ne híre ðú him ðæt ðú hine bedíglige *non audias eum ut occultes eum,* Deut. 13, 8. On gríne ða ðe hí bedíglodon *in laqueo quem absconderunt,* Ps. Spl. 9, 16. Bedíglod *occultus,* Ælfc. Gr. 28, 3; Som. 31, 5.

be-díhlian; *p.* -díhlode; *pp.* -díhlod *To hide.* v. be-díglian.

beding, e; *f.* *Bedding, covering of a bed, a bed,* Ps. Spl. 131, 3. v. bedding.

be-dipped, bedypt *dipped, dyed;* tinctus. v. be-dyppan.

bedling *a delicate person.* v. bædling.

be-dofen *drowned;* submersus, Homl. Th. ii. 472, 5; *pp. of* be-dúfan.

be-dolfen *buried,* Elen. Kmbl. 2159; El. 1081; *pp. of* be-delfan.

be-dón [be, dón *to do*] *To shut;* claudere:—Ðæt ðú ðíne doru mihtest bedón fæste *that thou mightest shut fast thy doors,* Ps. Th. 147, 2.

bēd-ræden, -ræðenn, e; *f.* *An assignment, ordinance* or *appointment;* assignatio, Som. v. ge-bēd-ræden.

be-dræf *drove,* Exon. 108 a; Th. 412, 5; Rä. 30, 9,=be-dráf; *p. of* be-drífan.

be-dráf *drove,* Ors. 3, 11; Bos. 72, 38; *p. of* be-drífan.

be-dragan; *p.* -dróg, -dróh, *pl.* -drógon; *pp.* -dragen *To draw aside, seduce;* seducere:—Ðe hie dearnenga bedróg *who seduced her secretly,* Cd. 29; Th. 38, 5; Gen. 602.

bed-reáf, es; *m. Bed-clothes, bedding;* lodix, fulcrum, lectisternia, Ælfc. Gl. 27; Som. 60, 109: 111; Som. 79, 62, 64; R. Ben. 55.

bed-reda, -rida, an; *m.* [bed *a bed,* reda=rida *from* riden *ridden, pp. of* rídan *to ride,* hence the *def. adj.* bedreda *bedridden, and the noun* bedreda, bedrida *one bedridden*] One BEDRIDDEN; clinicus:—Ðær læg be ðam wege án bedreda *there lay by the way one bedridden,* Homl. Th. ii. 422, 4. Arás se bedreda, and arn blissigende *the bedridden arose, and ran rejoicing,* ii. 422, 9. Ðá ðá se sunderhálga Iosias ðæt tácn geseah on ðam bedreðan [*def. adj.*] men, ðá feól he to ðæs apostoles fótum *when the pharisee Josias saw that miracle in the bedridden man, then fell he at the apostle's feet,* ii. 422, 11. Drihten cwæþ to sumum bedridan *the Lord said to one bedridden,* i. 472, 23.

bed-rest *a bed;* lectica, Ælfc. Gl. 66; Som. 69, 75: Judth. 10; Thw. 21, 26; Jud. 36. v. bedd-rest.

bed-rida *one bedridden,* Homl. Th. i. 472, 23. v. bed-reda.

be-drífan; *p.* -dráf, -dræf, *pl.* -drifon; *pp.* -drifen; *v. a.* I. *to drive, thrust on* or *upon, to compel, constrain* or *enforce one to do a thing, to pursue, follow;* cogere, compellere, agere, adigere:—Perdica hine bedráf into ánum fæstene *Perdiccas drove him into a fastness,* Ors. 3, 11; Bos. 72, 38. Hí him hám bedrifon [MS. bedrifan] and sige áhton *they drove them home and had a victory,* Bd. 1, 14; S. 482, 20. Wiht ða húðe hám bedræf *a creature drove the spoil home,* Exon. 108 a; Th. 412, 5; Rä. 30, 9. Ðú bedrifen [MS. bidrifen] wurde on ðas þeóstran worulde *thou wast driven into this dark world,* Exon. 28 b; Th. 86, 17; Cri. 1409. II. *to drive* or *beat against, to surround;* obruere, obducere, circumflare:—He geseah stapulas standan storme bedrifene *he saw columns standing driven by the storm,* Andr. Kmbl. 2987; An. 1496: Rood Kmbl. 123; Kr. 62. DER. drífan.

be-drincan; *p.* -dranc, *pl.* -druncon; *pp.* -druncen *To drink in* or *up, absorb;* imbibere:—Ðonne ðæt bedruncen sý, eft hit geniwa *when that is drunk up, renew it again,* Med. ex Quadr. 2, 10; Lchdm. i. 336, 4, MS. B.

bēd-ríp, e; *f. The cutting* or *reaping of corn on request;* ad preces messio, L. R. S. 5; Th. i. 436, 4, note. v. bēn-ríp.

be-dróg *seduced,* Cd. 29; Th. 38, 5; Gen. 602; *p. of* be-dragan.

be-droren; *pp. Deceived, deluded, bereaved, deprived;* deceptus, orbatus, Cd. 26; Th. 33, 31; Gen. 528: 93; Th. 120, 22; Gen. 1998; *pp. of* be-dreósan. v. dreósan, bi-droren.

be-druncen *drunk in, absorbed,* Med. ex Quadr. 2, 10; Lchdm. i. 336, 4, MS. B; *pp. of* be-drincan.

bed-ryda, an; *m. A bedridden man;* clinicus:—Se bedryda wearþ gehæled sóna; and eóde him ðá hám, hál on his fótum, se ðe ær wæs geboren on bære to cyrcan *the bedridden man was soon healed; and he then went home, whole on his feet, who before was borne on a bier to church,* Glostr. Frag. 10, 4, 15–18. v. bed-reda, drí, drían.

bed-stede, es; *m.* [bed *a bed;* stede *a place, station;* locus, situs] A BEDSTEAD; sponda. v. stede.

bed-þēn, es; *m.* [bed *a bed,* þēn *for* þegn *a servant*] *A chamberlain, a servant who has the care of a chamber;* lecti minister, camerarius, Ælfc. Gl. 27; Som. 60, 101.

bed-tíd, e; *f.* BEDTIDE, *bed time;* lecti adeundi tempus, serum, Ælfc. Gl. 95; Som. 76, 2.

bēdu *prayers;* orationes, Bd. 1, 7; S. 477, 43. v. bēd; *n.*

be-dúfan; *p.* -deáf, -dufon; *pp.* -dofen To bedive, *put under;* submergere, Homl. Th. ii. 392, 13. v. be-dofen. DER. dúfan.

bēdul; *adj. Prayerful, suppliant;* petitiosus, Ælfc. Gl. 101; Som. 77, 46.

be-dulfon *buried,* Ors. 3, 6; Bos. 58, 7; *p. pl. of* be-delfan.

bed-wahrift, es; *n. A curtain;* cortina, Cod. Dipl. A.D. 995; Kmbl. vi. 133, 9.

be-dyderian; *p.* ode; *pp.* od *To deceive;* decipere. v. be-didrian. DER. dyderian *to deceive.*

be-dydrung, e; *f. A deceit, deceiving;* deceptio. DER. dydrung.

be-dyppan; *p.* -dypte, *pl.* -dypton; *pp.* -dypped; *v. trans. To dip, immerse;* mergere, intingere, tingere:—Se ðe bedypþ on disce mid me his hand *qui intingit mecum manum in paropside,* Mt. Bos. 26, 23. Se ðe ic ræce bedyppedne hláf *is cui ego intinctum panem porrexero,* Jn. Bos. 13, 26. Híg bedypton his tunecan on ðam blóde *tinxerunt tunicam ejus in sanguine,* Gen. 37, 31. Ic bedyppe *mergo,* Ælfc. Gr. 28, 4; Som. 31, 36.

be-dyrnan, bi-dyrnan; *p.* de; *pp.* ed *To hide, conceal;* occultare:—Ne mihte him bedyrned wyrþan *it might not be hidden from him,* Cd. 14; Th. 17, 18; Gen. 261: Elen. Kmbl. 1201; El. 602: 1164; El. 584. v. dyrnan.

be-ebbian; *p.* ode, ade; *pp.* od, ad *To leave aground by ebbing;* aqua privare:—Scipu wæron be-ebbode [be-ebbade] *the ships were left aground by the ebb,* Chr. 897; Th. 176, 30. v. ebbian.

beel, es; *n. A pile;* rogus, Gl. E. 6, Lye. v. bæl.

be-eódon *dwelt, inhabited,* Bd. 1, 26; S. 488, 1; *p. of* be-gán.

beer *a bier, bed,* Cot. 23; Jn. Lind. War. 5, 8. v. bær.

be-fæstan, bi-fæstan; *p.* -fæste; *pp.* -fæsted. I. *to fasten, make fast, fix;* infigere:—Biþ se þridda dæl líge befæsted, in gléda grípe *the third part shall be fastened in fire, into the gripe of flames,* Elen. Kmbl. 2598; El. 1300. II. *to establish;* fundare, firmare:—Wæs se bisceophád fægere befæsted *the bishopric was fairly established,* Elen. Kmbl. 2423; El. 1213. III. *to commend, recommend, commit, deliver, put in trust, entrust;* commendare, tradere, committere:—He his geféran his freóndum wæs befæstende *socios amicis suis commendavit,* Bd. 4, 26; S. 602, 38. Ic him befæsted wæs *I was entrusted to him,* 5, 6; S. 618, 37: Ps. Th. 30, 5. Hyt gebyrede ðæt ðú befæstest feoh myneterum *oportuit te committere pecuniam numulariis,* Mt. Bos. 25, 27: L. C. S. 28; Th. i. 392, 10.

be-fæsting, e; *f. An entrusting.* DER. fæsting.

be-fæðman; *p.* ede; *pp.* ed *To embrace with the arms;* ulnis amplecti:—Befæðman, Cd. 163; Th. 204, 32; Exod. 428. v. fæðman.

be-fættian; *p.* ode; *pp.* od [be, fættian *to fatten*] *To make fat, anoint;* impinguare. v. ge-fættian.

be-falden *covered.* v. swegl-befalden.

be-fangen *taken,* Jos. 7, 15; *pp. of* be-fón.

be-faran; *p.* -fór, *pl.* -fóron; *pp.* -faren; *v. trans.* [be, faran *to go*] *To go round, to travel through, go all over, to traverse, to go, march, encompass, to surround;* peragrare, circumvenire:—Ne befaraþ ge Israhéla burga ǽrðan ðe mannes sunu cume *ye shall not go over the cities of the Israelites before the son of man come,* Mt. Bos. 10, 23. Rómáne on ungewis on án nyrewett befóran, óþ hý Somnite útan beföran *the Romans marched unwittingly into a narrow pass, till the Samnites surrounded them on the outside,* Ors. 3, 8; Bos. 63, 8: Cd. 167; Th. 209, 10; Exod. 497.

be-fealdan, bi-fealdan; *p.* -feóld, *pl.* -feóldon; *pp.* -fealden, -falden *To fold, infold, clasp, involve, surround, inwrap, cover, overwhelm;* implicare, involvere, amplecti, circumdare:—Ðú miht on ánre hand eáðe befealdan ealne middaneard *thou canst easily infold in one hand all the midearth,* Hy. 7, 119; Hy. Grn. ii. 289, 119. Ðá he ðá bóc befeóld *cum plicuisset librum,* Lk. Bos. 4, 20. He befeóld his handa mid ðæra tyccena fellum *pelliculas hædorum circumdedit manibus,* Gen. 27, 16. Mec hý-gedryht befeóld *a body of domestics surrounded me,* Exon. 94 b; Th. 353, 32; Reim. 21. DER. swegl-befalden.

be-feallan, ic -fealle, ðú -feallest, -fylst, he -fealleþ, -fylþ, *pl.* -feallaþ; *p.* -feól, -feóll, *pl.* -feóllon; *pp.* -feallen. I. *to fall;* cadere, incidere:—Ân of ðám ne befylþ on eorþan *unus ex illis non cadet super terram,* Mt. Bos. 10, 29. Hie oft befeallaþ on micel yfel *they often fall into great evil,* Past. 40, 3; Hat. MS. 53 b, 8: Cd. 18; Th. 21, 26; Gen. 330: Lk. Bos. 10, 36: Gen. 15, 12. II. *to fall off;* cadere ab aliquo; *pp.* befeallen *deprived, bereft;* orbatus, privatus:—Freóndum befeallen *bereft of friends,* Beo. Th. 2256; B. 1126: 4504; B. 2256. DER. feallan.

be-feastnian; *p.* ade; *pp.* ad *To betrothe;* desponsare:—Befeastnad *betrothed;* desponsatus, Mt. Lind. Stv. 1, 18. v. be-fæstan.

be-fêhþ *includes,* Bt. 24, 1; Fox 80, 14; *3rd pers. pres.* of be-fôn.

be-felan, -feolan; *p.* -fæl, *pl.* -fælon; *pp.* -feolen, -folen *To commit, commend, deliver, assign, allot;* committere, commendare, tradere, Leo 140. v. be-feolan.

be-felgan, bi-felgan; *p.* -fealg, -fealh, -felh, *pl.* -fulgon; *pp.* -folgen. I. *v. intrans. To stick* or *cling to, betake oneself;* inhærere, insistere:—Þilcum wordum heó him befelh ælce dæge *hujuscemodi verbis per singulos dies mulier molesta erat ei,* Gen. 39, 10. Æfter ðon ðe he ðær sum fæc hálgum leornungum befealh *after he had there for a while betaken himself to holy learning,* Bd. 4, 23; S. 594, 19. Ðæt he ðám hálwendan ongynnessum georne gefeole [befulge MS. B.] *ut cœptis salutaribus insisteret,* Bd. 5, 19; S. 637, 11, note. II. *v. trans. To deliver, transmit, consign;* tradere, committere:—He hine róde befealg *he delivered him to the cross,* Andr. Kmbl. 2654; An. 1328.

be-fellan; *p.* de; *pp.* ed *To fell;* cædere. v. be-fyllan.

be-fêng *concubuerit,* Gen. 19, 33. v. be-fôn.

be-feohtan; *p.* -feaht, *pl.* -fuhton; *pp.* -fohten *To deprive by fighting;* pugnando privare. v. bi-feohtan.

be-feól, -feóll *fell,* Lk. Bos. 10, 36; *p.* of be-feallan.

be-feolan, bi-feolan; *p.* -fæl, *pl.* -fælon; *pp.* -folen, -feolen *To commit, commend, deliver, grant;* committere, commendare, tradere:—Morðor under eorþan befeolan *to commit murder under the earth,* Exon. 90 b; Th. 340, 23; Gn. Ex. 115: Cd. 202; Th. 251, 7; Dan. 560. Ðú him for inwite yfel befæle *propter dolos disposuisti eis mala,* Ps. Th. 72, 14. Him wæs hálig gást befolen fæste *the holy spirit was fully granted to him,* Elen. Kmbl. 1870; El. 937: 391; El. 196. v. be-felan.

be-feóld *folded,* Lk. Bos. 4, 20; *p.* of be-fealdan.

BEFER, beofer, beofor, byfor, es; *m. A* BEAVER; castor, fiber:—Befer *fiber, castor, ponticus?* Ælfc. Gl. 19; Som. 59, 3; Wrt. Voc. 22, 47. Beofor, byfor *fiber,* Ælfc. Gr. 8; Som. 7, 13. [*Plat. Dut.* bever: *Ger. M. H. Ger.* biber: *O. H. Ger.* pipar, pipur: *Dan.* bäver: *Swed.* bäfver: *O. Nrs.* bifra, *f: Slav.* bobr. Grm. Wrtbch. i. 1806 connects the word with *Ger.* bauen *to build.*]

be-fêran; *p.* de; *pp.* ed *To go about, to go round, surround;* circumire, circumdare:—He lærende ða castel befêrde *circumibat castella in circuitu docens,* Mk. Bos. 6, 6. He befêrde ðæt Israhélisce folc *he surrounded the people of Israel,* Ex. 14, 9. DER. fêran.

be-fician *to deceive, to go round;* decipere, Off. Episc. 8.

be-fîlan; *p.* de; *pp.* ed *To befoul, defile:*—Ná mid meoxe befîled *not defiled with dung,* L. Ælf. P. 45; Th. ii. 384, 11. v. be-fýlan.

be-filgan; *p.* -filgde, *pp.* -filged *To follow after, pursue;* insequi:—Wolde me befilgende beón mid sáre *voluit me insequi cum dolore,* Bd. 4, 19; S. 589, 28, note. v. be-felgan.

be-flagen fláesc, es; *n.* [MS. flǽc] *The bowels;* viscera:—Beflagen fláec [= flǽsc] *vel* innopes innewearde *viscera,* Ælfc. Gl. 75; Som. 71, 99; Wrt. Voc. 45, 7. v. be-fleán.

be-fleán; *p.* -flóg, *pl.* -flógon; *pp.* -flagen *To flay, to skin,* or *take off the skin* or *bark;* decorticare, Cot. 62. v. beflagen flǽsc.

be-fleógan; *p.* -fleáh, *pl.* -flugon; *pp.* -flogen *To fly around* or *about;* circumvolare:—Ða spearcan beflugon on ðæs húses hróf *the sparks flew about on the roof of the house,* Bd. 3, 10; S. 534, 31, note.

be-fleón, to be-fleónne; *p.* -fleáh, *pl.* -flugon; *pp.* -flogen *To flee, flee away, escape;* fugere, effugere, evitare:—Hú he mihte befleón fram ðam toweardan yrre *quomodo posset fugere a ventura ira,* Bd. 4, 25; S. 599, 39. Hwider mæg ic ðinne andwlitan befleón *a facie tua quo fugiam?* Ps. Th. 138, 5: 61, 6. Nô ðæt ýðe byþ to befleónne *it is not easy to flee from that,* Beo. Th. 2010; B. 1003.

be-flówan; *p.* -fleów, *pl.* -fleówon; *pp.* -flówen *To overflow;* diffluere, redundare:—Wætre beflówen *overflowed with water,* Exon. 115 b; Th. 444, 19; Kl. 49.

be-fôh *contain;* complectere, Solil. 3; *impert.* of be-fôn.

be-folen *granted,* Elen. Kmbl. 1870; El. 937; *pp.* of be-felan, be-feolan.

be-fôn, bi-fôn, ic -fô, ðú -fêhst, he -fêhþ, *pl.* -fôþ; *p.* -fêng, *pl.* -fêngon; *impert.* -fôh; *pp.* -fangen, -fongen; *v. trans.* I. *to comprehend, grasp, seize, take hold of, catch;* comprehendere, apprehendere, capere:—Swá he ealle befêhþ ánes cræfte, heofon and eorþan *even as he comprehendeth all by his sole power, heaven and earth,* Andr. Kmbl. 653; An. 327. Habbaþ me helle clommas fæste befangen *the clasps of hell have firmly grasped me,* Cd. 19; Th. 24, 7; Gen. 374. Heó ánne hæfde befangen *she had seized one,* Beo. Th. 2594; B. 1295. Befangen on ðam fracodan gilte

deprehensus in hoc facinore, Jos. 7, 15. Ne mihton hîg his word befôn *non potuerunt verbum ejus reprehendere,* Lk. Bos. 20, 26. Gif mon forstolenne ceáp befêhþ *if a man seize stolen cattle,* L. In. 47; Th. i. 132, 4: L. Ath. i. 9; Th. i. 204, 10. Ðæt hîg woldon ðone Hælend on his sprǽce befôn *ut caperent eum in sermone,* Mt. Bos. 22, 15. II. *to surround, encompass, encircle, envelop, contain, clothe, case, receive, conceive, conceipe:*—He hafaþ ðam brídle bú tú befangen *he has encompassed both with the bridle,* Bt. Met. Fox 11, 58; Met. 11, 29. Befongen freáwrásnum *encircled with noble chains,* Beo. Th. 2906; B. 1451. Fýre befangen *enveloped in fire,* Beo. Th. 4540; B. 2274. Ne mihte ðes middaneard ealle ða béc befôn *non potest capere mundus omnes eos libros,* Jn. Bos. 21, 25: Bt. 24, 1; Fox 80, 14. Befôh hit mid feáum wordum *complectere hoc paucis verbis,* Solil. 3: Ps. Th. 74, 2. Ne hêt he ná etan ðone líchaman ðe he mid befangen wæs *he bade them not eat that body with which he was surrounded,* Homl. Pasc. Lisl. 9, 19: Soul Kmbl. 67; Seel. 34: Job 19, 26; Thw. 168, 2. Saglas, golde befongne *poles, cased in gold,* Past. 22, 2; Hat. MS. 33 a, 25. Ic hêr hǽlu calic hǽbbe befangen *calicem salutaris accipiam,* Ps. Th. 115, 4: Exon. 9 a; Th. 6, 7; Cri. 80.

be-fongen *encircled,* Beo. Th. 2906; B. 1451; *pp.* of be-fôn.

be-fôran, bi-fôran; *prep.* I. *dat.* II. *acc.* [be *by, proximity,* fôran *fore,* as æt fôran] BEFORE; ante, coram, prae:—I. *dat.* He swîðe oft befôran fremede folces rǽswum wundor æfter wundre *he very often performed before the princes of the people miracle after miracle,* Andr. Kmbl. 1237; An. 619. Ealdormen hêredon hîg befôran him *principes laudaverunt eam apud illum,* Gen. 12, 15. Hwá ne wáfaþ ðæs, ðæt ða steorran scínaþ befôran ðam mónan, and ne befôran ðære sunnan *who wonders not at this, that the stars shine before the moon, and not before the sun?* Bt. 39, 3; Fox 214, 30. II. *acc.* oft befôran him com *ante illum venire consueverat,* Bd. 5, 2; S. 614, 42, note. Sweord manige gesáwon befôran beorn beran *many saw a sword borne before the hero,* Beo. Th. 2052; B. 1024. III. befôran *frequently comes after the case:*—Him befôran fêreþ leóht *light goeth before him,* Cd. 222; Th. 288, 29; Sat. 389. Him bifôran *before them,* Exon. 47 a; Th. 160, 22; Gú. 947.

be-fôran; *adv. Before, at hand, openly;* ante, antea, prae, in conspectu, in conspectum:—He sceal befôran fêran *he shall advance before,* Bt. Met. Fox 4, 35; Met. 4, 18. Wundor on eorþan he befôran cýþde *he revealed miracles on earth openly,* Andr. Kmbl. 1212; An. 606. Wæs se atola befôran ðam wicked one was at hand, Cd. 224; Th. 295, 17; Sat. 487. He befôran gengde *he went before,* Beo. Th. 2829; B. 1412.

befôran-cwedan; *p.* -cwæþ, *pl.* -cwǽdon, -cwêdon; *pp.* -cweden *To foretell;* praedicere, Bd. 4, 19; S. 588, 15, note: 5, 2; S. 615, 13, note.

befôran-gestihtian; *p.* ode; *pp.* od *To fore-ordain;* praeordinare. DER. ge-stihtian.

Befor-leág *Beverley, in Yorkshire.* v. Beofer-lic.

be-fôtian, -fótigan; *p.* ode; *pp.* od [be, fótian, fôt a *foot*] *To befoot, to cut off the feet;* pedes abscindere, Som. v. be-heáfdian *to behead.*

be-freón; *p.* -freóde; *pp.* -freód *To free;* liberare, Ps. C. 50, 110; Ps. Grn. ii. 279, 110.

be-frinan, -frynan; *p.* -fran, *pl.* -frunon; *pp.* -frunen [be, frinan *to ask*] *To ask, inquire, learn;* interrogare, sciscitari, discere:—Ic befrine *sciscitor,* Ælfc. Gr. 25; Som. 27, 4. Herodes befran hî *Herodes didicit ab eis,* Mt. Bos. 2, 7.

beftan *after, behind, without;* post, sine, Som. Lye. v. bæftan.

be-fýlan, -fîlan; *p.* -fýlede; *pp.* -fýled, -filed, -fýld; *v. trans.* [be, fúl *foul*] *To* BEFOUL, *pollute, defile, make filthy;* inquinare, fœdare, contaminare:—Befîled, L. Ælf. P. 45; Th. ii. 384, 11: Basil. admn. 7; Norm. 48, 23: Lchdm. iii. 208, 7: Cot. 104.

be-fyllan; *p.* -fylde; *pp.* -fylled [be, fyllan *to fill*] *To fill, fill up;* adimplere:—Befyllan, Bd. 1, 27; S. 489, 26.

be-fyllan; *p.* -fylde, -fealde; *pp.* -fylled; *v. trans.* [be, fyllan, fellan *to fell*]. I. *to fell, strike down;* cædere, prosternere, projicere:—Hwæt befealdest ðú wærfæstne rinc *why didst thou fell the upright man?* Cd. 48; Th. 62, 6; Gen. 1010. He us hæfþ befylled *he has struck us down,* 19; Th. 23, 17; Gen. 361. II. *to deprive by felling, bereave;* cædendo orbare:—Secgum befylled *bereft of his warriors,* Cd. 97; Th. 128, 10; Gen. 2124.

befylþ *falls,* Mt. Bos. 10, 29; *3rd pers. pres.* of be-feallan.

bêg, es; *m. A bracelet, ring, crown;* armilla, corona:—Hie feredon brýd and bêgas *they conveyed bride and bracelets,* Cd. 90; Th. 112, 25; Gen. 1876. Hí on beorg dydon bêgas [MS. beg] and siglu *they placed in the mound rings and jewels,* Beo. Th. 6308, note; B. 3164. v. beáh.

be-galan; *p.* -gól, *pl.* -gólon; *pp.* -galen [be, galan *to sing, enchant*] *To enchant;* incantare:—Gyf hwylc yfel-dǽde man ôðerne begaleþ *if any ill-doing man enchants another,* Herb. 87, 4; Lchdm. i. 190, 10.

be-gan *began,* Gen. 9, 20. v. be-ginnan.

be-gân, bi-gân, ic -gá, ðú -gǽst, he -gǽþ, *pl.* -gáþ; *p.* -eóde, *pl.* -eódon; *pp.* -gân [be, gân *to go*]. I. *to go over, to surround, occupy, dwell, cultivate, till;* perambulare, circumdare, incolere, habitare, colere:—Ic fêrde geónd ðas eorþan and hí be-eóde *I walked through* [*over*]

the earth, and perambulated it, Job 1, 7; Thw. 164, 16. Se ðe æcer begæþ *he who goes over the land, a farmer*, Ælfc. Gr. 7; Som. 6, 44. Mid ðȳ Rômâne ðâ gyt Breotone be-eódan *dum adhuc Romani Brittaniam incolerent*, Bd. 1, 26; S. 488, 1. Hî ðone bûr ûtan be-eódon *they surrounded the dwelling without*, Chr. 755; Th. 83, 26, col. 1. II. *to go to. visit, attend, cherish, honour, worship;* obire, colere, excolere:—Plegan begân *to go to or attend plays*, Ors. 6, 2; Bos. 117, 9. Ðæt mynster seó ylce cwên swȳðe lufode and ârwyrþode and be-eóde eadem regina hoc monasterium multum diligebat, venerabatur, excolebat, Bd. 3, 11; S. 535, 15: 2, 13; S. 517, 1. III. *to commit, exercise, practise, observe;* committere, perficere, observare:—Synne, ða ic selfa be-eóde *sins, which I committed myself*, Ps. C. 50, 66; Ps. Grn. ii. 278, 66. He begæþ unmætas [MS. unætas] *he commits gluttonies*, Deut. 21, 20. Begâ ðê sylfne to ârfæstnysse *exercise thyself in or devote thyself to piety*, 1 Tim. 4, 7; Bt. Met. Fox 8, 33; Met. 8, 17: Ps. Th. 105, 12. Ða ðe be-eódon ídelnesse *observantes vanitatem*, 30, 6: 118, 23: 119, 5: 98, 4: Bd. 2, 13; S. 517, 4.

be-gân *tilled, cultivated:*—On begânum landum *in cultivated lands*, Herb. 5, 1; Lchdm. i. 94, 6; *pp.* of be-gân.

bêgan; he bêgþ; *p.* de; *pp.* ed. I. *to bow, bend, turn;* flectere, inflectere, deprimere:—Ðeáh ðû teó hwelcne bôh of dûne to ðære eorþan, swelce ðû bêgan mæge *though thou pull any bough down to the earth, such as thou mayest bend*, Bt. 25; Fox 88, 23. Se Ælmihtiga bêgþ ðider he wile mid his ânwealde *the Almighty bends them whither he will by his power*, Bt. Met. Fox 13, 6; Met. 13, 3: Cd. 221; Th. 288; 15; Sat. 381: Bd. 4, 11; S. 580, 10. II. *to bow to, to settle;* inflectere, insistere:—Ðara bearn swylce bêgaþ ædelum settum beámum, samed anlîce, standan on staðule stîðe wið geóguþe *quorum filii sicut novellæ plantationes stabilitæ a juventute sua*, Ps. Th. 143, 14. DER. a-bêgan, for-, ge-, ofge-. v. bygan.

be-gang, be-gong, bi-gang, bi-gong, bi-gencg, es; *m.* [be, gang *a step, proceeding*]. I. *a course, way, passage, circuit, district;* cursus, via, tenor, circuitus:—Ofer geofenes begang *over the course of ocean*, Beo. Th. 729; B. 362. Holma begang *the passage of the deeps*, Andr. Kmbl. 390; An. 195. Gârsecges begang *the circuit of ocean*, 1059; An. 530. II. *an undertaking, a business, exercise, service, religious worship;* negotium, exercitatio, cultus:—Ða willnode he hyne sylfne fram eallum begangum ðisse worulde fremde gedôn *cupivit se ab omnibus sæculi hujus negotiis alienare*, Bd. 3, 19; S. 549, 38. On bigange ðæs âncorlîfes *in exercenda vita solitaria*, 5, 1; S. 613, 9. Ðæt heó môste healdan ðone geleáfan and bigong hire æfestnysse *ut fidem cultumque suæ religionis servaret*, 2, 9; S. 510, 29: 1, 7; S. 477, 21: Jos. 23, 7. Bigencg *observatio, studium*, Scint. 7.

be-ganga, bi-gonga, bi-genga, bi-gengea, an; *m.* An *inhabitant, a dweller, cultivator, observer, benefactor, worshipper;* incola, cultor:—Be ærran bigengum [begangum MS. B.] *of the first inhabitants*, Bd. 1, 1; S. 473, 7. Þearfena bigenga *a benefactor of the poor;* cultor pauperum, Bd. 3, 14; S. 540, 23: 2, 15; S. 519, 8. DER. land-begenga.

be-gangan, -gongan, bi-gangan; -gongan; *pp.* -gangen [be, gangan *to go*]. I. *to go round, surround;* circumdare:—Cartaina wæs mid sǽ ûtan befangen [begangen Cot.] *Carthage was outwardly surrounded by sea*, Ors. 4, 13; Bos. 99, 39. II. *to go to or after, to attend, commit, practise, exercise, perform, observe, worship;* exercere, incumbere, procurare, colere:—Begangan his gebêdu *to attend his prayers*, Bd. 3, 16; S. 542, 34, col. 1. Begangan wæccan *to attend wakes*, Bd. 3, 17; S. 545, 11. Forligru ne begange *should not commit adultery*, L. C. E. 7; Th. i. 364, 24. Ðæt ðû his bebod georne begange *that thou shouldst gladly perform his command*, Elen. Kmbl. 2339; El. 1171: Ps. Th. 118, 48. Swȳðe ic begangen wæs *exercitatus sum*, Ps. Th. 76, 4: 54, 2. Gif ðû fremdu godu bigongest *if thou wilt worship strange gods*, Exon. 67 b; Th. 250, 3; Jul. 121.

begannes, -ness, e; *f.* [beginnan *to begin*] *The calends, the first day of the month;* calendæ, Cot. 202.

bêgaþ *shall settle*, Ps. Th. 143, 14; *pres. and fut. pl.* of bêgan II.

beg-beám, beig-beám, es; *m.* [begir *a berry*, beám *a tree*] *The mulberry-tree, the blackberry-bush, a tree bearing berries, a bramble;* morus, rubus:—Moyses æt-ȳwde wið ænne beigbeám *Moyses ostendit secus rubum*, Μωσῆς ἐμήνυσεν ἐπὶ τῆς βάτου, Lk. Bos. 20, 37.

begea *of both*, Judth. 11; Thw. 23, 19; Jud. 128; *gen.* of begen.

bêgean *to bow, bend:*—Cneó bêgean scolden *genua flectere deberent*, Bd. 3, 17; S. 544, 39, col. 2. v. bêgan.

be-geat, be-geáton *obtained*, Ors. 3, 11; Bos. 72, 6; *p.* of be-gytan.

be-gellan *to celebrate by song, to sing.* v. bi-gellan.

be-gêmed *taken care of, governed; pp.* of be-gȳman.

BEGEN; *nom. m. only, Both;* ambo; *adj. pron. pl:*—Hîg feallaþ begen on ænne pytt *ambo in foveam cadunt*, Mt. Bos. 15, 14. Wit wæron begen ðâ git on geógoðfeore *we [Beowulf and Breca] were both yet in youthful life*, Beo. Th. 1077; B. 536.—*Nom. m. f. n.* bâ, bû, bô *both;* ambo, ambæ, ambo:—Ða idesa, *f.* bâ *both the women*, Judth. 11; Thw. 23, 22; Jud. 133. Þrym, *m.* sceal mid wlenco, þriste, *m.* mid cênum; sceolon bû recene beadwe fremman *pomp shall be with pride,*

the confident with the bold; both shall quickly promote war, Exon. 89 b; Th. 337, 9; Gn. Ex. 62: Elen. Kmbl. 1225; El. 614. Blôd, *n.* and wæter, *n.* bû tû ætgædre eorþan sôhton *blood and water, both the two sought the earth together*, Exon. 70 a; Th. 260, 5; Jul. 292: Cd. 35; Th. 46, 29; Gen. 751.—*Nom. m. and f. or f. and n.* bâ, bû *both;* ambo et ambæ *vel* ambæ et ambo, *n:*—Sorgedon bâ twâ, Adam and Eue *both the two sorrowed, Adam and Eve*, Cd. 37; Th. 47, 24; Gen. 765: 39; Th. 52, 8; Gen. 840. Hî bû þêgon [MS. þegun] æppel *they both [Adam and Eve]* ate the apple, Exon. 61 b; Th. 226, 8; Ph. 402: Cd. 10; Th. 12, 18; Gen. 187. Wæron bû tû rihtwîse befôran Gode *both the two [Zacharias and Elizabeth] were righteous before God*, Lk. Bos. 1, 6, 7: Cd. 27; Th. 36, 20; Gen. 574. Wæter, *n.* and eorþe, *f.* sint on gecynde cealda bâ twâ *water and earth, both the two are by nature cold*, Bt. Met. Fox 20, 152; Met. 20, 76. Bû samod, lîc, *n.* and sâwl, *f.* both together, body and soul, Elen. Kmbl. 1775; El. 889: Exon. 27 a; Th. 81, 20; Cri. 1326. Niwe wîn, *n.* sceal beón gedôn on niwe bytta [*acc. pl.* of bytt, *f.*], ðonne beóþ bû tû gehealden *new wine shall be put into new bottles, then both the two shall be preserved*, Mk. Bos. 2, 22.—*Gen. m. f. n.* begra, begea, bega *of both;* amborum, ambarum, amborum:—Se Hâlga Gâst, ðe gǽþ of ðam Fæder and of ðam Suna, is heora begra lufu *the Holy Ghost, who proceedeth from the Father and the Son, is the love of them both*, Hexam. 2; Norm. 4, 22: Ælfc. T. 3, 4. Heora begra eágan wurdon ge-openode *the eyes of them both were opened*, Gen. 3, 7: Cd. 90; Th. 113, 27; Gen. 1893. Hyra begea nest *earum ambarum cibum*, Judth. 11; Thw. 23, 19; Jud. 128: Ps. Th. 86, 2. Engla and deófla, weorþeþ bega cyme *of angels and of devils, of both shall be a coming*, Exon. 21 a; Th. 56, 8; Cri. 897. Heora bega fæder *earum ambarum pater*, Cd. 123; Th. 157, 4; Gen. 2600.—*Dat. m. f. n.* bâm, bǽm *to both;* ambobus, ambabus, ambobus:—Se Hâlga Gâst, ðe gǽþ of ðam Fæder and of ðam Suna, is him bâm gemǽne *the Holy Ghost, who proceedeth from the Father and the Son, is common to them both*, Hexam. 2; Norm. 4, 22: Lk. Bos. 7, 42. He sceóp bâm naman *he gave names to both*, Cd. 6; Th. 8, 23; Gen. 128: Exon. 45 b; Th. 154, 14; Gû. 842.—*Acc. m. f. n.* bâ, bû *both;* ambos, ambas, ambo:—Bysmeredon uncit [*Inscription* Bismærede ungket] men, bâ ætgædre *they [men] reviled us two, both together*, Runic Inscrip. Kmbl. 354, 30. Ða beón beraþ, bû tû ætsomne, ârlîcne anleofan and ǽtterne tægel *the bees bear excellent food and a poisonous tail, both the two together*, Frag. Kmbl. 35; Leás. 19. On bâ healfa *on both sides*, Beo. Th. 2614; B. 1305: Ps. Th. 59, 5. Sceolde bû witan ylda æghwilc yfles and gôdes *each of men must know both of evil and good*, Cd. 24; Th. 31, 3; Gen. 479.—*Acc. m. and f. or f. and n.* bâ, bû *both;* ambos et ambas *vel* ambas et ambo:—Ðæt ðæt fȳr ne mæg foldan, *f.* and merestreám, *m.* forbærnan, ðeáh hit wið bâ twâ sîe gefeged *that the fire may not burn up earth and sea, though it be joined with both the two*, Bt. Met. Fox 20, 230; Met. 20, 115. Bringaþ Drihtne, bû ætsomne, wlite, *m.* and âre, *f.* bring to the Lord, both together, glory and honour, Ps. Th. 95, 7. Hât bû tû aweg Agar fêran and Ismael *command both the two to go away, Hagar and Ishmael*, Cd. 134; Th. 169, 12; Gen. 2798. Gehwylc hafaþ ætgædre bû lîc, *n.* and sâwle, *f.* each shall have together both body and soul, Exon. 23 a; Th. 64, 13; Cri. 1036.—*Instr. m. f. n.* bâm, bǽm *with or by both;* ambobus, ambabus, ambobus:—Mid bǽm handum *with both hands*, Elen. Kmbl. 1607; El. 805. [R. *Brun.* beie, *gen:* R. *Glouc.* beye, bey: *Laym.* beie, beine, beigene: *Orm.* beȝenn, *gen:* O. *Scot.* baith: O. *Sax.* bêdie, bêdea: *Frs.* bêthe: *Dut.* beide: M. *Dut.* bede: *Ger.* M. *Ger.* beide: N. L. *Ger.* beede: O. *Ger.* pêde, pêdo, pêdiu: *Goth.* bai *and* bayoþs; *n.* ba: *Dan.* baade: *Swed.* både: O. *Nrs.* bâðir, bâðar, bædi: *Lat.* ambo: *Grk.* ἄμφω: *Lith.* abbu; *f.* abbi; O. *Slav.* oba: *Sansk.* ubha; *dual* ubhau; *pl.* ubhe.]

be-geondan, be-iundan; *prep. acc.* [be *by*, geond, geondan *over*] BEYOND; per, trans:—Him fyligdon mycele menigu fram Iudea and fram begeondan Iordanem *secutæ sunt eum turbæ multæ de Judæa et de trans Jordanem*, Mt. Bos. 4, 25. Alîse me to farenne and to geseónne ðæt sêloste land begeondan Iordane *transibo et videbo terram hanc optimam trans Jordanem*, Deut. 3, 25. Begeondan sǽ *in transmarinis partibus*, Bd. 5, 19; S. 639, 10. Gewendon begeondan sǽ *went beyond sea*, Chr. 1048; Erl. 180, 16. Beiundan Iordane *trans Jordanem*, Deut. 1, 5.

be-geondan; *adv. Beyond;* ultra:—Feor begeondan *far beyond*, Ælfc. Gr. 38; Som. 41, 3. v. geond; *adv.*

be-geótan, bi-geótan; he -gȳt; *p.* -geát, *pl.* -guton; *pp.* -goten, -geten [be, geótan *to pour*]. I. *to pour out, to cast upon, to sprinkle, cover;* aspergere:—Ic wæs mid blôde bestêmed, begoten of ðæs guman sîdan *I was wet with blood, poured from the man's side*, Rood Kmbl. 97; Kr. 49. Mid blôde beón besprenged *sprinkled with blood*, Chr. 734; Th. 76, 18: Herb. 96, 4; Lchdm. i. 210, 3: Rood Kmbl. 13; Kr. 7. II. *to pour into;* infundere:—He me lâre on gemynd begeát *he poured knowledge into my head*, Elen. Kmbl. 2494; El. 1248.

be-geten, L. H. E. 2; Th. i. 28, 2; *for* be-gitan *to seize, obtain.*

be-getende *seeking out*, = be-gitende, Ps. Spl. T. 110, 2. v. be-gitan.

be-gêton *begot*, Cd. 223; Th. 294, 20; Sat. 474; *p.* of be-gitan.

beggen *both*, L. Ælf. P. 35; Th. ii. 378, 13, 15, 16; *nom. m.* = begen.

bēgian; *p.* ode; *pp.* od [bēg *a crown*] *To crown*; coronare:—Ðû bēgodest us *coronasti nos*, Ps. Spl. C. 5, 15. v. beágian.

be-gíetan *to get, obtain*, Exon. 65 b; Th. 242, 6; Ph. 669. v. begitan.

be-gíman *to guard*; custodire, Gen. 2, 15. v. be-gýman.

be-gímen *observation, care*; observatio, Wanl. Catal. 78, 24. v. begýmen.

be-gíming, e; *f. An invention, a device*; adinventio, Ps. Spl. 105, 36.

be-gínan; *p.* -gán, *pl.* -ginon; *pp.* -ginen *To open the mouth wide, gape, yawn?* oscitare in aliquem?—Ic begíne *I yawn*, Exon. 129 b; Th. 497, 19; Rä. 87, 3.

be-ginnan, ic -ginne, ðû -ginnest, -ginst, he -ginneþ, -gineþ, -ginþ, *pl.* -ginnaþ, -ginaþ; *p.* -gan, *pl.* -gunnon; *pp.* -gunnen; *v. a.* [be, ginnan, q. v.] *To* BEGIN; incipere:—Nóe ðá began tp wircenne ðæt land *Noe tunc cœpit exercere terram*, Gen. 9, 20: 18, 27: Hy. 10, 36; Hy. Grn. ii. 293, 36. v. on-ginnan.

be-giondan *beyond*, Past. Pref. MS. Hat. v. be-geondan.

be-girdan; *p.* -girde; *pp.* -girded *To begird*, Apol. Th. 12, 17. v. be-gyrdan.

be-gitan, -gietan, -gytan; *part.* -gitende; ic -gite, ðû -gytst, he -gyteþ, *pl.* -gytaþ; *p.* -geat, *pl.* -geáton; *pp.* -geten; *v. a.* [be, gitan *to get*] *To get, obtain, take, acquire, to seek out, receive, gain, seize, lay hold of, catch*; sumere, obtinere, assequi, acquirere, nancisci, capere, comprehendere, arripere:—Ǽlc mód wilnaþ sôþes gódes to begitanne *every mind wishes to get the true good*, Bt. 24, 2; Fox 82, 1. Hî ða burh mihton eáðe begitan *they might easily have taken the city*, Ors. 3, 4; Bos. 56, 10. He begeat ealle ða eást land *he obtained all the east country*, Ors. 3, 11; Bos. 72, 6. Hwæt begytst ðû of ðínum cræfte *quid acquiris de tua arte?* Coll. Monast. Th. 23, 3: Ps. Th. 83, 3: 68, 37. Ðe hý under Alexandre begeáton *which* [*riches*] *they had gained under Alexander*, Ors. 3, 11; Bos. 73, 27: Beo. Th. 4490; B. 2249. Fin sweord-bealo begeat *misery from the sword seized Fin*, Beo. Th. 2297; B. 1146.

be-gleddian, ic -gleddige; *p.* ode; *pp.* od *To dye, stain*; inficere:—Ic begleddige *inficio*, Ælfc. Gr. 28, 6; Som. 32, 37. And begleddod is eorþe on blôdum *et infecta est terra in sanguinibus*, Ps. Spl. 105, 36.

be-glídan; *p.* -glád, *pl.* -glidon; *pp.* -gliden *To glide* or *disappear from any one, to desert any one*; evanescere ab aliquo, derelinquere:—Unriht me eall beglíde *iniquitas a me omnis transeat*, Ps. Th. 56, 1.

be-gnagan; *p.* -gnóg, *pl.* -gnôgon; *pp.* -gnagen *To* BEGNAW, *gnaw*; corrodere, Martyrol. 9, Jul.

begne, an; *f. An ulcer, a carbuncle*; carbunculus:—Seó blace begne *the black ulcer*; carbunculus, Ælfc. Gl. 64; Som. 69, 21; Wrt. Voc. 40, 52.

be-gnornian; *p.* ode; *pp.* od *To deplore*; lugere:—Begnornodon deplored, Beo. Th. 6338; B. 3179.

be-gong, es; *m. A course*:—Under swegles begong *under the course of heaven*, Beo. Th. 1724; B. 860. v. be-gang.

be-gongan *to exercise*, Exon. 32 b; Th. 103, 24; Cri. 1693 [MS. bigongan]. v. be-gangan.

be-goten *covered*, Rood Kmbl. 13; Kr. 7; *pp.* of be-geótan.

begra *of both*:—He is heora begra lufu *he is the love of them both*, Hexam. 2; Norm. 4, 22. v. begen.

be-grafan, bi-grafan; *p.* -gróf, *pl.* -grófon; *pp.* -grafen [be, grafan *to dig*] *To bury*; defodere, sepelire:—Rôda greóte begrafene [MS. begrauene] *crosses buried in the sand*, Elen. Kmbl. 1666; El. 835.

be-grauen *buried*; = begrafen; *pp.* of be-grafan.

be-greósan; *p.* -greás, *pl.* -gruron; *pp.* -groren *To overwhelm fearfully*; horrore afficere, formidolose obruere?—Atole gâstas sûsle begrorene [MS. begrorenne] *the horrid spirits fearfully overwhelmed with torment*, Cd. 214; Th. 268, 9.

be-grétan, -grǽtan; *p.* -grét, *pl.* -gréton; *pp.* -gréten, -grǽten *To lament, bewail*; lamentare, deplorare:—Fǽmnan ne wǽran geonge begrétte *virgines eorum non sunt lamentatæ*, Ps. Th. 77, 63. v. grétan.

be-grindan; *p.* -grand, *pl.* -grundon; *pp.* -grunden. **I.** *to grind, polish*; perfricare, polire, exacuere:—Sindrum begrunden *ground with cinders*, Exon. 107 a; Th. 408, 3; Rä. 27, 6. **II.** *to deprive*; privare:—Ǽlc hine selfa begrindeþ gâsteȝ dugeþum *each deprives himself of his soul's happiness*, Cd. 75; Th. 91, 33; Gen. 1521. DER. grindan.

be-grípan; *p.* -gráp, *pl.* -gripon; *pp.* -gripen; *v. trans.* [be, grípan *to gripe*] *To* BEGRIPE, *chasten, chide*; increpare, Ps. Spl. T. 15, 7.

begrorene [MS. begrorennë] *fearfully overwhelmed*, Cd. 214; Th. 268, 9; *pp.* of be-greósan.

be-grornian *to lament, to grieve for*; mœrere, Cd. 13; Th. 16, 14; Gen. 243. v. gnornian.

be-grynian; *p.* ode; *pp.* od *To ensnare, entrap*; illaqueare, irretire:—Ðæt hig swâ beón begrynode *ut sic irretientur*, Coll. Monast. Th. 21, 17. v. grinian.

be-gunnon, be-gunnen *began, begun*, C. R. Ben. 22. v. be-ginnan.

---

be-gyldan; *p.* -gylde; *pp.* -gylded *To gild*; inaurare, deaurare:—Begylded fatu *vasa deaurata*, Lye. v. gyldan, gildan.

be-gýman, be-gíman; *p.* de; *pp.* ed; *v. trans. To take care of, to keep, govern, regard, serve, attend*; custodire, curare, servare, observare, attendere:—Godes þeówum ðe ðære cyrcan begýmaþ *to God's servants who serve the church*, L. Ælf. C. 24; Th. ii. 352, 11: Ps. Spl. 77, 63: Lk. Bos. 10, 35: Mt. Bos. 6, 1: Ps. Spl. 5, 2.

be-gýmen, be-gímen, e; *f. Care, regard, observation, shew, pomp*; observatio:—Mid begýmene = μετὰ παρατηρήσεως, *with shew* or *that it can be observed*, Lk. Bos. 17, 20.

be-gyrdan, -girdan; *p.* de; *pp.* ed, or be-gyrd; *v. trans.* [be, gyrdan *to gird*]. **I.** *to* BEGIRD, *surround*; cingere, præcingere, accingere:—Begyrdaþ eówer lendenu *renes vestros accingetis*, Ex. 12, 11. He ðæt eálond begyrde and gefæstnade mid dîce *he begirt and secured the island with a dike*, Bd. 1, 5; S. 476, 10. God se begyrde me of mihte *Deus qui præcinxit me virtute*, Ps. Spl. 17, 34: Ps. Th. 17, 37. He wæs begyrded mid wǽpnum ðæs gâstlîcan camphâdes *accinctus erat armis militiæ spiritalis*, Bd. 1, 7; S. 477, 24. **II.** *to clothe*; amicire:—Begyrded oððe bewǽfed leóhte swâ swâ mid hrægle *amictus lumine sicut vestimento*, Ps. Lamb. 103, 2.

be-gytan *to obtain*, Mt. Bos. 5, 7. v. be-gitan.

be-gytst *obtainest*, Coll. Monast. Th. 23, 3. v. be-gitan.

bêh *a crown*:—On ðone bêh *in coronam*, Bd. 5, 21; S. 643, 28. v. bêg.

be-habban, he -hæfeþ; *p.* -hæfde; *pp.* -hæfed, -hæft; *v. a.* [be *by, near*, habban *to have*]. **I.** *to compass, encompass, surround*; cingere, circumdare:—Ðîne fýnd behabbaþ ðê *inimici tui circumdabunt te*, Lk. Bos. 19, 43: Jos. 6, 20. Behǽfde heápa wyn Hǽlendes burg *the joy of bands surrounded the Saviour's tomb*, Exon. 120 a; Th. 460, 16; Hö. 18: Cd. 112; Th. 148, 9; Gen. 2454. **II.** *to comprehend*; comprehendere, continere:—Behabban hreðre or on hreðre *to comprehend in the mind*, Andr. Kmbl. 1633; An. 818: Exon. 92 b; Th. 347, 9; Sch. 10: Ps. Spl. 76, 9. **III.** *to restrain, detain, stay*; detinere:—Hî behæfdon hine *detinebant illum*, Lk. Bos. 4, 42.

be-hæfednes, -ness, e; *f. A detention, care*; conservatio:—Behæfednes fæsten *sparingness, parsimony*, Cot. 191. v. fæst-hafolnes.

be-hæftan; *p.* -hæfte; *pp.* -hæfted, contr. -hæft, -hæft *To betake, take, bind*; captare, vincire:—Be-hæft *held*; captus = gehæft, q. v. Gen. 22, 13. v. *pp.* of hæftan. v. ge-hæftan.

be-hǽs, e; *f.* [be *by, near*, hǽs *command*] *A self-command, vow, promise*. Hence our *behest*; votum:—He fela behǽsa behêt *he promised many vows*, Chr. 1093; Th. 359, 33. v. hǽs, behât.

be-hǽtst *vowest*, Gen. 38, 17. v. be-hâtan.

be-hangen *hung round*; *pp.* of be-hôn.

behât, es; *n. A promise, vow*; promissum, votum:—Ic sende on eów mînes fæder behât *ego mitto promissum Patris mei in vos*, Lk. Bos. 24, 49. Ðonne ðû behât behǽtst Drihtene *cum votum voveris Domino*, Deut. 23, 21. DER. be-hâtan, ge-hât.

be-hâtan, ic -hâte, ðû -hâtest, -hǽtst, he -hâteþ, *pl.* -hâtaþ; *p.* -hêt, *pl.* -hêton; *pp.* -hâten [be, hâtan *to call, promise*, vide **II**] *To promise, vow, threaten*; spondere, pollicere, vovere, comminari:—Ðæt ðû me behǽtst *quod polliceris*, Gen. 38, 17. Behêt he mid âþe *cum juramento pollicitus est*, Mt. Bos. 14, 7. Ðonne ðû behât behǽtst Drihtene *cum votum voveris Domino*, Deut. 23, 21. Drihten God behêt us wedd *Dominus Deus pepigit nobiscum fœdus*, 5, 2. Ǽlc yfel man him behêt *they threatened him every evil*, Chr. 1036; Ing. 209, 12; Ælf. Tod. 11.

be-hâwian; *p.* ode, ade; *pp.* od, ad *To see, see clearly*; videre:—Behâwa ðonne ðæt ðû ûtadô ðæt mot *see then clearly* [τότε διαβλέψεις] *that thou take out the mote*, Mt. Bos. 7, 5.

be-heáfdian; *p.* ode; *pp.* od; *v. trans.* [be, heáfod *head*] *To* BEHEAD; decollare:—He beheáfdode Iohannem *decollavit Iohannem*, Mt. Bos. 14, 10: Judth. 12; Thw. 25, 32; Jud. 290.

be-heáfdung, e; *f. A* BEHEADING; decollatio, L. Ath. i. prm; Th. i. 194, 21.

be-healdan, bi-healdan, ic -healde, ðû -healdest -hylst, he -healdeþ, -hylt, -hilt, *pl.* -healdaþ; *p.* ic, he -heóld, ðû -heólde, *pl.* -heóldon; *pp.* -healden; *v. trans.* [be *near*, healdan *to hold, observe*]. **I.** *to hold by* or *near, possess, observe, consider, beware, regard, mind, take heed, behave, to mean, signify*; tenere, inhabitare, servare, curare, gerere:—Heora ǽ to behealdenne *to observe their laws*, Ors. 3, 5; Bos. 57, 21. Adam sceal mînne stronglîcan stôl behealdan *Adam shall possess my strong seat*, Cd. 19; Th. 23, 28; Gen. 366. He gemetfæstlîce and ymbsceáwiendlîce hine sylfne on eallum þingum beheóld *se modeste et circumspecte in omnibus gereret*, Bd. 5, 19; S. 637, 5. Hwæt ðæt swefen beheóld *what the dream signified*, Gen. 41, 8. **II.** *to* BEHOLD, *see, look on*; observare, aspicere, videre:—Beheald ða tunglu *behold the stars*, Bt. 39, 13; Fox 232, 25. Loth ða beheóld geond eall, and geseah *elevatis itaque Lot oculis, vidit*, Gen. 13, 10.

be-heáwan, bi-heáwan; *p.* -heów; *pp.* -heáwen *To beat, bruise, hew* or *cut off, to separate from, deprive of*; tundere, cædendo privare, amputare:—Beheáwene mid swingellan *tunsi per flagella*, Past. 36, 5; Hat. MS. 47 b, 15. Heáfde beheáwan *to behead*, Bt. Met. Fox 1, 85; Met.

I, 43. Hwonne me wráþra sum aldre beheówe *when some enemy might deprive me of life*, Cd. 128; Th. 163, 21; Gen. 2701.

be-hédan; *p.* -hédde; *pp.* -héded *To watch, heed, guard;* cavere, curare, Leo 178. v. hédan.

be-héfe, es; *m:* be-héfnes, -ness, e; *f.* [be-hófen] *Gain, advantage, benefit*, BEHOOF; lucrum. v. be-hófian *to have need of.*

be-héfe; *adj. Necessary, behoveful;* necessarius:—Ðe behéfe synd *qui necessarii sunt,* Lk. Bos. 14, 28. Behéfe þing *necessary things, necessaries,* C. R. Ben. 46. DER. efn-behéfe.

be-hegian; *p.* ede; *pp.* ed *To* BEHEDGE, *hedge around;* circumsepire. v. hegian.

be-helan, bi-helan; *p.* -hæl, *pl.* -hǽlon; *pp.* -holen *To conceal, hill* or *cover over, hide;* occultare, Beo. Th. 833; B. 414: Bd. 4, 16; S. 584, 25, note. v. helan, be-helian.

be-heáld *availed,* Chr. 1123; Th. 374, 23. v. be-healdan.

be-heáldan [=be-healdan?] *To attend, intend;* attendere, intendere:—Wesan díne eáran gehýrende and beheáldende *fiant aures tuæ intendentes,* Ps. Th. 129, 2.

be-helian, bi-helian; *p.* ode, ede; *pp.* od, ed; *v. trans.* [be, helian *to cover*] *To cover, cover over, conceal, obscure, hide;* condere, sepelire:—Wurdon behelede ealle ða dúna *operti sunt omnes montes,* Gen. 7, 19. Se heofen mót ðæt leóht behelian *the heaven may obscure the light,* Bt. 7, 3; Fox 20, 21: Elen. Kmbl. 858; El. 429.

be-helman; *p.* ede; *pp.* ed *To cover over, to cover;* cooperire:—Heolstre behelmed *covered with darkness,* Salm. Kmbl. 209; Sal. 104. v. bi-helmian.

Behémas, *pl. m:* Béme, *nom. acc; gen.* a; *dat.* um; *pl. m.* The *Bohemians;* Bohémi:—Hí Maroaro habbaþ, be westan him Þyringas, and Behémas, and Bægware healfe *they, the Moravians, have, on their west, the Thuringians, Bohemians, and part of the Bavarians,* Ors. I, I; Bos. 18, 42.

be-héng, *pl.* -héngon *hung round;* *p.* of be-hón.

be-heófian; *p.* ode; *pp.* od *To bewail, lament;* lugere, lamentari:—Heora mǽdena ne synt beheófode *virgines eorum non sunt lamentatæ,* Ps. Lamb. 77, 63. v. heófian.

be-heóld *beheld,* Gen. 13, 10; *p.* of be-healdan.

be-heonan, -heonon; *adv.* [be *by,* heonan *hence*] *On this side, close by;* cis, citra:—Get beheonon *yet nearer;* citerius, Ælfc. Gr. 38; Som. 41, 4: Cot. 33.

be-heopian; *p.* ode; *pp.* od *To hew* or *cut off;* amputare, Cd. 125; Th. 160, 2, note a; Gen. 2644,=be-heáwan? *q.v.*

be-heówe *might deprive,* Cd. 128; Th. 163, 21; Gen. 2701. v. be-heáwan.

be-hét *promised,* Deut. 5, 2; *p.* of be-hátan.

be-hicgan *to confide, trust, rely, depend upon;* acquiescere, niti, inniti:—Ðe on Gode behicgaþ *qui in Deo acquiescunt,* R. Ben. 31. DER. hicgan.

be-hídan; *p.* -hídde *To hide;* abscondere:—Forðamðe ic eom nacod, ic behídde me *quod nudus essem, abscondi me,* Gen. 3, 10, 8. v. be-hýdan.

be-hidiglíce *carefully,* Bd. 3, 19; S. 547, 29. v. be-hydelíce.

be-hilt *beholds;* respicit, R. Ben. 8; *pres.* of be-healdan.

be-hindan; *prep. dat. Behind;* post, pone:—He lét him behindan ciólas *he left ships behind him,* Bt. Met. Fox 26,45; Met. 26, 23. Ligeþ him behindan hefig hrusan dǽl *behind it lies the heavy mass of earth,* 29, 106; Met. 29, 52. Ne ðé behindan nú lǽt mænige ðus micle *now leave not behind thee such a multitude of people,* Exon. 10 a; Th. 10, 19; Cri. 155.

be-hindan; *adv. Behind, back;* a tergo, pone, post:—Ac behindan beleác mid wǽge *but inclosed them behind with the wave,* Cd. 166; Th. 206, 24; Exod. 456. Ðú ðone héhstan heofon behindan lǽtst *thou shalt leave the highest heaven behind,* Bt. Met. Fox 24, 58; Met. 24, 29.

be-hionan *on this side,* Past. pref. v. be-heonan.

be-híring *a hiring,* Ælfc. Gl. 13; Som. 57, 123. v. be-hýring.

be-hlád *covered,* Ors. 3, 3; Bos. 56, 6; *p.* of be-hlídan.

be-hlǽman *to overwhelm with noise;* strepitu obruere. v. bi-hlǽman.

be-hlǽnan *to beset by leaning anything against another;* acclinando circumdare. v. bi-hlǽnan.

be-hlǽstan *to load a ship;* navem onerare. v. be, hlæstan.

be-hleápan; *p.* -hleóp, *pl.* -hleópon; *pp.* -hleápen *To leap upon* or *in, to fix;* insilire:—Ðæs monnes mód and his lufu biþ behleápen on ða lǽnan sibbe *the man's mind and his love are fixed on the fragile peace,* Past. 46, 5; Hat. MS. 67 a, 9.

be-hlehhan, bi-hlyhhan; *p.* -hlóh, *pl.* -hlógon; *pp.* -hlahen, -hleahen *To laugh at, deride;* ridere aliquid, exultare de aliqua re:—Ic ne þearf behlehhan *I need not deride,* Exon. 52 b; Th. 183, 22; Gú. 1331. DER. hlehhan.

be-hlídan; *p.* -hlád, *pl.* -hlidon; *pp.* -hliden [hlídan *to cover*] *To cover over, to cover, close;* tegere, claudere:—Híg awylton ðone stán, and ðone pytt eft behlidon *thei schulden turne awei the stoon, and thei schulden put it eft on the pit,* Wyc; Gen. 29, 3. Seó eorþe siððan togædere behlád *the earth then closed together,* Ors. 3, 3; Bos. 56, 6.

be-hlidenan = be-lidenan *the left* or *departed, the dead;* mortuos, Andr. Kmbl. 2179; An. 1091; *acc. pl. pp. from* be-lídan, *q.v.*

be-hlígan, he -hlíþ *To dishonour, defame;* infamare:—Oft hí mon wómmum behlíþ *man often defames her with vices,* Exon. 90 b; Th. 339, 29; Gn. Ex. 101.

be-hlýdan; *p.* de; *pp.* ed *To deprive;* privare, spoliare:—Ic sceal heáfodleás behlýded licgan *I must lie deprived of head,* Exon. 104 a; Th. 395, 20; Rä. 15, 10.

be-hófen *supplied, provided;* ornatus:—Ðæt ealle Godes cyricean sýn wel behófene *that all God's churches be well supplied* or *well provided* [*with all they have need of*], L. Edm. E. 5; Lambd. 58, 7; Wilk. 73, 13. v. be-hweorfan.

be-hófian, bi-hófian; *p.* ode; *pp.* od; *v. a. To have need of, to need, require;* egere, indigere. *Impersonally, it* BEHOVETH, *it concerns, it is needful* or *necessary;* oportet, interest:—Mycel wund behófaþ mycles lǽcedómes *a great wound has need of a great remedy,* Bd. 4, 25; S. 599, 40. He mægenes behófaþ gódra gúþrinca *he requires strength of good warriors,* Beo. Th. 5288; B. 2647: Exon. 98 a; Th. 367, 1; Seel. 1. Ðeáh ða scearpþanclan witan ðisse Engliscan geþeódnesse ne behófien *though the sharp-minded wise men may not have need of this English translation,* MS. Cot. Faust A. x. 150 b; Lchdm. iii. 440, 32. Behófaþ *oportet,* Jn. Lind. War. 3, 7. DER. a-behófian.

be-hóf-líc; *adj. Behoveful, needful;* necessarius:—Ðæt his líf him behóflíc wǽre *quia necessaria sibi esset vita ipsius,* Bd. 5, 5; S. 618, note 3. Behóflíc is *is necessary,* Mk. Skt. Lind. 11, 3.

be-hogadnes, -ness, e; *f. Use, custom, practice;* exercitatio, Cot. 114.

be-hogian *to be anxious, solicitous, wise, very careful;* solicitum esse, C. R. Ben. 58. v. hogian, hycgan.

be-hón; *p.* -héng, *pl.* -héngon; *pp.* -hangen, -hongen [be, hón *to hang*] *To* BEHANG, *to hang round;* circumpendere, circumdare, ambire:—Behongen beón mid bellum *to be behung* or *hung round with bells,* Past. 15, 4; Hat. MS. 19 b, 7.

be-hongen *hung round,* Past. 15, 4; Hat. MS. 19 b, 7; *pp.* of be-hón.

be-horsian; *p.* ode, ade, ude; *pp.* od, ad, ud *To deprive of a horse;* equo privare:—Ðá eóde se here to hyra scipum . . . and hí wurdon ðǽr behorsode *then the army went to their ships . . . and they were then deprived of their horses,* Chr. 886; Th. 152, 28, col. 3. DER. horsian.

be-hreósan, *pl.* -hreósaþ; *p.* -hreás, *pl.* -hruron; *pp.* -hroren *To rush down, fall;* ruere, corruere, incidere:—Behreósaþ on helle *incidunt in gehennam,* Lupi Serm. 5, 8.

be-hreówsian; *part.* -hreówsigende; ic -hreówsige, ðú -hreówsast, he -hreówsaþ, *pl.* -hreówsiaþ; *p.* ode; *pp.* od *To repent, feel remorse, make amends* or *reparation;* pœnitere, compungi, satisfacere:—Behreówsian *pœnitere,* Ælfc. Gr. 33; Som. 37, 22. Behreówsiaþ *compungimini,* Ps. Lamb. 4, 5. Ic behreówsige *satisfacio,* Ælfc. Gr. 37; Som. 39, 40. Behreówsigende *pœnitens,* Scint. 9. DER. hreówan, hreów.

be-hreówsung, e; *f. A lamenting, repentance, penitence;* pœnitentia:—Behreówsung oððe dǽdbót *pœnitentia,* Ælfc. Gr. 33; Som. 37, 22.

be-hríman; *p.* de; *pp.* ed [hrím *rime, hoar-frost*] *To cover with rime* or *hoar-frost;* pruinis circumfundere, Exon. 115 b; Th. 444, 17; Kl. 48.

be-hringed, be-hrincged; *part.* [be, hring *a ring*] *Inclosed in a ring, encircled, surrounded;* circumdatus:—Behringed beón *to be surrounded,* Past. 21, 5; Hat. MS. 32 a, 8.

be-hrópan; *p.* -hreóp, *pl.* -hreópon; *pp.* -hrópen [hrópan *to call* or *cry out*] *To scoff at, rail, trouble;* sugillare:—Ðe-læs heó cume me behrópende *ne veniens sugillet me,* Lk. Bos. 18, 5.

be-hroren; *p. part. Fallen off, deprived of;* a quo aliquid decidit, orbatus:—Fatu hyrstum behrorene *vessels deprived of their ornaments,* Beo. Th. 5517; B. 2762; *pp.* of be-hreósan, *q.v.*

be-hrúmig; *adj. Swarthy, sooty;* fuliginosus, Martyr. 3, April. v. hrúmig.

be-hrumod; *p. part. Bedaubed, dirtied;* cacabatum, Cot. 31: 189. v. besciten.

behþ, e; *f. A token, sign, proof;* signum, testimonium:—Heó hét hyre þinenne ðæs herewǽdan heáfod to behþe blódig ætýwan ðám burhleódum *she ordered her servant to shew the bloody head of the leader of the army to the citizens as a token,* Judth. 11; Thw. 24, 6; Jud. 174.

be-hwearf, es; *m. A change, an exchange;* commutatio:—On behwearfum heora *in commutationibus eorum,* Ps. Spl. 43, 14.

be-hweorfan; *p.*-hwearf,*pl.*-hwurfon; *pp.*-hworfen, -hweorfen. **I.** *to turn, spread about;* vertere, convertere:—Hleahtre behworfen *turned to laughter,* Andr. Recd. 3402; An. 1705. Híg behwurfon híg búton ðære wícstówe *they spread them about outside of the camp,* Num. 11, 32. **II.** *to turn* or *put in order, arrange;* disponere, parare:—Ðæt ealle Godes cyrcan sýn wel behworfene [behweorfene, H.] *that all God's churches be well put in order,* L. Edm. E. 5; Th. i. 246, 12. Ðæt ǽlc preost hæbbe eal mæsse-reáf wurþlíce behworfen *that every priest have all his mass-vestments worthily arranged,* L. Edg. C. 33; Th. ii. 250, 28. v. be-hweorfan.

be-hwerfan; *p.* de; *pp.* ed [be, hwerfan *to turn*] *To turn, prepare, instruct;* vertere, instruere:—Ðonne hió ǽrest síe útan behwerfed *when*

it is first turned round about, Bt. Met. Fox 13, 154; Met. 13, 77. Ic wolde mid sumre bîsne ðê behwerfan ûtan *I would instruct thee further* [ûtan *from without*] *by some example*, Bt. 34, 4; Fox 138, 27.

be-hwon *whence*; unde, Bd. 2, 2; S. 503, 2. v. hwonan.

be-hwurfon *spread about*, Num. 11, 32; *p. pl. of* be-hweorfan.

be-hwylfan; *p.* -hwylfde; *pp.* -hwylfed *To cover or vault over*; operire, obruere :—Ne behwylfan mæg heofon and eorþe his wuldres word *the word of his glory may not cover over heaven and earth*, Cd. 163; Th. 204, 28; Exod. 426. v. hwylfan.

be-hwyrfan *to treat, direct, exercise, practice*; tractare, exercere :—Behwyrf ðê sylfne *exerce temet ipsum*, Coll. Monast. Th. 31, 37 : R. Ben. 32. v. be-hweorfan.

be-hycgan, -hicgan *to think, consider, bear in mind, trust*; meditari, considerare, sollicitum esse de re, confidere, niti :—He sceal deópe behycgan þroht þeóden-gedál *he must deeply bear in mind the dire decease of his lord*, Exon. 52 b; Th. 183, 7; Gú. 1323. Ðe on Gode behicgaþ *qui in Deo acquiescunt*, R. Ben. 31. v. hycgan.

be-hýdan, bi-hýdan; *p.* -hýdde; *pp.* -hýded, -hýdd, -hýd *To hide, conceal, cover*; abscondere, occultare, operire :—Se ðe hine behýdde fram hǽton his *qui se abscondit a calore ejus*, Ps. Spl. 18, 7 : Salm. Kmbl. 604 : Sal. 301. Ðæt wæs lange behýded *which was long concealed*, Elen. Kmbl. 1582; El. 793. Heolstre behýdd *covered with darkness*, Elen. Kmbl. 2161; El. 1082. Behýdd *absconditum*, Mk. Bos. 4, 22.

be-hydelíce, -hidiglîce, big-hydilîce, big-hidiglîce; *adv. Carefully*; sollicite, sollerter, Bd. 1, 27; S. 489, 39 : 3, 19; S. 547, 29 : 4, 23; S. 595, 4.

be-hydig, bî-hidig; *adj. Careful, vigilant, wary, watchful, solicitous, anxious*; sollers :—He wæs se behydegesta [MS. behydegǽsta] *erat sollertissimus*, Bd. 5, 20; S. 642, 13 : 4, 7; S. 574, 33. v. hydig.

be-hýdignys, -nyss, e; *f.* [be, hýdan *to hide*] *A desert, a wilderness*; desertum :—Stefn Drihtnes tosceacende behýdignys *vox Domini concutientis desertum*, Ps. Spl. C. 28, 7.

be-hyldan *to put off, to flay, skin*; excoriare :—He hêt hý behyldan *he ordered to flay it*, Ors. 4, 6; Bos. 84, 45.

be-hýpan; *p.* -hýpte; *pp.* -hýped [hýpe *a heap*] *To heap or cover over, surround, encompass*; contegere, circumsepire, circumdare :—Wæs mid wǽpnum and mid feóndum eall ûtan behýped *cum armis et hostibus circumseptus erat*, Bd. 3, 12; S. 537, 28.

be-hýring, -hîring, e; *f. A hiring, letting out to hire*; locatio :—Behîring *vel* gehýred feoh *locatio*, Ælfc. Gl. 13; Som. 57, 123; Wrt. Voc. 20, 60. v. ge-hýran.

be-hýðelíce; *adv. More sumptuously*; sumptuosius, Cot. 186.

be-hýðlíc *sumptuous*. v. hyðelíc.

beig-beám, es; *m. A bramble*; rubus :—Moyses ætýwde wið ǽnne beigbeám *Moyses ostendit secus rubum*, Lk. Bos. 20, 37. v. begbeám.

be-innan *prep. dat. In, within*; in, intra :—Boëtius ðâ nânre frófre beinnan ðam carcerne ne gemunde *then Boethius thought of no comfort within the prison*, Bt. 1; Fox 4, 2.

be-irnan; *impert.* be-irn; *p.* -arn, *pl.* -urnon; *pp.* -urnen *To come or run into*; incurrere :—Ne be-irn ðû ou ða inwitgecyndo *do not run into their guilty nature*, Salm. Kmbl. 660 : Sal. 329. v. be-yrnan.

be-iundan *beyond*; trans, ultra :—Beiundan Iordane *trans Iordanem*, Deut. 1; S. 11, 30. v. be-geondan.

be-lácan; *p.* -lêc, -leólc, *pl.* -lêcon; *pp.* -lácen *To flow around, inclose*; circumfluere :—Ýþ mec lagufæðme beleólc *the wave inclosed me in its watery bosom*, Exon. 122 b; Th. 471, 26; Rä. 61, 7.

be-ládian, ic -ládige; *p.* de *To clear, excuse*; excusare :—Ðæt he wolde beládian his môdor *that he might clear his mother*, Ors. 3, 9; Bos. 65, 24 : Ælfc. Gr. 28, 6; Som. 32, 35. v. ládian.

be-ládigend, es; *m. One who makes excuses, a defender*; excusator, Ælfc. Gl. 23; Wrt. Voc. 83, 64.

be-ládung, e; *f. An excuse*; apologeticus, excusatio :—Beládung *apologeticus*, Ælfc. Gl. 106; Som. 78, 65; Wrt. Voc. 57, 44. v. ládung.

be-lǽdan; *p.* -lǽdde; *pp.* -lǽd, -lêd; *v. a. To bring, lead by, mislead, lead*; seducere, inferre, inducere, impellere :—Ðû belǽddest us on grin *thou hast mislead us into a snare*; induxisti nos in laqueum, R. Ben. 7. Belǽd beón mid unþeáwum *impelli vitiis*, R. Ben. 64. v. lǽdan.

be-lǽfan; *p.* de; *pp.* de *To remain, to be left*; remanere, superesse :—Ân of him ne belǽfde *unus ex eis non remansit*, Ps. Spl. C. 105, 11. v. lǽfan.

be-lǽg *surrounded*, Ps. Th. 118, 153; *p. of* be-licgan.

be-lǽndan *to deprive of land*, Chr. 1112; Th. 369, 39. v. be-landian.

be-lǽðed *part.* [láþ *evil*] *Loathed, detested*; exosus. v. láðian.

be-lǽwa, an; *m. A destroyer*; proditor, traditor. v. lǽwa.

be-lǽwan; *p.* -lǽwde; *pp.* -lǽwed; *v. a. To bewray, betray*; tradere, prodere :—Ðæt he hyne wolde belǽwan *ut traderet eum*, Mt. Bos. 26, 15, 16. Heó hine belǽwde *she betrayed him*, Jud. 16, 21. Ðæt Iohannes belǽwed wæs *quod Ioannes traditus esset*, Mt. Bos. 4, 12.

be-lǽwing, e; *f. A betraying, treason*; proditio, Homl. Th. ii. 244, 22. v. be-lǽwan, lǽwa *a betrayer*.

be-láf *remained*, Jos. 5, 1; *p. of* belífan.

be-lagen beón *to be oppressed*; opprimi, Past. 58, 1; Hat. MS.

be-lamp *happened, befell*, Beo. Th. 4928; B. 2468; *p. of* belimpan.

be-landian; *p.* ode, ede; *pp.* od, ed; *v. a. To deprive of land, to confiscate, disinherit*; terris privare :—Wearþ Eádgár belandod *Edgar was deprived of land*, Chr. 1091; Th. 359, 5. Hî hî ǽr belandedon *they had deprived them previously of their lands*, 1094; Th. 361, 12. v. be-lendian. Opposed to gelandian *to inherit*.

belced-sweora; *adj. Possessed of an inflated neck*; inflata cervice præditus :—Ic eom belced-sweora *I am neck-inflated*, Exon. 127 b; Th. 489, 24; Rä. 79, 1.

belcentan *to utter, give forth, belch, eructate*; eructare :—Se lǽcecræft biþ swíðe swête belcentan *the medicine is very sweet to eructate*, Bt. 22, 1, Bodl; Fox 76, note 17. v. belcettan.

belcettan; *p.* te; *pp.* ted *To utter, give forth*; eructare :—Nû míne weleras ðê wordum belcettaþ *ymnas elne eructabunt labia mea hymnum*, Ps. Th. 118, 171. v. bealcettan.

beld, beldo *boldness, rashness*; audacia. v. byld, byldo.

be-leác *shut in*, Ors. 4, 5; Bos. 81, 40; *p. of* belúcan.

be-leán; *p.* -lôh, *pl.* -lôgon; *pp.* -leahen *To hinder by blame, reprehend, reprove, forbid*; prohibere, reprobare, reprehendere :—We lǽraþ ðæt preostas oferdruncen beleán ôðrum mannum *we enjoin that priests reprehend drunkenness in other men*, L. Edg. C. 57; Th. ii. 256, 14. He him ðæt swýðe belôh *hoc multum illi prohibuit*, Bd. 5, 19; S. 638, 28, note : Beo. Th. 1027; B. 511. v. leán.

be-lecgan, bi-lecgan; *p.* -legde, -lêde, *pl.* -legdon; *pp.* -legd, -lêd; *v. a. To lay or impose upon, cover, invest, load, afflict, charge, accuse*; imponere, afficere, onerare, accusare :—Heó ðone hleóðor-cwyde husce belegde *she covered the revelation with scorn*, Cd. 109; Th. 143, 21; Gen. 2382. Papirius wæs mid Rômânum swylces dômes beleád *Papirius was invested with such authority by the Romans*, Ors. 3, 8; Bos. 63, 40. We hine clommum belegdon *we loaded him with chains*, Andr. Kmbl. 3119; An. 1562. Hî ðê wítum belecgaþ *they afflict thee with torments*, 2424; An. 1213. Gyf man sacerd belecge mid tyhtlan and mid uncræftum *if one charges a priest with an accusation and with evil practices*, L. C. E. 5; Th. i. 362, 8, 19, 21. Se ðe hine belecge *he who accuses him*, L. O. D. 6; Th. i. 354, 30 : 4; Th. i. 354, 15.

be-lêd *impelled*, R. Ben. 64; *pp. of* belǽdan.

be-lêd = be-legd *charged, accused*, L. O. D. 4; Th. i. 354, 15; *pp. of* be-lecgan.

be-lêgan, bi-lêgan; *p.* -lêgde; *pp.* -lêgd *To surround with flame*; circumflagrare flamma :—Lîge belêgde *surrounded with flame* [Ger. umlodert mit lohe], Cd. 188; Th. 234, 22; Dan. 296. v. lêgan.

be-legde *covered*, Cd. 109; Th. 143, 21; Gen. 2382; *p. of* be-lecgan.

be-lendan, be-lǽndan; *p.* de; *pp.* ed *To deprive of land*; terris privare :—Se cyng belǽnde ðone eorl *the king deprived the earl of his land*, Chr. 1112; Th. 369, 39, 41 : 1104; Th. 367, 11. Wearþ Eoda eorl and manege ôðre belende *earl Eudes and many others were deprived of their lands*, 1096; Th. 362, 36. v. be-landian.

belene, beolone, belone, an; *f. Henbell, henbane*; hyoscyamus niger :—Belenan meng wið rysele *mix henbane with lard*, L. M. 1, 31; Lchdm. ii. 72, 1. Dô belenan seáw *apply the juice of henbane*, 3, 2; Lchdm. ii. 310, 7. Genim beolonan sǽd *take the seed of henbane*, 1, 2; Lchdm. ii. 38, 1. v. beolone, henne-belle. [Henbane is so called from the baneful effects of its seed upon poultry, of which Matthioli says that 'birds, especially gallinaceous birds, that have eaten the seeds perish soon after, as do fishes also.' The *A. Sax.* belene and beolone, *Ger.* bilse, *O. Ger.* belisa, *Pol.* bielún, *Hung.* bilénd, *Rus.* belená are words derived (according to Zeuss, p. 34) from an ancient Celtic god Belenus, corresponding to the Apollo of the Latins : 'Dem Belenus war das Bilsenkraut heilig, das von ihm Belisa und Apollinaris hiess,' Prior 109.]

be-leógan; *p.* -leág, *pl.* -lugon; *pp.* -logen *To belie, deceive by lies*; fallere :—Belogen beón *falli*, Gr. Dial. 1, 14. DER. leógan.

be-leólc *flowed around, inclosed*, Exon. 122 b; Th. 471, 26; Rä. 61, 7; *the reduplicated p. of* be-lácan, v. lácan, *and Goth. cognates at the end of* lácan.

be-leóran *to pass over*. v. bi-leóran.

be-leósan, bi-leósan; *p.* -leás, *pl.* -luron; *pp.* -loren [be, leósan *to loose*] *To let go, to deprive of, to be deprived of, lose*; privare, orbare, privari, amittere :—Leóhte belorene *deprived of light*, Cd. 5; Th. 6, 9; Gen. 86 : Beo. Th. 2150; B. 1073; Andr. Kmbl. 2159; An. 1081. Ðær ic swíðe beleás hêrum, ðâm ðe ic hæfde *there I was much deprived of the hairs, which I had*, Exon. 107 a; Th. 407, 35; Rä. 27, 4. v. for-leósan.

be-lêwa, an; *m. A betrayer*; proditor. v. be-lêweda, lǽwa.

be-lêweda, an; *m. A betrayer*; proditor :—Mid Iudan úres Drihtenes belêwedan *with Judas the betrayer of our Lord*, Wanl. Catal. 137, 38, col. 1. v. belêwa, belǽwa.

bele-wite *simple*; simplex :—Se wer wæs swíðe belewite and rihtwís *erat vir ille simplex et rectus*, Job 1, 1; Thw. 164, 2. v. bile-wit.

bel-flȳs, es; *n.* [bell *a bell*, flȳs *a fleece*] *The* BELL-WETHER'S FLEECE, *the fleece of a sheep that carries the bell*; tympani vellus, *i. e.* ducis gregis

tintinnabulum gestantis vellus:—Bel-flýs *id est, tympani vellus*, L. R. S. 14; Th. i. 438, 23.

BELG, belig, bylg, bylig, bilig, bælg, bælig, es; *m. A* BULGE, *budget, bag, purse, bellows, pod, husk,* BELLY; bulga, follis, siliqua, uter :—Bylg *bulga,* Cot. 27. Bylig *follis,* Ælfc. Gl. 27; Wrt. Voc. 86, 15. Bilig *uter,* Ps. Spl. M. 118, 83. [*Dut.* balg, *m* : *Ger.* balg, *m* : *M. H. Ger.* balc, *m* : *O. H. Ger.* balg, *m. follis, uter* : *Goth.* balgs, *m* : *Dan.* bælg, *m* : *O. Nrs.* belgr, *m.*] DER. beán-belg, -bælg, blást-, mete-, wín-, v. ge-belg.

BELGAN, ic belge, ðu bilgst, bilhst, he bilgþ, bilhþ, bylgþ, *pl.* belgaþ; *p.* ic, he bealg, bealh, ðú bulge, *pl.* bulgon; *pp.* bolgen. I. *v. reflex. acc. To cause oneself to swell with anger, to make oneself angry, irritate oneself, enrage oneself;* ira se tumefacere, se irritare, se exasperare :—Nelle ðú on ǽcnesse ðé áwa belgan *non in æternum indignaberis,* Ps. Th. 102, 9. Ic bidde ðæt ðú ðé ne belge wið me *ne, quæso, indigneris,* Gen. 18, 30. Bealg hine swíðe folc-ágende *the people's lord irritated himself greatly,* Exon. 68 a; Th. 253, 25; Jul. 185. II. *intrans. To swell with anger, to be angry, to be enraged;* ira tumere, indignari, irasci :—Gé belgaþ wið me *mihi indignamini,* Jn. Bos. 7, 23. [*O. Sax.* belgan, *v. reflex.* ; *p.* balg; *pp.* bolgan *irasci, indignari* : *N. H. Ger.* balgen *pugnis certare* : *O. H. Ger.* belgan *tumere, irasci.*] DER. a-belgan, ge-, bolgen-mód.

bel-hringes beácn, es; *n. A sign by bell-ringing;* signum sonitu campanæ datum, R. Ben. 43.

bel-hús, bell-hús, es; *n. A* BELL-HOUSE, *a room or tower in the castle of a Thane, generally built between the kitchen and porter's lodge, where was a bell or bells to summon the inhabitants to prayers,* and for other purposes; campanile *vel* campanarium, turris in qua pendent tintinnabulum *vel* tintinnabula, Du Cange, fol. 1681, col. 712; CAMPANA, col. 708 :—Gif ceorl hæfde fíf hída ágenes landes cirican and cycenan, bell-hús . . . ðonne wæs he þegen-rihtes weorþe *if a freeman had five hides of his own land, a church and kitchen, a bell-house . . . then was he worthy of thane-right,* L. R. 2; Th. i. 190, 15.

be-libban; *p.* -lifde, *pl.* -lifdon; *pp.* -lifed, -lifd *To deprive of life;* vita privare :—Líc cólode belifd under lyfte *the corpse was lifeless cold in the air,* Exon. 51 b; Th. 180, 19; Gú. 1282. v. libban.

be-licgan, he -ligeþ, -líþ, *pl.* -licgaþ; *p.* -læg, *pl.* -lágon, -lágon; *pp.* -legen; *v. a.* [be *by,* licgan *to lie*] *To lie or extend by or about, to surround, encompass;* circumdare, cingere :—Hí belicgaþ us mid fyrde *circumdabunt nos exercitu,* Jos. 7, 9. Sió eá Etheopia land beligeþ úton *the river encompasseth the Ethiopian land,* Cd. 12; Th. 15, 7; Gen. 229. Me néd belæg *want surrounded me,* Ps. Th. 118, 153.

be-lidenes *of the left or departed,* Elen. Kmbl. 1752; El. 878; *gen. pp. from* be-líðan, *q. v.*

be-lífan, ic -lífe, ðú -lífest, -lífst, he -lífeþ, -lífþ; *p.* -láf, *pl.* -lifon; *pp.* -lifen *To remain, abide, to be left;* superesse, manere, remanere :—Ne se rysel ne belífþ óþ morgen *nec remanebit adeps usque mane,* Ex. 23, 18. He ána beláf ðær bæfta *mansit solus,* Gen. 32, 24; Ps. Spl. 105, 11. Hí námon ðæt of ðám brytsenum beláf, seofon wilian fulle *sustulerunt quod superaverat de fragmentis, septem sportas,* Mk. Bos. 8, 8. [*Plat.* bliven; *p.* bléf: *Dut.* blijven; *p.* bleef: *Ger.* bleiben; *p.* blieb: *M. H. Ger.* be-líben; *p.* be-leip: *O. H. Ger.* pi-lípan; *p.* pi-leip: *Dan.* blive; *p.* blev: *Swed.* blifva, bli; *p.* blef, ble: in *O. Nrs.* the word is wanting, as well as in *Goth.*] v. lífan.

be-lifd *-lifed deprived of life, lifeless, inanimate;* defunctus, Exon. 51 b; Th. 180, 19; Gú. 1282; *pp. of* be-libban.

belig *a bag.* v. belg.

be-ligeþ *encompasseth,* Cd. 12; Th. 15, 7; Gen. 229. v. be-licgan.

be-limp *an event;* eventus, Lchdm. iii. 202, 28. v. gelimp.

be-limpan; *p.* -lamp, *pl.* -lumpon; *sub.* -lumpe; *pp.* -lumpen [be, limpan *to appertain*] *To concern, regard, belong, pertain, appertain;* curare, pertinere :—Ne belimpþ to ðé *non ad te pertinet,* Mk. Bos. 4, 38. Hwæt ðæs to him belumpe *what of that concerned him?* Bd. 2, 12; S. 513, 39. Hwæt belimpþ his to ðé *what of it belongs to thee?* Bt. 14, 2; Fox 42, 35. Hit belimpþ tó ðære spræce *it appertains to the discourse,* Bt. 38, 2; Fox 198, 19. II. *to happen, occur, befall;* evenire, accidere, contingere :—Ðá him sió sár belamp *when that pain befell him,* Beo. Th. 4928; B. 2468.

be-lisnian, -listnian; *p.* ode; *pp.* od; *v. trans.* [be *from,* lystan *to desire*] *To evirate, emasculate, castrate;* castrare. *Part. p.* belisnod, belistnod *emasculated* :—Belisnod *spadatus, eunuchizatus,* Ælfc. Gl. 2; Som. 55, 53; Wrt. Voc. 16, 26. Used as a noun,—*A eunuch* :—Belisnod *spado, eunuchus,* Ælfc. Gr. 9, 3; Som. 8, 32. Sóþlíce synd belistnode, ðe of hyra módor innoðum cumaþ, and eft synt belistnode ða men ðe man belistnaþ, and eft synd belistnode ðe híg sylfe belistnodon for heofona ríce *sunt enim eunuchi, qui de matris utero sic nati sunt, et sunt eunuchi, qui facti sunt ab hominibus, et sunt eunuchi, qui se ipsos castraverunt propter regnum cælorum,* Mt. Bos. 19, 12. v. a-fýran.

be-lisnod, -listnod *a eunuch,* Ælfc. Gr. 9, 3; Som. 8, 32 : Ælfc. Gl. 2; Som. 55, 53. v. be-lisnian.

be-líþ *surrounds,* Cd. 12; Th. 15, 13; Gen. 232. v. be-licgan.

be-líðan; *p.* -láþ, *pl.* -liðon = -lidon; *pp.* -liðen = -liden [be *from,*

líðan *to go, sail*] *To go from, to leave;* effugere, relinquere :—Lífe belidenes líc *the body of the left by life,* i. e. *the body of the lifeless,* Elen. Kmbl. 1752; El. 878 : Exon. 52 a; Th. 182, 18, note; Gú. 1312: Judth. 12; Thw. 25, 26; Jud. 280. Ða belidenan [MS. behlidenan] *the dead;* mortuos, Andr. Kmbl. 2179; An. 1091.

BELL, e; *f*: belle, an; *f. A* BELL; campana, tintinnabulum, cymbalum :—Cyrice bell *the church-bell.* Hleóðor heora bellan *a sound of their bell,* Bd. 4, 23; S. 595, note 40. Belle *tintinnabulum,* Ælfc. Gr. 5; Som. 4, 39. Hériaþ hine on bellum *laudate eum in cymbalis,* Ps. Lamb. 150, 5. Seó lytle belle *the little bell.* Seó mycele belle *the large bell;* campana, Lye. [*Plat. Dut.* belle, bel.] v. bellan.

bell *a bellowing, roar, cry?* Cd. 148; Th. 185, 12; Exod. 121. v. bǽl-egesa.

BELLAN; *part.* bellende; ic belle, ðú bilst, he bilþ, *pl.* bellaþ; *p.* ic, he beal, ðú bulle, *pl.* bullon; *pp.* bollen *To* BELLOW, *to make a hollow noise, to roar, bark, grunt;* boare, latrare, grunnire :—Bearg bellende *a roaring [grunting] boar,* Exon. 111 b; Th. 428, 10; Rä. 41, 106. [*Ger.* bellen : *Swed.* böla : *O. Nrs.* belja.]

belle, an; *f. A bell;* tintinnabulum :—Hleóðor heora bellan *a sound of their bell,* Bd. 4, 23; S. 595, note 40 : Ælfc. Gr. 5; Som. 4, 39. v. bell.

bell-hús *a bell-house,* L. R. 2; Th. i. 190, 15. v. belhús.

be-locen *shut up, inclosed,* Cd. 209; Th. 259, 24; Dan. 696; *pp. of* be-lúcan.

be-logen *deceived,* Gr. Dial. 1, 14. v. be-leógan.

be-lóh *forbade,* Bd. 5, 19; S. 638, 28, note. v. be-leán.

belone, an; *f.* Henbane :—Henne-belone, óðrum naman belone *henbane, by another name bane,* Herb. 5, 1; Lchdm. i. 94, 5, note 9. v. hennebelle, belene.

be-loren *deprived,* Cd. 5; Th. 6, 9; Gen. 86; *pp. of* be-leósan.

BELT, es; *m. A* BELT, *girdle;* balteus, Cot. 25. [*O. H. Ger.* palz, balz, *m?* *a girdle* : *Ger.* Belt, *m. name of the narrow straits between the Danish isles* : *Dan.* belte *a belt* : *Swed.* bälte, *id* : *O. Nrs.* belti, *n. id* : *Lat.* balteus.] v. gyrdel.

be-lúcan, he -lýcþ; *p.* -leác, *pl.* -lucon; *pp.* -locen; *v. trans.* [be, lúcan *to lock*] *To lock up, inclose, surround, shut, shut up;* concludere, recludere, includere, circumcludere, amplecti, obserare, claudere :—Drihten hí beleác *Dominus conclusit eos,* Deut. 32, 30. Gif he ðone oxan belúcan nolde *si non recluserit bovem,* Ex. 21, 29. Ðá hét he hine gebringan on carcerne and ðær inne belúcan *he gave an order to take him to prison and therein lock him up,* Bt. 1; Fox 2, 26: Ors. 4, 5; Bos. 81, 40: Gen. 41, 49: Ps. Spl. C. T. 16, 11. Belocen leoðu-bendum *locked up in limb-bonds,* Andr. Kmbl. 327; An. 164. Wealle belocen *inclosed with a wall,* Cd. 209; Th. 259, 24; Dan. 696. Ðæt man belúce ǽlc deofulgyld-hús *that one should close every idol-temple,* Ors. 6, 30; Bos. 127, 36.

be-lumpe *concerned;* pertineret, Bd. 2, 12; S. 513, 39. v. be-limpan.

belune *henbane,* Som. Lye. v. belene.

be-lýcþ *locks,* Hexam. 5; Norm. 8, 27; *pres. of* belúcan.

be-lytegan; *p.* ade; *pp.* ad; *v. a.* [lyteg *crafty*] *To allure, inveigle, seduce;* procare :—He belytegade Crēce *he allured Greece,* Ors. 3, 7; Bos. 59, 39.

be-mǽnan, bi-mǽnan; *p.* de; *pp.* ed [be, mǽnan *to moan,* III. *q. v.*] *To* BEMOAN, *bewail, lament, mourn;* lugere, dolere, congemere :—Ða heófungdagas wǽron ðá gefyllede, ðe híg Moisen bemǽndon *completi sunt dies planctus lugentium Moysen,* Deut. 34, 8.

be-mǽtan = be-mǽton *measured, compared,* Ors. 3, 7; Bos. 60, 43; *p. pl. of* be-metan.

Bēme; *nom. acc; gen.* a; *dat.* um; *pl. m. The Bohemians;* Bohēmi :—Riht be eástan syndon Bēme *right to the east are the Bohemians,* Ors. 1, 1; Bos. 18, 33. v. Behēmas.

bēme, an; *f. A trumpet;* tuba, salpinx :—Bēman bláwan *to blow the trumpet,* Cd. 227; Th. 302, 19; Sat. 602. Bēme *barbita,* Cot. 27. v. býme.

be-mearn *mourned,* Cd. 106; Th. 139, 14; Gen. 2309. v. be-meornan.

be-meornan; *p.* -mearn, *pl.* -murnon; *pp.* -mornen [be, meornan *to mourn*] *To mourn,* BEMOURN, *bewail, deplore;* lugere :—Ðín ferhþ bemearn *thy soul mourned,* Cd. 106; Th. 139, 14; Gen. 2309. Nó ic ða stunde bemearn *I bemourned not the time,* Exon. 130 a; Th. 499, 12; Rä. 88, 14.

bēmere *a trumpeter,* Lye. v. býmere.

be-metan; *p.* -mæt, *pl.* -mǽton; *pp.* -meten; *v. trans.* [be, metan *to measure*] *To measure by, compare, estimate, consider;* metiri, commetiri, comparare, æstimare :—Ðæt hý ðá æt nihstan hý sylfe to nóhte bemǽtan *that they at last compared themselves to nought,* Ors. 3, 7; Bos. 60, 43. Ðæt hý ná siðdan nánes anwealdes hý ne bemǽtan, ne nánes freódōmes *that afterwards they did not consider themselves [possessed] of any power, nor of any freedom,* 3, 7; Bos. 62, 11. Ðæt hý heora miclan anwealdes and longsuman hý sylfe siðdan wið Alexander to náhte [ne] bemǽtan *that, in respect of their great and lasting power, they estimated themselves at nothing against Alexander,* 3, 9; Bos. 65, 39 : 4, 6; Bos. 86, 17.

be-míðan, bi-míðan; *p.* -máþ, *pl.* -miðon; *pp.* -miðen [be, míðan *to hide*] *To hide, conceal;* abscondere, occultare :—He ne mihte hit bemíðan

non potuit latere, Mk. Bos. 7, 24. Hí ne mágon heortan geþohtas fóre Waldende bemíðan *they cannot conceal their heart's thoughts before the Supreme*, Exon. 23 a; Th. 65, 4; Cri. 1049. He his mǽgwlite bemíðen hǽfde *he had concealed his shape*, Andr. Kmbl. 1712; An. 858.

be-murcnian; *p.* ode; *pp.* od [be, murcnian *to murmur*] *To murmur, murmur greatly;* obmurmurare :—Hú ungemetlíce, ge Rómware, bemurcniaþ *how immoderately, O Romans, do ye murmur*, Ors. 1, 10; Bos. 34, 9.

be-murnan, bi-murnan; *p.* -murnde; *pp.* -murned [be, murnan *to mourn*] *To bemoan, bewail, mourn, to care for;* lugere, curare, sollicitum esse de re :—Hwæt bemurnest ðú *why bemoanest thou?* Exon. 10 b; Th. 11, 26; Cri. 176. ' Síþ ne bemurneþ *he bewails not his lot*, 117 a; Th. 449, 31; Dóm. 79. Feorh ne bemurndon grǽdige gúþrincas *the greedy warriors cared not for the soul*, Andr. Kmbl. 308; An. 154.

be-mútian *to exchange for;* commutare. v. bi-mútian.

be-myldan [molde *mould*] *To cover with mould* or *earth, to bury, inter, hide* or *put under ground;* inhumare, humare, Cot. 101.

BEN, benn, e; *f.* [*connected with* bana *a slayer, murderer*] *A wound;* vulnus :—Ne ðǽr ǽnig com blód of benne *nor came there any blood from the wound*, Cd. 9; Th. 12, 6; Gen. 181. Heortan benne *the wounds of heart*, i. e. *sadness, grief*, Exon. 77 a; Th. 289, 17; Wand. 49. Blátast benna *the palest of wounds*, Exon. 19 a; Th. 48, 13; Cri. 771. Hí feóllon bennum seóce *they fell sick with wounds*, Cd. 92; Th. 118, 29; Gen. 1972. With this word the MSS. often confound the *pl.* of bend, as in Cd. 195; Th. 243, 12; Dan. 435, where *benne* stands for *bende*: and in Andr. Recd. 2077; An. 1040: Exon. 73 a; Th. 273, 21, note; Jul. 519, where *bennum* stands for *bendum.* v. bend. [*O. H. Ger.* bana, *f:* Goth. banya, *f: Icel.* ben, *f.*] DER. bennian, ge-.

BĚN; *gen. dat.* bēne; *acc.* bēn; *pl. nom.* bēna, bēne; *f. A praying, prayer, petition, an entreaty, a deprecation, supplication, demand.* Hence in *Chaucer* bone *and our* BOON; precatio, deprecatio, oratio, preces, postulatio :—Ðeáh ðe ðæs cyninges bēne mid hine swíðode and genge wǽren [wǽren, MS. T: wǽre, MSS. Ca. O.] *though the king's prayers were powerful and effectual with him*, Bd. 3, 12; S. 537, 18: 1, 4; S. 475, 32: 5, 1; S. 614, 15: 5, 21; S. 643, 6. Be ryhtes bēne *of praying for justice*, L. In. 8; Th. i. 106, 19. Ðín bēn ys gehýred *exaudita est deprecatio tua*, Lk. Bos. 1, 13. Ic underféng ðíne bēne *suscepi preces tuas*, Gen. 19, 21. Hí heom ðǽra bēna forwyrdnon *they gave to them a denial of their requests*, Ors. 2, 2; Bos. 40, 34. Micelra bēna dæg *litania major*, Martyr. 25, April. [*O. Nrs.* bón, *f. a petitioner.*]

bēn, benn *summoned; p. of* bannan.

bēna, an; *m. A petitioner, demander;* rogator, supplex :—Gehýr me helpys bēnan *exaudi me auxilii supplicem*, Ps. Th. 101, 2. Hý bēna wǽron *they were demanders*, or *they demanded*, Ors. 3, 11; Bos. 73, 36. Hence bēna wesan *to demand, request*, Beo. Th. 6272; B. 3140: Cd. 107; Th. 142, 6; Gen. 2357.

be-nacian; *p.* ode; *pp.* od, ed [be, nacian *nudare*] *To make naked;* denudare :—Ðú benacodest grundweall ôþ hneccan *denudasti fundamentum usque ad collum*, Cant. Abac. Lamb. fol. 190 a; 13.

be-nǽman, be-nēman; *p.* -nǽmde, -nēmde; *pp.* -nǽmed, -nēmed [be, niman *to take*] *To deprive, take away;* auferre, privare :—He ne meahte hí ðæs landes benǽman *he could not deprive them of their land*, Ors. 1, 10; Bos. 33, 35: Cd. 98; Th. 129, 32; Gen. 2152. Ealdre benǽman *to deprive of life*, Judth. 10; Thw. 22, 24; Jud. 76. Wuldre benēmed *deprived of glory*, Cd. 215; Th. 272, 18; Sat. 121.

BENC, e; *f.* A BENCH; scamnum, abacus :—Bugon to bence *they turned to a bench*, Beo. Th. 659; B. 327. On bence wæs helm *a helm was on the bench*, Beo. Th. 2491; B. 1243. [*Plat. O. Sax. Dut. Ger.* bank, *f: M. H. Ger.* banc, *m.f: O. H. Ger.* panch, *f: Dan. Swed.* bänk: *O. Nrs.* bekkr, *m.*] DER. ealu-benc, meodu-.

benc-sittende; *part. Sitting on a bench;* in scamno sedens, Judth. 10; Thw. 21, 20; Jud. 27: Exon. 88 a; Th. 332, 1; Vy. 78.

benc-swég, es; *m. A' bench-noise, noise from the benches, convivial noise;* clamor in scamnis ad convivium sedentium, Beo. Th. 2326; B. 1161.

benc-þel, es; *pl.* -þelu; *n. A bench-floor, a floor on which benches are put;* scamnorum tabulatum, Beo. Th. 976; B. 486: 2482; B. 1239.

bend, bænd, e; *f:* es; *m.* What ties, binds, or bends,—*A band, bond, ribbon, a chaplet, crown, ornament;* vinculum, ligamen, diadema :—Ðæt benda onlýseþ *that looseneth bonds*, Exon. 8 b; Th. 5, 12; Cri. 68. On láþne bend *in a loathsome bond*, Cd. 225; Th. 298, 27; Sat. 539. Heora bendas towearp *vincula eorum disrupit*, Ps. Th. 106, 13: 115, 7: 149, 8. Ða benda sumes gehæftes *vincula cujusdam captivi*, Bd. 4, 22; S. 590, 28. Ðá Iohannes on bendum gehýrde Cristes weoruc *Joannes cum audisset in vinculis opera Christi*, Mt. Bos. 11, 2. Bend agimmed and gesmiðed *diadema*, Ælfc. Gl. 64; Som. 69, 12; Wrt. Voc. 40, 46. Mid golde gesiwud bend *nimbus*, 64; Som. 69, 13. DER. ancorbend, fýr-, hell-, hyge-, íren-, searo-, wæl-, wíte-.

bendan; *p.* bende; *pp.* bended; *v. trans.* [bend *a band*]. I. *to* BEND; flectere, tendere, intendere :—He his bogan bendeþ *intendit arcum suum*, Ps. Th. 57, 6. He bende his bogan *arcum suum tetendit*, 7, 13. II. *to bind, fetter;* vincire :—Sume hí man bende *some they*

---

bound, Chr. 1036; Th. 294, 6, col. 2; Ing. 208, 28; Ælf. Tod. 4. DER. ge-bendan.

bend-feorm, e; *f. A feast for the reaping* [*binding*] *of corn, a harvest-feast;* firma ad congregandas segetes, firma messis :—On sumere þeóde gebyreþ bend-feorm [bēn-feorm] *for rípe in some one province a harvest-feast is due for reaping the corn*, L. R. S. 21; Th. i. 440, 26.

bēne; *gen. dat. s; nom. acc. pl.* of bēn *a prayer, q.v.*

be-neah *he requires*, Elen. Kmbl. 1233; El. 618. v. be-nugan.

be-neced *naked* :—Of hæftnede benecedes *de captivitate nudati*, Cant. Moys. Isrl. Lamb. 194 b, 42; *pp.* of be-nacian.

be-nēman; *p.* -nēmde; *pp.* -nēmed *To deprive;* privare :—Wuldre benēmed *deprived of glory*, Cd. 215; Th. 272, 18; Sat. 121. v. be-nǽman.

be-nemnan; *p.* -nemde; *pp.* -nemed [be, nemnan *to name*] *To affirm, declare, stipulate;* asserere, stipulari :—Áþe benemnan *to declare by oath*, Exon. 123 b; Th. 475, 18; Bo. 49. Fin Hengeste áþum benemde *Fin declared to Hengest with oaths*, Beo. Th. 2199; B. 1097: 6131; B. 3069: Ps. Th. 88, 3: 94, 11: 88, 42.

be-neótan, bi-neótan; *p.* -neát, *pl.* -nuton; *pp.* -noten [be, neótan *to enjoy, use*] *To deprive of the enjoyment* or *use of anything;* privare :—Aldre beneótan *to deprive of life*, Beo. Th. 1364; B. 680. Heáfde beneótan *to deprive of the head, to behead*, Apstls. Recd. 92; Ap. 46: Cd. 50; Th. 63, 32; Gen. 1041: 89; Th. 110, 1; Gen. 1831.

be-neoðan, be-nyðan; *prep. dat.* [be, neoðan *under*], BENEATH, *below, under;* infra :—Hió biþ swíðe fior hire selfre beneoðan *she is very far beneath herself*, Bt. Met. Fox 20, 444; Met. 20, 222. Gif se sconca biþ þyrel beneoðan cneówe *if the shank be pierced beneath the knee*, L. Alf. pol. 63; Th. i. 96, 16, 17: 66; Th. i. 96, 31. Nis nán wuht benyðan [him] *no creature is beneath him* [*beneath God's notice*], Bt. 36, 5; Fox 180, 18.

Benesing-tún *Bensington*, Chr. 571; Th. 33, 28, col. 1. v. Bensingtún.

bēn-feorm, e; *f. Food required from a tenant;* firma precum, L. R. S. 21; Th. i. 440, 26, for MS. bend-feorm, *q.v.*

ben-geat, es; *pl. nom. acc.* -geato; *n. A wound-gate, the opening of a wound;* vulneris porta :—Bengeato burston *the wound-gates burst open*, Beo. Th. 2246; B. 1121.

be-niman, bi-niman; *p.* -nam, *pl.* -námon; *pp.* -numen [be, niman *to take*] *To deprive, bereave;* privare :—Sceolde hine yldo beniman ellendǽda *age should deprive him of bold deeds*, Cd. 24; Th. 31, 12; Gen. 484. He hine his ríces benam *eum regno privavit*, Bd. 3, 7; S. 529, 31. He us hæfþ heofonríce benumen *he has bereft us of heaven's kingdom*, Cd. 19; Th. 23, 20; Gen. 362.

be-niðan; *adv.* [be, neoðan *under*] *Beneath, below, under;* infra, subter :—Ðú bist ǽfre bufan and ná beniðan *eris semper supra et non subter :* *thou shalt be above only, and thou shalt not be beneath*, Deut. 28, 13.

benn, e; *f. A wound;* vulnus, Cd. 9; Th. 12, 6; Gen. 181. v. ben.

bennian, bennegean; *p.* ode, ade; *pp.* od, ad [ben *a wound*] *To wound;* vulnerare :—Mec ísern bennade *iron wounded me*, Exon. 130 a; Th. 499, 7; Rä. 88, 12. Ic geseah winnende wiht wído bennegean [benne gean, Th.] *I saw a block* [*wood*] *wound* [lit. *to wound = wounding*] *a striving creature*, 114 a; Th. 438, 4; Rä. 57, 2. DER. ge-bennian.

be-nohte, *pl.* -nohton *enjoyed*, Andr. Kmbl. 3407; An. 1707; *p.* of be-nugan, *q.v.*

be-norþan, *adv. In the north;* partibus borealibus :—Ofer eall benorþan þan *everywhere in the north*, Chr. 1088; Th. 357, 10.

be-notian; *p.* ode; *pp.* od [be, notian *to use*] *To use, consume;* uti :—Hie hæfdan heora mete benotodne *they had consumed their provisions*, Chr. 894; Th. 166, 15, col. 2.

bēn-ríp, e; *f. The reaping of corn by request;* ad preces messio. Originally the tenant came to reap corn etc. at his lord's request: in time, it grew into a custom or duty, but its old designation bēn-ríp was still used :—Eác he sceal hwíltídum geára beón on manegum weorcum to hláfordes willan, to-eácan bēnyrþe and bēnrípe and mǽdmǽwecte *etiam debet esse paratus ad multas operationes voluntaris domini sui, et ad bēnyrþe, id est, araturam precum, et bēnrípe, id est, ad preces metere, et pratum falcare*, L. R. S. 5; Th. i. 436, 3–5.

bēnsian; *part.* ende; *p.* ode; *pp.* od [bēn *a prayer*, sian or sígan *to fall down*] *To fall down in prayer, to pray, entreat in prayer;* supplicare, deprecari, orare :—Drihten bēnsian *Dominum deprecari*, Bd. 4, 25; S. 601, 4. He wæs bēnsiende ða uplícan árfæstnesse mínra gesynta *supplicans erat supernæ pietati pro sospitate mea*, 5, 6; S. 619, 35: 3, 12; S. 537, note 20.

Bensing-tún, Benesing-tún, Bænesing-tún, es; *m.* BENSINGTON or *Benson in Oxfordshire;* Bensington in agro Oxoniensi :—Hēr Cuðulf feówer túnas genam, Liggeanburh, and Æglesburh, and Bensingtún, and Egoneshám *here, in 571, Cuthwulf took four towns*, LENBURY, *and* AYLESBURY, *and* BENSON, *and* ENSHAM, Chr. 571; Th. 32, 29, col. 2; 33, 28, col. 1; 32, 29, col. 1: 777; Th. 92, 12, col. 2.

benst, he benþ *summonest, summons; 2nd and 3rd pers. pres. of* bannan.

bēn-tíd, e; f. [bēn a prayer, tíd time] Prayer-time, rogation-days, time for supplication; rogationum dies :—Ðæt is heálíc dæg, bēn-tíd brēmu that is a high day, a celebrated time for supplication, Menol. Fox 148; Men. 75.

bēn-tíðe, bēn-tigðe, bēn-tiðige; adj. [bēn a prayer; tíða, tíðe possessing, having obtained; compos]. I. having obtained a prayer, benefitted, favoured, successful; precum vel supplicationis compos, fortunatus :—Hie ðær, Godes þances, swíðe bēntíðe [bēntiðige, col. 2; bēntigðe, p. 153, 10, cols. 1, 2] wurdon æfter ðam gehāte there, God be thanked, they were very successful after that vow, Chr. 883; Th. 152, 9, col. 3. II. accepting a prayer, exorable, gracious; deprecabilis :—Beó ðū bēntýðe vel gehlystfull ofer ðíne þeówan deprecabilis esto super servos tuos, Ps. Lamb. 89, 13.

be-nugan, he be-neah, pl. be-nugon; p. be-nohte; subj. pres. benuge [Goth. binauhan, binah; pp. binauht, δεῖ, oportet] To need, want, require, enjoy; indigere, frui :—Ðonne he bega beneah when he requires both, Elen. Kmbl. 1233; El. 618: Exon. 123 b; Th. 475, 12; Bo. 46. Gif hí ðæs wuda benugon if they enjoy [have enjoyment of] the wood, Bt. 25; Fox 88, 19. Wið ðan ðe mín wíf ðær benuge inganges dummodo uxor mea fruatur ingressu, Hick. Thes. ii. 55, 32. And sið nō frófre benohte and never since he enjoyed comfort, Andr. Kmbl. 3407; An. 1707: 2320; An. 1161. v. nugan.

be-numen deprived, Cd. 19; Th. 23, 20; Gen. 362; pp. of be-niman.

bēn-yrþ, e; f. Ploughed land; precum aratura :—Eác he sceal hwíltídum geára beón on manegum weorcum to hláfordes willan, to-eácan bēnyrþe and bēnrípe and mædmǽwecte etiam debet esse paratus ad multas operationes voluntatis domini sui, et ad bēnyrþe, id est, araturam precum, et bēnrípe, id est, ad preces metere, et pratum falcare, L. R. S. 5; Th. i. 436, 3-5.

be-nyðan beneath, under; infra, Bt. 36, 5; Fox 180, 18. v. be-niðan.

BEÓ; indecl. in s; pl. nom. acc. beón; gen. beóna; dat. beóum, beóm; f. A BEE; apis. The keeping of bees was an object of much care in the economy of the Anglo-Saxons. The great variety of expressions, taken from the flavour of honey, sufficiently account for the value they placed upon it. While the bee-masters [beó-ceorlas, v. beó-ceorl] enjoyed their own privileges, they had to pay an especial tax for the keeping of bees :—Swā swā seó beó sceal losian as the bee shall perish, Bt. 31, 2; Fox 112, 26. Sió wílde beó sceal forweorþan, gif hió yrringa awuht stingeþ the wild bee shall perish, if she angrily sting anything, Bt. Met. Fox 18, 9; Met. 18, 5. Ða beón beraþ ārlícne anleofan and ǽterne tægel the bees carry a delicious food and a poisonous tail, Frag. Kmbl. 34; Leás. 19. Be ðām ðe beón bewitaþ concerning those who keep bees, L. R. S. 5; Th. i. 434, 35. Ymbtrymedon me swā swā beón circumderunt me sicut apes, Ps. Spl. 117, 12: Ps. Th. 117, 12. [Dut. bij, bije, f: Ger. biene, bien, f: M. H. Ger. bíe, f: O. H. Ger. pía, f: Dan. Swed. bi, n: O. Nrs. bý, n; generally bý-fluga, f, a bee-fly.] DER. beó-bread, -ceorl, -gang, -þeóf, -wyrt.

beó I am or shall be; sum, ero: be thou; sis :—Gefultuma me fæste, ðonne beó ic fægere hál adjuva me, et salvus ero, Ps. Th. 118, 117. Ic beó ero, Ælfc. Gr. 32; Som. 36, 29. Beó ðū sis: Beó he sit, 32; Som. 36, 30: Beo. Th. 777; B. 386. v. beón.

beó-bread, bió-bread, bí-bread, es; n. I. BEE-BREAD, the pollen of flowers collected by bees and mixed with honey for the food of the larvæ; apum panis. ☞ Quite distinct from weax beeswax; cera = κηρός: and hunig-camb honey-comb :—Ic eom swētra ðonne ðú beóbread blénde mid hunige I am sweeter than if thou blendedst bee-bread with honey, Exon. 111 a; Th. 425, 20; Rä. 41, 59. Hí synt swētran ðonne hunig oððe beóbread they are sweeter than honey or bee-bread, Ps. Th. 18, 9. Þynceþ bíbread swētre, gif he ǽr bitres onbyrgeþ bee-bread seems sweeter, if he before has had a taste of bitter, Bt. Met. Fox 12, 17; Met. 12, 9. Hit is hunige micle and beóbreáde betere and swētre it is better and sweeter than much honey and bee-bread, Ps. Th.118,103. II. sometimes, from a deficient knowledge of natural history, beó-bread is used for hunig-camb honey-comb; favus :—Swētran [MS. swetra] ofer hunig and beóbreáde dulciora super mel et favum, Ps. Lamb. 18, 11. Híg brohton him ðǽl gebræddes fisces, and beóbreád illi obtulerunt ei partem piscis assi, et favum mellis; οἱ ἐπέδωκαν αὐτῷ ἰχθύος ὀπτοῦ μέρος, καὶ ἀπὸ μελισσίου κηρίου and from a honey-comb, Lk. Bos. 24, 42.

beóce a beech-tree. v. bēce, bóce, bóc.

beó-ceorl, beó-cere, es; m. A BEE-CEORL, bee-farmer or keeper; bocherus, apum custos :—Be ðām ðe beón bewitaþ. Beóceorle gebyreþ, gif he gafolheorde healt, ðæt he sylle ðonne lande geræd beó. Mid us is geræd ðæt he sylle v sustras huniges to gafole concerning those who keep bees. It behoves a keeper of bees, if he hold a taxable hive [stock of bees], that he then shall pay to the country what shall be agreed. With us it is agreed that he shall pay five sustras of honey for a tax; 'bochero, id est, apum custodi, pertinet, si gavelheorde, id est, gregem ad censum teneat, ut inde reddat sicut ibi mos [MS. moris] erit. In quibusdam locis est institutum, reddi v [MS. vi] mellis ad censum,' L. R. S. 5; Th. i. 434, 35-436, 2. Swā ic ǽr be beócere cwæþ sicut de custode apum dixi, L. R. S. 6; Th. i. 436, 17. [beócere = Barbarous Lat. bocherus = beó a bee, cherus = herus a master.] DER. þeów-beócere.

BEÓD, es; m. A table; mensa :—Ðā ða gebrōðru æt beóde sǽton sedentibus ad mensam fratribus, Bd. 3, 2; S. 525, 9. Ðū gearcodest beföran mínre gesihþe beód vel beódwyste vel mýsan parasti in conspectu meo mensam, Ps. Lamb. 22, 5. Beódas lances, Cot. 123. [O. Sax. biod: O. H. Ger. piot: Goth. biuds: O. Nrs. bjóðr.]

BEÓDAN, biódan; ic beóde, bióde, ðū beódest, býtst, býst, he beódeþ, být, pl. beódaþ; p. ic, he beád, ðū bude, pl. budon; pp. boden; v. trans. I. to command, BID, order; jubere, mandare :—Ðæs þing ic eów beóde hæc mando vobis, Jn. Bos. 15, 17. He beád Iosepe ðæt he bude his brōðrum dixit ad Joseph ut imperaret fratribus suis, Gen. 45, 17: Ors. 6, 7; Bos. 119, 38; Andr. Kmbl. 692; An. 346. II. to announce, proclaim, inspire, bode, threaten; nuntiare, annuntiare, nuntium vel mandatum deferre, prædicare, significare, inspirare, minari alicui aliquid :—He him friþ beódeþ he announces peace to them, Exon. 27 b; Th. 82, 20; Cri. 1341. Geácas geár budon cuckoos announced the year, 43 b; Th. 146, 27; Gū. 716. Him wæs hild boden to him was war proclaimed, Elen. Kmbl. 36; El. 18. Hwæt seó rūn bude what that mystery boded, Cd. 202; Th. 250, 6; Dan. 542. Geác monaþ geómran reorde, sorge beódeþ bitter in breósthord the cuckoo exhorts with mournful voice, inspires bitter sorrow to the heart, Exon. 82 a; Th. 309, 9; Seef. 54. Ðeáh him feónda hlóþ feorhcwealm bude though the band of fiends threatened death to him, 46 a; Th. 157, 6; Gū. 887: Mk. Bos. 10, 48. III. to offer, give, grant; offerre, præbere :—Beód him ǽrest sibbe offerres ei primum pacem, Deut. 20, 10. Hafa ārna þanc ðara, ðe ðū unc bude have thanks for the kindnesses, which thou hast offered us, Cd. 111; Th. 147, 7; Gen. 2435. [Plat. bēden to command, offer: O. Sax. biodan to offer: O. Frs. biada id: Dut. bieden id: Ger. bieten id: M. H. Ger. biuten id: O. H. Ger. biotan id: Goth. biudan id: Dan. byde to bid, offer: Swed. bjuda id: O. Nrs. bjóða id.] DER. a-beódan, be-, bi-, for-, ge-, on-.

beódas; pl. m. Dishes, plates, scales; lances, Cot. 123. v. beód.

beód-bolla, an; m. A table-bowl, a cup, bowl; cupa, Som.

beód-cláþ, es; m. A table-cloth, carpet, hanging; gausape = γαυσάπης, Ælfc. Gr. 9, 2; Som. 8, 28.

beódende commanding, R. Ben. 5; part. of beódan.

beódendlíc gemet the imperative mood. v. be-beódendlíc gemet.

beód-ern, es; n. [beód a table, ern a place] A refectory, a dining-room; refectorium, Ælfc. Gl. 107; Som. 78, 94; Wrt. Voc. 58, 9.

beód-fers, es; m. [beód a table, fers a verse] A song or hymn sung during meal-time; ad mensam carmen, hymnus, Dial. 1, 19.

beód-gæst, es; m. A guest at table; mensæ consors, convictor, Andr. Kmbl. 2177; An. 1090.

beód-geneát, es; m. A table-companion; mensæ socius, convictor, Beo. Th. 691; B. 343: 3431; B. 1713.

beód-gereordu; pl. n. [beód a table, gereord a feast] A table-meal, a feast; convivium, Cd. 74; Th. 91, 27; Gen. 1518.

beód-hrægl, es; n. [beód a table, hrægl clothing] A table-cloth; gausape = γαυσάπης, Ælfc. Gl. 30; Som. 61, 61; Wrt. Voc. 26, 60.

beód-sceát, es; m: beód-scýte, es; m. A table-cloth, table-napkin, hand-towel; mantile, mappa, Cot. 136.

beód-wist, beód-wyst, e; f. [beód a table, wist food] Food placed on a table, board, a table; mensa :—Ðū gearcodest beföran mínre gesihþe beód vel beód-wyste vel mýsan parasti in conspectu meo mensam, Ps. Lamb. 22, 5.

beofer, beofor, es; m. A beaver; castor, Ælfc. Gr. 8; Som. 7, 13. v. befer.

Beofer-lic, Beofor-lic, es; m. [beofer, lic? = lie, leá, leáh, q. v. Ric. A. D. 1184, Beverli: Brom. 1330, Beverlith] BEVERLEY, Yorkshire; Beverlea in agro Eboracensi :—Hēr forþférde se hālga biscop Iohannes, and his líc resteþ [MS. restad] in Beoferlíc here, A. D. 721, the holy bishop John died, and his body resteth at Beverley, Chr. 721; Erl. 45, 25; Th. 73, 15, col. 2; Beoforlíc, col. 1.

beofian; p. ode; pp. od To tremble, quake, be moved; tremere, contremere, commoveri :—Beofaþ eal beorhte gesceaft all the bright creation shall tremble, Exon. 116 b; Th. 448, 22; Dōm. 58. Seó eorþe beofode the earth trembled, 24 b; Th. 70, 27; Cri. 1145. Beofaþ middangeard the mid-earth shall quake, 20 b; Th. 55, 12; Cri. 882. For his ansýne sceal eorþe beofian commoveatur a facie ejus universa terra, Ps. Th. 95, 9: 103, 30. v. bifian.

beofung, e; f. A trembling, quaking; tremor. DER. eorþ-beofung an earthquake. v. bifung.

beó-gang, es; m. A swarm of bees; examen, Cot. 15, 164.

beógol, beógul; adj. Agreeing, consenting, bending wholly to; consentiens. v. ge-býgel.

beó-hāta? Cd. 156; Th. 193, 27. v. beót-hāta.

beolone, an; f. Henbane; hyoscyamus niger :—Genim beolonan sǽd take seed of henbane, L. M. 1, 6; Lchdm. ii. 50, 17: 1, 2; Lchdm. ii. 38, 1: 1, 3; Lchdm. ii. 42, 15: 1, 63; Lchdm. ii. 136, 26: 3, 37; Lchdm. ii. 328, 23. v. belene.

beóm am, Exon. 30 a; Th. 91, 13; Cri. 1491. v. beón.

beóm a beam, Chr. 1137; Erl. 262, 13. v. beám.

beó-móder; f. A BEE-MOTHER, *queen-bee;* chosdrus? *vel* castros? Ælfc. Gl. 22; Som. 59, 104; Wrt. Voc. 23, 61.

BEÓN [bión], to beónne; *part.* beónde; ic beó [beóm], ðú bist, byst, he biþ, byþ, *pl.* beóþ; *impert.* beó, *pl.* beóþ; *subj.* beó, *pl.* beón To BE, *exist, become;* esse, fieri :—Hí ne tweódon férende beón to ðam ēcan lífe *non dubitabant esse transituros ad vitam perpetuam,* Bd. 4, 16; S. 584, 38, 18. Ðe ðǽr beón noldon *who would not be there,* Byrht. Th. 137, 13; By. 185: Exon. 100a; Th. 376, 29; Seel. 162: Cd. 24; Th. 31, 15; Gen. 485: Mt. Bos. 19, 21: Bt. 5, 3; Fox 12, 12: Ælfc. Gr. 25; Som. 26, 48. Ic ðæs folces beó hyrde *I am the people's pastor,* Cd. 106; Th. 139, 24; Gen. 2314. Ic beó gearo sóna *I shall be soon ready,* Beo. Th. 3655; B. 1825: Exon. 71a; Th. 264, 17; Jul. 365: Andr. Kmbl. 144; An. 72. Ic beó hál *I shall be safe,* Mt. Bos. 9, 21: Mk. Bos. 5, 28: Ex. 3, 12. Ðonne ic stille beóm *when I am still,* Exon. 102b; Th. 387, 5; Rä. 4, 74: 72a; Th. 268, 26; Jul. 438: Mt. Lind. Rush. Stv. 9, 21. Ðú ána bist eallra dēma *thou alone art judge of all,* Hy. 8, 38; Hy. Grn. ii. 291, 38: Bt. Met. Fox 24, 53; Met. 24, 27: Exon. 8b; Th. 4, 24; Cri. 57: Cd. 26; Th. 34, 16; Gen. 538: Bd. 5, 19; S. 640, 43: Mk. Lind. War. 14, 70: Lk. Lind. Rush. War. 1, 76: Ðú yrre byst *tu terribilis es,* Ps. Th. 75, 5: 101, 24: Lk. Bos. 1, 76: Deut. 23, 22. Hiora birhtu ne biþ to gesettane *their brightness is not to be compared,* Bt. Met. Fox 6, 11; Met. 6, 6. Biþ ealles leás *he is void of all,* Cd. 217; Th. 276, 1; Sat. 182: 109; Th. 144, 19; Gen. 2392: Beo. Th. 604; B. 299: Ps. Th. 118, 142: Andr. Kmbl. 3383; An. 1695: Mt. Bos. 5, 19, 22, 37: Ors. 1, 1; Bos. 20, 18: Bt. 37, 3; Fox 190, 15. Fela biþ *many there are,* Exon. 78a; Th. 293, 14; Crä. 1: 26a; Th. 76, 5; Cri. 1235. Ne byþ lang *it shall not be long,* Elen. Grm. 433: Beo. Th. 3529; B. 1762. Sēlre biþ ǽghwǽm *it is better for every one,* Andr. Kmbl. 640; An. 320: Ps. Th. 111, 9: Beo. Th. 2009; B. 1002: Mt. Bos. 5, 14, 19, 21, 22. Yldo beóþ on eorþan ǽghwæs cræftig *age is on earth powerful of everything,* Salm. Kmbl. 583; Sal. 291: Exon. 36b; Th. 118, 27; Gū. 246. Ðǽr wit tū beóþ *where we two are,* Exon. 125a; Th. 480, 21; Rä. 64, 5: Beo. Th. 3681; B. 1838: Cd. 133; Th. 168, 20; Gen. 2785: Hy. 7, 88; Hy. Grn. ii. 289, 88: Ors. 1, 1; Bos. 20, 21: Bd. 4, 16; S. 585, 2: Bt. 10; Fox 30, 14: Nicod. 17; Thw. 8, 23: Mt. Rush. Stv. 26, 31. Beó ðú sunum mínum gedēfe *be thou gentle to my sons,* Beo. Th. 2457; B. 1226: Andr. Kmbl. 428; An. 214: Exon. 81a; Th. 305, 18; Fä. 90: Cd. 229; Th. 310, 25; Sat. 733: Jn. Bos. 3, 2. Ne beóþ ge tó forhte *be not ye too terrified,* Andr. Kmbl. 3216; An. 1611: Ps. Th. 104, 4. Ne beón ic gescynded *non confundar,* Ps. Th. 118, 6. Beón ða oferhydegan ealle gescende *confundantur superbi,* Ps. Th. 118, 78: 148, 12. [*Orm.* beon; *pres.* beo, best, beoþ, beþ; *subj.* beo, be, ben: *Laym.* beon; *pres.* beo, beost, bist, beoþ, beþ, biþ, biðe; *subj.* beo: *O. Sax.* bium, bist: *O. Frs.* bem, bim, ben, bin: *Dut.* ben: *O. Dut.* bem: *Ger. M. H. Ger.* bin: *O. H. Ger.* pim: *Slav.* byti: *Zend* bū: *Sansk.* bhū, bhavámi.] v. eom *I am,* wesan *to be.*

beón bees, Ps. Spl. 117, 12: L. R. S. 5; Th. i. 434, 35. v. beó.

beón, beónn *commanded, assembled;* p. of bannan.

beón-breád *bee-bread,* Ps. Spl. 18, 11. v. beó-breád.

beón-broþ, es; *n.* Perhaps *mead, a drink of water and honey mingled and boiled together;* melicratum, L. M. 2, 24; Lchdm. ii. 216, 12.

beónde *being,* Cot. 77; *part.* of beón.

be ongewyrhtum *freely;* gratis, Ps. Spl. C. 34, 8.

BEÓR, es; *m.* I. BEER, *nourishing* or *strong drink;* cerevisia, sicera. Beer, made from malted barley, was the favourite drink of the Anglo-Saxons. In their drinking parties, they pledged each other in large cups, round at the bottom, which must be emptied before they could be laid down, hence perhaps the name of a tumbler. We are speaking of the earliest times, for beer is mentioned in Beowulf :—Gebeótedon beóre druncne oret-mecgas, ðæt hie in beór-sele bídan woldon Grendles gūðe *the sons of conflict, drunk with beer, promised that they would await in the beer-hall the attack of Grendel,* Beo. Th. 965; B. 480. Æt beóre *at the beer,* 4088; B. 2041. ☞ Beer was the common drink of the Anglo-Saxons, hence *a convivial party* was called Gebeórscipe, *q. v : a place of entertainment,* beórsele *a beer-hall,* or beórtūn *a beer-enclosure.* Hence also the other compounds, as beór-scealc *a beer-server,* beór-setl *a beer-bench* or SETTLE, and beór-þegu *a beer-serving.* The following remark seems to be as applicable to the Anglo-Saxons as to the Icelanders,—Öl heitir með mönnum, en með Ásum bjór *ale is called, by men and by gods,* BEER, Alvismál.—Beóre druncen *drunk with beer,* Beo. Th. 1066; B. 531: Exon. 72b; Th. 271, 22; Jul. 486. He ne drincþ wín ne beór *vinum et siceram non bibet,* Lk. Bos. 1, 15: Deut. 14, 26. Ðæt mon geselle twelf seoxtres beóras *that they give twelve sesters of beer,* Th. Diplm. A. D. 901–909; 158, 22. II. *a beverage made of honey and water, mead;* metheglin, hydromeli, Ytis, *n.* = ὑδρόμελι, ydro-mellum, mulsum :—Beór ydromellum, Ælfc. Gl. 32; Som. 61, 114; Wrt. Voc. 27, 43. Beór mulsum, Ælfc. Gl. 32; Som. 61, 118; Wrt. Voc. 27, 46. [*Plat.* beer, *n.*: *Frs.* biar, *n.*: *Dut. Ger.* bier, *n.*: *Icel.* bjór, bjórr, *m.*: *O. H. Ger.* pier, *n.*: *Sansk.* pā *to drink.*] DER. beór-hyrde, -scealc, -scipe, -sele, -setl, -þegu, -tūn: gebeór, -scípe.

beora, an; *m. A grove;* lucus *vel* nemus, Ælfc. Gl. 110; Som. 79, 39; Wrt. Voc. 59, 11. v. bearo.

beoran *to bear* :—Ic sceal beoran *I shall bear,* Cd. 216; Th. 274, 22; Sat. 158: 217; Th. 277, 17; Sat. 206. v. beran.

beorc, e; *f.* I. *a birch-tree;* betula. v. birce, byrc. II. *the Anglo-Saxon Rune* ᛒ = b, the name of which letter in Anglo-Saxon is beorc *a birch-tree,* hence this Rune not only stands for the letter b, but for beorc *a birch-tree,* as,—ᛒ byþ blǽda leás *a birch-tree is void of fruit,* Hick. Thes. i. 135; Runic pm. 18; Kmbl. 342, 27.

BEORCAN, ic beorce, he byrcþ; *p.* bearc, *pl.* burcon; *pp.* borcen [*Icel.* barki, *m. guttur*]. I. *to make a sharp explosive sound;* latratum *vel* sonum edere. v. gebeorc. II. *to* BARK; latrare :—Ða dumban hūndas ne mágon beorcan. We sceolon beorcan and bodigan ðám lǽwedum *dumb dogs cannot bark. We ought to bark and preach to the laymen,* L. Ælfc. C. 23; Th. ii. 350, 34. Ic hwílum beorce swá hūnd *I sometimes bark as a dog,* Exon. 106b; Th. 406, 16; Rä. 25, 2. Hūnd byrcþ *canis latrat,* Ælfc. Gr. 22; Som. 24, 8. Ne mæg he fram hūndum beón borcen *he may not be barked at by dogs,* Herb. 67, 2; Lchdm. i. 170, 17. [*O. Nrs.* berkja.] DER. gebeorc, borcian.

beorcen *birchen;* tiliaceus [*Kil.* bercken]. v. bircen.

Beordan íg, e; *f.* [íg *an island,* beordan = bridan = bridum *with the young of birds*] BARDNEY *in Lincolnshire;* cœnobii locus in agro Lincolniensi, Som.

beorende *bringing forth;* *part.* of beoran.

beorg, beorh, biorg, biorh; *gen.* beorges; *dat.* beorge; *pl. nom. acc.* beorgas; *gen.* beorga; *dat.* beorgum; *m.* I. *a hill, mountain;* collis, mons :—On Sýne beorg *on Sion's hill,* Exon. 20b; Th. 54, 29; Cri. 876. Oþ ða beorgas ðe man hǽt Alpis *to the mountains which they call the Alps,* Ors. 1, 1; Bos. 18, 44; 16, 17. Ælc mūnt and beorh byþ genyðerod *omnis mons et collis humiliabitur,* Lk. Bos. 3, 5. Æt ðæm beorge ðe man Athlans nemneþ *at the mountain which they call Atlas,* Ors. 1, 1; Bos. 16, 6. II. *a heap,* BURROW or *barrow, a heap of stones, place of burial;* tumulus :—Worhton mid stānum ānne steápne beorh him ofer *congregaverunt super eum acervum magnum lapidum,* Jos. 7, 26. Bǽd ðæt ge geworhton in bǽlstede beorh ðone heán *he commanded [bade] that you should work the lofty barrow on the place of the funeral pile,* Beo. Th. 6186; B. 3097: 5606; B. 2807: Exon. 50a; Th. 173, 26; Gū. 1166: 119b; Th. 459, 31; Hö. 8. [*Laym.* berhȝe: *Piers* bergh; *still used in the dialect of Yorkshire: Plat.* barg: *O. Sax.* berg: *O. Frs.* berch, birg: *Ger.* berg: *M. H. Ger.* berc: *O. H. Ger.* perac: *Goth.* bairga-hei *a mountainous district: Dan.* bjærg, *n: Swed.* berg, *n: O. Nrs.* berg, *n :* derived from beorgan.] DER. ge-beorg, -beorh, heáh-, mund-, sǽ-, sand-, stān-.

beorg, berg *a protection, refuge;* præsidium, refugium. DER. heáfod-beorg, ge-beorg, scúr-beorg: cin-berg.

BEORGAN; ic beorge, ðú byrgst, byrhst, he byrgeþ, byrgþ, byrhþ, *pl.* beorgaþ; *p.* ic, he bearg, bearh, ðú burge, ðú burgon; *impert.* beorg, beorh, *pl.* beorgaþ, beorge ge; *pp.* borgen; *v. a.* I. *cum dat. To save, protect, shelter, defend, fortify, spare, preserve;* servare, salvare, custodire, tueri, parcere :—Beorh ðínum feore salva animam tuam, Gen. 19, 17. Woldon feore beorgan *they would save their lives,* Andr. Kmbl. 3075; An. 1540. Beorh me, Drihten, swá swá man byrhþ ðám ǽplum on his eágum mid his brǽwum *custodi me, Domine, ut pupillam oculi,* Ps. Th. 16, 8. Ðæt se bittra bryne beorgan sceolde ǽfæstum þrím *that the bitter burning should spare the pious three,* Exon. 53b; Th. 189, 10; Az. 57. II. *dat. of the pers. acc. of the thing or following wiđ,—To defend, secure, guard against, avoid;* defendere, arcere, cavere, vitare :—Hý him hryre burgon *they secured him from fall,* Exon. 43a; Th. 145, 30; Gū. 702: 55a; Th. 195, 21; Az. 159. Hý beorgaþ him bealoníþ *they guard themselves against baleful malice,* 44b; Th. 150, 19; Gū. 781. Druncen beorg ðē *from drunkenness guard thyself,* 80b; Th. 302, 10; Fä. 34. Ðæt preóstas beorgan wiđ ofer-druncen *that priests avoid [over-drinking] drunkenness,* L. Edg. C. 57; Th. ii. 256, 13. [*Orm.* berrȝhenn: *Plat.* bargen: *O. Sax.* gi-bergan: *M. H. Ger.* bergen: *O. H. Ger.* perkan, bergan: *Goth.* bairgan: *Dan.* bjærge: *Swed.* bærga: *O. Nrs.* biarga: *Grm. Wrtbch.* i. 1507 refers to *Grk.* φράγνυμι, φάργνυμι *to hedge round, to secure.*] DER. be-beorgan, ge-, ymb-.

beorgan *to taste;* gustare :—Fēnix of ðám wyll-gespryngum brimcald beorgeþ æt baða gehwylcum *the Phœnix tastes ocean-cold [water] from the well-springs at every bath,* Exon. 57b; Th. 205, 9; Ph. 110. v. byrgan.

Beorg-ford, Beorh-ford, es; *m.* [beorg *a hill,* ford *a ford;* collis ad vadum] BURFORD *in Oxfordshire* :—Hēr Cúþrēd, Wæst-Seaxna cining, gefeaht ðý xxii geára his ríces, æt Beorgforda [MS. Beorhforda], wiđ Æðelbald, Myrcena cing, and hine geflýmde *here, in 752, Cuthred, king of the West-Saxons, fought in the twenty-second year of his reign, at Burford, with Æthelbald, king of the Mercians, and conquered him,* Chr. 752; Erl. 49, 13.

beorg-hleoþ, es; *n. A mountain-brow;* montis fastigium :—Ofer beorghleoða *over the mountain-brows,* Exon. 114a; Th. 438, 27; Rä. 58, 2. v. beorh-hliþ.

beorg-seđel, es; *n. A mountain-dwelling;* habitaculum in monte:—He ongan beorgseđel bûgan *he began to inhabit a mountain-dwelling,* Exon. 34 a; Th. 108, 15; Gû. 73.

beorh; *gen.* beorges; *m. A hill, mountain;* collis, mons:—Ælc mûnt and beorh byþ genyđerod *omnis mons et collis humiliabitur,* Lk. Bos. 3, 5. v. beorg.

beorh *save,* Ps. Th. 16, 8; *impert. of* beorgan.

beorh-hlíþ, -hleoþ, es; *n. A mountain-height, mountain-brow;* montis clivus *vel* fastigium:—Under beorhhlíđe *under the mountain-height,* Elen. Kmbl. 1572; El. 788: 2015; El. 1009. Wǽron beorhhliđu blóde bestêmed *the mountain-brows were besteamed with blood,* Cd. 166; Th. 206, 7; Exod. 448. Under beorhhleođum *among the mountain-heights,* 98; Th. 130, 13; Gen. 2159.

beorh-stal, -stól, es; *m.* [beorh *a hill,* stal *a place, seat, dwelling*] *A hill-seat, dwelling on a hill;* sedes super collem *vel* clivum. v. burg-stal.

beorh-stede, es; *m. A mountain-place, place on a mountain, a mountain, mound;* locus in monte, mons, collis:—On beorhstede *on the mound,* Exon. 60 a; Th. 217, 22; Ph. 284.

beorht, es; *n. Brightness, a glistening, light, sight, glance, twinkling;* splendor, lumen, lux:—Đis leóhte beorht cymeþ morgna gehwâm *this pure brightness cometh each morn,* Exon. 93 a; Th. 350, 6; Sch. 59. Onfêng đam beorhte hire eágena *received the sight [full sight, sparkling] of her eyes,* Bd. 4, 10; S. 578, 2. Đæt biþ an eágan beorht *that is in the twinkling of an eye,* Bd. 2, 13; S. 516, note 20. v. bearhtm.

BEORHT, berht, byrht, bryht; *adj.* BRIGHT, *light, clear, lucid, splendid, excellent;* splendidus, lucidus, coruscus, clarus, formosus:—Eall đín lîchama biþ beorht *totum corpus tuum lucidum erit,* Mt. Bos. 6, 22. Beorht êđles wlite *the land's bright beauty,* Exon. 27 b; Th. 82, 32; Cri. 1347. Beorht sumor *bright summer,* 54 b; Th. 191, 29; Az. 95. To đære beorhtan byrg *to the bright tity,* 15 a; Th. 33, 1; Cri. 519. Beorhte burhwealías beorhte scînaþ *the lucid city-walls shine brightly,* Cd. 220; Th. 282, 31; Sat. 295. Đá cwom sunnan beorhtra líg *then came a fire, brighter than the sun,* Elen. Kmbl. 2218; El. 1110. Hí môdes eágan beorhtran gedón *they make the mind's eye clearer,* Bt. Met. Fox 21, 54; Met. 21, 27. Sum hafaþ beorhte stefne *one has a clear voice,* Exon. 79 b; Th. 298, 32; Crä. 94. II. *bright, brilliant, magnificent, noble, glorious, sublime, divine, holy;* clarus, præclarus, eximius, augustus, divus, sanctus:—In đa eástor-tíd, on đone beorhtan dæg *in the Easter-time, on that bright day,* Exon. 48 b; Th. 168, 17; Gû. 1079. Meotud ælmihtig, beorht cyning *Almighty God, noble king,* Andr. Kmbl. 1804; An. 905. Ne wolde him beorht fæder bearn ætniman *the glorious father [God] would not take the child from him,* Cd. 162; Th. 204, 4; Exod. 414. Se ân dêma is gestæđđig and beorht *the only judge is steadfast and sublime,* Bt. 36, 2; Fox 174, 20: Exon. 14 b; Th. 30, 22; Cri. 483. Mid đý beorhtan gebêde *with the holy prayer [the Lord's prayer],* Salm. Kmbl. 87; Sal. 43. [*Wyc.* bright: *Plat.* Brecht *a proper name, f: O. Sax.* berht, beraht: *Ger.* preserved in proper names as Bertha, Albrecht: *M. H. Ger.* berht: *O. H. Ger.* peraht: *Goth.* bairhts: *O. Nrs.* biartr: *Lat.* fulgeo, flagrare: *Grk.* φλέγειν *to burn,* from the *Sansk.* root bhrâj *to shine;* bhargas *splendour, brightness.*] DER. æl-beorht, eall-, efen-, gold-, heáfod-, heofon-, hîw-, rôdor-, sadol-, sigel-, sigor-, sun-, swegl-, þurh-, wlite-.

beorhtan, berhtan, byrhtan; *p.* te; *pp.* ed *To shine;* lucere, Ps. Th. 143, 7.

beorhte; *adv. Distinctly, clearly, lucidly, brightly;* clare:—He geseah Egypta heáhyrig beorhte blícan *he saw the Egyptians' cities brightly glitter,* Cd. 86; Th. 109, 13; Gen. 1822. Đonne seó sunne beorhtost scîneþ *when the sun shines brightest,* Bt. 39; Fox 26, 15: Beo. Th. 3039; B. 1517.

beorht-hwíl, e; *f. A glance;* ictus oculi, Lye. v. bearhtm-hwíl.

beorhtian, beorhtigan; *p.* ode; *pp.* od. I. *to shine, brighten;* clarere:—Đǽr his geearnunge oft miclum mægenum scînaþ and beorhtigaþ *there his earnings often shine and brighten with great virtues,* Bd. 3, 19; S. 550, 17. II. *to sound clearly* or *loudly;* clare sonare:—Beorhtode bencswêg *the bench-noise sounded loudly,* Beo. Th. 2326; B. 1161.

beorht-líc; *adj. Bright, light, clear, lucid, splendid;* lucidus, clarus, splendidus, Runic pm. 6; Hick. Thes. i. 135; Kmbl. 340, 19: Ps. Th. 67, 3.

beorht-líce; *adv. Clearly, distinctly, splendidly;* clare, splendide:—Đæt he beorhtlíce eall geseah *ut clare videret omnia,* Mk. Bos. 8, 25: Ps. Th. 118, 98: 147, 7.

beorhtm, *m. Tumult;* tumultus:—Hwǽr ahangen wæs heriges beorhtme rôdera waldend *where the Lord of glory was hung up by the tumult of the host,* Elen. Kmbl. 410; El. 205. v. breahtm *a noise,* brecan *to break.*

beorht-nes, byrht-nes, -ness, -nys, -nyss, e; *f.* [beorht *bright*] BRIGHT-NESS, *clearness, splendour;* splendor, claritas, nitor:—Godes beorhtnes him ymbsceán *claritas Dei circumfulsit illos,* Lk. Bos. 2, 9: Ælfc. Gr. 36; Som. 38, 54: Ps. Th. 118, 130. Eágena beorhtnes *brightness of the eyes,* Herb. 31, 2; Lchdm. i. 128, 13: Hy. 7, 31; Hy. Grn. ii. 287, 31.

beorht-rôdor, es; *m. The bright firmament, heaven;* æther, Cd. 146; Th. 183, 19; Exod. 94.

beorhtu, beorhto, birhtu, byrhtu, e; *f. Brightness, splendour;* claritas, splendor:—Gif hæleþa hwilc mæg ǽfre ofsión heofones leóhtes hlûtre beorhto *if any man may ever behold the clear brightness of heaven's light,* Bt. Met. Fox 21, 78; Met. 21, 39.

beór-hyrde, es; *m. A beer-keeper, butler;* cervisiæ custos, pincerna:—Sum biþ gewittig æt wínþege, beórhyrde gôd *one is witty at wine-bibbing, a good beer-keeper,* Exon. 79 b; Th. 297, 28; Crä. 75.

BEORMA, an; *m:* bearm, es; *m. Barm, leaven, yeast, froth;* fermentum:—Se beorma awent đa gesceafta of heora gecynde *barm changes creatures from their nature,* Homl. Th. ii. 278, 21. Wistfullian on yfelnysse beorman *to feast on the barm of evil,* ii. 278, 25. Heofena ríce is gelíc đam beorman *cœlorum regnum simile est fermento,* Mt. Bos. 13, 33: Lk. Bos. 13, 21. Nim ele and hunig and beorman *take oil and honey and barm,* Lchdm. i. 398, 6: Exon. 71 b; Th. 266, 11; Jul. 396. [*Plat. Dut.* barm, *m. fæx: Ger.* barme, bärme, *f: Dan. Swed.* bærme *dregs, lees, barm.*] v. and-, andbita.

Beormas; *gen.* a; *pl. m. The Biarmians.*—The Biarmians inhabited the country on the shores of the White Sea, north-west of the river Dwina. Alfred calls them Beormas. They were called Biarmians by Icelandic historians, and Permiaki by the Russians, and now Permians. In the Middle Ages, the Scandinavian pirates gave the name of Permia to the whole country between the White Sea and the Ural, Malte-Brun's Univer. Geog. vol. vi. p. 419. In an Icelandic MS. on geography, written in the 14th century, Beormia and two Cwenlands are located together. Kvenlönd II, ok ero þau norþr frá Bjarmalandi. Duæ Quenlandiæ, quæ ulterius quam Bjarmia boream versus extenduntur, Antiquitates Americanæ, p. 290.—Haldorson's Lexicon Islandico-Latino-Danicum, edited by Rask, has—'Biarmaland, Biarmia, quæ ob perpetuas nives albicatur, Bjarmeland, Permien. Biarmia ortum versus ad mare album vel gandvikam sita est:'—Fela spella him sǽdon đa Beormas, ǽgþer ge of hyra âgenum lande, ge of đǽm landum, đe ymb hý ûtan wǽran; ac he nyste hwæt đæs sóđes wæs, forđæm he hit sylf ne geseah. Đa Finnas, him þuhte, and đa Beormas sprǽcon neáh ân geþeóde *the Biarmians told him many stories, both about their own country and about the countries which were around them; but he knew not what was true, because he did not see it himself. The Finns and the Biarmians, as it seemed to him, spoke nearly the same language,* Ors. 1, 1; Bos. 20, 11—15. Đa Beormas hæfdon swíđe well gebûn hyra land *the Biarmians had very well inhabited their land,* 1, 1; Bos. 20, 7.

beorn *children,* Th. Diplm. A. D. 830; 466, 5. v. bearn.

beorn *for* bearn *burned,* Beo. Th. 3764, note; B. 1880; *p. of* beornan.

BEORN, biorn, es; *m.* [this word is only used by poets]. I. *a man;* vir:—Se beorn on waruþe scip gemêtte *the man found a ship on the strand,* Andr. Kmbl. 478; An. 239: 1203; An. 602. Boêtius wæs beorn bôca *Boethius was a man skilled in books,* Bt. Met. Fox 1, 103; Met. 1, 52: Exon. 83 a; Th. 313, 22; Môd. 4. Beornes blôde *with man's blood,* Bt. Met. Fox 8, 67; Met. 8, 34. Beornas Bađan nemnaþ *men name Bath,* Chr. 973; Erl. 124, 12; Edg. 5. Beornas geonge *young men,* Cd. 184; Th. 230, 13; Dan. 232. Beorna sêlost *the best of men,* 162; Th. 203, 10; Exod. 401: Bt. Met. Fox 21, 82; Met. 21, 41. II. *a prince, nobleman, chief, general, warrior, soldier;* princeps, vir nobilis, dux, miles:—Se beorn ageaf teóđan sceát *the prince gave a tenth portion,* Cd. 97; Th. 128, 1; Gen. 2120: 176; Th. 222, 3; Dan. 99. Þurh đæs beornes cyme *through the chief's coming,* Exon. 15 b; Th. 33, 24; Cri. 530. He đam beorne oncwæþ *he answered the warrior,* Byrht. Th. 138, 65; By. 245. Me on beáme beornas sticedon *soldiers pierced me on the cross,* Cd. 224; Th. 297, 1; Sat. 510. Beorna beáhgyfa *bracelet-giver of warriors* or *a rewarder of heroes,* Chr. 937; Erl. 112, 2; Edg. 30. III. *rich;* dives:—Beornum and þearfum *to rich and poor,* Runic pm. 12; Hick. Thes. i. 135; Kmbl. 341, 25. [*Dan. Swed. Icel.* björn, *m. a bear;* ursus.] DER. folc-beorn, gúþ-, sige-.

BEORNAN, byrnan; ic beorne, byrne, đú beornest, beornst, byrnest, byrnst, he beorneþ, beornþ, byrneþ, byrnþ, *pl.* beornaþ; *p.* ic, he bearn, barn, born, đú burne, đú burnon; *pp.* bornen. I. *v. n. To* BURN, *be on fire;* ardere, exardere, comburi:—Đonne beorneþ [byrneþ, Spl.] eorre his *cum exarserit ira ejus,* Ps. Surt. 2, 13. Se đe ǽfre nû beorneþ on bendum *he who now ever burns in bonds,* Cd. 222; Th. 290, 12; Sat. 414. Bearn [MS. beorn] breóstsefa [*their*] *spirit burned,* Exon. 15 b; Th. 34, 10; Cri. 540. Heofoncandel barn *the heavenly candle burnt,* Cd. 148; Th. 184, 31; Exod. 115. Hređer innan born *his spirit burned within,* Exon. 46 b; Th. 158, 18; Gú. 910. Him sorga burnon on breóstum *sorrows burned in their breasts,* Cd. 37; Th. 48, 17; Gen. 777. II. *v. trans. To* BURN; urere, comburere:—Swâ fýr wudu byrneþ *sicut ignis comburit silvas,* Ps. Th. 82, 10. [*O. Sax. M. H. Ger. O. H. Ger.* brinnan: *Ger.* brennen: *Swed. O. Nrs.* brenna.] DER. a-beornan, for-, ge-. v. bærnan, byrnan, on-brinnan.

beorn-cyning, es; *m. A king of men;* virorum rex:—Mâđmas ic đe, beorncyning, bringan wylle *I will bring thee treasures, king of men,* Beo. Th. 4302; B. 2148.

beorne, an; *f.* *A coat of mail;* lorica, Cod. Dipl. 716; A. D. 996–1006; Kmbl. iii. 351, 26. v. byrne.

Beornica ríce, es; *n:* mægþ, e; *f. The kingdom* or *province of the Bernicians, that part of Northumbria which lies between the river Tees and the Scottish sea* or *frith;* regnum *vel* provincia Berniciorum, a Tesi ad fretum Scoticum olim pertingens:—Oswio ðone óðerne dǽl Norþanhymbra ríces hæfde, ðæt is Beornica *Oswi possessed the other part of the Northumbrian kingdom, that is Bernicia,* Bd. 3, 14; S. 539, 35: 5, 14; S. 635, 6.

Beornice; *gen.* a; *dat.* um; *pl. m. The Bernicians;* Bernicii:—Man gehálgode twegen biscopas on his stal, Bosan to Derum and Eátan to Beornicum *two bishops were hallowed in his stead, Bosa over the Deirians and Eata over the Bernicians,* Chr. 678; Th. 61, 17, col. 1: Bd. 3, 24; S. 556, 45.

beorn-preát, es; *m. A band of men* or *warriors;* virorum turma:—Monig beornþreát *many a band of warriors,* Exon. 96 a; Th. 358, 24.

beorn-wiga, an; *m.* [wíga *a warrior*] *A soldier, hero;* loricatus bellator, Menol. Fox 447; Men. 225.

beór-scealc, es; *m. A beer-server, a butler;* cerevis·æ minister:—Beórscealca sum *some one of the beer-servers,* Beo. Th. 2485; B. 1240.

beór-scipe *a feast.* v. gebeór-scipe.

beór-sele, biór-sele, es; *m. A beer-hall, feasting-hall, hall, mansion, palace;* cerevisiæ aula, convivis recipiendis locus, aula, mansio, palatium:—In [on] beórsele *in the beer-hall,* Beo. Th. 968; B. 482: 988; B. 492: Runic pm. 14; Hick. Thes. i. 135; Kmbl. 342, 5. Gesittaþ beórselas beorna *they shall inhabit the beer-halls of chieftains,* Cd. 170; Th. 214, 2; Exod. 563.

beór-setl, es; *n. A* BEER-SETTLE or *bench;* scamnum cerevisiam bibentium:—Ofer beórsetle [MS. -sele] *on the beer-bench,* Exon. 75 b; Th. 283, 28; Jul. 687.

beor-swinig; *adj.* [= bǽr-synnig] *Openly-wicked, a publican,* Lk. Rush. War. 19, 2. v. bǽr-synnig.

beorþ, berþ, byrþ, e; *f:* es; *n?* [beorþ *bears, from* beoran, *as* byrþ birþ *from* beran] *A* BIRTH, *the act of coming into life, the thing born;* nativitas, partus, fetus, Cot. 87. Found in the compounds berþ-estre, berþ-ling: v. also beorþor, beorþor-cwelm, -þínen; hyse-beorþor. [O. Sax. gi-burd, *f:* O. Frs. berthe, *f:* O. H. Ger. burt, *f:* Goth. ga-baurþs, *f:* O. Nrs. burðr, *m.*] v. ge-byrd.

beór-þegu, e; *f. A beer-receiving, beer-serving, beer-drinking;* cerevisiæ acceptio vel ministratio, cerevisiæ potatio:—Ðæt wæs biter beórþegu *that·was a bitter beer-serving,* Andr. Grm. 1533; An. 1535. Æfter beórþege *after the beer-drinking,* Beo. Th. 234; B. 117: 1239; B. 617.

beorþor, byrþor, berþor, borþor, es; *n? Child-birth, that which is born, a fetus;* partus, fetus:—Æfter beorþre *after child-birth,* Med. ex Quadr. 4, 6; Lchdm. i. 344, 1: L. M. 3, 37; Lchdm. ii. 330, 1. Ðe him hyra beorþor losie *quibus fetus pereat,* Med. ex Quadr. 4, 4; Lchdm. i. 342, 21. Mid beorþre *fetu,* Cot. 87. DER. ge-beorþor, hyse-.

beorþor-cwelm, es; *m. A dead birth, an abortion, a miscarriage;* fetus mortuus *vel* abortivus, abortus, Cot. 11.

beorþor-þínen, e; *f. A midwife;* obstetrix [beorþor *child-birth,* þínen *a maid-servant*]. v. bróðor-þínen.

beór-tún, es; *m. A beer-hall;* convivis recipiendis locus *vel* aula, Mann. v. beór-sele.

Beorwíc [wíc *a village* or *residence,* Beornica *of the Bernicians;* Berniciorum vicus] BERWICK *on Tweed,* Som.

beosmriende *deceiving,* Bd. 5, 12; S. 628, 31, note, = bysmriende. v. bysmerian.

BEÓST, býst, býsting, es; *m?* BIESTINGS, *the first milk of a cow after calving;* colostrum:—Beóst biestings; obesta, Ælfc. Gl. 31; Som. 61, 102. Býst *colostrum,* Ælfc. Gl. 31; Som. 61, 102. Býsting, þicce meolc biest, biestings, *thick milk,* Ælfc. Gl. 33; Som. 62, 20. [*Plat.* beest, beest-melk: *Dut. Ger.* biest: *O. H. Ger.* biost: *Goth.* beist.]

BEÓT, es; *n.* I. *a threatening, threat, command, menace;* comminatio, minæ:—He ne wæs ondredende ða beótunge [beót, MSS. B. C.] ðæs ealdormannes *minas principis non metuit,* Bd. 1, 7; S. 477, 23: Exon. 68 a; Th. 253, 7; Jul. 176. II. *peril; periculum:*—Denden [ðen, MS.] in ðam beóte wǽron *while they were in that peril,* Cd. 187; Th. 232, 25; Dan. 265. III. *a boasting, boasting promise, promise;* jactantia, promissio gloriosa, promissum:—Wæs him gylp forod, beót forborsten *their vaunt was broken, their boasting shattered,* Cd. 4; Th. 5, 11; Gen. 70. He beót eal wið ðé sóðe gelǽste *he truly fulfilled all his promise to thee,* Beo. Th. 1051; B. 523: 160; B. 80. [*Ger. M.H.Ger.* butze, *m. larva, terriculamenta.*] DER. ge-beót, word-.

beót *beat, hurt,* Cd. 187; Th. 232, 24; Dan. 265; *p. of* beátan.

beóþ *is, are, shall be,* Exon. 44 a; Th. 149, 28; Gú. 768; 96 b; Th. 361, 20; Wal. 22: Ælfc. Gr. 25; Som. 26, 14: Th. Diplm. A.D. 743–745; 28, 27. v. beón.

beóðan *are,* Mt. Rush. Stv. 5, 11, = beóþ. v. beón.

beót-háta, an; *m.* [MS. beo = beót, gebeót *a command, decree,* háta *a caller, commander*] *A commander, leader;* imperator, dux:—Ahleóp ðá fór hæleðum hilde calla, bald beót-háta bord upahóf *then the herald of*

---

war leaped before the warriors, the bold commander [*Moses*] upraised his shield, Cd. 156; Th. 193, 27; Exod. 253.

beó-þeóf, es; *m. A thief* or *stealer of bees;* apum fur, L. Alf. pol. 9; Th. i. 68, 6.

beótian, beótigan; *p.* ode, ede; *pp.* od, ed [beót I. *a threatening*]. I. *to threaten;* minari, minitari:—Agustinus is sǽd, ðæt he beótigende fórecwǽde *Augustinus fertur minitans prædixisse,* Bd. 2, 2; S. 503, 29: Exon. 67 b; Th. 250, 35; Jul. 137. II. *to boast, vow, promise;* magna loqui, polliceri, spondere:—Swá he beótode ǽr wið his beáhgifan *as he boasted before towards his ring-giver,* Byrht. Th. 140, 18; By. 290. Ful oft wit beótedan, ðæt unc ne gedǽlde nemne deáþ ána *full oft we two vowed, that naught should part us save death alone,* Exon. 115 a; Th. 442, 32; Kl. 21.

beótian; *p.* ode; *pp.* od [*from* bót *a restoring, cure*] *To become* or *grow better;* melius fieri, convalescere:—Ða sóna gefélde ic me beótiende and wyrpende *then I felt myself soon getting better and turning;* confestim me melius habere sentirem, Bd. 5, 6; S. 620, 12.

beót-líce; *adv. In a threatening manner, threateningly;* minaciter, Jos. 8, 10: Num. 14, 44.

beótung, e; *f. A threatening, raging;* comminatio, minæ:—Beótunge dǽdum gefyldon [*they*] *followed the threatening with deeds,* Bd. 1, 15; S. 483, 39. Ða wæs his mód mid ðam beótungum gebreged *then was his mind frightened by the threatenings,* 2, 12; S. 513, 14: 1, 7; S. 477, 23. DER. ge-beótung.

beót-word, es; *n.* I. [beót I. *a threat*] *a word of threatening, threats;* minæ:—Beótwordum sprǽc folcágende *the people's lord spoke in words of threatening,* Exon. 68 a; Th. 253, 24; Jul. 185. II. [beót III. *a boasting*] *a word of boasting;* jactationis verbum:—Beówulf beótwordum sprǽc *Beowulf spake in words of boasting,* Beo. Th. 5014; B. 2510.

Beó-wulf, es; *m.* [= Beado-wulf *a war-wolf,* = Icel. Böðúlfr *a warwulf*] BEOWULF, a celebrated warrior of the Scyldings' race, a record of whose heroic deeds is given in the Anglo-Saxon poem bearing his name. It appears most probable that Beowulf was originally an Old Norse heathen Saga, written in the language common at the earliest age in Denmark, Sweden, and Norway, but now only spoken in Iceland. This Saga it is hoped may yet be found in some Swedish library. The story informs us that Hrothgar built a splendid palace at Heorot in the north of Jutland. This palace was soon made a scene of slaughter, in consequence of the nightly attacks of a monster called Grendel, who carried off at one time no less than thirty thanes, for the purpose of devouring them in his retreat. These dreadful visitations are continued during a period of twelve years. Intelligence of this calamity having reached the heroic Beowulf, a relation of Hrothgar, Beowulf resolves to rid the Danish land of this monster; and, in pursuance of this design, sails from home with a company of fifteen warriors. In terrific conflicts he kills Grendel and his mother.—It was the first heroic poem by any Germanic nation, and must have been translated into Anglo-Saxon by a Christian, as is evident by Grendel's mother being spoken of as a descendant of Cain, and numerous Christian allusions, when the Danish sovereignty in England was at its height, perhaps in the reign of Canute, about A.D. 1020. If it were originally written in the Old Norse or Icelandic the Saga would be called Böðúlfr, and the translator into Anglo-Saxon would naturally write it Beado-wulf contracted to Beó-wulf:—

Beówulf wæs brḗme,	*Beowulf was renowned,*
blǽd wíde sprang	*the glory of Scyld's offspring*
Scyldes eaferan	*widely spread*
Scede-landum in,	*in the Swedish lands.*
Beo. Th. 35-38; B. 18, 19.	
Heorot [Hróþgár] eardode	[*Hrothgar*] *occupied Heorot,*
sincfáge seld [MS. sel],	*the richly variegated seat.*
Beo. Th. 335; B. 166.	
[Grendel] atol æglǽca;	[*Grendel*] *the fell wretch;*
him on eaxla wearþ	*a deadly wound was manifest*
syndolh sweotol,	*in his shoulder,*
seonowa onsprungon,	*the sinews sprang asunder,*
burston bánlocan:	*the bone-inclosures burst:*
Beówulfe wearþ	*to Beowulf*
gúþhréþ gyfeðe;	*warlike fierceness was given;*
scolde Grendel ðonan	*Grendel, death-sick,*
feorhseóc fleón,	*must thence flee.*
Beo. Th. 1636-1644; B. 816-820.	
Geféng ðá be eaxla	*The War-Goths' lord*
Gúþ-Geáta leód	*seized then by the shoulder*
Grendles módor.	*Grendel's mother.*
Brægd ðá beadwe heard,	*Then the fierce warrior dragged*
feorhgenídlan,	*the mortal foe,*
ðæt heó on flet gebeáh:	*so that she bowed on the place:*
Beo. Th. 3078-3085; B. 1537-1540.	
- - oil eal þurhwód,	- - *the falchion passed through all*
fægne flǽschoman,	*her fated carcase,*
heó on flet gecrong.	*she sank on the ground.*
Beo. Th. 3139-3141; B. 1567, 1568.	

beó-wyrt, e; *f.* [beó *a bee*, wyrt *a plant*] BEE-WORT, *balm-mint*, *sweet flag*; apiastrum, acorus = ἄκορος, acorus calamus, Lin :—Beówyrt *apiastrum*, Cot. 12 : Ælfc. Gl. 39 ; Som. 63, 55 ; Wrt. Voc. 30, 9. Ðeós wyrt, ðe man on Léden *veneriam*, and on úre geþeóde beówyrt, nemneþ, heó biþ cenned on begánum stówum, and on wyrtbeddum, and on mǽdum *this plant, which in Latin is called veneria, and in our language bee-wort, is produced in cultivated places, and in wort-beds, and in meads*, Herb. 7, 1 ; Lchdm. i. 96, 21 : L. M. 1, 26 ; Lchdm. ii. 68, 4.

be-pǽcan; *part.* be-pǽcende ; *p.* be-pǽhte ; *pp.* be-pǽht ; *v. a.* [be *by*, pǽcan *to deceive*] *To deceive, entice, seduce, draw away*; decipere, pellicere, illudere, seducere :—Seó nǽddre bepǽhte me *serpens decepit me*, Gen. 3, 13 : Mt. Bos. 2, 16 : Ælfc. Gr. 28, 5 ; Som. 32, 1. Ic be-pǽce oððe forlǽde *seduco*, 47 ; Som. 48, 53 : Jud. 16, 5.

be-pǽcestre, an; *f. She who deceives, flatters*, or *entices, a harlot*; pellex, Ælfc. Gr. 28, 5 ; Som. 32, 1.

be-pǽcung, e; *f. Lewd practice*; lenocinium, Som. v. be-pǽcan.

be-pǽht *deceived*, Mt. Bos. 2, 16 ; *pp.* of be-pǽcan.

be-prenan, be-preðan *To wink*; nictare :—Tele nú ða lenge ðǽre hwíle, ðe ðú ðín eáge on beprenan [bepreðan, Cott.] mǽge *compare now the length of the time, wherein thou mayest wink thine eye*, Bt. 18, 3 ; Fox 66, 7.

bér, beer, e ; *acc.* bér, bére ; *f. A bed*; lectus, grabatus :—Nim bér ðín *tolle grabatum tuum*, Jn. Lind. War. 5, 12. Nim bére ðíne, Jn. Rush. War. 5, 12. v. bǽr.

BERA, an; *m. A* BEAR; ursus :—Dauid gewylde ðone wíldan beran *David subdued the wild bear*, Ælfc. T. 13, 26. Eofor oððe bera onginnan *to attack a boar or bear*, Exon. 92 a ; Th. 344, 21 ; Gn. Ex. 177. Sceall gyldan án beran fel *shall pay one bear's skin*, Ors. 1, 1 ; Bos. 20, 37. Bera *ursus*, Ælfc. Gl. 21 ; Som. 59, 69 ; L. Ecg. P. iv. 28 ; Th. ii. 212, 22. [*Laym.* beore : *Plat.* baar, *m* : *Dut.* beer, *m* : *Ger.* bär, *m* : *M. H. Ger.* ber : *O. H. Ger.* pero : *Dan.* biörn, *c* : *Swed.* biörn, *m* : *O. Nrs.* björn, *m.*]

be-récan *to cause to smoke*, Herb. 14, 2 ; Lchdm. i. 106, note 24. v. be-récan.

be-rédan; *p.* -rǽdde ; *pp.* -rǽd [be-*dis-*, rǽdan *to possess*] *To dispossess, deprive of*; privare :—He hine ríces berǽdde *he deprived him of his realm*, Andr. Kmbl. 2653 ; An. 1328 : 266 ; An. 133. Hie unscyldigne feore berǽddon *they deprived the guiltless of his life*, Elen. Kmbl. 993 ; El. 498. Earnulf hine berǽdde æt ðam ríce *Arnulf deprived him of the kingdom*, Chr. 887 ; Th. 156, 32, col. 1 ; 33, col. 2, 3 : Bt. tit. 1 ; Fox x. 3.

be-rǽsan; *p.* de ; *pp.* ed [be, rǽsan *to rush*] *To rush into*; irruere :—Ðá ðonne hie berǽsaþ on swelce weámódnesse *when they then rush into such anger*, Past. 40, 5 ; Hat. MS. 55 a, 25 ; Gen. 14, 15.

be-rafan; *p.* -róf, *pl.* -rófon ; *pp.* -rafen *To bereave*; spoliare :—Ða de Sodoma golde beròfon [MS. berofan] *those that had bereaved Sodom of gold*, Cd. 95 ; Th. 125, 13 ; Gen. 2078. v. be-reáfian, be-reófan.

BERAN, beoran, ic bere, beore, ðú birest, birst, byrst, he bireþ, byreþ, birþ, byrþ, *pl.* beraþ ; *p.* ic, he bær, ðú bǽre, *pl.* bǽron ; *pp.* boren ; *v. a.* I. *to* BEAR, *carry, bring, bear* or *carry a sacrifice, offer, bear off, carry out, extend, wear, support, endure, suffer*; ferre, portare, afferre, offerre, deferre, proferre, extendere, gerere, tolerare :—Ðú eall þing birest *thou bearest all things*, Bt. Met. Fox 20, 551 ; Met. 20, 276. Heó gár bireþ *she beareth the javelin*, Salm. Kmbl. 876 ; Sal. 437. Eft byreþ ofer lagustreámas leófne mannan *shall bear back over the water-streams the beloved man*, Beo. Th. 598 ; B. 296 : 4117 ; B. 2055. Se ðæt wicg byrþ *he whom the horse carries*, Elen. Kmbl. 2390 ; El. 1196. On handum hî beraþ ðe *in manibus portabunt te*, Ps. Spl. 90, 12. Secgas bǽron beorhte frætwa *the warriors bare bright arms*, Beo. Th. 347 ; B. 213. Ðe bǽron byrðena on ðises dæges hǽtan *qui portavimus pondus diei et æstus*, Mt. Bos. 20, 12 ; Lk. Bos. 11, 27. Ne bere ge sacc *nolite portare sacculum*, Lk. Bos. 10, 4 : Ex. 22, 13. Him wæs ful boren *to him the cup was borne*, Beo. Th. 2388 ; B. 1192 : Cd. 6 ; Th. 8, 7 ; Gen. 120. Deóflum onsǽgdnesse bær *dæmonibus hostias offerebat*, Bd. 1, 7 ; S. 477, 13. Byreþ blódig wæl *will bear off my bloody corpse*, Beo. Th. 900 ; B. 448. Ða wiccungdóm wídost bǽron *who carried the magic art furthest*, Cd. 178 ; Th. 223, 18 ; Dan. 121. Ðæt ða hætt beran móston *that they might wear [bear] a hat*, Ors. 4, 10 ; Bos. 96, 20, 18. Ic nelle beran eówre gýmeleáste *I will not endure your negligence*, L. Ælf. C. 1 ; Th. ii. 342, 10. II. *to* BEAR, *produce, bring forth*; facere, ferre, edere, parere :—Ælc gód treów byrþ góde wæstmas *every good tree produces* [facit] *good fruits*, Mt. Bos. 7, 17 : 7, 18. Ðæt wæs deáþes beám se bær bitres fela *that was the tree of death which bare much of bitter*, Cd. 24 ; Th. 31, 2 ; Gen. 479 : 30 ; Th. 40, 26 ; Gen. 645. Gif he to ðæm ríce wæs on rihte boren *if he to that kingdom was rightly born*, Bt. Met. Fox 26, 92 ; Met. 26, 46. [*O. Sax.* beran *ferre, portare* : *O. Frs.* bera : *O. H. Ger.* beran *ferre, parere, gignere, generare* : *Goth.* bairan ; *p.* bar, *pl.* berum ; *pp.* bairans *to bear, carry, bring, bear children* : *O. Nrs.* bera *ferre, portare, sustinere, tolerare* : *Grk.* φέρειν : *Sansk.* bhri *to bear, hence Goth.* barn *a child* : *A. Sax.* bearn *a child*.] DER. a-beran, æt-, be-, for-, fôr-, forþ-, ge-, in-, on-, ôþ-, to-, under-, up-, upa-, upge-, ymb- : berende, deáþ-, feorh-, gár-, helm-, leóht-, reord-,

sǽd-, sweord-, un-, wæstm- : berend, gár-, gást-, helm-, reord-, sáwl-, segn-, tácn- : berendnis, un- : bere, -ærn, -corn, -flór, -gafol, -grǽs, -hláf, -sǽd, -tún, -wíc : berie, berige, berge, blæc-, byrig-, hind-, streów-, wín- : brid : bearn, cyne-, dryht-, folc-, freó-, frum-, god-, hǽlu-, húsul-, steóp-, sweostor-, world-, þryþ- : -cennung, -eácen, -eácnung, -gebyrdo, -gestreón, -lést, -lufe, -myrþra, -teám : bearm, -cláþ, -rægl : beorma, bearm, gebyrman : byre : ge-byrd, -dæg, -tíd, -wiglǽre, -witega : byrde, ge-, in- : frum-byrdling, in-byrdling : beorþ, berþ, berþ-estre, berþ-ling : hyse- : beorþor, -cwelm, -pínen, hyse- : bǽr, bǽran, bǽr-disc : bǽre, æppel-, corn-, cwealm-, cwyld-, hlís-, horn-, leóht-, lust-, wæstm-, unwæstm- : bǽrnes, lust-, wæstm-, unwæstm- : byrðen, mægen-, sorg-, syn- : bora, cǽg-, horn-, mund-, rǽd-, rǽs-, segen-, sôþ-, sweord-, tácn-, wǽg-, wǽpen-, wíg-, wôþ-, wrôht- : boren, æðel-.

Beran burh; *gen.* burge ; *dat.* byrig ; *f.* [*Hunt.* Beranbiri : *Kni.* Banbyry] BANBURY, *Oxfordshire* :—Hér Cynríc and Ceawlin fuhton wið Brettas æt Beran byrig *here*, A. D. 556, *Cynric and Ceawlin fought with Britons at Banbury*, Chr. 556 ; Th. 30, 9, col. 1, 2, 3.

berbéna, æ ; *f. Latin* : berbéne, an ; *f. Vervain*; verbéna :—Berbéna [berbéne MS. H.] Ðeós wyrt, ðe man περιστερεών, and ôðrum naman berbénam, nemneþ, heó ys culfron swíðe híwcúþ. *Vervain. This plant, which they call vervain, and by another name verbena, in colour is very like to doves*, Herb. 67, 1 ; Lchdm. i. 170, 11–14. *Verbéna officinalis* is intended by the drawing in MS. V. and by περιστερεών in Dioskorides. v. æsc-þrote.

berc *a birch-tree*; betula :—Nim birc rinde *take birch-tree rind*, L. M. 3, 39 ; Lchdm. ii. 332, 9. v. birce.

bere, an ; *f. A female bear*; ursa. v. bera *ursus*.

BERE, es ; *m. Barley*; hordeum : — Ðá hét he him bere sǽd bringan *inde hordeum jussit afferri*, Bd. 4, 28 ; S. 605, 36 : Ælfc. Gr. 8 ; Som. 7, 63. Hira flex and hira beras [MS. bernas] wǽron fordône *eorum linum et hordea læsa sunt*, Ex. 9, 31. [*Scot. and North E.* bear, bere *barley*: *Goth.* barizeins, *adj. made of barley*; hordeaceus : *Swed. Norw. Icel.* barr, *n.* I. *spina abietis vel pinus*, II. *granum, semen, hordeum*.]

bére *a bed*; *acc. sing.* of bér.

bere-ærn, ber-ern, beren, bern, bearn, es ; *n. A barley-place, a corn-place, a barn*; horreum :—He gegaderaþ his hwǽte on his bern *congregabit triticum suum in horreum*, Mt. Bos. 3, 12 : 13, 30. He feormaþ hys berenes flóre *purgabit aream suam*, Lk. Jun. 3, 17. Ic towurpe míne berenu *destruam horrea mea*, 12, 18 : 12, 24 : Mt. Kmbl. Lind. 3, 12 : Leo 103 : 110.

be-reáfian, bi-reáfian, -reáfigean, ic -reáfige ; *p.* -reáfode ; *pp.* -reáfod ; *v. a. To* BEREAVE, *seize, spoil, take away*; eripere, spoliare, privare :—Heó hit ne mæg his gewittes bereáfian *she cannot bereave it of its faculty*, Bt. 5, 3 ; Fox 12, 25. Hú mæg man hys fata hyne bereáfian *quomodo potest quisquam vasa ejus diripere?* Mt. Bos. 12, 29 ; Mk. Bos. 3, 27. Ic ondréd, ðæt ðú me bereáfodest ðínra dôhtra *timui, ne violenter auferres filias tuas*, Gen. 31 : 43, 18 : 43, 14 : Ors. 3, 7 ; Bos. 61, 16 : Cd. 40 ; Th. 53, 11 ; Gen. 859.

be-récan, -rǽcan [rǽcan *to smoke*] *To cause to smoke*; facere ut fumet aliquid :—Beréc hit on hátum ahsum *make it smoke on hot ashes*, Herb. 14, 2 ; Lchdm. i. 106, 17.

be-reccan, -reccean ; *p.* -reahte, -rehte ; *pp.* -reaht, -reht. I. *to relate, recount, explain*; narrare, exponere :—Nú wille we sum þing scortlíce eów be him bereccan *now will we relate to you shortly something concerning him*, Nat. S. Greg. Els. 3, 2. II. *to explain one's conduct, justify one's self*; se excusare, se purgare, accusatorum criminibus respondere :—Hí simle séceaþ endleáse ládunga, hú hie bereccan [MS. C. bereccean] mægen *they always seek endless excuses, how they may justify themselves*, Past. 35, 2 ; Hat. MS. 45 a, 19. Him wæs lýfnesse seald ðæt he him môste scyldan and besecgan [MS. B. bereccan] *accepit locum se defendendi*, Bd. 5, 19 ; S. 640, 11, note. v. reccan.

bere-corn, es ; *n.* [bere *barley*, corn *a grain*] BARLEY-CORN, *a grain of barley*; hordei granum :—IX bere-corna *nine barley-corns*, L. Ath. iv. 5 ; Th. i. 224, 11.

bere-flór, es ; *m. A* BARLEY-FLOOR, *barn-floor*; hordei area, Lk. Lind. Rush. War. 3, 17.

bere-gafol, es ; *n. Barley-rent, a tribute of barley*; hordei tributum. One of the rents paid in kind, which, by the following enactment, is fixed at the rate of six pounds weight for every labourer employed in the barley harvest :—Mon sceal simle to bere-gafole agifan æt ánum wyrhtan six pund-wǽga *a man shall always give for barley-rent for every labourer six pounds weight*, L. In. 59 ; Th. i. 140, 5.

bere-grǽs, es ; *n.* BARLEY-GRASS, *a farrago*; hordei gramen :—Gréne beregræs *green fodder for cattle* [farrago], Ælfc. Gl. 59 ; Som. 67, 124.

bere-hláf, es ; *m. A* BARLEY-LOAF, *barley-bread*; hordeaceus panis. v. bere *barley*, hláf *a loaf*.

beren, es ; *n.* [bere-ærn, *q. v.*] *A barley-place, a barn*; horreum, Lk. Jun. 3, 17 : 12, 18, 24.

beren; *adj. Barley, made of barley*; hordeaceus :—Genim smæl beren mela *take fine barley-meal*, L. M. 1, 36 ; Lchdm. ii. 86, 24. Hæfþ fíf

berene hláfas *habet quinque panes hordeaceos*, Jn. Bos. 6, 9 : 6, 13.
v. bere.

beren, byren; *adj.* [bera *a bear*] *Belonging to a bear, ursine*; ursinus :—Se byrdesta sceall gyldan berenne cyrtel [kyrtel MS.] oððe yterenne *the richest must pay a bear- or otter-skin vest*, Ors. 1, 1; Bos. 20, 37.

berende; *part. Bearing, fruitful*; ferens, gerens, abundans, ferax :—Wíneard berende *vitis abundans*, Ps. Spl. 127, 3 : Cot. 85. Berende bóh *germen*, Ælfc. Gl. 60; Som. 68, 32. v. beran.

berendlíc; *adj. Bearable, tolerable.* v. a-berendlíc.

berendnis, -niss, e; *f. Fertility, fruitfulness*; fertilitas, Leo 110. v. un-berendnis.

be-rénian; *p.* ode; *pp.* od [regnian, rénian *to arrange*] *To cause, moliri* :—Heó wroht berénodon [berenedon MS.] *they caused strife*, Cd. 149; Th. 187, 6; Exod. 147.

be-reófan, bi-reófan; *p.* -reáf, *pl.* -rufon; *pp.* -rofen [be, reófan *to reave, rob*] *To bereave, deprive*; spoliare, privare :—Since berofene *deprived of treasure*, Cd. 144; Th. 179, 30; Exod. 36 : Beo. Th. 5855; B. 2931.

be-reótan; *p.* -reát, *pl.* -ruton; *pp.* -roten *To deplore*; deplorare :—Æðelinges deáþ bereótan *to deplore the death of the noble*, Exon. 119 b ; Th. 459, 27; Hö. 6.

ber-ern *a barley-place, a barn*; horreum, Mt. Kmbl. Lind. 3, 12. v. bere-ærn.

bere-séd, es; *n. Barley-seed, barley*; hordeum, Bd. 4, 28; S. 605, 36. v. bere.

bereþ *bears, brings forth, produces, 3rd pres. of* beran, Mt. Rush. Stv. 1, 21 : Hick. Thes. i. 135; Runic pm. 18; Kmbl. 342, 28.

bere-tún, es; *m.* [bere *barley, corn*, tún *an inclosure, a place shut in*] *A barley-inclosure, court-yard, threshing-floor, corn-farm, grange, corn-village,* BARTON; hordei area, villa frumentaria. 'BARTON, *Prædium dominicum,* vel *terræ* quas vocant *Dominicales,* hoc est, quas in distributione manerii dominus non elocavit hæreditarie, sed alendæ familiæ suæ causâ propriis manibus reservavit : *Dominicum, Gallice Domaine.* Vox in Devonia, inquit Spelmannus, et plaga Angliæ Occidentali bene nota,' Du Cange Glos :—Þerh-clænsade beretún his *permundavit aream suam,* Mt. Kmbl. Lind. 3, 12.

bere-wíc, es; *n. A barley-village, a corn-village*; hordeaceus *vel* frumentarius vicus, Th. Diplm. A. D. 1060; 382, 12 : A. D. 1093; 443, 31. v. bere-tún.

berg *a hill, mountain,* Som. DER. berg-ælfen. v. beorg.

berg-ælfen *mountain-elves*; oreades. v. ælf, -ælfen.

bergan *to surety*; gustare :—Ða ðe ne bergaþ deáþ *qui non gustabúnt mortem,* Mt. Kmbl. Rush. 16, 28. v. byrgan.

berge, an; *f. A berry, grape,* Deut. 23, 24. v. berie II.

bergels-leóþ, es; *n. A burial ode*; sepulcrale carmen, Leo 116. v. byrgen-leóþ.

bergel-song, es; *m. A burial song*; sepulcralis cantus, Leo 116. v. byrgen-song.

bergena *of berries,* Deut. 23, 24; *g. pl. of* berie.

Berghám-styde, es; *m.* BERHAM, *near Canterbury* :—In ðære stówe, ðý hátte Berghámstyde *in the place which is called Berham,* L. Wih. pref; Th. i. 36, 6.

bergyls, es; *m. A burial-place, a sepulchre*; sepulcrum, Coll. Monast. Th. 32, 33. v. byrgels.

berh *for* bearh *shunned*; vitavit, Bd. 2, 12; S. 513, 28; *p. of* beorgan.

berht; *adj. Bright*; splendidus, clarus, Bt. Met. Fox 22, 43; Met. 22, 22. v. beorht.

berhtan *to shine*; lucere. DER. ge-berhtan. v. beorhtan.

Berhte, an; *f. Bertha*; Bercta, *the daughter of Cariberht, king of Paris, and granddaughter of Clotaire, king of the Franks and Burgundians. In the year 570, she married Æðelbryht, king of Kent.* By the queen's Christian conduct, the heathen predilections of the king were removed, and the way made clear for the preaching of Augustine in 597. v. Æðelbryht :—Ær ðam, becom hlísa to him ðære cristenan æfestnysse, for ðon he cristen wíf hæfde, seó wæs him forgifen of Francena cyningcynne, Berhte wæs háten. Ðæt wíf he onféng fram hire yldrum ðære arédnesse, ðæt heó his leáfnysse hæfde ðæt heó ðone þeáw ðæs cristenan geleáfan, and hire æfestnysse, ungewemmedne healdan móste, mid ðý biscop, ðone ðe hí hire to fultume ðæs geleáfan sealdon, ðæs nama wæs Leodheard *before that, a report of the Christian religion had come to him [Æðelbryht] for he had a Christian wife, who was given to him from the royal kin of the Franks, her name was Bertha. He received his wife from her parents on condition, that she should have his leave that she might hold the manner of the Christian belief, and of her religion, unspotted, with the bishop, whose name was Liudhard, whom they gave her for the help of that faith,* Bd. 1, 25; S. 486, 30–36.

berhtm-hwæt; *adj. Swift as an eye-blink*; celer ut oculi nictus :—Ðec lígetu bláce, berhtmhwate ða ðec bletsige *the pale lightnings, swift as an eye-blink, these shall bless thee,* Cd. 192; Th. 240, 3; Dan. 381. v. bearhtm.

berhtra, *acc.* berhtre *brighter,* Bt. Met. Fox 22, 43; Met. 22, 22; *comp. of* berht, beorht, *q. v.*

berian *berries,* Ælfc. Gl. 47; Som. 65, 30; *pl. of* berie.

berian; *p.* ode, ede; *pp.* od [bær *bare*] *To bare, make naked, expose, exhibit, make a shew of*; nudare, denudare, in medium proferre, ostentare :—Benc-þelu beredon *they made bare the bench-floor,* Beo. Th. 2482 ; B. 1239. Ða ðe me fór werode wisdóm bereþ *who to me make a shew of wisdom before the people,* Cd. 179; Th. 224, 27; Dan. 142. v. barenian, a-barian.

berian *to taste.* v. bergan, byrgan, on-berian.

berian=byrian *to happen.* DER. ge-berian.

be-rídan, he -rít; *p.* -rád, *pl.* -ridon; *pp.* -riden; *v. a.* I. *to ride round, to surround, besiege*; perequitare, præcingere :—Ðæt he his gefán beríde *that he besiege his enemy,* L. Alf. pol. 42; Th. i. 90, 4. II. *to ride after, pursue*; persequi :—Ðá beríd ðone ðæt wíf *then they pursued the wife,* Chr. 901; Ing. 125, 14. He hine berád *he rode after him,* 755; Ing. 70, 1.

BERIE, berge, berige, berigie, an; *f.* I. *a* BERRY; bacca :—Berian *berries,* Cot. 36. Bergan *berries*; baccæ, Cot. 23. Nym wínberian, ðe beóþ acende æfter óðre berigian *take grapes, which are formed after other berries,* Lchdm. iii. 114, 5. II. *a grape*; uva. Though wín-berie, *q. v. a wine-berry, is generally used in Anglo-Saxon for a grape, yet* berge, berige *are sometimes found, as,* —Gif ðú gange binnan ðínes freóndes wíneard, et ðæra bergena swá fela, swá ðú wylle, and ne ber ðú ná má út mid ðé *if thou shalt go within thy friend's vineyard, eat as many of the grapes as thou wilt, and carry not out with thee any more,* Deut. 23, 24. Beóþ ðínes wífes wélan gelíce swá on wíngearde weaxen berigean, and on ðínes húses hwommum genihtsum *the riches of thy wife shall be like as grapes may grow in a vineyard, and abundant on the corners of thy house,* Ps. Th. 127, 3. [O. Sax. beri, *n* : Dut. bes, *f* : O. H. Ger. beri, *n* : Goth. basi, *n* : O. Nrs. ber, *n.* The Goth. Plat. and Dut., says Grimm [i. 1243], do not allow us to derive these words from the root of Goth. bairan, *A. Sax.* beran *to bear, but it is probably connected with* bær *bare, naked, signifying the bare fruit, which can be eaten immediately.* Bopp derives the *Teutonic* words and the *Lat.* bacca from *Sansk.* bhaksh *edere*; so the Goth. basi= bhakshya *cibus, eatable fruit.*] DER. blæc-berie, byrig-, hind-, streów-, streáw-, wín- [-berie, -berge, -berige, -berigie].

berig *to a city,* Wrt. Voc. 84, 45, = byrig; *dat. of* burh.

berig-drenc, es; *m.* [berige *a berry,* drenc *a drink*] *Drink made of mulberries*; diamoron, Wrt. Voc. 20, 23.

berige, an; *f. A berry, grape,* Ps. Th. 127, 3. v. berie II.

berigea, an; *m. A surety,* L. H. E. 6; Th. i. 30, 5. v. byriga.

berigean *berries, grapes,* Ps. Th. 127, 3; *nom. pl. of* berige. v. berie.

berigie *a berry,* Lchdm. iii. 114, 5. v. berie I.

be-rindan; *p.* de; *pp.* ed [be *off,* rind *the bark*] *To bark, peel* or *strip off the bark*; decorticare :—Berinde *decorticavit,* Cot. 62.

be-riówsian *to repent,* Ælfc. Gr. 33, MS. D; Som. 37, 22. v. behreówsian.

bern, es; *n. A barn*; horreum :—Nabbaþ ða hrefnas héddern ne bern *the ravens have not store-house nor barn* [cellarium neque horreum], Lk. Bos. 12, 24 : 12, 18 : 3, 17 : Mt. Bos. 3, 12 : 13, 30. Bern *horreum,* Ælfc. Gl. 109; Som. 78, 131. v. bere-ærn.

bernan *to burn*; ardere, Ælfc. Gr. 35; Som. 38, 5. v. beornan.

berne-lác, es; *n. A burnt offering*; holocaustum :—Ic ðé bernelác brengan móste *I must bring thee a burnt offering,* Ps. C. 50, 123; Ps. Grn. ii. 279, 123.

bernes *a burning,* Bd. 4, 21; S. 590, 21. v. bærnes.

bernet, bernett, es; *n. A burning*; incendium, R. Ben. interl. 28. v. bærnet.

berning, e; *f. A burning*; combustio, ustio, Som. Lye. v. bærning.

be-rofen *bereaved,* Beo. Th. 5855; B. 2931. v. be-reófan.

béron *might bear, carry, bring, for* bæren, *perf. subj. of* beran, Byrht. Th. 133, 49; By. 67.

be-rówan; *p.* -reów, *pl.* -reówon; *pp.* -rówen *To row round*; remigando circumnavigare, Chr. 897; Th. 176, 41.

berst *loss*; damnum, malum, ruina, Lupi Serm. i. 2 : Wulfstani Archiepiscopi Ebor. Admonitio sive Paraenesis, 8. *etc.* DER. berstan. v. byrst.

BERSTAN; *part.* berstende; ic berste, ðú birst, he birsteþ, biersteþ, birst, byrst, bierst, *pl.* berstaþ; *p.* ic, he bærst, ðú burste, *pl.* burston; *pp.* borsten. I. *to* BURST, *break, fail, fall*; cum fragore dissilire, corruere, rumpi, frangi :—Heofonas berstaþ *the heavens burst,* Exon. 21 b ; Th. 58, 10; Cri. 933. Burston bán-locan *the bone-inclosures burst,* Beo. Th. 1640; B. 818. Wægas burston *the waves broke,* Cd. 167; Th. 208, 15; Exod. 483. Ðá burston ða weallas *muri illico corruerunt,* Jos. 6, 20 : Ors. 1, 7; Bos. 29, 38. Gif him áþ burste *if an oath failed them,* L. Ed. 3; Th. i. 160, 20. II. *to make the noise of a bursting* or *breaking, to crash, dash, crack*; fragorem edere, sonare, crepare :—Brim berstende blód-egesan hweóp *the dashing sea threatened bloody horrors,* Cd. 166; Th. 208, 2; Exod. 477. Fingras burston *his fingers cracked,* Beo. Th. 1525; B. 760. [*Laym.* bersten : *Wyc.* berste, breste : *Plat.*

barsten: *O. Sax.* brestan: *O. Frs.* bersta: *Dut. Ger.* bersten: *M. H. Ger.* bresten: *O. H. Ger.* brestan: *Dan.* bröste: *Swed.* brista: *O. Nrs.* bresta.] DER. a-berstan, æt-, for-, ðþ-, to-, ût-.

bersting, e; *f.* A BURSTING, *rent;* ruptura. DER. mûþ-bersting, *q.v.*

berþ *a birth.* v. berþ-estre, berþ-ling, beorþ.

Berþa *Bertha;* Bercta, *Lat. f. the queen of Æðelbryht, king of Kent.* v. Berhte.

berðen, e; *f.* A *burthen, load;* sarcina:—Seám *vel* berðen *sarcina,* Wrt. Voc. 16, 27. v. byrðen.

berþ-estre, an; *f.* A *bearer of children;* genetrix, Leo 110. v. -estre.

berþ-ling, es; *m. Child-birth.* v. hyse-berþling.

berþor *child-birth.* v. beorþor, hyse-beorþor.

bert-hwíl *a moment;* momentum, R. Ben. 5. v. beorht-hwíl.

berwe; *dat. of* beara *a grove, q.v.*

be-rýfan [= be-reófan] *to bereave;* spoliare, privare:—Ðá hí þohton þeóden-stóles rícne berýfan *then they thought to bereave the powerful of his throne,* Exon. 84 a; Th. 317, 9; Mód. 63. DER. reófan *to reave, rob, bereave.*

be-rýpan; *p.* -rýpde, -rýpte, *pl.* -rýpton; *pp.* -rýped, -rýpt *To spoil;* spoliare:—Berýpton, Bt. Met. Fox 2, 23; Met. 2, 12. v. rýpan *to rip, tear.*

be-sacan; *p.* -sôc, *pl.* -sôcon; *pp.* -sacen *To dispute about anything;* in controversiam vocare. DER. un-besacen. v. sacan.

be-sæncan; *p.* -sæncte; *pp.* -sænct *to sink;* mergere, L. Ælf. P. 13; Th. ii. 368, 27. v. sencan.

be-sænct *sunk;* mersus; *pp. of* be-sæncan.

be-sæt, be-sæton *besieged,* Ors. 1, 14; Bos. 37, 15; *p. of* be-sittan.

be-sanc *sank;* submersit, Ors. 3, 11; Bos. 75, 32; *p. of* be-sincan.

be-sárgian; *p.* ode; *pp.* od *To lament, bewail, to mourn* or *be sorry for, to condole;* lamentari, condolere, compati, deflere:—Ic besárgige *compatior,* Ælfc. Gr. 29; Som. 33, 52: Ælfc. T. 42, 1: Scint. 45, 50. v. sárgian.

be-sárgung, e; *f.* A *sorrowing,* Hymn. Surt. 126, 24. v. sárgung.

be-sárigende *condoling.* v. be-sárgian, sárgian.

be-sáwan *to sow;* conserere. v. sáwan.

be-sáwe, *pl.* -sáwen *looked,* Bt. 35, 6; Fox 170, 9; *p. subj. of* be-seón.

be-scær, -scear, *pl.* -scæron, -sceáron *sheared, shaved;* *p. of* be-sceran.

be-sceadan; *p.* ede; *pp.* ed *To shadow;* obumbrare:—For hwám besceadeþ heó mûntas and môras *why shadoweth it mountains and moors?* Salm. Kmbl. 680; Sal. 339. v. sceadian, ofer-.

be-sceáden *separated,* L. E. I. 32; Th. ii. 430, 9; *pp. of* be-sceádan.

be-sceáf *cast,* Andr. Kmbl. 2384; An. 1193; *p. of* be-scúfan.

be-sceát *shot into, precipitated one's self,* Ors. 3, 3; Bos. 56, 5; *p. of* be-sceótan.

be-sceáwian; *p.* ode; *pp.* od *To look round upon, look on, consider, regard, watch;* circumspicere, intueri, considerare, respicere, perscrutari, providere:—Hí besceáwigende *circumspiciens eos,* Mk. Bos. 3, 5. Ic onlócige, oððe ic besceáwige *intueor,* Ælfc. Gr. 27; Som. 29, 60. Be-sceáwiaþ æcyres lílian *considerate lilia agri,* Mt. Bos. 6, 28. Ðû ne besceáwast nánes mannes hád *non respicis personam hominum,* Mt. Bos. 22, 16. Ðæt he Alexandres [wisan] besceáwode *that he might watch Alexander's conduct,* Ors. 4, 5; Bos. 82, 22: R. Ben. 55. DER. sceáwian.

be-sceáwigere, be-sceáwere *a beholder;* spectator, Som.

be-sceáwodnes, -ness, e; *f.* A *seeing, vision, sight;* visio, Ps. Spl. T. 9, 11.

be-scencan *to give to drink.* v. bi-scencan.

be-sceoren *shorn,* Bd. 5, 7; S. 621, 15,=be-scoren; *pp. of* be-sceran.

be-sceótan; he -sceóteþ, -scýt; *p.* -sceát, *pl.* -scuton; *pp.* -scoten *To shoot into, inject, precipitate one's self, to be sent, go;* injicere, se præcipitare, mitti, ire:—Ne bescýt se deófol næfre swá yfel geþóht in to ðám men *nunquam diabolus tam pravas cogitationes in hominem injicit,* Alb. resp. 40. Curtius besceát *Curtius se præcipitavit,* Ors. 3, 3; Bos. 56, 5. Ðæt hí on grûnd ne bescuton *ut in abyssum ne irent,* Lk. Bos. 8, 31.

be-sceran, bi-sceran, -sciran, -scyran; *p.* -scær, -scear, *pl.* -scæron, -sceáron; *pp.* -scoren *To shear off, to shave, cut off;* attondere, amputare, præcidere:—Hý eall heora heáfod besceáron *they all shaved their heads,* Ors. 4, 11; Bos. 96, 37; capitibus rasis, Ors. Hav. 4, 20; p. 270, 5. Ðæt he to preóste bescoren beón mihte *that he might be shorn as a priest,* Bd. 4, 1; S. 564, 24. Iulianus ðeáh to preóste bescoren wære *though Julian had been shorn for a priest,* Homl. Th. i. 448, 29. Ic næs næfre ge-efsod ne næfre bescoren, and gif ic beó bescoren, ðonne beó ic unmihtig ôðrum mannum *gelíc ferrum nunquam ascendit super caput meum, si rasum fuerit caput meum, recedet a me fortitudo mea et deficiam eroque sicut ceteri homines,* Jud. 16, 17. Man ne môt hine besciran *a man must not shear him,* Jud. 13, 5. Gif he hine to preóste bescire [-scyre MSS. B. H.], mid xxx scillinga gebête *if he shave him like a priest, let him make amends with thirty shillings,* L. Alf. pol. 35; Th. i. 84, 7, 9. Biscær, Reim. 26. v. sceran.

be-scerian, -scirian, -scyran, -scyrigan; *p.* ede; *pp.* ed *To deprive, separate, defraud;* privare, separare, fraudare:—Hér, A. D. 821. wærþ Ceolwulf his ríces bescered *here Ceolwulf was deprived of his kingdom,*

Chr. 821; Erl. 63, 10. Ðonne ic bescired beó fram tûnscíre *when I am deprived of my stewardship,* Lk. Bos. 16, 4. Ðone we sceoldan bescyrian ðære onfangenan ealdorlícnysse *quem nos privare auctoritate percepta debemus,* Bd. 1, 27; S. 492, 14. Ne syndon hí to bescyrianne gemænsumnysse Cristes líchoman and blôdes *non corporis ac sanguinis Domini communione privandi sunt,* 1, 27; S. 491, 27. He bescyraþ hine sylfne fram ðære écan méde *he separates himself from the everlasting reward,* Homl. Th. ii. 534, 34. Ná bescyreþ of gôdum hí ða gangendan on unscyldignysse *non privabit bonis eos qui ambulant in innocentia,* Ps. Spl. 83, 13. Mec bescyrede Scyppend eallum *the Creator deprived me of all,* Exon. 111 b; Th. 427, 34; Rä. 41, 101. He wæs eallra his lima þénunge bescyred *he was deprived of the use of all his limbs,* Bd. 5, 5; S. 617, 38. He hæfþ us ðæs leóhtes bescyred *he hath deprived us of the light,* Cd. 17; Th. 25, 12; Gen. 392: 21; Th. 25, 16; Gen. 394. Ðæt ic meahte ongitan Godes ágen bearn, scyldum bescyredne *that I might comprehend God's own child, separated from protections* [shields], Exon. 83 b; Th. 314, 2; Mód. 8. Wuldre bescyrede *from glory separated,* Andr. Kmbl. 3235; An. 1620: Cd. 221; Th. 285, 26; Sat. 343: Exon. 8 a; Th. 3, 7; Cri. 32: 45 b; Th. 155, 29; Gû. 867: Ps. Th. 77, 29. Syndon hí to bescyriganne Cristes líchoman and blôdes *corporis et sanguinis Domini privandi sunt,* Bd. 1, 27; S. 491, 34. Híg ne synt bepæhte oððe bescyrede fram heora gewilnunge *non sunt fraudati a desiderio suo,* Ps. Lamb. 77, 30; thei weren not defraudid of her desier, Wyc. v. bi-scirian.

be-scerwan *to deprive;* privare:—Ne ðínra árna me bescerwe *do not deprive me of thy mercy,* Ps. C. 50, 98; Ps. Grn. ii. 279, 98.

be-sciered *deprived,* Chr. 821; Erl. 62, 11,=be-scired; *pp. of* be-scirian.

be-scínan; *p.* -scán; *pp.* -scinen *To shine upon, illuminate;* collustrare, illuminare:—Mec heáðosigel bescíneþ *the glorious sun shines upon me,* Exon. 126 b; Th. 486, 18; Rä. 72, 17.

be-sciran *to shear, shave,* Jud. 13, 5: L. Alf. pol. 35; Th. i. 84, 7, 9. v. be-sceran.

be-scirian *to deprive,* Lk. Bos. 16, 4. v. be-scerian.

be-scítan; *p.* -scát; *pp.* -sciten *To bedaub;* cacare:—Besciten *caccabatum,* Cot. 189. v. scítan.

be-scofen *thrust off, precipitated,* Mk. Bos. 5, 13; *pp. of* be-scúfan.

be-scoren *shorn, shaved,* Jud. 16, 17; *pp. of* be-sceran.

be-screádian *to cut off;* descindere. DER. screádian.

be-screopan; *p.* -scræp, *pl.* -scræpon; *pp.* -screpen *To scrape,* BESCRAPE, *make level;* radere. v. screopan.

be-scrifen *part. Confessed, that hath undergone confession;* confessus. v. scrífan.

be-scúfan; *p.* -sceáf, *pl.* -scufon; *pp.* -scofen; *v. a. To shove, thrust, cast, hurl* or *throw, to precipitate;* intrudere, immittere, detrudere, præcipitare:—Hêt hine ðá niman, and ðær on bescúfan *then ordered to take him, and to shove him in there,* Ors. 1, 12; Bos. 36, 38. Wá biþ ðæm, ðe sceal sáwle bescúfan in fýres fæðm *woe shall be to him, who shall thrust a soul into the fire's embrace,* Beo. Th. 371; B. 184. Se mihtiga cyning niðer bescúfeþ in súsla grûnd *the mighty king casteth thee down into the abyss of sulphur,* Elen. Kmbl. 1883; El. 943. Ðé se Ælmihtiga heolstor besceáf *the Almighty cast thee into darkness,* Andr. Kmbl. 2384; An. 1193. Seó heord wearþ on sæ bescofen *grex precipitatus est in mare,* Mk. Bos. 5, 13. v. scúfan, sceófan.

be-scuton *went,* Lk. Bos. 8, 31; *p. pl. of* be-sceótan.

be-scyldigian; *p.* ode; *pp.* od *To accuse;* accusare, criminari. v. scyldigian, ge-.

be-scylian; *p.* ede; *pp.* ed *To look upon, to regard;* intueri:—Ðú bescylst mid ôðre eágan on ða heofenlícan þing, mid ôðre ðú lôcast on ðás eorþlícan *thou lookest with one eye on the heavenly things, and with the other thou lookest on these earthly* [things], Bt. 38, 5; Fox 206, 18.

be-scyran *to shave,* L. Alf. pol. 35; Th. i. 84, 7, 9, MSS. B. H. v. be-sceran.

be-scyre *should shave;* attonderet, L. Alf. pol. 35; Th. i. 84, 7, 9; *3rd pers. pres. subj. of* be-scyran.

be-scyred *deprived,* Bd. 5, 5; S. 617, 38; *pp. of* be-scyrian.

be-scyrednes, -ness, e; *f.* An *abdication, a casting off, depriving;* abdicatio, Cot. 14.

be-scyrian *to deprive, separate, defraud,* Bd. 1, 27; S. 492, 14: 1, 27; S. 491, 27: Homl. Th. ii. 534, 34: Ps. Spl. 83, 13: Exon. 111 b; Th. 427, 34; Rä. 41, 101: Bd. 5, 5; S. 617, 38: Cd. 21; Th. 25, 12; Gen. 392: 21; Th. 25, 16; Gen. 394: Exon. 83 b; Th. 314, 2; Mód. 8: Andr. Kmbl. 3235; An. 1620: Cd. 221; Th. 285, 26; Sat. 343: Exon. 8 a; Th. 3, 7; Cri. 32: 45 b; Th. 155, 29; Gû. 867: Ps. Th. 77, 29: Ps. Lamb. 77, 30. v. be-scerian.

be-scyrigan *to deprive,* Bd. 1, 27; S. 491, 34. v. be-scerian.

be-scyrþ *shaves; 3rd pers. pres. of* be-sceran.

be-scyrung, e; *f.* [be *from,* scerung *from* sceran *to tonsure* or *consecrate*] A *deposing, degrading, putting from holy orders;* exauctoratio, desecratio, exordinatio. DER. be-scyrian?

be-scýt *injects,* Alb. resp. 40; *3rd pers. pres. of* be-sceótan.

be-seah *looked about,* Gen. 24, 63 ; *p. of* be-seón.

be-seald *surrounded,* Cd. 2 ; Th. 3, 27 ; Gen. 42 ; *pp. of* be-sellan.

be-secgan ; *p.* -sægde, -sǽde, *pl.* -sægdon, -sǽdon ; *pp.* -sægd, -sǽd [be, secgan *to answer*] *To defend ;* defendere :—Him wæs lýfnesse seald, ðæt he him móste scyldan and besecgan on andweardnesse his gesacena *leave was given him, that he might shield and defend himself in the presence of his accusers,* Bd. 5, 19 ; S. 640, 11. v. be-reccan.

be-sellan ; *p.* -sealde, -salde, *pl.* -sealdon, -saldon ; *pp.* -seald [be *by, about,* sellan *to give*] *To surround, bring on ;* circumdare, obducere :—Sinnihte beseald *surrounded with perpetual night,* Cd. 2 ; Th. 3, 27 ; Gen. 42.

besema, an ; *m. A besom ;* scopæ :—He gemét hyt [hús] geclǽnsod mid besemum *invenit eam [domum] scopis mundatam,* Mt. Foxe 12, 44. v. besma.

be-sencan, bi-sencan ; *p.* -sencte ; *pp.* -senced *To sink, immerge ;* mergere, demergere :—Ic besence *mergo,* Ælfc. Gr. 28, 4 ; Som. 31, 36. Hreóhnys besencte me *tempestas demersit me,* Ps. Spl. 68, 3 : Ps. Th. 68, 2 : Menol. Fox 421 ; Men. 212. De-læs me besencen *ne me demergant,* Ps. Th. 68, 14. Sî besenced on sǽs grúnd *demergatur in profundum maris,* Mt. Bos. 18, 6 : Lk. Bos. 10, 15. Ðæt he gesáwe Satanan besencedne on ðám grúndum helle *that he saw Satan sunk in the depths of hell,* S. 5, 14 ; S. 634, 25. DER. sencan.

be-sengan ; *p.* -sengde ; *pp.* -senged, -sengd *To singe, scorch, burn ;* ustulare, urere, æstuare :—Beren ear beseng *singe a barley ear,* L. M. 1, 51 ; Lchdm. ii. 124, 18. Oðra wéron forberned oððe besenged [MS. besenced] *alia æstuaverunt,* Mt. Kmbl. Lind. 13, 6. Hí besáwon on ða besengdan burh and on ða wéstan *they looked on the burnt and wasted city,* Ors. 2, 8 ; Bos. 51, 42.

be-seón, -sión, bi-seón ; *ic* -seó, ðú -sihst, he -sihþ, -syhþ, *pl.* -seóþ ; *p. ic,* he -seah, ðú -sáwe, *pl.* -sáwon ; *impert.* -sih ; *pp.* -sewen [be *by, near, about ;* seón *to see*]. **I.** *to look about* or *around ;* circumspicere :—Sóna ðá hí besáwon hí, nánne hí mid him ne gesáwon *suddenly when they looked about them, they saw no one with him,* Mk. Bos. 9, 8. Ðá he beseah, ðá geseah he olfendas *when he looked about, then he saw the camels,* Gen. 24, 63. **II.** *to see, look, behold ;* videre, aspicere :—Abraham beseah upp and geseah þrí weras *Abraham looked up and saw three men,* Gen. 18, 2. Eágan his on þearfena beseóþ *oculi ejus in pauperem respiciunt,* Ps. Spl. 10, 5. Besih on me *aspice in me,* Ps. Lamb. 118, 132. **III.** *to go to see, visit ;* visere, visitare :—Beseoh wíngeard ðisne *visita vineam istam,* Ps. Th. 79, 14.

be-serian ; *p.* ode ; *pp.* od *To rob, plunder, deprive, deceive ;* spoliare, fraudare :—He hine feore [MS. fere] beserode *he deprived him of life,* Ps. C. 50, 22 ; Ps. Grn. ii. 277, 22. v. be-syrwan.

be-seten *beset ;* circumdatus, Ps. Th. arg. 19 ; *pp. of* be-sittan.

be-settan ; *p.* -sette, *pl.* -setton ; *pp.* -seted, -sett ; *v. a.* [be *by,* settan *to set*] *To* BESET, *set near, appoint, to place, own, possess ;* circumdare, collocare, ponere :—Seó cwén ða róde héht mid eorcnanstánum besettan [MS. besetton] *the queen commanded them to beset the cross with jewels,* Elen. Kmbl. 2049 ; El. 1026. Ic ðé mægene besette *I beset thee with strength,* Andr. Kmbl. 2866 ; An. 1435. Wæpna smiþ besette swínlícum hine *the armour-smith beset it with figures of swine,* Beo. Th. 2910 ; B. 1453. Se hálga wæs searoþancum beseted *the saint was beset with various thoughts,* Andr. Kmbl. 2511 ; An. 1257 : Exon. 60 a ; Th. 218, 19 ; Ph. 297. Domicianus ða rédan éhtnyssa besette on ðám cristenum *Domitian appointed the cruel persecutions of the Christians,* Ælfc. T. 32, 10. Sǽd þeówna his besetton ða *semen servorum ejus possidebit eam,* Ps. Spl. 68, 42.

be-sih *see, look, behold ;* aspice, Ps. Lamb. 118, 132 ; *impert. of* be-seón.

be-sincan ; *p.* -sanc, *pl.* -suncon ; *pp.* -suncen *To sink ;* submergere, demergere :—Seó burh besanc on eorþan *the city sank into the earth,* Ors. 3, 11 ; Bos. 75, 32. Twá byrig on eorþan besuncon *two cities sunk into the earth,* Ors. 3, 2 ; Bos. 54, 43. Wæs ic swíðe besuncen *I was deeply sunk,* Exon. 103 b ; Th. 392, 5 ; Rä. 11, 3. v. sincan.

be-singan ; *p.* -sang, -song, *pl.* -sungon ; *pp.* -sungen *To utter enchantments, to enchant, charm, bewail ;* excantare incantationibus, deplorare :—Ne sceal nán man mid galdre wyrte besingan *no man shall enchant a herb with magic,* Homl. Th. i. 476, 9. Besing *enchant,* Herb. 93, 2 ; Lchdm. i. 202, 13. Ge sceolon weán wópe besingan *ye shall bewail torment with weeping,* Exon. 41 b ; Th. 139, 3 ; Gú. 587.

besining, e ; *f. A bending ;* sinuatio :—Besining *sinuatio,* Ælfc. Gl. 100 ; Som. 77, 8 ; Wrt. Voc. 55, 11.

be-sión *to look about :*—Ðæt he hine ne besió *that he look not about him,* Bt. 35, 6 ; Fox 170, 17. v. be-seón I.

be-sittan, to be-sittanne ; *p.* -sæt, -sætt, *pl.* -sǽton ; *pp.* -seten [be *by, near,* sittan *to sit*]. **I.** *to sit round, surround, beset, besiege ;* circumdare, cingere, obsidere :—Ða Lácedemonian besǽton ða burh Mǽsiane tyn winter *the Lacedæmonians surrounded the city of Messene for ten years,* Ors. 1, 14 ; Bos. 37, 15. Se cyng lét [hí] besittan ðone castel *the king permitted [them] to beset the castle,* Chr. 1087 ; Erl. 226, 9. He besæt ða sinherge sweorda láfe *circumdedit magno exercitu*

ensium reliquias [superstites], Beo. Th. 5864 ; B. 2936. He fór to Hrofe ceastre, and besætt ðone castel *he went to Rochester, and beset the castle,* Chr. 1087 ; Erl. 226, 5. Hie hine besǽton on ælce healfe on ánum fæstenne *they beset it [the army] on every side in a fastness,* Chr. 894 ; Erl. 92, 23 : 918 ; Erl. 102, 35. He wæs beseten mid his feóndum on ðære byrig *he was beset by his enemies in the city,* Ps. Th. arg. 19 : Chr. 894 ; Erl. 92, 7. Ic eom beseten *obsideor,* Ælfc. Gr. 37 ; Som. 39, 8. Cassander hý hét ðǽr besittan *Cassander commanded to besiege them there,* Ors. 3, 11 ; Bos. 74, 16. Hí þohton [MS. þohtan] hine inne to besittanne *they thought to besiege him therein,* Chr. 1094 ; Erl. 230, 22. Antigones hine bedráf into ánum fæstenne and hine ðǽr besæt *Antigonus drove him into a fastness and besieged him there,* Ors. 3, 11 ; Bos. 73, 18 : Chr. 1106 ; Erl. 241, 8. Gif he ðæs mægenes ne hæbbe ðæt he hine inne besitte *if he have not sufficient power that he may besiege him within,* L. Alf. pol. 42 ; Th. i. 90, 11. **II.** *to be in session, to hold sessions, to be able to sit as master of, be in possession, to possess ;* considere, considere ad aliquid, possidere :—Fira bearn æht·besittaþ *filii hominum ad deliberationem considunt,* Andr. Kmbl. 820 ; An. 410. Ealdormen æht besǽton *princes sat in council,* Andr. Kmbl. 1216 ; An. 608 : 1254 ; An. 627 : Elen. Kmbl. 944 ; El. 473. Wálá wá ! ðæt is sárlíc, ðæt swá leóhtes andwlitan men sceolan ágan and besittan þýstra ealdor *alas ! it is a woful thing, that the prince of darkness should own and possess [have influence over by sitting or being near, hold, be in possession of] men of so bright a countenance ;* heu, proh dolor ! quod tam lucidi vultus homines tenebrarum auctor possidet, Bd. 2, 1 ; S. 501, 16.

be-siwian ; *p.* ede ; *pp.* ed *To sew together, to join ;* jungere :—Besiwed feðergeweorc *opus plumarium,* Cot. 145. v. siwian.

be-slægen *slain, cut off,* Chr. 937 ; Th. 205, 28, col. 2,=be-slagen ; *pp. of* be-sleán.

be-slépan ; *p.* -slép ; *pp.* -slápen [be, slǽpan *to sleep*] *To sleep ;* dormire :—He oft beslép *he often slept,* L. Pen. 16 ; Th. ii. 284, 3.

be-slagen *slain, taken away,* Chr. 937 ; Th. 204, 28, col. 1 ; *pp. of* be-sleán.

be-sleán ; *p.* -slóh, *pl.* -slógon ; *pp.* -slagen, -slægen, -slegen ; *instr. To beat, strike* or *cut off, take away, bereave ;* decollare, cædendo orbare, privare :—Ðǽr wæs heáfde beslagen se strengesta martyr sanct Albanus *decollatus itaque martyr fortissimus sanctus Albanus, there the bravest martyr, St. Alban, was beheaded,* Bd. 1, 7 ; S. 478, 33. He beslóh synsceaðan gewealde *he bereft the impious of power,* Cd. 4 ; Th. 4, 17 ; Gen. 55. Wuduwan freóððum beslægene *widows bereft of friends,* 94 ; Th. 121, 15 ; Gen. 2010.

be-slegen *slain,* Chr. 937 ; Th. 205, 28, col. 1,=be-slagen ; *pp. of* be-sleán.

be-slép *slept,* L. Pen. 16 ; Th. ii. 284, 3 ; *p. of* be-slǽpan.

be-slépan ; *p.* -slépte ; *subj. pl.* -slépen ; *pp.* -sléped, -slépt *To slip, lay, place, put,* and with the preposition *on* on, upon,—*to slip, put* or *lay on, to impose, clothe ;* ponere, imponere, induere :—Hú hefig geoc he beslépte on ealle *how heavy a yoke he laid on all !* Bt. 16, 4 ; Fox 58, 16. Beslépen hí on hý bysmor *induantur confusione !* Ps. Th. 34, 24. Beslépte mid gyldnum fnasum *in fimbriis aureis circumamicta,* 44, 15. v. slépan.

be-slítan ; *p.* -slát, *pl.* -sliton ; *pp.* -sliten *To slit, tear ;* findere, lacerare :—Ðec sculon moldwyrmas monige seonowum beslítan *many mouldworms shall tear thee from thy sinews,* Exon. 99 a ; Th. 371, 9 ; Seel. 73. Hér sculon abídan bán besliten seonwum *here shall abide the bones torn from the sinews,* Exon. 99 a ; Th. 370, 20 ; Seel. 62. v. slítan.

be-slógon *they beslóh bereft,* Cd. 4 ; Th. 4, 17 ; Gen. 55 ; *p. of* be-sleán.

BESMA, besema, an ; *m. A* BESOM, broom, *an instrument of punishment made of twigs, a rod ;* scopæ, virga :—Geclǽnsod mid besmum *scopis mundatam,* Mt. Bos. 12, 44 ; clensid with bismes, Wyc. He hit [hús] gemét mid besmum afeormod *invenit eam [domum] scopis mundatam,* Lk. Bos. 11, 25 ; he fyndith it [hous] clensid with beesmes, Wyc. He [Brutus] hý [his fíf suna] hét gebindan, and mid besman swingan *he [Brutus] gave orders to bind them [his five sons], and scourge them with rods [virgis cecidit,* Hav.], Ors. 2, 3 ; Bos. 42, 3. [*Frs.* besma, *m :* Dut. bézem, *m :* O. Dut. besem, bessem, *m :* Ger. besen, *m :* O. H. Ger. besamo : *Bret.* bezo, *m. a birch.*]

be-smítan ; *p.* -smát, *pl.* -smiton ; *pp.* -smiten [be, smitta *smut*] *To* BESMUT, *defile, dirty, pollute, contaminate ;* polluere, inquinare, coinquinare, contaminare :—Ðæt hine besmítan mǽge *quod possit eum coinquinare,* Mk. Bos. 7, 15 : Cd. 127 ; Th. 162, 14 ; Gen. 2681 : Judth. 10 ; Thw. 22, 12 ; Jud. 59 : Exon. 81 a ; Th. 305, 8 ; Fä. 85. Ic besmíte *polluo,* Ælfc. Gr. 28, 3 ; Som. 30, 49 : Ps. Spl. C. 88, 34. Ðis synt ða þing ðe ðone mann besmítaþ ; ne besmít ðone mann, ðeáh he unþwogenum handum ete *hæc sunt quæ coinquinant hominem ; non lotis autem manibus manducare, non coinquinat hominem,* Mt. Bos. 15, 20 : 15, 18 : Ps. Th. 54, 20. Besmiten mid synne *defiled with sin,* Cd. 74 ; Th. 91, 30 ; Gen. 1520 : Jos. 7, 12 : Bt. Met. Fox 8, 65 ; Met. 8, 33. Ðæt hýg nǽron besmitene *ut non contaminarentur,* Jn. Bos. 18, 28 : Ps. Th. 52, 1 : 106, 16 : Mk. Bos. 7, 2, 5.

be-smitenes, -ness, -nyss, e ; *f. Dirtiness,* SMUTTINESS, *filthiness,*

pollution, abomination, infection; sordes, inquinamentus, pollutio, coinquinatio :—Tilode se Drihtnes wer ða stöwe fram unsyfernyssum geclǽnsian ðara ǽrrena mǎna and besmitenessa *the man of God toiled to cleanse the place from the impurities of former misdeeds and abominations*, Bd. 3, 23; S. 554, 28. Ðæt of wyrtruman besmitenysse acenned biþ *quod ex pollutionis radice generatur*, Bd. 1, 27; S. 494, 38, 41. Bûtan ǽlcere besmitennysse *without any pollution*, Homl. Th. i. 538, 28.

be-smiðian; *p.* ode; *pp.* od; *v. trans.* *To forge, to make* or *work as a smith does*; excudere, fabricare, fabrefacere :—Innan and ûtan îrenbendum searoþoncum besmiðod *within and without, cunningly forged with iron bands*, Beo. Th. 1554; B. 775. DER. smiðian.

be-smyred; *pp. Besmeared*; interlitum, Cot. 108. DER. smyrian.

be-snédan; *p.* de; *pp.* ed *To cut, lop*; amputare :—Engel hêt besnédan *an angel commanded to cut it*, Cd. 200; Th. 248, 16; Dan. 514. Ðæt ðæt treów sceolde, telgum besnǽded, áfeallan *that the tree, lopped of its branches, should fall*, Cd. 202; Th. 250, 34; Dan. 556.

be-sníwod; *pp.* BESNOWED, *covered with snow, snowy*; nive tectus, ninguidus :—Besníwod *ninguidus*, Ælfc. Gl. 93; Som. 75, 94; Wrt. Voc. 52, 44. DER. sníwan.

be-snyðian; *p.* ede; *pp.* ed *To deprive*; privare :—Ongênþeów ealdre besnyðede Hǽþcyn *Ongentheow had deprived Hæthcyn of life*, Beo. Th. 5841; B. 2924; Andr. Kmbl. 2650; An. 1326: Exon. 107 a; Th. 407, 29; Rä. 27, 1.

be-solcen; *pp. Slow, inactive, dull, stupefied*; deses, torpidus :—Ðý læs he weorþe besolcen *lest he becomes stupefied*, Past. 35, 1; Hat. MS. 45 a, 15. v. solcen.

be-sóne; *adv. Soon, immediately*; mox, statim :—Cweðe se preóst besóne *let the priest immediately say*, L. Ælf. C. 36; Th. ii. 358, 24. v. sóna.

beso-reádian; *p.* ode; *pp.* od [*baso red, purple*, reádian *to redden*] *To make a reddish purple*; rubefacere :—Besoreáda ða rinda ealle ûtan *make all the rinds on the outside a reddish purple* [*by soaking in chalybiate water?*], L. M. 1, 47; Lchdm. ii. 116, 3.

be-sorg, -sorh; *adj. Anxious, careful, dear, beloved*; sollicitus, carus :—Ðǽr wǽron ofslægene hyre þægna feówer ðe hyre besorge wǽron *there were slain four of her thanes which were dear to her*, Chr. 917; Erl. 105, 25. Papinianus wæs ealra his deorlinga besorgost *Papinianus was the most beloved of all his favourites*, Bt. 29, 2; Fox 104, 25. Besorh carus, R. Ben. 72.

be-sorgian, bi-sorgian; *p.* ode; *pp.* od *To be sorry for, to care for, be anxious about, fear*; curare :—Gif ðú me lufodest, ðú hit besorgodest *if thou lovedst me, thou wouldst be sorry for it*, Apol. Th. 20, 27. Ne deáþ ne bisorgaþ *he cares not for death*, Exon. 61 a; Th. 223, 32; Ph. 368. Ðú hæfst gesúnd gehealden eall ðæt deórwyrðoste, ðætte ðú ðe besorgod hæfdest *thou hast kept entire everything most precious, which thou wast anxious about*, Bt. 10; Fox 28, 10. Ne bisorgaþ he synne to fremman *he feareth not to perpetrate sin*, Exon. 30 b; Th. 95, 12; Cri. 1556.

be-sorh *anxious, dear, beloved*, R. Ben. 72. v. be-sorg.

be-spanan, bi-spanan; *p.* -spón, -speón; *pp.* -spanen, -sponen; *v. trans. To allure, entice, incite, urge, induce, bring on any one*; allicere, illicere, incitare, provocare, inducere :—He deriende leóda bespeón to ðysan earde *he allured pernicious people to this land*, Chr. 959; Th. 219, 18. Gif he ǽnigne man on synne bespeóne *if he have enticed any man to sin*, L. Pen. 16; Th. ii. 284, 13. Ðæt gewin ðe hió hine on bespón mid manigfealdon firen-lustum *the war which she brought upon him by her manifold wicked desires*, Ors. 1, 2; Bos. 26, 40.

be-sparrad *shut*, Cot. 145. v. sparran.

be-speón, be-spón *allured, enticed*, Chr. 959; Th. 219, 18; *p.* of be-spanan.

be-spirian, -spirigan, -spyrigan; *p.* ode; *pp.* od *To inquire, trace*; inquirere, investigare :—Se ðe þrfe bespirige *of him who traces cattle*, L. Ath. iv. 2; Th. i. 222, 13. Se ðe bespyrige [bespirige, Wilk.] yrfe innan ôðres land, aspirige hit ût, se ðe ðæt land áge, gif he mæge *he who traces cattle into another's land, let him trace it out, who owns that land, if he can*, iv. 2; Th. i. 222, 14.

be-spræc, *pl.* -sprǽcon *spoke to, charged*; *p.* of be-sprecan.

be-sprǽcen *spoken to, charged*, L. Eth. ii. 9; Wilk. 105, 47, = besprecen; *pp.* of be-sprecan.

be-sprǽngan; *p.* de; *pp.* ed *To besprinkle*; aspergere, Herb. 86, 4; Lchdm. i. 190, 11, note. v. be-sprengan.

be-sprecan; *part.* -sprecende, ic -sprece, ðú -sprecest, -sprycst, he -sprecþ, -sprycþ, *pl.* -sprecaþ; *pp.* -sprecen, -sprǽcen [be *by*, sprecan *to speak*] *To speak to, to tell, pretend, plead, speak against, to complain, charge, accuse, impeach*; obloqui :—Fram stefne besprecendre oðða ofersprecendes *a voce obloquentis*, Ps. Lamb. 43, 17. Cristene Rôma besprycþ *Christian Rome complains*, Bos. 44, 45. Hû ge besprecaþ *how ye complain!* Ors. 1, 10; Bos. 34, 9. Hit besprecen biþ *it is charged*, L. Eth. ii. 8; Th. i. 288, 16; Ors. 1, 12; Bos. 36, 39.

be-sprengan; *p.* de; *pp.* ed *To besprinkle*; aspergere :—Bespreng me

mid ysopon, ðæt ic beó geclǽnsod *asperges me hyssopo, et mundabor*, Ps. Th. 50, 8. Besprengc hyne mid ðam wætere *besprinkle him with the water*, Herb. 86, 4; Lchdm. i. 190, 11.

be-sprycþ *tells, complains*, Ors. 2, 4; Bos. 44, 45; 3rd *pers. pres.* of be-sprecan.

be-spyrigan *to inquire, trace*, L. Ath. iv. 2; Th. i. 222, 14. v. bespirian.

best; *adv. sup.* BEST, *most*; optime :—Ðe helpes best behôfaþ *who most wants help*, L. C. S. 69; Th. i. 412, 3; MS. A. [Plat. Dut. Ger. best, beste.] The usual form is wel *well*, bet *better*, betst *best* = *most*. In the text the preceding passage has betst behôfaþ *most wants*.

besta; *m.* seó, ðæt beste *the* BEST; optimus :—Scipio, se þesta Rômâna witena *Scipio, the best of the Roman senators*, Ors. 5, 4; Bos. 104, 38; Cot. MS. Tib. B. I. fol. 85 b. v. betst; *adj.*

be-stæl, *pl.* -stǽlon *stole upon*, Ors. 1, 10; Bos. 33, 33: Chr. 876; Erl. 79, 13; *p.* of be-stelan.

be-stæpp *steps, steps upon, treads*, Jos. 1, 3; *pres.* of be-stapan.

be-standan; *p.* -stód, *pl.* -stódon; *pp.* -standen *To stand by* or *near, to stand around, surround, to stand on* or *upon, occupy, detain*; adstare, circumstare, circumdare, detinere :—Him bestande man *adstet quis ei*, L. Alf. P. 48; Th. ii. 384, 35. Abraham hîg bestód on ða ealdan wîsan *Abraham stood by her after the old custom*, Gen. 23, 2. Fæderas and môddru bestandaþ heora bearna lîc *fathers and mothers stand around the corpses of their children*, Homl. Th. ii. 124, 17. Ða bestódon ða Iudeas hyne ûtan *circumdederunt ergo eum Judæi*, Jn. Bos. 10, 24; Byrht. Th. 133, 51; By. 68. Ahrede me æt ðám ðe me habbaþ ûtan bestanden *redime me a circumdantibus me*, Ps. Th. 31, 8. Ðæs wîf wæs hû hugu xl daga mid grimre ádle bestanden *cujus conjux quadraginta ferme diebus erat acerbissimo languore detenta*, Bd. 5, 4; S. 617, note 6.

be-stapan; he -stæpþ; *p.* -stóp, *pl.* -stópon; *pp.* -stapen *To step, step upon, tread with the foot, go, enter*; gradi, calcare, ire, inire :—Eall ðæt rýmet, ðe eówer fôtswaðu on bestæpþ *omnem locum, quem calcaverit vestigium pedis vestri*, Jos. 1, 3. Se deófol into Iudan bestóp *the devil went* [*entered*] *into Judas*, Homl. Th. ii. 242, 14.

be-stelan, bi-stelan; *p.* -stæl, *pl.* -stǽlon; *subj. p.* -stǽle, *pl.* -stǽlen; *pp.* -stolen *To steal away* or *upon*; fugere, obrepere :—Gif hwá on ôðre scîre hine bestele *if any one steal himself away into another shire*, L. In. 39; Th. i. 126, 10. Bestelan on Theodosius hindan *to steal upon Theodosius behind*, Ors. 6, 36; Bos. 131, 25. Hannibal bestæl on Marcellus *Hannibal stole upon Marcellus*, Ors. 4, 10; Bos. 94, 19: Past. 28, 6; Hat. MS. 38 a, 6. Ðá he nihtes on ungearwe hî on bestæl, and hî swîðe forslôh and fordyde *then he stole upon them unawares by night, and grievously slew and destroyed them*, Ors. 1, 10; Bos. 33, 33. Hî nihtes bestǽlon ðǽre fyrde *they stole upon the army by night*, Chr. 876; Erl. 79, 13. Ðý-læs he on niht onweg fluge and bestǽle *lest he should have fled and stole away by night*, Bd. 4, 22; S. 591, 11.

be-stéman, -stýman; *p.* de; *pp.* ed *To* BESTEAM, *bedew, make damp, make wet*; humectare, madefacere, circumfundere :—Wǽron beorhhlidu blôde bestémed *the mountain-brows were besteamed with blood*, Cd. 166; Th. 206, 8; Exod. 448. Wæs ðæs hálgan lîc swǽte bestémed *the body of the saint was besteamed with blood*, Andr. Kmbl. 2480; An. 1241. Usses Dryhtnes rôd blôde bestémed *our Lord's rood bedewed with blood*, Exon. 23 b; Th. 67, 10; Cri. 1086. Hwîlum hit [beácen] wæs mid wǽtan bestémed *at times it* [*the beacon*] *was damped with wet*, Rood Kmbl. 44; Kr. 22. Ic wæs mid blôde bestémed begoten of ðæs guman sîdan *I was wet with blood poured from the man's side*, 96; Kr. 48. Hû ðú wǽgflotan wǽre bestémdan sünd wisige *how thou directest the sailing of the wave-floater* [*ship*] *wetted with the sea*, Andr. Kmbl. 974; An. 487. Dreóre bestémed *wet with blood*, 2949; An. 1477. DER. stéman.

be-stingan; *p.* -stang, *pl.* -stungon; *pp.* -stungen *To besting, thrust, push*; trudere, immittere, Med. ex Quadr. 5, 1; Lchdm. i. 348, 4.

be-stód, *pl.* -stódon *stood by* or *near, stood around, surrounded*, Gen. 23, 2; Byrht. Th. 133, 51; By. 68; *p.* of be-standan.

be-stolen *stolen*, Exon. 103 b; Th. 393, 7; Rä. 12, 6; *pp.* of be-stelan.

be-stóp *stepped, stepped into, entered*, Homl. Th. ii. 242, 14; *p.* of be-stapan.

be-streddon *heaped up*; aggeraverunt, Bd. 3, 2; S. 524, note 20. v. be-styrian.

be-stréowian; *p.* ode; *pp.* od *To* BESTREW; superspargere :—Hí mid duste heora heáfod bestréowodon *sparserunt pulverem super caput suum*, Iob Grn. 2, 12.

be-streðan, -stryðan; *p.* ede, de; *pp.* ed *To heap up, erect*; aggerare, obducere :—Stánum bestreðed *heaped up with stones*, Exon. 128 b; Th. 493, 28; Rä. 81, 38. Bestryðed fæste *firmly erected*, Exon. 93 b; Th. 351, 29; Sch. 87: Bd. 3, 2; S. 524, note 20.

be-strídan; he -strît, *pl.* -stridaþ; *p.* -strád, *pl.* -stridon; *pp.* -striden *To* BESTRIDE; ascendere :—Bestrídan hors *to bestride a horse*; equum ascendere, Lye. v. be, strídan *to stride*.

be-stroden *bespoiled, confiscated, robbed*, Cot. 108; *pp.* of bestrûdan.

be-strúdan; _p._ -streád, _pl._ -strudon; _pp._ -stroden To bespoil, spoil, confiscate, rob; spoliare, privare, confiscare:—Ða ðe Sodoma and Gomorra golde berófan bestrudon stigwitum _qui Sodoma et Gomorra auro spoliarunt, incolis privarunt_, Cd. 95; Th. 125, 14; Gen. 2079. Bestroden _confiscatus_, Cot. 108.

be-strýpan; _p._ -strýpte; _pp._ -strýped To strip, rob, spoil, bereave; exuere, spoliare:—Bestrýpan widuwan _viduas spoliare_ vel _exuere_, Off. Episc. 8. Ealle ða bestrýpte he æt lande _he bereaved all those of land_, Chr. 1065; Erl. 196, 11.

be-strýðan; _p._ ede, de; _pp._ ed To heap up, erect; aggerare, obducere:—Bestrýðed fæste _firmly erected_, Exon. 93 b; Th. 351, 29; Sch. 87: Bd. 3, 2; S. 524, note 20. v. be-streðan.

be-stungen _pushed:_—On næspyrl bestungen _pushed into the nostril_, Med. ex Quadr. 5, 1; Lchdm. i. 348, 4; _pp._ of be-stingan.

be-stýman; _p._ de; _pp._ ed To besteam, bedew, make damp, make wet; humectare, madefacere, circumfundere:—Drihtsele blôde bestýmed _the princely hall besteamed with blood_, Beo. Th. 977; B. 486. v. be-stêman.

be-styrian; _p._ ede; _pp._ ed [be, styrian _to move_] To heap up, pile up; aggerare:—His þegnas mid moldan hit bestyredon and gefæstnedon _his thanes heaped up with mould and fastened it_, Bd. 3, 2; S. 524, 20.

be-styrman; _p._ de; _pp._ ed To bestorm, storm, agitate; flatibus agere, agitare:—Ðonne hit bestyrmaþ ðisse worulde ungeþwærnessa _quando ipsam agitant hujus mundi inquietudines_, Bt. 3, 2; Fox 6, 8. DER. styrman.

be-suncen _sunk_, Exon. 103 b; Th. 392, 5; Rä. 11, 3; _pp._ of be-sincan.

be-swâc _deceived, enticed, seduced_, Andr. Kmbl. 1226; An. 613; _p._ of be-swícan.

be-swǽlan; _p._ de; _pp._ ed To burn, sweal, scorch, singe; adurere, ustulare:—Næs hyra feax fýre beswæled _nor was a hair of them burned by the fire_, Cd. 195; Th. 243, 18; Dan. 438. Glêdum beswæled _scorched by gleeds_, Beo. Th. 6075; B. 3041. DER. swelan.

be-swâpan; _p._ -sweóp, _pl._ -sweópon; _pp._ -swâpen [be, swâpan _to sweep_] To clear up, persuade, cover over, clothe, protect; suadere, cooperire, amicire, munire:—Gif hwylc Rædwolde on môd beswâpe _si qui Redualdo suadeat_, Bd. 2, 12; S. 514, 3. Hî hî mid scýtan besweóp _she covered herself over with a sheet_, 3, 9; S. 534, 13. Beswâpen [beswapyn MS.] leóhte swâ swâ of rægle _amictus lumine sicut vestimento_, Ps. Spl. C. 103, 2. Ðæt he bió wið ælce orsorgnesse beswâpen _that he shall be protected against every pleasure_, Past. 14, 3; Hat. MS. 17 b, 21.

be-swemman; _p._ -swemde; _pp._ -swemmed, -swemd To make to swim; natare facere:—Ðeáh hî beswemde weorþon _though they be made to swim_, Bt. 37, 4; Fox 192, 28.

be-sweóp _covered over, clothed_, Bd. 3, 9; S. 534, 13; _p._ of be-swâpan.

be-sweðian, bi-sweðian; _p._ ede; _pp._ ed To bind up, swathe; ligare:—Mid âcumban besweðe _bind up with tow_, L. M. 1, 1; Lchdm. ii. 22, 21. v. sweðian, ge-sweðian.

be-swíc, big-swíc, bî-swíc, es; _m._ [be, big, bî _intensive_; swíc _deceit_, swícan _to deceive_] Deceit, a deceiving, treachery, snare; fraus, deceptio, dolus = δόλος, decipula:—Bútan bræde and beswíce [bigswíce, bîswíce, Th. i. 160, 7, note 6] _absque figmento et fraude_, L. Ed. 1; Wilk. 48, 38. Bîswícum _deceptionibus_, Mone B. 1174. Philippus ealle ða cyninges mid bîswíce ofslôh _Philip slew all the kings by treachery_, Ors. 3, 7; Bos. 60, 13. To bîswíce his nýhstan _in dolo proximo suo_, Ps. Th. 23, 4. Beswíc _decipula_, Cot. 61. Ða woruldwélan synt gesceapene to bîswíce monnum _worldly riches are created for a snare to men_, Bt. 14, 1; Fox 42, 3.

be-swícan, bi-swícan; ic -swíce, ðú -swícest, -swícst, he -swíceþ, -swícþ, _pl._ -swícaþ; _p._ -swâc, _pl._ -swicon; _pp._ -swicen; _v. a._ [be by, swícan _to deceive_] To deceive, entice, seduce, delude, betray, offend, supplant, weaken, evade; decipere, illicere, seducere, illudere, prodere, scandalizare, supplantare, deficere, evadere:—He ongan sirwan hû he hine beswícan mihte _he began to plot how he might deceive him_, Ors. 1, 12; Bos. 35, 19: Cd. 23; Th. 29, 17; Gen. 451. Hý beswícaþ weardas _the guardians deceive them_, Exon. 116 a; Th. 446, 2; Dôm. 16: Ps. Th. 61, 9. Ne beswíc ðú ðinne nêxtan _deceive not thy neighbour_, Lev. 19, 11. Me nædre beswâc _the serpent deceived me_, Cd. 42: Th. 55, 20; Gen. 897: Exon. 61 b; Th. 226, 30; Ph. 413: Andr. Kmbl. 1226; An. 613. We beswicon [MS. beswican] us sylfe _we have deceived ourselves_, Exon. 121 a; Th. 464, 31; Hö. 96. Hycgaþ hû ge hî beswícen _think how ye may deceive them_, Cd. 22; Th. 28, 9; Gen. 433. Mid gedwolan beswicen _errore deceptus_, Deut. 30, 17: Exon. 97 a; Th. 363, 20; Wal. 56. Ge sind beswícene _ye are deceived_, Andr. Kmbl. 1489; An. 746. Ic beswíce _illicio_, Ælfc. Gr. 28, 5; Som. 31, 67. Gif hwá fǽmnan beswíce unbeweddode _if any one entice an unbetrothed woman_, L. Alf. 29; Th. i. 52, 5. Feóndas sôþfæstra sâwle willaþ beswícan _fiends will seduce the souls of the righteous_, Exon. 41 a; Th. 136, 12; Gû. 540. Wæs he beswicen fram his wífe _ab uxore sua seductus est_, Bd. 2, 15; S. 518, 29: Cd. 26; Th. 33, 32; Gen. 529. Hiora ealdormen wæron beswicene _principes eorum seducti sunt_, Ps. Th. 106, 39. Forðamðe ðú me

beswice _quia illusisti mihi_, Num. 22, 29. Sindon ge beswicene _ye are deluded_, Exon. 41 b; Th. 139, 22; Gû. 597. Us Godríc hæfþ beswicene _Godric has betrayed us_, Byrht. Th. 138, 51; By. 238. Ðæt eów beswícþ _hoc vos scandalizat?_ Jn. Bos. 6, 61: Mt. Bos. 18, 6. Ða ðe þohton beswícan færelde míne _qui cogitaverunt supplantare gressus meos_, Ps. Spl. 139, 5. Úton acræftan hû we heora mágon [MS. magan] beswícan _let us plan how we can weaken them_, Ors. 2, 5; Bos. 47, 20. Ne mæg hit wildeór beswícan _a wild beast cannot evade it_, Salm. Kmbl. 572; Sal. 285. Ðú hafast ðínra feónda handa beswicene _hostium manus evasisti_, Bd. 2, 12; S. 515, 23.

be-swícende, an; _f._ A deceiver, harlot; pellex, Cot. 170.

be-swícian; _p._ ode, ede, ade; _pp._ od, ed, ad [be, swícian _to wander_] To go from, evade, escape, be without, be free from; evadere, carere:—Ða ðe ðone deáþ beswícian myhton [myhtan MS.] _qui mortem evadere poterant_, Bd. 1, 12; S. 481, 1. Ðæt he ðoné deáþ beswícode _ut ipse mortem evaderet æternam_, Bd. 3, 23; S. 555, 36: 2, 12; S. 512, 36. Ðæt heó ðære langan untrumnesse beswícede _se infirmitate longa carere_, Bd. 3, 5, 4; S. 617, 24. Torhtgyþ ðære tungan onstyrenesse beswícade _Torctgyd linguæ motu caruit_, 4, 9; S. 577, 17.

be-swincan; _p._ -swanc, _pl._ -swuncon; _pp._ -swuncen To toil, labour, make with toil; laborare:—Ic sende eów to rípanne, ðæt ðæt ge ne beswuncon; óðre swuncon, and ge eódon on hyra geswinc _ego misi vos metere quod vos non laborastis; alii laboraverunt, et vos in labores eorum introistis_, Jn. Bos. 4, 38. Ðæt hrægl is beswuncen _laboratur vestis_, Ælfc. Gr. 19; Som. 22, 48.

be-swingan; _p._ -swang, _pl._ -swungon; _pp._ -swungen To scourge, beat; flagellare, verberare:—Ic wæs beswungen ealne dæg _fui flagellatus tota die_, Ps. Lamb. 72, 14: Bt. Met. Fox 25, 91; Met. 25, 46. Híg ne beóþ beswungene _non flagellabuntur_, Ps. Lamb. 72, 5: Ex. 5, 16. Ic eom beswungen _verberor_, Ælfc. Gr. 5; Som. 3, 32. Ic eom beswungen _I am beaten_; vapulo, 19; Som. 23, 3.

be-swuncen _made with toil_, Ælfc. Gr. 19; Som. 22, 48; _pp._ of beswincan.

be-swungen _beaten_, Bt. 37, 1; Fox 186, 20; _pp._ of be-swingan.

be-swylian; _p._ ede; _pp._ ed To soil, stain; polluere, inficere:—Hit wæs beswyled mid swâtes gange _it was soiled with running of blood_, Rood Kmbl. 45; Kr. 23.

be-sylfred; _pp._ [seolfer _silver_] Silvered, BESILVERED; deargentatus, Ps. 67, 14, Lye. v. ofer-sylfrian.

be-sylian; _p._ ede; _pp._ ed To soil, stain; maculare, inquinare:—Besyled _stained_; maculatus, Bt. 16, 4; Fox 58, 18: Elen. Kmbl. 1390; El. 697. v. selian.

be-syrewian; _p._ ede; _pp._ ed To ensnare, deceive; circumvenire, decipere, machinare:—Hî woldon hine besyrewian _they would deceive him_, Chr. 1002; Erl. 137, 34. v. be-syrwan.

be-syrian; _p._ ode, ede; _pp._ od, ed To rob, plunder, deprive, deceive; spoliare, fraudare, dejicere:—Ðæt hî mægon besyrian ðone earman _ut dejiciant inopem_, Ps. Th. 36, 13. Cirus hý besyrode _Cyrus ensnared them_, Ors. 2, 4; Bos. 45, 20. Hine Rodbeard besyrede _Robert deceived him_, Chr. 1093; Erl. 229, 5. Ða Scottas heora cyng Dunecan besyredon _the Scots ensnared their king Duncan_, 1094; Erl. 230, 40. v. be-syrwan.

be-syrwan, -syrewian, -syrian, -serian; _p._ -syrwde; _pp._ -syrwed To ensnare, deceive; circumvenire, decipere, machinari:—Ðæt hig woldon ðone Hǽlend mid fâcne besyrwan _ut Iesum dolo tenerent_, Mt. Bos. 26, 4. Mynte se mânscaða manna cynnes sumne besyrwan _the wicked spoiler expected to ensnare one of the race of men_, Beo. Th. 1430; B. 713: 1888; B. 942: Cd. 127; Th. 162, 13; Gen. 2680. v. syrwan.

BET, bett; _adv._ [? _from_ bet _well_; _comp._ betor _better?_ contracted _to_ bet; _sup._ betost _contracted to_ betst, _q. v._] BETTER; melius:—Ðá acsode he, to hwylcum tíman him wǽre _interrogabat ergo horam ab eis in qua melius habuerit_, Jn. Bos. 4, 52. Ðæt se hwǽte mǽge ðý bet weaxan _that the wheat may grow the better_, Bt. 23; Fox 78, 24. Hwoune his horse betst wurde _till his horse should be better_, Bd. 3, 9; S. 533, 34. [_Chauc. Piers:_ bet: _Scot._ bet: _O. Sax._ bet: _Frs._ bet: _M. Dut. N. Dut._ bet: _Ger._ basz: _M. H. Ger._ baz: _O. H. Ger._ baz: _O. Nrs._ betr.] DER. abet. v. wel _well_.

be-tǽcan; _p._ -tǽhte, _pl._ -tǽhton; _pp._ -tǽht; _v. a._ [be _by_, tǽcan _to teach, shew_]. **I.** _to shew_; ostendere:—He eów betǽcþ mycele heallan _ipse ostendet vobis cœnaculum magnum_, Lk. Bos. 22, 12. **II.** _to_ BETAKE, _impart, deliver, commit, put in trust_; impertire, assignare, tradere, commendare:—Ic betǽce híg ðam yrþlincge _adsigno eos aratori_, Coll. Monast. Th. 20, 31. Sum man clypode hys þeówas, and betǽhte hym hys ǽhta _homo vocavit servos suos, et tradidit illis bona sua_, Mt. Bos. 25, 14; a man clepide his seruauntis, and bitoke to hem his goodis, Wyc. Gen. 9, 2: Ps. Th. 104, 17: Ors. 2, 5; Bos. 48, 6. Swâ us betǽhton, ða ðe hit of frymþe gesâwon _sicut tradiderunt nobis, qui ab initio ipsi viderunt_, Lk. Bos. 1, 2: Elen. Kmbl. 1167; El. 585. Man hý ðære abedissan betǽhton _they committed her to the abbess_, Chr. 1052; Erl. 181, 28. Ðæt we môton ðé betǽcan sáwle úre _that we may commit our souls to thee_, Hy. 7, 82; Hy. Grn. ii. 289, 82: Runic pm. 20;

Kmbl. 343, 18; Hick. Thes. i. 135.   **III.** *to send, follow, pursue;* mittere, insequi, amandare:—Betǽcan [MS. betæcan] cildru on scôle *to send children to school;* mittere pueros in scholam, Obs. Lun. § 4; Lchdm. iii. 184, 28. Mid swiftum hûndum ic betǽce wildeór *with swift hounds I pursue wild beasts;* cum velocibus canibus insequor feras, Coll. Monast. Th. 21, 27. Ic betǽce fram me *amando,* Ælfc. Gr. 47; Som. 48, 35.

be-tǽcung, e; *f. A betaking;* traditio. v. be, tǽcung, be-tǽcan.

be-tǽhte, *pl.* -tǽhton *delivered, committed,* Gen. 9, 2: Chr. 1052; Erl. 181, 28; *p.* of be-tǽcan II.

be-tǽht *betrothed,* Mt. Lind. Stv. 1, 18,=be-tǽht; *pp.* of be-tǽcan II.

be-tǽhten, Chr. 654; Erl. 29, 11,=betǽhton *committed; p.* of betǽcan II.

bêtan, ic bête; *p.* bêtte; *pp.* bêted; *v. trans.* [Goth. ô=*A. Sax.* ô, ê, thus Goth. bôtyan=bôtan=*A. Sax.* bêtan].   **I.** *to make better, to improve, amend, repair, restore;* emendare, reparare, reficere, mederi, expiare:—Đæt he bêtte *that he should improve,* Bd. 5, 13; S. 632, 11: Ex. 21, 22. Hû đû meaht đîne æceras bêtan *how thou mayest improve thy fields,* Lchdm. i. 398, 1.   **II.** joined with fŷr *to mend* or *repair a fire, to light* or *make a fire, to kindle;* focum reparare. [In this sense böten is used in Low German at the present day:—Böt füer *make the fire.* So in Frs. fiûr boetsje *struere focum.*] Đâ hêt he bêtan micel fŷr *then he ordered a great fire to be lighted,* Ors. 6, 32; Bos. 129, 10.   **III.** *to remedy, compensate, make amends;* compensare:—Ic hit bête *I will remedy it,* Deut. 1, 17. [*Chauc.* bet: *Piers* bete: *R. Glouc.* bete: *Laym.* beten, beten: *Orm.* betenn: *O. Sax.* bôtean: *Plat.* betern *to repair:* böten *to mend the fire: Dut.* baten *to profit;* beteren *to amend: O. Frs.* beta, beteria *to repair: Ger.* bessern *to repair: Goth.* bôtyan: *Dan.* böde: *Swed.* böta: *Icel.* bæta; bet *better.*] DER. gebêtan, gebêtung, unbêted.

betast *best;* optimus:—Betast herefêđan blîcaþ *best martial bands shine,* Exon. 22 b; Th. 62, 36; Cri. 1012. v. betst; adj.

BÊTE, an; *f.* bête, an; *n? BEET, a root from which sugar is often extracted;* bêta=σεῦτλον, n:—Sindon eáþ begeátra bête and mealwe *beet and mallow are more easily procured,* L. M. 2, 30; Lchdm. ii. 226, 25: iii. 12, 26. Wyrc drænc of đære bêtan [MS. bêten] *work a drink of the beet,* Lchdm. iii. 22, 6. Beđe mid bêtan leáfum *foment with leaves of beet,* L. M. 1, 39; Lchdm. ii. 100, 12: iii. 2, 8: 44, 8: 114, 13. Nim đa bêtan, đe gehwǽr weaxaþ *take the beet, which groweth anywhere,* L. M. 2, 33; Lchdm. ii. 238, 3: iii. 22, 12. Nim bête [*acc. n.*] đe biþ ânsteallet *take beet, which is one-stalked,* iii. 70, 2. [*Dut.* beet, biet, *f: Ger.* beete, *f: O. H. Ger.* bieza, *f: Fr.* bette, *f: Ital.* bieta, *f: Lat.* beta, *f.*]

be-teáh *accused;* accusavit, Chr. 1096; Th. 362, 32; *p.* of be-teón.

be-teldan, bi-teldan; *p.* -teald, *pl.* -tuldon; *pp.* -tolden [be, teldan *to cover,* teld *a tent*]. *To cover, cover over, surround, overwhelm;* tegere, supertegere, circumdare, obruere:—He đæt wælreáf wyrtum biteldeþ he *covers the dead spoil with herbs,* Exon. 59 b; Th. 217, 1; Ph. 273. Lâme bitolden *covered with clay* [*buried*], 64 a; Th. 235, 11; Ph. 555: 64 b; Th. 238, 25; Ph. 609. Hæfde sigora weard betolden leófne leódfruman mid lofe sînum *the lord of triumphs had surrounded the dear chieftain with his praise,* Andr. Kmbl. 1976; An. 990. Fuglas hringe beteldaþ Fênix *the birds surround the Phœnix in a ring,* Exon. 60 b; Th. 221, 24; Ph. 339. Wæs wôpes hring torne bitolden *the weeping circle was overwhelmed with grief,* 15 b; Th. 34, 6; Cri. 538.

be-tellan; *p.* -tealde, -telede, *pl.* -tealdon, -teledon; *pp.* -teald, -teled, -tæled; *v. a.* [be, tellan *to tell*] *To speak about, to answer, excuse, justify, clear;* excusare:—Đæt he môste hine betellan *that he might answer him,* Chr. 1048; Erl. 180, 12. Godwine betealde hine *Godwin cleared himself,* 1052; Ing. 238, 20.

bêtende *part.* [from bêtan *to make better, atone*] *Amending, atoning;* reparans, expians:—Bêtende [MS. betend] crungon hergas to hrusan *the atoning bands sank to earth,* Exon. 124 a; Th. 477, 24; Ruin. 29.

be-teón; *p.* -teáh, *pl.* -tugon; *pp.* -togen.   **I.** *to draw over* or *round, cover, surround, inclose, protect;* obducere, superinducere, circumducere, concludere, munire:—Heora scyldas wǽron betogene mid hŷdum *their shields were covered with hides,* Ors. 5, 7; Bos. 107, 8. Betogen [betogan MS.] cræt *a covered carriage;* capsus, Ælfc. Gl. 49; Wrt. Voc. 34, 23. Hîg betugon mycele menigeo fixa *concluserunt copiosam multitudinem piscium,* Lk. Bos. 5, 6. Se reáda æppel biþ betogen mid ânfealdre rinde, and monig corn on-innan him hæfþ *in malo punico uno exterius cortice multa interius grana muniuntur,* Past. 15, 5; Hat. MS. 19 b, 22. v. teón I.   **II.** *to leave by law, bequeath;* legare, Th. Diplm. A. D. 1037; 567, 9.   **III.** *to bring a charge against any one, accuse;* criminari, accusare:—Beteáh Gosfrei Bainard Willelm of Ou *Geoffrey Bainard accused William of Eu,* Chr. 1096; Th. 362, 32. Se đe biþ betogen *he who is accused,* L. In. 54; Th. i. 136, 10: 71; Th. i. 148, 2. v. teón II.

betera, betra; *m:* betere, betre; *f. n. adj.* [*from* bet *good,* v. bet-lîc *good-like,* comp. betera, betra *better; sup.* betest, betst *best,* v. besta, gôd] BETTER; melior:—Đæt hŷ wǽron beteran þegnas *that they were better*

thanes, Ors. 4, 9; Bos. 92, 23. Đa betran tîda *the better times,* 4, 9; Bos. 92, 18. To beteran tîde *to a better time,* Bd. 3, 14; S. 539, 39. Wîtodlîce micle mâ mann ys sceápe betera? Mt. Bos. 12, 12; *hou moche more is a man betre than a sheep?* Wyc. Hit is betre *it is a better* [*thing*], Bt. 38, 7; Fox 210, 5: 29, 1; Fox 102, 6.

beterian; *p.* ode; *pp.* od [betera *better*] *To make better, ameliorate;* meliorari, emendare. v. ge-beterian.

betesta *best:*—Se betesta *the best,* Cot. 153. v. betst; adj.

beþ *is,* Chr. 675; Erl. 38, 8,=biþ; *3rd pers. pres.* of beón.

be-þæht *covered,* Bd. 3, 10; S. 534, 32,=be-þeaht; *pp.* of be-þeccan.

be-þærfeþ *expedit,* Mt. Rush. Stv. 19, 10,=be-þearfeþ. v. be-þearfan.

be-þeaht, -þeht *covered,* Exon. 117 a; Th. 451, 4; Dôm. 98: Elen. Kmbl. 2593; El. 1298; *pp.* of be-þeccan.

be-þearf, ic, he, đû be-þearft *I have, thou hast, he has need,* Elen. Kmbl. 1082; El. 543: Ps. Spl. 15, 1; *pres.* of be-þurfan.

be-þearfaþ *he needs, wants;* opus habet:—Hwæt helpeþ *vel* beþearfeþ [MS. beþearfaþ] menn *what does it help to a man* or *what needs a man* [*of what use is it to a man*]? quid prodest homini? Mt. Rush. Stv. 16, 26. v. þearfan, þurfan, be-þurfan.

be-þeccan, bi-þeccan; *p.* -þeahte, -þehte, *pl.* -þeahton, -þehton; *pp.* -þeaht, -þeht *To cover, cover over, conceal;* tegere, contegere, operire:—Đa rôde earme beþeahte *he covered the cross with his arm,* Elen. Kmbl. 2470; El. 1236: Cd. 185; Th. 230, 26; Dan. 239. Æghwæđer ôđerne earme beþehte *each covered the other with his arm,* [*each embraced the other*], Andr. Kmbl. 2030; An. 1017. Mec mon biþeahte mid wǽdum *one covered me with weeds,* Exon. 28 b; Th. 87, 10; Cri. 1423: 51 b; Th. 179, 1; Gû. 1255. Hie heora lîchoman leáfum beþeahton *they covered their bodies with leaves,* Cd. 40; Th. 52, 18; Gen. 845: Elen. Kmbl. 1669; El. 836. Se wæs beþeaht mid þæce *quod erat fœno tectum,* Bd. 3, 10; S. 534, note 32: Exon. 117 a; Th. 451, 4; Dôm. 98. Biþeaht *covered,* Exon. 47 b; Th. 163, 36; Gû. 1004. Heó helltregum wunodon þystrum beþeahte *they dwelt in hell-torments covered with darkness,* Cd. 4; Th. 5, 23; Gen. 76. Synfulle beóþ þrosme beþehte *the sinful shall be covered with foulness,* Elen. Kmbl. 2593; El. 1298. Me beþeahton [Spl. C. beþehton] þeóstru *contexerunt me tenebræ,* Ps. Th. 54, 5.

beđen, e; *f? A fomentation, embrocation;* fomentum:—Mid beđenum *with fomentations,* Bd. 4, 32; S. 611, 20. v. beđing.

be-þencan, bi-þencan; *p.* -þohte, *pl.* -þohton; *pp.* -þoht *To consider, bear in mind, BETHINK, remember, trust, confide, entrust;* considerare, recordari, in se reverti, meminisse, fidere, confidere:—Scyle gumena gehwylc georne biþencan, đæt us bicwom meahta Waldend *each man should well consider, that the Lord of might came to us,* Exon. 19 b; Th. 51, 27; Cri. 822. Đæt we gǽstes wlite biþencen *that we bear in mind the spirit's beauty,* Exon. 20 a; Th. 53, 14; Cri. 850: 51 b; Th. 179, 32; Gû. 1270. Đâ beþohte he hine he bethought himself; in se autem reversus, Lk. Bos. 15, 17: Ælfc. T. 35, 21. Hîg beþohton đæt hîg hym seofon weras gecuron *they bethought that they would choose them seven men,* Nicod. 20; Thw. 10, 4. He beþôhte swîđost to Arpelles *he trusted most in Harpalus,* Ors. 1, 12; Bos. 35, 34. Gif đû to sǽmran gode biþencest *if thou confidest in a worse god,* Exon. 66 b; Th. 245, 30; Jul. 52. Beþohton [MS. beþohtan] hŷ ealle heora wîgcræftas to Exantippuse *they entrusted all their military forces to Xantippus,* Ors. 4, 6; Bos. 85, 16. Cassander hæfde hys wisan beþoht to Seleucuse *Cassander had entrusted his affairs to Seleucus,* 3, 11; Bos. 74, 45.

be-þenede *served: substituted by Thorpe,* Beo. Th. 4077, *for* be-wenede. v. be-wenian.

be-þennan; *p.* ede; *pp.* ed *To cover;* obducere:—He mec beþenede *he covered me,* Exon. 107 a; Th. 408, 15; Rä. 27, 12.

beđian, beđigean *to bathe, foment;* fovere:—Beđa đa eágan *foment the eyes,* Herb. 1, 3; Lchdm. i. 72, 3: Med. ex Quadr. 4, 18; Lchdm. i. 346, 20. v. badian.

beđigean *to wash, foment;* fovere:—Đa eágan to beđigeanne *to foment the eyes,* Med. ex Quadr. 4, 18; Lchdm. i. 346, 20; MS. H. Ic beđige *foveo,* Ælfc. Gr. 26, 5; Som. 28, 66. v. badian.

beđing, e; *f. A fomentation, an assuaging* or *nourishing medicine;* fomentum, Bd. 4, 32; S. 611, 20; MS. B. v. beđen.

be-þoht, -þohte, *bethought, trusted, entrusted,* Ælfc. T. 35, 21: Ors. 3, 11; Bos. 74, 45; *p. and pp.* of be-þencan.

be-þorfte, -þorfton *did need,* Bt. 33, 4; Fox 128, 14; *p.* of be-þurfan.

be-þridian, -þrydian; *p.* ede; *pp.* ed [þrydian *from* þryþ *power, force*] *To force, overpower;* cogere, vi superare:—Đæt hine man wolde beþridian mid đam ilcan wrence *that they would overpower him by the same stratagem,* Ors.'6, 36; Bos. 132, 4. Đæt hŷ ân cyning swâ ŷđelîce on his geweald beþrydian sceolde *that one king should so easily force them under his power,* Ors. 3, 7; Bos. 59, 42: 2, 5; Bos. 47, 11.

be-þringan, bi-þringan; *p.* -þrang, *pl.* -þrungon; *pp.* -þrungen *To throng* or *press around, encompass, surround;* undique urgere, circumvenire, circumdare:—Ic wæs bîsgum beþrungen *I was encompassed with misery,* Elen. Kmbl. 2488; El. 1245: 1896; El. 950. Se sceal wesan wyrmum beþrungen *he shall be surrounded with worms,* Exon. 54 a;

Th. 316, 30; Môd. 56. Fēnix biþ on middum þreátum biþrungen *the phœnix is in the midst surrounded by multitudes*, 60 b; Th. 221, 27; Ph. 341.

be-þrungen *encompassed*, Elen. Kmbl. 2488; El. 1245; *pp. of* beþringan.

be-þryccan *to press on, impress.* v. bi-þryccan.

be-þuncan *To consider, look out;* consulere, prospicere, Exon. 113 a; Th. 432, 29; Rä. 49, 7.

be-þurfan, bi-þurfan, ic, he -þearf, ðú -þearft, *pl.* -þurfon; *p.* -þorfte, *pl.* -þorfton; *subj.* -þurfe, *pl.* -þurfen; *p.* -þorfte, *pl.* -þorften; *gen.* or *acc.* or *v. n. To need, have need, want, to be in want, to require;* opus habere, egere, indigere:—Wísdômes beþearf *I need wisdom,* Elen. Kmbl. 1082; El. 543. Ic árna biþearf *I need mercy,* Exon. 76 a; Th. 285, 17; Jul. 715: Ælfc. Gr. 26, 2; Som. 28, 48. Gôda mínra ðú ne beþearft *bonorum meorum non eges,* Ps. Spl. 15, 1. Ge beþurfon *indigetis,* Mt. Bos. 6, 32. We bicgaþ ða þing ðe we beþurfon *ememus necessaria,* Gen. 43, 4, 8. Máre ðonne he beþurfe *more than he has need of,* Bt. 14, 2; Fox 44, 21.

be-þweán, ic -þweá; *p.* -þwôh, *pl.* -þwôgon; *pp.* -þwegen *To wet, bedew, wash;* rigare:—Mid mínum teárum strecednysse míne oððe míne beddinge ic beþweá oððe ic gelecce *lacrimis meis stratum meum rigabo,* Ps. Lamb. 6, 7.

be-þwyr; *adj.* [be, þwir *wicked*] *Perverse, depraved;* depravatus, Cot. 63.

be-þýan; *p.* -þýde, -þýdde, *pl.* -þýddon; *pp.* -þýed, -þýd *To thrust;* trudere:—Hí beþýddon *they thrust,* Ors. 4, 1; Bos. 78, 8.

be-þýddon *thrust,* Ors. 4, 1; Bos. 78, 8; *p. pl. of* be-þýan.

be-tiénan *to shut, shut up;* concludere, Ps. Spl. T. 34, 3: Cot. 58. v. be-týnan.

betígean *to be* or *make better;* meliorare:—Sôna hý betigeaþ [MS. batigeaþ] *they will be better soon,* Lchdm. iii. 54, 33. v. beterian, betrian.

be-tíhan; *p.* -táh, *pl.* -tigon; *pp.* -tigen, -tygen [be, tíhan, II. *to bring a charge against any one*] *To accuse, impeach;* criminari, accusare:—Gif he oft betygen wǽre *if he has often been accused,* L. In. 18; Th. i. 114, 6: 37; Th. i. 124, 21: 52; Th. i. 134, 12. v. be-teón, III.

be-tíhtlian, -týhtlian; *p.* ode, ede, ade; *pp.* od, ed, ad *To accuse, charge;* accusare, criminari:—Gif he betíhtlod weorþe *if he be accused,* L. C. S. 31; Th. i. 396, 1. Ðe oft betíhtlede wǽron *who have often been accused,* L. Ath. i. 7; Th. i. 202, 25. Ælc mynetere ðe betíhtlad sî *every moneyer who is accused,* L. Eth. iii. 8; Th. i. 296, 15. Gif he betýhtlad wurðe *if he should be accused,* L. Eth. i. 1; Th. i. 280, 8, 16.

be-tilldon, be-teldon, Bd. 4, 26; S. 602, 19, *for* betældon *deceived.* v. tǽlan.

be-timbran; *p.* ede; *pp.* ed *To build, construct with timber;* ædificare, construere:—Hí betimbredon bēcn *they constructed a beacon,* Beo. Kmbl. 6312; B. 3160.

bêting *a cable,* Bt. 41, 3; Fox 250, 15. v. bǽting.

bêtl, es; *m. A* BEETLE; blatta:—Ða blacan bêtlas *blattæ nigro colore,* Cot. 141. v. bítel.

bet-líc; *adj. sup.* bet-lícast *Good-like, excellent;* eximius:—Bold wæs betlíc *the mansion was excellent,* Beo. Th. 3854; B. 1925. Betlícast, Exon. 8 b; Th. 5, 7; Cri. 66.

bet-nes, -ness, e; *f. [bet better] Satisfaction, amends, amendment, recompence;* satisfactio, compensatio:—Ðæt ic búton betnesse beó mínra synna *that I am without amendment of my sins,* L. De Cf. 10; Th. ii. 264, 16.

betoce *the herb betony,* L. M. 1, 39; Lchdm. ii. 104, 4. v. betonice.

be-togen. I. *drawn over, covered, inclosed,* Ors. 5, 7; Bos. 107, 8. II. *accused,* L. In. 54; Th. i. 136, 10; *pp. of* be-teón.

betogenes, -ness, -niss, e; *f. An accusation;* accusatio:—Be cierlisces monnes betogenisse [MS. H. betogenisse] *of a churlish man's accusation,* L. In. 37, titl; Th. i. 124, note 50.

be-tolden *surrounded,* Andr. Kmbl. 1976; An. 990; *pp. of* beteldan.

betonice, an; *f: also Lat.* betonīca, æ; *f. The herb* BETONY; betonīca officinālis. This species is the *common wood-betony,* the *betonīca officinālis* of Linnæus. It is a species of the genus *Stachys,* but it was formerly a species of the genus *Betonica.* It is very plentiful in Great Britain, and formerly much used in medicine. The leaves have a rough bitter taste, and are slightly aromatic. The roots are nauseous and very bitter, and when taken act as purgatives and emetics:—Genim betonican gôdne dǽl *take a good deal of betony,* Lchdm. iii. 22, 16. Nim betonican sǽd *take seed of betony,* iii. 72, 6. Wyl on ealaþ betonican *boil betony in ale,* L. M. 1, 16; Lchdm. ii. 58, 24. Wyrc betonican and pipores seofon and xx corna tosomne getrifulad *work betony and twenty-seven corns of pepper triturated together,* 1, 21; Lchdm. ii. 64, 6: 1, 22; Lchdm. ii. 64, 16. *Latin,* Betonīca, æ; *f:*—Ðis is seó grēne sealf,—betonīca, rude, etc. *this is the green salve,—betony, rue, etc.* Lchdm. iii. 6, 8. Genim ðás ylcan wyrte and betonīcam *take this same wort and betony,* Herb. 135, 3; Lchdm. i. 252, 4. [Betonīca, quæ et Vettonīca dicitur,

quod eam Vettones = Οὐέττονες, in Hispania invenerunt, Plin. 25, 8: Prior 20.]

betost; *adj. Best;* optimus:—Nû is ôfost betost, ðæt we þeódcyning ðǽr sceáwian *now is speed best, that we may see there the great king,* Beo. Th. 6007; B. 3007. v. betst.

be-trǽppan, -treppan [be, treppan *to trap*] *To* BETRAP, *to entrap;* circumvallare:—Meahton hí ðone here betrǽppan [betreppan, col. 1] *they might entrap the army,* Chr. 992; Th. 238, 40, col. 2.

betre *better:*—Hit is betre *it is better,* Bt. 38, 7; Fox 210, 5. v. betera.

be-tredan; *p.* -træd, *pl.* -trǽdon; *pp.* -treden *To tread upon, cover;* conculcare:—Þýstru betredaþ me *tenebræ conculcabunt me,* Ps. Spl. C. 138, 10.

be-treppan *to entrap,* Chr. 992; Th. 238, 40, col. 1. v. be-trǽppan.

betrian, betrigan; *p.* ode; *pp.* od [bet *well,* betra *better*] *To be better, to excel, to make better, to grow better;* meliorari, emendare:—Ic betrige melioror, Ælfc. Gr. 25; Som. 27, 13. v. beterian, gebeterian.

betrung, bettrung, e; *f.* [betrian *to be better*] *A* BETTERING, *amending;* emendatio:—Ðæt hit wǽre heora betrung *that it was their amendment,* Bt. 38, 7; Fox 210, 13.

be-trymian; *p.* ede; *pp.* ed [be, trymian *to fortify*] *To besiege, environ;* circumdare vallo:—Ðíne fýnd ðé betrymiaþ *circumdabunt te inimici tui vallo,* Lk. Bos. 19, 43. Ge geseóþ Hierusalem mid here betrymede *ye shall see Jerusalem besieged with an army,* 21, 20.

betst, betest; *adj. sup. def.* se betsta, betesta; seó, ðæt beteste; *pos.* gôd [bet *good;* v. bet-líc *good-like*] *Best, the best, first;* optimus, primus:—Ða þing ðe ge betstan gelífaþ [MS. betst ongelifaþ] *ea quæ vos optima credebatis,* Bd. 1, 25; S. 487, note 12. Scipio, se besta [Laud MS. betsta] Rômana witena *Scipio, the best of the Roman senators,* Ors. 5, 4; Bos. 104, 38; Cot. MS. Tib. B. I. fol. 85 b. Se betesta *the best,* Cot. 153. He scalde ðæt betste hors *he gave the best horse,* Bd. 3, 14; S. 540, 16. Ðara betstena sumes *of some one of the best,* Bt. 30, 1; Fox 110, 5. [*Goth.* bats? *good; comp.* batiza *better; sup.* batists *best:* O. Nrs. comp. betri *better; sup.* beztr *best.*]

betst; *adv. sup. of* wel [? bet *well, q. v.*] *Best, most;* optime:—Ðæt betst lícaþ *that pleases best,* Bt. 18, 2; Fox 64, 23. Ic him betst truwode *I most trusted them,* Bt. 2; Fox 4, 12. Albínus wæs betst gelǽred *Albinus was most learned,* Bd. pref; S. 471, 23.

betst-boren; *pp. Best-born, eldest;* major natu:—Moises clipode ða betstborenan *Moyses vocavit majores natu,* Lev. 9, 1: Gen. 50, 7: Deut. 5, 23.

bett *better,* Bd. 3, 9; S. 533, 34. v. bet.

bêtte *corrected,* Bd. 4, 25; S. 599, 25; *p. of* bêtan.

bettonice, an; *f. The herb betony;* betonīca officinālis:—Genim bettonican and pipor *take betony and pepper,* Lchdm. i. 380, 24. v. betonice.

bettrung, e; *f. A bettering, ameliorating;* emendatio:—To his bettrunge [Cot. betrunge] *to his amelioration,* Past. 31, 1; Hat. MS. 39 b, 8. v. betrung.

be-tugon *shut in, inclosed;* concluserunt, Lk. Bos. 5, 6; *p. pl. of* be-teón.

be-tuh; *prep. dat. acc. Between;* inter:—He bewîcode betuh ðǽm twâm hergum *he encamped between the two armies,* Chr. 894; Ing. 115, 4; Th. 164, 23, col. 2; 165, 22, col. 1; 23, col. 2. Betuh Arabia and Palestina *between Arabia and Palestine,* Ors. 1, 3; Bos. 27, 20: Cd. 37; Th. 47, 26; Gen. 766. v. be-tweoh.

be-tux *between.* v. betux-sittan, be-tweoh.

betux-sittan *to insert, interpose, to set, put* or *bring in;* interserere. v. betux, sittan.

be-tweoh, be-tweohs, be-tweox, be-twih, be-twyh, be-twyx, be-twyxt, be-twuh, be-twuht, be-twux, be-twuxt, be-tuh, be-tux; *prep. dat. acc.* [be by, with; twi, twihs, tweox, twux *duo*] *Between,* BETWIXT, *among, amid, in the midst;* inter, in medio. I. *dat:*—Men and nêtenu habbaþ andan betweoh him *men and beasts have enmity between them,* Bt. Met. Fox 28, 104; Met. 28, 52. Betweohs him *among them;* in cujus medio, Ex. 34, 10. Betwyh him *among them,* Bt. 39, 12; Fox 230, 27. Betwuh ðǽm wæs seó Magdalenisce Maria, and Maria Iacobes môder *inter quas erat Maria Magdalene, et Maria Iacobi mater,* Mt. Bos. 27, 56. Betwuht him *between them,* Bt. 39, 13; Fox 234, 5. Betwux wífa bearnum *inter natos mulierum,* Lk. Bos. 7, 28. Betwuxt ðǽm warum *among their wares* [merchandise], Nat. S. Greg. Els. 11, 14. Hǽðe stent betuh Winedum and Seaxum and Angle *Haddeby stands in the midst of the Winedi, Saxons and Angles,* Ors. 1, 19; Bos. 21, 30. II. *acc:*—Swá lamb betweox wulfas *sicut agnos inter lupos,* Lk. Bos. 10, 3. Betwih ða mægen *inter virtutes,* Bd. 4, 9; S. 576, 28. Ne byþ swá betweox eów *non ita erit inter vos,* Mt. Bos. 20, 26. III. *the case sometimes precedes the prep.* or *is separated from it:*—Hí him healdaþ betwuh sibbe *they keep peace between themselves,* Bt. Met. Fox 29, 8; Met. 29, 4. Him betuh *between them,* Cd. 37; Th. 47, 26; Gen. 766.

be-tweohs *among:*—Betweohs him *among them,* Ex. 34, 10. v. be-tweoh.

be-tweonan; *prep. dat. acc. Between;* inter :—Unc betweonan *between us two,* Cd. 91; Th. 114, 10; Gen. 1902. v. be-tweonum.

be-tweonum, be-tweonan, be-twinum, be-twinan, be-twynan, bi-tweon, bi-tweonum; *prep.* I. *dat.* II. *acc.* [be, bi *by, with,* tweo *two; dat.* tweonum, twinum, tweon, twin, twyn] BETWEEN, *betwixt, among, amid, in the midst;* inter, in medio. I. *dat.* Betweonan ðám *between them,* Ps. Th. 102, 12. Betweonum ðissum þingum *amid these things,* Bd. 1, 27; S. 488, note 26. Ðá Iudeas cwǽdon betweonan him sylfum *then the Jews said among themselves,* Jn. Bos. 7, 35. II. *acc.* Ðú hí betweonum wætera weallas lǽddest *thou ledest them between water-walls,* Ps. Th. 105, 9. Ðá seó cwén ongan lǽran ðæt hie sybbe swá same sylfra betweonum freóndrǽdenne gelǽston *then the queen began to teach that they should hold peace also amid their friendly band,* Elen. Kmbl. 2412; El. 1207. III. *sometimes* betweonum *follows its case, or is separated from it :*—Ðá gewearþ hí him betweonum *they then agreed between themselves,* Ors. 6, 30; Bos. 126, 24. Léton him ða betweonum tán wisian *they let the lot decide between them,* Andr. Kmbl. 2199; An. 1101. Ne sceólon unc betweonan teónan weaxan *injury shall not wax between us two,* Cd. 91; Th. 114, 10; Gen. 1902. Hluton hell-cræftum, hǽðengildum teledon betwinum *they cast lots, counted, with hellish arts, amid heathen gods,* Andr. Kmbl. 2207; An. 1105. Gif ge habbaþ lufe eów betwynan *si dilectionem habueritis ad invicem,* Jn. Bos. 13, 35. Friþ freóndum bitweon *peace between friends,* Exon. 32 a; Th. 101, 15; Cri. 1659. IV. *sometimes the case is placed between* be *and* tweonum, *as,*—Be sǽm tweonum *between the seas,* Cd. 163; Th. 205, 28; Exod. 442: 170; Th. 214, 1; Exod. 562. v. bi-tweonum.

be-tweonum; *adv. Between;* inter, in medio :—Ne sí lang fæc betweonum *ne sit longum spatium in medio,* Bd. 4, 9; S. 577, 27.

be-tweox *between;* inter :—Nú ic eów sende swá swá lamb betweox wulfas *ecce ego mitto vos sicut agnos inter lupos,* Lk. Bos. 10, 3 : 11, 51: Ps. Th. 87, 4 : 88, 5: Bt. Met. Fox 11, 90; Met. 11, 45 : 11, 168; Met. 11, 84 : 24, 25; Met. 24, 13. v. be-tweoh.

be-twih *between.* v. betwih-licgan, be-tweoh.

betwih-licgan, he -ligeþ *To lie between;* interjacere :—Gif mycel feornys síþfætes betwihligeþ *si longinquitas itineris magna interjacet,* Bd. 1, 27; S. 491, 39. v. be-tweoh.

be-twinan; *prep. dat. Within, among;* intra, inter :—Cwǽdon sume bóceras him betwinan *some scribes said among themselves,* Mt. Bos. 9, 3. v. be-tweonum III.

be-twinum *between, amid;* inter, in medio, Andr. Kmbl. 2207; An. 1105. v. be-tweonum.

be-twion; *adj.* [be *by, with;* twám, twǽm, *dat. of* twá *two] Double, folding, twofold;* duplex :—Mid betwion mentle *with a folding mantle;* diploide, Ps. Spl. T. 108, 28.

be-twuh; *prep. dat. acc. Between, among;* inter :—He gewícode be-twuh ðǽm twám hergum *he encamped between the two armies,* Chr. 894; Th. 164, 23, col. 1. He betwuh him wunaþ *he dwells among them,* Bt. 39, 13; Fox 234, 10: Bt. Met. Fox 29, 8; Met. 29, 4. v. be-tweoh.

be-twuht; *prep. dat. Between;* inter :—Betwuht him *between them,* Bt. 39, 13; Fox 234, 5. v. be-tweoh.

be-twux *between, among;* inter :—Nis betwux wífa bearnum, nán mǽrra wítega, ðonne Iohannes se Fulluhtere *major inter natos mulierum propheta nemo est Ioanne Baptista,* Lk. Bos. 7, 28: Gen. 3, 14. v. be-tweoh.

betwux-alegednes, -nyss, e; *f.* [betwux *between;* aleged, alegd *laid*] *What is laid or placed between, an interposition, interjection;* interjectio :—Interjectio mæg beón gecweden betwuxalegednyss on Englisc, forðanðe he líþ betwux wordum *an interjection may be called a laying between in English, because it lies between words,* Ælfc. Gr. 48; Som. 48, 61. v. betwyx-aworpennyss.

betwux-aworpennys *an interjection;* interjectio. v. betwyx-aworpennyss.

be-twuxt *among;* inter :—Ðá geseah Grēgŏrius betwuxt ðám warum, cýpecnihtas gesette *then Gregory saw among their wares, youths set for sale,* Nat. S. Greg. Els. 11, 14. v. be-tweoh.

be-twyh *between, among;* inter, in medio :—Betwyh ðás þing *between these things, in the mean while, whilst;* interea, Bd. 1, 27; S. 488, 26. Betwyh him *among them,* Bt. 39, 12; Fox 230, 27. v. be-tweoh.

betwyh-geset *interposed;* interpositus, Bd. 4, 9; S. 576, 42.

be-twynan; *prep. dat. Between, among;* inter :—Him betwynan *among them,* Mt. Jun. 9, 3: Jn. Bos. 16, 17. Ge habbaþ lufe eów betwynan *dilectionem habueritis ad invicem,* Jn. Bos. 13, 35. v. be-tweonum.

be-twyx *betwixt, between* :—Betwyx wífa bearnum *inter natos mulierum,* Mt. Bos. 11, 11: Chr. 1126; Th. 377, 10. v. betwyx-sendan, betweoh.

betwyx-aworpennyss, e; *f. An interjection;* interjectio :—Interjectio is betwyxaworpennyss. Se dǽl líþ betwyx óðrum wordum, and geswutelaþ ðæs módes styrunge. Heu geswutelaþ módes sárnesse *an interjection is a throwing between. This part of speech lieth between*

other words, *and denotes a stirring of the mind.* Heu *denotes a soreness of mind,* Ælfc. Gr. 5; Som. 3, 55. v: betwux-alegednes.

betwyx-sendan *to send between;* intermittere, R. Conc. Procem.

be-twyxt *betwixt, between;* inter, Hemm. p. 403. v. be-tweoh.

be-tygen *accused,* L. In. 14; Th. i. 110, 16; *pp. of* be-tíhan.

be-týhþ *accuses,* L. In. 46; Th. i. 130, 12; *pres. of* be-teón.

be-týhtlian *to accuse,* L. Eth. i. 1; Th. i. 280, 8, 16. v. be-tíhtlian.

be-týnan, -tiénan, bi-týnan; *p.* -týnde, *pl.* -týndon; *impert.* -týn, -tiéne; *pp.* -týned, -tiéned, -týnd; *v. a.* [be, týnan *to hedge in*]. I. *to inclose* or *surround with a hedge, inclose, close, shut, shut up;* sepem circumdare, sepire, intercludere, claudere, occludere, concludere :—Sum hírēdes ealdor wæs, se plantode wíngerd, and betýnde hyne *homo erat paterfamilias, qui plantavit vineam, et sepem circumdedit ei,* Mt. Bos. 21, 33: Mk. Bos. 12, 1. Ceorles weorþig sceal beón betýned *a churl's close ought to be surrounded with a hedge,* L. In. 40; Th. i. 126, 13. Hí hine betýndon in án nearo fæsten *they inclosed him in a narrow fastness,* Bd. 4, 26; S. 602, note 19. Háteþ heáhcyning helle betýnan *the mighty king shall command to close hell,* Salm. Kmbl. 348; Sal. 173. Ðæs heán biscopes leoma on ðysse byrigenne syndon betýnde [MS. be-tyned] *pontificis summi hoc clauduntur membra sepulchro,* Bd. 2, 1; S. 500, 22: Exon. 110 b; Th. 422, 25; Rä. 41, 11. Wearþ se hálga wong betýned *the holy plain was closed,* 61 b; Th. 227, 7; Ph. 419. He hine inne betýnan nolde *he would not shut it in,* L. Alf. 21; Th. i. 48, 31. He ðæt folc úte betýnde *he shut the people out,* Ors. 4, 5; Bos. 81, 40. Hý betýndon lanes duru *they shut the doors of Janus,* 6, 7; Bos. 120, 5 : 5, 14; Bos. 113, 42. Gif hwá wæterpyt betýnedne ontýne, and hine eft ne betýne, gelde swelc neát swelc ðǽron befealle *if any one open a water-pit [that is] shut up, and close it not again, let him pay for whatever cattle may fall therein,* L. Alf. 22; Th. i. 50, 6, 7. Betiéne togeánes híg *conclude adversus eos,* Ps. Spl. T. 34, 3. II. *to end, finish, conclude;* finire :—Heó ðus ðæt word betýnde *thus she ended the speech;* ita sermonem conclusit, Bd. 4, 9; S. 577, 28.

be-tyran [be, tyrwa *tar*] *To* BETAR, *to smear over, to stain a dark colour;* pice liquida inficere, *q. d.* pullo *vel* bætico colore imbuere, Ælfc. vern. 2.

be-ufan; *adv.* [be, ufan] *Above;* supra :—Swá we hér be-ufan cwǽdon *as we here have said above,* L. Ath. iv. 4; Th. i. 224, 4. v. búfan.

be-útan; *prep. dat.* [be, útan *out*] *Without;* extra :—Wundorlíc is geworden ðín wísdóm eall, se is be-útan me *mirabilis facta est scientia tua ex me,* Ps. Th. 138, 4. Gif ic míne fiðeru gefó, fleóge ǽr leóhte, oþ ðæt ic be-útan wese eallum sǽwum *si sumpsero pennas meas ante lucem, et habitavero in postremo maris,* 138, 7. Ða be-útan beóþ earce bordum *who shall be without the boards of the ark,* Cd. 67; Th. 81, 32; Gen. 1354. v. bútan.

be-waden; *part. p. A quo aliquid abiit?*—Of wombe bewaden, Exon. 130 b; Th. 499, 32; Rä. 88, 24. DER. be, wadan.

be-wǽfan; *p. de; pp.* ed [wǽfan *to cover*] *To befold, wrap round, cover, clothe;* obvolvere, amicire, operire, induere :—Mid ánre scýtan bewǽfed *amictus sindone,* Mk. Bos. 14, 51: Homl. Th. ii. 242, 24. Heó nam hyre wǽfels and bewǽfde híg *illa sustulit pallium et operuit se,* Gen. 24, 65. Martinus me bewǽfde mid ðyssere wǽde *Martin clothed me with this garment,* Homl. Th. ii. 500, 34. His cempan mid wolcnreádum wǽfelse hine bewǽfdon *his soldiers clothed him in a scarlet robe,* ii. 252, 25. Ic eom reáde bewǽfed *I am clothed with red,* Exon. 126 a; Th. 484, 2; Rä. 70, 1: Past. 14, 3; Hat. MS. 17 b, 19. [Goth. bi-waibyan *to wind, put round.*]

be-wǽg *surrounded,* Bt. 39, 4; Fox 216, 25; *p. of* be-wegan.

be-wǽgan; *p. de; pp.* ed *To deceive, disappoint;* frustrari :—Ne bewǽgde him *non frustratus est eum,* Ps. Spl. C. 131, 11. v. bi-wǽgan.

be-wǽgnan; *p.* ede; *pp.* ed *To offer;* offerre :—Him wæs freónd-laðu bewǽgned *a friendly invitation was offered to him,* Beo. Th. 2390; B. 1193.

be-wǽlan; *p.* de; *pp.* ed *To afflict;* undique vexare, affligere, cruciare :—Wítum bewǽled *afflicted with torments,* Andr. Kmbl. 2721; An. 1363.

be-wǽpnian, -wépnian; *p.* ede; *pp.* ed [be, wǽpen *a weapon*] *To take away arms, disarm;* armis spoliare :—Be ðam ðe óðerne bewǽpnaþ *de eo qui alium armis spoliaret,* L. C. S. 61, titl; Th. i. 408, 16. Gif man æt unlagum man bewǽpnige [bewepnie MS. B.] *if any one unlawfully disarm a man,* 61; Th. i. 408, 18.

be-wand *wrapped, enwrapped,* Bd. 3, 11; S. 536, 9: Lk. Bos. 2, 7; *p. of* be-windan.

be-warenian, -warnian; *p.* ode; *pp.* od *To guard, beware;* custodire, cavere :—He wel ne bewarenaþ wið ða unþeáwas *he does not well guard against the vices,* Bt. Met. Fox 16, 45; Met. 16, 23. Ða ðe hie wið ða lǽssan scylda bewareniaþ *those who guard themselves against the lesser sins,* Past. 57, 1; Hat. MS.

be-warian, -warigan; *p.* ode; *pp.* od *To keep, guard, preserve;* custodire, arcere :—Bisceopas godcunde heorda bewarian and bewerian sceolon *bishops ought to guard and defend their spiritual flocks,* L. C. E. 26;

Wilk. 133, 22; Th. i. 374, 24. Ðæt ðú meaht wîte bewarigan *that thou mayest ward off punishment*, Cd. 27; Th. 35, 31; Gen. 563. v. warian.

be-warnian *to beware*, R. Ben. 7. v. be-warenian.

be-weallan; *p.* -weóll, *pl.* -weóllon; *pp.* -weallen *To boil away*; decoquere:—Óþ-ðæt þrydda dǽl sý beweallen *till the third part be boiled away*, Med. ex Quadr. 1, 3; Lchdm. i. 328, 17: 8, 10; Lchdm. i. 360, 1.

be-wealwian; *p.* ode; *pp.* od *To wallow*; volutare:—Swîn on ða solu bewealwiaþ *swine wallow in the mire*, Bt. 37, 4; Fox 192, 29.

be-weardian, -weardigan; *part.* -weardigende; *p.* ode; *pp.* od *To ward, protect, keep*; custodire, protegere, observare:—Ðû, Drihten, beweardast us *tu, Domine, custodies nos*, Ps. Spl. 11, 8. Hálige englas ða dǽda beweardiaþ *holy angels protect the deeds*, L. C. E. 4; Th. i. 360, 31. Beweardigende *observantes*, Ps. Spl. 30, 7.

be-wearp *cast*, Bt. 7, 2; Fox 16, 25; *p.* of be-weorpan.

be-weaxan, bi-weaxan; *p.* -weóx, *pl.* -weóxon; *pp.* -weaxen *To overgrow, cover over*; obducere, obserere:—Sindon burgtúnas brérum beweaxene [MS. beweaxne] *the city-dwellings are overgrown with briers*, Exon. 115 b; Th. 443, 17; Kl. 31. Scyllun biweaxen *overgrown with scales*, 60 a; Th. 219, 21; Ph. 310.

be-weddian, -weddigan; *p.* ede, ode; *pp.* ed, od *To espouse, wed*; spondere, despondere:—Ic beháte oððe ic beweddige [MS. bewedige] *spondeo*, Ælfc. Gr. 26, 6; Som. 29, 10. Gif he hîg his suna beweddaþ *si filio suo desponderit eam*, Ex. 21, 9. v. weddian.

be-weddung, e; *f. A betrothal, wedding*; oppigneratio, connubium:—Be wîfmannes beweddunge *of a woman's betrothal*, L. Edm. B. titl; Th. i. 254, 1.

be-wefan; *p.* -wæf, *pl.* -wǽfon; *pp.* -wéfen *To cover over, envelope*; obtexere, obducere:—Biþ ðæt brægen mid reáman bewefen *the brain is covered over with a membrane*, Lchdm. iii. 146, 4.

be-wegan; *p.* -wæg, *pl.* -wǽgon; *pp.* -wegen *To cover, cover over, surround*; obducere, circumdare:—Bewegen wælmiste *covered with the mist of death*, Exon. 87 b; Th. 329, 30; Vy. 42. He hî bewæg mid wuda útan *he surrounded them with wood*, Bt. 39, 4; Fox 216, 25.

be-wendan; *p.* -wende; *pp.* -wended, -wend *To turn, turn round or about, convert*; vertere, convertere:—Bewend to ðǽre menigu *conversus ad turbam*, Mk. Bos. 5, 30. Se Hǽlend bewende hyne *the Saviour turned himself about*, Mt. Bos. 9, 22: Mk. Bos. 8, 33. Æt sumum cyrre bewend *aliquando conversus*, Lk. Foxe 22, 32. v. wendan.

be-wenian; *p.* ede; *pp.* ed [be, wenian *to accustom, draw to one's self, honour*] *To entertain, take care of*; hospitio accipere:—We wǽron hér tela bewenede *we were here kindly entertained*, Beo. Th. 3646, note; B. 1821. Dryht-bearn Dena duguþa bewenede [MS. *and* Thorpe's note, 4077; bî werede, B. 2035] *a noble offspring of the Danes entertained the knights*, 4077, note.

be-weópon *wept over, bewailed*, Num. 20, 30; *p. pl.* of be-wépan.

be-weorcean *to adorn*, Elen. Kmbl. 2045; El. 1024. v. be-wyrcan.

be-weorpan, -wyrpan; ic -weorpe, ðú -wyrpst, he -weorpeþ, -wyrpþ, *pl.* -weorpaþ; *p.* -wearp, *pl.* -wurpon; *pp.* -worpen. **I.** *to cast, cast down, throw*; projicere, dejicere:—Seó cwén hét [híg] ðam cyninge heáfod ofaceorfan, and bewyrpan on ânne cylle *the queen commanded [them] to cut off the king's head, and to cast it into a vessel*, Ors. 2, 4; Bos. 45, 33. Hwæt bewearp ðé on ðás gnornunga *what has cast thee into these lamentations?* Bt. 7, 2; Fox 16, 25. He hæfþ us beworpen on ealra wîta mǽste *he hath cast us down into the greatest of all torments*, Cd. 21; Th. 25, 13; Gen. 393. Ic wæs hér unscildig on pytt beworpen *I was thrown here innocent into a dungeon*, Gen. 40, 15. **II.** *to cast about or over, cover over, surround*; conjicere, supertegere, cingere:—Hláford, lǽt hine [fíctreów] gyt ðis geár, óþ ic hine bedelfe, and ic hine beweorpe mid meoxe *Lord, suffer it [the fig-tree] yet this year, till I dig about it, and cast it about [surround it] with dung*, Lk. Bos. 13, 8. Oft beweorpeþ ânre þecene wundrum gewlitegad *often casts over with a covering wondrously adorned*, Exon. 128 b; Th. 493, 20; Rä. 81, 34. Hafaþ fægerne eard wætre beworpen *it hath a fair dwelling surrounded with water*, Runic pm. 28; Kmbl. 345, 8; Hick. Thes. i. 135. DER. weorpan.

be-weotian; *p.* ode; *pp.* od *To observe, watch over*; observare, curæ habere:—Draca hord beweotode *a dragon watched over the hoard*, Beo. Th. 4431; B. 2212. v. be-witian.

be-wépan; *p.* -weóp, *pl.* -weópon; *pp.* -wópen *To weep, weep over, bewail*; flere, deflere, plorare:—Ic bewépe *defleo*, Ælfc. Gr. 26, 1; Som. 28, 28. Hî beweópon Aarones forþsîþ *they bewailed Adron's death*, Num. 20, 30. Wyduwan heora næron bewópene *viduæ eorum non plorabantur*, Ps. Lamb. 77, 64: Ors. 2, 8; Bos. 51, 41.

be-wépnian *to unweapon, disarm*, L. C. S. 61, titl; Th. i. 408, 16. v. be-wǽpnian.

be-werenes, -ness, e; *f.* [be-wered *forbidden*] *A forbidding*; prohibitio:—Óþ bewerenesse to onfónne ðam hálgan gerýne *usque ad prohibitionem percipiendi sancti mysterii*, Bd. 1, 27; S. 496, 43.

be-werian, bi-werian, -wergan; *p.* ede, ode; *pp.* ed, od *To defend, restrain*; defendere, prohibere, tueri:—Bisceopas godcunde heorda bewarian and bewerian sceolon *bishops ought to guard and defend [tueri debent] their spiritual flocks*, L. C. E. 26; Wilk. 133, 22; Th. i. 374, 25. Bewerede *coercuit*, Cot. 56. Bewered *prohibitus*, Bd. 1, 27; S. 493, 10. Bewerode *defendit*, Ex. 2, 17. Bewerod *prohibitus*, Ælfc. Gl. 63; Som. 68, 104. DER. werian.

be-werigend, es; *m. A defender*; protector, Ps. Spl. 27, 11.

be-werung, e; *f. A defence, fortification*; tutamen:—Bewerung strang *a strong defence*, Scint. 64.

be-wícian; *p.* ode; *pp.* od *To encamp*; castra metari:—Ælfréd cyning bewícode ðam twâm hergum *king Alfred encamped between the two armies*, Chr. 894; Gib. 92, 21.

be-wimman; *g.* -wimmannes; *f.* [be-wimmen, Wrt. Voc. 72, 36] *A niece*; neptis, Som. Lye. v. wimman.

be-windan, bi-windan; *p.* -wand, -wond, *pl.* -wundon; *pp.* -wunden; *v. a. To wind* or *bind around* or *about, entwine, wrap, enwrap, encircle, surround, wind, turn*; amplecti, involvere, cingere, circumdare, volvere:—Hí îsene næglas mid flexe bewundon *they wound iron nails round with flax*, Ors. 4, 1; Bos. 78, 8. Wæs bewunden *was wound round*, Andr. Kmbl. 38; An. 19. Wîrum bewunden *bound round with wires*, Beo. Th. 2066; B. 1031. Iosep bewand hyne mid clǽnre scýtan *Ioseph involvit illud in sindone munda*, Mt. Bos. 27, 59: Lk. Bos. 2, 7: Bd. 3, 11; S. 536, 9. Geseah heó monnes líchoman mid scýtan bewundenne *vidit corpus hominis sindone involutum*, Bd. 4, 9; S. 576, 32. Wæs Cristes lof on fyrhþlocan bewunden *Christ's praise was entwined within his breast*, Andr. Kmbl. 116; An. 58: Beo. Th. 6283; B. 3146. Biwunden *entwined*, Exon. 69 a; Th. 256, 20; Jul. 234. Sum gǽstes þearfe móde bewindeþ *one wraps his spirit's need in his mind*, 79 b; Th. 298, 18; Crä. 87: Ps. Th. 102, 12. Wæs feorh ædelinges flǽsce bewunden *the prince's soul was wrapped in flesh*, Beo. Th. 4840; B. 2424. Mec mon folmum biwond, and mec ðâ on þeóstre alegde biwundenne mid wonnum cláðum *one with hands enwrapped me, and then laid me in darkness enwrapped in dusky clothes*, Exon. 28 b; Th. 87, 9-12; Cri. 1422-1424. He wæs cláðum biwunden *he was enwrapped with clothes*, 18 b; Th. 45, 27; Cri. 725. Ðǽr is geat gylden wynnum bewunden *there is the golden gate encircled with joys*, Cd. 227; Th. 305, 21; Sat. 650: Beo. Th. 6097; B. 3052. He is wuldre biwunden *he is encircled with glory*, Exon. 65 b; Th. 241, 34; Ph. 666. Ða þreó wæter steápe stánbyrig streámum bewindaþ *the three waters surround lofty cities of stone with their streams*, Cd. 100; Th. 133, 18; Gen. 2212. Hwonne us líffreá tîre bewinde *when the Lord of life may surround us with honour*, Exon. 8 a; Th. 3, 1; Cri. 29. Ic eom bewunden mid wuldre *I am surrounded with glory*, 108 a; Th. 412, 18; Rä. 31, 2. He geseah Sennera feld sîdne bewindan *he saw Shinar's field wide winding*, Cd. 205; Th. 253, 28; Dan. 602. Abraham bewand ða hleóðorcwydas on hige sínum *Abraham turned the revelations in his mind*, 107; Th. 140, 34; Gen. 2337.

be-wiste *governed, presided*, Gen. 24, 2; *p.* of be-witan.

be-witan; ic. he -wát, ðú -wást, *pl.* -witon; *p.* -wiste, *pl.* -wiston; *pp.* -witen; *v. trans.* [be near, witan *to know, see, take care of*] *To overlook, watch over, superintend, preside, govern, administer*; præesse, administrare:—Ðe ealle his þing bewiste *qui præerat omnibus quæ habebat*, Gen. 24, 2. Ne miht ðú leng tún-scíre bewitan *jam non poteris villicare*, Lk. Bos. 16, 2: Ex. 3, 7: 5, 14: Ors. 2, 2; Bos. 41, 33: 2, 4; Bos. 43, 21: 6, 37; Bos. 132, 21. Fæder ealle gesceafte bewát *the father watches over all creatures*, Exon. 128 a; Th. 492, 5; Rä. 81, 9. To bewitanne, Gen. 39, 4.

be-witian, -witigan, -weotian; *p.* ode; *pp.* od *To observe, take care of, administer, perform*; observare, curæ habere, exsequi, peragere:—Ne mágon hî tunglu bewitian *they may not observe the heavenly bodies*, Exon. 89 b; Th. 335, 31; Gn. Ex. 40. Hí oft bewitigaþ sorgfulne sîþ *they often perform a sorrowful journey*, Beo. Th. 2861; B. 1428: Exon. 12 b; Th. 22, 18; Cri. 353.

be-wlát *looked, beheld*, Cd. 142; Th. 177, 6; Gen. 2925; *p.* of be-wlîtan.

be-wlátian; *p.* ode; *pp.* od *To see, look, behold*; videre, conspicere:—Eágan ðíne geséon oððe bewlátion [MS. bewlatiun] *ufnysse oððe rihtwîsnesse oculi tui videant æquitates*, Ps. Lamb. 16, 2. To gescyldnysse mínre beseoh oððe bewláta *ad defensionem meam conspice*, 21, 20.

be-wlátung, e; *f. Show, sight, pageant*; spectaculum. DER. be-wlátian.

be-wlítan; *p.* -wlát, *pl.* -wliton; *pp.* -wliten *To look, behold*; spectare, respicere:—Se eádega bewlát rinc ofer exle *the happy man looked over his shoulder*, Cd. 142; Th. 177, 6; Gen. 2925.

be-wópen *bewailed*, Ors. 2, 8; Bos. 51, 41; *pp.* of be-wépan.

be-worht *made, built, covered*, Jos. 2, 1; *pp.* of be-wyrcan.

be-worpen *cast, cast down, thrown, cast about, surrounded*, Cd. 21; Th. 25, 13; Gen. 393: Gen. 40, 15: Runic pm. 28; Kmbl. 345, 8; Hick. Thes. i. 135; *pp.* of be-weorpan.

be-wrǽcon *exiled, sent forth*, Cd. 189; Th. 235, 12; Dan. 305; *p. pl.* of be-wrecan.

H

be-wreáh *covered, covered over, protected,* Ps. Th. 104, 34; *p. of* be-wreón.

be-wrecan, bi-wrecan; *p.* -wræc, *pl.* -wrǽcon; *pp.* -wrecen. **I.** *to exile, send forth;* pellere, propellere :—Ðu úsic bewrǽce in ǽht-gewealda *thou hast exiled us into bondage,* Exon. 53 a; Th. 186, 25; Az. 25. Ða us bewrǽcon *they have sent us forth,* Cd. 189; Th. 235, 12; Dan. 305. **II.** *to strike* or *beat around, afflict;* circum pulsare :—We land gesóhton wære bewrecene *we sought the land beaten round* [*afflicted*] *with the sea,* Andr. Kmbl. 537; An. 269. **III.** *to drive* or *bring to;* appellere :—Ceólas léton sande bewrecene *they let the keels* [*ships*] *be driven to the sand* [*shore*], Elen. Kmbl. 502; El. 251. DER. wrecan.

be-wrencan; *p.* -wrencte; *pp.* -wrenced [be *about,* wrenc *deceit*] *To deceive;* occultis machinationibus circumvenire, Prov. Kmbl. 34.

be-wreón; *p.* -wreáh, *pl.* -wrugon; *pp.* -wrogen *To cover, cover over, protect, clothe;* tegere, contegere, operire, protegere, velare :—Bewrugon [bewreogon MS.] me þýstru *contexerunt me tenebræ,* Ps. Spl. 54, 5. Mid mínum bysmre ic eom bewrogen *confusio vultus mei operuit me,* Ps. Th. 43, 17. Ðú bewruge me fram gemétinge awyrgedra *protexisti me a conventu malignantium,* Ps. Spl. 63, 2. He hí wolcne bewreáh *he protected them with a cloud,* Ps. Th. 104, 34. Ic wæs nacod, and ge me noldon bewreón *I was naked, and ye would not clothe me,* Past. 44, 7; Hat. MS. 62 b, 21.

be-wrigen, -wrigon *covered, concealed,* Bt. Met. Fox 4, 93; Met. 4, 47; *pp. and p. pl. of* be-wríhan.

be-wrigennes, -ness, e; *f. A hiding, keeping close* or *concealing;* occultatio. DER. be-wríhan.

be-wríhan, bi-wríhan; *p.* -wráh, *pl.* -wrigon; *pp.* -wrigen *To cover over, conceal, wrap up;* velare, operire :—Se snáw bewríhþ wyrta cíþ *the snow covers over the germ of herbs,* Salm. Kmbl. 302. Ic goldwine mínne hrusan heolstre biwráh *I covered my bounteous patron in a cave of the earth,* Exon. 76 b; Th. 287, 32; Wand. 23. Bewrigen mid wrencum *concealed by frauds,* Bt. Met. Fox 4, 93; Met. 4, 47: Cd. 8; Th. 10, 14; Gen. 156. Bewrigenum *wrapped up, instr.* Cd. 77; Th. 95, 28; Gen. 1585. DER. be-wrigennes, wríhan.

be-wríhþ *covers over,* Salm. Kmbl. 605; Sal. 302; *3rd pers. pres. of* be-wríhan.

be-wrítan, bi-wrítan; *p.* -wrát, *pl.* -writon; *pp.* -writen *To write down, inscribe;* inscribere, Exon. 92 b; Th. 347, 27; Sch. 19.

be-wríðan, he -wríþ; *p.* -wráþ, *pl.* -wriðon *To bind, bind round, begird;* ligare, redimire :—Meotud bewríþ mid his wuldre eall eorþbüend *the Creator shall wreathe with his glory all earth's inhabitants,* Exon. 18 a; Th. 45, 12; Cri. 718. Duru wundurclommum bewriðen *the door bound with wondrous bands,* 12 a; Th. 19, 33; Cri. 310. DER. wríðan.

be-wrogen *covered, covered over;* opertus, Ps. Th. 43, 17; *pp. of* be-wreón.

be-wruge *hast protected;* protexisti, Ps. Spl. 63, 2; *2nd pers. sing. p. of* be-wreón.

be-wunden *wrapped, enwrapped,* Beo. Th. 4840; B. 2424: -wundon *wound* or *bound round,* Ors. 4, 1; Bos. 78, 8; *pp. and p. pl. of* be-windan.

be-wyddod *betrothed;* desponsatus, L. Ethb. 83; Th. i. 24, 5, = be-weddod; *pp. of* be-weddian.

be-wyrcan, -weorcean, bi-wyrcan; *p.* -worhte, *pl.* -worhton; *pp.* -worht *To work, work in, insert, make, build, cover, adorn;* elaborare, immittere, facere, ædificare, inducere, exornare :—Bewyrc us on heortan Háligne Gást *work the Holy Ghost into our hearts,* Hy. 7, 79; Hy. Grn. ii. 288, 79. Ne wát ic mec beworhtne wulle flýsum *I know not that I was made with fleeces of wool,* Exon. 109 a; Th. 417, 11; Rä. 36, 3. He lǽmen fæt biwyrcan hêt *he commanded to make an earthen vessel,* 74 a; Th. 277, 3; Jul. 575. Babylónia is mid stænenum wíghúsum beworht *Babylon is built with stone towers,* Ors. 2, 4; Bos. 44, 30: Jos. 2, 1. Se mid weaxe beworhte *he covered it with wax,* Ors. 2, 5; Bos. 46, 30. Ða têþ on golde bewyrc *cover the teeth with gold,* Med. ex Quadr. 1, 1; Lchdm. i. 326, 16. Seó cwén ða róde hêht golde beweorcean *the queen commanded to adorn the cross with gold,* Elen. Kmbl. 2045; El. 1024.

be-wyrpan *to cast, throw,* Ors. 2, 4: Bos. 45, 33. v. be-weorpan.

be-yrnan, -irnan; he -yrnþ; *p.* -arn, *pl.* -urnon; *pp.* -urnen [be *by,* yrnan *to run*] *To run by, to come in, occur, incur;* percurrere :—Be-arn me on móde *it occurred to my mind,* Homl. Th. i. 2, 6. Án wundor me nú on móde be-arn *one wonder now* [*runs by me into the mind*] *occurs to me,* Dial. 1, 10. He ne be-arn on leásunga synne *he incurred not the sin of* [*leasing*] *lying,* Dial. 1, 2 : Bd. de nat. rerum; Wrt. popl. science 7, 1; Lchdm. iii. 244, 20.

bezera, an; *m. The baptist* :—Se bezera, Mt. Rush. Stv. 3, 1. v. bædzere.

bi *by, near, concerning.* v. be, bí.

bí *a bee;* apis : *found in the compound* bí-breád.

bí; *prep. dat.* [Bí is more frequently shortened into be. In compounds it is generally written be- or bi-; but bí- is long where it is used for big, or is a contraction, thus,—bí-spell for big-spell, and as bí-breád for beó-breád. v. be.] **1.** *dat. By, near to, at, in, upon;* juxta, prope, apud, in :—Arás bí ronde oretta *the champion arose by his shield,* Beo. Th. 5069; B. 2538. He bí sesse geóng *he went by the seat,* 5506; B. 2756. Bí staðe fæste *fast by the shore,* Exon. 96 b; Th. 361, 11; Wal. 18. Hwearf bí bence *turned by the bench,* Beo. Th. 2380; B. 1188. **2.** *dat. Of, about;* de, quoad :—Ðæt bí ðé sóþfæst sægde Esaias *what Isaiah said truly of thee,* Exon. 12 a; Th. 19, 16; Cri. 301. Hýrde ic secgan gén bí sumum fugle *I have yet heard tell of a certain bird,* 97 b; Th. 365, 17; Reb. 1. Bí ðon se wítga song of whom *the prophet sang,* 17 a; Th. 41, 4; Cri. 650. **3.** *dat. By, through, because of, after, according to, in comparison with;* per, secundum, pro, ex :—Bí hwon scealt ðú lifgan *by what art thou to live?* Exon. 36 b; Th. 118, 23; Gú. 244. Bí noman gehátne *called by name,* 23 b; Th. 66, 16; Cri. 1072. Bí heofonwóman *through the crash of heaven,* 20 a; Th. 52, 18; Cri. 835. Leán cumaþ werum bí gewyrhtum *retribution shall come to men according to their works,* 27 b; Th. 84, 3; Cri. 1368: 76 a; Th. 286, 8; Jul. 728. Ðisses fugles gecynd fela gelíces bí ðám gecornum Cristes þegnum *the nature of this bird is much like to the chosen servants of Christ,* 61 b; Th. 225, 12; Ph. 388. **4.** *sometimes* bí *is separated from its case* :—Bí wædes ófre *by the shore of the sea,* Exon. 96 b; Th. 360, 22; Wal. 9.

biágian; *p.* ode; *pp.* od [beág *a crown*] *To crown;* coronare :—Ðú biágodyst hine *coronasti eum,* Ps. Spl. C. 8, 6. v. beágian.

biaþ *are; for* bióþ, Mt. Lind. Stv. 26, 31; *pl. pres. of* bión = beón.

bi-baðian; *p.* ode; *pp.* od *To bathe, wash;* lavare :—Se ædela fugel hine bibaðaþ in ðam burnan *the noble fowl bathes itself in the brook,* Exon. 57 b; Th. 205, 3; Ph. 107. v. be-baðian.

bi-beódan; *p.* -beád, *pl.* -budon; *pp.* -boden *To order, command, bid;* jubere, mandare, Exon. 56 a; Th. 200, 6; Ph. 36: 93 a; Th. 349, 13; Sch. 45. v. be-beódan.

biblio-þéce, an; *f.* [βιβλιοθήκη = βιβλίον *a book,* θήκη *repository, a library*]. **I.** *a library;* bibliotheca, C. R. Ben. 50. **II.** *a collection of books in one volume, hence,—The Bible;* biblia :—Hieronimus, se wurþfulla and se wísa bócere, úre Biblioþecan gebrohte to Lédene of Grēciscum bócum and of Ebréiscum *Jerome, the worthy and wise author, translated our Bible out of the Greek and Hebrew books into Latin,* Ælfc. T. Grn. 16, 6–8. Se saltere ys án bóc on ðære Biblioþecan *the psalter is one book in the Bible,* Ælfc. T. 14, 15. Iohannes awrát ða bóc, Apocalipsis gehâten, and ðeós bóc ys æftemyst on ðære Biblioþecan *John wrote the book called Revelation, and this book is the last in the Bible,* Ælfc. T. 31, 23.

bi-bod, es; *n. A command, decree, an order;* mandatum, jussum, Exon. 25 a; Th. 71, 22; Cri. 1159: Hy. 4, 34; Hy. Grn. ii. 283, 34. v. be-bod.

bí-breád, es; *n. Bee-bread;* apium panis :—Þynceþ bíbreád swêtre, gif he ǽr bitres onbyrgeþ *bee-bread seemeth sweeter, if he before has a taste of bitter,* Bt. Met. Fox 12, 17; Met. 12, 9. v. beó-breád **I.**

bi-búgan; *p.* -beág, *pl.* -bugon; *pp.* -bogen *To avoid;* avertere, Exon. 45 a; Th. 154, 9; Gú. 840. v. be-búgan.

bi-bycgong, e; *f.* [be, bycg *from* bycgan *to buy*] *A selling away;* venditio. v. bebycgean.

bi-byrgan; *p.* de; *pp.* ed *To bury,* Exon. 24 b; Th. 71, 21; Cri. 1159. v. be-byrgan.

BICCE, bice, bicge, an; *f. A* BITCH, *a female of the canine kind;* canicula :—Biccean [biccan MS. B.] meolc *bitch's milk,* Med. ex Quadr. 9, 8, 9; Lchdm. i. 362, 15, 18. [*Piers P.* bicche; *Ger.* bätze, betze, petze, *f.: Icel.* bikkja, *f.*]

biccen; *adj. Belonging to a bitch;* caninus; *the adj. of* bicce.

bi-cerran *to pass by;* prætérire, Mk. Lind. Rush. War. 6, 48. v. be-cerran.

bicgan *to buy, procure,* Jn. Bos. 4, 8: Beo. Th. 2615; B. 1305: Exon. 120 b; Th. 463, 11; Hö. 68: Salm. Kmbl. 403; Sal. 202: Exon. 114 a; Th. 436, 37; Rä. 55, 12. v. bycgan.

bicge *a bitch;* canicula, Ælfc. Gl. 21; Wrt. Voc. 23, 33. v. bicce.

bi-clyppan; *p.* -clypte; *pp.* -clypt *To clip, embrace, inclose, clasp;* amplecti, Exon. 59 b; Th. 217, 8; Ph. 277. v. be-clyppan.

bícnian, bícnigan; *part.* bícniende; he bícneþ; *p.* ode; *pp.* od; *v. a.* **I.** *to beckon, nod;* innuere :—Bícnode him bícniende him *erat innuens illis,* Lk. Bos. 1, 22. Bícnodon hí to his fæder *innuebant patri ejus,* 1, 62 : 5, 7. **II.** *to indicate, signify, announce, shew;* indicare, significare :—He sceal mid bellan bícnigan ða tída *he shall with bells announce the times,* L. Ælf. C. 11; Th. ii. 346, 29. v. beácnian.

bícnung *a sign;* signum, signatio. v. beácnung.

bi-cowen *gnawed,* Exon. 99 b; Th. 373, 20; Seel. 111; *pp. of* bi-ceówan. v. -ceówan.

bi-cweðan; *p.* -cwæþ, *pl.* -cwǽdon; *pp.* -cweden *To say;* dicere, Exon. 37 b; Th. 123, 32; Gú. 331. v. be-cweðan.

bí-cwide *a proverb,* Prov. 22. v. big-cwide.

bi-cwom, *pl.* -cwómon *came, entered* :—Ðá ic to hám bicwom *when I came home,* Exon. 86 a; Th. 324, 14; Wíd. 94 : 20 b; Th. 53, 32;

Cri. 859: 17 a; Th. 39, 33; Cri. 631: 48 b; Th. 168, 2; Gû. 1071. Ût bicwômon [MS. bicwoman], 24 a; Th. 69, 1; Cri. 1114. v. be-com, p. of be-cuman.

bíd, es; n. Delay, abiding; mora:—Wearþ on bíd wrecen was driven to delay [on delay], Beo. Th. 5917; B. 2962. On bíd wríceþ drives on delay, Exon. 101 b; Th. 382, 29; Rä. 4, 3. DER. an-bíd, on-: bíd-fæst, -steal.

bi-dǽlan; p. -dǽlde; pp. -dǽled To deprive, bereave of anything, to deliver, release, free from anything; privare, sejungere, expertem reddere:—Duguþum bidǽled bereft of honours, Exon. 16 a; Th. 35, 24; Cri. 563. v. be-dǽlan.

BÍDAN, ic bíde, ðû bídest, bítst, bíst, he bídeþ, bít, pl. bídaþ; p. ic, he bád, ðû bide, pl. bidon; pp. biden; acc. gen. To BIDE, abide, continue, remain, tarry, wait, await, expect, endure; manere, remanere, morari, habitare in aliquo loco, expectare, consequi, sustinere:—Ic in wîte sceal bídan in bendum I in torment must abide in bonds, Cd. 214; Th. 268, 2; Sat. 49. Seó eorþe gíniende bád the earth continued yawning, Ors. 3, 3; Bos. 56, 4. Ðonne ðæt he ðǽr leng bide than that he should abide there longer, Ors. 2, 5; Bos. 48, 4. Mere stille bád the sea remained still, Cd. 158; Th. 197, 2; Exod. 300. Bídaþ assan on þurste expectabunt onagri in siti sua, Ps. Th. 103, 11. Swá mín sáwl bád sicut expectavit anima mea, 55, 6. He geþyldum bád he waited patiently, Exon. 46 a; Th. 157, 4; Gû. 886. Ûtan we well ðǽre tîde bídan bene expectemus horam illam, Bd. 4, 24; S. 599, 5. Bídaþ Dryhtnes dómes they await the Lord's doom, Exon. 23 a; Th. 63, 17; Cri. 1021. Bád sôþra geháta he awaited the faithful promises, Cd. 71; Th. 86, 2; Gen. 1424. Hie ðæs bidon for this they waited, Exon. 10 a; Th. 10, 4; Cri. 147. In helle heó bryne welme bídau sceolden in hell they must abide [endure] scorching heat, Cd. 213; Th. 266, 25; Sat. 27. Ðá seó circe hér eahtnysse bád then the church here endured persecution, Exon. 18 a; Th. 44, 18; Cri. 704. [Laym. biden, ibiden; p. ibæd, ibad, pl. biden; pp. ibiden, ibede: O. Sax. bídan: N. Frs. bida: O. Frs. bidia: N. Dut. N. L. Ger. beiden: N. Ger. dial. beiten: M. H. Ger. bíten: O. H. Ger. bítan: Goth. beidan: Dan. bie: Swed. bida: O. Nrs. bíða [for bída]: Ir. Gael. feith.] DER. a-bídan, ge-, ofer-, on-.

BIDDAN, ic bidde, ðû biddest, bidst, bitst, he biddeþ, bit, byt, bitt, pl. biddaþ; impert. bide, pl. biddaþ; p. ic, he bæd, ðû bæde, pl. bǽdon; pp. beden; followed by an acc. of the person, or by the prep. to, and a gen. of the thing; v. trans. To ask, pray, intreat, beseech, BID, order, require; petere, poscere, orare, quærere, precari, deprecari, rogare, postulare, præcipere, requirere:—Ic bidde peto, Ælfc. Gr. 28, 1; Som. 30, 41. Eádréd, cyning, biddeþ and hálsaþ Eadred, king, prayeth and intreateth, Cod. Dipl. 433; A. D. 955; Kmbl. ii. 304, 24: Ælfc. Gr. 33; Som. 37, 31. Ic bidde precor, 25; Som. 27, 11. Andreas ongann merelíðendum miltsa biddan Andrew began to ask mercy for the sea-faring men, Andr. Kmbl. 706; An. 353. Hû hí hine bǽdon [MS. bædan] rihtes geleáfan and fullwihtes bædes how they had asked him the favour of a right belief and of a font of baptism, Ors. 6, 34; Bos. 130, 30. Ongunnon ealle biddan ðæs ðe he bæd all began to pray that which he prayed, Bt. 35, 6; Fox 168, 30. Hý him to eów árna bǽdun they prayed to you for compassion, Exon. 27 b; Th. 83, 9; Cri. 1353. Bide hine ora eum, Ps. Spl. 36, 6. Bidde ðé mín Drihten quæso Domine mi, Gen. 19, 18. We biddaþ quæsumus, Ælfc. Gr. 33; Som. 37, 41. Ðone alwaldan ára biddan to intreat the all-powerful for benefits, Cd. 217; Th. 277, 24; Sat. 209. Gehýr, God, gebéd mín ðon ic bidde exaudi, Deus, orationem meam cum deprecor, Ps. Spl. 63, 1. Biddaþ rogate, Ps. Th. 121, 6. He bit sibbe rogat ea quæ pacis sunt, Lk. Bos. 14, 32. Gif he bit æg si petierit ovum, 11, 12. Gif hit [cild] æges bitt if he ask for an egg, Homl. Th. i. 250, 9. Gif hit [cild] hine hláfes bitt if he ask him for bread, 250, 8. Gif he byt fisces if he ask for a fish, Lk. Bos. 11, 11. Bide me postula a me, Ps. Th. 2, 8. Hí dôþ swá ic bidde they do as I bid, Beo. Th. 2467; B. 1231. He bæd him hláfas wyrcan he bade him make loaves, Cd. 228; Th. 307, 1; Sat. 673. Ðû bitst me ðæt ic lǽde ût ðis folc præcipis ut educam populum istum, Ex. 33, 12. Bide his me eft de manu mea require illum, Gen. 43, 9. [Orm. biddenn: Laym. bidde, bidden; he biddeþ, pl. biddeþ; impert. bide, bid: O. Sax. biddean: Frs. bidde: O. Frs. bidda: Dut. bidden: N. Ger. M. H. Ger. bitten: O. H. Ger. bitjan: Goth. bidyan: Dan. bede: Swed. bedja: O. Nrs. poet. biðja petere, rogare.] DER. a-biddan, ge-, on-: v. bedd.

biddende praying, Ors. 2, 5; Bos. 47, 40; part. of biddan.

biddere, es; m. A petitioner; petitor vel petax, Ælfc. Gl. 114; Som. 80, 19. v. biddan.

bide pray, ora:—Bide ðinne fæder ora tuum patrem, Mt. Bos. 6, 6; sing. impert. of biddan.

bi-deáglian to hide, cover, conceal, keep close or secret, Exon. 51 a; Th. 177, 12; Gû. 1226. v. be-deáglian.

bi-déglad hidden, obscured:—Bidéglad on dægréd obscured at dawn, Exon. 57 a; Th. 204, 15; Ph. 98; pp. of bi-déglian. v. be-déglad.

bídende abiding, Elen. Kmbl. 966; El. 484; part. of bídan.

bíd-fæst; adj. [bíd an abiding, delay; fæst fast, firm] Stationary,

firm; stabilis:—Hyre fôta wæs bídfæst [biidfæst MS.] ôðer one of its feet was stationary, Exon. 114 a; Th. 438, 13; Rä. 57, 7.

bíding, es; f. A BIDING, abode; mansio, statio:—Ðǽr hý bídinge môstun tídum brúcan where they might at times enjoy a biding, Exon. 35 b; Th. 114, 30; Gû. 180.

bi-droren deprived; orbatus, Exon. 77 b; Th. 291, 8; Wand. 79; pp. of bi-dreósan. v. dreósan, be-droren.

bíd-steal, -steall, es; m. [bíd an abiding, delay; steal a stall, place] A stand, halt; statio, mora:—He, beald in gebéde, bídsteal gifeþ he, bold in prayer, maketh a stand, Exon. 71 a; Th. 265, 29; Jul. 388. Ic eofore eom cénra, ðonne he, gebolgen, bídsteal gifeþ I am bolder than a wild boar, when he, enraged, makes a stand, 110 b; Th. 423, 11; Rä. 41, 19.

bi-dyrnan; p. de; pp. ed To hide, conceal; occultare, Exon. 24 a; Th. 67, 16; Cri. 1089. v. be-dyrnan.

bie be, Mk. Lind. War. 10, 44, for bió; subj. of bión to be.

biéon a beacon, wonder, Ps. Spl. C. 104, 25. v. beácen.

Bieda, an; m. Bieda the son of Port:—Hér com Port on Brytene, and his twegan sunau, Bieda and Mægla here, A. D. 501, Port came to Britain, and his two sons, Bieda and Mægla, Chr. 501; Erl. 15, 14.

Biedan heáfod; gen. heáfdes; dat. heáfde; m. [Biedan Bieda's, heáfod head: Flor. Bidanheafod, A. D. 1114] BIEDA'S HEAD = Bedwin, Wilts?—Hér Wulfhere and Æscwine gefuhton æt Biedan heáfde here, A. D. 675, Wulfhere and Æscwine fought at Bedwin, Chr. 675; Erl. 36, 9; Th. 58, 15, col. 1, 3.

Biedan ford Bedford, Chr. 571; Th. 32, 26, col. 2. v. Bedan ford.

bién-codd beanpod, Lk. Foxe 15, 16. v. beán-belgas.

bi-eóde venerated, Exon. 68 b; Th. 255, 3; Jul. 208; p. of bi-gán.

biereþ bears, carries, Exon. 58 b; Th. 211, 18; Ph. 199; for bireþ; 3rd pres. of beran.

bierm a bosom, Ps. Spl. C. 73, 12. v. bearm.

biernende burning, for byrnende. v. byrnan.

biersteþ, bierst bursts, Exon. 102 a; Th. 386, 16; Rä. 4, 62; 3rd pres. of berstan.

bieþ are, for bióþ, Mk. Lind. War. 10, 43. v. bión.

bi-féerende part. Passing by, Lk. Lind. War. 18, 36. v. be-féran.

bi-fæstan; p. -fæste; pp. -fæsted To fasten, make fast, fix, commit, intrust; infigere, committere, tradere, Exon. 97 a; Th. 362, 2; Wal. 30: 50 a; Th. 173, 26; Gû. 1166. v. be-fæstan.

bi-fangen surrounded, Exon. 15 b; Th. 33, 18, note; Cri. 527; pp. of bi-fón. v. be-fón.

bi-fealdan; p. -feóld, pl. -feóldon; pp. -fealden To infold, involve, inwrap, cover, overwhelm; implicare, involvere, circumdare, Exon. 9 b; Th. 8, 14; Cri. 117. v. be-fealdan.

bi-felgan; p. -fealh, pl. -fulgon; pp. -folgen To deliver, transmit, consign; tradere, committere, Exon. 72 b; Th. 271, 13; Jul. 481. v. be-felgan.

bi-féng, pl. -féngon held, seized; apprehendit, Exod. 415; Grn. i. 88, 415; p. of bi-fón. v. be-fón.

bi-feohtan; p. -feaht, pl. -fuhton; pp. -fohten To deprive by fighting; pugnando privare:—Feore bifohten deprived of life, Exon. 101 b; Th. 384, 23; Rä. 4, 32.

bi-feolan; p. -fæl, pl. -fǽlon; pp. -folen To commit, commend, deliver; immittere, commendare, tradere:—Bifolen in foldan committed to earth, Exon. 71 b; Th. 267, 18; Jul. 417: 17 b; Th. 42, 5; Cri. 668. v. be-feolan.

bifgende, bifigende trembling, trembling with a fever:—Bifgende febricitantem, Mt. Rush. Stv. 8, 14. v. bifian.

BIFIAN, bifigan, byfian, beofian; p. ode; pp. od To tremble, shake, be moved; tremere, contremere, commoveri:—Drihten besihþ eorþan and déþ hýg bifian Dominus respicit terram et facit eam tremere, Ps. Lamb. 103, 32: Rood Kmbl. 72; Kr. 36. He, bifiende, feóll to Iohannes fôtum he, trembling, fell at John's feet, Ælfc. T. 10: Cd. 92; Th. 118, 25; Gen. 1970. Ic bifige tremo, Ælfc. Gr. 35; Som. 38, 8. Eorþe [eorþan MS.] bifode terra tremuit, Ps. Spl. 75, 8: Rood Kmbl. 83; Kr. 42. Ða wudas bifodon the woods shook, Bt. 35, 6; Fox 168, 8. [O. Sax. biðon: Frs. bibbe, bibje: O. Frs. beva: Dut. beven: Ger. beben: M. H. Ger. biben: O. H. Ger. bibén: Dan. bäve: Swed. bäfwa: O. Nrs. bifast: Lat. pavere: Grk. φέβομαι: Sansk. bhî to fear.] DER. a-bifian.

bifigan to tremble; tremere, Ælfc. Gr. 35; Som. 38, 8. v. bifian.

bi-fleón; part. -fleónde To escape, to pass by or under, to go away privately; subterfugere, Cot. 192. v. be-fleón.

bi-folen committed, commended; Exon. 71 b; Th. 267, 18; Jul. 417; pp. of bi-feolan.

bi-fón; p. -féng, pl. -féngon; pp. -fangen, -fongen. I. to comprehend, grasp, seize, take hold of, attach, catch, ensnare; comprehendere, apprehendere, reprehendere, deprehendere, capere:—Folm mec mæg bifón the hand may grasp me, Exon. 111 a; Th. 425, 6; Rä. 41, 52. II. to surround, encompass, encircle, envelop, contain, invest, clothe, case, receive, conceive; circumdare, amplecti, capere, cingere,

tegere, operire, accipere, concipere :—Flǽsce bifongen *surrounded with flesh*, Exon. 98 a ; Th. 368, 33 ; Seel. 34. v. be-fôn.

bi-fongen *surrounded*, Exon. 98 a ; Th. 368, 33 ; Seel. 34 ; *pp. of* bi-fôn. v. be-fôn.

bi-fôran; *prep. dat. Before* ; ante, coram :—Wineleás guma gesihþ him bifôran fealwe wegas *the friendless mortal sees before him seared ways*, Exon. 77 a ; Th. 289, 10 ; Wand. 46 : 47 a ; Th. 160, 22 ; Gú. 947. v. be-fôran ; *prep.*

bi-fôran; *adv. Before, of old* ; antea :—Swâ ǽr bifôran *as ere of old*, Exon. 14 b ; Th. 29, 26 ; Cri. 468. v. be-fôran ; *adv.*

bifung, beofung, e ; *f.* [bifian *to tremble*] *A trembling, shaking* ; tremor :—Fyrhto oððe bifung begrâp híg *tremor apprehendit eos*, Ps. Lamb. 47, 7 : 54, 6. DER. eorþ-bifung.

bî-fylc, es ; *n.* [bî *by, near to* ; fylc *a tribe, country, province*] *A neighbouring people, province,* or *region* ; provincia *vel* populus adjacens :—Of eallum ðyssum bífylcum *de cunctis prope provinciis*, Bd. 3, 14 ; S. 540, 11.

big; *prep. dat. Of, about, concerning* ; de, quoad :—Big ðam ðe ic ðê ǽr sægde *de qua tibi ante dixi*, Bd. 2, 12 ; S. 514, 35. v. be 2.

bi-gǽþ *commits*, Exon. 27 a ; Th. 80, 18 ; Cri. 1308 ; *pres. of* bi-gán.

bi-gán, he -gǽþ ; *p.* -eóde, *pl.* -eódon ; *pp.* -gán.   I. *to commit, exercise, observe, enjoy* ; committere, exercere, observare, frui, Exon. 27 a ; Th. 80, 18 ; Cri. 1308.   II. *to honour, worship, venerate* ; colere, Exon. 68 b ; Th. 255, 3 ; Jul. 208. v. be-gán.

bîgan; *p.* de ; *pp.* ed ; *v. trans. To bow, bend, bend down, turn, turn back* ; flectere, deflectere, incurvare, retorquere :—His cneów bîgde on eorþan *genua flexit in terram*, Bd. 5, 21 ; S. 643, 15 : 3, 2 ; S. 524, 14 : Mt. Bos. 27, 29 : Exon. 62 b ; Th. 229, 23 ; Ph. 459 : Bd. 3, 19 ; S. 548, 8 : Lev. 1, 15. v. bŷgan.

bi-gang, -gong, es ; *m.*   I. *a course, way, passage, circuit* ; cursus, via, tenor, circuitus :—Tîda bigong *the course of seasons*, Exon. 11 a ; Th. 15, 13 ; Cri. 235.   II. *an undertaking, business, exercise, religious worship* ; negotium, exercitatio, cultus, Bd. 5, 1 ; S. 613, 9. v. be-gang.

bi-gangan *to go round, go to, attend, commit, practise, exercise, worship* ; exercere, incumbere, colere, Bd. 1, 7 ; S. 477, 33. v. bi-gongan, be-gangan.

big-cwide, bí-cwide, es ; *m.* [be, big *by* ; cwide *a saying*] *A by-saying, by-word, proverb, fable, tale* ; proverbium, fabula :—Ge forwurðaþ þurh bigspell and bigcwidas *eris perditus in proverbium et fabulam*, Deut. 28, 37. Bícwide *proverbium*, Prov. 22.

bige, es ; *n*? [bycgan, bicgan *to buy*] *A buying, exchange, commerce, traffic* ; emptio, permutatio, commercium, mercatus :—Gif gebyrige ðæt for neóde heora hwilc wið úre bige habban wille, oððe we wið heora, mid yrfe and mid ǽhtum, ðæt is to þafianne *if it happen that from necessity any of them will have traffic with us, or we with them, with cattle and with goods, that is to be allowed*, L. A. G. 5 ; Th. i. 156, 2-4.

bige *buy*, Jn. Bos. 13, 29 ; *impert. of* bicgan.

bíge, es ; *m. A bending, turning, bend, an angle, a corner* ; flexus, sinus, angulus :—Se engel eóde into ânum nyrwette, ðe he ne mihte forbúgan on nâðere healfe, forðamðe ðǽr nân bíge næs *angelus ad locum angustum transivit, ubi nec ad dexteram nec ad sinistram poterat deviare*, Num. 22, 26. Bíge *limes fractura membri*, Fulg. 19. v. bŷge.

bîgean *to bow, bend* ; flectere :—His cneówu bîgean *genua flectere*, Bd. 4, 31 ; S. 610, 23 : 3, 2 ; S. 524, 21 : Ps. Th. 94, 6. v. bŷgan.

bi-geat *obtained, seized*, Exon. 81 b ; Th. 306, 12 ; Seef. 6 ; *p. of* bi-gitan. v. be-gitan.

bi-gegnes, bi-gegnys, -ness, e ; *f. A going about* or *applying one's self to anything, the pursuit* or *study of anything* ; studium :—Bigegnes *vel* smeágung *studium*, Ælfc. Gl. 90 ; Wrt. Voc. 51, 27 : Gr. Dial. 1, 10. DER. eorþ-bi-gegnys.

bi-gellan; *p.* -geal, *pl.* -gullon ; *pp.* -gollen *To celebrate by song, to scream* ; canendo celebrare, exclamare :—Ful oft ðæt earn bigeal *the eagle screamed that often*, Exon. 81 b ; Th. 307, 16 ; Seef. 24.

bígels, es ; *m. An arch, a vault, an arched roof* ; arcus, fornix, camera, Ælfc. Gl. 93 ; Som. 75, 91 ; Wrt. Voc. 52, 41 : Cot. 201.   DER. for-bígels.

bi-genog *worship, observation*, Scint. 7. v. be-gang II.

bi-geng, es ; *m. Observation, worship, service* ; cultuð :—Bigeng *cultus*, Ælfc. Gr. 11 ; Som. 15, 18, MSS. C. D.   He bæd híg ðâ georne, ðæt híg búgan ne sceoldon fram Godes bigengum *he bade them then earnestly, that they should not decline from the services of God*, Jos. 23, 7. v. be-gang.

bi-genga, an ; *m. An inhabitant, dweller, cultivator* ; incola, cultor :—Ðæt ðæt Eálond Wiht onféng Cristene bigengan *ut Vecta insula Christianos incolas susceperit*, Bd. 4, 16 ; S. 584, 2.   Se ârfæsta bigenga ðæs gástlícan landes *pius agri spiritalis cultor*, 2, 15 ; S. 519, 8 : Deut. Grn. 4. 3.   DER. land-bigenga. v. be-ganga.

bi-geongende, bi-gongende ; *part.* [*part. of* bi-gongan, v. be-gongan, be-gangan] *Passing by* ; præteriens, Mk. Lind. War. 15, 21 ; Mk. Rush. War. 15, 21.

bi-gerdel *a purse, public purse*, Ælfc. Gl. 65 ; Som. 69, 35 ; Wrt. Voc. 40, 63. v. big-gyrdel.

biggencere, es ; *m. A worker* ; operator :—Ic hæbbe smiþas... and manega ôðre mistlícra cræfta biggenceras *habeo fabros... et multos alios variarum artium operatores*, Coll. Monast. Th. 30, 3.

big-geng *observation, worship* ; cultus :—Biggeng [MS. biggend] *cultus*, Ælfc. Gr. 11 ; Som. 15, 18. v. begang.

big-gyrdel, bî-gyrdel, -gerdel ; *g.* -gyrdles, -gerdles ; *m.* [big, bî, gyrdel *a girdle, belt, purse*] *A belt, girdle*, and as girdles were used to carry money, hence, *a purse, public purse, treasury* ; zôna = ζώνη, saccus = σάκκος, fiscus :—Næbbe ge feoh on eówrum bígyrdlum *nolite possidere pecuniam in zonis vestris*, Mt. Bos. 10, 9.   Bígerdel *saccus*, Ælfc. Gl. 3 ; Som. 55, 68 ; Wrt. Voc. 16, 41.   Cyninges [MS. kinges] gafoles bígerdel *saccus vel fiscus*, 65 ; Som. 69, 35 ; Wrt. Voc. 40, 63. Biggyrdel *fiscus vel saccus publicus*, 17 ; Som. 58, 94 ; Wrt. Voc. 22, 11.

big-hydig; *adj. Careful, watchful, solicitous, anxious* ; solicitus, sollers :—Wæs seó môder ðære gesomnunge bíhydig [MS. B. byghydig = bighydig] *sollicita est mater congregationis*, Bd. 4, 7 ; Whel. 277, 27. v. be-hydig.

big-hydiglíce, -hydilíce, -hydlíce, -hidiglíce ; *adv. Carefully* ; sollicite, sollerter :—Ðe he bighydiglíce heóld *which he carefully held*, Bd. 4, 31 ; S. 611, 2.   Heó hine bighydilíce [bighydlíce, Whel. 324, 8] sôhte *she carefully sought him*, 4, 23 ; S. 595, 4.   Bighidiglíce *sollicite*, 1, 27 ; S. 489, note 39. v. be-hydelíce.

bi-gitan, -gytan *to get, obtain, seize* ; assequi, acquirere, arripere, corripere, Exon. 32 b ; Th. 103, 19 ; Cri. 1690. v. be-gitan.

big-leofa, bí-leofa, an ; *m.* [big, bí *for*, líf *life*, leofen *living, nourishment*].   I. provision by which life is maintained, *Food, victuals, nourishment* ; cibus, victus, alimentum :—Ðú nimst witodlíce of eallum mettum... ðæt híg beón ðé to bigleofan *tolles igitur ex omnibus escis... et erunt tam tibi quam illis in cibum*, Gen. 6, 21.   Hwæt begytst ðú of ðínum cræfte ? Bigleófan, and scrúd, and feoh *quid adquiris de tua arte* ? *Victum, et vestitum, et pecuniam*, Coll. Monast. Th. 23, 3-6.   Bigleofa *victus*, Ælfc. Gr. 28, 5 ; Som. 32, 6. Bíleofa *alimentum*, C. R. Ben. 49.   II. that by which food is procured, *Money, wages* ; stips, stipendium :—Scipe *vel* bigleofa *stipendium*, Ælfc. Gl. 12 ; Som. 57, 92 ; Wrt. Voc. 20, 33. v. an-leofa, and-leofen.

big-leofan; *part.* ende ; *p.* ede ; *pp.* ed *To nourish, feed, support* ; cibare. v. big-leofa.

bi-glídan *to glide* or *disappear from any one, to desert any one* ; evanescere ab aliquo, derelinquere, Exon. 94 a ; Th. 353, 18 ; Reim. 14.

bíg-nes, -ness, e ; *f. A bending, bowing* ; flexio :—Se earm nǽnige bígnesse on ðam elnbogan hæfde *brachium nihil prorsus in cubito flexionis habuit*, Bd. 5, 3 ; S. 616, 23. v. bŷgan.

bi-gong *a course*, Exon. 54 b ; Th. 193, 29 ; Az. 129. v. be-gang.

bi-gongan *to attend, practise, observe, worship*, Exon. 44 b ; Th. 150, 11 ; Gú. 777. v. be-gangan.

bi-grafan *to bury* ; sepelire, Exon. 29 a ; Th. 89, 33 ; Cri. 1466. v. be-grafan.

bigsen *an example*, Bd. 3, 28, MS. B ; S. 560, note 35. v. bŷsen.

big-sittan; *p.* -sæt, *pl.* -sǽton ; *pp.* -seten *To sit by* or *near* ; adsidere :—Se bisceop ðæt geseah ðe him bigsæt *the bishop who sat by him saw it* ; quo viso pontifex qui adsidebat, Bd. 3, 6 ; S. 528, 22.

big-spæc, e ; *f. A by-speech, deceiving* ; supplantatio.   DER. big. spǽc.

big-spell, bí-spell ; *g.* -spelles ; *pl. nom. acc.* -spell, -spellu ; *n.* [big, bí, spell *a history*] *A by-history, a parable, fable, example, proverb, story* ; parabola, fabula, exemplum, proverbium, narratio :—Gehýre ge ðæs sáwendan bigspell *vos audite parabolam seminantis*, Mt. Bos. 13, 18. Ic ahylde on bigspelle eáre mín *inclinabo in parabolam aurem meam*, Ps. Spl. 48, 4.   Ealle ðás þing se Hǽlend spræc mid bigspellum to ðâm weredum ; and nân þing ne spræc he bútan bigspellum *hæc omnia locutus est Iesus in parabolis ad turbas* ; *et sine parabolis non loquebatur eis*, Mt. Bos. 13, 34, 35 : Ps. Lamb. 48, 5.   Bigspellu, *acc. pl.* Lchdm. iii. 214, 15.   He him rehte bígspell be ðære sunnan *he related to him a parable of the sun*, Bt. titl. vi ; Fox x. 12.   Ðeáh we sculon manega and mistlíce bísna and bíspell reccan *though we should relate many and various examples and fables*, Bt. 35, 5 ; Fox 166, 13, 19.   Gehýr sum bíspell *hear an example*, 37, 3 ; Fox 190, 21 : 39, 6 ; Fox 220, 21. Þurh bigspell and bigcwidas *in proverbium et fabulam*, Deut. 28, 37. We sculon ðé sum bíspell reccan *we will relate a story to thee*, Bt. 35, 6 ; Fox 166, 27 : Bt. Met. Fox 23, 17 ; Met. 23, 9. [*Kil.* bijspel : *Ger.* beispiel, *n* : *M. H. Ger.* bîspel, *n.*]   DER. bigspell-bôc.

bigspell-bôc, e ; *f.* [bigspell *parabola, proverbium*, bôc *liber*] *A book of parables, the Book of Proverbs* ; proverbiorum liber :—Salomon gesette þreó béc þurh his snoternisse : ân ys bigspellbôc *Solomon wrote three books by his wisdom : one is the Book of Proverbs*, Ælfc. T. 14, 26.

big-standan; *p.* -stôd, *pl.* -stôdon ; *pp.* -standen [big = bî *by, near*, standan *to stand*] *To stand by* or *near one, to support* ; stare cum aliquo, adstare, adjuvare :—Bigstandaþ me, strange geneátas *stand by me, strong associates*, Cd. 15 ; Th. 18, 36 ; Gen. 284.   Ða ðe him bigstôdon *those*

who stood by him, Byrht. Th. 137, 7; By. 182: Beo. Th. 6086; B. 3047.

big-swíc, es; *m. Deceit, guile*; fraus:—Bútan brede and bigswíce *without fraud and guile*, L. Ed. 1; Th. i. 160, 7. v. be-swíc.

big-wist, bí-wist, e; *f.* [wist *subsistence, victuals, food*; wesan *to be, exist*] *Food, nourishment, provision*; pabulum, alimentum, commeatus:—Bigwist *alimentum, pabulum*, Abus. 4. We lǽrap, ꝺæt hí habban þreóra daga bíwiste *we enjoin, that they have provision for three days*, L. Edg. C. 3; Th. ii. 244, 12. He habban sceal ꝺám þrim gefērscipum bíwiste *he must have provisions for the three classes*, Bt. 17; Fox 60, 3, 4.

bí-gyrdel *a girdle, purse*, Mt. Bos. 10, 9. v. big-gyrdel.

bi-gytan *to get, obtain, seize*; assequi, acquirere, arripere, corripere, Exon. 32 b; Th. 103, 19; Cri. 1690. v. be-gitan.

bi-healdan; *p.* -heóld, *pl.* -heóldon; *pp.* -healden.   I. *to hold by or near, guard, observe, preserve*; tenere, inhabitare, custodire, servare, præservare:—Ꝺǽr se ánhaga eard bihealdeþ *there the lonely [bird] holds its dwelling*, Exon. 57 a; Th. 203, 21; Ph. 87. Mec sǽwelcund hyrde bihealdeþ *a spiritual shepherd guardeth me*, Exon. 37 a; Th. 121, 15; Gú. 289. Hine weard biheóld of heofonum *a guardian from heaven guarded him*, Exon. 34 a; Th. 108, 22; Gú. 76: 54 b; Th. 193, 22; Az. 125. Se sceal ꝺære sunnan-síþ bihealdan *he shall observe the sun's course*, Exon. 57 a; Th. 203, 27; Ph. 90: 57 b; Th. 205, 17; Ph. 114. Hâteþ mec heáh-cyning bihealdan *the high king commands [them] to preserve me*, Exon. 110 b; Th. 424, 15; Rä. 41, 39.   II. *to see, look on, behold*; videre, intueri, aspicere:—Freó ꝺæt bihealdeþ hû me of hrífe fleógaþ hylde pílas *my master beholds how the shafts of battle fly from my belly*, Exon. 105 a; Th. 399, 3; Rä. 18, 5. v. be-healdan.

bi-heáwan; *p.* -heów; *pp.* -heáwen *To hew or cut off, to deprive of*; cædendo privare:—Iohannes bibeád heáfde biheáwan *commanded to cut off John's head*, Exon. 70 a; Th. 260, 10; Jul. 295. v. be-heáwan.

bi-helan; *p.* -hæl, *pl.* -hǽlon; *pp.* -holen *To conceal*; occultare, Exon. 27 a; Th. 80, 23; Cri. 1311. v. be-helan.

bi-helian *to hide, conceal*, Exon. 52 b; Th. 183, 14; Gú. 1327. v. be-helian.

bi-helmian; *p.* ade; *pp.* ad *To cover over, to cover, shroud*; co-operire:—Heolstre bihelmad *shrouded with darkness*, Exon. 69 a; Th. 257, 2; Jul. 241. v. be-helman.

bi-heonan *on this side.* v. be-heonan.

bi-hlǽman *to overwhelm with noise, to fall upon*; strepitu obruere:—Ꝺonne foldbúende se micla dæg meahtan Dryhtnes mægne bihlǽmeþ *then the great day of the mighty Lord will fall with might upon the earth's inhabitants*, Exon. 20 b; Th. 54, 18; Cri. 870. [*O. Sax. O. H. Ger.* hlamon *crepitare.*]

bi-hlǽnan; *p.* de; *pp.* ed *To surround or beset by leaning anything against another*; acclinando circumdare:—Lǽmen fæt wudu-beámum, holte bihlǽnan [bilǽnan MS.] *an earthen vessel with forest trees, with wood beset*, Exon. 74 a; Th. 277, 7; Jul. 577.

bi-hlemman; *v. a.* [be, hlemman *to d' sh together*] *To dash together*; collidere cum strepitu:—He ꝺa grimman goman bihlemmeþ fæste togædre *he dashes the grim jaws [gums] fast together*, Exon. 97 b; Th. 364, 26; Wal. 76.

bi-hlyhhan; *p.* -hlóh, *pl.* -hlógon; *pp.* -hlahen, -hleahen *To laugh at, deride*; ridere aliquid, exultare de aliqua re, Exon. 73 b; Th. 274, 1; Jul. 526. v. be-hlehhan.

bi-hófian; *p.* ode; *pp.* od *To have need of, to need, require*; egere, indigere, Exon. 37 b; Th. 123, 33; Gú. 332. v. be-hófian.

bi-hongen *behung, hung round*, Exon. 81 b; Th. 307, 1; Seef. 17; *pp.* of bi-hón. v. be-hón.

bi-hreósan; *p.* -hreás, *pl.* -hruron; *pp.* -hroren *To rush down, cover*; ruere, obruere, incidere:—Hríme bihrorene *covered with rime*, Exon. 77 b; Th. 291, 4; Wand. 77.

bi-hroren *rushed*, Exon. 77 b; Th. 291, 4; Wand. 77. v. bi-hreósan.

bi-hýdan; *p.* -hýdde; *pp.* -hýded *To hide, conceal, cover*; abscondere, occultare, operire, Exon. 61 b; Th. 227, 4; Ph. 418. v. be-hýdan.

bí-hydig *careful*, Bd. 4, 7; S. 574, 33. v. be-hydig, big-hydig.

bii; *prep. dat.* [= big = bí = be] *By, near to*; juxta, prope:—Se eádiga ærcebiscop Sanctus Laurentius bii his fôregengan Sancte Agustine bebyrged wæs *beatus archiepiscopus Laurentius juxta prædecessorem suum Augustinum sepultus est*, Bd. 2, 7; S. 509, 6. v. be 1.

BIL, bill, es; *n. An old military weapon, with a hooked point, and an edge on the back, as well as within the curve, a* BILL *or a broad two-edged sword, a falchion. Whatever its shape, it must have had two edges; as, in the earliest poem, an envoy is attacked*, billes ecgum, *with the edges of a bill*; falx, marra, falcastrum, ensis curvus. Hitherto this word has only been found in poetry:—Ꝺá ic, on morgne, gefrægn mǽg ôꝺerne billes ecgum on bonan stǽlan *then on the morrow, I have heard of the other kinsman setting on the slayer with the edges of a bill*, Beo. Th. 4963; B. 2485. Geseah ꝺá sige-eádig bil, eald sweord eótenisc *then he saw a victorious bill, an old giant sword*, Beo. Th. 3119; B. 1557. Abrægd mid ꝺý bille *he brandished with his sword*, Cd. 142; Th. 177, 17; Gen. 2931. Billa ecgum *with the edges of swords*, Cd. 210; Th.

260, 14; Dan. 709. Billum abreótan *to destroy with swords*, Cd. 153; Th. 190, 14; Exod. 199. [*Laym.* bil *a falchion*: *O. Sax.* bil, *n*: *Dut.* bijl, *f*: *Ger.* beil, beihel, *n*: *M. H. Ger.* bîle, bîl, *n*: *O. H. Ger.* bihal, bial, *n*: *Sansk.* bil *to divide*; findere.] DER. gúþ-bil, hilde-, stán-, twí-, wíg-, wudu-.

bí-lage [bí *by, near*, lagu *a law*] *A* BYE-LAW; lex privata, Chr. W. Thorn. an. 1303.

bíle, es; *m? A* BILL, *beak of a bird, a proboscis, the fore part of a ship*; rostrum, proboscis = προβοσκίς:—Bíle *rostrum*, Wrt. Voc. 77, 26. Ylpes bíle *vel* wrót *an elephant's proboscis*, Ælfc. Gl. 18; Som. 58, 128; Wrt. Voc. 22, 42.

bíle *a bile, carbuncle, sore*; ulcus, Som. Lye. v. býl.

bi-leác *locked up, shut up*, Exon. 124 b; Th. 479, 1; Rä. 62, 1, = beleác; *p.* of be-lúcan.

bi-lecgan; *p.* -legde, -léde; *pp.* -legd, -léd *To lay or impose upon, to lay round, cover, load, afflict, charge*; imponere, afficere, onerare, accusare, Exon. 107 a; Th. 409, 6; Rä. 27, 25. v. be-lecgan.

bi-légan; *p.* -légde; *pp.* -légd *To surround with flame*; circumflagrare flamma:—Lége bilégde *surrounded with flame* [*Ger.* umlodert mit lohe], Exon. 53 a; Th. 186, 7; Az. 16. v. be-légan.

bile-hwít; *adj.* [bile *the beak*, hwít *white*, referring to the beaks of young birds, then to their nature, *Junius*] *Simple, sincere, honest, without fraud or deceit, meek, mild, gentle*; simplex, mitis:—Arnwi munec wæs swíꝺe gód man and swíꝺe bilehwít *monk Arnwi was a very good man and very meek*, Chr. 1041; Erl. 169, 12. v. bilewit.

bile-hwítlíce; *adv. Honestly, simply*; honeste, simpliciter:—Andswarede Dryhthelm bilehwítlíce, forꝺon he wæs bylehwítre gleáwnesse and gemetfæstre gecynde man *Drycthelme respondebat simpliciter, erat namque homo simplicis ingenii ac moderatæ naturæ*, Bd. 5, 12; S. 631, 30.

bí-leofa *food*, C. R. Ben. 49. v. big-leofa.

bí-leofen, -lifen, e; *f.* [leofen *living, livelihood*] *Food, provisions*; annona, pulmentum:—Bí-leofene [MS. bileouene] *annona*, C. R. Ben. 43. Bílifen *pulmentum*, Cot. 171. v. big-leofa.

bi-leóran; *p.* de, ade; *pp.* ed *To pass by or over*; transire, præterire:—Gif bileórade fram [MS. from] him seó [MS. ꝺio] tíd *si transiret ab eo hora*, Mk. Skt. Lind. 14, 35, 36. Se bileórde *qui præterivit*, Ps. Surt. 89, 4. v. leóran.

bi-leósan; *p.* -leás, *pl.* -luron; *pp.* -loren *To bereave, deprive*; orbare, privare:—Ꝺá afyrhted wearþ ár, elnes biloren *then the messenger was affrighted, bereft of courage*, Exon. 52 a; Th. 181, 30; Gú. 1301. v. be-leósan.

bile-wit, bele-wit, bil-wit; *adj.* [bile, wit *mind, wit*] *Merciful, mild, gentle, simple, honest*; æquanimus, mansuetus, mitis, simplex, honestus:—Bilewit Dryhten *merciful Lord*, Ps. C. 50, 99; Ps. Grn. ii. 279, 99: Bt. Met. Fox 20, 138; Met. 20, 69: 20, 510; Met. 20, 255: 20, 538; Met. 20, 269. We bletsiaþ bilewitne feder *we bless the merciful father*, Hy. 8, 8; Hy. Grn. ii. 290, 8. Gehýran ꝺa bilewitan [MS. bylewitan] *audiant mansueti*, Ps. Spl. 33, 2. Beóþ eornustlíce gleáwe swá nædran, and bilwite [MS. bilwyte] swá culfran *estote ergo prudentes sicut serpentes, et simplices sicut columbæ*, Mt. Bos. 10, 16: 11, 29.

bile-witness, bil-witness, e; *f. Mildness, simplicity, innocence*; simplicitas:—Se God wunaþ simle on ꝺære heán ceastre his ánfealdnesse and bilewitnesse *God dwells always in the high city of his unity and simplicity*, Bt. 39, 5; Fox 218, 19. Hý on bilwitnesse hyra líf alysdon *they passed their lives in simplicity*, Ors. 1, 2; Bos. 27, 5.

bil-gesleht, bill-gesliht, -geslyht, es; *n.* [bil, bill *a sword*, gesleht *a clashing, conflict, slaughter*; from sleán *to slay, kill*] *A clashing of swords, battle*; ensium concutio, pugna:—Gelpan ne þorfte beorn blandenfeax bilgeslehtes [billgeslyhtes, Cott. Tiber. A. vi; billgeslihtes, Cott. Tiber. B. i: Cott. Tiber. B. iv] *the grizzly-haired warrior needed not boast of the clashing of swords*, Chr. 937; Th. 204, 35, col. 1; Æðelst. 45.

bilgst, bilhst, he bilgþ, bilhþ *art angry, is angry*; 2nd *and* 3rd *pers. pres.* of belgan *to be angry*.

bil-hete, bill-hete, es; *m.* [bil, bill *ensis*, hete *odium*] *The hate of swords*; odium ope ensium manifestatum:—Æfter billhete *after the hate of swords*, Andr. Kmbl. 156; An. 78.

bí-libban; *p.* -lifde; *pp.* -lifed, -lifd [bí 1. *by, upon*, libban *to live*] *To live by or upon, to be sustained or supported*; vesci, sustentari:—Sciꝺꝺium wearþ emleóf, ꝺæt hý gesáwon mannes blód agoten, swá him wæs ꝺara nýtena meolc, ꝺe hý mǽst bílibbaþ *it was as agreeable to the Scythians to see* [lit. *that they saw*] *man's blood shed, as it was* [*to see*] *the milk of their cattle, upon which they mostly live*, Ors. 1, 2; Bos. 26, 31-33. God ꝺás eorþan, ꝺe ealle cwice wihta bílibbaþ, ealle hire wæstmbǽro gelytlade *God lessened this earth, all its fruitfulness, by which all living creatures are supported*, 2, 1; Bos. 38, 8.

bi-liden *left, departed*, Exon. 52 a; Th. 182, 18, = be-liden; *pp.* of be-líꝺan, *q. v.*

bí-lifen *food*, Cot. 171. v. bí-leofen.

bilig *a bag, bottle, skin*; uter, Ps. Spl. M. 118, 83. v. belg.

bi-lihþ *dishonours, defames,* Exon. 90 a; Th. 337, 16; Gn. Ex. 65,= be-hlíþ; *pres. of* be-hlígan, q. v.

bilíþ, es; *n. An image, a representation, resemblance, likeness, pattern, example;* imago, effigies:—Bilíþe wǽron eorlas Ebrēa *the men were the images* [*likenesses*] *of the Hebrews,* Cd. 187; Th. 232, 7, note a. [*O. Sax.* biliði, n: *Frs. O. Frs.* bilethe, byld, n: *Dut.* beeld, n: *Ger.* bild, n: *M. H. Ger.* bilde, n: *O. H. Ger.* biladi, bilidi, n: *Dan.* billed, billede, n: *Swed.* bild, m; belǽte, n: *O. Nrs.* bílldr, m. forma, aspectus; bílǽti, n. effigies, statua, Rask Hald.]

bill *a bill, falchion;* falcatus ensis:—Bill *falcastrum,* Ælfc. Gl. 51; Som. 66, 1; Wrt. Voc. 34, 61: Beo. Th. 5548; B. 2777. v. bil.

bill-gesliht, -geslyht *a clashing of swords,* Chr. 937; Th. 205, 35; Æðelst. 45. v. bil-gesleht.

bi-locen *locked up,* Exon. 26 a; Th. 77, 21; Cri. 1260,=be-locen; *pp. of* be-lúcan.

bilod *having a bill, nib* or *snout;* rostratus, Som. DER. bile *a bill, beak.* v. ge-bilod.

bi-loren *deprived,* Exon. 52 a; Th. 181, 30; Gú. 1301,=be-loren; *pp. of* be-leósan.

bilst, he biþ *bellowest, bellows; 2nd and 3rd pers. pres. of* bellan.

bil-swæþ; *gen.* -swædes, *pl. nom.* -swaðu; *n. A bill* or *sword track:*—Bilswaðu blödige *bloody sword tracks,* Cd. 160; Th. 198, 27; Exod. 329. v. swæþ, n.

bi-lúcan *to lock up, inclose, surround,* Exon. 31 b; Th. 99, 14; Cri. 1624. v. be-lúcan.

bil-wetnes *innocence,* Bd. 3, 27; S. 559, 28. v. bile-witness.

bil-wit *mild,* Cd. 40; Th. 53, 4; Gen. 856. v. bile-wit.

bil-witness *simplicity,* Ors. 1, 2; Bos. 27, 5. v. bile-witness.

bi-mǽnan; *p.* de; *pp.* ed *To bemoan, bewail, lament, mourn;* lugere:—Woldan wíf wôpe bimǽnan ædelinges deáþ *the women would with weeping bewail the noble's death,* Exon. 119 b; Th. 459, 24; Hö. 4. v. be-mǽnan.

bi-míðan; *p.* -máþ, *pl.* -miðon; *pp.* -miðen *To hide, conceal;* occultare, abscondere, Exon. 34 b; Th. 110, 33; Gú. 118: Ps. Th. 68, 6. v. be-míðan.

bi-murnan; *p.* -murnde; *pp.* -murned *To mourn, be troubled about, care for;* lugere, curare, sollicitum esse de re, Exon. 87 a; Th. 328, 7; Vy. 14: 34 a; Th. 110, 1; Gú. 101. v. be-murnan.

bi-mútian; *p.* ade; *pp.* ad [mútung *mutuum,* Cot. 136] *To exchange for;* commutare:—Swá ðás woruldgestreón on ða mǽran gód bimútad weorþaþ *so these world-treasures shall be exchanged for the greater good,* Exon. 33 b; Th. 106, 17; Gú. 42.

BIN, binn, e; *f. A manger, crib,* BIN, *hutch;* præsepe, præsepium:—Binn *præsepe,* Ælfc. Gr. 9, 2; Som. 8, 27. Heó hine on binne alède *reclinavit eum in præsepio,* Lk. Bos. 2, 7, 12, 16: Exon. 18 b; Th. 45, 25; Cri. 724. On heora assena binne *in the manger of their asses,* Homl. Th. i. 30, 13, 31. [*Chauc.* binn: *Dut.* ben, f: *Ger.* benne, binne, f.]

BINDAN, to bindenne; ic binde, ðú bindest, bintst, binst, he bindeþ, bint, *pl.* bindaþ; *p.* ic, he band, bond, ðú bunde, *pl.* bundon; *pp.* bunden; *v. a. To* BIND, *tie;* ligare, alligare:—Hió bindan þenceaþ cyningas *she thinks to bind kings,* Ps. Th. 149, 8. Fæste binde swearte wealas *I bind the swart strangers fast,* Exon. 103 b; Th. 393, 21; Rä. 13, 3. Hrusan [MS. hruse] bindeþ wintres wôma *the winter's violence binds the earth,* Exon. 78 a; Th. 292, 21; Wand. 102. Híg bindaþ hefige byrðyna *alligant onera gravia,* Mt. Bos. 23, 4. He band hine *he bound him,* Gen. 42, 24. Hrím hrusan bond *frost bound the earth,* Exon. 87 a; Th. 307, 31; Seef. 32. Úser Hǽlend [MS. hælendes] wæs bunden fæste *our Saviour was bound fast,* Exon. 116 b; Th. 449, 5; Dôm. 66. [*Chauc.* binde: *Laym.* binde, binden: *Orm.* bindenn: *O. Sax.* bindan: *Frs.* bynnen: *O. Frs.* binda: *Dut. Ger. M. H. Ger.* binden: *O. H. Ger.* bintan: *Goth.* bindan: *Dan.* binde: *Swed.* binda: *O. Nrs.* binda.] DER. an-bindan, be-, for-, ge-, in- [=un-], on- [=un-], un-, ymb-.

binde, an; *f.* [bindan *to bind*] *A band, wreath, head-band, fillet;* corolla, fascia:—Hió an Ceoldryþe hyre betstan [MS. betsðan] bindan *she gives to Ceoldryth her best band,* Cod. Dipl. 1290; A.D. 995; Kmbl. vi. 133, 18, 20.

bindele, byndele, byndelle, an; *f. A binding, tying, fastening with bands;* vinculis constrictio:—Be mannes bindelan *concerning* [*the*] *binding* [*putting in bands*] *of a man.* L. Alf. pol. 35; Þh. i. 84, 1, note 2.

bindere, es; *m.* [bindan *to bind*] *One who binds, a* BINDER; ligator:—Ic eom binder and swingere *I am a binder and a scourger,* Exon. 107 b; Th. 409, 25; Rä. 28, 6.

bi-neótan; *p.* -neát, *pl.* -nuton; *pp.* -noten *To deprive of the enjoyment* or *use of anything:*—On hyge hálge heáfde bineótan *to deprive the holy one in spirit of his head,* Exon. 74 b; Th. 278, 28; Jul. 604. He hine ealdre bineát *he deprived him of life,* Beo. Th. 4784; B. 2396. v. be-neótan.

bi-niman [*Goth.* bi-niman *auferre, furari;* κλέπτειν] *to deprive.* v. be-niman.

binn *a manger,* Ælfc. Gr. 9, 2; Som. 8, 27. v. bin.

binnan [be-innan]; *prep. dat. acc. Within, in, into;* intra, infra, in :—

Ðe binnan ðam fæstenne wǽran *who were within the fastness,* Ors. 4, 11; Bos. 97, 39: Mt. Bos. 2, 16. Gyt ne com se Hǽlend binnan ða ceastre *nondum Iesus venerat in castellum,* Jn. Bos. 11, 30. [*Northumb.* binna, bionna: *Frs.* binnen: *O. Frs.* binna, binnia: *Dut. Kil. Ger. M. H. Ger.* binnen.] DER. innan.

bi-nom, *pl.* bi-nômon *deprived,* Exon. 100 a; Th. 378, 15; Deór. 16: 37 b; Th. 122, 30; Gú. 313, = be-nam, -nâmon; *p. of* be-niman. v. niman.

bi-noten *deprived,* Exon. 45 b; Th. 156, 10; Gú. 872; *pp. of* bi-neótan.

bintst, binst, he bint *bindest, binds; 2nd and 3rd pers. pres. of* bindan *to bind.*

bið *I am* or *shall be,* Bt. 40, 5; Fox 240, 24; *pres. of* bión. v. beó, beón.

bið-breád *honey-comb,* Bt. 23; Fox 78, 25. v. beó-breád.

biódan *to command, announce, offer,* Beo. Th. 5777; B. 2892: Bt. 25; Fox 88, 18. v. beódan.

bióm *I am, shall be;* sum, ero:—Ic beóm hál *vel* gehǽled ic bióm *salva ero,* Mk. Lind. Rush. War. 5, 28: Jn. Rush. War. 7, 34; *1st pers. pres. of* bión. v. beón.

bión, ic bió, bióm, he bióþ, *pl.* bióþ, bieþ, biaþ; *subj.* bió, bie *to be;* esse, existere, fieri :—Ic bió swíðe fægn *I shall be very glad,* Bt. 40, 5; Fox 240, 24. Bióm, Jn. Rush. War. 7, 34. Hwæt iów ðý bet bió oððe þince *what is or appears to you the better?* Bt. Met. Fox 10, 130; Met. 10, 65; Beo. Th. 5487; B. 2747: Mk. Lind. War. 10, 44. Ne mæg hira ǽnig bútan ôðrum bión *nor can any of them exist without the others,* Bt. Met. Fox 20, 290; Met. 20, 145: 11, 102; Met. 11, 51: Bt. 33, 4; Fox 130, 26: Th. Diplm. A.D. 804; 459, 16. Ðonne bióþ brocene *then will be broken,* Beo. Th. 4132; B. 2063: Andr. Kmbl. 815; An. 408: Elen. Grm. 1289: Bt. Met. Fox 7, 46; Met. 7, 23: 24, 121; Met. 24, 61: Ps. C. 50, 80; Ps. Grn. ii. 278, 80: Mk. Lind. War. 10, 43: Mt. Lind. Stv. 26, 31. v. beón.

biór *beer,* Prov. 31. v. beór.

biorg *a hill, mountain;* collis, mons, Exon. 35 a; Th. 112, 20; Gú. 146. v. beorg.

biorhto *brightness,* Bt. 41, 1; Fox 244, 7. v. beorhtu.

biorn, es; *m. A warrior, soldier, hero;* bellator, miles, heros:—Biorn under beorge bordrand onswâf wið Geáta dryhten *the hero under the mount turned his shield's disc against the lord of the Goths,* Beo. Th. 5111, note; B. 2559. DER. folc-biorn. v. beorn II.

biór-sel·, es; *m. A beer-hall, feasting-hall,* Beo. Th. 5263; B. 2635. v. beór-sele.

bióþ *is, are,* Bt. Met. Fox 7, 46; Met. 7, 23: 24, 121; Met. 24, 61; *3rd pers. pres. of* bión. v. beón.

biótian *to threaten;* intentare, Cot. 108. v. beótian I.

biótul *a beetle, staff,* Cot. 28. v. býtl.

bið-wyrt *bee-wort;* apiastrum, Glos. Epnl. Recd. 153, 20. v. beó-wyrt.

BIRCE, ean; *f.* berc, beorc, byrc, e; *f. A birch-tree;* betula alba :—Genim bircean *take of the birch-tree,* L. M. 1, 36; Lchdm. ii. 86, 7: Wrt. Voc. 285, 22. [*Scot.* birk: *Plat.* barke, f: *Dut.* berke-boom, m: *Kil.* berck: *Ger.* birke, f: *M. H. Ger.* birke, birche, f: *O. H. Ger.* bircha, f: *Dan.* birk, m. f: *Swed. O. Nrs.* biörk, f. betula vel quæcunque arbor viridis.]

bircen, beorcen; *adj.* BIRCHEN, *belonging to birch;* betulaceus, Som. Lye. [*Kil.* bercken.]

birc-holt, es; *n. A birch holt* or *grove;* betuletum. v. byrc-holt.

bird *the young of any of the feathered tribe;* pullus:—Birdas *pullos,* Lk. Lind. Rush. War. 2, 24. v. brid.

bi-reáfian; *p.* ode; *pp.* od *To bereave;* privare, Exon. 87 b; Th. 328, 30; Vy. 25. v. be-reáfian.

bi-reófan; *p.* -reáf, *pl.* -rufon; *pp.* -rofen *To bereave, deprive;* spoliare, privare:—Rǽdum birofene *bereft of counsel,* Exon. 30 a; Th. 93, 14; Cri. 1526: 104 a; Th. 394, 22; Rä. 14, 7. v. be-reófan.

birest, he bireþ *bearest, bears,* Bt. Met. Fox 20, 551; Met. 20, 276: L. In. 57; Th. i. 138, 15; *2nd and 3rd pers. pres. of* beran.

birgan; *p.* de; *pp.* ed *To cover with a mound, to bury;* sepelire:—Birge man hine ðæs ilcan dæges *sepelietur in eadem die,* Deut. 21, 23: Gen. 49, 31. v. be-birgan. v. byrgan.

birgean *to bury:*—Iosue hēt hí birgean *Joshua ordered to bury them,* Jos. 10, 27. v. birgan.

birgels, es; *m. A burial-place, sepulchre;* sepulcrum :—Him sylfum to birgelse *in possessionem sepulcri,* Gen. 23, 9. v. byrgels.

birgen, birgenn, e; *f. A burying-place, sepulchre;* sepulcrum, Gen. 23, 4, 6: 49, 30: 50, 5: Num. 11, 34. v. byrgen.

birg-nes, -ness *a taste,* Cot. 97. v. býrignes.

birhtu *brightness, splendour;* claritas, splendor, Bt. Met. Fox 6, 11; Met. 6, 6: 20, 537; Met. 20, 269. v. beorhtu.

birig *to a city, for* byrig, Gen. 13, 12: Deut. 14, 27; *d. s. of* burh.

birigan *to bury.* v. be-birigan, byrigan.

birigh-man *a city officer;* ædilis, Ælfc. Gr. 9, 28; Som. 11, 29. v. byrig-man.

bí-rihte, -ryhte; *prep. dat. Near, close by;* juxta :—Geseh he on

greóte gingran síne bíryhte [Kmbl. birihte] him swefan on slǽpe *he saw his disciples near him slumbering in sleep on the sand*, Andr. Recd. 1699; An. 850.

birihto *brightness*, L. E. I. 20; Th. ii. 414, 11. v. beorhtu.

birilian, birlian, byrlian; *p.* ode, ade; *pp.* od, ad *To draw, bear*; haurire, Jn. Lind. Rush. War. 2, 8, 9.

bi-rinnan; *p.* -ran; *pp.* -runnen *to run as a liquid*, hence,—*To wet, bedew*; fluere, perfundere, irrigare:—Ða wearþ beám monig blódigum teárum birunnen, sæp wearþ to swáte *then many a tree became bedewed with bloody tears, their sap became [turned to] blood*, Exon. 25 a; Th. 72, 19–23; Cri. 1175–1177.

Birīnus, i; *m.* Latin: Birīne, Byrīne, es; *m. Birinus, the first bishop of Wessex, sent by pope Honorius to Britain in* A.D. 634:—Ðære tíde ða West-Seaxna þeód mid Cynigelse heora cyninge Cristes geleáfan onféng, bodade him and lǽrde Godes word Birīnus biscop, se mid Honorius geþeahte ðæs Papan com on Breotene .... He ða lǽrde ðǽr godcunde láre, and ðone cyning to Cristes geleáfan gecyrde, and hine gecristnade, and hine eft æfter fæce mid fulluhtbæðe aþwógh mid his þeóde West-Seaxum. Hit gelamp on ða sylfan tíd ðe mon ðone cyning fullade, ðæt ðǽr wæs se hálgesta and se sigefæstesta cyning Norþan Hymbra Oswald andweard .... Ða sealdon hí and geáfon ðam bisceope begen ða cyningas eardungstówe and biscopsetl on Dorcceastre, and he ðǽr, se biscop, Gode lifde and cyricean worhte and hálgode .... and he ðǽr his dagas ge-endode and to Drihtne férde, and in ðære ylcan ceastre bebyriged wæs, and eft æfter monigum geárum Hædde bisceop hét his líchoman up adón and lǽdan [MS. lædon] to Winton ceastre *eo tempore* [A.D. 634] *gens Occidentalium Saxonum, (qui antiquitus Gevissæ vocabantur,) regnante Cynigilso fidem Christi suscepit, prædicante illis verbum Birīno episcopo, qui cum consilio papæ Honorii venerat Brittaniam .... Itaque evangelizante illo in præfata provincia, cum rex ipse catechizatus, fonte baptismi cum sua gente ablueretur, contigit tunc temporis sanctissimum ac victoriosissimum regem Nordanhymbrorum Osualdum adfuisse .... Donaverunt autem ambo reges eidem episcopo civitatem quæ vocatur Dorcic [Dorchester], ad faciendum inibi sedem episcopalem; ubi factis dedicatisque ecclesiis ... migravit ad Dominum, sepultusque est in eadem civitate, et post annos multos Hædde episcopatum agente translatus inde in Ventam civitatem* [Winchester], Bd. 3, 7; S. 529, 4–6; 12–16; 18–21; 22–24. Hér forþférde Birīnus se biscop *here*, A.D. 650, *Birinus the bishop died*, Chr. 650; Th. 51, 1, col. 2. Hér Ægelbryht of Galwalum æfter Birīne [Byrīne, col. 2, 3] ðam Rómániscan bisceope onféng Wesseaxna biscopdóme *here*, A.D. 650, *Ægelbyrht of Gaul succeeded to the bishopric of the West-Saxons after Birinus the Roman bishop*, 650; Th. 50, 1–5, col. 1.

birst, he birsteþ, birst *burstest, bursts; 2nd and 3rd pers. pres. of* berstan.

birþ *bears; 3rd pers. pres. of* beran.

bi-sæc *a bag*, Mt. Rush. Stv. 10, 10. v. sæc, codd.

bi-sǽce, es; *n? m?* I. *a visit*; visitatio:—Bád bísǽce betran hyrdes *waited the visit of a better keeper*, Exon. 35 b; Th. 115, 11; Gú. 188. II. *persecution, dispute, litigation*; controversia, litigatio:—Bísǽce *in litigation*, L. Edg. C. 62; Th. ii. 258, 3. Gif ðǽr hwæt bísǽces sý, seme se biscop *if there be somewhat of dispute, let the bishop settle it*, Const. vii; Th. ii. 258, note a. DER. sǽcan, sécan *to seek, visit, persecute, dispute*.

bi-scær *sheared or cut off*, Reim. 26; *p. of* bi-sceran. v. be-sceran.

bi-scencan; *p.* -scencte, *pl.* -scencton; *pp.* -scenced [scencan *to give drink*, scenc *drink*] *To give to drink; ad potionem dare*:— Ge in wræcsíðe longe lifdon, lége biscencte *ye [fallen spirits] have long lived in exile, flame being given [you] to drink*, Exon. 41 b; Th. 139, 21; Gú. 596.

bisceop, biscop, biscep, es; *m.* I. *a* BISHOP, *prelate*; episcopus:—Se bisceop is gecweden *episcopus* and is *ofersceáwigend* on Englisc, ðæt he ofersceáwige symle his underþeóddan *the bishop is called episcopus, that is in English, overseer, because he constantly oversees his subordinates*, L. Ælf. P. 37; Th. ii. 378, 28. Nis ná máre betwyx mæsse-preóste and biscope, búton ðæt [Th. ii. 348, 24] se biscop is geset to máran bletsunge ðonne se mæsse-preóst sý; ðæt is, circan to hálgigenne, and to hádigenne preóstas, to bisceopgenne cild [Th. ii. 348, 26: MS. men to biscopienne], and to bletsigenne ele *there is no difference between a mass-priest and a bishop, but that the bishop is appointed for greater benediction [blessing] than is the mass-priest; that is, to hallow churches, and to ordain priests, to confirm children, and to bless oil*, 36; Th. ii. 378, 20; v. mæsse-preóst. Seó mægþ hafþ twegen bisceopas *the province has two bishops*, Bd. 4, 5; S. 573, 33. II. *a chief priest of the Jews*; pontifex:—Se forma biscop, ðe God silf gesette, wæs Aaron geháten *the first high priest, whom God himself appointed, was called Aaron*, L. Ælf. P. 38; Th. ii. 378, 32. Scrídde ðone biscop mid línenum reáfe *vestivit pontificem subucula linea*, Lev. 8, 7. Ða astyredon ða bisceopas ða menegu *pontifices autem concitaverunt turbam*, Mk. Bos. 15, 11. Se bisceop acsode ðone Hǽlend *pontifex interrogavit Iesum*, Jn. Bos. 18, 19, 22, 24. III. *a heathen priest of the Romans and Egyptians; the chief priest of the Romans was called Pontifex*

Maximus, which was a title assumed by the Consuls and Emperors, v. yldest-bisceop:—Sǽdon ða Égyptiscan bisceopas, ðæt ða Godes wundor hiora ágnum godum getealde wǽron, ðæt sint deófol-gild *the Egyptian priests said, that the godlike wonders were ascribed to their own gods, which are idols*, Ors. 1, 5; Bos. 28, 25. Bisceopas on Róme sǽdon, ðæt heora godas bǽdon ðæt him man worhte anfiteatra *the priests in Rome said, that their gods ordered them to build an amphitheatre*, Ors. 3, 3; Bos. 55, 26. Lucinius Crassus, se consul, wæs eác Rómána yldesta bisceop *Lucinius Crassus, the consul, was also the chief priest* [pontifex maximus] *of the Romans*, Ors. 5, 4; Bos. 104, 16. IV. *the rank of an Anglo-Saxon bishop was equal to that of the Ealdorman, or highest nobleman, being only inferior to the Ædeling or prince, for they had equal power as judges in civil courts of law,—and their burh-brice and wer-gyld were the same*:—Bisceope gebyreþ ǽlc rihting, ge on godcundan þingan ge on woruldcundan *to a bishop belongs every direction [righting] both in divine and worldly things*, L. I. P. 7; Th. ii. 312, 9. Sculon bisceopas, mid woruld-déman, dómas dihtan ðæt hí ne geþafian, gyf hí waldan magan, ðæt ǽnig unriht up-aspringe *bishops, with temporal judges, should so direct judgments that they never permit, if it be in their power, that any injustice spring up*, 7; Th. ii. 312, 35–37. And séce man hundred-gemót swá hit ǽr geset wæs; and hæbbe man þríwa on geáre burh-gemót; and túwa scír-gemót; and ðǽr beon ða geséte scíre bisceop and se ealdorman, and ðǽr ægðer tǽcan ge Godes riht ge woruld-riht *and let the hundred-moot be attended as it was before fixed; and thrice in the year let a city-moot be held; and twice a shire-moot; and let there be present the bishop of the shire and the ealdorman, and there each expound both God's law [right] and the world's law*, L. Edg. ii. 5; Th. i. 268, 2–5: L. C. S. 18; Th. i. 386, 4–8. Biscopes and ealdormannes burg-bryce biþ lx scillinga *a bishop's and an ealdorman's burh-bryce shall be sixty shillings*, L. Alf. pol. 40; Th. i. 88, 8, note 19, H. Biscopes and ealdormannes mund-brice gebéte mid ii pundum *recompense a bishop's and an ealdorman's mund-brice with two pounds*, L. Eth. vii. 11; Th. i. 332, 1. Biscopes and ealdormannes wér-gyld is viii þúsend þrymsa *a bishop's and an ealdorman's wer-gild is eight thousand thrymsas*, L. Wg. 3; Th. i. 186, 7. V. *the bishops were the best educated men of their age, and often the most energetic, their advice and assistance were, therefore, naturally sought in every case of emergency in the cabinet or in the field,—Hence Ealhstan, the bishop of Sherborne for fifty years* [Ealhstán hæfde ðæt biscopríce l wintra æt Scyreburnan, A.D. 817–867: Chr. 867; Ing. 98, 12–14], *became a general of Egbert and of his son Æthelwulf*:—Ecgbryht, West-Seaxna cyning, sende Æðelwulf his sunu of ðære fyrde, and Ealhstán his bisceop, to Cent micele werede, and hý Baldred ðone cyning norþ ofer Temese adryfon *Egbert, king of the West-Saxons, sent his son Æthelwulf, and Ealhstan his bishop, into Kent, with a large part of the army, and drove Baldred the king northward over the Thames*, Chr. 823; Ing. 87, 6–15: 845; Ing. 92, 1. Æt Mere-túne wearþ Heáhmund biscop ofslegen, and feala gódra monna *at Merton bishop Heahmund was slain, and many good men*, 871; Ing. 101, 1–9. [Orm. bisskopp, bisscopp, bisshopp: Laym. biscop, bissop: Wyc. bischop: O. Sax. biskop: Dut. bisschop: Ger. M. H. Ger. bischof: O. H. Ger. piscof: Goth. aipiskaupus: Dan. bisp: Swed. biskop: O. Nrs. biskup: Fr. évêque: Span. obispo: It. vescovo: Wel. esgob: Gael. easbuig: Ir. easbog: Arm. eskop: Slav. biskup: Lith. wyskupas. From the Lat. [e-piscop-us, hence O. H. Ger. piscof: A. Sax. biscop: Orm. bisshopp: Laym. biscop: Wyc. bischop: Eng. bishop]=Grk. ἐπίσκοπος *an overseer, guardian*, from ἐπί *upon, over*,—σκοπός *one who watches*,—σκοπέω *to look, watch, consider, contemplate*.] DER. arce-biscop, -biscop, ealdor-: bisceop-dóm, -gegyrelan, -hád, -hyrde, -líc, -ríce, -roc, -scír, -seld, -seðel, -setl, -stól, -þenung, -wíte, -wyrt: bisceopian.

bisceop-dóm, biscop-dóm, biscep-dóm, es; *m.* I. [bisceop *a bishop*, dóm *judgment*] *a bishop's doom, excommunication*; episcopi judicium, excommunicatio:—Sýn hí begen ðæs bisceopdómes scyldige *let them both be guilty of the bishop's doom [excommunication]*, Bd. 4, 5; S. 573, note 1. II. *the province of a bishop, a bishopric*; episcopi provincia, episcopatus:—He onféng biscopdóm Parisiace hátte *he received the bishopric called Paris*; accepto episcopatu Parisiacæ civitatis, Bd. 3, 7; S. 530, note 10. Ps. Lamb. 108, 8. Wine heóld ðone biscopdóm iii geár *Wine held the bishopric three years*, Chr. 660; Erl. 34, 7.

bisceop-gegyrelan *episcopal robes*. v. biscop-gegyrelan.

bisceop-hád, biscop-hád, es; *m.* [bisceop *a bishop*; hád *hood, condition, state*] BISHOPHOOD, *the office or state of a bishop, the episcopate, a bishopric*; munus episcopale, flaminium, episcopatus, episcopi provincia:—Wæs se bisceophád befæsted *the bishopric was established*, Elen. Kmbl. 2422; El. 1212. Biscophád *flaminium*, Cot. 86: 186. He biscopháde ge ǽr bisceopháde *in episcopatu et ante episcopatum*, Bd. 4, 6; S. 574, 2, 3: 5, 6; S. 620, 19. His biscophád [biscophád, Spl.] brúcan feóndas *let his enemies enjoy his episcopate*, Ps. Th. 108, 8.

bisceop-hyrde, biscop-hyrde, es; *m. A bishop's shepherd or clergyman*; episcopi clericus, Cot. 44. v. hyrde.

bisceopian, biscopgan; *p.* ode; *pp.* od *To exercise the office of a bishop*,

to oversee, visit, confirm; episcopali munere fungi, visitare, confirmare :— Se bisceop biþ gesett to hádigenne preóstas, and to bisceopgenne cild *the bishop is appointed for the ordaining of priests, and confirming of children,* L. Ælf. C. 17; Th. ii. 348, 26.

bisceop-líc, biscop-líc; *def.* se -líca, seó, ðæt -líce; *adj.* BISHOPLIKE, *episcopal, belonging to a bishop;* episcopalis, pontificalis :— He ðæt biscoplíce líf be-eóde *episcopalem vitam exercebat,* Bd. 5, 18; S. 635, 23. On bisceoplícum gerece *pontificali regimine,* 2, 15; S. 519, 13.

bisceop-ríce, biscop-ríce, es; *n.* [bisceop *a bishop,* ríce *a region*] A BISHOPRIC, *diocese, province of a bishop;* episcopi provincia, diœcesis = διοίκησις :—Mellitus féng to ðam bisceopríce *Mellitus succeeded to the bishopric,* Bd. 2, 7; S. 509, note 8. Seaxulf his biscopríce onféng *Saxulf succeeded to his bishopric,* 4, 6; S. 573, 35.

bisceop-roc, -rocc *a bishop's rochet.* v. biscop-roc.

bisceop-scír, biscop-scír, e; *f.* [bisceop *a bishop,* scír *a province*]. I. *the province of a bishop, a diocese;* episcopi provincia, diœcesis = διοίκησις, parochia = παροικία :—Bisceopscír *diœcesis* vel *parochia,* Ælfc. Gl. 68; Som. 69, 123; Wrt. Voc. 42, 4. Ðæt nænig bisceop ốðres bisceopscíre onswóge *ut nullus episcoporum parochiam alterius invadat,* Bd. 4, 5; S. 572, 32: 4, 13; S. 582, 1: 4, 6; S. 573, 39. He todǽlde on twá biscopscíre West-Seaxna mǽgþe *he divided the province of the West-Saxons into two dioceses,* 3, 7; S. 530, 6, 10. II. *the office of a bishop, episcopate;* episcopatus :—Seó bisceopscír Wihte ðæs eálondes belimpeþ to Daniele Wintan ceastre bisceope *episcopatus Vectæ insulæ ad Danihelem pertinet episcopum Ventæ civitatis,* 5, 23; S. 646, 22. Se forlét ða bisceopscíre *he left the episcopate;* relicto episcopatu, 3, 21; S. 551, 38.

bisceop-seld *a bishop's seat or residence, an episcopal see.* v. biscop-seld.

bisceop-seðel *a bishop's seat or residence, an episcopal see.* v. biscop-seðel.

bisceop-setl, biscop-setl, biscep-setl, es; *n.* [bisceop *a bishop,* setl *a seat*]. I. *a bishop's seat or residence;* sedes episcopalis :—Sæt he ðæt bisceopsetl xxxvii wintra and six mónaþ and feówertyne dagas *he occupied the episcopal residence thirty-seven [of] years [winters] and six months and fourteen days,* Bd. 5, 23; S. 646, 9. He ðam Wine gesealde bisceopsetl on Wintan ceastre *Vino in civitate Venta sedem episcopalem tribuit,* 3, 7; S. 530, 7, 14. Se eádiga Petrus se apostol gesæt biscepsetl on Róme *the blessed Peter the apostle occupied the episcopal residence in Rome,* Chr. 45; Erl. 6, 19. II. *a bishopric;* episcopatus :—Wine wæs adrifen of his bisceopsetle *Wine was driven from his bishopric;* pulsus est Vini de episcopatu, Bd. 3, 7; S. 530, 13.

bisceop-stól, biscop-stól, es; *m.* [stól *a stool, seat*] *A bishop's seat or residence, an episcopal see, bishopric;* sedes episcopalis, pontificatus :—He ne mihte ðone Rómaniscan bisceopstól eallunge forlǽtan *he could not altogether neglect the Roman episcopal see,* Nat. S. Greg. Els. 28, 8. Agefen to Wigorna cestre ðam bisceopstóle *given to the episcopal see at Worcester,* Th. Diplm. A. D. 883; 131, 27. Augustinus cyrde to his bisceopstóle *Augustine returned to his bishopric,* Nat. S. Greg. Els. 37, 5. Seó on setl biscopstóles wæs to ætýced *quæ in sedem pontificatus addita est,* Bd. 5, 23; S. 646, 32.

bisceop-þénung, e; *f.* [þénung *duty, office*] *The duty or office of a bishop;* episcopi officium :—Þegnode se árwurþa bisceop Willferþ on ðam dǽlum ða bisceopþénunge árwurþlíce fíf geár *the venerable bishop Wilfrith exercised the office of a bishop in those parts honourably five years,* Bd. 4, 13; S. 583, 15. Féng Eádulf to ðære bisceopþénunge *Eadulf succeeded to the bishop's office,* 5, 23; S. 645, 19.

bisceop-wíte *a bishop's fee for visiting.* v. biscop-wíte.

bisceop-wyrt, biscop-wyrt, biscep-wyrt, e; *f.* [wyrt *a wort, herb, plant*] BISHOP'S-WORT, *bishop's weed, betony, vervain, marsh-mallow;* ammi = ἄμμι [ammi majus, *Lin.*], betonica, verbena, hibiscum = ἰβίσκος :— Wyrc to drence æscþrotu, betonice, bisceopwyrt *make into a drink ash-throat, betony, bishop's-wort,* L. M. 1, 47; Lchdm. ii. 120, 10: 1, 23; Lchdm. ii. 66, 2, 10. Genim bisceopwyrt ða súðernan *take the southern bishop's-wort,* L. M. 2, 54; Lchdm. ii. 274, 27. To monnes stæmne nim biscopwyrt *for a man's voice take bishop's-wort,* Lchdm. iii. 46, 26: Ælfc. Gl. 40; Som. 63, 93; Wrt. Voc. 30, 43. Genim ða brádan biscopwyrt *take the broad bishop's-wort,* Lchdm. iii. 46, 2. Betonice, ðæt is, biscopwyrt *betony, that is, bishop's-wort,* Herb. cont. 1; Lchdm. i. 2, 1. Seó læsse biscopwyrt *betonica,* Ælfc. Gl. 43; Som. 64, 49; Wrt. Voc. 31, 59. Biscopwyrt [MS. biscopwyrtil] *verbena,* 41; Som. 64, 1; Wrt. Voc. 31, 14. Biscepwyrt *hibiscum,* Wrt. Voc. 286, 15.

biscep *a bishop,* Chr. 110; Erl. 8, 11: 636; Erl. 24, 14: 690; Erl. 42, 15. v. bisceop.

biscep-dóm *the province of a bishop, a bishopric,* Chr. 660; Erl. 34, 7. v. bisceop-dóm II.

biscep-setl *an episcopal see,* Chr. 45; Erl. 6, 19. v. bisceop-setl.

biscep-wyrt *marsh-mallow,* Wrt. Voc. 286, 15. v. bisceop-wyrt.

bi-scerian, -scírian, -scyrian; *p.* ede; *pp.* ed *To deprive, separate;* privare, separare :—Wilna biscírede *from desires separated,* Exon. 48 b; Th. 166, 24; Gú. 1047. Dreámum biscyred *from joys separated,* 88 a;

Th. 330, 23; Vy. 55. Faraþ nú, awyrgde, willum biscyrede engla dreámes, on éce fír go now, accursed, *wilfully deprived of the joy of angels, into eternal fire,* 30 a; Th. 93, 3; Cri. 1520: 95 a; Th. 355, 28; Reim. 84: 42 b; Th. 142, 17; Gú. 645. v. be-scerian.

bi-scirian *to separate,* Exon. 48 b; Th. 166, 24; Gú. 1047. v. bi-scerian.

biscop *a bishop,* Chr. 910; Erl. 100, 9, 10. v. bisceop.

biscop-dóm *the province of a bishop, a bishopric,* Bd. 3, 7; S. 530, note 10. v. bisceop-dóm II.

biscopgan *to confirm,* L. Ælf. C. 18; Wilk. 155, 51. v. bisceopian.

biscop-gegyrelan; *pl. m.* [gegyrela *a garment, robe*] *Episcopal robes;* indumenta episcopalia :—He sende him biscopgegyrelan *he sent him episcopal robes,* Bd. 1, 29; S. 498, 10.

biscop-hád *the office or state of a bishop, the episcopate,* Cot. 86: Ps. Spl. 108, 7. v. bisceop-hád.

biscop-heáfod-lín *a bishop's head linen, an ornament which bishops wore on their heads;* infula :—Biscop-heáfod-lín *infula,* Ælfc. Gl. 64; Som. 69, 10.

biscop-líc *episcopal,* Bd. 5, 18; S. 635, 23. v. bisceop-líc.

biscop-ríce *a bishopric,* Bd. 4, 6; S. 573, 35. v. bisceop-ríce.

biscop-roc, -rocc, es; *m.* [roc, rocc *a tunic*] *A bishop's rochet;* dalmatica :—Mid biscoprocce scrýdan *to clothe with a bishop's rochet,* Lchdm. iii. 202, 26.

biscop-scír *a diocese,* Bd. 3, 7; S. 530, 6, 10. v. bisceop-scír.

biscop-seld, es; *n.* [seld *a seat, residence*] *A bishop's seat or residence, an episcopal see;* sedes episcopalis :—Se cyning sealde him stówe and biscopseld on Lindesfearona eá *rex locum sedis episcopalis in insula Lindisfarnensi tribuit,* Bd. 3, 3; S. 525, 35.

biscop-seðel; *g.* -seðles; *n.* [seðel *a seat*] *A bishop's seat or residence;* sedes episcopalis :—Mellitus féng to ðam biscopseðle Contwara burge cirican *Mellitus succeeded to the episcopal residence of Canterbury church;* Mellitus sedem Doruvernensis ecclesiæ suscepit, Bd. 2, 7; S. 509, 8.

biscop-setl *an episcopal residence,* Chr. 604; Th. 38, 1. v. bisceop-setl.

biscop-stól *an episcopal seat,* Bd. 5, 23; S. 646, 32. v. bisceop-stól.

biscop-wíte, es; *n.* *A bishop's fee for visiting, procuration;* episcopo debita, Chr. 675; Erl. 38, 5.

biscop-wyrt *bishop's-wort, betony,* Lchdm. iii. 46, 26: Herb. cont. 1; Lchdm. i. 2, 1. v. bisceop-wyrt.

bi-scyrian *to deprive, separate,* Exon. 88 a; Th. 330, 23; Vy. 55: 30 a; Th. 93, 3; Cri. 1520: 95 a; Th. 355, 28; Reim. 84: 42 b; Th. 142, 17; Gú. 645. v. bi-scerian.

bi-seah *looked about,* Exon. 51 b; Th. 180, 8; Gú. 1276, = be-seah; *p.* of be-seón.

bisegu *occupation,* Bt. 33, 4; Fox 132, 28. v. býsgu.

bisen; *gen.* bísne, bísene; *f.* *An example, similitude, command, precept,* Bt. 22, 2; Fox 78, 13: 29, 1; Fox 102, 12: Exon. 40 a; Th. 133, 33; Gú. 499: Lk. Rush. War. 13, 6: Cd. 27; Th. 36, 13; Gen. 571. v. býsen.

bi-sencan *to sink,* Exon. 25 a; Th. 72, 8; Cri. 1169. v. be-sencan.

bi-seón; *p.* -seah *to see,* Exon. 23 b; Th. 67, 13; Cri. 1088. v. be-seón, seón.

bises; *indecl. m.* *A leap year;* bisextile, bisextus :—Bútan bises geboden weorþe, feorþan geáre *unless a leap year is appointed, [being] the fourth year,* Menol. Fox 64; Men. 32.

bi-settan; *p.* -sette, *pl.* -setton; *pp.* -seted, -sett *To set, beset, surround;* inserere, circumdare :—Ðonne gim in goldfate smiþa orþoncum biseted weorþeþ *when a gem has been set in a golden vessel by the artifice of smiths,* Exon. 60 a; Th. 219, 9; Ph. 304. Mid wyrtum se wilda fugel his nest biseteþ útan *the wild bird surrounds its nest without with herbs,* 63 b; Th. 233, 26; Ph. 530. v. be-settan.

bisgian *to occupy, busy,* Cd. 64; Th. 76, 29; Gen. 1264: Bt. procem; Fox viii. 6. v. býsgian.

bísgu, e; *f.* *Occupation, toil, affliction, care,* Bt. procem; Fox viii. 5, 6: Exon. 114 a; Th. 438, 14; Rä. 57, 7: 82 b; Th. 311, 6; Seef. 88: 74 b; Th. 280, 7; Jul. 625: Beo. Th. 3490; B. 1743: Bt. Met. Fox 22, 127; Met. 22, 64. v. býsgu.

bísgung, e; *f.* [= a-bísgung = a-býsgung] *Business, occupation;* negotium, occupatio :—Fint he ða ryhtwísnesse gehýdde mid his módes bísgunga *he will find the wisdom concealed by the occupation of his mind,* Bt. 35, 1; Fox 156, 12. Ne forlǽte se reccere ða inneran giémenne ðæs godcundan þiówdómes for ðære abísgunge ðara úterra weorca *let not the ruler forsake the inner care of the divine ministration for the occupation of outer works,* Past. 18, 1; Hat. MS. 25 a, 29, 27, 30. v. býsgu.

bísigu *occupation, labour,* Beo. Th. 567, note; B. 281. v. býsgu.

bisleásung, e; *f.* *Fiction;* figmentum, Ps. Spl. M. 102, 13. v. leásung.

bismǽrian; *p.* ede; *pp.* ed *To revile;* maledicere :—Bismǽredon uncit [*Inscription,* Bismǽrede ungket] men, bá ætgædre *they [men] reviled us two, both together,* Runic Inscrip. Kmbl. 354, 30. v. bysmerian.

bismǽr-word, es; *n.* [= bismer-word : bismer *opprobrium, contumelia;*

word *verbum*] *A disgraceful* or *abusive word, reproach, insult;* ignominiosum *vel* contumeliosum verbum, opprobrium, insultatio :—Mid bismærwordum *with insults,* L. H. E. 11; Th. i. 32, 5.

bismer, bismor, bysmer, bysmor; *gen.* bismeres, bysmres; *n.* [be, smeru *fat, grease*] *Filthiness, pollution, abomination, disgrace, infamy, mockery, reproach, contumely, blasphemy, calumny;* ludibrium, pollutio, abominatio, infamia, opprobrium, contumelia, blasphemia, calumnia :—Hí amyrdon heora folc on bysmore *they defiled their people with filthiness,* Ælfc. T. 15, 21. Seó stów gewearþ swíðe mǽre for Rómǎna bismere *the place became famous for the disgrace of the Romans,* Ors. 3, 8; Bos. 62, 44. His mód wæs mid ðam bismre ahwæt *his mind was whetted with that disgrace,* Ors. 6, 30; Bos. 126, 17. Hí mængdon eced and geallan togædere and hit, on his bismer, Criste gebudon *they mingled vinegar and gall together, and offered it to Christ, in mockery of him,* L. Edg. C. 39; Th. ii. 252, 17. Ðú hí, Drihten, dést deópe to bysmre *tu, Domine, deridebis eos,* Ps. Th. 58, 8. He hálge láre brygdeþ on bysmer *he turneth holy lore into mockery,* Exon. 117 a; Th. 449, 14; Dóm. 71. Hí gefremedan óðer bysmer *they made another reproach;* irritaverunt eum, Ps. Th. 105, 25 : 106, 10. Dracan ðú dysne geheowadest, héte syððan him bysmere bráde healdan *draco iste, quem formasti ad illudendum ei,* Ps. Th. 103, 25. Ðæt he dóþ to bysmore ðínum feóndum *he makes that for a reproach to thine enemies,* 8, 3. Ge gehýrdon his bysmer *audistis blasphemiam,* Mk. Bos. 14, 64. Ðæt ðú mǽge þolie bysmor on ǽlcne tíman *ut omni tempore calumniam sustineas,* Deut. 28, 29. [*O. Sax.* bismer, *n. opprobrium.*] DER. bismer-full, -leás, -leóþ, -líc, -líce, -nes, -spræc, -sprecan, -word : bismerian, ge- : bismerung : bismeriend.

bismer-full; *adj. Polluted, abominable, disgraceful;* pollutus, detestabilis, turpis. v. bysmor-full.

bismerian; *p.* ode, ede; *pp.* od, ed *To mock, deride, irritate, reproach, blaspheme, defame, revile;* illudere, deridere, irritare, irridere, blasphemare, calumniam facere, maledicere. DER. bismer. v. bysmerian.

bismeriend, es; *m. A deceiver;* illusor, Prov. 11, 4. DER. bismer.

bisme.-leás; *adj. Without pollution, spotless, blameless;* sine pollutione, immaculatus, irreprehensus. v. bysmer-leás.

bismer-leóþ, es; *n.* [bismer *mockery, reproach;* leóþ *a song*] *A reproachful song, an incantation;* carmen invectivum, nenia, Cot. 188.

bismer-líc, bismor-líc; *adj.* [bismer, bismor *disgrace,* -líc] *Disgraceful, ignominious, dirty, unpleasant;* turpis, ignominiosus, fœdus :— Mid ðam bismerlícestan áþe *with the most disgraceful oath,* Ors. 4, 3; Bos. 79, 39 : 1, 7; Bos. 29, 35. We lǽraþ, ðæt man geswíce bismerlícra efesunga *we enjoin, that a man abstain from ignominious tonsures,* L. Edg. C. 20; Th. ii. 248, 16. On ðone bismerlícostan eard *in the most unpleasant province,* Ors. 3, 11; Bos. 73, 34.

bismer-líce; *adv. Disgracefully, indecently, irreverently, contemptuously, reproachfully;* probrose, indecore, inverecunde, contumeliose. v. bismor-líce.

bismer-nes, -ness, e; *f.* [bismer *filthiness, pollution,* -nes] *A polluting, staining* or *defiling;* pollutio, Bd. 1, 27; S. 497, note 7.

bismer-spræc, -spæc, e; *f. A speaking blasphemy, blasphemy;* blasphemia. v. bysmor-spræc.

bismer-sprecan, -specan; *p.* -spræc, -spæc, *pl.* -sprǽcon, -spǽcon; *pp.* -sprecen, -specen [bismer *blasphemia,* sprecan, specan *loqui*] *To speak blasphemy, blaspheme;* blasphemiam loqui, blasphemare. v. bysmer-, specan.

bismerung, e; *f. Blasphemy;* blasphemia, Mk. Skt. Hat. 3, 28. v. bysmrung.

bismer-word, es; *n. A disgraceful* or *abusive word, reproach, insult;* ignominiosum *vel* contumeliosum verbum, opprobrium, insultatio. v. bismær-word.

bismiriende *deriding;* insultans, Greg. Dial. 2, 1, = bismeriende. v. bysmerian.

bismor *a disgrace,* Chr. 992; Erl. 131, 31. v. bismer, bismor-líc, -líce.

bismor-líc *disgraceful, ignominious,* L. Edg. C. 20; Th. ii. 248, 16. v. bismer-líc.

bismor-líce, bysmor-líce, bysmer-líce; *adv.* [bismer, bismor *disgrace,* -líce] *Disgracefully, indecently, irreverently, contemptuously, reproachfully;* probrose, indecore, inverecunde, contumeliose :—Bysmerlíce *disgracefully,* Judth. 10; Thw. 23, 2; Jud. 100. Hí willaþ, binnan Godes húse, bysmorlíce plegian *they will play irreverently within God's house,* L. Ælf. C. 35; Th. ii. 356, note 2, line 20. Worpaþ hine deófol on dómdæge bismorlíce *the devil shall cast him down contemptuously in the day of judgment,* Salm. Kmbl. 53; Sal. 27.

bismrian *to mock,* Ps. Spl. 103, 28. v. bysmerian.

bísnian *to give* or *set an example,* Bt. 33, 4; Fox 128, 20 : 39, 11; Fox 230, 2. v. býsnian.

bísnung *an example;* exemplum, Ælfc. T. 5, 15. v. býsnung.

bi-sorgian *to care for, fear,* Exon. 61 a; Th. 223, 32; Ph. 368 : 30 b; Th. 95, 12; Cri. 1556. v. be-sorgian.

bi-spanan; *p.* -spón, -speón; *pp.* -spanen, -sponen; *v. trans. To allure, entice, incite, urge;* allicere, illicere, seducere, incitare, impellere :—

Ic Herode in hyge bispeón, ðæt he Iohannes bibeád heáfde biheáwan *I Herod in mind incited, that he commanded John's head to be cut off,* Exon. 70 a; Th. 260, 8; Jul. 294. v. be-spanan.

bí-spell *a fable,* Bt. 35, 5; Fox 166, 19 : Ors. 1, 6; Bos. 29, 11. v. big-spell.

bissexte *a leap year;* bisextus, Bd. 5, 23; S. 648, 19. v. bises.

bist *art, shalt be;* es, eris, Bd. 5, 19; S. 640, 43 : Ælfc. Gr. 25; Som. 26, 28; *2nd pers. pres. and fut.* of beón.

bi-stelan; *p.* -stæl, *pl.* -stǽlon; *pp.* -stolen *To rob, deprive;* furari, privare :—Strengo bistolen *deprived of strength,* Exon. 107 b; Th. 410, 8; Rä. 28, 13. v. be-stelan.

bi-swác *deceived, seduced,* Exon. 70 a; Th. 260, 25; Jul. 302; *p.* of bi-swícan.

bi-sweðian; *p.* ede; *pp.* ed *To bind, wind round, inwrap;* ligare, involvere :—Hí biwundon oððe bisweðedon [biuundun ł bisueððun MS.] hine *ligaverunt eum,* Jn. Lind. War. 19, 40. Sibbum bisweðede, sorgum biwerede *inwrapt in peace, from cares protected,* Exon. 32 a; Th. 100, 19; Cri. 1644. v. be-sweðian.

bi-swíc, es; *m. Deceit;* fraus, Ors. 3, 7; Bos. 60, 13. v. be-swíc.

bi-swícan; *p.* -swác, *pl.* -swicon; *pp.* -swicen *To deceive, seduce;* decipere, seducere :—Ic Néron biswác [MS. bisweac] *I deceived Nero,* Exon. 70 a; Th. 260, 25; Jul. 302. v. be-swícan.

bi-swícol; *adj.* [bi-swíc *deceit;* dolus] *Deceitful;* dolosus :—We sculon geþencean ðæt ðis líf, ðæt we nú onlibbaþ, is bíswícol eallum ðǽm ðe hit lufiaþ *we ought to think that this life, in which we now live, is deceitful to all those who love it,* L. E. I. prm; Th. ii. 400, 16.

bit *asks, prays,* Lk. Bos. 11, 12; *3rd pers. pres.* of biddan.

bita, an; *m.* [biten; *pp.* of bítan *to bite*]. I. *a BIT, morsel, piece, fragment;* frustum, buccella :—Ne mihte hyra ǽlc ánne bitan of ðam gelæccan *every one of them could not get a morsel,* Homl. Th. i. 182, 10. Ǽfter ðam bitan *post buccellam,* Jn. Bos. 13, 27. II. *anything that bites, a biter, an animal;* ferus :—Ǽnlíce [ǽnlige MS.] bita *singularis ferus,* Ps. Spl. 79, 14.

BÍTAN; *part.* bítende; ic bíte, ðú bítest, bítst, he bíteþ, bítt, bít, *pl.* bítaþ; *p.* ic, he bát, ðú bite, *pl.* biton; *pp.* biten. I. *to BITE with the teeth;* mordere :—Ic bíte *mordeo,* Ælfc. Gr. 26, 6; Som. 29, 10. Monnan ic ne bíte nymþe he me bíte *I bite no man unless he bite me,* Exon. 125 a; Th. 482, 9, 10; Rä. 66, 5. Ǽghwá bíteþ mec on bær líc *every one bites me on the bare body,* 125 a; Th. 482, 7; Rä. 66, 4. Monige mec bítaþ *many bite me,* 125 a; Th. 482, 12; Rä. 66, 6. Ðæt mǽden bát and totær ǽlcne ðe heó gerǽcan mihte *the maiden bit and tore every one whom she could reach,* Homl. Th. i. 458, 14 : Beo. Th. 1488; B. 742. Biton [MS. byton] hine lýs *lice bit him,* Hexam. 17; Norm. 24, 30. Nim ðis ofæt, bít hit and byrge *take this fruit, bite it and taste,* Cd. 25; Th. 33, 12; Gen. 519. II. *used metaphorically of the biting* or *wounding by a sword,—to cut, wound;* cædere, vulnerare :—Se gist onfand ðæt se beadoleóma bítan nolde *the guest found that the war-beam [the sword] would not wound,* Beo. Th. 3051; B. 1523 : 2913; B. 1454. Sió ecg gewác, bát unswíðor *the edge [of the sword] failed, cut less sharply,* 5150; B. 2578. Ðeáh mec heard bite stíðecg stýle *though the stiff-edge steel wounded me greatly,* Exon. 130 a; Th. 499, 10; Rä. 88, 13. [*Chauc. Wyc.* bite : *R. Glouc.* byten : *Laym.* biten : *Orm.* bitenn : *Northumb.* bíta *discerpere* : *Plat.* biten : *O. Sax.* bítan : *O. Frs.* bíta : *Dut.* bijten : *Ger.* beiszen : *M. H. Ger.* bízen : *O. H. Ger.* bízan : *Goth.* beitan : *Dan.* bíde : *Swed.* bita : *Icel.* bíta : *Sansk.* bhid *findere, perforare.*] DER. a-bítan, on-.

bite, es; *m.* [bítan *to bite*] *A BITE, pain, the biting* or *pain of a wound, a biting disease* or *cancer;* morsus, cancri morbus *vel* cancer :—Hyt ða wédendan bítas gehǽleþ *it heals the maddening bites,* Med. ex Quadr. 13, 7; Lchdm. i. 370, 14. Wið apan bíte *for the bite of an ape,* 11, 7; Lchdm. i. 366, 24: L. Ethb. 35; Th. i. 12, 5 : Beo. Th. 2060; B. 2060. Þurh sweordes bíte *through the bite of the sword,* Apstls. Kmbl. 68; Ap. 34. Bíte írena *the bite of swords,* Beo. Th. 4511; B. 2259. Gnættas cómon ofer ðæt land mid fýrsmeortendum bítum *gnats came over the land with fire-smarting bites,* Ors. 1, 7; Bos. 29, 30. Wið cancerádle, ðæt is, bíte *against cancer-disease, that is, a biting disease,* L. M. 1, 44; Lchdm. ii. 108, 9. DER. láþ-bíte.

bítel, bítela, bétl; *m. A beetle;* blatta :—Ða blacan bétlas *the black beetles,* Cot. 141.

bi-teldan *to cover, surround, overwhelm,* Exon. 59 b; Th. 217, 1; Ph. 273 : 64 b; Th. 238, 25; Ph. 609. v. be-teldan.

bítende *biting;* mordax, Cot. 134; *part.* of bítan.

BITER, bitor, bitter, bittor; *g. m. n.* biteres, bitres, bittres; *f.* bitre; *sup.* biteresta, bitresta; *adj.* BITTER, *sharp, severe, dire;* amarus, acerbus, acer, dirus, atrox :—Ðæt bitereste [MS. biteroste] clyster *botri amarissimi,* Deut. 32, 32; *the clustre most bittir,* Wyc. Ðæt he bibúgan mǽge ðone bitran drync *that he may escape the bitter drink,* Exon. 45 a; Th. 154, 10; Gú. 840. Hí beheóldon bogan [MS. boga], þing [þingc MS.] biter *intenderunt arcum, rem amaram,* Ps. Spl. 63, 3; Ps. Th. 78, 5. Bittor, Exon. 82 a; Th. 309, 10; Seef. 55. Bittor, Exon. 47 a; Th. 161, 13; Gú. 958. Boda bitresta *the bitterest messenger,* Cd. 36; Th. 47, 19;

Gen. 763. Bittres; *g.* Salm. Kmbl. 658; Sal. 328. Biteres; *g.* Rood Kmbl. 225; Kr. 114. [*Orm.* bitterr: *O. Sax.* bittar: *Dut. Ger. M. H. Ger.* bitter: *O. H. Ger.* bittar: *Goth.* baitrs: *Dan. Swed.* bitter: *Icel.* bitr.] DER. þurh-biter, -bitter, winter-.

bitere *bitterly, sharply,* Ps. Th. 101, 18 : 128, 2. v. bitre.

biterian, biterigan; *p.* ode; *pp.* od *To embitter, make sharp;* acerbare :—Ðætte us biterige sió hreówsung *that the repentance may be bitter to us,* Past. 54, 5. DER. a-biterian, ge-.

biter-líce, bitter-líce;' *adv.* BITTERLY; amare :—He weóp biterlíce [Bos. bityrlíce] *he wept bitterly,* Mt. Jun. 26, 75. He ongan biterlíce [Smith, 600, 29, bitterlíce] wépan *he began to weep bitterly,* Bd. 4, 25 ; Whelc. 337, 43.

biter-nys, -nyss, e ; *f.* BITTERNESS; amaritudo :—Híg cómon to ðære stówe, ðe ys Mara genemned, ðæt ys on úre lýden biternys; ðá ne mihton híg drincan ðæt wæter, forðamðe hit wæs biter: ðá héton híg ealle his naman Mara, ðæt ys on úre lýden biternys *venerunt in Mara, nec poterant bibere aquas de Mara, eo quod essent amaræ, unde et congruum loco nomen imposuit vocans illum Mara, id est amaritudinem,* Ex. 15, 23. Heortan biternys *bitterness of heart,* Homl. Th. ii. 220, 18. Ðæs' múþ full is biternysse *cujus os plenum est amaritudine,* Ps. Spl. second 9, 8. Nolde his onbyrian for ðære biternysse *he would not taste it for its bitterness,* Homl. Th. ii. 254, 18, 19.

biter-wyrde; *adj.* Inclined to bitterness; ad amaritudinem pronus :— Ne he biterwyrde næs *he was not inclined to bitterness,* Homl. Th. i. 320, 15 : ii. 44, 22.

biþ *is, shall be;* est, erit, Bt. Met. Fox 6, 11; Met. 6, 6 : Cd. 217; Th. 276, 1; Sat. 182; *3rd pers. pres. and fut. of* beón.

bi-peahte, -peaht *covered over,* Exon. 96 a; Th. 359, 11; Pa. 61: 101 a; Th. 382, 10; Rä. 3, 9; *p. and pp. of* bi-þeccan. v. beþeccan.

bi-pearf ic *I need,* Exon. 76 a; Th. 285, 17; Jul. 715. v. biþurfan, be-.

bi-peccan *to cover,* Exon. 28 b; Th. 87, 10; Cri. 1423 : 51 b; Th. 179, 1; Gú. 1255. v. be-þeccan.

bi-pencan *to consider, bear in mind, confide,* Exon. 19 b; Th. 51, 27; Cri. 822: 20 a; Th. 53, 14; Cri. 850: 51 b; Th. 179, 32; Gú. 1270: 66 b; Th. 245, 30; Jul. 52. v. be-þencan.

bi-pringan *to surround,* Exon. 60 b; Th. 221, 27; Ph. 341. v. beþringan.

bi-pryccan; *p.* -þrycte, *pl.* -þrycton; *pp.* -þrycced [þryccan *to press*] *To press on;* imprimere :—Hí hwæsne beág ymb mín heáfod gebýgdon, þreám biþrycton *they bent a sharp crown around my head, pressed it on with reproaches,* Exon. 29 a; Th. 88, 26; Cri. 1446.

bi-purfan *to need, to have need,* Exon. 76 a; Th. 285, 17; Jul. 715. v. be-þurfan.

bítl *a mallet, hammer,* Past. 36, 5; Cott. MS. v. býtl.

bit-mælum; *adv.* [bit, mælum, *dat. pl. of* mæl, *n.*] *Piecemeal, by bits;* mordicus, Ælfc. Gr. 38; Som. 42, 5.

bitol, es; *n.* *A bridle;* frænum :—On gewealde and bitole ceácan heora gebind *in camo et fræno maxillas eorum constringe,* Ps. Spl. 31, 12.

bi-tolden *covered, overwhelmed,* Exon. 64 b; Th. 238, 25; Ph. 609; *pp. of* bi-teldan. v. be-teldan.

bitre, bitere, bittre; *adv.* [biter *bitter*] *Bitterly, sharply, cruelly;* amare, acriter, atrociter :—Ic eom bitre abolgen *I am bitterly vexed,* Exon. 119 b; Th. 458, 31; Hy. 4, 109 : 120 b; Th. 463, 4; Hö. 65 : Beo. Th. 4651; B. 2231. Unc he bitre forgeald *he bitterly requited us,* Cd. 222; Th. 290, 21; Sat. 418. Hí gebléndon bittre tosomne unswétne drync ecedes and geallan *they mingled bitterly together an unsweet drink of vinegar and gall,* Exon. 29 a; Th. 88, 11; Cri. 1438 : 119 a; Th. 457, 4; Hy. 4, 78.

bitst, he bit *askest, he asks,* Ex. 33, 12 : Homl. Th. i. 250, 8, 9; *2nd and 3rd pers. pres. of* biddan.

bitst, bist, bít *bidest, bides; 2nd and 3rd pers. pres. of* bídan.

bitt *a bottle;* uter. v. byt.

bitter *bitter;* amarus, Exon. 82 a; Th. 309, 10; Seef. 55. v. biter.

bitter-líce *bitterly,* Bd. 4, 25; S. 600, 29. v. biter-líce.

bitter-nes *bitterness,* Scint. 61. v. biter-nys.

bittor *bitter,* Exon. 47 b; Th. 163, 23; Gú. 998. v. biter.

bittre *bitterly, sharply, cruelly,* Exon. 94 b; Th. 354, 24; Reim. 50. v. bitre.

bi-tweon; *prep. dat.* *Between;* inter, Exon. 32 a; Th. 101, 15; Cri. 1659. v. be-tweonan III.

bi-tweonum; *prep. dat.* *Between;* inter :—Hornum bitweonum [horna abitweonum MS. Th.] *between the horns,* Exon. 107 b; Th. 411, 19; Rä. 30, 2. v. abi-tweonum, be-tweonum.

bi-týnan *to close, shut up,* Exon. 61 b; Th. 227, 7; Ph. 419. v. betýnan.

bityr-líce *bitterly;* amare :—Petrus weóp bityrlíce *Petrus flevit amare,* Mt. Bos. 26, 75. v. biter-líce.

bi-wǽgan; *p.* de; *pp.* ed; *v. a.* *To disappoint;* frustrari :—Ne biwǽgde hine *non frustratus est eum,* Ps. Surt. 131, 11. v. be-wǽgan.

bi-wǽrlan; *p.* de; *pp.* ed [v. bí- in be- II] *To pass by;* præterire, Lk. Lind. War. 10, 31 : 11, 42 : Lk. Rush. War. 11, 42. DER. wǽrlan.

bi-wǽwan; *p.* -weów; *pp.* -wǽwen *To blow against;* afflare :— Winde biwáwne [MS. biwaune] *waved or shaken by the wind,* Exon. 77 b; Th. 291, 2; Wand. 76. DER. wáwan.

bi-weaxan *to overgrow.* Exon. 60 a; Th. 219, 21; Ph. 310. v. beweaxan.

bi-weddian *to espouse, betrothe, wed;* desponsare :—Wæs sió fǽmne wélegum biweddad *the woman was betrothed to the rich one,* Exon. 66 a; Th. 244, 25; Jul. 33. v. be-weddian.

bi-werian, -wergan *to defend, restrain, forbid,* Exon. 87 b; Th. 329, 23; Vy. 38 : Exon. 45 a; Th. 153, 3; Gú. 820. v. be-werian.

bi-windan *to entwine, enwrap, encircle,* Exon. 69 a; Th. 256, 20; Jul. 234 : 28 b; Th. 87, 9, 12; Cri. 1422, 1424 : 18 b; Th. 45, 27; Cri. 725 : 65 b; Th. 241, 34; Ph. 666. v. be-windan.

bi-wist *food, provision,* Bt. 17; Fox 60, 4 : L. Edg. C. 3; Th. ii. 244, 12. v. big-wist.

bi-word, -wyrd, es; *n.* [be, bí *by,* word *a word*] *A* BYEWORD, *proverb;* proverbium :—Man segþ [seið MS.] to bíworde, 'hæge sitteþ ða æceras dæleþ' *man saith for a proverb,* '*the hedge abides which fields divides,*' Chr. 1130; Erl. 259, 13. Bíword, bíwyrd *proverbium,* Cot. 157.

bi-worpen *cast about, surrounded;* cinctus :—Is ðæt églond fenne biworpen *the island is surrounded with a fen,* Exon. 100 b; Th. 380, 9; Rä. 1, 5, = be-worpen; *pp. of* be-weorpan.

bi-wráh *covered,* Exon. 76 b; Th. 287, 32; Wand. 23; *p. of* biwríhan. v. be-wríhan.

bi-wrecan; *p.* -wræc, *pl.* -wrǽcon; *pp.* -wrecen *To strike or beat around, to surround;* circum pulsare, circumdare :—Hí sculon onfón in fýrbaðe wælmum biwrecene wráþlíc andleán *they must receive dire retribution in the fire-bath surrounded with flames,* Exon. 20 a; Th. 52, 11; Cri. 832. v. be-wrecan.

bi-wríhan; *p.* -wráh, *pl.* -wrigon; *pp.* -wrigen *To cover.* v. be-wríhan.

bi-wrítan; *p.* -wrát, *pl.* -writon; *pp.* -writen [be *by,* wrítan *to write*] *To write after, by,* or *out of, to copy;* postscribere, exscribere, Past. pref; Hat. MS. v. be-wrítan.

bi-wyrcan *to make,* Exon. 74 a; Th. 277, 3; Jul. 575. v. be-wyrcan.

bixen; *adj.* [box *the box-tree*] *Belonging to box,* BOXEN, *made of box-wood;* buxeus :—Bixen box *a box made of box-wood;* pyxis, Ælfc. Gl. 26; Som. 60, 96; Wrt. Voc. 25, 36.

blác; *adj.*   I. *bright, shining;* lucidus, splendidus :—On bryne blácan fýres *into the burning of the bright fire,* Cd. 186; Th. 231, 13; Dan. 246. Lígetta hérgen blác dýrne Dryhten *lightnings bright praise the beloved Lord,* Exon. 54 b; Th. 192, 16; Az. 107. Engel ða burh oferbrægd blácan lýge, hátan heaðowealme *an angel spread over the town a bright flame, hot warlike floods,* Andr. Kmbl. 3081; An. 1543. Blácum leóhte *with bright light,* Bt. Met. Fox 4, 15; Met. 4, 8. Lígetu blác *lightnings bright,* Cd. 192; Th. 240, 3; Dan. 381.   II. *bleak, pale, pallid, livid, as in death;* pallidus, de moribundis et mortuis :—Biþ his líf scæcen, and he blác his *life is departed, and he pale,* Exon. 87 b; Th. 329, 28; Vy. 41. Scylfing hreás blác *Scylfing fell pale,* Beo. Th. 4969; B. 2488: Runic pm. 29; Kmbl. 345, 16. Blácne *pale, acc.* Judth. 12; Thw. 25, 26; Jud. 278. He hæfde blæc feax and blácne andwlitan *he had black hair and a pale countenance,* Bd. 2, 16; S. 519, 34. Se móna mid his blácan leóhte *the moon with her pale light,* Bt. 4; Fox 6, 34. [*Prompt.* bleyke *pallidus, subalbus,* from blác, *p. of* blícan *to shine.*]   ☞ Observe the difference between blác *bright, shining,* bleak, *pale,* and blæc *black,* se blaca *the black.*

blác *shone,* Exon. 52 a; Th. 182, 4; Gú. 1305; *p. of* blícan.

blace berian *black berries;* mori, Ælfc. Gl. 47; Som. 65, 30. v. blæc-berie.

blác-ern, es; *n.* [blác *light,* ærn, ern *a place*] *A light place, a lamp, candlestick, lantern, light, candle;* lucerna :—Bæd ðæt hí ðæt blácern adwæscton *prayed that they would put out the light* [lucernam], Bd. 4, 8; S. 575, 40. Bærnaþ eówer blácern *light your candle,* Bd. 4, 8; S. 576, 6 : Ps. Th. 131, 18.

blác-hleór; *adj.* [blác II. *pale,* hleór *a face, cheek*] *Having a pale face, pale-faced, fair;* pallidus *vel* candidus genis :—Sceolde monig bláchleór ides bifiende gán *many a pale-faced damsel must trembling go,* Cd. 92; Th. 118, 23-25; Gen. 1969, 1970: Judth. 11; Thw. 23, 18; Jud. 128.

blácian, blácigan, to blácienne, blácigenne; *p.* ode; *pp.* od [blác *pallid, bleak, pale*] *To grow pale;* pallere, pallescere :—Ic blácige *palleo,* Ælfc. Gr. 26, 2; Som. 28, 42 : 35; Som. 38, 5. Ic onginne to blácigenne [blácienne MS. C.] *pallesco,* 35; Som. 38, 6. Onsýn blácaþ *his face grows pale,* Exon. 82 b; Th. 311, 13; Seef. 91. DER. a-blácian.

blácung, e; *f.* *Paleness, wanness;* pallor :—Blácung *pallor,* Ælfc. Gr. 9, 21; Som. 10, 27. On blácunge goldes *in pallore auri,* Ps. Lamb. 67, 14.

BLÆC, es; *n.* *Ink;* atramentum :—Ðæt hí habban blæc and bócfel *that they have ink and parchment,* L. Edg. C. 3; Th. ii. 244, 11. Blæc

atramentum, Wrt. Voc. 47, 3. [Plat. blak ink: O. H. Ger. blach ink: Dan. blæk, n. ink: Swed. blæck, n. ink: Icel. blek, n. atramentum.]

BLÆC; gen. m. n. blaces, f. blæcre; def. m. se blaca, f. n. blace: bleac; adj. BLACK, swarthy; niger, fuscus:—He hæfde blæc feax, and blácne andwlitan he had black hair, and a pale [lean, thin] countenance; nigro capillo, facie macilenta, Bd. 2, 16; S. 519, 33. Forðonðe ðú ne mæht ænne loc hwítne gewirce oððe blæcne quia non potes unum capillum album facere aut nigrum, Mt. Kmbl. Rush. 5, 36. Ofslógon ðone blacan Heawald they killed the black Heawald, Bd. 5, 10; S. 624, 40. Ða sind blace swíðe they are very black, Exon. 114 b; Th. 438, 28; Rä. 58, 2. Swearte wæron lástas, swaðu swíðe blacu swart were their footsteps, their tracks very black, 113 b; Th. 434, 19; Rä. 52, 3. [Icel. blakkr niger, Egils. v. A. Sax. blæc ink.] ☞ Observe the difference between blæc; gen. m. n. blaces, se blaca black, swarthy, and blác shining, pallid, bleak, pale, from blác; p. of blícan to shine; remark also blæc pale, livid, from blǽcan to bleach, whiten. v. blícan, blǽcan.

blǽc; adj. Shining, pale, livid; lucidus, pallidus, lividus:—Ís brycgade blǽce brimráde the ice bridged the pale water road, Grn. An. 1264. v. blǽcan to bleach.

blǽcan, blǽcean; p. de; pp. ed To BLEACH, whiten, fade; pallidum colorem inducere, albicare:—Blǽced bleached, Exon. 107 b; Th. 410, 27; Rä. 29, 5. Ne mæg ne sunne blǽcan no sun can bleach, Bd. 1, 1; S. 473, 20: blǽcean, note 20. note 20.

blǽc-berie, an; f. A BLACKBERRY, mulberry; vaccinium, morus:—Blace berian mori, Ælfc. Gl. 47; Som. 65, 30; Wrt. Voc. 33, 29. DER. blǽc.

blǽc-ern, es; n. An inkstand; atramentarium. DER. blæc ink, ærn a place.

blǽc-ern, es; n. [blǽc light, ærn a place] Literally a lamp or candle-stick, also the light itself; verbum de verbo, candelabrum, etiam candela, lucerna:—Bæd ðæt hí ðæt blǽcern acwencton prayed that they would put out the light [lucernam], Bd. 4, 8; S. 575, 40, note, MS. B. Ne menn blǽcern in beornaþ men do not light a candle [lucernam], Mt. Kmbl. Rush. 5, 15. Blǽcern fótum mínum lucerna pedibus meis, Ps. Th. 118, 105. v. blác-ern.

blǽc-fexed; adj. [blǽc black, feax, fex hair] Having black hair, black-haired; nigris capillis:—He is blæcfexed [MS. blæcfexede] he is black-haired, Homl. Th. i. 456, 16.

blǽc-gym; g. -gymmes; m. A black fossil, called jet; nigro-gem-meus, lapis gagates = γαγάτης, Bd. 1, 1; S. 473, 24.

blǽco, es; n. [blǽc pale, livid; blǽcan to bleach] Paleness, leprosy; pallor, lepra = λέπρα:—Blǽco pallor, Cot. 157. Lǽcedómas wið ðam yfian blǽce leechdoms against the evil leprosy, L. M. cont. 1, 32; Lchdm. ii. 8, 1. Wið blǽce genim góse smero for leprosy take goose-grease, L. M. 1, 32; Lchdm. ii. 76, 9, 1, 4, 7, 18. v. blǽcþa.

blǽc-teru; g. wes; n. Black-tar, tar, naphtha, a sort of bituminous fluid; pix fluida, naphtha, Som.

blǽcþa, an; m: blǽcþ-rust, es; m. Leprosy; vitiligo, Cot. 221. v. blǽco.

BLÆD; gen. blædes; nom. pl. blado, n. A leaf, BLADE; folium, palmula:—Brád blado broad leaves, Cd. 48; Th. 61, 8; Gen. 994. Róðres blæd the blade of an oar; palmula, Ælfc. Gl. 83; Som. 73, 77; Wrt. Voc. 48, 16. [O. Sax. blad, n: Frs. O. Frs. bled, n: Dut. blad, n: Ger. blatt, n: M. H. Ger. blat, n: O. H. Ger. blat, n: Dan. Swed. blad, n: Icel. blað, n. folium.] DER. ár-blæd.

blæd, e; f. A cup, bowl, goblet, vial; patera, phiala, Æthelflædæ Test. Lye. v. bledu.

blæd, es; m. I. a blast, blowing, breath, spirit, life, mind; flamen, flatus, inspiratio? spiritus, vita, animus:—Gif máre blæd windes astág [MS. astahg] if a stronger blast of wind arose; si flatus venti major adsurgeret, Bd. 4, 3; S. 569, 8. Þurh gǽstes blæd through the spirit's inspiration, Exon. 63 b; Th. 234, 33; Ph. 549. God ableów on his ansýne lífficne blæd God blew into his face the breath of life, Hexam. 11; Norm. 18, 26. His blæd forleósan to lose his life, Judth. 10; Thw. 22, 16; Jud. 63. Náh seó módor gewald bearnes blædes the mother hath not power over her child's life, Salm. Kmbl. 769; Sal. 384. Beorht on blǽde bright in life, Elen. Kmbl. 975; El. 489. II. enjoyment, prosperity, abundance, success, blessedness, gift, reward, benefit, glory, honour; fruitio, prosperitas, abundantia, successus, beatitudo, donum, præmium, beneficium, gloria, dignitas:—Hyra blæd leofaþ æt dómdæge their enjoyment shall exist [live] at doomsday, Exon. 31 b; Th. 100, 4; Cri. 1636. Blǽdes full full of enjoyment, Exon. 32 a; Th. 101, 13; Cri. 1658. Eorþan blǽdas the enjoyments of earth, 116 b; Th. 447, 28; Dóm. 46. He heóld blæd mid bearnum he possessed prosperity with his children, Cd. 79; Th. 97, 5; Gen. 1608. Hie ne meahton blǽdes brúcan they might not enjoy prosperity, 90; Th. 113, 26; Gen. 1893. On his blǽde in his prosperity, 205; Th. 253, 26; Dan. 601. Sý him wuldres blǽd may there be to him abundance of glory, Exon. 65 b; Th. 241, 27; Ph. 662. Ða feóndas ðæs blǽdes gebrocen hæfdon the fiends had enjoyed their success, Exon. 38 b; Th. 127, 28; Gú. 393. Écan lífes blǽd the blessedness of eternal life, Exon. 82 b; Th. 310, 24; Seef. 79.

Wæs his blæd mid God his reward was with God, 39 a; Th. 128, 27; Gú. 410: 20 b; Th. 55, 4; Cri. 878. Wæs heora blæd micel their glory was great, Cd. 1; Th. 2, 5; Gen. 14. Hie Iudéa blæd forbrǽcon billa ecgum they destroyed the Jews' glory with the edges of swords, Cd. 210; Th. 260, 13; Dan. 709. [O. H. Ger. blát flatus.] DER. fér-blæd, wuldor-: blǽd-ágende, -dæg, -fæst, -gifa, -horn, -wéla.

blǽd, bléd, e; f. What is produced,—A flower, blossom, fruit; flos, olus, fructus:—His leáf and his blǽda ne fealwiaþ its leaves and its flowers shall not fall; folium ejus non decidet, Ps. Th. 1, 4. Wudu sceal blǽdum blówan a wood shall blow with flowers, Menol. Fox 527; Gn. C. 34. Geseh he geblówene bearwas standan, blǽdum gehrodene he saw blowing groves stand, adorned with blossoms, Andr. Kmbl. 2896; An. 1451. Bléda wyrta olera herbarum, Ps. Spl. 36, 2. He déþ ǽlc twíg aweg on me, ðe blǽda ne byrþ; and he feormaþ ǽlc ðara, ðe blǽda byrþ, ðæt hyt bere blǽda ðe swíðor omnem palmitem in me non ferentem fructum, tollet eum; et omnem, qui fert fructum, purgabit eum, ut fructum plus afferat, Jn. Bos. 15, 2. Beorc biþ blǽda leás the birch-tree is fruitless [void of fruit], Runic pm. 18; Kmbl. 342, 27; Hick. Thes. i. 135. Hærfest bryngþ rípa bléda harvest brings ripe fruits, Bt. 39, 13; Fox 234, 15: 34, 10; Fox 150, 5. Balsames blæd fruit of balsam, Cot. 48.

blǽd-ágende; part. Possessing abundance, prosperous; abundantiam habens, prosper, Beo. Th. 2031; B. 1013.

blǽd-dæg; g. -dæges; pl. nom. -dagas; g. pl. -daga; m. A prosperous or happy day; prosperitatis dies, faustus dies:—Ðær we mótun brúcan blæddaga where we may enjoy prosperous days, Exon. 65 b; Th. 242, 16; Ph. 674: Cd. 60; Th. 73, 7; Gen. 1201.

blǽddre a blister, pimple, the bladder, Ex. 9, 9, 10: Ælfc. Gl. 75; Som. 71, 74; Wrt. Voc. 44, 56. v. blǽdre.

blǽd-fæst; adj. Prosperous; prosper:—Heó abreát blǽdfæstne beorn she destroyed a prosperous hero, Beo. Th. 2602; B. 1299. DER. ge-blǽdfæst.

blǽd-gifa, an; m. A giver of prosperity, happiness, or glory; prospe-ritatis, beatitudinis, vel gloriæ largitor:—Beorht blǽdgifa bright giver of prosperity, Andr. Kmbl. 167; An. 84: 1311; An. 656.

blǽd-horn, es; m. A blast-horn, a trumpet; classicum:—Blǽdhornas classica, Ælfc. Gl. 52; Som. 66, 44; Wrt. Voc. 35, 32.

blǽdre, blǽddre, an; f. [blǽwan to blow; flare] That which is blown out, hence I. an inflated swelling, blister, pimple, blain, pustule; pustula, papula:—Be ǽghwylcum uncúþum blǽdrum ðe on mannes nebbe sittaþ of all strange blisters which exist on a man's face, Herb. cont. 2, 19; Lchdm. i. 6, 10: Herb. 2, 19; Lchdm. i. 86, 5. Eall folc wæs on blǽdran, and ða wǽron swíðe hreówlíce berstende all the people had blisters [lit. was in blister], and they were very painfully bursting, Ors. 1, 7; Bos. 29, 37. On mannum and on nýtenum beóþ wunda and swel-lende blǽddran there shulen ben in men and yn beestis biles and bleynes swellynge, Wyc; Ex. 9, 9, 10. II. the BLADDER, receptacle for the urine; vesica:—Bǽres blǽdre a boar's bladder, Med. ex Quadr. 8, 12; Lchdm. i. 360, 8. Wið sáre ðære lifre and ðære blǽdran for sore of the liver and of the bladder, Herb. cont. 145, 2; Lchdm. i. 54, 27: Herb. 41, 2; Lchdm. i. 142, 8: 80, 1; Lchdm. i. 182, 12. Gif weaxan stánas on ðære blǽdran if stones grow in the bladder, L. M. 3, 20; Lchdm. ii. 320, 6. Genim eoferes blǽdran take a boar's bladder, Med. ex Quadr. 8, 11; Lchdm. i. 360, 5. Blǽddre vesica, Ælfc. Gl. 75; Som. 71, 74; Wrt. Voc. 44, 56. Wið ðære blǽddran sáre for sore of the bladder, Herb. 107; Lchdm. i. 220, 15: 126; Lchdm. i. 238, 10: Med. ex Quadr. 8, 11; Lchdm. i. 360, 4. [Chauc. Wyc. bladder: Piers P. bleddere: Dut. blaar, f: O. Dut. blaeder, blaere: Ger. blatter, f: M. H. Ger. bláter, f: O. H. Ger. blátara, f: Dan. blære, m. f: Swed. bläddra, f: Icel. blaðra, f.]

blǽd-wéla, an; m. Fruitful riches; opes uberes:—Ic ðé on ða fægran foldan gesette to neótenne neorxna wonges blǽdwélan I set thee on the fair earth to enjoy the fruitful riches of Paradise, Exon. 28 a; Th. 85, 16; Cri. 1392.

blǽge, an; f. A BLAY, bleak, the gudgeon; gobio = κωβιός:—Blǽge gobio, Ælfc. Gl. 101; Som. 77, 59; Wrt. Voc. 55, 64. [Ger. bleie, bleihe, f. a blay.]

blǽ-hǽwen, blǽ-hwen, blǽwen; adj. [bleó blue, hǽwen hued] Of a blue hue, bluish, violet or purple colour; cæruleus, perseus:—Moises scrídde ðone bisceop [Aaron] mid línenum reáfe, and girde hine, and dyde ymbe hine blǽhwene tunecan, and léde eaxlclaþ ofer hine Moses clothed the bishop [Aaron] with a linen garment, and girded him [with a girdle], and put around him a blue tunic, and laid a cope [lit. shoulder-cloth] upon him, Lev. 8, 7. Blǽwen perseus, Ælfc. Gl. 80; Som. 72, 94; Wrt. Voc. 46, 51.

blǽse, blase, an; f. I. a BLAZE, flame; ardor, flamma. v. bǽl-blæse. II. that which makes a blaze,—A torch, lamp; fax, facula, lampas = λαμπάς:—Blǽsefax, Greg. Dial. 2, 8: Glos. Prudent. Recd. 143, 33. Iudas com ðyder mid leóhtfatum, and mid blasum, and mid wǽpnum Iudas venit illuc cum laternis, et facibus, et armis, Jn. Bos. 18, 3. Blǽsum faculis, Mone B. 3487. Blase lampas, Ælfc. Gl. 30; Som. 61, 54; Wrt. Voc. 26, 53. [M. H. Ger. blas, n. fax, lampas.]

blæsere, blasere, blysıere, es; *m.* [blæse I. *a blaze, flame*] *A burner, incendiary;* incendiarius:—Be blæserum *of incendiaries,* L. Ath. i. 6; Th. i. 202, 18. We cwǽdon be ðám blaserum *we have ordained concerning incendiaries,* L. Ath. iv. 6; Th. i. 224, 13.

blæst, es; *m.* [blæse I. *a blaze, flame*] *A burning, blaze, flame;* ardor, flamma:—Ne mæg ðǽr, rěn ne snáw, ne fýres blæst, wihte gewyrdan *there rain nor snow, nor flame of fire can aught injure,* Exon. 56 a; Th. 198, 25; Ph. 15: Andr. Kmbl. 1674; An. 839. Ðæt he [Fěnix] onfón móte, þurh líges blæst, líf æfter deáþe *that it [the Phœnix] may, through the fire's flame, receive life after death,* Exon. 62 a; Th. 228, 6; Ph. 434. Léges blæstas weallas ymbwurpon *flames of fire overwhelmed the walls,* Andr. Kmbl. 3163; An. 1554.

blæst, es; *m.* [blˊáwan *to blow*] *flare*] *A blowing,* BLAST *or gust of wind, a breeze;* flatus:—Sǽgrundas súþwind fornam, bæþweges blæst *the south wind, the sea breeze, dried up the depths of the sea,* Cd. 158; Th. 196, 11; Exod. 290. [*Chauc.* blast: *Laym.* blæst: *Ger. M. H. Ger.* blast, *m:* O. H. Ger. blást, *m:* Icel. blástr, m.]

blæst-belg bellows, Wrt. Voc. 286, 76. v. blást-belg.

blæt, ðú blætst *is livid, thou art livid; 3rd and 2nd pers. pres. of* blátan.

blǽt, es; *m. A bleating, a* BLEAT *like a sheep;* balatus. DER. blǽtan.

BLǼTAN; *p.* blǽtte; *pp.* blǽtted; *v. n.* [blǽt *a bleat*] *To* BLEAT; balare:—Ic blǽte swá gát *I bleat as a goat,* Exon. 106 b; Th. 406, 17; Rä. 25, 2. Scép blǽt *ovis balat,* Ælfc. Gr. 22; Som. 24, 9. Hit biþ swíðe dyslíc ðæt se man beorce oððe blǽte *it is very foolish that the man bark or bleat,* 22; Som. 24, 12. [*Piers P.* blete: *Orm.* blætenn: *Dut.* bleeten, bláten: *M. Dut.* bleten: *Ger.* blaszen: *O. H. Ger.* blazan *to cry as a sheep* or *goat, to bleat.*]

blǽtesung, e; *f. A flaming, blazing, sparkling;* flagrantia, Ps. Spl. T. 76, 18.

blǽwen *light blue;* perseus; Ælfc. Gl. 80; Som. 72, 94; Wrt. Voc. 46, 51. v. blǽ-hǽwen.

blǽweþ, blǽwþ *blows,* Bt. Met. Fox 6, 15; Met. 6, 8: ðú blǽwest, blǽwst *thou blowest; 3rd and 2nd pers. pres. of* blǽwan.

blan *ceased,* Bd. 1, 8; S. 479, 17; *p. of* blinnan.

BLANC; *adj.* BLANK, *white, grey;* pallidus, albus, candidus:—Gewiton mearum rídan beornas on blancum *the warriors departed to ride on white horses,* Beo. Th. 1716; B. 856. [*Relq. Ant. W.* i, 37, 30, blonc *white: Dut.* blank *white, shining: Ger.* blank *albus: M. H. Ger.* blanc: *O. H. Ger.* blanch *candidus: Dan. Swed.* blank *bright: Q. Nrs.* blankr *albus,* Rask Hald: hence *Span.* blanco *white: Fr.* blanc: *It.* bianco.]

blanca, blonca, an; *m. A white* or *grey horse;* equus albus *vel* candidus:—On blancan *on a grey horse,* Elen. Grm. 1185. [*Laym.* blank, blonk *a horse, steed: O. Nrs.* blakkr, *m. cquus.*] DER. blanc.

bland, es; *n. A mixture, confusion;* mixtio:—Swég swíðrode and sanges [MS. sances] bland *sound prevailed and a confusion of song,* Cd. 158; Th. 197, 19; Exod. 309. [*Icel.* bland, *n.*] DER. ge-bland, -blond, wind-.

BLANDAN, blondan, ic blande, blonde, ðú blandest, he blandeþ, blent; *pl.* blandaþ; *p.* ic, he bleónd, blěnd, ðú bleónde, blěnde, *pl.* bleóndon, blěndon; *pp.* blanden, blonden *To mix,* BLEND, *mingle;* miscere:—Ic eom on gôman gena swětra ðonne ðú beóbreád blěnde mid hunige *I am yet sweeter on the palate than if thou blendedst bee-bread with honey,* Exon. 111 a; Th. 425, 21; Rä. 41, 59. [A strong verb in all the Teutonic dialects: *Goth.* blandan; *p.* baibland; *pp.* blandans: *O. Sax.* blandan: *O. H. Ger.* blandan: *Swed. O. Nrs.* blanda.] DER. ge-blandan: ge-blondan, on-: be-blonden: ge-bland: ge-bloud, ær-, ár-, ear-, earh-, sund-, ýþ-: wind-blond.

blanden-feax, blonden-feax, -fex; *adj.* [blanden; *pp. of* blandan *to mix;* feax, fex *hair*] *Having mixed* or *grizzly hair, grey-haired, old;* comam mixtam *vel* canam habens, senex. Blanden-feax is a phrase which in Anglo-Saxon poetry is only applied to those advanced in life; and is used to denote that *mixture* of colour which the hair assumes on approaching or increasing senility, Price's Warton i. xcvi. note 20:—Gelpan ne þorfte beorn blandenfeax [MS. blandenfex, col. 2] bilgeslehtes *the grizzly-haired warrior ought not to boast of the clashing of swords,* Chr. 937; Th. 204, 34, col. 1; Æðelst. 45. Abraham ne wěnde, ðæt him Sarra, brýd blondenfeax, bringan meahte on woruld sunu *Abram thought not that Sarah, his grey-haired wife, could bring a n into the world,* Cd. 107; Th. 141, 7; Gen. 2341: 123; Th. 157, 5; Gen. 2600: Beo. Th. 3586; B. 1791. Blondenfexa *the grizzly-haired,* 5916; B. 2962. Hruron teáras blondenfeaxum *tears fell from the grizzly-haired [prince],* 3750; B. 1873. Blondenfeaxe, gomele, ymb gôdne ongeador sprǽcon *the grizzly-haired, the old, spoke together about the good [warrior],* 3193; B. 1594.

blann *ceased, rested,* Bd. 3, 20; S. 550, 28; *p. of* blinnan.

blase *a torch, lamp,* Ælfc. Gl. 30; Som. 61, 54; Wrt. Voc. 26, 53. v. blæse II.

blasere, es; *m. An incendiary,* L. Ath. iv. 6; Th. i. 224, 13. v. blæsere.

blást-belg, es; *m. A blast-bag, bellows;* follis, Cot. 86.

BLÁT; *comp.* blátra; *superl.* blátast; *adj. Livid, pale, ghastly;* lividus, pallidus:—Þurh ðæs beornes breóst blát weóll waðuman streám *a livid stream bubbled in waves through the man's breast,* Andr. Kmbl. 2560; An. 1281. Hungres on wěnum, blátes beódgæstes *in expectation of hunger, of a pale table-guest,* 2177; An. 1090. Ðæt biþ frěcne wund, blátast benna *that is a dangerous wound, most ghastly of sores,* Exon. 19 a; Th. 48, 13; Cri. 771. [O. H. Ger. bleizza *livor.*]

blátan; *part.* blátende; ic bláte, ðú blátest, blátst, he bláteþ, blǽt, *pl.* blátaþ; *p.* bleót, blét, ðú bléte, *pl.* bléton; *pp.* bláten; *intrans. To be livid, pale,* or *dark as with envy;* livere:—Hygewælmas teáh beorne on breóstum blátende níþ *darkening [livid, pale] envy drew agitations of mind to the breast of the man,* Cd. 47; Gen. 981.

bláte; *adv. Lividly, pallidly;* livide, pallide:—Helle fýr bláte forbærnþ biteran lěge *the fire of hell lividly burns up with a dire [bitter] flame,* Bt. Met. Fox 8, 107; Met. 8, 54. Ðæt fýr ne mæg foldan and merestreám bláte forbærnan *the fire cannot pallidly burn up earth and sea,* 20, 229; Met. 20, 115.

blátende; *part. Darkening, making livid* or *pallid;* livens, Cd. 47; Th. 60, 14; Gen. 981. v. blátan.

BLÁWAN; *part.* bláwende; ic bláwe, ðú bláwest, bláwst, blǽwest, blǽwst, he bláweþ, bláwþ, blǽweþ, blǽwþ, *pl.* bláwaþ; *p.* bleów, blěw, *pl.* bleówon; *p.* bláwen *To blow, breathe;* flare, sufflare. I. *v. intrans:*—Ge geseóþ súþan bláwan *ye see the south [wind] blow,* Lk. Bos. 12, 55. Ic bláwe *flo,* Ælfc. Gr. 24; Som. 25, 41. Wind wráðe bláweþ *the wind fiercely blows,* Bt. Met. Fox 7, 104; Met. 7, 52: Ps. Th. 147, 7. Bǽwþ gást his and flówaþ wæteru *flabit spiritus ejus et fluent aquæ,* Ps. Lamb. 147, 18: Bt. Met. Fox 6, 15; Met. 6, 8. Swôgaþ windas, bláwaþ brecende, bearhtma mǽste *winds shall howl, crashing blow, with the greatest of sounds,* Exon. 21 b; Th. 59, 11; Cri. 951. Se wind súþan bleów *the wind blew from the south,* Bd. 3, 7; S. 509, 27. Bleów he on hí *he breathed on them,* Jn. Bos. 20, 22. Bleówon [MS. bleowun] windas *flaverunt venti,* Mt. Bos. 7, 25, 27. Bláwen is on smiððan *conflatur in conflatorio,* Prov. 27. II. *v. trans:*—Drihten háteþ hěh-englas běman bláwan *the Lord shall command the archangels to blow the trumpets,* Cd. 227; Th. 302, 19; Sat. 602. Englas bláwaþ býman *angels shall blow the trumpets,* Exon. 20 b; Th. 55, 10; Cri. 881. Ne bláwe man býman befóran ðě *let not a man blow a trumpet before thee,* Mt. Bos. 6, 2. [*Laym.* blæwen, blauwen, blawen, blowen: *Ger.* blähen: *M. H. Ger.* blæjen: *O. H. Ger.* blájan: *Lat.* flo.] DER. a-bláwan, for-, ge-, to-: bláwennys: blǽwere: bláwung.

bláwen-nys, -nyss, e; *f. A blowing* or *puffing up, a windy swelling;* inflatio, sufflatio. DER. bláwan.

blǽwere; *m.* [bláwan *to blow;* flare] *A* BLOWER; conflator:—Ídel wæs se bláwere *the blower was useless;* frustra conflavit conflator, Past. 37, 3; Hat. MS. 50 a, 24.

bláwung, e; *f.* [bláwan *to blow;* flare] *A* BLOWING; flatus:—Ðá hét Gedeon his geféran habban heora býman him mid tó ðære bláwunge *then Gideon commanded his companions to have their trumpets with them for the blowing,* Jud. 7, 16. DER. a-bláwung.

bleac; *def.* se bleaca; *adj. Black;* niger:—Wæs ðis gesceád ðæt for missenlíce heora feaxes hiwe, óðer wæs cweden se bleaca Heawold, óðer se hwíta Heawold *ea distinctione ut pro diversa capillorum specie, unus niger Hewald, alter albus Hewald diceretur,* Bd. 5, 10; S. 624, 16. v. blæc, blaca *black.*

BLEÁT; *def.* se bleáta, seó, ðæt bleáte; *adj. Wretched, miserable;* miser, miserabilis:—Ǽnig ne wæs mon on moldan ðætte meahte bibúgan ðone bleátan drync deópan deáþweges *there was not any man on earth that could avoid the miserable drink of the deep death-cup,* Exon. 47 a; Th. 161, 24; Gú. 963. [*Scot.* blait *nudus: Frs.* bleat *nudus: O. Frs.* blat *nudus; thi* blata *pauper, miser: Dut.* bloot: *M. Dut.* blôt: *Ger.* blosz: *M. H. Ger.* blôz: *Icel.* blautr.]

bleáte; *adv. Wretchedly, miserably;* misere, miserabile:—He geseah ðone leófestan lífes æt ende bleáte gebǽran *he saw his dearest [friend] bearing [himself] wretchedly at life's end,* Beo. Th. 5640, note; B. 2824.

BLEÁÞ; *adj. Gentle, timid, peaceful, inactive;* timidus, imbellis, ignavus:—Ic eom tó ðon bleáþ ðæt mec mæg gríma abrěgan *I am so timid that a phantom may frighten me,* Exon. 110 b; Th. 423, 4; Rä. 41, 16. Ne wæs him bleáþ hyge *his mind was not inactive,* Andr. Kmbl. 462; An. 231. [*Laym.* blæð *destitute: O. Sax.* blôði: *Dut.* bloode: *Ger.* blöde: *M. H. Ger.* blöde: *O. H. Ger.* blôdi: *Dan.* blöd: *Swed.* blöt: *Icel.* blauðr.] DER. here-bleáþ.

Blecinga ég, e; *f. Blekingey, the sea-coast of the Blekingians, a province on the south-west of Sweden; in A. Sax. times belonging to Denmark,* Ors. 1, 1; Bos. 22, 1.

bled, e; *f. A bowl, the dish of a balance, a scale.* v. helur-bled, bledu.

bléd, e; *f. A shoot, branch, flower, fruit;* germen, ramus, frons, flos, fructus:—Ðæt cymen [MS. cyme] grěne bléda *that green shoots come,* Cd. 200; Th. 248, 24; Dan. 518. On ðæs beámes blédum *on the branches of the tree,* Cd. 200; Th. 248, 5; Dan. 508. Ne

dreósaþ beorhte blēde *bright fruits fall not*, Exon. 56 a; Th. 200, 3; Ph. 35: 62 b; Th. 230, 2; Ph. 466. God lǣteþ hrusan syllan beorhte blēda beornum and þearfum *God lets earth give delightful fruits to rich and poor*, Hick. Thes. i. 135, 24. DER. wudu-blēd. v. blǣd.

blēdan; *p.* de; *pp.* ed [blōd *blood*] *To* BLEED, *emit blood;* sanguinem emittere :—Blēdaþ ǣdran *the veins shall bleed*, Salm. Kmbl. 290; Sal. 144. Se blēdenda fīc *the bleeding fig* or *disease*, Wanl. catal. 305, 4. Wið ðone blēdende fīc nim murran *for the bleeding fig* or *disease take myrrh*, Lchdm. iii. 8, 1. [*Dut.* bloeden: *Ger.* bluten: *O.H.Ger.* bluotan: *Dan.* blöde: *Swed.* blöda.]

blēd-hwæt; *g.* -hwates; *adj.* [blēd *a shoot*, hwæt *quick*] *A shoot growing quickly;* germen velox :—Ðonne ic hrēre bearwas blēd-hwate *then I shake the quick-growing groves*, Exon. 101 a; Th. 381, 10; Rä. 2, 9.

bledsian; *p.* ode; *pp.* od *To bless, consecrate;* benedicere, consecrare. DER. ge-bledsian. v. bletsian.

bledsung *a blessing*, Chr. 813; Erl. 60, 21. v. bletsung.

bledu, bled, blæd, e; *f. A bowl, vial, goblet, the dish of a balance, a scale;* patera, phiala, lanx. trutinæ, scala, Ælfc. Gl. 25; Wrt. Voc. 24, 44: Æthelflēdæ Test. Lye. DER. helur-bled.

BLĒGEN, e; *f. A* BLAIN, *blister, bile* or *ulcer;* pustula, ulcus :—Wið ða blēgene, genim nigon ægra and seóþ hig fæste *for blains, take nine eggs and boil them hard*, Lchdm. i. 380, 1. Wið ða blacan blēgene *against black blains*, L. M. 1, 58; Lchdm. ii. 128, 21. [*Tyndl.* blain: *Chauc.* blein: *Wyc.* bleines, *pl:* *Dut.* blein, *f:* *Dan.* blegn.]

blencan; *p.* blencte; *pp.* blenced *To deceive, cheat;* decipere, fallere :—He wrenceþ and blenceþ *he deceives and cheats*, Exon. 83 b; Th. 315, 18; Mōd. 33. [*Prov. Eng.* blench: *Icel.* blekkja *to impose upon*.]

blēnd *mixed, blended, mingled;* *p.* of blandan.

blendan, he blent; *p.* blende; *pp.* blended, blend; *v. trans.* [blind *cæcus*] *To* BLIND, *deprive of sight, darken;* cæcare, obscurare :—Se dæg blent and þióstraþ hiora eágan *the day blinds and darkens their eyes*, Bt. 38, 5; Fox 206, 5. Man hine blende, and hine swā blindne brohte to ðām munecum *they blinded him, and brought him thus blind to the monks*, Chr. 1036; Th. 294, 17, col. 2; Ælf. Tod. 14. [*Chauc. Piers* blende: *Laym.* a-blenden; *Orm.* blendenn: *O.Frs.* blenda, blinda: *Dut.* blinden: *Ger.* M.H.Ger. blenden: *O.H.Ger.* blentjan: *Goth.* gablindyan: *Dan.* for-blinde: *Swed.* för-blinda: *Icel.* blinda.] DER. a-blendan, ge-.

blent *blends;* *3rd pers. pres.* of blandan.

bleó *a colour, hue, complexion*, Ælfc. Gl. 79; Som. 72, 78; Wrt. Voc. 46, 35. v. bleoh.

bleó *blue* or *azure colour;* cæruleus, Som.

bleó-bord, es; *n.* [bleoh, bleó *colour*, bord *a table*] *A coloured table on which games of chess are played;* tabula colorata in qua prœlia latronum luduntur (Ettm. p. 311) :—Dryhten dǣleþ sumum gūþe blǣd, sumum tæfle cræft, bleóbordes gebregd *the Lord allots to one success in war, to another skill at the table, cunning at the coloured board*, Exon. 88 a; Th. 331, 20; Vy. 71.

bleó-brygd, es; *m? n?* [bleó *colour*, bregdan *to change*] *A variegated colour;* color variegatus :—Is se fugel fæger, bleóbrygdum fāg *the bird is fair, shining with variegated colours*, Exon. 60 a; Th. 218, 9; Ph. 292.

bleó-cræft, es; *m.* BLEE-CRAFT, *the art of embroidering;* ars plumaria, ars acupingendi :—Bleócræft *ars plumaria*, Cot. 17.

bleó-fæstnes, -ness, -nyss, e; *f. That which gives pleasure from its colour,—Pleasure, delight;* jucunditas, deliciæ :—In onleóhtnes oððe onlīhting on bleófæstnessum [bleófæstnyssum, Spl.] oððe ēstum mīnum *nox illuminatio est in deliciis meis*, Ps. Lamb. 138, 11.

bleó-fāg, -fāh; *adj.* [bleoh, bleó *color;* fāg, fāh *varius*] *Of various colours, party-coloured;* versicolor :—Byrne is mīn bleófāg *my byrnie is party-coloured*, Exon. 105 b; Th. 400, 18; Rä. 21, 3: Cot. 115. Oferslop bleófāh habban ǣrende fúllīc getācnaþ *to have a party-coloured overcoat betokens an unpleasant message*, Lchdm. iii. 200, 6.

BLEOH, bleóh, blioh, bliō; *gen.* bleós; *n. A colour, hue, complexion;* color, species :—Bleoh *color*, Ælfc. Gl. 79; Som. 72, 70; Wrt. Voc. 46, 27. Mislīc bleó *a mixed colour*, 79; Som. 72, 78; Wrt. Voc. 46, 35. Bliō *color*, Prov. 23. Ðæt wæs hwītes bleós swā cristalla *it was of a white colour like crystal*, Num. 11, 7. Ānes bleós *of one colour;* unicolor, concolor, Ælfc. Gl. 79; Som. 72, 76; Wrt. Voc. 46, 33.: Ælfc. Gr. 9, 21; Som. 10, 35. Hwī is se rēnboga mislīces bleós *why is the rainbow of a mixed colour?* Boutr. Scrd. 21, 25. Menn māgon cēpan be ðæs mōnan bleó hwylc weder toweard byþ *men may observe by the moon's colour what weather is at hand*, Bd. de nat. rerum; Wrt. popl. science 15, 9; Lchdm. iii. 268, 5. Hī brugdon on wyrmes bleó *they changed to a worm's hue*, Exon. 46 a; Th. 156, 32; Gū. 883: 71 a; Th. 264, 12; Jul. 363: Elen. Kmbl. 2210; El. 1106. Seolocenra hrægla mid mistlīcum bleowum hī ne gīmdon *they cared not for silken garments of various colours*, Bt. 15; Fox 48, 11. Mōnan bleoh habban hȳnþe getācnaþ *for the moon to have colours betokens humility*, Lchdm. iii. 206, 27. Hī habbaþ blioh and færbu ungelīce *they have different*

colours *and forms*, Bt. Met. Fox 31, 7; Met. 31, 4. Bleóum *with colours*, Exon. 94 a; Th. 352, 31; Reim. 4: Salm. Kmbl. 301; Sal. 150. Secgaþ guman ðæt Iosephes tunece wǣre bleóm bregdende *men say that Joseph's coat varied* [lit. *was varying*] *in colours*, Exon. 95 b; Th. 357, 3; Pa. 23: 87 a; Th. 327, 14; Vy. 3. Geseah ic ðæt beácen wendan bleóm *I saw the beacon change in colours*, Rood Kmbl. 43; Kr. 22: Elen. Kmbl. 1515; El. 759. [*Prov. Eng.* blee: *Chauc.* blee: *O.Sax.* blī, *n: North Frs.* bläy: *O.Frs.* blie, bli, *n.*] DER. ge-bleoh, wundorbleó.

bleóm *in colours*, Elen. Kmbl. 1515; El. 759; *inst. pl.* of bleoh.

bleónd, *pl.* bleóndon *mixed, blended;* *p.* of blandan.

bleónde *hast mixed, blended;* *p.* of blandan.

bleó-reád, -reód; *adj.* BLUE RED, *purple, myrtle-coloured;* cæruleoruber, myrteus :—Bleóreád *myrteus*, Cot. 135. Bleóreód *myrteus*, Ælfc. Gl. 79; Som. 72, 89; Wrt. Voc. 46, 46.

bleó-stǣning, e; *f. Coloured stone-work* or *pavement, Mosaic work;* opus musivum, pavimentum segmentatum, Som. Lye: Cot. 131.

bleót *was livid, pale;* *p.* of blātan.

bleót, ðū bleóte, *pl.* bleóton *sacrificed, sacrificedst, sacrificed;* *p.* of blōtan.

bleóum *in colours*, Salm. Kmbl. 301; Sal. 150; *inst. pl.* of bleoh.

bleów, bleówe, *pl.* bleówon *blew, breathed*, Jn. Bos. 20, 22; *p.* of blāwan.

bleów, ðū bleówe, *pl.* bleówon *flourished, hast flourished, flourished*, Ps. Surt. 27, 7; *p.* of blōwan.

bleowum *to* or *with colours*, Bt. 18; Fox 48, 11; *dat. pl.* of bleoh.

blere, es; *m? An onyx, gem;* onyx = ὄνυξ, *m.* a *nail* :—Blere *onyx*, Wrt. Voc. 288, 55.

blēt, blēte, *pl.* blēton *was livid, pale;* *p.* of blātan.

blēt *sacrifices;* *3rd pers. pres.* of blōtan.

bletsian, bletsigan; *part.* bletsiende, bletsigende; *p.* ode, ade; *pp.* od, ad; *v. a. To* BLESS, *wish happiness, consecrate;* benedicere, consecrare :—Ic Ismael ēstum wille bletsian *I will bless Ishmael with favours*, Cd. 107; Th. 142, 5; Gen. 2357: 191; Th. 238, 23; Dan. 359: Gen. 17, 16. He, bletsiende [bletsigende, Jun.], bræc ða hlāfas, and sealde his leorningcnihtum *he, blessing, brake the loaves, and gave to his disciples*, Mt. Bos. 14, 19. Ic bletsie ealle ða ðe hit healden *I bless all who may observe it*, Chr. 675; Erl. 39, 25. Ic bletsige oððe wel secge *benedico*, Ælfc. Gr. 37; Som. 39, 38. Ic bletsige ðē on mīnum līfe *benedicam te in vita mea*, Ps. Lamb. 62, 5: Exon. 41 b; Th. 138, 22; Gū. 580. Ðū geáres hring mid gyfe bletsast *benedices coronæ anni benignitatis tuæ*, Ps. Th. 64, 12. We ðec bletsiaþ, Fæder ælmihtig *we bless thee, Father almighty*, Cd. 192; Th. 241, 6; Dan. 400: Exon. 64 b; Th. 239, 12; Ph. 620: Ps. Lamb. 128, 8. Ðū bletsodest [bletsadest, Th.] Drihten eorþan ðīne *benedixisti Domine terram tuam*, Ps. Spl. 84, 1. He bletsode hī *benedicebat eos*, Mk. Bos. 10, 16: Ps. Spl. 106, 38. Mid heora mūþe hīg bletsodon, and mid heora heortan hīg wergdon *ore suo benedicebant, et corde suo maledicebant*, Ps. Lamb. 61, 5. Hī hine bletsadon meáglum wordum *they blessed him in strenuous words*, Exon. 43 a; Th. 146, 6; Gū. 705. Bletsa eálā ðū mīn sāwl Drihtne *benedic anima mea Domino*, Ps. Lamb. 103, 1. Bletsiaþ Drihtne ealle englas his *benedicite Domino omnes angeli ejus*, 102, 20. Neáta gehwilc naman bletsic *every* [*kind*] *of cattle bless* [*thy*] *name*, Cd. 192; Th. 240, 22; Dan. 390. Bletsien ðec, Dryhten, deór and nȳten *beasts and cattle bless thee, O Lord*, Exon. 55 a; Th. 194, 26; Az. 144. [*Chauc.* blisse, blysse: *Wyc.* blisse: *Laym.* bletseiȝen: *Orm.* blettcenn, blettsenn: *Northumb.* bletsia, bloetsia, bloedsia: *Icel.* bleza, bletza, blessa: *Goth.* bleiþs *merciful, kind*, bleiþyan *to have mercy*.] DER. ge-bletsian.

bletsing-bōc, e; *f. A blessing-book;* liber benedictionum formulas continens, Wanl. catal. 80, 33.

blētst *sacrificest;* *2nd pers. pres.* of blōtan.

bletsung, bledsung, e; *f. A* BLESSING; benedictio :—Sī bletsung Drihtnes ofer eów *sit benedictio Domini super vos*, Ps. Spl. 128, 7: Exon. 9 a; Th. 7, 12; Cri. 100. He onfón sceal mīnre bletsunge *he shall receive my blessing*, Cd. 106; Th. 140, 22; Gen. 2331. Cyn his on bletsunge byþ *semen illius in benedictione erit*, Ps. Lamb. 36, 26. Mid bletsunge [bledsunge, col. 1] ðæs pāpan *with the blessing of the pope*, Chr. 813; Th. 108, 22, col. 2, 3. Brohte him bletsunge, se ðe him ǣ sette *benedictionem dabit, qui legem dedit*, Ps. Th. 83, 7: 113, 21. Him se beorn bletsunga leán ageaf *the prince gave him the gift of his blessings*, Cd. 97; Th. 128, 2; Gen. 2120.

blēwþ, ðū blēwst *blows, thou blowest*, Ps. Spl. 102, 14; *3rd and 2nd pers. pres.* of blōwan.

BLÍCAN, ic blīce, ðū blícest, blícst, he blíceþ, blícþ, *pl.* blícaþ; *p.* ic, he blāc, ðū blice, *pl.* blicon; *pp.* blicen; *v. n.* I. *to shine, glitter, dazzle, sparkle, twinkle;* lucere, fulgere, coruscare, micare :—Ðū ðære gyldnan gesihst Hierusalem weallas blícan *thou seest the walls of the golden Jerusalem shine*, Salm. Kmbl. 469; Sal. 235: Exon. 57 a; Th. 204, 10; Ph. 95. Mōna swā seó Godes circe beorhte blíceþ *the church of God shines brightly like the moon*, 18 a; Th. 44, 11; Cri. 701: 58 b; Th. 210, 16; Ph. 186. Blícþ ðeós beorhte sunne *this bright sun glitters*,

Cd. 38; Th. 50, 19; Gen. 811. Hý fôre leódum leóhte blícaþ *they with light shall shine before the people*, Exon. 26 a; Th. 76, 14; Cri. 1239. Heofoncandel blác ofer lagoflôdas *the sun* [lit. *heaven's candle*] *shone over the water-floods*, Andr. Kmbl. 486; An. 243. Blicon bordhreóðan *bucklers glittered*, Cd. 149; Th. 187, 30; Exod. 160. Hwonne swegles tapur hǽdre blíce *when the sun* [lit. *heaven's taper*] *serenely shines*, Exon. 57 b; Th. 205, 20; Ph. 115. II. *to shine by exposure, as the bones;* denudando in conspectu dari :—Hí twigena ordum hine weallaþ ôþ ðæt him bân blícaþ *they shall vex him with points of twigs until his bones appear* [*shine*], Salm. Kmbl. 289; Sal. 144. [*Laym.* blikien: *O. Sax.* blîkan: *Frs.* blike *apparere:* *O. Frs.* blîka: *Ger.* er-bleichen *pallescere:* *M. H. Ger.* blîchen *fulgere:* *O. H. Ger.* ar-blîchan *pallescere:* *O. Nrs.* blika, blîkja: *Lat.* flag-ra-re: *Grk.* φλέγ-ω: *Lith.* blizg-ù *I shine:* *Sansk.* bhrâj *to shine.*] DER. a-blícan.

blíce, es; *m.* [blícan II. *to shine by exposure, as the bones*] *An exposure;* denudatio :—Gif bânes blíce weorþeþ, þrím scillingum gebête *if there be an exposure of the bone* [*by wounding*], *let amends be made with three shillings*, L. Ethb. 34; Th. i. 12, 4.

blícettan; *p.* blícette; *pp.* blícetted [blícan I. *to shine, glitter*] *To glitter, quiver;* vibrare :—Blícette *vibrabat*, Cot. 178. [*O. H. Ger.* blechazan *micare.*]

blícettung, e; *f.* [blícettan *to glitter*] *A coruscation, shining;* coruscatio :—Blícettunga *coruscationes*, Ps. Vos. 76, 18 : 143, 8. [*O. H. Ger.* blechazunga, *f. fulmen.*]

blícon *shone, glittered*, Cd. 149; Th. 187, 30; Exod. 160; *p. pl. of* blícan.

blícst, he blícþ *shinest, shines*, Cd. 38; Th. 50, 19; Gen. 811; *2nd and 3rd pers. pres. of* blícan.

blíds *joy, gladness*, Ps. C. 50, 99; Ps. Grn. ii. 279, 99. v. blíþs.

blin, blinn, e; *f.* [=be-lin; v. linnan *to cease*] *A ceasing, rest, intermission;* cessatio, intermissio :—Bútan blinne *without ceasing;* sine intermissione, Bd. 5, 12; S. 628, 20: Elen. Kmbl. 1648; El. 826. [*Old Eng.* blin, Ben. Jonson.] DER. un-ablínn. v. blinnan.

BLIND; *def.* se blinda, seó, ðæt blinde; *adj.* BLIND, *deprived of sight;* cæcus :—Ðâ wæs him broht ân deófolseóc man, se wæs blind and dumb *tunc oblatus est ei dæmonium habens, cæcus et mutus*, Mt. Bos. 12, 22: Mk. Bos. 10, 46: Cd. 115; Th. 150, 13; Gen. 2491. Ðæt ðú grâpie on midne dæg, swâ se blinda deþ on þîstrum *ut palpes in meridie, sicut palpare solet cæcus in tenebris*, Deut. 28, 29: Mt. Bos. 23, 26. Ǽt-hrân he ðæs blindan hand *he took the hand of the blind* [*man*], Mk. Bos. 8, 23. Hwâ geworhte mannes mûþ oððe hwâ geworhte dumne oððe deáfne and blindne oððe geseóndne *quis fecit os hominis aut quis fabricatus est mutum et surdum, cæcum et videntem?* Ex. 4, 11: Chr. 1036; Erl. 165, 29; Ælf. Tod. 15. Hîg synt blinde, and blindra lât-teówas: se blinda, gyf he blindne lǽt, hîg feallaþ begen on ǽnne pytt *cæci sunt, et duces cæcorum: cæcus si cæco ducatum præstet, ambo in foveam cadunt*, Mt. Bos. 15, 14: 9, 27: 20, 30: Lk. Bos. 7, 22: Andr. Kmbl. 1162; An. 581.' Blinde on geþoncum *blind in thoughts*, Exon. 24 b; Th. 69, 28; Cri. 1127: Bt. Met. Fox 19, 59; Met. 19, 30. Mæg wôd man blindra manna eágan ontýnan *numquid dæmonium potest cæcorum oculos aperire?* Jn. Bos. 10, 21. Manegum blindum he gesihþe forgeaf *cæcis multis donavit visum*, Lk. Bos. 7, 21: 4, 18. Ðonne ðû gebeórscype dô, clypa þearfan, and wanhâle, and healte, and blinde *cum facis convivium, voca pauperes, debiles, claudos, et cæcos*, Lk. Bos. 14, 13: Ps. Th. 145, 7. Drihten onleóhteþ ða blindan [MS. blinden] *Dominus illuminat cæcos*, Ps. Lamb. 145, 8. Eálá ge dysegan and blindan *O ye foolish and blind*, Mt. Bos. 23, 17, 19, 24. ¶ Blind slite *or* slyte *a blind or inward wound*, i. e. *a bite, the wound of which does not appear because of the swelling of the part affected;* morsus, cujus vulnus non apparet præ tumore partis affectæ, Herb. 4, 12; Lchdm. i. 92, 25. Seó blinde netele *or* netle *the blind or dead nettle;* archangelica [lamium album, *Lin.*], Ælfc. Gl. 43; Som. 64, 51; Wrt. Voc. 31, 61: L. M. 1, 23; Lchdm. ii. 66, 4. Blinda mann *a parasite;* palpo, Ælfc. Gr. 36; Som. 38, 46, 47. Blinde cweartern *a blind or dark prison;* cæcus vel tenebrosus carcer :—Gebrohton hí hine binnan ðam blindan cwearterne *they brought him into the dark prison*, Homl. Th. i. 416, 28. Engel scînende ðæt blinde cweartern mid leóhte afylde *a shining angel filled the dark prison with light*, ii. 382, 6. [*O. Sax.* blind: *O. Frs. Dut. O. Dut.* blind: *M. H. Ger. O. H. Ger.* blint, *gen.* blindes: *Goth.* blinds: *Dan. Swed.* blind: *Icel.* blindr.] DER. hyge-blind, môd-.

blindan *is not found, but the Gothic* ga-blindyan *to blind, exists; so also A. Sax.* blendan *to blind, q. v.*

blind-líce; *adv. In a blind manner,* BLINDLY, *rashly;* temere :—Hû blindlíce monige sprecaþ *how blindly* [*rashly*] *many speak*, Ors. i, 10; Bos. 34, 17.

blind-nes, -ness, -nyss, e; *f.* BLINDNESS; cæcitas :—Ðâ ge blindnesse bôte forsêgon *when ye renounced the remedy of blindness*, Elen. Kmbl. 777; El. 389: Exon. 41 b; Th. 139, 24; Gú. 600. Ofer hyra heortan blindnesse *super cæcitate cordis eorum*, Mk. Bos. 3, 5: Elen. Kmbl. 597; El. 299. Sende ðê Drihten on ungewitt and blindnysse *percutiat te Dominus amentia et cæcitate*, Deut. 28, 28.

blinnan; *part.* blinnende; ic blinne, ðû blinnest, blinst, he blinneþ, blinniþ, blinþ, *pl.* blinnaþ; *p.* ic, he blan, blon, blann, blonn, ðû blunne, *pl.* blunnon; *pp.* blunnen; *v. intrans.* [be, linnan *to cease*] *To cease, rest, leave off;* cessare, desinere :—Seó rêþnes ðæs stormes wæs blinnende *the fierceness of the storm ceased* [lit. *was ceasing*], Bd. 5, 1; S. 614, 9. Blǽd his blinniþ *his prosperity ceaseth*, Exon. 94 b; Th. 354, 29; Reim. 53. We Dryhten bletsigaþ, ne ðæs blinnaþ âwa to worulde *we bless the Lord, nor cease from this for ever*, Ps. Th. 113, 25. Seó êhtnes [MS. ehtnysse] blan *the persecution ceased*, Bd. 1, 8; S. 479, 17. Blann [blonn MS. T.] se bysceophâd eall geár and ðæs ôðres syx mônaþ *the bishopric was vacant* [lit. *rested*] *all one year and six months of the next*, 3, 20; S. 550, 28. Ic nôht ðon ǽr ðære ærninge blon *I naught the sooner left off from running*, 5, 6; S. 619, 15: Andr. Kmbl. 2532; An. 1267. Ðû wuldres blunne *thou forfeitedst glory*, 2760; An. 1382. Rômane blunnon [MS. blunnun] rícsian on Breotene *Romani in Brittania regnare cessarunt*, Bd. 1, 11; S. 480, 13. Blinn from eorre and forlêt hât-heortnisse *desine ab ira et derelinque furorem*, Ps. Surt. 36, 8. [*Chauc.* blinne.] DER. a-blinnan, ge-.

blinnende, an; *f.* [blinnende, *part. of* blinnan *to cease*] *A ceasing, rest, intermission;* cessatio, intermissio :—Bútan blinnendan *without ceasing;* sine intermissione, Bd. 5, 12; S. 628, note 20. v. blin.

blinnes, blinness, e; *f. Rest;* cessatio, Som. Ben. Lye. DER. blin, nes.

blinniþ *ceases*, Exon. 94 b; Th. 354, 29; Reim. 53; *3rd pers. pres. of* blinnan.

blió, blioh *a colour, hue, complexion*, Prov. 23: Bt. Met. Fox 31, 7; Met. 31, 4. v. bleoh.

bliótan *for* bleóton *sacrificed; 3rd pl. p. of* blôtan *to sacrifice* :—Hû ða burhleóde on Cartaina bliótan [=bleóton] *men hira godum how the towns-people in Carthage sacrificed men to their gods*, Ors. cont. 4, 4; Bos. 11, 32.

blis, bliss, blys, blyss, e; *f.* [contracted from blíþs, q. v.] I. BLISS, *joy, gladness, exultation, pleasure;* lætitia, gaudium, exultatio, beatitas :—Ne seó hêhste blis nis on ðâm flæsclícum lustum *the highest bliss is not in the fleshly lusts*, Bt. 33, 1; Fox 120, 5: Ps. Spl. 29, 6. On heofonum is singal blis *in heaven is eternal bliss*, Rood Kmbl. 280; Kr. 141: Exon. 18 b; Th. 47, 5; Cri. 750: 48 b; Th. 167, 5; Gú. 1055. Ðanon com ǽrest cristendôm and blis fôr Gode and fôr worulde *whence first came christianity and joy before God and before the world*, Chr. 1011; Erl. 146, 22. Ðû eart blis mîn *tu es exultatio mea*, Ps. Spl. 31, 9. Úre bliss on ânum ðê êce standeþ *our bliss eternally remaineth in thee alone*, Ps. Th. 86, 6. Gehýrde he of hrófe ðære ylcan cyricean upp astígan ðone ylcan blisse song *audivit ascendere de tecto ejusdem oratorii idem lætitiæ canticum*, Bd. 4, 3; S. 568, 2: Bt. 24, 4; Fox 86, 32: Andr. Kmbl. 2130; An. 1066. Stefn blisse *vox exultationis*, Ps. Spl. 117, 15. Þeówiaþ Drihtne on blisse, [and] insteppaþ oððe ingâþ on gesihþe his on blisse *servite Domino in lætitia,* [*et*] *introite in conspectu ejus in exultatione*, Ps. Lamb. 99, 2. Ðis is se dæg ðæne Drihten worhte eadigum to blisse *this is the day which the Lord made for joy to the blessed*, Menol. Fox 125; Men. 62: Exon. 15 b; Th. 35, 2; Cri. 552. Ðæt bearn bringeþ blisse ðê *that infant* [*Christ*] *bringeth bliss to thee*, Exon. 8 b; Th. 5, 11; Cri. 68: Chr. 975; Erl. 126, 30; Edg. 56. Ðû eart on heofonum blisse beorhtost *thou art the brightest of joys in heaven*, Hy. 7, 10; Hy. Grn. ii. 287, 10: Exon. 26 a; Th. 77, 15; Cri. 1257. Se burgstede wæs blissum gefylled *the city-place was filled with joys*, Exon. 52 a; Th. 181, 11; Gú. 1291: 27 b; Th. 82, 31; Cri. 1347. Blissum hrêmig *exulting in gladness*, Elen. Kmbl. 2273; El. 1138: Exon. 44 b; Th. 168, 18; Gú. 1079. II. *friendship, kindness, benevolence, grace;* comitas, benignitas, benevolentia, gratia :—Hí me to wendon heora bacu bitere, and heora blisse from *they turned their bitter backs on me, and* [*took*] *their friendship from* [*me*], Bt. Met. Fox 2, 30; Met. 2, 15. Þurh ðê eorþbúende ealle onfôþ blisse mînre and bletsunge *through thee all dwellers upon earth shall receive my grace and blessing*, Cd. 84; Th. 105, 30; Gen. 1761: 106; Th. 140, 21; Gen. 2331. [*Laym. Orm.* blisse.] DER. heáh-blis, -bliss, woruld-.

blisgere, es; *m. An incendiary;* incendii auctor :—Blisgeras *incendiaries*, L. Ath. i. 6; Th. i. 203, note 38. v. blǽsere.

blissian, blyssian, blissigan, blissigean; *part.* blissiende, blissigende; ic blissie, blissige, ðû blissast, he blissaþ, *pl.* blissiaþ; *p.* ode, ede, ade; *pp.* od, ed, ad [blis, bliss *bliss, joy*]. I. *v. intrans. To rejoice, exult, be glad or merry;* lætari, gaudere, exultare, ovare :—Heora láreówas blissigende hâm hwurfon *doctores eorum domum rediere lætantes*, Bd. 3, 30; S. 562, 20. Blissigende [blissiende MS. C.] *ovans*, Ælfc. Gr. 33; Som. 37, 46. Ic blissige [Spl. blissie] ofer spæce ðînre *lætabor ego super eloquia tua*, Ps. Lamb. 118, 162. Blissaþ se rihtwîsa on Drihtne *lætabitur justus in Domino*, Ps. Lamb. 63, 11: 57, 11: Andr. Kmbl. 1268; An. 634. Ða ðe ondrædaþ ðê, geseóþ me, and hí blissiaþ *qui timent te, videbunt me, et lætabuntur*, Ps. Lamb. 118, 74: Exon. 26 b; Th. 79, 51; Cri. 1287. Hyge blissode *their spirit rejoiced*, Andr. Kmbl. 1156; An. 578. Hí on ðon swýðe blissedon *they rejoiced very much at that*, Bd. 5, 12; S. 628, 34. Blissiaþ on Drihtne *lætamini in Domino*, Ps. Lamb. 31, 11. Blyssiaþ mid me *rejoice with me*, Lk. Bos. 15, 9.

Blissie [Lamb. blissige] heorte sēcendra Drihten *lætetur cor quærentium Dominum*, Ps. Spl. 104, 3. Blissian [blissien, Th. 66, 4] and fægnian hīg þeóda *lætentur et exultent gentes*, Ps. Lamb. 66, 5. Ðæt hī blission mid Criste *that they rejoice with Christ*, Chr. 1036; Erl. 165, 17; Ælf. Tod. 9.   II. *v. trans. dat. or acc. To make to rejoice, to gladden, delight, exhilarate;* lætificare :—Sum sceal on heápe blissian æt beóre bencsittendum *one shall in company delight the bench-sitters at beer*, Exon. 88 a; Th. 331, 34; Vy. 78. Ðá se hálga ongann hæleþ blissigean *then the saint began to gladden the man*, Andr. Kmbl. 3213; An. 1609. Ðû, God, eallum blissast *thou, O God, makest all to rejoice*, Hy. 7, 34; Hy. Grn. ii. 287, 34. Heortan manna wīndrinc blissaþ *vinum lætificet cor hominis*, Ps. Th. 103, 14. He sárig folc blissade *he gladdened the sorrowful people*, Ps. Th. 106, 32. Hyge wearþ mongum blissad *the mind of many was made to rejoice*, Exon. 24 b; Th. 71, 30; Cri. 1163. [Laym. blissien: Orm. blissen.] DER. ge-blissian: môd-blissiende.

blissung, blisung, e; *f.* [blis, bliss *exultatio*] *A triumphing, exultation;* exultatio :—Blisunga beorgas beóþ ymbgyrde *exultatione colles accingentur*, Ps. Spl. 64, 13. DER. ge-blissung.

blíð = blíðe *sweet, pleasant;* suavis, amœnus :—Ðis ofet is swēte, blíð on breóstum *this fruit is sweet, pleasant in the stomach*, Cd. 30; Th. 41, 13; Gen. 656.

BLÍÐE, *comp.* blíðra; *superl.* blíðost; *def.* se blíða, seó, ðæt blíðe; *adj.*   I. *joyful, glad, merry, cheerful, pleasant,* BLITHE; lætus, hilaris :—Beó blíðe, ðū gôda þeów *be joyful, thou good servant*, Mt. Bos. 25, 21. Wæs Iethro blíðe for eallum ðám þingum, ðe Drihten dyde Israhéla folce *Jethro was glad for all the things, which the Lord did for the people of Israel*, Ex. 18, 9. Wæs engla þreát hleahtre blíðe geworden *the host of angels became merry with laughter*, Exon. 18 b; Th. 46, 19; Cri. 739: 20 b; Th. 55, 3; Cri. 878: Cd. 178; Th. 223, 10; Dan. 117. Wæs se blíða gǣst fūs on forþweg *the blithe spirit was eager for departure*, Exon. 46 b; Th. 158, 30; Gú. 917. He bæd hine blíðne beón æt ðære beórþege *he bade him be merry at the beer-drinking*, Beo. Th. 1238; B. 617: Menol. Fox 193; Men. 98. Dô ðínes scealces sáwle blíðe *lætifica animam servi tui*, Ps. Th. 85, 3. Mid ðás blíðan gedryht [MS. gedryt] *with this joyful host*, Exon. 15 a; Th. 33, 2; Cri. 519. Ic God bletsige blíðe môde *I will bless God with a joyful mind*, 41 b; Th. 138, 23; Gú. 580: Ps. Th. 54, 11: 65, 7: Rood Kmbl. 242; Kr. 122. Wīgan wæron blíðe *the warriors were blithe*, Elen. Kmbl. 492: El. 246: Cd. 171; Th. 215, 12; Exod. 582: Th. 52, 8: 106, 41. Cyning wæs ðý blíðra *the king was the blither*, Elen. Kmbl. 192; El. 96: Bt. Met. Fox 9, 63; Met. 9, 32: Byrht. Th. 136, 5; By. 146. Híg blíðost [blíðust MS.] wǣron *they were most merry*, Jud. 16, 25.   II. *gentle, kind, friendly, clement, mild, sweet;* mansuetus, benignus, comis, clemens, mitis, suavis :—Him biþ engla Weard milde and blíðe *the Lord of angels will be mild and gentle to them*, Elen. Kmbl. 2631; El. 1317: Ps. Th. 118, 88: Beo. Th. 877; B. 436. Eallum is ūre Drihten milde and blíðe *suavis Dominus universis*, Ps. Th. 144, 9: 66, 6. Wese us beorhtnes ofer blíðan Drihtnes ūres *let the beauty [brightness] of our gentle Lord be over us*, 89, 19. Weorc ánra gehwæs beorhte blíceþ in ðam blíðan hám *the works of every one shall brightly shine in that sweet home*, Exon. 64 b; Th. 238, 5; Ph. 599. Fæder ongon, þurh blíðne geþoht, his bearn lǣran *a father began, through kind thought, to teach his son*, Exon. 80 b; Th. 302, 30; Fä. 44: Andr. Kmbl. 1941; An. 973: Ps. Th. 102, 19. Utan us biddan ðone blíðan gǣst ðæt he us gescilde wið sceáðan wǣpnum *let us pray the kind spirit [i. e. the Holy Ghost] that he shield us against the spoiler's weapons*, Exon. 19 a; Th. 48, 20; Cri. 774. Blíðe môde *with gentle mind*, Ps. Th. 89, 18: Exon. 14 b; Th. 467, 5; Hö. 134. Hý wǣron blíðe wið me on heora gebǣrum, and on heora môde hí blissedon on mínum ungelimpe *they were friendly with me in their manner, and in their mind they rejoiced for my misfortune*, Ps. Th. 34, 15. Swylce habban sceal blíðe gebǣro *such shall have gentle demeanour*, Exon. 115 b; Th. 444, 8; Kl. 44. Hý se æðeling grētte blíðum wordum *the chieftain greeted her with kind words*, 68 a; Th. 252, 19; Jul. 165.   III. *quiet, calm, peaceful;* tranquillus, placidus :—Lēton ðone hálgan swefan on sibbe under swegles hleó, blíðne bídan *they left the saint sleeping in peace, calm abiding under the vault of heaven*, Andr. Kmbl. 1665; An. 835. Ðæt he smylte môde and blíðe him eall forlēt *quod ille placida mente dimitteret*, Bd. 3, 22; S. 553, 21. Ða ýða swýgiaþ, blíðe weorþaþ *the waves grow silent, become calm*, Ps. Th. 106, 28. [Chauc. R. Glouc. blithe: Laym. blíðe, blíðen: Orm. bliþe: O. Sax. blíði: North Frs. blide in blid-skip *joy*: Dut. blijde: M. H. Ger. blíde: O. H. Ger. blídi: Goth. bleiþs: Dan. Swed. blid: Icel. blíðr.] DER. hyge-blíðe, ofer-, un-.

blíðe; *adv.*   I. *joyfully, gladly;* læte :—Bletsa, míne sáwle, blíðe, Drihten *bless the Lord joyfully, O my soul*, Ps. Th. 102, 1: Exon. 44 a; Th. 149, 9; Gú. 759.   II. *kindly, mildly;* benigne, clementer :—Ðú me, milde and blíðe, þurh ysopon ahluttra *asperges me hyssopo*, Ps. C. 50, 72; Ps. Grn. ii. 278, 72: Ps. Th. 54, 17.

blíðe-heortnys, -nyss, e; *f. Merry-heartedness;* lætitia, mansuetudo. DER. blíðe, heorte, -nes.

blíðe-líce; *comp.* -lícor; *adv. Gladly, joyfully,* BLITHELY, *merrily;* læte, hilariter :—He hine blíðelíce onfēng *he received him joyfully*, Lk. Bos. 19, 6: Gen. 46, 30. Ge māgon blíðelíce hlihhan *potestis hilariter ridere*, Ors. 3, 7; Bos. 62, 28. Ðæt he ðý blíðelícor þrôwode *that he the more gladly might suffer*, Bd. 5, 14; S. 634, 42.

blíðe-môd; *adj. Blithe of mind, glad, cheerful;* lætus animo, lætus, hilaris :—Wæs â blíðemôd bealuleás cyning [MS. kyng], ðeáh he lang ǣr, lande bereáfod, wunode wræclástum *the innocent king was ever blithe of mind [cheerful], though he long before, bereft of land, dwelt in exile*, Chr. 1065; Erl. 196, 34: Edw. 15: Cd. 72; Th. 88, 21; Gen. 1468: 86; Th. 108, 2; Gen. 1800: 210; Th. 260, 21; Dan. 713. Hyssas wǣron blíðemôde *the youths were cheerful [blithe of mind]*, 186; Th. 231, 26; Dan. 253.

blíð-heort; *adj.*   I. BLITHE *of* HEART, *merry, joyful;* lætus corde, hilaris :—Hræfn blaca, blíðheort, bodode cuman beorhte sunnan *the black raven, blithe of heart [merry], foretold the coming of the bright sun*, Beo. Th. 3608; B. 1802: Andr. Kmbl. 2526; An. 1264. Gefēgon beornas, blíðheorte, burhweardes cyme *the men, blithe of heart, rejoiced in the coming of the prince [lit. the city-guardian]*, Andr. Kmbl. 1319; An. 660.   II. *kind of heart, merciful;* benignus corde, misericors :—Gebletsode blíðheort Cyning, Metod alwihta, wíf and wǣpned *the merciful King, Lord of all things, blessed female and male*, Cd. 10; Th. 12, 28; Gen. 192.

blíð-nes, -ness, -nyss, e; *f. Joyfulness, enjoyment, a leaping for joy, exultation, mirth;* gaudium, exultatio, hilaritas :—Gif ðú nú atelan wilt ealle ða blíðnessa wið ðám unrôtnessum *if thou wilt now reckon all the enjoyments against the sorrows*, Bt. 8; Fox 24, 22. On blíðnysse *in exultatione*, Ps. Spl. 99, 2. Blíðnysse líf *vita hilaritatis*, Lchdm. iii. 212, 1.

blíþs, blíðs, e; *f. Joy, gladness;* lætitia :—Liódum to blíþse *to the gladness of the people*, Ps. C. 50, 118; Ps. Grn. ii. 279, 118. Sæle blíðse me *give me joy*, 50, 99; Ps. Grn. ii. 279, 99. [O. Sax. blíðsea, f.] DER. blíðe. v. blis.

blíþsian; *p.* ode; *pp.* od *To rejoice, be glad, blithe, merry;* lætari :—Hí tô swíðe blíþsodon *they rejoiced too much*, Past. 50, 2; Hat. MS. Blíþsa, cniht on ðínum gióguþhāde *rejoice, young man, in thy youth*, 49, 5; Hat. MS. [O. Sax. blídsean: Ger. blitzen *exsilire gaudio*: O. H. Ger. blíðen.] DER. blíþs.

blíðust *very merry*, Jud. 16, 25, = blíðost; *superl. of* blíðe, *adj.*

BLÔD, es; *n.* BLOOD, *gore;* sanguis, cruor :—Ðæt blôd eów byþ to tácne on ðám hūsum, ðe ge on beóþ: ðonne ic ðæt blôd geseó, ðonne forbūge ic eów *erit sanguis vobis in signum in ædibus, in quibus eritis, et videbo sanguinem et transibo vos*, Ex. 12, 13: Gen. 4, 10: Jn. Bos. 6, 55: Mt. Bos. 16, 17. Wæs ðæt blôd hât *the blood was hot*, Beo. Th. 3237; B. 1616: 3339; B. 1667: Cd. 9; Th. 12, 6; Gen. 181: Exon. 116 b; Th. 447, 15; Dôm. 40: Andr. Kmbl. 1907; An. 956. His swát wæs swylce blôdes dropan *est sudor ejus sicut guttæ sanguinis*, Lk. Bos. 22, 44: Mt. Bos. 27, 6, 8: Gen. 4, 11: Exon. 21 b; Th. 58, 15; Cri. 936. Lâ hwilc nýtwyrþnes on mínum blôde *quæ utilitas in sanguine meo?* Ps. Lamb. 29, 10: Lk. Bos. 22, 20: Beo. Th. 1698; B. 847. Hit biþ geworden to blôde *vertetur in sanguinem*, Ex. 4, 9: 7, 17: 29, 21. Swá hwá swá agít mannes blôd, his blôd biþ agoten *quicumque effuderit humanum sanguinem, fundetur sanguis illius*, Gen. 9, 6: Ps. Lamb. 13, 3: 49, 13: Andr. Kmbl. 46; An. 23. Gebletsode Romulus mid his brôðor blôde ðone weall, and mid ðara sweora blôde ða cyrican, and mid his eámes blôde ðæt ríce *Romulus blessed [consecrated] the wall [of Rome] with his brother's blood, the temples with the blood of their fathers-in-law, and the kingdom with his uncle's blood*, Ors. 2, 2; Bos. 41, 5-7. Meotud ðē gebohte blôde ðý hálgan *the Lord bought thee with his holy blood*, Exon. 98 a; Th. 368, 26; Seel. 30: Rood Kmbl. 96; Kr. 48. Blôde fáh *stained with blood*, Beo. Th. 1873; B. 934: 3192; B. 1594: 5940; B. 2974. Begleddod is eorþe on blôdum *infecta est terra in sanguinibus*, Ps. Spl. 105, 36. Deád blôd *clotted blood, gore;* cruor, Wrt. Voc. 283, 79. [Chauc. blod: Wyc. blood: Laym. Orm. blod: Scot. bloud: Plat. blod, n: O. Sax. blôd, n: Frs. bloed, n: North Frs. blot, blöt, n: O. Frs. blod, n: Dut. O. Dut. bloed, n: Ger. blut, n: M. H. Ger. bluot, n: O. H. Ger. bluot, n: Goth. bloþ, n: Dan. Swed. blod, n: Icel. blóð, n.] DER. blôd-dolg, -egesa, -fág, -geóte, -gíta, -gýte, -hreów, -hreówa, -lǣtan, -lǣtere, -leás, -reád, -reów, -ryne, -seax, -seten, -siht, -spíwing, -wyrt, -yrnende: blôdig, -tôþ: blôdeg: blôdegian, ge-.

blôd-dolg, es; *n. A bloody wound;* cruentum vulnus. DER. blôd, dolg, *q. v.*

blôd-dryncas; *pl. m. Blood-sheddings, blood-shed;* sanguinis effluvium :—Seó eorþbeofung tácnade ða miclan blôddryncas *the earthquake betokened the great blood-sheddings*, Ors. 4, 2; Bos. 79, 29.

blôd-egesa, an; *m.* [egesa, egsa *fear, terror*] *Bloody horror;* cruentus terror :—Brim berstende blôdegesan hweóp *the bursting sea threatened bloody horrors*, Cd. 166; Th. 208, 3; Exod. 477.

blôðegian; *p.* ode; *pp.* od [blôdig *bloody*] *To make bloody;* cruentare. DER. ge-blôdegian.

blódes flównyss, e; _f. A bloody flux, a flowing of blood;_ sanguinis fluxus:—Ðæt wíf wæs þrówiende blódes flównysse _mulier fluxum patiebatur sanguinis,_ Bd. 1, 27; S. 494, 4. v. blód-yrnende, flównes.

blód-fág; _adj._ [fág _tinctus_] _Stained with blood;_ sanguine tinctus:— Is me bánhús blódfág _my body_ [lit. _bone-house_] _is stained with blood,_ Andr. Kmbl. 2809; An. 1407: Beo. Th. 4127; B. 2060.

blód-forlǽtan; _p._ -forlét, _pl._ -forléton; _pp._ -forlǽten _To let blood, bleed;_ sanguinem emittere, phlebotomare:—Ðæt heó niwan blódforlǽten wǽre on earme _that she had been lately bled in the arm;_ quia phlebotomata est nuper in brachio, Bd. 5, 3; S. 616, 4.

blód-geótan _to pour out_ or _shed blood;_ sanguinem effundere. DER. blód, geótan.

blód-geóte, es; _m. Blood-shedding, a shedding of blood;_ sanguinis effusio:—Be blódgeóte _of blood-shedding,_ L. Edm. S. 4; Th. i. 248, 22, 24. v. blód-gýte.

blód-geótende; _part. Shedding blood, blood-thirsty;_ sanguinem effundens, sanguinolentus:—Weras blódgeótende _viri sanguinum,_ Ps.Spl. 54, 27.

blód-gíta, an; _m. A shedder of blood;_ sanguinis effusor:—Ðæne wer, ðe is blódgíta, gehiscþ Drihten _the Lord hates the man who is a blood-shedder,_ Ps. Lamb. 5, 8.

blód-gýte, es; _m._ [blód, gýte _a flowing, from_ gýt _flows out, pres. of_ geótan]. I. _a flowing_ or _running of blood;_ sanguinis profluvium:—Gif men blód út of nósum yrne tó swíðe, syle him drincan fífleáfan on wíne, and smyre ðæt heáfod mid ðam; ðonne óþstandeþ se blódgýte sóna _if blood run from a man out of his nostrils too much, give him to drink fiveleaf in wine, and smear the head with it; then the blood-running will soon staunch,_ Herb. 3, 5; Lchdm. i. 88, 8–10. II. _a blood-shedding, bloodshed;_ sanguinis effusio:—Ðér wæs se mæsta blódgýte _there was the greatest bloodshed,_ Ors. 4, 2; Bos. 79, 26. Wǽron ða mæstan blódgýtas _there were the greatest blood-sheddings,_ Ors. 3, 9; Bos. 67, 31. Bútan blódgýte _without bloodshed,_ Bd. 1, 3; S. 475, 11.

blód-hrǽcan; _p._ te; _pp._ ed _To retch_ or _spit blood;_ sanguinem excreare. DER. blód, hrǽcan _to retch._

blód-hrǽce, es; _m. A spitting of blood;_ sanguinis excreatio. v. blód, hrǽce.

blód-hreów; _def._ se blód-hreówa; _adj._ [hreów _cruel_] _Blood-thirsty, cruel;_ sanguinolentus, crudelis:—Me wið blódhreówes weres bealuwe gehǽle _save me from the wickedness of the blood-thirsty man,_ Ps. Th. 58, 2. Blódhreówe weras ge bebúgaþ me _viri sanguinum declinate a me,_ 138, 17. Se blódhreówa wer _sanguinum vir,_ Ps. Grn. 54, 24; Ps. Grn. ii. 153, 24.

blódig; _def._ se blódiga, seó, ðæt blódige; _adj._ BLOODY; sanguineus, cruentus:—Ne sý him bánes bryce, ne blódig wund _let there not be to him a breaking of bone, nor a bloody wound,_ Exon. 42 b; Th. 143, 33; Gú. 670: Andr. Kmbl. 2945; An. 1475. Se bræd of ðæm beorne blódigne gár _he plucked the bloody dart from the chief,_ Byrht. Th. 136, 21; By. 154. Geseoh nú swá ðín swát ageát, blódige stíge _behold now where thy blood poured forth, a bloody path,_ Andr. Kmbl. 2883; An. 1444. He byreþ blódig wæl _he will bear off my bloody corpse,_ Beo. Th. 900; B. 448. He his mǽg ofscét blódigan gáre _he shot his kinsman with a bloody arrow,_ 4872; B. 2440. Ealle him brimu blódige þuhton _all the waters seemed bloody to them,_ Cd. 170; Th. 214, 20; Exod. 572. Ða hwettaþ hyra blódigan téþ _who whet their bloody teeth,_ L. E. I. prm; Th. ii. 396, 6. Blódigum teárum _with bloody tears,_ Exon. 25 a; Th. 72, 20; Cri. 1175. Blódig útsiht _a dysentery;_ dysenteria, Ælfc. Gl. 11; Som. 57, 51; Wrt. Voc. 19, 53. [O. Sax. blódag: O. Frs. blodich: Dut. bloedig: Ger. blutig: M. H. Ger. bluotec: O. H. Ger. blótag: Dan. Swed. blodig: Icel. blóðigr.] DER. ge-blódegian.

blódig-tóþ; _adj. Bloody-toothed, cruel;_ cruentus dentibus, crudelis:—Bona blódigtóþ _the bloody-toothed murderer,_ Beo. Th. 4170; B. 2082.

blód-lǽswu, e; _f. A blood-letting;_ sanguinis emissio:—Frægn se biscop hwonne hire blódlǽswu ǽrest wǽre _the bishop asked when was first her blood-letting,_ Bd. 5, 3; S. 616, 12, 15. On ðære blódlǽswe _in the blood-letting,_ 5, 3; S. 616, 5.

blód-lǽtan; _p._ -lét, _pl._ -léton; _pp._ -lǽten _To let blood, bleed;_ sanguinem emittere, phlebotomare:—Blódlǽtan móna gód ys _it is a good moon for letting blood,_ Lchdm. iii. 184, 11: Bd. 5, 3; S. 616, 14.

blód-lǽtere, es; _m. A blood letter;_ phlebotomarius, Ælfc. Gl. 17; Som. 58, 93; Wrt. Voc. 22, 10.

blód-leás; _adj._ BLOODLESS; exsanguis, Ælfc. Gr. 9, 28; Som. 11, 58.

blód-mónaþ '_blood-month,_' i. e. _November._ v. blót-mónaþ.

blód-reád; _adj._ BLOOD-RED; sanguineus:—Ðæt þridde cyn ys _sanguineus, ðæt is blódreád _the third sort is sanguineus, that is blood-red,_ Herb. 131, 1; Lchdm. i. 242, 16.

blód-reów; _adj. Sanguinary;_ sanguinolentus:—Breóst-hord blódreów _a sanguinary heart_ [lit. _breast-hoard_ or _treasure_], Beo. Th. 3442; B. 1719.

blód-ryne, es; _m._ [ryne _a running, course_] _A running of blood, an issue;_ sanguinis fluxus:—Án wíf þolode blódryne twelf geár _mulier sanguinis fluxum patiebatur duodecim annis,_ Mt. Bos. 9, 20. On blódryne _in fluxu sanguinis,_ Lk. Bos. 8, 43.

blód-seax, blód-sex, es; _n. A blood-knife, a lancet;_ phlebotomus = φλεβοτόμον, Ælfc. Gl. 17; Som. 58, 91; Wrt. Voc. 22, 9. v. æder-seax.

blód-setenn, e; _f._ [blód _blood,_ setenn _from_ seten, _pp._ of sittan _to sit, stop_] _The stoppage of blood;_ sanguinis profluentis restrictio. v. sittan.

blód-siht, e; _f. A flowing of blood;_ sanguinis profluvium. DER. blód blood, siht _a flowing, flux._

blód-spíwing _a spewing of blood._ v. blót-spíung.

blód-wanian; _p._ ode; _pp._ od [wanian _to diminish_] _To diminish blood;_ sanguinem minuere:—Nys ná gód móna blódwanian _it is not a good moon for diminishing blood,_ Lchdm. iii. 184, 16.

blód-wíte, es; _n._ [blód, wíte _mulcta_] _Blood;_ sanguis:—Ná ic gegadrige gesamnunga heora of blódum oððe of blódwítum _non congregabo conventicula eorum de sanguinibus,_ Ps. Lamb. 15, 4.

blód-wyrt, e; _f._ BLOODWORT or _bloody-dock from its red veins and stems;_ rumex sanguineus, Lin. v. wyrt.

blód-yrnende; _part._ [blód, yrnende, _part._ of yrnan _to run, flow_] _Blood-flowing;_ sanguinans, sanguine fluens:—Ðæt wíf blódyrnende þrówaþ _the blood-flowing woman suffereth_ [_was suffering_], Bd. 1, 27; S. 494, note 8, B. v. blódes flównyss.

BLÓMA, an; _m._ [blów + am + a, Ettm. 314] _Metal, the metal taken from the ore,_ Wrt. Voc. 34, note 1: _a mass;_ metallum, massa = μᾶζα _that which adheres together like dough,_ Wht. Dict:—Ísenes blóma _a mass of iron;_ ferri massa, Som: Cot. 135. Blóma oððe ðáh _massa,_ Wrt. Voc. 85, 16; _Lye says truly, referring to this quotation,—' Ínter ea quæ pertinent ad metalla.'_ Blóma _is contained in one of our oldest glossaries:—_ Dáh [MS. dað] _vel_ blóma _massa,_ Ælfc. Gl. 51; Som. 66, 9; Wrt. Voc. 34, 68. Also in a Semi-Saxon glossary of the 12th century,—Blóma _vel_ dáh _massa,_ Wrt. Voc. 94, 63. DER. gold-blóma.

blon, blonn _ceased;_ cessavit, Bd. 5, 6; S. 619, 15: 3, 20; S. 550, note 27; _p. of_ blinnan.

blonca, an; _m. A grey horse;_ equus albus:—Beornas and bloncan mid _warriors and their grey horses with them,_ Exon. 106 a; Th. 405, 5; Rä. 23, 18. v. blanca.

blondan _to mix, blend, mingle;_ miscere. DER. ge-blondan. v. blandan.

blonden-feax; _part. Having mixed hair;_ comam mixtam habens, Cd. 107; Th. 141, 7; Gen. 2341: 123; Th. 157, 5; Gen. 2600: Beo. Th. 3586; B. 1791: 5916; B. 2962: 3750; B. 1873: 3193, B. 1594. v. blanden-feax.

blóstm, es; _m:_ e; _f? A blossom, flower;_ flos:—Blóstm _flos,_ Ælfc. Gl. 46; Som. 65, 10; Wrt. Voc. 33, 9. Blóstma hiwum _in hues of flowers,_ Exon. 94 a; Th. 352, 32; Reim. 4. v. blóstma.

BLÓSTMA, blósma, an; _m._ [= blóstm _a blossom_] A BLOSSOM, bloom, flower; flos:—Swá swá blósma æceres swá he blóweþ _tamquam flos agri sic efflorebit,_ Ps. Lamb. 102, 15. Ofer hine scír cymeþ mínra [minre MS.] sóþfæst blóstma _super ipsum florebit sanctificatio mea,_ Ps. Th. 131, 19. Ðeáh ðe lilie sý beorht on blóstman, ic eom betre ðonne heó _though the lily be bright in its blossom, I am better than it,_ Exon. 110 b; Th. 423, 26; Rä. 41, 28: Ps. Th. 102, 14. Ne feallaþ on foldan fealwe blóstman _the fallow blossoms fall not on earth,_ Exon. 57 a; Th. 202, 24; Ph. 74. Ic geseah ðone fægrestan feld full grówendra blóstma _I saw the most beautiful field full of growing flowers,_ Bd. 5, 12; S. 629, 20. Ellenes blósman genim _take blossoms of elder,_ L. M. 2, 59; Lchdm. ii. 288, 2. Ic geseah ðǽr on weaxende blósman litlum and litlum, and æfter ðám blósman wínbergean _I saw blossoms growing thereon by little and little, and after the blossoms grapes_ [lit. _wine-berries_], Gen. 40, 10. He dysegaþ se ðe wintregum wederum wile blósman [Cot. blostman] sécan _he is foolish who will seek flowers in wintry weather,_ Bt. 5, 2; Fox 10, 32. Ðænne wangas blóstmum blówaþ _then_ [i. e. _in summer_] _the fields bloom with flowers,_ Menol. Fox 179; Men. 91: Exon. 82 a; Th. 308, 31; Seef. 48. [_Tynd._ blossom: _Chauc. Piers P._ blosme: _Orm._ blostme: _Dut._ bloesem, _m: O. Dut._ blosem, _Kil: Dan._ blomst, _c: Swed._ blomster, _n: Icel._ blómstr, _m._]

blóstm-bǽrende; _part._ [blóstm, bǽran _to bear_] _Blossom-bearing;_ florifer:—Seó blóstmbǽrende stów is seó stów on ðære beúþ onfangene sóþfæstra sáula _the blossom-bearing place is the place to which are taken the souls of the righteous,_ Bd. 5, 12; S. 630, 14.

blóstmian; _part._ blóstmiende; _p._ ode; _pp._ od _To_ BLOSSOM, _blow;_ efflorere:—Seó beorhtnes ðæs blóstmiendan feldes wæs gesewen _the brightness of the blossoming field was seen,_ Bd. 5, 12; S. 629, 38.

BLÓT, es; _n. A sacrifice;_ sacrificium:—He ealle ða cuman to blóte gedyde _he gave all the strangers for a sacrifice,_ Ors. 1, 8; Bos. 31, 4. On blóte _by sacrifice,_ L. C. S. 5; Th. i. 378, 21. [_Icel._ blót, _n._] DER. ge-blót: blótan: blót-mónaþ.

blót = blód _blood;_ sanguis. v. blót-spíung.

blótan, ic blóte, ðú blótest, blétst, he blóteþ, blét, _pl._ blótaþ; _p._ ic, he bleót, ðú bleóte, _pl._ bleóton; _pp._ blóten; _v. a._ [blót _a sacrifice_] _To sacrifice, to kill for a sacrifice;_ immolare, sacrificare:—Ðæt hí hiora godum ðe ýð blótan meahton _that they might the more easily sacrifice to their gods,_ Ors. 2, 2; Bos. 40, 37: 4, 4; Bos. 80, 39: 5, 2; Bos. 102, 16. Ongunnon heora bearn blótan feóndum _immolaverunt filios suos dæmoniis,_ Ps. Th. 105, 27: Cd. 138; Th. 173, 5; Gen. 2856.

Úre yldran on ðam mónþe bleóton á *our forefathers always sacrificed in this month*, Hick. Thes. i. 219, 57. Ða burhleóde on Cartaina bleóton [bliotan MS.] men hira godum *the inhabitants of Carthage sacrificed men to their gods*, Ors. cont. 4, 4; Bos. 11, 32. Ðæt hine mon ænigum godum blóte *that a man sacrifice him to any gods*, Ors. 1, 8; Bos. 31, 11. Ðæt hí ða git swíðor blótten, ðonne hie ær dydon *that they should sacrifice still more than they had done before*, 4, 4; Bos. 80, 18. [*M. H. Ger.* bluoten: *O. H. Ger.* blozan, ploazzan, plozan: *Goth.* blotan: *O. Dan.* blothe: *Swed.* blota: *Icel.* blóta *sacrificare.*] DER. a-blótan, on-.

blót-mónaþ, es; *m.* [blót *a sacrifice*, mónaþ *month*] November, the month of sacrifice, so called because at this season the heathen Saxons made a provision for winter, and offered in sacrifice many of the animals they then killed. In an account of the Saxon months, it is thus described:—Se mónaþ is nemned on Léden *Novembris*, and on úre geþeóde blótmónaþ, forðon úre yldran, ða hý hæðene wæron, on ðam mónþe hý bleóton á, ðæt is, ðæt hý betæhton and benæmdon hyra deófolgyldum ða neát ða ðe hý woldon syllan *this month is called Novembris in Latin, and in our language the month of sacrifice, because our forefathers, when they were heathens, always sacrificed in this month, that is, that they took and devoted to their idols the cattle which they wished to offer*, Hick. Thes. i. 219, 56–58: Menol. Fox 387; Men. 195.

blót-spíung, e; *f.* [blót = blód *blood*, spíwing *spewing*] A throwing up of blood; hæmoptois, Ælfc. Gl. 10; Som. 57, 33; Wrt. Voc. 19, 38.

blótung, e; *f. A sacrificing, sacrifice*; sacrificium, immolatio:—þurh heora blótunge *per eorum sacrificium*, Ors. 3, 3; Bos. 55, 33. v. blót.

BLÓWAN; *part.* blówende; ic blówe, ðú blówest, blæwst, he blóweþ, blǽwþ, *pl.* ic blówe, ðú bleówe, *pl.* bleówon; *pp.* blówen; *v. n.* 1. *to* BLOW, *flourish, bloom, blossom*; florere, efflorere, reflorere:—Wudu sceal blædum blówan *the wood shall blow with flowers*, Menol. Fox 527; Gn. C. 34: Exon. 109 a; Th. 417, 6; Rä. 35, 9. Wæs Aarones gyrd gemétt blówende and berende hnyte *Aaron's rod was found blossoming and bearing nuts*, Homl. Th. ii. 8, 15. Ic eom bearu blówende *I am a blooming grove*, Exon. 108 a; Th. 412, 22; Rä. 31, 4. Ic blówe *floreo*, Ælfc. Gr. 26, 2; Som. 28, 44. Swá swá blósma æceres swá he blóweþ [blǽwþ, Spl.] *tamquam flos agri sic efflorebit*, Ps. Lamb. 102, 15. Hió gréwþ and blǽwþ and westmas bringþ *it grows and blossoms and produces fruits*, Bt. 33, 4; Fox 130, 6. Se rihtwísa swá palmtreów blǽwþ *justus ut palma florebit*, Ps. Lamb. 91, 13. Híg blówaþ swá swá gærs eorþan *florebunt sicut fœnum terræ*, 71, 16. Aarones gyrd greów and bleów and bær hnyte *Aaron's rod grew and blossomed and bare nuts*, Homl. Th. ii. 8, 18. Bleów flǽsc mín *refloruit caro mea*, Ps. Lamb. 27, 7. Ær ðon eówre treówu telgum blówen [MS. blówe] *ere your trees flourish with branches*, Ps. Th. 57, 8. 2. blówan *to blossom*, is sometimes used in Anglo-Saxon instead of bláwan *to blow*; and thus, blówan was occasionally used by the Anglo-Saxons as the present English *to blow*. We say *to blow as the wind*, and *to blow or blossom as a flower*. v. bláwan. [*Wyc. R. Glouc.* blowe: *Laym.* blowen: *O. Sax.* blôjan: *Frs.* bloeyen: *North Frs.* blôye: *O. Frs.* bloia: *Dut.* bloeijen: *Ger.* blühen: *M. H. Ger.* blüejen, blüen, bluon: *O. H. Ger.* bluohan, bluojan, bluon: *Lat.* florere: *Grk.* φλέω, φλοίω *to be in full vigour or bloom*: *Sansk.* phal *to burst, blossom*.] DER. ge-blówan.

blunne, *pl.* blunnon; *pp.* blunnen *hast been deprived, ceased, rested*, Andr. Kmbl. 2760; An. 1382: Bd. 1, 11; S. 480, 13; *p. and pp. of* blinnan.

blysa, blisa, an; *m. A torch*; fax:—Ðes blisa [blysa, D.] *hæc fax*, Ælfc. Gr. 9, 59; Som. 13, 37.

blysiere, es; *m. An incendiary*; incendii auctor:—Blysieras *incendiaries*, L. Ath. i. 6; Th. i. 202, 19. v. blæsere.

blysige, an; *f. A torch*; fax:—Þæcile, blysige *fax*, Wrt. Voc. 284, 20. v. þæcele.

blyssian *to rejoice*, Lk. Bos. 15, 9. v. blissian.

bó *both*; ambo, Ps. Th. 103, 9; *nom. pl.* = bú, bá. v. begen.

BÓC, e; *f.* bóc-treów, es; *n.* bócce, beóce, béce, bǽce, an; *f. A beech-tree*; fagus silvatica, fagus = φηγός, æsculus:—Bóc *fagus*; bóc æsculus, Ælfc. Gl. 45; Som. 64, 99, 100. [*Plat.* book, böke, *f.*: *Dut.* beuk, beuke, *f.*: *Kil.* boecke, buecke: *Ger.* buche, *f.*: *Icel.* bók, *f.*: *Lat.* fâgus, *f.* = Grk. φηγός, *f.*] DER. bóc-scyld, -treów, -wudu.

bóc; *g.* bóce? béc; *d.* béc; *acc.* béc; *pl. nom. acc.* béc; *g.* bóca; *d.* bócum, bócan; *f.* I. *a* BOOK; liber:—Seó bóc is on Englisc awend *the book is turned into English*, Homl. Th. ii. 358, 30. On fóre-werd ðære bóce [MS. bóc] oððe on heáfde béc awriten is be me *in capite libri scriptum est de me*, Ps. Lamb. 39, 9. On fórewardre ðyssere béc ys awriten be me, ðæt ic sceolde ðínne willan wyrcan, Ps. Th. 39, 8; *in the hed of the boc it is write of me, that I do thi wil*, Wyc. Ic wrát bóc *I wrote a book*, Bd. 5, 23; S. 648, note 37. Adilga me of ðínre béc *dele me de libro tuo*, Ex. 32, 32, 33. Swá he ða bóc unfeóld *so he unfolded the book*, Lk. Bos. 4, 17, 20: Deut. 31, 26. Ða béc befón *to contain the books*; capere libros, Jn. Bos. 21, 25. On ðæra cininga bócum *in the kings' books*, Ælfc. T. Lisle 21, 1: 23, 19: 40, 4. On ðære béc *in this book*, 24, 25. Bóca bedǽled *deprived of books*, 2, 3.

On fíf béc *in five books*, Bd. 5, 23; S. 648, 31. Ðis is seó bóc Adames mægrace *hic est liber generationis Adam*, Gen. 5, 1: Mt. Bos. 5, 31. Feówer Cristes béc *four books of Christ, the four gospels*, Ælfc. T. Lisle 24, 22. Bóca streón *a treasury of books, a library*, Bd. 5, 21; Whelc. 451, 30, MS. C. II. *a charter*; charta = χάρτης, *m.*:—Ðis is seó bóc, ðe Æðelstán cing gebócode Friþestáne bisceope *this is the charter, which king Æthelstan chartered to bishop Frithestane*, Th. Diplm. A. D. 938; 187, 18. Heó cýðaþ on ðisse béc *they declare by this charter*, Th. Diplm. A. D. 886–899; 137, 12. Ic him sealde ðæt lond on éce erfe, and ða béc *I gave him the land in perpetual heritage, and the charters*, Th. Diplm. A. D. 872–915; 168, 10. 2. *for the books which a priest ought to possess*, v. mæsse-preóst, 2; *for his canonical hours*, v. 3. [*Chauc.* booke: *Laym.* boc, bac, *f.*: *Orm.* boc: *Plat.* book, *n.*: *O. Sax.* bôk, *n. f.*: *Frs.* bok, *f.*: *boek, n.*: *O. Frs.* bok, *f. n.*: *Dut.* boek, *n.*: *Ger.* buch, *n.*: *M. H. Ger.* buoch, *f.*: *O. H. Ger.* bôh, *n.*: *Goth.* boka, *f.*: *Dan.* bog, *c.*: *Swed.* bok, *f.*: *Icel.* bók, *f.*: *O. Slav.* bukva, *f.* All these words have evidently the same origin. Wormius, Saxo, Junius, etc. suppose that as bóc denotes *a beech-tree*, as well as *a book*, in the latter case it was used in reference to the material from which the Northern nations first made their books. Wormius infers, that pieces of wood, cut from the beech-tree, were the ancient Northern books, *Lit. Run.* p. 6. Saxo Grammaticus states, that Fengo's ambassadors took with them letters engraved in wood [literas ligno insculptas], because that was formerly a celebrated material to write upon, *Lib.* iii. p. 52: *Turner's Hist.* App. b. ii. ch. 4, n. 25, vol. i. p. 238. Thus the *Latin* liber, and *Greek* βίβλος *a book*, took their origin from the materials of which books were made. *Liber* originally signified *the inner bark of a tree*, and βίβλος or βύβλος, *an Egyptian plant* [Cyperus papyrus, *Lin.*], which, when divided into lamina and formed into sheets to write upon, was called πάπυρος, hence papyrus *paper*. Martinius, Stiernhielmius, Wachter, Adelung, etc. rather derive buch, bóc, etc. from bügen *to bend* or *fold in plaits*, referring to the folded leaves of the parchment. Thus distinguishing these books from their folds. The ancient *volumina* were denominated from being in rolls, or rolled in the form of cylinders. At the Council of Toledo, in the 8th century, a book was denominated *complicamentum*, that which is folded. In still earlier times, even one fold of parchment was denominated a book, and *Ker.* calls a letter puah, and *Not.* brief puoch, lit. *a letter book*.] DER. æ-béc, ærend-bóc, bigspell-, bletsing-; Cristes bóc; dóm-, fór-, gódspell-, hand-, mæsse-, pistol-, ræding-, sang-, scrift-, síþ-, spel-, traht-, wís-: bóc-æceras, -cest, -cræft, -cræftig, -ere, -fel, -ge-streón, -hord, -hús, -ian, -land, -lár, -leáf, -léden, -líc, -rædere, -ræding, -reád, -riht, -scamel, -stæf, -tæcing, -talu, -ung.

bóc, *pl.* bócon *baked*; coxit, coxerunt, Ex. 12, 39; *p. of* bacan.

bóc-æceras, *pl. m. Booked acres, book-land, freehold.* v. bóc-land.

bócan = bócum *for books*, L. Eth. vi. 51; Th. i. 328, 8; *dat. pl. of* bóc.

bóca streón *a place for books, library*; bibliotheca, Bd. 5, 21; Whelc. 451, 30, MS. C.

bócce, beóce, béce, bǽce, an; *f. A beech-tree*; fagus = φηγός; æsculus. v. bóc, e; *f. a beech-tree*.

bóc-cest, e; *f.* [cest, cyst *a chest*] A book-chest, book-shop, tavern; taberna:—Bóccest *taberna*, Ælfc. Gl. 17; Som. 58, 89; Wrt. Voc. 22, 7.

bóc-cræft, es; *m.* [bóc *a book*, cræft *art, science*] Book-learning, learning, literature; literatura:—Boëtius wæs in bóccræftum se rihtwísesta Boëthius, *in book-learning, was the most wise*, Bt. 1; Fox 2, 13. Ðara bóccræfta *of the knowledge of letters, of literature*, Greg. Dial. pref. 2.

bóc-cræftig; *adj. Book-crafty or learned, learned in the Bible*; in libris literatus, in Bibliis doctus:—Hí breóton [MS. breotun] bóccræftige *they destroyed those learned in the Bible*, Exon. 66 a; Th. 243, 25; Jul. 16.

bócere, es; *m. A writer, scribe, an author, a learned man, instructor*; scriptor, scriba, interpres, vir doctus *vel* literatus:—Ða cwæþ se bócere, Láreów, well ðú on sóþe cwǽde *then the scribe said, Master, thou in truth hast well said*, Mk. Bos. 12, 32. Hwæt secgeaþ ða bóceras *why say the scribes?* Mt. Bos. 17, 10. Hieronimus se wurþfulla and se wísa bócere awrát be Iohanne *the worthy and the wise author Jerome wrote concerning John*, Ælfc. T. Lisle 32, 1. Ælc gelǽred bócere forlǽt ealde þing and niwe *every learned writer brings out old things and new*, 39, 5. Swá ðætte swá hwæt swá he of godcundum stafum þurh bóceras geleornode *ita ut quicquid ex divinis literis per interpretes disceret*, Bd. 4, 24; S. 596, 33. We witan ðæt, þurh Godes gyfe, þræl wearþ to þegene, and ceorl wearþ to eorle, sangere to sacerde, and bócere to biscope *we know that, by the grace of God, a slave has become a thane, and a ceorl [free man] has become an earl, a singer a priest, and a scribe a bishop*, L. Eth. vii. 21; Th. i. 334, 7–9.

bóc-fel, -fell, es; *n.* [fell *skin*] A skin prepared for books, parchment, vellum; charta pergamena, membrana:—Bócfel *membrana*, Ælfc. Gl. 80; Som. 72, 111; Wrt. Voc. 46, 68. Bócfel *bargina*, 16; Som. 58, 57; Wrt. Voc. 21, 44. Ðæt hí habban blæc and bócfel *that they have ink and vellum*, L. Edg. C. 3; Th. ii. 244, 11.

bóc-gestreón, es; *n. A book-treasury, library*; bibliotheca:—He ðider

I

micel bócgestreón and ǽðele begeat *he acquired there a great and noble library,* Bd. 5, 20; S. 642, 2.

bóc-hord, es; *n.* A BOOK-HOARD, *a library* or *receptacle for books, papers, etc;* bibliotheca, archivum:—Bóchord [MS. boochord] *bibliotheca* vel *armarium* vel *archivum,* Ælfc. Gl. 109; Som. 79, 4; Wrt. Voc. 58, 47.

bóc-hús, es; *n.* A BOOK-HOUSE, *library;* librarium:—Bóchús *librarium,* Ælfc. Gl. 109; Som. 79, 5; Wrt. Voc. 58, 48.

bócian; *p.* ode; *pp.* od *To give by charter, to charter;* libro vel charta dare:—Oswald biscop bócaþ Wihtelme his þegne *bishop Oswald charters to Wihthelm his thane,* Cod. Dipl. 531; A.D. 966; Kmbl. iii. 6, 9. DER. ge-bócian.

bóc-land, -lond, es; *n.* BOOK-LAND, *land held by a charter or writing, free from all fief, fee, service or fines.* Such was formerly held chiefly by the nobility, and denominated allodialis, which we now call *freehold;* ex scripto sive charta possessa terra, terra codicillaris:—Ðe on his bóclande cyricean hæbbe *who on his freehold has a church,* L. Edg. i. 2; Th. i. 262, 11: L. Ed. 2; Th. i. 160, 14. Se mon bócland hæbbe *the man has a freehold,* L. Alf. pol. 41; Th. i. 88, 16: Bd. 2, 3; S. 504, 29: 3, 24; S. 556, 4: Cod. Dipl. 317; A.D. 871-889; Kmbl. ii. 120, 6. Hæfde Rómánum to bóclande gesealde *Romanis per testamentum tradiderat,* Ors. 5, 4; Bos. 104, 18. Bóclandes, Cot. 83. v. folc-land *and* land.

bóc-lár, e; *f.* [lár *lore, learning*] *Book-learning, learning;* doctrina:—Blind biþ se láreów, gif he ða bócláre ne cann *blind is the teacher, if he know not book-learning,* L. Ælf. C. 23; Th. ii. 352, 6.

bóc-leáf, es; *n. The leaf of a book, a charter;* folium codicis, charta, instrumentum donationis. v. leáf.

bóc-léden *book-language, and as most books were written in Latin,* hence *Latin,* Chr. Erl. 3, 3. v. léden.

bóc-líc; *adj.* BOOK-LIKE, *biblical, bookish, relating to books;* biblicus:—Gregorius wæs fram cildháde on bóclícum lárum getýd *Gregory was from childhood instructed in book-learning,* Homl. Th. ii. 118, 16. On bóclícum gewritum *in book-writings,* 284, 24. Ðæt we consider *the book-lore,* 284, 24.

bócod *booked, chartered.* v. bócian, gebócian.

bócon *baked;* coxerunt, Ex. 12, 39; *p. pl.* of bacan.

bóc-rǽdere, es; *m. A reader of books, a reader;* lector, Cot. 126.

bóc-rǽding *book-reading, reading.* v. rǽding.

bóc-reád *Book-red, vermilion:* so named, because it was much used in ornamenting books; minium:—Of bócreáde *ex minio,* Cot. 75: 176.

bóc-riht, es; *n.* BOOK-RIGHT, *the right of a will* or *charter;* testamenti rectitudo *vel* jus:—Ðegenes lagu is, ðæt he sý his bócrihtes wyrðe *taini lex est, ut sit dignus rectitudine testamenti sui,* L. R. S. 1; Th. i. 432, 1.

bóc-scamel, es; *m. A reading-desk* or *seat;* pluteus, lectorium. DER. bóc, scamel *a bench.*

bóc-scyld, es; *m.* [bóc *a beech-tree,* scyld *a shield*] *A beechen shield;* fagineum scutum:—Ic ge-an [MS. geann] Siferþe mínes bócscyldes *I give to Siferth my beechen shield,* Th. Diplm. A.D. 938; 561, 5.

bóc-stæf, es; *pl. nom. acc.* -stafas; *g.* -stafa; *d.* -stafum; *m. A book-staff, a letter, character;* litera, character = χαρακτήρ:—Awrítaþ hie on his wǽpne wælnota heáp, bealwe bócstafas *they cut upon his weapon a heap of fatal marks, baleful letters,* Salm. Kmbl. 325; Sal. 162. Engel Drihtnes wrát in wáge worda gerýnu baswe bócstafas *the angel of the Lord wrote on the wall mysteries of words in crimson letters,* Cd. 210; Th. 261, 10; Dan. 724. Ðæt he him bócstafas arǽdde and arehte *that he [Daniel] should read and explain the characters to them,* 212; Th. 262, 7; Dan. 740. Hwá wrát bócstafas ǽrest *who first wrote letters?* Salm. Kmbl. 200, 23: 192, 6. Bócstafa *of letters,* Salm. Kmbl. 199; Sal. 99. Wæs se beám bócstafum awriten *the beam was inscribed with letters,* Elen. Kmbl. 182; El. 91. DER. stæf.

bóc-sum; *adj. Obedient, flexible,* BUXOM; obediens, flexibilis. [*Frs.* Halbert. p. 540, búchsom *flexibilis: Dut.* boogh-saem *flexibilis: Ger.* biegsam *flexibilis.*]

bóc-sumnes, -ness, e; *f. Obedience, pliantness,* BUXOMNESS; obedientia. [*Ger.* biegsamkeit *flexibilitas: Verst. Restitn.* buhsomnesse, bowsomenesse *pliableness.* Chaucer writes buxsomnesse, p. 211.]

bóc-tǽcing, e; *f.* v. bóc-talu; *f. Book-teaching, a book of decrees, writings, the scriptures, holy writ, the Bible;* Scripta Lambardo; Sacra Scriptura Bromto: rectius fortasse Sacri Canones, *vel* Liber Judicialis, Lye:—Be bóctǽcinge *ex scriptis,* L. C. S. 35; Wilk. 140, 3. Be bóctale *by scripture,* L. C. S. 38; Th. ii. 398, 21. v. dóm-bóc.

bóc-talu, e; *f. Book-story* or *narration, the Bible.* v. bóc-tǽcing.

bóc-treów, es; *n. A beech-tree;* fagus:—Bóc-treów *fagus,* Wrt. Voc. 79, 76. v. bóc *fagus.*

bóc-ung, e; *f. A* BOOKING, *a setting down in a book;* inscriptio. DER. bóc, ung.

bóc-wudu; *m.* BEECH-WOOD; locus fagis consitus:—On bócwuda *in the beech-wood,* Exon. 111 b; Th. 428, 11; Rä. 41, 106.

BOD, es; *pl.* u, o, a; *n. A command, commandment, precept, mandate, an edict, order, message;* jussum, mandatum, edictum:—Hwæt is ðæt bod micle [MS. micla] *in ǽ quod est mandatum magnum in lege?*

Mt. Lind. Stv. 22, 36: Mk. Lind. Stv. 12, 28, 29, 30, 31: Lk. Lind. Stv. 2, 1. Bod on cine *diploma,* Ælfc. Gl. 80; Som. 72, 110; Wrt. Voc. 46, 67. Hwá swá halt ðis bod [bode MS.] wurðe he éfre wunnende mid God *whosoever observes this command, may he ever dwell with God,* Cod. Dipl. 990; A.D. 680; Kmbl. v. 29, 23. We díne bodu brǽcon *we broke thy commandments,* Hy. 7, 109; Hy. Grn. ii. 289, 109. [*Laym.* bode, bod: *Orm.* bode: *Scot.* bode, bod: *Plat.* bod, ge-bodd, *n:* O. *Sax.* gibod, *n:* O. *Frs.* bod, *n:* Dut. ge-bod, *n:* Ger. bot, ge-bot, *n:* M.H.Ger. ge-bot, *n:* O. H. Ger. ga-bot, *n:* Goth. busns, *f.* in ana-busns: *Dan.* bud, *n:* Swed. bud, *n:* Icel. boð, *n. a commandment.*] DER. ǽ-bod, be-, bi-, for-, ge-.

boda, an; *m.* [bod *a message,* -a, *q. v.*] I. *a messenger, ambassador, herald, apostle, angel;* nuntius, legatus, prǽco, apostolus, angelus:—Eálá Wísdóm, ðú eart boda and fórrynel ðæs sóðan leóhtes *O Wisdom, thou art the messenger and forerunner of true light,* Bt. 36, 1; Fox 170, 28. Me ðes boda sǽgde wǽrum wordum *this messenger told me in cautious words,* Cd. 32; Th. 42, 30; Gen. 680: 32; Th. 43, 6; Gen. 686: 33; Th. 45, 11; Gen. 725. Heó ðæs láðan bodan lárum hýrde *she obeyed the advice of the loathsome messenger,* 33; Th. 44, 18; Gen. 711. Ða bodan us fǽrdon *nuntii nos terruerunt,* Deut. 1, 28: Exon. 27 a; Th. 80, 9; Cri. 1305. Sende he bodan beféran his ansýne *misit nuntios ante conspectum suum,* Lk. Bos. 9, 52: Gen. 32, 3: Exon. 24 b; Th. 71, 7; Cri. 1152. Cyninges bodan underfón *to receive a king's ambassador,* Lchdm. iii. 210, 15. Se sóða boda ðæs hean leóhtes Agustinus wæs fram him eallum bodad *verus summæ lucis prǽco ab omnibus prædicatur Augustinus,* Bd. 2, 2; S. 502, 32. Brimmanna boda *prǽco nautarum,* Byrht. Th. 133, 12; By. 49. Gefeohtes bodan *heralds of war;* præfeciales, Ælfc. Gl. 53; Som. 66, 81; Wrt. Voc. 36, 7. Ðú Drihtnes eart boda of heofnum *thou art the Lord's angel from heaven,* Cd. 26; Th. 34, 5; Gen. 533: Elen. Kmbl. 153; El. 77. Bodan hyrdum cýðdon sóþne gefeán *angels announced to the shepherds true joy,* Exon. 14 a; Th. 28, 20; Cri. 449. II. *a foreboder, prophet;* propheta, vates:—Gleáw bodan ǽrcwide *skilled in a prophet's prediction,* Exon. 83 a; Th. 313, 22; Mód. 4. [R. *Brunne* bode: *Laym.* boden, *pl:* O. *Sax.* bodo, *m:* Frs. bode, boade, *c:* O. *Frs.* boda, *m:* Dut. bode, *m:* Ger. M. H. Ger. bote, *m:* O. H. Ger. boto, *m:* Dan. bud: Swed. båd, *m:* Icel. boði, *m. a messenger.*] DER. ǽ-boda, eðel-, fór-, heáh-, nýd-, síþ-, spel-, wil-.

bodad *announced, proclaimed,* Andr. Kmbl. 2241; An. 1122, = bodod; *pp.* of bodian.

boden *ordered, offered, proclaimed,* Elen. Kmbl. 36; El. 18; *pp.* of beódan.

bodere, es; *m. A teacher, a master;* præceptor, Lk. Rush. War. 9, 33.

bodian, bodigan, bodigean; *part.* bodiende, bodigende; *p.* ode, ede, ade, ude; *pp.* od, ed, ad, ud; *v. a.* [bod *a message*]. I. *to tell, announce, proclaim, preach;* nuntiare, annuntiare, enuntiare, narrare, prædicare, evangelizare:—Ongan se Hǽlend bodian *cæpit Iesus prædicare,* Mt. Bos. 4, 17: Mk. Bos. 1, 45: Exon. 49 a; Th. 169, 2; Gú. 1088. He ongan bodigean on Decapolim *cæpit prædicare in Decapoli,* Mk. Bos. 5, 20: Cd. 169; Th. 210, 4; Exod. 510. Ic eom asend ðé ðis bodian *missus sum hæc tibi evangelizare,* Lk. Bos. 1, 19: Bd. 5, 9; S. 622, 13. To bodianne godcunde láre *ad prædicandum doctrinam divinam,* 5, 9, titl; S. 622, 4. Com se Hǽlend on Galileam Godes ríces gódspell bodigende *venit Iesus in Galilæam prædicans evangelium regni Dei,* Mk. Bos. 1, 14: Ps. Lamb. 2, 6. Ic bodie *annuntiabo,* Ps. Spl. 54, 17. Ðæt ic bodige oððe ðæt ic cýðe ealle hérunga oððe lofunga ðíne *on* geatum dóhter oððe dóhtra ðæs múntes [Siones] *ut annuntiem omnes laudationes tuas in portis filiæ Sion,* Ps. Lamb. 9, 15: Exon. 103 a; Th. 391, 3; Rä. 9, 10. Me ðes ár bodaþ frécne fǽrspell *this messenger announces to me a horrible unforeseen message,* Exon. 69 b; Th. 259, 3; Jul. 276: Bt. Met. Fox 29, 45; Met. 29, 23. Heofonas bodiaþ oððe cýðaþ wuldor Godes *cæli enarrant gloriam Dei,* Ps. Spl. 18, 1; Salm. Kmbl. 474; Sal. 237. Ðes apostol Iacobus bodode on Iudéa lande *this apostle James preached in Judea* [lit. *in the land of the Jews*], Homl. Th. ii. 412, 23. Ymb Bethleem bododon englas ðæt acenned wæs Crist on eorþan *angels announced about Bethlehem that Christ was born on earth,* Hy. 10, 23; Hy. Grn. ii. 293, 23. Bodedon heofonas rihtwísnysse his *annuntiaverunt cæli justitiam ejus,* Ps. Spl. 96, 6; Judth. 12; Thw. 25, 6; Jud. 244. Hý bodudon *annuntiaverunt,* Ps. Spl. 43, 1. Cýðaþ oððe bodiaþ betwux þeódum his gecneordnyssa oððe his ymbhoga *annuntiate inter gentes studia ejus,* Ps. Lamb. 9, 12: Andr. Kmbl. 669; An. 335. Wæs ðæt weátácen geond ða burh bodad *the fatal token was proclaimed throughout the town,* 2241; An. 1122. II. *to foretell, predict, prophesy, promise;* prædicere, promittere:—Him ðone dæg willan Drihten bodode *the Lord had foretold* [*promised*] *to him that day of desire,* Cd. 133; Th. 168, 2; Gen. 2776: Beo. Th. 3608; B. 1802. Ðæt wæs oft bodod ǽr beféran fram fruman worulde *it was often foretold long before from the beginning of the world,* Elen. Kmbl. 2280; El. 1141. [*Laym.* bodien: O. *Frs.* bodia: Icel. boða *to announce.*] DER. fóre-bodian, ge-, to-.

BODIG, es; *n.* I. *bigness* or *height of body, stature;* statura:—Ðæt se mon wǽre lang on bodige *quod esset vir longæ staturæ,*

Bd. 2, 16; S. 519, 33. Wæs Oswine se cyning on bodige heáh *king Oswine was tall in stature*, 3, 14; S. 540, 7. **II.** *the trunk, chest or parts of the chest, as the back-bone;* truncus corporis :—Bodig *truncus*, Wrt. Voc. 283, 26: *spina*, Cot. 177: 196. **III.** *the* BODY; corpus :—Ægðer ge his fêt ge his heáfod ge eác eall ðæt bodig *either his feet, or his head or even all the body*, Past. 35, 3; Hat. MS. 45 b, 12. [*Wyc.* body: *R. Glouc. Laym.* bodi: *Orm.* bodiȝ: *Ger.* bottech, *m:* *Bav.* bottich, *m: M.H.Ger.* botech, *m: O.H.Ger.* botah, *m: Gael.* bodhag, *f.*]

bodigean *to publish, preach,* Mk. Bos. 5, 20: Cd. 169; Th. 210, 4; Exod. 510. v. bodian.

bod-lâc, es; *n. A decree, ordinance;* decretum, Chr. 1129; Ing. 359, 21; Erl. 258, 13.

bod-scipe, es; *m.* [bod *a command,* scipe] *A message, an embassy, a commandment;* nuntium, mandatum :—Swâ ic him ðisne bodscipe secge *when I tell him this message,* Cd. 27; Th. 35, 10; Gen. 552. Ðâ hie Godes hæfdon bodscipe abrocen *when they had broken God's commandment,* 37; Th. 48, 29; Gen. 783. DER. ge-bodscipe.

bododon *announced;* annuntiaverunt, Ps. Spl. 43, 1, = bododon; *p. pl.* of bodian.

bodung, e; *f. A preaching, publishing, divulging;* prædicatio, pronuntiatio :—Niniuetisce men dædbôte dydon æt Ionam bodunge *viri Ninivitæ pœnitentiam egerunt ad prædicationem Ionæ,* Lk. Bos. 11, 32.

bodung-dæg, es; *m. An annunciation day;* annuntiationis dies :—Ðes dæg is gehâten Annuntiatio Sanctæ Mariæ, ðæt is Marian bodungdæg gecweden *this day is called* Annuntiatio Sanctæ Mariæ, *which is interpreted, the annunciation-day of Mary,* Homl. Th. i. 200, 25.

bóem *to both,* Th. Diplm. A.D. 830; 465, 22; *for* bâm; *dat.* of begen.

Boëties, Boötes; *m.* Boötes; Boötês, *e; m.* [*= Βοώτης,* ov; *m. a ploughman, from* βοῦς *an ox*]. The ancient constellation, the chief star of which is the bright Arcturus, v. arctos *the bear;* Ursa Major. The modern representation of Boötes is a man with a club in his right hand, and in his left a leash, which holds two dogs :—Hwâ ne wundraþ ðætte sume tunglu habbaþ scyrtran hwyrft ðonne sume habban? For ðý hí habbaþ swâ sceortne ymbhwyrft, for ði hí sint swâ neáh ðam norþende ðære eaxe, ðe eall ðes rôdor on hwerfþ, swâ nû Boëties dêþ *who wonders not that some constellations have a shorter course than others have? Therefore they have so short a course, because they are so near the north end of the axis, on which all the sky turns, as now Boötes does,* Bt. 39, 3; Fox 214, 17–24. Boötes beorhte scîneþ *Boötes shines brightly,* Bt. Met. Fox 28, 53; Met. 28, 27.

Boëtius; *nom. acc.; g.* Boëties, Boëtiuses; *d.* Boëtie; *m.* [*Βοηθόος warlike*] *Anicius Manlius Severinus* Boëthius, born in Rome between A.D. 470–475, was Consul in 510. He was so eminent for his integrity and talents that he attracted the attention and obtained the patronage of Theodoric the Great, king of the East or Ostrogoths. He was afterwards accused of treason, and cast into prison, where he wrote his celebrated work *De Consolatione Philosophiæ,* which king Alfred translated into *Anglo-Saxon* about A.D. 888. Being condemned to death, without a hearing, he was beheaded in prison about A.D. 524 :—Ðâ wæs sum consul, ðæt we heretoha hâtaþ, Boëtius wæs hâten. Se wæs, in bôccræftum and on worold-þeáwum, se rihtwîsesta *there was a certain consul, that we call heretoha, who was named Boëthius. He was, in book-learning and in worldly affairs, the most truly wise* [*=most righteous*], Bt. 1; Fox 2, 12–14. Se Boëtius wæs ôðre naman gehâten Seuerînus: se wæs heretoga Rômâna *Boëthius was by another name called Severinus: he was a consul of the Romans,* Bt. 21; Fox 76, 3–4. Hû Gotan gewunnon Rômâna rîce, and hû Boëtius hî wolde berǽdan, and Þeódríc ðâ ðæt anfunde and hine hêt on carcerne gebringan *how the Goths conquered the empire of the Romans, and how Boëthius wished to deliver them, and Theodoric discovered it, and gave orders to take him to prison,* Bt. title 1; Fox x. 2–4. Hû se Wîsdom com to Boëtie ǽrest inne on ðam carcerne *how Wisdom first came to Boëthius in the prison,* Bt. title 3; Fox x. 6: Fox xiv. 18. Hêr endaþ nû seó æftre frôfer-bôc Boëtiuses [Cot. MS. æftere frôfr-bôc Boëties] *here now endeth the second consolation-book of Boëthius,* Bt. 21; Fox 76, 2–3. Hêr endaþ nû seó þridde bôc Boëties *here now endeth the third book of Boëthius,* Bt. 35, 6; Fox 170, 23.

bog *the arm, shoulder,* Ælfc. Gl. 73; Som. 71, 16; Wrt. Voc. 44, 2. v. boh.

boga, an; *m.* [bogen; *pp.* of bûgan *to bow, bend*] Anything curved,—*A* BOW, *an arch, a corner;* arcus, angulus :—Æteówþ mîn boga on ðam wolcnum *apparebit arcus meus in nubibus,* Gen. 9, 14. Boga sceal stræle *a bow shall be for an arrow,* Exon. 91 b; Th. 343, 8; Gn. Ex. 154. Ðæt híg fleón fram ansýne bogan *ut fugiant a facie arcus,* Ps. Lamb. 59, 6. Híg aþenodon bogan heora *intenderunt arcum suum,* 36, 14: 57, 8: 63, 4. Hí lêton gâras fleógan, bogan wǽron bysige *they let the arrows fly, bows were busy,* Byrht. Th. 134, 66; By. 110. Bogan [MS. bogen] streng *a bow-string;* anquina, Ælfc. Gl. 52; Som. 66, 37; Wrt. Voc. 35, 26. [*Wyc.* bowe, bouwe: *Laym.* boȝe, bowe: *O.Sax.* bogo, *m: Frs.* boage: *O.Frs.* boga, *m: Dut.* boog, *m: Ger.* boge,

bogen, *m: M.H.Ger.* boge, *m: O.H.Ger.* bogo, *m: Dan.* bue, *c: Swed.* bâge, *m: Icel.* bogi, *m.* arcus.] DER. brægd-boga, flân-, horn-, hring-, rên-, scûr-, stân-, wîr-.

bôgan *to boast;* jactare, Scint. 46. v. bôn.

boga-net, boge-net, -nett, es; *n. A* BOW-NET, *weel, wicker-basket with a narrow neck for catching fish;* nassa :—Æwul *vel* boganet *nassa,* Ælfc. Gl. 102; Som. 77, 85; Wrt. Voc. 56, 9. Bogenet *vel* leáp *nassa,* 84; Som. 73, 90; Wrt. Voc. 48, 28. Bogenet *nassa,* 105; Som. 78, 41; Wrt. Voc. 57, 23.

boge-fôdder, es; *m.* [boga *a bow,* fôdder *fodder, from* fôd *food*] *A* BOW-FEEDER, *case for arrows, a quiver;* corýtos *= κωρυτός* :—Boge-fôdder *corytos* [MS. *coriti*], Ælfc. Gl. 53; Som. 66, 67; Wrt. Voc. 35, 53.

bogen *bowed, bent, gave way; pp.* of bûgan.

bogen *rosemary,* L. M. 3, 30; Lchdm. ii. 324, 25, = boðen, *q.v.*

bogen *a bow-net, weel,* Ælfc. Gl. 105; Som. 78, 41; Wrt. Voc. 57, 23. v. boga-net.

bogen streng, es; *m.* [bogen = bogan; *gen.* of boga *a bow;* streng *a string*] *The string of a bow, a* BOW-STRING; arcus chorda, anquina, Ælfc. Gl. 52; Som. 66, 37; Wrt. Voc. 35, 26. v. boga.

bogetung, e; *f.* [bogen; *pp.* of bûgan *to bend*] *A bending, crook;* anfractus, Cot. 18.

bôgian; *p.* ode; *pp.* od *To inhabit;* incolere :—Bôgodon *incoluerunt,* Ælfc. T. Lisle 21, 13. v. bûgian.

bogung, e; *f.* [bogen *bent; pp.* of bûgan *to bow, bend*] *Crookedness, perversity;* pravitas, perversitas :—Þurh heora upahefednysse and âgenre bogunge *through their arrogance and own perversity,* Homl. Th. ii. 428, 13.

boh, bog, es; *m.* [bogen *bent; pp.* of bûgan *to bow, bend*] *Anything curved or bent,—hence* **I.** *the arm, shoulder;* armus *= ἁρμός,* humerus, lacertus :—Se swîðra boh *armus dexter,* Lev. 7, 32: the riȝt schuldor, Wyc. Bog *lacertus,* Ælfc. Gl. 73; Som. 71, 16; Wrt. Voc. 44, 2. Eorl sceal on eós boge rîdan *a chief shall ride on a horse's back* [*lit. shoulder*], Exon. 90 a; Th. 337, 11; Gn. Ex. 63. Ðû nymst of ðam ramme ðone swýðran boh *tolles de ariete armum dextrum,* Ex. 29, 22. Mec se beaducâfa bogum bilegde *the battle-prompt man embraced me in his arms,* Exon. 100 b; Th. 380, 21; Rä. 1, 11. **II.** *the arm of a tree, a* BOUGH, *branch;* ramus, stipes, palmes :—Bôh *ramus,* Scint. 1. Boh *stipes,* Ælfc. Gr. 9, 26; Som. 11, 16. Berende boh *germen,* Ælfc. Gl. 60; Som. 68, 32; Wrt. Voc. 39, 18. Deáh ðû hwilcne boh ðæs treówes býge *though thou bendest any bough of a tree,* Bt. Met. Fox 13, 105; Met. 13, 53. Hit wearþ mycel treów, and heofenes fugelas reston on his bogum *factum est in arborem magnam, et volucres cœli requieverunt in ramis ejus,* Lk. Bos. 13, 19: Cd. 30; Th. 40, 2; Gen. 645. He astrehte his bogas ôþ ða sǽ *extendit palmites suos usque ad mare,* Ps. Lamb. 79, 12. **III.** *a branch of a family, offspring, progeny;* propago :—Tyddrung oððe boh *propago,* Ælfc. Gr. 36; Som. 38, 49. [*Chauc.* bow: *Piers P.* bowe: *Wyc.* boow, bouȝ, boȝ: *Orm.* boȝh: *Dut.* boeg, *m. the bow of a ship:* *Ger.* bug, *m. armus:* *M.H.Ger.* buoc, *m: O.H.Ger.* buoc, *m. armus:* *Dan.* bov, boug, *c. shoulder, bow of a ship:* *Swed.* bog, *m. the shoulder, haunch:* *O.Nrs.* bógr, *m. the shoulder of an animal.*] DER. wæter-boh, wîn-.

boh-scyld, es; *m. A shoulder shield;* ad humerum clypeus, Æthelst. Test. Mann.—bôc-scyld, *q.v.*

bohte, *pl.* bohton *bought;* emit, emerunt, Gen. 49, 30; *p.* of bycgan.

BOLCA, an; *m. The gangway of a ship;* forus navis :—Bolca *forus,* Cot. 86. Geseah weard beran ofer bolcan beorhte randas *the guard saw bright shields borne over the ship's gangway,* Beo. Th. 467; B. 231: Andr. Kmbl. 1203; An. 602. He on bolcan sæt *he sat on the gangway,* 610; An. 305. [*Icel.* búlki, *m. the cargo of a ship.*]

BOLD, es; *n.* **I.** *a building, dwelling, house;* ædificium, domicilium, domus :—Wæs ðæt bold tobrocen swîðe *the dwelling was much shattered,* Beo. Th. 1998; B. 997. Ðær ic wîc búge, bold mid bearnum *where I inhabit a dwelling, a house with children,* Exon. 104 b; Th. 396, 23; Rä. 16, 9. Bold wæs betlíc *the building was excellent* [*good-like*], Beo. Th. 3854; B. 1925. Nis ðæt betlíc bold [blod MS.] *that is no goodly dwelling,* Exon. 116 a; Th. 446, 16; Dôm. 23. **II.** *a superior house, hall, castle, palace, temple;* aula, palatium, ædes :—He him gesealde bold and bregostôl *he gave to him a habitation and a princely seat,* Beo. Th. 4398; B. 2196. Ne môt ic brúcan burga ne bolda *I may not enjoy towns nor palaces,* Cd. 216; Th. 273, 19; Sat. 139. Ðâ wæs Beówulfe gecýðed, ðæt his sylfes hâm, bolda sêlest, brynewylmum mealt *then it was made known to Beowulf, that his own home, the best of mansions, was consumed by flames of fire,* Beo. Th. 4641; B. 2326. Gewât beorht blǽdgifa in bold ôðer *the bright giver of glory departed into another temple,* Andr. Kmbl. 1312; An. 656. [*R. Glouc.* bold: *A. Sax.* bylda *a builder:* *Eng.* to build.. v. botl.] DER. feorh-bold, fold-: bold-âgende, -getæl, -getimber, -wéla.

bold-âgende; *part.* [bold *a house,* âgende *owning*] *House-owning, possessing a house;* domum possidens :—Hæleða monegum boldâgendra *to many of house-owning men,* Beo. Th. 6215; B. 3112: Exon. 90 b; Th. 339, 12; Gn. Ex. 93.

bold-getæl, es; *n.* [bold *a house*, getæl *a number, tribe, register*] *A dwelling-place, mansion, habitation, house;* domicilium, mansio, vicus, domus :—Gif mon wille of boldgetale [boldgetæle MS. B.] in ōðer boldgetæl hláford sēcan, dô ðæt mid ðæs ealdormonnes gewitnesse ðe he ǽr in his scíre folgode *if a man from one dwelling-place wish to seek a lord in another dwelling-place, let him do it with the knowledge of the alderman, whom he before followed in his shire*, L. Alf. pol. 37; Th. i. 86, 2; that is, *If a person who had* commended *himself, wished to take his name off the manor-roll of one lord, etc.* Thorpe's Laws, vol. i. p. 86, note a.

bold-getimber; *gen.* -getimbres; *pl. nom. acc.* -getimbru; *n. The timber of a house;* ædificii tignum :—Leóht [fýr] briceþ and bærneþ boldgetimbru *light [fire] breaketh and burneth the timbers of the house*, Salm. Kmbl. 826; Sal. 412.

bold-wēla, an; *m.* [bold *a house*, wēla *wealth*]. I. *a dwelling of wealth or happiness;* prædium, opes domesticæ :—Ne mæg ðē adón ðínne boldwélan *thou mayest not take thee thy dwelling of wealth or happiness*, Soul Kmbl. 118; Seel. 59. II. *paradise, heaven;* paradisus = παράδεισος, coelum :—Ðē is neorxna wang boldwēla fægrost *paradise is to thee the fairest dwelling of happiness*, Andr. Kmbl. 206; An. 103. Adam and Æue anforlēton beorhtne boldwēlan *Adam and Eve forsook bright paradisal happiness*, Exon. 73 a; Th. 272, 22; Jul. 503. He gesōhte swegle dreámas, beorhtne boldwélan *he sought the joys of heaven, the bright dwelling of happiness*, Apstls. Kmbl. 65; Ap. 33. He [God] sceal rǽdan, se ðe rôdor ahôf, wuldres fylde beorhtne boldwēlan *he [God] shall rule, who uplifted the firmament, with glory filled the bright dwelling of wealth*, Andr. Kmbl. 1047; An. 524.

bolgen *vexed, irritated, angry; pp.* of belgan.

bolgen-mód; *adj. Enraged in mind;* iratus animo :—Him bolgenmód yrre andswarode *enraged in mind, answered them angrily*, Cd. 183; Th. 228, 26; Dan. 209: Beo. Th. 1422; B. 709: Andr. Kmbl. 255; An. 128: Exon. 40 b; Th. 135, 25; Gū. 529.

BOLLA, an; *m. Any round vessel, cup, pot*, BOWL, *a measure;* vas, cyathus = κύαθος :—Bolla cyathus, Glos. Epnl. Recd. 156, 16. Cærenes gôdne bollan fulne meng togædere *mingle together a good bowl full of boiled wine*, L. M. 1, 1; Lchdm. ii. 24, 19. Ðǽr wǽron bollan steápe boren æfter bencum *there were carried deep bowls behind the benches*, Judth. 10; Thw. 21, 14; Jud. 17. [*Piers P. Laym.* bolle : *O. Frs.* bolla, *m.* in kne-bolla, strot-bolla : *Dut.* bol, *m*: *Kil.* bolle caput, globus : *Ger.* punsch-bole, *f.* a *punch-bowl* : *M. H. Ger.* hirn-bolle : *O. H. Ger.* hirni-polla *the brain-pan, skull* : *Dan.* bolle, *c* : *Swed.* bål, *n* : *O. Nrs.* bolli, *m.* a *bowl.*] DER. beód-bolla, heáfod-, þrot-.

bollen *bellowed, roared; pp.* of bellan.

BOLSTER; *gen.* bolstres; *m. A* BOLSTER, *a pillow for the head;* cervical :—He his heáfod onhylde to ðam bolstre, and medmycel fæc onslǽpte *reclinavit caput ad cervical, modicumque obdormivit*, Bd. 4, 24; S. 599, 7. He wæs on scipe, ofer bolster slápende *erat in puppi, super cervical dormiens*, Mk. Bos. 4, 38. Hit geondbrǽded wearþ beddum and bolstrum *it was overspread with beds and bolsters*, Beo. Th. 2484; B. 1240. [*Dut.* bolster, *m.* a *shell* : *Kil.* bolster culcita : *Ger.* polster, *m.* cervical : *M. H. Ger.* bolster, *m* : *O. H. Ger.* bolstar, *m* : *Swed.* bolster, *n.* a *mattress* : *O. Nrs.* bólstr. *m.* a *bolster*.] DER. heáfod-bolster, hleór-.

BOLT, es; *pl.* boltas; *m. A* BOLT, *a warlike engine to throw bolts, arrows;* catapulta, Cot. 45. [*Chauc.* bolt : *Dut.* bout, *m* : *Kil.* bolt sagitta : *Ger. M. H. Ger.* bolz, *m* : *O. H. Ger.* bolz : *Dan.* bolt, *c* : *O. Nrs.* bolti, *m.*]

bōn [bōgan *to boast*] *To boast;* jactare :—He bôþ his sylfes swíðor micle ðonne se sélla mon *he boasts of himself much more than a better man*, Exon. 83 b; Th. 315, 9; Mód. 28.

bona, an; *m. A killer;* interfector :—Se wites bona *the destroyer of the mind [the devil]*, Exon. 11 b; Th. 17, 3; Cri. 264. Fugel-bona *a bird killer*, 79 b; Th. 298, 5; Crä. 80. v. bana.

bond *bound;* ligavit, Exon. 42 b; Th. 143, 29; Gū. 668; = band; *p.* of bindan.

bonda, an; *m. A husband, an householder, a master of a family;* maritus :—Se bonda sæt *the husband dwelt*, L. C. S. 73; Th. i. 414, 21 : 77; Th. i. 418, 24. v. bunda.

bonde-land, es; *n. Bond or leased land, land held under restrictions, or on conditions expressed in writing;* tributaria terra :—Ān abbot, Beonne geháten, lēt Cūþbriht ealdorman x bonde-lande [x tributariorum terram, *vel* terram x manentium] æt Swinesheáfde, mid læswe and mid mǽdwe, and mid eal ðæt ðǽrto læi, and swá ðæt Cūþbriht geaf ðam abbote 1 punde ðǽrfore, and ilca geár ánes nihtes feorme, ouðer xxx scyllinge penega; swá eác ðæt eafter his dæi scolde ðæt land ongeán into ðam mynstre *an abbot, called Beonna, let to the aldorman Cuthbriht ten 'bonde-lands' at Swineshead, with leasow and with meadow, and with all lying thereto, and so that Cuthbriht should give to the abbot fifty pounds for it, and every year one night's entertainment, or thirty shillings in pennies; and also that after his day the land should come again to the monastery*, Chr. 777; Th. 92, note 1; Cod. Dipl. 165; A. D. 786–796; Kmbl. i. 201.

bon-gār, es; *m.* [bana, ban *a killer, death?* gār *a spear*] *A death-spear;* letifera hasta, Beo. Th. 4066; B. 2031.

bonnan; *p.* beónn, *pl.* beónnon; *pp.* bonnen *To summon, call together;* citare, convocare :—Sió býman stefen and se beorhta segn bonnaþ sáwla gehwylce *the voice of the trumpet and the bright sign shall summon every soul*, Exon. 23 b; Th. 66, 6; Cri. 1067. v. bannan.

booc-hord *a library*, Ælfc. Gl. 109; Som. 79, 4; Wrt. Voc. 58, 47. v. bóc-hord.

BŌR. I. *a borer, gimlet;* terebra, Leo 121. II. *a lancet, a surgeon's or barber's instrument, a burin, or graving tool;* scalprum rasile, Cot. 63. [*Plat.* baar : *Dut.* boor, *f* : *Dan.* bor, *n* : *Swed.* borr, *m* : *O. Nrs.* bor, *m.* terebra, Rask Hald.]

bora, an; *m.* [boren, *pp.* of beran *to bear*] *One who bears* or *sustains the charge of anything, a ruler;* qui rem aliquam gerit, gestor :—Ríces boran *the rulers of the state*, Cd. 224; Th. 296, 10; Sat. 500.

-bora, an; *m.* [*from* boren; *pp.* of beran] Often used as a termination to denote *A bearer, bringer, supporter;* is qui fert, gerit; as, Cǽg-bora, horn-, mund-, rǽd-, rǽs-, segen-, sweord-, tācn-, wǽg-, wæpen-, wíg-, wôþ-, wrôht-. v. -bǽre.

borcian; *p.* ade, ode *To bark;* latrare :—Hió borcade : þancode willum *it barked: thanked willingly*, Exon. 129 a; Th. 495, 11; Rä. 84, 6. v. beorcan.

BORD, es; *n.* I. *a* BOARD, *plank;* tabula sectilis, tabula :—Bord tabula, Wrt. Voc. 63, 80. Borda gefēg *a joining of boards;* commissura, R. 62. Hwílum ic bordum sceal heáfodleás behlýðed licgan *sometimes I must lie on boards deprived of head*, Exon. 104 a; Th. 395, 18; · Rä. 15, 9. Wirc ðē ǽnne arc of aheáwenum bordum *make thee an ark of planed planks*, Gen. 6, 14; fac tibi arcam de lignis levigatis, Vulg. II. *what is made of a board,—A table, shield;* mensa, clypeus :—Ic on wuda stonde, bordes on ende *I stand upon wood, at the end of the table*, Exon. 129 a; Th. 496, 15, 18; Rä. 85, 15, 16. Geweorþe bord oððe mēse heora befóran him on grine *fiat mensa eorum coram ipsis in laqueum*, Ps. Surt. T. 68, 27. Scip sceal genægled, scyld gebunden, leóht bord *a ship shall be nailed, a shield bound, the light shield* [lit. *board*], Exon. 90 b; Th. 339, 16; Gn. Ex. 95: Byrht. Th. 134, 67; By. 110: Fins. Th. 58; Fin. 29. He fýsde forþ flán genehe: hwílon he on bord sceát, hwílon beorn tǽsde *he poured forth his arrows abundantly: sometimes he shot on the shield, sometimes he pierced the warrior*, Byrht. Th. 139, 46; By. 270: Beo. Th. 5041; B. 2524: Cd. 156; Th. 193, 28; Exod. 253. Ðǽr wæs borda gebrec *there was clash of shields*, Elen. Kmbl. 227; El. 114: Beo. Th. 4510; B. 2259. Beraþ bord fór breóstum *bear shields before their breasts* Judth. 11; Thw. 24, 16; Jud. 192: 12; Thw. 26, 9; Jud. 318. He mid bordum hēt wyrcan ðone wíhagan *he commanded to raise with the shields the fence of war*, Byrht. Th. 134, 49; By. 101: Andr. Kmbl. 2412; An. 1207. III. *the board, covering* or *deck of a ship, the ship itself;* tabulatum, stega = στέγη, constratum, navis :—Hý twegen sceolon habban gomen on borde, in sídum ceóle *they two shall have pastime on board, in the spacious ship*, Exon. 92 a; Th. 345, 5; Gn. Ex. 183. He drugaþ his ār on borde *he draws his oar on board*, 92 a; Th. 345, 15; Gn. Ex. 188. Ofer ceóles bord *from the vessel's deck*, Exon. 20 b; Th. 54, 2; Cri. 862. Lǽd under earce bord eaforan ðíne *lead thy children under the covering of the ark*, Cd. 67; Th. 80, 23; Gen. 1333: 67; Th. 82, 4; Gen. 1357. Bord oft onfēng ýða swengas *the ship often received the blows of the waves*, Elen. Kmbl. 476; El. 238. Ic wille eall acwellan ða be-ūtan beóþ earce bordum *I will destroy all who shall be without the boards of the ark or all who are not in the ark or ship*, Cd. 67; Th. 81, 33; Gen. 1354. IV. *with the prepositions* innan *and* ūtan *governing the genitive case, at home and abroad;* domi et foris :—Hie sibbe innan bordes gehióldon *they preserved peace at home* [lit. *inside the boundary*], Past. pref; Hat. MS. Man ūtan bordes wísdôm hieder on lond sôhte *one from abroad* [lit. *outside the boundary*] *sought wisdom in this land*, Past. pref; Hat. MS. [*Wyc.* boord : *R. Brun.* bord : *R. Glouc.* bord, borde : *Laym.* bord, beord, burd : *Orm.* bord, borde : *O. Sax.* bord, *m*: *Frs.* boerd, bord, *m*: *O. Frs.* bord, *m*: *Dut.* boord, *m*: *Ger.* boord, *m. and n*: *M. H. Ger.* bort : *O. H. Ger.* bort, borti, borto, *m*: *Goth.* fotu-baurd, *n.* a *foot-stool* : *Dan.* bord, *n*: *Swed.* bord, *m*: *Icel.* borð, *n*: *Fr.* bord, *m*: *Span. It.* bordo, *m*: *M. Lat.* bordus : *Wel.* bwrdh, *m*: *Corn.* bord, *f*: *Ir. Gael.* bord, *m*: *Armor.* bourz.] DER. bleó-bord, fāmig-, gūþ-, hilde-, hleó-, nægled-, þryþ-, wǽg-, wíg-, ýþ-.

borde, an; *f. A board, table;* tabula, mensa :—Fǽmne æt hyre bordan gerîseþ *it becomes a damsel to be at her board*, Exon. 90 a; Th. 337, 14; Gn. Ex. 64.

bord-gelāc, es; *n.* [lācan *to play, sport, fly*] *What flies against a shield, hence,—A missile, dart;* telum :—Ðý-læs ingebúge biter bordgelāc under bánlocan *lest the bitter dart enter in under the skin*, Exon. 19 a; Th. 48, 9; Cri. 769. v. bord II.

bord-hæbbende; *part.* [bord *scutum, clypeus*; habban *habere, vel* hebban, hæbban *levare, tollere*] *Shield-bearing;* scutum ferens, scutifer, Beo. Th. 5782; B. 2895.

bord-haga, an; *m.* [bord II. *a shield*, haga *a hedge*] *The cover of*

shields; clypeorum sepimentum:—Gefeallen under bordhagan *fallen under the cover of shields*, Elen. Kmbl. 1300; El. 652.

bord-hreóða, -hréða, an; *m.* [bord II. *a shield,* hreóðan *to cover, protect*]. I. *the cover* or *protection of the shield;* clypei tegmen vel tutela:—Hǽðne heápum þrungon under bordhreóðan *the heathens thronged in heaps under the cover of shields,* Andr. Kmbl. 256; An. 128: Beo. Th. 4412; B. 2203: Cd. 154; Th. 192, 23; Exod. 236. II. *a shield, buckler;* clypeus:—Blicon bordhreóðan *shields glittered,* Cd. 149; Th. 187, 30; Exod. 160. Hǽfdon hie ofer bordhreóðan beácen arǽred *they had a signal reared over their bucklers,* 160; Th. 198, 9; Exod. 320. Brǽcon bordhréðan *they broke through the bucklers,* Invent. Crs. Recd. 242; El. 122.

bord-rand, es; *m.* [bord II. *a shield,* rand *a rim, margin*] The margin or disc of a shield; scúti margo:—Biorn bordrand onswáf *the hero turned his shield's disc,* Beo. Th. 5112; B. 2559.

bord-stæp, es; *pl. nom. acc.* -staðu; *n.* [stæp *a shore, bank*] The sea-shore; litus:—Eágorstreámas beóton bordstaðu [bordstǽdu MS.] *the ocean-streams beat the sea-shores,* Andr. Kmbl. 883; An. 442.

bord-þaca, an; *m.* Board thatch, *a warlike engine, a cover or roof of a house, a snare;* testudo, laquearium:—Bordþacan *laquearii,* Cot. 119.

bord-weall, es; *m.* A board-wall, *a shield;* scutorum agger, testudo, clypeus:—He brǽc ðone bordweall *he broke through the board-wall,* Byrht. Th. 139, 60; By. 277: Beo. Th. 5952; B. 2980.

bord-wudu; *m.* Shield-wood, *a shield;* clypei lignum, clypeus, Beo. Th. 2490; B. 1243. v. bord II.

boren *borne, carried, born,* Bt. Met. Fox 26, 92; Met. 26, 46; *pp. of* beran.

boren-nes, -ness, e; *f.* [boren *born,* -nes] *Birth, nativity;* partus, nativitas. DER. æðel-borennes.

borg *a surety* or *pledge,* L. Alf. pol. 3; Th. i. 62, 8. v. borh.

borgas *sureties, debtors,* L. Eth. i. 1; Th. i. 280, 21; *pl. of* borh.

borgen *saved, protected, sheltered; pp. of* beorgan.

borges bryce *a breaking* or *breach of a suretyship* or *pledge,* L. Alf. pol. 3; Th. i. 62, 9, 10, 12. v. borh-bryce.

borg-gylda, an; *m.* A usurer; fœnerator, Ps. Spl. C. 108, 10.

borgian, he borgaþ; *p.* ode, ede; *pp.* od, ed [borh *a pledge, loan*] To take or *give a loan,* BORROW, *lend;* mutuari, commodare:—Ðam ðe wylle æt ðé borgian, ne wyrn ðú him *volenti mutuari a te, ne avertas,* Mt. Bos. 5, 42. Borgaþ se synfulla and ná gefillþ oððe he ne agylt *mutuabitur peccator et non solvet,* Ps. Lamb. 36, 21. Borgedon [MS. borgedan] *commodarunt,* Cot. 38. [Chauc. R. Glouc. borwe: Piers P. borwen: Laym. burȝen: Plat. borgen: O. Frs. borga: Dut. Ger. M. H. Ger. borgen: O. H. Ger. borgên cavere: Dan. borge: Swed. borga: O. Nrs. borga fidejubere.] DER. a-borgian.

borgiend, es; *m.* [part. of borgian *to lend*] A usurer; fœnerator:—Smeáge borgiend [MS. borgiende] ealle spéda his *scrutetur fœnerator omnem substantiam ejus,* Ps. Spl. 108, 10.

borg-wed, -wedd, es; *n. Anything given in pledge, a promise;* vadimonium. v. wed, wedd.

BORH; *g.* borges; *d.* borge; *acc.* borh; *pl. nom. acc.* borgas; *g.* a; *d.* um; *m.* I. *a security, pledge, loan, bail;* fœnus:—Ic wille, ðæt ǽlc mann sý under borge ge binnan burgum ge bútan burgum *I will that every man be under security both within cities and without cities,* L. Edg. S. 3; Th. i. 274, 6. Abere se borh ðæt he aberan sciíde *let the borh bear that he ought to bear,* L. Edg. ii. 6; Th. i. 268, 9. On his ágenon borge *on his own security,* L. Eth. i. 1; Th. i. 282, 10. Gif ðú feoh to borge selle *if thou give money on loan,* L. Alf. 35; Th. i. 52, 21. Be borges andsæce *concerning a denial of a bail,* L. In. 41; Th. i. 128, 1, note 1. II. *a person who gives security, a surety, bondsman, debtor;* fidejussor, debitor.—Bail was taken by the Saxons from every person guilty of *theft, homicide, witchcraft, etc:* indeed, every person was under bail for his neighbour. It is generally thought, that the borh originated with king Alfred, but the first time we find it clearly expressed, is in the Laws of Ine, v. *Turner's Hist. of A. S.* Bk. vi. Append. 3, ch. 6, vol. ii. p. 499:—Sette getreówe borgas *shall appoint true sureties,* L. Eth. i. 1; Th. i. 280, 21: 280, 6, 7, 8: L. Edg. 6; Th. i. 162, 19, 20. Ge asécaþ eówre borgas *ye shall search out your debtors,* L. E. I. 42; Th. ii. 438, 35. [Chauc. Wyc. borwe: R. Glouc. borewes, *pl:* Piers P. borgh: Laym. borh: Frs. borch, m: O. Frs. borch, borch, m: Dut. borg, m. and f: Ger. borg, m: M. H. Ger. borc, m.]

borh-bryce, borg-bryce, es; *m.* [borh *a pledge,* bryce *a breaking*] A pledge-breaking, *violation of a bail;* fidejussionis violatio:—Be borh-bryce *concerning a pledge-breaking,* L. Alf. pol. 3; Th. i. 62, 7, note 10. Borh-bryce, L. In. 31; Th. i. 122, note 20. Borg-bryce, L. Alf. pol. 1; Th. i. 60, 19.

borh-fæstan, geborh-fæstan; *p.* -fæste; *pp.* -fæsted [borh *a surety,* fæstan *to fasten*] To fasten or *bind by pledge* or *surety;* fidejussione obligare:—Man borhfæst ðam cyninge [MS. kyninge] ealle ða þægnas *they bound by pledge all the thanes to the king,* Chr. 1051; Ing. 228, 33; Erl. 181, 5.

borh-hand, borhond, e; *f. A pledge by the hand, a pledger, surety,*

security; sponsor, fidejussor:—Borh-hand *sponsor, fidejussor,* Ælfc. Gl. 114; Som. 80, 15; Wrt. Voc. 60, 50: Ælfc. Gr. 9, 25; Som. 10, 66: 9, 35; Som. 12, 32.

borhigenda, an; *m.* [borh *a loan,* ágenda *a possessor*] A usurer; fœnerator:—Ascrudnige borhigenda ealle spéde oððe ǽhte his *scrutetur fœnerator omnem substantiam ejus,* Ps. Lamb. 108, 11.

borh-leás; *adj. Void of security;* fidejussore carens:—Gif hwá borh-leás orf habbe . . . agife ðæt orf, and gilde xx oran *if any one have cattle borhless* [i. e. *for which no borh has been given*] . . . *let him give up the cattle, and pay twenty oran* [which at 1s. 4d. each, would make £1. 6s. 8d. in our money, v. púnd], L. Eth. iii. 5; Th. i. 296, 1.

borh-wed, -wedd, es; *n. Anything given in pledge;* vadimonium. v. wed, wedd.

BÓRIAN; *p.* ode; *pp.* od To BORE, *to make a hole, perforate;* tere-brare, perforare:—Wyrm ðe bóraþ treów *a worm that perforates wood;* termes *vel* teredo, Ælfc. Gl. 23; Som. 60, 4; Wrt. Voc. 24, 8. [Tynd. bore: *Dut.* boren: *Ger.* bohren: *M. H. Ger.* born: *O. H. Ger.* borjan, borôn: *Dan.* bore: *Swed.* borra: *Icel.* bora: *Lat.* for-are: *Zend* bar *to cut, bore.*]

born *burnt; p. of* beornan:—Forðonðe se Godes wer stronglíce innon born mid ðý fýre godcundre lufan *quia vir Dei igne divinæ caritatis fortiter ardebat,* Bd. 2, 7; S. 509, 30.

bornen *burnt; pp. of* beornan.

borsten *burst; pp. of* berstan.

borþor *child-birth.* v. beorþor, hyse-beorþor.

Boruchtuari, -orum; *pl. m. Lat. A people of ancient Germany, conquered by the Old-Saxons;* Boructuari:—Ðá Swýþbyrht hæfde bisceop-háde onfongen, he gewát to ðære þeóde Boruchtuarorum; . . . ac ðá æfter noht langre tíde seó ylce þeód wæs oferwunnen fram Eald-Seaxum, and ða wǽron wíde todrifene *Suidberct, accepto episcopatu, ad gentem Boructuarorum secessit; . . . sed expugnatis non longo post tempore Boructuaris, quolibet hi, a gente Antiquorum Saxonum, dispersi sunt,* Bd. 5, 11; S. 626, 6-11. v. Boruct-ware.

Boruct-ware; *gen.* a; *dat.* um; *pl. m:* Boructuari, -orum; *pl. m. A people of ancient Germany, occupying the country between the Rhine, the Lippe, Ems, and Weser;* Bructĕri = Βρούκτεροι:—Wǽron Frysan, Rugine, Dene, Hune, Eald-Seaxan, Boructware *sunt Fresones, Rugini, Danai, Hunni, Antiqui Saxones, Boructuari,* Bd. 5, 9; S. 622, 16. Tacitus always mentions the Bructeri with the Tencteri,—Bructeri et Tencteri, Ann. xiii. 56: Hist. iv. 21, 77. Zeuss supposes they may have inhabited the country near the Lippe, which was called *Boroctra* or *Borhtergo,* Deut. Nachbarst. 353.

Bosan-hám, Bosen-hám, es; *m.* [Flor. A.D. 1114; Sim. Dunelm. 1164 Bosanham: Hovd. 1204 Boseham] BOSEHAM or BOSHAM *in Sussex;* in agro Sussexiensi:—Ðá gewende Swegen to his scypum [MS. scypon] to Bosanhám *Swegen then went with his ships to Bosham,* Chr. 1049; Erl. 172, 34. Gewende ðá Swegen eorl to Bosenhám *earl Swegen then went to Bosham,* 1048; Erl. 180, 15.

BÓSG, bósig, bósih, es; *m?* *n?* An ox or cow-stall, *where the cattle stand all night in winter; a* BOOSE, as it is now called by the common people, in the Midland and Northern counties. It is now [1874] more generally used for the upper part of the stall where the fodder lies,—They say, ' you will find it in the *cow's boose,*' that is, in the place for the *cow's food;* præsepium:—Of bósge *a præsepio,* Lk. Rush. War. 13, 15. Of bósih a *præsepio,* Lk. Lind. War. 13, 15. [Frs. bos *a cottage:* Ger. banse, *m.* or *f:* Goth. bansts, *m. a barn:* Dan. baas, *c:* Swed. bås, *n:* Icel. bás. *m. stabulum, præsepium bovis,* Rask Hald.]

BÓSUM, bósm, es; *m. The space included by the folding of the arms, the* BOSOM, *lap, breast, interior parts;* sinus, gremium, pectus, interna:—Ðæt ic híg bære on mínum bósume, swá fóstormódor déþ cyld *ut portarem eos in sinu meo, sicut portare solet nutrix infantulum,* Num. 11, 12. Mín gebéd on bósme mínum byþ gecyrred *oratio mea in sinu meo convertetur,* Ps. Lamb. 34, 13: 73, 11: 78, 12: 88, 51. Ic winde sceal swelgan of sumes bósme *I* [i. e. *a horn*] *shall swell with wind from some one's bosom,* Exon. 104 a; Th. 395, 30; Rä. 15, 15: 109 b; Th. 419, 17; Rä. 38, 7: 127 a; Th. 489, 11; Rä. 78, 6. Gescype scylfan on scipes bósme *make shelves in the interior* [lit. *bosom*] *of the ship,* Cd. 65; Th. 79, 5; Gen. 1306: 67; Th. 80, 21; Gen. 1332: 71; Th. 85, 6; Gen. 1410: Chr. 937; Erl. 112, 27; Æðelst. 27. Of brimes bósme *from the sea's bosom,* Andr. Kmbl. 887; An. 444. Dó ðíne hand on ðínne bósum. Ðá he híg dyde on his bósum *mitte manum tuam in sinum tuum: cum misisset in sinum,* Ex. 4, 6, 7. Án man mihte faran ofer his ríce, mid his bósum full goldes, ungederad *a man might go over his kingdom, with his bosom full of gold, unhurt,* Chr. 1086; Erl. 222, 4. Ðú ðínre módor bósm sylfa gesóhtes *thou thyself soughtest thy mother's bosom,* Exon. 121 b; Th. 465, 27; Hö. 110. Ðú wuldres þrym bósme gebǽre *thou barest the majesty of glory* [Christ] *in thy breast,* 9 a; Th. 6, 14; Cri. 84. [Wyc. bosum: Laym. bosm: Orm. bosemm: Plat. bussen, bossen: O. Sax. bósom, m: O. Frs. bosm, m: Dut. boezem, m: Ger. busen, m: M. H. Ger. buosem, buosen: O. H. Ger. bósam, buosam, m. sinus.] DER. fǽmig-bósm, swegl-.

BÔT, e; *f.*   I. *help, assistance, remedy, cure*; auxilium, remedium, emendatio, sanatio :—Hér ys seó bôt, hû ðu meaht ðîne æceras bêtan *here is the remedy, how thou mayest improve thy fields,* Lchdm. i. 398, 1.   Findest ðû ðær æt bôte and ælteowe hǽlo *thou shalt find therein a remedy and perfect healing,* Herb. 1, 29; Lchdm. i. 80, 6. Byþ hræd bôt *the cure will be quick,* Med. ex Quadr. 6, 15; Lchdm. i. 354, 11.   II. *a* BOOT, *compensation due to an injured person as damages for the wrong sustained, redressing, recompense, an amends, a satisfaction, correction, reparation, restoring, renewing, repentance, an offering*; compensatio, emendatio, reparatio, oblatio :—Gif feaxfang geweorþ, L scætta to bôte *if there be a taking hold of the hair, let there be 50 sceats for compensation,* L. Ethb. 33; Th. i. 12, 3.   For bôte his synna *for a redressing of his sins,* Bd. 4, 25; S. 599, 32: 5, 13; S. 632, 13.   Bringaþ ânne buccan to bôte *bring a kid for an offering,* Lev. 4, 23, 28: L. Alf. pol. 2; Th. i. 62, 6: Bd. 1, 27; S. 489, 9. ¶ To-bôte *to-boot, with advantage, moreover, besides.* [*Piers P.* bote: *Laym. Orm.* bote: *Plat.* bote, *f:* O.*Sax.* bôta, *f:* O.*Frs.* bote, *f: Dut.* boete, *f: Ger.* busze, *f:* M.H.*Ger.* buoz, buoze: O.H.*Ger.* bôza, *f: Goth.* bota, *f: Dan.* bod, *c:* Swed. bot, *m: Icel.* bót, *f.*]   DER. bric-bôt, bricg-, burh-, hâd-, weofod-.

bôþ *boasts* :—He bôþ *he boasts,* Exon. 83 b; Th. 315, 9; Môd. 28; *pres. of* bôn.

boðen, es; *m? n? Rosemary, darnel*; rosmarinus, rosmarinus officinalis, Lin. lolium :—Deós wyrt, ðe man *rosmarinum* [MS. *rosmarim*] and ôðrum naman boðen, nemneþ, byþ cenned on sandigum landum *this herb, which is called* rosmarinus, *and by another name rosemary, is produced in sandy lands,* Herb. 81, 1; Lchdm. i. 184, 5.   Ceów boðenes moran *chew roots of rosemary,* L. M. 3, 4; Lchdm. ii. 310, 17.   Deós wyrt ys boðene gelíc *this herb is like rosemary,* Herb. 149, 1; Lchdm. i. 274, 6. Boðen *lolium,* Ælfc. Gl. 101; Som. 77, 30; Wrt. Voc. 55, 35.

botl, es; *n. An abode, a dwelling, mansion, house, hall*; domus, ædes, domicilium, atrium :—Gif he him nân botl ne selþ *if he do not give him an abode,* L. In. 67; Th. i. 146, 5.   Fordrífe ðý botle *let him be driven from the abode,* 68; Th. i. 146, 5.   Wæs Gûþláce botles neód *Guthlac was in need of a dwelling* [lit. *there was need to Guthlac of a dwelling*], Exon. 37 a; Th. 122, 4; Gû. 300.   Pharao eóde in to his botle *Pharao ingressus est domum suam,* Ex. 7, 22.   Mîn se êca dǽl in gefeán fareþ, ðǽr he fægran botles brúceþ *my eternal part* [i. e. *the soul*] *shall go into joy, where it shall enjoy a beautiful mansion,* Exon. 38 a; Th. 125, 14; Gû. 354.   To ðæra sacerda ealdres botle *in atrium principis sacerdotum,* Mt. Bos. 26, 3, §8.   Cynelíc botl *a kingly dwelling, a palace*; palatium, Ælfc. Gl. 81; Som. 73, 9; Wrt. Voc. 47, 16.   DER. ealdor-botl, heáfod-.

bôt-leás; *adj.* [bôt *boot,* leás *less*] BOOTLESS, *unpardonable, what cannot be remedied, recompensed or expiated*; inexpiabilis :—Ðonne síg ðæt bôtleás *then is that unpardonable,* L. C. E. 2; Th. i. 358, 24.   Hûsbryce is bôtleás *housebreaking is unpardonable,* L. C. S. 65; Th. i. 410, 6.

botl-gestreón, es; *n.* [gestreón *riches, wealth*] *Household property, goods, or treasure*; domesticæ opes :—Chus wæs brytta brôðrum sínum botlgestreóna *Cush was a dispenser of household treasures to his brothers,* Cd. 79; Th. 97, 32; Gen. 1621.   Lameh onféng æfter fæder dæge botlgestreónum *Lamech succeeded to the household goods after his father's day,* 52; Th. 65, 32; Gen. 1075: 91; Th. 116, 3; Gen. 1930.

botl-weard, -werd, es; *m.* [weard *a keeper, guardian*] *A housesteward*; ædilis :—Hófweard *vel* byriweard *vel* botlweard *ædilis,* Ælfc. Gl. 8; Som. 56, 105; Wrt. Voc. 18, 54.   Botlwerd *ædilis,* Ælfc. Gr. 9, 28; Som. 11, 29.

botl-wéla, an; *m.* [botl *a house,* wéla *weal, wealth*] *House-wealth, a collection of houses, village*; domesticæ opes, vicus :—Ðér is botlwéla Bethlem hâten *there is a village called Bethlem,* Cd. 86; Th. 107, 34; Gen. 1799.

BOTM, es; *m. A* BOTTOM; fundus :—Scipes botm *a ship's bottom, the keel*; carina, Ælfc. Gl. 83; Som. 73, 64; Wrt. Voc. 48, 3: 103; Som. 77, 112; Wrt. Voc. 56, 32.   Satan on botme [ðære helle] stôd *Satan stood at the bottom* [*of hell*], Cd. 229; Th. 310, 5; Sat. 721: 18; Th. 21, 27; Gen. 330: 19; Th. 23, 18; Gen. 361.   Heó to [ðæs fennes] botme com *she came to the bottom* [*of the fen*], Beo. Th. 3017; B. 1506. [*Chauc.* botome: *Wyc.* botme: O.*Sax.* bodom, *m: Frs.* boyem, *c:* O.*Frs.* boden, *m: Dut.* bódem, *m: Ger.* M.H.*Ger.* bodem, boden, *m:* O.H.*Ger.* bodam, *m: Dan.* bund, *c:* Swed. botten, *m: Icel.* botn, *m: Lat.* fundus, *m: Grk.* πυθμήν, *m: Ir.* bonn, *m: Gael.* bonn, buinn, *m: Sansk.* budhna, *m. the bottom,* from the root budh *to fathom a depth, penetrate to the bottom.*]   DER. byden-botm, tunne-.

bôt-wyrþe; *adj. Pardonable, expiable, that may be atoned for*; emendabilis :—Æt bôtwyrþum þingum *among pardonable things,* L. C. E. 3; Th. i. 360, 16.

BOX, es; *m? n? The* BOX-*tree*; buxus = πύξος, buxus sempervirens, Lin :—Box *buxus,* Ælfc. Gl. 47; Som. 65, 39; Wrt. Voc. 33, 36: 79, 71.   Æt ðam boxe, of ðam boxe *at the box-tree, from the box-tree,* Cod. Dipl. 1102; A.D. 931; Kmbl. v. 195, 14. [*Chauc.* box-tree: *Dut.* box-boom: *Ger.* buchs, *m:* M.H.*Ger.* buhs, *m:* O.H.*Ger.* buhs-boum:

Dan. bux-bom: *Swed.* bux-bom: *Lat.* buxus: *Grk.* πύξος *the box-tree* or *box-wood.*]   DER. bixen.

box, es; *m? n?* [box *the box-tree*] *A wooden case made of box-wood, a* BOX; buxum, pyxis = πυξίς :—Bixen box *a box made of box-wood*; pyxis, Ælfc. Gl. 26; Som. 60, 96; Wrt. Voc. 25, 36.   Forcorfen [MS. forcaruen] box *a carved box*; buxum, Ælfc. Gr. 6, 9; Som. 5, 59.   Seó hæfde box mid deórwyrþre sealfe *she had a box of precious ointment,* Mt. Bos. 26, 7.   Ellenes blôsman gedô on box *put blossoms of elder into a box,* L. M. 2, 59; Lchdm. ii. 288, 3.   Hundteontig boxa *a hundred* [*of*] *boxes,* Jn. Bos. 19, 39. [*Chauc. R. Glouc.* box: *Dut.* bus, *f: Ger.* büchse, *f:* M.H.*Ger.* bühse, *f:* O.H.*Ger.* buhsa, *f: Lat.* buxum, *n;* pyxis, *f: Grk.* πυξίς, *f. a box.*]   DER. sealf-box.

box-treów, es; *n. The* BOX-TREE; buxus = πύξος :—Ðis boxtreów *hæc buxus,* Ælfc. Gr. 6, 9; Som. 5, 59.   v. box.

bracan; *p.* bróc, *pl.* brócon; *pp.* bracen *To break, bruise* or *bray in a mortar, to beat up*; conterere, contundere :—Ðâ sceolon beón ele bracene *then shall they be beaten up with oil,* Lev. 6, 21.   v. brecan.

braccas, *pl. m. Breeches*; bracæ :—Braccas on swefnum geseón *to see breeches in dreams,* Lchdm. iii. 198, 28.   v. bróc; *pl.* brêc, brǽc.

brac-hwíl *a glance while, a moment.*   v. bearhtm-hwíl.

bracigean *to dress, mingle* or *counterfeit with brass*; ærare.   v. brǽsian.

BRÁD; *def.* se brâda, seó, ðæt brâde; *comp. m.* brâdra, *f. n.* brâdre, brǽdre; *superl.* brâdost; *adj.* BROAD, *open, large, spacious, copious*; latus, expansus, amplus, spatiosus, copiosus :—Ðæt eálond on Wiht is twelf míla brâd *the isle of Wight is twelve miles broad,* Bd. 1, 3; S. 475, 19: Ors. 1, 1; Bos. 21, 4, 5, 6.   Wæs his ríce brâd *his kingdom was broad,* Exon. 65 b; Th. 243, 10; Jul. 8: Elen. Kmbl. 1831; El. 917: Beo. Th. 6296; B. 3158.   Brâd is bebod ðín *latum est mandatum tuum,* Ps. Lamb. 118, 96.   Se brâda sǽ *the broad sea,* Exon. 24 b; Th. 70, 28; Cri. 1145: Chr. 942; Erl. 116, 11; Edm. 5: Ps. Th. 79, 10.   Beówulfe brâde ríce on hand gehwearf *the broad realm passed into the hand of Beowulf,* Beo. Th. 4421; B. 2207.   Beorn monig seah on ðâs beorhtan burg brâdan ríces *many a chief looked on this bright city of a broad realm,* Exon. 124 b; Th. 478, 9; Ruin. 38.   Ofer Babilône brâdum streáme we sittaþ *we sit over the broad stream of Babylon,* Ps. Th. 136, 1.   On ðam brâdan brime *on the broad ocean,* Exon. 55 a; Th. 194, 20; Az. 142.   Se hearda þegn lêt brâdne mêce brecan ofer bordweal *the fierce thane caused his broad sword to break over the shield,* Beo. Th. 5948; B. 2978.   Ðû scealt ðínum breóstum tredan brâde eorþan *thou shalt tread the broad earth on thy breast,* Cd. 43; Th. 56, 5; Gen. 907: 83; Th. 105, 12; Gen. 1752: Ps. Th. 118, 32: Exon. 22 b; Th. 61, 29; Cri. 992.   He him brâd syleþ lond *he will give him broad land,* Exon. 88 a; Th. 331, 29; Vy. 75.   On brâd wæter *on the broad water,* Ps. Th. 105, 8: Salm. Kmbl. 52: Sal. 275.   Ðâ he healdan mihte brâd swurd *when he could hold his broad sword,* Byrht. Th. 132, 12; By. 15: 136, 38; By. 163: Beo. Th. 3096; B. 1546.   Brâde synd on worulde grêne geardas *in the world there are broad green regions,* Cd. 25; Th. 32, 29; Gen. 510. Of ðâm brâd blado sprytan ongunnon *thence broad leaves began to spring,* 48; Th. 61, 8; Gen. 994.   Engle and Seaxe ofer brâde brimu Brytene sôhton *the Angles and Saxons sought Britain over the broad seas,* Chr. 937; Erl. 115, 20, note; Æðelst. 71: Exon. 13 a; Th. 22, 25; Cri. 357.   Sceolde he ða brâdan lígas sêcan *he must seek the broad flames,* Cd. 36; Th. 47, 20; Gen. 763.   Hit mæg bión syxtig míla brâd, oððe hwene brǽdre; and middeweard þrîtig oððe brâdre *it may be sixty* [*of*] *miles broad, or a little broader; and midway thirty or broader,* Ors. 1, 1; Bos. 21, 1, 2.   Ðeáh hit ǽlce geáre sý brâdre and brâdre *though it is broader and broader every year,* 2, 6; Bos. 50, 22.   Ic eom brǽdre ðonne ðes wong grêna *I am broader than this green plain,* Exon. 111 a; Th. 425, 3; Rä. 41, 50: 111 b; Th. 426, 32; Rä. 41, 82.   Ðæt býne land is easteweard brâdost *the inhabited land is broadest eastward,* Ors. 1, 1; Bos. 20, 45.   Sume hyne slôgon on his ansýne mid hyra brâdum handum *some smote him on his face with their open hands,* Mt. Bos. 26, 67.   Brâd *amplus,* Ælfc. Gr. 37; Som. 39, 35.   Seó sunne is swâ brâd swâ eall eorþan ymbhwyrft, ac heó þincþ [MS. þingþ] us swýðe unbrâd, forðamðe heó is swíðe feorr fram ûrum gesihþum *the sun is as large as the whole compass of the earth, but he* [lit. *she*] *appears to us very small* [lit. *un-broad*], *because he is very far from our sight,* Bd. de nat. rerum; Wrt. popl. science 3, 8–11; Lchdm. iii. 236, 6–9.   Ða steorran, ðâ us lyttle þinceaþ [MS. þingeaþ], synd swýðe brâde *the stars, which seem little to us, are very large,* 3, 16; Lchdm. iii. 236, 14.   Se deófol brohte him brâde stânas *the devil brought large stones to him,* Cd. 228; Th. 306, 31; Sat. 672.   Byþ se niwa môna brâdra [MS. braddra] gesewen *the new moon appears* [lit. *is seen*] *larger,* Bd. de nat. rerum; Wrt. popl. science 14, 14; Lchdm. iii. 264, 26.   Ðær is brâde lond in heofoníce *there is a spacious land in heaven's kingdom,* Cd. 218; Th. 278, 2; Sat. 215.   Hí bebúgaþ brâdne hwyrft *they shall inhabit the spacious orb,* 190; Th. 236, 16; Dan. 322: Exon. 53 b; Th. 187, 29; Az. 38.   Ðû gearwodest beföran me brâdne beód *thou preparedst a copious table before me,* Ps. Th. 22, 6.   Ge onsceáwiaþ beágas and brâd gold *ye will behold bracelets and ample gold,* Beo. Th. 6201; B. 3105.   Ic his cynn gedô

brád and bresne *I will make his race large and powerful*, Cd. 134; Th. 169, 17; Gen. 2801. Brád earmbeáh *a broad or large arm-bracelet*; dextrocherium, Ælfc. Gl. 114; Som. 80, 30; Wrt. Voc. 61, 10. [*Chauc. Wyc.* brod, brood: *R. Glouc.* brod: *Laym.* braed, brad, brod: *Orm.* brad: *Scot.* braid, brade: *Plat.* breed: *O. Sax.* brêd: *Frs.* bred: *O. Frs.* bred, breid: *Dut.* breed: *Ger. M. H. Ger.* breit: *O. H. Ger.* breit: *Goth.* braids: *Dan. Swed.* bred: *Icel.* breiðr: *Lat.* latus for platus: *Grk.* πλατύς: *Lith.* platùs: *Zend* frath-anh *breadth: Sansk.* prithu *broad, wide;* prith *to extend.*] DER. un-brád, wíd.

brád-æx, e; *f. A broad axe, an axe;* dolatura, dolabrum :—Brádæx dolatura, Cot. 68 : dolabrum, Ælfc. Gl. 51; Som. 65, 131; Wrt. Voc. 34, 59.

Brádan ē; *indecl. f.* [i. e. latus fluviuś, *Hist. Eccl. Petroburg.* Bardanea, *Gib. Chr. explicatio* 15] *Broadwater;* Bradanea :—Þurh ân scŷr wæter, Brádan ē hátte *through a clear water called Broadwater,* Chr. 656; Erl. 31, 17; per unam pulcram aquam, Bradanea nomine, Cod. Dipl. 984; A. D. 664; Kmbl. v. 5, 3.

Brádan-ford, es; *dat.* -forde, -forda; *m.* [brád *broad,* ford *a ford*] BRADFORD *in Wilts;* loci nomen vadum amplum *vel* latum significans, hodie *Bradford* in agro Wiltoniensi :—Cénwalh gefeaht æt Brádanforda be Afne *Kenwealh fought at Bradford near the Avon,* Chr. 652; Erl. 26, 22.

Brádan-relic, Brádun-reolic, es; *m :* Brádan-reíg, -eíg = íg, e; *f.* [eíg, íg *an island, broad island*] *Flat Holme, an island in the mouth of the Severn :*—Sǽton hie úte on ðam íglande, æt Brádanrelice *they sat outward on an island, Flat Holme,* Chr. 918; Ing. 132, 19.

bráddra *broader, larger,* Bd. de nat. rerum; Wrt. popl. science 14, 14; Lchdm. iii. 264, 26, = brádra; *comp. def. m. of* brád.

bráde; *adv. Broadly, widely;* late :—Físon bráde bebúgeþ *Pison widely encompasses it,* Cd. 12; Th. 14, 23; Gen. 223: Exon. 13 a; Th. 24, 5; Cri. 380: Ps. Th. 106, 37.

Bráden, Brǽden, es; *m. [Flor.* Bradene: so called from its *size,* from brád, brǽd *broad, open, spacious;* dene, es; *m. vallis, locus silvestris,* v. denu] BREDON *Forest, near Malmesbury, Wiltshire;* silvæ nomen in agro Wiltoniensi :—Hie cômon to Creccagelâde, and fôron ðǽr ofer Temese, and nâmon, ǽgðer ge on Brádene, ge ðǽr ymbútan, eall ðæt hie gehentan mehton *they came to Cricklade, and there they went over the Thames, and took, both in Bredon, and thereabout, all that they could carry off,* Chr. 905; Th. 180, 22, col. 1, 2.

brád-hláf, es; *m.* [brádan *to roast,* hláf *bread*] *A biscuit, pârched or baked bread;* paximatium = παξαμάδιον, panis torrefactus :—Brádhláf paximatium, Wrt. Voc. 288, 66.

brádiende; *part.* [brád *broad, spread out*] *Stretching out, extending, reaching;* amplificans, extendens, tendens :—Fram ðam heofone brádiende niðer ôþ ða eorþan *reaching from the heavens down to the earth,* Ors. 5, 10; Bos. 108, 25. v. brǽdan.

brád-nes, -ness, -nis, -niss, -nys, -nyss, e; *f.* [brád *broad, large,* -nes, -nis, -nys -*ness*] BROADNESS, *extent, largeness, surface;* latitudo, amplitudo, facies, superficies :—Se rôdor belýcþ on his bôsme ealle eorþan brádnysse *the firmament incloses in its bosom all the extent of the earth,* Hexam. 5; Norm. 8, 27. Se wǽta, gyf hit sealt byþ of ðære sǽ, byþ þurh ðære lyfte brádnysse to ferscum wǽtan awend *the moisture, if it is salt from the sea, is turned into fresh water through the extent of the atmosphere,* Bd. de nat. rerum; Wrt. popl. science 19, 3, 27; Lchdm. iii. 278, 11; 280, 14. Gehêrde me on tobrǽdednesse oððe on brádnesse Drihten exaudivit me in latitudine Dominus, Ps. Lamb. 117, 5. Salomone forgeaf God brádnysse heortan *God gave Solomon largeness* [or *liberality*] *of heart,* Homl. Th. ii. 576, 29. Þeóstru wǽron ofer ðære niwelnisse brádnisse *tenebræ erant super faciem abyssi,* Gen. 1, 2. Ðære eorþan brádnis wæs adrúwod *exsiccata esset superficies terræ,* 8, 13. Byþ ðære eorþan brádnys betweox us and ðære sunnan *the surface of the earth is between us and the sun,* Bd. de nat. rerum; Wrt. popl. science 5, 8; Lchdm. iii. 240, 14. Sumes þinges brádnyss *the surface of something;* superficies, Ælfc. Gr. 47; Som. 48, 47. Ân wyll asprang of ðære eorþan, wǽtriende ealre ðære eorþan brádnysse *fons ascendebat e terra, irrigans universam superficiem terræ,* Gen. 2, 6.

brádost *broadest,* Ors. 1, 1; Bos. 20, 45; *superl. of* brád.

brádre *broader,* Ors. 2, 6; Bos. 50, 22; *comp. f. n. of* brád.

brád-þistel; *gen.* -þistles; *m. A thistle with long leaves, sea-holm, sea-holly;* eryngium = ἠρύγγιον, eryngium maritimum, Lin :— Brádþistel eryngion, Cot. 212.

brǽc, ðú brǽce, *pl.* brǽcon *broke, didst break,* Mt. Bos. 14, 19: Exon. 28 a; Th. 85, 20; Cri. 1394: Cd. 32; Th. 43, 5; Gen. 686; *p. of* brecan.

brǽc, es; *n.* [brǽc; *p. of* brecan *to break*] *A breaking, flowing, rheum, catarrh;* rheuma = ῥεῦμα :—Brǽc *rheuma,* Ælfc. Gl. 10; Som. 57, 21; Wrt. Voc. 19, 27. DER. ge-brǽc, fȳr-ge-, hrǽc-ge-, neb-ge-.

brǽc *breeches;* braccæ, Som. femoralia, Wrt. Voc. 81, 63, = brêc; *pl. of* brôc, *f.*

brǽc-côðu, e; *f.* [brǽc *a breaking,* côðu *a disease*] *The breaking or falling disease, epilepsy;* epilepsia = ἐπιληψία :— Brǽc-côðu, fylle-seóc

epilepsia vel *caduca* vel *larvatio* vel *commitialis,* Ælfc. Gl. 10; Som. 57, 20; Wrt. Voc. 19, 26.

brǽce; *adj. Breaking;* violans. DER. ǽw-brǽce, un-. v. brecan.

brǽc-seóc; *adj.* [brǽc, seóc *sick, diseased*] *Troubled with the falling sickness, epileptic, frantic, lunatic;* epilepticus, phreneticus, lunaticus :— Sum brǽcseóc man becom ðyder *phreneticus devenit ibi,* Bd. 4, 3; Whelc. 267, 45, MSS. B. C. DER. ge-brǽcseóc.

brǽc-seócnes, -ness, e; *f. Epilepsy;* epilepsia. DER. brǽc-seóc *epileptic, frantic;* -nes -*ness.*

brǽd, bred, es; *m.* [= brǽgd, bregd from bregdan *to braid, weave, twist*] *Fraud, deceit;* fraus, dolus :—He hit dyde bútan brede [brǽde MS. B.] and bigswíce *he did it without fraud and guile,* L. Ed. 1; Th. i. 160, 6. Ic spæce drífe bútan brǽde and bútan bíswíce *I prosecute my suit without fraud and without guile,* L. O. 2; Th. i. 178, 13. Bred *fucus, fraus, astus,* Cot. 10.

brǽd *plucked, drew out,* Byrht. Th. 136, 20; By. 154; *p. of* bredan.

brǽd, e; *f.* brǽdo, brǽdu; *indecl. f.* [brád *broad*; latus] BREADTH, *width, latitude;* latitudo, amplitudo :—Biþ se arc fíftig fǽðma on brǽde *the ark shall be fifty fathoms in breadth;* quinquaginta cubitorum erit latitudo arcæ, Gen. 6, 15. On brǽdo his stealles *latitudine sui status,* Bd. 1, 1; S. 474, 29. Ic on brǽdu [brǽde, Spl.] gange *ambulabam in latitudine,* Ps. Th. 118, 45. Drihten me gehȳrde on heáre [= heáhre, MS. hearr] brǽdu *exaudivit me in latitudine Dominus,* 117, 5. Hí habbaþ ingang swá mycelre brǽdo swá mon mæg mid liðeran geworpan *habet ingressum amplitudinis quasi jactus fundæ,* Bd. 4, 13; S. 583, 11. [*Chauc.* brede: *Wyc.* breede: *O. Frs.* brede, *f:* *Dut.* breedte, *f:* *Ger. M. H. Ger.* breite, *f:* *O. H. Ger.* breiti, *f:* *Goth.* braidei, *f:* *Dan.* brede, *c :* *Swed.* bredd, *f:* *Icel.* breidd, *f. breadth.*] DER. hand-brǽd.

brǽd *broad;* latus, Beo. Th. 4421, note. v. brád.

brǽdan, brédan; *to brǽdanne,* brédanne; *part.* brǽdende; he brǽdeþ, brǽd; *p.* brǽdde, *pl.* brǽddon; *pp.* brǽded, brǽd [brád *broad;* latus]. I. *v. trans. To make broad,* BROADEN, *extend, spread, stretch out;* dilatare, propalare, expandere :—Hí heora stôwe brǽddon *they broadened their places,* Bd. 1, 8; S. 479, 24. He gesihþ brimfuglas brǽdan feðra *he sees sea-fowls spread their wings,* Exon. 77 a; Th. 289, 13; Wand. 47. Ge wilniaþ eówerne hlísan to brǽdanne *ye wish to spread your fame,* Bt. 18, 1; Rawl. 38, 33, MS. Cot. Se wallenda lêg hine brǽdde to ðam biscope *the raging flame spread itself to the bishop,* Bd. 2, 7; S. 509, 22. Brǽddon æfter beorgum flotan feldhúsum *the sailors spread* [*themselves*] *amongst the hills with their tents,* Cd. 148; Th. 186, 1; Exod. 132. Ðæt hí his naman brǽden [MS. brǽdan] *that they spread his name,* Bt. 30, 1; Fox 108, 11. Se cyning his handa wæs uppweardes brǽdende wið ðæs heofenes *the king stretched* [lit. *was stretching*] *out his hands upwards towards heaven,* Ors. 4, 5; Bos. 81, 36. II. *v. intrans. To be extended or developed, grow or rise up;* dilatari, adolescere :—Leáf and gærs brǽd geond Bretene *leaves and grass are extended* [lit. *leaf and grass is extended*] *over Britain,* Bt. Met. Fox 20, 197; Met. 20, 99. Treó sceolon brǽdan *trees shall rise up,* Exon. 91 b; Th. 343, 22; Gn. Ex. 160. [*Laym.* breden: *Scot.* brade: *Plat.* breden, bredden: *O. Sax.* brêdian, brêdôn: *Kil.* breeden: *Ger. M. H. Ger.* breiten: *O. H. Ger.* breitan: *Goth.* braidyan: *Dan.* brede: *Swed.* breda: *Icel.* breiða *to broaden.*] DER. ge-brǽdan, geond-, ofer-, to-.

BRÆDAN, brêdan, to brǽdenne; *part.* brǽdende; *p.* brǽdde; *pp.* brǽded, brǽd; *v. a. To roast, broil, warm;* assare, fovere :—We mâgon brǽdan ða þing [þingc MS.] ðe to brǽdene synd *non possumus assare quæ assanda sunt,* Coll. Monast. Th. 29, 21. Brêdan, weormian *fovere,* Cot. 86. Brǽdende *assans,* Cot. 195. [*Laym.* breden: *Scot.* brade: *Plat.* braden, braën: *Frs.* briede: *O. Frs.* breda: *Dut.* braden: *Ger.* braten: *M. H. Ger.* brâten: *O. H. Ger.* brâtan *assare.*] DER. ge-brǽdan.

brǽde, es; *m.* [brǽdan *to roast*] *Roasted meat;* assatura :—Brǽde *assura* vel *assatura,* Ælfc. Gl. 31; Som. 61, 85; Wrt. Voc. 27, 15. [*Dut.* ge-braad, *n :* *Ger.* brate, *m. caro assa :* *M. H. Ger.* brâte, *m :* *O. H. Ger.* brâto, *m. assatura.*]

brǽde, an; *f. The breadth;* latum. v. lenden-brǽde.

brǽded-nes, -ness, e; *f.* [brǽded; *pp. of* brǽdan *to broaden,* -nes -*ness*] *Broadness, breadth, width, latitude;* amplitudo, latitudo. DER. to-brǽdednes. v. brádnes.

brǽdels, es; *m?* [brǽdan *to spread* or *stretch out*] *Anything spread or stretched out, a carpet, covering, garment, dress;* palla, stragulum, velamentum, opertorium :—Brǽdels *stragulum,* R. 4, Lye. DER. ofer-brǽdels.

Brǽden *Bredon Forest :*—On Brǽdene ge ðǽr onbútan *in Bredon and thereabout,* Chr. 905; Th. 181, 23, col. 1, 2. v. Bráden.

brǽding, e; *f.* [brǽdan *to spread, extend*] *A spreading;* ampliatio :— Mæg hine scamian ðære brǽdinge his hlísan *he may be ashamed of the spreading of his fame,* Bt. 19; Fox 68, 24.

brǽding-panne, an; *f.* [brǽdan *to roast, broil,* panne *a pan*] *A frying-pan;* sartago, Cot. 173. v. brǽd-panne.

brǽd-ísen, bred-ísern, es; *n.* [brǽd, *p. of* bredan; ísen, ísern *iron*] *A scraping or graving tool, file;* scalprum, scalpellum :—Brǽdísen *scal-*

prum, *scalpellum*, Cot. 173. Bredîsern *scalpellum*, Glos. Epnl. Recd. 162, 28.

bræd-nys, -nyss, e; *f. Broadness;* latitudo. DER. to-brǽdnys. v. brádnes.

brǽdo *breadth, width,* Bd. 1, 1; S. 474, 29: 4, 13; S. 583, 11. v. brǽd.

bræd-panne, an; *f.* [brǽdan *to roast,* panne *a pan*] *A frying-pan;* sartago, frixorium, Cot. 115. v. brǽding-panne.

brǽdre *broader,* Ors. 1, 1; Bos. 21, 2, = brádre; *comp. f. n. of* brád.

brǽdu *breadth, width,* Ps. Th. 117, 5 : 118, 45. v. brǽd.

brægd, bregd, es; *m.* [brægd, *p. of* bregdan *to twist, braid, weave*] *Deceit, fraud;* dolus, fraus. DER. ge-brægd, -bregd, nearo-. v. bræd.

brægd *bent,* Beo. Th. 1593; B. 794; *p. of* bregdan.

brægdan *to modulate;* modulari:—Hî gehêraþ hleódrum brægdan ódre fugelas *they hear other birds modulate their songs,* Bt. Met. Fox 13, 94; Met. 13, 47.

brægd-boga, an; *m.* [brægd, *p. of* bregdan *to draw, bend,* brægd *deceit;* boga *a bow*] *A drawn* or *bent bow, a deceitful* or *fraudulent bow;* arcus incurvatus *vel* fraudulentus:—He in folc Godes forþ onsendeþ of his brægdbogan biterne strǽl *he* [*the devil*] *sendeth forth, amongst God's people, the bitter arrow from his deceitful bow,* Exon. 19 a; Th. 48, 1; Cri. 765.

brægden; *adj.* [= bregden; *pp. of* bregdan] *Deceitful, cunning, crafty;* dolosus:—Sendon [sendan MS.] hî Marium, ðone consul, ongeán Geoweorþan, â swá lytigne, and â swá brægdenne, swá he wæs *they sent Marius, the consul, against Jugurtha, as he was always so cunning, and always so crafty,* Ors. 5, 7; Bos. 106, 29.

brægd-wís; *adj.* [brægd *deceit,* wîs *wise*] *Wise in deceit, crafty, fraudulent;* astutus, fraudulentus, dolosus:—Brægdwís bona *a crafty murderer,* Exon. 33 b; Th. 107, 13; Gû. 58.

BRÆGEN, brægn, bragen, es; *n. The* BRAIN; cerebrum, cerebellum :—Wið tobrocenum heáfde, and gif ðæt brægen ûtsíge, genim æges ðæt geoluwe *for a broken head, and if the brain appears, take the yolk of an egg,* L. M. 1, 1; Lchdm. ii. 22, 19. Brægen *cerebrum vel cerebellum,* Ælfc. Gl. 69; Som. 70, 38; Wrt. Voc. 42, 46. Brægn *cerebrum,* Wrt. Voc. 64, 25. Bragen *cerebrum,* 70, 25. Brægenes ádl *the disease of the brain,* L. M. 2, 27; Lchdm. ii. 222, 3. On his brægn astíge his unriht *in verticem ipsius iniquitas ejus descendet,* Ps. Th. 7, 16. [*Chauc.* brain: *R. Glouc.* brayn: *Laym.* brain, braȝen: *Plat.* brägen: *O. Frs.* brein, brin, *n : Dut.* brein, *n. cerebrum.*]

Brægent-ford *Brentford in Middlesex,* Chr. 1016; Th. 280, 26, col. 2: 1016; Th. 282, 5, col. 2. v. Brent-ford.

brǽhtm *a glimpse, glittering, twinkling,* Bd. 2, 13; Whelc. 142, 23, MS. B. v. bearhtm.

brǽmbel *a bramble,* Herb. 89, 1; Lchdm. i. 192, note 6. v. brêmel.

brǽmbel-brǽr, es; *m.* [brǽmbel *a bramble,* brǽr, brêr *a brier*] *A bramble-brier;* tribulus, Wrt. Voc. 285, 64. v. brêmel.

brǽmbel-leáf, es; *n.* [brǽmbel *a bramble,* leáf *a leaf*] *The leaf of a bramble;* rubi folium:—Nim brǽmbel-leáf *take bramble-leaves,* Lchdm. iii. 40, 26. v. brêmel.

brǽmel *a bramble.* DER. brǽmel-berie. v. brêmel.

brǽmel-berie, an; *f.* [brǽmel = brêmel *a bramble,* berie *a berry*] *A bramble-berry;* rubi bacca:—Drince seóca óf brǽmelberian gewrungene *let the sick man drink of wrung bramble-berries,* Lchdm. iii. 8, 17.

brǽr *a brier;* tribulus. v. brǽmbel-brǽr. v. brêr.

BRÆS, es; *n.* BRASS; æs:—Brǽs oððe ár æs, Ælfc. Gr. 5; Som. 4, 59. [*O. Nrs.* bras, *n. ferumen, soldering of iron,* Rask Hald.] v. ár.

bræsen, bresen; *def.* se brǽsna, seó, ðæt brǽsne, bresne; *adj.* I. BRAZEN, *made of brass;* æreus, æneus:—Bræsen oððe æren *æneus,* Ælfc. Gr. 5; Som. 4, 59. Ðû gesettest swá swá bogan brǽsenne earmas mîne *posuisti ut arcum æreum brachia mea,* Ps. Lamb. 17, 35. II. *strong, powerful, bold, daring;* validus, fortis, potens, procax:—Gebeád ðá se brǽsna Babilône weard *then the bold lord of Babylon proclaimed,* Cd. 196; Th. 244, 16; Dan. 449.

bræsian, brasian, ic brǽsige, ðû brǽsast, he brǽsaþ, *pl.* brǽsiaþ; *p.* ode; *pp.* od *To cover* or *furnish with brass, to make of brass;* ærare:—Ic brǽsige [MSS. C. D. brasige] *æro,* Ælfc. Gr. 36; Som. 38, 39.

brǽsna *strong, bold,* Cd. 196; Th. 244, 16; Dan. 449. v. brǽsen.

BRÆÞ, breþ, es; *m. An odour, a scent, smell good* or *bad, a savour;* odor, odoramen:—God underfêng ðære wynsumnysse brǽþ *odoratus est Dominus odorem suavitatis,* Gen. 8, 21. Ongan se cealc mid ungemete stincan, ðá wearþ ungemete mid ðam brǽþe ofsmorod *the plaster* [*lit. chalk*] *began to smell excessively, and Jovian was smothered with the smell,* Ors. 6, 32; Bos. 129, 12. Brêþ *odor,* Ælfc. Gl. 70; Wrt. Voc. 42, 58. [*Chauc. Piers P.* breeþ: *Ger.* bradem, *m : M. H. Ger.* bradem, *m : O. H. Ger.* bradam, *m.*] v. ǽdm.

BRÆW, breáw, breág, brêg, brêgh, brêhg, es; *m. An eye-lid;* palpebra:—Wið þiccum brǽwum *for thick eye-lids,* L. M. 1, 2; Lchdm. ii. 38, 9. Ðæt biþ swíðe gód sealf ðam men ðe hæfþ þicce brǽwas *that will be a very good salve for a man who has thick eye-lids,* 1, 2; Lchdm. ii. 38, 22, 12. Unwlîtig swile and atelíc his eágan brêgh [brêg MS. C.]

wyrde and wemde an unsightly and fearful swelling harmed and corrupted his eye-lid, Bd. 4, 32; S. 611, 18. Ðá he ðá ðam feaxe onfêng ðæs hálgan heáfdes, he togesette ðam untruman brêhge *cum accepisset capillos sancti capitis, adposuit palpebræ languenti,* 4, 32; S. 611, 40. Ðá gehrán he his eágan, gemêtte he hit swá hál mid ðý brǽwe *contingens oculum, sanum cum palpebra invenit,* 4, 32; S. 612, 7. Brǽwas [brêgas, Surt.] his axiaþ oððe befrinaþ bearn manna *palpebræ ejus interrogant filios hominum,* Ps. Lamb. 10, 5; the eȝelidis of hym asken the sones of men, Wyc. Brǽwas *palpebræ,* Wrt. Voc. 70, 41: 282, 50. Breáwas *palpebræ,* Ælfc. Gl. 70; Som. 70, 63; Wrt. Voc. 42, 71. Gif ic selle swefnu oððe slǽp eágum mînum, and breáwum [brǽwum, Spl: brêgum, Surt.] mînum hnappunga *si dedero somnum oculis meis, et palpebris meis dormitationem,* Ps. Lamb. 131, 4; I shal not ȝiue slep to myn eȝen, and to my eȝe lidis napping, Wyc. Ic eom wide calu, ne ic breága ne brúna brúcan môste *I am very bald, nor can I make use of eye-lids nor eye-lashes,* Exon. 111 b; Th. 427, 32; Rä. 41, 100. Betwux oferbrúan and brǽwum *intercilium* [= *intercilia*], Ælfc. Gl. 70; Som. 70, 70; Wrt. Voc. 43, 4. [*O. Sax.* brâha, brâwa, *f : O. Frs.* ag-bre, *n. an eye-lid: M. H. Ger.* brâ, *f : O. H. Ger.* brâ, *n : Icel.* brá, *f. an eye-lid: Lat.* frons, *f. the forehead, brow : Grk.* ὀφρύς, *f. the eye-brow : Sansk.* bhrû, *f. an eye-brow, the brow.*] DER. ofer-brǽw. v. brú.

bragen *the brain,* Wrt. Voc. 70, 25. v. brægen.

BRAND, brond, es; *m.* I. *a* BRAND, *fire-brand, torch;* titio, torris:—Brand *titio vel torris,* Ælfc. Gl. 30; Som. 61, 76; Wrt. Voc. 27, 6. Brand *titio,* Wrt. Voc. 82, 55; Glos. Epnl. Recd. 163, 42. Bǽron brandas on bryne blácan fŷres *they bare fire-brands into the burning of the bright flame,* Cd. 186; Th. 231, 12; Dan. 246. Se ád wæs ǽghwonan ymb-boren mid brondum *the funeral pile was heaped around on every side with fire-brands,* Exon. 74 a; Th. 277, 15; Jul. 581. II. *a burning, flame, fire;* incendium, flamma, ignis:—Brond þeceþ hús *the burning covers the house,* Exon. 59 a; Th. 212, 27; Ph. 216. Hǽfde landwara líge befangen, bǽle and bronde *he had enveloped the land-inhabitants in flame, with fire and burning,* Beo. Th. 4633; B. 2322. Reóteþ meówle, seó hyre bearn gesihþ brondas þeccan *the woman weeps, who sees the flames covering her child,* Exon. 87 b; Th. 330, 7; Vy. 47. Ðá beágas sceal brond fretan *fire shall consume the rings,* Beo. Th. 6021; B. 3014: Exon. 19 b; Th. 51, 7; Cri. 812. He his sylfes ðǽr bán gebringeþ, ðá ǽr brondes wylm on beorhstede forþylmde *it* [*the Phœnix*] *brings its own bones there, which the fire's rage had before encompassed on the mound,* Exon. 60 a; Th. 217, 21; Ph. 283. Ðá fŷnd þoliaþ helle to-middes brand and brǽde lígas *the fiends suffer fire and broad flames in the midst of hell,* Cd. 18; Th. 21, 16; Gen. 325. Hŷ hine ne môston bronde forbǽrnan *they could not consume him with fire,* Beo. Th. 4258; B. 2126. Brondas lácaþ on ðam deópan dæge *fires shall flare on that awful day,* Exon. 116 b; Th. 448, 23; Dôm. 58. Bronda *of fires,* Beo. Th. 6302; B. 3161: Exon. 116 a; Th. 445, 25; Dôm. 13. Bronda beorhtost *brightest of fires* or *lights, the sun,* 93 b; Th. 350, 17; Sch. 65. III. metaphorically from its shining, *A sword* [hence the *Eng. to* BRANDISH] ; ensis:—Ic gean Eádmunde mînum [minon MS.] brêðer ánes brandes *I give to Edmund my brother one sword,* Th. Diplm. 559, 24. Ðæt hine nó brond ne beádomêcas bîtan ne meahton *that no sword nor battle-falchions might bite it,* Beo. Th. 2912; B. 1454. [*Chauc.* bronde *a torch: Laym.* brond, brand *a sword: Plat.* brand, *m : Frs.* brân, *c. gladius: O. Frs.* brond, brand, *m. a fire-brand : Dut.* brand, *m. a burning, fire : Ger.* brand, *m. titio, torris, ensis : M. H. Ger.* brant, *m : O. H. Ger.* brant, *m. titio, torris : Dan.* brand, *m. f : Swed.* brand, *m. a fire-brand, fire : Icel.* brandr, *m.* I. *a brand, fire-brand;* II. *the blade of a sword.*]

brand? Beo. Th. 2045, note; B. 1020, note; *an error of the copyist for* bearn *a son.*

brand-hát, brond-hát; *def.* se -háta, seó, ðæt -háte; *adj.* [brand II. *a burning,* hát *hot*] *Burning hot, very hot, ardent, passionate;* ardentissimus, vehemens, fervidus:—Brandháta níþ weóll on gewitte *ardent malice boiled in their mind,* Andr. Kmbl. 1536; An. 769. Born in breóstum brondhát lufu *ardent love burned in his breast,* Exon. 46 b; Th. 160, 2; Gû. 937.

brand-hord *ardent treasure;* ardens thesaurus. v. brond-hord.

brand-ísen, es; *n.* [brand II. *a burning,* ísen *iron*] *A* BRANDING-IRON, *a tripod;* andena, tripes:—Brandîsen *andena* Ælfc. Gl. 30; Som. 61, 77; Wrt. Voc. 27, 7: 82, 54. [*Dut.* brandijzer, *n : O. Dut.* brandijzer *fulcrum focarium,* Kil: *Ger.* brandeisen, *n. cauterium.*] v. Du Cange, vol. i. col. 187, Andena.

brand-rád, e; *f.* [ród I. *a rod*] *A branding-rod;* andena, Glos. Epnl. Recd. 153, 4. [*O. Frs.* brondrad: *O. Dut.* brandroede.]

brand-stæfn *the shining prowed.* v. brond-stæfn.

brang, brong; *pl.* brungon *brought; p. of* bringan.

brant, bront; *adj. High, deep, steep, difficult;* altus, arduus:—Ðæt ðú us gebrohte brante ceóle, heá hornscipe, ofer hwæles êðel, on ðære mægþe *that thou wouldst bring us with the steep keel, the high pinnacled ship, over the whale's home, to that tribe,* Andr. Kmbl. 545–549; An. 273–275. Ðe brontne ceól ofer lagustrǽte lǽdan cwômon *who came leading*

a *high keel over the water-street*, Beo. Th. 482; B. 238. Ymb brontne ford *about the deep ford*, 1140; B. 568. Lêton ofer fífelwǽg scrîdan bronte brimþísan *they let the high ships go over the ocean-wave*, Elen. Kmbl. 475; El. 238. [*Wrt. Provncl.* brant *steep: Dan.* brat *steep: Swed.* brant *precipitous: Icel.* brattr *steep.*]

bran-wyrt, e; *f. A bilberry shrub;* vaccinium :—Branwyrt *vaccinium*, Ælfc. Gl. 39; Som. 63, 73; Wrt. Voc. 30, 25. v. brûn-wyrt II.

brasian, brasigan, ic brasige *I cover with brass;* æro, Ælfc. Gr. 36; Som. 38, 39, Bodleian copy, C. D. v. bræsian.

brassica, an; *m. Colewort, cabbage;* brassica, æ, *f.* :—Wyrta sindon betste bēte and mealwe and brassica *beet and mallow and cabbage are the best herbs,* L. M. 2, 30; Lchdm. ii. 228, 1.

BRASTL, es; *m. A noise, brustle, rustle, creak, crackle, burning?* crepitus, strepitus, fractio, arsio? Som. [*Ger.* brassel, prassel, geprassel, *n. a crackling noise.*] v. brastlung.

brastlian, brastligan, to brastlienne, brastligenne; *part.* brastliende, brastligende; he brastlaþ; *p.* ode; *pp.* od [berstan *rumpi, frangi*] To BRUSTLE, *rustle, crackle, make a noise, murmur;* crepare, crepitare, strepere, murmurare :—Began to brastligenne þunor *thunder began to crackle,* Homl. Th. ii. 196, 23. Ðæt treów brastliende sáh to ðam hálgan were *the tree fell crackling towards the holy man,* ii. 508, 33. Brastligende mid brandum *crackling with fire-brands,* ii. 140, 16. Se begeáton þeósterfulle wununga afyllede mid brastligendum lígum *ye have obtained dark dwellings filled with crackling flames,* i. 68, 5. Se þuner oft egeslíce brastlaþ *thunder often crackles fearfully,* Bd. de nat. rerum; Lchdm. iii. 280, 13. [*Laym.* brastlien : *Ger. M.H.Ger.* brasteln : *Swed.* prassla *to crackle.*]

brastlung, e; *f.* A BRUSTLING, *rustling, creaking, breaking, crashing;* strepitus, crepitus, fractio :—Híg tobrǽcon ða bûcas mid micelre brastlunge *they broke the pitchers with great crashing,* Jud. 7, 20. Brastlung treówa *rustling of trees,* Ælfc. Gr. 1; Som. 2, 35 : Greg. Dial. 1, 2.

bratt *A cloak;* pallium :—Forlēt hrægl' oððe bratt *remitte pallium,* Mt. Kmbl. Lind. 5, 40. [*Prov. Eng.* brat *a child's pinafore : Chauc.* bratt *a coarse mantle, rag : Wel.* brat *a rag : Gael.* brat *a mantle, apron, cloth.*]

breác *enjoyed,* Exon. 77 a; Th. 289, 7; Wand. 44; *p. of* brûcan.

BREÁD, breód, es; *n. A bit, fragment, morsel,* BREAD; buccella, panis :—Æfter ðæt breád *post buccellam,* Jn. Lind. War. 13, 27, 30. Hí ge-ēton ðæt breád *manducaverunt panem,* 6, 23. [*Chauc.* brede: *Wyc.* breed, brede: *Piers P.* breed: *R. Brun. R. Glouc.* brede: *Laym.* bred: *Orm.* brǽd: *Plat.* brood, *n: O. Sax.* brôd, *n: Frs.* braed, *n: O. Frs.* brad, *n: Dut. O. Dut.* brood. *n: Ger.* brot, *n: M. H. Ger.* brôt, *n: O. H. Ger.* brôt, *n: Dan. Swed.* bröd, *n: Icel.* brauð, *n.* Breád is first used in a compound word in Anglo-Saxon, v. beó-breád. It was first used as a separate word in the Lindisfarne Gospels, about A. D. 946-968, and breód in the Rushworth, John 13, 27, A. D. 901-1000. Breád and breód there signify *a morsel.* In John 6, 23, Lindisfarne and Rushworth, it signifies *bread,* panis.] DER. beó-breád.

breág *an eye-lid :*—Breága *palpebrarum,* Exon. 111 b; Th. 427, 32; Rä. 41, 100. v. bræw.

breahtm, brehtm, bearhtm, beorhtm, byrhtm, es; *m. A noise, tumult, sound, cry;* fragor, strepitus, tumultus, clamor, vociferatio :—Ðá wearþ breahtm hæfen *then a noise was raised,* Exon. 36 a; Th. 118, 1; Gû. 233. Breahtem stîgeþ *a tumult rises,* 83 b; Th. 314, 25, note; Môd. 19. Breahtmum hwurfon ymb ðæt hâte hûs hǽþne leóde *the heathen people surrounded that hot house with cries,* 55 a; Th. 195, 25; Az. 161: 57 b; Th. 206, 29; Ph. 134. [*O. Sax.* brahtum, braht, *m: M. H. Ger. O. H. Ger.* braht, *m: Dan.* brag, *n: Swed. Icel.* brak, *n.*] DER. brecan to *break.*

breahtm *a shining, moment, glance, an atom;* scintillatio, atomus :—Breahtm *atomus,* Cot. 36: 100. v. bearhtm *brightness.*

breahtum-hwǽt; *adj. Swift as the twinkling of an eye;* celer ut oculi nictus :—Ðec lígetta hērgen, blâce, breahtum-hwate *may the lightnings praise thee, pale, swift as the twinkling of an eye,* Exon. 54 b; Th. 192, 16; Az. 107. v. berhtm-hwǽt.

breard, es; *m. A brim, margin, rim, the highest part of anything :*—To brearde heofnes *ad summum cœli,* Mk. Lind. War. 13, 27. v. brerd.

breát *destroyed,* Beo. Th. 3430; B. 1713; *p. of* breótan.

breátan, ic breáte, ðú breátest, brýtst, he breáþ, brýt, *pl.* breátaþ; *p.* breót, *pl.* breóton; *pp.* breáten To *break, demolish, destroy, kill;* frangere, conterere, necare :—Hí hálge cwelmdon, breóton [breotun MS.] bôccræftige [bôccræftge MS.] bærndon gecorene *they slew the holy, destroyed the book-learned, burned the chosen,* Exon. 66 a; Th. 243, 25; Jul. 16. DER. a-breátan. v. breótan.

breáw, *pl.* bruwon *brewed; p. of* breówan.

breáw *an eye-lid,* Ælfc. Gl. 70; Som. 70, 63; Wrt. Voc. 42, 71 : Ps. Lamb. 131, 4. v. bræw.

breáw-ern, es; *n. A brewing-place, brew-house;* coquina cerevisiæ, Grm. ii. 338, 3 :—Breáwern *aporleriterium,* forte *apolyterium,* Ælfc. Gl. 55; Som. 67, 17.

brec, es; *n. A breaking, crash, noise;* fractio, fragor, strepitus. DER. ge-brec, bān-ge-, cumbol-ge-. v. brecan.

brēc *the breech, breeches,* L. M. 1, 71; Lchdm. ii. 146, 3 : R. Ben. 55; *acc. s. and nom. pl. of* brôc, *f.*

breca, an; *m. A breaker;* violator. DER. ǽw-breca, wiðer-. v. brecan.

BRECAN; ic brece, ðú bricest, bricst, he briceþ, bricþ, *pl.* brecaþ; *p.* ic, he bræc, ðú brǽce, *pl.* brǽcon; *pp.* brocen. I. *v. trans.* 1. to ˌBREAK, *burst, violate, break* or *burst through;* frangere, confringere, rumpere, perfringere, perrumpere :—Lēt se hearda Higeláces þegn brádne mēce brecan ofer bordweal *the fierce thane of Higelac caused his broad sword to break over the shield,* Beo. Th. 5952; B. 2980: Exon. 102 b; Th. 387, 10; Rä. 5, 3 : Andr. Kmbl. 1007; An. 504 : Salm. Kmbl. 202; Sal. 100. Hit þurh hróf wadeþ, briceþ boldgetimbru *it goeth through the roof, breaketh the timbers of the house,* 825; Sal. 412: Exon. 125 a; Th. 482, 8; Rä. 66, 4. Se Hǽlend bræc ða hláfas *Iesus fregit panes,* Mt. Bos. 14, 19: 15, 36: Beo. Th. 3027; B. 1511: 3138; B. 1567. Ne brǽcon hí nâ his sceancan *non fregerunt ejus crura,* Jn. Bos. 19, 33. Swâ swâ fæt tigelan ðú bricst hí *tanquam vas figuli confringes eos,* Ps. Spl. 2, 9. Seó wiht, gif hió gedýgeþ, dûna briceþ *the creature, if it escape, will burst the hills,* Exon. 109 b; Th. 420, 6; Rä. 39, 6. Him egsa becom ðá dēma duru in helle brǽc *dread came over them when the judge burst the doors in hell,* Cd. 221; Th. 288, 15; Sat. 381. Gif hie brecaþ his gebodscipe, he him abolgen wurþeþ *if they break* [*violate*] *his commandment, he will be incensed against them,* 22; Th. 28, 3; Gen. 430. Ðú mín bibod brǽce *thou didst break my commandment,* Exon. 28 a; Th. 85, 20; Cri. 1394. Bræc se here ðone friþ *the army broke* [*violated*] *the peace,* Chr. 911; Erl. 100, 16: 921; Erl. 106, 6. Heó Alwaldan bræc willan *she broke* [*violated*] *the Almighty's will,* Cd. 29; Th. 37, 34; Gen. 599. Yldran usse in oferhygdum ðín bibodu brǽcon *our forefathers in pride broke thy commandments,* Exon. 53 a; Th. 186, 13; Az. 19: Cd. 188; Th. 234, 28; Dan. 299. Gif hwâ his âþ brece, bēte swâ dômbōc tǽce *if any one break his oath, let him make amends as the doom-book may teach,* L. Ed. 8; Th. i. 164, 2. Ðæt ǽnig mon wǽre ne brǽce *that any man should not break the compact,* Beo. Th. 2205; B. 1100. Bióþ brocene âþsweord eorla *the oaths of the warriors will be broken,* 4132; B. 2063. He lǽteþ inwitflân brecan ðone burgweal *he lets the shafts of treachery break through the town-wall,* Exon. 83 b; Th. 315, 28; Môd. 38. Ic hwîlum ēðelfæsten brece *sometimes I break through a land-fastness,* Exon. 126 b; Th. 487, 4; Rä. 72, 23. Se storm and seó stronge lyft brecaþ brâde gesceaft *the storm and the strong blast shall break through the broad creation,* Exon. 22 b; Th. 61, 29; Cri. 992. Eádweard bræc ðone bordweall *Edward broke through the wall of shields,* Byrht. Th. 139, 60; By. 277. Brǽcon bordhreóðan [*they*] *broke through the wall of shields,* Elen. Kmbl. 243; El. 122. Leóht lyftedoras bræc *the light burst through the aerial dwellings,* Cd. 155; Th. 193, 24; Exod. 251. 2. *to press, force, urge;* urgere :—Lufian hine fyrwet bræc Iulianan *desire urged him to love Juliana,* Exon. 66 a; Th. 244, 14; Jul. 27 : Salm. Kmbl. 493; Sal. 247: Beo. Th. 470; B. 232: 5562; B. 2784. 3. *to rush into a place, take a place by storm;* in locum irrumpere, expugnare :—Siððan he for wlence beorgas brǽce *since he for pride rushed into the mountains,* Exon. 35 b; Th. 114, 29; Gû. 180. Cwom [MS. cuom] feorþe healf hund scipa on Temese mûþan, and brǽcon Contwara burg and Lundenburg *three hundred and fifty ships came to the mouth of the Thames, and took Canterbury and London by storm,* Chr. 851; Erl. 66, 34. II. *v. intrans.* 1. *to break* or *burst forth, make a noise* or *crash;* erumpere, crepare, fremere :—Geseah streám brecan of beorge [*he*] *saw a stream burst forth from the mount,* Beo. Th. 5085; B. 2546. Wæter wynsumu of ðære moldan tyrf brecaþ *pleasant waters burst forth from the turf of the earth,* Exon. 56 b; Th. 202, 9; Ph. 67. Swôgaþ windas, blâwaþ brecende, bearhtma mǽste *winds shall howl, crashing blow, with greatest of sounds,* Exon. 21 b; Th. 59, 11; Cri. 951. 2. *to sail;* navigare :—Scealtú ceól gestígan, and brecan ofer bæþweg *thou shalt ascend a ship, and sail over the sea* [lit. *bath-way*], Andr. Kmbl. 445; An. 223 : Elen. Kmbl. 487; El. 244. We brecaþ ofer bæþweg brimhengestum *we sail over the sea in ships* [lit. *sea-horses*], Andr. Kmbl. 1025; An. 513. III. *v. reflex. To retch;* screare :—Gebrǽd he hine seócne, and ongan hine brecan to spíwenne *he feigned himself sick, and began retching to spew,* Chr. 1003; Erl. 139, 9. [*Wyc.* breke, breek : *Piers* breken : *R. Glouc.* breke : *Laym.* breken : *Orm.* brekenn : *Plat.* broeken, breken : *O. Sax.* brekan : *Frs.* brekke : *O. Frs.* breka : *Dut.* breken : *Ger.* brechen : *M. H. Ger.* brëchen : *O. H. Ger.* brechan : *Goth.* brikan : *Dan.* bräkke : *Swed.* bråka, bräcka : *Icel.* braka *to creak.*] DER. a-brecan, be-, for-, ge-, ofer-, on-, to-, þurh-, ge-, -mǽlum, -ung ; ǽ-, ge-, bān-ge-, cumbol-ge- : breca, breoca, ǽw-, wiðer- : brece, hláf-ge- : brecendlíc, una- : brecþ, edor- : bræc, -côðu, -seóc, -seócnes ; ge-, fýr-ge-, hrǽc-ge-, neb-ge- : brice, bryce, ǽw-, aþ-, bān-, borh-, burh-, ciric-, cyric-, eodor-, fæsten-, freóls-, ful-, ge-, griþ-, hād-, hûs-, lah-, mund-, sām-, wed- : breahtm : broc, scip-ge-, un-.

brece, es; *n. A bit, morsel, piece;* frustum, buccella. DER. hláf-gebrece. v. brecan.

Brecenan-mere, es; *m.* [*Bd.* Britannemere : *Flor.* Bricenanmere : *Hunt.* Brecanammere : *Hovd.* Bricenamere] *Brecknock, the capital of*

Brecknockshire in South Wales; Brechinia. Gibson says,—Ad secundum circiter milliare a Brecknock in Wallia conspicitur Brecknockmere. Arx autem quam in nostris Annalibus Æthelfleda dicitur expugnasse, fuit, opinor, apud ipsum *Brecknock*, Chr. explicatio, p. 16, col. 1 :—Sende Æðelflǽd fyrd on Wealas, and abræc Brecenanmere *Æthelfled sent a force into Wales, and took Brecknock by storm*, Chr. 916; Th. 190, 35.

brecendlíc; adj. [brecende, *part.* of brecan *to break*, -líc] *Breakable*; fragilis. DER. un-abrecendlíc.

bréc-hrægl, -hrægl, es; *n.* [bréc *breeches, pl.* of bróc, *f.*; hrægl *a garment*] *A sort of garment;* lumbare, diplois = διπλοΐς :—Him sí abrogden, swá of bréchrægle [mid twýfealdum mentle, Spl.], hiora sylfra sceamu *operiantur* [*aperiantur?*] *sicut diploide confusione sua*, Ps. Th. 108, 28.

brec-mǽlum; adv. [brece *a bit, piece;* mǽlum, *dat. pl.* of mǽl, n.] *By bits, piecemeal;* minutatim, Mone B. 1819.

brecþ, e; *pl. nom.* brecþa; *f.* [brecan *to break*] *A broken state, fracture,* used figuratively *of mental contrition, grief;* fractio, ærumna :—Ðæt wæs wræc micel wine Scyldinga, módes brecþa *that was great wretchedness to the friend of the Scyldings, his mind's griefs,* Beo. Th. 344; B. 171. DER. edor-brecþ.

brecung, e; *f.* [brecan *frangere*] *A* BREAKING; fractio :—On brecunge breódes *in fractione panis*, Lk. Rush. War. 24, 35.

bred, es; *pl. nom. acc.* bredu; *n. A surface, plank, board, table, tablet;* superficies, tabula, tabella :—Ðisse eorþan ymbhwyrft is, wið ðone heofon to mettanne, swilce án lytel pricu on brádan brede *the circumference of this earth is, compared with the heaven, like a little point on a large surface,* Bt. 18, 1; Fox 62, 4. Breda þiling *vel* flór on to þerscenne *a joining of planks* or *a floor to thrash on;* area, Ælfc. Gl. 57; Som. 67, 73; Wrt. Voc. 37, 59. Hí bǽron anlícnysse Drihtnes on brede afægde and awritene *they bore the likeness of the Lord figured and drawn on a board;* ferebant imaginem Domini in tabula depictam, Bd. 1, 25; S. 487, 3. Lytle hús of bredan [=bredum] *small houses with tables, eating-houses, taverns;* tabernæ *vel* gurgustia, Ælfc. Gl. 55; Som. 67, 12; Wrt. Voc. 37, 7. Ic bær ða stǽnenan bredu, on ðám wæs ðæt wedd, ðe Drihten wið eów gecwæþ *acciperem tabulas lapideas, tabulas pacti, quod pepigit vobiscum Dominus,* Deut. 9, 9. [*Dut.* berd, *n : O. Dut.* bred, *n : Ger.* bret, brett, *n : M. H. Ger.* brët, *n : O. H. Ger.* bret, *n.*] DER. wex-bred.

bred *deceit,* L. Ed. 1; Th. i. 160, 6. v. bræd.

bréd *broad,* Chr. 189; Erl. 9, 25. v. brád.

bredan; ic brede, ðú britst, brist, he brit, bret, *pl.* bredaþ; *p.* bræd, *pl.* brudon; *pp.* broden, breden. I. *to weave,* BRAID, *knit, join together, draw, pluck;* plectere, nectere, vibrare, gladium stringere :—Ic brede nett *plecto,* Ælfc. Gl. 28, 5; Som. 32, 8. Ic brede me max *plecto mihi retia,* Coll. Monast. Th. 21, 13. Beadohrægl broden on breóstum læg *the armour* [lit. *war-garment*] *joined together lay on my breast,* Beo. Th. 1108; B. 552: 3100; B. 1548. Byrhtnóþ bræd bill of sceðe *Byrhtnoth drew his battle-axe from its sheath,* Byrht. Th. 136, 36; By. 162. Híg brudon up heora ancran *they drew up their anchors,* Chr. 1052; Erl. 184, 23. Sweord ǽr gemealt, forbarn broden mǽl, wæs ðæt blód to ðæs hát *the sword had already melted, the drawn brand was burnt, so hot was the blood,* Beo. Th. 3236; B. 1616. Se bræd of ðæm beorne blódigne gár *he plucked the bloody dart from the chief,* Byrht. Th. 136, 20; By. 154. II. *to change, vary, transform;* vertere, variare, transformare :—Simon bræd his hiw ætforan ðam cásere, swá ðæt he wearþ fǽrlíce geþuht cnapa, and eft hárwenge *Simon changed his appearance before the emperor, so that he suddenly seemed a boy, and again a hoary man,* Homl. Th. i. 376, 11. Hǽðen cild biþ gefullod, ac hit ne bret ná his hiw wiðútan, ðeáh ðe hit beó wiðinnan awend *a heathen child is baptized, but it varies not its aspect without, although it be changed within,* Homl. Th. ii. 268, 30. DER. a-bredan, æt-, for-, ge-, ofer-, on-, óþ-, to-, upa-, úta-, wið-. v. bredan.

brédan *to roast, broil, warm,* Cot. 86. v. brǽdan.

brédan *to make broad,* Bt. 18, 1; Rawl. 38, 33, MS. Cot. v. brǽdan.

bréd-búr *a bed-chamber,* Hymn Surt. 34, 30: 103, 17. v. brýd-búr.

bréden; adj. *Broad;* latus :—Seuerus geworhte weall of turfum, and brédenne [breden MS: bred weal, col. 1: bred weall, col. 2] ðǽr on ufon, fram sǽ to sǽ *Severus made a wall of turfs, and a broad wall thereupon, from sea to sea,* Chr. 189; Th. 15, 22, col. 3. v. brád.

bredende; adj. [*part.* of bredan] *Deceitful, cunning, crafty;* dolosus :—Sendon [MS. sendan] hí Marius, ðone consul, ongeán Geoweorþan, á swá lytigne, and á swá bredende, swá he wæs *they sent Marius, the consul, against Jugurtha, as he was always so cunning, and so crafty,* Ors. 5, 7; Bos. 106, 29; notes, p. 24.

bréd-guma *a bridegroom,* Mt. Kmbl. Hat. 9, 15. v. brýd-guma.

bréding-panne, an; *f.* [brédan *to roast,* panne *a pan*] *A frying-pan;* sartago, Wrt. Voc. 288, 38. v. brǽding-panne.

brédi-panne, bréding-panne, an; *f.* [brédan *to roast,* panne *a pan*] *A frying-pan;* sartago :—Brédipanne [MS. bredipannæ] *sartago,* Glos. Epnl. Recd. 11 30. Bréding-panne *sartago,* Wrt. Voc. 288, 38. v. brǽd-panne.

---

bred-ísern *a graving iron,* Glos. Epnl. Recd. 162, 28. v. brǽd-ísen.

brég *an eye-lid,* Ps. Surt. 131, 4: Bd. 4, 32; S. 611, note 18. v. brǽw.

brega; *m. A governor, ruler, prince;* imperator, princeps :—Ðá se brega mǽra geladade leóf weorud *when the great prince assembled the dear company,* Exon. 14 a; Th. 29, note 1; Cri. 456, note. v. brego.

brégan, brégean; *p.* de; *pp.* ed; *v. a.* [bróga *fear, terror*] *To give fear, frighten, make afraid, terrify, astonish;* terrere, pavefacere, stupefacere :—Hí sǽ-ýða swýðe brégaþ *the sea-waves greatly frighten them,* Runic pm. 21; Kmbl. 343, 24; Hick. Thes. i. 135. Ðeáh hí me swá brégdon, ne dorston hí me gehrínan *though they frightened me so, they durst not touch me,* Bd. 5, 12; S. 628, 45. Ne beó ge brégede fram ðám ðe ðone líchaman ofsleáþ *be ye not afraid of those who slay the body,* Lk. Bos. 12, 4: 21, 9. Hý hine brégdon *they terrified him,* Exon. 40 b; Th. 136, 4; Gú. 536. Ne biþ he bréged mid ænigum ógan *he will not be terrified with any dread,* Herb. 73, 2; Lchdm. i. 176, 4. We hí scylen manian and brégean *we should admonish and frighten them,* Past. 53, 8; Hat. MS. Sume wíf us brégdon *some women astonished us,* Lk. Bos. 24, 22. DER. a-brégan, ge-.

brégd, brégda *fear, terror, dread.* v. bróga, brégnes.

BRÉGDAN, bredan, ic bregde, ðú bregdest, he bregdeþ, *pl.* bregdaþ; *p.* brægd, *pl.* brugdon; *pp.* brogden, bregden. I. *v. a. To move to and fro, vibrate, cast, draw, drag, change, bend, weave;* vibrare, vibrare gladium, jactare, stringere, trahere, nectere, plectere :—Git mundum brugdon *ye vibrated with your hands,* Beo. Th. 1033; B. 514. Ðæt hie ne móste se synscaða bregdan *that the sinful spoiler might not draw them,* 1419; B. 707: Exon. 42 b; Th. 142, 23; Gú. 648. Ic underbæc bregde nebbe *I draw my face backwards,* Exon. 130 a; Th. 498, 6; Rä. 87, 8. Bócstafa brego bregdeþ feónd be ðam feaxe *the prince of letters shall draw the fiend by his hair,* Salm. Kmbl. 200; Sal. 99. Saga, hwá mec bregde of brimes fæðmum *say, who drew me from the bosom of the ocean,* Exon. 101 a; Th. 382, 18; Rä. 3, 13. Sǽ-rófe árum bregdaþ ýþbord [MS. yþborde] neáh *brave seamen draw the vessel near with oars,* 79 a; Th. 296, 26; Crä. 57. Brægd beadwe heard feorh-geníðlan *the fierce warrior dragged the mortal foe,* Beo. Th. 3082; B. 1539: 1593; B. 794. Brugdon hæleþ of scæðum sweord *the warriors drew their swords from their sheaths,* Cd. 93; Th. 120, 8; Gen. 1991: Judth. 11; Thw. 24, 38; Jud. 229. Nǽfre hie ðæs sellíce bleóum bregdaþ *let them never so strangely change with colours,* Salm. Kmbl. 301; Sal. 150. Bleóm bregdende *changing in colours,* Exon. 95 b; Th. 357, 3; Pa. 23. Sceal mǽg nealles inwitnet óðrum bregdan *a kinsman should not weave a net of treachery for another,* Beo. Th. 4341; B. 2167. Ic gefrægn sunu Wihstánes beran brogdne beadu-sercean *I heard that Wihstan's son bore his weaved war-sark,* 5503; B. 2755. Ðær wæs on eorle brogden byrne *there was on the man the woven mail-shirt,* Elen. Kmbl. 513; El. 257: Exon. 64 b; Th. 238, 11; Ph. 602. Bregden feðrum *woven with feathers,* 60 a; Th. 219, 13; Ph. 306: Ps. Th. 138, 9. II. *v. n. to turn into;* se vertere in aliquid :—Hí brugdon on wyrmes bleó *they turned into the hue of a worm,* Exon. 46 a; Th. 156, 30; Gú. 882. [*Wyc. R. Glouc.* breide: *Scot.* brade: *O. Sax.* bregdan: *O. Frs.* brida: *L. Ger.* breiden: *O. H. Ger.* brettan: *Icel.* bregða.] DER. a-bregdan, be-, for-, ge-, ofer-, on-, to-, upa-, úta-.

Bregent-ford *Brentford in Middlesex,* Chr. 1016; Th. 280, 28, col. 1. v. Brent-ford.

brégh *an eye-lid,* Bd. 4, 32; S. 611, 18. v. brǽw.

brég-nes, -less, e; *f.* [brégan *to give fear*] *Fear, terror, dread;* terror :—Brégnessa [MS. bregnes] díne hý gedréfdon me *terrores tui conturbaverunt me,* Ps. Spl. T. 87, 17.

BREGO, bregu, brega, breogo; *indecl. m.* A word chiefly used by poets, denoting *A leader, governor, ruler, prince, king, Lord;* imperator, princeps, rex, Dominus :—Se beorna brego *a leader of men,* Judth. 12; Thw. 25, 11; Jud. 254. Norþmanna bregu *the leader of North men,* Chr. 937; Erl. 112, 33; Æðelst. 33. Brego engla *the ruler of angels,* Cd. 9; Th. 12, 7; Gen. 181. Brego moncynnes *ruler of mankind,* Bt. Met. Fox 20, 86; Met. 20, 43. Babilóne brego *the king of Babylon,* Cd. 187; Th. 232, 6; Dan. 256. Se brega mǽra *the great prince,* Exon. 14 a; Th. 29, note 1; Cri. 456, note. Beorna breogo *the king of men,* Andr. Kmbl. 609; An. 305. [*Icel.* bragr, *m. vir primarius, princeps.*]

brego-ríce, es; *n.* [brego *a governor, ruler, king;* ríce *a region, kingdom*] *A kingdom;* regnum :—Se wæs Babylónes bregoríces fruma *he was the founder of the kingdom of Babylon,* Cd. 79; Th. 98, 21; Gen. 1633.

brego-stól, breogo-stól, es; *m.* [brego *a ruler, prince, king;* stól *a stool, seat, throne*] *A prince's stool* or *chair, a throne, a prince's dominion, kingdom;* principis sella, thronus, regnum :—He him gesealde bold and bregostól *he gave him a habitation and a princely seat,* Beo. Th. 4398; B. 2196: 4729; B. 2370. He hámes niósan lét ðone bregostól *he left the kingdom to visit his home,* 4767; B. 2389. Breogostól, Andr. Kmbl. 417; An. 209.

brego-weard, es; *m.* [brego *a ruler, prince;* weard *a guard, keeper*] *A royal guard, prince, lord;* princeps, dominus, Cd. 131; Th. 166, 13; Gen. 2747: 106; Th. 140, 26; Gen. 2333.

bregu *a leader, ruler, prince*, Chr. 937; Erl. 112, 33; Ædelst. 33. v. brego.

brēgyd *made afraid, frightened*, Lk. Foxe 12, 4,=brēged; *pp.* of brēgan.

brēhg *an eye-lid*, Bd. 4, 32; S. 611, 40. v. bráew.

brehtm, es; *m.* *A noise, tumult, sound, cry*; fragor, strepitus, tumultus, clamor :—Đa com hæleþa þreát..... weorodes brehtme *then came the troop of heroes..... with the tumult of a host*, Andr. Kmbl. 2544; An. 1273. v. breahtm *a noise*.

brehtnian *To make a noise* or *crackling*; crepare, Cot. 202.

brehtnung, e; *f.* *A noise, clattering, cracking*; crepitus, Cot. 49.

brēman; *part.* brēmende; *p.* de; *pp.* ed; *v. a.* [brēme *celebrated*] *To celebrate, solemnise, make famous, have in honour*; celebrare, honorare :—Đæt hie đæt hālige gerȳne brēman mǽgen *that they may celebrate the holy mystery* [i.e. *the sacrament*], L. E. I. 4; Th. ii. 404, 27. Á brēmende *ever celebrating*, Exon. 13 a; Th. 24, 20; Cri. 387. We đec brēmaþ *we celebrate thee, holy Lord, in our prayers*, hālig Drihten, gebēdum Cd. 192; Th. 241, 17; Dan. 406: Menol. Fox 186; Men. 94. Bodiaþ and brēmaþ beorhtne geleáfan *preach and make famous bright belief*, Exon. 14 b; Th. 30, 21; Cri. 483. DER. ge-brēman.

brēmbel *a bramble*, L. M. 2, 65; Lchdm. ii. 296, 23. v. brēmel.

brēmbel-æppel, es; *m.* *Bramble-fruit, blackberry*; rubi pomum, L. M. 1, 64; Lchdm. ii. 138, 26: 3, 41; Lchdm. ii. 334, 12.

brēmbel-rind, e; *f.* [brēmbel *a bramble*, rind *rind, bark*] *Bramble-rind*; rubi cortex :—Genim brēmbel-rinde *take bramble-rind*, L. M. 3, 47; Lchdm. ii. 338, 11. v. brēmel.

brēmber *a bramble*, Cd. 142; Th. 177, 12; Gen. 2928. v. brēmel.

brēmblas *brambles*, Homl. Th. i. 18, 17; *pl.* of brēmbel. v. brēmel.

BRĒME, brȳme; *def.* se brēma, seó, đæt brēme; *comp.* brēmra; *sup.* brēmest, brȳmust; *adj.* *Celebrated, renowned, illustrious, famous, notable*, BRIM, *glorious, esteemed*; celeber, clarus, illustris, famosus, notus, cognitus :—Og wæs brēme cyning on Basane *Og was a celebrated king in Basan*, Ps. Th. 135, 21: Menol. Fox 80; Men. 40. Đæt is heálic dæg, bēntīd brēmu *that is a high day, a celebrated time for supplication*, 148; Men. 75. Đis is anlícnes đæs brēmestan mid đām burgwarum in đære ceastre *this is the image of the most celebrated amongst the inhabitants in the city*, Andr. Kmbl. 1435; An. 718. Beówulf wæs brēme *Beowulf was renowned*, Beo. Th. 35; B. 18: Cd. 177; Th. 222, 13; Dan. 104. Đá wearþ se brēma on mōde blíđe *then was the illustrious one blithe in mind*, Judth. 10; Thw. 22, 10; Jud. 57. Ne hȳrde ic bisceop brēmran *I have not heard a more illustrious bishop*, Menol. Fox 205; Men. 104. Bēc syndon brēme *books are famous*, Salm. Kmbl. 473; Sal. 237. Salomon wæs brēmra, đeáh đe Saturnus sumra hæfde bóca cǽga *Salomon was the more famous, though Saturn had the keys of some books*, 366; Sal. 182. Fram gebyrdtíde brēmes Cyninges *from the birth-time of the glorious King* [*Christ*], Chr. 973; Erl. 124, 20; Edg. 12. Hí Rómána brȳmuste wǽron *they were the most esteemed of the Romans*, Ors. 2, 2; Bos. 41, 30. [*Northumb.* brōeme *clarus*.]

brēme; *adv.* *Famously, notably, gloriously*; famose, solemniter, gloriose :—Is his miht ofer middangeard brēme gebledsod *his might is gloriously blessed throughout the earth*, Andr. Kmbl. 3434; An. 1721.

BRĒMEL, brēmbel, brǽmbel, brēmber, es; *m.* *A* BRAMBLE, *brier, blackberry bush*; tribulus, vepres, rubus fruticosus, Lin :—*Herba rubus* [*erusti* MS.=*rubus fruticosus*], đæt is brēmel [brēmbel MS. H.] *the herb rubus, that is bramble*, Herb. cont. 89; Lchdm. i. 34, 21. Genim đás wyrte đe man brēmel [brǽmbel MS. H.] nemneþ *take this herb which a man calls bramble*, Herb. 89, 1; Lchdm. i. 192, 9. Brēmelas *vepres*, Wrt. Voc. 80, 23: Brēmlas *vepres*, Ælfc. Gr. 13; Som. 16, 15: Gl. 48; Som. 65, 52; Wrt. Voc. 33, 48. Abraham geseah ánne ramm betwux đām brēmelum be đām hornum gehæft *Abraham vidit arietem inter vepres hærentem cornibus*, Gen. 22, 13. Þornas and brēmelas heó asprīt đē *spinas et tribulos germinabit tibi*, 3, 18: Homl. Th. i. 432, 34. Wiđ ūtwærce, brēmbel đe sīen begen endas on eorþan *for dysentery, a bramble of which both ends are in the earth*, L. M. 2, 65; Lchdm. ii. 290, 30. Seó eorþe sylþ đē þornas and brēmblas *the earth shall give thee thorns and brambles*, Homl. Th. i. 18, 17. He rom geseah brēmbrum fæstne *he saw a ram fast in the brambles*, Cd. 142; Th. 177, 12; Gen. 2928. [*Chauc.* brember: *Wyc.* brembil, brimbil: *Plat.* brummel-beere, *f*: *Dut.* braam, *m.* *a bramble*; braam-bézie, *f.* *a blackberry*: *Kil.* braeme, breme *rubus*: *Ger.* brom-beere, *f.* *a blackberry*: *O. H. Ger.* brāma, *f*; brāmo, *m*; brāmal, *n*: *Dan.* brambær, *n*: *Swed.* brombär, *m.*] DER. heop-brēmel.

brēmel-æppel *bramble-fruit, blackberry*. v. brēmbel-æppel.

brēmel-berie *a bramble-berry*. v. brǽmel-berie.

brēmel-brǽr *a bramble-brier*. v. brǽmel-brǽr.

brēmel-leáf *the leaf of a bramble*. v. brǽmbel-leáf.

brēmel-rind *bramble-rind*. v. brēmbel-rind.

brēmel-þyrne, an; *f.* [brēmel *a bramble*, þyrne *a thorn*] *A bramble-thorn, bramble-bush*; rubus :—On middan ánre brēmelþyrnan *de medio rubi*, Ex. 3, 2, 4.

brēmen; *adj.* *Illustrious, glorious*; illustris, gloriosus :— Brēmen

Dryhten *the glorious Lord*, Exon. 54 b; Th. 193, 4; Az. 116: 55 a; Th. 194, 21; Az. 142. v. brēme.

Bremes burh; *gen.* burge; *dat.* byrig; *f.* BRAMSBURY or *Bramsby, Lincolnshire*; urbis *vel* arcis nomen in agro Lincolniensi :—Hēr, A. D. 909, Æđelflǽd getimbrode Bremes burh *in this year*, A. D. 909, Æthelfled built Bramsbury, Chr. 909; Th. 183, 30, col. 2. Hēr, A. D. 910, Æđelflǽd getimbrede đa burh æt Bremes byrig *in this year*, A. D. 910, Æthelfled built the fortress at Bramsbury, 910; Th. 184, 11, col. 2.

brēmlas *brambles*, Ælfc. Gr. 13; Som. 16, 15; *pl. nom.* of brēmel.

bremman; *part.* bremmende; *p.* de; *pp.* ed *To rage, roar*; rudere, fremere :— Bremman *rudere*, Cot. 192. Bremmende *rudens*, 192. Bremmde *fremuit*, Jn. Lind. War. 11, 33, 38. [*Frs.* brimje, brimme : *Dut.* brommen : *Kil.* bremmen : *Ger.* brummen : *M. H. Ger.* brimmen : *O. H. Ger.* breman : *Lat.* fremere : *Grk.* βρέμειν.]

brēmra *more illustrious*, Salm. Kmbl. 366; Sal. 182; *comp.* of brēme.

brencþ *brings*, Bt. Met. Fox 13, 120; *3rd pers. pres.* of brengan.

brenneþ *burns*, Runic pm. 15; Kmbl. 342, 11; Hick. Thes. i. 135,= berneþ; *3rd sing. pres.* of bernan.

brengan; ic brenge, đū brengest, brengst, he brengeþ, brengþ, brencþ, *pl.* brengaþ; *p.* ic, he brohte, đū brohtest, *pl.* brohton; *pp.* broht; *v. a.* *To bring, adduce, lead, produce, bear, carry*; ferre, afferre, offerre, proferre :—Đæt geár mōt brengan blōsman *the year may bring blossoms*, Bt. 7, 3; Fox 20, 22. He brengeþ æfter swegeltorht sunne *he brings after him the heavenly-bright sun*, Bt. Met. Fox 29, 46; Met. 29, 23. Eorþe sió cealde brengþ wæstma fela *the cold earth bringeth many fruits*, 20, 201; Met. 20, 101. Brencþ *brings*, 13, 120; Met. 13, 60. Wæter and eorþe wæstmas brengaþ *water and earth produce fruits*, 20, 150; Met. 20, 75. Nū scíneþ đe leóht, đæt ic from Gode brohte *now the light shineth, which I brought from God*, Cd. 29; Th. 38, 32; Gen. 615. Đū brohtest *thou broughtest*, Exon. 121 a; Th. 463, 34; Hö. 80: 121 a; Th. 464, 12; Hö. 86. Gabriel brohte *Gabriel brought*, Exon. 12 b; Th. 21, 18; Cri. 336: Cd. 156; Th. 194, 12; Exod. 259. Áras brohton *the messengers brought*, Elen. Kmbl. 1989; El. 996. Đa he hæfde ǽr him to wīfe broht *whom he had formerly married* [lit. *he had formerly taken to himself for a wife*], Bd. 3, 7; S. 529, 30. DER. ætgebrengan: forþ-brengan, ge-, ofer-, onge-, ongeán-.

brengnes, -ness, e; *f.* *An offering*; oblatio :—Onsægednissa and brengnesse đū nolde *sacrificia et oblationem noluisti*, Ps. Spl. T. 39, 9.

brenning *a burning*; crematio, Som. Lye. v. bærning.

Brent-ford, Bregent-ford, Brægent-ford; *gen.* -fordes; *dat.* -forde, -forda; *m.* [Brent *the river Brent*, ford *a ford* : Brenford, *Sim. Dun*: Brendeford, *Hunt.*] BRENTFORD *in Middlesex, situate where the river Brent flows into the Thames*; oppidum in agro Middlesexiæ, in sinu quodam ubi se in Tamesin effundit Brent fluvius :—Eádmund cyng fērde ofer Temese æt Brentforda *king Edmund went over the Thames at Brentford*, Chr. 1016; Th. 282, 4, col. 1: 281, 26, col. 1.

brenting, es; *m.* *A ship*; navis :—Hí brentingas ofer flōda genīpu feorran drífaþ *they drive ships from afar over the mists of floods*, Beo. Th. 5607; B. 2807.

breód *a bit, morsel, bread*, Jn. Rush. War. 13, 27. v. breád.

breodian; *p.* ode; *pp.* od *To cry out*; vociferari :—He breodaþ *he cries out*, Exon. 83 b; Th. 315, 8; Mōd. 28.

breodwian; ic breodwige, đū breodwast, he breodwaþ, *pl.* breodwiaþ; *p.* ode; *pp.* od *To prostrate*; prosternere ?—Beóþ đa gebolgne, đa đec breodwiaþ, tredaþ đec and tergaþ *they are enraged, they will prostrate thee, will tread and tear thee*, Exon. 36 b; Gl. 258. DER. a-bredwian.

breogo *a ruler, prince, king*, Andr. Kmbl. 609; An. 305. v. brego.

breogo-stōl *a throne, kingdom*, Andr. Kmbl. 417; An. 209. v. brego-stōl.

BREÓST, es; *n.* I. *the breast of man* or *beast*; pectus :—Đæt míne breóst wereþ *that defends my breast*, Beo. Th. 911; B. 453. On breóstum læg *lay on my breast*, 1109; B. 552. He beót his breóst *percutiebat pectus suum*, Lk. Bos. 18, 13. Blíđ on breóstum *mild in the breast* [*stomach*], Cd. 30; Th. 41, 13; Gen. 656. Đū gǽst on đínum breóste *super pectus tuum gradieris*, Gen. 3, 14. II. *the breasts*; ubera :—Đa breóst đe đū suce *ubera quæ suxisti*, Lk. Bos. 11, 27. Đa breóst đe ne sīcton *ubera quæ non lactaverunt*, 23, 29. Đǽr wearþ Alexander þurhscoten mid ánre flán underneođan ōđer breóst *there Alexander was shot through with an arrow underneath one breast*, Ors. 3, 9; Bos. 68, 27. III. *the breast as the seat of the vital powers, of the feelings, and of the affections, The heart, mind, thought*; pectus, cor, mens :—Drihtnes wæs bám on breóstum byrnende lufu *in both their breasts there was the burning love of the Lord*, Cd. 10; Th. 12, 25; Gen. 191. Hwæđre he in breóstum đa git hērede—in heortan—heofonríces weard *nevertheless he still in his breast—in his heart—honoured the guardian of heaven's kingdom*, Andr. Kmbl. 102; An. 51. Mæg đín mód wesan blíđe on breóstum *thy mind may be blithe in thy breast*, Cd. 35; Th. 46, 28; Gen. 751. Beoran on breóstum geþohtas *to bear in our breasts blithe thoughts*, 217; Th. 277, 17; Sat. 206. Adame innan breóstum his hyge hwyrfde *Adam within his breast changed*

his mind, 33; Th. 44, 27; Gen. 715. Ðū ūra breósta āna aspyrigend eart *tu nostrorum pectorum solus investigator es*, Hymn. Surt. 33, 21. Dēma ðū ǣtbist smēgan dǣda breóstes *judex aderis rimari facta pectoris*, 36, 20. Gefyll mid heofonlīcre gyfe ðe ðū gesceópe breóst *imple superna gratia quæ tu creasti pectora*, 92, 9. [*Chauc. Wyc.* brest : *R. Glouc.* breste : *Laym.* breoste : *Orm.* brest : *Plat.* borst, bost, *f* : *O. Sax.* briost, breost, *n* : *Frs.* boarst, *m. f* : *O. Frs.* brust : *Dut. Kil.* borst, *f* : *Ger. M. H. Ger. O. H. Ger.* brust, *f* : *Goth.* brusts, *f* : *Dan.* bryst, *n* : *Swed.* bröst, *n* : *Icel.* brjóst, *n*.] DER. byled-breóst, fōre-.

breóst-bán, es; *n.* [breóst *the breast*, bán *a bone*] The BREAST-BONE; *pectoris os, pectusculum,* Ælfc. Gl. 73; Som. 71, 25; Wrt. Voc. 44, 11.

breóst-bedern, es; *n. The breast-chamber, the inmost thoughts, the mind, the breast, chest;* pectoris conclave *vel* cubile, *i. e.* pectus intimum, thorax = θώραξ :—Fōran-bodig *vel* breóstbedern [MS. beden] *thorax* [MS. *torax*], Ælfc. Gl. 73; Som. 71, 26; Wrt. Voc. 44, 12.

breóst-beorh, -beorg, es; *m. A breast-defence, breast-plate;* pectoris tutamen. DER. breóst, beorg.

breóst-cearu, e; *f.* [breóst II. *the heart, mind,* cearu *care*] *The care of the heart, anxiety, grief, sorrow;* ægritudo, mæror :—Ic bitre breóstceare gebiden hæbbe *I have suffered bitter grief,* Exon. 81 b; Th. 306, 7; Seef. 4: 115 b; Th. 444, 9; Kl. 44.

breóst-côfa, an; *m.* [breóst *the breast, the heart, mind,* côfa *a cave, chamber*] *The breast-chamber, breast, heart, mind;* pectoris cubile, pectus, uber, cor, animus :—Under breóstcôfan *sub pectore,* Wanl. Catal. 48, 43. Ðū eart hiht mīn fram breóstcôfan mōdor mīnre *tu es spes mea ab uberibus matris meæ,* Ps. Lamb. 21, 10. He wæs ðe blīðra on breóstcôfan *he was the blither in his heart,* Bt. Met. Fox 9, 64; Met. 9, 32 : Cd. 27; Th. 36, 19; Gen. 574: Exon. 76 b; Th. 287, 22; Wand. 18.

breóst-gebeorh, -geborh; *gen.* -gebeorges; *m.* [breóst, gebeorh *a defence*] *A defence for the breast,* hence *a defence generally, bulwark, tower;* propugnaculum, Cot. 152.

breóst-gehygd, e; *f.:* es; *n.* [breóst II. *the heart, mind,* gehygd *thought, meditation*] *The thought of the heart* or *mind, a thought;* cordis *vel* animi cogitatio, cogitatio :—Ðæt wæs gingeste word breóstgehygdum *that was the last word from his mind's thoughts,* Beo. Th. 5628; B. 2818: Andr. Kmbl. 1994; An. 999.

breóst-geþanc, -geþonc, es; *m.* [breóst II. *the heart, mind,* geþanc *thought*] *The thought of the heart* or *mind, a thought;* cordis *vel* animi cogitatio, cogitatio :—Annanias ðec, and Adzarias and Misaél, Metod, dōmige, breóstgeþancum *Hananiah and Azariah and Mishael glorify thee, O God, in their minds' thoughts,* Cd. 192; Th. 241, 5; Dan. 400. Breóstgeþoncum, Exon. 80 b; Th. 302, 8; Fä. 33.

breóst-gewǣdu; *pl. n.* [breóst I. *the breast,* gewǣde *a garment, clothing*] *A covering for the breast, corselet;* pectoris vestimentum, lorica :—Gehwearf in Francna fǣðm feorh cyninges, breóstgewǣdu, and se beáh somod *the king's life fell into the power of the Franks, his corselet, and his collar also,* Beo. Th. 2426; B. 1211: Beo. Th. 4330; B. 2162.

breóst-hord, es; *n. m.* [breóst II. *the heart, mind,* hord *a hoard, treasure*] *The breast's treasure, the thought, mind, heart;* pectoris thesaurus, cogitatio, mens, cor :—Oþ-ðæt wordes ord breóst-hord þurhbræc *until the point* [or *issue*] *of the word broke through his mind,* Beo. Th. 5577; B. 2792. Him on ferhþe greów breóst-hord blōdreów *in his mind there grew a bloodthirsty thought,* Beo. Th. 3442; B. 1719 : Exon. 82 a; Th. 309, 10; Seef. 55.

breóst-hyge, es; *m.* [breóst, hyge, hige *the mind*] *The breast-thought;* pectoris cogitatio, Andr. Elen. Grm. xxxix. v. hyge, hige.

breóst-lín, e; *n.* [breóst, lín *linen*] *A breast-linen* or *bandage, breast-cloth;* pectoralis fascia, Cot. 89.

breóst-loca, an; *m.* [breóst, loca *an inclosure*] *The breast-inclosure, the mind;* pectoris clausura, mens :—Swefen he onfōn ne meahte in his breóstlocan *he could not contain the dream in his mind,* Cd. 180; Th. 226, 7; Dan. 167: Elen. Kmbl. 2498; El. 1250.

breóst-net, -nett, es; *n.* [breóst, net *a net*] *A breast-net, covering for the breast, breast-plate;* pectorale reticulatum, thorax :—Him on eaxle læg breóstnet broden *on his shoulder lay the braided breastplate.* Beo. Th. 3100; B. 1548: Cd. 154; Th. 192, 24; Exod. 236.

breóst-rocc, es; *m.* [breóst, rocc *clothing*] *Breast-cloth;* thorax :—Breóstrocc *thorax,* Cot. 163. Stīðe and ruge breóstroccas [MS. breóstrocces] *stiff and rough breast-clothes;* renones, Ælfc. Gl. 63; Som. 68, 114; Wrt. Voc. 40, 24.

breóst-sefa, an; *m.* [breóst *the breast,* sefa *the mind*] *The mind* or *heart in the breast, the mind, heart;* mens *vel* cor in pectore, mens, cor :—Árēred wearþ beornes breóstsefa *the mind of the man was exalted,* Elen. Kmbl. 1606; El. 805 : Exon. 15 b; Th. 34, 10; Cri. 540. Ic onsende in breóstsefan bitre geþoncas *I send into his mind bitter thoughts,* 71 b; Th. 266, 28; Jul. 405.

breóst-toga, an; *m. A breast-leader;* pectoris dux :—Sumra hæfde bald breóst-toga bôca cǣga *the bold chief had the keys of some books,* Salm. Kmbl. 369; Sal. 184.

breóst-wǣrc, es; *n?* *A breast-pain, the asthma, short windedness;*

---

pectoris dolor *vel* morbus, forsan asthma, Lye, = ἄσθμα *short breath, a panting.* v. wærc.

breóst-weall, es; *m.* [breóst, weall *a wall*] *A wall as high as the breast, a rampart, defence;* structura in muris ad pectus alta, munimentum, propugnaculum, Cot. 199.

breóst-weorþung, e; *f.* [breóst, weorþung *a honouring*] *A breast-decoration, an ornament;* pectoris decoratio, ornamentum :—Nalles he Fres-cyninge breóstweorþunge bringan môste *he could not bring the ornament to the Frisian king,* Beo. Th. 5001; B. 2504.

breóst-wylm, es; *m. The fountain of the breast, a breast, teat, emotion of the breast, grief;* pectoris fons, uber, pectoris æstuatio, ærumna :—Ðū eart hiht mīn fram breóstwylmum mōdor mīnre *tu es spes mea ab uberibus matris meæ,* Ps. Spl. 21, 8. He ðone breóstwylm forberan ne mihte *he could not restrain the emotion of his breast,* Beo. Th. 3758; B. 1877.

BREÓTAN; ic breóte, ðū breótest, breótst, brýtest, brýtst, he breóteþ, brýteþ, brýt, *pl.* breótaþ; *p.* ic, he breát, ðū brute, *pl.* bruton; *pp.* broten; *v. a. To bruise, break, demolish, destroy;* conterere :—Hergas breótaþ *break idols,* Exon. 14 b; Th. 30, 26; Cri. 485. Heremōd breát bolgen-mōd eaxlgesteallan *Heremod in angry mood destroyed his bosom friends,* Beo. Th. 3430; B. 1713. [*O. H. Ger.* bretôn *cædere : Dan.* bryde: *Swed.* bryta : *Icel.* brjóta.] DER. a-breótan. v. breátan.

Breoten, e; *f. Britain;* Britannia, Bd. 1, 17; S. 484, 26. v. Bryten.

breóðan; ic breóðe, ðū breóðest, brýst, he breóðeþ, brýþ, *pl.* breóðaþ; *p.* breáþ, *pl.* bruðon; *pp.* broðen *To ruin, destroy;* perdere. DER. a-breóðan. v. breótan.

Breoton *Britain,* Bd. 1, 1; S. 473, 8. v. Bryten.

breótun *destroyed,* Exon. 66 a; Th. 243, 25; Jul. 16, = breóton; *p. pl.* of breátan.

BREÓWAN; ic breówe, ðū breówest, brýwst, he breóweþ, brýwþ, *pl.* breówaþ; *p.* breáw, *pl.* bruwon; *pp.* browen, ge-browen *To* BREW; cerevisiam coquere :—Ne biþ ðǣr nænig ealo gebrowen mid Estum *there is no ale brewed by the Esthonians,* Ors. 1, 1; Bos. 22, 17. Ne dranc he nānes gemencgedes wǣtan, ne gebrowenes he drank not *of any mixed or brewed fluid,* Homl. Th. i. 352, 7. [*Dut.* brouwen : *Ger.* brauen : *M. H. Ger.* briuwen : *O. H. Ger.* briuwan : *Dan.* brygge : *Swed.* brygga : *Icel.* brugga.] DER. twy-browen.

BRÉR, es; *m. A* BRIER, *the bramble;* tribulus, rubus fruticosus :—Genim brēr ðe hiopan on weaxaþ *take a brier on which hips grow,* L. M. 1, 38; Lchdm. ii. 96, 15. Sindon burgtūnas brērum beweaxene [MS. beweaxne] *the city-dwellings are overgrown with briers,* Exon. 115 b; Th. 443, 17; Kl. 31. [*Chauc. Wyc.* brere : *Orm.* breress, *pl* : *Northumb.* breer, *m* : *Fr.* bruyère *heather* : *O. Fr.* bruière : *M. Lat.* bruarium *a heath, barren land rough with brambles and bushes,* Du Cange.] DER. brǣmbel-brǣr, hind-brēr.

BRERD, breord, breard, briord, es; *m. A brim, margin, rim, top of a pot* or *vessel, a shore, bank, brink;* labrum, ora, margo, summitas, summum :—Hīg gefyldon ða ðone brerd *impleverunt eas usque ad summum,* Jn. Bos. 2, 7. Ofer brūnne brerd *over the dark brim,* Exon. 107 a; Th. 408, 8; Rä. 27, 9. Brerd *vel* ôfer *crepido,* Ælfc. Gl. 98; Som. 76, 81; Wrt. Voc. 54, 25. Stæþ *vel* brerd *labrum, margo, vel crepido,* 106; Som. 78, 44; Wrt. Voc. 57, 25. To bearde heofnes *ad summum cæli,* Mk. Lind. War. 13, 27. [*Wyc.* brerde : *Laym.* breorde : *Orm.* brerd : *O. H. Ger.* brart, brort, *m. prora, ora, labrum, margo, limbus : Icel.* broddr, *m. a spike : Sansk.* bhṛishṭi, *f. a spike.*]

bresne; *adj. Strong, powerful, bold;* potens :—Ic his cynn gedô brād and bresne *I will make his race wide-spread and powerful,* Cd. 134; Th. 169, 17; Gen. 2801: 180; Th. 226, 18; Gen. 173. v. brǣsen II.

bret *varies, changes;* 3rd pres. of bredan :—Hǣðen cild biþ gefullod, ac hit ne bret nā his hiw widūtan, ðeáh ðe hit beó wiðinnan awend *a heathen child is baptized, but it varies not its aspect without, although it be changed within,* Homl. Th. ii. 268, 30. v. bredan II.

Bret-, **Bryt-** *a Welshman.* v. Bret-walas, Bret-walda, Bryt-land.

Breten *Britain,* Bt. Met. Fox 20, 197; Met. 20, 99. v. Bryten.

Bretenan-mere, es; *m. The British mere* or *lake, Welshpool, Montgomeryshire;* loci nomen apud Cambrenses, Som. v. Brecenan-mere.

brêþ *breath,* Wrt. Voc. 42, 58. v. brǣþ.

brêðer *to a brother;* fratri, Lk. Bos. 12, 13; *dat. of* brôðor.

Bret-land, es; *n. Britain* :—On Bretlande *in Britain,* Ors. 6, 30; Bos. 126, 2. v. Bret-, Bryt-land.

bretta, an; *m. A steward, lord, the Lord;* dispensator, dominus, Deus :—Lífes Bretta *Lord of life,* Ps. C. 50, 122; Ps. Grn. ii. 279, 122. v. brytta.

Brettas *Britons,* Chr. Th. 4, 4, col. 1; also *Bretons,* Chr. 890; Th. 160, 10, col. 1. v. Bryttas.

brettnere *a steward;* dispensator. v. brytnere.

Bret-walas; *pl. m. The Britons of Wales;* Walli :—Cynríc ða Bret-walas geflíémde *Cynric routed the Welsh,* Chr. 552; Th. 28, 39, col. 1.

Bret-walda, an; *m. A ruler of the Saxons in Britain, the chief Saxon king in England;* Saxonum in Britannia rex supremus. Turner and Lappenberg suppose that the Bretwalda was elected by the other Saxon kings and by the collected nobility and other electors in Britain, because

Hunt. lib. ii, about A. D. 1148, says, 'Omnia jura regni Anglorum, reges scilicet et proceres et tribunos in ditione sua tenebat:'—Ecgbryht wæs se eahteða cyning, se ðe Bretwalda wæs *Egbert was the eighth king, who was the Bretwalda*, Chr. 827; Th. 112, 21, col. 1.—There does not appear to be any historical evidence that the Bretwalda denoted any special title or office. The word is given in this alphabetical order because it occurs once in the Chronicle, and is thus written by historians; however, its more correct form appears to be brýten-walda, *q. v.*

bric- *a bridge* [= bricg], *found in the compound* bric-bót, *q. v.*

brica, an; *m. A breaker;* ruptor. DER. æw-breca, L. M. I. P. 16; Th. ii. 268, 30.

bric-bót, e; *f. A repairing* or *restoring of a bridge;* pontis restitutio *vel* instauratio :—Bricbóta aginne man georne *let a man diligently begin the repairings of bridges*, L. Eth. vi. 32; Th. i. 322, 31: v. 26; Th. i. 310, 24.

brice, bryce, es; *m.* [*from* briceþ, brycþ, *pres. of* brecan *to break*] *A breaking, rupture, fracture, fragment, violation, breach;* fractio, ruptura, fractura, fragmentum, violatio :—Híg hine oncneówon on hláfes brice *cognoverunt eum in fractione panis*, Lk. Bos. 24, 35. We witon ful georne, ðæt to miclan bryce sceal micel bót nýde *id compertum est nobis, immanis ubi facta est ruptura, ibi opus esse, ut large resarciatur*, Lupi Serm. i. 3; Hick. Thes. ii. 99, 30. Ne sý bānes bryce *let there not be a fracture of a bone*, Exon. 42 b; Th. 143, 32; Gū. 670. Gefēg ðǽs bricas to ānsúndnysse *join these fragments to soundness*, Homl. Th. i. 62, 7, 9. Hí gegaderodon ða bricas *they gathered the fragments*, i. 182, 22. Wǽron seofan spyrtan afyllede mid ðǽm bricum *seven baskets were filled with the fragments*, ii. 396, 9: i. 190, 4, 11. Ðæs borges bryce *a violation* or *infraction of the pledge* or *security*, L. Alf. pol. 3; Th. i. 62, 9, 10, 12. [*Plat.* bräk, *m.*: *Frs.* brek, *m. f.*: *O.Frs.* breke, *m. f.*: *Dut.* breuk, *f.*: *Dan.* bræk, brök: *Swed.* brak, *n.*: *Icel.* brek, *n. a fraudulent purchase of land :* like *Ger.* ge-brechen, *n. vitium ;* bruch, *m. a breaking, breach, from Ger.* brechen, *A. Sax.* brecan *to break.*] DER. æw-brice, -bryce, āþ-, bān-, borh-, burh-, ciric-, cyric-, eodor-, fæsten-, freóls-, ful-, ge-, griþ-, hād-, hús-, lah-, mund-, sām-, wed-.

bríce *use, service* :—God híg gesceóp eallum mannum to brícе *God created them for the use of all men*, Deut. 4, 19. v. brýce.

bríce; *adj. Useful;* utilis :—Dæg byþ eallum brícе *day is useful to all*, Runic pm. 24; Kmbl. 344, 14; Hick. Thes. i. 135. v. brýce.

brícest, he bríceþ *breakest, he breaks*, Exon. 63 a; Th. 232, 10; Ph. 504; *2nd and 3rd pers. pres. of* brecan.

bricg, e; *f. A bridge;* pons :—He hēt ða ofermetan bricge mid stāne gewyrcan *he ordered a very large bridge to be built with stone*, Ors. 2, 5; Bos. 48, 11. v. brycg.

Bricg, Brycg, e; *f.* [*Sim. Dun.* Brige: *Hovd.* Briges: *Matt. West.* Brigges.] **I.** *Bridgenorth in Shropshire;* oppidum in agro Salopiensi :—Æðelflǽd ða burh getimbrede æt Bricge *Æthelfled built the fortress at Bridgenorth*, Chr. 912; Th. 186, 10, col. 2; 187, 10, col. 1. **II.** *Bruges in Belgium;* Brugæ, Flandriæ emporium :—Heó com to Bricge begeondon sǽ *she came to Bruges beyond the sea*, Chr. 1037; Erl. 166, 7. Fērde Swegen ūt to Baldewines lande to Brycge *Sweyn went out to Baldwin's land to Bruges*, 1045; Erl. 170, 11: 1046; Erl. 175, 6: 1052; Erl. 181, 20: 1052; Erl. 182, 4.

bricg-bót, e; *f. A repairing of a bridge;* pontis instauratio :—Bricgbóta aginne *let the repairings of bridges be begun*, L. C. S. 10; Th. i. 380, 27. v. brycg-bót.

bricg-geweorc, es; *n.* BRIDGE-WORK, *the construction* or *reparation of a bridge;* pontis opus, pontis exstructio *vel* instauratio :—Brycg-geweorc, Heming. 104, Lye. *Turner's Hist. of A. S.* App. No. 4, c. 3, vol. ii. p. 539, 8vo. 1823. v. brycg-geweorc.

Bricg-stów, e; *f.* [Bricstowa, *Flor:* Brigestou, Bristou, *Hunt:* Brycstoue, *Sim. Dun:* Brikestow, Bristohw, *Hovd:* Bristow, *Kni:* brycg *a bridge*, stów *a place*] BRISTOL *in Gloucestershire and Somersetshire;* Bristova in finibus agrorum Glocestriensis et Somersetensis :—Híg fērdon to Bricgstōwe *they went to Bristol*, Chr. 1087; Erl. 224, 18.

bricg-weard, es; *m.* [bricg *a bridge*, weard *a keeper, guardian*] *A keeper* or *defender of a bridge;* pontis custos *vel* defensor :—Hí ðǽr bricgweardas bitere fundon *they found there the stern defenders of the bridge*, Byrht. Th. 134, 16; By. 85.

bricsian; *p.* ade *To profit;* prodesse, Bd. 5, 13; S. 632, 6. v. brýcsian.

bricst, he bricþ *thou shalt break, he shall break;* confringes, confringet, Ps. Spl. 2, 9; *2nd and 3rd pers. pres. and fut. of* brecan.

bricst *shalt eat;* edes, Gen. 3, 19; *pres. and fut. of* brūcan.

brid, bridd, es; *m. The young of any of the feathered tribe;* pullus :—Earnes brid *an eagle's young*, Exon. 59 a; Th. 214, 7; Ph. 235. þurh briddes hād *through the state of a young bird*, 61 a; Th. 224, 7; Ph. 372. ðæt híg offrunge sealdon twegen culfran briddas *ut darent hostiam duos columbæ pullos*, Lk. Bos. 2, 24: Lev. 1, 14: Ps. Spl. 83, 3. On swealwan bridda magan *in the maw of the young ones of a swallow*, L. M. 3, 1; Lchdm. ii. 306, 7. Hit sculon beón micle briddas *it should be big young ones*, L. M. 3, 1; Lchdm. ii. 306, 14. Hrefnes briddum

corvi pullis, Ps. Th. 146, 10. [*Chauc.* brid, bryd: *Wyc. Piers P.* brid: *Orm.* bridd: *O. Nrs.* burdr, *m. Rask*, burðr, *m. Vigf. partus.*]

bríd *a bride;* sponsa. v. brýd.

bríd-bletsung, e; *f. A marriage-blessing;* nuptialis benedictio :—Man ne mót sillan him brídbletsunge *they [priests] may not give them the marriage-blessing*, L. Ælf. P. 43; Th. ii. 382, 33.

bríd-būr *a bedchamber.* v. brýd-būr.

briddas *the young of any of the feathered tribe;* pulli. v. brid.

BRIDEL; *gen.* bridles; *m. A* BRIDLE; frenum :—Bridel *bagula?* Ælfc. Gl. 15; Som. 58, 46; Wrt. Voc. 21, 35. Bridles midl *a bridle's middle, a bit;* camus, 21; Som. 59, 61; Wrt. Voc. 23, 22: Runic pm. 21; Kmbl. 343, 26; Hick. Thes. i. 135. On hælftre and bridle ceácan heora gewríþ *in camo et freno maxillas eorum constringe*, Ps. Lamb. 31, 9. He ðǽne bridel of ateáh *he took the bridle off [his horse]*, Bd. 3, 9; S. 533, note 34. Se gemetgaþ ðone bridel *he regulates the bridle*, Bt. 36, 2; Fox 174, 18. Mid his bridle *with his bridle*, Bt. 21; Fox 74, 6: Bt. Met. Fox 11, 45, 57, 157; Met. 11, 23, 29, 79: 24, 73; Met. 24, 37. He ðæt gewealdleðer forlǽt ðara bridla *he shall let go the rein [lit. governing leather] of the bridles*, Bt. 21; Fox 74, 31: Bt. Met. Fox 11, 151; Met. 11, 76. Drihten welt eallra gesceafta mid ðām bridlum *his anwealdes the Lord governs all creatures with the bridles of his power*, Bt. 25; Fox 88, 3: Bt. Met. Fox 13, 5; Met. 13, 3. [*Chauc.* bridel, bridle: *Wyc.* brydil, bridel: *Dut.* breidel, *m:* Kil. breydel: *O. H. Ger.* brittil, *m. a bridle.*]

bridels, es; *m. A bridle;* frenum :—On bridels dôn *to put on a bridle*, Elen. Kmbl. 2348; El. 1175: 2367; El. 1185: 2396; El. 1199 v. bridel.

bridels-hring, es; *m. A bridle-ring;* in freno annulus :—Ðæs cyninges sceal mearh midlum geweorþod, bridelshringum *the king's horse shall be adorned with bits, with bridle-rings*, Elen. Kmbl. 2385; El. 1194.

bridel-þwangas; *pl. m. Bridle-thongs* or *reins;* freni :—Ic wyrce bridelþwangas [MS. bridel-þwancgas] *facio frenos*, Coll. Monast. Wrt. 9, 9.

bríd-gifu, e; *f.* [bríd = brýd *a bride*, gifu *a gift*] *A marriage-portion, dowry;* dos :—Ðeós brídgifu *hæc dos*, Ælfc. Gr. 9, 31; Som. 12, 1.

bridles *of a bridle*, Ælfc. Gl. 21; Som. 59, 61; Wrt. Voc. 23, 22; *gen. of* bridel.

bridlian; *p.* ode; *pp.* od [bridel *a bridle*] *To* BRIDLE, *curb, rule;* frenare. DER. ge-bridlian.

brig *a bridge*, Chr. 1125; Erl. 254, 19. v. bricg, brycg.

brigd, es; *n.* [bregdan *to change*] *A change, variety;* varietas :—Ðæs deóres hiw brigda gehwæs wundrum lixeþ *the animal's hue of every variety wondrously shines*, Exon. 95 b; Th. 357, 9; Pa. 26. [*Icel.* brigði, *n. a change.*]

briht *bright*, Lk. Hat. 11, 34, Lye. v. bryht, beorht.

brihtan; *p.* brihte; *pp.* brihted [briht = beorht *bright*] *To brighten;* illuminare. DER. ge-brihtan. v. beorhtian.

briht-líce; *adv. Clearly, brightly;* clare, splendide :—Ðæt he brihtlíce eall geseah *ut videret clare omnia*, Mk. Skt. Hat. 8, 25. v. beorht-líce.

BRIM, brym, es; *n. m. Surf, the sea, ocean, surface of the sea;* æstus aquæ, mare, pelagus = πέλαγος, æquor :—Brim sceal sealt weallan *the salt sea shall foam*, Menol. Fox 552; Gn. C. 45: Andr. Kmbl. 884; An. 442: 3147; An. 1576: Cd. 166; Th. 208, 2; Exod. 477: Exon. 95 b; Th. 356, 6; Pa. 7. Beáteþ [MS. beataþ] brim staðo [MS. stæðo] *the sea beats the shores*, Andr. Kmbl. 991; An. 496. Wæs brim blóde fáh *the sea's surface was stained with blood*, Beo. Th. 3192; B. 1594: 1699; B. 847. Ic of fǽðmum cwom brimes *I came from the bosom of the sea*, Exon. 103 b; Th. 392, 13; Rä. 11, 7: Andr. Kmbl. 884; An. 442: Beo. Th. 5599; B. 2803. On ðam brādan brime *on the broad ocean*, Exon. 55 a; Th. 194, 20; Az. 142: Elen. Kmbl. 505; El. 253: Menol. Fox 423; Men. 213. Brimo fædmaþ [MS. fædmeð] *in ceastra gehwǽre the seas surround [them] in every city*, Elen. Kmbl. 1941; El. 972. Ealle him brimu blódige þuhton *all the waters seemed bloody to them*, Cd. 170; Th. 214, 20; Exod. 572: Ps. Th. 106, 28: Beo. Th. 1145; B. 570. Cealde [MS. ceald] brymmas *cold seas*, Chr. 1065; Erl. 196, 31; Edw. 12. Engle and Sexe becōmon ofer brāde brimu *Angles and Saxons came over the broad seas*, Chr. 937; Th. 208, 5; Æðelst. 71: Andr. Kmbl. 1037; An. 519. [*Icel.* brim, *n. surf, the sea:* Sansk. bhram *to agitate, fluctuate.*]

brim-ceald, -cald; *adj.* [brim, ceald *cold*] *Cold as the water of the sea, ice-cold;* frigidus ut aqua maris, frigidissimus, gelidus :—Fēnix brimcald beorgeþ *the Phoenix tastes the ocean-cold [water]*, Exon. 57 b; Th. 205, 9; Ph. 110. Wæter wynsumu of ðære moldan tyrf brimcald brecaþ *pleasant waters, sea-cold, break forth from the turf of the earth*, 56 b; Th. 202, 9; Ph. 67.

brim-clif, es; *n.* [brim, clif *a cliff, rock*] *A sea-cliff;* marinus scopulus :—Ða liðende land gesáwon, brimclifu blícan, beorgas steápe *the voyagers saw land, the sea-cliffs shine, steep mountains*, Beo. Th. 449; B. 222.

brim-faroþ? es; *n.* [brim, faroþ *the shore*] *The sea-shore;* maris litus :—Bebūgaþ brādne hwyrft ôþ ðæt brimfaroþ [MS. brimfaro] *they*

shall inhabit the spacious orb unto the sea-shore, Cd. 190; Th. 236, 17; Dan. 322.

brim-flód, brym-flód, es; *m.* [brim, flód *a flowing, flood*] *The sea's flowing, the ocean-flood, sea*; maris fluctus, cataclysmus = κατακλυσμός, mare :—Heofonsteorran búgaþ brádne hwearft oþ brimflódas *the stars of heaven encircle the spacious orb unto the ocean floods*, Exon. 53 b; Th. 187, 30; Az. 38. Brymflód *cataclysmus*, Ælfc. Gl. 115; Som. 80, 45; Wrt. Voc. 61, 23: Cot. 50.

brim-fugel; *gen.* -fugles; *m.* [brim, fugel *a bird, fowl*] *A sea-fowl, sea-gull*; marina avis :—He gesihþ baðian brimfuglas *he sees sea-fowls bathe*, Exon. 77 a; Th. 289, 12; Wand. 47.

brim-gæst, -giest, es; *m.* [brim, gæst *a guest*] *A sea-guest, sailor*; marinus hospes, nauta :—Biþ hlúd brimgiesta breahtm *the sailors' noise is loud*, Exon. 101 b; Th. 384, 9; Rä. 4, 25.

brim-hengest, es; *m.* [brim, hengest *a horse*] *A sea-horse, ship*; marinus equus, navis :—Hí brimhengest bringeþ to lande *the ship brings them to land*, Runic pm. 16; Kmbl. 342, 19; Hick. Thes. i. 135. We brecaþ ofer bæþweg brimhengestum *we sail over the sea in ships*, Andr. Kmbl. 1026; An. 513.

brim-hlæst, e; *f.* [brim, hlæst *a burden*] *The sea's burden, fishes*; maris onus, pisces :—Brúcaþ brimhlæste and heofonfugla *enjoy fishes and fowls of heaven*, Cd. 10; Th. 13, 10; Gen. 200.

brim-lád, e; *f.* [brim, lád *a way, path*] *The path of the sea, sea-way*; maris via :—Ic in brimláde bídan sceolde *I must remain on the sea's path*, Exon. 81 b; Th. 307, 27; Seef. 30. Đe brimláde teáh *who came the sea-way*, Beo. Th. 2107; B. 1051.

brim-líðende; *part.* [brim, líðende; *part.* of líðan *to go, sail*] *Sea-faring*; per æquora navigans :—Se beót abeád brimlíðendra *he declared the threats of the sea-faring [men]*, Byrht. Th. 132, 37; By. 27. Hie ymb brontne ford brimlíðende ne letton *they have not hindered sea-faring [men] about the deep ford*, Beo. Th. 1141; B. 568.

brim-man, -mann, es; *m.* [brim, man *a man*] *A seaman, sailor*; nauta :—Brimmen wódon *the seamen proceeded*, Byrht. Th. 140, 29; By. 295. Brimmanna, *gen. pl.* 133, 12; By. 49.

brim-nesen, e; *f.* [brim, nesan *to be saved from*] *A safe sea-passage*; per æquora iter salvum :—Gif hie brimnesen settan mósten *if they should make a safe sea-passage*, Elen. Kmbl. 2006; El. 1004.

brim-rád, e; *f. The sea-road, the sea*; maris cursus, mare :—Geofon swaðrode, brimrád gebád *the ocean subsided, the sea-road stopped*, Andr. Kmbl. 3172; An. 1589: 2525; An. 1264.

brim-stréam, brym-stréam, es; *m.* [brim, stréam *a stream, river*]. **I.** *the sea's current, ocean-stream, the sea, ocean*; maris fluctus, mare, oceanus :—Ic on brimstreáme spræc worda worn *I spake many words on the ocean-stream*, Andr. Kmbl. 1806; An. 905. Beóton brimstreámas *the sea-streams dashed*, 477; An. 239. Ic eów ferian wille ofer brimstreámas *I will convey you over the seas*, 695; An. 348: Beo. Th. 3825; B. 1910. **II.** *a rapid stream, river*; fluvius rapidus, amnis :—Humbran eá, bráda brimstreám *Humber's river, broad rapid stream*, Chr. 942; Th. 208, 38, col. 1, 2, 3.

brim-þisa, an; *m.* -þise, an; *f.* [brim, -þisa, -þise *a noise*] *A ship*; navis :—He brimþisan æt sǽs faroþe sécan wolde *he would seek a ship on the sea-shore*, Andr. Kmbl. 3313; An. 1659. Léton ofer fífelwæg scríðan bronte brimþisan *they let the high ships go over the ocean*, Elen. Kmbl. 475; El. 238.

brim-wísa, an; *m.* [brim, wísa *a leader, guide*] *A sea-leader, leader of sailors*; per maris æstum dux, nautarum dux :—Abreót brimwísan, brýd aheorde *he slew the sea-leader, set free his bride*, Beo. Th. 5852; B. 2930.

brim-wudu; *m.* [brim, wudu *wood*] *Sea-wood, a ship*; maris lignum, navis :—Brimwudu scynde leóht to hýðe *the light ship hastened to the port*, Exon. 52 a; Th. 182, 5; Gú. 1305. Meahte gesión brecan ofer bæþweg brimwudu *he could see the ship sail over the sea*, Elen. Kmbl. 488; El. 244.

brim-wylf, e; *f.* [brim, wylf *a she-wolf*] *A sea-wolf*; marina lupa. An epithet applied to Grendel's mother :—Hine seó brimwylf abroten hæfde *the sea-wolf had destroyed him*, Beo. Th. 3202; B. 1599.

brim-wylm, es; *m.* [brim, wylm *æstus*] *The sea's surge*; maris æstus :—Brimwylm onféng hilde rince *the sea's surge received the man of war*, Beo. Th. 2993; B. 1494.

bring; *m.* [bringan *to bring*] *That which is brought, an offering, a sacrifice*; sacrificium, holocaustum :—Đú onféhst bringas *acceptabis holocausta*, Ps. Trin. Camb. 50, 20. DER. on-bring.

BRINGAN; *part.* bringende; ic bringe, bringce, ðú bringst, he bringeþ, bringcþ, bringþ, *pl.* bringaþ; *p.* ic, he brang, brong, ðú brunge, *pl.* brungon; *pp.* brungen; *v. a. To* BRING, *adduce, lead, produce, bear, carry*; ferre, adducere, ducere, producere, offerre, proferre :—Hwær is ðæt tiber, ðæt ðú bringan þencest *where is the gift which thou thinkest to bring?* Cd. 140; Th. 175, 7; Gen. 2891: Exon. 23 b; Th. 65, 23; Cri. 1059. Ic ðé þúsenda þegna bringe *I will bring thee thousands of warriors*, Beo. Th. 3663; B. 1829: Exon. 103 a; Th. 390, 22; Rä. 9, 5. Winter bringeþ weder ungemetcald *winter brings weather*

excessively cold, Bt. Met. Fox 11, 117; Met. 11, 59: 11, 125; Met. 11, 63. Regn wolcen bringceþ *a cloud brings rain*, Ps. Th. 67, 10. Seó eorþe westmas bringþ *the earth produces fruits*, Bt. 33, 4; Fox 130, 7. His bodan bringaþ *his angels bring*, Cd. 25; Th. 32, 28; Gen. 510: 231; Th. 286, 24; Sat. 357. Bring us hǽlo líf *bring us a life of health*, Exon. 10 a; Th. 10, 11; Cri. 150. He ða býsene from Gode brungen hæfde *he had brought the mandates from God*, Cd. 30; Th. 41, 4; Gen. 651: 176; Th. 221, 3; Dan. 82. [*Chauc. R. Brun. R. Glouc.* bringe: *O. Sax.* brengian, bringan: *Frs.* bringe: *O. Frs.* branga, bringa: *Dut.* brengen: *Kil.* brenghen: *Ger. M. H. Ger.* bringen: *O. H. Ger.* bringan: *Goth.* briggan.] DER. ge-bringan, onge-, to-, þurh-.

brinnan; *p.* bran, *pl.* brunnon; *pp.* brunnen *To burn*; ardere. DER. on-brinnan. v. beornan.

briord, es; *m. A brim, margin, rim, the highest part of anything*; labrum, ora, margo, summitas, summum :—Gefyldon ða to briorde *impleverunt eas ad summum*, Jn. Lind. War. 2, 7. v. brerd.

briosa; *m. A* BREESE, *gad-fly*; asilus, tabānus, Cot. 160; Wrt. Voc. 281, 32.

brist *supportest*; vehis; *for* birst, *2nd pres. s. of* beran *to bear, support* :—Đú birst [MS. brist] ealle þing búton geswince *thou supportest all things without labour*, Bt. 33, 4; Fox 132, 36.

bristl *a bristle*; seta. v. byrst.

brit *knits*; plectit. v. bredan.

Briten, Britten, e; *f. Britain*; Britannia :—Britene ígland ys eahta hund míla lang *the island of Britain is eight hundred miles long*, Chr. Th. 3, 1, col. 3. Brittene ígland *the island of Britain*, Chr. Th. 3, 1, col. 2. v. Bryten.

Brittas; *pl. m. The Britons*; Britones, Chr. Th. 3, 31, col. 2. v. Brytas, Bryttas.

brittian *to dispense* :—Gold brittade *dispensed gold*, Cd. 59; Th. 72, 4; Gen. 1181. v. bryttian.

Brittisc *British*, Chr. Erl. 3, 3; Th. 3, 5, col. 2. v. Bryttisc.

brittnere *a steward*; dispensator, Past. 63, Lye. v. brytnere.

BRÍW, es; *m. A thick pottage made of meal, pulse, etc*, BREWIS; puls; *gen.* pultis = πόλτος *porridge* :—Đes bríw *this pottage*; Ælfc. Gr. 9, 46; Som. 13, 9; Wrt. Voc. 290, 38. Swá þicce swá bríw *as thick as pottage*, L. M. 1, 36; Lchdm. ii. 88, 18: 2, 51; Lchdm. ii. 266, 25. Ete ðone bríw *let him eat the pottage*, 1, 36; Lchdm. ii. 88, 2: 2, 51; Lchdm. ii. 264, 19. Bríwas niman *pultes accipere*, Lchdm. iii. 210, 4. [*Plat.* brij, *m: Frs.* bry: *Dut.* brij, *m: Ger.* brei, *m: M. H. Ger.* brî, brîe, *m: O. H. Ger.* brî, brío, *m.*] DER. calwer-bríw.

bríwan; *p.* de; *pp.* ed *To cook, dress food*; coquere :—Bríw his mete wið ele *dress his meat with oil*, L. M. 2, 51; Lchdm. ii. 264, 22; 266, 29. v. breówan.

BROC, es; *m? A* BROCK, *badger*; taxo = tassus [= tasso *It:* taisson *Fr.*], meles :—Broc *taxo vel melus*, Wrt. Voc. 78, 4: Ælfc. Gl. 19; Som. 59, 10; Wrt. Voc. 22, 53. Sum fyðerféte nýten is, ðæt we nemnaþ taxonem, ðæt ys broc on Englisc *there is a four-footed animal, which we name taxonem, that is brock in English*, Med. ex Quadr. 1, 2; Lchdm. i. 326, 12. [*Wyc.* brok: *Laym.* brockes, *pl: Dan.* brok: *Icel.* brokkr, *m: Wel. Corn.* broch: *Ir.* broc, *m: Gael.* broc, bruic, *m: Manx* broc, *m: Armor.* broc'h, *m.*]

BRÓC; *gen.* bróce; *dat.* bréc; *acc.* bróc, bréc; *pl. nom. acc.* bréc, bræc; *gen.* bróca; *dat.* brócum; *f.*   **I.** *the* BREECH; nates :—Under ða bréc *under the breech*. L. M. 1, 71; Lchdm. ii. 146, 3.   **II.** *a covering for the breech, pl.* BREECHES, *trousers, pantaloons*; braca, bracæ, femoralia :—Bréc *femoralia*, R. Ben. 55. Bréc *femoralia*, Wrt. Voc. 81, 63. [*Chauc.* brech, *pl: Wyc.* brechis, *pl: Piers P.* brech, *pl: R. Brun.* breke, *pl: R. Glouc.* brych, *pl: Laym.* brechen, *dat. s;* breches, *pl: Scot.* breek, breik; *pl.* breeks, breiks: *Plat.* brook, broke, *f: Frs.* broek, *f. pudendorum tegumentum: O. Frs.* brok, *pl.* brek, *f: Dut.* broek, *f: Kil.* broecke *bracha: Ger.* bruch, *f. n. femorale: M. H. Ger.* bruoch, *f: O. H. Ger.* bruoh, bruoch, bróch, *n;* bruocha, *f: Dan.* brog, *c:* Swed. bracka, *f: Icel.* brók, *pl.* brækr, *f: Fr.* braie, *f: Span. Port.* braga: *Lat.* brācæ, *pl. f: Grk.* βράκαι, *pl. f: Ir.* broages: *Armor.* bragez, *m.*] DER. bréc-hrægel: wǽd-bréc.

bróc, es; *m.* [bróc, *perf. of* bracan *to break, purl, ripple*] *A* BROOK; latex, torrens :—Se bróc *the brook*, Bt. 6; Fox 14, 27. Burna oððe bróc *latex*, Wrt. Voc. 80, 69. Bróc *torrens*, Ælfc. Gl. 98; Som. 76, 78; Wrt. Voc. 54, 22. Bróc biþ onwended *the brook is turned aside*, Bt. Met. Fox 5, 38; Met. 5, 19. [*Laym.* broc: *Plat.* brook, *f: Ger.* bruch, *m. n. palus: M. H. Ger.* bruoch, *n: O. H. Ger.* bruoh.]

bróc, es; *pl.* brócu; *n:* bróc, gebróc, metaphorically, that which violently breaks from the body or mind; hence, *Affliction, misery, tribulation, trouble, labour, adversity, a disease, malady, sickness*; afflictio, miseria, tribulatio, labor, adversitas, morbus, ægritudo :—God nyle náu unaberendlíce bróc him ansettan *God wishes not to put on them any unbearable affliction*, Bt. 39, 10; Fox 228, 4. Mid heardum bróce *with severe [hard] affliction*, Bt. 39, 11; Fox 228, 25. He on ðǽm bróce nyle alǽtan ðás eorþlican wilnunga *in affliction he will not give up these earthly desires*, Past. 37, 3; Hat. MS. 50 a, 18, 21, 22: 36, 4; Hat. MS. 47 b, 7. On

ðám brócum *in these afflictions*, Th. Diplm. A. D. 880–885; 485, 24. Ðæt hit sý gefreód æghwylcere uneáþnesse ealles woroldlíces bróces *that it be freed from every annoyance of all worldly trouble*, 1061; 389, 30: 864; 125, 13: Past. 37, 3; Hat. MS. 50 a, 7. Ðæt biþ swíðe hefig bróc *it is a very severe labour*; gravis labor est, 61, 1; Hat. MS. Eucharius wæs þearle geswenct mid langsumum bróce *Eucharius was much afflicted with a protracted disease*, Homl. Th. ii. 24, 16: 176, 32. Brócu *miseriæ*, Lye. DER. ge-bróc.

bróc, es; *m?* [bróc, *p.* of bracan] *An inferior horse, a shaking horse, jade*; caballus, equus vilior :—Ðæt hie sécen him bróc on onráde, and on wǽne, oððe on ðon ðe hie â þrówian mǽgen *that they look for themselves to ride on a horse, and in a wain, or in that which they can ever endure*, L. M. 2, 6; Lchdm. ii. 184, 13. [*Chauc*. brok: *Icel*. brokkr, *m*.]

broccen *vel* gǽten roc, es; *m*. [broc *a badger*, gǽten *goaten*, *caprine*, roc *a garment*] *A garment made of badger or goat-skins, extending from the shoulders to the loins*; melotes, Ælfc. Gl. 63; Som. 68, 117; Wrt. Voc. 40, 27.

bróce *use*, Bd. 3, 22; Whelc. 221, 39, note B. C. v. brýce.

brocen *enjoyed*, = gebrocen, Exon. 38 b; Th. 127, 29; Gú. 393; *pp.* of brúcan, gebrúcan.

brocen *broken*, Beo. Th. 4132; B. 2063; *pp.* of brecan.

brócian; *part*. brócigende; ic brócie, ðú brócast, he brócaþ, *pl.* bróciaþ; *p.* ode; *pp.* ge-brócod; *v. a.* [bróc *affliction*] *To oppress, vex, afflict, break up, injure, blame*; opprimere, vexare, affligere, confringere, nocere, accusare :—Ic beóde ðæt hý nán man ne brócie *I command that no man oppress them*, Th. Diplm. A.D. 880–885; 492, 10. Ða manigfealdan yrmþa ða wérigan burh brócigende wǽron *manifold miseries afflicted [lit. were afflicting] the weary city*, Ors. 2, 4; Bos. 42, 36. Ða gebétan ðe hí bróciaþ *to amend those whom they afflict*, Bt. 39, 11; Fox 230, 8. Se synfulla biþ gebrócod *for his unrihtwísnysse the sinful is afflicted for his unrighteousness*, Homl. Th. i. 472, 3: 474, 19. Ðæt gebrócode flǽsc gelǽrþ ðæt upahæfene môd *the afflicted flesh teaches the proud mind*, Past. 36, 7; Hat. MS. 48 a, 22. We for ûrum synnum gebrócode beóþ *we are afflicted for our sins*, Homl. Th. i. 476, 19. Nǽfde se here Angelcyn gebrócod *the army had not broken up the English race*, Chr. 897; Erl. 94, 30. Hí gefeóllon of ânre upflóran and sume swíðe gebrócode wǽron *they fell from an upper floor and some were much injured*, 978; Erl. 127, 12. Gif ðé mon brócie for rihtre scylde, geþola hit wel *if a man blame thee for a just cause, bear it well*, Prov. Kmbl. 45. DER. wiðer-brócian.

bróc-líc; *adj. Sick, grieved, miserable*; æger. DER. bróc.

bróc-líce; *adv. Sickly, grievously*; ægre. DER. bróc.

bróc-minte, an; *f*. bróc-mint, e; *f*. BROOKMINT; horsemint; mentha sylvestris, Lin. Σισύμβριον sisymbrium officinale :—Brócminte. Genim ðysse wyrte wôs, ðe man sisymbrium, and ôðrum naman brócminte nemneþ *Brookmint. Take the juice of this plant, which men call* σισύμβριον, *and by another name, brookmint*, Herb. 107; Lchdm. i. 220, 17.

brócu *troubles*; *pl.* of bróc, es; *n*.

brócung, e; *f*. [bróc *affliction, sickness*] *Sickness*; ægritudo :—þurh his brócunge *through his sickness*, Homl. Th. i. 472, 7.

bród, e; *f*. I. *a growing together, congealing, waxing hard*; concretio, Cot. 55. II. *a* BROOD; proles. v. bródig. [*R. Glouc*. brod: *Scot*. brod: *Dut*. ge-broed, *n*: *Ger*. brut, *f. a brood*: *M. H. Ger*. bruot, *f*.]

bród; *adv. Freely, of free cost*; gratis :—Bród *gratis*, Wrt. Voc. 284, 71.

broddetan, brodettan *To tremble, quake, to pant for fear*; tremere, trepidare, palpitare, Greg. Dial. 2, 25; Cot. 154, Som. Lye.

broden *woven, braided*, Beo. Th. 1108; B. 552; *pp.* of bredan.

bróðer *a brother* :—Bróðer sune *a brother's son*, Ælfc. Gl. 91; Som. 75, 27; Wrt. Voc. 51, 71. v. bróðor.

brodetung, e; *f. A work, workmanship, fashion, forged tale, a lie*; figmentum :—He oncneów brodetunge [MS. brogdetunge] ûre *ipse cognovit figmentum nostrum*, Ps. Spl. C. 102, 13.

bródig; *adj.* BROODY, *brooding*; incubans :—Bródige henne *a broody hen*, Bridf.

broel, brogel, es; *n*. [corrupted from the *Mid. Lat*. brolium or briolium] *A park, warren stored with deer*; hence the BROYL, *a wood in Sussex, belonging to the Archbishop of Canterbury*; vivarium, hortus cervorum, Som. [*O. H. Ger*. brogil, broil.]

BRÓGA, an; *m. A prodigy, monster, trembling, fear, terror, horror, dread*; monstrum, tremor, terror, horror :—Ǽnig ôðer bróga *any other prodigy*, Bt. 36, 1; Fox 172, 17. Iówer ege and bróga sie ofer ealle eorþan nítenu *terror vester ac tremor sit super cuncta animalia terræ*, Past. 17, 2; Hat. MS. 22 a, 14. Brógan ðíne gedréfdon me *terrores tui conturbaverunt me*, Ps. Spl. 87, 17. Bûtan brógan *without dread*, Lev. 26, 6. Hine se bróga angeat *terror laid hold of him*, Beo. Th. 2587; B. 1291. Ne con he ðæs brógan dǽl *he knoweth not a portion of the terror*, Exon. 117 a; Th. 449, 15; Dôm. 71. Ðǽr is brógna [= brógena] hýhst *there is the greatest of terrors*, 116 a; Th. 446, 17; Dôm.

23. [*O. H. Ger*. brôgo, *m*.] DER. bryne-brôga, gryre-, here-, spere-, wæter-, wíte-.

brogden *woven, cast*, Elen. Kmbl. 513; El. 257; *pp.* of bregdan.

brogden-mǽl, es; *n*. [brogden, *pp.* of bregdan, mǽl *a spot, mark*] *Turned or marked with a spot or sign*; tortum *vel* curvatum signum :—Beofaþ brogden-mǽl *what is marked by signs [the sword] trembles or glitters*, Elen. Kmbl. 1514; El. 759.

brohte, ðú brohtest, *pl.* brohton; *pp.* broht *Brought, broughtest, brought*, Cd. 29; Th. 38, 32; Gen. 615: Exon. 121 a; Th. 463, 34; Hö. 80: Elen. Kmbl. 1989; El. 996: Bd. 3, 7; S. 529, 30; *p. and pp.* of brengan.

brôh-þreá; *m. f. n.* indecl. *but in dat. and inst. pl.* [brôh = brôg *terror*, þreá *calamitas*] *Terrific calamity*; calamitas terroris plena :—Ðæt brôh-þreá Cananêa wearþ cynne getenge *the terrific calamity was grievous to the Canaanites' race*, Cd. 86; Th. 108, 29; Gen. 1813. v. þreá.

BRÓM, es; *m. The well-known shrub from which besoms are made, hence* BROOM; genista :—Bróm *genista*, Ælfc. Gl. 46; Som. 64, 130; Wrt. Voc. 32, 64: L. M. 1, 55; Lchdm. ii. 126, 12: 1, 32; Lchdm. ii. 78, 19: Wrt. Voc. 80, 16: 285, 69. Genim brómes ahsan *take ashes of broom*, L. M. 1, 2; Lchdm. ii. 32, 12. [*Chauc. Wyc*. bromes, *pl*: *Dut*. brem, *f*: *Kil*. brem *genista*.]

Bróm-dún, e; *f*. [bróm *broom*, dún *a hill*] BRUMDON, *Dorset*; hodie opinor Brumdon in agro Dorsetensi :—Ðæt gemót wæs on Brómdúne *the meeting was at Brumdon*, L. Eth. iii. 4; Th. i. 294, 14: Cod. Dipl. 1322; A. D. 1035; Kmbl. vi. 186, 13, 14.

bróm-fæsten; *n*. [bróm *broom*, fæsten *an inclosed place*] *A broomfield, a field, close or wood of broom*; myricæ campus, myricetum, genesteium, Cot. 97.

brond *a fire-brand, fire, sword*, Exon. 74 a; Th. 277, 15; Jul. 581: Beo. Th. 6021; B. 3014: 2912; B. 1454. v. brand.

brond-hát *ardent*, Exon. 46 b; Th. 160, 2; Gú. 937. v. brand-hát.

brond-hord, es; *n*. [brand II. *a burning*, hord *a hoard, treasure*] *A burning or ardent treasure, a treasure exciting ardent desires*; ardens thesaurus :—Se ǽr in dæge wæs dýre, scríðeþ nú deóp feor, brondhord geblówen, breóstum in forgrówen *copper was dear in [that] day, now it circulates wide and far, an ardent treasure flourishing, grown up in the hearts*, Exon. 94 b; Th. 354, 15; Reim. 46.

Brondingas; *nom. acc*; *gen.* a; *dat.* um; *pl. m. The Brondings*, *supposed to be the inhabitants of the island Brännö, lying off the coast of West Gothland in the Cattegat*; populi nomen :—Breca gesôhte swǽsne êðel, lond Brondinga *Breca sought his own country, the land of the Brondings*, Beo. Th. 1047; B. 521. Breoca weóld Brondingum *Breca ruled the Brondings*, Scóp Th. 51; Wíd. 25.

brond-stæfn; *adj. The shining prowed*; proram spuma fulgentem habens :—Storm ne mæg brecan brondstæfne *a storm cannot break the shining [foaming] prowed [ship]*, Andr. Kmbl. 1007; An. 504.

brong *brought*; *p.* of bringan.

bront *high, deep, steep, difficult*, Beo. Th. 482; B. 238: 1140; B. 568: Elen. Kmbl. 475; El. 238. v. brant.

BRORD, es; *m? A prick or point, a lance, javelin, the first blade or spire of grass or corn, etc*; punctus, cuspis, frumenti spica, herba :—Brord *punctus*, Cot. 157. Ne furðan brordas *not even blades*; ne herbæ quidem, Bd. 4, 28; S. 605, 35. Brord *herba*, Mt. Lind. Rush. Stv. 13, 26. Ðæt brord *natum*, Lk. Lind. War. 8, 6. [*Orm*. brodd: *Dan*. brodd, brodde, *m. f*: *Swed*. brodd, *m*: *O. Nrs*. broddr, *m. aculeus, telum, frons aciei vel agminis*.]

brosnian; *part*. brosniende; ic brosnige, ðú brosnast, he brosnaþ, *pl.* brosniaþ; *p.* ode, ade; *pp.* od *To corrupt, decay, rot, perish*; corrumpi, deficere, dissolvi, perire :—Ðære fǽmnan líchoma brosnian ne mihte *the body of the maiden could not corrupt*; feminæ caro corrumpi non potuit, Bd. 4, 19; S. 587, 36. Him hyge brosnaþ *his mind corrupts*, Exon. 81 a; Th. 304, 11; Fä. 68. Brosnaþ enta geweorc, hrófas sind gehrorene *the work of giants is decaying, the roofs are fallen*, Exon. 124 a; Th. 476, 4; Ruin. 2: Beo. Th. 4512; B. 2260. Ða beámas â grêne stondaþ, nǽfre brosniaþ *the trees always stand green, never decay*, Exon. 56 a; Th. 200, 10; Ph. 38. Cristene Rôma besprycþ, ðæt hyre weallas for ealdunge brosnian *Christian Rome complains, that her walls decay with age*, Ors. 2, 4; Bos. 44, 45. Ðes brosnienda wêla *this perishing wealth*, Bt. 16, 1; Fox 50, 33. Brosnade burgsteal *the city-place has perished*, Exon. 124 a; Th. 477, 23; Ruin. 29. DER. ge-brosnod, unge-: brosniendlíc, brosnigendlíc, un-: brosnung, ge-, un-.

brosniend-líc, brosnigend-líc; *adj. Corruptible, perishable*; corruptibilis :—Ðæt wæter is brosniendlíc wǽta *water is a corruptible fluid*, Homl. Th. ii. 270, 5, 8, 13, 33. Geneálǽhþ ðam brosniendlícum wætere *he approaches the corruptible water*, ii. 270, 1. DER. un-brosniendlíc.

brosnung, e; *f. Corruption, decay*; corruptio, defectio :—Ic niðerastíge on brosnunge *descendo in corruptionem*, Ps. Lamb. 29, 10: Homl. Th. ii. 206, 2: 268, 35: 536, 20. Wæs ne wêlan brosnung *there was no decay of wealth*, Exon. 44 b; Th. 151, 25; Gú. 800. DER. ge-brosnung, un-.

brot, es; *n*. [broten; *pp.* of breótan *to break*] *A fragment*; fragmentum. [*Icel*. brot, *n*.] DER. ge-brot.

Broten *Britain*, Bd. 3, 29; S. 561, 15. v. Bryten.

broten *bruised, broken*; *pp. of* breótan.

BROÞ, es; *n.* BROTH; jus:—Broþ *jus*, Wrt. Voc. 82, 60. Fætt broþ ge mágon habban *pingue jus potestis habere*, Coll. Monast. Th. 29, 13. [*M. H. Ger. Bav.* brod, *n.: O. H. Ger.* bród, bröt, *n.*]

bróðar *a brother*, Th. Diplm. A. D. 830; 466, 3. v. bróðor.

bróðer *a brother* :—Ne ic hýrde wæs bróðer mínes *nor was I keeper of my brother*, Cd. 48; Th. 62, 2; Gen. 1008: Mt. Bos. 5, 24. v. bróðor.

BRÓÐOR, bróðer, bróðer, bróður; *d.* bréðer; *but often indecl. in sing*; *pl. nom. acc.* bróðor, bróðer, bróður, bróðru, bróðro, ge-bróðor, er, ru, ro, ra; *g.* bróðra, ge-bróðra; *d.* bróðrum, ge-bróðrum; *m. A* BROTHER; frater :—Úre bróðor *noster frater, nom. s*; *g.* úres bróðor *nostri fratris*; *dat.* úrum bréðer *nostro fratri*; *acc.* úrne bróðor *nostrum fratrem*; *voc.* eálá ðú úre bróðor *O noster frater* ! *abl.* fram úrum bréðer *a nostro fratre*: *pl. nom.* úre gebróðra *nostri fratres*; *g.* úra gebróðra *nostrorum fratrum*; *dat.* úrum gebróðrum *nostris fratribus*; *acc.* úre gebróðra *nostros fratres*; *abl.* fram úrum gebróðrum *a nostris fratribus*, Ælfc. Gr. 15; Som. 19, 18-23. Hwær is ðín bróðor *ubi est frater tuus?* Gen. 4, 9: Mt. Bos. 5, 23. Bróðor Arones *Aaron's brother*, Cd. 124; Th. 158, 21; Gen. 2620: 47; Th. 60, 19; Gen. 984. Geboren bróðer *germanus frater*, Greg. Dial. 2, 13. Ðínes bróðor blód clypaþ *fratris tui sanguis clamat*, Gen. 4, 10. His bróðor bearn *his brother's child*, Beo. Th. 5231; B. 2619. Sege mínum bréðer *dic fratri meo*, Lk. Bos. 12, 13. Cain gewearþ to ecg-banan ángan bréðer *Cain became a murderer to his only brother*, Beo. Th. 2529; B. 1262: Ps. Th. 34, 14: Mk. Bos. 12, 19. Bróðor þrý *the three brothers*, Cd. 94; Th. 122, 28; Gen. 2033. His bróðru fóron *fratres ejus ascenderunt*, Jn. Bos. 7, 10. His bróðro cwædon *fratres ejus dixerunt*, 7, 3. For míne bróðru *propter fratres meos*, Ps. Th. 121, 8. Ðe ne onfó bróðru and swustra *qui non accipiat fratres et sorores*, Mk. Bos. 10, 30. Hyre bróðra deáþ *the death of her brothers*, Exon. 100 a; Th. 377, 24; Deór. 8. Gemang bróðrum *inter fratres*, Jn. Bos. 21, 23. [*Plat.* broder, *m: O. Sax.* bróthar, *m: O. Frs.* bróther, broder, *m: Dut.* broeder, *m: M. H. Ger.* bruoder, *m: O. H. Ger.* bruodar, bróðar, *m: Goth.* broþar, *m: Dan. Swed.* broder, *m: O. Nrs.* bróðir, bróðir, *m: Lat.* frater, *m: Grk.* φράτηρ: *Ir.* brathair, *m: Wel.* brawd; *pl.* brodyr, *m: Sansk.* bhrátṛi, *from root* bhṛi [*A. Sax.* beran] *to bear, support, a brother being the natural supporter of sisters who have lost their father.*] DER. fæderen-bróðor, freó-, ge-, sige-. v. ge-bróðor.

bróðor-bana, an; *m. A brother-slayer, fratricide*; fratricida :—Ic monnes feorh seðe to bróðorbanan *I will avenge man's life on the fratricide*, Cd. 75; Th. 92, 9; Gen. 1526.

bróðor-cwealm, es; *m. Brother-murder, fratricide*; fratricidium :—Se me gemonige bróðorcwealmes *who shall remind me of my fratricide*, Cd. 49; Th. 63, 10; Gen. 1030.

bróðor-gefæðred *a brother by the same father*; frater ex eodem patre ortus, Ors. 3, 7; Bos. 60, 19. v. ge-fæðrian.

bróðor-gemédred *a brother by the same mother*; frater ex eadem matre ortus, Gen. Grn. 43, 29. v. ge-médrian.

bróðor-gyld, es; *n. Brother-retribution, vengeance for brothers*; fratrum cædis retributio :—On hyra bróðorgyld [bróðra gyld, *Thorpe*] *in vengeance for their brothers*, Cd. 153; Th. 190, 15; Exod. 199.

bróðor-leás; *adj.* BROTHERLESS; fratrem non habens, Exon. 129 a; Th. 496, 17; Rä. 85, 16.

bróðor-líc, bróðer-líc; *adj.* BROTHERLY; fraternus :—þurh ða bróðorlícan þingunge *per fraternam intercessionem*, Bd. 4, 22; S. 592, 21: Ælfc. Gr. 5; Som. 4, 57.

bróðor-lícnes, -nys, -nyss, e; *f.* BROTHERLINESS; fraternitas :—Ðín bróðorlícnys is on Mynstres reogolum getýd and geláered *tua fraternitas Monasterii regulis erudita est*, Bd. 1, 27; S. 489, 10.

bróðor-ræden, bróðer-ræðenn, e; *f. Brotherhood*; fraternitas, Ælfc. Gr. 5; Som. 5, 21.

bróðor-sib, -sibb, -syb, -sybb, e; *f.* **I.** *brotherhood, the relationship between brothers*; cognatio fraternalis, germanitas :—Syndon him on æðelum óðere twegen beornas, geborene bróðorsybbum [Kmbl. 1380, -sibbum] *to him in his family are other twain men, born in brotherly-relationship*, An. 690: Cot. 100. **II.** *brotherly love*; fraternus amor :—Hí bróðorsibbe georne bigongaþ *they earnestly cultivate brotherly love*, Exon. 44 b; Th. 150, 10; Gú. 776.

bróðor-slaga, an; *m. A* BROTHER-SLAYER; fratricida, Wrt. Voc. 85, 47. v. bróðor *a brother*, slaga *a slayer*.

bróðor-þínen, -þínenu, e; *f. A midwife at the birth of twin-brothers*; fratres geminos parturienti obstetrix, Gen. 38, 28.

bróðor-wyrt, e; *f.* BROTHER-WORT, *the herb pennyroyal*; mentha pulegium, Wrt. Voc. 68, 61.

bróður; *m. A brother*; frater :—His ágen bróður *his own brother*, Ps. Th. 107, 7: 132, 1. He geseh Iacobum Zebedei and Ioannem his bróður *vidit Iacobum Zebedæi et Ioannem fratrem ejus*, Mt. Bos. 4, 21. v. bróðor.

browen *brewed, cooked*; *pp. of* breówan. v. ge-browen, twy-.

BRÚ; *gen. dat. acc.* brúwe; *pl. nom. acc.* brúa, brúwa; *gen.* brúwena;

brúena, brúna; *dat.* brúwum; *f. A* BROW, *an eye-brow, eye-lash*; cilium, supercilium, tauto :—Brúa *cilia*, Ælfc. Gl. 70; Som. 70, 62; Wrt. Voc. 42, 70. Brúwa *cilium* [=*cilia*], Wrt. Voc. 64, 35: 282, 49. Brúwa *tautones*, Wrt. Voc. 64, 28. Ic eom wíde calu, ne ic breága ne brúna [=brúena] brúcan *móste I am very bald, nor can I make use of eye-lids nor eye-lashes*, Exon. 111 b; Th. 427, 32; Rä. 41, 100. Betweoh brúwum *intercilium* [=*intercilia*], Wrt. Voc. 64, 34: 282, 48. [*Wyc.* browe, brewe: *Laym.* breowe, bruwe, brouwe: *Prompt.* browe *supercilium*: *Scot.* bre, bree: *Plat.* brane: *Dut.* wenk-braaw, *f. the brow, eye-brow*: *O. Dut. Kil.* brauwe, brouwe, *f. cilium, supercilium*: *Ger.* braue, braune, *f. supercilium*: *M. H. Ger.* bráwe, *f: O. H. Ger.* bráwa, *f: Dan. Swed.* bryn, *f. n. a border, brink, eye-brow*: *Icel.* brún, *f. the eye-brow*: *Lat.* frons, *f. the forehead, brow*: *Grk.* ὀφρύς, *f. the eye-brow*: *Sansk.* bhrú, *f. an eye-brow, the brow.*] DER. ofer-brú. v. bræw.

BRÚCAN, to brúcanne; ic brúce, ðú brúcest, brýcst, brícst, he brúceþ, brýcþ, *pl.* brúcaþ; *p.* ic, he breác, ðú bruce, *pl.* brucon; *pp.* brocen; *v. a. gen. To use, make use of, to pass, spend, enjoy, have enjoyment of, to eat, bear, discharge*; utî, frui, possidere, habere, gaudere aliqua re, edere :—Ðæt he beáh-hordes brúcan móste *that he might have enjoyment of the ring-hoard*, Beo. Th. 1793; B. 894. Ne benohton beornas to brúcanne *needed not men to enjoy*, Andr. Kmbl. 2321; An. 1162. Sáwla móton lífes brúcan *souls may enjoy* [*have enjoyment of*] *life*, Andr. Kmbl. 458; An. 229. Brúceþ fódres *enjoys* [*has an enjoyment of*] *food*, Runic pm. 28; Kmbl. 345, 5; Hick. Thes. i. 135. Brúc ðisses beáges *make use of this collar*, Beo. Th. 2436; B. 1216. He giefstólas breác *he enjoyed gifts*, Exon. 77 a; Th. 289, 7; Wand. 44. Ðe hyra lífes þurh lust brucon [MS. brucan] *who have spent their life in pleasure*, Exon. 38 b; Th. 127, 19; Gú. 388. Ne brícst *usest not*, Deut. 28, 30. Ðú brícst ðínes hláfes *thou shalt eat of thy bread*, Gen. 3, 19. Brúcaþ, Jn. Bos. 4, 9. [*Piers P.* brouke: *Laym.* bruken: *Orm.* brukenn: *Plat.* bruken: *O. Sax.* brúkan: *Frs.* bruke: *O. Frs.* bruka: *Dut.* gebruiken: *Ger.* brauchen: *M. H. Ger.* brúchen: *O. H. Ger.* brúchan: *Goth.* brukyan: *Dan.* bruge: *Swed. Icel.* brúka.] DER. þurh-brúcan: ge-brúcan.

brúcing, e; *f. A function, an occupation, enjoyment*; functio, fruitio, occupatio, usus, Som. Lye. DER. brúcan.

brudon *spread*; dilatarunt, Cd. 154; Th. 191, 29; Exod. 222; *p. pl. of* bredan.

brugdon *laid hold of, drew*; strinxerunt, Cd. 93; Th. 120, 8; Gen. 1991; *p. pl. of* bregdan.

BRÚN; *adj.* BROWN, *dark, dusky*; fuscus, subniger, rufus, furvus :—Sum brún *part brown*, Exon. 60 a; Th. 218, 17; Ph. 296. Brúne leóde *brown people*; Æthiopes, Cd. 146; Th. 182, 4; Exod. 70. Sió brúne ýþ *the dusky wave*, Bt. Met. Fox 26, 58; Met. 26, 29. [*Chauc.* browne: *R. Glouc.* broune: *Frs.* brun: *O. Frs.* brun: *Dut.* bruin: *Ger.* braun: *M. H. Ger. O. H. Ger.* brún: *Dan.* braun: *Swed.* brun: *Icel.* brúnn.] DER. sealo-brún.

brúna *of eye-brows*, Exon. 111 b; Th. 427, 32; Rä. 41, 100, = brúena; *gen. pl. of* brú.

Brunan burh; *gen.* Brunan burge; *dat.* Brunan byrig; *f. Brunanburh, about five miles south-west of Durham, or on the plain between the river Tyne and the Browney, Dr. Guest properly writes 'round Brunanburh*;' v. example 1; Brunæ castellum. [Brunan burh is a pure Anglo-Saxon word, and signifies *the castle of Bruna*, though in a charter of Athelstan, dated 978, the year after the battle, it is called Bruninga feld, *the plain of the Brunings*, or *the descendants of Bruna*, as -ing denotes, v. -ing,—'Acta est hæc præfata donatio anno ab incarnatione Domini nostri Jesu Christi DCCCCXXXVIII, in quo anno bellum factum est in loco qui *Bruninga feld* dicitur, ubi Anglis victoria data est de cælo,' Th. Diplm. 186, 34-37; Cod. Dipl. 374; A. D. 938; Kmbl. ii. 210, 33-37. *Brunanburh* was written by *Ingulf*, in A. D. 1109, Brunford: *Hunt.* in 1148, Brumesburh, Brunesburih, Brunesburh, Bruneburh: *Hovd.* in 1204, Brunnanbyrg, Brumenburh: *Brom.* in 1330, Brunneburyh.] As the exact place cannot be determined by the name of any large town now existing, it is necessary to enter into the history of the battle, and thus ascertain its most probable locality.—Sihtric, king of Northumbria, which then extended from the Humber to the Frith of Forth [v. Angle], was son of Ingwar, and grandson of Ragnar Lodbrog. Sihtric was baptized and married Athelstan's sister in A. D. 925. He soon put away his wife, and renounced Christianity. Athelstan prepared to attack him for rejecting his sister, but Sihtric died, when Anlaf his son fled to Ireland, and Athelstan added Northumbria to his dominions. All the leaders of the Anglo-Danes and the Welsh were jealous of the increasing power of Athelstan, and combined against him. Anlaf, king of Dublin, commenced the fray by sailing from Ireland with 615 ships, containing about 100 men each, making more than 61,000 men: with this force he entered the Humber. He was joined by the Anglo-Danes, by the Welsh, and by Constantine, his father-in-law, the king of the Scots. Athelstan completely routed the immense army brought against him about Brunanburh, and became the first king of England. Alfred the Great was king over all the Anglo-Saxons, but by this complete victory Athelstan became

the undisputed king over all England [Engla land, q. v.]—The locality of Brunanburh has not yet been determined. It appears to me, it must be north of Beverley, as Athelstan is reported by Ingulf to have visited the tomb of St. John at Beverley, and to have placed his dagger on the altar, making a vow that if victory was granted to him, he would redeem it at a worthy price. The credibility of this story has been questioned; but, whatever doubt may remain, it proves that in the time of Ingulf, A. D. 1109, there was a general impression that Athelstan marched north of Beverley to oppose his invaders, and that, after the victory in the north, on returning to the south, he redeemed his pledge at Beverley by granting many privileges. Anlaf, collecting the remnant of his conquered army, could have no difficulty in returning to his ships in the Humber, as he had to pass through the country of the Anglo-Danes, his friends, and subjects of his late father.—Now all this history indicates that Anlaf marched north to unite his army with that of his father-in-law, Constantine, king of the Scots. Athelstan followed him, and their forces met about Brunanburh. I think it was on the west of Durham. I am led to this conclusion by these facts relating to the battle, and by the *Feodarium Prioratus Dunelmensis, published by the Surtees Society,* vol. lviii, in 1872. There is a plain between the rivers Wear and Browney [Brunan eá], and west of Durham, well adapted for a great battle. We find, in the present day, east and west Brandon [Brunan dún] and Brandon castle, the property of Viscount Boyne. There is still the river Browney [Brunan eá]. In the *Feod. Dunelmen.* compiled about A. D. 1430, we find the name of a river, of persons, and of places mentioned on the west of Durham. We have 'Ultra aquam de Wer usque ad aquam de Brun,' pref. p. lv: p. 192, note. 'De Brune,' 192, 193, note: 194, note. 'Petro de Brandone,' p. 180, note. 'Petrus de Brandone,' 200, note. On looking at the map of the learned Bishop Gibson, in his Anglo-Saxon Chronicle, 4to. 1692, I find he is of my opinion, that *Brunanburh* was north of Beverley. I cannot, however, discover why he places it to the north of Northumbria. For the reasons I have stated, I believe it was to the south-west of Durham.—Dr. Guest, Master of Caius College, Cambridge, in his excellent work, *A History of English Rhythms,* 8vo. 1838, gives the following account of this battle,—'In the year 937, was fought the battle of *Brunanburh*—a battle, that involved more important interests than any, that has ever yet been fought within this Island. It was indeed a battle between races.... Round the banner of Athelstan were ranged one hundred thousand Englishmen, and before them was the whole power of Scotland, of Wales, of Cumberland, and of Ireland under Anlaf, king of Dublin, led on by sixty thousand Northmen. The song, which celebrated the victory, is worthy of the effort that gained it. This song is found in all the copies of the Chronicle, but with considerable variations. Price collated three of them: *The Dunstan MS.* Tib. A. VI; the Abingdon, Tib. B. I; and the Worcester, Tib. B. IV. I have taken copies from all these MSS, and also from the Plegmund MS. in Ben'et Library. The Dunstan MS. appears to be by far the most correct transcript of the four. Price formed a text, so as best to suit the convenience of translation. The result might have been foreseen, and is such as little encourages imitation. I shall rather give the text, as it is found in *one* of these copies—*the Dunstan MS.* v. Chr. 937; Th. 200, col. 2. Not a word need be altered, to form either good sense or good poetry,' vol. ii. pp. 60, 61. In Mr. Earle's Chronicle, 8vo. 1865, p. 113, note x, are some excellent remarks on this song.—Dr. Guest has arranged the lines according to his system of Rhythm. I have arranged them according to the Anglo-Saxon punctuation, as in the article Beówulf. Dr. Guest's text is given within brackets, when the general orthography, of the word, seemed to require alteration:—

Hér, DCCCCXXXVII,	*Now,* A. D. 937,
Æðelstán cing,	*Athelstan king,*
eorla drihten,	*of earls the lord,*
beorna beág-gifa,	*of barons the bracelet-[beigh-] giver,*
and his bróðor eác,	*and his brother also [eke],*
Eádmund æðeling,	*Edmund the prince [etheling],*
ealdor langne tír	*elders a long train [tire]*
geslógan æt sæcce [sake],	*slew in battle,*
sweorda ecggum,	*with sword-edges,*
embe Brunan burh.	*round Brunanburh.*
Gst. Rthm. ii. 60, 26-62, 3.	
Ðǽr læg secg manig,	*There lay many a soldier,*
gárum forgrunden,—	*by the darts brought low,—*
guman norþerne,	*northern men,*
ofer scyld sceoten,	*over shield shot,*
swylce Scyttisc eác	*so also [eke] the Scotchman's*
wérig wigges sæd.	*wretched war-spawn.*
Gst. Rthm. ii. 64, 1-4.	
Fífe lágon	*Five lay*
on dæm campstede—	*on that battle-field [war-stead]—*
ciningas geonge	*youthful kings*
sweordum aswefede;	*sword-silenced;*

swilce seofone eác	*so also seven*
eorlas Ánláfes,	*earls of Anlaf,*
unrím herges—	*a host of the robber-band—*
flotan and Scotta.	*shipmen and Scots.*
Gst. Rthm. ii. 64, 14-18.	
Gewitan him ðá Norþmen	*Went [gan] then the Northmen*
nægled-cnearrum—	*in their nailed barks—*
[dreórig daroða láf	*[the darts' sad leavings*
on dynges mere]	*on the noisy sea]*
ofer deóp wæter,	*over deep water,*
Dyflen sécean	*Dublin [Dyflen]*
eft Iraland.	*Ireland [the land of the Ire] to seek once more.*
Gst. Rthm. ii. 66, 19-22.	
Ne wearþ wæl máre	*Was no greater carnage*
on ðisum [ðys] églande	*ever yet,*
ǽfre gyta, ...	*within this island, ...*
syððan eástan, hider	*since from the east, hither*
Engle and Sexan	*up came*
upp becóman.	*Angles and Saxons [Engle and Sexe].*
Gst. Rthm. ii. 68, 10-15.	

Hér, A. D. 937, Æðelstán cyning lǽdde fyrde to Brunan byrig *in this year,* A. D. 937, *king Athelstan led an army to Brunanburh,* Chr. 937; Th. 201, 25–27, col. 2. Hér, A. D. 937, Æðelstán [Ædestan MS.] cing and Eádmund his bróðer lǽdde fyrde to Brunan byrig [MS. Brunan byri]; and ðár gefeht wið Ánláfe [MS. Anelaf]; and, Criste fultumegende, sige hæfde *in this year,* A. D. 937, *king Athelstan and Edmund his brother led an army to Brunanburh; and there fought against Anlaf; and, Christ aiding, they had victory,* Chr. 937; Erl. 113, 2–4.

brún-basu, -baso; *adj.* [brún *brown,* basu *purple*] *Dark-purple, purple, purple-red, scarlet;* purpureus, ostriger, coccineus, puniceus:— Brúnbasere reádnysse *purpureo ostro,* Mone B. 6102. Brúnbasewum [MS. -bæsewum], reádum *purpureis,* 2087. Brúnbasum *purpureis,* 189. Brúnbaso *ostriger,* Cot. 145. Brúnbasne *coccineum,* Mone B. 6153. Ðý brúnan oððe ðý brúnbasewan *puniceo,* Cot. 183.

brún-ecg; *adj.* [brún *brown,* ecg *an edge*] *Brown-edged;* nigra acie præditus:—Byrhtnoþ brǽd bill of scéðe, brád and brúnecg *Byrhtnoth drew his battle-axe from its sheath, broad and brown of edge,* Byrht. Th. 136, 38; By. 163: Beo. Th. 3096; B. 1546.

brúnéða, an; *m.* A disease called brunella or pruna; morbus quidam, idem forte, qui Belgis *bruyne,* id est, Erysipelas [= ἐρυσίπελας] cerebri. Oris vitium, cum linguæ tumore, exasperatione, siccitate, et nigredine, vulgo, inquit *Kilianus,* brunella, *Som* :—Ðæt biþ strang sealf and gód wið swelcre abláwunge and brúnéðan, and wið ðara ceácna geswelle, oððe asmorunge *that is a strong salve and good for such inflation and brunella, and for swelling of the jaws, or smothering,* L. M. 1, 4; Lchdm. ii. 48, 10–12.

brún-fág; *adj.* [brún *brown,* fág *coloured, dyed*] *Of a brown colour, brown-hued;* fulvi coloris:—Ætbær brúnfágne helm *he bore away the brown-hued helmet,* Beo. Th. 5223; B. 2615.

brunge, *pl.* brungon; *pp.* brungen *broughtest, brought,* Cd. 30; Th. 41, 4; Gen. 651; *p. and pp. of* bringan.

brún-wann; *adj.* [brún *fuscus,* wan, wann *ater*] *Dark-brown, dusky;* fusco-ater:—Niht helmade brúnwann beorgas steápe *dusky night covered over the steep mountains,* Andr. Kmbl. 2613; An. 1308.

brún-wyrt, brúne-wyrt, e; *f.* I. BROWNWORT or *water-betony;* scrofularia aquatica:— Genim bánwyrt and brúnwyrt *take bonewort and brownwort,* L. M. 1, 25; Lchdm. ii. 66, 18. Brúne wyrt, 1, 61; Lchdm. ii. 132, 7. Genim brúne wyrt *take brownwort,* 2, 51; Lchdm. ii. 268, 9, 13: 1, 39; Lchdm. ii. 100, 5: 1, 48; Lchdm. ii. 122, 16. II. *wood-betony* or *brownwort;* scrofularia nodosa:—[Genim] ða brúnan wyrt brádleáfan, sió weaxeþ on wuda *take the broad-leafed brownwort, which grows in woods,* L. M. 1, 38; Lchdm. ii. 92, 23.

brute; *pl.* bruton *bruisedst, broke; p. of* breótan.

brúwa *brows, eye-brows,* Wrt. Voc. 64, 35, = brúa; *pl. nom. of* brú.

bryc *a bridge;* pons:—Ðæt he dó bryc-geweorc *that he do bridge-work,* L. R. S. 1; Th. i. 432, 2. v. brycg.

bryce *a violation, infraction,* L. Alf. pol. 3; Th. i. 62, 9. v. brice.

bryce; *adj.* [brycþ, *pres. of* brecan *to break*] *Breakable, worthless, frail, fleeting;* fragilis, futilis, caducus:—Mín bigengea gewát bryce on feorweg *incolatus meus prolongatus est,* Ps. Th. 119, 5. DER. un-bryce.

BRÝCE, bríce, es; *m.* [brýcst, brícst, *pres. of* brúcan *to use, enjoy*] *Use, service, the occupation* or *exercise of a thing, profit, advantage, fruit;* usus, ministerium, commodum:—Gif ðæt ówiht brýce wæs *if that was any use;* si hoc aliquid prodesset, Bd. 5, 14; S. 634, 8, note. Láfe on hwylc hugu fatu gehiwade wǽron mennisces brýces *recisuræ in vasa quælibet humani usus formarentur,* 3, 22; S. 552, 14. Bríce oððe gewuna *usus,* Ælfc. Gr. 11; Som. 15, 16. Ealle werþeóde lifgaþ bí ðám lissum, ðe éce Dryhten gesette sínum bearnum to bríce *all tribes of men live by the blessings, which the eternal Lord bestowed on his children for their use,* Exon. 54 b; Th. 193, 3; Az. 116. We sceoldon ða hwílendlícan þing to úrum brícum habban *we should have transitory things for*

K

our use, Homl. Th. ii. 460, 28. God híg gesceóp eallum mannum to bríce *quæ creavit Deus in ministerium cunctis gentibus,* Deut. 4, 19. Brýce *commodum,* Cot. 59. Lǽnes landes brýce *fructus,* Cot. 92. [Plat. bruuk: *Dut.* ge-bruik, *n: Kil.* bruyk: *Ger.* brauch, *m: O.H. Ger.* brûh, *m: Dan.* brug, *c: Swed.* bruk, *n: O. Nrs.* brûk, *n. usus, mos,* Rask Hald.]

brýce, bríce; *adj.* [brýcst, brícst, *pres. of* brúcan *to use*] *Useful, profitable;* utilis :—He monegum on Godes cyricum brýce wæs *multis in ecclesia utilis fuit,* Bd. 3, 23; S. 555, 33. He monegum brýce lifde *vitam multis utilem duxit,* 4, 26; S. 602, 41: Ps. Th. 118, 35. DER. un-brýce.

BRYCG, bricg, e; *f. A* BRIDGE; pons :—Ðeós brycg *hic pons,* Ælfc. Gr. 9, 39; Som. 12, 59. Ðæt he ne myhte tǽre brycge cuman *that he could not come to the bridge,* Ors. 2, 5; Bos. 48, 14. Eádweard cyning hêt gewyrcan ða brycge ofer Treontan *king Edward commanded the bridge over the Trent to be built,* Chr. 924; Erl. 110, 10: 887; Erl. 84, 30: 1071; Erl. 210, 17: Ors. 2, 5; Bos. 46, 7. [*Chauc.* brigge: *Piers P.* brugg: *R. Brun.* brigge: *R. Glouc.* brugg: *Plat.* brugge, brügge, *f: Frs.* bregge: *O. Frs.* bregge, brigge, *f: Dut.* brug, *f: Ger.* brücke, *f: M.H. Ger.* brucke, brücke, brügge, *f: O.H. Ger.* brucca, *f: Dan.* brygge, bro, *m. f: Swed.* brygga, bro, *f: Icel.* bryggja, brú, *f.*] DER. stân-bricg.

Brycg *Bruges in Belgium,* Chr. 1052; Erl. 182, 4. v. Bricg.

brycg-bót, bricg-bót, e; *f.* [brycg *a bridge,* bót *a repairing*] *A repairing* or *restoring of a bridge;* pontis restitutio *vel* instauratio :— Brycgbóta aginne man georne *let a man diligently begin the repairing of bridges,* L. C. S. 10; Th. i. 380, 27, note 65: 66; Th. i. 410, 8, note 11.

brycg-geweorc, es; *n.* BRIDGE-WORK; pontis opus :—Brycg-geweorc, Heming 104. v. bricg-geweorc.

brycgian; *p.* ade; *pp.* ad [brycg *a bridge*] *To bridge, bridge over, make a bridge;* pontem trajicere *vel* construere :—Sceal îs brycgian *ice shall bridge over* [*water*], Exon. 90 a; Th. 338, 4; Gn. Ex. 73. Ofer eástreámas îs brycgade *the ice bridged over the water-streams,* Andr. Kmbl. 2524; An. 1263. DER. ofer-brycgian.

Brycg-stów *Bristol,* Chr. 1052; Th. 314, 27. v. Bricg-stów.

brycg-weard *a keeper* or *defender of a bridge.* v. bricg-weard.

brýcian, brícsian; *p.* ode, ade; *pp.* od, ad [brýce, bríce *use*] *To be of use, profit, benefit, do good;* prodesse, proficuum esse :—He his geférum brýcian gýmde *he took care to do good to his companions,* Bd. 5, 9; S. 623, 33. Hî brýcaþ monigra hǽlo *multorum saluti proficuum erit,* Bd. 4, 22; S. 590, 32. Him sylfum brícsade *benefited himself,* Bd. 5, 13; S. 632, 6.

brýcþ, ðú brýcst *uses,* thou usest; 3rd and 2nd *pres. of* brúcan.

bryd, es; *n. A drawing, drawing out;* extractio :—Mid wǽpnes bryde *by the drawing of a weapon,* L. Alf. pol. 38; Th. i. 86, 16.

BRŸD, bríd, e; *f.* One owned or purchased,—*A* BRIDE, *woman about to be married* or *newly married, a wife, spouse, woman;* sponsa, nupta, uxor, mulier :—Seó geladung is gecweden Cristes brýd and clǽne mǽden *the church is called Christ's bride and a pure maiden,* Boutr. Scrd. 19, 39. Brýd *sponsa,* Ælfc. Gl. 87; Som. 74, 57; Wrt. Voc. 50, 39. Brýde láste *with the step of a bride,* Cd. 129; Th. 164, 15; Gen. 2715. Tyn fǽmnan fêrdon ongên ðone brýdguman and ða brýde *decem virgines exierunt obviam sponso et sponsæ,* Mt. Bos. 25, 1. Se ðe brýde hǽþ, se ys brýdguma *qui habet sponsam, sponsus est,* Jn. Bos. 3, 29. Ðá wæs Adames brýd gáste gegearwod *then Adam's bride was endued with soul,* Cd. 10; Th. 12, 16; Gen. 186. Him brýd sunu brohte *his wife brought to him a son,* Cd. 58; Th. 71, 16; Gen. 1171. Lothes brýd underbæc beseah *Lot's wife looked backwards,* 119; Th. 154, 27; Gen. 2562: Beo. Th. 4067; B. 2031. Adam ongan ôðres striénan bearnes be brýde, Cd. 55; Th. 68, 18; Gen. 1119: Th. 108, 28; Gen. 1813. Loth gelǽdde brýd mid bearnum in Sǽgor *Lot led his wife with their children into Zoar,* 118; Th. 153, 11; Gen. 2537: 129; Th. 164, 22; Gen. 2718. Nerôn his brýde ofslóg self mid sweorde *Nero himself slew his wife with a sword,* Bt. Met. Fox 9, 60; Met. 9, 30: Beo. Th. 5904; B. 2956: Cd. 125; Th. 159, 21; Gen. 2638. Him brýda twá eaforan fêddon *two wives brought forth offspring to him,* 52; Th. 65, 33; Gen. 1075. Feóllon wergend brýda, bennum seóce *the defenders of the wives fell, sick with wounds,* 92; Th. 118, 28; Gen. 1972. God me ðæs brýd forgeaf *God gave me this woman,* 26; Th. 33, 27; Gen. 526. [*Piers P.* burde: *Laym.* brude: *Orm.* brid: *O. Sax.* brûd: *Frs. O. Frs.* breid: *Dut.* bruid: *Ger.* braut: *M. H. Ger. O. H. Ger.* brût: *Goth.* bruþs νύμφη *nurus: Dan. Swed.* brud: *Icel.* brúðr.] v. wíf.

orýd-bed, es; *n. A bride-bed;* genialis torus, Ælfc. Gl. 66; Som. 69, 72; Wrt. Voc. 41, 26.

brýd-bletsung, e; *f. A bride's blessing;* nuptialis benedictio. v. bríd-bletsung.

brýd-búr, es; *n. A bedchamber;* thalamus :—Of brýdbúre his *de thalamo suo,* Ps. Spl. 18, 5: Beo. Th. 1846; B. 921.

brýd-ealo, -eala; *gen.* -ealowes; *n.* [ealu *ale*] *A bride-ale, bride* or *marriage feast;* nuptiale convivium :—Ðær wæs ðæt brýdealo [Laud.

MS. -eala], ðæt wæs manegra manna bealo *there was the bride-ale, which was many men's bale,* Chr. 1076; Erl. 213, 26.

brýd-ealoþ; *indecl. n.* [ealaþ *ale*] *A bride-ale, bride* or *marriage feast;* nuptiale convivium :—Hí wǽron æt ðam brýdealoþ *they were at the marriage feast,* Chr. 1075; Erl. 214, 15.

brydel; *gen.* brydles; *m. A bridle;* frenum, lupatum :—Brydel *bagula, salivare,* Ælfc. Gl. 21; Wrt. Voc. 23, 23. v. bridel.

brýde *with conjugal footstep,* Cd. 129; Th. 164, 16; Gen. 2715. v. brýd, lást.

brýdelíc gewrit, es; *n. A bride-like writing, a play;* drama, Cot. 66.

brýdel-þwang, -twancg, es; *m. A bridle rein;* frenum. v. bridel-þwang, -twancg, Coll. Monast. Th. 27, 35.

brýden wah *a broad wall,* Bd. Whelc. 1, 8; p. 48, 27. v. brêden, wah *a wall.*

brýd-gifa *espousals;* sponsalia, Ælfc. Gl. 87; Som. 74, 53. v. bríd-gifu.

brýd-guma, brýdi-guma, an; *m.* [brýd, guma *a man*] *A bride-man, bridegroom;* sponsus :—Swá swá brýdguma of his brýdbúre *tamquam sponsus procedens de thalamo suo,* Ps. Th. 18, 5. Cweðe ge sceolun ðæs brýdguman cnihtas wêpan, ða hwíle ðe se brýdguma mid him byþ *numquid possunt filii sponsi lugere quamdiu cum illis est sponsus?* Mt. Bos. 9, 15: 25, 1. Se ðe brýde hǽfþ, se ys brýdguma *qui habet sponsam, sponsus est,* Jn. Bos. 3, 29: Ælfc. Gl. 87; Som. 74, 55.

brýdi-guma *a bridegroom,* Ælfc. Gl. 87; Som. 74, 55. v. brýd-guma.

brýd-lác, es; *n. A marriage gift* or *feast, the celebration of a marriage;* nuptiale offertorium, nuptiarum celebritates :—Ne nán preóst môt beón æt ðam brýdlácum áhwǽr ðǽr man eft wífaþ, oððe wíf eft ceorlaþ *nor may any priest be at the celebration of a marriage anywhere where a man marries a second wife, or a woman marries again,* L. Ælfc. C. 9; Th. ii. 346, 18.

brýd-leóþ, es; *n. A marriage song;* epithalamium = ἐπιθαλάμιον, Mone B. 3121: 3123.

brýd-líc; *adj. Bridal;* nuptialis :—Reáf brýdlíc *vestem nuptialem,* Mt. Lind. Stv. 22, 12.

brýd-loca, an; *m.* [loca *a place shut in*] *A bride-chamber;* sponsæ cubile :—On ðæm brýdlocan *in the bride-chamber,* Homl. Blick. 9, 10.

brýd-lufe, an; *f.* [lufe *love, favour*] *A bride's love;* sponsæ amor :— He ða brýdlufan sceal sécan *he must seek a bride's love,* Exon. 67 b; Th. 249, 20; Jul. 114.

brýd-ræst *a bride-bed;* genialis lectus, Cot. 99. v. brýd-bed.

brýd-reáf, es; *n. A nuptial garment;* nuptialis vestis :—Mid brýdreáf *veste nuptiali,* Mt. Lind. Stv. 22, 11.

brýd-sang, es; *m. A marriage song;* hymenæus = ὑμεναῖος, epithalamium = ἐπιθαλάμιον, Ælfc. Gl. 87; Som. 62, 40; Wrt. Voc. 28, 22.

brýd-þing, es; *n. A bride-thing, what relates to marriage, in pl. nuptials;* nuptiæ :—Gabriel wæs ðissa brýdþinga ǽrendwreca *Gabriel was the messenger of these nuptials,* Homl. Blick. 3, 13.

brydyls *a bridle,* Ps. Spl. C. 31, 12. v. bridels.

brygc *a bridge,* Wrt. Voc. 80, 50. v. brycg.

brygdan, he brygdeþ *To turn;* vertere :—He hálge láre brygdeþ on bysmer *he turneth holy lore to mockery,* Exon. 117 a; Th. 449, 14; Dôm. 71. DER. on-brygdan. v. bregdan.

bryht *bright,* Ps. Spl. T. 15, 6. v. beorht.

bryhtm *a glance* :—Eágan bryhtm *an eye's glance, a moment,* Bd. 2, 13; S. 516, 20. v. bearhtm.

bryidan; *p.* ede; *pp.* ed *To take;* tollere, sumere :—Ðæs áþ ðe his ǽhte bryideþ *the oath of him who takes* [Th. *discovers*] *his property,* L. O. 4; Th. i. 180, 8. v. bregdan. DER. æt-bryidan, ge-.

brym *the sea,* Cd. 100; Th. 132, 12; Gen. 2192: Chr. 1065; Erl. 196, 31; Edw. 12. v. brim.

brýme *famous,* Ors. 2, 2; Bos. 41, 30. v. brême.

brym-flód *a deluge,* Ælfc. Gl. 115; Som. 80, 45; Wrt. Voc. 61, 23. v. brim-flód.

brymme, es; *m. A* BRIM, *brink, an edge, a border, lip of a pot, and such like;* ora, margo :—Brymmas sǽs *the borders* or *shores of the sea, a strait,* Hymn. Lye. [*Chauc.* brimme: *Laym.* brimme, *dat: Kil.* breme: *Ger.* bram, *n;* bräme, *f. margo, fimbria.*]

brym-streám *the sea, a river,* Mt. Rush. Stv. 8, 18: Chr. 942; Th. 209, 38, col. 1; Edm. 5. v. brim-streám.

brýmuste *most famous,* Ors. 2, 2; Bos. 41, 30. v. brême.

bryne, byrne, es; *m.* [byrnan *to burn*] *A burning, fire, flame, heat;* ustio, ardor, incendium, ignis, flamma, fervor :—Ne se bryne beót mǽcgum *the burning did not hurt the youths,* Cd. 187; Th. 232, 24; Dan. 265: Exon. 59 a; Th. 213, 24; Ph. 229: 53 b; Th. 189, 9; Az. 57. Mid ðý me of sweoran forþlifaþ seó reádnes and bryne ðæs swyles *dum mihi de collo rubor tumoris ardorque promineat,* Bd. 4, 19; S. 589, 31: Exon. 32 a; Th. 101, 22; Cri. 1662. On bryne ge gremedon Drihten *in incendio provocastis Dominum,* Deut. 9, 22: Cd. 186; Th. 231, 12; Dan. 246. Þurh fýres bryne *through the fire's burning,* 197; Th. 245, 11; Dan. 461: Exon. 64 a; Th. 236, 16; Ph. 575. Hie sceolon þrówian biterne bryne *they shall suffer bitter burning,* Andr. Kmbl. 1231; An. 616. Ǽr ðam ðe ðæt mynster mid byrne fornumen wǽre *priusquam*

monasterium esset incendio consumptum, Bd. 4, 25; S. 599, 18. Se biscop ða brynas ðara húsa gebiddende adwæscte *episcopus incendia domorum orando restinxerit*, 1, 19; S. 484, 36. Brego Caldêa gewât to ðam bryne *the prince of the Chaldeans went to the fire*, Exon. 55 b; Th. 196, 27; Az. 180. Hie ðone bryne fandedon *they proved the fire*, Cd. 196; Th. 244, 29; Dan. 455: Exon. 72 b; Th. 270, 31; Jul. 473. Bryne stígeþ heáh to heofonum *the flame rises high to heaven*, Exon. 63 a; Th. 233, 6; Ph. 520: 55 b; Th. 196, 23; Az. 178. Beóþ amerede monna gæstas þurh bryne fýres *the souls of men will be proved through the fire's heat*, 63 b; Th. 234, 25; Ph. 545: Salm. Kmbl. 124; Sal. 61. [*Laym.* brune: *O. Sax.* brunni, *m*: *Goth.* brunsts, *f*: *Icel.* bruni, *m*.] DER. fær-bryne, helle-, lêg-, líg-, mân-, sun-.

bryne BRINE, *salt liquor*; salsugo, muria, Ælfc. Gl. 33; Som. 62, 14; Wrt. Voc. 27, 67. [*Kil.* brijn *muria*.] DER. fisc-bryne.

bryne-ádl, e; *f.* [ádl *a disease*] *A burning disease, a fever*; æstuans morbus, febris, Cot. 92.

bryne-bróga, an; *m.* [bróga *fear, dread*] *Fear or dread of fire*; incendii terror:—Wið brynebrógan *against the fire's dread*, Exon. 55 a; Th. 195, 24; Az. 161.

bryne-gield, es; *n. A burnt-offering, burnt-sacrifice*; holocaustum, Cd. 140; Th. 175, 6; Gen. 2891: 142; Th. 177, 18; Gen. 2931.

bryne-hát; *adj. Burning hot*; ardentissimus:—Ǽr se wlonca dæg bodige brynehátne lêg *ere the awful day proclaim the burning hot flame*, Exon. 116 b; Th. 448, 9; Dóm. 51.

bryne-leóma, an; *m.* [leóma *a ray of light, beam*] *A fire-beam, flame*; flamma:—Bryneleóma stód *the flame stood*, Beo. Th. 4616; B. 2313.

bryne-teár, es; *m.* [teár *a tear*] *A burning tear*; fervida lacrima:—Bitrum bryneteárum *with bitter burning tears*, Exon. 10 a; Th. 10, 14; Cri. 152.

bryne-welm, -wylm, es; *m. A burning flame, flame of fire, burning heat*; incendii fervor *vel* æstus:—Brynewylmum mealt gifstól Geáta *the gift chair of the Goths was consumed by flames of fire*, Beo. Th. 4642; B. 2326: Exon. 42 a; Th. 142, 14; Gú. 644. In helle heó brynewelme bídan sceolden sáran sorge *in hell they must endure great sorrow from the burning heat*, Cd. 213; Th. 266, 24; Sat. 27.

bryngaþ *bring*; afferte, Ps. Spl. 28, 1, = bringaþ; *impert. pl. of* bringan.

bryrdan; he bryrdeþ, bryrdþ, bryrþ; *p.* bryrde; *pp.* bryrded, bryrd [brord *stimulus, cuspis*] *To prick, goad, incite, urge, constrain*; compungere, stimulare, instigare, urgere, compellere:—Se Ælmihtiga ealle gesceafta bryrþ mid his bridlum *the Almighty constrains all creatures with his bridles*, Bt. Met. Fox 13, 5; Met. 13, 3. DER. a-bryrdan, an-, in-, on-.

bryrd-dæg, es; *m. Passion-day*; passionis dies, Som.

bryrding, e; *f. Compunction, instigation*; compunctio, impulsio. v. on-bryrding.

bryrdnys, -nyss, e; *f. A pricking, goading, stimulation, instigation*; compunctio, stimulatio, instigatio:—Mid bryrdnysse ðæs upplícan êðles *by stimulation from the country above*, Bd. Whelc. 173, 16. DER. a-bryrdnes, an-, in-, on-.

bryrþ *urges, constrains*, Bt. Met. Fox 13, 5; Met. 13, 3; *3rd pres. of* bryrdan.

brýsan; he brýsþ; *p.* brýsde; *pp.* brýsed, ge-brýsed *To* BRUISE; conterere. [*Wyc.* brisse: *Tynd. pp.* brosed: *Plat.* brusen *to make a rushing noise: Dut.* bruisen *to foam* or *roar as the sea: Ger.* brausen *to ferment: Dan.* bruse *to roar: Swed.* brusa *to roar: O. Nrs.* brúsa *æstuare*.] DER. to-brýsan: ge-brýsed.

Bryt- *A Welshman; Wallus*: used in compounds. v. Bryt-land.

brýt; e; *f. A nymph, bride*; nympha [= νύμφη *a bride*], Ælfc. Gl. 88; Som. 74, 64; Wrt. Voc. 50, 45. v. brýd.

brýt *breaks*; *3rd pers. pres. of* breótan.

bryta, an; *m. A lord*:—Swegles brytan *lords of heaven*, Cd. 213; Th. 266, 17; Sat. 23. v. brytta.

brytan *to break*, Herb. 1, 3; Lchdm. i. 72, note 8, B: 13, 1; Lchdm. i. 104, 20: 32, 1; Lchdm. i. 130, note 12. v. bryttian.

Brytas, Bryttas, Brittas; *pl. m. The Britons*; Britones:—Hit hafdon Brytas *the Britons had it*, Chr. Th. 3, 29, col. 3. Bryttas, 3, 8, col. 1, 3: 4, 4, col. 2, 3.

brytednys, -nyss, e; *f. A breaking, bruising*; contritio. DER. to-brytednys.

Bryten, Bryton, Briten, Breoten, Breoton, Broten, Brittan, Britten, Brytten; *gen. dat. acc.* e; *f. acc. also as nom.* BRITAIN; Britannia, Cambria:—Brytene ígland is ehta hund míla lang *the island of Britain is eight hundred miles long*, Chr. Th. 3, 1, col. 1: 3, 10, col. 1, 3. Syxtigum wintrum ǽr ðam ðe Crist wǽre acenned, Gaius Iulius, Rômâna cásere [MS. kasere], mid hund-eahtatigum scipum, gesôhte Brytene *sixty years before Christ was born, Caius Julius, emperor of the Romans, with eighty vessels, sought Britain*, Chr. Th. 5, 17–21, col. 3, 1, 2. Breoton [Brytene C] is eálond ðæt wæs iú geára Albion hâten *Britain is an island that was formerly called Albion*, Bd. 1, 1; S. 473, 8: 2, 1;

S. 501, 10. On Breotone *into Britain*, Bd. 1, 15; S. 483, 2. Bryten, *acc.* Exon. 45 b; Th. 155, 5; Gú. 855.

brýten-cyning, es; *m. A powerful king*; rex præpotens, Exon. 88 a; Th. 331, 28; Vy. 75.

brýten-grúnd, es; *m. The spacious earth*; terra spatiosa, Exon. 13 a; Th. 22, 25; Cri. 357.

Bryten-lond, es; *n. The land of Britain*; Britanniæ terra:—Maximus, se cásere, wæs on Bryten-londe geboren *Maximus, the emperor, was born in the land of Britain*, Chr. 381; Ing. 11, 9.

brýten-ríce, es; *n. A spacious kingdom*; regnum spatiosum, Exon. 54 b; Th. 192, 17; Az. 107.

brýten-walda, brýten-wealda, brêten-ânwealda, an; es; *m. A powerful ruler or king*; præpotens rex. It is affirmed [*Kmbl. Sax. Eng.* ii. 21, *and note* 1] that the true meaning of brýten-walda, compounded of walda *a ruler*, and the *adj.* brýten, is totally unconnected with Brettas or Bretwalas, the name of the British aborigines; for brýten is derived from breótan *to bruise, break, to break into small portions, to disperse*; and, when coupled with walda, wealda *a ruler, king*, means no more than *an extensive* or *powerful king, a king whose power is widely extended*. Many similar compounds are found, thus in Exon. 88 a; Th. 331, 28; Vy. 75 we have brýten-cyning *a powerful king* exactly equivalent to brýten-walda. Brýten-grúnd *the wide expanse of earth*, 13 a; Th. 22, 25; Cri. 357. Brýten-ríce *a spacious realm*, 54 b; Th. 192, 17; Az. 107. Brýten-wong *the spacious world*, 13 a; Th. 24, 6; Cri. 380. The uncompounded *adj.* is used in the same sense. Breóton bold *a spacious dwelling*, Cd. 228; Th. 308, 3; Sat. 687. Turner thinks that the Bret-walda [*Hist. of A. Sax.* bk. iii. ch. 5, vol. i. pp. 318 and 378] was *a war-king*, elected by the other Anglo-Saxon kings and their nobility, as their leader in the time of war. Lappenberg [*Th. Lapbg.* i. 125–129] takes the same view; while Kemble [*Sax. Eng.* ii. 8–21] opposes both Turner and Lappenberg, asserting that there was not any general ruler or superior *war-king* elected by the Anglo-Saxons, and that even Bret-walda [*q. v.*] does not refer to the Britons, that it is so written in only one MS. of the Chr. while each of the five others has the word brýten-, and therefore the word ought to be written as above, brýten-walda. Of these Brýten-waldan the Chronicle names the following eight,—Ðý geáre ge-eóde Ecgbriht cing Myrcna ríce, and eal ðæt be súþan Humbre wæs, and he wæs eahtoða cing, ðe brýtenwalda wæs. Ǽrest wæs Ælle, [Súþ-Seaxna] cing, se ðus mycel ríce hæfde. Se æftera wæs Ceawlin, West-Sexna cing. Se þridda wæs Æðelbriht, Cantwara cing. Se feórþa wæs Rædwald, East-Engla cing: fifta wæs Eádwine, Norþhymbra cing: syxta wæs Oswald, ðe æfter him ríxode: seofoða wæs Osweo, Oswaldes bróðor: eahtoða Ecgbriht, West-Seaxna cing *in this year* [A.D. 827] *king Ecgbriht subdued the kingdom of the Mercians, and all that was south of the Humber, and he was the eighth king, who was Brýtenwalda. The first was Ælle* [A.D. 477–514], *king of the South-Saxons, who had thus much sway. The second was Ceawlin* [A.D. 560–593], *king of the West-Saxons. The third was Æthelbriht* [A.D. 593–616], *king of the men of Kent. The fourth was Rædwald* [A.D. 617?–625], *king of the East-Angles: the fifth was Eadwine* [A.D. 625–635], *king of the Northumbrians: the sixth was Oswald* [A.D. 635–642], *who reigned after him: the seventh was Oswiu* [A.D. 642–670], *Oswald's brother: the eighth was Ecgbriht* [A.D. 800–836], *king of the West-Saxons*, Chr. 827; Th. 112, 16–34, col. 2, 3: Palgrv. Eng. Com. pp. ccxxxiv–v.

brýten-wong, es; *m.* [brýten, wang, wong *a plain, field*] *A spacious plain* or *field, in pl. the world*; spatiosus campus, mundus:—Geond brýtenwongas *throughout the spacious world*, Exon. 13 a; Th. 24, 6; Cri. 380.

brýtest, brýtst, he brýteþ, brýt *breakest, breaks*; *2nd and 3rd pers. pres. of* breótan.

Bryt-ford, es; *m.* [Bryt *a Briton*, ford *a ford*] BRITFORD, *near Sarum, Wiltshire*:—Tostig wæs ða æt Brytforda [MS. Brytfordan] mid ðam cinge [MS. kinge] *Tostig was then at Britford with the king*, Chr. 1065; Erl. 194, 38.

bryðen, es; *n? A drink, brewing*; potus:—Bryðen wæs ongunnen, ðætte Adame Eue gebyrmde æt fruman worulde *the drink was prepared, which Eve fermented for Adam at the beginning of the world*, Exon. 47 a; Th. 161, 4; Gú. 953: L. M. 1, 67; Lchdm. ii. 142, 15. Ân bryðen mealtes *one brewing of malt*, Wulfgeat's Will.

brytian *to dispense, distribute*, Past. 44, 1; Hat. MS. 61 a, 13. v. bryttian.

brýtian *to profit*, Bd. 5, 9; S. 623, note 32, 33, T. v. brýcian.

Bryt-land, Bryt-lond, es; *n. The land of Britain, Wales*; Britannia, Cambria:—Ðá fór Harold mid scipum of Brycgstôwe abútan Brytland *then Harold went with his ships from Bristol about Wales*, Chr. 1063; Ing. 251, 21. Into Brytlande *in Walliam*, Chr. 1063; Gib. 170, 41, note 1. v. Bryten.

brytnere, es; *m. A distributor, steward*; dispensator:—Hwá sí [MS. sie] wís brytnere *who can be a wise steward?* Past. 63. v. brytta.

brytnian; *p.* ode, ede, ade; *pp.* od, ed, ad *To dispense, distribute*;

administer ; dispensare, administrare :—He sinc brytnade *he dispensed treasure,* Beo. Th. 4756 ; B. 2383. Hí weolan brytnodon *they dispensed wealth,* Chr. 1065 ; Erl. 197, 40 ; Edw. 21. Æðelingas wēlan brytnedon *the nobles distributed riches,* Cd. 209 ; Th. 259, 14 ; Dan. 691. v. bryttian.

brýtofta *espousals ;* sponsalia, Ælfc. Gl. 87 ; Som. 74, 53 ; Wrt. Voc. 50, 35. v. brýd-gifa, bríd-gifu.

Bryton *Britain,* Bd. 1, 7 ; S. 476, 34. v. Bryten.

Bryton-land ; es ; *n. British land, Britain,* Chr. 979 ; Th. 233, 7, col. 1.

brytsen ; *gen. dat. acc.* brytsene ; *pl. nom. gen. acc.* brytsena ; *dat.* brytsenum ; *f.* [brytan *to break] A broken part, fragment ;* fragmentum :— Hí nāmon ða lāfa, twelf wilian fulle ðæra brytsena *tulerunt reliquias, duodecim cophinos fragmentorum plenos,* Mt. Jun. 14, 20 : Jn. Bos. 6, 13. Of ðām brytsenum *de fragmentis,* Mk. Bos. 8, 8. Gaderiaþ ða brytsena *colligite fragmenta,* Jn. Bos. 6, 12.

brytta, bryta, bretta, an ; *m. A bestower, dispenser, distributor, prince, lord, God?* largitor, dispensator, administrator, princeps, dominus, Deus?— Sinces brytta *a dispenser of treasure,* Cd. 89 ; Th. 111, 18 ; Gen. 1857 : Judth. 10 ; Thw. 21, 22 ; Jud. 30 : Beo. Th. 1219 ; B. 607 : 3849 ; B. 1922 : Exon. 76 b ; Th. 288, 3 ; Wand. 25. Goldes brytta *a distributor of gold,* Cd. 138 ; Th. 173, 26 ; Gen. 2867 : 93 ; Th. 120, 20 ; Gen. 1997. Beága brytta *a distributor of rings or bracelets,* Beo. Th. 69 ; B. 35 : 709 ; B. 352 : 2978 ; B. 1487. Synna brytta *the prince of sins, the devil,* Elen. Kmbl. 1913 ; El. 958. Morðres brytta *the prince of murder, the devil,* Andr. Kmbl. 2342 ; An. 1172. Boldes brytta *the lord of a house,* Elen. Kmbl. 323 ; El. 162. Lifes brytta *the Lord of life* = God, Cd. 6 ; Th. 8, 10, 24 ; Gen. 122, 129 : Exon. 12 b ; Th. 21, 14 ; Cri. 334 : Andr. Kmbl. 1644 ; An. 823. Swægles brytta *the Lord of heaven* = God, Cd. 215 ; Th. 272, 24 ; Sat. 124 : Exon. 12 a ; Th. 18, 10 ; Cri. 281. Tíres brytta *the Lord of power* = God, 14 b ; Th. 29, 14 ; Cri. 462. [Icel. bryti, *m. a steward, bailiff.*]

Brytta *of the Britons,* Bd. 1, 34 ; S. 499, 20 ; *gen. pl.* of Bryttas.

Bryttas, Brittas, Brettas, Breotas, Brytas, Britas ; *pl. m.* I. *Britons ;* Britones :—Ǽrest wǽron búend ðyses landes Bryttas *the first inhabitants of this land* [England] *were the Britons,* Chr. Th. 3, 8, col. 1, 3. Mód and mægen Bryttas onfēngon *the Britons took heart and power,* Bd. 1, 16 ; S. 484, 19 : 1, 15 ; S. 483, 17. Ðætte Angel-þéod wæs gelaðod fram Bryttum on Breotone *that the Angle-nation was invited by the Britons into Britain,* 1, 15 ; S. 483, 2. II. *Bretons ;* Armoricani :—Ðý ilcan geáre fōr se here of Sigene to Sant Laudan, ðæt is betweoh Brettum [Bryttum, col. 2, 3] and Francum *in the same year the army went from the Seine to St. Lô, which is between the Bretons and the Franks,* Chr. 890 ; Th. 160, 10, col. 1. Hí speónan ða Bryttas heom to *they enticed the Bretons to them,* 1075 ; Th. 349, 26.

Brytten, e ; *f. Britain,* Chr. Th. 3, 11, col. 2. v. Bryten.

bryttian, brittian, bryttigan, brytian ; *pl.* bryttigaþ ; *p.* bryttade ; *v. a. To divide into fragments, dispense, rule, use ;* dispensare frustatim, gubernare :—Hí hit him bryttian sceoldon *they should dispense it to them,* Past. 44, 1 ; Hat. MS. 61 a, 13. Land bryttade *ruled the land,* Cd. 62 ; Th. 75, 6 ; Gen. 1236. Worulde mægen bryttigan *might use force,* Cd. 4 ; Th. 4, 12 ; Gen. 52. [Icel. brytja *to chop, cut in pieces.*]

brýttian ; *p.* ode, ade ; *pp.* od *To possess, enjoy ;* possidere, frui :— Sculon wēlan bryttian *shall enjoy wealth,* Cd. 99 ; Th. 131, 19 ; Gen. 2178. Woruld bryttade *enjoyed the world,* Cd. 62 ; Th. 74, 22 ; Gen. 1226. v. brýtian.

Bryttisc, Brittisc ; *adj. British ;* Britannicus :—He wæs Bryttisc *he was British,* Chr. 1075 ; Erl. 213, 3.

Brytt-wealas, Bryt-walas ; *pl. m. The Brito-Welsh, Britons ;* Britanni :—Cynríc ða Bryttwealas geflýmde *Cynric routed the Britons,* Chr. 552 ; Gib. 20, 2. Brytwalas, 167 ; Erl. 9, 20 : 443 ; Erl. 11, 33 : 571 ; Erl. 19, 15.

bú, bý, es ; *n?* [ic búe, he býþ, *pres. of* búan *to dwell] A dwelling, habitation ;* habitatio, habitaculum :—Bearn hēr bú nāmon, and ðǽr eardedon *here children obtained a dwelling, and there settled,* Ps. Th. 101, 25. Stanford and Deóra bý wǽron under Norþmannum *Stamford and Derby* [Deóra bý *habitation of deer or animals*] *were under the Northmen,* Chr. 942 ; Th. 210, 4 ; Edm. 8. Se ðe húsoðde bý hæfde *qui domicilium habebat,* Mk. Skt. Lind. 5, 3. [Plat. buw, *m :* O. Sax. bú, *n :* Dut. bouw, *m :* Ger. bau, *m :* M. H. Ger. bú, bou, *m :* O. H. Ger. pú, *m :* Dan. bo, *m. f :* Swed. bo, *m :* Icel. bú, *n. domus :* Sansk. bhú, *f. the earth, site, place.*]

bú *both, nom. m. f.* or *n : acc. m. f. n.* of begen ; ambæ, ambo :—Hí bú þēgon [MS. þegun] æppel *they both* [Adam and Eve] *ate the apple,* Exon. 61 b ; Th. 226, 8 ; Ph. 402 : Cd. 10 ; Th. 12, 18 ; Gen. 187 : 82 ; Th. 102, 13 ; Gen. 1699. v. bá.

BÚAN, búgan ; ic búe, ðú búst, he býþ ; *p.* búde, *pl.* búdon ; *pp.* gebún ; *v. anom.* I. *intrans. To dwell, live ;* habitare, versari aliquo loco :—He búde on Eást-Englum *he dwelt among the East-Angles,* Chr. 890 ; Erl. 86, 29 : Ors. 1, 1 ; Bos. 19, 26. Gif he weard onfunde búan [MS. buon] on beorge *if he found the keeper dwelling in the mount,*

Beo. Th. 5676 ; B. 2842. II. *v. a. acc. To inhabit, occupy ;* inhabitare, colere, incolere :—He lēt heó ðæt land búan *he let them inhabit the land,* Cd. 13 ; Th. 16, 6 ; Gen. 239. Ðæt ðú búst eorþan *ut inhabites terram,* Ps. Th. 36, 33. Ðæt hēr men bún ðone heán heofon *that here men inhabit the high heaven,* Cd. 35 ; Th. 45, 32 ; Gen. 735. Ne mæg mon meduseld búan *a man may not occupy the mead-bench,* Beo. Th. 6123 ; B. 3065. [Plat. buwen, bouen, buen, bujen : O. Sax. búan : Frs. bouwje : O. Frs. buwa, bowa : Dut. bouwen : Ger. bauen : M. H. Ger. buwen, biuwen, bouwen : O. H. Ger. búan, búwan : Goth. bauan : Dan. boe : Swed. bo : Icel. búa : Lith. bu-ti *to be :* Slav. by-ti *to be :* Zend bú *to be, become :* Sansk. bhú *to become, spring up, be, exist, live.*] DER. ge-búan : ân-búende : bú, bý : búgan, búgend : búgian, búian, búwian.

BÚC, es ; *m. A BUCK, a male deer ;* cervus ; Ælfc. Gl. 19 ; Som. 59, 22 : Wrt. Voc. 22, 63. v. dá *a doe.*

BÚC, es ; *m.* I. *the belly, stomach ;* venter, alvus :—Hit is betwux tóðum tocowen and into ðam búce asend *it is chewed between the teeth and sent into the stomach,* Homl. Th. ii. 270, 34. II. *a vessel that bulges out, as a bottle, jug, pitcher ;* lagena, hydria :—Búc *lagena,* Wrt. Voc. 83, 24. Þurh heora blāwunge and ðæra búca swēg *through the sound of their blowing and of the pitchers,* Jud. 7, 21. Hí tobrǽcon ða búcas mid micelre brastlunge *they broke the pitchers* [hydrias confregerunt] *with great crashing,* 7, 20. [Chauc. bouke *bulk, body :* Plat. buuk, *m. venter :* O. Sax. bûk, *m. uter :* Frs. buk, *m. f :* O. Frs. buk, buch, *m. venter :* Dut. buik, *m. belly :* Kil. buyck *corporis truncus :* Ger. bauch, *m. venter, alveus :* M. H. Ger. bûch, *m. venter :* O. H. Ger. bûh, *m. venter :* Dan. bug, *m. f. the stomach, belly* or *middle of a vessel :* Swed. buk, *m. belly :* Icel. búkr, *m. the trunk, body.*] DER. æscen, hrygile-búc.

bucc *a cheek, part of a helmet ;* buccula, Cot. 25.

BUCCA, an ; *m.* [buc *a buck] A he-goat,* BUCK ; caper, hircus :—Bucca *caper vel hircus,* Wrt. Voc. 78, 32. Bucca *hircus,* Ælfc. Gr. 8 ; Som. 7, 30. Bucca *caper vel hircus vel tragos* [= τράγος], Ælfc. Gl. 20 ; Som. 59, 36 ; Wrt. Voc. 22, 77. Gif se ealdor syngaþ, bringeþ ánne buccan to bóte *si peccaverit princeps, offerat hircum immaculatum,* Lev. 4, 23 : 9, 3. He asyndrode twáhund gáta and twentig buccena *separavit capras ducentas et hircos viginti,* Gen. 32, 14 : Ps. Lamb. 49, 13. Ic ne underfó of eowedum ðínum buccan *non accipiam de gregibus tuis hircos,* 49, 9 : Deut. 32, 14. Buccan horn *a buck's horn, one of the twelve signs of the zodiac, Capricorn,* Bd. de nat. rerum ; Wrt. popl. science 7, 8 ; Lchdm. iii. 246, 3. Buccan beard *a goat's beard,* Wrt. Voc. 289, 10. [Chauc. buck : Orm. bucc : Plat. buk, *m :* O. Sax. buc, *m :* Frs. bok, *m. f :* Dut. bok, *m :* Ger. bock, *m :* M. H. Ger. boc, *m :* O. H. Ger. boch, *m :* Dan. buk, *m. f :* Swed. bock, *m :* Icel. bokki, *m.*] DER. firgen-bucca, stán-, wudu-.

Bucc-inga ham ; *gen.* hammes ; *m.* [Hunt. Bukingeham : Brom. Bukyngham : Bucc, -inga ham, q. v.] BUCKINGHAM ; oppidum primarium agri Buccinghamensis :—Fōr Eádweard cyning to Buccinga hamme *king Edward went to Buckingham,* Chr. 918 ; Erl. 104, 18.

Buccinga ham-scír, e ; *f.* BUCKINGHAMSHIRE ; ager Buccinghamensis :—Hí wendon ðanou on Buccinga hamscíre *they turned thence to Buckinghamshire,* Chr. 1010 ; Th. 264, 11 : 1011 ; Erl. 144, 35 : 1016 ; Erl. 154, 6, 24.

búc-ful, -full, e ; *f. A pitcherful :*—Him wearþ ðá geboren to búcful wæteres *a pitcherful of water was then borne to him,* Homl. Th. ii. 422, 29.

bude *hast offered,* Cd. 111 ; Th. 147, 7 ; Gen. 2435 : budon *offered,* Beo. Th. 2175 ; B. 1085 ; *p. s. and pl.* of beódan.

búde *dwelt ;* habitavit, Ors. 1, 1 ; Bos. 19, 26 ; *p.* of búan.

búende ; *part.* búend, es ; *m. Inhabiting* or *dwelling ;* inhabitans :— Búendra leás *void of those inhabiting* [Cd. 5 ; Th. 6, 16 ; Gen. 89] or *inhabitants, thus used as a noun,* though sometimes in composition declined as a *m.* noun, búend, es ; *m. :* it is *often declined as a m. part.* that is an *adj.* ending in e. It would then be declined nom. s. -búende ; *gen.* -búendes ; *d.* -búendum ; *acc.* -buendne ; but most frequently as an *adj. pl ;* nom. acc. -búende ; *gen.* -búendra [*as a noun,* búenda] ; *d.* -búendum :—Mid búendum *cum habitantibus,* Ps. Lamb. 82, 8. DER. ân-búende, ceaster-búend, ēg-, eorþ-, feor-, fold-, grúnd-, hēr-, íg-, land-, neáh-, sund-, þeód-, woruld-.

búfan, búfon ; *prep. dat.* [be-ufan] *Above ;* super ; used in opposition to *under :*—God totwǽmde ða wæteru, ðe wǽron *under* ðǽre fæstnisse fram ðǽm ðe wǽron búfan ðǽre fæstnisse *Deus divisit aquas, quæ erant sub firmamento ab his quæ erant super firmamentum,* Gen. 1, 7. Búfan ðam māran wealle *above the greater wall,* Ors. 2, 4 ; Bos. 44, 28. Twentig míla búfan Lundenbyrig *twenty miles above London,* Chr. 896 ; Th. 172, 25. DER. *prep.*

búfan, búfon [be-ufan] ; *adv. Above, before ;* supra :—Be ðære búfan sǽd wæs *de qua supra dictum est,* Bd. 4, 22 ; S. 592, 13 : Mt. Rush. Stv. 2, 9. [Plat. baven : Dut. boven supra.] DER. ufan ; *adv.*

búgan ; *p.* ede ; *v. a. acc. To inhabit ;* inhabitare, incolere :—Þenden git mōston ân lond búgan *while ye might inhabit one land,* Exon. 123 a ; Th. 473, 20 ; Bo. 17. Ðǽr ic wíc búge *there I inhabit a dwelling,* 104 b ;

Th. 396, 22 ; Rä. 16, 8 : 103 a ; Th. 389, 23 ; Rä. 8, 2. Đær nô men búgaþ *eard where men inhabit not a home*, 58 a ; Th. 208, 18 ; Ph. 157. Búgede *habitavit*, Aldh. Gl. Grn. v. búan, búgian.

BÚGAN ; *part.* búgende ; ic búge, ðu búgest, býhst, býgst, he búgeþ, býhþ, býgþ ; *p.* ic, he beág, beáh, ðu buge, *pl.* bugon ; *imp.* búge, búh ; *pp.* bogen ; *v. intrans. To* BOW *or bow down oneself, bend, swerve, give way, submit, yield, turn, turn away, flee* ; *se flectere vel inclinare, curvare, declinare, desistere, cedere, vertere, divertere, fugere* :—Hí noldon búgan to nánum deófolgilde *they would not bow down to any idol*, Homl. Th. ii. 18, 29 : Rood Kmbl. 71 ; Kr. 36 : Num. 25, 2. Ne eom ic wyrðe ðæt ic his sceóna þwanga búgende uncnytte, Mk. Bos. 1, 7 ; *I knelinge am not worthi for to vndo the thwong of his schoon*, Wyc. Seó eá, norþ búgende, ût on ðone Wendel-sǽ *the river, bending northward,* [*flows*] *out into the Mediterranean sea*, Ors. 1, 1 ; Bos. 17, 33 : Exon. 103 a ; Th. 390, 24 ; Rä. 9, 6. Seó eorþe nǽfre ne býhþ ne ufor ne nyðor ðonne se ælmihtiga Scyppend hí gestaðelode *the earth never swerves neither higher nor lower than the almighty Creator established it*, Bd. de nat. rerum ; Wrt. popl. science 10, 19 ; Lchdm. iii. 254, 18. Hí bugon and flugon *they gave way and fled*, Chr. 999 ; Erl. 135, 25. Ic sceolde on bonan willan búgan *I must submit to a murderer's will*, Exon. 126 b ; Th. 486, 4 ; Rä. 72, 7 : Beo. Th. 5829 ; B. 2918. Him beág gód dǽl ðæs folces *a good part of the people submitted to him*, Chr. 913 ; Erl. 102, 7 : 921 ; Erl. 108, 1. He to fulluhte beáh *he submitted to baptism*, Homl. Th. i. 386, 32 : Ex. 32, 26. Hí bugon to ðam *they submitted to that*, Jos. 9, 27 : Chr. 975 ; Erl. 125, 24. Ælc burhwaru wæs búgende to him *every city was yielding to him*, Jos. 11, 19. Búge ic to eówerum hǽðenscipe *I will turn to your heathendom*, Homl. Th. i. 70, 28. Híg bugon of ðam wege *they have turned out of the way*, Ex. 32, 8. Ðæt ge ne bugon eft to woruldþingum *that ye turn not again to worldly things*, Boutr. Scrd. 22, 46. Se Hǽlend beáh fram ðære gegaderunge *the Saviour turned away from the company*, Jn. Bos. 5, 13 : Beo. Th. 5905 ; B. 2956. Búh fram yfele and dô oððe wyrc gód *diverte a malo et fac bonum*, R. Ben. in procem. He sceal búgan *fugere debeat*, Ex. 21, 13 : Gen. 19, 21 : Byrht. Th. 139, 58 ; By. 276. Hí bugon fram beaduwe *they fled from the fight*, 137, 12 ; By. 185 : Beo. Th. 5190 ; B. 2598. [*Laym.* buʒen, buwen : *Orm.* buʒhenn : *Plat.* bögen : *Dut.* buigen : *Kil.* buyghen : *Ger. M. H. Ger.* biegen : *O. H. Ger.* biugan : *Icel.* boginn *bent* : *Sansk.* bhuj *to bend.*] DER. a-búgan, an-, be-, bi-, for-, ge-, in-, on-, under-, ymb-.

búgend, es ; *m.* [búgende, *part. of* búgan, búan *to dwell*] *A dweller, an inhabitant* ; *habitator* :—Ærost wǽron búgendas [MS. búgend] ðyses landes Bryttas *at first the inhabitants of this land* [*England*] *were Britons*, Chr. Th. 3, 7, col. 3.

búgende *bowing, kneeling*, Mk. Bos. 1, 7. v. búgan *to bow down*.

búgian, búian, búwian, to búgianne ; *p.* ode ; *pp.* od.   **I.** *intrans. To dwell* ; *habitare* :—Ge ðǽr búgiaþ *ye dwell there*, Bt. 18, 1 ; Fox 62, 22.   **II.** *v. a. acc. To inhabit, occupy* ; *inhabitare, incolere* :—Ðis is land to búgianne *this is to inhabit land*, Bt. 17 ; Fox 60, 4. v. búan.

búh *turn* :—Búh fram yfle *diverte a malo*, R. Ben. in procem. *impert. of* búgan *to bow, turn.*

búh-somnes, -ness ; *f.* BOWSOMENESS, *pliableness* ; *obedientia*, Verst. Restitn. p. 211. v. bócsumnes.

búian *to dwell, inhabit* ; *habitare, incolere* :—Ðæt we môston búian *that we should dwell*, Ps. Th. 28, 8. Ðe on eorþan búiaþ *who dwell on earth*, Ps. Th. 32, 7. Búiaþ *inhabit*, Ps. Th. 32, 12. v. búgian.

bule *a stud, boss, brooch* ; bulla, Cot. 26. [*Ger.* bulle ; *f.*]

bulentse, an ; *f. The name of a plant, which, from not knowing its Latin or English name, I call* bulentse :—Nime bulentsan ða smalan *take the small bulentse*, L. M. 1, 47 ; Lchdm. ii. 118, 1.

bulge *wast angry* ; *p. of* belgan.

bulgon *made angry, were angry* ; *p. pl. of* belgan.

bulle *bellowedst, roaredst* ; bullon *bellowed, roared* ; *p. of* bellan.

bulluca, an ; *m. A male calf, a* BULLOCK ; *vitulus*, Scint. 54.

bulot, bulut *Ragged robin or cuckoo-flower* ; *lychnis, flos cuculi* ; Lin :—Bulot-niðeweard *the nether part of cuckoo-flower*, L. M. 1, 58 ; Lchdm. ii. 128, 15. Nim bulut *take cuckoo-flower*, 3, 48 ; Lchdm. ii. 340, 1.

bunda, bonda, an ; *m.*   **I.** *a wedded or married man, a husband* ; *maritus, sponsus* :—Ne mæg nán wíf hire bondan [bundan MS. B, note 57] forbeódan, ðæt he ne môte into his cotan gelogian ðæt ðæt he wille *no wife may forbid her husband, that he may not put into his cot what he will*, L. Cnut. pol. 74 ; Wilk. 145, 41 ; Th. i. 418, 23-25 ; Schmd. 312, 76, § 1. Sé hit bonda, sé hit wíf *sive maritus sit, sive uxor*, Hick. Diss. Ep. 18, 40.   **II.** *the father or head of a family, a householder* ; *paterfamilias, œconomus* :—Swá ymbe friðes bóte swá ðam bondan [bundan MS. A. L. C. S. 8] sí sélost and ðam þeófan sí láðost *so concerning frithes-bót as may be best for the householder* [*patrifamilias*] *and worst for the thief*, L. Ænh. Wilk. 122, 40 ; Eth. vi. 32 ; Th. i. 322, 27 ; Schmd. 232, § 32 : L. Cnut. pol. 8 ; Wilk. 134, 40 ; Th. i. 380, 14 ; Schmd. 274, 8. And ðær se bonda [MS. B, bunda] sæt uncwyd and unbecrafod sitte ðæt wíf and ða cild on ðam ylcan unbesacen. And

gif se bonda [MS. B, bunda] beclypod wǽre, etc. *and where the householder dwelt without claim or contest, let the wife and the children dwell in the same, without litigation. And if the householder had been cited, etc.* L. Cnut. pol. 70 ; Wilk. 144, 39 ; Th. i. 414, 21 ; Schmd. 310, 72. *The early Latin version is*, Et ubi bonda [bunda, L. Th. í. 526, 3], i. e. paterfamilias manserit, sine compellatione et calumpnia, sint uxor et pueri in eodem, sine querela. Et si [bunda, i. e. paterfamilias] compellatus fuerat, etc. L. Cnut. 73 ; Th. ii. 542, 13-15.   **2.** every word has its history by which its introduction and use are best ascertained. Bede tells us [Bk. i. 25, 2] that Ethelbert, king of Kent, married a Christian wife Bertha, a Frankish princess. The queen prepared the way for the friendly reception of Augustine and his missionary followers by Ethelbert in A. D. 597, who was the first to found a school in Kent, and wrote Laws which are said to be asette on Augustines dæge *established in the time of Augustine*, between A. D. 597 and 604. The cultivation and writing of Anglo-Saxon [Englisc] began with the conversion of Ethelbert. Marriage, and the household arrangements depending upon it, were regulated by the law of the church, and indigenous compound words were formed to express that law,—thus ǽ *law, divine law* ; Cristes ǽ *Christi lex. Rihte ǽ legitimum matrimonium*, Bd. 4, 5 ; S. 573, 17. *Ǽw wedlock, marriage,* ǽw-boren *lawfully born, born in wedlock* : ǽw-breca, -brica, *m. wedlock breaker, an adulterer* : ǽw-fæst-man *marriage-fast-man, a wedded man, a husband* : ǽw-nian *to wed, take a wife*.   **3.** Hús-bunda, -bonda *a wedded man, husband, householder*. This compound is one of the oldest in the language. It is found in the interpolated passage of Matt. xx. between vers. 28 and 29. The passage is in all the Anglo-Saxon MSS. of the Gospels, except the interlineary glosses. The Anglo-Saxon is a literal version of the Augustinian MS. in the Bodleian Library, Oxford [Codex August. 857 D. 2. 14], the Old Italic version, from which the text of the Latin vulgate of the Gospels was formed by St. Jerome about A. D. 384. Though we do not know the exact dates when the Gospels were translated from Latin into Anglo-Saxon, Cuthbert assures us that Bede finished the last Gospel, St. John, on May 27, 735, [see Pref. to Goth. and A. Sax. Gos. Bos. pp. ix-xii.] As the three preceding Gospels were most likely translated before St. John, then the following sentence was written before 735. Se hús-bonda [hús-bunda in MS. Camb. Ii, 2, 11] háte ðé arísan and rýman ðam óðrum *the householder bid thee rise and make room for the other*, Notes to Bosworth's Goth. and A. Sax. Gos. Mt. xx. 28, p. 576. Hús-bonda is also used by Ælfric in his version of the Scriptures about 970, Ex. 3, 22.   **4.** Bunda, bonda *one wedded or bound, a husband*, from bindan ; *p.* band, bundon ; *pp.* bunden *to bind* must have been of earlier origin than the compound hús-bunda. It is a well-known rule that in Anglo-Saxon *a person or agent* is denoted by adding *a*, as býtl *a hammer*, býtla *a hammerer* ; ánweald *rule, government*, ánwealda *a ruler, governor* ; bunden, bund *bound*, bunda, bonda *one bound, a husband*. Bunda might be banda as well as bonda, for *a* is often used for *o*, as mon for man *a man*. The early use of hús-bunda, -bonda would at once indicate that it was not likely to be of Norse or Icelandic origin. It could not be derived from the Norse búa *to dwell* ; *part.* búandi, bôandi *dwelling* ; nor even from the *A. Sax.* búan *to dwell*, because the *ú* and *ô* are long in the *Norse* búa *to dwell*, búandi, bôandi *dwelling*, and in the *A. Sax.* búan *to dwell*, búende *dwelling*, búend *a dweller* ; while the *u* and *o* are always short in bunda and bonda. So, in other compounds, from bindan *to bind*, as bonde-land *bond or leased land, land let on binding conditions*. Bunda then is a pure Anglo-Saxon word derived from bindan *to bind*. Búan *to dwell*, with the *part.* búende *dwelling*, and the noun búend, es ; *m. a dweller*, is quite a distinct word with its own numerous compounds. v. búende, búend, es ; *m.*

bunden *bound, tied* ; bundon *bound*, Beo. Th. 3805 ; B. 1900 ; *pp. and p. of* bindan.

bunden-stefna, an ; *m.* [bunden *bound*, stefna *the prow of a ship*] *A bound prow* ; *ligata prora* :—Sǽgenga fleát ofer ýðe, bundenstefna ofer brimstreámas *the ship* [lit. *sea-goer*] *floated over the wave, the bound prow over the ocean-streams*, Beo. Th. 3824 ; B. 1910.

bune, an ; *f. A sort of cup* ; *carchesium* = καρχήσιον, *poculi genus*, Judth. 10 ; Thw. 21, 14 ; Jud. 18 : Beo. Th. 5544 ; B. 2775 : Exon. 77 b ; Th. 338, 23 ; Gn. Ex. 83.

Bune, Bunne, an ; *f? Boulogne in France* ; *Bononia* :—Se micla here férde to Bunan [Bunnan, Th. 162, 20, col. 1] *the great army went to Boulogne*, Chr. 893 ; Th. 163, 20, col. 3.

buoptalmon ; *n.* [βούφθαλμον = βοῦς, ὀφθαλμός] *Ox-eye, chamomile* ; *anthemis nobilis*, Lin :—Buoptalmon ... heó hafaþ geoluwe blôstman eal swylce eáge, ðanon heó ðone naman onféng *Ox-eye ... it has yellow blossoms all like an eye, whence it took the name*, Herb. 141, 1 ; Lchdm. i. 262, 4.

BÚR, es ; *n. A* BOWER, *cottage, dwelling, an inner room, storehouse* ; *tabernaculum, conclave, casa* :—Wiht wolde hyre on ðære byrig búr atimbran *a creature would construct a bower for itself in the town*, Exon. 108 a ; Th. 411, 26 ; Rä. 30, 5. On búre, ahôf brýd Abrahames hleahtor *in the inner room, Abraham's wife raised a laugh*, Cd. 109 ; Th. 144, 7 ;

Gen. 2386. Cumena bûr *a guest-house*, Bd. 4, 31; S. 610, 11. Bedcófa *vel* bûr *cubiculum*, Ælfc. Gl. 27; Som. 60, 99; Wrt. Voc. 25, 39. Wæs to bûre Beówulf fetod *Beowulf was fetched to his dwelling*, Beo. Th. 2624; B. 1310. On his suna bûre *in his son's dwelling*, Beo. Th. 4902; B. 2455. Æfter bûrum *along the dwellings*, Beo. Th. 282; B. 140. [*Chauc.* boure: *Piers P.* bour: *R. Glouc.* boures, *pl:* Laym. bur: *Orm.* bure: *Plat.* bur, buur, *m:* Ger. bauer, *m:* O. H. Ger. bûr: *Dan.* buur, *n:* *Swed.* bur, *m:* *Icel.* búr, *n.*] DER. brýd-bûr.

burcg, e; *f.* *A city:*—Ðære burcge *of the city*, Bt. 18, 2; Fox 64, 18. v. burh.

bûr-cote, an; *f.* [bûr *a bower*, cote *a couch*] *A bed-chamber; cubiculum:*—On hira bûrcotum, and on hiera beddum *in their bed-chambers, and in their beds*, Past. 16, 2; Hat. MS. 20 b, 15.

burg, e; *f.* *A city; urbs:*—Sceal seó burg bídan *the city shall remain*, Exon. 121 b; Th. 466, 30; Hö. 129. v. burh.

burg- = beorg- *a hill, in some compounds, as in* burg-stal, *q. v.*

burga *cities, of cities*, Mt. Bos. 11, 20: Salm. Kmbl. 613. v. burh.

burg-ágende; *part.* *Possessing a fortress* or *palace; arcem vel palatium possidens*, Elen. Kmbl. 2347; El. 1175.

burga man, es; *m.* *A citizen; civis:*—Sí hit burga man *sive civis sit ille*, Deut. 1, 16. v. burh-man.

burgan = burgen, Ors. 2, 5; Bos. 47, 15; *p. pl. subj. of* beorgan *to save*.

burgat, es; *pl.* burgatu; *n.* [burg *a city*, gat, geat *a gate*] *A city-gate; urbis porta:*—Ðá Samson genam ða burggatu [MS. burgatu] and gebær on his hricge *then Samson took the city-gates and bore them on his back*, Jud. 16, 3.

burg-bryce, burh-bryce, -brice, es; *m.*   I. *a breaking into a castle* or *dwelling; castelli vel domus violatio*, L. In. 45; Th. i. 130, 7.   II. *the fine to be paid for this burglary; mulcta ob castelli vel domus violationem*, L. Alf. pol. 40; Th. i. 88, 7.

burgen, e; *f.* *A burying-place, sepulchre*, Ps. Th. 29, 9. v. byrgen.

Burgenda land, es; *n.* *The land of the Burgundians, an island in the west of the Baltic sea; Boringia.* Burgenda land is the Icelandic Burgundarhólmr, of which the present Danish and Swedish name Bornholm is a contraction:—Burgenda land *the land of the Burgundians*, Ors. 1, 1; Bos. 21, 44.

Burgendan; *pl. m.* *The Burgundians; Burgundiones:*—Burgendan habbaþ ðone ylcan sæs earm be westan him *the Burgundians have the same arm of the sea to the west of them*, Ors. 1, 1; Bos. 19, 19. v. Burgendas.

Burgendas; *gen.* a; *pl. m:* Burgendan; *pl. m.* *The Burgundians;* Burgundiones. These, in Alfred's time, dwelt to the north-west of the Osti. We find them at another period on the east bank of the Oder. They have given them name to the island of Bornholm in the Baltic:—Osti habbaþ be norþan him Winedas and Burgendas *the Esthonians have to the north of them the Wends and the Burgundians*, Ors. 1, 1; Bos. 19, 18. Wine Burgenda *friend of the Burgundians*, Wald. 85; Vald. 2, 14. Weóld Burgendum Gifica *Gifica ruled the Burgundians*, Scóp Th. 40; Wíd. 19: 131; Wíd. 65.

Burgende; *gen.* a; *dat.* um; *m.* *The Burgundians, inhabitants of Burgundy, an old province in the east of France;* Burgundiones:—Profentse hæfþ be norþan hyre ða beorgas, ðe man Alpis hæt, and be sûþan hyre is Wendel-sæ, and be norþan hyre and eástan synd Burgende, and Wascan be westan *Provence has on the north of it the Burgundians, which people call the Alps, and on the south of it is the Mediterranean sea, and on the north and east of it are the Burgundians, and on the west the Gasconians*, Ors. 1, 1; Bos. 24, 2.

bûr-geteld, es; *n.* [bûr *a bower*, geteld *a tilt, cover*] *A tilt* or *covering of a tent, a tent; tentorium:*—He in ðæt bûrgeteld neðde *he ventured into the tent*, Judth. 12; Thw. 25, 24; Jud. 276: 10; Thw. 22, 10; Jud. 57: 12; Thw. 25, 8; Jud. 248.

burg-geat *a city-gate*, Andr. Kmbl. 1679; An. 842. v. burh-geat.

burg-hleoþ, es; *n.* *A fortress-height*, Exon. 107 b; Th. 409, 17; Rä. 28, 2. v. burh-hleoþ.

burg-loca, an; *m.* *A city-inclosure, city-barrier*, Andr. Kmbl. 2075; An. 1040: 2132; An. 1067: 1879; An. 942. v. burh-loca.

burg-lond, es; *n.* *City-land; urbis solum:*—Eálá sancta Hierusalem, Cristes burglond *O holy Jerusalem, city-land of Christ!* Exon. 8 b; Th. 4, 12; Cri. 51.

burgon *preserved*, Elen. Kmbl. 268; El. 134; *p. pl. of* beorgan.

burg-ræced, es; *nom. acc. pl.* -ræced; *n.* *A city-dwelling, house surrounded by a wall* or *rampart of earth; urbanæ ædes, circumvallata domus:*—Beorht wæron burgræced *bright were the city-dwellings*, Exon. 124 a; Th. 477, 9; Ruin. 22.

burg-rúnan *the fates, furies, fairies.* v. burh-rúnan.

burg-sæl, es; *nom. acc. pl.* -salu, -salo; *n.* *A castle-hall, city-dwelling; arcis aula, urbana domus:*—Ofer burgsalu *over the city-dwellings*, Exon. 51 b; Th. 179, 7; Gú. 1258: 52 a; Th. 182, 4; Gú. 1305: 96 a; Th. 358, 23; Pa. 50.

burg-sele, es; *m.* *A castle-hall, city-dwelling; arcis aula, urbana*

---

domus:—Burgsele beofode *the castle-hall trembled*, Exon. 94 b; Th. 353, 49; Reim. 30.

burg-sittend *a city-dweller, citizen*, Bt. Met. Fox 27, 34; Met. 27, 17: Elen. Kmbl. 552; El. 276. v. burh-sittend.

burg-sittende *city-dwelling, inhabiting a city*, Cd. 52; Th. 66, 24; Gen. 1089: Exon. 12 b; Th. 21, 20; Cri. 337: 53 a; Th. 186, 14; Az. 19: 106 b; Th. 407, 10; Rä. 26, 3. v. burh-sittende.

burg-stal, -stól, es; *m.* [burg = beorg, beorh *a hill*, stal *a place, seat, dwelling*] *A hill-seat, dwelling on a hill; sedes super collem vel clivum*, Cot. 209. *The name of places built on a hill, as* Burstall *in Suffolk*, Borstall *in Kent and Oxfordshire, etc.*

burg-steal, es; *m.* [burg *a fortress, city*, steal *a place*] *A city-place; arcis locus, arx:*—Brosnade burgsteal *the city-place has perished*, Exon. 124 a; Th. 477, 23; Ruin. 29. [*Ger. M. H. Ger.* burgstall.]

burg-stede *a city-place, city*, Exon. 52 a; Th. 181, 10; Gú. 1291: 124 a; Th. 476, 3; Ruin. 2. v. burh-stede.

burg-tún, es; *m.* *A borough-town, city-inclosure, city-dwelling; urbis septum, urbana domus:*—Sindon burgtúnas brérum beweaxne *the city-dwellings are overgrown with briers*, Exon. 115 b; Th. 443, 16; Kl. 31.

burg-waran, burh-waran, *gen.* -warena; *pl. m.* *Inhabitants of a city, citizens; urbis incolæ, cives:*—Ealle burgwaran *all the city-inhabitants*, Exon. 121 b; Th. 467, 6; Hö. 134: 120 b; Th. 462, 23; Hö. 56. Burgwarena fruma *the chief of the city-warriors*, Scóp Th. 182; Wíd. 90.

burg-ware *inhabitants of a city, citizens*, Andr. Kmbl. 3164; An. 1585: Chr. 919; Th. 192, 25: Exon. 18 b; Th. 46, 25; Cri. 742. v. burh-ware.

burg-waru *the inhabitants of a city as in a body*, Andr. Kmbl. 2189; An. 1096. v. burh-waru.

burg-weall, -weal *a city-wall*, Exon. 83 b; Th. 315, 28; Mód. 38: 22 a; Th. 61, 1; Cri. 978. v. burh-weall.

burg-wígende; *part. pl.* *City-warring;* used substantively, *city-warriors; ex arce belligerentes, cives belligeri:*—Swylce Húna cyning meahte abannan to beadwe burgwígendra *whomsoever the king of the Huns might summon to the fight*, Elen. Kmbl. 68; El. 34.

BURH, burg; *gen.* burge; *dat.* byrig, byrg; *acc.* burh, burg; *pl. nom. acc.* burga; *gen.* burga; *dat.* burgum; *f.* [beorh, beorg = burh, burg *the impert. of* beorgan *to defend*].   I. *the original signification was arx, castellum, mons, a castle for defence. It might consist of a castle alone; but as people lived together for defence and support, hence a fortified place, fortress, castle, palace, walled town, dwelling surrounded by a wall* or *rampart of earth; arx, castellum, mons, palatium, urbs munita, domus circumvallata:*—Se Abbot Kenulf macode fyrst ða wealle abútan ðone mynstre, [and] geaf hit ðá to nama Burh [Burch MS.], ðe ær hét Medeshámstede *the Abbot Kenulf first made the wall about the minster, and gave it then the name Burh* = Burg [Petres burh *Peter's burg* = Peterborough], *which before was called Meadow-home-stead*, Chr. 963; Erl. 123, 27–34; Th. 221, 34–39. ☞ The style of the Anglo-Saxon indicates a late date, perhaps about 1100 or 1200. Burg *arx*, Cot. 10. Stíþlíc stán-torr and seó steápe burh on Sennar stód *the rugged stone-tower and the high fortress stood on Shinar*, Cd. 82; Th. 102, 15; Gen. 1700. Oþ ðæt hie on Sodoman weall-steápe burg wlítan meahton *till they on Sodom's lofty-walled fortress might look*, 109; Th. 145, 7; Gen. 2402. Ðær se hálga heáh, steáp reced, burh timbrede *there the holy man built a high, steep dwelling, a walled town*, 137; Th. 172, 6; Gen. 2840. Burge weall *the wall of a city; murus*, Ps. Th. 17, 28. Ðæt hie geseón mihten ðære wlitegan byrig weallas *that they might see the walls of the beautiful city*, Judth. 11; Thw. 23, 24; Jud. 137: Ps. Th. 44, 13: 47, 11. On leófre byrig and háligre *in montem sanctificationis suæ*, 77, 54: 77, 67. Ðá férdon híg þurh ða burhga *egressi circuibant per castella*, Lk. Bos. 9, 6. Eádweard cyng fór mid fierde to Bedan forda, and beget ða burg *king Edward went with an army to Bedford, and gained the walled town*, Chr. 919; Th. 192, 24, col. 1. Ge binnan burgum, ge búton burgum *both within walled towns, and without walled towns*, L. Edg. S. 3; Th. i. 274, 7. Ðone æðeling on ðære byrig métton, ðær se cyning ofslægen læg *they found the atheling in the inclosure of the dwelling, where the king lay slain*, Chr. 755; Th. 84, 19, col. 1: L. Edm. S. 2; Th. i. 248, 16: L. Eth. iii. 6; Th. i. 296, 5.   II. *a fortress* or *castle being necessary for the protection of those dwelling together in cities* or *towns,—a city, town, burgh, borough; urbs, civitas, oppidum:*—Rôma burh *the city Rome*, Bd. 1, 11; S. 480, 10, 12. Ða ðe in burh móton gongan, in Godes ríce *those they may go into the city*, [*may go*] *into God's kingdom*, Cd. 227; Th. 303, 16; Sat. 613. Ðonne hý hweorfaþ in ða hálgan burg *when they pass into the holy city*, Exon. 44 b; Th. 150, 26; Gú. 784. Ðæt he gesáwe ða burh *ut videret civitatem*, Gen. 11, 5. Ða burh ne bærndon *they burnt not the city*, Ors. 2, 8; Bos. 52, 8. Burge weard *the guardian of the city*, Cd. 180; Th. 226, 19; Dan. 173: Ps. Th. 9, 13. Ðonne hí eów êhtaþ on ðysse byrig *cum persequentur vos in civitate ista*, Mt. Bos. 10, 23: Exon. 15 b; Th. 34, 14; Cri. 542. Binnan ðære byrig *within the city*, Ors. 2, 8; Bos. 52, 4. Beóþ byrig mid Iudéum

getimbrade *ædificabuntur civitates Judæ*, Ps. Th. 68, 36. Byrig fægriaþ *towns appear fair*, Exon. 82 a; Th. 308, 32; Seef. 48. Ðá ongan he hyspan ða burga *tunc cœpit exprobrare civitatibus*, Mt. Bos. 11, 20. On burgum *in the towns*, Beo. Th. 105; B. 53. [*Piers P. Chauc.* burghe: *R. Brun.* burgh: *R. Glouc.* borȝ: *Laym.* burh: *Orm.* burrh: *Plat.* borch, *f.*: *O. Sax.* burg, *f. urbs, civitas*: *Frs.* borge, *m. f.*: *O. Frs.* burch, burich, *f.*: *Dut.* burgt, *f.*: *Kil.* borg, borght: *Ger.* burg, *f. arx, castellum*: *M. H. Ger.* burc, *f.*: *O. H. Ger.* buruc, burg, *f. urbs, civitas*: *Goth.* baurgs, *f.*: *Dan.* borg, *m. f.*: *Swed.* borg, *m.*: *O. Nrs.* borg, *f.*] DER. ealdor-burh [-burg], fóre-, freó-, freoðo-, gold-, heáfod-, heáh- [heá-], hleó-, hord-, in-, leód-, mæg-, medo-, meodu-, rand-, rond-, sceld-, scild-, scyld-, stán-, under-, weder-, wín-, wyn-.

burh-ágende; *part. Possessing a fortress.* v. burg-ágende.

burh-bót, e; *f. The repairing of fortresses, which was one of the burdens on all landed property; urbium vel castrorum instauratio,* L. Eth. v. 26; Th. i. 310, 23: vi. 32; Th. i. 322, 31: L. C. S. 10; Th. i. 380, 27: L. R. S. 1; Th. i. 432, 2.

burh-brece *a breaking into a castle,* L. In. 45; Th. i. 130, 6, note 9. v. burh-bryce.

burh-bryce, -brice, es; *m. A breaking into a castle or dwelling,—the fine for this burglary,* L. In. 45; Th. i. 130, 6, note 9: L. Alf. pol. 40; Th. i. 88, 7, note 16. v. burg-bryce.

burh-ealdor, -ealder; *gen.* -ealdres; *m. A ruler of a city, mayor, citizen; urbis præfectus, municeps,* Ælfc. Gr. 14; Som. 16, 55: 9, 55; Som. 13, 24.

burh-fæsten, es; *n. A city-fastness, fortress, citadel; arx munita, castellum*:—Com God sceáwigan beorna burhfæsten *God came to view the chieftains' city-fastness,* Cd. 80; Th. 101, 10; Gen. 1680.

burhg, e; *f. A fortress, city, walled-town*:—Férdon híg þurh ða burhga *egressi circuibant per castella,* Lk. Bos. 9, 6: Bd. 4, 1; S. 563, 12. v. burh.

burh-gata *city-gates,* Jos. 2, 5. v. burh-geat.

burh-geat, -gat, burg-, es; *pl. nom. acc.* u, a, o; *n. A city-gate; urbis porta*:—Æt burhgeate *at the city-gate,* Cd. 111; Th. 146, 22; Gen. 2426. Mid ðam ðe ða burhgata belocene wurdon *cum portæ clauderentur,* Jos. 2, 5. Fóre burg-geatum *before the city-gates,* Andr. Kmbl. 1679; An. 842: Exon. 120 a; Th. 461, 20; Hö. 38.

burh-geat-setl, es; *n. A town-gate-seat, where a court was held for trying causes of family and tenants; ad urbis portam sedes,* L. R. 2; Th. i. 190, 15.

burh-gemót, es; *n. A BURGMOTE, city-moot, meeting of townsmen, corporation; urbis comitia*:—Hæbbe man þriwa on geáre burhgemót *thrice in a year let a city-moot be held,* L. Edg. ii. 5; Th. i. 268, 3: L. C. S. 18; Th. i. 386, 4.

burh-geréfa, an; *m. A BOROUGH-REEVE, city-reeve, the governor and chief magistrate of a city or town; urbis prætor, præfectus, præpositus, quæstor, curialis,* Wrt. Voc. 18, 7: 18, 42.

burh-geþingþ, -geþincgþ, e; *f. The city council or assembly,* L. Eth. iii. 1; Th. i. 292, 7. v. ge-þingþ.

burhge weardas; *pl. m.* [=burge weardas] *The guardians of the city,* Cd. 212; Th. 262, 6; Dan. 740.

burh-hleoþ, burg-hleoþ, es; *n. A fortress-height, the hill on which a city is built; clivus montis, in quo arx vel urbs sita est*:—Forbærned burhhleoðu *scorched fortress-heights,* Cd. 146; Th. 182, 3; Exod. 70. Ic eom brungen of burghleoðum *I am brought from fortress-heights,* Exon. 107 b; Th. 409, 17; Rä. 28, 2. v. beorh-hliþ.

burh-land, es; *n. City-land; urbis solum.* v. burg-lond.

burh-leóde; *nom. acc; gen.* -leóda; *dat.* -leódum; *pl. m. Town-people, citizens; cives*:—Him ða burhleóde wiðcwædon *the citizens withstood him,* Ors. 3, 7; Bos. 61, 6: Cd. 226; Th. 300, 7; Sat. 561: Judth. 11; Thw. 24, 14; Jud. 187: 11; Thw. 24, 6; Jud. 175. [*O. Sax.* burg-liudi *incolæ, cives.*]

burh-loca, burg-loca, an; *m. A city-inclosure, city-barrier or defence, as—a wall, mound or moat; urbis septum, arcis claustrum vel clausura*:—He geládde brýd mid bearnum under burhlocan, in Ságor *he led his wife with the children within the city-inclosure, into Zoar,* Cd. 118; Th. 153, 12; Gen. 2537; Andr. Kmbl. 2132; An. 1067: Beo. Th. 3860; B. 1928. He nænige forlét under burglocan bendum fæstne *he left not one under the city-barriers fast in bonds,* Andr. Kmbl. 2075; An. 1040: 1879; An. 942.

burh-man, -mann, es; *m. A townsman, citizen; urbanus, civis*:—Burhman *vel* burhsita *urbanus,* Ælfc. Gl. 50; Som. 65, 103; Wrt. Voc. 34, 32: Nathan. 1.

burh-ræced, es; *n. A city-dwelling.* v. burg-ræced.

burh-ræden, -rædenn, e; *f. Citizenship; municipatus,* Cot. 128.

burh-riht, es; *n. The civil law; jus civile,* Som. v. riht *law.*

burh-rúnan; *pl. f.* [-rúne, an; *f.*] *The fates, furies, fairies; parcæ, furiæ, oreades*:—Burhrúnan *furiæ,* Cot. 92.

burh-sæl, es; *m. A castle-hall, city-dwelling.* v. burg-sæl.

burh-sǽta, an; *m. A dweller in a city, citizen; civis.* v. burh-séta.

burh-scipe, es; *m. A township,* [BOROUGH-SHIP], *free borough, an*

incorporated city or town; municipium, Ælfc. Gr. 10; Som. 14, 50: Ælfc. Gl. 54; Som. 66, 104. DER. ge-burh-scipe.

burh-soir, e; *f. A city-boundary, city-liberty; urbis territorium*:—Ða yfelan leóda fíf burhscíra ðæs Sodomítisces eardes *the evil people of the five city-boundaries of the Sodomitish land,* Ælfc. T. 7, 20: Jos. 13; Thw. 152, 9: Cot. 148.

burh-sele, es; *m. A castle-hall, city-dwelling.* v. burg-sele.

burh-séta, es; *m. A city-dweller, townsman, citizen; civis, oppidanus,* Wrt. Voc. 18, 36. v. burh-sǽta.

burh-síta, an; *m. A city-dweller, citizen*:—Burhsíta *urbanus,* Wrt. Voc. 34, 32. v. burh-sǽta.

burh-sittend, burg-sittend, es; *m. A city-dweller, an inhabitant of a city, citizen; urbis incola, civis*:—Ðú scealt sunu ágan, ðone sculon burhsittende Isaac hátan *thou shalt have a son, whom the city-dwellers shall call Isaac,* Cd. 106; Th. 140, 12; Gen. 2326: 136; Th. 172, 2; Gen. 2838. Ðá wurdon blíðe burhsittende *then the citizens became merry,* Judth. 11; Thw. 23, 37; Jud. 159: Cd. 188; Th. 235, 1; Dan. 299. Ðæt is wíde cúþ burhsittendum *that is widely known to the city-dwellers,* Cd. 135; Th. 170, 18; Gen. 2815: 210; Th. 261, 11, 23; Dan. 724, 730. His gebídan ne mágon burgsittende *citizens cannot wait for him,* Bt. Met. Fox 27, 34; Met. 27, 17: Elen. Kmbl. 552; El. 276. v. burh-sittende.

burh-sittende, burg-sittende; *part. City-dwelling, inhabiting a city; urbem incolens*:—He folgode ánum burhsittendum men ðæs ríces adhæsit uni civium regionis illius, Lk. Bos. 15, 15. Folca bearn burgsittende *the sons of men dwelling in cities,* Cd. 52; Th. 66, 24; Gen. 1089: Exon. 12 b; Th. 21, 20; Cri. 337. Burgsittendra, *gen. pl.* 106 b; Th. 407, 10; Rä. 26, 3.

burh-spræc, -spæc, e; *f. Civil or courtly speech, polite behaviour, urbanity; urbanus sermo, urbanitas,* Cot. 202.

burh-staðol, es; *m. A dwelling in a city, a mansion, house; urbana sedes, mansio, habitaculum.* v. burh, staðol in staðel.

burh-steal, es; *m. A city-place; arcis locus, arx.* v. burg-steal.

burh-stede, burg-stede, es; *m. A city-place, city; urbis locus, urbs*:—On ðam burh-stede *in that city,* Cd. 52; Th. 65, 7; Gen. 1062: 174; Th. 218, 31; Dan. 47. Hí ágon beorhtne burhstede *they shall have a bright city-place,* 221; Th. 287, 6; Sat. 363: Beo. Th. 4522; B. 2265. Æfter burhstedum *through the cities,* Andr. Kmbl. 1161; An. 581. Se burgstede wæs blissum gefylled *the city-place was filled with joys,* Exon. 52 a; Th. 181, 10; Gú. 1291: 124 a; Th. 476, 3; Ruin. 2.

burh-þelu, e; *f. A castle-floor.* v. buruh-þelu.

burh-tún, es; *m. A city-inclosure, city-dwelling; urbis septum, urbana domus.* v. burg-tún.

burh-waran; *gen.* -warena; *pl. m. Inhabitants of a city; cives*:—Wearþ eal here burhwarena blind *all the multitude of the city-inhabitants became blind,* Cd. 115; Th. 150, 13; Gen. 2491. v. burg-waran.

burh-ware, burg-ware; *gen.* a; *dat.* um; *pl. m. Inhabitants of a city, citizens; urbis incolæ, cives*:—Him cyrdon to mǽst ealle ða burhware *almost all the inhabitants of the city turned to him,* Chr. 919; Ing. 133, 15. Se geháten wæs mid ðǽm burhwarum Brutus *he was called Brutus by the citizens,* Bt. Met. Fox 10, 93; Met. 10, 47. Ofer burhware *over the inhabitants,* Cd. 181; Th. 226, 31; Dan. 179. Wurdon burgware blíðe on móde *the citizens were blithe in mood,* Andr. Kmbl. 3164; An. 1585. Ðá wearþ burgwarum éce gefeá *then was to the citizens everlasting joy,* Exon. 18 b; Th. 46, 25; Cri. 742.

burh-waru, burg-waru; *gen. dat.* e; *acc.* e, u; *f. The inhabitants of a city considered as a community, the whole body of citizens; civitas, civitatis populus*:—Ǽlc burhwaru wæs búgende to him *non fuit civitas quæ se traderet illis,* Jos. 11, 19. Wearþ eall seó burhwaru onstyred *commota est universa civitas,* Mt. Bos. 21, 10: Chr. 1013; Th. 271, 28, col. 1. Wæs mycel nienegu ðære burhware mid hyre *erat turba civitatis multa cum illa,* Lk. Bos. 7, 12. Ic gefrægn leóde tosomne burgwaru bannan *I learnt that the people, the body of citizens, were summoned together,* Andr. Kmbl. 2189; An. 1096.

burh-waru-man, -mann, es; *m. A citizen; civis,* Bd. 1, 7; S. 479, 12.

burh-wealda, an; *m. A city-ruler, citizen; urbis rector, civis,* Bd. 1, 7; S. 479, 12, note 12.

burh-weall, burg-weall, -weal, es; *m. A city-wall; urbis vallum, mœnia*:—Burhweall *mœnia,* Ælfc. Gl. 55; Som. 66, 116; Wrt. Voc. 36, 36. Léton ðone hálgan burhwealle néh *they left the saint near the city-wall,* Andr. Kmbl. 1666; An. 835. Beorhte burhweallas *bright city-walls,* Cd. 220; Th. 282, 31; Sat. 295. Brecan ðone burgweal *to break through the city-wall,* Exon. 83 b; Th. 315, 28; Mód. 38: 22 a; Th. 61, 1; Cri. 978.

burh-weard, es; *m. A city-ward or guardian, city-defender; urbis custos vel defensor*:—Hæfde abrocene burhweardas *had slain the city-guardians,* Cd. 144; Th. 180, 2; Exod. 39; Andr. Kmbl. 1320; An. 660.

burh-wéla, an; *m. City-wealth; urbis opes*:—Þenden he burh-wélan

brúcan môste *while he might have the enjoyment of city-wealth*, Beo. Th. 6191; B. 3100.

burh-wered, es; *n. A city-multitude*; urbis multitudo :—Heánra burhwered *vulgus vel plebs*, Wrt. Voc. 18, 37.

burh-wígende; *part. pl. City-warring*. v. burg-wígende.

burh-wita, an; *m. A knowing and polished man of the city, city-counsellor, citizen*; urbanus, homo civilis, urbis consiliarius, municeps :—Portgeréfa *vel* burhwita *municeps*, Wrt. Voc. 18, 41.

burig=byrig *to a city*, Ors. 6, 23, MS. C; *the dat. of burh a city.*

BURN, e; *f*: burne, an; *f*: burna, an; *m*. [*from* burnon, *p. pl. of* beornan *to boil, bubble*; fervere] *A bubbling or running water, a* BOURN, *brook, stream, river*; torrens, rivus :—Hefe upp ðíne hand ofer burna and ofer môras *extende manum tuam super rivos et super paludes*. Ex. 8, 5. v. burne, burna. ☞ As a prefix or termination to the names of places, burn or burne denotes that they were near a stream; as, Burnham, Burnley, Bornemouth, Radburne, Swanburne, Sherborne. [*Piers P.* bourn: *Scot.* burn: *Plat.* born, *m*: *O. Sax.* brunno, *m. a source*: *O. Frs.* burna, *m*: *Dut.* born, bron, *f*: *Kil.* borne: *Ger.* brunne, born, *m*: *M. H. Ger.* brunne, burne, *m*: *O. H. Ger.* brunno: *Goth.* brunna, *m*: *Dan.* brönd, *m. f*: *Swed.* brunn, *m*: *Icel.* brunnr, *m*.]

burna, an; *m. A stream, bourn*; torrens, latex :—Burna oððe brôc *latex*, Wrt. Voc. 80, 69. Scîr burna bið gedréfed : brôc bið onwended *the clear stream is disturbed : the brook is turned aside*, Bt. Met. Fox 5, 37; Met. 5, 19. He hine bibaðaþ in ðam burnan *he bathes himself in the stream*, Exon. 57 b; Th. 205, 3; Ph. 107. Burna *latex*, Wrt. Voc. 54, 21. v. burn, burne.

burne, an; *f. Running water, a stream, brook, river*; torrens, rivus :—Burnan flôweþ *aquæ fluent*, Ps. Th. 147, 7. He of stán-clife stearce burnan lædde *he drew a strong stream from the stony rock*, Ps. Th. 135, 17. Se Hælend eóde ofer ða burnan Cedron *Iesus egressus est trans torrentem Cedron*, Jn. Bos. 18, 1. Burnan unrihtwísnysse gedréfdon me *torrentes iniquitatis conturbaverunt me*, Ps. Spl. 17, 5. Aþene ðíne hand ofer ealle flôdas, ge ofer burnan, ge ofer meras, and ofer ealle wæter-pyttas *extende manum tuam super omnes fluvios, et rivos, ac paludes, et omnes lacus aquarum*, Ex. 7, 19. Wit unc in ðære burnan baðodan ætgædre *we two bathed together in the brook*, Exon. 121 b; Th. 467, 1; Hö. 132. v. burn. DER. wylle-burne.

burne *hast burnt, wast on fire*; *p. of* beornan.

burne *burned*, Ors. 4, 7; Bos. 88, 45; *subj. p. of* beornan.

burn-sele, es; *m*. [burn *a spring, brook*; sele *a dwelling, mansion*] *A bath-house*; balneum, Exon. 124 a; Th. 477, 10; Ruin. 22.

búr-reáf, es; *n*. [búr *a chamber*, reáf *a garment*] *Hangings for a chamber, tapestry*; tapete, Th. Diplm. 530, 36.

burste *hast burst, broken, failed*; burston *burst, broken*, Beo. Th. 1640; B. 818; *p. of* berstan.

búr-þegen, -þên, es; *m*. [búr *a chamber*, þegen *a servant, attendant*] *A chamber-servant, chamberlain, chancellor, secretary*; cubicularius, cancellarius, scriniarius :—Búrþên *cubicularius*, Ælfc. Gl. 27; Som. 60, 100; Wrt. Voc. 25, 40. His þeóden þanc gesæde ðam búrþêne *his chief gave thanks to the chamberlain*, Byrht. Th. 135, 20, note; By. 121. Búrþên *cancellarius vel scriniarius*, Ælfc. Gl. 114; Som. 80, 22; Wrt. Voc. 61, 3.

burþre, an; *f. A birth, issue*; natus, partus :—Þurh ða burþran we wæron gehælde, and þurh ðæt gebeorþor we wurdon alýsde *through the issue we were saved, and through the birth we were redeemed*, Homl. Blick. 105, 2b.

burug *a city*, Mt. Kmbl. Lind. 5, 14. v. burh.

buruh *a castle, city*, Fins. Th. 72; Fin. 36: Ors. 5, 5; Bos. 105, 24: Mt. Foxe 10, 11. v. burh.

buruh-þelu, e; *f*. [burh *a castle*, þelu *a plank, board*] *A castle-floor*; arcis tabulatum :—Buruhþelu dynede *the castle-floor sounded*, Fins. Th. 61; Fin. 30.

búta=[be, út *out*] *Without*; extra :—Búta ðæt lond *extra regionem*, Mk. Lind. Rush. War. 5, 10. Búta ðæm wîngeard *extra vineam*, Mt. Lind. War. 21, 39. v. bútan; *prep.*

búta; *adv. Without*; foras, foris :—He eóde búta *exiit foras*, Mk. Lind. War. 14, 68. Petrus stôd to dura búta *Petrus stabat ad ostium foris*, Jn. Rush. War. 18, 16.

búta, búte; *conj. Unless*; nisi :—Ænig mon wât ðone sunu búta ðe Fæder *nemo novit filium nisi Pater*, Mt. Lind. War. 11, 27. Búta ðes ûtacunda *nisi hic alienigena*, Lk. Lind. War. 17, 18. v. bútan; *conj.*

bútá *both*; ambo :—Swelton hîg bútá *they both shall die*, Deut. 22, 22: Exon. 113 b; Th. 436, 25; Rä. 55, 6. v. bútú.

bútan, búton, bútun; *prep.* [be, útan *out*]. I. *with the dative*; cum dativo. 1. *out of, against*; extra, contra :—Forbærn ðæt celf bútan ðære wicstôwe *ipsum vitulum comburet extra castra*, Lev. 4, 21. Bútan leódrihte *against the law of the land*, Andr. Kmbl. 1357; An. 679. 2. *without, except*; sine, absque, præter :—Bútan leahtre *sine crimine*, Mt. Bos. 12, 5. Bútan ânum cnihte *excepto uno puerulo*, Bd. 3,

---

23; S. 555, 26. Bútan geþeahte *without thought*, 3, 1; S. 523, 31. Bútan ende *without end*, Exon. 11 b; Th. 17, 16; Cri. 271: L. E. I. prm; Th. ii. 400, 28. II. *with the accusative*; cum accusativo. 1. *out of*; extra :—He lædde hine bútan ða wíc *eduxit eum extra vicum*, Mk. Bos. 8, 23. 2. *without, except*; sine, præter :—Bútan sealm *præter psalmodiam*, Bd. 3, 27; S. 559, 10. III. sometimes *bútan* is separated from its case :—Ðæt wæs geworden bútan weres frigum *that came to pass without the favours of man*, Exon. 8 b; Th. 3, 17; Cri. 37. [*Chauc.* but: *R. Brun.* bot: *R. Glouc.* bote: *Laym.* bute, bote: *Orm.* buttan, butt: *O. Sax.* bútan, bôtan: *Frs.* buten: *O. Frs.* buta: *Dut.* buiten: *Kil.* buyten: *Ger.* bauszen.]

bútan, búton, bútun; *conj.* [be, útan *out*]. I. *with the subj. Unless, save that*; nisi :—Bútan ðú [eorþan spéde] gedælde Dryhtne sylfum *unless thou hadst bestowed [the riches of the earth] for the Lord himself*, Exon. 99 a; Th. 371, 19; Seel. 78. Búton ðæt hit sý útaworpen *nisi ut mittatur foras*, Mt. Bos. 5, 13. Bútan ær wyrce êce Dryhten ende worlde *save ere the eternal Lord shall work an end of the world*, Exon. 98 a; Th. 367, 24; Seel. 12. II. *with the ind. Save or except that*; nisi :—Egorhere eall acwealde búton ðæt earce bord heóld heofona freá *the water-host destroyed all save that the Lord of heaven held the ark board*, Cd. 70; Th. 84, 26; Gen. 1403. III. *without a dependent verb, Except, save, besides, but*; nisi :—Ond eallum dagum bútan sunnan dagum *diebus cunctis excepta dominica*, Bd. 3, 23; S. 554, 32. Ic ne gehýrde bútan hlimman sæ *I heard nought save the sea roaring*, Exon. 81 b; Th. 307, 4; Seef. 18. Sume men sædon ðæt ðær næran bútan twegen dælas *some men said that there were but two parts*, Ors. 1, 1; Bos. 15, 6.

búte *without*; foris, Jn. Lind. War. 18, 16. v. búta; *adv.*

búte; *conj. Unless, but*; nisi, sed :—Nán þing wyrþe [geweorþe Cot.] búte hit God wille *nothing comes to pass unless God wills it*, Bt. 41, 2; Fox 244, 18: Bt. Met. Fox 18, 20; Met. 18, 10. Búte ic nât *but I know not*, Bt. 34, 10; Fox 148, 16. Búte ge to him gecyrren *nisi convertimini*, Ps. Th. 7, 12. v. bútan; *conj.*

búte *both*; ambo :—Búte ða þinc *ambæ res*, R. Ben. interl. 5. v. bútú.

butere, an; *f.* BUTTER; butyrum [= βούτυρον, βοῦς *a cow*, τυρός *cheese*] :—Butere *butyrum*, Wrt. Voc. 82, 27. Dô ðonne mele fulne buteran *add then a basin full of butter*, L. M. 1, 36; Lchdm. ii. 86, 17, 19, 22. On ðære buteran *in the butter*, 1, 36; Lchdm. ii. 88, 1. On gôdre buteran *in good butter*, 3, 32; Lchdm. ii. 326, 18 : 3, 41; Lchdm. ii. 334, 14². Ahlyttre ða buteran *purify the butter*, 3, 2; Lchdm. ii. 308, 28: Coll. Monast. Th. 34, 27. [*Wyc.* butter: *Plat.* botter, *f*: *Frs.* buter: *O. Frs.* butera, botera: *Dut.* bóter, *f*: *Ger.* butter, *f*: *M. H. Ger.* buter: *O. H. Ger.* butere, *f*: *Fr.* beurre, *m*: *It.* butirro, burro, *m*: *Lat.* butyrum: *Grk.* βούτυρον.]

buter-flége *a butterfly*; papilio. v. buttor-fleóge.

buter-geþweor, es; *n. Butter-curd, what is coagulated, butter*; butyri coagulum, butyrum :—Buter-geþweor ælc and cýsgerunn losiaþ [MS. losaþ] eów *butyrum omne et caseus pereunt vobis*, Coll. Monast. Th. 28, 19.

buteric *a bottle*, Coll. Monast. Th. 27, 35. v. buteruc.

buter-stoppa, an; *m*. [butere *butter*, stoppa *a vessel*] *A butter-vessel, butter-dish*; butyri vas, Wrt. Voc. 290, 24.

buteruc, buteric, buturuc, butruc, es; *m. A leathern bottle*; flasco, uter :—Buteruc *flasco*, Ælfc. Gl. 26; Som. 60, 76; Wrt. Voc. 25, 16. Ic bicge hýda and fell, and wyrce of him butericas *ego emo cutes et pelles, et facio ex iis utres*, Coll. Monast. Th. 27, 35. Ðæt wæter asceortode, ðe wæs on ðam buturuce *consumpta esset aqua in utre*, Gen. 21, 15. Butruc *flasco*, Wrt. Voc. 85, 83. [*O. Sax.* buteric, *m*: *O. H. Ger.* butrih *uter*.]

búton *without*; sine :—Búton ælcum eorþlicum fæder *without any earthly father*, Homl. Th. i. 24, 30. Búton synne ânum *without any sin*, i. 24, 35. v. bútan; *prep.*

búton; *adv. Gratuitously, without a cause*; gratis :—Forðan ðe búton hí behíddon me onforwyrde *quoniam gratis absconderunt mihi interitum*, Ps. Spl. 34, 8.

butruc *a bottle*, Wrt. Voc. 85, 83. v. buteruc.

butsa-carlas [bâtes carlas, *i. e.* bât-sæ carlas] *Seamen, sailors*; nautæ, Chr. 1066; Ing. 259, 4.

Butting-tún, es; *m.* BODDINGTON, *Gloucestershire* :—Offóron hie ðone here hindan æt Buttingtúne on Sæferne staðe *they followed after the army to Boddington on the bank of the Severn*, Chr. 894; Erl. 92, 22. Mr. Earle has the following pertinent note on the locality :—Two places have hitherto contended for this site, viz. Boddington near Cheltenham, and Buttington in Montgomeryshire, near Welshpool. But Mr. Ormerod [*Archæologia*, vol. xxix; and *Strigulensia*, p. 60] has put forward a claim for Buttinton in Tidenham, on the peninsula formed by the Severn and the Wye. There are traces of works here, though less considerable than those at Buttington in Montgomeryshire. Mr. Ormerod grounds his claim mainly upon Matthew of Westminster's 'paganos tam navali quam terrestri exercitu circumcinxit.' No such thing appears in the text before us, but to the opposite effect. One is almost tempted to suspect that

this 'Verwirrer der Geschichte' [as Lappenberg calls Matthew of West-minster] caught sight of 'sciphere' in the next line, and imagined the rest. But it must be allowed, Mr. Ormerod's position has its advantages. It does not, however, suit 'ðā up þe Sæferne,' if this means that they went up stream, which would seem to be its meaning, though not in Florence, Chr. Erl. notes, p. 318.

buttor-fleóge, an; f. [butere *butter*, fleóge *a fly*] A BUTTERFLY; papilio, Ælfc. Gl. 22; Som. 59, 115; Wrt. Voc. 23, 70. [*Ger.* butter-fliege, f.] DER. niht-buttorfleóge.

bútū [bū = bā *both*, tū = twā *two*] *Both*; ambo :—Đonne beóþ bútū gehealden *then both [the two] shall be preserved*, Mk. Bos. 2, 22. Đā bútū abulgon Isaace and Rebeccan *then both [the two] were a grief to Isaac and Rebecca*, Gen. 26, 35 : Lk. Bos. 1, 6, 7. Wit him bútū sprecaþ *we both* [lit. *we two both*] *speak to him*, Cd. 27; Th. 36, 20; Gen. 574: 39; Th. 52, 4; Gen. 838: 40; Th. 52, 22; Gen. 847. Đǽr hie sǽton bútū *where they both* [lit. *they two both*] *sat*, 133; Th. 168, 8; Gen. 2779. v. bātwā.

bútun *without* :—Bútun geongum litlingum, and heordum *absque parvulis, et gregibus*, Gen. 50, 8. v. bútan; *prep.*

bútun *unless, save* ; nisi, Mt. Bos. 11, 27 : 12, 4. v. bútan; *conj.*

buturuc *a bottle*, Gen. 21, 15. v. buteruc.

búwian; *p.* ode; *pp.* od *To inhabit*; inhabitare :—Búwa eorþan *inhabita terram*, Ps. Th. 36, 3. v. búgian.

bý, es; n? *A dwelling, habitation*; habitatio :—Se ðe hús oððe lytel [MS. lytelo] bý hæfde in byrgenum [MS. byrgennum] *qui domicilium habebat in monumentis*, Mk. Skt. Lind. 5, 3. Hence, by and bye in the termination of the names of places. v. bū.

BYCGAN, bicgan, bycgean; ic bycge, bicge, ðū bygest, he bygeþ, *pl.* bycgaþ, bicgaþ; *p.* bohte, *pl.* bohton; *impert.* byge, bige, *pl.* bycgaþ; *pp.* boht; *v. a. To buy, procure*, redimere :—Hí woldon mete bicgan *cibos emerent*, Jn. Bos. 4, 8. Đæt hie bicgan sceoldon *which they must buy*, Beo. Th. 2615; B. 1305 : Exon. 120 b; Th. 463, 11; Hö. 68. Đā híg férdon bycgean *dum irent emere*, Mt. Bos. 25, 10. Ic bicge *I buy*, Salm. Kmbl. 403; Sal. 202. Mete bygeþ he *he buys meat*, Exon. 90 b; Th. 340, 14; Gn. Ex. 111. Hí bycgaþ *they buy*, 33 b; Th. 106, 27, note; Gū. 47. Đæt góde men mid feó bicgaþ *which good men buy with money*, 114 a; Th. 436, 37; Rä. 55, 12. Đæt bohte Abraham *quam emit Abraham*, Gen. 49, 30: Chr. 963; Erl. 123, 27. Menn heora land bohton [MS. bohtan] *men bought their land*, Chr. 1066; Erl. 203, 10. Bige us to ðæs cynges þeówette *eme nos in servitudinem regiam*, Gen. 47, 19. Bige ða þing *eme ea*, Jn. Bos. 13, 29. Bycgaþ eów ele *emite oleum vobis*, Mt. Bos. 25, 9. [*Wyc.* bigge, bye, biȝe : *R. Brun.* bie : *Laym.* bugge : *Orm.* biggenn : *O. Sax.* buggean : *Frs.* bikje : *Goth.* bugyan.] DER. a-bycgan, -bicgan, be-, ge- : un-boht, unbe-, unge-.

bycgean *to buy, procure*; emere :—Híg woldon bycgean *they would buy*, Mt. Bos. 25, 10. v. bycgan.

bycgen, bycgenn, e ; f. *A buying, selling*; emptio, Som. Ben. Lye.

bȳcnend-líc; *adj. Allegorical, mystical*; allegoricus :—Býcnendlíc racu *allegorica expositio*, Bd. 5, 23; S. 647, 42. v. beácniend-líc.

bȳcnian, býcnan; *p.* ode; *pp.* od *To beckon, shew, signify*; indicare :—Niht niht býcneþ *nox nocti indicat*, Ps. Spl. 18, 2. Đe býcnaþ [gehiwode *finxit*, Lamb : býcnaþ *fixit?*] eáge *qui finxit oculum*, 93, 9. v. beácnian.

bȳcniend-líc gemet, es ; n. *The indicative mood*; indicativus modus, Ælfc. Gr. 21; Som. 23, 18, MS. C.

bȳcnung, e; f. *A figure, trope*; figura :—Under býcnunge ðæs bíges *sub figura coronæ*, Bd. 5, 22; S. 644, 10. v. beácnung.

bȳd = beád? *commanded*, bid, Gen. 50, 5 ; *p.* of beódan.

bȳdel, es; m. [beódan *to bid, order, proclaim*] I. *one who bids or cries out, a herald, proclaimer, minister*; præco, nuncius :—Býdel *præco*, Ælfc. Gr. 47; Som. 48, 41 : Wrt. Voc. 84, 40. Se Godes býdel *a messenger of God, minister*, Homl. Th. ii. 530, 2. Se Godes Sunu sende his býdel tofóran him *the Son of God sent his proclaimer before him*, ii. 36, 25, 27. Bisceopas sindon býdelas Godes lage *bishops are proclaimers of God's law*, L. C. E. 26; Th. i. 374, 15. Biscopas sind to býdelum gesette *bishops are ordained to be ministers*, Homl. Th. ii. 320, 8. Drihten sende his býdelas ætfóran him *the Lord sent his messengers [prophets] before him*, ii. 530, 9. II. *one who bids or summons to appear in a court of law, a* BEADLE; apparitor, exactor, bedellus :—Đē sylle se dēma ðam býdele, and se býdel ðē sende on cwertern *judex tradat te exactori, et exactor mittat te in carcerem*, Lk. Bos. 12, 58. Býdele gebýraþ, ðæt he for his wycan sý weorces frigra ðonne óðer man *bedello pertinet, ut pro servitio suo libertior sit ab operatione quam alii homines*, L. R. S. 18; Th. i. 440, 6. He þurh his býdelas his gafoles myngaþ *he reminds him of his tribute by his messengers* [lit. *beadles*], L. Edg. S. 1; Th. i. 270, 19. Aaron hēt býdelas beódan,—to morgen biþ simbeldæg *Aaron commanded beadles to proclaim,—to-morrow is a feast day*, Ex. 32, 5. [*Piers P.* bedele : *Dut.* beul, *m*: *Ger.* büttel, *m*: *M.H.Ger.* bütel : *O.H.Ger.* butil, *m*.]

BYDEN, bydenn, e; f. I. *a bushel*; modius :—Cwyst ðū cymþ ðæt leóhtfæt ðæt hit beó under bydene aset *numquid venit lucerna*

ut sub modio ponatur? Mk. Bos. 4, 21: Lk. Bos. 11, 33. II. *a barrel, tun, butt*; dolium, cupa :—Hí mec baðedon in bydene *they bathed me in a tub*, Exon. 107 b; Th. 409, 24; Rä. 28, 6. Byden *cupa*, Ælfc. Gl. 49; Som. 65, 94; Wrt. Voc. 34, 24. [*O. H. Ger.* butin *cupa*.]

byden-botm, es ; *m. The bottom of a vessel*; fundus, Ælfc. Gl. 25; Som. 60, 49; Wrt. Voc. 24, 49.

bȳe *to a habitation*; dat. of bý.

byffan *to mutter*; mutire, Cot. 154. DER. a-byffan.

byfian; *p.* ode; *pp.* od *To tremble*; tremere :—Eorþe ondréd oððe byfode and heó geswāc oððe heó wæs stille *terra tremuit et quievit*, Ps. Lamb. 75, 9. v. bifian.

byfor, es; *m. A beaver*, Ælfc. Gr. 8; Som. 7, 13, MS. T. v. befer.

bȳgan, bígan, bígean, bégan; he bȳgeþ; *p.* de; *pp.* ed; *v. trans. To bow, bend, turn, turn back, bow down, humble, abase;* flectere, inflectere, incurvare, retorquere, deflectere, humiliare :—Bȳgdest ðū ðē for hæleðum *thou bowedst thyself before men*, Exon. 100 a; Th. 376, 11; Seel. 153. Đeáh ðū hwilcne boh bȳge wið eorþan *though thou bend any bough towards the earth*, Bt. Met. Fox 13, 106; Met. 13, 53. Bȳgaþ hine, ðæt he on hinder gāþ *they shall turn him back, so that he shall go backward*, Salm. Kmbl. 252; Sal. 125. He herm-cweðend hýneþ and bȳgeþ *humiliabit calumniatorem*, Ps. Th. 71, 5. [*Dan.* böje, boie : *Swed.* böja : *O. Nrs.* beygja.] DER. for-bȳgan, -bígan, ge-, on-. v. búgan.

bȳge, bíge, es; *m.* [bȳgan *to bow*] *A bowing, bending, turning, a corner, an angle, a bay, bosom, the apex of a helmet*; flexus, ancon, angulus, sinus, conus :—Đā gestóp he to ānes wealles bȳge *then he stepped to a bend of a wall*, Ors. 3, 9; Bos. 68, 23 : Num. 22, 26. Helmes bȳge *conus galeæ*, Wrt. Voc. 36, 3.

bȳgend-líc; *adj. Flexible, pliable*; flexilis, flexibilis :—Bȳgendlíc on ðām geþeódnessum his liða *flexibilibus artuum compagibus*, Bd. 4, 30; S. 608, 37. v. bȳgan.

bȳgest, he bygeþ *buyest, he buys*, Exon. 90 b; Th. 340, 14; Gn. Ex. 111 ; *2nd and 3rd pers. pres.* of bycgan.

byggan *to build*; ædificare, Som. Ben. Lye. v. bȳtlian.

bȳgnes, -ness, e ; f. *A bending, bowing*; flexio. v. bígnes.

byg-spǽc, e ; f. *A beguiling in speech*; supplantatio, Ps. Spl. 40, 10.

bȳgþ, bȳhþ, ðū bȳgst, bȳhst *bows, thou bowest;* *3rd and 2nd pers. pres.* of búgan *to bow.*

byht, es; *m.* [bȳgan *to bend*] *A bending, corner, dwelling, an abode, bay,* BIGHT ; habitatio, dominium, sinus :—Andlang norþgeardes ðæt hit cymþ in ðone byht *along the north yard till it comes to the corner*, Cod. Dipl. 538; A.D. 967; Kmbl. iii. 18, 29 : Cod. Dipl. Apndx. 308; A.D. 875; Kmbl. iii. 399, 25, 32. Eall ðæt sculon āgan eaforan ðíne, þeódlanda gehwilc, folcmægþa byht *thy sons shall own all that, each country, the dwelling of nations*, Cd. 100; Th. 133, 20; Gen. 2213. Mec ahebbaþ ofer hæleða byht ðeós heá lyft *this lofty air raises me above the dwellings of men*, Exon. 103 a; Th. 389, 26; Rä. 8, 3. Ofer wætres byht to lande *over the water's abode [bay] to the land*, Exon. 106 a; Th. 404, 23; Rä. 23, 12. [*Dut.* bogt, *f*: *Ger.* bucht, *f*: *Dan.* bugt, *m.f*: *Swed.* bugt, *m*: *Icel.* bygð, *f*.]

bȳing, e; *f. A habitation*; domus, Mk. Skt. Rush. 5, 3. v. bý.

BȲL, bȳle, bíle, es; *m. A* BILE, *blotch, sore*; carbunculus, Cot. 183. [*O. Frs.* bel, beil : *Dut.* buil, *f*: *Kil.* buyll : *Ger.* beule, *f*: *M.H.Ger.* biule, *f*: *Dan.* bule, *m.f*: *Swed.* bula, *f*: *O. Nrs.* beyla, *f*.]

BYLD, e ; *f*: byldo ; *f. indecl. in s. Constancy, boldness*; constantia :—Bídeþ þurh byldo *awaiteth with constancy*, Exon. 9 b; Th. 8, 5 ; Cri. 113. He sceolde ða byldo anescian *poterat emollire constantiam*, Bd. 1, 7; S. 477, note 43. [*O.H.Ger.* baldî, *f*: *Goth.* balþei, *f.* boldness.] DER. ge-byld.

bylda, an; *m.* [bold *a house*] *A* BUILD⁓R.; ædificator :—Sum biþ bylda til hām to habbanne *one is a good builder to raise a house*, Exon. 79 b; Th. 297, 29; Crä. 75.

byldan; *p.* bylde; *pp.* bylded; *v. trans.* [beald *bold*; v. byld] *To make bold, to animate, instigate, exhort, encourage, confirm*; animare, instigare, hortari, confirmare :—He Fresena cyn byldan wolde *he would encourage the race of the Frisians*, Beo. Th. 2193; B. 1094. Geongne æðeling sceolon góde gesíðas byldan *good companions should encourage a young prince*, Menol. Fox 488; Gn. C. 15. Hí bylde bearn Ælfríces *the son of Ælfric encouraged them*, Byrht. Th. 137, 60 ; By. 209. Swā hí ealle bylde Godríc to gúþe *so Godric encouraged them all to the war*, Byrht. Th. 141, 11; By. 320. Bǽdon hí Sigebyrht ðæt he mid him to ðam gefeohte fóre and hyra fultum trymede and bylde *rogaverunt Sigberctum ad confirmandum militem secum venire in prælium*, Bd. 3, 18; S. 546, 20, col. 1.

bȳle *a bile, blotch, sore.* v. bȳl.

byled-breóst; *adj.* [byled, breóst *a breast*] *Puff-breasted*; rostrato pectore præditus :—Ic eom byled-breóst *I am puff-breasted*, Exon. 127 b; Th. 489, 23; Rä. 79, 1. v. gebilod.

byle-wit *merciful*; æquanimus, mansuetus :—Gehýran ða bylewitan *audeant mansueti*, Ps. Spl. 33, 2. v. byly-wit, bile-wit.

bylg *a bulge, bag*, Cot. 27. v. belg.

bylgan; p. de; pp. ed To offend, anger, vex; offendere, irritare, vexare. DER. a-bylgan. v. belgan.

bylgean to bellow; mugire, Martyr. 17, Jan. v. bellan.

Bylges leg, es; n. [Flor. Bililesleaga = Sim. Dun. Byligesleage: Hovd. Biligesleage] BISLEY, in Gloucestershire:—Hí cômon to Bylges lege they came to Bisley, Chr. 1055; Erl. 190, 15.

bylgþ is angry; 3rd pers. pres. of belgan.

bylig bellows; follis, Wrt. Voc. 86, 15. v. belg.

byllinc a cake; collyris, collyrida, Cot. 208.

bylwet, bylwit simple. v. bile-wit.

bylwet-lîce; adv. Simply; simpliciter, Ors. 1, 2; Bos. 26, 29. v. bile-hwîtlîce.

byly-wit merciful, kind; æquanimus, mitis :—Bylywit fæder merciful father, Cd. 191; Th. 238, 32; Dan. 363. v. bile-wit.

BŶME, bême, an; f. A trumpet; tuba, salpinx = σάλπιγξ :—Bŷme sang the trumpet sounded [lit. sang], Cd. 148; Th. 186, 2; Exod. 132. Dære bŷman swêg ŵeóx sonitus buccinæ crescebat, Ex. 19, 19: 20, 18: Ps. Spl. 46, 5: Exon. 23 b; Th. 65, 29; Cri. 1062. Bŷmiaþ oððe hlyriaþ on niwum mônþe mid bŷman buccinate in neomenia tuba, Ps. Lamb. 80, 4. Bŷman sungon the trumpets sounded [lit. sung], Elen. Kmbl. 218; El. 109. Drêmaþ Drihtne on bŷman psallite Domino in tubis, Ps. Lamb. 97, 6. Seofon sacerdas blâwon mid bŷmon septem sacerdotes clangent buccinis, Jos. 6, 4, 13. [Laym. bemen, beomen; pl. trumpets.] DER. heofon-bŷme, here-, sige-.

bŷmere, es; m. [bŷme a trumpet] A trumpeter; tubicen, salpista = σαλπιστής :—Bŷmere tubicen, Ælfc. Gr. 9, 12; Som. 9, 24; Wrt. Voc. 73, 57. Bŷmere salpista, Ælfc. Gl. 114; Som. 80, 11; Wrt. Voc. 60, 47.

bŷme-sangere, es; m. [bŷme a trumpet, sangere a singer] A trumpeter; salpicta = σαλπιγκτής, Ælfc. Gl. 114; Som. 80, 13; Wrt. Voc. 60, 48.

bŷmian; p. ode; pp. od [bŷme a trumpet] To sound or play on a trumpet; tuba canere, buccinare :—Ic bŷme salpizo vel buccino, Ælfc. Gl. 114; Som. 80, 14; Wrt. Voc. 60, 49. Bŷmiaþ oððe hlyriaþ on niwum mônþe mid bŷman buccinate in neomenia tuba, Ps. Lamb. 80, 4.

bŷn; def. se bŷna, seó, ðæt bŷne; adj. [bŷþ; pres. of bûan to inhabit, occupy] Inhabited, occupied; habitatus :—Ðæt bŷne land is eásteweard brâdost the inhabited land is broadest eastward, Ors. 1, 1; Bos. 20, 45. Licgaþ wilde môras on emnlange ðæm bŷnum lande wild mountains lie along the inhabited land, 1, 1; Bos. 20, 44.

byndele, byndelle a binding, L. Alf. pol. 35; Th. i. 84, 1, MS. H. v. bindele.

byóþ are, shall be, = bióþ; pres. pl. of bión.

byro, e; f. A birch-tree; betula :—Byrc betula [MS. betulus], Ælfc. Gl. 47; Som. 65, 20; Wrt. Voc. 33, 20. v. birce.

byro-holt, es; n. A birch holt or grove; betuletum, Ælfc. Gl. 47; Som. 65, 21.

byroþ barks, Ælfc. Gr. 22; Som. 24, 8; pres. of beorcan.

byrd birth; nativitas. v. ge-byrd.

byrd-dæg, es; m. A birth-day; natalis dies. v. ge-byrd-dæg.

byrde; sup. byrdest, def. se byrdesta; adj. Born, well-born, noble, rich; natus, natu vel genere præstans, nobilis, opulentus :—Se byrdesta sceall gyldan the richest must pay, Ors. 1, 1; Bos. 20, 36. DER. ge-byrde, in-. v. ge-byrd.

byrden a burden, Som. Ben. Lye. v. byrðen.

byrdest, se byrdesta the highest born, most noble, richest, Ors. 1, 1; Bos. 20, 36; sup. of byrde.

byrdicge a weaver's tool; plumaria, N. Som. Wrt. Voc. 282, 3.

byrdnys, -nyss, e; f. Quality, state, condition; qualitas, status, conditio. DER. an-byrdnys, in-. v. ge-byrd.

byrd-scype, es; m. [byrd, ge-byrd birth, scype state, condition] Birthship, child-bearing; gestatio, partus :—Ic tô fela hæbbe ðæs byrdscypes bealwa onfongen I have received too many injuries from this child-bearing, Exon. 10 b; Th. 12, 7; Cri. 182.

byrd-tîd, e; f. Birth-tide, time of birth; natale tempus. v. ge-byrd-tîd.

byrdu-scrûd, es; n. [byrdu = bord a shield, scrûd a garment, clothing] The covering of a shield, a shield; clypei tegmen, clypeus :—Unc sceal sweord and helm, byrne and byrduscrûd bâm gemæne sword and helmet, armour and shield, shall be common to us both, Beo. Th. 5313; B. 2660.

byre; gen. byres; dat. byre; acc. byre: pl. nom. acc. byras, byre; gen. byra; dat. byrum; m. A son, child, descendant; natus, filius, soboles, proles :—Ðonne æfre byre monnes hŷrde under heofonum than ever child of man heard under heaven, Exon. 57 b; Th. 206, 18; Ph. 128: Beo. Th. 4113; B. 2053. Ðær hyre byre wæron where her sons were, 2381; B. 1188. Dæs ða byre siððan gyrne onguldon, ðe hî ðæt gyfl þêgun for which their children since with grief have paid, that they ate that fruit, Exon. 61 b; Th. 226, 22; Ph. 409. Mæru cwên bædde byras geonge the illustrious queen solicited her young sons, Beo. Th. 4040; B. 2018. Lamech bearna strŷnde; him byras wôcan eafora and idesa; he ðone yldestan Noæ nemde Lamech begat children; to him descendants were

born of sons and daughters; the eldest he named Noah, Cd. 62; Th. 75, 1; Gen. 1233. [Goth. baur, m. one born, a son: O. Nrs. burr, borr, m.] v. bearn.

bŷre, es; m. An event, the time at which anything happens, a favourable time, an opportunity; eventus, tempus quo accidit aliquid, opportunitas, occasio, = καιρός :—Wæs ðær mid him ôþ ðone bŷre ðæt Swegen wearþ deád was there with him until the time that Sweyn was dead, Chr. 1013; Th. 272, 22. Ðá he bŷre hæfde when he had opportunity, Byrht. Th. 135, 21; By. 121. DER. ge-bŷre. v. ge-bŷrian.

byrele a cup-bearer, butler, Wrt. Voc. 290, 51: Beo. Th. 2327; B. 1161. v. byrle.

byrelian to pour out, give to drink, serve, Exon. 45 b; Th. 154, 13; Gû. 842. v. byrlian.

byren; adj. Belonging to a bear; ursinus, Som. Ben. Lye. v. beren.

byrene, an; f. A she-bear; ursa, Ælfc. Gl. 21; Som. 59, 70; Wrt. Voc. 23, 29. v. bera.

byreþ bears, Beo. Th. 598; B. 296; 3rd pers. pres. of beran.

byreþ it pertains to, it is lawful; pertinet ad, licet, Jn. Lind. War. 10, 13. v. bŷrian.

byrg to a city, Exon. 15 a; Th. 33, 1; Cri. 519; dat. of burh.

byrga of cities or inclosed dwellings, for burga; gen. pl. of burh, Runic pm. 8; Kmbl. 341, 3.

byrga a pledger, creditor, Cot. 37. v. byrgea.

BYRGAN, birgan, byrigan, birigan, birgean, byrigean, byrian; p. de; pp. ed [beorg tumulus]; v. trans. To raise a mound, to BURY; tumulare, tumulo condere, sepelire :—Hí his lîchaman on cyrican neáh weofode byrgan woldon they would bury his body in the church near the altar, Bd. 3, 19; S. 550, 10: Exon. 82 b; Th. 311, 27; Seef. 98. Birge man hine ðæs ilcan dæges sepelietur in eadem die, Deut. 21, 23. Ðær hine man birgde ibi sepelierunt eum, Gen. 49, 31. Alŷf me ærest byrigan mînne fæder permitte mihi primum sepelire patrem meum, Lk. Bos. 9, 59 : 9, 60. Hine man byrigde swâ him wel gebŷrede they buried him as well became him, Chr. 1036; Th. 294, 21; Hy. 10, 29; Hy. Grn. ii. 293, 29. [Wyc. birie: Piers P. yburied, pp: Chauc. buried: R. Glouc. ybured: Laym. burien; Orm. birrȝenn: Dut. bergen: O. Dut. berghen condere, abscondere, servare, tueri: Ger. M. H. Ger. bergen: O. H. Ger. bergan, ga-bergan condere, recondere: Goth. bairgan tueri, conservare: O. Nrs. byrgja includere.] DER. be-byrgan, bi-, ge-: byrgen.

BŶRGAN, bŷrian, bŷrigan, bŷrgean, bŷrigean, beorgan; p. de; pp. ed To taste, eat; gustare, manducare :—Ðú dînes gewinnes wæstme bŷrgest labores fructuum tuorum manducabis, Ps. Th. 127, 2. Nympe ðú æppel ænne bŷrgdest unless thou hast tasted an apple, Cd. 42; Th. 54, 21; Gen. 880. Hî bû þêgun æppel, bŷrgdon forbodene they both ate the apple, tasted the forbidden [fruit], Exon. 61 b; Th. 226, 11; Ph. 404. Nim ðê ðis ofet on hand, bît hit and bŷrge take to thee this fruit in hand, bite it and taste, Cd. 25; Th. 33, 12; Gen. 519. [O. Nrs. bergja to taste; gustare.] DER. a-bŷrgan, ge-, on-.

byrgea, byrigea, byriga, berigea, an; m. [borh, borg a pledge, security] A person who gives a pledge, a surety; fidejussor :—Gif ðú hæbbe byrgean, mana ðone ðæs ângyldes if thou have a surety, admonish him of the recompense, L. In. 22; Th. i. 116, 11. Mid lx scillinga gebête ðam byrgean let amends be made to the surety with sixty shillings, L. Alf. pol. 18; Th. i. 72, 12, 15, 16: L. In. 31; Th. i. 122, 6. Se man ðam ôðrum byrigean geselle let the man give surety to the other, L. H. E. 8; Th. i. 30, 12. Gif he byrigan forwærne if he refuse surety, 9, 10; Th. i. 30, 15, 17. Him man wilsumne berigean geselle [MS. gefelle] let a man give him a sufficient surety, 6; Th. i. 30, 5. DER. leód-gebyrgea.

bŷrgean to taste; gustare :—He byreþ blôdig wæl, bŷrgean þenceþ, eteþ unmurnlîce he will bear off my bloody corpse, will resolve to taste it, will eat it without repugnance, Beo. Th. 901; B. 448. DER. a-bŷrgan. v. bŷrgan.

byrged buried. v. byrgan.

byrgels, birgels, bergels, es; m. A BURIAL-place, sepulchre, tomb; sepulcrum, bustum :—Byrgels bustum, Cot. 183. To birgelse in possessionem sepulcri, Gen. 23, 9. v. byrgen.

byrgen, byrgenn, birgen, byrigen, burgen, e; f. [beorg tumulus] A burying, grave, sepulchre, tomb; sepulcrum, monumentum, tumba :—Byrgen sepulcrum, Ps. Th. 48, 9 : Ps. Surt. 13, 3. Hât nû healdan ða byrgene jube ergo custodire sepulcrum, Mt. Bos. 27, 64 : 27, 66. On ðam wyrt-tûne wæs niwe byrgen in horto erat novum monumentum, Jn. Bos. 19, 41 : 19, 42. Com to ðære byrgene venit ad monumentum, Jn. Bos. 20, 1 : 20, 3, 4, 6, 8, 11. Ðŷ þriddan dæge of byrgenne, of deáðe, arâs Dryhten on the third day the Lord arose from the sepulchre, from death, Elen. Kmbl. 371; El. 186 : 965; El. 484: Exon. 18 b; Th. 45, 34; Cri. 729: Ps. Th. 29, 8. Byrgenum sepulcris, 13, 5: Salm. Kmbl. 445; Sal. 223. On his byrgenne is awriten byrgen-leóþ scriptum est in tumba ipsius epitaphivm, Bd. 2, 1; S. 500, 17.   2. in the districts of England first occupied by the Angles, Saxons, and Jutes, numerous extensive cemeteries of the heathen period have been examined. In these cemeteries the graves are usually arranged in rows, and are dug exactly in the same manner and form as our modern church-

yard graves, which are probably copied from them. After the burial, a low circular mound was raised over the grave. From their contents we learn that the body of the deceased was buried in the full dress worn when living,—the men with their arms and military equipments,—the women with their personal ornaments and jewelry. The body was generally laid on its back, on the floor of the grave; but in the wealthier classes, it was frequently inclosed in a wooden coffin, for in A. D. 679, it is said—Æðeldryþ on treówene þruh wæs bebyriged Æthelrith was buried in a wooden coffin, Bd. 4, 19; S. 588, 21; or in the Latin of Bede—Ædilthryd ligneo in locello sepulta, S. 163, 15. 3. the belief in a future life is shewn by the care with which the relatives and friends of better condition, placed in the grave of the dead objects which it was supposed would be necessary or useful in the next world: even mere personal ornaments, or articles to which the deceased had been attached, or which can only have been placed there as tokens of affectionate remembrance. Evidence is also found of the sentiments of tenderness which followed them to their last resting-place. It was believed that the dead were exposed to evil spirits, for amulets are usually found interred with them,—especially beads of amber, which were thought to be protective against such influences. The frequent occurrence, among the earth in the grave, of bones of animals, which were commonly eaten by the Anglo-Saxons, would seem to shew that there were both sacrifices and feasting at the burial. Human bones have been found in such a position as to justify a supposition, that a slave had been slain and thrown into the grave, perhaps in the belief that he would continue to serve his master in the spiritual world. 4. in the districts which were occupied by the Angles in Britain, and Old Saxons on the continent, νεκροκαυστία, cremation or the burning of the bodies before burial, appears to have been almost universal, among rude nations, from the age of Homer to that of Alfred. The interment, therefore, consists of an urn filled with the burnt bones. It has been supposed that cremation was originally the mode of burial in use among the Angles; and that the Saxons and Jutes buried the body entire, or that they had adopted this mode of burial when they came into Britain. See Kemble in the Archæological Journal, No. 48. It is recorded of the Esthonians and Old Saxons, who were a very warlike and powerful people, once occupying the whole north-west corner of Germany,—And ðæt is mid Estum þeáw, ðæt ðær sceal ælces geþeódes man beón forbærned; and gyf ðár man án bán findeþ unforbærned, hí hit sceolon miclum gebētan it is also a custom with the Esthonians, that there men of every tribe must be burned; and if any one find a single bone unburnt, they shall make a great atonement, Ors. 1, 1; Bos. 23, 3-5. It is certain that in Beowulf, which is supposed to be an Old Norse poem, the body of the hero is described as being burnt:—Hit sǽ-líðend syððan hātan Biówulfes biorh sea-farers may afterwards call it Beowulf's mound [barrow], Beo. Th. 5604-5606; B. 2806, 2807. Him ðā gegiredon Geáta leóde ád unwáclícne, helm-behongen, hilde bordum, and beorhtum byrnum the people of the Goths raised for him a mighty funeral pile, hung with helmets, shields, and bright breast-plates, 6265-6271; B. 3137-3140. Ongunnon ðā bæl-fýra mǽst wígend weccan: wudu-rēc astāh sweart of Swió-þole then the warriors began to kindle the greatest of bale-fires: the wood-smoke ascended black from the Swedish pine, 6277-6281; B. 3143-3145. Hí on beorg dydon beágas and siglu, eall swylce hyrsta on the mound they placed rings and jewels, also ornaments, 6307-6309; B. 3164, 3165. Ðā ymbe hlǽw ridon æðelingas . . . cyning mǽnan, word-gyd wrecan then nobles rode round the mound . . . their king bewail, a verbal lay recite, 6319-6325; B. 3170-3173. Swā begnornodon Geáta leóde thus the people of the Goths deplored, 6338, 6339; B. 3179. 5. it is probable that down to a very late period the people adhered to many of their ancient burial customs. Charlemagne, so late as the year 789, ordered his Christian Saxon subjects to bury their dead in the Christian cemeteries, and not in the tumuli of the pagans, in these words,—' Jubemus ut corpora Christianorum Saxonum ad cœmeteria ecclesiæ deferantur, et non ad tumulos paganorum,' Capit. Carl. Mag. Walter, tom. ii. p. 107. In England, the ordinary converts appear to have been drawn reluctantly from the burial places of their forefathers by the establishment of Christian cemeteries attached to the churches, and even there they seem long to have continued many of their old rites. A few of these ceremonies are mentioned in the Anglo-Saxon ecclesiastical laws and constitutions relating to funerals. 6. it appears from a regulation, which, though only preserved in the laws of Henry I, evidently belonged to the Anglo-Saxon period, that as soon as any person was dead, the body was laid out, with the feet to the east and the head to the west. This law enjoins any one who, either in revenging a feud or defending himself, should kill a man, not to take anything belonging to him, whether his horse, or his helmet, or his sword, or any money he may have, but to lay out his body in the manner usually observed with the dead, the head to the west and the feet to the east, upon his shield, if he have one; and to fix his lance, and place his arms round, and attach his horse by the reins; and to go to the nearest town and give information to the first person he meets: the Latin of the law is,—' Si quis in vindictam vel in se defendendo

occidat aliquem, nihil sibi de mortui rebus aliquis usurpet, non equum, non galeam, vel gladium, vel pecuniam prorsus aliquam; sed ipsum corpus solito defunctorum more componat, caput ad occidens, pedes ad oriens versum, super clipeum, si habeat; et lanceam suam figat, et arma circummittat, et equum adregniet; et adeat proximam villam, et cui prius obviaverit denunciet,' L. H. 83, § 6; Th. i. 591. 7. during the time that the dead body remained unburied, the relations and friends assembled to watch or wake over it [this watching or waking is mentioned under the word líc a body, see líc II], and this proceeding was evidently accompanied with feasting and drinking carried to a very great excess. So late as the end of the tenth century, archbishop Ælfric addressed the following injunction to his clergy:—Ge ne scylan fægnigan forþ-farenra manna, ne ðæt líc gesécan, būton eów mann laðige ðǽr-to: ðænne ge ðǽr-to gelaðode sýn, ðonne forbeóde ge ða hǽðenan sangas ðæra lǽwedra manna, and heora hlūdan cheahchetunga; ne ge sylfe ne eton, ne ne drincon ðǽr ðæt líc inne líþ, ðe-læs ðe ge syndon efen-lǽce ðæs hǽðenscypes ðe hý ðǽr begáþ ye shall not rejoice on account of men deceased, nor attend on the corpse, unless ye be thereto invited: when ye are thereto invited, then forbid ye the heathen songs of the laymen, and their loud cachinations; nor eat ye, nor drink, where the corpse lieth therein, lest ye be imitators of the heathenism which they there commit, L. Ælf. C. 35; Th. ii. 356, 23-358, 5. The clergy gave little attention to these injunctions, for they are warned against being ' hunters of funerals,' and Ælfric tells us how some priests ' Fægniaþ ðonne men forþfaraþ, and unbedene gaderiaþ hí to ðam líce, swā swā grǽdige ræmmas, ðár ðár hí hold geseóþ; ac heom gebíraþ mid rihte to bestandenne ða men, ðe híraþ into heora mynstre; and ne sceal nán faran on óðres folgoþ to nánum líce būton he gebeden sý rejoice when men depart hence, and unbidden gather about the corpse, like greedy ravens, wherever they see a dead carcase; whereas it properly becomes them to bury those men, who belong to their minster; and no one ought to go in another's following to any corpse unless he be invited,' L. Ælf. P. 49; Th. ii. 386, 2-6. 8. we have no reason for supposing that people who were not rich were buried in coffins, but the body, having been wrapped up in its winding-sheet, appears to have been merely laid in the grave, and then covered with earth. The first coffins used by the converted Anglo-Saxons were undoubtedly of wood [vide 2], and it was the ecclesiastics who introduced the stone sarcophagi for eminent personages of their own order. Sebbi, king of the East-Saxons, was buried in a coffin of stone:—Gearwodan hí his líchoman to bebyrigeanne on stǽnenre þruh cujus [Sebbi] corpori tumulando præparaverant sarcofagum lapideum, Bd. 4, 11; S. 580, 4. 9. at every funeral a payment, called a soul-sceát [v. sāwel-sceát], was made to the church where the interment took place, and a legacy was also expected. A mancus of gold, or even a much higher sum, was usually paid in the case of a king or bishop, or of a person of high rank. 10. the graves were no doubt arranged in rows and covered with small mounds, as in the older pagan cemeteries, except that the mounds were elongated instead of being circular, and had head-stones. They seem, at an early period, to have been laid north and south, like many of those in the pagan cemeteries, and not east and west, as was the position of the bodies of the nuns of Hartlepool, buried towards the end of the seventh century, which were uncovered about thirty years ago. Small flat stones, the largest less than a foot square, had been laid over the graves at Hartlepool, each bearing a cross, and the name of the person it commemorated; some engraved in Anglo-Saxon runes, and some in the Roman letters of the seventh century, for to the latter end of that period they evidently belonged. v. Thrupp's Anglo-Saxon Home, 8vo. 1860, pp. 397-405. A very valuable paper by George Rolleston, Esq. M. D. F. R. S. On the modes of sepulture in early Anglo-Saxon times in this country, reprinted from the Translations of the International Congress of Prehistoric Archæology, Third Session: Douglas's Nenia Britannica: Faussett's Inventorium Sepulchrale: Akerman's Remains of Pagan Saxondom: Wylie's Fairford Graves: Braybrooke's Saxon Obsequies: and Mr. C. Roach Smith's Collectanea Antiqua.

byrgend, es; m. A burier; sepultor:—Náhtan byrgendas non erat qui sepeliret, Ps. Th. 78, 3.

byrgen-leóþ, es; n. A tomb-elegy, an epitaph; sepulcrale carmen, epitaphium:—On his byrgenne is awriten byrgen-leóþ scriptum est in tumba ipsius epitaphium, Bd. 2, 1; S. 500, 18.

byrgen-song, es; m. A burial song; cantus sepulcralis, Leo 116. v. bergel-song.

byrgen-stów, byrigen-stów, e; f. A burying-place, cemetery; sepulcri locus, cœmeterium, Cot. 75: Bd. 5, 23; S. 645, 19.

byrgere, es; m. A burier, corpse-bearer; vespillo, Cot. 155.

byrging [byrgung, Ettm.], e; f. A burying, the act of burying; sepultura, Jn. 20, 1, 4, Lye.

býrging, e; f. Taste, tasting; gustus, Scint. 12, Lye. [O. Nrs. berging, f. gustus, sacra synaxis vel participatio divinæ Eucharistiæ.] v. on-býrging.

byrgst, byrhst, he byrgeþ, byrgþ, byrhþ protectest, he protects, Ps. Th. 16, 8; 2nd and 3rd pers. pres. of beorgan.

byrht *bright, clear, lucid, loud;* clarus, splendidus, clarisonus, Beo. Th. 2402; B. 1199: Cd. 217; Th. 275, 15; Sat. 172. v. beorht.

byrhtan *to shine;* lucere, Exon. 24 a; Th. 67, 18; Cri. 1090. v. beorhtan.

byrhtm, es; *m. Noise, tumult;* fragor, tumultus, Apstls. Kmbl. 42; Ap. 21. v. breahtm.

byrhtm-hwŷl *a moment.* v. bearhtm-hwîl.

byrht-nes *brightness,* Ps. Spl. 118, 130. v. beorht-nes.

byrhtu, e; *f. Brightness, splendour,* Exon. 26 a; Th. 76, 15; Cri. 1240. v. beorhtu.

byrht-word; *adj.* [byrht = beorht *bright,* word *a word*] *Bright of word, clear in words* or *speech;* clarus voce :—Byrhtword arás engla ordfruma *the creator of angels, bright of words,* arose, Cd. 218; Th. 279, 15; Sat. 238.

byri = byrig *to a city.* v. byri-weard.

byrian; *p.* ede, ide; *pp.* ed *To bury* :—Ðǽr hí mon byride *where they buried her,* Ors. 3, 6; Bos. 58, 9. DER. be-byrian. v. byrgan.

byrian, *3rd s.* byreþ; *p.* ede; *pp.* ed [byre *an event, a favourable time, an opportunity*] *To happen, pertain to, belong to;* evenire, contingere, pertinere ad [v. ge-byrian]: *found as v. impers.: it pertains to, it concerns, it belongs to, it is lawful;* pertinet ad, oportet, licet :—Ne byreþ to him from scipum *non pertinet ad eum de ovibus,* Jn. Lind. War. 10, 13: Mk. Lind. War. 4, 38. Ðe ne byrede him to etanne *quem non licebat ei edere,* Mt. Kmbl. Rush. 12, 4. DER. ge-byrian.

byrian *to taste;* gustare. v. a-byrian *under* a-byrgan.

byrig *to a city,* Ps. Th. 44, 13 : 47, 11; *dat. of* burh.

byrig, e; *f: acc. s.* byrig, byrige *A city;* urbs, civitas :—Hér Cúþa gefeaht wið Brytwalas æt Biedcan forda, and genam Lygeanbyrig and Ægles byrig *in this year Cutha fought against the Brito-Welsh at Bedford, and took Lenbury and Aylesbury,* Chr. 571; Th. 33, 28. Cantwara byrig forbarn ðÿ geáre *Canterbury was burnt down in this year,* 754; Th. 81, 36, col. 2. v. burh.

byrig, es; *n. A mulberry-tree;* morus :—He ofslôh byrig heora on hagule *occidit moros eorum in pruina,* Ps. Spl. 77, 52 : L. M. 2, 53; Lchdm. ii. 274, 17.

byriga, an; *m. A surety;* fidejussor :—He him byrigan gesealdne hæbbe *he has given him surety,* L. H. E. 10; Th. i. 30, 17. v. byrgea.

byrigan, birigan; *p.* de; *pp.* ed *To bury;* sepelire :—Alýf me ǽrest byrigan mînne fæder *permitte mihi primum sepelire patrem meum,* Lk. Bos. 9, 59 : 9, 60: Chr. 1036; Th. 294, 21; Hy. 10, 29; Hy. Grn. ii. 293, 29: Nicod. 21; Thw. 10, 30: 21; Thw. 11, 4. DER. be-byrigan. v. byrian, byrgan.

byrigan; *p.* de *To taste;* gustare :—Deáþ he ðǽr byrigde *he there tasted death,* Rood Kmbl. 199; Kr. 101. Ðæt he hire sealde ðæt wæter to byrigenne *ut gustandam illi daret eam aquam,* Bd. 5, 4; S. 617, 21. DER. on-byrigan. v. byrian, byrgan.

byrig-berge, an; *f. A mulberry* :—Byrigbergena seáw selle drincan *give him to drink juice of mulberries,* L. M. 2, 30; Lchdm. ii. 230, 12.

byrigea *a surety,* L. H. E. 8; Th. i. 30, 12. v. byrgea.

byrigean *to bury.* v. byrgan, be-byrigan.

byrigean *to taste.* v. on-byrigean, byrgan.

byrigen, byrigenn, e; *f.* [beorg *tumulus*] *A burying-place, a sepulchre, tomb, burying;* sepulcrum, monumentum, tumba, sepultura, Bd. 4, 19; S. 588, 37 : 3, 8; S. 532, 15, 17 : 3, 11; S. 535, 32 : 1, 33; S. 499, 7. v. byrgen.

byrigen-stôw, e; *f. A burying-place* :—He sylfa byrigenstôwe worhte *sibi ipse in locum sepulcri fecerat,* Bd. 5, 23; S. 645, 19. v. byrgen-stôw.

byrig-leóþ, es; *n. An epitaph;* epitaphium, Bd. 2, 1, Lye. v. byrgen-leóþ.

byrig-man, -mann, es; *m.* [byrig *a city,* man *a man*] *A city officer;* ædilis, Ælfc. Gr. 9, 28, MS. D; Som. 11, 29. v. burh-man.

byrignes, -ness, -nyss, e; *f. A burying, burial;* sepultura, Bd. 4, 11; S. 580, 8. DER. be-byrignes.

byrignes, bîrgnes, -ness, e; *f. A tasting, a taste;* gustus :—Mid byrignesse ðæs wæteres *by the tasting of the water,* Bd. 5, 18; S. 635, 29. Bîrgness *gustus,* Cot. 97. DER. an-byrignys. v. byrgan.

Byrîne, es; *m. Birinus, the first bishop of Wessex,* Chr. 649; Th. 50, 3, col. 2, 3; 51, 2, col. 1. v. Birînus.

byris, e; *f? A graving-iron, file;* scalprum, scalpellum :—Byris *scalprum,* Glos. Epnl. Recd. 162, 36 : *scalpellum,* 162, 51. [*O. H. Ger.* bursa, *f.*]

byri-weard, es; *m.* [byrig, *dat. of* burh *a city,* weard *a guard*] *A city-guardian;* urbis custos, ædilis, Wrt. Voc. 18, 54. v. burh-weard.

BYRLE, byrele, es; *m. A cup-bearer, butler;* pocillator, calicum magister, pincerna :—Byrle *pincerna,* Ælfc. Gl. 113; Som. 80, 1; Wrt. Voc. 60, 37 : 74, 16. Egipta cynges byrle *pincerna regis Ægypti,* Gen. 40, 1. Byrele *pincerna,* Wrt. Voc. 290, 51. Þurh byreles hond *through the cup-bearer's hand,* Exon. 88 a; Th. 330, 15; Vy. 51. Byrlas ne gǽldon *the cup-bearers delayed not,* Andr. Kmbl. 3065; An. 1535. Geleornedon

his byrelas him betweonum *his cup-bearers planned among themselves,* Ors. 3, 9; Bos. 69, 10 : Beo. Th. 2327; B. 1161. Geþohte he ðǽra byrla ealdor *recordatus est magistri pincernarum,* Gen. 40, 20, 21, 23. Yldest byrla *a caliculis, magister calicum,* Ælfc. Gl. 113; Som. 79, 130; Wrt. Voc. 60, 34. Ðara ôðer bewiste his byrlas, ôðer his bæcestran *alter pincernis præerat, alter pistoribus,* Gen. 40, 2. [*Laym.* birle, borle : *Orm.* birrless, *pl : Icel.* byrli, byrlari, *m.*]

byrlian, byrlian; *p.* ade; *pp.* ad [byrle, byrele *a cup-bearer*] *To pour out, give to drink, serve;* propinare :—Ic him byrlade wrôht of wêge *I poured out complaint to them from the cup,* Exon. 72 b; Th. 271, 23; Jul. 486. Feónd byrlade ðære idese bittor bǽdewǽg *the fiend gave the woman the bitter cup to drink,* 47 a; Th. 161, 8; Gû. 955. Ðone bitran drync Eue Adame byrelade *Eve served to Adam the bitter drink,* 45 b; Th. 154, 13; Gû. 842.

byrman; *p.* de; *pp.* ed [beorma *barm*] *To ferment with barm, to leaven;* fermentare.

byrnan; *part.* byrnende; he byrneþ.   **I.** *v. intrans. To burn, to be on fire;* ardere :—Sín eówer leóhtfatu byrnende *sint vestræ lucernæ ardentes,* Lk. Bos. 12, 35 : Deut. 9, 15. Ðonne byrneþ gramen his *cum exarserit ira ejus,* Ps. Spl. 2, 13: Bd. 5, 3; S. 616, 36.   **II.** *v. trans. To burn;* urere, comburere :—Swá fÿr wudu byrneþ *sicut ignis comburit sylvas,* Ps. Th. 82, 10. v. beornan.

BYRNE, an; *f. A corslet, coat of mail;* lorica, thorax :—Môt he gesellan monnan and byrnan and sweord *he may give a man a corslet and a sword,* L. In. 54; Th. i. 138, 1. Ðǽr wæs on eorle brogden byrne *there was on the man the twisted coat of mail,* Elen. Kmbl. 513; El. 257. Ætbær hringde byrnan *he bore away the ringed coat of mail,* Beo. Th. 5224; B. 2615. Ongan wyrcan síde byrnan *he began to make a large coat of mail,* Salm. Kmbl. 906; Sal. 453: Judth. 12; Thw. 26, 15; Jud. 328. [*Laym.* burne, brunie: *Ger.* brünne, *f: M. H. Ger.* brünje, brünne, *f: O. H. Ger.* brunja, brunna, *f: Goth.* brunyo, *f: Dan.* brynie, *m. f: Swed. Icel.* brynja, *f: O. Slav.* brunija.] DER. gúþ-byrne, heaðo-, heaðu-, here-, îren-, îsern-.

byrne, es; *m. A burning;* incendium :—Ǽr ðam ðe ðæt mynster mid byrne fornumen wǽre *priusquam monasterium esset incendio consumptum,* Bd. 4, 25; S. 599, 18. v. bryne.

byrne, an; *f. Running water, a stream;* torrens, rivus :—Ofer byrnan bôsm *over the stream's bosom,* Exon. 102 a; Th. 386, 15; Rä. 4, 62. v. burne.

byrnendra *more burning,* Bd. 5, 3; S. 616, 36. v. byrnan.

byrn-hom, es; *m.* [byrne *a coat of mail,* hom *a covering, garment*] *A coat of mail;* lorica :—Beraþ bord fôr breóstum and byrnhomas *bear shields before your breasts and coats of mail,* Judth. 11; Thw. 24, 17; Jud. 192.

byrn-wiga, an; *m. A soldier clothed in armour;* loricatus miles :—Se byrnwîga bûgan sceolde *the mailed warrior must submit,* Beo. Th. 5828; B. 2918: Exon. 77 b; Th. 292, 5; Wand. 94. Byrnwîgena brego *the chief of mailed soldiers,* Judth. 9; Thw. 21, 28; Jud. 39.

byrn-wigende, -wiggende; *part. Clothed in armour, mailed;* loricatus :—Swá hire weoruda helm byrnwiggendra beboden hæfde *as the prince of the mailed armies had commanded her,* Elen. Kmbl. 447; El. 224. Gehlôdon byrnwîgendum werum wǽghengestas *they loaded the ships with men covered with armour,* Elen. Kmbl. 470; El. 235.

byrn-wiggend, es; *m. A soldier clothed in armour, a mailed warrior;* loricatus miles *vel* bellator :—Bealde byrnwîggende *bold warriors,* Judth. 9; Thw. 21, 13; Jud. 17.

byrst, es; *n. A bristle;* seta :—Byrst *seta,* Wrt. Voc. 286, 57: Glos. Epnl. Recd. 162, 49. Hyre twigu beóþ swylce swînene [MS. swinen] byrst *its twigs are like swine bristles,* Herb. 52, 2; Lchdm. i. 156, 3. [*Frs.* boarstel, *m. f: Dut.* borstel, *m: Ger.* borste, *f: O. H. Ger.* burst, *n;* bursti, pursta, *f: Dan.* börste, *m. f: Swed.* borst, *m: Icel.* burst, *f.*]

byrst *bursts, breaks, fails; 3rd pers. pres.* of berstan.

byrst, he byrþ *bearest, he bears, produces;* facit, Mt. Bos. 7, 17; *2nd and 3rd pers. pres.* of beran.

byrst, berst, es; *m. A loss, defect;* damnum, calamitas :—Gylde ðone byrst, ðe ðæt fÿr ontende *reddet damnum, qui ignem succenderit,* Ex. 22, 6, 12 : Ps. Th. 108, 18. We habbaþ fela byrsta gebiden *multas calamitates sumus perpessi,* Lupi Serm. i. 2; Hick. Thes. ii. 99, 21.

byrþ *a birth.* v. beorþ, byrþ-ling.

BYRÐEN, berðen, byrðyn; *gen.* byrðenne; *f. A BURTHEN, load, weight, bundle;* onus, sarcina, fascis :—Hefig byrðen *onus grave,* Ps. Th. 37, 4. Sorh biþ swǽrost byrðen *sorrow is the heaviest burthen,* Salm. Kmbl. 623; Sal. 311. Seám oððe byrðen *onus,* Ælfc. Gr. 9, 32; Som. 12, 14. Byrðen *fascis,* 9, 28; Som. 11, 44: Mt. Lind. Stv. 13, 30. [*O. Sax.* burðinnia, *f: O. Frs.* berthe, berde, *f: Ger. M. H. Ger.* bürde, *f: O. H. Ger.* burdi, *f: Goth.* baurþein, *f: Dan.* byrde, *f: Swed.* börda, *f: Icel.* byrðr, byrði, *f.*] DER. mægen-byrðen, sorg-, syn-.

byrðene dǽl, es; *m. A share of a burthen, a portion;* portio, Ps. Spl. 49, 19.

byrðen-mǽlum; *adv.* [byrðen, mǽlum, *dat. pl.* of mǽl, *n.*] *By burdens;* oneribus :—Se dêma hǽt his englas gadrian ðone coccel byrðen-

mǽlum *the judge will command his angels to gather the tares by burdens*, Homl. Th. i. 526, 22.

byrðen-meto; *indecl*; *f?* *An excessive burden*; oneris excessus, onerosa mensura, Prov. 27, Ettm.

byrðen-strang; *adj. Burthen-strong, strong to bear burdens*; oneribus portandis robustus :—Assa is stunt nýten, and byrðenstrang *an ass is a foolish beast, and strong for burdens*, Homl. Th. i. 208, 13.

byrþere; *gen.* byrþres; *m.* [beran *to bear, carry*] *A bearer, carrier, supporter*; portarius, vespillo, fulcimen :—Crist ðone wácan assan geceás him to byrþre *Christ chose the mean ass for his bearer*, Homl. Th. i. 210, 16. Ða byrþeras hine to byrgenne féredon *the bearers bare him to the grave*, i. 492, 27. Seó untrumnys his gecyndes behófode sumes byrþres *the infirmity of his nature had need of some supporter*, i. 308, 12.

byrþ-ling, beorþ-ling, es; *m. A born image, birthling, child*. v. beorþ, hyse-berþling.

byrþor, es; *n?* *Child-birth, a fetus*; partus, fetus :—Bútan byrþres intingan *sine partus causa*, Bd. 1, 27; S. 493, 40. v. beorþor.

byrþor-cwelm, es; *m. An abortion, a miscarriage*. v. beorþor-cwelm.

byrþor-þinen, e; *f. A midwife*. v. beorþor-þinen.

byrðyn, e; *f. A burthen*; onus :—Mín byrðyn ys leóht *meum onus est leve*, Mt. Bos, 11, 30. v. byrðen.

Byr-tún, es; *m.* [*Hovd.* Burhtun : *Brom.* Burton super Trent : *Stub. Kni.* Burton] BURTON *on Trent, Staffordshire*; oppidum ad ripam fluminis Trentæ, in agro Staffordiensi :—Se cyng geaf him ðæt abbotríce on Byrtúne *the king gave him the abbacy at Burtòn*, Chr. 1066; Erl. 203, 16.

býsegu *occupation*, Bt. Met. Fox 20, 509; Met. 20, 255. v. býsgu.

BÝSEN, bísen, býsn, e; *f.* I. *a pattern, an example, model, resemblance, similitude, parable*; norma, exemplum, modellum, similitudo, parabola :—Ðú bútan býsne, Ælmihtig God, eall geworhtest þing þearle gód [good, MS.] *thou, Almighty God, madest all things very good, without a pattern*, Bt. Met. Fox 20, 85; Met. 20, 43. Seó býsen ðæs rihtan geleáfan Angel cyricean to Róme geléæded wæs *exemplum catholicæ fidei Anglorum Romam perlatum est*, Bd. 4, 18; S. 587, 11: 2, 1; S. 590, 26: 4, 23; S. 595, 10. Gúþlác mongum wearþ býsen on Brytene *Guthlac was an example to many in Britain*, Exon. 35 a; Th. 112, 19; Gú. 146. Ðiós óðru býsen *this other similitude*, Bt. Met. Fox 12, 13; Met. 12, 7. Æfter heora býsne *after their example*, Ps. Th. arg. 28: Cd. 217; Th. 276, 29; Sat. 196. On býsene ðære frymþelícan cyricean *in exemplum primitivæ ecclesiæ*, Bd. 4, 23; S. 593, 40. Be sumere bísene *by some example*, Bt. 22, 2; Fox 78, 13. Ðæt hí ealle gemyndige wǽron hyre býsene *that they all should be mindful of her example*, Bd. 4, 23; S. 595, 20. He býsene gegearwode *he gave an example*, 4, 23; S. 594, 24. He us býsene sealde his árfæstnysse *he gave us an example of his piety*, Homl. Th. i. 492, 23. Wolde ic eów býsne onstellan *I would give you an example*, Andr. Kmbl. 1942; An. 973: Bd. 4, 27; S. 604, 1. Secgan Dryhtne lof ealra ðara bísena ðe us his wísdóm cýðaþ *let us speak to the Lord praise for all the examples which manifest his wisdom*, Exon. 40 a; Th. 133, 33; Gú. 499. Ealle béc sint fulle ðara bísna ðara monna, ðe ǽr us wǽron [MS. wǽran] *all books are full of examples of the men, who were before us*, Bt. 29, 1; Fox 102, 12. Onlícnesse oððe bísene *a parable*; similitudinem, Lk. Rush. War. 13, 6. II. *a command, precept, admonition*; mandatum, præceptum, admonitio :—Ic gelýfe ðæt hit from Gode cóme, broht from his býsene *I believe that it came from God, brought by his command*, Cd. 32; Th. 42, 29; Gen. 680. Ðæt he ða býsene from Gode brungen hæfde *that he had brought those commands from God*, 30; Th. 41, 3; Gen. 651. Hwylce ðú selfa hæíst bísne on breóstum *what precepts thou thyself hast in thy breast*, 27; Th. 36, 13; Gen. 571. Ic ðínra býsna ne mæg wuht oncnáwan *I cannot understand aught of thy commands*, 26; Th. 34, 6; Gen. 533. [*Laym.* bisne, bysne, *dat. a pattern, example*: *Orm.* bisne *example*: *O. Sax.* busan, *f. in* am-busan, *f. a commandment*: *Goth.* ana-busns, *f. a command*.] DER. fóre-býsen, lár-: býsnian, ge-, mis-: býsnung, ge-.

býsenian *to give an example*, C. R. Ben. 2. v. býsnian.

býsenung *an example*, C. R. Ben. 61. v. býsnung.

býsgian, bísgian, býsigan; *p.* ode, ade; *pp.* od, ad *To occupy, busy, fatigue, trouble, afflict*; occupare, fatigare, affligere, tribulare :—Se man biþ hérigendlíc, ðe mid gódum weorcum hine sylfne býsgaþ *the man is praiseworthy, who busies himself with good works*, Homl. Th. ii. 406, 16. For ðæm manigfealdum bísgum, ðe hine oft ægðer ge on móde ge on líchoman bísgodon [MS. bisgodan] *on account of the manifold occupations, which often busied him* [*king Alfred*] *both in mind and in body*, Bt. procem; Fox viii. 6: Cd. 64; Th. 76, 29; Gen. 1264. Ic eom býsgod on sange *occupatus sum cantu*, Coll. Monast. Th. 18, 25; Wrt. Voc. 2, 11. Ðeáh ðæs líchoman leahtras and hefignes and unþeáwas oft býsigen monna módsefan *though the sins and heaviness and vices of the body may often trouble the minds of men*, Bt. Met. Fox 22, 60; Met. 22, 30. Hine hungor býsgaþ *hunger afflicts him*, Exon. 97 a; Th. 363, 10; Wal. 51. Ðē untrymnes on ðisse nýhstan niht býsgade *infirmity afflicted thee in this last night*, 47 b; Th. 163,

10; Gú. 991. [*Frs.* bisgje, bysgje *occupare*.] DER. a-býsgian, ge-: býsgung, a-, woruld-.

BÝSGU, bísgu, býsigu, bísigu, býsegu, bísegu; *gen. e*; *dat. e*; *acc. u, o*: *nom. acc. pl.* u; *gen. a*; *dat.* um; *f. Occupation, business, labour, care, toil, difficulty, trouble, affliction*; occupatio, negotium, labor, cura, opus, difficultas, dolor, tribulatio :—Ða bísgu us sint swíðe earfoþ ríme *the occupations are to us very difficult to be numbered*, Bt. procem; Fox viii. 6. For ðǽm manigfealdum bísgum, ðe 'hine oft ægðer ge on móde ge on líchoman bísgodon [bisgodan MS.] *on account of the manifold occupations, which often busied him* [*Alfred*] *both in mind and in body*, Bt. procem; Fox viii. 5. Of ðisum býsegum *from these occupations*, Bt. Met. Fox 20, 509; Met. 20, 255. Of ðissum bísegum *from these occupations*, Bt. 33, 4; Fox 132, 28. Býsigum gebǽded *oppressed with labours*, Beo. Th. 5153; B. 2580. Biþ se slǽp tó fæst bísgum gebunden *the sleep is bound too fast by cares*, Beo. Th. 3490; B. 1743: Bt. Met. Fox 22, 117; Met. 22, 64. Óðer bísgo dreág *the other suffered toil*, Exon. 114 a; Th. 438, 14; Rä. 57, 7: 82 b; Th. 311, 6; Seef. 88. Oþ-ðæt he ða býsgu oferbiden hæfde *until he had surmounted the trouble*, Exon. 40 b; Th. 135, 2; Gú. 518. Mec his býsgu gehreáw *his affliction grieved me*, Exon. 43 a; Th. 144, 31; Gú. 686. Bísigu, Beo. Th. 567, note; B. 281. Ic bísga unrím dreág *I suffered numberless [of] afflictions*, Exon. 74 b; Th. 280, 7; Jul. 625. Méðe for ðǽm miclum [miclan MS.] býsgum *weary on account of the great afflictions*, 49 a; Th. 168, 25; Gú. 1083. [*Dut.* bézig-heid, *f. occupation*.] DER. nýd-býsgu.

býsgung, e; *f. Business, occupation, care*; negotium, occupatio, cura. DER. a-býsgung, woruld-. v. bísgung.

býsig; *adj. Occupied, diligent, laborious, BUSY, industrious*; occupatus, sedulus, laboriosus, negotiosus, industrius :—Býsig æfter bócum *occupied over books*, Salm. Kmbl. 123; Sal. 61. Bogan wǽron býsige *bows were busy*, Byrht. Th. 134, 66; By. 110: Ps. Th. 58, 3. [*Chauc.* besy, bisy, bysy: *Laym.* bisi, bisie.] DER. líc-býsig, líf-, nýd-, þrag-.

býsigan *to occupy, trouble*, Bt. Met. Fox 22, 60; Met. 22, 30. v. býsgian.

býsigu *labour*, Beo. Th. 5153; B. 2580. v. býsgu.

bysmer *mockery, reproach, blasphemy*, Exon. 117 a; Th. 449, 14; Dóm. 71: Ps. Th. 58, 8: 103, 25: 105, 25: 106, 10: Mk. Bos. 14, 64. v. bismer.

bysmerian, bysmrian, bismrian, bismǽrian, bysmorian, bysmrígan, to bismerienne, bysmrigenne; *p.* ode, ede; *pp.* od, ed [bismer, bysmer *mockery, blasphemy*] *To mock, deride, irritate, reproach, blaspheme, defame, revile*; illudere, deridere, irritare, irridere, blasphemare, calumniam facere, maledicere :—Draca ðes, ðone ðú ýwodest to bismrienne him *draco iste, quem formasti ad illudendum ei*, Ps. Spl. 103, 28. Hí sellaþ hine þeódum to bysmrigenne *tradent eum gentibus ad illudendum*, Mt. Bos. 20, 19. Ðæt he me bysmrode *ut illuderet mihi*, Gen. 39, 17. Ðæt he bysmorode us *ut illuderet nobis*, 39, 14. Ðú, Drihten, bysmrast hí tu, Domine, deridebis eos, Ps. Spl. 58, 9. Us fýnd bysmriaþ *enemies deride us*, Ps. Th. 79, 6. Ealle bysmrodon me *omnes deriserunt me*, Ps. Spl. 21, 6. Hí bysmeredon hí on ðone reádan sǽ *irritaverunt eos in rubrum mare*, Ps. Th. 105, 8. Ongunnon hí on ðám wícum Moyses bysmrian *they began to irritate Moses in the camps*, 105, 14. Hú lange bysmraþ se wiðerwearda naman ðínne *usquequo irritat adversarius nomen tuum?* Ps. Spl. 73, 11. Se ðe eardaþ on heofonum bysmeraþ hý *qui habitat in cælis irridebit eos*, Ps. Spl. 2, 4. Se ðone Hálgan Gást bysmeraþ, se næfþ on écnysse forgyfenesse *qui blasphemaverit in Spiritum Sanctum, non habebit remissionem in æternum*, Mk. Bos. 3, 29. Hí bysmeriaþ *they blaspheme*, 3, 28. Ða wegférendan hyne bysmeredon *prætereuntes blasphemabant eum*, Mt. Bos. 27, 39, 41. Ne bysmra ðú ðínne mǽg *non facies calumniam proximo tuo*, Lev. 19, 13. Ne lǽt bysmrian banan mancynnes ða ðín lof beraþ *let not the murderers of men revile those who bear thy praise*, Andr. Kmbl. 2587; An. 1295. Bysmeredon hie bútu ætgædere *they reviled us both together*, Rood Kmbl. 95; Kr. 48: Andr. Kmbl. 1923; An. 964. Uton gangan ðæt we bysmrigen him *let us go that we may revile him*, 2713; An. 1359. DER. gebysmerian.

bysmer-leás; *adj.* [bismer, bysmer *pollution, abomination, disgrace*; -leás *-less*] *Without pollution, spotless, blameless*; sine pollutione, immaculatus, irreprehensus :—Ðæt he mǽge éðles mid monnum brúcan bysmerleás *that he may enjoy the world blameless with men*, Exon. 27 a; Th. 81, 19; Cri. 1326.

bysmer-líce *disgracefully*, Judth. 10; Thw. 23, 2; Jud. 100. v. bismor-líce.

bysmer-spǽc, e; *f. Blasphemy*; blasphemia :—For ðínre bysmer-spǽce *for thy blasphemy*, Jn. Bos. 10, 33. v. bysmor-sprǽc.

bysmer-specan, ic -spece, ðú -spicst, -spycst, he -speceþ, -spicþ, -spycþ, *pl.* -specaþ; *p.* -spæc, *pl.* -spǽcon; *pp.* -specen [bismer, bysmer *blasphemia*, specan *loqui*] *To speak blasphemy, to blaspheme*; blasphemiam loqui, blasphemare :—Ðú bysmerspycst *blasphemas*, Jn. Bos. 10, 36.

bysmer-spycst *blasphemest*; blasphemas, Jn. Bos. 10, 36. v. bysmer-specan.

bysmerung *blasphemy;* blasphemia, Mk. Bos. 3, 28. v. bysmrung.

bysmor *filthiness, reproach, calumny,* Ælfc. T. 15, 21: Ps. Th. 8, 3: Deut. 28, 29. v. bismer.

bysmor-full; *adj.* [bismer, bysmor *pollution, abomination, disgrace;* full *full*] *Polluted, abominable, disgraceful;* pollutus, detestabilis, turpis :—Ðæt híg búgan ne sceoldon to ðam bysmorfullum hǽðengilde *that they should not bow to the abominable heathen idol,* Jos. 23, 7.

bysmorian *to mock,* Gen. 39, 14. v. bysmerian.

bysmor-líce *disgracefully, irreverently,* L. Ælf. C. 35; Th. ii. 356, note 2, line 20. v. bismor-líce.

bysmor-spræc, bysmur-spræc, bysmer-spæc, e; *f.* [bismer, bysmer *blasphemy;* spræc, spæc *a speaking, word, speech*] *A speaking blasphemy, blasphemy;* blasphemia :—Ðes sprycþ bysmorspræce *this* [*man*] *speaketh blasphemy;* hic blasphemat, Mt. Bos. 9, 3. Ælc synn and bysmurspræc byþ forgyfen mannum, sóþlíce ðæs Hálgan Gástes bysmurspræc ne byþ forgyfen *omne peccatum et blasphemia remittetur hominibus, Spiritus Sancti autem blasphemia non remittetur,* 12, 31. Ðis ys bysmorspræc *this is blasphemy,* 26, 65. For ðínre bysmerspæce *on account of thy blasphemy,* Jn. Bos. 10, 33.

bysmrian; *p.* ode; *pp.* od *To deride, irritate, reproach, defame, revile,* Gen. 39, 17: Ps. Spl. 58, 9: Ps. Th. 105, 14: Lev. 19, 13: Andr. Kmbl. 1923; An. 964. v. bysmerian.

bysmrigan *to mock, revile,* Mt. Bos. 20, 19: Andr. Kmbl. 2713; An. 1359. v. bysmerian.

bysmrung, bysmerung, e; *f.* [bismer, bysmer *infamy, blasphemy*] *Deceit, infamy, blasphemy;* illusio, infamia, blasphemia :—Ðeós bysmrung nis to ondrǽdanne *hæc illusio non est timenda,* Bd. 1, 27; S. 496, 39, 41: 497, 6. Is on ðære ylcan bysmrunge swýðe nýðþearflíc gesceád *est in eadem illusione valde necessaria discretio,* 1, 27; S. 496, 34, 21. Hió hyre firenluste fulgan ne môste bútan manna bysmrunge *she could not fulfil her wicked desire without the infamy of mankind,* Ors. 1, 2; Bos. 27, 14. Ealle sinna synd manna bearnum forgyfene, and bysmerunga, ðám ðe hí bysmeriaþ *omnia dimittentur filiis hominum peccata, et blasphemiæ, quibus blasphemaverint,* Mk. Bos. 3, 28.

bysmur-spræc *blasphemy,* Mt. Bos. 12, 31. v. bysmor-spræc.

býsnian, bísnian, býsnigan, býsenian; *p.* ode; *pp.* od [býsen, býsn *an example*] *To give or set an example;* exemplum dare :—We lǽraþ, ðæt preóstas aa wel býsnian *we enjoin that priests always set a good example,* L. Edg. C. 52; Th. ii. 254, 28. Gif ða láreówas wel tǽcaþ, and wel býsniaþ, beóþ hí gehealdene *if the teachers teach well, and give good example, they shall be saved,* Homl. Th. ii. 50, 3. Ne bísnode ðé nán man, forðamðe nán ǽr ðé næs *no man set thee an example, for no one was before thee,* Bt. 33, 4; Fox 128, 20. Ða bísnodon hiora æftergengum *they set an example to their successors,* 39, 11; Fox 230, 2. Gif he yfel býsnige *if he give evil example,* Homl. Th. ii. 48, 35: L. Edg. C. 66; Th. ii. 258, 17. DER. ge-býsnian, mis-.

býsnigan *to give or set an example,* Homl. Th. ii. 48, 35: L. Edg. C. 66; Th. ii. 258, 17. v. býsnian.

býsnung, bísnung, býsenung, e; *f.* [býsen, býsn *an example*] *An example;* exemplum :—For ðære miclan bísnunge *for the great example,* Ælfc. T. 5, 15. DER. ge-býsnung.

býst *art, shalt be,* Lk. Bos. 1, 76: Ælfc. Gr. 25; Som. 26, 12. v. beón.

býst *biestings,* Ælfc. Gl. 31; Som. 61, 102. v. beóst.

býst *commandest, offerest; 2nd pers. pres. of* beódan.

býsting, es; *m.* BIESTINGS, *the first milk of a cow after calving;* colostrum :—Býsting, þicce meolc *biestings, thick milk,* Ælfc. Gl. 33; Som. 62, 20; Wrt. Voc. 28, 3. v. beóst.

BYT, bytt, e; *f.: pl.* bytta *A bottle, flagon,* BUTT, *tun;* uter, dolium :—Byt *uter,* Wrt. Voc. 85, 82. Bytt *uter,* Ælfc. Gr. 9, 18; Som. 9, 58. Ne híg ne dóþ niwe wín on ealde bytta; gýf hí dóþ, ða bytta beóþ tobrocene, and ðæt wín agoten, and ða bytta forwurðaþ. Ac híg dóþ niwe wín on niwe bytta, and ǽgðer byþ gehealden *neque mittunt vinum novum in utres veteres; alioquin rumpuntur utres, et vinum effunditur, et utres pereunt. Sed vinum novum in utres novos mittunt, et ambo conservantur,* Mt. Bos. 9, 17: Jos. 9, 4: Ps. Lamb. 32, 7. [*Ger.* butte, bütte, *f.: M. H. Ger.* büte, bütte, *f.: Dan.* bötte, *m. f.: Swed.* bytta, *f.: Icel.* bytta, *f.*]

byt *asks, prays,* Lk. Bos. 11, 11: Ex. 5, 16,=bit; *3rd pers. pres. of* biddan.

být *commands, bids, offers,* Ex. 5, 10; *3rd pers. pres. of* beódan.

byþ *is, shall be,* Mt. Bos. 5, 14. v. beón.

byþ *inhabits; 3rd pers. pres. sing. of* búan.

byþne *a keel.* v. bytne.

býtl, bítl, es; *n. m?* [být, *pres. of* beátan *to beat, strike*] *A* BEETLE, *hammer;* malleus :—Seó wífman án ðæra teldsticcena geslóh mid ánum býtle búfan his þunwengan *the woman struck one of the tent-nails with a hammer above his temples,* Jud. 4, 21. Nán mon ne gehiérde bítles swég *no man heard the sound of hammer,* Past. 36, 5; Cott. MS. [*Plat.* bötel.]

býtla, an; *m.* [býtl *a hammer,* -a *q. v.*] *A hammerer, builder;* ædificator :—Se býtla ðǽr háligne hám arǽrde *the builder raised up a holy home there,* Exon. 34 b; Th. 110, 36; Gú. 119.

býtlian; *p.* ode, ede; *pp.* od, ed [býtla *a builder*] *To build;* ædificare :—Hí ongunnon býtlian heora burh *they began to build their town,* Cd. 90; Th. 112, 33; Gen. 1880: 99; Th. 131, 15; Gen. 2176. He ne býtlaþ of ðam grúndwealle *he builds not from that foundation,* Homl. Th. i. 368, 25. Býtlode *ædificavit,* R. Ben. in procem. Hí worhton ðæt geweorc æt Tæmeseforda, and hit búdon, and býtledon *they wrought the work at Tempsford, and inhabited it, and built,* Chr. 921; Erl. 106, 18. DER. ge-býtlian.

býtlung, e; *f.* [býtl, ung] *A building, edifice;* structura, ædificium :—Seó býtlung is ofer Criste gelogod *the building is founded on Christ,* Homl. Th. i. 368, 22.

bytne *the keel* or *bottom of a ship;* carina, Cot. 32.

býtst *commandest, offerest; 2nd pers. pres. of* beódan.

býtt *ordains,* Homl. Th. i. 358, 31,=být, *q. v.*

bytta *bottles,* Mt. Bos. 9, 17; *pl. of* byt.

bytte-hlid, es; *n. A lid of a butt;* dolii opertorium, Cot. 208: Mann.

bytt-fylling, e; *f. A filling of butts;* doliorum impletio, L. Ath. v. § 8, 1; Th. i. 236, 4.

býwan; *p.* de; *pp.* ed *To prepare, adorn;* parare, ornare :—Ða ðe beadogrímman býwan sceoldon *those who should prepare the war-helmet,* Beo. Th. 4507, note; B. 2257. [*O. Nrs.* búa *parare.*] DER. a-býwan.

C

In Gothic and Icelandic ᴄ is entirely wanting, being always represented by *k*. It is remarkable that the Anglo-Saxons have seldom made use of *k*; but, following the Latin, have preferred the use of *c*. **1.** the letter *c* is found as an initial, medial, and final.—As an initial letter it corresponds to the Gothic and Icelandic *k*; as,—*A. Sax.* corn *corn,* Goth. kaurn, *Icel.* korn; *A. Sax.* ceósan *to choose,* Goth. kiusan, *Icel.* kjósa. As a medial and final letter *c* corresponds to the Gothic and Icelandic *k*,—thus *A. Sax.* æcer *a field,* Goth. akrs, *Icel.* akr; *A. Sax.* eác *also,* Goth. auk, *Icel.* ok [og]. **2.** *c* and *cc* are often changed into *h* or *hh* before *s* or *þ*, and especially before *t;* as, strehton *they stretched,* for strecton *from* streccan. Ahsian *for* acsian *or* axian *to ask;* séhþ *for* sécþ *seeks, from* sécan *to seek.* In words immediately derived from Anglo-Saxon, *k* is frequently substituted for the Anglo-Saxon *c;* as, cyning *a king;* cyn *kin* or *kindred.* Sometimes *q* or *ch;* as, cwén *queen;* cild *a child;* cin *a chin.* **3.** the Runic letter ᚻ not only stands for the letter ᴄ, but also for the name of the letter in Anglo-Saxon cén *a torch.* v. cén and RÚN.

cac, es; *m?* Dung, *excrement;* stercus, foria, merda, Som. Ben. Lye. [*Plat.* kak, kakk: *Dut.* kak, *m: Kil.* kack: *Ger.* kack, *m: Dan.* kag, *m. f: Grk.* κάκκη : *Lat.* cacare: *Grk.* κακκάω.]

cac-hús, es; *n. A privy;* latrina, Som. Ben. Lye. [*Kil.* kack-huys.]

cæd, ced, es; *m. A boat;* linter, Mone B. 120, Ettm.

cæder-beám, es; *m. A cedar-tree;* cedrus :—Hériaþ Drihten, múntas and ealle beorgas, treówu wæstmbǽru, and ealle cæder-beám *laudate Dominum, montes et omnes colles, ligna fructifera, et omnes cedri,* Ps. Spl. 148, 9. v. ceder-beám.

Cædmon, es; *m.* [Cædmon, MS. C. C. C. Oxford: Cædmon, Bd. 4, 24; S. 170, 50; Cedmon, S. 597, 12 : Ceadmon, MS. B, S. 597, note 12: Cadmon, Runic Monmnts. by Prof. Stephens, fol. Cheapinghaven, 1868, p. 419, 11 : cæd *linter,* mon *homo*] *A* man employed by the monks of Whitby in the care of their cattle in the early part of the seventh century. He is the first person of whom we possess any metrical composition in our vernacular language. So striking and similar are some of his thoughts to Paradise Lost, it has been supposed that Milton had read his Poems. He became a monk of Whitby, and died in the monastery about A. D. 680. A full account is given of him in Bede's History, bk. iv. ch. 24. The origin of his Poem is thus recorded in king Alfred's Anglo-Saxon version of Bede :—Ðá stód him sum mon æt þurh swefen, and hine hálette and grétte, and hine be his naman nemde, Cædmon [Cedmon, Bd. 4, 24; S. 597, 12], sing me hwæt-hwegu. Ðá andswarede he and cwæþ, ne con ic nán þing singan ... Eft he cwæþ, ac ðú meaht me sprecende wæs, hwæðere ðú meaht me singan. Cwæþ he, hwæt sceal ic singan? Cwæþ he, sing me frumsceaft. Ðá he ðá ðás andsware onféng; ðá ongan he sóna singan, in hérenesse Godes scyppendes, ða fers and ða word ðe he nǽfre ne gehýrde ... Ðá arás he from ðam slǽpe and eall ðæt he slǽpende song fæste on gemynde hæfde ... Song he ǽrest be middangeardes gesceape, and be fruman moncynnes, and eall ðæt stǽr Genesis, and eft be útgonge Israhéla folces of Ægypta lande, and be ingonge ðæs gehátlondes, and be óðrum monigum spellum ðæs hálgan gewrites Canones bóc; and be Cristes menniscnesse, and be his þrówunge, and be his uppastígnesse on heofonas; and big ðæs hálgan Gástes cyme, and ðæra Apostola láre; and eft big ðam ege ðæs toweardan dómes, and be fyrhto ðæs tintreglícan wítes, and be swétnesse ðæs heofonlícan ríces : he monig

leóþ geworhte *then stood some man by him in a dream, and hailed and greeted him, and named him by his name,* 'Cædmon, canta mihi aliquid,' = *Cædmon, sing me something. Then he answered and said, I cannot sing anything . . . Again, he who was speaking with him said, Yet thou must sing to me. Said he, What shall I sing? Said he, Sing me the origin of things. When he received this answer, then he began forthwith to sing, in praise of God the Creator, the verses and the words which he had never heard . . . Then he arose from sleep, and had fast in mind all that he sleeping had sung . . . He first sang of earth's creation, and of the origin of mankind, and all the history of Genesis, and then of the departure of the people of Israel from the Egyptians' land, and of the entrance of the land of promise, and of many other histories of the canonical books of Holy Writ; and of Christ's incarnation, and of his passion, and of his ascension into heaven; and of the coming of the Holy Ghost, and the doctrine of the Apostles; and also of the terror of the doom to come, and the fear of hell-torment, and the sweetness of the heavenly kingdom: he made many poems,* Bd. 4, 24; S. 597, 11–18, 25, 26– 598, 9–17. **2.** Cædmon was first published by Junius, from the Bodleian MS. the only one in existence. Junius published the Anglo-Saxon text only at Amsterdam in 1655, without a translation, in very small 4to, pp. 116. It was again published by B. Thorpe, F.S.A. in large 8vo. 1832, with an English translation, notes, and a verbal index, pp. 341. **3.** Bouterwek, with German translation and notes, an excellent vocabulary, Lateinischangelsächsisches Wörter-verzeichniss, in 2 vols. 8vo. 1854. Gütersloh bei C. Bertelsmann. **4.** Grein in 2 vols. 8vo. 1857, Text, vol. i. pp. 148.

cæfester, es; *m?* *A halter, head-stall;* capistrum, Cot. 31: 33. DER. ge-cafstrian.

cæfian, cefian; *p.* ede; *pp.* ed *To embroider;* acu pingere. DER. be-cæfian, ymb-.

CÆG; *gen.* cǽge; *pl. nom. acc.* cǽga, cǽgia; *f:* cǽge, an; *f. A* KEY; clavis:—Stæfcræft is seó cǽg ðe ðæra bóca andgytt unlýcþ *grammar is the key that unlocketh the sense of books,* Ælfc. Gr. pref; Som. 1, 23: 9, 28; Som. 11, 54: Past. 15, 2; Hat. MS. 19 a, 17. Ge ætbrudon ðæs ingehýdes cǽge *tulisti clavem scientiæ,* Lk. Bos. 11, 52. Saturnus sumra hæfde bóca cǽga *Saturn had the keys of some books,* Salm. Kmbl. 370; Sal. 184. Ðé ic sylle heofona ríces cǽgia *tibi dabo claves regni cælorum,* Mt. Bos. 16, 19. Gástes cǽgum [MS. cǽgon] *with the keys of the spirit,* Cd. 169; Th. 211, 11; Exod. 524. Cǽgan, Exon. 112 a; Th. 429, 29; Rä. 43, 12. [*Chauc.* key: *Wyc.* keie, keye: *R. Glouc.* keyen, *pl:* *Frs.* cay, cayce *a small key:* O. *Frs.* kei, kai, *m:* *Wel.* can *to shut, inclose.*] DER. lioðu-cǽge, searo-cǽg.

cǽg-bora, an; *m. A key-bearer;* claviger, Ælfc. Gr. 8; Som. 7, 19.

cǽge, an; *f. A key;* clavis:—Cǽgan, Exon. 112 a; Th. 429, 29; Rä. 43, 12. v. cǽg.

cǽggian; *p.* ode; *pp.* od *To lock, shut fast;* obserare. DER. cǽg.

cǽg-hyrde, es; *m.* [hyrde *a keeper, guardian*] *A keeper of keys, gaoler;* clavicularius. DER. cǽg.

cǽg-loca, an; *m. The action of locking up, a key-locking, any repository locked up;* clavis et loculamentum:—Búton hit under ðæs wífes cǽglocan [cǽglocum MS. A.] gebroht wǽre, sý heó clǽne, ac ðæra cǽgean heó sceal weardian *that is, hire hordern, and hire cyste, and hire tege unless it has been brought under his wife's 'lock and key,' let her be clear; for it is her duty to keep the keys of them; namely, her 'hord-ern,' and her chest, and her cupboard,* L. C. S. 77; Th. i. 418, 19–22. The Latin version reads: 'Sed suum hordern quod dicere possumus dispensam, et cistam suam, et teage, id est scrinium suum, debet ipsa custodire.' A similar provision is found in the old Scottish law: 'Tamen uxor in certis casibus respondere tenebitur; videlicet, si furtum inveniatur sub clavibus suis quas ipsa habet sub custodia et cura sua, utpote spensæ, arcæ suæ vel scrinii sui. Et si aliquod furtum sub clavibus suis inveniatur, uxor cum viro suo tamquam ei consentaneus erit culpabilis et punietur,' *Quon. Attachi.* xii. c. 7. There is a republication of the same law in the Stat. Willielmi Regis, with this variation: 'Spensa et arca robarum et jocalium suorum et de scrinio seu coffero,' xix. c. 3. We may therefore, perhaps, render the terms in the quotation above, '*locked up in her store-room, her chest, and her cupboard,*' L. Th. i. 418, note b.

cǽlan; *p.* de; *pp.* ed *To make cold or cool, to cool;* infrigidare, Cot. 113. DER. ge-cǽlan. v. calan.

cælc, es; *m. A cup, chalice, goblet;* calix:—Cælc oððe scenc *calicem,* Mt. Lind. Rush. Stv. 10, 42. v. calic.

cæle *A* KEEL *or bottom of a ship;* carina, Som. Ben. Lye.

cælic, es; *m. A cup, chalice, goblet;* calix:—Cælic hǽle ic onfó *calicem salutaris accipiam,* Ps. Spl. 115, 4. v. calic.

cælþ *is cold,* Hexam. 20; Norm. 28, 22; *3rd pres. of* calan.

cæmban *to comb;* pectere, Ælfc. Gr. 28, 3; Som. 30, 61, MS. D. v. cemban.

cæmpa, an; *m. A soldier;* pugnator:—Wer cæmpa *vir pugnator,* Cant. Moys. Lamb. 186 b, 3. v. cempa.

cænnan *to clear, prove;* manifestare:—Mynstres aldor hine cænne in

preóstes canne *let the chief of a monastery clear himself with a priest's clearance,* L. Wih. 17; Th. i. 40, 13: 22; Th. i. 42, 3: L. Edg. S. 11; Th. i. 276, 12. v. cennan *to declare,* II.

cænnan; *p.* cænde; *pp.* cænned *To bring forth, produce;* parere:—Ðeós wyrt biþ cænned abúton dícum *this herb is produced about ditches,* Herb. 13, 1; Lchdm. i. 104, 18, MSS. H. B. v. cennan *to beget,* I.

cænnestre, an; *f. One who has borne, a mother, dam;* genitrix. v. cynnestre.

cæn-ryn, es; *n. A generation,* Ps. Spl. 47, 12. v. cyn-ren.

cǽpe-hús, es; *n.* [cépa *a merchant,* hús *a house*] *A storehouse;* armarium:—Ælces cynnes cǽpe-hús *armarium,* Ælfc. Gl. 109; Som. 79, 19; Wrt. Voc. 58, 59.

CÆPPE, an; *f. A* CAP, *cape, cope, hood;* cappa, pileus, cucullus, planeta:—Cǽppe *cappa,* Wrt. Voc. 81, 67. Cǽppe *planeta,* Ælfc. Gl. 27; Som. 60, 114; Wrt. Voc. 25, 54: 81, 45. Gerénod cæppe *an adorned hood;* penula, Ælfc. Gl. 27; Som. 60, 115; Wrt. Voc. 25, 55. [*Piers P.* cope: *Chauc.* cappe, cope: *Laym.* cape, cope: *Plat.* kappe: *Frs.* kæpe: O. *Frs.* kappe: *Dut.* kap, *f:* *Kil.* kappe: *Ger. M. H. Ger.* kappe, *f:* O. *H. Ger.* kappa, *f:* *Dan.* kaabe, kappe, *m. f:* *Swed.* kappa, kåpa, *f:* *Icel.* kápa, *f:* from *M. Lat.* cappa, 'quia capitis ornamentum est,' Isidorus.]

cærc-ærn *a prison;* carcer, Som. Ben. Lye. v. carc-ærn.

cærcian *to chirk, chirp,* Ælfc. Gr. 26, 5; Som. 29, 7, MS. C. v. cearcian.

cæren *a sort of wine, boiled wine;* defrutum, carenum, Cot. 66: L. M. 1, 1; Lchdm. ii. 24, 19. v. ceren.

cærfille, an; *f. Chervil;* cerefolium:—Cærfille *cerefolium,* Ælfc. Gl. 43; Som. 64, 45; Wrt. Voc. 31, 55. v. cerfille.

CÆRSE, cerse, an; *f.* CRESS, *watercress;* nasturtium, cardǎmum = κάρδαμον:—Man nasturcium, and óðrum naman cærse [cerse B.] nemneþ *one nameth nasturtium, and by another name, cress,* Herb. 21, 1; Lchdm. i. 116, 17. Ðeós wyrt, cærse, ne biþ sáwen, ac heó of hyre sylfne cenned biþ on wyllon and on brócen *this herb, cress, is not sown, but it is propagated of itself in wells and in brooks,* i. 116, 15. [*Piers P.* kerse: *Dut.* kers, *f:* *Ger. M. H. Ger.* kresse, *m. f:* O. *H. Ger.* kresso, *m.* cressa, *f.*] DER. eá-cærse, -cerse, fen-, tún-, wylle-.

cǽs *chose,* Chr. 963; Erl. 123, 35, = ceás; *p. of* ceósan.

cæster, e; *f. A city;* civitas, Mt. Rush. Stv. 5, 14: 8, 34. v. ceaster.

CÁF; *comp.* ra, re; *sup.* est, ost; *adj. Quick, sharp, prompt, nimble, swift;* acer, celer, præceps:—Ðá geseah Iohannes sumne cniht swíðe glæd on móde and on anginne cáf *there John saw a certain youth very cheerful in mind and quick in design,* Ælfc. T. 33, 17: R. Ben. 7: Fulg. 9. Cáf *præceps,* Glos. Prudent. Recd. 143, 32. Hét ðá hæleða hleó healdan ða bricge wígan wíghearde cáfne *then the defence [the chief] of the soldiers commanded a warrior, hardy in battle and nimble, to defend the bridge,* Byrht. Th. 133, 66; By. 76. Ðæt hí sceoldon beón cáfe [MS. caue] to Godes willan *that they might be prompt for God's will,* Homl. Th. ii. 44, 31. Sume earniaþ ðæt hie síen ðý cáfran *some merit that they may be the more nimble,* Bt. 34, 7; Fox 144, 8. [*R. Brun.* kof *boisterous:* *Relq. Ant. W.* i. 212, 8, cof: *Orm.* kafe *bold:* O. *Nrs.* á-kafr *promptus, velox.*] DER. beadu-cáf. v. cífan.

cáfe; *adv. Quickly, promptly;* celeriter, prompte:—Mægen samnode cáfe to ceáse *he promptly collected his strength for the fight,* Elen. Kmbl. 111; El. 56. DER. cífan.

cáfer-tún, es; *m. A hall, inclosure, court, vestibule;* atrium, vestibulum:—Mycel and rúm heall *vel* cáfertún *atrium,* Ælfc. Gl. 109; Som. 79, 21; Wrt. Voc. 58, 59: Lk. Bos. 11, 21: Jn. Bos. 18, 15: Bt. 18, 1; Rawl. 38, 30. Seó fǽmne geneálǽhte ðam cáfertúne ðyses húses *the maiden came nigh the court of this house,* Bd. 3, 11; S. 536, 36: 5, 2; S. 615, 2: Ps. Lamb. 95, 9. For ðí ðe is betere án dæg on ðínum cáfertúnum ofer þúsenda hér *quia melior est dies una in atriis tuis super milia,* Ps. Lamb. 83, 11: 95, 8: 115, 8: 121, 2: 134, 2: Ps. Th. 121, 2: 133, 2: 134, 2. Infaraþ on cáfertúnas his on ymnum *introite atria [courts] ejus in hymnis,* Ps. Spl. 99, 4: Ps. Lamb. 99, 4. DER. cífan.

cáf-líce; *adv. Quickly, hastily, stoutly, manfully, valiantly;* velociter, viriliter:—Ðám gemettum wæs beboden ðæt hí sceoldon cáflíce etan *the partakers were commanded to eat quickly,* Homl. Th. ii. 282, 3: i. 494, 11: Glos. Prudent. Recd. 146, 38: Byrht. Th. 136, 19; By. 153: Num. 31, 6. DER. cífan.

cáf-scype, es; *m. A quickness;* velocitas, R. Ben. 5. DER. cífan.

cál, es; *m. A herb, wild cole-wort;* arboracia, lapsana?—Cál arboracia *vel* lapsana? Ælfc. Gl. 44; Som. 64, 73; Wrt. Voc. 32, 9. v. cawel.

CALAN, ic cale, ðú calest, cælst, he caleþ, cælþ, *pl.* calaþ; *p.* cól, *pl.* cólon; *pp.* calen; *v. intrans. To be or become cool or cold;* algere, frigescere:—Ðonne him cælþ, he ceþþ him hlywþe *when he is cold, he betakes himself to shelter,* Hexam. 20; Norm. 28, 22. Hwæðer ða welgan ne cale ða ríc never become cold? Bt. 26, 2; Fox 92, 34. [*Wyc.* kele, koole: *Orm.* kelenn: *Plat.* kölen: O. *Sax.* kôlon: O. *Frs.* kela: *Dut.* koelen: *Ger.* kühlen: *M. H. Ger.* kuolen *to become cold:* O. *H. Ger.* kuoljan: *Dan.* koele: *Swed.* koela: *Icel.* kala; *p.* kól, *pp.* kalit *algere: Lat.* gelare.] DER. a-calan, of-: calian: célan, a-, ge-

cēle, cȳle, fǽr-; -gicel, -wyrt: cēlnes, ge-: cēling; cēlung, ge-: cōl, -nes: cōlian, a-: ceald, cald, æl-, brim-, eal-, hrím-, ís-, morgen-, ofer-, sin-, snāw-, wæl-, winter-: caldu, sin-: cald-heort: cealdian, a-: cīlian: cǽlan, ge-.

calc, es; *m. A shoe, little shoe, sandal;* calceus, sandalium:—Gesceóde mid calcum *calceatos sandaliis,* Mk. Bos. 6, 9: Cot. 209.

calc-rond; *adj. Round of hoof;* calceis *vel* soleis ferreis marginatus:—Calcrondes, Exon. 91 a; Th. 342, 15; Gn. Ex. 143.

cald *cold;* gelidus, frigidus:—Đonne cymþ forst fyrnum cald *then cometh bitter cold frost,* Cd. 17; Th. 20, 28; Gen. 316: 227; Th. 304, 29; Sat. 637: Andr. Kmbl. 619; An. 310. Caldra *colder,* Exon. 111 a; Th. 425, 10; Rä. 41, 54. Caldast *coldest,* 81 b; Th. 308, 1; Seef. 33. v. ceald, calan.

cald, es; *n. Cold, coldness,* Exon. 81 b; Th. 306, 16; Seef. 8. v. ceald *frigus.*

cald-heort; *adj. Cold-hearted, unfeeling, cruel;* frigidus cordis, inhumanus, crudelis:—Cirmdon caldheorte *the cold-hearted cried out,* Andr. Kmbl. 275; An. 138. v. calan.

caldu, e; *f. Cold, coldness;* gelu, frigus. DER. sin-caldu. v. calan.

calend, es; *m.* I. *a month;* mensis:—Calend [kalend MS.] Martius rēđe *the fierce month of March,* Menol. Fox 62; Men. 31. II. *the appointed time* or *day of life;* dies, terminus vitæ:—Ǽr þæg cyme, đæt sȳ his calend arunnen *ere the day come, when his appointed time be run out,* Salm. Kmbl. 959; Sal. 479.

calf *a calf,* Ps. Spl. 49, 10. v. cealf.

calferu, *acc. pl. Calves;* vitulos, Ps. Surt 49, 9. v. cealf.

calfian *to* CALVE; vitulum edere, Som. Ben. Lye. v. cealfian.

calfru *calves,* Ps. Th. 21, 10. v. cealf.

calfur *calves;* vituli:—Ymb-saldon me calfur *circumdederunt me vituli,* Ps. Surt. 21, 13: 50, 21. v. cealf.

cālian; *p.* ode; *pp.* od; *v. intrans. To be* or *become cold;* algere, frigescere. v. calan.

CALIC, cælic, cǽlc, calc, es; *m. A cup,* CHALICE, *goblet;* calix:—Se calic mínre blisse *the cup of my joy,* Ps. Th. 15, 5: Ps. Spl. 22, 7. Dæl calices mínes *pars calicis mei,* Ps. Spl. 15, 5. He genam đone calic *accepit calicem,* Mt. Bos. 26, 27, 28: Ps. Th. 115, 4: Ps. Surt. 115, 13. [*Plat.* kelk: *O. Sax.* kelik, *m: O. Frs.* tzielk, tzilik, *m: Dut.* kelk, *m: Ger.* kelch, *m: M. H. Ger.* kelich, kelch, *m: O. H. Ger.* kelih, *m: Dan.* kalk, *m. f: Swed. Norw.* kalk, *m: Icel.* kalkr, *m;* from *Lat.* calix: *Grk.* κύλιξ.]

calla, an; *m.* [ceallian *to call*] *A herald,* found in the phrase,—hilde calla [*q. v.*] *war's herald* or *a herald of war,* Cd. 156; Th. 193, 26; Exod. 252.

CALU, caluw; *adj.* CALLOW, *bald, without hair;* calvus, glaber:—Calu ođđe hnot *glaber* [MS. *glabrio*], Ælfc. Gr. 9, 3; Som. 8, 36: Exon. 111 b; Th. 427, 31; Rä. 41, 99. Monig man weorþ fǽrlíce caluw *many a man becomes bald suddenly,* Prov. Kmbl. 42. [*Wyc.* calu: *Plat.* kaal: *Frs.* keal: *Dut.* kaal: *Kil.* kael: *Ger.* kahl: *M. H. Ger.* kal: *O. H. Ger.* chalo, chalaw: *Lat.* calvus: *Ir. Gael.* calbh: *O. Slav.* golu.]

caluw *bald,* Prov. Kmbl. 42. v. calu.

calwa, an; *m. A disease which causes baldness, the mange;* alopecia = ἀλωπεκία, Cot. 12.

calwer, es; *m. Pressed curds;* calmaria? gabalacrum?—Calwer [MS. caluuær] *calmaria?* Glos. Epnl. Recd. 157, 21: *gabalacrum?* 157, 26. Calwer *gabalacrum?* Cot. 96. v. cealre.

calwer-bríw, cealer-bríw, es; *m. A thick pottage made of curds;* calviale, Wrt. Voc. 290, 37. v. bríw.

calwere, es; *m?* [calu *bald*] *A bald place on the top of the head, a skull, place of skulls, place for burial;* calva, calvaria, Som. Ben. Lye.

camal *a camel,* Lk. Lind. War. 18, 25. v. camel.

camb, es; *m.* [camb *joined*] *p.* of cimban]. I. *a comb for cleaning hair, wool, flax, etc;* pecten, Wrt. Voc. 86, 11. v. bannuc-camb, fleđe-camb, wulfes camb. II. *the crest of a cock, the crest* or *top of a helmet, etc;* crista:—Helmes camb *the helmet's crest;* crista, Ælfc. Gl. 53; Som. 66, 75; Wrt. Voc. 36, 2. Camb on hætte, *vel* on helme *a crest on the hat* or *helmet;* crista, Cot. 46. [*Orm.* camb: *Scot.* kaim: *O. Sax.* camb, *m: Frs.* kaem: *Dut. Kil.* kam, *m: Ger.* kamm, *m: M. H. Ger.* kam, *m;* kambe, *f: O. H. Ger.* kamp, kampo, *m: Dan.* kam, *m. f: Swed.* kam, *m: Icel.* kambr, *m: Sansk.* jambha, *m. tooth.*]

camb, e; *f. A comb, an assemblage of cells in which bees store their honey;* favus:—Hí ymbþrungon me swā swā beón camba *they surrounded me as bees* [*surround*] *the combs,* Ps. Lamb. 117, 12.

cambiht [camb, iht]; *adj. Combed, having a crest;* cristatus. v. camb II.

camel, camell, camal, es; *m. A camel;* cămēlus =κάμηλος, גָמָל:—Wæs Iohannes gegerelad mid hērum cameles [camelles, Lind.] *erat Iohannes vestitus pilis cameli,* Mk. Skt. Rush. 1, 6. Iohannes hæfde gewēde of hērum đæra camella *Iohannes habebat vestimentum de pilis camelorum,* Mt. Kmbl. Lind. 3, 4. Se camal, Lk. Lind. War. 18, 25.

cammoc, cammuc, commuc, es; *n. m? The cammoc, kex, an umbel-*

liferous plant, brimstone wort, hog's fennel, cow weed, cow parsley. Kambuck *is still a name of the kexes in Suffolk,* Prior 36, 126; peucedănum officinale, = πευκέδᾰνον, n; πευκέδᾰνος, f. *sulphur wort, hog's fennel:*—Đás wyrte man peucedanum, and ōđrum naman cammoc [cammuc MS. H.] nemneþ *this wort is called peucedanum, and by another name cammoc,* Herb. 96, 1; Lchdm. i. 208, 17. Wyrc gōdne drenc, elenan iii snǽda, commuces viii *make a good drink, three portions of elf dock, eight of cammoc,* L. M. 3, 30; Lchdm. ii. 324, 20.

camp, es; *m. A bond, fetter, chain;* compes:—Hió bindan þenceaþ cyningas on campum *ad alligandos reges eorum in compedibus,* Ps. Th. 149, 8. v. cops.

CAMP, comp, es; *m. A contest, war, battle;* certamen, pugna, bellum:—Ic ne gȳme đæs compes *I care not for the contest,* Exon. 105 b; Th. 402, 26; Rä. 21, 35. Drihten tǽcþ handa míne to gefeohte, and fingras míne to slehte ođđe to campe *Dominus docet manus meas ad prælium, et digitos meos ad bellum,* Ps. Lamb. 143, 1: Bd. 3, 24; S. 556, 21: Judth. 11; Thw. 24, 21; Jud. 200: Beo. Th. 5003: B. 2505: Chr. 937; Th. 202, 2, col. 1, 2; Æđelst. 8: Andr. Kmbl. 2651: An. 1327. Mec gesette Crist to compe *Christ has placed me in battle,* Exon. 102 b; Th. 389, 3; Rä. 7, 2: Andr. Kmbl. 468; An. 234. He ofercom campe feónda folcriht *he overcame the liberty of enemies in battle,* Cd. 143; Th. 178, 33; Exod. 21. [*Laym.* comp *a conflict:* *Plat.* kamp: *O. Frs.* kamp, komp, *m: Dut.* kamp, *m. a battle: Ger. M. H. Ger.* kampf, *m. a fight: O. H. Ger.* champh, *m: Dan.* kamp, *m. f: Swed.* kamp, *m: Norw. Icel.* kapp, *n: Wel.* camp, *f.*] DER. camp-dōm, -hād, -rǽden, -stede, -wǽpen, -wered, -weorud, -wíg, -wudu: comp-wǽpen, -weorod, -wíg.

camp-dōm, es; *m. Warfare;* militia, Scint. 29, 1. DER. camp.

camp-hād, es; *m. Warfare;* militia:—Hí synd bigongende woruldlícne camphād *they are exercising worldly warfare,* Bd. 5, 24; S. 647, 9. DER. camp.

campian, compian; *p.* ode; *pp.* od [camp *war*] *To fight, contend against;* militare, pugnare:—Sceal oretta ā Gode campian *a champion shall ever fight for God,* Exon. 37 b; Th. 123, 1; Gū. 316: Bd. 1, 15; S. 483, 12. Se deófle campaþ [compaþ, Ps. Lamb. fol. 183 b, 18] *he fights for the devil,* Hy. 2, 5; Hy. Grn. ii. 281, 5. Ic longe Gode campode *I have long fought for God,* Exon. 42 a; Th. 140, 25; Gū. 615. He for his ēđle mid his leódum compode *he fought for his country with his men,* Bd. 3, 9; S. 533, 17. [*Scot.* kemp: *Dut.* kampen: *Ger.* kämpfen: *M. H. Ger.* kempfen: *O. H. Ger.* chamfan, chemfan: *Dan.* kämpe: *Swed.* kämpa: *Icel.* keppa.] DER. wiđ-compian.

camp-rǽden, -rǽdenn, e; *f. State* or *condition of contest, contest, war;* certandi modus, certamen, pugna:—Nō hyra þrym alæg camprǽdenne *their vigour did not fail in the contest,* Andr. Kmbl. 7; An. 4. DER. camp.

camp-stede, es; *m. The place of battle, battle-field;* locus pugnæ:—On đam campstede *on the battle-field,* Chr. 937; Th. 204, 2, col. 1; Æđelst. 29: 937; Th. 206, 1, col. 1; Æthelst. 49. Fōr campstede [MS. campsted] sēcan *he went forth to seek the place of battle,* Bt. Met. Fox 26, 28; Met. 26, 14. DER. camp.

camp-wǽpen *a battle-weapon, military weapon.* v. comp-wǽpen.

camp-weorud, es; *n. Fighting-men, soldiers;* militia, exercitus, Bd. 3, 24; S. 556, 33. v. camp-wered.

camp-wered, -weorud, comp-weorod, es; *n.* [werod, es; *n. an army*] *Warriors, soldiers, fighting-men, army;* militia, exercitus:—Hí sceoldan for heora campwered gebiddan and to Gode þingian *they should pray and make intercession to God for their warriors,* Bd. 2, 2; S. 503, 39. Æđelhere mon slōh mid ealle his campweorude đe he mid him brohte *Ethelhere was slain with all the fighting-men whom he had brought with him,* 3, 24; S. 556, 33. Đa ārleásan cyningas ofslegene wǽron mid heora compweorode *the wicked kings were slain with their army,* 2, 5; S. 507, 40. DER. camp.

camp-wíg *a battle.* v. comp-wíg.

camp-wudu; *gen.* -wuda; *m.* War-wood, *a shield;* lignum pugnæ, clipeus:—Đonne rand dynede, campwudu clynede *then rang the shield, the war-wood sounded,* Elen. Kmbl. 101; El. 51. DER. camp.

can, cann, e; *f. A knowledge, clearance.* v. cann.

can, ic he *I know, he knows:*—Ic ođđe he can, Elen. Kmbl. 1363; El. 683; Ps. Th. 88, 13. He can *he can,* Bt. 39, 2; Fox 214, 10. v. cunnan.

Cananēisc; *adj. Canaanitish;* Chananæus:—Cham ys fæder đære Cananēiscre þeóde *Ham is the father of the Canaanitish people,* Gen. 9, 18.

canceler, es; *m. A chancellor;* cancellarius:—Se cyng Willelm betǽhte Rodbeard his cancelere đæt biscoprīce on Lincolne *the king William transferred the bishopric of Lincoln to Robert his chancellor,* Chr. 1093; Ing. 306, 7.

cancer; *gen.* cancres; *m?* I. *a cancer, an eating* or *spreading disease;* cancer, morbus:—Gif đu wille cancer ablendan, genim đonne fífleáfan đa wyrte: seóþ on wíne *if thou desire to stop a cancer, then take the herb fiveleaf: boil it in wine,* Herb. 3, 9; Lchdm. i. 88, 20. Ealne đone bíte đæs cancres heó aformaþ *it clears away all the pain* [*bite*] *of the cancer,* 167, 3; Lchdm. i. 296, 22. Wiđ cancre, nim gāte geallan

and hunig *against cancer, take goat's gall and honey*, L. M. 3, 36; Lchdm. ii. 328, 13 : Herb. 32, 3; Lchdm. i. 130, 12, MS. O, note 24. Wið cancre *for cancer*, Med. ex Quadr. 6, 21; Lchdm. i. 354, 25. **II.** *a crab;* cancer, animal. v. cancer-hæbern.

cancer-ádl, e; *f. A cancer-disease, a canker;* cancer, carcinoma = κορκίνωμα:—Wið canceráðle, ðæt is, bíte *against cancer-disease, that is, a biting disease*, L. M. 1, 44; Lchdm. ii. 108, 9.

cancer-hæbern, es; *n.* [cancer *a crab*, hæbern = hæb-ærn *a place, dwelling-place*] *A crab-hole;* caverna, cavernula D.

cancettan; *part.* cancettende; *p.* cancette; *pp.* cancetted *To laugh aloud* or *in a cackling manner;* cachinnare :—Mæssepreóst ne sceal lufigean micelne and ungemetlícne cancettende hleahtor *nor shall a mass-priest love great and immoderate cackling laughter*, L. E. I. 21; Th. ii. 416, 36. v. ceahhetan.

cancetung, e; *f. A laughing in a cackling manner;* cachinnus, Cot. 58. v. ceahhetung.

CANDEL, candell, condel, condell, e; *f:* candel, es; *n. A* CANDLE; candela, lampas = λαμπάς :—Hádre scíneþ ródores candel *the sun* [*the candle of the firmament!*] *serenely shines*, Beo. Th. 3148; B. 1572. Candeles leóma *the light of a candle;* lampas, Ælfc. Gl. 67; Som. 69, 88; Wrt. Voc. 41, 41. Glád ofer grúndas Godes condel beorht *God's bright candle glided over the grounds*, Chr. 937; Th. 202, 16, col. 1; Ædelst. 15 : Exon. 51 b; Th. 179, 20; Gú. 1264 : 72 a; Th. 269, 23; Jul. 454. Se sceal ðære sunnan síþ bihealdan, Godes condelle *he shall observe the sun's course, God's candle*, 57 a; Th. 204, 2; Ph. 91. [*Chauc. Laym.* candel : *Pers.* قنديل kandeel *a candle :* *Fr.* chandelle; *Span. It.* candela, from the *Lat.* candela, from candēre *to shine*.] DER. dæg-candel, friþ-, heofon-, mére-, swegel-, weder-, woruld-, wyn- : candel-bora, -leóht, -mæsse, -snytels, -stæf, -sticca, -treów, -twist, -weoc, -wyrt.

candel-bora, an; *m. A* CANDLE-BEARER, *a subdeacon, a clerk;* acolythus = ἀκόλουθος, Cot. 203.

candell, e; *f. A candle;* candela, lampas. v. candel.

candel leóht, es; *n. Candle-light;* lucernæ lumen, C. R. Ben. 53. DER. candel.

Candel-mæsse, an; *f.* CANDLEMAS, the mass at the feast of purification which, in the Romish church, *is celebrated with many lighted candles;* festum purificationis beatæ Mariæ :—Æt Candelmæssan *at Candlemas*, L. Eth. ix. 12; Th. i. 342, 32. Hér, A.D. 1014, Swegen ge-endode his dagas to Candelmæssan *here, A.D. 1014, Sweyn ended his days at Candlemas*, Chr. 1014; Th. 272, 25, col. 1. DER. candel.

candel-snytels, es; *m?* *Candle-snuffers;* emunctorium :—Candel-snytels *emunctorium*, Ælfc. Gl. 30; Som. 61, 56. DER. candel.

candel-stæf, es; *m. A candle-staff* or *stick;* candelabrum :—Ne hí ne ælaþ hyra leóhtfæt, and hit under cyfe settaþ, ac ofer candelstæf *neque accendunt lucernam, et ponunt eam sub modio, sed super candelabrum*, Mt. Bos. 5, 15.

candel-sticca, an; *m. A* CANDLESTICK; candelabrum, Chr. 1102; Th. 366, 20. DER. candel.

candel-treów, es; *n. A candlestick with branches, a candlestick;* candelabrum :—Ne menn blæcern in beornaþ and settaþ hine under mytte, ah on candeltreów *neque accendunt lucernam et ponunt ea sub modio, sed super candelabrum*, Mt. Kmbl. Rush. 5, 15. DER. candel.

candel-twist, es; *n. A pair of snuffers;* emunctoria :—Candel-twist *emunctoria*, Ælfc. Gl. 82; Som. 73, 50; Wrt. Voc. 47, 54. DER. candel.

candel-weoc, e; *f. A wick of a candle, a torch;* funale, funis :—Candelweoca *funalia vel funes*, Ælfc. Gl. 67; Som. 69, 87; Wrt. Voc. 41, 40. DER. candel.

candel-wyrt, e; *f.* [candel *a candle*, wyrt *a herb, plant*] CANDLEWORT, *hedge-taper, mullein;* lucernaria, phlomos = φλόμος, verbascum : thapsus, lapa. A plant useful for wicks of lamps :—Candelwyrt *phlomos* [MS. *fromos*] *vel lucernaria* [MS. *lucernaris*], Ælfc. Gl. 44; Som. 64, 90; Wrt. Voc. 32, 25.

cann *know, knows;* scio, scit, Ps. Th. 91, 5 : 93, 11. v. cunnan.

cann, e; *f. A knowledge, cognizance, averment* or *positive assertion, clearance;* notitia, cognitio, assertio :—Mynstres aldor hine cænne in preóstes canne *let the chief of a monastery clear himself with a priest's cognizance*, L. Wih. 17; Th. i. 40, 13. Mid rihtre canne *by lawful averment*, L. H. E. 16; Th. i. 34, 12. Ðanne is cirican canne riht *then is the church clearance right*, L. Wih. 21; Th. i. 42, 1. [*Kil.* konne, kunne : *Ger.* kunde, *f.*]

CANNE, an; *f. A* CAN, *cup;* crater :—Canne *crater vel canna*, Ælfc. Gl. 24; Som. 60, 38; Wrt. Voc. 24, 38. [*Wyc.* cannes, *pl :* *Plat.* kanne : *Dut.* kan, *f:* *Ger. M. H. Ger.* kanne, *f:* *O. H. Ger.* channa, *f:* *Dan.* kande, *m. f:* *Swed. Icel.* kanna, *f.*]

CANON, es; *m. A* CANON, *rule;* regula, canon = κανών :—Se canon cwæþ *the canon said*, L. Ælf. P. 31; Th. ii. 376, 26. Se canon awriten is *the canon is written*, Bd. 5, 23; S. 648, 43. Ða canonas openlíce beódaþ *the canons openly command*, L. Ælf. P. 31; Th. ii. 376, 20. Canones bóc *the book of the canon*, Bd. 4, 24; S. 598, 13.

canon-dóm, es; *m. A canonship, office of a canon;* canonicatus. v. canon, -dóm *office, state, condition*.

canonec-líc; *adj. Canonical;* canonicus :—Æfter canoneclícan gewunan *according to canonical custom*, Canon. Hrs. 359, 8.

canonic, es; *m. A canon, prebendary;* canonicus :—Ðæt Godes þeówas, biscopas and abbodas, munecas and mynecena, canonicas and nunnan, to rihte gecyrran *that God's servants, bishops and abbots, monks and mynchens, canons and nuns, turn to right*, L. Eth. vi. 2; Th. i. 314, 17 : vi. 4; Th. i. 316, 1 : v. 7; Th. i. 306, 13.

canst *knowest, canst*, Andr. Kmbl. 135; An. 68 : March 176; *2nd pers. sing. pres. of* cunnan.

cantel-cap, es; *m.* CANTEL-COPE, *a sort of priest's garment;* caracalla, Chr. 1070; Ing. 274, 1.

cantere, es; *m. A singer;* cantor, Som. Ben. Lye.

cantic, es; *m. A canticle, song;* canticum :—Hafaþ se cantic ofer ealle Cristes béc wídmærost word *the canticle hath the greatest repute over all Christ's books*, Salm. Kmbl. 99; Sal. 49. Ðæt ic sí gebrydded þurh ðæs cantices cwide *that I may be touched through the word of the canticle*, 33; Sal. 17. Moises wrát ðone cantic and lærde Israéla folc *scripsit Moyses canticum et docuit filios Israel*, Deut. 31, 22 : 31, 19 : Salm. Kmbl. 47; Sal. 24: Ps. Th. 143, 10.

Cantwara burg, Cantware-burg, Cantwar-burg, -burh; *gen.* burge; *f:* Cantwara byrig, e; *f.* [Cant-wara, *gen. pl.* of Cant-ware *Kentish men*, burh *a city*] *A city* or *fortress of the men of Kent;* Cantuariorum urbs *vel* castellum. **I.** CANTERBURY; Durovernensis civitas :—Cantwara burg forbærn ðý geáre *Canterbury was burnt in that year*, Chr. 754; Th. 80, 35, col. 1. Brǽcon Cantwara burh *they took Canterbury by storm*, 853; Th. 120, 28, col. 3. Ðá sealde Ædelbyrht him wununesse and stówe on Cantwara byrig, seó wæs ealles his ríces ealdorburh *dedit ergo Ædilberctus eis mansionem in civitate Durovernensi* [*Canterbury*], *quæ imperii sui totius erat metropolis*, Bd. 1, 25; S. 487, 18 : 4, 5; S. 572, 9. To Cautwarebyrig *to Canterbury*, Chr. 1009; Th. 260, 37. He wæs bebyrged innan Cantwarbyrig *he was buried within Canterbury*, 690; Th. 65, 23, col. 1 : 754; Th. 81, 36. **II.** *Rochester;* Roffensis civitas, Roffa :—Putta Cantwara burhge bisceop, seó is cweden æt Hrofesceastre *Putta Episcopus castelli Cantuariorum, quod dicitur Rofecester*, Bd. 4, 5; Whelc. 272, 35.

Cantwara mægþ, e; *f. The county of Kent, men of Kent;* Cantianorum provincia :—On Cantwara mægþe *in the county of Kent*, Bd. pref; S. 471, 26.

Cant-ware; *gen.* a; *dat.* um; *acc.* e; *pl. m. Kentish men, inhabitants of Kent;* Cantuarii :—Of Geáta fruman syndon Cantware and Wihtsætan *de Jutarum origine sunt Cantuarii et Victuarii*, Bd. 1, 15; S. 483, 22. Cantwara cyningas *kings of Kentish men*, L. H. E; Th. i. 26, 4, 5 : 34, 3 : 36, 2. Agustinus nú on Brytene rest, on Cantwarum *Augustine now rests in Britain, among the inhabitants of Kent*, Menol. Fox 207; Men. 105.

capelein, capellan *A chaplain;* capellanus, Chr. 1099; Ing. 318, 14.

capian; he capaþ; *p.* ode; *pp.* od *To turn, incline oneself;* vertere, se inclinare :—Capaþ he up *he turns upwards*, Bd. de nat. rerum; Wrt. popl. science 15, 3; Lchdm. iii. 266, 23.

capitol, capitul, es; *m:* capitula, an; *m. A chapter;* capitulum :—Hér onginþ se forma capitul *here begins the first chapter*, L. Ecg. P. cont. i. 1; Th. ii. 170, 3 : iii. 1; Th. ii. 194, 2. On ðam ende dises capitulan *in the end of this chapter*, Bt. 32, 2; Fox 116, 33.

capitol-mæsse, an; *f. Early* or *morning mass, first mass;* prima *vel* matutinalis missa :—We sungon capitol-mæssan *cantavimus primam missam*, Coll. Monast. Th. 33, 29.

cappa *a cap, cope, priest's garment;* capitulum :—Heáfod-cláþ *vel* cappa *capitulum vel capitularium*, Ælfc. Gl. 64; Som. 69, 15. v. cæppe.

CAPÚN, es; *m. A* CAPON; gallinaceus, capo = κάπων :—Capún *gallinaceus*, Wrt. Voc. 63, 9 : Ælfc. Gl. 39; Som. 63, 48: Wrt. Voc. 30, 3. Capún *capo*, 39; Som. 63, 46; Wrt. Voc. 30, 1. [*Plat.* kappuun : *Dut.* kapoen, *m :* *Kil.* kappuyn, kaphoen : *Ger.* kapaun, *m :* *M. H. Ger.* kapûn, *m :* *Dan. Swed.* kapun, *m :* *O. Nrs.* kapún, *m.* Rask Hald : from the *Lat.* capo : *Grk.* κάπων.]

cara *care*, Ælfc. Gl. 89; Som. 74, 96; Wrt. Voc. 51, 9. v. cearu.

carc CARK, *care;* cura, Som. Ben. Lye. v. carc-ern.

carc-ern, carc-ærn, es; *n.* [carc *care*, or *Lat.* carcer *a prison;* ærn, ern *a place*] *A prison, a house of correction;* carcer, latomiæ :—Aléd of carcernes clúse míne sáwle *educ de carcere animam meam*, Ps. Th. 141, 8. Ðonne þincþ him ðæt he síe on carcerne gebroht *then it seems to him that he is brought into prison*, Bt. 37, 1; Fox 186, 15. Ic wæs on cearcerne [MS. Cot. carcærne] *eram in carcere*, Past. 44, 7 : Hat. MS. 62 b, 22. To ðam carcerne *to the prison*, Andr. Kmbl. 179; An. 90 : Exon. 8 a; Th. 2, 27; Cri. 25 : Cd. 227; Th. 304, 28; Sat. 637 : Cot. 124 : 191.

car-clífe, an; *f. Agrimony;* agrimonia, Wrt. Voc. 79, 62. v. gar-clífe.

care *care*, Ps. Th. 143, 18; *acc. of* caru. v. cearu.

care-líce; *adv. Sorrowfully, miserably, wretchedly;* misere :—Me
L

deorc earfoðe carelíce cnyssedan *dark troubles wretchedly weakened me,* Ps. Th. 85, 6.

Carendre, an; *f. A province of Germany,* now the duchy of *Carinthia* or *Kärnthen,* a crown land of the Austrian empire :—On óðre healfe Donua ðære eá is ðæt land Carendre, súþ óþ ða beorgas ðe man hæt Alpis *on the other side of the river Danube is the country Carinthia,* [*lying*] *south to the mountains which are called the Alps,* Ors. 1, 1; Bos. 18, 43. Be eástan Carendran is Pulgara land *to the east of Carinthia is the country of the Bulgarians,* 1, 1; Bos. 19, 1.

car-ful; *adj.* CAREFUL, *anxious, curious;* sollicitus, curiosus :—Drihten carful oððe ymhydig is mínes *Dominus sollicitus est mei,* Ps. Lamb. 39, 18. Carful *curiosus,* Ælfc. Gl. 89; Som. 74, 112; Wrt. Voc. 51, 25. v. cear-ful.

carful-líce; *adv.* CAREFULLY, *diligently;* sollicite, diligenter :—Se sacerd sceal dón carfullíce Godes þenunga *the priest shall carefully do God's services,* L. Ælf. C. 36; Th. ii. 360, 25. Twá þing sind ðe we sceolon carfullíce scrutnian *there are two things that we should diligently attend to,* Homl. Th. ii. 82, 25.

carful-nys, -nyss, e; *f.* CAREFULNESS, *curiosity;* sollicitudo, curiositas :—Godes cwydas sind to smeágenne mid micelre carfulnysse *the words of God are to be considered with great carefulness,* Homl. Th. ii. 280, 18 : Lchdm. iii. 210, 5.

carian; *p.* ode; *pp.* od *To take care, regard, heed, to be anxious;* curare, sollicitum esse :—Ðæt abbodas næfre idele wlænca carian *that abbots should never regard vain pomps,* L. I. P. 13; Wilk. 150, 25. Se morgenlíca dæg caraþ ymb hyne sylfne *crastinus dies sollicitus erit sibi ipsi,* Mt. Bos. 6, 34: Homl. Th. i. 66, 9. Carian *to take heed, care,* L. I. P. 14; Th. ii. 322, 5. Ða cariaþ mid wacelum móde *they care with watchful mind,* Homl. Th. ii. 78, 2. v. cearian.

carited *charity;* caritas :—Heóld mycel carited in ðe hús *held much charity in the house,* Chr. 1137; Erl. 263, 6.

carl, es; *m.* [=ceorl *a churl*] *A churl, rustic;* rusticus, colonus :—Carles wæn *the churl's wain* or *waggon,* Æqu. Vern. 30, 5; Wrt. popl. science 16, 5; Lchdm. iii. 270, 11, 12; Boutr. Scrd. 29, 31. v. carles wæn.

carl; *adj. Male, masculine;* masculus. Used in compounds, as carl-cat, -fugel, -man.

carl-cat, es; *m. A male* or *he cat;* masculus cattus, Som. Ben. Lye.

car-leás; *adj.* [caru *care,* leás *less*] CARELESS, *reckless, void of care, free;* improvidus, securus :—Wulfas sungon, carleásan deór *wolves howled, reckless beasts,* Cd. 151; Th. 188, 10; Exod. 166. He on ðam dóme freoh and carleás biþ *in judicio liber erit,* R. Ben. 2.

carleás-nes, -ness, e; *f. Freedom from care, security,* CARELESSNESS; securitas, Ælfc. Gl. 89; Som. 74, 113; Wrt. Voc. 51, 26. v. car-leás.

car-leást, e; *f. Freedom from care, security, carelessness;* securitas :—Ring on swefuum underfón carleáste getácnaþ *to receive a ring in dreams betokens freedom from care,* Lchdm. iii. 198, 21, 29: 210, 5.

carles wæn [*gen.* of carl] *the churl's wain, the constellation of the Great Bear;* Ursa Major :—Carles wæn ne gæþ næfre adúne under ðyssere eorþan, swá swá óðre tunglan dóþ *the churl's wain never goes down under this earth, as other constellations do,* Bd. de nat. rerum; Wrt. popl. science 16, 5; Lchdm. iii. 270, 11, 12. v. arctos.

carl-fugel, es; *m. A male* or *cock bird;* mas avis, Som. Ben. Lye.

carl-man, -mann, es; *m. A male, man;* masculus, homo :—Ða námen hí carlmen and wimmen *then took they men and women,* Chr. 1137; Ing. 366, 7.

CARR, es; *m.* **I.** *a stone, rock,* SCAR; petrus =πέτρος, petra =πέτρα :—Ðæt is getrahtad carr *quod interpretatur petrus,* Jn. Lind. War. 1, 42. Ðæt wæs geheáwen of carre oððe stáne *quod erat excisum de petra,* Mk. Skt. Lind. 15, 46. Se ðe gesette ða grundas ofer carr oððe stán *qui posuit fundamenta supra petram,* Lk. Lind. War. 6, 48: Mt. Kmbl. Lind. 7, 24. **II.** *Charmouth, in Dorsetshire, at the mouth of the river* Carr, = the Norman Charr, or *Charmouth;* 'in agri Dorsætensis parte maritima, post *c* literam addito *h,* ad morem Normannorum, *Gib* :—Æðelwulf cyning gefeaht æt Carrum wið xxxv sciphlæsta *king Æthelwulf fought at Charmouth against the crews of thirty-five ships,* Chr. 840; Th. 120, 3, col. 1, 2, 3; 121, 3, col. 1, 2, 3: 833; Th. 116, 4, col. 1, 2, 3; 117, 4, col. 1, 2, 3. [*North Eng.* carrock : *Scot.* cairn : *Wel.* carn : *Corn.* carn, *m* : *Ir.* carn : *Gael.* carr, *m* : *Manx* carn, *m.*]

Carrum *the place of a naval engagement, near Charmouth, Dorsetshire,* Chr. 840; Erl. 67, 12. v. Carr II.

Cartaina; *indecl* : Cartaine, an; *f. Carthage;* Carthago :—Cartaina toworpen wæs *Carthage was overthrown,* Ors. 5, 2; Bos. 101, 18. Scipia hæfde gefaren to ðære niwan byrig Cartaina *Scipio had gone to the new city Carthage,* 4, 10; Bos. 93, 41: 4, 13; Bos. 99, 27. Ðæt mon ealle Cartaina towurpe *that one would overthrow all Carthage,* 4, 13; Bos. 99, 25. He þohte Cartainan toworpan *he wished to overthrow Carthage,* 4, 13; Bos. 100, 3.

Cartaine; *nom. acc;* *gen.* a; *dat.* um; *pl. m. The Carthaginians;* Carthaginienses :—Wilnedon Cartaine friðes to Rómánum *the Cartha-*

ginians *sued for peace to the Romans,* Ors. 4, 6; Bos. 87, 12. Terrentius, se mæra Cartaina sceóp, bær hætt on his heáfde *Terence, the great poet of the Carthaginians, wore a hat on his head,* 4, 10; Bos. 96, 18 : 4, 11; Bos. 97, 11 : 4, 13; Bos. 99, 24. Wearþ Cartainum friþ alýsed fram Scipian *peace was granted to the Carthaginians by Scipio,* 4, 10; Bos. 96, 11 : 4, 6; Bos. 86, 32. Rómáne wunnon on Cartaine *the Romans fought against the Carthaginians,* 4, 7; Bos. 87, 37 : 4, 6; Bos. 86, 37.

carte, an; *f.* [*Lat.* charta] *Paper, a piece of paper, a deed;* chárta = χάρτης :—Híg hym tosendon áne cartan, seó wæs ðus awriten [MS. awryten] *they sent a paper to him, which was thus inscribed,* Nicod. 20; Thw. 10, 5. Alecge ða sealfe on hátne cláþ oððe cartan *lay the salve on a hot cloth or on paper,* L. M. 2, 19; Lchdm. ii. 202, 10. Cartan wrítan [MS. wirtan] oððe rædan *to write or read a paper,* Lchdm. iii. 200, 35.

caru *care, sorrow, grief,* Lk. Bos. 10, 40 : Ps. Th. 60, 1 : 78, 11. v. cearu.

cáser-dóm, es; *m. An emperor's rule;* imperium :—Ða wæs syxte geár Constantínes cáserdómes *then was the sixth year of Constantine's imperial rule,* Elen. Kmbl. 16; El. 8.

Cásere, es; *m.* [= *Lat.* Cæsar; *gen.* Cæsáris] *Cæsar, an emperor;* imperator :—Wearþ Gaius Gallica cásere *Caius Caligula was emperor,* Ors. 6, 3; Bos. 117, 18: Elen. Kmbl. 84; El. 42 : 1995; El. 999. For þingum ðæs ærran cáseres *for the deeds of the former emperor,* Ors. 6, 4; Bos. 118, 15: Exon. 65 a; Th. 240, 6; Ph. 634 : Elen. Kmbl. 524; El. 262 : 1098; El. 551 : 1335; El. 669. Ðæs [MS. ðes] cáseres cwén *the woman* or *wife of the emperor;* imperatrix *vel* augusta, Wrt. Voc. 72, 58. Cáseres wíf *the emperor's wife;* imperatrix *vel* augusta, Ælfc. Gl. 68; Som. 70, 1; Wrt. Voc. 42, 10. Aulixes under hætde ðæm cásere cynerícu twá *Ulysses had two kingdoms under the emperor,* Bt. Met. Fox 26, 11; Met. 26, 6. Ða gesettan Rómáne twegen cáseras *then the Romans appointed two emperors,* Ors. 6, 24; Bos. 124, 18. Hí hæfdon *Cæsares* ofer híg, ðæt we cweðaþ cáseras, ða beóþ cyninga yldest *they had Cæsares over them, that we call emperors, who are the greatest of kings,* Jud. Thw. 161, 29. DER. heáh-cásere.

cásering, e; *f. A cæsaring, a coin with an emperor's image, a coin;* drachma = δραχμή, didrachma :—Gif wíf losaþ cásering *si mulier perdiderit drachmam,* Lk. Lind. Rush. War. 15, 8. Ne unband cásering *non solvit didrachma,* Mt. Lind. Stv. 17, 23.

cáser-líc; *adj. Cæsar-like, imperial;* imperialis, Cot. 115.

Cásern; *f.* [Cásere + en, *f.* termin. Cáseren, Cásern] *An empress;* augusta :—Æfter ðam ðe Róme burh getimbred wæs DCCC wintra and LXVII, féng Adriánus to Rómána ánwealde. He [Cásere] wearþ Rómánum swá leóf, and swá weorþ, ðæt wí hine nánuht ne héton búton fæder; and, him to weorþscype, hí héton his wíf, cásern [cásere + en, *the f.* termin.] *eight hundred and sixty-seven years after the building of Rome, Hadrian succeeded to the government of the Romans. He became so dear to the Romans, and so honoured, that they never called him anything but father; and, in honour of him, they called his wife, empress,* Ors. 6, 11; Bos. 121, 5–15.

cassoc *hassock, hassock-grass,* Lchdm. iii. 24, 3. v. cassuc.

cassuc, cassoc, e; *f. Hassock, hassock-grass, rushes, sedge* or *coarse grass;* aira cæspitosa, carex paniculata, Lin :—Dó him ðis to læcedóme : eoforþrote, cassuc, etc. *give him for this a leechdom : everthroat, hassock,* etc. L. M. 3, 63; Lchdm. ii. 350, 23 : 1, 63; Lchdm. ii. 136, 30 : 3, 67; Lchdm. ii. 354, 24. To háligre sealfe sceal cassoc *hassock shall be for a holy salve,* Lchdm. iii. 24, 3. Dó in gléde finol and cassuc and récels : bærn eal tosomne *put fennel and hassock and incense upon a fire : burn all together,* iii. 56, 5 : L. M. 1, 62; Lchdm. ii. 134, 30 : 3, 62; Lchdm. ii. 350, 6 : 3, 64; Lchdm. ii. 352, 13. Weorc Cristes [MS. Criste] mæl of cassuce fífo *make five crosses of hassock-grass,* Lchdm. iii. 56, 8.

cassuc-leáf; *pl. n. Hassock-leaves* :—Wið eárum [earon MS.] genim ða brádan biscopwyrt and cassucleáf *for the ears take the broad bishopwort and hassock-leaves,* Lchdm iii. 46, 2.

CASTEL, castell, es; *n. m. A town, village,* CASTLE; villa, oppidum, castellum :—Faraþ on ðæt castel [to ðam castelle, *Hat.* in ðas cæstre, *Rush.*], ðæt fóran ongeán eów ys *ite in castellum, quod contra vos est,* Mt. Bos. 21, 2. He ða lærende ða castel beférde *et circuibat castella in circuitu docens,* Mk. Bos. 6, 6. His wíf wæs innan ðam castele *uxor sua fuit in castello,* Chr. 1075; Gib. 183, 3 : 1053; Erl. 187, 9. Ða castelas gewunnon *castella expugnarunt,* 1069; Gib. 174, 28. [*Lat.* castellum, *dim.* of castrum *a camp, fortified place;* akin to casa *a hut,* and caveo *to guard, protect.*] DER. castel-men, -weorc.

castel-men; *gen.* -manna; *pl. m. Castle-men;* castellani :—Ða casteltelmen ðe wæron on Engla lande him togeánes cómon [MS. comen] *the castle-men who were in England came against him,* Chr. 1075; Erl. 213, 18.

castel-weorc, es; *n. Castle-work;* castellorum opus :—Hí suencten ðe men of ðe land mid castelweorces [*for* castelweorcum] *they oppressed the men of the land with castle-works* [castellis ædificandis], Chr. 1137; Th. 382, 20.

casul, e; f? A cassock, short cloak; birrhus, casŭla, lacerna, sacrum pallium [Ger. kasel; f.], Som. Ben. Lye.

cāsus; gen. cāsūs; m. [Lat. cāsus, from cădo to fall; as the Grk. πτῶσις a fall, case, from πίπτω to fall] A case, falling or change to denote the relation of nouns, adjectives, and pronouns to other words in a sentence:—Mid ðam casu with the case, Ælfc. Gr. 7; Som. 6, 16, 17, 20, 22, 25, 28. Ðás six cāsus these six cases, Som. 6, 32. Cāsus, ðæt is fyll oððe gebígednis a case, that is, a declining or inflection, Ælfc. Gr. 14; Som. 17, 23. Ða pronomina, ðe habbaþ vocativum, ðá habbaþ six casus the pronouns which have a vocative, then have six cases, Ælfc. Gr. 18; Som. 20, 54. v. ge-bígednys.

CAT, catt, es; m. A CAT; cătus, murĭceps:—Cat cattus vel murilĕgus aut murĭceps, Wrt. Voc. 78, 20. Cat murĭceps vel musio, murilĕgus, Ælfc. Gl. 21; Som. 59, 71; Wrt. Voc. 23, 30. [Piers P. Chauc. cat: Plat. katte, f: O. Frs. katte, f: Dut. kat, f: Kil. katte: Ger. M. H. Ger. kater, m: katze, f: O. H. Ger. kazza, f: Dan. kat, m. f: Swed. katt, m: Icel. köttr, m: Fr. chat, m: Span. gato, m: Ital. gatto, m: Lat. cătus, m: Grk. κάττα, f: Wel. câth: Corn. cath, f: Ir. cat: Gael. cat, cait, m: Manx cayt: Armor. kaz, m.]

cattes mint, e; f. Cat's mint, cat-mint; felina mentha, nepeta cataria, Lin. Som. Ben. Lye.

caul a basket, Cot. 45: 196. v. cawl.

CAWEL, cawl, caul, es; m. COLE, colewort, cabbage; caulis, magudăris = μαγύδαρις, brassica, Lin :—Caul caula [= caulis] vel magudaris, Wrt. Voc. 79, 44. Befeald on caules [cawles MS. H.] leáf fold it in the leaf of a cabbage, Herb. 14, 2; Lchdm. i. 106, 17: L. M. 1, 46; Lchdm. ii. 114, 22: 2, 24; Lchdm. ii. 214, 23. Sele him etan gesodenne cawel on gódum broþe give him colewort to eat sodden in good broth, L. M. 3, 12; Lchdm. ii. 314, 15: 3, 44; Lchdm. ii. 336, 18. Wild cawel wild cole; brassica silvatica, Herb. 130, 1; Lchdm. i. 240, 17. Se bráda cawel the broad colewort, cabbage, L. M. 1, 33; Lchdm. ii. 80, 9. [Scot. kail, kale: Frs. koal, kool: Dut. kool, f: Ger. kohl, m: M. H. Ger. köle, kol, m: O. H. Ger. kôl: Dan. kaal, m. f: Swed. kal, m: Icel. kál, n: Fr. chou, m: Span. col, m: Ital. cavolo, m: Lat. caulis, m: Grk. καυλός, m: Wel. cawl: Corn. caul, m: Ir. câl: Gael. câl, m: Manx kail, f: Armor. kaol, m.]

cawel-leáf, es; n. A cabbage-leaf; brassicæ folium :—Nim cawel-leáf take cabbage-leaves, Lchdm. iii. 40, 24.

cawel-sǽd, es; n. Cabbage-seed; brassicæ semen :—Nim cawel-sǽd take cabbage-seed, Lchdm. iii. 72, 5.

cawel-stela, an; m. [stela a stalk] A cabbage-stem; brassicæ caudex :—Nim cawelstelan take a cabbage-stem, Lchdm. iii. 102, 7.

cawel-wyrm, -wurm, es; m. A cabbage-worm, caterpillar; curculio, eruca :—Cawelwurm gurgulu [= curculio], Ælfc. Gl. 23; Som. 59, 127; Wrt. Voc. 24, 2.

cawl, caul, ceawl, ceaul, es; m. A basket; sporta, corbis, cophĭnus = κόφĭνος :—Cawl sporta, Ælfc. Gl. 50; Som. 65, 118; Wrt. Voc. 34, 47. Hý heora cawlas afylled hæfdon they had filled their baskets, Ors. 4, 8; Bos. 90, 34. Caul corbis, Cot. 45: 196. Ceawlas cophinos, Mt. Kmbl. Lind. 14, 20. Ceaulas cophinos, Mk. Skt. Lind. 6, 43.

ceác, es; m. A pitcher, jug, basin, laver; urceus, caucus = καῦκος, luter = λουτήρ :—Ceác urceus, Wrt. Voc. 85, 67 : Ælfc. Gl. 26; Som. 60, 80; Wrt. Voc. 25, 20. Calica fyrmþa and ceáca baptismata calicum et urceorum, Mk. Bos. 7, 4, 8. Ðæt he hét ðǽr ǽrene ceácas onhón ut ibi æreos caucos suspendi juberet, Bd. 2, 16; S. 520, 6. Befóran ðæm temple stód ǽren ceác, onuppan twelf ǽrenum oxum … Se ceác wæs swá micel ðæt he oferhelede ða oxan ealle, búton ða heáfudu totodon út a brazen laver stood before the temple, upon twelve brazen oxen … The laver was so large that it covered the oxen entirely, save that the heads projected out, Past. 16, 5; Hat. MS. 21 b, 3, 4. On ðæm ceáce in the laver, 16, 5; Cot. MS.

ceác-bán, es; n. The cheek-bone, jaw; mandibula :—Ceác-bán vel ceácan vel cin-ban mandibula, Ælfc. Gl. 71; Som. 70, 81; Wrt. Voc. 43, 14. v. ceáce.

ceác-bora, an; m. A jug or pitcher-bearer; anhilus? Cot. 13; anthevilus? Wrt. Voc. 285, 14.

ceace a trial, proof; exploratio, tentamentum, experientia, N. Som. Ben. Lye.

CEÁCE, an; f. The jaw, CHEEK; maxilla, mala, mandibula, gena :—Ðæt tácen ðære bærnesse he on his ceácan bær signum incendii in maxilla portavit, Bd. 3, 19; S. 549, 16. He gehrán his ceácan contigit maxillam ejus, 3, 19; S. 549, 1. Ceácan malæ, maxillæ, Wrt. Voc. 282, 58, 59. On hælftre and bridle ceácan heora gewríþ in camo et freno maxillas eorum constringe, Ps. Lamb. 31, 9. Ceácan mandibulæ, Wrt. Voc. 64, 46. Ceác-bán vel ceácan vel cin-bán mandibula, Ælfc. Gl. 71; Som. 70, 81; Wrt. Voc. 43, 14. Ðæt biþ gód sealf wið ðara ceácna [= ceácena] geswelle that is a good salve for swelling of the cheeks, L. M. 1, 5; Lchdm. ii. 48, 11. [Wyc. cheek-boon the jaw; Piers P. R. Brun. cheke: Chauc. cheeke, cheke: Plat. käkel: O. Frs. keke, tziake, f: Dut. kaak, f: Kil. kaecke: Swed. kek, m: Icel. kjálki, m.]

ceác ful; adj. A pitcher full, jug full :—Brohte Romanus ceác fulne

---

wæteres Romanus brought a jug full of water, Homl. Th. i. 428, 1. Gedó on ceác fulne wínes put [it] into a jug full of wine, L. M. 1, 2; Lchdm. ii. 30, 23.

CEAF, cef, es; pl. nom. acc. ceafu; n. CHAFF; palea :— Ceaf palea, Ælfc. Gl. 59; Som. 68, 1; Wrt. Voc. 38, 52. Ðæt ceaf he forbærnþ on unacwencedlícum fýre paleas comburet igni inextinguibili, Lk. Bos. 3, 17. Ða ceafu he forbærnþ on unadwæscendlícum fýre paleas comburet igni inextinguibili, Mt. Bos. 3, 12. Ðæt folc wæs todrifen ofer eall Egipta land cef to gadrienne dispersus est populus per omnem terram Ægypti ad colligendas paleas, Ex. 5, 7, 10, 12, 16, 18. [R. Brun. Chauc. Laym. chaf: Orm. chaff: Plat. kaff: Dut. kaf, n: Ger. kaff, n: M. H. Ger. kaf, n.]

CEAFER, ceafor, es; m. A beetle, CHAFER; brúchus = βροῦχος :—Ceafor bruchus, Ælfc. Gl. 23; Som. 59, 118; Wrt. Voc. 23, 72 : 77, 50 : 281, 45. He cwæþ and com gærshoppa, and ceaferas ðæs næs gerím oððe getel dixit et venit locusta. et bruchus cujus non erat numerus, Ps. Lamb. 104, 34. [O. Sax. Dut. kever, m: Ger. käfer, m: M. H. Ger. kevere, m: O. H. Ger. kĕvar, kĕvaro, m.]

ceafer-tún a hall; atrium. v. cäfer-tún.

ceafes a harlot; pellex, concubina, L. C. S. 55; Th. i. 406, 16, note 26 A. v. cyfes.

CEAFL, es; m. A bill, beak, snout, jaw, cheek; rostrum, rictus, fauces, maxilla :—Se wîda ceafl gefylled biþ the wide jaw is filled, Exon. 97 b; Th. 363, 26; Wal. 59: Andr. Kmbl. 3403; An. 1705. Blódigum ceaflum with bloody jaws, 318; An. 159: Exon. 26 a; Th. 77, 5; Cri. 1252. Dauid gewylde ðone wildan beran, and his ceaflas totær David subdued the wild bear, and tore apart his jaws, Ælfc. T. 13, 26: 14, 2. [Wyc. chaul: Laym. cheuel, chæfl, choul: O. Sax. kaflôs, pl. m: Dut. kevels, pl. f: Ger. kiefel, kifel, kiffel, m.] DER. helle ceafl.

ceahhetan; p. te; pp. ed To laugh loud or in a cackling manner; cachinnare :—Ceahhetton they laughed in a cackling manner, Bd. 5, 12; S. 628, 34 [= ceachetan : Dut. kakelen : Kil. gachelen : Ger. M. H. Ger. kachen: O. H. Ger. kachazzen, chahhazen : Lat. cachinnare : Grk. καχάζω: Sansk. kakh to laugh]. v. cancettan.

ceahhetung, e; f. A loud or cackling laughter; cachinnus, cachinnatio :—Ðá gehýrde ic mycel gehlýd and ceahhetung, swá swá ungelǽredes folces then heard I a great noise and a cackling laughter, as of rude folk, Bd. 5, 12; S. 628, 30. Ceahhetung vel cincung cachinnatio, Ælfc. Gl. 88; Som. 74, 86.

CEALC, es; m. Plaster, cement, CHALK; calx arenata, calx :—Iuuinianus wæs sume niht on ánum niwcilctan húse: ðá hét he ðæt he ðǽr-inne mycel fýr, forðon hit wæs ceald weder. Ðá ongan se cealc mid ungemete stincan, ðá wearþ Iuuinianus mid ðam brǽþe ofsmorod Jovian was one night in a newly-plastered house: then he ordered a great fire to be lighted therein, because it was cold weather. Then the plaster began to fume excessively, and Jovian was smothered with the vapour, Ors. 6, 32; Bos. 129, 9–12. [Dut. kalk, f: Kil. kalck: Ger. kalk, kalch, m: M. H. Ger. kalc, m: O. H. Ger. calc, chalch: Dan. kalk, m. f: Swed. Norw. kalk, m: Icel. kalk, n: Lat. calx, m. and f: Grk. χάλιξ, m. and f: Wel. Corn. calch, m: Ir. calc: Gael. cailc, f: Manx kelk, m.] DER. niw-cilct.

Cealca ceaster; gen. ceastre; f. The chalk city. Camden thinks it is Tadcaster, in Yorkshire; idem, ut opinatur clarus Camdenus, quod hodie Tadcaster in agro Eboracensi, sic olim vocatum a calce ibidem copiose effossa, Som. Ben. Lye.

Cealc-hýþ, e; f. The name of a place, Challock, Chalk, in Kent :—Hér wæs gefitfullíc sinoþ æt Cealc-hýþe here [in A. D. 785] there was a contentious synod at Chalk, Chr. 785; Erl. 57, 13.

cealc-stán, es; m. Chalk-stone, chalk; calculus, Ælfc. Gl. 25; Wrt. Voc. 85, 25. v. mealm-stán 2.

CEALD, cald; comp. ra; sup. ost; adj. [ceald = cald, q. v.] Cool, COLD; frigidus, gelidus :—Hú ðone cealdan magan ungelíclíce mettas lyste how various meats please the cool stomach, L. M. cont. 2, 16; Lchdm. ii. 160, 7. Forst se biþ fyrnum ceald frost which is intensely cold, Cd. 38; Th. 50, 16; Gen. 809. Dú ðæm wætere wǽtum and cealdum foldan fæste gesettest thou firmly settest the earth to the water wet and cold, Bt. Met. Fox 20, 180; Met. 20, 90: 20, 152; Met. 20, 76. Wedera cealdost the coldest of tempests, Beo. Th. 1097; B. 546. [Laym. cald: Plat. koold, kold, kolt: O. Sax. O. Frs. kald: Dut. koud: Kil. koud, kaud: Ger. M. H. Ger. kalt: O. H. Ger. calt: Goth. kalds, m: kald, n: Dan. kold: Swed. kall: Icel. kaldr: Lat. gelidus: Lith. száltas: Lett. salts: Sansk. jala.] DER. æl-ceald, brim-, eal-, hrím-, ís-, morgen-, ofer-, sin-, snáw-, wæl-, winter-. v. calan.

ceald, cald, es; n. Cold, coldness; frigus :—Somod hát and ceald heat also and cold, Cd. 192; Th. 239, 29; Dan. 377: Cd. 216; Th. 273, 5; Sat. 132. Hátes and cealdes of heat and of cold, Exon. 117 b; Th. 451, 20; Dóm. 106. Hý ðeóþ cealde geclungene they are shrivelled with cold, Salm. Kmbl. 609; Sal. 304. Calde geþrungen wǽron míne fét my feet were pierced with cold, Exon. 81 b; Th. 306, 16; Seef. 8. v. calan.

cealdian; p. ode; pp. od; v. intrans. To become cold; frigescere :—Eorþmægen ealdaþ, ellen cealdaþ [MS. côlaþ] earthly power grows old,

L 2

courage becomes cold, Exon. 95 a; Th. 354, 62; Reim. 69, Grn. Gl. DER. a-cealdian. v. calan.

cealer-bríw, es; *m. A thick pottage made of curds;* calviale, Gl. Lchdm. ii. 375, 18. v. calwer-bríw.

CEALF, celf, calf, es; *pl.* cealfru, calfru; *n. m. A* CALF; vitulus, vitula :— He genam án fætt cealf *tulit vitulum tenerrimum*, Gen. 18, 7. He ofslóh án fæt celf *occidit vitulum saginatum*, Lk. Foxe 15, 27. Ne onfó ic ná of eówrum húse cealfas *non accipiam de domo tua vitulos*, Ps. Th. 49, 10. Ðæt hálige cealf *the holy calf*, Ps. C. 50, 137; Ps. Grn. ii. 280, 137. Me ymbhringdon mænige calfru *circumdederunt me vituli multi*, Ps. Th. 21, 10. Ic ne on-foo of húse ðínum calferu *non accipiam de domo tua vitulos*, Ps. Surt. 49, 9. On-settaþ ofer wi-bed ðín calfur *acc. pl. imponent super altare tuum vitulos*, 50, 21. [*Orm.* callf: *Plat.* kalf, kalv, *n: O. Sax.* calf, *n: Dut.* kalf, *n: Ger.* kalb, *n: M. H. Ger.* kalp, *n: O. H. Ger.* kalb, *n: Goth.* kalbo, *f. a young cow, heifer: Dan.* kalv, *m. f: Swed.* kalf, *m: Icel.* kálfr, *m.*]

cealf-ádl, e; *f.* [ádl *a disease, pain] A calf-disease, a sort of disease;* morbi genus, L. M. 35, Lye.

cealfa hús, es; *n. A house for [of] calves;* vitularius, Ælfc. Gl. 1; Som. 55, 24; Wrt. Voc. 15, 24.

cealfian; *p.* ode; *pp.* od *To calve;* vitulum parere. v. cealf.

ceallian; *p.* ode; *pp.* od [calla *a caller, herald] To* CALL, *cry out, shout;* clamare :—Ongan [MS. ongean] ceallian ofer cald wæter Byrhthelmes bearn *the son of Byrhthelm began to shout across the cold river*, Byrht. Th. 134, 28; By. 91. [*Chauc. R. Brun.* calle : *Piers P.* callede, *p: O. Frs.* kaltia, kella : *Dut. Kil. Ger. M. H. Ger.* kallen : *O. H. Ger.* challôn : *Dan.* kalde : *Swed. Norw. Icel.* kalla : *Lat.* calare : *Grk.* καλεῖν.] DER. hilde calla.

cealre, calwer, es; *m. Pressed curds, a jelly made of curds* or *sour milk;* calmaria, gabalacrum ?—Cealre [MS. cealfre] calmaria, Wrt. Voc. 290, 33. Nim súr molcen, wyrc to cealre, and beþ mid ðý cealre *take sour curds, work them to a jelly, and foment with the jelly*, L. M. 1, 39; Lchdm. ii. 98, 25, 26. Súr meolc wyrce cealre, and beðe mid cealre *work sour milk into jelly, and foment with the jelly*, Lchdm. iii. 42, 26. Gewirc niwne cealre *make new jelly*, L. M. 1, 44; Lchdm. ii. 108, 13. Nim ða wyrta and wyrce togadere swá micel swá cealras [MS. celras] *take the herbs and work them together as thick as curds*, Lchdm. iii. 118, 14. Calwer *gabalacrum*, Cot. 96. DER. cealer-bríw.

ceaol *a basket;* cophinus, Lk. Lind. War. 9, 17. v. cawl.

CEÁP, es; *m.*　　　　**I.** *cattle;* pecus :—Ðǽm landbúendum is beboden ðæt ealles ðæs ðe him on heora ceápe geweaxe, híg Gode ðone teóðan dǽl agyfen *to farmers it is commanded, that of all which increases to them of their cattle, they give the tenth part to God*, L. E. I. 35; Th. ii. 432, 29. Ceápas *cattle*, Cd. 83; Th. 105, 2; Gen. 1747. His neáhgebúres ceáp *his neighbour's cattle*, L. In. 40; Th. i. 126, 15. Ceápes cwild *murrain of cattle*, Chr. 897; Erl. 94, 31.　　　　**II.** *as* cattle were the chief objects of sale, hence,—*Saleable commodities, price, sale, bargain, business, market;* pretium, negotium, pactio, venditio, forum :—Ceápas *saleable commodities, goods*, Cd. 85; Th. 106, 16; Gen. 1772: 90; Th. 112, 28; Gen. 1877. Deópum ceápe gebohte *redeemed us at a great [deep] price*, L. C. E. 18; Th. i. 370, 28. Sume wǽron to ceápe gesealde *some were sold at a price*, Nathan. 8: Gen. 41, 56. Awyrigende ceáp [MS. cep] *malignum negotium*, Lchdm. iii. 206, 32. Ic gange to ceápe *I go to market;* veneo, Ælfc. Gr. 32; Som. 36, 23. [*Laym.* cheap, chep *value, purchase: Plat.* koop, *m: O. Sax.* kop, *m. purchase, money: O. Frs.* káp, *m. purchase, sale: Dut.* koop, *m. bargain: Ger.* kauf, *m: M. H. Ger.* kouf, *m. purchase: O. H. Ger.* chouf, kóuf, *m. negotium: Dan.* kjöb, *n: Swed.* köp, *n. purchase: Icel.* kaup, *n. bargain.*] DER. land-ceáp, orleg-, searo-.

ceáp-cniht, es; *m. A hired servant, a slave;* emptitius, Cot. 72.

ceáp-dæg; *gen.* -dæges; *pl. nom. acc.* -dagas; *m. A bargaining* or *market-day:*—Ceáp-dagas *the Nones* or *stated times when the common people came to market;* nonæ, Ælfc. Gl. 96; Som. 76, 27; Wrt. Voc. 53, 36: Cot. 142.

ceáp-eádig; *adj. Rich in goods, rich in cattle :*—Nefne him hafaþ ceápeádig mon *unless a man rich in cattle retains him*, Exon. 90 b; Th. 340, 8; Gn. Ex. 108.

ceáp-ealeðel, -ealoþ, es; *n. The ale-selling place, an ale-house;* taberna, popina, cervisiarium :—Ne sceolon mæsse-preóstas æt ceáp-ealeðelum ne etan ne drincan *mass-priests should not eat nor drink at ale-houses*, L. E. I. 13; Th. ii. 410, 18.

ceáp-gyld, es; *n.*　　　　**I.** *bargain money;* justum rei venditæ pretium :—Þolige ðæs ceápgyld *perdat pretium emptionis*, L. Ath. i. 24; Wilk. 61, 25; Th. i. 212, 16, note 33.　　　　**II.** *price or market-price of what is stolen;* rei furto ablatæ pretium :—Gilde man ðam teónde his ceápgyld *let a man pay to the accuser the market-price* [pretium], L. C. S. 2; Th. i. 390, 23.

ceápian; *p.* ode; *pp.* od [ceáp II] *To bargain, chaffer, trade, to contract for the purchase* or *sale of a thing, to buy, to bribe;* negotiari, emere, comparāre :—Ceápiaþ óþ-ðæt ic cume *negotiamini dum venio*, Lk. Bos. 19, 13. He adráf út ealle ða ðe ceápodon inuan ðam temple

ejiciebat omnes ementes et vendentes in templo, Mt. Bos. 21, 12. Gyfum ceápian *to bribe with gifts*, Cd. 212; Th. 262, 5; Dan. 739. Mid ðám hí útwǽpnedmonna freóndscipes him ceápiaþ *quibus externorum sibi virorum amicitiam comparent*, Bd. 4, 25; S. 601, 18. Mihte ýþ geceápian, gif ǽnig man ceápode *might easily buy, if any one bargained*, Ors. 5, 7; Bos. 106, 17. DER. a-ceápian, be-, ge-, ofa-.

ceáping; e; *f. A buying, marketing;* emptio :—Ðæt nán ceáping ne sý Sunnan dagum *that no marketing be on Sundays*, L. Ath. i. 24; Th. i. 212, 15, note 31. v. ceápung.

ceáp-man, cýp-man, cýpe-man; *gen.* -mannes; *dat.* -men; *pl. nom. acc.* -men; *gen.* -manna; *dat.* -mannum; *m. A* CHAPMAN, *merchant, market-man;* mercator, negotiator, nundinator :—Gif ceápman uppe on folce ceápie, dó ðæt beóran gewitnessum *if a chapman traffic up among the people, let him do it before witnesses*, L. In. 25; Th. i. 118, 12, note 32: Obs. Lun. § 14; Lchdm. iii. 190, 23. Ða cýpmen binnon ðam temple getácnodon unrihtwíse láreówas on Godes geladunge *the chapmen within the temple betokened unrighteous teachers in God's church*, Homl. Th. i. 410, 35: ii. 120, 15. Cýpemen monig cépeþing to ceápstówe brohte *chapmen brought many saleable things to market*, Bd. 2, 1; S. 501; 4.

ceáp-sceamul, -sceamel, es; *m.* [scamel *a bench, seat] A toll-booth, custom-house, treasury;* mercatorium scabellum, telonium = τελώνιον, gazophylacium = γαζοφυλάκιον :— He geseah Leui, æt ceápsceamule sittende *vidit Levi, sedentem ad telonium*, Lk. Bos. 5, 27. Ðás word he spæc æt ceápsceamele *hæc verba locutus est in gazophylacio*, Jn. Bos. 8, 20.

ceáp-scip, es; *n. A merchant ship, trading ship;* navis mercatoria :— Hí wícingas wurdon, and æt ánum cyrre án c and eahtatig ceápscipa geféngon *they became pirates, and took, at one time, one hundred and eighty trading ships*, Ors. 3, 7; Bos. 61, 2.

ceáp-setl, cép-setl, es; *n.* [setl *a seat] A toll-booth, custom-house;* telonium = τελώνιον :—He geseah Leuin sittende æt hys cépsetle *vidit Levi sedentem ad telonium*, Mk. Bos. 2, 14.

ceáp-stów, e; *f. A market-place, a market;* forum, emporium :—Lundenceaster is monigra folce ceápstów of lande and of sǽ-cumendra *Lundonia civitas est multorum emporium populorum terra marique venientium*, Bd. 2, 3; S. 504, 19. Cýpemen monig cépeþing to ceápstówe brohte *chapmen brought many saleable things to market*, 2, 1; S. 501, 5: Cot. 138.

ceáp-strǽt; e; *f.* [ceáp II. *saleable commodities*, strǽt *a street, public place, market] A street* or *place for merchandise, a market;* vicus mercatorius, forum, mercatus, Som. Ben. Lye.

ceápung, e; *f. Business, trade, traffic, commerce;* negotium, negotiatio :—Be ceápunge *concerning traffic* or *commerce*, L. Ed. 1; Th. i. 158, 8. Fram ceápunge þurhgangende on þýstrum *a negotio perambulante in tenebris*, Ps. Spl. C. 90, 6. Ic ne ongeat grame ceápunga *non cognovi negotiationes*, Th. 70, 15.

ceápung-gemót, es; *n. A meeting for trade, a market;* mercatus, Cot. 133.

ceápung-þing, es; *n. A buying, setting a price;* mercatus, Som. Ben. Lye.

cear; *adj. Sorrowful, anxious, sollicitous;* angore plenus, anxius, sollicitus :—On cearum cwidum *with anxious words*, Cd. 214; Th. 269, 2; Sat. 67 : 134; Th. 169, 3; Gen. 2794.

cearc, es; *m. n? Care, anxiety;* cura, sollicitudo :—Iudas ne meahte oncyrran cearces [MS. rex, = crex, = cerx, = cearx, = cearces] genídlan *Judas could not avert the pressure of anxiety*, El. 610. v. carc.

cearc-ern, es; *n. A prison;* carcer :—Ic wæs on cearcerne *eram in carcere*, Past. 44, 7; Hat. MS. 62 b, 22. v. carc-ern.

cearcetung, e; *f. A gnashing, grinding, crashing noise, as of the teeth;* stridor, Som. Ben. Lye.

cearcian, cearcigan; *part.* cearciende; *p.* ode; *pp.* od *To chatter, creak, crash, gnash;* strídere, strídere, crepitare :—Cearciende teþ *gnashing the teeth;* strídentes dentes, Som. Ic cearcige oððe gristbítige strídeo vel strído, Ælfc. Gr. 26, 5; Som. 29, 7.

cearg *sorrowful*, Andr. Kmbl. 2218; An. 1110. v. cearig.

ceare-líce *sorrowfully, miserably, wretchedly.* v. care-líce.

cearena *of cares* or *sorrows*, Exon. 22 a; Th. 59, 33; Cri. 962; *gen. pl.* of ceáru.

cearf *carved*, Solil. in præf; *p.* of ceorfan.

cear-ful; *adj. Careful, full of care, sad;* sollicitus :—Cleopaþ swá cearful se gǽst to ðam duste *the spirit so sad shall call to the dust*, Exon. 98 a; Th. 368, 1; Seel. 15. Cwǽdon cearfulle, Criste láðe, to Gúþláce *the foes of Christ, full of care, said to Guthlac*, 41 a; Th. 136, 30; Gú. 549: 8 a; Th. 2, 26; Cri. 25.

cearful-líce *carefully, diligently.* v. carful-líce.

cearful-nes, -ness, e; *f. Carefulness, curiosity.* v. carful-nys.

cear-gǽst, es; *m. A spirit of anxiety, fearful ghost;* terribilis spiritus :—In lyft astág ceargǽsta [MS. ceargesta] cirm *in the air arose a cry of fearful ghosts* or *spirits*, Exon. 38 a; Th. 125, 34; Gú. 364.

cear-gealdor; *gen.* -gealdres; *n.* [galdor *an incantation, charm] A dire* or *horrible enchantment;* cantio *vel* loquela mæsta :—Helle gǽst

cleopade fôr corþre ceargealdra full *the spirit of hell cried before the multitude, full of dire enchantments,* Exon. 74 b; Th. 279, 24; Jul. 618.

ceari *anxious,* Exon. 100 a; Th. 376, 29; Seel. 162. v. cearig.

cearian, cearigan, carian; ic cearige, ðú cearast, he cearaþ, *pl.* ceariaþ; *p.* ode; *pp.* od [cearu *care*] *To take care, heed, to be anxious* or *sorry;* cúrare, sollicitum esse :—Hwæt bemurnest ðú cearigende *why mournest thou sorrowing?* Exon. 10 b; Th. 11, 27; Cri. 177. He æt gúþe ná ymb his líf cearaþ *he cares not about his life in battle,* Beo. Th. 3077; B. 1536. Ne ceara ðú fleáme dælan somwist incre *care not thou to part your fellowship by flight,* Cd. 104; Th. 137, 25; Gen. 2279: 130; Th. 165, 16; Gen. 2732.

cearig, ceareg, ceari; *adj.* [cearu *care, sorrow*] *Careful, sorrowful, pensive, wary,* CHARY, *anxious, grieving, dire;* sollicitus, cautus, querens, mente turbatus, dirus :—Hie bidon hwonne bearn Godes cwôme to cearigum *they waited till the child of God should come to the sorrowful,* Exon. 10 a; Th. 10, 6; Cri. 148. Cearegan reorde *in a sorrowful voice,* Andr. Kmbl. 2218; An. 1110. Wæs Meotud on beám bunden fæste cearian clomme *the Creator was bound fast on the tree with dire bond,* Exon. 116 b; Th. 449, 6; Dôm. 67. Ne þurfon wyt beón cearie æt cyme Dryhtnes *we need not be anxious at the Lord's coming,* Exon. 100 a; Th. 376, 29; Seel. 162. DER. earm-cearig, ferhþ-, gnorn-, hreów-, môd-, sorg-, winter-.

cear-leás *void of care, careless, reckless, free.* v. car-leás.

cearleás-nes *freedom from care, security, carelessness.* v. carleás-nes.

cear-leást *freedom from care, security, carelessness.* v. car-leást.

cearo *care, sorrow, grief,* Exon. 32 a; Th. 101, 23; Cri. 1663. v. cearu.

cear-seld, es; *n. A place of sorrow;* habitaculum mæroris, Exon. 81 b; Th. 306, 10; Seef. 5.

cear-síþ, es; *m.* [síþ *fortune, fate*] *A sorrowful fate, sad fortune;* curæ sors, fortuna tristis :—Cealdum cearsíþum *with cold sad fortunes,* Beo. Th. 4783; B. 2396.

cear-sorg, e; *f. Sorrowful care, anxious sorrow;* cura sollicita :—Me cearsorge of môde asceáf þeóden usser *our Lord removed anxious care from my mind,* Cd. 55; Th. 68, 9; Gen. 1114.

CEARU, caru, cearo, e; *f.* CARE, *sorrow, grief;* cura, dolor, mæror :—Cearu wæs geniwod geworden in wícum *care was become renewed in the dwellings,* Beo. Th. 2611; B. 1303: Exon. 22 b; Th. 62, 7; Cri. 998: 119 b; Th. 459, 10; Hy. 4, 114. Nis ðé nán caru *non est tibi curæ,* Lk. Bos. 10, 40; Ps. Th. 60, 1. Ðonne biþ þearfendum cwíðende cearo *then there shall be wailing care to the miserable,* Exon. 26 b; Th. 79, 5; Cri. 1286: 77 a; Th. 289, 29; Wand. 55. Gehýr me, ðonne ic.to ðé bidde ceare full *hear me, when I, full of care, pray to thee,* Ps. Th. 140, 1. Ic sceolde ána míne ceare cwíðan *I must bewail my care alone,* Exon. 76 b; Th. 287, 4; Wand. 9: Ps. Th. 118, 145, 147. Ne cleopigaþ hí care *they speak not their care,* 113, 16: 143, 18. Ða ceare seófedun ymb heortan *sorrows sighed round my heart,* Exon. 81 b; Th. 306, 20; Seef. 10. Cearena full *full of sorrows,* Exon. 22 a; Th. 59, 33; Cri. 962. Hý in cearum cwíðaþ *they mourn in sorrows,* 35 b; Th. 115, 23; Gú. 194. Ðe-læs eówer heortan gehefegode sýn on ðises lífes carum *ne forte graventur corda vestra in curis hujus vitæ,* Lk. Bos. 21, 34: 8, 14. Mid cearum hí cwíðdun sorhfully [lit. *with sorrows*] *they mourned,* Exon. 24 b; Th. 69, 35; Cri. 1131: 21 a; Th. 55, 31; Cri. 892. [*Piers P.* kare; *Chauc.* care; *Laym. Orm.* care, kare; *O. Sax.* kara, *f.; M. H. Ger.* kar, *f.; O. H. Ger.* chara, *f.; Goth.* kara, *f.*] DER. aldor-cearu, breóst-, gúþ-, líf-, mæl-, môd-, sorg-, úht-, woruld-.

cearung, e; *f.* [cearu *care*] *Pensiveness, anguish of mind, a complaint;* sollicitudo, Som. Ben. Lye.

cear-wylm, -welm, -wælm, es; *m.* [wylm *heat of mind, emotion*] *Sorrowful* or *anxious emotion, agitation;* sollicita perturbatio, agitatio :—Ða cearwylmas côlran wurþaþ *the anxious emotions become cooler,* Beo. Th. 569; B. 282. Á wæs sæc cnyssed cearwelmum *the contest was ever tossed with waves of sorrow,* Elen. Kmbl. 2513; El. 1258. Æfter cearwælmum *after anxious emotions,* Beo. Th. 4138; B. 2066.

CEÁS, e; *f.* es; *n. A quarrel, strife;* lis :—Gif man mannan wæpnum bebyreþ ðær ceás weorþ *if a man supply another with weapons where there is strife,* L. Ethb. 18; Th. i. 6, 19. On ceáse *in strife,* L. Alf. 18; Th. i. 48, 17. Mearh mægen samnode to ceáse *the horse collected his strength for the strife,* Elen. Kmbl. 111; El. 56. [*O. Frs.* kase, *f. quarrel: O. H. Ger.* kôsa, *f. eloquium, fabula.*] DER. un-ceás.

ceás *chose,* Chr. 975; Th. 226, 21; Edg. 22; *p. of* ceósan.

ceásan? *p.* ceós, *pl.* ceóson; *pp.* ceásen [ceás *strife*] *To strive, fight;* contendere. v. be-ceásan.

ceásega, an; *m. A chooser;* elector. DER. wæl-ceásega, *q. v.*

ceásnes, -ness, e; *f. Election, choice;* electio, Som. Ben. Lye.

ceást, e; *f?* es; *n? Strife, contention, murmuring, sedition, scandal;* lis, rixa, seditio :—On ceáste *in strife,* L. Alf. 18; Th. i. 48; note 34. Gif he þurh unnytte ceáste man ofsleá fæste x geár *si in inutili rixa hominem occiderit, x annos jejunet,* L. Ecg. P. iv. 68, § 22; Th. ii. 230, 29. Ne he ceáste ne astirige *he shall not stir up strife,* L. Ælf. P. 50; Th. ii. 386, 12. Folcslíte *vel* æswícung, sacu, ceást *seditio,* Ælfc. Gl. 15; Som. 58, 39; Wrt. Voc. 21, 30. [*Piers P.* cheeste, cheste.] v. ceás *strife.*

ceaster, cæster, cester; *gen. dat.* ceastre; *acc.* ceastre, ceaster, *pl.* ceastra; *f.* The names of places ending in -caster and -chester were probably sites of a castrum *a fortress,* built by the Romans; the Saxon word is burh, Gen. 11, 4, 5. I. generally *f.* but sometimes *n.* vide II. *A city, fort, castle, town;* urbs, civitas, castellum :—Ne mæg seó ceaster beon behýd *non potest civitas abscondi,* Mt. Bos. 5, 14. On ðære heán ceastre *in the high city,* Bt. 39, 5; Fox. 218, 18. Ða cômon ða weardas on ða ceastre *then the keepers came into the city,* Mt. Bos. 28, 11. Ðú in ða ceastre gong *go thou into the city,* Andr. Kmbl. 1878; An. 941. Ælla and Cissa ymbsæton ceaster *Ella and Cissa besieged the city,* Chr. 491; Erl. 15, 6. Se Hælend ymbfôr ealle burga and ceastra *circuibat Iesus omnes civitates et castella,* Mt. Bos. 9, 35. II. ceaster; *gen.* ceastres; *n. A city, etc :* it is thus declined in the termination of Exan-cester, -ceaster :—Ymsæton Exancester *besieged Exeter,* Chr. 894; Erl. 91, 9; Th. 166, 30, col. 1. Ymbsæton Exanceaster, Th. 167, 26, col. 1, 2. Ða wende he hine west wið Exanceastres *then he turned west towards Exeter* [versus Exanceaster], Chr. 894; Erl. 91, 10; Th. 166, 31, col. 1; 29, col. 2; 167, 28, col. 1, col. 2. Se cyning hine west wende mid ðære fierde wið Exanceastres *the king turned west with the army towards Exeter,* 168, 26, col. 1; 24, col. 2; 169, 21, col. 1; 18, col. 2. III. *the name of a particular place, as* CHESTER, CAISTOR, CASTOR, *the city;* hæc civitas :—He him sende scipon æfter, and Hugo eorl of Ceastre *he sent ships after him, and Hugh earl of Chester,* Chr. 1094; Erl. 230, 28: 1120; Erl. 248, 8.

ceaster-æsc, es; *m. Black hellebore;* helleborus niger :—Wyrc gôdne drenc ceasteræsces *make a good drink of black hellebore,* L. M. 3, 30; Lchdm. ii. 324, 20. Nim ceasteræsc *take black hellebore,* Lchdm. iii. 28, 20: 30, 14: 56, 15.

ceaster-búend, es; *m. City-dweller;* urbem habitans :—He áteáh ceasterbúendum *he came to the city-dwellers,* Beo. Th. 1540; B. 768.

ceaster-hlid, es; *n.* [hlid *a cover;* tegmen] *Cover of a city, gate;* urbis tegmen, porta :—Ðæt ænig meahte ðæs ceasterhlides clustor unlúcan *that any one might unlock the inclosure of the city-gate,* Exon. 12 a; Th. 20, 7; Cri. 314.

ceaster-hof, es; *n.* [hof *a house, dwelling*] *A city-dwelling;* urbis ædes :—Storm upp arás æfter ceasterhofum *a storm arose along the city-dwellings,* Andr. Kmbl. 2475; An. 1239.

Ceaster-scír, e; *f.* [ceaster III. *Chester,* scír *a shire*] *Cheshire;* ager Cestrensis :—Rodbeard wæs gecoren to bisceope to Ceasterscíre *Robert was chosen bishop of Cheshire,* Chr. 1085; Erl. 218, 21.

ceaster-ware; *gen.* -wara; *dat.* -warum; *pl. m. City-inhabitants, citizens;* cives :—Wearþ Húna cyme cúþ ceasterwarum *the coming of the Huns was known to the citizens,* Elen. Kmbl. 83; El. 42: Andr. Kmbl. 3290; An. 1648.

ceaster-waru, e; *f. Townsmen as a body, the citizens* or *city;* cives, civitas :—Ða eóde eall seó ceaster-warú *then the whole city* [citizens as a body] *came out,* Mt. Bos. 8, 34.

ceaster-wyrhta, an; *m. An embroiderer, damask-weaver;* polymitarius, Cot. 156.

ceaster-wyrt, e; *f. Black hellebore;* helleborus niger, Lchdm. ii. 375, 24.

ceást-full; *adj. Full of contention, tumultuous;* tumultuosus, contentiosus, Scint. 28: Fulg. 23.

ceastra *cities,* Mt. Bos. 9, 35; *pl. of* ceaster.

ceat *a thing;* res, Cot. 100 :—Ceatta *cheats;* circumventiones, Som. Ben. Lye.

ceáw, *pl.* cuwon *chewed;* *p. of* ceówan.

Ceawan hlǽw, es; *m. Cheawan low,* CHALLOW :—To Ceawan hlǽwe [MS. læwe] *to Challow,* Chron. Abing. i. 138, 5: Cod. Dipl. v. 310, 33.

ceawl, ceaul *a basket;* cophinus, Mt. Lind. Stv. 14, 20: Mk. Skt. Lind. 6, 43. v. cawl.

ced *a boat;* linter, Mone B. 120. v. cæd.

cedelc, e; *f. The herb mercury;* mercurialis perennis, Lin :—Cedelc *mercurialis,* Glos. Brux. Recd. 41, 44. Herba mercurialis, ðæt is, cedelc *the herb mercurialis, that is, mercury,* Herb. cont. 84; Lchdm. i. 34, 3. Wið ðæs innoþes heardnysse genim ðás wyrte, ðe man *mercurialis,* and ôðrum naman cedelc nemneþ *for hardness of the inwards take this herb, which is called* mercurialis, *and by another name mercury,* Herb. 84, 1; Lchdm. i. 186, 23.

ceder; *gen.* cedre; *f. The cedar;* cedrus = κέδρος :—God brycþ ða heán ceder on Libano *confringet Dominus cedros Libani,* Ps. Th. 28, 5. On eallum cedrum *to all cedars,* 148, 9.

ceder-beám, cæder-beám, es; *m. A cedar-tree;* cedrus = κέδρος :—Cederbeám *cedrus,* Ælfc. Gl. 47; Som. 65, 41; Wrt. Voc. 33, 38: 80, 17. Libanes cederbeámas ða ðú gesettest *cedri Libani quas plantasti,* Ps. Th. 103, 16. Ic geseah árleásne geuferodne swá swá cedertrýw ðæs wuda oðde cederbeámas ðæs holtes *vidi impium elevatum sicut cedros Libani,* Ps. Lamb. 36, 35.

ceder-treów, -trýw, es; *n. A cedar-tree;* cedrus = κέδρος :—Ic geseah árleásne geuferodne swá swá cedertrýw ðæs wuda oðde cederbeámas ðæs holtes *vidi impium elevatum sicut cedros Libani,* Ps. Lamb. 36, 35.

cef *chaff*, Ex. 5, 7, 10, 12, 16, 18. v. ceaf.

cefes, e; *f. A concubine*, L. C. S. 55; Th. i. 406, 16, note 26 B. v. cyfes.

cēgan, cēgean *to call, call upon, invoke*, Ps. Spl. 137, 4: Ps. Lamb. 74, 2: Chr. 974; Th. 224, 27, col. 2, 3; Edg. 7. v. cīgan.

cehhettung, e; *f. A laughing in a cackling manner, a laugh of scorn, scorn;* cachinnus, contemptus:—Hwelce cehhettunge ge woldon ðæs habban, and mid hwelcum hleahtre ge woldon beón astyred *what scorn ye would have at this, and with what laughter ye would be moved*, Bt. 16, 2; Fox 52, 4. v. ceahhetung.

cel, *pl.* celas *a basket*, Mt. Lind. Stv. 15, 37. v. cawl.

cēlan; *p.* de; *pp.* ed; *v. intrans. To be* or *become cold;* algere, refrigerari:—Cēlan is of untrumnysse ðæs gecynnes *algere ex infirmitate naturæ est*, Bd. 1, 27; S. 494, 15. DER. a-cēlan. v. calan.

cēle, es; *m. A cold, coldness;* frigus:—Fōr andwlītan cēles *ante faciem frigoris*, Ps. Th. 147, 6: Bt. Met. Fox 20, 219; Met. 20, 110: 20, 225; Met. 20, 113: 20, 315; Met. 20, 158. v. cȳle.

celender, cellender, es; *n. The herb coriander;* coriandrum, L. M. 1, 4; Lchdm. ii. 44, 17: 1, 35; Lchdm. ii. 82, 6. v. celendre.

celendre, cellendre, an; *f:* celender, cellender, es; *n. The herb coriander;* coriandrum = κορίαννον, coriandrum sativum, Lin:—Celendre *coriandrum*, Ælfc. Gl. 43; Som. 64, 44; Wrt. Voc. 31, 54: 286, 16. Genim ðās wyrte, ðe man *coliandrum*, and, ōðrum naman ðam gelīce, cellendre nemneþ *take this herb, which is called* coriandrum, *and, by another name like that, coriander*, Herb. 104, 1; Lchdm. i. 218, 16. Genim celendran seáw grēnre *take juice of green coriander*, L. M. 1, 3; Lchdm. ii. 42, 4: 1, 31; Lchdm. ii. 72, 12: 3, 3; Lchdm. ii. 310, 5. Nim cellendran *take coriander*, 3, 47; Lchdm. ii. 338, 6, 7: 2, 39; Lchdm. ii. 248, 3. Genim celender and beána togædere gesodene *take coriander and beans sodden together*, 1, 4; Lchdm. ii. 44, 17. Celendres sǣd gegnīd *rub seed of coriander*, 2, 48; Lchdm. ii. 262, 21. Cellendres sǣd gedō on scearp wīn *put seed of coriander into sour wine*, 2, 33; Lchdm. ii. 236, 30. Mid cellendre *with coriander*, 1, 35; Lchdm. ii. 82, 6.

celeþonie, an; *f. The herb celandine* or *swallow-wort;* chelidonium = χελιδόνιον, chelidonium majus, Lin:—Celeþonie *celandine*, L. M. 1, 45; Lchdm. ii. 110, 21. Nim celeþonian moran *take roots of celandine*, 3, 41; Lchdm. ii. 334, 26: 3, 42; Lchdm. ii. 336, 9: 3, 60; Lchdm. ii. 344, 2. Genim celeþonian *take celandine*, 1, 2; Lchdm. ii. 38, 14: 1, 32; Lchdm. ii. 78, 27: 1, 39; Lchdm. ii. 102, 1: 1, 48; Lchdm. ii. 122, 16: 3, 2; Lchdm. ii. 306, 23.

celf *a calf*, Lk. Foxe 15, 27. v. cealf.

cēling, cēlung, e; *f. A cooling, refreshing;* refrigerium, refrigeratio, Som. Ben. Lye.

cellendre *coriander*, Herb. 104, 1; Lchdm. i. 218, 16. v. celendre.

celmert-mon, -monn, es; *m. A hired servant, hireling;* mercenarius:—He celmertmon is *mercenarius est*, Jn. Rush. War. 10, 12, 13. Celmertmonn *mercenarius*, Jn. Lind. War. 10, 12. Ða celmertmenn *mercenarii*, Lk. Lind. War. 15, 17. From celmertmonnum ðīnum *de mercenariis tuis*, 15, 19: Mk. Skt. Lind. 1, 20.

cēl-nes, cōl-nes, -ness, e; *f. Coolness, cool air, a breeze;* refrigerium, aura:—Ðū lǣddest us on cēlnesse *eduxisti nos in refrigerium*, Ps. Spl. C. T. 65, 11. To sēcanne wið hǣto cēlnes *quærere contra æstum auras* [*breezes*], Bd. 1, 27; S. 494, 17. DER. ge-cēlnes. v. calan.

cēlod, cēllod; *part.* [ceól *the keel of a ship*] *Formed like a keel* or *boat;* scaphiformis:—Cēlod bord *a shield shaped as a boat*, Fins. Kmbl. 57; Fin. 29. Cēllod bord, Byrht. Th. 140, 4; By. 283.

celras *curds*, Lchdm. iii. 118, 14. v. cealre.

cemban, cæmban; *p.* de; *pp.* ed [camb *a comb*, I. *q. v.*] *To* COMB; pectere:—Ic cembe *pecto*, Ælfc. Gr. 28, 3; Som. 30, 61.

cemes, e; *f. A linen night-gown, chemise;* camisia, Cot. 31.

cempa, an; *m.* [camp *war, battle*, -a, *q. v.*] *A soldier, warrior*, CHAMPION; miles, bellator, athleta = ἀθλητής:—Cempa *miles vel athleta*, Wrt. Voc. 72, 68. Se cempa oferwon frēcnessa fela *the champion overcame many perils*, Exon. 35 a; Th. 113, 2; Gū. 151: Andr. Kmbl. 922; An. 461: Byrht. Th. 135, 17; By. 119: Beo. Th. 2629; B. 1312. Ðā ða cempan hine ahēngon, hī nāmon his reáf, and worhton feówer dǣlas, ǣlcum cempan ānne dǣl *milites cum crucifixissent eum, acceperunt vestimenta ejus, et fecerunt quatuor partes, unicuique militi partem*, Jn. Bos. 19, 23. Scyld sceal cempan *a shield shall be for a soldier*, Exon. 91 a; Th. 341, 22; Gn. Ex. 130: Beo. Th. 3901; B. 1948: Andr. Kmbl. 460; An. 230. Woldun hȳ geteón in orwēnnysse Meotudes cempan *they would draw God's soldier into despair*, Exon. 41 a; Th. 136, 28; Gū. 548: Salm. Kmbl. 279; Sal. 139. Hūslfatu hālegu cempan genāmon *the warriors took the holy vessels of sacrifice*, Cd. 210; Th. 260, 9; Dan. 707: Fins. Th. 29; Fin. 14. We his þegnas sind, gecoren to cempum *we are his thanes, chosen to* [*be his*] *warriors*, Andr. Kmbl. 647; An. 324. Alǣten cempa *a soldier who has served his time, a veteran;* emeritus, Ælfc. Gl. 7; Som. 56, 62; Wrt. Voc. 18, 15. Gecorene cempan *chosen soldiers, adjutants;* optiones, 7; Som. 56, 64; Wrt. Voc. 18, 17. Cempena yldest *a chief of soldiers, a commander;* militum tribunus, Ors. 4, 9; Bos. 91, 18. Twā hund cempna [= cempena] *two hundred* [*of*] *soldiers: manipulus*, Ælfc. Gl. 7; Som. 56, 75; Wrt. Voc. 18, 27. Fīf hund cempena ealdor *a commander of five hundred soldiers;* cohors, 7; Som. 56, 61; Wrt. Voc. 18, 14. DER. fēðe-cempa, sige-.

CĒN, es; *m.* I. *the Anglo-Saxon Rune* ᚳ = the letter *c*, the name of which letter in Anglo-Saxon is cēn *a torch;* pinus, tæda; hence this Rune not only stands for the letter *c*, but for cēn *a torch*, as,—ᚳ byþ cwicera gehwām cūþ on fȳre *torch on fire is well known to all living*, Hick. Thes. vol. i. p. 135; Runic pm. 6; Kmbl. 340, 17: Exon. 76 a; Th. 284, 28; Jul. 704. II. this Rune appears sometimes to stand for the *adj.* cēne *bold*, II. *q. v.* [*Plat.* keen: *Ger. M. H. Ger.* kien, *m. n. a fir* or *pine saturated with the gum of turpentine: O. H. Ger.* kien, kēn *pinus, fax, tæda.*]

CĒNE, cȳne; *adj.* I. KEEN, *fierce, bold, brave, warlike;* acer, audax, animosus, bellicosus:—Se wæs ūþwita cēne and cræftig *who was a philosopher keen and profound*, Bt. Met. Fox 10, 101; Met. 10, 51. Stōp ūt cēne collenferþ *he stept out bold* [*and*] *firm of mind*, Andr. Kmbl. 3154; An. 1580. Eofore eom ǣghwǣr cēnra *than a wild boar I am everywhere bolder*, Exon. 110 b; Th. 423, 9; Rä. 41, 18. Cende cneow-sibbe cēnra manna *he begat a race of brave men*, Cd. 161; Th. 200, 14; Exod. 356. Þriste mid cēnum *the confident with the brave*, Exon. 89 b; Th. 337, 8; Gn. Ex. 61: Beo. Th. 1541; B. 768. II. this word is sometimes expressed by the Rune ᚳ:—Ðonne ᚳ cwacaþ *then the bold shall quake*, Exon. 19 b; Th. 50, 8; Cri. 797: Elen. Grm. 1258. [*Piers P. R. Brun. Chauc. R. Glouc. Laym.* kene: *Dut.* koen: *Ger.* kühn: *M. H. Ger.* küene, kuon: *O. H. Ger.* kōn, kōni, kuon, kuoni.] DER. dǣd-cēne, gār-.

cēnlīce; *adv. Keenly, boldly, courageously, notably;* animose, audacter, insigniter, Ælfc. T. 15, 17.

CENNAN, cænnan, cynnan; *part.*-nende; *p.* de; *pp.* ed; *v. trans.* I. *to beget, conceive, create, bring forth;* gignere, creare, facere, parere:—Ic to-dæg cende ðē *ego hodie genui te*, Ps. Spl. 2, 7. Sceal, ic nū eald wīf, cennan *shall I, now an old woman, conceive?* Gen. 18, 13. Iob sunu Waldendes freónoman cende *Job gave* [*created, made*] *a noble name to the Lord's son*, Exon. 17 a; Th. 40, 9; Cri. 636. Ðam wæs Judas nama cenned *to him was given* [*created, made*] *the name Judas*, Elen. Kmbl. 1170; El. 587: Ps. Th. 73, 7. Heó cende hyre frumcennedan sunu *peperit filium suum primogenitum*, Mt. Bos. 1, 25. II. *to bring forth from the mind, to declare, choose, ascribe, clear, prove;* advocare, confiteri, adscribere, purgare, manifestare:—Gif he cynne ðæt he hit bohte *if he declare that he bought it*, L. Edg. S. 11; Th. i. 276, 12, MS. F. Ic me to cyninge cenne Iudas *I chose Judah to me for a king*, Ps. Th. 107, 8. We deórwyrþne dǣl Dryhtne cennaþ *we ascribe the precious lot to the Lord*, Exon. 35 a; Th. 113, 8; Gū. 154. Cenne he hwanon hit him cōme *let him declare whence it came to him*, L. Eth. ii. 8; Th. i. 288, 14, 21, 22, 23, 25. Gif he cenþ ðæt he hit bohte *if he declare that he bought it*, L. Edg. S. 10; Th. i. 276, 6. Mynstres aldor hine cænne in preóstes canne *let the chief of a monastery clear himself with a priest's clearance*, L. Wih. 17; Th. i. 40, 13: 22; Th. i. 42, 3; L. Edg. S. 11; Th. i. 276, 12. [*Piers P.* kennen, kenne *to teach: Chauc.* kennen *to know: R. Brun.* ken *to know: Laym.* kenne, kennen *to know, make known, acknowledge: Orm.* kennedd *begotten: O. Sax.* kennian *gignere, cognoscere: Frs.* kinnen: *O. Frs.* kanna, kenna *to know: Dut. Ger. M. H. Ger.* kennen *to know: O. H. Ger.* kannjan: *Goth.* kannyan *to make known: Dan.* kjende: *Swed.* känna: *Icel.* kenna *to know, teach.*] DER. a-cennan, ge-, on-.

cennend-līc; *adj. Begetting, genital;* gignens, genitalis:—Ða cennendlīcan *genitalia*, Wrt. Voc. 283, 53. v. cennan.

cennestre *one who has borne, a mother*. v. cynnestre.

cenning, e; *f. Birth, a producing;* partus:—Ðære cenninge tīma *tempus pariendi*, Gen. 25, 24. DER. ed-cenning.

cenning-tīd, e; *f. The time of bringing forth, birth-time;* pariendi tempus, puerperii hora:—Ðā wæs gefylled Elizabethe cenningtīd, and heó sunu cende *Elisabeth autem impletum est tempus pariendi, et peperit filium*, Lk. Bos. 1, 57. On ðære cenningtīde *instante partu*, Gen. 38, 27.

cennynde *producing*, Bd. 1, 27; S. 493, 23, = cennende; *part. of* cennan.

cēnost *keenest, bravest, boldest*, Cd. 160; Th. 198, 14; Exod. 322; *sup. of* cēne.

Cēnrēd, es; *m.* [cēne, rēd *counsel*] *Cenred, son of Ceolwald, and father of Ine, king of Wessex:*—Cēnrēd wæs Ceolwalding *Cenred was the son of Ceolwald*, Chr. Th. 2, 2. Ingeld wæs Īnes brōðor, and hī, begen brōðra, wǣron [MS. wareon] Cēnrēdes suna: Cēnrēd wæs Ceoldwalding *Ingeld was Ine's brother, and they, both brothers, were Cenred's sons: Cenred was son of Ceolwald*, Text. Rof. 61, 12-18. v. Īne.

Cent; *indecl. n. The county of* KENT; Cantium = Κάντιον:—Wæs he sended to Cent *he was sent into Kent*, Bd. 3, 15; S. 541, 24: Chr. 823; Erl. 62, 19. Se cyning wæs on Cent *the king was in Kent*, Chr. 911; Erl. 101, 37: 1009; Erl. 143, 14. Se mūþa Limene is on easteweardre Cent *the mouth of the Limen is in the east of Kent*, 893; Erl. 88, 26.

centaurie, an; *f. The herb centaury;* centaureum = κενταύριον,

erythræa centaureum, Lin :—Nim centaurian *take centaury*, L. M. 2, 8 ; Lchdm. ii. 186, 26 : 2, 39 ; Lchdm. ii. 248, 13.

cênþu, e ; *f. Boldness;* audacia :—Cræft and cênþu *strength and boldness*, Beo. Th. 5385 ; B. 2696.

Centingas ; *pl. m. Men of Kent, Kentish men ;* Cantiani :—Hî forneáh ealle west Centingas fordýdon *they ruined nearly all the west Kentish men*, Chr. 999 ; Th. 248, 12, col. 2 : 1011 ; Th. 267, 7, col. 1.

Centisc ; *adj.* KENTISH, *belonging to Kent ;* Cantianus :—Seó Centisce fyrd com ongeán hî *the Kentish force came against them*, Chr. 999 ; Th. 249, 6, col. 2. Ætsǽton ða Centiscan ðǽr *the Kentish [men] remained there*, 905 ; Erl. 98, 23.

Cent-land, -lond, es ; *n. Kentish land, Kent ;* Cantium :—Eást-Seaxe syndon Temese streáme tosceádene fram Centlande *the East-Saxons are divided from Kent by the river Thames*, Bd. 2, 3 ; S. 504, 17 : 3, 15 ; S. 541, note 24. Æðelréd oferhergode Centland [Centlond, col. 1] *Æthelred ravaged Kent*, Chr. 676 ; Th. 60, 8, col. 2, 3. Ða Brettas forléton Centlond *the Britons forsook Kent*, 457 ; Erl. 12, 19.

Cent-ríce, es ; *n. The kingdom of Kent ;* Cantii regnum :—Hér Eádberht fêng to Centrîce here, A.D. 725, *Eadberht succeeded to the kingdom of Kent*, Chr. 725 ; Erl. 44, 31.

CEÓ, ció ; *indecl. f. A* CHOUGH, *a bird of the genus corvus, a jay, crow, jackdaw ;* cornix, gracculus, monedula :—Ðeós ceó *hæc cornix*, Ælfc. Gr. 9, 64 ; Som. 13, 58. Ceó gracculus vel *monedula*, Ælfc. Gl. 37 ; Som. 63, 13 ; Wrt. Voc. 29, 36. [*Scot.* keaw : *Dut.* kauw, *f : M.H.Ger.* kouch, *m. a horned owl : O.H.Ger.* kaha, *f : Dan.* kaa, kaje, *m. f :* Swed. kaja, *f : Icel.* kjói, *m. a sea-bird.*]

ceóce *a cheek-bone, cheek*, Wrt. Voc. 64, 44, = ceáce, *q.v.*

ceofl *a basket ;* côphīnus = κόφινος, Lk. Rush. War. 9, 17. v. cawl.

ceol *a basket ;* sporta, Mt. Kmbl. Lind. 15, 37 : Mk. Skt. Lind. 8, 20. v. cawl.

CEÓL, ciól, es ; *m. The* KEEL *of a ship, a ship ;* carina, celox, navis :—Ðe brontne ceól ofer lagustrǽte lǽdan cwômon *who came leading a high keel over the water-street*, Beo. Th. 482 ; B. 238. Ðæt ðû us gebrohte brante ceóle, heá hornscipe, ofer hwæles êðel, on ðǽre mǽgþe *that thou wouldst bring us with the steep keel, the high pinnacled ship, over the whale's home, to that tribe*, Andr. Kmbl. 545–549 ; An. 273–275. Ceól *celox*, Glos. Epnl. Recd. 156, 12 : Wrt. Voc. 288, 30. Ceól on lande stôd *the ship stood on land*, Beo. Th. 3829 ; B. 1912 : Exon. 90 b ; Th. 339, 20 ; Gn. Ex. 97. Ofer ceóles bord *from the vessel's deck*, 20 b ; Th. 54, 2 ; Cri. 862 : Andr. Kmbl. 620 ; An. 310. In ðam ceóle wæs cyninga wuldor *the glory of kings was in the ship*, 1707 ; An. 856 : Exon. 81 b ; Th. 306, 9 ; Seef. 5. He ceól gesôhte *he sought the ship*, Andr. Kmbl. 759 ; An. 380. Hî cômon on þrim ceólum to Brytene *they came in three ships to Britain*, Chr. 449 ; Erl. 13, 3 : Bt. Met. Fox 21, 22 ; Met. 21, 11. Ceólas lêton on brime bîdan *they let the ships abide in the sea*, Elen. Kmbl. 500 ; El. 250. Hwanon cômon ge ceólum lîðan *whence came ye sailing in ships?* Andr. Kmbl. 512 ; An. 256 : Exon. 20 a ; Th. 53, 18 ; Cri. 852. [*Plat.* keel : *Dut.* kiel, *f : Ger. M.H.Ger.* kiel, *m : O.H.Ger.* chiol, cheol, chiel, *m : Dan.* kiöl, *m. f :* Swed. köl, *m : Icel.* kjóll, *m.*] DER. þríerêþre-ceól.

ceola *a little cottage, a cabin ;* stega, Som. Ben. Lye.

ceolas ; *pl. m. Cold winds, cold ;* auræ frigidæ, frigus :—Ðec ceolas weorþian Fæder, forst and snäw *thee, O Father, cold winds adore, frost and snow*, Exon. 54 b ; Th. 192, 9 ; Az. 103.

CEOLE, ciole, an ; *f. The throat,* JOWL ; *guttur, fauces :—Ðý-læs sió ceole sîe aswollen *lest the throat be swollen*, L. M. 1, 4 ; Lchdm. ii. 48, 26. Wið ceolan swile *for swelling of throat*, 1, 12 ; Lchdm. ii. 54, 23 ; 56, 2. Wið sweorcôðe, rîges seofoþa seóþ on geswêttum wætere, swille ða ceolan mid ðý gif se sweora sâr sîe *for quinsy, seethe the siftings of rye in sweetened water, swill the throat with it if the neck be sore*, 1, 4 ; Lchdm. ii. 48, 21. Hû swête ceolum mînum spræce ðîne, ofer hunig mûþe mîne *quam dulcia faucibus meis eloquia tua, super mel ori meo*, Ps. Spl. 118, 103. Ne cleopigaþ hî, ðeáh ðe hî ceolan habban *they [i.e. idols] cry not, though they have throats*, Ps. Th. 113, 16. [*Plat.* kele : *Dut.* keel, *f : Kil.* keele, kele : *Ger.* kehle, *f : M.H.Ger.* kël, *f : O.H.Ger.* këla, *f : Lat.* gula, *f : Sansk.* gala, *m.*]

ceoler ; *gen.* ceolre ; *f. The* COLLAR *or throat ;* guttur :—Sind gefægnunga Godes on ceolre oðﬆe þrote heora *sunt exaltationes Dei in gutture eorum*, Ps. Lamb. 149, 6. v. ceole.

Ceóles íg, e ; *f.* [ceól *a ship*, íg *an island*] CHELSEA, *'on the bank of the Thames, Middlesex ; Somner* says, 'Insularis olim et navibus accommodata, ut nomen significat.'

Ceóles íg, e ; *f.* CHOLSEY, *Berks, near Wallingford*, Chr. 1006 ; Th. 256, 27.

ceól-þelu, e ; *f. The deck of a ship, a ship ;* navis tabulatum, navis :—Ic eom hér cumen on ceólþele *I am come here in a ship*, Exon. 123 a ; Th. 473, 1 ; Bo. 8.

Ceolwald, es ; *m.* [ceol, -wald, es ; *m. power*] *Ceolwald, son of Cuthwulf, an ancestor of the West-Saxon kings*—Ceolwald wæs Cûþwulfing *Ceolwald was the son of Cuthwulf*, Chr. Th. 2, 3. v. Cênrêd, Íne.

ceorf-æx, e ; *f. A cutting axe, executioner's axe ;* securis :—Wǽran ða

heáfda mid ceorfæxum ofacorfena *their heads were cut off with axes*, Ors. 4, 1 ; Bos. 79, 7.

CEORFAN ; *part.* ceorfende ; ic ceorfe, ðû ceorfest, cyrfst, he ceorfeþ, cyrfþ, *pl.* ceorfaþ ; *p.* ic, he cearf, ðû curfe, *pl.* curfon ; *pp.* corfen ; *v.a. To cut, cut down, hew, rend, tear,* CARVE, *engrave ;* secare, concidere, succidere, excidere, conscindere, incidere, infindere :—He wæs hine sylfne mid stânum ceorfende *erat concidens se lapidibus*, Mk. Bos. 5, 5. He cearf of heora handa and heora nosa *he cut off their hands and their noses*, Chr. 1014 ; Erl. 151, 10. Hîg curfon ðone ram eall to sticceon *they cut the ram all to pieces*, Lev. 8, 20. Corfen *cut*, Exon. 107 b ; Th. 410, 24 ; Rä. 29, 4. Treówa ceorfan *to hew trees*, Obs. Lun. § 11 ; Lchdm. iii. 188, 24 : Cd. 200 ; Th. 248, 11 ; Dan. 511. On wuda treówa mid æxum hî curfon dura *in silva lignorum securibus exciderunt januas*, Ps. Spl. 73, 7. Curfon hie ðæt moldern of beorhtan stâne *they hewed the sepulchre out of bright stone*, Rood Kmbl. 132 ; Kr. 66. Ðû toslite oððe curfe hǽran mîne *thou hast rent my sackcloth ;* conscidisti saccum meum, Ps. Spl. 29, 13. Îsene ceorfan *to carve* or *engrave with iron*, Past. 37, 3 ; Hat. MS. 50 b, 5. Ceorfende *infindens*, Cot. 111. [*R. Glouc.* carf *cut : Chauc.* corven, *pp : Scot.* kerf : *Plat.* karven : *Frs.* kerven : *O.Frs.* kerva : *Dut.* kerven : *Ger. M.H.Ger.* kerben : *Dan.* karve : *Swed.* karfva.] DER. a-ceorfan, be-, for-, of-, ofa-, to-, ymb-.

ceorfincg-îsen, es ; *n. A marking* or *searing-iron ;* cauterium = καυτήριον, Scint. 9.

CEORIAN, ceorigan, ciorian, cerian ; *part.* ceorigende ; *p.* ode ; *pp.* od ; *v. intrans. To murmur, complain ;* murmurare, queri :—Ne underfêhþ nân ceorigende sáwul Godes rîce, ne nân ceorian ne mæg, se ðe to ðam becymþ *no murmuring soul receives God's kingdom, nor may any one murmur who comes to it*, Homl. Th. ii. 80, 11. We ne ceoriaþ *we murmur not*, ii. 80, 16. Hîg ceorodon ongeán God and Moysen *they murmured against God and Moses*, Num. 21, 5 : Homl. Th. i. 338, 11 : ii. 472, 1. Ic ceorige oððe cîde *queror*, Ælfc. Gr. 29 ; Som. 33, 52. [*Dut.* korren *to coo, as pigeons : Kil.* karien, koeren, koerien *gemere, instar turturis : Ger.* kerren *stridere : M.H.Ger.* kërren, kirren : *O.H.Ger.* kerren *garrire ;* queran *gemere : Lat.* garrio : *Grk.* γηρύω : *Zend* gar *to sing : Sansk.* gṝī *sonare.*] DER. be-ceorian.

CEORL, es ; *m.* I. *a freeman of the lowest class,* CHURL, *countryman, husbandman ;* homo liber, rusticus, colonus :— Ceorles weorþig sceal beón betýned *a churl's close must be fenced*, L. In. 40 ; Th. i. 126, 13. Se ceorl, 60 ; Th. i. 140, 8. Swâ we eác settaþ be eallum hâdum, ge ceorle ge eorle *so also we ordain for all degrees, whether to churl or earl [gentle or simple]*, L. Alf. pol. 4 ; Th. i. 64, 3. Twelfhyndes mannes aþ forstent vi ceorla aþ *a twelve hundred man's oath stands for six churls' oaths*, L. O. 13 ; Th. i. 182, 19. Be ceorles gærstûne *of a husbandman's meadow*, L. In. 42 ; Th. i. 128, 4, 5. Landes [MS. londes] ceorl *a land's man*, Bt. Met. Fox 12, 54 ; Met. 12, 27. II. *a man, husband ;* vir, maritus :—Ceorla cynge *king of the commons*, Chr. 1020 ; Erl. 160, 23. Ealdan ceorlas wilniaþ *old men wish*, Bt. 36, 5 ; Fox 180, 7. Clypa ðînne ceorl *voca virum [husband] tuum*, Jn. Bos. 4, 16, 17. Ðû hæfdest fîf ceorlas *thou hast had five husbands*, 4, 18. III. *a free man*, as opposed to þeów, and to þræl *a slave ;* or as opposed to þegen *a thane* or *nobleman*, as we say, 'gentle or simple :'—We witan ðæt, þurh Godes gyfe, þræl wearþ to þegene, and ceorl wearþ to eorle, sangere to sacerde, and bôcere to biscope *we know that, by the grace of God, a slave has risen to a thane, and a ceorl [free man] has risen to an earl, a singer to a priest, and a scribe to be a bishop*, L. Eth. vii. 21 ; Th. i. 334, 7–9. Gif ceorl gepeáh, ðæt he hæfde fullîce fîf hîda âgenes landes, cirican and cycenan [MS. kycenan], bell-hûs and burh-geat-setl, and sunder-note on cynges healle, ðonne wæs he ðonon-forþ þegen-rihtes weorþe *if a free man thrived, so that he had fully five hides of his own land, church and kitchen, bell-house and a city-gate-seat, and special duty in the king's hall, then was he thenceforth worthy of thane-right*, L. R. 2 ; Th. i. 190, 14–17. [*Chauc.* cherl : *Wyc.* cherl, churl : *Laym.* cheorl : *Orm.* cherl *a young man : Plat.* kerel : *Frs.* tzierl : *O.Frs.* tzerle, tzirle : *Dut.* karel, *m : Ger. M.H.Ger.* kerl, *m : O.H.Ger.* charal, charl, *m : Icel.* karl, *m.*] DER. ceorl-boren, -folc, -ian, -isc, -iscnes, -lîc, -lîce, -strang : æcer-ceorl, hûs-.

ceorl-boren ; *part. Country* or *free-born, common, low-born*, opposed to þegen-boren *noble-born :—Ne þearf he hine gyldan mâ, sý he þegen-boren, sý he ceorl-boren *he need not pay more for him, be he born a thane, be he born a churl*, L. O. D. 5 ; Th. i. 354, 20.

ceorl-folc, es ; *n. Common people, the public ;* vulgus :—Ðis ceorlfolc [ceorle folc MS.] *hoc vulgus*, Ælfc. Gr. 8 ; Som. 7, 35. Ceorlfolc *vulgus*, 13 ; Som. 16, 7 : Wrt. Voc. 72, 73.

ceorlian ; *p.* ode ; *pp.* od [ceorl *a husband*] *To take a husband, to marry ;* nubere. Spoken of a woman, and opposed to wîfian *to take a wife :—Ne wîfiaþ hîg, ne hîg ne ceorliaþ *they take not a wife, nor do they take a husband*, Mt. Bos. 22, 30. Ne nân preóst ne môt beón æt ðam brýdlâcum âhwǽr, ðǽr man eft wîfaþ, oððe wîf eft ceorlaþ *no priest may be at a marriage anywhere, where a man marries a second wife, or a woman a second husband*, L. Ælf. C. 9 ; Th. ii. 346, 19.

ceorlisc, ciorlisc, cierlisc, cirlisc, cyrlisc; *adj.* [ceorl, -isc, *q. v.*] CHURLISH, *rustic, common;* rusticus, vulgaris :—Ceorlisc *rusticus,* Cot. 188. Ceorlisc hláf *common bread;* cibarius [panis], Ælfc. Gl. 66; Som. 69, 61; Wrt. Voc. 41, 17. Ceorlisc folc *common people;* vulgus vel plebs, 87; Som. 74, 45; Wrt. Voc. 50, 27. Gif cierlisc [ciorlisc MS. H; cyrlisc B.] mon betygen wǽre *if a common man has been accused,* L. In. 18; Th. i. 114, 6. Se cierlisca [ceorlisce MS. B; ciorlisca H.] mon *the common man,* 37; Th. i. 124, 21. Be cierlisces [cyrlisces MSS. B. G.] monnes ontŷnesse *of the accusing of a common man,* 37; Th. i. 124, 20. Be cirliscum [ceorliscum MS. B; cyrliscum G; cierliscum H.] þeófe *of a common thief,* 18; Th. i. 114, 5. Sǽton feáwa cirlisce [cyrlisce, col. 2, 3; 165, col. 1, 2] men *a few countrymen remained,* Chr. 893; Th. 164, 4, col. 1.

ceorlisc-nes, -ness, e; *f.* CHURLISHNESS, *rudeness, vulgarity;* rusticitas, sordes. v. cyrliscnys.

ceorl-líc, ceorlíc; *adj.* CHURL-LIKE, *rustic, common;* rusticus, vulgaris :—Ceorlíc ǽhta *common property;* peculium, Ælfc. Gl. 13; Som. 57, 122; Wrt. Voc. 20, 59. v. ceorlisc.

ceorl-líce, ceorlíce; *adv. Commonly;* vulgariter, Bridf.

ceorl-strang; *adj. Strong as a man, manlike;* fortis, virilis :—Ceorlstrang fǽmne *a manlike woman;* virago, Ælfc. Gl. 5; Som. 56, 10; Wrt. Voc. 17, 18.

Ceortes íg, Certes íg, e; *f.* [*Hovd. Matt. West.* Certesie] *Cerot's island,* CHERTSEY, *in Surrey, on the bank of the Thames;* Ceroti insula, Certesia, in agro Surriensi, ad ripam Tamesis fluminis :—Ercenwold getimbrede mynster on Sūþrigena lande, þe Temese streáme, on þǽre stówe ðe is nemned Ceortes íge *Earconvaldus monasterium construxerat in regione Sudergeona, juxta fluvium Tamensem, in loco qui vocatur Cerotæsei, id est, Ceroti insula,* Bd. 4, 6; S. 574, 15. Hér drǽfde Eádgár cyng ða preóstas of Ceortes íge [Certes ige, 223, col. 3] *in this year,* A. D. 964, *king Edgar drove the priests from Chertsey,* Chr. 964; Th. 222, 5, 10.

ceorung, e; *f.* [ceorian *to murmur*] *A murmuring, complaint, grudging;* murmuratio, querimonia, querela :—Sum ceorung mihte beón gif he his behát ne gelǽste *there might be some murmuring if he performed not his promise,* Homl. Th. ii. 80, 26, 12. Æfter ceorunge *after murmuring,* ii. 80, 9. Módignys aceñþ ceorunge *pride begets murmuring,* ii. 222, 8. Ic gesylle fram me Israhéla ceorunge *cohibebo a me querimonias filiorum Israel,* Num. 17, 5. Beóþ cumlíðe eów betwŷnan búton ceorungum *be hospitable among yourselves without grudging,* Homl. Th. ii. 286, 14.

CEÓSAN, ciósan, ic ceóse, ðú ceósest, cýst, he ceóseþ, cýst, císt, *pl.* ceósaþ; *p.* ic, he ceás, cés, ðú cure, *pl.* curon; *impert.* ceós, *pl.* ceósaþ; *pp.* coren; *v. a.* **I.** *to* CHOOSE, *select, elect;* legere, seligere, eligere :—Ðæt hí woldon óðerra wera ceósan *that they would make a choice of other husbands,* Ors. 1, 10; Bos. 32, 32. He hêht him wine ceósan *he commanded him to choose friends,* Cd. 90; Th. 112, 8; Gen. 1867: Runic pm. 29; Kmbl. 345, 15; Hick. Thes. i. 135. Drihten ðé císt *the Lord will choose thee,* Deut. 28, 9. Hí leófne ceósaþ ofer woruldwélan *they choose the beloved above worldly wealth,* Exon. 62 b; Th. 230, 29; Ph. 479. Bebodu ðíne ic ceás *mandata tua elegi,* Ps. Spl. 118, 173. Hér Eádgár, Engla cyning, ceás him óðer leóht, and ðis wáce forlêt líf *here,* A. D. 975, *Edgar, king of the Angles, chose him another light, and left this frail life,* Chr. 975; Erl. 124, 30; Edg. 22: 1041; Erl. 169, 10. Æfæste men him ðá wíc curon *the pious men chose them a dwelling there,* Cd. 86; Th. 108, 9; Gen. 1803: Andr. Kmbl. 808; An. 404. Ceós ðé geféran and feoht ongén Amalech *elige viros et pugna contra Amalec,* Ex. 17, 9: Deut. 17, 15. Ðæt ic neóbed ceóse *that I may choose a death-bed,* Exon. 63 b; Th. 235, 7; Ph. 553. Ðæt se cyning him ceóse sumne wísne man *ut provideat rex virum sapientem,* Gen. 41, 33: Ps. Th. 105, 5. Ceósan us eard in wuldre *may we choose us a dwelling in glory,* Cd. 217; Th. 277, 14; Sat. 204. Ðæt he óðer líf cure *that he chose another life,* Bd. 5, 19; S. 638, 6. Ǽr he bǽl cure *ere he chose the funeral pile,* Beo. Th. 5629; B. 2818: Exon. 100 a; Th. 376, 20; Seel. 157. Ðæt hí him cyning curan *ut regem sibi eligerent,* Bd. 1, 1; S. 474, 22. Ðéh ðe fell curen synnigra cynn *though the race of sinners chose death,* Andr. Kmbl. 3217; An. 1611. **II.** *to accept by choice or what is offered, to accept;* oblatum accipere, accipere :—Ðæt he ðone cynedóm ciósan wolde *that he would accept the kingdom,* Beo. Th. 4742; B. 2376. Hie curon æðelinges ést *they accepted the chieftain's bounty,* Cd. 112; Th. 147, 20; Gen. 2442. [*Wyc.* Piers P. *Chauc. R. Glouc.* chese: *Laym.* cheosen: *Orm.* chesenn: *Plat.* kösen, kören: *O. Sax.* kiosan, keosan: *Frs.* kiezjen, tziezjen: *O. Frs.* kiasa, tziesa: *Dut.* kiezen: *Ger.* kiesen: *M. H. Ger.* kiusen, kiesen: *O. H. Ger.* kiusan, kiosan: *Goth.* kiusan: *Dan.* keise: *Icel.* kjósa: *Lat.* gustare: *Grk.* γεύω: *Sansk.* jush *to like, be fond of, choose.*] DER. a-ceósan, forþ-, ge-, on-, wið-, wiðer-.

CEOSEL, ceosol, cisil, cysel, es; *m? Gravel, sand;* glarea, sabulum. Hence the sand-hill in Dorsetshire is called CHESSIL :—Cisil *glarea,* Glos. Epnl. Recd. 157, 12. [*Kil.* kijsel, kesel: *Ger.* kiesel, m: *M. H. Ger.* kisel, m: *O. H. Ger.* kisil, m.] DER. sǽ-ceosel, sand-.

ceosel-stán, cysel-stán, es; *m. Sand-stone, gravel;* glarea, calculus :—Ceoselstán *glarea,* Wrt. Voc. 63, 70. Cyselstán *calculus,* Ælfc. Gl. 11; Som. 57, 46; Wrt. Voc. 19, 48.

ceosol, cesol, es; *m? n? A hut, cottage;* gurgustium :—Cesol *gurgustium,* Glos. Epnl. Recd. 157, 8.

ceósung, e; *f. A choosing;* electio, Som. Ben. Lye. DER. a-ceósung. v. ceósan.

ceoul *a basket;* cophinus, Jn. Lind. War. 6, 13. v. cawl.

CEÓWAN, to ceówenne, ic ceówe, ðú ceówest, cýwst, he ceóweþ, cýwþ, *pl.* ceówaþ; *p.* ceáw, *pl.* cuwon; *pp.* cowen *To* CHEW, *gnaw, eat, consume;* ruminare, manducare :—He hêt hine ceówan mid tóþum his fingras *he commanded him to gnaw his fingers with his teeth,* Homl. Th. ii. 510, 34. Ongunnon ða nǽddran to ceówenne heora flǽsc and heora blód súcan *the serpents began to chew their flesh and suck their blood,* ii. 488, 34, 27. Ðæt híg eton ða nýtenu ðe hira clawe todǽlede beóþ and ceówaþ *omne quod habet divisam ungulam, et ruminat in pecoribus, comedetis,* Lev. 11, 3, 4. Hí cuwon heora girdlas, and gærs ǽton *they chewed their own girdles, and ate grass,* Ælfc. T. 42, 9: Homl. Th. i. 404, 5. Ðec sculon mold-wyrmas monige ceówan *many mould-worms shall consume* [*chew, eat*] *thee,* Exon. 99 a; Th. 371, 8; Seel. 72. [*Chauc.* chewe; *Orm.* chewwenn: *Scot.* chaw, chow: *Plat.* kaujen, kauwen, kawwen: *Dut.* kaauwen: *Kil.* kauwen, kouwen, kuwen: *Ger.* käuen, kauen: *M. H. Ger.* kiuwen: *O. H. Ger.* kiuwan: *Dan.* tygge: *Swed.* tugga: *Icel.* tyggja, tyggva.] DER. be-ceówan, for-, to-.

ceowl *a basket;* sporta, Mk. Skt. Rush. 8, 8. v. cawl.

ceówung, e; *f. A chewing;* ruminatio, Som. Ben. Lye. v. cýwung.

cép, es; *m. A sale, bargain, business;* negotium :—Awyrigende cép *malignum negotium,* Somn. 159; Lchdm. iii. 206, 32. Sellan to cépe *to give for sale, sell,* Deut. 28, 68. v. ceáp II.

cépa, an; *m. A chapman, merchant;* mercator :—Nǽnig cépa ne seah ellendne wearod *no merchant saw a foreign shore,* Bt. Met. Fox 8, 58; Met. 8, 29. Ne geseah nán cépa eáland *no merchant visited an island,* Bt. 15; Fox 48, 13. Cépena þinga gewrixle *the interchange of merchants' goods, commerce;* commercium, Ælfc. Gl. 16; Som. 58, 53; Wrt. Voc. 21, 41. v. cýpa.

CÉPAN, to cépanne; *p.* cépte, *pl.* cépton; *pp.* céped, cépt; *v. a. gen. acc. To observe, keep, regard, await, desire, take, betake oneself to, meditate, bear;* observare, tenere, manere, appetere, captare, se conferre, meditari, portare :—Menn mágon cépan be his bleó hwylc weder toweard byþ *men may observe by his hue what weather is coming,* Bd. de nat. rerum; Wrt. popl. science 15, 9; Lchdm. iii. 268, 5. Híg mínne hó oððe hóhfót cépaþ oððe begemaþ *ipsi calcaneum meum observabunt,* Ps. Lamb. 55, 7: Homl. Th. ii. 324, 16: Ælfc. T. 28, 3. Ðe willaþ ðysre deópnysse cépan *who will keep this precept,* Homl. Th. ii. 94, 7. Ðæt folc his cépte *the people regarded him,* Homl. Th. ii. 506, 7. Hí brycge ne cépton *they regarded not the bridge,* Chr. 1013; Erl. 148, 11. Ða sceoldon cépan Godwines eorles *they were to lay in wait for earl Godwine,* 1052; Erl. 183, 34. Ða munecas ðæs ándagan cépton *the monks awaited the day appointed,* Homl. Th. ii. 172, 13. He dysigra manna hérunga céþþ *he desires the praises of foolish men,* i. 412, 7. Ðæt hí cépaþ ðæs ydelan hlýsan *that they desire vain renown,* ii. 566, 2. Swá hwilcne swá ic cysse, cépaþ his sóna *whomsoever I kiss, take him forthwith,* ii. 246, 11. He nolde him nánes fleámes cépan *he did not wish to betake himself to flight,* Ælfc. T. 36, 18. Ðonne him cælþ, he céþþ him hlywþe *when he is cold, he betakes himself to shelter,* Hexam. 20; Norm. 28, 22. Ic gylpes cépte *I have persevered in boasting;* jactantiæ insistebam, Mod. confitendi 1. Nele he him hearmes cépan *he will not meditate harm against him,* Homl. Th. ii. 522, 20. He me hearmes céþþ *he meditates harm against me,* i. 56, 3. Ðe cépton heora deáþes *who meditated their death,* L. Ælf. C. 2; Th. ii. 342, 20. Ðæt ðú cépe [MS. kepe] him hearmes *that thou meditate harm against him,* Basil admn. 5; Norm. 46, 4. Ne cép [MS. kep] ðú dínum néxtan fácnes *devise not deceit against thy neighbour,* 5; Norm. 46, 10. Geþyldelíce synd to cépanne *patienter portandi sunt,* R. Ben. interl. 36. [*Chauc. R. Glouc. Laym.* kepe: *Kil.* kepen.]

cépe-cniht, es; *m. A bought servant, slave;* venalis puer, servus :—Gregorius geseah cépecnihtas ðǽr gesette *Gregory saw slaves placed there,* Bd. 2, 1; S. 501, 7. v. ceáp-cniht.

cépe-man, es; *m. A chapman, merchant;* mercator :—Gif man feormaþ cépeman *if a man entertain a chapman,* L. H. E. 15; Th. i. 32, 17. Hit cépemen ne gefaraþ *merchants do not visit it,* Bt. 18, 2; Fox 64, 1. v. ceáp-man.

cépe-stów *a market-place, market;* forum, emporium, Som. Ben. Lye. v. ceáp-stów.

cépe-þing; *pl. n. Saleable things, goods, ware, merchandise;* venalia, merces :—Secgeaþ hí ðæt cýpemen monig cépeþing to ceápstowe brohte *dicunt quia mercatoribus multa venalia in forum fuissent conlata,* Bd. 2, 1; S. 501, 4. Cépeþing [MS. cepeþingc] *merces,* Ælfc. Gl. 16; Som. 58, 52; Wrt. Voc. 21, 40.

céping, e; *f. Traffic, merchandise;* negotiatio :—Hús cépinge *domum negotiationis,* Jn. Rush. War. 2, 16. To cépinge his *ad negotiationem*

suam, Rtl. 107, 25   Betre is tosocnung his cēpinge seolferes and goldes *melior est acquisitio ejus negotiatione argenti et auri*, 81, 14.

cēp-man, -mann, es; *m. A chapman, merchant;* mercator :—Hīg fōron mid ōðrum cēpmannum *they went with other merchants*, Gen. 42, 5. v. ceáp-man.

cēp-sceamol, es ; *m. A toll-booth, seat of custom, treasury ;* telonium = τελώνιον, gazophylacium = γαζοφυλάκιον :—Ðás word he spræc æt cēpsceamole *hæc verba locutus est in gazophylacio*, Jn. Foxe 8, 20. v. ceápsceamul.

cēp-setl, es; *n. A toll-booth, seat of custom;* telonium = τελώνιον :— He geseah Leuin sittende æt hys cēpsetle *vidit Levi sedentem ad telonium*, Mk. Bos. 2, 14. v. ceáp-setl.

cer *a turn.* v. cerr, cyrr.

Cerdic, es ; *m. Cerdic, the founder of the West-Saxon kingdom ;* Cerdīcus :—Ðȳ geáre ðe wæs agan fram Cristes acennesse cccc wintra and xcv [MS. xciiii] wintra, ðá Cerdic and Cynríc his sunu cwom up æt Cerdices ōran mid v scipum. Ond ðæs ymb vi geár, ðæs ðe hie up cwōmon, ge-eódon West-Seaxna ríce; and ðæt wǽron ða ǽrestan cyningas ðe West-Seaxna lond on Wealum ge-eódon; and he hæfde ðæt ríce xvi geár; and ðá he gefōr, ðá fēng his sunu Cynríc to ðam ríce, and heóld xxvii [MS. xvii] winter. Ðá he gefōr, ðá fēng Ceol to ðam ríce and heóld vii geár. Ðá he gefōr, ðá fēng Ceolwulf to his brōður, and he rícsode xvii geár; and hiera cyn gǽþ to Cerdice. Ðá fēng Cynegils, Ceolwulfes brōður sunu, to ríce and rícsode xxxi wintra ; and he onfeng ǽrest fulwihte Wesseaxna cyninga; and ðá fēng Cēnwalh to and heóld xxxi wintra; and se Cēnwalh wæs Cynegilses sunu *in the year that was past from the birth of Christ 495, then Cerdic and Cynric his son landed at Cerdic's shore from five ships. And six years after they landed, they subdued the West-Saxons' kingdom ; and they were the first kings, who conquered the West-Saxons' land from the Welsh ; and he had the kingdom sixteen years ; and when he died, then his son Cynric succeeded to the kingdom, and held it twenty-seven winters. When he died, then Ceol succeeded to the kingdom, and held it seven years. When he died, then Ceolwulf his brother succeeded, and he reigned seventeen years ; and their kin reaches to Cerdic. Then Cynegils, Ceolwulf's brother's son, succeeded to the kingdom, and reigned thirty-one winters ; and of the West-Saxons' kings, he first received baptism ; and then Cenwalh succeeded, and held it thirty-one winters ; and Cenwalh was the son of Cynegils*, Chr. Erl. 2, 1–20. Hér, A. D. dxxxiv, Cerdic forþfērde, and Cynríc his sunu rīxode xxvii wintra and hie gesealdon heora twám nefum, Stufe and Wihtgāre, Wihte eálond *here*, A. D. 534, *Cerdic died, and Cynric his son reigned twenty-seven years, and they gave their two nephews, Stuf and Wihtgar, the isle of Wight*, Chr. 534; Th. 26, 40. v. Cerdices ford, Cerdices leáh, Cerdices ōra, Birīnus, Cynegils.

Cerdices ford, es ; *m. Cerdic's ford, the ford of a little river in the south of Dorsetshire on Cerdices ōra, q.v ;* Cerdīci vadum :—Hér Cerdic and Cynríc West-Sexena ríce onfēngun ; and ðȳ ilcan geáre hie fuhton wið Brettas, ðær mon nū nemneþ Cerdices ford *in this year Cerdic and Cynric took the kingdom of the West-Saxons ; and in the same year they fought against the Britons, where it is now named Cerdic's ford*, Chr. 519; Th. 26, 21–26, col. 1.

Cerdices leáh ; *gen. leáge; f. Cerdic's ley, in the south of Dorset-shire ;* Cerdīci campus :—Hér Cerdic and Cynríc [MS. Cinric] fuhtan wið Bryttas on ðære stōwe ðe is gecweden Cerdices leág [MS. Laud ford] *in this year Cerdic and Cynric fought against the Britons at the place which is called Cerdic's ley*, Chr. 527; Th. 26, 30–33, col. 3.

Cerdices ōra, Certices ōra, an ; *m. Cerdic's shore, on the south of Dorsetshire,* v. Cerdices ford ; Cerdīci lítus :—Ðá Cerdic and Cynríc his sunu cwom up æt Cerdices ōran mid v scipum *then*, A. D. 495, *Cerdic and Cynric his son came up to Cerdic's shore with five ships*, Chr. Erl. 2, 3. Hér cwōmon Cerdic and Cynríc his sunu on Breteue, mid v scipum, in ðone stede ðe is gecweden Cerdices [Certices 25, 29, col. 1, 2] ōra *here*, A. D. 495, *Cerdic and Cynric his son came to Britain, with five ships, at the place which is called Cerdic's shore*, Chr. 495; Th. 24, 31, col. 1, 2, 3 : 514; Th. 26, 16, col. 1.

ceren, cæren, cyren, es ; *n? New wine boiled down one third or one half, sweet wine;* carenum = κάροινον :—Hī, ða sylfe betweónum, indrencton mid ðam cerenum ðære gōdspellícan swētnysse *betweenthemselves, they pledged with the wines of gospel sweetness*, Guthl. 17; Gdwin. 72, 7. Cærenes gōdne bollan fulne meng togædere *mingle together a good bowl full of boiled wine*, L. M. 1, 1; Lchdm. ii. 24, 19. Cyren *vel* awilled wīn *dulcisapa*, Cot. 62.

CEREN, cyrin, e; *f. A CHURN;* vas in quo lac agitatur et butyrum cogitur, fidelia, sinum :—Cyrin *sinum*, Wrt. Voc. 290, 31. [*Prompt.* chyrne : *Scot.* kirn : *Plat.* karne : *Ger. dial.* kerne, *f : Dan.* kjerne, *m. f : Swed.* kärna, *f : Icel.* kirna, *f.*]

cerfe *shall separate;* secabit :—Ne cerfe *non secabit*, Lev. 1, 17. v. ceorfan.

CERFILLE, cærfille, cyrfille, an ; *f.* CHERVIL; cærefolium = χαιρέφυλλον, chærophyllum sylvestre, Lin :—Genim ðysse wyrte ðe man *cerefolium*, and ōðrum naman ðam gelíce cerfille nemneþ þrý croppas

take three heads of this herb, which is named cerefolium, *and by the other like name chervil*, Herb. 106; Lchdm. i. 220, 9 : Lchdm. ii. 72, 6. To monnes stemne nim cerfillan *for a man's voice take chervil*, 1, 83; Lchdm. ii. 152, 15 : 2, 52; Lchdm. ii. 272, 10. [*Plat.* karwel : *Dut.* kervel, *f : Ger.* kerbel, *m : M. H. Ger.* kërvele, *f : O. H. Ger.* kerfola, *f : Dan.* kiörvel, *m. f : Swed.* kyrfvel, *m : Icel.* kerfill, *m.* Rask Hald : *Lat.* cærefolium ; from. *Grk.* χαιρέφυλλον.] DER. wudu-cerfille.

cerg; *adj.* [= cearig, *q. v.*] *Sad, dire, wicked;* tristis, sollicitus, dirus, malus :—Cerge reóðaþ fōre onsýne ēces dēman *the wicked shall wail before the face of the eternal judge*, Exon. 20 a ; Th. 52, 20; Cri. 836.

cerian *to murmur*, Wanl. Catal. 4, 6. v. ceorian.

cerlic, es ; *m ? n ? The herb* CARLOCK *or* CHARLOCK ; rapum sylvestre :— Nim cerlices sǽd *take seed of charlock*, L. M. 1, 39; Lchdm. ii. 102, 2 : 2, 34; Lchdm. ii. 238, 30.

cernan; *p. de; pp.* ed [ceren *a churn*] *To churn;* agitare butyrum, Som. Ben. Lye.

cerr, es; *m. A turn, time;* versio, temporis spatium :—Æt ōðrum cerre *alio tempore*, Bt. 35, 2; Fox 156, 17. v. cyrr.

cerran; *p. de; pp.* ed *To turn, return;* verti, reverti :—On wōh cerde *turned to wrong, deviated;* deviavit, Cot. 61. Cer ðe on bæcling *turn thee behind*, Cd. 228; Th. 308, 26; Sat. 698. Hió cerrende Criste hērdon *they returning obeyed Christ*, Ps. C. 50, 56; Ps. Grn. ii. 278, 56. Cerreþ on upróðor leóht *light returns to the sky*, Bt. Met. Fox 29, 102; Met. 29, 50. v. cyrran.

cerrednes, -ness, e; *f.* [cerred, *pp.* of cerran; -nes] *A turning;* versio, Ben. Lye. DER. a-cerrednes. v. cyrrednes.

cerse, an; *f. Cress;* nasturtium, Herb. 21; Lchdm. i. 116, 17, MS. B : L. M. 1, 26; Lchdm. ii. 68, 4: 1, 31; Lchdm. ii. 74, 10 : 128, 13 : ii. 182, 15 : 188, 8 : ii. 340, 24. v. cærse.

Certes íg, es ; *f.* CHERTSEY; Certesia :—Hér [MS. hier] wurþan ða canonicas gedrifen ūt of ealdan mynstre fram Eádgāre cynge, and eác of niwan [MS. niwen] mynstre and of Certes íge, and of Mideltūne, and he sette ðarto munecas and abbodas : to niwan [MS. niwen] mynstre Æþelgārum, to Certes íge Ordberhtum, to Mideltūne Cyneward *here the canons were driven out of the old monastery [at Winchester] by king Edgar, and also from the new monastery, and from Chertsey, and from Milton, and he placed thereto monks and abbots : Æthelgar to the new monastery, Ordberht to Chertsey, [and] Cyneward to Milton*, Chr. 964; Th. 223, 1–11. v. Ceortes íg.

Certices ōra, an; *m. Cerdic's shore;* Cerdīci lítus :—On ðone stede ðe is gehāten Certices ōra *at the place which is called Cerdic's shore*, Chr. 495; Th. 25, 29, col. 1, 2 : 514; Th. 27, 15, col. 1, 2. v. Cerdices ōra.

ceruille *chervil*, Lchdm. iii. 106, 19. v. cerfille.

cēs *chose, elected;* *p.* of ceósan.

cēse *a cheese*, L. In. 70; Th. i. 146, 19. v. cȳse.

cēse-lib *rennet or runnet;* coagulum, Som. Ben. Lye. v. cȳs-lib.

cesol *a cottage*, Glos. Epnl. Recd. 157, 8. v. ceosol.

cest, e; *f. A chest;* cibotium = κιβώτιον, cistella, loculus, Ælfc. Gl. 3; Som. 55, 64 : Jn. Rush. War. 13, 29. v. cyst.

cester *a city*, Chr. 491; Erl. 14, 6. v. ceaster.

cete, an; *f. A cabin, cellar;* cella, Ælfc. Gl. 108; Som. 78, 99; Wrt. Voc. 58, 14. v. cote, cyte.

cetel, cetil, es; *m. A* KETTLE; cācābus = κάκκάβος :—Cetil *cacabum*, Glos. Epnl. Recd. 155, 26. v. cytel.

cetel-hrūm, es; *m. Kettle-soòt;* cacabi fuligo :—Genim cetelhrūm *take kettle-soot*, L. M. 1, 61; Lchdm. ii. 134, 2.

Cetrehta, an; *m. Catterick, near Richmond, Yorkshire;* Cataracta, oppidi nomen in agro Richmondensi :—Tūn, ðe he oftust oneardode wel neáh Cetrehtan, gyt to-dæg mon his naman cneódeþ *cujus nomine vicus in quo maxime solebat habitare, juxta Cataractam, usque hodie, cognominatur*, Bd. 2, 20; S. 522, 24.

cewl *a basket*, Mt. Kmbl. Lind. 16, 9: Mk. Skt. Lind. 8, 8. v. cawl.

chor, es ; *m? A dance, chorus, choir ;* chŏrus = χορός :—Chor *chorus*, Wrt. Voc. 81, 21.

chor-gleów, es; *n.* [gleó, gleów *glee, joy, music*] *A musical dance, dance ;* chorus = χορός :—Hērian híg naman his on chorgleówe *laudent nomen ejus in choro*, Ps. Lamb. 149, 3 : 150, 4.

cicel; *gen.* cicles ; *m. A morsel, little mouthful, cake ;* buccella, placenta :—Cicel *buccella*, Cot. 26 : 126. Se cicel *the cake*, Lchdm. iii. 30, 21. Gemenged wið meolowe and to cicle abacen *mingled with meal and baked to a cake*, Med. ex Quadr. 9, 17; Lchdm. i. 364, 14. Bac hym ánne cicel *bake him a cake*, Lchdm. iii. 134, 20 : L. M. 1, 46; Lchdm. ii. 114, 25 : Lchdm. iii. 30, 19, 26 : 96, 17.

CICEN, es; *pl. nom. acc.* cicenu; *gen.* a ; *dat.* um ; *n. A* CHICKEN; pullus :—Cicen *pullus*, Ælfc. Gl. 39; Som. 63, 49; Wrt. Voc. 30, 4: 281, 24. Cicen oððe brid oððe fola *pullus*, Wrt. Voc. 77, 37. Henne mid cicenum gesihþ ceápas eácan getácnaþ *a dream of a hen with chickens betokens trade to be increasing*, Lchdm. iii. 204, 31. Seó henn hyre cicenu under hyre fyðeru gegaderaþ *gallina congregat pullos suos sub alas*, Mt. Bos. 23, 37. Cicena mete *chickens' meat, chick-weed ;* modera,

alsÿne = ἀλσίνη, Ælfc. Gl. 44; Som. 64, 66; Wrt. Voc. 32, 3 : 69, 27 : 79, 39 : L. M. 3, 8; Lchdm. ii. 312, 16 : Lchdm. iii. 6, 14 : 118, 29 : 134, 1. [*Wyc.* chykenys, *pl* : *Piers P.* chicknes, *pl* : *Chauc.* chike : *Prompt.* chekyn : *Plat.* kiken, küken : *Dut.* kieken, kuiken, *n* : *Kil.* kiecken : *Ger.* küch-lein, *n* : *Dan.* kylling, *m. f* : *Swed.* kyckling, *m* : *Icel.* kjúk-lingr, *m* : *O. Nrs.* kyk-lingr, *m.* Rask Hald.]

cicene, an; *f. A* KITCHEN; coquina, culina :—Cicene [MS. cicen] *coquina vel culina*, Ælfc. Gl. 107; Som. 78, 77; Wrt. Voc. 57, 55. v. cycene.

cicle *to a cake*, Med. ex Quadr. 9, 17; Lchdm. i. 364, 14; *dat.* of cicel.

cíd, cýd, es; *m? Strife, chiding, contention;* contentio, jurgium, rixa, Somn. 305. DER. ge-cíd.

CÍDAN, to cídenne; *p.* cídde, *pl.* cíddon, cídon; *pp.* cíded, cídd [cíd *strife, chiding*] *To* CHIDE, *rebuke, blame, contend, strive, quarrel, complain;* increpare, rixari, altercari, queri :—Cídan on swefnum ceápes eácan getácnaþ *to chide in dreams betokens increase of trade,* Lchdm. iii. 208, 3 : 204, 32. Rihtwís cídeþ me *justus increpabit me,* Ps. Spl. 140, 6. Cídde him se Hǽlend *increpavit illum Jesus,* Lk. Bos. 4, 35 : Mk. Bos. 1, 25 : 8, 33 : Homl. Th. i. 300, 24 : ii. 44, 21. His leorningcnihtas cíddon him *discipuli ejus increpabant illos,* Lk. Bos. 18, 15. Cíde he wið God *let him blame God,* Homl. Th. i. 96, 1. Gif men cídaþ *si rixati fuerint viri,* Ex. 21, 18. Begunnon hí to cídenne *they begun to quarrel,* Homl. Th. ii. 158, 13. Ic cíde *altercor,* Ælfc. Gr. 25; Som. 27, 12. Ic cíde oððe ceorige *queror,* 29; Som. 33, 53. [*Wyc.* chide, chiden : *Piers P.* chiden : *Chauc.* chide : *Laym.* chiden : *Ger.* kiden, kyden *to sound.*] DER. ge-cídan.

cídde *told,* Gen. 9, 22; = cýdde; *p.* of cýdan.

CIDER, es; *m?* CIDER; vinum pomarium, Lye. [*Wyc.* sydur, sidir : *Dut.* cider, *f* : *Ger.* cider, *m* : *Fr.* cidre : *Span.* cidra : *It.* cidro, sidro.]

cíding, cýdung, e; *f. A* CHIDING, *reproving, rebuke;* increpatio :—For his cídinge *for his chiding,* Ors. 4, 12; Bos. 99, 8. Of cýdunge dínre hí fleóþ *ab increpatione tua fugient,* Ps. Spl. T. 103, 8.

ciefes, e; *f. A concubine;* concubina, Ors. 6, 30; Bos. 126, 41. v. cyfes.

cíegan *to call, call upon, invoke,* Ps. Th. 52, 5 : 74, 1. v. cígan.

cíele, es; *m. Cold;* frigus :—For cíele nele se sláwa erian *propter frigus piger arare nonvult,* Past. 39, 2; Hat. MS. 53 a, 14, 16, 18. v. cíle.

cielf *a calf,* Ps. Spl. C. 105, 20. v. cealf.

ciellan; *pl. m. Vessels for drink, wooden tankards, leather bottles;* obbæ, Dial. 1, 5. v. cyll.

ciepe *an onion;* cǽpe :—Genim ciepan *take an onion,* L. M. 1, 3; Lchdm. ii. 40, 6. v. cipe.

ciépe-mon *a merchant,* Som. Ben. Lye. v. ceáp-man.

cier, cierr, es; *m. A turn, time, business, affair;* versio, temporis spatium, negotium :—Æt ánum cierre *uno eodemque tempore,* Past. 61, 2. Mid óðrum cierrum *with other affairs,* Past. 4, 1; Hat. MS. 9 b, 7. v. cyrr.

cierlisc *churlish, rustic,* L. In. 37; Th. i. 124, 20, 21. v. ceorlisc.

ciern, es; *n? Must or new wine boiled thick;* sapa, Cot. 170 : 184. v. ceren.

CÍFAN? *p.* cáf, *pl.* cifon; *pp.* cifen *To quarrel;* litigare. [*Dut.* kijven *to quarrel* : *Ger.* keifen *to scold* : *Icel.* kífa *to strive, quarrel.*] DER. cáf, cáf-líce, -scype; un-cáf-scipe; cáfer-tún.

cifes *a harlot;* pellex, Alb. resp. 64 : Cot. 150 : 190. v. cyfes.

cifes-gemána, an; *m. Fornication;* concubinatus :—We lǽraþ, ðæt man geswíce cifesgemánan [MS. cifesgemanna] *docemus, ut cessent concubinatus,* L. Edg. C. 21; Wilk. 84, 1.

CÍGAN, cígean, cýgan, cýgean, ciégan, cǽgan, cégean; *part.* cígende; *p.* de; *pp.* ed. **I.** *v. trans. To call, name, call upon, invoke, call together, summon;* vocare, nominare, invocare, convocare :—Drihten mæg steorran be naman cígean ealle *the Lord can call all the stars by name,* Ps. Th. 146, 4. Ealle gewunedon hí móder cýgean *all were accustomed to call her mother,* Bd. 4, 23; S. 594, 39. Swá hine cígþ Engle and Seaxe *as the Angles and Saxons call it,* Menol. Fox 366; Men. 184. Ðone [MS. þonne] niða bearn nemnaþ and cígaþ Pentecostenes dæg *which children of men name and call the day of Pentecost,* Chr. 973; Erl. 124, 15; Edg. 7. He cígde hungor ofer eorþan *vocavit famem super terram,* Ps. Spl. 104, 15. Ufan engla sum Abraham cýgde *an angel from above called Abraham,* Cd. 141; Th. 176, 9; Gen. 2909. Ðú eart líðe eallum cígendum ðé *tu es mitis omnibus invocantibus te,* Ps. Lamb. 85, 5 : Ps. Spl. 146, 10. Swá hwylce daga ic ðé cíge, gehýr me *in quacumque die invocavero te, exaudi me,* Ps. Th. 137, 4. Ðínne naman we cígaþ *nomen tuum invocabimus,* Ps. Lamb. 79, 19. Ðe cígaþ naman his *qui invocant nomen ejus,* Ps. Spl. 98, 6. Abraham wordum God torhtum cígde *Abraham called upon God with fervent words,* Cd. 86; Th. 108, 16; Gen. 1807 : Ps. Th. 90, 15. God híg ne cígdon *Deum non invocaverunt,* Ps. Lamb. 52, 6 : 78, 6 : Ps. Spl. 98, 7. Us gehýr swilce we ðé daga, Drihten, cígen *hear us, O Lord, on whatever day we may call upon thee,* Ps. Ben. 19, 9; Ps. Grn. ii. 148, 19, 9. Moyses bebeád eorlas cígean sweot sande neár *Moses bade his men summon the multitude near to the sand,* Cd. 154; Th. 191, 24; Exod. 219. **II.**

v. intrans. To cry, call; clamare, vocare :—Abeles blód to me cígeþ *Abel's blood crieth to me,* Cd. 48; Th. 62, 12; Gen. 1013. Ic cígde to Dryhtne *I called to the Lord,* Ps. Th. 117, 5. DER. a-cígan.

cignis, niss, e; *f. A name, naming;* nomen, Som. Ben. Lye.

cilct; *part.* [cealc *chalk*] *Chalked;* calce illitus. DER. niw-cilct.

CILD; *gen.* cildes, *pl.* cild, *sometimes* cildru, cildra; *n. A* CHILD, *infant;* infans, puer :—Arís and nim ðæt cild *surge et accipe puerum,* Mt. Bos. 2, 13, 14. Ðæt cild wixþ and gewurþ eft cnapa and eft syððan cniht *the child grows, and then becomes a boy, and afterwards a young man,* Hom. Sax. Þurh cildes hád *in the state of childhood,* Exon. 65 a; Th. 240, 15; Ph. 639. Eálá cild, hú eów lícaþ ðeós spǽc *O pueri, quomodo vobis placet ista locutio?* Col. Monast. Th. 32, 7. Eálá ge cildra *O pueri,* 35, 33. Mid cilde beón, weorþan, or wesan *to be with child,* Bd. Whelc. 487, 22. [*Chauc. Laym. Orm.* child : *O. Sax. O. Frs.* kind, *n* : *Ger.* kind, *n* : *M. H. Ger.* kint, *n* : *O. H. Ger.* kind, kint, *n. proles* : *Goth.* kilþei, *f. fœtus* : *Icel.* kind, *f.*] DER. módor-cild, steóp-.

cilda hyrde, oððe láreów, es; *m. A herder or teacher of children, schoolmaster;* pædagogus = παιδαγωγός, Ælfc. Gl. 80; Som. 72, 103; Wrt. Voc. 46, 60.

cilda mæsse-dæg, es; *m. Childermas* [*Innocents'*]-*day;* festum innocentium :—Ðys Gódspel sceal on cilda [MS. cylda] mæsse-dæg *this Gospel must be on Childermas* [*Innocents'*]-*day, Dec. 28th,* Rubc. Mt. Bos. 2, 13–18; Notes, p. 574.

cilda trog, es; *m.* [cild, trog *a trough, cradle*] *A child's cot, cradle;* cunæ, arum, *pl. f.* Som. Ben. Lye.

cild-cláþ, es; *n. A child-cloth, a swaddling-cloth;* infantilis pannus :—Hine mid cildcláðum bewand *pannis eum involvit,* Lk. Bos. 2, 7.

cild-cradol, es; *m. A child's cradle;* cunabula, *pl.* Ælfc. Gr. 13; Som. 16, 23. On cildcradole *in a child's cradle,* Homl. Th. i. 82, 29.

cild-faru, e; *f. A carrying of children.* v. cyld-faru.

cild-fostre, -festre, an; *f. A child-fosterer, nurse;* nutrix :—Mót he habban mid him his cildfostran [-festran, Roff.] *debet habere secum nutricem infantis sui,* L. In. 64; Wilk. 25, 4.

cild-geong; *adj. Young as a child;* infans, Andr. Kmbl. 1369; An. 685.

cild-hád, es; *m.* CHILDHOOD, *infancy;* infantia :—Of cildháde *ab infantia,* Mk. Bos. 9, 21 : Elen. Kmbl. 1826; El. 915.

cild-hama, an; *m. The womb;* matrix, uterus, Ælfc. Gl. 74; Som. 71, 57; Wrt. Voc. 44, 39.

cild-isc; *adj.* CHILDISH, *puerile;* puerilis :—Cildisc wesan *to be childish,* Cd. 106; Th. 139, 32; Gen. 2318. v. cild-líc.

cildiung-wíf, es; *n. A child-bearing woman;* puerpera, Wrt. Voc. 17, 17.

cild-líc, cildisc; *adj. Childish;* infantilis, puerilis :—Cildlíc *puerilis,* Ælfc. Gr. 5; Som. 5, 23 : 9, 28; Som. 11, 38. For ðære cildlícan yldo *propter infantilem ætatem,* Bd. 4, 8; S. 575, 28.

cildru *children,* Homl. Th. i. 80, 20; *acc. pl.* of cild.

cild-sung, e; *f. Childishness;* puerilitas, Som. Ben. Lye.

cíle, es; *m. A cold;* frigus :—Cíle wið hǽto *cold with heat,* Bt. Met. Fox 29, 101; Met. 29, 50 : Gen. 8, 22. v. cýle.

cilfer-lamb, cilfor-lamb, es; *n. A female lamb;* agna femina :—Bringe án cilforlamb *offerat agnam,* Lev. 5, 6.

cílian, ic cílige; *p.* ode; *v. intrans. To be cold;* algere :—Ic cílige *algeo,* Ælfc. Gr. 26, 3; Som. 28, 55. v. calan.

cilic, es; *m. Hair-cloth;* cilicium, Mt. Kmbl. Lind. 11, 21.

cille *a leather bag;* ascopera = ἀσκοπήρα, Wrt. Voc. 288, 37. v. cyll.

Cilt-ern, es; *n.* [ceald *cold,* ærn *place*] *The* CHILTERN, *high hills in Buckinghamshire and Oxfordshire;* montes quidam excelsi in agris Bucingamiensi et Oxoniensi :—Námon hí [Þurkilles here] ǽnne upgang út þuruh Ciltern, and swá to Oxena forda, and ða buruh forbærndon *they* [*Thorkell's army*] *took an upward course out through Chiltern, and so to Oxford, and burned that town,* Chr. 1009; Th. 262, 21, col. 1.

cim, cim-stanas; *pl. m. The bases of a pillar;* bases, Som. Ben. Lye.

cimbal, es; *m:* cimbala, an; *m. A cymbal;* cymbalum :—Cimbal *cymbalum,* Ælfc. Gl. 20; Wrt. Voc. 82, 17. Cimbalan oððe psalteras æt-hrínan [MS. ætrínan] *saca hit getácnaþ to touch cymbals or psalteries betokens a lawsuit,* Somn. 74; Lchdm. iii. 202, 14 : Greg. Dial. 1, 9.

cimban? *p.* camb, *pl.* cumbon; *pp.* cumben *To join;* jungere. DER. camb; bannuc-camb, ðeðe-, wulfes-.

cimbing, e; *f. A joint, conjunction;* commissura, Som. Ben. Lye.

cime, es; *m. A coming,* Cd. 29; Th. 39, 1; Gen. 618. v. cyme.

cimþ *comes,* Ps. Th. 15, 11; *3rd pres.* of cuman.

CIN, cyn, e; *f. The* CHIN; mentum :—Cin *mentum,* Wrt. Voc. 71, 1. [*Chauc.* chinne : *Piers P.* chyn : *Laym.* chin : *O. Sax.* kinni, *n* : *O. Frs.* kin, ken : *Dut.* kin, *f* : *Ger. M. H. Ger.* kinn, *n* : *O. H. Ger.* kinni, *n* : *Goth.* kinnus, *f. the cheek* : *Dan.* kind, *m. f* : *Swed.* kind, *f* : *Icel.* kinn, *f* : *Lat.* gena : *Grk.* γένυς : *Sansk.* hanu, *m. f. the jaw.*] DER. cin-bán.

cin *a kind;* genus. v. cinn, cyn, cynn.

cínan *a chink,* Bt. 35, 3; Fox 158, 28, note; *acc.* of cíne.

cínan; *p.* cán, *pl.* cinon; *pp.* cinen *To gape, to break into chinks;* hiare, dehiscere, Som. Ben. Lye. DER. to-cínan.

cin-bán, es; *n. The* CHIN-BONE; mandibula, mentum :—Cin-bán man-

dibula, Ælfc. Gl. 71; Som. 70, 81; Wrt. Voc. 43, 14. Cin-bán *mentum*, Text. Rof. 40, 1. Se ðe cin-bán forslæhþ, mid xx scillingum forgelde *let him who breaks the chin-bone pay for it with twenty shillings*, L. Ethb. 50; Th. i. 16, 1.

cin-berg, e; *f. That part of the helmet which protects the chin*; menti protectio :—Grímhelm gespeón cining, cinberge *the king clasped his grim helmet, the protection of his chin*, Cd. 151; Th. 188, 28; Exod. 175.

cincg *a king*, Th. Diplm. A. D. 743–745; 28, 21. v. cyning.

cincung, e; *f. A loud* or *cackling laughter*; cachinnatio :—Ceahhetung, *vel* cincung *cachinnatio*, Ælfc. Gl. 88; Som. 74, 86.

cind *a kind, nature.* v. cynd.

cine, es; *m.* I. *a commander of four men*, or *a fourth part of an army*; quaternio :—Cine oððe feówer manna ealdor *quaternio*, Ælfc. Gr. 9, 3; Som. 8, 34. II. *a sheet of parchment folded into four parts, a quarto sheet*; quaternio :—Cine *quaternio*, Ælfc. Gl. 80; Som. 72, 108; Wrt. Voc. 46, 65: 75, 10. Bod on cine *a command in folded parchment*; diploma = δίπλωμα, Ælfc. Gl. 80; Som. 72, 110; Wrt. Voc. 46, 67.

CÍNE, cýne, an; *f. A chink, fissure, vault*; rima, caverna :—Ic geseah áne lytle cýnan [Cott. cínan] *I saw a little chink*, Bt. 35, 3; Fox 158, 28. Cínan *rimas*, Glos. Prudent. Recd. 149, 5. Cínum *cavernis*, 148, 81. [*Wyc.* chyne : *Dut.* keen, *f.*]

cine-líc; *adj.* [cyn *fit, suitable*] *Of a like kind, agreeable, suitable, adequate*; congruus, competens :—Ðæt we wilnian to heom fultum be swá manegum mannum swá us cinelíc þince æt swá micelere spræce *that we desire aid from them of so many men as may seem to us adequate for so great a suit*, L. Ath. v. § 8, 3; Th. i. 236, 16.

cinen, cínende *gaping*; *pp.* and *pres. part. of* cínan.

cing *a king*, Deut. 11, 3; Chr. 894; Erl. 92, 17. v. cyning.

Cinges tún, es; *m.* [cinges tún *the king's town*] KINGSTON; regia villa :—Æðelstán wæs to cinge æt Cinges túne gehálgod *Athelstan was consecrated king at Kingston*, Chr. 925; Th. 198, 7, col. 3; 8, col. 2: 979; Th. 234, 9, col. 1; 235, 6, col. 2. v. Cynges tún.

cining *a king*, Cd. 151; Th. 188, 28. v. cyning.

cín-líc *gaping.* v. cíne.

cinn, es; *n. A kind*; genus :—Fleógende cinn *flying kind*; volatile, Gen. 1, 20. Creópende cinn *creeping kind*; reptilia, 1, 24. Æfter his cinne *after its kind*, 1, 11. v. cyn, cynn.

cinnan, ic cinne, ðú cinnest, he cinneþ, cinniþ, *pl.* cinnaþ; *p.* ic, he can, ðú cunne, *pl.* cunnon; *pp.* cunnen *To generate, procreate*; generare, procreare :—Sorgum cinniþ *brings forth with sorrows*, Exon. 94 b; Th. 354, 28; Reim. 52. From this verb, the *p.* ic, he can are taken as a present tense. Hence it is called one of the twelve præterito-præsentia, enumerated under ágan. For cúðe the weak *p.* of cunnan, v. the *inf.* cunnan. DER. for-cinnan.

cin-tóþ, es; *m. A front tooth, grinder*; molaris, Prov. 30, Lye.

cínu, e; *f. A chink, fissure*; rima, fissura :—Cínu *rima vel fissura*, Wrt. Voc. 85, 18. Gemétte he ðæt fæt swá gehál ðæt ðær nán cínu on næs gesewen *he found the vessel so whole that there was no chink seen in it*, Homl. Th. ii. 154, 22. v. cíne, an; *f.*

ció *a chough, sort of crow*; cornicula, Wrt. Voc. 281, 2. v. ceó.

ciól, es; *m. A ship*; navis :—He lét him behindan ciólas nigon and hundnigontig *he left behind him ninety-nine ships*, Bt. Met. Fox 26, 46; Met. 26, 23. v. ceól.

ciole, an; *f. The throat*; guttur :—Sting finger on ciolan *thrust a finger into the throat*, L. M. 1, 59; Lchdm. ii. 130, 5.

ciorian *to complain*, Ælfc. Gr. 29, MS. D; Som. 33, 52. v. ceorian.

ciorl *a rustic*, L. In. 40; Th. i. 126, 12, note 28. v. ceorl.

ciorlisc *churlish, rustic, common*, L. In. 18; Th. i. 114, 6, note 8. v. ceorlisc.

ciósan *to choose, accept*, Beo. Th. 4742; B. 2376. v. ceósan.

cípan; *p.* cípte, *pl.* cípton, cíptun; *pp.* cípt *To sell*; vendere :—Híg cíptun *vendiderunt*, Gen. 47, 20. v. cýpan.

cipe, ciepe, an; *f. An onion*; cæpa, allium cæpe, Lin :—Cipe *an onion*, L. M. 1, 39; Lchdm. ii. 102, 24. Genim garleac and cipan *take garlic and onion*, 1, 3; Lchdm. ii. 40, 15. Twá cipan oððe þreó gebræd on ahsan *roast two or three onions in ashes*, 1, 69; Lchdm. ii. 144, 14.

cipe-leac, es; *n. A leek*; cipus, Cot. 55.

cipp, es; *n? A coulter, ploughshare*; dentale :—Cipp *dentale*, Ælfc. Gl. 1; Som. 55, 71; Wrt. Voc. 15, 7.

Cippan-ham, -hamm, es; *m.* [*Hunt.* Cipenham : *Brom.* Chipenham] CHIPPENHAM, *Wilts*; villæ nomen in agro Wiltoniensi :—Hér hine bestæl se here on midne winter ofer twelftan niht to Cippanhamme *in this year* [A. D. 878], *at mid-winter, after twelfth night, the army stole itself away to Chippenham*, Chr. 878; Erl. 79, 29. Hér fór se here to Cirenceastre of Cippanhamme, and sæt ðær án geár *in this year* [A. D. 879] *the army went from Chippenham to Cirencester, and remained there one year*, Chr. 879; Erl. 80, 26; 81, 23.

cipresse, an; *f. The cypress-tree*; cupressus, Som. Ben. Lye. v. cypresse.

cíptun *bought*, Gen. 47, 20; *p. pl. of* cípan. v. cýpan.

cir *a turn, time* :—Æt ðam fiftan cire *at the fifth turn* or *time*, Lchdm. i. 214, 6, MS. B. note 8. v. cirr, cyrr.

circe, an; *f. A church*; ecclesia = ἐκκλησία :—Circe *ecclesia*, Ælfc. Gl. 107; Som. 78, 82; Wrt. Voc. 57, 58. We læraþ, ðæt man innan circan ænigne man ne birige *we enjoin that they do not bury any man within a church*, L. Edg. C. 29; Th. ii. 250, 15: Bd. 2, 7; S. 509, 5. v. cyrice.

Circe, Kirke, an; *f. Circe the sorceress*; Circe, es; *f.* = Κίρκη, ης; *f.* :—Cyninges dóhtor sió Circe wæs *Circe was the king's daughter*, Bt. Met. Fox 26, 112; Met. 26, 56.

circe-weard, es; *m. A churchwarden*; ecclesiæ custos, Chr. 1131; Erl. 260, 12. v. cyric-weard.

circe-wíca, an; *m. A church-dwelling, sacristy*; sacrarium :—To ðe circewícan *to the sacristy*, Chr. 1137; Erl. 263, 13.

circ-líc; *adj.* [circe *a church*] *Like a church, ecclesiastical*; ecclesiasticus :—Mid circlícum þénungum *with ecclesiastical services*, Wanl. Catal. 118, 4, col. 2. v. cyric-líc.

circ-nyt, -nytt, e; *f.* [nyt *duty, service*] *Church-duty* or *service*; ecclesiæ ministerium *vel* officium :—Sum cræft hafaþ circnytta fela *one has skill in many church-services*, Exon. 79 b; Th. 298, 27; Crä. 91.

circol-wyrde, es; *m. A calculator, reckoner*; computator :—Feówer síðon syx byþ feówer and twentig : ða syx tíða sind genemned þurh ðæra circolwyrda gleáwnysse quadrantes *four times six are four-and-twenty : the six hours are called by the wisdom of calculators quadrants*, Bridf. 63.

circul, es; *m. A circle, the zodiac*; circulus, zodiacus = ζωδιακός :—Ðær ðæs emnihtes circul is geteald *where the circle of the equinox is reckoned*, Bd. de nat. rerum; Wrt. popl. science 4, 18; Lchdm. iii. 238, 23. Ætýwdan feówer circulas onbútan ðære sunnan *four circles appeared round the sun*, Chr. 1104; Erl. 239, 17. For ðam brádan circule ðe is zodiacus geháten, under ðam circule yrnþ seó sunne *on account of the broad circle which is called zodiacus, under which circle the sun runs*, Bd. de nat. rerum; Wrt. popl. science 5, 20, 21; Lchdm. iii. 242, 2. Ðæt heó be-yrne ðone miclan circul zodiacum *that she runs through the great circle the zodiac*, Bd. de nat. rerum; Wrt. popl. science 7, 1; Lchdm. iii. 244, 21.

circul-ádl, e; *f. Circle-disease, the shingles*; zona, circínus :—Læcedómas wið ðære ádle ðe mon hæt circuládl *leechdoms for the disease, which man calls the circle-disease* or *shingles*, L. M. Cont. 1, 36; Lchdm. ii. 8, 18 : L. M. 1, 36; Lchdm. ii. 86, 5.

circul-cræft, es; *m. Circle-craft, the zodiac*; sphæræ cognitio :—Sceal on circule cræfte findan hálige dagas *shall by circle-craft* [or *the zodiac*] *find out holy days*, Menol. Fox 134; Men. 67.

cire-bald; *adj. Bold in decision*; arbitrii strenuus :—Ðá him cirebaldum Meotud mancynnes módhord onleác *then the Lord of mankind unlocked the treasure of words to him bold in decision*, Andr. Kmbl. 341; An. 171.

Ciren-ceaster, Cyren-ceaster, Cyrn-ceaster; *gen.* ceastre; *f.* [*Asser.* Cirrenceastre : *Hunt.* Cirecestere : *Brom.* Circestre] CIRENCESTER, *Cicester, Gloucestershire*; Cirencestria in agro Glocestriensi :—Hie genámon iii ceastra, Gleawanceaster, and Cirenceaster [Cyrenceaster, col. 2, 3], and Baðanceaster *they took three cities, Gloucester, and Cirencester, and Bath*, Chr. 577; Th. 32, 41, col. 1. Æt Cirenceastre [Cyrenceastre, col, 2, 3] *at Cirencester*, 628; Th. 44, 13, col. 1. Hér fór se here to Cirenceastre [Cyrenceastre, col. 2, 3] of Cippanhamme, and sæt ðær án geár *in this year* [A. D. 879] *the army went from Chippenham to Cirencester, and remained there one year*, 879; Th. 148, 38, col. 1: 880; Th. 150, 8, col. 1. Hér, on Eastron, wæs micel gemót æt Cyrenceastre *in this year* [A. D. 1020], *at Easter, there was a great council at Cirencester*, 1020; Th. 286, 12, col. 2. Him eóde on hand se cyning and ða burhware ðe wæron on Cyrnceastre *the king came into his hands and the townspeople who were in Cirencester*, Ors. 5, 12; Bos. 110, 22.

ciric-belle, an; *f.* [cirice *a church*] *A church-bell*; ecclesiæ campana :—Of ciricbellan *from a church-bell*, L. M. 1, 63; Lchdm. ii. 136, 29.

ciric-bryce, cyric-bryce, es; *m.* [cirice *a church*, brice, bryce *a breaking, violation, breach*] *Church-breach, a breaking into a church*; in ecclesiam irruptio :—Be ciricbryce *of church-breach*, L. Ath. i. 5; Th. i. 202, 5, 6.

ciric-dór, es; *n. A church-door*; ecclesiæ porta :—Se ðe man ofslehþ binnan ciricdórum [MS. -derum] sylle ðære cirican cxx scillinga *let him who slays a man within church-doors give to the church 120 shillings*, L. Eth. vii. 13; Th. i. 332, 9.

cirice, an; *f. A church*; ecclesia = ἐκκλησία :—We læraþ, ðæt preóstas cirican healdan to godcundre þénunge *we enjoin that priests keep their churches for divine service*, L. Edg. C. 26; Th. ii. 250, 3 : 30; Th. ii. 250, 19. v. cyrice, circe.

ciric-frith *church-peace*, L. Alf. pol. 2; Th. i. 62, 5. v. cyric-frith.

ciric-fultum, es; *m.* [fultum *help, aid*] *Church-help, ecclesiastical support*; ecclesiæ auxilium :—We læraþ, ðæt preóstas geóguþe geornlíce læran ðæt hí ciricfultum habban *we enjoin that priests diligently teach youth that they may have ecclesiastical support*, L. Edg. C. 51; Th. ii. 254, 26.

ciric-griþ, cyric-griþ; es; *n. Church-peace, right of sanctuary;* ecclesiæ pax :—Stande ǽlc ciricgriþ [cyric- MS. A.] swá swá hit betst stód *let every church-peace stand as it has best stood,* L. Edg. i. 5 ; Th. i. 264, 25 : L. E. G. 1 ; Th. i. 166, 20. Gif ǽnig man Godes ciricgriþ swá abrece, ðæt he binnon ciricwagum mannslaga weorþe, ðonne síg ðæt bótleás *if any man so break God's church-peace, that he be a homicide within church-walls, then let that be bootless,* L. C. E. 2 ; Th. i. 358, 22 : 2 ; Th. i. 360, 4 : L. Eth. vi. 14 ; Th. i. 318, 24 : ix. 1 ; Th. i. 340, 1, 5.

ciriclec *ecclesiastical,* Chr. 716 ; Erl. 44, 19. v. cyriclíc.

ciric-mangung, e ; *f. Church-móngering, the sale or purchase of ecclesiastical offices, simony;* sacrorum nundinatio :—Ænig man ciricmangunge né macie *let no man commit simony,* L. Eth. v. 10 ; Th. i. 306, 28 : vi. 15 ; Th. i. 318, 27.

ciric-mitta, an ; *m.* [mitta *a measure, bushel*] *A church measure;* ecclesiastica mensura :—VI ciricmittan ealaþ *six church measures of ale,* Th. Diplm. A. D. 900; 144, 33.

ciric-ragu, e ; *f. Church-lichen or moss;* ecclesiæ muscus, L. M. 1, 63; Lchdm. ii. 138, 1.

ciric-sceat, es ; *m. Church-scot, church-money, tax or rate;* ecclesiæ census. v. cyric-sceat.

ciric-sócn, cyric-sócn, e ; *f. Church-privilege;* ecclesiæ immunitas :—Be ciricsócnum *of church-privileges,* L. In. 5 ; Th. i. 104, 12.

ciric-þén, es ; *m.* [þén *a servant, minister*] *A church-minister, clergyman;* ecclesiæ minister, clericus :—Ænig man ciricþén ne útige búton biscopes geþehte *let no man turn out a church-minister without the bishop's counsel,* L. Eth. v. 10 ; Th. i. 306, 29 : vi. 15 ; Th. i. 318, 27.

ciric-þénung, e ; *f.* [þénung *duty, service*] *Church-duty or service;* ecclesiæ ministerium :—We lǽraþ ðæt preóstas on ciricþénungum ealle án dreógan, and beón efenweorþe on geáres fæce on eallum ciricþénungum *we enjoin that priests in church-duties all perform service at the same time, and, in the space of a year, be like worthy in all church-duties,* L. Edg. C. 50 ; Th. ii. 254, 22-24.

ciric-tún, es ; *m.* [tún *an inclosure*] *A church-inclosure, church-yard, cemetery;* ecclesiæ sepimentum, cœmetérium = κοιμητήριον :—Ne binnan cirictúne ǽnig hund ne cume *let not any dog come within the church-yard,* L. Edg. C. 26 ; Th. ii. 250, 7.

ciric-wæcce, an ; *f. A church-watch or wake;* vigilia :—We lǽraþ ðæt man, æt ciricwæccan, swíðe gedreóh sí *we teach that a man, at the church-wakes, be very sober,* L. Edg. C. 28 ; Th. ii. 250, 12.

ciric-wag, es ; *m. A church-wall;* ecclesiæ murus :—Se ðe ofslehþ man binnan ciricwagum biþ feorhscyldig *he who slays a man within church-walls is life-guilty,* L. Eth. viii. 13 ; Th. i. 332, 8 : ix. 1 ; Th. i. 340, 5 : L. C. E. 2 ; Th. i. 358, 23.

ciris-beám, es ; *m. A* CHERRY-*tree;* cěrǎsus = κεράσύς :—Cirisbeám cerasus, Wrt. Voc. 285, 44. Cirisbeám [MS. cisirbeam] cerasus, Glos. Epnl. Recd. 156, 19.

cirlisc *rustic,* Chr. 893 ; Erl. 88, 33. v. ceorlisc.

CIRM, cyrm, es ; *m. A noise, shout, clamour, uproar;* strepitus, clamor, fragor, clangor :—Hlynn wearþ on ceastrum, cirm árleásra cwealmes on óre *din was in the cities, the clamour of the shameless at the point of death,* Cd. 119 ; Th. 153, 31 ; Gen. 2547. In the following references it is written cirm, Exon. 20 a ; Th. 52, 19 ; Cri. 836 : 22 b ; Th. 62, 7 ; Cri. 998 : 36 a ; Th. 118, 5 ; Gú. 235 : 38 a ; Th. 125, 34 ; Gú. 364 : 83 b ; Th. 314, 26 ; Mód. 20 : Andr. Kmbl. 82 ; An. 41 : 2476 ; An. 1239. Cyrm, dyne *fragor,* Mone B. 4413. Cyrm *clangor,* Ælfc. Gr. 5 ; Som. 4, 40. Wæs on eorþan cyrm *a noise was on the earth,* Byrht. Th. 134, 61 ; By. 107 : Andr. Kmbl. 2252 ; An. 1127. Hlúd herges cyrm *loud was the shout of the host,* Cd. 148 ; Th. 184, 14 ; Exod. 107. Ic gehýre synnigra cyrm swíðe hlúdne *I hear the uproar of sinners very loud,* 109 ; Th. 145, 17 ; Gen. 2407. Cyrmum *clangoribus,* Mone B. 6276. DER. here-cirm, wíg-.

cirman, cyrman ; *p.* de ; *pp.* ed ; *v. intrans.* [cirm *a noise, shout*] *To make a noise,* CHIRM, *cry out, shout;* strepere, clamare, exclamare :—Hí ongunnon cirman hlúde *they began to cry out aloud,* Judth. 12 ; Thw. 25, 20 ; Jud. 270. Ic hlúde cirme *I cry out aloud,* Exon. 103 a ; Th. 390, 18 ; Rä. 9, 3. Ða hlúde cirmaþ *they loudly cry out,* 114 b ; Th. 439, 4 ; Rä. 58, 4. He hlúde cirmde *he cirmde he did not cry out with a loud voice,* 113 a ; Th. 432, 20 ; Rä. 49, 3. Swá wilde deór cirmdon *they cried out as wild beasts,* 46 a ; Th. 156, 25 ; Gú. 880. Herewópa mǽst láðe cyrmdon *the enemies shouted the loudest of army-cries,* Cd. 166 ; Th. 207, 3 ; Exod. 461. [*Scot.* chirm : *Dut. Kil.* kermen : *Ger. M. H. Ger.* karmen *to wail.*]

Cirn-ceaster *Cirencester,* Chr. 628 ; Erl. 25, 14. v. Ciren-ceaster.

cirnel *a kernel,* Som. Ben. Lye. v. cyrnel.

cirpsian; *p.* ede ; *pp.* ed *To crisp, curl;* crispare, Som. Ben. Lye. v. cyrpsian.

cirps-loccas *crisped* or *curled locks,* Som. Ben. Lye. v. crisp, cyrps.

cirr *a turn, business, affair;* versio, negotium :—Mid óðrum cirrum *with other affairs,* Past. 4, 1 ; Swt. 36, 23. v. cir, cyrr.

cirran; *p.* de ; *pp.* ed *To turn;* vertere :—Him cirde to þurferþ eorl *earl Thurferth turned to him,* Chr. 921 ; Erl. 107, 27 : Invent. Crs. Recd. 1833 ; El. 915. v. cyrran.

cís; *adj. Choice, nice in eating;* fastidiosus in edendo :—Gyf hwá sý cís *if any one be choice,* Herb. 8, 2 ; Lchdm. i. 98, 15.

cisil *sand, gravel;* glarea, Glos. Epnl. Recd. 157, 12. v. ceosel.

cisil-stán *sand-stone.* v. ceósol-stán.

císnes, -ness, e ; *f. Choiceness, niceness;* fastidium, curiositas, R. Ben. 39 : L. M. 2, 1 ; Lchdm. ii. 174, 21. v. ceásnes.

Cisse-ceaster; *gen.* -ceastre ; *f.* [*Flor.* Cissaceaster : *Sim. Dun.* Cissacestre] *Cissa's city,* CHICHESTER, *Sussex;* Cissæ castellum, Cicestria in agro Sussexiensi :—Hergodon hie upon Súþ-Seaxum neáh Cisseceastre *they harried on the South-Saxons near Chichester,* Chr. 895 ; Erl. 93, 27. To Cisseceastre *at Chichester,* L. Ath. i. 14 ; Th. i. 208, 3.

cist, e ; *f. A band, company;* cohors :—On folcgetæl fíftig cista : hæfde cista gehwilc x hund tíreádigra *in the number of the people were fifty bands : each band had ten hundred illustrious warriors,* Cd. 154 ; Th. 192, 9-16 ; Exod. 229-232. DER. eóred-cist, here-.

cist *goodness, bounty,* Ælfc. T. 9, 1. v. cyst.

cist, e ; *f. A chest;* cista, Wrt. Voc. 288, 31. v. cyst.

cist *chooses,* Deut. 28, 9 ; *3rd sing. pres. of* ceósan.

cisten-beám, es ; *m. A chesnut-tree;* castanea = κάστανον :—Cistenbeám [MS. cistenbean] *castanea,* Wrt. Voc. 285, 46. v. cyst-beám.

cist-mǽlum *earnestly;* certatim, Som. Ben. Lye.

citel *a kettle,* Wrt. Voc. 288, 35. v. cytel.

CITELIAN; *p.* ode ; *pp.* od *To tickle;* titillare, Ettm. [*Scot.* kittle : *Plat.* kiddeln, keddeln, kitteln, ketteln : *Dut.* kittelen, ketelen : *Ger.* kitzeln : *O. H. Ger.* kizilón, kuzilón : *Dan.* kildre : *Swed.* kittla : *Icel.* kitla.]

citelung, e ; *f. A tickling;* titillatio :—Citelung [MS. kitelung] titillatio, Wrt. Voc. 289, 21.

CÍÞ, cýþ, es ; *m.*   I. *a young shoot of a herb or tree, a* CHIT, *sprout, germ, sprig, mote;* germen, festuca :—Swá dropan ofer gærsa cíþas *quasi stillæ super graminum germina,* Deut. 32, 2. Forhwí ǽlc sǽd to cíþum and wyrtrumum weorþe *why should every seed turn to germs and roots?* Bt. 34, 10 ; Fox 148, 32. On eallum cedrum cíþ alǽded [MS. cuþ, ciiþ=cíþ alǽded] *the germ formed on all cedar trees,* Ps. Th. 148, 9. Eall eorþan cíþ *every shoot of the earth,* 103, 12. se snáw bewríhþ wyrta cíþ *the snow covers the germ of herbs,* Salm. Kmbl. 605 ; Sal. 302. Seó eorþe cýþ mid hire cíþum, ðæt se tíma is geáres anginn *the earth makes known by her plants, that the time is the beginning of the year,* Homl. Th. i. 100, 16. Forst sceal lúcan eorþan cíþas *frost shall lock up the germs of the earth,* Exon. 90 a ; Th. 338, 7 ; Gn. Ex. 75. Genim wegbrǽdan þrý cýþas *take three sprouts of plantain,* Herb. 2, 14 ; Lchdm. i. 84, 14. Ðú meaht gesión lytelne cíþ on ðínes bróður eágan *thou canst see a little mote in thy brother's eye,* Past. 33, 6 ; Cot. MS. 42 b, 32. Se smala cíþ *the small mote,* 33, 6 ; Hat. MS. 43 a, 2, 3. Cunna hwæðer ðú mæge adón ðone cíþ of ðínes bróður eágan *try if thou canst remove the mote from thy brother's eye,* 33, 6 ; Hat. MS. 43 a, 6.   II. *seed;* crementum :—Cýþ *crementum,* Glos. Brux. Recd. 38, 7 ; Wrt. Voc. 64, 16. Cíþ, *vel* weres sǽd *crementum, vel hominis semen vel crementum,* Ælfc. Gl. 74 ; Som. 71, 73 ; Wrt. Voc. 44, 55. [*O. Sax.* kîd, *m* : *O. H. Ger.* kîdi, *n.*] DER. gærs-cíþ.

cíþ-fæst; *adj. Rooted, growing;* radicatus, crescens :—Se man ðe plantaþ treówa oððe wyrta he hí wæteraþ óþ-ðæt hí beóþ cíþfæste *the man who plants trees or herbs waters them until they are rooted,* Homl. Th. i. 304, 26.

citil *a kettle,* Som. Ben. Lye. v. cytel.

CLÁ, cleó, clawu ; *gen. dat. acc.* clawe ; *pl. nom. acc.* cleó, clawa, clawe, clawe ; *gen.* clawena ; *dat.* clám, clawum ; *f. A nail,* CLAW, *hoof;* unguis, ungula :—Fénix fýres láfe clám biclyppeþ *the Phœnix seizes the relics of the fire with its claws,* Exon. 59 b ; Th. 217, 8 ; Ph. 277. Nægl oððe clawu *unguis,* Ælfc. Gr. 9, 28 ; Som. 11, 46. Wurdon forþaborene ísene clawa *iron claws were brought forth,* Homl. Th. i. 424, 19. Sume wǽron mid ísenum clawum totorene *some were torn with iron claws,* Homl. Th. i. 542, 30. Hóf oððe clawu *ungula,* Wrt. Voc. 71, 66. Ðe clawe ne todǽlaþ *qui ungulam non dividunt,* Lev. 11, 4. Hearde cleó *hard hoofs,* Ps. Th. 68, 32. Hira clawe todǽlede beóþ *their hoofs are divided,* Lev. 11, 3. Gelícaþ Gode ofer cealf iungne forþbringende clawu [clawa, Spl.] *placebit Deo super vitulum novellum producentem ungulas,* Ps. Lamb. 68, 32. [*Wyc.* cle, clee *a hoof : Wrt. Gl.* 12th cent. p. 87, 26 clau *unguia : O. Sax.* clâuua, *f. a claw, hoof : Frs.* klauwe : *O. Frs.* klewe *a claw : Dut.* klaauw, *m* : *Ger.* klaue, *f. unguis, ungula : M. H. Ger.* klâ, *f* : *O. H. Ger.* klawa, kloa, *f. unguis, ungula : Dan.* klo, *m. f* : *Swed.* klo, *m* : *Icel.* kló, *f.*] DER. clawan, clawung, cleweða.

clæc-leás, clac-leás ; *adj. Free;* immunis, Cot. 104. Clacleás [clacles MS.] *free,* Hick. Thes. i. 149, 51, 57.

clæfer-wyrt, e ; *f. Clover-wort, clover;* trifolium minus :—Nim ða smalan clæfer-wyrt nioðowearde *take the netherward part of the small clover-wort,* L. M. 1, 39 ; Lchdm. ii. 102, 26.

CLÆFRE, an ; *n. f.* CLOVER ; trifolium pratense :—Ðysse wyrte man crision and óðrum naman clæfre nemneþ *a man names this herb* κίρσιον,

and by another name clover, Herb. 70; Lchdm. i. 172, 16. Clæfre nom. 172, 14. Hwîte clæfran wyrc clame *work white clover to a paste*, L. M. 1, 21; Lchdm. ii. 64, 4. Clæfre calta vel trifillon, Ælfc. Gl. 41; Som. 64, 3; Wrt. Voc. 31, 15. Nim reád clæfre *take red clover*, L. M. 3, 8; Lchdm. ii. 312, 20. Clæfran seáwes *of juice of clover*, 2, 24; Lchdm. ii. 214, 11. Nim clæfran wyrttruman *take roots of clover*, 2, 40; Lchdm. ii. 250, 12. [*Plat.* klever, klewer: *Dut.* klaver, *f*: *Ger.* klee, *m*: *M. H. Ger.* klê; *gen.* klêwes, *m*: *O. H. Ger.* klê, chlêo; *gen.* chlêwes: *Dan.* klöver, *n*: *Swed.* klöfver, *m*.]

CLÆG, es; *m?* CLAY; Samia terra, Ælfc. Gl. 56; Som. 67, 36; Wrt. Voc. 37, 26. [*Wyc.* cley: *Chauc.* clei: *Plat.* klei: *Frs.* klaey: *O. Frs.* klai: *Dut.* klei, *f*: *Kil.* kleye: *Ger.* klei, klai, *m*: *Dan.* klåg, kleg, *m. f. n*: *O. Nrs.* kleggi, *m. massa compacta*, Rask Hald. The fundamental idea is *slimy, tenacious*.]

clǽig; *def.* se clǽiga, clǽia; *adj.* CLAYEY; argillaceus:—On ða clǽian lane, *of ðære clǽian lane to the clayey lane, from the clayey lane*, Cod. Dipl. 741; A. D. 1024; Kmbl. iv. 31, 8, 9.

Clǽig-hangra, an; *m.* [clǽig = clǽg *clay*] *Clay-hanger* or *Claybury*, *Essex*:—Eádmund cyning gegaderede fyrðe and fêrde to Lundene, eal be norþan Temese, and swá ût þuruh Clǽighangran *king Edmund gathered a force and went to London, all north of the Thames, and so out through Clayhanger*, Chr. 1016; Erl. 156, 24.

CLÆMAN; *p.* de; *pp.* ed *To* CLAM, *smear, anoint*; linere:—Ic clæme lino, Ælfc. Gr. 28, 1; Som. 30, 35. Ðû wircst wununge binnan ðam arce and clæmst wiðinnan and wiðútan mid tyrwan *mansiunculas in arca facies et bitumine linies intrinsecus et extrinsecus*, Gen. 6, 14. Clæm on ðone cancer *smear it on the cancer*, L. M. 1, 44; Lchdm. ii. 110, 4: 3, 45; Lchdm. ii. 336, 22. Clæme on ðæt geswel *smear it on the swelling*, Lchdm. iii. 38, 23. [*Wyc.* clemede *smeared*: *Kil.* kleemen: *O. H. Ger.* kleimjan, chleimen: *Icel.* kleima.] DER. ge-clǽman.

clǽmende *hardening*; obfirmans, Cot. 145.

clǽmming, e; *f.* A *blotting, daubing, smearing, hardening*; litura, oblimatio, Ælfc. Gr. 47, Som. Ben. Lye.

CLǼNE, clêne; *def.* se clǽna, seó, ðæt clǽne; *comp. m.* clǽnra, *f. n.* clǽnre; *sup.* clǽnest; *adj.* I. CLEAN, *pure, clear*; mundus, purus, merus, serenus:—Ðonne ân unclǽne gást biþ adrifen of ðæm men, ðonne biþ ðæt hús clǽne *when an unclean spirit is driven out of a man, then the house is clean*, Past. 39, 1; Hat. MS. 53 a, 8. Swá swá clǽne nýten eodorcende in ðæt swêteste leóþ gehwyrfde *quasi mundum animal ruminando in carmen dulcissimum convertebat*, Bd. 4, 24; S. 598, 6: Homl. Th. i. 138, 20. Clǽne oflete, and clǽne wîn, and clǽne wæter *a pure oblation, and pure wine, and pure water*, L. Edg. C. 39; Th. ii. 252, 13. Wæs seó lyft swîðe clǽne *the air was very clear*, Chr. 1110; Erl. 243, 1. Se clǽna ôþscûfeþ scearplîce *the pure [bird] flies quickly away*, Exon. 58 a; Th. 209, 8; Ph. 167. Ðæt land ic selle Cynulfe for syxtigum mancessa clǽnes goldes *I sell the land to Cynulf for sixty mancuses of pure gold*, Cod. Dipl. 313; A. D. 883; Kmbl. ii. 111, 21. Calic on handa Drihtnes wînes [MS. win] clǽnes [MS. clǽnis] full is *calix in manu Domini vini meri plenus*, Ps. Spl. 74, 7. Forbærne hit man on clǽnum fire *let a man burn it in a pure fire*, L. Edg. C. 38; Th. ii. 252, 8: Exon. 55 a; Th. 194, 11; Az. 137: Bt. Met. Fox 12, 9; Met. 12, 5. Clǽnre heortan *mundo corde*, Ps. Spl. 23, 4. Gebærnedne hláf clǽnne seóþ on ealdum wîne *seethe pure toasted bread in old wine*, L. M. 2, 2; Lchdm. ii. 180, 26. Cyning [MS. kynincg] seeal on Drihtne clǽne blisse habban *a king shall have pure bliss in the Lord*, Ps. Th. 62, 9. Ne acyr ðú fram ðínum cnihte ðîn clǽne gesihþ *ne avertas faciem tuam a puero tuo*, 68, 17. Gewât him se hálga sêcan ðone clǽnan hám *the holy one departed to seek the pure home*, Andr. Kmbl. 1956; An. 980. Húslfatu Caldéas clǽne genámon *the Chaldeans took the clean vessels of sacrifice*, Cd. 210; Th. 260, 10; Dan. 707. Clǽnum stefnum *with pure voices*, Elen. Kmbl. 1496; El. 750. God ðone ǽrestan ǽlda cynnes of ðære clǽnestan foldan geworhte *God made the first of the race of men from the purest earth*, Exon. 44 b; Th. 151, 12; Gú. 794. II. *chaste, innocent*; castus, innoxius:—Clǽne *castus*, Ælfc. Gl. 90; Som. 74, 121; Wrt. Voc. 51, 34. Clǽne [MS. cleane] oððe heofonlíc [MS. -lice] *cælebs*, Ælfc. Gr. 9, 49; Som. 13, 13. Ðú byst clǽne *absque peccato eris*, Deut. 23, 22: Chr. 1066; Erl. 198, 4; Edw. 23. Gif heó clǽne sýj *if she be innocent*, L. Ath. v. § 1, 1; Th. i. 228, 17: L. Eth. iii. 7; Th. i. 296, 9. On háligra clǽnre cyricean *in ecclesia sanctorum*, Ps. Th. 149, 1. Ic onfêng fǽmnan clǽne *I received a chaste damsel*, Exon. 10 b; Th. 12, 18; Cri. 187. Ðone clǽnan sacerd *the pure priest*, 9 b; Th. 9, 18; Cri. 136. Beón ða ôðre clǽne *let the others be innocent*, Gen. 44, 10. Sint spræcu Drihtnes spræcu clǽne *sunt eloquia Domini eloquia casta*, Ps. Lamb. 11, 7. Seó clǽneste cwên *the most chaste woman*, Exon. 11 b; Th. 17, 26; Cri. 276. [*Piers P.* clene: *Laym.* clæne, clene, clane: *Orm.* cleane: *Plat.* kleen *parvus*: *Frs.* klien *parvus*: *O. Frs.* klen *parvus*: *Dut.* kleen *little*: *Kil.* kleyn *exilis, minutus*: *Ger.* klein *parvus*: *M. H. Ger.* kleine *subtilis, parvus*: *O. H. Ger.* kleini *subtilis*: *Dan.* klein: *Swed.* klen *thin, slight*: *Icel.* klénn *snug, tiny*.] DER. hyge-clǽne, un-.

clǽne, clâne, clêne; *adv.* CLEAN, *entirely*; penitus, omnino:—Ne rîpe

ge ðæt land tô clǽne *reap not the land too clean*, Lev. 23, 22: Ors. 4, 1; Bos. 76, 30: Bd. 3, 10; S. 534, 35. Clǽne biþ beorhtast nesta bǽle forgrunden *the brightest of nests is entirely destroyed by the fire*, Exon. 59 a; Th. 213, 18; Ph. 226: Ps. Th. 88, 37. Ðæt mín cynn clǽne [MS. clane] gewíte *that my race be clean gone*, Cod. Dipl. 235; A. D. 835; Kmbl. i. 311,'16. Clêne *entirely*, Cd. 213; Th. 265, 14; Sat. 7.

clǽn-georn; *adj. Yearning after purity*; puritatis amans:—Clǽngeorn and cystig *yearning after purity and bountiful*, Exon. 128 a; Th. 492, 25; Rä. 81, 21. Ne mágon ná swilce men macian wununge ðam clǽngeornan Gode on clǽnre heortan *no such men can make a dwelling in a pure heart for a God desirous of purity*, Basil. admn. 7; Norm. 48, 19.

clǽn-heort; *def.* se clǽn-heorta; *adj. Clean-hearted, pure in heart*; mundo corde:—Eádige synd ða clǽnheortan, forðamðe hî God geseóþ *beati mundo corde, quoniam ipsi Deum videbunt*, Mt. Bos. 5, 8: Homl. Th. ii. 580, 33.

clǽn-lîc; *adj. Pure*, CLEANLY; purus, mundus:—Mid clǽnlîcre lufe *with pure love*, Bt. 21; Fox 74, 38: Bt. Met. Fox 11, 183; Met. 11, 92.

clǽn-lîce; *adv. Purely, cleanly*; purè, L. Ælfc. C. 36; Th. ii. 360, 25.

clǽnnes, -ness, -niss, -nyss, e; *f.* CLEANNESS, *chastity, purity, modesty*; puritas, castimonia:—Clǽnnesse riht *castimoniæ jura*, Bd. 2, 5; S. 507, 1. Heó on clǽnnesse Gode þeówode *she served God in chastity*, 4, 9; S. 576, 21: L. Eth. v. 9; Th. i. 306, 20. Mid clǽnnesse *with purity*, L. Eth. v. 7; Th. i. 306, 15: vi. 4; Th. i. 316, 2: Ps. Th. 88, 37. Ðæt he healdan wille his clǽnnisse *that he will keep his chastity*, L. Eth. v. 6; Th. i. 306, 8. Þurh ða heálícan clǽnnysse *through exalted purity*, Homl. Th. i. 346, 1: L. Edg. S. 1; Th. i. 272, 16: Ps. Spl. 17, 22, 26. DER. un-clǽnnes.

clǽnsend, es; *m.* [*part. of* clǽnsan = clǽnsian] A *cleanser*; purgator. DER. eár-clǽnsend.

clǽnsere, es; *m.* A *cleanser, purifier, priest*; purgator, Som. Ben. Lye.

clǽnsian, clênsian, to clǽnsianne; *part.* clǽnsiende; *p.* ode, ade; *pp.* od, ad [clǽne, *pure*] *To* CLEANSE, *purify, chasten, clear oneself*; mundare, purgare, castigare, se liberare:—Gif man eard wille clǽnsian *if a man wishes to cleanse the land*, L. Eth. ix. 40; Th. i. 348, 25: L. C. S. 7; Th. i. 380, 7. Sió wamb biþ tô clǽnsianne *the stomach is to be cleansed*, L. M. 2, 46; Lchdm. ii. 260, 12. Clǽnsie man ða þeóde *let a man cleanse the people*, L. E. G. 11; Th. i. 174, 2. Hî tiliaþ hî selfe to clǽnsianne mid ðý wôpe *they strive to purify themselves with mourning*, Past. 54; Hat. MS. Ðis wæter cristnaþ and clǽnsaþ cwicra menigo *this water cristeneth and purifieth a multitude of men*, Salm. Kmbl. 791; Sal. 395. Heó ða iungran lǽrde and clǽnsade ge mid hire lâre ge mid lifes bysne *she taught and purified the younger ones both by her doctrine and by the example of her life*, Bd. 4, 9; S. 576, 23. Clǽnsa me *munda me*, Ps. Spl. 18, 13. Clǽnsiende clǽnsode me Drihten *castigans castigavit me Dominus*, Ps. Spl. 117, 18. Gif he mid ða ádle clǽnsad beón sceolde *if he must be chastened by disease*, Bd. 4, 31; S. 610, 26. Gif hwá þeóf clǽnsian wylle *if any one will clear a thief*, L. Eth. iii. 7; Th. i. 296, 7. Preóst hine clǽnsie sylfes sôþe *let a priest clear himself by his own truth*, L. Wih. 18; Th. i. 40, 14, 16: 19; Th. i. 40, 17: 20; Th. i. 40, 19: L. Eth. ii. 8; Th. i. 288, 19: ii. 9; Th. i. 290, 10. Hine geréfa clǽnsie *let the reeve clear him*, L. Wih. 22; Th. i. 42, 4. [*Wyc.* Piers P. clense: *Orm.* clennsenn.] DER. a-clǽnsian, be-, ge-, un-: un-geclǽnsod.

clǽnsnian, clǽnsnigan; *p.* ode; *pp.* od *To cleanse, clear oneself*; se purgare:—Clǽnsnaþ [MS. clǽnsnoþ] he ðone he clears him, L. Eth. ii. 8; Th. i. 288, 20. Clǽnsnige hine sylfne *let him clear himself*, ii. 9; Th. i. 290, 11. Bûton he frînd hæbbe ðe hine clǽnsnian *unless he have friends who may clear him*, ii. 9; Th. i. 290, 13. v. clǽnsian.

clǽnsung, e; *f.* A CLEANSING, *purifying, chastening, expiation, chastity*; emundatio, purificatio, castigatio, expiatio, castitas:—Ðú towurpe hine fram clǽnsunge *destruxisti eum ab emundatione*, Ps. Lamb. 88, 45: Mk. Bos. 1, 44. Wæs Rômâna gewuna ðæt hî clǽnsunge þweáles and bæþes sôhton *Romanorum usus fuit lavacri purificationem quærere*, Bd. 1, 27; S. 495, 15. Wæs he mid clǽnsunge forhæfednesse weorþ and mǽre *erat abstinentiæ castigatione insignis*, 4, 28; S. 606, 39. Biþ heó fremiende to his clǽnsunge *erit in expiationem ejus proficiens*, Lev. 1, 4. Ðe belumpon to ðære mynsterlícan clǽnsunge *quæ monasticæ castitatis erant*, Bd. 5, 19; S. 637, 14. DER. ge-clǽnsung, mynster-, un-.

clǽppettan; *p.* tte; *pp.* ted *To* palpitate, *have a palpitation*; palpitare:—Gif sino clǽppette *if a sinew have palpitation*, L. M. 1, 26; Lchdm. ii. 68, 8. v. clappan.

clǽppetung, e; *f. The pulse*; pulsus, Ælfc. Gl. 76; Som. 71, 109; Wrt. Voc. 45, 15. Ǽdra clǽppetung *the pulse of the veins*, L. M. 2, 46; Lchdm. ii. 258, 16.

clǽsnian; *p.* ode; *pp.* od *To cleanse*; mundare, purgare:—Sceal mon clǽsnian ða yflan wǽtan *one must cleanse the evil humours*, L. M. 2, 30; Lchdm. ii. 228, 14, note 2: 2, 32; Lchdm. ii. 234, 25, note 2: 2, 35; Lchdm. ii. 240, 23, note 4: 2, 48; Lchdm. ii. 262, 17, note 2. v. clǽnsian.

clǽþ *a cloth*:—Dô on clǽþ *put on a cloth*, L. M. 2, 47; Lchdm. ii. 260, 28. v. clâþ.

clǽweða *a clawing, scratching*, Past. 11, 6; MS. Oth. v. cleweða.

olǎf, *pl.* clifon *clave, adhered; p. of* clífan.

clam; *gen.* clammes; *m. n?*   I. *what is clammy, mud, clay; malagma, lutum* :—Wyrc swā to clame *so work to clam* [*a clammy substance*], Herb. 2, 11; Lchdm. i. 84, 3. Mid heardum weorcum clames *operibus duris luti*, Ex. 1, 14.   II. *a bandage, what holds or retains, as a chain, net, fold, prison;* vinculum :—He ðē clamme belegde *he loaded thee with a chain*, Andr. Kmbl. 2386; An. 1194. Of ðǽm clammum *with tnose chains*, Bt. Met. Fox 1, 165; Met. 1, 83: Exon. 112 a; Th. 429, 30; Rä. 43, 12. Gebindan ǽrenum clammum *to bind with brazen bands*, Cd. 200; Th. 248, 28; Dan. 520: Beo. Th. 2675; B. 1335: 1931; B. 963. v. clom; *gen.* clommes.

clǎm *with claws*, Exon. 59 b; Th. 217, 8; Ph. 277; *dat. of* clǎ.

clamb, clomm, *pl.* clumbon *climbed; p. of* climan, climban.

clǎne *clean, clear*, L. M. 2, 65; Lchdm. ii. 296, 6. v. clǽne.

clang *shrunk*, Andr. Kmbl. 2522; An. 1262; *p. of* clingan.

clappan *to* CLAP, *move, palpitate;* palpitare, Som. Ben. Lye.

CLÁTE, an; *f. The herb* CLOT-*bur, a bur that sticks to clothes, burdock, goose-grass, clivers;* philanthropos = φιλάνθρωπος, lappa, arctium lappa, galium aparine, Lin :—Ðás wyrte man *philanthropos* nemneþ, ðæt ys on úre geþeóde menlufigende, forðȳ heó wyle hrædlíce to ðam men geclyfian : ða man eác óðrum naman clǎte nemneþ *this herb is called philanthropos, that is in our language men-loving, because it will readily cleave to a man: it is also named by another name clivers*, Herb. 174, 1; Lchdm. i. 306, 2–5: Ælfc. Gl. 40; Som. 63, 105; Wrt. Voc. 30, 53: 41; Som. 63, 108; Wrt. Voc. 30, 56: 66, 67. Clǎte *lappa*, Wrt. Voc. 75, 79, 41: Ælfc. Gl. 40; Som. 63, 91; Wrt. Voc. 30, 41. Wið ceolan swile clǎtan wyl on ealaþ *for swelling of throat boil burdock in ale*, L. M. 1, 12; Lchdm. ii. 56, 3: 1, 45; Lchdm. ii. 110, 13: 2, 53; Lchdm. ii. 274, 3. Nim ða smalan clǎtan *take the small burdock*, 1, 39; Lchdm. ii. 100, 23. Genim doccan oððe clǎtan, ða ðe swimman wolde *take dock or clote, such as would swim*, 1, 50; Lchdm. ii. 122, 22. [*Wyc.* clote, cloote : *Chauc.* clote-lefe *a leaf of the clot-bur: Ger. M. H. Ger.* klette. *f.: O. H. Ger.* kletta, kledda, *f.*]

CLÁÞ; *gen.* clǎðes; *m.* CLOTH; *pannus:* in the plural, *clothes;* vestimenta :—Ne dēþ nān man niwes clǎðes scyp on eald reáf *nemo immittit commissuram panni rudis in vestimentum vetus*, Mt. Bos. 9, 16. Heó ða moldan on clǎðe bewand *she wound the mould in a cloth*, Bd. 3, 11; S. 536, 8. Dó on clǎþ *put on a cloth*, L. M. 2, 2; Lchdm. ii. 180, 5, 10, 28: 2, 47; Lchdm. ii. 262, 2. Awring þurh clǎþ *wring through a cloth*, 2, 53; Lchdm. ii. 274, 7. Hig bewundon hine mid línenum clǎðe *ligaverunt illud linteis*, Jn. Bos. 19, 40. Ðæt is heora bíwist; wǽpnu, and mete, and ealo, and clǎðas *this is their provision; weapons, and meat, and ale and clothes*, Bt. 17; Fox 60, 5. Him wyrþ oftohen ðara clǎða *he is deprived of the clothes*, 37, 1; Fox 186, 14: Bt. Met. Fox 25, 46; Met. 25, 23. Of ðínum clǎðum *a vestimentis tuis*, Ps. Th. 44, 10: Exon. 18 b; Th. 45, 27; Cri. 725: 28 b; Th. 87, 12; Cri. 1424. Ruben tær his clǎðas *Reuben tore his clothes*, Gen. 37, 29: Bt. 37, 1; Fox 186, 10. [*R. Glouc.* cloth : *Laym.* claðe, cloð, claed : *Orm.* claþ : *Scot.* claith, clayth : *Plat.* kleed : *Frs.* klaed : *O. Frs.* klath, klad, kleth, *n : Dut. Kil.* kleed, *n : Ger.* kleid, *n : M. H. Ger.* kleit, *n : Dan. Swed.* kläde, *n : Icel.* klæði, *n.*] DER. bearm-clǎþ, cild-, feax-, heáfod-, sǎr-, swát-.

clǎþ-scear *a pair of shears.* v. scear IV.

clatrung, e; *f. Anything that makes a clattering, a drum, rattle;* crepitaculum, Cot. 51.

clauster; *gen.* claustres; *n. An inclosed place, a cloister;* claustrum :—Eálá ge cildra, gǎþ ūt, būtan hygeleáste, to claustre, oððe to leorninge *O vos pueri, egredimini, sine scurrilitate, in claustrum, vel in gymnasium*, Coll. Monast. Th. 36, 9. Fæsten *vel* clauster *claustrum*, Ælfc. Gl. 109; Som. 79, 15; Wrt. Voc. 58, 56. v. clūstor.

clawan, ic clawe; *p.* ede; *pp.* ed [clǎ *a nail, claw*] *To* CLAW; scalpere :—Ic clawe *scalpo*, Ælfc. Gr. 28, 4; Som. 31, 20. [*Dut.* klaauwen : *Ger.* klauen : *O. H. Ger.* klawjan : *Dan.* klöe : *Swed.* klå : *Icel.* klá *to scratch*, klóask *to fight with claws*.]

clawu *a nail, claw, hoof*, Ælfc. Gr. 9, 28; Som. 11, 46; Wrt. Voc. 71, 66. v. clǎ.

clawung, e; *f.* [clǎ *a claw*] *A pain, the gripes;* tormina :—Lǽcedómas wið clawunga *leechdoms for the gripes*, L. M. cont. 2, 32; Lchdm. ii. 164, 16: 2, 32; Lchdm. ii. 236, 1.

cleacian; *p.* ode; *pp.* od *To go nimbly, hurry;* festinare, trepidare :—He cleacode swíðe earhlíce to porte *he hurried very timidly to town;* in via totus trepidabat, M. H. 115 a.

cleadur *a clatter, drum, rattle;* crepitaculum, Som. Ben. Lye.

cleáf, *pl.* clufon *clove, separated; p. of* cleófan.

cleáfa, an; *m. A cellar;* cellarium :—Hwá gefylþ cleáfan his *quis replet cellaria sua?* Coll. Monast. Th. 28, 17. v. cleófa.

Clede-mūþa, an; *m.* [*the mouth of the river Cleddy*] GLADMOUTH, CLEDMOUTH, *South Wales* :—Hér Eádweard cyning getimbrede ða burh æt Cledemūþan in A.D. 921, *king Edward built the burgh at Cledmouth*, Chr. 921; Th. 194, 1–3, col. 3; Th. 195, 1–3, col. 1.

olemman; *p.* de; *pp.* ed [clam II. *a chain*] *To fetter, bind, inclose;* vincire, includere. DER. be-clemman.

clencan; *p.* te; *pp.* ed *To* CLINCH, *hold fast;* prehendere, prensare. v. be-clencan, *Supl.*

clǽne *clean, pure, clear*, Ps. C. 50, 88; Ps. Grn. ii. 278, 88: Chr. 1110; Erl. 243, 1. v. clǽne; *adj.*

clǽne *cleanly, entirely;* penitus :—Deópne ymblyt clǽne ymbhaldeþ meotod *the Lord entirely upholdeth the deep expanse*, Cd. 213; Th. 265, 14. v. clǽne; *adv.*

clengan; *p.* de; *pp.* ed *To exhilarate;* exhilarare :—Dreám clengeþ *joy exhilarates*, Exon. 107 b; Th. 411, 6; Rä. 29, 8.

clǽnsian *to cleanse, clear oneself*, L. Wih. 22; Th. i. 42, 4. v. clǽnsian.

cleó *a claw, hoof*, Ps. Th. 68, 32. v. clǎ.

cleof *a cliff, rock*, Exon. 101 b; Th. 384, 15; Rä. 4, 28. v. clif.

cleófa, cleáfa, cliófa, an; *m. That which is cloven, a cleft, chasm, den, cell, chamber;* cubile, cellarium, cubiculum :—On heora cleófum oððe holum híg beóþ gelogode *in cubilibus suis collocabuntur*, Ps. Lamb. 103, 22. Unriht he byþ smeágende on his cliófan *iniquitatem meditatus est in cubili suo*, Ps. Th. 35, 3. Sinewealt cleófa *vel portic absida*, Ælfc. Gl. 108; Som. 78, 122; Wrt. Voc. 58, 34. Deós sweoster wæs ūtgangende of hire cleófan *hæc soror egressa est de cubiculo*, Bd. 4, 9; S. 576, 31. DER. clūstor-cleófa, ferhþ-, hord-, in-, nȳd-. v. clyfa.

CLEÓFAN, ic cleófe, ðū clýfst, he clýfþ, *pl.* cleófaþ; *p.* cleáf, *pl.* clufon; *pp.* clofen *To* CLEAVE, *separate, split;* findere, dissecare :—Cleófan *secare*, Glos. Prudent Recd. 149, 54: *scindere*, 150, 9. Bordweall clufon aforan Eádweardes *Edward's sons clove the board-wall*, Chr. 937; Th. 200, 38, col. 3; Ædelst. 5. Clufon, Byrht. Th. 140, 4; By. 283. [*Piers P.* cleven: *Chauc.* cloven, *pp : Orm.* clofenn, *pp : Plat.* klöwen, klöven : *O. Sax.* klioðan: *Dut.* klieven, klooven: *Ger.* klieben: *M. H. Ger.* kliuben, klieben: *O. H. Ger.* kliuban: *Dan.* klöve: *Swed.* klyfva: *Icel.* kljúfa.] DER. to-cleófan: cleófa, cleáfa, clýfa, clífa, bed-, clūstor-, ferhþ-, gebed-, hord-, in-, nȳd-.

Cleofes hoo *Cliff, near Rochester*, Chr. 822; Th. 110, 14, col. 3. v. Clofes hoo.

cleofian, he cleofaþ, *pl.* cleofiaþ; *p.* ode; *pp.* od *To cleave, adhere, stick;* adhærere :—Ða ðe him on cleófiaþ *those who cleave to him*, Exon. 97 b; Th. 364, 20; Wal. 73. v. clifian.

cleopian; *p.* ode; *pp.* od *To cry, call;* clamare :—Ic nū wille geornlíce to Gode cleopian *I will now earnestly call upon God*, Bt. 3, 4; Fox 6, 28: Andr. Kmbl. 2796; An. 1400. Ic cleopode to ðē *clamavi ad te*, Ps. Th. 118, 146, 147. v. clypian.

cleopigend, cleopend, es; *m. A vowel;* vocalis, Som. Ben. Lye.

cleopung, e; *f. A cry;* clamor, Mt. Rush. Stv. 25, 6. v. clypung.

cleót *a clout*, Som. Ben. Lye. v. clūt.

cleóða, an; *m. A plaster, salve, poultice;* malagma :—Ðone hālwendan cleóðan *malagma*, Mone B. 2976. v. clíða.

cleowen *a clew, ball of thread or yarn, ball*, Wrt. Voc. 59, 37: Exon. 59 a; Th. 213, 17; Ph. 226. v. cliwen.

clepian; *p.* ode; *pp.* od *To cry, call;* clamare, vocare :—Ic clepode forðanðe ðū gehȳrdest me eálá ðū God *ego clamavi quoniam exaudisti me Deus*, Ps. Lamb. 16, 6. v. clypian.

clepung, e; *f. A calling;* vocatio, clamor :—Se nān clepunge ðǽrto nā hafde māre *he had not any more calling thereto*, Chr. 1129; Erl. 258, 9. Clepung mín on ansȳne oððe on gesihþe his ineóde to his eárum *clamor meus in conspectu ejus introivit in aures ejus*, Ps. Lamb. 17, 7. v. clypung.

clerc, cleric, clerec, es; *m.* [*Lat.* clericus = κληρικός *belonging to the clergy, clerical*] A CLERK, *clergyman, generally a deacon* or *priest;* clericus :—Gregorius wæs clerc *Gregory was a priest*, Chr. 1129; Erl. 258, 25 : 1123; Erl. 250, 20. He drāf ūt ða clerca of ðe biscopríce *he drove the clergy out of the bishopric*, 963; Erl. 121, 13. Preóst oððe cleric *clericus*, Wrt. Voc. 71, 77. We lǽraþ ðæt preósta gehwilc to sinoþe hæbbe his cleric *we enjoin that every priest at a synod have his deacon*, L. Edg. C. 4; Th. ii. 244, 14. Hí wǽron ealle ðæs cynges clerecas *they were all the king's clergy*, Chr. 1085; Erl. 218, 22.

clerc-hād, cleric-hād, cleroc-hād, es; *m. The clerical office, priesthood;* sacerdotium, clericatus :—Clerchādes man *a man of the clerical order*, Chr. 1123; Erl. 250, 11. Clerichād *clericatus*, C. R. Ben. 60. Clerochād *clericatus*, Cot. 45.

cleweða, clǽweða, an; *m. A clawing, scratching;* scalpturigo, scalpurigo :—Se giecþa [gicþa MS. Cot.] biþ swíðe unsǎr, and se cleweða [MS. Oth. clǽweða] biþ swíðe rów, and ðeáh-hwǽðere gif him mon tó longe fylgþ, he wundaþ, and wund sǎraþ *the itch is very free from pain, and the scratching is very comfortable, and yet if it be kept up too long, it produces a wound, and the wound is painful*, Past. 11, 6; Hat. MS. 15 b, 23. DER. clawu, clǎ *a nail, claw*.

CLIBBOR; *adj.* [clifian *to cleave, adhere*] *Sticky, adhesive;* tenax :—Weá biþ wundrum clibbor *grief is wonderfully adhesive*, Menol. Fox 485; Gn. C. 13. [*M. H. Ger.* klēber: *O. H. Ger.* klebar *adhesive*.]

cliewe *a clew*, Som. Ben. Lye. v. clywe.

CLIF, clyf, cleof, es; *n. A* CLIFF, *rock, steep descent, promon ory;*

clivus, rupes, promontorium :—Ða Iudéi lǽddon Crist to ánum clife, and woldon hine niðerascúfan *the Jews led Christ to a cliff, and would cast him down,* Homl. Th. ii. 236, 33. Æt Eádwines clife *at Edwin's cliff,* Chr. 761; Th. 89, 24, col. 1. Ðæt hí ne hlipen on ðæt scorene clif *that they leap not down the abrupt cliff,* Past. 33, 1; Hat. MS. 41 a, 9. Be clifum on the ʼcliffs, Exon. 81 b; Th. 306, 15; Seef. 8. Ðæt hie Geáta clifu ongitan meahton *that they might perceive the cliffs of the Gauts,* Beo. Th. 3826; B. 1911. Ofer cald cleofu *over the cold cliffs,* Andr. Kmbl. 619; An. 310: Exon. 101 b; Th. 384, 15; Rä. 4, 28. Ðú hluttor lǽtest wæter of clife clǽnum *thou lettest forth clear waters from the pure rock,* Exon. 55 a; Th. 194, 11; Az. 137: Bt. Met. Fox 5, 25; Met. 5, 13. Se ðe gecyrde clyf on w͵yllan wætera *qui convertit rupem in fontes aquarum,* Ps. Spl. M. C. 113, 8. God clifu cyrreþ on wæteres wellan *God turneth rocks into wells of water,* Ps. Th. 113, 8. Clif *promontorium,* Ælfc. Gl. 67; Som. 69, 117; Wrt. Voc. 41, 67. Nílus seó eá, hyre ǽwylme, is neáh ðæm clife ðære Reádan Sǽs *the spring of the river Nile is near the promontory of the Red Sea,* Ors. 1, 1; Bos. .7, 19, 29. [*O. Sax.* klif, *n. a rock :* *Dut.* klip, *f. a rock, cliff :* *Kil.* kleppe, klippe *rupes, petra :* *Ger.* klippe, *f. rupes :* *O. H. Ger.* clep *promontorium :* *Dan.* klippe, *m. f. a rock, cliff :* *Swed.* klippa, *f :* *Icel.* klif, *n. a cliff.*] DER. brim-clif, ég-, heáh-, holm-, stán-, weal-.

clífa, an; *m. A den, cave ;* cubile, spelunca, Bd. 3, 23; S. 554, 22. v. clýfa.

CLÍFAN, ic clífe, ðú clífest, clífst, he clífeþ, clífþ, *pl.* clífaþ; *p.* cláf, *pl.* clifon; *pp.* clifen To CLEAVE, *adhere ;* adhǽrere. [*Piers P.* clyven : *Plat.* kleeven : *O. Sax.* bi-klíban : *Frs.* be-klieuwen : *O. Frs.* bi-kliva : *M. H. Ger.* klíben : *O. H. Ger.* klíban : *Dan.* klæbe : *Swed.* klibba.] DER. óþ-clífan ; clifian, cleofian, cliofian.

clife, an; *f.* I. *the greater burdock ;* arctium lappa :—Dó clifan *use burdock,* L. M. 1, 67; Lchdm. ii. 142, 16. II. *the small burdock :*—Seó smæle clife *the small burdock,* CLIVERS; galium aparine, L. M. 1, 50; Lchdm. ii. 124, 2. DER. gar-clife.

clifer; *gen.* clifres; *m. A claw, talon ;* ungula :—Clifras [MS. cifras] *ungulas,* Glos. Prudent. Recd. 150, 37. Clífra *ungularum,* 149, 7. DER. clifrian.

clif-hlép, clif-hlýp *right down, under foot ;* pessum, Cot. 155, Som. Ben. Lye.

clifian, cleofian, cliofian, clyfian; *p.* ode; *pp.* od To *cleave, adhere ;* adhǽrere :—Hí willaþ clifian on ðǽm monnum *they will cleave to him,* 16, 3; Fox 56, 10: L. M. 1, 2; Lchdm. ii. 38, 20. His flǽsces lima clifaþ ǽlc on óðrum *each of the limbs of his flesh cleaves to another,* Past. 47; Hat. MS. Ðín tunge clifaþ to ðínum goman *thy tongue cleaveth to thy gums,* Homl. Th. ii. 530, 28. To ðære lifre clifiaþ ad-hǽrent jecori, Lev. 1, 8. Ðæt dust, ðæt of eówre ceastre on úrum fótum clifode, we drígeaþ on eów *pulverem, qui adhæsit nobis de civitate vestra, extergimus in vos,* Lk. Bos. 10, 11. [*Wyc.* cleuyde *cleaved :* *Laym.* cleouieþ *cleaveth :* *O. Sax.* klibón : *Dut.* kleeven : *Ger.* kleben, kleiben : *O. H. Ger.* kleben, klebjan.] DER. æt-clifian, ge-, on-, to-, to-ge-.

clifig, clifiht; *adj.* CLIFFY, *steep ;* clivosus, Ælfc. Gl. 9; Som. 56, 120; Wrt. Voc. 19, 4: Cot. 34: 209.

clifon *cleaved, adhered ;* adhǽserunt; *p. pl. of* clífan.

clifrian, ic clifrige; *p.* ode; *pp.* od [clifer *a claw*] To *claw, scratch ;* scabere :—Ic clifrige *scabo,* Ælfc. Gr. 28, 6; Som. 32, 25. DER. to-clifrian.

clif-stán, es; *m. A rough stone, rock ;* cautes :—Clifstánas *cautes,* Cot. 44.

clif-wyrt, e; *f. Maiden-hair, water-wort, fox-glove ;* agrimonia :—Clifwyrt, sume men hataþ foxes clife, sume eá-wyrt *cliff-wort, some men call fox-glove, some water-wort,* L. M. 1, 15; Lchdm. ii. 58, 3.

climan, ðú climst, he climþ; *p.* clomm *to climb.* v. climban and ofer-climm.

CLIMBAN, ic climbe, ðú climst, he climþ, *pl.* climbaþ; *p.* clamb, *pl.* clumbon; *pp.* clumben; *v. a.* To CLIMB; scandere, ascendere :—Clumbon [MS. Clumben] upp to ðe stépel *climbed up to the steeple,* Chr. 1070; Erl. 209, 9. Clumbon [MS. Clumben] upp to ðe halge róde *climbed up to the holy cross,* Erl. 209, 6. [*Laym.* climben *to climb,* he climbeth; *p.* cluombe, *pl.* clumben; *pp.* iclumben : *Orm.* climbenn *to climb :* *Dut.* klimmen *scandere :* *O. H. Ger.* klimban : *M. H. Ger.* klimmen, klam, klummen, geklummen : *Sansk.* kram *incedere, ascendere.*] DER. ofer-climan, ofer-climban : climan, clymmian.

climmian *to climb.* v. clymmian, climan, climban.

climst, he climþ *climbest, climbs ;* 2nd *and* 3rd *pers. pres. of* climan, climban.

CLINGAN, ic clinge, ðú clingst, he clingþ, *pl.* clingaþ; *p.* clang, *pl.* clungon; *pp.* clungen, geclungen. I. *to wither, pine, to* CLING [*in this sense, rarely used in English*] *or shrink up ;* se contrahere, marcescere :—Clang wæteres þrym ofer eástreámas : ís brycgade blǽce brimráde *the glory of water shrank over river streams : ice bridged a pale water-road,* Andr. Kmbl. 2522; An. 1262. Ic clinge *marcesco,* Ælfc. Gr. 35; Som. 38, 7. [*Piers P.* clyngen *to shrink, wither, pine.*] v. for-clingan,

ge-clungen. II. *to* CLING, *stick close ;* circumcludere, includere. v. be-clingan.

cliof *a cliff, rock, pointed rock, crag ;* cautes, Cot. 30. v. clif.

cliófa *a den, chamber,* Ps. Th. 35, 3. v. cleófa.

cliofian, he cliofaþ, *pl.* cliofiaþ; *p.* ode; *pp.* od To *cleave ;* adhǽrere :—Hí willaþ cliofian on ðǽm monnum *they will cleave to the men,* Bt. 16, 3; Fox 54, 19, note 9. v. clifian.

cliofung, e; *f. A* CLEAVING; sectio :—Cliofung *sectio,* Ælfc. Gl. 62; Som. 68, 83; Wrt. Voc. 39, 66.

cliopian; *part.* clioppende; *p.* ode; *pp.* od To *cry, call ;* clamare :—Se Hǽland ongann cliopian [MS. cliopia] *the Saviour began to cry,* Mk. Skt. Lind. 10, 47. Clioppende, 9, 36: 15, 39: Mt. Kmbl. Lind. 14, 26. v. clypian, clipian.

cliowen *a clew, ball,* Mone B. 1662. v. cliwen.

clipian, clipigan, *pl.* clipiaþ; *p.* ode; *pp.* od To *make a vocal sound, call, address, invoke ;* vocare, alloqui :—We clipiaþ to ǽlcum þinge *we address everything,* Ælfc. Gr. 7; Som. 6, 25. v. clypian, clipigendlíc.

clipigendlíc; *adj.* I. *calling, vocative ;* vocativus :—Vocativus is clipigendlíc oððe gecígendlíc : mid ðam casu we clipiaþ to ǽlcum þinge, Eálá ðú man cum hider *O! homo veni huc :* Eálá ðú man sprec to me *O! homo loquere ad me :* Eálá ðú láreów tǽce me sum þing *O! magister doce me aliquid : vocative is calling or invoking : with this case we address everything, as*—O! thou man come hither : O! thou man speak to me : O! thou master teach me something, Ælfc. Gr. 7; Som. 6, 24–27. II. *making a vocal sound ;* vocalis. v. clypiendlíc, clypigendlíc.

clipur, es; *m. A* CLAPPER *of a bell ;* tintinnabuli *vel* campanæ malleus :—Se bend ðe se clipur ys mid gewriðen, ys swylce hyt sý sum gemetegung ðæt ðære tungan clipur mæge styrian, and ða lippan æt-hwega beátan. Sóþlíce mid ðæs rápes æt-hríne se bend styraþ ðone [MS. ðæne] clipur *the band with which the clapper is tied, is as it were a method for moving the clapper of the tongue, and beating more or less the lips. So with the touch of the rope the band moves the clapper,* Wanl. Catal. 109, col. 2, 16–20. [*Dut.* klepel, *f : M. H. Ger.* klepfel, *m. tubillus ;* klepfer, *m. clapper.*]

cliroc, es; *m. A clerk, priest ;* clericus :—Cliroc hine clǽnsie *let a clerk clear himself,* L. Wih. 19; Th. i. 40, 17. v. clerc.

Clistún, es; *m.* CLIST *or* CLYST, *near Exeter, Devon,* Chr. 1001; Gib. 132, 16; Ing. 175, 7. v. Glistún.

clite, an; *f. The herb colt's foot ;* tussilago :—Genim ða langan clitan [MS. lancge cliton] *take the long colt's foot,* Lchdm. iii. 22, 16.

clíða, clýða, an; *m. A plaster, salve, poultice ;* emplastrum, malagma = μάλαγμα :—Se wítega Isaias worhte ðam cyninge Ezechie clíðan to his dolge *the prophet Isaiah made for king Hezekiah a plaster for his sore,* Homl. Th. i. 476, 1. Clíða *malagma,* Wrt. Voc. 74, 9 : Ælfc. Gr. 9, 1; Som. 8, 22. Man sceal him wyrcean clíðan toföran his heáfde *one must make him a poultice for his forehead,* Lchdm. iii. 8, 13, 16. Swylce ðǽr clýða togelǽd wǽre *as if a poultice were laid there,* Herb. 51, 2; Lchdm. i. 154, 18. Ðyssa wyrta genim ða læssan, wyrc to clýðan *take the lesser of these herbs, make it into a poultice,* 143, 5; Lchdm. i. 266, 15 : 173, 4; Lchdm. i. 304, 15. Genim ðyssa wyrta wyrttruman, gecnucude mid ele, and mid hwǽtenan meluwe, and mid sápan, ðam gemete ðe ðú clýðan wyrce *take roots of these herbs, pounded with oil, and with wheaten meal, and with soap, in the manner in which thou wouldst make a poultice,* 184, 4; Lchdm. i. 322, 14 : 130, 1; Lchdm. i. 240, 21 : 125; Lchdm. i. 236, 21.

cliwen, clywen, cleowen, cliowen, es; *n.* [cliwe = clywe] *A clew, anything that is globular, a ball of thread, ball ;* glomus, globus :—Cliwen *glomus,* Wrt. Voc. 66, 18 : 82, 8 : 282, 1. Clywen *glomus,* Ælfc. Gl. 28; Som. 61, 5; Wrt. Voc. 26, 4. Cleowen *glomer, globellum,* Ælfc. Gl. 111; Som. 79, 68; Wrt. Voc. 59, 37. Án cliwen gódes nettgernes *one ball of good net-yarn,* Cod. Dipl. Apndx. 461; A. D. 956; Kmbl. iii. 451, 7. Cliwenes *globi,* Mone B. 560. Mintan wel getrifulade meng wið hunig, wyrc to lytlum cliwene *mingle mint, well triturated, with honey, make it into a little ball,* L. M. 1, 48; Lchdm. ii. 122, 11. Ða ýslan onginnaþ lúcan togædere geclungen to cleowenne *the ashes begin to combine together shrunk up into a ball,* Exon. 59 a; Th. 213, 17; Ph. 226. Arǽfaþ ðæt cliwen ðære twífaldan heortan *unravels the clew of the double heart,* Past. 35, 5; Hat. MS. 46 b, 2. Men gesáwon scínan æt his hnolle swilce fýren clywen *men saw shining on his crown as it were a fiery circlet,* Homl. Th. ii. 514, 2. Cliwene *glomere,* Mone B. 3713. Cleóne [= cleowene] *glomere,* 526. Cliowena *globos,* 1662.

CLOCCIAN; *p.* ode; *pp.* od To CLUCK, *sigh ;* glocire, glocitare, singultire, bombum *sive* sonitum edere :—Ðeáh seó bródige henn sárlíce cloccige *though the brooding hen sorely cluck,* Bridf. 76. [*Scot.* clock : *Plat.* klukken : *Dut.* klokken : *Kil.* klocken : *Ger. M. H. Ger.* klucken, glucken : *Dan.* klukke : *Swed.* klokka, klukka : *Icel.* klökkva : *Lat.* glocíre : *Grk.* κλωσσω.]

clod-hamer, es; *m? A field-fare ?* turdus piláris ?—Clodhamer *vel* feldefare *a field-fare ;* scorellus ? [turdus piláris ? Lin.], Wrt. Voc. 63, 27.

Clod-hangra, an; *m.* [clod, hangra *a meadow*] *Clodhanger :*—Þurh

ût Clodhangran; of ðan hangran andlang rôde ût on Mules dene *out through Clodhanger; from the meadow along the road out to Mule's dean,* Cod. Dipl. 1198; A.D. 956; Kmbl. v. 374, 28.

clofen *cloven, separated; pp.* of cleófan.

Clofes hoo = Clofes hô; *gen.* hôs; *pl. nom. acc.* hôas; *gen.* hôa; *dat.* hôum; *m. Cliff, near Rochester :*—Hêr sinoþ wæs æt Clofes hoo [æt Clofes hô, col. 2] *in this year* [A.D. 822] *there was a synod at Cliff,* Chr. 822; Th. 111, 14, col. 1; 110, 14, col. 1, 2. Æt Clofes hôum *at Cliff,* Th. Diplm. A.D. 803; 52, 32: A.D. 825; 73, 12. Ðâ wæs sionoþlíc gemôt on ðære mæran stôwe ðe mon hâteþ Clofes hôas *then there was a synodal meeting in the famous place which is called Cliff,* Th. Diplm. A.D. 825; 70, 11.

clof-þung, -þunc, e; *f. The herb crow-foot,* Herb. 9, 1; Lchdm. i. 98, 23, 25, MS. B: Lchdm. iii. 54, 21. v. cluf-þung.

clof-wurt *the herb buttercup,* Herb. 10; Lchdm. i. 100, 14, MS. B. v. cluf-wyrt.

CLOM; *gen.* clommes; *m:* clam; *gen.* clammes; *m. A band, bond, clasp, bandage, chain, prison;* vinculum, carcer :—Habbaþ me swâ helle clommas fæste befangen *the clasps of hell have so firmly grasped me,* Cd. 19; Th. 24, 6; Gen. 373. Ðes wîtes clom *this bond of torture,* 215; Th. 271, 10; Sat. 103. Ðysne wîtes clom *this bond of torment,* 216; Th. 274, 21; Sat. 157 : 223; Th. 293, 11; Sat. 453. On ðissum fæstum clomme *in this fast bondage,* 21; Th. 26, 17; Gen. 408. Clommum fæste *fast in bonds,* Andr. Kmbl. 260; An. 130. Cealdan clommum *with cold bands,* 2425; An. 1214. DER. bealu-clom, fýr-, hæfte-, helle-, wæl-, wîte-, wundor-. v. clam; *gen.* clammes; *m.*

clomm *climbed;* scandit; *p.* of climan.

clough *a cleft of a rock,* or *down the side of a hill,* Som. Ben. Lye.

CLÛD, es; *m. A stone, rock, hill;* saxum, rupes, collis :—Clûdas feóllan of muntum *stones fell from the mountains,* Ors. 6, 2; Bos. 117, 12. Clûd *rupes,* Ælfc. Gr. 9, 27; Som. 11, 24. Mid clûdum ymbweaxen *surrounded with rocks,* Ors. 3, 9; Bos. 67, 22. Sumra wyrta eard biþ on clûdum *the soil of some herbs is on rocks,* Bt. 34, 10; Fox 148, 24. Beorh oððe clûd *collis,* Ælfc. Gr. 9, 28; Som. 11, 46. [Laym. clude, chlud *a cliff, rock:* Orm. cludess *hills:* Plat. kluut, klute, kloot : Dut. kluit, *f.:* kloot, *m:* Kil. klot : Ger. klosz, *m. gleba:* M. H. Ger. klôz, *m. a lump:* O. H. Ger. kloz, *m. massa:* Dan. klode, *m. f. a ball:* Swed. klot, *n:* Icel. klót, *n.* knob on a sword's hilt: hence the Eng. CLOD.] DER. stân-clûd.

clûdig; *adj. Stony, rocky;* saxeus :—Ðæt Norþ-manna land is on sumum stôwum swýðe clûdig *the country of the Northmen is in some places very rocky,* Ors. 1, 1; Bos. 20, 42.

clufe *an ear of corn, a clove of garlic;* spica, Som. Ben. Lye. Clufe? *f. pl.* in e, *A clove,* the bulb or tuber of a plant, Glos. of Lchdm. ii. Twâ clufe *two cloves,* L. M. 3, 41; Lchdm. ii. 336, 3. Garleaces iii clufe *three cloves of garlic,* 3, 62; Lchdm. ii. 350, 8.

clufeht, clufiht; *adj. Bulbed;* bulbosus :—Nim clufehte wenwyrt *take the bulbed wenwort,* L. M. i. 58; Lchdm. ii. 128, 17. Gegnîd on twâ clufe ðære clufehtan wenwyrte *rub them upon two bulbs of the bulbed wenwort,* 3, 41; Lchdm. ii. 336, 3.

clufon *clove, separated,* Chr. 937; Th. 200, 38, col. 3; Æðelst. 5; *p. pl.* of cleófan.

cluf-þung, e; *f.* cluf-þunge, an; *f.* [clufe, þung *monkshood, hellebore;* aconitum = ἀκόνῑτον] *The herb crow-foot;* ranunculus sceleratus, Lin :—Clufþung *crow-foot,* L. M. i, 1; Lchdm. ii. 20, 4: 1, 24; Lchdm. ii. 66, 14: 1, 28; Lchdm. ii. 70, 2: 1, 47; Lchdm. ii. 120, 1: 3, 8; Lchdm. ii. 312, 20: iii. 12, 27. Ðeós wyrt ðe man *sceleratam,* and ôðrum naman clufþunge nemneþ, biþ cenned on fuhtum and on wæteregum·stôwum *this herb which is called scelerata, and by another name crow-foot, is produced in damp and watery places,* Herb. 9, 1; Lchdm. i. 98, 24–26. Genim clufþungan wôs *take juice of crow-foot,* 110, 3; Lchdm. i. 224, 7.

cluf-wyrt, e; *f. The herb buttercup;* batrachion = βατράχιον, ranunculus acris, Lin :—Ðeós wyrt ðe man *batrachion,* and ôðrum naman clufwyrt nemneþ, biþ cenned on sandigum landum and on feldum : heó biþ feáwum leáfum and þynnum *this herb which is called batrachion, and by another name buttercup, is produced on sandy lands and in fields: it is of few and thin leaves,* Herb. 10, 1; Lchdm. i. 100, 15–17 : L. M. 3, 8; Lchdm. ii. 312, 13.

CLUGGE, an; *f. A bell, small bell;* campana :—Hleoðor heora cluggan, ðære hî gewunedon to gebêdum gecîgde and awehte beón, ðonne heora hwylc of weorulde geféred wæs *the sound of their bell, by which they were wont to be called and awaked to prayers, when any of them had gone out of the world,* Bd. 4, 23; S. 595, 40. [Plat. klokke *a bell, clock:* O. Frs. klokke : Dut. klok, *f. a clock, bell:* Ger. glocke, *f:* M. H. Ger. glogge, *f:* O. H. Ger. glokka, *f:* Dan. klokke, *m. f. a bell, clock:* Swed. klocka, *f. a bell, clock:* Icel. klukka, klokka, *f.*]

clumbon; *pp.* clumben *climbed,* Chr. 1070; Erl. 209, 9; *p. pl.* and *pp.* of climban.

clumian; *p.* ode; *pp.* od *To murmur, mutter;* mussitare :—Hî clu-

---

miaþ mid ceaflum ðær hî scoldon clypian *they mutter with their jaws where they ought to speak aloud,* Wanl. Catal. 30, 14.

clungon; *pp.* clungen *withered, pined; p. pl.* and *pp.* of clingan.

CLÛS, e; *f:* clûse, an; *f. An inclosure, a narrow passage, close, bond, prison;* claustrum, carcer :—Ðeáh he hie mid fîftigum clûsum beclemme *though he surround it with fifty bonds,* Salm. Kmbl. 143; Sal. 71. Alæd of carcernes clûse mîne sâwle *educ de carcere animam meam,* Ps. Th. 141, 8. He fram ðære clûsan afaren wæs wið ðara scipa *he was gone from the pass towards the ships,* Ors. 6, 36; Bos. 131, 26, 22. Ðâ hæfdon hý heora clûsan belocene *when they had closed their passes,* 3, 7; Bos. 60, 4. Annas and Caiphas wæron forþgangende to ðære clûsan *Annas and Caiaphas were going forth to the prison,* Nicod. 14; Thw. 7, 10 : 16; Thw. 8, 6, 9. [Plat. kluse : Dut. kluis, *f:* Kil. kluyse : Ger. klause, *f:* M. H. Ger. klôse, klûs, klûse, *f:* O. H. Ger. klûsa, *f:* M. Lat. clusa, clausa : Lat. clausus, *pp.* of claudēre *to shut, inclose.*]

clûse, an; *m. An inclosure;* claustrum, Ors. 6, 36; Bos. 131, 26. v. clûs.

cluster, es; *n. A* CLUSTER, *bunch;* botrus = βότρυς, *f:*—Cluster ðæt bitereste *botrus amarissima,* Cant. Moys. Isrl. Lamb. 193 b, 32. v. clyster.

CLÛSTOR, clûster, clauster; *gen.* clûstres; *pl. nom. acc.* clûstor, clûstro; *n. A lock, bar, barrier, cell;* claustrum, clausura :—Meahte ðæs ceasterhlides clûstor onlûcan *might unlock the lock of the city-gate,* Exon. 12 a; Th. 20, 8; Cri. 314. Wæs mid clûstre carcernes duru behliden *the door of the prison was shut with a lock,* Exon. 69 a; Th. 256, 23; Jul. 236. Ða locu feóllon [feollan MS.], clûstor of ðâm ceastrum *the locks fell, the barriers from that city,* 120 a; Th. 461, 23; Hô. 40. Ðæt he mihte cuman þurh ðâs clûstro *that he might pass through these barriers,* Cd. 22; Th. 27, 11; Gen. 416. He hine hêht on carcernes [MS. carcerne] clûster behûsan *he commanded him to be locked in a prison's cell,* Bt. Met. Fox 1, 146; Met. 1, 73. [O. Sax. klûstar, *n:* Frs. klooster, kleaster : O. Frs. klaster, *n:* Dut. klooster, *n:* Kil. klooster : Ger. kloster, *n:* M. H. Ger. O. H. Ger. klôster, *n:* Dan. Swed. kloster, *n:* Icel. klaustr, *n:* Lat. claustra, *pl. n. a lock, bar, bolt.*]

clûstor-cleófa, an; *m. A prison-chamber, cell;* carceris cubiculum :—On clûstorcleófan *in the prison-chamber,* Andr. Kmbl. 2041; An. 1023.

clûstor-loc, clûster-loc, es; *pl. nom.* -loca; *n. A prison-lock, lock, bar;* claustellum, claustrum :—Clûstor-loca [MS. -locæ] *claustella,* Glos. Epnl. Recd. 156, 2. Cluster-loc *claustellum,* Cot. 34 : *claustrum,* 181.

CLÛT, es; *m. A small piece of cloth,* CLOUT, *patch, piece of metal, plate;* pittacium, commissura, lamina :—Clût *pittacium,* Glos. Epnl. Recd. 161, 19 : *commissura,* Ælfc. Gl. 28; Som. 61, 4; Wrt. Voc. 26, 3 : 82, 2. Wurdon forþaborene îsene clûtas *iron plates were brought forth,* Homl. Th. i. 424, 19. Lecgaþ ða îsenan clûtas hâte glôwende to his sîdan *lay the iron plates glowing hot to his side,* Homl. Th. i. 424, 35. [Wyc. Piers P. clout : Chauc. cloutes *rags:* Orm. clutesse, *pl:* Dan. klud, *m. f:* Swed. klut, *m:* Icel. klútr, *m:* Wel. clwt, *m:* Gael. clùd, clùid, *m. a clout, rag, patch.*] DER. ge-clûtod. v. clûd.

clyf *a cliff, rock,* Ps. Spl. M. C. 113, 8. v. clif.

clýfa, clifa, an; *m.* [cleófa, cleófan *to cleave, divide, separate*]. **I.** *a separate place for man,*—*A chamber;* cubiculum, cubile :—Ne mæge we hreppan ǽnne wyrm binnon ðînum clýfan *we may not touch a worm in thy chamber,* Homl. Th. ii. 416, 23. On diglum oððe on incôfan, oððe on clýfum *in cubilibus,* Ps. Lamb. 4, 5. On his incôfan oððe on his clýfan *in cubīli suo,* 35, 5. **II.** *a separate place for wild beasts,*—*A cave, den;* antrum, caverna, cubile :—On ðâm clîfum ðe dracan oneardedon *in the dens which dragons dwelt in;* in cubilibus, in quĭbus dracōnes habitābant, Bd. 3, 23; S. 554, 22. DER. bed-clýfa, gebed-, hord-, in-. v. cleófa.

clyfer-fête; *adj.* [clifer *a claw, talon*] *Claw-footed, talon-footed, cloven-footed;* fissipes :—Ða fugelas ðe be flǽsce lybbaþ syndon clyferfête *the birds which live by flesh are cloven-footed,* Hexam. 8; Norm. 14, 19.

clyfian, clyfigan; *p.* ode; *pp.* od *To cleave, adhere;* adhǽrere :—Ðæt feax ðe on ðâm cambe clyfige somnige *let her collect the hair that cleaveth to the comb,* Med. ex Quadr. 1, 7; Lchdm. i. 332, 21, MS. B.

clyfigende âdl *a joint-disease, the gout,* Som. Ben. Lye.

clýfst, he clýfþ *cleavest, cleaves; 2nd* and *3rd pers. pres. sing.* of cleófan.

clyf-wyrt *clivers, fox-glove,* Ælfc. Gl. 40; Som. 63, 91; Wrt. Voc. 30, 41 : 79, 41. v. clif-wyrt.

clymmian, he clymmaþ, *pl.* clymmiaþ; *p.* ode; *pp.* od [climan *to climb*] *To climb;* scandere :—Leóht clymmaþ *light ascends* [climbeth], Salm. Kmbl. 829; Sal. 414.

CLYMPRE, an; *n? A lump* or CLUMP *of metal, metal;* massa metalli, metallum :—Hefigere ic eom ðonne unlytel leádes clympre *I am heavier than a huge clump of lead,* Exon. 111 b; Th. 426, 18; Rä. 41, 75. Wyrc greáte clympran [MS. clymppan] feówur *make four great lumps,* Lchdm. iii. 134, 31. Clympre *metallum,* Wrt. Voc. 286, 73. [Plat. klump : Dut. klomp, *m:* Kil. klompe : Ger. klump, klumpen, *m:*

Dan. klump, m. f: Swed. klump, m: O. Nrs. klumbr, klumpr, m. Rask Hald.]

clynan; p. ede; pp. ed [clyne metal] To ring, sound; clangere:—Rand dynede, campwudu clynede the shield rang, the war-wood sounded, Elen. Kmbl. 101; El. 51.

clyne, es; m? n? clyna, clyne, clyno; indecl. f. A mass, lump, ball, metal; massa, sphæra = σφαῖρα, metallum:—Clynes, trendles sphæræ, Mone B. 3491. Ælces cynnes wecg, vel ôra oððe clyna metallum, Ælfc. Gl. 51; Som. 66, 8; Wrt. Voc. 34, 67. Clyne, clyno massa, metallum, Cot. 132: 182. Sile hym ǽne clyne give him one lump, Lchdm. iii. 134, 33. Trendel, clyne sphæra, Mone B. 3465. Clyne, clottum massa, 3478.

clypenes, -ness an embrace, Bd. 3, 24; S. 557, 6, note. v. clypnys.

CLYPIAN, clypigan, clipian, cleopian, clepian; part. clypiende, clypigende; ic clypie, clypige, ðû clypast, he clypaþ, pl. clypiaþ; p. ode, ade; impert. clypa, pl. clypiaþ; pp. od, ad To make a vocal sound, speak, speak aloud, to cry out, call, say; loqui, clamare, vocare, dicere:—He ongan clypian cœpit clamare, Mk. Bos. 10, 47. Ne com ic rihtwîse clypian I came not to call the righteous, Lk. Bos. 5, 32: 19, 15. Hlûddre stæfne clypigan to cry with a loud voice, Bd. 4, 19; S. 589, 12, note. Clypiende dicens, R. Ben. 44. Mid micelre stemne clypigende crying with a loud voice, Homl. Th. i. 48, 5. Ic clypie to Gode clamabo ad Deum, Ps. Lamb. 56, 3. Drihten gehŷrþ me ðonne ic clypige to him Dominus exaudiet me cum clamavero ad eum, Ps. Lamb. 4, 4. Ðû clypast thou callest, Hy. 7, 45; Hy. Grn. ii. p. 288, 45. Hwî clypaþ Dauid hyne Drihten quomodo David vocat eum Dominum? Mt. Bos. 22, 43, 45. Ge clypiaþ me lâreów vos vocatis me magister, Jn. Bos. 13, 13. To ðê ic clypode ad te clamavi, Ps. Lamb. 60, 3: 65, 17. Ic to ðê, Drihten, clypade ego ad te, Domine, clamavi, Ps. Th. 87, 13. He clypode mid micelre stemne he cried with a loud voice, Homl. Th. i. 596, 5: Bd. 3, 2; S. 524, 21: Byrht. Th. 132, 33; By. 25: 139, 19; By. 256. Israéla folces prafostas clypodon to Pharaone præpositi filiorum Israel vociferati sunt ad Pharaonem, Ex. 5, 15: Homl. Th. i. 72, 28. Clypa ða wyrhtan voca operarios, Mt. Bos. 20, 8: Lk. Bos. 14, 12, 13: Jn. Bos. 4, 16. Clypiaþ hyne vocate eum, Ex. 2, 20. [Wyc. Piers P. Chauc. clepe: Laym. clepie, clepien, cleopie, cleopien: Orm. clepenn: Scot. clep, clepe to call, name.] DER. be-clypián, forþ-, of-, on-, to-, toge-: healf-clypiende.

clypiendlíc, clypigendlíc, clipigendlíc; adj. Making a vocal sound; vocalis [from vox, vocis the voice]:—Syndon fîf vocales, ðæt synd clypigendlíce, a, e, i, o, u. Ðâs fîf stafas æteówiaþ heora naman þurh hî silfe, and bûton ðâm stafum ne mæg nân word beón awriten, and forðî hîg sind quinque vocales gehâtene there are five vocales, a, e, i, o, u, which are vocal [sounds]. These five letters indicate their names by themselves, and without these letters no word can be written, and therefore they are called the five vocal sounds, Ælfc. Gr. 2; Som. 2, 44–46. Consonantes, ðæt is samod-swêgende, forðanðe hî swîgaþ mid ðâm fîf clypigendlícum consonants, that is, sounding together, because they are made articulate by the five vocal sounds, Som. 2, 50. v. sylf-swêgend.

clypnys, clypenes, -nyss, -ness, e; f. An embrace; complexus:—To clypnysse ðæs heofonlícan brýdguman eádig fæmne ineóde ad complexum sponsi cælestis virgo beata intraret, Bd. 3, 24; S. 557, 6.

clypol; adj. Vocal; vocalis, Bridf. 101.

clypola, an; m. A vowel; vocalis, Bridf. 101.

CLYPPAN; p. clypte; pp. clypt To embrace, clasp, CLIP, cherish; complecti, amplexari:—Ðæt he his mondryhten clyppe and cysse that he his embrace and kiss his lord, Exon. 77 a; Th. 289, 2; Wand. 42. Nâwuht ðes woruldgielp is ðe hie clyppaþ and lufiaþ this worldly glory is worthless which they embrace and love, Past. 41, 1; Hat. MS. 56 a, 3. Ðâ Laban gehîrde ðæt Iacob wæs cumen his swustor sunu, ðâ arâs he to-geánes and clypte hine cum audisset Laban venisse Iacob filium sororis suæ, cucurrit obviam ei complexusque eum, Gen. 29, 13. Ioseph clypte hira ælcne and cyste hîg and weóp amplexatus et osculatus est Ioseph et ploravit super singulos, 45, 15. Ongan seó abbudisse clyppan and lufian ða Godes gife abbatissa amplexata gratiam Dei, Bd. 4, 24; S. 598, 1. Hine sybbe and lufu swylce clyppeþ justitia et pax complexæ sunt se, Ps. Th. 84, 9. Clyppende amplexans, Prœm. R. Conc. Hý hî lufan fæste clyppaþ they firmly clasp them with love, Exon. 107 a; Th. 409, 8; Râ. 27, 26. Heáfodswîma heortan clypte insensibility seized his heart, Cd. 76; Th. 94, 30; Gen. 1569. Æghwæðer óðerne earme beþehte, cyston hie and clypton each embraced the other with his arm, they kissed and clasped each other, Andr. Kmbl. 2031; An. 1018. [Wyc. Piers P. Chauc. clippe: Laym. clappe: Orm. clippenn: O. Frs. kleppa: Dan. klippe: Swed. Icel. klippa.] DER. be-clyppan, bi-, ymb-.

clypung, clepung, e; f. Articulation, speaking out, the forming of words, a cry; eloquium, clamor:—Se mûþ drýfþ ût ða clypunge, and seó lyft biþ geslagen mid ðære clypunge the mouth produces [driveth out] the articulation, and the air is struck in the articulation, Ælfc. Gr. 1; Som. 2, 31. Clypung mîn infærþ [ineóde, Lamb.] on eárum his clamor meus introivit in aures ejus, Ps. Spl. 17, 8. Clypunga the kalends; kalendæ, Ælfc. Gr. 13; Som. 16, 19.

clýsan; p. de; pp. ed To close, shut; claudere. DER. be-clysan: clýsing.

clýsing, clýsung, e; f. A CLOSING, inclosure, conclusion of a sentence, period; claustrum, periodus = περίοδος:—Seó fæstnung ðære hellícan clýsinge ne geþafaþ ðæt ða wiðercoran ǽfre ûtabrecon the fastening of the hellish inclosure never allows the wicked to break out, Homl. Th. i. 332, 20. Hî on hellícere clýsunge andbídodon they waited in the hellish inclosure, Homl. Th. ii. 80, 6. Clýsunga claustra, R. Ben. Interl. 67. Periodos is clýsing oððe ge-endung ðæs ferses a period is the conclusion or ending of a sentence [lit. verse], Ælfc. Gr. 50, 14; Som. 51, 18. DER. be-clýsing.

CLYSTER; gen. clystres; pl. nom. acc. clystru; gen. clystra; dat. clystrum; n. A CLUSTER, bunch, branch; botrus = βότρυς, f. racemus, propago:—Clyster botrus, Ælfc. Gl. 47; Som. 65, 32; Wrt. Voc. 33, 31. Hira wînberie ys gealla and ðæt biteroste clyster uva eorum uva fellis et botri amarissimæ, Deut. 32, 32. Clystru botros, Mone B. 2548. Clystrum racemis, 3835. Ic geseah wîneard, on ðam wǽron þreó clystru videbam vitem in qua erant tres propagines, Gen. 40, 10, 12. [Prompt. clustyr: Plat. kluster: Kil. klister.]

clýsung an inclosure, Homl. Th. ii. 80, 6. v. clýsing.

clýða a poultice; emplastrum, malagma, Herb. 51, 2; Lchdm. i. 154, 18. v. clîða.

CLYWE, an; f. n? A CLEW, ball of thread or yarn, ball; globus, glomus:—Clywe globus, Ælfc. Gl. 111; Som. 79, 66; Wrt. Voc. 59, 35. [Plat. kluwe, klouwen: Dut. kluwen, klouwen, n: Kil. klouwe, kluwe: Ger. kläuel, kleuel, knäuel, n. m: M.H.Ger. kliuwel, n: O.H.Ger. kliuwa, f. cliuwi, n.] v. cliwen.

clywen a clew, ball of thread or yarn, ball, circlet, Ælfc. Gl. 28; Som. 61, 5; Wrt. Voc. 26, 4: Homl. Th. ii. 514, 2. v. cliwen.

cnæd, ðû cnæde, pl. cnædon kneaded, hast kneaded, fermented; p. of cnedan.

CNÆP, cnæpp, cnep, es; m. A top, cop, KNOP; vertex, jugum, supercilium:—Uppan ðæs muntes cnæp in montis vertice, Ex. 19, 20. Hîg astigon to ðæs muntes cnæppe ascenderunt in verticem montis, Num. 14, 44. Ofer cneppas trans juga, Glos. Prudent. Recd. 149, 55. Hîg læddon hine ofer ðæs muntes cnæpp duxerunt illum ad supercilium montis, Lk. Bos. 4, 29. [Piers P. knappe: Chauc. knoppes, pl: Plat. knoop: O. Frs. knop, knap, m: Dut. knop, m: Kil. knoppe: Ger. M.H.Ger. knopf, m. nodus, globulus: O.H.Ger. knoph, m: Dan. knap, m. f: Swed. knapp, m: Icel. knappr, m: Wel. Ir. cnap: Gael. cnap, cnaip, m.]

cnæpling, es; m. A stripling, youth, boy; adolescens, puer:—Eom ic cnæpling I am a boy, Homl. Th. ii. 576, 14: Mone B. 2514.

cnǽwe, cnâwe; adj. Knowing, conscious, aware; cognoscens, conscius. DER. ge-cnǽwe, or-.

cnǽwst he cnǽwþ knowest, knows; 2nd and 3rd pers. pres. of cnâwan.

CNAPA, cnafa, an; m. I. a boy, young man, KNAVE; puer, juvenis, adolescens:—He betǽhte hys cnapan and se cnapa hit ofslôh he gave it [a calf] to his young man and the young man slew it, Gen. 18, 7. Heó sealde ðam cnapan drincan dedit puero bibere, 21, 19: 22, 19: 42, 22: 48, 16: Homl. Th. i. 186, 14. Ic hæbbe sumne cnapan habeo quemdam puerum, Coll. Monast. Th. 19, 27. Abraham fêrde mid twâm cnapum to fyrlenum lande Abraham ducens secum duos juvenes abiit in locum, Gen. 22, 3, 5. Syle cnapan [cnafan C.] ðînum da puero tuo, Ps. Spl. 85, 15. Ðæt wîf wearþ wrâþ ðam cnapan mulier molesta erat adolescenti, Gen. 39, 10. II. a servant; servus:—He hêt his cnapan behealdan to ðære sǽ he ordered his servant to look towards the sea, Bd. de nat. rerum; Wrt. popl. science 18, 23; Lchdm. iii. 276, 24. [Wyc. knaue-child a male child: Piers P. Chauc. knave: Laym. cnaue: Orm. cnapess, gen: Plat. knape, knawe: O. Sax. knapo, m: Frs. knape: O. Frs. knapa, knappa, m: Dut. knaap, m: Kil. knape: Ger. M.H.Ger. knabe, m: O.H.Ger. knabo, m: Swed. knape, m: Icel. knapi, m.] DER. þeów-cnapa.

CNÁWAN; ic cnâwe, ðû cnâwest, cnǽwst, he cnâweþ, cnǽwþ, pl. cnâwaþ; p. cneów, pl. cneówon; pp. cnâwen To KNOW; noscere:—Ða byþ cnæwene noscuntur, Mone B. 169. [Wyc. Piers P. Chauc. knowen, knowe: Laym. i-cnawen: Orm. cnawenn: O.H.Ger. knâjan: Icel. knâ: Lat. novi, old form gnovi I came to know: Grk. γι-γνώ-σκω: Sansk. jñâ.] DER. an-cnâwan, be-, ge-, on-, to-.

cnâwing, e; f. Knowledge, a knowing; cognitio, Som. Ben. Lye. DER. on-cnâwing.

CNEAR, cnearr, es; m. A small ship, galley used for ships of the Northmen; navis, septentrionalium naves:—Cnear on flot the ship on float, Chr. 937; Erl. 114, 1, notes, p. 326; Æðelst. 35. [Icel. knarri, m. navis, id. qu. knörr, m. navis, in specie mercatoria; Ólafs Saga hins helga, 27, 1, ubi promiscue ponuntur knörru et kaupskipum, Egils. sub knörr.] DER. nægled-cnear.

cneátian; p. ode; pp. od To argue, dispute, contend; disceptare, contendere:—Cneátian disceptare, Mone B. 967. Cneátiaþ contendunt, 1867.

cneátung, e; *f. A debate, an inquiry, a search;* disputatio, scrutinium, Scint. 14.

CNEDAN; ic cnede, ðú cnidest, cnist, he cnit, *pl.* cnedaþ; *p.* ic, he cnæd, ðú cnǽde, *pl.* cnǽdon; *pp.* cneden To KNEAD, *ferment;* subigere, fermentare :—Cnede to ðam [MS. ðan] hláfe *to knead bread,* Lchdm. iii. 134, 21. Óþ-ðæt sie cneden *donec fermentaretur,* Lk. Skt. Rush. 13, 21. [*Chauc.* knede: *Orm.* knednen: *Dut. Kil.* knéden: *Ger.* kneten: *M. H. Ger.* knëten: *O. H. Ger.* knetan: *Dan.* knede: *Swed.* knåda: *Icel.* knoða.] DER. ge-cnedan.

CNEÓ, cneów, es; *n.*   **I.** *a* KNEE; genu :—Ðæt he on cneó lecge honda and heáfod *that he lays his hands and head on his knee,* Exon. 77 a; Th. 289, 3; Wand. 42. Me synt cneówu unhále *genua mea infirmata sunt,* Ps. Th. 108, 24. Cneówa *genua,* Wrt. Voc. 283, 68. Hie on cneówum sǽton *they sat on their knees,* Cd. 181; Th. 227, 2; Dan. 180: Chr. 979; Erl. 129, 22: Ors. 3, 9; Bos. 68, 35: Exon. 48 a; Th. 164, 19; Gú. 1014. Cneó bígeþ *bends the knees,* Exon. 62 b; Th. 229, 23; Ph. 459. Cneó bégean scolden *genua flectere deberent,* Bd. 3, 17; S. 544, 39, col. 2: Elen. Kmbl. 1693; El. 848: Exon. 63 a; Th. 232, 29; Ph. 514: 112 b; Th. 431, 9; Rä. 45, 5.   **II.** *a generation, relationship;* generatio, propinquitatis gradus :—On ánum cneówe *in generatione una,* Ps. Th. 108, 13. Óþ hund cneówa [MS. cnea] *to a hundred generations,* Exon. 124 a; Th. 476, 16; Ruin. 8. Binnan cneówe *within relationship,* L. E. G. 12; Th. i. 174, 25. In ðam þriddan cneówe mid Crécum mót man wíf niman, in fiftan mid Rómánum *in tertio propinquitatis gradu apud Græcos viro licet uxorem ducere,* in quinto apud Romanos, L. Ecg. C. 28; Th. ii. 152, note h. Binnan ðam feórþan cneówe *within the fourth degree of relationship,* L. Eth. vi. 12; Th. i. 318, 15. [*Piers P.* knowes *knees:* Laym. cneo: *Orm.* cnewwe: *Plat.* knee *knee, generation:* O. Sax. knio, kneo, *n. knee:* O. Frs. kni, kne, *n. knee, degree of relationship:* Dut. Kil. knie, *f. knee:* M. H. Ger. knie, *n:* O. H. Ger. kniu, kneo, *n:* Goth. kniu, *n: Dan.* knæ, *n:* Swed. knä, *n: Icel.* kné, *n: Lat.* genu, *n: Grk.* γόνυ, *n: Sansk.* jānu, *m. n.*]

cneódan; he cneódeþ; *p.* cneád, *pl.* cnudon; *pp.* cnoden To *give;* tribuĕre, cognominare :—He naman cneódeþ *he gives a name,* Bd. 2, 20; S. 522, 24. v. cnódan.

cneóeht; *adj.* [cneó *a knee,* -eht = -iht, *adj. termination,* q. v.] *Knotty;* geniculatus :—Sió cneóehte wenwyrt *the knotty wenwort,* L. M. 1, 64; Lchdm. ii. 140, 8.

Cneoferis burh, burg, e; *f. Burghcastle, Suffolk;* villæ nomen in agro Suffolciensi :—Ðá wæs fæger mynster getimbred on wuda neáh sǽ on sumre ceastre, seó is nemned on Englisc Cneoferis burh *erat monasterium silvanum, et maris vicinitate amœnum, constructum in castro quodam, quod lingua Anglorum Cnobheres burg, id est, urbs Cnobheri vocatur,* Bd. 3, 19; S. 547, 22. v. Cnobheres burh.

cneoht *a boy,* Bd. 2, 6; S. 508, 18 : 3, 18; S. 545, 45, col. 2. v. cniht.

cneó-mǽgas, cneów-mǽgas, -mágas; *pl. m.* [cneó **II.** *generation,* mǽg *relation*] *Relations of the same sex or the same generation;* consanguinei :—Cneówmǽgas *relations,* Cd. 83; Th. 104, 11; Gen. 1733. From cneómǽgum *from their relations,* Chr. 937; Erl. 112, 8; Æðelst. 8. Enos ongon, mid ðám cneómágum, ceastre timbran *Enoch began, with his kinsmen, to build a city,* Cd. 50; Th. 64, 28; Gen. 1057: Andr. Kmbl. 1370; An. 685: Elen. Kmbl. 1170; El. 587.

cneord; *adj. Diligent, intent;* sollers, intentus. DER. ge-cneord.

cneord-lǽcan; *p.* -lǽhte; *pp.* -lǽht *To be diligent, study;* studere, M. H. 14 a. DER. ge-cneordlǽcan.

cneordnys, -nyss, e; *f. Diligence, study, learning;* studium, disciplina :—Cneordnysse *studio,* Mone B. 2464: *disciplina,* 1034. DER. ge-cneordnys.

cneóres, cneórys, cneóris, cneórnis, -ress, e; *f. A generation, posterity, race, tribe, family;* generatio, posteritas, gens, tribus, familia :—Cneóres *generatio,* Ælfc. Gl. 91; Som. 75, 18; Wrt. Voc. 51, 63: Mt. Bos. 1, 18. Ðeós cneórys is mánfull cneórys *generatio hæc generatio nequam est,* Lk. Bos. 11, 29. Hwí sécþ ðeós cneóris tácen *quid generatio ista signum quærit?* Mk. Bos. 8, 12 : Ps. Lamb. 23, 6: Bd. 1, 27; S. 491, 9. Cneóresse *generationis,* Mone B. 896. Mid ðisse cneórysse mannum *cum viris generationis hujus,* Lk. Bos. 11, 31. Cneórisse bóc *liber generationis,* Mt. Bos. 1, 1: Ps. Th. 94, 9. Ne gesihþ nán man of ðisse wirrestan cneóresse ðæt góde land *non videbit quispiam de hominibus generationis hujus pessimæ terram bonam,* Deut. 1, 35: Ps. Th. 44, 18. On ðære þriddan cneórisse *in the third generation,* Bd. 1, 27; S. 491, 8: Mk. Bos. 8, 12: Lk. Bos. 11, 30. Fram cynrene on cneórisse *a generatione in generationem,* Ps. Lamb. 89, 1: 101, 19. Mid ðisse cneórysse *cum generatione hac,* Lk. Bos. 11, 32: 17, 25. Ealle cneóressa *omnes generationes,* Mt. Bos. 1, 17. Ðás sind ðære heofenan and ðære eorþan cneórnisse *istæ sunt generationes cæli et terræ,* Gen. 2, 4. Ðás sind Noes cneórnissa *hæ sunt generationes Noe,* Gen. 6, 9. Ða on cneóressum cýðed syndan *they are known to generations,* Ps. Th. 101, 16. Sie gefeá gehwám ðe in cneórissum cende weorþen *let there be joy to each one who*

in their generations shall be born, Exon. 11 a; Th. 15, 6; Cri. 232: Cd. 190; Th. 236, 10; Dan. 319: Ps. Th. 144, 13. Cneóresse *posteritatem,* Mone B. 648. Ðære cneórisse wæs Cainan weard *Cainan was guardian of that race,* Cd. 57; Th. 70, 18; Gen. 1155 : 106; Th. 139, 31; Gen. 2318. Hine weorþiaþ wera cneóressa *races of men worship him,* Ps. Th. 71, 15. Ealle wera cneórissa ðé weorþiaþ *omnes gentes adorabunt te,* 85, 8 : 74, 6. Com God wera cneórissa weorc sceáwigan *God came to behold the work of the races of men,* Cd. 80; Th. 101, 8; Gen. 1679. Secgaþ on cneórissum *dicite in gentibus,* Ps. Th. 95, 9 : Cd. 64; Th. 77, 12; Gen. 1274. Cneóres *tribus,* Ælfc. Gl. 49; Som. 65, 73; Wrt. Voc. 34, 8. Cneórisse cende wǽron *ascenderunt tribus,* Ps. Th. 121, 4. Se biþ wiðerbreca wera cneórissum *he shall be an adversary to the tribes of men,* Cd. 104; Th. 138, 8; Gen. 2288: Exon. 44 b; Th. 151, 7; Gú. 791. Mín awóc on ðære cneórisse cynebearna rím *one raised up in that family a number of princely children,* Cd. 82; Th. 102, 22; Gen. 1704. Of Cames cneórisse wóc wermǽgþa fela *from Ham's family arose many tribes of men,* 79; Th. 98, 29; Gen. 1637.

cneó-rím, cneów-rím, es; *n. The number of kin, progeny, family;* cognatorum numerus, progenies, familia :—Of ðam wíd folc, cneórím micel, cenned wǽron *from whom a wide-spread people, a great progeny, were born,* Cd. 79; Th. 98, 32; Gen. 1639. Cneórím [MS. cneorisn] Caines *the family of Cain,* 63; Th. 76, 12; Gen. 1256. He his cynnes cneórím ícte *he increased the progeny of his race,* 59; Th. 72, 22; Gen. 1190. Ða ðæs cynnes cneówrím ícton *they increased the progeny of the race,* 52; Th. 65, 13; Gen. 1065.

cneóris *a generation, race, tribe, family,* Mk. Bos. 8, 12 : Ps. Th. 74, 6 : 121, 4: Cd. 79; Th. 98, 29; Gen. 1637. v. cneóres.

cneórnis, -niss, e; *f. A generation,* Gen. 2, 4 : 6, 9. v. cneóres.

cneórys *a generation,* Lk. Bos. 11, 29, 31, 32 : 17, 25. v. cneóres.

cneó-sib *a race, generation.* v. cneów-sib.

cneów, es; *n.*   **I.** *a knee;* genu :—Cneów *genu,* Ælfc. Gl. 75; Som. 71, 87; Wrt. Voc. 44, 69 : 71, 52. Heó on cneów sette *she knelt down,* Elen. Kmbl. 2270; El. 1136: Ps. Th. 94, 6. Hí bígdon heora cneów befóran him *they bowed their knees before him,* Mt. Bos. 27, 29.   **II.** *a generation;* generatio :—In ðære þeóde awóc his ðæt þridde cneów *in that nation rose the third generation from him,* Cd. 209; Th. 258, 16; Dan. 676. v. cneó.

cneów, *pl.* cneówon *knew;* *p.* of cnáwan.

cneó-wǽrc, cneów-wǽrc, es; *n? A pain in the knees;* genuum dolor :—Wið cneówwærce *for a pain in the knees,* Lchdm. iii. 16, 16. Wið cneów-wærce, L. M. 1, 24; Lchdm. ii. 66, 11.

cneów-holen, cneó-holen, es; *m. n?* KNEEHOLM, *knee-hulver, kneeholly, butcher's broom;* ruscum, victoriola, ruscus aculeatus, Lin :—Genim twegen scenceas fulle wóses ðysse wyrte, ðe man *victoriola,* and óðrum naman cneówholen, cenneþ *take two cups full of the juice of this herb, which is called* victoriola, *and by another name knee-holly,* Herb. 59; Lchdm. i. 162, 6. Genim cneówholen *take knee-holly,* L. M. 1, 36; Lchdm. i. 86, 10 : 1, 39; Lchdm. ii. 102, 9 : 2, 51; Lchdm. ii. 266, 15 : iii. 4, 29 : 30, 14. Wyrc to drence twá cneówholen *make into a drink the two knee-hollies,* L. M. 1, 47; Lchdm. ii. 120, 8.

cneówian, cneówigan; *part.* cneówigende; *p.* ode; *pp.* od [cneó, cneów *a knee*] *To bow the knee, to kneel;* genuflectere :—Benedictus on his gebédum cneówode *Benedict knelt down in prayer,* Homl. Th. ii. 154, 20 : 178, 33. Cneówigende *genuflectens,* Procem. R. Conc. DER. ge-cneówian.

cneówlian; *p.* ode; *pp.* od To KNEEL; genuflectere, MS. Tib. A. iii. fol. 94. v. cneówian.

cneów-mǽgas, -mágas *relations,* Cd. 83; Th. 104, 11; Gen. 1733: Elen. Kmbl. 1372; El. 688. v. cneó-mǽgas.

cneów-rím *progeny,* Cd. 52; Th. 65, 13; Gen. 1065. v. cneó-rím.

cneów-sib; *gen.* -sibbe; *f. A race, generation;* generatio :—Cende cneówsibbe cénra manna *he begat a race of brave men,* Cd. 161; Th. 200, 13; Exod. 356.

cneówung, cnéwung, e; *f. A kneeling;* genuflectio, Bd. 3, 17; S. 544, 39, note.

cneów-wǽrc *a pain in the knees,* L. M. 1, 24; Lchdm. ii. 66, 11. v. cneó-wǽrc.

cneów-wyrste; *pl. f.* [wrist, wyrst *the wrist*] *Knee-joints;* genicula, Ælfc. Gl. 75; Som. 71, 88; Wrt. Voc. 44, 70.

cnep *a top, summit,* Glos. Prudent. Recd. 147, 55. v. cnæp.

cnídan; *p.* cnád, *pl.* cnidon; *pp.* cniden To *beat;* cædere :—Ða sume cnidon [MS. cnidun] *they beat some;* alium ceciderunt, Mt. Kmbl. Rush. 21, 35. DER. for-cnídan.

cnidest, cnist, he cnit *kneadest, kneads; 2nd and 3rd pers. pres. of* cnedan.

CNÍF, es; *m. A* KNIFE; culter, cultellus, artavus, *Low Latin* = cultellus :—Cníf *artavus,* Wrt. Voc. 82, 40. [*Chauc.* knyfes, *pl :* Laym. Orm. cnif : *Plat.* knief, kniiv : *Frs.* knyf : *Kil.* knijf : *Ger.* kneif, *m :* Dan. kniv, *m. f :* Swed. knif, *m : Icel.* knífr, *m. a knife or dirk.*] v. seax.

CNIHT, cneoht, cnyht, es; *m. A boy, youth, attendant, servant,* KNIGHT : hence the modern knights of a shire are so called because they

serve the shire; puer, juvenis, adolescens, servus:—Sum lytel sweltende cniht *a little dying boy*, Bd. 4, 8; S. 575, 23: Ors. 3, 7; Bos. 58, 43. Tyn wintra cniht *a boy of ten years*, L. In. 7; Th. i. 106, 18: Lk. Bos. 7, 7: Bd. 5, 19; S. 637, 4: Byrht. Th. 136, 18; By. 153. Fram ðínum cnihte *a puero tuo*, Ps. Th. 68, 17. Heó cwæþ to ðam cnihte *ait ad puerum*, Gen. 24, 65. Cwicne abregd cniht of áde *take the boy alive from the pile*, Cd. 141; Th. 176, 20; Gen. 2914: 162; Th. 203, 20; Exod. 406. Ðú ðone cnyht to us brohtest in Bethlem *thou broughtest the boy to us in Bethlehem*, Exon. 121 a; Th. 463, 33; Hö. 79. He scôle gesette in ðære cneohtas and geonge menn lærde wǽron *he set up a school in which boys and young men were taught*, Bd. 3, 18; S. 545, 45, col. 2. Ðyssum cnyhtum wes líðe *be gentle to these boys*, Beo. Th. 2443; B. 1219. Ðæt hie ðæs cnihtes cwealm gesôhton *that they should seek the young man's death*, Andr. Kmbl. 2243; An. 1123: 1824; An. 914. Ða cnihtas cræft leornedon *the youths learned science*, Cd. 176; Th. 221, 4; Dan. 83: 182; Th. 228, 2; Dan. 196. To cwale cnihta *for the destruction of the youths*, Cd. 184; Th. 229, 32; Dan. 226. Cnyhta *of the youths*, Exon. 55 a; Th. 195, 32; Az. 165. Wundor Godes on ðam cnihtum gecýðed wæs *the miracle of God was manifest on the youths*, Cd. 197; Th. 245, 32; Dan. 472. Moises sende cnihtas *Moyses misit juvenes*, Ex. 24, 5: Cd. 176; Th. 221, 16; Dan. 89: Cd. 195; Th. 243, 5; Dan. 431. Cnihtas wurdon ealde ge giunge ealle forhwerfde to sumum dióre *the attendants [of Ulysses], old and young, were all transformed to some beast*, Bt. Met. Fox 26, 170; Met. 26, 85. Agynþ beátan ða cnihtas and ða þínena *cœperit percutere servos et ancillas*, Lk. Bos. 12, 45. Ic, Oswold bisceop, landes sumne dǽl sumum cnihte ðæm is Osulf nama, for uncre sybbe, forgeaf *I, bishop Oswald, have given a portion of land to a knight named Osulf, for our kinship*, Cod. Dipl. 557; A.D. 969; Kmbl. iii. 49, 32: 612; A.D. 977; Kmbl. iii. 159, 25. [*Wyc.* kniȝt, knyȝt: *R. Brun.* knyght: *Chauc.* knight, knyght: *R. Glouc.* knygt: *Laym.* cniht: *Orm.* cnihtess, *pl:* *Scot.* knecht, knycht: *Plat.* knecht, knekt: *Frs.* knecht: *O. Frs.* kniucht, knecht, *m:* *Dut. Kil. Ger.* knêht, *m:* *M.H. Ger.* kneht, *m:* *O. H. Ger.* kneht, *m:* *Dan.* knegt, *m.f:* *Swed.* knekt, *m.*] DER. in-cniht, leorning-.

cniht-cild, es; *n. A male child, boy; puer*:—Wæs on ðam ylcan mynstre cnihtcild sum, ne wæs yldre ðonne þrý-wintre *there was in the same monastery a boy, he was not older than three years*, Bd. 4, 8; S. 575, 27.

cniht-gebeorþor, *gen.* -gebeorþres; *n. A boy-bearing, child-bearing; pueri partus*:—On ðæm cnihtgebeorþre heó â clǽne þurhwunode *in child-bearing she continued ever immaculate*, Homl. Blick. 3, 12.

cniht-geong; *adj. Young as a child; puerilis*, Elen. Kmbl. 1276; El. 640.

cniht-hád, es; *m. The period between childhood and manhood, youth, boyhood*, KNIGHTHOOD; *pubes*:—Cnihthád *pubes*, Ælfc. Gr. 9, 28; Som. 11, 50. Oþ cnihtháde *to youth; pube tenus*, 47; Som. 48, 8.

cniht-iugoþ, e; *f. Youth, boyhood; juventus*:—Cnihtiugoþ and sumor beóþ gelíce *youth and summer are alike*, Bridf. 11: 12.

cniht-leás; *adj.* KNIGHTLESS, *without an attendant; sine servo*, M. H. 113 b.

cniht-líc; *adj. Boyish, childish; puerilis*:—Ne he cnihtlíce gálnysse næs begangende *nor was he [Guthlac] addicted to boyish levity*, Guthl. 2; Gdwin. 12, 16. Swâ oft swâ cnihtlícu yldo begǽþ *as childish age is often wont*, 2; Gdwin. 12, 19.

cniht-wesende; *part. Being a boy or youth, while a youth; dum puer est*:—On ðam mynstre on ðam cnihtwesendum *in monasterio tunc puero*, Bd. 3, 12; S. 537, 17: 2, 15; S. 518, 36. Cnihtwesende *being a youth*, Exon. 85 a; Th. 320, 34; Wíd. 39: Beo. Th. 750; B. 372: 1075; B. 535.

cniht-wíse, an; *f. Youthwise, boy's manner; pueri mos*:—Sprecan æfter cnihtwísan *to speak after the manner of a boy*, Guthl. 2; Gdwin. 12, 13.

cnittan *to knit*, Ælfc. Gr. 36; Som. 38, 22, MS. C. v. cnyttan.

Cnobheres burh; *gen.* burge; *f.* [MS. Cneoferis burh] *Burghcastle, Suffolk;* Cnobheri urbs, in agro Suffolciensi ad ostia Garionis fluvii:—Ceaster, seó is nemned on Englisc Cneoferis burh. *In his original Latin, Bede says,*—Castrum, '*quod lingua Anglorum Cnobheres burg, id est,* urbs Cnobheri *vocatur*,' Bd. 3, 19; S. 547, 22.

cnocian *to knock*. DER. ge-cnocian. v. cnucian.

cnódan, cneódan; ic cnóde, ðú cnódest, he cnódeþ, cneódeþ; *pl.* cnódaþ; *p.* cneád, *pl.* cnudon; *pp.* cnoden, gecnoden *To give, assign, call, carry out, exalt; tribuěre, attribuěre, efferre*:—Gyt mon his naman cneódeþ *yet man calls by his name*, Bd. 2, 20; S. 522, 24. Gif hwæt welgedónes biþ, ðonne cnódaþ him ealle mid hêrenesse *if anything be well done, then all exalt him with praise; si qua bene gesta sunt, omnes laudibus efferunt*, Past. 17, 3; Hat. MS. 22 b, 3.

CNOLL, es; *m. A* KNOLL, *hill-top, cop, summit; cacumen, vertex*:—On ðam teóðan mónþe æteówodon ðæra munta cnollas *decimo mense apparuerunt cacumina montium*, Gen. 8, 5. Garganus hine gemêtte standan uppon ðam cnolle ðære heálícan dúne *Garganus found him standing on the knoll of the high hill*, Homl. Th. i. 502, 13. Heá dúne,

hyllas and cnollas *high downs, hills and knolls*, Exon. 18 a; Th. 45, 11; Cri. 717. On cnolle *in vertice*, Mone B. 927. To ufeweardum ðam cnolle *ad verticem montis*, Jud. 16, 3. He hit ne sette upon ðone hêhstan cnoll *he should not set it upon the highest hill-top*, Bt. titl. xii; Fox xii. 15. On ðam lytlan cnolle ðe Ermon hâtte *Hermonis a monte modico*, Ps. Th. 41, 7. [*Prompt.* knolle: *Plat.* knúlle: *Dut.* knol, *m:* *Kil.* knolle: *Ger.* knolle, knollen, *m:* *M.H. Ger.* knolle: *Dan.* knold, *m.f:* *Swed.* knöl, *m.*]

CNÓSL, es; *n. A race, progeny, offspring, kin, family; proles, genus, generatio*:—Gewît ðú nú fêran, and ðíne fare lǽdan, ceápas to cnósle *begin thou now to depart, and lead thy family, thy cattle for progeny*, Cd. 83; Th. 105, 2; Gen. 1747. Mínes cnósles *of my progeny*, Exon. 105 a; Th. 399, 22; Rä. 19, 4: 112 a; Th. 430, 15; Rä. 44, 9. Gôdes and yfles ðǽr ic cunnade, cnósle bidǽled *there I tried good and evil, separated from my offspring*, 85 b; Th. 321, 27; Wíd. 52. Bearn vel cnósl *soboles vel proles*, Ælfc. Gl. 91; Som. 75, 19; Wrt. Voc. 51, 64. Cnósle *genere*, Mone B. 1608. Hêht from hweorfan mânscyldigne cnósle sínum *he bade the crime-guilty depart from his kindred*, Cd. 50; Th. 64, 12; Gen. 1049. On cnósle oððe on cynne *in generatione*, Ps. Lamb. 32, 11. Gewât him mid cnósle *he departed with his family*, Cd. 83; Th. 104, 4; Gen. 1730. [*O. Sax.* knôsal, *n:* *Ger.* knôsel, *m. a little man:* *O. H. Ger.* knuosli, knôsli, *n.*] DER. fæderen-cnôsl, geóguþ-.

cnossian, he cnossaþ; *p.* ode; *p.* od *To beat, strike, dash; tundi, quassari, illidi*:—Ýða gewealc mec oft bigeat, æt nacan stefnan, ðonne he be clifum cnossaþ *the rolling of the waves has often caught me, at the vessel's prow, when it strikes on rocks*, Exon. 81 b; Th. 306, 15; Seef. 8.

CNOTTA, an; *m. A* KNOT, *fastening, knitting; nexus*:—Cnotta *nexus*, Ælfc. Gr. 11; Som. 15, 10. Gyt hêr is óðer cnotta ealswâ earfoðe *there is yet another knot equally difficult*, Homl. Th. ii. 386, 22. To onlýsanne [MS. onlýsenne] ða fæstan cnottan [MS. cnotten] *to loosen the fast knots*, Th. Diplm. A.D. 1035; 334, 9: Wanl. Catal. 42, 23. Mid cnottum *nexibus*, Mone B. 3128: Homl. Th. ii. 28, 26. [*Prompt. Chauc.* knotte: *Plat.* knutte: *Frs.* knotte: *Dut.* knot, *f:* *Kil.* knutte: *Ger.* knoten, knote, *m:* *M.H. Ger.* knode, knote, *m:* *O. H. Ger.* knodo, *m:* *Dan.* knude, *m.f:* *Swed.* knut, *m:* *Icel.* knútr, *m.*]

CNUCEL; *gen.* cnucles; *m. A* KNUCKLE, *joint; articulus*, Som. Ben. Lye. [*Prompt.* knokylle: *Relq. Ant. W.* i. 190, 30, knokelys, *pl:* *Plat.* knukkel, knúchel: *Frs.* kneukel: *O. Frs.* knokele, knokle: *Dut.* kneukel, *m:* *Kil.* knokel: *Ger.* knöchel, *m:* *Dan.* knogle, *m.f:* *Swed.* knoge, *m:* *Icel.* knúi, *m.*]

CNUCIAN, cnucigan; *p.* ode; *pp.* od *To* KNOCK, *beat, pound; pulsare, tundere, pertundere*:—Cnuciaþ and eów biþ ontýned *pulsate et aperietur vobis*, Mt. Bos. 7, 7: Lk. Bos. 11, 9. Ðam cnuciendum biþ ontýned *pulsanti aperietur*, Mt. Bos. 7, 8: Lk. Bos. 11, 10. He cnucode æt ðære dura *he knocked at the door*, Homl. Th. ii. 382, 17, 22. Ic cnucige *tundo, pertundo*, Ælfc. Gr. 28, 7; Som. 32, 56, 65. Ða leáf cnuca on ânum mortere *pound the leaves in a mortar*, Herb. 41, 4; Lchdm. i. 142, 18: 57, 1; Lchdm. i. 158, 20: 63, 7; Lchdm. i. 166, 29: 64; Lchdm. i. 168, 5: 65; Lchdm. i. 168, 11. Cnucige ealle ða wyrta *pound all the herbs*, Lchdm. i. 382, 15. [*Prompt.* knokkyñ: *Wyc.* Piers P. knocken: *Chauc.* knocke: *Plat.* knukken *to utter a deep sound:* *Icel.* knoka: *Wel.* cnociaw: *Corn.* cnoucye.] DER. ge-cnucian.

cnuian; *p.* ode; *pp.* od *To pound*, Lchdm. ii. 340, 15. v. cnuwian.

Cnut, es; *m. Cnut was the Danish king of England for twenty-one years, from A.D. 1014-1035*:—Hêr, on ðissum geáre, Swegen ge-endode his dagas to Candelmæssan iii ñ Feb'. And se flota ða eal gecurón Cnut to cyninge *here, in this year, A.D. 1014, Sweyn ended his days at Candlemas, on the 3rd of the Nones of February [Feb. 3rd]. And then all the fleet chose Cnut for king*, Chr. 1014; Erl. 150, 20-22. Hêr forþfêrde Cnut cing, on ii Id' Novemb' æt Sceftes byrig, and hine man ferode ðanon to Winceastre, and hine ðǽr bebyrigde *here departed king Cnut, on the 2nd of the Ides of November [= Nov. 12] at Shaftesbury, and they bore him thence to Winchester, and buried him there*, 1035; Erl. 164, 17-19. Hêr man drǽsde út Ælfgife, Cnutes cynges láfe, seó wæs Hardacnutes cynges môdor *here, A.D. 1037, they drove out Ælfgifu, widow of king Cnut, who was mother of king Hardacnut*, 1037; Erl. 167, 1. [Knúta, os, ossis. Leggja mót wið marga prúða knútu *cum multis splendidis* [nitidis] *artubus congredi*, Hh. 83, 1, i. e. *cum multis militibus, prædæ destinatis*. Raskius, F. vi. 403, *pro nom. propr. accipit, a* Knútr, *aut de principibus viris aut bellatoribus*, Egils.]

cnuwian, cnuian; *p.* ode; *pp.* od *To knock, pound; pinsere*:—Genim lǽfre neoðowearde, cnuwa and wring *take the netherward part of a bulrush, pound it and wring*, Lchdm. i. 382, 18. Cnua beolenan *pound henbane*, L. M. 3, 50; Lchdm. ii. 340, 15. DER. ge-cnuwian. v. cnucian.

cnyht *a boy, youth*, Exon. 121 a; Th. 463, 33; Hö. 79: 55 a; Th. 195, 32; Az. 165: Beo. Th. 2443; B. 1219. v. cniht.

CNYLL, es; *m. A* KNELL, *sound of a bell; signum campanæ*:—Hwîlon ic gehýre cnyll and ic arîse *aliquando audio signum et surgo*, Coll. Monast. Th. 35, 29. [*Prompt.* knyll-ynge *tintillacio: Relq. Ant. W.* ii. 31, cnul *sound of a bell: Ger.* knall, *m.fragor, crepitus: Dan.* knald, *n.*

sound : Swed. knall, *m. a loud noise : Wel.* cnul, cnull, *m. a passing bell.*]

CNYLLAN, cnyllsan; *p.* de; *pp.* ed *To* KNELL, *sound a bell ;* pulsare, campanâ signum dare :—Ðæm cnyllende ontýned biþ *pulsanti aperietur,* Lk. Skt. Rush. 11, 10. Cnyllaþ [cnyllsaþ, Lind.] and ontýned biþ iów *pulsate et aperietur vobis,* 11, 9 : 12, 36 : R. Ben. 48. Cnylled *pulsatus,* R. Conc. 1. [*Ger.* knallen, knellen *crepare, fragorem edere : M. H. Ger.* knillen, knüllen *to beat : Dan.* knalde *fragorem edere : Swed.* knalla *to make a noise : Icel.* knylla *to beat with a blunt weapon.*]

cnyllsan *to knell, sound a bell,* Lk. Skt. Lind. 11, 9 : 12, 36. v. cnyllan.

CNYSSAN, cnysan; *part.* cnyssende; *p.* cnyssede, cnysede, cnyste; *pp.* cnyssed *To* press, *trouble, toss, strike, dash, beat, overcome ;* premere, tribulare, pulsare, contundere, vincere :—Ic wæs hearde cnyssed *I was hard pressed,* Ps. Th. 117, 13. Ne læt úsic costunga cnyssan tô swíðe *let not temptations trouble us too much,* Exon. 122 a ; Th. 469, 7 ; Hy. 5, 9. Me costunge [MS. costunce] cnyssaþ *trials trouble me,* Ps. Th. 63, 1 : Exon. 81 b ; Th. 308, 2 ; Seef. 33. Me costunge cnyssedan *trials troubled me,* Ps. Th. 65, 13 : 85, 6 : 114, 4. Cnysedon, 58, 17. Cnysdon, 119, 1. Cnysdan, 118, 143 : 137, 7. Se storm biþ cnyssende ðæt scip *the storm is tossing the ship,* Past. 9, 2 ; Hat. MS. 13 b, 10. Ne mec sceal ámas cnyssan *the weaver's reeds shall not strike me,* Exon. 109 a ; Th. 417, 22 ; Rä. 36, 8. Cnysseþ ðæt sár on ða rib *the sore striketh upon the ribs,* L. M. 2, 46 ; Lchdm. ii. 258, 3. Ne se hearda forst cnyseþ ǽnigne *the hard frost strikes not any,* Exon. 56 b ; Th. 201, 21 ; Ph. 59. He cnyste Petres sídan *he struck Peter's side,* Homl. Th. ii. 382, 7. Ðás stánhleoðu stormas cnyssaþ *storms dash these stony rocks,* Exon. 78 a ; Th. 292, 19 ; Wand. 101. Gaius Iulius se Cásere Brettas mid gefeohte cnysede *Caius Julius Cæsar beat the Britons in battle,* Chr. Erl. 4, 24. Ahteniense hí mid gefeohte cnysedon *the Athenians beat them in battle,* Ors. 3, 1 ; Bos. 53, 5. Ðæt hine ne cnysse sió wilnung *lest desire overcome him,* Past. 19, 1 ; Hat. MS. 28 a, 6. [*Scot.* knuse *to press down with the knees : Plat.* knusen *to squeeze : Frs. Japx.* kniesen *to bruise : Dut.* kneuzen *to bruise : Kil.* knisschen *terere, quassare : Ger.* knüssen *to push, beat : M. H. Ger.* knüsen, knüssen *to press, push, beat : O. H. Ger.* knusjan, knussan *concutere : Goth.* knussyan *to press down : Dan.* knuse *to bruise : Swed.* knusa *to bruise : Icel.* knosa *to bruise, beat.*] DER. a-cnyssan, ge-, on-, to-, úta-.

cnyssung, e; *f. A striking, stroke ;* ictus :—Of ðære lyfte cnyssunge *from the striking of the air,* Ælfc. Gr. 1 ; Som. 2, 30. Sweng oððe cnyssung *ictus,* 43 ; Som. 44, 55.

CNYTTAN, cnittan; *p.* cnytte; *pp.* cnytted, cnytt, cnyt *To tie, bind,* KNIT; nectere, nexere, ligare :—Ic cnytte *necto,* Ælfc. Gr. 36 ; Som. 38, 22. Ic cnytte [MS. C. cnitte] *nexo,* 36; Som. 38, 23 : 28, 3 ; Som. 51, 61. Genim ðysse ylcan coliandran sǽd, endlufon corn oððe þreóttyne, cnyte mid ánum þrǽde *take seed of this same coriander, eleven or thirteen grains, knit them with a thread,* Herb. 104, 2 ; Lchdm. i. 218, 20. [*Prompt.* knyttyñ *nodo, confedero : Wyc.* knyt, knyttide, *pp : Piers P.* knytte : *R. Brun.* knytte : *Chauc.* knitte : *Laym.* icnutten, *p. pl. knotted : Plat.* knutten *nodare : Dut.* knotten *to tie : Kil.* knodden *nodare : Ger.* knoten, knöten *nodare : Dan.* knytte *to knit : Swed.* knyta *to knit, tie : Icel.* knytja *to knit together : Lat.* nodare *to tie : Sansk.* nah *to bind, tie.*] DER. be-cnyttan, ge-, un-.

cnyttels, es; *m? A knitting thread, string, thong ;* nervus :—Strenga, cnyttelsa *nervorum,* Mone B. 2858.

COC, cocc, es; *m. A* COCK, *a male fowl or bird ;* gallus, pullus :—Coc *gallus,* Ælfc. Gl. 39 ; Som. 63, 47 ; Wrt. Voc. 30, 2 : 63, 8 : 77, 34. Creów se cocc *gallus cantavit,* Mt. Bos. 26, 74, 34 : Jn. Bos. 13, 38. Cocca *pullorum,* Mone B. 4913. Ðonne coccas cráwan *when cocks crow,* Lchdm. iii. 6, 5. [*Prompt.* cok : *Chauc.* cok, cock : *Kil.* kocke : *Dan.* kok, *m: Icel.* kokkr, *m: Fr.* coq, *m: O. Fr.* coc.] DER. sǽ-coc, wudu-.

CÓC, es; *m. A* COOK; coquus :—Cóc *coquus,* Ælfc. Gr. 28, 5 ; Som. 32, 7 : Wrt. Voc. 82, 50. Hwæt secgaþ we be cóce *quid dicimus de coquo?* Coll. Monast. Th. 29, 5. Hí cócas gehyrstan *cooks roasted them,* Ps. Th. 101, 3. [*Prompt.* cooke : *Piers P.* coke : *Chauc.* coke : *Laym.* coc : *Plat.* kokk : *O. Sax.* kok, *m: Dut.* kok, *m: Kil.* kock : *Ger. M. H. Ger. O. H. Ger.* koch, *m: Dan.* kok, *m. f: Swed.* kock, *m: Icel.* kokkr, *m: Ital.* cuóco, *m: Lat.* co, coquus, *m: Wel.* côg : *Corn.* cog, *m: Ir. Gael.* coca : *Armor.* cok : *O. Slav.* kuchari.]

COCCEL, es; *m.* COCKLE, *darnel, tares ;* zizania :—Æteówde se coccel hine *apparuerunt zizania,* Mt. Bos. 13, 26. He oferseów hit mid coccele on middan ðam hwǽte *superseminavit zizania in medio tritici,* 13, 25 : Homl. Th. i. 526, 20. Se sóða Déma hǽt his englas gadrian ðone coccel *the true Judge shall bid his angels gather the cockle,* 526, 21 : Mt. Bos. 13, 27, 29, 30. Coccela *zizaniorum,* Mone B. 2332. [*Prompt.* cokylle : *Wyc.* cockil, cokil : *Chauc.* cockle.]

COCER, cocor, cocur, es; *m.* I. *a quiver for arrows, a case ;* pharetra = φαρέτρα :—Cocer *pharetra,* Wrt. Voc. 84, 31. Hý gyrdon flána heora on cocere *paraverunt sagittas suas in pharetra,* Ps. Spl. 10, 2. Nim ðín gesceót, ðínne cocur and ðínne bogan, and gang út *sume*

arma tua, pharetram et arcum, et egredere foras, Gen. 27, 3. II. *a sword, spear ;* framea :—Ageót cocor *effunde frameam,* Ps. Spl. 34, 3. Genera fram cocore míne sáwle *erue a framea animam meam,* 21, 19. [*Prompt.* cocur *cothurnus : Piers P.* cokeres *stockings : Laym.* koker, *m : Plat.* köker, käker : *O. Sax.* cocǽre, *m : Frs. O. Frs.* koker : *Dut. Kil.* kóker : *Ger.* köcher, *m : M. H. Ger.* kochǽre, kocher, *m : O. H. Ger.* kochar : *Dan.* kogger, *n : Swed.* koger, *n.*]

cócer-panne, cócor-panne, an; *f.* [cóc *a cook,* panne *a pan*] *A cooking-pan, frying-pan ;* sartago, frixorium :—On cócerpannan *in frixorio,* Ps. Th. 101, 3. Cócorpanne *sartago,* Mone B. 4694.

cócnunga, *pl. f.* [cóc *a cook*] *Things cooked, pies* :—Metegearwa and cócnunga sint to forbeódanne *meat-preparations and things cooked must be forbidden,* L. M. 2, 23 ; Lchdm. ii. 210, 26 : 2, 32 ; Lchdm. ii. 236, 10.

cocor, es; *m. A sword ;* framea, Ps. Spl. 21, 19. v. cocer II.

cócor-mete, es; *m.* [cóc *a cook,* mete *meat, food*] *Meat divided into four parts?* quadripartitum, Wrt. Voc. 290, 41.

cocur *a quiver,* Gen. 27, 3. v. cocer I.

cod-æppel, es; *m. A quince-pear, quince ;* malum cydoneum *vel* cotoneum, Cot. 93.

CODD, es; *m. A bag, sack,* COD, *husk ;* pera = πήρα, folliculus, siliqua :—Codd *folliculus,* Ælfc. Gl. 59 ; Som. 67, 128 ; Wrt. Voc. 38, 50. Ne nime ge nán þing on wege, ne gyrde, ne codd *nihil tuleritis in via, neque virgam, neque peram,* Lk. Bos. 9, 3 : 22, 36 : Mt. Bos. 10, 10 : Mk. Bos. 6, 8. Nim wínberian coddas [MS. coddes] *take husks of the grape,* Lchdm. iii. 112, 13. [*Prompt.* codde : *Wyc.* coddes, coddis *pods : Chauc.* cod : *Scot.* cod *a pillow : Kil.* kodde *a bag, sack : Swed.* kudde, *m. a cushion : Icel.* koddi, *m. a pillow.*] DER. bién-codd, sceát-.

coelnes *coolness,* Wanl. Catal. 304, 49. v. cólnes.

coerin *boiled wine,* Cot. 61. v. ceren.

CÓFA, an; *m. A* COVE, *cave, repository, inner room, chamber, ark ;* cubile, cubiculum, arca :—On cófan *in a chamber,* Exon. 125 a ; Th. 480, 18 ; Rä. 64, 4. Wæs culufre eft of cófan sended *the dove was sent again from the ark,* Cd. 72 ; Th. 88, 13 ; Gen. 1464. On cyninga cófum *in cubilibus regum,* Ps. Th. 104, 26. DER. bán-cófa, bed-, breóst-, ferhþ-, gást-, heolstor-, hord-, hreðer-, in-, mearh-, morþor-, nýd-, rún-, þeóster-: cóf-godas.

Cofan-treó, Cofen-treó, Couen-tré, es; *n.* [a monachorum conventu sic dictum putant quidam] COVENTRY, *Warwickshire ;* Coventria in agro Warwicensi :—Leófwine abbod on Cofantreó feng to ðam bisceopríce *Leofwine, abbot at Coventry, succeeded to the bishopric,* Chr. 1053; Erl. 188, 7. Leofríc liþ æt Cofentreó *Leofric lieth at Coventry,* 1057 ; Erl. 192, 30. Of Couentré *at Coventry,* 1066 ; Erl. 203, 16 : 1130 ; Erl. 258, 37.

Cofer-flód, Cofor-flód, es; *n. m. The sea of Galilee ;* Galilæum mare :—Ic fare on wæteres hricg ofer Coferflód, Caldéas sécan *I depart upon the water's back over the sea of Galilee, to seek the Chaldeans,* Salm. Kmbl. 39 ; Sal. 20. Ðú gewítest on Wendelsǽ, ofer Coforflód, cýððe sécean *thou goest on the Mediterranean sea, over the sea of Galilee, to seek thy country,* 407 ; Sal. 204.

cóf-godas, *pl. m. Household-gods ;* penates, Ælfc. Gl. 113 ; Som. 79, 113 ; Wrt. Voc. 60, 20 : Glos. Prudent. Recd. 152, 28.

cófincel, es; *n. A hand-mill ;* pistrilla, Cot. 155.

cóf-líce *quickly,* Som. Ben. Lye. v. cáf-líce.

cóf-scipe *quickness,* Som. Ben. Lye. v. cáf-scype.

cohhetan ; *p.* te; *pp.* ed *To bluster ;* tumultuári :—Hí ongunnon cohhetan *they began to bluster,* Judth. 12 ; Thw. 25, 20 ; Jud. 270.

CÓL ; *gen.* cóles; *pl. nom. acc.* cóla, cólu ; *gen.* cóla; *dat.* cólum ; *n.* COAL; carbo :—Cól *carbo,* Wrt. Voc. 86, 20 : 286, 79. Swá sweart swá cól *as black as coal,* L. M. 3, 39 ; Lchdm. ii. 332, 19. Cól [MS. coll] *carbo,* Ælfc. Gl. 30 ; Som. 61, 75 ; Wrt. Voc. 27, 4. On hát cól *upon a hot coal,* L. M. 1, 50 ; Lchdm. ii. 124, 6. Cóla onælde synd fram him *carbones succensi sunt ab eo,* Ps. Spl. 17, 10, 15. Feallaþ ofer hí cólu *cadent super eos carbones,* Ps. Spl. C. 139, 11. Þurh ða cólu ðæs alteres *by the coals of the altar,* Past. 7, 1; Hat. MS. 12 a, 10. Ða twegen drýmen wurdon awende to cóla gelícnyssum *the two wizards were turned to the likeness of coals,* Homl. Th. ii. 496, 28. [*Prompt.* cole *carbo : Wyc.* colis, *pl : Chauc.* cole : *Laym.* col : *Scot.* coill, coyll : *Plat.* kole : *O. Frs.* koal : *Dut.* kool, *m. f : Kil.* kole : *Ger.* kohle, *f : M. H. Ger.* kol, *m : O. H. Ger.* kolo, *m ;* kol, *n : Dan.* kul, *n : Swed.* kol, *n : Icel.* kol, *n.*] DER. heofon-cól.

CÓL ; *comp.* ra; *sup.* ost; *adj.* COOL, *cold ;* frigidus :—Oft æspringe útaweallep of clife hárum cól and hlutor *a fountain often springs out of a hoar rock cool and clear,* Bt. Met. Fox 5, 26 ; Met. 5, 13. Hrér mid sticcan óþ ðæt hit cól síe *stir it about with a spoon till it be cool,* L. M. 3, 26 ; Lchdm. ii. 242, 1 : 2, 51 ; Lchdm. ii. 270, 2 : 3, 30 ; Lchdm. ii. 326, 6 : 3, 31 ; Lchdm. ii. 326, 15. Wyrc him leage of ellenahsan, þweah his heáfod mid cólre *make him a ley of elder ashes, wash his head with this cold,* 3, 47 ; Lchdm. ii. 338, 26. Ða cearwylmas cólran wurþaþ *the anxious emotions become cooler,* Beo. Th. 570 ; B. 282 : 4139; B. 2066. [*Prompt.* cole *algidus : R. Glouc.* cole : *Plat.* kölig, köl :

Dut. koel; *Kil.* koel: *Ger.* kühl, kühle: *M. H. Ger.* küele: *O. H. Ger.* kuol: *Dan.* kölig, köl: *Swed.* kylig.]

côledon *cooled, became cold,* Andr. Kmbl. 2514; An. 1258; *p. pl. of* côlian.

côlian; *p.* ode, ede; *v. intrans. To* COOL, *to be or become cold;* algere, refrigerari:—Lêt ðonne hyt côlian *then let it cool,* Herb. 94, 4; Lchdm. i. 204, 23. Flæsc onginneþ côlian *the flesh begins to cool,* Runic pm. 29; Kmbl. 345, 14. Côlaþ Cristes lufu *the love of Christ cooleth,* Exon. 33 a; Th. 104, 17; Gû. 9. Sumur-hât côlaþ *summer-heat becomes cold,* Exon. 95 a; Th. 354, 58; Reim. 67. Lîc côlode *the corpse became cold,* Exon. 51 b; Th. 180, 18; Gû. 1281. Weder côledon *the storms were cold,* Andr. Kmbl. 2514; An. 1258. Leomu côlodon *the limbs became cold,* Elen. Grm. 882. DER. a-côlian, ge-. v. calan.

coliandre, an; *f. The herb coriander;* coriandrum = κορίαννον:—Cnuca coliandran sædes nigon corn *pound nine grains of coriander seed,* Herb. 52, 2; Lchdm. i. 156, 3: 104, 2; Lchdm. i. 218, 19. v. celendre.

colla, an; *m. Rage, strife;* ardor, furor. DER. morgen-colla.

collen-ferhtan; *p.* -ferhte; *pp.* -ferhted *To make empty or void, render desolate;* exinanire:—Ða ðe cweðaþ, ge collenferhtaþ oððe aîðliaþ ôþ grundweal oððe to staðolfæstnunga on hire *qui dicunt, exinanite, exinanite usque ad fundamentum in ea,* Ps. Lamb. 136, 7.

collen-ferhþ, -ferþ, -îyrhþ; *adj.* [collen, *pp. of* cellan *to swell? p.* ceall, *pl.* cullon; *pp.* collen, Ettm: ferhþ *mind*] *Fierce-minded, bold of spirit, bold;* animi ferox, audax:—Cleopode collenferhþ cearegan reórde *the fierce-minded cried out in a sorrowful voice,* Andr. Kmbl. 2217; An. 1110. Wîgan wæron blîðe, collenferhþe *the warriors were blithe, bold of spirit,* Elen. Kmbl. 493; El. 247: Judth. 11; Thw. 23, 22; Jud. 134. Ðonne he beót spriceþ collenferþ *when he bold of spirit utters a promise,* Exon. 77 b; Th. 290, 26; Wand. 71: Apstls. Kmbl. 107; Ap. 54. In ceól stigon collenfyrhþe *the bold of spirit stept into the ship,* Andr. Kmbl. 698; An. 349. Collenferþ *bold of spirit,* Exon. 96 b; Th. 361, 9; Wal. 17. Eódon mid collenferhþe *the bold went together,* Elen. Kmbl. 755; El. 378: 1694; El. 849. Hwæðer collenferþ cwicne gemêtte *whether he should find the bold [warrior] living,* Beo. Th. 5563; B. 2785. Cuma collenferhþ *the bold guest,* 3616; B. 1806. Hleóþrade cempa collenferhþ *the bold warrior spake,* Andr. Kmbl. 1075; An. 538. Stôp ût hræde, collenferþ *he quickly stept out, firm of mind,* 3154; An. 1580.

collon-crôh, -crôg, es; *m. A water-lily;* nymphæa = νυμφαία:—Colloncrôh nymphæa, Wrt. Voc. 68, 20: Mone A. 461. Colloncrôg *nymphæa,* Cot. 140.

côl-mâse, an; *f.* [côl *coal,* mâse *a titmouse*] *A coal-titmouse, coal-tit;* parus ater:—Côlmâse parra, Wrt. Voc. 62, 39: parula, 281, 11: bardioriolus, Ælfc. Gl. 39; Som. 63, 52; Wrt. Voc. 30, 7. [*Dut.* koolmees, *f. a titmouse.*]

Coln, e; *f? The river* COLNE, *Essex;* Colnius, in agro Essexiensi:—Hie flugon ofer Temese, ðá up be Colne on ânne îggaþ *they fled over the Thames, then up by the Colne to an island,* Chr. 894; Erl. 90, 28.

coln *a pebble stone;* calculus, Som. Ben. Lye.

côlne *pertaining to coals;* carbonarius, Som. Ben. Lye.

Colne-ceaster; *gen.* -ceastre; *f.* COLCHESTER, *Essex, so called from the river Colne;* Colcestria, in agro Essexiæ, ad ripam Colnii fluvii:—Hî fôron to Colneceastre *they went to Colchester,* Chr. 921; Erl. 107, 9; 108, 5.

côl-nes, -ness, e; *f.* COOLNESS, *cool air, a breeze;* refrigerium, aura:—On côlnesse *in refrigerium,* Ps. Th. 65, 11. v. cêl-nes, calan.

côlode *cooled,* Exon. 51 b; Th. 180, 18; Gû. 1281; *p. of* côlian.

côl-pyt, -pet; *gen.* -pyttes, -pettes; *m. A* COAL-PIT; carbonis fossa:—Fram Hlypegete to ðam côlpytte: fram côlpette *from Lipgate to the coal-pit: from the coal-pit,* Cod. Dipl. 1322; A.D. 1035; Kmbl. vi. 186, 9.

COLT, es; *m. A* COLT; pullus:—He asyndrode þrítig gefolra olfendmyrena mid heora coltum, and twentig assmyrena mid heora coltum [MS. coltun] *separavit camelos fœtas cum pullis suis triginta, et asinas viginti et pullos earum,* Gen. 32, 15. [*Prompt.* colte: *Wyc. Chauc.* colt.]

colt-græig, e; *f?* [græg, grig *grey?*] *The herb colt's foot;* tussilago farfara, Lin. v. Prior 51:—Coltgræig caballopodia vel ungula caballi, Ælfc. Gl. 44; Som. 64, 63; Wrt. Voc. 31, 73.

côl-þræd, -þrêd, es; *m. A coal or blackened thread, plumb-line;* perpendiculum:—Côlþrêd perpendiculum, Glos. Epnl. Recd. 160, 73.

coltræppe, an; *f? Ram, whin or Christ's thorn;* rhamnus = ῥάμνος, Cot. 156.

Coludes burh, burhg; *gen.* burge; *dat.* byrig; *f. Colud's city, Coldingham, Berwickshire, Scotland;* Coludi vel Coludana urbs, Colania, in agro Barovici:—Eóde Ædeldryþ on Æbban mynstre ðære Abbudissan, seó wæs Ecfriþes faðu ðæs cyninges, ðæt is geseted on ðære stôwe ðe mon nemneþ Coludes burh *Ædiltryda intravit monasterium Æbbæ abbatissæ, quæ erat amita regis Ecgfridi, positum in loco quem Coludi urbem nominant,* Bd. 4, 19; S. 587, 42. Ærðamðe ðæt mynster æt Coludes byrig mid byrne fornumen wære *priusquam monasterium Coludanæ urbis esset*

incendio consumptum, 4, 25; S. 599, 18. Hêr Coludes burh forbarn mid godcundum fýre *in this year* [A.D. 679] *Coldingham was burnt with divine fire,* Chr. 679; Erl. 41, 12. Ðæt nunmynster ðæt mon nemneþ Coludes burhg þurh ungýmenne synne fýres lîge wæs fornumen *monasterium virginum quod Coludi urbem cognominant per culpam incuriæ flammis absumptum est,* Bd. 4, 25; S. 599, 19.

Columba, an; *m. An Irish priest, the Apostle of the Highlands,* born about A. D. 520, and arrived in Scotland in 565. He preached to the Picts, whose king gave him the Western Isle, Iona, in which he founded his abbey and college. Columba was abbot 32 years, and died there, at the age of 77, on the 9th of June, 597 [Bd. 3, 4; S. 106, 107: *it is not in king Alfred's A. Sax. version*]. Columba is thus spoken of in the Chr. A. D. 565:—Columba, messapreóst, com to Pyhtum, and hí gecyrde to Cristes geleáfan; ðæt sind ðonne [ðone MS.] wærteras [MS. wærteres] be norþum môrum; and heora cyning him gesealde ðæt êgland ðe man nemnaþ Iî, ðær sindon v hîda, ðæs ðe men cweðaþ. Ðær se Columba getymbrade mynster; and he ðær wæs abbot xxxii wintra; and ðær forþférde, ðá ðá he wæs lxxvii wintra. Ða stôwe habbaþ nû git his erfewærdas [MS. erfewærdes]…. Nû, sceal beón æfre on Iî abbod, næs biscoep; and ðam sculon beón underþædde ealle Scotta biscoepas, forðam ðe Columba wæs abbod, nes biscoep *Columba, mass-priest, came to the Picts, and converted them to the faith of Christ; who are now dwellers by the northern mountains; and their king gave him the island which men name Iona, where there are five hides, from what men say. There Columba built a monastery; and he was abbot there thirty-two years, and there died when he was seventy-seven years. His inheritors yet have the place…. Now, in Iona, there must ever be an abbot, not a bishop; and to him must all the bishops of the Scots be subject, because Columba was an abbot, not a bishop,* Chr. 565; Th. 31, 29, col. 1–33, 7, col. 1.

com, *pl.* cômon came, Beo. Th. 865; B. 430: Cd. 160; Th. 199, 20; Exod. 341; *p. of* cuman.

comb, es; *m. A low place inclosed with hills, a valley;* vallis, Som. Ben. Lye. v. cumb.

combol, es; *n. An ensign, military standard.* DER. here-combol.

cométa, an; *m. A comet;* cométa, cométes, æ; *m.* = κομήτης, ου; *m. long-haired:*—Higegleáwe hátaþ cométa be naman *the wise-minded call a comet by name,* Chr. 975; Th. 228, 38, col. 1, 2, 3; Edg. 52.

commuc, es; *n. m? The cammoc, kex, brimstone wort;* peucedănum officinale, Lin, L. M. 3, 30; Lchdm. ii. 324, 20. v. cammoc.

comp, es; *m. A battle, contest;* certamen, pugna, Exon. 105 b; Th. 402, 26; Rä. 21, 35: 102 b; Th. 389, 3; Rä. 7, 2: Andr. Kmbl. 468; An. 234. v. camp.

comp-dôm *warfare,* Rtl. 8, 15. v. camp-dôm.

comp-gim *gen.* -gimmes; *m. A precious gem;* pretiosa gemma:—Mid ðam neorxna wonges compgimmum astæned *stoned with the gems of paradise,* Salm. Kmbl. 150, 10.

comp-hâd *warfare,* Som. Ben. Lye. v. camp-hâd.

compian *to fight, contend against;* militare, pugnare, Exon. 37 b; Th. 123, 1; Gû. 316: Bd. 1, 15; S. 483, 12: 3, 9; S. 533, 17: Ps. Lamb. fol. 183 b, 18. v. campian.

compung, e; *f. A combating, fighting, contest;* pugna, concertatio, Cot. 49.

comp-wæpen, es; *n. A battle-weapon, military weapon;* arma:—Oft ic gæstberend cwelle compwæpnum *I often kill the living with battle-weapons,* Exon. 105 b; Th. 401, 9; Rä. 21, 9. v. camp-wæpen.

comp-weorod, es; *n. An army;* exercitus, Bd. 2, 5; S. 507, 40. v. camp-wered.

comp-wîg, es; *m. n. A battle;* pugna:—Compwîge *in battle,* Judth. 12; Thw. 26, 18; Jud. 333.

con *I know, he knows; I, he can,* Cd. 227; Th. 304, 13; Sat. 629: Bd. 3, 24; S. 556, 16. v. cunnan.

côn, coon *bold,* Som. Ben. Lye. v. coon, cêne.

condel, condell, e; *f. A candle;* candela, lampas, Chr. 937; Th. 202, 16, col. 1; Ædelst. 15: Exon. 51 b; Th. 179, 20; Gû. 1264: 72 a; Th. 269, 23; Jul. 454. v. candel.

Cone-ceaster; *gen.* -ceastre; *f. Caster, a town seven miles from Newcastle;* oppidum septimo a Novo-castro milliario, N. Som. Ben. Lye.

conned *proved;* probatus, Lye. v. cunnian.

consolde, an; *f. The herb comfrey;* consolida:—Dô him ðis to læcedôme, streáwbergean leáf, consolde, etc. *give him this for a remedy, strawberry leaves, comfrey, etc.* L. M. 3, 63; Lchdm. ii. 350, 27.

const *knowest, canst,* Beo. Th. 2759; B. 1377; *2nd pers. pres. of* cunnan.

Constantînus, as *Lat. gen.* i; *dat.* o; *acc.* um; *m: also gen.* es; *dat.* e; *m. Constantine the Great, Roman Emperor,* A.D. 306–337. He is said to have been converted to Christianity, about 312, by the vision of a luminous cross in the sky, on which was the inscription ἐν τούτῳ, νίκα *by this, conquer.* In 330 he removed the seat of empire to Byzantium, which he called after his own name Κωνσταντίνου πόλις, *the city of Constantine,* CONSTANTINOPLE:—Fêrde Constantius forþ on Breotone, and Constantînus his sunu, ðam gôdan Câsere, his rîce forlêt.

Wríteþ Eutropius ðæt Constantínus, se Cásere, wǽre on Breotene acenned *Constantius died in Britain* [A.D. 306], *and left his kingdom to his son Constantine, the good emperor. Eutropius writes that the emperor Constantine was born in Britain,* Bd. 1, 8; S. 479, 30–32. Constantius, se mildesta man, fôr on Bryttanie, and ðǽr gefôr; and gesealde his suna ðæt ríce, Constantínuse, ðone he hæfde be Elenan his wífe *Constantius, the most merciful man, went into Britain, and died there; and gave the empire to Constantine, his son, whom he had by Helena his wife,* Ors. 6, 30; Bos. 126, 39–41. Notes and various readings, p. 28, col. 2, § 4, 41 h, MS. C. wífe; L. ciefese. Ðá wæs syxte geár Constantínes cáserdómes *then was the sixth year of Constantine's imperial power,* Elen. Kmbl. 15; El. 8. Ðá sige forgeaf Constantíno cyning ælmihtig þurh his rôde *then the king Almighty gave victory to Constantine through his cross,* 289; El. 145. Mid Constantíne *with Constantine,* Ors. 6, 31; Bos. 127, 42. *Also dat.* Constantínuse, 6, 30; Bos. 127, 7, 17, 23. v. Elene.

consul, es; *m. A consul; one of the two chief magistrates of the Romans chosen annually after the expulsion of their kings,* geár-cyning, *q. v;* consul:—Him ða Rômane æfter ðǽm [cyningum] látteówas gesetton, ðe hí consulas hêton, ðæt hiora ríce heólde ân geár ân man *after them [the kings] the Romans appointed over themselves leaders, whom they called consuls, that one man of them should hold power one year,* Ors. 2, 2; Bos. 41, 36. Brutus wæs se forma consul *Brutus was the first consul,* Ors. 2, 3; Bos. 41, 40, 41: 2, 4; Bos. 42, 27. Ân consul forðon [MS. þæne] triumphan *one consul [Fabius] declined the triumph,* 2, 4; Bos. 42, 43. Senátas cômon ongeán hyra consulas *the senators came to meet their consuls,* 2, 4; Bos. 43, 5, 20, 26. Under ðǽm twám consulum *under the two consuls,* 2, 4; Bos. 42, 33, 39: 2, 4; Bos. 43, 10, 16. Hæfdon him consulas, ðæt we cweðaþ rǽdboran *they had consuls, that we call counsellors,* Jud. Thw. 161, 22. [Consul, consul-ere *to consult, take counsel,* hence *counsellor.*]

consula bêc, cyninga bêc, *pl. f. Books of consuls, or kings' annals, calendars;* fastorum libri, fasti, Cot. 92.

Contwara burg *Canterbury,* Chr. 851; Erl. 66, 34. v. Cantwara burg.

Cont-ware *inhabitants of Kent,* Chr. 616; Erl. 20, 38. v. Cantware.

coon *bold,* Som. Ben. Lye. v. côn, cêne.

coorta, an; *m. A band of soldiers, cohort;* cohors:—He hæfde eahta ond hund-eahtatig coortena [MS. coortana], ðæt we nû truman hátaþ, ðæt wæs, on ðǽm dagum, fíf hund manna, and ân þúsend *he had eighty-eight cohorts, which we now call bands, each of which was, in those days, one thousand five hundred men,* Ors. 5, 12; Bos. 111, 14, 17.

cop; *gen.* coppes; *m. A top,* COP, *summit;* vertex, summitas:—Coppe summitate, Mone B. 1576.

côp, es; *m? A cope, an outer garment worn by priests;* ependytes = ἐπενδύτης:—Côp *vel* hoppada *vel* ufrescrúd *ependeton* [=ependytes], Ælfc. Gl. 112; Som. 79, 83; Wrt. Voc. 59, 52.

cope-man *a merchant,* Som. Ben. Lye. v. ceáp-man.

copenere, es; *m. A lover;* amator:—Ðú eart forlegen wið manigne copenere *tu fornicata es cum amatori multo,* Past. 52, 3; Hat. MS.

copest *chiefest, most precious;* pretiosissimus, Som. Ben. Lye. v. cop *a summit.*

copian; *p.* ode, ade; *pp.* od, ad *To plunder, pillage, steal;* compilare:—Copade and stæl *compilabat,* Cot. 53.

cop-líc *fit;* coplíce *fitly, well;* apte, Gr. Dial. 1, 1, Lye.

copor, es; *n? Copper;* cuprum:—Nim hwetstán brádne and gníd ða buteran on ðæm hwetstáne mid copore *take a large whetstone and rub butter on the whetstone with copper,* Lchdm. iii. 16, 22.

copp, es; *m. A cup, vessel;* calix, vas:—Calic oððe copp wætres *calicem aquæ,* Mk. Skt. Lind. 9, 41. Copp *vas,* Cot. 175. v. cuppe.

copped; *part.* [cop *a top*] *Having the top cut off, topped, polled;* capite recisus, decacuminatus:—To ðan coppede þorne *to the topped thorn,* Cod. Dipl. 1121; A.D. 939; Kmbl. v. 240, 28, 29. Andlang weges on ða coppedan âc *along the way to the polled oak,* Th. Diplm. A.D. 900; 145, 29.

COPS, cosp, es; *m. A rope, cord, fetter;* funis, anquina, compes:—Cops anquina [anguina, MS.], Ælfc. Gl. 104; Som. 78, 10; Wrt. Voc. 56, 56. Hí sǽdon ðæt hió sceolde sleán on ða raccentan and on cospas *they said that she should throw them into chains and fetters,* Bt. 38, 1; Fox 194, 32. [O. Sax. cosp, *m: Lat.* compes *a fetter.*] DER. fôt-cops, hand-, swur-.

corcíp, es; *m. An increase;* incrementum:—Loc hine geseon corcíp getácnaþ *capillum se videre incrementum significat,* Lchdm. iii. 212, 9. v. cíp.

coren *chosen, elected,* Chr. 675; Th. 58, 34; *pp.* of ceósan.

corenes, -ness, e; *f.* [coren, *pp.* of ceósan *to choose*] *An election, a choice;* electio, C. R. Ben. Lye. DER. ge-corenes, wið-, wiðer-.

corfen *cut, carved,* Exon. 107 b; Th. 410, 24; Rä. 29, 4; *pp.* of ceorfan.

Corfes geat, Corf-geat, es; *n.* [*Sim. Dun.* Coruesgeate: *Hovd.* Coruesgate] *Corfgate, Purbeck, Dorsetshire:*—Hér wæs Eádweard cyning ofslæ-

gen æt Corfes geate [Corfgeate, Th. 233, 2, col. 2] *in this year* [A.D. 979] *king Edward was slain at Corfgate,* Chr. 979; Th. 232, 3, col. 2.

corflian; *p.* ode; *pp.* od [ceorfan *to cut*] *To cut up small, mince;* concidere:—Ðás wyrta sý swýðe smæl corflode *let these herbs be minced very small,* Lchdm. iii. 292, 5.

coríon, es; *n?* [= κόριον for κορίαννον = κορίανον, Anac. 138] *The herb coriander;* coriandrum [ὑπέρικον hypericon, Diosc. 3, 171], Som. Ben. Lye. v. celendre.

CORN, es; *n.*   I. CORN, *a grain, seed, berry;* frumentum, granum, bacca:—Corn *frumentum,* Ælfc. Gl. 59; Som. 67, 122; Wrt. Voc. 38, 44. Wæs corn swá dýre, swá nán man ǽr ne gemunde *corn was so dear, as no man before remembered it,* Chr. 1044; Erl. 168, 21: Homl. Th. ii. 68, 17. Hie wǽron benumene ǽgðer ge ðæs ceápes ge ðæs cornes *they were deprived both of the cattle and of the corn,* Chr. 895; Erl. 93, 18: Bd. de nat. rerum; Wrt. popl. science 10, 8; Lchdm. iii. 254, 4. Se Dêma gegadaraþ ðæt clǽne corn into his berne *the Judge will gather the pure corn into his barn,* Homl. Th. ii. 68, 18: Chr. 894; Erl. 93, 11. Hý heora corn ripon *they reaped their corn,* Ors. 4, 8; Bos. 90, 33: Chr. 896; Erl. 94, 6: Past. 52; Hat. MS. Corn *granum,* Wrt. Voc. 83, 16. Ðæt hwætene corn wunaþ âna *granum frumenti solum manet,* Jn. Bos. 12, 24; Bt. 35, 1; Fox 156, 2, 4. Senepes corn *granum sinapis,* Lk. Bos. 17, 6. Heofena ríce is geworden gelíc senepes corne, ðæt seów se man on hys æcre *simile est regnum cælorum grano sinapis, quod homo seminavit in agro suo,* Mt. Bos. 13, 31: Lk. Bos. 13, 19. Hægl byþ hwítust corna *hail is the whitest of grains,* Runic pm. 9; Kmbl. 341, 4; Hick. Thes. i. 135. Se æppel monig corn oninnan him hæfþ *the apple has many seeds inside it,* Past. 15, 5; Hat. MS. 19 b, 23. Ífig byrþ corn golde gelíce *ivy bears berries like gold,* Herb. 121, 1; Lchdm. i. 234, 4. Genim ðysse wyrte twentig corna *take twenty grains of this herb* [*ivy*], 121, 2; Lchdm. i. 234, 6.   II. *a hard* or *cornlike pimple, a corn, kernel on the feet;* pustula, clavus:—Ðis mæg horse wið ðon ðe him biþ corn on ða fêt *this may be for a horse which has corns on his feet,* Lchdm. iii. 62, 22. [*Prompt.* corne: *Wyc. Chauc. R. Glouc.* corn: *Laym.* corn, *n: Orm.* corn: *Plat.* koren, koorn: *O. Sax.* korn, korni, kurni, *n: O. Frs.* korn: *Dut.* kóren, *n: Ger. M. H. Ger. O. H. Ger.* korn, *n: Goth.* kaurno, *n. a grain of corn; Dan. Swed. Icel.* korn, *n. a grain of corn.*] DER. giþ-corn, mete-, sand-, sund-.

corn-æsceda *Corn-sweepings, chaff;* quisquiliæ:—Æppelscreáda *vel* cornæsceda *quisquiliæ,* Ælfc. Gl. 17; Som. 58, 97; Wrt. Voc. 22, 13.

corn-appla, *pl. n. Pomegranates;* mala Punica, Mone B. 3822.

corn-bǽre; *adj. Corn-bearing;* graniger:—Corn-bǽre *graniger,* Ælfc. Gr. 8; Som. 7, 20: Homl. Th. i. 450, 11. Cornbǽrum *granigera,* Mone B. 1435.

corn-gesǽlig; *adj.* [gesǽlig *fortunate, rich*] *Wealthy in corn;* frumento opulentus:—Cild corngesǽlig biþ *a child will be wealthy in corn,* Obs. Lun. § 9; Lchdm. iii. 188, 11.

corn-gesceót, es; *n? A payment* or *contribution of corn;* frumenti solutio *vel* munus:—Se wudu beó gelǽst binnan þrým dagum æfter ðam corngesceóte *let the wood be supplied within three days after the contribution of corn,* Cod. Dipl. 942; Kmbl. iv. 278, 10.

corn-hrycce, an; *f. A* CORN-RICK; frumenti acervus:—Wearþ gemêt ðæt feoh uppon ânre cornhryccan *the money was found upon a corn-rick,* Homl. Th. ii. 178, 8.

corn-hús, es; *n. A corn-house, granary;* granarium, Ælfc. Gl. 109; Som. 78, 130; Wrt. Voc. 58, 42.

corn-hwæcca, an; *m. A corn-chest, bin;* arca frumentaria. v. hwæcca, Som. Ben. Lye.

cornoch, es; *m. A crane;* grus, Som. Ben. Lye.

corn-treów, es; *n. A cornel-tree;* cornus:—Corntreów *cornus,* Ælfc. Gl. 46; Som. 64, 124; Wrt. Voc. 32, 58: Cot. 49.

corn-troh -trog, es; *m.* [troh *a trough*] *A corn-trough, bin, a vessel for cleansing grains of corn;* cista frumentaria, capisterium:—Corntroh *capisterium,* Ælfc. Gl. 3; Som. 55, 62; Wrt. Voc. 16, 35.

Corn-weal, es; *m.* CORNWALL, Cornubia, Som. Ben. Lye.

Corn-wealas; *gen.* -weala; *dat.* -wealum; *pl. m. Cornishmen, the inhabitants of Cornwall in a body, Cornwall;* Cornubienses, Cornubia:—Cômon hí to lande on Cornwealum *they came to land in Cornwall,* Chr. 892; Th. 160, 39, col. 3: 997; Erl. 134, 8. v. Wealh.

corn-wurma, an; *m. A corn-worm, weevil;* vermiculus, Ælfc. Gl. 17; Som. 58, 84; Wrt. Voc. 22, 2.

cors, es; *m. A curse;* execratio, Ben. Lye. v. curs.

corsian *to curse,* Ben. Lye. v. cursian.

cor-snǽd, e; *f.* [cor, cer, cyrr *a choice;* snǽd *a bit, piece*] *A choice or trial piece;* panis conjurátus, offa consecráta. A sort of ordeal in which the person accused had placed in his mouth an ounce of bread or cheese. If he ate it freely and without hurt, he was considered innocent; but guilty, if he could not swallow it, or had a difficulty in doing so. The Host was used for this purpose in Christian times:—Gif man freónd-leásne weofod-þên mid tihtlan belecge, gá to corsnǽde *if a friendless servant of the altar be charged with an accusation, let him go to the*

corsnǽd, L. Eth. ix. 22; Th. i. 344, 23: L. C. E. 5; Th. i. 362, 19. To corsnǽde *to the* corsnǽd, Th. i. 362, 25 : Th. i. 344, 29.

corþer; *gen.* corþres; *n* : corþer; *gen.* corþre; *f. A band, multitude, company, troop, body, train, pomp;* multitudo, cohors, copia, pompa :—Cirmdon caldheorte, corþer óðrum getang *the cold-hearted cried out, troop thronged on troop,* Andr. Kmbl. 276; An. 138. Cyning corþres georn *a king desirous of pomp,* Cd. 176; Th. 221, 28; Dan. 95. Wǽron ealle ætgædere cyningas on corþre *the kings were altogether in a body,* 151; Th. 189, 27; Exod. 191: 166; Th. 207, 11; Exod. 465: Exon. 15 a; Th. 31, 11; Cri. 494: 46 a; Th. 156, 25; Gú. 880. Stígeþ cirm on corþre *clamour arises in the company,* 83 b; Th. 314, 26; Mód. 20. Cyning on corþre *a king amid his train,* Beo. Th. 2310; B. 1153: Ps. Th. 54, 16. On wera corþre *in the company of men,* Elen. Kmbl. 608; El. 304: 1081; El. 543: 140; El. 70. Heó cleopode fór corþre *she cried before the assemblage,* Exon. 74 b; Th. 279, 23; Jul. 618: Bt. Met. Fox 26, 169; Met. 26, 85: Andr. Kmbl. 3428; An. 1718. Se sunu Wihstánes acígde of corþre cyninges þegnas *the son of Wihstan called the king's thanes from the band,* Beo. Th. 6233; B. 3121. Mid corþre *with a troop,* Andr. Kmbl. 2151; An. 1077: 2244; An. 1123: 2410; An. 1206: Elen. Kmbl. 1379; El. 691. Corþre *ne lytle with no little train,* Exon. 16 a; Th. 36, 19; Cri. 578. Hér Eádgár wæs Engla waldend corþre micelre *in this year* [A. D. 973] *Edgar became ruler of the Angles with much pomp,* Chr. 973; Erl. 124, 10; Edg. 2. Hí cwómon in ða ceastre corþra mǽste *they came to the city with the greatest of companies,* Elen. Kmbl. 548; El. 274: Exon. 58 a; Th. 209, 7; Ph. 167. Corþrum miclum *in large bands,* Cd. 80; Th. 99, 27; Gen. 1652: 112; Th. 148, 7; Gen. 2453. [*O. H. Ger.* kortar, *n. grex : Lat.* cohors, *gen.* cohortis = cors, *gen.* cortis *a company.*] DER. hilde-corþer, mægen-.

cor-wurma, an; *m. A purple colour;* mūrex :—Corwurmum *mūrĭcĭbus,* Mone B. 6170.

COS, coss, es; *m. A* KISS; osculum :—Cos *osculum,* Wrt. Voc. 72, 44. Ic hine to mínum cosse arǽrde *I raised him to my kiss,* Homl. Th. ii. 32, 11. Coss ðú me ne sealdest *osculum mihi non dedisti,* Lk. Bos. 7, 45. Mannes sunu ðú mid cosse sylst *osculo filium hominis tradis,* 22, 48. Betwux ðám cossum *between the kisses,* Homl. Th. i. 566, 19. Cossas syllan hearm getácnaþ *to give kisses betokens harm,* Lchdm. iii. 208, 27. [*Wyc.* cos, coss, cosse : *Laym.* coss : *Plat.* kuss : *O. Sax.* kus, *m* : *O. Frs.* kos, *m* : *Dut. Kil.* kus, *m* : *Ger.* kuss, *m* : *M. H. Ger.* kus, *m* : *O. H. Ger.* kus, *m* : *Dan.* kys, *n* : *Swed.* kyss, *m* : *Icel.* koss, *m* : *Wel.* cusan, *m* : *Corn.* cussin, *m* : *Sansk.* kus *to embrace.*]

Coshám, es; *m.* COSHAM *or* CORSHAM, *Wilts*; loci nomen in agro Wiltoniensi :—Læg se cyng seóc æt Coshám *the king lay sick at Corsham,* Chr. 1015; Erl. 152, 13.

cosp, es; *m. A fetter;* compes :—On cospas *into fetters,* Bt. 38, 1; Fox 194, 32. v. cops.

cossas *kisses,* Lchdm. iii. 208, 27; *acc. pl. of* cos.

cossian; *p.* ode; *pp.* od [cos *a kiss*] *To kiss;* osculari :—Heó hit cossode *she kissed it,* Homl. Th. i. 566, 19. v. cyssan.

cost, es; *m? The herb costmary;* costus = κόστος, balsamita vulgaris, Lin :—Cost *costus,* Ælfc. Gl. 39; Som. 63, 71: Wrt. Voc. 30, 23 : 79, 21. Costes gódne dǽl gebeát smǽle and gegníd to duste *beat small a good deal of costmary and rub to dust,* L. M. 2, 55; Lchdm. ii. 276, 6 : 2, 24; Lchdm. ii. 212, 26. Genim pipor and cymen and cost *take pepper and cummin and costmary,* 1, 17; Lchdm. ii. 60, 15 : 1, 23; Lchdm. ii. 66, 9 : 1, 47; Lchdm. ii. 120, 9. Ænglisc [MS. Æncglisc] cost *English costmary, tansy;* [tanacetum vulgare, Lin.], Lchdm. iii. 24, 8.

cost; *adj.* [costian *to tempt, try, prove*] *Tried, proved;* probatus :—Cempan coste cyning weorþodon *the tried champions glorified the king,* Andr. Kmbl. 2111; An. 1057. DER. ge-cost.

costere, costnere, es; *m. A tempter;* tentator :—Manna cynnes [MS. manna kynnes] costere hafaþ acenned on ðé ða unablinnu ðæs yfelan geþohtes *the temptᵒʳ of mankind* [lit. *of the race of men*] *hath begotten in thee the unrest of this evil thought,* Guthl. 7; Gdwin. 46, 9. Se costere cwæþ to him *tentator dixit ei,* Mt. Kmbl. Rush. Lind. 4, 3.

costere, es; *m? A digging tool, spade;* fossorium :—Costere *vel* delfísen *vel* spadu *vel* pal *fossorium,* Ælfc. Gl. 2; Som. 55, 40; Wrt. Voc. 16, 14.

COSTIAN, costigan, costnian; *p.* ode, ade, ede; *pp.* od, ad, ed *To tempt, try, prove;* probare, tentare. **I.** *v. trans. gen. acc.* **1.** *with the genitive;* cum genitivo :—Ðæs rinces se ríca ongan cyning costigan *the powerful king began to tempt the chief,* Cd. 137; Th. 172, 18; Gen. 2846. Ðú mín costadest, Drihten *Domine, probasti me,* Ps. Th. 138, 1. He mín costode *he tried me,* Beo. Th. 4175; B. 2084. Úre costade, God *probasti nos, Deus,* Ps. Th. 65, 9. Costodon mín *tentaverunt me,* Ps. Spl. C. M. 94, 8. Hí Godes costodon [MS. costodan] *tentaverunt Deum,* Ps. Th. 77, 41. Hí on wéstenne heora Godes costedon [MS. costedan] *tentaverunt Deum in inaquoso,* 105, 12, 31. Costa mín, God *proba me, Deus,* 138, 20. **2.** *with the accusative;* cum accusativo :—He ðæt folc costian lét *he let* [*them*] *try the people,* Ors. 6, 3; Bos. 118, 6. He costode cyning alwihta *he tempted the king of all*

creatures, Cd. 228; Th. 306, 28; Sat. 671: Homl. Blick. 29, 24, 34. Hí costodon God *tentaverunt Deum,* Ps. Spl. 105, 14: Mt. Bos. 16, 1. Ne costa ðú ðínne Drihten God *tempt not the Lord thy God,* Homl. Blick. 29, 33 : Ps. Spl. C. T. 25, 2. **II.** *v. intrans* :—Ðonne bryne costaþ hú gehealdne sind sáwle wið synnum *when the burning proveth how abstinent are souls from sins,* Exon. 23 b; Th. 65, 24; Cri. 1059. Feówertig daga he wæs fram deófle costod *diebus quadraginta tentabatur a diabolo,* Lk. Bos. 4, 2: Homl. Blick. 29, 14. [*Laym.* i-costned, *pp. proved, tried : O. Sax.* koston *to try, tempt : Ger.* kosten *to taste, try by tasting;* tentare, gustare : *O. H. Ger.* koston *tentare : Goth.* kausyan *to taste : Icel.* kosta *to try, tempt.*] DER. fore-costian, ge-.

costigan *to tempt,* Cd. 137; Th. 172, 18; Gen. 2846. v. costian.

costigend, costnigend, es; *m. A tempter;* tentator :—Se costigend eóde to him *the tempter went to him,* Homl. Blick. 27, 4. Se costnigend *tentator,* Mt. Bos. 4, 3.

costing *a temptation,* Exon. 33 a; Th. 104, 18; Gú. 9. v. costnung.

costnere, es; *m. A tempter;* tentator :—Swá swá se geleáfa strengra biþ, swá biþ ðæs costneres miht lǽsse *as the faith is stronger, so is the might of the tempter less,* Homl. Th. ii. 392, 20. v. costere.

costnes, -ness, e; *f. A temptation;* tentatio, Som. Ben. Lye. DER. ge-costnes.

costnian; *part.* costnigende; *p.* ode; *pp.* od; *v. trans. gen. acc. To tempt;* tentare :—Hyne costnigende *tentantes eum,* Mt. Bos. 19, 3. Ic hys costnode *I tempted him,* Nicod. 26; Thw. 14, 15. Costnodon me *tentaverunt me,* Num. 14, 22 : Ps. Lamb. 94, 9. Afanda me Drihten, and costna me *proba me Domine, et tenta me,* Ps. Spl. 25, 2. Ne costna ðú Drihten ðínne God *non tentabis Dominum Deum tuum,* Mt. Bos. 4, 7 : Lk. Bos. 4, 12. v. costian.

costnigend, es; *m. A tempter;* tentator, Mt. Bos. 4, 3. v. costigend.

costnung, costung, costing, e; *f.* [costnian, costian *to tempt, try*] *A temptation, trying, trial, tribulation;* tentatio, probatio, tribulatio :—Ðeós costnung is of ðam níþfullan deófle *this temptation is from the malicious devil,* Boutr. Scrd. 23, 10, 8. Wæs seó ǽreste costung ofercumen *the first temptation was overcome,* Exon. 39 a; Th. 128, 24; Gú. 409: Homl. Th. ii. 156, 26: Ex. 17, 7. On ðære costnunge tíman *in tempore tentationis,* Lk. Bos. 8, 13. Æfter dæge costunge *secundum diem tentationis,* Ps. Spl. 94, 8. Ne gelǽd ðú us on costnunge *ne nos inducas in tentationem,* Mt. Bos. 6, 13 : 26, 41: Mk. Bos. 14, 38: Lk. Bos. 11, 4 : 22, 40, 46: Homl. Th. ii. 596, 9: 600, 16. On costunge *in tentatione,* Deut. 9, 22. Sindan costinga monge arisene *many temptations are arisen,* Exon. 33 a; Th. 104, 18; Gú. 9. Ðæt he us gescylde wið ða þúsendlícan crǽftas deófles costunga *that he shield us from the thousand crafts of the devil's temptations,* Homl. Blick. 19, 17. Micle costnunge ge gesáwon *tentationes magnas viderunt oculi tui,* Deut. 29, 3. Drecþ se deófol mancynn mid mislícum costnungum *the devil vexes mankind with various temptations,* Boutr. Scrd. 19, 44. Seó costnung ðære éhtnesse gestilled wæs *the trial of the persecution was stilled,* Bd. 1, 8; S. 479, 19. Me costung and sár cnyssedan *tribulation and sorrow troubled me,* Ps. Th. 114, 4. Hí on costunge cleopedan to Drihtne *clamaverunt ad Dominum cum tribularentur,* 106, 12, 18, 27 : 117, 5 : 142, 12. Ðonne me costunge cnysedon *in die tribulationis meæ,* Ps. Th. 58, 17 : 65, 13. Me costunga cnysdan *tribulatio et angustia invenerunt me,* Ps. Th. 118, 143 : 119, 1 : 137, 7. DER. nýd-costing.

costung, e; *f. A temptation, trying;* tentatio, tribulatio, Ex. 17, 7 : Ps. Spl. 94, 8 : Deut. 9, 22 : Ps. Th. 114, 4. v. costnung.

COT, cott, es; *pl. nom. acc.* cotu; *gen.* cota; *dat.* cotum, cottum; *n. A* COT, *cottage, house, bed-chamber, den;* casa, domus, cubiculum, cubile, spelunca :—Onbútan ða cotu *about the cots,* Cod. Dipl. 551; A. D. 969; Kmbl. iii. 35, 6. Ongeán ða cotu *towards the cots,* 559; A. D. 969; Kmbl. iii. 52, 16. We witan ðæt hý ne durran hý selfe æt hám æt heora cotum werian *we know that they dare not defend themselves at home in their own houses,* Ors. 3, 9; Bos. 69, 26. Ingá in cotte ðínum *intra in cubiculum tuum,* Mt. Kmbl. Lind. 6, 6. In cotum [Lind. cottum] *in cubiculis,* Lk. Skt. Rush. 12, 3 : 11, 7. Ge worhton ðæt to þeófa cote *fecistis illam speluncam latronum,* Mt. Bos. 21, 13. [*Prompt.* coote: *Wyc.* Piers P. cotes, *pl* : *Chauc.* cote : *Plat.* kate, katen : *Dut.* kot, *n* : *Ger.* kot, *n* : *Dan.* koje, *m. f* : *Swed.* kette, *m* : koja, *f* : *Icel.* kot, *n* : *Wel.* cwt : *Gael.* cot, *m.*]

cote, an; *f. A cot, cottage, house;* casa, domus :—Gif hwilc man forstolen þingc hám to his cotan bringe *if any man bring a stolen thing home to his house,* L. C. S. 77; Th. i. 418, 18. v. cyte.

cóða *diseases;* nom. gen. acc. pl. of cóðu.

cóð-líce; *adv.* [cóða, cóðu *a disease*] *Badly, miserably;* male, misere :—Cóðlíce racentan geræped *miserably bound in chains,* Bt. Met. Fox 25, 72; Met. 25, 36.

cóðu, e; *f:* cóðe, an; *f:* cóða, an; *m. A disease, sickness, pestilence;* morbus :—Mycel orfes wæs ðæs geáres forfaren þurh mistlíce cóða *much cattle was destroyed this year through various diseases,* Chr. 1041; Erl. 169, 9. Swylc cóðe com on mannum ... ðæt mænige swulton *such a disease came on men ... that many died,* Chr. 1087; Th. 353, 37. Seó miccle cóðu *the great disease, leprosy;* elephantinus morbus, Homl. Th. ii. 480, 10.

Seó côðu ðe lǽcas hâtaþ paralisin *the disease which physicians call palsy*, ii. 546, 29. He fram ðǽre côðe hine gehǽlde *he healed him from the disease*, i. 400, 10. Wið wambe côðum *for diseases of the stomach*, L. M. 2, 32; Lchdm. ii. 234, 1. DER. ban-côða, -côðu, bræc-, eár-, fǽr-, fôt-, heort-, in-, mûþ-, sweor-, un-.

cot-líf, es; *pl. nom. acc. -líf; gen. -lífa; n. [cot a cot, cottage; líf*, II. *a place to live in] A village;* villa:—Ðæt cotlíf *the village*, Cod. Dipl. 828; A.D. 1066; Kmbl. iv. 191, 13: 845; Kmbl. iv. 204, 31: 855; Kmbl. iv. 211, 25: 859; Kmbl. iv. 214, 6: 864; Kmbl. iv. 217, 7. He bohte feola cotlíf *he bought many villages*, Chr. 963; Erl. 121, 24. Hý forbærndon ôðra cotlífa fela *they burned many other villages*, 1001; Erl. 136, 32.

cot-sǽta, an; *m. An inhabitant of a cottage, a cottager;* casæ habitator, Som. Ben. Lye.

cot-setla, cote-setla, an; *m.* [MS. kot-setla, kote-setla] *A cottager;* casárius:—Cotsetlan [MS. kotsetlan] riht *a cottager's right*, L. R. S. 3; Th. i. 432, 15. Cotesetlan [MS. kotesetlan] riht, be ðam ðe on lande stent. On sumon he sceal ælce Môndæge ofer geáres fyrst his láforde wyrcan, ôðð iii dagas ælcre wucan on hærfest: ne þearf he landgafol syllan. Him gebýriaþ v æceras to habbanne, mâre gyf hit on lande þeáw sý, and tô lytel hit biþ beó hit ǽ læsse, forðan his weorc sceal beón oft rǽde. Sylle his heorþ-pænig on hálgan þunres dæg, eal swâ ælcan frigean men gebýreþ, and werige his hláfordes inland, gif him man beóde æt sǽ-wearde and æt cyniges deór-hege, and æt swilcan þingan swilc his mǽþ sý, and sylle his ciric-sceát to Martinus mæssan *cotsetle rectum est juxta quod in terra constitutum est. Apud quosdam debet omni die Lunae, per anni spatium, operari domino suo, et tribus diebus unaquaque septimana in Augusto. [Apud quosdam, operatur per totum Augustum, omni die, et unam acram avene metit pro diurnale opere. Et habeat garbam suam quam praepositus vel minister domini dabit ei.] Non dabit landgablum. Debet habere quinque acras ad perhabendum, plus si consuetudo sit ibi, et parum nimis est si minus sit quod deservit, quia saepius est operi illius. Det super heorþpenig in sancto die Jovis, sicut omnis liber facere debet, et adquietet inland domini sui, si submonitio fiat de sewarde, id est, de custodia maris, vel de regis deorhege, et ceteris rebus quae suae mensurae sunt: et det suum cyricsceatum in festo sancti Martini*, L. R. S. 3; Th. i. 432, 16–434, 2.

cot-stôw, e; *f.* [stôw *a place] A place of cottages;* casarum situs:—On ða ealdan cotstôwa *to the old cot-places*, Cod. Dipl. 578; A. D. 973; Kmbl. iii. 97, 30.

cott *a bed-chamber*, Mt. Kmbl. Lind. 6, 6: Lk. Skt. Lind. 11, 7: 12, 3. v. cot.

cottuc, es; *m. Mallow;* malva:—Cottuc wyl on wætere *boil mallow in water*, L. M. 1, 32; Lchdm. ii. 78, 19: 1, 60; Lchdm. ii. 130, 23. Nim niðeweardne cottuc *take the netherward part of mallow*, 1, 68; Lchdm. ii. 144, 5.

cowen *chewed, eaten; pp. of* ceówan.

coxre *a quiver*, Som. Ben. Lye. v. cocer.

CRABBA, an; *m.*   I. *A* CRAB, *crayfish;* cancer:—Crabba cancer, Ælfc. Gl. 102; Som. 77, 74; Wrt. Voc. 55, 78: 77, 68. Hwæt fêhst ðú on sǽ? Crabban and lopystran *quid capis in mari? Cancros et polypodes*, Coll. Monast. Th. 24, 11.   II. *a sign of the zodiac, cancer;* signum zodiaci, cancer:—Feórþa ðæra tâcna ys gehâten *cancer*, ðæt is crabba *the fourth of the signs is called* cancer, *that is, a crab*, Bd. de nat. rerum; Wrt. popl. science 7, 5; Lchdm. iii. 244, 25. [Dut. krab, *f: Kil.* krabbe: *Ger.* krabbe, *f;* krebs, *m: M. H. Ger.* krebez, *m: O. H. Ger.* chrêpazo, *m: Dan.* krabbe, *m. f: Swed.* krabba, *f: Icel.* krabbi, *m: Lat.* karabus, *m: Grk.* κάραβος, *m. a crab: Sansk.* śarabha, *m. a grasshopper, crab.*]

cracettan *to* CROAK; crocitare, Gr. Dial. 2, 8, Som. Ben. Lye.

Crac-gelád *Cricklade*, Chr. 905; Th. 180, 21, col. 2. v. Crecca-gelád.

CRACIAN; *part.* craciende; *p.* ode; *pp. od To* CRACK, *quake;* crepare:—Craciendum *crepante*, Mone B. 123. Sió eorþe eall cracode *the whole earth quaked*, Ps. Th. 45, 3. [Piers P. craked *broke: Chauc.* crakke: *Laym.* crakeden, chrakeden, *p. pl: Plat. Dut.* kraken: *Ger. M. H. Ger.* krachen: *O. H. Ger.* krachjan, krachôn: *Gael.* crac *crepare.*]

CRADEL, cradol, es; *m. A* CRADLE; cunabula:—Cradel *cunabula*, pl. [MS. *cunabulum*], Ælfc. Gl. 27; Som. 60, 112; Wrt. Voc. 25, 52. On cradele [MS. B. cradole] *in a cradle*, L. C. S. 77; Th. i. 420, 1. [*Prompt.* credel, cradel: *R. Brun.* credille: *Chauc. R. Glouc.* cradel: *Gael.* creathail, *f. a cradle.*] DER. cild-cradol.

cradol *a cradle*, L. C. S. 77; Th. i. 420, 1, MS. B. v. cradel.

cradol-cild, es; *n. A cradle-child, infant;* e cunabulis infans:—Syndon cradolcild geþeówode *infantes e cunabulis sunt mancipati*, Lupi Serm. 1, 5; Hick. Thes. ii. 100, 30.

cræcetung, e; *f. A croaking;* crocitatio:—Cræcetung hræfena *the croaking of ravens*, Guthl. 8; Gdwin. 48, 4.

Cræcilád *Cricklade*, Chr. 1016; Erl. 153, 5. v. Crecca-gelád.

cræfian *to crave*, Cod. Exon. 5 b, Lye. v. crafian.

CRÆFT, es; *m.*   I. *power, might, strength as of body or*

externals; vis, robur, potentia:—On ðam gefeohte Mêda cræft gefeól *in that battle the power of the Medes fell*, Ors. 1, 12; Bos. 35, 43. He cwæþ ðæt ðín abal and cræft mâra wurde *he said that thy strength and power would become greater*, Cd. 25; Th. 32, 9; Gen. 500: 155; Th. 193, 13; Exod. 245: 212; Th. 262, 3; Dan. 738: Beo. Th. 2571; B. 1283. His ágnes cræftes *of his own strength*, Bt. 16, 2; Fox 54, 5. Þurh his cræftes miht *by the might of his power*, Andr. Kmbl. 1170; An. 585: Elen. Kmbl. 1112; El. 558: Exon. 24 b; Th. 70, 29; Cri. 1146. He cræft mâran hæfde *he had greater power*, Cd. 14; Th. 18, 6; Gen. 269: 22; Th. 27, 12; Gen. 416: 23; Th. 266, 3; Jul. 392: Exon. 33 b; Th. 107, 14; Gú. 58: Beo. Th. 1402; B. 699. Nýdaþ cræfte tíd *the tide forces it with power*, Salm. Kmbl. 790; Sal. 394: Cd. 23; Th. 29, 13; Gen. 449: Exon. 71 b; Th. 266, 3; Jul. 392: Beo. Th. 1969; B. 982. Mid eallum hiora cræftum *with all their forces*, Ors. 1, 13; Bos. 37, 4: Exon. 109 a; Th. 417, 24; Rä. 36, 9. He his dryhtne hýrde þurh dýrne cræftas *he obeyed his lord through secret powers*, Salm. Kmbl. 904; Sal. 451: Cd. 184; Th. 230, 1; Dan. 226: Exon. 88 b; Th. 332, 33; Vy. 94: 92 b; Th. 346, 27; Sch. 5.   II. *an art, skill,* CRAFT, *trade, work;* ars, peritia, artificium, occupatio, opus:—Se cræft ðæs lareówdômes biþ cræft ealra cræfta *the art of teaching is the art of all arts*, Past. 1, 1; Hat. MS. 6 b, 8. Cræft ars, Wrt. Voc. 73, 35. Wolde ic ânes to ðê cræftes neósan *I would inquire of one art from thee*, Andr. Kmbl. 968; An. 484. He byþ forlǽten fram ðam cræfte *ipse dimittetur ab arte*, Coll. Monast. Th. 31, 35. Ic gearcie hig mid cræfte mínum [MS. minon] *praeparo eas arte mea*, 27, 31: Bt. 39, 4; Fox 216, 24. Seó þeód ðone cræft ne cûðe ðæs fiscnôþes *the people knew not the art of fishing*, Bd. 4, 13; S. 582, 43. Betweoh ðâs cræftas *inter istas artes*, Coll. Monast. Th. 30, 17. On his mycclum cræfte *by his great skill*, Hexam. 1; Norm. 4, 3. Nân mon ne mæg nænne cræft cýðan bûtan tôlum *no man can shew any skill without tools*, Bt. 17; Fox 58, 29: Boutr. Scrd. 17, 8. Wundorlíce cræfte ðú hit hæfst gesceapen *with wonderful skill thou hast made it*, Bt. 33, 4; Fox 130, 11: Ors. 1, 12; Bos. 35, 35. Cræft biþ betere ðonne ǽhta *a craft [=trade] is better than wealth*, Prov. Kmbl. 20: Coll. Monast. Th. 27, 27: 28, 5, 7, 9: 30, 11. Ælces cræftes andweorc *the materials of any craft*, Bt. 17; Fox 58, 30. Hwæt begytst ðú of ðínum cræfte *what gettest thou by thy trade?* Coll. Monast. Th. 23, 3: 28, 3, 31. Ðeáh ðê ðíne sǽlþa forlǽton, ne forlǽt ðú ðínne cræft *though thy wealth desert thee, desert not thou thy trade*, Prov. Kmbl. 57: Coll. Monast. Th. 21, 1, 11: 22, 35, 37: Bt. 17; Fox 58, 31: 17; Fox 60, 2. Mistlícra cræfta biggenceras *workers of various trades*, Coll. Monast. Th. 30, 1. To cræftum [MS. cræftan] *to educate in trades*, L. Edg. C. 51; Th. ii. 254, 26. Gif ðú bearn hæbbe, lǽr ða cræftas, ðæt hí mægen be ðám libban *if thou have children, teach them trades, that they may live by them*, Prov. Kmbl. 20: 57. Seó cwên bebeád cræftum getýde girwan Godes tempel *the queen commanded men skilled in crafts [=trades] to make a temple of God*, Elen. Kmbl. 2034; El. 1018. Wæs ǽfre unbegunnen Scyppend, se ðe gemacode swylcne cræft *the Creator, who made such a work, was ever without beginning*, Hexam. 1; Norm. 4, 5.   III. *craft of mind, cunning, knowledge, science, talent, ability, faculty, excellence, virtue;* astutia, machinatio, scientia, facultas, praestantia, virtus:—Þurh deófles cræft *through the devil's craft*, Cd. 25; Th. 31, 29; Gen. 492. Ðeáh Eue on deófles cræft bedroren wurde *though Eve had been deceived by the devil's craft*, 38; Th. 51, 7; Gen. 823: Exon. 17 b; Th. 43, 7; Cri. 685: Andr. Kmbl. 2590; An. 1296: Frag. Kmbl. 56; Leás. 30. Feóndes cræfte *by a fiend's craft*, Andr. Kmbl. 2394; An. 1198: Exon. 71 a; Th. 264, 5; Jul. 359. Mínum cræftum *by my devices*, 72 b; Th. 271, 11; Jul. 480. Beald biþ se ðe onbýrigeþ bôca cræftes *he is bold who tasteth of book-knowledge*, Salm. Kmbl. 484; Sal. 242. On bôclícum cræfte *in book-knowledge*, Boutr. Scrd. 17, 5. Ða cnihtas cræft leornedon *the youths learned science*, Cd. 176; Th. 221, 5; Dan. 83. Ic wilnode ðæt míne cræftas ne wurden forgitene *I was desirous that my talents should not be forgotten*, Bt. 17; Fox 60, 9. Ða yfelan ǽfre habbaþ nænne cræft *the wicked never have any ability*, 36, 3; Fox 174, 35. Seó gesceádwísnes is synderlíc cræft ðære sáwle *reason is a peculiar faculty of the soul*, 33, 4; Fox 132, 10: 32, 1; Fox 116, 3. Ða cræftas ðe we ǽr ymbe sprǽcon ne sint to wiðmetanne wið ðære sáwle cræfta ǽnne *the faculties which we have before spoken about are not to be compared with any one of the faculties of the soul*, 32, 1; Fox 116, 1, 2, 4. Omêrus on his leóþum swíðe hêrede ðære sunnan cræftas *Homer in his poems greatly praised the sun's excellences*, 41, 1; Fox 244, 7. Sint ða cræftas betran ðonne ða unþeáwas *the virtues are better than the vices*, 36, 5; Fox 180, 15. Simmachus is wísdômes and cræfta full *Symmachus is full of wisdom and virtues*, 10; Fox 28, 17. Se eorþlíca ânweald næfre ne sǽwþ ða cræftas *earthly power never sows the virtues*, 27, 1; Fox 94, 25: 30, 1; Fox 110, 5. Nân man for his ríce ne cymþ to cræftum, ac for his cræftum he cymþ to ríce *no man by his authority comes to virtues, but by his virtues he comes to authority*, 16, 1; Fox 50, 21, 23, 24.   IV. *a* CRAFT, *any kind of ship;* navis qualiscunque:—Gif massere geþeah ðæt he fêrde þrige ofer wíd-sǽ be his âgenum cræfte, se wæs ðonne syððan þegenrihtes weorþe *if a merchant thrived, so that*

he fared thrice over the wide sea in his own craft, then was he thenceforth worthy of thane-right, L. R. 6; Th. i. 192, 10. Ic æfre ne geseah on sǽ leódan syllícran cræft *I never saw a more wonderful craft sailing on the sea*, Andr. Recd. 1004; An. 500. [*Wyc. Piers P. Chauc.* craft: *Laym.* cræft, craft: *Orm.* crafft: *Plat.* kraft, kracht: *O. Sax.* kraft, m. *and f*: *Frs. O. Frs.* kreft: *Dut.* kracht, *f: Kil.* kracht: *Ger. M. H. Ger. O. H. Ger.* kraft, *f: Dan.* kraft, m. *f: Swed.* kraft, *m: Icel.* kraptr, kraftr, m.] DER. aclǽc-cræft, ǽ-, átor-, beadu-, bealo-, bóc-, deófol-, dreám-, drý-, dwol-, ellen-, firen-, flíter-, galdor-, gleó-, gúþ-, hell-, hyge-, lǽce-, lár-, leornung-, leóþ-, leoðo-, leóðu-, mód-, morþor-, nearo-, ofer-, rím-, sang-, sceóp-, scín-, scip-, scóp-, searo-, snytro-, stæf-, sundor-, swinsung-, tungel-, tungol-, un-, wæl-, wic-, wicce-, wíg-, word-, woruld-, wóþ-, wundor-.

cræfta, an; *m.* [cræft *art*] *An artist, a craftsman, workman*; artifex:—Cræfta *artifex*, Ælfc. Gl. 10; Som. 14, 43. v. cræftiga.

cræftan; *p.* te; *pp.* ed [cræft *art*] *To exercise a craft, to build*; architectari:—Ic cræfte *architector*, Ælfc. Gr. 36; Som. 38, 35. DER. a-cræftan, ge-.

cræftca *a workman*; artifex, opifex, Wrt. Voc. 73, 36, 38. v. cræftiga.

cræftega *a workman*, Past. 37, 3; Hat. MS. 50 b, 6. v. cræftiga, cræfta.

cræftga *an artificer*, Bt. Met. Fox 11, 184; Met. 11, 92. v. cræftiga.

cræftgast *most skilful*, Bt. Met. Fox 30, 4; Met. 30, 2; *sup. of* cræftig.

cræftgian *to strengthen, make powerful*. DER. ge-cræftgian.

cræft-gleáw; *adj. Sage-minded, science-learned*; animi prudens:—Cræft-gleáwe men *sage-minded men*, Chr. 975; Erl. 126, 26; Edg. 52.

cræftica *a workman*, Ælfc. Gl. 81; Som. 73, 2; Wrt. Voc. 47, 9. v. cræftiga.

cræftig; *adj. Ingenious, skilful*, CRAFTY, *cunning, virtuous, powerful*; ingeniosus, peritus, astutus, probus, potens:—Sum biþ fugelbona hafeces cræftig *one is a fowler skilful with the hawk*, Exon. 79 b; Th. 298, 6; Crä. 81: 91; Th. 361, 24; Wal. 24: Ps. C. 50, 11; Ps. Grn. ii. 277, 11. Án reordode, ðam wæs Iudas nama, wordes cræftig *one spake, whose name was Judas, crafty in word*, Elen. Kmbl. 837; El. 419: Exon. 97 b; Th. 364, 18; Wal. 72: Beo. Th. 2936; B. 1466. He sende cræftige wyrhtan *misit architectos*, Bd. 5, 21; S. 643, 7. Móde ðæs cræftig *with a mind so cunning*, Exon. 79 b; Th. 299, 6; Crä. 98. Cêne and cræftig *brave and virtuous*, Bt. Met. Fox 10, 101; Met. 10, 51: Bt. 36, 6; Fox 182, 10, 11. Sume men bióþ cræftige *some men are virtuous*, 39, 10; Fox 228, 7. Yldo beóþ on eórþan æghwæs cræftig *age is powerful over everything on earth*, Salm. Kmbl. 584; Sal. 291: Beo. Th. 3929; B. 1962: Chr. 1066; Th. 334, 1; Edw. 5. Weras wîsfæste, wordes cræftige *wise men, powerful of speech*, Elen. Kmbl. 628, 630; El. 314, 315. Nán cræftigra is ðonne ðú *no one is more skilful than thou*, Bt. 33, 4; Fox 128, 18. Omêrus wæs leópa cræftgast *Homer was most skilful in poems*, Bt. Met. Fox 30, 4; Met. 30, 2. Elþeódge wíf hæfdon gegán ðone cræftgestan dǽl *strange women had overcome the most powerful part*, Ors. 1, 10; Bos. 33, 41. DER. ǽ-cræftig, æl-, ár-, beadu-, bóc-, eácen-, hyge-, lagu-, leóþ-, leoðu-, má-, mód-, rím-, rún-, searo-, sundor-, un-, wíg-.

cræftiga, cræftega, cræftica, cræftca, cræftga, an; *m. A craftsman, workman, artificer, architect*; artifex, opifex, architectus:—Se micla cræftiga *the great craftsman*, Past. 8, 1; Hat. MS. 12, 15: Andr. Recd. 3264; An. 1635. Cræftica [MS. D. cræftca] *artifex*, Ælfc. Gr. 10; Som. 14, 43, MS. C: Ælfc. Gl. 81; Som. 73, 2; Wrt. Voc. 47, 9. Cræftca *artifex*, Wrt. Voc. 73, 36. Ðyssera cræftcena *horum artificum*, Ælfc. Gr. 10; Som. 14, 44. Se cræftega wyrcean mæg to ðæm ðe he wile *the workman can make what he likes of it*, Past. 37, 3; Hat. MS. 50 b, 6. Cræftiga *opifex*, Ælfc. Gl. 9; Som. 56, 128; Wrt. Voc. 19, 11. Cræftca *opifex*, Wrt. Voc. 73, 38. Swá swá ǽlc cræftega þencþ his weorc *as every artificer considers his work*, Bt. 39, 6; Fox 220, 4. Se cræftga geférscipas fæste gesamnaþ *the artificer firmly unites societies*, Bt. Met. Fox 11, 184; Met. 11, 92: Exon. 8 a; Th. 1, 22; Cri. 12. Cræftiga [MS. C. cræftca] *architectus*, Ælfc. Gr. 36; Som. 38, 35, MS. D.

cræftig-líce; *adv. Workmanlike*, CRAFTILY; fabre, artificiose:—Cræftiglíce *fabre*, Cot. 84. Seó heáfodstôw cræftiglíce geworht ætýwde *locus capitis fabrefactus apparuit*, Bd. 4, 19; S. 590, 1.

cræftigra *more skilful*, Bt. 33, 4; Fox 128, 18; *comp. of* cræftig.

cræft-leás; *adj. Artless, unskilful, innocent, simple, inexpert*; iners, indoctus, innocens:—Cræftleás *iners*, Wrt. Voc. 73, 50. Dǽl-leás· *vel* cræftleás *expers, indoctus*, Ælfc. Gl. 18; Som. 58, 123; Wrt. Voc. 22, 36.

cræft-líc; *adj. Artificial*; artificialis, Bridfr. Som. Ben. Lye.

cræft-líce; *adv. Cunningly, craftily*; affabre:—Cræftlíce *vel* smícere *affabre*, Ælfc. Gl. 99; Som. 76, 113; Wrt. Voc. 54, 55: Ælfc. Gr. 38; Som. 41, 32. v. cræftig-líce.

cræft-searo; *gen.* -searowes; *n. An instrument of war, a device, stratagem*; machina, Som. Ben. Lye.

cræft-wyrc; *es*; *n. Workmanship*; artificium, Scint. 29.

cræn *a crane*, Som. Ben. Lye. v. cran.

cræsta, an; *m. A* CREST, *tuft, plume*; crista, Som. Ben. Lye.

CRÆT, crat, es; *pl. nom. acc.* cratu, crætu; *gen.* cræta; *dat.* cratum, crætum; *n. A chariot*, CART; currus, pilentum:—Cræt *currus*, Ælfc. Gl. 49; Som. 65, 91; Wrt. Voc. 34, 22: 85, 71. Betogen [MS. betogan] cræt *capsus*, 49; Som. 65, 93; Wrt. Voc. 34, 23. Wǽrun Godes cræta gegearwedra tyn þúsendo *currus Dei decem millibus*, Ps. Th. 67, 17. On horsum and on cratum *equis ac curribus*, Deut. 11, 4. Mid gebeótlícum crætum and gilplícum riddum *with threatening chariots and proud horsemen*, Homl. Th. ii. 194, 23: Ps. Spl. C. 19, 8. He hæfde cratu and rídende men *habuit currus et equites*, Gen. 50, 9: Ex. 14, 27. Heó oferarn Pharao, and ealle his crætu and riddan it [*the sea*] *overwhelmed Pharaoh, and all his chariots and horsemen*, Homl. Th. ii. 194, 27. Crat *pilentum vel petorrítum*, Ælfc. Gl. 49; Som. 65, 95; Wrt. Voc. 34, 25. [*Prompt.* cart *biga, rheda, quadriga*: *Wyc.* cart, carte: *Piers P.* cartwey: *Chauc.* carte: *R. Glouc.* carte-staf: *Laym.* carte, *dat*: *Dut.* krat, *n*: *Ger.* krätze, kretze, m. *f*: *M. H. Ger.* kretze, m.*f*: *O. H. Ger.* cratto, *m*: *Icel.* kartr, *m*: *Wel.* cart: *Ir.* cairt: *Gael.* cairt, cartach, *f*.]

cræte-hors; *es*; *n.* [cræt *a cart*, hors *a horse*] *A cart-horse*; veredus, Ælfc. Gl. 5; Som. 56, 17; Wrt. Voc. 17, 21.

cræt-wǽn; *es*; *m.* [wǽn *a waggon*] *A chariot, wain*; currus:—Crætwǽn mid seolfre gegyred *a chariot mounted with silver*, Ors. 2, 4; Bos. 43, 14. Mid crætwǽne *with a chariot*, 2, 4; Bos. 43, 6. Sceoldon senátas rídan on crætwǽnum *the senators must ride in chariots*, 2, 4; Bos. 43, 9.

créwst, he créwþ *crowest, crows*, Lk. Bos. 22, 34; *2nd and 3rd pers. pres. of* cráwan.

CRAFIAN, crafigan; *p.* ode, ede; *pp.* od, ed *To ask*, CRAVE, *implore, demand, summon*; petere, postulare, in jus vocare:—Gif hwá wíte crafige *if any one crave a fine*, L. C. S. 70; Th. i. 412, 24. Se man crafode inne on hundrede *the man summoned him before the hundred court* Lchdm. iii. 288, 4. He mid rihte crafede ðás ða he crafede *he with right craved those things which he craved*, Chr. 1070; Erl. 208, 18, 23. [*Piers P.* craven: *Dan.* kræve: *Swed.* kräfva: *Icel.* krefja.] DER. be-crafian: un-crafod, unbe-.

crammian, ic crammige; *p.* ode; *pp.* od *To* CRAM, *stuff*; farcire:—Ic crammige oððe fylle *farcio*, Ælfc. Gr. 30, 2; Som. 34, 36. [*Wyc.* crammyd, *pp*: *Piers P.* ycrammed, *pp*.] DER. under-crammian.

CRAN, es; *m*: e; *f. A* CRANE; grus:—Cran *grus*, Ælfc. Gr. 9, 33; Som. 12, 20: Ælfc. Gl. 38; Som. 63, 34; Wrt. Voc. 29, 53: 62, 20: 77, 16: 280, 25. [*Prompt.* crane *grus*: *Laym.* cron, crane: *Plat.* kraan: *O. Sax.* krano, *m*: *Dut.* kraan, *f*: *Kil.* kraene: *Ger.* kranich, *m*: *M. H. Ger.* kranech, *m*: *O. H. Ger.* kranuh, *m*: *Dan.* trane, *m. f*: *Swed.* trana, *f*: *Icel.* trani, *m*; trana, *f*: *Lat.* grus, *f*: *Grk.* γέραν-ος, *m. and f*: *Wel.* Corn. garan, *f*: *Ir.* Gael. garan, *m*: *Armor.* garan, *f*.]

cranc, *pl.* cruncon *yielded*; *p. of* crincan.

cranc-stæf; *es*; *m. A weaver's instrument*; instrumenti genus ad textores pertinentis, Som. Ben. Lye.

crang, *pl.* crungon *fell, perished, died*; *p. of* cringan.

crang *dead, killed*; mortuus, occisus, Mann.

crano-hawc [cran *a crane*; hafoc, es; *m. a hawk*] *A crane-hawk*; accipiter, qui gruem mordet, Spelm. Gl. Ben. Lye.

cráp *should creep*, Chr. 1131; Erl. 260, 3, = creápe; *p. subj. of* creópan.

crat *a waggon*, Ælfc. Gl. 49; Som. 65, 95; Wrt. Voc. 34, 25. v. cræt.

CRÁWAN, ic cráwe, ðú cráwest, créwst, he cráweþ, créwþ; *p.* creów, *pl.* creówon; *pp.* cráwen *To* CROW *as a cock*; cantare instar galli:—Ne cráwþ se hana to-dæg *non cantabit hodie gallus*, Lk. Bos. 22, 34. Ne cráwþ se cocc, ǽr ðú wiðsæcst me þríwa, Jn. Bos. 13, 38; *the koc schal not crowe, til thou schalt denye me thries*, Wyc. Ærðamðe cocc cráwe, þríwa ðú me wiðsæcst, 26, 75; *bifore the cok crowe, thries thou shalt denye me*, Wyc. Ær hana cráwe *priusquam gallus vocem dederit*, Mk. Bos. 14, 30. Ær se hana cráwe, 14, 72; *bifore the cok synge*, Wyc: Lk. Bos. 22, 61. Sôna se cocc creów *statim gallus cantavit*, Jn. Bos. 18, 27. Hrædlíce ðá creów se cocc, Mt. Bos. 26, 74; *anon the cok crew*, Wyc. Se hana creów *gallus cantavit*, Mk. Bos. 14, 68: Lk. Bos. 22, 60. Ðá eftsôna creów se hana; Mk. Bos. 14, 72; *anon eftsones the cok song*, Wyc. [*Wyc.* crowe: *Plat.* kreien, kreijen: *Dut.* kraaijen: *Kil.* kraeyen: *Ger.* krähen: *M. H. Ger.* kræjen: *O. H. Ger.* krájan, kráhan.]

CRÁWE, an; *f.* I. *a* CROW; cornix:—Cráwe *cornix*, Ælfc. Gl. 10; Som. 63, 8; Wrt. Voc. 29, 31: 62, 29: 280, 34. II. *a raven*; corvus:—Se selþ nýtenum mete heora, and briddum cráwan cígendum hine *qui dat jumentis escam ipsorum, et pullis corvi invocantibus eum*, Ps. Spl. T. 146, 10. [*Chauc.* crow: *Plat.* kreie, kraie: *O. Sax.* krâia, *f: Frs.* Japx. krie: *Dut.* kraai, *f: Kil.* kraeye: *Ger.* krähe, *f: M. H. Ger.* krâ, *f: O. H. Ger.* krâa, *f: Lat.* corvus, cornix: *Grk.* κόραξ, κορώνη: *Sansk.* kârava, *m. a crow*.]

cráw-leác, es; *n.* [cráwe *a crow*, leác *a leek*] *Crow-garlic*; allium vineale, Lin:—Nim hermodactylos = ἑρμο-δάκτυλος [MS. datulus] ða wyrt ... ðæt is on úre geþeóda ðæt greáte [MS. greáta] cráwleác

[MS. crauleac] *take the wort allium vineale . . . that is in our language the great crow-garlic,* Lchdm. i. 376, 3. Cráwan leác *hermodactylus,* Ælfc. Gl. 44; Som. 64, 84; Wrt. Voc. 32, 20.

Creacan ford *Crayford,* Chr. 456; Th. 22, 5, col. 2, 3. v. Crecgan ford.

Creácas; *gen.* Creáca; *pl. m. The Greeks;* Græci :—Mid eallan Creáca cræftum *with all the arts of the Greeks,* Ors. 1, 10; Bos. 33, 29, 31 : Bos. 34, 6. v. Grēcas.

Creacc-gelád *Cricklade,* Chr. 905; Th. 181, 21, col. 1. v. Creccagelád.

Creácisc; *adj. Greek, Grecian;* Græcus, Ors. 1, 10; Bos. 33, 12. v. Grēcisc.

creád *pressed,* Chr. 937; Th. 204, 14, col. 1; Æðelst. 35; *p. of* creódan.

creáp, *pl.* crupon *crept, crawled,* Glostr. Frag. 6, 7: Ors. 1, 7; Bos. 29, 33; *p. of* creópan.

Creca-lád *Cricklade,* Chr. 1016; Erl. 153, 38. v. Crecca-gelád.

Crēcas; *gen.* Crēca; *pl. m. The Greeks;* Græci :—Fôr on Crēcas *he went against the Greeks,* Ors. 2, 5; Bos. 46, 15, 31. Ymbe Crēca land *about the land of the Greeks,* Ors. 1, 1; Bos. 23, 11: 23, 12, 13, 17, 22: 1, 6; Bos. 29, 6. Perseus of Crēca lande in Asiam fôr *Perseus went from the land of the Greeks into Asia,* 1, 8; Bos. 31, 14. v. Grēcas.

crecca, an; *m. A* CREEK, *bay, wharf;* crepido, Som. Ben. Lye.

Crecca-gelád, Cre-gelád, e; *f.* [gelád *a road, way:* Flor. Criccelade: *Hunt.* Crikelade: *Sim. Dun.* Criccelad: *Brom.* Criklade] CRICKLADE, *Wiltshire;* oppidi nomen in agro Wiltoniensi :—Hie hergodon ofer Mercna land ôþ hie cômon to Creccagelâde, and fôron ðær ofer Temese *they harried over the Mercians' land until they came to Cricklade, and there they went over the Thames,* Chr. 905; Erl. 98, 15. On ðissum geáre com Cnut mid his here ofer Temese into Myrcum æt Cregelâde *in this year* [A. D. 1016] *Cnut came with his army over the Thames into Mercia at Cricklade,* 1016; Erl. 153, 23.

Creccan ford *Crayford,* Chr. 456; Th. 23, 4, col. 2. v. Crecgan ford.

Crēce; *gen.* a; *dat.* um; *pl. m. The Greeks;* Græci :—He belytegade ealle Crēce on his geweald *he allured all the Greeks into his power,* Ors. 3, 7; Bos. 59, 39, 40. Philippus alýfde eallum Crēcum *Philip gave leave to all the Greeks,* 3, 7; Bos. 61, 42. v. Crēcas, Grēcas.

Creogan ford, Creccan ford, es; *m.* [*Hunt.* Creganford: *the ford of the river Cray*] CRAYFORD, *Kent;* loci nomen in agro Cantiano :—Hēr Hengest and Æsc fuhton wið Brettas in ðære stówe ðe is gecweden Crecgan ford *in this year* [A. D. 457] *Hengest and Æse fought against the Britons at the place which is called Crayford,* Chr. 457; Erl. 12, 18.

Crēcisc *Grecian,* Bt. Met. Fox 26, 55; Met. 26, 28. v. Grēcisc.

crēda, an; *m.* [*Lat.* crēdo *I believe*] *The creed, belief;* symbolum fidei :—Se læssa crēda *the less or Apostles' creed,* Homl. Th. ii. 596, 11. We andettaþ on úrum crēdan ðæt Drihten sitt æt his Fæder swiðran *we confess in our creed that the Lord sits at the right hand of his Father,* i. 48, 28 : 274, 23. Ælc cristen man sceal æfter rihte cunnan his crēdan . . . mid ðam crēdan he sceal his geleáfan getrymman *every christian man by right ought to know his creed . . . with the creed he ought to confirm his faith,* 274, 20, 21. DER. mæsse-crēda.

Cre-gelád *Cricklade,* Chr. 1016; Erl. 153, 23. v. Crecca-gelád.

crencestre, crencistre, an; *f. A female weaver, a spinster;* textrix, Cod. Dipl. 1290; Kmbl. vi. 131, 32.

Creocc-gelád *Cricklade,* Chr. 905; Erl. 99, 20. v. Crecca-gelád.

CREÓDAN, ic creóde, ðú creódest, crýtst, crýst, he creódeþ, crýdeþ, crýt, *pl.* creódaþ; *p.* ic, he creád, ðú crude, *pl.* crudon; *pp.* croden *To* CROWD, *press, drive;* premere, premi, pellere, pelli :—Ðonne heáh geþring on cleofu crýdeþ *when the towering mass on the cliffs presses,* Exon. 101 b; Th. 384, 15; Rä. 4, 28. Creád cnear on flot *the bark drove afloat,* Chr. 937; Th. 204, 14; col. 1; Æðelst. 35. [*Prompt.* crowdyñ' *impello:* Chauc. croude, crowde *push:* Kil. kruyen, kruyden *trudere, propellere.*]

CREÓPAN; *part.* creópende; ic creópe, ðú crýpest, crypst, creópest, creópst, he crýpeþ, crýþ, creópeþ, creópþ, *pl.* creópaþ; *p.* creáp, *pl.* crupon; *pp.* cropen *To* CREEP, *crawl;* repere, serpere :—He næfþ his fóta geweald and onginþ creópan *he has not the use of his feet and begins to creep,* Bt. 36, 4; Fox 178, 14, Cott. MS. Him cômon to creópende fela næddran *many serpents came creeping to them,* Homl. Th. ii. 488, 21. Mægen creópendra wyrma biþ on heora fótum *the power of reptiles* [lit. *creeping worms*] *is in their feet,* Ors. 4, 6; Bos. 84, 44: Gen. 7, 21. Nán wilde deór, ne on fyðerfótum ne on creópendum, nis to wiðmetenne yfelum wífe *no wild beast, neither among the four-footed nor the creeping, is to be compared with an evil woman,* Homl. Th. i. 486, 29. Læde seó eorþe forþ creópende cinn æfter heora hiwum *producat terra reptilia secundum species suas,* Gen. 1, 24, 25, 26. Ic creópe *repo,* Ælfc. Gr. 28, 4; Som. 31, 23. Se biþ mihtigra se ðe gæþ ðonne se ðe crýpþ *he is more powerful who goes than he who creeps,* Bt. 36, 4; Fox 178, 16. Hí creópaþ and snícaþ *they creep and crawl,* Bt. Met. Fox 31, 12; Met. 31, 6. Heó creáp betwux ðám mannum *she crept among the men,*

Homl. Th. ii. 394, 11: Glostr. Frag. 6, 7. Ða munecas crupon under ðam weofode *the monks crept under the altar,* Chr. 1083; Erl. 217, 22 : Ors. 1, 7; Bos. 29, 33. [*Piers P.* crepen: *Chauc. R. Glouc.* crepe: *Laym.* crepen: *Plat.* krupen: *O. Sax.* criepan: *Frs.* krippen: *O. Frs.* kriapa: *Dut.* kruipen: *Kil.* kruypen: *Ger.* kriechen: *M. H. Ger.* kriuchen: *O. H. Ger.* kriuchan: *Dan.* krybe: *Swed.* krypa: *Icel.* krjúpa.] DER. be-creópan, þurh-, under-.

creópere, es; *m. A* CREEPER, *cripple;* serpens, clinicus :—Seó ealde cyrce wæs eall behangen mid criccum and mid creópera sceamelum *the old church was all hung around with crutches and with cripples' stools,* Glostr. Frag. 12, 17.

creópung, e; *f. A* CREEPING, *stealing;* obreptio, Cot. 144.

creów, *pl.* creówon *crew,* Jn. Bos. 18, 27; *p. of* cráwan.

crépel, es; *m. A burrow;* cuniculum, Mone B. 2774.

cresse *cress,* Glos. Epnl. Recd. 162, 61. v. cærse.

CRICC, crycc, e; *f. A* CRUTCH, *staff;* baculus :—Gird ðín and cricc ðín me frēfredon *virga tua et baculus tuus me consolata sunt,* Ps. Spl. C. 22, 5. He, mid his cricce wreðiende, on cyricean eóde *baculo sustentans intravit ecclesiam,* Bd. 4, 31; S. 610, 28. He, mid his crycce hine awreðiende, hâm becom *baculo innitens domum pervenit,* 4, 31; S. 610, 17. He mid criccum his fēðunge underwreðode *he supported his gait with crutches,* Homl. Th. ii. 134, 24. [*Laym.* crucche, *dat:* *Plat.* krukke, krükke: *Dut.* kruk, *f:* *Kil.* krucke: *Ger.* krücke, *f:* *M. H. Ger.* krücke, krucke, *f:* *O. H. Ger.* krucka, *f:* *Dan.* krykke, *m. f:* *Swed.* krycka, *f.*]

Cric-gelád *Cricklade,* Chr. 1016; Th. 276, 29, col. 2. v. Crecca-gelád.

Cridian tûn, es; *m.* [tûn *a town:* Flor. Cridiatun] CREDITON, *Devonshire, formerly the seat of the bishops of Devonshire, so called because it is situated on the banks of the river Creedy;* oppidi nomen in agro Devoniensi :—Hēr æt Kyrtlingtûne forþfērde Sideman bisceop, on hrædlícan deáþe: se wæs Defnascíre bisceop, and he wilnode ðæt his lícræst sceolde beón æt Cridian tûne, æt his bisceopstóle *in this year* [A. D. 977] *bishop Sideman died at Kirtlington, by sudden death : he was bishop of Devonshire, and he desired that his body's resting-place might be at Crediton, at his episcopal see,* Chr. 977; Erl. 127, 35–38: Cod. Dipl. 1334; A.D. 1046; Kmbl. vi. 196, 15.

crimman; *p.* cramm, cram, *pl.* crummon; *pp.* crummen *To crumb, crumble, mingle;* friare, inserere :—Hornes sceafoðan crim on ðæt dolh *crumble shavings of horn on the wound,* L. M. 1, 61; Lchdm. ii. 132, 12. Cram inseruit, Glos. Prudent. Recd. 151, 33. DER. a-crimman.

crincan, ic crince, ðú crincst, he crincþ, *pl.* crincaþ; *p.* cranc, *pl.* cruncon; *pp.* cruncen *To yield;* occumbere :—Wígend cruncon, wundum wērige *the fighters yielded, oppressed with wounds,* Byrht. Th. 140, 43; By. 302. DER. ge-crincan.

crincgan *to fall,* Byrht. Th. 140, 23; By. 292. v. cringan.

cringan, crincgan; ic cringe, crincge, ðú cringest, cringst, he cringeþ, cringþ, *pl.* cringaþ, crincgaþ; *p.* crang, crong, *pl.* crungon; *pp.* crungen *To yield,* CRINGE, *fall, perish, die;* occumbere, mori :—Sume on wæl crungon *some had fallen in the slaughter,* Beo. Th. 2231; B. 1113. Hí sceoldon begen crincgan on wælstówe *they should both fall on the battle-field,* Byrht. Th. 140, 23; By. 292: Andr. Kmbl. 2062; An. 1033: Chr. 937; Th. 202, 6; col. 2; Æðelst. 10. Crungon *they perished,* Exon. 124 a; Th. 477, 17; Ruin. 26: 124 a; Th. 477, 24; Ruin. 29. Fæge crungon *the fated died,* Cod. 167; Th. 208, 11; Exod. 481: Beo. Th. 1275; B. 635. DER. ge-cringan. v. gringan.

crisma, an; *m.* [chrisma, ātis, *n.* = χρῖσμα, ᾰτος; *n. an unction,* from χρίω [*fut.* χρίσω] *I touch the surface of a body, I rub or anoint*]. I. *the chrism, unction or holy oil, used for anointing by the Roman Catholic church after baptism;* oleum chrismātis :—Eálá ge mæsse-preóstas, mîne gebróðra, we secgaþ eów nú ðæt we ǽr ne sædon, forðonðe we to-dæg sceolan dǽlan úrne ele, on þreó wîsan gehâlgodne, swá swá us gewissaþ seó bôc; *i. e. oleum sanctum, et oleum chrismatis, et oleum infirmorum,* ðæt is on Englisc, hâlig ele, óðer is crisma, and seóccra manna ele: and ge sceolan habban þreó ampullan gearuwe to ðâm þrým elum; forðanðe we ne durran dôn hi togædere on ânum elefate, forðanðe hyra ǽlc biþ gehâlgod on sundron to synderlícre þênunge. Mid ðam hâligan ele, ge scylan ða hǽðeran cild mearcian on ðam breóste, and betwux ða gesculdru, on middeweardan, mid rôde tâcne, forðanðe ge hit fullian on ðam fantwætere; and ðonne hit of ðæm wætere cymþ, ge scylan wyrcan rôde tâcen uppon ðæm heáfde mid ðam hâligan crisman. On ðam hâligan fante, ǽrðanðe ge hý fullian, ge scylon dôn crisman on Cristes rôde tâcne; and man ne môt besprengan men mid ðæm fantwætere, syððan se crisma biþ ðæron gedôn *O ye mass-priests, my brethren, we will now say to you what we have not before said, because to-day we are to divide our oil, hallowed in three ways, as the book points out to us; i. e. oleum sanctum, et oleum chrismatis, et oleum infirmorum, that is, in English, holy oil, the second is chrism, and sick men's oil: and ye ought to have three flasks ready for the three oils; for we dare not put them together in one oil vessel, because each of them is hallowed apart for a particular service. With holy oil, ye shall mark heathen children on*

the breast, and between the shoulders, in the middle, with the sign of the cross, before ye baptize it in the font water; and when it comes from the water, ye shall make the sign of the cross on the head with the holy chrism. In the holy font, before ye baptize them, ye shall pour chrism in the figure of the cross of Christ; and no one may be sprinkled with the font water, after the chrism is poured in, L. Ælf. E. Th. ii. 390, 1-17. Mid crysman smyreþ his breóst *chrismate pectus eorum unguet,* L. Ecg. C. 36; Th. ii. 162, 1. Ðonne he crisman fecce *when he fetches chrism,* L. Edg. C. 67; Th. ii. 258, 20: L. N. P. L. 9; Th. ii. 292, 3. II. *the white vesture, called chrisom, which the minister puts upon the child immediately after dipping it in water, or pouring water upon it in baptism;* chrismale, id est, vestis candida, quæ super corpus baptizati ponitur. *In the Liturgy of Edward VI,* 1549, *it is said,* 'Then the minister shall put upon the child the white vesture, commonly called the *Chrisom;* and say, Take this white vesture for a token of the innocency, which, by God's grace, in this holy sacrament of baptism, is given unto thee,' p. 112. This white vesture was worn for a month after the child's birth, and if it died before the expiration of that time, it had the chrisom for its shroud. A child, thus dying, was called a Chrisom-child :—Wǽron eác gefullade æfter-fyligendre tíde óðre his [Eádwines] bearn of Ædelburhge dǽre cwéne acende, Ædelhún, and Ædeldriþ his dóhter, and óðer his suna Wuscfreá hátte, ac ða ǽrran twegen under crisman forþgeférdon, and on cyrican in Eoferwícceastre bebyrigde wǽron *baptizati sunt tempore sequente et alii liberi ejus [Æduini] de Ædilberga regina progeniti, Ædilhun, et Ædilthryd filia, et alter filius Vuscfrea quorum primi albati adhuc rapti sunt de hac vita* [lit. the former two died under chrism], *et Eburaci in Ecclesia sepulti,* Bd. 2, 14; S. 518, 1 : 5, 7; S. 620, 40. Under crysmum *baptizatus in albis,* Mone B. 2096.

crism-hálgung, e; *f. The consecration of the oil of chrism;* chrismatis consecratio, Wanl. Catal. 121, col. 2, 57.

crism-lýsing, -lísing, e; *f. A leaving off the baptismal vest;* chrismatis solutio :—His crismlýsing [crismlising MS. A.] wæs æt Wedmor *the leaving off his baptismal vest was at Wedmore,* Chr. 878; Erl. 81, 20. v. crisma.

crisp; *adj.* CRISP, *curly;* crispus :—He hæfde crispe loccas *he had curly locks,* Bd. 5, 2; S. 615, 30. v. cyrps.

Crist, Krist, es; *m.* CHRIST; Christus = Χριστός *the anointed one, as a translation of the Heb.* מָשִׁיחַ *Messiah :*—Se Hǽlend, ðe is genemned Crist *Iesus, qui vocatur Christus;* Ἰησοῦς, ὁ λεγόμενος Χριστός, Mt. Bos. 1, 16. Crist wæs acenned, Hǽlend geháten *Christ was born, called Jesus* [*Saviour*], Menol. Fox 1-7. Hér is on cneórisse bóc Hǽlendes Cristes *liber generationis Iesu Christi,* Mt. Bos. 1, 1. Hér ys gódspelles angyn Hǽlendes Cristes, Godes suna *initium evangelii Iesu Christi, filii Dei,* Mk. Bos. 1, 1. Beseoh onsýne cristes ðínes *behold the face of thine anointed,* Ps. Th. 83, 9 : 88, 32, 44. Feówer Cristes béc *the four Gospels,* Ælfc. T. Grn. 12, 27 : Bd. 5, 19; S. 638, 16. Seó Cristes bóc *the Gospel,* Ælfc. T. 30, 1. Feoh bútan gewitte ne can Crist gehérian *cattle without understanding cannot praise Christ,* Salm. Kmbl. 48; Sal. 24. Ofer ealle Cristes béc *over all Christ's books* [*Gospels*], 100; Sal. 49. On Cristes onlícnisse *in Christ's likeness,* Salm. Kmbl. 146, 15.

cristalla, an; *m:* cristallus, i; *m. Lat.* I. *crystal;* crystallus = κρύσταλλος :—Ðæt wæs hwítes bleós swá cristalla *it was of a white colour like crystal,* Num. 11, 7. Cristallan *crystallum,* Glos. Prudent. Recd. 140, 49. II. *the herb crystallium, flea-bane, flea-wort;* crystallion = κρυστάλλιον, psyllion = ψύλλιον :—Nim cristallan and disman *take crystallium and tansy,* Lchdm. iii. 10, 29.

cristen; *def.* se cristena; *sup.* se cristenesta; *adj.* [Crist *Christ*] *Christian;* christianus :—Ælc cristen man hæfde sibbe *every christian man had peace,* Ors. 6, 13; Bos. 122, 7 : 6, 30; Bos. 127, 22. Cristnu gesamnung *the christian church,* Ps. Th. 44, 11. Gif hwá cristenes mannes blód ageóte *if any one shed a christian man's blood,* L. Edm. E. 3; Th. i. 246, 2 : Ps. Th. 106, 31. He forbeád ðæt man nánum cristenum men ne abulge *he forbade men to annoy any christian man,* Ors. 6, 11; Bos. 121, 10; L. Edm. E. 2; Th. i. 244, 16: Elen. Kmbl. 1974; El. 989. Hí bebudon ðæt man ælcne cristenne man ofslóge *they commanded men to slay every christian man,* Ors. 6, 13; Bos. 121, 32. Him sealde Iustinus áne cristene bóc *Justin gave him a christian book,* 6, 12; Bos. 121, 24. Godes þeówas for eall cristen folc þingian *let the servants of God intercede for all christian people,* L. Eth. v. 4; Th. i. 304, 25 : vi. 2; Th. i. 314, 18: L. C. E. 6; Th. i. 364, 7. Cristene men secgaþ *christian men say,* Bt. 39, 8; Fox 224, 14 : Ors. 6, 11; Bos. 121, 8. Nero wæs ǽrest éhtend cristenra manna *Nero was the first persecutor of christian men,* 6, 5; Bos. 119, 22 : 6, 9; Bos. 120, 18 : Elen. Kmbl. 1956; El. 980. Fram óðrum cristenum mannum *from other christian men,* Ors. 6, 9; Bos. 120, 22 : 6, 12; Bos. 121, 25. Hí cristene men pinedon *they tormented christian men,* 6, 11; Bos. 121, 17 : 6, 19; Bos. 123, 16. Oswig se cristena cyning to his ríce féng *Oswy the christian king succeeded to his kingdom,* Bd. 3, 21; S. 551, 30. Se cristena dóm *christianity,* Bt. 1; Fox 2, 15. Bǽdon [MS. bædan] hí ða cristenan

men they asked the christian men, Ors. 6, 13; Bos. 121, 41 : 6, 30; Bos. 127, 14. Se mon wæs se cristenesta and se gelǽredesta *the man was most christian and most learned,* Bd. 2, 15; S. 518, 43 : 3, 1; S. 523, 7 : 3, 9; S. 533, 6.

cristen, es; *m:* cristena, an; *m. A christian;* christianus :—He wæs cristen *he was a christian,* Bt. 1; Fox 2, 7 : Chr. 167; Erl. 8, 16 : Bd. 3, 21; S. 551, 4. He hét ealle ða cristenan *he ordered all the christians,* Ors. 6, 30; Bos. 127, 10.

Cristen-dóm, es; *m. Christianity,* CHRISTENDOM, *the christian world;* christianitas :—Se cristendóm weóx on heora tíman *christianity increased in their time,* Jud. Grn. Epilog. 264, 7 : Jud. Thw. 161, 21. Æghwylc cristen man gýme his cristendómes georne *let every christian man strictly keep his christianity,* L. Eth. v. 22; Th. i. 310, 5 : vi. 27; Th. i. 322, 5 : L. C. E. 19; Th. i. 370, 32 : Ælfc. T. 28, 3. Gif hwá cristendóm wyrde *if any one violate christianity,* L. E. G. 2; Th. i. 168, 1 : L. Eth. v. 1; Th. i. 304, 4, 7 : L. C. S. 11; Th. i. 382, 7. On cristendóm *in christendom,* Chr. 1129; Erl. 258, 29.

cristenest, se. cristenesta *the most christian, pious, holy,* Bd. 3, 9 S. 533, 6 : 2, 15; S. 518, 43; *sup. of* cristen.

Cristes bóc, e; *f.* CHRIST'S BOOK, *the Gospel;* Christi liber, evangelium, Ælfc. T. 30, 1: Salm. Kmbl. 100; Sal. 49. v. Crist.

cristlíc; *adj. Christlike, christian;* christianus :—We lǽraþ, ðæt æghwilc cristen man cristlíce lage rihtlíce healde *we direct, that every christian man rightly observe the christian law,* L. Eth. vi. 11; Th. i. 318, 11, note 4.

cristnian; *p.* ode; *pp.* od *To christianize, catechize;* catechizare :—Ðæt Paulínus ðǽr ðæt folc cristnode and fullode [MS. cristnade ꞇ fullade] *that Paulinus might there christen and baptize the people,* or as the original Latin of Bede has it, with greater precision,—*ut Paulinus cum eis catechizandi et baptizandi officio deditus moraretur,* Bd. 2, 14; S. 518, 7, 8; Latin 95, 34.

croc, crocc, crogg, crohh, es; *m. A crock, pitcher, waterpot, flagon, a little jug* or *lentil-shaped vessel;* urceus, lagena, lenticula, legythum :—Croccas, Cot. 209: Grm. iii. 458, 15. DER. croc-wyrhta.

CROCCA, an; *m. A* CROCK, *pitcher, earthenware pot* or *pan;* vas fictile, testa, olla :—Mín mægen ys forseárod, swá swá lǽmen crocca *exaruit velut testa virtus mea,* Ps. Th. 21, 13. Crocca *olla,* Ps. Lamb. 59, 10 : Ælfc. Gr. 7; Som. 6, 53 : Wrt. Voc. 82, 56. Wyl wæter on croccan *boil water in a crock,* L. M. 1, 40; Lchdm. ii. 104, 19. On ǽnne croccan ðone ðe sie gepicod útan *in a crock that is pitched on the outside,* I, 2; Lchdm. ii. 26, 23. Ic gedó ðæt ðú hí miht swá eáðe abrecan, swá se croccwyrhta mæg ǽnne croccan *tamquam vas figuli confringes eos,* Ps. Th. 2, 9 : Herb. 126, 2; Lchdm. i. 238, 6. [*Piers P.* krokke : *Plat.* kruke: *O. Sax.* crúka, *f:* Frs. kruwch : *O. Frs.* krocha, *m:* Dut. kruik, *f:* Kil. kruycke: *Ger.* krug, *m:* M. H. Ger. kruoc, *m: O. H. Ger.* krôg, *m:* Dan. krukke, *m. f:* Swed. kruka, *f: Icel.* krukka, *f.*]

croc-hwǽr, es; *m.* [hwer *an ewer*] *A kettle;* cacabus, Som. Ben. Lye.

croc-sceard, es; *n.* [sceard *a shred, fragment*] *A shred* or *fragment of a crock* or *pot, a* POTSHERD; testa, testu :—Adruwode oððe forseárode swá swá blýwnys oððe crocsceard mægen mín *aruit tamquam testa virtus mea,* Ps. Lamb. 21, 16. Mid ánum crocscearde *with a potsherd,* Job Thw. 166, 34: Homl. Th. ii. 452, 29. Crocsceard *testu,* Ælfc. Gr. 11; Som. 15, 29.

croc-wyrhta, crocc-wyrhta, -wirhta, an; *m. A crockworker, potter;* figulus, luti figulus :—Crocwyrhta *figulus vel luti figulus,* Ælfc. Gr. 28, 5; Som. 31, 62. Ic gedó ðæt ðú hí miht swá eáðe abrecan, swá se crocwyrhta mæg ǽnne croccan *tamquam vas figuli confringes eos,* Ps. Th. 2, 9. Fæt crocwirhtan *vel* tygelwirhtan *vas figuli,* Ps. Lamb. 2, 9.

croda, an; *m.* [croden, *pp.* of creódan *to crowd, press, drive*] *A crowd, press;* collisus. DER. lind-croda.

croden *crowded, pressed; pp. of* creódan.

croft, es; *m. A* CROFT, *a small inclosed field;* prædiolum, agellulus septus :—Æt ðæs croftes heáfod *at the top of the croft,* Cod. Dipl. 553; A. D. 969; Kmbl. iii. 37, 23. In ðone croft, of ðæm crofte *to the croft, from the croft,* 681; A. D. 972; Kmbl. iii. 261, 11 : 679; A. D. 972-992; Kmbl. iii. 258, 27, 28.

crog, crogg, crohh, es; *m. A small vessel, chrismatory, bottle;* legythum, lenticula, lagena :—Crog oððe ampella *lenticula,* Cot. 124. v. croc.

croh, es; *m? Saffron;* crocus = κρόκος, crocus sativus, Lin :—Meng mid [MS. wið] croh *mingle it with saffron,* L. M. 2, 37; Lchdm. ii. 244, 27: Herb. 118, 2; Lchdm. i. 232, 7: Med. ex Quadr. 5, 4; Lchdm. i. 348, 14.

crohh *a pitcher;* legythum, lagena *vel* ampulla, Cot. 119. v. crog.

croma *a crumb,* Mt. Kmbl. Rush. 15, 27. v. cruma.

crompeht; *adj. Full of crumples, wrinkled;* folialis, Cot. 91.

crong *killed, perished; p. of* cringan.

CROP, cropp, es; *m.* I. *a sprout* or *top of a herb, flower, berry, an ear of corn, a bunch of berries* or *blooms, cluster;* cyma = κῦμα, thyrsus = θύρσος, spica, corymbus = κόρυμβος, racemus, uva :—Crop *cyma,* Ælfc. Gl. 60; Som. 68, 18; Wrt. Voc. 39, 4. Crop *tursus, cimia* [= *thyrsus, cyma*], 42; Som. 64, 28; Wrt. Voc. 31, 38. Dó him

merscmealwan crop *give him a sprout of marsh mallow,* L. M. 3, 63;
Lchdm. ii. 350, 25.   Genim ðysse wyrte þrý croppas *take three sprouts
of this herb,* Herb. 106; Lchdm. i. 220, 10.   Genim ðysse wyrte croppas *take the tops of this herb,* 110, 4; Lchdm. i. 224, 9: 130, 1;
Lchdm. i. 240, 18.   Genim ðysse wyrte croppas *take berries of this herb*
[*ivy*], 100, 3; Lchdm. i. 214, 3.   Þegnas his ða croppas eton *discipuli
ejus spicas manducabant,* Lk. Skt. Lind. 6, 1.   Wið ðon biþ gód lustmocan crop *a bunch of 'lustmock' is good for that,* L. M. 1, 38;
Lchdm. ii. 92, 9.   Genim lustmocan crop *take a bunch of 'lustmock',* 1,
38; Lchdm. ii. 98, 16.   Croppas *racemos,* Mone B. 2572.   Croppum
uvis, 3836.   **II.** *the* CROP *or craw of a bird;* vesicula gutturis:—
Wurp ðone cropp and ða feðera widæftan ðæt weofod *vesiculam gutturis
et plumas projiciet prope altare,* Lev. 1, 16.   **III.** *a kidney;*
rien:—Crop *rien,* Ælfc. Gl. 76; Som. 71, 107; Wrt. Voc. 45, 13.
[*Prompt.* croppe *cyma:* Piers P. crop: *Chauc.* crop, croppe: *Plat.*
kropp: *Dut.* krop, *m: Kil.* krop, kroppe: *Ger. M.H. Ger.* kropf, *m:
O.H. Ger.* kroph, *m: Dan.* krop, *m. f: Swed.* kropp, *m: Icel.*
kroppr, *m.*]   DER. ifig-crop.

cropen *crept, crawled; pp. of* creópan.

crop-leác, es; *n. Garlic;* allium satīvum, Lin:— Genim cropleác
take garlic, L. M. 1, 3; Lchdm. ii. 42, 14: 3, 68; Lchdm. ii. 356, 5.

croppa, an; *m. The top or flower of a herb;* corymbus, pluma:—
Bānwyrt hæbbe croppan *bonewort hath clusters of flowers,* L. M. 2, 51;
Lchdm. ii. 266, 6.   v. crop **I.**

croppiht; *adj.* [crop **I.** *a bunch, cluster;* -iht, *adj. termination,* q. v.]
Croppy, full of clusters; racemosus, L. M. 1, 39; Lchdm. ii. 102, 12.

cruce, an; *f. A cruse, pitcher, waterpot;* urceus, urceolus:—Cruce
viciolum [=*urceolus*], Wrt. Voc. 290, 67.

crucet-hús, es; *n. A torment house;* afflictionis domus:—Sume hí
diden in crucet-hús, ðæt is in ān ceste ðæt was scort, and nareu, and undeþ,
and dide scærpe stānes ðērinne, and þrengde ðe man ðærinne, ðæt him
bræcon alle ðe limes *some they put into a crucet-house, that is into a chest
that was short, and narrow, and undeep, and put sharp stones therein,
and pressed the man therein, so that they brake all his limbs,* Chr. 1137;
Th. 382, 28.

crudon *crowded, pressed; p. pl. of* creódan.

cruft, es; *m?* crufte, an; *f. A vault, crypt, hollow place under the
ground;* crypta:—Cruftan, cruftes *cryptæ,* Mone B. 2017.   Crufte
crypta, 4931.   Cruftan *crypta,* 3298.   [*Ger.* gruft, *f. a crypt.*]

Crúland, Crúwland, es; *n.* [Interpreto Ingulpho *crúda et cœnosa
terra,* Gib. Chr. explicatio, p. 22, col. 1] CROWLAND OR CROYLAND,
Lincolnshire; loci nomen in agro Colnolniensi. St. Guthlac, hermit of
Crowland, passed a great part of his life and died here in A. D. 714.
After his death, king Æthelbáld of Mercia founded a monastery at
Crowland in A. D. 716:—Ðæt abbotríce of Crúlande *the abbacy of
Crowland,* Chr. 1066; Erl. 203, 17: 963; Erl. 123, 5.   Hér wæs
Walþeóf eorl beheáfdod on Wincestre, and his líc wearþ geléd to
Crúlande, and he ðær is bebyrged *in this year* [A. D. 1077] *earl Waltheof
was beheaded at Winchester, and his body was taken to Crowland, and
he is there buried,* 1077; Th. 350, 10.   Hí cómon to ðære stówe ðe
man hāteþ Crúwland *they came to the place which is called Crowland,*
Guthl. 3; Gdwin. 22, 1: 12; Gdwin. 58, 12.   Ðā wæs se eahtoða dæg
ðæs kalendes Septembres, ðā se eádiga wer, Gūþlāc, com to ðære fóre-
sprecenan stówe, Crúwlande... hæfde he ðā on ylde six and twentig
wintra *it was the eighth day before the kalends of September* [Aug. 24th,
A. D. 699], *when the blessed man, Guthlac, came to the aforesaid place,
Crowland... he was then twenty-six years of age,* Guthl. 3; Gdwin. 22,
25-24, 3: 22; Gdwin. 96, 21.   v. Gūþ-lāc.

CRUMA, an; *m. A* CRUMB, *fragment;* mica:—Cruma *mica,* Wrt.
Voc. 83, 1.   We hēdaþ ðæra crumena ðæs hláfes *we take care of the
crumbs of the bread,* Homl. Th. ii. 114, 33.   Ða hwelpas etaþ of ðām
crumum *catelli edunt de micis,* Mt. Bos. 15, 27: Lk. Bos. 16, 21. Lege
on ðone magan hláfes cruman *lay crumbs of bread on the stomach,*
L. M. 2, 12; Lchdm. ii. 190, 15: Homl. Th. ii. 114, 29. [*Prompt.*
crumme *mica:* Wyc. crummes, *pl:* Chauc. Piers P. cromes, *pl:* Orm.
crummess, *pl: Scot.* crum: *Plat.* kröme, kroom: *Dut.* kruim, *f: Kil.*
kruyme: *Ger.* krume, *f: Dan.* krumme, *m. f: Swed.* krumma, *f.*]

CRUMB, crump; *adj. Bent down, stooping;* cernuus, obuncus:—
Crump *obuncus,* Cot. 144.   Ða crumban *obunca,* 185.   [*Prompt.* crombe,
crome *bucus: Orm.* crumb: *Scot.* crummet: *O. Sax.* O. Frs. crumb:
Dut. krom: *Ger.* krumm: *M.H. Ger.* krump: *O.H. Ger.* krumb: *Dan.*
krum: *Swed.* krum: *Wel.* crwm *bent: Corn.* crom *crooked: Ir. Gael.* crom
bent.]

cruncon; *pp.* cruncen *yielded,* Byrht. Th. 140, 43; By. 302; *p. pl.
and pp. of* crincan.

crundel, crundol, crundul; *gen.* crundeles, crundles; *dat.* crundle,
crundelle; *m.*   **I.** *a barrow, mound raised over graves to protect
them;* tumulus:—On ðone durnan [MS. durnen] crundel; of ðam durnan
crundelle on ðone þorn *to the retired barrow; from the retired barrow to
the thorn,* Cod. Dipl. 1053; A.D. 854; Kmbl. v. 105, 26.   Ðonan on
morþcrundle; of morþcrundle on ðone brádan herpæþ [MS. herpaþ]

---

thence to the death-barrow [*to the tumulus of the dead*]; *from the tumulus
of the dead to the broad military road,* Cod. Dipl. 543; A.D. 968;
Kmbl. iii. 23, 34, 35.   Ðēr þwyres ofer þrý crundelas *there across over
three barrows,* Cod. Dipl. 985; Kmbl. v. 13, 32.   **II.** in later times
crundel is *n:*—On ðæt crundel *to the barrow,* Cod. Dipl. 1283; Kmbl. vi.
120, 8.   [Kemble, in his Glossary Cod. Dipl. iii. pref. p. xxi, says,—'*It
seems to denote a sort of water-course, a meadow through which a stream
flows.*' Yet the following example in this same vol. proves that **a**
crundel could not be *a meadow through which a stream flows,* as it was
on a hill:—Crāwan crundul on Wereðan hylle *Crow's crundle on Weretha's
hill,* Cod. Dipl. 698; A.D. 997; Kmbl. iii. 301, 35. Professor Leo
says,—' A crundel or crundwel is *a spring* or *well, with its cistern,
trough,* or *reservoir,*' and cites,—Ðonon eft on crundwylle *then again
to crund-spring,* Cod. Dipl. 1188; Kmbl. v. 354, 20, 28.   The crundle
on *Weretha's hill* militates against Dr. Leo's view, as well as Kemble's;
Mr. Thorpe therefore concludes,—' My belief is, that the word is not
Anglo-Saxon, nor Germanic, but British, and signifies *a tumulus* ŏr *barrow,*
and is akin to the Welsh carneddaw *a cairn or heap of stones,*' Th. Diplm.
Glossary, p. 654.]   DER. morþ-crundel, stān-.

crungon; *pp.* crungen *yielded, perished,* Exon. 124 a; Th. 477, 17;
Ruin. 26; *p. pl. and pp. of* cringan.

crupon *crept, crawled,* Ors. 1, 7; Bos. 29, 33: Chr. 1083; Erl. 217,
22; *p. pl. of* creópan.

crusene, crusne, an; *f. A robe made of skins;* mastruga:—Crusene
oððe deórfellen roc *crusen or a beastfelt or skin garment,* Wrt. Voc. 82, 4.
Crusne *mastruga,* Ælfc. Gl. 65; Som. 69, 39; Wrt. Voc. 40, 66.

crup *a crowd;* multitudo, turba confertissima, Som. Ben. Lye. v.
creódan.

Crúwland *Crowland, Lincolnshire,* Guthl. 12; Gdwin. 58, 12. v.
Crúland.

CRYB; *gen.* crybbe; *f. A* CRIB, *bed, stall;* stratum, præsepe:— Ic
læg cildgeong on crybbe *I lay as a young child in a crib,* Exon. 28 b;
Th. 87, 16; Cri. 1426. [*Prompt.* crybbe *præsepe:* Orm. cribbe: *Scot.*
crufe, cruife, crofe: *Plat.* kribbe, krubbe: *O. Sax.* cribbia, *f: Frs.
O. Frs.* kribbe, *f: Dut.* krib, kribbe, *f: Kil.* krippe: *Ger. M.H. Ger.*
krippe, *f: O.H. Ger.* krippa, kripha, *f: Dan.* krybbe, *m. f: Swed. Icel.*
krubba, *f: Fr.* crèche, *f: Prov.* crepcha: *It.* gréppia, *f: Slav.* kripa, *f.
a basket.*]

crycce *a crutch, staff,* Bd. 4, 31; S. 610, 17.   v. cricc.

crýdeþ *presses,* Exon. 101 b; Th. 384, 15; Rä. 4, 28; *3rd pers. pres.
of* creódan.

crýfele *a den, passage under ground;* spelunca, meatus subterraneus,
Som. Ben. Lye. v. crýpele.

crymbig *crooked,* Som. Ben. Lye. v. crumb.

crymbing, e; *f. A bending;* curvatura, Cot. 56.

crýpan; *p.* crýpte; *pp.* crýped *To creep;* repere:—He næfþ his fóta
geweald and onginþ crýpan *he has not the use of his feet and begins to
creep,* Bt. 36, 4; Fox 178, 14.   v. creópan.

crýpele, es; *m? A den, burrow;* cuniculum, Mone B. 2774.

crýpest, crýpst, he crýpeþ, crýpþ *creepest, creeps; 2nd and 3rd pers.
pres. of* creópan.

crysma *chrism,* L. Ecg. C. 36; Th. ii. 162, 1.   v. crisma.

crysum-lýsing *a leaving off the baptismal vest,* Chr. 879; Th. 148,
32, col. 3.   v. crism-lýsing.

crýt = crýdeþ *crowdeth:* ðu crýtst, crýst *thou crowdest; 3rd and
2nd pers. pres. of* creódan.

CÚ; *nom. acc; gen.* cúe, cú, cuus, cús; *dat.* cý; *pl. nom. acc.* cý;
gen. cúa, cúna; *dat.* cuum, cúm; *f. A* cow; vacca, bucula:—Cú *vacca,*
Wrt. Voc. 287, 56.   Cú *vacca vel bucula,* Ælfc. Gl. 21; Som. 59, 82;
Wrt. Voc. 23, 40: 78, 42. Iung cú *a young cow;* juvenca, Ælfc. Gl.
22; Som. 59, 89; Wrt. Voc. 23, 46.   An cú wearþ gebroht to ðam
temple *a cow was brought to the temple,* Homl. Th. ii. 300, 33: Chr.
1085; Erl. 218, 36. Gesomna cúe mesa *collect the dung of a cow,*
L. M. 1, 38; Lchdm. ii. 98, 5.   On ðære cú hricge *on the cow's back,*
M. H. 194 a.   Be cuus horne *of a cow's horn,* L. In. 59; Th. i. 140, 1,
3.   Cús eáge *a cow's eye,* 59; Th. i. 140, 4.   Of ðære cý *from the cow,*
M. H. 194 a.   Gif mon cú forstele *if a man steal a cow,* L. Alf. pol. 16;
Th. i. 70, 24: L. In. 38; Th. i. 126, 5: L. Ath. v. § 6, 2; Th. i. 234,
1: L. O. D. 7; Th. i. 356, 5.   Cúa *of cows,* Cod. Dipl. 201; A.D. 814;
Kmbl. i. 253, 28.   Feówertig cúna *vaccas quadraginta,* Gen. 32, 15:
Cod. Dipl. 732; A.D. 1016-1020; Kmbl. iv. 10, 23: 949; A.D.
1049-1052; Kmbl. iv. 284, 8.   On cuum *in vaccis,* Ps. Lamb. 67, 31.
Ðū wāst, ðæt ic hæbbe hnesce litlingas and ge-eáne eówa and gecelfe cý
mid me *nosti quod parvulos habeam teneros et oves et boves fœtas mecum,*
Gen. 33, 13: Cod. Dipl. 235; A.D. 835; Kmbl. i. 310, 18, 25, 27:
675; A.D. 990; Kmbl. iii. 255, 13. [*Prompt.* cowe *vacca:* Piers P.
kow, cow: R. Brun. kie, *pl:* Plat. ko, *pl.* koie: *O. Sax.* kô, *f: Frs.*
kw, *pl.* ky, *f: O. Frs.* ku, *f: Dut.* koe, *f: Kil.* koe, koeye: *Ger.* kuh,
f: M.H. Ger. kuo, *f: O.H. Ger.* kua, kô, *f: Dan.* kô, koe: *Swed.*
ko, *f: Icel.* kýr, *f. dat. and acc.* kú: *Lat.* cēva *a heifer: Sansk.* go,
gaus *bos, vacca.*]   DER. folc-cú, mete-.

cualme-stów, e; f. A place of burial; calvariæ locus, Som. Ben. Lye. v. cwealm-stów.

cú-butere, an; f. Cow's butter, butter made of cow's milk; vaccæ butyrum:—Reáde netlan awylle on hunige and on cúbuteran boil red nettles in honey and in cow's butter, L. M. 2, 51; Lchdm. ii. 268, 18: iii. 16, 20.

cuc quick, alive; vivus:—He lêt cucne he left alive, Ors. 6, 2; Bos. 116, 41: Gen. 1, 20: Ælfc. Gl. 35; Som. 62, 90. v. cwic.

cú-cealf, es; n. A cow's calf; vaccæ vitulus:—Gif man of myran folan adrífþ oððe cúcealf if a man drives off a mare's foal or a cow's calf, L. Alf. pol. 16; Th. i. 70, 23.

cuceler, cuculer, cucler, es; m. A spoon, half a drachm; cochlear:—Fíf cuceleras fulle five spoonsful, Herb. 26, 3; Lchdm. i. 122, 23. Þrý cuculeras three spoons, 26, 3; Lchdm. i. 122, 24. [Lat. cochlear, áris; n.]

cucen alive; vivus, Wanl. Catal. 3, 12. v. cucon.

cucian; p. ode; pp. od To quicken, make alive; vivificare, Som. Ben. Lye. v. cwician.

cucler, es; m. A spoon; cochlear:—Ðæt seáw sele on cuclere give the juice in a spoon, L. M. 1, 48; Lchdm. ii. 120, 19. Genim celeþonian [MS. cileþonian] seáwes cucler fulne take a spoon full of juice of celandine, L. M. 1, 2; Lchdm. ii. 28, 3. The following are examples of cucler:—2, 1; Lchdm. ii. 178, 6: 2, 4; Lchdm. ii. 182, 23: 2, 7; Lchdm. ii. 186, 5: 2, 24; Lchdm. ii. 214, 5, 25. v. cuceler.

cucler-mǽl, es; n. [mǽl a measure] A spoon measure; cochlearis mensura:—Án cuclermǽl one spoon measure, L. M. 2, 7; Lchdm. ii. 186, 10. Tú cuclermǽl two spoon measures, 1, 2; Lchdm. ii. 28, 3.

cucon, cucun alive, quick; vivus:—Ðæt he Wulfnóþ cuconne oððe deádne begytan sceolde that he should take Wulfnoth alive or dead, Chr. 1009; Erl. 142, 3. v. cuc, cwic.

cuculer, es; m. A spoon; cochlear:—Þrý cuculeras three spoons, Herb. 26, 3; Lchdm. i. 122, 24. v. cuceler.

cucumis; gen. eris; m. Lat. A cucumber; cucumis:—Cucumeres, ðæt synd eorþæppla cucumbers, which are earth-apples, Num. 11, 5.

cud, cudu, es; n? A cud, what is chewed; rumen:—Ðe heora cudu ne ceówaþ: ða clǽnan nýtenu ðe heora cudu ceówaþ which chew not the cud: the clean beasts which chew their cud, M. H. 138 b. v. cwudu.

cudele a cuttlefish; sepia = σηπία:—Cudele vel wasescite sepia, Ælfc. Gl. 102; Som. 77, 82; Wrt. Voc. 56, 6.

cú-eáge, an; f. A cow's eye; vaccæ oculus:—Cúeáge biþ scillinges weorþ a cow's eye is worth a shilling, L. In. 59; Th. i. 140, 4, note 11.

cuellan to kill, Som. Ben. Lye. v. cwellan.

cúe mesa, an; m. Cow's dung; lætámen:—Gesomna cúe mesa collect cow's dung, L. M. 1, 38; Lchdm. ii. 98, 5.

cuén a queen, Chr. 672; Erl. 34, 35: 737; Erl. 46, 22: 836; Erl. 64, 33: 855; Erl. 68, 30: 885; Erl. 84, 5: 888; Erl. 86, 18. v. cwén.

cuffle, an; f. A cap, coif, hood, head dress; pileus, cucullus, capitis tegmen:—Hió an Æðelfléde hyre cuffian she gives to Æthelfled her hood, Cod. Dipl. 1290; A.D. 995; Kmbl. vi. 133, 20.

cugele, cugle, cuhle, an; f. A COWL, monk's hood; cuculla:—Twá cugelan two cowls, R. Ben. 55. Cugle cuculla, Wrt. Voc. 81, 71. Seó cuhle the cowl, R. Ben. 55. [Ger. kogel, gugel, f: M. H. Ger. gugele, f: O. H. Ger. cucula, f: M. Lat. cuculla: Span. cogúlla, f.]

cú-horn, cuu-horn, es; m. A cow's horn; vaccæ cornu:—Cuuhorn [cú- MSS. B. H.] biþ twegea pæninga wurþ a cow's horn shall be worth two pence, L. In. 59; Th. i. 140, 2.

cú-hyrde, es; m. [hyrde a keeper, guardian] A cowherd, person who has the charge of cows; vaccarius, bubulcus:—Cúhyrde gebýreþ ðæt he hæbbe ealdre cú meolc vii niht, syððan hed nige cealfod hæþ, and frymetlinge býstinge xiv niht; and gá his metecú mid hláfordes cú vaccarii rectum est, ut habeat lac vaccæ veteris vii noctibus, postquam enixa erit, et primitivarum bistinguium xiv noctibus; et eat ejus vacca cum vaccis domini, L. R. S. 13; Th. i. 438, 18–20. Cúhyrdas bubulcos, Mone B. 2408.

cuic living, Jn. Lind. War. 4, 10. v. cwic.

cuic-beám, es; m. A juniper-tree; juniperus. v. cwic-beám.

cuide a saying, Past. 35, 5; Hat. 46 b, 4. v. cwide.

cúle a cowl, Wanl. Catal. 131, 74, col. 1. v. cugele.

CULFRE, culfre, culefre, an; f: culfer, e; f. A dove, CULVER, pigeon; columba:—Se hálega Gást astáh swá án culfre descendit Spiritus sanctus sicut columba, Lk. Bos. 3, 22; Wrt. Voc. 77, 20: 280, 31. Wæs culufre of cófan sended a dove was sent from the ark, Cd. 72; Th. 88, 12; Gen. 1464. Culfer columba, Ælfc. Gl. 37; Som. 63, 2; Wrt. Voc. 29, 25. Ðæt híg offrunge sealdon, twegen culfran briddas ut darent hostiam, duos columbæ pullos, Lk. Bos. 2, 24; Ps. Th. 67, 13. On culfran hiwe in likeness of a dove, Homl. Th. i. 104, 21. Fyðeras culefran oferseolfrade pennæ columbæ deargentatæ, Ps. Lamb. 67, 14. He asende út áne culfran emisit columbam, Gen. 8, 8, 10, 12. He forlêt háswe culufran he let out a livid dove, Cd. 72; Th. 87, 20; Gen. 1451: 72; Th. 89, 8; Gen. 1477. Ða hálgan apostolas wǽron swilce culfran the holy apostles were as doves, Homl. Th. i. 586, 1: Homl. Blick. 23,

27: Bilwyte swá culfran simplices sicut columbæ, Mt. Bos. 10, 16: Ps. Th. 54, 6. [Wyc. culver, culvere: Chauc. culver: Piers P. colvere: R. Glouc. colfren, pl: Orm. cullfre: Laym. culveren, pl: Lat. columba.] DER. wudu-culfre.

culmille, an; f. The lesser centaury; erythræa centaurium, Lin:—Genim ða lytlan culmillan take the small centaury, L. M. 1, 16; Lchdm. ii. 58, 20. v. curmealle.

culpa, an; m. A fault; culpa:—Ne ic culpan in ðê ǽfre onfunde I have never found any fault in thee, Exon. 10 b; Th. 11, 28; Cri. 177.

culpian; p. ode; pp. od To humiliate, cringe; humiliare:—Hú ne is ðæt ðonne sum dǽl ermþa, ðæt mon scyle culpian to ðam ðe him gifan scyle is not this then somewhat of misery, that a man must cringe to him who can give to him? Bt. 32, 1; Fox 114, 15.

CULTER, cultur; gen. cultres; m? A COULTER or CULTER, dagger; culter, sica:—Hwanon ðam yrþlinge culter, búton of cræfte mínon unde aratori culter, nisi ex arte mea? Coll. Monast. Th. 30, 31; Wrt. Voc. 74, 73. Cultur sica, 287, 5. Gefæstnodon sceare and cultre mid ðære syl confirmato vomere et cultro aratro, Coll. Monast. Th. 19, 21. [Prompt. culter: Wyc. culter, cultre: Piers P. cultour, kultour: Fr. coutre: It. coltro: Lat. culter: Sansk. kṛit to cut.]

culufre a dove, Cd. 72; Th. 88, 12; Gen. 1464. v. culfre.

cum come:—Nú ðú cum now come thou, Exon. 10 a; Th. 10, 9; Cri. 149; imp. of cuman.

cuma, an; m. [cum, imp. of cuman to come; -a, termination, q. v.] A comer, guest, stranger; advena, hospes:—Ic wæs cuma eram hospes, Mt. Bos. 25, 35, 38, 43; Wrt. Voc. 86, 43. Mon cýðe cynewordum, hú se cuma hátte let a man make known in fitting words, how the guest is called, Exon. 112 b; Th. 430, 30; Rä. 44, 16: Beo. Th. 3616; B. 1806. Gúþlác swýðe blíðe wæs ðæs heofonlícan cuman Guthlac was right glad of the heavenly guest, Guthl. 4; Gdwin. 30, 2. Fram eallum ðám cumum a cunctis hospitibus, Bd. 4, 31; S. 610, 6. Metodes þeów grétan eóde cuman the Lord's servant went to meet the guests, Cd. 111; Th. 146, 32; Gen. 2431. Ðæt he wolde ǽlcne cuman swíðe árlíce underfón that he would very honourably receive every stranger, Bt. 16, 2; Fox 52, 31. Cuman árfæste righteous strangers, Cd. 114; Th. 150, 3; Gen. 2486. Cómon Sodomware cuman acsian the inhabitants of Sodom came to demand the strangers, 112; Th. 148, 8; Gen. 2453: Ors. 1, 8; Bos. 31, 4. Cumena árþegn an attendant of guests, Bd. 4, 31; S. 610, 4. Cumena búr a guest-chamber, 4, 31; S. 610, 11. Cumena hús a guest-house, an inn, Lk. Bos. 2, 7: 22, 11. Cumena inn a guest-house, an inn, Greg. Dial. 2, 22. Cumena wícung a guest-dwelling, an inn, Ælfc. Gl. 58; Som. 67, 85; Wrt. Voc. 58, 11. DER. cwealm-cuma, wil-.

CUMAN; part. cumende; ic cume, ðú cymst, cymest, he cumeþ, cymþ, cymeþ, cimþ, pl. cumaþ; p. ic, he com, cwom, ðú cóme, pl. cómon, cwómon; imp. s. cum, cym, pl. cumaþ; subj. indef. ic cume, cyme, pl. cumon, cumen, cymen; p. cóme, pl. cómen; pp. cumen, cymen. I. to COME, go, happen; venire, ire, accidere, evenire:—Sceal se gást cuman the spirit shall come, Soul Kmbl. 17; Seel. 9. Cuman ongunnan they attempted to come, Beo. Th. 494; B. 244. Cum to ðam lande, ðe ic ðê geswutelige come to the land, which I will shew thee, Gen. 12, 1. Ne cumon eów ðás worde of gemynde let not these words depart out of your mind, Deut. 4, 9. Ðonne wíg cume when war happens, Beo. Th. 46; B. 23. Ðonne his fyll cóme when his fall has happened, Cd. 200; Th. 248, 15; Dan. 513. Cumaþ ðonne mid cumendum venientes autem venient, Ps. Th. 125, 6. II. cuman is used with the infinitive expressing manner or purpose; as, Com féran came walking or happened to walk, Cd. 40; Th. 52, 31; Gen. 852. Com lǽdan came leading or came to lead, 85; Th. 106, 19; Gen. 1773. Sunnan leóma cymeþ scýnan a sunbeam shall come shining or begin to shine, Exon. 21 a; Th. 56, 17; Cri. 902. Secgan cymeþ shall come to say, Cd. 22; Th. 28, 20; Gen. 438. Com grétan came to greet, 97; Th. 126, 31; Gen. 2103. Com weorc sceáwigan came to view the work, 80; Th. 101, 7; Gen. 1678. [Prompt. cum, come: Wyc. Chauc. Piers P. come: Laym. come, cumen, cummen, kumen: Orm. cumenn: Plat. kamen: O. Sax. kuman: Frs. kommen: O. Frs. kuma, coma: Dut. komen: Ger. kommen: M. H. Ger. komen: O. H. Ger. queman: Goth. qiman: Dan. komme: Swed. komma: Icel. koma: Lat. venire: Grk. βαίνειν: Sansk. gam.] DER. a-cuman, an-, aweg-, be-, fór-, fóre-, forþ-, ge-, in-, of-, ofer-, oferbe-, onbe-, ongeán-, þurh-, to-, tobe-, up-.

CUMB, es; m. I. a hollow among hills, narrow valley, COMB; caverna inter colles, vallis angusta:—Andlang cumbes along the valley, Cod. Dipl. Apndx. 354; A. D. 931; Kmbl. iii. 406, 10: 489; A. D. 962; Kmbl. iii. 457, 29. In cumb, of ðam cumbe to a valley, from the valley, Cod. Dipl. Apndx. 118; A. D. 770; Kmbl. iii. 380, 5. II. a liquid measure; mensura quædam liquidorum: hence, perhaps, our dry measure COMB or COOMB = four bushels:—Cumb fulne líðes aloþ, and cumb fulne Weïsces aloþ a comb full of mild ale and a comb full of Welsh ale, Th. Diplm. A. D. 791–796; 40, 5: Lchdm. iii. 28, 9. [Dut. kom, f. a basin: Ger. kumpf, kump, m. 1. a dry measure for corn and fruit; II. a cup, basin: M. H. Ger. kumpf a vessel, dry measure: O. H. Ger. chumph cimpus? O. Fr. combe a deep valley: Grk. κύμβος

the hollow of a vessel, cup, bowl; κύμβη *a basin : Wel.* cwm, *m. a hollow, deep valley : Sansk.* kumbha, *m. a pot, jug.*] DER. fild-cumb.

cumbel-gehnád, es; *n.* [cumbel = cumbol, gehnád *a conflict*] *A conflict of ensigns* or *banners, a battle;* signorum conflictus, proelium, Chr. 937; Erl. 114, 15; Ædelst. 49, note.

Cumber-land, Cumbra-land, Cumer-land, es; *n.* [*Sim. Dun.* Cumbreland : *Hunt. Hovd. Brom.* Cumberland] CUMBERLAND; Cumbria :—Hér Eádmund cyning oferhergode eal Cumbraland *in this year* [A. D. 945] *king Edmund overran all Cumberland,* Chr. 945; Th. 212, 10; 213, 10, col. 1, 2 : Cumberland, 213, 10, col. 3. On ðisum geáre se cyning férde into Cumerlande [Cumberlande, col. 2] *in this year the king went into Cumberland,* 1000; Th. 248, 29, col. 1; 249, 29.

CUMBOL, cumbl, cuml, es; *n.* **I.** *a sign, image, military standard, ensign, banner;* signum, imago, signum militare, vexillum :—In campe gecrong cumbles hyrde *the standard's guardian fell in battle,* Beo. Th. 5004; B. 2505. Hie fôr ðam cumble on cneówum sæton *they sat on their knees before the image,* Cd. 181; Th. 227, 1; Dan. 180. Cumbol lixton wîges on wênum *ensigns glittered in hopes of battle,* 151; Th. 188, 29; Exod. 175; Andr. Kmbl. 8; An. 4. To weallgeatum wîgend þrungon, cêne under cumblum *the warriors thronged to the wallgates, bold beneath their ensigns,* Andr. Kmbl. 2409; An. 1206 : Judth. 12; Thw. 26, 18; Jud. 333. **II.** *a sign* or *evidence of disease, a wound;* morbi signum, vulnus :—Se læce, ðonne he cymþ ðone untruman to snîðanne, ærest [MS. æresð] he sceáwaþ ðæt cumbl [cuml MS. Oth.] *the surgeon, when he comes to cut the patient, first examines the wound;* ad ægrum medicus venerat, secandum vulnus videbat, Past. 26; Hat. MS. 36 a, 7. [*O. Sax.* kumbal, *n. a heavenly sign : O. H. Ger.* cumpal *cohortes : Swed.* kummel, *n. tessera, signum : Icel.* kuml, kumbl, kubl, *n. a sign, badge, mark, war-badge.*]

cumbol-gebrec *a crash* or *clashing of banners.* v. cumbul-gebrec.

cumbol-gehnád *a conflict of ensigns* or *banners, a battle.* v. cumbel-gehnád.

cumbol-gehnást, es; *n.* [cumbol I. *an ensign, banner;* gehnást *a conflict*] *A conflict of ensigns* or *banners, a battle;* signorum conflictio, bellum :—Ðæt hie beadoweorca beteran wurdon on campstede, cumbolgehnástes *that they were better in works of war on the battle-field, at the conflict of banners,* Chr. 937; Th. 206, 2, col. 2; 207, 2.

cumbol-haga, an; *m.* [haga *a hedge*] *A compact rank, phalanx;* phalanx :—Ic sceal sêcan ôðerne under cumbolhagan cempan *I must seek another soldier in the rank,* Exon. 71 b; Th. 266, 8; Jul. 395.

cumbol-hete, es; *m.* [hete *hate*] *Warlike hate;* bellicum ôdium :—þurh cumbolhete *through warlike hate,* Exon. 75 a; Th. 280, 30; Jul. 637.

cumbol-wîga, an; *m.* [wîga *a warrior*] *A warrior, soldier;* bellator, miles, Judth. 12; Thw. 25, 5; Jud. 243 : 12; Thw. 25, 14; Jud. 259.

cumbor; *gen.* cumbres; *n.* [=cumbol, *q. v.*] *A banner, standard, ensign;* signum militare :—Hroden hilte cumbor *a banner adorned on the hilt,* Beo. Th. 2048.

Cumbra-land *Cumberland,* Chr. 945; Erl. 116, 29. v. Cumber-land.

cumbul-gebrec, es; *n.* [cumbul=cumbol I, gebrec *a noise, crashing*] *A crashing of banners* or *ensigns;* signorum fragor, Ps. C. 50, 11; Ps. Grn. ii. 277, 11.

cumen *come,* Gen. 48, 2; *pp.* of cuman.

cumende *coming,* Ps. Lamb. 125, 6; *part.* of cuman.

cú-meoluc, e; *f.* [meolc *milk*] *Cow's milk;* vaccæ lac :—Gâte geallan meng wið cûmeoluc *mingle goat's gall with cow's milk,* L. M. 1, 3; Lchdm. ii. 40, 19.

Cumer-land *Cumberland,* Chr. 1000; Erl. 137, 1. v. Cumber-land.

cum-feorm, e; *f.* [cuma *a stranger,* feorm *food, support, hospitality*] *Entertainment of strangers;* hospitium, Th. Diplm. A. D. 848; 102, 30.

cú-migoþa, an; *m.* [migþa, migoþa *urine*] *Cow's urine;* vaccæ urina :—Gesomna cûmigoþan [MS. -migoþa] *collect cow's urine,* L. M. 1, 38; Lchdm. ii. 98, 5.

cumin *the herb cummin,* Som. Ben. Lye. v. cymen.

cuml *a wound, swelling,* Past. 26; MS. Oth. v. cumbol **II.**

cum-lîðe; *adj.* [cuma *a comer,* lîðe *mild, gentle*] *Kind to comers* or *strangers, hospitable;* hospitalis :—Cumlîðe hospitalis, Ælfc. Gr. 9, 28; Som. 11, 37. Cild cumlîðe *a child will be hospitable,* Obs. Lun. § 15; Lchdm. iii. 192, 1 : 16; Lchdm. iii. 192, 8. Beóþ cumlîðe eów betwŷnan bûton ceorungum *be hospitable among yourselves without grudging,* Homl. Th. ii. 286, 14.

cum-lîðian [cuma *a guest,* lîðian *to nourish*] *To lodge, to receive as a guest;* hospitari, R. Ben. Interl. 1.

cum-lîðnys, -nyss, e; *f. Hospitableness, hospitality;* hospitalitas :—Cumlîðnys is swîðe hlîsful þing *hospitality is a very excellent thing,* Homl. Th. ii. 286, 16. þurh ða cumlîðnysse *by hospitality,* 286, 2, 7; 8, 11, 13, 17, 27.

cummáse *a coal-titmouse, coal-tit,* Wrt. Voc. 281, 10. v. côl-máse.

cum-pæder, es; *m. A godfather;* compater :—ðe Æðeréd his cumpæder healdan sceolde *which Æthelred his godfather had to defend,* Chr. 894; Erl. 92, 2.

cumul, es; *pl. nom. acc.* cumulu; *n. A glandular swelling;* tumor glandulósus :—Wið cyrnlu and wið ealle yfele cumulu *for kernels and for all evil lumps,* Herb. 158, 5; Lchdm. i. 286, 17. v. cumbol **II.**

cúna *of cows,* Gen. 32, 15; *gen. pl.* of cú.

-cund, an adjective termination, denoting KIND, *sort,* or *origin, likeness;* as, ædel-cund, deóful-, engel-, eorþ-, feor-, feorran-, gæst-, god-, heofon-, hîw-, in-, sâwel-, ufan-, up-, woruld-. [*O. Sax.* -kund *oriundus,* in god-kund *divine : O. H. Ger.* -kund : *Goth.* -kunds : *Grk.* -γενής : *Lat.* -gena.]

cune-glæsse, an; *f. The herb hound's* or *dog's tongue;* cynoglossos =κυνόγλωσσον, cynoglossum officinale, Lin :—Wið cancerâdle, cuneglæsse nioðoweard *for cancer, the netherward part of hound's tongue,* L. M. 1, 44; Lchdm. ii. 110, 1.

cunelle, an; *f. Thyme;* thymus [=θύμος] vulgaris :—Wyl cunellan [MS. cunille] *boil thyme,* L. M. 1, 31; Lchdm. ii. 74, 22. DER. wudu-cunelle.

cuning *a king,* Greg. Dial. MS. Hat. Bodl. fol. 9 a, 7. v. cyning.

CUNNAN, ic can, con, ðú canst, const, he can, con, *pl.* cunnon; *p.* ic, he cûðe, ðú cûðest, *pl.* cûðon; *subj.* cunne, *pl.* cunnen; *p.* cûðe, *pl.* cûðen; *pp.* [on]-cunnen, cûþ; *v. a.* **I.** *to be* or *become acquainted with, to know;* noscĕre, scire :—Ic ða stôwe ne can *I know not the place,* Elen. Kmbl. 1363; El. 683 : 1267; El. 635. Ic eów ne con *I know you not,* Cd. 227; Th. 304, 13; Sat. 629. Ðú canst *thou knowest,* Andr. Kmbl. 135; An. 68. Const, Beo. Th. 2759; B. 1377. Cann, Ps. Th. 91, 5 : 93, 11. Conn, Exon. 43 a; Th. 145, 12; Gú. 693. Ge ne cunnon *ye know not,* Cd. 179; Th. 224, 25; Dan. 141. Ðæt ðú cunne *that thou knowest,* 228; Th. 308, 34; Sat. 702 : Elen. Kmbl. 748; El. 374. Ic cûðe *I knew,* Cd. 216; Th. 273, 26; Sat. 142 : 19; Th. 24, 30; Gen. 385 : Ors. 1, 2; Bos. 26, 34. Hwanon cûðest ðú me *unde me nosti?* Jn. Bos. 1, 48. Cûðon, Cd. 18; Th. 23, 10; Gen. 357 : Andr. Kmbl. 1504; An. 753 : Gen. 29, 5. Heó weán cûðon *they became acquainted with woe,* Cd. 4; Th. 5, 20; Gen. 74. Men ne cunnon *men know not,* Beo. Th. 327; B. 162. Ic ne conn þurh gemæcscipe monnes ôwêr *I know not anywhere of a man through cohabitation,* Exon. 10 b; Th. 13, 6; Cri. 198. **II.** *with inf. To know how to do', to have power, to be able,* CAN; scire, posse :—Ic can eów læran *I can teach you,* Cd. 219; Th. 280, 3; Sat. 250. Ðe can naman ðînne neóde hêrigean *qui scit jubilationem,* Ps. Th. 88, 13. Hêrian ne cûðon wuldres waldend' *they knew not how to praise the ruler of glory,* Beo. Th. 367; B. 182. Dydon swâ hie cûðon *they did as they could,* Cd. 187; Th. 232, 11; Dan. 368. [Cunnan is the second of the twelve Anglo-Saxon verbs, called *præterito-præsentia,* given under âgan, *q. v.* The *inf.* cunnan and the *pres.* can, *pl.* cunnon, retaining preterite inflections, are taken from the *p.* of the strong verb cinnan, ascertained from can, *pl.* cunnon, which shews the ablaut or internal change of the vowel in the *p.* tense of the twelfth class of Grimm's division of strong verbs [Grm. i. edn. 2, p. 898; Koch, i. p. 252], and requires, by analogy with other verbs of the same class, the *inf.* cinnan, *q. v.* and the *pp.* cunnen. Thus we find the original verb cinnan, *p.* can, *pl.* cunnon; *pp.* cunnen. The weak *p.* cûðe, *pl.* cûðon, *for* cunde, cundon, is formed regularly from the *inf.* cunnan. The *pp.* generally takes the weak form, in Anglo-Saxon as well as in the cognate words; but strong and weak forms are both found, in *A. Sax.* the strong on-cunnen, and the weak cûþ, and in *M. H. Ger.* the strong ver-kunnen, and the weak kunt. The same *præterito-præsens* may be generally observed in the following cognate words :—

	inf.	*pres.*	*pl.*	*p.*	*pp.*
Eng.		can,		could,	
Laym.	cunne,	can,	cunnen,	cuðe, couðe,	cuþ.
Wyc.	kunne,	can, kan,	cunnen, kunnen,	koude, kouthe,	cunde, koud.
Plat.	könen,	kann,	könen,	kunden, kunnen,	kunt.
O. Sax.	kunnan,	kan,	kunnun,	consta,	kuþ.
O. Frs.	kunna,	kan,	kunnon,	kunda,	kuth, kud.
Ger.	können,	kann,	können,	konnte,	gekonnt.
M.H.Ger.	kunnen,	kan,	kunnen,	kunde,	-kunnen, kunt.
O.H.Ger.	kunnan, kan,		kunnumês,	kunda, kunsta,	kund.
				konda, konsta,	
Goth.	kunnan, kann,		kunnum,	kunþa,	kunþs.
O. Nrs.	kunna, kann,		kunnum,	kunna,	kunnat.]

DER. for-cunnan, on-.

cunne, *pl.* cunnen *know, can,* Cd. 228; Th. 308, 34; Sat. 702 : Elen. Kmbl. 748; El. 374; *subj. pres.* of cunnan.

cunnere, es; *m. A tempter;* tentator, Mt. Lind. Stv. 4, 3.

cunnian; *p.* ode, ade, ede; *pp.* od, ad, ed; *v. a.* **I.** *to prove, try, inquire, search into, seek for, explore, examine, investigate, tempt, venture;* probare, tentare, explorare, requirere, experiri, periclitari :—Woldon cunnian, hwæðer . . . *they would prove, whether . . . ,* Andr. Kmbl. 257; An. 129. Môt ic nú cunnian *may I now inquire?* Bt. 5, 3; Fox 10, 34. Uncûþne eard cunnian *to seek for an unknown home,* Exon. 28 b; Th. 87, 1; Cri. 1418 : Beo. Th. 2893; B. 1444. Se cunnaþ Dryhtnes meahta *he tempteth the Lord's might,* Salm. Kmbl. 454; Sal. 227. He ðín cunnode *he has proved thee,* Cd. 163; Th. 204, 16; Exod. 420 : Bd. 3, 2; S. 525, 15. **II.** *with gen. To have experience of,*

to make trial of; periclitari, experiri:—Gódes and yfles ðǽr ic cunnade there I had experience of good and evil, Exon. 85 b; Th. 321, 26; Wïd. 52. Git wada cunnedon *ye made a trial of the fords,* Beo. Th. 1021; B. 508. [*Orm.* cunnenn *to try, attempt:* O. H. Ger. kunnēn *experiri, tentare.*] DER. a-cunnian, be-, ge-.

cunning, e; f. *Experience,* CUNNING; experientia, Som. Ben. Lye. v. on-cunning.

cunnung, e; f. *Probation;* probatio, tentatio, Exon. 118 a; Th. 453, 33; Hy. 4, 24.

cuopel; *gen.* cuople; f? *A coble, small ship;* navicula:—Ofstīgende hine oððe he ofstäg in lytlum scipe oððe in cuople *ascendente eo in naviculam,* Mt. Kmbl. Lind. 8, 23.

CUPPE, an; f. *A small drinking vessel,* CUP; poculum, obba:—Cuppe obba, Ælfc. Gl. 24; Som. 60, 43; Wrt. Voc. 24, 43. Nime ǽne cuppan let him take a cup, L. M. 2, 64; Lchdm. ii. 290, 2: Lchdm. iii. 72, 17: Cod. Dipl. 492; Kmbl. ii. 380, 35. Ic ge-an mīnum hláforde iv cuppan *I give four cups to my lord,* Th. Diplm. A. D. 972; 519, 24. [*Prompt. Wyc.* cuppe: *Piers P.* coppe, coupe: *Chauc.* cuppe: *R. Glouc.* coupe: *Orm.* cuppess, pl: *Laym.* cuppe: *Plat.* kop-jen, kop-ken *a little basin:* *Frs. O. Frs. Dut.* kop, m: *Dan.* kop, m. f: *Swed.* kopp, m: *Icel.* koppr, m: *Fr.* coupe, f: *It.* cóppa, f: *Span.* cópa, f: *Lat.* cupa, f. *a tub, cask:* *Grk.* κύπ-ελλον *a cup, goblet:* *Wel.* cwpan, f; cwb, m: *Ir.* cupa: *Sansk.* kūpa, kumbha, m. *a vessel for water.*] DER. scencing-cuppe, sop-.

curfon *carved,* Lev. 8, 20; p. pl. *of* ceorfan.

curmealle, curmelle, curmille, an; f. *Centaury;* centaurēum = κενταύρειον:—Wið ūtsihtādle; curmealle, etc. *for diarrhœa; centaury, etc.* L. M. 3, 22; Lchdm. ii. 320, 11: 1, 32; Lchdm. ii. 76, 20. Curmille *centaury,* 1, 32; Lchdm. ii. 78, 21. Wring curmeallan seáw *wring juice of centaury,* 3, 1; Lchdm. ii. 310, 9: Lchdm. iii. 38, 26: 58, 10. Genim grēne curmeallan *take green centaury,* 10, 19: 18, 23: 28, 28: L. M. 3, 26; Lchdm. ii. 322, 21: 3, 30; Lchdm. ii. 324, 21. Wyl on ealaþ twā curmeallan *boil in ale the two centauries,* L. M. 3, 38; Lchdm. ii. 330, 14. The centaury may be spoken of as, **I.** *the greater centaury;* chlora perfoliata, Lin:—Genim ðäs wyrte ðe Grēcas *centauria major* and Angle curmelle seó mǽre nemnaþ *take this herb which the Greeks name* centaurea major *and the English the greater centaury,* Herb. 35, 1; Lchdm. i. 134, 3. Curmelle *centaurea major,* Ælfc. Gl. 42; Som. 64, 29; Wrt. Voc. 31, 39. **II.** *the lesser centaury;* erythræa centaurium, Lin:—Ðeós wyrt ðe man *centauriam minorem* and ōðrum naman curmelle seó læsse nemneþ, biþ cenned on fæstum landum *this herb which is named* centaurea minor *and by another name the lesser centaury, is produced on stiff lands,* Herb. 36, 1; Lchdm. i. 134, 17. v. eorþ-gealla.

curn-stān *a mill-stone,* Glos. Prudent. Recd. 149, 79. v. cweorn-stān.

curon *chose,* Cd. 86; Th. 108, 9; Gen. 1803; p. pl. *of* ceósan.

CURS, es; m. *A* CURSE; maledictio:—On ænigne man curse asettan *to set a curse on any man,* Offic. Episc. 3. Gif hïg ǽnig man ūtabrede, hæbbe he Godes curs *if any man take them away let him have God's curse,* Wanl. Catal. 81, 5: Cod. Dipl. 310; A. D. 871–878; Kmbl. ii. 107, 5: 1057; Kmbl. v. 114, 25: Chr. 656; Erl. 33, 12: 675; Erl. 39, 20, 21, 27, 28: 963; Erl. 123, 14. [*Prompt.* curce: *Wyc.* curs: *Chauc.* cursing: *R. Brun.* cursyng.]

cursian; p. ode, ede; pp. od, ed *To* CURSE; maledicere:—Cursiende [MS. cursiynde] *maledicentes,* Ps. Spl. C. 36, 23. Ðe biscopes and lēred men heó cursede *the bishops and clergy cursed them,* Chr. 1137; Erl. 262, 37.

cursung, e; f. A CURSING, *curse, torment, hell;* maledictio, damnatio, gehenna = γέεννα:—He lufode cursunge, and heó cume him *dilexit maledictionem, et veniet ei,* Ps. Spl. C. 108, 16: Mt. Kmbl. Lind. 5, 29: 10, 28: Lk. Skt. Lind. Rush. 20, 47.

cūs *of a cow:*—Cūs eáge biþ scillinges weorþ *a cow's eye shall be worth a shilling,* L. In. 59; Th. i. 140, 4; gen. *of* cū.

CŪSC; adj. *Chaste, modest, pure, clean;* castus, purus:—Þurh cūscne siodo *through modest conduct,* Cd. 29; Th. 39, 2; Gen. 618. [*Plat.* küsk: *Dut.* kuisch: *Kil.* kuysch: O. *Sax.* kūsko, *adv:* *Frs.* kuwsch: *O. Frs.* kusk: *Ger.* keusch: *M. H. Ger.* kiusche, kiusch: *O. H. Ger.* kiuski, kūski *sobrius, pudicus:* *Dan.* kydsk: *Swed.* kysk.]

cusceote, cuscote, cuscute, an; f. [*Lancashire, cowshot*] *A ringdove, wood-pigeon;* palumbes, palumbä:—Cusceote *palumba,* Wrt. Voc. 280, 32. Cuscote, wuduculfre *palumbes,* 62, 27. Cuscutan *palumbes,* Glos. Epnl. Recd. 161, 58.

cūslyppe, cūsloppe, an; f. *A* COWSLIP; primula veris, Lin:—Nim wudubindes leáf and cūslyppan *take leaves of woodbine and cowslip,* L. M. 3, 30; Lchdm. ii. 326, 4: 3, 31; Lchdm. ii. 326, 10: iii. 30, 8: 46, 22. Cūsloppe *britannica,* Ælfc. Gl. 42; Som. 64, 30; Wrt. Voc. 31, 40.

cūsnis *choiceness;* fastidium, Glos. Epnl. Recd. 156, 40. v. cīsnes.

cū-tægel, -tægl, es; m. *A cow's tail;* vaccæ cauda:—Cūtægl biþ fíf penega weorþ *a cow's tail shall be worth five pence,* L. In. 59; Th. i. 140, 3, MS. B.

cuter *resin;* mastix, resina:—Cuter *mastix* vel *resina,* Ælfc. Gl. 48; Som. 65, 53; Wrt. Voc. 33, 49.

cūþ; *comp.* -ra; *sup.* -ost, -est; adj. [cūþ *known, pp. of* cunnan]. **I.** *known, clear, plain, evident, manifest;* notus, cognītus, manifestus:—Ðæt wæs monegum cūþ *that was known to many,* Exon. 100 b; Th. 378, 21; Deór. 19: Lk. Bos. 8, 17. Cūþ is wīde *it is widely known,* Exon. 40 b; Th. 134, 14; Gū. 507. Cūþ is, ðæt *it is manifest, that,* Cd. 198; Th. 246, 20; Dan. 482. Cūþ standeþ, ðæt he gescylded wæs *quem esse servatum constat,* Bd. 3, 23; S. 555, 27: 1, 27; S. 492, 38. Ðæt wæs ðara fæstna folcum cūþost *that was of those fastnesses most known to nations,* Cd. 209; Th. 259, 16; Dan. 692. **II.** *known, well known, sure, safe, noted, known as excellent, famed, celebrated;* notus, certus, præstans, egregius:—Cūþe ærenddracan *nuntii certi,* Bd. 4, 1; S. 564, 40. Cūþran gewitnesse *certiori notitia,* Bd. 4, 19; S. 588, 40. Se cūþesta gewita *certissimus testis,* 4, 19; S. 587, 27. Cūþes werodes *of the famed host,* Cd. 154; Th. 192, 14; Exod. 230: Beo. Th. 1738; B. 867: 4362; B. 2178: Cd. 226; Th. 302, 9; Sat. 596. **III.** *familiar, intimate, related, friendly;* notus, familiāris, amīcus, benevŏlus:—Swā swā he cūþre stæfne wæs to me sprecende *quasi familiari me voce alloquens,* Bd. 4, 25; S. 600, 43. Ne sint me winas cūþe eorlas elþeódige *the strange men are no affable friends to me,* Andr. Kmbl. 396; An. 198. Feor ðū me dydest freóndas cūþe *longe fecisti notos meos a me,* Ps. Th. 87, 8. Mīne cūþe *notos meos,* 87, 18: 54, 13: 131, 18. [*Wyc.* koud, kowd *known, pp. of* kunne: *Chauc.* couth, kouth, *pp. of* conne: *Orm.* cuþ, *pp. of* cunnenn: *Laym.* cuð, coð, icuð *known, renowned, pp. of* cuðe *to make known:* O. *Sax.* kuð *known:* O. *Frs.* kuth, kund, kud: *Dut.* kond: *Ger.* kund: M. H. *Ger.* kunt: O. H. *Ger.* kund: *Goth.* kunþs *known, pp. of* kunnan: *Icel.* kunnr, kuðr *known.*] DER. folc-cūþ, for-, hīw-, hīw-, in-, un-, unfor-, wīd-: cýþig, on-, un-.

cūða, an; m. [cūþ *known, pp. of* cunnan: -a, *termination,* q. v.] *One known, an acquaintance, a familiar friend, a relation;* notus, cognātus:—Ðū cūða mīn *tu notus meus,* Ps. Spl. 54, 14: Lk. Bos. 2, 44. Ne clypa ðū ðīne frýnd ne ðīne cūðan *noli vocare amīcos tuos neque cognātos,* 14, 12: 1, 58. v. cūþ.

cūðe; adv. *Clearly;* manifeste:—Ic cūðe gesette *I have clearly set,* Ps. Th. 88, 3.

cūðe, pl. cūðon *knew, could,* Ors. 1, 2; Bos. 26, 34; p. *of* cunnan.

cūþe-līc, cūþ-līc; adj. *Known, certain;* notus, Som. Ben. Lye. DER. un-cūþlīc.

cūðe-līce; adv. *Certainly:*—Ac we ðæt cūþelīce oncneówan *but that we certainly have known,* Bd. 1, 27; S. 491, 4. v. cūþlīce.

cūðe-men; pl. m. *Relations;* cognati:—Ða cūðemen *cognati,* Lk. Skt. Rush. 1, 58.

cūðen *knew, could,* Exon. 25 a; Th. 73, 6; Cri. 1185; subj. p. *of* cunnan.

cūðest *knewest, couldst;* 2nd pers. p. *of* cunnan.

cūþice; adv. = cūþlīce *Clearly;* manifeste:—Forðon ic cūþlīce [MS. cupice] on ðæm, hēr nū cwicu lifige *quia in ipsis vivificasti me,* Ps. Th. 118, 93.

cūþ-lǽtan [cūþ = cýþ *relationship,* lǽtan *to admit*] *To enter into friendship;* societatem facere, Som. Ben. Lye.

cūþ-līce, cūþe-līce; comp. or; adv. **I.** *certainly, manifestly;* certo, aperte:—Ic cūþlīce wât *scio certissime,* Bd. 2, 12; S. 513, 42: 4, 19; S. 589, 25. Ðæt his līf ðe cūþlīcor claresceat certius, 5, 1; S. 613, 14, note. Acyrred cūþlīce from Cristes ǽ *turned manifestly from Christ's law,* Exon. 71 b; Th. 267, 6; Jul. 411: Ps. Th. 103, 16: 106, 6: 121, 1: 146, 4: 149, 8. **II.** *for, indeed, therefore;* nempe, igitur:—Cweðaþ cūþlīce *for indeed they said,* Ps. Th. 70, 10: 82, 4: Hy. 10, 20; Hy. Grn. ii. 293, 20. **III.** *familiarly, courteously, kindly;* familiariter, civiliter, comiter:—Ðæt he ðe cūþlīcor from ðām hālgum ge-earnode in heofonum onfongen beón *quo familiarius a sanctis recipi mereretur in cœlis,* Bd. 5, 7; S. 621, 12: Cd. 111; Th. 146, 32; Gen. 2431. Ðæt he eáþmēdum ellorfūsne oncnāwe cūþlīce *that he should with affability kindly treat the ready to depart,* Andr. Kmbl. 643; An. 322: Ps. Th. 118, 146, 154: 54, 16: 90, 15. DER. for-cūþlīce, in-, un-.

cūþ-nes, -ness, e; f. *Knowledge, acquaintance;* scientia, Scint. 38, Som. Ben. Lye. DER. cūðe *knew;* p. *of* cunnan *to know.*

cūþ-noma, an; m. *A surname;* cognomen, Mt. Kmbl. Præf. p. 8, 13.

cūðo-menn; pl. m. *Relations;* cognati:—Cūðomen *cognatos, acc. m.* Lk. Skt. Lind. 14, 12. v. cūðe-men.

cūðon *knew, could,* Cd. 18; Th. 23, 10; Gen. 357; p. pl. *of* cunnan.

cūþra *more sure,* Bd. 4, 19; S. 588, 40; comp. *of* cūþ.

cūðudyst = cýddest *innotuisti,* Ps. Spl. C. 143, 4; 2nd pers. p. *of* cýðan.

Cūþ-wulf, es; m. *Cuthwulf:*—Cūþwulf wæs Cūþwining *Cuthwulf was the son of Cuthwin,* Chr. Th. 2, 3. Hēr DLXXI Cūþwulf feaht wið Bretwalas æt Bedcan forda *in this year,* A. D. 571, *Cuthwulf fought with the Brito-Welsh at Bedford,* Chr. 571; Th. 32, 25, col. 1.

cuu; gen. cuus; f. A *cow;* vacca:—Be cuus horne *of a cow's horn,* L. In. 59; Th. i. 140, 1, 3: Ps. Lamb. 67, 31. v. cū.

cuwon *chewed,* Ælfc. T. 42, 9; *p. pl.* of ceówan.

CWACIAN, cwacigan; *part.* cwaciende, cwacigende; *p.* ode; *pp.* od *To* QUAKE, *shake, tremble;* tremere, contremere :—Seó eorþe wæs cwaciende *the earth was quaking,* Ors. 2, 6; Bos. 49, 41. Seó cwacigende swustor *the quaking sister,* Homl. Th. ii. 32, 26, 31. Heó gemētte ealle hire bearn cwacigende eallum limum *she found all her children quaking in every limb,* 30, 20. Heard ecg cwacaþ *the hard edge shaketh,* Elen. Kmbl. 1513; El. 758. Cēne cwacaþ *the bold shall quake,* Exon. 19 b; Th. 50, 8; Cri. 797. Ða tēþ cwaciaþ on swīðlícum cýle *their teeth shall quake in the intense cold,* Homl. Th. i. 132, 27 : 530, 35. Ic cwacode eal on fefore *I quaked all in a fever,* ii. 312, 19. Cwacode eorþe contremuit terra, Ps. Spl. C. 17, 9. Cwacode he sóna *he instantly quaked,* Homl. Th. ii. 312, 15 : 32, 3, 19. [*Prompt.* quakyñ *tremere: Wyc. Piers P.* quaken: *R. Brun. Chauc. R. Glouc.* quake: *Laym.* quakien, cwakie.]

cwacung, e; *f. A* QUAKING, *trembling;* tremor :—Sóna biþ ætstilled sió · cwacung *the quaking will soon be stilled,* L. M. 1, 26; Lchdm. ii. 68, 11. Cwacung gegrāp hīg *tremor apprehendit eos,* Ps. Spl. C. 47, 5. On cwacunge *in tremore,* Ps. Spl. C. 2, 11. Wæs se mūnt Garganus bifigende mid ormǣtre cwacunge *the mount Garganus was trembling with immense quaking,* Homl. Th. i. 504, 28. Būton cwacunge *without quaking,* ii. 32, 18.

cwǣde, *pl.* cwǣdon *said,* Ps. Th. 89, 3 : Cd. 191; Th. 238, 28; Dan. 361; *2nd sing. p.* and *p. pl.* of cweðan.

cwǣl, *pl.* cwǣlon *died; p.* of cwelan.

cwǣlm *death,* Som. Ben. Lye. v. cwealm.

cwǣlu *a violent death,* Som. Ben. Lye. v. cwalu.

cwēman *to please,* Som. Ben. Lye. v. cwēman.

cwēn *a queen* :—Æðelfriþ cwēn, seó wæs Ælfrēdes swuster, forþfērde, and hire líc liþ æt Pauian *queen Æthelfrith, who was Alfred's sister, died, and her body lies at Pavia,* Chr. 888; Erl. 87, 16–18. v. cwēn.

cwært-ern *a prison,* Mt. Kmbl. Rl. 25, 43, 44. v. cweart-ern.

cwǣstednys *a trembling,* Som. Ben. Lye. DER. to-cwæstednys.

cwæþ QUOTH, *said, spoke,* Deut. 32, 26: Bd. 3, 5; S. 527, 30, 31; *p.* of cweðan.

cwæðst *sayest,* Ælfc. Gr. 2; Som. 3, 7, = cwedst; *2nd pres. sing.* of cweðan.

CWALU, e; *f. A quelling with weapons, torment, a violent death, slaughter, destruction;* nex, cædes, exitium :—Se cyning Eádwine mid ārleásre cwale ofslegen wæs *rex Æduini impia nece occisus,* Bd. 2, 14; S. 517, 32 : 2, 12; S. 513, 9, 12, 16. þurh ānes engles cwale, on Cristes cwale *through an angel's death, by Christ's death,* Boutr. Scrd. 17, 38. Hū nyt is ðe mín slæge, oððe mín cwalu *slaughter,* oððe mín rotung on byrgenne? Ps. Th. 29, 8. To cwale cnihta *for the destruction of the youths,* Cd. 184; Th. 229, 32; Dan. 226. To cwale syllan *to give to death,* Exon. 70 a; Th. 259, 29; Jul. 289. To cwale lǣdan *to lead to death,* 74 b; Th. 279, 14; Jul. 613. [*Laym.* quale *murrain;* quale-huse, cwal-huse *a torture-house :* O.Sax. quala, f: Dut. kwaal *malum, morbus :* Kil. quaele *languor, ægritudo :* Ger. qual, f: M.H.Ger. quël, f. *torment:* O.H.Ger. quâla *nex, pernicies :* Dan. qwal, m. f: Swed. qual, n. *anguish, agony :* Icel. kwal- in compounds, *pain, torment.*] DER. deáþ-cwalu, feorh-, gāst-, hearm-, hell-, līg-, níþ-, swylt-, sylf-.

cwanc, *pl.* cwuncon *disappeared; p.* of cwincan.

CWĀNIAN; *part.* cwāniende; *p.* ode, ede; *pp.* od, ed *To bewail, deplore, lament, mourn;* plorare, deplorare, queri, lugere. I. *v. trans* :—Sum sceal, leómena leás, sār cwānian *one, void of light, shall bewail his pain,* Exon. 87 b; Th. 328, 18; Vy. 19 : 73 b; Th. 274, 23; Jul. 537. II. *v. intrans* :—Cwāniendra cirm *the cry of mourning men,* Exon. 20 a; Th. 52, 19, note; Cri. 836. Weras cwānedon *the men lamented,* Andr. Kmbl. 3071; An. 1538. [*Plat.* kwinen *'to languish:* Dut. kwijnen *to linger, pine :* Kil. quenen, quynen *tabescere :* M.H.Ger. quinen *to languish :* Goth. qainon *lamentari, lugere :* Icel. kweina *to wail, lament.*]

cwānig; *adj.* [cwānian *to bewail, lament, mourn*] *Complaining, bewailing, sad;* querulus, tristis. DER. mōd-cwānig.

Cwanta-wíc, es; *n.* [wíc *a dwelling*] *St. Josse-sur-Mer* or *Estaples, the ancient name of which was Quantovic* or *Quentawich* :—Hēr wæs micel wælsliht on Lundenne and on Cwanta-wíc and on Hrófes ceastre *in this year* [A. D. 839] *there was a great slaughter at London and at Estaples and at Rochester,* Chr. 839; Erl. 66, 17.

cwart-ern *a prison* :—Ic wæs on cwarterne *eram in carcere,* Mt. Kmbl. Hat. 25, 36, 39. v. cweart-ern.

Cwat-brycg, -bricg, e; *f.* [*Ethelw.* Cantbricge: *Flor.* Quatbrig: *Hunt.* Quadruge: *Matt. West.* Quantebridge] *Bridgenorth in Shropshire;* oppidi nomen in agro Salopiensi :—Hī gedydon æt Cwatbricge *be Sæfern they arrived at Bridgenorth on the Severn,* Chr. 896; Th. 173, 43, col. 1 : col. 2 has Brygce. Æt Cwatbrycge, Th. 174, 1, col. 1, 2. Sǣton hie ðone winter æt Cwatbrycge [Bricge, Th. 174, 10, col. 2 : 175, 9, col. 1 : Brygcge, 175, 10, col. 2] *they remained that winter at Bridgenorth,* Chr. 896; Th. 174, 11, col. 1. v. Bricg.

CWEAD, es; *n. Dung, filth, ordure;* stercus :—Sume nimaþ wearm cwead *some take warm dung,* L. M. 1, 50; Lchdm. ii. 124, 8 : 2, 48; Lchdm. ii. 262, 18. Of cweade *de stercore,* Ps. Spl. 112, 6. [*Wyc.* quad, quade, *adj. bad: Piers P.* queed *the evil one, devil : Plat.* quaad, *adj. bad, evil :* O.Frs. quad, qwad, *adj. bad, evil :* Dut. kwaad, *n. evil, mischief :* Kil. quaed, quaet, quat, kat *stercus, oletum :* Ger. koth, *m.* merda, lutum : M.H.Ger. kât, kôt, quat, *m. n. stercus :* O.H.Ger. chot *stercus :* Zend gūtha, *m. dirt :* Sansk. gūtha, *m. n. excrement.*]

cweahte, *pl.* cweahton *quaked, vibrated; p.* of cweccan.

cwealde, *pl.* cwealdon *slew,* Exon. 65 b; Th. 243, 3; Jul. 5 : Ors. 4, 4; Bos. 80, 41; *p.* of cwellan.

cwealm, cwēlm, es; *m. n.* [cwelan *to die*] *Death, destruction, a violent death, slaughter, murder, torment, plague, pestilence, contagion,* QUALM; mors, pernicies, nex, cædes, homicidium, cruciatus, lues, pestis, pestilentia, contagium :—Hine se cwealm ne þeáh *death profited him not,* Exon. 74 b; Th. 278, 30; Jul. 605 : Cd. 79; Th. 98, 1; Gen. 1623 : Elen. Kmbl. 1349; El. 676. Him cwelm gesceód *death destroyed him,* Cd. 208; Th. 257, 36; Dan. 668. Ylda cwealm *a slaughter of men,* Andr. Kmbl. 363; An. 182. Cwealmes wyrhta *a worker of murder, a murderer,* Cd. 48; Th. 61, 29; Gen. 1004. Ðider sōþfæstra sāwla mótun cuman æfter cwealme *thither the souls of the just may come after death,* Exon. 32 b; Th. 103, 14; Cri. 1688 : Cd. 166; Th. 207, 18; Exod. 468. To wera cwealme *for the destruction of men,* Andr. Kmbl. 3013; An. 1509. Ic honda gewemde on Caines cwealme mīne *I have polluted my hands in Cain's murder,* Cd. 52; Th. 67, 4; Gen. 1095. In Caines cynne ðone cwealm gewrǣc Drihten *the Lord avenged the death [of Abel] on Cain's race,* Beo. Th. 215; B. 107 : Exon. 28 b; Th. 87, 17; Cri. 1426 : Andr. Kmbl. 2243; An. 1123. Ðū wāst cwealm hātne in helle *thou knowest hot torment in hell,* 2374; An. 1188 : 562; An. 281. þurh deáþes cwealm *through pain of death,* Exon. 35 b; Th. 115, 26; Gū. 195 : Cd. 224; Th. 296, 9; Sat. 499. Mid morþes cwealme *with pain of death,* 35; Th. 47, 9; Gen. 758. Cwealma mǣst *the greatest of torments, hell,* Exon. 31 b; Th. 99, 20; Cri. 1627. Micel cwealm wearþ ðæs folces *the mortality of the people was great,* Homl. Th. ii. 122, 18. Cwealm *pestilentia* vel *contagium* vel lues, Ælfc. Gl. 9; Som. 57, 8; Wrt. Voc. 19, 18. Ðæt us cwealm on ne becume *ne forte occidat nos pestis,* Ex. 5, 3. To ðam swíðe awēdde se cwealm ðæt hundeahtatig manna of lífe gewiton *the plague raged to that degree that eighty men departed from life,* Homl. Th. ii. 126, 18 : Exon. 89 a; Th. 335, 7; Gn. Ex. 30. On ðissum geáre com micel māncwealm on Brytene īgland, and on ðam cwealme forþfērde Tuda biscop *in this year* [A. D. 664] *there was a great plague in the island of Britain, and bishop Tuda died of the plague,* Chr. 664; Erl. 35, 19 : Homl. Th. ii. 124, 2. Godes miltsung ðone rēdan cwealm gestilde *God's mercy stilled the cruel pestilence,* ii. 126, 22. Beóþ mycele eorþan styrunga geond stōwa, and cwealmas *terræmotus magni erunt per loca, et pestilentiæ,* Lk. Bos. 21, 11. *In the following example* cwealm *is neuter* :—Sume ic þurh mislíc cwealm mínum hondum slóg *some I slew by my hands through various deaths,* Exon. 73 a; Th. 272, 2; Jul. 493. [*Chauc.* qualm *sickness : Laym.* qualm *mortality, plague : Plat.* qualm *vapour, smoke : O.Sax.* qualm, m. *violent death, murder : Dut.* kwalm, m. *reek, moist : Ger.* qualm, m. *vapour, smoke : M.H.Ger.* qualm, m. *anguish : O.H.Ger.* qualm, m. *nex : Dan.* qwalm, m. f. *vapour, smoke : Swed.* qwalm, n. *sultriness.*] DER. beadu-cwealm, bealo-, brōðor-, deáþ-, feorh-, gār-, mān-, morþor-, níþ-, orf-, ūt-, wæl-, yrf-.

cwealm-bǣre, cwylm-bǣre; *adj.* [-bǣre, an *adj.* termination; *producing, bearing*] *Death-bearing, deadly;* mortiférus :—Ðeáh ðe he cwealmbǣre wǣre *though he was death-bearing,* Wanl. Catal. 164, 48, col. 1. Drenc mid ðam cwealmbǣrum āttre gemenged *a drink mingled with deadly poison,* Homl. Th. ii. 158, 17 : 260, 11. Cwealmbǣrne *mortiférum,* Mone B. 4905. Cōmon ða cempan mid cwylmbǣrum tōlum *the soldiers came with deadly tools,* Homl. Th. ii. 260, 7.

cwealm-bǣrnes, -ness, e; *f. Destruction, ruin, deadliness, mortality;* pernicies, mortalitas. v. cwelm-bǣrnys.

cwealm-bealu; *gen.* -bealuwes; *n.* [bealo, bealu *bale, evil*] *Deadly evil;* cædis malum :—Ðæt hit mōste cwealmbealu cȳðan *that it must make known the deadly evil,* Beo. Th. 3884; B. 1940.

cwealm-cuma, an; *m.* [cuma, *q. v. a comer, guest*] *A deadly guest;* advena cædem parans :—Ne wolde eorla hleó ðone cwealmcuman cwicne forlǣtan *the refuge of the earls would not leave the deadly guest living,* Beo. Th. 1588; B. 792.

cwealm-dreór, es; *m.* [dreór *blood*] *Slaughter-gore;* sanguis cæde profusus, Cd. 47; Th. 60, 22; Gen. 985.

cwealmnes, cwylmnes, -ness, -nyss, e; *f. Torment, pain, anguish;* cruciatus :—Ða wǣron missenlícum cwealmnyssum þrēste *qui diversis cruciatibus torti,* Bd. 1, 7; S. 479, 13. Fram swā myclum cwylmnessum *a tamque diutinis cruciatibus,* 4, 9; S. 577, 10.

cwealm-stede, es; *m.* [stede *a place*] *A death-place;* mortis locus :—To cwealmstede *ad palæstram,* Glos. Prudent. Recd. 148, 46.

cwealm-stów, e; *f.* [stów *a place*] *A place of execution;* patibuli vel supplicii locus :—He to ðǣre cwealmstōwe lǣded wæs *he was led to the place of execution,* Bd. 1, 7; S. 478, note 38.

cwealm-þreá; *indecl; m. f. n.* [cwealm, þreá *a vexing, terror*] *Deadly terror; letalis terror* :—Mid cwealmþreá *with deadly terror*, Cd. 116; Th. 151, 12; Gen. 2507.

cwearn *a mill-stone*, Mk. Skt. Rush. 9, 42. v. cwyrn, cweorn-stán.

cweart-ern, cwert-ern, es; *n. A guard-house, prison;* custodia, carcer :—Ðæs cwearternes hirde híg betæhte Iosepe *custos carceris tradidit eos Ioseph*, Gen. 40, 4. Ic wæs on cwearterne *eram in carcere*, Mt. Bos. 25, 36, 39: Lk. Bos. 3, 20: Jn. Bos. 3, 24: Ælfc. Gr. 9, 18; Som. 9, 59. [*Prompt.* qwert, whert *incolumis, sanus, sospes.*]

cweartern-líc; *adj. Of or belonging to a prison;* carceralis :—Þurh cwearternlíce cyp *per carceralem stipitem*, Glos. Prudent. Recd. 150, 38.

CWECCAN; *part.* cweccende; ic cwecce, ðú cwecest, cwecst, he cweceþ, cwecþ, *pl.* cweccaþ; *p.* cwehte, cweahte, *pl.* cwehton, cweahton; *pp.* cweaht *To vibrate, move;* torquēre, quatēre, vibrāre, movēre :—Cweccende torquens, Glos. Prudent. Recd. 147, 49. He cwecþ his sweord *gladium suum vibrabit*, Ps. Th. 7, 12. Þegn Hróþgáres, þrymmum cwehte *Hrothgar's thane, violently quaked*, Beo. Th. 476; B. 235. Iohannes cwehte his heáfod *John shook his head*, Ælfc. T. 36, 9. Hí cwehton [MS. cwehtun] heora heáfod *moverunt caput*, Ps. Lamb. 21, 8. Ða wegférendan cwehton heora heáfod *the passers-by shook their heads*, Mt. Bos. 27, 39: Mk. Bos. 15, 29. [*Laym.* quecchen *to shake, move: Icel.* kwika *to move, stir.*] DER. a-cweccan.

cweccung, e; *f. A moving, wagging;* commotio :—Ðú gesettest us on cweccunge heáfdes on folcum *posuisti nos in commotionem capitis in populis*, Ps. Lamb. 43, 15.

cwede *a saying*, Som. Ben. Lye. v. cwide.

cweden *spoken, said, called*, Exon. 15 b; Th. 34, 24; Cri. 547: Chr. 455; Erl. 13, 23: Bd. 5, 19; S. 636, 45; *pp.* of cweðan.

cwehte, *pl.* cwehton *shook, moved, quaked*, Beo. Th. 476; B. 235: Ælfc. T. 36, 9: Ps. Lamb. 21, 8: Mt. Bos. 27, 39: Mk. Bos. 15, 29; *p.* of cweccan.

CWELAN, ic cwele, ðú cwilst, he cwelþ, cwilþ, cwylþ, *pl.* cwelaþ; *p.* cwæl, *pl.* cwǽlon; *pp.* cwolen *To die;* mori :—Cwele ic *I die*, Exon. 125 a; Th. 482, 2; Rä. 66, 1. Swá swá fixas cwelaþ gyf hí of wætere beóþ, swá eác cwelþ [cwylþ MSS. R. L.] ǽlc eorþlíc líchama gyf he byþ ðære lyfte bedǽled *as fishes die if they are out of water, so also every earthly body dies if it be deprived of the air*, Bd. de nat. rerum; Wrt. popl. science 17, 9–11; Lchdm. iii. 272, 25 and note 36. [*Laym.* quelen *to die: O. Sax.* quelan *to die from a violent death* or *as a martyr: Dut.* quelen *languore tabescere: O. H. Ger.* quelan *cruciari, pati, mori.*] DER. a-cwelan, óþ-: cwild, -bǽre, -bǽrlíce, -tíd: cwalu: cwellan, a-: cwellere: a-cwelledness: cwealm, -bǽre, -bǽrness, -bealu, -cuma, -dreór, -ness, -stede, -stów, -þreá: cwelman, cwylmaṅ, ge-: cwylming.

cweldeht; *adj.* [cweid = cwyld *destruction*, -eht = -iht *adj. termination*, q. v.] *Mortified;* corruptionis plenus :—Wið wyrmǽtum líce and cweldehtum *for a worm-eaten and mortified body*, L. M. 1, 54; Lchdm. ii. 126, 4.

CWELLAN, ic cwelle, ðú cwelest, cwelst, he cweleþ, cwelþ, *pl.* cwellaþ; *p.* cwealde, *pl.* cwealdon; *pp.* cwelled, cweled, cweald; *v. a. To kill, slay* = QUELL? necare, trucidare, occidere, mactare :—Ða cwelleras ne woldan hine cwellan *the executioners would not kill him*, Bd. 5, 19; S. 638, 30: Cd. 140; Th. 176, 2; Gen. 2905: Hy. 7, 105; Hy. Grn. ii. p. 289, 105. Oft ic cwelle compwæpnum *often I kill with battle-weapons*, Exon. 105 b; Th. 401, 9; Rä. 21, 9. Ðú ramm cwelst *thou shalt kill the ram*, Ex. 29, 16. We cwellaþ *we kill*, Ex. 8, 26. Cwealde *had killed*, Andr. Kmbl. 3247; An. 1626. Hí stearcferþe cwellan þohtun *the stern of mind resolved to slay her*, Exon. 75 a; Th. 280. 31; Jul. 637. Ðú Grendel cwealdest *thou didst slay Grendel*, Beo. Th. 2673; B. 1334. Árleás cyning cwealde cristne men *the impious king slew christian men*, Exon. 65 b; Th. 243, 3; Jul. 5. [*Prompt.* qwellyṅ *suffocare: Wyc.* quellere *a killer: Piers P.* quellan *to kill: Chauc. R. Glouc.* quelle: *Laym.* quelle-n: *Orm.* cwellenn: *O. Sax.* quellian: *Dut.* kwellen *to vex: Kil.* quellen *molestare: Ger.* quälen *to vex: M. H. Ger.* queln, quellen, kellen *to press, vex: O. H. Ger.* queljan *necare: Dan.* qwäle *to quell, torture: Swed.* qwälja *to torment: Icel.* kwelja *to torment.*] DER. a-cwellan.

cwellend, es; *m.* [cwellende, *part.* of cwellan *to kill*] *A killer, slayer;* interfector :—Cwellend *sector*, Glos. Prudent. Recd. 150, 27.

cwellere; *m. A killer, man-slayer, executioner*, QUELLER, *tormentor;* lanio, interfector, spiculator? carnifex :—Se cwellere *the executioner*, Bd. 1, 7; S. 478, 15. 35. Ða cwelleras *the executioners;* carnifices, 5, 19; S. 638, 29. Herodes sende ǽnne cwellere, and bebeád ðæt man his heáfod on ánum disce brohte *Herod sent an executioner, and commanded that they should bring his 'John Baptist's' head on a dish*, Mk. Bos. 6, 27. Hyldere, oð... [MS. flæctawere] *lanio, vel lanista, vel carnifex, vel macellarius*, Ælfc. Gl. 113; Som. 79, 122; Wrt. Voc. 60, 27.

cwelm *destruction, death*, Cd. 208; Th. 257, 36; Dan. 668. v. cwealm.

cwelman, cwylman, cwilman; *part.* -ende; *p.* de; *pp.* ed [cwealm, cwelm *death, destruction, torment*] *To torture, torment, destroy, kill;* trucidare, cruciare :—Cwelmende fýr *destroying fires*, Exon. 22 a; Th. 59,

28; Cri. 959. He wæs ðæt folc cwilmende *he tortured the people*, Ors. 1, 12; Bos. 36, 25. He eorþ-cyningas yrmde and cwelmde *he oppressed and slew the kings of the earth*, Bt. Met. Fox 9, 94; Met. 9, 47. Mæssepreóstas wǽron cwylmde *sacerdotes trucidabantur*, Bd. 1, 15; S. 484, 1: 4, 13; S. 582, note 29. Hí hálge cwelmdon *they slew the holy*, Exon. 66 a; Th. 243, 24; Jul. 15. Ðæt hí cwylmen rihte heortan *ut trucident rectos corde*, Ps. Spl. 36, 15. Ðú hungre scealt cwylmed weorþan *thou shalt be put to death with hunger*, Elen. Kmbl. 1373; El. 688. [*O. Sax.* quelmian *to kill.*] DER. ge-cwelman, -cwylman.

cwelm-bǽrnys, -nyss, e; *f.* [cwealm, cwelm *death, destruction*] *Destruction, ruin, deadliness, mortality;* pernicies, mortalitas :—Cwelmbǽrnyss *pernicies*, Ælfc. Gr. 12; Som. 15, 52. Þurh myrran is gehíwod cwelmbǽrnys úres flǽsces *by myrrh is typified the mortality of our flesh*, Homl. Th. i. 118, 3.

cwelþ *dies*, Bd. de nat. rerum; Wrt. popl. science 17, 10; *3rd pres. sing.* of cwelan.

cwéman; *part.* cwémende; *p.* de; *pp.* ed; *v. a. dat. To give pleasure, please, delight, propitiate, satisfy;* placere, satisfacere :—Sum sceal on heápe hæleðum cwéman *one shall in company give pleasure to men*, Exon. 88 a; Th. 331, 33; Vy. 77. Ic mínum Criste cwéman þence leófran láce *I purpose to please my Saviour with a dearer gift*, 37 a; Th. 120, 26; Gú. 277: Ors. 1, 12; Bos. 36, 27: Cd. 220; Th. 283, 16; Sat. 305. Se ðe ne þenceþ Meotode cwéman *he who thinketh not to propitiate the Creator*, 217; Th. 276, 5; Sat. 184: Exon. 69 a; Th. 257, 25; Jul. 252: Ps. Th. 91, 3: 94, 1. God tostencþ bán heora ða ðe mannum cwémendra *Deus dissipavit ossa eorum qui hominibus placent*, Ps. Spl. 52, 7. Ic cwéme Drihtne on ríce lýfigendra *placebo Domino in regione vivorum*, 114, 9: Ps. Th. 53, 6. Esne his hláforde cwémeþ *a servant gives pleasure to his master*, 122. 2. Martiras Meotode cwémaþ *martyrs give delight to the Creator*, Cd. 228; Th. 305, 31: Sat. 655: Exon. 39 a; Th. 130, 5; Gú. 433: Ps. Th. 71, 10. Nǽnig man scile orþances útabredan wǽpnes ecgge, ðeáh ðe him se wífte cwéme *no man should draw forth the weapon's edge without a cause, although its beauty please him*, Salm. Kmbl. 332; Sal. 165. Ðæt we cwéman Criste *that we please Christ*, Cd. 226; Th. 302, 8; Sat. 596. Ðam ic georne cwémde *whom I have earnestly propitiated*, Exon. 48 b; Th. 167, 11; Gú. 1058. Him lofsangum cwémdon [MS. cwemdan] *cantaverunt laudes ejus*, Ps. Th. 105, 11. [*Laym.* queme, cweme, iquemen, icweme *to please: Orm.* cwemenn: *Ger.* bequemen *to accommodate.*] DER. ge-cwéman.

cwéme; *adj.* [cwéman *to please*] *Pleasant, pleasing, grateful, acceptable, fit;* gratus, acceptus, congruus. DER. ge-cwéme.

cwéming, e; *f. A pleasing, satisfying;* placentia, satisfactio, Greg. Dial. 4, 28.

cwémnys, -nyss, e; *f. A satisfaction, an appeasing, a mitigation;* satisfactio :—Cwémnys uncysta *satisfactio vitiorum*, Bd. 1, 27; S. 495, 32.

CWÉN; *gen. dat.* cwéne; *acc.* cwén, cwénn, cwéne; *pl. nom. acc.* cwéne, cwéna; *gen.* cwéna; *dat.* cwénum; *f*: cwéne, cwýne; *gen. dat. acc.* cwénan, cwýnan; *pl. nom. acc.* cwénan; *gen.* cwénena; *dat.* cwénum; *f.* I. *a woman;* femina :—Seó clǽneste cwén ofer eorþan *the purest woman upon earth*, Exon. 12 a; Th. 17, 27; Cri. 276. Þurh ða ǽdelan cwénn *through the noble woman*, 25 b; Th. 73, 34; Cri. 1199. Cwéna sélost *the best of women*, Menol. Fox 334; Men. 168. Ealdra cwéna spell *old women's talk;* anilis fabula, Ælfc. Gl. 100; Som. 77, 20; Wrt. Voc. 55, 24. Ic wæs feaxhár cwéne *I was a hoary-headed woman*, Exon. 126 b; Th. 487, 13; Rä. 73, 1. On cwénena bróce, of cwénena bróce, *from the women's brook*, Cod. Dipl. Apndx. 426; A. D. 949; Kmbl. iii. 429, 34. II. *a wife;* uxor :—Abrahames cwén *Abraham's wife*, Cd. 103; Th. 136, 17; Gen. 2259. Hæleða cwénum *to the wives of the warriors*, 169; Th. 210, 7; Exod. 511. Gif preóst cwénan forlǽte, and óðre nime, anaþema sit *if a priest forsake his wife, and take another, let him be excommunicated*, L. N. P. L. 35; Th. ii. 296, 1. Gif man mid esnes cwýnan geligeþ, be cwicum ceorle, ii gebéte *if a man lie with an 'esne's' wife, her husband still living, let him make twofold amends*, L. Ethb. 85; Th. i. 24, 9. III. *a king's or emperor's wife, a* QUEEN, *empress;* regina, imperatrix, augusta :—Cwén *regina*, Ælfc. Gl. 68; Som. 69, 128; Wrt. Voc. 42, 8: 72, 56: Mt. Bos. 12, 42: Lk. Bos. 11, 31: Ors. 1, 10; Bos. 33, 23: 3, 11; Bos. 73, 37: Chr. 672; Erl. 35, 37: 722; Erl. 45, 26: Beo. Th. 1851; B. 923: Elen. Kmbl. 494; El. 247. Ðæs [MS. ðes] cáseres cwén *imperatrix vel augusta*, Wrt. Voc. 72, 58. Oft on ánre tíde acenþ seó cwén and seó wyln *the queen and the slave often bring forth at one time*, Homl. Th. i. 110, 27: Elen. Kmbl. 832; El. 416: 1113; El. 558: Beo. Th. 2311; B. 1153. Seo ylce cwén Saméramis *the same queen Semeramis*, Ors. 1, 2; Bos. 27, 6. Ðǽr wearþ Marsepia, sió cwén, ofslagen *Marpesia, the queen, was slain there*, 1, 10; Bos. 33, 22, 24: Elen. Kmbl. 756; El. 378: Bt. Met. Fox 26, 178; Met. 26, 89. Deós cwén *this queen*, Elen. Kmbl. 1064; El. 533: 1099; El. 551. He wæs on ðære cwéne gewealdum *he was in the queen's power*, 1217; El. 610: 2269; El. 1136. Ðone hie ðære cwéne

N

agēfon *they gave him up to the queen*, 1171; El. 587: 2257; El. 1130. Aðelwulf cyng Carles dôhtor hæfde to cwēne *king Æthelwulf had the daughter of Charles for his queen*, Chr. 885; Erl. 85, 3: 1017; Erl. 161, 10: 1048; Erl. 180, 21. Mid ða æðelan cwēn *with the noble queen*, Elen. Kmbl. 550; El. 275: Beo. Th. 1334; B. 665: Exon. 86 a; Th. 324, 29; Wíd. 102. Ofslóh ge ðone cyning, ge ða cwēne *slew both the king and the queen*, Ors. 3, 11; Bos. 74, 4: Homl. Th. i. 438, 21: Exon. 90 a; Th. 338, 22; Gn. Ex. 82. Cyningas and cwēne *kings and queens*, 113 a; Th. 433, 15; Rä. 50, 8. Hiora twā wǣron heora cwēna, Marsepia and Lampida wǣron hātene *two of them, called Marpesia and Lampeto, were their queens*, Ors. 1, 10; Bos. 33, 14, 35. Se wæs Melcolmes sunu cynges and Margarite ðære cwēnan *he was the son of king Malcolm and queen Margaret*, Chr. 1097; Erl. 234, 37. [*Prompt.* quene *regina;* quen, womann of lytylle price: *Wyc.* queene: *Piers P.* queyne, quene: *R. Brun. R. Glouc.* quene: *Laym.* quen-e, *f: Orm.* cwen: *Scot.* queyn, quean *a young woman: Plat.* quene: *O. Sax.* cwān, cwēna, *f. uxor: Dut.* kween, *f. a married woman: Kil.* quene *uxor, mulier: Ger.* königin, *f: M. H. Ger.* kone, kon, *f. uxor: O. H. Ger.* quena, chena, chone, *f. mulier, conjux, uxor: Goth.* qens, *f. mulier, uxor: Dan.* qwinde, kone *mulier, uxor: Swed.* qwinna, *f. mulier, uxor:* kåna, *f. a low woman: Icel.* kona, kuna, kván, kwǽn *a woman, wife, queen: Grk.* γυνή *femina, genitrix: Slav.* shena: *Sansk.* gnā, jani, *f. a woman, wife, mother.*] DER. dryht-cwēn, folc-, gūþ-, sige-, þeód–.

Cwēna land *the land* or *country of the Quaines*, Ors. 1, 1; Bos. 21, 10. v. Cwēnas, Cwēn-land.

Cwēnas; *gen.* a; *pl. m. The Quaines;* Cayani. *The inhabitants of Cwēn-land, q. v:*—Is to-emnes ðǣm lande sūþeweardum, on ôðre healfe ðæs môres, Sweóland, ôþ ðæt land norþeweard; and to-emnes ðǣm lande norþeweardum, Cwēna land. Ða Cwēnas hergiaþ hwílum on ða Norþmen ofer ðone môr; hwílum ða Norþmen on hý; and ðǣr sint swíðe micle meras fersce geond ða môras; and beraþ ða Cwēnas hyra scypu ofer land on ða meras, and ðanon hergiaþ on ða Norþmen. Hý habbaþ swýðe lytle scypa, and swýðe leóhte *over against the land* [*Finland*] *southward, on the other side of the waste, is Sweden, northward up to the land; and over against the land northward is the land of the Quaines. The Quaines sometimes make war on the Northmen over the waste; sometimes the Northmen on them; and there are very large fresh lakes beyond the wastes; and the Quaines carry their boats over land into the lakes, and thence make war on the Northmen. They have very little boats, and very light,* Ors. 1, 1; Bos. 21, 8–15.

cwencan; *p.* cwencte; *pp.* cwenced, cwenct *To extinguish*, QUENCH; extinguere. DER. acwencan.

cwēne, cwýne, an; *f. A woman, wife, queen, common woman, harlot;* femina, uxor, regina, meretrix:—Ic wæs feaxhār cwēne *I was a hoary-headed woman*, Exon. 126 b; Th. 487, 13; Rä. 73, 1. Cwēnan forlǣtan *to forsake a wife*, L. N. P. L. 35; Th. ii. 296, 1. Mid esnes cwýnan *with an 'esne's' wife*, L. Ethb. 85; Th. i. 24, 9. Margarite ðære cwēnan *of queen Margaret*, Chr. 1097; Erl. 234, 37. Wið āne cwēnan fylþe adreógaþ *cum una meretrice spurcitiem exercent*, Lupi Serm. 1, 11; Hick. Thes. ii. 102, 26. v. cwēn.

cwēn-fugol, es; *m. A female* or *hen bird; avis feminea*, Som. Ben. Lye.

Cwēn-land, es; *n. Cwēn-land lies between the White Sea* [Cwēn Sǣ] *and Norway, north of the Gulf of Bothnia. The country east and west of the Gulf of Bothnia, from Norway to the Cwēn or White Sea, including Finmark on the north. Malte-Brun says that the inhabitants of Cwēn-land were a Finnish race. They were called Quaines, and by Latin writers Cayani. Gerchau maintains, in his history of Finland, 1810, that the Laplanders only were called Finns, and that they were driven from the country by the Quaines. 'They settled in Lapland, and on the shores of the White Sea, which derived from them the name of Quen Sea or Quen-vik.' ... Adamus Bremensis happened to be present at a conversation, in which king Swenon spoke of Quen-land or Quena-land, the country of the Quaines, but as the stranger's knowledge of Danish was very imperfect, he supposed the king had said Quinna-land, the country of women or Amazons; hence the absurd origin of his Terra Feminarum, mistaking the name of the country, for quinna a woman. Malte-Brun's Universal Geog. Edin. 1827, vol. vi. p. 495.—Dr. Latham's Germania of* Tacitus, 174, 179:—Sweón habbaþ be sūþan him ðone sǣs earm Osti; and be eástan him Sermende; and be norþan. him ofer ða wēstennu is Cwēn-land *the Swedes have, to the south of them, the Esthonian arm of the sea; and to the east of them the Sermende; and to the north of them, over the wastes, is Cwēn-land*, Ors. 1, 1; Bos. 19, 21–23: 21, 10.

cwēn-líc; *adj.* QUEENLY, *feminine;* muliebris:—Ne biþ swylc cwēnlíc þeáw *such is not a feminine custom*, Beo. Th. 3885; B. 1940.

cwēnn *a woman*, Exon. 25 b; Th. 73, 34; Cri. 1199; *acc. s. of* cwēn.

Cwēn-sǣ; *gen.* -sǣs; *m. The White Sea;* hyperboreus oceanus:—Fram ðære eá Danais, west ôþ Rín ða eá ... and eft sūþ ôþ Donua ða eá ... and norþ ôþ ðone gársecg, ðe man Cwēnsǣ hǣt: binnan ðǣm syndon manega þeóda; ac hit man hǣt eall, Germania *from the river Don, westward to the river Rhine ... and again south to the river Danube ... and north to the ocean, which is called the White Sea:*

within these are many nations; but they call it all, Germania, Ors. 1, 1; Bos. 18, 21–28. v. Cwēnas, Cwēn-land.

cweoc *quick, alive*, Symb. Athan. Lye. v. cwic.

cweodo *a cud, quid*, L. M. 2, 14; Lchdm. ii. 192, 6. v. cwudu.

cweorn, e; *f:* cweorne, an; *f. A mill, hand-mill, quern*, Mt. Kmbl. Hat. 24, 41: Ex. 11, 5. v. cwyrn.

cweorn-bill, es; *n.* [bil *a bill, falchion*] *A stone chisel for dressing querns;* lapidaria, Cot. 125.

cweorn-stān *a mill-stone*, Mk. Bos. 9, 42: Lk. Bos. 17, 2. v. cwyrn-stān.

cweorn-tēþ; *pl. m. Molar teeth, grinders;* molares, Wrt. Voc. 282, 75. v. cweðan.

cwert-ern, es; *n. A prison:*—Ðe-læs ðú sý on cwertern send *ne forte in carcerem mittaris*, Mt. Bos. 5, 25: Lk. Bos. 12, 58. v. cweart-ern.

cweþ *says*, Ælfc. Gr. 15; Som. 18, 45, =cweðeþ; *3rd pres. sing. of* cweðan.

cweþ ðú *say thou*, cweðe he *let him say*, cweðaþ, cweðe ge *say ye*, Ælfc. Gr. 33; Som. 37, 33, 39; Mt. Bos. 3, 9: Gen. 50, 19; *impert. of* cweðan.

CWEÐAN, to cweðanne; *part.* cweðende; ic cweðe, ðú cweðest, cweðst, cwæðst, cwidst, cwyðst, cwîst, cwýst, he cweðeþ; cweþ, cwiþ, cwyþ, *pl.* cweðaþ; *p.* ic, he cwæþ, ðú cwǣde, *pl.* cwǣdon; *impert.* cweþ, cweðe, *pl.* cweðaþ, cweðe; *subj.* cweðe, *pl.* cweðen; *p.* cwǣde, *pl.* cwǣden; *pp.* cweden *To say, speak, call, proclaim;* dicere, loqui, vocare, indicere. **I.** *v. trans:*—Ic ðé wolde lofsang cweðan *laudem dixi tibi*, Ps. Th. 118, 164: Rood Kmbl. 230; Kr. 116. For ðam worde ðe se Wealdend cwyþ *for the word which the Lord shall speak*, Rood Kmbl. 220; Kr. 111. Gehýraþ hwæt se unrihtwísa dēma cwyþ *audite quid judex iniquitatis dicit*, Lk. Bos. 18, 6. Him ða word hí cweðaþ *they say the words to him*, Exon. 13 b; Th. 25, 15; Cri. 401. Ne cwæþ ic wiht *I spake not aught*, 125 a; Th. 482, 1; Rä. 66, 1: Bt. Met. Fox 10, 69; Met. 10, 35. Drihten cwæþ word to Noe *the Lord spake words to Noah*, Cd. 74; Th. 91, 11; Gen. 1510: Beo. Th. 5318; B. 2662: Andr. Kmbl. 658; An. 329. Arríus se gedwola cwæþ gemôt ongeán ðone bisceop *Arius the heretic proclaimed a synod against the bishop*, Homl. Th. i. 290, 12. Alýs mîne sáwle of ðām welerum ðe wom cweðen *deliver my soul from the lips which may speak evil*, Ps. Th. 119, 2. Hí geornlíce smeádon hwæt he cwǣde *they earnestly considered what he said*, Bd. 3, 5; S. 527, 37. On ðære stôwe ðe is cweden Ægeles þrep *at the place which is called Aylesthorpe*, Chr. 455; Erl. 13, 23: Exon. 11 a; Th. 13, 32; Cri. 211. **II.** *v. intrans:*—Hwæt mágon we cweðan ongén úrne hláford *what can we say to our lord?* Gen. 44, 16: Cd. 229; Th. 310, 24; Sat. 732. Hú hie cweðan woldon *how they would speak*, 201; Th. 249, 17; Dan. 531: Exon. 28 a; Th. 84, 22; Cri. 1377. Ðæt is wundor to cweðanne *quod mirum dictu est*, Bd. 3, 6; S. 528, 10. Ðus cweðende, he forþférde *hæc dicens, expiravit*, Lk. Bos. 23, 46: Homl. Th. i. 380, 2, 21: Ps. Th. 104, 10. Ic cweðe to ðysum, and ic cweðe to ôðrum *dico huic et alii*, Mt. Bos. 8, 9; Ælfc. Gr. pref; Som. 1, 39: 5; Som. 3, 27: 15; Som. 17, 36: 18; Som. 21, 26, 27, 29, 59, 61, 63. Ic cweðe aio, inquio, 33; Som. 37, 31, 37. Ðú cwedst ais, 33; Som. 37, 31: Ps. Lamb. 87, 11. Gif ðú cwǣdst *if thou sayest*, Ælfc. Gr. 2; Som. 3, 7. Ðú cwidst inquis, 33; Som. 37, 38. Ðú cwyðst *thou sayest*, 2; Som. 3, 8: 5; Som. 3, 27, 32, 33, 36: 15; Som. 17, 36: 18; Som. 21, 62. Ðú cwîst ðæt ic ðé andwyrdan scyle *thou sayest that I must answer thee*, Bt. 5, 3; Fox 12, 16: Num. 11, 22, 23: 23, 12: Ps. Th. 87, 12. Ðú cwýst ðæt ic me gebiddan sceole to dumbum stānum *thou sayest that I must pray to dumb stones*, Homl. Th. i. 424, 9: Ælfc. Gr. 5; Som. 3, 29: Ps. Th. 88, 16. Man cweðeþ *dicet homo*, Ps. Th. 57, 10. He cweþ *he says*, Ælfc. Gr. 5; Som. 3, 50: 15; Som. 18, 45. He cweþ *ait*, 33; Som. 37, 31. Ðonne cwiþ se engel *then the angel shall speak*, Exon. 32 b; Th. 102, 7; Cri. 1669: Beo. Th. 4088; B. 2041. Swā hwylc swā cwyþ to ðisum munte *quicumque dixerit huic monti*, Mk. Bos. 11, 23: Mt. Bos. 7, 21: Jn. Bos. 4, 10: 16, 18. He cwyþ *inquit*, Ælfc. Gr. 33; Som. 37, 38. We cweðaþ *we say*, Ælfc. Gr. 18; Som. 21, 67. Ge cweðaþ *ye say*, Deut. 28, 67. Sume men cweðaþ on Englisc ðæt hit sié feaxede steorra *some men say in English that it* [a comet] *is a long-haired star*, Chr. 891; Erl. 88, 18. Híg cweðaþ *they say*, Deut. 31, 17: Exon. 12 a; Th. 18, 14; Cri. 283: Cd. 63; Th. 75, 13; Gen. 1239. Hí cweðaþ *aiunt, inquiunt*, Ælfc. Gr. 33; Som. 37, 32, 38. Ic cwæþ *dixi*, Deut. 32, 26: Ps. Lamb. 29, 7: 39, 8: Jn. Bos. 11, 42. Ðú cwǣde, ðæt ðú me woldest wel dôn *tu locutus es, quod benefaceres mihi*, Gen. 32, 12: Andr. Kmbl. 2822; An. 1413: Ps. Th. 89, 3. Ðú cwǣde inquisti, Ælfc. Gr. 33; Som. 37, 39. He cwæþ sylf to me *ipse dixit mihi*, Gen. 20, 5: Ex. 1, 15: Lev. 6, 19, 24: Num. 10, 36: Deut. 1, 34: Jos. 3, 6: Jud. 4, 18: Mt. Bos. 8, 4: Mk. Bos. 2, 5: Lk. Bos. 2, 48: Jn. Bos. 5, 8: Fins. Th. 48; Fin. 24. Híg cwǣdon *him betwýnan mutuo loquebantur*, Gen. 37, 19: Num. 16, 3: Cd. 191; Th. 238, 28; Dan. 361: Beo. Th. 6342; B. 3181: Elen. Kmbl. 1138; El. 571. Hí cwǣdon *aiebant*, Ælfc. Gr. 33; Som. 37, 33. Ðus cweþ *thus say*, Ex. 19, 3. Cweþ ðú *ai, inque*, Ælfc. Gr. 33; Som. 37, 33, 39. Cweðe he *inquiat*, 33; Som. 37, 39. Ne cweðaþ betwux eów *say not among yourselves*, Mt. Bos. 3, 9. Cweðe

ge *say ye*, Gen. 50, 19. Ðý-læs ðú cweðe *lest thou shouldest say*, Cd. 98; Th. 129, 18; Gen. 2145: Ælfc. Gr. 7; Som. 6, 16: 21; Som. 23, 28, 38. Gif se þeówa cweðe ðæt he nelle fram ðé faran *if the servant should say that he will not go from thee*, Deut. 15, 16. Ðý-læs cweðen [MS. cweðan] óðre þeóda *lest other nations should say*, Ps. Th. 78, 10. Gif ic cwæde *if I said*, 72, 12. Hú wunda cwæden to hæleðum *how the wounds spake to men*, Exon. 114 b; Th. 441, 13; Rä. 60, 17. Ðæt is wel cweden *that is well spoken*, 15 b; Th. 34, 24; Cri. 547. [Piers P. quod *quoth: Chauc.* quethe: *Orm.* cwaþþ said: *Laym.* queð, i-queð, quaeð, quað *quoth*; iqueðen, *pp.* said: *O. Sax.* queðan, quethan: *O. Frs.* quetha, queda, quan: *M. H. Ger.* quiden, kiden: *O. H. Ger.* quedan: *Goth.* qiþan: *Dan.* qwaede: *Swed.* kwaeda: *Icel.* kweða: *Lat.* in-quit *quoth: Sansk.* root kath *to converse with any one*.] DER. a-cwedan *to say, tell*, æfter-, be-, bi-, for-, fóre-, ge-, hearm-, on-, onbe-, onge-, to-, wið-.

cweðs ðú lá = cwýst ðú lá *O! sayest thou?* numquid? Ps. Lamb. 7, 12. v. cwýst ðú, cweðan.

cweðst *sayest, speakest*, Ps. Lamb. 87, 11; *2nd pres. sing.* of cweðan.

CWIC, cwyc, cwuc, cuc; *def.* se cwica, seó, ðæt cwice; *adj. Alive*, QUICK; vivus, vivax:—Enoch cwic gewát mid cyning engla *Enoch departed alive with the king of angels*, Cd. 60; Th. 73, 25; Gen. 1210: Exon. 16 b; Th. 37, 8; Cri. 590: Ps. Th. 118, 57. Cwyc *alive*, 104, 8. Ne biþ se cwuca nyttra ðe se deáda, gif him his yfel ne hreówþ *the quick [living] is not better than the dead, if he repent not of his evil*, Bt. 36, 6; Fox 182, 20. Se iunga wæs cwices módes *the youth was of a quick mind;* erat adolescens animi vivacis, Bd. 5, 19; S. 637, 37. He nó ðær ðit cwices læfan wolde *he would have naught alive there*, Beo. Th. 4618; B. 2314. Ælc wuht cwices [cwuces Cot.] biþ innanweard hnescost *everything alive is inwardly softest*, Bt. 34, 10; Fox 150, 5. Ne ofsleá ic ælc þing cuces *non percutiam omnem animam viventem*, Gen. 8, 21: Wrt. Voc. 85, 51. On cwicum ceápe *in live stock*, L. Ath. i. prm; Th. i. 194, 6: Homl. Blick. 39, 18. Æt cwicum [cwicon MS.] menn *for a living man*, L. Eth. iii. 1, 2; Th. i. 292, 10, 13. Be cwicum ceorle *the husband being alive*, L. Ethb. 85; Th. i. 24, 9. On cucum [MS. cucan] ceápe *in live stock*, Cod. Dipl. 1201; A. D. 956; Kmbl. v. 378, 20. Seó sealf ðone wyrm ðæron deádne gedéþ, oððe cwicne ofdrífþ *the salve will make the worm therein dead, or drive it away alive*, L. M. 3, 39; Lchdm. ii. 332, 26. Hie ænigne cwicne ne métton *they found not any alive*, Andr. Kmbl. 2166; An. 1084: Elen. Kmbl. 1378: El. 691. Abraham leófa, ne sleah ðín ágen bearn, ac ðú cwicne abregd cniht of áde, eaforan ðínne *beloved Abraham, slay not thine own child, but take thou the boy, thy son, alive from the pile*, Cd. 141; Th. 176, 19; Gen. 2914: Beo. Th. 1589; B. 792: Exon. 90 b; Th. 340, 21; Gn. Ex. 114: Ps. Th. 118, 154. Ic hyne eft cwycne ageaf *I gave him back again alive*, Nicod. 26; Thw. 14, 28, 38. Tiberius forneáh nænne ðæra senátussa ne lét cucne *Tiberius left hardly any of the senators alive*, Ors. 6, 2; Bos. 116, 41: L. C. S. 25; Th. i. 390, 21. Cwice, *acc. f. alive*, Glos. Prudent. Recd. 148, 51. Gif hió cwic bearn gebyreþ *if she bare a live child*, L. Ethb. 78; Th. i. 22, 4. Sníþ ðæt cwice líc *cut the body alive*, L. M. 1, 35; Lchdm. ii. 84, 29. Cwicre stæfne *with the living voice;* viva voce, Bd. 4, 18; S. 586, 39. Cwice *quick, alive, pl. nom. m.* Ps. Th. 105, 5: Andr. Kmbl. 258; An. 129. Híg in to helle cuce síðodon *descenderunt vivi in infernum*, Num. 16, 33: Chr. 794; Erl. 59, 23. Ðe ealle cwice wihta bílibbaþ *by which all creatures alive are supported*, Ors. 2, 1; Bos. 38, 8. Se Ælmihtiga líf gesceóp cynna ðara ðe cwice hwyrfaþ *the Almighty created life for each of the kinds that go to and fro alive*, Beo. Th. 197; B. 98. Cwyce secgeaþ his wundorweorc *his wondrous works alive shall speak*, Ps. Th. 104, 1. Ða cwican nó genihtsumedon ðæt hí ða deádan bebyrigdan *those alive were not enough to bury the dead*, Bd. 1, 14; S. 482, 31. Cwicera manna *of men alive*, Judth. 11; Thw. 24, 41; Jud. 235: Runic pm. 6; Kmbl. 340, 17; Hick. Thes. i. 135. Ðær biþ cwicra gewin *there shall be strife of the quick*, Exon. 22 b; Th. 62, 8; Cri. 998: 51 a; Th. 177, 7; Gú. 1223: Salm. Kmbl. 792; Sal. 395. Ðú bist déma cwucra ge deádra *thou art the judge of quick and dead*, Hy. 8, 39; Hy. Grn. ii. 289, 39. He is God cwucera gelwelces *he is the God of each of those alive*, Bt. Met. Fox 29, 160; Met. 29, 80. Blis astíhþ cwicera cynna cyninge *the joy of quick kinds ascends to the king*, Menol. Fox 183; Men. 93: Andr. Kmbl. 1823; An. 914: Judth. 12; Thw. 26, 12; Jud. 324. Cwicra wihta *of beings alive*, Exon. 107 b; Th. 411, 5; Rä. 29, 8. His is mycel sæ, ðær is unrim cwycra *his is the great sea, where is a countless number of things alive*, Ps. Th. 103, 24. Ic wille mid flóde acwellan cynna gehwilc cucra wuhta *with a flood I will destroy every kind of creatures alive*, Cd. 61; Th. 78, 23; Gen. 1297. Be cwicum mannum *the men being alive*, L. Eth. ix. 4; Th. i. 340, 18: L. C. E. 3; Th. i. 360, 9. Cwycum and deádum *to quick and dead*, Hy. 7, 117; Hy. Grn. ii. 289, 117. Wylle on glédum cwicum *boil on live coals*, L. M. 2, 28; Lchdm. ii. 224, 20. On cwicum wædum *in living garments*, Salm. Kmbl. 280; Sal. 139. To démenne ægðer ge ðám cucum ge ðám deádum *to judge both the quick and the dead*, Homl. Th. ii. 596, 20: 598, 6: Num. 16, 48. Seó wiht bindeþ cwice *the creature will bind the*

quick, Exon. 109 b; Th. 420, 8; Rä. 39, 7. Ðe ðær cwice méteþ fýr *who shall find there fires alive*, 22 a; Th. 59, 27; Cri. 959. Déman ða cucan and deádan *judicare vivos et mortuos*, Ps. Lamb. fol. 199 a, 25: 202 a, 27. [*Wyc.* quyk: *Piers P. R. Brun.* quik: *Chauc.* quik, quick: *R. Glouc.* quyc: *Laym.* cwic, cwik, quic, quike: *Orm.* cwicc, cwike: *Plat.* quik, qwik: *O. Sax.* quik, quic: *Frs.* quick: *O. Frs.* quik: *Dut.* kwik: *Kil.* quick: *Ger.* keck gay, *brisk;* quecksilber *mercury: M. H. Ger.* quëc, këc: *O. H. Ger.* quek, quik, chuech: *Goth.* qius, *gen.* qiwis *vivus: Dan.* quik: *Swed.* kwick: *Icel.* kwikr, kykr: *Lat.* vivus *alive;* victum, *supine of* vivere *to live: Grk.* βίος *life: Sansk.* jíva *vivus*.] DER. healf-cwic, sám-.

cwic-æht, cwyc-æht, e; *f.* [æht *cattle*] *Live stock, cattle;* pecus:— Gebéte on cwicæhtum [cwyc- MS. B.] *let amends be made in live stock*, L. Alf. pol. 18; Th. i. 72, 12.

cwic-beám, es; *m. The* QUICKBEAM, *a sort of poplar?* forte populus tremula? cariscus, juniperus:—Genim cwicbeám *take quickbeam*, L. M. 1, 23; Lchdm. ii. 66, 1. Cwicbeám *cariscus*, Ælfc. Gl. 46; Som. 64, 119; Wrt. Voc. 32, 53.

cwicbeám-rind, e; *f. Bark of quickbeam:*—Wyl on wætere cwicbeámrinde *boil bark of quickbeam in water*, L. M. 1, 32; Lchdm. ii. 78, 12: 1, 36; Lchdm. ii. 86, 5.

cwice, an; *f. Quick-growing grass, couch-grass, quitch-grass;* gramen:— Cwice *gramen*, Ælfc. Gl. 42; Som. 64, 24; Wrt. Voc. 31, 34. Genym ðysse wyrte leáf, ðe man *gramen*, and ðórum naman cwice nemneþ *take leaves of this herb, which is named* gramen, *and by another name quitch*, Herb. 79; Lchdm. i. 182, 8: Lchdm. iii. 12, 28: 16, 8. Genim cwican *take quitch*, L. M. 2, 51; Lchdm. ii. 268, 10. [*Plat.* qwäk, queek, quek, quik *viticum repens: Dut.* kweek-gras, *n. dog's grass: Ger.* quecke, *f. any grass with creeping roots: Dan.* qwik-græs *couch-grass: Swed.* qwick-hwete, *n. dog's grass growing among wheat*.]

cwicen, cwucen, cucen, cucon, cucun; *adj.* [cwic *alive*, -en *adj.* termination] *Alive, quick;* vivus:—Hwá cwicenne me on ðysum ealdre fréfrade *who comforted me quick [living] in this life*, Ps. Th. 118, 82. We ne mágon hátan deádne mon for cwucene *we cannot call a dead man quick [living]*, Bt. 36, 6; Fox 182, 20. Ðone cyning hí brohton cucenne to Iosue *regem viventem obtulerunt Iosue*, Jos. 8, 23: Homl. Th. i. 294, 15. Gewylde man hine swá cucenne [cucunne MS. D: cwicne G.] swá deádne *let them seize him whether alive or dead*, L. Edg. ii. 7; Th. i. 268, 18. Ðæt he Wulfnóþ cuconne oððe deádne begytan sceolde *that he should take Wulfnoth alive or dead*, Chr. 1009; Erl. 142, 3. Genim cucne hrefn *take a live crab*, L. M. 3, 2; Lchdm. ii. 306, 20, 21.

cwic-feoh; *gen.* -feós; *n. Living property, cattle;* vivum munus, pecus, Som. Ben. Lye.

cwic-fýr, es; *n. Living fire, fire of brimstone, sulphur;* ignis vivus, sulphur:—Gifeóll ðæt fýr and cwicfýr of heofne *pluit ignem et sulphur de cælo*, Lk. Skt. Rush. 17, 29.

Cwichelmes hléw, Cwicchelmes hlǽw, Cwicelmes hlǽw, es; *m.* [hlǽw *a heap, barrow, small hill: Flor.* Cuiccelmeslawe: *Hunt.* Chichelmeslaue: *Hovd.* Cwichelmelow: *Cwichelm's hill;* Cwichelmi agger] CUCKHAMSLEY *hill* or *Cuchinslow, Berkshire, a large barrow on a wide plain overlooking White Horse Vale;* Cwichelmi agger in agro Berchensi:—Wendon to Wealingæforda, and ðæt eall forswælldon; and wæron him ðá áne niht æt Ceóles ége, and wendon him ðá andlang Æsces dúne to Cwichelmes [Cwicelmes, Th. 256, 28, col. 1: Cwicchelmes, 257, 27, col. 1] hlǽwe, and ðær onbídedon beótra gylpa, forðan oft man cwæþ, gif hí Cwichelmes [Cwicelmes, col. 1] hlǽwe gesóhton, ðæt hí nǽfre to sǽ gangan [gangen MS.] ne sceoldan *they went to Wallingford, and burned it all down; and were then one night at Cholsey, and then went along Ashdown to Cuckhamsley hill, and there tarried out of threatening vaunt, because it had often been said, if they came to Cuckhamsley hill, that they would never go to the sea* Chr. 1006; Th. 256, 25-32, col. 2. Æt Cwicelmes hlǽwe *at Cuckhamsley hill*, Th. Diplm. A. D. 995; 288, 24. On Cwicelmes hlǽw *to Cuckhamsley hill*, 291. 28.

cwic-hrérende; *part.* [hréran *to move*] *Quick-moving?*—Wilt ðú biddan ðé gesecge sídra gesceafta cræftas cwichrérende *wilt thou desire that he tell thee the quick-moving powers of wide-spread creatures?* Exon. 92 b; Th. 346, 28; Sch. 5.

cwician, cwycian, cucian; *p.* ode, ade; *pp.* od, ad [cwic *alive, quick*]. I. *v. intrans. To come to life*, QUICKEN; vívère et spírare:— Wǽron ða leoma cwiciende *the limbs were quickening*, Greg. Dial. 4, 36. Smire mid ða sáran limu, hie cwiciaþ sóna *smear the sore limbs therewith, they will soon quicken*, L. M. 3, 47; Lchdm. ii. 338, 25. Se synfulla mid godcundre onbryrdnysse cucaþ *the sinful quickens with divine stimulation*, Homl. Th. i. 494, 15. II. *v. trans. To make alive*, QUICKEN; vivificare:—Me ðín spræc cwycade *eloquium tuum vivificavit me*, Ps. Th. 118, 50. Ðú us cwica *quicken thou us*, 79, 17. [*Prompt.* qwycchyñ *movère: Wyc.* quikene, quykne, quycken *to revive: Piers P.* quykne *to bring to life: Chauc.* quiken *to become or make alive: Plat.* queken, *v. n. and a. to grow, cultivate: O. Sax.* -quikôn, -quiccôn: *Dut.* kweeken *to foster, manure, cultivate: Kil.* quicken, quecken *nutrire, alere, educare: Ger.* er-quicken *to refresh: M. H. Ger.* quicken,

kücken *to make alive* : O. H. Ger. quikjan *vivificare* : Dan. qwæge : Swed. qwicka : Icel. kweykja, kweykwa.] DER. a-cwician, ed-, ge-, ge-ed-.

cwic-lífian, -lífigan ; *p.* -lifode ; *pp.* -lifod *To live* ; vivere :—Cwiclifigende *living*, Salm. Kmbl. 840 ; Sal. 419. Ðǽr sceal fǽsl wesan cwiclifigendra cynna gehwilces *there shall be food for each of living kinds*, Cd. 65 ; Th. 79, 14 ; Gen. 1311.

cwic-seolfor ; *gen.* -seolfres ; *dat.* -seolfre ; *n.* QUICKSILVER ; vivum argentum :—Wið magan wærce ; rudan sǽd and cwicseolfor *for pain of stomach ; seed of rue and quicksilver*, L. M. 3, 69 ; Lchdm. ii. 356, 19. Cwicseolfor *argentum vivum*, Cot. 16.

cwic-súsl, cwyc-súsl, es ; *n* : e ; *f.* [súsl *sulphur, brimstone, torment, punishment*] *Living punishment, hell-torment* ; sempervivum tormentum, infernum, barathrum = βάραθρον :—Cwicsúsl *vel* helelíc deópnes barathrum, vorago profunda, Ælfc. Gl. 54 ; Som. 66, 96 ; Wrt. Voc. 36, 20. Satanas ðæs cwicsúsles ealdor ðære helle *Satan the chief of the living torment of hell*. Nicod. 26 ; Thw. 14, 12. On ðam cwicsúsle *in hell-torment*, 25 ; Thw. 13, 30 : Exon. 16 a ; Th. 35, 21 ; Cri. 561 : 97 a ; Th. 362, 18 : Wal. 38. Of ðysse cwycsúsle *from this hell-torment*, Nicod. 30 ; Thw. 17, 28. Faraþ ða unrihtwísan into écere cwicsúsle, mid deófle and his awyrigedum englum *the unrighteous will go into everlasting torment, with the devil and his accursed angels*, Homl. Th. ii. 108, 31.

cwic-treów, es ; *n.* ₄ *The asp* or *aspen-tree* ; populus tremula, Lin :—Cwictreów *cresis?* tremulus, Ælfc. Gl. 47 ; Som. 65, 26. v. cwic-beám.

cwicu, cwico, cucu = cuc ; *nom. acc. m. f. n :* *pl. nom. acc. m. f. n.* cwicu, cwico, cucu ; *adj. Alive, quick* ; vivus :—Cwicu *alive*, *nom. m.* Ps. Th. 118, 93. Cwico wæs ic *I was living*, Exon. 125 a ; Th. 482, 1 ; Rä. 66, 1 : Beo. Th. 6178 ; B. 3093. Cucu *vivus*, Wrt. Voc. 85, 56. Samson miccle má on his deáþe acwealde, ðonne he ǽr cucu dyde *Samson multo plures interfecit moriens, quam ante vivus occiderat*, Jud. 16, 30 : Boutr. Scrd. 18, 11 : Homl. Th. i. 52, 20 : ii. 212, 33 : Cod. Dipl. 897 ; Kmbl. iv. 233, 5, 13. Ne sécþ seó cucu [turtle] nǽfre hire óðerne gemacan *the quick [living turtle-dove] never seeks to itself another mate*, Homl. Th. i. 142, 14. Heó sóna cucu arás *she instantly arose alive*, ii. 26, 32. Gif hit cucu [cwicu MS. G.] feoh wǽre *if it were live cattle*, L. Alf. 28 ; Th. i. 52, 1. Ǽlc þing ðe cucu byþ *everything which is alive* ; animal, Wrt. Voc. 78, 50. Ic hæfde ferþ cwicu *I had a soul alive*, Exon. 126 b ; Th. 487, 21 ; Rä. 73, 5. Ic hæfde feorh cwicu *I had a soul alive*, 103 b ; Th. 392, 11 ; Rä. 11, 6 : 104 a ; Th. 394, 14 ; Rä. 14, 3. Teón ða wæteru forþ swimmende cynn cucu on lífe *producant aquæ reptile animæ viventis*, Gen. 1, 20 : Ex. 22, 4. Hí cwico nǽron *they were not alive*, Exon. 24 b ; Th. 69, 36 ; Cri. 1131. Cwicu *quick [living]*, *pl. nom. n.* Ps. Th. 108, 24. Cwicu *quick [living]*, *pl. acc. m.* 87, 18. He clifu cyrreþ on cwicu wæteres wellan *he turneth the rocks to quick [living] springs of water*, 113, 8. v. cwic.

cwicu-líce ; *adv. In a living manner, vigorously* ; vivide :—Me on weg ðínne lǽde cwiculíce *in via tua vivifica me*, Ps. Th. 118, 37.

cwid-bóc, e ; *f. The Book of Proverbs* ; proverbiorum liber :—Be ðæm is awriten on Salomonnes cwidbócum *about which it is written in the Proverbs of Solomon*, Past. 36, 8 ; Cot. MS.

cwiddung, cwyddung, e ; *f. A saying, tale, report, speech* ; dictum, sermunculus :—Manegra manna cwyddung is *it is a saying of many men*, Bd. de nat. rerum ; Wrt. popl. science 10, 28 ; Lchdm. iii. 256, 4. Æt frǽmdra monna cwiddunge *from the report of strangers*, Bt. 18, 4 ; Fox 66, 25. Ná swilce he nyste manna cwyddunga be him *not as though he knew not the sayings of men concerning him*, Homl. Th. i. 366, 7.

cwide, cwyde, cwyðe, es ; *m.*   **I.** *the expression of a thought, a sentence, period* ; sententia :—We todǽlaþ ða bóc to cwydum, and siððan ða cwydas to dǽlum, eft ða dǽlas to stæfgefégum, and siððan ða stæfgefégu to stafum ; ðon beóþ ða stafas untodǽledlíce, forðonðe nán stæf ne biþ náht, gif he gǽþ on twá. Ǽlc stæf hæfþ þreó þing, nomen, figura, potestas, ðæt is nama, and hiw, and miht *we divide the book into sentences, and then the sentences into words [parts], again the words into syllables, and then the syllables into letters ; now the letters are indivisible, because a letter is nothing if divided into two [if it go in two]. Every letter has three properties*, nomen, figura, potestas, *that is a name, and a form, and a sound [power]*, Ælfc. Gr. 2 ; Som. 2, 37–41.   **II.** *a saying, proverb, speech, discourse, sermon, will* ; dictum, dictio, sermo, homilia, testamentum :—Eówer cwide stande *may your saying stand*, Jos. 2, 21. Singende ðone ealdan cwide *singing the old adage*, Bt. 14, 3 ; Fox 46, 29. Þurh ryhtlícne cwide [MS. cuide] and dóm *through a righteous sentence and judgment*, Past. 35, 5 ; Hat. MS. 46 b, 4. On ǽgðer ðæra bóca sind feówertig cwyda, búton ðǽre fórespræce *in each of these books there are forty discourses, without the preface*, Homl. Th. ii. 2, 14 : i. 28, 20. Ætfóran ǽlcum cwyde we setton ða swutelunge on Léden *before each discourse we have set the argument in Latin*, ii. 2, 17. Ðes [MS. ðis] is Byrhtríces níhsta cwide *this is Byrhtric's last will*, Th. Diplm. A. D. 950 ; 500, 24 : A. D. 958 ; 509, 3 : A. D. 998 ; 544, 25 : A. D. 1002 ; 543, 33. Ðæt se cwyde standan móste *that the will might stand*, A. D. 950 ; 501, 11 : A. D. 972 ; 519, 17 : A. D. 997 ; 539, 22 : A. D. 996–1006 ; 549, 11. Cwydas dón *to make wills*, Lchdm. iii. 210,

30.   **III.** *a legal enactment, decree* ; edictum, decretum :—Swa hit ǽr Eádmundes cwide wæs *as it was formerly the enactment of Edmund*, L. Edg. H. 2 ; Th. i. 258, 9. Swá úre ealra cwide is *as is the decree of us all*, L. Eth. i. 4 ; Th. i. 284, 5 : L. C. S. 33 ; Th. i. 396, 19. [Laym. cwide, quide-n *a testament* ; *pl.* quides, cwides *speeches, words* : O. Sax. quidi, *m.* speech, saying : O. H. Ger. quidi, *f. n.* dictum, verbum : Goth. qiss, *f.* speech : Icel. qwiðr, *m.* a saying ; *word, speech.*] DER. ǽr-cwide, big-, ed-, ge-, gegn- [geagn-, gén-], galdor-, gilp-, heard-, hearm-, hleóðor-, hosp-, lár-, leahtor-, mǽdel-, meðel-, sár-, sib-, sóþ-, teón-, torn-, wiðer-, wom-, word- : cwidian.

cwide-gied, -giedd, es ; *n.* [gid, gied *a song, lay*] *A song, ballad* ; carmen :—Fela cúþra cwidegiedda *many [of] known songs*, Exon. 77 a ; Th. 289, 28 ; Wand. 55.

cwide-leás *speechless, intestate.* v. cwyde-leás.

cwidian, cwiddigan, cwydian, cwyddian ; *p.* ode ; *pp.* od [cwide, cwyde *a saying*] *To speak, say* ; dicere :—Ongan hine hyspan and hearm cwiddigan [cwidian, Cot.] *he began to revile and speak ill of him*, Bt. 18, 4 ; Fox 66, 33.

cwid-rǽden *an agreement* ; pactum. v. gecwid-rǽden.

cwidu *what is chewed, a cud*, QUID, L. M. 2, 3 ; Lchdm. ii. 182, 3 : 2, 4 ; Lchdm. ii. 182, 17. v. cwudu.

cwiert-ern *a prison*, Mt. Kmbl. B. 25, 36, 39. v. cweart-ern.

cwiferlíce ; *adv. Anxiously* ; sollicitè, C. R. Ben. 64.

cwild *a plague, pestilence, murrain, destruction*, Wrt. Voc. 75, 54 : Ælfc. Gr. 9, 27 ; Som. 11, 25 : Chr. 897 ; Erl. 94, 31 : Ps. Spl. C. 28, 9 : 31, 8. v. cwyld.

cwild-bǽre ; *adj. Pestilence-bearing, deadly* ; pestiferus, Scint. 53 : 63.

cwild-bǽrlíce ; *adv. Pestilentially, destructively* ; pestifere, Scint. 8.

cwilde flód, es ; *n. m. The destruction's flood, deluge* ; diluvium, Ps. Spl. C. 28, 9. v. cwyld.

cwild-tíd *a dead time.* v. cwyld, cwyl-tíd.

cwilman *to torture, kill*, Ors. 1, 12 ; Bos. 36, 25. v. cwelman.

cwilst, he cwilþ *diest, dies* ; *2nd and 3rd pers. pres. of* cwelan.

cwiman *to come* ; venire, *the supposed infin. of* cwom, *q. v.*

cwínan, *p.* cwán, *pl.* cwinon ; *pp.* cwinen *To waste* or *dwindle away* ; tabescere. DER. a-cwínan.

cwincan, ic cwince, ðú cwincst, he cwincþ, *pl.* cwincaþ ; *p.* cwanc, *pl.* cwuncon ; *pp.* cwuncen *To disappear, vanish, decrease* ; evanescere, diminuere, deficere, Leo A. Sax. Gl. 209. DER. a-cwincan.

cwínod *wasted*, Bt. 10 ; Fox 28, 29. v. cwánian.

cwis, cwiss, e ; *f.* [cweðan *to say, speak*] *A saying, speaking* ; locutio. DER. and-cwis, ge- : un-cwis.

cwíst *sayest, speakest*, Bt. 5, 3 ; Fox 12, 13 : Ps. Th. 87, 12, = cweðst ; *2nd pres. sing. of* cweðan.

CWIÞ, es ; *m* : cwiða, an ; *m. The womb* ; matrix, uterus :—Beðe mid ðone cwiþ *bathe the womb therewith*, L. M. 3, 37 ; Lchdm. ii. 330, 2 : 3, 38 ; Lchdm. ii. 330, 19. Cwiþ *matrix*, Ælfc. Gl. 76 ; Som. 71, 118. Wið ðæs cwiðan sáre *for soreness of the womb*, Herb. 165, 2 ; Lchdm. i. 294, 11. [O. H. Ger. quiti : Goth. qiþus, *m* : Swed. qwed : Icel. kwiðr.]

cwiþ *saith, speaks*, Exon. 14 a ; Th. 28, 28 ; Cri. 453 : 30 a ; Th. 92, 35 ; Cri. 1519, = cweðeþ ; *3rd pres. sing. of* cweðan.

cwiðan, cwýðan ; he cwíðeþ ; *p.* de ; *pp.* ed *To speak* or *moan in grief, mourn, lament* ; lamentáre, plangére :—Wópe cwíðan *with weeping to lament*, Cd. 48 ; Th. 61, 13 ; Gen. 996. Ic sceolde ána míne ceare cwíðan *I must alone mourn my care*, Exon. 76 b ; Th. 287, 4 ; Wand. 9. We cwíðdon [MS. cwiðdun] *lamentavimus*, Mt. Bos. 11, 17. Fǽmnan ne synd cwýðede [cwyðde MS.] *virgines non sunt lamentatæ*, Ps. Spl. C. 77, 69. Adames cyn cwíðeþ *Adam's race lamenteth*, Exon. 22 a ; Th. 59, 34 ; Cri. 962. Hý in cearum cwíðaþ *they mourn in sorrows*, Exon. 35 b ; Th. 115, 23 ; Gú. 194. Ðonne biþ þearfendum cwíðende cearo *then shall be wailing care to the miserable*, 26 b ; Th. 79, 5 ; Cri. 1286. [O. Sax. quíðean : Swed. quida : Icel. kwíða *to feel anxiety about.*]

cwiðend-líc ; *adj. Proper, peculiar, natural* ; genuínus, Cot. 96, Som. Ben. Lye.

cwið-nes, -ness, e ; *f. A wailing, lamentation* ; lamentum, Greg. Dial. 3, 15, 37.

cwíðst *sayest, speakest*, Ælfc. Gr. 33 ; Som. 37, 38, = cweðst ; *2nd pres. sing. of* cweðan.

cwoellan *to kill* ; necare, interficere :—Sóhton hine Iudéas to cwoellanne *quærebant eum Judæi interficere*, Jn. Lind. War. 5, 18. v. cwellan.

cwolen *died* ; *pp. of* cwelan.

cwolstan *to swallow.* DER. for-cwolstan, *q. v.*

cwom, *pl.* cwómon *came* ; venit, venerunt ; *have the same meanings as the contracted forms* com, *pl.* cómon, *p. of* cuman, *q. v.* The *p. indic.* cwom, *pl.* cwómon, -an, -un ; *p. subj.* cwóme :—Ðá hleóðor cwom *when the sound came*, Cd. 181 ; Th. 226, 29 ; Dan. 178. Ðá ðú ǽrest cwóme *when thou first camest*, Exon. 39 a ; Th. 129, 25 ; Gú. 426. Hwonne bearn Godes cwóme *when the child of God should have come*, 10 a ; Th. 10, 6 ; Cri. 148. To Hierusalem cwómon *they came to Jerusalem*, Elen. Kmbl. 547 ; El. 274. Cwóman englas *angels came*, Exon. 15 b ; Th. 34, 21 ; Cri. 545. Wuldres áras cwómun *messengers of glory came*,

15 a; Th. 31, 11; Cri. 494. Cwom, *pl.* cwômon, *seem to be from* cwiman, *which I have not found in A. Sax. It is in Goth.* qiman [*pronounced* kwiman = cwiman]; *p.* qam, *pl.* qemum; *pp.* qumans *to come*; venire. *Goth.* Ni mag qiman [kwiman = cwiman]. *A. Sax.* Ic ne mæg cuman *I cannot come*, Lk. Bos. 14, 20. v. cwiman, cuman.

cwuc; *def.* se cwuca *alive, quick*, Bt. 36, 6; Fox 182, 20. v. cwic.

cwucen *alive, quick*, Bt. 36, 6; Fox 182, 20. v. cwicen.

cwuda *a cud, quid*, L. M. 2, 2; Lchdm. ii. 178, 26: 2, 52; Lchdm. ii. 270, 28. v. cwudu.

CWUDU, cwuda, cweodo, cwidu, cudu; *gen.* ues, wes; *n. What is chewed, a cud, quid*; manducatum, rumen :—Ðe heora cudu ne ceówaþ: ða clǽnan nýtenu ðe heora cudu ceówaþ *which chew not their cud: the clean beasts which chew their cud*, M. H. 138 b. ¶ Hwît cwudu *white cud, mastich*; an odoriferous gum from the mastich-tree, which was called by Lin. *pistacia lentiscus.* This gum was used for chewing in the East; *mastiche* = μαστίχη :—Hwît cwudu *mastich*, L. M. 1, 23; Lchdm. ii. 66, 3. Gedô gódne dǽl ðæron hwîtes cweodwes *put a good deal of mastich therein*, 2, 14; Lchdm. ii. 192, 6. Ofersceade mid hwîtes cwidues duste *sprinkle over with dust of mastich*, 2, 3; Lchdm. ii. 182, 3. Of hwîtum cwidue and wîne *with mastich and wine*, 2, 4; Lchdm. ii. 182, 17. Hwît cwudu gecnuwa swîðe smale *pound mastich very small*, 1, 13; Lchdm. ii. 56, 5: 1, 8; Lchdm. ii. 54, 3 : 1, 47; Lchdm. ii. 118, 29: 3, 2; Lchdm. ii. 308, 24. Genim ele and gedô hwît cwuda on ðone ele *take oil and put mastich into the oil*, 2, 2; Lchdm. ii. 178, 26: 2, 52; Lchdm. ii. 270, 28. Nim hwît cudu *take mastich*, Lchdm. iii. 72, 15: 124, 25: 134, 10. [*Prompt.* cudde = *Wyc.* code, quede, quide, kude: *Orm.* cude.]

cwuncon; *pp.* cwuncen *disappeared, vanished*; *p. pl.* and *pp.* of cwincan.

cwyc *alive, quick* :—Cwyc *alive*, Ps. Th. 104, 8 : Nicod. 26; Thw. 14, 28, 38. v. cwic.

cwyc-ǽht *live stock* :—On cwycǽhtum *in live stock*, L. Alf. pol. 18; Th. 1, 72, 12. note 28. v. cwic-ǽht.

cwycian *to make alive, quicken*, Ps. Th. 118, 50. v. cwician II.

cwyc-sûsl *hell-torment*, Nicod. 30; Thw. 17, 28. v. cwic-sûsl.

cwyddian; *p.* ode; *pp.* od *To speak, say*; dicere :—Ðæt me oferhydige ǽfre ne môtan hearm cwyddian *that the proud may never speak evil of me*, Ps. Th. 118, 122. Crist hî befran hû men cwyddodon be him *Christ asked them how men spake concerning him*, Homl. Th. ii. 388, 31. v. cwidian.

cwyddung *a saying*, Homl. Th. i. 366, 7. v. cwiddung.

cwyde. I. *a sentence*; sententia, Ælfc. Gr. 2; Som. 2, 38. II. *a discourse, sermon* :—Smeágaþ ðysne cwyde *consider this sermon*, Homl. Th. i. 28, 20 : ii. 2, 14: 2, 17. v. cwide.

cwydele, an; *f. An inflamed swelling*; pustula, varix :—Cwydele *pustula*, Ælfc. Gl. 9; Som. 57, 10; Wrt. Voc. 19, 19. Cwydele *vel* hwylca *varix*, 76; Som. 71, 129; Wrt. Voc. 45, 32.

cwyde-leás; *adj. Speechless, intestate*; mutus, intestatus :—He læg cwydeleás, bûtan andgite *he lay speechless, without sense*, Homl. Th. i. 86, 26. Gif hwá cwydeleás of ðyssum lîfe gewîte *if any one depart this life intestate*, L. C. S. 71; Th. i. 412, 27.

cwydian; *p.* ode; *pp.* od *To speak, say*; dicere :—Menn cwydodon *men said*, Chr. 1085; Erl. 217, 38. v. cwidian.

cwydol; *adj.* [cweðan *to say, speak*] *Speaking, saying*; dicens, loquens. DER. wyrig-cwydol, *q. v.*

cwyd-rǽden *an agreement*; pactum. v. gecwid-rǽden.

cwyld, cwild, es; *m. n.* cwyld, cwild, e; *f.* [cweald, *pp. of* cwellan *to kill*] *A plague, pestilence, murrain, destruction*; pestis, pestilentia, clades :—Boreas ealne ðone cwyld *m.* aflîgþ *Boreas [the north wind] drives every plague away*, Bd. de nat. rerum; Wrt. popl. science 18, 9; Lchdm. iii. 276, 7. Cwilde *f.* flôd *the flood of destruction, deluge*; diluvium, Ps. Spl. C. 28, 9: 31, 8. Auster mistlîce cwyld *n.* blǽwþ geond ðas eorþan *auster [the south wind] blows various plagues through this earth*, Bd. de nat. rerum; Wrt. popl. science 17, 26; Lchdm. iii. 274, 17. Cwild [cwyld MSS. C. D.], *m. f.* or *n.* clades, Ælfc. Gr. 9, 27; Som. 11, 25. Cwild, *m. f.* or *n.* pestis, Wrt. Voc. 75, 54. Mid ceápes cwylde *m. f.* or *n.* with a murrain of cattle, Chr. 897; Th. 174. 22, col. 2; 175, 20. Se ðe on þrymsetle cwyldes *m.* or *n.* nâ sæt *qui in cathedra pestilentiæ non sedit*, Ps. Spl. C. 1, 1 : Mone B. 2711. Cwyld-tîd or cwyl-tîd *evening time*; conticinium :—Cwyl-tîd *vel* gebed-giht *conticinium*, Ælfc. Gl. 16; Som. 58, 63; Wrt. Voc. 21, 50. v. cwyld-seten. DER. mon-cwyld.

cwyld-bǽre; *adj. Pestilence-bearing, deadly.* v. cwild-bǽre.

cwyld-bǽrlíce; *adv. Pestilentially.* v. cwild-bǽrlíce.

cwyld-full; *adj. Destructive, pernicious*; perniciosus :—Cwyldfulle wæfersêne *perniciosum spectaculum*, Mone B. 1259.

cwyld-rôf; *adj. Devoted to slaughter*; necandi strenuus :— Deór cwyldrôf = wulfas *the beasts devoted to slaughter = wolves*, Cd. 151; Th. 188, 10, 11 = 7; Exod. 166 = 164.

cwyld-seten, cwyl-seten, e; *f.* [cwyld, cwyl = cweald, *pp. of* cwellan *to kill*: *Icel.* kweld, *n. evening*: as if the night *quelled* or *killed* daylight]

A setting in of the evening, the first part of the night; conticinium :— Cwylseten *conticinium*, Mone B. 3747. Cwylsetene *conticinio*, 3748. Cwyldsetene *galli cantu*, 4677.

cwylla, an; *m. A well, spring*; fons :—Riht sûþ be eástan ðam cwyllan ôþ ða wýde strǽte *right south by east of the spring as far as the wide road*, Cod. Dipl. 409; A. D. 946; Kmbl. ii. 265, 32. [*Ger.* quelle, *f. a spring, source, fountain.*]

cwylm *destruction, slaughter*, Glos. Prudent. Recd. 152, 12. v. cwealm.

cwylman; *p.* ede; *pp.* ed *To kill, torment*, Ps. Spl. 36, 15 : Elen. Kmbl. 1373; El. 688. v. cwelman.

cwylm-bǽre; *adj. Death-bearing, pernicious*; mortiférus :—Cômon ða cempan mid cwylmbǽrum tôlum *the soldiers came with deadly tools*, Homl. Th. ii. 260, 7. v. cwealm-bǽre.

cwylmd = cwylmed *killed*, Bd. 1, 15; S. 484, 1; *pp. of* cwylman.

cwylmende, cwilmende; *part. Tormenting*; crucians, Ors. 1, 12; Bos. 36, 25. v. cwelman.

cwylmian; *part.* cwylmigende; *p.* ode; *pp.* od [cwealm *pain, torment*] *To suffer, suffer torment* or *pain*; cruciāri :—Heó sceal ǽcelîce cwylmian *it [the soul] shall suffer eternally*, Homl. Th. ii. 232, 29. Ða mánfullan beóþ ǽfre cwylmigende on helle sûsle *the sinful shall ever be suffering pain in hell torment*, 608, 11. We cwylmiaþ *we suffer torment*, 416, 5. Gehwylce mánfulle geféran on ðám ěcum tintregum cwylmiaþ *all wicked associates shall suffer in everlasting torments*, i. 526, 27.

cwylming, e; *f.* [cwylmian *to suffer*] *Torture, trouble, suffering, a cross*; cruciātus, crux :—Cwylminge [MS. cwylmingce] *cruciātu*, Mone B. 3178. Se ðe ne nimþ hys cwylminge, and fyligþ me, nys he me wyrðe *qui non accipit crucem suam, et sequitur me, non est me dignus*, Mt. Bos. 10, 38 : Lk. Bos. 9, 23.

cwylmnes *torment*, Bd. 4, 9; S. 577, 10. v. cwealmnes.

cwylþ *dies*, Bd. de nat. rerum; Lchdm. iii. 272, note 36; *3rd pres. sing. of* cwelan.

cwyl-tíd *dead time*, Ælfc. Gl. 16; Som. 58, 63; Wrt. Voc. 21, 50. v. cwyld.

cwýne *a wife*, L. Ethb. 85; Th. i. 24, 9. v. cwên, cwêne.

CWYRN, cweorn, e; *f.*: cweorne, an; *f. A mill, hand-mill*, QUERN; mola :—Twá beóþ æt cwyrne grindende: ân byþ genumen, and ôðer byþ lǽfed *duæ molentes in mola: una assumětur, et una relinquětur*, Mt. Bos. 24, 41. Ðæt hîg grundon on cwyrne *populus illud frangēbat mola*, Num. 11, 8. Æt ðære cweornan *ad molam*, Ex. 11, 5. [*Prompt.* querne *mola manualis: Wyc. Chauc.* querne: *Plat.* queern, qwern *a handmill: O. Sax.* querna, *f.: O. Frs.* quern: *Dut. Kil.* querne: *M. H. Ger.* kürne, kürn, kurn, *f.: O. H. Ger.* quirn, *f.: Goth.* qairnus, *m.* or *f.: Dan.* qwærn, *m. f.: Swed.* qwarn, *f.: Icel.* kwern, kwörn, *f.*] DER. esul-cwyrn, hand-.

cwyrn-bill *a stone chisel for dressing querns.* v. cweorn-bill.

cwyrn-burne, an; *f. A mill-stream*; molāris torrens, Som. Ben. Lye.

cwyrn-stán, cweorn-stán, es; *m. A mill-stone*; molaris lapis, mola :— Cwyrnstán *mola*, Wrt. Voc. 83, 8. Ðæt him wǽre getiged ân ormǽte cwyrnstán to his swuran, and he swá wurde on deóppre sǽ besenced *that an immense mill-stone was tied to his neck, and he was so sunk in the deep sea*, Homl. Th. i. 514, 17 : Mt. Bos. 18, 6. Ân cweornstán *lapis molaris*, Lk. Bos. 17, 2 : Mk. Bos. 9, 42.

CWYSAN; *p.* de; *pp.* ed *To crush*, QUASH, *shake, bruise, dash against*; quassare, terere, allidere :—Se ðe forgnîdeþ oððe cwysþ lytlungas ðîne to stáne *qui allidet parvulos tuos ad petram*, Ps. Lamb. 136, 9. Ðû genýðeredest oððe ðú cwysdest me *allisisti me*, 101, 11. [*Prompt.* quaschyñ *quassāre: R. Brun.* quassed, *p. quashed: Plat.* quesen, questen *to crush: O. Sax.* quetsan *to push, squeeze: Frs.* quetsen *vulnerare: O Frs.* quetsene *a bruise: Dut.* kwetsen *to bruise, wound, injure: Kil.* quetsen *quassare, lædere: Ger.* quetschen *to squeeze: M. H. Ger.* quetzen *to squeeze: Goth.* qistyan *to destroy: Dan.* quæste *to squeeze: Swed.* qwäsa *to squash, bruise, wound: Icel.* kwista *to destroy, cut down: Fr.* casser *to break: Lat.* quassare, quatere *to batter, break in pieces.*] DER. for-cwysan, to-.

cwýst *sayest, speakest*, Homl. Th. i. 424, 9, = cweðst; *2nd pres. sing. of* cweðan.

cwýst ðû, cwýst ðú lá, cwýst tû lá *sayest thou?* used in questions, as interrog. adv. numquid? — Cwýst ðú eom ic hyt? Mt. Bos. 26, 22 *whether it am I?* Wyc. note 11; numquid ego sum? Vulg: Ps. Spl. 29, 12 : 7, 12. v. cweðan.

cwyð, e; *f.* [= cwide, cwyde] *A word, saying*; verbum, dictum :— Him ða cwyðe frecne scódon *these words overwhelmed him with woe*, Cd. 78; Th. 96, 18; Gen. 1596. v. cwide.

cwyþ *saith, speaks*, Jn. Bos. 16, 18 : Rood Kmbl. 220; Kr. 111, = cweðeþ; *3rd pres. sing. of* cweðan.

cwýðan *to lament*, Ps. Spl. C. 77, 69. v. cwîðan.

cwyðe *a saying*, S. Greg. Hom. 23, 104, Lye. v. cwide.

cwyðele *an inflamed swelling.* v. cwydele.

cwyðst *sayest, speakest*, Ælfc. Gr. 18; Som. 21, 62, = cweðst; *2nd pres. sing. of* cweðan.

cȳ cows, Gen. 33, 13; acc. pl. of cū.

CYCENE, cicene, an; f. A kitchen; coquīna, culīna :— Cycene coquīna, Wrt. Voc. 82, 49 : culīna, Mone B. 3731. Đæt seó cycene [MS. kycene] eal forburne that the kitchen was all burning, Homl. Th. ii. 166, 5, 11. Wurpon hí ða anlícnysse inn to heora cycenan [MS. kycenan] they cast the image into their kitchen, ii. 166, 3. Gif ceorl hæfde cirican and cycenan [MS. kycenan] if a free man had a church and a kitchen, L. R. 2; Th. i. 190, 15. [Piers P. kytchen: Chauc. kichen: Plat. köke, käke: Dut. keuken, f: Kil. kokene, keuckene: Ger. küche, f: M. H. Ger. küche, küchen, kuche, kuchen, f: O. H. Ger. kuchina, f: Dan. kjökken, n : Swed. kök, n : Icel. kock-hús : Fr. cuisine, f: Prov. cozina: Span. cocina, f: It. cucina, f: Lat. coquīna, f: Wel. cegin, f: Corn. cegin, keghin, f: Ir. cucann: Armor. kegin: Lith. kukne: Russ. kuchnja.]

cȳdde said, told, Chr. 1066; Th. 336, 21, = cýðde; p. of cýðan.

cȳdung a chiding, Ps. Spl. T. 103, 8. v. cíding.

CȲF, e; f: cýfe, an; f. A vessel, vat, cask, bushel; dolium, modius :— Cýf dolium, Ælfc. Gl. 25; Som. 60, 48; Wrt. Voc. 24, 48. Stód ðær án æmtig cýf an empty cask stood there, Homl. Th. ii. 178, 34. Cýfe dolium, Wrt. Voc. 83, 25. Se hét afyllan áne cýfe mid ele he commanded a vat to be filled with oil, Homl. Th. i. 58, 25. Under cýfe sub modio, Mt. Bos. 5, 15. [Prompt. kowpe crater : Plat. kope dolium: O. Sax. côpa, f: dolium: Dut. kuip, f. a tub: Kil. keuwe, kuype cupa, dolium: Ger. kufe, f. a vessel: M. H. Ger. kuofe, f. cupa: O. H. Ger. kuofa, f. dolium, tunna: Dan. kippe, kyper, m. f. a dyer's tub: Swed. kyp, m. a dyer's tub: kupa, f. a case, box: Icel. kúpa, f. a bowl, basin, box: Fr. cuve, f: Span. cuba, f. cask for wine or oil: M. Lat. cuppa, f: Lat. cupa, f. a tun: Grk. κύπ-ελλον a tub, cask: Sansk. kūpa a cistern; kumbha vessel for water.]

CYFES, cyfys, cifes, ciefes, e; f: cyfese, an; f. A concubine, handmaid; concubina, pellex, ancilla :—Cyfes pellex, Wrt. Voc. 86, 73. Of cifise ex pellĭce, Mone B. 4553. Se ðe hæbbe riht wif, and eác cifese [MS. A. ceafese] B. cefese] ne dô him nán preóst nán ðara gerihta, ðe man cristenum men' dôn sceal he who has a right wife, and also a concubine, let no priest do for him any of those rites, which ought to be done for a christian man, L. C. S. 55; Th. i. 406, 16, and note 26. Cyfys [= cyfes] oððe bepǽcystre [MSS. C. D. bepǽcestre] pellex, Ælfc. Gr. 28, 5; Som. 32, 1. Constantius gesealde his suna ðæt ríce, Constantinuse, ðone he hæfde be Elenan his ciefese Constantius gave the empire to Constantine, his son, whom he had by Helena his concubine [wife, v. notes to Ors. Bos. p. 28, col. 2], Ors. 6, 30; Bos. 126, 41. Gif he cyfesan hæbbe, and náne riht æwe, he áh ðæs to dônne swá him geþincþ; wite he ðeáh ðæt he beó on ánre gehealden, beó hit cyfes, beó hit æwe si concubinam habeat, et nullam legitimam uxorem, erit ei proinde quod ipsi videbitur faciendum; sciat tamen ut cum una ei manendum sit, sit concubina, sit uxor, L. Ecg. P. ii. 9; Th. ii. 186, 2–5: L. M. I. P. 17; Th. ii. 270, 6, 9: Boutr. Scrd. 22, 22. Be ðinre cyfese super ancilla tua, Gen. 21, 12. [Laym. chevese, chivese a concubine: Plat. keves: Dut. kevis, f. a concubine: Kil. kevisse, kiese pellaca, concubina: Ger. kebse, f. concubina, pellex: M. H. Ger. kebes, kebese, kebse, f. concubina: O. H. Ger. kebis, kebisa, f. pellex, concubina: Icel. Vigf. kefsir, m. concubitor, concubinus: O. Nrs. Rask Hald. képsi, kéffir servus molestus, oblocutor.]

cyfes-boren; def. se cyfes-borena; part. Born in concubinage, base-born; e concubina genĭtus :—His cyfesborena bróðor siððan ríxode, se ðe wende to Scottum his base-born brother afterwards reigned, who had gone to the Scots, Homl. Th. ii. 148, 17.

cyfes-hád, es; m. Whoredom, adultery, concubinage; pellicātus, Cot. 186.

cyfys pellex, Ælfc. Gr. 28, 5; Som. 32, 1. v. cyfes.

cȳgan, cȳgean to call, call upon, invoke, Bd. 4, 23; S. 594, 39: Cd. 141; Th. 176, 9: Gen. 2909: Ps. Spl. 78, 6. v. cígan.

cȳging, e; f. A calling, naming; appellatio, Som. Ben. Lye. v. cȳgan.

cȳgling, e; f. A relation; cognātus :—Cȳgling his cognātus ejus, Jn. Rush. War. 18, 26. v. cȳðling.

cyld, es; n. Cold, coldness; frigus :—For cylde præ frigŏre, Coll. Monast. Th. 19, 29. v. ceald frigus.

cyld, es; m. A child, Bt. 36, 5; Fox 180, 6: Mt. Jun. 2, 13, in the title. v. cild.

cyld-faru; e; f. A carrying of children; parvulōrum subvectio :—Đæt híg nymon wǽnas to hira cyldfare ut tollant plaustra ad subvectiōnem parvulōrum, Gen. 45, 19.

CȲLE, cíle, cēle, es; m. A cold, coldness, CHILL; frigus :—Ne mæg fýres feng ne forstes cýle somod eardian the grasp of fire and chill of frost cannot dwell together, Salm. Kmbl. 708; Sal. 353. Beóforan ansíne cýles ante faciem frigŏris, Ps. Spl. 147, 6. Nabbaþ we to hyhte nymþe cýle and fýr we have nought in hope, save chill and fire, Cd. 220; Th. 285, 10; Sat. 335. Hý wyrcaþ ðone cýle hine on they bring the cold upon him, Ors. 1, 1; Bos. 23, 6, 8. [Prompt. cole algor: Piers P. Laym. Orm. chele chill, cold: Plat. köle, f. pain: Ger. kühle, f: M. H. Ger. küele, f: O. H. Ger. kuolî, f: Dan. köle, m. f. coolness of

the air: Swed. kyla, f. a chill: Icel. kylr, m. a gust of cold air: Lat. gelu.] DER. fǽr-cȳle.

cȳle-gicel, es; m. An icicle; frigŏris stiria :—Land wǽron freórig cealdum cýlegicelum the lands were frozen with cold icicles, Andr. Kmbl. 2521: An. 1262: Exon. 56 b; Th. 201, 20; Ph. 59. v. gicel.

CYLEN, cyln, e; f. A kiln, an oven; fornācŭla, siccatōrium :—Cylene fornācŭlæ, Cot. 86. Cyln vel ast siccatōrium, Ælfc. Gl. 109; Som. 78, 132; Wrt. Voc. 58, 44. [Prompt. kylne: Icel. kylna, f: Wel. kylyn, m.]

cylenisc; adj. Like a kiln; fornāceus, Som. Ben. Lye.

cyleþenie, an; f. The herb celandine; chelidonium majus :—Cyleþenie, Herb. 75; Lchdm. i. 176, 15, 18. v. celeþonie.

cylew, cylu; adj. Spotted, speckled; guttātus :—Cylew guttātus, Cot. 99. Cylu guttātus, Ælfc. Gl. 80; Som. 72, 92; Wrt. Voc. 46, 49.

cȳle-wyrt, e; f. Sour-sorrel; oxylapăthum, Cot. 216.

cylin, cyline heorþ a kiln; fornācŭla. v. cylen.

CYLL, e; f: cylle, cille, an; f: cylle, es; m. A leather bottle, flagon, vessel; uter, ascopēra = ἀσκοπήρα :—Gesomnigende swá swá on cylle wætera sǽs congregans sicut in utrem aquas maris, Ps. Spl. C. 32, 7. Đas cylle istum utrem, Greg. Dial. 3, 37. Swá ðú on hríme setest hlance cylle sicut uter in pruina, Ps. Th. 118, 83. Flaxe oððe cylle asscopa [= ascopēra], Ælfc. Gl. 5; Som. 56, 27; Wrt. Voc. 17, 32. Æmtige cillan vacuum utrem: ða cillan istum utrem, Greg. Dial. 3, 37. Gefylde he ðære cyrcan cyllan implevit lampades ecclesiæ, 1, 5. He gegaderode eall sǽwætru tosomne, swylce hí wǽron on ánum cylle congregans sicut in utrem aquas maris, Ps. Th. 32, 6. Seó cwén [Tomyris] hét ðæt heáfod bewyrpan on ánne cylle se wæs afylled mannes blódes the queen [Tomyris] ordered the head to be thrown into a vessel which was filled with man's blood, Ors. 2, 4; Bos. 45, 34. Se ðe fæstne hider cylle [MS. kylle] brohte ... gif hwelc þyrelne cylle [kylle MS.] brohte to ðys burnan who has brought hither a water-tight bottle ...if any has brought to this spring a leaky bottle, Past. 65; Hat. MS. [Icel. kyllir, m. a bag or pouch.] DER. stór-cylle, -cille.

cyln a kiln :—Cyln vel ast siccatōrium, Ælfc. Gl. 109; Som. 78, 132; Wrt. Voc. 58, 44. v. cylen, ast.

cylu spotted, Ælfc. Gl. 80; Som. 72, 92; Wrt. Voc. 46, 49. v. cylew.

cym come, Exon. 13 a; Th. 23, 22; Cri. 372; impert. of cuman.

cymast most beautiful, Ps. Th. 86, 2; superl. of cyme, adj.

cyme, cime, es; m. [cuman to come] A coming, an approach, advent; adventus :—Me is ðín cyme on myclum þonce gratus mihi est multum adventus tuus, Bd. 4, 9; S. 577, 21: Exon. 21 a; Th. 56, 8; Cri. 897: 21 a; Th. 57, 10; Cri. 916: 44 b; Th. 152, 2; Gū. 802: 56 b; Th. 201, 9; Ph. 53: 69 b; Th. 258, 3; Jul. 259. Wearþ Húna cyme cúþ ceasterwarum the approach of the Huns was known to the citizens, Elen. Kmbl. 82; El. 41. He ongeat ðone intingan heora cymes he understood the cause of their coming, Bd. 2, 2; S. 504, 1. He wítgode hú his ealdormenn sceoldon fægnian his cymes of his wrǽcsíðe he prophesied how his chief men should rejoice at his coming from his banishment, Ps. Th. arg. 23. Sixtygum wintra ǽr Cristes cyme sixty [of] years [winters] before the coming of Christ, Bd. 1, 2; S. 475, 4: Exon. 23 a; Th. 64, 1; Cri. 1031: 100 a; Th. 376, 30; Seel. 162: 57 b; Th. 205, 4; Ph. 107: 59 b; Th. 214, 27; Gū. 945: 56 b; Th. 200, 28; Ph. 47: 63 a; Th. 231, 16; Ph. 490: Cd. 151; Th. 189, 4: Exod. 179: Elen. Kmbl. 2170; El. 1086. Þurh mínne cime through my coming, Cd. 29; Th. 39, 1; Gen. 618. Gefégon beornas burhweardes cyme the men rejoiced at the coming of the prince, Andr. Kmbl. 1320; An. 660: Menol. Fox 62; Men. 31. Ic ne wât hwonan his cymas [MS. cyme] sindon I know not whence his comings are, Exon. 50 b; Th. 175, 18; Gū. 1196: Beo. Th. 520; B. 257. DER. be-cyme, eft-, forþ-, from-, geán-, hér-, hider-, hleoðor-, ofer-, ongeán-, seld-, þrym-, to-, up-, ymb-, ym-.

cyme; adj. Becoming, convenient, suitable, lovely, beautiful, splendid; commŏdus, conveniens, aptus, splendidus :—Cumaþ nú and geseóþ, hú cyme weorc Drihten worhte come now and see what lovely works the Lord has wrought, Ps. Th. 65, 4. Đe on Chananéa cymu worhte wundur qui fecit mirabilia in terra Chanaan, 105, 18. Gif ic míne gewǽda on wítehrægl cyme cyrde if I turned my beautiful garments into sackcloth, Ps. Th. 68, 11. Đæt ðú sí cymast ceastra Drihtnes that thou may be the most beautiful of the cities of the Lord, Ps. Th. 86, 2. DER. un-cyme.

cymed, es; n. The plant wall-germander; forte chamædrys = χαμαίδρυς, teucrium chamædrys, Lin :—Genim cymed take germander, L. M. 1, 16; Lchdm. ii. 58, 20: 1, 15; Lchdm. ii. 58, 16. Nim cymed take germander, 1, 39; Lchdm. ii. 102, 20.

cymen, es; m. n. The herb cummin; cŭmīnum = κύμῐνον, cŭmīnum, cyminum, Lin :—Ge tiógoðiaþ eówre mintan and eówerne dile and eówerne cymen [MS. kymen] ye tithe your mint and your dill and your cummin, Past. 57; Hat. MS. Dô ðæt cymen on.eced put the cummin into

vinegar, L. M. 2, 44; Lchdm. ii. 256, 6. Cymen *cymīnum*, Ælfc. Gl. 44; Som. 64, 64; Wrt. Voc. 32, 1: Herb. 155, 1; Lchdm. i. 280, 23: L. M. 2, 39; Lchdm. ii. 246, 23: iii. 6, 16: 24, 9. Cymenes *of cummin*, Herb. 152, 1; Lchdm. i. 276, 21: L. M. 2, 2; Lchdm. ii. 180, 20: 2, 15; Lchdm. ii. 192, 15: 2, 30; Lchdm. ii. 228, 26: 2, 44; Lchdm. ii. 256, 6. Wyrc *sealfe of cymene make a salve with cummin*, 2, 22; Lchdm. ii. 206, 20. Genim cymen *take cummin*, Herb. 94, 2; Lchdm. i. 204, 16: 376, 5: L. M. 1, 2; Lchdm. ii. 36, 11: 1, 17; Lchdm. ii. 60, 15: 1, 48; Lchdm. ii. 120, 24: 2, 6; Lchdm. ii. 184, 15: 2, 24; Lchdm. ii. 214, 17: iii. 28, 11: 72, 14. Cymenes sǽd *seed of cummin*, L. M. 3, 12; Lchdm. ii. 314, 21. Cymenes dust *dust of cummin*, 3, 23; Lchdm. ii. 322, 3.

cymen *come*, Exon. 8 b; Th. 5, 8; Cri. 66; *pp. of* cuman.

Cymēn, es; *m.* Cymen, *son of Ælle, who was the first Bretwalda* [v. Bret-walda, brýten-walda]; Cymēnus:—For example, v. Cymēnes óra.

Cymēnes óra, an; *m.* Cymen's *shore, near Wittering, Sussex*; Cymēni litus, qui ibi naves ad terram appulit. Nunc nomen amisit, sed fuisse prope *Wittering*, in agro Sussexiensi, Charta Donatiōnis quam Cedwalla Rex Ecclesiæ Selsiensi fecit, planissĭme convincit, Camd. Camden and, after him, Gibson say, in the preceding Latin, this place was near Wittering on the coast of Sussex. They rely on a Charter which Kemble [Cod. Dipl. 992] has marked as spurious, but which was no doubt constructed with a regard for probability. In this Charter [Cod. Dipl. 992; A. D. 683; Kmbl. v. 33, 22] the name occurs as Cumeneshora, a form which countenances Ingram's guess that Shoreham is the place; quasi *Cymene*shoreham, v. Chr. Erl. 281, A. D. 477:—Hér, A. D. 477, com [MS. cuom] Ælle on Bretonlond, and his iii suna, Cymen, and Wlencing, and Cissa, mid iii scipum, on ða stówe ðe is nemned Cymēnes óra, and ðǽr ofslógon monige Wealas, and sume on fleáme bedrifon on ðone wudu ðe is genemned Andredes leáge *in this year*, A. D. 477, *Ælle came to Britain, and his three sons, Cymen, and Wlencing, and Cissa, with three ships, at the place which is named Cymen's shore, and there slew many Welsh, and drove some in flight into the wood which is named Andredsley*, Chr. 477; Erl. 12, 28–32.

cym-líc; *adj. Comely, convenient, lovely, beautiful, splendid*; aptus, commŏdus, splendidus:—Hierusalem, ðū wǽre swá swá cymlíc ceaster getimbred *Jerusalem, thou wert built as a beautiful city*, Ps. Th. 121, 3: Exon. 108 b; Th. 415, 24; Rä. 34, 2.

cym-líce; *comp.* -lícor; *adv. Conveniently, fitly, beautifully, splendidly*; commode, apte, Andetaþ Drihtne, and his ēcne naman cīgaþ cymlíce *confitemini Domino et invocate nomen ejus*, Ps. Th. 104, 1: 98, 7. Cymlícor ceól gehládenne *a more fitly laden ship*, Andr. Kmbl. 721; An. 361: Beo. Th. 75; B. 38.

cym-lícor *more aptly* or *fitly*, Andr. Kmbl. 721; An. 361: Beo. Th. 75; B. 38; *comp. of* cym-líce.

cymst, cymest *comest*, Cd. 203; Th. 252, 28; Dan. 585: Beo. Th. 2769; B. 1382; *2nd pres. sing. of* cuman.

cymþ, cymeþ *comes*, Cd. 17; Th. 20, 26; Gen. 315: Beo. Th. 4123; B. 2058; *3rd sing. pres. of* cuman.

cyn *the chin*; mentum. v. cin.

CYN, cynn, es; *n.* **I.** *every being of one kind, a kindred, kind, race, nation, people, tribe, family, lineage, generation, progeny*, KIN; genus, gens, natio, populus, stirps, tribus, familia, natales, origo, generatio, proles, progenies:—Ðæt hie ne mōton ǽgnian mid yrmþum Israhēla cyn *that they may not hold in misery the race of Israel*, Cd. 156; Th. 194, 24; Exod. 265: 170; Th. 213, 21; Exod. 555. Monna cynn *hominum genus*, Exon. 20 b; Th. 55, 23; Cri. 888: 98 b; Th. 370, 1; Seel. 50: Cd. 212; Th. 261, 33; Dan. 735. Eorþan cynn *terræ tribus*, Ps. Th. 71, 18. Eal engla cynn *all the race of angels*, Exon. 75 a; Th. 281, 10; Jul. 644. Eall gimma cynn *all kinds of gems*, Andr. Kmbl. 3037; An. 1521. Fór cynn æfter cynne *tribe went after tribe*, Cd. 161; Th. 200, 3; Exod. 351. Ðis cynn ne byþ ūtadryfen *hoc genus non ejicitur*, Mt. Bos. 17, 21. Ðæt wíf wæs hǽðen, Sirofenisces cynnes *erat mulier gentīlis, Syrophœnissa genere*, Mk. Bos. 7, 26. Lā nǽddrena cyn *progenies viperarum*, Mt. Bos. 3, 7. Of cynne on cynn *from generation to generation*; a progenie in progeniem, Ps. Th. 84, 5: 88, 1. Adames cyn *the race of Adam*, Cd. 222; Th. 289, 35; Sat. 408: Exon. 22 a; Th. 59, 33; Cri. 961. Ymb fisca cynn *de piscium genere*, Exon. 96 b; Th. 360, 6; Wal. 1. DER. cyn-recen, cynn-recenis, -ren, -ryn: ælf-cyn, -cynn, átor-, cyne-, deór-, earfoþ-, engel-, eormen-, eorþ-, fæderen-, feorh-, fífel-, fisc-, fleóh-, from-, frum-, fugel-, fugol-, gim-, gum-, hǽðen-, helle-, heoloþ- [= hæleþ-], hwǽte-, lǽce-man-, médren-, óm-, orf-, sigor-, treó-, wǽpned-, wer-, wyrm-, wyrt-. **II.** *in grammar,—Gender*; genus:—Syndon twā cynn,—masculinum, ðæt is werlíc, and *femininum*, wíflíc. Werlíc cynn biþ ðes wer *hic vir: there are two genders*,—masculine, *that is manlike, and* feminine, *womanlike. Masculine gender is* ðes wer *this man*, Ælfc. Gr. 6; Som. 5, 27, 28. Ælc nýten biþ oððe he, oððe heó *every animal is either* he, *or* she, 6; Som. 5, 34. Neutrum is náðor cynd, ne werlíces, ne wíflíces *neuter is neither kind, neither of male nor of female*, 6; Som. 5, 32. Ðis gebýraþ oftost to náðrum cynne, swá swá is ðis word *hoc*

verbum: this oftest belongeth to the neuter gender, as is ðis word *this word*, 6; Som. 5. 35. Twíflíces cynnes ðæt is *dubii generis*, 6; Som. 5, 46. Sume naman synd óðres cynnes on ánfealdum getele, and óðres cynnes on mænigfealdum getele *some nouns are of one gender in the singular number, and of another gender in the plural number*, 13; Som. 16, 25. The *m. f. n.* occur in the following sentence, indicated by the articles se, seó, ðæt :—Seó sáwel ys má ðonne se líchama, and se líchama mā ðonne ðæt reáf *anima plus est quam esca, et corpus plus quam vestimentum*, Lk. Bos. 12, 23. **III.** *a sex*; sexus:—Hwæðeres cynnes bearn heó cennan sceal *of which sex she shall bear a child*, Lchdm. iii. 144, 6. [Wyc. kyn *family, generation*: Chauc. kin: Piers P. kynne: R. Glouc. R. Brun. kyn: Laym. cun, kun *race, progeny, kind*: Orm. kin: O. Sax. kunni, cunni, *n. race*: Dut. kunne, *f. gender*: Kil. konne, kunne *genus, species, sexus*: O. Frs. ken, kin, kon, *n. genus*: M. H. Ger. künne, *n. family*: O. H. Ger. kunni, *n. genus, gens*: Dan. kjön, *n. genus*: Swed. kön, *n. sex*; kynne, *n. disposition*: Icel. kyn, *n. a kind, kin*: Lat. genus, gens: Grk. γένος: Sansk. janus *gens*.]

cyn, cynn; *adj. Akin, suitable, fit, proper*; congruus, condignus:—Ðæt is cyn *that is proper* or *reasonable*, Bt. 33, 1; Fox 122, 4. Swá hit cynn [cyn Cot.] was *as was suitable* or *fit*, 35, 4; Fox 162, 24. Swylce hit kyn [cyn MS. B; cynn H.] sié *as it may be right*, L. In. 42; Th. i. 128, 11. Hit ys cyn *it is proper*, Ps. Th. 29, 11: 9, 34: 138, 20.

cyncg *a king*, L. E. G. pref; Th. i. 166, 3. v. cyning.

CYND, es; *n.* **I.** *nature*, KIND; natura:—Gif hió hire cynd healdan wile *if she desire to retain her nature*, Bt. 35, 4; Fox 160, note 21, MS. Cot. **II.** *a sort, gender*; natura, genus:—Neutrum is náðor cynd, ne werlíces, ne wíflíces *neuter is neither sort* [gender], *neither of male nor of female*, Ælfc. Gr. 6, 3; Som. 5, 32. [Prompt. keende, kyynde *genus*: Wyc. kynde *nature*: Piers P. kynde *nature, race, kind*: Laym. i-cunde *nature, kind, race*: Orm. kinde *nature, kind, race*: O. Sax. kind, *n. a child*: Dut. kind, *n. a child*: Ger. kind, *n. a child*: M. H. Ger. kint, *gen.* kindes, *n. a child*: O. H. Ger. kind, kint, *n. proles*: Icel. kind, *f. species, race, kind*: Lat. gent-em, *acc. of* gens.] DER. ge-cynd.

cynde; *adj. Natural, innate, inborn*; naturalis, innatus, ingenitus:—Cniht weóx and þág swá him cynde wǽron *the boy waxed and thrived as to him was natural*, Cd. 132; Th. 167, 26; Gen. 2771. DER. ge-cynde, un-, unge-.

cynde-líc; *adj. Natural*, KINDLY; naturalis, ingenitus:—Sídra gesceafta cræftas cyndelíce *the kindly powers of wide-spread creatures*, Exon. 92 b; Th. 346, 27; Sch. 5. DER. ge-cyndelíc, unge-.

cyne-, used in compounds, signifying *kingly, royal, special*; regius, præ-. v. cyne-bǽnd, -bearn, -boren, -bót, -botl, -cyn, -dóm, etc.

cyne; *adj. Bold, brave*; audax:—Cyninga cynost *bravest of kings*, Ps. C. 50, 3; Ps. Grn. ii. p. 276, 3. DER. searo-cýne. v. cēne.

cýne, an; *f. A chink, fissure*; rima:—Ðæs leóhtes scíma þurh ða cýnan ðære dura ineóde *the glare of the light came through the chinks of the door*, Bd. 4, 7; S. 575, 19.

cyne-bǽnd, es; *m.* [bend, bǽnd *a band, chaplet, crown*] *A royal crown, a diadem*; regia corona, diadema = διάδημα, Som. Ben. Lye.

cyne-bearn, es; *n. A kingly child, royal offspring*; regius puer, regia proles:—Ne mihton oncnáwan ðæt cynebearn *they might not acknowledge the royal child*, Andr. Kmbl. 1131; An. 566. Wuldres cynebearn *the royal child of glory*, Menol. Fox 316; Men. 159: Cd. 82; Th. 102, 23; Gen. 1704.

cyne-boren; *part. Of royal birth*; regia stirpe natus, M. H. 12 a.

cyne-bót, e; *f.* [bót *boot, compensation*] *A king's compensation* or *recompense*; regis compensatio:—Gebírаþ seó cynebót ðām leódum *the king's compensation belongs to the people*, L. Wg. 1; Th. i. 186, 4: L. M. L; Th. i. 190, 8.

cyne-botl, es; *n.* [botl *a dwelling*] *A kingly dwelling, a palace*; palatium, Wrt. Voc. 86, 27.

cyne-cyn, -cynn, es; *n.* [cyne *regius, regalis*; cyn, cynn, *gens, stirps, familia*] *A royal race, royal lineage, royal offspring* or *family*; gens regia, proles regia, stirps *vel* familia regia:—Of Francena cynecynne *de gente Francorum regia*, Bd. 1, 25; S. 486, note 32: 2, 14; S. 518, 3. He wæs hiora cynecynnes *he was of their royal race*, Bt. Met. Fox 26, 83; Met. 26, 42. He wæs cynecynnes *he was of royal lineage*, Bt. 38, 1; Fox 194, 14: Bd. 3, 18; S. 546, 39, col. 1: L. Wg. 1; Th. i. 186, 18.

cyne-dóm, es; *m.* [dóm *power, dominion*] *A royal dominion* or *power, kingdom, realm*; imperium, regnum, sceptrum, potestas :— Cynedóm *sceptrum*, Ælfc. Gl. 69; Som. 69, 127; Wrt. Voc. 42, 7. We willaþ ðæt án cynedóm fæste stande æfre on þeóde *we will that one kingship stand fast for ever in the nation*, L. N. P. L. 67; Th. ii. 302, 8. Hanna wæs mid ungemete ðæs cynedómes gyrnende *Hanno had an immoderate longing for the kingdom*, Ors. 4, 5; Bos. 81, 43: L. Wg. 1; Th. i. 186, 4: Ps. C. 50, 149; Ps. Grn. ii. 280, 149. Rūmes cynedómes *augustæ potestatis*, Mone B. 3931. For ðam cynedóm *for the kingdom*, L. M. L; Th. i. 190, 6. Claudius Orcadas eáland to Rōmwara cynedóme geþeódde *Claudius Orcadas insulas Romano adjecit imperio*, Bd. 1, 3; S. 475, 7: Chr. 47; Erl. 6, 26. He ðone cynedóm ciósan wolde *he would choose*

the kingdom, Beo. Th. 4741; B. 2376: L. Eth. ix. 42; Th. i. 350, 3. Ðætte ryhte cynedómas þurh úre folc gesæstnode wǽron *that just royal governments might be settled throughout our people*, L. In. pref; Th. i. 102, 9.

cyneg *a king*, Jos. 10, 5: Homl. Th. ii. 540, 17. v. cyning.

cyne-geard *a royal wand, sceptre*, Ælfc. Gl. 68; Som. 69, 127; Wrt. Voc. 42, 7. v. cyne-gyrd.

cyne-gerd *a sceptre*, Ælfc. Gl. 6; Som. 56, 47; Wrt. Voc. 18, 2. v. cyne-gyrd.

cyne-gerela, an; m. [gerela *a robe*] *A kingly robe*; regius vestitus:—Gif mon wolde him awindan of ðæs cynegerelan [MS. -gerelum] *if any one would strip off from him these kingly robes*, Bt. Met. Fox 25, 45; Met. 25, 23.

cyne-gewǽdu; pl. n. [gewǽde *a garment, robe*] *Royal robes*; regiæ vestes:—He onféng cynegewǽdum *he took the royal robes*, Bd. 1, 6; S. 476, 19.

cyne-gild, -gyld, es; n. [gild *compensation*] *A king's compensation*; regis compensatio:—To bóte on cynegilde [-gylde MS. H.] *as offering for the king's compensation*, L. M. L. Th. i. 190, 7.

Cynegils, es; m. *Cynegils, sixth king of the West Saxons*; Cyne-gilsus:—Cynegilses, gen. Chr. Erl. 2, 20: Chr. 688; Erl. 42, 10. Hér, A.D. 611, Cynegils féng to ríce on Wesseaxum, and heóld xxxi wintra *here, Cynegils succeeded to the kingdom of the West Saxons, and held it thirty-one years*, 611; Erl. 20, 33. Hér, A.D. 635, Cynegils [MS. Kynegils] wæs gefullod fram Byríne ðam biscope on Dorcaceastre, and Oswold Norþhymbra cining his onféng *here, Cynegils was baptized by Birinus the bishop of Dorchester, and Oswold, king of Northumbria, was his sponsor*, 635; Erl. 25, 33. Cynegils onféng ǽrest fulwihte Wesseaxna cyninga *Cynegils was the first of the West Saxon kings who received baptism*, Erl. 2, 16.

cyne-gód; adj. *Excellent, noble*; præstans, nobilis:—Him cynegódum *to him excellent*, Cd. 78; Th. 96, 5; Gen. 1590. Him ða cynegóde on Carran ædelinga bearn eard genámon *then the noble children of men took them a dwelling in Harran*, 83; Th. 104, 16; Gen. 1736: 182; Th. 228, 2; Dan. 196: 195; Th. 243, 8; Dan. 433: Exon. 85 b; Th. 321, 34; Wíd. 56.

cyne-gold, es; n. *Royal gold, a crown*; diadema = διάδημα, corona:—Þeódnes cynegold sóþfæstra gehwone glengeþ *the Lord's crown shall adorn each of the just*, Exon. 64 b; Th. 238, 17; Ph. 605.

cyne-gyrd, -geard, -gerd; f. [gyrd *a rod, wand*] *A royal wand, sceptre*; sceptrum:—Cynegyrd *sceptrum*, Wrt. Voc. 72, 55. Cynegeard *sceptrum*, Ælfc. Gl. 68; Som. 69, 127; Wrt. Voc. 42, 7. Cynegerd *sceptrum*, 6; Som. 56, 47; Wrt. Voc. 18, 2. Hí to ðæs caseres cyne-gyrde gebugon *they submitted to the emperor's sceptre*, Homl. Th. ii. 502, 16.

cyne-hád, es; m. [hád *form, condition*] *A royal personage* or *condition, dignity, kinghood*; regia persona *vel* dignitas:—Ðæt se cynehád [MS. cynehade] ðæs hálgan weres éce gemynd hæfde *ut regia viri sancti persona memoriam haberet æternam*, Bd. 3, 11; S. 535, 30, note. Ic Ælfréd, gifendum Criste, mid cynehádes mǽrnesse geweorþaþ hæbbe cúþlíce ongiten *I Alfred, adorned, by the grace of Christ, with the dignity of a king have well perceived*, Greg. Dial. MS. Hat. fol. 1, 1.

cyne-hám, es; m. [hám *a house, dwelling, home*] *A royal residence*; regia villa:—On ðam cyneháme ðe is gecýged Bearwe *at the royal residence which is called Barrow*, Cod. Dipl. 90; A.D. 716-743; Kmbl. i. 109, 15. On his ágenum cynehámum *in his own royal residences*, 598; A.D. 978; Kmbl. iii. 138, 7.

cyne-helm, -healm, es; m. [helm *a crown*] *A crown, diadem*; corona, diadema:—Cynehelm *corona, diadema*, Ælfc. Gl. 51; Som. 66, 14; Wrt. Voc. 35, 5: Mone B. 2166. Cynehealm *diadema*, Wrt. Voc. 74, 56. Wundon cynehelm of þornum, and asetton ofer hys heáfod *plectentes coronam de spinis posuerunt super caput ejus*, Mt. Bos. 27, 29; Jn. Bos. 19, 2, 5. Cynehelme *corona*, Mone B. 3019. For cynehelme *for a royal diadem*, Homl. Blick. 23, 34.

cyne-hláford, es; m. [hláford *a lord*] *A royal lord, sovereign lord, king*; regius *vel* supremus dominus, rex:—Be his cynehláfordes geþafunge *with the permission of his royal lord*, Cod. Dipl. 593; A.D. 965-975; Kmbl. iii. 127, 8. Æt his leófan cynehláforde Eádgáre cyninge *from his dear sovereign lord king Edgar*, 583; A.D. 963-975; Kmbl. iii. 111, 26: 598; A.D. 978; Kmbl. iii. 138, 22: Chr. 1016; Erl. 158, 5, 17, 29. Ðæt we ealle ánum cynehláforde holdlíce hýran *that we all faithfully obey one sovereign lord*, L. Eth. vi. 1; Th. i. 314, 10. Utan ǽnne cynehláford holdlíce healdan *let us faithfully support one sovereign lord*, v. 35; Th. i. 312, 21: ix. 44; Th. i. 350, 12.

cynelec; adj. *Royal*; regalis:—In ðæm cynelecan túne *in the royal town*, Bd. 3, 17; S. 543, 21, col. 2. v. cyne-líc.

cyne-líc, cynelíc, cynelec; adj. *Kingly, royal, regal, belonging to the state, public*; regius, regalis, publicus:—Eádward cyng man bebyrigde bútan ǽlcum cynelícum wurþscipe *king Edward was buried without any kingly honour*, Chr. 979; Erl. 129, 3. Ðæt is cynelíc þing *that is a royal thing*, Exon. 124 b; Th. 478, 26; Ruin. 48. Wæs ðæs ylcan

mynstres abbudisse on ða tíd seó cynellíce fǽmne Ælflǽd *præerat quidem tunc eidem monasterio regia virgo Ælbflæd*, Bd. 4, 26; S. 603, 3. Ðæt se cynelíca hád ðæs hálgan weres éce gemynd hæfde *ut regia viri sancti persona memoriam haberet æternam*, 3, 11; S. 535, 30. In ðæm cyne-lecan túne *in the royal town*, Bd. 3, 17; S. 543, 21, col. 2. Cynelícre *publica*, Glos. Prudent. Recd. 145, 30. Cynelíco getimbro and ánlípie *publica ædificia et privata*, Bd. 1, 15; S. 483, 45. Chaldéas cynelícan getimbro mid fýre fornámon [MS. fornaman] *the Chaldeans destroyed the royal buildings with fire*, 1, 15; S. 483, 42. He onféng cynelícum gewǽdum and com on Breotone *he took the royal robes and came into Britain*, 1, 6; S. 476, 19, note. Wið ða cynelícan ádle ðe man *auriginem* nemneþ *ad morbum regium, hoc est, auriginem* [= auruginem], Herb. 87, 1; Lchdm. i. 190, 14. Cynelíc reáf *trabea*, Ælfc. Gl. 63; Som. 68, 122; Wrt. Voc. 40, 30. Cynelíc [MS. kyne-] botl *palatium*, 81; Som. 73, 9; Wrt. Voc. 47, 16.

cyne-líce; adv. *Royally*; regie:—Ðú miltse on us gecýþ cynelíce *shew mercy royally on us*, Exon. 10 a; Th. 10, 24; Cri. 157.

cynelíc-nys, -nyss, e; f. *Royalty, as shewn in the deportment, a kingly likeness*; regia dignitas:—For his cynelícnysse ge módes geonsýnes *for his kingliness both of his mind and appearance*, Bd. 3, 14; S. 540, 9.

cynellíc *kingly, royal*, Bd. 4, 26; S. 603, 3. v. cyne-líc.

Cyne-mǽres ford, es; m. [*Flor.* Kimeresford: cyne *royal*; mǽre *a mere*; ford *a ford*] KEMPSFORD, *Gloucestershire*:—Rád Æðelmund alderman ofer æt Cynemǽresforda *alderman Æthelmund rode over at Kempsford*, Chr. 800; Erl. 60, 6.

cyne-ríce, -rýce, es; n. *A royal region* or *possession, a kingdom, realm*; regnum:—Secg monig wyscte ðæt ðæs cyneríces ofercumen wǽre *many a warrior wished that there was an end of that kingdom*, Exon. 100 b; Th. 378, 34; Deór. 26. Féng his bearn to cyneríce *his child succeeded to the kingdom*, Chr. 975; Erl. 126, 5; Edg. 31: 1066; Erl. 201, 1: 1076; Erl. 215, 2. On ðý cyneríce be súþan Temese *in the kingdom south of the Thames*, 871; Erl. 76, 9. On cynerýce *in the realm*, Exon. 53 b; Th. 187, 23; Az. 35. He ge-eóde ealle ða cynerícu ðe on Crécum wǽron *he over-ran all the kingdoms which were in Greece*, Ors. 3, 7; Bos. 58, 39. Cyneríca mǽst *greatest of kingdoms*, Exon. 85 a; Th. 321, 1; Wíd. 39. Ðæt he ealdordóm ágan sceolde ofer cynerícu *that he should possess eldership over the kingdoms*, Cd. 158; Th. 198, 5; Exod. 318: Bt. Met. Fox 26, 12; Met. 26, 6.

cyne-róf; adj. [róf *famous*] *Royally famous, noble*; nobilis:—Wolde ic ánes to ðé, cyneróf hæleþ, cræftes neósan *I would inquire of thee of one art, noble hero*, Andr. Kmbl. 967; An. 484: 1169; An. 585. Cirdon cynerófe *the noble ones turned*, Judth. 12; Thw. 26, 6; Jud. 312: 11; Thw. 24, 21; Jud. 200.

cyne-scipe, es; m. *Kingship, royalty, honour*; regia dignitas:—Hæbbe ic mínes cynescipes gerihta *I may have my rights of royalty*, L. Edg. S. 2; Th. i. 272, 27. Me to fullum cynescipe *to my perfect royalty*, 2; Th. i. 272, 25. Him sylfum to cynescipe *in honour of himself*, L. Edg. i. prm; Th. i. 262, 4: L. C. E. prm; Th. i. 358, 6.

cyne-setl, es; n. [setl *a seat*] *A royal seat, throne*; imperii sedes, solium:—Constantinopolis is nú ðæt heáhste cynesetl ealles eástríces *Constantinople is now the chief royal seat of all the eastern empire*, Ors. 3, 7; Bos. 61, 11. Ðe sit on his cynesetle *qui sedet in solio ejus*, Ex. 11, 5.

cyne-stól, es; m. [cyne *royal*, stól *a seat, stool*] *A royal throne* or *dwelling, chief city, capital*; thronus, urbs regia, arx, metropolis:—On his cynestóle *on his kingly throne*, Exon. 25 b; Th. 75, 6; Cri. 1217: Elen. Kmbl. 659; El. 330. Of cynestólum *from royal seats*, Exon. 96 a; Th. 358, 22; Pa. 49. Constantinopolis is Créca cynestól *Constantinople is the royal dwelling-place of the Greeks*, Bt. 1; Fox 2, 22: Ors. 3, 9; Bos. 65, 45. Cynestóle Creácas wióldon *the Greeks possessed the metropolis*, Bt. Met. Fox 1, 95; Met. 1, 48: Menol. Fox 208; Men. 105. We becómon to ðam cynestóle, ðǽr getimbred wæs tempel Dryhtnes *we came to the royal city, where the temple of the Lord was built*, Andr. Kmbl. 1332; An. 666. Ðǽr heó ǽfre forþ wunian móten cestre and cynestól *where they may evermore possess cities and a kingly throne*, Cd. 220; Th. 283, 1; Sat. 298: Chr. 975; Erl. 125, 31. Sancta Hierusalem, cynestóla cyst *holy Jerusalem, choicest of royal cities*, Exon. 8 b; Th. 4, 11; Cri. 51.

cyne-strǽt, e; f. *A royal street* or *road*; regia via, publicum, Cot. 153.

Cynete, an; f.   **I.** *the river* KENNET *which rises in Wiltshire*; fluvii nomen qui originem suam habet in agro Wiltoniensi:—Ǽrest on Cynetan, ðæt up andlang strémes...ðæt eft innan Cynetan strém *first to the Kennet, then up along the stream...then again to the river Kennet*, Cod. Dipl. 792; A.D. 1050; Kmbl. iv. 122, 21, 26: Cod. Dipl. Apndx. 378; A.D. 939; Kmbl. iii. 413, 22, 30: Cod. Dipl. 1120; A.D. 939; Kmbl. v. 238, 17, 25, 35: 1152; A.D. 944; Kmbl. v. 300, 16, 18: 1199; A.D. 956; Kmbl. v. 376, 6, 16: 1282; A.D. 984; Kmbl. vi. 118, 1, 6.   **II.** KENNET, *a village on the river Kennet in Wiltshire*; villæ nomen in agro Wiltoniensi:—Wæs fyrd gesomnod æt Cynetan *a force was assembled at Kennet*, Chr. 1006; Erl. 140, 23.

cyne-þrym; *gen.* -þrymmes; *m.* [þrym *a multitude, majesty, glory*] *A kingly host, royal majesty* or *glory*; regia multitudo, regis majestas :— Mid cyneþrymme *with a kingly host*, Cd. 209; Th. 260, 8; Dan. 706: Exon. 120 b; Th. 462, 12; Hö. 51. He cwom on cyneþrymme *he came in royal majesty*, Ps. Th. 95, 12. Ryhtfremmende cyneþrym cýðaþ *the righteous doers shall proclaim the royal majesty*, Exon. 65 a; Th. 240, 5; Ph. 634: Andr. Kmbl. 2645; An. 1324. Ðu me gecýðdest cyneþrymma wyn *thou declaredst to me joy of kingly glories*, Exon. 120 b; Th. 463, 23; Hö. 74.

cyne-wîse, an; *f.* [wîse *an affair*] *The state, republic, commonwealth*; respublica :— Se næht freomlîces ongan on ðære cynewîsan *he began nothing profitable in the state*, Bd. 1, 3; S. 475, 21. Rehte ða cynewîsan *rempublicam rexit*, 1, 5; S. 476, 8.

cyne-widðe, an; *f. A royal wreath, diadem*; redimiculum :— Cynewiddan *redimicula*, Mone B. 6270: Cot. 185.

cyne-word, es; *n.* [word *a speech*] *A proper speech* or *word*; proprium verbum :— Mon cýðe cynewordum, hû se cuma hâtte *let a man make known in fitting words, how the guest is called*, Exon. 112 b; Th. 430, 29; Rä. 44, 16.

Cynewulf, es; *m. An Anglo-Saxon poet, who has preserved his name in Runes, in his poem on Elene's Recovery of the Cross.* Mr. Kemble will best describe his own discovery.—In the Vercelli MS. is contained a long poem on the finding of the Cross by the Empress Helena [=Elene]. After the close of the poem, and apparently intended as a tail-piece to the whole book, comes a poetical passage, in which the author principally refers to himself, and after a reference to his own increasing age and the change from the strength and joyousness of youth, he breaks out, in the 15th Canto, into a moralizing strain, in which he concludes his work. The following thirty lines, containing Runes, form a portion of this Canto :—

Â wæs sæc ôþ-ðæt,	*Ever was contest till then,*
cnyssed cearwelmum	*with waves of sorrow tossed*
ᚻ [cên] drûsende,	C [*the torch*] *sinking,*
ðeáh he, in medohealle	*though he, in meadhall*
mâþmas, þege	*treasures, handled*
æplede gold,	*appled gold,*
ᛡ [yr] gnornode,	Y [*sorrow*] *he mourned,*
ᚾ [nýd] gefêra,	N [*need*] *his consort,*
nearu sorge dreáh,	*narrow sorrow he suffered,*
enge rûne,	*a close rune,*
ðær him ᛖ [êh] fôre	*where* E [*the horse*] *before him*
mîlpaðas mæt,	*measured the mile-paths,*
môdig þrægde	*proudly hastened*
wîrum gewlenced.	*with wires adorned.*
ᚹ [wên] is geswîþrad,	W [*hope*] *is overpowered,*
gomen æfter gearum,	*my joy in my old age,*
geógoþ is gecyrred	*youth is turned back*
ald onmedla.	*my old pride.*
ᚢ [ûr] wæs geára	U *I was of old*
geógoþhâdes glæm,	*a gleam of youth,*
nû synt geárdagas	*now are the days of my life*
æfter fyrstmearce	*after the appointed space*
forþgewitene,	*departed,*
lîfwynne geliden,	*the joy of life flowed away,*
swâ ᛚ [lagu] toglîdeþ,	*as* L [*lake or water*] *glideth,*
flôdas gefýsde.	*the floods that hasten.*
ᚠ [feoh] æghwam biþ	F [*wealth*] *will be for every man*
læne under lyfte,	*failing under the heaven,*
landes frætwe	*the ornament of the land*
gewîtaþ under wolcnum.	*will depart under the welkin.*

Elen. Kmbl. 2512-2541; El. 1257-1272.

The extreme rudeness and abruptness of these lines, and the apparent uselessness of the Runes, led me to suspect that there was more in them than merely met the eye. This I found to be the case; for, on taking the Runes out of the context, using them as single letters and uniting them in one word, they supplied me with the name **CYNEWULF**, undoubtedly no other than the author of the poems. I cannot here bestow space upon a long argument to shew who this *Cynewulf* was. I believe him to have been the Abbot of Peterborough of that name, who flourished in the beginning of the eleventh century, who was accounted in his own day a celebrated poet, both in Latin and Anglo-Saxon, whose works have long been lost, but whose childish ingenuity has now enabled us with some probability to assign to him the authorship of the Vercelli and Exeter Codices, Archæologia, vol. xxviii. 1840, by Kemble, pp. 327–372. The Reverend Jn. Earle, M. A. etc. Rector of Swanswick, with some pertinent remarks, supposes Cynewulf to be the same person as Cyneweard. v. Chr. Erl. Introduction, pp. xx–xxii.

cyng *a king*, Chr. 664; Erl. 34, 20: 894; Erl. 91, 32: L. Ath. iv. pref; Th. i. 220, 1. v. cyning.

cynge *a king*, L. Edg. S. 1; Th. i. 270, 7. v. cyning.

Cynges tûn, es; *m.* [cynges tûn *king's town*] KINGSTON; regia villa :— Aðelstân wæs to cynge æt Cynges tûne gehâlgod *Athelstan was consecrated king at Kingston*, Chr. 924; Th. 199, 8, col. 1: 979; Th. 234, 10, col. 2. Æt Cyninges tûn *at Kingston*, Chr. 979; Th. 235, 9, col. 1. v. Cinges tûn, Cyninges tûn.

cyning, cyng, es; *m.* [cyn *people*, -ing *originating from, son of*]. **I.** *a king, ruler, emperor*; rex, imperator. He is the representation of the people, and springs from them, as a son does from his parents. The Anglo-Saxon king was elected from the people; he was, therefore, the king of the people. He was the chosen representative of the people, their embodiment, the child, not the father of the people. He was not the lord of the soil, but the leader of his people. He completed the order of freemen, and was the summit of his class. As the freeman [ceorl] was to the noble [æðele], so was the noble to the king. The Anglo-Saxon king was the king of a tribe or of a people, but never of the land. We read of kings of the West Saxons or of the Mercians, but not of Wessex or of Mercia. The king was, in truth, essentially one with the people, by them and their power he reigned; but his land was like theirs, private property. It was not the feudal system, and was never admitted that the king was owner of all the land in a country :—Se cyning mildelîce onfêng *the king received* [*him*] *gladly*, Ors. 1, 8; Bos. 30, 44. Se Iudêa cyning *the king of the Jews*; ὁ βασιλεὺς τῶν Ἰουδαίων, Mt. Bos. 2, 2. Saul wæs gecoren ærest to cyninge on Israhêla þeóde *Saul was first chosen king of the people of Israel*, Ælfc. T. 13, 3. Eart ðû wîtodlîce cyning *ergo rex es tu?* οὐκοῦν βασιλεὺς εἶ σύ; Jn. Bos. 18, 37. Cyninges botl *a king's dwelling, palace*, Bd. 2, 14; S. 518, 18. Cyninga [MS. cininga] bôc *the book of kings*, Ælfc. T. Grn. 6, 38: 8, 3. Cyninga [MS. kyninga] byrgen *a burying-place of kings*; mausoleum, bustum, Ælfc. Gl. 85; Som. 74, 3; Wrt. Voc. 49, 27. Maximian, ârleás cyning *Maximian, the wicked emperor*, Exon. 65 b; Th. 243, 1; Ju'. 4. **2.** *a spiritual King, God, Christ*; Deus, Christus :—Heofona Cyning *the King of heaven*, Andr. Kmbl. 3008; An. 1507: 3017; An. 1511: Cd. 137; Th. 172, 18; Gen. 2846. Crist is ealra cyninga Cyning *Christ is King of all kings*, Homl. Th. ii. 588, 9: Exon. 9 b; Th. 9, 17; Cri. 136: 11 a; Th. 14, 6; Cri. 215: Andr. Kmbl. 1955; An. 980. **3.** *the devil*; diabôlus, satânas :—Hellwarena cyning *the king of hell's inhabitants*, Exon. 70 a; Th. 261, 28; Jul. 322. Se ofermôda cyning, Satan *the haughty king, Satan*, Cd. 18; Th. 22. 9; Gen. 338. **II.** *Anglo-Saxon kings were at first elected from a family or class, by Witena gemôt the assembly of the wise.* **2.** *fidelity was sworn to them by the people,* in the following words :—Ðus man sceal swerigean hyld-âþas. 'On ðone Drihten, ðe ðes hâligdôm is fôre hâlig, ic wille beón N. hold and getrîwe, and eal lufian ðæt he lufaþ, and eal ascûnian ðæt he ascûnaþ, æfter Godes rihte and æfter woroldgerysnum, and næfre, willes ne gewealdes, wordes ne weorces, ôwiht dôn ðæs him lâþre biþ; wið ðam ðe he me healde swâ ic earnian wille, and eall ðæt læste ðæt uncer fôrmæl wæs, ðâ ic to him gebeáh and his willan geceás *thus shall a man swear oaths of fidelity* [or *homage*]. *By the Lord, before whom this relic is holy, I will be to N. faithful and true, and love all that he loves, and shun all that he shuns, according to God's law, and according to the world's principles, and never, by will nor by force, by word nor by deed, do aught of what is loathful to him; on condition that he keep me as I am willing to deserve, and all that fulfil that our agreement was, when I submitted to him and chose his will,* L. O. 1; Th. i. 178, 2–9. If this was taken in A. D. 924, it was not long before the power of the king was limited, for we have the following oath administered to Æðelréd, when he was consecrated king at Kingston in A. D. 978, as is stated in the Chronicle,—On ðys geáre wæs Æðelréd to cininge gehâlgod æt Cinges tûne *in this year Æthelred was consecrated king at Kingston*, Chr. 978 [MS. 979]; Th. 234, 9, col. 1. **3.** *the king took a corresponding oath to his people.* The words of the king's oath are,—Ðis gewrit is gewriten, stæf be stæfe, be ðam gewrite ðe Dûnstân arcebisceop sealde ûrum hlâforde æt Cinges tûne â on dæg ðâ hine man hâlgode to cinge, and forbeád him ælc wedd to syllanne bûtan ðysan wedde, ðe he up on Cristes weofod lêde, swâ se bisceop him dihte. 'On ðære hâlgan Þrýnnesse naman, Ic þreó þing behâte cristenum folce, and me underþeóddum :—*Ân ærest,* ðæt ic Godes cyrice and eall cristen folc mînra gewealda sôðe sibbe healde. *Oðer* is, ðæt ic reáflâc and ealle unrihte þing eallum hâdum forbeóde. *Þridde,* ðæt ic behâte and bebeóde on eallum dômum riht and mildheortnisse, ðæt us eallum ærfæst and mildheort God þurh ðæt his êcean miltse forgife, se lifaþ and rîxaþ' *this writing is copied, letter for letter, from the writing which archbishop Dunstan delivered to our lord at Kingston on the very day when he was consecrated king, and he forbade him to give any other pledge but this pledge which he laid upon Christ's altar, as the bishop instructed him. 'In the name of the Holy Trinity, three things do I promise to this christian people, my subjects. First, that I will hold God's church and all the christian people of my realm in true peace. Second, that I will forbid rapine and all injustice to men of all conditions. Third, that I promise and enjoin justice and mercy in all judgments, whereby the just and merciful God may give us all his eternal favour who liveth*

and reigneth,' Relq. Ant. W. ii. 194. **4.** from the freedom with which the educated spoke of the Doom's Day Survey of William the Conquerer, indicating their love of freedom, *we have no reason to suppose this oath was the first oath taken by kings in our limited monarchy.* The spirit of the monks may be seen in the following extract from the Chronicle:—Willelm, Engla landes cyng, ðe ða wæs sittende on Normandige, forðig he áhte ægðer ge Engla land ge Normaudige ... sende ða ofer eall Engla land into ælcere scíre his men ... Swá swýðe nearwelíce he hit lett út aspyrian, ðæt næs án ælpig híde, ne án gyrde landes, ne, furðon, hit is seeame to tellanne, ac hit ne þuhte him nán sceame to dónne, án oxa [MS. oxe], ne án cú, ne án swín næs belyfon, ðæt næs gesæt on his gewrite, and ealle ða gewrita wæron gebroht to him syððan *William, king of England, who was then resident in Normandy, for he owned both England and Normandy ... then sent his men over all England into each shire ... So very narrowly did he commission them to trace it out, that there was not one single hide, nor a rood of land, nay, moreover, it is shameful to tell, though he thought it no shame to do it, not an ox, nor a cow, nor a swine was left, that was not set down in his writ, and all the recorded particulars were afterwards brought to him,* Chr. 1085; Erl. 218, 2–4 ... 24, 25 ... 33–38. **5.** the Anglo-Saxon king *had royal power to pardon transgressors;*—Gif hwá in cyninges healle gefeohte, oððe his wæpn gebrede, and hine mon gefó; sié ðæt on cyninges dóme, swá deáþ, swá líf, swá he him forgifan wille *if any one fight in the king's hall, or draw his weapon, and he be taken, be it in the king's power, either death or life, or pardon,* L. Alf. pol. 7; Th. i. 66, 8, 9. Sié on cyninges dóme hwæðer he líf áge ðe náge *be it in the king's power whether he shall or shall not have life,* L. In. 6; Th. i. 106, 3, 4. Búton him cyning [MS. kyning] árian wille *unless the king will be merciful to him,* 36; Th. i. 124, 19. Ðæt he wære his feores scyldig, búton he cyng gesóhte, and he him his feorh forgifan wolde; eall swá hit ær æt Greátan leá and æt Exan ceastre and æt þunres felda gecweden wæs *that he should be liable in his life, unless he should flee to the king, and he should give him his life; all as it was before ordained at Greatley and at Exeter and at Thundersfield,* L. Ath. v. § 1, 4; Th. i. 230, 6–9: L. Edm. S. 6; Th. i. 250, 11: L. Edg. ii. 7; Th. i. 268, 24, 25: L. Eth. iii. 16; Th. i. 298, 14: vii. 9; Th. i. 330, 24. **6.** *of all forfeits the king had one half*—to healfum:—Fó se cyng to healfum,—to healfum ða men ðe on ðære ráde beón *let the king take possession of half, of [the other] half the men, who may be in the riding [shall take possession],* L. Ath. i. 20; Th. i. 210, 6, 7. **7.** *treasure-trove, or treasure or money found, of which the owner was unknown, belonged to the king.* It is designated in Anglo-Saxon charters by the words—ealle hordas búfan eorþan, and binnan eorþan *all hoards above the earth, and within the earth.* As we learn from Beowulf, in early and heathen times, much treasure was buried in the mound raised over the ashes of the dead, besides what was burned with the body:—Hí on beorg dydon bégas [MS. beg] and siglu, forléton eorla gestreón eorþan healdan, gold on greóte, ðær hit nú gén lífaþ yldum swá unnyt swá hit ær wæs *they placed rings and jewels in the mound, they left the treasure of earls to the earth to hold, gold in the dust, where it now yet remains as useless to men as it was before,* Beo. Th. 6307–6318; B. 3164–3169. The legend of Guthlac [about A.D. 700, v. Crúland] supplies a very early instance of the search for gold and silver in the mounds :—Wæs ðær on ðam eálande sum hláw mycel ofer eorþan geworht, ðone ylcan men iú geára for feós wilnunga gedulfon and brǽcon: ða wæs ðær on ðóðre sídan ðæs hláwes gedolfen swylce mycel wæterseáþ wære *there was on the island a great mound raised upon the earth, which some men of yore had dug and broken up in hopes of treasure: then there was dug up on the other side of the mound as it were a great water-pit,* Guthl. 4; Gdwin. 26, 4–8. **8.** *Pastus or Convivium*=Cyninges feorm. The king visited different districts personally or by deputy to see that justice was done to all his subjects. In these periodical journeys the king received support and entertainment wherever he went. Hence perhaps the privileges of our judges. In A.D. 814 Cénwulf released the bishop of Worcester from a *pastus* of twelve men, whom he was bound to find. This was so great an expense that the exemption was worth an estate of thirteen hides, v. Cod. Dipl. 203; A.D. 814; Kmbl. i. 256. **9.** *Vigilia*= heáfodweard *head ward, or a proper watch set over the king, which he claimed when he came into any district.* The sǽweard or *coast guard* was also a regal right, performed by the tenants of those land owners whose estates lay contiguous to the sea. **10.** *the mint or coinage of money.* The king exercised a superintendence over the circulating medium. Æðelrǽd not only enacted that there should be no moneyers besides the king's, but that their number should be diminished:—Nán man ne áge nænne mynetere búton cyng *let no man have a moneyer except the king,* L. Eth. iii. 8; Th. i. 296, 15. Ut monetarii pauciores sint quam antea fuerint, iv. 9; Th. i. 303, 2. **11.** *the grant of a market, with power to levy tolls, was also a royalty,* Cod. Dipl. 1075; A.D. 873–899; Kmbl. v. 142: 1084; A.D. 904; Kmbl. v. 157. v. The Rights of Anglo-Saxon Kings, explained more fully in *Kemble's Saxons in England,* 2 vols. 8vo. 1849. Bk. ii. chap. 2; vol. ii. pp. 29–103. [*Prompt.* kynge: *Wyc.* kyng:

Piers P. Chauc. king: *R. Glouc.* kyng: *Laym. Orm.* king: *Plat.* kŏnig: *O. Sax.* kuning, cunig, m: *Frs.* kening: *O. Frs.* kining, kinig, kening, keneng, koning: *Dut.* koning, m: *Kil.* koningh, m: *Ger.* kŏuig, m: *M. H. Ger.* künic, künec, künc, m: *O. H. Ger.* kuning, m: *Dan.* konning, konge, m: *Swed.* konung, kong, kung, m: *Icel.* konungr, kóngr, m: *Lett.* kungs *dominus.*] DER. æðel-cyning, Angel-, beorn-, brýten-, eorþ-, eðel-, folc-, gást-, geár-, gúþ-, hæðen-, heáh-, heofon-, leód-, mægen-, ródor-, sǽ-, segn-, self-, sige-, sóþ-, swegl-, þeód-, þrym-, þryþ-, woruld-, wuldor-.

cyning-bald; *adj. Kingly or nobly bold;* nobiliter audax:—Férdon forþ cyningbalde men *the nobly bold men went forth,* Beo. Th. 3273; B. 1634.

cyning-cynn, es; *n.* [cynn *a sort, race,* v. cynn] *A royal race;* regium genus :—Of ðæs strýnde monigra mægþa cyningcynn fruman lǽdde *the royal race of many tribes drew its beginning from his stock,* Bd. 1, 15; S. 483, 30. Eanfriþ wæs ðære mægþe cyningcynnes *Eanfrith was of the royal race of that province,* 3, 1; S. 523, 14. Penda wæs se fromesta esne of Mercna cyningcynne *Penda was the boldest man of the royal race of the Mercians,* 2, 20; S. 521, 9. v. cyne-cyn.

cyning-dóm, es; *m.* [-dóm *dominion, power*] *Kingly power, a* KINGDOM; regimen, regnum:—Cyningdóm habban to have kingly power, Cd. 173; Th. 216, 7; Dan. 3. Metod ðec aceorfeþ of cyningdóme *the Lord will cut thee off from thy kingdom,* 202; Th. 251, 24; Dan. 568. Caldéas cyningdóm áhton *the Chaldeans held the kingdom,* 209; Th. 258, 24; Dan. 680. v. cyne-dóm.

Cyninges tún *Kingston,* Chr. 979; Th. 235, 9, col. 1. v. Cynges tún.

cyninges wyrt, e; *f. The herb marjoram;* sampsuchum = σάμψυχον, origanum majorana, Lin :—Cyninges wyrt *sampsuchum,* Mone A. 529.

cyning-feorm, cyninges feorm, e; *f.* [feorm *food, support*] *Royal purveyance, tribute for the royal household;* regis firma :—Ic heó gefreóge ðcelíce ðæs gafoles, ðe hió nú get to cyninges handa ageofan sceolan of ðam dǽle ðe ðær ungefreód to láfe wæs ðære cyningfeorme, ge on hlutrum alaþ, ge on beóre, ge on hunige, ge hrýðrum, ge on swýnum, ge on sceápum *I free them for ever from the impost which they have still to pay into the king's hand, from that portion, which was there left unfreed of the royal purveyance, whether in pure ale, or in beer, or in honey, or in oxen, or in swine, or in sheep,* Cod. Dipl. 313; A.D. 883; Kmbl. ii. 111, 4–9. Ðe cyninges feorm to belimpe *to which the royal purveyance belongs,* L. Alf. pol. 2; Th. i. 60, 24.

cyning-gereord, -gereorde, es; *n.* [gereord *food, a repast, feast*] *A royal feast;* regis convivium :—Cyning-gereorde *fercula,* Cot. 93.

cyning-gierela, an; *m. A royal crown, diadem;* regalis tænia [= ταινία] diadema = διάδημα, Som. Ben. Lye.

cyning-ríce *a kingdom,* Som. Ben. Lye. v. cyne-ríce.

cyn-líc; *adj.* [cyn *suitable, fit*] *Becoming, fitting;* décorus :—Suilce iów cynlíc þynce *as to you may seem fitting,* Th. Diplm. A.D. 804–829; 401, 36. Swá him rihtlíc and cynlíc þince *as to them may seem just and becoming,* Th. Diplm. A.D. 905; 493, 12.

cyn-líce; *adv. Becomingly, fitly;* congruenter :—Hi cynlíce to ðe cleopiaþ *they fitly call upon thee,* Ps. Th. 64, 14 : 118, 57, 82, 145, 147: 126, 2.

cynn, es; *n. A sort, kind;* genus, Ps. Th. 144, 13. v. cyn.

cynn *suitable, fit,* Bt. 35, 4; Fox 162, 24: L. In. 42; Th. i. 128, 11, MS. H. v. cyn.

cynnan *to declare, clear, prove;* advocáre, purgáre, manifestáre :—Gif he cynne ðæt he hit bohte *if he declare that he bought it,* L. Edg. S. 11; Th. i. 276, 12, note 7. v. cennan II.

cynnestre, an; *f.* [cennan *to bring forth,* -estre *a female* termination, q. v.] *One who brings forth, a mother;* genitrix, mater :—Ðæt cild oncneów Marian stemne, cynnestran *the child knew the voice of Mary, the mother,* Homl. Th. i. 352, 27.

cynning-stán, es; *m.* [cennan II. *to try, prove;* stán *a stone*] *A trying-stone;* tessera :—Cynning-stán on tæfle *a little wooden tower on the side of a gaming-board, hollow and having steps inside, through which the dice were thrown upon the board;* pyrgus [= πύργος], turricula, Ælfc. Gl. 61; Som. 68, 65; Wrt. Voc. 39, 48.

cynn-recceniss, e; *f.* [reccenys *a narration, history*] *A reckoning of relationship, a genealogy;* genealogia, Mt. Kmbl. Lind. 1, title.

cyn-recen; gen. -recenne; *f. A pedigree, genealogy, parentage;* generatio, genealogia, parentela, Som. Ben.

cyn-ren, -ryn, es; *n.* [cyn *a kindred, race, nation, family, generation;* ren, ryn *a course*] *A family course, family, generation, kind, nation, posterity;* generatio, genus, natio, progenies, propago :—He forlét his ríce and his cynren *he left his country and his family,* Bt. 38, 1; Fox 194, 27. Cynren *generatio,* Wrt. Voc. 72, 49. Ðis ys Thares cynryn *this is the generation of Terah,* Gen. 11, 27. On cynrynum cynrena [MS. kynrynum kynrena] *in generationes generationum,* Ps. Lamb. 71, 5. On ðam fíftan dæge úre Drihten gesceóp ða mycelan hwalas on heora cynrynum *on the fifth day our Lord created the great whales with their kinds,* Hexam. 8; Norm. 14, 8. Fisc sceal on wætere cynren cennan [MS. cynran cennen] *a fish shall propagate his kind in the water,*

Menol. Fox 515; Gn. C. 28. Cynrenu *genera*, Scint. 53. Ic andette ðē on cynrenum [cynrenon MS.], Drihten *confitebor tibi in nationibus, Domine*, Ps. Spl. 17, 51. Lā ge nædrena cynryn *progenies viperarum*, Mt. Bos. 12, 34. Cynren *propago*, Ælfc. Gl. 91; Som. 75, 17; Wrt. Voc. 51, 62.

Cynríc, es; *m. Cynric, the second king of the West Saxons, son of Cerdic*, q. v; Cynrícus:—Hēr, A. D. ccccxcv, cōman twegen ealdormen on Brytene, Cerdic and Cynríc his sunu, mid v scipum on ðone stede ðe is gecweden Cerdices ōra, and ð‎ӯ ilcan dæge hie gefuhtan wið Wealum *here*, A. D. 495, *came two aldormen to Britain, Cerdic and Cynric his son, with five ships, at the place which is called Cerdic's shore* [on the south of Dorsetshire, v. Cerdices ōra], *and on the same day they fought against the Welsh*, Chr. 495; Th. 24, 26-33. Hēr Cerdic forþférde, and Cynríc his sunu rícsode forþ xxvi wintra *in this year* [A. D. 534] *Cerdic died, and Cynric his son reigned for twenty-six years*, 534; Erl. 14, 32.

cyn-ryn, es; *n. A family course, generation*; generatio, progenies, Gen. 11, 27: Ps. Lamb. 71, 5: Hexam. 8; Norm. 14, 8: Mt. Bos. 12, 34. v. cyn-ren.

CÝP; *gen.* cyppes; *m. A* CHIP, *beam, log, trunk of a tree*; festuca, trabs, stipes:—Cyppes *stipĭtis*, Glos. Prudent. Recd. 148, 80. Cyp *stipitem*, 150, 39. [*Prompt.* chyppe *assula*: *Chauc.* chippes, *pl*: *R. Brun.* chip: *Kil.* kippen *cudere*: *Icel.* kippa *to pull, snatch*; kippr, *m. a pull, shock, spasm*.]

cýp, e; *f. A measure, bushel*; modius, dolium:—Under cýpe *sub modio*, Mt. Kmbl. Hat. 5, 15. Cýpe *dolium*, Mone B. 3630. v. cýf.

cýpa, cēpa, an; *m.* [ceáp II.] I. *a factor, merchant, trader*; negotiator, mercator:—Ðā ðær fōron Madianisce cýpan *then there passed Midianitish merchants*, Gen. 37, 28. Cýpa *mercator*, Glos. Prudent. Recd. 140, 38. Ðās hálgan cýpan, Petrus and Andreas, mid heora nettum and scipe him ðæt ēce líf geceápodon *these holy traders, Peter and Andrew, with their nets and ship bought for themselves everlasting life*, Homl. Th. i. 580, 19. Drihten adræfde ðillíce cýpan of ðam hálgan temple *the Lord drove such chapmen from the holy temple*, 406, 24. II. *what a merchant has his goods in,—A basket*; cofinus = κόφινος:—Man nam ða gebrotu ðe ðār belifon, twelf cýpan fulle *sublatum est quod superfuit illis, fragmentorum cophĭni* [κόφινοι] *duodecim*, Lk. Bos. 9, 17. [*Scot.* couper, coper *one who buys and sells*: *O. Frs.* kapere, *m. a purchaser*: *Dut.* kooper, *m*: *Ger.* käufer, *m*: *M. H. Ger.* koufer, *m*: *O. H. Ger.* koufâri, *m*: *Dan.* kjöber: *Swed.* köpare, *m*: *Lat.* caupo *a merchant*: *Grk.* κάπηλος *one who sells provisions*: *Lith.* kupczus *mercator*.] DER. mynet-cýpa.

cýpan, cīpan; ic cýpe, ðū cýpest, cýpst, he cýpeþ, cýpþ, *pl.* cýpaþ; *p.* cýpte, ðū cýptest, *pl.* cýpton, cīptun *To sell*; vendere:—Ic wylle cýpan *volo vendere*, Coll. Monast. Th. 27, 19. Ic cýpe míne þingc *ego vendo meas res*, 26, 33. Hwær cýpst ðū fixas ðíne *ubi vendis pisces tuos?* 23, 21. Ðū sældest *vel* cýptest folc ðín *vendidisti populum tuum*, Ps. Spl. T. 43, 14. Sēde ðam ðe ða culfran cýpton *dixit his qui columbas vendebant*, Jn. Bos. 2, 16. Gāþ to ðam cýpendum and bycgaþ eów ele *ite ad vendentes et emite vobis oleum*, Mt. Bos. 25, 9: Gen. 47, 20. [*Prompt.* chepyn' *licitari*: *Chauc.* chepe *to buy, market*: *Piers P.* chepen *to buy*: *Scot.* coup *to buy and sell*: *Plat.* kopen, köpen *to buy*: *O. Sax.* kōpôn *to bargain*: *Frs.* keapjen: *O. Frs.* kapia *to buy*: *Dut.* koopen *to buy*: *Ger.* kaufen: *M. H. Ger.* koufen: *O. H. Ger.* koufen, koufôn *mercari*: *Goth.* kaupon *to bargain*: *Dan.* kjöbe *to buy*: *Swed.* köpa *to buy*: *Icel.* kaupa, *p.* keypti *to bargain*.] DER. be-cýpan, ge-. v. ceápian.

cýpe-cniht, es; *m. A bought servant, slave*; venalis puer, servus:—Ða geseah he cypecnihtas *he then saw slaves*, Homl. Th. ii. 120, 18.

cýpe-man, -mann, es; *m. A merchant*, Bd. 2, 1; S. 501, 4. v. ceáp-man.

cypera, an; *m. A* KIPPER, *salmon in the state of spawning*; salmo ova gignens:—Ðonne eów fōn lysteþ leax oððe cyperan *when you desire to catch a salmon or a kipper*, Bt. Met. Fox 19, 23; Met. 19, 12.

cyperen; *adj. Coppery, belonging to copper*; æreus:—Seóþ on cyperenum citele *seethe it in a copper kettle*, L. M. 1, 15; Lchdm. ii. 56, 19. Dô on cyperen fæt *put it into a copper vessel*, 1, 2; Lchdm. ii. 36, 1. Gemultan ealle ða anlícnessa togædere, ðe ðær binnan wǽran, ge gyldene, ge sylfrene, ge ǽrene, ge cyperene *all the statues, which were in it, of gold, and of silver, and of brass, and of copper, were melted together*, Ors. 5, 2; Bos. 101, 22. Forðonðe he forgnāþ gatu cyperene *quia contrivit portas æreas*, Ps. Spl. 106, 16. Cyperen hwer *a copper ewer or vessel*; cucuma, Ælfc. Gl. 26; Som. 60, 83; Wrt. Voc. 25, 23.

cýpe-þing; *pl. n. Saleable things, merchandise*; merces, Cot. 133. v. cēpe-þing.

cýping, cýpingc, cíping, e; *f.* [ceáping, ceáp *a price*, q. v. II]. I. *a bargaining, setting a price, marketing, chapping, traffic*; negotiatio, nundina:—Ðæt nān cýping ne s‎ӯ Sunnan dagum *that no marketing be on Sundays*, L. Ath. i. 24; Th. i. 212, 15: v. 10; Th. i. 240, 9. Ða ealdorbiscopas geþafedon ðæt ðær cýping binnan gehæfd *were the high-priests allowed chapping to be held therein*, Homl. Th. i. 406, 6. Cýpingc *negotiatio*, Ælfc. Gl. 81; Som. 73, 18; Wrt. Voc. 47, 25.

Sunnan dæges cýpinge we forbeódaþ ǽghwār *we forbid Sunday's traffic everywhere*, L. N. P. L. 55; Th. ii. 298, 21. Cýpingc, L. C. E. 15; Th. i. 368, 15. Ne fortruwige he hiene æt ðære cípinge *let them not be too confident of their bargain*, Past. 44, 6; Hat. MS. 62 b, 9. Cýpinga *nundinæ*, Ælfc. Gr. 13; Som. 16, 21. Ðæt hí Sunnan dæges cýpinga georne geswícan *that they strictly abstain from Sunday marketings*, L. Eth. vi. 44; Th. i. 326, 21: vi. 22; Th. i. 320, 12: v. 13; Th. i. 308, 11: ix. 17; Th. i. 344, 7. II. *a market-place, market*; forum:—Ðæs tūnes cýping and seó innung ðara portgerihta gange into ðære hálgan stōwe *let the market of the town and the revenue of the port dues go to the holy place*, Cod. Dipl. 598; A. D. 978; Kmbl. iii. 138, 10. To-middes ðære cýping *in the midst of the market*, M. H. 117 a. Andlang stræte ūt on ða cýpinge, swā up anlang cýpinge *along the road out to the market-place, so up along the market-place*, Cod. Dipl. 720; A. D. 1012; Kmbl. iii. 359, 12, 13.

cýp-man; *gen.* -mannes; *m. A chapman, merchant*; mercator:—Ða cýpmen binnon ðam temple getācnodon unrihtwíse lāreówas on Godes geladunge *the chapmen within the temple betokened unrighteous teachers in God's church*, Homl. Th. i. 410, 35: ii. 120, 15. Drihten adræfde of ðam temple ða cýpmen *the Lord drove the chapmen from the temple*, i. 406, 1. Sume synt cýpmenn *alii sunt mercatores*, Coll. Monast. Th. 19, 7. Be cýpmanna fōre *of the journeying of chapmen*, L. In. 25; Th. i. 118, 11, note 27, B. G. v. ceáp-man.

Cyppan-ham, -hamm *Chippenham, Wilts*:—Hēr hine bestæl se here to Cyppanhamme *here the army stole itself away to Chippenham*, Chr. 878; Th. 146, 21, col. 2, 3: 880; Th. 148, 39, col. 3. v. Cippan-ham.

cypresse, an; *f. The cypress*; cupressus [= κυπάρισσος], cupressus sempervirens, Lin:—Of cypressan *from the cypress*, Lchdm. iii. 118, 21.

cypsed; *pp. Bound, fettered*; compeditus. DER. ge-cypsed. v. cyspan.

cýp-strǽt, e; *f.* [cýp = ceáp II, strǽt *a street*] *A street or place for merchandise, cheap street*; vicus mercatorius:—Andlang cýpstrǽte *along cheap street*, Cod. Dipl. 1291; A. D. 996; Kmbl. vi. 135, 17.

cyrc, e; *f. A church*; ecclesia:—Cristes cyrc *Christ's church*, Chr. 1066; Erl. 202, 1. In ðære cyrce *in the church*, 1070; Erl. 209, 40. Ða cyrce *the church*, 1070; Erl. 209, 36. v. cyrce, cyrice.

cyrc-brǽce, es; *m. Church-breach, a breaking into a church*; in ecclesiam irruptio:—Ða heáfodleahtras sind, mansliht, cyrcbrǽce, etc. *the chief sins are, murder, church-breach, etc.* Homl. Th. ii. 592, 4. v. ciricbryce.

cyrce; *gen.* cyrcan, cyrcean; *f. A church*; ecclesia:—Seó cyrce mid hire portice mihte fíf hund manna eaðelíce befōn on hire r‎ӯmette *the church with her porch could easily contain in its space five hundred men*, Homl. Th. i. 508, 13: ii. 584, 3: 592, 22. Cyrcan duru *a church's door*, i. 64, 31. Crist is se grundweall ðære gāstlícan cyrcan *Christ is the foundation of the spiritual church*, ii. 588, 22. Ne sceal cyrcean timber to ænigum ōðrum weorce, būton to ōðre cyrcean *ligna ecclesiæ non debent ad aliud opus poni, nisi ad aliam ecclesiam*, L. Ecg. P. A. 16; Th. ii. 234, 16, 17. v. cyrice.

cyrce weard *a warden of the church, sacristan*, Chr. 1070; Erl. 207, 33. v. cyric-weard, cyrc-weard.

cyrc-hálgung *hallowing or consecrating a church*, Homl. Th. ii. 582, 27. v. cyric-hálgung.

cyrc-líc *ecclesiastical*, Chr. 716; Th. 70, 35, col. 3: L. Ælf. C. 33; Th. ii. 356, 13: Homl. Th. i. 600, 8. v. cyric-líc.

cyrc-þénung *church-service*, Glos. Prudent. Recd. 145, 81. v. ciricþénung.

cyrc-þingere, es; *m. A priest*; sacerdos:—Sacerd *vel* cyrcþingere *sacerdos*, Ælfc. Gl. 68; Som. 70, 14; Wrt. Voc. 42, 23. v. þingere II, cyric-þingere.

cyrc-weard, cyric-, -werd, es; *m. A churchwarden, sacristan*; ecclesiæ custos, sacri scriniarius:—Cyrcweardes þénung *a churchwarden's duty*, Greg. Dial. 1, 5. Æðelstān cyric-weard [MS. -wyrd] fēng to ðam abbodríce æt Abban dūne *Æthelstan, warden of the church, succeeded to the abbacy at Abingdon*, Chr. 1044; Th. 300, 26. Cyrcweard *sacri scriniarius*, Ælfc. Gl. 114; Som. 80, 23; Wrt. Voc. 61, 4. Cyrcwerd *ædituus*, R. Conc. 1. Se bisceop befran ðone cyrcweard hwær ðæs hálgan wǽpnu wǽron *the bishop asked the sacristan where the weapons of the saint were*, Homl. Th. i. 452, 2. Ðā wæs ān cyrce weard Ywarn wæs gehāten *there was a sacristan called Ywarn*, Chr. 1070; Erl. 207, 33.

cyrde, *pl.* cyrdon *turned, returned*, Lk. Bos. 14, 21: Jn. Bos. 6, 66; *p.* of cyrran.

cyre, es; *m.* [ceósan *to choose*] *Choice, free choice, free will*; electio, hærēsis = αἵρεσις, optio, arbitrium:—Cyre [MS. kyre] *hæresis*, Ælfc. Gl. 3; Som. 55, 84; Wrt. Voc. 16, 55. Cyre *optio*, Glos. Prudent. Recd. 146, 52. God forgeaf him āgenne cyre, forðanðe ðæt is rihtwísnys ðæt gehwylcum s‎ӯ his āgen cyre geþafod *God gave them their own free will, for it is righteousness that to every one be allowed his own free will*, Homl. Th. i. 112, 4, 5, 8, 11, 12, 14: 110, 35: 292, 32: ib. 490, 16. Ic wylle ðæt hý s‎ӯn heora freólses wyrðe and hyra cyres *I will that they be worthy of their freedom and their free will*, Cod. Dipl. 314; A. D. 880-885; Kmbl. ii. 116, 30. Hwí wæs se man betǽht to

his ágenum cyre *why was the man [Adam] committed to his own free will?* Boutr. Scrd. 17, 25. Mid cyre *arbitrio,* Mone B. 1344: 2616. [*Laym.* cure, *m. choice:* *Plat.* köre *election:* *Dut.* keur, *f. choice:* *Kil.* keur, kore *optio, electio, arbitrium:* *Ger.* kür, kur, chur, *f. election:* *M. H. Ger.* kür, küre, *f. examination, election:* *O. H. Ger.* churi, *f. deliberatio, electio:* *Dan.* kaar, *n. choice:* *Swed.* kor *electio:* *Icel.* kjörr, keyr, *n. choice, decision.*]

cyre-áþ, es; *m.* [cyre *a choice,* áþ *an oath*] *The select oath, the oath sworn by the accused, together with a certain number of consacramentals selected by him out of a fixed number of persons named to him by the judge;* juramentum electum, quod quis præstabat cum aliquot conjuratoribus ab ipso selectis e quibusdam a judice nominatis [Schmd. 566] :—Nemne him man x men and begite ðara twegen and sylle ðone áþ . . . and stande ðæs cyre-áþ ofer xx peninga *let there be named ten men to him and let him get two of them and give the oath . . . and let his select oath stand for over twenty pence,* L. Ath. i. 9; Th. i. 204, 15. v. ungecoren áþ.

cyre-bald *bold in decision;* arbitrii strenuus. v. cire-bald.

cyre-líf, es; *n. A choice of life, where on decease of a lord, the cultivators choose a lord for themselves;* optio vitæ, ubi, mortuo domino, villani sibi dominum eligunt:—Ic bidde, on Godes naman, and on his háligra, ðæt mínra maga nán ne yrfewearda ne geswence nán nænig cyrelíf ðara ðe ic foregeald, and me West-Seaxena wítan to rihte gerehton, ðæt ic hí mót lætan swá freó swá þeówe, swæðer ic wille; ac ic, for Godes lufan and for mínre sáwle þearfe, wylle ðæt hý sýn heora freólses wyrðe and hyra cyres; and ic, on Godes lífiendes naman, beóde ðæt hý nán man ne brócie, ne mid feós manunge, ne mid nænigum þingum, ðæt hý ne mótan ceósan swylcne mann swylce hý wyllan *I pray in the name of God, and his saints, that no one of my kinsmen nor heirs molest any choice of life of those for whom I have paid, and the witan of the West Saxons have rightly confirmed to me, that I might leave them either free or servile, as I will; but I, for love of God and for my soul's need, will that they be entitled to their freedom and their choice; and I, in the name of the living God, command that no man oppress them, either by exaction of money, or in any other way, so that they may not choose whatever lord they will,* Cod. Dipl. 314; A. D. 880–885; Kmbl. ii. 116, 24–33.

cyren *must, wine boiled down;* dulcisapa :—Awilled wín *vel* cyren *dulcisapa,* Cot. 62. v. a-willan, ceren.

Cyren-ceaster, Cyrn-ceaster *Cirencester, Cicester, Gloucestershire:*— Æt Cyrenceastre *at Cirencester,* Chr. 1020; Th. 286, 12, col. 2: Ors. 5, 12; Bos. 110, 22. v. Ciren-ceaster.

cyrf, e; *f? A cutting off, an instrument to cut with;* abscissio, ferrum abscissionis :—Cyrf *abscissio,* R. Ben. 28. Be ðisum cyrfe *of this cutting,* Homl. Th. ii. 406, 33. Cyrf *ferrum abscissionis,* C. R. Ben. 40. DER. æ-cyrf, of-.

CYRFÆT, cyrfet, es; *m? A gourd;* cucurbita :—Cyrfæt *cucurbita,* Ælfc. Gl. 43; Som. 64, 38; Wrt. Voc. 31, 48. Hwerhwettan oððe cyrfet gesihþ on swefnum untrumnysse getácnaþ *to see in dreams a cucumber or a gourd betokens ailment,* Somn. 43; Lchdm. iii. 200, 16. Wylde cyrfet *wild gourd,* colocynthis = κολοκυνθίς, Ælfc. Gl. 39; Som. 63, 58; Wrt. Voc. 30, 12. Wild cyrfet *vel* hwit wíngeard *bryonia* = βρυωνία, 44; Som. 64, 81; Wrt. Voc. 32, 17. [*Plat.* körbs, körwitz, kürwes, *m: Dut.* kauwoorde, *f. a gourd: Kil.* kauwoorde, kouworde: *Ger.* kürbiss, *m: M. H. Ger.* kürbez, *m: O. H. Ger.* kurbiz, *m: Fr.* gourde, *f: O. Fr.* gougourde: *Lat.* cucurbita.]

cyrfel, es; *m.* [cyrf *a cutting off*] *A little stake, a peg;* paxillus :— Cyrfel *vel* litel stigul [= sticel?] *paxillus,* Ælfc. Gl. 29; Som. 61, 46; Wrt. Voc. 26, 45.

cyrfille, an; *f. Chervil;* cærefolium :—Nim cyrfillan *take chervil,* Lchdm. iii. 12, 13: 46, 25. v. cerfille.

cyrfst, he cyrfþ *carvest, carves; 2nd and 3rd pers. pres. of* ceorfan.

cyric *a church.* v. *in the compounds* cyric-æwe, -belle, -bóc, -bót, -bryce, -burh, -dór, -friþ, -fultum, -georn, -geriht, -griþ, *etc.*

cyric-æwe, ciric-æwe, es; *m. An ecclesiastical marriage;* ecclesiasticum matrimonium :—Hí, þurh heálicne hád, ciricæwe underféngan *they, through holy orders, have entered into an ecclesiastical marriage,* L. I. P. 23; Th. ii. 334, 14. v. cyric; æw, æwe.

cyric-belle *a church-bell;* ecclesiæ campana. v. ciric-belle.

cyric-bóc, e; *f. A church-book;* liber continens ritus et ceremonias ecclesiæ :— To æghwælcre neóde man hæfþ on cyricbócum mæssan gesette *masses for every necessity have been placed in church-books,* Lupi Serm. 2, 3; Hick. Thes. ii. 107, 32.

cyric-bót, ciric-bót, e; *f. Church-repair;* ecclesiæ reparatio :—To cyricbóte *for church-repair,* L. Eth. vi. 51; Th. i. 328, 6. To ciricbóte sceal eall folc fylstan mid rihte *all people must lawfully give assistance to church-repair,* L. C. S. 66; Th. i. 410, 12: L. Eth. ix. 6; Th. i. 342, 8.

cyric-bryce *church-breach, a breaking into a church,* L. Ath. i. 5; Th. i. 202, 6, MSS. B. L. v. ciric-bryce.

Cyric-burh; *gen.* -burge; *dat.* -byrig; *f.* [*Hunt.* Cereburih: *Brom.*

Cyrebury: *the church city*] *Chirbury, Shropshire;* loci nomen in agro Salopiensi :—Æðelflæd ða burh getimbrede æt Cyricbyrig *Æthelfled built the fortress at Chirbury,* Chr. 913; Th. 186, 35, col. 2; 187, 35, col. 1.

cyric-dór *a church-door;* ecclesiæ porta. v. ciric-dór.

CYRICE, cirice, cyrce, circe; *gen.* an, ean; *f: cyric, ciric, in the compound* cyric-æwe, *etc.* q. v. cyrc, e; *f.* circ, *in the compound* circ-líc, *etc.* q. v. **I.** *the* CHURCH *as a temporal and spiritual body;* ecclesia = ἐκκλησία :—Seó cyrice on Breotone hwæt hwugu fæc sibbe hæfde *the church in Britain for some time had peace,* Bd. 1, 8; S. 479, 17. Seó Godes circe, seó circe æfyllendra *the church of God, the church of the faithful,* Exon. 18 a; Th. 44, 8, 16; Cri. 699, 703. To ðære ánnesse ðære hálgan Cristes cyrican *to the unity of Christ's holy church,* Bd. 1, 26; S. 488, 13. Agustinus on Cent ðære frymþelícan cyrican líf and láre wæs onhýrigende *Augustine in Kent imitated the life and lore of the early church,* 1, 26; S. 487, 27. Gregorius féng to biscopháde ðære Rómániscan cyrican *Gregory succeeded to the bishopric of the Roman church,* 1, 23; S. 485, 23: 1, 4; S. 475, 29. Ongunnon hí ðæt apostolíce líf ðære frymþelícan cyrican onhýrigean *they began to imitate the apostolic life of the early church,* Bd. 1, 26; S. 487, 32. Fram ðam biscope ðære Rómániscan cyrican *by the bishop of the Roman church,* 1, 13; S. 481, 38. On Norþanhymbra þeóde and cyrican *in the nation and church of the Northumbrians,* 2, 20; S. 521, 19. On ðære hálgan Rómánisce cyrican *in the holy Roman church,* 1, 27; S. 489, 33, 38. Hælend Crist is se grundweall ðære gástlícan cyrcan *Jesus Christ is the foundation of the spiritual church,* Homl. Th. ii. 588, 22. Ealle Godes cyrcan sind getealde to ánre cyrcan, and seó is geháten geladung *all God's churches are accounted as one church, and that is called a congregation,* ii. 580, 22. On ciricean Crist Drihten God bletsige *in ecclesiis benedicite Dominum Deum,* Ps. Th. 67, 24. Hí hýndon and hergedon Godes cyrican *they oppressed and harried God's church,* Bd. 1, 6; S. 476, 21. Crist getimbrode ða gástlícan cyrcan, ná mid deádum stánum ac mid lybbendum sáwlum *Christ built the spiritual* [lit. *ghostly*] *church, not with dead stones but with living souls,* Homl. Th. ii. 580, 12. **II.** *a church, the material structure;* ecclesia :—Ðær wæs cyrice geworht *a church was built there,* Bd. 1, 7; S. 479, 6: 1, 26; S. 487, 42. Wæs cirice gehálgod *a church was consecrated,* Andr. Kmbl. 3291; An. 1648. Ðæt seó cyrce afealle *that the church may fall down,* Homl. Th. i. 70, 27. Godes cyrce is úre gebédhús *God's church is our prayer-house,* ii. 584, 3. Circe *ecclesia,* Ælfc. Gl. 107; Som. 78, 82; Wrt. Voc. 57, 58. Awriten mid ðám bróðrum ðære cyricean æt Lindesfarena *written by the brethren of the church at Lindesfarne,* Bd. pref; S. 472, 29. Nim úre cyrcan máðmas *take our church's treasures,* Homl. Th. i. 418, 14, 17. Nis ná alýfed ðæt ðæs mynstres hláford sylle ðære cyrcean land to óðre cyrcean *non licet monasterii domino terram ecclesiæ alii assignare ecclesiæ,* L. Ecg. P. A. 25; Th. ii. 236, 15, 16. Ceadwala cining wæs gebyrged innan Sče Petres cyrican *king Ceadwalla was buried in St. Peter's church* [*at Rome*], Chr. 688; Erl. 43, 7. Hí on cyrican in Eoferwícceastre bebyrigde wæron *they were buried in the church at York,* Bd. 2, 14; S. 518, 2. Æðelbyrht cyning on cyricean ðara eádigra apostola Petrus and Paulus bebyriged wæs *king Æthelbert was buried in the church of the blessed apostles Peter and Paul,* 2, 5; S. 506, 22. On eorþlícere cyrcan líþ stán ofer stáne *in an earthly church stone lies over stone,* Homl. Th. ii. 582, 17: i. 452, 2: 504, 8: 506, 11, 18. Se Cénwalh hét atimbrian ða cyrican on Wintan-ceastre *Cenwalh commanded the church at Winchester to be built,* Chr. 641; Erl. 27, 13. Eádwine cyning wæs gefullod fram Pauline ðam bisceope on Eoferwícceastre, ðý hálgestan Eásterdæge, on sancti Petres cyricean ðæs apostoles, ðá he ðær hæfde geweorce of treówe cyricean getimbrede, syððan he gecristnad wæs . . . and sóna ðæs ðe he gefullad wæs, he ongan, mid ðæs bisceopes láre, máran cyrican and hýhran stænene timbrian, and wyrcean ymb ða cyrican útan ðe he ær worhte *king Edwin was baptized by bishop Paulinus on the most holy Easter day, in the church of St. Peter the apostle at York, when he had there built a church of wood, with hasty work, after he was christened . . . and soon after he was baptized, he began, by the bishop's advice, to build a larger and higher church of stone, and to construct it about the church which he had formerly wrought,* Bd. 2, 14; S. 517, 22–30: Chr. 626; Erl. 23, 40; 25, 2: Bd. 2, 3; S. 504, 23, 27: 2, 14; S. 518, 18: 2, 16; S. 519, 22. Hió cirican getimbrede, tempel Drihtnes, on Caluarie *she built a church, a temple of the Lord, on Calvary,* Elen. Kmbl. 2014; El. 1008. Se hét ciricean getimbran, Godes tempel *he commanded a church to be built, a temple of God,* Andr. Kmbl. 3265; An. 1635. Hí ðærofer cyrcan arærdon and weofod *they raised a church and altar thereover,* Homl. Th. i. 506, 15, 19, 25, 35. Ne wæron cyrican getimbrede *churches were not built,* Bd. 2, 14; S. 518, 16. Ða menigfealdan cyrcan ateoriaþ *the manifold churches will decay,* Homl. Th. ii. 582, 6. Ða cyrcean, ðe beóþ fram ðam bisceopum gehálgode, sceolon mid hálig wætere beón geondstrédde *ecclesiæ, ab episcopis illis consecratæ, aqua benedicta debent aspergi,* L. Ecg. P. A. 5; Th. ii. 232, 20. On éhtnysse Godes cyrcena *in the persecution of God's churches,* Bd. 1, 6; S. 476, 22. On ðám lácum geleáfsumra ðe hí to

Godes cyricum bringaþ *of the gifts of the faithful which they bring to* God's churches, 1, 27 ; S. 488, 39. On Cristes cyrican ða ðe on Brytene wǽron *in Christ's churches which were in Britain*, 1, 8; S. 479, 26. Constantīnus hēt ðæt man cyricean timbrede, and ðæt man belūce ǽlc deófulgyldhūs *Constantine ordered churches to be built, and every heathen temple to be closed*, Ors. 6, 30 ; Bos. 127, 36 : Bd. 1, 8; S. 479, 22, 23. Maximian, ārleás cyning, cwealde cristne men, circan fylde *Maximian, the wicked emperor, slew christian men, overthrew churches*, Exon. 65 b ; Th. 243, 4; Jul. 5. On ðison geáre barn Cristes cyrc *in this year* [A.D. 1066] *Christchurch* [Canterbury] *was burnt*, Chr. 1066; Erl. 202, 1. Cyrice weard, cyrce weard *a warden of a church*, 1043 ; Erl. 169, 33 : 1070; Erl. 207, 33. In ðære cyrce *in the church*, 1070; Erl. 209, 40. Ða cyrce *the churches*, 1070; Erl. 209, 36. **III.** *a heathen temple* ; templum paganum :—Gebletsode Romulus mid ðara sweora blōde ða cyrican *Romulus consecrated the temples with the blood of their fathers-in-law*, Ors. 2, 2 ; Bos. 41, 7. [*Prompt.* chyrche : *Wyc.* cherche : *Piers P.* kirk : *Chauc.* chirche : *R. Glouc.* chirches, *pl* : *Laym.* chirche, chireche, *f* : *Scot.* kirk : *Plat.* karke, kerke : *O. Sax.* kirika, *f* : *Frs.* tjercke : *O. Frs.* kerke, sthereke, sziurke, tsiurike, *f* : *Dut.* kerk, *f* : *Kil.* kercke : *Ger. M. H. Ger.* kirche, *f* : *O. H. Ger.* kiricha, *f* : *Dan.* kirke, *m. f* : *Swed.* kyrka, *f* : *Icel.* kirkja, *f* : *Grk.* κυριακή [οἰκία] *the Lord's* [house].] DER. cyric-ǽwe, -belle, -bōc, -bōt, -bryce, -burh, -dōr, -friþ, -fultum, -georn, -geriht, -griþ, -hād, -hālgung, -līc, -mangung, -mitta, -neód, -nyt, -pæþ, -ragu, -rēna, -sang, -sangere, -sceat, -sōcn, -stīg, -þēn, -þēnung, -þingere, -tīd, -tūn, -wæcce, -wǽd, -wag, -waru, -weard.

cyric-friþ, ciric-friþ, es ; *m. n. Church-peace, right of sanctuary* ; ecclesiæ pax :—Cyricfriþ *church-peace*, L. Ethb. 1 ; Th. i. 2, 6. Ciric-friþes [cyric- MS. H.] to bōte *as compensation for the church-peace*, L. Alf. pol. 2 ; Th. i. 62, 5.

cyric-fultum *church-help, ecclesiastical support.* v. ciric-fultum.

cyric-georn ; *adj. Diligent in attending church* ; ad ecclesiam libenter frequens, L. Ecg. C. prm ; Th. ii. 132, 15.

cyric-geriht, es ; *n. A church-due* ; ecclesiæ debitum :—Hī gyrnaþ heora sceatta on teoðungum, and on eallum cyricgerihtum *they desire their monies for tithes, and for all church-dues*, L. I. P. 19 ; Th. ii. 328, 1.

cyric-griþ, es ; *n. Church-peace* ; ecclesiæ pax :—Stande ǽlc cyricgriþ swā swā hit betst stōd *let every church-peace stand as it has best stood*, L. Edg. i. 5 ; Th. i. 264, 25, MS. A. v. ciric-griþ.

cyric-hād, es ; *m.* [hād II. *degree, order*] *A church-degree, order of the church* ; ecclesiæ ordo :—For ðām seofon cyrichādum [-hādan MS.] ðe se mæssepreóst, þurh Godes gife, geþeáh ðæt he hæfde, he biþ þegen-rihtes wyrðe *for the seven orders of the church, which the mass-priest, through the grace of God, has acquired, he is worthy of thane-right*, L. O. 12 ; Wilk. 64, 41.

cyric-hālgung, cyrc-hālgung, e ; *f. Church-hallowing, consecration of a church* ; encænia = ἐγκαίνια, ecclesiæ consecratio :—Ðys sceal to cyric-hālgungum *this shall be for the consecration of a church*, Rubc. Jn. Bos. 10, 22 ; Notes, p. 580. Æt ðære ealdan cyrchālgunge *at the old church-hallowing*, Homl. Th. ii. 582, 27.

cyric-līc, circ-līc, cyrc-līc; *adj. Like a church, ecclesiastical* ; ecclesiasticus :—Cyriclīc wer *vir ecclesiasticus*, Bd. 2, 20 ; S. 522, 21. Magister cyriclīces sanges *magister ecclesiasticæ cantionis*, 2, 20 ; S. 522, 27. Fram ǽlcere cyriclīcre gesamnunge *a quaque ecclesiastica congregatione*, L. Ecg. P. A. 30 ; Th. ii. 236, 35. Hie heóldan ða cyriclīcan sceare *they observed the ecclesiastical tonsure*, Chr. 716 ; Th. 70, 34, col. 2. Ðæt cyriclīce stǽr ūres eálondes and þeóde ic wrāt on fīf bēc *I* [Bede] *wrote the ecclesiastical history of our island and nation in five books*, Bd. 5, 24 ; S. 648, 31. Cyriclīce preóstas *ecclesiastici presbyteri*, L. Ecg. P. A. 5 ; Th. ii. 232, 17. Monad mid gelomlīcre smeáwunge and leornunge cyriclīcra gewrita *admonitus ecclesiasticarum frequenti meditatione scripturarum*, Bd. 5, 21 ; S. 642, 26 ; 5, 23 ; S. 645, 15. Mid ōðrum cyriclīcum bōcum *cum cæteris ecclesiasticis voluminibus*, 5, 20 ; S. 642, 1.

cyric-mangung *church-mongering, simony*, L. Eth. vi. 15 ; Wilk. 121, 19. v. ciric-mangung.

cyric-mitta *a church-measure.* v. ciric-mitta.

cyric-neód, e ; *f. Church-need* ; ecclesiæ necessitas :—Riht is ðæt man betǽce ǽnne dǽl preóstum, ōðerne dǽl to cyricneóde, þriddan dǽl ðām þearfum *it is right that one part* [of the alms] *be delivered to the priests, a second part for the need of the church, a third part for the poor*, L. Edg. C. 55, note 4 ; Th. ii. 256, 30.

cyric-nyt, -nytt *church-duty* or *service.* v. circ-nyt.

cyric-pæþ, es ; *m. A church-path* ; ad ecclesiam semita :—Of ðære dīce on ðæne cyricpæþ *from the ditch to the church-path*, Cod. Dipl. 736 ; A.D. 1021–1023; Kmbl. iv. 19, 9.

cyric-ragu *church-lichen* or *moss.* v. ciric-ragu.

cyric-rēna, an ; *m.* [rān *robbery*] *Church-robbery, sacrilege* ; sacrilegium :—On cyricrēnan *in sacrileges*, L. Eth. vi. 28 ; Th. i. 322, 20.

cyric-sang, -song, es ; *m. A church-song* ; ecclesiasticum carmen :—He ða cyricsangas lǽrde, ðe hī ǽr ne cūðan *quæ illi non noverant,*

carmina ecclesiastica doceret, Bd. 5, 20 ; S. 642, 8. He wæs on cyricsonge se gelǽredesta *qui cantandi in ecclesia erat peritissimus*, 2, 20 ; S. 522, 25.

cyric-sangere, es ; *m. A church-singer* ; ecclesiæ cantator :—He sumne ædele cyricsangere begeat, se wæs Mafa hāten *he got a famous church-singer, who was named Mava*, Bd. 5, 20 ; S. 642, 5.

cyric-sceat, ciric-sceat, es ; *m. Church-scot, church-money, tax* or *rate* ; ecclesiæ census. Church-scot was at first a certain measure of corn paid to the church. In a charter of Bishop Werfrith, those to whom it was granted, agreed,—Ðæt hī agefen ǽlce gēre þreó mittan hwǽtes to ciric-sceatte to Cliffe *that they should give yearly to Cliff three measures of wheat as church-scot*, Bd. S. 772, 8. Be cyric-sceattum. Cyric-sceattas sīn agifene be Scē Martines mæssan. Gif hwā ðæt ne gelǽste, sié he scyldig lx scill. and be xii fealdum agife ðone ciric-sceat *of church-scots. Let church-scots be given at Martinmas. If any one do not perform that, let him forfeit sixty shillings, and give the church-scot twelvefold*, L. In. 4 ; Th. i. 104, 8–11. Ðæt neád-gafol ūres Drihtnes ; ðæt sȳn, ūre teoðunga and cyric-sceattas *the necessary tribute of our Lord ; that is, our tithes and church-scots*, L. Edg. S. 1 ; Th. i. 270, 25. Cyric-sceat was also a general word, and included not only corn, but poultry or any other provision, that was paid in kind to the church. So in the Inquisition of the Rents of the Abbey of Glastonbury, A.D. 1201 :—In church-scet lx gallinas et semen frumenti ad tres acras, *Chartul. de Glaston. MS.* f. 38 : L. In. 61 ; Th. i. 140, 12–14 : L. Ath. i. prm ; Th. i. 196, 7–10 : L. Edm. E. 2 ; Th. i. 244, 15–18 : L. Edg. i. 2 ; Th. i. 262, 10–17 : L. Eth. vi. 18 ; Th. i. 320, 1–2 : L. Eth. ix. 11 ; Wilk. 114, 19–22 ; Th. i. 342, 27–29.

cyric-sōcn *a church-privilege*, Cod. Dipl. 870 ; Kmbl. iv. 220, 19. v. ciric-sōcn.

cyric-stīg, e ; *f.* [stīg *a way, path*] *A church-path* ; ad ecclesiam callis :—Of ðam hylle on cyricstīge, of cyricstīge on ða blacan þyrnan *from the hill to the church-path, from the church-path to the black-thorn*, Cod. Dipl. 1368 ; Kmbl. vi. 220, 19, 20.

cyric-þēn *a minister of the church*, L. I. P. 25 ; Th. ii. 340, 13. v. ciric-þēn.

cyric-þēnung *church-service*, L. I. P. 23 ; Th. ii. 334, 30. v. ciric-þēnung.

cyric-þingere *a priest.* v. cyrc-þingere.

cyric-tīd, e ; *f. Church-time, time of service in a church* ; in ecclesia ministerii tempus :—His cyrictīda on rihtlīcne tīman *his church-hours at the right time*, L. I. P. 8 ; Th. ii. 314, 20.

cyric-tūn *a church-inclosure, church-yard.* v. ciric-tūn.

cyric-wæcce *a church-watch* or *wake*, L. Edg. C. 28 ; Wilk. 84, 30. v. ciric-wæcce.

cyric-wǽd, e ; *f. A church-garment* ; ecclesiæ vestimentum :—To cyricwǽdum [MS. -wǽdan] *for church-garments*, L. Eth. vi. 51 ; Th. i. 328, 8.

cyric-wag *a church-wall*, L. Eth. vii. 13 ; Wilk. 111, 17. v. ciric-wag.

cyric-waru, e ; *f. A church-congregation* ; in ecclesia congregatio :—On cyricware *in a church-congregation*, L. O. 13 ; Th. i. 184, 12.

cyric-weard, -wyrd *a churchwarden*, Chr. 1044 ; Th. 300, 26, col. 1. v. ciric-weard.

cyrin *a churn* ; sinum, Wrt. Voc. 290, 31. v. ceren.

Cyring-ceaster *Cirencester* :—Æt Cyringceastre *at Cirencester*, Chr. 1020; Th. 286, 13, col. 1. v. Ciren-ceaster.

cyrlisc *rustic, rural* ; rusticus, L. In. 18 ; Th. i. 114, 6, note 8, B. v. ceorlisc.

cyrliscnys, -nyss, e ; *f.* CHURLISHNESS, *clownishness, rudeness* ; rusticitas, Som. Ben. Lye.

cyrm *a noise, shout, uproar*, Andr. Kmbl. 2313 ; An. 1158 : Scint. 55 : Cot. 86. v. cirm.

cyrman *to cry out, shout*, Cd. 166 ; Th. 207, 3 ; Exod. 461. v. cirman.

cyrn *a churn* ; sinum. v. cyrin.

Cyrn-ceaster *Cirencester* :—On Cyrnceastre *in Cirencester*, Chr. 1020; Th. 287, 12, col. 1. v. Ciren-ceaster.

cyrnel, cyrnl ; *gen.* es ; *dat.* cyrnele ; *pl. nom. acc.* cyrnlu ; *gen.* cyrnla ; *n. m?* **I.** *a* KERNEL, *grain* ; nucleus, granum :—Men geseóþ oft ðæt of ānum lytlum cyrnele cymþ micel treów ; ac we ne māgon geseón on ðam cyrnele nāðor ne wyrtruman, ne rinde, ne bogas, ne leáf ; ac God forþtīhþ of ðam cyrnle treów, and wæstmas, and leáf *men often see that of one little kernel comes a great tree ; but in the kernel we can see neither root, nor rind, nor boughs, nor leaves ; but from the kernel God draws forth tree, and fruits, and leaves*, Homl. Th. i. 236, 16–20. Cyrnel *granum*, Ælfc. Gl. 46 ; Som. 65, 8 ; Wrt. Voc. 33, 7. Nim ðone cyrnel ðe byþ innan ðan persogge *take the kernel which is within the peach*, Lchdm. iii. 102, 6. Genim of pīnhnyte xx geclǽnsodra cyrnela *take twenty* [of] *cleansed kernels of the nuts of the stone pine*, L. M. 2, 2 ; Lchdm. ii. 180, 19. Sele ða cyrnlu ðæs eorþ-ifiges on hātum wætre drincan *give him the grains of the ground ivy in hot water to drink*, 2, 39 ; Lchdm. ii. 248, 26. **II.** *a hard*

concretion in the flesh, an indurated gland or *strumous swelling*; toles, glandulæ duriores, quæ succrescunt in isto tumore, quem strumam dicimus:—Wið cyrnlu *for kernels* [or *swelled glands*], Herb. 14, 2; Lchdm. i. 106, 13, 19: Herb. cont. 4, 3; Lchdm. i. 8; 4, 3: 14, 2; Lchdm. i. 12; 14, 2: Herb. 4, 3; Lchdm. i. 90, 8: Med. ex Quadr. 3, 7; Lchdm. i. 340, 14. Lege ofer ða cyrnlu *lay it over the kernels* or *swelled glands*, Herb. 14, 2; Lchdm. i. 106, 19. Wið cyrnla sáre *for sore of kernels* or *swelled glands*, Med. ex Quadr. 6, 3; Lchdm. i. 352, 1. Lege tó ðam cyrnlum [MS. -lun] *lay to the kernels* or *swelled glands*, Herb. 75, 5; Lchdm. i. 178, 13. [*Prompt.* kyrnel: *Plat.* karn: *Dut.* kern, *f.*: *Kil.* kerne: *Ger.* kern, *m.*: *M. H. Ger.* kërne, kërn, *m.*: *O. H. Ger.* kerno, *m.*: *Dan.* kjerne, *m. f.*: *Swed.* kärna, *f.*: *Icel.* kjarni, *m.*] DER. æppel-cyrnel.

cyrps; *adj. Curly*; crispus, tortus:—He is blæcfexede and cyrps *he is black-haired and curly*, Homl. Th. i. 456, 17. Cyrpsum loccum *with curly locks*, Mone B. 1236.

cyrpsian; *p.* ode; *pp.* od *To crisp, curl*; crispare, asperare:—Cyrpsiendum [MS. cyrpisiendum] *crispantibus*, Mone B. 1239. Cyrpsaþ [MS. cypsaþ] *asperat*, Glos. Prudent. Recd. 144, 61.

cyrr, cerr, cirr, cierr, es; *m. A turn, space of time, an occasion, affair*; versio, vices, temporis spatium, negotium:—Æt ðam feórþan cyrre [sǽle, *q. v.*] *at the fourth turn* or *time*, Herb. 100, 3; Lchdm. i. 214, 5, 6, 7, 8: Gen. 38, 18. Æt sumum cyrre *at some turn* or *time, when*; aliquando, Lk. Bos. 22, 32. Se biþ abísgod, on færelde mid óðrum cierrum *who is busied, in a journey with other affairs*, Past. 4, 1; Hat. 9 b, 7. [*Laym.* chærre, cherre: *Plat.* keer, kere, *f.*: *Dut.* keer, *m.*: *Ger.* kehr, kehre, *f.*: *M. H. Ger.* kêre, *f.* kêr, *m.*: *O. H. Ger.* kêra, *f.* kêr, *m.*] DER. ed-cyrr, frum-, ofer-, on-, sǽ-.

cyrran, ic cyrre, ðú cyrrest, he cyrreþ, *pl.* cyrraþ; *p.* cyrde, *pl.* cyrdon; *pp.* cyrred. I. *to turn*; vertere:—He clifu cyrreþ on wæteres wellan *he turneth rocks into wells of water*, Ps. Th. 113, 8. Gif ic míne gewǽda on wíte-hrægl cyme cyrde *et posui vestimentum meum cilicium*, Ps. Th. 68, 11. Cyrred, *pp. turned*, Exon. 107 b; Th. 410, 25; Rä. 29, 4. II. *to be turned, to turn himself, to go, return*; verti, se vertére, ire, reverti:—Ðú wille cyrran *thou wilt be turned*, Cd. 91; Th. 115, 13. Nú cyrrest *now turnest thyself*, Elen. Kmbl. 1329; El. 666. Hí cyrraþ *they return*, Ps. Th. 69, 3. Cyrdon *returned*, Cd. 195; Th. 243, 8; Dan. 433. [*Laym.* charten: *Scot.* cair, kair *to drive backwards and forwards*: *Plat.* keren: *O. Sax.* kêran: *Frs.* keeren: *O. Frs.* kera: *Dut.* keeren: *Kil.* keren, kerien *verrere*: *Ger.* kehren *verrere, vertere*: *M. H. Ger.* kêren: *O. H. Ger.* kerjan *verrere, vertere*: *Dan.* kjöre: *Swed.* köra *to drive*: *Icel.* keyra *to whip, lash, drive*.] DER. a-cyrran, -cerran, be-, for-, ge-, mis-, ofer-, on-. ongeán-, to-, under-, ymb-.

cyrredness, -ness, e; *f. A turning, conversion*; versio, conversio. v. a-cyrredness, ge-.

cyrse, an; *f. Cress*; nasturtium, Lacn. 89; Lchdm. iii. 58, 22. v. cærse.

cyrs-treów, es; *n. A cherry-tree*; cerasus = κεράσος, Ælfc. Gl. 46; Som. 64, 123; Wrt. Voc. 32, 57. v. ciris-beám.

CYRTEL, kyrtel; *gen.* cyrtles; *m. A KIRTLE, vest, garment, frock, coat*; palla, tunica:—Cyrtel *vel* oferbrǽdels *palla*, Ælfc. Gl. 4; Som. 55, 86; Wrt. Voc. 16, 56. Ic gean sancte Æðelþryþe ánes wullenan cyrtles [kyrtles MS.] *I give tó saint Æthelthryth one woollen kirtle*, Cod. Dipl. 782; A.D. 1046; Kmbl. iv. 107, 7. Bicgaþ cyrtlas *buy kirtles*, Homl. Th. i. 64, 13. Ðam ðe wylle on dóme wið ðé flitan, and niman ðíne tunecan [cyrtel oððe hrægl, Mt. Kmbl. Lind.] læt him tó ðinne wǽfels *ei qui vult tecum judicio contendere et tunicam tuam tollere, dimitte ei et pallium*, Mt. Bos. 5, 40; to hym that wole stryue with thee in dome, and take awey thi coote, leeue thou to hym and thin ouer clothe, Wyc. Næbbe ge ne twá tunecan [cyrtlas, Mt. Kmbl. Lind.] *nolite possidere neque duas tunicas*, 10, 10; nyl ȝe welden nether two cootis, Wyc: Lk. Lind. War. 3, 11. Berenne cyrtel [kyrtel MS.] *a bear-skin vest*, Ors. 1, 1; Bos. 20, 38. [*Prompt.* kyrtyl *tunica*: *Piers P.* kirtel: *R. Brun.* kirtelle: *Chauc.* kirtel: *Laym.* curtel: *Orm.* kirrtell: *Plat.* kiddel: *Dut.* kiel, *m.*: *Kil.* kedel, kele: *Ger.* kittel, *m.*: *M. H. Ger.* kitel, kittel, *m.*: *Dan.* kjortel, *m. f.*: *Swed.* kjortel, *m.*: *Icel.* kyrtill, *m.*]

cyrten; *adj. Beautiful, elegant*; venustus:—Hlísful and cyrten *famous and beautiful*, Homl. Th. ii. 220, 29. Ful cyrtenu ceorles dóhtor *a churl's very beautiful daughter*, Exon. 106 b; Th. 407, 16; Rä. 26, 6.

cyrten-lǽcan; *p.* -lǽhte; *pp.* -lǽht *To make lovely, to beautify*; venustare:—Ic cyrtenlǽce *venusto*, Ælfc. Gl. 99; Som. 76, 115; Wrt. Voc. 54, 57.

cyrten-líce; *adv. Notably, solemnly, cunningly*; notabiliter, solemniter, subtiliter, Scint. 38.

CYSE, cése, *m.*: cýsa, an; *m. A CHEESE*; caseus:—Cýse *caseus*, Wrt. Voc. 82, 26: 290, 32. Niwe gáte cýse *new goat's cheese*, Med. ex Quadr. 6, 5, 6, 7; Lchdm. i. 352, 5, 7, 9. Ferscne cýse on lege *lay on fresh cheese*, L. M. 1, 39; Lchdm. ii. 102, 14, 1, 53; Lchdm. ii. 126, 1: Lchdm. iii. 96, 24. Nim cýsan *take cheese*, 96, 21. Tyn cýse [cýsas B. H.] *ten cheeses*, L. In. 70; Th. i. 146, 19. [*Prompt.* chese: *Plat.* kese: *O. Sax.* kêsi, *m.*: *Dut.* kaas, *f.*: *Kil.* kaese, kese: *Frs.* tzys:

O. Frs. kise, tzise, *m.*: *Ger.* käse, *m.*: *M. H. Ger.* kæse, *m.*: *O. H. Ger.* kasi, *m.*: *Lat.* caseus: *Wel.* caws, *m.*: *Corn.* caus, cos, ces, *m.*: *Ir.* cais: *Gael.* caise: *Manx* caashey, *m.*: *Armor.* caouz.]

cýse-fæt, es; *n. A cheese-vat*; vas pro caseo asservando, calăthus = κάλαθος, Cot. 53.

cýse-hwæg, es; *n. Cheese-whey*; siringia:—Ða rinda wyl on cýsehwæge *boil the rinds in cheese-whey*, L. M. 3, 39; Lchdm. ii. 332, 9.

cysel *gravel, sand*; glarea. v. ceosel.

cysel-stán *gravel*, Ælfc. Gl. 11; Som. 57, 46; Wrt. Voc. 19, 48. v. ceosel-stán.

cýs-gerunn, es; *n*? [ge-runnen *coagulatus*] *Rennet* or *runnet, a substance used to produce curd*; lactis coagulum:—Butergeþweor ǽlc and cýsgerunn losaþ eów *butyrum omne et caseus pereunt vobis*, Coll. Monast. Th. 28, 19.

cýs-lyb, -lybb, es; *pl. nom. acc.* -lybbu; *n.* [cýse *cheese*, lyb, lib *a drug*] *Cheese-drug, rennet* or *runnet*; casei coagulum:—Haran cýslybb syle drincan ðam wífe *give the woman a hare's runnet to drink*, Med. ex Quadr. 4, 14; Lchdm. i. 346, 4. Ða meolc geren mid cýslybbe *turn the milk with rennet*, Lchdm. iii. 18, 11. Cýslybbu *coagula*, Glos. Prudent. Recd. 141, 25.

cyspan; *p.* ede; *pp.* ed [cosp *a fetter*] *To bind, fetter*; compedibus constringére:—Sǽdon ðæt hió sceolde cyspan mænigne *they said that she would bind many*, Bt. Met. Fox 26, 154; Met. 26, 77.

cyssan; *p.* cyste; *pp.* cyssed; *v. a.* [cos *a kiss*] *To KISS*; osculari:—Ic cysse ðé *osculor te*: ic eom fram ðé cyssed *osculor a te*, Ælfc. Gr. 19; Som. 22, 51, 52. Ic cysse, ðú cyst, he cyst *osculor, oscularis, osculatur*, 25; Som. 26, 58, 59. Swá hwæne swá ic cysse, se hyt is *quemcumque osculatus fuero, ipse est*, Mt. Bos. 26, 48. Hwílum mec on cófan cysseþ *sometimes he kisses me in a chamber*, Exon. 125 a; Th. 480, 19; Rä. 64, 4. Mec weras cyssaþ *men kiss me*, 108 a; Th. 412, 27; Rä. 31, 6: 104 a; Th. 395, 6; Rä. 15, 3. Ic cyste *osculatus sum*, Ælfc. Gr. 25; Som. 26, 60. He hine cyste *he kissed him*, Homl. Th. ii. 422, 34: ii. 426, 12: Bd. 3, 6; S. 528, 23. He cyste hyne *osculatus est eum*, Mt. Bos. 26, 49: Gen. 48, 10. Æghwæðer óðerne cyston hie *they kissed each other*, Andr. Kmbl. 2031; An. 1018. Ðæt he his mondryhten clyppe and cysse *that he embrace and kiss his lord*, Exon. 77 a; Th. 289, 2; Wand. 42. [*Prompt.* kissin: *Wyc.* kisse: *Piers P.* kissen: *R. Brun.* kisse: *Chauc.* kisse: *R. Glouc.* cussede, *þ*: *Laym.* cusseþ: *O. Sax.* kussian: *O. Frs.* kessa: *Dut.* kussen: *Ger. M. H. Ger.* küssen: *O. H. Ger.* kussjan, kussan: *Goth.* kukyan: *Dan.* kysse: *Swed.* kyssa: *Icel.* kyssa: *Grk.* κυνεῖν, *inf. aor.* κύσαι *to kiss*: *Sansk.* kus *amplecti*.] DER. ge-cyssan.

CYST, cist, cest, e; *f. A CHEST, coffer, coffin, sheath, casket*; capsa, capsella, cista, cistella, loculus:—Hire cyste *cistam suam*, L. C. S. 77; Th. i. 418, 21. He ða cyste æt-hrán *tetigit loculum*, Lk. Bos. 7, 14. On cyste dyde *condidit in capsella*, Bd. 3, 11; S. 536, 9. Ðæt hí woldan his bán on niwe cyste gedón *ut ossa illius in novo recondita loculo locarent*, 4, 30; S. 608, 30: 3, 6; S. 528, 29. Cist *cista*, Wrt. Voc. 288, 31. Cest *cistella*, Ælfc. Gl. 3; Som. 55, 64; Wrt. Voc. 16, 37. [*Chauc.* cheste: *Scot.* kist, kyst: *Dut.* kist, kast: *Kil.* kiste: *O. Frs.* kiste: *Ger. M. H. Ger.* kiste, *f.*: *O. H. Ger.* kista, *f.*: *Dan.* kiste, *m. f.*: *Swed. Icel.* kista, *f.*: *Lat.* cista: *Grk.* κίστη *a chest, box*: *Manx* kishtey, *m. a chest*: *Armor.* kest, *f. a basket*.] DER. bóc-cest.

cyst, cist, e; *f.* [ceósan *to choose*]. I. *choice, election*; optio, electio:—Ic ðé cyst abeád *I have offered thee a choice*, Cd. 91; Th. 115, 14; Gen. 1919. Ðonne beóþ gesomnad, on ða swíðran hond, ða clǽnan folc, Criste sylfum gecorene bi cystum *then shall be assembled, on the right hand, the pure people, chosen by election by Christ himself*, Exon. 25 b; Th. 75, 19; Cri. 1224: Ps. Th. 64, 4. II. *with gen. pl. What is chosen*; æstimatio:—Írena cyst *what is chosen of swords*, Beo. Th. 1350; B. 673: 1609; B. 802: 3398; B. 1697. Wǽpna cyst *what is chosen of weapons*, 3123; B. 1559. Symbla cyst *what is chosen of feasts*, 2469; B. 1232. Him gewát Abraham eástan eágum wlítan on landa [MS. lande] cyst *Abraham departed from the east to look with his eyes on what is chosen of lands* [*Canaan*], Cd. 86; Th. 107, 26; Gen. 1795. Wedera cyst *what is chosen of weathers*, 191; Th. 238, 6; Dan. 350. Sancta Hierusalem, cynestóla cyst *holy Jerusalem, what is chosen of royal thrones*, Exon. 8 b; Th. 4, 11; Cri. 51. Folgoþa cyst *what is chosen of services*, 13 b; Th. 24, 27; Cri. 391. Godwebba cyst, ðæs temples segl *what is chosen of textures, the veil of the temple*, 24 b; Th. 70, 8; Cri. 1135. Eardríca cyst *what is chosen of habitations* [*the garden of Eden*], 45 a; Th. 153, 14; Gú. 825. Eardwíca cyst *what is chosen of dwellings*, 98 a; Th. 366, 21; Reb. 15. Ic swefna cyst segan wylle *I will relate what is chosen of dreams*, Rood Kmbl. 1; Kr. 1. Burga cyst, Róm *what is chosen of cities, Rome*, Bt. Met. Fox 1, 35; Met. 1, 18. III. *excellence, virtue, munificence, goodness*; præstantia, virtus, largitas, bonitas:—Þíonde on eallum cystum and cræftum *flourishing in all excellencies and virtues*, Bt. 38, 5; Fox 206, 23: Exon. 79 b; Th. 299, 22; Crä. 106. Hí hêton heom seggan ðæs landes cysta *they bade them be told of the excellencies of the land*, Chr. 449; Erl. 12, 6. Fród fæder freóbearn lǽrde cystum eald *a wise father,*

old in excellencies, taught his dear son, Exon. 80 a; Th. 300, 7; Fä. 2. Wēnaþ menn ðæt he hit dō for cystum [kystum MS.] men think that he does it for virtue, Past. 20, 1; Hat. MS. 29 a, 27. Ðæt ðū ðīne cysta cýðe that thou mayest shew thy virtues, Prov. Kmbl. 46. Cystum gōd good in virtues, Chr. 1065; Erl. 199, 6; Edw. 23: Beo. Th. 1738; B. 867: 1850; B. 923. Seó gitsung gedēþ gitseras lāðe, and ða cysta gedóþ ða leóftæle covetousness makes misers loathsome, and munificence makes them estimable, Bt. 13; Fox 38, 16. Hū me cynegōde cystum dohten how the noble munificently treated me, Exon. 85 b; Th. 322, 1; Wīd. 56. Þurh Godes micclan cyste through the great goodness of God, Homl. Th. ii. 468, 14. For his micclan ciste of his great goodness, Ælfc. T. 9, 1. [Laym. custe manner, quality: O. Sax. kust, f. choice: Frs. O. Frs. kest, f. choice: Ger. kurst=kur, f. election: M. H. Ger. kust, f. mánner of choosing: O. H. Ger. kust, f. æstimatio, electio, virtus: Goth. ga-kusts, f. what has been tried, a trial; kustus, m. examination: Icel. kostr, m. trial, choice.] DER. gum-cyst, hilde-, un-.

cyst; adj. Desirable; desiderabilis:—Ne hī for âwyht eorþan cyste ða sēlestan geseón woldan pro nihilo habuerunt terram desiderabilem, Ps. Th. 105, 20.

cýst choosest, chooses; 2nd and 3rd pers. pres. of ceósan.

cyst-beám, es; m. [beám a tree] A chestnut-tree; castānea=κάσ-ταvos:—Cystel vel cystbeám castānea, Ælfc. Gl. 46; Som. 65, 6; Wrt. Voc. 33, 5.

cystel, e; f? A chestnut-tree, Ælfc. Gl. 46; Som. 65, 6; Wrt. Voc. 33, 5. v. cyst-beám.

cyste-líce; adv. [cyst munificence] Munificently; largiter:—Sý wuldor and lof ðam wēlegan Drihtne, se ðe his gecorenan swā cystelíce wurþaþ be glory and praise to the bounteous Lord, who so munificently honours his chosen, Homl. Th. ii. 154, 2. Cystelíce largiter, Ælfc. Gr. 38; Som. 41, 42. Ic gife cystelíce largior, 31; Som. 35, 54. Cystelíce dǽlan to distribute bountifully, Homl. Th. ii. 228, 18.

cysten=cystan to get, procure, get the value of; acquirere, æquiparare facere=Se man ðe hafde ân pūnd he ne mihte cystan [MS. cysten] ænne peni at ânne market the man who had a pound could not get the value of a penny at a market, Chr. 1125; Erl. 253, 28: 1124; Erl. 252, 39.

cystig; adj. Munificent, benevolent, bountiful, liberal, generous, good; munificus, largus, probus, bonus:—Cystig largus, Ælfc. Gr. 38; Som. 41, 41; Wrt. Voc. 76. 4. Ðæt he sié cystig that he be benevolent, Past. 20, 2; Cot. MS. Seó mōdor clǽngeorn biþ and cystig the mother is pure and bountiful, Exon. 128 a; Th. 492, 25; Rä. 81, 21. Cystig largus vel dapsilis, Ælfc. Gl. 82; Som. 73, 34; Wrt. Voc. 47, 38: larga, Glos. Prudent. Recd. 145, 51. Bióþ ðǽm to ungemetlíce cystige they are immoderately generous to them, Past. 44, 6. DER. un-cystig.

cystignes, cystines, -ness, -nyss, e; f. Bountifulness, goodness, munificence; liberalitas, largitas, munificentia:—Cystignesse, cystignysse libe-ralitatis, Mone B. 2511. Cystines liberalitas, 2494. We sceolon ofer-winnan woruldlíce gytsunge mid cystignysse úres clǽnan mōdes we must overcome worldly covetousness by the bounty of our pure mind, Homl. Th. ii. 222, 20.

cyst-leás; adj. Fruitless, reprobate; reprōbus:—Him [God] ðā se cystleása [Cain] cwealmes wyrhta andswarode then the reprobate [man] Cain, the worker of murder, answered God, Cd. 48; Th. 61, 28; Gen. 1004.

cystlíc; adj. Munificent; munifícus, Som. Ben. Lye.

cystlíce; adv. Munificently; largiter, Ælfc. Gr. 38. v. cystelíce.

cýs-wuce, an; f. [cýse cheese, wuce a week] Cheese-week, the last week of eating cheese before Lent; septimana dominicæ quinquagesimæ. In the Greek church quinquagesima Sunday is the last day on which cheese may be eaten till Easter. The same rule prevailed in monasteries of the Benedictine order, which only were known in England before the Conquest. 'Abstinentiam ovorum et casei incipimus feria secunda post quinquagesi-mam:'—Ðis sceal on Wōdnes dæg, on ðære syxteóðan wucan ofer Pentecosten; and on Fríge dæg innan ðære cýs-wucan this [Gospel] must be on Wednesday, in the sixteenth week after Pentecost; and on Friday within the cheese-week, Rubc. Mt. Bos. 5, 43, Notes, p. 575.

CÝTA, an; m. A KITE, bittern; milvus, būteo, Ælfc. Gl. 37; Som. 63, 9; Wrt. Voc. 29, 32: Glos. Brux. Recd. 37, 3; Wrt. Voc. 63, 17. [Piers P. kytte: Chauc. kyte: Wel. cud, m.]

cyte, cote, an; f. A cot, cottage, bedchamber, cell; casa, cubiculum, cella:—Tær ðæt hors ðæt þæc of ðære cytan hrófe the horse tore the thatch off the roof of the cottage, Homl. Th. ii. 136, 17. Hī hine lǽddon ūt of ðære cytan they led him out of the cottage, Guthl. 5; Gdwin. 36, 8. Gecyrde he to sumes hyrdes cytan he turned into a shepherd's cottage, Homl. Th. ii. 136, 14. In ðæm he hæfde cirican and cytan in hac habuit ecclesiam et cubiculum, Bd. 3, 17; S. 543, 24, col. 2. Cyte cella, Wrt. Voc. 85, 75. Wæs sum munuc on nēhnesse his cytan eardiende in vicinia cellæ illius habitabat quidam monachus, Bd. 5, 12; S. 630, 42. Leóht of heofenum gefylde ða cytan a light from heaven filled the cell, Homl. Th. ii. 546, 34.

CÝTEL, citel, cetel, es; m. A kettle, brazen or copper pot, cauldron;

cācābus=κάκκᾰβος, lēbes=λέβης:—Hwer vel cytel lebes: cytel cacābus, Ælfc. Gl. 26; Som. 60, 84, 85; Wrt. Voc. 25, 24, 25. Cytel cacābus, Wrt. Voc. 82, 57. On niwum cytele in a new kettle, L. M. 1, 3; Lchdm. ii. 44, 2. On cyperenum citele in a copper kettle, 1, 15; Lchdm. ii. 56, 19. On micelne citel, on læssan citel in a large kettle, in a smaller kettle, 1, 38; Lchdm. ii. 98, 10, 12. Ceteles hrūm kettle-soot, 1, 72; Lchdm. ii. 148, 10. Genim tyn-ámberne cetel take a kettle holding ten ambers, L. M. 1, 36; Lchdm. ii. 86, 13. [Prompt. ketyl, chetyle: Wyc. ketels, cheteles, pl: Plat. ketel: O. Sax. ketil, m: Dut. ketel, m: Frs. tjettel: O. Frs. ketel, szetel, tsetel, m: Ger. kessel, m: M. H. Ger. kezzel, m: O. H. Ger. kezil, m: Goth. katils, m: Dan. kjedel, kedel, m. f: Swed. kittel, m: Icel. ketill, m.]

cytel-hrūm kettle-soot. v. cetel-hrūm.

cytere, an; f. A harp; cithāra=κιθάρα:—Arîs saltēre and cytere exsurge, psaltērium et cithāra, Ps. Spl. C. 56, 11.

CÝÞ, cýþþ, e; f. I. knowledge; notitia, cognitio, scientia:—Cýþþe notitiæ, Mone B. 4214. Of mīnre sylfre cýþþe from my own knowledge, Bd. 5, 24; S. 647, 18. Ðe náne cýþþe to Gode næfdon who have had no knowledge of God, Homl. Th. i. 396, 28. Ðære godcundan cýþþe divinæ cognitionis, Bd. 5, 22; S. 644, 13, 16. II. re-lation, relationship, KITH; familiaritas, munus:—Gif he to ðam cyninge furðor cýþþe hæbbe if he have further relation to the king, L. C. S. 72; Th. i. 414, 17. III. a known land, native country, region, place; situs naturalis, natale solum, patria regio:—Ðis is mīn ágen cýþ this is my own country, Bt. Met. Fox 24, 98; Met. 24, 49. On heora ágenre cýþþe in their own country, Bt. 27, 4; Fox 100, 11. Eorlas on cýþþe men in the country, Andr. Kmbl. 1467; An. 735. Cniht of cýþþe a boy from his country, Cd. 134; Th. 169, 15; Gen. 2800. Ðū meaht to heora cýþþe becuman thou mayest come to their country, Bt. Met. Fox 12, 47; Met. 12, 24. Gif ðū gewītest cýþþe sēcean if thou goest to seek thy country, Salm. Kmbl. 408; Sal. 204. Cýþ region, Bt. 33, 4; Fox 130, 14. Ðær úre cýþþ wæs there was our place, Ps. Th. 121, 2: 119, 5. [Piers P. kith, kyth relationship: Laym. cuðde, f. country, race, kin: Orm. cuþe acquaintance: Plat. kunde, kunne knowledge: O. Frs. kethe, kede news: Dut. kunde, f. knowledge, kindred: Kil. konde notitia: Ger. kunde, f. knowledge, news: M. H. Ger. künde, kunde, f. knowledge, acquaintance, home: O. H. Ger. kundi, f. in un-kundi fraus: Goth. kunþi, n. knowledge: Dan. kunde, f: Swed. kund, m. a customer: Icel. kynni, n. acquaintance.] DER. eald-cýþ, -cýþþ, feor-, ge-, on-.

cýþ, es; m. I. a sprout, germ; germen:—Genim wegbrǽdan þrý cýþas take three sprouts of plantain, Herb. 2, 14; Lchdm. i. 84, 14. II. seed; crementum:—Cýþ crementum, Glos. Brux. Recd. 38, 7; Wrt. Voc. 64, 16. v. cíþ.

CÝÐAN; p. ic, he cýðde, cýdde, ðú cýðdest, cýddest; pp. cýðed. I. to make known, tell, relate, proclaim, announce; nuntiare, annuntiare, narrare, referre, effari, prædicare:—Wordum cýðan to make known in words, Cd. 102; Th. 135, 14; Gen. 2242: Exon. 12 a; Th. 19, 7; Cri. 297. Ongan Dryhtnes ǽ georne cýðan he began the Lord's law gladly to proclaim, Elen. Kmbl. 398; El. 199: 2510; El. 1256. Cýþ narra, Lk. Bos. 8, 39: Mt. Bos. 2, 8: Gen. 37, 14: Bd. 2, 9; S. 511, 32. Cýðdon Cristes gebyrd they announced Christ's birth, Exon. 8 b; Th. 5, 5; Cri. 65: Ps. Th. 77, 7: 101, 16. Cýðe his neáhgebūrum let him tell to his neighbours, L. Edg. S. 7; Th. i. 274, 20. II. to declare, reveal, manifest, shew, perform, confess, confirm, testify, prove; notum facere, revelare, manifestare, ostendere, perhibere, confiteri, testari, probare:—Ic him cýðde ðīnne naman notum feci eis nomen tuum, Jn. Bos. 17, 26. Wīsdom sceoldon weras Ebrēa wordum cýðan [MS. cyðdon] the Hebrew men must reveal wisdom by words, Cd. 176; Th. 221, 33; Dan. 97. Ellen cýðan to manifest valour, Beo. Th. 5384; B. 2695. Wundor cýðan to perform a miracle, Elen. Kmbl. 2222; El. 1112: Andr. Kmbl. 1142; An. 571. Ðe me cýþ befóran mannum qui confitebitur me coram hominibus, Mt. Bos. 10, 32: Jn. Bos. 1, 20. Cýðde, Bd. 4, 25; S. 600, 30. Ðú cýddest tu innotuisti, Ps. Spl. 143, 4. He cýþ testatur, Jn. Bos. 3, 32: 1, 15. Mid áþe cýðan to prove on oath, L. C. S. 15; Th. i. 384, 10. Eallra heora dóme wæs cýðed [MS. kyþed] omnium judicio probatum est, Bd. 5, 19; S. 640, 13. [Piers P. couthen: Chauc. kithe, kythe: Laym. cuðe, cuðen: Orm. kiþenn: O. Sax. kúðian, kundan: O. Frs. ketha, keda: Ger. M. H. Ger. künden: O. H. Ger. kundjan, kundan: Goth. kunþyan: Dan. kynde: Swed. kunna: Icel. kynna.] DER. a-cýðan, for-, ge-, of-, ofer-.

cýðere, es; m. I. a witness; testis:—Oñarison on me cýðeras unrihtwíse insurrexerunt in me testes iniqui, Ps. Spl. 26, 18. Cýðras testes, 34, 13. Hwî gewilnige we gyt cýðera quid adhuc desideramus testes? Mk. Bos. 14, 63. II. a martyr, one who bears witness by his death; martyr=μάρτυρ a witness:—Stephānus is se forma cýðere Stephen is the early martyr, Homl. Th. ii. 34, 13. Þurh ðæs hālgan cýðeres þingunge through the pleading of the holy martyr, 28, 33. Eallum cýðerum to all martyrs, 34, 23.

-cýðig -known? notus? Only used in the compounds on-cýðig, un-, q. v. In German, however, kündig known, is used as a simple word, and as a compound.

cýð-lǽcan; _p._ -lǽhte; _pp._ -lǽht _To become known;_ innotescere :—Cýðlǽce _innotescat,_ Mone B. 4286.

cýð-líc, cýðe-líc; _adj. Manifest;_ manifestus. v. ge-cýðelíc.

cýþling _a relation;_ cognatus, Jn. Lind. War. 18, 26. v. cúða.

cýð-nes, -nys, -ness, -nyss, e; _f. A witness, testimony, testament;_ testimonium, testamentum :—Sume sǽdon leáse cýðnesse agén hine _quidam falsum testimonium ferebant adversus eum,_ Mk. Bos. 14, 57. Cýðnys, 14, 59 : Jn. Bos. 3, 32, 33 : Bd. 2, 7 ; S. 509, 17. Cýðnys _testamentum,_ Ps. Spl. 24, 15. DER. ge-cýðnes.

cýþþe; _gen. dat. acc._ of cýþ, Bt. 27, 3 ; Fox 100, 1, Cott. note 1.

cýððu, e; _f. A native country, home;_ situs natalis :—Fugel his cýððu sécep _the bird seeks its home,_ Exon. 59 b ; Th. 217, 9 ; Ph. 277 : Exon. 119 b; Th. 459, 9 ; Hy.ʼ4, 114. v. cýþ.

cyt-wér, es; _m._ [wér _a weir_] _A weir with a kiddle_ or _a cut for a fish trap;_ kidellus, machina piscatoria in fluminibus ad salmones, aliosque pisces intercipiendos :—On Sæuerne xxx cytwéras _thirty ' cyt-wérs ' on the Severn,_ Cod. Dipl. Apndx. 461; A. D. 956 ; Kmbl. iii. 450, 13, 15, 20, 21, 23.

cýwst, he cýwþ _chewest, chews; 2nd and 3rd pers. pres._ of ceówan.

cýwung, cíwung, e; _f. A chewing;_ ruminatio, Ælfc. Gl. 99 ; Som. 76, 121; Wrt. Voc. 54, 62. v. ceówung, ceówan.

D

D is sometimes changed into ð, as Ic wurde, _or_ Ic wurðe: snídan, snídan _to cut._ 2. _d_ and _t_ are often interchanged, as métte _met,_ for métde. 3. nouns ending in _d_ or _t_ are generally feminine, as Gebyrd, e ; _f. birth:_ Miht, e ; _f. might, power._ 4. a word terminating with ed, d [_Icel._ at, t : _Ger._ et, t] indicates that a person or thing is furnished or provided with that which is expressed by the root, and is usually considered as a participle, although no verb may exist to which it can be assigned ; such words have, therefore, generally ge prefixed to them ; as gehyrned _horned;_ gesceód _shod,_ Rask's Gr. by Thorpe, § 326. 5. the perfect participle ends in ed, od, but when the letters _t, p, c, h, x,_ and _s,_ after another consonant, go before the infinitive an, the vowel before the terminating _d_ is not only rejected, but _d_ is changed into _t;_ as from dyppan _to dip_ would be regularly formed dypped _dippcd,_ contracted into dyppd, dyppt, and dypt _dipped._ 6. the Rune Ⓜ not only represents the letter _d,_ but stands for dæg _a day._ v. dæg III. and RÚN.

DÁ ; _gen._ dán ; _f._ [_that is_ dae = dá ; _gen. dat. acc._ daan = dán ; _pl. nom. acc._ daan = dán ; _gen._ daena = dána ; _dat._ daaum = dáum] _A_ DOE; _f._ :—Dá _damma vel dammula,_ Ælfc. Gl. 13 ; Wrt. Voc. 78, 28. [_Prompt._ doo _dama:_ Wyc. doo : _Chauc._ does, _pl :_ Dan. daa _a doe._] v. buc, bucca; _m. a buck._

daag _anything that is loose, dagling, dangling;_ sparsum, Wrt. Voc. 288, 67. v. dǽg.

DÆD; _gen. dat._ dǽde; _acc._ dǽde, dǽd; _pl. nom. acc._ dǽda, dǽde; _f. A_ DEED, _action;_ actio, actus, factum :—Dǽd _actio,_ Ælfc. Gr. 9, 3; Som. 8, 38 : actus, 11 ; Som. 15, 12. Be ðam ðe seó dǽd sý _according as the deed may be,_ L. Eth. v. 31; Th. i. 312, 10 : vi. 38; Th. i. 324, 23 : L. C. E. 3; Th. i. 360, 13. Seó árfæste dǽd _the goodly deed,_ Bd. 3, 6 ; S. 528, 22 : Cd. 28; Th. 37, 24; Gen. 594 : 226 ; Th. 301, 4; Sat. 576 : Bt. Met. Fox 9, 36; Met. 9, 18 : Chr. 1036; Erl. 165, 11 ; Ælf. Tod. 6. Gesǽton land unspédigran ðonne se frumstól wæs, ðe hie, æfter dǽde, ofadrifen wurdon _they inhabited a land more barren than the first settlement was, which they, after their deed, were driven from,_ Cd. 46; Th. 59, 15 ; Gen. 964. For ðære dǽde _for that deed,_ 125; Th. 159, 23 ; Gen. 2639 : 126 ; Th. 161, 24 ; Gen. 2670. Hió speón hine on ða dimman dǽd _she urged him to that dark deed,_ 32 ; Th. 43, 3 ; Gen. 685. Sceolde he dǽd ongyldan _he must expiate the deed,_ 15 ; Th. 19, 23 ; Gen. 295 : 17; Th. 20, 15 ; Gen. 309 : 25 ; Th. 32, 23 ; Gen. 507 : Beo. Th. 5772 ; B. 2890 : Elen. Kmbl. 772 ; El. 386. Ða alecgendlícan word getácniaþ dǽde _the deponent verbs signify action,_ Ælfc. Gr. 19 ; Som. 22, 56. Ic wraxlige _I wrestle;_ luctor, hér is dǽd _here is action,_ 19 ; Som. 22, 57. Mid ðisre dǽde _with this deed,_ Homl. Th. i. 218, 7 : Exon. 103 b ; Th. 393, 8; Rä. 12, 7. Ne sindon him dǽda dyrne _deeds are not dark to him,_ 23 a ; Th. 65, 5 ; Cri. 1050 : 39 b ; Th. 130, 12 ; Gú. 437. Ðæt his góde dǽda swýðran wearþan ðonne misdǽda _that his good deeds be more prevailing than his misdeeds,_ Chr. 959; Erl. 121, 5. Opene weorþaþ monna dǽde _men's deeds shall be open,_ Exon. 23 a ; Th. 64, 34; Cri. 1047. Ðú scealt þrówian ðinra dǽda gedwild _thou shalt expiate the error of thy deeds,_ Cd. 43 ; Th. 57, 2 ; Gen. 922 : 188 ; Th. 233, 27 ; Dan. 282: Bd. pref ; S. 471, 13 : Exon. 34 a ; Th. 185, 16 ; Az. 8. Ðeáh ðe he dǽda gehwæs dyrstig wære _although he were daring in every deed,_ Beo. Th. 5668 ; B. 2838 : Elen. Kmbl. 2563 ; El. 1283. In his dǽdum _in his deeds,_ Exon. 82 a ; Th. 308, 17 ; Seef. 41: 76 a ; Th. 284, 34 ; Jul. 707 : Cd. 29 ; Th. 38, 6 ; Gen. 602 : Chr. 755 ; Erl. 49, 21. Wile Dryhten sylf dǽda gehýran

the Lord himself will hear of the deeds, Exon. 99 b ; Th. 372, 14 ; Seel. 91: Beo. Th. 393 ; B. 195. Dǽda his hí ongeáton _facta ejus intellexerunt,_ Ps. Spl. 63, 10. Ðæt we ǽfæstra dǽde démen _that we consider the deeds of the pious,_ Exon. 40 a ; Th. 133, 31 ; Gú. 498 : 44 a ; Th. 148, 13 ; Gú. 744 : Ps. Th. 118, 17, 43. Gódum dǽdum _by good deeds,_ Cd. 74 ; Th. 91, 5 ; Gen. 1507 : 91; Th. 116, 14 ; Gen. 1936 : Exon. 53 a ; Th. 185, 5 ; Az. 3 : Ps. Th. 104, 7 : 124, 1 : 135, 3. [_Prompt._ dede _factum:_ Wyc. dedis, _pl :_ R. Brun. dedes, _pl :_ Chauc. R. Glouc. dede : Laym. Orm. dede, _f :_ O. Sax. dâd, _f :_ Frs. diede, dead : O. Frs. diede, _f :_ Dut. daad, _f :_ Kil. dæd : Ger. that, _f :_ M. H. Ger. tat, _f :_ O. H. Ger. tât, _f :_ Goth. deds, _f :_ Dan. daad, _m. f :_ Swed. dåd, _f :_ Icel. dáð, _f._] DER. ǽr-dǽd, bealu-, deófol-, ellen-, fácen-, firen-, gleó-, gód-, gu-, iu-, lof-, mægen-, mis-, oncýþ-, syn-, weá-, wel-, wom-, won-, yfel-.

dǽd-bana, an; _m._ [dǽd _a deed,_ bana _a killer_] _An evil-doer, a perpetrator of murder;_ homicida :—Gif man gehádodne mid fǽhþe belecge, and secge ðæt he wǽre dǽdbana _if any one charge one in holy orders with enmity, and say that he was a perpetrator of homicide,_ L. Eth. ix. 23 ; Th. i. 344, 26.

dǽd-béta, an ; _m. A deed amender, penitent;_ maleficii compensator :—Se dǽdbéta _the penitent,_ L. M. I. P. 3; Th. ii. 266, 16.

dǽd-bétan; _part._ -ende; _p._ -bétte; _pp._ -béted _To make amends, give satisfaction, to be penitent, to repent;_ maleficium compensare, malum bono pensáre, pœnitere :—His sáwle wúnda dǽdbétende gelácnian _to heal the wounds of his soul by making amends,_ Homl. Th. i. 124, 14. Dǽd-béte _shall make amends,_ L. C. S. 41 ; Th. i. 400, 16 : L. Eth. ix. 26 ; Th. i. 346, 6. Ðæt he sealde sóðe gebýsnunge eallum dǽdbétendum, ðe to Drihtene gecyrraþ _that he should give a true example to all, who shall turn to the Lord by doing amend deeds,_ Ælfc. T. 38, 4.

dǽd-bót, e; _f. An amends-deed, repentance, penitence;_ pœnitentia, maleficii compensatio :—Behreówsung oððe dǽdbót _pœnitentia,_ Ælfc. Gr. 33 ; Som. 37, 22. Deóplíc dǽdbót biþ, ðæt lǽwede man swá æscǽre beó, ðæt íren ne cume on hǽre, ne on nægle _it is a deep penitence, that a layman be so untrimmed, that scissors [iron] come not on hair, nor on nail,_ L. Pen. 10 ; Th. ii. 280, 17 : 3; Th. ii. 278, 8. Eornostlíce dóþ médemne weastm ðære dǽdbóte _facite ergo fructum dignum pœnitentiæ,_ Mt. Bos. 3, 8 : Lk. Bos. 3, 3, 8. Búton hý to rihtre dǽdbóte gecyrran _unless they turn to right repentance,_ L. Edm. E. 6 ; Th. i. 246, 16 : Chr. 963; Erl. 123, 15, 21. Dóþ dǽdbóte : sóþlíce genéálǽceþ heofona ríce _pœnitentiam agite: appropinquavit enim regnum cœlorum,_ Mt. Bos. 3, 2 : L. M. I. P. 1 ; Th. ii. 266, 5. Þurh dǽdbóte _through penance,_ L. Pen. 7 ; Th. ii. 278, 19 : L. Edm. E. 3 ; Th. i. 246, 3. Dǽdbóta sind gedihte on mislíce wísan _penances are devised in various ways,_ L. Pen. 13 ; Th. ii. 282, 3.

dǽd-bótnys, -nyss, e ; _f. Penitence;_ pœnitentia, Scint. 9.

dǽd-céne ; _adj. Deed-bold;_ agendo fortis, audax :—Com ingán ealdor þegna, dǽdcéne mon _the prince of thanes, the deed-bold man, came entering,_ Beo. Th. 3294 ; B. 1645.

dǽd-from ; _adj. Deed-strong;_ agendo strenuus :—Hí beóþ ðý dǽd-fromran _they are so much the more energetic,_ Ps. Th. 109, 8.

dǽd-fruma, an ; _m._ [dǽd _a deed,_ fruma II. _an author, inventor_] _A deed-doer, perpetrator, labourer;_ facinoris _vel_ facinorum auctor, actor :—Eádmund cyning, ðýre dǽdfruma _king Edmund, the dear deed-doer,_ Chr. 942 ; Erl. 116, 9 ; Edm. 3 : Andr. Kmbl. 149; An. 75. Grendel, diór dǽdfruma _Grendel, the dire perpetrator,_ Beo. Th. 4186 ; B. 2090. Cain and Abel, ða dǽdfruman, dugeþa strýndon, wélan and wiste _Cain and Abel, the original labourers, acquired goods, wealth and food,_ Cd. 46 ; Th. 59, 27 ; Gen. 970.

dǽd-hata, an ; _m._ [hatian _to hate_] _A deed-hater;_ facinorum osor :—Deógol dǽdhata _a secret deed-hater,_ Beo. Th. 555 ; B. 275.

dǽd-hwæt, _pl._ -hwate, -hwatan ; _adj. Deed quick_ or _active, strenuous, bold;_ promptus et expeditus ad agendum, acer, strenuus :—Hæleþ dǽdhwate _men prompt of deed,_ Exon. 65 b ; Th. 242, 26 ; Jul. 2. Ge wǽron dǽdhwæte _ye were bold of deeds,_ Elen. Kmbl. 584 ; El. 292 : Exon. 23 a ; Th. 24, 15 ; Cri. 385. Ða dǽdhwatan geond ðone ofen eódon _the bold of deed went through the oven,_ Cd. 191 ; Th. 238, 12 ; Dan. 353.

dǽd-leán; _n. A deed-loan_ or _reward, a recompence;_ factorum præmium :—Him eallum wile mihtig Drihten dǽdleán gyfan _the mighty Lord will give them all a recompence,_ Cd. 156 ; Th. 194. 20 ; Exod. 263.

dǽd-líc; _adj. Deedlike, active;_ activus :—Twegen dǽlnimende cumaþ of ðam dǽdlícum worde _duo participia veniunt a verbo activo,_ Ælfc. Gr. 24 ; Som. 25, 30. Dǽdlíce word _activa verba,_ Ælfc. Gr. 19 ; Som. 22, 28. Ðás and ðýlíce synd _activa,_ ðæt synd dǽdlíce gehátene, forðanðe hí geswuteliaþ dǽda _these and the like are activa, which are called active, because they declare actions,_ 19 ; Som. 22, 30, 37.

dǽdon, dǽdun _did, made:_—Ðæt hie to mete dǽdon _that they made for food,_ Cd. 33 ; Th. 45, 6 ; Gen. 722 ; _p. pl._ of dón.

dǽd-róf; _adj. Deed-famed, illustrious, valiant;_ agendo celeber _vel_ strenuus :—Abraham andswarode, dǽdróf, Drihtne sínum _Abram the deed-famed answered his Lord,_ Cd. 99 ; Th. 131, 8 ; Gen. 2173 : 121 ; Th. 156, 16 ; Gen. 2589.

dǽd-scúa, an; *m.* [scúa *a shade*] *One who acts in the dark*; in tenebris agens, diabolus:—Deorc dǽd-scúa *a dark deed actor* [*the devil*], Exon. 11 b; Th. 16, 22; Cri. 257. v. deáþ-scúa.

dǽd-weorc, es; *n. A work of works, great work*; facinus egregium:—Hereþreátas for ðam dǽdweorce Drihten hēredon *the army-bands praised the Lord for that great work*, Cd. 170; Th. 214, 26; Exod. 575.

dǽftan; *p.* dǽfte; *pp.* dǽft *To make convenient* or *ready, put in order*; apparare, sternere:—Ðæt he sceolde gearcian and dǽftan his weg [MS. weig] *that he might prepare and make ready his way*, Homl. Th. i. 362, 8. Menn dǽftaþ heora hús *men put their houses in order*, ii. 316, 7. Dǽfte *straverat*, Glos. Prudent. Recd. 149, 73. DER. ge-dǽftan.

dǽft-líce; *adv.* DEFTLY, *aptly, fitly*; commode, opportune. DER. ge-dǽftlíce, unge-.

DÆG; *gen.* dæges; *pl. nom. acc.* dagas; *m:* daga, an; *m.* I. *a* DAY; dies:—Se dæg segþ ðam óðrum dæge Godes wundru *one day to another tells of God's wonders*, Ps. Th. 18, 2. God hét ðæt leóht, dæg *God called the light, day*, Gen. 1, 5. Se þridda dæg *the third day*, Gen. 1, 13. Emnihtes dæg *the day of equinox*; æquinoctium, Menol. Fox 347; Men. 175. Wintres dæg *the winter's day* or *beginning of winter*, Menol. Fox 401; Men. 202. II. *the time of a man's life*; tempus vitæ humanæ:—On midle mínra dagena *in the midst of my days*, Ps. Th. 101, 21. Heora dagena tíd *dies eorum*, 77, 32. On þreóra monna dæg *in three men's days* or *lives*, Bd. App. S. 771, 45. III. *the Anglo-Saxon Rune* ᛞ = *the letter d, the name of which letter in Anglo-Saxon is dæg a day*; hence this Rune not only stands for the letter *d*, but for dæg *a day*, as,—ᛞ byþ Drihtnes sond, deóre mannum *day is the Lord's messenger, dear to men*, Hick. Thes. vol. i. p. 135; Runic pm. 24; Kmbl. 344, 9. IV. *the daily service of the early English church is recorded, referring to the example of the Psalmist*, thus,—Dauid cwæþ seofon síðon on dæg ic sang ðē, Drihten, to lofe,—Ðæt is Ǽrst on ǽrne morgen;—Eft on undern-tíde; and 3 on midne dæg,—and 4 on nón,—and 5 on ǽfen,—and 6 on fóran niht,—and 7 on úhtan tíman *David said,—seven times in a day, O Lord, I sang to thee in praise, that is,—First, in early morning [at break of day]*;—*Next at nine o'clock*;—and *3ly at midday*;—and *4ly at the nones*, 3 *o'clock*;—and *5ly at even, at 6 o'clock, the 12th or an even or equal part of the 24 hours from 6 a.m. to 12 p.m*;—and *6ly at the fore night [at 9 o'clock]*;—and *7ly at midnight, that is from 12 o'clock at night, to 3 or later in the morning* ǽr dægrēde *before dawn*, Canon. Hrs. 361, 7-362, 6: Ælfc. Gl. 95; Som. 75, 126-76, 1; Wrt. Voc. 53, 7-15. v. tíd-sang. ¶ On dæg *in the day, by day*. To dæg *to-day*. Dæg ǽr *the day before*. On ǽrran dæg *on a former day*. Óðre dæg *another day*. [Laym. dæi, dai: Orm da33: Plat. dag: O. Sax. dag, *m*: Frs. dey: O. Frs. di, dei, dach, *m*: Dut. dag, *m*: Ger. tag: M. H. Ger. O. H. Ger. tac, tag, *m*: Goth. dags, *m*: Swed. Dan. dag, *m*: Icel. dagr, *m*: Lat. dies: Sansk. div, dyaus, *m. f. day*.] DER. ǽr-dæg, blǽd-, deáþ-, dóm-, eald-, ealdor-, earfoþ-, ende-, feorh-, freóls-, fyrn-, gang-, geár-, gebéd-, gebyrd-, gefeoht-, geheald-, geld-, gemynd-, geswinc-, gewin-, gyrstan-, lǽn-, líf-, mǽl-, mid-, ræst-, síþ-, swylt-, symbel-, tíd-, weder-, weorc-, wíl-, win-, winter-, wyn-: heó-dæg: án-dæge: daga, án-daga.

dæg-candel, -condel, -candell, e; *f. Day-candle, the sun*; diei candela, sol:—Dægcondel, Exon. 130 b; Th. 499, 34; Rä. 88, 26. Dryhten forlét dægcandelle scínan *the Lord permitted the sun [the day-candle] to shine*, Andr. Kmbl. 1670; An. 837. DER. candel.

dæges; *adv.* [*from gen. of* dæg] *Daily*; die:—Dæges and nihtes *die ac nocte*, Ps. Th. 1, 2: Bt. 35, 6; Fox 168, 7.: Chr. 894; Erl. 93, 5. DER. ig-dæges, y-dæges.

dæges eáge, ége, an; *n.* [dæges, *gen. of* dæg *a day*; eáge, ége *an eye*: *a day's eye*] *A* DAISY; bellis perennis, Lin:—Dæges eáge consolida, Wrt. Voc. 79, 14. Dæges ége consolida, Ælfc. Gl. 42; Som. 64, 26; Wrt. Voc. 31, 36: Lchdm. iii. 292, 8.

dæg-fæsten, es; *n.* [fæsten *a fast*] *A day's fast*; diei jejunium:—Is se ǽresta lǽcedóm dægfæsten, ðæt mon mid ðý ða wambe clǽnsige, ðæt hió ðý ðe leóhtre sié *the first remedy is a day's fast, that, with that, a man may cleanse the stomach, that it may be the lighter*, L. M. 2, 25; Lchdm. ii. 216, 25.

dæg-feorm, e; *f.* [feorm *food, sustenance*] *Food for a day*; unius diei victus:—Áne dægfeorme *a day's sustenance*, Cod. Dipl. 477; A.D. 958; Kmbl. ii. 355, 5.

dæg-hluttre; *adv.* [hluttre *brightly, clearly*] *Brightly as day*; clare instar diei:—Dýre Dryhtnes þegn dæghluttre scán *the Lord's dear minister shone brightly as day*, Exon. 42 b; Th. 143, 23; Gú. 665.

dæg-hwam; *adv. Daily*; quotidie:—Nim cneówholen dæghwam *take knee holly daily*, L. M. 1, 39; Lchdm. ii. 102, 10. Lufiaþ ða ðe dæghwam Dryhtne þeówiaþ *they love those who daily serve the Lord*, Exon. 33 b; Th. 106, 34; Gú. 51: 38 a; Th. 125, 20; Gú. 357.

dæg-hwamlíc, -hwomlíc; *def.* se -líca, seó, ðæt -líce; *adj. Daily*; diurnus, quotidianus:—Hit ealle beorhtnysse dæghwamlíces leóhtes ofer-swýðde *it overshone all the brightness of the daily light*, Bd. 4, 7; S. 575, 20. Syle us to-dæg úrne dæghwamlícan hláf *panem nostrum quotidianum da nobis hodie*, Lk. Bos. 11, 3: Mt. Bos. 6, 11: Homl. Th. i. 264, 31.

Betwyh gehald regollíces þeódscipes and ða dæghwamlícan gýmenne to singanne on cyricean, ne symble swéte and wynsum wæs ðæt ic oððe leornode, oððe lǽrde, oððe wríte *inter observantiam disciplinæ regularis et quotidianam cantandi in ecclesia curam, semper aut dicëre, aut docëre, aut scribëre dulce habui*, Bd. 5, 23; S. 647, 26 28. To dæghwomlícum bigleófan *for their daily subsistence*, Homl. Th. ii. 118, 30.

dæg-hwamlíce; *adv. Daily*; quotidie:—Ic dæghwamlíce mid eów wæs *quotidie eram apud vos*, Mk. Bos. 14, 49. Se bróðor dæghwamlíce wæs wyrse and wyrse *the brother was daily worse and worse*, Bd. 4, 32; S. 611, 24.

dæg-hwíl, e; *f.* [dæg *day*, hwíl *time*] *Day-time, time of life*; diei hora vel tempus:—Ðæt he dæghwíla gedrogen hæfde, eorþan wynne *that he had finished his days, his joy of earth*, Beo. Th. 5445; B. 2726.

dæg-hwomlíc *daily*, Homl. Th. ii. 118, 30. v. dæg-hwamlíc.

dægian *to dawn, become day*, Som. Ben. Lye. v. dagian.

dǽglan *secret, hidden, unknown*, Bt. 25; Fox 88, 26; *acc. pl. def. of* dǽgol = dígol.

dæg-lang, -long; *adj. Lasting a day*:—Dæglounge fyrst *per totam diem*, Salm. Kmbl. 1000; Sal. 501.

dæg-langes; *adv. During one day, for a day*; per unam diem:—Beó ðé stille dæglanges ðínre fyrdinge *be still for a day from thy march*, Homl. Th. ii. 482, 29. v. dæg-lang.

dæg-líc; *adj. Daily*; quotidianus:—Twá dæglíc fæsten oððe þreó dæglíc is genóh to healdenne *biduanum vel triduanum sat est observare jejunium*, Bd. 4, 25; S. 600, 8.

dæg-mǽl, es; *n.* [mǽl *a mark*] *A day-mark, an instrument for telling the hour, a dial, clock*; horologium = ὡρολόγιον = ὥρα *an hour*; λόγιον *a telling, an announcement*, Ælfc. Gl. 30; Som. 61, 58; Wrt. Voc. 26, 57.

dæg-mǽls-pílu [*for* dæg-mǽles píl], e; *f. The style of a dial*; horologii gnomon, Ælfc. Gl. 30; Som. 61, 59.

dæg-mǽl-sceáwere, es; *m. Who* or *what shews the time of day*; horoscopus, Ælfc. Gl. 112; Som. 79, 103: 4; Som. 56. 2.

dæg-mete, es; *m.* [dæg *a day*, mete *meat, food*] *Daily food*; quotidianus cibus:—Dæg-mete *agapis*, Cot. 15, Som. Ben. Lye.

dæg-rǽd, -rǽd, es; *n. Dawn, daybreak, early morning*; dilúculum, matutīnum, auróra:—Dægréd *dilúculum*, Ælfc. Gl. 95; Som. 75, 127; Wrt. Voc. 53, 8. Syxta is *matutīnum vel auróra* ðæt is dægréd [-ræd MS. R.] *the sixth is matutínum vel auróra that is dawn*, Bd. de nat. rerum; Wrt. popl. science 6, 18; Lchdm. iii. 244, 5. Ðis wæs eall geworden ǽr dægréde *this was all performed ere daybreak*, Cd. 223; Th. 294, 4; Sat. 466: Homl. Th. i. 508, 32: 592, 22. Betweox ðam dægréde [-ræde MS. R.] and sunnan upgange *between dawn and sunrise*, Bd. de nat. rerum; Wrt. popl. science 6, 19; Lchdm. iii. 244, 6. Cwom María on dægréd *Mary came at dawn*, Exon. 119 b; Th. 459, 34; Hö. 9: 57 a; Th. 204, 15; Ph. 98: Cd. 222; Th. 289, 27; Sat. 404: Salm. Kmbl. 429; Sal. 215. Se Hǽlend com on dægréd to ðam temple *Iesus dilúculo venit in templum*, Jn. Bos. 8, 2: Lk. Bos. 24, 1: Ex. 8, 20. To ǽfenne þurhwunaþ wóp and on dægréd blisse *ad vesperum demorábitur fletus et ad matutínum lætítia*, Ps. Lamb. 29, 6: Gen. 32, 22. Ðæt leóht, ðe we hátaþ dægréd, cymþ of ðære sunnan *the light, which we call dawn, cometh from the sun*, Bd. de nat. rerum; Wrt. popl. science 2, 29; Lchdm. iii. 234, 29. Ic-gá út on dægréd *exeo dilúculo*, Coll. Monast. Th. 19, 13: Ælfc. T. 24, 11.

dægréd-líc; *adj. Of* or *belonging to the morning, early*; matutinus, matutinalis:—Fram heordnesse dægrédlíce *a custodia matutina*, Ps. Lamb. 129, 6. We sungon dægrédlíce lofsangas *cantavimus matutinales laudes*, Coll. Monast. Th. 33, 27.

dægréd-sang, es; *m. Morning song*; matutīna cantio, C. R. Ben. 20.

dægréd-wóma, an; *m.* [dægréd *daybreak, dawn*, wóma *a noise, rushing*] *Rush* or *noise of dawn*; auroræ strepitus:—Oþ-ðæt eástan cwom ofer deóp gelád dægrédwóma, wedertácen wearm *until there came from the east over the deep way the rush of dawn, a warm weather-token*, Exon. 51 b; Th. 179, 24; Gú. 1266: Andr. Kmbl. 249; An. 125.

dæg-rím, es; *n.* [dæg *day*, rím *a number*] *A number of days, a course of days*; dierum numerus:—Wiste ðe geornor ðæt his aldres wæs ende gegongen, dógora dægrím *he knew the better that his life's end was passed, his days' number*, Beo. Th. 1650; B. 823. Upon ðæt ígland ðǽr Apollines dóhtor wunode dægrímes worn *upon the island where Apollo's daughter dwelt a number of days*, Bt. Met. Fox 26, 66; Met. 26, 33: Cd. 47; Th. 60, 1; Gen. 975: 67; Th. 80, 20; Gen. 1331. On his dægríme *in his number of days*, Exon. 83 b; Th. 314, 10; Mód. 12. Dægríme fród *wise in number of days*, 130 a; Th. 498, 15; Rä. 88, 2: Cd. 99; Th. 131, 9; Gen. 2173.

dæg-rima, an; *m.* [dæg *day*, rima *a rim, edge*] *Daybreak, morning*; aurora:—Hwæt is ðeós ðe astíhþ swilce arísende dægrima *what is this which ascends like the rising morn?* Homl. Th. ii. 442, 33. Dægríma *aurora*, Ælfc. Gl. 95; Som. 75, 128; Wrt. Voc. 53, 9: Hymn. Surt. 8, 21.

Dægsan stán, Degsa-stán, Dæg-stán, es; *m.* [*Flor. Hunt.* Degsastan: *the stone of Degsa*] DAWSTON cr *Dalston, Cumberland*; loci nomen in

agro Cumbriæ:—Hér Ægþan Scotta cyng feaht wid Dælreoda, and wid Æðelferþe, Norþhymbra cynge, æt Dægstáne [Dægsan stáne, Th. 37, 26], and man ofslóh mæst ealne his here *in this year* [A.D. 603] *Ægthan king of the Scots fought against the Dalreods, and against Æthelfrith, king of the Northumbrians, at Dawston, and almost all his army was slain,* Chr. 603; Th. 36, 24–29, col. 1. Wæs ðis gefeoht geworden on ðære mæran stówe ðe cweden is Degsastán *this battle was fought in the famous place which is called Dawston,* Bd. 1, 34; S. 499, 32.

dæg-sceald, es; *m.* [dæg *day*, sceald = scild, scyld *a shield*] *A day shield* or *screen;* diei velamen:—Dægscealdes hleó wand oter wolcnum *the day shield's shade* [i.e. *the pillar of cloud*] *rolled over the clouds,* Cd. 146; Th. 182, 22; Exod. 79.

dæg-steorra, an; *m.* [dæg *a day*, steorra *a star*] *The day star;* lucifer, aurora:—Seó sunne and se móna, and æfensteorra and dægsteorra, and óðre þrý steorran, ne synd ná fæste on ðam firmamentum *the sun and the moon, and the evening star and the day star, and three other stars, are not fast in the firmament,* Bd. de nat. rerum; Wrt. popl. science 15, 28; Lchdm. iii. 270, 3: Ælfc. T. 24, 11. Upasprungen scínþ dægsteorra *ortus refulget lucifer,* Hymn. Surt. 27, 23. Nú gæþ dægsteorra up *jam ascendit aurora,* Gen. 32, 26. Ær dægsteorran ic cende ðé *ante luciferum genui te,* Ps. Spl. 109, 4.

dæg-ðerlíc [= dæg-hwæðer-líc]; *adj. Daily, present;* diurnus, hodiernus:—Ðis dægðerlíce gódspel sprecþ ymbe ðæra Iudéiscra þwyrnysse *this daily gospel speaks of the perversity of the Jews,* Homl. Th. ii. 224, 29. On ðisre dægðerlícan rædinge *in this daily lecture,* i. 194, 24. Se gódspellere Lucas beleác ðis dægðerlíce gódspel mid feáwum wordum *the evangelist Luke concluded the gospel of this day with few words,* i. 90, 8. Ðás dægðerlícan þénunga *these daily services,* ii. 86, 24. Hí þeónde þurhwunodon óþ ðisum dægðerlícum dæge *they have continued prospering to this present day,* ii. 132, 14: i. 28, 28: 32, 8.

dæg-þern, e; *f. A day's space;* diei spatium:—Læt simle dægþerne betweonum *leave always a day's space between,* L. M. 2, 39; Lchdm. ii. 248, 20: 2, 51; Lchdm. ii. 268, 1.

dæg-tíd, e; *f.* [dæg *day*, tíd *time*] *Day-time, time;* diei tempus:—On ðære dægtíde *at that time,* Cd. 80; Th. 100, 4; Gen. 1659. On dægtídum *in the day-time,* Exon. 105 a; Th. 398, 26; Rä. 189, 3: 126 a; Th. 484, 23; Rä. 71, 6.

dæg-tíma, an; *m.* [tíma *time*] DAY-TIME, *day;* diurnum tempus, dies:—þurh dægtíman oððe geond dæg sunne ne forswæle ðé ne móna *per diem sol non uret te, neque luna,* Ps. Lamb. 120, 6.

dæg-wæccan, *pl. f.* [wæcce *a watching*] *Day-watchings;* excubiæ, Ælfc. Gl. 7; Som. 56, 68; Wrt. Voc. 18, 20.

dæg-weard, es; *m.* [weard *a watchman*] *A day-watchman;* excubitor, vigil, Ælfc. Gl. 7; Som. 56, 69; Wrt. Voc. 18, 21.

dæg-weorc, es; *n.* [weorc *work*] *A day's work;* diei opus:—Him mihtig God ðæs dægweorces deóp leán forgeald *the mighty God recompensed to him a high reward for that day's work,* Cd. 158; Th. 197, 30; Exod. 315: 167; Th. 209, 28; Exod. 506: Byrht. Th. 136, 8; By. 148. Æt ðam dæg-weorce *at that day's work,* Elen. Kmbl. 291: El. 146. Ðætte he ðæt dægweorc dreóre gebohte *that he bought that day's work with blood,* Cd. 149; Th. 187, 14; Exod. 151: 169; Th. 210, 21; Exod. 518.

dæg-weorþung, e; *f.* [weorþung *an honouring, celebration*] *A commemoration* or *celebration of a feast-day;* diei festi celebratio:—Ðe on gemynd nime ðære deórestan dægweorþunga róde under róderum *who may bear in remembrance the honouring of the day of the most precious cross under the firmament* [i.e. *the feast of the Invention of the Cross*], Elen. Kmbl. 2466; El. 1234.

dæg-wine, es; *n? A day's pay;* diarium:—Dægwine *diarium,* Ælfc. Gl. 33; Som. 62, 32; Wrt. Voc. 28, 15. Dægwine *pensum vel diarium,* 64; Som. 69, 9; Wrt. Voc. 40, 43. v. wine.

dæg-wist, e; *f.* [wist *food*] *A day's food;* diei victus:—Ðæt he him dægwistes tiðode *that he would give him a day's food,* Homl. Th. ii. 134, 30.

dæg-wóma, an; *m.* [wóma *a noise*] *The rush of day, the dawn;* diei apparitio, aurora:—Dægwóma becwom, morgen mæretorht *the dawn came, the beautiful morning,* Cd. 160; Th. 199, 26; Exod. 344. Dægwóman bitweon and ðære deorcan niht *between dawn and the dark night,* Exon. 50 b; Th. 175, 7; Gú. 1191.

DÆL; *gen.* dæles; *dat.* dæle; *pl. nom. acc.* dalu, dalo; *n. A* DALE, *den, gulf;* vallis, barathrum:—Ðæs dæles se dæl *the part of the dale,* Ors 1, 3; Bos. 27, 29. In deóp dalu *into the deep dales,* Exon. 130 a; Th. 498, 21; Rä. 88, 5: 56 a; Th. 199, 11; Ph. 24. We synd aworpene on ðás deópan dalo *we are cast into these deep dens* [*hell*], Cd. 22; Th. 27, 21; Gen. 421. On ðæt deópe dæl deófol gefeallaþ *devils shall fall into the deep gulf,* Exon. 30 b; Th. 93, 26; Cri. 1532. [*Prompt.* dale *vallis:*—Piers P. Chauc. Laym. Orm. dale: *Plat.* daal: *O. Sax.* dal, *n: Frs.* dalle, deil: *O. Frs.* del, deil: *Dut.* dal, *n: Ger.* thal, *n: M. H. Ger. O. H. Ger.* tal, *n: Goth.* dal, *n: Dan.* dal, *m. f: Swed.* dal, *m: Icel.* dalr, *m: Wel.* dôl: *Corn.* dol, *f: Ir. Gael.* dail: *Manx* dayll, *f.*] DER. of-dæl.

DÆL, es; *m.* I. *a part, portion,* DEAL; pars, portio:—Ðæs dæles se dæl *the part of the dale,* Ors. 1, 3; Bos. 27, 29. Ðú offrast teóðan dæl smedeman *thou shalt offer a tenth deal of flour;* offeres decimam partem similæ, Ex. 29, 36, 40. Hí heora gód on swá manige dælas todælaþ *they divide their goods into so many parts,* Bt. 33, 2; Fox 122, 26. Micel dæl bewylledes wæteres on huniges gódum dæle *a great deal of boiled water in a good deal of honey,* L. M. 2, 20; Lchdm. ii. 202, 27. Gódne dæl *a good deal,* L. M. 2, 55; Lchdm. ii. 276, 6. Ðæs íglandes mycelne dæl *a great deal of the island,* Chr. 189; Ing. 9, 11. Fæder, syle me mínne dæl mínre æhte, ðe me to gebýreþ, Lk. Bos. 15, 12; *fadir, gyue to me the porcioun of substaunce, that byfallith to me,* Wyc. Be dæle *in part, partly,* Chr. 1048; Erl. 178, 5. Sume dæle *in some part, partly,* Cot. 154. II. *a part of speech in grammar;* pars orationis:—Eahta dælas sind *partes orationis sunt octo,* Ælfc. Gr. 5; Som. 3, 22. Interjectio is betwyxaworpennyss. Se dæl líþ betwux óðrum wordum, and geswutelaþ ðæs módes styrunge *an interjection is a throwing between. This part of speech lieth between other words, and denotes a stirring of the mind,* 5; Som. 3, 55. III. *a part of a sentence, a word;* verbum:—We todælaþ ða bóc to cwydum, and siððan ða cwydas to dælum, eft ða dælas to stæfgefégum *we divide the book into sentences, and then the sentences into words* [*parts*], *again the words into syllables,* Ælfc. Gr. 2; Som. 2, 37–39. [*Prompt.* dele: Wyc. deel: Piers P. del, deel: Chauc. del, delle: Laym. dæle, dal, del: Orm. dæl, dale, del: Scot. dail: Plat. deel: O. Sax. dél, deil, m: Frs. deel: O. Frs. del, n: Dut. deel, n: Kil. deel, deyl: Ger. theil, m: M. H. Ger. teil, m: O. H. Ger. teil, m. n: Goth. dails, f: Dan. deel, m. f: Swed. del, m: Icel. deill, m: Sansk. dal *findere.*] DER. eást-dæl, niðer-, norþ-, súþ-, west-: or-dæle.

dælan; *p.* de; *pp.* ed; *v. a* [dæl *a part, deal*] *To divide, separate, distribute, bestow, spend, dispense,* DEAL, DOLE; dividěre, distribuěre, separáre ab aliquo:—Israélas ongunnon dælan ealde mádmas *the Israelites began to divide old treasures,* Cd. 171; Th. 215, 17; Exod. 584. Onfóþ and dælaþ betwux eów *accipite et dividite inter vos,* Lk. Bos. 22, 17: Ps. Spl. 21, 17: 111, 8. Mathusal mágum dælde gestreón *Mathuselah distributed the treasure to his brethren,* Cd. 52; Th. 65, 21; Gen. 1069. Dælde eall ðæt heó áhte *she had spent all that she had,* Mk. Bos. 5, 26. [*Prompt.* delyñ: Wyc. delen: Piers P. delen, dele, deelen: Chauc. dele: R. Brun. daile: R. Glouc. dele: Laym. dælen, dalen, delen: Orm. dælenn: Plat. delen: O. Sax. délian, deilan: Frs. deelen: O. Frs. dela: Dut. deelen: Kil. deelen, deylen: Ger. theilen: M. H. Ger. teilen: O. H. Ger. teiljan: Goth. dailjan: Dan. dele: Swed. dela: Icel. deila.] DER. a-dælan, be-, bi-, for-, ge-, to-.

dæledlíce *by itself, apart,* Som. Ben. Lye. DER. ge-dæledlíce, to-.

dælend, es; *m.* [dælende, *part. of* dælan *to divide*] *A dealer, divider, distributor;* divísor:—Hwá sette me déman, oððe dælend, ofer inc *quis me constituit judicem, aut divisórem, super vos?* Lk. Bos. 12, 14.

dælere, es; *m. A* DEALER, *divider, distributor, agent;* divísor, sequester:—Dælere *divísor,* Ælfc. Gl. 33; Som. 62, 28; Wrt. Voc. 28, 11: 74, 15. Ic wæs dælere betwix Gode and eów *ego sequester et medius fui inter Dominum et vos,* Deut. 5, 5. Ðam wædlan gedafenaþ ðæt he gebidde for ðane dælere *on the indigent it is incumbent that he pray for the distributor,* Homl. Th. i. 256, 33. God gesette ðone wélegan dælere on his gódum God *appointed the wealthy a distributor of his goods,* ii. 102, 28.

dælf, es; *n?* [delfan *to dig*] *Anything dug out, a* DELF, *ditch;* fossa, scrobis:—Eástweard to cynges dælf *eastward to the king's delf,* Chr. 963; Erl. 122, 17: 963; Erl. 123, 6.

dæling, e; *f. A dividing, parting;* partitio, Som. Ben. Lye. DER. to-dæling.

dæl-leás; *adj.* [dæl *a part, portion*] *Without a part, portionless, deficient;* expers:—Dælleás *vel* cræftleás *expers, indoctus,* Ælfc. Gl. 18; Som. 58, 123; Wrt. Voc. 22, 36: 90; Som. 75, 2; Wrt. Voc. 51, 47.

dæl-mælum; *adv.* [mælum, *dat. pl. of* mæl, *n.*] *By parts* or *pieces;* partim, Ælfc. Gr. 38; Som. 41, 59: *particulátim,* Mone B. 148: 3549: *paulátim,* 2635.

dæl-neomend *a sharer, partaker,* Ps. Th. 118, 63. v. dæl-nimend.

dæl-niman; *p.* -nam, *pl.* -námon; *pp.* -numen *To take part, to participate;* participáre. v. dæl-nimend, *etc.*

dæl-nimend, -nymend, -neomend, es; *m.* [nimende, *part. of* niman *to take*]. I. *a taker of a part, a sharer, partaker, participator;* particeps:—Ðæt se Hælend dælnimend wære úre deádlícnysse *that the Saviour was a partaker of our mortality,* Homl. Th. i. 36, 33. Se níþfulla is ðæra deófla dælnimend *the envious is a par icipator with devils,* i. 606, 5. Ic eom dælneomend ðe heom ondrædaþ ðé *particeps ego sum omnium timentium te,* Ps. Th. 118, 63. Toforan eallum his dælnymendum on ðære menniscnysse *before all his participators in humani'y,* Homl. Th. ii. 230, 26. II. *in grammar,—A participle;* participium:—Participium is dælnimend: he nimþ ænne dæl of naman, and óðerne of worde *a participle is a taker of parts: it takes one part from a noun, and the o'her from a verb,* Ælfc. Gr. 5; Som. 3, 40. Sume

adverbia cumaþ of dǽlnimendum *some adverbs come from participles*, 38; Som. 41, 11.

dǽl-nimendlíc *sharing, partaking, participial*, Som. Ben. Lye.

dǽl-nimendnes, -ness, e; *f. A sharing, participation*; participatio:—Ðǽre dǽlnimendnes [-nimendes] his on ðæt sylfe *cujus participatio ejus in id ipsum*, Ps. Spl. 121, 3.

dǽl-nimung, e; *f. A share, portion*; portio:—Dǽlnimung oððe spéde mín on lande lyfigendra *portio mea in terra viventium*, Ps. Lamb. 141, 6.

dǽl-numelnes, -ness, e; *f.* [numol *taking, receiving*] *A sharing, partaking, participation*; participatio:—Ðǽre dǽlnumelnes is hire on ðæt sylfe *cujus participatio est ejus in id ipsum*, Ps. Lamb. 121, 3.

dǽl-nymend *a sharer, participator*, Homl. Th. ii. 230, 26. v. dǽlnimend.

dǽma, an; *m. A judge*; judex, arbiter:—Bête swá mycel swá dǽman tǽcan *subjacebit damno quantum arbitri judicaverint*, Ex. 21, 22. v. dêma.

Dǽne; *pl. nom. acc; gen.* Dǽna; *m. The Danes*; Dani:—Dǽna lagu *the law of the Danes*, L. C. S. 15; Th. i. 384, 3, note 4. v. Dene.

dǽne-land, es; *n.* [dǽnu *a valley*] *A valley*; convallis:— Dǽneland getelda ic amete convallem *tabernaculorum metibor*, Ps. Lamb. 59, 8.

dǽnn, es; *n. A den*; cubile:—Godwine geann Leófwine ðæs dǽnnes æt Swiðrǽdingdǽnne *Godwine gives to Leofwine the den at Surrenden*, Cod. Dipl. 1315; A.D. 1020; Kmbl. vi. 178, 8, 13. v. denn.

dǽnnede *became slippery*, Chr. 937; Erl. 112, 12,=dennode; *p. of* dennian.

dǽnu, e; *f. A vale, valley*; convallis:—On ðisse sárgan dǽne *in convalle lacrymarum*, Ps. Th. 83, 6. v. denu.

Dǽrenta-mûþa, Derta-mûþa, an; *m.* [mûþa *the mouth of a river*] *Dartmouth, Devonshire*; Tremunda, in agro Devoniæ:—Hí férdon to Dǽrentamûþan [Dertamûþan, Th. 310, 5, col. 2] *they went to Dartmouth*, Chr. 1049; Th. 310, 6, col. 1.

dǽrst, es; *m. Leaven*; fermentum:—Ongelíc is dǽrste *simile est fermento*, Lk. Lind. War. 13, 21. *Rush. has dat. pl.* Gelíc is dǽrstum, Lk. Rush. War. 13, 21: 12, 1: 22, 1. DER. ge-dǽrsted.

dǽrstan, derstan; *pl. f?* Dregs, lees; fæx:—Nyle he ða dǽrstan him dôn unbrýce *verumtamen fæx ejus non est exinanita*, Ps. Th. 74, 8. Ða derstan beóþ gôde *the dregs will be good*, L. M. 1, 2; Lchdm. ii. 38, 18, 19. Wið ecedes derstan *with lees of vinegar*, 1, 39; Lchdm. ii. 98, 24.

dǽru, e; *f. Harm*; damnum:—His brôðer to dǽre and to lættinge *to the harm and hindrance of his brother*, Chr. 1101; Erl. 237, 18. v. daru.

dafen; *adj. Becoming, fit, suitable*; decens, congruus, conveniens. DER. ge-dafen: dafenian, ge-: dafenigendlíce, ge-: dafenlíc, ge-, unge-: dafenlíce, ge-, unge-: dafenlícnes, ge-, unge-.

dafenian, dafnian; *p.* ode; *pp.* od *To be seemly or becoming*; decere:—Swá swá dafnaþ munuce *as becomes a monk*; sicut decet monacho, Coll. Monast. Th. 35, 5. DER. ge-dafenian, -dafnian.

dafenigendlíce *suitably, conformably*. DER. ge-dafenigendlíce.

dafenlíc, dafnlíc; *adj. Becoming, fit, suitable*; decens, congruus, conveniens:—Dafnlícum *congruis*, Mone B. 1359. DER. ge-dafenlíc, unge-.

dafenlíce *becomingly, properly, fitly*. DER. ge-dafenlíce, unge-.

dafenlícnes, -nys, -nyss, e; *f. A fit time, opportunity*; opportunitas:—Ðú forsihst on dafenlícnyssum gedréfednysse *despicis in opportunitatibus in tribulatione*, Ps. Spl. C. second 9, 1. DER. ge-dafenlícnes, unge-.

dafnaþ *becomes*, Coll. Monast. Th. 35, 5; *3rd pres. sing. of* dafenian.

dǽg, es; *n? What is dangling*; sparsum:—Dǽges hlæfpe *sparsio*, Wrt. Voc. 288, 68. v. daag.

daga, an; *m. A day*; dies,—found in the compound word ân-daga, *q. v;* also v. dæg II.

dagas *days*, Bd. 1, 1; S. 473, 32: 474, 31; *pl. nom. acc. of* dæg.

dagena *of days*; dierum:—Him bebeád seofon dagena fæsten *enjoined them a fast of seven days*, Homl. Th. i. 434, 21: Exon. 31 a; Th. 97, 8; Cri. 1587: Menol. Fox 128; Men. 64; *gen. of* daga.

dages *daily*; die, Ps. Lamb. 1, 2. v. dæges, dæg.

dagian, dagigan; *p.* ode; *pp.* od [dagas *days, pl. of* dæg *a day*] *To* DAWN, *to become day, be day*; lucescere:—Mín leóht me tocymeþ ðonne hit dagian ongynneþ *mea lux, incipiente aurora, mihi adventura est*, Bd. 4, 8; S. 576, 7: 4, 9; S. 576, 30. Ne ðis ne dagaþ eástan *this dawns not from the east*, Fins. Th. 4; Fin. 3. Swylce hit ealle niht dagie [dagige MSS. P. S.] *as though it were day all night*, Bd. de nat. rerum; Wrt. popl. science 12, 9; Lchdm. iii. 260, 1. [*Prompt.* dagyñ': *Piers P. Chauc.* dawe: *Laym.* dæʒen, daiʒen, daʒiʒen: *Plat.* dagen: *Dut.* dágen: *Kil.* daghen: *Ger. M.H.Ger.* tagen: *O.H.Ger.* tagên: *Dan.* det daget *it dawns: Swed. Icel.* daga.] DER. ân-dagian, ge-ân-.

dagung, e; *f. A dawning, dawn, day-break*; aurora, tempus matutinum, diluculum:—Betwux hancrêd and dagunge *between cock-crowing and dawn*, Chr. 795; Erl. 59, 26: 802; Erl. 61, 19. Eóde he ût on dagunge *of ðam hûse egressus est tempore matutino de cubiculo*, Bd. 3, 27; S. 559, 1. On dagunge he eft acwicode and semninga uppasæt *diluculo reviviscens ac repente residens*, 5, 12; S. 627, 13: 4, 8; S. 576, 9: 4, 23; S. 596, 17. On dagunge ðæs fiftan dæges *quinta inlucescente die*, 5, 19; S. 640, 26.

DÁH, dôh; *gen.* dáges; *m?* DOUGH; farina subacta, massa = μάζα:—Blôma oððe dáh *massa*, Wrt. Voc. 85, 16: 94, 63. Dáh [MS. dað] *vel* blôma *massa*, Ælfc. Gl. 51; Som. 66, 9; Wrt. Voc. 34, 68. Cned hyt ðæt hit sí swá þicce swá dôh *knead it that it may be as thick as dough*, Lchdm. iii. 88, 17. Wyrc clam of dáge *make a paste of dough*, L. M. 3, 59; Lchdm. ii. 342, 18. [*Prompt.* dowe *pasta: Wyc.* dough: *Plat.* deeg: *Dut.* deeg. *n: Kil.* deegh *massa: Ger.* teig, *m: M.H.Ger.* teic, *gen.* teiges: *O.H.Ger.* teig, *m:* Goth. daigs, *m.* dough; deigan *to make dough: Dan.* deig, *m. f:* Swed. deg, *m: Icel.* deig, *n: Sansk.* dih *to smear, plaster.*]

dáhle *hid*, Bd. 4, 27; S. 604, 24,=dígle; *pl. nom. acc. of* dígol.

dahum *to days*, Bt. 4; Fox 8, 5,=dagum; *dat. pl. of* dæg.

dál, es; *n. A division, allotment, portion*, DOLE; discrimen, divisio, portio:—Ic sette dál betwux ðín folc and mín folc *ponam divisionem inter populum meum et populum tuum*, Ex. 8, 23. Is ðes middangeard dálum gedǽled *this earth is divided into parts*, Exon. 33 a; Th. 105, 18; Gú. 25. Swá beóþ môd-sefan dálum gedǽled, sindon dryht-guman ungelíce *dispositions are distributed by parts, while people are unlike*, 83 b; Th. 314, 29-32; Môd. 21-23. DER. ge-dál, to-.

Dalamensan; *gen.* -ena; *pl. m. The Dalamensan*; Dalamensæ: a Slavonic race, who dwelt in Misnia on both sides of the river Elbe:—Be norþan eástan Maroara syndon Dalamensan, and be eástan Dalamensan [MS. Dalamensam] sindon Horithi, and be norþan Dalamensan [MS. Dalomensam] sindon Surpe *to the north-east of the Moravians are the Dalamensan, and to the east of the Dalamensan are the Horithi, and to the north of the Dalamensan are the Surpe*, Ors. 1, 1, § 12; Bos. 19, 4-6.

dalc, dolc, es; *m. A clasp, buckle, brooch, bracelet*; fibula, spinther, regula:—Preón *vel* oferfeng *vel* dalc *fibula*, Ælfc. Gl. 64; Som. 69, 22; Wrt. Voc. 40, 53. Dalc *spinther*, Ælfc. Gr. 9, 18; Som. 9, 63. Ic geseah sumne gildenne dalc on fíftigum entsum *vidi regulam auream quinquaginta siclorum*, Jos. 7, 21. Dolc oððe preón *spinther*, Wrt. Voc. 74, 59.

dalf *dug*, Mt. Kmbl. Lind. 21, 33, = dealf; *p. of* delfan.

dap-fugel *the dip-fowl or diver, a gull*; mergus, mergulus, Som. Ben. Lye. v. dop-fugel.

daraþ, dareþ *a dart, spear, javelin*, Exon. 66 b; Th. 246, 27; Jul. 68: Beo. Th. 5689; B. 2848. v. daroþ.

dareþ-lácende, deareþ-lácende; *part.* [daroþ, dareþ *a dart, spear*; lácende, *part. of* lácan *to play*] *Playing with a dart, dart-brandishing*; telo ludens:—Beornþreát monig ófestum gefýsde, dareþlácende *many a band of nobles hurried with haste, dart-brandishing*, Exon. 96 a; Th. 358, 29; Pa. 53. Dareþlácendra *of the dart-players*, Elen. Kmbl. 1298; El. 651. Deareþlácende stæde wícedon *the dart-players bivouacked on the shore*, 73; El. 37.

daro *hurt, harm*, Bd. 3, 2; S. 525, 17. v. daru.

daroþ, daraþ, dareþ, es; *m.* [derian *to hurt*] *A* DART, *spear, javelin, weapon*; telum, jaculum, hasta:—Daroþ sceal on handa *the spear shall be in the hand*, Menol. Fox 502; Gn. C. 21. Forlét daroþ of handa fleógan *let a dart fly from the hand*, Byrht. Th. 136, 11; By. 149: 139, 17; By. 255. Reórdode ríces hyrde, daraþ hæbbende *the realm's guardian spake, raising his spear*, Exon. 66 b; Th. 246, 27; Jul. 68. Daroþas wǽron weó ðǽre wihte *darts were an affliction to the creature*, 114 a; Th. 438, 8; Rä. 57, 4. Þurh daroþa gedrep *through the stroke of darts*, Andr. Kmbl. 2886; An. 1446. Dareþa *of darts*, Chr. 937; Th. 207, 11; Æðelst. 54. Ða ne dorston dareþum lácan *who durst not play with javelins*, Beo. Th. 5689; B. 2848. [*Prompt.* darte: *Wyc.* dartis, *pl: R. Brun.* darte: *Chauc.* dart: *O.H.Ger.* tart *lancea: Swed.* dart, *m. a dagger: Icel.* darraðr, *m.* hasta.]

daroþ-æsc, es; *n? An ash-dart*; jaculum fraxineum:—Daroþæsc flugon *ash-darts flew*, Elen. Kmbl. 280; El. 140.

DARU, daro, e; *f. Hurt, harm, damage*; damnum, noxa:—Hwelc is mǽre daru *what is a greater hurt?* Bt. 29, 2; Fox 106, 14. Gemétte he his earm and his hand swá hále and swá gesúnde swá him nǽfre bryce ne daro gedôn wǽre *he found his arm and his hand so hale and so sound, as if breach or hurt had never been done to them*, Bd. 3, 2; S. 525, 17. Him to dare *to his harm*, Exon. 42 b; Th. 144, 2; Gú. 672. Ne astrece ðú ðíne hand búfon ðam cilde, ne him náne dare ne gedô *stretch thou not thine hand over thy son, nor do him any harm*, Homl. Th. ii. 60, 35. Búton ǽlcere dare *without any hurt*, i. 102, 8. Ðæt môd mid þwyrlícum geþohtum hogaþ ôðrum dara *the mind will meditate harm to others with perverse thoughts*, i. 412, 28. [*Kil.* dere, deyre *nocumentum: O.H.Ger.* tara, f.]

Daðan, es; *m. Dathan*, one of the sons of Eliab, Num. 26, 9:—Æfter ðam arison Chore and Hon, Daðan and Abiron ongeán Moisen *after that Korah and On, Dathan and Abiram rose up against Moses*, Num. 16, 1: 16, 27, 32: Deut. 11, 6. [דָּתָן *Dáthán*.]

Datia, Ors. 1, 1, § 12; Bos. 19, 3,=Datie; *gen.* Datia; *pl. m. The* DACIANS; Dáci; *gen.* ôrum; *m.*=Δακοί *A celebrated warlike people in Upper Hungary, in Transylvania, Moldavia, Wallachia, and in*

Bessarabia. They were originally of the same race as the Getæ. Trajan crossed the Danube and conquered the country in A D. 106, and colonised it with Romans. At a later period Dacia was invaded by the Goths; and as Aurelian considered it more prudent to make the Danube the boundary of the Empire, he resigned Dacia to the barbarians, removed the Roman inhabitants to Mœsia, and gave to the Dacians the name of the Aureliani, who inhabited that part of the province along the Danube in which they were settled:—And be eástan dǽm sind Datie [MS. Datia] ðá ðe iú wǽron Gotan *and to the east of them [the Wisle] are the Dacians who were formerly Goths,* Ors. 1, 1, § 12 ; Bos. 19, 3.

Dauid, es ; *m. David ;* Dávid, ídis ; *m :*—Dauid sang ðysne syxtan sealm *David sang this sixth psalm,* Ps. Th. arg. 6. Dauides sealm *the psalm of David,* Ps. Th. arg. 4. Dauides sunu *David's son,* Homl. Blick. 15, 18, 20. Crist onwráh, in Dauides dýrre mǽgan, ðæt is Euan scyld eal forpynded *Christ revealed that, in David's dear kinswoman, the sin of Eve is all turned away,* Exon. 9 a ; Th. 7, 4 ; Cri. 96. [דָוִד, Dávid, from דּוּד *dúd affection.* We have, in the same meaning, the classical name *Erasmus,* from ἐράσμιος *lovely, affectionate.*]

deácon *a levite, deacon;* levítes:—Aaron ðín bróður, deácon, hæfþ góðe sprǽce *Aaron frater tuus, levites, eloquens est,* Ex. 4, 14. v. diácon.

deácon-hád *deaconhood, deaconship;* diaconátus, Bd. 5, 23 ; S. 647, 29. v. diácon-hád.

DEÁD; *def.* se deáda ; seó, ðæt deáde ; *adj.* DEAD ; mortuus:—Lazarus ys deád *Lazarus mortuus est,* Jn. Bos. 11, 14 : Mt. Bos. 9, 24 : Jud. 3, 25 : Elen. Kmbl. 1761 : El. 882 : Beo. Th. 939 ; B. 467 : Exon. 126 b ; Th. 487, 19 ; Rä. 73, 4. Næs ðǽr nán þing deád of ðám *nec erat quidquam mortuum de his,* Ex. 9, 7 : 21, 34. Me hátran sind Dryhtnes dreámas ðonne ðis deáde líf *the Lord's joys are more exciting to me than this dead life,* Exon. 24 a ; Th. 309, 31 ; Seef. 65. Sceal yrfe gedǽled deádes monnes *the inheritance of a dead man shall be divided,* 90 a ; Th. 338, 19 ; Gn. Ex. 81. Grǽf deádum men heófeþ *the grave shall groan for the dead man,* 91 b ; Th. 342, 29 ; Gn. Ex. 149. Mec deádne ofgeáfun fæder and móder *father and mother gave me up as dead,* 103 a ; Th. 391, 7 ; Rä. 10, 1 : Beo. Th. 2623 ; B. 1309. Brihtríc þohte ðæt he Wulfnóþ cuconne oððe deádne begytan sceolde *Brihtric thought that he would get Wulfnoth alive or dead,* Chr. 1009 ; Erl. 142, 3. Ealle synd deáde *mortui sunt omnes,* Ex. 4, 19 : Mt. Bos. 28, 4 : Th. 113, 24. Deáde of duste arísaþ þurh Drihtnes miht *the dead shall rise from the dust through power of God,* Cd 227 ; Th. 302, 24 ; Sat. 605 : Exon. 25 a ; Th. 72, 30 ; Cri. 1180. Hí ǽton deádra lác *manducavērunt sacrificia mortuōrum,* Ps. Th. 105, 22 : Mt. Bos. 23, 27. Land dryrmyde deádra hrǽwum *the land mourned over the corpses of the dead,* Cd. 144 ; Th. 180, 6 ; Exod. 41 : Elen. Kmbl. 1299 ; El. 651 : 1887 ; El. 945. Be deádum *for the dead,* Exon. 82 b ; Th. 311, 27 ; Seef. 98. Mid ðám deádum fellum *with the dead skins,* Boutr. Scrd. 20, 29. Ne dó hý tó deádan *ne occidēris eos,* Ps. Th. 58, 10 : 61, 3 : Ex. 21, 35, 36. Ne willaþ eów andrǽdan deáde fēðan *dread ye not dead bands,* Cd. 156 ; Th. 194, 26 ; Exod. 266 : Exon. 24 b ; Th. 71, 21 ; Cri. 1159 : Andr. Kmbl. 2156 ; An. 1079. Lǽt deáde bebyrigean hyra deádan *let the dead bury their dead,* Mt. Bos. 8, 22. Ne húru wundur wyrceaþ deáde *numquid mortuis facies mirabīlia ?* Ps. Th. 87, 10. [*Prompt.* dede: *Wyc.* ded : *Piers P.* deed : *Chauc.* dede : *R. Glouc.* ded : *Plat.* dood : *O. Sax.* dôd : *Frs.* dea : *O Frs.* dad, dath : *Dut.* dood : *Ger.* todt : *M. H. Ger. O. H. Ger.* tôt : *Goth.* dauþs : *Dan. Swed.* död : *Icel.* dauðr.] DER. woruld-deád.

deád-bǽre; *def.* se deád-bǽra, seó, ðæt deád-bǽre ; *adj. Death-bearing, deadly ;* mortifer, lethālis, lethifer :—Deádbǽre *lethāle,* Mone B. 1859. Se drenc deádbǽra wæs *the drink was deadly,* Homl. Th. ii. 158, 22. Ðæt ðín heorte forhtige for ðam deádbǽrum drence *that thy heart may fear the deadly drink,* i. 72, 16. Deádbǽre sprancan *lethiferas labruscas,* Mone B. 1993.

deád-bǽrende; *part. Death-bearing, deadly;* mortifer:—Se Arrianisca gedwola ðæt deádbǽrende áttor his getreówleásnysse on eallum middangeardes cyricum strégde *the Arian heresy spread the death-bearing venom of its truthlessness in all the churches of the earth,* Bd. 1, 8 ; S. 479, 34. v. deáþ-berende.

deád-bǽrlíc; *adj. Deadly;* mortifer :—Him ne deraþ, deáh hí hwæt deádbǽrlíces drincon *si mortifērum biberint, non eis nocēbit,* Mk. Bos. 16, 18.

deád-bǽrnes, -ness, e ; *f. A killing, mortification;* mortificātio, Mone B. 3934.

deád blód *dead blood, congealed blood,* Wrt. Voc. 283, 79. v. blód.

deád-boren; *part. Dead-borne;* mortuus fœtus :—Deádboren tuddur *mortuus fœtus,* Herb. 63, 2 ; Lchdm. i. 166, 3.

deád-líc; *def.* se deád-líca, seó, ðæt deád-líce ; *adj.* DEADLY, *mortal ;* mortālis, morticīnus :—Ðæt án deádlíc man mihte ealne middaneard oferseón *that a mortal man could see over all the world,* Homl. Th. ii. 186, 5. Rómāne deádlícne sige gefóran *the Romans gained a deadly victory,* Ors. 3, 8 ; Bos. 63, 33. Se cyning and monige of his folce lufodon ðis deádlíce líf *the king and many of his people loved this deadly*

life, Bd. 3, 30 ; S. 561, 41 : Boutr. Scrd. 20, 29. We onlybbaþ on ðisum deádlícum lífe *we live in this deadly life,* 30, 12. Deádlíce *morticinas,* Glos. Prudent. Recd. 145, 23. DER. un-deádlíc.

deád-líce; *adv. Mortally;* lethalíter, Cot. 123.

deád-lícnys, -nyss, e ; *f. Deadliness, mortality;* mortalītas :—Ðæt he dælnimend wǽre úre deádlícnysse *that he was a partaker of our mortality,* Homl. Th. i. 36, 34. He becóm on ða tíde ðǽre myclan deádlícnysse *tempŏre mortalitātis adveniens,* Bd. 3, 23 ; S. 555, 9 : 3, 30 ; S. 561, 38. Ealle his geféran on ðǽre deádlícnysse ðæs wǽles of worulde genumene wǽron *omnes sŏcii ipsōrum mortalitāte [cædis] de sæcŭlo rapti,* 3, 27 ; S. 558, 36. He hæfde ealle deádlícnyssa aworpen *he had cast off all mortalities,* Homl. Th. ii. 290, 1. DER. un-deádlícnys.

deád-rægl *clothing of the dead, a shroud;* pallium sepulchrāle, Som. Ben. Lye.

deád-spring, es ; *m.* [spring *an ulcer*] *A malignant ulcer, carbuncle ;* carbuncŭlus :—Wið wúnda and wið deádspringas *for wounds and ulcers,* Herb. 4, 2 ; Lchdm. i. 90, 5 : 9, 2 ; Lchdm. i. 100, 1 : 87, 3 ; Lchdm. i. 190, 24 : 91, 7 ; Lchdm. i. 200, 17.

DEÁF; *adj.* DEAF ; surdus :—Deáf *surdus vel surdaster,* Ælfc. Gl. 77 ; Som. 72, 21 ; Wrt. Voc. 45, 54. Ic swá swá deáf ne gehýrde *ego tamquam surdus non audiēbam,* Ps. Lamb. 37, 14. Eart ðú dumb and deáf *thou art dumb and deaf,* Exon. 99 a ; Th. 370, 26 ; Seel. 65. Nǽddran deáfre *aspĭdis surdæ,* Ps. Lamb. 57, 5. Hwá geworhte dumne oððe deáfne *quis fabricātus est mutum et surdum ?* Ex. 4, 11. Hí lǽddon him ǽnne deáfne and dumbne *addūcunt ei surdum et mut:m,* Mk. Bos. 7, 32 : Exon. 113 a ; Th. 433, 3 ; Rä. 50, 2. Anlíc nǽdran seó hí deáfe déþ *like an adder which makes herself deaf,* Ps. Th. 57, 4. Eálá deáfa and dumba gást *surde et mute spīrĭtus,* Mk. Bos. 9, 25. Deáfe gehýraþ *surdi audiunt,* Mt. Bos. 11, 5 : Mk. Bos. 7, 37 : Lk. Bos. 7, 22 : Andr. Kmbl. 1154 ; An. 577. Ðæt ic dumbum and deáfum deófolgieldum gaful onhāte *that I promise tribute to dumb and deaf idols,* Exon. 68 a ; Th. 251, 24 ; Jul. 150. Ne wirige ðú deáfe *curse not the deaf,* Lev. 19, 14. Deáf corn *deaf or barren corn,* Past. 52, 9 ; Hat. MS. [*Prompt.* deffe *surdus :* *Wyc.* def : *Piers P.* deef, *pl.* deve : *Chauc.* deef : *R. Glouc.* deve : *Plat.* doov : *O. Sax.* douf : *O. Frs.* dâf : *Dut.* doof : *Ger.* taub : *M. H. Ger.* toup : *O. H. Ger.* toup, doup : *Goth.* daubs, daufs *hardened, obdurate : Dan.* döv : *Swed.* döf : *Icel.* daufr.] DER. a-deáf : deáfian, a- : deáfu : a-deáfung.

deáf *dived,* Exon. 126 b ; Th. 487, 18 ; Rä. 73, 4 ; *p.* of dúfan *to dive,* q v.

deáfian *to become* or *wax deaf.* v. a-deáfian.

deáf-líc; *adj.* [deáf = défe *fitting, proper*] *Suitable, fitting, proper ;* conveniens :—Deáflíc to gehírenne on heálícum gemóte *fitting to be heard at a public assembly,* Ælfc. T. 15, 4. v. ge-défe.

deáfu, e ; *f.* [deáf *deaf*] *Deafness ;* surditas :—Wið eárwærce and wið deáfe *for ear-ache and for deafness,* L. M. 1, 3 ; Lchdm. ii. 40, 8. Wið eárena deáfe *for deafness of ears,* 1, 3 ; Lchdm. ii. 40, 20.

deag, es ; *m. A day ;* dies :—Æfter feáum deagum *after a few days,* Bd. 5, 9 ; S. 623, 7. v. dæg.

deág *is of use, is good, avails,* Exon. 8 a ; Th. 2, 19 ; Cri. 21 : 10 b ; Th. 12, 22 ; Cri. 189 ; *pres.* of dugan.

deágan; ic deáge, ðú deágest, deágst, dýhst, he deágeþ, deágþ, dýgþ, dýhþ, *pl.* deágaþ ; *p.* deóg, *pl.* deógon ; *pp.* deágen *To dye, colour ;* tingĕre :—Heoro-dreóre deáþfæge deóg *the death-doomed dyed it with fatal gore,* Beo. Th. 1704 ; B. 850.

deáge *of a colour* or *dye,* Homl. Th. ii. 254, 5 ; *gen.* of deáh.

deággede *gouty,* Ælfc. Gl. 77 ; Som. 72, 12 ; Wrt. Voc. 45, 46. v. deág-wyrmede.

deágian, dégian ; *p.* ode ; *pp.* od [deáh *a colour, dye*] *To colour,* DYE ; fucāre, inficĕre, tingĕre :—Deágian *fucāre,* Mone B. 1245 : *inficĕre,* 6225. Dégian *tingĕre,* 6251. DER. ge-deágod, twí-gedeágod.

deáglenes *solitariness,* Cot. 18. v. dígolnes.

deágol *secret,* Exon. 110 b ; Th. 424, 14 ; Rä. 41, 39 : L. M. 2, 66 ; Lchdm. ii. 298, 8 : Bd. 3, 16 ; S. 542, 34, MS. T. v. dígol.

deágollíce, deágolíce *secretly,* L. E. I. 45 ; Th. ii. 440, 33 : L. M. 2, 66 ; Lchdm. ii. 298, 6. DER. un-deágollíce. v. dígollíce.

deágolnes *hiding-place,* Bd. 4, 27 ; S. 604, 22. v. dígolnes.

deágung, e ; *f. A dyeing, colouring ;* tinctūra :—Deágung *tinctūra,* Ælfc. Gl. 28, 5 ; Som. 31, 59. Ne mihte nán eorþlíc cyning swá wlítige deágunge his hræglum begytan swá swá róse hæfþ *no earthly king could get such beautiful dyeing for his garments as the rose has,* Homl. Th. ii. 464, 10.

deág-wyrmede, deággede ; *part.* [deág = deáw *dew,* wyrm *a worm*] *Dew-wormed, gouty ;* podagricus = ποδαγρικός :—Deágwyrmede *vel* deággede *podagricus,* Ælfc. Gl. 77 ; Som. 72, 12 ; Wrt. Voc. 45, 46.

deáh *is of use, is good* or *virtuous, avails,* Herb. 2, 22 ; Lchdm. i. 86, 18 : Bt. 27, 2 ; Fox 98, 15 : Exon. 80 b ; Th. 303, 5 ; Fä. 48 : Beo. Th. 1151 ; B. 573 ; *pres.* of dugan.

deáh; *gen.* deáge ; *f. A colour,* DYE ; tinctūra, fucus, stĭbium, murex :—Deáh *tinctūra :* reád deáh *coccus,* Ælfc. Gl. 64 ; Som. 69, 5, 6 ; Wrt. Voc. 40, 39, 40. Deáge *tinctūræ,* Mone B. 6226. Mid ðǽre deáge

hiwe *with the colour of the dye*, Homl. Th. ii. 254, 5. Deáge *fuco*, Mone B. 1080: 6224. Twí-gedeágadre deáge *bis tincto cocco*, 1094. Deáge *stibio*, 4649 : *rubenti*, 6235 : *murīce*, 6268. Reádre deáge *rubro stibio*, 1242.

deáhl [= deágol] ; *def.* se deáhla ; seó, đæt deáhle ; *adj. Dark, secret ;* obscūrus, secrētus :—Đære deáhlan neahte *of the dark night*, Bd. 2, 6 ; S. 508, 13. v. dígol.

deal, deall ; *adj. Proud, exulting, eminent ;* superbus, clarus :—Fugel feđrum deal *a bird proud of feathers*, Exon. 59 b ; Th. 216, 10 ; Ph. 266. Bǽr-beágum deall *proud of bearing rings*, 108 b ; Th. 414, 18 ; Rä. 32, 22. Sum sceal wildne fugel atemian, fiđrum dealne *one shall tame the wild bird, exulting in his plumes*, 88 b ; Th. 332, 21 ; Vy. 88. Wíggendra þreát cômon, æscum dealle *a troop of warriors came, proud with their spears*, Andr. Kmbl. 2195 ; An. 1099 : Exon. 106 a ; Th. 404, 22 ; Rä. 23, 11. Đær swíþferhþe sittan eódon, þryþum dealle *the strong of soul went to sit there, proud of their strength*, Beo. Th. 992 ; B. 494. Sprǽcon wlonce monige, dugeþum dealle *many proud ones spake, eminent with virtues*, Cd. 89 ; Th. 111, 1 ; Gen. 1849.

dealf *dug*, Mt. Bos. 25, 18 ; *p.* of delfan.

deapung, e ; *f. A dipping ;* immersio, Som. Ben. Lye. v. dyppan.

dear ; ic, he *I dare, he dares*, Gen. 44, 34 : Beo. Th. 1373 ; B. 684 ; *pres.* of durran.

dearf, *pl.* durfon *laboured ; p.* of deorfan.

dearnunga, dearnenga, dearninga ; *adv.* [dyrne *secret, obscure*] *Secretly, privately, clandestinely ;* clam, occulte, clandestino :—He wolde dearnunga mid mándǽdum menn beswícan *he would secretly deceive men with wicked deeds*, Cd. 23 ; Th. 29, 14 ; Gen. 450. Gif đín bróđor đé lǽre dearnunga *si tibi voluerit persuadere frater tuus clam*, Deut. 13, 6 : Jn. Bos. 19, 38. Ođđe eáwunga ođđe dearnunga *either publicly or privately*, L. Edg. ii. 8 ; Th. i. 270, 5 : L. Ath. v. § 1, 2 ; Th. i. 228, 21. Be đon ic mon dearnenga [dearnunga MSS. G. H.] bearn gestriéne *in case a man beget a child clandestinely*, L. In. 27 ; Th. i. 120, 1, 2 : L. Alf. 6 ; Th. i. 44, 17. Đeáh heó dearnenga fordón wurde *though she was secretly seduced*, Cd. 30 ; Th. 39, 21 ; Gen. 629 : 29 ; Th. 38, 5 ; Gen. 602. Hwæt he dearninga on hyge hogde *what he secretly meditated in his mind*, Exon. 51 a ; Th. 177, 13 ; Gú. 1226. DER. un-dearnunga.

dearr-líc ; *adj. Daring, rash ;* temerārius, Som. Ben. Lye.

dearr-scipe, es ; *m. Rashness, presumption ;* temerītas, Som. Ben. Lye.

dearst đú *thou darest*, Beo. Th. 1061 ; B. 527 ; *2nd pres. sing.* of durran.

DEÁÞ, es ; *m.* DEATH ; mors :—Đeáh đe him se bitera deáþ geboden wǽre *though bitter death were announced to them*, Cd. 183 ; Th. 229, 26 ; Dan. 223 : Exon. 31 b ; Th. 98, 6 ; Cri. 1603 : Beo. Th. 899 ; B. 447 : 5773 ; B. 2890. Se deáþ cymþ *death comes*, Bt. 8 ; Fox 26, 6 : Chr. 1065 ; Erl. 198, 7 ; Edw. 26. Hí ofercume unþinged deáþ *vēniat mors super illos*, Ps. Th. 54, 14. Nis me đæs deáþes sorg *there is no fear of death to me*, Exon. 38 a ; Th. 125, 7 ; Gú. 350 : 40 a ; Th. 133, 25 ; Gú. 495 : Cd. 25 ; Th. 31, 28 ; Gen. 492 : Elen. Kmbl. 1165 ; El. 584 : Bt. 8 ; Fox 26, 6. Đú đe upahefst me of geatum deáþes *qui exaltas me de portis mortis*, Ps. Lamb. 9, 15. He is deáþes scyldig *reus est mortis*, Mt. Bos. 26, 66 : Ps. Th. 54, 4 : 72, 3. Gif hwá sié deáþes scyldig *if any one be guilty of death*, L. In. 5 ; Th. i. 104, 13 : 27 ; Th. i. 120, 3. He men of deáþe worde awehte *he woke men from death with his word*, Andr. Kmbl. 1166 ; An. 583 : Exon. 14 b ; Th. 29, 23 ; Cri. 467 : 41 b ; Th. 139, 25 ; Gú. 598. Gif he man to deáþe gefylle beó he útlah *if he fell a man to death let him be an outlaw*, L. E. G. 6 ; Th. i. 170, 10 : L. C. S. 2 ; Th..i. 376, 18 : Chr. 979 ; Erl. 129, 10 : Boutr. Scrd. 17, 25 : 18, 11. Eall đæt gemót sôhte leáse saga ongén đone Hǽlend, đæt híg hyne to deáþe sealdon *omne consilium quærēbat falsum testimōnium contra Iesum, ut eum morti traderent*, Mt. Bos. 26, 59 : 20, 18 : Ps. Th. 114, 8 : 117, 18. Fram deáþe to lífe *a morte in vitam*, Jn. Bos. 5, 24. Deáþ he đǽr býrigde *he there tasted death*, Rood Kmbl. 199 ; Kr. 101 : Cd. 228 ; Th. 306, 17 ; Sat. 665 : Exon. 119 b ; Th. 459, 25 ; Hö. 5. Þurh fǽrlícne deáþ *through sudden death*, L. C. S. 71 ; Th. i. 412, 29. Unrôt ys mín sáwl ôþ deáþ *tristis est anima mea usque ad mortem*, Mt. Bos. 26, 38 : 16, 28 : Ex. 10, 17 : Deut. 30, 15. He sceal deáþe sweltan *he shall perish by death*, L. Alf. 14, 15 ; Th. i. 48, 3, 7, 8. Đæt đú deáþe sweltest *that thou shalt perish by death*, Exon. 67 b ; Th. 250, 11 ; Jul. 125. Deáþe cwylman *mortificāre*, Ps. Spl. 108, 15. Đæt he deáþa gedál dreógan sceolde *that he should undergo death*, Exon. 36 a ; Th. 116, 12 ; Gú. 206. Gegang đa deáþa bearn đe hí dêmaþ nú *possīde filios morte punītōrum*, Ps. Th. 78, 12. Deáþas *spirits, ghosts ;* manes, Cot. 134. [*Wyc.* deeth : *Chauc.* deth : *Laym.* dæd, dæað, deað, deð, *m* : *Orm.* dæþ : *O. Sax.* dôð, *m* : *Frs.* dead, dea : *O. Frs.* dad, dath, *m* : *Dut.* dood, *m* : *Ger.* tod, *m* : *M. H. Ger.* tôt, *m* : *O. H. Ger.* tôd, *m* : *Goth.* dauþus, *m* : *Dan.* död, *m. f* : *Swed.* död, *m* : *Icel.* dauði, *m.*] DER. 'ǽr-deáþ, ende-, gúþ-, mere-, swylt-, wæl-, wundor-.

deáþ-bǽre *death-bearing, deadly*, Som. Ben. Lye. v. deád-bǽre.

deáþ-beám, es ; *m. A death-tree, tree of death ;* mortis arbor, mortifera :—Deáþbeámes ofet *fruit of the tree of death*, Cd. 30 ; Th. 40, 13 ; Gen. 638.

deáþ-bed, -bedd, es ; *n. A death-bed, grave ;* mortis stratum, sepulcrum :—Nú is wilgeofa deáþbedde fæst *the kind giver is now fast in his death-bed* [= *grave*], Beo. Th. 5795 ; B. 2901.

deáþ-berende ; *part. Death-bearing, deadly ;* mortifer :—Eue sealde deáþberende gyfl *Eve gave the deadly fruit*, Exon. 45 a ; Th. 153, 8.

deáþ-bérnis, -niss, e ; *f. Death, destruction, pestilence ;* pernícies, pestilentia :—Deáþbérnisse ođđe uncúþo ádlo *pestilentiæ*, Lk. Skt. Lind. 21, 11.

deáþ-cwalu, e ; *f. A deadly pain* or *plague, agony ;* mortis dolor :—Sió wērge sceólu hreósan sceolde in wíta forwyrd, đǽr hie in wylme nú dreógaþ deáþcwale *the wretched crew were compelled to fall into the ruin of punishment, where they now suffer deadly pains in flame*, Invent. Crs. Recd. 1533 ; El. 766. Ne geweóx he him to willan, ac to deáþcwalum Deniga leódum *he waxed not for their benefit, but for a deadly plague to the Danes' people*, Beo. Th. 3428 ; B. 1712.

deáþ-cwealm, es ; *m.* [cwealm *a violent death, slaughter*] *Slaughter ;* nex :—Ic wræc deáþcwealm Denigea *I avenged the slaughter of the Danes*, Beo. Th. 3344 ; B. 1670.

deáþ-cwylmende, -cwylmmende ; *part.* [cwelman, cwylman *to destroy, kill*] *Put to death, destroyed, killed ;* mortificātus :—Geáhna bearn adýdra ođđe deáþcwylmmendra *possīde filios mortificātōrum*, Ps. Lamb. 78, 11.

deáþ-dæg, es ; *m. Death-day, day of death ;* mortis dies :—Æfter deáþdæge *after the day of death*, Beo. Th. 376 ; B. 187 : Menol. Fox 581 ; Gn. C. 60. To đínum deáþdæge *to thy death-day*, Exon. 98 a ; Th. 369, 6 ; Seel. 37.

deáþ-denu, e ; *f. The valley of death ;* mortis vallis :—In đisse deáþdene *in this valley of death*, Exon. 12 b ; Th. 21, 33 ; Cri. 344. In đas deáþdene *in this death-vale*, Exon. 61 b ; Th. 226, 35 ; Ph. 416.

deáþ-drepe, es ; *m. Death-stroke ;* letālis ictus :—Đý deáþ-drepe *in the death-stroke*, Cd. 167 ; Th. 209, 6 ; Exod. 495. v. drepe.

deáþ-fǽge ; *adj.* [deáþ *death*, fǽge *fated, doomed*] *Death-doomed ;* morti addictus :—Deáþfǽge deóg *the death-doomed had dyed it*, Beo. Th. 1704 ; B. 850.

deáþ-gedál, es ; *n.* [gedál *a separation*] *A deathly separation, separation of body and soul in death ;* letālis separātio :—Næs egle [MS. engle] on môde deáþgedál *the deathly separation was not oppressive to his soul*, Exon. 46 b ; Th. 159, 33 ; Gú. 936.

deáþ-godas ; *pl. m. Death-gods, spirits, ghosts ;* manes, Cot. 134.

deáþ-lég, es ; *m.* [lég *a flame*] *A death-flame ;* letālis flamma :—Wihta gehwylce deáþlég nimeþ *the death-flame shall seize each creature*, Exon. 22 a ; Th. 61, 12 ; Cri. 983.

deáþ-líc ; *adj. Deadly, mortal, good and bad angels ;* mortālis :—Đis is bísen đara sôþena gesælþa, đara wilniaþ ealle deáþlíce men to begitanne *this is an example of the true goods, which all mortal men desire to obtain*, Bt. 24, 2 ; Fox 80, 30. Híg gesetton hrǽwas ođđe đa deáþlícan đínra þeówana mettas fugelum heofonan *posuērunt morticina servōrum tuōrum escas volatilĭbus cœli*, Ps. Lamb. 78, 2.

deáþ-lícnes *mortality*, Som. Ben. Lye. v. deád-lícnys.

deáþ-mægen ; *gen.* -mægnes ; *n. A deadly power* or *band ;* letifera caterva, Exon. 45 b ; Th. 155, 28 ; Gú. 867.

deáþ-ræced, es ; *n.* [ræced, reced *a house*] *A death-house, sepulchre ;* mortis domus, sepulcrum :—Deáþræced onhliden weorþaþ *the death-houses shall be opened*, Exon. 56 b ; Th. 200, 30 ; Ph. 48.

deáþ-rǽs, es ; *m.* [rǽs *a rush*] *Death-rush, rushing of death ;* mortis impĕtus :—Ealle deáþrǽs forfêng *the death-rush clutched them all*, Andr. Kmbl. 1990 ; An. 997.

deáþ-reów ; *adj.* [reów *cruel*] *Deadly cruel, savage ;* atrox :—Com seofona sum to sele geogan deóful deáþreów *a savage devil came with seven others unto the hall*, Andr. Kmbl. 2629 ; An. 1316.

deáþ-scúa, an ; *m.* [scúa *a shade*] *The shadow of death, death ;* mortis umbra, mors, Beo. Th. 322 ; B. 160.

deáþ-scúfa, an ; *m.* [scúfa = scúwa *a shade*] *The shadow of death, death ;* mortis umbra, mors :—Forđanđe nis on deáþe ođđe on deáþscúfan, đe gemyndig sý đín *quoniam non est in morte, qui memor sit tui*, Ps. Lamb. 6, 6.

deáþ-scyld, e ; *f.* [deáþ *death ;* scyld *sin, crime*] *A death-fault, capital crime ;* capitāle crīmen :—Gif gehádod man hine forwyrce mid deáþscylde *if a man in orders ruin himself with capital crime*, L. E. G. 4 ; Th. i. 168, 22 ; L. C. S. 43 ; Th. i. 400, 27.

deáþ-scyldig ; *adj.* [deáþ *death*, scyldig *guilty*] *Death-guilty, condemned ;* damnātus :—Gif deáþscyldig man scriftspræce gyrne *if a man guilty of death desire confession*, L. E. G. 5 ; Th. i. 168, 24 : L. C. S. 44 ; Th. i. 402, 3.

deáþ-sele, es ; *m.* [deáþ *death ;* sele *a dwelling, hall*] *A death-hall ;* mortis aula :—In đam deáþsele *in the death-hall*, Exon. 48 b ; Th. 166, 25 ; Gú. 1048. On wítehús, deáþsele deófoles *into the house of torment, the death-hall of the devil*, 30 b ; Th. 94, 8 ; Cri. 1537 : 97 a ; Th. 362, 1 ; Wal. 30.

deáþ-slege, es ; *m.* [slege *a blow, stroke*] *A death-blow ;* letālis ictus :—

þurh deáþslege *through deadly stroke*, Exon. 102 b; Th. 388, 27; Rä. 6, 14.

deáþ-spere, es; *n.* [spere *a spear*] *A deadly spear*; letális hasta :—Dol him ne ondrædeþ deáþsperu *the foolish will not dread the deadly spears*, Exon. 102 a; Th. 385, 32; Rä. 4, 53.

deáþ-stede, es; *m.* [deáþ *death*, stede *a place*] *A death-place*; mortis campus :—Lágon on deáþstede drihtfolca mæst *the greatest of people lay on their death-place*, Cd. 171; Th. 216, 1; Exod. 589.

deáþ-þénunga, *pl. f.* [þénung *a service*] *Funeral services, funerals*; exséquiæ, Cot. 74.

deáþ-wang, es; *m.* [deáþ *death*, wang *a field, plain*] *A death-plain*; mortis campus :—Hí swæfon dreóre druncne, deáþwang rudon *they slept drunken with blood, made the death-plain red* or *bloody*, Andr. Recd. 2009; An. 1005.

deáþ-wége, es; *n.* [deáþ *death*, wége *a cup*] *A deadly cup*; mortis póculum :—Ænig ne wæs mon on moldan ðætte meahte bibúgan ðone bleátan drync deópan deáþwéges *there was not any man on earth that could avoid the miserable drink of the deep deadly cup*, Exon. 47 a; Th. 161, 25; Gú. 964.

deáþ-wérig, *adj. Death-weary, dead*; mortuus :—Ne móston deáþwérigne Deniga leóde bronde forbærnan *the Danes' people could not consume the death-weary one with fire*, Beo. Th. 4256; B 2125.

deáþ-wíc, es; *n.* [deáþ *death*, wíc *a mansion*] *A mansion of death*; mortis mansio :—He gewát deáþwíc seón *he departed to see the mansion of death*, Beo. Th. 2555; B. 1275.

deáþ-wyrda; *pl. f.* [wyrd *fate*] *Death-events, fates*; fata, Cot. 89.

DEÁW, es; *m. n.* DEW; ros :—Swá swá deáw ðære dúne ðætte [se, Th; se ðe, Spl.] niðerastáh on munte oððe to dúne *sicut ros Hermon qui descendit in montem Sion*, Ps. Lamb. 132, 3. On morgen wæs ðæt deáw abútan ða fyrdwíc *mane ros jacuit per circuitum castrórum*, Ex. 16, 13: Num. 11, 9. Deáw and deór scúr ðec dómige *the dew and heavy rain exalt thee*, Cd. 192; Th. 239, 18; Dan. 372: Exon. 16 b; Th. 38, 19; Cri. 609: 108 a; Th. 412, 11; Rä. 30, 12: Deut. 32, 2. þurh dropunge deáwes and rénes *through the dropping of dew and rain*, Ps. Th. 64, 11. Syle ðé God of heofenes deáwe *det tibi Deus de rore cœli*, Gen. 27, 28, 39. [*Prompt.* dewe: Piers P. *Chauc.* dewes, *pl: Orm.* dæw: *Plat.* dau, *m: Frs.* dauwe, douwe: *O. Frs.* daw, *m: Dut.* dauw, *m: Kil.* dauw, dauwe: *Ger.* thau, tau, *m: M. H. Ger. O. H. Ger.* tou, *n: Dan.* dug, dugg, *m. f: Swed.* dagg, *m: Icel.* dögg, *f.*] DER. mele-deáw, sun-.

deáw-driás, es; *m?* [dreósan *to fall*] *A fall of dew, dew-fall*; róris cásus :—Deáwdriás on dæge weorþeþ winde geondsáwen *the dew-fall in day is scattered by the wind*, Cd. 188; Th. 233, 17; Dan. 277.

deáwian *to* DEW, *bedew*; rorare, Som. Ben. Lye.

deáwig; *adj.* DEWY; roscídus :—Gúþcyste onþrang deáwig-sceaftum *the war-tribe pressed onwards with dewy shafts*, Cd. 160; Th. 199, 25; Exod. 344. Ðara breósta biþ deáwig wǽtung *there is a dewy wetting of the breasts*, L. M. 2, 46; Lchdm. ii. 258, 17.

deáwig-feðere; *def.* se -feðera, seó, ðæt -feðere; *adj. Dewy-feathered*; roscídus pennis :—Sang se wanna fugel, deáwigfeðera *the sad fowl sang, dewy of feathers*, Cd. 93; Th. 119, 24; Gen. 1984. Hwreópon herefugolas, deáwigfeðere *the fowls of war screamed, dewy-feathered*, 150; Th. 188, 4; Exod. 163.

deáw-wyrm, es; *m. A ringworm, tetter*; impetígo :—Wið deáwwyrmum genim doccan *for ringworms take dock*, L. M. 1, 50; Lchdm. ii. 122, 21: 124, 5, 7.

deccan; *impert.* dec *To cover*; tegére :—Dec ánne cláþ ðær of *cover a cloth therewith*, Herb. 47, 1; Lchdm. i. 150, 19. DER. ge-deccan. v. þeccan.

Decem-ber; *gen.* -bris; *m.* [décem *ten : Sansk.* vára : *Pers.* bár *time, space : the tenth month of the Romans, beginning with March, and as we begin with January, it is our twelfth month*] *The month of December*; Décember, bris, *m* :—Mónaþ Decembris, ærra iúla [geóla] *the month of December, the former yule*, Menol. Fox 437; Men. 220; *January being after yule* or *Christmas is called Se æftera geóla the after yule*, Cott. Tibérius. B. i; Hick. Thes. i. 212, 57.

declínigendlíc; *adj. Declinable*; declinábilis :—Feówer synd declinabilia, ðæt is declinigendlíce *four are declinabilia, that is declinable*, Ælfc. Gr. 50, 3; Som. 51, 7.

declínung, e; *f. A declension*; declinátio :—Seó forme declínung *the first declension*, Ælfc. Gr. 7; Som. 6, 3. On fíf declínungum *in five declensions*, 6. 2.

déd *dead*, Chr. 1129; Erl. 258, 22. v. deád.

déda *of deeds*, Ps. C. 50, 147; Grn. ii. 280, 147,=dæda; *gen. pl. of* dæd.

defe; *adj. Becoming, fit, suitable*; décens, congruus, convéniens. DER. ge-défe, lær-ge-, un-ge-: défelíc, ge-: défelíce, ge-, un-ge-.

défe-líc *becoming, fit*. DER. ge-défelíc.

défe-líce *becomingly, fitly, suitably*. DER. ge-défelíce, un-ge-.

Defénas, Defnas; *gen.* a; *dat.* um; *pl. m. Devonians, the inhabitants of Devonshire in a body, Devonshire*; Devonienses, Devónia :—Hé wæs

Weala gefeoht and Defena [Defna, Th. 110, 16] *in this year* [A. D. 823] *there was a fight of the Welsh and Devonians*, Chr. 823; Th. 111, 16, col. 1, 2. Ægðer ge on Defenum [Defnum, col. 2] ge welhwær be ðæm særiman *both in Devon and elsewhere on the sea-shore*, Chr. 897; Th. 176, 8, col. 1: 981; Th. 234, 31: 997; Th. 246, 5. Forþférde Ælfgar on Defenum *Ælfgar died in Devonshire*, Chr. 962; Th. 218, 38.

Defena scír, Defna scír, e; *f.* [*Hunt.* Deuenesire, Dauenescyre : *Hovd.* Daveneshire : *Brom.* Devenescire : *Kni.* Devenchire, Devenschyre] DEVONSHIRE; Devónia :—He wæs on Defena scíre *he was in Devonshire*, Chr. 878; Th. 146, 33, col. 1: 851; Th. 120, 20, col. 1. Hí ymbsæton án geweorc on Defna scíre *they besieged a fortress in Devonshire*, 894; Th. 166, 28. Sideman wæs Defna scíre bisceop *Sideman was bishop of Devonshire*, 977; Th. 230, 16.

Defenisc; *adj. Of* or *belonging to Devonshire*; Devóniensis :—Gesomnede man ormæte fyrde Defenisces folces *an immense force of Devonshire people was collected*, Chr. 1001; Th. 250, 5.

défre; *adj. Timely, seasonable*, Som. Ben. Lye; *comp. of* défe?

deg *a day*, Th. Diplm. A. D. 830; 465, 21: A. D. 972; 520, 7. v. dæg.

dég *profits*; prodest, Mt. Kmbl. Lind. 16, 26,=deág, deáh; *pres. of* dugan.

dég *a colour, dye*, Som. Ben. Lye. v. deáh.

dégelíce [dégel-líce] *secretly*, Bt. Met. Fox 1, 127; Met. 1, 64. v. dígollíce.

dégelnis *solitude*, Mt. Lind. Stv. 6, 4, 6. v. dígolnes.

dégian *to colour, dye*, Mone B. 6251. v. deágian.

dégle *secret, hidden*, Lk. Lind. War. 8, 17; *nom. n. of* dégol. v. dígol.

déglíce *secretly*, Mt. Lind. Stv. 20, 11. v. dígollíce.

dégol *obscurity, mystery*, Elen. Grm. 340: Exon. 46 b; Th. 159, 11; Gú. 925. v. dígol.

dégol *secret, unknown*, Exon. 8 b; Th. 3, 24; Cri. 41: 104 b; Th. 397, 17; Rä. 16, 21. v. dígol; *adj.*

dégol-ful; *adj. Full of secret, mysterious*; secréti plénus, mystícus :—Ic míðan sceal dégolfulne dóm mínne *I must conceal my mysterious power*, Exon. 127 b; Th. 491, 14; Rä. 80, 14.

dégollíce *secretly*, Mk. Rush. War. 9, 28. v. dígollíce.

dégolnis *solitude*, Mt. Lind. Stv. 6, 6. v. dígolnes.

dégullíce *secretly*, Mt. Rush. Stv. 1, 19. v. dígollíce.

dégulnes *solitude*, Mt. Rush. Stv. 6, 4, 6. v. dígolnes.

déhter *to a daughter*, Exon. 67 b; Th. 251, 7; Jul. 141; *dat. of* dóhtor.

dehtnung *a disposing*, Prov. 24. v. dihtnung.

Deira ríce *the kingdom of the Deirians*, Som. Ben. Lye. v. Dera ríce.

delan; *p.* dæl, *pl.* dælon; *pp.* dolen *To fall, sink*; lábi :—Ærðon engla weard for oferhygde dæl on gedwilde *ere the angels' guardian for pride sank into error*, Cd. 1; Th. 2, 22; Gen. 23.

délan *to divide*, Cant. Moys. Ex. 15, 10; Thw. notes, p. 29, 10. v. dælan.

delf, es; *n. A delving, the act of digging*; fossio, Th. Anlct. DER. ge-delf, stán-ge-. v. dælf.

DELFAN; ic delfe, ðú delfest, dilfst, he delfeþ, dilfþ, *pl.* delfaþ; *p.* ic, he dealf, ðú dulfe, *pl.* dulfon; *subj.* delfe, *pl.* delfen : *p.* dulfe, *pl.* dulfen; *pp.* dolfen; *v. a. To dig, dig out*, DELVE : fódere, effódere :—Ne mæg ic delfan *fódere non váleo*, Lk. Bos. 16, 3. Ongan he eorþan delfan *he began to dig the earth*, Elen. Kmbl. 1655; El. 829. Ic delfe fódio, Ælfc. Gr. 28, 6; Som. 32, 45. Ðær þeófas hit delfaþ *ubi fures effódiunt*, Mt. Bos. 6, 19, 20: Exon. 111 b; Th. 427, 27; Rä. 41, 97. Ic dealf ðisne pytt *ego fodi púteum istum*, Gen. 21, 30. Se dealf deópe *qui fodit in altum*, Lk. Bos. 6, 48. Wæterpyttas ðe ge ne dulfon *wells which ye dug not*, Deut. 6, 11. Hí dulfon áne mycle díc *they dug a great ditch*, Chr. 1016; Erl. 155, 22: Ex. 7, 24: Ps. Lamb. 21, 17: Ps. Th. 56, 8. Swelce hwá delfe eorþan *as if any one should dig the earth*, Bt. 40, 6; Fox 242, 5. Gif se delfere ða eorþan nó ne dulfe *if the digger had not dug the earth*, 40, 6; Fox 242, 7. [*Prompt.* delvyn' fódere: *Wyc.* delue: Piers P. delven: *Chauc.* delve: *Laym.* dælfen, deluen: *Orm.* dellfeþþ burieth: *Plat.* dölben: *O. Sax.* bi-delban *to bury*: *Frs.* dollen: *O Frs.* delva, dela: *Dut.* delven: *Ger.* delben: *M. H. Ger.* télben: *O. H. Ger.* bi-telban *sepelíre*.] DER. a-delfan, be-, ge-, of-, þurh-, under-, upa-, úta-.

delfere, es; *m. A digger*; fossor :—Gif se delfere ða eorþan nó ne dulfe *if the digger had not dug the earth*, Bt. 40, 6; Fox 242, 7.

delfing, es; *m. A* DELVING, *digging, laying bare, exposing*; ablaqueátio :—Niderwart treówes delfing, bedelfing *ablaqueátio*, Ælfc. Gl. 60; Som. 68, 11; Wrt. Voc. 39, 2. DER. be-delfing.

delf-ísen, es; *n. A digging-iron, spade*; fossórium :—Costere *vel* delfísen *vel* spadu *vel* pal *fossórium*, Ælfc. Gl. 2; Som. 55, 40; Wrt. Voc. 16, 14: Cot. 90.

delu, e; *f: pl. nom. gen. acc.* dela; *dat.* delum *A teat, nipple*; mamma :—Wæron forbrocene ða dela hiora mægdenhádes ... bióþ forbrocene ða wæstmas ðæra dela *fractæ sunt mammæ pubertátis eárum ... pubertátis mammæ franguntur*, Past. 52; Hat. MS. [*O. H. Ger.* tila, tili, *f. mamma*.]

dem, demm, es; *m. Damage, mischief, harm, injury, loss, misfortune*:

damnum, mălum, noxa, injūria, detrĭmentum, calămĭtas:—Ðǽr wæs ân swá mĭcel dem *there was so great a loss*, Ors. 6, 14; Bos. 122, 21. Be ðæs demmes ehte *pro damni æstimatiōne*, Ex. 22, 5. He ðone demm his giémelíéste gebétan ne mæg *he cannot remedy the mischief of his neglect*, Past. 36, 3; Hat. MS. 47 a, 22. Ne wéne ic ðæt ænig man atellan mǽge ealne ðone dem ðe Rómánum gedón wearþ *I do not think that any man can tell all the harm which was done to the Romans*, Ors. 2, 8; Bos. 51, 28. Hit oft gebýraþ ðæt seó leáse wyrd ne mæg ðam men dôn nǽnne dem *it often happens that deceitful fortune can do no injury to a man*, Bt. 20; Fox 70, 23. He geman ðone demm oððe ðæt bismer, ðæt him ǽr gedôn wæs *he remembers the injury or the disgrace that was formerly done to him*, Past. 33, 7; Hat. MS. 43 b, 2: Ors. 2, 4; Bos. 43, 29. Ôðrum monnum þyncþ ðæt hie mǽstne demm [dem MS. Cott.] þrôwigen *it seems to other men that they suffer the greatest misfortune*, Past. 14, 5; Hat. MS. 18 a, 26.

DÉMA, an; *m.* [déman *to deem, judge, think*]. **I.** *a deemer, thinker, judge, an umpire*; censor, consul, jūdex, arbĭter:—Ic eom se déma *I am the judge*, Exon. 42 b; Th. 144, 8; Gū. 675: 69 a; Th. 257, 19; Jul. 249: Judth. 10; Thw. 22, 12; Jud. 59. Se Déma gegaderaþ ðæt clǽne corn into his berne *the Judge shall gather the pure corn into his barn*, Homl. Th. ii. 68, 17: i. 526, 21. Gehýraþ hwæt se unrihtwísa déma cwyþ *audite quid judex iniquitātis dicit*, Lk. Bos. 18, 6, 2: Ps. Lamb. 74, 8: Ps. Th. 67, 6; Wrt. Voc. 72, 66. Déma *judex*, vel *censor*, vel *arbĭter*, Ælfc. Gl. 68: Som. 70, 9; Wrt. Voc. 42, 18: 86; Som. 74, 21; Wrt. Voc. 50, 5: *consul*, Ælfc. Gr. 9, 10; Som. 9, 16. Sceall ǽghwylc ðǽr riht gehýran dǽda gehwylcra, þurh ðæs déman múþ *there shall every one hear the right of all his deeds, through the judge's mouth*, Elen. Kmbl. 2564; El. 1283: Exon. 69 b; Th. 257, 33; Jul. 256. Him egsa becom fôr déman *dread came over them before their judge*, Cd. 221; Th. 288, 13; Sat. 380: 175; Th. 220, 15; Dan. 71. Ic ðone déman in dagum mínum wille weorþian *I will worship the judge in my days*, Exon. 41 b; Th. 139, 8; Gū. 590. Besencte syndon wið stán déman heora *absorpti sunt juxta petram judices eōrum*, Ps. Lamb. 140, 6. Déman *censōres*, vel *judīces*, vel *arbitri*, Ælfc. Gl. 8; Som. 56, 87; Wrt. Voc. 18, 39. Ealra déman ðam gedéfestan *to the most benevolent of all judges*, Exon. 93 a; Th. 350, 3; Sch. 58. Ærmorgenes gancg wið ǽfentíd ealle ða déman Drihten healdeþ *exĭtus matutīni et vespĕre delectabĕris*, Ps. Th. 64, 9. **II.** *the judge, who gave a wrong judgment, was subject to a fine of one hundred and twenty shillings; and if a man could not obtain justice, the judge to whom he applied was fined thirty shillings. As the judge represented the king, he was at the king's disposal*:—Se déma, ðe ôðrum wôh déme, gesylle ðam cynge hundtwelftig scillinga *to* bôte, bútan he mid áþe gecýðan durre, ðæt he hit ná rihtor ne cúðe, and þolige á his þegenscipes, bútan he hine æt ðam cynge gebicge, swá swá he him geþafian wille, and amanige ðære scíre bisceop ða bôte *to* ðæs cynges handa *let the judge, who judges wrong to another, pay to the king one hundred and twenty shillings for a fine, unless he dare to prove on oath, that he knew it not more rightly, and let him forfeit for ever his thaneship, unless he will buy it of the king, so as he is willing to allow him, and let the bishop of the shire exact the fine [and pay it] into the king's hands*, L. Edg. ii. 3; Th. i. 266, 15–20. Gif hwá him ryhtes bidde befóran hwelcum scírmen oððe ôðrum déman, and abiddan ne mǽge, and him wedd, mon sellan nelle, gebéte xxx scillinga, and binnan vii nihton gedô hine ryhtes wierþne *if any one demand justice before a sheriff or other judge, and cannot obtain it, and the man will not give him a promise, let him make compensation with thirty shillings, and within seven days do him justice*, L. In. 8; Th. i. 106, 20–108, 2. [*Laym.* deme *a judge*: *Orm.* deme *a chief, ruler, judge*: *O. H. Ger.* tuomo, *m. judex, dux.*] DER. ealdor-déma, heofon-, sige-.

déman; *to* démanne, démenne; *part.* démende; ic déme, ðú démest, démst, he démeþ, démþ, *pl.* démaþ; *p.* démde, *pl.* démdon; *impert.* dém, déme, *pl.* démaþ, déme ge; *pp.* démed; *v. trans. dat. acc.* [dóm *judgment, opinion*] *To* DEEM, *judge, think, consider, estimate, reckon, determine, examine, prove, doom, condemn*; judicāre, arbitrāri, æstimāre, censēre, recensēre, decernĕre, sancīre, examināre, condemnāre:—He com déman eorþan *venit judicāre terram*, Ps. Lamb. 95, 13: Elen. Kmbl. 621; El. 311: Exon. 63 a; Th. 231, 25; Ph. 494. Nellen ge déman, ðæt ge ne sýn fordémede *nolīte judicāre, ut non judicēmini*, Mt. Bos. 7, 1. Eorþan to démanne *judicāre terram*, Ps. Th. 97, 8; Bd. 4, 3; S. 569, 27. To démenne ǽgðer ge ðam cucum ge ðám deádum *to judge both the quick and the dead*, Homl. Th. ii. 596, 20: 598, 6. Démende *judging*, Past. 15, 6; Hat. MS. 20 a, 19. Ic rihtwísnessa déme *justĭtias judicābo*, Ps. Lamb. 74, 3. Ðæs ðe ic déme *ut arbĭtror*, Bd. 1, 27; S. 497, 5. Ic déme oððe asmeáge *censeo*, Ælfc. Gr. 26, 2; Som. 28, 51. Ic déme oððe ic gefette oððe ic hálgige *sancio*, 30, 1; Som. 34, 33. Ðú démst [Th. démest] folctruman on emnysse *judĭcas popŭlos in æquitāte*, Ps. Lamb. 66, 5. Nǽfre God démeþ ðæt ǽnig ðæs earm geweorþe *God never deems* [= *decrees*] *that any should become so poor*, Exon. 78 b; Th. 294, 17; Crä. 16. Ðis fýr æfter weorca ge-earnunge ânra gehwylcum démeþ and bærneþ *iste rogus juxta merĭta opĕrum singŭlos examĭnat*, Bd. 3, 19; S. 548, 27. He démþ folcum mid rihte *judicābit*

popŭlos cum justĭtia, Ps. Th. 9, 9: Ps. Lamb. 95, 13. Ðam ylcan dôme ðe ge démaþ, eów biþ gedémed *in quo judĭcio judicātis, judicabimĭni*, Mt. Bos. 7, 2. He monige démde to deáþe *he doomed many to death*, Elen. Kmbl. 997; El. 500. Moises and Aaron gegaderodon ealle ðás and démdon him *quos Moyses et Aaron congregavĕrunt recensentes eos*, Num. 1, 18. Ne dém náu unriht... déme rihte ðínum néxtan *non injuste judicābis... juste judĭca proxĭmo tuo*, Lev. 19, 15. Rihtlíce démaþ eálá ge suna manna *rec'e judicāte filii homĭnum*, Ps. Lamb. 57, 2: 81, 3. Hú lange déme ge unrihtwísnesse *usquequo judicātis iniquitātem?* 81, 2. Ne wæs sôna his hálgung [MS. halgunge] démed *nec statim ordinātio decrēta*, Bd. 4, 28; S. 606, 22. Beóþ his dagas swýʹce démde gelíce swá ðú on scimiendre sceade lócige *dies ejus sicut umbra prætĕreunt*, Ps. Th. 143, 5. [*Wyc. Piers P. Chauc.* demen; *R. Glouc.* ydemd, *pp.*: *Laym.* demen: *Orm.* deme, demen: *O. Sax.* dôman, duomian: *O. Frs.* déma: *M. H. Ger.* tücmen: *O. H. Ger.* tuomian: *Goth.* domyan: *Dan.* dömme: *Swed.* döma: *Icel.* dæma.] DER. a-déman, for-, ge-, to-.

démend, es; *m. A judge, an umpire*; jūdex, arbĭter:—God sceal on heofenum dǽda demend *God shall be in the heavens judge of actions*, Menol. Fox 531; Gn. C. 36: Exon. 76 a; Th. 286, 1; Jul. 725: Andr. Kmbl. 173; An. 87: 2379; An. 1191.

démere, es; *m. A* DEEMER, *judge*; jūdex, L. Alf. 18; Th. i. 48, note 38.

demm *damage, mischief, harm*, Ex. 22, 5: Ors. 2, 4; Bos. 43, 29. *v.* dem.

demman; *p.* de; *pp.* ed *To* DAM, *stop water*; obturāre flūmen, Som. Ben. Lye. [*O. Frs.* demma, damma: *Dut.* dammen: *Ger.* dämmen: *M. H. Ger.* temmen: *O. H. Ger.* bi-temman *occupāre*: *Goth.* faurdammyan *to dam*: *Dan.* dämme: *Swed.* dämma: *Icel.* demma.] DER. fordemman.

Dena lagu, lag, lah, e; *f. The law of the Danes, Danish law*; Danōrum lex, Danisca lex. *v.* Dene.

Dena mearc *the land of the Danes, Denmark*, Ors. 1, 1; Bos. 21, 33. *v.* Dene-mearc.

den-bera; *pl. n. Lat.* [beara *a grove, wood*] *Swine-pastures, places yielding mast for the fattening of hogs*; pascua porcōrum:—Pascua porcōrum quæ nostra lingua Saxonĭca denbera nomināmus, Cod. Dipl. 288; A.D. 863; Kmbl. ii. 75, 27: 281; A.D. 858; Kmbl. ii. 65, 6. Adjectis denberis in commūni saltu, 160; A.D. 765–791; Kmbl. i. 194, 34: 179; A.D. 801; Kmbl. i. 216, 26: 198; A.D. 811; Kmbl. i. 248, 17: 239; A.D. 838; Kmbl. i. 317, 20.

denegan *to knock, ding*; tundĕre, Som. Ben. Lye.

dene, an; *f. A valley*; vallis:—Dene *vallis*, Ælfc. Gr. 9, 28; Som. 11, 55: Wrt. Voc. 80, 44: Ælfc. Gl. 97; Som. 76, 64; Wrt. Voc. 54, 8. Ælc dene biþ gefylled *every valley shall be filled*, Homl. Th. i. 360, 33. Seó dene ðe ðú geseáh *vallis illa quam aspexisti*, Bd. 5, 12; S. 630, 3. Seó dene wæs afylled mid manna sáwlum *the valley was filled with men's souls*, Homl. Th. ii. 350, 9. Seó nicele byrnende dene *the great burning valley*, ii. 352, 20. *v.* denu.

dene, es; *m. A valley*; vallis:—Abram com and eardode wið ðone dene Mambre *Abram venit et habitāvit juxta convallem Mambre*, Gen. 13, 18. *v.* denu.

Dene; *nom. acc.; gen. a; dat.* um; *pl. m. The Danes*; Dāni:—Ðá ða Engle and Dene to friþe and to freóndscipe fullíce féngon *when the English and Danes fully took to peace and to friendship*, L. E. G; Th. i. 166, 7. Gif hláford his þeówan freóls-dæge nýde to weorce, gylde lahslihte inne on Dena lage, and wíte mid Englum *if a lord oblige his servant to work on a festival-day, let him pay penalty within the Danish law, and fine among the English*, L. E. G. 7; Wilk. 53, 1. Sunnan dæges cýpinge gif hwá agynne þolie ðæs ceápes, and twelf ôrena mid Denum, and xxx scillingas mid Englum *if any one engage in Sunday marketing, let him forfeit the chattel, and twelve ores among the Danes, and thirty shillings among the English*, L. E. G. 7; Th. i. 170, 16.

Dene-mearc, -mearc, e; *f.* -marce, -mearce, -merce, an; *f.* DENMARK; Dānia, Cimbrica Chersonēsus = Χερσόνησος, *f. a land island, peninsula; from* χέρσος, χέρρος *land, and* νήσος, ου; *f. an island* [Dene *the Danes*,—denu *a plain, vale, valley; and* mearc *a boundary*. The Saxon Chronicle, in 1005, 1023, 1036, has Denemearc; Denmearc, in 1019, 1075; Denmarc, in 1070 and 1119. In Danish mark signifies *a country*; hence Denmark *the low country of the Danes*: so Finmark *the low country of the Finns*. Wulfstan [Alfred, A.D. 892] is the most early writer hitherto known, who mentions Denmark]:—Wulfstán sǽde ðæt he geföre of Hǽðum. Weonoðland him wæs on steór-bord, and on bæc-bord him wæs Langa land, and Lǽland, and Falster, and Scon-ég; and ðás land eall hýraþ to Denemearcan *Wulfstan said that he went from Haddeby. He had Weonodland on the right, and Langland, Laaland, Falster, and Sconey on his left; and all these lands belong to Denmark*, Ors. 1, 1; Bos. 21, 39, 41–43. Ða ígland in Denemearce hýraþ *these islands belong to Denmark*, 1, 1; Bos. 21, 38.

Denisc; *def.* se Denisca; *adj.* DANISH; Dānicus:—Gif man ofslagen weorþe, ealle we lǽtaþ efen dýrne, Engliscne and Deniscne *if a man be slain, we estimate all equally dear, English and Danish*, L. A. G. 2;

Th. i. 154, 1. Wið Deniscne here *against the Danish army*, Chr. 837; Erl. 66, 7 : 845; Erl. 66, 23. Næron hí náwðer ne on Frysisc gesceapen ne on Denisc *they were shapen neither as the Frisian nor as the Danish*, Chr. 897; Th 177, 3, col. 2. Hér, A.D. 872, Ælfréd cyning gefeaht wið feówer sciphlæstas Deniscra monna *here*, A.D. 872, *king Alfred fought against four ship-crews of Danish men*, 872; Th. 150, 28, col. 1. Ða com ðæm Deniscum scipum flód to *then the tide came to the Danish ships*, 897; Th. 176, 37, col. 1. Com ðá se Denisca flóta to Sandwíc *then*, A.D. 1006, *the Danish fleet came to Sandwich*, 1006; Th. 257, 4, col. 1.

Deniscan; *gen.* ena; *pl. m.* [Denisca, *def.* of Denisc; *adj.*] *The Danish men, the Danes*; Dänici viri, Dáni :—Hér, A.D. 835, Ecgbryht, Westseaxna cing, geflýmde ge ða Wealas ge ða Deniscan *here*, A.D. 835, *Ecgbryht, king of the West Saxons, routed both the Welsh and the Danes*, Chr. 835; Th. 116, 13–23, col. 1, 2. Ða Deniscan áhton wælstówe ge-weald *the Danes obtained power of the battle-place*, Chr. 833; Erl. 65, 19 : 837; Erl. 67, 8 : 840; Erl. 67, 13 : 871; Erl. 75, 15 : 871; Erl. 77, 6 : 999; Erl. 134, 26.~ On ðæra Deniscena healfe wæs ofslægen Eoric cyning *king Eric was slain on the side of the Danes*, Chr. 905; Erl. 99, 32 : 910; Erl. 100, 15. v. Denisc.

Denisses burna, an; *m.* DENISESBURN, *the river Denis*; Denisi rivus :— On ðære stówe ðe Engle nemnaþ Denisses burna *in loco qui lingua Anglórum Denises burna, id est rivus Denisi vocátur*, Bd. 3, 1; S. 524, 10.

DENN, es; *n. A* DEN; cubile, lustrum? [lustra MS.] :—Denn *cubile*, Ælfc. Gr. 9, 2; Som. 8, 27. Wild-deóra holl and denn *lustrum ferárum* [MS. *lustra*], Ælfc. Gl. 110; Som. 79, 38; Wrt. Voc. 59, 10. Se lég-draca gewát ðennes :—Geseah he wundur on ðæs wyrmes denn *he saw wonders in the dragon's* [lit. *worm's*] *den*, 5512; B. 2759. [*Prompt.* deñ *specus*: *Wyc.* den: *Chauc.* dennes *caves*: *Laym.* denne: *Dut.* denne, *f. deck of a ship*: *Kil.* denne *área*, *antrum*: *Ger.* tenne, *f. area*: *M. H. Ger.* tenne, *n. area*: *O. H. Ger.* tenni, *n. area*.]

dennian; *p.* ode, ade; *pp.* od, ad *To become slippery*; lubrícum fíeri :—Feld dennode [dennade, col. 1] secga swäte *the plain became slippery with the blood of soldiers*, Chr. 937; Th. 203, 10, col. 2; Ædelst. 12.

den-séte; *m. pl. Dwellers in valleys or plains*; vallicólæ. v. sǽte.

denu, e; *f*: dene, an; *f*: dene, es; *m. A plain, vale, dale, valley*; vallis, convallis :—Seó denu ðe ðú gesáwe weallendum lígum *vallis illa quam aspexisti flammis ferventíbus*, Bd. 5, 12; S. 630, 3, note, MS. B. Seó stów ðær seó denu wæs *the place where the valley was*, 5, 12; S. 630, note 3, MS. T. Ælc denu biþ gefylled *omnis vallis implebítur*, Lk. Bos. 3, 5. Ða becóme wit to ánre dene, seó wæs ormǽtlíce deóp and wíd, and forneán on lenge unge-endod *we two then came to a valley, which was immensely deep and wide, and in length almost endless*, Homl. Th. ii. 350, 6 : Bd. 5, 12; S. 627, 36; Ps Lamb. 83, 7 : Bt. Met. Fox 7, 73; Met. 7, 37 : Salm. Kmbl. 458; Sal. 229. From Ebron dene *de valle Hebron*, Gen. 37, 14. He gebirgde hine on ðære dene Moab landes ongeán Phogor *sepelivit eum in valle terræ Moab contra Phogor*, Deut. 34, 6. Dene getelda ic mete *convallem tabernaculórum dimétiar*, Ps. Spl. 107, 7. Dena genihtsumiaþ of hwǽte *valles abundábunt frumento*, 64, 14 : Exon. 115 b; Th. 443, 14; Kl. 30. Dene, *nom. pl.* Exon. 56 a; Th. 199, 11; Ph. 24. Ðú ðe asendst wyllas on denum *qui emittis fontes in convallibus*, Ps. Lamb. 103, 10 : Exon. 107 b; Th. 409, 18; Rä. 28, 3. [It is often used as a termination of the names of places situate in a plain or valley, as *Tenterden*, etc.] DER. deáp-denu.

deófel-líc; *adj. Diabolical, devilish*; diabólicus :—Mid deófellícum wiglungum *with devilish incantations*, Homl. Th. i. 102, 11.

deófel-seócnys, -nyss *devil-sickness*, Mt. Bos. 4, 24. v. deófol-seócnes.

deófles *of the devil*, Andr. Kmbl. 86; An. 43; *gen.* of deófol.

deóflic, *adj.* deófel-líc; *adj. Devilish, diabolical*; diabólicus :—Úre heo-fenlíca Hláford ðone deóflícan deáþ nyðeratræd *our heavenly Lord trod down the diabolical death*, Nicod. 29; Thw. 16, 40. Undergeat se apostol ðás deóflícan fácn *the apostle perceived these diabolical wiles*, Homl. Th. i. 62, 31. Mid deóflícum wiglungum *with diabolical incantations*, i. 102, 15.

DEÓFOL, deóful, dióful; contracted to deófl; *gen.* es; *dat.* e; *nom. pl.* deóflu, deófol; *gen.* deófla. *m. n. The* DEVIL; diabólus. **I.** *m.* Nú þencþ menig man and smeáþ hwanon deófol cóme ? Ðonne wite he ðæt God gesceóp, to mǽran engle, ðone ðe nú is deófol ; ac God ne gesceóp hine ná to deófle ; ac ðá ðá he wæs mid ealle fordón and for-scyldgod þurh ða miclan upahefednysse and wiðerweardnysse, ðá wearþ he to deófle awend, se ðe ǽr wæs mǽre engel geworht *now many a man will think and inquire whence the devil came? Then let him know that God created, as a great angel, him who is now the devil ; but God did not create him as the devil ; but when he was wholly done for and guilty towards God, through his great haughtiness and enmity, then became he changed to the devil, who before was created a great angel*, Homl. Th. i. 12, 18–23. Se deófol ne wunode ná on sóþfæstnysse, forðamðe seó

sóþfæstnyss nis náteshwon on him *the devil abided not in the truth, because the truth is not in any wise in him*, Hexam. 10; Norm. 16, 18. Ðæt he ðone deófol adrífe *ut dæmónium ejicéret*, Mk. Bos. 7, 26. **II.** *n.* Him biþ ðæt deófol láþ *the devil is loathly to them*, Salm. Kmbl. 246; Sal. 122. Hyre ðæt deófol oncwæþ *the devil addressed her*, Exon. 72 b; Th. 270, 5; Jul. 460. Heó ðæt deófol genom *she took the devil*, 69 b; Th. 259, 27; Jul. 288. Heó ðæt deófol teáh bendum fæstne *she drew the devil fast in bonds*, 73 b; Th. 274, 17; Jul. 534. On deófla ealdre he drífþ út deóflu *in princípe dæmóniórum ejicit dæmónes*, Mt. Bos. 9, 34. Deófol, *nom. pl.* Exon. 30 b; Th. 93, 27; Cri. 1532: *acc. pl.* Exon. 118 b; Th. 455, 18; Hy. 4, 51. [*Prompt.* dewle, devylle: *Wyc.* deuel: *Piers P.* deovel: *Chauc.* deuill: *Laym.* deauel, deouel: *Orm.* deofell, defell: *Plat.* düvel, düwel, *m*: *O. Sax.* diuðal, diobol, diabol, diuvil, *m*: *Frs.* deal, dijvel, *m*: *O. Frs.* diovel, divel, *m*: *Dut.* duivel, *m*: *Ger.* teufel, *m*: *M. H. Ger.* tiuvel, tievel, *m*: *O. H. Ger.* tiufal, *m*: *Goth.* diabaulus, *m*: *Dan.* dævel, djævel, *m*: *Swed.* djefvul, *m*: *Icel.* djöfull, *m*: *Lat.* diabólus, *m*: *Grk.* διάβολος *an accuser* or *slanderer, m*; from διαβάλλω *to cast* or *dart through* or *against*; figuratively, *to stab with an accusation* or *slander*; διά *through, against*, and βάλλω *to cast*. Διάβολος = ἀντίδικος *an opponent, adversary* = שָׂטָן *m.* Satan, *q.v.*] DER. helle-deófol, hilde-.

deófol-cræft, es; *m. Devil-craft, the black art, witchcraft*; dæmoniáca ars :—þurh dígolnesse deófolcræftes *per dæmoniacæ artis arcána*, Bd. 4, 27; S. 604, 9. Hí nalæs mid deófolcræfte ac mid godcunde mægene gewélgade cóman *illi non dæmoniaca sed divina virtúte prædíti veniébant*, Bd. 1, 25; S. 487, 1.

deófol-cund *diabolical*. v. deóful-cund.

deófol-dǽd, e; *f. A devil-deed, diabolical deed*; diabóli machinátio, diabólícum facinus :—Hie wlenco anwód deófoldǽdum *pride invaded them with diabolical deeds*, Cd. 173; Th. 217, 5; Dan. 18.

deófol-gild, deóful-gild, diófol-gild, -geld, -gield, -gyld, es; *n.* [deófol, gild *tribute, worship*] *Devil-worship, sacrifice to devils, idolatry, an idol, an image of the devil*; diabóli *vel* dæmónum cultus, idololatría = εἰδωλολατρεία, idólum, simulacrum :—Ðæt man mihte dón heora deófolgyld *that they might do their devil-worship*, Ors. 3, 3; Bos. 55, 29, 33, 37 : Andr. Kmbl. 3372; An. 1690: Exon. 66 b; Th. 245, 29; Jul. 52: Bd. 1, 7; S. 477, 4 : L. Ecg. C. 38; Th. ii. 162, 22, note 6. Betwih deófolgyldum *lísdon inter idóla vivérent*, Bd. 3, 30; S. 562, 19: Exon. 68 a; Th. 251, 25; Jul. 150. Beóþ deófolgyld dysigra þeóda gold and seolfur *simulacra gentium argentum et aurum*, Ps. Th. 134, 15 : 113, 12 : Bd. 3, 30; S. 561, 43: Cd. 145; Th. 180, 18; Exod. 47: Elen. Grm. 1041: Cot. 118.

deófol-gylda, an; *m.* [gild = gyld *a worship*, with -a *a worshipper*] *A worshipper of the devil, an idolater*; idololatres = εἰδωλολάτρης :—Ða deófolgyldan gecwǽdon ðæt hí woldon ðone apostol to heora hǽðenscipe geneádian *the idolaters said that they would force the apostle to their heathenship*, Homl. Th. i. 70, 23.

deófolgyld-hús *a heathen temple*. v. deófulgyld-hús.

deófol-scín, es; *pl. nom. acc.* -scínnu; *n.* [scín *a vision, phantom, demon*] *A diabolical vision, phantom, demon*; dæmoniácus visus, dæmon :—Deófolscín *dæmoniácus visus*, M. H. 106 b. Deófolscínnu *dæmónia*, Scint. 7.

deófol-seóc; *def.* se deófol-seóca; *adj.* [seóc *sick*] *Devil-sick, possessed with a devil*; dæmónium hábens, dæmoniácus :—Ða wæs him broht án deófolseóc man *tunc oblátus est ei dæmónium habens*, Mt. Bos. 12, 22 : 9, 32. Híg brohton him manege deófolseóce *obtulérunt ei multos dæmónia habentes*, 8, 16. Deófolseóc *dæmoniácus*, Ælfc. Gl. 78; Som. 72, 34; Wrt. Voc. 45, 66. Hí ða ofsettan deófolseócan forléton *they forsook the possessed demoniacs*, Homl. Th. i. 64, 26.

deófol-seócnes, deóful-seócnes, deófel-seócnes, -ness, -nyss, e; *f. Devil sickness, possession with the devil*; dæmónium = δαιμό ιον :—Deófolseóc-nessa us synd on dínum naman underþeódde *dæmónia subjiciuntur nobis in nomine tuo*, Lk. Bos. 10, 17. Sumue we gesáwo on dínum naman deófolseócnessna útadrífende *vidimus quemdam in non ne tuo ejicientem dæmónia*, Mk. Bos. 9, 38: 16, 17: Lk. Bos. 9, 49: ? 5, 32. He sealde him mihte ofer ealle deófolseócnessa *dedit illis virt.i'em super omnia dæmónia*, Lk. Bos. 9, 1. De hæfdon deófolseócnesse *habentes dæmónia*, Mt. Bos. 8, 28. Deófolseócnysse he hæfþ *dæmónium habet*, Lk. Bos. 7, 33. Deófulseócnysse *dæmónium*, Mt. Bos. 11, 18. De ða deófulseóc-nyssa hæfdon *qui dæmónia habúérant*, 8, 33. Deófelseócnyssa *dæmónia*, 4, 24.

deófol-wítga, an; *m. A devil-prophet, soothsayer, wizard*; vates diabólicus, magus :—Him andswaredon deófolwítgan *the soothsayers answered him*, Cd. 178; Th. 223, 31; Dan. 128.

deóful *the devil*, Mt. Bos. 13, 19. v. deófol.

deóful-cund; *adj. Devil-kind* or *similar, diabolical*; diabólicus :—Gewát se deófulcunda *the diabolical departed*, Judth. 10; Thw. 22, 14; Jud. 61.

deóful-gild, -gyld *idolatry, an idol*, Andr. Kmbl. 3372; An. 1690: Ors. 6, 36; Bos. 131, 41: Bd. 3, 1; S. 523, 23 : 3, 30; S. 562, 15. v. deófol-gild.

deófulgyld-hús, es; *n.* *A heathen temple;* paganōrum templum :—Constantinus hĕt ðæt man cyricean timbrede, and ðæt man belúce ælc deófulgyldhús *Constantine ordered churches to be built, and every heathen temple to be closed,* Ors. 6, 30; Bos. 127, 36.

deóful-seócnys, -nyss *devil-sickness,* Mt. Bos. 8, 33: 11, 18. v. ðeóful-seócnes.

deóg, *pl.* deógon *dyed, coloured,* Beo. Th. 1704; B. 850; *p. of* deágan.

deógol *secret,* Beo. Th. 555; B. 275: Elen. Grm. 1093. v. dígol.

deógollíce *secretly :*—Deógollíce folcrǽd fremede *secretly did public benefits.* Andr. Kmbl. 1241; An. 621. v. dígollíce.

DEÓP, díóp; *adj.* DEEP, *profound, stern, awful, solemn;* prōfundus, grăvis, sōlemnis :—Ðes pytt is deóp *this well is deep,* Jn. Bos. 4, 11. Deóp wæter *the deep water,* Exon. 54 b; Th. 193, 19; Az. 124. Fíftena stód deóp ofer dúnum flód elna *the flood stood fifteen ells deep over the hills,* Cd. 69; Th. 84, 15; Gen. 1398. Noe oferláþ ðone deópestan drencflóda *Noah sailed over the deepest of drowning floods,* 161; Th. 200, 29; Exod. 364. Hú héh and deóp hell seó *how high and deep hell is!* 228; Th. 309, 9; Sat. 707. Deópra dolga *of deep wounds,* Exon. 114 a; Th. 438, 7; Rä. 57, 4. Wǽrun díne geþancas þearle deópe *nimis profundæ factæ sunt cogitatiōnes tuæ,* Ps. Th. 91, 4. Deóp leán *a deep requital,* Cd. 167; Th. 209, 29; Exod. 506. Þurh deópne gedwolan *through profound error,* Exon. 70 a; Th. 260, 22; Jul. 301. Onguldon deópra firena *they atoned for their deep crimes,* 45 a; Th. 153, 23; Gú. 830. Þurh deópne dóm *through stern doom,* 42 a; Th. 142, 8; Cú. 641. On ðam deópan dæge *on that awful day,* 116 b; Th. 448, 24; Dóm. 59. Ðú míne sáwle ofer deópum deáþe gelǽddest *eripuisti animam meam de morte,* Ps. Th. 114, 8. Deópne áþ Drihten aswór *jurāvit Dominus solemne jurāmentum,* 131, 11. Moyses sægde hálige sprǽce, deóp ǽrende *Moses delivered a holy speech, a solemn message,* Cd. 169; Th. 210, 20; Exod. 518. [*Prompt. Wyc.* depe: *Piers P.* dupe: *Chauc. R. Glouc.* depe: *Laym.* deop, deap: *Orm.* deope, deep, deop, dep: *Plat.* deep, deip: *O. Sax.* O. *Frs.* diop, diap: *Dut.* diep: *Kil.* duyp: *Ger. M. H. Ger.* tief: *O. H. Ger.* tiuf: *Goth.* diups: *Dan.* dyb: *Swed.* djup: *Icel.* djúpr.] DER. un-deóp.

deóp, dýp. díóp, es; *n.:* dýpe, an; *f. Depth, the deep, abyss;* prōfundum :—Ne me forswelge sǽ-grundes deóp *ne me absorbeat profundum,* Ps. Th. 68, 15. Adó me of deópe deorces wæteres *libera me de profundo aquārum,* 68, 14. Ic slóh gársecges deóp *I struck the ocean's deep,* Cd. 147; Th. 195, 24; Exod. 281: Beo. Th. 5091; B. 2549: Exon. 93 b; Th. 351, 21; Sch. 83.

deópe, díópe; *comp.* -or; *sup.* -ost; *adv. Deeply, profoundly, thoroughly, entirely, earnestly;* prōfunde, grăvĭter, subtīlĭter, penĭtus, solemnĭter :—He wearþ deópe gedolgod *he became deeply wounded,* Exon. 113 b; Th. 435, 25; Rä. 54, 6. Gedréfede ða deópe syndan *turbāti sunt grăvĭter,* Ps. Th. 106, 26. Se ðis líf deópe geond þenceþ *who profoundly contemplates this life,* Exon. 77 b; Th. 291, 29; Wand. 89. Búton he ðe deóppor hit gebéte *unless he amend it the more earnestly,* Cod. Dipl. 773; A. D. 1044; Kmbl. iv. 87, 13. Ðæt ðú deópost cunne *what thou most thoroughly knowest,* Exon. 88 b; Th. 333, 10; Gn. Ex. 2. Nis mín bán wið ðé deópe behýded *non est [penĭtus] occultātum os meum abs te,* Ps. Th. 138, 13. Nú ic ðé halsie deópe *now I beseech thee earnestly,* Exon. 121 a; Th. 465, 22; Hö. 108.

deóp-hycgende, *part. Deeply meditating;* contemplābundus, Exon. 49 a; Th. 168, 29; Gú. 1085: Elen. Grm. 353: 881.

deóp-hydig; *adj. Deeply meditating, thoughtful;* contemplābundus :—Cwicra gehwylc deóp-hydigra *each thoughtful being,* Exon. 117 a; Th. 450, 31; Dóm. 96: 47 a; Th. 162, 12; Gú. 974.

deóplíc; *adj. Deep;* prōfundus :—Deóplíc dǽdbót biþ *it is a deep penitence,* L. Pen. 10; Th. ii. 280, 17: Exon. 98 a; Th. 367, 5; Seel. 3: 49 a; Th. 169, 32; Gú. 1103.

deóp-líce, díóp-líce; *comp.* -lícor; *sup.* -lícost; *adv.* DEEPLY, *profoundly, thoroughly;* prōfunde, subtīlĭter :—Þearle deóplíce ðú sprycst *valde profunde loqueris,* Coll. Monast. Th. 32, 9: Exon. 49 a; Th. 169, 13; Gú. 1094: Bt. Met. Fox 22, 5; Met. 22, 3. Díóplíce spirigan æfter ryhte *to search deeply after truth,* Bt. 35, 1; Fox 154, 19. Wit sculon deóplícor ymbe ðæt *we two must inquire more deeply about it,* 5, 3; Fox 12, 12. Ðe deóplícost Drihtnes geryno reccan cúðon *who most profoundly could relate the Lord's mysteries,* Elen. Kmbl. 559; El. 280.

deópnes, díópnes, -ness, -nys, -nyss, -niss, e; *f.* DEEPNESS, *depth, an abyss;* prōfundum, altĭtūdo, ăbyssus = ἄβυσσος, vŏrāgo :—Onafǽstnod ic eom on líme deópnesse . . . ic com on deópnysse sǽ *infixus sum in limo profundi . . . vēni in altitūdinem măris,* Ps. Lamb. 68, 3. Ǽuig ne wát ða deópnesse Drihtnes mihta *no one knows the depth of the Lord's might,* Hy. 3, 33; Hy. Grn. ii. 282, 33. Is neowelnes oððe deópnes swá swá scrúd oððe hrægl gegyrlu oððe wǽfels his *est abyssus sicut vestimentum amictus ejus,* Ps. Lamb. 103, 6. Deópnys abyssus, Ælfc. Gl. 98; Som. 76, 91; Wrt. Voc. 54, 35. Nywelnes oððe deópnys deópnissa gecígd *abyssus abyssum invŏcat,* Ps. Lamb. 41, 8. On ðære hellican deópnysse *in the hellish abyss,* Nicod. 24; Thw. 12, 20. Gesettende on goldhordum díópnyssa oððe nywelnyssa *ponens in thesauris abyssos,* Ps. Lamb.

32, 7. Cwicsúsl *vel* helelíc deópnes *barathrum, vorāgo profunda,* Ælfc. Gl. 54; Som. 66, 97; Wrt. Voc. 36, 20.

deóp-þancol; *adj. Deep-thinking, contemplative;* cogitābundus, contemplātivus, Som. Ben. Lye. DER. un-deópþancol.

DEÓR, diór, es; *n. An animal, any sort of wild animal, a wild beast,* DEER; *mostly in contrast to domestic animals;* fĕra, bestia :—Is ðæt deór pandher háten *the animal is called panther,* Exon. 95 b; Th. 356, 16; Pa. 12. Ðæt is wrǽtlíc deór, hiwa gehwylces *that is a curious beast, of every hue,* 95 b; Th. 356, 29; Pa. 19. God geworhte ðære eorþan deór æfter hira hiwum, and ða nítenu on heora cynne *fēcit Deus bestias terræ juxta spĕcies suas, et jumenta in genĕre suo,* Gen. 1, 25. Uton wircean man to andlícnisse, and to úre gelícnisse, and he sig ofer ða deór *faciāmus hominem ad imaginem, et similitudinem nostram, et præsit bestiis,* 1, 26. Lǽde seó eorþe forþ cuce nítena on heora cinne, and deór æfter heora hiwum *prodūcat terra animam viventem, jumenta in genĕre suo, et bestias terræ secundum spĕcies suas,* 1, 24. Ohthere hæfde, ðá he ðone cyninge sóhte, tamra deóra unbebohtra syx hund. Ða deór hí hátaþ hránas *Ohthere had, when he came to the king, six hundred of tame DEER unbought* [*non emptus untrafficked* or *traded in*]. *These DEER they call reins,* Ors. 1, 1; Bos. 20, 25–27. Rēðe deór *a fierce beast;* bellua, Ælfc. Gl. 18; Som. 58, 126. Ánhyrne deór *a one-horned beast, unicorn, rhinoceros;* unicornis *vel* monocĕros *vel* rinocĕros, μονόκερως *vel* ῥινόκερως, 18; Som. 58, 130; Wrt. Voc. 22, 43: 78, 1. [*R. Brun. Chauc. R. Glouc.* der: *Laym. Orm.* deor, der: *Plat.* deert, n: *O. Sax.* dier, n: *O. Frs.* diar, dier, n: *Dut.* dier, n: *Ger.* thier, n: *M. H. Ger.* tier, n: *O. H. Ger.* tior, tier, n: *Goth.* dius, n: *Dan.* dyr, n: *Swed.* djur, n: *Icel.* dýr, n: *Grk.* θήρ *a wild beast.*] DER. heá-deór, mere-, ráh-, sǽ-, wǽg-, wild-.

deór, diór, dýr; *adj.* [deór *an animal*]. I. *brave, bold,* as a wild beast; fortis, strēnuus :—Se hálga wæs to hofe lǽded, deór and dómgeorn *the holy one was led to the house, bold and virtuous,* Andr. Kmbl. 2617; An. 1310: Exon. 108 b; Th. 414, 6; Rä. 32, 16. Nis mon in his dǽdum to ðæs deór *there is not a man so bold in his deeds,* Exon. 82 a; Th. 308, 17; Seef. 41. Ðæt wæs se deóra, Didimus wæs háten *that was the bold one, he was called Didymus,* Cd. 225; Th. 299, 1; Sat. 543. Georne gewyrcan deóres dryhtscipes *to zealously labour for bold rulership,* Salm. Kmbl. 775; Sal. 387. Deórum dǽdum *by bold deeds,* Exon. 82 b; Th. 310, 17; Seef. 76. Wǽron mancynnes dugoþa dýrust *they were of mankind the bravest of people,* Cd. 174; Th. 218, 10; Dan. 37. II. *heavy, severe, dire, vehement;* grăvis, dīrus, vehĕmens :—Deór scúr *heavy rain,* Cd. 192; Th. 239, 18; Dan. 372. Diór dǽdfruma *the dire perpetrator, Grendel,* Beo. Th. 4186; B. 2090. Ðone deóran síþ *the severe journey,* Salm. Kmbl. 723; Sal. 361. Swenga ne wyrnaþ deórra dynta *they are not sparing of strokes, severe blows,* Salm. Kmbl. 245; Sal. 122. DER. deór-líc, -mód: heaðo-deór, hilde-.

Deóra bý, Deór-bý, es; *n?* [*Hunt.* Dereby, Derebi: *Ethel.* Derebi: deór *an animal, deer;* bý *a dwelling, habitation; a habitation of deer* or *animals*] DERBY; Derbia :—Hér Ædelflǽd, Myrcna hlǽfdige, begeat ða burh ðe is geháten Deóra bý *in this year* [A. D. 917] *Æthelfled, lady of the Mercians, obtained the burgh which is called Derby,* Chr. 917; Erl. 105, 24: 942; Erl. 116, 14; Edm. 8. Hér wæs eorþstyrung on Deórbý *in this year* [A. D. 1049] *there was an earthquake at Derby,* 1049; Erl. 173, 18.

Deora mǽgþ, Deora ríce *the province* or *kingdom of the Deirians,* Som. Ben. Lye. v. Dera mǽgþ, Dera-ríce.

deóran, dýran; *p.* ede; *pp.* ed *To hold dear, love;* cārum habēre :—Heó deóraþ míne wísan *they love my ways,* Exon. 103 b; Th. 393, 9; Rä. 12, 7. Dýran sceolde he his dreámas on heofonum *he should hold dear his joys in heaven,* Cd. 14; Th. 17, 9; Gen. 257.

deór-boren, diór-boren; *comp.* -ra; *sup.* -est; *adj. Noble-born, noble;* nātu nōbilis :—Ða ilcan riht dó man be ðam deórborenran *let the same rights be done with respect to the nobler-born,* L. In. 34; Th. i. 124, 3.

Deór-bý *Derby,* Chr. 1049; Erl. 195, 35. v. Deóra bý.

Deórbý-scír, Deórbí-scír, e; *f.* [*Brom.* Derbyshire] DERBYSHIRE; ager Derbiensis :—He fór súþ mid ealre ðære scíre, and mid Snotinghamscíre, and Deórbýscíre [Deorbíscíre, Erl. 194, 20] *he went south with all the shire, and with Nottinghamshire, and Derbyshire,* Chr. 1065; Erl. 195, 35.

DEORC; *def. se* deorca, seó, ðæt deorce; *adj.* DARK, *obscure, gloomy, sad;* tenebrōsus, obscūrus :—Niht-helm geswearc, deorc ofer dryhtgumum *the helm of night grew murky, dark o'er the vassals,* Beo. Th. 3584; B. 1790: Exon. 30 b; Th. 95, 22; Cri. 1561: 101 b; Th. 384, 2; Rä. 4, 21. Hí me asetton on seáþ [MS. sceaþ] hinder, ðær wæs deorc þeóstru, and deáþes scúa *posuērunt me in lacu inferiōri, et in tenĕbris, et in umbra mortis,* Ps. Th. 87, 6: Lk. Bos. 11, 34. Biþ se deorca deáþ ge-endad *the dark death shall be ended,* Exon. 63 a; Th. 231, 34; Ph. 499: Ps. Th. 101, 9. Seó deorce niht gewíteþ *the dark night departs,* Exon. 57 a; Th. 204, 16; Ph. 98. Adó me of deópe deorces wæteres *libera me de profundo aquārum,* Ps. Th. 68, 14. He hí of ðám þýstrum ðanon alǽdde, and of deáþes scúan deorcum generede *eduxit eos de tenĕbris, et umbra mortis,* 106, 13. On ðære deorcan niht *in the dark night,* Andr. Kmbl. 2922; An. 1464: Exon. 50 b; Th. 175, 8; Gú. 1191. Drihten sealde him dimne and deorcne deáþes scúwan *the Lord gave him death's*

shadow dim and dark, Cd. 223; Th. 293, 14; Sat. 455: Exon. 61 a; Th. 225, 2; Ph. 383. Ðú dæg settest, and deorce niht *tuus est dies, et tua est nox*, Ps. Th. 73, 16: 142, 4. Wæs ðæs fugles flyht dyrne and dégol ðám ðe deorc gewit hæfdon on hreðre *the bird's flight was hidden and secret to those who had a dark understanding in their breasts*, Exon. 17 a; Th. 40, 18; Cri. 640: Cd. 5; Th. 7, 19; Gen. 108. Se ðis deorce líf deópe geondþenceþ *he profoundly contemplates this dark life*, Exon. 77 b; Th. 291, 28; Wand. 89. Feónd seondon réðe, dimme and deorce *our foes are fierce, dim and dark*, Cd. 215; Th. 271, 13; Sat. 105: Ps. Th. 73, 19: 113, 12. Gebrecu féraþ deorc ofer dreohtum [MS. dreontum] *the crashes go dark over multitudes*, Exon. 102 a; Th. 385, 15; Rä. 4, 45: 48 b; Th. 168, 1; Gú. 1071. Cwíst ðú oncnáwaþ híi wundru ðíne, on ðám dimmum deorcan þýstrum *numquid cognoscentur in tenebris mirabilia tua?* Ps. Th. 87, 12. He wát deorce grundas *he knows the dark places*, 134, 6: 145, 6. Ðú scealt andettan hwæt ðú þurhtogen hæbbe deorcum gedwildum *thou shalt confess what thou hast accomplished by dark errors*, Exon. 72 b; Th. 270, 4; Jul. 460: Beo. Th. 556; B. 275. Þurhdrifon híi me mid deorcan næglum *they pierced me with dark nails*, Rood Kmbl. 91; Kr. 46. [*Prompt. dele: Wyc.* derk-: *Chauc.* dark-: *Piers P.* derk: *R. Glouc.* derk: *O.H. Ger.* tarni *latens*, tarhnjan *occultāre*: *Icel.* dökkr: *Gael.* dorch *dark, black, dusky*.] DER. deorce: deorcian, a-: deorcung.

deorce; *adv.* Darkly, sadly; obscure:—Ðú his dagena tíd deorce gescyrtest *minorasti dies tempŏris ejus*, Ps. Th. 88, 38. Næfre ge heortan geþanc deorce forhyrden *nolite obdurāre corda vestra*, 94, 8.

deorc-full; *adj.* Darksome, dark; tenebrōsus:—Deorcfull wæg *via tenebrōsa*, Scint. 59.

deorcian; *p.* ode; *pp.* od To darken, to grow dark; obscurāre, obscūre facĕre. DER. a-deorcian. v. deorc.

deorc-líce; *adv.* Darkly, horridly; tetrum, Glos. Prudent. Recd. 142, 7.

deorcung, e; *f.* Twilight; crepuscŭlum:—Tweóne leóht *vel* deorcung *crepuscŭlum*, Ælfc. Gl. 94; Som. 75, 122; Wrt. Voc. 53, 3. Deorcunge, æfnunge *crepuscŭlo*, Mone B. 178.

deór-cynn, es; *n. Animal-kind, beast-kind*; anĭmālium *vel* bestiārum gĕnus:—Sume wurdon to ðam deórcynne ðe mon hát tigris *some were turned to the kind of beast which man calls tiger*, Bt. 38, 1; Fox 196, 1. On ðam syxtan dæge God gescóp eall deórcynn *on the sixth day God created all kinds of animals*, Bd. de nat. rerum; Wrt. popl. science 2, 16; Lchdm. iii. 234, 14: Hexam. 9; Norm. 14, 27. To mistlícum deórcynnum *to various kinds of beasts*, Bt. 38, 1; Fox 196, 2.

DEÓRE, dióre; *adj.* **I.** *DEAR, beloved*; cārus, dilectus, familiāris:—Deóre wæs he Drihtne úrum *he was dear to our Lord*, Cd. 14; Th. 17, 17; Gen. 261: 214; Th. 269, 32; Sat. 82: Exon. 105 a; Th. 399, 13; Rä. 18, 10. Dæg byþ deóre mannum *day is dear to men*, Runic pm. 24; Hick. Thes. i. 135; Kmbl. 344, 10. His se deóra sunu *his dear son*, Cd. 218; Th. 279, 25; Sat. 243: Exon. 76 a; Th. 286, 2; Jul. 725. Âhte ic holdra ðý lǽs, deórre duguþe *I owned the less of faithful ones, of dear attendants*, Beo. Th. 980; B. 488. He æfter deórum men dyrne langaþ *he longs secretly after the dear man*, Beo. Th. 3762; B. 1879: Ps. Th. 119, 1. Ic me on mínne Drihten deórne getreówige *ego in te sperābo, Domĭne*, Ps. Th. 54, 24; 77, 69: 88, 17. He gedǽlde him deóre twá *he separated two dear to him*, Cd. 131; Th. 166, 8; Gen. 2744. Deórast ealra *dearest of all*, Exon. 76 a; Th. 284, 15; Jul. 697. Ðín mildheortnes standeþ deórust *thy mercy is most dear*, Ps. Th. 102, 16. Aldorþegn ðone deórestan *the dearest chief*, Beo. Th. 2622; B. 1309. **II.** *dear of price, precious, of great value, desirable, excellent, glorious, magnificent, noble, illustrious*; pretiōsus, magni æstimandus, desiderābilis, exīmius, gloriōsus, magnificus, nobĭlis, illustris:—Deóre [MS. deor] hit is *pretiōsum est*, Ælfc. Gl. 35; Som. 62, 82; Wrt. Voc. 28, 60. Sege me hwæðer se ðín wéla deóre seó ðé *tell me whether thy wealth is precious to thee*, Bt. 13; Fox 38, 6. Ðeáh gold gód seó and deóre [dióre MS. Cot.] *though gold is good and precious*, 13; Fox 38, 11. Deórum máðme *for the precious treasure*, Beo. Th. 3060; B. 1528. On Dryhtnes naman deórum *in the Lord's precious name*, Ps. Th. 117, 10. Gesáwon dryncfæt deóre *they had seen the precious drinking vessel*, Beo Th. 4500; B. 2254. Deóran since *with precious metal*, Exon. 124 a; Th. 19, 31; Cri. 309. Deóre máþmas *precious treasures*, Beo. Th. 4464; B. 2236. Gód hlísa biþ betera and deórra [diórra MS. Cot.] ðonne ænig wéla *good fame is better and more precious than any wealth*, Bt. 13; Fox 38, 24: Exon. 128 b; Th. 493, 16; Rä. 81, 31. Ða me synd golde deórran *they are dearer to me than gold*, Ps. Th. 118, 127. Sinc biþ deórost *treasure is most precious*, Menol. Fox 480; Gn. C. 10. Hwæt ðé deórast [diórust MS. Cot.] þince: hwæðer ðe gold ðe hwæt? *what seems to thee most precious: whether gold or what?* Bt.: 13; Fox 38, 10: Exon. 103 b; Th. 393, 13; Rä. 12, 9. In ðam deóran hám *in that desirable home*, Exon. 45 b; Th. 154, 15; Gú. 843: Cd. 218; Th. 278, 10; Sat. 219. On getýnum ðe ymb Dryhtnes hús deóre syndan *in the courts which are glorious about the Lord's house*, Ps. Th. 115, 8. Ðǽr seó deóre scólu leófne lofiaþ *where the glorious assemblage praise the beloved*, Exon. 64 a; Th. 235, 21;

Ph. 560. Ðeáh hwá æðele sié, duguþum dióre *though any be noble, magnificent in riches*, Bt. Met. Fox 10, 57; Met. 10, 29. Deóre ríce Engla landes *in the glorious kingdom of England*, Chr. 1065; Erl. 196, 38; Edw. 19. Is mín módor mægþa cynnes ðæs deórestan *my mother is of the noblest race of women*, Exon. 109 a; Th. 416, 11; Rä. 34, 10. [*Prompt. Wyc. Piers P. R. Brun. Chauc. R. Glouc.* dere: *Laym.* deore, dure: *Orm.* deore, dere: *Plat.* dür: *O. Sax.* diuri: *Frs.* djoer: *O. Frs.* diore, diure: *Dut.* dier: *Ger.* theuer: *M. H. Ger.* tiure: *O. H. Ger.* tiuri: *Dan. Swed.* dyr: *Icel.* dýrr *dear, precious*.] DER. deór-boren, -líce, -ling, -wurþe, -wyrþe, -wurþnes, -wyrþnes: un-deóre. v. dýre.

deóre, dióre; *adv. Dearly, with great price*; cāre, magno:—Deóre he hit bohte *vel* sealde *he bought* or *sold it dearly*; care vendidit, Ælfc. Gl. 35; Som. 62, 84; Wrt. Voc. 28, 62. Dióre gecépte drihten Créca Troia burh *the lord of the Greeks dearly bought the city of Troy*, Bt. Met. Fox 26, 37; Met. 26, 19. DER. un-deóre.

deóren; *adj.* [deór *an animal, wild beast*] *Of* or *belonging to a wild beast*; bestiālis:—Mid deórenum ceaflum *bestialĭbus rictĭbus*, Mone B. 3289.

deoreþ-sceaft, es; *m.* [deoreþ=daroþ *a dart*, sceaft *a shaft, handle*] *A dart-shaft, a spear*; hasta:—Under deoreþsceaftum *amid the dart-shafts*, Cd. 93; Th. 119, 23; Gen. 1984.

deorf, es; *n. Labour, trouble, tribulation*; lăbor, tribulātio. DER. ge-deorf.

deór-fald, es; *m. A deer-fold, a park, an inclosure for deer*; cervōrum hortus, vivārium, saltus, Som. Ben. Lye.

DEORFAN, ic deorfe, ðú dyrfst, he dyrfþ, *pl.* deorfaþ; *p.* dearf, *pl.* durfon; *pp.* dorfen To labour; laborāre:—Ne wiðcweðe ic to deorfenne gyt, gif ic nýdbehéfe eom gyt ðínum folce *I refuse not to labour still, if I am yet needful to thy people*, Homl. Th. ii. 516, 26. Þearle ic deorfe *I labour very much*, Coll. Monast. 19, 13. [*O. Sax.* far-dervan *to perish*: *Ger. M. H. Ger.* ver-derben *to destroy, perish*.] DER. ge-deorfan.

deór-fellen; *adj.* [fell *a skin*] *Made of beast-skins*; ex pellibus ferārum:—Crusene oððe deórfellen roc *crusen* or *a beast-skin garment*; mastruga, Wrt. Voc. 82, 4.

deór-friþ, es; *n. Deer-protection, game-protection*; cervōrum tūtēla:—Se cyng Willelm sætte mycel deórfriþ, and he lægde laga ðærwið, ðæt swá hwá swá slóge heort oððe hinde, ðæt hine man sceolde blendian *king William constituted much protection to game, and he laid down laws therewith, that whosoever should slay hart or hind should be blinded*, Chr. 1086; Erl. 222, 25–27.

Deór-hám, es; *m.* [deór *a wild beast*, hám *home, dwelling*] DERHAM, *Gloucestershire*, DEREHAM, *Norfolk*; lŏcōrum nōmen in agris Glocestriæ et Norfolciæ:—Híi iii ciningas ofslógon in ðære stówe ðe is gecweden Deórhám *they slew three kings at the place which is called Derham*, Chr. 577; Erl. 19, 21. On ðysum geáre Wihtburge líchama wearþ gefunden eal gehál and unformolsnod æt [MS. a] Deórhám, æfter líf and fífti geáran ðæs [MS. þas] ðe heó of ðysum lífe [MS. liue] gewát *in this year* [A.D. 798] *the body of Wihtburh was found at Dereham, all whole and uncorrupted, five and fifty years after she had departed from this life*, Chr. 798; Th. 105, 15–21, col. 3.

deór-hege, es; *m.* [hege *a hedge, fence*] *A deer-fence*; cervōrum sepīmentum:—Deórhege to cyniges háme *the deer-fence for the royal mansion*, L. R. S. 1; Th. i. 432, 4: 2; Th. i. 432, 11: 3; Th. i. 432, 24.

Deór-hyrst, es; *m.* [hyrst *a hurst, copse, wood*] DEERHURST, *Gloucestershire*; lóci nōmen in agro Glocestriæ:—Æt Olaníge wið Deórhyrste *at Olney near Deerhurst*, Chr. 1016; Th. 282, 40, col. 2. On Deórhyrste *at Deerhurst*, Chr. 1053; Th. 322, 13, col. 2.

deoriende *hurting*, Chr. 959; Erl. 121, 4,=deriende; *part.* of derian.

deór-líc; *adj.* [deór I. *brave, bold*] *Bold*; fortis:—Breca næfre git swá deórlíce dæd gefremede *Breca never yet performed such a bold deed*, Beo. Th. 1174; B. 585.

deór-líce; *adv. Preciously, worthily*; prĕtiōse, digne:—To hwan hió ða næglas sélost and deórlícost gedón meahte *to what she might best and most worthily employ the nails*, Elen. Kmbl. 2315; El. 1159.

deór-ling, diór-ling, dýr-ling, es; *m. A dearling*, DARLING, *minion, favourite*; unīce dilectus, dēlĭciæ:—Gif ðé lícode his dysig, swá wel swá his dysegum deórlingum dyde *if his folly had pleased thee, as well as it did his foolish favourites*, Bt. 27, 2; Fox 96, 23: Wanl. Catal. 127, 49, col. 2. Se godcunda ánweald gefriþode his diórlingas [deórlingas MS. Cot.] *the divine power saved his darlings*, Bt. 39, 10; Fox 228, 11. He his diórlingas duguþum stépte *he decked his favourites with honours*, Bt. Met. Fox 15, 15; Met. 15, 8. Iohannes se Godspellere, Cristes dýrling *John the Evangelist, Christ's darling*, Homl. Th. i. 58, 1: Menol. Fox 230; Men. 116.

deór-mód; *adj.* [deór I. *brave, bold*; mód *mood, mind*] *Bold of mind, brave*; fortis anĭmi:—Wearþ adræfed deórmód hæleþ *the brave hero was driven away*, Chr. 975; Erl. 126, 18; Edg. 44: Exon. 46 b; Th. 159, 11; Gú. 925: 79 b; Th. 298, 22; Crä. 89: Andr. Kmbl. 1251; An. 626: Fins. Th. 46; Fin. 23. On felda ðam ðe deórmóde Diran héton *in the plain which the brave men called Dura*, Cd. 180;

Th. 226, 14; Dan. 171. Deórmódra síþ *the march of the brave*, 147; Th. 183, 25; Exod. 97.

deór-net, -nett, es; *n. A beast-net, hunting-net*; rēte venātĭcum, cassis:—Deórnet *cassis*, Ælfc. Gl. 84; Som. 73, 91; Wrt. Voc. 48, 29.

deornunga *secretly*, L. In. 27; Wilk 19, 12. v. dearnunga.

deór-tún, es; *m.* [tún *an inclosure*] *A deer-inclosure*; cervōrum sepīmentum, Som. Ben. Lye.

Deorwente, an; *f.* [deor = Celt. dwr *water*; went *turned, bent*; v. wendan] *The river* DERWENT, *in Yorkshire, Derbyshire, Cumberland, and Durham*; quatuor fluviōrum nomen in agris Eboracensi Derbiensi Cumbriensi et Dunholmensi:—Be Deorwentan dære eá *by the river Derwent* [*Yorkshire*], Bd. 2, 9; S. 511, 18: 2, 13; S. 517, 16: Of dam ðe ða fruman aweallaþ Deorwentan streámes *from which the beginnings of the river Derwent spring*, 4. 29; S. 607, 11.

deór-wyrþe, -wurþe; *adj.* [deóre *dear*, weorþe *worth*] *Precious, dear, of great worth* or *value*; prētiōsus:—Ðá he funde dæt án deórwyrþe meregrot *inventa autem una prētiōsa margarīta*, Mt. Bos. 13, 46. Deórwurþe *prētiōsus*, Wrt. Voc. 85, 61. Ealra gecorenra hálgena deáþ is deórwurþe on Godes gesihþe *the death of all the chosen saints is precious in the sight of God*, Homl. Th. i. 48, 34. Ofer gold and stáne deórwyrþum *super aurum et lapĭdem prētiōsum*, Ps. Lamb. 18, 11: 20, 4. We deórwyrþne dæl Dryhtne cennaþ *we ascribe the precious lot to the Lord*, Exon. 35 a; Th. 113, 7; Gú. 154. Hí wurdon gehwyrfede to deórwurþum gymmum *they were turned to precious gems*, Homl. Th. i. 64, 5. Hí nǽfre swá deórwurþe gymstánas ne gemétton *they have never before met with such precious gems*, i. 64, 10. Ðæt is git deórwyrþre ðonne monnes líf *it is even more valuable than man's life*, Bt. 10; Fox 28, 38. Ðú hæfst gesund gehealden eall dæt deórwyrþoste *thou hast kept entire everything most precious*, Bt. 10; Fox 28, 9. Mid dam deórwurþustan reáfe *with the most valuable raiment*, Gen. 27, 15.

deór-wyrþnes, -wurþnes, -ness, e; *f. Preciousness, a precious thing, treasure*; res prētiōsa:—Mid eallum deórwyrþnessum *with all precious things*, Bt. 7, 4; Fox 22, 31. Ðe ða frécnan deórwurþnessa funde *who found the dangerous treasures*, 15; Fox 48, 24.

dépan; *p.* te; *pp.* ed *To dip, baptize*; baptizāre:—Dépiþ *vel* dyppeþ *baptizābit* = βαπτίσει, Mt. Rush. Stv. 3, 11. v. dyppan.

Déprobane; *indecl. f. An island in the Indian ocean, Ceylon*; Taprŏbăna = Ταπροβάνη:—Be súþan eástan dam porte is dæt ígland Déprobane *to the south-east of the port* [*Calymere*] *is the island Ceylon*, Ors. ☧, 1; Bos. 16, 16. v. Táprabane.

Dera mǽgþ, e; *f.* [Dere *the Deirians*, mǽgþ *a province, region, country*] *The country of the Deirians, Deira, being part of Northumbria, situate between the Tyne and Humber*; Deirōrum provincia:—In Dera mǽgþe *in provincia Deirōrum*, Bd. 2, 14; S. 518, 14. v. Dera ríce.

Dera ríce, es; *n.* [Dere *the Deirians*, ríce *a kingdom*] *The kingdom of the Deirians, Deira*; Deirōrum regnum:—Fēng to Dera ríce *suscēpit regnum Deirōrum*, Bd. 3, 1; S. 523, 9. Se hæfde Dera ríce *qui in Deirōrum partĭbus regnum habēbat*, 3, 23; S. 554, 8.

Dere; *gen.* Dera; *pl. m.* The Deirians, *inhabitants of Deira between the rivers Tyne and Humber*; Deiri:—Andswarede him mon and cwæþ dæt hí Dere nemde wǽron *responsum est quod Deiri vocārentur*, Bd. 2, 1; S. 501, 21, 22: Homl. Th. ii. 120, 34, 35. Mid dysses cyninges geornesse ða twá mǽgþa Norþan Hymbra Dere and Beornice on áne sibbe geteáh *hujus industria regis Deirōrum et Berniciōrum provinciæ in unam sunt pācem*, Bd. 3, 6; S. 528, 30. He wæs vii winter Dera cyning *he was king of the Deirians seven years*, 3, 14; S. 539, 32. Man gehálgode ii biscopas on his stal, Bosan to Derum, and Eatan to Beornicum *two bishops were consecra¹ed in his stead, Bosa to Deira* [lit. *to the Deirians*], *and Eata to Bernicia*, Chr. 678; Erl. 41, 7. v. Dera mǽgþ.

deregaþ *injure*, Bt. 4; Fox 8, 16, = deriaþ; *pres. pl.* of derian.

DERIAN, derigan; *part.* deriende, derigende; *ic* derige, ðú derast, derest, he deraþ, dereþ; *pl.* deriaþ, deregaþ; *p.* ode, ede; *pp.* od, ed; *v. trans. dat. To injure, hurt, harm, damage*; nocēre, lædĕre, obesse:—Him ða stormas derian ne máhan [derigan ne mǽgon MS. Cot.] *the storms cannot hurt him*, Bt. 7, 3; Fox 22, 6: Bt. Met. Fox 12, 8; Met. 12, 4. He ne forlét mannan derian heom *non reliquit homĭnem nocēre eis*, Ps. Lamb. 104, 14. Derigende *nocĕns*, Ælfc. Gr. 9, 38; Som. 12, 51. Dém Driht derigende [deriende MS. T; ða deriendan, Lamb.] me *judĭca Domĭne nocentes me*, Ps. Spl. 34, 1. Ic derige *noceo*, Ælfc. Gr. 43; Som. 44, 41: Ps. Lamb. 88, 34. Hit me ne deraþ *it shall not hurt me*, Homl. Th. i. 72, 13: Boutr. Scrd. 31, 18. Hió oft dereþ unscyldegum *she often injures the guiltless*, Bt. Met. Fox 4, 71; Met. 4, 36: 26, 221; Met. 26, 111. On woru!de monnum ne deriaþ máne áþas *wicked oaths inflict no injury on men in the world*, 4, 95; Met. 4, 48: Past. 59; Hat. MS. Náuht ne deregaþ monnum máne áþas *wicked oaths in no wise injure men*, Bt. 4; Fox 8, 16. He derode manna gesihþum *he injured men's sight*, Homl. Th. i. 454. 21: Hexam. 16; Norm. 24, 3: Chr. 1032; Erl. 164, 2: Boutr. Scrd. 18, 3. Gif ðú ðínum cristenum bréðer deredest *if thou injuredst thy christian brother*, Homl. Th. i. 54, 22. Him ówiht ne derede *naught harmed them*, Cd. 188; Th. 233, 11; Dan. 274: 23; Th. 30, 24; Gen. 471. Ðæt ðú me ne derige *ne nŏceas*

mihi, Gen. 21, 23. Swá hwæt swá mannum derige, dæt is eall for úrum synnum *whatsoever is injurious to men, is all for our sins*, Homl. Th. i. 16, 25. [*Piers P.* dere: *Chauc.* dere: *Laym.* derede, *p:* O. Sax. derian: *Frs.* deare, derre: *O.Frs.* dera: *Dut.* deren: *O.H.Ger.* terjan, terran *nocēre.*] DER. ge-derian: un-deriende.

deriendlíc, derigendlíc; *def.* se -líca, seó, dæt -líce; *adj. Injurious, noxious, hurtful*; nocīvus, noxius, nŏcens:—Deriendlíc *nocīvus*, Fulg. 20: noxius, Hymn. Surt. 5, 7. Hit ne biþ dam men derigendlíc *it will not be injurious to a man*, Boutr. Scrd. 20, 18. Ðæt we forbúgan ǽlc þing derigendlíces *vitēmus omne noxium*, Hymn. Surt. 14, 13: 37, 16: 93, 3. Afyrsa hǽtan derigendlíce *aufer calōrem noxium*, 10, 31. Him wǽron derigendlíce dracan and næddran *serpents and adders were noxious to them*, Hexam. 17; Norm. 24, 32. Híg swíðe gedrehton ða deriendlícan *the hurtful greatly afflicted them*, Ælfc. T. Grn. 11, 35. Ðæt ðú derigendlíce ætbrede *ut noxia subtrăhas*, Hymn. Surt. 133, 7. Us he gehealde fram derigendlícum *nos servet a nocentĭbus*, 9, 7.

dér-ling *a darling:*—Dérling mín *dilectus meus*, Mt. Kmbl. Lind. 12, 18. v. deór-ling.

derne *secret, hidden*, Ps. C. 50, 70; Ps. Grn. ii. 278, 70. v. dyrne.

dern-geliger, dern-geligr; *f:* dern-geliger-scipe, es; *m. A secret lying, adultery*; clandestīnus concubĭtus, adultĕrium:—In derngeligerscipe [MS. derne-gilegerscipe] *in adultĕrio*, Jn. Rush. War. 8, 3. v. ge-liger.

dern-unga; *adv.* [derne, unga *a termination*] *Secretly*; clam:—Dernunga *clam*, Mt. Kmbl. Rush. 2, 7. v. dearnunga.

derodine? *scarlet dye*, Past. 14, 4; Hat. MS. 18 a, 3. v. dyrodine.

derstan *dregs, lees*. L. M. 1, 2; Lchdm. ii. 38, 18, 19: 1, 39; Lchdm. ii. 98, 24. v. dærstan.

derung, e; *f. An injuring, harming*; læsio, injúria, nocumentum, Greg. Dial. 3, 16.

dést *doest, dost*, Jn. Bos. 6, 30; déþ *does*, Basil admn. 4; Norm. 40, 29; *2nd and 3rd sing. pres. of* dón.

diácon, deácon, es; *m. A deacon, minister of the church, levite*; diácŏnus = διάκονος *a servant, waiting man* = Lat. minister, levīta, levītes = λευίτης:—Diáconus is þén, ðe þénaþ dam mæsse-preóste, and ða offrunga sett uppon dæt weofod, and gódspell eác rǽt æt Godes þénungum. Se mót fulligan cild, and dæt folc húsligan [i. e. he mót eác hláf sillan, gif þearf biþ *he may also give the bread, if need be*, L. Ælf. P. 34; Th. ii. 378, 12] *deacon is a minister, who ministers to the mass-priest, and sets the offerings upon the altar, and also reads the gospels at God's services. He may baptize children, and housel the people*, L. Ælf. C. 16; Th. ii. 348, 12. [Gif frigman] diácones feoh [stele], vi gylde [forgylde] *if a freeman steal the property of a deacon, he must repay sixfold*, L. Ethb. 1, 4; Th. i. 2, 5: 4, 3; *about* A.D. 599. Swylce diácon hine clǽnsie *so let a deacon clear himself*, L. Wih. 18; Th. i. 40, 16: L. Eth. ix. 20; Th. i. 344, 15: L. C. E. 5; Th. i. 362, 12, 17: Bd. 3, 20; S. 550, 21. We nú gehýrdon of dæs diácones múþe *we have now heard from the mouth of the deacon*, Homl. Th. i. 152, 3. Ða Iudéas sendon diáconas *misērunt Iudæi levītas* [Wyc. *dekenys*], Jn. Bos. 1, 19. Diácon *levīta* [Wyc. *dekene*], Lk. Bos. 10, 32. Ða apostolas gehádodon seofon diáconas... Ðæra diácona wæs se forma Stephănus... Hí mid gebédum and bletsungum to diáconum gehádode wurdon *the apostles ordained seven deacons... The first of the deacons was Stephen... They were ordained deacons with prayers and blessings*, Homl. Th. i. 44, 10, 13, 20: 416, 9, 11. DER. arce-diácon, erce-, under-. v. hád II.

diácon-hád, es; *m. The office of a deacon, deaconship*; diaconātus:—On diáconháde *in deaconship*, Homl. Th. ii. 120, 13.

diácon-þénung, e; *f.* [þénung *duty, office*] *The duty* or *office of a deacon*; diaconātus officium:—He diáconþénunge mycelre tíde brúcende wæs *diaconātus officio non pauco tempŏre fungebātur*, Bd. 4, 3; S. 570, 28.

díc, es; *m. A* DIKE, *a bank formed by throwing the earth out of the ditch*; vallum, id est tumŭlus, qui terra effossa exstructus est:—Andlang díces *along the dike*, Cod. Dipl. Apndx. 442; A.D. 956; Kmbl. iii. 438, 18. Ondlong ridiges on ðone díc *along the ridge to the dike*, 620; A.D. 978; Kmbl. iii. 169, 2: iii. 168, 35. On ánne micelne díc *to a great dike*, iii. 169, 7. Of ðæm díc *from the dike*, iii. 169, 2. To dæm ealdan díc *to the old dike*, Th. Diplm. A.D. 905; 494, 17. On ðone díc *to the dike*, 494, 37. [O. Sax. díc, *m. a dike, dam:* O.Frs. dik, *m. a dike, dam:* Dut. dijk, *m. a dike:* Ger. deich, *m. a mound:* Sansk. dehī, *f. a mound, bank, rampart.*] DER. ýlen-díc [éaland-díc].

díc, e; *f.* I. *a ditch, the excavation* or *trench made by throwing out the earth, a channel for water*; fossa, excavātio *vel* scrŏbis unde terram fodĕrant:—Ðonne to dære díce hyrnan *then to the corner of the ditch*, Th. Diplm. A.D. 905; 495, 21. Ðonne on ðone weg, ðe scýt ofer ða díc *then to the way, that leads over the ditch*, Th. Diplm. A.D. 900; 145, 27. On ða díc *to the ditch*, Cod. Dipl. Apndx. 441; A.D. 956; Kmbl. iii. 437, 11, 15, 27. Of dam bróc on ða ealdan díc *from the brook to the old ditch*, 556; A.D. 969; Kmbl. iii. 48, 21. On ða reádan díc *in the reedy ditch*, Cod. Dipl. 1172; A.D. 955; Kmbl. v. 332, 13. Binnon lytlum fæce wendon to Lundene; and dulfon ða áne mycele díc, on ða súþ-healfe, and drógon heora scipa [scypo MS. Cot. Tiber. B. i; scipo MS. Cot. Tiber. B. iv] on

west-healfe ðære brycge *within a little space they went to London; and they then dug a great ditch, on the south side, and dragged their ships to the west side of the bridge,* Chr. 1016; Th. 281, 4-7, col. 1.   II. *sometimes* díc, es; *m. is found to denote*—*a ditch* or *channel for water :*— Ymbútan ðone weall [Babilónes] is se mæsta díc, on ðam is yrnende se ungeféglecesta streám; and, wiðútan ðam díce, is geworht twegra elna heáh weall *round the wall [of Babylon] is a very great ditch, in which runs the deepest stream; and, outside the ditch, a wall is built two ells high,* Ors. 2, 4; Bos. 44, 26, 27. [*Prompt.* dyke *fossa :* Piers P. dyk, dych *a ditch :* Chauc. dich *a ditch :* Laym. dic, dich, *f. a ditch :* Plat. diek, dík, *m. a pond :* Frs. dijck, *w. vallum :* Ger. teich, *m. a pond :* M. H. Ger. tîch, *m. a pond :* Dan. dige, *n. a ditch :* Swed. dike, *n. a ditch, trench :* Icel. díki, dík, *n. a ditch.*]

dícere, es; *m. A ditcher, digger;* fossor, Ælfc. Gl. 60; Som. 68, 21; Wrt. Voc. 39, 7.

dícian; *p.* ode; *pp.* od *To* DIKE, *bank, mound;* aggĕráre, cingĕre :— Ðær Severus hēt dícian and eorþwall gewyrcan *there Severus commanded to raise a bank and to make an earth wall,* Bd. 1, 12; S. 481, 9.   DER. be-dícian, ge-.

dícung, e; *f. A ditching, digging;* fossio, Ælfc. Gl. 60; Som. 68, 20; Wrt. Voc. 39, 6.

dide *did,* Chr. 616; Erl. 23, 5, = dyde; *p. of* dón.

didon *did,* Hy. 7, 107; Hy. Grn. ii. 289, 107, = dydon; *p. pl. of* dón.

diégel *hidden, obscure,* Past. 43, 2; Hat. MS. 59 a, 17.   v. dígol.

diégel-líce *secretly,* Som. Ben. Lye.   v. dígol-líce.

diégelnes *solitude, recess,* Bt. 13; Fox 38, 26.   v. dígolnes.

dielf *dug.*   v. be-dielf.

dielgian *to destroy,* Past. 55, 2.   v. dilgian.

dierne *hidden, secret,* Elen. Kmbl. 2160; El. 1081.   v. dyrne I.

Difelin, Dyflen, Dyflin, es; *m?* [Hovd. Diveline] *Dublin;* Dublána :— Gewiton hie ða Norþmen ofer deóp wæter Difelin [Dyflen, Th. 206, 14, col. 2 : Dyflin, 207, 14, col. 1] sēcan *the Northmen departed over the deep water to seek Dublin,* Chr. 937; Th. 206, 14, col. 1; Æðelst. 56.

dígel *hidden, secret,* Greg. Dial. Hat. MS. fol. 1 a, 20; Homl. Th. ii. 314, 17.   v. dígol.

dígelan *to hide,* Som. Ben. Lye.   v. díglian.

dígel-líce *secretly,* Ors. 6, 21; Bos. 123, 29.   v. dígollíce.

dígelnes, dígelnys *solitariness, recess,* Ps. Spl. second 9, 10 : Ors. 2, 1; Bos. 39, 40.   v. dígolnes.

dígle, dígele *secret, hidden,* Mk. Bos. 4, 22 : Ælfc. Gr 33; Som. 37, 24; *nom. n. of* dígol.

dígle; *adv. Secretly;* secrēto, clam :—Ic to ðé, Drihten, dígle cleopode *clamávi ad te, Domine, secrēto,* Ps. Th. 141, 5.   Dígle *furtim,* Glos. Prudent. Recd. 144, 30.

díglian; *p.* ede, ode; *pp.* od *To hide;* occulĕre, occultáre :—Hí on wudum and on wēstenum and on scræfum hí hýddon and dígledon *se silvis, ac desertis abdĭtisve speluncis occulĕrant,* Bd. 1, 8; S. 479, 22.   DER. be-díglian, ge-deigelian.

díglíce *secretly,* Mt. Bos. 17, 19.   v. dígollíce.

díglod *hidden,* Fulg. 16; *pp. of* díglian.

dígneras, dýneras; *pl. m. Small pieces of money;* folles, dēnárii :— Dígneras *folles,* Cot. 93.   Dýneras *folles,* Ælfc. Gl. 106; Som. 78, 55; Wrt. Voc. 57, 35.

DÍGOL, dýgol, dēgol, es; *n. Concealment, a secret place, secret, darkness, the grave, mystery;* secrētum, absconditum, mystērium :— Ðæt ðín ælmesse sý on díglum *ut sit eleemosýna tua in abscondĭto,* Mt. Bos. 6, 4.   He wæt díglu heortan *ipse nōvit abscondĭta cordis,* Ps. Spl. 43, 24 : 50, 7.   Mægen he cýðde on dígle *he revealed his power in secret,* Andr. Kmbl. 1251; An. 626.   He ðý þriddan dæge of dígle arás *he rose the third day from the secret place [the grave],* Exon. 96 a; Th. 359, 13; Pa. 62.

dígol, dýgol, diógol; *gen. m. n.* dígles, *f.* dígolre; *def. nom. m.* dígla; *f. n.* dígle; *adj. Secret, hidden, private, dark, obscure, profound, abstruse, unknown;* secrētus, occultus, obscūrus, ignōtus :—Se þeóden gewát sēcan dígol land *the king departed to seek a secret land,* Andr. Kmbl. 1396; An. 698.   He ána gesæt on dígolre stówe *he sat alone in a secret place,* Bd. 3, 27; S. 559, 2.   Sóþlíce nis nán þing dígle, ðæt ne sý geswutelod *non est enim occultum, quod non manifestētur,* Lk. Bos. 8, 17.   He ðær wolde dígol beón *he would there be hidden,* Bd. 3, 14; S. 539, 44.   On dígle, deorce stówe *in an obscure, dark place,* Ps. Th. 142, 4.   Is seó forþgesceaft dígol and dyrne *the future condition is dark and secret,* Menol. Fox 585; Gn. C. 62.   Me Daniel dýglan swefnes sóðe gesæde *Daniel said soothly to me of the dark dream,* Cd. 198; Th. 246, 21; Dan. 482.   Ðæt wit mægen smeálícor sprecan and diógolran wordum *that we two may argue more closely and with profounder words,* Bt. 13; Fox 36, 32. [Laym. digelliche *secretly :* O. H. Ger. tougal *opācus, obscūrus, occultus.*]

dígol-líce, dígolíce; *adv. Secretly;* secrēto, clam :—His leorning-cnihtas hine dígollíce ahsodon *discipŭli ejus secrēto interrogābant eum,* Mk. Bos. 9, 28 : Ps. Th. 9, 29.   Albánus hæfde ðone Cristes andettere dígollíce mid him *Alban had Christ's confessor secretly with him,* Bd. 1,

7; S. 477, 7.   Se dígolíce lácnod wæs fram his wundum *who was secretly healed of his wounds,* 4, 16; S. 584, 30.   DER. un-deágollíce.

dígolnes, dígolnys, -ness, -nyss, e; *f. Solitariness, solitude, privacy, secrecy, mystery, hiding-place, recess;* solitūdo, absconditum quid, secrētum, arcāna, latebra :—He to dígolnesse and to stilnesse becom ðære godcundan sceáwunge *he came to the privacy and stillness of the divine contemplation,* Bd. 4, 28; S. 605, 10.   Se cyning his geþohte ðære cwēne on dígolnysse onwreáh *rex cogitatiōnem suam reginæ in secrēto revēlávit,* 2, 12; S. 514, 36.   Him Dryhten synderlíce his dígolnysse onwreáh *Domĭnus ei specialiter sua revēlābat arcāna,* 4, 3; S. 567, 20.   Nænig ðara andweardra his heortan deágolnesse him helan dorste *nullus præsentium latebras ei sui cordis celāre præsumpsit,* 4, 27; S. 604, 22.

dígul *secret,* Ps. Th. 106, 23.   v. dígol.

díhglum, díhlum = díglum *secret, retired, shady :*—On díhglum stówum *in shady places,* Herb. 38; Lchdm. i. 138, 22.   Ón díhlum *in secret,* Mt. Bos. 6, 6; *dat. pl. of* dígol, *q. v.*

díhlíce *secretly,* Mt. Bos. 1, 19 : 24, 3.   v. dígollíce.

díhlum *in secret,* Mt. Bos. 6, 6.   v. díhglum.

DIHT, es; *n?*   I. *a setting in order, disposing, contriving, disposition, conduct, consultation, deliberation, purpose;* dispositio, excogitātio, consilium, propŏsitum :—God gefylde on ðam seofoðan dæge his weorc ðe he worhte on wunderlícum dihte, and he on ðam seofoðan dæge geswác ðæs dihtes ðæs deóplícan cræftes *God completed on the seventh day his works which he had wrought with wondrous contriving, and on the seventh day he ceased from the disposition of the profound art,* Hexam. 12; Norm. 20, 10, 14.   Hit stent on úrum ágenum dihte hú us biþ æt Gode gedēmed *it stands by our own conduct how we shall be judged before God,* Homl. Th. i. 52, 32.   Ða mágas ðe æt ðam dihte wæron þolian ðone ylcan dóm *cognāti qui illi consilio interfuĕrint patiantur eandem sententiam,* L. M. I. P. 16; Th. ii. 270, 4.   Ic eom unscyldig, ægðer ge dæde ge dihtes, æt ðære tihtlan *I am guiltless, both in deed and purpose, of the accusation,* L. O. 5; Th. i. 180, 16.   II. *a dictating, direction, order, command;* dictātio, directio, jussum, mandātum :—Moyses underfēng of Godes sylfes dihte ealle ða deópnyssa ðe he on fíf bócum syððan afæstnode *Moses received from the dictating of God himself all the mysteries which he afterwards inscribed in five books,* Hexam. 1; Norm. 2, 17.   Saul wearþ Gode ungehýrsum and nolde faran be his dihte *Saul was disobedient to God and would not walk by his direction,* Homl. Th. ii. 64, 3 : L. E. G. pref; Th. i. 166, 19 : L. C. S. 71; Th. i. 412, 30.   Ealle ða þing ðe he dyde, he dyde be his dihte *all the things which he did, he did by his [God's] command,* Gen. 39, 3. [Dut. dicht, *n. poetry :* Ger. dicht, ge-dicht, *n. a poem :* M. H. Ger. tihte, *f. a composing :* tihte, *n. a poem, fiction :* O. H. Ger. dihta, *f. dictation, fiction :* Dan. dight, *n. a poem, fiction :* Swed. dikt, *m. a fable, poem :* Icel. dikt, *n. a composition :* Lat. dictum *a saying, order.*]

dihtan, ic dihte; *p.* ic, he [dihtde =] dihte, dyhte, *pl.* dihton; *pp.* dihted; *v. a.*   I. *to set in order, dispose, arrange, appoint, direct, compose;* parāre, dispōnĕre, instruĕre, constituĕre, compōnĕre :—Abram ðá dyde, swá swá him dyhte Sarai *Abraham then did as Sarah arranged,* Gen. 16, 3 : Jn. Bos. 18, 14.   Ic eów dihte, swá mín Fæder me ríce dihte *ego dispōno vobis, sicut dispŏsuit mihi pater meus regnum,* Lk. Bos. 22, 29.   Ðær se Hælend heom dihte *ubi constituĕrat illis Iesus,* Mt. Bos. 28, 16 : 25, 19.   II. *to order, dictate, indite;* dirigĕre, dictāre :—Hí didon ðá, swá swá him dihte Iosue *then they did as Joshua ordered them,* Jos. 8, 8.   Drihten dihte him hwæt he dōn sceolde *Domĭnus omnia opĕra ejus dirĭgēbat,* Gen. 39, 23.   [Wyc. diting *an inditing, writing :* Piers P. Chauc. dighte *to dispose :* Laym. dihte, dihten *to rule, dispose, indite :* Plat. tichten *to fix, appoint, dispose :* Dut. Ger. dichten *carmĭna compōnĕre :* Kil. dichten *dictāre :* M. H. Ger. tihten *fingĕre :* O. H. Ger. dihtôn *dictāre :* Dan. digte *to make poems :* Swed. dikta *to fable, feign :* Icel. dikta *to compose, feign :* Lat. dictāre *to dictate.*]   DER. a-dihtan, ge-.

dihtaþ *dictates,* Bd. 1, 27; S. 490, 21, = dihteþ; *3rd pres. sing. of* dihtan.

dihtere, dihtnere, es; *m. An informant, expounder, disposer, manager, steward;* auctor, commentātor, expŏsĭtor, dispensātor :—Ic wríte swá me ða dihteras sædon ðe his líf geornost cúðon *I write as the informants who knew his life most accurately told me,* Guthl. prol; Gdwin. 4, 23; 6, 8.   Dihtere *commentātor, expŏsĭtor,* Ælfc. Gl. 49; Som. 65, 86; Wrt. Voc. 34, 18.   Dihtnere *dispensātor,* 33; Som. 62, 29; Wrt. Voc. 28, 12.   Hwá ys getrýwe and gleáw dihtnere, ðæne se hláford geset ofer his híred *quis est fidēlis dispensātor, et prudens, quem constītuet Domĭnus supra familiam suam?* Lk. Bos. 12, 42 : Homl. Th. ii. 344, 5.

dihtig; *adj. Doughty;* valĭdus, Cd. 93; Th. 120, 11; Gen. 1993.   v. dyhtig.

dihtnere *an arranger, a steward;* dispensātor, Lk. Bos. 12, 42.   v. dihtere.

dihtnung, e; *f. A disposing, ordering;* dispositio, condĭtio :—Ealle ðínre synd dihtnunge underóðde *omnia tuæ sunt condĭtiōni subjecta,* Wanl. Catal. 293, 50, col. 1.   DER. ge-dihtnung.

DILE, dyle, es; *m.* DILL, *anise;* anēthum = ἄνηθον, anēthum gravéölens, Lin :—Genim diles blóstman *take blossoms of dill,* L. M. 1, 1: Lchdm. ii. 20, 7.   Genim diles sædes áne yntsan *take one ounce of seed*

of dill, L. M. 2, 12; Lchdm. ii. 190, 9: 2, 15; Lchdm. ii. 192, 14.
Selle him mon dile gesodenne on ele let a man give him dill sodden in
oil, **2**, 23; Lchdm. ii. 236, 15. Ge tiogoðiaþ eówre mintan and eówerne
dile and eówerne cymen ye tithe your mint and your dill and your
cummin, Past. 57; Hat. MS: Mt. Bos. 23, 23. Genim ðas wyrte, ðe
man anēthum, and ŏðrum naman dyle, nemneþ take this herb, which is
named anēthum, and by another name dill, Herb. 123, 1; Lchdm. i. 234,
20: Wrt. Voc. 79, 9. [Dut. dille, f: Ger. dill, m; dille, f: M. H. Ger.
tille: O. H. Ger. tilli anēthum: Dan. dild, m. f: Swed. dill, m.]

DILEGIAN, dilgian, dielgian; p. ode; pp. od To destroy, abolish,
blot out, erase; delēre, abōlēre:—Gif se wrītere ne dilegaþ ðæt he ǽr
wrāt if the scribe does not erase what he wrote before, Past. 54, 5; Hat. MS.
Swā swā fenn strǽta ic dilgie hīg ut lutum platēārum delēbo eos, Ps. Spl.
17, 44. To dielgianne hira synna to blot out their sins, Past. 55, 2;
Hat. MS. [Orm. dillghenn: O. Sax. far-diligôn delēre; Frs. dylgjen:
O. Frs. diligia: Ger. tilgen: M. H. Ger. tīligen, tilgen: O. H. Ger. tili-
gōn.] DER. a-dilegian, -dilgian, for-: un-dilegod.

dilfst, he dilfþ diggest, digs; 2nd and 3rd pers. pres. of delfan.

dilgian to destroy; delēre, Ps. Spl. 17, 44. v. dilegian.

DIM; def. se dimma, seó, ðæt dimme; adj. DIM, dark, obscure, hidden;
obscūrus, tenebrōsus:—Ðes wīda grund stōd deóp and dim this wide
abyss stood deep and dim, Cd. 5; Th. 7, 12; Gen. 105: 24; Th. 30,
36; Gen. 478. Nǽnegum þuhte dæg on þonce, gif sió dimme niht ǽr
ofer eldum egesan ne brohte the day would seem delightful to none, if
the dark night did not bring terror over men, Bt. Met. Fox 12, 32;
Met. 12, 16. Com hæleða þreát to ðǽre dimman ding the troop of
heroes came to the dark dungeon, Andr. Kmbl. 2541; An. 1272: Cd. 215;
Th. 271, 27; Sat. 111. On ðǽre dimman ádle in the hidden malady, Exon.
49 b; Th. 171, 31; Gū. 1135. Drihten sealde him dimne and deorcne
deáþes scūwan the Lord gave them death's shadow, dim and dark, Cd. 223;
Th. 293, 14, note; Sat. 455. Nabbaþ we to hyhte nymþe ðone dimman
hām we have nought in hope save this dim home, Cd. 221; Th. 285, 14;
Sat. 337. Hió speón hine on ða dimman dǽd she urged him to that
dark deed, 32; Th. 43, 3; Gen. 685. On ðis dimme hol in this dim
hole, Bt. 2; Fox 4, 11: Andr. Kmbl. 2618; An. 1310. Sindon dena
dimme the dells are dim, Exon. 115 b; Th. 443, 14; Kl. 30: Cd. 215;
Th. 271, 13; Sat. 105: Ps. Th. 108, 8. Cwīst ðū oncnāwaþ hī wundru
ðīne on ðām dimmum deorcan þystrum numquid cognoscentur in tenēbris
mirabilia tua? 87, 12. [Piers P. dymme: Chauc. dim: O. Frs. dim:
Ger. dial. dimmer: M. H. Ger. timber, timmer: O. H. Ger. timbar:
Icel. dimmr dark.]

dim-hofe, dym-hofe, an; f. A lurking-place, hiding-place; latībūlum,
lātēbra:—He gesette þystru dymhofan oððe dymnes oððe behȳdednesse
his pōsuit tenēbras latībūlum suum, Ps. Lamb. 17, 12. Dimhofan latēbræ,
Ælfc. Gr. 13; Som. 16, 21. Dimhofum latībūlis, Mone B. 85. Gregorius on dymhofum [MS. -hofon] ætlūtode Gregory concealed himself
in hiding-places, Homl. Th. ii. 122, 33.

dimlíc, dymlíc; adj. Dim, secret, hidden, concealed; obscūrus, clandestīnus:—Of dimlícum clandestīnis, Mone B. 872. Nā swylce he
todrǽfe ða dymlícan þeóstra not as if he dispelled the dim darkness,
L. Ælf. C. 14; Th. ii. 348, 7.

dimman to dim, darken, obscure; obscūrāre. DER. a-dimmian, for-.

dimnes, dymnys, -ncss, -nyss, e; f. DIMNESS, darkness, obscurity;
cālīgo, obscūritas:—Dimnes cālīgo, Ælfc. Gl. 94; Som. 75, 120; Wrt.
Voc. 53, 1. Ðis biþ gōd lǽcedōm wið eágna dimnesse this is a good
remedy for dimness of eyes, L. M. 1, 2; Lchdm. ii. 26, 9. Wolcnu and
dimnys on his ymbhwyrfte nubes et cālīgo in circuitu ejus, Ps. Lamb. 96,
2: Mone B. 3240. Se dæg is þeóstra dæg and dimnysse the day is a
day of darkness and dimness, Homl. Th. i. 618, 17. Dymnys cālīgo,
Ælfc. Gr. 9, 3; Som. 8, 56.

dim-scūa, an; m. [scūwa, scūa a shade, shadow] Dimness, darkness;
tenēbræ:—Oft hira mōd onwōd under dimscūan deófles lārum their mind
often went under darkness by the devil's lore, Andr. Kmbl. 281; An. 141.

dincge, dyncge, an; f. Ploughed land, fallow land; novāle:—Dincge
nŏvāle, Wrt. Voc. 66, 56. Dyncgum novālibus, Mone B. 1434: 2326.

ding, e; f. A dungeon, prison; carcer:—Com hæleða þreát to ðǽre
dimman ding the troop of heroes came to the dark dungeon, Andr. Kmbl.
2541; An. 1272.

dingiung, e; f. A dunging, manuring; stercŏrātio:—Dingiung stercŏrātio, Ælfc. Gl. 1; Som. 55, 5; Wrt. Voc. 15, 5.

dinig, dingc, e; f? Dung; fimus:—Dinig fimus, Ælfc. Gl. 1; Som.
55, 6; Wrt. Voc. 15, 6. Dingc [MS. dingce] thymiāma, Mone B. 4795.
v. dung.

dinne, es; m. A storm, tempest; procella:—On dinnes mere on a stormy
sea, Chr. 938; Ing. 144, 24; Whel. 556, 44.

diófol-gild, es; n. Devil-worship, an image of the devil, an idol, Ors.
1, 5; Bos. 28, 27. v. deófol-gild.

diógol secret, obscure, profound, Bt. 13; Fox 36, 32. v. dígol; adj.

dióhlu secrets, Prov. 11. v. dígol.

dióp deep, Prov. 22. v. deóp; adj.

dióp depth, Ps. Spl. T. 64, 7. v. deóp.

diópe deeply, solemnly, Beo. Th. 6131; B. 3069. v. deópe.

dióplíce deeply, Bt. 35, 1; Fox 154, 19. v. deóplíce.

diópnys, -nyss deepness, depth, an abyss; ăbyssus = ἄβυσσος, Ps. Lamb.
32, 7. v. deópnes.

diór heavy, severe, dire, Beo. Th. 4186; B. 2090. v. deór; adj. II.

diór a beast, animal, Bt. Met. Fox 26, 183; Met. 26, 92: 27, 21;
Met. 27, 11. v. deór.

diór-boren noble-born, noble:—Apollines dōhtor diórboren Apollo's
noble-born daughter, Bt. Met. Fox 26, 103; Met. 26, 52. v. deór-boren.

dióre dear, precious, glorious, magnificent, Bt. 13; Fox 38, 10, MS.
Cott: Bt. Met. Fox 10, 57; Met. 10, 29. v. deóre.

dióre dearly, with great price, Bt. Met. Fox 26, 37; Met. 26, 19.
v. deóre.

diór-ling a darling, Bt. Met. Fox 15, 15; Met. 15, 8. v. deórling.

diór-wyrþe precious, costly, Bt. 15; Fox 48, 5. v. deór-wyrþe.

dippan; p. de, te; pp. ed, d, t To dip, Ps. Spl. 67, 25: Ex. 12, 22.
v. dyppan.

DISC, es; m. A plate, bowl, DISH; discus, cătīnus, păropsis:—Eallswā
se disc also the dish, L. Ælf. C. 22; Th. ii. 350, 23. Disc discus, Wrt.
Voc. 82, 22: 290, 20. Clǽnsa ǽryst ðæt wiðinnan ys calices and discas
munda prius quod intus est calīcis et paropsĭdis = παροψίς, ίδος; f. Mt.
Bos. 23, 26. Þweah ðæt gewrit of ðam disce wash the writing off the
dish, L. M. 1, 62; Lchdm. ii. 136, 9. Syle me on ānum disce Iohannes
heáfod ðæs Fulluhteres da mihi in disco caput Ioannis Baptistæ, Mt.
Bos. 14, 8, 11: Mk. Bos. 6, 25, 27. Se ðe his hand on disce mid me
dypþ qui intingit mecum manum in cătīno, 14, 20. On disce in părop-
sĭde, Mt. Bos. 26, 23. Bebeád ðæt mon ðone disce tobrǽce to styccum
and ðām þearfum gedǽlan discum confringi, atque paupĕribus minūtātim
dīvidi præcēpit, Bd. 3, 6; S. 528, 21. Discas lāgon dishes lay [there],
Beo. Th. 6088; B. 3048. Ic gefrægn ānne mannan him on bearm hlādan
bunan and discas I heard that one man loaded in his bosom cups and
dishes, 5544; B. 2775. Ge clǽnsiaþ ðæt wiðūtan ys, caliceas and discas
mundātis quod deforis est călĭcis et paropsĭdis, Mt. Bos. 23, 25.
[Prompt. dysshe: Wyc. disch, dishe a disc, quoit: Piers P. dissh:
Chauc. dish: Laym. disc: Plat. disch, m. table: O. Sax. disk, disc, m. a
table: Dut. disch, m. a dining-table: Ger. M. H. Ger. tisch, m. a table:
O. H. Ger. tisc, m. discus, mensa, fercŭlum: Dan. disk, m. f. a table,
dish: Swed. disk, m. a counter: Icel. diskr, m. a plate: Lat. discus:
Grk. δίσκος a round plate, quoit, dish.] DER. bǽr-disc, hlǽd-, hūsel-.

disc-berend, es; m. A dish-bearer; discĭfer, Cot. 65.

discipul, es; m. A disciple, scholar; discĭpŭlus:—Se wæs iu on
Brytene Bosles discipul discĭpŭlus quondam in Brittania Boisili, Bd. 5, 9;
S. 622, 28. Crist cwæþ to his discipulum Christ said to his disciples,
Boutr. Scrd. 22. 45: Homl. Th. ii. 266, 33: 320, 13.

discipul-hád, es; m. DISCIPLEHOOD, pupilage; discĭpŭlātus:—Ðysses
discipulhāde Cūþberht wæs eádmōdlíce underþeóded hujus discipŭlātui
Cudberct humĭlĭter subdĭtus, Bd. 4, 27; S. 603, 39.

disc-þén, es; m. [þegen, þēn a minister, servant] A dish-servant,
dish-bearer, minister of food, sewer; discĭfer, discophŏrus, cibi minister:—
Discþén discĭfer vel discophŏrus, Ælfc. Gl. 30; Som. 61, 68; Wrt. Voc.
26, 65. Godes engel gebrohte ðone discþén ðǽr he hine ǽr genam ihe
angel of God brought the minister of food where he had before taken him,
Homl. Th. i. 572, 9.

disg foolish, Deut. 32, 21. v. dysig.

disig folly, Hy. 7, 107; Hy. Grn. ii. 289, 107. v. dysig.

disme, an; f? The herb tansy? tanacētum?—Nim cristallan and
disman take crystallium and tansy, Lchdm. iii. 10, 29.

distæf, es; m. [dis = Gael. dos a bush, tuft; stæf a staff] A DISTAFF;
colus:—Distæf colus, Ælfc. Gl. 28; Som. 61, 15; Wrt. Voc. 26, 14:
82, 9.

dŏ do, Elen. Kmbl. 1078; El. 541; impert. of dôn.

DOCCE, an; f. DOCK, sorrel; lăpăthum = λάπαθον, rumex:—Ðeós
wyrt ðe man lăpăthum, and ŏðrum naman docce nemneþ, biþ cenned on
sandigum stōwum, and on ealdum myxenum this herb which is called
lăpăthum, and by another name dock, is produced in sandy places, and
on old dunghills, Herb. 14, 1; Lchdm. i. 106, 10–12, note 14: L. M.
3, 63; Lchdm. ii. 350, 26: Wrt. Voc. 67, 54. Doccan moran dust dust
of root of dock, L. M. 1, 54; Lchdm. ii. 126, 6. Sume seóðaþ bētan
oððe doccan on geswēttum wīne some seethe beet or dock in sweetened
wine, L. M. 2, 25; Lchdm. ii. 218, 7: 1, 38; Lchdm. ii. 80, 1, 76;
Lchdm. ii. 150, 10. Seó fealwe docce the fallow dock; rumex marītima
vel palustris, L. M. 1, 49; Lchdm. ii. 122, 19. Seó reáde docce the red
dock; rumex sanguīnea, L. M. 1, 49; Lchdm. ii. 122, 19: 1, 50;
Lchdm. ii. 124, 2. Seó scearpe docce the sharp or sour dock, sorrel;
oxylăpăthum = ὀξυλάπαθον, rumex acētōsa, Som. Ben. Lye. Docce seó
ðe swimman wille the dock which will swim, the water-lily; nymphæa,
L. M. 3, 71; Lchdm. ii. 358, 8: 2, 65; Lchdm. ii. 292, 11: 1, 50;
Lchdm. ii. 122, 21. [Chauc. docke a sour herb: Kil. docke, blæderen
the herb colt's foot.] DER. eá-docce, sūr-, wudu-.

DOCGA, an; m. A DOG; canis:—Docgena canum, Glos. Prudent.
Reed. 148, 23. [Piers P. R. Glouc. dogge: Chauc. dogges, pl: Plat.

dogge *a big dog: Dut.* dog, *m. a bull-dog: Ger.* dog, dogge, docke, *m. f. canis molossus Anglicus: Dan.* dogge, *m. f: Swed.* dogg, *m. a mastiff.*]

dôchtor *a daughter,* Ælfc. Gl. 91; Som. 75, 22; Wrt. Voc. 51, 66. v. dôhtor.

doefe *perfect,* Mt. Kmbl. Rush. 19, 21. v. dêfe.

doeg *a day,* Mt. Kmbl. Lind. 27, 62. v. dæg.

doema *a judge,* Mt. Kmbl. Lind. 5, 25. v. dêma.

doeman *to judge,* Mt. Kmbl. Lind. 7, 1. v. dêman.

dôende *doing,* Ps. Spl. 102, 6, = dônde; *part. of* dôn.

dôere, es; *m. A doer, worker;* opifex:—Dôere, ðæt is Gâst se hâlga *opifex, id est Spiritus sanctus,* Rtl. 198, 13.

doeþ-bérnis, -niss *a pestilence,* Lk. Skt. Rush. 21, 11. v. deáþ-bérnis.

dofen *dived, dipped;* mersus, immersus; *pp. of* dûfan.

Dofere, Dofre, an; *f.* [*Hunt.* Douere, Doure: *Sim. Dun. Kni.* Dovere: *Houd.* Dowere: *Brom.* Dover: *Thorn.* Dovore: *Wel.* dwfr *water*] DOVER; Dubris, Dofris, is; *f.:*—His men cômon to Doferan *his men came to Dover,* Chr. 1050; Th. 313, 20, col. 2: 1051; Th. 317, 25, col. 2. On ðam ylcan geáre com Eustatius up æt Doferan *in the same year Eustace landed at Dover,* 1052; Th. 312, 26, col. 2: 1095; Th. 361, 21. He to Dofran gewende *he went to Dover,* 1048; Th. 313, 32, 34, 35, col. 1; 315, 18, col. 1: 1052; Th. 319, 26, col. 1.

dofung, e; *f. Dotage;* deliramentum:—Dofunga *deliramenta,* Cot. 69: Mone B. 1621: 4192. Dofunga *insidias,* Mone B. 2721.

dôger *a day;* dies:—Dôgera *of days,* Bd. 4, 3; 569, 4. v. dôgor.

dôgian; *p.* ode; *pp.* od *To bear, suffer;* pati?:—Ic dôgode *I suffered,* Exon. 100 b; Th. 380, 17; Rä. 1, 9.

DÔGOR, dôger, es; *m. n. A day;* dies:—Ymb ântîd ôðres dôgores *about the first hour of the second day,* Beo. Th. 444; B. 219: 1215; B. 605. He to ðam ŷtemæstan dôgore becom *he came to his last day,* Bd. 4, 8; S. 575, 30, 39. Ðys dôgor ðû geþyld hafa weána gehwylces *do thou have patience this day for every woe,* Beo. Th. 2794; B. 1395. Ðý dôgore *in that day,* 3599; B. 1797: Judth. 9; Thw. 21, 10; Jud. 12. Uferan dôgore *at a later day,* Past. 38, 8; Hat. MS. 52 b, 7: Ors. 4, 5; Bos. 82, 15. Dôgor beoþ mîn forþscriðen *my days will be departed,* Exon. 48 a; Th. 164, 14; Gû. 1011. He dôgora gehwám dreám gehŷrde hlûdne in healle *he heard loud merriment each day in the hall,* Beo. Th. 176; B. 88: Bt. Met. Fox 13, 42; Met. 13, 21: 22, 122; Met. 22, 61. His dôgora wæs rîm aurnen *the number of his days was run out,* Cd. 79; Th. 98, 5; Gen. 1625: 119; Th. 155, 12; Gen. 2571. Emb ahta dôgera rîmes *after the number of eight days,* Menol. Fox 189; Men. 96. He wæs his ðara nŷhstana dôgera gemyndig *he was mindful of his last days,* Bd. 4, 3; S. 569, 4. His forgifnesse gumum to helpe dæleþ dôgra gehwám Dryhten weoroda *the Lord of hosts dealeth his forgiveness each day in help to men,* Exon. 14 a; Th. 27, 9; Cri. 428: 33 a; Th. 105, 23; Gû. 27: Beo. Th. 2184; B. 1090. Ic mâna fela æfter dôgrum dyde *I did many evils during my days,* Hy. 4, 51; Hy. Grn. ii. 284, 51. Þrió dôgor *for the space of three days;* triduo, Mt. Kmbl. Lind. 15, 32. Uferan dôgrum *in later days,* Beo. Th. 4407; B. 2200. [*Icel.* dægr, dœgr, *n. a day: Goth.* -dogs; *adj.* in ahtau-dogs *on the eighth day;* fidur-dogs *on the fourth day.*] DER. dôgor-gerîm, -rîm: ende-dôgor. v. dæg.

dôgor-gerîm, es; *n.* [gerîm *a number*] *Number of days, allotted time of life;* diêrum numerus, vitæ spatium:—Wæs eall sceacen dôgorgerîmes *all the number of his days was departed,* Beo. Th. 5449; B. 2728. Næfre he sôþra swâ feala wundra gefremede dôgorgerîmum *he could never have performed so many true miracles during his life,* Elen. Kmbl. 1556; El. 780.

dôgor-rîm, es; *n.* [rîm *a number*] *Number of days, time of life;* diêrum numerus, vitæ spatium:—Oþ-ðæt ende cymeþ dôgorrîmes *till the end of the number of days cometh,* Exon. 62 b; Th. 231, 6; Ph. 485. Nâne forlêt deáþ dôgorrîme *death lets none escape after a number of days,* Bt. Met. Fox 10, 133; Met. 10, 67. Is ðæs þroht to ðæs heard dôgorrîmum *this suffering is so hard in the days of my life,* Elen. Kmbl. 1406; El. 705.

dôh *dough,* Lchdm. iii. 88, 17. v. dâh.

dôhtar *a daughter,* Th. Diplm. A. D. 830; 466, 4. v. dôhtor.

dohte *benefited,* Chr. 1006; Erl. 140, 13: dohtest *shouldst benefit,* Deut. 15, 11; *p. of* dugan.

dôhter *a daughter:*—Lothes dôhter *Lot's daughter,* Cd. 123; Th. 157, 22; Gen. 2610. v. dôhtor.

dohtig; *def.* se dohtiga; *adj.* [dohte, *p. of* dugan *to avail*] DOUGHTY, *valiant, good;* fortis, validus, probus:—Forþférde Hacun, se dohtiga eorl, on sæ Hakon, *the doughty earl, died at sea,* Chr. 1030: Erl. 162, 40. Ðyssa þinga is gecnæwe ælc dohtig man on Cent [MS. Kænt] and on Sûþ-Seaxum [MS. -Sexan] *every good man in Kent and in Sussex is cognizant of these things,* Th. Diplm. A. D. 1016-1020; 313, 19. v. dyhtig.

dohton *benefited, were honest,* Bt. 18, 3; Fox 64, 37; *p. pl. of* dugan.

DÔHTOR, dôhtur, dôhter; *indecl. in sing. but the dat.* dêhter *is found: pl. nom. acc.* dôhtor, dôhtra, dôhtru, dôhter; *gen.* dôhtra; *dat.*

instr. dôhtrum; *f. A* DAUGHTER; filia:—Mîn dôhtor is deád *filia mea defuncta est,* Mt. Bos. 9, 18. Gelŷf, dôhtor *confide, filia,* 9, 22. Ðû fram mînre dôhtor onwôce *thou from my daughter wast born,* Cd. 223; Th. 292, 11; Sat. 439. Ðâ wæs ellen-wôd fæder wið dêhter *then was the father furious with his daughter,* Exon. 67 b; Th. 251, 7; Jul. 141: Gen. 29, 18: Mk. Bos. 7, 26, 29: Homl. Th. ii. 26, 33. Ðæm forgeaf Hrêðel ângan dôhtor *to whom Hrethel gave his only daughter,* Beo. Th. 755; B. 375. Cynincga dôhtor *regum filiæ,* Ps. Th. 44, 10. Fægnigan dôhtra *exultent filiæ,* Ps. Spl. 47, 10: Ps. Th. 44, 14. Heora dôhtru *eorum filiæ,* 143, 15. Ðæt ðû me bereáfodest ðînra dôhtra *ne violenter auferres filias tuas,* Gen. 31, 31. Fyllaþ eorþan sunum and dôhtrum *fill the earth with sons and daughters,* Cd. 10; Th. 13, 5; Gen. 198. Ðû scealt cennan sunu and dôhtor *thou shalt bring forth sons and daughters,* 43; Th. 57, 7; Gen. 924. Suna and dôhter *filios et filias,* Ps. Th. 105, 27. [*Wyc.* douȝtir: *Piers P.* doughtres, *pl: Chauc.* doughter, doughtre: *R. Brun.* doughter: *R. Glouc.* dogtren, *pl: Laym.* dohter, douter, doȝter: *Orm.* dohhterr: *Plat.* dogter, dochter, *f: O. Sax.* dohtar, dohtor, dohter, *f: Frs.* dochter, doayter: *O. Frs.* dochter, *f: Dut.* dochter, *f: Ger.* tochter, *f: M. H. Ger.* tohter, *f: O. H. Ger.* tohtar, *f: Goth.* dauhtar, *f: Dan.* datter, *f: Swed.* dotter, *f: Icel.* dóttir, *f: Grk.* θυγάτηρ, *f: Lith.* dukte: *Zend* dughdhar: *Sansk.* duhitṛi, *f. a daughter,* properly *a milkmaid,* from duh *to milk.*] DER. steóp-dôhtor.

dôhtur *a daughter:*—Ðære Herodiadiscean dôhtur *Herodiadis filia,* Mt. Bos. 14, 6. v. dôhtor.

**DOL; *def.* se dola, seó, ðæt dole; *adj.* DULL, *foolish, erring, heretical;* stolidus, stultus, hæreticus = αἱρετικός:—Dol biþ se ðe him his Dryhten ne ondrædeþ *foolish is he who dreads not his Lord,* Exon. 83 a; Th. 312, 7; Seef. 106: 89 a; Th. 335, 17; Gn. Ex. 35: Salm. Kmbl. 447; Sal. 224. Ge weorþmyndu in dolum dreáme Dryhtne gieldaþ *ye pay reverence to the Lord in foolish joy,* Exon. 39 a; Th. 130, 8; Gû. 435. Oþ hie to dole wurdon *until they became foolish,* Cd. 18; Th. 22, 14; Gen. 340. Ne ondrædaþ ða dolan *the foolish are not afraid,* Past. 7, 2; Hat. MS. 12 a, 25. Ða dolan rædas *stolida consulta,* Cot. 189. Ic dole hwette *I excite the dull,* Exon. 103 b; Th. 393, 1; Rä. 12, 3: 107 b; Th. 410, 16; Rä. 28, 17: Ps. Th. 118, 126. [*Chauc.* dul: *Orm.* dill *sluggish: Plat.* dul mad: *O. Sax.* dol *stultus: Frs.* dol, mad: *Dut.* dol *insânus: Ger.* toll mad: *M. H. Ger.* tol, dol mad: *O. H. Ger.* tol *stultus: Goth.* dwals: *Icel.* dulr *silent, close.*]

dolc *a buckle,* Wrt. Voc. 74, 59. v. dalc.

dolc-swaðu *scars,* Ps. Lamb. 37, 6, = dolh-swaðu; *pl. nom. of* dolh-swæþ.

dolfen *dug;* *pp. of* delfan. v. a-dolfen.

dolg *a wound, scar,* L. M. 1, 45; Lchdm. ii. 114, 1: Exon. 24 a; Th. 68, 24; Cri. 1108. v. dolh.

dolg-ben, -benn, e; *f.* [ben *a wound*] *A wound;* vulnus:—Dolgbennum þurhdrifen *pierced through with wounds,* Andr. Kmbl. 2793; An. 1399.

dolg-bôt *compensation for a wound,* L. Alf. pol. 23; Th. i. 78, 7. v. dolh-bôt.

dolgian; *p.* ode; *pp.* od [dolg = dolh *a wound*] *To wound;* vulnerâre:—Dolgdon, *p. pl.* Exon. 114 b; Th. 441, 2; Rä. 60, 11. DER. ge-dolgian.

dol-gilp, es; *m.* [dol *foolish;* gilp *pride, haughtiness*] *Foolish pride, vain-glory;* vana glôria:—Git wada cunnedon for dolgilpe *ye both made trial of the fords for foolish vaunt,* Beo. Th. 1022; B. 509.

dolg-rune *pellitory,* L. M. 1, 25; Lchdm. ii. 66, 16. v. dolh-rune.

dolg-sealf *a wound salve, poultice for a wound,* L. M. cont. 1, 38; Lchdm. ii. 8, 26, 29. v. dolh-sealf.

dolg-slege, es; *m.* [slege *a blow*] *A wounding blow;* vulnerans ictus:—Þurh dolgslege *through a wounding blow,* Andr. Kmbl. 2948; An. 1477. Deáh he sâres swâ feala deópum dolgslegum dreógan sceolde *although he must suffer so much pain through deep wounding blows,* 2489; An. 1246.

DOLH, dolg, es; *n. A wound, scar of a wound, cut, gash, sore;* vulnus, cicatrix, ulcus:—Cnua grêne betonican and lege on ðæt dolh gelôme, oþ-ðæt ðæt dolh [sŷ] gebâtod *pound green betony and lay it on the wound frequently, until the wound is bettered,* L. M. 3, 33; Lchdm. ii. 328, 2, 3: 1, 38; Lchdm. ii. 96, 9, 15, 16: 1, 72; Lchdm. ii. 148, 21. Gyf yfele dolh oððe wunda on heáfde sŷn, genim ðas ylcan wyrte *if evil cuts or wounds be on the head, take this same herb,* Herb. 1; Lchdm. i. 234, 15. Me ecga dolg eácen weorþaþ *to me the edges' sores become increased,* Exon. 102 b; Th. 388, 25; Rä. 6, 13. Deópra dolga *of deep gashes,* 114 a; Th. 438, 7; Rä. 57, 4. To deópum dolgum *for deep wounds,* L. M. 1, 45; Lchdm. ii. 114, 1. Wið ða sweartan dolh, genim ðas ylcan wyrte *for black scars, take this same herb,* Herb. 10, 3; Lchdm. i. 100, 23: Homl. Blick. 91, 1. Ðám biþ grorne dolg sceáwian *it shall be sad to them to behold the scars.* Exon. 25 b; Th. 74, 16: Cri. 1207: 24 a; Th. 68, 24; Cri. 1108. Blôd-dolh *a blood-letting wound,* L. M. 1, 72; Lchdm. ii. 148, 12, 15. [*Frs.* dolge *vulnus: O. Frs.* dolch, dulg, dolech, dulich, *n. vulnus: O. H. Ger.* tolg, *n. vulnus:*

Goth. dulgs, *m. culpa :* Icel. dólg, *n. direful enmity.*] DER. feorh-dolh, heoru-, seono-, syn-.

dolh-ben, -benn *a wound.* v. dolg-ben.

dolh-bót, dolg-bót, e ; *f.* [bót *compensation*] *A wound-fine* or *compensation for a wound ;* vulnĕris compensātio :—Bête dolgbóte [dolhbôte MS. H.] *let him make compensation for the wound,* L. Alf. pol. 23 ; Th. i. 78, 7.

dolh-drenc, es ; *m.* [drenc *a drink*] *A wound-drink, potion for a wound ;* vulnĕrāria pōtio :—Dolhdrenc : ribbe niođeweard aud ufeweard cnuwa smale *a wound-drink : pound small the netherward and upward part of ribwort,* L. M. 1, 38 ; Lchdm. ii. 98, 1 : 1, 38 ; Lchdm. ii. 96, 19, 22.

dolh-rune, dolg-rune, dulh-rune, an ; *f.* The herb *pellitory, which grows upon walls ;* perdīcium = περδίκιον, parietāria officinālis, Lin :— Wiđ lungen-ádle ; dolhrune, etc. *for lung-disease ; pellitory, etc.* L. M. 2, 52 ; Lchdm. ii. 268, 16 : Herb. 83, 1 ; Lchdm. i. 186, 12, 13 : Lchdm. iii. 16, 9. Dulhrune *pellitory,* L. M. 3, 8 ; Lchdm. ii. 312, 16. To sealfe wiđ springe, nim dolhrunan *for a salve against a pustule, take pellitory,* 1, 33 ; Lchdm. ii. 80, 8 : 1, 38 ; Lchdm. ii. 96, 11 : 3, 65 ; Lchdm. ii. 354, 1 : Lchdm. iii. 4, 10 : 38, 26. Genim dolgrunan *take pellitory,* L. M. 1, 25 ; Lchdm. ii. 66, 16 : 1, 47 ; Lchdm. ii. 120, 5.

dolh-sealf, dolg-sealf, e ; *f.* [sealf *a salve, poultice*] *A wound-salve, poultice for a wound ;* vulnĕrārium emplastrum :—Dolhsealf ; genim wegbrædan sǽd, getrifula smale, scead on đa wunde, sôna biþ sêlre *a wound-salve ; take seed of waybroad, bray it small, put [shed] it on the wound, soon it will be better,* L. M. 1, 38 ; Lchdm. ii. 90, 27 : 1, 38 ; Lchdm. ii. 96, 2, 7, 10, 13. Grundeswelge đa đe weaxaþ on worþigum biþ gód to dolhsealfe *the groundsel which grows in highways is good for a wound-salve,* 1, 38 ; Lchdm. ii. 92, 27. Hêr sindon dolhsealfa to eallum wundum *here are wound-salves for all wounds,* 1, 38 ; Lchdm. ii. 90, 23. Dolgsealf wiđ lungen-ádle *a wound-salve for lung-disease,* L. M. cont. 1, 38 ; Lchdm. ii. 8, 29. Dolgsealfa wiđ eallum wundum *wound-salves for all wounds,* L. M. cont. 1, 38 ; Lchdm. ii. 8, 26.

dolh-slege *a wounding blow.* v. dolg-slege.

dolh-smeltas ; *pl. m.* Linen *bandages ;* tæniæ = ταινίαι :— Tæppan *vel* dolhsmeltas [MS. dolsmeltas] *tæniæ* [MS. tenia], Ælfc. Gl. 4 ; Som. 55, 93 ; Wrt. Voc. 16, 64. v. tæppan, from tæppa, *m.*

dolh-swæþ ; *gen.* -swæđes ; *pl. nom. acc.* -swađu, -swađo ; *n :* dolhswađu, e ; *f :* -swađo ; *indecl. f.* [swæþ, swađu *a trace, vestige*] *A trace of a wound, a scar ;* cicatrīcis vesīigium, cicātrix :—Dolhswæþ [MS. -swađ] *cicātrix,* Ælfc. Gl. 85 ; Som. 73, 115 ; Wrt. Voc. 49, 22. Forrotodon gewemmede and hîg synt dolhswađu [dolcswaþu MS : dolhswađo, Spl.] mîne *putruērunt et corruptæ sunt cicatrices meæ,* Ps. Lamb. 37, 6. Dolhswađu *cicātrix,* Wrt. Voc. 85, 50. Đæt seó þynneste dolhswađo and seó læste ætýwde *that the thinnest and the least scar was to be seen,* Bd. 4, 19 ; S. 589, 19.

dolh-wund ; *adj.* [wund *wounded*] *Wounded ;* vulnĕrātus :—He on swîman læg druncen and dolhwund *he lay in stupor drunk and wounded,* Judth. 10 ; Thw. 23, 6 ; Jud. 107.

dol-líc, dol-líg ; *adj. Foolish, rash ;* stultus, temĕrārius :—He manna mǽst mǽrþa gefremede, dǽda dollícra *he of men had achieved most glories, rash deeds,* Beo. Th. 5285 ; B. 2646. Druncen beorg đê and dollíg word *guard thyself from drunkenness and foolish words,* Exon. 80 b ; Th. 302, 11 ; Fä. 34.

dollíce ; *adv. Foolishly, rashly ;* stulte, insāne :—Sprǽc heálíg word dollíce wiđ Drihten sînne *he spake proud words foolishly against his Lord,* Cd. 15 ; Th. 19, 22 ; Gen. 295 : Homl. Th. ii. 330, 26. Ne man ne sceal drincan, ođđe dollíce etan binnan Godes húse *nor may any one drink, nor foolishly eat within God's house,* L. Ælf. C. 35 ; Th. ii. 356, note 2, line 10 : Past. 20, 1 ; Hat. MS. 29 b, 4.

dol-sceađa, an ; *m.* [dol *foolish ;* sceađa *a robber*] *A foolish* or *rash robber ;* temĕrārius spoliātor :—God eáđe mæg đone dolsceađan dǽda getwǽfan *God may easily sever the doltish robber from his deeds,* Beo. Th. 962 ; B. 479.

dol-scipe, es ; *m.* [dol *foolish ;* scipe *termination,* q. v.] *Foolishness, folly, error ;* stultītia, error :—Giongra monna dolscipe hî ofslihþ *the folly of young men kills them,* Past. 50, 2 ; Hat. MS.

dol-sprǽc, e ; *f.* [sprǽc *a speaking, talk*] *Foolish* or *vain talk, loquacity ;* fātuus sermo :—Ðýlæs we, for dolsprǽce, tô wîdgangule weorþen *lest, from loquacity, we wander too far,* Past. 49, 4 ; Hat. MS.

dol-willen, es ; *n. Rashness, madness ;* temĕrĭtas, dementia :—Ðú þurh dîn dolwillen gedwolan fylgest *thou followest error through thy rashness,* Exon. 68 b ; Th. 254, 24 ; Jul. 202.

dol-willen ; *adj. Rash, mad ;* temĕrārius, dēmens :—Ic đec gedyrstig and đus dolwillen gesôhte *I have sought thee thus daring and rash,* Exon. 72 a ; Th. 269, 17 ; Jul. 451.

dol-wîte, es ; *n.* [dol *foolish, audacious* = Ger. toll-kühn ; wîte *a punishment*] *Punishment for audacity, temerity* or *fool-hardiness ;* temerĭtātis pœna :—Nales dolwîte *no punishment for audacity,* Exon. 107 a ; Th. 408, 25 ; Rä. 27, 17.

DÓM, es ; *m.* I. DOOM, *judgment, judicial sentence, decree,*

ordinance, *law ;* jūdicium, sententia, decrētum, jus, lex :—Hit ys Godes dôm *Dei jūdicium est,* Deut. 1, 17 : Jn. Bos. 11, 31. Dômes dæg *jūdicii dies,* Mt. Bos. 10, 15 : 11, 22, 24. Đam ylcan dôme đe ge dêmaþ, eów biþ gedêmed *in quo jūdicio judicavĕrĭtis, judicabimĭni,* Mt. Bos. 7, 2 : Ex. 6, 6 : 23, 6. Æfter eówrum ágnum dôme *according to your own judgment,* Bt. 14, 2 ; Fox 44, 35. Sýn hî bisceopes dôme scyldig *let them be liable to the bishop's sentence,* Bd. 4, 5 ; S. 573, 1. Đone ryhtan dôm *the righteous sentence,* Exon. 27 b ; Th. 84, 6 ; Cri. 1369 : 42 a ; Th. 142, 8 ; Gú. 641. Hie noldon hyra þeódnes dôm þafigan *they would not obey their lord's decree,* Cd. 181 ; Th. 227, 21 ; Dan. 190 : Exon. 65 a ; Th. 240, 21 ; Ph. 642. On gewritum findaþ dôma gehwilcne đara đe him Drihten bebeád *they find in the scriptures each of the ordinances which the Lord commanded him* [Moses], Cd. 169 ; Th. 211, 2 ; Exod. 520. Ðis syndon đa dômas đe Æđelbirht cyning asette on Agustinus dæge *these are the laws which king Ethelbert established in Augustine's day,* L. Ethb. pref ; Th. i. 2, 2 : L. H. E. pref ; Th. i. 26, 3. Be Ínes dômum *of Ine's laws,* L. In. pref ; Th. i. 102, 1. II. *a ruling, governing, command ;* rectio, gubernātio, impĕrium :—Dôme Drihten eorþan ymbhwyrft ealle gesette *Dŏmĭnus correxit orbem terræ,* Ps. Th. 95, 9 : Exon. 39 a ; Th. 129, 3 ; Gú. 415 : Beo. Th. 5708 ; B. 2858. III. *might, power, dominion, majesty, glory, magnificence, honour, praise, dignity, authority ;* potentia, potestas, majestas, glôria, splendor, honor, laus, dignĭtas, auctôrĭtas :—Đær wearþ Læcedemonia ánweald and heora dôm alegen *there was the dominion of the Lacedæmonians and their power laid low,* Ors. 3, 1 ; Bos. 53, 30. Hí on dryhtlícestum dôme lifdon *they lived in most lordly majesty,* Exon. 82 b ; Th. 311, 1 ; Seef. 85. Sigemunde gesprong dôm unlytel *no little glory sprang to Sigemund,* Beo. Th. 1775 ; B. 885 : 1913 ; B. 954. Hæfde Daniel dôm micelne in Babilônia *Daniel had much honour in Babylon,* Cd. 180 ; Th. 225, 33 ; Dan. 163. Eów Dryhten geaf dôm unscyndne *the Lord gave you shameless glory,* Elen. Kmbl. 730 ; El. 365. Se đe wile dôm árǽran *who desires to exalt his dignity,* Exon. 87 a ; Th. 327, 2 ; Wíd. 140. Drihten á dôm áge, leóhtbǽre lof *may the Lord ever have glory, bright praise,* Exon. 80 a ; Th. 299, 33 ; Crä. 111. Dôme gewurþad *honoured with glory,* Beo. Th. 3295 ; B. 1645. Dôma sêlast *best of dignities,* Exon. 122 a ; Th. 467, 20 ; Alm. 4. IV. *will, free will, choice, option ;* arbitrium, optio :—On eówerne ágenne dôm *in your own will,* Andr. Kmbl. 677 ; An. 339. Ðæt he beáh-hordes brúcan môste selfes dôme *that he might enjoy the ring-hoard of his own free will,* Beo. Th. 1794 ; B. 895 : 5545 ; B. 2776. V. *sense, meaning, interpretation ;* significātio, interprĕtātio :—Ge sweltaþ deáþe nymþe ic dôm wite sôþan swefnes *ye shall perish by death unless I know the interpretation of my true dream,* Cd. 179 ; Th. 224, 29 ; Dan. 143. [*Prompt.* dome : *Wyc.* dom, dome, doom : *Piers P.* doom, dome : *Chauc.* dome : *Laym. Orm.* dom : *O. Sax. O. Frs.* dôm, *m. jūdicium, arbitrium, honor :* Dut. doeming, *f. condemnation :* Kil. doeme *judicium :* Ger. *in the termination* -tum, -thum -dom : M. H. Ger. O. H. Ger. tuom, *m. n. jūdicium :* Goth. doms, *m. judgment :* Dan. đom, *m. f :* Swed. dom, *m :* Icel. dómr, *m :* Sansk. dhāman, *n. a dwelling-place, state, condition, law, from dhā to put.*]

-dóm, es ; *m.* as the termination of nouns is always masculine, and denotes *Dominion, power, authority, property, right, office, quality, state, condition ;* as Cyne-dôm *a king's power, office, etc. a kingdom ;* freó-dôm *freedom ;* hálig-dôm *holiness ;* wís-dôm *wis-dom ;* i. e. *the state* or *condition of being free, holy, wise.*

dóm-bóc ; *f.* [bóc *a book,* q. v.] DOOM-BOOK, *a book of decrees* or *laws ; liber judiciālis :*—Bête be đam đe seó dôm-bóc secge *let him pay a fine according as the doom-book may say,* L. Ath. 1, 8 ; Th. i. 202, 7 : L. Edg. i. 3 ; Th. i. 262, 23 : i. 5 ; Th. i. 264, 20. Swá hit on đǽre dôm-bêc stande *as it stands in the doom-book,* L. Ed. prm ; Th. i. 158, 4. Ne þearf he nânra dômbóca ôđerra cêpan *he need not heed any other doom books,* L. Alf. 49 ; Th. i. 56, 30. Ôþ-đæt he com to đám dômbócum, đe se heofenlíca Wealdend his folce gesette *until he came to the doom-books, which the heavenly Ruler appointed for his people,* Homl. Th. ii. 198, 18.

dóm-dæg, es ; *m.* [dómes dæg *doom's day,* L. E. I. 25 ; Th. ii. 422, 10 : Salm. Kmbl. 649 ; Sal. 324] DOOMSDAY, *judgment-day ;* dies jūdicii :—Ǽr he dômdæges dyn gehýre *before he shall hear doomsday's din,* Salm. Kmbl. 545 ; Sal. 272. Æt dômdæge, Exon. 31 b ; Th. 99, 3 ; Crl. 1619. On dômdæge, 99 b ; Th. 372, 19 ; Seel. 95 : Cd. 227 ; Th. 302, 15 ; Sat. 600. On đam micclan dômdæge *in die jūdicii,* L. Ælf. P. 40 ; Th. ii. 380, 39. Ðæt he dômdæg [dômes dæg MS. B.] ondrǽde *that he dread doomsday,* L. C. E. 25 ; Th. i. 374, 13.

dóm-eádig ; *adj. Blessed with power ;* pŏtens, nôbilis, beātus, glôria abundans :—Wæs đǽre fǽmnan ferþ geblissad dômeádigre [-eadigra MS.] *the damsel's soul, the noble one's was rejoiced,* Exon. 69 b ; Th. 259, 26 ; Jul. 288 : 32 a ; Th. 101, 11 ; Cri. 1657 : 43 a ; Th. 145, 23 ; Gú. 699 : Cd. 63 ; Th. 75, 29 ; Gen. 1247.

dómere, es ; *m. A judge ;* jūdex :—Swá him dómeras [dêmeras MS. H.] gereccen *as the judges may prescribe to him,* L. Alf. 18 ; Th. i. 48, 18. Heretogan and dômeras hæfdon mǽstne weorþscipe *consuls and judges*

had most honour, Bt. 27, 4; Fox 100, 13. Settaþ ða to dómerum *appoint them judges*, Past. 18, 2; Hat. MS. 26 a, 6.

Domer-hám, Domar-hám, es; *m.* DAMERHAM, *Wiltshire;* loci nomen in agro Wiltoniensi:—Æðelflǽd æt Domerháme, Ælfgáres dóhter ealdormannes, was his cwén *Æthelfled at Damerham, daughter of Ælfgar the alderman, was his* [*king Edmund's*] *queen,* Chr. 946; Erl. 117, 25. Ic geán ðæs landes æt Domarháme into Glæstinga byrig *I give the land at Damerham to Glastonbury,* Th. Diplm. A. D. 972; 519, 30.

dóm-ern, es; *n. A judgment-place, a court-house;* forum judiciāle, tribūnal, prætōrium:—Dómern *tribūnal,* Glos. Prudent. Recd. 143, 70. Ðá underféngon ðæs déman cempan ðone Hǽlend on ðam dómerne, and gegaderodon ealne ðone þreát to heom *tunc milites præsidis suscipientes Iesum in prætorium, congregāvērunt ad eum universam cohortem,* Mt. Bos. 27, 27: Jn. Bos. 18, 28, 33: 19, 9: Homl. Th. ii. 422, 1. Wyðútan hys dómern *outside his judgment-hall,* Nicod. 10; Thw. 5, 9.

dóm-fæst; *adj.* [fæst *fast, firm*] *Firm in judgment, just, firm, powerful; justus, pŏtens:*—Noe wæs dómfæst and gedéfe *Noah was just and meek,* Cd. 64; Th. 78, 2; Gen. 1287: 108; Th. 143, 8; Gen. 2376: Exon. 54 b; Th. 192, 1; Az. 99. Syle us to-dæg dómfæstne blǽd *give us to-day firm prosperity,* 122 a; Th. 469, 1; Hy. 5, 6. Twelfe wǽron dǽdum dómfæste *the twelve were powerful in deeds,* Apstls. Kmbl. 9; Ap. 5. Ic séce swegelcyning, dómfæstra dreám *I seek the King of heaven, the joy of the just,* Exon. 48 b; Th. 167, 6; Gú. 1056.

dóm-fæstnes, -ness, es; *f.* [fæstnes *firmness*] *Firmness of judgment, judgment;* jūdicii integritas, jūdicium:—Mildheortnessa and dómfæstnes ic singe *misericordiam et jūdicium cantābo,* Ps. Lamb. 100, 1.

dóm-georn; *adj.* [georn *desirous, eager*] *Eager for justice, ambitious, just, virtuous;* justitiæ appětens, justus:—Se hálga wæs to hofe lǽded, deór and dómgeorn *the holy one was led to the house, dear and virtuous,* Andr. Kmbl. 2617; An. 1310. Hleóþrodon dugoþ dómgeorne *the ambitious rulers spake,* 1385; An. 693: Exon. 76 b; Th. 287, 20; Wand. 17: Elen. Kmbl. 2579; El. 1291.

dóm-hús, es; *n.* [hús *a house*] *A judgment-house;* cūria, epicaustērium, capitōlium:—Dóm-hús *cūria,* Ælfc. Gl. 55; Som. 67, 1; Wrt. Voc. 36, 44. Dóm-hús vel mót-hús *epicaustērium,* 107; Som. 78, 74; Wrt. Voc. 57, 52. Dóm-hús *capitōlium,* 107; Som. 78, 97; Wrt. Voc. 58, 12.

dóm-hwæt; *adj.* [hwæt *quick, strenuous*] *Strenuous in judgment;* in jūdicio strēnuus:—We hine dómhwate, dǽdum and wordum hērgen hold-líce *we strenuous, may praise him faithfully in deeds and words,* Exon. 14 a; Th. 27, 11; Cri. 429.

dómian; *p.* ode; *pp.* od [dóm *justice, glory*] *To praise, glorify;* celebrāre, gloriam tribuěre:—Annanias ðec and Adzarias and Misael Metod dómige *Hananiah and Azariah and Mishael may glorify thee, O Lord,* Cd. 192; Th. 241, 4; Dan. 399: 192; Th. 239, 19; Dan. 372.

dóm-leás; *adj. Inglorious, powerless, hapless;* inglōrius, impŏtens, infortūnātus:—Æðelingas gefricgean dómleásan dǽd *nobles shall hear of your inglorious deed,* Beo. Th. 5772; B. 2890. Sceolon nú ǽfre dreógan dómleáse gewinn *now we shall ever wage powerless war,* Cd. 218; Th. 279, 3; Sat. 232. Ealle swylt fornam, druron dómleáse *death tore them all away, hapless they fell,* Andr. Kmbl. 1989; An. 997.

dóm-líc; *adj. Judicial, glorious;* judiciālis, gloriōsus:—Dómlíc *judiciālis,* Ælfc. Gr. 9, 28; Som. 11, 36. Wǽron hwæðre monge ða ðe Meotude gehýrdun dǽdum dómlícum *there were yet many who obeyed the Creator with glorious deeds,* Exon. 62 a; Th. 228, 28; Ph. 445: 62 a; Th. 229, 8; Ph. 452.

dóm-líce; *adj. Judicially, powerfully, gloriously;* judiciāliter, potenter, gloriōse:—Sýn me ðíne handa on hǽlu nú, and ðæt dómlíce gedón weorþe *fiat manus tua et salvum me facias,* Ps. Th. 118, 173: Exon. 54 b; Th. 193, 19; Az. 124: Judth. 12; Thw. 26, 10; Jud. 319.

Dommoc-ceaster *Dunwich, Suffolk,* Bd. 2, 15; S. 519, 12. v. Domuc.

domne, es; *m. A lord;* dŏminus:—Hér resteþ domne Agustinus, se ǽresta ærcebisceop Cantwarena burge *here resteth lord Augustine, the first archbishop of Canterbury,* Bd. 2, 3; S. 504, 43. Mín domne bisceop *my lord bishop,* 3, 14; S. 540, 25: 3, 19; S. 548, 23.

dóm-setl, es; *n.* [dóm *judgment,* setl *a seat*] *A judgment-seat, tribunal;* tribūnal:—Ðis dómsetl *hoc tribūnal,* Ælfc. Gr. 9, 5; Som. 9, 2. Se geréfa hét Iulianan út gelǽdan to his dómsetle *the count bade Juliana be led out to his judgment-seat,* Exon. 73 b; Th. 274, 16; Jul. 534: 68 a; Th. 252, 12; Jul. 162. On his dómsetle *pro tribūnāli,* Mt. Bos. 27, 19.

dóm-settend, es; *m. One sitting in judgment, a judge, a lawyer;* jurisconsultus, Cot. 113.

Domuc, e; *f? Dommoc-ceaster; gen.* -ceastre; *f. Dunwich, on the sea coast of Suffolk, the seat of the first East Anglian bishopric, which was subsequently fixed at Norwich;* loci nomen in agri Suffolciensi ora maritima:—Alfhun bisceop forþférde on Sudberi, and he wearþ bebyrged in Domuce, and Tídfriþ wearþ gecoren æfter him *bishop Alfhun died at Sudbury, and he was buried at Dunwich, and Tidfrith was chosen after him,* Chr. 798; Th. 105, 9–13, col. 3. Felix se bisceop, se com of Burgundana ríces dǽlum, onféng bisceopsetl on Dommocceastre, and mid ðý he seofontyne winter on bisceoplícum gerece fóre wæs, ðær he on

sibbe his líf ge-endode *Felix episcŏpus, qui de Burgundiōrum partĭbus venit, accēpit sedem episcopātus in civitāte Domnoc, et cum decem ac septem annos eidem provinciæ pontificāli regimĭne præesset, ibĭdem in pace vitam finīvit,* Bd. 2, 15; S. 519, 12.

DÓN, to dónne; *part.* dóende, dónde; ic dó, ðú dést, he déþ, *pl.* dóþ; *p.* ic, he dyde, ðú dydest, *pl.* dydon; *impert.* dó, *pl.* dóþ; *subj.* dó, *pl.* dón, dó; *p.* dyde, *pl.* dyden; *pp.* dón, dén *To* DO, *make, cause;* agěre, facěre:—Ne mót ic dón ðæt ic wylle *non licet mihi quod volo facěre?* Mt. Bos. 20, 15: Chr. 876; Erl. 79, 12: 994; Erl. 133, 17: Cd. 10; Th. 12, 23; Gen. 189: Beo. Th. 2349; B. 1172: Bt. Met. Fox 19, 78; Met. 19, 39. Alýfþ on restedagum wel dón, oððe yfele *licet sabbătis benefacěre, an male?* Lk. Bos. 6, 9. He sǽde ðæt he hit náhte to dónne *he said that he ought not to do it,* Chr. 1070; Erl. 208, 5: 1091; Erl. 227, 13: Mt. Bos. 12, 2: Exon. 26 b; Th. 79, 11; Cri. 1289. Hyt ys alýfed on restedagum wel to dónne *licet sabbătis benefacěre,* Mt. Bos. 12, 12. Dóende [dónde, Lamb.] *faciens,* Ps. Spl. 102, 6. Eádig ys se þeów, ðe hys hláford hyne gemét ðus dóndne, ðonne he cymþ *beatus ille servus, quem cum veněrit dominus ejus, inveněrit sic facientem,* Mt. Bos. 24, 46: Lk. Bos. 12, 43. Ic dó *ago,* Ælfc. Gr. 28, 6; Som. 32, 12: *fácio,* 28, 6; Som. 32, 36. Ic dó oððe wyrce *fácio,* 33; Som. 37, 47. Ic dó gyt *faxo,* 33; Som. 37, 43. Ic dó ðæt gyt beóþ manna fisceras *fáciam vos fǐeri piscatōres homĭnum,* Mt. Bos. 4, 19. Ðú dést *faxis,* Ælfc. Gr. 33; Som. 37, 44. Hwí dést ðú wið me swá *why doest thou with me so?* Gen. 12, 18: Jn. Bos. 6, 30. Se ðe hit déþ, se biþ mycel *he who does it shall be great,* Mt. Bos. 5, 19: 13, 23: 18, 35: Boutr. Scrd. 19, 41: Ælfc. Gr. 33; Som. 37, 44: Salm. Kmbl. 364; Sal. 181: Ps. Th. 139, 12: Bt. Met. Fox 9, 123; Met. 9, 62: Beo. Th. 2121; B. 1058. Se árleása déþ ðæt fýr cymþ ufan *the impious one will cause fire to come from above,* Homl. Th. i. 6, 7: Mt. Bos. 5, 32. Gyf ge ðæt dóþ *if ye do that,* Mt. Bos. 5, 47. Ne winne ge ongén ða ðe eów yfel dóþ *strive not against those who do you wrong,* Mt. Bos. 5, 39: 12, 2. Ðæt cild weóx swá swá óðre cild dóþ *the child grew as other children do,* Homl. Th. i. 24, 35: 18, 26: Boutr. Scrd. 18, 13: Cd. 60; Th. 73, 18; Gen. 1206: Exon. 34 a; Th. 109, 35; Gú. 100. Ne dyde ic for fácne *I did it not for fraud,* Cd. 128; Th. 162, 34; Gen. 2691. Ðú ondsæc dydest *thou madest denial,* Andr. Kmbl. 1854; An. 929. Ðæt dyde unhold mann *inimĭcus homo hoc fecit,* Mt. Bos. 13, 28: Boutr. Scrd. 20, 2: Cd. 33; Th. 44, 12; Gen. 708: Exon. 24 a; Th. 68, 4. Iosep dyde swá Drihtnes engel him bebeád *Joseph fecit sicut præcēpit ei angēlus Domini,* Mt. Bos. 1, 24: Ps. Th. 93, 7. Se wilnode ðæs westdæles, swá se óðer dyde ðæs eástdǽles *he wished for the west part, as the other did for the east part,* Ors. 3, 9; Bos. 66, 26: Boutr. Scrd. 18, 2: Cd. 215; Th. 272, 10; Sat. 117: Rood Kmbl. 226; Kr. 114: Beo. Th. 893; B. 444: Exon. 8 a; Th. 2, 11; Cri. 17. He ne cúðe hwæt ða cynn dydon *he knew not what the people did,* Cd. 92; Th. 116, 23; Gen. 1944: Exon. 53 a; Th. 186, 10; Az. 17. Híg dydon swá hwæt swá híg woldon *fecērunt quæcumque voluērunt,* Mt. Bos. 17, 12: Chr. 1001; Erl. 137, 9. Reced weardode unrím eorla, swá hie oft ǽr dydon *countless warriors guarded the mansion, as they had often done before,* Beo. Th. 2481; B. 1238: Cd. 227; Th. 304, 6; Sat. 625: Exon. 14 a; Th. 28, 32; Cri. 455. Dó fac, Ælfc. Gr. 33; Som. 37, 47. Dó swá ic ðé bidde do *as I pray thee,* Cd. 101; Th. 134, 16; Gen. 2225: Elen. Kmbl. 1078; El. 541. Dóþ wel ðám ðe eów yfel dóþ *benefacíte his qui odērunt vos,* Mt. Bos. 5, 44: Cd. 106; Th. 140, 6; Gen. 2323: Exon. 41 a; Th. 137, 24; Gú. 564: Beo. Th. 2467; B. 1231: Ps. Th. 30, 28. Dóþ his síðas rihte *make his paths straight,* Mt. Bos. 3, 3: Ps. Th. 61, 8: 67, 4. Beheald ðæt ðú ðas dǽde ne dó *see that thou do not this deed,* Homl. Th. i. 38, 25. Ðæt he dó máre *ut salvos facēret omnes,* Ps. Th. 75, 6: 118, 126. Ðæt heó dó ðæt ðæt heó ǽr dyde *that she may do that which she before did,* Bt. 25; Fox 88, 35, 36. Hwæt dó we ðæt we wyrceon Godes weorc *quid faciēmus ut operēmur opěra Dei?* Jn. Bos. 6, 28: Exon. 49 b; Th. 372, 28; Seel. 99. Hwæt dó ge máre *quid amplius facĭtis?* Mt. Bos. 5, 47. Ðeáh hí wom dón *though they commit sin,* Exon. 81 a; Th. 304, 15; Fä. 70: Cd. 109; Th. 145, 26; Gen. 2411: Ps. Th. 95, 7. ¶ Dón dǽdbóte *to do penance, repent,* Mt. Bos. 3, 2: 4, 17: 11, 20, 21: 12, 41. Dón edleán *to give a reward,* Boutr. Scrd. 22, 37. Dón fram *to depart,* Ps. Lamb. 17, 22. Dón in *to put in* or *into,* Bd. 2, 3; S. 504, 33: L. M. 1, 1; Lchdm. ii. 22, 13: Cd. 100; Th. 248, 31; Dan. 521. Dón neóde *to supply want,* Basil. admn. 4; Norm. 40, 29. Dón preóste *to give to a priest,* L. Edg. 1, 2: Th. i. 262, 15. Dón of *to take off, doff,* L. M. 1, 36; Lchdm. ii. 86, 15: Beo. Th. 5610; B. 2809. Dón on *to put on, in,* or *into, to don,* L. M. 1, 1; Lchdm. ii. 18, 13; 24, 1: 1, 2; Lchdm. ii. 30, 5; 32, 14, 15, 17, 21; Herb. 1, 7; Lchdm. i. 72, 21: 2, 7; Lchdm. i. 82, 12: 13, 2; Lchdm. i. 104, 23: Beo. Th. 2293; B. 1144: 6307; B. 3164: Elen. Kmbl. 2348; El. 1175: Exon. 88 b; Th. 332, 19; Vy. 87: Hy. 9, 55; Hy. Grn. ii. 292, 55: Hy. Bos. 9, 16, 17. Dón to *to put to,* Past. 49, 2; Hat. MS.: L. M. 1, 2; Lchdm. ii. 28, 15. Dón to witanne *to do to wit, to make to know* or *understand,* Past. 46, 8; Hat. MS. 68 a, 12: Prov. Kmbl. 11. Betre dón *to prefer,* Bd. 2, 2; S. 502, 15. For náuht dón *to consider as naught,* Past. 38, 1; Hat. MS. 50 b, 19: Lev. 26, 15:

Deut. 31, 16. Furðor dôn *to prefer, esteem,* Past. 17. 7; Hat. MS. 23 b, 14. Gifta dôn *to keep nuptials,* Somn. 186; Lchdm. iii. 208, 21. Huntaþ dôn *to be hunting,* 239; Lchdm. iii. 212, 3. Gode dôn *to render to God,* L. Edg. C. 54; Th. ii. 256, 2. Gŷmen [MS. gyman] dôn *to take care, regard,* Ors. 3, 9; Bos. 68, 25. Munuclíf dôn *to lead a monastic life,* Bd. 4, 23; S. 593, 19. On wôh dôn *to pervert,* Past. 2, 1; Cot. MS. To cyninge dôn *to make a king,* Ors. 6, 4; Bos. 118, 25: Bt. Met. Fox 15, 26; Met. 15, 13. Wrace dôn *to take revenge,* L. In. 9; Th. i. 108, 4. [*Prompt.* dooñ': *Wyc.* don, doon: *Piers P.* doon: *Chauc.* do *to cause:* *Laym.* don, do: *Orm.* don: *Plat.* doon: *O. Sax.* dôn, duôn, duan, dôan: *Frs.* dwaen, dien: *O. Frs.* dua: *Dut.* doen: *Ger.* thuen, thun: *M. H. Ger.* tuon: *O. H. Ger.* tuoan, tuon: *Sansk.* dhâ *ponĕre.*] DER. a-dôn, be-, for-, ge-, iu-ge-, of-, of-a-, ôfer-, on-, on-ge-, ôþ-, to-, to-ge-, un-, under-, up-a-, ût-a-.

Dona-feld; *gen.* -feldes; *dat.* -felde, -felda; *m.* TANFIELD, *near Ripon, Yorkshire;* Campodônum in agro Eboracensi:—On Donafelda, ðǽr wæs ðâ cyninges botl, hêt Eádwine ðǽr cyricean getimbrian *in Campodōno, ubi tunc etiam villa rēgia erat, Æduïni rex fecit basīlicam,* Bd. 2, 14; S. 518, 17.

dôn-líc; *adj. Active;* practicus =πρακτικός, Cot. 149.

Donua; *indecl. f. The river Danube;* Danūbius =Δανούβιος:—Sûþ ôþ Donua ða eá, ðǽre ǽwylme is neáh ðǽre eá Rînes *south to the river Danube, whose spring is near the river Rhine,* Ors. 1, 1; Bos. 18, 24, 29. On ôðre healfe ðǽre eá Donua *on the other side of the river Danube,* 1, 1; Bos. 18, 31, 43.

dooc *the south wind;* notus, auster, Som. Ben. Lye.

dop-enid, -ænid, e; *f.* [ened *a duck*] *A dipping-duck, a moorhen, fen-duck, coot;* fulica, fulix:—Dop-enid *fulĭca,* Ælfc. Gl. 38; Som. 63, 30; Wrt. Voc. 29, 50. Ganot, dop-ænid *fulix,* Glos. Epnl. Recd. 156, 53.

dop-fugel, es; *m. A dipping-fowl, a water-fowl, a moorhen;* mergus, mergŭlus:—Dop-fugel *mergus,* Wrt. Voc. 280, 12. Dop-fugel *mergŭlus,* Glos. Brux. Recd. 36, 6; Wrt. Voc. 62, 6.

doppettan; *p.* te; *pp.* ed *To dip often, dip in, immerse;* mersāre:—Geseah he swymman scealfran on flôde, and gelôme doppettan adûne to grunde, êhtende þearle ðǽre eá fixa *he saw gulls swimming on the water, and frequently dipping down to the bottom, eagerly pursuing the fishes of the river,* Homl. Th. ii. 516, 7. Ic doppette *merso,* Ælfc. Gr. 36; Som. 38, 21. v. dyppan.

Dor, es; *m.* DORE, *Derbyshire;* loci nomen in agro Derbiensi:—Ecgbryht Wesseaxna cyning lǽdde fierd to Dore wið Hymbre *Egbert king of the West Saxons led an army to Dore against the Northumbrians,* Chr. 827; Erl. 64, 7.

DÔR, es; *pl. nom. acc.* dôr, dôru, dûru; *n. A large door;* porta:—Ðæt ðû ðíne dôru mihtest bedôn fæste *that thou mightest shut fast thy doors,* Ps. Th. 147, 2. Gâþ nû on his dôru *intrāte portas ejus,* Ps. Th. 99, 3. Hôh ða wyrte on ðam [MS. ðan] dôre *hang the herbs on the door,* Lchdm. iii. 56, 29. Forðon he ǽren dôr eáde gescêneþ [MS. gesceeneþ] *quia contrīvit portas æreas,* Ps. Th. 106, 15. Dûru *doors,* Exon. 97 b; Th. 364, 29. [*Prompt.* dore: *Wyc. Piers P. Chauc.* dore: *Laym.* dure, dore: *Plat.* döre *a door;* door *a gate: O. Sax.* dor, *n. a door, gate: Frs.* doare, doar: *O. Frs.* dore, dure *a door: Ger.* thüre, *f. a door;* thor, *n. a gate: Goth.* daúr, *n;* daúro, *f: Dan.* dör, *n: Swed.* dörr, *f: Icel.* dyrr, *f: O. Nrs.* dyrr, *n: Grk.* θύρα: *Sansk.* dvâr, *f;* dvâra, *n.*] DER. Fífel-dôr, hel-, helle, weall-. v. dûru, *f.*

dora, an; *m. A humble-bee, dumble-*DORE; bombus terrestris, attăcus =ἄττάκος:—Dora *atticus* [=attăcus] vel *burdo* [=*Fr.* bourdon], Ælfc. Gl. 22; Som. 59, 112; Wrt. Voc. 23, 68. Doran hunig *dumbledore's honey,* L. M. 1, 2; Lchdm. ii. 28, 20. Celeþenian seáw gemeng wið dorena hunig *mingle juice of celandine with dumbledores' honey,* 1, 2; Lchdm. ii. 26, 7. Ða ahsan gemenge wið dorena hunig *mix the ashes with dumbledores' honey,* Lchdm. ii. 28, 26.

Dorce-ceaster, Dorces ceaster, Dorca-ceaster, Dorceaster; *gen.* -ceastre; *f.* [*Bd.* Dorcinca, Dorcic: *Hunt.* Dorecestre: *Brom.* Dorke-cestre: *Matt. West.* Dorcestre] DORCHESTER, *Oxfordshire, the episcopal seat of the first bishop of the West Saxons, which was subsequently removed to Lincoln;* Durocastrum, in agri Oxoniensis parte Berceriensi finitima:—Hêr Cynegils [MS. Kynegils] wæs gefullod fram Byríne ðam biscope on Dorcaceastre *in this year* [A. D. 635] *Cynegils was baptized at Dorchester by bishop Birinus,* Chr. 635; Th. 47, 4, col. 1. Hêr wæs Cwichelm gefullod on Dorcceastre [Dorces ceastre, Th. 46, 10, col. 1] *in this year* [A. D. 636] *Cwichelm was baptized at Dorchester,* 636; Th. 47, 9, col. 1: 639; Th. 46, 18, col. 2; 47, 17, col. 1. Æt Dorceceastre [Dorceastre, Th. 175, 28, col. 2] *at Dorchester,* 897; Th. 174, 31, col. 1, 2; 175, 27, col. 1. Geáfon ðam bisceope begen ða cyningas eardungstôwe and biscopsetl on Dorceceastre *both the kings* [*Cynegils of the West Saxons and Oswald of the Northumbrians*] *gave the bishop* [*Birinus*] *a dwelling-place and episcopal see at Dorchester,* Bd. 3, 7; S. 529, 20. Ætla wæs on Dorcceastre to bisceope gehâlgod *Ætla was consecrated bishop of Dorchester,* 4, 23; S. 594, 11. Hêr Wulstan arcebiscop onfêng eft biscopríces, on Dorceceastre *in this year* [A. D. 954]

archbishop *Wulfstan again received a bishopric, at Dorchester,* Chr. 954; Th. 215, 26, col. 1.

dorfen *laboured, perished;* *pp. of* deorfan. v. ge-deorfan.

Dorm-ceaster; *gen.* -ceastre; *f.* [*by the Britons called Cair-Dorm, by Antonïnus* Durobrivæ, *from the passage over the water; and the Anglo-Saxons, for the same reason, called it also Dornford*] *Dornford* or *Dorgford, in Huntingdonshire, on the river Nen,* Som. Ben. Lye.

Dorn-sǽte, Dor-sǽte; *gen.* -sǽta; *dat.* -sǽtum, -sǽton, -sǽtan; *pl. m.* [dor =*Celt.* dwr, dur *water;* -sǽte *dwellers, inhabitants: dwellers by water*] *Inhabitants* or *men of Dorsetshire, people of Dorsetshire in a body,* DORSETSHIRE; Dorsetenses, Dorsetia:—Ðŷ ilcan geáre gefeaht Æðelhelm wið Deniscne here mid Dornsǽtum [Dorsǽtan, Th. 118, 17, col. 2; Dorsǽton, 119, 17, col. 1; Dorsǽtum, 119, 16, col. 2] *in the same year* [A. D. 837] *Æthelhelm fought against the Danish army with the Dorset-men,* Chr. 837; Th. 118, 17, col. 1. Mid Dornsǽtum [Dorsǽtum, Th. 120, 12, col. 2, 3; Dorsǽton, 121, 11, col. 1, 2, 3] *with the Dorset-men,* 845; Th. 120, 12, 36. Alfwold wæs bisceop on Dorsǽtum *Alfwold was bishop of Dorset,* 978; Th. 232, 7, col. 1: 982; Th. 234, 38: 236, 8: 1015; Th. 276, 13; 277, 13. Hí up eódon into Dorsǽton [Dorsǽtan, Th. 247, 19] *they went up into Dorsetshire,* 998; Th. 246, 19: Cod. Dipl. 1302: A. D. 1006; Kmbl. vi. 155, 6: 1334; A. D. 1046; Kmbl. vi. 195, 31. On Dorsǽtan *in Dorsetshire,* Cod. Dipl. 841; Kmbl. iv. 200, 26: 871; Kmbl. iv. 221, 5: Chr. 1078; Th. 350, 17.

Dornwara ceaster; *gen.* ceastre; *f.* [*the city of the inhabitants of Dorsetshire*] DORCHESTER, *the chief town of Dorsetshire;* Dorcestria, agri Dorsetensi caput:—Ðis wæs gedôn in ðam cynelícan setle on ðǽre stôwe ðe is genǽmned Dornwara ceaster *this was done in the royal residence in the place which is named Dorchester,* Th. Diplm. A. D. 864; 126, 8: Cod. Dipl. 1061; A. D. 868; Kmbl. v. 119, 26.

dorste, *pl.* dorston *durst,* Ors. 1, 10; Bos. 33, 30: 4, 11; Bos. 97, 14; *p. of* durran.

Dorwit-ceaster; *gen.* -ceastre; *f. Canterbury;* Dorobernia:—Hrôfes ceaster is xxiv míla fram Dorwitceastre *Rochester is twenty-four miles from Canterbury,* Chr. 604; Erl. 21, 24.

dott, es; *m. A* DOT, *small spot, speck;* punctum:—Geopenige mon ðone dott, and binde ðone clîðan to ðan swyle *let the speck* [*at the head of a boil*] *be opened, and the poultice be bound to the swelling,* Lchdm. iii. 40, 14.

drabbe *dregs, lees,* DRAB; fæces, Som. Ben. Lye. [*Prompt.* draffe *segestārium, drascum: Wyc.* draf *dreg, refuse;* draffis *dregs: Piers P. Chauc. Laym.* draf *dregs: Dut.* draf, *m.*]

DRACA, an; *m.* **I.** *a dragon;* draco:—Draca ðes ðone ðû ywodest *draco iste quem formasti,* Ps. Spl. 103, 28. Tredan león and dracan *conculcāre leonem et dracōnem,* Ps. Th. 90, 13. Ðû fortrydst leóna and dracena *thou shalt be a treader down of lions and dragons,* Ps. Spl. 90, 13: Ps. Th. 148, 7. **II.** *a serpent;* serpens:—Is ðæt deór pandher, se is æt-hwǽm freónd, bûtan dracan ânum *the beast is the panther, which is to each a friend, save to the serpent only,* Exon. 95 b; Th. 356, 24; Pa. 16. **III.** *the serpent = the devil;* diabŏlus:—Worpaþ hine deófol, draca egeslíce *the devil, the fearful dragon, shall cast him down,* Salm. Kmbl. 52; Sal. 26: Exon. 96 a: Th. 359, 4; Pa. 57. [*R. Glouc.* dragon: *Laym.* drake, *m: Orm.* drake: *Plat.* drake, *m: Dut.* draak, *m: Ger.* drache, *m: M. H. Ger.* trache, tracke, *m: O. H. Ger.* tracho, *m: Dan.* drage, *m. f: Swed.* drake, *m: Icel.* dreki, *m: Fr.* dragon, *m: Span.* dragón, *m: Ital.* dragóne, *m: Lat.* draco: *Grk.* δράκων *a dragon, from* δέρκομαι *to flash, gleam.*] DER. eorþ-draca, fŷr-, lêg-, líg-, níþ-, sǽ-.

dracan blôd, es; *n. Dragon's blood, a pigment obtained from the dragon's blood-tree;* cinnabăris =κιννάβᾰρι, Cot. 210. v. dracentse.

dracentse, dracente, dracanse, draconze, an; *f. Dragon-wort, dragons;* dracontea =δρακόντιον, arum dracuncŭlus, Lin:—*Herba dracontea, ðæt ys* dracentse, Herb. Cont. 15, 1; Lchdm. i. 12; 15, 1. Deós wyrt, ðe man dracontea *and ôðrum naman* dracentse *nemneþ, ys sǽd ðæt heó of dracan blôde acenned beón sceolde this herb, which is named* dracontea, *and by another name dragons, is said to be produced from dragon's blood,* Herb. 15, 1; Lchdm. i. 106, 22. Nim dracentan wyrtruman [MS. wyrtruma] *take roots of dragons,* Lchdm. iii. 114, 8. Dracanse *dragons,* iii. 24, 3. Draconzan, *acc. dragons,* L. M. 3, 62; Lchdm. ii. 350, 7.

DRÆDAN; *ic* drǽde, *ðû* drǽdest, drǽtest, drǽst, *he* drǽdeþ, drǽt, *pl.* drǽdaþ; *p.* drêd, dreórd, *pl.* drêdon; *pp.* drǽden *To* DREAD, *fear;* timēre, pavēre: *found in the compounds* a-drǽdan, an-, on-drǽdan, on-drǽd-endlíc, on-drǽd-ing: of-drǽd. [*Wyc.* drede, dreed: *Piers P. Chauc.* drede: *Laym.* drede: *Orm.* drædenn, dredenn: *O. Sax.* ant-drâdan, an-drâdan: *M. H. Ger.* en-trâten: *O. H. Ger.* an-trâtan.]

drǽf, drâf, e; *f. A driving out, an expulsion;* expulsio:—Be drǽfe [drâfe MS. B.] *of expulsion,* L. In. 68; Th. i. 146, 6. DER. ût-drǽf.

drǽfan; *p.* de; *pp.* ed *To drive;* agĕre, pellĕre. DER. a-drǽfan, ge-, to-, ge-drǽfnes, to-drǽfednes, ût-drǽfere. v. drífan.

drǽfend, es; *m. A hunter;* venātor:—Sum biþ deóra drǽfend *one is a hunter of beasts,* Exon. 78 b; Th. 295, 24; Crä. 38.

drǽge, es; *n? A* DRAG, *drag-net;* tragŭla, verricŭlum:—Drǽg-net

P

vel dræge *tragŭla*, Ælfc. Gl. 1; Som. 55, 13; Wrt. Voc. 15, 13. **Dræge** *tragŭla* vel *verricŭlum*, 105; Som. 78, 40; Wrt. Voc. 57, 22.

drægeþ, ðú drægest *drags, thou draggest; 3rd and 2nd pers. pres. of* dragan.

dræg-net, -nett, es; *n. A drag-net;* tragum, verricŭlum :—Dræg-net *vel* dræge *tragŭla*, Ælfc. Gl. 1; Som. 55, 13; Wrt. Voc. 15, 13. Dræg-net *verricŭlum*, 84; Som. 73, 89; Wrt. Voc. 48, 27.

drægþ, ðú drægst *drags, thou draggest*, Past. 56, 2; Hat. MS; *3rd and 2nd pers. pres. of* dragan.

dræhþ, ðú dræhst *drags, thou draggest; 3rd and 2nd pers. pres. of* dragan.

dræn *a drone*, Wrt. Voc. 77, 48. v. drán.

drænc *a drink*, L. M. I. P. 10; Th. ii. 268, 6. v. drinc.

dræp, ðú dræpe, *pl.* dræpon *struck; p. of* drepan.

drǽtest, drǽst, he drǽt *dreadest, dreads; 2nd and 3rd pers. pres. of* drǽdan.

drǽf, e; *f.* [dráf *drove, p. of* drífan] *A* DROVE, *herd, band;* armenta, grex, agmen :—Ðá ðá seó ormǽte micelnyss his orfes on ðære dúne læswede, sum módig fearr wearþ ángencga, and ðære heorde dráfe oferhogode *when the immense multitude of his cattle was grazing on the mountain, an unruly bull wandered alone, and despised the companionship of the herd,* Homl. Th. i. 502, 10. Oft twegen sǽmen oððe þrý hwílum drífaþ ða dráfe cristenra manna fram sǽ to sǽ *sæpe duo tresve e pirātis christiānorum agmen congregātum a mari usque ad mare compellunt,* Lupi Serm. i. 15; Hick. Thes. ii. 103, 34. Hí drífon heora dráfa into Medewæge *they drove their herds into the Medway,* Chr. 1016; Erl. 157, 4, 16.

dráf *drove,* Chr. 1099; Ing. 318, 16; *p. of* drífan.

DRAGAN, ic drage, ðú drægest, drægst, drǽhst, he drægeþ, drægþ, dræhþ, *pl.* dragaþ; *p.* dróg, dróh, *pl.* drógon; *pp.* dragen. I. *v. a.* *To* DRAG, *draw;* trahĕre :—Eall ðæt ða beón dragen toward ða dráne dragaþ fraward *all that the bees draw towards them the drones draw from them,* Chr. 1127; Th. 378, 24, 25. Simon Petrus dróg ðæt nett on eorþe *Simon Petrus traxit rete in terram,* Jn. Lind. War. 21, 11. Hí me drógon, and ic hit nyste... hit mon drægþ swá hit ne gefret *traxērunt me et ego non sensi... trahĭtur et nequaquam sentit,* Past. 56, 2; Hat. MS. Hí drógon heora scipa on west-healfe ðære brycge *they dragged their ships to the west side of the bridge,* Chr. 1016; Erl. 155, 9, 23. II. *v. intrans. To draw oneself, to draw, go;* se conferre, ire :—Drógon swá wíde swá wegas to lǽgon *they went as far as the roads lay before them,* Andr. Kmbl. 2465; An. 1234. Ongon dragan Dryhtnes cempa *the Lord's champion began to go,* Exon. 43 a; Th. 145, 23; Gú. 699. [*Wyc.* drow, droȝ, drowȝ *drew: Laym.* draȝen, drawe *to draw: Orm.* draghenn *to draw: Plat.* drägen *to bear, endure: O. Sax.* dragan *to bear: Frs.* dreagjen, dreagen, dreyn: *O. Frs.* drega, dreaga *to bear: Dut.* dragen *to bear: Ger. M. H. Ger.* tragen *to bear, endure: O. H. Ger.* tragan *portāre: Goth.* dragan *to carry: Dan.* drage *to draw, carry: Swed.* draga *to wear: Icel.* draga *to drag, carry: Lat.* trahĕre *to pull.*] DER. be-dragan, út-.

DRÁN, drǽn, e; *f. A* DRONE; fucus :—Drán *fucus,* Ælfc. Gl. 22; Som. 59, 106; Wrt. Voc. 23, 62. Drǽn *fucus,* Wrt. Voc. 77, 48. Ðǽr he wunede eall riht swá dráne dóþ on híue: eall ðæt ða beón dragen toward ða dráne dragaþ fraward *he abode there just as drones do in a hive: all that the bees draw towards them the drones draw from them,* Chr. 1127; Erl. 256, 20, 21. [*Piers P.* drane: *Plat.* drone: *O. Sax.* drán, *f. fucus: Ger.* drone, thräne, *f;* dran, *m. fucus: M. H. Ger.* tren, *m. fucus: O. H. Ger.* treno, *m. attăcus, fucus: Dan.* drone, *m. f: Swed.* drönje, drön-are, *m: Grk.* ἀν-θρήν-η, *f. a hornet, bee: Sansk.* druṇa, *m. a bee;* dhraṇ *to sound.*]

dranc *drank,* Gen. 9, 21; *p. of* drincan.

dreá *a magician, wizard,* Salm. Kmbl. 89, MS. A; Sal. 44. v. drý.

dreág, dreáh *did, suffered,* Exon. 74 b; Th. 280, 9; Jul. 626: Cd. 145; Th. 180, 22; Exod. 49; *p. of* dreógan.

dreahnian; *p.* ode; *pp.* od *To strain out, drain;* excolāre :—Dreahna út þurh wyllene cláþ *drain [it] out through a woollen cloth,* Lchdm. iii. 72, 23. v. drehnigean.

dreahte, ðú dreahtest, *pl.* dreahton; *pp.* dreaht *Vexed, vexedst, troubled,* Exon. 98 a; Th. 368, 6; Seel. 17; *p. and pp. of* dreccan.

DREÁM, es; *m.* I. *joy, pleasure, gladness, mirth, rejoicing, rapture, ecstasy, frenzy;* jūbĭlum, lætĭtia, gaudium, delīrium :—Ðǽr biþ drincendra dreám se micla *there is the great joy of drinkers,* Exon. 88 a; Th. 332, 3; Vy. 79: Beo. Th. 999; B. 497: Cd. 169; Th. 211, 25; Exod. 531. Ðǽr biþ engla dreám *there [in heaven] is joy of angels,* Exon. 32 b; Th. 102, 22; Cri. 1676: Elen. Kmbl. 2461; El. 1232: Apstls. Kmbl. 96; Ap. 48. Ic eam ealles leás ecan dreámes *I am bereft of all eternal joy,* Cd. 216; Th. 275, 8; Sat. 168: 217; Th. 276, 2; Sat. 182: Exon. 27 b; Th. 82, 24; Cri. 1343: Rood Kmbl. 285; Kr. 144. In dolum dreáme *in foolish joy,* Exon. 39 a; Th. 130, 8; Gú. 435. In ðam uplícan engla dreáme *in the exalted joy of angels,* 9 a; Th. 7, 17; Cri. 102. He dreám gehýrde hlúdne in healle *he heard loud mirth in the hall,* Beo. Th. 177; B. 88. Sorh cymeþ in manna dreám *sorrow cometh into the joy of men,* Frag. Kmbl. 3; Leás. 2: Exon. 35 a;

Th. 114, 2; Gú. 166. Heó móton ágan dreáma dreám mid Gode *they may possess joy of joys with God,* Cd. 220; Th. 283, 32; Sat. 314: Exon. 16 a; Th. 36, 22; Cri. 580: Apstls. Kmbl. 163; Ap. 82. Eart ðú dumb and deáf, ne sindan ðíne dreámas wiht *thou art dumb and deaf, thy pleasures are naught,* Exon. 99 a; Th. 370, 27; Seel. 65. Dreáma leás *void of joys, joyless,* Beo. Th. 1705; B. 850: Cd. 2; Th. 3, 23; Gen. 40: 5; Th. 7, 18; Gen. 108. Ic dreáma wyn sceal ágan mid englum *I shall possess joy of joys with angels,* Exon. 42 b; Th. 142, 31; Gú. 652. Hie forþ heónon gewiton of worulde dreámum *they have departed hence from the world's joys,* Rood Kmbl. 263; Kr. 133: Exon. 43 b; Th. 146, 19; Gú. 712. Hér ge-endode eorþan dreámas Eádgár Engla cyning *in this year* [A. D. 975] *Edgar, king of the Angles, ended the pleasures of earth,* Chr. 975; Erl. 124, 29; Edg. 21: Exon. 32 b; Th. 102, 5; Cri. 1668. Sécan mid sibbe swegles dreámas *to seek in peace the joys of heaven,* Andr. Kmbl. 1618; An. 810: Cd. 14; Th. 17, 9; Gen. 257: Exon. 26 a; Th. 76, 28; Cri. 1246: Judth. 12; Thw. 26, 31; Jud. 350. On swylcum wódum dreáme *in such insane ecstasy or frenzy,* Ors. 3, 6; Bos. 58, 14: Homl. Th. i. 524, 34 : 526, 1: ii. 50, 28: 110, 18, 31. II. *what causes mirth,—An instrument of music, music, rapturous music, harmony, melody, song;* organum = ὄργανον, musĭca, concentus, harmŏnia = ἁρμονία, modulātio, modus, melōdia = μελῳδία, cantus :—Ne mágon ðam breahtme býman ne hornas, ne hearpan hlyn, ne organan swég, ne ǽnig ðara dreáma ðe Dryhten gescóp gumum to gliwe in ðas geómran woruld *trumpets nor horns can [equal] that sound, nor sound of harp, nor organ's tone, nor any of those kinds of music which the Lord hath created for delight to men in this sad world,* Exon. 57 b; Th. 206, 29–207, 10; Ph. 134–139. On saligum we ahófon oððe ahéngon dreámas úre *in salicibus suspendĭmus orgăna nostra,* Ps. Lamb. 136, 2. Sǽde se engel ðæt se dreám wǽre of ðam upplícum werode *the angel said that the melody was from the celestial host,* Homl. Th. ii. 342, 10: Exon. 52 a; Th. 181, 9; Gú. 1290. Werhádes men ongunnon symle ðone dreám, and wífhádes men him sungon ongeán andswariende *men always begun the melody, and women answering sung in turn,* Homl. Th. ii. 548, 12: Cd. 220; Th. 284, 28; Sat. 328. Iohannes gehýrde swylce býmena dreám *John heard, as it were, the sound of trumpets,* Homl. Th. ii. 86, 35. Dreáme *harmŏnia, modulatiōne,* Mone B. 2528, 2529. Dreámas *concentus,* 4940. Dreámum *modis,* Glos. Prudent. Recd. 143, 9. [*Laym,* dræm, dream, drem, *m. joy, rejoicing: Orm.* dræm *sound.*] DER. dreám-cræft, -ere, -hæbbende, -healdende, -leás, -líc, -nes, -swinsung: dréman, drýman, freá-: dréme, drýme, ge-, unge-: éðel-dreám, gleó-, god-, gum-, heofon-, man-, medu-, sele-, sin-, swegl-, woruld-, wuldor-, wyn-.

dreám-cræft, es; *m. The art of music, music;* musĭca :—Gedeþ se dreámcræft ðæt se mon biþ dreámere *the art of music causes the man to be a musician,* Bt. 16, 3; Fox 54, 31.

dreámere, es; *m. A musician;* musĭcus, Bt. 16, 3; Fox 54, 31.

dreám-hæbbende; *part.* [dreám I. joy, hæbbende *having, possessing*] *Possessing bliss, joyful;* lætābundus :—Þrymmas weóxon dreámhæbbendra *the glories of the possessors of bliss increased,* Cd. 4; Th. 5, 34; Gen. 81.

dreám-healdende; *part.* [healdende *holding*] *Holding joy, joyful;* lætābundus :—Beó ðú sunum mínum gedéfe, dreámhealdende *be thou gentle to my sons, holding them in joy,* Beo. Th. 2459; B. 1227.

dreám-leás; *adj. Joyless, sad;* mæstus :—Dreámleás gebád *he continued joyless,* Beo. Th. 3445; B. 1720: Cd. 202; Th. 251, 4; Dan. 558. Ðis is dreámleás hús *this is a joyless house,* Exon. 31 b; Th. 99, 22; Cri. 1628.

dreám-líc; *def.* se -líca, seó, ðæt -líce; *adj. Joyous, musical;* jucundus, musĭcus :—Dreámlíc oððe wynsum sý him spæc [MS. spæce] mín *jucundum sit ei eloquium meum,* Ps. Lamb. 103, 34. Ða dreámlícan musĭca, Cot. 133.

dreámnes, -ness, e; *f. A singing;* cantio :—Word dreámnessa oððe sanga *verba cantiōnum,* Ps. Lamb. 136, 3.

dreám-swinsung *mirth-harmony, harmony,* Cot. 4. v. swinsung.

dreáp, *pl.* drupon *dropped; p. of* dreópan.

dreápian *to drop,* Ps. Surt. 67, 9. v. dreópian.

dreárung, e; *f. A falling;* destillātio, Cd. 191; Th. 238, 3; Dan. 349. v. dreórung.

dreás *rushed, fell; p. of* dreósan.

dreás *soothsayers;* hariŏli, Prov. 23, = drýas; *pl. nom. of* drý.

DRECCAN, dreccean, drecan, ic drecce, drece, ðú drecest, drecst, he dreceþ, drecþ, *pl.* dreccaþ, drecceaþ; *p.* [drechede = drehde =] drehte, dreahte, *pl.* drehton, dreahton; *pp.* [dreched = drehed = dreht, dreaht] dreht, dreaht *To vex, afflict, trouble, torture, torment;* vexāre, afflīgĕre, tribulāre, turbāre, cruciāre :—Mec sorg dreceþ *sorrow vexeth me,* Cd. 99; Th. 131, 21; Gen. 2179. Drecþ se deófol mancynn mid mislicum costnungum *the devil vexes mankind with various temptations,* Boutr. Scrd. 19, 44. Me Agar drehte dógora gehwam *Hagar hath vexed me each day,* Cd. 102; Th. 135, 27; Gen. 2249. Yrfweardnysse ðíne hí drehton *hæreditātem tuam vexavērunt,* Ps. Spl. 93, 5: Chr. 897; Erl. 95, 7. Ic drece *vexo,* Ælfc. Gr. 24; Som. 25, 44. Ðeáh hine se ymbhoga ðyssa woruldsǽlþa wráðe drecce *though the anxious care of*

these worldly goods severely afflicts him, Bt. Met. Fox 7, 108; Met. 7, 54: Homl. Th. i. 156, 21. Ne wendaþ hine wyrda, ne hine wiht dreceþ *fates change him not, nor doth aught afflict him*, Exon. 88 b; Th. 334, 1; Gn. Ex. 9: Bt. Met. Fox 7, 50; Met. 7, 25. Ðonne mîne fýnd me drecceaþ *dum affligit me inimicus*, Ps. Th. 42, 2. Ic ðé bebeóde ðæt ðû nânum men ne drece *I command thee that thou afflict no man*, Homl. Th. ii. 296, 5. On ðam ecan lîfe ðǽr ne cymþ nán deófol ne nán yfel mann, ðe us mǽge dreccan *in the eternal life there will come no devil nor evil man who may trouble us*, i. 272, 10. Hwî drecst ðû leng ðone lâreów *why troublest thou the master longer?* Mk. Bos. 5, 35. Hî hine dreccaþ *they trouble him*, Ps. Th. arg. 25: Homl. Th. ii. 540, 34. To hwon dreahtest ðû me *for what [why] hast thou tortured me?* Exon. 98 a; Th. 368, 6; Seel. 17. Gif hine dreccean môt ðissa yfla hwæðer *if either of these evils can torment it*, Bt. Met. Fox 5, 80; Met. 5, 40. [*Piers P.* drecchen *to vex: Chauc.* drecche: *Laym.* i-dræcched, -dracched, -drecched, *pp. injured, disturbed.*] DER. ge-dreccan.

dreccednys, -nyss, e; *f. Vexation, affliction, tribulation;* vexātio, afflictio, tribulātio:—He ðære dreccednysse geswác *he ceased the affliction*, Homl. Th. i. 454, 28. DER. ge-dreccednys.

dreccing, e; *f. Tribulation;* vexātio, Som. Ben. Lye.

drêd, *pl.* drêdon *dreaded, feared; p. of* drǽdan.

DRÉFAN; *part.* drêfende; *p.* drêfde; *pp.* drêfed *To disturb, agitate, disquiet, vex, trouble;* commovēre, turbāre, conturbāre, tribulāre, contristāre:—Uparǽr môd ûre drêfende *erige mentes nostras turbidas*, Hymn. Surt. 127, 6. Ðonne ic wado drêfe *when I disturb the waters*, Exon. 103 a; Th. 389, 24; Rä. 8, 2. Ðû drêfst hî *turbābis eos*, Ps. Spl. 82, 14. For-hwý unrôt eart sâwle mîn, and for-hwon drêfst me *quare tristis es anima mea, et quare conturbas me?* Ps. Spl. 41, 6, 15: 42, 5. Drêfaþ *conturbant*, Mone B. 2613. Ne lagu drêfde *it disturbed not the water*, Exon. 106 a; Th. 404, 31; Rä. 23, 16. Ðæt ðû lagu drêfde *that thou mightest disturb the water*, Exon. 123 a; Th. 473, 26; Bo. 20. Gewât him on nacan, drêfan deóp wæter *he departed in the bark, to agitate the deep water*, Beo. Th. 3812; B. 1904. Hwý ge scylen eówer môd drêfan *why should ye trouble your mind?* Bt. Met. Fox 27, 3; Met. 27, 2. He to nâhte gelǽdeþ ða drêfendan us *ipse ad nihilum deducet tribulantes nos*, Ps. Spl. 59, 13. To-hwý gemænigfylde synd ða ðe drêfaþ me *quid multiplicāti sunt qui tribūlant me?* Ps. Spl. 3, 1. For-hwî drêfe ge eówru môd *why vex ye your minds?* Bt. 39, 1; Fox 210, 24. For-hwý drêfed ic gange, ðonne swencþ me feónd *quare contristātus incēdo, dum affligit me inimicus?* Ps. Spl. 41, 13. [*Laym.* i-drêfed, *pp. disturbed;* to-drefed, -dreved *oppressed: Orm.* dræsedd, dreofedd, drefedd *disturbed, troubled: Plat.* dröven: *O. Sax.* drôbian, druovan *turbāri, conturbāre: Kil.* droeven *tristāri, turbāre: Ger.* trüben: *M. H. Ger.* trüeben: *O. H. Ger.* truobian: *Goth.* drobyan *to trouble, confound: Dan.* be-dröve: *Swed.* be-dröfva.] DER. ge-drêfan, to-: un-drêfed. v. dróf.

drêfednes, -ness, -nyss, e; *f. Vexation, affliction, tribulation;* vexātio, afflictio, tribulātio:—Syððon cômon [comen MS.] ealle drêfednysse [MS. dræuednysse] and ealle ifele to ðone mynstre *after that all troubles and all evils came to the monastery*, Chr. 1066; Erl. 203, 31. DER. ge-drêfednes.

drêfing, e; *f. A disturbing;* conturbātio, Ælfc. Gl. 5; Som. 56, 24; Wrt. Voc. 17, 29.

drêfliende, *part. Troubled with rheum;* rheumaticus = ῥευματικός:—Saftriende *vel* drêfliende *rheumaticus*, Ælfc. Gl. 77; Som. 72, 14; Wrt. Voc. 45, 48.

drêfre; *adj. Agitated, disturbed;* turbulentus, C. R. Ben. 64. v. dróf.

drege *dry*, Prov. 16. v. drige.

drehnigean, drehnian, dreahnian; *p.* ode; *pp.* od *To strain out,* DRAIN; excolāre, percolāre:—Lâ blindan lâtteówas, ge drehnigeaþ ðone gnæt aweg *duces cæci, excolantes culicem*, Mt. Bos. 23, 24.

drehte, *pl.* drehton; *pp.* dreht *Vexed, afflicted*, Cd. 102; Th. 135, 27; Gen. 2249: Ps. Spl. 93, 5; *p. of* dreccan.

drêman, drýman; *p.* de; *pp.* ed [dream *joy, music*] *To rejoice, to play on an instrument;* jubilāre, psallēre:—Drêmaþ Gode Iacobes *jubilāte Deo Iacob*, Ps. Spl. 80, 1. Drêmaþ oððe fægniaþ on gesihþe cyninges *jubilāte in conspectu regis*, Ps. Lamb. 97, 7. We drêmaþ mægnu ðînum *psallēmus virtūtes tuas*, Ps. Spl. 20, 13. Drêmaþ oððe singaþ cyninge ûrum *psallīte regi nostro*, Ps. Lamb. 46, 7: 97, 5. [*Laym.* dremen, dreomen *to revel, resound: O. Sax.* drômian *jubilāre.*] DER. freá-drêman.

drême, drýme; *adj.* [dream II. *music, melody, harmony*] *Melodious, harmonious;* canōrus:—Mid dremere stefne *canōra voce*, Mone B. 2538. DER. ge-drême, -drýme, unge-.

drenc, es; *m.* I. *a* DRENCH, *dose, draught, drink;* pōtus, pōtio:—Wið ûtsiht-âdle drenc *a dose for diarrhœa*, L. M. cont. 3, 22; Lchdm. ii. 300, 23. Drenc *pōtus*, Ælfc. Gr. 11; Som. 15, 16: Wrt. Voc. 82, 46: *pōtio*, 74, 7. Se drenc deáðbǽra wæs *the drink was deadly*, Homl. Th. ii. 158, 22. Wîn nys drenc cilda *vinum non est pōtus puerōrum*, Coll. Monast. Th. 35, 19: Homl. Th. ii. 158, 17. Wið sîdan sâre ðære swiðran hwîte clæfran wyrc to drence *for sore of right side make white clover to a drink*, L. M. 1, 21; Lchdm. ii. 64, 4: 1, 23;

Lchdm. ii. 64, 27: Homl. Th. ii. 158, 16. Wyrc drenc wið hwôstan *make a dose for cough*, L. M. 1, 15; Lchdm. ii. 56, 18. Sele him oft styrgende drenc *give him often a stirring drink*, 1, 42; Lchdm. ii. 106, 25. Se yrþling sylþ us hlâf and drenc *arātor dat nobis panem et potum*, Coll. Monast. Th. 31, 3. Hî ðone gâstlîcan drenc druncon *they drank the spiritual drink*, Homl. Th. ii. 202, 3. Drenc wyð ættre *a dose* or *antidote against poison;* theriāca = θηριακή, Ælfc. Gl. 12; Som. 57, 78; Wrt. Voc. 20, 20. Swylfende drenc *a dose to be gulped* or *swallowed down, a pill;* catapōtium = καταπότιον, 12; Som. 57, 80; Wrt. Voc. 20, 22. II. *a drowning;* demersio, submersio:—Sume drenc fornam on lagostreáme *drowning took off some in the water-stream*, Elên. Kmbl. 272; El. 136. Gæst in deáþ-sele drence bifæsteþ scipu mid scealcum *the guest commits ships and crews to the death-hall by drowning*, Exon. 97 a; Th. 362, 2; Wal. 30. DER. berig-drenc, dolh-, dust-, ofer-, wyrt-.

DRENCAN; *part.* drencende; *p.* ic, he drencte, ðû drenctest, *pl.* drencton; *pp.* drenced; *v. a.* I. *to give to drink, to* DRENCH, *make drunk;* potum *vel* potiōnem dāre, potāre, inebriāre:—Of burnan willan ðînes ðû drenctest [Th. drencst] hî *torrente voluntātis tuæ potābis eos*, Ps. Spl. 35, 9. Ðû drenctest us mid wîne *potasti nos vino*, 59, 3. On þurste mînum hî drencton me mid ecede *in siti mea potavērunt me acēto*, 68, 26. Drencende *inebrians*, 64, 11. Se inwida dryht-guman sîne drencte mid wîne *the wicked one made his people drunk with wine*, Judth. 10; Thw. 21, 21; Jud. 29. II. *to drown;* submergēre, Ps. Th. 106, 17. [*Wyc.* drenche: *Piers P.* drenchen, drenche: *Chauc.* drenche: *Plat.* drenken: *O. Sax.* drenkan: *Frs.* drinssen: *O. Frs.* drenka, drinka, drinsa *to drown: Dut.* drenken *to drench: Ger.* tränken *to give to drink: M. H. Ger.* trenken: *O. H. Ger.* trankjan, trenkjan *potāre: Goth.* dragkyan *to give to drink: Swed.* dränka *to drown: Icel.* drekkja *to drown.*] DER. a-drencan, for-, ge-, in-, ofer-, ofge-, on-. v. drincan.

drenc-cuppe, an; *f. A drinking-vessel, a cup;* pocūlum, Wrt. Voc. 82, 42.

drenc-fæt, es; *n.* [fæt *a vessel*] *A drinking-vessel, cup;* calix = κύλιξ:—Gâst ýsta oððe storma is dæl drencfætes heora oððe heora calices *spiritus procellārum est pars calicis eorum*, Ps. Lamb. 10, 7: 15, 5: 22, 5. v. drinc-fæt.

drenc-flôd, drence-flôd, es; *m.* [drenc II. *a drowning,* flôd *a flood*] *A drowning-flood, deluge;* dilūvium:—Noe oferlâþ ðone deópestan drencflôda [MS. dren-flôda] *Noah sailed over the deepest of deluges*, Cd. 161; Th. 200, 30; Exod. 364. Fîftena stôd deóp ofer dûnum se [MS. sæ] drenceflôd elna *the deluge stood fifteen ells deep over the hills*, 69; Th. 84, 16; Gen. 1398.

drenc-horn, es; *m. A drinking-horn;* potōrium cornu:—Ic geann into ðære stôwe ðone drenc-horn ðe ic ǽr [MS. êr] æt ðam hîrêde gebohte *I give to that place the drinking-horn which I formerly bought from the brotherhood*, Cod. Dipl. 722; Kmbl. iii. 361, 31.

drenc-hûs, es; *n. A drinking-house;* potiōnārium:—Ælces cinnes drenc-hûs *potiōnārium*, Ælfc. Gl. 110; Som. 79, 30; Wrt. Voc. 59, 4.

DRENG, es; *m. A warrior, soldier;* bellātor, miles:—Forlêt drenga sum daroþ of handa fleógan *one of the warriors let fly a dart from his hand*, Byrht. Th. 136, 10; By. 149. [*Laym.* dring *a thane, warrior, servant: Dan.* dreng *a boy, youth: Swed.* dreng, dräng, *m. a man, servant, soldier: Icel.* drengr, *m. a youth, valiant man.*]

drenge *a drink:*—Drenge ðû sylst us *potum dabis nobis*, Ps. Spl. 79, 6. v. drenc.

dreó-cræft, es; *m. Magical art, magic;* magica ars:—Simon se drý þurh dreócræft worhte ærene næddran, and ða hie styredan *Simon the sorcerer made brazen serpents by magic, and they moved of themselves*, Homl. Blick. 173, 21. v. drý-cræft.

DREÓGAN, to dreóganne; *part.* dreógende; ic dreóge, ðû dreógest, drýhst, he dreógeþ, drýhþ, drîhþ, *pl.* dreógaþ; *p.* ic, he dreáh, dreág, ðû druge, *pl.* drugon; *pp.* drogen; *v. trans.* I. *to do, work, perform, to pass life, to fight;* ágěre, făcěre, perficěre, patrāre, vitam ágěre, militāre:—To dreóganne wordum and dǽdum willan ðînne *to do thy will by words and deeds*, Cd. 107; Th. 141, 23; Gen. 2349. Ðe he dreógan sceolde *which he had to do*, Exon. 37 b; Th. 122, 28; Gû. 312. Hwæt dreógest ðû *what doest thou?* Exon. 69 a; Th. 257, 14; Jul. 247. Þeódnes willan dreógeþ *he does the will of the Lord*, Exon. 38 a; Th. 125, 20; Gû. 357. Gif mæsse-preóst oððe munuc hæmed-þincg drîhþ, fæste x geát se presbýter *vel* monăchus fornicatiōnem commisěrit, x annos jejūnet, L. M. I. P. 28; Th. ii. 272, 22. Drugon ðæt dæges and nihtes *fecērunt hoc die ac nocte*, Ps. Th. 54, 8. Gewin drugon *they fought*, Beo. Th. 1601; B. 798. Drugon wǽpna gewin *they fought the strife of arms, they wáged war*, Exon. 92 b; Th. 346, 7; Gn. Ex. 201. Hû manega gefeoht he ðǽr dreógende wæs *how many battles he was there fighting*, Ors. 1, 11; Bos. 35, 9. II. *to bear, suffer,* ENDURE, *endure;* ferre, pati, sustinēre, tolerāre:—Mán ne cûðon dôn ne dreógan *they knew not to do nor suffer crime*, Cd. 10; Th. 12, 23; Gen. 190. Ðe ða earfeða oftost dreógeþ *who oftenest suffers those afflictions*, Exon. 52 b; Th. 183, 19; Gû. 1329. Earfeða dreág *suffered hardships*, Exon. 74 b; Th. 280, 9; Jul. 626. Swâ ðæt fæsten dreáh *who endured that bondage*, Cd. 145; Th. 180, 22; Exod. 49. We lǽraþ ðæt man ǽnig gedrinc, and.

ǽnig unnit ðǽr ne dreóge *we teach that man suffer not there any drinking,
nor any vanity*, L. Edg. C. 28; Th. ii. 250, 14.   **III.** *to enjoy;*
frui :—He sibbe dreáh *he enjoyed peace*, Cd. 130; Th. 165, 28; Gen.
2738. Symbel-wynne dreóh *enjoy the pleasure of the feast!* Beo. Th.
3569; B. 1782.   **IV.** *v. intrans. To be employed, be busy;*
ágére, negótiósum esse :—Nǽnig manna wât hû mín hyge dreógeþ, býsig
æfter bócum *no man knows how my mind is employed, busy over books*,
Salm. Kmbl. 122, MS. B; Sal. 60. Dreógan, *inf.* Cd. 104; Th. 137,
31; Gen. 2282. Dreág, *p.* Exon. 53 a; Th. 185, 5; Az. 3. [*Chauc.*
drye *to suffer, endure:* *Laym.* drizen, drigen, drien *to suffer, do:* *Orm.*
dreghenn *to suffer, endure:* *Scot.* dre, dree, drey *to suffer:* *Goth.* driugan
to do military service.] DER. a-dreógan, ge-.

dreóh-lǽcan *magicians, sorcerers;* magi, Som. Ben. Lye. v. drý.

DREÓPAN; ic dreópe, ðû drýpst, he drýpþ, *pl.* dreópaþ; *p.* dreáp,
pl. drupon; *pp.* dropen *To drop;* stillāre, Prov. 19. [*Chauc.* drope:
Piers P. droppeiṛ: *Plat.* drüppen: *Dut.* druipen: *Kil.* droppen, druppen
manāre: Frs. drippen: *O. Frs.* driapa: *Ger.* tropfen, triefen: *M. H. Ger.*
triufen: *O. H. Ger.* triufan: *Dan.* dryppe: *Swed.* drypa: *Icel.* drjúpa *to
drip.*] DER. a-dreópan.

dreópian, dreápian, dropian, drupian; *p.* ode, ede; *pp.* od, ed *To drop;*
stillāre, distillāre :—Swá dropa on ðas eorþan dreópaþ *as a drop, which
droppeth on this earth*, Ps. Th. 71, 6. Heofonas [MS. Heofenas] dreápe-
dun *cæli distillāvērunt*, Ps. Surt. 67, 9. Myrre and cassia dropiaþ of
ðínum cláðum *myrrh and cassia drop from thy clothes*, Ps. Th. 44, 10.
Heofanas drupodon *cæli distillāvērunt*, Ps. Spl. 67, 9.

DREÓR, es; *m. Blood;* cruor :—Ic his blôd ageát, dreór on eorþan
I shed his blood, his gore on earth, Cd. 49; Th. 63, 12; Gen. 1031.
Dreóre fâhne *stained with gore*, Beo. Th. 898; B. 447. Dreóre druncne
drunk with blood, Andr. Kmbl. 2005; An. 1005. [*O. Sax.* drôr, *m.
cruor, sanguis: M. H. Ger.* trôr, *m. n. a dripping, blood: O. H. Ger.*
trôr *cruor: Icel.* dreyri, drôri, *m. blood.*] DER. cwealm-dreór, heoru-,
sâwel-, wæl-. v. dreósan.

dreórd, *pl.* dreórdon, dreórdun *dreaded, feared*, Mt. Kmbl. Rush. 9, 8:
19, 25, = drēd, *pl.* drēdon; *p.* of drǽdan.

dreór-fâh; *adj. Stained with gore;* cruentātus, Beo. Th. 974;
B. 485.

dreórgian; *p.* ode; *pp.* od [dreór *blood*] *To be dreary, to fall, to
perish;* mǽrēre, cadēre, corruēre :—Ðás hofu dreórgiaþ *these courts are
dreary*, Exon. 124 a; Th. 477, 26; Ruin. 30.

dreórig, dreóreg, dreórg, driórig; *def.* se dreóriga, dreórega, seó, ðæt
dreórige; *adj.*   **I.** *bloody, gory, glorious;* cruentus, cruentātus,
gloriōsus :—Wæter stôd dreórig and gedréfed *water stood gory and troubled*,
Beo. Th. 2838; B. 1417: Ps. Th. 135, 20: Exon. 72 b; Th. 211, 14;
Jul. 482. Hwæt druh ðû dreórega *lo thou gory dust!* Soul Recd. 33;
Seel. 17.   **II.** *sad, sorrowful, pensive,* DREARY; mœstus :—Híg
wurdon swíðe dreórige *they became very sorrowful*, Gen. 44, 13: Mk.
Bos. 14, 19. On ðas dreórgan tíd *in this sorrowful tide*, Exon. 48 b;
Th. 167, 10; Gú. 1058. [*Wyc.* drerg, dreri, drury *sad: Chauc.* drery
sad: Laym. druri, dreri *sad: Orm.* dreorig, drerig *sad: O. Sax.* drôrag
cruentus: Dut. treurig *sad: Ger.* traurig *sad: M. H. Ger.* trûrec *sad:
O. H. Ger.* trûrag *mœstus: Icel.* dreyrigr, dreyrugr *bloody.*] DER. heoru-
dreórig. v. dreósan.

dreórig-ferþ; *adj. Sad in soul;* tristis anīmo :—Dreórig-ferþe *sad in
soul*, Exon. 24 a; Th. 68, 26; Cri. 1109.

dreórig-hleór; *adj. Sad of countenance;* tristis facie :— Sumne
dreórighleór in eorþ-scræfe eorl gehýdde *a man sad of countenance has
hidden one in an earth-grave*, Exon. 77 b; Th. 291, 17; Wand. 83.

dreórig-líce; *adv. Drearily, mournfully;* mœste, Anlct. v. dreór-líc.

dreórig-môd; *adj. Sad of mind;* tristis anīmo :— Abraham dráf
dreórig-môd tû of earde *Abraham drove the two sad of mind from his
habitation*, Cd. 134; Th. 169, 24; Gen. 2804.

dreórignys, dreórinys, -nyss, e; *f.* DREARINESS, *sadness;* mœstitia :—
Gif he ne gehulpe hire sárlícan dreórinysse *if he might not relieve her
painful dreariness*, Greg. Dial. MS. Hat. fol. 5 a, 8.

dreór-líc, dreórilíc; *adj.*   **I.** *bloody;* sanguinolentus :—Ne
wearþ dreórlíce [dreórilícre, col. 2] dǽd gedôn syððan Dene cômon *no
bloodier deed was done since the Danes came*, Chr. 1036; Th. 294, 9;
Ælf. Tod. 6.   **II.** *mournful, sad;* mœstus, tristis :—Dreórilíc
frécednys *triste periculum*, Glos. Prudent. Recd. 151, 83.

dreór-sele, es; *m. A dreary, desolate-looking hall;* domus mœstitiæ :—
On dreórsele in the dreary hall, Exon. 115 b; Th. 444, 20; Kl. 50.

dreórung, dreárung, e; *f. A falling;* destillātio :—Ðonne on sumeres
tíd sended weorþeþ dropena dreórung *when a falling of drops is sent in
summer's :ime*, Exon. 54 a; Th. 189, 23; Az. 64. v. dreósan.

DREÓSAN; ic dreóse, ðû drýst, he dreóseþ, drýst, *pl.* dreósaþ;
p. dreás, *pl.* druron; *pp.* droren *To rush, fall, perish;* cadēre, ruēre :—
Wæstmas ne dreósaþ *the fruits do not fall*, Exon. 56 a; Th. 200, 2;
Ph. 34. Dreóseþ deáw and rēn *dew and rain fall*, 16 b; Th. 38, 19; Cri.
609. Druron dômleáse *they fell ingloriously*, Andr. Kmbl. 1989; An.
997. Swylge seó gitsung ða dreósendan wélan ðisses middangeardes
avarice swallows the perishable riches of this earth, Bt. 12; Fox 36, 13:

Bt. Met. Fox 7, 32; Met. 7, 16.   [*Laym.* drese *to fall down:* *O. Sax.*
driosan *cadēre:* *Goth.* driusan *to fall.*] DER. a-dreósan, ge-.

DREPAN; ic drepe, ðû drepest, dripest, dripst, he drepeþ, dripeþ,
dripþ, *pl.* drepaþ; *p.* ic, he drep, dræp, ðû dræpe, *pl.* drǽpon; *pp.* drepen,
dropen *To strike;* percutēre :—Ic sweorde drep ferhþgeníþlan *I struck
the deadly foe with my sword*, Beo. Th. 5753; B. 2880. Ðonne biþ on
hreðre, under helm drepen biteran strǽle *then he will be stricken with the
bitter shaft in the breast, beneath the helmet*, Beo. Th. 3495; B. 1745.
Wæs him feorh dropen *his life was stricken*, Beo. Th. 5955, note;
B. 2981. [*Plat.* drēpen *to hit:* *Dut. Ger.* treffen: *M. H. Ger.* triffen:
O. H. Ger. trefan *percutēre, pulsāre: Dan.* dræbe *to slay:* *Swed.*
dræpa *to kill, slay:* *Icel.* drepa *to hit.*]

drepe, drype, es; *m. A slaying, stroke, violent death;* occīsio :—He
drepe þrôwade *he suffered the stroke* [*death-stroke*], Beo. Th. 3183;
B. 1589. DER. deáþ-drepe.

drepen, drepenn, e; *f. A stroke;* percussio. v. gemynd-drepen.

dresten = drestan; *pl. f?* Dregs, *lees;* fæx :—Dresten his nys aídlude
fæx ejus non est exinanita, Ps. Spl. T. 74, 8. v. dærstan.

drettan *to consume.* v. ge-drettan.

drí, es; *m. A sorcerer, magician;* magus :— Be drían = dríum *by
sorcerers*, Glostr. Frag. 10, 30. v. dríian.

dríian = dríum = drýum *with sorcerers*, Glostr. Frag. 10, 30 : as fisceran
and fugeleran = fiscerum and fugelerum, Ors. 1, 1; Bos. 20, 5; *the dative
plural of* drí, drý, fiscere, and fugelere, *q. v.*

dríás, es; *m?* [dreósan *to fall*] *A falling, fall;* casûs. DER. deáw-
dríás.

drican [= drincan] *to drink*, Somn. 112, 113; Lchdm. iii. 204, 22, 23 :
Ps. Spl. 77, 49. v. drincan.

drí-cræfteg *skilful in magic*, Ex. 7, 11. v. drý-cræftig.

dríe *dry*, Ex. 14, 21: Bt. 5, 2; Fox 10, 31. v. drige.

drif, e; *f.*   **I.** *a fever;* febris :—Seó drif [sio drif MS.] *febris*,
Mt. Kmbl. Rush. 8, 15.   **II.** but drif, es; *m. or n. in the following
example* :—Full-neáh ǽfre ðe ôðer man wearþ on ðam wyrrestan yfele,
ðæt [MS. þet] is on ðam drife *almost every other man was in the worst
evil, that is with fever*, Chr. 1087; Th. 353, 38. DER. ge-drif.

DRÍFAN, drýfan, ic drífe, ðû drífest, drífst, he drífeþ, drífþ, drîft,
pl. drífaþ; *p.* ic, he dráf, ðû drife, *pl.* drifon, dreofon; *pp.* drifen.   **I.**
v. trans. To DRIVE, *force, pursue;* pellēre, mināre, impellēre, persēqui :—
Se geréfa hie wolde drífan to ðæs cyninges tûne *the reeve would drive them
to the king's vill*, Chr. 787; Erl. 56, 13. Se Hǽlend ongan drífan of ðam
temple syllende and bicgende *Iesus cæpit ejicēre vendentes et ementes in
templo*, Mk. Bos. 11, 15. Sum mæg ofer sealtne sǽ sundwudu drífan *one
can drive a vessel over the salt sea*, Exon. 17 b; Th. 42, 24; Cri. 677.
For hwan ðû us, God, woldest fram ðē drífan *ut quid repulisti nos, Deus?*
Ps. Th. 73, 1. Ic drífe sceáp mîne to heora lease *mino oves meas ad
pascua*, Coll. Monast. Th. 20, 11. Ic ða of Drihtnes drífe ceastre *I will
drive them from the Lord's city*, Ps. Th. 100, 8. Ða wēregan neát, ðe
man daga gehwam drífeþ and þirsceþ, ongitaþ hira gôddénd *the brute
animals, which man drives and beats every day, understand their benefactors*, Elen. Kmbl. 716; El. 358. Flinte ic eom heardra, ðe ðis fýr
drífeþ of ðissum strongan stýle *I am harder than flint, which this fire
drives from this strong steel*, Exon. 111 b; Th. 426, 24; Rä. 41, 78.
Hwîlum ðæt drige drife ðone wǽtan *sometimes the dry drives away the
wet*, Bt. Met. Fox 29, 98; Met. 29, 48. Us drífaþ ða ællreordan to sǽ
the barbarians drive us to sea, Bd. 1, 13; S. 481, 44 : Beo. Th. 5609;
B. 2808. Ôðerne he dráf mid sticele, ôðrum he wiðteáh mid bridle *the
one he drove with a goad, the other he restrained with a bridle*, Past. 40,
3; Hat. MS. 54 b, 12. Abraham dráf dreorig-môd tû of earde *Abraham
drove the two sad of mind from his dwelling*, Cd. 134; Th. 169, 23;
Gen. 2804. Ne eart ðû se sylfa God, ðe us swá drife *nonne tu, Deus,
qui repulisti nos?* Ps. Th. 59, 9. Hî drifon scipu into Medwæge *they
drove the ships into the Medway*, Chr. 1016; Erl. 157, 16. Hîg hyne
drifon ût *ejēcērunt eum foras*, Jn. Bos. 9, 35. Ða hîg eów drifon *cum
vos persequerentur*, Deut. 11, 4. Hî dreofon hine onweg *they drove him
away*, Bd. 2, 5; S. 507, 27. Ge fleóþ, ðeáh eów man ne drífe *fugiētis,
nemine persēquente*, Lev. 26, 17. Ðæt he on wræc drífe his selfes sunu
that he should drive into exile his own son, Cd. 134; Th. 168, 32; Gen.
2791. Drífan drýcræft *to exercise magic*, Bt. Met. Fox 26, 107; Met.
26, 54. Ceáp drífan *to drive* or *transact a bargain*, R. Ben. 57.
Mangunge drífan *to follow a trade*, Homl. Th. ii. 94, 34. Spæce or
spræce drífan *to prosecute a suit, urge a cause*, L. O. 2; Th. i. 178, 13:
L. Ælf. C. 35; Th. ii. 356, note 2, 4: Th. Diplm. 376, 11. Wôh
drífan *to practise wrong*, L. I. P. 11; Th. ii. 320, 4.   **II.** *v. intrans.*
To drive, rush with violence; ruēre :—Ic com mid ðý heáfde and mid
handa on ðone stân drífan *I came driving on the stone with my head and
hands*, Bd. 5, 6; S. 619, 23. [*Wyc.* dryue: *Piers P.* dryven: *Chauc.*
drife, drive: *Laym.* driuen, driue: *Orm.* drifenn: *Plat.* drîwen, drîben:
O. Sax. drîban *ágēre, pellēre: Frs.* drieuwen: *O. Frs.* driva: *Dut.*
drijven: *Ger.* treiben: *M. H. Ger.* trîben: *O. H. Ger.* trîban: *Goth.*
dreiban: *Dan.* drive: *Swed.* drifva: *Icel.* drífa.] DER. a-drífan, be-, for-,
ge-, in-, of-, ofa-, ofer-, þurh-, to-, ût-, ûta-, wið-.

Driffeld; *gen.* es; *dat.* a, e; *m.* [*in* A. D. 1360 *it was written* Dyrffeld] *Great* DRIFFIELD, *in the East Riding of Yorkshire;* oppĭdi nomen in agro Eboracensi:—Hēr.Aldfriþ Norþan Hymbra cining forþfērde, on xix kl' Jan. on Driffelda *in this year* [A.D. 705] *Alfred, king of the Northumbrians, died at Driffield, on the 19th of the kalends of January* [*December 14th*], Chr. 705; Erl. 43, 33.

drigan, drygan, drigean; *p.* de; *pp.* ed; *v. a.* [drige *dry*] *To* DRY, *make dry, rub dry, wipe;* siccāre, tergēre, extergēre:—Se hâta sumor giereþ and drigeþ sǽd and blēda *the hot summer prepares and dries seeds and fruits,* Bt. Met. Fox 29, 120; Met. 29, 60. Đæt dust, đæt of eówre ceastre on ūrum fótum clifode, we drigeaþ on eów *pulvĕrem, qui adhǽsit nobis de civitāte vestra, extergĭmus in vos,* Lk. Bos. 10, 11. Heó ongan mid hyre teárum his fēt þweán, and drigde mid hyre heáfdes feaxe *lacrўmis cœpit rigāre pedes ejus, et capillis capĭtis sui tergēbat,* Lk. Bos. 7, 38, 44. Seó drigde his fēt mid hyre loccum *extersit pedes ejus capillis suis,* Jn. Bos. 11, 2: 12, 3. DER. a-drigan, -drygan, ge-, ofa-, ūta-.

DRIGE, dryge, drīe; *def.* se driga, dryga, drīa; seó, đæt drige, dryge, drīe; *adj.* DRY; siccus, arĭdus:—Se wind blǽwþ norþan and eástan, heálíc, and ceald, and swīđe drige [drīe MSS. P. L.] *the wind blows from the north-east, violent, and cold, and very dry,* Bd. de nat. rerum; Wrt. popl. science 18, 8; Lchdm. iii. 276, 6. Drige wudu *dry wood, firewood;* ligna, Wrt. Voc. 80, 31. Adrugode se stream swâ đæt he mihte dryge ofergangan *the stream dried up so that he might go over dry,* Bd. 1, 7; S. 478, 14: Exon. 111 b; Th. 426, 22; Rä. 41, 77. Tunge biþ drige *the tongue is dry,* L. M. 2, 46; Lchdm. ii. 258, 8. Seó [MS. sie] eorþ is dryge *the earth is dry,* Bt. 33, 4; Fox 128, 34: Andr. Kmbl. 3161; An. 1583. Læg ân drīe strǽt þurh đa sǽ *a dry road lay through the sea,* Ex. 14, 21. Đæs fўres gecynd is hât and drīe *the nature of fire is hot and dry,* Boutr. Scrd. 18, 22, 23. Hwîlum đæt drige drîft đone wǽtan *sometimes the dry drives away the wet,* Bt. Met. Fox 29, 97; Met. 29, 48. Seó sǽ, ûtflówende, gerўmde þreóra mîla drīes færeldes *the sea, flowing out, made room for a dry passage of three miles,* Homl. Th. i. 564, 18. Đa sacerdas ætstódon on đam grunde on drigre moldan on middan đære eá be drīum grunde *sacerdōtes stābant per siccam humum in medio Iordānis,* Jos. 3, 17. Gif híg on grēnum treówe đæs þing dóþ, hwæt dóþ híg on đam drigum *si in virĭdi ligno hæc faciunt, in arĭdo quid fiet?* Lk. Bos. 23, 31: Ps. Th. 105, 9. Drihten gewende đa sǽ to drīum *mare Domĭnus vertit in siccum,* Ex. 14, 21. Betwux đære drygan and đære cealdan eorþan and đam hâtan fўre *between the dry and the cold earth and the hot fire,* Bt. 33, 4; Fox 128, 37. Đæt seó sǽ drigne grund đam folce gegearcige *that the sea should prepare dry ground for the people,* Homl. Th. i. 564, 24. In drygne seáþ *into a dry pit,* Invent. Crs. Recd. 1388; El. 693. Worhte his tolme foldan drige *arĭdam fundavērunt manus ejus,* Ps. Th. 94, 5: Cd. 8; Th. 10, 29; Gen. 164. Uppan drīe eorþan *super arĭdam,* Ex. 4, 9. Se đe gecyrde sǽ on drige land *qui convertit mare in arĭdam,* Ps. Spl. 65, 5. Dó drige pic to *add dry pitch,* L. M. 2, 38; Lchdm. ii. 246, 14. Đa drigan eorþan *the dry earth,* Bt. 33, 4; Fox 130, 2. Hwîlum flîht se wǽta đæt dryge *sometimes the wet drives away the dry,* Bt. 39, 13; Fox 234, 11. Wǽron đa wareþas drige *the shores were dry,* Ps. Th. 105, 9. Wegas syndon dryge *the ways are dry,* Cd. 157; Th. 195, 28; Exod. 283. Drîra *arentum,* Glos. Prudent. Recd. 151, 22. Dysegaþ se đe wile sǽd ðþfæstan đam drīum [drygum, Cot.] furum *he does foolishly who will sow seed in the dry furrows,* Bt. 5, 2; Fox 10, 31. Hí fērdon ođđe fóron on drigum flódum *abiērunt in sicco flumĭna,* Ps. Lamb. 104, 41. He gǽþ geond drige stówa *ambŭlat per loca arĭda,* Mt. Bos. 12, 43: Ps. Th. 65, 5. Se wyrcþ drige [drīe MSS. P. L.] wolcnu *it makes dry clouds,* Bd. de nat. rerum; Wrt. popl. science 18, 2; Lchdm. iii. 274, 24. Fram đære burnan đe he drigum fótum ofereóde *from the brook which he went over with dry feet,* Bd. 1, 7; S. 478, 32. Mid drīum handum *with dry hands,* L. M. 2, 3; Lchdm. ii. 182, 8. Đæt Israhēlisce folc gâ drīum fótum innan đa sǽ *ut gradiantur filii Israel in medio mari per siccum,* Ex. 14, 16, 29. [*Wyc.* drie: *Piers P.* drye: *Chauc.* drey: *Orm.* driȝȝe: *Plat.* dröge, drüge, dræge: *Dut.* droog: *Ger.* trocken: *M. H. Ger.* trucken: *O. H. Ger.* trukan *siccus: Dan.* dröi *solid: Swed.* dryg *heavy: Icel.* drjúgr *solid, substantial.*]

drigian, đú drígast; *p.* ode; *pp.* od [dreógan *to suffer, endure*] *To suffer, endure;* tolerāre, pati:—Đú on đisum andweardan lífe mâ earfođa drígast *thou sufferest more troubles in this present life,* Guthl. 5; Gdwin. 32, 13.

drig-nes, dryg-nes, -ness, -nis, -niss, -nyss, e; *f.* DRYNESS; siccĭtas:—Đære drignesse ne sceal he huniges onbîtan ac eald wîn *for the dryness he must not taste of honey but old wine,* L. M. 2, 27; Lchdm. ii. 222, 19. Æteówige drignis *let dryness appear;* appāreat ārĭda, Gen. 1, 9. God gecîgde đa drignisse eorþan *vŏcāvit Deus ārĭdam terram,* 1, 10. On drignysse in *inăquōso,* Ps. Spl. 77, 20. Drygnessa his handa gescópan *siccam mănus ejus formāvērunt,* Ps. Lamb. 94, 5.

Driht', Driht *the Lord,* used with or without the apostrophe in Spelman's Psalms for all the cases of Drihten. v. Dryht'.

driht, e; *f. A multitude, an army,* Cd. 146; Th. 182, 21; Exod. 79: Cd. 47; Th. 61, 6; Gen. 993. v. dryht.

driht-ealdor, drihte ealdor; *gen.* ealdres; *m. The lord of a feast;* architriclīnus:—Se drihtealdor cwæþ to đam brўdguman *the lord of the feast said to the bridegroom,* Homl. Th. ii. 70, 25, 28. Se drihte ealdor đæs wînes onbўrgde *gustāvit architriclīnus vinum,* Jn. Bos. 2, 9. Beraþ đære drihte ealdre *ferte architriclīno,* 2, 8. v. dryht-ealdor.

drihten; *gen.* drihtnes, drihtenes; *m. A ruler, lord, the Lord:*—Gumena drihten *lord of men,* Cd. 205; Th. 254, 18; Dan. 613. Eorla drihten *lord of earls,* Beo. Th. 2105; B. 1050. Drihten Crēca *lord of the Greeks,* Bt. Met. Fox 26, 38; Met. 26, 19. Drihten mîn *my lord,* Cd. 101; Th. 134, 15; Gen. 2225. Witig Drihten, rōdera Rǽdend *the wise Lord, Ruler of the skies,* Beo. Th. 3113; B. 1554. Drihten wereda *the Lord of hosts,* Beo. Th. 4378; B. 2186. Ēce Drihten wið Abrahame sprǽc *the Lord eternal spake with Abraham,* Cd. 106; Th. 139, 1; Gen. 2303. Ic eom Drihten đîn God *ego sum Domĭnus Deus tuus,* Ex. 20, 2. Þurh ûrne Drihten Crist *through our Lord Christ,* L. Ælf. P. 39; Th. ii. 380, 3. On đæm naman Drihtnes ūres Godes *in nomĭne Domĭni Dei nostri,* Ps. Th. 19, 7. Se seofođa ys Drihtnes restedæg đînes Godes *septĭmo die sabbătum Domĭni Dei tui est,* Ex. 20, 10. Eálā Drihtenes þrym *O majesty of the Lord,* Cd. 216; Th. 274, 34; Sat. 164: Ps. Lamb. 26, 13: Ps. Th. 68, 37. v. dryhten.

Drihten-líc; *def.* se -líca, seó, đæt -líce; *adj. Belonging to the Lord, Lordly;* Domĭnicus:—Drihtenlíces Domĭnici, Mone B. 429. Angelþeóde đæs Drihtenlícan geleáfan gife geleornode *gens Anglōrum Domĭnicæ fidei et dona discēret,* Bd. 3, 3; S. 525, 29. He nǽfre mete onfēng bûtan đў Drihtenlícan dæge *he never took meat except on the Lord's day,* 4, 25; S. 599, 30.

Drihten-líce; *comp.* -lícor; *adv. According to the Lord, by the Lord;* secundum Domĭnum, a Domĭno:—Đæt he Drihtenlícor mæge beón hâlig genemned *that he may be called holy by the Lord,* L. E. I. 21; Th. ii. 418, 9.

drihten-weard, es; *m.* [weard *a keeper, guardian*] *A guardian lord, king;* domĭnus custos, rex:—On đam drihtenweard deópne wisse sefan sîdne geþanc *in whom the guardian lord knew [to exist] deep ample thought of mind,* Cd. 201; Th. 249, 24; Dan. 535.

driht-folc *a nation,* Cd. 144; Th. 179, 26; Exod. 34. v. dryht-folc.

driht-gesíþ, es; *m.* [gesíþ *a companion*] *An associate, attendant;* satelles:—Nân ne feól drihtgesíþa *none of the associates fell,* Fins. Th. 84; Fin. 42.

driht-guma, an; *m. A popular man, man of the people, a warrior, retainer,* Beo. Th. 2781; B. 1388: 198; B. 99. v. dryht-guma.

drihþ *does, performs, commits,* L. M. I. P. 28; Th. ii. 272, 22; 3rd *pres. sing.* of dreógan.

drihtin-beáh; *gen.* -beáges; *dat.* -beáge; *m.* [drihtin = drihten *a lord,* beáh *a ring, bracelet*] *A lord-ring* or *money paid for slaying a freeman.* In the laws of Edward the Confessor it is called Manbóte:—Manbóte in lege Anglōrum, regi et archiepiscŏpo, iii marc de homĭnibus suis; episcŏpo comitātus, comĭti comitātus, et dapifĕro regis, xx sol; barōnibus cetēris, x solid, L. Ed. C. 12; Th. i. 447, 28–31. Gif man frigne mannan ofsleahþ, cyninge l scillinga to drihtin-beáge *if any one slay a freeman,* [*let him pay*] *fifty shillings to the king, as 'drihtin-beah,'* L. Ethb. 6; Th. i. 4, 6, 7.

driht-líc, driht-lec *lordly,* Menol. Fox 511; Gn. C. 26: Cd. 33; Th. 168, 12; Gen. 2781. v. dryht-líc.

driht-líce *in a lordly manner,* Cd. 98; Th. 129, 4; Gen. 2138. v. dryht-líce.

driht-né; *pl. nom. acc.* -néas; *m. A dead body of a host;* cadāver agmĭnis:—Ofer drihtnéum *over the bodies of the slain,* Cd. 150; Th. 188, 5; Exod. 163. v. né.

Drihtnes *of the Lord;* Domĭni, Ex. 20, 10; *gen.* of Drihten. v. dryhten.

driht-scipe *rulership,* Cd. 24; Th. 31, 14; Gen. 485. v. dryht-scipe.

driht-sele *a princely hall,* Beo. Th. 974; B. 485. v. dryht-sele.

driht-weras; *pl. m.* [wer *a man*] *Men, chieftains;* popŭlares viri:—Oþ-đæt drihtweras duguþum gefóran đær is botlwéla Bethlem hâten *till that the fellow men journeyed to where there is a village called Bethel,* Cd. 86; Th. 107, 32; Gen. 1798. Đú móst heonon hûþe lǽdan ealle, bûton dǽle đissa drihtwera *thou mayest lead all the spoil hence, save the part of these chieftains,* 98; Th. 129, 27; Gen. 2150.

dríme *joy;* jubĭlum, Cot. 109. v. dreám.

DRINC, drync, es; *m.* drinca, an; *m.* drince, an; *f.* DRINK, *a drink, draught;* potus, haustus:—Mín blôd ys drinc *sanguis meus est potus,* Jn. Bos. 6, 55. Ic ofþyrsted wæs gæstes drinces *I was thirsty for the soul's drink,* Exon. 98 a; Th. 369, 15; Seel. 41. Hēr gefór Harþacnut swâ đæt he æt his drince stód *in this year* [A.D. 1042] *Harthacnut died as he stood at his drink,* Chr. 1042; Erl. 166, 34. Ic mínne drinc mengde mid teárum *potum meum cum fletu tempĕrābam,* Ps. Th. 101, 7. Swâ hwylce swâ sylþ ánne drinc cealdes wæteres ánum đyssa lytylra manna *quicumque potum dedĕrit uni ex minĭmis istis calĭcem aquæ frigĭdæ,* Mt. Bos. 10, 42. We đē drinc sealdon *dedĭmus tibi potum,* 25, 37: Bt. Met. Fox 8, 43; Met. 8, 22. Nǽron đa mistlíce drincas *there were not then various drinks,* Bt. 15; Fox 48, 5: Bt. Met. Fox 8, 18; Met. 8, 9.

[*Wyc.* drynk: *Piers P.* drenke: *Chauc.* drinke: *Laym.* drænc, drench; drinc: *Orm.* drinnc, drinnch: *Plat.* drunk, drank, *m*: *O. Sax.* drank, *m. n*: *Frs.* dranck: *O. Frs.* drank in compounds: *Dut.* dranc, dronc, *m*: *Ger.* trank, trunk, *m*: *M. H. Ger.* tranc, *n. m*; trunc, *m*: *O. H. Ger.* trank, *n. potus*; trunk, *m. haustus*: *Goth.* draggk, dragk, *n. drink*: *Dan.* drik, *m. f*: *Swed.* drick, dryck, *m*: *Icel.* drekka, *f. beverage*.] DER. ātor-drinc, ge-, mān-, medo-, ofer-, wīn-, wīuge-.

drinca, an; *m*: drince, an; *f*. [drinc *drink*] Drink; potus:—Eáðe we māgon geseón hwǽr se drinca is *we can easily see where the drink is*, Ors. 5, 8; Bos. 107, 30. He wolde beran drincan his gebróðrum *he would bear drink to his brethren*, Homl. Th. ii. 180, 5. He bæd him drincan and heó him blīðelīce sealde *he asked for drink and she gave it him gladly*, Jud. 4, 19: Basil admn. 4; Norm. 42, 24. He bæd God ðæt he him asende drincan *he prayed God to send him drink*, Jud. 15, 18. Drince mylsce drincan sió gebēt ða biternesse *let him drink a mulled drink which will amend the bitterness*, L. M. 1, 42; Lchdm. ii. 108, 2. DER. āttor-drinca, on-.

drincan, to drincenne, ic drince, ðū drincst, he drincþ, dryncþ, *pl.* drincaþ; *p.* dranc, *pl.* druncon; *pp.* druncen [drinc *drink*]. I. *to* DRINK, *imbibe*; bibēre, potāre, imbibēre:—He dranc of ðam wīne, ðā wearþ he druncen *bibens vinum inebriātus est*, Gen. 9, 21: Lev. 10, 9. We ǽton and druncon beforan ðe *manducāvimus coram te, et bibīmus*, Lk. Bos. 13, 26. Ðonne hīg druncene beóþ *cum inebriāti fuērint*, Jn. Bos. 2, 10. II. the Anglo-Saxons often drank to excess, as is evident by the exhortation of Abbot Ælfric to his friend Sigferd, to whom he dedicated his Treatises on the Old and New Testaments:—Ðū woldest me laðian, ðā ðā ic wæs mid ðē ðæt ic swīðor drunce, swilce for blisse. Ac wite ðū, leóf man, ðæt se ðe óðerne neádaþ ofer his mihte to drincenne ðæt se mót aberan heora begra gild, gif him ǽnig hearm of ðam drence becymþ. Úre Hǽlend forbeád ðone oferdrenc. Ða láreówas alēdon ðone unþeáw þurh heora láreówdóm and tǽhton ðæt se oferdrenc fordēþ untwī-līce ðæs mannes sāwle and his gesúndfullnysse. Unhǽl becymþ of ðam drence *when I was with thee, thou wouldest urge me to drink very much, as it were for bliss. But know thou, dear friend, that he who forces another man to drink more than he can bear, shall answer for both, if any harm come thereof. Our Saviour hath forbidden over drinking. The learned fathers have also put down that bad habit by their wise teaching, and taught that the over drinking surely destroys a man's soul and soundness. Unhealthiness cometh after [over] drinking*, Ælfc. T. 43, 6–17. [*Piers P.* drinken: *Chauc.* dronken, *pp.*: *Laym.* drinchen, drinken: *Orm.* drinnkenn: *Plat.* drinken: *O. Sax.* drinkan: *Frs.* drincken: *O. Frs.* drinka: *Dut.* drinken: *Ger. M. H. Ger.* trinken: *O. H. Ger.* trinkan: *Goth.* drigkan: *Dan.* drikke: *Swed.* dricka: *Icel.* drekká.] DER. a-drincan, be-, for-, ge-, ofa-, ofer-, on-.

drince-fæt, es; *n. A cup*; calix:—Ic geseah Pharaones drincefæt on mīnre handa *vidēbam cālicem Pharaōnis in manu mea*, Gen. 40, 11, 13. v. drinc-fæt.

drince-leán, es; *n. Tributary drink, scot-ale, the contribution of tenants to purchase ale for the entertainment of their lord or his steward on the fee*, Glos. to Th. Laws, vol. ii. *Or, perhaps, the ale given by the seller to the buyer on concluding a bargain*; retrĭbūtio potus *vel* præmium bibendi:—Drinceleán and hláfordes riht gifu stande ǽfre unawend *let the tributary drink and the lord's rightful gift ever stand unchanged*, L. C. S. 82; Th. i. 422, 2: L. N. P. L. 67; Th. ii. 302, 7.

drincere, es; *m. A* DRINKER; potātor:—Drincere wīnes *potātor vini*, Mt. Kmbl. Lind. 11, 19.

drinc-fæt, drync-fæt, drync-fæt, drenc-fæt; *gen.* -fætes; *pl. nom. acc.* -fatu; *n.* [fæt *a vessel*] *A drinking-vessel, cup*; pōcŭlum, calix = κύλιξ:—Beóþ heora drincfatu gefyldu *their drinking-vessels shall be filled*, Ps. Th. 10, 7.

drinc-lagu, e; *f. Drinking-law*; assisa potus:—Statūtum, scilĭcet edictum, lex, *vel* constĭtūtio de potus vendendi mensūris, Som. Lye.

drinc-wērig, adj. *Drink weary, satisfied with drinking*; potu defessus, temŭlentus, Cot. 124.

driórig *bloody*; cruentātus, gloriōsus:—Driórigne, *acc.* Beo. Th. 5572; B. 2789. v. dreórig.

dripest, dripst, he dripeþ, dripþ *strikest, strikes*; *2nd and 3rd pers. pres. of* drepan.

dris-līc *fearful.* DER. on-dris-līc. v. dryslīc.

drisn, e; *f? A wig, false hair*; capillāmentum, galerĭcŭlum:—Rupe *vel* drisne *capillāmenta*, Ælfc. Gl. 35; Som. 62, 96; Wrt. Voc. 28, 73. v. rupe.

DRÓF; adj. *Draffy, dreggy, dirty, troubled*; sordĭdus, turbŭlentus, turbĭdus:—Se ðe his bróðor hataþ, he hæfþ unstilnesse, and swyðe dróf [MS. drofi] mód *he that hateth his brother has disquietude, and a very troubled mind*, Basil admn. 4; Norm. 44, 16. Flód dróf *a turbid flood*, Somn. 102; Lchdm. iii. 204, 11. [*Laym.* drof *disturbed, grieved*: *O. Sax.* dróbi, druoƀi *turbĭdus, nubĭlus*: *Kil.* droef *turbĭdus, turbŭlentus, fecŭlentus*: *Ger.* trübe *troubled, obscure, dark, dull, sad*: *M. H. Ger.* trüebe: *O. H. Ger.* truobi *turbĭdus, turbātus*.] DER. ge-dróf.

dróf-denu, e; *f. A den* or *valley where droves of cattle feed*; armen-

tōrum cubile. Locus nemorōsus armentōrum receptui accommŏdus, Som. Ben. Lye. v. dráf.

dróf-līc; adj. *Agitated, disturbed, troublesome, irksome, sad*; turbŭlentus, molestus:—Him biþ fȳr ongeán, dróflīc wīte *before them shall be fire, sad punishment*, Exon. 116 a; Th. 446, 8; Dóm. 19.

dróf-man, -mann, es; *m. A drove-man, cattle-keeper*; būbulcus, Som. Ben. Lye. v. dráf.

drófnys, -nyss, e; *f. Dirtiness, sedition*; turbulentia, Som. Ben. Lye.

dróg *drew*, Jn. Lind. War. 21, 11; *p. of* dragan.

drogan = drugon *suffered*; tolerārunt, Bt. 38, 1; Card. 302, 21; *p. pl. of* dreógan.

droge, an; *f? Dung*, DRAUGH; stercus:—Nim monnes drogan *sume stercus humānum*, L. M. 3, 36; Lchdm. ii. 328, 16.

drogen *done, worked*; *pp. of* dreógan.

drógon *drew*, Andr. Kmbl. 2465; An. 1234; *p. pl. of* dragan.

dróh *dragged, drew*; *p. of* dragan.

droht, es; *m? Manner* or *condition of life*; vitæ condĭtio:—Hū he his wīsna trūwade, drohtes, on ðære dimman ādle *how he trusted in his morals, his manner of living, in that hidden malady*, Exon. 49 b; Th. 171, 31; Gū. 1135. v. drohtaþ.

droht *drawn, draught*; tractus, haustus, Cot. 202, Som. Ben. Lye.

drohtaþ, drohtoþ, es; *m*. [dreógan *to do, suffer, pass life, live*] *Conversation, manner* or *way of life, condition, conduct, society*; condĭtio vitæ, stătio, conversātio:—Is se drohtaþ strang ðam ðe lagolāde cunnaþ *severe is the way of life for him who trieth a sea-journey*, Andr. Kmbl. 626; An. 313: 2770; An. 1387: Exon. 20 a; Th. 53, 28; Cri. 857. Duguþ and drohtaþ *virtue and converse*, Exon. 42 b; Th. 143, 4; Gū. 656. Ne wæs his drohtoþ swylce he on ealderdagum ǽr gemētte *his condition was not such as he had before found in his life-days*, Beo. Th. 1517; B. 756. Ðæt hie ðe eáþ mihton ofer ȳða geþring drohtaþ adreógan *that they might the easier endure their way of life over the clash of waves*, Andr. Kmbl. 737; An. 369: 2564; An. 1283: Exon. 103 a; Th. 389, 20; Rä. 7, 10. Hī mā lufedon dióra drohtaþ *they loved more the society of beasts*, Bt. Met. Fox 26, 183; Met. 26, 92. Drohtaþ sēcan *to seek a sojourn*, Cd. 86; Th. 109, 6; Gen. 1818: Exon. 61 b; Th. 227, 1; Ph. 416.

drohtian *to converse, live*, Bd. 1, 27; S. 488, 37: 5, 6; S. 618, 28: Salm. Kmbl. 894; Sal. 446. v. drohtnian.

drohtigen *that ye converse*; *pl. pres. subj. of* drohtian. v. drohtnian.

drohtnian, drohtian; *part.* drohtniende, drohtende; *p.* ode, ade; *pp.* od, ad *To converse, dwell* or *keep company with, pass life, live*; versāri, conversāri, dēgĕre, vitam ăgĕre:—Bī bisceopum, hū hī mid heora geférum drohtian and lifigean scylon *de episcŏpis, quălĭter cum suis clerĭcis conversentur*, Bd. 1, 27; S. 488, 37: Hy. 4, 89; Hy. Grn. ii. 285, 89. Cild ic eom under gyrde drohtniende *puer sum sub virga dēgens*, Coll. Monast. Th. 34, 23. Wæs he on his geférscipe drohtiende *in clero illius conversātus*, Bd. 5, 6; S. 618, 28. Hī drohtende duguþe beswīcaþ *they by converse deceive the virtuous*, Exon. 97 a; Th. 362, 6; Wal. 32. Ic drohtnige *conversor*, Ælfc. Gr. 37; Som. 39, 15. Drohtnaþ on temple God *versātur in templo Deus*, Hymn. Surt. 44, 7. To hwām drohtaþ heó mid us *why dwelleth she with us?* Salm. Kmbl. 894; Sal. 446: Exon. 57 a; Th. 203, 22; Ph. 88. We drohtniaþ *degīmus*, Hymn. Surt. 113, 17. Ða ungeleáffullan, ðe būton Godes gelaðunge dwollīce drohtniaþ *the unbelieving, who live in error without the church of God*, Homl. Th. ii. 60, 14. Se in ðam mynstre eardode and drohtnade *qui in illo monastērio degēbat*, Bd. 4, 25; S. 601, 32. Fela wītegan under ðære ǽ Gode gecwēmelīce drohtnodon *many prophets under the old law passed their days acceptably to God*, Homl. Th. ii. 78, 34. Ðæt mid Suna Meotudes drohtigen dæghwamlīce *that ye converse daily with the Son of God*, Andr. Kmbl. 1363; An. 682.

drohtnung, drohtung, e; *f*. [droht *vitæ condĭtio*] *Conversation, condition, conduct, life, actions*; conversātio, condĭtio, stătio, actio:—Hira drohtnung sī afandud *quorum conversātio sit probāta*, Deut. 1, 13. Manega hālige bēc cȳðaþ his [Gregoriuses] drohtnunge and his hālige līf *many holy books manifest his [Gregory's] conduct and his holy life*, Homl. Th. ii. 116, 29. Of ðære munuclīcan drohtnunge *from the monastic life*, 120, 12. Sume on mynsterlīcre drohtnunge on reogollīcum līfe getreówlīce Drihtne þeówdon *some served the Lord truly in monastic conversation in regular life*, Bd. 3, 27; S. 558, 24: Bd. de nat. rerum; Wrt. popl. science 4, 5; Lchdm. iii. 238, 4. On micelre drohtnunge *in great renown*, L. Ælf. P. 40; Th. ii. 380, 33. He his līf in Gode mid wyrþre drohtunge gefylde *vitam in Deo digna conversatiōne complēvit*, Bd. 5, 6; S. 620, 24. On ðæra Apostola drohtnunge *in the Acts of the Apostles*, R. Ben. 33. Oþ-ðæt he full hāl sȳ on his drohtnungum *until he be full sound in his conditions*, Homl. Th. i. 126, 2.

DROPA, an; *m*. I. *a* DROP; stilla, gutta, stillicĭdium:—Dropa *gutta vel stilla*, Ælfc. Gl. 97; Som. 76, 70; Wrt. Voc. 54, 14. Yrnþ dropmǽlum swīðe hluttor wæter, ðæt gecīgdon ða ðe on ðære stōwe wunodon *stillam, ðæt is dropa very pure water runs [there] drop by drop, which those who dwelt in the place called* stilla, *that is drop*, Homl. Th. i. 510, 1. Flōwe min spræc swā dropan ofer gærsa cīþas *fluat elŏquium*

meum quasi stillæ super gramĭna, Deut. 32, 2. Snáw cymþ of ðam þynnum wǽtan, ðe byþ upatogen mid ðǽre lyfte, and byþ gefroren ǽr ðan ðe he to dropum geurnen *sӯ snow comes of the thin moisture, which is drawn up with the air, and is frozen before it be run into drops*, Bd. de nat. rerum; Wrt. popl. science 19, 14; Lchdm. iii. 278, 25. His swǽt wæs swylce blŏdes dropan on eorþan yrnende *est sudor ejus sĭcut guttæ sanguĭnis decurrentis in terram*, Lk. Bos. 22, 44. Swá dropa, ðe on ðas eorþan dreópaþ *as a drop which droppeth on this earth*, Ps. Th. 71, 6. Heó ŏðerne dropan on ðæt ŏðer eáge dyde *she put [did] another drop on the other eye*, Guthl. 22; Gdwin. 98, 3. Nime ǽnne eles dropan *take a drop of oil*, Orþ. 4, 7; Bos. 88, 11: L. M. 1, 2; Lchdm. ii. 34, 26. Swá swá dropan dropende ofer eorþan *sĭcut stillicĭdia stillantia super terram*, Ps. Spl. 71, 6. Dropan stígaþ *the drops shall rise*, Salm. Kmbl. 90; Sal. 44. Dropena dreorung *a fall of drops*, Exon. 54 a; Th. 189, 23; Az. 64: Cd. 191; Th. 238, 3; Dan. 349: 213; Th. 265, 23; Sat. 12. **II.** *a disease, paralysis?* morbus, parălȳsis = παράλυσις :— Wið fŏt-ádle, and wið ðone dropan *against gout [foot disease] and against the paralysis [the drop]*, Lchdm. i. 376, 1. Wið ðone dropan *against the paralysis [the drop]*, Herb. 59; Lchdm. i. 162, 4, 7. Heó ǽlc yfel blŏd and ðæne dropan gewyldeþ *it subdues all evil blood and the paralysis [the drop]*, 124, 1; Lchdm. i. 236, 13. [*Wyc.* droppes, *pl :* Laym. drope: *Plat.* droppen, drüppen, *m : O. Sax.* dropo, *m : O. Frs.* dropta *dropping : Dut.* drop, *m : Kil.* droppe : *Ger.* tropfen, *m : M. H. Ger.* tropfe, *m : O. H. Ger.* trofo, tropfo, *m. gutta : Dan.* dryp, *n ;* draabe, *m. f : Swed.* droppe, *m : Icel.* dropi, *m.*] DER. hleór-dropa, rēn-, spēd-, wǽg-, wŏp-, wrŏht-.

dropan, droppan; *pres. part.* ende; *p.* ede; *pp.* ed *To drop;* stillāre :— Swá swá dropan dropende ofer eorþan *sĭcut stillicĭdia stillantia super terram*, Ps. Spl. 71, 6. Droppende, Ps. Lamb. 71, 6. DER. dropa *a drop.*

dropen *stricken :*—Wæs feorh dropen *life was stricken*, Beo. Th. 5955, note; B. 2981; *pp. of* drepan.

dropen *dropped; pp. of* dreópan.

drop-fág *stronius?* Wrt. Voc. 289, 27.

drop-fáh, -fág; *adj.* [dropa *a drop*, fáh *coloured, stained*] *Drop-coloured, variegated in spots, spotted;* stillātus :—Stillātus, ðæt is on úre geþeóde, dropfáh *stillātus, that is in our language, spotted*, Herb. 131, 1; Lchdm. i. 242, 14. Wið dropfágum andwlatan *for a spotted face*, Med. ex Quadr. 5, 6; Lchdm. i. 348, 21.

dropian *to drop*, Ps. Th. 44, 10. v. dreópian.

drop-mǽlum; *adv. By drops, drop by drop;* guttātim :—Yrnþ dropmǽlum swiðe hluttor wæter *very clear water runs drop by drop*, Homl. Th. i. 508, 34. v. mǽl III.

droppan *to drop :*—Droppende *stillans*, Ps. Lamb. 71, 6. v. dropan.

droppetian, droppetan; *p.* ode, ede; *pp.* od, ed *To drop, fall by drops, distil ;* distillāre :—Heofonas droppetodon fram ansӯne Godes *cæli distillāvērunt a facie Dei*, Ps. Lamb. 67, 9. Fór ansӯne Drihtnes heofonas droppetaþ *the heavens drop before the face of the Lord*, Ps. Th. 67, 9.

droppetung, e; *f. A dropping, falling by drops, drop by drop ;* stillicĭdium :—Swá swá niðer astīhþ droppetung droppende ofer eorþan *as falling [rain] comes down, dropping over the earth*, Ps. Lamb. 71, 6.

dropung, e; *f. A dropping ;* stillicĭdium :—þurh dropunge deáwes and rēnes *through dropping of dew and rain*, Ps. Th. 64, 11: Ps. Vos. 71, 6. v. droppetung.

droren *fallen, perished; pp. of* dreósan.

dros DROSS, *filth, lees ;* sordes, fæx, aurĭcŭla, Cot. 14. [*Kil.* droes *fæx.*] v. drosna.

drosen-líc ; *adj. Brittle, weak ;* frăgĭlis, Som. Ben. Lye.

DROSNA, drosne, *nom. acc ; gen.* drosna; *dat.* drosnum; *pl. f. Grounds, sediment, lees, dregs ;* fæx, fæces :—Ðǽs drosna *hæc fæx*, Ælfc. Gr. 9, 70 ; Som. 14, 14; Wrt. Voc. 83, 22. His drosna [drosne, Ps. Spl. T. 74, 8] nis aĭdlad *fæx ejus non est exinānīta*, Ps. Lamb. 74, 9. Drosna *fæces*, Ælfc. Gl. 33; Som. 62, 25; Wrt. Voc. 28, 8. He gelǽdde me of fenne drosna *edūxit me de luto fæcis*, Ps. Spl. 39, 2. Of ðam drosnum *from the dregs*, Ps. Th. 39, 1. Hī druncon ŏþ ða drosna *usque ad fæces bibērunt*, Ælfc. Gr. 47; Som. 47, 45. Eles drosna *dregs of oil ;* amurca = ἀμόργη, Ælfc. Gl. 47; Som. 65, 18; Wrt. Voc. 33, 18. [*Kil.* droessem *fæx : Ger.* drusen, *f. fæx : M. H. Ger.* truosen, *f. barm, yeast : O. H. Ger.* truosana, trōsana *fæx, amurca.*]

drugaþ, drugoþ, e; *f.* [drige *dry*] *A* DROUGHT, *dryness ;* siccĭtas, arĭdĭtas :—Drugaþ [MS. drugaþe] *siccĭtas vel arĭdĭtas*, Ælfc. Gl. 96; Som. 76, 35; Wrt. Voc. 53, 43. Drugaþ oððe hǽþ *siccĭtas*, Wrt. Voc. 76, 77. Bearn Israēla eódon þurh drugoþe *filii Israel ambulāvērunt per siccum*, Ps. Lamb. fol. 189 a, 21.

drugian, he drugaþ, *pl.* drugiaþ; *p.* ode; *pp.* od; *v. n.* [drige *dry*] *To become dry, wither ;* arescĕre :—Drugaþ his ár on borde *his oar becomes dry on board*, Exon. 92 a; Th. 345, 15; Gn. Ex. 188. On mergen swá wyrt gewíteþ, on mergen blŏweþ and fareþ, on ǽfen afylþ, astĭðaþ, and drugaþ *mane sĭcut herba transeat, mane flŏreat et transeat, vespĕre decĭdat, indūret, et arescat*, Ps. Spl. 89, 6. Gif ðæt wæter hī ne geþwǽnde, ðonne drugode hió *if the water moistened it [the earth] not, then it would*

become dry, Bt. 33, 4; Fox 130, 8. DER. a-drugian, for-, ge-: unadrugod.

drugon *suffered, endured*, Beo. Th. 1601; B. 798; *p. pl. of* dreógan.

drugung, e; *f. A dryness, a dry place ;* siccĭtas, ināquōsus lŏcus :—Hī costadon God in drugunge *temptāvērunt Deum in siccĭtāte*, Ps. Surt. 105, 14: 77, 17.

druh, e; *m. Dust ;* pulvis :—Hwæt! druh ðū dreórega *lo! thou gory dust!* Soul Recd. 33; Seel. 17.

druncaþ *drink*, Exon. 99 b; Th. 373, 23; Seel. 114, = drincaþ; *pres. pl. of* drincan.

druncen *drunken*, Gen. 9, 21; *pp. of* drincan. DER. un-druncen, wíndruncen.

druncen, es; *n? e; f? Drunkenness ;* ēbriĕtas :—Ðæt he ne onbíte ǽniges þinges ðe druncen ofcume *that he taste not anything from which drunkenness may come*, L. Pen. 11; Th. ii. 280, 23. Druncen beorg ðē and dollíg word *guard thyself from drunkenness and foolish words*, Exon. 80 b; Th. 302, 10; Fä. 34. Gif hit þurh druncen gewurþe, bēte ðe deóppor *si ex ebriĭtāte accidĕrit, eo grăvius emendet*, L. M. I. P. 41; Th. ii. 276, 12. Gif ðū hwæt on druncen misdō, ne wít ðū hit ðam ealoþe *if thou have misdone in drunkenness, blame not the drink*, Prov. Kmbl. 39. ofer-druncen.

druncen-georn ; *adj. Drink-desirous, drunken ;* bĭbax, ebriōsus, R. Ben. 4.

druncen-hád, es ; *m.* [MS. -hed] *Drunkenness ;* ebriĕtas :—þurh heora druncenhád [MS. -hed] *through their drunkenness*, Chr. 1070; Th. 345, 42.

druncen-lǽt ; *adj. Slow ;* lentus, Cot. 124.

druncennes, druncennys, druncenys, -ness, e ; *f.* DRUNKENNESS ; ebriĕtas :—Warniaþ eów, ðe-læs eówer heortan gehefegode sӯn on druncenesse *attendĭte autem vobis ne forte graventur corda vestra in ebriĕtāte*, Lk. Bos. 21, 34. Ða hūs ða ðe on to gebiddenne geworhte wǽron syndon nū on hūs gehwyrfed oferǽta and druncennesse *the houses which were built to pray in are now turned into houses of gluttony and drunkenness*, Bd. 4, 25; S. 601, 13. Mid druncennysse *by drunkenness*, Ors. 1, 6; Bos. 29, 17. For ðære druncenysse *because of the drunkenness*, Gen. 19, 33, 35. On druccennysse and on wiste hiora wombe þeówiaþ, nas Gode *in drunkenness and feasting they minister to their belly, not to God*, L. Eccl. 45 ; Wilk. 195, 25; L. E. I. 45; Th. ii. 440, 38. v. drincan II.

druncen-scipe, es; *m. Drunkenness ;* ebriĕtas, Som. Ben. Lye.

druncen-wille ; *adj. Drunken ;* ebrius :—Drincþ mid ðam druncenwillum [drucen-willum MS.] monnum *bibit cum ebriis*, Past. 17, 8; Hat. MS. 24 a, 23.

drunc-mennen, es ; *n. A drunken maid-servant ;* ebria ancilla, Exon. 103 b; Th. 393, 32; Rä. 13, 9.

druncne *drunken*, Beo. Th. 965; B. 480; *nom. pl. of* druncen, *pp.*

druncnian ; *p.* ode; *pp.* od. **I.** *to be or become drunk;* inebriāri :—Iohannes se Fulluhtere ne dranc nǽðor ne wín, ne beór, ne ealu, ne nán ðære wǽtan ðe menn of druncniaþ *John the Baptist drank neither wine, nor beer, nor ale, nor of the liquor from which men become drunk*, Homl. Th. ii. 38, 7. Ðonne ða gebeóras druncniaþ *when the guests are drunk*, ii. 70, 27. **II.** *to sink, drown;* mergi :—Mid [MS. mid] ðӯ ne ongan druncnian [MS. druncnia] *cum cæpisset mergi*, Mt. Kmbl. Lind. 14, 30. DER. on-druncnian.

druncning, e; *f. A drinking ;* ebriĕtas :—Drencfæt oððe calic mín drincende oððe on druncninge lā hū scínende oððe hū beorht is *calix meus inebrians [in ebriĕtāte] quam præclārus est*, Ps. Lamb. 22, 5.

druncon *drank*, Lk. Bos. 13, 26; *p. pl. of* drincan.

drupian *to drop*, Ps. Spl. 67, 9. v. dreópian.

drupon *dropped; p. pl. of* dreópan.

druron *fell*, Kmbl. 1989; An. 997; *p. pl. of* dreósan.

drūsan, drūsian; *part.* drūsende; *p.* ode, ade; *pp.* od, ad; *v. intrans. To sink, become low, slow, inactive, to* DROWSE; cadĕre, lentum *vel* segnem esse :—Cēn drūsende *the sinking flame*, Elen. Kmbl. 2514; El. 1258. Lagu drūsade, wældreóre fág *the stream became slower, stained with deadly gore*, Beo. Th. 3265; B. 1630. He drūsende deáþ ne bisorgaþ *he cares not for death when he becomes inactive [by age]*, Exon. 61 a; Th. 223, 31; Ph. 368: 52 b; Th. 184, 33; Gū. 1353. v. drīsan.

druwian *to become dry, wither.* DER. a-druwian, for-. v. drugian.

DRÝ, drí; *gen.* drӯs; *dat. acc.* drӯ; *pl. nom. acc.* drӯas; *gen.* drīra? *dat.* drӯum, drīum; *m. A magician, sorcerer, wizard ;* magus, malĕfĭcus :—Drý *magus*, Wrt. Voc. 74, 41. Petres wiðerwinna wæs sum drý, se wæs Simon geháten: ðes drý wæs mid ðam awyrgedum gáste afylled *Peter's adversary was a certain sorcerer, who was called Simon: this sorcerer was filled with the accursed spirit*, Homl. Th. i. 370, 32: 374, 18: 376, 3: 380, 16: Homl. Blick. 173, 8, 18, 28, 32: 175, 6, 17, 31: 183, 17: 187, 32. He getengde wið ðæs drӯs *he hastened towards the magician*, Homl. Th. i. 374, 5. Petrus cwæþ to ðam drý *Peter said to the magician*, i. 372, 6: 380, 21: Homl. Blick. 173, 2, 9, 33: 175, 25. Hí woldon forbærnan ðone drý *they would burn the magician*, Homl. Th. i. 372, 30: 374, 22: 376, 10: 380, 23: Homl. Blick. 173, 11, 30: 175, 1: 181, 33. Ðū miht mid ðӯ gebēde blŏd

onhǽtan ðæs deófles drý *thou mayest with prayer heat the blood of the devil's wizard*, Salm. Kmbl. 89; Sal. 44. Hý drýas wǽron *they were sorcerers*, Exon. 70 a; Th. 260, 23; Jul. 301: Andr. Kmbl. 67; An. 34. Hý getrymedon hyra drýas *their magicians encouraged them*, Ors. 1, 7; Bos. 30, 21. Cwǽdon ða drýas to Pharaone *dixērunt malefīci ad Pharaōnem*, Ex. 8, 19: 9, ¶11. Drīra [drīa?] *magōrum*, Mone B. 4018. Herodes biswicen wæs from drýum oððe tungulcræftgum *Herōdes insulsus erat a magis*, Mt. Kmbl. Lind. 2, 16. Ðýlæs-ðe se deófol us be drīum [MS. drian] mǽge *lest the devil have power over us by sorcerers*, Glostr. Frag. 10, 30. [*Orm.* drig-menn *magicians*: Gael. draoi, draoidh, druidh, *m. a druid, magician*.]

drý-cræft, es; *m.* [cræft *craft, art*] *Magical art, magic, sorcery*; ars magica *vel* malefīca :—Hī sǽdon ðæt hió scolde mid hire drýcræft ða men forbredan *they said that she should overthrow the men by her sorcery*, Bt. 38, 1; Fox 194, 30. Gif hī hwylcne drýcræft hæfdon *si quid malefīcæ artis habuissent*, Bd. 1, 25; S. 486, 40: Ex. 7, 11. Sum man wæs mid drýcræfte bepǽht *some man was deceived by magic*, Homl. Th. i. 448, 13. Warna ðé ðæt ðū ne gíme drýcræfta ne swefena ne hwatena *nec inveniātur in te, qui ariolos sciscītētur et observet ·somnia atque augūria*, Deut. 18, 10. Drīfan drýcræftas *to exercise magical arts*, Bt. Met. Fox 26, 107; Met. 26, 54. Mid drýcræftum *by sorceries*, Ors. 1, 7; Bos. 30, 22.

drý-cræftig, drī-cræfteg; *adj.* [cræftig *crafty, skilful*] *Skilful or crafty in magic or sorcery, magical*; magicæ artis perītus, magīcus :—Sió, hī sǽdon, sceolde bión swīðe drýcræftigu *she, they said, would be very skilful in sorcery*, Bt. 38, 1; Fox. 194, 20. Pharaon gegaderude ealle ða drīcræftegustan men *vocāvit Pharao sapientes et malefīcos*, Ex. 7, 11.

drýfan *to drive*; pellēre :—Sceoldon drýfan *should drive*, Ors. 2, 4; Bos. 43, 10. v. drīfan.

drýfan; *p.* de, *pl.* don; *pp.* ed *To trouble, vex*; vexāre :—Mǽst hine drýfdon his ágene men [MS. mæn] *his own men vexed him most*, Chr. 1118; Erl. 246, 34. v. drēfan.

drygan; *p.* de; *pp.* ed *To dry, make dry, rub dry, wipe*; siccāre, tergēre, extergēre :—Se hāta sumor drygþ and gearwaþ sǽd and blēda *the hot summer dries and prepares seeds and fruits*, Bt. 39, 13; Fox 234, 14. Cómon twegen seolas of sǽlícum grunde, and hī mid heora flýse his fēt drygdon *two seals came from the sea-ground, and they dried his feet with their fur*, Homl. Th. ii. 138, 12. Hie beóþ oft drygde *they are often dried*, Past. 11, 4; Hat. MS. 15 a, 19. v. drigan.

dryge *dry*, Exon. 111 b; Th. 426, 22; Rä. 41, 77: Andr. Kmbl. 3161; An. 1583: Cd. 157; Th. 195, 28; Exod. 283. v. drige.

drygge *dry*, Bt. Met. Fox 7, 31; Met. 7, 16. v. drige.

dryg-nes, -ness *dryness*, Ps. Lamb. 94, 5. v. drig-nes.

dryht, driht, e; *f. A people, multitude, army*, in *pl.* men; pŏpŭlus, multitūdo, cāterva, fāmīlia, hŏmīnes :—Dryhtum to nytte *for use to people*, Exon. 113 a; Th. 433, 25; Rä. 51, 2. Ic dryhtum þeówige *I serve multitudes*, 104 a; Th. 394, 9; Rä. 13, 15: Cd. 146; Th. 182, 21; Exod. 79. Ðæt ðý deáþ-drepe drihta [MS. drihte] swǽfon *that the armies slept in the swoon of death*, Cd. 167; Th. 209, 7; Exod. 495: 217; Th. 275, 26; Sat. 177. Drihta bearnum *to the children of men*, 47; Th. 61, 6; Gen. 993: Exon. 95 b; Th. 357, 7; Pa. 25. [*Laym.* drihte *retinue*: O. Sax. druht, *only in composition, as* druht-folc *comĭtātus*, pŏpŭlus: *Frs.* dregte: O. Frs. dracht, drecht: M. H. Ger. O. H. Ger. truht, trut, *f. multitude*: Icel. drótt, *f.* pŏpŭlus. v. Goth. ga-drauhts, *m. a soldier*, from driugan *to do military service*: A. Sax. dreógan.] DER. folc-dryht, -driht, mago-; gedriht, gedryht, hī-, hý-, sib-, wil-.

Dryht', Driht', *or without the apostrophe* Dryht, Driht *The Lord*; Dŏmĭnus : *chiefly used in the interlinear Psalms, published by Spelman and by the Surtees' Society, for all the cases of* Dryhten, Drihten.

dryht-bearn, es; *n. A child of the people, a noble child*; puer pŏpŭlāris, nŏbilis :—Dryhtbearn Dena *the Danes' princely child*, Beo. Th. 4076; B. 2035.

dryht-cwēn, e; *f. A noble queen*; dŏmĭna et rēgīna :—Dryhtcwēn duguþa *a noble queen of chieftains*, Exon. 86 a; Th. 324, 21; Wíd. 98.

dryht-ealdor, driht-ealdor, driht ealdor, es; *m. The ruler of a household, meeting*, or *feast, a bridesman*; dŏmĭnus, archi-triclīnus, parănymphus = παράνυμφος :—Brýdguma *vel* dryhtealdor *parănymphus*, Ælfc. Gl. 87; Som. 74, 60; Wrt. Voc. 50, 42.

dryhten, drihten; *gen.* dryhtnes, dryhtenes; *m.*   **I.** *a ruler, lord, prince*; dŏmĭnus, princeps :—Geáta dryhten *the Goths' lord*, Beo. Th. 2973; B. 1484. Eorla dryhten *lord of earls*, Beo. Th. 4666; B. 2338. Dryhten Higelāc *lord Higelac*, Beo. Th. 4005; B. 2000. In gemynd his dryhtnes naman brohte *it brought his lord's name into his mind*, Exon. 114 b; Th. 440, 25; Rä. 60, 8.   **II.** *the supreme ruler, the Lord*; *chiefly used for God and Christ*; Dŏmĭnus :—Him Dryhten sylf, heofona heáhcyning, hlyt getǽhte *the Lord himself, high king of heaven, assigned a lot to them*, Andr. Kmbl. 10; An. 5. Dryhtna Dryhten *the Lord of lords*, Andr. Kmbl. 1747; An. 876. Dryhten God *the Lord God*, Exon. 96 a; Th. 358, 33; Pa. 55. Dryhten Crist *the Lord Christ*, Exon. 41 a; Th. 137, 25; Gú. 564. Ðe in Dryhtnes noman cwōme *who camest in the Lord's name*, Exon. 13 b; Th. 26, 5; Cri. 413. We fōr Dryhtene iu dreámas hefdon *we formerly had joys before the Lord*, Cd. 214; Th. 267, 26; Sat. 44. [*Laym.* drihten: *Orm.* drihtin: O. Sax. drohtin: O. Frs. drochten *Lord, only used for God and Christ*: O. H. Ger. truhtīn dŏmĭnus: Icel. dróttinn *princeps*.] DER. freá-dryhten, freó-, gum-, hleó-, man-, sige-, weoruld-, wine-.

dryhten-beáh *a lord-ring*. v. drihtin-beáh.

dryhten-bealo, -bealu; *gen.* -bealowes; *n.* [bealo *evil*] *Profound misery, extreme evil*; permagna calămĭtas :—He sceal dreógan dryhtenbealo *he shall suffer profound misery*, Exon. 88 a; Th. 330, 22; Vy. 55. Ellen biþ sēlast ðām ðe sceal dreógan dryhtenbealu *courage is best for those who must suffer extreme evil*, 52 b; Th. 183, 6; Gú. 1323.

dryhten-dōm, es; *m.* [-dōm *termination, q. v.*] *Sovereignty, majesty*; dŏmĭnātus, majestas :—Se hālga hērede on hēhþo heofoncyninges dryhtendōm *the saint praised the majesty of heaven's king on high*, Andr. Kmbl. 1997; An. 1001.

Dryhten-līc *belonging to the Lord, Lordly*. v. Drihten-līc.

Dryhten-līce *according to the Lord, by the Lord*. v. Drihten-līce.

dryhten-weard *a guardian-lord, king*. v. drihten-weard.

dryht-folc, driht-folc, es; *n.* [folc *a people*] *A nation, multitude*; pŏpŭlus, multĭtūdo :—Micel arīseþ dryhtfolc to dōme *a great multitude shall arise to judgment*, Exon. 23 a; Th. 64, 23; Cri. 1042. Dryhtfolca helm *a protector of nations*, 107 a; Th. 408, 24; Rä. 27, 17. Wæs deápe gedrenced drihtfolca mǽst *the greatest of nations was drenched with death*, Cd. 144; Th. 179, 26; Exod. 34: 160; Th. 198, 13; Exod. 322: 171; Th. 216, 2; Exod. 589.

dryht-gesíþ *an associate, attendant*. v. driht-gesíþ.

dryht-gestreón, es; *n.* [gestreón *a treasure*] *A nation's or people's treasure*; pŏpŭli ŏpes :—Eodor gefylled dryhtgestreóna *an inclosure filled with people's treasures*, Exon. 105 a; Th. 398, 25; Rä. 18, 3.

dryht-guma, driht-guma, an; *m. A popular man, man of the people, warrior, retainer, follower*;—*pl.* men, *people*; vir popŭlāris *vel* nōbilis, mīles, sătelles,—hōmĭnes :—Semninga biþ, ðæt ðec, dryhtguma, deáþ oferswýðeþ *suddenly it will be, that thee, warrior, death overpowers*, Beo. Th. 3540; B. 1768. Druncne dryhtguman dóþ swā ic bidde *the drunken retainers do as I bid*, 2466; B. 1231. Weccaþ of deáþe dryhtgumena bearn, eall monna cynn *the sons of men, all mankind, shall wake from death*, Exon. 20 b; Th. 55, 22; Cri. 887. Beóþ mōdsefan dālum gedǽled, sindon dryhtguman ungelíce *dispositions are by parts distributed, people are unlike*, 83 b; Th. 314, 31; Mōd. 22: 79 a; Th. 297, 23; Crä. 72.

drýhþ, ðū drýhst *does, thou doest*; *3rd and 2nd pers. pres. of* dreógan.

dryht-leóþ, es; *n.* [leóþ *a song*] *A lordly song*; nōbile carmen :—Be ðam Dauid cyning dryhtleóþ agōl *king David sang a lordly song of him*, Elen. Kmbl. 684; El. 342.

dryht-līc, driht-līc, driht-lec; *comp.* -līcra; *sup.* -līcest; *adj. Lordly, noble, distinguished*; principālis, nōbilis, eximius :—We gehýrdon ðæt mid Sigelwarum yppe wearþ dryhtlíc dōm Godes *we have heard that the lordly doom of God was revealed among the Ethiopians*, Apstls. Kmbl. 129; Ap. 65: Exon. 94 b; Th. 354, 1; Reim. 39. Sweord sceal on bearme, drihtlíc īsern *the sword shall be in the bosom, lordly iron*, Menol. Fox 511; Gn. C. 26. Him drihtlícu mǽg þuhte *she seemed a noble damsel to them*, Cd. 89; Th. 111, 2; Gen. 1849. Cwæþ drihtlecu mǽg, brýd to beorne *his noble mate, his wife, spake to the chief*, 133; Th. 168, 12; Gen. 2781. Drihtlíce cempan hyra sweord getugon *the noble warriors drew their swords*, Fins. Th. 29; Fin. 14: Beo. Th. 2320; B. 1158. Hí on dryhtlícestum dōme lifdon *they lived in the most lordly power*, Exon. 82 b; Th. 310, 35; Seef. 85.

dryht-líce, driht-líce; *adv. In a lordly manner, divinely*; nōbĭlĭter :—God leóht and þýstro gedǽlde dryhtlíce *God divinely parted light and darkness*, Exon. 11 a; Th. 14, 32; Cri. 228. Abraham fōr eorlum drihtlíce sprǽc *Abram spake in a lordly manner before the people*, Cd. 98; Th. 129, 4; Gen. 2138.

dryht-māþm, es; *m.* [māþm *a treasure*] *A noble or lordly treasure*; nōbĭles ŏpes :—Wearþ dryhtmāþma dǽl forgolden *his share of noble treasures was paid for*, Beo. Th. 5678; B. 2843.

dryht-né *a dead body of a host*. v. driht-né.

dryhtnes *of a lord*, Exon. 114 b; Th. 440, 25; Rä. 60, 8; *gen. of* dryhten.

dryht-scipe, driht-scipe, es; *m.* [-scipe *termination*] *Rulership, lordship, domination, dignity*; dŏmĭnātus, dignĭtas :—Ðara dōm leofaþ and hira dryhtscipe *their dignity and their lordship shall live*, Elen. Kmbl. 899; El. 451. For hwam nele mon him on giógoþe georne gewyrcan deóres dryhtscipes *why will not man in youth zealously work for himself bold rulership*? Salm. Kmbl. 775; Sal. 387. Sceolde hine yldo beniman ellendǽda dreámas and drihtscipes *age must take from him the joys of bold deeds and of rulership*, Cd. 24; Th. 31, 14; Gen. 485. Nalles feallan lēt dōm and drihtscipe *he let not his power and domination sink*, Cd. 60; Th. 73, 4; Gen. 1199. Ne lǽt ðīn dryhtscipe feallan *let not thy mighty rule fall*, Wald. 12; Vald. 1, 7.

dryht-sele, driht-sele, es; *m.* [sele *a dwelling, hall*] *A princely dwelling, hall*; aula :—Draca hord eft gesceát, drihtsele dyrnne *the*

dragon darted back to his hoard, his secret hall, Beo. Th. 4629; B. 2320: 1538; B. 767. Wæs drihtsele dreórfáh *the princely hall was stained with blood*, 974; B. 485.

dryht-sib, -sibb, e; *f.* [sib *peace, kinship*] *Peace between two nations, lordly kinship*; pax *vel* amícítia inter duas gentes:—Ic Heaðobeardna ne talige dryhtsibbe dæl Denum unfæcne *I esteem not part of the Heathobeards' lordly kinship to the Danes guileless*, Beo. Th. 4142; B. 2068.

dryht-weras *men, chieftains.* v. driht-weras.

dryht-wuniende; *part.* [wuniende, *part. of* wunian *to dwell*] *Dwelling among people*; in pŏpŭlo dēgens:—Ðara æghwylc mót dryhtwuniendra dæl onfón *each of those dwelling among people may receive a share*, Exon. 78 a; Th. 293, 26; Crä. 7.

drýman; *part.* drýmende; *p.* de; *pp.* ed *To rejoice, be joyful*; jubĭláre:—Hí mótun drýman mid Dryhten *they may rejoice with the Lord*, Exon. 32 b; Th. 102, 27; Cri. 1679. Him gefylgan ne mæg drýmendra gedryht *the multitude of the joyful cannot follow him*, Exon. 60 b; Th. 222, 13; Ph. 348. Eall druncon and drýmdon *all drunk and rejoiced*, Cd. 133; Th. 168, 11; Gen. 2781. Drýmaþ Gode eall eorþe *jubĭláte Deo omnis terra*, Ps. Spl. 97, 5, 7: 46, 1. v. drēman.

drýme *a song*, Som. Ben. Lye. v. dreám.

drý-men *magicians, sorcerers*, Homl. Th. ii. 472, 14. v. drý.

drýming, e; *f. A soft* or *murmuring noise*; sŭsurrus, Som. Ben. Lye. v. dreám.

drync, es; *m. Drink, a drink, draught*; potus, haustus:—Ðǽr wæs ælcum genóg drync *there was enough drink for each*, Andr. Kmbl. 3069; An. 1537. Ic ofþyrsted wæs gǽstes drynces *I was thirsty for the soul's drink*, Soul Kmbl. 82; Seel. 41. Drync ðú selst us *potum dabis nobis*, Ps. Lamb. 79, 6; Andr. Kmbl. 44; An. 22: Exon. 29 a; Th. 88, 12; Cri. 1439. Of mistlícum dryncum *from various drinks*, Bt. 37, 1; Fox 186, 17. DER. heoru-drync, ofer-. v. drinc.

drync-fæt, es; *n. A drinking-vessel*; pōcŭlum:—Gesáwon dryncfæt deóre *they saw the precious drinking-vessel*, Beo. Th. 4500; B. 2254: 4601; B. 2306. v. drinc-fæt.

dryncþ *drinks*, Ps. Spl. 74, 8; *3rd pres. sing. of* drincan.

drynge, es; *m. Drink*; potus:—Drynge mínne [MS. min] mid wópe ic gemengde *potum meum cum fletu tempĕrābam*, Ps. Spl. 101, 10. v, drinc.

drynge *I drink*, Ps. Spl. 49, 14; *for* drince. v. drincan.

drypan; *p.* de, te; *pp.* ed *To drop, mòisten*; stillāre, humectāre:—Nime ánne eles dropan, and drype on án mycel fýr *take a drop of oil, and drop it on a large fire*, Ors. 4, 7; Bos. 88, 11: L. M. i, 3; Lchdm. ii. 40, 5, 7, 24, 28, 30. Heó drypte in ða eágan *she dropped it on the eyes*, Guthl. 22; Gdwin. 98, 2. Míne handa drypton myrran *my hands dropped myrrh*, Homl. Th. i. 118, 4. He bæd ðæt Lazarus móste his tungan drypan *he prayed that Lazarus might moisten his tongue*, i. 330, 29. DER. ge-drypan. v. dropa.

drype, es; *m. A stripe, blow*; ictus:—Ðéh ðú drype þolie *though thou suffer a stripe*, Andr. Kmbl. 1910; An. 957: 2436; An. 1219. v. drepe.

drýpst, he drýpþ *droppest, drops; 2nd and 3rd pers. pres. of* dreópan.

dryre, es; *m. Fall, decline, ceasing*; cāsus, lapsus, cessātio:—Hrímes dryre *a fall of rime*, Exon. 56 a; Th. 198, 27; Ph. 16. Ðær wæs ne dreámes dryre *there was no ceasing of joy*, 44 b; Th. 152, 1; Gú. 802. DER. fær-dryre. v. dreósan.

dryrmian *to make sad, to be made sad, to mourn*; lugēre:—Dryrmyde, Cd. 144; Th. 180, 5; Exod. 40. v. drysmian.

drys-lic, dris-líc; *adj. Fearful, terrible*; terrĭbĭlis:—Ahwilc *vel* egeslíc *vel* dryslíc *terrĭbĭlis*, Ælfc. Gl. 116; Som. 80, 65; Wrt. Voc. 61, 43. v. on-drislíc, an-drysenlíc, an-drysne, drysne.

drysmian, dryrmian; *p.* ode; *pp.* od *To become dark, gloomy, to be made sad, to mourn*; calĭgāre, obscūrāri, mœstĭtia affici, lugēre:—Óþ-ðæt lyft drysmaþ *until the air grows gloomy*, Beo. Th. 2755, note; B. 1375.

drysnan; *p.* ede; *pp.* ed *To put out, quench, extinguish*; extinguĕre:—Ðæt fýr ne biþ drysned *ignis non extinguĭtur*, Mk. Skt. Rush. 9, 46. DER. ge-drysnan, un-drysnende, un-adrysnendlíc.

drysne *terrible*; reverendus. v. on-drysne.

drýst *rushest, rushes; 2nd and 3rd pers. pres. of* dreósan.

DUBBAN; *p.* ade; *pp.* ad *To strike*, DUB, *create*; percutĕre, creāre:—Se cyng dubbade his sunu Henric to rídere *the king dubbed* [or *created*] *his son Henry a knight*, Chr. 1085; Erl. 219, 1. [*R. Brun.* dubbid, *p*: *Chauc.* dubbed: *Laym.* dubben: *Swed.* dubba: *Icel.* dubba, dybba: *Fr.* dauber *to strike*.]

DUCE, an; *f. A* DUCK; anas:—On ducan seáþe, of ducan seáþe *to the duck's pond, from the duck's pond*, Cod. Dipl. 538; A.D. 967; Kmbl. iii. 18, 16, 17: Apndx. 308; A.D. 875; Kmbl. iii. 399, 18. [*Piers P. Chauc.* doke: *Plat.* dūker: *Kil.* duycker mergus.]

dúfan, ic dúfe, ðú dýfst, he dýfþ, *pl.* dúfaþ; *p.* ic, he deáf, ðú dufe, *pl.* dufon; *pp.* dofen *To* DIVE, *sink*; mergi:—Ic deáf under ýðe *I dived under the wave*, Exon. 126 b; Th. 487, 18; Rä. 73, 4: 113 b; Th. 434, 23; Rä. 52, 5. Dúfe seo hand æfter ðam stáne óþ ða wriste *let the hand*

dive after the stone up to the wrist, L. Ath. iv. 7; Th. i. 226, 16. Gif ðú dýfst *if thou sinkest*, Homl. Th. ii. 392, 35. Mid ðam ðe he deáf *when he was sinking*, ii. 392, 2: 390, 21. DER. be-dúfan, ge-, onge-, þurh-: dýfan.

dúfe-doppa, an; *m. A pelican*; pelĭcánus = πελεκανος:—Gelíc geworden ic eom niht-hræfne oððe dúfedoppan wéstennes *simĭlis factus sum pelĭcáno solĭtúdinis*, Ps. Lamb. 101, 7.

dúfian; *p.* ode, ede; *pp.* od, ed *To sink, immerge*; immergĕre, Ben. Lye.

DUGAN; *part.* dugende; ic, he deah, deag; ðú duge, *pl.* dugon; *p.* dohte, *pl.* dohton *To avail, to be of use, able, fit, strong, vigorous, good, virtuous, honest, bountiful, kind, liberal*; valēre, prödesse, frūgi esse, bŏnum esse, munĭfĭcum, *vel* libĕrālem se præbēre:—Ðonne his ellen deah *when his valour avails*, Beo. Th. 1151; B. 573: Andr. Kmbl. 920; An. 460: Bt. 29, 2; Fox 106, 1. Se ðe his heorte deah *he whose heart is good*, Cd. 219; Th. 282, 8; Sat. 283. Húru se aldor deah [Th. þeáh, Beo. 744], se ðǽm heaðorincum hider wísade *the chief is able indeed, who has led the warriors hither*, B. 369. Ðeáh ðú heaðoræsa gehwǽr dohte, grimre gúþe *though thou hast everywhere been vigorous in martial onslaughts, in grim war*, Beo. Th. 1057; B. 526. Gif he ǽr ne dohte *if he were not before virtuous*, Bt. 27, 2; Fox 98, 14. Dó á ðætte duge *do ever what is virtuous*, Exon. 80 a; Th. 300, 10; Fä. 4. Ðet him náðor ne dohte ne innhere ne úthere *so that neither the in-army nor the out-army was of use to them*, Chr. 1006; Th. 257, 15, col. 1. Swá swá hí sceoldon, gif hí dohton *as they ought, if they were honest*, Bt. 18, 3; Fox 64, 37. Ðæt ðú dohtest ðínum brêðer and wædlan and þearfan *that thou be bountiful to thy brother, to the poor, and to the needy*, Deut. 15, 11. Ðú us wel dohtest *thou wast truly kind to us*, Beo. Th. 3647; B. 1821: 2693; B. 1344. Hú me cyne-góde cystum dohten *how the good by race were munificently liberal to me*, Exon. 85 b; Th. 322, 1; Wíd. 56: 86 a; Th. 324, 4; Wíd. 89. Ða sceolon eall dugende beón swá swá hit gedafenaþ ðam háde *they shall all be virtuous so as is befitting the order*, L. Ælf. C. 16; Th. ii. 348, 16. [Dugan is the third of the twelve Anglo-Saxon verbs called *præterito-præsentia*, and given under ágan, *q. v.* The *inf.* dugan and the *pret.* deah, *pl.* dugon, retaining preterite inflections, are taken from the *p.* of a strong verb deogan, *p.* deah, *pl.* dugon; *pp.* dogen, ascertained from deah; *pl.* dugon, which shews the ablaut or internal change of the vowel in the *p.* of the twelfth class of Grimm's division of strong verbs [Grm. i. p. 898; Koch, i. p. 252], and requires by analogy with other verbs of the same class the *inf.* deogan and the *pp.* dogen; thus we find the original verb deogan, *p.* deah, *pl.* dugon; *pp.* dogen. The weak *p.* dohte, *pl.* dohton [= duhte, duhton], is formed regularly from the *inf.* dugan. The same *præterito-præsens* may be generally observed in the following cognate words:—

	inf.	*pres.*	*pl.*	*p.*
Piers P. Orm.		degh, dægh,		
O. Sax.	dugan,	dóg,	dugun,	
O. Frs.	duga,	duch,		
M. H. Ger.	tugen,	touc,		tohte,
O. H. Ger.	tugen,	touc,	tugun, *3rd pers. pl.*	tohta,
Goth.	dugan,	dáug,	dugum,	daúhta.]

dugeþ, dugoþ *good, virtuous, honourable*; bonus, probus, Mann. v. duguþ; *adj.*

dugoþ-gifu, e; *f.* [dugoþ = duguþ, gifu *a gift*] *Liberality, munificence*; largítas, munificentia:—Ic Wulfstán Lundeniscra manna bisceop mínes hláfordes dugoþgife ǽfre geþwǽrige *I Wulfstan, bishop of the London men, ever consent to my lord's munificence*, Cod. Dipl. 715; A. D. 1006; Kmbl. iii. 350, 36.

duguþ, dugoþ, e; *f.* [dugan valēre]. **I.** *manhood and all who have reached manhood*; ætas virĭlis [O. H. Ger. an dero tugende *in virĭli ætáte*, tugent daz ist die metilscaft des menniskinen alteris vĭres, hoc est mēdia virĭlis ætas, Graff's Sprch. v. 372]:—Todǽlan duguþe and geógoþe *to distribute to old and young*, Andr. Kmbl. 304; An. 152. Ymb-eóde ða ides Helminga duguþe and geógoþe dǽl æghwylcne *then the Helmings' dame went round every part* [*group*] *of old and young*, Beo. Th. 1246; B. 621: 323; B. 160: 3352; B. 1674: Andr. Kmbl. 2245; An. 1124. **II.** *multitude, troops, army, people, men, attendants, the nobles, nobility, the heavenly host*; cōpiæ, exercĭtus, pŏpŭlus, hŏmĭnes, comĭtátus, prócĕres, mīlĭtia cœlestis:—Duguþ samnade *the multitude collected*, Andr. Kmbl. 250; An. 125: 2542; An. 1272. Áhte ic holdra ðý læs, deórre duguþe *I owned the less of faithful ones, of dear attendants*, Beo. Th. 980; B. 488. Dugoþ Israhēla *the army of Israel*, Cd. 146; Th. 183, 13; Exod. 91: 167; Th. 209, 17; Exod. 500. Duguþe ðínre *to thy people*, Hy. 7, 69; Hy. Grn. ii. 288, 69. Ðæt is duguþum cúþ *that is known to men*, Andr. Kmbl. 1364; An. 682. Ðú ðe in Dryhtnes noman dugeþum cwóme *thou who camest in the Lord's name to men*, Exon. 13 b; Th. 26, 6; Cri. 413. Be ðám hringum mon mihte witan hwæt Rómána duguþe gefeallen wæs *by the rings one might know how many of the nobility of the Romans had fallen*, Ors. 4, 9; Bos. 91, 11: 3, 11; Bos. 74, 30: 1, 12; Bos. 35, 43. Se cining wæs

gefullod mid eallum his dugoþe *the king was baptized with all his nobility*, Chr. 626; Th. 43, 29: 1016; Th. 283, 30. He spræc mid duguþe ealdrum *lŏcūtus est cum magistrātĭbus*, Lk. Bos. 22, 4: 12, 11. Dugoþ Drihten hērigaþ *the heavenly host praises the Lord*, Cd. 170; Th. 213, 2; Exod. 546: Exon. 23 b; Th. 65, 32; Cri. 1063. *God and Christ are called* duguþa helm, dryhten, dēmend, etc. *helmet, lord, ruler, etc. of the hosts or heavenly hosts*, Cd. 216; Th. 274, 35; Sat. 164: Exon. 19 a; Th. 49, 7; Cri. 782: Andr. Kmbl. 173; An. 87.   III. *majesty, glory, magnificence, power, virtue, excellence, ornament*; majestas, magnificentia, potentia, virtus, dĕcus :—Ealra duguþa duguþ, Drihten Hǽlend *majesty of all majesties, Lord Saviour*, Hy. 3, 24; Hy. Grn. ii. 282, 24. He sōhte Drihtnes duguþe *he sought [entered into] the Lord's glory*, Cd. 60; Th. 73, 15; Gen. 1205. Wuldre benēmed, duguþum bedēled *bereft of glory, deprived of power*, Cd. 215; Th. 272, 19; Sat. 122: 212; Th. 263, 21; Dan. 765: Exon. 16 a; Th. 35, 24; Cri. 563. Seó duguþ ðæs wlītes ðe on gimmum biþ *the excellence of the beauty, which is in gems*, Bt. 13; Fox 40, 3. On ðǽm is swiotol sió gifu and ealla ða duguþa hiora fæder *in whom is manifest the ability and all the virtues of their father*, 10; Fox 28, 32. Simmachus seó duguþ ealles moncynnes *Symmachus the ornament of all mankind*, 10; Fox 28, 12.   IV. *advantage, gain, good, happiness, prosperity, riches, blessings, salvation*; commŏdum, lucrum, bŏnum, prospērĭtas, divĭtiæ, ōpes, sālus :—Hwæt ðú us to duguþum gedôn wille *what thou wilt do to our advantage*, Andr. Kmbl. 683; An. 342. Adrifen from duguþum *driven from good*, Cd. 106; Th. 140, 5; Gen. 2323. Gifa ðe him to duguþe Drihten scyrede *the gifts which the Lord had bestowed on him for his happiness*, 176; Th. 221, 12; Dan. 87. He him duguþa blǽd forgeaf *he gave them abundance of prosperity*, 121; Th. 156, 2; Gen. 2582. On ðǽre dægtíde duguþe wǽron *there were riches at that time*, 80; Th. 100, 5; Gen. 1659. Eallum bidǽled duguþum *and* dreámum *deprived of all blessings and joys*, Exon. 28 b; Th. 86, 16; Cri. 1409: Cd. 43; Th. 57, 18; Gen. 930.   V. *benefit, gift*; benefĭcium, mūnus, dōnum :—Secgan Drihtne þonc duguþa gehwylcre *to say thanks to the Lord for all benefits*, Exon. 16 b; Th. 38, 4; Cri. 601: 96 a; Th. 359, 3; Ps. 57: Cd. 74; Th. 91, 10; Gen. 1510.   VI. *that which is seemly, suitable, seemliness*; dĕcōrum :—He cúðe duguþe þeáw *he knew the usage of decorum [decorous usage]*, Beo. Th. 724; B. 359: 6330; B. 3175. Æfter duguþum *according to seemliness*, Cd. 104; Th. 137, 31; Gen. 2282. [*Laym.* duȝeðe *nobles* : *Plat.* dögt, *f. solidiness* : *O. Frs.* duged, *f. power* : *Ger.* tugend, *f. virtus* : *M. H. Ger.* tugent, *f* : *O. H. Ger.* tugad, *f. vis, rōbur, virtus* : *Dan.* ayd, *f* : *Swed.* dygd, *f* : *Icel.* dygð, *f. virtue*.] DER. ǽðel-duguþ, ealdor-, heofon-, woruld-.

duguþ, dugoþ, dugeþ; *adj. Good, honourable*; bonus, probus, Mann. v. dugeþ; *adj.*

duguþ-gifu, e; *f. Liberality*; munificentia, Som. Ben. Lye. v. dugoþgifu.

duhte *did good*, Chr. 1013; Erl. 149, 5, = dohte; *p. of* dugan.

dulfon *dug*, Ps. Th. 56, 8; *p. pl. of* delfan.

dulh-rune *pellitory*, L. M. 3, 8; Lchdm. ii. 312, 16. v. dolh-rune.

dulmūnus; *gen. pl.* dulmūna; *m. The war-ship of the Greeks, which king Alfred assures us would hold a thousand men*; longa nāvis. These ships were the μακρὰ πλοῖα or νῆες μακραί, generally called in Greek 'ὁ δρόμων, ωνος, *m. the light war-vessel of the Greeks*. They were the longæ nāves *the long war-ships* of the Romans, which had often more than fifty rowers. The Romans called their vessel drōmo, ōnis, defining it as *a fast rowing vessel*, evidently deriving their word from the Greek δρόμων, Cod. Just. 1, 27, 1, § 8; Cassiod. Var. 5, 17, *init.* where it is described as 'trīrēme vehĭculum rēmōrum tantum nŭmĕrum prōdens, sed hŏmĭnum fácies dilĭgenter abscondens.' Some suppose that Alfred derived his word dulmūnus from the *Icel.* drómundr, *m.* which Egilsson in his Lexĭcon Poëtĭcum, Hafniæ, 8vo. 1860, explains 'nāvis grandior, cūjus gĕnĕris tantum extra regiōnes septemtrionāles, ut in māri mediterrāneo, mentio fit,' S.E. i. 582, 3, Orkn. 82, 1, 3. Vigfusson, in his Icelandic-English Dictionary, 4to. Oxford, 1869-1874, in drómundr gives only the Latin and Greek, and O. H. Ger. drahemond as cognates. What Orosius calls longas nāves, Alfred translates dulmūnus in Anglo-Saxon. As we read in the Anglo-Saxon Chronicle of A. D. 897; Th. i. 174, 41,—Hēt Ælfrēd cyng timbrian lang-scipu ongēn ða æscas *king Alfred commanded to build long-ships against those ships*, v. æsc **IV**.—Alfred, in his translation of Orosius, says :—Ær he [Ercol] ongan mid Creáca scypum, ðe mon dulmūnus hǽt, ðe man segþ ðæt ǽn scip mǽge ǽn þúsend manna *before he [Hercules] began with Grecian ships, which are called dulmunus, of which it is said that one ship can hold a thousand men*, Ors. 1, 10; Bos. 33, 31-33. He [Xersis] hæfde scipa ðæra mycclena dulmūna ǽn M and ii hund *he [Xerxes] had one thousand two hundred of the large ships*, dulmunus, Ors. 2, 5; Bos. 46, 32, 33. v. Glossārium ad scriptōres mĕdiæ et infĭmæ Latinĭtātis Dōmĭni Du Cange, Dufresne; Francofurti ad Mœnum, 3 vols. fol. 1681, Dromōnes.

DUMB, *def.* se dumba, seó, ðæt dumbe; *adj.* DUMB, *speechless, mute*; mūtus, e-linguis :—Eart ðú dumb and deáf *thou art dumb and deaf*, Exon. 99 a; Th. 370, 26; Seel. 65: 108 b; Th. 414, 7; Rä. 32, 16. Beó

ðú dumb ôþ-ðæt ðæt cild beó acenned *be thou dumb until the child shall be born*, Homl. Th. i. 202, 7: L. Alf. pol. 14; Th. i. 70, 14. Dumb *mūtus*, Wrt. Voc. 75, 36: Mt. Bos. 12, 22: Lk. Bos. 11, 14. Se dumba fæder *the dumb father*, Homl. Th. i. 354, 27: Salm. Kmbl. 457; Sal. 229. Se dumba spræc *lŏcūtus est mūtus*, Mt. Bos. 9, 33: Lk. Bos. 11, 14. Dumbes *elinguis*, Glos. Prudent. Recd. 143, 1. Híg brohton him dumbne man *obtŭlērunt ei hŏmĭnem mūtum*, Mt. Bos. 9, 32: Mk. Bos. 9, 17: Ex. 4, 11. Gesēgun ða dumban gesceaft *they saw the dumb creation*, Th. Bos. 69, 30; Cri. 1128: 113 a; Th. 433, 3; Rä. 50, 2. Ða ôðre nigon consonantes synd gecwedene mūtæ, ðæt synd dumbe *the other nine consonants are called mūtæ, which are dumb*, Ælfc. Gr. 2; Som. 3, 1, 2. He dyde ðæt deáfe gehȳrdon, and dumbe sprǽcon *surdos fēcit audīre, et mūtos lŏqui*, Mk. Bos. 7, 37: Mt. Bos. 15, 31. Ic sceal dǽda fremman swā ða dumban neát *I shall do deeds such as the dumb cattle*, Andr. Kmbl. 134; An. 67. Dumbra *of the dumb*, Salm. Kmbl. 158; Sal. 78. Be dumbera manna dǽdum *of dumb men's deeds*, L. Alf. pol. 14; Th. i. 70, 13. Hí forgeáfon dumbum sprǽce *they gave speech to the dumb*, Homl. Th. i. 544, 33: 424, 10: Andr. Kmbl. 1153; An. 577: Exon. 68 a; Th. 251, 24; Jul. 150. [*Piers P.* dombe : *Wyc.* doumbe : *Chauc.* dombe : *Laym.* dumbe : *Orm.* dumb : *O. Sax.* dump *stultus* : *Frs.* domme, dom : *O. Frs.* dumbe, dume *stultus, mūtus* : *Dut.* dom *stupid* : *Ger.* dumm *stupid* : *M. H. Ger.* tump *stupid* : *O. H. Ger.* tumb *mūtus, stultus* : *Goth.* dumbs *mute* : *Dan.* dum *stupid* : *Swed.* dum *stupid* : dumb *mute* : *Icel.* dumbr *mute*.] DER. dum-nys : a-dumbian.

dumle? *the pelican*; onocrŏtălus = ὀνοκρόταλος, Cot. 23.

dumnys, -nyss, e; *f.* DUMBNESS, *speechlessness*; loquendi impotentia, Som. Ben. Lye.

DUN; *adj.* DUN, *a colour partaking of brown and black*; fuscus, aquĭlus :—Dun *fuscus*, Cot. 141, 147: natius [= nātīvus?], Ælfc. Gl. 79; Som. 72, 86; Wrt. Voc. 46, 43. Dunn *balidus* [= βαλιός?], Wrt. Voc. 289, 28. On ðone [MS. ðonne] dunnan stān *to the dun stone*, Cod. Dipl. 1120; A. D. 939; Kmbl. v. 238, 32. [*Chauc.* dunne, donne *dark-coloured* : *Ir.* dunn *a dun colour* : *Wel.* dwn *dun, swarthy, dusky* : *Gael.* donn *brown-coloured*.] DER. ásse-dun.

DÚN; *pl. nom. acc.* dúna, dúne; *f. A mountain, hill*, DOWN; mons, collis :—Seó dún, ðe se Hǽlend ofastāh, getácnode heofenan ríce *the mountain, from which Jesus descended, betokened the kingdom of heaven*, Homl. Th. i. 120, 21; 502, 2, 7: Exon. 101 b; Th. 384, 1; Rä. 4, 21. Deós dún *hic mons*, Ælfc. Gr. 9, 39; Som. 12, 58: 5; Som. 4, 8: Ps. Lamb. 67, 16: Wrt. Voc. 80, 42. Hie be hliðe heáre dúne eorþscræf fundon *they found an earth-cavern by the slope of a high hill*, Cd. 122; Th. 156, 26; Gen. 2594: Homl. Th. i. 502, 13. Betwux ðære dúne Sion, and ðam munte Oliueti *between mount Sion and the mount of Olives*, i. 440, 15: 502, 2, 9: 120, 10. Genôh lange ge wunodon on ðisse dúne *sufficit vobis, quod in hoc monte mansistis*, Deut. 1, 6: Gen. 31, 54: Mt. Bos. 24, 3. Stôpon stíðhycgende on ða dúne up *the stout-hearted went aloft upon the hill*, Elen. Kmbl. 1430; El. 717: Bt. Met. Fox 19, 20; Met. 19, 10: Cd. 228; Th. 307, 21; Sat. 683. Ðec heá duna hergen *high downs praise thee*, Exon. 54 b; Th. 193, 6; Az. 117. Of denum and of dúnum *from dells and from downs*, 107 b; Th. 409, 18; Rä. 28, 3: Cd. 69; Th. 84, 15; Gen. 1398: 71; Th. 85, 28; Gen. 1421. Seó wiht dúna briceþ *the creature will burst the hills*, Exon. 109 b; Th. 420, 6; Rä. 39, 6. Wurdon behelede ealle ða hēhstan dúna under ealre heofenan. And ðæt wæter wæs fíftyne fæðma deóp ofer ða hēhstan dúna *operti sunt omnes montes excelsi sub universo cœlo. Quindĕcim cŭbĭtis altior fuit aqua super montes, quæ operuĕrat*, Gen. 7, 19, 20. He gehleápeþ heá dúne *he shall leap the high downs*, Exon. 18 a; Th. 45, 10; Cri. 717. Seó stôw is on Oliuetes dúne ufeweardre *the place is on the high mount of Olives*, Homl. Blick. 125, 19. [*R. Glouc.* dounes *hills* : *Laym.* dune, *f* : *Orm.* dun *a hill* : *Plat.* dünen *sandhills on the seashore* : *Dut.* duin, *n* : *Kil.* duyne *agger mărinus* : *Ger.* düne, *f* : *O. H. Ger.* dún, dúna *mons* : *Fr.* dune, *f* : *Span.* dúnas, *pl. f* : *Ital.* dúna, *f. an elevation of sand thrown up by the sea* : *Ir.* dun, *m. a fortified hill, fortress* : *Corn.* dun, din, *f. a hill*.] DER. a-dún, -dúne, of-.

dún; *adj. Mountainous, hilly*; montānus :—To dún-landum *to hilly lands*, Deut. 1, 7. v. dún-land.

dún-elfen, e; *f.* [-ælfen *a fairy*] *A down or mountain-fairy*; castālis, ĭdis; *f. one of the muses*; castālides, um, *f.* Ælfc. Gl. 113; Som. 79, 112; Wrt. Voc. 60, 19.

dúne-ward, dúne-weard *downward*, Som. Ben. Lye. v. a-dúnweard.

dun-falu, dun-fealu; *adj.* [dun *dun*, fealu *fallow-coloured*] *Dun or tawny colour*; color cervŏrum :—Dun-fealu [MS. -falu] *cervīnus*, Ælfc. Gl. 79; Som. 72, 88; Wrt. Voc. 46, 45. v. fealo.

DUNG, e; *f.* DUNG; fimus, stercus :—Ic hine bedelfe, and ic hine beweorpe mid dunge *fŏdiam circa illam et mittam stercŏra*, Lk. Skt. Hat. 13, 8. [*Wyc.* dong, dung : *Piers P. Chauc.* donge : *Frs.* dong : *O. Frs.* dung : *Ger.* dung, *m. manure* : *M. H. Ger.* tunc, *f* : *O. H. Ger.* tunga, *f* : *Dan.* dynge, *m. f. a heap of dung* : *Swed.* dynga, *f* : *Icel.* dyngja, *f. a heap, dung*.]

Dún-holm, es; *m.* [*Flor.* Dunhelm : *Brom.* Durem, Durham : dún

a hill, holm *wa'er, an island*] DURHAM ; Dunelmia :—Hēr forlēt Ægelrîc bisceop his bisceoprîce æt Dûnholm *in this year* [A. D. 1056] *bishop Ægelric left his bishopric at Durham,* Ch. 1056 ; Erl. 191, 14. Đa menn hine befôron innan đære burh æt Dûnholme *the men surrounded him in the burgh at Durham,* Chr. 1068 ; Erl. 205, 34 : 1072 ; Erl. 211, 9, 29 : 1075 ; Erl. 212, 35 : 1080 ; Erl. 216, 12 : 1087 ; Erl. 224, 6, 32 : 1087 ; Erl. 226, 9 : 1096 ; Erl. 232, 39.

dûn-land, es ; *n.* Down *or hilly land ;* terra montāna : *it is opposed to* feld-land *plain or level land* :—Faraþ to Amorrēa' dûne and to ôðrum feld-landum and dûn-landum and to unhēheran landum *venîte ad montem Amorrhæōrum et ad cētĕra campestria atque montāna et humĭliōrđ lŏca,* Deut. 1, 7.

dûn-lendisc ; *adj.* Hilly, *mountainous land ;* montānus :—Sume sind *derivatîva,* swâ dûn-lendisc *montānus,* Ælfc. Gr. 5 ; Som. 4, 10.

Dunnan tûn, es ; *m. Dunna's town* = Dunnington.

dunnian, he dunnaþ, *pl.* dunniaþ ; *p.* ode ; *pp.* od *To make of a dun* or *a dark colour, to obscure, darken ;* obscūrāre :—Se môna đa beorhtan steórran dunnaþ [MS. dunniaþ] *the moon obscures the bright stars,* Bt. 4 ; Fox 6, 35.

dûn-sǣte ; *gen.* -sǣta ; *dat.* -sǣtum, -sǣtan ; *pl. m.* [dûn *a mountain,* -sǣte *dwellers, inhabitants*] Mountaineers, *inhabitants of the mountains of Wales ;* monticŏlæ Walliæ :—Đis is seó gerǣdnes đe Angelcynnes witan and Wealhþeóde rǣdboran betweox Dûnsǣtum [MS. Dûnsǣtan] gesetton *this is the ordinance which the witan of the English race and the counsellors of the Welsh nation established among the inhabitants of the mountains of Wales,* L. O. D. pref ; Th. i. 352, 2. Be Wentsǣtum and Dûnsǣtum. Hwîlon Wentsǣte hýrdon into Dûnsǣtan, ač hit gebýreþ rihtor into West-Sexan : đyder hý scylan gafol and gislas syllan. Eác Dûnsǣte beþyrfan, gif heom se cyning an, đæt man hûru friþgislas to heom lǣte *of the Gwents* [i. e. *the people of West Wales, in Carmarthenshire, Pembrokeshire, and Cardiganshire*] *and the Dûnsǣte. Formerly the Gwents belonged to the Dûnsǣte, but more properly they belong to the West Saxons : thither they shall give tribute and hostages. The Dûnsǣte also need, if the king grant it to them, that at least peace-hostages be allowed them,* L. O. D. 9 ; Th. i. 356, 16-20.

dûn-scrǣf ; *gen.* -scrǣfes ; *pl. nom. acc.* -scrafu ; *gen.* -scrafa ; *dat.* -scrafum, -scrǣfum ; *n.* [dûn *a mountain,* scræf *a den, cave*] *A mountain-cave ;* montāna caverna :—Dûnscrafu, *nom. pl. mountain-caves,* Exon. 56 a ; Th. 199, 12 ; Ph. 24. He sêceþ dýgle stôwe under dûnscrafum *he seeks a secret place among the mountain-caves,* 96 a ; Th. 357, 32 ; Pa. 37. Weras woldon to dûnscrǣfum drohtoþ sêcan *the men would seek a refuge in mountain-caves,* Andr. Kmbl. 3076 ; An. 1541.

Dûn-stân, es ; *m.* Dunstan ; Dunstānus :—Hēr S. Dûnstân wearþ geboren *in this year* [A. D. 925] *St. Dunstan was born,* Chr. 925 ; Th. 199, 4, col. 3. Hēr Eádmund cing betæhte Glæstinga beri S. Dûnstâne, đǣr he siđđan ǣrest abbod wearþ *in this year* [A. D. 943] *king Edmund delivered Glastonbury to St. Dunstan, where he afterwards first became abbot,* 943 ; Th. 211, 17-21, col. 3. On đam ylcan geáre wæs Dûnstân abbod adrǣfed ofer sǣ *in the same year* [A. D. 957] *abbot Dunstan was driven away over sea,* 957 ; Th. 217, 2-4, col. 1. Hēr Eádgâr sende æfter S. Dûnstâne, and geaf [MS. gif] him đæt bisceoprîce on Wigarceastre, and syđđan đæt bisceoprîce on [MS. an] Lundene *in this year* [A. D. 959] *Edgar sent after St. Dunstan, and gave him the bishopric of Worcester, and afterwards the bishopric of London,* 959 ; Th. 219, 25-29, col. 3. Hēr Sêe Dûnstân fêng to arcebisceoprîce *in this year* [A. D. 961] *St. Dunstan succeeded to the archbishopric* [*of Canterbury*], 961 ; Th. 218, 34, col. 1. On đissum geáre ealle đa yldestan Angelcynnes witan gefeóllon æt Calne of ânre upflôran, bûton se hâlga Dûnstân arcebisceop âna ætstôd uppan ânum beáme ; and sume đǣr swîđe gebrôcode wǣron, and sume hit ny [= ne] gedydon mid đam lîfe *in this year* [A. D. 978] *all the chief witan of the English race fell at Calne from an upper floor, but the holy archbishop Dunstan alone stayed upon a beam ; and some there were very much maimed, and did not escape with life,* Chr. 978 ; Th. 231, 30-39, col. 1. Hēr Dûnstân se hâlga arcebisceop forlēt đis lîf, and gefêrde đæt heofonlîce *in this year* [A. D. 988] *the holy archbishop Dunstan departed this life, and passed to the heavenly* [*life*], 988 ; Th. 239, 9-11, col. 1.

dûn-strǣt, e ; *f. A hilly road ;* via montāna, Som. Ben. Lye.

dunung, e ; *f. A noise ;* crêpitus, Som. Ben. Lye.

dûr, es ; *n. A door.* v. dûru, *pl. nom. n.* v. dôr, *n.*

dûre, es ; *f. A door ;* ostium, jânua :—To đære dûran *at the door,* Mk. Bos. 1, 33. v. dûru.

dûre-leás ; *adj. Doorless ;* sine jânua :—Dûreleás is đæt hûs *the house is doorless,* Anlct. 153, 24, col. 2.

dûreras ; *m. Folding doors ;* valvæ, Cot. 183.

dûre-þinen *a female door-keeper,* Jn. Bos. 18, 16. v. dûru-þinen.

dûre-weard, -werd, es ; *m. A door-ward, door-keeper,* Mk. Bos. 13, 34 ; Wrt. Voc. 81, 12 : L. Ælf. C. 11 ; Th. ii. 346, 28. v. dûru-weard.

durfon *laboured, perished ; p. pl. of* deorfan.

durne ; *adj. Retired, secret ;* reclûsus, secrētus :—On đone durnan [MS. durnen] crundel ; *of đam durnan crundelle on đone* þorn *to the*

retired barrow ; from the retired barrow to the thorn, Cod. Dipl. 1053 ; A. D. 854 ; Kmbl. v. 105, 26. v. dyrne.

DURRAN, ic, he dear, đû dearst, *pl.* durron, durran ; *p.* dorste, *pl.* dorston, dorstan ; *pp.* dorren *To* DARE, *presume ;* audēre :—Ne dear ic hâm faran *I dare not go home,* Gen. 44, 34 : Ex. 32, 30 : Cd. 40 ; Th. 54, 1 ; Gen. 870. Gif đû Grendles dearst neán bîdan *if thou darest abide near Grendel,* Beo. Th. 1059 ; B. 527 : Andr. Kmbl. 2700 ; An. 1352. Gif he gesêcean dear *if he dares to seek,* Beo. Th. 1373 ; B. 684. Ne durran we ôwêr gefêran *we dare not go anywhere,* Exon. 70 b ; Th. 262, 10 ; Jul. 330. Hî durron, Bd. 1, 27 ; S. 491, 33. Hwæđer đû durre gilpan *whether thou dare boast,* Bt. 14, 1 ; Fox 40, 22 : Bt. Met. Fox 11, 107 ; Met. 11, 54. Sêc gif đû dyrre *seek it if thou durst,* Beo. Th. 2763 ; B. 1379. Hwæđer he winnan dorste *whether he durst fight,* Ors. 4, 1 ; Bos. 97, 14 : Cd. 121 ; Th. 156, 15 ; Gen. 2589. Hî dorston, Beo. Th. 5688 ; B. 2848 : dorstan, Bd. 3, 11 ; S. 536, 41. Gif hî dorsten *if they durst,* Bt. Met. Fox 1, 54 ; Met. 1, 27. [Durran is the fourth of the twelve Anglo-Saxon verbs, called *prætĕrito-præsentia,* and given under âgan, *q. v.* The *inf.* durran and the *pres.* dear, *pl.* durron, retaining preterite inflections, are taken from the *p.* of the verb, ascertained from dear, *pl.* durron, which shews the ablaut or internal change of the vowel in the *p.* tense of the twelfth class of Grimm's division of strong verbs [Grm. i. p. 898 ; Koch, i. p. 252], and requires by analogy with other verbs of the same class the *inf.* deorran = deorşan [*Goth.* daursan] and the *pp.* dorren. Thus we find the original verb *inf.* deorran = deorsan ; *p.* dear, *pl.* durron ; *pp.* dorren. The weak *p.* dorste, *pl.* dorston [= durște, durston], is formed regularly from the *inf.* durran = dursan. The same *prætĕrito-præsens* may be generally observed in the following cognate words :—

	inf.	*pres.*	*pl.*	*p.*
Engl.	dare,	dare,	dare,	durst,
Wyc.	dore,	dar,	durn,	
Laym. Orm.		der, darr,	durren,	durste,
O. Sax.	gi-durran,	gi-dar,	durren,	gi-dorsta,
O. Frs.	thura,	thur, dur,	thuron,	thorste,
M. H. Ger.	turren,	tar,	turren,	torste,
O. H. Ger.	turran,	tar,	turrumês,	torsta,
Goth.	daursan,	dars,	daursum,	daursta.]

durste *durst,* Chr. 1154 ; Erl. 266, 4, = dorste ; *p. of* durran.

dûr-stodl, es ; *n. A door-post ;* postis :—Dûr-stodl *postes,* Wrt. Voc. 290, 15. v. dûru-stod.

dûru ; *gen.* e ; *dat.* e, a ; *acc.* e, a, u ; *pl. nom.* a ; *gen.* ena ; *dat.* um ; *acc.* a, u ; *f* : dûre, an ; *f. An opening, a door, the door of a house ;* ostium, jânua, fŏris :—Dûru ymbstandennesse welerum mînum '*keep the door* [*opening ostium*] *of my lips,*' Eng. versn. Ps. Lamb. 140, 3. Seó dûru wæs belocen *clausa erat jânua,* Mt. Bos. 25, 10. Dûru sôna on arn *soon he rushed on the door,* Beo. Th. 1447 ; B. 721. Dûra, Andr. Kmbl. 1998 ; An. 1001. Đâ đa dûra wǣron belocene *cum fores essent clausæ,* Jn. Bos. 20, 19. Of đære dûra *from the door,* Mt. Bos. 26, 71. Belocenum dûrum *januis clausis,* Jn. Bos. 20, 26. DER. eág-dûru, fôre-, helle-, hlîn- : dûru-leás, -stod, -þegn, *m.* -þînen = þignen, *f.* -weard : dýr : ge-dýre, ofer-gedýre. v. dôr, *n.*

dûru ; *pl. n.* Doors, Exon. 97 b ; Th. 364, 29 ; Wal. 78, = dôru. v. dôr, *n.*

dûru-leás *doorless ;* sine janua. v. dûre-leás.

dûru-stod, e ; *f.* [stod = studu *a post*] *A door-post ;* ostii postis, Cot. 157. v. dûr-stodl.

dûru-þegn, es ; *m.* [þegen *a servant*] *A door-keeper ;* jânîtor :—Dûruþegnum wearþ hildbedd stýred *the death-bed was spread for the door-keepers,* Andr. Kmbl. 2182 ; An. 1092.

dûru-þînen, dûre-þînen, e ; *f. A female door-keeper ;* ancilla ostiâria :—Cwæþ seó dûruþînen to Petre *dîcit Petro ancilla ostiâria,* Jn. Bos. 18, 17. Se leorningcniht cwæþ to đære dûreþînene *discîpulus dixit ostiâriæ,* 18, 16.

dûru-weard, dûre-weard, -werd, es ; *m. A door-keeper ;* jânîtor, ostiârius :—Se man beóde đam dûrewearde, đæt he wacige *homo janîtōri præcēpit ut vigĭlet,* Mk. Bos. 13, 34. Ostiârius is dûruweard se đe circan cǣgan healt *ostiārius is the door-keeper who holds the keys of the church,* L. Ælf. P. 34 ; Th. ii. 378, 5. Dûreweard *ostiārius,* Wrt. Voc. 81, 12. Ne sceal nân dûruwerd forsecgan nânne rǣdere mid nânre wrohte *non licet ostiārio ulli accûsāre lectōrem ullum ulla accusātiōne,* L. Ecg. C. 41 ; Th. ii. 168, 1, 3. Ostiārius [MS. Hostiārius] is đære cyrcean dûrewerd, se sceal mid bellan bîcnigan đa tîda, and đa cyrcan unlûcan geleáffullum mannum, and đam ungeleáffullum belûcan wiđûtan *ostiārius is the door-keeper of the church, who shall announce the hours with bells, and unlock the church to believing men, and shut the unbelieving without,* L. Ælf. C. 11 ; Th. ii. 346, 28-30. v. hâd II.

DUST, es ; *n.* DUST ; pulvis :—Hwæđer đē đæt dust hêrige *numquid confitĕbĭtur tibi pulvis?* Ps. Th. 29, 9 : Ps. Lamb. 77, 27. Ligeþ dust đǣr hit wæs *the dust shall lie where it was,* Exon. 99 b ; Th. 373, 8 ; Seel. 105 : 108 a ; Th. 412, 10 ; Rä. 30, 12. Hió wǣre fordrugod to

duste *it would be dried to dust*, Bt. Met. Fox 20, 207; Met. 20, 104: Salm. Kmbl. 630; Sal. 314: Exon. 98 a; Th. 368, 4; Seel. 16: Bd. 4, 30; S. 608, 30. Hí beóþ duste gelícran, ðonne hit wind toblǽwþ *tamquam pulvis, quem projícit ventus a fácie terræ*, Ps. Th. 1, 5: 89, 6. Asceacaþ ꝺæt dust of eówrum fótum *excútíte pulvĕrem de pĕdibus vestris*, Mk. Bos. 6, 11: Lk. Bos. 10, 11. [*Wyc. Chauc.* dust: *R. Glouc.* douste: *R. Brun.* doste: *Laym.* dust, doust, *n: Orm.* dusst: *Plat.* dust, *m: O. Frs.* dust: *Dut.* duist: *Ger.* dust, *m. pulvis: Dan.* dyst, *m. f: Icel.* dust, *n: Sansk.* dhû-li, *m;* from dhû *to shake,* Willms. 457.]

dust-drenc, es; *m.* *A drink made of the seeds of herbs rubbed to dust;* pótio ex herbārum quārumdam semĭnibus, in pulvĕrem redactis, compŏsíta:—Wyrc gódne dustdrenc: nim merces sǽd, and finoles sǽd, dilesǽd, etc. . . . gegníd ealle wel to duste: dó ꝺæs dustes gódne cuclerfulne on strang hluttor eala *make a good dust-drink* [*thus*]: *take seed of marche, and seed of fennel, dill-seed, etc. . . . rub all well to dust: put a good spoonful of the dust into strong clear ale,* L. M. 3, 12; Lchdm. ii. 314, 17–23.

dust-sceáwung, e; *f.* [sceáwung *a beholding, contemplation*] *A dust-viewing, contemplation of dust;* pulvĕris spectātio *vel* contemplātio:—He gewát from ꝺære dustsceáwunga *he departed from the contemplation of the dust,* Homl. Blick. 113, 29.

duþhamor, dyþhomar, es; *m. Papýrus* = πάπῡρος:—Duþhamor *papýrus,* Ælfc. Gl. 43; Som. 64, 39; Wrt. Voc. 31, 49.

DWǼS; *adj. Dull, foolish, stupid;* hĕbes, stultus, fátuus:—Dwǽs *vel* sott *hĕbes,* Ælfc. Gl. 88; Som. 74, 79; Wrt. Voc. 50, 59: 74, 35: Ælfc. Gr. 9, 26; Som. 11, 5. Abroten *vel* dwǽs *váfer vel fátuus vel* sócors, Ælfc. Gl. 9; Som. 56, 115; Wrt. Voc. 18, 62. Dwǽs *indŏcĭlis,* Glos. Prudent. Recd. 152, 26. [*Plat.* dwas: *O. Frs.* dwes: *Dut.* dwaas.] DER. ge-dwǽs.

dwǽscan; *p.* dwǽscede, dwǽscte; *pp.* dwǽsced, dwǽsct *To extinguish, put out;* extinguĕre:—Dryhten lǽnan lífes leahtras dwǽsceþ *the Lord extinguishes the crimes of this frail life,* Exon. 62 b; Th. 229, 17; Ph. 456: 128 b; Th. 493, 19; Rä. 81, 33. Feóndscype dwǽscaþ, sibbe sáwaþ on sefan manna *extinguish enmity, sow peace in the minds of men,* 14 b; Th. 30, 28; Cri. 486. DER. a-dwǽscan, to-: un-adwǽscendlíc.

dwǽs-nys, -nyss, e; *f. Dulness, foolishness, stupidity;* hĕbĕtūdo, stultĭtia, stūpĭdĭtas:—Dwǽsnys *vel* sotscipe *hĕbĕtūdo,* Ælfc. Gl. 88; Som. 74, 80; Wrt. Voc. 50, 60.

dwala *an error, doubt;*—Nánnes dwala is *non dubium est,* Mt. Kmbl. Præf. p. 2, 13. v. dwola.

dwalian; *p.* ede; *pp.* ed *To err;* errāre:—Híg dwaledon errāvĕrunt, Ps. Lamb. 57, 4. v. dwelian I.

dwán, *pl.* dwinon *pined, dwindled;* *p.* of dwínan.

dwás-líht, es; *n.* [dwǽs *dull,* líht *light*] *A false light;* ignis fatuus, Som. Ben. Lye.

dwealde, *pl.* dwealdon *deceived,* Bt. 35, 5; Fox 164, 32; *p.* of dwellan.

DWELAN, ic dwele, ðú dwelest, dwilst, he dweleþ, dwilþ, *pl.* dwelaþ; *p.* ic, he dwæl; ðú dwǽle, *pl.* dwǽlon; *pp.* dwolen; *v. n. To be led into error, err;* in errôrem dúci, errāre. [*O. Sax.* far-dwelan *to neglect: O. H. Ger.* twelan *torpēre.*] DER. ge-dwelan: dwelian, a-, ge-, ofa-: dwellan, ge-.

DWELIAN, dweligan, dweoligan, dwalian, dwolian, dwoligan; *part.* dweliende, dweligende; ic dwelige, ðú dwelast, he dwelaþ, *pl.* dwelíaþ, dwelígaþ, dwelígeaþ; *p.* ode, ede; *pp.* od, ed. I. *v. n. To be led into error, err;* in errôrem dúci, errāre:—Dwelian he dyde híg on wæglǽste oðꝺe bútan wege, and ná on wege *errāre fecit eos in invio, et non in via,* Ps. Lamb. 106, 40. Wæs ꝺæt dweligende sceáp ongeán fered *the wandering sheep was brought back,* Homl. Th. i. 340, 4. Dysige men, dweligende, sécaþ ꝺæt hêhste gód on ꝺa sámran gesceafta *foolish men, erring, seek the highest good in the worse creatures,* Bt. 33, 1; Fox 120, 12. Ꝺa seofon dweligendan steorran *the seven wandering stars, the planets,* Boutr. Scrd. 18, 26, 29. Ge dweliaþ *errātis,* Mt. Bos. 22, 29. Hú ne dweligaþ ge *nonne ídeo errātis?* Mk. Bos. 12, 24. Swýde ge dweligeaþ *multum errātis,* 12, 27. Ꝺa ongunnon clypian ꝺæt se rihtwísa dwelode *they begun to say that the righteous man erred,* Homl. Th. ii. 300, 17. He dyde ꝺæt ge dwelodon of ꝺam wege *ut errāre te facĕret de via,* Deut. 13, 5. Hí dwelodon on þwyrlícum dǽdum *they erred in perverse actions,* Homl. Th. ii. 398, 7: 46, 26. Hí dwelodon errāvĕrunt, Ps. Spl. 57, 3. II. *v. a. To lead into error, mislead, deceive;* in errôrem dúcĕre, decípĕre:—Ꝺæt folc dweliende *misleading the people,* Homl. Th. ii. 492, 35. Ic ꝺé ne dwelode *I have not deceived thee,* Bt. 35, 5; Fox 166, 1; 164, 32, MS. Bod. Me þincþ ꝺæt ðú me dwelige *methinks that thou misleadest me,* 35, 5; Fox 164, 12. [*O. Sax.* duelan *errāre: Frs.* dwæljen, dwyljen *to err: O. Frs.* dwela, dwila *to err: Dut.* dwálen *to err.*] DER. a-dwelian, ge-, ofa-.

DWELLAN, ic dwelle, ðú dwelest, dwelst, he dweleþ, dwelþ, *pl.* dwellaþ; *p.* dwealde, dwelede; *pp.* dweald, dweled. I. *v. a. To lead into error, deceive, mislead;* in errôrem dúcĕre, decípĕre:—Ic ꝺé ne dwelle *I do not deceive thee,* Bt. 35, 5; Fox 166, 1, MS. Cot. Ðú sǽdest ꝺæt ic ꝺé dwealde *thou saidst that I deceived thee,* 35, 5; Fox 164,

32. Me þincþ ꝺæt ðú me dwelle *methinks that thou misleadest me,* 35, 5; Fox 164, 12, MS. Cot. II. *v. a. To prevent, hinder, delay;* impēdíre, tardāre:—Ic dysge dwelle *I delay the foolish,* Exon. 103 b; Th. 392, 27; Rä. 12, 3. Ne hine wiht dweleþ, ádl ne yldo *nothing prevents him, disease nor age,* Beo. Th. 3475, note; B. 1735. Se ealda dweleþ miltse mid máne *the old one* [*the devil*] *prevents mercy with wickedness,* Frag. Kmbl. 62; Leás. 33. III. *v. n. To continue, remain, DWELL;* mănēre, habitāre:—Nero on ꝺam holte on cýle and on hungre dwelode, óþ-ꝺæt hine wulfas totǽron *Nero remained in the wood, in cold and hunger, until wolves tore him to pieces,* Homl. Th. i. 384, 10. [*Piers P.* dwelle *to inhabit: Chauc.* dwell *to inhabit: Orm.* dwellenn *to dwell, delay: O. Sax.* bi-dwelian *to delay, prevent: M. H. Ger.* twelen *morāri: O. H. Ger.* twălôn, twaljan, tweljan *morāri, impēdíre: Dan.* dwæle *to tarry, delay, dwell: Swed.* dwäljas *to dwell: Icel.* dvala *to delay;* dwelja *to dwell, wait, stay.*] DER. ge-dwellan.

dweola, dweolda *error, heresy.* DER. ge-dweola, -dweolda. v. dwola.

dweoligan; *part.* dweoligende *To err;* errāre:—Hí to ꝺam dweoligendum lǽcedómum deófolgylde efeston *they hastened to the erring cures of idolatry,* Bd. 4, 27; S. 604, 7. v. dwelian I.

DWEORG, dweorh, es; *m. A dwarf;* nānus:—Dweorg *pygmæus* vel *nānus* vel *pūmílio,* Ælfc. Gl. 114; Som. 80, 20; Wrt. Voc. 61, 1. Dweorh *nānus,* Wrt. Voc. 73, 53. [*Plat.* dwark, dwarf, *m: Frs.* dwirg: *O. Frs.* dwirg: *Dut.* dwerg, *m. f: Ger.* zwerg, *m: M. H. Ger.* twerc, *n: O. H. Ger.* twerg, *m: Dan.* dvárg, dverg, *m. f: Swed.* dverg, *m: Icel.* dvergr, *m.*]

dweorge-dwosle, -dwostle, an; *f.* [dweorg *a dwarf*] *The herb pennyroyal;* mentha pulēgium, Lin:—Herba pollēgion [= pulēgium], ꝺæt is dweorge-dwosle, Herb. cont. 94, 1; Lchdm. i. 38, 12. Ꝺeós wyrt, ꝺe man pollēgium [= pulēgium], and óðrum naman dweorge-dwosle nemneþ *this herb, which is called* pulēgium, *and by another name pennyroyal* [*dwarf dwosle*], Herb. 94, 1; Lchdm. i. 204, 6, 7: 156, 2; Lchdm. i. 282, 23: iii. 6, 19. Nim dweorge-dwoslan *take pennyroyal,* Herb. 106; Lchdm. i. 220, 10: iii. 6, 12. Dweorge-dwostle *pennyroyal,* L. M. 1, 48; Lchdm. ii. 120, 23: 2, 53; Lchdm. ii. 274, 9, 13. Lege dweorge-dwostlan gecowene on *lay on chewed pennyroyal,* 2, 30; Lchdm. ii. 228, 19: 2, 32; Lchdm. ii. 236, 10: 3, 1; Lchdm. ii. 304, 29: iii. 74, 5.

dwés *dull,* Som. Ben. Lye. v. dwǽs.

dwild, dwyld, es; *n. Error, heresy, a prodigy, spectre; error,* hærĕsis = αἵρεσις, prodigium, spectrum:—Wærþ mycel dwyld on Cristendóm *there was much error in Christendom,* Chr. 1129; Erl. 258, 29. On Engla land feole dwild weáren geseogen and geheórd *many prodigies were seen and heard in England,* 1122; Erl. 249, 13. DER. ge-dwild, -dwyld, mis-gedwield.

dwilman *to confuse, perplex, confound.* DER. for-dwilman.

dwimor, dwimer, dwymer, es; *n. An illusion, delusion, apparition, phantom; error,* fallācia, phantasma = φάντασμα. DER. ge-dwimor.

dwimor-líc; *adj. Visionary;* tamquam per visum, Som. Ben. Lye. DER. ge-dwymorlíc.

DWÍNAN, ic dwíne, ðu dwínest, dwínst, he dwíneþ, dwínþ, *pl.* dwínaþ; *p.* dwán, *pl.* dwinon; *pp.* dwinen *To pine, fade,* DWINDLE, *waste away;* tabescĕre:—Ðonne dwíneþ seó wamb sóna *then soon will the belly dwindle,* Herb. 2, 4; Lchdm. i. 82, 2. Dwinon *tabuĕrunt,* Cot. 190. [*Wyc.* dwyne, *p.* dwynede *to pine, waste away: Chauc.* dwined, *pp.* wasted, shrunk: *Plat.* dwinen *to vanish: Kil.* dwijnen *extenuāre, perīre: Dan.* tvine *to weep, vanish: Swed.* twina *to languish, pine away: Icel.* dvína, dvina *to dwindle, pine away.*] DER. a-dwínan, for-, ge-.

dwola, an; *m.* [dwolen, *pp.* of dwelan *to err*] *Error, heresy; error,* hærĕsis = αἵρεσις:—Seó mǽgþ on dwolan wæs lifigende *provincia in errôre versāta est,* Bd. 2, 15; S. 518, 42. Se [Arrianisca] dwola on ꝺam niwan sinoþe geníðerad wæs *Arriāna hærĕsis in Nicæna synôdo damnāta erat,* 1, 8; S. 479, 36, MS. B. DER. ge-dwola.

dwol-cræft, es; *m.* [cræft *a craft*] *Foolish craft, magic;* prāva *vel* magíca ars:—Him geblendon drýas þurh dwolcræft drync unheórne *the wizards mixed for them through magic a fatal drink,* Andr. Kmbl. 67; An. 34.

dwolema *darkness, chaos,* Bt. Met. Fox 5, 86; Met. 5, 43. v. dwolma.

dwolian, dwoligan; *part.* dwoliende, dwoligende; *p.* ede; *pp.* ed *To wander out of the way, err;* errāre:—Þurh monige stówe dwoliende *wandering through many places,* Bd. 4, 3; S. 570, 11. Dysige men, dwoliende, sécaþ ꝺæt hêhste gód on ꝺa sǽmran gesceafta *foolish men, erring, seek the highest good in the worse creatures,* Bt. 33, 1; Fox 120, 12, MS. Cot. Hider and ꝺider dwoligende *wandering hither and thither,* 36, 5; Fox 180, 12. To ꝺam dwoligendum deófolgyldum *to the erring idolatry,* Bd. 4, 27; S. 604, 7, MS. B. Híg dwolíaþ on heortan *hi errant corde,* Ps. Lamb. 94, 10. Ꝺa synfullan dwoledon *peccātôres errāvĕrunt,* Ps. Surt. 57, 4. DER. ge-dwolian. v. dwelian I.

dwol-líc; *def.* se -líca, seó, ꝺæt -líce; *adj. Foolish, erring, heretical;* stultus, hærĕticus:—Nis ꝺis nán dwollíc sagu *this is not a foolish saying,* Jud. 15, 19. Hý adwǽscdon ꝺa dwollícan lára *they extinguished the heretical doctrines,* L. Ælf. C. 33; Th. ii. 356, 11.

dwol-líce; *adv. Foolishly, heretically;* stulte, hærĕtíce:—Ne man ne

mót drincan, ne dwollíce plegan, ne etan innan tyrican *no one may drink, nor foolishly play, nor eat in a church*, L. Ælf. E; Th. ii. 392, 16: L. Ælf. C. 33; Th. ii. 356, 12. Ðe dwollíce leofaþ *who lives in heresy*, Hexam. 20; Norm. 28, 17.

dwolma, dwolema, an; *m. Chaos, a chasm, gulph*; chaos, *n.* = χάος, τό, hiātus:—Dwolma *chaos*, Cot. 40: 204. Betweox us and eów is mycel dwolma getrymed *inter nos et vos chaos magnum firmātum est*, Lk. Bos. 16, 26. Ða twegen tregan teóþ to-somne wið ðæt mód fóran mistes dwoleman *the two vexations draw together before the mind a chaos of darkness*, Bt. Met. Fox 5, 86; Met. 5, 43. DER. dwilman, for-.

dwolung, e; *f. Dotage*; delirāmentum, Cot. 69.

dworge-dwostle, an; *f. Pennyroyal*; pulēgium:—Nim dworge-dwostlan *take pennyroyal*, Lchdm. iii. 100, 25, 27. v. dweorge-dwoslè.

dwyld *error, heresy*, Chr. 1129; Erl. 258, 29. DER. ge-dwyld. v. dwild.

dwyrge-dwysle, an; *f. Pennyroyal*; pulēgium:—Hylwyrt oððe dwyrge-dwysle *pollēgia* [= *pulēgium*], Wrt. Voc. 79, 54. v. dweorge-dwosle.

dyd, e; *f. A deed*; actum:—Se consul [Fauius] gedyde ða bysmer-lícestan dyde *the consul* [*Fabius*] *did the most disgraceful deed*, Ors. 5, 2. Barrington, A. D. 1773, 180, 15. v. dæd.

dýdan; *p.* dýdde, *pl.* dýddon; *pp.* dýded, dýdd, dýd; *v. a.* [deád *dead*] *To put to death, kill*; morti trādĕre, occīdĕre:—Ne dýde man æfre on Sunnan dæges freólse ǽnigne forwyrhtne man *let not a man ever put any condemned man to death on the festival of Sunday*, L. C. S. 45; Th. i. 402, 9. DER. a-dýdan.

dyde *did*, Bt. 25; Fox 88, 36; *p. of* dón.

dyde; *acc. sing. of* dyd [dyde *what was done, p. of* dón *to do*] *a deed*; actum, Ors. 5, 2; Barrington, 180, 15,= dæde, Ors. 5, 2; Bos. 102, 21.

dyderian, dydrian; *p.* ode; *pp.* od; *v. trans. To deceive, delude*; illūdĕre:—Me þincþ ðæt ðú me dwelige and dyderie [dwelle and dydre, Cot.], swá mon cild dêþ *methinks that thou misleadest and deludest me, as any one does a child*, Bt. 35, 5; Fox 164, 12. DER. be-dyderian, be-didrian.

dyderung, dydrung, e; *f. An illusion, delusion, pretence*; delūsio, simūlātio:—Ðæs hálgan andwerdnyss acwencte ðæs deófles dyderunge *the presence of the saint quenched the delusion of the devil*, Homl. Th. ii. 140, 19. Hit wæs ðæs deófles dydrung *it was an illusion of the devil*, ii. 166, 6. He nys wís ðe mid dydrunge hyne sylfne beswícþ *non est sapiens qui simulātiōne semet ipsum decīpit*, Coll. Monast. Th. 33, 3. DER. be-dydrung.

dydest *didst, didst put*, Hy. 9, 55; Hy. Grn. ii. 292, 55: dydon *they did*, Lk. Bos. 10, 13; *p. of* dón.

dydrin, es; *m? A yolk*; vitellus:—Nim æges dydrin *take the yolk of an egg*, L. M. 1, 38; Lchdm. ii. 92, 20. [*Bav.* dottern, *m.*]

dýfan; *p.* de; *pp.* ed *To dip, immerse*; immergĕre:—Mec feónda sum dýfde on wætre *some enemy dipped me in water*, Exon. 107 a; Th. 407, 32; Rä. 27, 3. He hine on ðam streáme sencte and dýfde *he sank and immersed himself in the stream*, Bd. 5, 12; S. 631, 22. [*Icel.* dýfa *to dip*.] v. dúfan.

dýfen, e; *f. Desert, reward*; merĭtum:—Æft heora ge-earnungum and dýfene *juxta eorum merīta*, C. R. Ben. 2.

dýfing, e; *f. A diving*; immersio, urīnātio, Som. Ben. Lye.

Dyflen, Dyflin *Dublin*, Chr. 937; Th. 206, 14, col. 2; 207, 14, col. 1; Ædelst. 55. v. Difelin.

dýfst, he dýfþ *divest, dives*; *2nd and 3rd pers. pres. of* dúfan.

dýgan; *p.* dýgde; *pp.* dýged [dugan *vălēre*] *To do good, benefit*; prodesse, vălēre:—Ic secge ðæt sió fóresprǽc ne dýge náuðer ne ðam scyldigan, ne ðam ðe him fore þingaþ *I say that the defence does no good either to the guilty or to him who pleads for him*, Bt. 38, 7; Fox 210, 6. Ðæt ys to gelýfenne ðæt hit dýge *it is to be believed that it may benefit*, Herb. 2, 15; Lchdm. i. 84, 19. DER. ge-dýgan, -dígan, -dégan.

dýgel *secret, unknown*, Beo. Th. 2719; B. 1357: *gen. pl.* dýgelra, Exon. 92 b; Th. 347, 26; Sch. 18. v. dígol.

dýgle *secret, hidden*, Exon. 35 b; Th. 115, 7; Gú. 186: Cd. 178; Th. 224, 2; Dan. 130; *def. nom. f. n. of* dýgol. v. dígol.

dýgol *darkness*, Exon. 39 b; Th. 130, 13; Gú. 437. v. dígol.

dýgol *secret, hidden*. v. dýgle, dígol.

dýhst, he dýgþ, dýhþ *dyest, dyes*; *2nd and 3rd pers. pres. of* deágan.

dyht *a direction*, Chr. 1097; Erl. 234, 18. v. diht.

dyhte *arranged*, Mt. Bos. 25, 19: Gen. 16, 3; *p. of* dyhtan. v. dihtan.

DYHTIG, dihtig; *adj.* [dugan *vălēre*] *Doughty, strong*; vălĭdus:—Sweord ecgum dyhtig *a sword doughty of edges*, Beo. Th. 2578; B. 1287. Dihtig, Cd. 93; Th. 120, 11; Gen. 1993. [*Piers P.* douthy, doghty: *Chauc.* douhty: *Laym.* duhti: *Orm.* duhtig: *Plat.* dugtig: *Ger.* tüchtig: *M. H. Ger.* tühtic *able, strong, fit: O. H. Ger.* tugad-ig *virtuous: Dan.* dygtig: *Swed.* dugtig: *Icel.* dygðugr.] v. dohtig.

dyle *dill*, Wrt. Voc. 79, 9. v. dile.

dylsta? *pl.* dylstan *Matter, corruption, mucus*; tabum, mucus:—Fleó

ða mettas ða ðe him dylsta on innan wyrcen *let him avoid the meats which may work mucus in his inside*, L. M. 2, 29; Lchdm. ii. 226, 10. Ðǽr dylstan on synd *whereon the mucus is*, 1, 31; Lchdm. ii. 72, 20.

dylstiht; *adj.* [dylsta *matter*; -iht, *adj. termination*, q. v.] *Mattery, mucous*; mucōsus:—Gif hie dylstihte sién *if they be mucous*, L. M. 1, 29; Lchdm. ii. 70, 9.

dym-hofe *a lurking-place, hiding-place*, Ps. Lamb. 17, 12: Homl. Th. ii. 122, 33. v. dim-hofe.

dymlíc *dim, obscure*:—Ða dymlícan þeóstra *the dim darkness*, L. Ælf. C. 14; Th. ii. 348, 7. v. dimlíc.

dymnys *dimness, darkness*:—Dymnys cālīgo, Ælfc. Gr. 9, 3; Som. 8, 56. v. dimnes.

dyncge *ploughed land*, Mone B. 1434: 2326. v. dincge.

DYNE, dyn, es; *m. A* DIN, *noise*; sonus, fragor, strepitus:—Se dyne becom hlúd of heofonum *the din came loud from heaven*, Cd. 223; Th. 294, 5; Sat. 466. Cyrm, dyne *fragor*, Mone B. 4413: Cd. 221; Th. 288, 13; Sat. 380: 222; Th. 289, 7, 27; Sat. 394, 404. Ær he dóm-dæges dyn gehýre *ere he shall hear doomsday's din*, Salm. Kmbl. 546; Sal. 272: 650; Sal. 324. Dyne *fragŏre*, Mone B. 4425. [*Chauc.* dinne: *Dan.* dön, *n. a loud noise: Swed.* dän, *n. a* din, *noise: Icel.* dynr, *m. a* din, *noise*.] *Dan.* eorþ-dyne: ge-dyn, swég-.

dýneras *small pieces of money*, Ælfc. Gl. 106; Som. 78, 55; Wrt. Voc. 57, 35. v. digneras.

dyngan; *p.* ede; *pp.* ed [dung *dung*] *To* DUNG, *manure*; stercŏrāre. [*Piers P.* dongen: *Wyc.* dunge: *Frs.* dongjen: *O. Frs.* donga, denga: *Ger.* düngen: *M. H. Ger.* tungen: *Dan.* dynge *to heap up*.] DER. ge-dyngan.

dynge, dinge, dynige, es; *m? A noise, dashing, storm*; sonus, stre-pĭtus, procella:—On dynges mere *on the sea of noise*, Gst. Rthm. ii. 66, 20; Chr. 937; Th. 206, 12, col. 2. v. dyne.

dynian, he dyneþ; *p.* ede; *pp.* ed; *v. intrans.* [dyne *a din, noise*] *To make a noise,* DIN, *resound*; fragōrem edĕre, sŏnāre, perstrĕpĕre, clan-gĕre:—Gif eáran dynien *if the ears din*, L. M. 1, 3; Lchdm. ii. 40, 1: 42, 24. Dyneþ upheofon *heaven above shall resound*, Exon. 116 b; Th. 448, 25; Dóm. 59: 21 b; Th. 58, 5; Cri. 931. Hleóðor dynede *the noise resounded*, Andr. Kmbl. 1478; An. 740: Beo. Th. 1538; B. 767: Fins. Th. 61; Fin. 30: Judth. 10; Thw. 21, 18; Jud. 23: Exon. 94 b; Th. 353, 46; Reim. 28. Dynedon scildas *the shields rang*, Judth. 11; Thw. 24, 24; Jud. 204. [*O. Sax.* dunian *fragōrem edĕre: Swed.* dåna *to make a noise, ring: Icel.* dynja *to gush, shower: Lat.* tonāre *to make a loud noise, to thunder: Sansk.* dhan, dhvan *to sound, to cause a sound*.]

dynige *mountainous places*; montāna, L. M. 3, 8; Lchdm. ii. 312, 21, Som. Ben. Lye.

DYNT, es; *m.* **I.** *a stroke, stripe, blow*; ictus, plaga, per-cussio:—He, mid ðam dynte, nyðer astáh *he, with the blow, fell down*, Chr. 1012; Th. 268, 29, col. 2: Jn. Lind. Rush. War. 18, 22. On-drǽden him ðone dynt *let them fear the stroke*, Past. 45, 2; Hat. MS. 64 b, 23. **II.** *the mark or noise of a blow,—A bruise,* DINT, *noise, crash*; contusio, impressio, sonus:—Gif dynt sie, scilling; gif he heáhre handa dyntes onfêhþ, scilling forgelde *if there be a bruise, a shilling; if he receive a right hand bruise, let him* [*the striker*] *pay a shilling*, L. Ethb. 58; Th. i. 18, 1. Ne wyrnaþ deórra dynta *they are not sparing of severe dints*, Salm. Kmbl. 245; Sal. 122. Wyrcþ hlúdne dynt *makes a loud crash*, Bt. 38, 2; Fox 198, 9. [*Piers P. Chauc.* dint *a blow, knock: R. Brun.* dynt: *R. Glouc.* dunt, *pl.* dyntes: *Orm.* dinnt *a blow, stroke: Icel.* dyntr, dyttr, *m;* dynta, *f. a dint*.]

dýp, es; *n. The deep*; profundum:—Ofer dýpe, Exon. 101 b; Th. 384, 1; Rä. 4, 21. v. deóp.

dýpan; *p.* dýpte = dýpte [dýp *deep*] *To make deep, deepen, increase, augment*; profundius reddĕre, augēre:—We cwǽdon be ðám blaserum, ðæt man dýpte ðone áþ be þryfealdum *we have ordained concerning incendiaries that the oath be augmented threefold*, L. Ath. iv. 6; Th. i. 224, 14. [*Laym.* ideoped, *pp. deepened: Frs.* djepjen: *O. Frs.* diupa: *Dut.* diepen: *Ger.* tiefen in ver-tiefen *to make deeper: M. H. Ger.* tiefen *to deepen: Goth.* ga-diupyan: *Dan.* for-dybe: *Swed.* för-djupa: *Icel.* dýpka *to become deeper, to deepen*.]

dýpe, an; *f:* dýp, es; *n. Depth, the deep, sea*; profundum, altītūdo, altum:—Híg næfdon ðære eorþan dýpan *non habēbant altitūdĭnem terræ*, Mt. Bos. 13, 5. Ascúfaþ hine út on middan ðære dýpan *thrust him out into the middle of the deep*, Homl. Th. i. 564, 8. Teóh hit on dýpan *duc in altum*, Lk. Bos. 5, 4. v. deóp.

DYPPAN, dippan; ic dyppe, he dypþ, dyppeþ, *pl.* dippaþ; *p.* dypte; *pp.* dypped, dypd = dypt; *v. a. To* DIP, *immerge, baptize*; immergĕre, intingĕre, tingĕre, baptizāre:—Se ðe his hand on disce mid me dypþ *qui intingit mecum manum in cătīno*, Mk. Bos. 14, 20. Dyppe his finger ðǽron *let him dip his finger therein*, Lev. 4, 17. Biþ dipped fót ðín on blóde *ut intingātur pes tuus in sanguine*, Ps. Spl. 67, 25. Dippaþ ysopan sceaft on ðam blóde *fascĭculum hyssōpi tingĭte in sanguĭne*, Ex. 12, 22. Ic eówic dêpu oððe dyppe, se eówic dêpiþ oððe dyppeþ *ego baptizo vos, ipse baptīzābit vos*, Mt. Rush. Stv. 3, 11. Dyppende *baptīzantes*, 28, 19.

[*Wyc.* dippe: *Orm.* dippesst *dippest*: *Plat.* döpen *to baptize*: *O. Sax.* dōpian *baptizāre*: *Dut.* doopen *tp baptize, immerge*: *Ger.* taufen *to baptize*: *M. H. Ger.* toufen *to baptize*: *O. H. Ger.* toufen *baptizāre*: *Goth.* daupyan *to baptize*.] DER. be-dyppan, ge-, onbe-.

dȳr, es; *n. A door*; ostium, jānua :—We lǽraþ, ðæt mæssepreósta oððe mynsterpreósta ǽnig ne cume binnan circan dȳre būton his oferslipe *we enjoin, that no mass-priest, or minster-priest, come within the church-door without his upper vestment*, L. Edg. C. 46; Th. ii. 254, 9. v. dór.

dȳr *brave, bold*, Cd. 174; Th. 218, 10; Dan. 37. v. deór I.

dȳran *to hold dear, love* :—Dȳran *to hold dear*, Cd. 14; Th. 17, 9; Gen. 257. v. deóran.

Dyra wudu, Dera wudu; *gen. dat.* wuda; *m.* [Dere *the Deirians*, wudu *a wood: the wood of the Deirians*] *Beverley, Yorkshire*; oppĭdi nomen in agro Eboracensi :—Se sōþfæsta Berhthun eft wæs abbud ðæs mynstres ðæt ys gecȳged on Dyra wuda *veracissimus Bercthun nunc abbas monastērii quod vocātur in Derauuda, id est, in silva Derōrum*, Bd. 5, 2; S. 614, 29. - He wæs bebyriged in Sce' Petres portice on his mynstre ðæt is cweden in Dera wuda *sepultus est in portĭcu sancti Petri, in monastērio suo, quod dicĭtur in silva Derōrum*, 5, 6; S. 620, 21. Iohannes fōr to his mynstre on Dera wuda *John went to his monastery at Beverley*, Chr. 685; Erl. 41, 35.

dȳre; *adj.* **I.** *dear, beloved*; cārus, dilectus :—Se wæs him dȳre *he was dear to him*, Lk. Bos. 7, 2: Gen. 44, 5: L. Eth. vii. 22; Th. i. 334, 12: Chr. 942; Erl. 116, 9; Edm. 3: Cd. 63; Th. 75, 28; Gen. 1247: Exon. 42 b; Th. 143, 22; Gū. 665: Runic pm. 26; Kmbl. 344, 24; Hick. Thes. i. 135, 51: Ps. Th. 87, 1: Exon. 32 a; Th. 100, 33; Cri. 1651: Ps. Th. 88, 3: Exon. 9 a; Th. 7, 5; Cri. 96: Cd. 25; Th. 32, 22; Gen. 507: Exon. 54 b; Th. 192, 18; Az. 108: Ps. Th. 131, 5: Exon. 120 b; Th. 463, 14; Hö. 70: Menol. Fox 381; Men. 192: Elen. Kmbl. 583; El. 292. **II.** *dear of price, precious, costly*; prĕtiōsus, magni æstimandus :—Ou ðisum gēre wæs corn swā dȳre swā nán man ǽr ne gemunde, swā ðæt se sester hwǽtes eóde to lx penega, and eác furðor *in this year* [A. D. 1044] *corn was so dear as no man before remembered it, so that the sester of wheat went for sixty pence, and even more*, Chr. 1044; Erl. 168, 21: Exon. 94 b; Th. 354, 13; Reim. 45: Exon. 113 a; Th. 433, 12; Rä. 50, 6: Beo. Th. 4106; B. 2050: 4601; B. 2306: Beo. Th. 6089; B. 3048: 6253; B. 3131: Wanl. catal. 32, 16. v. deóre.

dyrfst, he dyrfþ *labourest, labours*; *2nd and 3rd pers. pres. of* deorfan.

dȳrling *a darling* :—Þeódnes dȳrling Iohannes *John, the Lord's darling*, Menol. Fox 230; Men. 116. v. deórling.

dyrnan; *p.* de; *pp.* ed; *v. a.* [dyrne *hidden, secret*] *To hide, secrete, restrain*; occultāre, celāre, obscurāre, cohibēre :—Ðeáh hí hit ǽr swíðe dyrndon *though they had before quite hidden it*, Ors. 5, 10; Bos. 108, 15. Ne mihte Iosep hyne leng dyrnan *non se potĕrat ultra cohibēre Ioseph*, Gen. 45, 1. DER. be-dyrnan, bi-, ge-.

dyrne, es; *n. A secret*; secrētum :—Nelle ic ðē mín dyrne gesecgan *I will not tell thee my secret*, Exon. 88 b; Th. 333, 11; Gn. Ex. 2.

DYRNE, dierne; *def.* se dyrna, seó, ðæt dyrne; *adj.* **I.** *close, hidden, secret, obscure*; occultus, secrētus, latens, obscūrus :—Ðá ðæt wíf geseah, ðæt hit [wíf] him næs dyrne *when the woman saw that she [the woman] was not hid from him*, Lk. Bos. 8, 47: Elen. Kmbl. 1443; El. 723: Menol. Fox 585; Gn. C. 62. Ne sceal dyrne sum wesan *nothing shall be secret*, Beo. Th. 548; B. 271. Dȳlæs ða smyltnesse ðæs dōmes gewemme oððe se dierna [dyrna MS. Cot.] æfst oððe tō hræd ierre *lest secret envy or too hasty anger corrupt the calmness of judgment*, Past. 13, 2; Hat. MS. 17 a, 12. Draca hord eft gesceát, dryhtsele dyrnne *the dragon darted back to his hoard, his secret hall*, Beo. Th. 4629; B. 2320. Hie hafaþ in siofan innan dyrne wūnde *they have within their mind a secret wound*, Frag. Kmbl. 57; Leás. 30. Ne sindon him dǽda dyrne *deeds are not hidden from him*, Exon. 23 a; Th. 65, 5; Cri. 1050: 39 b; Th. 130, 12; Gū. 437: 39 b; Th. 131, 32; Gū. 464. Ne dō ðú ne dyrne ðíne ða deóran bebodu *non abscondas a me mandāta tua*, Ps. Th. 118, 19: 134, 6. **II.** *dark, deceitful, evil*; tenebricōsus, subdŏlus :—Dyrne deófles boda wearp hine on wyrmes líc *the devil's dark messenger changed himself into a worm's body*, Cd. 24; Th. 31, 24; Gen. 490. Ðú mid lignum fare þurh dyrne geþanc *thou mayest come with lies through evil design*, 26; Th. 34, 3; Gen. 532: Exon. 115 a; Th. 442, 13; Kl. 12. Sceal mæg nealles inwit-net ōðrum bregdan dyrnum cræfte *a kinsman should not braid a net of treachery for another with deceitful craft*, Beo. Th. 4342; B. 2168. He to forþ gestóp dyrnan cræfte *he had stept forth with evil craft*, 4569; B. 2290. Ides sceal dyrne cræfte hire freónd gesēcan *the woman shall with deceitful art seek her friend*, Menol. Fox 547; Gn. C. 43. Dyrnra gǽsta *of evil spirits*, Beo. Th. 2718; B. 1357: Exon. 71 a; Th. 264, 22; Jul. 368. [*Piers P. Chauc.* derne *secret*: *Laym.* deorne, derne *secret*: *Orm.* dærne *secret, hidden*: *O. Sax.* derni *secret*: *O. Frs.* dern, dren in compounds *occultus*: *O. H. Ger.* tarni *latens*.] DER. un-dyrne.

dyrne-geliger; *gen.* -geligre; *f.* [dyrne *secret*, geliger *a lying*] *A secret lying, adultery*; adultĕrium :—Heó hæfde dyrne-geligre *she [Eurydice] had secret adultery*, Ors. 3, 11; Bos. 73, 39: Ps. Spl. C. 72, 26. v. geliger.

dyrn-gewrit, es; *n.* [dyrne *secret*, gewrit *a writing*] *A secret writing, in the pl. books whose authors are not known, the apocryphal books*; occulta scripta, apocrypha, Cot. 10.

dyrn-liogan; *part.* -licgende, -licgynde [dyrne *secret*, licgan *to lie*] *To lie secretly, to fornicate*; fornicāri :—Dyrnlicgynde *fornicāti sunt*, Ps. Spl. C. 105, 36.

dyrodine, derodine? *Scarlet dye or colour*; coccus = κόκκος :—On ðæs sacerdes hrægle wæs dyrodine twegera bleó *on the priest's raiment there was twice-dyed scarlet*, Past. 14, 6; Hat. MS. 18 b, 1. Ðæt hrægl wæs beboden ðæt scolde bión geworht of purpuran and of tweóbleóm derodine *superhumerale ex purpura et bis tincto cocco fieri præcipitur*, 14, 4; Hat. MS. 18 a, 3.

dyrre *durst*, Beo. Th. 2763; B. 1379; *subj. pres. of* durran.

dȳrre *dearer, more precious*; prĕtiōsior, carior, *comparative of* dȳre **II** :—Forðonðe hí sint dȳrran ðonne ǽnige ōðre *because they are dearer than any others*, Ors. 5, 2; Bos. 101, 25.

dȳrsian, to ; *f. Tribulation*; tribŭlātio. DER. ge-dyrst.

dyrst, e; *f. Tribulation*; tribŭlātio. DER. ge-dyrst.

dyrste-líce; *adv. Boldly*; audacter :—Dyrstelíce *audācĭter*, Ælfc. Gr. 38; Som. 41, 66. Iosep dyrstelíce in to Pilate eóde *Ioseph audacter introivit ad Pilātum*, Mk. Jun. 15, 43. DER. ge-dyrstelíce.

dyrstig; *adj. Daring, bold, rash*; audax, ausus :—Ðeós and ðis dyrstige *audax*, Ælfc. Gr. 9, 60; Som. 13, 41. Dyrstig oððe gedyrst-lǽht *ausus*, 41; Som. 43, 29. Hú wǽre ðú dyrstig ofstician bār *quomōdo fuisti ausus jugŭlare aprum*? Coll. Monast. Th. 22, 13: Bd. 2, 6; S. 508, 25, note: Nicod. 12; Thw. 6, 23. Ðeáh ðe he dyrstig wǽre *though he were daring*, Beo. Th. 5669; B. 2838. DER. ge-dyrstig, unge-.

dyrstigan; *p.* ede; *pp.* ed *To dare*. v. ge-dyrstigan.

dyrstig-líce; *adv. Boldly*; audacter, Mk. Bos. 15, 43. v. dyrste-líce.

dyrstignes, dyrstnes, -nyss, e; *f. Boldness, presumption, arrogance, rashness*; audācia, temerĭtas :—Sió gedyrstignes [MS. Cot. dyrstignes] his mōdes *præsumptio spīrĭtus*, Past. 13, 2; Hat. MS. 17 a, 15. Ðæt ðín mōd ne beó ahafen mid dyrstignysse [dyrstnysse, Nat. S. Greg. Els. p. 39, note 1] *that thy mind be not lifted up with arrogance*, Homl. Th. ii. 132, 4. DER. ge-dyrstignes.

dyrsting-panne, an; *f. A frying-pan*; sartāgo, frixōrium, Ælfc. Gl. 25; Som. 60, 59; Wrt. Voc. 25, 1. v. hyrsting-panne.

dyrst-lǽcan; *p.* -lǽhte; *pp.* -lǽht *To dare*; audēre :—Ðæt nán ne dyrstlǽce ceósan hláfordas of lǽwedan mannan *that none dare to choose lords of laical men*, Chr. 796; Ing. 82, 26. DER. gedyrst-lǽcan, lǽcan.

dyrst-líc; *adj. Bold*; audax. v. un-dyrstlíc.

dyrstnys, -nyss *arrogance*, Nat. S. Greg. Els. p. 39, note 1. v. dyrstignes.

dȳr-wurþe; *comp. m.* -wurþra; *f. n.* -wurþre; *adj. Of great worth or value, precious*; prĕtiōsus :—Seó ðe dȳrwurþre wǽre eallum māþmum *quæ omnĭbus ornamentis prĕtiōsior est*, Bd. 2, 12; S. 514, 40. v. deór-wyrþe.

dyseg *foolish*, Bt. Met. Fox 19, 57; Met. 19, 29. v. dysig.

dysegian, dysgian, dysian; *part.* dysigende, dysiende; he dysegaþ; *p.* ede, ode; *pp.* ed, od; *v. intrans.* [dysig *foolish*]. **I.** *to be foolish, act foolishly, err*; ineptīre, errāre :—Ða, dysiende, wēnaþ ðætte ðæt þing sié ælces weorþscipes wyrþe *they, foolish, think that the thing is worthy of all estimation*, Bt. 24, 4; Fox 86, 9. He dysegaþ, se ðe wile sǽd ōpfæstan ðám driúm furum *he does foolishly, who will sow seed in the dry furrows*, 5, 2; Fox 10, 30. Ðæt ða dysegien *that they are foolish*, 24, 4; Fox 86, 9, MS. Bod. Ðæt hí on heortan hyge dysegedon hí errant corde, Ps. Th. 94, 10. **II.** *to talk foolishly, blaspheme*; blasphēmāre :—Manega ōðre þing híg him to cwǽdon dysigende *alia multa blasphĕmantes dicēbant in eum*, Lk. Bos. 22, 65. He dysegaþ *blasphēmat*, Mk. Bos. 2, 7.

dyselíc *foolish*, Bd. 4, 27; S. 604, 2. v. dys-líc.

dysg; *adj. Foolish, weak, ignorant*; stultus, ignorans :—Dysgum monnum *by ignorant men*, Bt. 33, 4; Fox 130, 28. v. dysig.

dysgung, e; *f. Silliness, foolishness*; stultĭtia :—Wið dysgunge *against foolishness*, L. M. 1, 66; Lchdm. ii. 142, 1.

dysi *folly*, Bt. 36, 1; Fox 172, 8. v. dysig.

dysi *stupid*, Bt. Met. Fox 28, 130; Met. 28, 65. v. dysig; *adj.*

dysian; *part.* dysiende *to be foolish*, Bt. 24, 4; Fox 86, 9. v. dysegian.

DYSIG, dyseg, dysg, disig, disg, dysi; *adj.* DIZZY, *foolish, unwise, stupid*; stultus, insĭpiens, insānus :—Dysig ná ongyt ðás ðing *stultus non intellĭgit hæc*, Ps. Spl. 91, 6. He biþ swá dysig and swá ungewiss *he is so foolish and so ignorant*, Bt. 11, 2; Fox 34, 25. Ða dysige men *foolish men*, 33, 3; Fox 126, 8. Ða dysegan sint on gedwolan wordene *the foolish are in error*, Bt. Met. Fox 19, 57; Met. 19, 29. Híg sint dysegran *they are more foolish*, 19, 82; Met. 19, 41. Cyninga dysegast *the most foolish of kings*, 15, 22; Met. 15, 11. Dysegum neátum *jumentis insipientĭbus*, Ps. Th. 48, 11. Dysgum monnum *by unwise men*, Bt. 33, 4; Fox 130, 28: Bt. Met. Fox 28, 130; Met. 28, 65: Deut. 32, 21. [*Plat.* dusig, dösig, düsig *giddy*: *O. Frs.* dusig *giddy*: *Dut.* duizelig *giddy*: *Ger.* dusig, däsig *stupid*; duselig *giddy*: *O. H. Ger.* tusig *stultus, hĕbes*.]

dysig, disig, dysi, es; *n. An error, ignorance, folly, foolishness*; error, stultĭtia, insānia, insĭpientia :—Đæt is hefig dysig *that is a grievous folly*, Bt. Met. Fox 19, 1; Met. 19, 1: Bt. 32, 3; Fox 118, 7. Đē lícode his dysig and his unrihtwísnes *his folly and his injustice pleased thee*, 27, 2; Fox 96, 22. Dysi and unrihtwísnes nū rícsaþ ofer ealne middaneard *folly and wickedness now reign over all the mid-earth*, 36, 1; Fox 172, 8. Fægniaþ irmingas hiera ägnes dysiges and hearmes *the wretches rejoice at their own folly and sorrow*, Past. 35, 4; Hat. MS. 46 a, 14: Bt. 36, 5; Fox 180, 6. Ulcinienses and Thrusci đa folc forneáh ealle forwurdon for heora ägnum dysige *the Volscians and the Etruscans nearly all perished through their own folly*, Ors. 4, 3; Bos. 79, 43: Bt. 18, 2; Fox 64, 4. Ne lócaþ näfre to ídelnesse, ne to leásungum, ne to dysige *non respexit in vanĭtātes, et insānias falsas*, Ps. Th. 39, 4. Mīne wŭnda rotedan and fūledon for mínum dysige *computŭērunt et deterŏrāvērunt cicatrīces meæ, a fācie insĭpientiæ meæ*, 37, 5. Abigall forswigode đæt dysig hiere fordruncnan hláfordes *Abigail concealed the folly of her drunken lord*, Past. 40, 4; Hat. MS. 55 a, 12, 15: 45, 2; Hat. MS. 64 b, 25. Lífes weard of mōde abrit đæt micle dysig *the guardian of life removes from his mind that great ignorance*, Bt. Met. Fox 28, 156; Met. 28, 78: 19, 77; Met. 19, 39: Bt. 39, 3; Fox 216, 5: Past. 30; Hat. MS. 39 a, 5. Đeáh ic mid dysige þurhdrifen wäre *though I was thoroughly penetrated with folly*, Elen. Kmbl. 1410; El. 707: Ps. Th. 75, 4. We sinna fela didon for ūre disige *we committed many sins through our foolishness*, Hy. 7, 107; Hy. Grn. ii. 289, 107.

dysig-dóm, es; *m. Foolishness, ignorance*; impērĭtia, Pref. R. Conc.

dysig-nes, dysi-nes, -ness, e; *f.* DIZZINESS, *blasphemy*; stultĭtia, blasphēmia :—Wäron heó mid elreordre dysignesse onbläwne *inflāti erant barbăra stultĭtia*, Bd. 2, 5; S. 507, 13. Of manna heortan yfele geþancas cumaþ, dysinessa *de corde hŏmĭnum malæ cogĭtātiōnes procēdunt*, blasphēmia, Mk. Bos. 7, 22.

dys-líc, dyse-líc; *def.* se -líca, seó, đæt -líce; *adj. Foolish, stupid*; stultus :—Hit biþ swíđe dyslíc đæt se man beorce odđe bläte *it is very foolish that the man bark or bleat*, Ælfc. Gr. 22; Som. 24, 11: Bd. 1, 27; S. 493, 11. Oft ge dyslíce dæd gefremedon *often ye have done a foolish deed*, Elen. Kmbl. 771; El. 386. From đæm lífe đæs dyselícan gewunon *a vita stultæ consuetūdinis*, Bd. 4, 27; S. 604, 2. On dyslícum geswincum *in foolish labours*, Past. 18, 2; Hat. MS. 26 a, 11.

dys-líce; *adv. Foolishly*; stulte :—Se Godes cunnaþ ful dyslíce *he tempteth God very foolishly*, Salm. Kmbl. 455; Sal. 228. Dyslíce đū dydest *stulte opĕrātus es*, Gen. 31, 28.

dystig; *adj.* DUSTY; pulvĕrŭlentus, Cot. 183.

dyphomar *papўrus* = πάπῦρος :— [Nim] dyphomar [*take*] *papўrus*, L. M. 1, 41; Lchdm. ii. 106, 17. v. duþhamor.

DYTTAN; *p.* de; *pp.* ed *To* DIT, *close or shut up*; opprĭmĕre, occlūdĕre, obtūrāre :—Ongunnon đa Fariséi his mūþ dyttan *cæpērunt Pharisæi os ejus opprĭmĕre*, Lk. Bos. 11, 53. Anlíc nädran seó dytteþ hyre eáran *secundum similitūdĭnem serpentis obtūrantis aures suas*, Ps. Th. 57, 4. [*Laym.* dutte, *p. pl.* stopt: *Orm.* dittenn *to shut, stop:* O. Nrs. ditta *rimas occlūdĕre*, Rask Hald.] DER. for-dyttan.

dyxsas *dishes, platters*, Mt. Foxe and Jun. 23, 25, = discas; *pl. acc. of* disc.

E

A. Anglo-Saxon words, containing the short or unaccented vowel *e*, are often represented by modern English words of the same meaning, having the sound of *e* in *net, meṭ*; as, Nett, bedd, weddian, hell, well, denn, fenn, webb, ende. **2.** the short *e* in Anglo-Saxon generally comes (1) before a double consonant; as, Nebb, weccan, tellan, weddian: (2) before any two consonants; as, Twentig, sendan, bernan: (3) before one or two consonants, when followed by a long *o* by a final vowel; as, Selǝ, henne. **3.** e is often contracted from ea; as, Ceaster and cester *a burgh, fortified town;* eahta and ehta *eight*.

B. Words containing the long or accented Anglo-Saxon *ē* are very frequently represented by English terms of the same signification, with the sound of *e* in *heel*; as, Réc, méd, hēl, cwēn, gēs, fēt, tēþ, hēdan, fēdan, mētan *to meet.* Some remarks on the accented *ē* in Grimm's Deutsche Grammatik, 2nd Edit. Göttingen, small 8vo. 1822, vol. i. pp. 229, 230: 3rd Edit. small 8vo. 1840, vol. i. pp. 361, 362, may be found useful, and are especially recommended to the student of Anglo-Saxon. **2.** it is, however, difficult to say when the *ē* is long in Anglo-Saxon, but it may be useful to remember, the *ē* is often long before the single consonants *l, m, n, r, c, d, f, g, s, t,* and þ; as, in hēl *a heel*, fēlan *to feel*, dēman *to deem, think*, fēnix *a phœnix*, hēr *here*, gēs *geese*, fēt *feet*, fēdan *to feed*, tēþ *teeth*, bēc *books*, blēgen *a blain*, drēfan *to trouble.*

C. The Runic ᛗ not only stands for the vowel *e*, but also for the name of the letter in Anglo-Saxon, eh *a war-horse.* v. eh *a war-horse*, and RŪN.

-e, in the termination of nouns, denotes a person; as, Hyrde, es; *m. A shepherd*, from hyrdan *to guard.* The vowel *-e* is also used to form

nouns denoting inanimate objects; as, Cȳle, es; *m. Cold:* cwide, es; *m. A saying, testament:* brice, es; *m. A breach:* wlite, es; *m. Beauty.* These are mostly derived from verbs, and are masculine, but when derived from adjectives they are feminine; as, Rihtwíse, an; *f. Justice.*

-e, is the termination of derivative adjectives; as, Wyrđe *worthy*, from wyrþ *worth:* forþgenge *forthcoming, increasing.*

-e, is also the usual letter by which adverbs are formed from adjectives ending in a consonant; as, Rihte *rightly*, sóþlíce *truly*, yfele *badly.*

ē; *dat.* or *inst. to* or *from a river* :—Of đære ē Indus *from the river Indus*, Ors. 1, 1; Bos. 16, 25; *dat. sing.* v. eá.

EÁ; *often indeclinable in the sing, but eás is sometimes found in gen;* and *ē, ǽ, eǽ in dat; pl. nom. acc.* eá, eán; *gen.* eá; *dat.* eáum, eám, eán; *f.: ǽ; indecl. f. Running water, a stream, river, water;* flūvius, flūmen, torrens, aqua :—Eá of dūne *water from the hill*, Menol. Fox 520; Gn. C. 30. Seó feorþe eá ys gehäten Eufrates *flūvius quartus est Euphrātes*, Gen. 2, 14: Bd. 3, 24; S. 556, 34, 46. On twá healfe đære eás *on the two sides of the river*, Chr. 896; Th. 172, 39, col. 1. On ōđre healfe đære eá [MS. L. eás] *on the other side of the river*, Ors. 1, 1; Bos. 20, 3. Be đære eá ōfrun *by the banks of the river*, Gen. 41, 3: Ors. 1, 3; Bos. 27, 28: 2, 4; Bos. 44, 13. Be đære eá *by the river*, Chr. 896; Th. 172, 35, col. 2. Đa eá oferfaran wolde *would go over the river*, Ors. 2, 4; Bos. 44, 2. On đæm lande syndon twá mycele eá *Idaspes and Arbis in the country are two great rivers, Hydaspes and Arabis*, Ors. 1, 1; Bos. 16, 34. Lät streámas weallan, eá in flēđe *let streams well out, a river in flood*, Andr. Kmbl. 3006; An. 1506. Đás synd đa feówer eán of đnum wyllspringe *these are the four streams from one well-spring*, Ælfc. T. 25, 19. He hí upforlēt on feówer hund eá and on syxtig *he divided it into four hundred and sixty streams*, Ors. 2, 4; Bos. 44, 9. Betweox đäm twám eáum *between the two rivers*, Ors. 5, 2; Bos. 102, 34. Ofer đäm eám *super flūmĭna*, Ps. Th. 23, 2. Betweoh đæm twám eán *between the two rivers*, Ors. 1, 1; Bos. 16, 28. On feówer eán *into four streams*, Gen. 2, 10. [*Laym.* æ, *f:* Orm. æ: O. Sax.* aha, *f:* O. Frs. a, *f:* Ger. aa, *f. name of rivers or brooks;* -ach *suffix of river-names:* M. H. Ger. ahe, *f:* O. H. Ger. aha, *f:* Goth. ahwa, *f: Dan.* aa, *m. f:* Swed. å, *f:* Icel. á, *f:* Lat. aqua.] v. ǽg-, ēg-, ēh-, íg-.

eá, eáw *oh! alas! commonly* eá-lá; *interjec.* q. v.

eác; *prep. dat.* With, *in addition to, besides;* cum, præter :—Gif đū sunu áge, odđe swäsne mäg, odđe freónd ænigne eác đissum idesum, alæde of đysse leód-byrig *if thou have a son, or beloved kinsman, or any friend with* [*in addition to*] *these damsels, lead* [*them*] *from this city*, Cd. 116; Th. 150, 31; Gen. 2500. Đæt gér wäs đæt sixte eác feówertigum *that year was the six and fortieth*, i. e. *the sixth with the fortieth*, or *the sixth increased with forty*, Bd. 1, 3; S. 475, 16: 1, 13; S. 481, 35, 39; Bt. Met. Fox 1, 87; Met. 1, 44. DER. to-eác. v. eác; *conj.*

EÁC; *conj.* **I.** EKE, *also, likewise, moreover, and;* etiam, quoque, et :—Abeád eác Adame ēce Drihten *the Lord eternal beseeched also to Adam*, Cd. 43; Th. 57, 8; Gen. 925. Eác we đæt gefrugnon *we also have heard that*, Exon. 12 a; Th. 19, 15; Cri. 301: Cd. 174; Th. 220, 8; Dan. 68: Beo. Th. 195; B. 97. Hondum slógun, folmum areahtum and fystum eác *struck with their hands, with outstretched palms and with fists also*, Exon. 24 a; Th. 69, 24; Cri. 1125: 9 b; Th. 9, 18; Cri. 136: Cd. 69; Th. 82, 35; Gen. 1372. And ge sceolon eác þweán eówer ælc ōđres fēt *and likewise ye ought to wash one another's feet*, Jn. Bos. 13, 14, 9. Ic eów secge, eác máran đonne wítegan *I say unto you, and more than a prophet*, Mt. Bos. 11, 9. Adam häfde nigen hund wintra and þrítig eác *Adam had nine hundred winters, and thirty also*, Cd. 55; Th. 68, 31; Gen. 1126: 58; Th. 71, 3; Gen. 1165. Fíf and syxtig wintra häfde and eác þreó hund *he had five and sixty winters, and also three hundred*, 62; Th. 74, 4; Gen. 1217: 74, 34; Gen. 1232. Ne his wordum eác woldan gelýfan *et non credĭdērunt in verbis ejus*, Ps. Th. 105, 20. **II.** eác hwäđre, hwæđre eác *Nevertheless, however;* nihilōmĭnus :—Eác hwæđre ceald lyft is gemenged *the cold air nevertheless is mingled*, Bt. Met. Fox 20, 156; Met. 20, 78. Wäs me hwæđre eác láþ *nevertheless it was to me unpleasant*, Exon. 100 b; Th. 380, 23; Rä. 1, 12. **2.** eác swilce, swylce eác *So also, also, moreover, very like, even so, as if;* parimŏdo, tamquam :—Đa apostoli gesetton eác swilce lárspell to đäm leódscipum đe to geleáfan bugon *the apostles moreover gave instructions to the nations submitting to the faith*, Ælfc. T. 27, 20. Đá wäs eác swilce se scucca him betwux *there was also the devil between them*, Th. Anlct. 37, 9: Ps. Th. 55, 4: 108, 29. Eác swylce heó sprecende sý to eallum mancynne *as if it spoke to all mankind*, Ors. 2, 4; Bos. 44, 34. Wíte þoliaþ swilce eác đa biteran rēcas *they suffer torments, so also the bitter reeks*, Cd. 18; Th. 21, 17; Gen. 325: Judth. 12; Thw. 26, 20, 25, 30; Jud. 338, 344, 349: Exon. 120 b; Th. 462, 5; Hö. 47: 34 b; Th. 112, 1; Gü. 137. Swylce grúndas eác *so also the abyss*, 10 a; Th. 9, 35; Cri. 145. **3.** ge eác swylce *Quin et* :—Eall đæt he on änweald onféng ge eác swylce monige Brytta eáland Angelcynnes ríce underþeódde *quæ omnia sub ditiōne accēpit quin et Mevanias insŭlas impērio subjŭgāvit Anglōrum*, Bd. 2, 9; S. 510, 16. **4.** eác swá *So also, even so, likewise* :—Swá đeós world eall

gewîteþ, and eác swâ some, ðe hire on wurdon atydrede *so all this world goes away, and even so those who were born upon it,* Elen. Grm. 1278. Se is eác wealdend ealra ðara ðe ðǽr in wuniaþ ungesewenlîcra, and eác swâ same ðara ðe we eágum on lôciaþ *he is also the ruler of all those creatures which therein dwell invisible, and even so of those that we behold with our eyes,* Bt. Met. Fox 11, 10; Met. 11, 5: 11, 19; Met. 11, 10: 11, 171; Met. 11, 86. Sió gesceádwîsnes sceal ðǽre wilnunge waldan and irsunge eác swâ *the reason ought to govern the will and the anger likewise,* 20, 398; Met. 20, 199: 20, 384; Met. 20, 192. [*Wyc.* eke: *Chauc.* eek, eke: *R. Glouc.* ek: *Laym.* æc, ac, ec, eke, æke: *Plat.* ook: *O. Sax.* ôk *etiam, quoque: Frs.* ak, eak: *O. Frs.* ak, oke *also, and: Dut.* ook: *Ger.* auch *etiam, quoque: M. H. Ger.* ouch: *O. H. Ger.* ouh *etiam: Goth.* auk *because: Dan.* og *and: Swed.* och *and; ock also: Icel.* og *atque, et: O. Nrs.* auk, ôk *etiam.*] v. ĕc, ǣc.

EÁCA, an; *m. An addition,* EEKING, *increase, usury, advantage;* additámentum :—Ðeáh mîn bân and blôd bûtū geworþen eorþan to eácan *though my bones and blood both become an increase to the earth,* Exon. 38 a; Th. 125, 10; Gû. 352. Þincþ ðē lytel eáca ðînra gesǣlþa *does it seem to thee little addition to thy felicities?* Bt. 20; Fox 72, 12. Is witena gehwām wôpes eáca *there is increase of weeping to every man,* Salm. Kmbl. 922; Sal. 460. Ic [Ælfríc Abbod] geset hæbbe feówertig lârspella, and sumne eácan ðǽrto *I [Abbot Ælfric] have composed forty sermons, and some addition thereto,* Ælfc. T. 27, 18. Gif he hæfþ sumne eácan yfeles *if he has some addition of evil,* Bt. 38, 3; Fox 200, 19. For ðæs yfles eácan *for the addition of evil,* 200, 21. Ne gehêne ðū hine mid ðŷ eácan *oppress him not with the usury,* L. Alf. 35; Th. i. 52, 23; *neque humília illum ūsúra tua,* Wilk. 31, 45. ¶ To eácan *besides, moreover :*—Ðæt wæs to eácan ôðrum unarîmedum yflum *that was besides other innumerable evils :* literally, *in or for, addition to, etc.* Bt. 1; Fox 2, 11. To eácan himselfum *besides himself :* literally, *in addition to,* Bt. 26, 2; Fox 92, 20. Ôðer is to eácan andgete *the second is moreover manifest,* Exon. 26 a; Th. 76, 21; Cri. 1243. DER. mægen-eáca, ofer-.

EÁCAN; *p.* eóc, *pl.* eócon; *pp.* eácen, ēcen *To be increased, augmented, enlarged, indued;* augēri, increscĕre :—Adam wearþ gāste eácen *Adam was with spirit indued,* Cd. 48; Th. 61, 23; Gen. 1001: Exon. 102 b; Th. 388, 26; Rä. 6, 13. Eácen feoh increased cattle, Cd. 74; Th. 91, 25; Gen. 1517. Heó wæs mago-timbre be Abrahame eácen worden *she had been increased with offspring by Abraham,* Cd. 102; Th. 135, 2; Gen. 2236: 123; Th. 157, 14; Gen. 2606: 132; Th. 167, 15; Gen. 2766. Ðæt þurh bearnes gebyrd brŷd eácen wearþ *that through child-bearing the bride was increased,* Exon. 8 b; Th. 3, 19; Cri. 38. Heó ongieten hæfde ðæt heó eácen wæs *she had discovered that she was pregnant,* Exon. 100 a; Th. 378, 4; Deór. 11. Ælmihtig eácenne gāst in sefan sende *the Almighty sent an enlarged spirit into his soul,* Cd. 198; Th. 246, 27; Dan. 485. 'Is dôhtor mîn eácen, upliden *my daughter is magnified, exalted,* Exon. 109 a; Th. 416, 13; Rä. 34, 11. [*Wyc.* echen, eche, eeche: *Chauc.* eche: *R. Glouc.* eche: *Scot.* eik: *O. Sax.* ôkian, ôcon: *O. Frs.* aka: *O. H. Ger.* auhôn: *Goth.* aukan: *Dan.* öge: *Swed.* öka: *Icel.* auka: *Lat.* aug-eo: *Grk.* αὔξ-ω: *Lith.* aug-u *to increase.*]

eácen; *adj.* [*pp.* of eácan] *Increased, great, vast, powerful;* auctus, magnus, pŏtens, grăvidus :—Eácne fuglas *the teeming fowls,* Cd. 98; Th. 130, 12; Gen. 2158. Se wæs æðele and eácen *who was noble and vigorous,* Beo. Th. 398; B. 198: Exon. 10 b; Th. 13, 20; Cri. 205. Eald sweord eácen *an old, powerful sword,* Beo. Th. 3330; B. 1663: 4286; B. 2140. Eácne eardas *the vast dwellings,* 3246; B. 1621. Insende eácne egesan *he sent in mighty terror,* Salm. Kmbl. 947; Sal. 473. Cræfte eácen *great in skill,* Exon. 128 a; Th. 492, 26; Rä. 81, 21: 14; Rä. 81, 15: 103 a; Th. 391, 21; Rä. 10, 8. Nǣron ge swâ eácne môd-geþances *ye were not so powerful in mental thought,* Cd. 179; Th. 224, 14; Dan. 136. DER. feorh-eácen, mægen-. v. ēcen.

eácen-cræftig; *adj. Exceedingly strong;* vălidus, pollens, ingens :—Wæs ðæt yrfe eácencræftig *that heritage was exceedingly strong,* Beo. Th. 6095; B. 3051: 4549; B. 2280.

eá-cerse, an; *f. Water-*CRESS :— Eácersan getrifula oððe geseóþ on buteran *bruise or seethe water-cress in butter,* L. M. 1, 38; Lchdm. ii. 94, 4.

eácnian, eácnigan, eánian; *part.* -iende, -igende; *p.* ode, ade *To increase, to be augmented, to become pregnant, to bring forth;* augēri, concipĕre, parturíre :—Ellen eácnade *the fortitude increased,* Exon. 94 b; Th. 353, 51; Reim. 31: Ps. Spl. 7, 15. Eácniende wîf *muliĕrem prægnantem,* Ex. 21, 22. DER. ge-eácnian, to-ge-. v. eánian.

eácnigende; *part. Bringing forth;* partŭriens :—Ðær sárnessa swâ swâ eácnigendes wîfes *ibi dolōres ut partŭrientis,* Ps. Lamb. 47, 8; *part. of eácnigan.* v. eácnian.

eácnung, e; *f. Increase, a conception;* conceptio :—Hū ðū eácnunge onfēnge bearnes þurh gebyrde *how thou didst receive increase through child-bearing,* Exon. 9 a; Th. 5, 26; Cri. 75. DER. bearn-eácnung, ge-.

EÁD, es; *n. A possession, riches, prosperity, happiness, bliss;* possessio, ŏpes, dīvĭtiæ, prospĕritas, felîcitas, beatitúdo :—Se him ðæt eád gefēþ *who gives the happiness to it,* Exon. 60 b; Th. 220, 13; Ph. 319.

Se rinc ageaf eorþcunde eád *the prince gave up earthly happiness,* Cd. 79; Th. 98, 8; Gen. 1627. Niótan ðæs eádes *to have enjoyment of the bliss,* Cd. 21; Th. 26, 5; Gen. 402. [*O. Sax.* ôd, *n. estate, wealth: O. H. Ger.* ôt, *n. prædium: Icel.* auðr, *m. riches, wealth.*]

eád; *adj. Rich, wealthy, blessed, happy;* dīves, opŭlentus, beātus :—Ic ðē eád mæg gecŷðe *I will shew thee the blessed virgin,* Exon. 70 b; Th. 263, 19; Jul. 352: Cd. 151; Th. 189, 17; Exod. 186.

eádan; *p.* eód, *pl.* eódon; *pp.* eáden *To give, concede, grant;* dăre, concēdĕre :—Swâ him eáden wæs *as was granted to them,* Bt. Met. Fox 31, 18; Met. 31, 9. Is æfestum eáden *it [the soul] is given to envy,* Exon. 118 b; Th. 455, 7; Hy. 4, 46. Ac me eáden wearþ *but it was granted to me,* 10 b; Th. 13, 10; Cri. 200.

Eádbald, -bold, es; *m.* [eád *happy,* bald *bold*] *Eadbald, son of Ethelbert, king of Kent. He succeeded his father to the kingdom of Kent in A.D. 616, and died in A.D. 640 :*—Hēr Æðelbryht Contwara cyning forþférde, and Eádbald his sunu fēng to rîce, se forlēt his fulluht and leofode on hǣðenum þeáwe, swâ ðæt he hæfde his fæder lāfe to wîfe *in this year [A.D. 616] Ethelbert, king of the Kentish people, died, and Eadbald his son succeeded to the kingdom, who disregarded his baptism, and lived in heathen manner, so that he had his father's widow to wife,* Chr. 616; Th. 40, 2-9: Bd. 2, 5; S. 506, 36. Hēr Eádbald [Eádbold, col. 2] Cantwara cining forþférde, se wæs cining xxiv wintra *in this year [A.D. 640] Eadbald, king of the Kentish people, died, who was king twenty-four years,* Chr. 640; Th. 47, 20, col. 1: Bd. 3, 8; S. 531, 6.

eádeg *happy,* Cd. 72; Th. 89, 6; Gen. 1476. v. eádig.

eadesa *an adze,* Ps. Surt. 73, 6. v. adesa.

Eádes burh; *gen.* burge; *dat.* byrig; *f.* [*Hunt.* Edesbirh: *Brom.* Edesbury] EDDESBURY, *Cheshire;* loci nomen in agro Cestriensi :—Æðelflǣd Myrcna hlǣfdige ða burh getimbrede æt Eádes byrig *Æthelfled, lady of the Mercians, built the fortress at Eddesbury,* Chr. 913; Th. 186, 30, col. 2.

eád-fruma, an; *m. Author of happiness;* beatitūdinis auctor :—Ēce eádfruma *the eternal author of happiness,* Exon. 15 b; Th. 33, 27; Cri. 532: Andr. Kmbl. 2585; An. 1294.

eádga, eádge *happy,* Cd. 90; Th. 113, 10; Gen. 1885: Exon. 67 a; Th. 249, 1; Jul. 105. v. eádig.

Eádgár, es; *m.* [eád *happy,* gár *spear*] *Edgar, second son of Edmund, and grandson of Alfred the Great. Edgar, in A.D. 955, succeeded to the kingdom of Mercia; and, at the death of his brother Eadwig, in A.D. 959, to the kingdoms of Wessex and Northumbria, over which he reigned sixteen years. He was, therefore, king for twenty years, from A.D. 955-975 :*—Hēr, A.D. 955, Eádgár fēng to Myrcena rîce *here Edgar succeeded to the kingdom of Mercia,* Chr. 955; Erl. 119, 32. Hēr, A.D. 959, forþférde Eádwíg cing, and Eádgár his brôðor fēng to rîce, ægðer ge on West-Seaxum, ge on Myrcum, ge on Norþhymbrum *here king Eadwig died, and Edgar his brother succeeded to the kingdom, as well of the West-Saxons as of the Mercians, and of the Northumbrians,* Chr. 959; Th. 216, 10-15, col. 2. Hēr, A.D. 975, Eádgár cing forþférde *here king Edgar died,* Chr. 975; Th. 227, 19, col. 3.

eádgian; *p.* ode; *pp.* od [eád *bliss*] *To bless, enrich;* beatǐfǐcāre, Exon. 8 a; Th. 2, 16; Cri. 20.

eád-giefu, e; *f. Gift of blessedness;* beatitūdinis dōnum :—Ðæt hí ēce eádgiefe ānforlēton *that they forsook the eternal gift of blessedness,* Exon. 73 a; Th. 272, 20; Jul. 502: 74 a; Th. 276, 8; Jul. 563. v. eád-gifu.

eád-gifa, -giefa, an; *m. Giver of prosperity* or *happiness;* prosperǐtātis vel beatitūdinis dātor :—Engla eádgifa *bliss-giver of angels,* Andr. Kmbl. 147; An. 74: 901; An. 451: Exon. 15 b; Th. 34, 22; Cri. 546.

eád-gifu, -giefu, e; *f. Blessed grace, gift of blessedness;* beāta grātia, beatitūdinis dōnum :—Ðæt ðū me ne lǣte of lofe hweorfan ðînre eádgife *that thou let me not turn from the praise of thy blessed grace,* Exon. 69 b; Th. 259, 2; Jul. 276.

eád-hrēdig; *adj. Happy, blessed;* beātus :—Eádhrēdig mæg *O blessed maiden!* Exon. 69 b; Th. 257, 34; Jul. 257. Eádhrēdige mægþ *the blessed maidens,* Judth. 11; Thw. 23, 22; Jud. 135. v. eáþ-hrēdig.

eádi- *happy.* v. Ps. Th. 64, 14, *in* eádig-lîc, eádi-lîc.

eádig, eádeg; *adj.* [eád *happiness, prosperity,* ig] *Happy, blessed, prosperous, fortunate, rich, perfect;* beātus, fēlix, gaudii plēnus, faustus, abundans, opŭlentus, dīves :—Se eádega wer *the happy man,* Cd. 72; Th. 89, 6; Gen. 1476. Se eádga *the blessed [man],* 90; Th. 113, 10; Gen. 1885. Seó eádige *the blessed [maid],* Elen. Grm. 618. Seó eádge *the blessed [maid],* Exon. 67 a; Th. 249, 1; Jul. 105. Forðon se biþ eádig *therefore he shall be blessed,* Cd. 220; Th. 283, 13; Sat. 304. Eádig on eorþan *rich on earth,* 98; Th. 129, 21; Gen. 2147: Exon. 22 b; Th. 63, 3; Cri. 1014. Ôðer biþ unlǣde on eorþan, ôðer biþ eádig *the one is miserable on earth, the other fortunate,* Salm. Kmbl. 365. Earm ic wæs on ēðle ðînum ðæt ðū wurde eádig on mînum *I was poor in thy residence that thou mightest be rich in mine,* Exon. 29 b; Th. 91, 25; Cri. 1497: 30 b; Th. 95, 8; Cri. 1554. Æðeling eádig *a prosperous noble,* Beo. Th. 2454; B. 1225. Eádig and ânmôd *blessed and steadfast,* Andr. Kmbl. 107; An. 54: Exon. 43 b; Th. 146, 29;

Gú. 717. Eádig on elne *perfect in courage*, 47 b; Th. 163, 25; Gú. 999. To ðissum eádigan hám *to this happy home*, Cd. 228; Th. 306, 7; Sat. 660. Habbaþ eádigne bearn ealle ymbfangen *all have encircled the blessed child*, 216; Th. 273, 29; Sat. 144. Eádigra gedryht *the company of the blessed*, Exon. 32 a; Th. 101, 26; Cri. 1664. Eádgest, *superl* :—Ðær he to ðám eádgestum ǽrest mæþleþ *where he first shall speak to those most blessed*, Exon. 27 b; Th. 82, 13; Cri. 1338. [Laym. ædie, eædi, eadi, edi *blessed, beautiful* : *Orm.* ædig *blessed* : *O. Sax.* ôdag *rich, happy* : *O. H. Ger.* ôtag *dives* : *Goth.* audags *blessed* : *Icel.* auðigr, auðugr *rich, opulent*.] DER. ceáp-eádig, dôm-, efen-, eft-, hréþ-, hwæt-, sige-, sigor-, tír-.

eádigan *to bless, enrich*; beatificáre, App. Scint. Lye. v. eádgian.

eádig-líc, eádi-líc; *adj. Happy, prosperous*; prosper, abundans, faustus:—Biþ ðæt ǽrende eádiglícre *that errand will be more prosperous*, Exon. 100 a; Th. 375, 1; Seel. 131. Cunnaþ eádilíc wæstm on wangas *convalles abundábunt frumento*, Ps. Th. 64, 14.

eádig-líce; *adv. Happily*; felíce:—Ða drihtguman lifdon eádiglíce *the retainers lived happily*, Beo. Th. 200; B. 100.

eádignes, -ness, e; *f. Happiness*; beatítúdo, opúlentia:—Ic sceal ýcan eádignesse *I shall increase happiness*, Exon. 108 a; Th. 413, 4; Rä. 31, 9 : 83 a; Th. 313, 7; Seef. 120 : Bt. 40, 4; Fox 240, 8.

eádi-líc; *adj. Happy*, Ps. Th. 64, 14. v. eádiglíc.

eád-leán *a reward*, Som. Ben. Lye. v. edleán.

eád-leánnung, e; *f. Proper recompense, remuneration, retribution*; retribútio, Ps. Spl. 54, 22. v. ed-leánung.

eád-lufe, an; *f. Happiness of love*; beatífícans ámor:—Éce eádlufan *the eternal happiness of love*, Exon. 67 a; Th. 248, 31; Jul. 104.

eád-méd, es; *n. Humility*; humílítas, generally found in the pl:—Ic eádmédu efnan þence *humiliátus sum*, Ps. Th. 118, 107. On mínum eádmédum *in humilítáte mea*, 118, 92. v. eáþ-méd.

eád-médan; *p. de To humble*; humiliáre, Ps. Spl. 74, 7 : 38, 3. DER. ge-eádmédan. v. eáþmédan.

eád-méde; *adj. Humble*; humílis anÍmi:—Ic eom eádméde *humiliátus sum*, Ps. Th. 115, 1 : 118, 75. v. eáþ-méde.

eád-médlíc *humble, respectful*, Anlct.

eád-méd, eáþ-mód; *adj. Humble, meek, mild*; hŭmÍlis, Mt. Bos. 11, 29.

eád-módan *to humble*; humíliáre, Ps. Spl. T. 17, 29. v. eáþ-módian.

eád-módlíce; *adv. Humbly, submissively*; humÍlÍter, Ps. Spl. 130, 3 : Ps. Th. 114, 2. v. eáþ-módlíce.

eád-módnes, eád-módnys, -ness, -nyss, e; *f. Humbleness, humility, humanity*; humÍlÍtas:—Crist eardaþ on dǽre dene eádmódnesse *Christ dwells in the vale of humility*, Bt. 12; Fox 36, 23 : Ps. Spl. 9, 13. v. eáþmódnis.

Eádmund, es; *m.* [eád *happy*, mund *protection*]. 1. Edmund the Martyr, king of East Anglia, was of the Old-Saxon race. He began to reign in A.D. 855. 'Anno DomÍnícæ incarnatiónis DCCCLV,—Eádmundus Orientálium Anglórum gloriosissÍmus cœpit regnáre VIII. Kalend. Januárii, id est die natális DômÍni, anno ætátis suæ decÍmo quarto,' Asser. p. 7, 26–30. He reigned fifteen years, and his death is thus recorded,—Hér, A.D. 870, fór se here ofer Myrce innon Eást-Ængle;—and, on ðam geáre, Scé Eádmund [MS. Æðmund] cíning him wið gefeaht, and ða Deniscan sige náman, and ðone cíning ofslógon, and ðæt land eall gé-eódon *here the army went over Mercia into East-Anglia;—and, in that year, St. Edmund the king fought with them, and the Danes gained the victory, and slew the king, and overran all that land*, Chr. 870; Erl. 73, 29–75, 1. 2. Edmund Atheling, second son of Edward the Elder, and younger brother of Athelstan, whom he succeeded. Edmund was king of Wessex for six years and a half, from A.D. 940-946 :—Hér, A.D. 940, Æðelstán cyning forþférde, and Eádmund Æðeling féng to ríce *here king Athelstan died, and Edmund Atheling succeeded to the kingdom*, Chr. 940; Th. 209, 13-20, col. 1. Hér, A.D. 946, Eádmund cyning forþférde, on Scs AgustÍnus mæssedæge, and he hæfde ríce seofoðe healf geár; and ðá féng Eádréd Æðeling, his bróðor, to ríce *here king Edmund died, on St. Augustine's mass-day [May 26th], and he held the kingdom six years and a half; and then Eadred Atheling, his brother, succeeded to the kingdom*, Chr. 946; Erl. 116, 33-36. 3. Edmund Ironside, son of Æthelred Atheling. Edmund began to reign in A.D. 1016, and died the same year :—A.D. 1016, ðá gelamp hit ðæt se cyning Æðelréd forþférde, and ealle ða witan ða on Lundene wǽron, and seó burhwaru gecuron Eádmund to cyninge *then it happened that king Æthelred died, and all the witan that were in London, and the townsmen chose Edmund for king*, Chr. 1016; Erl. 155, 15-19. A.D. 1016, ðá to Scé Andreas mæssan, forþférde Eádmund cyng *then, on St. Andrew's mass-day [Nov. 30th], king Edmund died*, Chr. 1016; Th. 284, 12, col. 2.

Eádmundes burh; *gen.* burge; *dat.* byrig; *f.* [Eádmundes *Edmund's*, burh *the town*] St. Edmundsbury, Bury St. Edmunds, Suffolk :—Hér, A.D. 1046, forþférde Æðelstán abbot on Abban dúne and féng Spearhafoc munuc to of Scé Eádmundes byrig *here died Æthelstan, abbot of*

Abingdon, and monk Spearhawk of St. Edmundsbury succeeded, Chr. 1046; Erl. 170, 15.

eád-nes, -nys, -ness, -nyss, e; *f. Happiness, prosperity*; beatítúdo :—Ós byþ eorla gehwám eádnys *mind is to every man prosperity*, Hick. Thes. vol. i. 135, 8; Runic pm. 4; Kmbl. 340, 10. Ongan he wurþigan eádnysse and hýrsumnysse *he began to esteem happiness and obedience*, Guthl. 2; Gdwin. 18, 16. v. éþnes.

eá-docce, an; *f. A water-dock*; rŭmex aquatíca, Lchdm. ii. 379.

eador; *adv. Together*; una, simul :—Eall eador *all together*, Cd. 119; Th. 154, 18; Gen. 2557. Ða wæs eall eador [geador, Kmbl.] *there was all together*, Andr. Recd. 3253; An. 1629. v. geador.

eador *a hedge, dwelling*. v. edor.

eador-geard, es; *m. The inclosure of arteries, the body*; dómus vénárum, corpus; aula septa, Grm. Andr. Elen. 129, 4. Lǽtaþ spor eadorgeard [ealdorgeard, Kmbl.] sceoran, fǽges feorhhord *let the spur raze the dwelling [of arteries? or of life?], the soul-hoard of the mortal*, Andr. Recd. 2362; An. 1183. v. ǽdre.

Eádréd, es; *m.* [eád *happy*, réd = rǽd *counsel*] Eadred Atheling, third son of Edward the Elder. Eadred was king of Wessex and Northumbria, for nine years and a half, from A.D. 946-955 :—Hér, A.D. 946, féng Eádréd Æðeling to ríce *here Eadred Atheling succeeded to the kingdom*, Chr. 946; Erl. 116, 35. Hér, A.D. 955, Eádréd [MS. Æðréd] cyning forþférde, and féng Eádwíg to ríce, Eádmundes sunu *here king Eadred died, and Eadwig, Edmund's son, succeeded to the kingdom*, Chr. 955; Erl. 119, 8.

Eadulfes næs, Ealdulfes næs, næss, es; *m. Eadulf's ness, Walton-on-the-Naze?* Æðulphi promontórium in agro Essexiensi :—Ða ôðre fôron on Eást-Seaxon to Eadulfes næsse *the others went on to Essex, to Eadulf's ness*, Chr. 1049; Ing. 220, 24 : 1051; Th. 319, 2, col. 2 : 1052; Th. 321, 10.

eád-wacer, es; *m. A watchman of property*; bonórum custos, Exon. 101 a; Th. 380, 30; Rä. 1, 16.

Eádweard, -ward, es; *m.* [eád *happy*, weard *ward, guardian*]. 1. Edward the Elder, the eldest son of Alfred the Great. Edward was king of Wessex for twenty-four years, from A.D. 901-925 :—Hér, A.D. 901, gefór Ælfréd cyning, and féng Eádweard his sunu to ríce *here king Alfred died, and Edward his son succeeded to the kingdom*, Chr. 901; Erl. 97, 8-10. Hér, A.D. 925, Eádweard cyning [MS. cing] forþférde, and Æðelstán his sunu féng to ríce *here king Edward died, and Æthelstan his son succeeded to the kingdom*, Chr. 925; Erl. 110, 19. 2. Edward the Martyr, son of Edgar. Edward was king of Wessex, Mercia, and Northumbria, for three years, from A.D. 975-978 :—Hér, A.D. 975, Eádweard, Eádgáres sunu, féng to ríce *here king Edward, Edgar's son, succeeded to the kingdom*, Chr. 975; Th. 227, 37, col. 1. Hér, A.D. 978, wearþ Eádweard cyning gemartyrad *here king Edward was martyred*, Chr. 978; Th. 232, 1-3, col. 1. 3. Edward the Confessor, son of Æthelred. Edward was king of England for twenty-four years, from A.D. 1042-1066 :—Hér, A.D. 1042, wæs Eádward gehálgod to cinge on Wincestre *here Edward was consecrated king at Winchester*, Chr. 1042; Erl. 168, 2. Hér, A.D. 1066, forþférde Eádward [MS. Eáduuard] cyning [MS. king], and Harold eorl féng to ðam ríce *here king Edward died, and earl Harold succeeded to the kingdom*, Chr. 1066; Erl. 198, 1.

eád-wéla, an; *m. Happy weal, riches, happiness, blessedness*; divítiæ, opúlentia, felícítas, beatítúdo :—Sumum eádwélan dǽleþ *to some he dispenses riches*, Exon. 88 a; Th. 331, 12; Vy. 67 : 59 b; Th. 215, 10; Ph. 251 : 80 a; Th. 301, 17; Fä. 20. Sáwul fundaþ to ðam longan gefeán in eád-wélan *the soul tendeth to that lasting joy into happiness*, 48 b; Th. 167, 22; Gú. 1064 : 64 a; Th. 237, 6; Ph. 586.

Eádwíg, es; *m.* [eád *happy*, wíg *war*] Eadwig, son of Edmund. Eadwig was king of Wessex and Northumbria for four years, from A.D. 955-959 :—Hér, A.D. 955, féng Eádwíg to ríce, Eádmundes sunu *here Eadwig, Edmund's son, succeeded to the kingdom*, Chr. 955; Erl. 119, 8. Hér, A.D. 959, Eádwíg cyning forþférde, and féng Eádgár his bróðor to ríce *here king Eadwig died, and Edgar his brother succeeded to the kingdom*, Chr. 959; Erl. 119, 11.

eǽ; *dat.* or *abl. To or by a river* :—Be ðære eǽ *by the river*, Chr. 896; Th. 172, 35, col. 1. v. eá.

eæd-leænian *to reward*; retríbuére, Ps. Spl. T. 17, 22. DER. ge-eædleænian. v. edleænian.

eældian *to grow old*; inveterascére, Ps. Spl. T. 17, 47 : 31, 3. v. ealdian.

eællenge; *interj. Behold*; en, ecce, Ps. Spl. T. 53, 4. v. eallenga.

eærdung, e; *f. A tabernacle*; tabernáculum, Ps. Spl. T. 59, 6. v. eardung.

eærfoðian *to trouble*; tribuláre, Ps. Spl. T. 12, 5 : 41, 14.

eærfoþnes, -ness, e; *f. Difficulty, trouble*; diffícultas, tribulátio, Ps. Spl. T. 33, 19 : 65, 10 : 117, 5. v. earfoþnes.

eærpung, e; *f. A harping, harp*; cíthára, Ps. Spl. T. 32, 2. v. earpa.

eæþ-mód; *adj. Mild*; nútis, Ps. Spl. T. 24, 10. v. eáþ-mód.

eafera *a son*, Beo. Th. 2374; B. 1185. v. eafora.

Q

eá-fisc, -fix, es; m. A river-fish; flŭviālis piscis:—Iór byþ eáfixa [sum] eel [?] is a river-fish, Runic pm. 28; Kmbl. 345, 4. Eáfiscas sēcan to seek river-fishes, Bt. Met. Fox 19, 48; Met. 19, 24.

eafor, es; m. A boar, wild boar; aper:—Sume wǽron eaforas some were wild boars, Bt. Met. Fox 26, 161; Met. 26, 81. v. eofor.

eafora, eafera, eafra, eofera, afora, afera, afara, an; m. An offspring, successor, heir, son; prōles, successor, fīlius:—Wearþ Adame eafora fēded a son was born to Adam, Cd. 55; Th. 67, 23; Gen. 1105: 82; Th. 103, 3; Gen. 1712: Bt. Met. Fox 26, 69; Met. 26, 35. Ne wearþ Heremód swā eaforum Ecgwēlan Heremod was not so to Ecgwela's successors, Beo. Th. 3424; B. 1710. Ðæt we on Adame and on his eafrum andan gebētan that we repair our wrongs on Adam and his offspring, Cd. 21; Th. 25, 24; Gen. 399. [O. Sax. aƀaro, m. prōles, fīlius.]

eafor-heáfod-segn, es; m. A boar-head banner; signum ad capĭtis aprīni similitūdĭnem fabrĭcātum, vel signum apri præcĭpuum:—Hēt in beran eaforheáfodsegn he bade the boar-head banner to be borne in, Beo. Th. 4311; B. 2152.

eafoþ, es; n. Strength, violence, might; vis:—Wæs seó mæg ānrǽd and unforht, eafoða gemyndig the maid was resolved and fearless, of her strength mindful, Exon. 74 b; Th. 278, 22; Jul. 601. Him Geáta sceal eafoþ and ellen gebeódan a Goth shall offer him strength and valour, Beo. Th. 1208; B. 602. Heremódes hild sweþrode, eafoþ [MS. earfoþ] and ellen Heremod's war had ceased, his strength and energy, 1808; B. 902: 4687; B. 2349. Hie unlǽdra eafoðum gelýfdon they believed in the might of savage spirits, Andr. Kmbl. 284; An. 142. Unlǽdra eafoþ the violence of the wretched men, 59; An. 30. v. eofoþ.

eág-æppel, es; m. The apple of the eye; pupilla, Som. Ben. Lye.

eágan beorht, es; n. An eye's glance, a moment; ocŭli micātio, momentum, Bd. 2, 13; S. 516, 20, MSS. C. B. v. eágan bryhtm.

eágan brégh, e; f. An eyebrow; palpebra, Bd. 4, 32; S. 611, 18. v. brǽw.

eágan bryhtm, es; m. An eye's twinkle, a moment; ocŭli micātio, momentum, Bd. 2, 13; S. 516, 20. v. eágan beorht.

eá-gang, es; m. A water-course; flumĭnis cursus:—On ðære eágang in the water-course, Ors. 2, 4; Bos. 44, 13.

eágan weán, wenn A ringworm, tetter; impetīgo:—Eágan weán vel wearhbrǽde impetīgo, Ælfc. Gl. 73; Som. 71, 9; Wrt. Voc. 43, 62.

eága-swind the eyelid, the cheek; gēna, Som. Ben. Lye; Grm. Gr. iii. 401 proposes eágan-spind.

eág-dúru; e; f. An eye-door, a window; fenestra, Martyr. 12, Jan. Lye.

EÁGE, ēge; gen. dat. -an; acc. -e; pl. nom. acc. -an, -on; gen. -ena, -na; dat. -um, -on; n. I. an EYE; ocŭlus:—Gyf ðīn swýðre eáge ðē ǽswīcie si ocŭlus tuus dexter scandalīzat te, Mt. Bos. 5, 29. Mínra eágna leóht light of my eyes, Exon. 67 a; Th. 248, 14; Jul. 95. Eágena gesihþ the sight of the eyes, Andr. Kmbl. 60; An. 30. Eágum to wynne to their eye's delight, Exon. 26 a; Th. 76, 26; Cri. 1245. II. the eye of a needle; forāmen:—Þurh nǽdle eáge per forāmen acus, Mt. Bos. 19, 24: Lk. Bos. 18, 25. [Piers P. eighe, pl. eighen: Wyc. eiʒe, eʒe, iʒe, yʒe, pl. eiʒen: Chauc. R. Glouc. eye, pl. eyen: Laym. eʒe, eʒ-hen, æʒene: Orm. eghe, pl. eghne, ehhne, ehne: Scot. ee, pl. een: Plat. ooge, pl. aagen: O. Sax. ōga, n; pl. ōgun: O. Frs. age, ag, ach, oge, n; pl. agon: Dut. ooge, n: Ger. auge, n: M. H. Ger. ouge, n: O. H. Ger. ouga, auga, n: Goth. augo, n: Dan. öie, n: Swed. öga, n; pl. ögon: Icel. auga, n: Lat. oc-ulus, m: Grk. ὄκος, ὄκκος, m: Lith. akis, f: Sansk. aksha, n.]

eág-ece, es; m. Eye-ache; ocŭlōrum dŏlor, Som. Ben. Lye.

eáge-spring, -sprinc, es; n. [eáge an eye; spring a spring] A spring or twinkling of the eye; ocŭli ictus, Som.

eág-fleá A spot in the eye; albūgo, Ælfc. Gl. 73; Som. 71, 10.

eág-gebyrd; e; f. The nature or power of the eye; ocŭli nātūra, Exon. 60 a; Th. 219, 3; Ph. 301.

eág-hill, es; m. An eyebrow; supercĭlium, Mann.

eág-hringas; pl. m. The eyebrows, eyelids; palpebræ, genæ? Som. Ben. Lye.

eágh-þyrl a window, Bd. 4, 3; S. 568, 6. v. eág-þyrl.

eágor-streám, es; m. A water-stream, ocean; māre, Andr. Kmbl. 882; An. 441: Bt. Met. Fox 20, 244; Met. 20, 122. v. ēgor-stream.

eág-sealf, e; f. Eye-salve; colliria, Ælfc. Gl. 12; Som. 57, 82.

eág-seoung, -sioung, e; f. An eye-disease; glaucōma, Cot. 97: 170, Lye.

eág-séung, e; f. Eye-seeing, eye-sight; ocŭlōrum acies, Som. Ben. Lye.

eág-sýne; adj. Visible to the eye; ocŭlis conspĭcuus, Andr. Kmbl. 3099; An. 1552.

eág-þyrl, ēg-þyrl, ēh-þyrl, es; n. An eye-hole, a window; fenestra:—Ontŷnde se bisceop ðæt eág-þyrl ðære cyrican apĕruit episcŏpus fenestram oratōrii, Bd. 4, 3; S. 568, 6: 5, 12; S. 629, 15.

eág-wræc, es; n. A pain of the eyes; ocŭlōrum dŏlor, Med. ex Quadr. 9, 4; Lchdm. i. 362, 1.

eág-wyrt, e; f. Eye-wort, eye-bright; ocŭlāria, L. M. 3, 30; Lchdm. ii. 324, 19.

eáh- eye-, = eág-, in compounds, q. v.

eáh-mist, es; m. Eye-mist or dimness; ocŭlōrum calīgātio, Som. Ben. Lye.

eáh-streám a water-stream, Exon. 25 a; Th. 72, 6; Cri. 1168. v. eá-stream.

eaht, æht, eht, e; f. Deliberation, council; delibĕrātio, consĭlium, Exon. 80 a; Th. 301, 24; Fä. 24.

EAHTA, ahta, æhta, ehta eight; octo:—Eahta dagas dies octo, Lk. Bos. 9, 28. Būton ðām eahta mannum except eight men, Ælfc. T. 6, 26. To eahta geára fyrste for a space of eight years, Jud. 3, 8. Æfter eahta dagum post dies octo, Jn. Bos. 20, 26. He hēht eahta mearas on flet teón he commanded eight steeds to be led into the court, Beo. Th. 2075; B. 1035. [Wyc. eighte: Laym. æhte, æhten, eahte, ehte: Orm. ehhte: O. Sax. ahto: O. Frs. achta, achte, acht: Dut. Ger. acht: M. H. Ger. aht, eht: O. H. Ger. ahtō: Goth. ahtau: Dan. otte: Swed. åtta: Icel. átta: Fr. huit: Span. ocho: Ital. otto: Lat. octo: Grk. ὀκτώ: Sansk. ashṭan.] DER. eahta-teóða, -toða, -tyne: hund-eahtatig.

eáhtan, ēhtan, iehtan. I. to observe, judge; observāre, æstimāre, reputāre:—We māgon eáhtan and sōþe secgan ðæt we may judge and soothly say that, Exon. 30 b; Th. 94, 34; Cri. 1550. Wile fæder eáhtan hū suna bringen sáwle the father will judge how his sons bring their minds, 23 b; Th. 66, 20; Cri. 1074. II. c. gen. To watch any one, pursue, persecute; persĕqui:—Bona eáhteþ ānbüendra the murderer persecutes lone dwellers, Exon. 33 b; Th. 107, 15; Gū. 59: 37 b; Th. 123, 4; Gū. 317: Ps. Th. 118, 150. [O. Sax. ahtian persĕqui: O. Frs. achta, echta, achtia damnāre, judicāre: Ger. æchten proscrībere: M. H. Ger. āhten, æhten: O. H. Ger. āhtian, āhtōn, ahtēn persĕqui.] v. ōht.

eahta-teóða; m: eahta-teóðe; f. n. adj. The eighteenth; duodevicē-sĭmus:—On ðam eahtateóðan geáre in the eighteenth year, Ors. 6, 2; Bos. 117, 10. Ðysne eahtateóðan sealm Dafid sang David sang this eighteenth psalm, Ps. Th. arg. 18.

eahtatig eighty. v. hund-eahtatig.

eahta-tyne, ehta-tyne; adj. EIGHTEEN; octōdĕcim:—Hig him þeówodon eahtatyne geár they served him eighteen years, Jud. 3, 14: 10, 8.

eahteða, eahteoða eighth, Exon. 47 b; Th. 164, 11; Gū. 1010: Menol. Fox 6: Men. 3. v. eahtoða.

eahtian, eahtigan, ehtian; p. ode, ade, ede; pp. od. I. to meditate, devise, deliberate; meditāre, reputāre, deliberāre:—Eahtade hū wynna þorfte brūcan he meditated how he might enjoy delights, Exon. 37 b; Th. 122, 17; Gū. 307. Sum dōmas con, ðǽr dryhtguman rǽd eahtiaþ one understands dooms, where people devise counsel, 79 a; Th. 297, 24; Crä. 73: 74 b; Th. 279, 6; Jul. 609; Andr. Kmbl. 2325; An. 1164: Dan. 7; B. 1407: 347; B. 172. II. to esteem; æstimāre:—Eahtodon eorlscipe and his ellen-weorc they esteemed his bravery and his valiant works, Beo. Th. 6327; B. 3174.

eáhtnes, ēhtnes, -nys, -ness, -nyss, e; f. Persecution; persecūtio:—Se eáhtnysse ahóf who raised persecution, Exon. 65 b; Th. 243, 2; Jul. 4: 18 a; Th. 44, 18; Cri. 704.

eahtoða, eahteða, ehteoða, ehtuða; m: -ðe; f. n: adj. The eighth; octāvus:—Eahtoðan sīðe an eighth time, Exon. 80 b; Th. 303, 26; Fä. 59.

eahtung, æhtung, e; f. A price, an estimation; æstimātio, Som. Ben. Lye. v. ehtung.

eáhum with eyes; = eágum; pl. dat. or inst. of eáge, Bt. 5, 1; Fox 8, 25, MS. Bod.

EAL, eall; gen. m. n. ealles; f. ealre, eallre; dat. m. n. eallum; f. ealre, eallre; acc. m. ealne, eallne, f. ealle, n. eal; inst. ealle; pl. nom. acc. ealle, ealla; gen. ealra, eallra; dat. eallum; sometimes used indecl; adj. I. ALL; tōtus, omnis, cunctus, unĭversus:—Eal ða earfeðu all the pains, Exon. 25 b; Th. 74, 5; Cri. 1202: 118 a; Th. 452, 25; Hy. 4, 7: Andr. Kmbl. 1889; An. 947. Eal here ðe whole host, Cd. 114; Th. 150, 12; Gen. 2490: Salm. Kmbl. 645; Sal. 322. Eal ic I all, Exon. 115 a; Th. 443, 13; Kl. 29. Ealles ðæs gafoles of all the tribute, Exon. 16 a; Th. 35, 16; Cri. 559. Ealre worlde of all the world, Hy. 7, 57: 11, 20. Ealles ðæs of all that, Exon. 119 a; Th. 456, 19; Hy. 4, 69. Ealne ðisne ymbhwyrft all this orb, 110 b; Th. 423, 1; Rä. 41, 14. Ealne ðone egesan all the terror, Cd. 202; Th. 250, 3; Dan. 541. Geond ealne middangeard tōto orbe, Bd. 2, 4; S. 505, 26. Ealne weg always, Bt. 38, 4; Fox 204, 10, 11. Ealle ða gesceaft all the creation, Bt. Met. Fox 20, 37; Met. 20, 19. Ealle ǽ unĭversam legem, Deut. 4, 8. Ðīne ealle gebann omni mandāta tua, Ps. Th. 118, 86. Ealle gesceafte all creatures, Andr. Kmbl. 2997; An. 1501. Ealle ða þing omnia, Gen. 1, 31: Deut. 4, 3. Ealle þing cuncta, Bd. 1, 26; S. 487, 34: Mk. Bos. 9, 23. Ealle ða gelǽredestan men plūres vĭri doctissĭmi, Bd. 2, 2; S. 502, 38. Ealle his bigengan omnes cultōres ejus, Deut. 4, 3. Ealla gesceafta all creatures, Bt. Met. Fox 13, 14; Met. 13, 7: 20, 105; Met. 20, 53: Bt. 39, 13; Fox 234, 24. Ealle mægne with all power, Bt. Met. Fox 26, 128; Met. 26, 64. Ealle gemete omni mŏdo, Bd. 1, 27; S. 496, 39. Ealra ðara gifena for all the gifts, Exon. 41 b; Th. 138, 18; Gū. 578. Earmost ealra wihta poorest of all creatures, 110 a; Th. 421, 7; Rä. 40, 14. On eallum biþ ðæm līchoman it is in

all the body, Bt. Met. Fox 20, 360; Met. 20, 180. Eallum heora eaforum to all their offspring, Cd. 26; Th. 35, 5; Gen. 550. Eal wæs ðæt mearcland the border-land was all, Andr. Kmbl. 37; An. 19. Ealles ðû ðæs wîte awunne for all this thou hast obtained suffering, Exon. 39 b; Th. 130, 18; Gû. 440. Ealra we healdaþ sancta symbel we keep the feast of all the saints, Menol. Fox 396; Men. 199. Ealle wyrd forsweóp mîne mágas fate has swept away all my kinsmen, Beo. Th. 5621; B. 2814. Deáh hit wið ealle sié eft gemenged weoruld-gesceafta though it is still mixed with all worldly creatures, Bt. Met. Fox 20, 255; Met. 20, 128. Þreó eal on ân all three in one, Exon. 22 a; Th. 60, 16; Cri. 970. Ðæs ealles nôwîht nothing of all that; nil omnîmôdis, Bd. 4, 11; S. 579, 21. Fram him eallum by them all, 2, 2; S. 502, 32. On woruld ealle through the whole world, Cd. 32; Th. 42, 16; Gen. 674. His earfoðo ealle ætsomne all his woes at once, 216; Th. 272, 30; Sat. 127. We ealle we all, Exon. 120 b; Th. 463, 12; Hö. 69. Feówer eallum to all four, 113 b; Th. 434, 28; Rä. 52, 7. Me ealne, Ps. C. 50, 98. Hit eal it all, Beo. Th. 3220; B. 1608. Iob sæt ðâ sârlîce eal on ânre wûnde Job sat there doleful all [covered with] a wound, Job Thw. 166, 32. Wæs ðæt bold tobrocen swîðe eal inneweard all the dwelling was much shattered within, Beo. Th. 2000; B. 998. He lîfes gesteald in ðam êcan hâm eal sceáwode he saw all the dwelling-place of life in the eternal home, Exon. 12 a; Th. 19, 24; Cri. 305. Ðæs we ealles sculon secgan þonc for all that we ought to give thanks, 16 b; Th. 38, 24; Cri. 611. Sió his rîces wæs ealles êðel-stôl it was the metropolis of his whole empire, Bt. Met. Fox 9, 21; Met. 9, 11. Hie ðâ ânmôde ealle cwædon they all said then unanimously, Andr. Kmbl. 3201; An. 1603. Nîðða bearna ǽrest ealra first of all the children of men, Cd. 56; Th. 69, 15; Gen. 1136. Us is eallum neód to us all it is needful, Exon. 11 b; Th. 15, 33; Cri. 245. II. without substantive, and sometimes governing the genitive :—Eal [acc. n.] ic recce I govern all, Exon. 110 b; Th. 424, 2; Rä. 41, 33. We oncnâwaþ eal [acc. n.] ðæt we geworhton we acknowledge all that we have done, Hy. 7, 91. Hæfde unlifgendes eal gefeormod he had devoured all the lifeless, Beo. Th. 1493; B. 744. Him ealles þonc ǽghwâ secge let each give thanks to him for all, Exon. 88 b; Th. 333, 4; Vy. 97. Ðê sié ealles þonc meorda and miltsa thanks be to thee for all, for the rewards and mercies, 118 b; Th. 456, 14; Hy. 4, 66. Sindon ealle nyt all are useful, 114 a; Th. 437, 20; Rä. 56, 10. Ealle ætsomne omnes pariter, Bd. 2, 13; S. 515, 38. Ofer ealle over all, Elen. Grm. 386. Ealra aldor chief of all, Cd. 228; Th. 306, 14; Sat. 664: Elen. Grm. 372. Âna wið eallum alone against all, Beo. Th. 292; B. 145: Cd. 218; Th. 279, 28; Sat. 245. Metod eallum weóld gumena cynnes the Creator ruled over the whole of the race of men, Beo. Th. 2119; B. 1057. III. ealles, ealle, ealra are sometimes used, almost adverbially :—Ealles gelîcost most like of all, Cd. 188; Th. 233, 13; Dan. 275. Ealles mǽst maxime, Bd. 2, 4; S. 505, 7: Ps. Th. 119, 3. Ealles edgiong quite young again, Exon. 64 a; Th. 236, 28; Ph. 581: Ps. Th. 138, 14. Ealles tô swîðe all too readily, L. C. S. 3; Th. i. 376, 22: Nicod. 17; Thw. 8, 18: Bt. Met. Fox 5, 59; Met. 5, 30. Ealles swâ swîðe all so readily, 4, 70; Met. 4, 35: 12, 64; Met. 12, 32. Sille ic ðê ealle xxx pûnda I will give thee thirty pounds in all, Salm. Kmbl. 25; Sal. 13. Mid ealle altogether; pěnitus, Bd. 1, 12; S. 480, 38: Ors. 2, 4; Bos. 45, 21: Chr. 893; Th. 162, 24: Exon. 22 a; Th. 60, 28; Cri. 976. Ealra swîðost maxime, Bd. 2, 4: S. 505, 22: Cd. 18; Th. 22, 8, 36; Gen. 337, 351. Ealra wǽron fîfe in all they were five, Exon. 112 b; Th. 432, 1; Rä. 47, 6. [Wyc. Piers P., al, pl. alle : Chauc. al : Laym. al : Orm. all, alle : O. Sax. al : Frs. O. Frs. al, ol : Dut. al, alle, alles : Ger. all, aller, alle, alles : M. H. Ger. al, inflected aller, alliu, alleȝ, elliu, elle, alle : O. H. Ger. al, all : Goth. alls : Dan. Swed. al : Icel. allr, öll, allt, alt : Grk. ὅλος.]

eal, e; f. An awl; subûla :—Þurhþyrlige his eáre mid eale [mid âne eale, Roff.] perföret aurem illîus subûla, L. Alf. 11; Wilk. 29, 12. v. al.

eala ale :—Eala cervisia, celia, Ælfc. Gl. 32; Som. 61, 106; Wrt. Voc. 27, 35. v. ealu.

eálá, æálá, ǽlá, hêlá; interj. O! alas! Oh! eheu! euge! proh :—Eálá ge næddran O! ye serpents, Mt. Bos. 23, 33 : 23, 37. Eálá, eálá euge, euge, Ps. Spl. 69, 4. Eálá eálá! oðđe wel wel! ahah ahah! or well well! euge euge! vel bene bene! Ps. Lamb. 34, 25. Ǽálá, ðû Scippend O, thou Creator, Bt. Met. Fox 4, 1; Met. 4, 1. Ǽlá Drihten leóf, Hy. 1, 1 : 2, 1. Eálá! gif he wolde O that he would, Bt. Met. Fox 9, 105; Met. 9, 53. Eálá hwæt se forma wǽre alas! that the first should have been, 8, 109; Met. 8, 55. Eálá! ðæt hit wurde O! that it might be, 8, 77; Met. 8, 39. Eálá! ðǽr we mágon geseón alas! there we may see, Exon. 27 a; Th. 80, 27; Cri. 1313. v. eáw.

eá-lâd, e; f. A water-way; aquôsa via :—Frêcne þuhton egle eálâda the fearful water-ways appeared terrible, Andr. Kmbl. 881; An. 441.

eala-hûs, eal-hûs, es; n. An ale-house; taberna :—On eala-hûse in an ale-house, L. Eth. iii. 11; Th. i. 292, 9.

eá-land, -lond, es; n. Water-land, an island; insûla [eás land island, lit. water's land, land of water, v. gen. eás in eá] :—Ne geseah nân cêpa eáland no merchant visited the island, Bt. 15; Fox 48, 13. Ðæs fægerne gefeán habbaþ eálanda mænig lætentur insûlæ multæ, Ps. Th. 96, 1. Cumaþ hî of eálandum ûtan they shall come forth from the islands, 71, 10. Swylce he eác Orcadas ða eálond to Rômwara rîce geþeódde Orcâdas ětiam insûlas Rômâno adjêcit impěrio, Bd. 1, 3; S. 475, 13: Beo. Th. 4657; B. 2334: Exon. 52 a; Th. 181, 27; Gû. 1299 : 96 b; Th. 360, 27; Wal. 12 : 361, 17; Wal. 21 : 60 a; Th. 217, 28; Ph. 287. v. îg-land.

eala-scóp, es; m. An ale-poet, L. N. P. L. 41; Th. ii. 296, 12. v. ealu-scóp.

ealaþ, ealoþ, alaþ, alþ, aloþ, eoloþ; n : indecl. in s. but gen. alþes, Rtl. 116, 42, Ale; cervisia :—Twelf ambra Wilisces ealaþ [MS. B. ealoþ] twelve ambers of Welsh ale, L. In. 70; Th. i. 146, 17 : Ors. 5, 3; Bos. 103, 33. v. ealu.

ealaþ-wyrt, e; f. Ale-wort; cervisia mustea, nova, Som. Ben. Lye.

eal-beorht all-bright. v. eall-beorht.

ealc each :—He ofslôh ða hǽðenan on ealcum gefeohte he slew the heathen in every fight, Ælfc. T. 13, 18. v. ǽlc.

eal-ceald all-cold. v. æl-ceald.

eal-cræftig all-powerful, all-mighty. v. æl-cræftig.

eal-cyn of every kind, universal. v. eall-cyn.

EALD, adj; adj. comp. yldra, eldra, eoldra; sup. yldest. I. old, ancient; větus, ætâte provectus, priscus, antîquus :—Ic eom nû eald I am now old, Lk. Bos. 1, 18. Eald ǽfensceóp an old evening-bard, Exon. 103 a; Th. 390, 21; Rä. 9, 5 : Beo. Th. 4426; B. 2210. Ealde ŷþmearas old horses of the waves, Exon. 20 b; Th. 54, 5; Cri. 864. Geongum and ealdum to young and old, Beo. Th. 144; B. 72. Hwæt niwes oðđe ealdes what of new or old, Exon. 115 a; Th. 441, 24; Kl. 4. Se ealda ðe old one [Satan], Frag. Kmbl. 61; Leás. 32. Eald enta geweorc the old work of giants, Exon. 77 b; Th. 291, 24; Wand. 87 : 60 b; Th. 220, 16; Ph. 321 : 86 b; Th. 326, 1; Wîd. 122. Of ðære ealdan moldan hâtaþ hŷ upp-astandan he bids them to arise up from the old mould, 21 a; Th. 55, 25; Cri. 889. Ða ealdan wûnde the old wounds, 24 a; Th. 68, 23; Cri. 1108. Mid ðŷ ealdan lîge with the ancient flame, 30 b; Th. 94, 28; Cri. 1547. Ða ealdan race the old story, 28 a; Th. 85, 26; Cri. 1397. Wrecaþ ealdne nîþ avenge your ancient grudge, 74 b; Th. 280, 3; Jul. 623. II. eminent, great, exalted; eminens, præstans, excelsus : it has the same meaning in compounds, v. eald-wîta :—Nâ ðæt ǽlc eald sŷ, ac ðæt he eald sŷ on wîsdôme not that every one is old, but that he is old in wisdom, L. Ælf. C. 17; Th. ii. 348, 21. [Wyc. eld, elde, olde : Chauc. elde, olde : Laym. æld, alde, olde : Orm. ald : O. Sax. ald : Frs. aod, aud, oad : O. Frs. ald : Dut. oud : Ger. M. H. Ger. O. H. Ger. alt : Goth. alþeis old.] DER. efen-eald, ofer-, or-.

eald-a-wered worn, wasted with age; vetustus, R. Ben. 51, Lye. v. eald-wêrig.

eald-owén, e; f. An old wife, an old crone; větula :—Ealdra cwêna spell větulârum fâbûla, R. 100.

eald-oŷþ, eald-cŷðđu, e; f. The old country; prisca patria :—Ðæt he his ealdcŷðđu sêcan môte that he may seek its old country, Exon. 62 a; Th. 228, 9; Ph. 435 : 61 a; Th. 222, 19; Ph. 351 : 18 b; Th. 46, 16; Cri. 738.

eald-dagas; pl. m. Ancient days, days of old; prisci dies :—In ealddagum, Exon. 12 a; Th. 19, 19; Cri. 303: Ors. 3, 7; Bos. 61, 44: Bd. 4, 27; S. 604, 41, MS. B.

eald-dôm, es; m. Age; větustas :—Hyre ânweald is hreósende for ealddôme her power is decreasing from age, Ors. 2, 4; Bos. 45, 4.

ealde men; homines, Ps. Th. 93, 9. v. ylde.

eald-ealdfæder a great-grandfather; proâvus, Som. Ben. Lye.

ealder an elder, R. Ben. 4. v. ealdor.

ealdermen aldermen, Jud. Thw. 157, 32. v. ealdorman.

eald-fæder, ealde-fæder; indecl. in s. but sometimes gen. -fæderes and dat. -fædere are found; pl. nom. acc. -fæderas; gen. a; dat. um; m. A grandfather, ancestor; âvus, antecessor :—Ealdefæder âvus, Ælfc. Gl. 91; Som. 75, 6; Wrt. Voc. 51, 51. Ðû forþfærst tô dînum ealdfæderum tu îbis ad patres tuos, Gen. 15, 15 : Beo. Th. 751; B. 373. v. fæder 2.

eald-feónd, eald-fînd, es; m. An ancient foe, arch-fiend, Satan; antîquus inîmicus, diâbôlus :—Ealdfeónda cyn the tribe of ancient foes, Cd. 174; Th. 219, 20; Dan. 57 : 196; Th. 244, 26; Dan. 454: Exon. 16 a; Th. 35, 32; Cri. 567. Ðæt he ne lête him ealdfeónd oncyrran môd from his Meotude that he did not let the ancient fiend turn his mind from his Creator, 37 b; Th. 124, 7; Gû. 336: 62 a; Th. 229, 2; Ph. 449 : 121 a; Th. 464, 18; Hö. 89. v. eald-geníþla, eald-gewinna, eald-hettende.

eald-gecynd, es; n. Old or original nature; antîqua nâtûra vel indôles :—Wudu-fuglas on treówum ealdgecynde wuniaþ the wood-birds live in the trees in their old nature, Bt. Met. Fox 13, 79; Met. 13, 40 : 25, 114; Met. 25, 57: Exon. 54 b; Th. 193, 26; Az. 127.

eald-geneát, es; m. An old companion; větus cômes :—Se wæs eald-geneát he was an old companion, Byrht. Th. 140, 58; By. 310. v. eald-gesîþ.

eald-geníþla, an; m. An ancient foe, arch-fiend, Satan; antîquus inîmicus, diâbôlus :—Ðe-læs him ealdgeníþlan scyðđan cômon lest the old

foes might come to injure him, Andr. Kmbl. 2098; An. 1050 : Judth. 11; Thw. 24, 37. Ealdgeníþla, helle hæftling *the old fiend, hell's captive*, Andr. Kmbl. 2682; An. 1343. v. eald-feónd.

eald-gesegen, e; *f. An old saga*; antīqua narrātio :—Se ðe ealdgesegena worn gemunde *who remembered a great number of old sagas*, Beo. Th. 1743; B. 869.

eald-gesíþ, es; *m. An old companion*; vĕtus cŏmes :—Gewiton ealdgesíþas *the old companions departed*, Beo. Th. 1711; B. 853 : Andr. Kmbl. 2210; An. 1106. v. eald-geneát.

eald-gestreón, es; *n. An old treasure*; antīquus thēsaurus :—Ic ðé ða fǽhðe leánige ealdgestreónum *I will recompense thee for the strife with old treasures*, Beo. Th. 2766; B. 1381 : Beo. Th. 2921; B. 1458 : Exon. 31 a; Th. 96, 8; Cri. 1571.

eald-geweorc, es; *n. An ancient work, the world*; priscum ŏpus, mundus :—Freán ealdgeweorc *the ancient work of the Lord*, Bt. Met. Fox 11, 80; Met. 11, 40 : 20, 232; Met. 20, 116.

eald-gewin, -gewinn, es; *n. An ancient conflict*; antīquum bellum :—Ðæt wæs eald-gewinn *that was an ancient conflict*, Elen. Kmbl. 1290; El. 647 : Beo. Th. 3566; B. 1781.

eald-gewinna, an; *m. An old foe*; antīquus inĭmīcus :—Grendel wearþ, eald-gewinna, ingenga mín *Grendel, my old foe, became my invader*, Beo. Th. 3556; B. 1776. v. eald-feónd.

eald-gewyrht, es; *n. An ancient action*; prisca actio :—He þrówode for Adames ealdgewyrhtum *he suffered for Adam's ancient actions*, Rood Kmbl. 198; Kr. 100 : Beo. Th. 5307; B. 2657.

eald-hád, es; *m. Old age*; senectus. v. ald-hád.

Ealdhelm *Aldhelm*, Chr. 731; Th. 74, 31, col. 2, 3; Aldhelm, 74, 31, col. 1. v. Aldhelm.

eald-hettende; *pl. m. Old foes*; antīqui inĭmīci, Judth. 12; Thw. 26, 11; Jud. 321. v. eald-feónd.

eald-hláford, es; *m.* [eald *old*, *ancient*; hláford *a lord*] *An old* or *ancient lord*; pristīnus dŏmĭnus :—Ecg wæs íren eald-hláfordes *the sword of the old lord was iron*, Beo. Th. 5550; B. 2778. He hæfde heora eald-hláfordes sunu on his gewealde *he had the son of their old lord in his power*, Ors. 3, 11; Bos. 74, 25. Se Cásere wæs heora eald-hláford cynnes *the Cæsar was of the kin of their ancient lords*, Bt. 1; Fox 2, 22. He sende ǽrend-gewrit eald-hláfordum *he sent letters to the ancient lords*, Bt. Met. Fox 1, 126; Met. 1, 63.

eald-hryter-flǽsc, es; *n. A side of meat cut off*; succīdia, Ælfc. Gl. 31; Som. 61, 101; Wrt. Voc. 27, 29. *Mann.* suggests eald-hryðer-flǽsc *adulti bŏvis cāro.* v. hrysel.

ealdian; *p.* ode; *pp.* od *To grow* or *wax old*; senescĕre, inveterascĕre :—Syððan ic ealdode *postquam consĕnui*, Gen. 18, 12 : Jn. Bos. 21, 18 : Exon. 33 a; Th. 104, 27; Gú. 14. for-ealdian.

eald-líc; *adj. Old, senile, venerable*; sĕnīlis, grăvis :—Ealdlíc sĕnīlis, Ælfc. Gr. 9, 28; Som. 11, 38 : grăvis, Off. Episc. 1.

eald-m.dér, ealde-mōder; *f. A grandmother*; avia :—Ealde-mōder *avia*, Ælfc. Gl. 91; Som. 75, 9; Wrt. Voc. 51, 54.

ealdnys, -nyss, e; *f.* OLDNESS, *age*; vĕtustas :—Ealdnyss *vĕtustas*, Ælfc. Gr. 5; Som. 5, 21. We awurpon ða derigendlícan ealdnysse *we have cast off pernicious age*, Homl. Th. i. 194, 25.

ealdor, ealdur, aldor; *gen.* ealdres; *dat*. ealdre; *pl. nom. acc.* ealdras; *m.* I. *an* ELDER, *parent, head of a family, author*; părens, paterfamīlias, auctor :—Úre ealdras ða ǽrestan menn *primi părentes nostri*, Bd. 1, 27; S. 493, 3. Ðæt unriht ðe his ealdras ǽr gefremedon *inĭquĭtas patrum ejus*, Ps. Th. 108, 14. Sum híredes ealdor wæs hŏmo *erat paterfamīlias*, Mt. Bos. 21, 33. Þýstra ealdor *tenebrārum auctor*, Bd. 2, 1; S. 501, 16. II. *an elder, chief, governor, prince*; sĕnior, præpŏsĭtus, princeps :—Ðæs folces ealdoran *seniōres pŏpŭli*, Lev. 4, 15. Hundredes ealdor *centŭrio*, Mt. Bos. 8, 5 : Ælfc. Gl. 6; Som. 56, 58. Ðæra byrla ealdor *the chief butler*, Gen. 40, 9. Cwæþ se Hǽlend to ðæs temples ealdrum *dixit Iesus ad magistrātus templi*, Lk. Bos. 22, 52 : C. R. Ben. 25. Ðæt wæs ealdor heora *that was their chief*, Cd. 221; Th. 287, 27; Sat. 373. Heofna ealdor *the prince of the heavens*, Cd. 226 ; Th. 300, 20; Sat. 567. Ealdor þegna *the prince of thanes*, Beo. Th. 3293; B. 1644. Egesful ealdor *a dreadful prince*, Exon. 70 b; Th. 262, 7 : Jul. 329. He ofer his ealdre gestód *he stood opposite his sovereign*, 55 b; Th. 196, 1; Az. 167. Ealdras of Zabulone *princĭpes Zabulon*, Ps. Th. 67, 25 : 82, 9. [*Wyc.* eldren, eldres *fathers, seniors* : *Laym.* ældere, aldere *a chieftain;* ældere, ælderen, alderen *ancestors, parents* : *Plat.* elder, *m. senior* ; in *pl. parents* : *O. Sax.* aldiro, aldro, *m. ancestor;* *pl.* eldiron *parents* : *Dut.* ouder *a parent;* *pl.* ouders, ouderen, *m. parents* : *O. Frs.* alder *a parent* : *Ger.* eltern, ältern *parents* : *M. H. Ger.* altern *parents* : *O. H. Ger.* altiron, eltiron *parents* : *Dan.* äldre *elder, older;* for-ældre *parents* : *Swed.* äldre *elder, older;* for-äldrar *parents.*] v. yldra.

EALDOR, aldor, es; *n* : e; *f?* I. *life*; vita :—Ealdres æt ende *at life's end*, Beo. Th. 5573; B. 2790. Ðe him wolde ealdres geunnan *which would grant him life*, Andr. Kmbl. 2263; An. 1133. On ðissum ealdre *in this life*, Ps. Th. 87, 14. Deáþ geþrydeþ ealdor ánra gehwæs *death expels the life of every one*, Exon. 62 b; Th. 231, 10; Ph. 487. Nalles for ealdre mearn *he cared not for life*, Beo. Th. 2889; B. 1442.

He æt wíge gecrang ealdres scyldig *he succumbed in battle, his life forfeiting*, 2680; B. 1338 : 4128; B. 2061. Ne wæs me feorh ðá gén ealdor in innan *there was as yet no soul, no life within me*, Exon. 103 a; Th. 391, 10; Rä. 10, 3 : Andr. Kmbl. 2276; An. 1139 : Salm. Kmbl. 711; Sal. 355. Swá biþ geóguþe þeáw, ðǽr ðæs ealdres egsa ne stýreþ *so is the wont of youth, where fear of life checks not*, Exon. 38 b; Th. 127, 24; Gú. 391. On ealdre ealre *in the whole life*, Ps. Th. 126, 6. II. *age, in the expressions*, on ealdre, on aldre *ever;* unquam *and* to ealdre *always;* semper, *which are used not only with regard to the duration of life, but also in general for an unlimited period of time, independently or with the addition of* á, áwa, ǽfre, ǽfter, éce *as well in positive as in negative sentences* :—Ne mæg hine on ealdre ǽnig onhrēran *non commovēbitur in æternum*, Ps. Th. 124, 1 : 79, 15. Ne weorþe ic on ealdre ǽfre gescended *non confundar in æternum*, 70, 1 : 118, 80. Ic ǽr ǽfre on ealdre ne wolde melda weorþan *I never before would be the narrator*, Exon. 50 b; Th. 175, 29; Gú. 1202. Him gewearþ yrmþu to ealdre *misery was to them for ever*, 73 a; Th. 272, 24; Jul. 504. Á to ealdre, 116 a; Th. 446, 28; Dóm. 29. Ǽfre to ealdre, 56 b; Th. 200, 13; Ph. 40. Áwa to ealdre, 14 b; Th. 30, 13; Cri. 479. Éce to ealdre, 18 a; Th. 43, 17; Cri. 690 : Meuol. Fox 303; Men. 153. [*O. Sax.* aldar, *n. ætas* : *O. Frs.* alder *age in alderlong* : *Dut.* ouder in ouder-dom *lifetime* : *Ger. M. H. Ger.* alter, *n. age* : *O. H. Ger.* altar, *n. ætas, ævum, vĕtustas, sĕnectus* : *Goth.* alds, *f. ævum* : *Dan.* alder, *m. f. age* : *Swed.* ålder, *m. age* : *Icel.* aldr, *m. age, life, period, everlasting life.*]

ealdor-apostol, aldor-apostol, es; *m. The chief apostle, the chief of the apostles*; princeps apostŏlōrum :—He mynster getimbrede on áre Sce Petres ðæs ealdorapostoles *he built a monastery in honour of St. Peter, the chief apostle*, Bd. 4, 18; S. 586, 26.

ealdor-bana *a life-destroyer*; vitæ destructor. v. aldor-bana.

ealdor-bealu, aldor-bealu; *gen.* -bealuwes, -bealwes; *n. Vital evil;* malum vitæ afficiens :—Fá þrówiaþ ealdor-bealu egeslíc *the hostile shall suffer terrific vital evil*, Exon. 31 b; Th. 98, 31; Cri. 1616.

ealdor-biscop, es; *m. An elder* or *chief bishop, an archbishop;* sĕnior episcŏpus, archiepiscŏpus; *the Pope is so called by king Alfred* :—Ðá wæs Vitalianus Papa ðæs apostolícan setles ealdorbiscop *then Pope Vitalian was the chief bishop of the apostolic seat;* sedi apostŏlīcæ præerat, Bd. 4, 1; S. 563, 23 : 2, 13; S. 516, 1 : 5, 8; S. 621, 39. v. biscop.

ealdor-botl, es; *n. A royal house* or *villa;* rēgālis villa :—Ðǽr wæs ðá cyninges ealdorbotl *ubi tunc erat villa rēgālis*, Bd. 2, 9; S. 511, 18.

ealdor-burh, -burg; *gen.* -burge; *f. A royal city, metropolis;* rēgia arx, metrŏpŏlis :—On Cantwara byrig, seó wæs ealles his ríces ealdorburh *in civĭtāte Doruvernensi, quæ impĕrii sui tōtīus erat metrŏpŏlis*, Bd. 1, 25; S. 487, 19 : 1, 13; S. 482, 6. Godes ealdorburg *God's royal city*, Exon. 114 b; Th. 441, 8; Rä. 60, 15.

ealdor-cearu *life-care, care for life, life-long care.* v. aldor-cearu.

ealdor-dæg, ealder-dæg, aldor-dæg; *gen.* -dæges; *pl. nom. acc.* -dagas; *m. Life-day, day of life;* vitæ dies :—On ealderdagum *in the days of his life*, Beo. Th. 1518; B. 757 : 1440; B. 718.

ealdor-déma *a supreme judge, a prince.* v. aldor-déma.

ealdor-dóm, ealdur-dóm, aldor-dóm, alder-dóm, es; *m.* [ealdor *an elder, a chief;* dóm *dominion, power*] *Eldership, authority, magistracy, principality;* auctōritas, magistrātus, princĭpātus, prīmātus, dŭcātus :—He his ealdordóm synnum aswefede *he [Reuben] had destroyed his eldership by sins*, Cd. 160; Th. 199, 8 : Exod. 335. Is heora ealdordóm gestrangod *confortātus est princĭpātus eōrum*, Ps. Th. 138, 15 : Cd. 60; Th. 73, 1; Gen. 1197 : Exon. 58 a; Th. 208, 20; Ph. 158 : 66 a; Th. 244, 10; Jul. 28. Theodor ealdordóm hæfde *Theodōrus prīmātum hăbēbat*, Bd. 4, 28; S. 606, 26, 6. Ealdordóm dŭcātus, Ælfc. Gl. 6; Som. 56, 48; Wrt. Voc. 18, 3. Ealdordómas vel ða hēhstan wurþscipas *fasces*, 112; Som. 79, 85; Wrt. Voc. 59, 53 : 68; Som. 70, 4; Wrt. Voc. 42, 13.

ealdor-duguþ, aldor-duguþ, e; *f. The chief nobility;* procēres, Judth. 12; Thw. 26, 5; Jud. 310.

ealdor-freá *a chief lord.* v. aldor-freá.

ealdor-gedál, aldor-gedál, es; *n. Separation from life, death;* vitæ divortium, mors :—Óþ his ealdorgedál *until his death*, Cd. 92; Th. 118, 2; Gen. 1959.

ealdor-gesceaft, e; *f. Condition of life;* vitæ condītio, Exon. 110 a; Th. 421, 24; Rä. 40, 23.

ealdor-gewinna, an; *m. Vital adversary;* adversārius qui vitæ insĭdiātur, Beo. Th. 5799; B. 2903 : Exon. 40 b; Th. 134, 10; Gú. 505.

ealdor-lang; *adj. Life-long;* sempiternus :—Hí ealdorlangne tír geslógon æt sæcce *they won life-long glory in the battle*, Chr. 937; Erl. 112, 3; Æðelst. 3.

ealdor-leás, aldor-leás; *adj. Lifeless;* vita prīvātus :—Hie gefricgeaþ freán úserne ealdorleásne *they shall hear our lord [is] lifeless*, Beo. Th. 5998; B. 3003.

ealdor-leás *deprived of parents.* v. aldor-leás.

ealdor-leg, aldor-leg, -læg, es; *n.* [ealdor, læg; *p. of* licgan] *Life-law,*

fate, death; fātum, mors:—Æfter ealdorlege *after death,* Exon. 51 a; Th. 177, 29; Gū. 1234.

ealdorlíc, aluorlíc; *adj. Principal, chief, excellent;* princĭpālis, magnĭfĭcus:—Ealdorlíc *principālis,* Ælfc. Gr. 9, 28; Som. 11, 37. Ealdorlíc ánnyss *principālis unĭtas,* Hymn. Surt. 1, 5. Ealdorlíce Gáste *Spīrĭtu principāli,* Ps. Grn. 50, 13; ii. 149, 13.

ealdor-líce; *adv. Excellently.* v. aldor-líce.

ealdorlícnes, -ness, -nys, -nyss, e; *f. Principality, authority;* auctōrĭtas:—Ne syllaþ we ðé ænige ealdorlícnysse *nullam tibi auctōrĭtātem trĭbuĭmus,* Bd. 1, 27; S. 492, 12, 15, 22, 26. Mid máran ealdorlícnysse *mājōre auctōrĭtāte,* 3, 22; S. 553, 3, 35.

ealdor-man, -mann, -mon, ealdur-, aldor-, eldor-, es; *m.* [eald *old, not only in age, but in knowledge,* v. eald, hence ealdor *an elder;* man *hŏmo*]. **I.** *an elderman,* ALDERMAN, *senator, chief, duke, a nobleman of the highest rank, and holding an office inferior only to that of the king;* mājor nātu, sēnātor, prŏcer, princeps, prīmas, dux, præfectus, trĭbūnus, quīcunque est aliis grādu aut nātu mājor. The title of Ealdorman or Aldorman denoted civil as well as military pre-eminence. The word *ealdor* or *aldor* in Anglo-Saxon denotes princely dignity: in Beowulf it is used as a synonym for cyning, þeóden, and other words applied to royal personages. Like many other titles of rank in the various Teutonic languages, it, strictly speaking, implies age, though practically this idea does not survive in it any more than it does in the word *Senior,* the original of the feudal term *Seigneur.* Every shire had its ealdorman, who was the principal judicial officer of the shire, and also the leader of its armed force. The internal regulations of the shire, as well as its political relation to the whole kingdom, were under his immediate guidance and supervision,—the scír-geréfa, or sheriff, being little more than his deputy, and under his control. The dignity of the ealdorman was supported by lands within his district, which appear to have passed with the office,—hence the phrases, ðæs ealdormonnes lond, mearc, gemǽro, etc. which so often occur. The ealdorman had also a share of the fines and other monies levied to the king's use; though, as he was invariably appointed from among the higher nobles, he must always have possessed lands of his own to the extent of forty hides, v. *Hist. Eliens.* ii. 40. The ealdormen of the several shires seem to have been appointed by the king, with the assent of the higher nobles, if not of the whole witena gemót, and to have been taken from the most trustworthy, powerful, and wealthy of the nobles of the shire. The office and dignity of ealdorman was held for life,—though sometimes forfeited for treason and other grave offences; but it was not strictly hereditary :—Fram ðám bróðrum and ðám ealdormannum *a fratrĭbus ac majōrĭbus,* Bd. 5, 14; S. 634, 10: 5, 19; S. 637, 6. Ofslógon Rómāna ealdorman *slew a Roman noble,* Ors. 5, 10; Bos. 108, 30. Ealdormen, *nom. pl. princĭpes,* Ps. Th. 67, 24: Gen. 12, 15. Ðæt he his ealdormen lǽrde *ut erudīret princĭpes suos,* Ps. Th. 104, 18. Án ealdormann *unus de princĭpĭbus,* 81, 7. Ealdormenn Iudan *princĭpes Jŭda,* 67, 25: 82, 9: Mt. Bos. 20, 25: Mk. Bos. 6, 21. His ealdormannum and his þegnum *suis dŭcĭbus ac ministris,* Bd. 3, 3; S. 526, 1: 4, 15; S. 583, 27. Arbatus his ealdorman, ðe he gesct hæfde ofer Méðas ðæt land *Arbaces, his chief officer, whom he had set over the country of the Medes,* Ors. 1, 12; Bos. 35, 17: 2, 1; Bos. 38, 35: Bd. 4, 12; S. 580, 34: 1, 13; S. 481, 40. Ðæt se ylca ða dóhter ðæs ealdormannes blinde onlíhte *ut idem fīliam trĭbūni cæcam inlumināvĕrit,* 1, 18; S. 484, 30: Bt. 10; Fox 28, 31. **II.** the new constitution introduced by Cnut, who reigned in England from A. D. 1014 to 1035, reduced the ealdorman to a subordinate position,—one *eorl, Nors.* jarl, being placed over several shires. The Danish kings ruled by their eorlas or jarls, and the ealdormen disappeared from the shires. Gradually the title ceased altogether, except in the cities, where it denoted an inferior judicature, much as it now does among ourselves :—Ðis is ðonne seó woruldcunde gerǽdnes, ðe ic [Cnut] wille, mid mínan witenan rǽde, ðæt man healde ofer eall Engla land *this is then the secular ordinance which I [Cnut], with the counsel of my witan, will, that it be observed over all the land of the English,* L. C. S. pref; Th. i. 376, 3, 4. Ðæt is ðonne ǽrest ðæt ic wylle; ðæt man rihte laga upp-atǽre, and ǽghwilce unlaga georne afylle, and ðæt man aweódige and awyrtwalige, ǽghwylc unriht, swá man geornost mǽge, of ðissum earde *this is then the first that I will; that right laws be established, and all unjust laws carefully suppressed, and that every injustice be weeded out and rooted up, with all possible diligence, from this land,* L. C. S. 1; Th. i. 376, 5–8. And habbe man þríwa on geára burh-gemót, and twá scír-gemót and thrice a year let there be a borough meeting, and twice a shire meeting, L. C. S. 18; Th. i. 386, 4, 5. v. eorl, scírgeréfa, and húscarl.

ealdor-mon, -monn, es; *m. An elderman, alderman, nobleman, chief;* mājor nātu, princeps :—Ebrinus se ealdormon *Ebrinus mājor dŏmus rēgiæ,* Bd. 4, 1; S. 564, 33: 2, 13; S. 515, 32. v. ealdor-man.

ealdor-ner, aldor-ner, es; *n. A life-salvation, life's safety, refuge, asylum;* vitæ servātio, refūgium :—Cwom him to áre and to ealdor-ner *he came to them for mercy and for their life's salvation,* Exon. 53 b; Th. 189, 4; Az. 54. v. ner.

ealdor-sacerd, es; *m. A high priest;* summus sacerdos :—Ongan

ealdorsacerd hyspan *the high priest began to revile,* Andr. Kmbl. 1340; An. 670.

ealdor-scype, es; *m. Eldership, supremacy;* principātus, prīmātus :—Ða on þeódum ealdorscype habbaþ *they have eldership among the nations,* Mk. Bos. 10, 42. Ealdorscype healdan *prīmātum tĕnēre,* Coll. Monast. Th. 30, 17.

ealdor-stól, es; *m. The lord's seat;* domĭni sēdes :—Áhte ic ealdorstól *I possessed the lord's seat,* Exon. 94 b; Th. 353, 36; Reim. 23.

ealdor-þegn, aldor-þegn [-þægn], es; *m. The principal thane* or *servant;* princĭpālis minister :—Ealdorþegnas *principal servants,* Menol. Fox 257; Men. 130. Hie ðæt ðám ealdorþegnum cýðan eódon *they went to announce it to the principal thanes,* Judth. 12; Thw. 25, 4; Jud. 242.

ealdor-wísa, *a chief ruler.* v. aldor-wísa.

eald-riht, es; *n. An ancient right;* vĕtus jus *vel* privĭlēgium :—He him gehét ðæt hý ealdrihta ǽlces mósten wyrðe gewunigen *he promised them that they should remain possessed of each of their ancient rights,* Bt. Met. Fox 1, 71; Met. 1, 36: 1, 114; Met. 1, 57. Bǽdon hine ðæt he him to heora ealdrihtum gefultumede *they prayed him that he would succour them with respect to their ancient rights,* Bt. 1; Fox 2, 24.

Eald-Seaxe, Ald-Seaxe; *gen.* -Seaxa; *dat.* -Seaxum; *pl. m:* Eald-Seaxan; *pl. m. The Old-Saxons;* antĭqui Saxōnes; *the German* or *continental Saxons occupying the territory between the Eyder and the Weser:*—Hér Eald-Seaxe [Ald-Seaxe, Th. 92, 29, col. 1] and Francan gefuhton *in this year* [A. D. 779] *the Old-Saxons and the Franks fought,* Chr. 779; Th. 93, 29, col. 1, 2. Gegadrode mycel sciphere on Eald-Seaxum [Ald-Seaxum, col. 1] *a large naval force assembled among the Old-Saxons,* 885; Th. 154, 20, col. 2, 3: 449; Th. 20, 20, 26: 924; Th. 199, 10: Bd. 5, 10; S. 624, 12, 22. Be norþan Þyringum syndon Eald-Seaxan and be norþan westan him syndon Frysan, and be westan Eald-Seaxum is Ælfe miþa ðǽre eá and Frysland *to the north of the Thuringians are the Old-Saxons, and to the north-west of them are the Friesians, and to the west of the Old-Saxons is the mouth of the river Elbe and Friesland,* Ors. 1, 1; Bos. 18, 34: Bos. 19, 14.

eald-spell, es; *n. An old story;* antīqua narrātio :—Ælfréd us ealdspell reahte *Alfred told us an old story,* Bt. Met. Fox introduc. 2; Met. Einl. 1. On ealdspellum *in old tales,* Bt. 39, 4; Fox 216, 19.

eald-sprǽc, e; *f. An old speech, history,* Leo A. Sax. Gl. 149.

Ealdulfes nǽs, Chr. 1052; Th. 321, 10. v. Eádulfes nǽs.

ealdung, e; *f. Age;* sĕnectus :—Róma besprycþ ðæt hyre weallas for ealdunge brosnian *Rome complains that her walls decay from age,* Ors. 2, 4; Bos. 44, 45. DER. ealdian.

ealdur *a prince,* Jn. Foxe 16, 11. v. ealdor.

ealdur-dóm *authority, principality,* Ps. Th. 113, 2. v. ealdor-dóm.

ealdur-man, -mann, es; *m. An elderman, alderman, nobleman;* mājor nātu, princeps :—Nelle ge on ealdurmenn áne getreówian *nōlīte confīdere in princĭpĭbus,* Ps. Th. 145, 2: 118, 161. v. ealdor-man.

eald-wérig; *adj. Vile of old;* jampridem malignus :—Ealdwérige Egypta folc *the folk of Egypt vile of old,* Cd. 145; Th. 180, 24; Exod. 50.

eald-wíf, es; *n. An old woman;* anus, anŭla, vĕtŭla :—Sceal ic nú ealdwíf cennan *num vere parĭtūra sum anus,* Gen. 18, 13: Ælfc. Gl. 88; Som. 74, 67; Wrt. Voc. 50, 48.

eald-wita, an; *m.* [eald *old,* wita *one who knows*] *One old* or *eminent in knowledge, a priest;* presbyter :—Presbiter is mæsse-preóst oððe ealdwita; ná ðæt ǽlc eald sý, ac ðæt he eald sy on wísdóme *presbyter is the mass-priest* or *one eminent in knowledge: not that every one is old, but that he is old in wisdom,* L. Ælf. C. 17; Th. ii. 348, 20: Bd. 2, 16; S. 519, 29.

eald-writere, es; *m. An antiquarian, one that writes of old* or *ancient matters;* antīquārius, Som. Ben. Lye.

ealeðe-tún, es; *m. An ale-house;* taberna, Som. Ben. Lye.

eal-fela *Very much, full many;* permultum :—Se mæg ealfela singan and secgan *he can sing and say very much,* Exon. 17 b; Th. 42, 2; Cri. 666: Beo. Th. 1742; B. 869: 1770; B. 883.

eal-felo *All-fell, very baleful;* omnīno pernĭciōsus :—Eal-felo áttor *very baleful venom,* Exon. 106 b; Th. 405, 28; Rä. 24, 9. v. æl-fǽle, fell.

eal-fremd *foreign;* aliēnus. v. æl-fremd.

eal-geador, eall-geador; *adv. Altogether;* omnīno :—Ðǽr wæs ealgeador Grendles grápe *there was altogether Grendel's grasp,* Beo. Th. 1675; B. 835. v. geador.

eal-gearo, eall-gearo; *adj. All ready* or *prepared;* omnīno promptus *vel* parātus :—Beorh ealgearo wunode on wonge *the mountain stood all ready on the plain,* Beo. Th. 4475; B. 2241: 155; B. 77: 2465; B. 1230.

eal-geleáflíc *believed by all;* catholĭcus. v. eall-geleáflíc.

ealgian, algian; *p.* ode; *pp.* od *To defend;* defendĕre :—Nemne we mǽgen feorh ealgian þeódnes *unless we may defend the life of the prince,* Beo. Th. 5304; B. 2655: 5329; B. 2668. Hí æt campe wið láþra gehwǽne land ealgodon *they defended the land in conflict against every foe,* Chr. 937; Th. 202, 4; Æðelst. 9: Andr. Kmbl. 20; An. 10: Beo. Th. 2413; B. 1204: R. Ben. 64: 69. DER. ge-ealgian, -algian.

eal-gréne, eall-gréne, æl-gréne; *adj. All-green*; omnīno vĭrĭdis:—Gesēgun eorþan ealgréne *they saw the earth all-green*, Exon. 24 b; Th. 69, 31; Cri. 1129.

eal-gylden, eall-gylden; *adj. All-golden*; omnīno aurĕus:—Swŷn eal-gylden *the all-golden swine*, Beo. Th. 2227; B. 1111.

ealh *a residence, temple.* v. alh, healh.

eal-hálig *all-holy*; omnīno sanctus. v. eall-hálig.

ealh-stede, alh-stede, eolh-stede, es; *m. A protecting* or *sheltering place, city, temple*; lŏcus qui præbet tūtēlam, arx, templum:—In ðære wīdan byrig, ealhstede eorla *in the wide city, the sheltering place of men*, Cd. 208; Th. 258, 11; Dan. 674.

eal-hús *an ale-house*, Som. Ben. Lye. v. eala-hús.

eal-hwít *all-white.* v. eall-hwít.

eá-lifer, e; *f.* [eá *water*, lifer *liver*] *Liverwort?* eupătōrium cannăbĭnum, Lin:—Eálifer hätte wyrt gníd on ealaþ *rub in ale the herb called liverwort*, L. M. 1, 22; Lchdm. ii. 64, 21: 2, 24; Lchdm. ii. 216, 14.

eal-íren *all of iron.* v. eall-íren.

eal-ísig *all-icy*; omnīno glaciālis. v. eall-ísig.

eá-lĭðend, es; *m. A wave-sailor, sailor*; qui æquor navĭgat:—Wǣron eorlas onlíce eálĭðendum *the men were like sailors-over-the-wave*, Andr. Kmbl. 502; An. 251.

eall; *adj. All*; tōtus:—Eall tōtus, Ælfc. Gr. 18; Som. 21, 10. Eall ðín líchama *all thy body*, Mt. Bos. 6, 22. Eall ðeós woruld *all this world*, Cd. 29; Th. 38, 9; Gen. 604: Exon. 20 a; Th. 52, 34; Cri. 843: Lk. Bos. 23, 18: Jn. Bos. 11, 50: Mk. Bos. 4, 34: Andr. Kmbl. 652; An. 326: 2294; An. 1148: 2867; An. 1436: Bt. Met. Fox 26, 121; Met. 26, 61: 28, 9; Met. 28, 5: Beo. Th. 4091; B. 2042: 4181; B. 2087: Exon. 22 a; Th. 60, 5; Cri. 965: Salm. Kmbl. 2; Sal. 1: Bt. 38, 4; Fox 204, 9: Bd. 1, 12; S. 480, 35: 1, 26; S. 487, 37: Ors. 2, 4; Bos. 45, 15. v. eal.

ealla, an; *m. Gall, bile;* fel:—Hym man drincan mengde myd eallan and myd ecede *one mingled him a drink with gall and with vinegar*, Nicod. 26; Thw. 14, 18. Wið ðæs eallan [geallan MS. H.] togotennysse *for effusion of the bile*, Herb. 146, 2; Lchdm. i. 270, 4: 141, 2; Lchdm. i. 262, 12. v. gealla.

eall-beorht, æll-beorht, æl-beorht; *adj. All-bright;* pĕnītus splendĭdus, fulgentissĭmus:—Englas eall-beorhte *angels all-bright*, Cd. 224; Th. 297, 23; Sat. 522.

eall-cyn; *adj. Of every kind, universal*; omnĭgĕnus, unĭversus:—Eallcyn sǽd getreówfulra [Iacobes MSS. C. T.] *universum semen Iacob*, Ps. Spl. 21, 22.

eallenga, eællenge; *adv. Altogether, utterly*; prorsus, omnīno:—Ðonne wæs se óðer eallenga sweart *then was the other utterly black*, Cd. 24; Th. 30, 35; Gen. 477.

Ealleríca, an; *m. Alaric, king of the Goths* :—Rædgota and Ealleríca Rómāne burig abrǽcon *Rhadgast and Alaric broke into the city of Rome*, Bt. 1; Fox 2, 2. v. Alríca.

eall-geador; *adv. Altogether*, Andr. Kmbl. 2196; An. 1099. v. eal-geador.

eall-gearo; *adj. All-ready*; omnīno promptus:—Ic beóm eall-gearo *I am all-ready*, Exon. 106 b; Th. 405, 19; Rä. 24, 4. v. eal-gearo.

eall-geleáflíc; *adj.* [geleáflíc *believed*] *Believed by all, catholic*; cathŏlĭcus = καθολικός:—Ðæt monega cyricean on Hibernia, lǽrendum Athamnano, ða eallgeleáflícan Eástran onféngon *ut plurĭmæ Scottōrum ecclesiæ, instante Adamnano, cathŏlĭcum Pascha suscēpĕrint*, Bd. 5, 15; S. 635, 10.

eall-gréne; *adj. All-green*; omnīno vĭrĭdis:—Hwá furðum teóde eorþan eall-gréne *who first produced the earth all-green*, Andr. Recd. 1599; An. 799. v. eal-gréne.

eall-gylden; *adj. All-golden*; omnīno aurĕus:—He geseah segn eall-gylden *he saw an ensign all-golden*, Beo. Th. 5528; B. 2767: Judth. 10; Thw. 22, 3; Jud. 46. v. eal-gylden.

eall-hálig; *adj. All-holy*; omnīno sanctus:—Drihten, ðú earce eart eall-háliga *O Lord, thou art the ark of the all-holy*, Ps. Th. 131, 8.

eall-hwít; *adj. All-white*; omnīno albus:—On eallhwítre þryh *in an all-white coffin*, Th. Diplm. A. D. 970; 241, 11.

eal-líc *universal, general, catholic*, Som. Ben. Lye. v. al-líc.

ealling; *adv. Always;* semper:—Ealling byþ, ymb tyn niht ðæs, tiid [=tíd] geweorþad Barþolomeus *the time of Bartholomew is always honoured about ten nights from hence*, Menol. Fox 304; Men. 153: 344; Men. 173. v. ealneg.

eallinga; *adv. Altogether, wholly*; prorsus, omnīno:—He eallinga ne adíligaþ eów *he will not altogether destroy you*, Deut. 4, 31; Salm. Kmbl. 835: Sal. 417. v. eallunga.

eall-íren; *adj. All of iron*; omnīno ferrĕus:—He hêht gewyrcean eall-írenne wígbord wrǣtlíc *he commanded a wondrous battle-shield, all of iron, to be made*, Beo. Th. 4665; B. 2338.

eall-ísig; *adj. All-icy*; omnīno glaciālis:—Saturnus is se cealda eall-ísig tungel *Saturn is the cold all-icy star*, Bt. Met. Fox 24, 45; Met. 24, 23. Se is eall-ísig *it is all icy*, Bt. 36, 2; Fox 174, 13.

eall-mægen, al-mægen, es; *n. All-power, all-might*; omnis vis:—Gif hī, eall-mægene, ne þiówoden þeódne mǣrum *if they, with all might, served not the illustrious Lord*, Bt. Met. Fox 29, 193; Met. 29, 98.

eall-mihtig, -meahtig, -mehtig, -mihteg [el-, æl-, æll-]; *adj. All-mighty*; omnĭpŏtens:—Drihten eallmihtig *Dominus Deus*, Ps. Th. 93, 22.

eall-nacod; *adj. Entirely naked*; omnīno nūdus:—Ic eom eallnacod *I am all naked*, Cd. 42; Th. 54, 3; Gen. 871.

eallneg; *adv. Always*; semper:—Ðú eallneg siófodest ðæt hí eallneg nǣron on wíte *thou always didst lament that they were not always punished*, Bt. 38, 4; Fox 204, 10, 11, MS. Cot. v. ealneg.

eall-niwe; *adj. All-new*; omnīno nŏvus:—He wearþ gebunden mid eallniwum rápum *nŏvis funĭbus vinctus est*, Jud. 16, 11.

eallnunge; *adv. Altogether*; omnīno, C. R. Ben. 55. v. eallunga.

eall-reord *foreign speaking, barbarous*, Bd. 1, 23; S. 485, 32. v. el-reord.

Eallríca, an; *m. Alaric, king of the Goths*:—Eallríca Gotona cyning *Alaric, king of the Goths*, Ors. 2, 1; Bos. 39, 37. v. Alríca.

eall-rúh; *adj. All-rough*; omnīno hirsūtus:—Se wæs reád and eall rúh *he was red and all hairy*, Gen. 25, 25.

eall-swá; *adv. Also, so, so as, likewise, even as, even so*; sīcut:—Eallswá he sǽde *sīcut dixit*, Mk. Bos. 14, 16. v. eal-swá.

eall-tela; *adv. Quite well*; omnīno bĕne, Cd. 91; Th. 114, 17; Gen. 1905.

eallunga, allunga, eallenga, eællenge, eallinga, eallnunge; *adv. Altogether, entirely, quite, indeed, at all, assuredly, utterly*; prorsus, omnīno, profecto:—Ðæt ge eallunga ne swerion *non jurāre omnīno*, Mt. Bos. 5, 34: Exon. 21 b; Th. 57, 23; Cri. 923: Bt. Met. Fox 25, 131; Met. 25, 66: Bt. 10; Fox 30, 3. Eallunga Godes ríce on eów becymþ *profecto pervēnit in vos regnum Dei*, Lk. Bos. 11, 20.

eall-wealda; *adj. All-ruling, almighty*; omnĭpŏtens:—Eallwealdan Gode *to almighty God*, Andr. Recd. 414; An. 205. v. eal-wealda.

eall-wihta, al-wihta, æl-wihta; *pl.* [eall *all, every*; wiht *creature*] *All beings*; omnia creāta:—Cyning eall-wihta *king of all creatures*, Andr. Kmbl. 3204; An. 1605: Cd. 47; Th. 60, 7; Gen. 978: 5; Th. 7, 28; Gen. 113. v. wiht I, *for* wihta, *nom. pl.*

eall-wundor; es; *n.* [wundor *a wonder*] *A very wonderful thing*; res omnīno mirābĭlis:—Weras fyrdleóþ gólon [MS. galan] eall-wundra fela *the men sung a martial song of many very wonderful things*, Cd. 171; Th. 215, 5; Exod. 578.

eal-mægen *all power, all might.* v. eall-mægen.

eal-mǽst, æl-mǽst; *adv.* ALMOST; totum fere, pene:—Hit is eal-mǽst mid hāligra manna naman geset *sanctōrum hŏmĭnum nŏmĭnĭbus totum fere obsĭtum est*, Bd. Whelc. 448, 18; Homl. Th. ii. 466, 22.

eal-mihtig; *adj. All-mighty*; omnĭpŏtens:—On Godes ealmihtiges naman *in the name of almighty God*, Th. Diplm. A. D. 886–899; 138, 34. v. eall-mihtig.

eal-myrca *an Ethiopian.* v. æl-myrca.

eal-nacod *entirely naked.* v. eall-nacod.

ealneg, ealnig, eallneg; *adv.* [ealne weg, Bt. 38, 4; Fox 204, 10, 11] *Always, quite*; semper, prorsus:—Ýþ wið lande ealneg winneþ *the wave contends always against the land*, Bt. Met. Fox 28, 114; Met. 28, 57: Ors. 3, 7; Bos. 62, 36. Ðe ǽfre biþ ealnig smylte *which ever is quite calm*, Bt. Met. Fox 21, 30; Met. 21, 15.

ealning; *adv. Always*; semper:—Swá he ealning dyde æt Saltwíc *as he always did at Saltwich*, Th. Diplm. A. D. 886–899; 138, 15. v. ealling, ealneg.

eal-niwe *all-new, quite new.* v. eall-niwe.

eal-nósu, eall-nósu, eall seó násu, e; *f. All nose* or *all the nose, a swelling of the uvula*; columella, columna nasi:—Eal ufweard nósu *tota ascendens columna nasi*, Ælfc. Gl. 71; Som. 70, 86; Wrt. Voc. 43, 18. Eall-nósu, Mann: eal-nósu *the swelling of the uvula*; columella. v. Som. Eall seó násu *columna*, Wrt. Voc. 282, 64.

ealo *ale*, Ors. 1, 1; Bos. 22, 17: Bt. 17; Fox 60, 5. v. ealu.

ealo-benc, e; *f. An ale-bench* :—In ealo-bence *on the ale-bench*, Beo. Th. 2062; B. 1029. v. ealu-benc.

ealo-fæt, es; *n. An ale-vat, vessel in which ale was left to ferment*; lăcus:—Under ðæt ealo-fæt *under the ale-vat*, L. M. 1, 67; Lchdm. ii. 142, 12.

ealo-gál; *adj. Ale-drunk*; cervīsia inebriātus:—Ic gehýre ealogálra gylp *I hear the boast of the ale-drunken*, Cd. 109; Th. 145, 19; Gen. 2408.

ealo-geweorc, es; *n. Ale-work, brewing*; cervīsiæ coctio:—On ðære byrig wæs ǽrest ealo-geweorc ongunnen *in that city ale-brewing was first begun*, Ors. 5, 3; Bos. 103, 35.

eálond, es; *n. An island*; insŭla:—Breoton is gārsecges eálond *Brittania est oceăni insŭla*, Bd. 1, 1; S. 473, 8: 1, 3; S. 475, 13. v. eáland.

ealoþ *ale*, L. In. 70; Th. i. 146, 17, MS. B. v. ealaþ.

ealo-wǽge, es; *n. The ale-cup* :—Se ðe bær hroden ealowǽge *who bare the ornamented ale-cup*, Beo. Th. 995; B. 495. Ofer ealowǽge *over the ale-cup* [*during a drinking*], Beo. Th. 966; B. 481. v. ealu-wǽge.

ealo-wósa, an; *m. Ale-wetter* or *drinker*; cervīsiæ inebriātor *vel* pōtor:—

Sumum yrrum ealowôsan, were wînsadum *from one irritated as an ale-drinker, a wine-sated man*, Exon. 87 b ; Th. 330, 10 ; Vy. 49.

eal-riht; *adj. All-right*; pĕnĭtus rectus, R. Ben. 72, Lye.

eal-sealf, e ; *f.* [eal *all*, sealf *salve*] *The herb called the oak of Jerusalem* or *the oak of Cappadocia*; ambrŏsia, Som. Ben. Lye := ἀμβροσία *a perfumed salve, a plant*; ambrŏsia mărĭtima, Diosc. 3, 129, L. S. Lex. under ἀμβροσία.

eal-seolcen; *adj. All-silken*; holosērĭcus = ὁλοσηρικός, Ælfc. Gl. 62 ; Som. 68, 92 ; Wrt. Voc. 40, 3.

eal-swâ, **eall-swâ**; *adv.* ALSO, *so, so as, likewise, even as, even so*; sĭmĭlĭter, sĭcut :—Cristenum cyninge gebýreþ ðæt he sý ealswâ hit riht is *it is the duty of a Christian king to be as it is right*, L. I. P. 2 ; Th. ii. 304, 8, 22. Ðâ cwæþ he ealswâ to ðâm ôðrum *dixit sĭmĭlĭter ad altĕrum*, Mt. Bos. 21, 30. Gewurþe ðê, ealswâ ðû wylle *fiat tĭbi sĭcut vis*, Mt. Bos. 15, 28. [*Piers P. Chauc.* als *also : Laym.* alse, al so, al swa, al swo *as, so, also, thus, as if : Orm.* allse, alls, allswa, all swa *also, as, so : O. Sax.* alsô sĭmĭlĭter, tanquam, sĭcut, quăsi, quum : *Frs.* als, az, alsa sĭcut, cum, ita, si : *O. Frs.* alsa, olsa *ita, cum : Dut.* als *when, if;* alzoo *thus, so : Ger.* also *thus;* als as, *when : M.H.Ger.* als, alsô, alse *thus, when : O.H.Ger.* al sô *ut, sĭcut, vĕlut, sic.*]

eal-téaw; *adj.* [eal *all*, teaw = tǽw *good*] *Entirely good ;* omnīno bŏnus :—Gif he ealteawne ende gedreógeþ *if he enjoys a very good end*, Cd. Jun. 110, 16 ; Hy. 2, 13 ; Hy. Grn. ii. 281, 13.

EALU, ealo, es ; *n: generally indecl. in sing.* ALE ; cervĭsia, sĭcĕra :—Ne he ealu ne drince nǽfre oððe wîn *let him never drink ale nor wine*, Jud. 13, 4. Iohannes se Fulluhtere ne dranc nâðor ne wîn, ne beór, ne ealu *John the Baptist drank neither wine, nor beer, nor ale*, Homl. Th. ii. 38, 7 : Bt. 17 ; Fox 60, 5, MS. Cot : L. M. 1, 47 ; Lchdm. ii. 120, 15 : Beo. Th. 1542 ; B. 769. [*Chauc. Laym.* ale : *O. Sax.* alo in alo-fat, *n. an ale-cup : Dan. Swed. Icel.* öl, *n.*]

ealu-benc, ealo-benc, e ; *f. An ale-bench ;* scamnum cervĭsiam bibentium :—On ealu-bence *on the ale-bench*, Beo. Th. 5726 ; B. 2867.

ealu-clýfe *an ale-house*, Som. Ben. Lye. v. eala-hûs.

ealu-fæt *an ale-vat*, Som. Ben. Lye. v. ealo-fæt.

ealu-gafol, es ; *n.* [gafol *tax, tribute*] *Tribute* or *excise paid for ale ;* cervĭsiæ trĭbûtum :—On sumen lande gebûr sceal syllan hunig-gafol, on suman mete-gafol, on suman ealu-gafol *in one place a boor shall give honey-tribute, in another meat-tribute, in another ale-tribute*, L. R. S. 4 ; Th. i. 434. 32.

ealu-gâl *ale-drunken.* v. ealo-gâl.

ealu-geweorc *ale-brewing.* v. ealu.

ealu-malt *malt used for making ale.* v. ealu, alo-malt.

ealu-sceop, es ; *m. An ale-brewer, a brewer ;* cervĭsiārius, Som. Ben. Lye.

ealu-scóp, eala-scóp, es ; *m. An ale-poet :*—We lǽraþ, ðæt ǽnig preóst ne beó ealu-scóp *we teach that no priest be an ale-poet*, L. Edg. C. 58 ; Th. ii. 256, 15.

ealu-wǽge, es ; *n. An ale-cup ;* pătĕra, scyphus :—Dôhtor Hrôþgâres eorlum ealuwǽge bær *Hrothgar's daughter bore the ale-cup to the earls*, Beo. Th. 4047 ; B. 2021.

ealu-wôsa *ale-wetter* or *drinker.* v. ealo-wôsa.

Eal-walda, an ; *m. All-ruler, the Almighty ;* omnium rector, Cd. 14 ; Th. 16, 20 ; Gen. 246. v. Eal-wealda.

eal-weald; *adj. All-powerful, almighty ;* omnĭpŏtens :—Ǽrende eal-wealdan Gode wæs sprecen *a message was spoken to the all-powerful God*, Andr. Kmbl. 3239 ; An. 1622.

Eal-wealda, an ; *m. All-ruler, God, the Almighty ;* omnium rector, Deus, omnĭpŏtens :—For ðam ealwealdan [MS. alwealdan] *for the all-ruler* [*God*], Cd. 19 ; Th. 23, 13 ; Gen. 359. Noldon ealwealdan [MS. alwealdan] word weorþian *they would not revere the all-ruler's* [*the Almighty's*] *word*, 18 ; Th. 21, 23 ; Gen. 328.

eal-werlíce; *adv. All-manly, liberally, freely ;* prorsus virĭlĭter, benigne :—Ealwerlíce [MS. ealwerlíc] dô Driht *benigne fac Domine*, Ps. Spl. 50, 19.

eal-wihta *all beings.* v. eall-wihta.

eal-wundor *a very wonderful thing.* v. eall-wundor.

eam *am :*—Ic eam biddende Drihten *ad Deum deprĕcâtus sum*, Ps. Th. 141, 1. Ic eam leás ĕcan dreámes *I am bereft of eternal joy*, Cd. 216 ; Th. 275, 7 ; Sat. 168 : Exon. 10 a ; Th. 11, 8 ; Cri. 167 : Exon. 36 a ; Th. 116, 34 ; Gû. 217 : Mt. Rush. Stv. 11, 29. v. eom.

EÁM, es ; *m. An* EAM, *uncle chiefly on the mother's side ;* avuncŭlus :—Eám *avuncŭlus*, Wrt. Voc. 72, 42 : Beo. Th. 1766 ; B. 881 : Exon. 112 b ; Th. 431, 35 ; Rä. 47, 6 : Chr. 1066 ; Erl. 203, 17. Nim ðê wîf of Labanes dôhtrum ðînes eámes *accĭpe tĭbi uxôrem de fĭliābus Labāni avuncŭli tui*, Gen. 28, 2 : 29, 10 : Ors. 1, 12 ; Bos. 35, 32 : 2, 2 ; Bos. 41, 7 : Bd. 5, 19 ; S. 637, 33. Romŭlus slôh his eám *Romulus slew his uncle*, Ors. 2, 3 ; Bos. 41, 43 : Chr. 1046 ; Erl. 175, 5, 23. Mîn eám *avuncŭlus meus :* mînes eámes fæder *avuncŭlus meus magnus :* mînes eámes yldre fæder *proavuncŭlus meus :* mînes eámes þridde fæder *abavuncŭlus meus*, Ælfc. Gl. 93 ; Som. 75, 65–71 ; Wrt. Voc. 52, 21–24. [*Chauc.* eem, eme : *Laym.* æm, eam, æem, hem : *Plat.* oom, *m : Dut.* oom, *m : Frs.* yem, yeme : *O.Frs.* em, *m : Ger.* ohm, oheim, *m :*

M.H.Ger. ôheim, oeheim, *m : O.H.Ger.* ôheim, *m.*] For an uncle on the father's side, v. fædera.

eám, eán *to waters :*—Ofer ðam eám *sŭper flūmĭna*, Ps. Th. 23, 2. Betweoh ðâm twâm eán *between the two waters*, Ors. 1, 1 ; Bos. 16, 28 ; *dat. pl.* of eá.

eánian, eánigan ; *part.* eánigende ; *p.* eánode ; *pp.* eánod [eáw = eówu *a female sheep, a ewe*] *To* YEAN, *bring forth as a ewe ;* enīti, parturīre :—He genam hine of eówedum sceápa, fram eánigendum he genam hine *sustŭlit eum* [*Dāvĭdem*] *de grĕgĭbus ovium, de post fetantes* [*oves*] *accĕpit eum*, Ps. Lamb. 77, 70. DER. ge-eán. [*Prompt.* enyñ', *brynge forthe kyndelyngys* [*A. Sax.* litlingas]. The verb *to ean* or *yean*, which is commonly applied only to the bringing forth of lambs, here appears to have had anciently the more general signification of the word from which it is derived, *A. Sax.* eánian enīti, *parturīre : Wyc.* ene, eene, ȝeene, ȝene, yeene *sheep with lambs*, Ps. 143, 13 : Is. 40, 11 : *Dut. dial.* oonen *to produce young.*]

eá-ôfer, es ; *m. A river-bank ;* rīpa flūmĭnis :—Be sǽwaroþe, and be eá-ôfrum *by the sea-shore, and by river-banks*, Bt. Met. Fox 19, 43.

eapl *an apple*, Cd. 222 ; Th. 290, 7 ; Sat. 411. v. æppel.

ear, ær, es ; *m. Sea, ocean ;* măre, oceānus :—Hyre [dûne] deorc on lâst eare geblonden ôðer fereþ *dark on its* [*the down's*] *track goes another mixed with the ocean*, Exon. 101 b ; Th. 384, 3 ; Rä. 4, 22. v. ear-gebland, ear-grund.

EAR, es ; *n. An* EAR *of corn ;* spīca :—Seó eorþe wæstm beraþ, ǽrest gærs, syððan ear, syððan fulne hwǽte on ðam eare *terra fructĭfĭcat, primum herbam, deinde spīcam, deinde plēnum frumentum in spīca*, Mk. Bos. 4, 28. Ða seofon fullan ear getácniaþ seofon wæstmbǽre geár and wēlige *septem spīcæ plēnæ septem ubertātis anni sunt*, Gen. 41, 26, 27. Pharao rehte Iosepe be ðâm oxum and be ðâm earum *Pharaoh told Joseph of the oxen and of the ears* [*of corn*], Gen. 41, 17. Hig onguunun pluccian ða ear cœpērunt vellĕre spīcas, Mt. Bos. 12, 1 : Mk. Bos. 2, 23 : Lk. Bos. 6, 1. Him þuhte, ðæt he gesâwe seofon ear weaxan on ânum healme fulle and fægre *septem spīcæ pullŭlābant in culmo uno plēnæ atque formôsæ*, Gen. 41, 5 : Lev. 23, 22 : Deut. 23, 25. [*Wyc.* eere, ere : *R. Glouc.* eres, *pl : Plat.* aar, aare : *Dut.* aar, *f : Ger.* ähre, *f ;* äher, *n : M.H.Ger.* äher, eher, *n : O.H.Ger.* ahir, eher, *n : Goth.* ahs, *n : Dan. Swed. Icel.* ax, *n.* Grimm supposes the root of these words to be ak *sharp*, and refers to *Lat.* acus, acies, acidus : *Ger.* ecke *a corner.*]

eár *before*, Chr. 1041 ; Th. 299, 15, col. 1. v. ǽr.

EÁR, es ; *m. The Anglo-Saxon Rune* ᛠ, *which stands for the letters eá :* v. Steph. Runic Monmnts. p. 100, 11 ; 117, col. 7 : and p. 137 : *the earth, the ground ;* hūmus :—ᛠ byþ egle eorla gehwylcum, ðonne fæstlíce flǽsc onginneþ hrâw côlian, hrusan ceósan to gebeddan *the ground is hateful to every man, when surely the flesh beginneth to cool as a corpse, to choose the earth for a consort*, Runic pm. 29 ; Kmbl. 345, 10 : Hick. Thes. i. 135, 57. [*Icel.* aurr, *m. hŭmus.* Hylja auri *hŭmo condĕre*, Kormak's Saga.]

earan *are*, Th. Diplm. A. D. 804–829 ; 463, 1. v. eom.

earbe, an ; *f? A tare ;* ervum :—Dô earban to *add tares*, L. M. 1, 26 ; Lchdm. ii. 68, 4. v. earfe.

earc, e ; *f :* earce, an ; *f.* I. *the ark of Noah ;* arca :—Noe on ða earce eóde *Noah went into the ark*, Mt. Bos. 24, 38 : Lk. Bos. 17, 27. Under earce bord *under the boards of the ark*, Cd. 67 ; Th. 80, 23 ; Gen. 1333. Earce bordum *with the boards of the ark*, 67 ; Th. 81, 33 ; Gen. 1354. II. *a chest, the ark of the covenant ;* cista, cistella :—Cest *vel* earc *cibôtium* = κιβώτιον, vel cistella, Ælfc. Gl. 3 ; Som. 55, 64 ; Wrt. Voc. 16, 37. On earce *in the chest*, Exon. 124 b ; Th. 479, 3 ; Rä. 62, 2. Æt Godes earce *to the ark of God*, Cd. 212 ; Th. 262, 30 ; Dan. 752 : Ps. Th. 131, 8. v. earce II ; arc, *m.*

earce, an ; *f. The ark ;* arca :—Ðû earce eart eall-hâligra *tu arca sanctĭfĭcātiōnis tuæ*, Ps. Th. 131, 8. v. earc II.

eár-clǽnsend, es ; *m.* [eáre, clǽnsian *to cleanse*] *An ear-cleanser, the little finger ;* dĭgĭtus aurĭcŭlāris :—Eárclǽnsend [MS. earclæsnend] aurĭcŭlāris, Wrt. Voc. 283, 24.

earcnan-stân, es ; *m. A precious stone, gem ;* gemma, lăpis prĕtiôsa :—Se earcnanstân *the precious stone*, Exon. 25 a ; Th. 73, 27 ; Cri. 1196. v. eorcnan-stân.

eár-côðu, e ; *f.* [eáre, côðu *a disease*] *An ear-disease ;* parôtis = παρωτίς :—Eár-côðu *parôtĭdes*, Ælfc. Gl. 11 ; Som. 57, 57 ; Wrt. Voc. 20, 1.

EARD, es ; *m.* I. *native soil* or *land, country, province, region, place of residence, dwelling, home ;* sŏlum nātīvum, patria, rĕgio, dŏmĭcĭlium :—Sumra wyrta oððe sumes wuda eard biþ on dûnum, sumra on merscum ... on ðære stôwe ðe his eard biþ *the native soil of some herbs* or *of some wood is on hills, some in marshes ... in the place which is its native soil*, Bt. 34, 10 ; Fox 148, 22–26. He com to his earde *vēnit in patriam suam*, Mt. Bos. 13, 54. Nys mán wîtega bûtan wurþscype, bûton on hys earde *non est prophēta sine hŏnôre, nĭsi in patria sua*, 13, 57. Eard *patria*, Ælfc. Gl. 97 ; Som. 76, 57 ; Wrt. Voc. 54, 1. Ðis is mîn âgen cýþ, eard and êðel *this is my own country, dwelling, and home*, Bt. Met. Fox 24, 99 ; Met. 24, 50. Ðû gebunde ðæt fýr ðæt hit ne mæg cuman to his âgenum earde *thou hast bound the fire, that it may not come to its own region*, Bt. 33, 4 ; Fox 130, 32, 24. Ne ðær elþeó-

dige **eardes** brúcaþ *strangers enjoy no dwelling there*, Andr. Kmbl. 560; An. 280. Earda leás *deprived of dwellings*, Cd. 128; Th. 163, 29; Gen. 2705. Earda sélost *happiest of dwellings* [*heaven*], Hy. 7, 29; Hy. Grn. ii. 287, 29: Exon. 42 a; Th. 141, 16; Gú. 628: 36 b; Th. 120, 7; Gú. 268. Fífel-cynnes eard *the dwelling of the Fifel race*, Beo. Th. 209; B. 104. Of ðam heofon-fugelas healdaþ eardas *super ea volucres cæli habĭtābunt*, Ps. Th. 103, 11. Eard gemunde *he remembered his home*, Beo. Th. 2263; B. 1129. II. *earth* or *land, in contrast to water, as a firm place on earth* or *on land*; terra, terra firma :—He gefæstnude foldan staðelas, eorþan eardas *he made fast foundations of the ground, the firm places of the earth*, Ps. Th. 103, 6. Eard git ne const frécne stówe, ðǽr ðú findan miht secg *thou dost not yet know the land, perilous place, where thou mayest find the man*, Beo. Th. 2759; B. 1377: Exon. 38 b; Th. 128, 4; Gú. 399 : 129 a; Th. 495, 20; Rä. 85, 6. Lǽt nú gebídan on earde *let us now abide on land*, Andr. Kmbl. 799; An. 400. From hróf eardes *a summo terræ*, Mk. Lind. War. 13, 27. Gǽst and líc geador síðedan on earde *soul and body journeyed together on earth*, Exon. 76 a; Th. 285, 16; Jul. 715. III. *state, station, condition*; sĭtus, condĭtio :—Fundiaþ ǽlc gesceaft ðider swíðost, ðider his eard and his hǽlo swíðost bíoþ *every creature chiefly tends thither, where its station, and its health especially is*, Bt. 34, 11; Fox 150, 22. Man us tyhhaþ twegen eardas, Drihtnes áre oððe deófles þeówet *two conditions are appointed to us, the glory of God or bondage of the devil*, Hy. 7, 97; Hy. Grn. ii. 289, 97. [Orm. ærd *place, region*: Laym. ærde, ard *land, earth*: O. Sax. ard, m. *habĭtātio*: Dut. aard, m. *nature, temper*: Kil. ærd: Ger. art, f. *nātūra, indōles, mōdus, spĕcies, gĕnus*: M. H. Ger. art, gen. ardes, m; art, gen. arte, f. *nātūra, indōles*: O. H. Ger. art, f. arātio; der. of erian *to plough?*] DER. éðel-eard, herh-, middan-, somud-, wíc-.

eard-begenga, -begænga, -begenda, an; m. [beganga, begenga *a dweller*] *An inhabitant, dweller*; incŏla :—Eardbegenga wæs sáwle mín *incŏla fuit anĭma mea*, Ps. Lamb. 119, 6. Ðá ðá híg wǽron eardbegendan *cum essent incŏlæ*, Ps. Lamb. 104, 12. Eardbegǽngan *incŏlæ*, Ps. Spl. M. 104, 11.

eard-begengnes, -biggengnes, -ness, e; f. *An abode, habitation*; habĭtātio, incŏlātus :—Eardbegengnes oððe elþeódignys mín afeorrad oððe gelængd is *incŏlātus meus prolongātus est*, Ps. Lamb. 119, 5. Eardbiggengnes [MS. eardbiggendes] mín aforseorsode is *incŏlātus meus prolongātus est*, Ps. Spl. 119, 5.

eard-éðel-riht, es; n. *Land-inheritance right, patrimonial right*; patrium jus, Beo. Th. 4402; B. 2198.

eard-éðel-wyn, -wynn, e; f. *Joy of an estate*; prædii gaudium :—He me lond forgeaf, eardéðelwyn *he gave me land, joy of property*, Beo. Th. 4979; B. 2493. v. éðel-wyn.

eard-fæst; adj. *Earth fast, settled, established in a place, abiding*; sŏlo fixus, habĭtans :—Ðe eardfæst byþ on Hierusalem *qui habĭtat in Hierusalem*, Ps. Th. 124, 1: Exon. 44 2; Th. 149, 8; Gú. 758: Cd. 136; Th. 171, 27; Gen. 2834: Bt. Met. Fox 7, 76; Met. 7, 38 : Ors. 5, 4; Bos. 105, 11: 6, 33; Bos. 129, 33.

eard-geard, es; m. *A dwelling-place, the earth*; habĭtātiōnis lŏcus, terra :—In ðam eardgearde *in that dwelling-place* [*in Jerusalem*], Exon. 8 b; Th. 4, 19; Cri. 55. Ydde ðisne eardgeard ælda Scyppend *the Creator of men overwhelmed this world*, 77 b; Th. 291, 20; Wand. 85.

eard-gyf, es; n. *A gift from one's native land*; patrium dōnum :—Kyningas eard-gyfu bringaþ *Spl.* has, cyningas gyfa togelǽdaþ : *rēges dōna addúcent*, Ps. Th. 71, 10.

eard-hæbbendra [= eard, hæbbendra], Ps. Th. 86, 6; *gen. pl.* of eard-hæbbende; *part. pres.* of eard-habban = habban *to have.*

eardian, eardigan, eardigean, ærdian; *part.* eardiende, eardigende, eardende; ic eardige, ðú eardast, he eardaþ, *pl.* eardiaþ, eardigaþ; *p.* ode, ade, ede; *pp.* od, ad, ed. I. *v. intrans. To dwell, live, feed*; habĭtāre :—Heofenes fugelas eardian mágon under his sceade *possunt sub umbra ejus aves cæli habĭtāre*, Mk. Bos. 4, 32 : Exon. 129 b; Th. 496, 24; Rä. 85, 19 : Ps. Th. 67, 6 : Ps. Spl. 2, 4 : 5, 5. Eardigan, Bt. 33, 4; Fox 130, 10. Loth ne dorste on ðam fæstenne leng eardigean *Lot might not longer dwell in that fastness*, Cd. 121; Th. 156, 19; Gen. 2591 : Ps. Spl. C. 112, 8. Ic eardige, Ps. Th. 60, 3. Ðú eardast, Hy. 5, 1; Hy. Grn. ii. 289, 5. Ðǽr his hírêd eardaþ *where his flock feeds*, Cd. 226; Th. 302, 2; Sat. 592. Æt helle dúru dracan eardigaþ *dragons dwell at the gate of hell*, 215; Th. 270, 30; Sat. 98. On earda eorðan *dwell on earth*, Ps. Spl. 36, 3. Ðeáh hí somod eardien *though they dwell together*, Bt. Met. Fox 20, 292; Met. 20, 146. For yfelnesse ðara eardiendra ðǽr on ðære byrig *a malĭtia inhabĭtantium in eo*, Bd. 4, 25; S. 599, 22: Ps. Th. 82, 6 : 135, 27 : Ps. Spl. 16, 13. Eardendra, Ps. Th. 106, 33. Abram eardode on ðam lande Chanaan *Abram habĭtāvit in terra Chanaan*, Gen. 13, 12. Eardodon, Beo. Th. 6093; B. 3050. Se me be healfe eardade *who dwelled by my side*, Exon. 129 b; Th. 496, 26; Rä. 85, 20. Eardedon, 9 b; Th. 8, 30; Cri. 125. II. *v. trans. To inhabit*; inhabĭtāre. incŏlĕre :—Peohtas ongunnon eardigan ða norþ-dǽlas ðysses eálondes *Picti habĭtāre per septentriōnāles insŭlæ partes cœpĕrunt*. Bd. 1, 1; S. 474, 18. Sceolde wíc eardian elles hwergen

he should inhabit a dwelling elsewhere, Beo. Th. 5172; B. 2589: Ps. Th. 104, 19. DER. ge-eardian, on-, on-eardiend.

eardigendlíc; *adj. Inhabitable*; habĭtābĭlis :—Seó stów eardigendlíc wæs geworden *lŏcus habĭtābĭlis factus est*, Bd. 4, 28; S. 605, 21.

earding, e; f. *A habitation, dwelling*; habĭtācŭlum :—Ðǽr we mótun ágan eardinga *where we may possess dwellings*, Exon. 65 b; Th. 242, 14; Ph. 673. v. eardung.

eard-land, es; n. *Country*; patria :—Sealde heora eardland eall Isrāhēlum *dĕdit terram eōrum hærĕdĭtātem Israel*, Ps. Th. 134, 12.

eard-ríce, es; n. *A dwelling-land*; terra habĭtātiōnis :—Eardríce cyst *the best of habitations* [*Paradise*], Exon. 45 a; Th. 153, 14; Gú. 825.

eard-stapa, an; m. *A land-stepper, wanderer*; terras peragrans, peregrīnātor :—Swá cwæþ eard-stapa *so said a wanderer*, Exon. 76 b; Th. 286, 25; Wand. 6.

eard-stede, es; m. *A dwelling-place*; lŏcus habĭtātiōnis :—Ða swétestan somnaþ and gædraþ wyrta wynsume and wudubléda to ðam eardstede it [*the Phœnix*] *collects and gathers pleasant herbs and forest leaves to that dwelling-place*, Exon. 58 b; Th. 211, 9; Ph. 195.

eardung, eardŭng, ærdung, e; f. *A habitation, a dwelling, tabernacle*; habĭtātio, habĭtācŭlum :—Is geworden eardung his on Sion *facta est habĭtātio ejus in Sion*, Ps. Spl. 75, 2 : 32, 14 : Ps. Spl. T. 77, 32 : Ps. Th. 106, 3 : Hy. 6, 11; Hy. Grn. ii. 286, 11 : Bd. 4, 28; S. 605, 20. v. earding.

eardung-burh; *gen.* -burge; f. *A dwelling-city, city of tabernacles*; tabernacŭlōrum urbs :—Híg getimbrodun Pharaones eardungburga Phiton and Rameses *ædificāvērunt urbes tabernacŭlōrum Pharaōni Phithon et Ramesses*, Ex. 1, 11.

eardung-hús, es; n. *A habitation*; habĭtācŭlum :—Gemǽne eardunghús *commūne habĭtācŭlum*, Bd. 4, 28; S. 605, 26.

eardung-stów, e; f. *A dwelling-place, a tent, tabernacle*; habĭtātiōnis lŏcus, tabernācŭlum :—On eallum eówrum eardungstówum *in cunctis habĭtācŭlis vestris*, Ex. 12, 20 : Ps. Th. 106, 6 : Jn. Bos. 14, 2 : Bd. 4, 28; S. 605, 19.

eard-wíc, es; n. *A dwelling-place*; habĭtātiōnis lŏcus :—Ðonne ic sceal eardwíc uncúþ gesécan *when I shall seek the uncouth dwelling-place*, Apstls. Kmbl. 185; Ap. 93. He getimbreþ eardwíc niwe *it builds a new dwelling-place*, Exon. 62 a; Th. 228, 1; Ph. 431.

eard-wrecca, -wreca, an; n. [eard I. *native country*; wrecca = wræcca *an exile*] *One banished from his native country, an exile*; exsul :—Þurh eardwrecena feormunge *by harbouring of exiles*, L. Alf. pol. 4; Th. i. 62, 16, note 24.

EÁRE, an; n : *nom. acc. sing.* eáre; *nom. acc. pl.* eáran *The* EAR *of man or an animal*; auris :—Ðæs eáre slóh Petrus of *cujus abscĭdit Petrus aurĭculam*, Jn. Bos. 18, 26 : Mk. Bos. 7, 33, 35 : 14, 47 : Ælfc. Gl. 71; Som. 70, 92; Wrt. Voc. 43, 23 : Ps. Th. 140, 8 : Exon. 128 b; Th. 494, 19; Rä. 83, 3 : Cd. 216; Th. 275, 13; Sat. 171. [Wyc. eer, eere, ere : Piers P. ere : Chauc. ere : Orm. ære : Plat. oor, n : O. Sax. óra, n : Frs. ær, ear, eare : O. Frs. are, ar, n : Dut. oor, n : Ger. ohr, n : M. H. Ger. ôre, n : O. H. Ger. ôra, n : Goth. auso, n : Dan. öre, n : Swed. öra, n : Icel. eyra, n : Lat. auris, f : Grk. οὖς, n : Lith. ausis, f.]

eáre-finger, es; m. *An ear-finger, the little finger*; auricŭlārius digĭtus, minĭmus digĭtōrum :—Eárefinger *auricŭlārius*, Wrt. Voc. 71, 34.

eáre-lippric, eár-lipric, e; f : eór-lippric, es; n. *A flap of the ear*; auricŭla :—In eárlipricum, *dat. pl.* Mk. Lind. War. 7, 33. Eárliprica, *acc. pl.* Mk. Lind. War. 7, 33 : Jn. Rush. War. 18, 26. Ða eárelipprica, *acc. pl.* Mk. Lind. War. 14, 47.

earendel, earendil, es; m? *A shining light, ray*; jûbar :—Leóma, earendil *jûbar*, Glos. Epnl. Recd. 158, 25. Eálá earendel! eala beorhtast! ofer middangeard monnum sended *O ray! brightest of angels! sent to men over mid-earth*, Exon. 9 b; Th. 7, 20; Cri. 104. [O. H. Ger. Orendel, nn. pr.]

EARFE, earbe, an; f? *A tare*; ervum, orŏbus = ὄροβος :—Earfan wyl on wætere *boil tares in water*, L. M. I, 8; Lchdm. ii. 52, 16. [Dut. erwt, f. pea : Kil. erwete, erte, f : Ger. erbse, f. a pea : M. H. Ger. areweiȝ, erweiȝ, f : O. H. Ger. araweiȝ, arawiȝ, erbiȝ pisum : Dan. ært, ert, m. f. a pea : Swed. ärt, f. a pea : Icel. ertr, f. pl. peas.]

earfednyme *an heir*; hēres, Lk. Skt. Hat. 20, 14. v. yrfenuma.

EARFEÐE, earfoþ, es; *pl. nom. acc.* u, o, a; n. *Hardship, labour, difficulty, trouble, suffering, woe*; lābor, molestia, tribŭlātio :—Ic ðæt earfeðe wonn *I suffered the hardship*, Exon. 28 b; Th. 87, 21; Cri. 1428. Earfoðes feala *tribŭlātiōnis multum*, Ps. Th. 70, 19. Earfoðu, 21, 9 : 24, 15 : 68, 27. Ðe ða earfeða dreógeþ *who suffers those afflictions*, Exon. 52 b; Th. 183, 18; Gú. 1329. Earfeðum, Ps. Th. 106, 5, 27. Earfeða dǽl *a deal of sufferings*, Cd. 9; Th. 12, 4; Gen. 180. [Plat. arbeed, f : Hel. arabêd, arbed, f : arabêdi, arbêdi, n : O. Sax. arbeit, f : arbeithi, arbeidi, arvit, n : Frs. aerbeyde : O. Frs. arbeid, arbed, n : Dut. arbeid, m : Ger. arbeit, f : M. H. Ger. arbeit, arebeit, f : O. H. Ger. arabeit, arbeit, f : Goth. arbaiþs, f : Dan. arbeid, arbeide, n : Swed. arbete, n : Icel. erfiði, ervíði, n. toil, labour, distress.] DER. firen-earfeðe, -earfoþ, ge-, mægen-, mód-, would-.

earfeðe, earfoþ; *adj. Hard, difficult, troublesome*; difficilis, mŏlestus :—

Nis me earfeðe to geþolianne willan Dryhtnes mînes *it is not hard for me to endure the will of my Lord*, Exon. 48 a; Th. 166, 6; Gú. 1038. Ða bisgu us sint swîðe earfoþ *the occupations are to us very difficult*, Bt. prœem; Fox viii. 7. Earfoðest *most difficult*, Bt. 39, 4; Fox 216, 15. [*Orm.* arrfeþþ *difficult*.]

earfeþ-mæcg, es; *m. An unhappy* or *unfortunate man:* infortûnâtus hŏmo :—Se endestæf earfeþmæcgum weálîc weorþeþ *the end to the unfortunate is miserable*, Exon. 87 a; Th. 328, 3; Vy. 11. v. earfoþ-mæcg.

earfeþ-sîþ *a misfortune, calamity :*—Earfeþsîðas *calamities*, Andr. Kmbl. 2568; An. 1285. v. earfoþ-sîþ.

earfoþ, es; *n. Hardship, trouble;* lâbor, tribûlâtio :—Mâ earfoða *more of troubles*, Guthl. 5; Gdwin. 32, 13. v. earfeðe.

earfoþ; *adj. Hard, difficult*, Bt. prœem; Fox viii. 7. v. earfeðe; *adj.*

earfoþ-cyn, -cynn, es; *n. A violent generation;* prâva gens :—Ðæt wæs earfoþcynn yrre and rêðe *gĕnus prâvum et peramârum*, Ps. Th. 77, 10.

earfoþ-dæg, es; *m. A trouble-day, day of trouble;* tribûlâtiônis dies :—Ic on earfoþ-dæge Drihten sôhte *in die tribûlâtiônis Deum exquîsîvi*, Ps. Th. 76, 2.

earfoþ-fere; *adj. Difficult to pass;* difficilis transîtu, Scint. 10.

earfoþ-hâwe; *adj. Difficult to be seen;* difficilis vîsu :—Earfoþhâwe is *it is difficult to be seen*, Bt. Met. Fox 20, 303; Met. 20, 152: Bt. 33, 4; Fox 130, 30.

earfoþ-hwîl, e; *f. A time of hardship;* mŏlestum tempus :—Ic earfoþhwîle þrôwade *I suffered a time of hardship*, Exon. 81 b; Th. 306, 5; Seef. 3.

earfoþ-hylde; *adj. Ill-inclined, ill-disposed, ill-natured;* malĕvŏlus, malignus :—Se ðe earfoþhylde biþ, and gyrnþ ðæra þinga ðe he begitan ne mihte, bûton twŷn him geneálæhþ se hreófla Giezi *he who is ill-inclined, and yearns for the things which he could not obtain, without doubt to him approximates the leper Gehazi*, Homl. Th. i. 400, 1.

earfoþ-lǽre; *adj. Difficult to be taught, dull;* difficilis doctu :—Earfoþlǽran brôðru *indôciles fratres*, Greg. Dial. 2, 3.

earfoþ-lǽte; *adj. Difficult to be sent forth;* difficilis emissu :—Earfoþlǽte micga *a painful discharge of urine, strangury;* strangûria = στραγγουρία, Ælfc. Gl. 11; Som. 57, 47; Wrt. Voc. 19, 49.

earfoþ-lîc; *adj. Irksome;* laboriôsus :—Eall is earfoþlîc eorþan rîce *the realm of earth is all irksome*, Exon. 78 a; Th. 292, 28; Wand. 106. Gif eów ǽnig þing þince earfoþlîce *si difficile vôbis vîsum ălîquid fuĕrit*, Deut. 1, 17.

earfoþ-lîce; *adv. With difficulty, reluctantly, sorely, hardly;* difficile, invîte, ægre :—Earfoþlîce wæs gûþ getwǽfed *the contest had been parted with difficulty*, Beo. Th. 3318; B. 1657: 3276; B. 1636: Mk. Bos. 10, 23. Se ellen-gǽst earfoþlîce þrage geþolode *the potent ghost reluctantly endured for a time*, Beo. Th. 173; B. 86: Exon. 98 a; Th. 369, 8; Seel. 38. Ða wæs gegongen earfoþlîce *then it befel sorely*, Beo. Th. 5636; B. 2822: Andr. Kmbl. 1028; An. 514. Hî ôþ-eódon earfoþlîce *they hardly escaped*, Beo. Th. 5861; B. 2934.

earfoþlícnes, -ness, -nyss, e; *f. Difficulty, pain;* difficultas :—Heó earfoþlícnysse [-nesse MS. B.] ðæs migþan astyreþ *it stirreth a difficulty of the urine* [*strangury*], Herb. 143, 1; Lchdm. i. 266, 3. Wið ðæs migþan earfoþlícnyssa [-nysse MS. H : -nesse MS. B.] *for difficulties of the urine*, 156, 3; Lchdm. i. 284, 4.

earfoþ-mæcg, earfɔþ-mæcg, es; *m. An unhappy man;* infortûnâtus hŏmo :—Se earfoþmæcg up lôcode *the afflicted man looked up*, Cd. 206; Th. 255, 12; Dan. 623.

earfoþnes, -ness, -niss, -nyss, e; *f. Difficulty, hardship, anxiety, tribulation, misfortune;* difficultas, lâbor, angustiæ, tribûlâtio, infortûnium :—God ealle þing gediht bûton earfoþnysse *God regulates all things without difficulty*, Bd. de nat. rerum; Wrt. popl. science 19, 5; Lchdm. iii. 278, 13. Wið wîfa earfoþnyssum [-nessum MS. B.] *for the difficulties of women*, Med. ex Quadr. 2, 7; Lchdm. i. 334, 18. He geheóld his rîce mid myclum geswince and earfoþnessum [-nyssum, Th. 278, 40, col. 2 : -nissum, 279, 41, col. 1] *he held his kingdom with much labour and hardships*, Chr. 1016; Th. 278, 41, col. 1. Bûtan micelre earfoþnysse *without much tribulation*, Homl. Th. i. 476, 13 : Boutr. Scrd. 20, 35. Bûtan earfoþnyssum *without tribulations*, Homl. Th. i. 476, 11. Mihte we ðŷ êþ geþolian swâ hwæt earfoþnessa swâ us on becôme *we might the more easily bear whatsoever misfortunes come upon us*, Bt. 10; Fox 30, 12.

earfoþ-recce; *adj.* [reccan *to relate*] *Difficult to be told;* difficilis narrâtu, Lupi Serm. 5, 3, Lye.

earfoþ-rîme; *adj. Difficult to be numbered;* difficilis numĕrâtu :—Ða bisgu us sint swîðe earfoþrîme *the occupations are to us very difficult to be numbered*, Bt. prœem; Fox viii. 7.

earfoþ-sǽlig; *adj. Unblessed;* infêlix :—Ne biþ ǽnig ðæs earfoþsǽlig mon on moldan *there is not any man on earth so unblessed*, Exon. 78 b; Th. 294, 1; Crä. 8.

earfoþ-sîþ, earfeþ-sîþ, es; *m. A laborious journey, misfortune, calamity;* mŏlestum iter, infortûnium, calâmitas :—Weoru geferaþ earfoþsîða *ye travel plenty of laborious journeys*, Andr. Kmbl. 1355; An. 678: Cd. 72;

Th. 89, 5; Gen. 1476. Se folc-toga findan sceolde earfoþsîðas *the nation's leader should find calamities*, 208; Th. 257, 13; Dan. 657: Exon. 88 a; Th. 330, 30; Vy. 59. Ðû wâst ânra gehwylces earfeþsîðas *thou knowest every man's calamities*, Andr. Kmbl. 2568; An. 1285.

earfoþ-tǽcne; *adj. Difficult to be shewn;* difficilis demonstrâtu :—Eorþe and wæter earfoþtǽcne wuniaþ on fŷre *earth and water dwell in fire difficult to be shewn*, Bt. Met. Fox 20, 294; Met. 20, 147.

earfoþ-þrag, e; *f. Time of tribulation;* mŏlestum tempus :—Á syððan earfoþþrage þolaþ *ever after will suffer a time of tribulation*, Beo. Th. 572; B. 283.

EARG, earh; *comp.* eargra, earhra; *sup.* eargost; *adj.* I. *inert, weak, timid, cowardly;* iners, ignâvus, segnis, tîmidus :—Se earga fêðe Brytta *ácies* segnis Brittŏnum, Bd. 1, 12; S. 481, 19. Ful oft mon wearnum tîhþ eargne *full oft one urges the inert with threats*, Exon. 92 a; Th. 345, 14; Gn. Ex. 188. Ne biþ swylc earges sîþ *such is not the path of the cowardly*, Beo. Th. 5076; B. 2541: Ors. 6, 36; Bos. 131, 27. II. *evil, wretched, vile;* prâvus, imprŏbus :—Ða cyningas, ðe æfter Romuluse rîcsedan, wǽran eargran ðonne he wǽre *the kings who reigned after Romulus, were more vile than he was*, Ors. 2, 2; Bos. 41, 24. Tarcuinius hiora eallra eargost wæs *Tarquin was the most vile of them all*, 2, 2; Bos. 41, 26. Swâ fela eargra worda *so many evil words*, Cd. 27; Th. 36, 32; Gen. 580: Exon. 26 b; Th. 79, 29; Cri. 1298. [*Chauc.* erke *indolent, indisposed : Laym.* earȝh *timid : Scot.* arch, argh, ergh *averse : Frs.* erg *bad, wicked : O. Frs.* erch, erg, arg *bad : Dut.* erg *bad : Ger.* arg *bad, wicked : M. H. Ger.* arc *mălus, prâvus : O. H. Ger.* arg *avârus, prâvus : Dan.* arg, arrig *bad, wicked, passionate : Swed.* arg *angry : Icel.* argr *emasculate, effeminate.*] DER. un-earg.

earge; *adv. Inertly, badly;* segniter, mâle :—Earge gê ðæt lǽstun *ye performed that badly*, Exon. 30 a; Th. 92, 3; Cri. 1503.

ear-gebland; *n. Wave-mingling;* oceâni turbâtio, undârum commixtio :—Ofer eargebland [æra gebland, col. 1] land gesôhtan *they sought the land over the ocean* [lit. *the wave-mingling*], Chr. 937; Th. 202, 38, col. 2: Th. 203, 38, col. 1, 2: Bt. Met. Fox 8, 59; Met. 8, 30.

eár-gespeca, eár-gespreca, an; *m. An ear-speaker, a whisperer;* auricŭlârius, susurro, Cot. 14.

earg-faru, e; *f. A flight* or *shooting of an arrow*, Exon. 71 b; Th. 266, 26; Jul. 404. v. earh-faru.

eargian *to be slothful, dull, idle;* torpescĕre. DER. a-eargian.

eargra *weaker*, Bt. 26, 2; Fox 92, 27, = *comp. of* earg.

ear-grund, es; *m. The ocean's ground;* oceâni fundus, Exon. 53 b; Th. 188, 3; Az. 40.

eargscipe, earhscipe, es; *m. Idleness, sloth;* ignâvia, Lye.

earh *ocean.* DER. earh-gebland. v. ear.

earh; *adj. Swift, fleeing through fear, timorous, weak;* fûgax, Ælfc. Gr. 9, 60; Som. 13, 43 : Byrht. Th. 138, 50; By. 238. v. earg.

EARH, e; *f.* arewe, an; *f. An ARROW;* sagitta :—Earh ǽttre gemǽl *the arrow stained with poison*, Andr. Recd. 2661; An. 1333. [*Laym.* Chauc. arwe : *Piers P.* arwe, *pl.* arewes : *Wyc.* arewe, arwe : *Goth.* arhwazna, *f. telum : O. Nrs.* ör; *gen.* örvar, *f. sagitta.*]

earh-faru, e; *f.* [earh *an arrow;* faru *a going, journey, passage*] *A flight of arrows;* sagittârum vŏlâtus :—Habbaþ scearp speru, atole earhfare *they have sharp spears, a terrible flight of arrows*, Salm. Kmbl. 259; Sal. 129. Mid earhfare *with a flight of arrows*, Andr. Kmbl. 2097; An. 1050. Ða us gescildaþ wið sceððendra eglum [MS englum] earhfarum *they shall shield us against the enemies' noxious flights of arrows*, Exon. 19 a; Th. 47, 28; Cri. 762.

earh-geblond *wave-mingling*, Elen. Kmbl. 477; El. 239. v. ear-gebland.

earhlíce; *adv. Fearfully, timidly, disgracefully, basely;* trepide, remisse, ignâve, turpiter :—Earhlíce *timidly*, Gen. 20, 4. Hî hine earhlíce ofslógon *they basely slew him*, Chr. 1086; Erl. 223, 9. v. earh.

earhra *weaker*, Bt. 26, 2; Fox 92, 27, MS. Bod. v. earg.

eár-hring, eár-ring, es; *m. An ear-ring;* inauris :—Nymaþ gyldene eár-hringas of eówer wîfa eáron *tollîte inaures aureas de uxôrum vestrârum auribus*, Ex. 32, 2; Ælfc. Gl. 4; Som. 55, 91. v. eár-ring.

eá-risc, e; *f. A water-rush, bulrush;* scirpus, juncus, Cot. 219: R. 42 ? Lye. v. ǽ-risc.

eá-rîþ, es; *m. A water-stream;* aquæ rîvus :—Ðǽr synd fûle eáriþas yrnende *there are foul running water-streams*, Guthl. 3; Gdwin. 20, 5.

eá-rixe, an; *f. A water-rush :*—Nim eárixena wyrtruman *take roots of water-rushes*, Lchdm. iii. 122, 8. v. eá-risc.

eár-læppa, an; *m.* [eáre *an ear,* læppa *a lap*] *An ear-lap;* pinnûla :—Eár-læppa *vel* ufweard [MS. ufwaard] eáre *pinnûla :* flêran *vel* eár-læppan *pinnûlæ*, Ælfc. Gl. 71; Som. 70, 83, 84; Wrt. Voc. 43, 15, 16.

eár-loccas; *pl. m.* [eár = ǽr *before*] *Forelocks;* antiæ, Ælfc. Gl. 64; Som. 69, 16; Wrt. Voc. 40, 49.

EARM, es; *m.* I. *an* ARM, *the limb extending from the shoulder to the hand;* brachium :—Gif se earm biþ forad bûfan elmbogan, ðǽr sculon xv scillinga tô bôte *if the arm be broken above the elbow, there shall be fifteen shillings for compensation*, L. Alf. pol. 54; Th. i. 94, 24: 66; Th. i. 96, 28. Earm *brachium*, Wrt. Voc. 64, 69: 71, 22: 283, 7:

Ps. Lamb. 88, 22 : 97, 1. On mycelnysse earmes ðínes *in magnitūdine brachii tui*, Cant. Moys. Lamb. 187 b, 16 : Ps. Th. 70, 17 : 78, 12. He worhte mægne on hys earme *fēcit potentiam in brachio suo*, Lk. Bos. 1, 51 : Ex. 6, 6 : Ps. Lamb. 76, 16 : 135, 12 : Beo. Th. 4711 ; B. 2361. Se ðe earm þurhstinþ vi scillingum gebête : gif earm forbrocen weorþ, vi scillingum gebête *let him who stabs [another] through the arm make amends with six shillings : if the arm be broken, let him make amends with six shillings*, L. Ethb. 53 ; Th. i. 16, 7, 8 : Byrht. Th. 136, 43 ; By. 165. Ánra gehwylc wið earm gesæt, hleonade wið handa *each one rested on his arm, leaned on his hand*, Cd. 223 ; Th. 291, 18 ; Sat. 432 : Beo. Th. 1503 ; B. 749. Æghwæðer óðerne earme beþehte *each embraced the other with his arm*, Andr. Kmbl. 2030 ; An. 1017 : Elen. Kmbl. 2470 : El. 1236. Forðande earmas synfulra beóþ tobrocene oððe beóþ tobrytte *quóniam brachia peccatōrum contĕrentur*, Ps. Lamb. 36, 17 : 43, 4. Næfde sēllícu wiht exle ne earmas *the wonderful thing had not shoulders nor arms*, Exon. 108 b ; Th. 415, 4 ; Rä. 33, 6 : 129 a ; Th. 494, 24 ; Rä. 83, 6. Ðe me mid hys earnum worhte *who made me with his arms*, Cd. 26 ; Th. 34, 28 ; Gen. 544 : Ps. Th. 90, 11. Muscl ðæs earmes *the muscle of the arm*; tōrus *vel* muscŭlus *vel* lăcertus, Ælfc. Gl. 72 ; Som. 70, 123 ; Wrt. Voc. 43, 48. II. *anything projecting from a main body, as an inlet of the sea or ocean, etc*; sínus, rāmus :—Ðæs sǽs earm *an arm of the sea*, Ors. 1, 1 ; Bos. 19, 10, 15, 19, 21. Earmes, 23, 20 : 24, 16, 17. Gársecges earm, Ors. 1, 1 ; Bos. 18, 23 : 19, 9. [*Wyc.* arm : *Chauc.* arme : *Laym.* ærm, arm : *Orm.* arrmess, *pl* : Plat. O.Sax. arm, *m* : *Frs.* earm : O.Frs. erm, arm, *m* : *Dut. Ger. M.H.Ger.* arm, *m* : O.H.Ger. aram, arm, *m* : *Goth.* arms, *m* : *Dan.* arm, *m.f* : *Swed.* arm, *m* : *Icel.* armr, *m* : *Lat.* armus, *m* : *Grk.* ἁρμός, *m. the shoulder-joint* : *Sansk.* îrma, *m. the arm.*] DER. sǽ-earm : earm-beáh, -boga, -gegyrela, -hreád, -scanca, -slífe, -strang, -swíþ.

EARM, ærm, arm ; *comp.* earmra ; *sup.* earmost ; *adj.* I. *poor, miserable, helpless, pitiful, wretched* ; pauper, miser :—Ðá com án earm wuduwe *cum vēnisset vĭdua una pauper*, Mk. Bos. 12, 42, 43 : Bt. 39, 2 ; Fox 212, 16. Nú eart tú earm sceaða *now art thou a miserable wretch*, Cd. 214 ; Th. 268, 19 ; Sat. 57 : 226 ; Th. 301, 9 ; Sat. 579 : Ps. Th. 136, 8. Earm biþ se him his frýnd geswícaþ *miserable is he whom his friends betray*, Exon. 89 a ; Th. 335, 22 ; Gn. Ex. 37. Se wæs ord-fruma earmre láfe *who was the chief of the poor remnant*, Cd. 179 ; Th. 225, 11 ; Dan. 152. Gé sindon earme ofer ealle menn *you are wretched above all men*, Andr. Kmbl. 1351 ; An. 676. Nó ic gefrægn earmran mannan *I have not heard of a more miserable man*, Beo. Th. 1159 ; B. 577. Ic wolde cweðan ðæt hí wǽron earmoste *I should say that they were most miserable*, Bt. 38, 2 ; Fox 198, 13 : Exon. 110 a ; Th. 421, 6 ; Rä. 40, 14. II. *the poor and destitute for whom the church made a provision* ; pauperes :—Be teóðunge. Se cyng and his witan habbaþ gecoren and gecweden, ealswá hit riht is,—ðæt þridda [MS. þriddan] dǽl ðare teóðunge, ðe to circan gebýrige, gá to ciric-bóte ;—and óðer dǽl ðám Godes þeówum ;—þridde Godes þearfum, and earman þeówetlingan *concerning tithe. The king and his witan have chosen and decreed, as is just,—that a third part of the tithe, which belongs to the church, go to church-repair ;—and a second part to the servants of God ;—a third to God's poor, and the needy in thraldom*, L. Eth. ix. 6 ; Th. i. 342, 6–9. v. þearfa. [*Laym.* ærm : *Plat. O.Sax.* arm : *Frs.* earm : erm : *Dut. Ger. M.H.Ger.* arm : O.H.Ger. arm, aram : *Goth.* arms : *Dan. Swed.* arm : *Icel.* armr.]

earm-beáh ; *gen.* -beáges ; *dat.* -beáge ; *m. An arm-ring, bracelet* ; armilla :—Brád earmbeáh *a broad* or *large arm-bracelet* ; dextrochĕrium, Ælfc. Gl. 114 ; Som. 80, 30 ; Wrt. Voc. 61, 10. Earmbeága fela *many bracelets*, Beo. Th. 5520 ; B. 2763.

earm-boga, an ; *m. An arm-bow, elbow* ; brachii curvātūra, Som. Ben. Lye.

earm-cearig ; *adj. Miserable and sad* ; miser et tristis :—Hú ic, earmcearig, íscealdne sǽ, winter wunade *how I passed a winter, miserable and sad, on the ice-cold sea*, Exon. 81 b ; Th. 306, 27 ; Seef. 14 : 76 b ; Th. 287, 26 ; Wand. 20.

earmè ; *adv. Wretchedly, badly* ; mĭsĕre, măle :—He lyt ongeat ðæt him swá earme gelamp *he little knew that it would fall out to him so badly*, Cd. 76 ; Th. 9ɔ, 26 ; Gen. 1567.

earm-gegyrela, -gegirela, an ; *m.* [gegyrela *clothing, apparel*] *A bracelet to be worn on the right arm* ; dextrāle :—Earmgegirelan dextrālia, Cot. 63.

earm-heort ; *adj. Tender-hearted, merciful* ; misĕricors, Greg. Dial. 1, 2.

earm-hreád ; *adj. f. An arm-ornament* ; brachii ornāmentum :—Earmhreáda [MS. earm reade] twá *two arm-ornaments*, B. 1194. v. hreóðan.

earmian ; *p.* ode ; *pp.* od ; *v. reflex. To commiserate, feel pity* ; misĕrēri :—Hwam ne mæg earmian swylcere tíde *who cannot feel pity for such a time?* Chr. 1087 ; Th. 354, 2.

earming, erming, yrming, es ; *m. A wretched* or *miserable being* ; miser :—Earming miser, Ælfc. Gr. 8 ; Som. 7, 18 : Ælfc. Gl. 77 ; Som. 72, 17 ; Wrt. Voc. 45, 50 : 75, 33. Syle ðín eáre ðínum earminge *give thy ear to thy wretched one*, Ps. Lamb. fol. 183 b, 17. Ne ondrǽd ðé, lá earming. git ðú hæfst lífes hiht *dread not, O wretched man, thou hast yet*

hope of life, Ælfc. T. 37, 2. Ða ðe ðæs wélan gitsiaþ, hí biþ symle wædlan and earmingas on hyra móde *they who covet wealth are always poor and miserable beings in their mind*, Prov. Kmbl. 50.

earmlíc ; *sup.* earmlícost ; *adj. Miserable, wretched* ; miser :—Ðér sceal earmlíc ylda cwealm æfter wyrþan *then must afterwards miserable slaughter of men take place*, Andr. Kmbl. 363 ; An. 182. Wǽs gehýred earmlíc ylda gedræg *the wretched tumult of men was heard*, Andr. Kmbl. 3108 ; An. 1557 : Beo. Th. 1618 ; B. 807 : Bd. 5, 13 ; S. 632, 29. Ðæt is earmlícost ealra þinga *this is the most wretched of all things*, Bt. Met. Fox 19, 55 ; Met. 19, 28 : 27, 32 ; Met. 27, 16 : 28, 148 ; Met. 28, 74.

earmlíce ; *adv. Miserably, wretchedly* ; mĭsĕre :—He wæs earmlíce beswicen *he was wretchedly beguiled*, Bd. 5, 13 ; S. 632, 26 : 1, 12 ; S. 481, 21 : Cd. 81 ; Th. 101, 35 ; Gen. 1692 : Exon. 88 a ; Th. 330, 20 ; Vy. 54. Earmlícor *more miserably*, Bd. 5, 14 ; S. 635, 3.

earm-scanca, an ; *m. An arm-bone* [=shank] ; crus :—Gif ða earmscancan beóþ begen forade *if the arm-bones be both broken*, L. Alf. pol. 55 ; Th. i. 94, 26.

earm-sceapen ; *adj. Miserable, wretched* ; miser :—Ne mihte earmsceapen áre findan *the poor wretch might not find pity*, Andr. Kmbl. 2259 ; An. 1131 : 2689 ; An. 1347 : Beo. Th. 2707 ; B. 1351 : Cd. 206 ; Th. 255, 30 ; Dan. 632.

earm-slífe, an ; *f. An arm-sleeve* ; brachĭle, R. Ben. Interl. 55.

earm-strang ; *adj. Arm-strong, muscular* ; tŏrōsus, Ælfc. Gl. 72 ; Som. 70, 124 ; Wrt. Voc. 43, 49.

earm-swíþ ; *adj. Arm-powerful, muscular, strong* ; lacertōsus, Cot. 123 : 200.

earmþu, e ; *f. Misery, poverty* ; misĕria :—Gif ða earmþa ealle sóðe sint *if the miseries are all true*, Bt. 38, 2 ; Fox 198, 14, 16. v. yrmþu.

earmung ; e ; *f. Misery, poverty* ; misĕria :—Hió biþ eádgum leóf, earmunge tǽse [earmum getǽse, Grn.] *she is dear to the rich, benevolent to poverty*, Exon. 128 a ; Th. 492, 28 ; Rä. 81, 22.

Ear-múþa, an ; *m.* [ear *the sea, the river Yare*, múþa *the mouth*] *Great* YARMOUTH, *Norfolk* ; oppĭdum in agro Norfolciensi, et in insŭla Vecti, Lye.

EARN, es ; *m. An eagle* ; aquĭla :—Se earn *the eagle*, Herb. 31, 2 ; Lchdm. i. 128, 10. Earn aquĭla, Ælfc. Gl. 36 ; Som. 62, 107 ; Wrt. Voc. 29, 5 : 62, 1 ; 77, 12 : 280, 1. Swá earn his briddas spænþ to flihte and ofer híg flíceraþ, swá he tobrǽdde his feðeru *sĭcut aquĭla prŏvŏcans ad vŏlandum pullos suos et super eos vŏlĭtans expandit ălas suas*, Deut. 32, 11. Ŭrigfeðera earn sang ahóf *the dewy-feathered eagle raised his song*, Elen. Kmbl. 58 ; El. 29 : 222 ; El. 111 : Judth. 11 ; Thw. 24, 27 ; Jud. 210 : Byrht. Th. 134, 60 ; By. 107 : Exon. 111 a ; Th. 426, 1 ; Rä. 41, 67. Biþ ge-edniwad swylce earnes geógeþ ðín *renŏvābĭtur ut aquĭlæ juventus tua*, Ps. Lamb. 102, 5. Earnes brid *an eagle's young*, Exon. 59 a ; Th. 214, 7 ; Ph. 235. Earnes mearh *an eagle's marrow*, Lchdm. iii. 14, 24. Se wonna hrefn fela earne secgan *the dark raven [shall] say much to the eagle*, Beo. Th. 6044 ; B. 3026 : Exon. 59 a ; Th. 214, 12 ; Ph. 238 : Ps. Th. 102, 5. Ic onhyrge ðone haswan earn *I imitate the dusky eagle*, Exon. 106 b ; Th. 406, 21 ; Rä. 25, 4 : Chr. 937 ; Erl. 115 ; Æðelst. 63. Swá hwǽr swá hold byþ, ðæder beóþ earnas gegaderode *ubicumque fŭĕrit corpus, illic congrĕgābuntur et aquĭlæ*, Mt. Bos. 24, 28. Cōmon earnas on flyhte *eagles came in flight*, Andr. Kmbl. 1725 ; An. 865. He sende blódige earnas *he sent bloody eagles*, Salm. Kmbl. 943 ; Sal. 471. [*Chauc.* earn : R. Glouc. ern : *Laym.* ærn, erne : *Orm.* ærn : *Scot.* ern, erne, eirne, earn : *Plat.* arend, aarn, aarnd : *Dut.* arend, *m* : *Ger.* aar, *m* : *M.H.Ger.* arn, *m* : O.H.Ger. arn, aro, *m* : *Goth.* ara, *m* : *Dan.* örn, *m.f* : *Swed. Icel.* örn, *m*.]

earn, es ; *n. A house, cottage* ; căsa :—On ðære stówe ðe is gecíged æt hwítan earne *in the place which is called [at] Whitern [white house*, candĭda căsa], Bd. 5, 24 ; S. 646, 31. v. ærn.

earn-cynn, -cynn, es ; *n. Eagle-kind* ; gĕnus aquĭlæ :—Ne ete ge nán þing earncynnes *do not eat anything of the eagle-kind*, Lev. 11, 13.

earne *active*, Exon. 101 a ; Th. 380, 31 ; Rä. 1, 16 ; *acc. of* earu.

earn-geáp ? [earn *an eagle*, geáp *shrewd, cunning*] *A vulture, species of falcon* ; vultur, harpe :—Earn-geáp ? *vultur*, Ælfc. Gl. 38 ; Som. 63, 32 ; Wrt. Voc. 29, 51. Earn-geáp ? *arpa* [=*harpe*], Glos. Brux. Recd. 36, 2 ; Wrt. Voc. 62, 2, Ben. Lye. v. earn-geát.

earn-geát, e ; *f.* [gĕát, gāt *a goat*] *The goat-eagle, vulture* ; harpe = ἅρπη, vultur, Glos. Epnl. Recd. 153, 40 : Mone A. 2.

EARNIAN ; *p.* ode, ade, ede ; *pp.* od, ad ; *v. trans. gen. acc. To EARN, merit, deserve, get, attain, labour for* ; mĕrēri :—Byþ geseald ðære þeóde ðe hys earnaþ *it shall be given to the nation which deserves it*, Mt. Bos. 21, 43. Hú monna gehwylc earnode éces lífes *how every man merited eternal life*, Exon. 23 a ; Th. 65, 9 ; Cri. 1052. Ða he ne earnade elles wuhte *when he did not earn anything else*, Bt. Met. Fox 9, 39 ; Met. 9, 20. Gé ðæs earnedon *ye merited this*, Th. 83, 2 ; Cri. 1350. Uton we friþes earnian *let us merit peace*, 98 a ; Th. 366, 17 ; Reb. 13. He hæfþ ðæt ðæt he earnaþ *he has that which he earns*, Bt. 37, 2 ; Fox 188, 6. [*Plat.* arnen, arnden *to reap* : O.Frs. arn, *f. messis* : *Kil.* arnen, ernen *mĕtĕre sĕgĕtem* : *Ger.* ernten, ärnten *to reap, harvest* :

M. H. Ger. arnen *to reap: O. H. Ger.* arnên *mĕrēri;* arnôn *mĕtĕre: Goth.* asans, *f. harvest.*] DER. ge-earnian.

earning, e; *f. A merit;* mĕrītum:—Nô ðæs earninga ǽnige wǽron *for this were not any merits,* Exon. 118 b; Th. 456, 17; Hy. 4, 68. v. earnung.

earning-land, es; *n. Land earned* or *made freehold* = bôc-land, Cod. Dipl. 679; A.D. 972-992; Kmbl. iii. 259, 10; Sax. Engl. i. 312, note 2.

Earnulf, Arnulf, es; *m. Arnulf, emperor of Germany from A.D. 887 to 899, nephew of Charles le Gros = Ger. Karl der Dicke:*—Ð̄ȳ ilcan geáre, forþférde Carl, Francna cyning; and Earnulf, his brôður sunu, hine vi wicum ǽr he forþférde, berǽdde æt ðam rîce *in the same year, Charles, king of the Franks, died; and six weeks before he died, Arnulf, his brother's son, bereft him of the kingdom,* Chr. 887; Th. 156, 30. Mid Earnulfes geþafunge *with the consent of Arnulf,* 887; Th. 156, 36.

earnung, earning, e; *f. An* EARNING, *desert, reward, good turn, compassion;* mĕrītum, mĭsĕrātio, compassio:—For earnunge ēcan lifes *for the reward of eternal life,* Hy. 6, 26; Hy. Grn. ii. 286, 26. Hwylce earnunga uncre wǽron *such deserts have been ours,* Exon. 100 a; Th. 377, 3; Seel. 166. Se gewuldorbeágaþ ðē on earnunga *qui cŏrōnat te in mĭsĕratiōnĭbus,* Ps. Spl. 102, 4. DER. ge-earnung.

earon are, Ps. Th. 101, 21: Th. Diplm. A.D. 887; 133, 37; 134, 1; *3rd pres. pl. of* eom.

earp; *adj. Dark, dusky;* fuscus:—Earpan gesceafte, fûs ofer folcum, fȳre swǽtaþ *the dark creatures* [*clouds;* nubes], *hurrying over the people, sweat fire,* Exon. 102 a; Th. 385, 10; Rä. 4, 42. v. eorp.

earpa a harp, Ps. Spl. 107, 2. v. hearpa.

eár-plǽttan; *p.* -plǽtte; *pp.* -plǽtted [eáre *an ear,* plǽttan *to strike*] *To strike on the ear, to box the ear;* cōlăphum incŭtĕre:—Se byrle ðone apostol eár-plǽtte *the cup-bearer struck the apostle on the ear,* Homl. Th. ii. 520, 12.

eár-preón, es; *m. An ear-pin, ear-ring;* inauris:—Eárpreón *vel* eár-ring inauris, Ælfc. Gl. 65; Som. 69, 50; Wrt. Voc. 41, 7. Earpreónas *vel* eár-hringas inaures, 4; Som. 55, 91; Wrt. Voc. 16, 61.

eár-ring an ear-ring, Ælfc. Gl. 65; Som. 69, 50; Wrt. Voc. 41, 7. v. eár-hring.

EARS, ærs, es; *m. The breech, the buttocks, the hind part;* ānus, pōdex. [*Piers P.* ers: *Chauc.* ers, erse: *Plat.* aars, ars, eers, *m: Frs.* earse, earz: *O. Frs.* ers: *Dut.* aars, *m: Ger.* arsch, *m: M. H. Ger. O. H. Ger.* ars, *m. cŭlus, pōdex: Dan.* ars, arts, *m. f: Swed.* ars, *m: Icel.* ars, rass, *m.*] DER. open-ærs: ears-ende, -gang, -ling, -lȳre, -ode, -þerl.

eár-scrypel, es; *m. An ear-scraper, ear-finger;* dĭgĭtus auricŭlāris:—Eár-scrypel auricŭlāris, Glos. Brux. Recd. 38, 75; Wrt. Voc. 65, 3.

eár-sealf, e; *f. An* EAR-SALVE, L. M. 1, 3; Lchdm. ii. 40, 1.

ears-ende, es; *m. The breech, the buttocks;* nātes:—Ears-ende [MS. -endu] nātes, Wrt. Voc. 65, 36: [MS. -enda], 283, 61.

ears-gang, es; *m. Āni fŏrāmen, ānus.* v. ars-gang.

ears-ling; *adv. Only used adverbially with* on,—*On the back, backwards;* retrorsum:—Sȳn hî gecyrde on earsling *be thei turned awey bacward,* Wyc; avertantur retrorsum, Ps. Th. 34, 5. Gân hȳ on earsling *avertantur retrorsum,* 6, 8. v. bæcling, hinderling.

ears-lȳre, es; *m?* [lȳre = līra *muscle*] *The breech-muscle, the breech;* nātes:—Earslȳre nātes, Ælfc. Gl. 74; Som. 71, 71; Wrt. Voc. 44, 53.

earsode; *part. Having a breech, breeched;* tergōsus, Ælfc. Gl. 77; Som. 72, 4; Wrt. Voc. 45, 38.

eár-spinl, e; *f.* [spinl = spindel *a spindle*] *An ear-ring;* inauris, Prov. 25. v. eár-hring.

ears-þerl, es; *n.* [þerl = þyrel *a hole*] *Fŏrāmen āni, ānus:*—Ears-þerl ānus *vel* verpus, Ælfc. Gl. 74; Som. 71, 72; Wrt. Voc. 44, 54.

eart art:—Ðú eart ðē selfa ðæt hêhste good *thou thyself art the highest good,* Bt. Met. Fox 20, 90; Met. 20, 45: Bt. 10; Fox 26, 23: Ælfc. Gr. 32; Som. 36, 26: Beo. Th. 710; B. 352: 1016; B. 506: Andr. Kmbl. 2378; An. 1190: Elen. Grm. 808: Exon. 13 b; Th. 25, 19; Cri. 403: Ps. Th. 51, 8: Salm. Kmbl. 658; Sal. 328: Cd. 26; Th. 34, 4; Gen. 532: 214; Th. 268, 19; Sat. 57: Nicod. 4; Thw. 2, 34: Mk. Bos. 14, 70; *2nd pers. sing. of* eom.

earþ art, Cd. 205; Th. 254, 9; Dan. 609: Bt. 33, 4; Fox 128, 4. v. eom.

eárðan *before that;* antĕquam, Chr. 1041; Th. 299, 15, col. 1. v. ǽr; *adv.*

earþling a farmer. v. yrþling.

earu; *adj. Quick, active, ready;* cĕler, alăcer, parātus:—Gehȳrest ðú uncerne earne hwelp *hearest thou our active whelp?* Exon. 101 a; Th. 380, 31; Rä. 1, 16. [*Sansk. ara quick.*] v. arod.

earun are, Ps. Th. 104, 7; *pl. pres. of* eom.

eár-wǽrc, es; *n. Ear-ache, a pain in the ear;* auris dŏlor:—Wið eárwærce *for ear-ache,* L. M. 1, 3; Lchdm. ii. 40, 1.

eár-wicga, eór-wicga, an; *m. An* EARWIG *or worm;* vermis *vel* forfĭcŭla auricŭlāris:—Wið eárwicgan *against earwigs,* L. M. cont. 1, 3; Lchdm. ii. 2, 14: L. M. 1, 3; Lchdm. ii. 40, 1: 1, 3; Lchdm. ii. 44, 4. v. wicga.

earwunga [earnunga?]; *adv. Without cause;* grātis:—Afuhton me earwunga *expugnāvērunt me grātis,* Ps. Th. 108, 2: 68, 4: 118, 161: 119, 6. v. arwunga, earnung, bûtan ge-earnungum *grātis,* s. v. ge-earnung.

eás *of a river:*—On twâ healfe ðære eás *on the two sides of the river,* Chr. 896; Th. 172, 39, col. 1; *gen. of* eá, q. v.

eá-spring, ǽ-spring, es; *n. A water-spring, fountain;* ǽquæ fons, fons:—Ðæt Cûþbyrhtus ân eáspring of drigre eorþan up gelǽdde *ut Cudberct fontem de arente terra produxĕrit,* Bd. 4, 28; S. 605, 6.

EÁST; *m. The* EAST; ŏriens:—Ôþ Indéas eáste wearde *unto the Indies towards the east,* Bt. Met. Fox 16, 36; Met. 16, 18. Sió sunne norþ eft and eást otēweþ *the sun appears again in the north and east,* i. e. *in the north-east,* 13, 118; Met. 13, 59. Ðæt eálond on Wiht is þrittiges mîla lang eást and west; and twelf mîla brâd sûþ and norþ *Vecta insŭla hăbet ab ŏriente in occāsum triginta circĭter mîlia passuum; ab austro in bŏ̆ream duodĕcim,* Bd. 1, 3; S. 475, 19. [*Wyc.* est, eest: *Piers P.* eest: *Chauc.* est: *Laym.* æst: *Orm.* æst: *Plat.* oost: *O. Sax.* ôst-ar *towards the east: Frs.* æst, east: *O. Frs.* asta, ost: *Dut.* oost, oosten, *n: Ger.* ost, osten, *m: M. H. Ger.* ôsten, *n: O. H. Ger.* ôst, ôstan, *m: Dan.* öst, östen, öster: *Swed.* öster, *m: Icel.* austr, *m.*] DER. eástan, eástan-sûþan: Eást-Centingas, -dǽl, -ende, -Engle, -folc, -Francan, -healf, -land, -lang, -rîce, -rihte, -sǽ, -Seaxe: eáster, eáster-ǽfen, -dæg, -fæsten, -feorm, -lîc, -mônaþ, -niht, -þénung, -tîd, -wuce.

Eást; *adj.* EAST, *easterly;* ŏrientālis: EAST *used mostly in composition as a noun,* Eást-Engle *East-Angles,* Bd. 5, 24; S. 646, 19. Eást-Seaxe *East-Saxons,* 5, 24; S. 646, 19. Eást-Francan *East-Franks,* Ors. 1, 1; Bos. 18, 30: Chr. 891; Erl. 88, 3. v. eást, es; *m.*

eásta, an; *m. The east;* ŏriens:—He férde syððan to ðam mŭnte, be eástan Bethel *inde transgrĕdiens ad montem, qui erat contra ŏrientem Bethel,* Gen. 12, 8. Be eástan Rîne syndon Eást-Francan *to the east of the Rhine are the East-Franks,* Ors. 1, 1; Bos. 18, 29, 31, 33, 39, 45. Be eástan *in the east,* Bt. Met. Fox 29, 65; Met. 29, 33: Chr. 878; Erl. 80, 9: 894; Erl. 92, 19. v. eást, es; *m.*

eá-stæþ a river-bank; flūmĭnis rīpa. v. eá. stæþ *a shore, bank.*

eástan, eásten; *adj. East;* ŏrientālis:—Eástan sûþan wind *south-east wind;* vulturnus, Ælfc. Gl. 54; Som. 66, 86; Wrt. Voc. 36, 12. Norþan eástan wind eurus, euroauster, 54; Som. 66, 87; Wrt. Voc. 36, 13. Eásten wind subsōlānus, 54; Som. 66, 82; Wrt. Voc. 36, 8.

eástan, eásten, ēstan; *adv. From the east, easterly;* ab ŏriente:—Gif wind cymþ westan oððe eástan *if the wind come westerly or easterly,* Cd. 38; Th. 50, 10; Gen. 806: 80; Th. 99, 20; Gen. 1649: 86; Th. 107, 24; Gen. 1794. Æðeltungla wyn eástan lîxeþ *the delight of the noble stars shines easterly,* Exon. 60 a; Th. 218, 6; Ph. 290: 57 a; Th. 204, 24; Ph. 102: 20 b; Th. 55, 19; Cri. 886. Eásten hider *from the east hither,* Cd. 27; Th. 35, 16; Gen. 555. Hwonne up cyme ǽdelast tungla ēstan lîxan *when the noblest of stars riseth up shining easterly,* Exon. 57 a; Th. 204, 8; Ph. 94.

eástan-sûþan south-eastern, Ælfc. Gl. 54; Som. 66, 86; Wrt. Voc. 36, 12. v. eástan; *adj.*

Eást-Centingas; *pl. m. The East Kentians, men of East Kent;* Cantii ŏrientis habitātōres:—Ealle Eást-Centingas friþ wið ðone here genāmon *all the men of East Kent made peace with the army,* Chr. 1009; Th. 260, 39.

eást-dǽl, es; *m. The eastern part, the east;* terræ pars orientālis, ortus:—Cirus, Persa cyning, hæfde mǽst eallne ðæne eást-dǽl awêst *Cyrus, king of the Persians, had laid waste almost all the east,* Ors. 2, 4; Bos. 43, 43: Exon. 55 b; Th. 197, 20; Ph. 2. Eást-dǽl ortus, Ps. Lamb. 102, 12.

eásten; *adj. East;* orientālis. v. eástan; *adj.*

eásten; *adv. From the east, easterly;* ab ŏriente. v. eástan; *adv.*

eást-ende, es; *m. The east-end;* pars orientālis:—Æt ðæs wuda eást-ende *at the east-end of the wood,* Chr. 893; Th. 162, 28.

Eást-Engle; *pl. m. The East-Angles;* ŏrientes Angli:—Of Engle côman Eást-Engle and Middel-Engle *from Angeln came the Angles of the east and the middle Angles,* Bd. 1, 15; S. 483, 24.

eásten-wind, es; *m. The east wind;* subsōlānus. v. eástan; *adj.*

eáster, eástor; *gen.* eástres; *pl. nom. acc.* eástro; *gen.* eástrena; *dat.* eástron, eástran [= eástrum]; *n:* eástre, an; *n.* **I.** *Easter, the feast of Easter;* pascha = πάσχα:—On dæge symbeles eástres *in die solemni paschæ,* Lk. Lind. War. 2, 41. Wæs ðære ylcan nihte ðara hālgan Eástrena, ðæt seó cwên cende dôhtor ðæm cyninge *it was on that same holy night of Easter, that the queen bore to the king a daughter,* Bd. 2, 9; S. 511, 28. Æfter twǽm dagum beóþ eástro *post biduum pascha fiet,* Mt. Bos. 26, 2. Freóls-dæg, se is gecweden Eástre *a feast day which is called Easter,* Lk. Bos. 22, 1. **II.** *the passover, paschal lamb;* pascha:—To eástron *for the Easter lamb,* Mt. Bos. 26, 17. Ðá hî eástron offrodon ... ðæt ðú eástron ete *quando pascha immŏlābant ... ut manducēs pascha,* Mk. Bos. 14, 12. [*Ger. M. H. Ger.* ostern, *f: Ker.* ôstarun, ôstrun: *Otfr.* ôstarâ, ôstorun *dea, pascha: A. Sax.* Eástre, *the goddess of the rising sun, whose festivities were in April. Hence used by Teutonic christians for the rising of the sun of righteousness, the feast of the resurrection,* Bd. de Temp. Rat. Works, vol. ii. p. 81: Grimm's Deut. Mythol. 8vo. 1855, pp. 180-183.]

eáster, eástor; *adj. Easter;* paschālis :— Ðys sceal on eáster-æfen *this belongs to easter-even,* Rubc. Mt. Bos. 28, 1 ; Notes, p. 577, 28, 1 a. Eáster-tíd *easter-tide* or *time,* Homl. Th. ii. 266, 15, 19, 21. Eáster-mōnaþ *easter-month, April,* Menol. Fox 142 ; Men. 72.

eáster-æfen, eástor-æfen, es ; *m. Easter-even;* dies ante festum paschæ :—Ðys sceal on eáster-æfen *this [gospel] must be on easter-even,* Rubc. Mt. Bos. 28, 1 ; notes, p. 577, 28, 1 a.

eáster-dæg, eástor-dæg, es ; *m. Easter-day;* dies paschālis :—Com he to ðam cyninge ðy ǽrestan eáster-dæge *pervēnit ad rēgem prīmo die paschæ,* Bd. 2, 9 ; S. 511, 17.

eáster-fæsten, es ; *n. Easter-fast;* quadrāgēsĭma, jejūnium paschāle :— On fōreweard eáster-fæsten *in the beginning of the easter-fast;* incĭpiente quadrāgēsĭma, Bd. 5, 2 ; S. 614, 37.

eáster-feorm, eástor-feorm, e ; *f. Easter-feast* or *repast;* paschālis firma :—On sumere þeóde gebȳreþ winter-feorm [and] eáster-feorm *in quĭbusdam lŏcis dătur firma nātālis Dŏmĭni, et firma paschālis,* L. R. S. 21 ; Th. i. 440, 26.

eáster-líc, eástor-líc ; *adj. Easter, paschal;* paschālis :—Hý fōron to Hierusalem to ðam eásterlícan freólse *they went to Jerusalem to the paschal feast,* Lk. Bos. 2, 42 : Homl. Th. ii. 32, 15 : 284, 1.

eáster-mōnaþ, es ; *m. Easter-month; Aprīlis* mensis :—Eáster-mōnaþ cymeþ *easter-month comes,* Menol. Fox 142 ; Men. 72.

eást-ern, -erne ; *adj.* [ern *a place*] EASTERN, *oriental;* orientālis :— Ðonne cymþ eásterne wind *then comes the eastern wind,* Cd. 17 ; Th. 20, 27 ; Gen. 315. Se wer wæs mǽre betwux eallum eásternum *erat vir ille magnus inter omnes orientāles,* Job Thw. 164, 7.

eáster-niht, e ; *f. Easter-night;* nox paschālis :—In ðære eáster-niht *in the easter-night,* Exon. 120 a ; Th. 460, 10 ; Hö. 15.

eáster-þenung, e ; *f. The paschal feast, paschal lamb, the passover;* pascha :—Híg gegearwodon him eáster-þenunge *parāvērunt ei pascham,* Mt. Bos. 26, 19.

eáster-tíd, eástor-tíd, e ; *f. Easter-tide;* paschæ tempus :—Se Hælend geheóld ða eáster-tíde *the Saviour kept the easter-tide,* Homl. Th. ii. 242, 21 : 266, 15, 19, 21.

eáster-wuce, eástor-wice, an ; *f. Easter-week;* paschālis septimāna :— Ðys sceal on Sæternes dæg, on ðære eáster-wucan *this [gospel] must be on Saturday in easter-week,* Rubc. Jn. Bos. 20, 1, 11 ; Notes, p. 580, 20, 1 a, 11 a : 21, 1 ; Notes, p. 580, 21, 1 a.

eá-stæþ, eá-stæþ, es ; *n. A river-bank;* flūmĭnis rīpa :—Hí on ðam eástede ealle stódon *they all stood on the river-bank,* Byrht. Th. 133, 40 ; By. 63.

eásteweard *eastward,* Bt. 18, 1 ; Fox 60, 31. v. east ; *m.*

eást-folc, es ; *n. Eastern people;* pŏpŭlus orientālis, Som. Ben. Lye.

Eást-Francan ; *pl. m. East-Franks;* Franci orientāles :—Wyð norþan Donua ǽwylme. and be eástan Ríne, syndon Eást-Francan *to the north from the spring of the Danube, and to the east of the Rhine, are the East-Franks,* Ors. 1, 1 ; Bos. 18, 30. Mid Eást-Francum *with the East-Franks,* Chr. 891 ; Erl. 88, 3.

eást-healf, e ; *f. The east-side ;* orientāle lătus, plăga orientālis :—Ðe on eást-healfe ðære eá wǽron *who were on the east side of the river,* Chr. 894 ; Th. 170, 9, col. 2. On eást-healfe Iericho *contra orientālem plăgam urbis Iericho,* Jos. 4, 19 : Lev. 1, 16.

Eást-land, es ; *n. The east country, Esthonia [Eastland], the country of the Osti* or *Estas ;* orientālis terra, terra Esthonia :—Iacob com to ðam eástlande *Iacob vēnit in terram orientālem.* Gen. 29, 1. Eástland is swȳðe mycel *Esthonia is very large,* Ors. 1, 1 ; Bos. 22, 12.

eást-lang ; *adv. Along the east ;* orientem versus :—Se wudu is eástlang and westlang hund twelftiges míla lang oððe lengra *the wood, from east to west* [lit. *along the east and along the west*], *is one hundred and twenty miles long, or longer,* Chr. 893 ; Th. 162, 30.

eástor-æfen, es ; *m. Easter-even ;* dies ante festum paschæ :—On eástor-æfen *on easter-even,* L. E. I. 41 ; Th. ii. 438, 24. v. eáster-æfen.

eástor-dæg, es ; *m. Easter-day ;* dies paschālis :—Ðy sylfan eástor-dæge *on the same easter-day,* Bd. 5, 23 ; S. 645, 36. v. eáster-dæg.

eástor-feorm, e ; *f. Easter-feast* or *repast ;* firma paschālis :—Eallum ǽhte-mannum gebȳreþ mid-wintres feorm and eástor-feorm *omnĭbus ehtemannis jūre compĕtit nātālis firma et paschālis firma,* L. R. S. 9, 1 ; Th. i. 436, 33. v. eáster-feorm.

eástor-líc ; *adj. Easter, paschal ;* paschālis :—On ðære sylfan eástor-lícan symbelnesse *on the same easter-feast,* Bd. 4, 28 ; S. 606, 23 : 3, 24 ; S. 557, 40. v. eáster-líc.

eástor-tíd, e ; *f. Easter-tide ;* paschæ tempus :—In ða eástor-tíde *in the easter-tide,* Exon. 48 b ; Th. 168, 10 ; Gú. 1075 ; Bd. 5, 23 ; S. 645, 36. v. eáster-tíd.

eástor-wice, an ; *f. Easter-week ;* septimāna paschālis :—Ealle ða dagas ðære eástor-wican *all the days of the easter-week,* L. E. I. 41 ; Th. ii. 438, 25. v. eáster-wuce.

eástran, eástron ; *dat. pl.* of eáster ; *gen.* -tres, *q. v.* Eástron *seems to be used for other cases in the pl.*

eástre, an ; *n. Easter. the feast of easter ;* pascha, Lk. Bos. 22, 1. v. eáster.

---

eá-stream, es ; *m. A water-stream, a river ;* rīvus :—Heóldon forþryne eástreámas heora *the river-streams held their onward course,* Cd. 12 ; Th. 14, 9 ; Gen. 216. Ofer eástreámas ís brycgade blāce brimrāde *over the river-streams the ice bridged a pale water-road,* Andr. Kmbl. 2523 ; An. 1263. v. ēg-stream, eáh-stream.

eá-stream-ȳp, e ; *f. A river-stream-flood ;* rīvi fluctus, Cd. 192 ; Th. 240, 11 ; Dan. 385.

eást-ríce, es ; *n. East kingdom, eastern country, eastern part of a country ;* orientāle regnum, orientālis rĕgio, Chr. 893 ; Th. 162, 19, col. 1, 3 : Ors. 2, 1 ; Bos. 39, 21, 27.

eást-rihte ; *adv. East right, towards* or *in the east ;* contra ortum sōlis :—We witan ōðer eálond eást-rihte *nŏvīmus insŭlam aliam contra ortum sōlis,* Bd. 1, 1 ; S. 474, 15.

eástro *easter,* Mt. Bos. 26, 2 ; *nom. acc. pl.* of eáster.

eást-rōdor, es ; *m. The eastern part of heaven ;* pars orientālis cœli, ortus :—Ðes eást-rōdor *ortus,* Ps. Th. 102, 12.

eástron ; *dat. pl.* of eáster, eástor.

eást-sǽ, es ; *f. The east sea, sea on the east side of a country ;* orientāle mǽre, Bd. 1, 12 ; S. 481, 8 : 1, 15 ; S. 483, 40.

Eást-Seaxe ; *gen.* -Seaxa ; *dat.* -Seaxum ; *pl. m :* -Seaxan ; *gen.* -Seaxena, -Seaxna ; *dat.* -Seaxum ; *pl. m. The East-Saxons, people of Essex ;* orientāles Saxōnes :—Hēr Eást-Seaxe onfēngon geleáfan and fulwihtes bæþ *in this year* [A. D. 604] *the East-Saxons received the faith and bath of baptism,* Chr. 604 ; Th. 36, 33, col. 2, 3 : 823 ; Th. 110, 31, col. 1 : 894 ; Th. 170, 19, col. 1 : 904 ; Th. 181, 16, col. 2. Of Seaxum cōman Eást-Seaxan and Súþ-Seaxan and West-Seaxan *from the Saxons came the East-Saxons and the South-Saxons and the West-Saxons,* Bd. 1, 15 ; S. 483, 23. To-ætecte ðisse gedrēfnisse storm Sæberhtes deáþ Eást-Seaxna cyninges *the death of Saberht, king of the East-Saxons, increased the storm of this disturbance,* 2, 5 ; S. 507, 6. Mellitum Agustinus sende Eást-Seaxum to bodigenne godcunde lāre *Augustine sent Mellitus to preach divine doctrine to the East-Saxons,* 2, 3 ; S. 504, 16 : Chr. 604 ; Th. 36, 37, col. 1 : 921 ; Th. 194, 34 : 994 ; Th. 242, 10. Eást-Seaxena, -Seaxna land, ríce, þeód *the country, kingdom* or *nation of the East-Saxons,* Chr. 895 ; Th. 173, 7, col. 2 : 836 ; Th. 118, 6, col. 1 : 855 ; Th. 128, 15, col. 1 ; 129, 20 : Bd. 4, 11 ; S. 579, 4 : 2, 3 ; S. 504, 21.

eást-weard, eást-werd *eastward, in the east,* Ælfc. Gr. 28 ; Som. 40, 7. v. eást.

eást-weg, es ; *m. East-way ;* orientālis via :—On eást-wegas *in the east-ways,* Cd. 174 ; Th. 220, 11 ; Dan. 69 : Elen. Kmbl. 509 ; El. 255.

eáþ ; *adv. Easily ;* facĭliter :—Dryhten mæg gehwone eáþ gescildan *the Lord may easily shield each,* Exon. 40 b ; Th. 135, 23 ; Gú. 528 : Cd. 95 ; Th. 124, 6 ; Gen. 2058. Hie ðe eáþ mihton adreógan *they the easier might endure,* Andr. Kmbl. 735 ; An. 368. v. ēþ, ȳþ. v. eáðe ; *adj.*

eáþ-bēde ; *adj. Exorable ;* deprĕcābilis :—Wes ðínum scealcum wel eáþbēde *deprĕcābilis esto super servos tuos,* Ps. Th. 89, 15.

eáþ-bēne ; *adj. Exorable ;* deprĕcābilis :—Eáþ-bēne *deprĕcābilis,* Som. Ben. Lye ; Ps. Grn. ii. 200, 15, note.

EÁÐE, ēðe, ȳðe ; *comp. m.* eáðera, eáðra ; *f. n.* eáðere, eáðre ; *sup.* eáðost ; *adj. Easy, smooth ;* facĭlis, lēvis :—Gode þancedon ðæs ðe him ȳþ-lāda eáðe wurdon *they thanked God for that the wave-paths had been easy* [= *smooth*] *to them,* Beo. Th. 462 ; B. 228. Eáðere ys olfende to farenne þurh nædle þyrel, ðonne se ríca and se wēlega on Godes ríce gā *it is an easier [thing] for a camel to go through a needle's eye than a powerful and wealthy man to go into God's kingdom,* Mk. Bos. 10, 25. Eáðre is ðæt heofen and eorþe gewīton, ðonne án stæf of ðære ǽ fealle *it is an easier [thing] that heaven and earth pass away than one letter of the law fail,* Lk. Bos. 16, 17. [*Chauc.* ethe, eythe *easy ;* esy *light, gentle :* R. Glouc. eþ : *Laym.* æðe, eð : *Orm.* æþ : *Scot.* eith, eyth, eth : *O. Sax.* ōði : *Icel.* auð, adverbial prefix, *easy.*] DER. un-eáðe.

eáðe ; *sup.* eáðost, -ust ; *adv. Easily, readily, soon, perhaps ;* facĭliter :— Ða burh mihton eáðe begitan *they might easily have taken the city,* Ors. 3, 4 ; Bos. 56, 10 : Beo. Th. 961 ; B. 478. Ic eáðe forbær rūme regulas *I readily preferred the lax rules,* Exon. 39 b ; Th. 131, 22 ; Gú. 459. We ðē eáðe gecȳðaþ síþ úserne *we readily proclaim our adventure to thee,* Andr. Recd. 1721 ; An. 861. Hwā mæg eáðost [eáðust MS. B.] ða dūru ontȳnan *who may most easily open the door?* Salm. Kmbl. 71 ; Sal. 36 : Cd. 174 ; Th. 219, 6 ; Dan. 50 : Ps. Th. 76, 10. DER. un-eáðe. v. ēðe.

eáðelic, ǽðelíc ; *comp. m.* -lícra ; *f. n.* -lícre : *adj. Easy, possible ;* facĭlis :—Ealle þing synt mid Gode eáðelíce *with God all things are possible,* Mt. Bos. 19, 26. Hwæt is eáðelícre *what is easier ?* 9, 5. DER. un-eáðelíc. v. ǽðe-líc.

eáðelíce, ēðelíce, ȳðelíce ; *comp.* or ; *sup.* ost, ust ; *adv. Easily ;* facĭle :—Eáðelícor mæg se olfend gān þurh ánre nædle eáge *it is easier for a camel to go through the eye of a needle,* Lk. Bos. 18, 25. He sōhte hū he eáðelícost hine gesealde *he sought how he might most easily betray him,* 22, 6. DER. un-eáðelíce.

eáþ-fere ; *adj. Easily trod, easy ;* facĭlis ĭtu :—Eáþfere weg *ĭter vel ĭtus,* Ælfc. Gl. 56 ; Som. 67, 48 ; Wrt. Voc. 37, 35.

eáþ-fynde ; *adj. Easy to be found ;* facĭlis inventu :—Ða wæs eáþfynde

then was easy to be found, Beo. Th. 276; B. 138: Cd. 93; Th. 120, 12; Gen. 1993. v. êþ-fynde, ŷþ-.

eáþ-gesýne *easy to be seen, visible.* v. êþ-gesýne = ŷþ-gesêne.

eáþ-gete; *adj. Easily got, got ready, prepared;* făcĭlis adeptu, parātus:—Him wæs eáþgete ele to đam baþe *oil was made ready for his bath,* Ælfc. T. 32, 14. v. êþ-begete.

eáþ-hrêđig; *adj. Blessed;* beātus:—Seó eáþhrêđige Elene *the blessed Elene,* Elen. Kmbl. 531; El. 266; *for* eáđ-hrêđig, *q. v.*

eáþ-hylde *satisfied, contented.* v. êþ-hylde.

eáþ-lǽre; *adj. Easily taught, teachable;* dŏcĭbilis:—Ealle eáþlǽre beóþ Godes ĕrunt omnes dŏcĭbiles Dei, Jn. Bos. 6, 45.

eáþ-mêd, es; *n. Humility, affability, kindness;* humĭlĭtas, humānĭtas, generally found in the *pl:*—Ac mîne [MS. min] eáþmêdu geseah *vĭde humĭlĭtātem meam,* Ps. Th. 118, 153: 135, 24. On mĭnum eáþmêdum *in humĭlĭtāte mea,* 118, 50. For eáþmêdum *in humility,* Exon. 53 a; Th. 186, 5; Az. 15: 13 a; Th. 22, 29; Cri. 359. v. eáđ-mêd.

eáþ-mêdan *To adore;* adŏrāre:—Eáþmêdaþ feorr adŏrābĭtis prŏcul, Ex. 24, 1. DER. ge-eáþmêdan. v. eáđmêdan.

eáþ-mêde; *adj. Of an easy mind, humble;* mītis, hŭmĭlis:—He gebêtte mid eáþmêde ingeþance *he expiated with humble mind,* Ps. C. 50, 152; Ps. Grn. ii. 280, 152. v. eáđ-mêde.

eáþ-mêdum; *adv. [dat. pl. of* eáþmêd] *Humbly, kindly;* humĭlĭter, benigniter:—Eáþ-mêdum *humbly,* Exon. 46 a; Th. 157, 15; Gû. 892. Đæt he eáþmêdum oncnáwe *that he should treat [him] kindly,* Andr. Kmbl. 641; An. 321. Gewát him se hálga eáþmêdum *the holy one departed kindly,* 1957; An. 981.

eáþ-metto; *indecl. sing; pl. nom. acc.* -metta; *f. Humility;* humĭlĭtas:—Geseóh mîne eáþmetto *vĭde humĭlĭtātem meam,* Ps. Th. 9, 13: 24, 16. On đam stáne eáþmetta *on the rock of humility,* Bt. 12; Fox 36, 22: Bt. Met. Fox 7, 65; Met. 7, 33.

eáþ-mód; *adj. Humble, lowly, obedient;* hŭmĭlis, obēdiens:—Gif đú eáþmódne eorl gemête *if thou meet a lowly person,* Exon. 84 b; Th. 318, 5; Mód. 78. He eáþmóde him eorlas funde *he found men obedient to him,* Menol. Fox 195; Men. 99. His ætgiefan eáþmód weorþeþ *he becomes obedient to his feeder,* Exon. 88 b; Th. 332, 27; Vy. 91. v. eáđ-mód.

eáþ-módian *to obey;* obēdîre. v. ge-eáþ-módian.

eáþ-módlíce; *adv. Humbly;* humĭliter:—Abiddaþ hine eáþmódlíce *pray to him humbly,* Bt. 42; Fox 258, 21. v. eáđmódlíce.

eáþ-módnis, -nys, -niss, -nyss, e; *f. Humility;* humĭlĭtas:—Mid micelre eáþmódnisse *with great humility,* Th. Diplm. A. D. 804–829; 459, 15. On eáþmódnysse mîne *in humĭlĭtāte mea,* Ps. Spl. 118, 50. v. eáđmódnes.

eáþnes, -ness, e; *f. Easiness;* făcĭlĭtas. v. êþnes, eáđ-nes.

eatogeđa *eighth:*—Seó eatogeđe *the eighth,* Bd. 4, 5; S. 573, note 10. v. eahtođa.

eatol; *adj. Dire, terrible;* dīrus, terrĭbĭlis:—Gæst yrre cwom, eatol *the guest came angry, terrible,* Beo. Th. 4154; B. 2074: 4949, note; B. 2478. v. atol.

Eatole *Italy;* Itălia, Som. Ben. Lye.

Eatol-ware; *pl. m. Italians;* Ităli, Som. Ben. Lye.

eáu-fæstnys, -nyss, e; *f.* [eáu = ǽw, ǽ *law;* fæstnys *firmness] Firmness in the law, religion, devotion;* relĭgio:—Be eáufæstnysse and wundorlícre árfæstnysse Óswaldes cyninges *de relĭgiône ac piĕtāte miranda Osualdi rēgis,* Bd. 3, 6; S. 528, 2. v. ǽ-fæstnes.

eáum *to rivers,* Ors. 5, 2; Bos. 102, 34; *dat. pl. of* eá.

eáw, eáw-lá *oh! alas! O! eheu!* Bt. Met. Fox 9, 109; Met. 9, 55. v. eálá.

eáwan; *p.* de; *pp.* ed *To shew, manifest;* ostenděre, manĭfestāre:—Hí þenceaþ þreú þearle þeódum eáwan *they intend to shew a severe chiding to the nations,* Ps. Th. 149, 7. He eáweþ him egsan *he shews them terror,* Exon. 33 b; Th. 107, 11; Gû. 57: Beo. Th. 557; B. 276. Ne sindon đíne æhta wiht, đa đú monnum eáwdest *thy possessions are nought, which thou didst shew to men,* Exon. 99 a; Th. 371, 14; Seel. 75. Nǽfre wommes tácn eáwed weorþeþ *the sign of crime shall never be manifested,* 8 b; Th. 4, 20; Cri. 55: 22 a; Th. 59, 22; Cri. 956. [O. Frs. auwa, awa.] DER. ge-eáwan, óþ-. v. ŷwan.

eáwesclíce; *adv.* [eáwan *to shew, manifest] Openly;* pălam:—Đætte seó sáwl in deágolnisse þrówiende wæs, đætte se líchoma eáwesclíce fôretácnode *quod anĭma in occulto passa sit, căro pălam præmonstrābat,* Bd. 3, 19; S. 549, 17.

eáw-fæst; *adj.* [eáw = ǽw, ǽ *law;* fæst *fast, fixed] Firm in observing the law, religious, pious;* relĭgiósus, pius:—Gregorius wæs ædelborenre mægþe and eáwfæstre acenned *Gregory was born of a noble and pious family,* Homl. Th. ii. 118, 7. Se eáwfæsta papa *the pious pope,* ii. 118, 8. Mid eáwfæstum þeáwum relĭgiósis mōrĭbus, Bd. 3, 23; S. 555, 4. v. ǽ-fæst.

eáw-fæstnys, -nyss, e; *f.* [eáw = ǽw, ǽ *law;* fæstnys *firmness] Firmness in the law, religion, piety;* relĭgio, piĕtas:—Mid gelícere eáwfæstnysse *with similar piety,* L. E. I. 41; Th. ii. 438, 26. v. ǽ-fæstnes.

eáwu, e; *f. A ewe;* ovis fēmĭna:—Agefe mon to Liming 1 eáwa

and v cý *let fifty ewes and five cows be given to Lyming,* Th. Diplm. A. D. 835; 470, 29, 32. v. eówu.

eáwunga, eáwunge; *adv.* [eáwan *to shew, manifest] Openly, publicly;* mănifeste, pălam, cōram:—God eáwunga cymeþ *Deus mănĭfeste vēniet,* Ps. Spl. 49, 3. He wearþ dîgellíce cristen, forđon he eáwunga ne dorste *he was secretly a christian, because he durst not openly,* Ors. 6, 21; Bos. 123, 29: Exon. 126 b; Th. 487, 2; Rä. 72, 22. Ođđe eáwunga ođđe dearnunga *either publicly or privately,* L. Edg. ii. 8; Th. i. 270, 5. Eáwunge cōram, Ælfc. Gr. 38; Som. 41, 55.

eá-wylm *a welling* or *boiling up of water, spring,* Lye. v. ǽwelm.

eá-wyrt; e; *f. River-wort, burdock;* arctium lappa, Lin:—Genim clífwyrt, sume men hátaþ foxes clífe, sume eá-wyrt *take cliff-wort, some men call [it] fox-glove, some river-wort,* L. M. 1, 15; Lchdm. ii. 58, 4: iii. 74, 10. Nim eáwyrte niođowearde *take the netherward [part] of burdock,* L. M. 1, 87; Lchdm. ii. 154, 14.

eax *an axe;* secūris:—Seó eax *the axe,* L. In. 43; Th. i. 128, 23, note 65, MS. B. v. æx.

EAX, ex, æx, e; *f. An axis, axle-tree;* axis:—Neáh đam norþende đære eaxe *near the north end of the axis,* Bt. 39, 3; Fox 214, 20: 39, 13; Fox 232, 33: Bt. Met. Fox 28, 44; Met. 28, 22: 29, 36; Met. 29, 18. On đære ilcan eaxe hwerfeþ eall rûma rôdor *all the spacious sky turns on the same axis,* 28, 30; Met. 28, 15. Ymb đa eaxe *about the axis,* Bt. 39, 3; Fox 214, 23. On wǽnes eaxe hwearfaþ đa hweó1, and sió eax stent stille *the wheels turn on the waggon's axle-tree, and the axle-tree stands still,* 39, 7; Fox 220, 27, 30, 31: 39, 8; Fox 224, 5. Sió nafu ferþ nêhst đære eaxe *the nave goes nearest to the axle-tree,* 39, 7; Fox 222, 2, 12, 20, 21, 22, 28. Twegen steorran synd gehátene *axis,* đæt is ex, forđamđe se firmamentum went on đám twám steorran, swá swá hweogel tyrnþ on eaxe, and forđí hî standaþ symle stille *two stars are called* axis, *that is axle-tree, because the firmament turns on the two stars, as a wheel turns on an axle-tree, and because they always stand still,* Bd. de nat. rerum; Wrt. popl. science 16, 12–15; Lchdm. iii. 270, 20–23. [Wyc. ax-tre, ex-tre *an axle-tree:* Plat. asse: Dut. as, *f:* Ger. achse, axe, *f:* M. H. Ger. ahse, *f:* O. H. Ger. ahsa, *f:* Dan. axe, m. *f:* Swed. axel, m: Icel. axull, öxull, m; öxul-tré, n: Lat. axis, m: Grk. ἄξων, m: Lith. aszis, *f:* Sansk. aksha *the axle of a wheel, a wheel,* car.]

Eaxan ceaster, e; *f:* es; n. v. ceaster *Exeter, Devon:*—Wende he hine wið Eaxan ceastres *he turned towards Exeter,* Chr. 894; Th. 167, 28, col. 2: 894; Th. 169, 17, col. 2: 895; Th. 173, 10, col. 2. v. Exan ceaster.

Eaxan minster; *gen.* -minstres; *n. The minster on the river Ex, Axminster, Devon;* oppĭdum in agro Devōniensi, Som. Ben. Lye. v. Acsan mynster.

Eaxan múþa, an; *m. The mouth of the river Ex, Exmouth, Devon:*—To Eaxan múþan *to Exmouth,* Chr. 1001; Ing. 174, note a. v. Exan múþa.

EAXEL, eaxl, exl, e; *f:* eaxle, an; *f. The shoulder;* hŭmĕrus:—Standeþ [MS. standaþ] me hêr on eaxelum *stands here on my shoulders,* Wald. 92; Vald. 2, 18. Gefêng he be eaxle Grendles módor *he seized Grendel's mother by the shoulder,* Beo. Th. 3078; B. 1537. He forlêt earm and eaxle *he left arm and shoulder,* 1948; B. 972. He gewêrgad sæt freán eaxlum neáh *he sat wearied near his lord's shoulders,* 5699; B. 2853: 722; B. 358. Hæfde earmas and eaxle *it had arms and shoulders,* Exon. 129 a; Th. 494, 24; Rä. 83, 6. Gif eaxle gelæmed weorþeþ *if a shoulder be lamed,* L. Ethb. 38; Th. i. 14, 2. He hit set on his exla impōnit eam in hŭmĕros suos, Lk. Bos. 15, 5: Andr. Kmbl. 3148; An. 1577. [Laym. exle, dat: O. Sax. ahsla, *f:* O. Frs. axle, axele, *f:* Ger. achsel, *f:* M. H. Ger. ahsel, *f:* O. H. Ger. ahsala, *f:* Goth. amsa, m: Dan. axel, m. *f:* Swed. axel, m: Icel. öxl, *f:* Lat. axilla, *f.*]

eaxl-cláþ, es; *m. A shoulder-cloth, scapular;* hŭmĕrāle:—Lêde eaxlcláþ ofer hine desŭper hŭmĕrāle ei impŏsuit, Lev. 8, 7.

eaxle, an; *f. A shoulder;* hŭmĕrus:—Gif eaxle gelæmed weorþeþ *if a shoulder be lamed,* L. Ethb. 38; Th. i. 14, 2. v. eaxel, eaxl.

eaxle-gespan; *gen.* -gespannes; *n. The shoulder-span:*—Fífe gimmas wǽron on đam eaxlegespanne *five gems were on the shoulder-span,* Rood Kmbl. 17; Kr. 9.

eaxl-gestealla, an; *m. A shoulder companion, nearest friend, bosom friend, comrade;* cōmes qui est a lătĕre, sŏcius intĭmus, commĭlĭto:—Deád is Æschere, mîn eaxlgestealla *Æschere is dead, my bosom friend,* Beo. Th. 2656; B. 1326. Hæfde wígena tô lyt, eaxlgestealna *he had too few of warriors, comrades,* Elen. Kmbl. 127; El. 64. Ic eom ædelinges eaxlgestealla *I am a noble's bosom friend,* Exon. 127 a; Th. 489, 27; Rä. 78, 1. Heremód breát eaxlgesteallan *Heremod destroyed his bosom friends,* Beo. Th. 3432; B. 1714.

EBBA, an; *m? An* EBB *or receding of water;* rĕcessus măris:—Nêpflód vel ebba ledona, Ælfc. Gl. 105; Som. 78, 29; Wrt. Voc. 57, 11. Ebba [MS. ebbe] *recessus,* 105; Som. 78, 36; Wrt. Voc. 57, 18. Ebba [MS. ebbe] *vel gyte-streám rheuma,* 105; Som. 78, 38; Wrt. Voc. 57, 20. Gewrixle đæs flódes and đæs ebban *change of the flood and the ebb,*

Bt. 21; Fox 74, 30. Com flówende flód æfter ebban *the flowing flood came after the ebb*, Byrht. Th. 133, 45; By. 65 : Bt. Met. Fox 11, 138; Met. 11, 69. [*Chauc.* ebbe : *Plat.* ebbe, *f* : *O. Frs.* ebba, *n* : *Dut.* eb, *f* : *Kil.* ebbe : *Ger. M. H. Ger.* ebbe, *f* : *O. H. Ger.* ebba, *f* : *Dan.* ebbe, *m.f* : *Swed.* ebb, *m.*]

ebbian; *p.* ode, ade; *pp.* od, ad [ebba *an ebb*] *To ebb*; recēdĕre, refluĕre :—Will-flód ongan lytligan eft, lago ebbade sweart under swegle *the well-flood began again to lessen, the water ebbed dark under the firmament*, Cd. 71; Th. 85, 12; Gen. 1413. DER. a-ebbian, be-,. ge-: æbbung, sǽ-.

ĕbere-morþ, es; *n.* [*ǽber clear, manifest*; morþ *murder*] *Open murder, manslaughter*; homicĭdium manĭfestum, L. H. 12, § 1; Th. i. 522, 27, Som. Ben. Lye.

Ebreisc; *adj. Hebrew, belonging to Jews*; Hebræus :—Nychodēmus awrāt eall mid Ebreiscum stafum *Nicodemus wrote all in Hebrew letters*, Nicod. pref; Thw. 1, 4. Of Seme com ðæt Ebreisce folc *from Shem came the Hebrew people*, Ælfc. T. 7, 25.

ebur-þring, es; *m. The celestial sign Orion*, Som. Ben. Lye. v. eofor-þring.

ebylgan *to be angry*; īrasci, Ben. Lye. v. a-belgan.

ebylgnes, -ness, e; *f. Anger, indignation*; īra :—On ebylgnesse his *in indignātiōne ejus*, Ps. Spl. T. 29, 5. v. æbylignes.

ēc; *conj. EKE, also*; etiam :—Ða us ēc bewrǣcon *who also have sent us forth*, Cd. 189; Th. 235, 12; Dan. 305 : 151; Th. 190, 5; Exod. 194. Ēc sceoldon his þegnas ðǽr gewunian *his followers must also inhabit there*, 220; Th. 284, 23; Sat. 326 : Beo. Th. 6254, note; B. 3131 : Ps. Th. 131, 17. v. eác.

ēcan, ǽcan, īcan, iécan, ȳcan, ȳcean, ic ēce, ðū ēcest, he ēcþ, *pl.* ēcaþ; *p.* ēcte, *pl.* ēcton, ēhton; *pp.* ēced [eáca *an addition*] *To EKE, increase, prolong, add*; augēre, appōnĕre :—Dū scealt ēcan dīne yrmþu *thou shalt increase thy wretchedness*, Andr. Kmbl. 2767; An. 1386. Gē ēcaþ eówre ermþe *ye increase your poverty*, Bt. 26, 2; Fox 94, 9. Ðæt ēcþ his ermþa *that augments his misery*, 29, 1; Fox 102, 19. Ēcte ðæt spell mid leóþe *he prolonged the speech with verse*, 12; Fox 36, 6 : Ps. Th. 104, 20. Hī hira firena furdur ēhton *appŏsuērunt adhuc peccāre ei*, 77, 19. Ðæt se awyrgeda ne ēce, ðæt he hine leng myclie ofer eorþan *ut non appōnat ultra magnĭfĭcāre se hŏmo sŭper terram*, 9, 38. Hwæt biþ dē ealles seald oððe ēced swā from ðǽre inwitfullan ȳfian tungan *quid dētur tĭbi aut quid appōnātur tĭbi a lingua dŏlōsa?* 119, 3. DER. æt-ēcan, ge-, to-, to-æt-, to-ge- : to-æt-ȳcnys.

ēcce-lĭc; *adj. Eternal, perpetual, everlasting*; æternālis :—Upahebbaþ gatu ēccelīce *elevāmĭni portæ æternāles*, Ps. Spl. 23, 7. v. ēce-lĭc.

ece, æce, ace, es; *m. An AKE, pain*; dŏlor :—Efne swā se bisceop ðone ece and ðæt sār mid him.ūt bǽre *as if the bishop had borne the ake and the sore out with him*, Bd. 5, 3; S. 616, 37 : 5, 4; S. 617, 22. DER. acan.

ĒCE, ǽce; *gen. m. n.* ēces; *gen. f.* ēcre, ēcere; *dat. m. n.* ēcum; *f.* ēcre, ēcere; *def.* se ēcă, ēcea, seó, ðæt ēce; *gen.* ēcan, ēcean; *adj. Eternal, perpetual, everlasting*; sempĭternus, æternus :—Ðis ys sŏþlīce ēce líf *hæc est autem vīta ætern*a, Jn. Bos. 17, 3. Onwōd ēce feónd folcdriht wera *the eternal foe pervaded the nation of men*, Cd. 64; Th. 76, 23; Gen. 1261. Dē síe ēce hērenis *eternal praise be to thee*, Exon. 13 b; Th. 26, 10; Cri. 415. Ðæt is ēcu rest *that is eternal rest*, Bt. Met. Fox 13, 142; Met. 13, 71. Godes ēce bearn *God's eternal child*, Exon. 18 b; Th. 46, 29; Cri. 744. Swā him se ēca bebeád *as the Eternal bade him*, Cd. 107; Th. 142, 28; Gen. 2368. Ēces word *the Eternal's word*, Exon. 61 b; Th. 225, 33; Ph. 398. Fóre onsýne ēcan Dryhtnes *before the face of the eternal Lord*, 64 b; Th. 238, 7; Ph. 600. To ēcre gemynde *for a continual remembrance*, Homl. Blick. 127, 22. Wæs me andſencge ēcere hǽlu *tu es susceptor salūtis meæ æternæ*, Ps. Th. 88, 23. Ic þanc secge ēcum Dryhtne *I say thanks to the eternal Lord*, Beo. Th. 5584; B. 2796. Andetaþ ðam ēcean Gode *confĭtēmĭni Deo æterno*, Ps. Th. 135, 27. Cēgaþ his ēcne naman *invocāte nōmen ejus æternum*, 104, 1. On ðone ēcan eard ussa sāwla *to the eternal region of our souls*, Bt. Met. Fox 23, 21; Met. 23, 11. He him ēce meaht geceás *he chose to himself eternal power*, Exon. 45 b; Th. 154, 34; Gū. 852. He us sealde ēce staðelas *he gave us eternal seats*, 17 b; Th. 41, 26; Cri. 661. Se ðe ða ēcan āgan wille sóþan gesǽlþa *he who will possess the eternal true felicities*, Bt. Met. Fox 7, 57; Met. 7, 29. Ðæt he walde ēcra gestealda *that he shall rule the eternal mansions*, Elen. Kmbl. 1601; El. 802. Eorþan ðū gefyllest eceum wæstmum *thou fillest the earth with eternal fruits*, Ps. Th. 64, 9. Se mec āna mæg ēcan meahtum geþeón þrymme *who alone by his eternal powers can tame me with power*, Exon. 111 b; Th. 427, .3; Rä. 41, 90. [*Orm.* eche : *O. Sax.* ēwig : *O. Frs.* ewch, ewig, iowich, iowigh : *Dut.* eeuwig : *Ger. M. H. Ger.* ēwic, ēwec : *O. H. Ger.* ēwîg : *Goth.* ayuk-duþs *eternity* : *Dan. Swed.* evig.] DER. efen-ēce.

ēce; *adv. Ever, evermore, eternally, perpetually*; in æternum, semper, contĭnuo, perpĕtuo :—Hie on friþe lifdon ēce mid heora aldor *they lived ever in peace with their chief*, Cd. 1; Th. 2, 16; Gen. 20. Ðǽr he ēce sceal hāmfæst wesan *where he shall for ever sojourn*, Exon. 30 b; Th.

95, 9; Cri. 1554. Ðe wunaþ ēce *qui mănet in æternum*, Ps. Th. 54, 19. Ēce standeþ Godes hand-geweorc *God's handywork standeth evermore*, Canon. Hrs. 369, 17. Ðǽr is help gelong ēce to ealdre *there is our help for evermore at hand*, Exon. 75 a; Th. 281, 14; Jul. 646. Wunaþ symble ēce mănet in sēcŭlum sēcŭli, Ps. Th. 110, 2. Wunaþ ēce forþ *mănet in sēcŭlum sēcŭli*, Ps. Th. 118, 90.

ECED, æced, æcced, es; *n. m. ACID, vinegar*; acētum :—Ðā stód ān fæt full ecedes *vas ergo ērat pŏsĭtum acēto plēnum*, Jn. Bos. 19, 29. Se Hǽlend onfēng ðæs ecedes *the Saviour received the vinegar*, Jn. Bos. 19, 30. Onfēng ðe Hǽlend ðæt æced, Jn. Rush. War. 19, 30. Drync ecedes *a drink of vinegar*, Exon. 29 a; Th. 88, 13; Cri. 1439. Mid ecede *with vinegar*, Ps. Th. 68, 22. Wyl niðewearde netelan on ecede, dō oxan geallan on ðæt eced *boil the netherward [part] of nettle in vinegar, add ox gall to the vinegar*, L. M. 3, 7; Lchdm. ii. 312, 8, 9. Lege hit in ðone eced *lay it in the vinegar*, Lchdm. iii. 18, 2. [*Plat.* etik, *m* : *O. Sax.* ekid, *n* : *Dut.* edik, eek, *m* : *Ger.* essich, essig, *m* : *M. H. Ger.* ezzich, *m* : *O. H. Ger.* ezih, *m* : *Goth.* akeit, *n* : *Dan.* eddike, *m. f* : *Swed.* ättika, *f* : *Icel.* edik, *n.*] DER. eced-fæt, æced-fæt, -wín.

eced-fæt, æced-fæt, es; *n. An acid-vat, a vinegar-vessel*; acetābŭlum, Ælfc. Gl. 114; Som. 80, 32; Wrt. Voc. 61, 12.

eced-wín, es; *n. Acid-wine*. v. æced-wín.

ēce-lĭc, ēcce-lĭc; *adj. Eternal, perpetual, everlasting*; æternālis :—Upahebbaþ gatu ēcelīce *elevāmĭni portæ æternāles*, Ps. Spl. 23, 9. Ēccelíc *eternal*, 23, 7.

ēce-lĭce; *adv. Eternally, ever*; perpĕtuo, Ælfc. Gr. 38; Som. 42, 1. Ic ðas tíde Eástrena ēcelīce healdan wille *vŏlo hoc tempus Paschæ perpĕtuo observāre*, Bd. 5, 21; S. 643, 20.

ēcen *great, powerful*; magnus, pŏtens, Andr. Kmbl. 1271; An. 636: 1763; An. 884; =eácen; *pp.* of eácan *augēri*.

ecer *an acre*, Som. Ben. Lye. v. æcer.

ECG, e; *f. An EDGE, a sharpness, blade, sword*; ăcies, acūmen, glădius, ferrum :—On sweordes ecge *on the edge of the sword*, Lk. Bos. 21, 24. Hyne ecg fornam *the sword had destroyed him*, Beo. Th. 5538; B. 2772. Ecg wæs íren *the edge was iron*, 5549; B. 2778. Ecg grymetode *the blade rang*, Cd. 162; Th. 203, 24; Exod. 408. Ecga [MS. ecge] mihton helpan æt hilde *swords might help in battle*, Beo. Th. 5360; B. 2683 : 5649; B. 2828. Mid gryrum ecga *with terrors of swords*, 971; B. 483. Æscum and ecgum *with spears and swords*, 3548; B. 1772. Billa ecgum *with edges of bills*, Cd. 210; Th. 260, 14; Dan. 709. [*Wyc.* egge : *Laym.* egge, agge : *Orm.* egge : *Plat.* egge, *f* : *O. Sax.* eggia, *f* : *Frs.* ig : *O. Frs.* eg, ig, *f* : *Kil.* egghe, *f* : *Ger. M. H. Ger.* ecke, *f* : eck, *n* : *O. H. Ger.* ekka, *f* : *Dan.* eg, *m. f* : *Swed.* egg, *m* : *Icel.* egg, *f* : *Lat.* ăcies, acūmen : *Grk.* ἀκή, ἀκίς, ἀκμή : *Sansk.* aśri, *f. ăcies, ensis*.] DER. brūn-ecg, heard-, stíþ-, stȳl-, twȳ-.

ecgan; *p.* de; *pp.* ed; *v. trans.* [ecg *an edge*] *To give an edge, to sharpen*; acuĕre. Ecged *edged, sharpened, only found in compositions*, as twig-ecged *two-edged*; bĭceps, *q. v.*

ecg-bana, -bona, an; *m. A sword-killer, murderer*; glădio cædens, occīsor :—Cain gewearþ to ecgbanan āngan brēder *Cain became the murderer of his only brother*, Beo. Th. 2528; B. 1262. Ecg-bona, 5006; B. 2506.

Ecg-bryht, -briht, -berht, -byrht, es; *m.* [ecg *edge, sword*; bryht *bright, excellent*] *Egbert*; Ecgbryhtus; *king of Wessex for thirty-seven years and seven months, from A. D. 800–837. Egbert chose Swithun* [v. Swíþhūn] *for the preceptor to his son Æðelwulf, the heir to the throne of Wessex* :—Hēr, A. D. 800, Ecgbryht fēng to Wesseaxna ríce *here, A. D. 800, Egbert succeeded to the kingdom of the West-Saxons*, Chr. 800; Erl. 60, 4. Hēr, A. D. 837 [MS. 836], Ecgbryht cyning forþfērde, se rícsode xxxvii wintra and vii mōnþas *here, A. D. 837, king Egbert died, who reigned thirty-seven years and seven months*, Chr. 836; Th. 117, 25, col. 1.

Ecg-bryhtes stān, es; *m. Brixton Deverill, Wilts?*—He gerād to Ecgbryhtes stāne be eástan Sealwyda *he rode to Egbert's stone, on the east of Selwood*, Chr. 878; Th. 148, 3, col. 1.

ecg-clif *a sea cliff or shore*, B. 2893; =ēg-clif, *q. v.* Beo. Th. 5778.

ecg-heard; *adj. Hard of edge*; ăcie dūrus :—Lǽtaþ spor, íren ecg-heard, ealdorgeard sceoran *let the spur, the iron hard of edge, raze the dwelling of life*, Andr. Kmbl. 2363; An. 1183.

ecg-hete, es; *m. Sword-hate, hostile hate*; ŏdium glădiis manifestātum, bellum :—Ne gesacu ōhwǽr ecghete eóweþ *nor strife shews anywhere hostile hate*, Beo. Th. 3480; B. 1738.

ecg-plega, an; *m. A play of swords, sword-fight, battle*; pugna :—Hie ðam ealdorþegnum cýðan eódon atolne ecgplegan *they went to inform the principal thanes of the cruel sword-fight*, Judth. 12; Thw. 25, 6; Jud. 246.

ecg-þræc; *gen.* -þræce; *pl. nom. gen. acc.* -þraca; *f. Sword-strength, war or savage courage*; glădiōrum impĕtus :—He ne þearf atole ecg-þræce *he needs not the cruel sword-strength*, Beo. Th. 1196; B. 596.

ecg-wæl, es; *n. Sword's wail, slaughter*; străges glădio cæsōrum :—

On ecgwæle [MS. ecgwale] *amid the slaughter of swords*, Cd. 96; Th. 126, 2; Gen. 2089.

ecilma, an; *m. A chilblain*; pernio, Som. Ben. Lye. v. æcelma.

écne *great*; *acc. of* ēcen.

éc-nes, -nis, -nys, -ness, -niss, -nyss, e; *f. Eternity, everlasting*; æternĭtas:—Ðæt we wuldres eard in ēcnesse āgan mōsten *that we for ever might possess the abode in glory*, Exon. 25 b; Th. 74, 9; Cri. 1204: Ps. Th. 118, 152. On ēcnisse *for ever*, Cd. 23; Th. 30, 18; Gen. 469. On ēcnysse *for ever*, Mk. Bos. 3, 29: Ps. Th. 110, 6: 118, 44.

écra *of eternal*, Elen. Kmbl. 1601; El. 802; *gen. pl. of* ēce.

écre *for continual*, Homl. Blick. 127, 22; *dat. f. of* ēce.

éc-sóþ, ēc-sōþlīce *but truly, but also*; sed autem, vēre, Som. Ben. Lye.

éc-sóþlíce *but truly*; v. ēc-sōþ.

éd [eád *happiness*] *Safety, security, happiness*; sălus, asȳlum:—Ēd monne *safety of men, the ark*, Cd. 70; Th. 84, 30, Mann. Some think ed signifies *a renewing, restoration, regeneration*; renōvātĭo: ·then ed monne might be translated, *regeneration of men*. Grn. corrected ed monne *into* edniowne *renewed*:—Ðā he hine [ēgor-here] upp forlēt edniowne [*acc. referring to* hine=ēgor-here] strēamum stīgan *when he allowed it [the water-flood-' host'] renewed to mount up in streams*, Gen. 1405.

ed-, prefixed to words, denotes *anew, again*, as the Latin re- *meaning* rursus, dēnuo, itĕrum. Edniwian *to renew, to make new again*; renōvāre. [*Wyc.* ed-: *Plat. O. Frs.* et- in etmal: *M. H. Ger.* ite-: *O. H. Ger.* it-, ita-: *Goth.* id-: *O. Nrs.* ið-.]

-ed used as a termination of *pp.* v. D 4, 5.

éd- = ād *a funeral pile*. v. ēd-wylm.

ed-cenning, e; *f. Regeneration*; regenĕrātĭo:—On edcenninge *in regenĕrātĭōne*, Mt. Bos. 19, 28.

ed-cer, -cir, -cyr, -cerr, -cirr, -cyrr, es; *m. A return*; reversĭo, rĕdĭtus:—Ne hī edcerres ǣfre mōton wēnan *they may never think of return*, Cd. 223; Th. 293, 7; Sat. 451. Edcir ðære ādle *a return of the disease*, Past. 33, 7; Cot. MS. Edcyr of wræcsīþe [MS. spræc-sīðe] *postlĭmĭnium*, Ælfc. Gl. 15; Som. 58, 28; Wrt. Voc. 21, 22. DER. cerr.

ed-cǽlness, e; *f. A recooling, pleasant coolness*; refrigĕrātĭo, Ps. Spl. 65, 11? Lye.

ed-cucian, -cwician; *p.* ode, ade; *pp.* od, ad *To re-quicken, revive*; reviviscĕre, Greg. Dial. 1, 12, Lye. DER. ge-edcucian, -cwician.

ed-cwide, es; *m. A relation, retelling*; relātĭo, Lye.

ed-cyr, -cyrr, es; *m. A return*; rĕdĭtus, Wrt. Voc. 21, 22. v. ed-cer.

éde, es; *n. A flock*; grex:—Wæs ðǣr ēde *erat ibi grex*, Lk. Lind. War. 8, 32: 12, 32. v. eówde.

eder, es; *m. A hedge, house*; sēpes, dōmus:—Hryðge ða ederas *the houses [are] ruinous*, Exon. 77 b; Th. 291, 5; Wand. 77. v. eodor.

eder-gong, es; *m. A home-seeking*; desĭdērĭum dōmus:—Ðǽr nǣfre cymeþ edergong *there never comes a home*, Exon. 32 b; Th. 102, 21; Cri. 1676.

edesc-hen *an edish hen, a quail*; cŏturnix, Ps. Surt. 104, 40. v. edisc-hen.

ed-geong, ed-giong; *adj. Growing young again*; rejuvĕnescens:—Of ascan edgeong weseþ *from ashes he becomes young again*, Exon. 61 a; Th. 224, 10; Ph. 373.

ed-gifan; *p.* -geaf, *pl.* -geáfon; *pp.* -gifen *To give again, restore*; reddĕre, Leo, A. Sax. Gl. 108.

ed-gift, e; *f. A re-giving, restitution*; restĭtūtĭo, Lye.

ed-gild, es; *n. A re-payment*; rĕ-sŏlūtĭo, Leo, A. Sax. Gl. 250. v. gild.

ed-giong; *adj. Growing young again*; rejuvĕnescens, Exon. 64 a; Th. 236, 28; Ph. 581. v. ed-geong.

ed-grówung, e; *f. A re-growing*; recĭdīva, Ælfc. Gl. 60; Som. 68, 26; Wrt. Voc. 39, 12.

ed-gyldend, es; *m. A remunerator, rewarder*; remunĕrātor, Scint. 33, Som. Ben. Lye.

ed-hwyrft, es; *m. A returning, return*; rĕditio, rĕdĭtus:—He ne wēneþ, ðæt him ðæs edhwyrft cyme *he will not hope that its return may come*, Exon. 89 b; Th. 336, 3; Gn. Ex. 42: Beo. Th. 2566; B. 1281.

edisc, es; *n.* [ed-, *Lat.* te- *again*; isc a termination, generally an adj. but also es; *n.*] I. EDISH or *aftermath, pasture*; pascua:—Wǣrun we his sceáp, ða he on his edisce afēdde *we were his sheep, which he fed in his pasture*, Ps. Th. 94, 7; 99, 3. II. *a park*; vivārĭum, Cot. 207, Lye.

edisc-hen, -henn; e; *f. An* EDISH HEN, *quail*; cŏturnix:—Hī bǣdon, and com edischen *petiĕrunt, et vēnit cŏturnix*, Ps. Spl. 104, 38. Edeschen ' the edisse-henne,' Ps. Surt. 104, 40. v. ersc-hen.

edisc-weard, es; *m. The keeper of edish, of a park, warren, etc*; vivarii custos, Wrt. Voc. 288, 12, Som. Ben. Lye. v. edisc.

ed-lǽcan; *p.* -lǣhte; *pp.* -lǣht *To repeat, renew*; repĕtĕre, renŏvāre, Som. Ben. Lye.

ed-lǽcung, e; *f. A repetition*; repetĭtĭo:—He sceal God biddan ðæt he hyne gehealde wið ðara ǣrgedônra yfla edlǣcunge *he shall pray to God to preserve him against a repetition of the evils before committed*, L. E. I. 21; Th. ii. 416, 42.

ed-leánian, ed-leánian; *p.* ode; *pp.* od *To reward, recompense, renew,*

remit; retrĭbuĕre:—He edleánaþ me *retrĭbuit mihi*, Ps. Spl. T. 17, 26. DER. leánian.

ed-leánung, e; *f. A rewarding*; retrĭbūtĭo:—For edleánunge *propter retrĭbūtĭōnem*, Ps. Spl. T. 118, 112. v. ed-leánung.

ed-leán, ead-leán, æd-leán, es; *n.* [ed or ead; leán *a loan*] *A reward, recompense, requital, retribution*; præmium, retrĭbūtĭo:—Edleánes dæg *retrĭbūtĭōnis dies*, Lk. Bos. 4, 19. Ðæt edleán, Bt. 3, 4; Fox 6, 19: Andr. Kmbl. 2457; An. 1230. For edleáne *propter retrĭbūtĭōnem*, Ps. Spl. 118, 112.

ed-leánian *to reward*; remunĕrāre, Som. Ben. Lye. DER. leánian.

ed-leánung, e; *f. A rewarding, recompense*; retrĭbūtĭo:—Nylle ðū forgytan ealle edleánunga *vel* edleán his *nōli oblivisci omnes retrĭbūtĭōnes ejus*, Ps. Lamb. 102, 2. v. ed-leánung.

ed-lesende, ed-lesendlĭc; *adj. Reciprocal, relative*; relātīvus:—Gif ic cweðe, ðū wāst hwā ðys dyde *tu scis quis hoc fēcit*, ðon biþ se [hwā] *quis relātīvum*, ðæt is edlesendlĭc, Ælfc. Gr. 18; Som. 21, 30: 38; Som. 40, 62.

ed-lesung, e; *f. A relation, relating*; relātĭo, Ælfc. Gr. 18; Som. 21, 58.

éd-mód; *adj. Mild, obedient*; obēdiens, mītis, Ben. Lye.

éd-módian, -mōdigan; *p.* ode; *pp.* od *To be humble, to obey*; obēdīre:—Hi ēdmōdigaþ him *obēdiunt ei*, Mk. Lind. War. 1, 27.

ed-neowe; *adj. Renewed*; renŏvātus:—Eart ðū edneowe *renŏvātus es*, Ps. Th. 102, 5: Cd. 17; Th. 20, 25; Gen. 314. v. ed-niwe.

ed-niowunga; *adv. Anew*; dēnuo:—Ðe eów eágena leóht bôte gefremede edniowunga *who healed anew the light of your eyes*, Elen. Kmbl. 599; El. 300.

ed-niwan; *adv. Anew, again*; de nŏvo, dēnuo:—Eów gebȳraþ ðæt gē beón acennede edniwan *oportet nos nasci dēnuo*, Jn. Bos. 3, 7: 3, 3.

ed-niwe, ed-neowe; *adj. New, again new, renewed*; renŏvātus:—Eft cymeþ feorh edniwe *renewed life returns*, Exon. 59 a; Th. 213, 12; Ph. 223: 61 a; Th. 224, 4; Ph. 370: Bt. Met. Fox 11, 77; Met. 11, 39.

ed-niwe; *adv. Anew, again*; dēnuo:—Swá ´se fugel weorþeþ gomel æfter geárum geong edniwe *thus the bird becomes old after years and young again*, Exon. 59 b; Th. 215, 25; Ph. 258.

ed-niwian; *part.* igende; *p.* ode, ede; *pp.* od, ed *To make new, to renew*; renŏvāre:—Ðū edniwast ansíne eorþan *renŏvābis făciem terræ*, Ps. Spl. 103, 31. Hȳ fǣringa eald æfþoncan edniwedon [MS. edniwedan] *they suddenly renewed the old grudge*, Exon. 72 b; Th. 271, 21; Jul. 485. DER. ge-ed-niwian.

ed-niwinga; *adv. Anew*; dēnuo:—Se fugel líf eft onfēhþ edniwinga *the bird receives again life anew*, Exon. 63 b; Th. 234, 2; Ph. 534: Andr. Recd. 1569; An. 784.

ed-niwung, e; *f. A renewing, reparation, renovation*; repărātĭo:—Seó feórþe dǣl sceal beón to edniwunge Godes cyricean *the fourth part shall be to a renewing of God's church*, Bd. 1; 27; S. 489, 9.

édo *a flock*; grex:—Ge-eode [MS. ge-eáde] all suner *vel* ēdo in sǣ *ābiit tōtus grex in măre*, Mt. Kmbl. Lind. 8, 32. v. eówde.

edor, eder, es; *m. A hedge, fence, place inclosed by a hedge, fold, dwelling, house*; sēpes, dōmus, tectum:—Gif fríman edor gegangeþ *if a freeman forcibly enter a dwelling*, L. Ethb. 29: Th. i. 10, 3. Under edoras *under dwellings*, Cd. 112; Th. 147, 25; Gen. 2445: 114; Th. 150, 5; Gen. 2487. Ederas *houses*, Exon. 77 b; Th. 291, 5; Wand. 77. v. eodor.

edor-brecþ, e; *f.* [edor, brecþ *fractio*] *A fence-breaking, house-breaking*; sēpis fractĭo, dŏmus fractĭo:—Gif fríman edorbrecþe gedēþ *if a freeman commit house-breaking*, L. Ethb. 27; Th. i. 8, 15. v. eodor-brice.

edor-brice, -bryce *a fence-breaking*, L. Alf. pol. 40; Th. i. 88, 10, note 25. v. eodor-brice.

edre; *adv. Immediately, at once, forthwith*; stătim, prōtĭnus, illĭco:—Edre him ða eorlas agēfon ondsware *the earls gave answer to him immediately*, Andr. Kmbl. 801; An. 401: 1285; An. 643: 1900; An. 952: Invent. Crs. Recd. 1300; El. 649. v. ǽdre.

édre *an artery, vein*; artēria, vēna, Som. Ben. Lye. v. ǽdre.

ed-recan; *p.* te; *pp.* ed *To ruminate*; rumĭnāre, Som. Ben. Lye. v. eodorcan.

ed-recedroc, -rocc, es; *m. The belching thing*; rūmen, Cot. 169, Som. Ben. Lye.

ed-rine, es; *m. A meeting*; occursus, Ps. Spl. T. 18, 7.

edring, e; *f. A refuge, return*; refŭgium:—Dust ne mæg him edringe ǽnge gehâtan *the dust may not promise any refuge to him*, Exon. 99 b; Th. 373, 11; Seel. 107. v. edor.

ed-roc, es; *m. A chewing again, chewing the cud, considering*; rūmen, rumĭnātĭo:—Wasend *vel* edroc rūmen, Ælfc. Gl. 72; Som. 70, 116; Wrt. Voc. 43, 43. Cíwung *vel* edroc, *vel* aceócung rumĭnātĭo, Ælfc. Gl. 99; Som. 76, 121, 122; Wrt. Voc. 54, 62.

ed-sceaft, æd-sceaft, e; *f. A new creation, new birth*; regĕnĕrātĭo:—Com swefnes wôma, hū woruld wǣre wundrum geteód ungelíc yldum ôþ edsceafte *the terror of a dream came, how the world was wondrously framed unlike to men until regeneration*, Cd. 177; Th. 222, 30; Dan. 112: Bt. 34, 10; Fox 150, 14, 16.

ed-staðelian; _p._ ode ; _pp._ od [ed _again,_ staðelian _to establish,_ staðol _a foundation_] _To establish again, re-establish, restore ;_ restĭbĭlĭre, Som. Ben. Lye.

ed-staðelig; _adj. Firm, strong ;_ firmus :—Beó se awirged, ðe æfre eft gedô edstaðelige ðas burh Hiericho _mălĕdictus vir qui suscitāvĕrit et ædĭfĭcāvĕrit_ [_restĭbĭlĭ:am fĕcĕrit_] _cīvĭtātem Jĕrĭcho,_ Jos. 6, 26.

ed-staðelung, e ; _f. An establishing again, re-establishment, renewing ;_ repărātio, R. Ben. 36.

ed-þingung, e ; _f. A reconciliation ;_ reconcĭlĭātio :—Edþingung _reconcĭlĭātio,_ Ælfc. Gl. 90 ; Som. 74, 127 ; Wrt. Voc. 51, 40.·

ĕdulf-stæf, es ; _m. A family staff_ or _support, stay of the house ;_ prædii sustentācŭlum, Cd. 55 ; Th. 68, 16. v. ĕðyl-stæf.

ed-wendan; _p._ -wende ; _pp._ -wended ; _v. intrans. To return, desist from, cease ;_ reverti, cessāre :—Gyf him edwendan æfre scolde bealuwa bīsigu _if ever the tribulation of evils should return to him,_ Beo. Th. 565 ; B. 280.

ed-wenden, e ; _f. A reverse, alteration, end ;_ mūtātio, āversio, cessātio :—Edwenden cwom _a reverse came,_ Beo. Th. 4383, note ; B. 2188. Ǽr ðon edwenden worulde geweorþe _ere that an end shall be to the world,_ Exon. 56 b ; Th. 200, 14 ; Ph. 40.

ed-wendu, e ; _f. An alteration, change, end ;_ mūtātio, cessātio :—Ǽghwylc ðissa earfoða ēce standeþ, būtan edwende _all these sufferings are eternal, without a change,_ Salm. Kmbl. 951 ; Sal. 475.

ed-wielle _A whirlpool, dizziness ;_ vortex āquæ, Cot. 86.

ed-wihte; _pron. Anything, something ;_ ǎlĭquid :—Næfre, hleówlora [MS. hleor-lora] æt edwihtan mon weorþeþ _a man is never deprived of protection in anything,_ Cd. 92 ; Th. 117, 15 ; Gen. 1954. [Ed = _A. Sax._ æt _in_ æt-hwæga _somewhat;_ ǎlĭquantum : æt-hwôn _almost ;_ fēre : _Ger._ et : _M. H. Ger._ ete : _O. H. Ger._ etta, eta, ede.]

ed-winde _A winding again, a vortex ;_ vortex :—Edwinde _vortex,_ Ælfc. Gl. 98 ; Som. 76, 92 ; Wrt. Voc. 54, 36.

ed-wist, e ; _f._ [ed _re-, anew, again ;_ wist _support_] _Being, subsistence, existence, essence, substance ;_ substantia :—Ic adilegie ealle ða edwiste, ðe ic geworhte _dēlĕbo omnem substantiam, quam fēci,_ Gen. 7, 4. v. ætwist.

edwistfull; _adj._ [edwist _substance,_ full _full_] _Existing, substantial, substantive ;_ substantiālis, Som. Ben. Lye. v. edwistlíc.

edwistlíc; _adj. Existing, subsisting, substantial, substantive ;_ substantiālis :—Ic eom, is edwistlíc word _I am is the substantive_ [_existing_] _verb,_ Ælfc. Gr. 32 ; Som. 36, 24. DER. efen-edwistlíc.

ed-wít, æd-wít, es ; _n. A reproach, disgrace, blame, contumely, scorn ;_ opprobrium, prŏbrum, ignōmĭnia, cavillātio :—Wæs him on gemynde yfel and edwít _the evil and contumely was in his mind,_ Bt. Met. Fox 1, 109 ; Met. 1, 55. Ealle beóþ aweaxen of edwittes ýða heáfdum _all shall be grown over by the heads of the waves of scorn,_ Salm. Kmbl. 57 ; Sal. 29. Ne þearf ðê on edwít Abraham settan _Abraham need not put thee in reproach,_ i. e. _reproach to thee,_ Cd. 130 ; Th. 165, 7 ; Gen. 2728. And me eác fela dīnra edwíta on gefeóllon _et opprobria exprobrantium tĭbi cecĭdĕrunt sŭper me,_ Ps. Th. 68, 9 : 73, 21.

ed-wítan; _p._ -wāt, _pl._ -witon ; _pp._ -witen _To reproach, blame, upbraid ;_ exprobrāre :—Hosp edwítendre ðê hruron ofer me _opprobria exprobrantium tĭbi cecĭdĕrunt sŭper me,_ Ps. Spl. 68, 12. v. æt-wítan.

ed-wítfullíce; _adv. Disgracefully ;_ prŏbrōse, Cot. 195, Lye.

edwít-líf, es ; _n. A disgraceful life ;_ prŏbrōsa vīta :—Deáþ biþ sēlla eorla gehwylcum ðonne edwítlíf _death is better for every man than a disgraceful life,_ Beo. Th. 5775 ; B. 2891.

edwít-scype, es ; _m. Cowardice ;_ ignāvia, ignōmĭnia :—Þurh edwítscype _ignōmĭniōse,_ Wald. 23 ; Vald. 1, 14.

edwít-spræc, e ; _f. Contemptuous speech, scorn ;_ opprobrium, imprŏpĕrium, cavillātio :—Ðý-læs ic scyle þrōwian edwítspræce _lest I shall suffer contemptuous speech,_ Andr. Kmbl. 161 ; An. 81 : Ps. Th. 88, 43 : 101, 6.

edwít-spreca, an ; _m. A blame-speaker, scoffer, caviller ;_ cavillātor :—Him edwítsprecan ermþu gehēton _the cavillers threatened him with affliction,_ Exon. 39 a ; Th. 129, 8 ; Gú. 418.

edwít-stæf, es ; _m. A disgraceful letter, reproach, scandal, disgrace, dishonour ;_ opprobrium :—Eom ic to edwít-stæfe eallum geworden _factus sum opprobrium omnĭbus,_ Ps. Th. 108, 24 : 78, 4 : 118, 42.

ĕd-wylm, es ; _m._ [= ād a _funeral pile,_ wylm _heat, fire_] _Heat of fire, burning heat ;_ flammæ æstuātio :—Se fǽcna gebroht hafaþ æt ðam ēdwylme ða ðe him oncleófiaþ _the beguiler has brought into that burning heat those who cleave to him,_ Exon. 97 b ; Th. 364, 19 ; Wal. 73.

ed-wyrpan; _p._ -wyrpte ; _pp._ -wyrped _To recover, become better ;_ meliōrāri, Ben. Lye. DER. ge-edwyrpan.

ed-wyrping, e ; _f. Recovery, a growing better, recovering ;_ recŭpĕrātio :—Án eáwfæst mynecenu læg swíðe geswenct, orwēne ælcere edwyrpinge _a pious mynchen lay greatly afflicted, hopeless of any recovery,_ Homl. Th. ii. 26, 29.

Éfe, an ; _f. Eve :_—Éfe _Eve,_ Cd. 222 ; Th. 290, 1 ; Sat. 408. v. Éua.

efel; _adj. Evil, bad ;_ prāvus, mălus, Som. Ben. Lye. v. yfel.

efe-lang; _adj._ [= efen _even,_ lang _long_] _Even-long, equally long, oblong ?_ [_Wrt. Provncl._ evelong = _oblong_] ; æque longus, oblongus :—

ᚩ Ðæt hol ðæt he efe-lang ǽr gefylde _the oblong hole which he filled before,_ Exon. 112 b ; Th. 431, 13 ; Rä. 45, 7.

efe-láste, efen-láste, an ; _f._ [lǽstan _to last, continue, endure_] _The everlasting ;_ gnaphālium, Lin :—Genim efelāstan _take everlasting,_ L. M. 1, 1 ; Lchdm. ii. 20, 3 : 1, 32 ; Lchdm. ii. 78, 19 : 1, 47 ; Lchdm. ii. 120, 2 : 2, 65 ; Lchdm. ii. 292, 4. Nim efelāstan ufewearde _take the upper_ [_part_] _of everlasting,_ L. M. 2, 56 ; Lchdm. ii. 276, 20. Efelāste _herba mercŭriālis,_ Som. Ben. Lye.

efeleác; _n. An onion, a scallion ;_ cæpa, Som. Ben. Lye.

EFEN, efn, æfen ; _adj._ EVEN, _equal ;_ æquus, plānus, æquālis :—Môdes gecynde grēteþ grorn efen winde _the disposition of his mind approached sadness equal to the wind,_ Exon. 94 b ; Th. 354, 22 ; Reim. 49. On efen, _adv. together ;_ simul, una :—Englas on efen blāwaþ býman _angels shall blow the trumpet together,_ Exon. 20 b ; Th. 55, 10 ; Cri. 881 : Ps. Th. ₁16, 1. On efen, _prep. On even ground, on a level, by, near, aside with ;_ in æquāli, juxta :—Him on efn ligeþ ealdor-gewinna _by him lies his vital adversary,_ Beo. Th. 5798 ; B. 2903. [_Wyc. Piers P. Chauc._ even : _Laym._ æfne, efne : _Orm._ efenn : _Plat._ even, ewen, effen : _O. Sax._ eban : _Frs._ even : _O. Frs._ ivin, even : _Dut._ even, effen : _Ger._ eben : _M. H. Ger._ eben, ëbene : _O. H. Ger._ eban : _Goth._ ibns : _Dan._ jävn : _Swed._ jemn : _Icel._ jafn, jamn.] DER. un-efen.

efen, efne ; _adv. Evenly, equally, just so ;_ æque :—Wunedon ætsomne efen swā lange swā him lýfed wæs _they dwelled together just so long as was permitted to them,_ Bt. Met. Fox 20, 487 ; Met. 20, 244 : Exon. 41 a ; Th. 137, 24 ; Gú. 564.

efen, es ; _n. Evening ;_ vesper :—Éfna gehwām _each evening,_ Exon. 50 b ; Th. 176, 27 ; Gú. 1216. v. æfen.

efen-, efn-, efne-, in composition, denotes _even, equal,_ represented by co-, con-, com-, _as_

efen-æðele _equally noble._ v. emn-æðele.

efen-behéfe _equally useful_ or _necessary._ v. efn-behéfe.

efen-beorht; _adj. Equally bright ;_ æque splendĭdus :—Heofonsteorran ealle efen-beorhte æfre ne scīnaþ _the stars of heaven do not ever shine all equally bright,_ Bt. Met. Fox 20, 465 ; Met. 20, 233 : 20, 461 ; Met. 20, 231.

efen-bisceop, efn-biscop, es ; _m. A co-bishop ;_ co-episcŏpus :—Mid Laurentio and Justo his efenbisceopum _cum Laurentio et Justo co-episcŏpis,_ Bd. 2, 5 ; Whelc. 122, 38.

efen-blissian; _part._ -blissiende ; _p._ ode ; _pp._ od [blissian _to rejoice_] _To rejoice with, to rejoice equally ;_ congrātŭlāri :—Efenblissiende Breotone on his geleáfan, monige eálond blissiaþ _Britain equally rejoicing in his belief, many isles shall rejoice ;_ congrātŭlante in fide ejus Brittania, lætentur insulæ multæ, Bd. 5, 24 ; S. 647, 14.

efen-ceaster-wearan; _gen. ena ; pl. m. Fellow-citizens ;_ concīves :—Efenceasterwearan ðæs heofonlícan ríces _concīves regni cœlestis,_ Bd. 1, 26 ; S. 488, 16.

efen-cuman; _p._ -com, _pl._ -cōmon ; _pp._ -cumen ; _v. intrans. To come together, convene, assemble together, agree ;_ convēnīre :—Līcode us efencuman _plăcuit convēnīre nos,_ Bd. 4. 5 ; S. 572, 5. Efencumendum monegum bisceopum _convenientĭbus plūrĭmis episcŏpis,_ Bd. 3, 28 ; S. 560, 11.

efen-dýre; _adj. Equally dear ;_ æque cārus :—Ða syndon efen-dýre _they are equally dear,_ L. A. G. 2 ; Th. i. 154, 3.

efen-eádig; _adj. Equally blessed ;_ æque beātus :—Efeneádig bearn _equally blessed child,_ Hy. 8, 21 ; Hy. Grn. ii. 290, 21.

efen-eald, efn-eald ; _adj. Co-eval, of the same age ;_ co-ævus, co-ætāneus :—Ic æt efenealdum æfre ne mētte māran snyttro _I never met with greater prudence among those of his age,_ Andr. Kmbl. 1105 ; An. 553 : Bd. 5. 19 ; S. 637, 19. Nǽnig efen-eald him _no one of like age with him,_ Exon. 85 a ; Th. 321, 2 ; Wīd. 40. Plegende mid his efen-ealdum _playing with his co-evals,_ Homl. Th. ii. 134, 4.

efen-eardigende _Dwelling together ;_ cohăbĭtans :—Ðæt ðú sunu wǽre efen-eardigende mid ðinne ēngan Freán _that thou his son shouldst be dwelling together with thy sole Lord,_ Exon. 11 a ; Th. 15, 16 ; Cri. 237.

efen-ēce, emn-ēce ; _adj. Co-eternal ;_ co-æternus :—Ǽr ðon up-stíge efenēce bearn _agnum fæder ere that the co-eternal child ascended to his own father,_ Exon. 14 b ; Th. 29, 19 ; Cri. 465.

efen-edwistlíc; _adj. Consubstantial, of the same substance ;_ consubstantiālis :—Se Hālga Gāst is ðæs Fæder Gāst and ðæs Suna, him bām efenedwistlíc _the Holy Ghost is the Spirit of the Father and of the Son, consubstantial with them both,_ Homl. Th. ii. 362, 27. Ic gelýfe on ænne Crist, ðone āncennedan Godes Sunu, acennedne nā geworhtne, efen-edwistlícne ðam Fæder _I believe in one Christ, the only begotten Son of God, begotten not made, consubstantial with the Father,_ ii. 596, 30.

efen-ehþ, -nehþ, e ; _f. A plain ;_ plānĭties :—On ǽlcre efen-ehþe _on every plain,_ Chr. 894 ; Th. 170, 36.

efen-esne, es ; _m. A fellow-servant ;_ conservus. v. efue-esne.

efen-etan _to eat as much as any one._ v. efn-etan.

efen-fela, -feola ; _indecl. So many, as many ;_ tŏtĭdem, tot :—Eardas rūme Meotud arǽrde efen-fela bega þeóda and þeáwa _the Creator_

established spacious lands, as many of both nations and manners, Exon. 89 a ; Th. 334, 17 ; Gn. Ex. 17. Hilde abbudisse efen-feola wintra in munuclífe Drihtne gehálgode *Hild abbatissa tŏtĭdem annos in Monastĭca vīta Dŏmĭno consecrāvit,* Bd. 4, 23 ; S. 592, 42.

efen-gedǽlan *to share alike.* v. efngedǽlan.

efen-gefeón ; *p.* -gefeah, *pl.* -gefǽgon ; *pp.* -gefǽgen *To rejoice together ;* congaudēre :—Efengefeóndum eallum ðam folce *congaudente ūniverso pŏpŭlo,* Bd. 3, 22 ; S. 553, 13.

efen-gelíc ; *adj. Like, co-equal ;* sĭmĭlis, consĭmĭlis, co-æquālis :—Cweðaþ to hyra efengelícon *dīcunt co-æquālĭbus,* Mt. Bos. 11, 16.

efen-gemæcca, an ; *m. A companion, husband ;* consors, consortii jūre æqualis, Som. Ben. Lye. v. efn-gemæcca.

éfen-gereord, e ; *f. An evening repast, supper ;* cœna, Som. Ben. Lye.

éfen-gereordian *To sup ;* cœnāre, Som. Ben. Lye.

efen-háda-bisceop, es ; *m. A co-bishop ;* co-episcŏpus, Greg. Dial. 1, 5.

efen-hǽfdling, es ; *m. An equal, fellow, fellow-mate ;* co-æquālis, co-ætāneus :—Gesomnode miccle scóle his gehoftena and hys efen-hǽfdlingas *he collected a great troop of his companions and equals,* Guthl. 2 ; Gdwin. 14, 3.

efen-heáh ; *adj. Equally high ;* æque altus, Salm. Kmbl. 85, 28.

efen-heáp, es ; *m. A fellow-soldier, soldier of the same band ;* commănĭpŭlāris, Som. Ben. Lye.

efen-hérenis, -niss, e ; *f. A praising together ;* collaudātio, Ps. Spl. C. 32, 1.

efen-hérian ; *v. trans. To praise together ;* collaudāre, Som. Ben. Lye.

efen-hleóðor, -hleóðres ; *m. A sounding together, concordance of voices* or *sounds, united voice ;* concentus :—Bletsiaþ Bregu sēlestan efenhleóðre ðus *they bless the most excellent Lord thus with united voice,* Exon. 64 b ; Th. 239, 15 ; Ph. 621.

efen-hleta, -hlytta, an ; *m. A consort, companion, fellow ;* consors :— Hæfde Oswio efenhletan ðære cynelícan wurþnysse *hăbuit Oswiu consortem rēgiæ dignĭtātis,* Bd. 3, 14 ; S. 539, 29 : 5, 8 ; S. 621, 27. Ðæt we beón efenhlyttan his wuldres *that we be companions of his glory,* Homl. Th. i. 34, 1. Smyrode ðē God ðín mid ele blisse toföran ðínum efenhlyttum *unxit te Deus tuus ŏleo lætĭtiæ præ consortĭbus tuis,* Ps. Lamb. 44, 8.

efen-hlytta, an ; *m. A consort, companion ;* consors, Ælfc. Gr. 9, 44 ; Som. 13, 6, MSS. C. D. v. efen-hleta.

efen-lǽcan ; *p.* -lǽhte ; *pp.* -lǽht *To be equal, like, to imitate ;* imĭtāri, Lye. v. ge-efenlǽcan.

efen-lǽcend, es ; *m. An imitator ;* imĭtātor, Scint. 2, Lye.

efen-lǽcestre, an ; *f. A female imitator ;* imĭtatrix, Som. Ben. Lye.

efen-lǽcung, e ; *f. A matching* or *making like* or *equal ;* imĭtātio, æquĭpărātio, Som. Ben. Lye.

efen-láste, an ; *f. The everlasting ;* gnaphālium :—Genim efenlástan nyðowearde *take the netherward [part] of everlasting,* Lchdm. iii. 2, 2. v. efe-láste.

efen-líc ; *adj. Even, equal ;* æquālis :—Nǽnig efenlíc ðam in worlde gewearþ *wīfes gearnung a woman's desert was in the world not equal to that,* Exon. 8 b ; Th. 3, 20 ; Cri. 39 : Bd. 4, 17 ; S. 585, 38.

efen-líca, an ; *m. An equal ;* æquālis. v. efn-líca.

efen-líce ; *adv.* EVENLY, *alike ;* æque :—Efenlíce Godes man *æque Deo devōtus,* Bd. 3, 23 ; S. 554, 16.

efen-lícnes, -ness, e ; *f. Evenness, equality ;* æquālĭtas. v. efn-lícnes.

efen-ling, es ; *m. A consort, an equal.* v. efn-ling.

efen-mǽre *equally great.* v. efnmǽre.

efen-metan ; *v. trans. To make equal, to compare ;* compărāre, Som. Ben. Lye.

éfen-męte, es ; *m. Even-meat, supper ;* cœna, Som. Ben. Lye.

efen-micel ; *adj. Equally great ;* æque magnus :—Ðú meahte spéd efen-micle Gode ágan ne móste *thou mightest not possess abundance of power equally great with God,* Exon. 28 b ; Th. 86, 4 ; Cri. 1403.

efen-mid ; *adj. Middle ;* mĕdius, plāne mĕdius :—On ðisse eorþan efen-midre *in mĕdio terræ,* Ps. Th. 73, 12.

efen-neáh ; *adv. Equally near ;* æque vĭcīne :—Strĭceþ ymbūtan efen-neáh gehwæðer *it holds its course around equally near everywhere,* Bt. Met. Fox 20, 282 ; Met. 20, 141.

efen-niht, e ; *f. Even-night, equinox ;* æquinoctium, Bd. Whelc. 493, 38.

efen-nys, efyn-nis, -niss, -nes, e ; *f.* EVENNESS, *equality ;* æquālĭtas :— Efennys gecýdnys ðín on ēcnysse *æquĭtas testĭmōnia tua in æternum,* Ps. Spl. 118, 144 : 10, 8.

efen-ríce ; *adj. Equally mighty, of equal power ;* æquālis potentiæ, æquipollens :—Wǽron hí eft efenríce *they were again of equal power,* Bd. 5, 10 ; S. 624, 27.

efen-sárig ; *adj. Even* or *equally sorry ;* æque tristis, compassus :—He wearþ hyre sáre efensárig *ille ĕrat ejus dŏlōri compassus,* Greg. Dial. 2, 1, Lye.

efen-sárignyss, e ; *f. Compassion ;* compassio, Lye.

efen-scearp ; *adj. Equally sharp ;* æque acūtus :—Hí heora tungan

teóþ sweorde efen-scearpe *exăcuĕrunt ut glădĭum linguas suas,* Ps. Th. 63, 3.

efen-scyldig ; *adj. Equally guilty,* L. C. S. 77 ; Th. i. 420, 2.

efen-spédiglíc, efne-spédelíc ; *adj.* [efen, spédiglíc *substantiam hăbens] Consubstantial ;* consubstantiālis :—Þrýnnesse in ánnesse efenspédiglíce *Trinĭtātem in unĭtāte consubstantiālem,* Bd. 4, 17 ; S. 585, 37.

efen-swíþ ; *adj. Equally strong.* v. efn-swíþ.

éfen-þénung, e ; *f. Even-food, supper ;* vespertīna refectio, Fulg. 42, Mann.

efen-þeówa, an ; *m :* efen-þeów, efn-þeów, es ; *m. A fellow-servant ;* conservus :—Astrehte hys efen-þeówa hyne and bæd hyne *prŏcĭdens conservus ejus rŏgābat eum,* Mt. Bos. 18, 29. Hú ne gebýrede ðē gemiltsian ðínum efen-þeówan *nonne ergo oportuit te misĕrēri conservi tui?* Mt. Bos. 18, 33. He gemétte hys efen-þeówan *he found his fellow-servant,* 18, 28. Gesáwon hys efen-þeówas ðæt *his fellow-servants saw that,* 18, 31.

efen-þrówian ; *p.* ode ; *pp.* od *To suffer together, to compassionate, commiserate ;* compăti, commĭsĕrāri, Past. 16, 1 ; Hat. MS. 20 a, 25, 26.

efen-þrówung, e ; *f. A suffering together, compassion ;* compassio, Som. Ben. Lye.

efen-þwær ; *adj. Agreeing ;* concors, Procem. R. Conc. Lye.

efen-towistlíc ; *adj. Consubstantial ;* consubstantiālis, Som. Ben. Lye.

efen-wǽge, an ; *f. Even-weight ;* æquĭpondium, Som. Ben. Lye.

efen-wel *even, well, equally ;* æque, sĭmĭlĭter, Off. Regum 10, Lye.

efen-weorcan ; *v. trans. To co-operate ;* co-opĕrāri, Som. Ben. Lye.

efen-weorþ ; *adj. Even worth, equivalent ;* æque dignus, æquĭvălens, L. Edg. C. 50 ; Th. ii. 254, 23.

efen-werod, es ; *n. A soldier of the same company, a fellow-soldier ;* commănĭpŭlāris, Som. Ben. Lye.

efen-wesende *co-existent ;* co-existens :—Ðú mid Fæder ðínne gefyrn wǽre efenwesende *thou wast co-existent with thy Father of old,* Exon. 12 b ; Th. 22, 11 ; Cri. 350.

efen-wiht *even-weight ;* æquĭpondium, Som. Ben. Lye.

efen-wyrcan, -weorcan ; *v. trans. To co-operate ;* co-opĕrāri :—Efenwyrcend *co-opĕrātor,* Bd. 5, 20 ; S. 641, 27.

efen-wyrcung, e ; *f. A co-operating ;* co-opĕrātio, Som. Ben. Lye.

efen-wyrhta, an ; *m. A fellow worker ;* co-opĕrātor :—Com he to Róme mid hys efenwyrhtan and geféran ðæs ylcan weorces Ceólferþ *vēnit Rōmam cum co-opĕrātōre ac sŏcio ejusdem opĕris Ceolfrido,* Bd. 4, 18 ; S. 586, 28.

efen-wyrðe ; *adj. Equally worthy ;* condignus :—Mid efenwyrðum dǽdum *condignis actĭbus,* Bd. 3, 27 ; S. 559, 24 : 4, 6 ; S. 574, 18.

efen-yrfe-weard, es ; *m. A co-heir ;* cŏhēres :—Sibba, his geféra and efenyrfeward ðæs ylcan ríces *Sebbe, sŏcius ejus et cŏhēres regni ejusdem,* Bd. 3, 30 ; S. 562, 2. Swylce gedafenaþ ðæt hí engla efenyrfeweardas on heofonum sín *tāles angĕlōrum in cœlis dĕcet esse cŏhērēdes,* 2, 1 ; S. 501, 19.

éfeostlíce ; *adv. Quickly, hastily ;* cĕlĕrĭter :—He bebeád him ðæt he éfeostlíce sceolde to him cuman *he commanded him that he should quickly come to him,* Chr. 1114 ; Th. 370, 19.

efer, es ; *m. A wild boar ;* ăper, Anlct. v. eofor.

efer-fearn *fĭlix arbŏrătĭca,* Ælfc. Gl. 42 ; Som. 64, 14 ; Wrt. Voc. 31, 25. v. eofor-fearn.

Efer-wíc *York,* Chr. 188 ; Th. 15, 25, col. 3. v. Eofor-wíc.

EFES, e ; *f. Eaves of a house, a brim, brink, edge, side ;* margo, lātus :—Geworden ic eom swá spearwa ánhoga oððe ánwuniende on efese oððe on þecene *factus sum sĭcut passer solĭtārius in tecto,* Ps. Lamb. 101, 8. To ðære efese *to the edge,* Cod. Dipl. 353 ; A. D. 931 ; Kmbl. ii. 172, 22. Bí swá hwæðerre efese [MS. efes] *on whichever side,* Chr. 894 ; Erl. 90, 13. [*Wyc.* evese *brow of a hill : Laym.* eovesen, *dat. pl.* eaves *: Plat.* oese, ese *: O. Frs.* ose *edges of the roof : Ger. Bav. dial.* obesen *porch of a church : M. H. Ger.* obese, *f.* vestĭbŭlum *: O. H. Ger.* opasa *atrium,* vestĭbŭlum *: Goth.* ubizwa, *f. a hall, porch : Icel.* ups, *f.* eaves.]

efes-drypa, an ; *m. Eaves-drip ;* stillĭcidium. v. yfes-drypa, -dropa.

efesian, efosian, efsian ; *p.* ode ; *pp.* od [efes *the eaves,* q. v.] *To cut in the form of eaves, to round, to shear ;* in rŏtundum attondēre, tondēre :—Ne gē eów ne efesion ne beard ne sciron *neque in rŏtundum attondēbĭtis cŏmam nec rādētis barbam,* Lev. 19, 27. Ic efesige oððe ic scere scép oððe hors *tondeo ŏves aut ĕquos,* Ælfc. Gr. 26, 6 ; Som. 29, 9. DER. g-efesian, -efsian.

éfest, e ; *f. A hastening ;* festĭnātio. v. ófost.

éfestan, to éfestanne ; *p.* éfeste, *pl.* éfeston ; *impert.* éfest, *pl.* éfestaþ ; *pp.* éfested *To hasten, make haste, be quick ;* propĕrāre, concurrĕre, festĭnāre :—Hwylcum wegum to éfestanne sý to ingange his ríces *quĭbus sit viis ad ingressum regni illīus propĕrandum,* Bd. 2, 2 ; S. 502, 20. He éfeste norþweard *he hastened northward,* Chr. 1016 ; Erl. 154, 10. Hí to ðám dweoligendum lǽcedómum deófolgylde éfeston and scyndon *ad errātĭca idolatriæ medicāmĭna concurrēbant,* Bd. 4, 27 ; S. 604, 7. To gefultumianne me éfest *ad adjŭvandum me festīna,* Ps. Lamb. 69, 2. v. éfstan.

efesung, e ; *f. A polling, rounding, shearing, compassing ;* tonsūra, Som. Ben. Lye.

R

efesung-sceara, an; _f. A pair of scissors_ or _shears;_ forfex, Som. Ben. Lye.

efete, an; _f. An_ EFT, _a newt, lizard;_ lăcerta :—Efete lăcerta vel stīlio, Ælfc. Gl. 24; Som. 60, 18; Wrt. Voc. 24, 22. [_Wyc._ euete _a lizard._] v. áðexe.

efn; _adj. Even, equal;_ æquus, plānus, æquālis :—On efn, _adv. Together;_ simul, ūna, Ps. Th. 116, 1. On efn, _prep. On even ground, by, near, aside with;_ in æquāli, juxta, Beo. Th. 5798; B. 2903. v. efen.

efnan; _p._ ede, de; _pp._ ed; _v. trans._ I. _to throw down, prostrate, level, lay low;_ prosternĕre :—Ic efne to eorþan ealdne ceorl _I throw down the old churl to earth,_ Exon. 107 b; Th. 409, 28; Rä. 28, 8. II. _to perform, execute, labour, achieve;_ patrāre, perpetrāre, facĕre, præstāre :—Ic æ ðíne efne and healde _custōdiam lēgem tuam,_ Ps. Th. 118, 44 : 118, 131, 143. Óþ-ðæt his byre mihte eorlscipe efnan _until his son might achieve a valorous deed,_ Beo. 5237; B. 2622. De ær eorlscipe efndon _who before performed valorous deeds,_ 6006; B. 3007. Hie efndon unrihtdóm _they executed unrighteousness,_ Cd. 181; Th. 227, 7; Dan. 183. Hie ðat efnedon sóna _they performed that soon,_ Elen. Kmbl. 1423; El. 713. Efn elne ðis _perform this boldly,_ Exon. 80 a; Th. 300, 18; Fä. 8. DER. ge-efnan.

efn-behéfe; _adj._ [behéfe _necessary_] _Equally useful_ or _necessary;_ æque ūtilis _vel_ necessārius :—Is ðiós óðru býsen efnbehéfu _this other similitude is equally necessary,_ Bt. Met. Fox 12, 14; Met. 12, 7.

efn-biscop, es; _m. A co-bishop;_ co-episcŏpus :—Mid Laurentio and Justo his efnbiscopum _cum Laurentio et Justo co-episcŏpis,_ Bd. 2, 5; S. 507, 30. v. efen-bisceop.

efne, an; _f? Alum;_ alūmen, styptēria = στυπτηρία :—Efne _alūmen_ vel _stiptūra_ [=_styptēria_], Ælfc. Gl. 41; Som. 63, 126; Wrt. Voc. 31, 12. Efne _alūmen,_ 56; Som. 67, 38; Wrt. Voc. 37, 28.

efne [= efen] _adv. Even, exactly, precisely, just, alike, likewise, just now;_ plāne, æque, omnino, mŏdŏ, jam prīdem :—He wintra hæfde efne hund-seofontig ær him sunu wóce _he had just seventy winters ere a son was born to him,_ Cd. 57; Th. 70, 24; Gen. 1158. De we willaþ ferigan efne to ðam lande _we will convey thee even to the land_ [_to the very land;_ in eandem terram], Andr. Kmbl. 587; An. 294: Bt. Met. Fox 8, 95; Met. 8, 48. On witte weallende byrnþ efne sió gitsung _even the covetousness_ [i.e. _the just-mentioned covetousness_] _burns raging in his mind,_ 8, 91; Met. 8, 46. Gif ic on helle gedó hwyrft ænigne, ðú me æt-byst efne rihte _si descendĕro in infernum, părĭter ades,_ Ps. Th. 138, 6. He hæfde eorþan and up-ródor efne gedæled _he had divided the earth and firmament alike,_ Cd. 146; Th. 182, 16; Exod. 76. [v. efn-gedǽlan.] Ic æ ðíne efnast healde _I keep thy law most exactly,_ Ps. Th. 118, 77. Efne swá _even so, even as_ :—And efne swá he ðec gemétte meahtum gehrodene _and even so he found thee adorned with virtues,_ Exon. 12 b; Th. 21, 5; Cri. 330. Deór efne swá some æfter ðære stefne on ðone stenc faraþ _just so goes the beast after the voice in that odour,_ 96 a; Th. 358, 30; Pa. 53. Lixte se leóma efne swá of heofene scíneþ ródores candel _the beam shone even as from heaven shines the candle of the firmament,_ Beo. Th. 3146; B. 1571. He Hengestes heáp hringum ðenede efne swá swíðe swá he Fresena cyn byldan wolde _he should serve Hengest's band with rings even as abundantly as he would encourage the Frisian race,_ Beo. Th. 2188; B. 1092. He efne swá swíðe hí lufode, ðæt ... _he loved her even so greatly, that_ ... [adeo ut], Bt. Met. Fox 26, 129; Met. 26, 65. v. efen; _adv._

efne; _interj. Lo! behold! truly! indeed;_ en, ecce, certe, prŏfectŏ :—Ðá se tán gehwearf efne ofer ænne ealdgesíða _then indeed went the lot over one of the old comrades,_ Andr. Kmbl. 2209; An. 1106. And efne! ðá ætýwde Moyses and Helias et _ecce apparuērunt Moyses et Elias,_ Mt. Bos. 17, 3. Efne swá biþ gebletsad beorna æghwylc _ecce sic benedicētur hŏmo,_ Ps. Th. 127, 5. Efne me God fultumeþ _ecce Deus adjŭvat me,_ 53, 4 : 54, 7 : 86, 3 : 118, 40 : 138, 3.

efn-eald co-eval :—Efneald _æquævus_ vel _coætāneus,_ Ælfc. Gl. 9; Som. 56, 119; Wrt. Voc. 19, 3. v. efen-eald.

efne-cuman; _v. intrans. To convene;_ convĕnīre :—Efne-cômon to him _conveniēbant ad eum,_ Mk. Rush. War. 1, 45. v. efen-cuman.

efne-esne, es; _m. A fellow-servant;_ conservus :—Efne-esne ðín ic eom [MS. am] _conservus tuus sum,_ Rtl. 70, 41 : Mt. Kmbl. Lind. 18, 33.

efne-nú; _interj. Behold now;_ ecce :—Efnenú ge-eácnode unrihtwísnesse _ecce partŭriit injustĭtia,_ Ps. Lamb. 7, 15.

efnes, -ness, -nyss, e; _f. Evenness, equity, justice;_ æquĭtas :—Efnes syndon dómas ðíne _æquitas sunt jŭdicia tua,_ Ps. Spl. 118, 75 : Ps. Lamb. 118, 144. He dêmþ ymbhwyrft eorþan on efnesse _ipse judĭcābit orbem terræ in æquitāte,_ Ps. Lamb. 9, 9. Dú gelíffæst me on efnesse oððe emnesse ðínre _vivĭfĭcābis me in æquitāte tua,_ 142, 11. Eágan ðíne geseón oððe bewlátiun efnysse oððe rihtwísnesse _ocŭli tui vĭdeant æquĭtātes,_ 16, 2. v. efen-nys.

efne-spédelíc; _adj. Of the same substance, consubstantial;_ consubstantiālis :—On þrým hádum efenspédelícum _in tribus persōnis consubstantiālibus,_ Bd. 4, 17; S. 585, 38. v. efen-spédiglíc.

efn-etan _to eat as much as any one?_ or _to become equal, to equal?_ par esse alĭcui ēdendo? æquāre, æmŭlāri :—Ic mésan mæg meahtelícor and

---

efn-etan ealdum þyrse _I can feast more heartily and eat as much as the old giant,_ Exon. 111 a; Th. 425, 28; Rä. 41, 63.

efn-éðe; _adj. Equally easy;_ æque făcĭlis :—Is efnéðe up and of dúne to feallanne foldan ðisse _it is equally easy for this earth to fall up and down,_ Bt. Met. Fox 20, 333; Met. 20, 167.

efn-gedǽlan; _p._ de; _pp._ ed; _v. trans. To share alike;_ in æquāles partes divīdĕre :—Beámas twegen ðara æghwæðer efngedǽlde heáhþegnunga hāliges gástes _two pillars, of which shared alike the high services of the holy spirit,_ Cd. 146; Th. 183, 22; Exod. 95.

efn-gemæcca, an; _m._ [gemæcca _a companion_] _A fellow-companion, associate, fellow;_ consors :—Ða beóþ hira gelícan and hira efngemæccan on hira gecynde _they are their equals and their fellows in their nature,_ Past. 29; Hat. MS. 38 b, 16.

efn-líc; _adj. Equal;_ æquus. v. efenlíc.

efn-líca, an; _m. An equal;_ æquālis :—Nis nán efnlíca ðín _there is no one thine equal,_ Bt. Met. Fox 20, 38; Met. 20, 19. v. efen-líca.

efn-lícnes, -ness, e; _f. Evenness, equality;_ æquālītas :—Hie healdaþ má geférrǽdenne and efnlícnesse ðonne ealdordóm _they observe companionship and equality more than authority,_ Past. 17, 9; Hat. MS. 24 b, 6.

efn-ling, es; _m. A consort, an equal, a fellow;_ consors, Ps. Spl. T. 44, 9.

efn-mǽre; _adj. Equally great, illustrious, renowned;_ æstĭmātus, æque illustris, conspĭcuus :—He ðone wělegan wǽdlum efn-mǽrne gedéþ _he makes the rich equally great to the poor,_ Bt. Met. Fox 10, 63; Met. 10, 32.

efn-swíþ; _adj. Equally strong;_ æque vălĭdus :—Manigu óðru gesceaft efn-swíðe him _many other creatures equally strong with them,_ Bt. Met. Fox 11, 88; Met. 11, 44.

efn-þeów, es; _m. A fellow-servant;_ conservus :—Dǽm hláforde is to cýðanne, ðæt he ongiete ðæt he is efnþeów his _it is to be made known to the master, that he understand that he is his fellow-servant,_ Past. 29; Hat. MS. 38 b, 18. v. efen-þeówa.

efor, es; _m. A wild boar;_ áper :—Hiene ofslóg án efor _a wild boar slew him,_ Chr. 885; Erl. 82, 34. Sume sceoldan bión eforas _some should be wild boars,_ Bt. 38, 1; Fox 194, 34. v. eofor.

efor-fearn, es; _m. A species of fern, polypody;_ rădiŏlus, polýpŏdium = πολυπόδιον :—Herba rădiŏla ðæt is efor-fearn, Herb. cont. 85; Lchdm. i. 34, 7. Deós wyrt, ðe man rădiŏlum, and óðrum naman efor-fearn, nemneþ, ys gelíc fearne, and heó byþ cenned on stānigum stówum, and on ealdum hús-stedum, and heó hæfþ on æghwylcum leáfe twá endebyrdnyssa fægerra pricena, and ða scínaþ swá gold _this plant, which is named rădiŏlus, and by another name everfern, is like fern, and it is produced in stony places, and in old homesteads, and has on each leaf two rows of beautiful spots, and they shine like gold,_ Herb. 85, 1; Lchdm. i. 188, 10-14 : L. M. 1, 17; Lchdm. ii. 60, 13. v. eofor-fearn.

Eforwíc-ingas _inhabitants of York,_ Chr. 918; Th. 193, 9, col. 1. v. Eoforwíc-ingas.

efosian _to cut in the form of eaves, to round, shear;_ tondēre :—Hine man efosode _eum totondērunt,_ Gen. 41, 14. v. efesian.

éfre _ever, always,_ Chr. 675; Erl. 38, 26. v. ǽfre.

efsian, efsigean _to cut in the form of eaves, to round, shear;_ tondēre :—Man ne môt hine efsian _no one shall shear him,_ Jud. 13, 5 : Past. 18, 7; Hat. MS. 27 b, 11, 24. v. efesian.

éfstan, éfestan; _p._ éfstte, éfste, _pl._ éfston, éfstun; _impert._ éfst, éfste, _pl._ éfstaþ; _pp._ éfsted, éfst; _v. intrans._ [ófest, ófost, ófst _haste_] _To hasten, draw near, approach, make haste, be quick;_ festīnāre, propĕrāre, concurrĕre, appropinquāre, accelĕrāre :—Uton nú éfstan seón wundur _let us now hasten to see the wonders,_ Beo. Th. 6193; B. 3101 : Rood Kmbl. 67; Kr. 34. He éfste [éfstte, Th. 278, 23, col. 2] norþweard _he hastened northward,_ Chr. 1016; Th. 278, 22, col. 1. Abraham éfste in to ðam getelde _festĭnāvit Abraham in tabernācŭlum,_ Gen. 18, 6 : Lk. Bos. 19, 6 : Beo. Th. 2990; B. 1493 : Cd. 139; Th. 174, 2; Gen. 2872. Hí éfston óþ to gatum deáþes _appropinquāvērunt usque ad portas mortis,_ Ps. Spl. 106, 18 : Byrht. Th. 137, 55; By. 206. Hí genealǽhton oððe éfstun [éfston, Ps. Spl. 15, 3] _accelĕrāvērunt,_ Ps. Lamb. 15, 4. Éfst ardlíce ðyder _festīna ĭbi,_ Gen. 19, 22 : Lk. Bos. 19, 5. Éfst [Th. éfste] oððe neálǽce ðæt ðú generige me _accelĕra ut ēruas me,_ Ps. Spl. 30, 2. Éfstaþ and lǽdaþ hine to me _festĭnāte et addūcĭte eum ad me,_ Gen. 45, 13 : Boutr. Scrd. 22, 42 : Homl. Th. ii. 88, 32. Éfstaþ ðæt ge gangon þurh ðæt nearwe geat _hasten that ye go through the narrow gate,_ Lk. Bos. 13, 24. Ðæt we to ðé mid ealre heortan éfston _that we may hasten to thee with all our heart,_ Homl. Th. ii. 600, 3. To ðam ðe hit éfst wæs _ad quam festīnātum erat,_ Prov. 20. DER. ge-éfstan.

eft; _adv. Again, second time, then, afterwards;_ ĭtĕrum, dēnuo, rursus, re-, deinde, ĭtem :—Eft lufigende God _ĭtĕrum ămans Deum,_ Ælfc. Gr. 43; Som. 44, 58. Asende Noe út eft culfran _Noe rursus dīmīsit cŏlumbam,_ Gen. 8, 10 : Mt. Bos. 4, 7, 8 : Ælfc. Gr. 38; Som. 40, 51, 52 : Chr. 790; Erl. 56, 38 : 828; Erl. 64, 10 : 1046; Erl. 170, 17 : 797; Erl. 58, 16 : Chr. 838; Erl. 66, 13. Eft _ĭtem,_ Bd. 4, 8; S. 575, 38 : 5, 5; S. 617, 34. Eft on Cent forbærnde _afterwards burned in Kent,_ Chr. 685;

Erl. 40, 20. [*Piers P.* eft *again:* *Wyc.* eft, efte *again:* *Laym.* æft, afte, eft, efte *afterwards:* *Orm.* efft *afterwards, again:* *O.Sax.* eft *again:* *O.Frs.* eft, efta *behind, afterwards, then:* *Goth.* afta *behind, back.*] v. æft.

eft-agyfan *To give back;* reddĕre, i. e. re-dăre, Bd. 2, 1; S. 500, 19.

eft-betǽht, æft-betǽht *Re-assigned, re-delivered, given back;* re-consignātus, R. Ben. 4. v. be-tǽcan.

eft-cerran *To return;* redīre:—Eftcerdon *reversi sunt,* Lk. Skt. Lind. 10, 17.

eft-cuman *To come back;* revenīre:—He hēt ealle eftcuman *he commands all to come again,* Bt. 39, 13; Fox 234, 25. Eft-cymeþ *comes again,* Bd. 2, 13; S. 516, 21.

eft-cyme, es; *m. A coming again, return;* rĕdītus, reversio:—Ðæt eorlwerod sæt on wēnum eftcymes leófes monnes *the warrior band sat in expectation of the return of the dear man,* Beo. Th. 5785; B. 2896: Exon. 121 b; Th. 466, 33; Hö. 130. Treófugla tuddor tācnum cýðdon eádges eftcyme *the tree-fowls' offspring by signs made known the blessed man's return,* Exon. 43 a; Th. 146, 11; Gū. 708.

eft-eádig; *adj. Rich:*—Efteádig [ēst-, Th: sēft-, Grn.] secg *the favoured mortal,* Exon. 82 a; Th. 309, 12; Seef. 56.

eft-edwītan *To reprove, upbraid again;* re-probāre, Mt. Kmbl. Lind. 21, 42.

eft *after,* Cod. Dipl. 1073; A. D. 896; Kmbl. v. 140, 7; Th. Diplm. A. D. 896; 139, 8. v. æfter.

eft-gecígan, eft-gecígean *To recall, call back;* re-vocāre:—Sende he ðone biscop hí to sóþfæstnysse geleáfan eft-gecígean *he sent the bishop to call them again to the belief of the truth,* Bd. 3, 30; S. 562, 10.

eft-hweorfan *To turn back, return;* rĕ-vertĕre:—Æfter tīde eft-hweorfende to heofonum *after a time returning again to the heavens,* Bd. 4, 3; S. 568, 29. Eft-hwurfon *returned again,* 5, 6; S. 619, 9.

eft-leán, es; *n.* [leán *a reward*] *A recompense;* retrībūtio:—He eft-leán wile ealles gĕnomian *he will surely take a recompense,* Exon. 24 a; Th. 68, 8; Cri. 1100.

eft-lésing, e; *f. Redemption;* redemptio, Mt. Kmbl. Lind. 20, 28.

eft-ongén-bígan *To untwist again, to unwreathe;* re-torquēre:—Eft-ongén-bígde *retorsit,* Cot. 189.

eft-síþ, es; *m. A journey back, return;* rĕdītus:—Ár wæs on ófoste, eftsíðes georn *the messenger was in haste, desirous of return,* Beo. Th. 5560; B. 2783. Landweard onfand eftsīþ eorla *the land-warden perceived the return of the warriors,* Beo. Th. 3786; B. 1891: 2669; B. 1332.

eft-sittan; *p.* -sæt, *pl.* -sǽton; *pp.* -seten *To sit again, reside;* re-sīdĕre:—Ic eftsitte oððe ic uppsitte *resīdeo,* Ælfc. Gr. 26, 5; Som. 29, 6.

eft-sóna; *adv.* [eft *again,* sóna *soon*] EFTSOONS, *soon after, again, a second time;* itĕrum:—He hí lǽrde eftsóna *he taught them again,* Mk. Bos. 10, 1.

eft-spellung, e; *f. A recapitulation;* re-capĭtŭlātio, Cot. 171.

eft-swā-micel *Even so much;* tantundem:—Eft-swā-miceles *for so much, at that price;* tantĭdem, Som. Ben. Lye.

eft-wyrd, e; *f. Future fate, day of judgment;* futūrum fātum, judĭcii dies, Cd. 169; Th. 212, 15; Exod. 539.

eftyr *after;* post, Lye. v. æfter.

efyn-gelíc; *adj.* [efen *even,* gelíc *like*] *Even-like, alike, equal, co-equal;* co-æqualis, Som. Ben. Lye.

efynnis *Evenness, equity;* æqualĭtas, equĭtas, Ps. Spl. C. 110, 7. v. efennys.

efyr *a boar,* Ps. Spl. C. 79, 14. v. eofor.

ég, e; *f. Water, sea;* aqua, măre. Used to denote,—*The sea coast:*—Blecinga ég *Blekingley, the coast of the Blekingians,* Ors. 1, 1; Bos. 22, 1. Scon-ég *Sconey.* v. ég-.

ég-. Used in composition:—*water, sea;* aqua, măre. DER. ég-búende, -clif, -land, -streám. v. íg-.

égan *to fear, dread.* DER. on-égan, q. v.

ég-búende; *pl. m. adj.* Used as a noun, *An island dweller;* ad aquam *vel* in insŭla hăbĭtans:—On ðǽre ealdan byrig Acemannes ceastre; hie égbúendas [MS. egbuend] Baðan nemnaþ *in the old town Akemansceaster* [*the pained man's city*]; *the islanders call it Bath,* Chr. 974; Th. 224, 20, col. 2, 3; Edg. 4. Gehwæm égbúendra *to each of the islanders,* 975; Th. 230, 5; Edg. 57. v. íg-búende.

ég-clif, es; *n. A water-cliff* or *shore;* scŏpŭlus [= σκόπελος *a look-out place*] măris, lĭtus:—Ofer égclif [MS. ecgclif] ðæt eorl-werod sæt *the warrior band sat on the ocean's shore,* Beo. Th. 5778; B. 2893.

EGE, æge, eige, es; *m. Fear, terror, dread,* AWE; tĭmor, terror, formīdo:—Eorþcynincgum se ege standeþ *terrĭbĭli ăpŭd rēges terræ,* Ps. Th. 75, 9. On ðǽm dagum wæs mycel ege fram ðǽm wífmannan *in those days there was a great dread of these women,* Ors. 1, 10; Bos. 33, 26: Bt. Met. Fox 1, 143; Met. 1, 72. Ege Drihtnes *tĭmor Domĭni,* Ps. Spl. 18, 10. Beó eówer ege and óga ofer ealle nītenu *terror vester ac trĕmor sit sŭper cuncta anĭmālia terræ,* Gen. 9, 2. Nis me ege mannes for áhwæðer *non tĭmēbo quid făciat mĭhi hŏmo,* Ps. Th. 55, 4: 117, 6. Wearþ hit swā mycel æge fram ðam here *there was so great*

awe of the army, Chr. 1006; Erl. 140, 31. Gefeallaþ [MS. gefeællæþ] ofer hī eige and fyrhto *fear and dread shall fall upon them,* Cant. Moys. Ex. 15, 19; Thw. 30, 19. Ða Bryttas mid mycclum ege flugon to Lunden-byrig *the Britons fled to London in great terror,* Chr. 456; Erl. 13, 29: 823; Erl. 63, 24. Nā ðū ondrǽdst fram ege nihtlícum *non timēbis a tĭmōre nocturno,* Ps. Spl. 90, 5: Ps. Th. 118, 38: Bd. 5, 13; S. 632, 24. Ðū hæfdest eorþlícne ege *thou hadst earthly awe,* Homl. Th. i. 596, 8: Ors. 3, 9; Bos. 64, 9. Syleþ eallum mete, ðām ðe his ege habbaþ *escam dĕdit timentĭbus se,* Ps. Th. 110, 3: 59, 4. Ðe him Metodes ege, on his dǽdum, Drihten forhtaþ *qui tĭmet Dŏmĭnum,* 127, 5. [*Laym.* eie, eiȝe, eie, æie, *m.* awe, dread, anger: *Orm.* eȝȝe: *M. H. Ger.* ege, *f.: O. H. Ger.* egi, agi, *m. terror: Goth.* agei, *f.: Dan.* ave, *m. f.: Icel.* agi, *m. terror, discipline.*] DER. tíd-ege.

ége; *gen. dat. acc. of* ég *water,* Chr. 47; Th. 11, 6, col. 3. v. ég.

ége; *n. An eye:*—Mid ēgum *with eyes,* Cd. 229; Th. 310, 18; Sat. 728. Gif ðín ége *if thine eye,* Mt. Rush. War. 5, 29. v. eáge; *n.*

egean *To harrow* or *break clods;* occāre, Som. Ben. Lye.

ege-full; *adj. Fearful, terrible;* terrĭbĭlis:—Mǽre God, and mihtig and egefull *Deus magnus, et pŏtens et terrĭbĭlis,* Deut. 10, 17. Hit wæs swíðe egefull *it was very terrible,* Bt. 18, 2; Fox 64, 14. v. eges ful.

ege-healdan *To hold in fear, correct;* corrĭpĕre, Ps. Spl. T. 93, 10.

ege-láf; *f. What had escaped horror;* horrōris resĭduum:—Ege-lāfe [MS. ece-láfe], *acc.* Exod. 370.

ege-leás; *adj. Fearless;* impăvĭdus, Past. 36, 1, Lye.

egeleás-líce; *adv. Fearlessly;* impăvĭdē:—Hie nū egeleás-lícor and unnytlícor brúcaþ ðære mildheortlícan Godes giefe *they now enjoy the merciful gifts of God the more fearlessly and uselessly,* Past. 36, 1; Hat. MS. 46 b, 9.

Egeles ford, es; *m. Ailsford:*—Eádríc gewende ðone cyning ongeán æt Egeles forda *Eadric went to meet the king at Ailsford,* Chr. 1016; Th. 282, 10, col. 1. v. Ægeles ford.

egen *fear;* tĭmor, Wanl. Catal. p. 14, line 7, note z. DER. ege.

egenu *a little round heap;* glŏmŭlus, Som. Ben. Lye.

egenwirht *Hire, wages, a gift;* merces, Ps. Spl. T. 126, 4.

ege-nys, eges ful-nes, -ness, e; *f. Fearfulness, fear;* tĭmor, Ps. Spl. T. 88, 39.

egesa, egsa, ægsa, an; *m.* [ege *fear*] *Fear, horror, dread;* tĭmor, horror, terror, formīdo:—Him gásta weardes egesa on breóstum wunode *fear of the guardian of spirits dwelt in his breast,* Cd. 138; Th. 173, 24; Gen. 2866: Beo. Th. 1572; B. 784: Andr. Kmbl. 789; An. 445: Rood Kmbl. 170; Kr. 86: Judth. 12; Thw. 25, 10; Jud. 252. Būtan Godes egsan [MS. B. egesan] *without fear of God,* Bd. 4, 12; S. 581, 1: Cd. 178; Th. 223, 23; Dan. 124: Andr. Kmbl. 914; An. 457. Sió dimme niht ofer eldum egesan ne brohte *the dim night did not bring terror over men,* Bt. Met. Fox 12, 34; Met. 12, 17: Cd. 202; Th. 250, 3; Dan. 541: Ps. Th. 66, 6. Egesan geaclod *terrified with fear,* Andr. Kmbl. 1609; An. 806: Beo. Th. 5465; B. 2736. [*O. Sax.* egiso, *m: M. H. Ger.* egese, eise, *f. horror: O. H. Ger.* ekiso, *m;* egis, agis, *n. horror: Goth.* agis, *n. fear, terror, horror.*] DER. bǽl-egsa, blōd-egesa, flōd-, folc-, glēd-, hild-, líg-, niht-, þeód-, wæter-.

égessa, ēgsa, an; *m.* [ēkso; *m. possessor: O. Sax. Heli.* āgan *to own*] *An owner; possessor:*—Égesan ne gýmeþ *heeds not the owner,* Beo. Th. 3519; B. 1757.

eges ful, ege-ful, -full; *adj.* [eges ful *full of fear* =] *Fearful, terrible, wonderful;* tĭmōre plēnus, terrĭbĭlis, admīrābĭlis:—Ðū [God] eart egesful *tu* [*Deus*] *terrĭbĭlis es,* Ps. Lamb. 75, 8: Cd. 177; Th. 222, 17; Dan. 106: Exon. 30 a; Th. 93, 20; Cri. 1529. Bera sceal on hǽþe, eald and egesfull *the bear shall be on the heath, old and terrible,* Menol. Fox 519; Gn. C. 30: Beo. Th. 5850; B. 2929. Drihten ys mǽre God and mihtig and egefull *Dŏmĭnus est Deus magnus et pŏtens et terrĭbĭlis,* Deut. 10, 17: Bt. 18, 2; Fox 64, 14. Eálá Drihten, lā hū egesful oððe hū wundorlíc is ðín nama *Dŏmĭne, quam admīrābĭle est nōmen tuum!* Ps. Lamb. 8, 2, 10.

eges fullíc; *adj. Full of fear, fearful, awful;* terrĭbĭlis:—Hū egesfullíc he is in geþeahtingum ofer monna bearn *quam terrĭbĭlis est in consĭliis sŭper fĭlios homĭnum,* Bd. 4, 25; S. 601, 36. Egesfullícran, *nom. pl. more full of terror,* Salm. Kmbl. 93; Sal. 46.

eges ful-nes, -ness, e; *f. Fulness of fear, formidableness;* formīdŏlōsĭtas:—Eges fulnes, L. I. P. 3; Th. ii. 306, 21. v. egenys [= ege, -nys, -nes.]

eges grime, grimme, an; *f. A witch, sorceress;* vĕnēfĭca, malēfĭca, Som. Ben. Lye.

egesian; *p.* ode; *pp.* od *To affright;* terrēre, Som. Ben. Lye. v. egsian.

egesig *terrible, horrible.* v. eiseg.

eges líc; *def. se* eges líca, seó, ðæt eges líce; *adj.* [eges líc *a likeness of fear* =] *Fearful, terrible, dreadful, terrific, horrible, awful;* terrĭbĭlis, terrĭficus, horrĭbĭlis, horrendus:—Eorþscræf egeslíc *a fearful cavern,* Andr. Kmbl. 3174; An. 1590. Egeslíc æled eágsýne wearþ *the terrible fire was visible to the eye,* 3098; An. 1552: Rood Kmbl. 148; Kr. 74. Eálá hū egeslíc ðeós stów ys *quam terrĭbĭlis est lŏcus iste!* Gen. 28, 17.

He is egeslíc God, ofer ealle godu eorþbûendra *Domĭnus terribĭlis est sŭper omnes deos*, Ps. Th. 95, 4 : 88, 6 : Ps. Spl. 46, 2. Wæs đær swíđe egeslíc geatweard *there was a very horrible gatekeeper*, Bt. 35, 6 ; Fox 168, 18. Đæs egeslícan đæt đû dô feóndes aîđlian awyrgeđe syrwunga *horrendi fácias hostis văcuisse (?) malignas insĭdias*, Hymn. Surt. 47, 24. Egeslícne cwide sigora Weard ofer đæt fǽge folc forþ forlǽteþ *the Lord of victories shall send forth a dreadful utterance over the fated folk*, Exon. 30 a ; Th. 92, 30 ; Cri. 1516. Fâ þrôwiaþ ealdor-bealu egeslíc *the hostile shall suffer terrific vital evil*, 31 b ; Th. 98, 31 ; Cri. 1616. Đæt he monig þing ge egeslíce ge willsumlíce geseah *that he saw many things both awful and delightful*, Bd. 5, 12 ; S. 627, 29. Se đe worhte egeslícu on sǽ đære reádan *qui fecit terribĭlia in mâri rubro*, Ps. Lamb. 105, 22. Wæs heora sum đâm ôđrum egeslícra *one of them was more dreadful than the others*, Bd. 5, 13 ; S. 633, 3. Daga egeslícast *most terrible of days*, Exon. 23 a ; Th. 63, 20 ; Cri. 1022.

eges líce; *adv.* [eges líce *in likeness of fear* =] *Fearfully*; terrĭbĭlĭter :—Hî nâht ne belimpaþ to đam þunere đe on đyssere lyfte oft egeslíce brastlaþ *they do not appertain to the thunder which in this atmosphere often crackles fearfully*, Bd. de nat. rerum ; Wrt. popl. science 19, 26 ; Lchdm. iii. 280, 13. Worpaþ hine deófol on dômdæge egeslíce *the devil shall fearfully cast him down in the day of doom*, Salm. Kmbl. 52 ; Sal. 26.

egesung, e ; *f. A threatening, fear, dread*; commĭnātio, R. Ben. interl. 27, Som. Ben. Lye. v. egsung.

egeđe *a rake, harrow*; rastrum, Som. Ben. Lye.

egeđere, es ; *m. A raker*; occātor, Som. Ben. Lye.

eggian; *p.* ode; *pp.* od *To* EGG, *excite*; excĭtāre, Ben. Lye.

égh-þyrl, es ; *n. An eye hole, a window*; fenestra :—Đæs leóhtes scíma þurh đa cýnan đære dúra and þurh đa éghþyrla ineóđe *the glare of the light entered through the chinks of the door and through the windows*, Bd. 4, 7 ; S. 575, 20. v. eág-þyrl.

ég-hwelc *all, every*, Jn. Rush. War. 8, 34. v. ǽg-hwilc.

Egipte, Egypte ; *gen.* a ; *dat.* um ; *pl. m. The Egyptians, the people of Egypt in a body, Egypt*; Ægyptii :—Đæt Egipte ne forwurþon *that the Egyptians perish not*, Gen. 41, 36. Egipta land, Egypta land *the land of the Egyptians, Egypt*, Gen. 12, 10, 11, 14, 20 : 13, 10 : 21, 21 : 37, 25, 28, 36 : 39, 1. Egipta cyng, Egypta cyng *the king of Egypt*, Gen. 40, 1 : Ex. 3, 18, 19 : 5, 4. Egypta ealdor *a prince of the Egyptians*, Gen. 42, 6. Egipta here *the host of the Egyptians*, Deut. 11, 4. Fôron Iosepes tyn gebrôđru to Egiptum *Joseph's ten brothers went to Egypt*, Gen. 42, 3 : 45, 9. Hunger fornam swíđust Egipte *famine oppressed the Egyptians most*, Gen. 47, 13.

Egiptisc, Egyptisc ; *def.* se Egiptisca, Egiptiscea ; seó, đæt Egiptisce ; *adj. Belonging to Egypt, Egyptian*; Ægyptius :—Hér is ides Egyptisc *here is an Egyptian woman*, Cd. 101 ; Th. 134, 19 ; Gen. 2227. Fram đære Egiptiscan eá *from the Egyptian river*, Gen. 15, 18. Hine gebohte Egiptisc man *an Egyptian man bought him*, 39, 1 : Ex. 2, 11, 19. Đisra Egiptiscra manna *of these Egyptian men*, Gen. 50, 11. Se Egiptiscea cyng *the Egyptian king*, Ex. 1, 17. Befôran đam Egiptiscean folce *before the Egyptian people*, 3, 22. Þurh Egiptisce galdru *through Egyptian enchantments*, 7, 11. Đæt Egiptisce folc *the Egyptian people*, 11, 7. Đa Egyptiscan *the Egyptians*, Ex. 14, 18, 31. Iosep sealde hwǽte đâm Egiptiscan mannum *Joseph sold corn to the Egyptian men*, Gen. 41, 56.

egiđe *a rake*, Som. Ben. Lye. v. egeđe.

EGL, e ; *f. A mote*; festūca :—Hwí gesihst đú đa egle on đínes brôđor eágan *quid vides festūcam in ocŭlo fratris tui?* Lk. Bos. 6, 41, 42. [*Ger.* egel, achel, *f.* festūca, arista.]

eglan *to trouble*, Judth. 11 ; Thw. 24, 12 ; Jud. 185. v. eglian.

ég-land, ég-lond, es ; *n. Water-land, an island*; insŭla :—We witan ôđer égland *we know another island*, Chr. Erl. 3, 10. Geond đis égland *throughout this island*, Chr. 641 ; Erl. 27, 11. In đæt églond *on the island*, Exon. 96 b ; Th. 361, 7 ; Wal. 16. Églond monig *many an island*, 89 a ; Th. 334, 12 ; Gn. Ex. 15 : 100 b ; Th. 380, 8 ; Rä. 1, 5 : Bt. Met. Fox 1, 31 ; Met. 1, 16. v. íg-land.

EGLE; *adj. Troublesome, hateful, loathsome, horrid*; mŏlestus, odiōsus, infestus, turpis :—He him sylfum byþ egle *he is loathsome to himself*, Basil admn. 8 ; Norm. 50, 24 : Cd. 209 ; Th. 258, 21 ; Dan. 679. Gif egle wǽron *if they were troublesome*, Exon. 126 a ; Th. 485, 20 ; Rä. 71, 16. Đý-læs sceađan mihton egle ondsacan *lest the horrid apostates might injure* [*him*], Andr. Kmbl. 2297 ; An. 1150 : 2916 ; An. 1461. Eglum ättor-sperum *with horrid venomed spears*, Exon. 105 a ; Th. 399, 10 ; Rä. 18, 9. [*Goth.* agls *shameful, disgraceful*; aglus *difficult, troublesome*.]

eglian, eglan, ĕlan ; hit egleþ, eleþ ; *p.* ode, ade ; *pp.* od, ad ; *v. trans.* chiefly used impersonally with dat. of person. *To trouble, pain, grieve*, AIL ; molestare, dŏlēre :—Đæt he us eglan môste *that he could trouble us*, Judth. 11 ; Thw. 24, 12 ; Jud. 185. Me egleþ [eleþ, MS. H.] swýđe it *grieves me much*, L. Edm. S. proœm ; Th. i. 246, 22. Him næfre syđđan seó ádl ne eglode *the illness never ailed him afterwards*, Guthl. 12 ; Gdwin. 60, 8 : 13 ; Gdwin. 60, 19. Đæt him stranglíce eglade *it afflicted him severely*, Chr. 1086 ; Erl. 220, 33. Gif men innan wyrmas eglen [eglien MS. B.] *if worms trouble a man within*, Herb. 2, 10 ;

Lchdm. i. 82, 22. [*Piers P. Chauc.* eylen, eilen *to ail: Orm.* eʒʒlenn : *Plat.* echeln, öcheln *to be vexed, grieved at anything: Ger.* ekeln : *Goth.* aglyan *to molest*, in us-aglyan.] DER. æt-eglan, ge-.

Egones hám, Egnes hám, es ; *m.* [*Ethelw.* Egnesham: *Flor.* Eignesham : *Hunt.* Aeguesham : *Gerv.* Egenesham] ENSHAM or EYNSHAM, *Oxfordshire*; lôci nômen in agro Oxoniensi :—Hér Cûþwulf feaht wiđ Bretwalas and genom Egones hám *in this year* [A.D. 571] *Cuthwulf fought against the Britons and took Eynsham*, Chr. 571 ; Erl. 18, 14. Into Egnes hám *at Eynsham*, Cod. Dipl. 714 ; A.D. 1005 ; Kmbl. iii. 344, 16.

egor *nine ounces or inches, a span*; dodrans, Cot. 64, Som. Ben. Lye.

egor- *water, the sea*; aqua, mǎre. [*Icel.* ægir, *m.*] DER. egor-here, -stream.

égor-here, es ; *m. The water-host, the deluge*; undārum exercĭtus, dĭlŭvium :—Se égorhere eorþan tuddor eall acwealde *the water-host destroyed all the earth's progeny*, Cd. 69 ; Th. 84, 23 ; Gen. 1402 : 75 ; Th. 92, 31 ; Gen. 1537.

égor-stream, eágor-stream, es ; *m. A water-stream, water, the sea*; unda, flŭvius, mǎre :—Điós eorþe mæg and égorstream cræfta nâne adwæscan đæt đæt him on innan sticaþ *this earth and sea can by no means extinguish that which in them remains*, Bt. Met. Fox 20, 236 ; Met. 20, 118. Égorstreámas swôgan *the water-streams sounded*, Cd. 69 ; Th. 83, 4 ; Gen. 1374.

egsa, ægsa, an ; *m. Fear, horror, dread*; timor, horror, terror :—Egsa com ofer me *timor vēnit sŭper me*, Ps. Spl. 54, 5 : Exon. 20 a ; Th. 52, 26 ; Cri. 839 : Cd. 221 ; Th. 288, 12 ; Sat. 379. Beóþ egsan of heofene *ĕrunt terrōres de cœlo*, Lk. Bos. 21, 11 : Cd. 148 ; Th. 186, 10 ; Exod. 136. v. egesa.

egsa, an ; *m. An owner; possessor* :—Égsan wyn *the owner's pleasure*, Exon. 90 b ; Th. 340, 7 ; Gn. Ex. 107. v. égesa.

egsian; *p.* ode; *pp.* od [egsa *fear*] *To frighten*; terrēre :—Oft Scyld egsode eorl *Scyld often frightened man*, Beo. Th. 11 ; B. 6. DER. ge-egsian.

ég-stream, éh-stream, es ; *m. A water-stream, a river, the sea*; aquæ fluctus, flūmen, mǎre :—Hæfde Metod égstreám eft gecyrred *the just Creator had averted the stream*, Cd. 71 ; Th. 85, 15 ; Gen. 1415. Here wícode égstreáme neáh *the host encamped near the river*, Elen. Kmbl. 132 ; El. 66 : Beo. Th. 1158 ; B. 577. v. eá-stream.

egsung, e ; *f.* [egsa *fear*] *A terrible act, frightening, threatening*; terrĭbĭle, commĭnātio :—Strencþe egsunga ođđe egesfulra þinga đínra hî cweđaþ *virtūtem terribĭlium tuōrum dĭcent*, Ps. Lamb. 144, 6. Mid egsunge *by threatening*, Jud. Thw. 161, 37.

egþa, an ; *m. An instrument to beat out corn*; trĭbŭla, Ælfc. Gl. 2 ; Som. 55, 52 ; Wrt. Voc. 16, 25.

egþe *a rake*, Som. Ben. Lye. v. egeđe.

égđer *either* :—Égđer ge—ge botht—and, Gen. 4, 22. v. ægđer.

ég-þyrl *a window* :—Þurh đæs húses égþyrl *through the window of the house*, Jos. 2, 15. v. eág-þyrl.

égum *with eyes*, Cd. 229 ; Th. 310, 18 ; Sat. 728 ; *dat. pl.* of ēge = eáge ; n. q. v.

é-gylt *a fault*, Ps. Spl. T. 31, 5. v. ǽ-gylt.

Egypte; *pl. m. The Egyptians*, Ors. 1, 7 ; Bos. 30, 21. v. Egipte.

Egyptisc *Egyptian*, Ex. 6, 5. v. Egiptisc.

egys *full fearful*, Ps. Spl. C. 46, 2. v. eges ful.

eh, es ; *n.*   I. *a war-horse, charger*; equus bellātor :—Đa đa hors ôþbær, eh and eorlas *which bore away the horses, the chargers, and chiefs*, Exon. 106 a ; Th. 404, 21 ; Rä. 23, 11.   II. *the Anglo-Saxon Rune* M *= e, the name of which letter in Anglo-Saxon is eh a war-horse,—hence, this Rune not only stands for the letter e, but for eh a war-horse, charger, as,—* M *[eh] byþ for eorlum the war-horse is for chiefs*, Hick. Thes. i. 135, 37 ; Runic pm. 19 ; Kmbl. 343, 3. v. eoh.

éh- *water*, used in composition. v. íg.

ehennys, -nyss, e ; *f. Modesty*; pŭdor, Som. Ben. Lye.

eher *an ear of corn*, Mk. Lind. War. 4, 28. v. ear.

eh-heólođe, an ; *f. The plant elecampane* or *horseheal*; inŭla hĕlēnium, Lin, L. M. 1, 32 ; Lchdm. ii. 76, 20.

éhst *highest*, Ps. Spl. 49, 15, = heáhst, hêhst ; *superl.* of heáh.

éh-stream, es ; *m. A water-stream, ocean* :—Heliseus éhstreám sôhte, leólc ofer lagu-flôd *Heliseus sought the ocean, bounded over the water-flood*, Exon. 75 b ; Th. 283, 1 ; Jul. 673. v. ég-stream, eá-stream.

ehsýne *a face, countenance*; fǎcies, Som. Ben. Lye. v. an-sýn.

eht *value, estimation* :—Be đæs demmes ehte *pro damni æstimātiōne*, Ex. 22, 5. v. eaht.

ehta *eight* :—Ehta dagas gefylleđe wǽron *consummāti sunt dies octo*, Lk. Bos. 2, 21. v. eahta.

éhtan; he éht, *pl.* éhtaþ; *p.* éhte, *pl.* éhton; *pp.* ehted *To follow after, chase, pursue, persecute, annoy, afflict*; persĕqui, trĭbŭlāre, afflī-gĕre,—*followed by gen.* or *acc* :—Ne éht he nânre wuhte *he pursues not anything*, Bt. Fox 258, 3. Húndas míne wildeór éhton *cǎnes mei fěras persĕquebantur*, Coll. Monast. Th. 21, 15. Đonne hî eów éhtaþ on đysse byrig *cum persĕquentur vos in civĭtāte ista*, Mt. Bos. 10, 23 ;

5, 11: Ælfc. Gr. 29; Som. 33, 48: 25; Som. 26, 63. Ðara ðe mîn ëhtaþ *tribŭlantium me*, Ps. Th. 26, 14: 118, 157. Ðū us ahreddest æt ðām ðe ûre ëhton *libĕrasti nos ex affligentĭbus nos*, Ps. Th. 43, 9. v. eáhtan, ôht.

ehta-tyne; *adj. Eighteen*; octō-dĕcim:—Wēne gē ðæt ða ehtatyne wæron scyldige *pŭtātis quia illi dĕcem et octo dēbĭtōres fuĕrint?* Lk. Bos. 13, 4, 16. v. eahta-tyne.

ëhtend, es; *m. A persecutor*; persĕcūtor:—Domicianus wearþ ëhtend cristenra manna *Domitian was a persecutor of christian men*, Ors. 6, 9; Bos. 120, 18. He dreág ëhtendra nîþ *he endured the persecutors' malice*, Exon. 40 a; Th. 133, 28; Gū. 496. Ic his ëhtendas ealle geflýme *I will put all his persecutors to flight*, Ps. Th. 88, 20.

ehteoða, ehteða *eighth*:—On ðam ehteoðan dæge *on the eighth day*, Lk. Bos. 1, 59. v. eahtoða.

ëhtere, ëhtre, es; *m. A persecutor*; persĕcūtor:—Ëhtere *persĕcūtor*, Wrt. Voc. 74, 44. Of ëhtere *ex persecūtōre*, Bd. 1, 7; S. 478, 19. Sanctus Albanus cýðde ðām ëhterum Godes geleáfan ðæt he cristen wære *Saint Alban told the persecutors of God's truth that he was a christian*, 1, 7; S. 477, 22. Gebiddaþ for eówre ëhteras *pray for your persecutors*, Mt. Bos. 5, 44: Bd. 1, 7; S. 476, 37.

ëh-þyrl *eye-hole, a window*:—Ðū wircst ëhþirl *thou makest a window*, Gen. 6, 16: Jos. 2, 18. v. eág-þyrl.

ehtian *to esteem, deem, value*; æstĭmāre:—Swā monnum riht is to ehtienne *quantum homĭnĭbus æstĭmāre fas est*, Bd. 5, 6; S. 618, 30; Beo. Th. 2449; B. 1222: Cd. 193; Th. 241, 25; Dan. 410. v. eahtian.

ëhting, e; *f. Persecution*; persĕcūtio, Ors. 6, 23; Bos. 124, 11, notes, p. 28. 1.

ëhtnes, ëhtnys, -ness, -nyss, e; *f. Persecution*; persĕcūtio:—Seó ëhtnes ðara cristenra manna *the persecution of christian men*, Ors. 6, 23; Bos. 124, 11: Ps. Th. 118, 139. Fram Iudēa ëhtnesse *from the persecution of the Jews*, Ps. Th. arg. 17: Mt. Bos. 13, 21: Chr. 2; Erl. 4, 30. Eádige synd ða ðe ëhtnysse þoliaþ for rihtwîsnysse *beati qui persecūtiŏnem patiuntur propter justĭtiam*, Mt. Bos. 5, 10: Bd. 1, 6; S. 476, 22: 1, 8; S. 479, 19, 21. v. eáhtnes.

ehtoða, ehtuða *eighth*, Ex. 22, 30. v. eahtoða.

ëhtre *a persecutor*, Bd. 1, 7; S. 476, 37, MS. B. v. ëhtere.

ehtung, e; *f. Deliberation, council*; delĭbĕrātio:—Ehtunga ealle hæfdon *cogĭtāvērunt*, Ps. Th. 82, 3. v. eahtung.

eíg, e; *f. An island*; insŭla:—Wið eíge *near the island*, Chr. 878; Th. 148, 29, col. 1. v. îg.

eíge *fear*, Cant. Moys. Ex. 15, 19; Thw. 30, 19. v. ege.

eíg-land, es; *n.* [eá *water* = eíg, land *land*] *Water-land, an island*; insŭla:—On ðis eíglande *in this island*, Chr. 937; Erl. 115, 15. v. îg, eá, îg-land = eá-land.

eiseg; *adj.* [= egseg, egeseg, from egesa *fear*] *Terrible, horrible*; terrĭbĭlis:—Cleopaþ ðonne se alda ût of helle, wriceþ word-cwedas wēregan reorde, eisegan stefne *then the chief calleth out of hell, uttereth words with accursed speech, with horrible voice*, Cd. 213; Th. 267, 6–10; Sat. 34–36.

el-, ele- *foreign, strange*; peregrīnus, externus. v. el-land, ele-land.

él *an eel*, Som. Ben. Lye. DER. Él-îg. v. æl.

-el, -ol, a termination denoting persons, as, Fôrrîdel *an outrider*; býdel *a herald*. It denotes also inanimate objects; as, Gyrdel *a girdle*; stýpel *a steeple*.

elan *to trouble, pain, grieve*, L. Edm. S. proœm; Th. i. 246, 22, note 33. v. eglian.

élas *hedgehogs*, Ps. Spl. T. 103, 19. v. îl.

el-boga, ele-boga, eln-boga, an; *m. An* ELBOW; cŭbĭtum, ulna:—Elboga *cŭbĭtum*, Ælfc. Gl. 72; Som. 70, 125; Wrt. Voc. 43, 50. Fæðm betwux elbogan [MS. elboga] and hand-wyrste *a cubit, between the elbow and wrist*; cŭbĭtum, 72; Som. 70, 126; Wrt. Voc. 43, 51. [*Chauc. R. Glouc.* elbowe: *Plat.* ellbagen, m: *Dut.* elleboog. m: *Ger.* elbogen, elnbogn, ellenbogen, m: *M.H.Ger.* ellenboge, elenboge, m: *O.H.Ger.* elinbogo: *Dan.* albue, m. f: *Icel.* albogi, alnbogi, olbogi, ölbogi, ölnbogi, m. *elbow*.]

élc *each*, Mk. Skt. Rush. 16, 15. v. ǽlc.

elch, es; *m. The* ELK; alces, cervus alces, Lin. Som. Ben. Lye. [*Dut.* ellend, elland, eland, m. *tragĕlaphus, hircocervus, anĭmal septentrionālis regiōnis*: *Ger.* elk, m; commonly elen, elend, n. m; elen-thier, n: *M.H.Ger.* ëlch, ëlhe, m: *O.H.Ger.* elaho, eliho, elho, elocho, elch, m: *Dan.* els-dyr, n: *Swed.* elg, m: *Icel.* elgr, m: *Lat.* alces, f: *Grk.* ἄλκη, f. *an elk*.]

elcian; *part.* elcigende; *p.* ode; *pp.* od; *v. n. To put off, delay*; mŏrāri, differre, cunctāri, tempus trăhĕre:—Ic latige on sumere stôwe, oððe ic elcige *mŏror*, Ælfc. Gr. 25; Som. 27, 14. Ðæt he leng ne elcode to his geleáfan *that he no longer delayed his belief*, Homl. Th. ii. 26, 1. v. latian.

elcor, elcur, ælcor; *adv. Elsewhere, otherwise, besides, except*; alias, alĭter, præter, nisi:—Gif hit hwæt elcor biþ *sin alias*, Bd. 4, 28; S. 605, 17. [*O. Frs.* ekker, elker, elkes *alias*: *O. Sax.* elcor *alias*:

O.H.Ger. elichor, elicor, elihor *prorsus, ultra, amplius*: *Dan.* ellers: *Swed.* eljest *else*: *Icel.* ellegar, elligar, ellar, ella *alias*.]

elcra; *comp. adj.* [elcian *to put off, delay*; elcung *lateness, delay*] *Latter*; postērior:—Gif hie cumaþ of ôðrum biterum and yfelum wǽtum, ða ðe wyrceaþ ôman, ðonne beóþ ða elcran to stillanne ôþ-ðæt ðe hie unstrangran weorþan *if they come from other bitter and evil humours, which cause inflammations, then are the latter to be stilled until they become less strong*, L. M. 2, 1; Lchdm. ii. 178, 12–15.

elcung, e; *f. A delay*, R. Ben. 5, 71, Lye. v. eldung.

elcur; *adv. Otherwise*; alias, Mt. Rush. Stv. 6, 1: 9, 17: Lk. Lind. War. 5, 37. v. elcor.

eld *age*:—Sió forme eld *the first age*, Bt. Met. Fox 8, 7; Met. 8, 4: Bt. 15; Fox 48, 2. v. yldu.

elde *to tarry*, Som. Ben. Lye. v. yldan.

eldcung *delay*, Bt. 38, 3; Fox 202, 17, MS. Cot. v. eldung.

elde *men*, Elen. Kmbl. 949; El. 476: Beo. Th. 5215; B. 2611: Andr. Kmbl. 2115; An. 1059: Bt. Met. Fox 20, 199; Met. 20, 100. v. ylde.

eldendlíc; *adj. Slow, slack*; pĭger, tardus, Som. Ben. Lye.

eldo *old age*, Beo. Th. 4229; B. 2111. v. yldu.

eldor *a prince*, Som. Ben. Lye. v. ealdor.

eldor-man *an alderman*, Th. Diplm. A. D. 883; 129, 25. v. ealdor-man.

eldra *elder, older*; sĕnior:—Heora eldran fæder *of their older father, of their grandfather*, Bt. 10; Fox 28, 32, MS. Cot; *comp. of* eald.

eldran *elders, parents*, Bt. Met. Fox 1, 115; Met. 1, 58: 13, 55; Met. 13, 28. v. yldra.

eldre; *comp? omne*:—Ne dyde he âhwǽr swā eldran cynne *non fēcit tālĭter omni nātiōni*, Ps. Th. 147, 9.

eldung, eldcung, elcung, e; *f. Delay*; mŏra:—Hit is eldung and anbîd ðæs hëhstan dēman *it is the delay and waiting of the highest judge*, Bt. 38, 3; Fox 202, 17. v. ylding.

ELE, es; *m.* OIL; ŏleum:—Eles gecynd is ðæt he wile oferstîgan ǽlcne wǽtan: ageót ele uppon wæter oððe on ôðrum wǽtan, se ele flýt bufon: ageót wæter uppon ðone ele, and se ele abrecþ up and swimþ bufon *it is the nature of oil that it will rise above every fluid: pour oil upon water or on another fluid, the oil will float above: pour water upon the oil, and the oil will break through and swim above*, Homl. Th. ii. 564, 11–14. Ele *ŏleum*, Ælfc. Gl. 32; Som. 61, 109; Wrt. Voc. 27, 38: Ps. Lamb. 108, 18: 140, 5. Hund sestra eles *centum cădos ŏlei*, Lk. Bos. 16, 6: Ps. Lamb. 4, 8. Of eówrum ele *de ŏleo vestro*, Mt. Bos. 25, 8. Mid mînum hālgan ele *ŏleo sancto meo*, Ps. Lamb. 88, 21. Ðū amǽstest oððe ðū gefætnodest on ele heáfod mîn *impinguasti in ŏleo căput meum*, Ps. Lamb. 22, 5: 103, 15. Genexode synt his sprǽcu ofer ele *mollīti sunt sermōnes ejus sŭper ŏleum*, 54, 22. Hî ne nāmon nānne ele mid hym *non sumpsērunt ŏleum secum*, Mt. Bos. 25, 3, 4, 9: Lk. Bos. 10, 34: Gen. 28, 18: Lev. 2, 1, 6. Ðū nymst ânne holne hlâf mid ele gesprengedne *tolles tortam pānis unius crustŭlam conspersam ŏleo*, Ex. 29, 23: Lev. 2, 4: Ps. Lamb. 44, 8: Lk. Bos. 7, 46. Eles drosna *dregs of oil*; amurca = ἀμόργη, Ælfc. Gl. 47; Som. 65, 18; Wrt. Voc. 33, 18. [*Wyc. Chauc.* oile: *Orm.* ele: *Scot.* olye: *Plat.* oelje: *O. Sax.* olig, n: *Frs.* oalje: *O. Frs.* olie: *Dut.* olie, f: *Ger.* öl, n: *M. H. Ger.* ol, öl, n: *O. H. Ger.* olei, n: *Goth.* alew, n: *Dan.* olie, m. f: *Swed.* olja, f: *Icel.* olea, olía, f: *Lat.* oleum, n: *Grk.* ἔλαιον, n. *olive oil*; ἐλαία, f. *olive-fruit*.] DER. wyn-ele.

éle *a lamprey*, Som. Ben. Lye. v. ǽl.

ele-bacen; *pp. Oil-baked, baked in* or *with oil*; ŏleo coctus, ŏleātus:—Manna hîg gadredon and grundon on cwyrne oððe britton and sudon on croccan and worhton hlâfas ðærof: ða wǽron hîg swilce hîg wǽron elebacene *pŏpŭlus colligens Man frangēbat mŏla sive tĕrēbat in mortārio, cŏquens in olla et făciens ex eo tortŭlas săpŏris quăsi pānis ŏleāti*, Num. 11, 8.

ele-beám, es; *m.* [ele *oil*, beám *a tree*] *An olive-tree*; ŏlea, ŏliva:—Elebeám *ŏlea vel ŏliva*, Ælfc. Gl. 32; Som. 61, 111; Wrt. Voc. 27, 40: 47; Som. 65, 18; Wrt. Voc. 33, 17: 80, 25: 285, 70: Ps. Lamb. 51, 10. Heó brohte elebeámes twig *she brought a twig of olive-tree*, Cd. 72; Th. 88, 30; Gen. 1473. Heó brohte ân twig of ânum elebeáme *illa portāvit rāmum ŏlivæ*, Gen. 8, 11. Ealle eówre elebeámas forwurþaþ *all your olive-trees shall perish*, Deut. 28, 40: Ps. Th. 127, 4. Syndon bearn ðîne swā swā nywlîcra elebergena oððe guógaþ elebeáma *sunt filii tui sîcut novellæ ŏlivārum*, Ps. Lamb. 127, 3. Dô swā on ðînum wîn-earde and on ðînum elebeámon *ĭta făcies in vinea et in ŏlivēto tuo*, Ex. 23, 11. He eów sylþ elebeámas ðe gē ne plantudon *dĕdĕrit tibi ŏlivēta quæ non plantasti*, Deut. 6, 11. Unwæstmbǽre elebeám *an unfruitful or wild olive-tree*; ŏleaster, Ælfc. Gl. 47; Som. 65, 19; Wrt. Voc. 33, 19: Ælfc. Gr. 8; Som. 7, 15.

ele-beámen; *adj. Of or belonging to the olive-tree*, ŏleăgĭnus, Ælfc. Gl. 32; Som. 61, 110; Wrt. Voc. 27, 39.

ele-berge, an; *f.* [ele *oil*; berge = berie *a berry*] *An olive, the fruit of an olive-tree*; ŏliva:—Swā swā eleberge wæstmbǽra *sîcut ŏliva fructĭfĕra*, Ps. Spl. 51, 8. Syndon bearn ðîne swā swā nywlîcra elebergena oððe guógaþ elebeáma *sunt filii tui sicut nŏvellæ ŏlivārum*, Ps. Lamb. 127, 3.

He ûteóde on ðæne mûnt Oliuarum, ðæt is Elebergena *egressus ibat in montem Ólivārum,* Lk. Bos. 22, 39.

ele-boga, an; *m. An elbow;* ulna, Wrt. Voc. 71, 24. v. el-boga.

ele-byt, -bytt, e; *f.* [ele *oil,* byt *a bottle*] *An oil vessel or cruet, a chrismatory;* lentĭcŭla :—Elebyt ǽrenu *lentĭcŭla,* Cot. 121.

ele-fǽt, es; *n. An oil-vat, cruise or pot;* emĭcādium :—Elefæt *emĭcŭdium* [=*emĭcādium,* v. Du Cange, vol. ii. 238], Ælfc. Gl. 26; Som. 60, 79; Wrt. Voc. 25, 19. Stǽnen elefæt *ălăbastrum,* 24; Som. 60, 40; Wrt. Voc. 24, 40.

elehtre, eluhtre, an; *f. The plant lupine;* lŭpīnus albus, Lin :—Elehtre *lupine,* L. M. 2, 34; Lchdm. ii. 238, 30: 2, 65; Lchdm. ii. 296, 24: 3, 22; Lchdm. ii. 320, 12. Hafa clam geworht of elehtran *have a plaster made of lupine,* L. M. 3, 39; Lchdm. ii. 332, 21. Genim elehtran *take lupine,* Herb. 46, 3; Lchdm. i. 148, 22: L. M. 1, 33; Lchdm. ii. 80, 16: 1, 62; Lchdm. ii. 134, 13: 1, 64; Lchdm. ii. 138, 27: 1, 66; Lchdm. ii. 142, 2: 3, 41; Lchdm. ii. 334, 5: iii. 56, 26.

ele-lǽnde; *adj. Strange, foreign;* peregrīnus :—Elelænde eorþbigennys *cŏlōnia, ṗeregrīnōrum cultūra,* Ælfc. Gl. 54; Som. 66, 102; Wrt. Voc. 36, 25.

ele-land, es; *n. A foreign country;* externa terra :—Ðǽr ic on elelande ǽhte stôwe *there I owned a place in a foreign country,* Ps. Th. 118, 54.

ele-lendisc; *adj. Strange, foreign;* advĕna, aliēnus :—Elelendisc ic eom mid ðe advĕna *ĕgo sum ăpud te,* Ps. Lamb. 38, 13. Bearn elelendisce ferealdodon *fīlii aliēni invetĕrāti sunt,* Ps. Lamb. 17, 46.

elene, an; *f. The herb elecampane;* inŭla hĕlēnium, Lin :—Genim nioðowearde elenan *take the netherward part of elecampane,* L. M. 3, 26; Lchdm. ii. 322, 15: 3, 47; Lchdm. ii. 338, 14. v. eolone.

Elene, an; *f. Helena;* Helĕna =Ἑλένη: *The wife of the Roman emperor Constantius, and mother of Constantine the Great :*—Constantius gesealde his suna ðæt rîce, Constantinuse, ðone he hæfde be Elenan, his wîfe *Constantius gave the empire to his son Constantine whom he had by Helena his wife,* Ors. 6, 30; Th. 496, 33. His [Constantínes] môdor wæs cristen, Elena gehâten, swîðe gelýfed mann, and þearle eáwfæst his [Constantíne's] *mother was a christian, called Helena, a very faithful person, and very pious,* Homl. Th. ii. 306, 3. ☞ See Ors. Bos. Notes and Various Readings, p. 28, col. 2, in proof that *Helena was the lawful wife of Constantius :*—Ðâ him Elene forgeaf sincweorþunga *then Helĕna gave him treasures,* Elen. Kmbl. 2434; El. 1218. Fôr Elenan cneó *before the knee of Helĕna,* 1693; El. 848: 1903; El. 953. Se Câsere [Constantínus] hêht Elenan *the emperor Constantine told Helena,* Elen. Kmbl. 2003; El. 1003: 2124; El. 1063. Elene, 438; El. 219: 1204; El. 604: 1236; El. 620.

ele-sealf, e; *f. Oil-salve, sweet balm;* nardus =νάρδος, ambrŏsia = ἀμβρŏσία, Cot. 3: 146.

ele-seocche, an; *f?* [ele *ŏleum,* seocche =seohhe *colātōrium] A vessel for straining oil, an oil-strainer?*—Eleseocche *fisclum?* Ælfc. Gl. 66; Som. 69, 85; Wrt. Voc. 41, 38.

eleþ, es; *m. A man;* hŏmo :—Witon ðæt se eleþ êce bîdeþ *they know that the man eternally abideth,* Exon. 33 b; Th. 106, 8; Gû. 38. v. hæleþ.

ele-treów, es; *n. An oil-tree, olive-tree;* ŏlīva :—Swâ swâ eletreów wæstmbǽra *sicut ŏlīva fructĭfĕra,* Ps. Spl. C. T. 51, 8. Swâ niwe planta eletreówa *sicut nŏvellæ ŏlīvārum,* Ps. Spl. C. 127, 4.

ele-twîg, es; *n. An olive twig, a small branch of olive,* Cot. 146.

-elfen, e; *f.* [ælf *an elf,* en *a feminine termination] A fairy, nymph;* nympha : *used only as a termination :*—Dûn-elfen *castălĭdes;* feld-elfen *moĭdes(?);* wudu-elfen *dryădes;* wylde-elfen *hamadryădes;* sǽ-elfen *naĭădes,* Ælfc. Gl. 112, 113; Som. 79, 108–112; Wrt. Voc. 60, 15–19. v. -ælfen.

el-hygd, e; *f. Strange thought, distraction;* perturbātio :—Môdes elhygd *distraction of the mind,* L. M. 2, 46; Lchdm. ii. 258, 18.

Élig, e; *f.* [êl =æl *an eel,* íg *an island] The isle of* ELY, *Cambridgeshire;* insŭla Eliensis *in agro Cantabrigiensi :*—Is Élíg ðæt land on Eást-Engla mǽgþa, hû hugu syx hund hîda, on eálondes gelícnesse; is eall mid fenne and mid wætere ymbseald, and fram genihtsumnesse ǽla ða ðe on ðám ylcan fennum foṅgene beóþ hit naman onfeṅg *the land Ely is in the province of the East-Angles, of about six hundred hides, in the likeness of an island; it is all encompassed with a fen and with water, and took its name from the abundance of eels which are caught in the same fen,* Bd. 4, 19; S. 590, 3–6. Hér Sće Æðeldryht ongon ðæt mynster æt Élíge *in this year* [A.D. 673] *St. Ætheldryth began the monastery at Ely,* Chr. 673; Th. 58, 4. Æðeldryþ wæs abbudisse geworden on ðam þeódlande ðe is gecýged Élíge, ðǽr heó mynster getimbrade *Ætheldryth became abbess in the country which is called Ely, where she built a monastery,* Bd. 4, 19; S. 588, 1.

Élig-burh, Élí-burh; *gen.* -burge; *dat.* -byrig; *f. The city of Ely, Cambridgeshire;* urbs Eliensis *in agro Cantabrigiensi :*—Man hine lǽdde to Élígbyrig [Élíbyrig, Th. 294, 15, col. 2] *they led him to Ely,* Chr. 1036; Th. 294, 16, col. 1.

éliŏtrŏpus, éliŏtrŏpos *the turnsole,* Herb. 137; Lchdm. i. 254, 10, 16. v. héliotropus.

el-land, es; *n. A foreign country, strange land;* externa terra :—Mægþ sceal, geómormôd, elland tredan *a maiden, sad of mind, shall tread a strange land,* Beo. Th. 6031; B. 3019.

ellarn *an elder-tree,* Som. Ben. Lye. v. ellen.

ellefne; *adj. Elevĕn;* undĕcim :—Ellefne orettmæcgas *eleven champions,* Andr. Recd. 1331; An. 664. v. endleofan.

ELLEN; *gen.* elnes; *m. n. Strength, power, vigour, valour, courage, fortitude;* vis, rôbur, vîgor, virtus, fortĭtūdo :—Wîsdôm hæfþ on him feówer cræftas, ðara is ân wærscipe, ôðer metgung, þridde is ellen, feorþe rihtwîsnes *wisdom has in it four virtues, of which one is prudence, another temperance, the third is fortitude, the fourth justice,* Bt. 27, 2; Fox 96, 34: Beo. Th. 1151; B. 573: Cd. 64; Th. 78, 5; Gen. 1288: Exon. 52 b; Th. 183, 4; Gû. 1322: Andr. Kmbl. 920; An. 460: Menol. Fox 491; Gn. C. 16. Ðâ him wæs elnes þearf *when he had need of valour,* Beo. Th. 5745; B. 2876: Cd. 47; Th. 59, 32; Gen. 972: Exon. 45 b; Th. 156, 3; Gû. 869: Andr. Kmbl. 2002; An. 1003: Elen. Kmbl. 1446; El. 725: Salm. Kmbl. 21; Sal. 11: Ps. Th. 118, 23. Wæs Gûþlâc on elne strong *Guthlac was strong in courage,* Exon. 36 b; Th. 119, 34; Gû. 262: 62 b; Th. 231, 4; Ph. 484: Beo. Th. 5624; B. 2816. Hæfde him on innan ellen untweódne *he had within him unwavering courage,* Andr. Kmbl. 2485; An. 1244: Beo. Th. 5384; B. 2695. Ic gefremman sceal eorlîc ellen *I shall perform a manly deed of valour,* Beo. Th. 1278; B. 637. Ne lǽt ðîn ellen gedreósan *let not thy strength sink,* Wald. 10; Vald. 1, 6: Beo. Th. 1208; B. 602: Exon. 120 b; Th. 463, 7; Hö. 66: Apstls. Kmbl. 6; Ap. 3. Ellen fremman *to do a deed of valour,* Andr. Kmbl. 1210; An. 1210: Beo. Th. 6; B. 3. Moyses bebeád folc hycgan on ellen *Moses bade the people think on valour,* Cd. 154; Th. 191, 22; Exod. 218: Fins. Th. 21; Fin. 11: Ps. Th. 93, 2. Engel hine elne trymede *an angel strengthened him with courage,* Exon. 35 a; Th. 113, 21; Gû. 161: Cd. 98; Th. 129, 2; Gen. 2137: Beo. Th. 5715; B. 2861: Andr. Kmbl. 1966; An. 985: Rood Kmbl. 67; Kr. 34: Ps. Th. 128, 5. Elne *with strength, power,* or *courage, strongly, powerfully, courageously;* strēnue, fortĭter, Beo. Th. 3938; B. 1967: Exon. 80 a; Th. 300, 18; Fä. 8: Th. 52, 5: 59, 4: 118, 4, 176. [*Orm.* ellennlæs *powerless: O. Sax.* ellien, ellen, *n. strength, manhood: M. H. Ger.* ellen, *n. strength, manhood: O. H. Ger.* ellan, *m. zēlus, rōbur, virtus: Goth.* alyan, *n. zeal: Icel.* eljan, elja, *f. endurance, energy.*] DER. mægen-ellen.

ELLEN, es; *n. The elder-tree;* sambûcus nigra, *a small tree whose branches are filled with a light spongy pith. The fruit is a globular, purplish-black berry, of which wine is often made, called elder-berry wine. It is quite distinct from* alor *the alder-tree,* q. v :—Ellenes blôsman genim *take blossoms of elder,* L. M. 2, 59; Lchdm. ii. 288, 2. Genim ðas wyrte, ðe man sambûcus =σαμβύκη [MS. samsuchon = σάμψυχον] and ôðrum naman ellen, hâteþ *take this wort, which is named sambucus, and by another name elder,* Hcrb. 148, 1; Lchdm. i. 272, 14. Genim ellenes leáf *take leaves of elder,* L. M. 1, 27; Lchdm. ii. 68, 23: 2, 30; Lchdm. ii. 228, 4. [*Plat.* elloorn, *m: Ger. M. H. Ger.* holder, holunder, *m: O. H. Ger.* holder, holuntar, *m: Dan.* hyld, *m. f;* hyldetræ, *n: Swed.* hyll, *f.*]

ellen; *adj. Of elder, elder-;* sambûceus :—Genim ellenne sticcan *take an elder-stick,* L. M. 1, 39; Lchdm. ii. 104, 7.

ellen-campian; *p.* ode, ede; *pp.* od, ed *To contend vigorously;* fortĭter pugnāre :—Ellencampian *pugĭlāre,* Cot. 4. Ellencampedon *pugĭlāvērunt,* Cot. 40.

ellen-cræft, es; *m. Strength, power;* virtus, pŏtentia :—Ahebbaþ hâligne heofena Drihten, usserne God ellencræfta *exalt the holy Lord of heaven, our God of powers,* Ps. Th. 98, 5.

ellen-dǽd, e; *f.* [ellen *valour,* dǽd *a deed] A deed of valour, bold* or *valiant deed;* virtūtis factum :—Sceolde hine yldo beniman ellendǽda *age should deprive him of bold deeds,* Cd. 24; Th. 31, 13; Gen. 484: Judth. 12; Thw. 25, 22; Jud. 273. He secgan hýrde ellendǽdum *he heard tell of valiant deeds,* Beo. Th. 1756; B. 876: 1804; B. 900.

el-lende, ele-lænde; *adj. Strange, foreign;* extērus, peregrīnus :—Nǽnig cêpa se neáh ellendne wearod *no merchant saw a foreign shore,* Bt. Met. Fox 8, 60; Met. 8, 30. In ellende *in foreign land, afar,* Mt. Kmbl. Rush. 21, 33: 25, 14.

Ellen-dûn, Ellan-dûn, e; *f.* [*Flor.* Ellandun, i. e. mons Eallæ: *Will. Malm.* Hellendune: *Hunt.* Elendune] *Allington, near Amesbury, Wilts;* lôci nōmen in agro Wiltoniensi :—Hér gefeaht Ecgbryht cyning and Beornwulf cyning on Ellendûne [Ellandûne, Th. 111, 21, col. 2] *in this year Egbert and Beornwulf fought at Allington,* Chr. 823; Th. 110, 20.

ellen-gǽst, es; *m. A bold* or *powerful spirit;* pŏtens spîritus :—Se ellengǽst *the powerful spirit* [Grendel], Beo. Th. 172; B. 86.

ellen-gôdnes zeal, Bd. 3, 3; S. 525, 32, note. v. ellen-wôdnes.

ellen-heard; *adj. Hard of courage, bold, courageous;* fortis, strēnuus :—Wæs eorl ellenheard searoþancum beseted *the courageous warrior was beset with various thoughts,* Andr. Kmbl. 2509; An. 1256: Exon. 49 b; Th. 172, 3; Gû. 1138.

ellen-léca, an; *m. A champion;* pŭgil, agōnista, Cot. 15.

ellen-leás; *adj. Lacking courage;* fortĭtūdĭne cărens :—Ic sceal sêcan

óðerne ellenleásran cempan *I must seek another less courageous soldier,* Exon. 71 b; Th. 266, 7; Jul. 394.

ellen-líce; *adv. Boldly, daringly;* fortíter, strēnue, pŏtenter :—Wíf beorn acwealde ellenlíce *the woman daringly slew a warrior,* Beo. Th. 4250; B. 2122.

ellen-mǣrþ, e; *f.* [mǣrþ *greatness, glory*] *Glory of valour* or *courage;* fortitūdĭnis glōria :—Grendel nihtweorce gefēh, ellenmǣrþum *Grendel rejoiced in his night-work, his valour-glories,* Beo. Th. 1660; B. 828.

ellen-rind, e; *f. Elder-rind* or *bark;* sambūci cortex :—Well ellenrinde niðewearde *boil the nether part of elder-rind,* L. M. 1, 32; Lchdm. ii. 78, 5 : 1, 54; Lchdm. ii. 126, 5 : 1, 68; Lchdm. ii. 128, 14.

ellen-róf; *adj. Remarkably strong, powerful, daring, brave;* rŏbustus, strēnuus, fortis :—Strang oððe ellenróf *robustus,* Ælfc. Gr. 9, 22; Som. 10, 52. Beó ðū gestrangod and ellenróf *confortāre et esto rŏbustus,* Jos. 1, 7, 9 : Cd. 89; Th. 110, 26; Gen. 1844 : Beo. Th. 685; B. 340 : Exon. 96 a; Th. 358, 3; Pa. 40 : Judth. 10; Thw. 23, 7; Jud. 109 : Wald. 79; Vald. 2, 11. Hí woldon āninga ellenrófes mōd gemiltan *they would entirely subdue the bold man's mind,* Andr. Kmbl. 2784; An. 1394. Gif ic ǽnigne ellenrófne gemēte *if I find any brave man,* Exon. 71 a; Th. 265, 17; Jul. 382. Ellenrófe weras *the bold men,* Exon. 106 b; Th. 405, 9; Rä. 23, 20 : Cd. 94; Th. 122, 33; Gen. 2036 : Andr. Kmbl. 2284; An. 1143.

ellen-sióc; *adj.* [sióc = seóc *sick, diseased, infirm, languid*] *Infirm* or *languid from want of strength;* invālidus, dēbĭlis :—Hwæðer he cwicne gemētte in ðam wongstede Wedra þeóden ellensiócne *whether he should find the languid prince of the Goths alive on the field,* Beo. Th. 5567; B. 2787.

ellen-sprǽc, e; *f. Powerful speech;* pŏtens sermo :—He ne meahte ellensprǽce, hleóðor ahebban *he could not raise his voice, his powerful speech,* Exon. 49 b; Th. 171, 18; Gú. 1128.

ellen-þríst; *adj. Bold in courage, bold;* audax :—Ða idesa ellenþríste *the bold women,* Judth. 11; Thw. 23, 22; Jud. 133.

ellen-weorc, es; *n. A work of valour, valiant* or *powerful act;* fortitūdĭnis ŏpus, res fortíter gesta :—He wæs ānrǽd ellenweorces *he was steadfast in his work of valour,* Andr. Kmbl. 464; An. 232. Gif ðū ðæt ellenweorc aldre gedígest *if thou escapest with life from that work of valour,* Beo. Th. 1326; B. 661 : 5279; B. 2643 : Exon. 42 a; Th. 140, 20; Gú. 613. Ellenweorca *of valiant acts,* Beo. Th. 4789; B. 2399. Ellenweorcum *by valiant acts,* Andr. Kmbl. 2740; An. 1372.

ellen-wód, e; *f?* [wód *mad*] *Zeal;* zēlus = ζῆλος :—Me ðínes húses heard ellenwód æt zēlus dŏmus tuæ cŏmēdit me, Ps. Th. 68, 9.

ellen-wód; *adj.* [wód *mad*] *Raging, furious;* fŭriōsus :—Wæs ellenwód fæder wið dēhter *the father was furious with his daughter,* Exon. 67 b; Th. 251, 4; Jul. 140.

ellen-wódian; *p.* ode; *pp.* od [ellen-wód *zeal*] *To strive with zeal, emulate;* æmŭlāri :—Nylle ðū ellen oððe ellenwódian [MS. ellenwondian] on yfelwillendum *nóli æmŭlāri in malignantĭbus,* Ps. Spl. C. 36, 1.

ellen-wódnes, -ness, e; *f. Zeal, envy, emulation, ardour;* zēlus = ζῆλος, fervor :—Swindan me dyde ellenwódnes mín *tabescĕre me fēcit zēlus meus,* Ps. Spl. T. 118, 139; 78, 5. Aidanus hæfde Godes ellenwódnesse and his lufan micle *Aidan had much zeal and love for God,* Bd. 3, 3; S. 525, 32. He wæs mid wylme mycelre ellenwódnesse onbærned *zēlo magni fervōris accensus est,* 4, 24; S. 598, 22.

ellen-wyrt, e; *f. Elderwort, wallwort, danewort, dwarf-elder;* sambūcus ĕbŭlus, Lin :—Genim ðas wyrte, ðe man ĕbŭlum, and ōðrum naman ellenwyrte nemneþ, and eác sume men wealwyrt hātaþ *take this herb, which is named ĕbŭlum, and by another name, elderwort, and some men also call it wallwort,* Herb. 93, 1; Lchdm. i. 202, 5 : Wrt. Voc. 67, 12, 64 : 69, 17.

ELLES; *adv.* ELSE, *otherwise, in another manner;* ălíter, ăliōquin, ăliunde, sĕcus :—Elles ălíter, ăliōquin, Ælfc. Gr. 38; Som. 41, 7, 67. Elles næbbe ge mēde mid eówrum fæder ăliōquin mercēdem non hăbēbĭtis ăpud patrem vestrum, Mt. Bos. 6, 1 : Mk. Bos. 2, 21. Gif hit elles sý *sin autem,* Lk. Bos. 10, 6. He stýhþ elles ofer *ascendit ăliunde,* Jn. Bos. 10, 1. Hí ne mihton elles bión *they could not else exist,* Bt. 39, 13; Fox 234, 30 : Bt. Met. Fox 9, 104; Met. 9, 52 : Chr. 1044; Erl. 168, 17 : Beo. Th. 5034; B. 2520 : Exon. 67 b; Th. 249, 18; Jul. 113. Hwá aríst elles of Syon búton ðú *who else shall arise out of Sion but thou?* Ps. Th. 13, 11. Hwæt elles is *quid est ăliud?* Bd. 1, 27; S. 494, 15. Nyton hwæt hý elles sprecon *they know not what they else speak,* Ps. Th. 43, 16. Áhwǽr or ǽghwǽr elles *anywhere else,* Ps. Th. 71, 12 : 102, 15. Ná elles, ná hú elles *not otherwise, no how else;* haud sĕcus, Ælfc. Gr. 38; Som. 42, 3 : Bt. 32, 1; Fox 114, 8. Nówiht elles *nothing else;* nil ăliud, Bd. 2, 14; S. 518, 8. Elles áwiht, ówiht *or* wuht *anything else;* ăliud quid, 32; Th. 42, 33; Gen. 682 : 91; Th. 114, 16; Gen. 1905 : Exon. 82 a; Th. 308, 27; Seef. 46 : 115 a; Th. 443, 1; Kl. 23 : Bt. Met. Fox 9, 40; Met. 9, 20. Elles hwæt *anything else,* Bd. 4, 3; S. 569, 8. Elles hwǽr, hwár, hwergen *or* hwider *elsewhere;* ăliorsum, L. Eth. v. 12; Th. i. 308, 5 : L. C. E. 13; Th. i. 308, 6 : Beo. Th. 277; B. 138 : 5173; B. 2590 : Ælfc. Gr. 38; Som. 40, 7. [Wyc. Piers P. ellis : Chauc. R. Glouc. elles : Orm. elless : Scot. els,

ellis : O. Frs. elles, ellis : M. H. Ger. alles *ăliter :* O. H. Ger. alles, elles, ellies *ăliōquin :* Goth. allis *at all :* Swed. eljest : Lat. ălias.]

elles hwá *any;* ali-quis, March. § 136, 5 a.

ellicor; *adv. Elsewhere, otherwise,* Ælfc. Gr. 38, Lye, Ettm. v. elcor.

ellm, es; *m. An elm;* ulmus :—On ellmum *in ulmis,* L. Edg. C. 16; Wilk. 83, 47. v. elm.

ellnung, e; *f. Emulation, zeal;* æmŭlātio :—Hí hæfdon Godes ellnunge *æmŭlātiōnem Dei hăbēbant,* Bd. 5, 22; S. 644, 8. v. elnung.

ellor; *adv. Elsewhere;* ălias, ăliorsum :—Heó ðæt leóht geseah ellor scríðan *she saw the light depart elsewhere,* Cd. 37; Th. 48, 9; Gen. 773 : 133; Th. 168, 17; Gen. 2784 : Judth. 10; Thw. 23, 9; Jud. 112 : Beo. Th. 110; B. 55.

ellor-fús; *adj.* [fús *ready, quick*] *Desirous* or *ready to go elsewhere, ready to depart;* pĕregre eundi cŭpidus, ăliorsum íre părātus :—Oþ-ðæt gást, ellorfús, gangan sceolde to Godes dóme *until his spirit, ready to depart, must go to God's judgment,* Cd. 79; Th. 97, 7; Gen. 1609. He his hláford geseah ellorfúsne *he saw his lord ready to depart [about to die],* Exon. 48 a; Th. 165, 11; Gú. 1027 : Andr. Kmbl. 375; An. 188.

ellor-gást, -gǽst, es; *m. A spirit living* or *going elsewhere, a departing spirit;* spíritus ălĭbi dēgens :—Scolde se ellorgást on feónda geweald síðian *the departing spirit must go into the power of fiends,* Beo. Th. 1619; B. 807. Ellorgǽst *a departing spirit,* 3238; B. 1617. Hie gesáwon twegen ellorgǽstas *they saw two spirits living elsewhere,* 2702; B. 1349.

ellor-síþ, es; *m. A journey elsewhere, departure, death;* ălĭbi iter, mors :—Symble biþ gemyndgad eaforan ellorsíþ *his offspring's death will always be remembered,* Beo. Th. 4893; B. 2451.

ell-reord; *adj. Foreign-speaking, barbarous;* barbărus :—Eallum ellreordum cynnum *cunctis barbăris nātiōnĭbus,* Bd. 4, 2; S. 565, 31. v. el-reord.

ell-reordig; *adj. Foreign-speaking, barbarous;* barbărus :— Óðer [heretoga] wæs ðam hǽðenan rēþra and grimra fordon he ellreordig wæs alter [dux] *quia barbărus ĕrat, pāgāno sævior,* Bd. 2, 20; S. 521, 21, 24 : 3, 6; S. 528, 10. v. el-reord.

ell-þeód, e; *f. A strange people, foreign nation;* pĕregrīna gens :—Hý fóron on ellþiéde *they went into a foreign land,* Ors. 4, 4; Bos. 81, 6. v. el-þeód.

ell-þeódig -þiódig; *adj. Strange, foreign, a stranger, a foreigner :—* Ellþeódigra *of the foreigners,* Cd. 89; Th. 110, 8; Gen. 1835 : Lk. Lind. War. 17, 18 : 24, 18. v. el-þeódig.

ellyn *zeal,* Ps. Spl. C. 118, 139. v. ellen.

ELM, ellm, es; *m. An* ELM, *elm-tree;* ulmus :—Genim elmes rinde *take bark of elm,* L. M. 1, 6; Lchdm. ii. 52, 9. [Chauc. elmes, *pl :* Dut. olm, *m :* Ger. ulme, *f :* M. H. Ger. ëlm, *f :* O. H. Ger. elm, helmboum : Dan. alm, älm, *m. f :* Swed. alm, *f :* Icel. almr, álmr, *m :* Lat. ulmus, *f.*] DER. elm-rind.

elm-boga, an; *m. An elbow;* cŭbĭtum :—Gif se earm biþ forad búfan elmbogan *if the arm be broken above the elbow,* L. Alf. pol. 54; Th. i. 94, 24. v. el-boga.

el-mehtig *almighty,* Ps. C. 77 [Pfr. Germ. 10, 427]. v. eal-mihtig.

elmestlíc; *adj. Charitable;* mĭsĕricors :—Swǽ hit him bóem rehtlícast and elmestlícast wére *as might be most righteous and most charitable for both,* Th. Diplm. A. D. 830; 465, 23.

el-mihtig *almighty :*—God elmihtiga *almighty God,* Chr. 1086; Th. 353, 32. v. eal-mihtig.

elm-rind, e; *f.* ELM-RIND *or bark;* ulmi cortex :—Elmrind *bark of elm,* L. M. 1, 47; Lchdm. ii. 116, 2. Well elmrinde *boil elm-rind,* 1, 32; Lchdm. ii. 78, 5. Nim elmrinde *take elm-rind,* 1, 38; Lchdm. ii. 98, 8 : 3, 29; Lchdm. ii. 324, 15. Genim elmrinde grēne *take elm-rind grene,* 1, 56; Lchdm. ii. 126, 15. Mid elmrinde *with elm-rind,* 1, 25; Lchdm. ii. 66, 23.

ELN, e; *f.* I. *an* ELL, *a measure of length, the space from the point of the elbow to the end of the middle finger, eighteen inches.* This is the Heb. אמה [amma] *a cubit :* the Lat. cŭbĭtus *a cubit,* ulna *an ell.* Liddell and Scott say πῆχυς = cŭbĭtus, and ulna *an ell* properly contain twenty-four δάκτυλοι [δάκτυλος, the breadth of a finger, about ¾ of an English inch] :—Tís dè éx ὑμῶν μεριμνῶν δύναται προσθεῖναι ἐπὶ τὴν ἡλικίαν αὐτοῦ πῆχυν ἕνα; Mt. 6, 27; ỉþ whas izwara maurnands mag anaaukan ana wahstu seinana aleina aina? Mt. Bos. Goth. 6, 27; *quis autem vestrum cōgĭtans pŏtest adjĭcĕre ad stătūram suam cŭbĭtum ūnum?* Mt. Vulg. 6, 27; hwylc eówer mæg sóþlíce geþencan ðæt he ge-eácnige áne *elne* to hys anlícnesse? Mt. Bos. 6, 27; Wycl. says *cubite;* Tynd. *cubit.* It is therefore presumed that the Grk. πῆχυς = Heb. אמה was eighteen inches; for twenty-four δάκτυλοι x by ¾ = [.75] = eighteen inches. In the parallel passage, Lk. Bos. 12, 25, there is not any Gothic; the Grk. Lat. and A. Sax. are the same as in the preceding verse. Lk. Bos. 12, 25, is, therefore, not quoted. Hí wǽron unfeor fram lande, swylce hit wǽre twá hund elna *non longe ĕrant a terra, sed quăsi cŭbĭtis dŭcentis* [18 in. x 200 ÷ 12 = 300 ft.], Jn. Bos. 21, 8. Fíftena stód deóp ofer dúnum se drenceflód monnes elna *the deluge stood deep over the downs, fifteen ells of man,* Cd. 69; Th. 84, 17; Gen. 1399. Eln *ulna,* Glos. Brux. Recd. 38, 62; Wrt. Voc. 64, 71. II. *the Royal*

Persian ell, or cubit, is very nearly 20½ *inches;* for Herodotus says that the πῆχυς βασιλήιος, bk. i. § 178, is 3 δάκτυλοι longer than the common Grk. πῆχυς = *cubit* or *ell :* 24 δάκτυλοι, i. e. 24 + 3 = 27 δάκτυλοι ; 27 x [¾ of an inch and $\frac{1}{12}$ of an 8th, δάκτυλος *a finger's breadth* = about $\frac{75}{100}$ of an inch, that is ¾ of an inch and $\frac{1}{12}$ of an 8th = ¾ + $\frac{1}{98}$ = $\frac{72}{98} + \frac{1}{98} = \frac{73}{98}$ = .76] .76 = 20$\frac{5}{90}$ [= 20½ inches, and $\frac{2}{90}$ or $\frac{1}{40}$ of an inch] :—Se weall Babilônes is fîftig elna brâd, and twâ hund elna heáh .... and ymbûtan ðone weall is se mæsta dîc .... and wiðûtan ðam dîce is geworht twegra elna heáh weall *the wall of Babylon is fifty ells broad, and two hundred ells high .... and round the wall is a very great dike .... and outside the dike a wall is built two ells high,* Ors. 2, 4; Bos. 44, 23–28. **III.** *the ell in A. Sax. was sometimes about 24 inches, or 2 feet :*—Se hwæl biþ micle læssa ðonne ôðre hwalas: ne biþ he lengra ðonne syfan elna lang; ac, on his âgnum lande, ða beóþ eahta and feówertiges elna lange, and ða mæstan, fîftiges elna lange; ðara, he sæde, ðæt he syxa sum ofslôge syxtig on twâm dagum *this whale is much less than other whales: it is not longer than seven ells; but, in his own country* [Norway], *they are eight and forty ells long, and the largest, fifty ells long; of these, he said, that he was one of six, who killed sixty in two days,* Ors. 1, 1; Bos. 20, 18–23. In giving the size of the Horse-whale or Walrus, and of the Whale, Ohthere, a Norwegian, would most probably calculate by the measure of Scandinavia, the ell of Norway, Sweden, and Denmark. Molbeck, in his Dansk Ordbog, thus defines it :—'Alen, et vist længdemaal, som deles i 24 tommer ... Tomme een 12te fod, og een 24de alen,' ... That is, *Ell, a certain measure of length, which is divided into 24 inches .... An inch one 12th of a foot, and one 24th of an ell.* King Alfred, in his Anglo-Saxon version of Orosius, followed the calculation of Ohthere, who says that the Horse-whale or Walrus is 7 ells long, that is 14 feet, and the Whales 48 ells, and the largest 50, that is 96 feet, and the largest 100 feet long. These calculations approach very nearly to those given by Mr. Broderip, who says the length of the Walrus is from 10 to 15 feet, and Dr. Scoresby, who gives the length of the Physalus to be about 100 feet, Ors. Eng. p. 43, note 45. **2.** ells of different lengths were used in Anglo-Saxon times; and, even in the present day, 3 sorts of ells are known in England :—*The Flemish ell* is 3 quarters of a yard or 27 inches; *the English* 5 quarters or 45 inches; and *the French* 6 quarters or 54 inches. [Early English, *Wrt. spec.* 35, ân elne long: R. Glouc. 429, 3, elnen, *pl :* Plat. eel, *f :* Frs. jelne: O. Frs. ielne, elne, *f :* Dut. el, elle, *f :* Ger. elle, *f :* M. H. Ger. elne, elline, ellen, *f :* O. H. Ger. elina, elna, elle, *f :* Goth. aleina, *f :* Dan. aln, *f :* Swed. aln, *f :* Icel. alin, *f :* Lat. ulna, *f :* Grk. ὠλένη, *f.* Eln *the ell* is found in *A. Sax.* eln-boga, el-boga *the elbow :* Dut. elle-boog : Ger. ellen-boge.] Ell is an old Teutonic word being used in the oldest German, in the Gothic translation of Ulphilas about A. D. 360: in Anglo-Saxon about 895. The date of its use in other parts of Europe may be ascertained by referring to the languages quoted above, and in the list of contractions where the names and dates of the authors are given.

eln-boga, an; *m. An elbow;* cŭbĭtum :—Se earm nænige bígnesse on ðam elnbogan hæfde *the arm had no bending at the elbow,* Bd. 5, 3; S. 616, 23. v. el-boga.

elnes *of strength,* Beo. Th. 3063; B. 1529; *gen. of* ellen.

elnes = ellenes *of elder :*—Elnes rinde sele *give elder-rind,* L. M. 2, 30; Lchdm. ii. 230, 14. v. elne *an elder-tree.*

eln-gemet, es; *n. An ell-measure, the length of an ell, two feet?* cŭbĭtālis mensûra, ulnæ mensûra :—Ðæt fær gewyrc fîftiges wîd, þrittiges heáh, þreó hund lang elngemeta *make the vessel fifty wide, thirty high, three hundred long, of ell measures,* Cd. 65; Th. 79, 10; Gen. 1309. v. eln I. and III.

elnian; *part.* elnende; *p.* ode, ade; *pp.* od, ad [ellen *strength*]. **I.** *to make strong, strengthen;* confortâre :—Elnode he hine and sæt upp confortâtus sêdit in lectŭlo, Gen. 48, 2. **II.** *to strive with zeal after another, endeavour to be equal, emulate;* æmŭlâre, zēlâre :—Nyl ðû elnian betwih awergde, ne elnende ðû sié dônde unrehtwîsnisse *nôli æmŭlâre inter mâlignantes, neque æmŭlâtus fŭeris făciêntes iniquĭtâtem,* Ps. Surt. 36, 1, 7. Ic elnode [elnade, Ps. Th. 72, 2] ofer ða unrihtwîsan *zēlâvi sŭper iniquos,* Ps. Spl. C. 72, 3. Ne elna ðû *ne æmŭlêris,* Ps. Surt. 36, 8. DER. ge-elnian.

elnung, ellnung, e; *f. Zeal, hot emulation, envy;* zēlus, æmŭlâtio :—Elnung *zēlus,* Rtl. 192, 5. Elnung oððe æfista hûses ðînes ge-et [= geæt] mec *zēlus dŏmus tuæ cŏmêdit me,* Jn. Lind. War. 2, 17.

elone *the herb* elecampane, L. M. 1, 23; Lchdm. ii. 66, 2. v. eolone.

elp *an elephant,* Som. Ben. Lye. v. ylp.

elpen-bǽnen; *adj. Made of ivory;* ēburnĕus :—Fram hûsum elpenbǽnenum *a dŏmĭbus ēburnĕis,* Ps. Lamb. 44, 10.

elpen-bân, es; *n. An elephant's bone, ivory;* ēbur :—Elpenbânum *with ivory,* Ps. Spl. 44, 10: Cot. 71. v. ylpen-bân.

elpend, es; *m. An elephant;* ēlephas = Ἐλέφας :—Hwæðer ge seón mâran on eówrum lîchoman ðonne elpend *if ye were greater in your body than the elephant,* Bt. 32, 1; Fox 114, 25. Elpendes hŷd wyle drincan wǽtan gelîce and spinge dêþ *an elephant's hide will drink wet*

---

like a sponge, Ors. 5, 7; Bos. 107, 10. He genêþde under ânne elpend *he went boldly under an elephant,* Ors. 4, 1; Bos. 77, 20: 78, 9. He hæfde xx elpenda *he had twenty elephants,* 4, 1; Bos. 77, 5: 5, 7; Bos. 107, 8. To ðam elpendum [MS. elpendan] *to the elephants,* 4, 1; Bos. 77, 26. Hêt Pirrus dôn ða elpendas on ðæt gefeoht *Pyrrhus ordered the elephants to be brought into the battle,* 77, 16, 23: 78, 5, 28.

elpend-tôþ, es; *m. An elephant's tooth;* ēlephantis dens, Cot. 78.

elra; *comp? Stranger :*—He ne mêtte on elran men mundgripe mâran *he did not find a stronger hand-gripe in a stranger man,* Beo. Th. 1509; B. 752.

el-reord, ell-reord, æl-, æll-, eall-; *adj. Foreign-speaking, barbarous;* barbârus :—Wǽron heó mid elreordre dysignesse onblâwne *inflâti ĕrant barbâra stultĭtia,* Bd. 2, 5; S. 507, 13: 1, 5; S. 476, 11: 1, 14; S. 482, 12: Ps. Surt. 113, 1.

el-reordig, ell-reordig; *adj. Foreign-speaking, barbarous;* barbârus, pĕregrînus :—Of gramum folce ða elreordige ealle wǽron *de pŏpŭlo barbâro,* Ps. Th. 113, 1.

el-reordignes, -ness, e; *f. Barbarousness, outlandishness;* barbăries, Som. Ben. Lye.

el-riord; *adj. Barbarous;* barbărus :—Mid elriordre dysignesse *barbâra stultĭtia,* Bd. 2, 5; Whelc. 122, 3. v. el-reord.

el-þeód, æl-þeód, æl-þiód [ell-]; *gen.* e; *pl. nom. acc.* a, e; *f. A foreign nation, strange people;* gens pĕregrîna, alienĭgĕna, pĕregrîni :—Êhton elþeóda *they pursued the strange nations,* Elen. Kmbl. 277; El. 139. Fôre elþeódum *before strange nations,* Exon. 27 b; Th. 82, 12; Cri. 1337: 23 b; Th. 67, 6; Cri. 1084. On elþeóde *among a strange people,* Andr. Kmbl. 1943; An. 974: Exon. 123 b; Th. 474, 25; Bo. 36.

el-þeódian *foreigners;* barbăros, = el-þeódigan; *acc. pl. def. of* el-þeódig, Bd. 1, 14; S. 482, 12, note.

el-þeódig, æl-þeódig, el-þiódig [ell-]; *adj. Strange, foreign, barbarous, one who is abroad;* pĕregrînus, barbârus, advĕna, alienĭgĕna, qui pĕregre est :—Eorlas elþeódige *strange men,* Andr. Kmbl. 397; An. 199. Þearfum and elþeódigum symble eáþmôd *pauperĭbus et pĕregrînis semper hŭmĭlis,* Bd. 3, 6; S. 528, 10, note. Ðæt Bryttas ða elþeódian of heora gemǽrum adrîfan *ut Brittônes barbâros suis e fînĭbus pĕpŭlĕrint,* 1, 14; S. 482, 12, note. Nû cwom elþeódig *now a stranger has come,* Elen. Kmbl. 1813; El. 908: Cd. 124; Th. 159, 3; Gen. 2629. Hwonne me wrâþra sum elþeódigne aldre beheówe *when some enemy might bereave me, a stranger, of life,* 128; Th. 163, 20; Gen. 2701: Exon. 82 a; Th. 308, 11; Seef. 38: 87 b; Th. 329, 5; Vy. 29. Ða elþeódigan ealle Drihten lustum healdeþ *Dŏmĭnus custôdit advĕnam,* Ps. Th. 145, 8: 110, 4. v. el-þeód.

el-þeódglîce, æl-þeódiglîce; *adv. In foreign parts, among foreigners;* pĕregre. v. æl-þeódiglîce.

el-þeódignes, -þeódines, æl-þeódignes, -ness, -nyss, e; *f. A being or living abroad, pilgrimage;* pĕregrînâtio :—Ferde on elþeódignysse *pĕregre prŏfectus est,* Mt. Bos. 21, 33. On elþeódinysse, 25, 14. Elþeódignys oððe eardbegengnes mîn afeorrad oððe gelængd is *incŏlâtus meus prŏlongâtus est,* Ps. Lamb. 119, 5. v. eard-begengnes.

el-þeódisc; *adj. Foreign, strange;* pĕregrînus :—To bebyrgenne elþeódisce men *in sĕpultûram pĕregrînôrum,* Mt. Bos. 27, 7. v. el-þeódig.

el-þiód, e; *f. A foreign nation :*—In elþióde *pĕregre,* Mt. Rush. Stv. 21, 33. v. el-þeód.

el-þiódgian, -þiódigian; *p.* ode; *pp.* od [el, þeód *a people*] *To live in foreign parts, to lead a pilgrim's life;* pĕregrînâri :—Wilnode he on neáweste ðara hâligra stôwe to tîde elþiódgian on eorþan *cŭpîvit in vicînia sanctôrum lŏcôrum ad tempus pĕregrînâri in terris,* Bd. 5, 7; S. 621, 12.

el-þiódig *strange, foreign,* Bt. 39, 2; Fox 212, 17, note 3: Ors. 3, 7; Bos. 62, 35: Mt. Lind. Stv. 25, 14. v. el-þeódig.

eltst *eldest;* nâtu maxĭmus :—Seó mǽgþ asprang of Noes eltstan suna, se wæs gehâten Sem *that family sprang from Noah's eldest son who was called Shem,* Homl. Th. i. 24, 7, = yldest; *sup. of* eald.

eluhtre, an; *f. The plant lupine;* lŭpînus albus, Lin :—Wyl eluhtran on ealaþ *boil lupine in ale,* L. M. 1, 41; Lchdm. ii. 106, 11: 1, 63; Lchdm. ii. 136, 26. v. elehtre.

ŏlys *hedgehogs,* Ps. Spl. C. 103, 19. v. îl.

em-, in composition, denotes *even, equal :* v. efen *even,* emb *about.* v. em-lîcnes, em-niht, etc.

emb, *even about, round, around :*—Emb eahta niht *about eight nights,* Menol. Fox 418; Men. 210: 76; Men. 38: 188; Men. 95: 109; Men. 54: 259; Men. 131: 449; Men. 226. Embe fîf niht *about five nights,* Menol. Fox 21; Men. 11: 30; Men. 15: 38; Men. 19: 82; Men. 41: 385; Men. 194. v. ymb.

embe-fær, es; *n.* [fær *a going, journey*] *A going round, circuit;* circuĭtus :—Embefær tûna *circuĭtus villârum,* Procem. R. Conc.

embe-gân *to go round,* Lye. v. ymb-gân.

embe-gang, es; *m. A going round, circuit;* circuĭtus :—Se embegang ðara landa *the circuit of the lands,* Cod. Dipl. Apndx. 402; A. D. 944;

Kmbl. iii. 421, 6. Se mōna hæfþ læstne embegang *the moon has the least circuit*, Boutr. Scrd. 18, 38. Embegang dōn *processiōnem făcĕre*, R. Conc. 3. v. ymbe-gang.

embe-gyrdan; he -gyrt; *p.* -gyrde; *pp.* -gyrded *To surround, begird*; circumcingĕre:—Gârsecg embegyrt gumena rîce *the ocean surrounds the kingdoms of men*, Bt. Met. Fox 9, 81; Met. 9, 41. v. ymb-gyrdan.

embeht, es; *n. An office, serving*; ministĕrium:—Ymb oft embehte *circa frĕquens ministĕrium*, Lk. Skt. Lind. 10, 40. v. ambeht.

embehtian; *p.* ode, ade; *pp.* od, ad *To minister, serve*; ministrāre:—Heó embehtade odðe gehērde him *ministrābat eis*, Mt. Kmbl. Lind. 8, 15. Embehtaþ *ministrābit*, Lk. Skt. Lind. 12, 37.

embeht-mon, -monn, es; *m. A servant-man, servant, minister*; servus, minister:—Allra embehtmonn *omnium minister*, Mk. Lind. War. 9, 35. v. ambiht-man.

embehtsumnes, -nis, -niss, e; *f. A compliance, kind attention*; obsĕquium:—He dēmeþ embehtsumnisse [MS. embehtsumise] odðe hérnisse *arbitrētur obsĕquium*, Jn. Lind. War. 16, 2.

embe-hydignes, -ness, e; *f. Solicitude*; sollicitūdo, C. R. Ben. 43, Lye. v. ymb-hydignys.

Embene; *pl. m. The inhabitants of Amiens, Amiens, in Picardy, France*; Ambiānum:—Hér fór se here up on Sunnan to Embenum, and dær sæt ân geár *in this year* [A.D. 884] *the army went up the Somme to Amiens, and remained there one year*, Chr. 884; Erl. 82, 17.

embe-smeágung, e; *f. A considering about, experience*; empīria = ἐμπειρία:—Manega embesmeágunga *empīria*, Ælfc. Gl. 82; Som. 73, 49; Wrt. Voc. 47, 53.

embe-þencan; *part.* -þencende; *p.* -þohte; *pp.* -þoht *To think about, to be anxious for, careful*; sollicītus esse:—Ne beó ge embeþencende hû odðe hwæt ge sprecon, odðe andswarion *nōlīte sollicīti esse quāliter aut quid respondeātis, aut quid dīcātis*, Lk. Bos. 12, 11. v. ymbe-þencan.

embe-ûton; *adv. About*; circum:—Ân of dâm de dâr embe-ûton stôdon *one of those who stood there about*, Mk. Bos. 14, 47. v. ymbe-ûtan; *adv.*

emb-feran; *p.* -ferde; *pp.* -fered *To go round, surround*; circuīre:—Híg geond feówertig daga embferdon done eard *they went round the country for forty days*, Num. 13, 26.

embiht, es; *m. A servant*; minister:—Da embihtas *ministri*, Jn. Rush. War. 7, 46. v. ambeht; *m.*

embiht, es; *n. An office*; officium:—Gefylde wæron da dagas embihtes his *implēti sunt dies officii ejus*, Lk. Skt. Lind. 1, 23. v. ambeht; *n.*

embiht-mon, -monn, es; *m. A servant-man, servant, minister*; servus, minister:—Allra embihtmon *omnium minister*, Mk. Rush. War. 9, 35. v. ambiht-man.

emb-long *at length*, Som. Ben. Lye.

emb-rin, es; *n.* [= emb-îren *an encircling iron*] *A fetter*; compes:—Embrin *balus*? Cot. 203; Wrt. Voc. 288, 1.

emb-ryne, es; *m. A running round, a course, revolution, anniversary*; revōlūtio, circuĭtus:—Tyn embrynas *quinquennia jam dĕcem*, Glos. Prudent. Recd. 139, 1. v. ymb-rene.

emb-sittan; *p.* -sæt, *pl.* -sæton; *pp.* -seten *To sit round* or *about, surround, beset, besiege*; circumsēdĕre, obsīdĕre:—Porsenna and Tarcuinius embsæton Rôme burh *Porsenna and Tarquin surrounded Rome*, Ors. 2, 3; Bos. 42, 11. Fe besirede dæt folc de hí embsæten hæfdon *he deceived the people who had besieged them*, Ors. 4, 5; Bos. 83, 3. v. ymb-sittan.

emb-snîdan; *p.* -snáþ, *pl.* -snidon; *pp.* -sniden, -snyden *To cut round, circumcise*; circumcīdĕre:—Dæt dæt cild embsnyden wære *ut circumcīdĕrĕtur puer*, Lk. Bos. 2, 21. v. ymb-snîdan.

emb-stemn; *adv. By turns*; vĭcissim:—Embstemn *vel* dær gemang *vĭcissim*, Glos. Prudent. Recd. 140, 2.

emb-ûtan *about, round*; circum, circa:—Guton [MS. geoton; dæs celfes blôd] embûtan dæt weofod *they poured* [*the blood of the calf*] *round the altar*, Lev. 1, 5, 11. v. ymb-ûtan; *prep.*

emb-wlâtian; ic -wlâtige; *p.* ode; *pp.* od *To look about, contemplate*; contemplāri:—Ic embwlâtige *contemplor*, Ælfc. Gr. 25; Som. 27, 5, MS. D. v. ymb-wlâtian.

emb-wlâtung, e; *f. A viewing, contemplation*; contemplātio:—Hí brûcaþ dære incundan embwlâtunge his godcundnysse *they enjoy the closest contemplation of his divinity*, Homl. Th. i. 348, 7. v. ymb-wlâtung.

em-cristen *a fellow-christian*, L. Ed. C. 36; Th. i. 461, 1. v. emne-cristen.

eme *deceit, fraud*; fraus, Som. Ben. Lye.

emel, e; *f. A canker-worm, caterpillar, weevel*; ērūca, brūchus = βροῦχος:—He sealde emele odðe treówyrme wæstm heora *dĕdit ērūcæ fructus eōrum*, Ps. Spl. C. 77, 51. He sæde and com gærshoppe and emel dæs næs nâ gerîm *dixit et vēnit lŏcusta, et brūchus cūjus non ĕrat nŭmĕrus*, 104, 32. v. ymel.

emertung, e; *f. A tickling, an itching*; prūrīgo:—Emertung *prūrīgo*, Ælfc. Gl. 11; Som. 57, 61; Wrt. Voc. 20, 5.

emetig; *adj. Empty, vacant*; văcuus, văcans:—He gemēteþ hit [hûs]

emetig *invēnit eam* [*dŏmum*] *văcantem*, Mt. Kmbl. Rush. 12, 44. v. æmtig.

em-fela; *adj. Equally many*; tŏtīdem:—Gân inn emfela manna of ægdre healfe *let equally as many men of either side go in*, L. Ath. iv. 7; Th. i. 226, 20. v. efen-fela.

em-hydig; *adj. Anxious about, solicitous*; sollicītus, C. R. Ben. 33. v. ymb-hydig.

emitte, an; *f. An emmet, ant*; formīca:—Emittan *formīcæ*, Prov. 30. v. æmete.

em-lang; *adj. Equally long*; ejusdem longitūdinis, L. M. 2, 36; Lchdm. ii. 242, 15.

em-leóf; *adj. Equally dear*; æque cārus:—Him wearþ emleóf, dæt hý gesâwon mannes blôd agoten, swâ him wæs dara nýtena meolc *it was equally dear to them to see man's blood shed, as it was* [*to see*] *the milk of their cattle*, Ors. 1, 2; Bos. 26, 32.

em-lîce; *adv. Even-like, evenly, equally, patiently*; æquāliter, æquanĭmiter:—Hû emlîce hit gelamp *how evenly it happened!* Ors. 2, 1; Bos. 39, 25 : 3, 6; Bos. 57, 41. He done eard emlîce dælde betwux twelf mægþum *he divided all the country equally among the twelve tribes*, Homl. Th. ii. 214, 12 : Boutr. Scrd. 29, 11; Lchdm. iii. 266, 22. He forbær Godes swingele swîde emlîce *he bare God's scourging very patiently*, Homl. Th. ii. 98, 12. v. efen-lîce.

em-lícnes, -ness, e; *f. Evenness, equality, equity*; æquitas:—He dēmþ folc on emlícnesse *judicābit pŏpŭlos in æquitāte*, Ps. Spl. T. 95, 10: 110, 7 : 118, 75. v. efen-lícnes.

em-micel; *adj. Equally much*; æque multus:—Em-micel ealra *equally much of all*, L. M. 1, 2; Lchdm. ii. 30, 5. v. emn-micel, efen-micel.

emn; *adj. Even, equal, plain, level, just*; æquus, plānus, æquālis:—Dæs wîsan monnes môd biþ swîde emn *the wise man's mind is very even*, Past. 42, 1; Hat. MS. 58 a, 16 : 17, 5; Hat. MS. 23 a, 7 : Ps. Th. 10, 8. Næs ic næfre swâ emnes môdes *I was never of so even a mind*, Bt. 26, 1; Fox 90, 25. Seó burh wæs getimbred on swîde emnum lande *the city was built on very level land*, Ors. 2, 4; Bos. 44, 20 : Past. 4, 2; Hat. MS. 10 a, 14. Habbaþ emne wæga and emne gemetu and sestras *stătēra justa et æqua sint pondĕra, justus mōdius æquusque sextārius*, Lev. 19, 36. On emn *on even ground, by, near*; in æquāli, juxta, Gen. 16, 12 : 21, 19 : Jos. 10, 5 : Homl. Th. i. 30, 16; Byrht. Th. 137, 9; By. 184. To emnes *over against, opposite*; adversus, contra, Ors. 1, 1; Bos. 21, 8 : 3, 9; Bos. 68, 25 : Cod. Dipl. 1102; A.D. 931; Kmbl. v. 194, 32; 195, 2. v. efen.

emn-, emne-, in composition, *even, equal*, as efen:—Emne-cristen *a fellow-christian*. Emn-sceólere *a school-fellow*.

emn-ædele; *adj. Equally noble*; æque nōbīlis:—Ealle sint emn-ædele *all are equally noble*, Bt. 30, 2; Fox 110, 17 : Bt. Met. Fox 17, 27; Met. 17, 14.

emne; *comp.* emnor, emnar; *adv. Equally, even, exactly, precisely, just*; æquāliter, æque, omnīno:—Sió sunne and se mōna habbaþ todæled betwuht him done dæg and da niht swîde emne *the sun and the moon have divided the day and the night very equally between them*, Bt. 39, 13; Fox 234, 6 : Bt. Met. Fox 29, 72; Met. 29, 35 : Ps. Th. 9, 8. Crist hiene selfne ge-eáþmēdde emne óþ done deáþ *Christ humbled himself even unto death*, Past. 41, 1; Hat. MS. 56 a, 22 : 50; Hat. MS : Cd. 92; Th. 116, 28; Gen. 1943: Bt. Met. Fox 9, 76; Met. 9, 38 : 13, 89; Met. 13, 45 : Andr. Kmbl. 227; An. 114 : 441; An. 221: 665; An. 333. Ne wēne ic dæt ænige twegen lâtteówas emnar gefuhton *I do not think that any two leaders fought more equally*, Ors. 3, 1; Bos. 53, 32. v. efen.

emn-éce; *adj. Co-eternal*; coæternus:—Is emnéce mægenþrymnes *est coæterna majestas*, Ps. Lamb. fol. 200, 25. Ealle þrý hâdas emnéce him sylfum synt *totæ tres personæ coæternæ sibi sunt*, 201, 27. v. efen-éce.

emne-cristen, em-cristen, es; *m. A fellow-christian*; co-christiānus:—His emnecristen *fratrem suum in Christo*, L. Ed. C. 36; Wilk. 209, 18.

emne-lîce *evenly, equally*, Som. Ben. Lye. v. efen-lîce.

emnes, -ness, -niss, -nyss, e; *f. Evenness, equity, justice*; æquitas:—Drihten dû gelíffæst me on efnesse odðe emnesse dîne *Dŏmine vivificābis me in æquitāte tua*, Ps. Lamb. 142, 11. Emnesse geseah anwlita his *æquitātem vīdit vultus ejus*, 10, 8. He dēmþ ymbhwyrft eorþan on emnisse *ipse judicābit orbem terræ in æquitāte*, Ps. Spl. 9, 8. Eágan dîne geseón emnyssa *ŏcŭli tui vīdeant æquitātes*, 16, 3 : 51, 3 : 110, 7. v. efen-nyss.

emnett, es; *n? Level ground, a plain*; plānĭties, campus:—He hæfde on ham emnette gefaren *he had marched on the level ground*, Ors. 4, 8; Bos. 89, 38.

emnettan, emnyttan, to emnettenne; *p.* te; *pp.* ed *To make even* or *equal, to regulate*; æquāre, coæquāre:—Synt to emnettenne be dissere emnihte *they are to be regulated by this equinox*, Bd. de nat. rerum; Wrt. popl. science 11, 15; Lchdm. iii. 256, 24. Ic emnytte *coæquo*, Ælfc. Gr. 47; Som. 48, 56. DER. ge-emnettan.

emn-gód; *adj. Equally good*; æque bŏnus:—Nân wuht nis betere donne God ne emngód him *no creature is better than God nor equally good with him*, Bt. 34, 3; Fox 138, 7. Nyton nâuht emngód *they know nothing equally good*, 34, 2; Fox 136, 4.

emnian *to equal, to make alike*, Som. Ben. Lye.

em-niht, es; *n.* [em, emn *equal*; niht *night*] *Equal day and night, equinox*; æquǐnoctium:—On emnihtes dæg, ꝥæt is ꝺonne se dæg and seó niht gelíce lange beóþ *on the day of the equinox, that is when the day and night are equally long*, Bd. de nat. rerum; Wrt. popl. science 12, 19; Lchdm. iii. 260, 13. Ver is lencten tíd, seó hæfþ emnihte *spring is the lenten tide, which hath an equinox*, 8, 28; Lchdm. iii. 250, 10. Autumnus is hærfest, ꝺe hæfþ óꝺre emnihte *Autumn is harvest, which hath the other equinox*, 9, 1; Lchdm. iii. 250, 11. On ꝺæs hærfestlícan emnihtes ryne *in the course of the harvest [autumnal] equinox*, Lchdm. iii. 238, 27. To hærfestes emnihte *at the autumnal equinox*, Th. Diplm. A. D. 902; 151, 11.

emnis, -niss *evenness, equity*, Ps. Spl. 9, 8. v. emnes.

emn-land *even land, a plain*, Som. Ben. Lye.

emn-líce; *adv. Equally, evenly;* æquǎlǐter, æque:—Ðæt hine ealle emnlíce hérian *that all praise him equally*, Ps. Th. 32, 1: Bt. 13; Fox 38, 34, MS. Cot. v. efen-líce.

emn-micel, em-micel; *adj. Equally great;* æque magnus:—Habbaþ emnmicelne willan to cumenne *they have equally great desire to come*, Bt. 36, 4; Fox 178, 10: 42; Fox 256, 10. v. efén-micel.

emn-neáh; *prep. Equally near;* æque prǒpe:—On ælcere stówe he is hire emn-neáh *it is in every place equally near it*, Bt. 33, 4; Fox 130, 23. v. efen-neáh.

emn-réꝺe; *adj.* [réꝺe *cruel*] *Equally cruel;* æque sævus:—Romulus and Brutus wurdon emnréꝺe *Romulus and Brutus were equally cruel*, Ors. 2, 3; Bos. 41, 42.

emn-sár, es; *n. Equal sorrow* or *contrition;* æquǎlis dǒlor:—Hie ne mágon ealneg ealla on áne tíd emnsáre hreówan *they cannot always repent of all at one time with equal sorrow*, Past. 53, 3; Hat. MS.

emn-sárian *to be alike sorry, to condole;* condǒlére, Som. Ben. Lye.

emn-sárig *equally sorry*, Som. Ben. Lye. v. em-sárig, efen-sárig.

emn-sceólere, es; *m. A fellow-scholar;* condiscǐpǔlus:—He ofslóh his emnsceólere *he slew his fellow-scholar*, Ors. 3, 9; Bos. 67, 12.

emnys, -nyss *evenness, equity*, Ps. Spl. 16, 3: 51, 3: 110, 7. v. emnes.

emnyttan *to make equal*, Ælfc. Gr. 47; Som. 48, 56. v. emnettan.

empire *an empire;* impěrium, Lye.

em-rene, es; *m. A circle;* circǔlus, C. R. Ben. 18. v. ymb-rene.

em-sárig; *adj. Equally sorry;* æque tristis:—Hí woldon ꝺæt ꝺa óꝺre wíf wæran emsárige heom *they wished the other women to be equally sorry with themselves*, Ors. 1, 10; Bos. 33, 1. v. efen-sárig.

em-sníꝺan; *p.* -snáþ, *pl.* -snidon; *pp.* -sniden *To circumcise;* circumcídêre:—Ge emsníꝺaþ ꝺæt flǽsc eówres fylmenes *circumcídêtis carnem præpūtii vestri*, Gen. 17, 11. v. ymb-sníꝺan.

em-swápen *clothed;* amictus, Som. Ben. Lye. v. ymb-swápan.

emta, an; *m. Leisure;* ōtium:—On emtan to smeágeanne *to study at leisure*, Bd. pref. S. 471, 10. Ic get emtan næbbe *I have not leisure yet*, Bt. 38, 2; Fox 196, 24. v. æmta.

emtig; *adj. Empty, idle;* vacuus, ōtiōsus:—Híg synt emtige *they are idle*, Ex. 5, 8. v. æmtig.

em-trymming, e; *f. A fortress, fence;* mūnīmentum, Som. Ben. Lye. v. ymb-trymming.

em-twá *two even parts, halves;* dīmǐdia:—Ne dǽlaþ on emtwá heora dagas *non dīmǐdiābunt dies suos*, Ps. Lamb. 54, 24. He tobærst on emtwá *he burst asunder into halves*, Homl. Th. ii. 250, 26.

-en.   I. *m.* forms only a few masculine *terminations* of nouns; as, þeóden; *gen.* þeódnes; *m. a king*, from þeód *people:* dryhten; *gen.* dryhtnes; *m. a lord*, from dryht *people, subjects.* **II.** *f.* -en forms many feminine nouns = the *Ger.* -in, *Dan.* -inde; as, þínen, e; *f. a maid-servant* [*Ger.* dienerin], from þén [*Ger.* diener]; þeówen, e; *f. a female slave*, from þeów: wylen; *gen.* wylne; *f. the same*, from weal *a slave:* mennen, e; *f. a maid-servant*, from manna: gyden, e; *f. a goddess*, from god: munecen, e; *f. a nun*, from munec: cǽsern [= cǽsere + en], e; *f. an empress*, from cǽsere: fyxen, e; *f. a she-fox*, from fox. Also -en forms many nouns of the *f.* gender [corresponding to the *Icel.* -n, -in]; as, Segen; *gen.* segne; *f. tradition, saying*, Icel. sögn: gýmen, e; *f. heed, care:* byrgen, e; *f. a tomb:* sylen, e; *f. a gift:* byrꝺen, e; *f. a burden:* hiwrǽden; *gen.* hiwrǽdenne; *f. a family, house:* and several others in -ræden; as, Gecwyd-rǽden, e; *f. an agreement, contract:* mǽg-rǽden, e; *f. relationship:* gefér-rǽden, -rǽdenn, e; *f. a train, company, congregation.* **III.** some nouns in -en are neuters [corresponding to the *Icel.* -in, -en]; as, Mægen, es; *n. strength, might* = Icel. megin, magn: mǽden, es; *n. a maiden:* wésten, es; *n. a waste, desert:* swefen, es; *n. a dream:* midlen, es; *n. a middle:* fæsten, es; *n. a fortress, fastness.*

-en is a termination of adjectives,—hence from fýr *fire* is fýren *fiery;* stǽn *a stone;* stǽnen *stony:* -en is also the termination of *pp.* in strong verbs; arisen *risen*, from arísan *to rise;* dolfen *digged*, from delfan *to dig;* witen *known*, from witan *to know.*

én = ǽn = án- one, as,—ǽn-líc, *q. v.* = án-líc, *q. v.* én-wintre *one winter, q. v;* én-líc = án-líc, *q. v;* én-lípig = án-lípig, *q. v.*

encgel, es; *m. An angel;* angělus:—Hálig encgel *a holy angel*, Cd. 226; Th. 301, 24: Sat. 586, = engel *an angel.*

énd; *adv. Formerly, of old;* prius, ōlim:—Ic adreág fela siꝺꝺan ꝺú énd to me in síꝺadest *I have suffered much since thou didst come to me of old*, Exon. 120 b; Th. 463, 16; Hö. 71.

-end, es; *m.* the ending of nouns, denoting the agent:—Wegferend, es; *m. a way-faring man.*

ENDE, es; *m.*   **I.** *an* END; fīnis, termǐnus:—Ac nys ꝺonne gyt se ende *sed nondum est fīnis*, Mt. Bos. 24, 6. Á bútan ende *ever without end*, L. E. I. prm; Th. ii. 400, 28. Ðæt hí ꝺæs gewinnes sumne ende gedyden *that they would make an end of the war*, Ors. 2, 2; Bos. 41, 1. Ðú eart eallra þinga fruma and ende *thou art the beginning and end of all things*, Bt. Met. Fox 20, 549; Met. 20, 275: Andr. Kmbl. 1112; An. 556.   **II.** *a corner, part, sort;* angǔlus, pars, spěcies:—Ðæt sylfe wæter ꝺæt hí ꝺa bán mid þwógan, gutan in ænne ende ꝺære cyricean *the selfsame water that they washed the bones with, they poured into one corner of the church* [in angǔlo sacrārii], Bd. 3, 11; S. 535, 33. Harold of-slóh ꝺǽr mycelne ende ꝺæs folces *Harold slew there a great part of the people*, Chr. 1052; Gib. 166, 22; Th. 319, 14, col. 1. On feówer endum ꝺyses middangeardes *in the four parts of this world*, Ors. 2, 1; Bos. 38, 21. Ofer ealle eorþan endas *over all parts of the earth*, Ps. Th. 18, 4. Ne hæfde wit óꝺer uncymran hors and óꝺres endes *numquid non hǎbuǐmus ěquos vīliōres, vel ālias spěcies*, Bd. 3, 14; S. 540, 27. [*Chauc. Wyc.* ende; *O. Sax.* endi, *m. n: Frs.* eyn, eyne: *O. Frs.* enda, einde, eind, ein, *m: Dut.* einde, *n: Ger.* ende, *n. m: M. H. Ger.* ende, *n. m: O. H. Ger.* anti, enti, *m. n: Goth.* andeis, *m: Dan.* ende, *m. f: Swed.* ände, *m: Icel.* endi, endir, *m: Sansk.* anta, *m.*] DER. eást-ende, norþ-, west-, woruld-.

-ende, the termination forming the active participle:—Wegfer-ende *way-faring:* also found for -enne. v. -anne.

ende-byrd, e; *n? An arranging, arrangement, order;* ordo:—Se Ælmihtiga ealra gesceafta endebyrd wundorlíce gemetgaþ *the Almighty wonderfully regulates the arrangement of all creatures*, Bt. Met. Fox 13, 8; Met. 13, 4.

ende-byrdan; *p.* de; *pp.* ed *To set in order, adjust, dispose;* dispō-něre, Ps. Spl. 49, 6.

ende-byrdes; *adv. Orderly, for order;* per ordǐnem, ordǐnātim:—Ðe him róꝺera Weard endebyrdes gesette *which the Guardian of the skies has orderly appointed for them*, Bt. Met. Fox 11, 41; Met. 11, 21. Ðú ꝺysne middangeard todældest swá hit getǽsost wæs endebyrdes *thou hast divided this middle earth as it was most suitable for order*, 20, 23; Met. 20, 12.

ende-byrdlíc; *adj. Belonging to order, ordinal;* ordǐnālis:—Ende-byrdlíce naman *ordǐnālia nōmǐna*, Ælfc. Gr. 49; Som. 49, 53.

ende-byrdlíce; *adv. Orderly, in order, in succession;* successīve:—Ealle ꝺás wǽron endebyrdlíce bisceophǎda brúcende on Myrcna þeóde *all these in succession enjoyed the bishopric of Mercia*, Bd. 3, 24; S. 558, 4. Endebyrdlíce in order, Bt. 33, 4; Fox 128, 7.

ende-byrdnes, -byrnes, -ness, e; *f. Order, disposition, method, way, manner, means;* ordo:—Óþ endebyrdnesse *ex ordǐne*, Lk. Bos. 1, 3. On endebyrdnesse *in ordǐne*, 1, 8.

ende-dæg; *gen.* -dæges; *pl. nom. acc.* -dagas; *gen.* -daga; *dat.* -dagum; *m. The last day, the day of one's death;* dies suprēmus, dies mortis:—Ðá wæs endedæg ꝺæs ꝺe Caldéas cyningdóm áhton *then was the last day that the Chaldeans held the kingdom*, Cd. 209; Th. 258, 22; Dan. 679. Ic sceal endedæg mínne gebídan *I shall await my last day*, Beo. Th. 1279; B. 637. Án endedæg *one ending day*, Apstls. Kmbl. 157; Ap. 79.

ende-deáþ, es; *m.* [ende *an end*; deáþ *death*] *Final death;* mors vītam fīniens:—Líf bútan endedeáþe *life without final death*, Exon. 32 a; Th. 101, 4; Cri. 1653.

ende-dógor, es; *m. n. The final day, day of one's death;* fīnālis dies, mortis dies:—Wæs endedógor neáh geþrungen *the final day was near at hand*, Exon. 46 a; Th. 158, 8; Gú. 905: 49 b; Th. 171, 11; Gú. 1125: 50 a; Th. 174, 7; Gú. 1174. Ðæt eorlwerod sæt on wénum ende-dógores *the warrior band sat in expectation of the final day* [death], Beo. Th. 5784; B. 2896. Nis nú swíꝺe feor ꝺam ýtemestan endedógor *it is now not very far to the utmost final day*, Exon. 49 b; Th. 172, 8; Gú. 1140. Bád se endedógor *he awaited* [his] *final day*, 51 b; Th. 179, 10; Gú. 1259.

ende-láf, e; *f.* [ende *an end*; láf *a remainder, remnant*] *The last remnant;* extrēmum relǐquum:—Ðú eart endeláf usses cynnes *thou art the last remnant of our race*, Beo. Th. 5618; B. 2813.

ende-leán, es; *n.* [leán *a reward*] *A final reward;* fīnālis retrǐbūtio:—Him ꝺæs æfter becwom yfel endeleán *for this an evil final reward came on him afterwards*, Cd. 181; Th. 227, 15; Dan. 187. Him endeleán þurh wæteres wylm Waldend sealde *the Almighty gave to them a final reward through the water's rage*, Beo. Th. 3389; B. 1692.

ende-leás; *adj.* ENDLESS, *infinite, eternal;* infīnītus, perpětuus, æter-nus:—Ðæt is endeleás wundor *that is an endless wonder*, Bt. 36, 1; Fox 172, 18: Exon. 100 b; Th. 379, 8; Deór. 30: Andr. Kmbl. 1389; An. 695. Hý sceolon sár endeleás forþ þrówian *they must thenceforth suffer endless pain*, Exon. 31 b; Th. 99, 30; Cri. 1632: 69 a; Th. 257, 22;

Jul. 251. Ða earmþa beóþ endeleáse ðe éce bióþ *those miseries are endless which are eternal*, Bt. 38, 2; Fox 198, 16.

ende-leáslíce; *adv.* ENDLESSLY, *eternally;* infīnīte, Som. Ben. Lye.

ende-leásnys, -nyss, e; *f.* ENDLESSNESS, *eternity;* infīnītas, Ælfc. Gr. 18; Som. 21, 58.

ende-líf, es; *n. An end of life, death;* vīta fīnīta, mors:—Wurdon hie deáþes on wénan, áðes and endelīfes *they were in expectation of death, of the funeral pile and end of life,* Elen. Kmbl. 1166; El. 585.

ende-mæst *endmost, last;* extrēmus, Som. Ben. Lye.

ende-mes, endemes, ændemes, ændemest; *adv. Equally, likewise, in like manner, together;* parīter:—Forðon ic ne mæg eal ða monigfealdan yfel endemes areccan *because I cannot equally reckon all the manifest evils,* Ors. 2, 5; Bos. 49, 11: 3, 10; Bos. 69, 36. Ne mæg hió ealle endemest gescīnan *nor can she equally shine upon all,* Bt. 41, 1; Fox 244, 9.

endemestnes, -ness, e; *f. An extremity;* extrēmĭtas, R. Ben. interl. 6.

ende-néhst, -nýhst, ende-nēxta, ende-níhsta; *adj. The nighest end, the last, uttermost;* ultĭmus:—Drihten, ðū oncneówe ealle ða nywestan oððe ða endenīhstan [MS. ændenihstan] *Dŏmĭne, tu cognōvisti omnia novissĭma,* Ps. Lamb. 138, 5. Februārius se mōnaþ is ealra scyrtst and endenýhst *February is the shortest and last month of all,* Bd. de nat. rerum; Wrt. popl. science 13, 28; Lchdm. iii. 264, 8.

ende-rím, es; *n. The final number, the number;* fīnālis nŭmĕrus:—Daga enderīm he gesette *he set the number of days,* Cd. 213; Th. 265, 24; Sat. 12.

ende-sǽta, an; *m. An end* or *border inhabitant, one stationed at the extremity of a territory;* līmĭtis incŏla, Beo. Th. 487; B. 241.

ende-spǽc, e; *f. An end-speech, epilogue;* epĭlŏgus, Reg. Conc. in Epílŏgo.

ende-stæf; *pl. nom. acc.* -stafas; *m. An epilogue, conclusion, destruction;* epĭlŏgus, perorātio:—Heó endestæf gesceáwiaþ *they shall behold their end,* Cd. 225; Th. 298, 30; Sat. 541.

endian, ændian; *p.* ode; *pp.* od *To* END, *make an end;* fīnīre, dēsĭnĕre:—Hí hit endian sceoldon *they should end it,* Ps. Th. 9, 6. v. ge-endian.

endleofan, endlufon, endlyfun, *inflected cases of* endleof, endluf, endlyf [end=ān *one;* unus; leof=líf, *from* līfan *to leave;* relinquĕre, Grm. ii. 947, *or* end=ān *one;* lif *ten;* dĕcem; *existing in Teutonic languages only in the words for* 11 *and* 12; *A. Sax.* end-líf *and* twē-lf=twā-lf= twā-líf, Grm. Gsch. § 246] ELEVEN; undĕcim=ἕν-δεκα:—Ōsrēd ðæt ríce hæfde endleofan wintra *Osred held the kingdom for eleven years,* Bd. 5, 18; S. 635, 20. Mid hīra endlufon sunum *cum undĕcim fīliis,* Gen. 32, 22. Endleofan steorran *eleven stars,* Gen. 37, 9: Chr. 71; Th. 13, 3, col. 3. [*Wyc.* enleuene, enleuen, enleue: *R. Glouc.* endleve: *Laym.* elleoue, elleouen: *Plat.* elv, elwen: *O. Sax.* ellevan: *Frs.* alve, alue: *O. Frs.* andlova, elleva: *Dut.* elf: *Ger.* eilf, elf: *M. H. Ger.* einlif, einlef: *O. H. Ger.* einlif: *Goth.* ainlif: *Dan.* elleve: *Swed.* elfva: *Icel.* ellifu.] v. twelf.

endlyfta, ændlyfta, ællyfta, seó, ðæt, -e; *adj. The eleventh;* undĕcĭnus:—On ðam endlyftan mōnþe *undĕcimo mense,* Deut. 1, 3. Endlyfta ðæra tácna ys gehāten āquārius *the eleventh of the signs is called āquārius,* Bd. de nat. rerum; Wrt. popl. science 7, 9; Lchdm. iii. 246, 3.

endung, e; *f. An* ENDING, *end;* fīnis, consummātio:—Ðæt ríp is worulde endung *messis consummātio sæcŭli est,* Mt. Bos. 13, 39. DER. ge-endung.

end-werc, es; *n.* [werc=wærc *pain*] *A pain in the buttocks;* nātium dŏlor:—Ðes drænc is gód wið endwerce *this drink is good for pain in the buttocks,* Lchdm. iii. 50, 11.

ENED, e; *f.* I. *a duck;* ănas, gen. ănătis; *f.* ăněta:—Óþ enede mére *to the duck's mere,* Cod. Dipl. 204; A. D. 814; Kmbl. i. 258, 5. Ened ăněta, Ælfc. Gr. 7; Som. 6, 52: Wrt. Voc. 77, 22: 280, 8. II. ened, m. *A drake;* ănas, ănětārius, mascŭlus istīus ăvis:—Ened *a drake?* ănas, gen. ănătis; *m.* Ælfc. Gl. 36; Som. 62, 122; Wrt. Voc. 29, 18. Ened *a drake?* larax? Wrt. Voc. 280, 9. [*Dut.* eend, end, *f. a duck;* *m. a drake:* *Ger.* ente, *f. a duck;* enterich, *m. a drake:* *M. H. Ger.* ant, *f. a duck;* *m. a drake:* *O. H. Ger.* anut, anit ănas: *Dan.* and, *m. f.:* *Swed.* and, *f. a wild duck:* *Icel.* önd, *f. pl.* endr, andir *a duck:* *Lat.* ănas, gen. ănătis, *m. f.:* *Grk.* νῆττα, νῆσσα, *f. a duck.*]

eneleác, es; *n. An onion;* cæpe:—We hæfdon porleác and eneleác *in mentem nōbis vĕniunt porri et cæpe,* Num. 11, 5. v. enneleác.

enetere, ænitre; *adj. Of a year old;* annĭcŭlus:—Ðú dést ælce dæg on ðæt weofod twá enetere lamb *fácies in altāri agnos annĭcŭlos duos per singŭlos dies,* Ex. 29, 38. v. ān-wintre.

enga *sole:*—Mid ðinne engan Freán *with thy sole Lord,* Exon. 11 a; Th. 15, 17; Cri. 237. v. ānga.

enge *from confinement,* Cd. 71; Th. 86, 23; Gen. 1435. v. engu.

enge; *def.* se enga; *adj. Narrow, anxious;* angustus, anxius:—Ufan hit is enge *it is narrow above,* Exon. 116 a; Th. 446, 14; Dóm. 22: 47 a; Th. 162, 3; Gú. 970. Of ðam engan hofe *from that narrow house,* 73 b; Th. 274, 12; Jul. 532: 8 a; Th. 3, 6; Cri. 32. Enge ānpaðas *narrow passes,* Cd. 145; Th. 181, 8; Exod. 58: Beo. Th. 2824;

B. 1410. Helle wisceþ, ðæs engestan éðel-ríces *shall wish for hell, the narrowest realm,* Salm. Kmbl. 213; Sal. 106. v. ange.

ENGEL, ængel, angel, engyl; *gen.* engles; *dat.* engle; *pl. nom. acc.* englas, engel; *gen.* engla; *dat.* englum; *m. An* ANGEL, *a messenger;* angĕlus=ἄγγελος:—Se engel him to cwæþ *dixit illis angĕlus,* Lk. Bos. 2, 10: 1, 13: Mt. Bos. 28, 5: Gen. 22, 12. Godes engel stód on emn hī *the angel of God stood before them,* Homl. Th. i. 30, 15, 17: Mt. Bos. 1, 20, 24: Jn. Bos. 5, 4. Ðæt mæg engel ðin eáþ geféran *that thine angel may more easily travel,* Andr. Kmbl. 387; An. 194. Þurh ðæs engles word *through the angel's word,* Exon. 20 a; Th. 51, 31; Cri. 824: 34 b; Th. 110, 11; Gú. 166: Salm. Kmbl. 901; Sal. 450: Homl. Th. i. 30, 22. He ðam engle oncwæþ *he spake to the angel,* Cd. 141; Th. 176, 12; Gen. 2910: Lk. Bos. 2, 13. God sent his engel beforan ðē *Dŏmĭnus mittet angĕlum suum cōram te,* Gen. 24, 7: 16, 7. Māran cýðde habbaþ englas to Gode ðonne men *angels are more like God than men,* Homl. Th. i. 10, 3. Englas blāwaþ býman *angels shall blow the trumpet,* Exon. 20 b; Th. 55, 9; Cri. 881: 14 a; Th. 28, 17; Cri. 448. Cōmon twegen englas venērunt duo angĕli, Gen. 19, 1, 12, 15. Beheóldon ðæt [MS. ðær] engel Dryhtnes ealle *all the angels of the Lord beheld it,* Rood Kmbl. 18; Kr. 9. Hér sindon nigon engla werod *here are nine hosts of angels,* Homl. Th. i. 10, 14: 12, 8: Elen. Kmbl. 2559; El. 1281. Engla ríce *the kingdom of angels,* 2460; El. 1231. Engla beorhtast *brightest of angels,* Exon. 9 b; Th. 7, 21; Cri. 104. Gif ðú in heofonríce habban wille eard mid englum *if thou wilt have in heaven's realm a dwelling with angels,* Elen. Kmbl. 1240; El. 622: Andr. Kmbl. 1197; An. 599: 3440; An. 1724. Mid hys englum *cum angĕlis suis,* Mt. Bos. 16, 27. Englas God worhte, ða sind gāstas, and nabbaþ nǽnne līchaman *God created angels, which are spirits, and have no body,* Homl. Th. i. 276, 1. Mannes sunu sent his englas *mittet fīlius hŏmĭnis angĕlos suos,* Mt. Bos. 13, 41: Mk. Bos. 13, 27. [*Wyc.* aungel: *Chauc.* aungel: *Laym.* engles, *pl:* *Orm.* enngell: *O. Sax.* engil, *m:* *Frs.* ingel: *O. Frs.* angel, angl, engel, *m:* *Dut.* Ger. M. H. Ger.* engel, *m:* *O. H. Ger.* engil, *m:* *Goth.* aggilus, *m:* *Dan.* engel, *m:* *Swed.* engel, *m:* *Icel.* engill, *m:* *Lat.* angĕlus, *m:* *Grk.* ἄγγελος, *m. f. a messenger, angel.*] DER. heáh-engel, heofon-, up-.

Engel; *gen.* Engle; *f. Anglen in Denmark, the country from which the Angles came into Britain;* Angŭlus, terra quam Angli ante transĭtum in Britanniam cŏluērunt:—Of Engle cōman Eást-Engle, and Middel-Engle, and Myrce, and eall Norþhembra cynn *from Anglen came the East-Angles, and Middle-Angles, and Mercians, and all the race of the Northumbrians,* Bd. 1, 15; S. 483, 24. v. Angel.

engel-cund; *adj. Angelic;* angĕlĭcus:—God him giefe sealde engelcunde *God gave him angelic grace,* Exon. 34 a; Th. 108, 13; Gú. 72.

engel-cyn, -cynn, es; *n.* [engel angĕlus; cyn, cynn gĕnus] *The angel race* or *order;* gĕnus *vel* ordo angĕlōrum:—Wæs ðæt engelcyn [MS. encgelcyn] genemnod *the angel race was named,* Cd. 221; Th. 287, 12; Sat. 366. Ðú sitest ofer ðam engelcynne *thou sittest above the angel race,* Elen. Kmbl. 1463; El. 733. Hæfde se Ealwalda engelcynna tyne getrymede *the Almighty had ten established orders of angels,* Cd. 14; Th. 16, 21; Gen. 246: Andr. Kmbl. 1434; An. 717.

engel-líc, engle-líc; *adj. Angelic;* angĕlĭcus:—He ge-earnode ðæt he wæs brúcende engellícre gesihþe *angĕlica mĕruit vīsiōne perfrui,* Bd. 3, 19; S. 547, 13.

Engla feld; *gen.* feldes; *dat.* felda, felde; *m.* [Hovd. Englefeld: Brom. Englefelde: Matt. West. Anglefeld: *Angles' field, the field of the English*] ENGLEFIELD *or* INGLEFIELD, *near Reading, Berkshire;* lŏci nōmen in agro Berkeriensi:—Hér cwom se here to Reádingum on West-Seaxe, and ðæs ymb iii niht ridon ii eorlas up: ða gemétte hie Æðelwulf aldorman on Engla felda, and him ðær wið gefeaht, and sige nam *in this year* [A. D. 871] *the army came to Reading in Wessex, and three nights after two earls rode up: then alderman Æthelwulf met them at Inglefield, and there fought against them, and gained the victory,* Chr. 871; Erl. 74, 5-8.

Engla land, es; *n. The land of the Angles* or *Engles,* ENGLAND; Anglōrum terra. It extended in the time of Bede, A. D. 731, from the present Lincolnshire to the Frith of Forth, on the south of which Æbbercurnig is located:—Ðæt mynster Æbbercurníg, ðæt is geseted on Engla lande *the minster Abercorn, that is seated in the land of the Angles,* or Engla land=England, Bd. 4, 26; S. 602, 36.

Englan; *gen.* ena; *dat.* um; *acc.* an; *pl. m. The Angles;* Angli:—Ða Wealas flugon ða Englan [=Engle, Th. 22, 27, col. 2, 3] *the Welsh fled from the Angles,* Chr. 473; Th. 23, 26, col. 2; 23, 27, col. 1. Betweox Wealan and Englan *between the Welsh and Angles,* L. O. D. 2; Th. i. 354, 2: 3; Th. i. 354, 10. v. Engle, Angle *the Angles.*

englas *angels,* Homl. Th. i. 276, 1. v. engel.

Engle, Angle; *pl. nom. acc;* *gen.* a; *dat.* um; *pl. m.* Englan; *gen.* ena; *pl. m. The Angles;* Angli *The inhabitants of Anglen in Denmark.* Anglen was the province from which the English derived their being and name. Anglen [v. Engel] lies on the south-east part of the Duchy of Sleswick, in Denmark. **The majority of settlers in Britain**

were from Anglen and the neighbourhood, hence this country and people derived their name *England* and *English,* England being derived from Engla land *the land* or *country of the Angles:*—On ðæm landum eardodon Engle, ǽr hȳ hider on land cómon *the Angles [Engles] dwelt on these lands before they came hither on land* [i. e. *before they came to England*], Ors. 1, 1; Bos. 21, 36. Engla cyningas *kings of the Angles,* Bd. 2, 15; S. 518, 38. Betweox Wealum and Englum *between the Welsh and English,* L. O. D. 2; Th. i. 352, 14.

Engle *of Anglen,* Bd. 1, 15; S. 483, 24; *gen. dat. acc. of* Engel *Anglen,* q.v.

engle-líc; *adj. Angelic;* angēlicus:—Engelíce ansȳne hí habbaþ *angēlicam hăbent făciem,* Bd. 2, 1; S. 501, 18. v. engel-líc.

Englisc, Ænglisc, *adj.* ENGLISH; Anglĭcus:—Hér syndon on ðis íglande [Britene] fíf geþeóda [MS. þeóda], Englisc, and Brytisc, ... and Scyttisc, and Pihtisc, [and Bóc-Lǽden] *here are in this island* [*Britain*] *five languages, English, and British,* ... *and Scottish, and Pictish,* [*and Book-Latin*], Chr. Th. 3, 3–6, col. 3, 2. Ðæt is on Englisc, mín God *that is in English, my God,* Mt. Bos. 27, 46. On Englisc *in English,* Bd. 3, 19; S. 547, 22. On Englisc land, ne Englisc on Wilisc *in England* [*English land*], *nor English in Welsh,* L. O. D. 6; Wilk. 126, 3. Awendan of Lédene on Englisc *to translate from Latin into English,* Ælfc. pref. Gen. 1, 4. Seó bóc is on Englisc awend *the book is turned* [*translated*] *into English,* Homl. Th. ii. 358, 30. Ic [Ælfríc Abbod] gesett hæbbe wel feówertig lárspella on Engliscum gereorde *I* [*Abbot Ælfríc*] *have composed about forty sermons in the English tongue,* Ælfc. T. 27, 17. Ðeáh ða scearpþanclan witan ðisse Engliscan geþeódnesse ne behófien *though the sharp-minded wise men need not this English translation,* MS. Cot. Faust. A. x. 150 b; Lchdm. iii. 440, 31.

Englisc-man, -mon, es; *m. An Englishman;* Anglĭcānus:—Ic wille ðæt gē fédaþ ealle wæga ān earm Engliscmon *I will that ye entirely feed one poor Englishman,* L. Ath. i. prm; Th. i. 198, 5.

engu, e; *f. Narrowness, confinement, a narrow place;* angustiæ:—Of enge *from confinement,* Cd. 71; Th. 86, 23; Gen. 1435: Exon. 101 b; Th. 383, 17; Rä. 4, 12. On enge, Th. 383, 3; Rä. 4, 5. [*Ger. M. H. Ger.* enge, *f.* angustiæ: *O. Nrs.* öngum, *dat. pl.* angustiis.]

engyl, es; *m. An angel;* angĕlus:—His engyl ongan ofermód wesan *his angel began to be presumptuous,* Cd. 14; Th. 17, 19; Gen. 262: 15; Th. 19, 18; Gen. 293: Mt. Bos. 11, 10. v. engel.

enid *a duck, drake, coot, water-fowl;* ănăs, ăneta, fulĭca, Som. Ben. Lye. v. ened.

énig *any,* Th. Diplm. A. D. 830; 466, 1. v. ænig.

énitre; *adj. Of a year old;* annĭcŭlus:—Gif seó offrung beó of sceápon oððe of gátum, bring énitre offrunge *if the offering be of sheep or of goats, bring an offering of a year old,* Lev. 1, 10. v. énetere.

én-líc *only;* ūnĭcus, Lye. v. án-líc.

en-líhtan *to enlighten,* Som. Ben. Lye. v. on-líhtan.

én-lípig *each;* singŭlāris, Ælfc. Gr. 49, Lye. v. án-lípig.

-enne *the termination of the declinable infinitive in the dat. governed by* to, *as,*—To farenne *to go,* Mt. Bos. 8, 21. v. -anne.

enneleác, ennelēc, eneleác, ynneleác, yneleác, es; *n.* [leác *a leek,* onion] *An onion;* cæpe, ūnio:—Enneleác *an onion,* Glos. Brux. Recd. 41, 19; Wrt. Voc. 67, 34. Ennelēc *cæpe,* Ælfc. Gl. 40; Som. 63, 106; Wrt. Voc. 30, 54.

ent, es; *m. A giant;* gĭgas = γίγας:—He geblissode swá swá se mǽsta oððe swá swá ent to ge-yrnanne weg his *exultāvit ut gĭgas ad currendam viam ejus,* Ps. Lamb. 18, 6: Ps. Spl. 32, 16: Wrt. Voc. 73, 52. Nembroþ se ent *Nimrod the giant,* Boutr. Scrd. 21, 35: Ors. 2, 4; Bos. 44, 17. Dauid eóde to ánwíge ongeán ðone ent Goliam *David went in single combat against the giant Goliath,* Ælfc. T. 14, 3: Ors. 1, 10; Bos. 33, 29. Entas wǽron ofer eorþan on ðám dagum *gĭgantes ĕrant sŭper terram in dĭebus illis,* Gen. 6, 4: Homl. Th. i. 318, 15. He seah on enta geweorc *he looked on the work of giants,* Beo. Th. 5428; B. 2717: Exon. 77 b; Th. 291, 24; Wand. 87: Andr. Kmbl. 2988; An. 1497: Menol. Fox 463; Gn. C. 2. v. eten, eóten.

ent-cyn, -cynn, es; *n. Giant-kind, giant-race;* gĭgantum gĕnus:—We gesáwon of ðám entcynne Enachis bearna micelra wæstma *vidīmus monstra quædam fīliōrum Enac prŏcērae stătūrae,* Num. 13, 34.

entisc *belonging to* or *made by a giant, giant;* gĭgantēus:—Lét entiscne helm brecan *he caused the giant helmet to break,* Beo. Th. 5951; B. 2979. v. eótenisc.

entse, an; *f. A shekel, Jewish money;* sĭclus:—Ic geseah twáhund entsena hwítes seolfres and sumne gildenne dalc on fíftigum entsum *vīdi dūcentos sĭclos argenti regulamque auream quinquāginta sĭclōrum,* Jos. 7, 21. v. yntse.

én-wintre; *adj. Of a year old;* annĭcŭlus:—Énwintre *vecta?* Wrt. Voc. 287, 60. v. án-wintre.

eo.   I. unaccented, *generally* stands before *two consonants* lc, ld, lf, rc, rd, rf, rg, rh, rl, rm, rn, rp, rr, rt. rþ, x; as, Geolca *a yolk,* sceolde *should,* seolfor *silver,* deorc *dark,* sweord *a sword,* ceorfan *to carve,* beorgan *to protect,* beorht *bright,* eorl *earl,* beorma *barm,* ĕornost *earnest,* weorpan *to throw,* steorra *a star,* heorte *the heart,* eorþe *the earth,* meox *dung.*   II. eó accented, the diphthong, generally stands before the

consonants c, d, f, g, h, l, m, n, p, r, s, st, t, w; as, Seóc *sick,* beódan *to bid,* þeóf *a thief,* fleógan *to fly,* hreóh *rough,* hweól *a wheel,* leóma *a ray of light,* beón *to be,* deóp *deep,* beór *beer,* ceósan *to choose,* breóst *the breast,* fleótan *to float,* leóþ *a song,* ceówan *to chew.*   2. eó is also the termination of many words, and then the ó in eó is always accented; as, Beó *a bee;* ic beó *I shall be;* freó *free;* gleó *glee;* seó *the;* seó *sim, sis, sit;* treó *a tree;* þreó *three,* etc

eó the Runic character for these letters is **Z.** v. eóh = íw *a yew-tree.*

eóc, eócon *increased;* p. of eácan.

eóc *safety, help, succour,* Wald. 45; Vald. 1, 25. v. geóc.

eóde, es; *n. A flock;* grex:—Ðæt lytle eóde *pŭsillus grex,* Lk. Skt. Rush. 12, 32. v. eówde.

eóde, ðú eódest, *pl.* eódon *went, delivered,* Ps. Th. 60, 4: 67, 21: 94, 11; *p. of* gán.

EODOR, eoder, eodur, edor, eder, es; *m.*   I. *a hedge, fence, enclosure, dwelling, house;* sēpes, sēpīmentum, dŏmus, tectum:—Héht ða eahta mearas on flet teón in under eoderas *he commanded then eight steeds to be led into the court under the enclosures,* Beo. Th. 2078; B. 1037.   II. *a limit, end, region, zone;* ōra, margo, extrēmĭtas, plăga, rĕgio:—Gescóp heofon and eorþan and holma bigong eodera ymbhwyrft [*he*] *created heaven and earth and the seas' expanse, the circuit of zones,* Exon. 67 b; Th. 249, 17; Jul. 113.   III. *a prince, sovereign, protector;* princeps, tūtor:—Ic ðē biddan wille, eodor Scyldinga, ánre béne *I will entreat of thee, sovereign of the Scyldings, one boon,* Beo. Th. 860; B. 428: 2092; B. 1044: Exon. 90 a; Th. 339, 6; Gn. Ex. 90. [*O. Sax.* edor, *m*: *M. H. Ger.* ĕter, *m. n*: *O. H. Ger.* ĕtar: *Icel.* jaðarr, jöðurr, *m.*] DER. edor-brecþ, -brice, eder-gong, eodor-brice, -wír.

eodor-brice, edor-brice, -bryce, es; *m.* [eodor, edor *a hedge, fence;* brice, bryce *a breach, breaking*] *A fence-breaking;* sēpis fractio *vel* violātio:—Ceorles eodorbryce [Th. i. 88, 10, note 25, edorbryce, edor-brice] biþ fíf scillinga *for breaking a churl's fence shall be five shillings,* L. Alf. pol. 36; Lambd. 31, 31.

eodorcan, edorcan; *part.* eodorcende; *p.* te; *pp.* ed *To chew, ruminate;* rūmĭnāre:—He eall mid hine gemynegode and swá swá clǽne nýten eodorcende [*Whelc.* oðer cende] in ðæt swēteste leóþ gehwyrfde *ipse cuncta rĕmĕmōrando sēcum et quăsi mundum ănĭmal rūmĭnando in carmen dulcissĭmum convertēbat,* Bd. 4, 24; S. 598, 7.

eodor-wír, es; *m. A wire-enclosure;* cingŭlum, sēpiens fīlum mĕtallĭcum, Grn:—Ic eom mundbora mínre heorde, eodorwírum fæst *I am the protector of my flock, fortified by wire-enclosures,* Exon. 105 a; Th. 398, 23; Rä. 18, 2.

eodur, es; *m. A prince, sovereign, protector;* princeps, tūtor:—Him Hróþgár gewát, eodur Scyldinga *Hrothgar departed, the Scyldings' protector,* Beo. Th. 1330; B. 663. v. eodor.

eofel *evil,* Bt. 7, 3; Fox 22, 19. v. yfel.

eofer *a boar,* Ps. Th. 79, 13: Beo. Th. 2228; B. 1112: 2660; B. 1328. v. eofor.

eofera, an; *m. A successor;* successor:—Æfter Eorpwalde Rædwaldes eoferan *post Earpualdum Reualdi successōrem,* Bd. 3, 18; S. 545, 35, col. 1. v. eafora.

eofer-spreót, es; *m. A boar-spear;* contus ad vēnātiōnem ūsĭtātus:—Mid eoferspreótum *with boar-spears,* Beo. Th. 2879; B. 1437. v. eoforspreót.

Eofer-wíc *York,* Chr. 189; Th. 15, 28, col. 2. v. Eofor-wíc.

Eofes-ham, Eues-ham; *gen.* -hammes; *m.* [*Flor.* Eouesham: *Hovd.* Heuesham: *Brom.* Euesham: *Kni.* Evisham, Evysham, Ewesham, Evesham] EVESHAM, *Worcestershire;* oppĭdi nomen in agro Vigorniensi:—Ðæs gēres forþférde Æfic se ædela decanus on Eofesham *in this year* [A. D. 1037] *died Æfic the noble dean at Evesham,* Chr. 1037; Th. 294, 36, col. 2. Ælfward wæs abbad on Eofeshamme ǽrest *Ælfward was first abbot of Evesham,* Chr. 1045; Th. 303, 2. Ðæs ylcan geáres man hálgode ðæt mynster on Eofeshamme on vi id' Octobris *in the same year* [A. D. 1054] *was consecrated the monastery at Evesham, on the 6th of the Ides of October* [*October* 10th], Chr. 1054; Th. 322, 34, col. 1; 324, 3, col. 2: 1078; Th. 350, 15.

eofet *a debt,* L. Alf. pol. 22; Wilk. 39, 35. v. eofot.

eofne; interj. Behold! ecce!—Eofne! ða ðe fyrsiaþ híg fram ðē losiaþ *ecce! qui elongant se a te pĕrĭbunt,* Ps. Lamb. 72, 27: 82, 3. v. efne.

EOFOR, eofer, eafor, efor, efer, efyr, ofor, es; *m.*   I. *a boar, a wild boar;* ăper:—Fornam hine eofor of wuda *extermĭnāvit eam ăper de silva,* Ps. Spl. 79, 14; Ps. Th. has,—Hine útan of wuda eoferas wrótaþ 79, 13: Exon. 110 b; Th. 423, 8; Rä. 41, 18: 92 a; Th. 344, 20; Gn. Ex. 176. Sele ðú him flǽsc eofores *give him boar's flesh,* L. M. 2, 4; Lchdm. ii. 182, 14.   II. *the figure of a boar on a helmet;* signum apri sŭper găleam:—Swýn eal-gylden, eofer íren-heard *the swine all-golden, the boar iron-hard,* Beo. Th. 2228; B. 1112: 2660; B. 1328. [*Ger.* eber, *m*: *M. H. Ger.* ĕber, *m*: *O. H. Ger.* ebur, *m*: *Icel.* jöfurr, *m.*] DER. eofor-cumbol, -fearn, -líc, -spreót, -swín, -þring, -þrote, -wíc, -wíc-ceaster, -wícingas, -wíc-scír: eoforen, eoforen-denu.

eofora *a successor.* v. eafora.

eofor-cumbol, eofur-cumbol, -cumbul, es; *n.* [cumbol *a banner*]

A boar-banner; signum ad apri similĭtūdĭnem fabrĭcātum :—Ðǽr wæs on eorle ǽnlíc eoforcumbul there was on the man a beauteous boar-shaped ensign, Elen. Kmbl. 517; El. 259.

eoforen; adj. Belonging to a boar; aprīnus, Som.

eoforen-denu, e; f. A boar-vale; aprīna vallis, Som. Ben. Lye.

eofor-fearn, efor-fearn, efer-fearn, es; n. [fearn a fern] A species of fern, polypody; polypōdium vulgāre, Lin :—Eoforfearn fĭlix mĭnūta, polypŏdium, Glos. Brux. Recd. 41, 36; Wrt. Voc. 67, 51. Eoforfearn fĭlicīna, fĭlix arbŏrātĭca, 41, 66; Wrt. Voc. 68, 1. Wĭð ðon sceal eoforfearn polypody shall [do] for that, L. M. 1, 12; Lchdm. ii. 56, 1: 1, 63; Lchdm. ii. 138, 15: 2, 51; Lchdm. ii. 266, 16. Genim eofor-fearnes mǽst take most of polypody, L. M. 1, 15; Lchdm. ii. 56, 20: 1, 59; Lchdm. ii. 130, 9: iii. 74, 4. Eoforfearn dô on hunig put poly-pody into honey, L. M. 1, 60; Lchdm. ii. 130, 24: 1, 87; Lchdm. ii. 154, 17: iii. 56, 19.

eofor-líc, es; n. A boar-likeness; apri sĭmŭlacrum :—Eoforlíc scionon boar's likenesses shone, Beo. Th. 612; B. 303.

eofor-spreót, eofer-spreót, es; m. A boar-spear; vēnābŭlum, Cot. 200. v. eofer-spreót.

eofor-swín, es; n. A boar pig, male swine; verres :—Eoforswínes cwead verris stercus, L. M. 2, 48; Lchdm. ii. 262, 18.

eofor-þring, es; m. Orion? v. ebur-þring.

eofor-þrote, an; f. [eofor a boar, þrote the throat] The carline thistle; carlina acaulis, Lin :—Eoforþrote colucus? colicus? Glos. Brux. Recd. 41, 64; Wrt. Voc. 67, 79: 291, 7. Wĭð heáfodece sceal eofor-þrote carline thistle shall [serve] for head-ache, Lchdm. iii. 12, 25: 24, 7: L. M. 1, 31; Lchdm. ii. 74, 18: 1, 48; Lchdm. ii. 122, 13: 1, 62; Lchdm. ii. 134, 19, 28: 3, 8; Lchdm. ii. 312, 16. Nim eofor-þrotan sǽd take seed of carline thistle, 3, 12; Lchdm. ii. 314, 18. Eofor-þrotan awyl on ealaþ boil carline thistle in ale, 1, 45; Lchdm. ii. 110, 12, 23: 2, 53; Lchdm. ii. 274, 2: 3, 26; Lchdm. ii. 322, 24: 3, 48; Lchdm. ii. 340, 1.

Eofor-wíc, Eofer-wíc, Efer-wíc, Euer-wíc, es; n. [Hunt. Eouerwíc, Eouorwíc, Euerwíc: Dun. Eworwíc: Hovd. Eboracum] YORK; Ebŏrā-cum :—Seuerus ge-endode on Eoforwíc Severus ended [his days] at York, Chr. 189; Th. 15, 28, col. 1.

Eofor-wíc-ceaster; gen. -ceastre; f. York :—On ðǽre cyricean Eoforwícceastre in Eboracensi ecclēsia, Bd. 5, 24; S. 646, 29: Chr. 644; Th. 48, 20.

Eofor-wícingas, pl. m. Yorkists, people of York; Eboracenses :—Hǽfdon Eoforwícingas gehāten ðæt hie on hire rǽdenne beón woldan the people of York had promised that they would be at her disposal, Chr. 918; Th. 192, 9.

Eofor-wíc-scír, e; f. YORKSHIRE; comĭtātus Eboracensis :—Fóran ða þegnas ealle on Eoforwícscíre to Eoferwíc all the thanes in Yorkshire went to York, Chr. 1065; Th. 332, 7.

eofot, eofut, eofet, es; n. A debt, crime; dēbĭtum, culpa :—Be eofotes andetlan. Gif mon on folces gemōte ge-yppe eofot of confession of debt. If a man declare a debt at a folk-mote, L. Alf. pol. 22; Th. i. 76, 6. Reht oððe eofut oððe scyld dēbĭtum, Mt. Kmbl. Lind. 18, 25. Godes ágen bearn, unscyldigne eofota gehwylces, hēngon on heáne beám fæderas usse our fathers hung up God's own son on a high tree, guiltless of every crime, Elen. Kmbl. 846; El. 423.

eofoþ, es; n. Strength, violence, might, Beo. Th. 5062, note; B. 2534. v. eafoþ.

eoful-sæc, es; n? [eoful = yfel evil, sacan to accuse] Evil accusation, blasphemy; blasphēmia :—Ðæt ðú eofulsæc ǽfre ne fremme wið Godes bearne that thou never make blasphemy against God's son, Elen. Kmbl. 1045; El. 524.

eofur-cumbol, es; n. A boar-banner, Elen. Kmbl. 151; El. 76. v. eofor-cumbol.

eógoþ, e; f. Youth; jŭventus :—Duguþe and eógoþe with old and young, Andr. Kmbl. 2245; An. 1124. v. geóguþ.

eoh; nom. acc: gen. eohes = eos = eós; m. A war-horse, charger; ĕquus bellātor :—He gehleóp ðone eoh he mounted the charger, Byrht. Th. 137, 20; By. 189. Eorl sceal on eós bōge a chief shall [ride] on horse-back, Exon. 90 a; Th. 337, 11; Gn. Ex. 63. [O. Sax. ¹ ehu-scalc servus ĕquārius, compos: a scalc servus et ehu ĕquus, quod et nōmen cūjus-dam lĭtĕræ rūnĭcæ Saxŏnĭcæ est;' Heli. Schmel: O. Nrs. jó-r vel ió-r,—'sŏnus hūjus lĭtĕræ ĭdem fuit, atque hŏdie, in lingua vĕtĕre, sed ad figūram et nōmen quod attĭnet, non distinguēbātur ab i:' Egils.—gen. jó-s, ió-s; dat. jó, ió; acc. jó, ió, ó; pl. gen. acc. jóa, íoa.] v. eh.

eóh = íw; m. The Anglo-Saxon Rune Z = eó, the name of which letters in Anglo-saxon is eóh = íw a yew-tree; taxus,—hence this Rune not only stands for the diphthong eó, but for eóh a yew-tree, as,— Z [Eóh] biþ útan unsmēðe treów, heard, hrusan fæst yew is outwardly an unsmooth tree, hard, fast in the earth, Hick. Thes. i. 135, 25; Runic pm. 13; Kmbl. 341, 26. v. íw and RŪN.

eoldra, eolldra older, Bt. 16, 1; Fox 50, 7. Eolldra fæder grand-father, Bt. 10; Fox 28, 32; comp. of eald.

eolet, es; n. The sea, ocean; māre, ŏceănus :—Ðá wæs sundliden eoletes æt ende then was the sea-voyage at the end of the ocean, Beo. Th. 453, note; B. 224.

EOLH, eolc; gen. eolhes, eolces, eolcs, eolx; m. [eolx vĭdētur genĭtīvus ab eolc, eolh, Ettmül. Poet. 288, 15, note] An ELK; alces. The Rune Y = x seems to stand for the genitive of this word in the Runic poem,—hence, this Rune not only stands for the letter x, but for eolhx = eolcx = eolcs = eolces of an elk, as,— Y [eolhx = eolces] secg eard [seccard MS.] hæfþ oftust on fenne, wexeþ on wætere elk's sedge hath its place [earth] oftest in fen, waxeth in water, Hick. Thes. i. 135, 29; Runic pm. 15; Kmbl. 342, 7. Eolx secg papilluum, Wrt. Voc. 286, 36. [O. H. Ger. elaho: M. H. Ger. elch: O. Nrs. elgr: Lat. alces: Grk. ἄλκη.] v. RŪN.

eolh-sand amber; electrum, Cot. 75.

eolh-stede a sheltering-place, a temple, An. 1644. v. ealh-stede.

eolhx, eolx; gen. sing. of eolh, eolc an elk.

eolone, eolene, elone, elene, an; f. The plant elecampane; ĭnŭla hĕ-lĕnium, Lin :—Genim eolonan take elecampane, L. M. 1, 15; Lchdm. ii. 58, 18: 1, 32; Lchdm. ii. 76, 4: 1, 36; Lchdm. ii. 86, 11. Wyrc sealfe of eolonan make a salve of elecampane, L. M. 1, 28; Lchdm. ii. 70, 5. Eolene elecampane, L. M. 1, 23; Lchdm. ii. 66, 9.

eoloþ ale, L. In. 70; Th. i. 146, 17, MS. H. v. ealaþ.

eom [eam, am], ðú eart [earþ, art, arþ], he is, ys; I am, thou art, he is; sum, es, est: pl. sind, sindon [synd, sint, synt, sient, sindan, sindun, syndon, syndan, syndun, siendon, seondon, seondan, siondon, siondan, syondon; earon, earun, earan, aron]: pl. we, ye, they are; sŭmus, estis, sunt: subj. sí, sý, [sig, sige, síe, sýe, seó, sió] if I, if thou, if he be; sim, sis, sit: pt. sín, sýn [síe, síen, seón] if we, if ye, if they be; sīmus, sītis, sint :—Ic eom, sum, is edwistlíc word and gebýraþ to Gode ánum synder-líce, forðande God is ǽfre unbegunnen, and unge-endod on him sylfum, and þurh hine sylfne wunigende 'Sum,' I am, is the substantive verb, and belongs exclusively to God alone, because God is ever without beginning, and without end in himself, and existing by himself, Ælfc. Gr. 32; Som. 36, 24–26. Ic eom weg, and sōþfæstnys, and líf ĕgo sum via, et vērĭtas, et vīta, Jn. Bos. 14, 6. Ic sylf hit eom ego ipse sum, Lk. Bos. 24, 39. Ic eom I am, Beo. Th. 676; B. 335: Fins. Th. 49; Fin. 24: Exon. 102 b; Th. 388, 1; Rä. 8, 4; Cd. 19; Th. 24, 4; Gen. 372: Cd. 215; Th. 270, 28: Sat. 97; Ps. Th. 68, 6: Bd. 5, 19; S. 640, 40. [Orm. amm, arrt, iss, pl. arrn, sinndenn; subj. sí: Laym. eam, am, æm, em; eart, art, ært; his; pl. sunden, sundeþ, senden, sonden; subj. seo, sí; pl. seon, seoþ: O. Sax. is, ist, pl. sind, sint, sindon, sindun; subj. sí, sín: O. Frs. is, send; subj. se, sie; Ger. ist, sind; subj. sei, seien: M. H. Ger. O. H. Ger. ist, sint; subj. sí, sín: Goth im, is, ist, pl. sijum, sijuþ, sind; subj. sijau, sijais, sijai; pl. sijaima, sijaiþ, sijaina: O. Nrs. em, ert, er, erum, eruþ, eru; subj. sē, sér, sē, pl. sēim, sēiþ, sēi: Grk. εἰμί, ἐστί: Slav. jesmi, jesti: Sansk. asmi, asti.] DER. neom. v. wesan.

eom = heom to them; illis, Gen. 20, 8.

eond yond, beyond; ultra, per, Nicod. 19; Thw. 9, 28. v. geond.

eonde a species; spĕcies, Bd. 3, 14; S. 540, 16, note. v. ende.

eond-lýhtan; p. -lýhtde = -lýhte; pp. -lýhted = -lýhtd = -lýht [eond = geond through; lýhtan to shine] To shine through, enlighten; perlū-mĭnāre, illūmĭnāre :—We ealle eondlýhte wǽron we were all enlightened, Nicod. 24; Thw. 12, 21. Swylce gylden sunna wǽre ofer us ealle eondlýhte a golden sun as it were shone over us all, 24; Thw. 12, 23.

eond-send overspread, Nicod. 27, Lye. v. geond-sendan.

eonu moreover; porro, Som. Ben. Lye.

eorcnan-stán, eorcan-stán, eorclan-stán, earcnan-stán, es; m. A precious stone, pearl, topaz; lăpis prētiosus, gemma, tŏpāzion = τοπάζιον, τόπαζος; m. the yellow or oriental topaz, Ps. Spl. M. C. 118, 127: Elen. Kmbl. 2048; El. 1025: Exon. 64 b; Th. 238, 12; Ph. 603. Eorcan-stán, 124 b; Th. 478, 7; Ruin. 37. Eorclanstán, Beo. Th. 2420, note; B. 1208. [O. Nrs. iarknasteinn, m. lăpis pellūcĭdus: Goth. airknis; adj. good, holy: O. H. Ger. erchan egrēgius, summus.]

eord the earth, ground, Som. Ben. Lye. v. eorþe, eard.

eordian; p. ode; pp. od To dwell, inhabit; hăbĭtāre :—Ða on lífes hús eordiaþ they dwell in the house of life, Ps. Th. 134, 21. v. eardian.

eóred, eórod, es; n. Cavalry, a band, legion, troop; equĭtātus, lĕgio, turma :—Hie gesáwon eóred lixan they saw the band glittering, Cd. 149; Th. 187, 28; Exod. 157. Eórod sceal getrume rídan a troop shall ride in a body, Exon. 90 a; Th. 337, 12; Gn. Ex. 63. Legio, ðæt is on úre geþeóde, eóred legion, that is in our tongue, a troop, Lk. Bos. 8, 30. v. weorod, weorud.

eóred-cist, eórod-cist, -cyst, -cest, -ciest, e; f. [eóred a band, troop; cist a company] A company, troop; turma, lĕgio :—Wesseaxe eóred-cistum [eóredcystum, Th. 202, 28, col. 2; 203, 28] on lást legdun láðum þeódum the West-Saxons in troops followed the footsteps of the hostile nations, Chr. 937; Th. 202, 28, col. 1. Eóredcystum in troops, Exon. 96 a; Th. 358, 27; Pa. 52. Fór fyrda mǽst eóredcestum the greatest of armies marched in bands, Elen. Kmbl. 71; El. 36. Eóredciestum faraþ they go in bands, Exon. 60 b; Th. 220, 25; Ph. 325.

eóred-geatwe; pl. f. Military trappings; armāmenta :—Se eów geaf eóred-geatwe who gave to you military trappings, Beo. 5724; B. 2866.

eóred-mæcg, es; m. [mæcg a man] A horseman; ĕques :—Hæfdon

xi eóredmæcgas fríd-hengestas *the horsemen had eleven war-horses*, Exon. 106 a; Th. 404, 6; Rä. 23, 3.

eóred-man *a horseman;* ĕques, Som. Ben. Lye. v. eórod-man.

eóred-þreát, es; *m.* [þreát *a host, troop*] *A band, company;* turma, lĕgio:—Atol eóredþreát *a horrid band*, Exon. 102 a; Th. 385, 23; Rä. 4, 49.

eored-wered, es; *n.* [werod, wered *a company, multitude*] *A band, company, multitude;* exercĭtus, lĕgio:—Eóredweredu ðara deófla lĕgiónes *sive exercĭtus dæmónum*, Greg. Dial. 1, 10.

eorendel *the first dawn.* v. earendel.

eorfeðe *difficult;* diffĭcĭlis, Mt. Kmbl. Rush. 7, 14. v. earfeðe.

eorg *weak;* segnis:—Ðam eorgan Sisaran *to the weak Sisera*, Jud. 5; Thw. 156, 8. v. earg.

eó-risc *a bulrush;* scirpus. v. eá-risc.

EORL, es; *m.*   I. *an Anglo-Saxon nobleman of high rank, the yarl of the Danes, about the same as an ealdorman.* He who was in early times styled *ealdorman*, was afterwards denominated *an earl;* cŏmes, sătelles princĭpis. This title, which was introduced by the Jutes of Kent, occurs frequently in the laws of the kings of that district, the first mention of it being:—Gif on eorles túne man mannan ofslæhþ xii scillinga gebête *if a man slay a man in an earl's town, let him make compensation with twelve shillings*, L. Ethb. 13; Th. i. 6, 9, 10. Its more general use among us dates from the later Scandinavian invasions, and though originally only a title of honour, it became in later times one of office, nearly supplanting the older and more Saxon one of 'ealdorman:'—Swá we eác settaþ be eallum hádum, ge ceorle ge eorle *so also we ordain for all degrees, whether to churl or earl*, L. Alf. pol. 4; Th. i. 64, 3. Se eorl nolde ná geþwærian *the earl would not consent*, Chr. 1051; Ing. 227, 13, 23: 228, 4, 28, 35, 36: 229, 10, 21, 25, 26.   II. *a man, brave man, hero, general, leader, chief;* vir, pŭgil, vir fortis, dux:—Eorlas on cýþþe *men in the country*, Andr. Kmbl. 1467; An. 735. Him se Ebrisca eorl wísade *the Hebrew man* [*Lot*] *directed them*, Cd. 112; Th. 147, 24; Gen. 2444. Ða eorlas þrý, *nom. pl. the three men*, 95; Th. 123, 16; Gen. 2045. Eorlas wênaþ *men think*, 86; Th. 109, 22; Gen. 1826. Fór eorlum *before the people*, 98; Th. 129, 1; Gen. 2137. Þegna and eorla *of thanes and earls*, Bt. Met. Fox 25, 15; Met. 25, 8. Geared gumum gold brittade, se eorl wæs æðele *Jared dispensed gold to the people, the man was noble*, Cd. 59; Th. 72, 5; Gen. 1182. [*Piers P.* eerl: *Chauc.* erl: *R. Glouc.* erles *noblemen: Laym.* eorl: *Orm.* eorless, *pl : O. Sax.* Hel. erl, *m. a man, nobleman, male offspring, boy : Icel.* jarl, earl, *m. a gentleman, nobleman, warrior, chief.*]

eorl-cund; *adj. Earl kind, noble;* nŏbĭlis:—Gif mannes esne eorlcundne mannan ofslæhþ þreóm hundum scillinga gylde se ágend *if a man's servant slay a man of an earl's degree, let the owner pay three hundred shillings*, L. H. E. 1; Th. i. 26, 8.

eorl-dóm, es; *m. An* EARLDOM, *the province or dignity of an earl*, the same as ealdor-dóm, v. *Turner's Hist.* b. viii. c. 7; cŏmĭtis múnus:—Ælfgár eorl fêng to ðam eorldôme ðe Harold ær hæfde *earl Ælfgar succeeded to the earldom which Harold had before*, Chr. 1053; Erl. 189, 14.

eorl-gebyrd, e; *f.* [gebyrd *birth*] *Noble birth, nobility;* nŏbĭlĭtas:—Eorlgebyrdum *by noble birth*, Bt. Met. Fox 9, 52; Met. 9, 26: 10, 54; Met. 10, 27.

eorl-gestreón, es; *n.* [gestreón *treasure*] *Noble treasure, riches;* dīvĭtiæ:—Nis him gád eorlgestreóna *he lacks not noble treasures*, Exon. 123 b; Th. 475, 10; Bo. 45: Beo. Th. 4481; B. 2244.

eorl-gewǽde, es; *n.* [gewǽde *clothing*] *Manly clothing, armour;* vĭrīlis vestĭtus:—Gyrede hine Beówulf eorlgewǽdum *Beowulf clad himself in armour*, Beo. Th. 2888; B. 1442.

eorlíc [=eorl-líc]; *adj. Manly;* vĭrīlis:—Eorlíc ellen *manly strength*, Beo. Th. 1278; B. 637. v. eorlisc, eorl-líc.

eorlíce [=eorl-líce]; *adv. Manfully, strongly, greatly;* vĭrīlĭter, vĕhĕmenter, multum:—Gebealh heó swíðe eorlíce wið hire suna *she was very greatly incensed against her son*, Cod. Dipl. 755; Kmbl. iv. 54, 30.

eór-lippric, es; *n. A flap of the ear*, Jn. Lind. War. 18, 26. v. eárelippric.

eorlisc, eorl-líc; *adj.* EARLISH, *earl-like, like an earl;* nŏbĭlis:—Eorlisc, L. Ath. v. prm; Th. i. 228, 8. Eorllíc [MS. eorlíc], Beo. Th. 1278; B. 637.

eorl-mægen, es; *n. A host of men;* vĭrôrum turma:—Sió cwên bebeád ofer eorlmægen áras fýsan *the queen commanded messengers to hasten throughout the mass of the people*, Elen. Kmbl. 1958; El. 981.

eorl-riht, es; *n. An earl's right or privilege;* cŏmĭtis jus *vel* privĭlêgium:—Gif þegen geþeáh, ðæt he wearþ to eorle, ðonne wæs he syððan eorlrihtes weorþe *if a thane thrived, that he became an earl, then he was thenceforth worthy of an earl's right*, L. R. 5; Th. i. 192, 8.

eorl-scipe, -scype, es; *m. Manliness, bravery, courage, supremacy, nobility;* vĭrīlĭtas:—Hí eahtodon eorlscipe and his ellenweorc *they valued his manliness and his valiant works*, Beo. Th. 6327; B. 3174: Scóp. Th. 283; Wíd. 141: Beo. Th. 3458; B. 1727: 4272;

B. 2133. Eorlscipes, Salm. Kmbl. 22; Sal. 11. He eorlscype fremede *he effected supremacy*, Exon. 85 a; Th. 320, 31; Wíd. 37.

eorl-werod, es; *n.* [werod *a company, troop*] *A band of men, warrior band;* vĭrôrum turma:—Ðær ðæt eorlwerod sæt *the warrior band sat there*, Beo. Th. 5779; B. 2893.

Eorman-ríc, Eormen-ríc, es; *m. The celebrated king of the Ostrogoths or East Goths, the Alexander of the Goths;* Eormanrīcus, v. Gota III, Alríca, and þeód-ríc:—Eormanríc áhte wíde folc Gotena ríces *Ermanric possessed the wide nations of the kingdom of the Goths*, Exon. 100 a; Th. 378, 25; Deór. 21. Weóld Eormanríc Gotum *Ermanric ruled the Goths*, Scóp. Th. 38; Wíd. 18. Ic wæs mid Eormanríce *I was with Ermanric*, 178; Wíd. 88. Ðæt wæs inn-weorud Eormanríces *that was the household band of Ermanric*, 224; Wíd. 111. He searo-níðas fealh Eormenríces *he fell into the guileful enmity of Ermanric*, Beo. Th. 2406; B. 1201. *For the anachronisms and inconsistences I would refer to* W. Grimm's *Deutsche Heldensage, where may be found the particulars of this celebrated hero.*

eormen, eorman; *adj. Universal, immense, whole, general;* universālis, immensus, permagnus, tótus, ūnĭversus. *Used in composition, as in* eormen-cyn, -grund, -láf, -ríc, -strýnd, -þeód.

eormen-cyn, -cynn, es; *n. The human race;* hūmānum gĕnus:—God gesceapo ferede æghwylcum on eorþan eormencynnes *God has borne his decrees to every one of the human race on earth*, Exon. 88 b; Th. 333, 3; Vy. 96: Beo. Th. 3918; B. 1957.

eormen-grund, es; *n.* [grund *ground, earth*] *The spacious earth;* immensa terra:—Ofer eormengrund *over the spacious earth*, Beo. Th. 1722; B. 859.

Eormen-ríc *Ermanric*, Beo. Th. 2405; B. 1200. v. Eorman-ríc.

eormen-strýnd, e; *f. The great generation;* permagna gĕnĕrātio:—Ðú eart eorre eormenstrýnde *thou art of an angry, great* [*heathen*] *generation*, Salm. Kmbl. 659; Sal. 329.

eormen-þeód, e; *f. A great people;* permagnus populus. v. yrmenþeód.

eormþu *poverty, calamity:*—Eormþa, Bt. 7, 4; Fox 22, 29. Eormþum, 23; Fox 78, 31. v. yrmþu.

eornan *to run;* currĕre, Ps. Surt. 57, 8. v. yrnan.

eornende *running; part. of* eornan = yrnan.

eornes, eornest *a duel, combat;* duellum, Som. Ben. Lye.

eornest *earnest, earnestness*, Exon. 24 a; Th. 68, 9; Cri. 1101. v. eornost.

eorneste *earnest, serious*, Exon. 20 a; Th. 51, 32; Cri. 825: Homl. Th. i. 386, 20. v. eornoste; *adj.*

eorneste *in earnest, earnestly*, Bt. Met. Fox 13, 56; Met. 13, 28: 16, 44; Met. 16, 22. v. eornoste; *adv.*

eornestlíce *earnestly;* stŭdiôse. v. eornostlíce.

eornfullíce; *adv. Earnestly;* stŭdiôse. v. eornostlíce.

eornfullnes, -ness, e; *f. Earnestness, anxiety;* dīlĭgentia, sollĭcĭtudo:—Eornfullness ðisse worulde *sollĭcĭtudo istius sæcŭli*, Mt. Bos. 13, 22. v. geornfulnes.

eornigende *murmuring;* murmŭrans, L. E. I. 21; Th. ii. 416, 16.

eornlíce; *adv. Diligently;* dīlĭgenter:—Genim ðas wyrte eornlíce gecnucude mid ecede *take this herb diligently pounded with vinegar*, Herb. 87, 2; Lchdm. i. 190, 21. v. geornlíce.

EORNOST, eornust, eornest, e; *f.* EARNEST, *earnestness, zeal;* sērium, stŭdium:—Mid swelcum eorneste [eornoste MS. Cot.] *with such zeal*, Past. 15, 1; Hat. MS. 18 b, 27. On eornost, eornust *or* eornoste *in earnest, earnestly*, Ælfc. T. 12, 8: Homl. Th. ii. 250, 30: Mt. Bos. 5, 18: 13, 17: Gen. 14, 15. Þurh eorneste *in earnest, sternly*, Exon. 24 a; Th. 68, 9; Cri. 1101. [*Wyc.* ernes, eernes, ernest *earnest, pledge: Chauc.* erneste *zeal: Laym.* eornest *conflict: Frs.* ernste: *O. Frs.* ernst: *Dut.* ernst, *m : Ger.* ernst, *m : M. H. Ger.* ërnest, ërnst, *m : O. H. Ger.* ërnust, ërnost, ërnest, *m. f. vĭgor, sērium.*]

eorneste; *adj. Earnest, serious;* sērius, stŭdiôsus:—On eornostne hige *with earnest intention*, Cod. Dipl. 942; Kmbl. iv. 278, 15. Biþ eorneste ðonne eft cymeþ, rêðe and ryhtwís *he will be earnest when he comes again, stern and just*, Exon. 20 a; Th. 51, 32; Cri. 825. Mid eornestum móde *with earnest mind*, Homl. Th. i. 386, 20.

eornoste, eorneste; *adv. In earnest, earnestly, seriously, courageously, strongly;* sêrio, strênue, sēdŭlo, vĕhĕmenter:—He feaht eornoste *he fought earnestly*, Byrht. Th. 140, 1; By. 281: Judth. 11; Thw. 24, 39; Jud. 231. Hió onginþ eornoste racentan slítan *she will begin in earnest to sever her chains*, Bt. Met. Fox 13, 56; Met. 13, 28: 16, 44; Met. 16, 22.

eornostlíce; *adv.* EARNESTLY, *strictly, truly;* sēdŭlo:—Sunnan dæges cýpincge we forbeódaþ eornostlíce *we strictly forbid marketing on Sunday*, L. C. E. 15; Th. i. 368, 15.

eornostlíce, eornustlíce; *conj. Therefore, but;* ergo, ĭgĭtur, ĭtăque:—Abram ðá eornostlíce astírode his geteld *mŏvit ĭgĭtur tabernācŭlum suum Abram*, Gen. 13, 18. Eornostlíce ealle cneóressa fram Abraham oþ Dauid synd feówertyne cneóressa *omnes ĭtăque genĕrātiônes ab Abraham*

usque ad David, gĕnĕrātiōnes quatuordĕcim, Mt. Bos. 1, 17. Beóþ eornustlíce gleáwe *estōte ergo* [*oûn*] *prūdentes*, Mt. Bos. 10, 16, 26 : 2, 1 : 13, 40.

eornust *earnest, earnestness*, Mt. Bos. 13, 17. v. eornost.

eornustlíce *therefore, but*, Mt. Bos. 2, 1 : 10, 16, 26 : 13, 40. v. eornostlíce.

eórod, es ; *n. A band, legion, troop ;* turma, lĕgio :—Wíse men tealdon án eórod to six þúsendum, and twelf eórod sind twá and hundseofontig þúsend *wise men have reckoned a legion at six thousand, and twelve legions are seventy-two thousand,* Homl. Th. ii. 246, 28, 29, 25 : Jud. Thw. 161, 36. v. eóred.

eórod-man, -mann, es ; *m. A horseman ;* ĕques :—Líhte se eórod-man *desīluit ĕques,* Bd. 3, 9 ; S. 533, 33.

eorp, *earp ; adj. Dark, dusky, brown, swarthy ;* fuscus, badius :—Eorp werod *the swarthy host [the Egyptians],* Cd. 151 ; Th. 190, 4 ; Exod. 194 : Exon. 113 a ; Th. 433, 21 ; Rä. 50, 11. [*Icel.* jarpr *brown.*]

eorre, es ; *n. Anger, wrath ;* íra :—Warniaþ eów ðæs Drihtenes eorres and mínes *beware of the Lord's anger and of mine,* L. Ath. i. prm ; Th. i. 196, 33 : Ps. Lamb. 101, 11. v. yrre.

eorre ; *adj. Angry, enraged, fierce ;* írātus, īrācundus :—He us eorre gewearþ *he has become angry with us,* Cd. 219 ; Th. 280, 27 ; Sat. 261 : Elen. Kmbl. 801 ; El. 401. Þurh eorne hyge *through angry mind,* 1367 ; El. 685. Nalæs late wǣron eorre æscberend to ðam orlege *the fierce spear-bearers were not slow to the onset,* Andr. Kmbl. 93 ; An. 47 : 2153 ; An. 1078. v. yrre ; adj.

eorringa ; *adv. Angrily ;* írāte :—Hine eorringa gesēceþ bócstafa brego *the prince of letters shall angrily seek him,* Salm. Kmbl. 198 ; Sal. 98. v. yrringa.

eorsian *to be angry,* Ps. Spl. 4, 5 : Mt. Kmbl. Rush. 5, 22. v. yrsian.

eorsung *anger,* Cant. Moys. Ex. 15, 8 ; Thw. 29, 8. v. yrsung.

eorþ, e ; *f. The earth ;* terra :—Seó [MS. sie] eorþ is dryge and ceald, and ðæt wæter wǣt and ceald *the earth is dry and cold, and the water wet and cold,* Bt. 33, 4 ; Fox 128, 34. v. eorþe.

eorþ-æppel, es ; *m : nom. acc. pl. n.* -æppla *An earth-apple, a cucumber ;* cŭcŭmis :—Cŭcŭmeres, ðæt synd eorþæppla *cucumbers, which are earth-apples,* Num. 11, 5. Eorþæppel mandrăgŏra, Ælfc. Gl. 44 ; Som. 64, 79 ; Wrt. Voc. 32, 15.

eorþ-ærn, es ; *n. An earth-place, a tomb, sepulchre ;* spēlunca, sĕpulcrum :—Open wæs ðæt eorþærn *the sepulchre was open,* Exon. 120 a ; Th. 460, 18 ; Hö. 19. In ðæt eorþærn *in the sepulchre,* 119 b ; Th. 460, 4 ; Hö. 12 : Exon. 119 b ; Th. 459, 22 ; Hö. 3.

eorþ-beofung, e ; *f. An earthquake ;* terræ mōtus :—Seó eorþbeofung tácnade ða miclan blód-dryncas *the earthquake betokened the great blood-sheddings,* Ors. 4, 2 ; Bos. 79, 28. v. eorþ-bifung.

eorþ-bifung, -beofung, e ; *f.* [bifung *a trembling, shaking*] *An earthquake ;* terræ mōtus :—Ðǣr wearþ geworden micel eorþbifung *terræ mōtus factus est magnus,* Mt. Bos. 28, 2. Híg gesáwon ða eorþbifunge *vīdērunt terræ mōtum,* 27, 54.

eorþ-bigengnys, -bigennys, -nyss, e ; *f. Earth-cultivation, attention to agriculture ;* terræ cultūra, agricultūræ stŭdium : — Elelǣndra eorþbigennys cŏlōnia, id est peregrīnōrum cultūra, Ælfc. Gl. 54 ; Som. 66, 103 ; Wrt. Voc. 36, 25. v. eard-begengnes, el-þeódignes.

eorþ-bigenga, an ; *m.* [bigenga *an inhabitant, dweller*] *An inhabitant of the earth ;* terrĭcŏla, terrĭgĕna :—Ðæt he eorþbigengan awecce hine to ondrǣdanne *ut terrĭgĕnas ad tĭmendum se suscĭtet,* Bd. 4, 3 ; S. 569, 22.

eorþ-búend, es ; *m. An earth dweller, inhabitant ;* terrĭcŏla :—Eorþbúend, Ps. Th. 65, 1 : 101, 13 : 118, 4. v. búend, búende.

eorþ-burh ; *gen.* -burge ; *dat.* -byrig ; *f. An earth mound* or *burying place ;* agger, hŭmātio :—To ðare eorþbyrig *to the earth mound,* Cod. Dipl. Apndx. 335 ; A. D. 903 ; Kmbl. iii. 403, 31.

eorþ-byrig, e ; *f. An earth mound ;* agger :—Eorþ-byrig [MS. -byre] Ælfc. Gl. 56 ; Som. 67, 45 ; Wrt. Voc. 37, 33.

eorþ-cafer, es ; *m. An earth-chafer, a cock-chafer ;* taurus :—Eorþcaferas *tauri,* Ælfc. Gl. 24 ; Som. 60, 23 ; Wrt. Voc. 24, 26. v. ceafer.

eorþ-cend ; *pp.* [cend = cenned *born*] *Earth-born ;* terrĭgĕna :—Eorþcende *terrĭgĕnæ,* Ps. Spl. C. 48, 2.

eorþ-crypel, -cryppel ; *gen.* -crypeles, -cryples, -crypples ; *m. A creeper on the earth, one having the palsy, a paralytic person ;* părălytĭcus = παραλυτικός :—In ðære ðe eorþcrypel [se eorþcryppel, Lind.] læg *in quo părălytĭcus jăcēbat,* Mk. Skt. Rush. 2, 4 : Lk. Skt. 5, 18. Se Hǣlend cwæþ to ðæm eorþcrypele [eorþcrypple, Lind.] *Iēsus ait părălytĭco,* Mk. Skt. Rush. 2, 5. To cweðanne ðæm eorþcryple *dĭcĕre părălytĭco,* Mk. Skt. Rush. Lind. 2, 9. Brengende to him ðone eorþcrypel *ferentes ad eum părălytĭcum,* Mk. Skt. Lind. Rush. 2, 3 : Mt. Kmbl. Lind. 9, 2. Gebrohtun him eorþcryplas *obtŭlērunt ei părălytĭcos,* Mt. Kmbl. Lind. 4, 24.

eorþ-cund ; *adj. Earthly, terrestrial ;* terrestris :—Se rinc ageaf eorþcunde eád *the prince gave up earthly happiness,* Cd. 79 ; Th. 98, 8 ; Gen. 1627.

eorþ-cyn, -cynn, es ; *n. The earth-kind, terrestrial species ;* gĕnus

terricŏlārum :—Eallum eorþcynne *for each terrestrial species,* Cd. 161 ; Th. 201, 10 ; Exod. 370.

eorþ-cyning, es ; *m.* [cyning *a king*] *An earthly king, king of the land ;* terræ rex :—Sceótend Scyldinga to scypum feredon eal ingesteald eorþcyninges *the Scyldings' warriors conveyed all the house chattels of the king of the land to their ships,* Beo. Th. 2315 ; B. 1155. Ðam æðelestan eorþcyninga *for the noblest of earthly kings,* Elen. Kmbl. 2346 ; El. 1174 : Cd. 162 ; Th. 202, 23 ; Exod. 392 : 189 ; Th. 235, 14 ; Dan. 306. Eorþcyningum [MS. -cynincgum] se ege standeþ *terrĭbĭli ăpud rēges terræ,* Ps. Th. 75, 9. He eorþcyningas yrmde and cwelmde *he oppressed and slew the kings of the earth,* Bt. Met. Fox 9, 93 ; Met. 9, 47 : Ps. Th. 88, 24.

eorþ-draca, an ; *m. An earth-dragon ;* drăco in antro dēgens :—Sió wund ongon, ðe him se eorþdraca geworhte, swelan and swellan *the wound, which the earth-dragon had made in him, began to burn and swell,* Beo. Th. 5417 ; B. 2712 : 5642 ; B. 2825.

eorþ-dyne, es ; *m. Earth din, an earthquake ;* terræ mōtus :—On ðisan gēre wæs micel eorþdyne *in this year* [A. D. 1060] *was a great earthquake,* Chr. 1060 ; Erl. 193, 31 : 1122 ; Erl. 249, 14.

EORÞE, an ; *f :* eorþ, e ; *f.* **I.** *the* EARTH *in opposition to the sea, the ground, soil ;* terra, hŭmus, sŏlum :—God gecígde ða drignisse eorþan, and ðæra wætera gegaderunga he hēt sǣs *vŏcāvit Deus ārĭdam terram, cóngrĕgātiōnesque ăquārum appellāvit māria,* Gen. 1, 10. Spritte seó eorþe grówende gærs and sǣd wircende and æppelbǣre treów wæstm wircende æfter his cinne, ðæs sǣd sig on him silfum ofer eorþan *germĭnet terra herbam vĭrentem et făcientem sēmen et lignum pōmĭfĕrum făciens fructum juxta gĕnus suum, cujus sēmen in sēmetipso sit sŭper terram,* Gen. 1, 11, 12, 24, 25, 28, 29 : Cd. 57 ; Th. 69, 32 ; Gen. 1144 : Exon. 62 b ; Th. 231, 11 ; Ph. 487 : Beo. Th. 3069 ; B. 1532 : Elen. Kmbl. 1655 ; El. 829 : Bt. Met. Fox 8, 118 ; Met. 8, 59. Ic ðec ofer eorþan geworhte, on ðære ðú scealt yrmþum lifgan and to ðære ilcan scealt eft geweorþan *I made thee on earth, on which thou shalt live in misery and shalt become the same again,* Exon. 16 b ; Th. 39, 12–19 ; Cri. 621–624 : 38 a ; Th. 125, 10 ; Gú. 352. Cain wæs eorþan tilia *fuit Cain agrĭcŏla* [lit. *a tiller of the earth*], Gen. 4, 2. **II.** *the* EARTH, *terrestrial globe ;* tellus :—On anginne gesceóp God heofenan and eorþan *in the beginning God created heaven and earth,* Gen. 1, 1, 2, 17, 20, 26 : 2, 1, 4 : Cd. 98 ; Th. 129, 9 ; Gen. 2141 : Exon. 16 b ; Th. 38, 18 ; Cri. 608. Se Ælmihtiga eorþan worhte *the Almighty made the earth,* Beo. Th. 185 ; B. 92. Drihtnes is eorþe and fulnysse oððe gefyllednes hyre *the earth is the Lord's and the fulness thereof,* Ps. Lamb. 23, 1 : Ex. 9, 29 : Deut. 10, 14. Ðæt gē ne swerion þurh eorþan, forðamðe heó ys Godes fótscamul *that ye swear not by the earth, because it is God's foot-stool,* Mt. Bos. 5, 35. [*Piers P. Wyc.* erthe : *Laym.* eorðe, eorðen, earþe, erþe : *Orm.* eorþe, erþe : *Plat.* eerde, *f : O. Sax.* erða, *f : Frs.* yerd : *O. Frs.* irthe, erthe, erde, *f : Dut.* aarde, *f : Ger. M. H. Ger.* erde, *f : O. H. Ger.* erda, erada, *f : Goth.* airþa, *f : Dan.* jord, *m. f : Swed.* jord, *f : Icel.* jörð, *f. earth, land, estate.*]

eorþ-fæst, -fest ; *adj. Earth-fast, fixed in the earth ;* in terra firmus :—To ánum [MS. ane] eorþfestum treówe *to a tree firm in the earth,* Th. Anlct. 122, 10.

eorþ-fæt, es ; *n. An earthen vessel, the body ;* vas terrā factum, corpus :—Se gǣst nimeþ swá wíte swá wuldor, swá him in worulde ðæt eorþfæt ǣr geworhte *the spirit receives either punishment or glory, as the body has worked for him before in the world,* Exon. 98 a ; Th. 367, 15 ; Seel. 8.

eorþ-gealla, an ; *m.* [gealla *gall*] *The herb* EARTH-GALL, *the lesser centaury ;* fel terræ, erythrǣa centaurium, Lin :—Eorþgealla [MS. -gealle] *fel terræ* vel centauria, Wrt. Voc. 79, 50 : Ælfc. Gl. 41 ; Som. 64, 5 ; Wrt. Voc. 31, 17. Eorþgealla *centauria,* Mone A. 373. Nim centaurian, ðæt is *fel terræ,* sume hátaþ eorþgeallan *take centaury, that is fel terræ, some call it earth-gall,* L. M. 2, 8 ; Lchdm. ii. 186, 27.

eorþ-gemet, es ; *n. Earth-measure, geometry ;* geometria = γεωμετρία, Cot. 95.

eorþ-gesceaft, e ; *f.* [gesceaft *a creature*] *An earthly creature ;* terrestris creatūra :—Men habbaþ [MS. habbæþ] geond middangeard eorþgesceafta ealle oferþungen *men have all surpassed earthly creatures throughout the middle earth,* Bt. Met. Fox 20, 387 ; Met. 20, 194.

eorþ-græf, es ; *n. A hole dug in the earth, a ditch, well ;* fossa, pŭteus :—Isernes dǣl eorþgræf pæðeþ *a part of iron passes the well,* Exon. 114 b ; Th. 439, 26 ; Rä. 59, 9.

eorþ-gráp, e ; *f. Earth's grasp, the hold of the grave ;* terræ comprĕhensio :—Eorþgráp hafaþ waldend wyrhtan *earth's grasp* [i. e. *the grave*] *holdeth its mighty workmen,* Exon. 124 a ; Th. 476, 12 ; Ruin. 6.

eorþ-hele, es ; *m. A heap ;* tŭmŭlus :—Wæs ðæt deáw abútan ða fyrdwíc, swilce hit hagoles eorþhele wǣre *the dew was about the camp, as it were a heap of hail,* Ex. 16, 14.

eorþ-hnutu, -nutu, e ; *f. An earth-nut ;* būnium flexuōsum :—Of ðam cumbe in eorþnutena þorn *from the combe to the earth-nut thorn,* Cod. Dipl. Apndx. 308 ; A. D. 875 ; Kmbl. iii. 399, 7.

eorþ-hús, es ; *n. An earth-house, den, cave ;* hypŏgæum = ὑπόγαιον,

subterrāneum :—Eorþhûs *hypŏgæum* vel *subterrāneum*, Ælfc. Gl. 110; Som. 79, 37; Wrt. Voc. 59, 9. Rômâne him worhton eorþhûs for ðære lyfte wylme *the Romans built for themselves earth-houses because of the boiling heat of the air*, L. M. 1, 72; Lchdm. ii. 146, 16.

eorþ-ifi, es ; *n. Ground ivy;* hĕdĕra nigra :—Eorþ-ifies *of ground ivy;* hĕdĕr̆æ nigræ, L. M. 1, 2; Lchdm. ii. 30, 17. v. eorþ-ifig.

eorþ-ifig, -yfig, -ifi, -iui, es; *n. Earth* or *ground ivy;* hĕdĕra nigra, hĕdĕra terrestris, glechoma hĕdĕrācea :—Genim hederan nigran, ðe man ôðrum naman eorþifig nemneþ *take* hĕdĕra nigra, *which one calleth by another name ground ivy*, Herb. 100, 1; Lchdm. i. 212, 20.

eorþ-iui [=ivi], es; *n. Ground ivy;* hĕdĕra nigra :—Eorþ-iui [MS. eorðiuí], Herb. 100; Lchdm. i. 212, 18, note 17, MS. B. v. eorþ-ifig.

eorþ-líc; *def.* se -líca; seó, ðæt -líce; *adj.* EARTHLY, *terrestrial;* terrēnus, terrestris :—He wæs eorþlíc cing *he was an earthly king*, Chr. 979; Erl. 129, 9. Hí eorþlíces âuht ne haldeþ *nothing earthly holds them*, Bt. Met. Fox 20, 331; Met. 20, 166. Ða twelf bôcland him gefreóde eorþlíces camphâdes and eorþlícere hĕrenysse to bigongenne ðone heofonlícan camphâd *dōnātis duodĕcim possessiuncŭlis terrārum, in quĭbus ablāto stŭdio mĭlĭtiæ terrestris, ad exercendam mĭlĭtiam cælestem*, Bd. 3, 24; S. 556, 41. Ðû meahte ǽlc eorþlíc þing forsión *thou mayest look down upon every earthly thing*, Bt. Met. Fox 24, 13; Met. 24, 7. Hine nolden his eorþlícan mâgas wrecan *his earthly kinsmen would not avenge him*, Chr. 979; Erl. 129, 11, 15. Gif ic eów eorþlíce þing sǽde *si terrēna dixi vobis*, Jn. Bos. 3, 12. Hió ðæs lǽnan lufaþ eorþlícu þing *she loves these transitory earthly things*, Bt. Met. Fox 20, 447; Met. 20, 224. Ðû gegæderast ða hiofonlícan sâwla and ða eorþlícan lîchoman *thou bringest together the heavenly souls and the earthly bodies*, Bt. 33, 4; Fox 132, 23. He forsihþ ðas eorþlícan gôd *he despises these earthly goods*, 12; Fox 36, 25 : 33, 4; Fox 132, 14, 18.

eorþ-líce; *adv. In an earthly manner;* terrŭlenter :—Eorþlíce *terrŭlenter*, Glos. Prudent. Recd. 145, 20.

eorþ-ling, es; *m. A farmer;* terræ cultor :—Eorþling *birbicaliolus?* Glos. Brux. Recd. 36, 50; Wrt. Voc. 63, 4; Mone A. 50. v. yrþ-ling.

eorþ-mægen, es; *n. Earthly power;* terræ vis :—Eorþmægen ealdaþ *earthly power grows old;* terræ vīres invĕtĕrascunt, Exon. 95 a ; Th. 354, 61; Reim. 69: Ettmül. Poet. pref. xviii. 59; p. 223, 69.

eorþ-mistel, es; *m. The plant basil;* clīnŏpŏdium =κλινοπόδιον :—Genim eorþmistel *take basil*, L. M. 1, 36; Lchdm. ii. 86, 21.

eorþ-nafela, -nafola, -nafala, -nafla, an; *m. Earth-navel, asparagus;* aspărăgus officinālis :—Nim eorþnafelan *take asparagus*, Lchdm. iii. 40, 23. Genim eorþnafolan wyrttruman *take roots of asparagus*, Herb. 126, 2; Lchdm. i. 238, 5. Wyll miclan eorþnafolan *boil the great asparagus*, Lchdm. iii. 18, 7. Eorþnafala *asparagus*, iii. 6, 15. Genim eorþnaflan [MS. B. -nafelan] *take asparagus*, Herb. 97, 1; Lchdm. i. 210, 8.

eorþ-reced, es; *n.* [reced *a house*] *An earth-house, a cave;* subterrānea dŏmus, antrum :—Hû ða stânbogan ĕce eorþreced healde *how the stone arches held the eternal earth-house*, Beo. Th. 5431; B. 2719.

eorþ-rest, e; *f. A resting* or *lying on the ground;* chămeunia = χαμευνία, Cot. 31.

eorþ-ríce, es; *n. A kingdom of the earth, earth's kingdom, the earth;* terræ regnum, terra :—Geond ealle eorþrícu *per omnia regna terræ*, Deut. 28, 25 : Bt. Met. Fox 4, 74; Met. 4, 37. He eorþrícum eallum weold *regnum ipsĭus omnĭbus domĭnābĭtur*, Ps. Th. 102, 18. On eorþríce *on earth's kingdom, on earth*, Cd. 22; Th. 27, 18; Gen. 419 : 23; Th. 29, 22; Gen. 454: 26; Th. 35, 1; Gen. 548.

eorþ-rima, an; *m. A kind of plant, dodder?* herbæ gĕnus, cuscuta? L. M. 3, 41; Lchdm. ii. 334, 12.

eorþ-scræf, es; *n. An earth-cavern, a grave;* căverna, antrum, sĕpulcrum :—Hie be hlíðe heáre ʾdûne eorþscræf fundon, ðǽr Loth wunode *they found by the slope of a high hill an earth-cavern, where Lot dwelt*, Cd. 122; Th. 156, 27; Gen. 2595: Exon. 115 a; Th. 443, 11; Kl. 28. Eardiaþ on eorþ-scræfum *hăbĭtant in sĕpulcris*, Ps. Th. 67, 7: Andr. Kmbl. 1605; An. 804.

eorþ-sele, es; *m.* [sele *a hall*] *An earth-hall, cave;* subterrānea aula, antrum :—Eald is ðes eorþsele *this earth-hall is old*, Exon. 115 a; Th. 443, 12; Kl. 29. Mec me mânsceaða eorþsele ût gesêceþ *the atrocious spoiler will seek me out from his earth-hall*, Beo. Th. 5023; B. 2515. He eorþsele âna wisse *he alone knew the earth-hall*, 4811; B. 2410.

eorþ-slihtes; *adv.* [slihtes, *old gen.* of sliht *destruction, slaughter,* like nihtes *of* niht] *In an earth-destroying manner;* in mŏdo vastante terram :—Swâ swâ oxa gewunaþ to awêstenne gærs, ôþ ða wirttruman, eorþslihtes mid tôðum *as an ox is accustomed to consume grass with his teeth, even to the roots, in an earth-destroying manner*, Num. 22, 4.

eorþ-stede, es; *m.* [stede *a place*] *An earth-place;* terræ lŏcus :—Ðâ hí ðæt ðín fægere hûs on eorþstede gewemdan [MS. gewemdaþ] *in terra pollŭērunt tabernaculum*, Ps. Th. 73, 7.

eorþ-styrennis, -niss, e; *f.* [styrenes *motion*] *An earthquake;* terræ mōtus :—Eorþstyrennis gewarþ micelu *terræ mōtus factus est magnus*, Mt. Kmbl. Rush. 28, 2.

eorþ-styrung, -stirung, e; *f. An earth-stirring, earthquake;* terræ mōtus :—Eorþstyrung fela burhga ofhreás ... þreóttyne byrig þurh eorþ-

styrunge afeóllon *an earthquake has overthrown many cities ... thirteen cities fell through an earthquake*, Homl. Th. i. 608, 27, 29 : 244, 17. Micele eorþstyrunga beóþ gehwǽr *great earthquakes shall be everywhere*, i. 608, 18. Ðâ wearþ mycel eorþstirung *there was a great earthquake*, Nicod. 15; Thw. 7, 17.

eorþ-tilia, an; *m. An earth-tiller, husbandman;* agricŏla :—Mîn fæder ys eorþtilia *păter meus agricŏla est*, Jn. Bos. 15, 1.

eorþ-tilþ, e; *f.* [tilþ *culture*] *Earth-tillage, agriculture;* agricultūra :—Eorþtilþ *agricultūra*, Coll. Monast. Th. 30, 27.

eorþ-tudor; *gen.* -tudres; *n.* [tuddor *progeny*] *Progeny of earth, men;* terrestris prŏgĕnies, hŏmines :—Ðis ys se dæg ðe Drihten geworhte eallum eorþtudrum eádgum to blisse *this is the day which the Lord made for bliss to all happy men*, Ps. Th. 117, 22.

eorþ-tyrewa, an; *m.* [tyrwa *tar*] *Earth-tar, asphalte;* bītūmen :—Se weall is geworht of tigelan and eorþtyrewan *the wall* [*of Babylon*] *is built with bricks and earth-tar*, Ors. 2, 4; Bos. 44, 25.

eorþ-wæstm, e; *f. Fruit of the earth;* terræ frux :—Mycel eorþwæstm frūgum cōpia, Bd. 1, 14; S. 482, 13. Eorþwæstme grówaþ *fruits grow*, Ps. Th. 103, 12. Wæs seó stôw wædla eorþwæstma *ĕrat lŏcus frūgis inops*, Bd. 4, 28; S. 605, 18. On eorþwæstmum genôh þuhte *abundance appeared in the fruits of the earth*, Bt. Met. Fox 8, 12; Met. 8, 6.

eorþ-waru, e; *f.* -ware; *gen.* -wara; *pl. m.* -waran; *gen.* -warena; *pl. m. Inhabitants* or *population of the earth;* terrĭcŏlæ, terrĭgĕnæ :—Heofonwaru and eorþwaru *cælĭcŏlæ et terrĭcŏlæ*, Hy. 7, 95; Hy. Grn. ii. 289, 95. Crist sibb is heofonware and eorþware *Christ is the peace of the inhabitants of heaven and of the inhabitants of the earth*, Ors. 3, 5; Bos. 57, 27. Dêm eorþware *jūdĭca terram*, Ps. Th. 81, 8 : 98, 1 : 144, 13. Gehýraþ ðâs, eorþware *audīte hæc, terrĭgĕnæ*, Ps. Spl. 48, 2 : Exon. 13 a; Th. 24, 9; Cri. 382. Ðæt cynebearn acenned wearþ eallum eorþwarum *the royal child was born for all the inhabitants of the earth*, Andr. Kmbl. 1135; An. 568: Exon. 41 b; Th. 138, 21; Gû. 579 : Bt. Met. Fox 13, 120; Met. 13, 60: Menol. Fox 124; Men. 62. Hér wynnaþ earme eorþwaran *miserable inhabitants of earth strive here*, Bt. Met. Fox 4, 113; Met. 4, 57: 17, 1; Met. 17, 1. Ofer ealle eorþwaran *over all the inhabitants of earth*, Past. 43, 9; Hat. MS. 60 b, 7.

eorþ-weall, es; *m. An earth-wall, mound;* agger :—Under eorþweall *under the earth-wall*, Beo. Th. 6171; B. 3090. Mid eorþwealle *with an earth-wall*, Bd. 1, 5; S. 476, 10 : 4, 28; S. 605, 24.

eorþ-weard, es; *m. An earth-guard;* terræ custos :—Hæfde lîgdraca eorþweard forgrunden *the fire-dragon had destroyed the earth-guard*, Beo. Th. 4658; B. 2334.

eorþ-weg, es; *m. An earth-way;* terrestris via :—Hió me woldan ðisses eorþweges ende gescrîfan *consummāvērunt me in terra*, Ps. Th. 118, 87. Þurh ða rôde sceal ríce gesêcan of eorþwege ǽghwylc sâwl *every soul shall seek the kingdom away from earth through the cross*, Rood Kmbl. 237; Kr. 120: Exon. 58 b; Th. 209, 29; Ph. 178: Ps. Th. 71, 11. Of eorþwegum *from the earthly ways*, Elen. Kmbl. 1468; El. 736.

eorþ-wéla, an; *m. Earth-wealth, fertility;* terrestres dīvitiæ, fertĭlitas :—Mid Egyptum wearþ syfan geár se ungemetlíca eorþwéla *for seven years there was very great fertility in Egypt*, Ors. 1, 5; Bos. 28, 3. Biþ him eorþwéla ofer ðæt ĕce líf *earthly wealth to them is above the eternal life*, Exon. 33 a; Th. 105, 34; Gû. 33. Ne ic me eorþwélan ôwiht sinne *I care naught for earth's wealth*, Exon. 37 a; Th. 121, 17; Gû. 290. Sum him Metudes êst ofer eorþwélan ealne geceóseþ *one chooses his Creator's favour above all earthly wealth*, 79 b; Th. 298, 20; Crä. 88.

eorþ-weorc, es; *n. Earth-work;* terræ ŏpus :—Híg on eorþweorcum gehýnede wǽron *in terræ ŏpĕribus premĕbantur*, Ex. 1, 14.

eorþ-yfig, es; *n. Ground ivy;* hĕdĕra terrestris, Herb. 100; Lchdm. i. 212, 18. v. eorþ-ifig.

eór-wicga *an earwig;* blatta, Ælfc. Gl. 24; Som. 60, 20; Wrt. Voc. 24, 24. v. eár-wicga.

eóryd *a legion*, Mt. Foxe 26, 53. v. eóred.

eós *of a war horse*, Exon. 90 a; Th. 337, 11; Gn. Ex. 63; *gen. sing.* of eoh.

eosol *an ass*, Wrt. Voc. 287, 50. v. esol.

eóster *easter*, Lk. Skt. Lind. Rush. 22, 8, 15. v. eáster.

eosul *an ass*, Mt. Kmbl. Rush. 21, 5. v. esol.

eosul-cwearn, e; *f. An ass-mill, a mill turned by asses;* ăsĭnāria mŏla, Cot. 16.

Eota land, es; *n. The land of the Jutes, Jutland;* Jūtia :—Mægð seó is gecýd Eota land *a province which is called Jutland*, Bd. 4, 16; S. 584, 24. v. Iotas.

EÓTEN, es; *m.* **I.** *a giant, monster, Grendel;* gĭgas, monstrum, Grendel :—Wæs se grimma gǽst Grendel, Caines cyn,—ðanon untydras ealle onwôcon, eótenas and ylfe and orcnēas, swylce gigantas *Grendel was the grim guest, the race of Cain,—whence unnatural births all sprang forth, monsters, elves, and spectres, also giants*, Beo. Th. 204-226; B. 102-113. Eóten, *nom. sing.* Beo. Th. 1526; B. 761. Eótena, *gen. pl.* Beo. Th. 846; B. 421. **II.** Eótenas, *gen.* a; *dat.* um; *pl. m. the Jutes, Jutlanders, the ancient inhabitants of Jutland in the north of Denmark;* Jūtæ :—Eótena treówe *the faith of the Jutes*, Beo.

Th. 2148; B. 1072: 2180; B. 1088: 2286; B. 1141: 2294; B. 1145. [O. Nrs. jötunn, m.] v. ent, eten.

eótenisc, eótonisc; adj. Belonging to or made by a giant; giant; giganteus, a gigante factus:—Geseah ðā eald sweord eótenisc then he saw an old giant sword, Beo. Th. 3120; B. 1558. Ætbær eald sweord eótonisc bore away the old giant sword, 5225; B. 2616. v. entisc.

Eotol-ware; gen. -wara; dat. -warum; pl. m. Inhabitants of Italy, Italians, Italy; Itáli, Itália:—He sinoþ gesomnade Eotolwara biscopa cōgēret synōdum episcōpōrum Itāliæ, Bd. 2, 4; S. 505, 33.

eóton ate, Chr. 998; Erl. 135, 20, = æton; p. pl. of etan.

eótonisc, Beo. Th. 5225; B. 2616: 5950; B. 2979. v. eótenisc.

eóton-weard, e; f. Giant-protection; contra gigantem protectio:—Seleweard eótonweard abéad the hall-guard offered protection against the giant [Grendel], Beo. Th. 1341, note; B. 668.

eow, es; m? A griffin; gryps = γρύψ, gryphus:—Eow, fiðerfóte fugel griffin, a four-footed bird; griffes [= gryphus], Ælfc. Gl. 18; Wrt. Voc. 22, 44. v. giw.

eow, es; m. I. the yew; taxus, L. M. 3, 63; Lchdm. ii. 350, 24. v. íw. II. the mountain ash; ornus? Ælfc. Gl. 47; Som. 65, 40; Wrt. Voc. 33, 37.

eów to you, vou; vōbis, vos; ὑμῖν, ὑμᾶς; pers. pron; dat. acc. pl. of ðū, Ex. 6, 8: Mt. Bos. 6, 16: 5, 46: Lk. Bos. 12, 28. v. gē.

eów; interj. Wo! alas! væ! heu!—Eów me! heu mihi! Ps. Spl. T. 119, 5. v. wā.

eówa ewes, female sheep; pl. nom. acc. of eówu.

eówan; p. de; pp. ed; v. trans. To shew, manifest, confer; ostendēre, manifestāre, conferre:—Ne gesacu ōhwær ecghete eóweþ nor strife anywhere shews hostility, Beo. Th. 3480; B. 1738. Ðā gēn Abrahame eówde selfa hālige spræce then he himself shewed again to Abraham a holy speech, Cd. 98; Th. 130, 24; Gen. 2164. Ealne ðone egesan, ðe him eówed wæs all that terror which was shewn to him, 202; Th. 250, 4; Dan. 541. v. eáwan, ȳwan.

eów-berge, an; f. A yew-berry; taxi bacca, L. M. 3, 63; Lchdm. ii. 350, 24.

eówcig; adj. Of or belonging to a ewe; ad ŏvem fēmĭnam pertĭnens:—Mid eówcigre wulle with ewe's wool, L. M. 1, 31; Lchdm. ii. 74, 5. v. eówocig.

eówd a flock, herd, sheepfold, Ælfc. Gr. 9, 2, 61; Som. 8, 27; 13, 47. v. eówde.

eówde, eówede, eówode, es; n: eówd, eówod, e; f. A flock, herd; grex:—Neuter, Ðæt Drihtnes eówde the Lord's flock, Bd. 1, 14; S. 482, 25: 2, 6; S. 508, 15. We wærun sceáp eówdes ðínes nos ŏves grēgis tui, Ps. Th. 78, 14. He genam hine æt eówde, ūte be sceápum tūlit eum de grēgĭbus ŏvium, 77, 69. Ne scealt ðū ðæt eówde ānforlætan thou shalt not desert the flock, Andr. Kmbl. 3334; An. 1671. Hafaþ se awyrgda wulf tostenced, Dryhten, ðín eówde hath the accursed wolf scattered thy flock, O Lord? Exon. 11 b; Th. 16, 23; Cri. 257. Ofer ðín āgen eówde sceápa sŭper ŏves grēgis tuæ, Ps. Th. 73, 1: 118, 111. He gelædde hí swā swā eówde [eówode, Ps. Lamb. 77, 52] on wēstne perduxit eos tanquam grēgem in deserto, Ps. Spl. 77, 57. Of eówdum [eówedum, Ps. Lamb. 77, 70] sceápa de grēgĭbus ŏvium, Ps. Spl. 77, 76. Feminine, Ðeós eówd hic grex, Ælfc. Gr. 9, 61; Som. 13, 47. He ðæt sceáp bær on his exlum to ðære eówde he bare the sheep on his shoulders to the flock, Homl. Th. i. 340, 2. Ic wylle ahreddan mīne eówde wið eów I will deliver my flock from you, i. 242, 13. 2. eówd, e; f. A sheepfold, fold; ŏvīle:—Eówd ŏvīle, Ælfc. Gr. 9, 2; Som. 8, 27. Sceal beón ān eówd and ān hyrde there shall be one fold and one shepherd, Homl. Th. i. 244, 1, 3. Ic hæbbe ōðre scēp ðe ne sind nā of ðisre eówde I have other sheep which are not of this fold, Homl. Th. i. 242, 35: 244, 6: ii. 114, 21.

eówe, es; m. f. A sheep, L. In. 55; Th. i. 138, 6, MSS. G. H. v. ewe, ēs.

eówe of a ewe, L. In. 55; Th. i. 138, 6, note 11, MS. B. v. eówu.

eówede a flock, Ps. Lamb. 77, 70. v. eówde.

eówena of ewes:—Twāhund eówena two hundred sheep, Gen. 32, 14. v. eówu.

eowend membrum vĭrīle, L. Alf. pol. 25; Th. i. 78, 15.

eowendende; part. Returning; rĕdiens, Ps. Spl. 77, 44. v. awendan.

eówer of you; vestrûm vel vestri, ὑμῶν; gen. pl. of pers. pron. ðū:—A.. eówer ūnus vestrûm, Mt. Bos. 26, 21. Eówer sum one of you, Beo. Th. 502; B. 248. Eówer ǽnig any of you, Cd. 22; Th. 27, 34; Gen. 427. v. gē.

eówer YOUR; vester, vestra, vestrum, ὑμέτερ-os, -α, -ον; adj. pron:—Biþ eówer blæd micel your prosperity shall be great, Cd. 170; Th. 214, 3; Exod. 563. Sceal eall ēðel-wyn eówrum cynne leófum alicgean all joy of country shall fail to your beloved kindred, Beo. Th. 5763; B. 2885.

eówer-lendisc; adj. Of your land or country; vestras:—Eówerlendisc vestras, Ælfc. Gr. 15; Som. 17, 45.

eówes a sheep's, L. In. 55; Th. i. 138, 6, note 11, MSS. G. H. v. ēwes.

eówestras sheepfolds, Som. Ben. Lye. v. ēwestre.

eówian; p. ode; pp. od To shew; ostendēre:—Hí eówodon me ða

wunde monstrāvērunt mihi vulnus, Bd. 4, 19; S. 589, 17. Ðā hēt he his tungan forþdōn of his mūþe, and him eówian linguam prōferre ex ōre, ac sŭbi ostendēre jussit, Bd. 5, 2; S. 615, 6. v. eáwan, ȳwan.

eówic you; acc. pl. of pers. pron. ðū:—Fæder alwalda mid ār-stafum eówic gehealde may the all-ruling Father with honour hold you, Beo. Th. 640; B. 317. Eówic grētan hēt bade to greet you, 6182; B. 3095. v. gē.

eówih = eówic you; acc. pl. of pers. pron. gē ye.

Eowland, es; n. Oeland, an island on the coast of Sweden; Oelandia:—Wæron us ðās land, ða synd hātene Blecinga ēg, and Meore, and Eowland, and Gotland, on bæcbord we had, on our left, those lands which are called Blekingey, and Meore, and Oeland, and Gothland, Ors. 1, 1; Bos. 22, 1.

eówocig, eówcig; adj. Of or belonging to a ewe; ad ŏvem fēmĭnam pertĭnens:—Mid eówocigre wulle with ewe's wool, L. M. 1, 3; Lchdm. ii. 42, 25.

eówod, e; f. A flock, herd; grex, Homl. Th. ii. 514, 23. v. n. and f. in eówode.

eówode, es; n: eówod, e; f. A flock, herd; grex:—Neuter, He gebrohte hig swylce eówode on wēstene perduxit eos tamquam grēgem in deserto, Ps. Lamb. 77, 52. Feminine, He nȳtenum lǽcedōm forgeaf, ahredde fram wōdnysse, and hēt faran aweg to ðære eówode ðe hí ofadwelodon he gave medicine to animals, saved them from madness, and bade them go away to the herd from which they had strayed, Homl. Th. ii. 514, 21–23. v. eówde.

eówo-humele, an; f. The female hop-plant; humŭlus fēmĭna:—Genim eówohumelan take the female hop-plant, L. M. 3, 61; Lchdm. ii. 344, 8.

eówre your, Deut. 32, 11; acc. of eówer.

EÓWU; gen. eówe; pl. nom. acc. eówa; gen. eówena; dat. eówenum; f: ēwe, an; f. A EWE, female sheep; ŏvis fēmĭna:—Ewes were milked by the Anglo-Saxons. The milk was used for domestic purposes: butter and cheese were made from it; for Ælfric teaches the shepherd [sceáp-hyrde] to say, 'On fōrewerdne morgen ic drífe sceáp mīne to heora lease, and ic agēnlæde hig to heora loca, and melke hig tweówa on dæg, and cȳse and buteran ic dō in prīmo māne mīno ŏves meas ad pascua, et rĕdūco eas ad caulas, et mulgeo eas bis in die, et cāseum et butyrum făcio,' Coll. Monast. Th. 20, 11-19. Twāhund eówena, and twentig rammena two hundred ewes, and twenty rams, Gen. 32, 14. Eówu biþ, mid hire geonge sceápe, scilling weorþ a ewe, with her young sheep, shall be worth a shilling, L. In. 55; Th. i. 138, 7, MS. B. Be eówe weorþe of a ewe's worth; de ŏvis prētio, L. In. 55; Th. i. 138, 6, note 11, MS. B. Wyl on eówe meolce hindhioloðan boil water agrimony in ewe's milk, L. M. 1, 70; Lchdm. ii. 144, 22. v. ram, the m. of eówu. [Plat. ouwe, ouw a female sheep: Frs. eij, ei, n. ŏvis fēmĭna: Dut. ooi, f. a ewe-lamb: Ger. Swiss Dial. au, auw, ow, f. a female sheep: M. H. Ger. owe, f. a female sheep: O. H. Ger. awi, owi, au, f. ovīcŭla, agna: Goth. in the words awēþi, n. a herd of sheep; awistr, n. a sheepfold: Lat. ŏvis, f: Grk. ŏïs, m. f. a sheep: Lith. awis, f. a sheep: Sansk. āvi, m. f. a sheep.]

eówunga; adv. Openly; pălam, Mk. Rush. War. 8, 32. v. eáwunga.

epegitsung, e; f. Avarice, covetousness; avārĭtia, Ps. Spl. T. 118, 36.

epiphania = ἐπιφάνια the Epiphany, the manifestation of Christ to the Gentiles. v. twelfta dæg.

epistol, e; f. A letter; ĕpistŏla:—Eall heora gewinn awacnedon ǽrest fram Alexandres epistole all their wars arose first from a letter of Alexander, Ors. 3, 11; Bos. 72, 20. [Ger. epistel, f: M. H. Ger. epistole, f: O. H. Ger. epistula, f: Goth. aipistaule, f: Lat. ĕpistŏla, f: Grk. ἐπιστολή, f.] v. pistol.

epl, eppl an apple, Ps. Spl. 78, 1. v. æppel.

epse an asp-tree, Som. Ben. Lye. v. æps.

ér ere, before, Th. Diplm. A. D. 830; 465, 30. v. ǽr.

éran a shrill sound, the ears; tinnulus, aures, Som. Ben. Lye. v. eáre.

er-bleadd, es; n. [er = ear an ear of corn, bleadd = blæd a blade, leaf] A stalk, stem, blade, haulm, straw, stubble; stĭpŭla:—Ðū asendest yrre ðīn and hit æt hí swā swā erbleadd mīsisti īram tuam, quæ devŭrābit eos sīcut stĭpŭlam, Cant. Moys. Ex. 15, 8; Thw. 29, 8.

erc an ark, a chest:—Erc gehālgunge ðīnre arca sanctĭficātiōnis tuæ, Ps. Surt. 131, 8: Lk. Rush. War. 17, 27. v. earc II.

erce-biscop an archbishop, Bd. 2, 20; S. 521, 42. v. arce-bisceop.

erce-diácon an archdeacon; archidiāconus, Wrt. Voc. 71, 80: Homl. Th. i. 416, 29: 418, 16. v. arce-diācon.

erce-hád, es; m. Archhood, an archbishop's pall, his dignity, of which the pall was a sign; pallium:—Ðæt his æftergengan symle ðone pallium and ðone erchád æt ðam apostolīcan setle Rōmānisce geladunge feccan sceoldon that his successors should always fetch the pall and the archiepiscopal dignity from the apostolic seat of the Roman church, Homl. Th. ii. 132, 10.

Ercol, es; m: Erculus, i; m. Lat. Hercules; Hercŭles:—Hȳ Ercol ðær gebrohte Hercules brought them there, Ors. 3, 9; Bos. 68, 6. Erculus wæs Iobes sunu Hercules was the son of Jove, Bt. 39, 4; Fox 216, 23.

-ere, -er, es; *m.* as the termination of many nouns, signifies a person or agent. v. fulwer *and* fullere *a fuller, bleacher,* Mk. Bos. 9, 3 : from wer *a man;* plegere *a player;* sǽdere *a sower;* wrítere *a writer.*

erede *ploughed, eared,* Ors. 1, 1 ; Bos. 20, 31 ; *p. of* erian.

eregende *ploughing,* Lk. Bos. 17, 7, = erigende ; *part. of* erian, erigan.

éren; *adj. Brazen;* æreus, Ps. Spl. T. 17, 36 : 106, 16. v. ǽren.

ǽrest *first;* imprīmis, C. R. Ben. 4. v. ǽrest.

eretic; *adj. Heretical;* hæretĭcus, Bd. 4, 13, Lye.

erfe, es; *n. An inheritance;* hērēdĭtas :—Freólsgefa áge his erfe *let the freedom-giver have his heritage,* L. Wih. 8 ; Th. i. 38, 16. v. yrfe.

erfe-gewrit, es; *n. A charter of donation;* dōnātiōnis charta, Heming, p. 120, Lye.

erfeðe; *adj. Difficult, troublesome;* diffĭcĭlis, mŏlestus :—For hwon erfeðo sindon gē dǽm wífe *quid mŏlesti estis mŭlieri?* Mt. Kmbl. Lind. 26, 10. v. earfeðe.

erfe-weard, es; *m. An heir;* hēres :—Ðú eart erfeweard ealra þeóda *tu hērēdĭtābis in omnĭbus gentĭbus,* Ps. Th. 81, 8. Forleórt he ðæs hwílewendlícan ríces erfeweardas his suna þrié *tres fĭlios suos regni tempŏrālis hērēdes relīquit,* Bd. 2, 5 ; Whelc. 121, 41. v. yrfe-weard.

erfe-weardnis, -niss, e; *f. An inheritance;* hērēdĭtas :—Erfeweardnis mín *hērēdĭtas mea,* Rtl. 3, 34. v. yrfe-weardnes.

ergende *ploughing,* Chr. 876 ; Th. 144, 32, col. 1, = erigende ; *part. of* erian, erigan.

erhe, erhlíce *fearfully,* R. Ben. Interl. 5. v. earh-líce.

ERIAN, erigan, erigean, to erianne, eriganne, erigenne; *part.* erigende ; *p.* ede ; *pp.* ed ; *v. a. To plough,* EAR ; ărāre :—For cíele nele se sláwa erian [erigan MS. Cot.] *propter frĭgus pĭger ărāre nonvult,* Past. 39, 2 ; Hat. MS. 53 a, 14, 15. Nylle erigean [erian MS. Cot.] *nonvult ărāre,* 39, 2 ; Hat. MS. 53 a, 18. Míne æceras ic erige *mei agros ăro,* Ælfc. Gr. 15 ; Som. 19, 44. Ðú erast *thou ploughest,* Homl. Th. i. 488, 24. Ðǽr yrþling ne eraþ *where husbandman ploughs not,* i. 464, 25. Ðæt lytle ðæt he erede, he erede mid horsan *the little that he ploughed, he ploughed with horses,* Ors. 1, 1 ; Bos. 20, 31. Era mid ðínum oxan *plough with thine ox,* Prov. Kmbl. 67. Hit is tíma to erigenne [eriganne MS. D.] *tempus est ărandi,* Ælfc. Gr. 24 ; Som. 25, 17. Me is to erigenne [erianne MS. D.] *ărandum est mĭhi,* 24 ; Som. 25, 19. Hæfst ðú æceras to erigenne [eriganne MS. D.] *hăbes agros ad ărandum?* 24 ; Som. 25, 20. Erigende ic geþeó *ărando prŏfĭcio,* 24 ; Som. 25, 18. Hwylc eówer hæþ eregendne þeów *quis vestrum hăbet servum ărantem?* Lk. Bos. 17, 7. Ergende *ploughing,* Chr. 876 ; Th. 144, 32, col. 1. [*Wyc.* ere, eren, eeren *to plough :* Piers P. erien, erie, erye : *Chauc.* ere : *Laym.* ærien : *O. Frs.* era : *Dut. Kil.* erien, eren, eeren, æren : *Ger.* ären, eren : *M. H. Ger.* ern : *O. H. Ger.* aran, erran ărāre : *Goth.* aryan *to plough :* *Swed.* ärja : *Icel.* erja : *Lat.* ărāre : *Grk.* ἀροῦν *to plough, till.*] DER. ge-erian, on-.

ering-lond, es ; *n. Arable land;* arvum, Cod. Dipl. 1339 ; Kmbl. vi. 200, 7.

eriung, e ; *f. A ploughing, earing;* ărātio, Ælfc. Gl. 1 ; Som. 55, 3 ; Wrt. Voc. 15, 3.

erk, e ; *f. The ark;* arca :—Noe on erke eóde *Noe in arcam intrāvit,* Lk. Skt. C. C. 17, 27. v. arc.

Ermanríc, es ; *m. The celebrated king of the Ostro-Goths or East-Goths.* v. Eormanríc.

erming, es ; *m. A miserable or wretched being;* mīser :—Ðæt is sió án frófer erminga æfter ðám ermþum ðisses lífes *that is the only comfort of the wretched after the calamities of this life,* Bt. 34, 8 ; Fox 144, 29. v. earming.

Erming-strǽt, e ; *f.* [here-man-strǽt *via strāta mĭlĭtāris,* Som.] *Erming-street. One of the four great Roman roads in Britain,* Som. Lye. v. Wætlinga-strǽt.

ermþu, e ; *f. Misery, calamity;* mĭsĕria :—Cwom ofer eorþan ermþu *misery came upon the earth,* Ps. Th. 104, 14 : Exon. 11 b ; Th. 17, 17 ; Cri. 271 : Andr. Kmbl. 2325 ; An. 1164 : Bt. Met. Fox 16, 15 ; Met. 16, 8. Æfter ermþum *after calamities,* Bt. 34, 8 ; Fox 144, 30 : Elen. Kmbl. 1533 ; El. 768. v. yrmþu.

ern *a place,* Som. Ben. Lye. v. ærn.

ern, es ; *m. An eagle;* ăquĭla, Lye. v. earn.

-ern; *def. m.* -erna ; *f. n.* -erne ; an adjective termination from ærn, ern *a place,* denoting, *as -ern* in English, *Towards a place* :—Godrum se Norþerna cyning forþférde *Godrum, the Northern king, died,* Chr. 890 ; Th. 160, 1. He forþbrohte Súþerne wynd *transtūlit austrum,* Ps. Spl. 77, 30. Fram deófle Súþernum *a dæmŏnio mĕrĭdiāno,* Ps. Spl. 90, 6. Betwux eallum Eásternum *inter omnes orientāles,* Job Thw. 164, 7. Þurh ðone smyltan Súþan Westernan wind *through the mild South-western wind,* Bt. 4 ; Fox 8, 8.

ernþ, e ; *f. Standing corn, the crop;* sĕges :—Hí swá swa rípe ernþ fortreddon hí ealle *they trod them all down like ripe corn,* Bd. 1, 12 ; S. 480, 35, note. DER. earnian.

érra *the former,* Som. Ben. Lye. = ǽrra ; *comp. of* ǽr.

ersc, es ; *n. A park, preserve;* vīvārium, Ben. Lye. v. edisc.

ersc-hen, ærsc-hen, -hæn, -henn, e ; *f. A quail;* cŏturnix, perdix :—

Erschen *cŏturnix,* Wrt. Voc. 77, 36. Hí bǽdon and com erschen *pĕtíêrunt et vēnit cŏturnix,* Ps. Spl. M. C. 104, 38. Erschæn *cŏturnix,* Wrt. Voc. 63, 22. Drihten gesende swá micel fugolcyn on hira wícstówe swilce erschenna, ðæt is on Lýden cŏturnix *ascendens cŏturnix co-opĕruit castra,* Ex. 16, 13. v. edisc-hen.

-es is the termination of the genitive case singular, in the greater part of Anglo-Saxon nouns.—Cyninges botl *a king's palace.*—Abrahames God *Abraham's God.* In English e is omitted, but its place is denoted by an apostrophe.

-es is the termination of adverbs in many cases where the noun is not so formed ; as nihtes *by night, nightly;* nēdes *of necessity, necessarily.*

Esau; *gen.* Esawes [Esaues] ; *dat.* Esawe ; *m.* [Esau עֵשָׂו *hairy,* from עָשָׂה *to be hairy*] *Esau* :—Sôþlíce Iacob sende bodan to Esawe his brēðer *mĭsit autem Jacob nuntios ad Esau fratrem suum,* Gen. 32, 3. Esau, 32, 8 : 33, 4, 8, 15, 16. Alíse me of Esawes handa mínes brôður *ērue me de mănu fratris mei Esau,* 32, 11. Ða handa synd Esaues handa *the hands are the hands of Esau,* 27, 22. Esauwe *to Esau,* 32, 18. Esau, 32, 17, *acc.* Esauw, 33, 1, *acc.*

Escan ceaster *Exeter;* Exonia, Chr. 876, 877 ; Erl. 78, 13, 16. v. Exan ceaster.

Esces dún *Ashdown* :—In Esces dúne *at Ashdown,* Cod. Dipl. 998 ; Kmbl. v. 41, 15. v. Æsces dún.

esl, e ; *f. A shoulder;* hŭmĕrus :—He on esle ahôf *he raised* [*him*] *on his shoulder,* Cd. 228 ; Th. 307, 18 ; Sat. 681. v. eaxl.

ESNE, es ; *m. A man of the servile class, a servant, retainer, man, youth;* mercēnārius, servus, vir, jŭvĕnis. The esne was probably a poor freeman from whom a certain portion of labour could be demanded in consideration of his holdings, or a certain rent [gafol, *q. v.*] reserved out of the produce of the hives, flocks or herds committed to his care. He was a poor mercenary, serving for hire, or for his land, but was not of so low a rank as the þeów or wealh :—Ánan esne gebýreþ to metsunge xii pūnd gódes cornes, and ii scípæteras and i gód mete-cū, wudurǽden be landsíde *ūni æsno, id est, inŏpi, contingunt ad victum xii pondia bŏnæ annōnæ, et duo scæpeteras, id est, ŏvium corpŏra, et una bŏna convictuālis vacca, et sartĭcāre juxta sĭtum terræ,* L. R. S. 8 ; Th. i. 436, 26-28. Gif man mid esnes cwýnan gelígeþ, be cwicum ceorle, ii gebéte *if a man lie with an ' esne's' wife, her husband still living, let him make twofold amends,* L. Ethb. 85 ; Th. i. 24, 9. Gif man mannes esne gebindeþ, vi scillinga gebéte *if a man bind* [*another*] *man's esne, let him make amends with six shillings,* 88 ; Th. i. 24, 15. Gif esne ofer dryhtnes hǽse þeówweorc wyrce an Sunnan ǽfen, efter hire setlgange, ôþ Mónan ǽfenes setlgang, lxxx scillinga se dryhtne gebéte. Gif esne dêþ, his ráde, ðæs dæges, vi se wið dryhten gebéte, oððe síne hýd *if an esne do* *theow labour, contrary to his lord's command, from sunset on Sunday-eve till sunset on Monday-eve* [*that is, from sunset on Saturday till sunset on Sunday*], *let him make amends to his lord with eighty shillings. If an esne do* [*servile work*] *of his own accord on that day* [*Sunday*], *let him make amends to his lord with six shillings, or his hide,* L. Wih. 9, 10 ; Th. i. 38, 18-22. Ic eom ðín ágen esne, Dryhten *O Dŏmĭne, ĕgo sum servus tuus,* Ps. Th. 115, 6 : Gen. 24, 61, 66 : Exon. 112 a, 112 b ; Th. 430, 9, 17, 31 ; Rä. 44, 5, 9, 17. On ðínes esnes gebéd *in orātiōnem servi tui,* Ps. Th. 79, 5. Ic Dauide, dýrum esne, on áþsware ǽr benemde *jŭrāvi David servo meo,* Ps. Th. 88, 3. He him Dauid geceás, deórne esne *ēlēgit David servum suum,* 77, 69. Wæs se ofen onhǽted, hine esnas mænige wurpon wudu on innan *the oven was heated, many servants cast wood into it,* Cd. 186 ; Th. 231, 9 ; Dan. 244 : Ps. Th. 68, 37. Twá hund-teontig and fíftig ðara monna esna and mennena he gefullode *servos et ancillas dūcentos quinquāginta baptīzāvit,* Bd. 4, 13 ; S. 583, 20 : Ps. Th. 78, 11. Án esne of Leuies híwrǽdene *vir de dŏmo Levi,* Ex. 2, 1 : 11, 2. Se hwata esne *the brave man,* Bt. 40, 3 ; Fox 238, 10. Penda, se fromesta esne *Penda, vir strēnuissĭmus,* 32, 20 ; S. 521, 8. Ealle we synd ánes esnes suna *omnes fĭlii ūnius vĭri sŭmŭs,* Gen. 42, 11, 13. Uton agifan ðæm esne his wíf *let us give to the man his wife,* Bt. 35, 6 ; Fox 170, 7. [*O. H. Ger.* asni, *m. mercēnārius :* *Goth.* asneis, *m. a hireling.*] DER. fyrd-esne.

esne-wyrhta, an ; *m. A hireling, mercenary;* mercēnārius :—Esnewyrhta *mercēnārius,* Greg. Dial. 2, 3. Eallum fríóum monnum ðás dagas sién forgifene bútan þeówum mannum and esnewyrhtum *to all freemen let these days be given, but not to slaves and hirelings,* L. Alf. pol. 43 ; Th. i. 92, 3.

esn-líce; *adv. Manfully, valiantly;* vĭrīlĭter :—Onginnaþ esnlíce and beóþ staðulfæste *vĭrīlĭter ăgĭte et confortāmĭni,* Deut. 31, 6. Hwæt dó gē, bróður, dôþ esnlíce *what ye do, brother, do manfully,* Past. 47 ; Hat. MS. Hopa nú to Drihtne, and dó esnlíce *expecta Dŏmĭnum, et vĭrīlĭter áge,* Ps. Th. 26, 16 : 30, 28.

ESOL, esul, es ; *m. An ass;* ăsĭnus :—His ēstfulnesse widteáh se esol ðe he onuppan sæt *the ass, upon which he* [*Balaam*] *sat, opposed his zeal,* Past. 36, 7 ; Cot. MS. Gif ðǽr befeólle on oððe oxa oððe esol *if an ox or an ass fell into it,* Past. 63 ; Hat. MS. Ongan ðá his esolas bǽtan *began then to bridle* [*bit*] *his asses,* Cd. 138 ; Th. 173, 25 ; Gen. 2866.

[*O.Sax.* esil, *m*: *Dut.* ezel, *m*: *Ger. M.H.Ger.* esel, *m*: *O.H.Ger.* esil, *m*: *Goth.* asilus, *m*: *Slav.* osilu.] v. assa, asse.

essian; *p.* ode; *pp.* od *To waste, consume*; tābescĕre :—Essian me dyde æfþanca mîn *tābescĕre me fĕcit zēlus meus*, Ps. Spl. M. 118, 139.

ÉST, es; *m*: ĕst, e; *f.*   I. *will, consent, grace, favour, liberality, munificence, bounty*; bĕnĕplăcĭtum, consensus, grātia, bĕnĕvŏlentia, mū-nificentia :—Ofer mîne ĕst *against my will*, Andr. Kmbl. 2438; An. 1217. Ofer ĕst Godes *against God's consent*, Exon. 61 b; Th. 226, 10; Ph. 403. þurh ĕst Godes *through grace of God*, 44 b; Th. 151, 21; Gû. 798: Elen. Kmbl. 1968; El. 986. Hie on þanc curon æðelinges ĕst *they accepted thankfully the chieftain's bounty*, Cd. 112; Th. 147, 21; Gen. 2443. He gearwor hæfde ágendes ĕst ǽr gesceáwod *he had previously more fully experienced the owner's favour*, Beo. Th. 6142; B. 3075: Andr. Kmbl. 965; An. 483.   II. *delicacies*; dēlĭciæ:—Ðá ðe synd on ĕstum *qui sunt in dēlĭciis*, Lk. Bos. 7, 25. Éstas *dēlĭciæ*, Ælfc. Gr. 13; Som. 16, 16. Cyninga wist *vel* ĕstas *dăpes*, Ælfc. Gl. 65; Som. 69, 56; Wrt. Voc. 41, 13. [*Orm.* esstess, *pl. dainties: O.Sax.* anst, *f. favour, grace: O.Frs.* enst, est *favour: Ger.* gunst, *f. favour: M.H.Ger. O.H.Ger.* anst, *f. grātia: Goth.* ansts, *f. favour: Dan.* yndest, *m. f: Swed.* ynnest, *m. favour: Icel.* ást, *f. love, affection.*]

-est, the termination of the superlative degree, perhaps from ĕst *abundance.*

éstan *from the east, easterly*, Exon. 57 a; Th. 204, 8; Ph. 94. v. eástan; *adv.*

éste; *adj. Gracious, bountiful*; bĕnignus :—Ðæt he him ealra wæs ára éste *that he was bountiful to him in all gifts*, Cd. 74; Th. 91, 8; Gen. 1509. Ðæt hyre eald Metod éste wǽre bearngebyrdo *that the Lord of old was gracious to her in her child-bearing*, Beo. Th. 1895; B. 945.

Éste, Éstas; *nom. acc: gen.* Ésta; *dat.* Éstum; *pl. m. The Esthonians* or *Osterlings* are a Finnish race,—the Éstas of Wulfstan and the Osterlings of the present day. They dwelt on the shores of the Baltic on the east of the Vistula :—Ðæt Witland belimpeþ to Éstum *Witland belongs to the Esthonians*, Ors. 1, 1; Bos. 22, 5. Ne biþ nǽnig ealo gebrowen mid Éstum, ac ðǽr biþ medo genóh *no ale is brewed by the Esthonians, but there is mead enough*, Bos. 22, 17, 19: 23, 3.

éste-líce, ĕst-líce; *adv. Kindly, gladly, delicately, daintily*; bĕnigne, libenter, dēlĭcāte :—Éstelíce bĕnigne, Ps. Spl. T. 50, 19: R. Ben. 71. Éstelíce *dēlĭcāte*, Scint. 27: Prov. 29. We ðe ĕstlíce mid us willaþ ferigan *we will gladly convey thee with us*, Andr. Kmbl. 583; An. 292.

éster *easter*, Som. Ben. Lye. v. eáster.

ĕst-ful; *adj.* [ĕst *bounty*] *Full of kindness, devoted to, ready to serve*; dēvōtus, vōtĭvus, offĭciōsus :—Éstful *dēvōtus*, Greg. Dial. 1, 3, 11. Éstful *vel* gehýrsum *offĭciōsus*; éstful *vōtĭvus*, Ælfc. Gl. 115; Som. 80, 54, 56; Wrt. Voc. 61, 32, 34.

ĕstful-líce; *adv. Kindly, devotedly*; dēvōte, Greg. Dial. 2, 16.

ĕstfulnes, -ness, e; *f. Fulness of liberality, devotion, zeal*; dēvōtio :—Hí leorniaþ mid fulre éstfulnesse ða sóðan gód to sécanne *they learn to seek the true good with full devotion*, Past. 58, 1; Hat. MS. His éstfulnesse wiðteáh se esol ðe he onuppan sæt *the ass, on which he* [*Balaam*] *sat, opposed his zeal*, Past. 36, 7; Cot. MS.

ĕstig; *adj. Gracious, bounteous*; bĕnignus :—Duguþa éstig *bounteous in benefits*, Exon. 95 b; Th. 356, 23; Pa. 16.

ĕstines, -ness, e; *f. Benignity, kindness, bounteousness*; bĕnignĭtas :—Drihten selþ éstinesse *Dŏmĭnus dăbit bĕnignĭtātem*, Ps. Spl. T. 84, 13: 64, 12.

ĕst-land, es; *n. East-land, east country, the east*; terra ŏrientālis, Som. Ben. Lye. v. Eást-land.

ĕst-líce; *adv. Gladly*; libenter, bĕnigne :—We ðe ĕstlíce mid us willaþ ferigan *we will gladly convey thee with us*, Andr. Kmbl. 583; An. 292. v. éste-líce.

Ĕst-mere, es; *m.* [ĕst = eást *east*, mere *a lake*] *The Frische Haff, or fresh water lake which is on the north of east Prussia.* Hav or Haf signifies a sea, in Danish and Swedish. It is written Haff in German, and it is now used to denote all the lakes connected with the rivers on the coast of Prussia and Pomerania. The Frische Haff is about sixty miles long, and from six to fifteen broad. It is separated by a chain of sand banks from the Baltic Sea, with which, at the present time, it communicates by one strait called the Gat. This strait is on the north-east of the Haff, near the fortress of Pillau, *Malte Brun's Univ. Geog.* vol. vii. p. 14. This Gat, as Dr. Bell informs me, 'seems to have been formed, and to be kept open by the superior force of the Pregel stream.' This gentleman has a perfect knowledge of the Frische Haff and the neighbourhood, as he received his early education in the vicinity, and matriculated at the University of Königsberg, near the west end of the Haff. I am indebted to Dr. Bell for the map of the celebrated German Historian, Professor Voigt, adapted to his 'Geschichte Preussens von den ältesten Zeiten,' 9 vols. 8vo, Königsberg, 1827-1839.' In this map three or four openings from the Frische Haff to the Baltic. 'It is certain,' says Malte Brun, 'that in 1394 the mouth of one strait was situated at Lochsett, six or eight miles north of the fortress of Pillau.' Voigt's map gives the year 1311. Id. vol. vii. p. 15. The next is the Gat of Pillau, at

present the only opening to the Baltic, with the date 1510. The third Gat, marked in the map with the date 1456, is about ten or twelve miles south-west of Pillau; and the fourth, without any date, is much nearer the west end of the Frische Haff :—Seó Wisle líþ ût of Weonodlande, and líþ in Éstmere. and se Éstmere is hûru fíftene mîla brâd. Ðonne cymeþ Ilfing eástan in Éstmere of ðæm mere, ðe Truso standeþ in staðe *the Vistula flows out of Weonodland and runs into the Frische Haff* [*Estmere*]; *and the Frische Haff is, at least, fifteen miles broad. Then the Elbing comes from the east into the Frische Haff, out of the lake* [*Drausen*] *on the shore of which Truso stands*, Ors. 1, 1; Bos. 22, 5-8.

ĕst-mete, es; *m. Delicate meat, dainties, luxuries*; dēlĭcātus cĭbus, daps, dēlĭciæ :—Ðeós sand oððe ĕstmete *hæc daps*, Ælfc. Gr. 9, 54; Som. 13, 20. Seó wuduwe ðe lyfaþ on éstmettum, heó ne lyfaþ nâ, ac heó is deád. Ðeós Anna, ðe we embe sprecaþ, ne lufude heó nâ ést-mettas, ac lufude fæstenu *the widow who liveth in luxuries, she liveth not, but she is dead. This Anna, of whom we speak, loved not luxuries, but loved fasts*, Homl. Th. i. 146, 34-148, 1.

-estre, -istre, -ystre, an; *f.* are the feminine terminations of nouns of action, same as the Latin -ix and English -ess; *as* Fiðelestre *a female fiddler*, Wrt. Voc. 73, 62; hleápestre *a female dancer*, 73, 71; lǽrestre *an instructress*; myltestre *meretrix* vel *scortum*, Wrt. Voc. 86, 72; rǽdistre *a female reader*, Wrt. Voc. 72, 7; sangestre [MS. sangystre] *a songstress*, Wrt. Voc. 72, 5; seámestre *a seamstress*, 74, 13.

éstum; *adv.* [*dat.* or *inst. pl.* of ĕst, *q.v.*] *Willingly, gladly, kindly, bounteously*; libenter, bĕnigne, mūnĭfĭcenter :—He Freán hýrde éstum *he obeyed the Lord willingly*, Cd. 92; Th. 117, 11; Gen. 1952: Ps. Th. 140, 3. Him wæs wunden gold éstum ge-eáwed *twisted gold was kindly offered to him*, Beo. Th. 2392; B. 1194. Ic Ismael éstum wille bletsian *I will bless Ishmael bounteously*, Cd. 107; Th. 142, 4; Gen. 2356.

esul *an ass*, Som. Ben. Lye.   v. esol.

esul-cweorn, e; *f. A mill-stone turned by an ass*; mŏla asĭnāria, Cot. 16.

é-swíc, e; *f. Disgrace, offence*; scandălum :—Nis in him éswíc *non est in illis scandălum*, Ps. Surt. 118, 165. In éswíc *in scandălum*, 68, 23.   v. ǽ-swíc.

é-swíca, an; *m. A hypocrite, heathen*; hypocrĭta, ethnĭcus :—Ðû éswíca *hypocrĭta*, Mt. Kmbl. Lind. 7, 5. Éswíca *ethnĭcus*, 18, 17. v. ǽ-swíca.

ETAN, to etanne; *part.* etende; ic ete, ðû etest, etst, itst, ytst, ætst, he, heó, hit, yt, ytt, et, ett, eteþ, ieteþ, iteþ, yteþ, *pl.* etaþ; *p.* ic, he æt, ðû ǽte, *pl.* æton; *subj. indef.* ic ete, æte, *pl.* eten; *p.* ǽte, *pl.* ǽten; *pp.* eten; *v. a. To* EAT, *consume, devour*; ĕdĕre, cŏmĕdĕre, mandūcāre, vescēre :—Ðû scealt greót etan *thou shalt eat dust* [*grit*], Cd. 43; Th. 56, 9; Gen. 909: 43; Th. 57, 28; Gen. 935. Seó leó bringþ hungregum hwelpum hwæt to etanne *the lioness brings to hungry whelps somewhat to eat*, Ors. 3, 11; Bos. 71, 38. Rýnde him manna [mete] to etanne *pluit illis manna ad mandūcandum*, Ps. Spl. 77, 28. Ðæt treów wæs gód to etanne *quod bŏnum esset lignum vescendum*, Gen. 3, 6. Etende *eating*, Ps. Th. 105, 17. Ic ete *ĕdo*, ðû etst [ytst MS. D.] *es*, he et [ett MS. C; ytt D.] *est*; we etaþ *ĕdĭmus*, gê etaþ *ĕdĭtis*, hí etaþ *ĕdunt*, Ælfc. Gr. 32; Som. 36, 18, 19. Ðû itst oððe drincst *thou eatest or drinkest*, Bt. 14, 1; Fox 42, 14. Ðû ytst wyrta *thou shalt eat herbs*, Gen. 3, 18. Ðû ætst *thou shalt eat*; cŏmĕdes, Gen. 3, 17. Ðe ytt hláf *qui mandūcat pānem*, Jn. Bos. 13, 18. Se tô seldan ieteþ *he too seldom eats*, Exon. 90 b; Th. 340, 16; Gn. Ex. 112. Ne wiht iteþ *nor eats a thing* [*creature*], 114 b; Th. 439, 28; Rä. 59, 10. Gê etaþ *ye eat*, Gen. 3, 5. Ðû ǽte of ðam treówe *thou hast eaten of the tree*; cŏmĕdisti de ligno, Gen. 3, 17. He æt ða offring-hláfas *pānes prŏpŏsĭtiōnis cŏmĕdit*, Mt. Bos. 12, 4. He æt he ate, Gen. 3, 6. Fuglas ǽton ða vŏlucres cŏmĕdērunt ea, Mt. Bos. 13, 4. Deáh ðe gê of ðam treówe eten [MS. eton] *though ye should eat of the tree*, Gen. 3, 4. Ðæt gê ne ǽton *ut non cŏmĕdērĕtis*, 3, 1, 3. [*Tynd.* eat: *Wyc. Chauc.* ete: *Piers P.* eten, ete: *R. Glouc.* ete: *Laym.* æten, eten: *Orm.* etenn: *Northumb.* eta: *Plat.* eten: *O.Sax.* etan: *Frs.* ytten: *O.Frs.* eta, ita: *Dut.* eten: *Ger.* essen: *M.H.Ger.* ĕzzen: *O.H.Ger.* ezan, ezzan: *Goth.* itan; *p.* at, etum; *pp.* itans: *Dan.* äde: *Swed.* äta: *Icel.* eta: *Lat.* ĕd-o: *Grk.* ἔδ-ω: *Sansk.* ad *to eat.*] DER. fretan [= for-etan], ge-etan, of-, ofer-, þurh-, under-.

ete-lond, es; *n. Pasture land*; pascua terra :—Ægðer ge etelond ge yrþlond [MS. eyrðlond] *both pasture land and arable land*, Cod. Dipl. 299; A.D. 869; Kmbl. ii. 95, 14.

eten, es; *m. A giant*; gigas, Ps. Spl. T. 32, 16: 18, 6.   v. ent.

eten = eton *should eat*, Gen. 3, 4; *subj. of* etan.

etere, es; *m. An* EATER, *a consumer, devourer*; vŏrax :—Etere *vŏrax*, Mt. Kmbl. Lind. 11, 19. Eteras *commessātōres*, Prov. 18.

etest *shalt eat*; cŏmĕdes, Ps. Th. 127, 2; *2nd fut. of* etan.

et-felgan; *p.* -fealh, *pl.* -fulgon; *pp.* -folgen *To cleave* or *stick to, adhere*; adhærēre :—Nâ etfilgþ me heorte þweor *a wicked heart cleaves not to me*, Ps. Spl. T. 100, 4.   v. æt-felgan.

éþ; *adv. More easily* :—Ðæt ic ðý éþ mǽge ðæt sóþe leóht on ðe gebringan *that I may the more easily bring upon thee the true light*,

Bt. 5, 3; Fox 14, 20: 19; Fox 70, 3. Ðú meaht éþ gecnáwan *thou mightest more easily know*, Bt. Met. Fox 12, 43; Met. 12, 22: 10, 75; Met. 10, 38. v. eáþ.

éþ, e; *f. A wave*; unda :—Éþ unda, Ælfc. Gl. 98; Som. 76, 79; Wrt. Voc. 54, 23. v. ýþ.

éðan, *p.* de; *pp.* ed *To overflow, lay waste*; vastāre :—Ðá eác éðan gefrægn eald-feónda cyn win-burh wera *then also I heard that the tribe of ancient foes laid waste the people's beloved city*, Cd. 174; Th. 219, 19; Dan. 57. v. ýðan.

Eðan-dún, e; *f.* [*Hunt.* Edendune: *Matt. West.* Ethendune] EDDINGTON, *near Westbury, Wiltshire*; lŏci nōmen in agro Wiltonensi :—He fōr to Eðandúne *he went to Eddington*, Chr. 878; Erl. 81, 12.

éþ-begete; *adj. Easily got, got ready, prepared*; fǎcilis adeptu, pǎrātus :—Ðá wæs grim andswaru éþbegete *there was a fierce answer ready*, Beo. Th. 5714; B. 2861. v. eáþ-gete.

éþ-cwide, éþ-cwíde *a rehearsal*, Som. Ben. Lye. v. ed-cwide.

éðe; *adj.* [éðan *to lay waste, desert, desolate*; vastātus :—Ðæt he geheólde éðne éðel *that he might hold the desert land*, Cd. 175; Th. 220, 28; Dan. 78.

éðe; *comp.* éðre; *sup.* éðost; *adj. Easy, ready, mild, soft*; fǎcilis, mītis :—Ne wæs ðæt éðe síþ *that was no easy enterprise*, Beo. Th. 5166; B. 2586. Eall ðú ðín yrre éðre gedydest mītīgasti omnem īram tuam, Ps. Th. 84, 3: Mk. Bos. 2, 9: Elen. Kmbl. 2586; El. 1294. v. eáðe; *adj.*

éðe; *sup.* éðest; *adv. Easily*; fǎcíliter, Hy. 1, 6; Hy. Grn. ii. 280, 6. v. eáðe; *adv.*

ÉÐEL, æðel, éðel; *gen.* éðles; *dat.* éðle, éðele; *m. n.* **I.** *one's own residence* or *property, inheritance, country, realm, land, dwelling, home*; prædium ǎvītum, fundus hereditārius, patria, terra, sēdes, domĭcilium, tabernāculum :—Ðis is mín ágen cýþ, eard and éðel *this is my own country, dwelling and home*, Bt. Met. Fox 24, 99; Met. 24, 50. Hér sceal mín wesan eorþlic éðel *here shall be my earthly country*, Exon. 36 a; Th. 117, 30; Gú. 232. Ic ealne geondhwearf éðel Gotena *I traversed all the country of the Goths*, 86 b; Th. 325, 10; Wíd. 109. Nán wítega nis andfenge on his éðele *nēmo prophēta acceptus est in patria sua*, Lk. Bos. 4, 24. Se éðel úþgenge wearþ Adame and Éuan *the country became alien to Adam and Eve*, Exon. 45 a; Th. 153, 11; Gú. 824: Th. 152, 29; Gú. 816. Onfóþ mínes Fæder ríce, beorht éþles wlite *receive my Father's realm, the land's bright beauty*, 27 b; Th. 82, 32; Cri. 1347. Ic ferde to foldan ufan from éþle *I went to earth from the realm above*, Cd. 224; Th. 296, 2; Sat. 496. Engla éðel *the dwelling of angels*, Andr. Kmbl. 1049; An. 525. Hæleða éðel *the dwelling of heroes*, 41; An. 21. Ðæt he síþ tuge eft to éþle *that he would go his way again home*, Exon. 37 b; Th. 123, 21; Gú. 326: 36 b; Th. 119, 1; Gú. 248. Éþles neósan *to visit their home*, Andr. Kmbl. 1660; An. 832: 32; An. 16. On heora éðele *in tabernāculis eōrum*, Ps. Th. 68, 26. **2.** *the following three examples are neuter* :—Ðæt earme éðel *mīsĕra patria*, Bd. 1, 12; S. 480, 37. He wolde eft ðæt ðæt sēcan his hwīlendlīcan ríces *tempŏrālis sui regni sēdem repĕtiit*, 3, 22; S. 552, 33. His ríces éðel ðæt he hæfde *sēdem regni quam tĕnuit*, 4, 1; S. 563, 14. **II.** *the Anglo-Saxon Rune* ᛟ = œ, *the name of which letter in Anglo-Saxon is* œ́ðel, ǽðel, éðel *one's native country*,—hence, this Rune not only stands for the letters œ, but for éðel = éðel *one's native country*, as,—ᛟ [éðel] byþ oferleóf ǽghwylcum men *a native country is over-dear to every man*, Hick. Thes. i. 135, 45; Runic pm. 23; Kmbl. 344, 3: Beo. Th. 1045; B. 520: 1830; B. 913. [*O. Sax.* ôðil, *m.* domĭcilium, patria, prædium avītum: *O. Frs.* ēthel, *m:* *O. H. Ger.* uodal, *n.* prædium: *Icel.* óðal, *n.* fundus avītus.] DER. fæder-éðel.

éðel-boda, an; *m. A native preacher, the apostle of a country*; indĭgĕnus prædicātor, patriæ apostŏlus :—He éðelbodan wiste *he knew the native preacher*, Exon. 47 a; Th. 162, 15; Gú. 976.

éðel-boren; *adj. Noble-born*; nōbilis natu, Prov. 31. v. æðel-boren.

éðel-cyning, es; *m. A country's king, king of the land*; patriæ *vel* terræ rex :—Eall ǽr-gestreón éðelcyninga *all ancient treasure of the kings of the land* [*earth*], Exon. 22 b; Th. 62, 6; Cri. 997.

éðel-dreám, es; *m. Domestic pleasure, joy from one's country*; dŏmesticum gaudium, patriæ gaudium :—He heóld á éðeldreámas *he ever possessed domestic joys*, Cd. 78; Th. 97, 4; Gen. 1607.

éðele; *adj. Noble, famous, excellent*; nōbilis, egrĕgius :—Syle us on earfoðum éðelne fultum *da nōbis auxilium de trĭbŭlātiōne*, Ps. Th. 107, 11. v. æðele.

éðel-eard, es; *m. A native dwelling*; patrium domĭcilium :—Abraham wunode éðeleardum *Abraham abode in the native dwellings*, Cd. 92; Th. 116, 33; Gen. 1945.

éðel-fæsten, es; *n. Land-fastness, a country's fortress*; patriæ mūnīmentum :—Ic éðelfæsten brece *I break through a land-fastness*, Exon. 126 b; Th. 487, 3; Rä. 72, 22.

éðelíce; *adv. Easily*; facīliter :—Ðú eall þing birest éðelíce búton geswince *thou bearest all things easily without labour*, Bt. Met. Fox 20, 552; Met. 20, 276. Ðæt ðú mæge cumon éðelícost *that thou mayest most easily come*, Bt. 41, 5; Fox 254, 17. v. eáðelíce.

éðelícnes, -ness, e; *f. Easiness*; facilítas, Cot. 82. DER. un-éðelícnes. v. eáþnes.

éðeling *a noble, prince*, Chr. 617; Erl. 23, 17: 972; Erl. 125, 7. v. æðeling.

Éðelinga íg *the island of nobles, the island of Athelney*, Som. Ben. Lye. v. Æðelinga ígg.

éðel-lǽnd, -lond, es; *n. A native land, a country*; patria, terra :—Ðá wæs gúþ-hergum wera éðelland geond-sended *then with hostile bands was the people's native land overspread*, Cd. 92; Th. 118, 20; Gen. 1968: 69; Th. 83, 14; Gen. 1379. On éðelland ðǽr Salem stód *into the country where Salem stood*, 174; Th. 218, 15; Dan. 39. Sēceþ eádig éðellond *seeks [its] happy native land*, Exon. 59 b; Th. 217, 12; Ph. 279: 42 a; Th. 141, 17; Gú. 628.

éðel-leás; *adj. Countryless, homeless*; patria *vel* dŏmo cǎrens, extorris, exul :—Ðæt ðú éðelleásum dēman wille *that thou art willing to adjudge to me homeless*, Andr. Kmbl. 148; An. 74. Éðel-leáse ðysne gyst-sele gihþum healdaþ [healdeþ MS.] *the homeless hold this guest-hall in memory*, Cd. 169; Th. 212, 3; Exod. 533.

éðel-mearc, e; *f. One's country's boundary*; patriæ limes :—Him ðá Abraham gewát of Egipta éðelmearce *Abraham then departed from the Egyptians' country's boundary*, Cd. 85; Th. 106, 9; Gen. 1768: 90; Th. 112, 22; Gen. 1874: 100; Th. 133, 8; Gen. 2207.

éðel-ríce, es; *n. A native-realm, native-country*; patrium regnum, patria :—Ðæt ðú móste mínes éðelríces neótan *that thou mightest enjoy my native realm*, Exon. 29 a; Th. 89, 24; Cri. 1462: Andr. Kmbl. 239; An. 120: 864; An. 432: Salm. Kmbl. 214; Sal. 106.

éðel-riht, -rieht, es; *n. A land* or *country's right*; patrium jus :—Wǽron orwénan éðelrihtes *they were hopeless of country's right*, Cd. 114; Th. 191, 8; Exod. 211. Stód seó dýgle stów ídel and ǽmen éðelriehte feor *the secret spot stood void and desolate, far from patrial-right*, Exon. 35 b; Th. 115, 10; Gú. 187. DER. eard-éðel-riht.

éðel-seld, es; *n. A native seat, settlement*; patria sēdes, dōmĭcilium :—Sceoldon ða rincas sēcan ellor éðelseld *the chieftains must seek a settlement elsewhere*, Cd. 90; Th. 113, 32; Gen. 1896.

éðel-setl, es; *n. A native seat, a settlement*; patria sēdes, dŏmĭcilium :—Him ðá eard geceás and éðelsetl *chose him then a dwelling and a settlement*, Cd. 91; Th. 115, 30; Gen. 1927. v. éðel-seld.

éðel-stæf, es; *m. A family staff* or *support, stay of the house*; prædii sustentācŭlum. v. éðyl-stæf.

éðel-staðol, es; *m. A native settlement*; patrium habĭtācŭlum :—Hú he éðelstaðolas eft gesette, swegel-torhtan seld *how he might replenish the native settlements, heaven-bright seats*, Cd. 5; Th. 6, 25; Gen. 94.

éðel-stól, es; *m.* **I.** *a paternal-seat, native-seat, country, habitation*; patria sēdes, patria, dōmĭcilium :—Eafora æfter yldrum éðelstól heóld *the son after his parents held the paternal-seat*, Cd. 56; Th. 69, 2; Gen. 1129. He éðelstólas healdan cúðe *he could hold [his] paternal-seats*, Beo. Th. 4732; B. 2371. Engla éðelstól *native-seat of angels*, Exon. 8 b; Th. 4, 13; Cri. 52: 86 b; Th. 326, 1; Wíd. 122. Ðé is éðelstól eft gerýmed *to thee a habitation is again assigned*, Cd. 73; Th. 89, 23; Gen. 1485: 74; Th. 91, 19; Gen. 1514. **II.** *a chief city, metropolis*; urbs primāria, metrŏpŏlis = μητρόπολις :—He hēt forbærnan Rōmāna burig, sió his ríces wæs ealles éðelstól *he ordered to burn up the city of the Romans, which was the metropolis of his whole empire*, Bt. Met. Fox 9, 21; Met. 9, 11.

éðel-stów, e; *f. A dwelling-place*; habĭtātiōnis lŏcus :—Ðé wíc geceós, éðelstówe *choose thee a habitation, a dwelling-place*, Cd. 130; Th. 164, 33; Gen. 2724: 50; Th. 64, 19; Gen. 1052.

éðel-þrym, -þrymm, es; *m. One's country's dignity*; dignĭtas *vel* glōria patriæ :—He éðelþrym onhóf *he exalted his country's dignity*, Cd. 79; Th. 98, 23; Gen. 1634.

éðel-turf, éðyl-turf; *gen.* -turfe; *dat.* -tyrf; *f. Native turf* or *soil, native country, country*; patrium sŏlum, patria, territōrium :—On mínre éðeltyrf *on my native turf*, Beo. Th. 824; B. 410. Ðá com leóf Gode on ða éðelturf *then came the friend of God into that country*, Cd. 85; Th. 106, 20; Gen. 1774: 127; Th. 162, 6; Gen. 2677: Exon. 60 b; Th. 220, 17; Ph. 321.

éðel-weard, es; *n. A country's guardian* or *ruler, a king*; patriæ custos *vel* dŏmĭnus, rex :—Wæs ðæt fród cyning, eald éðelweard *that was a wise king, an old country's guardian*, Beo. Th. 4426; B. 2210. Gíomonna gestrión sealdon unwillum éðelweardas *the wealth of men of old their country's guardians unwillingly gave up*, Bt. Met. Fox 1, 48; Met. 1, 24.

éðel-wyn, -wynn, e; *f. Joy of country*; patriæ gaudium :—Nú sceal eall éðelwyn eówrum cynne leófum alicgean *now shall all joy of country to your beloved kindred fail*, Beo. Th. 5762; B. 2885. DER. eard-éðelwyn.

éðer *a hedge*; sēpes, Som. Ben. Lye. v. eodor.

éþfynde; *adj. Easily found*, Cd. 171; Th. 215, 6; Exod. 579. v. eáþ-fynde, ýþ-fynde.

éþ-gesýne; *adj. Easy to be seen, visible*; fǎcilis vīsu, visĭbĭlis :—Ðær biþ éþgesýne þreó tácen *there shall be easy to be seen three signs*, Exon. 26 a; Th. 76, 6; Cri. 1235: Beo. Th. 2225; B. 1110. v. ýþ-gesýne.

édgiende *breathing;* anhēlans, Cot. 1. v. éðian.

édgung, e; *f. A breathing, inspiration;* inspīrātio:—Of ēdgunge gāstes graman dīnes *ab inspīrātiōne spīrĭtus īræ tuæ,* Ps. Spl. T. 17, 18. v. éðung.

éþ-hylde; *adj. Easily inclined, satisfied, contented;* contentus:—On ánum were éþhylde heó ne biþ *she will not be contented with one man,* Obs. Lun. § 19; Lchdm. iii. 194, 1. Beóþ éþhylde on eówrum andlyfenum *contenti estōte stĭpendiis vestris,* Lk. Bos. 3, 14.

éðian, éðigean; *p.* ode; *pp.* od. **I.** *to breathe, inspire;* hālāre, spīrāre, inspīrāre:—He leórt tācen forþ, þurh fýres bleó, up éðigean *he let a token forth breathe up, through colour of fire,* Elen. Kmbl. 2211; El. 1107. Se gāst éðaþ *the spirit breathes,* Greg. Dial. 2, 21. Hý ealle éðiaþ *they all breathe,* 4, 3. Éðode him on ðone mūþe *inspīrāvit ei in os,* Martyrol. ad 28 April. **II.** *to smell;* ŏdōrāre:—Habbaþ opene nōse, ne mǎgon éðian *nares hǎbent et non ŏdōrābunt,* Ps. Th. 113, 14.

éðiende *abounding.* v. ýðian.

éðle *to a home,* Exon. 37 b; Th. 123, 21; Gū. 326: éðles *of a home,* Andr. Kmbl. 1660; An. 830; *dat. and gen. of* éðel *a home;* domĭcĭlium.

éðm, es; *m. Breath, steam, vapour;* hālĭtus, spīrĭtus, vǎpor:—Hū sīd se swarta éðm seó *how wide the black vapour is,* Cd. 228; Th. 309, 4; Sat. 704. Ne lǽte on ðone éðm *let him not allow the vapour on* [it], L. M. I, 32; Lchdm. ii. 78, 24. v. éðm.

éþnes, -ness, e; *f. Easiness, facility, favour;* facĭlĭtas:—He gemunde ðara éþnessa and ðara ealdrihta ðe hī under ðǽm Cǽserum hæfdon *he remembered the favours and the ancient rights which they had under the Cæsars,* Bt. 1; Fox 2, 16. v. eáðnes.

éðode *breathed, inspired,* Martyrol. ad 28 April; *p. of* éðian.

éðre *more easy,* Mk. Bos. 2, 9; *comp. of* éðe. v. eáðe; *adj.*

et-hrínan *to touch,* Som. Ben. Lye. v. æt-hrínan.

edða; *conj. Or;* aut:—Hū se cuma hātte, edða se esne *how the guest is called, or the servant,* Exon. 112 b; Th. 430, 31; Rä. 44, 17: Mt. Kmbl. Rush. 5, 18. v. oððe.

éðung, éðgung, e; *f. Breath, a breathing, inspiration;* hālĭtus, spīrātio, inspīrātio:—He lǽg swā swā deád mon, nemne þynre éðunge ætýwde *quási mortuus jǎcēbat, hālĭtu tantum pertĕnui quia vivĕret demonstrans,* Bd. 5, 19; S. 640, 24. Éðung *spīrātio,* Ælfc. Gl. 79; Som. 72, 63; Wrt. Voc. 46, 21. Of éðunge gāstes graman dīnes *ab inspīrātiōne spīrĭtus īræ tuæ,* Ps. Spl. C. 17, 18.

éðyl, es; *m. A native country, country;* patria, terra:—Gesǽton eard and éðyl unspēdĭgran ðonne se frumstōl wæs *they inhabited a dwelling and a country more barren than was the first settlement,* Cd. 46; Th. 59, 11; Gen. 962: 73; Th. 90, 9; Gen. 1492. v. éðel.

éðyl-stæf, édulf-stæf, es; *m. A family staff* or *support, stay of the house;* prædii sustentǎcŭlum:—Ic eom orwēna ðæt unc se [seó MS.] éðylstæf ǽfre weorþe gifeðe *I am hopeless that to us two the staff of the family will ever be by lot,* Cd. 101; Th. 134, 11; Gen. 2223. v. éðel-stæf.

éðyl-turf; *gen.* -turfe; *dat.* -tyrf; *f. Native turf* or *soil, native country, country;* patrium sŏlum, patria, terrĭtōrium, Cd. 12; Th. 14, 26; Gen. 224: 129; Th. 163, 33; Gen. 2707. v. éðel-turf.

Etna; *indecl?* Etne, Ætne, es; *m. Etna, the volcano of Sicily;* Ætna, æ; *f.* = Αἴτνη, ης; *f.* **1.** Etna [MS. Eðna] ðæt sweflene fýr tācnode, ðā hit upp of helle geate asprang on Sicilia ðam lande, and fela ofslōh mid bryne and mid stence [Ors. B.C. 458] *Etna betokened the brimstone fire, when it sprang up from the door of hell in the island of the Sicilians and slew many by burning and stench,* Ors. 2, 6; Bos. 50, 16–19. This is much abridged from Ors. 2, 14; Hav. 123–127. Though Alfred has given the impression of his age, respecting volcanoes, Orosius only speaks thus of Etna,—*Ætna ipsa, quæ tunc cum excĭdio urbium atque agrōrum crebris eruptiōnibus æstuābat, nunc tantum innoxia spĕcie ad prætĕrĭtōrum fĭdem fūmat,* Hav. 124, 2–4. On ðam geáre, asprang up Etna fýr on Sicilium, and mǎre ðǽs landes forbærnde ðonne hit ǽfre ǽr dyde *in that year* [B.C. 135], *fire sprang up from Etna among the Sicilians, and burnt more of the land than it ever did before,* Ors. 5, 2; Bos. 103, 16. Etna fýr afleów up swā brād and swā mycel, ðæt feáwa ðara manna mihte beón eardfæste, ðe on Lipara wǽron ðam íglande, ðe ðǽr níhst wæs, for ðære hǽte and for ðam stence *the fire of Etna flowed up so broad and so great, that few of the men, who were in the island Lipara, which was next to it, could abide in their dwellings, for the heat and for the stench,* 5, 4; Bos. 105, 9–12. **2.** Etne, Ætne, es; *m:*—Se mŭnt, ðe nū monna bearn Etne hātaþ, on íglonde Sicilia swefle byrneþ, ðæt mon helle fýr hāteþ wīde, fordæm hit simle biþ sinbyrnende *the mountain, which now the children of men call Etna, burns in the island of Sicily with sulphur, that men widely call fire of hell, because it ever is perpetually burning,* Bt. Met. Fox 8, 96–104; Met. 8, 48–52. Nū manna gitsung is swā byrnende, swā ðæt fýr on ðam munte ðe Ætne hātte *now the covetousness of men is as burning as the fire in the hell, which is in the mountain that is called Etna,* Bt. 15; Fox 48, 20. Se byrnenda swefl ðone mŭnt bærnþ, ðe we hātaþ Ætne *the*

burning brimstone burneth the mountain, which we call Etna, 16, 1; Fox 50, 5.

Etne, Ætne, es; *m. Etna:*—Monna bearn Etne hātaþ *the children of men call Etna,* Bt. Met. Fox 8, 97; Met. 8, 49. Ðe Ætne hātte *which is called Etna,* Bt. 15; Fox 48, 20. v. Etna.

et-néhstan; *adv. At nighest, at last, lastly;* postrēmo, novissĭme, Som. Ben. Lye. v. æt-nýhstan.

etol; *adj. Voracious, gluttonous;* ĕdax:—Etol *ĕdax,* Ælfc. Gr. 9, 60; Som. 13, 44. v. ettul-man.

eton *should eat,* Gen. 3, 4, = eten; *subj. of* etan *to eat.*

et-somne; *adv. Together;* conjuncte, sĭmul:—Et-somne cwom lx monna *sixty men came together,* Exon. 106 a; Th. 404, 1; Rä. 23, 1. v. æt-somne.

etst, he et *eatest, eats;* es, est, Ælfc. Gr. 32; Som. 36, 18; *2nd and 3rd pers. pres. of* etan.

ettan *to pasture land;* depascĕre:—Eal ðæt land ðæt man áðer oððe ettan oððe erian mæg *all the land that they could either pasture or plough,* Ors. 1, 1; Bos. 20, 41.

ettul-man, es; *m. A gluttonous man;* vŏrax hŏmo:—Hér ys ettulman *ecce hŏmo vŏrax,* Mt. Bos. 11, 19.

ettulnys, -nyss, e; *f. Greediness, gluttony;* edācĭtas, Som. Ben. Lye.

Éua, æ; *f. Lat:* Éve, Éfe, an; *f. Eve;* Hēva:—Éua, ðæt is lif; fordanðe heó is ealra libbendra mōdor *Hēva, id est vīta; eo quod mǎter esset cunctōrum viventium,* Gen. 3, 20. Be Éuan his gemæccan *by Eve* [Hēvam] *his wife,* 4, 1. Éua, Homl. Th. i. 16, 27. Éuan scyld *Eve's sin,* Exon. 9 a; Th. 7, 6; Cri. 97. [Heb. חַוָּה from חָיָה *to live.*]

euen *even,* Som. Ben. Lye. v. efen.

Euer-wíc *York,* Chr. 189; Th. 14, 23, col. 1. v. Eofor-wíc.

Eues-ham; *m. Evesham,* Chr. 1077; Erl. 215, 15. v. Eofes-ham.

eufæstnys, e; *f. Sincerity, religion;* relĭgio, Ælfc. T. 28, 11. v. ǽ-fæstnes.

eúwu *a ewe,* Heming. p. 129. v. eówu.

éw-bryce *adultery,* Som. Ben. Lye. v. ǽw-bryce.

éwe, an; *f. A ewe;* ŏvis fēmĭna:—Éwe biþ, mid hire giunge sceápe, scilling weorþ *a ewe, with her young sheep, shall be worth a shilling,* L. In. 55; Th. i. 138, 7. v. eówu.

éwe, es; *common gender A sheep, generally as* ŏvis:—Be éwes weorþe *of a sheep's worth;* de ŏvis prētio, L. In. 55; Th. i. 138, 6. v. eówu.

éwede *a flock,* Ps. Spl. T. 77, 57. v. eówde.

ewerdla *damage.* v. æf-werdla.

éwes *a sheep's,* L. In. 55; Th. i. 138, 6: *also* eówes *in MSS.* G, H; *gen. of* éwe, es. v. eówu.

éwestre, es; *m. A sheepfold;* ŏvīle, Cot. 7. v. eówestras.

éwiscnes, -ness, e; *f. Disgracefulness, impudence, shamelessness;* impŭdentia, Som. Ben. Lye. v. ǽwiscnys.

éwyde *a flock,* Ps. Spl. C. 77, 57. v. eówde.

ewyrdlu *damage.* v. æf-werdla.

ex, e; *f. An axe;* secūris. v. æx.

ex *an axis;* axis, Som. Ben. Lye. v. eax.

Ex, es; *m:* Exa, an; *m. The river Ex;* Isca, *in Devon.* v. Exan ceaster, Exan mūþa.

exámeron, es; *n. A work on the six days of creation;* hexæmĕron = ἑξά-ἡμερον = ἐξ *six,* ἡμέριος, ον *relating to a day:*—Exámeron, ðæt is be Godes six daga weorcum *Hexameron, that is concerning the six days' works of God,* Hexam. Norm. 1. Basilius awrāt áne wunderlīce bōc, be eallum Godes weorcum, ðe he geworhte on six dagum, Exámeron gehāten *Basil wrote a wonderful book about all the works of God, which he wrought in six days, called Hexameron,* Basil prm; Norm. 32, 12.

Exan ceaster, Exan ceaster, Exe cester, es; *n.* [*Flor.* Exancestre, Excestre: *Hovd.* Excester; Ex, Exa *the river Ex:* ceaster; *gen.* ceastres; *n. v.* ceaster *a city*] EXETER, *Devon;* cīvĭtas Exoniæ in agro Devōniensi, ad ripam Iscæ flūmĭnis:—Se here Exan ceaster beseten hæfde *the army had beset Exeter,* Chr. 895; Th. 172, 12. He wende hine wið Exan ceastres *he turned towards Exeter,* Chr. 894; Th. 166, 31. Wið Exan cestres *towards Exeter,* Chr. 894; Th. 168, 26, col. 1. Exacester, Chr. 1003; Th. 252, 14, col. 1. Eaxeceaster, Execiester, Th. 253, 14, col. 1, 2. v. ceaster II.

Exan mūþa, Eaxan mūþa, Axa-mūþa, an; *m:* Exan mūþ, es; *m. The mouth of the river Ex,* EXMOUTH, *Devon:*—Se here com to Exan mūþan *the army came to the mouth of the Ex,* Chr. 1001; Th. 249, 36. To Exan mūþe *to Exmouth,* Th. 249, col. 2, 36. To Axa-mūþan *to Exmouth,* Chr. 1049; Th. 307, 37.

exl, e; *f. Shoulder;* hŭmĕrus:—He hit set on his exla *impōnit in humeros suos,* Lk. Bos. 15, 5: Andr. Kmbl. 3148; An. 1577. v. eaxel.

ex-odus, i; *m.* [*Lat.* exodus = *Grk.* ἐξ *out;* ὁδός, *f. way, path, travelling*] *A going out;* exĭtus:—Exodus on Grēcisc, Exitus on Lýden, Útfæreld on Englisc *Exodus* [Ἔξοδος, *f.*] *in Greek, exĭtus in Latin, a going out in English,* Ex. Thw. Title. v. út-færeld.

exorcista, an; *m. A caster out of spirits,* L. Ælf. P. 34; Th. ii. 378, 6. v. hád II, hālsigend.

F

AT the end of syllables, and between two vowels, the Anglo-Saxon *f* is occasionally represented by *u*, the present English *v*; it is, therefore, probable that the Anglo-Saxon *f* in this position had the sound of our present *v*, as Luu, luf=lufu *love*; fíf *five*; hæuþ, hæfþ *haveth*; Euen, efen *even*. In the beginning of Anglo-Saxon words, *f* had the sound of the English *f*, as Fíf *five*, finger *finger*, finn *fin*, fisc *fish*. The Rune Ᵽ not only stands for the letter *f*, but for Feoh, which, in Anglo-Saxon, signifies *money*, *wealth*. v. feoh **IV** and **RÚN**.

fá *hostile*; *hostiles*:—Fá þrówiaþ bealu egeslíc *the hostile shall suffer fearful evil*, Exon. 31 b; Th. 98, 30; Cri. 1615; *pl. nom. acc. of* fáh.

faag *of a varying colour.* v. fág.

faca *of spaces*, Andr. Kmbl. 2741; An. 1373; *gen. pl. of* fæc.

facade *acquired*, Ors. 3, 11; Bos. 75, 28; *p. of* facian.

FÁCEN, fácn, es; *pl. nom. acc.* fácnu; *gen.* fácna; *n. Deceit, fraud, guile, treachery, malice, wickedness, evil, crime*; dolus, fraus, nēquĭtia, mălĭtia, inĭquĭtas, prævārĭcātio:—Eádig wer ðam ðe ná ætwíteþ Drihten synna, and nys on gáste his fácen *beātus vir cui non impūtābit Dŏmĭnus peccātum, nec est spīrĭtu ejus dŏlus*, Ps. Spl. 31, 2 : Ps. Lamb. 35, 4. Hér is Israhēlisc wer, on ðam nis nán fácn *ecce vēre Israēlīta, in quo dŏlus non est*, Jn. Bos. 1, 47. Ðis fácn *hæc fraus*, Ælfc. Gr. 9, 36; Som. 12, 34. Ne ætfyligeþ ðé ǣhwǣr fácn ne unriht *numquid adhǣret tĭbi sēdes inĭquĭtātis*, Ps. Th. 93, 19. Fácnes cræftig *skilled in guile*, Exon. 97 a; Th. 361, 24; Wal. 24 : 62 a; Th. 229, 4; Ph. 450. He ðæs fácnes fintan sceáwaþ *he sees the sequel of treachery*, 83 b; Th. 315, 16; Mód. 27. Gif heó ðæs fácnes gewíta nǣre *if she were not privy to the crime*, L. Ath. v. 1, § 1, 2; Th. i. 228, 17, 21. Ic feóde fácnes wyrcend *făcientes prævārĭcātiōnes odīvi*, Ps. Th. 100, 3 : 139, 10. Ne dyde ic for fácne *I did it not for fraud*, Cd. 128; Th. 162, 34; Gen. 2691 : Exon. 73 a; Th. 272, 10; Jul. 497. Bútan ǣghwylcum fácne *without any guile*, L. O. 2; Th. i. 178, 14. He hí ðonne bútan fácne fédeþ syððan *păvit eos sine mălĭtia cordis sui*, Ps. Th. 77, 71 : 93, 22. He lǣdige ða hand mid ðe man týhþ ðæt he ðæt fácen mid worhte *let him clear the hand therewith with which he is charged to have wrought the fraud*, L. Ath. i. 14; Th. i. 206, 24. Fácen ne dó ðú *ne fraudem fĕcĕris*, Mk. Bos. 10, 19. Eorl óðerne spreceþ fægere beśóran, and ðæt fácen swá ðeáh hafaþ in his heortan *one man speaks another fair before his face, and nevertheless hath evil in his heart*, Frag. Kmbl. 9; Leás. 5: Menol. Fox 574; Gn. C. 56. Hí fácen and unriht acwǣdon *lŏcūti sunt nēquĭtiam*, Ps. Th. 72, 6 : 94, 9. Ðæt he him nán fácn mid nyste *that he knew of no guile in him*, L. C. S. 29; Th. i. 392, 16: L. O. 9; Th. i. 182, 3. Se Hǣlend hyra fácn gehýrde *cognĭta Iesus nēquĭtia eōrum*, Mt. Bos. 22, 18. Him yfle ne mæg fácne sceððan *evil may not injure them by guile*, Exon. 64 b; Th. 237, 25; Ph. 595 : 70 b; Th. 263, 15; Jul. 350. Nóðer he ðý fácne mæg biwergan *nor may he defend himself from that evil*, 87 b; Th. 329, 22; Vy. 38. Innan of manna heortan yfele geþances cumaþ, fácnu *ab intus ēnim de corde hŏmĭnum mălæ cōgĭtātiōnes prŏcēdunt, dŏlus*, Mk. Bos. 7, 22. Ðú tó fela fácna gefremedes in flǣschoman *thou hast perpetrated too many guiles in the body*, Exon. Th. 137, 12; Gú. 558: Cd. 125; Th. 160, 16; Gen. 2651. [*Orm.* fakenn: *Plat.* faxen, *pl. fun*: *O. Sax.* fēkn, *n. a fraud, deceit*: *M. H. Ger.* veichen, *n*: *O. H. Ger.* feihan, *n*: *Icel.* feikn, *f. a token, an omen.*]

fácen-dǽd, e; *f. A wicked deed, sin*; peccātum:—For fyrenfulra fácendǣdum *pro peccātōrĭbus derelinquentĭbus*, Ps. Th. 118, 53.

fácen-ful, fácn-ful, -full; *def. se* -fulla, seó, ðæt -fulle; *adj. Deceitful, crafty*, fraudŭlentus, dŏlōsus:—Se fácenfulla [MS. fakenfulla] fægere word sprecþ *the deceitful man speaks fair words*, Basil admn. 5; Norm. 46, 5. Múþ ðæs fácenfullan ofer me geopened is *os dŏlōsi sŭper me ăpertum est*, Ps. Lamb. 108, 2. Fram menn fácenfullum [MS. fakenfullum] genera me *ab hŏmĭne dŏlōso ērue me*, 42, 1. Drihten alés sáwle míne fram tunge fácenfulre *Dŏmĭne lībĕra ănimam meam a lingua dŏlōsa*, 119, 2: 108, 3. On fácnfulre tungan *lingua dŏlōsa*, 51, 6. Ðæne wer ðe is blódgitol and géotende oððe wer blóda and fácenfulne gehiscþ oððe onscunaþ Drihten *virum sanguĭnum et dŏlōsum abŏmĭnābĭtur Dŏmĭnus*, 5, 8. Dó ðú feorr fram ðé ða fácenfullan [MS. fakenfullan] hiwunge *make far from thee deceitful dissimulation*, Basil admn. 5; Norm. 46, 9. Weras [MS. weres] blóda and fácnfulle ne dǣlaþ [MS. dǣla] on emtwá heora dagas *viri sanguĭnum et dŏlōsi non dimĭdiābunt dies suos*, Ps. Lamb. 54, 24.

fácen-fulnes, -ness, e; *f. Deceitfulness, deceit*; fraudŭlentia, Som. Ben. Lye.

fácen-gecwis, e; *f. A wicked consent, conspiracy*; conspĭrātio, Cot. 46.

fácen-geswipere, es; *n. Deceitful counsel, deceit*; consĭlium astūtum, dŏlus:—Hí on ðínum folce fácengeswipere syredan *in plēbem tuam astūte cōgĭtāvērunt consĭlium*, Ps. Th. 82, 3.

fácen-leás; *adj. Without deceit, simple, innocent*; simplex, Som. Ben. Lye.

fácen-líc; *adj. Deceitful*; dŏlōsus, R. Ben. in prœem: Ors. 3, 1? Lye.

fácen-líce; *adv. Deceitfully, fraudulently*; dŏlōse, fraudŭlenter:—Ðín bróðor com fácenlíce and nam ðíne bletsunga *vēnit germānus tuus fraudŭlenter et accēpit benedictiōnem tuam*, Gen. 27, 35. Ða leásan men fácenlíce þencaþ *false men think treacherously*, Frag. Kmbl. 49; Leás. 26.

fácen-searu, fácn-searu; *gen.* -searwes; *n. A treacherous wile, treachery*; machĭnātio dŏlōsa:—Þurh fácnsearu *by treachery*, Ps. Th. 55, 1. Gefylled fácensearwum *filled with treacherous wiles*, Exon. 83 b; Th. 315, 7; Mód. 27.

fácen-stæf, fácn-stæf, es; *pl. nom. acc.* -stafas; *m. A deceitful* or *treacherous deed*; nēquĭtia:—Nalles fácnstafas fremedon *they perpetrated no treacherous deeds*, Beo. Th. 2041; B. 1018.

fácen-tácen, es; *n. A false sign, sign of crimes*; scĕlĕrum signum:—Hafaþ fácentácen feores *they shall have the false sign of life*, Exon. 30 b; Th. 95, 32; Cri. 1566.

facg, fagc, es; *n? A flat-fish, plaice*; plătesia, Ælfc. Gl. 102; Som. 77, 64; Wrt. Voc. 55, 69.

facian; *p.* ode, ade; *pp.* od, ad *To acquire*; acquīrĕre:—Ðe he him sylfum facade Mæcedonia onweald *because he wished to get the government of the Macedonians for himself*, Ors. 3, 11; Bos. 75, 28.

fácn *deceit*, Jn. Bos. 1, 47. v. fácen.

fácne; *def. se* fácna; seó, ðæt fácne; *adj. Deceitful, fraudulent, factious*; subdŏlus, dŏlōsus, factiōsus:—Fácna *dŏlōsus*, Cot. 85 : factiōsus, 198. Gif hit fácne is *if it be fraudulent*, L. Ethb. 77; Th. i. 22, 2. Fácnum wordum *with factious words*, Cd. 214; Th. 268, 35; Sat. 65. v. fæcne; *adj.*

fácne; *adv. Deceitfully, fraudulently*; dŏlōse, fraudŭlenter:—Ic his feóndas fácne gegyrwe mid scame *inĭmĭcos ejus induam confūsiōne*, Ps. Th. 131, 19 : 138, 18. v. fæcne; *adv.*

fácn-ful, -full *deceitful*, Ps. Lamb. 51, 6 : 54, 24. v. fácen-ful.

fácon *deceit*, Jn. Lind. War. 1, 47. v. fácen.

fácyn-full *deceitful*, Prov. 14. v. fácen-ful.

fadian; *p.* ode; *pp.* od *To set in order, dispose, direct, guide*; ordĭnāre, dispōnĕre, dirĭgĕre:—Word and weorc freónda gehwylc fadige mid rihte *let every friend guide his works and words aright*, L. C. E. 19; Th. i. 372, 1. **DER.** ge-fadian, mis-.

fadung, e; *f. A setting in order, disposing, dispensation*; ordo, ordĭnātio, dispŏsítio:—Fadung ordo, ordĭnātio, R. Ben. 65: dispŏsítio, 18. Swá swá hit ðǣre godcundlícan fadunge gelícode *as it seemed good to the divine dispensation*, Homl. Th. i. 274, 31. **DER.** ge-fadung, mis-.

FÆC, es; *pl. nom. acc.* facu; *gen.* faca; *n. Space, interval, distance, portion of time*; spătium, intervallum, tempŏris intervallum:—On wæs lytlum fæce *in so short a space*, Elen. Kmbl. 1917; El. 960. Ðæt wæs on fæce syxtig furlanga fram Hierusalem *quod ĕrat in spătio stădiōrum sexăginta ab Ierūsălem*, Lk. Bos. 24, 13. Hí binnon lytlan fæce gewendon to Lundene *they within a little space went to London*, Chr. 1016; Erl. 155, 22. Myccle fæce *multo intervallo*, Bd. 1, 1; S. 473, 10. Ymb lytel fæc *after a little time*, Elen. Kmbl. 543; El. 272 : 765; El. 383. Þurh lytel fæc *for a little space*, Exon. 35 b; Th. 115, 6; Gú. 185. Se þeódwíga þreónihta fæc swífeþ on swefote *the noble creature is dormant in slumber a three nights' space*, 96 a; Th. 357, 34; Pa. 38. Geseah he ánre stówe fæc *vīdit ūnīus lŏci spătium*, Bd. 3, 10; S. 534, 19. Unfyrn faca *in a little time*, Andr. Kmbl. 2741; An. 1373. Twegra dæga fæc *two days' space*; duārum dīērum spătium, R. Ben. 53. Fífwintra fæc *five years' space*; olympias, Ælfc. Gl. 16; Som. 58, 69; Wrt. Voc. 21, 56. Lytel fæc *a little time, interval*; intervallum, Ælfc. Gr. 47; Som. 48, 35 : Beo. Th. 4472; B. 2240. Æfter fæce *after a while, afterwards*; postmōdum, Bd. 3, 5; S. 527, 16 : 5, 23; S. 645, 33. [*Plat.* fak: *Frs.* feck cāmĕra, *spătium, intervallum*: *O. Frs.* fek, fak: *Dut.* vak, *n. an empty place* or *space*: *Ger.* fach, *n. any inclosed space*: *M. H. Ger.* vach, *n*: *O. H. Ger.* fah mænia: *Dan.* fag, *n. a department, office*: *Swed.* fack, *n. a compartment.*]

fæccan *to fetch*, L. E. G. 3; Th. i. 168, 11, note 13. v. feccan.

fæcele, an; *f. A torch*; fax:—Fæcele stánes *fax scŏpŭli*, Cot. 169. v. þæcele.

fæcne, fácne; *adj. Deceitful, fraudulent, guileful, wicked*; subdŏlus, dŏlōsus, mălignus, nēquam:—Swá oft sceaða fæcne forfēhþ eorlas *as oft the guileful robber surprises men*, Exon. 20 b; Th. 54, 20; Cri. 871. Hæfde fæcne hyge *he had a crafty soul*, Cd. 23; Th. 29, 1; Gen. 443. Of firenfulra fæcnum handum *from the deceitful hands of the wicked*, Ps. Th. 81, 4 : 105, 10 : 136, 3. **DER.** fela-fæcne, un-.

fæcne, fácne; *adv. Maliciously, disgracefully*; măligne, turpĭter:—Gif me mín feónd fæcne wyrgeþ *si inĭmĭcus meus mălĕdixisset mĭhi*, Ps. Th. 54, 11 : 55, 2 : 65, 2 : 111, 7, 9.

fédde *fed*, Chr. 994; Erl. 133, 26,=fédde; *p. of* fédan.

FÆDER, feder; *indecl. in sing. but gen.* fæderes *and dat.* fædere *are sometimes found*; *pl. nom. acc.* fæderas; *gen. a*; *dat.* um; *m. A FATHER*, păter:—Fæder and módor *a father and mother*; hic et hæc parens, Ælfc. Gr. 9, 38; Som. 12, 48. On Fæder geardas *in the dwellings of the Father*, Salm. Kmbl. 832; Sal. 415. Mid fæder ðínne *with thy father*, Exon. 12 b; Th. 22, 9; Cri. 349. We bletsiaþ bilewitne Feder *we bless*

the merciful Father, Hy. 8, 8; Hy. Grn. ii. 290, 8. Sunu his fæderes *son of his father*, Cd. 226; Th. 301, 12; Sat. 580. Ðis is se ilca God, ðone fæderas cúðon *this is the same God, whom your fathers knew*, Andr. Kmbl. 1504; An. 753: Elen. Kmbl. 796; El. 398. Ne sleá man fæderas for suna gylton, ne suna for fædera gilton *non occidentur patres pro filiis, nec filii pro patribus*, Deut. 24, 16. Bebeád fæderum ussum *mandâvit patribus nostris*, Ps. Th. 77, 7. **2.** I Fæder *pāter :* 2 ealda [MS. ealde] fæder *ăvus :* 3 þridda [MS. þridde] fæder *proăvus :* 4 feówerþa [MS. feówerþe] fæder *ăbăvus :* 5 fífta [MS. fífte] fæder *ătăvus :* 6 sixta fæder *sextus pater*, trītăvus, Ælfc. Gl. 90, 91; Som. 75, 4–14; Wrt. Voc. 51, 49–59: 72, 18–23: Nat. S. Greg. Els. p. 4, note. [*Wyc.* fader, fadir : *Piers P. Chauc.* fader : *Laym.* fæder, fader, uader : *Orm.* faderr : *Plat.* vader, *m :* *O. Sax.* fader, fadar, *m :* *Frs.* faer : *O. Frs.* feder, fader, feider, *m :* *Dut.* vader, *m :* *Ger.* *M. H. Ger.* vater, *m :* *O. H. Ger.* fatar, fater, *m :* *Goth.* fadar ; *gen.* fadrs ; *dat.* fadr, *m :* *Dan. Swed.* fader, *m :* *Icel.* faðir, *m :* *Lat.* păter, *m :* *Grk.* πατήρ, *m :* *Sansk.* pi-tṛi *from* pā *to guard, preserve.*] DER. ǽr-fæder, eald-, forþ-, fóster-, god-, heáfod-, heáh-, sóþ-, steóp-, wealdend-, wuldor-: fædera, ge-fædera, suhter-.

fædera, fædra, an; *m.* *An uncle, a father's brother ;* patruus :—Mín fædera *patruŭs meus*, Wrt. Voc. 52, 13. · Bân hire fæderan *patrui sui ossa*, Bd. 3, 11; S. 535, 16: 3, 24; S. 556, 28: Cd. 90; Th. 114, 7; Gen. 1900. Mînes fæderan þridda fæder *my uncle's great grandfather*, Wrt. Voc. 52, 16. [*O. Frs.* federia, *m :* *O. H. Ger.* fataro, *m.*] DER. suhtor-fædra, suhter-ge-fædera. v. eám *an uncle on the mother's side.*

fæder-æðelo ; *indecl. n.* [æðelo *nobility, origin*] *Fatherly nobility, origin, ancestry, fatherly honours ;* gĕneălŏgia păterna, nŏbĭlĭtas hērēdĭtāria :—Ða ðe mægburge mǽst gefrunon, fæderæðelo gehwæs *those who most understand kinship, the ancestry of each*, Cd. 161; Th. 200, 24; Exod. 361. He scolde fæderæðelum onfón *he should succeed to his father's honours*, Beo. Th. 1826; B. 911.

fæderen, fædern, fædren; *adj.* *Paternal, belonging to a father ;* păternus, Cd. 79; Th. 98, 10; Gen. 1628.

fæderen-bróðor, es; *m.* *A brother from the same father ;* frāter ex eōdem patre ŏriundus :—Ic fram ðé wearþ fæderenbróðrum *exter factus sum fratribus meis*, Ps. Th. 68, 8.

fæderen-cnôsl, fædren-cnôsl, es; *n.* [cnôsl *a race, kin*] *A paternal race, father's kin ;* păterna prŏgĕnies, părentēla :—Be ðæs fædrencnôsles wére *according to the 'wer' of the father's kin*, L. Alf. pol. 9; Th. i. 68, 2.

fæderen-cyn, fædren-cyn, -cynn, es; *n.* [cyn *a race, kin*] *A paternal kin* or *race ;* păternum gĕnus :—Hiera ryht fæderencyn [fædrencynn, Th. 87, 14, col. 1] gæþ to Cerdice *their direct paternal kin goes to Cerdic*, Chr. 755; Th. 86, 14, col. 1. We areccan ne mágon ðæt fædrencynn *we cannot tell the paternal kin*, Exon. 11 b; Th. 16, 4; Cri. 248.

fæderen-healf, fædren-healf, e; *f.* *The father's side ;* păterna pars :—Hira nǽn næs on fædrenhealfe togeboren, búton him ânum *none of them on the paternal side was born thereto, except him alone*, Chr. 887; Erl. 86, 5.

fæderen-mǽg, fædern-mǽg, fædren-mǽg, -mág, es; *m.* [mǽg *a relation*] *A relation on the father's side, paternal relative ;* a patre cognātus, agnātus :—Cain gewearþ to ecgbanan fæderenmǽge *Cain became the murderer of his father's son*, Beo. Th. 2530; B. 1263. Fædrenmǽga mǽgleás *kinless of paternal relatives*, L. Alf. pol. 27; Th. i. 78, 20. Fædrenmǽgum hiora dǽl mon agife *let their share be given to the paternal kindred*, 8; Th. i. 66, 22. Fædernmágas *agnāti*, Ælfc. Gl. 92; Som. 75, 37; Wrt. Voc. 51, 79.

fæderen-mǽgþ, e; *f.* *Paternal kindred ;* păterna cognātio :—VIII fæderenmǽgþe *eight of the paternal kindred*, L. E. G. 12; Th. i. 174, 19.

fæder-éðel ; *gen.* -éðles; *m.* [éðel *a country, home*] *Father-land, paternal home ;* păterna rĕgio, patria :—Scipia swór ðæt him leófre wǽre, ðæt he hine sylfne ácwealde ðonne he forléte his fæderéðel *Scipio swore that he would rather kill himself than leave his father-land*, Ors. 4, 9; Bos. 91, 20. He bebeád, ðæt ǽlc cóme to his fæderéðle *he gave orders that every one should come to his father's home*, 5, 14; Bos. 114, 18, 22.

fæder-éðel-stól, es; *m.* *Father-land, paternal-seat ;* patria, sēdes patria :—Carram ofgif, fæderéðelstól *renounce Harran, thy father-land*, Cd. 83; Th. 105, 4; Gen. 1748: Exon. 15 a; Th. 32, 22; Cri. 516.

fæder-feoh, -fioh; *gen.* -feós; *n.* *A father-fee,—the marriage portion which reverted to the father, if his daughter became a widow, and returned home ;* Fæder-feum, dos a patre accepta, L. Ethb. 81; Th. i. 24, 1, note a. v. Du Cange in voce.

fæder-geard, es; *m.* *A paternal habitation ;* păternum dŏmĭcĭlium :—Fædergeardum feor *far from his paternal habitations*, Cd. 50; Th. 64, 20; Gen. 1053.

fæder-gestreón, es; *n.* *A father's property, patrimony ;* patrĭmōnium, Cot. 152.

fædering-mǽg, es; *m.* *A paternal relation ;* a patre cognātus, agnātus, L. Ethb. 81; Th. i. 24, 1. v. fæderen-mǽg.

fæderleás ; *adj.* FATHERLESS ; orbus patre, orphănus, Ps. Vos. 93, 6.

fæder-líc ; *def.* se -líca, seó, ðæt -líce ; *adj.* *Of* or *belonging to a*

father, FATHERLY, *paternal, ancestral ;* patrius, păternus, patrōnymĭcus :—Wæs he to ðǽre fæderlícan healle gelǽdd *he was led to his father's hall*, Guthl. 2; Gdwin. 12, 11. Ðýlæs toworpen síen fród fyrngewritu and ða fæderlícan láre forléten *lest the wise old scriptures should be overturned and our ancestral lore deserted*, Elen. Kmbl. 862; El. 431. Sume syndon patronimica, ðæt synd fæderlíce naman *some are patronymics, which are fatherly nouns*, Ælfc. Gr. 5; Som. 4, 52.

fædern-mǽg, -mág *a paternal relative*, Ælfc. Gl. 92; Som. 75, 37; Wrt. Voc. 51, 79. v. fæderen-mǽg.

fæder-ríce, es; *n.* *A paternal kingdom ;* păternum regnum :—In heora fæderríce *in their paternal kingdom*, Cd. 220; Th. 283, 22; Sat. 308.

fæder-slaga, an; *m.* *A father-slayer ;* parrĭcīda, Ælfc. Gl. 85; Som. 73, 113; Wrt. Voc. 49, 20.

fæderyn-cyn, -cynn, es; *n.* *A paternal kindred* or *race*, Cd. 170; Th. 213, 29; Exod. 559. v. fæderen-cyn.

fædra, an; *m.* *A paternal uncle*, Chr. 901; Th. 178, 22. v. fædera.

fædren *paternal, belonging to a father ;* păternus. v. fæderen.

fædren-cnôsl *father's kin*, L. Alf. pol. 9; Th. i. 68, 2. v. fæderen-cnôsl.

fædren-cyn, -cynn *a paternal kin*, Exon. 11 b; Th. 16, 4; Cri. 248. v. fæderen-cyn.

fædren-healf *the paternal side*, Chr. 887; Erl. 86, 5. v. fæderen-healf.

fædren-mǽg *a paternal relative*, L. Alf. pol. 27; Th. i. 78, 20. v. fæderen-mǽg.

fædrunga, an; *m.* *A paternal relation, any parental relation ;* cognātus a patre, părens :—Feóndes fædrunga *the fiend's parent* [i. e. Grendel's mother], Beo. Th. 4262; B. 2128. [*O. H. Ger.* fatarungo, *m.* v. Grm. ii. 363.]

fædyr *a father*, Mt. Foxe 23, 9. v. fæder.

FǼGE ; *def.* se fǽga, seó, ðæt fǽge ; *comp.* -ra ; *sup.* -est ; *adj.* **I.** *fated, doomed, destined ;* prŏpĕræ morti dēvōtus, cui mors imminet :—Æt fótum feóll fǽge cempa *the fated warrior fell at his feet*, Byrht. Th. 135, 17; By. 119: Exon. 89 a; Th. 335, 2; Gn. Ex. 27. Næs ic fǽge ðá gyt *I was not yet doomed*, Beo. Th. 4289; B. 2141: 5943; B. 2975. Pharaon gefeól, and his fǽge werud, on ðam Reádan Sǽ *excussit Pharaōnem, et exercĭtum ejus, in Mări Rubro*, Ps. Th. 135, 15. Lǽtaþ gáres ord ingedúfan in fǽges ferþ *let the javelin-point pierce the life of the doomed one*, Andr. Kmbl. 2665; An. 1334: Salm. Kmbl. 318; Sal. 158. Hogodon georne hwá ðǽr mid orde mihte on fǽgean men feorh gewinnan *they were earnestly anxious who there might first take life with a spear from the doomed man*, Byrht. Th. 135, 28; By. 125. Wyrd ne meahte in fǽgum leng feorg gehealdan *fate might not longer preserve life in the destined*, Exon. 48 a; Th. 165, 19; Gú. 1031. Bil eal þurhwód fǽgne flǽschoman *the falchion passed through all her fated carcase*, Beo. Th. 3140; B. 1568. On ðæt fǽge folc in the fated band, Elen. Kmbl. 233; El. 117. Wrǽce bísgodon fǽge þeóda *the fated people were busied in evil*, Cd. 64; Th. 76, 30; Gen. 1265. Fǽge swulton on geofene *the destined perished in the ocean*, Andr. Kmbl. 3059; An. 1532. Scipflotan fǽge feóllan *the death-doomed shipmen fell*, Chr. 937; Erl. 112, 12; Æðelst. 12. Ádl fǽgum feorh óþ-þringeþ *disease will expel life from the fated*, Exon. 82 b; Th. 310, 7; Seef. 71: Judth. 11; Thw. 24, 27; Jud. 209. Nó ðý fǽgra wæs *that was not the more fated*, Cd. 162; Th. 203, 6; Exod. 399. **II.** *dead, killed, slain ;* mortuus, occīsus :—Tódǽlan werum to wiste fǽges flǽschoman *to distribute the flesh of the slain to the men for food*, Andr. Kmbl. 307; An. 154. Ofer ðæt fǽge hús *over the dead house*, Elen. Kmbl. 1759; El. 881. Hirdas lǽgon gǽsne on greóte, fǽgra flǽschaman *the keepers lay lifeless on the sand, the carcases of the slain*, Andr. Kmbl. 2171; An. 1087. Fǽgum stæfnum *with dead bodies*, Cd. 166; Th. 207, 5; Exod. 462. **III.** *accursed, condemned ;* execrātus, damnātus :—Egeslícne cwide sylf sigora Weard ofer ðæt fǽge folc forþ forlǽteþ *the Lord of victories himself shall send forth a dreadful utterance over the condemned folk*, Exon. 30 a; Th. 92, 33; Cri. 1518. On ðæt deópe dæl gefeallaþ synfulra here, fǽge gǽstas *the band of the sinful shall fall into the deep gulf, accursed spirits*, 30 b; Th. 94, 3; Cri. 1534. **IV.** *feeble, timid ;* imbēcillus, tĭmĭdus :—Nis mín breóstsefa forht ne fǽge *my mind is not afraid nor feeble*, Exon. 37 a; Th. 120, 33; Gú. 281. Ne willaþ eów andrǽdan deáde féðan, fǽge ferhþlócan *dread ye not dead bands, feeble carcases*, Cd. 156; Th. 194, 27; Exod. 267. [*Laym.* feie: *O. Sax.* fégi : *Dut.* veeg : *Ger.* feig tĭmĭdus, ĭgnāvus : *M. H. Ger.* veige : *O. H. Ger.* feigi : *Icel.* feigr.] DER. deáþ-fǽge, slege-, un-: un-fǽglíc.

FǼGEN, fægn; *comp.* fægenra; *sup.* fægnost; *adj.* FAIN, *glad, joyful, rejoicing, elate ;* lætus, gaudens, hĭlāris, elātus :—Fægen fylle *joyful in slaughter*, Exon. 96 a; Th. 357, 27; Pa. 35. Wîta ne sceal tó fægen *the sagacious must not be too elate*, 77 b; Th. 290, 20; Wand. 68: Cd. 100; Th. 131, 26; Gen. 2182. Ic bió swíðe fægn [Cott. gefægen] gif ðú me lǽdest ðider ic ðé bidde *I shall be very glad if thou leadest me whither I desire thee*, Bt. 40, 5; Fox 240, 25. He, on ferþe fægn fácnes and searuwa, wælhríow wunode *he, rejoicing in his mind in stratagem and frauds, remained a tyrant*, Bt. Met. Fox 9, 73; Met. 9,

37. Ferdon forþ ðonon, ferhþum fægne *they went forth thence, rejoicing in their minds*, Beo. Th. 3270; B. 1633. Wǽron ealle fægen in firnum *they were all glad in their sufferings*, Cd. 223; Th. 292, 3; Sat. 435: Andr. Kmbl. 2084; An. 1043. Lyt monna wearþ lange fægen ðæs ðe he ððerne bewrencþ *few men rejoice long in what they have got by deceiving others*, Prov. Kmbl. 34. Fægenra *more joyful*, Bt. Met. Fox 12, 24; Met. 12, 12. Fægnost *most joyful*, Exon. 81 b; Th. 306, 26; Seef. 13. [*Piers P.* fayn: *Chauc.* fain, fawe: *R. Glouc.* fawe, fayn: *Laym.* fæin, fain: *O. Sax.* fagan: *Icel.* feginn.] DER. ge-fægen, on-, wil-.

fægenian; *p.* ode; *pp.* od *To rejoice*; gaudēre:—Ceruerus ongan fægenian mid his steorte *Cerberus began to wag [rejoice with] his tail*, Bt. 35, 6; Fox 168, 17. v. fægnian.

FÆGER, e; *f. Beauty, fairness*; pulchrĭtūdo:—Ðæs līchoman fæger *the body's beauty*, Bt. 32, 2; Fox 116, 30. [*O. H. Ger.* fagarī, *f.*]

fæger; *comp. m.* fægerra; *f. n.* fægerre; *sup.* -est, -ost, -ast, -ust; *adj.* [fæger *beauty, fairness*] FAIR, *beautiful, joyous, pleasant, pleasing, sweet*; pulcher, dĕcōrus, lætus, jucundus, dulcis:—Swā fæger swā swā Alcibiades wæs *as fair as Alcibiades was*, Bt. 32, 2; Fox 116, 18, 24, 25. Seó wæs fæger *which was fair*, Bd. 1, 7; S. 478, 22, 23. On hrusan ne feól fæger foldbold *the fair earthly dwelling fell not on the ground*, Beo. Th. 1550; B. 773: 2278: B. 1137. Biþ swā fæger fugles gebǽru *the bird's bearing is so pleasing*, Exon. 57 b; Th. 206, 11; Ph. 125. Hió dumb wunaþ, hwæðre hyre is on fôte fæger hleoðor *it continues dumb, yet there is in its foot a sweet voice*, 108 b; Th. 414, 9; Rä. 32, 17. Wæs geforþad ðīn fægere weorc *thy beautiful work was done*, Hy. 9, 24; Hy. Grn. ii. 291, 24. Mīn se ēca dæl fægran botles brūceþ *my eternal part shall enjoy a fair mansion*, Exon. 38 a; Th. 125, 13; Gū. 353. Is mīn flǽsc swylce, for fægrum ele, frēcne onwended *cāro mea immūtāta est propter ŏleum*, Ps. Th. 108, 24. Us wuldres weard þurh lāre speón to ðam fægeran gefeán *the Lord of glory drew us by his teaching to fair joy*, Andr. Kmbl. 1195; An. 598. Forht ic wæs for ðære fægran gesyhþe *I was terrified at the beautiful sight*, Rood Kmbl. 41; Kr. 21. Segnas stôdon on fægerne swēg *the banners rose at the joyous sound*, Cd. 170; Th. 214, 8; Exod. 566. Wīte ðū ðæt ðū ānforlēte Dryhtnes ðīne fægran gefeán *know thou that thou didst lose the Lord's fair joy*, Elen. Kmbl. 1894; El. 949: Exon. 33 a; Th. 105, 6; Gū. 19. Gif ðū gesihst ansīne ðīne fægere blisse getācnaþ *if you see your face fair it betokens bliss*, Lchdm. iii. 212, 30, 31. Oþ-ðæt heó reste stôwe fægere funde *until she found a joyous resting-place*, Cd. 72; Th. 88, 18; Gen. 1467. Se æðela geaf giestliðnysse fægre on flette *the noble gave a fair entertainment in his abode*, 112; Th. 147, 29; Gen. 2447: Exon. 123 b; Th. 474, 27; Bo. 37. Cyning wæs ðý blīðra on fyrhþsefan þurh ða fægeran gesihþ *the king was blither in his mind through the joyous vision*, Elen. Kmbl. 196; El. 98. Ic ðē on ða fægran foldan gesette *I set thee on the pleasant earth*, Exon. 28 a; Th. 85, 12; Cri. 1390: 41 b; Th. 139, 30; Gū. 601. He wīc āhte fæger and freólīc *he had a dwelling fair and goodly*, Cd. 83; Th. 103, 22; Gen. 1722. Gimmas stôdon fægere æt foldan sceátum *beautiful gems stood at the extremities of the earth*, Rood Kmbl. 14; Kr. 8. Folcstede fægre wǽron *the towns were pleasant*, Cd. 91; Th. 116, 9; Gen. 1933: Exon. 26 b; Th. 79, 23; Cri. 1295. Ðeáh he fæger word ūtan ætýwe *although it outwardly shew fair words*, Frag. Kmbl. 31; Leás. 17. Swā beóþ gelīce ða leásan men ða ðe mid tungan treówa gehātaþ fægerum wordum *such resemble false men who with the tongue promise fidelity in fair words*, 48; Leás. 26: Ps. Th. 89, 17. Wyllan onspringaþ fægrum foldwylmum *wells spring forth with pleasant bubblings from earth*, Exon. 56 b; Th. 202, 3; Ph. 64: 64 b; Th. 238, 26; Ph. 610. Heofon is betera and heálīcra, and fægerra ðonne eall his innung, būton monnum ānum *the heaven is better, and higher, and fairer than all which it includes, except men alone*, Bt. 32, 2; Fox 116, 10: Exon. 43 b; Th. 147, 2; Gū. 720. Ne hýrde ic sīþ ne ǽr on ēgstreáme idese lǽdan mægen fægerre *I never heard before or since that a female led on the ocean-stream a fairer power*, Elen. Kmbl. 484; El. 242. Ðǽr hī sceáwiaþ frætwe fægerran [MS. fægran] *where they behold a fairer decoration*, Exon. 60 b; Th. 221, 5; Ph. 330. Hī to ðam fægrestan heofonrīces gefeán hweorfan môstan *they might depart to the fairest joy of heaven's realm*, Exon. 45 a; Th. 152, 14; Gū. 808. Wlitig is se wong eall mid ðām fægrestum foldan stencum *all the plain is beauteous with the sweetest odours of earth*, 56 a; Th. 158, 10; Ph. 8. Ðē is neorxna wang boldwēla fægrost *paradise is to thee the fairest dwelling of happiness*, Andr. Kmbl. 206; An. 103. Oþ-ðæt he Adam gearone funde, and his wīf somed, freó fægroste *until he found Adam ready, and his wife also, fairest woman*, Cd. 23; Th. 29, 28; Gen. 457. Se biþ gefeána fægrast *that shall be the sweetest of joys*, Exon. 32 b; Th. 102, 1; Cri. 1666. Fægerust mægþa sôhte weroda God *the fairest of virgins sought the dear God of hosts*, Menol. Fox 294; Men. 148: 226; Men. 114. [*Chauc.* faire: *Laym.* fæiʒer, fæire, fære, faire, feier, ueir: *O. Sax.* fagar: *M. H. Ger.* fager: *O. H. Ger.* fagar: *Goth.* fagrs *adapted, fit*: *Dan.* fager, fauer, faver: *Swed.* fager: *Icel.* fagr.] DER. un-fæger.

fægere, fægre, fegere; *adv. Pleasantly, softly, gently, fairly, beautifully*;

suāvĭter, bĕnigne, cōmĭter, dĕcenter, pulchre:—Fægere leohte ðæt land lago yrnende *the running water pleasantly washed the land*, Cd. 12; Th. 13, 30; Gen. 210: Ps. Th. 125, 1: Menol. Fox 283; Men. 143: Elen. Kmbl. 2423; El. 1213. He fægere mid wætere oferwearp wuldres cynebearn *he gently sprinkled with water the royal child of glory*, Menol. Fox 314; Men. 158. Him fægere ēce Drihten andswarode *the eternal Lord answered him fairly*, Cd. 107; Th. 141, 27; Gen. 2351: Frag. Kmbl. 8; Leás. 5. Fægere he syngþ *pulchre cantat*, Ælfc. Gr. 38; Som. 40, 32: Elen. Kmbl. 1483; El. 743: Runic pm. 18; Kmbl. 342, 32; Hick. Thes. i. 135, 36: Ps. Th. 60, 3: 62, 7: 118, 117. DER. un-fægere.

fægernes, fægernys, -ness, -nyss, e; *f.* FAIRNESS, *beauty*; pulchritūdo:—On heofona wuldres fægernesse *with the beauty of heaven's glory*, Homl. Blick. 159, 16. Mid dīnum hiwe oððe wlite and fægernysse ðīnre begēm *spĕcie tua et pulchritūdine tua intende*, Ps. Lamb. 44, 5.

fæger-wyrde; *adj. Fair in word, fairly speaking*; suāvĭlŏquus, dĕcenter lŏquens:—Wes ðū dīnum yldrum ārfæst symle, fægerwyrde *be thou ever dutiful to thy parents, fair in word*, Exon. 80 a; Th. 300, 26; Fä. 12.

fægir; *adj. Fair*; pulcher:—Þurh fægir word *with fair words*, Cd. 42; Th. 55, 24; Gen. 899. v. fæger.

fægn glad, joyful:—Ic bió fægn *I shall be glad*, Bt. 40, 5; Fox 240, 25. v. fægen.

fægnian, fægenian, fagnian, fagenian, fahnian; *p.* ode; *pp.* od [fægen, fægn glad, joyful] *To rejoice, be glad, exult, applaud, to be delighted with, to wish for*; gaudēre, jubĭlāre, lætāri, exultāre, plaudēre, appĕtĕre:—Ne sceal he tô ungemetlīce fægnian ðæs folces worda *he ought not to rejoice immoderately at the people's words*, Bt. 30, 1; Fox 108, 9: 108, 7, 10, MS. Cott. Onginnaþ fægnian mid folmum *plaudent mănĭbus*, Ps. 97, 8. Ic afētige oððe fægnige [MS. fegnige] *plaudo*, Ælfc. Gr. 28, 4; Som. 31, 28. Fægnaþ Israhēla *lætābĭtur Israel*, Ps. Spl. 13, 11. We fægniaþ smyltre sǽ *we rejoice at the serene sea*, Bt. 14, 1; Fox 40, 18. Fægniaþ fealdas *gaudēbunt campi*, Ps. Spl. 95, 11: Bt. Met. Fox 29, 187; Met. 29, 95. Fægnode mīn cild on mīnum innoþe *exultāvit in gaudio infans in ūtero meo*, Lk. Bos. 1, 44. Fægnodon ealle *all rejoiced*, Bt. Met. Fox 1, 66; Met. 1, 33. Fægniaþ Gode ealle eorþe *jubĭlāte Deo omnis terra*, Ps. Spl. 65, 1. Fægniaþ rihtwīse *exultāte justi*, 31, 14. Hwæðer ðū fægerra blôstmena fægnige *dost thou rejoice in fair blossoms*? Bt. 14, 1; Fox 40, 25. Ðeáh he ðæs fægnige *though he rejoice at this*, 30, 1; Fox 108, 11. DER. ge-fægnian, on-.

fægnung, e; *f. A rejoicing, exultation*; jubĭlātio, exultātio:—Is eádig folc ðæt can wyndreámas oððe fægnunge *est beātus pŏpulus qui scit jubĭlātiōnem*, Ps. Lamb. 88, 16. On fægnunga hī rīpaþ *in exultatiōne mĕtent*, Ps. Spl. 125, 6, 8. Fægnunga Godes *exultātiōnes Dei*, 149, 6. DER. ge-fægnung.

fǽg-nys, -nyss, e; *f. Difference, diversity, variety*; vărietas:—Ymbgyrd oððe ymbwǽfd mid missenlīcum oððe mid fǽgnyssum *circumamicta varietātibus*, Ps. Lamb. 44, 15.

fægr *fair*, Bd. 3, 14, Lye. v. fæger.

fægre; *adv. Pleasantly, slowly, fairly, beautifully*; suāvĭter, pĕdetentim, pulchre:—Ðæt on foldan fægre stôde wudubeám *that a forest-tree pleasantly stood on earth*, Cd. 199; Th. 247, 17; Dan. 498: Exon. 59 b; Th. 217, 2; Ph. 274. Fægre *pĕdetentim*, Ælfc. Gr. 38; Som. 40, 30. v. fægere.

fægrian; *p.* ode; *pp.* od [fæger *fair*] *To become fair* or *beautiful*; pulchrescĕre:—Byrig fægriaþ *towns become fair*, Exon. 82 a; Th. 308, 32; Seef. 48. DER. a-fægrian.

fǽgþ, e; *f. Hostility*; hostilitas:—On ða fǽgþe *in that hostility*, Andr. Kmbl. 567; An. 284, = fǽhþ, *q. v.*

FǼHÞ, fǽgþ, e; *f.* fǽhþo, an; *f.* fǽhþo, fǽhþu; *indecl. f. Feud, vengeance, enmity, hostility, deadly feud, that enmity which the relations of the deceased waged against the kindred of the murderer*; capĭtālis inimĭcĭtia, vindĭcātio, hostilĭtas, factio ob hŏmĭnem interemptum:—Sió fǽhþ gewearþ gewrecen wrāþlīce *the feud was wrathfully avenged*, Beo. Th. 6115; B. 3061: 4798; B. 2403. Ne gefeáh he ðære fǽhþe *he rejoiced not in the enmity*, 218; B. 109: Exon. 29 a; Th. 88, 17; Cri. 1441. He nó mearn fore fǽhþe and fyrene *he mourned not for his enmity and crime*, Beo. Th. 274; B. 137: 3079; B. 1537. Gif man gehādodne mid fǽhþe belecge *if a man in holy orders be charged with deadly feud*, L. C. E. 5; Th. i. 362, 21: L. Eth. ix. 23; Th. i. 344, 25. Fǽhþe ic wille on weras stælan *I will place vengeance on men*, Cd. 67; Th. 81, 27; Gen. 1351: 227; Th. 305, 2: Sat. 641. Gif hwā ǽnigne man ofsleá, ðæt he wege sylf ða fǽhþe *if any one slay any man, that he himself bear the feud*, L. Edm. S. 1; Th. i. 248, 3, 9: L. In. 74; Th. i. 150, 2. He geþingade þeodbüendum wið Fæder swǽsne fǽhþa mǽste *he appeased for mankind the greatest feud with his dear Father*, Exon. 16 b; Th. 39, 5; Cri. 617. On ða fǽgþe *in that hostility*, Andr. Kmbl. 567; An. 284. Wæs seó fǽhþo open on ühtan *the deadly feud was open at early morn*, Cd. 222; Th. 289, 30; Sat. 405. Ðæt ys sió fǽhþo *that is the feud*, Beo. Th. 5990; B. 2999: 4971; B. 2489. Sceal ic fǽhþu dreógan *I must endure enmity*, Exon. 115 a; Th. 443, 7; Kl. 26. [*Plat.* vede, fede, veide: *O. Frs.* feithe, faithe, feythe, faythe, *f*: *Dut.*

veete, f: Ger. fehde, f: M. H. Ger. vêhede, vêde, f: Dan. feide, m. f. feud, war.] DER. wæl-fæhþ.

fæhþ-bót, e; f. Feud-amends, compensation for engaging in a feud or quarrel; inimicitiārum compensātio :—Ne þearf ænig mynster-munuc mid rihte fæhþbóte biddan, ne fæhþbóte bétan no minster-monk may lawfully demand feud-amends, nor pay feud-amends, L. Eth. ix. 25 ; Th. i. 346, 2 : L. C. E. 5 ; Th. i. 362, 27.

fæhþe, an ; f. Deadly feud; capitālis inimīcitia :—Wæs seó fæhþe open úhtan the deadly feud was open at early morn, Cd. 222 ; Th. 289, 30 ; Sat. 405. v. fǽhþ.

fæhþo, fæhþu; indecl. f. Feud, enmity; capitālis inimīcitia :—Ðæt is sió fæhþo that is the feud, Beo. Th. 5990 ; B. 2489. Sceal ic fæhþu dreógan I must endure enmity, Exon. 115 a ; Th. 443, 7 ; Kl. 26. v. fǽhþ.

fæiger; adj. Fair, beautiful ; pulcher :—Fægrestan heowes of the most beautiful colour, Bd. 3, 14 ; Whelc. 199, 34, MS. Cantab. v. fæger.

fæla many, Nicod. 17 ; Thw. 8, 18. v. fela.

fæ-lǽcan, fá-lǽcan ; p. -lǽhte ; pp. -lǽht To be at deadly enmity, to be at feud ; inimīcitiam capitālem movēre :—Gif hwá heora ænigne fælǽce [fálǽce MS. L.] if any one be at feud with any of them, L. Ath. i. 20 ; Th. i. 210, 10.

fæle; adj. Fell. DER. æl-fæle. v. felo.

fǽle; adj. Faithful, true, dear, good ; fidēlis, constans, cārus, bŏnus :—Wes us fǽle freónd be a faithful friend to us, Cd. 130 ; Th. 165, 1 ; Gen. 2725 : 135 ; Th. 170, 26 ; Gen. 2819 : Exon. 35 a ; Th. 112, 15 ; Gú. 144 : Elen. Kmbl. 175 ; El. 88 : Ps. Th. 66, 3 ; Th. i. 7 : 77, 34 : 94, 7. Se fǽla fugel the faithful bird, Exon. 17 a ; Th. 40, 27 ; Cri. 645. Wese áwa friþ on Israhéla fǽlum folce let peace ever be with the faithful people of Israel, Ps. Th. 148, 14. Mid Ealhhilde, fǽlre freoðuwebban with Ealhild, the faithful peace-weaver, Exon. 84 b ; Th. 319, 2 ; Wíd. 6 : Ps. Th. 76, 3 : 118, 155. Nafaþ æt gefeohte fǽlne helpend he has not a faithful helper in battle, Ps. Th. 88, 36 : 113, 18 : 120, 1. Ðone fǽlan geþanc the true thought, 138, 20. Ne afyr ðú me fǽle spræce take not away from me true speech, 118, 43. Ðín fǽle hús thy dear house, 78, 1. Onfóh me fǽle Drihten accept me dear Lord, 118, 116. Spræcon fǽle freoðoscealcas to Lothe the faithful ministers of peace spake to Lot, Cd. 115 ; Th. 150, 25 ; Gen. 2497. He his folc genam swá fǽle sceáp abstulit sicut ŏves pŏpŭlum suum, Ps. Th. 77, 52 : 78, 14 : 99, 3. DER. un-fǽle.

fǽle; adv. Faithfully, truly, well ; fidēliter, apte, bĕne :—Ðú míne fét fǽle beweredest thou faithfully protectedst my feet, Ps. Th. 55, 11 : 84, 1 : 90, 4.

fælg, e ; f: fælge, an ; f. A felly, a part of the circumference of a wheel ; canthus, Som. Ben. Lye. v. felg.

fælging a harrow ; occa, Som. Ben. Lye. v. fealga.

fællan; p. de ; pp. ed To offend ; scandălizāre :—Gif ðín ége aswícaþ ðé oððe fælle ðec si ŏcŭlus tuus scandălizat te, Mt. Kmbl. Rush. 5, 29, 30 : 18, 8.

fælniss, e ; f. An offence ; scandălum :—From fælnissum ab scandălis, Mt. Rush. Stv. 18, 7.

fælsian; p. ode ; pp. od To cleanse, purify ; lustrāre :—Ðæt ic móte Heorot fælsian that I may purify Heorot, Beo. Th. 869 ; B. 432. He Hróþgáres sele fælsode he had purified Hrothgar's hall, Beo. Th. 4694 ; B. 2352. DER. ge-fælsian.

fǽm foam, Som. Ben. Lye. v. fám.

fǽman; p. de ; pp. ed [fám foam] To foam or froth ; spūmāre :—Fæmþ spūmat, Lk. Bos. 9, 39. Fǽmende spūmans, Mk. Bos. 9, 20. DER. a-fǽman.

fǽmig; adj. Foamy ; spūmōsus :—Ðæt ceól scyle fǽmig rídan ýða hrycgum that the foamy vessel shall ride on the waves' backs, Exon. 101 b ; Th. 384, 24 ; Rä. 4, 32. v. fámig.

fǽmnan of a virgin, Exon. 66 b ; Th. 246, 10 ; Jul. 59 ; gen. of fǽmne.

fǽmnan hád, fǽmn-hád, es ; m. [fǽmne a virgin, woman] Virginity, maidenhood, womanhood ; virginitas :—Ic fǽmnan hád mínne geheóld I preserved my maidenhood, Exon. 9 a ; Th. 6, 31 ; Cri. 92. Þurh fǽmnan hád through womanhood, Cd. 224 ; Th. 296, 1 ; Sat. 495. On fǽmnan háde in virginity, Ors. 3, 6 ; Bos. 58, 5. Heó lyfode mid hyre were seofen geár of hyre fǽmnháde vixĕrat cum viro suo annis septem a virginitāte sua, Lk. Bos. 2, 36.

FÆMNE, fémne, an ; f. [fēmĭna a woman] A virgin, damsel, maid, woman ; virgo, puella, fēmĭna :—Wæs ðæs ylcan mynstres abbudisse on ða tíd seó cynelíce fǽmne Ælflǽd præĕrat quidem tunc eidem monastērio rēgia virgo Ælbflǽd, Bd. 4, 26 ; S. 603, 3, 6 : 4, 8 ; S. 575, 34 : Gen. 2, 23 : Mt. Bos. 1, 23. Seó fǽmne wæs Sarra háten the damsel was called Sarah, Cd. 83 ; Th. 103, 23 ; Gen. 1722 : 101 ; Th. 134, 17 ; Gen. 2226. Sceal fémne hire freónd geséccan the damsel shall seek her lover, Menol. Fox 548 ; Gn. C. 44. Geseah ic líchoman ðære hálgan Godes fǽmnan vidi corpus sacræ Deo virginis, Bd. 4, 19 ; S. 589, 15, 43 : 4, 19 ; S. 588, 30. Wæs ðære fǽmnan ferþ geblissad the damsel's soul was rejoiced, Exon. 69 b ; Th. 259, 24 ; Jul. 287 : 66 b ; Th. 246,

10 ; Jul. 59 : 67 a ; Th. 247, 15 ; Jul. 79. Be ðære grimman untrumnysse ðære fǽmnan de acerba puellæ infirmitāte, Bd. 3, 9 ; S. 534, 7 : 4, 8 ; S. 576, 11. Cirliscre fǽmnan of a churlish woman, L. Alf. pol. 11 ; Th. i. 68, 14 : L. Alf. 29 ; Th. i. 52, 7 : Apstls. Kmbl. 57 ; Ap. 29. Ðære fǽmnan líchoma brosnian ne mihte fēmĭnæ căro corrumpi non pŏtuit, Bd. 4, 19 ; S. 587, 36. Hæfde Nērgend fægere fóstorleán fǽmnan forgolden, éce to ealdre the Saviour had repaid the fair reward of fostering to the virgin, in eternal life, Menol. Fox 302 ; Men. 152. Gif hwylc man hine wið fǽmnan forlicge si hŏmo quis cum puella fornicātus fuĕrit, L. Ecg. P. 4, 68 ; Th. ii. 228, 10. He mid fǽmnan on flet gǽþ he walks with the woman in the court, Beo. Th. 4074 ; B. 2034. Ic of ðam torhtan temple Dryhtnes onféng freólíce fǽmnan clǽne I joyfully received a pure damsel from the bright temple of the Lord, Exon. 10 b ; Th. 12, 18 ; Cri. 187 : 66 a ; Th. 244, 13 ; Jul. 27. Gemétte he ðær sume fǽmnan invēnit puellam ibi, Bd. 3, 9 ; S. 534, 4, 9 : L. Ecg. P. 4, 68 ; Th. ii. 230, 15. Worhte God freólícu fǽmnan God wrought a goodly woman, Cd. 9 ; Th. 12, 12 ; Gen. 184 : L. Alf. 29 ; Th. i. 52, 5. Aryson ealle ða fǽmnan surrexērunt omnes virgĭnes illæ, Mt. Bos. 25, 7, 11 : Ps. Spl. 44, 16 : Ps. Th. 77, 63 : Ps. Lamb. 148, 12 : Bd. 4, 19 ; S. 589, 39. Síðedon fǽmnan and wuduwan the damsels and widows departed, Cd. 94 ; Th. 121, 14 ; Gen. 2010. Heó mynster getimbrade Gode willsumra fǽmnena constructo monastērio virgĭnum Deo devotārum, Bd. 4, 19 ; S. 588, 2. Fela fǽmnena many damsels, Exon. 120 b ; Th. 462, 8 ; Hö. 49. Byþ heofena ríce gelíc ðám týn fǽmnum sĭmĭle ĕrit regnum cælōrum dĕcem virgĭnibus, Mt. Bos. 25, 1. Onfóþ ðæm fǽmnum receive the damsels, Cd. 113 ; Th. 149, 7 ; Gen. 2471. [O. Sax. fēmea, fēhmia, f: Frs. fæm, f: O. Frs. famne, fomne, femne, fovne, fone, f: Icel. feima, f: Lat. fēmĭna, f. a female, woman.]

fæmnenlíc; adj. Virginlike ; virginālis, Som. Ben. Lye.

fæmn-hád virginity ; virgĭnĭtas, Lk. Bos. 2, 36. v. fǽmnan hád.

fæn, fænn, es ; n. m. A fen, mud ; pălus, lŭtum :—Mid fænne with a fen, Bt. 18, 1 ; Fox 62, 26. Swá swá fænn strǽtena ic adilgige hí ut lŭtum plăteārum delēbo eos, Ps. Lamb. 17, 43. v. fen.

fæna a vane, standard, Som. Ben. Lye. v. fana.

fæng-tóþ, es ; m. [fang, q. v ; tóþ a tooth] A fang tooth ; dens cănīnus, Text. Roff. p. 39, 26.

fæniht; adj. [fæn a fen, iht an adj. termination] FENNY, marshy, dirty, muddy; pălustris, Som. Ben. Lye.

fænn a fen, Ps. Lamb. 17, 43. v. fæn, fen.

fær; nom. acc : gen. færes ; dat. fære ; pl. nom. acc. faru ; gen. fara ; dat. farum, n : fær ; gen. dat. acc. fære ; pl. nom. gen. acc. fara ; dat. farum ; f? [from faran to go]. I. a going, journey, way, journeying, expedition ; iter, expeditio bellica :—Ánes dæges fær iter diei, Lk. Bos. 2, 44. Gódige folces fær facilitate the people's journeying, L. Pen. 15 ; Th. ii. 282, 9. Ðæt wæs fær micel that was a great expedition, Invent. Crs. Recd. 1295 ; El. 646. II. that in which a journey or voyage is made,—a vehicle, vessel, ship ; vehĭcŭlum, nāvis :—Ðú ðær [Th. Grn. ðæt that] fær gewyrc make thou that vessel, Cd. 65 ; Th. 79, 6 ; Gen. 1307. Fær Noes Noah's ark, Cd. 66 ; Th. 80, 4 ; Gen. 1323. [Piers P. Chauc. fare : Laym. fære, fare, uare : Plat. foore, foor, f: Dut. voer, n : Ger. fuhre, f: M. H. Ger. var, f: O. H. Ger. fuora, f: far, n : Dan. före, n : Swed. fora, f: Icel. för, f. a journey.] DER. ád-fær, ge-, in-, ofer-, ongeán-, út-, þurh-.

FǢR, fér, es ; m. Fear, danger, peril ; timor, terror, pĕrīcŭlum :—Hie se fǽr begeat the peril overwhelmed them, Beo. Th. 2141 ; B. 1068. Fǽr ongéton they felt fear, Cd. 166 ; Th. 206, 16 ; Exod. 452. [Wyc. R. Glouc. fere : Plat. vare, f. danger : O. Sax. fár, m. insĭdiæ : Dut. gevaar, n. danger : Kil. vaer mĕtus : Ger. fahr, ge-fahr, f. pĕrīcŭlum : M. H. Ger. vár, vâre, m. snares : O. H. Ger. fâra, f. insĭdiæ, pĕrīcŭlum : Dan. fare, m. f. danger : Swed. fara, f. peril : Icel. fár, n. harm, plague.] v. fǽr ; adj. sudden.

fǽr, fér, es ; m. A fever ; febris :—Wið þriddan dæges fére and feórþan dæges fére for a third day's fever and a fourth day's fever, L. M. cont. 1, 62 ; Lchdm. ii. 12, 27. v. fefer.

fǽr; adj. Fair, beautiful ; pulcher :—Hors ðæs fǽrestan heowes a horse of the most beautiful colour, Bd. 3, 14 ; S. 540, 16, note. v. fæger.

fǽr; adj. Sudden, intense, terrible, horrid ; sŭbĭtus, terrībĭlis, horrĭdus. Used in the compounds:—Fǽr-bifongen, -bryne, -cóðu, -cwealm, -cýle, -deáþ, -dryre, -fyll, -gripe, -gryre, -haga, -inga, -líc, -líce, -níþ, -sceaða, -scyte, -searo, -slide, -spel, -unga, -wundor, -wyrd.

fǽran to go ; íre :—Ic fǽre eo, Ælfc. Gr. 30, 5 ; Som. 34, 67. v. faran.

fǽran; p. de ; pp. ed [fǽr fear] To terrify, frighten ; terrēre :—Bodan us fǽrdon nuntii nos terruĕrunt, Deut. 1, 28. DER. a-fǽran.

fǽr-béna, an ; m. A husbandman, peasant, churl ; rusticus :—Gif hit sí fǽrbéna, gilde xii ór if it be a churl, let him pay twelve ores, L. N. P. L. 50 ; Th. ii. 298, 6.

fǽr-bifongen; adj. With perils encompassed ; pĕrīcŭlis vel terrōrĭbus circumventus :—Fǽrbifongen ic ðær furðum cwom I had just come there encompassed with perils, Beo. Th. 4022 ; B. 2009.

fǽr-bryne, es ; m. A terrible heat ; terrībĭle incendĭum :—Hálig God

wið færbryne folc gescylde *the holy God shielded the people against the intense heat,* Cd. 146; Th. 182, 7; Exod. 72.

FÆRBU, e; *f. Colour;* cŏlor :—Habbaþ færbu ungelíce and mǽgwlitas *they have colour and species unlike,* Bt. Met. Fox 31, 7; Met. 31, 4. [*Ger.* farbe, *f.*]

færcodon *brought,* Chr. 1009; Th. 261, 30, = fercodon; *p. pl. of* fercian, *q. v.*

fǽr-cóðu, e; *f. Sudden sickness* or *death, apoplexy;* repentīna ǽgrĭtūdo *vel* mors, apoplexia = ἀποπληξία, Som. Ben. Lye.

fǽr-cwealm, es; *m. A sudden pestilence;* repentīna pestĭlentia :—Æt ðæm færcwealme ðe his leódscipe swýðe drehte and wanode *in the pestilence which much afflicted and decreased his people,* L. Edg. S. 1; Th. i. 270, 8.

fǽr-cýle, es; *m. A terrible cold;* terrībĭle frīgus :—Geondfolen fýre and fǽrcýle *filled with fire and intense cold,* Cd. 2; Th. 3, 30; Gen. 43.

fǽrd *an army, expedition;* exercĭtus, expĕdītio mĭlĭtāris, Som. Ben. Lye. v. fyrd.

fǽr-deáþ, es; *m. Sudden death;* repentīna mors, Cot. 14.

fǽr-dryre, es; *m. A sudden* or *pernicious fall;* repentīnus *vel* pernĭciōsus lapsus :—Con he sīdne ræced fæste gefégan wið fǽrdryrum *he can firmly compact the spacious dwelling against sudden falls,* Exon. 79 a; Th. 296, 9.

fǽreld, fareld, fǽrelt, es; *n.* [fǽr *a going,* faran *to go*]. I. *a way, going, motion, journey, course, passage, progress, expedition, company, one who accompanies in the journey of life, a relation;* via, ĭter, cursus, gressus, expĕdītio, cognāta :—Hwá ne wundrige wolcna fǽreldes *who does not express a wonder of the way of the clouds?* Bt. Met. Fox 28, 4; Met. 28, 2. Wǽnes sió eax welt ealles ðæs fǽreldes *the axle-tree of a waggon regulates all its going,* Bt. 39, 7; Fox 220, 29. Á byþ on færylde *it is ever in motion,* Runic pm. 17; Kmbl. 342, 24; Hick. Thes. i. 135, 33. On ðissum geáre næs nān fǽreld to Rôme *in this year there was no journey to Rome,* Chr. 889; Th. 158, 33, col. 1. On fǽrelde *in itĭnĕre,* Past. 4, 1; Hat. MS. 9 b, 6. Ða habbaþ fǽreld *they have a course,* Bt. Met. Fox 28, 22; Met. 28, 11. Ne beó gē afyrhte þurh geswince ðæs langsuman fǽreldes, oððe þurh yfelra manna ymbe-spræce *be ye not afraid through the toil of the tedious journey, or through the conversation of evil men,* Homl. Th. ii. 128, 2. Se esne rehte ðā Isaace eall hys fǽreld *then the servant told Isaac all his journey,* Gen. 24, 66 : Ps. Spl. 36, 33 : 139, 5. On fǽrelde *in the expedition,* Runic pm. 27; Kmbl. 345, 2; Hick. Thes. i. 135, 54. On ðam fǽrelde *in the progress,* Bt. 39, 7; Fox 222, 19. On ðam fǽrelde *in the company,* Ors. 4, 6; Bos. 84, 36. Fǽreld ðīn *cognāta tua,* Lk. Rush. War. 1, 36. Fǽreldu [MS. fǽreldtu] *lustra, meātus,* Cot. 125 : 134. II. *a particular passage,—The passover of the Jews;* transĭtus, phase, id est transĭtus, *Vulg.* [= τὸ πάσχα, *indecl.*] :—Gáþ and nymaþ nýten þurh eówer hīwrǽdene, and offriaþ phase, ðæt ys fǽreld *īte tollentes ănĭmal per fămĭlias vestras, et immŏlāte phase,* Ex. 12, 21; go ȝe, and take a beeste by ȝoure meynees, and offre ȝe fase [*passover*], Wyc. Hit ys Godes fǽreldes offrung *victĭma transĭtus Dŏmĭni est; it is the sacrifice of the Lord's passover,* Ex. 12, 27. Biþ Drihtnes fǽreld *phase Dŏmĭni est,* Lev. 23, 5; is pask [*the passover*] of the Lord, Wyc. DER. an-fǽreld, fyrd-, in-, ofer-, on-, út-, ymb-.

fǽreld-freóls, es; *m. The passover feast;* transĭtûs *vel* paschæ festum, phase :—Híg worhton phase, ðæt ys fǽreld-freóls *they kept the passover, that is the passover feast;* fēcērunt phase, id est paschæ festum, Jos. 5, 10.

fǽreldtu? *passages;* meātus, lustra, Cot. 125 : 134. v. fǽreld.

fǽrelt, es; *n. A going, progress, expedition;* ĭter, gressus, expĕdītio :—Wǽnes sió eax welt ealles ðæs fǽreltes *the axle-tree of a waggon regulates all its going,* Bt. 39, 7; Fox 220, 29, note 26. On ðæm fǽrelte *in the progress,* 39, 7; Fox 222, 19, note 18. On fǽrelte *in itĭnĕre,* Past. 4, 1; Swt. 36, 22. He ðæt fǽrelt swīðost þurhteáh *he most chiefly undertook that expedition,* Ors. 4, 10; Bos. 93, 31. Ðæt Scipia ðæs fǽreltes consul wǽre *that Scipio was the leader of the expedition,* 4, 10; Bos. 95, 2 : 4, 10; Bos. 93, 34. Æt ðam ǽrran fǽrelte *in the former expedition,* 4, 10; Ors. 92, 31 : 4, 10; Bos. 93, 37. v. fǽreld.

fǽreng, e; *f. A swooning, trance;* dēlĭquium, Cot. 79.

fǽre-sceat, -sceatt, es; *m. Fare-scot, passage-money;* naulum, prĕtium transĭtus, Som. Ben. Lye.

fǽrest, fǽreþ *goest, goeth,* Bt. Met. Fox 24, 56; Met. 24, 28 : Elen. Kmbl. 2546; El. 1274; *2nd and 3rd pers. pres. and fut. of* faran.

fǽr-fyll, e; *f. A sudden* or *pernicious fall, a precipice;* repentīnus cāsus, prǽceps :—On fǽrfyll *in præceps,* Cot. 112.

fǽr-gripe, es; *m. A sudden* or *pernicious grasp;* sŭbĭtanea *vel* pernĭciōsa arreptio :—Him hrīnan ne mihte fǽrgripe flōdes *the flood's sudden grasp could not touch him,* Beo. Th. 3036; B. 1516. Under fǽrgripum *during his sudden grasps,* Beo. Th. 1480; B. 738.

fǽr-gryre, es; *m. A perilous horror;* terror perĭcŭlōsus :—Ða hyssas þrý fǽrgryre fýres oferfaren hæfdon *the three youths had passed through the fire's dire horror,* Cd. 197; Th. 245, 14; Dan. 463. Wið fǽrgryrum *against perilous horrors,* Beo. Th. 350; B. 174.

færh *a little pig;* porcellus, Glos. Epnl. Recd. 161, 40. v. fearh.

fǽr-haga, an; *m. A peril-hedge;* perĭcŭlōrum sēpes :—He his mōdsefan wið ðam fǽrhagan fæste trymede *he firmly strengthened his mind against the peril,* Exon. 46 b; Th. 159, 27; Gú. 933.

fǽringa, fǽrincga, fǽrunga, fǽrunge; *adv.* [fǽr *sudden,* -inga, -unga *dverbial terminations*] *Suddenly, quickly, by chance;* sŭbĭto, repente, forte :—Fǽringa hí geteorodon *sŭbĭto defēcērunt,* Ps. Spl. C. 72, 19. Ðú fǽringa gehogodest sæcce sēcean *thou suddenly resolvedst to seek conflict,* Beo. Th. 3980; B. 1988 : Exon. 46 b; Th. 158, 20; Gú. 911 : Bt. Met. Fox 28, 82; Met. 28, 41. Ðonne he fǽringa cymþ *cum vēnĕrit repente,* Mk. Bos. 13, 36. Fǽrincga fýr wudu byrneþ *fire quickly burneth a wood,* Ps. Th. 82, 10.

fǽrlíc, feárlíc; *def. se* fǽrlíca, seó, ðæt fǽrlíce; *adj. Sudden, unexpected, quick;* sŭbĭtus, repentīnus :—Him becom fǽrlíc yfel *a sudden plague came upon them,* Ors. 4, 5; Bos. 81, 22 : Gen. 19, 19. Fǽrlíc geþoht *a sudden thought,* Hexam. 14; Norm. 22, 5. Fǽrlíc rén *sudden rain;* imber, Ælfc. Gl. 94; Som. 75, 113; Wrt. Voc. 52, 63. Þurh fǽrlícne [feárlícne MS. A.] deáþ *through sudden death,* L. C. S. 71; Th. i. 412, 28. Se fǽrlíca dæg *repentīna dies,* Lk. Bos. 21, 34. Se fǽrlíca deáþ *sudden death,* Homl. Th. ii. 22, 19.

fǽrlíce, fērlíce, feárlíce; *adv. Suddenly, immediately, by chance;* sŭbĭto, repente, forte :—Cometæ synd gehátene ða steorran ðe fǽrlíce and ungewunelíce æteówiaþ *the stars are called comets which appear suddenly and unusually,* Bd. de nat. rerum; Wrt. popl. science 16, 20; Lchdm. iii. 272, 3 : Gen. 14, 15 : 19, 32 : Job Thw. 165, 23 : Bt. 38, 2; Fox 198, 8 : Exon. 77 a; Th. 290, 6; Wand. 61. He fǽrlíce hrýmþ *sŭbĭto clāmat,* Lk. Bos. 9, 39 : Ps. Lamb. 63, 6 : Coll. Monast. Th. 22, 17.

fǽrm *a supper, feast,* Mt. Kmbl. Lind. 22, 2, 3, 4. v. feorm.

fǽr-níþ, es; *m. A sudden* or *pernicious hostility, mischief;* pernĭciōsa hostĭlitas :—Sorh is me to secganne hwæt Grendel hafaþ fǽrnīða gefremed *it is sorrow for me to say what sudden mischiefs Grendel has perpetrated,* Beo. Th. 956; B. 476.

fǽrnys, -nyss, e; *f. A passage, fare;* transĭtus :—Ðær monna fǽrnys mǽst wæs *juxta publicos viārum transĭtus,* Bd. 2, 16; S. 520, 5.

fǽrr, es; *n. A passing;* transĭtus :—Nis faru oððe fǽrr *non est transĭtus,* Ps. Lamb. 143, 14. v. fær; *n.*

fǽrs *verse;* versus, Ælfr. præf. p. 3, Lye. v. fers.

fǽr-sceaða, an; *m. A sudden* or *dangerous enemy;* sŭbĭtum damnum ínferens hostis :—Ðæt he on ðam fǽrsceaðan feorh gerǽhte *that he might reach the life of the dangerous enemy,* Byrht. Th. 135, 62; By. 142.

fǽr-scyte, es; *m. A sudden* or *pernicious shot;* imprŏvīsus *vel* fātālis jactus :—We fæste sculon wið ðam fǽrscyte wearde healdan *we should firmly hold ward against that sudden shot,* Exon. 19 a; Th. 48, 4; Cri. 766 : 35 a; Th. 113, 13; Gú. 157.

fǽr-searo; *gen.* -searwes; *n. An insidious artifice;* insĭdiōsa machĭnātio :—Feónda fǽrsearo *the sudden artifice of foes,* Exon. 19 a; Th. 48, 11; Cri. 770.

fǽr-slide, es; *m. A sudden fall;* imprŏvīsus lapsus :—Ðú geheólde fēt mīne wið fǽrslide *thou keptst my feet from sudden fall,* Ps. Th. 114, 8.

fǽr-spel, -spell, es; *n. A sudden message, sudden news, horrible message;* imprŏvīsus *vel* terrībilis nuncius :—Hie him fǽrspel bodedon *they announced to them the sudden news,* Judth. 12; Thw. 25, 5; Jud. 244. On fyrd hyra fǽrspell becwom *the sudden tidings came in their tent,* Cd. 148; Th. 186, 8; Exod. 135. He ðæs fǽrspelles mōdsorge wæg *hefige æt heortan he bare mental sorrow heavy at heart at the sudden news,* Exon. 48 a; Th. 165, 4; Gú. 1023. For ðam fǽrspelle *at the sudden news,* Andr. Kmbl. 2173; An. 1088. Wæs seó fǽmne for ðam fǽrspelle egsan geaclad *the damsel was chilled with terror at the horrible message,* Exon. 69 b; Th. 258, 19; Jul. 267. Me ðes ǽr bodaþ frēcne fǽrspell *this messenger announces an impious horrible message to me,* 69 b; Th. 259, 4; Jul. 277.

fǽrst, fǽrsþ *goest,* Gen. 4, 12; fǽrþ *goes,* Bt. Met. Fox 20, 432; Met. 20, 216; *2nd and 3rd pres. sing. of* faran.

fǽrþ, es; *m. n. The mind;* mens :—On fǽrþe *in the mind,* Bt. Met. Fox 27, 47; Met. 27, 24. v. ferþ.

fǽrunga, fǽrunge; *adv. Suddenly, quickly, by chance;* sŭbĭto, repente, forte :—Fǽrunga forte, Ælfc. Gr. 38; Som. 41, 28 : Jos. 9, 7. Fǽrunge astorfen *sĭdĕrātus vel ictuatus,* Ælfc. Gl. 114; Som. 80, 29; Wrt. Voc. 61, 9. v. fǽringa.

fǽr-wundor; *gen.* -wundres; *n. A sudden* or *stupendous wonder;* inŏpīnātum et stŭpendum mīrācŭlum :—Gē onlóciaþ fǽrwundra *sum ye behold a stupendous wonder,* Cd. 157; Th. 195, 20; Exod. 279.

fǽr-wyrd, e; *f. A terrible fate, destruction, perdition;* terrībĭle fātum, intĕrĭtus, perdĭtio :—He wēnþ ðæt ðone mon ǽr mǽge gebrengan on fǽrwyrde *that he thinks may bring the man earlier to a terrible fate,* Past. 62; Hat. MS.

fǽryld, es; *n. A motion, journey;* via, Runic pm. 17; Kmbl. 342, 24; Hick. Thes. i. 135, 33. v. fǽreld.

fǽs, fǽss, fas, es; *pl. nom. acc.* fasu; *n. A fringe;* fimbria :—On fǽsum

gyldenum *in fimbriis aureis*, Ps. Spl. C. 44, 15. Wíf gehrán fas [fæss, Rush.] oððe wlóh wēdes his *mŭlier tĕtĭgit fimbriam vestĭmenti ejus*, Mt. Kmbl. Lind. 9, 20 : 14, 36. Micclaþ fasu hiora *magnĭficant fimbrias*, Mt. Kmbl. Rush. 23, 5.

FÆSL, es ; *m ? n ? Offspring, progeny ; fētus, prōles, sŭbŏles :*—Ðǽr sceal fæsl wesan cwiclifigendra cynna gehwilces *there shall be offspring of every living kind*, Cd. 65 ; Th. 79, 13 ; Gen. 1310 : 67 ; Th. 80, 17 ; Gen. 1330. To fæsle *for progeny*, 67 ; Th. 82, 8 ; Gen. 1359. [Plat. fasel *sŭbŏles* : Dut. Kil. fasel, vasel *fētus in ŭtĕro* : Ger. fasel, *m. fētus, sŭbŏles* : M.H.Ger. vasel, *n. fētus* : O.H.Ger. fasal, *f. fētus* : Icel. fösull, *m. a brood.*]

FÆST ; *adj.* FAST, *fixed, firm, stiff, solid, constant, fortified ; fixus, firmus, sŏlĭdus, constans, mūnītus :*—Ealle mǽst steorran synd fæste on ðam firmamentum *almost all stars are fixed in the firmament*, Bd. de nat. rerum ; Wrt. popl. science 15, 26 ; Lchdm. iii. 268, 23 : Andr. Kmbl. 2983 ; An. 1494. Fæste mōde *fixa mente*, Bd. 4, 3 ; S. 569, 14 : Exon. 8 a ; Th. 1, 10 ; Cri. 6. Se wille fæst hús timbrian *he will build a firm house*, Bt. 12 ; Fox 36, 7, 10 : Cd. 151 ; Th. 189, 1 ; Exod. 178. Mid fæstum geleáfan *with firm faith*, Boutr. Scrd. 20, 27 : Cd. 21 ; Th. 26, 17 ; Gen. 408. Ðeós wyrt biþ cenned on fæstum stówum *this herb is produced on solid places*, Herb. 20, 1 ; Lchdm. i. 114, 12 : 45, 1 ; Lchdm. i. 148, 5. On fæstum landum *on stiff lands*, 36, 1 ; Lchdm. i. 134, 18. On ðam weorce fæste *in ŏpĕre isto constantes*, Jos. 9, 27. Seó burh wæs fæst *the city was fortified*, Bd. 3, 16 ; S. 542, 19. Micle burga ōþ heofun fæste *urbes magnæ ad cælum usque mūnītæ*, Deut. 1, 28. Fæst innoþ *restricta alvus*, Herb. 1, 12 ; Lchdm. i. 74, 11. [*Laym.* faste, feste : Orm. fasst : Plat. fast : O.Sax. fast : Frs. O.Frs. fest : Dut. vast : Ger. fest : M.H.Ger. vast, veste : O.H.Ger. fasti, festi : Dan. Swed. fast : Icel. fastr.]

-fæst, as a termination, denotes *fast, very, perfectly, effectually*, as the English *fast asleep, perfectly asleep*; Ǽ-fæst *fast in the law, firm, religious* ; Sōþ-fæst *fast in truth, true, just* ; Staðol-fæst *steadfast, steady* ; Unstaðol-fæst *unsteady, unsteadfast.* DER. ǽ-fæst, ǽw-, ár-, bǽd-, blǽd-, cíþ-, dóm-, eard-, gemet-, gif-, gin-, gryre-, hals-, hám-, heáh-, hróf-, hyge-, leoðu-, líf-, mægen-, rǽd-, rægol-, sige-, sigor-, somod-, sóþ-, stæþ-, staðol-, stede-, þeáw-, þrym-, tír-, treów-, un-, unstaðol-, wǽr-, wís-, wlitig-, wuldor-.

fæstan, -nian ; *p.* fæste ; *pp.* fæsted [fæst *fast, firm*]. **I.** *to fasten, make fast or firm, entrust, commit, commend ;* firmāre, commendāre, Lk. Lind. War. 23, 46. **II.** some have taught and now teach that he who fasts properly, fastens or secures his salvation, hence, perhaps,—*To* FAST ; jējūnāre :—Ne mǽgon hí fæstan *non possunt jējūnāre*, Mk. Bos. 2, 19. [*Wyc.* fastiden, *p. pl.* fastened, made firm ; fasten = *to fast :* Piers P. festnen *to fasten* ; fasten *to fast :* Orm. fesstnenn *to fix* ; fasstenn *to fast :* Plat. vesten *to fasten* ; fasten *to fast :* O.Sax. festian, festan *to fasten :* Frs. festgjen *to fasten :* O.Frs. festigia *to fasten* ; festia *to fast :* Dut. vesten *to fasten* ; vasten *to fast :* Ger. festen *commonly* be-festigen *to fasten* ; fasten *to fast :* M.H.Ger. vesten *to fasten* ; vasten *to fast :* O.H.Ger. fastjan, festan *firmāre* ; fastēn *to fast :* Goth. fastan *to fasten, fast :* Dan. fæste *to fasten* ; faste *to fast :* Swed. fästa *to fasten* ; fasta *to fast :* Icel. festa *to fasten* ; fasta *to fast.*] DER. æt-fæstan, a-, be-, bi-, ge-, gelíf-, gesige-, líf-, óþ-.

fæste, feste ; *comp.* fæstor ; *adv.* **I.** *fast, firmly ;* fixe, firme :—Sceát he mid his spere ðæt hit sticode fæste on ðam hearge *he shot with his spear that it stuck fast in the temple*, Bd. 2, 13 ; S. 517, 12 : Cd. 8 ; Th. 10, 14 ; Gen. 156 : Jos. 6, 1. Swíðe fæste tosomne gelímed *very firmly cemented together*, Bt. 35, 2 ; Fox 156, 35 : Exon. 22 a ; Th. 61, 5 ; Cri. 980. He heóld hyne fæstor *he held him more firmly*, Beo. Th. 288 ; B. 143. **II.** *fastly, quickly ;* cĕlĕrĭter :—Fæste geþúfe *cĕlĕrĭter frŭtĭcans, luxŭrians*, Cot. 123 : 198.

fæsten, es ; *n.* [fæstan **II.** *to fast*]. **I.** *a fast, fasting ;* jējūnium :—Ðis feówertigfealde fæsten wæs asteald on ðære ealdan gecýðnysse *this fortyfold fast was established in the old testament*, Homl. Th. ii. 100, 1. Nis ðæs mannes fæsten náht, ðe hine sylfne on forhæfednysse dagum fordrencþ *the man's fasting is naught, who inebriates himself on days of abstinence*, 608, 23 : Homl. Blick. 37, 31. Twá dæglíc fæsten oððe þreó dæglíc is genóh to healdenne *bĭduānum vel trĭduānum sat est observāre jējūnium*, Bd. 4, 25 ; S. 600, 8. Ðes gearlíca ymryne un gebrincþ efne nú ða clǽnan tíd lenctenlíces fæstenes *this yearly course just now brings us the pure time of the lenten fast*, Homl. Th. ii. 98, 25 : Homl. Blick. 27, 23. Ðæs feówertiglícan fæstenes *quadrāgēsĭmæ*, Bd. 3, 23 ; S. 554, 38. Gif mæsse-preóst folc miswyssige æt fæstene *if a mass-priest misdirect the people about a fast*, L. E. G. 3 ; Th. i. 168, 9 : L. N. P. L. 11 ; Th. ii. 292, 11. Búton þurh gebédu and on fæstene *nisi in orātiōne et jējūnio*, Ps. Lamb. 34, 13. Hí fæsten lufiaþ *they love fasting*, Exon. 44 b ; Th. 150, 18 ; Gú. 780. Gif mon his heówum in fæsten flǽsc gefe *if a man during a fast give flesh-meat to his family*, L. Wih. 14 ; Th. i. 40, 9 ; L. E. G. 8 ; Th. i. 172, 6. Þurh gebéd and fæsten *per orātiōnem et jējūnium*, Mt. Bos. 17, 21 : Ps. Th. 68, 10. We úrne líchoman clǽnsiaþ mid fæstenum and mid gebédum *we cleanse our bodies with fastings and prayers*, Homl. Blick. 39, 2. On

fæstenum and on hálsungum *jējūniis et obsecrātiōnibus*, Lk. Bos. 2, 37 : Ps. Th. 108, 24. Freólsa and fæstena healde man rihtlíce *let festivals and fasts be rightly kept*, L. Eth. vi. 22 ; Th. i. 320, 10. **II.** *a fastness, fortress, bulwark, place of strength, a castle, wall ;* mūnīmentum, arx, castellum :—Ealle hire fæstenu híg fordilegodon mid fýre *all her strongholds they destroyed with fire*, Jos. 11, 12. Nearo fæsten *narrow fastness*, Bd. 4, 26 ; S. 602, 20. **III.** *an inclosed place, cloister ;* claustrum :—Fæsten *vel* clauster *claustrum*, Ælfc. Gl. 109 ; Som. 79, 15 ; Wrt. Voc. 58, 56. [O.Sax. festî, *f. fortress, strength* : O.Frs. fest *junction* : Dut. vest, *f. a city wall, fortress* : Ger. feste, *f. a fortress* : M.H.Ger. veste, *f. firmness, solidity, fortress* : O.H.Ger. fasti, festi, *f. firmĭtas, rŏbur, arx* : Dan. fæste, *n. a handle* : Swed. fäste, *n. firmament, castle* : Icel. festa, *f. a pledge* ; festr, festi, *f. that by which a thing is fastened.*] DER. burh-fæsten, ēðel-, lagu-, sǽ-, þell-, weall-, wudu-.

fæsten-behæfednes, -ness, e ; *f. Parsimony, niggardliness ;* parsĭmōnia, Cot. 191.

fæsten-brice, -bryce, es ; *m.* [fæsten *a fast*, brice, bryce *a breaking, breach*] *A breach of a fast, fast-breaking,* BREAKFAST ; jējūnii violātio, jentācŭlum :—On fæstenbricum [MS. fæstenbricon] *in breaches of fasts*, L. Eth. vi. 28 ; Th. i. 322, 19.

fæsten-dæg, es ; *m. Fast-day ;* jējūnii dies, C. R. Ben. 54.

fæsten-díc, es ; *m. A castle-ditch ;* arcis fossa :—Andlang riþe ōþ ðone fæstendíc *along the stream to the castle-ditch*, Cod. Dipl. 204 ; A. D. 814 ; Kmbl. i. 257, 32. v. díc ; *f.* **II.**

fæsten-geat, es ; *n. A fortress or city gate ;* arcis *vel* urbis porta :—Wið ðæs fæstengeates folc onette *the people hastened to the city gate*, Judth. 11 ; Thw. 23, 38 ; Jud. 162.

fæsten-gewerc, es ; *n. Fortification work, fortification ;* fortĭfĭcātio, arcium mūnīmentum, Heming, p. 104.

fæstennes, -ness, e ; *f. Fastness, a walled town ;* castellum, Som. Ben. Lye. v. fæstnes.

fæsten-tíd, e ; *f. Fast-tide or time ;* jējūnii tempus :—Man sceal freólstídum [MS. -tidan] and fæstentídum [MS. -tidan] geornlícost beorgan *one ought most earnestly to take care at festival-times and fast-times*, L. C. S. 38 ; Th. i. 398, 17. Yfel biþ ðæt man riht fæstentíde ǽr mǽle ete *it is bad that any one, at a lawful fast-time, eat before the time*, 47 ; Th. i. 402, 23 : L. Edg. C. 25 ; Th. ii. 250, 2.

fæster-mōdor *a foster-mother*, Bt. 3, 1 ; Fox 4, 30, MS. Cot. v. fóster-mōdor.

fæstes ; *adv. By chance ;* forte, Cot. 88.

fæst-gongel ; *adj. Firm and sure going, faithful, constant ;* sēcūrus progressus, fĭdēlis :—Sum geþyld hafaþ, fæstgongel ferþ *one has patience, a faithful soul*, Exon. 79 b ; Th. 298, 4 ; Crä. 80.

fæst-hafol, -hafel, -hafod ; *adj. Fast-having, sparing, miserly ;* tĕnax, parcus, sordĭdus :—Fæsthafol tĕnax, Ælfc. Gr. 9, 60 ; Som. 13, 44. Fæsthafol strengþ manna vĭgor, Hymn. Surt. 11, 2. Fæsthafel tĕnax, Ælfc. Gl. 82 ; Som. 73, 42 ; Wrt. Voc. 47, 46. Sint to manianne ða fæsthafolan *the miserly are to be admonished*, Past. 45, 2 ; Cot. MS. Fæsthafod oððe uncystig tĕnax, Wrt. Voc. 76, 5.

fæst-hafolnes, -ness, e ; *f. Fast-havingness, sparingness, economy ;* parcĭtas :—Fæsthafolnesse *parcĭtātem*, Past. 60 ; Hat. MS.

fæst-hydig ; *adj. Steadfast in mind ;* constans anĭmo :—Ic ðe wát fæsthydigne *I know thee steadfast in mind*, Cd. 67 ; Th. 81, 18 ; Gen. 1347 : Exon. 90 b ; Th. 339, 30 ; Gn. Ex. 102.

fæsting, e ; *f. An entrusting, act of confidence ;* commendātio :—Gif hwá óðrum his unmagan ōþfæste, and he hine on ðære fæstinge forferie *if any one commit his infant to another's keeping, and he die during such keeping*, L. Alf. pol. 17 ; Th. i. 72, 5. DER. be-fæsting.

fæstingan *to fasten, make firm ;* firmāre :—Ic fæstinge mín wedd mid eów *firmābo pactum meum vobiscum*, Lev. 26, 9. v. fæstnian.

fæsting-men, festing-men, -menn ; *pl. m.* [fæsting *an entrusting*, men, v. man *a man*] *Servants of the king entrusted to the keeping of the monasteries while going from place to place ;* servi rēgii ad cūram monastēriōrum commendāti in regno obeundo :—Terram līberābo ab refectione et hābĭtu illōrum omnium qui dīcuntur fæstingmen, Th. Diplm. A. D. 822 ; 65, 17 : A. D. 821 ; 64, 11 : A. D. 841 ; 92, 19. Festingmenn, A. D. 823 ; 67, 2 : A. D. 828 ; 79, 30.

fæstlíc ; *adj.* FASTLIKE, *firm ;* firmus :—Wæs se fruma fæstlíc *the man was firm*, Exon. 44 a ; Th. 148, 15 ; Gú. 745 : Cd. 220 ; Th. 284, 22 ; Sat. 325. Eálá ! ðæt on eorþan áuht fæstlíces weorces ne wunaþ ǽfre *alas ! that on earth aught of permanent work does not ever remain*, Bt. Met. Fox 6, 32 ; Met. 6, 16. Gehyge ðú fæstlícne rǽd *devise firm counsel*, Cd. 203 ; Th. 252, 30 ; Dan. 586. Fæstlíce fórescyttelsas *firm bars*, Exon. 12 a ; Th. 20, 3 ; Cri. 312.

fæstlíce ; *comp.* or ; *sup.* ost ; *adv. Firmly, constantly, fast, quickly ;* firmĭter, constanter, cĕlĕrĭter :—Híg fæstlíce weóxon *they constantly increased*, Jud. 4, 24. Færþ micle fæstlícor *goes much more firmly*, Bt. 39, 7 ; Fox 220, 30. DER. un-rǽd-fæstlíce, wuldor-fæstlíce.

fæst-mód ; *adj. Constant in mind ;* constans anĭmo :—He wiste hú fæstmód he wæs on his geleáfon *he knew how constant in mind he was in his belief*, Ors. 6, 33 ; Bos. 129, 28.

fæstmód-staðol, es; *m. A state of constancy of mind, constancy;* constantis animi stătus, constantia, Off. Episc. 1.

fæstn *a fasting;* jejūnium :—Mid fæstnum *with fastings,* Nat. S. Greg. Els. 34, 28. v. fæsten I.

fæstn *a fortification;* mūnīmentum :—Ðara fæstna *of those fortifications,* Cd. 209; Th. 259, 15; Dan. 692. v. fæsten II.

fæst-nes, -niss, -ness, -nyss, e; *f. Firmament, firmness, stability, fastness, fortification;* firmāmentum, firmĭtūdo, mūnĭmen, propugnā-cŭlum :—Firmamentum [fæstnes] is ðeós róderlíce heofen, mid manegum steorrum amett ... Seó [fæstnes] firmamentum tyrnþ symle onbútan us under ðyssre eorþan and búfan, ac ðǽr is ungerím ðæs betweox hyre and ðære eorþan *the firmament is this ethereal heaven, adorned with many stars ... The firmament always turneth about us under this earth and above it, but there is an immeasurable space between it and the earth,* Lchdm. iii. 254, 8–13. Gewurþe nú fæstnis tomiddes ðám wæterum ... And God geworhte ða fæstnisse, and totwǽmde ða wæteru, ðe wǽron under ðære fæstnisse, fram ðám, ðe wǽron búfan ðære fæstnisse ... And God hét ða fæstnisse, heofenan *fiat firmāmentum in mĕdio aquārum ... Et fēcit Deus firmāmentum, divisitque aquas, quæ erant sub firmāmento, ab his, quæ erant sŭper firmāmentum ... Vŏcāvitque Deus firmāmentum, cælum,* Gen. 1, 6–8. Behealdaþ nú ða wídgilnesse, and ða fæstnesse heofenes *behold now the immensity, and the firmness of heaven,* Bt. 32, 2; Fox 116, 5. Ymbtrymming oððe fæstnyss mūnimen, Ælfc. Gr. 9, 12; Som. 9, 32. DER. rǽd-fæstnes, sóþ-, staðol-. v. ródor.

fæstnian, festnian; *p.* ode, ed *To* FASTEN, *secure, confirm, bind;* firmāre, vincīre :—Hie handa fæstnodon *they fastened his hands,* Andr. Kmbl. 97; An. 49: Ps. Th. 47, 11. We willaþ griþ fæstnian *we will confirm the peace,* Byrht. Th. 132, 53; By. 35. DER. a-fæstnian, ge-.

fæstnung, e; *f. A* FASTENING, *confirmation;* fixūra :—Búton ic geseó ðæra nægla fæstnunge on his honda *nisi vĭdēro in manĭbus ejus fixūram clavōrum,* Jn. Bos. 20, 25.

fæst-rǽd; *def.* se fæst-rǽda; *adj. Firm in purpose, steadfast, constant, inflexible;* firmus consilii, constans :—Se fæstrǽda Cato *the steadfast Cato,* Bt. 19; Fox 70, 7: Bt. Met. Fox 10, 97; Met. 10, 49. Gehýrde fæstrǽdne geþoht *he heard a steadfast resolution,* Beo. Th. 1225; B. 610: Ps. Th. 134, 3. DER. un-fæst-rǽd.

fæst-rǽdlíce; *adv. Boldly, constantly;* constanter, Wulfst. Par. 5.

fæst-rǽdnes, -ness, e; *f. Fixed state of mind, fortitude, resolution;* fortĭtūdo :—Mót ic nú cunnian hwón ðíne [MS. ðinne] fæstrǽdnesse *may I now inquire a little concerning thy fortitude?* Bt. 5, 3; Fox 10, 35. DER. un-fæstrǽdnes.

fæst-steall; *adj. Fast-standing;* firmĭter stans :—Wǽron fæststealle fótas míne on ðínum cáfertúnum *stantes erant pĕdes nostri in atriis tuis,* Ps. Th. 121, 2.

FÆT, es; *pl. nom. acc.* fatu, fata; *gen.* fata; *dat.* fatum; *n. A vessel, cup,* VAT; *vas,* călix :—Swá swá fæt crocwirhtan oððe tygelwirhtan ðú tobrytst híg *tamquam vas fĭgŭli confringes eos,* Ps. Lamb. 2, 9. Fætes botm *the bottom of a vessel;* vāsis fundum, Cot. 92. Mid ðam fǽte *with the vessel,* Homl. Th. ii. 158, 19. He oferwríhð nán man mid fǽte his onǽlede leóhtfæt *nēmo autem lucernam accendens, opĕrit eam vāse,* Lk. Bos. 8, 16. In seolfren fæt *in a silver vessel,* Elen. Kmbl. 2050; El. 1026. He mid róde tácne ðæt fæt bletsode *he blessed the vessel with the sign of the cross,* Homl. Th. ii. 158, 19. On ðæt fæt *in călicem,* Gen. 40, 11. Geseah he fyrnmanna fatu *he saw vessels of men of yore,* Beo. Th. 5515; B. 2761. Gecuron híg ða gódan on hyra fatu *elēgērunt bŏnos in vāsa,* Mt. Bos. 13, 48. Adrifene fatu *graven* or *embossed vessels,* Ælfc. Gl. 67; Som. 69, 99; Wrt. Voc. 41, 49. Ne mæg man ðone strangan his ǽhta and his fatu bereáfian, and on his hús gán *nēmo pŏtest vāsa fortis ingressus in dŏmum dirĭpĕre,* Mk. Bos. 3, 27. Hú mæg man ingán on stranges hús, and hys fatu hyne bereáfian *quōmŏdo pŏtest quisquam intrāre in dŏmum fortis, et vāsa ejus dirĭpĕre,* Mt. Bos. 12, 29. [*Prompt.* fate *cupa : Scot.* fat *a cask, barrel : O. Sax.* fat, *n : Plat.* vat, fat, *n : Dut.* vat, *n : Ger.* fass, *n : M. H. Ger.* vaȝ, *n : O. H. Ger.* faz, *n : Dan.* fad, *n : Swed. Icel.* fat, *n.*] DER. ár-fæt, bán-, drinc-, eorþ-, gold-, hord-, húsel-, lám-, leóht-, líc-, lyft-, máðum-, sealm-, sinc-, sync-, -stán-, wǽg-, wæter-.

fæt, es; *m. A journey, going, path;* meátus, passus, gressus, ĭter, *used only in compound words.* v. fæt-hengest, síþ-fæt.

fæt; *adj. Fat;* pinguis :—Fæt *pinguis,* Wrt. Voc. 83, 45. Mid fǽtre lynde *with fat grease,* Ps. Th. 80, 15. v. fætt.

fæt, fætt, es; *n? A thin plate of metal, gold-leaf, ornament;* lāmĭna, bractea :—Sceal se hearda helm, hyrsted golde, fætum, befeallen *the hard helmet, adorned with gold, with ornaments, shall be fallen off,* Beo. Th. 4504, note; B. 2256. To ðæs ðe he goldsele gumena wisse, fættum fáhne *until he perceived the golden hall of men, variegated with ornaments,* 1436; B. 716.

fæted, fætt; *part. Covered with gold, gilt, golden, ornamented;* bracteātus :—Ðæt sweord fáh and fæted *the sword coloured and ornamented,* Beo. Th. 5395; B. 2701. Gesáwon fæted wǽge, dryncfæt deóre *they saw the golden cup, the precious drinking vessel,* Beo. Th. 4499; B. 2253: 4553; B. 2282: Exon. 113 b; Th. 434, 27; Rä. 52, 7: Andr. Kmbl. 601; An. 301.

fæted-hleór, es; *n. Ornamented cheek;* phălĕrāta gĕna :—He hêht ðá eahta mearas fætedhleóre on flet teón *then he commanded to lead into court eight steeds with ornamented cheek,* Beo. Th. 2076; B. 1036.

fæted-sinc, es; *n. Gilded treasure;* bracteātus thēsaurus = θησαυρός :—Ðeáh ic ðé lyt syllan mihte fætedsinces *though I might give to thee a little of gilded treasure,* Andr. Kmbl. 955; An. 478.

fætels, fetels, es; *pl. nom. acc.* fætelsas, fætels; *m. n. A vessel, vat, sack, bag, pouch;* vas, saccus, pēra = πήρα, marsūpium = μαρσύπιον :—Dó on swylc fætels swylce ðú wille *put [it] into whatever vessel thou wilt,* Lchdm. iii. 16, 26. Ðeáh man asette twegen fætels full ealaþ oððe wæteres, hý gedóþ ðæt óðer biþ oferfroren *if a man set two vats full of ale or of water, they cause that either shall be frozen over,* Ors. 1, 1; Bos. 23, 8. Seó mǽgþ gebrohte heáfod blódig on ðam fætelse *the woman brought the bloody head in the bag,* Judth. 11; Thw. 23, 18; Jud. 127. Ic bicge hýda and fell, and wyrce of him pusan and fætelsas *ĕgo ĕmo cŭtes et pelles et fācio ex iis pĕras et marsūpia,* Coll. Monast. Th. 28, 1. DER. mete-fætels.

fætelsian; *p.* ode; *pp.* od *To put into a vessel;* in vas infundĕre :—Fætelsa and heald hyt *put it into a vessel and preserve it,* Med. ex Quadr. 1, 3; Lchdm. i. 328, 17.

fætere *light, negligent;* levis, remissus, Som. Ben. Lye.

fæt-fellere; *m. Abatis;* aliter *abax?* Ælfc. Gl. 113; Som. 79, 118; Wrt. Voc. 60, 25.

fæt-gold, es; *n. Gold drawn out into thin plates;* in lāmĭnas dēductum aurum, B. 1921.

fǽdem, es; *m. Bosom, lap;* sĭnus, grĕmium :—In fǽdem *in sĭnu,* Jn. Lind. War. 1, 18. v. fǽdm.

fæt-hengest, es; *m. A road horse;* itĭnĕris ĕquus :—Ne fæt-hengest *nor a road horse,* Exon. 106 a; Th. 404, 27; Rä. 23, 14.

fǽder *a feather,* Deut. 32, 11. v. feðer.

fǽder-homa *a feather-covering, the wings,* Cd. 22; Jun. 11, 1. v. feðer-hama.

FǼÐM, es; *m: also in prose* fædm, e; *f.* I. *the embracing arms;* brachia amplexa, circumdăta :—Hí fǽdmum clyppaþ *they will clasp them in their arms,* Exon. 107 a; Th. 409, 8; Rä. 27, 25. He wæs upphafen engla fǽdmum *he was upraised in the arms of angels,* Exon. 17 a; Th. 41, 6; Cri. 651. Wæs Gúþláces gǽst gelǽded engla fǽdmum *the spirit of Guthlac was led in the arms of angels,* Exon. 44 a; Th. 148, 33; Gú. 754. Ðá hêt lífes brytta englas síne fǽdmum ferigean leófne *then the giver of life commanded his angels to bear the dear one in their arms,* Andr. Kmbl. 1647; An. 825. II. *what embraces* or *contains,—A lap, bosom, breast;* quicquid complectītur *vel* comprehendit alĭquid, sĭnus, grĕmium, interna, pectus :—Me on fǽdme sticaþ *places me in the bosom,* Exon. 103 b; Th. 394, 1; Rä. 13, 11. On fǽder fǽdme *in the bosom of the father,* Menol. Fox 583; Gn. C. 61. He lǽdeþ in his ánes fǽdm ealle gesceafta *he leadeth into the bosom of himself alone all creatures,* Exon. 93 a; Th. 349, 34; Sch. 56. Deáþ in eorþan fǽdm sendaþ lǽne líchoman *death sends frail bodies into earth's bosom,* Exon. 62 b; Th. 231, 11; Ph. 487. Heó losaþ ne on foldan fǽdm *she shall not escape into earth's bosom,* Beo. Th. 2790; B. 1393. To Fæder fǽdmum *in his Father's bosom,* Beo. Th. 378; B. 188. Uppastód of brimes bósme on bátes fǽdm egesa ofer ýþlid *terror uprose from the bosom of the sea on the lap of the boat over our wave-ship,* Andr. Kmbl. 888; An. 444. Ðara ðe lífes gást fǽdmum þeahte *of those who covered in their breasts the spirit of life,* Cd. 64; Th. 77, 28; Gen. 1282. In fǽdm fýres *into the bosom of the fire,* Cd. 184; Th. 230, 16; Dan. 234. Astág mægna gold-hord in fǽmnan fǽdm *the treasury of might [Christ] descended into a virgin's womb,* Exon. 19 b; Th. 49, 19; Cri. 788. III. *that part of the arm on which one leans, hence—A cubit, the length from the elbow to the wrist, said to be estimated at one foot six inches or 18 inches;* cūbĭtus. v. eln :—Fǽdm betwux elbogan and handwyrste *a cubit is betwixt the elbow and wrist,* Ælfc. Gl. 72; Som. 70, 126; Wrt. Voc. 43, 51. Þreó hund fǽdma biþ se arc on lenge *trĕcentōrum cŭbĭtōrum ĕrit longĭtūdo arcæ,* Gen. 6, 15. And ðú getíhst his heáhnisse togædere on ufeweardum to ánre fǽdme *et in cŭbĭto consummābis summĭtātem ejus,* Gen. 6, 16. IV. *both the arms extended, now a* FATHOM = *six feet;* spătium utriusque brachii extensione contentum, Cot. 162? Lye. V. *the arms extended for embracing* or *protecting,—An embrace, protection;* amplexus, complexus, protectio :—Wæs wíf Abrahames lǽded on fremdes fǽdm *the wife of Abraham was led to the embrace of a stranger,* Cd. 124; Th. 159, 7; Gen. 2631. Sceolde monig ides bifiende gán on fremdes fǽdm *many a damsel trembling must go into the embrace of a stranger,* Cd. 92; Th. 118, 26; Gen. 1971. Þurh flódes fǽdm *through the embrace of the flood,* Andr. Kmbl. 3230; An. 1618. Hæfde wederwolcen wídum fǽdmum eorþan and upródor gedǽled *the storm-cloud had divided with wide embraces the earth and firmament above,* Cd. 146; Th. 182, 14; Exod. 75. Hwá mec bregde of brimes fǽdmum *who drew me from the embrace of ocean?* Exon. 101 a; Th. 382, 19; Rä. 3, 13. VI. *in the hands* or *power of,—Grasp, power;* pŏtestas, dĭtio :—Gehwearf ðá in Francna fǽdm feorh cyninges *the life of the king then departed into the power*

[grasp] of the Franks, Beo. Th. 2424; B. 1210. Gê of feónda sǽðme weorþen ye escape from the power of enemies, Cd. 158; Th. 196, 20; Exod. 294. Ðe ic alýsde feóndum of fæðme which I released from the power of foes, Exon. 29 b; Th. 91, 2; Cri. 1486. VII. what is extended,—An expanse, abyss, deep; expansum, tractus, superficies, abyssus, profundum:—Siððan leóhtes weard ofer ealne foldan fæðm fýr onsendeþ after that the guardian of light shall send fire over all the expanse of earth, Exon. 116 b; Th. 448, 14; Dôm. 54. Bodiaþ beorhtne geleáfan ofer foldan fæðm preach the bright faith throughout the expanse of the earth, Andr. Kmbl. 671; An. 336. Se brâda sǽ bræc on eorþan fæðm the broad sea broke on to the tract of earth, Exon. 24 b; Th. 70, 32; Cri. 1147. Swâ hie wið eorþan fæðm þusend wintra ðǽr eardodon as if they had rested there on the plain of earth a thousand winters, Beo. Th. 6091; B. 3049. On flódes fæðm ceólum lâcaþ they sail in ships on the expanse of the flood, Andr. Kmbl. 503; An. 252. [Chauc. fadmen, pl. fathoms: Laym. ueðme fathom: Plat. fadem, faem a thread, cubit: O. Sax. faðmôs, pl. m. the hands and arms: Dut. vadem, vaam, f. a fathom: Kil. vadem filum quod intra mānus extensas contīnētur, mensūra mānuum expensārum, ulna, passus: Ger. faden, fadem, m. a thread, cubit: M. H. Ger. vadem, vaden, m: O. H. Ger. fadam, fadum, m. n. filum: Dan. favn, m. f: Swed. famn, m: Icel. faðmr, m. a fathom.] DER. heoru-fæðm, lagu-, wæl-.

fæðmian, fæðman; p. ade, ede; pp. ad, ed To FATHOM, embrace, contain, envelope, clasp, devour; amplecti, complecti, contĭnēre, comĕdĕre:—Hie lêton flôd fæðmian frætwa hyrde they let the flood embrace the treasures' guardian, Beo. Th. 6257; B. 3133; Andr. Kmbl. 3176; An. 1591. Feorhcynna fela fæðmeþ êglond an island contains many of mortal kinds, Exon. 89 a; Th. 334, 11; Gn. Ex. 14. Wæter fæðmedon the waters enveloped them, Andr. Kmbl. 3143; An. 1574. Ðæt mínne líchaman glêd fæðmie that fire should clasp my body, Beo. Th. 5298; B. 2652. Heora geóguþe fýr fæðmade jŭvĕnes eōrum comēdit ignis, Ps. Th. 77, 63. DER. fæðm-, ofer-: sîd-fæðmed.

fæðm-lîc; adj. Bending, winding; sĭnuōsus, Cot. 202.

fæðm-rîm, es; n. Fathom-measure; cŭbĭtōrum vel ulnârum nŭmĕrus:— Is ðæt torhte lond twelfum hêrra fæðmrîmes that glorious land is higher by twelve of fathom-measure, Exon. 56 a; Th. 199, 21; Ph. 29.

fætian to fetch; addūcĕre, Lye. v. fetian.

fætnes, -ness, -nyss, e; f. [fæt fat] FATNESS; pinguēdo, adeps:—Hî habbaþ fætnesse they have fatness, Ps. Th. 16, 9. Of fætnysse hwætes ex adĭpe frūmenti, Ps. Lamb. 80, 17. Fætnysse heora hî beclýsdon thei han closide togidere her fatnesse, Wyc; ădipem suum conclūsērunt, Ps. Spl. 16, 11. Mid ungle oððe mid fætnysse lamba cum ădĭpe agnōrum, Cant. Moys. Isrl. Lamb. 192 a, 14.

FÆTT, fett, fæt; adj. FAT, fatted; pinguis, săgīnātus, crassus:—Seó fætte gelynd the fat grease, Ps. Th. 62, 5. Ðîn fæder ofsloh ân fætt cealf occīdit păter tuus vĭtŭlum săgīnātum, Lk. Bos. 15, 27, 23, 30: Gen. 18, 7. Ðonne hîg etaþ and fulle beóþ and fætte cum comēdĕrint et sătūrāti crassique fuĕrint, Deut. 31, 20: Gen. 41, 2: Ps. Spl. 21, 30: Ors. 4, 13; Bos. 100, 25, 26: Ps. Lamb. 21, 13. Ða fættan fearas me ofsǽton tauri pingues obsēdērunt me, Ps. Th. 21, 10. He ofslôh heora fættan occīdit pingues eōrum, Ps. Lamb. 77, 31: Gen. 41, 4. Mâra ic eom and fættra ðonne amǽsted swîn I am larger and fatter than a fattened swine, Exon. 111 b; Th. 428, 8; Rä. 41, 105. Bringon eall ðæt ðêrinne fættest sî offĕrent quidquid pinguēdinis est intrinsĕcus, Lev. 3, 3. [Piers P. Chauc. fat: Laym. uatte, fatte, pl: Frs. fet: O. Frs. fat: O. Sax. feit: Dut. vet: Ger. fett, feist: M. H. Ger. veiȝ, veiȝt, veiȝet: O. H. Ger. feizt: Dan. feed, fed: Swed. fet: Icel. feitr.]

fætt; part. Covered with gold, gilt, golden, ornamented; bracteātus:— Sincgestreónum fættan goldes with precious treasures of rich gold, Beo. Th. 2190; B. 1093: 4484; B. 2246. Fættan golde with rich gold, 4210; B. 2102. Hwanon ferigeaþ gê fætte scyldas whence bear ye your ornamented shields? 672; B. 333. v. fæted.

fættian; p. ode; pp. od To FATTEN; pinguēfăcĕre, pinguescĕre:— Fættiaþ wlitige wêstenes the feire thingis of desert schulen wexe fatte, Wyc; pinguescent spĕciōsa deserti, Ps. Spl. 64, 13. v. ge-fættian, ge-fætnian.

fæx deceit; fūcus, Cot. 91, Lye.

fæx hair, Jn. Lind. War. 11, 2. v. feax.

FÁG, fâh; def. se fâga, seó, ðæt fâge; adj. Coloured, stained, dyed, tinged, shining, variegated; tinctus, cŏlōrātus, vărius, versĭcŏlor, discŏlor:—Wæter wældreóre fâg water stained with deadly gore, Beo. Th. 3267; B. 1631. Ðæt sweord fâh and fæted the sword blood-stained and ornate, 5395; B. 2701: 2576; B. 1286. Bleóbrygdum fâg shining with variegated colours, Exon. 60 a; Th. 218, 9; Ph. 292. Gâr golde fâh a weapon shining with gold, Menol. Fox 503; Gn. C. 22. Fýrmǽlum fâg variegated with marks of fire, Andr. Kmbl. 2269; An. 1136. Fâh vărius vel discŏlor, Ælfc. Gl. 79; Som. 72, 79; Wrt. Voc. 46, 36: 77, 3. Fultum ðû him afyrdest fâgan sweordes avertisti adjūtōrium glădii ejus, Ps. Th. 88, 36. Ic geann Ælmǽre ânes fâgan stêdan I give to Ælmær one pied steed, Th. Diplm. 560, 38. Ofer næddran and fâgum wyrme ðû gæst sŭper aspĭdem et basiliscum ambŭlābis, Ps. Spl. C. 90, 13.

He me habban wile dreóre fâhne he will have me stained with gore, Beo. Th. 898; B. 447. He geseah steápne hrôf golde fâhne he saw the steep roof shining with gold, 1858; B. 927. On fâgne flôr feónd treddode the fiend trod on the variegated floor, 1454; B. 725. Slôh ðone feóndsceaðan fâgum mêce slew the enemy with a blood-stained sword, Judth. 10; Thw. 23, 4; Jud. 104. He geseah since fâge he saw variegated treasures, Beo. Th. 3234; B. 1615. Fâgum sweordum with shining swords, Judth. 11; Thw. 24, 18; Jud. 194. [Laym. fæh: O. Sax. fêh: Ger. fech: M. H. Ger. vêch: O. H. Ger. fêh: Goth. faihs in filu-faihs many-coloured.] DER. ban-fâh, bleó-fâg, blôd-, brûn-, dreór-, gold-, haso-, reád-, searo-, sinc-, stân-, swât-, tigel-, wæl-, won-, wyrm-.

fâg guilty, criminal, outlawed, hostile, Beo. Th. 2531; B. 1263. v. fâh.

fâge A plaice, flounder; platesia, Coll. Monast. Th. 24, 12. v. facg.

fagen; adj. Glad; lætus:—Wǽron ða burhware fagene the citizens were glad, Ors. 5, 3; Bos. 103, 32. v. fægen.

fagenian; p. ode; pp. od To rejoice, to be glad; gaudēre:—He fagenode ðæs he rejoiced at it, Bt. 16, 4; Fox. 58, 9. Hîg fagenodon gāvisi sunt, Lk. Bos. 22, 5. v. fægnian.

fâgettan, fâgetan, fâggetan; p. te; pp. ed To turn colour, change, vary; văriāre:—Se môna fâggeteþ [fâgetteþ MS. R; fâgeteþ MS. P] oððe asweartaþ the moon turns colour or becomes dark, Bd. de nat. rerum; Lchdm. iii. 240, 23; Wrt. popl. science 5, 15.

fâgetung, e; f. A changing, change; vărietas, dīversītas:—Hêr is ðære lyfte fâgetung here is a changing of the air, Homl. Th. ii. 538, 33.

fâgian; p. ode; pp. od To shine, glitter, vary; vărĭāre:—Swâ hit nû fâgaþ so it now varies, Bt. Met. Fox 11, 79; Met. 11, 40. Hî fâgiaþ they vary, Bt. 21; Fox 74, 13.

fagnian; p. ode; pp. od To rejoice, be delighted with, wish for; gaudēre, appĕtĕre:—Fagnian to rejoice, Bt. 30, 1; Fox 108, 7, 10. Herodes fagnode, ðâ he ðone Hǽlend geseah Hērōdes, viso Jēsu, gāvisus est, Lk. Bos. 23, 8. To hwon fagnast ðû ðæs ðe ðû ǽr hæfdest why dost thou long for what thou formerly hadst? Bt. 14, 2; Fox 42, 32.

fâgnys, -nyss, e; f. A scab, ulcer, eruption; scăbies, ulcus, eruptio:— Lâþlic biþ ðæs hreóflian lîc mid mislîcum fâgnyssum loathsome is the body of the leper with divers scabs, Homl. Th. i. 122, 22. Ðæt Crist ûre sâwle fram synna fâgnyssum gehǽlan mǽge that Christ may heal our soul from the ulcers of sins, 122, 25. Seó fâgnys aweg gewât the eruption went away, Homl. Th. ii. 178, 15. Unlybba awende his hiw to wunderlîcere fâgnysse poison turned his appearance to a wonderful eruption, 178, 12.

fâgung, e; f. Difference, diversity, variety; vărietas, Gr. Dial. 2, 27.

fâh coloured; tinctus, colōrātus:—Blôde fâh coloured with blood, Beo. Th. 1873; B. 934. v. fâg; adj. coloured.

FÁH, fâg; pl. nom. acc. fâ; gen. fâra; dat. fâum; adj. Guilty, criminal, proscribed, outlawed, inimical, hostile; sons, reus, proscriptus, inīmĭcus, infensus, infestus:—Dǽdum fâh guilty of [wicked] deeds, Cd. 216; Th. 274, 19; Sat. 156. Mid dǽdum fâh, Th. 105, 28. Firendǽdum fâh guilty of sinful deeds, Exon. 22 b; Th. 62, 13; Cri. 1001: 66 b; Th. 246, 9; Jul. 59. Fyrendǽdum fâg, Beo. Th. 2006; B. 1001. Firendǽdum fâ, nom. pl. Exon. 31 b; Th. 99, 31; Cri. 1633. Leahtrum fâh guilty of crimes, Exon. 97 b; Th. 364, 6; Wal. 66. Leahtrum fâ, nom. pl. Exon. 20 a; Th. 52, 7; Cri. 830: 30 b; Th. 94, 12; Cri. 1539. Mâne fâh guilty of crime, Beo. Th. 1960; B. 978. Mâne fâ, nom. pl. Andr. Kmbl. 3196; An. 1601. Synnum fâh guilty of sins, Frag. Kmbl. 28; Leás. 16: Exon. 118 b; Th. 456, 9; Hy. 4, 64. Mid synnum fâh, Cd. 217; Th. 275, 32; Sat. 180. Weorcum fâh guilty of [wicked] works, Elen. Kmbl. 2484; El. 1243. Ðeáh ðû from scyle freómagum feor fâh gewîtan though thou, outlawed, shalt depart far from thy kindred, Cd. 50; Th. 63, 29; Gen. 1039: Exon. 31 b; Th. 98, 34; Cri. 1617: Andr. Kmbl. 3406; An. 1707: Elen. Kmbl. 1535; El. 769. He fâg gewât he outlawed departed, Beo. Th. 2531; B. 1263. Beó he fâh wið ðone cyng let him be hostile to the king, L. Ath. i. 20; Th. i. 210, 11: Cd. 215; Th. 270, 28; Sat. 97: Wald. 101; Vald. 2, 22. Me beswâc fâh wyrm þurh fægir word the hostile serpent deceived me with fair words, Cd. 42; Th. 55, 24; Gen. 899: Cd. 166; Th. 207, 31; Exod. 475: Exon. 127 b; Th. 490, 22; Rä. 80, 5. Fâgum wyrme to the hostile serpent, Cd. 42; Th. 55, 35; Gen. 904. Nemne we mǽgen fâne gefyllan unless we may fell the foe, Beo. Th. 5303; B. 2655. Fâ þrôwiaþ ealdorbealu egeslîc the hostile shall suffer terrific vital evil, Exon. 31 b; Th. 98, 30; Cri. 1615. Fâra monna of hostile men, Andr. Kmbl. 2045; An. 1025: Beo. Th. 1160; B. 578. Fâum folmum with hostile hands, Cd. 4; Th. 4, 31; Gen. 62: 114; Th. 149, 33; Gen. 2484. [Chauc. foo a foe: R. Glouc. fon foes: Laym. i-fa, i-fo, fo a foe: M. H. Ger. vêch, ge-vêch hostile: O. H. Ger. fêh, ga-fêh inīmĭcus: Goth. fayan to be hostile, to reproāch.] DER. gryre-fâh, nearo-, syn-.

fâh-man, -mon, es; m. A foeman, an enemy; inīmĭcus:—Gif hie fâhmon [fâhman MS. H.] geierne if a foeman flee to it, L. Alf. pol. 5; Th. i. 64, 9.

fahnian; p. ode; pp. od To rejoice; gaudēre:—Hî fahnodon gāvisi sunt, Mk. Bos. 14, 11. v. fægnian.

fahnys *a rejoicing;* jūbĭlātio, Som. Ben. Lye.

faht *fought,* Chr. 1122; Erl. 249, 23,=feaht; *p. of* feohtan.

fā-lǽcan *to be at deadly enmity, to be at feud,* L. Ath. i. 20; Th. i. 210, 10, MS. L.   v. fǽ-lǽcan.

fald, e; *f? A* FOLD, *a sheepfold, an ox-stall, stable;* septum, ŏvīle, būcētum, bŏvīle, stăbŭlum :—Into sceápa falde *in ŏvīle ovium,* Jn. Bos. 10, 1: L. R. S. 4; Th. i. 434, 13. Hryðra fald *būcētum,* Ælfc. Gl. 1; Som. 55, 22; Wrt. Voc. 15, 22: Gen. 18, 7. Scēpen steal *vel* fald *bŏvīle, stăbŭlum,* Ælfc. Gl. 1; Som. 55, 23; Wrt. Voc. 15, 23. Fald oððe hūs *be* wege *stăbŭlum,* Wrt. Voc. 85, 72. [*Wyc.* fold: *Orm.* faldes, *pl.*]   DER. rĭþ-fald.

fald-gang, es; *m. Fold-going, putting sheep in fold to manure the land;* secta faldæ, servitium, quo tĕnēbātur vassallus ŏves ipsīus ad ŏvīle dŏmĭni perdūcĕre, fundi dŏmĭnĭcālis stercŏrandi grātia. v. Spelm. Glos. Lye.

fald-gang-penig, es; *m. Fold-going money, money paid by a vassal to be free from sending sheep to fold on his lord's land;* nummus dŏmĭno sŏlūtus a vassallo, ut a secta faldæ lĭbĕrārētur, Som. Ben. Lye.

fald-wurþ; *adj. Fold-worthy, liberty of folding;* falda, sive libertāte faldagii dignus, dōnātus, Som. Ben. Lye.

falewe *fallow* or *pale yellow,* Som. Ben. Lye.   v. fealo.

falewende *yellow coloured;* flavescens, Cot. 191.

fallende *falling,* Bd. 5, 6; S. 618, 24,=feallende; *part. of* feallan.

FALS, es; *n. A* FALSE*hood, fraud, counterfeit;* falsum :—Būtan ǽlcon false *without any fraud,* L. Eth. vi. 32; Th. i. 322, 29: L. C. S. 8; Th. i. 380, 16. Se ðe ofer ðis fals wyrce, þolige ðæra handa ðe he ðæt fals mid worhte *he who after this shall make a counterfeit* [*coin*], *let him forfeit the hands with which he made the counterfeit,* L. C. S. 8; Th. i. 380, 16, 17, 20, 22. Hwí tíhþ ūre hláford us swā micles falses *why doth our lord accuse us of so great a fraud?* Gen. 44, 7. [*Orm.* fals: *O. Frs.* falsk, falsch: *Ger.* falsch, m. n: *M. H. Ger.* valsch, m: *Icel.* fals, n: *Lat.* falsum, n.]

Falster *an island in the Baltic,* Ors. 1, 1; Bos. 21, 43.

FĀM, es; *n.* FOAM; spūma :—Ðæt fām of ðam mūþe eóde *the foam went out of the mouth,* Bd. 3, 9; S. 533, 32: 3, 11; S. 536, 14: Ælfc. Gl. 98; Som. 76, 89; Wrt. Voc. 54, 33: Exon. 101 a; Th. 382, 1; Rä. 3, 4.   [*Ger.* feim, m: *M. H. Ger.* veim, m: *O. H. Ger.* feim, faim, m: *Sansk.* phena, m. n. *foam, froth, scum.*]   v. fǽman.

fām-bláwende; *def.* se -bláwenda; *part. Foam-blowing, emitting foam;* spūmam efflans :—Se lég fāmbláwenda seáþ and se fūla ðone ðū gesāwe, ðæt wæs helle tintreges mūþ *pŭteus ille flammĭvŏmus ac pŭtĭdus quem vĭdisti, ipsum est os gehennæ,* Bd. 5, 12; S. 630, 12, note, MS. T.

fāmgian; *p.* ode; *pp.* od *To foam;* spūmāre :—Flód fāmgode *the flood foamed,* Cd. 167; Th. 208, 10; Exod. 481.

fāmig, fǽmig; *adj.* FOAM*y;* spūmōsus :—Fāmig sǽ *the foamy sea,* Cd. 72; Th. 87, 22; Gen. 1452. Fāmige flōdas *foamy floods,* 100; Th. 133, 19; Gen. 2213: Exon. 101 b; Th. 383, 32; Rä. 4, 19: Salm. Kmbl. 315; Sal. 157.

fāmig-bord, es; *n. A foaming bank;* spūmōsa margo :—On streám fāmigbordum [MS. -bordon] *on a stream with foamy banks,* Bt. Met. Fox 26, 52; Met. 26, 26.

fāmig-bōsm, es; *m. A foamy bosom;* spūmōsus sĭnus, Cd. 167; Th. 209, 2; Exod. 493.

fāmig-heals; *adj. Foamy-necked;* spūmōsus in collo :—Sǽ-genga fōr, fleát fāmigheals *the sea-goer went, the foamy-necked floated,* Beo. Th. 3822; B. 1909: 441; B. 218: Andr. Kmbl. 993; An. 497.

fāmwǽstas *molles,* Cot. 131.

fan *a fan.* v. fann, fon.

FANA, an; *m. A standard, flag,* VANE; vexillum :—Fana hwearfode, scīr on sceafte *the standard waved, bright on the shaft,* Bt. Met. Fox 1, 20; Met. 1, 10: Cd. 155; Th. 193, 18; Exod. 248.   [*Chauc.* fane *a vane:* Plat. fane, *f: O. Sax.* fano, *m: O. Frs.* fona, fana, *m: Dut.* vaan, *f: Ger.* fane, fahne, *f: M. H. Ger.* vane, van, *m: O. H. Ger.* fano, *m: Goth.* fana, *m: Dan.* fane, *m. f: Swed.* fana, *f: Icel.* fáni, *m: Lat.* pannus, *m: Grk.* πῆνος, *m.*]   fāð-fana.

fand *found,* Cd. 72; Th. 87, 30; Gen. 1456; *p. of* findan.

fandere, es; *m. A tempter, trier;* tentātor, Som. Ben. Lye.

fandian, fandigan; *to* fandienne; *p.* ede, ode; *pp.* ed, od; *v. trans. gen. dat. acc. To try, tempt, prove, examine, explore, seek, search out;* tentāre, prŏbāre, exāmĭnāre, expĕrīri, inquīrĕre, vestĭgāre :—Gif ðē ǽfre geweorþeþ ðæt ðū wilt oððe mōst weorolde þióstro eft fandian *if it should happen that thou wilt or must again explore the world's darkness,* Bt. Met. Fox 24, 113; Met. 24, 57.   Ic bohte ān getýme oxena, nū wille ic faran and fandian hyra jŭga boum ēmi quinque, et eo prŏbāre illa, Lk. Bos. 14, 19. Ic wille fandigan nū hwæt ða men dōn *I will now seek to know what those men do,* Cd. 109; Th. 145, 24; Gen. 2410. Ðǽm weorce to fandienne *to prove the work,* Ors. 1, 12; Bos. 36, 37. He gārsecg fandaþ *he tempteth the ocean,* Runic pm. 25; Kmbl. 344, 20; Hick. Thes. i. 135, 50. Ðū fandodest us God *prŏbasti nos Deus,* Ps. Spl. 65, 9. Ferdon ða Pharisēi, and hí fandedon *exiērunt Pharisæi, tentantes eum,* Mk. Bos. 8, 11. Hý fandodon mín *tentāvērunt me,* Ps. Th. 34, 16: 40, 6. Ne fanda ðīnes Drihtnes *tempt not thy Lord,* Homl. Th. i. 166, 21. Fanda fonden: *Chauc.* fonde: *Laym.* fondien: *Orm.* fandenn: *O. Sax.* fandôn: *Frs.* fanljen: *O. Frs.* fandia, fandlia: *Dut. Kil.* vanden: *Ger.* fanden, fahnden: *M. H. Ger.* venden: *O. H. Ger.* fantôn *tentāre, explōrāre.*]   DER. a-fandian, ge-.

fandlíc *hostile;* hostīlis.   DER. a-fandelíc.

fandung, e; *f. A temptation, trial, proof;* tentātio, prŏbātio, inquĭsītio :—Ōðer is seó fandung ðe Iacob se apostol embe sprǽc *the other is the temptation of which the apostle James spoke,* Boutr. Scrd. 23, 8. Scearplícu and smeálícu fandung ðæs mōdes *the sharp and searching temptation of the mind,* Past. 21, 3; Hat. MS. 30 a, 26. Ðære lufe fandung is ðæs weorces fremming *the proof of love is the performance of work,* Homl. Th. ii. 314, 28. On ðære fandunge *in temptation,* Boutr. Scrd. 23, 8. He of earce forlét hāswe culufran on fandunga *he let out a livid dove from the ark on trial,* Cd. 72; Th. 87, 21; Gen. 1452. DER. a-fandung.

fang, es; *m.* [fangen; *pp. of* fôn *to take,* q. v.] *what is taken, A booty;* captūra, præda :—Hí fang woldon fôn *they would take booty,* Chr. 1016; Th. 281, 30.   [*Laym.* feng, ueng *booty: Scot.* fang *a capture: O. Frs.* fang, feng, *m: Dut.* vang, *m: Ger.* fang, *m: M. H. Ger.* vanc, *m: O. H. Ger.* fang, *m. captūra: Dan.* fang, *n: Swed.* fång, *n: Icel.* fang, *n. a catching.*]   DER. feax-fang, feoh-, fore-, for-, under-.

fangen *taken;* captus :—Hér beóþ fangene seólas and hrônas *here are caught seals and whales,* Bd. 1, 1; S. 473, 16; *pp. of* fôn *to take.*

fangen-nes, -ness, e; *f. A taking.* DER. on-fangenness, under-.

FANN, e; *f? A* FAN, *implement for winnowing grain;* vannus, ventĭlābrum :—Fann *vannus,* Ælfc. Gl. 50; Som. 65, 114; Wrt. Voc. 34, 43. Ðæs fann ys on his handa, and he afeormaþ his þyrscelflōre *cujus ventĭlābrum in mănu sua, et permundābit āream suam,* Mt. Bos. 3, 12: Lk. Bos. 3, 17.   [*Chauc.* fan: *Dut.* wan, wanne, *f: Ger. M. H. Ger.* wanne, *f: O. H. Ger.* wanna, *f: Swed.* vanna, *f: Lat.* vannus, *f.*]

fant, font, es; *m. Fountain, spring;* fons, tis, *m; pure water, that which holds pure or holy water, The font for baptism;* baptistērium = βαπτιστήριον :—Ne dō man nǽnne ele to ðam fante *let no one put any oil into the font,* L. Ælf. C. 36; Th. ii. 358, 35; Wilk. 159, 32. v. fant-fæt, fant-wæter, font-wæter.

fant-fæt; *gen.* fant-fætes; *pl. nom. acc.* fant-fatu; *n. A font vessel, the font for baptism;* baptistērii vas :—Hǽðen cild biþ gebroht synfull þurh Adames forgǽgednysse, to ðam fant-fæte, ac hit biþ aþwogen fram eallum synnum wiðinnan, ðeáh ðe hit wiðūtan his hiw ne awende *a heathen child is brought to the font-vessel, sinful through Adam's transgression, but it is washed from all sins within, though without it change not its appearance,* Homl. Th. ii. 268, 29-33.

fant-wæter, font-wæter, es; *n. Font-water, baptismal water;* baptistērii aqua :—Ðæt hālige fant-wæter, ðe is gehāten lífes wyl-spring, is gelíc on hiwe ōðrum wæterum *the holy font-water, which is called the well-spring of life, is in appearance like other waters,* Homl. Th. ii. 268, 34.

fara, an; *m. A farer, traveller;* viātor. v. ge-fara, mere-, nýd-, tíd-.

fāra, Andr. Kmbl. 2045; An. 1025; *gen. pl. of* fāh *hostile.*

FARAN, to farenne; ic fare, ðū farest, færest, færst, færsþ, he fareþ, færeþ, færþ, *pl.* faraþ; *p.* fôr, fôr, fôron; *pp.* faren, A word expressing every kind of going from one place to another, hence   I. *to go, proceed, travel, march, sail;* īre, vādĕre, incēdĕre, transīre, migrāre, nāvĭgāre :—Faran ofer feldas *to go over fields,* Exon. 108 b; Th. 415, 8; Rä. 33, 8.   Nū wylle ic faran *now I will go,* Lk. Bos. 14, 19, 31. We fōron *transīvĭmus,* Ps. Spl. 65, 11.   Ic fôr fram ðē *I went from thee,* Gen. 31, 31. Constantius, se mĭldesta man, fôr on Bryttanie, and ðǽr gefôr *Constantius, the mildest man, went into Britain, and there died,* Ors. 6, 30; Bos. 126, 39.   Fôr fāmig scip *the foaming ship sailed,* Cd. 71; Th. 85, 19; Gen. 1417.   II. *to* FARE, *happen, to be in any state;* versāri in ălĭqua re, se hăbēre ălĭquo mŏdo, Cd. 26; Th. 34, 2; Gen. 531.   Ic fare būtan bearnum *I have no children* [lit. *I go without children*], Gen. 15, 2. Hū mæg se man wel faran *how can the man fare well?* Ælfc. T. 40, 3. [*Piers P.* faren, fare: *Wyc. Chauc.* fare: *Laym.* fære, færen, faren, varen: *Orm.* farenn: Plat. faren: *O. Sax.* faran: *Frs.* farren: *O. Frs.* fara: *Dut.* váren: *Ger.* fahren, faren: *M. H. Ger.* varn: *O. H. Ger.* faran: *Goth.* faran: *Dan.* fare: *Swed.* fara: *Icel.* fara: *Sansk.* pṛi *to bring over.*]   DER. a-faran, be-, for-, forþ-, ge-, geond-, in-, of-, ofer-, on-, ôþ-, þurh-, tô-, ūt-, wið-, ymbe-.

faraþ-lácende; *part. Swimming;* nătans :—Fiscas faraþlácende *swimming fishes,* Exon. 97 b; Th. 364, 34; Wal. 80.   v. faroþ-lácende.

fare *in a journey,* Gen. 8, 1.   v. faru.

fareld *a journey* :—Þurh geswinc ðæs fareldes *through fatigue of the journey,* Nat. S. Greg. Els. 29, 10; and MS. at foot of plate facing Title. v. færeld.

fareþ-lácende; *part. Sailing;* nāvĭgans :—Fareþlácenduṃ nāvĭgantibus, Exon. 96 b; Th. 360, 14; Wal. 5.   v. faroþ-lácende.

Fariseisc; *def.* se Fariseisca; *adj. Pharisean;* Phărīsǽus :—Bæd hine sum Fariseisc man ðæt he ǽte mid him *rŏgāvit illum quĭdam Phărīsǽus ut prandēret ăpud se,* Lk. Bos. 11, 37. Ongan se Fariseisca on him smeágan and cweðan *Phărīsǽus cœpit intra se repūtans dīcĕre,* 11, 38.

Cômon to him ða bôceras and Fariseisce *accessērunt ad eum Scribæ et Phărĭsæi*, Mt. Bos. 15, 1. Ða Fariseiscan synt gedrēfede *Phărĭsæi scandălizāti sunt*, 15, 12.

Farnea eálond, es; *n. Farn island, on the coast of Northumberland, near Lindisfarne;* Farnensis insŭla, Som. Ben. Lye.

faroþ, es; *n?* *The floating of the waves, a billow, the shore;* fluctuātio mãris, unda, lĭtus:—Hî hyne ætbǣron to brimes faroþe *they bore him away to the sea's shore*, Beo. Th. 56; B. 28. Fûs on faroþe *ready on the shore*, Andr. Kmbl. 509; An. 255. DER. brim-faroþ, mere-, sǣ-, waroþ-.

faroþ-hengest *a sea-horse, ship.* v. fearoþ-hengest.

faroþ-lácende, faraþ-lācende, fareþ-lācende; *part.* [lācan *to sail*] *Sailing, swimming;* nāvĭgans, nătans:—Faroþlācende *sailing*, Andr. Kmbl. 1014; An. 507. Gewîciaþ faroþlācende on ðam eálonde *the seafaring [men] encamp on that island*, Exon. 96 b; Th. 361, 15; Wal. 20.

faroþ-rídende; *part. Wave-riding, sailing;* nāvĭgans:—We on sǣbâte wada cunnedon, faroþrídende *we in the sea-boat made a trial of the fords, riding over the waves*, Andr. Kmbl. 879; An. 440.

faroþ-strǣt, e; *f. The sea-street, the sea;* marĭtima via, mǎre:—Ic ongiten hæbbe ðæt ðû on faroþstrǣte feor ne wǣre *I have understood that thou wert not far from us upon the sea*, Andr. Kmbl. 1795; An. 900: 622; An. 311.

FARU, e; *f.* I. *a going, journey, passage;* ĭter, profectio, ĭtio, transĭtus:—Hit ys Godes faru *est transĭtus Dōmĭni [passover]*, Ex. 12, 11. II. *family, what is movable;* fămĭlia, cōmĭtātus:—God ðâ gemunde Noes fare *God then remembered Noah's family*, Gen. 8, 1. Mid ealre fare, and mid eallum ǣhtum *with all his family, and with all his possessions*, 12, 5. Abram ðâ ferde of Egipta lande mid ealre his fare *Abram then went from the land of the Egyptians with all his family*, 12, 20. Gewît ðû nû feran and dîne fare lǣdan ceápas *begin thou now to depart and lead thy family and thy cattle*, Cd. 83; Th. 105, 1; Gen. 1746. III. *expedition, march;* expĕdĭtio, agmen migrantium:—He ðas fare lǣdeþ *he leadeth this expedition*, Cd. 170; Th. 213, 19; Exod. 554. v. fǣr; *n.* and *f.* [Piers P. Chauc. Laym. fǣre, fare, uare, faren: *O. Frs.* fare, fera, fere, fer, *f:* Ger. far, fahr, *f. res mōbĭlis:* M. H. Ger. var, *f. ĭter:* O. H. Ger. fuora, *f. ĭtio: Icel.* för, *f. a journey, expedition.*] DER. earh- [earg-] faru, forþ-, fyrd-, gâr-, hægl-, man-, streám-, wǣg-, wolcen-, ŷþ-.

fas *a fringe*, Som. Ben. Lye. v. fæs.

fast *fast*, Som. Ben. Lye. v. fæst *fast, firm.* v. fæstan II.

fastitocalon [= ἀσπιδοχελώνη: Dietrich ἄστυ τὸ καλόν] *A large whale;* bālæna = φάλαινα:—Ic wille cŷðan bî ðam miclan hwale, ðam is noma cenned fastitocalon *I will make known concerning the great whale, to which the name Fastitocalon is given*, Exon. 96 b; Th. 360, 18; Wal. 7.

fatan; *p.* fôt, *pl.* fôton; *pp.* faten. *To go;* ĭre, volvi, volvĕre. v. fetan, fetian.

faðu, e; *f:* faðe, an; *f. A father's sister, paternal aunt;* ămĭta:—Faðu ămĭta, Ælfc. Gr. 6; Som. 5, 55: Wrt. Voc. 72, 43. Mîn faðu ămĭta mea; mînra faða môder ămĭta mea magna; mînre faðan yldre môder *proamĭta mea;* mînre [MS. mînra] faðan þridde môder abămĭta mea, Ælfc. Gl. 92, 93; Som. 75, 60–64; Wrt. Voc. 52, 17–20. Seó wæs Ecfriþes faðu ðæs cyninges *quæ erat ămĭta rēgis Ecgfridi*, Bd. 4, 19; S. 587, 41. Bûton hit sŷ his môder, oððe sweoster, oððe faðu, oððe môddrie *unless it be his mother, or sister, or father's sister, or mother's sister*, Homl. Th. ii. 94, 32. Ic gean mînre faðan Leófware ðæs heáfodbotles on Purleá *I give to my aunt Leofware the chief dwelling at Purley*, Cod. Dipl. 1293; A. D. 998; Kmbl. vi. 138, 23. v. môddrie *a maternal aunt.*

fatu, fata *vats, vessels*, Mk. Bos. 3, 27: Mt. Bos. 12, 29. v. fæt.

Faul; a word used as a charm against the bite of an adder:—Sume än word wið nǣdran bîte lǣraþ to cweðenne, ðæt is, Faul *some teach us against bite of adder to speak one word, that is, Faul*, L. M. 1, 45; Lchdm. ii. 114, 2.

feá; *indecl. n.* FEE, *money, goods;* pĕcūnia:—Gif ðû ðisses mannes feá in his synnum deádes ne onfênge *si hujus vĭri in peccātis suis mortui pĕcūniam non accēpisses*, Bd. 3, 19; S. 549, 10. v. feoh.

feá, an; *m. Joy;* gaudium:—Him he gehêt êcne feán *he promised him everlasting joy*, Bd. 1, 25; Whelc. 76, 1. v. ge-feá.

feá; *adj. Few;* pauci:—Ðis feá âna doþ a *few only do this*, Bd. 4, 25; S. 601, 8. Ðæt hêr wǣre mycel rîp [MS. riip] and feá wyrhtan *that a great harvest was here and few workmen*, 1, 29; S. 498, 5. Feá ðæt gedŷgaþ *few escape from that*, Exon. 102 a; Th. 386, 6; Rä. 4, 57. Feá worda cwæþ *he said few words*, Beo. Th. 5318; B. 2662. He feára sum beforan gengde *he with a few went before*, Beo. Th. 2828; B. 1412. Ealle nemne feáum ânum *all save a few only*, Beo. Th. 2167; B. 1081. Nales feáum sîþum *not a few times*, Elen. Kmbl. 1633; El. 818: Andr. Kmbl. 1210; An. 605. v. feáwa.

feá; *adv. Even a little, ever so little;* părum:—Ne mâgon feá gangan *they cannot walk even a little*, Ps. Th. 134, 18.

feágan, to feágenne [feá, gefeá *joy*] *To rejoice;* lætāri, plaudĕre:—To feágenne on blisse þeóde ðînre *ad lætandum in lætĭtia gentis tuæ*, Ps.

Lamb. 105, 5. Flôdas feágaþ oððe hafetiaþ mid handa *flūmĭna plaudent mănu*, 97, 8.

feaht *fought*, Byrht. Th. 139, 14; By. 254; *p. of* feohtan.

feala; *adj. Many, much;* multum, multa:—Ne sprǣc ic worda feala *non lŏcūtus sum verbōrum multa*, Ps. Th. 4: 77, 43: 105, 27. On feala wîsan *multis mŏdis*, Coll. Monast. Th. 25, 11. v. fela.

feala-fôr, feale-fôr, e; *f? A fieldfare?* turdus pĭlāris?—Fealafôr *torax?* Cot. 174, Som. Ben. Lye. v. feolu-fôr, felde-fare.

feala-hiw, es; *n. A varied colour:*—Feala-hiwes hrægel *pŏlymĭta*, Ælfc. Gl. 63; Wrt. Voc. 40, 14. v. hiw.

feald *a field*, Ps. Spl. 77, 15: 64, 12. v. feld.

feald, es; *n. A fold, inclosure, field;* septum, ăger, Som. Ben. Lye. DER. ge-feald.

-feald, the termination of numerals, as ân-feald *one-fold, single;* twî-feald or twŷ-feald *two-fold, double;* þreó-feald or þrŷ-feald *three-fold, treble;* seofon-feald *seven-fold;* manig-feald *manifold.* [*O. Sax.* -fald: *O. Frs.* -fald: *M. H. Ger.* -valt: *O. H. Ger.* -falt: *Goth.* -falþs.]

FEALDAN, ic fealde, ðû fealdest, fylst, he fealdeþ, fylt, *pl.* fealdaþ; *p.* feóld, *pl.* feóldon; *pp.* fealden [feald *a fold*] *To* FOLD *up, wrap;* plĭcāre:—Gôd scipstŷra hǣt fealdan ðæt segl *a good pilot gives order to furl the sail*, Bt. 41, 3; Fox 250, 14. Ic fealde *plĭco;* ic feóld *plĭcui* vel plĭcāvi, Ælfc. Gr. 24; Som. 25, 50. He feóld his fêt uppan his bedd *collēgit pĕdes suos sŭper lectŭlum*, Gen. 49, 32. Fingras feóldon [MS. feóldan] *mec fingers folded me*, Exon. 107 a; Th. 408, 4; Rä. 27, 7. Ðæt he hine fealde swâ swâ bôc *that it fold itself like a book*, Ps. Th. 49, 5. [*Wyc.* folden, falt, *pp.* bent, bowed: *Chauc.* folden: *Dut.* vouwen: *Ger.* falten: *M. H. Ger.* valten, valden: *O. H. Ger.* faldan: *Goth.* falþan: *Dan.* folde: *Swed.* fålla: *Icel.* falda.] DER. be-fealdan, bi-, ge-, onbe-, ongeán-, tobe-, to-, un-.

feale, *pl. nom. acc.* fealewe *fallow, pale yellow, dusky*, Chr. 937; Th. 204, 16, col. 1: Andr. Kmbl. 3177; An. 1591. v. fealo.

fealewe, *yellow;* flāvus, Cot. 81. v. fealo.

fealewian *to grow yellow, ripen, wither as leaves*, Salm. Kmbl. 627; Sal. 313. v. fealwian.

fealga *harrows*, Glos. Epnl. Recd. 160, 24; *pl. nom. acc. of* fealh.

FEALH; *gen.* fealge; *f. A harrow;* occa:—Fealh *occa*, Cot. 197. Fealga *occas*, Glos. Epnl. Recd. 160, 24. [*Ger.* felge: *M. H. Ger.* vëlge, *f: O. H. Ger.* fëlga, *f. flexūra, rădius, canthus, occa.*]

fealh *underwent*, Beo. Th. 2405; B. 1200; *p. of* felgan.

feall, e; *f? A trap, pitfall;* decĭpŭla, Lye, Ettm.

FEALLAN, to feallanne; *part.* feallende; ic fealle, ðî feallest, fealst, felst, fylst, he fealleþ, fealþ, felþ, fylþ, *pl.* feallaþ; *p.* feól, feóll, *pl.* feóllon; *pp.* feallen; *v. intrans. To* FALL, *fall down, fail;* cădĕre, decĭdĕre, procĭdĕre, defĭcĕre:—Hî sceolon raðe feallan on grimne grund *they shall fall rapidly into the grim abyss*, Exon. 30 a; Th. 93, 15; Cri. 1526: Beo. Th. 2145; B. 1070: Ps. Th. 87, 4: Rood Kmbl. 85; Kr. 43. Enoch nalles feallan lêt dôm *Enoch let not his power fail*, Cd. 60; Th. 73, 3; Gen. 1198. To feallanne *to fall*, Bt. Met. Fox 20, 335; Met. 20, 168. Gyf ðû feallende to me ge-eádmêtst *si cădens adorāveris me*, Mt. Bos. 4, 9: Lk. Bos. 10, 18. Heofones steorran beóþ feallende *stellæ cœli ērunt decĭdentes*, Mk. Bos. 13, 25. Ðis lîf is lǣnlic and feallende *this life is transitory and failing*, L. E. I. prm; Th. ii. 400, 16. Ic fealle *cădo*, Ælfc. Gr. 28, 7; Som. 32, 54. Se rên fealleþ *the rain falls*, Ps. Th. 71, 6: Exon. 56 b; Th. 201, 25; Ph. 61: Salm. Kmbl. 662; Sal. 330. Se hagol fealþ *the hail falls*, Ex. 9, 19: Bt. 6; Fox 14, 29: Boutr. Scrd. 18, 25. Him on innan felþ muntes mægenstân *a huge mountain-stone falls into it*, Bt. Met. Fox 5, 30; Met. 5, 15. Se ðe fylþ uppan ðysne stân, he byþ tobrŷsed *qui cecĭdĕrit sŭper lăpĭdem istum, confringētur*, Mt. Bos. 21, 44: Bd. de nat. rerum; Wrt. popl. science 19, 15; Lchdm. iii. 278, 25. Hîg feallaþ begen on ænne pytt *ambo in fŏveam cădunt*, Mt. Bos. 15, 14, 27: Bd. de nat. rerum; Wrt. popl. science 15, 21, 22: Exon. 57 a; Th. 213, 21: Ph. 74: Salm. Kmbl. 628; Sal. 313: Ps. Th. 57, 7. He on hrusan ne feól *he fell not on the earth*, Beo. Th. 1549; B. 772: Fins. Th. 83; Fin. 41: Byrht. Th. 135, 31; By. 126: Bt. Met. Fox 1, 161; Met. 1, 81: Exon. 108 a; Th. 412, 11; Rä. 30, 12. Ic feóll beforan Drihtne *procĭdi ante Dōmĭnum*, Deut. 9, 18. Feóll Abram astreht to eorþan *cĕcĭdit Abram prōnus in făciem*, Gen. 17, 3: Beo. Th. 5830; B. 2919: Byrht. Th. 135, 16; By. 119: Andr. Kmbl. 1835; An. 920: Ps. Th. 77, 27. Feónda beorn feóllon þicce *the lives of the foes fell thickly*, Cd. 95; Th. 124, 20; Gen. 2065: Beo. Th. 2089; B. 1042: Byrht. Th. 135, 1; By. 111: Elen. Kmbl. 253; El. 127. Ðæt heó feólle ðwâl rif, Boutr. Scrd. 18, 25. [*Piers P.* fallen: *Wyc.* falle: *Chauc.* falle: *Laym.* falle, fallen, fællen, uallen: *Orm.* fallenn: *O. Sax.* Frs. fallan: *O. Frs.* falla: *Dut.* vallen: *Ger.* fallen: *M. H. Ger.* vallen: *O. H. Ger.* fallan: *Dan.* falde: *Swed. Icel.* falla.] DER. a-feallan, be-, ge-, of-, onbe-, on-, oþ-, to-.

FEALO, fealu, feale; *def. se* fealwa; *adj.* FALLOW, *pale yellow* or *red coloured as withered grass* or *leaves, dusky, bay?* flāvus, gilvus, fuscus:—Fealo lîg feormaþ and fênix byrneþ *the yellow flame consumes and burns the Phœnix*, Exon. 59 a; Th. 213, 1; Ph. 218: 104 b; Th. 396, 8; Rä. 16, 1. Fealu *busĭus?* [=*fuscus?*], Ælfc. Gl. 79; Som. 72, 81;

Wrt. Voc. 46, 38. Se fealwa holen *the fallow* or *withered holly leaf,* Exon. 114 a; Th. 437, 19; Rä. 56, 10. Cing ût gewât on fealone [fealene, col. 1] flôd *the king departed on the dusky flood,* Chr. 937; Th. 204, 16, col. 2; Ædelst. 36: Beo. Th. 3904; B. 1950. Sum fealone wæg stefnan steórep *one steers the prow* [*on*] *the dusky wave,* Exon. 79 a; Th. 296, 19; Crä. 53. Fleón fealone streám *to escape the dusky stream,* Andr. Kmbl. 3074; An. 1540. Lang is deós sîpfæt ofer fealuwne flôd *this journey is long over the dusky flood,* 841; An. 421. Sindon fealwe fôtas *the feet are yellow,* Exon. 60 a; Th. 219, 22; Ph. 311. Ne feallap dær fealwe blôstman *fallow blossoms fall not there,* 57 a; Th. 202, 24; Ph. 74. Fealwe mearas *bay horses,* Beo. Th. 1735; B. 865. Se beorg tohlâd and in forlêt fealewe wægas *the hill opened and let in the dusky waves,* Andr. Kmbl. 3177; An. 1591. Meahte æghwylc wegan fealwe linde *each could bear the yellow shields,* Cd. 94; Th. 123, 14; Gen. 2044. Wineleás guma gesihp him biforan fealwe wegas *the friendless mortal sees before him seared ways,* Exon. 77 a; Th. 289, 11; Wand. 46: Beo. Th. 1837; B. 916. [*Chauc.* falwe: *Laym.* falewe, *pl* : *O. Sax.* falu : *Dut.* vaal : *Kil.* vael, vaeluwe : *Ger.* fal, fahl, falb : *M. H. Ger.* val : *O. H. Ger.* falo, falw : *Icel.* fölr *pale, fallow* : *Lat.* pallĭdus *pale* : *Sansk.* palita *grey.*] DER. æppel-fealu.

fealo *many,* Beo. Th. 5508, note; B. 2757, note. v. fela.

feá-lôg; *adj. Destitute;* destĭtûtus :—Ne eam ic swâ feálôg monna weorudes *I am not so destitute of men,* Exon. 36 a; Th. 116, 34; Gû. 217.

fealo-hilte; *adj. Having a yellow* or *golden handle;* căpŭlo flāvo *vel* aureo instructus :—Feóll tó foldan fealohilte swurd *the golden-hilted sword fell to the earth,* Byrht. Th. 136, 45; By. 166.

fealp *falleth, falls,* Bt. 6; Fox 14, 29; *3rd pers. pres. of* feallan.

fealu *fallow, pale yellow, dusky,* Ælfc. Gl. 79; Som. 72, 81; Wrt. Voc. 46, 38: Andr. Kmbl. 841; An. 421. v. fealo.

fealu; *gen.* fealuwes, fealwes; *n. Fallow ground, ground ploughed lying fallow after a crop;* nŏvāle :—Andlang weges óp done brôc, de scýt to fealuwes leá *along the way to the brook, which shoots to the field of fallow ground,* Cod. Dipl. 399; A. D. 944; Kmbl. ii. 251, 1. DER. fealo *a yellowish light red, like marly ground recently ploughed.*

fealuwian *to wither,* Bt. Met. Fox 11, 116; Met. 11, 58. v. fealwian.

fealvor, es; *m. A species of water-fowl, the sultana-hen;* porphyrio = πορφυρίων :—Fealvor *porphyrio,* Wrt. Voc. 280, 17. v. felofor.

fealwa *fallow,* Exon. 114 a; Th. 437, 19; Rä. 56, 10; *def. m. nom. sing. of* fealo.

fealwe *fallow, pale yellow, dusky, bay,* Exon. 57 a; Th. 202, 24; Ph. 74: 60 a; Th. 219, 22; Ph. 311: Beo. Th. 1735; B. 865: 1837; B. 916; *nom. acc. pl. of* fealo.

fealwian, fealewian, fealuwian; *p.* ode; *pp.* od *To grow yellow, ripen, to wither as leaves;* flāvescĕre :—On hærfest hit fealwap *in harvest it ripens,* Bt. 21; Fox 74, 23. His leáf ne fealwiap *its leaves shall not wither,* Ps. Th. 1, 4. Lytle hwîle leáf beóp grêne, donne hý eft fealewiap, feallap on eorpan *a little while the leaves are green, then they grow yellow again, fall to the earth,* Salm. Kmbl. 627; Sal. 313. Fealuwap *withers,* Bt. Met. Fox 11, 116; Met. 11, 58.

feán *joy,* Bd. 1, 25; Whelc. 76, 1; *acc. of* feá.

feánes, -ness, e; *f. Fewness;* paucĭtas :—Seó feánes nýdde dara sacerda dæt ân bisceop beón sceolde ofer tú folc *paucĭtas sacerdôtum côgêbat ûnum antistĭtem duôbus pŏpŭlis præfĭci,* Bd. 3, 21; S. 551, 33. v. feáwnes.

fear, es; *m. A bull, an ox;* taurus, bos :—Gif he hrîderu offrian wille, bringe unwemme fear odde heáfre *si de bobus võluĕrit offerre, marem sive fēminam immaculāta offĕret,* Lev. 3, 1. v. fearr.

feára *of a few,* Beo. Th. 2828; B. 1412. v. feá *few,* feáwa.

fearh, færh, ferh, es; *pl.* fearas; *m. A little pig, a* FARROW, *litter;* porcellus :—Fearh *porcellus,* Wrt. Voc. 78, 40. Fearas suilli vel *porcelli* vel *nefrendes,* Ælfc. Gl. 20; Som. 59, 35; Wrt. Voc. 22, 76.

fearh-hama, an; *m. A little stem;* caulĭcŭlus :—Fearh-hama *caulĭcŭlus,* Ælfc. Gl. 76; Som. 71, 117; Wrt. Voc. 45, 22.

feárlíc *sudden,* L. C. S. 71; Th. i. 412, 28, MS. A. v. færlíc.

feárlíce; *adv. Suddenly, quickly;* sŭbĭto :—He ódre fyrde hêt feárlíce abannan *he commanded another army to be quickly summoned,* Chr. 1095; Erl. 232, 6: 1120; Erl. 248, 12. v. færlíce.

fearm, es; *m. A freight, cargo, load;* ŏnus nāvis :—Ofer holmes hrincg hof sêleste fôr mid fearme *the most excellent house* [*the ark*] *sailed over the ocean's orb with its freight,* Cd. 69; Th. 84, 7; Gen. 1394. [*Icel.* farmr, *m. a fare, freight, cargo.*]

FEARN, FERN, es; *n. A* FERN; filix :—Fearn *filix,* Ælfc. Gl. 42; Som. 64, 10; Wrt. Voc. 31, 21: 67, 45: 79, 64. Genim dysse wyrte wyrt-truman, de man *filicem* and ódrum naman fearn nemnep *take a root of this plant, which is named* filix, *and by another name fern,* Herb. 78; Lchdm. i. 180, 25. Atió ǽrest of da pornas, and da fyrsas, and dæt fearn *draw out first the thorns, and the furze, and the fern,* Bt. 23; Fox 78, 22: Bt. Met. Fox 12, 5; Met. 12, 3. Dæt micle fearn *the large fern;* aspidium filix, L. M. i, 56; Lchdm. ii. 126, 14: Lchdm. i. 380, 19. [*Chauc.* ferne : *Dut.* váren, *n* : *Kil.* væren : *Ger.* farn, farren, *m* :

M. H. Ger. varm, varn, *m* : *O. H. Ger.* farm, farn, *n* : *Sansk.* parṇa, *n. a leaf, plant, tree.*] DER. eofor-fearn, fen-.

fearn-bed, es; *n. A fern-bed;* filĭcêtum, R. 85, Lye.

Fearn-dûn, e; *f.* [*Hunt.* Ferandune : *Brom.* Farandon : fearn *fern,* dûn *a hill*] *Faringdon, Berkshire?* or *Farndon, Northamptcnshire?*—Hêr Eádweard cing gefôr on Myrcum æt Fearndûne *in this year* [A. D. 924] *kind Edward died in Mercia at Farndon,* Chr. 924; Th. 198, 1, col. 2, 3.

Fearn-ham, -hamm, es; *m.* FARNHAM, *in Surrey;* lŏci nōmen in agro Surreiensi :—Sió fierd him wid gefeaht æt Fearnhamme *the army fought against them at Farnham,* Chr. 894; Erl. 90, 26.

fearn-leás, -lês; *adj. Fernless, without fern;* sine fĭlice, Hem. p. 86.

fearop-hengest, es; *m.* [fearop=farop, *q. v.*] *A sea-horse, ship;* mārīnus equus, nāvis :—Fearophengestas gearwe stôdon *the ships stood ready,* Elen. Kmbl. 452; El. 226.

FEARR, es; *m.*   I. *a bull, an ox;* taurus, bos :—Fearr *taurus,* Ælfc. Gr. 8; Som. 7, 30. He geworhte ânes fearres anlícnesse of âre *he made an image of a bull with brass,* Ors. 1, 12; Bos. 36, 29. Fearras fætte ofsettun odde ymbsæton me *tauri pingues obsēdērunt me,* Ps. Lamb. 21, 13 : Mt. Bos. 22, 4. Ete ic flǽscmettas fearra *mandūcăbo carnes taurôrum,* Ps. Lamb. 49, 13 : 67, 31 : Gen. 32, 15.   II. *the Bull, one of the twelve signs of the zodiac;* taurus :—Oder dæra tâcna ys gehâten *taurus,* dæt is fearr *the second of the signs is called* taurus, *that is a bull,* Bd. de nat. rerum; Wrt. popl. science 7, 4; Lchdm. iii. 244, 24. [*Dut.* var, varre, *m* : *Ger.* farre, farr, *m* : *M. H. Ger.* var, varre, *m* : *O. H. Ger.* farri, farro, far, *m* : *Icel.* farri, *m. a bullock.*]

feá-sceaft; *adj. Having few things, poor, naked, destitute;* mĭser, pauper, destitūtus :—Freónda feásceaft *destitute of friends,* Cd. 97; Th. 126, 24; Gen. 2100: 112, 14; Th. 149, 23; Gen. 2479: Andr. Kmbl. 2257; An. 1130. Ic feásceaft eom *I am destitute,* Cd. 99; Th. 131, 13; Gen. 2175: Beo. Th. 13; B. 7. Feásceaft guma *the miserable man,* Beo. Th. 1950; B. 973: Andr. Kmbl. 3110; An. 1558: Exon. 119 b; Th. 459, 5; Hy. 4, 112. Wæs bên getídad feásceaftum men *the prayer was granted to the poor man,* Beo. Th. 4559; B. 2285: 4775; B. 2393. God eáde mæg afrêfran feásceaftne *God may easily comfort the poor* [*one*], Exon. 10 b; Th. 11, 23; Cri. 175: Andr. Kmbl. 733; An. 367. Hwider fundast dû, feásceaft ides *whither art thou hastening, poor damsel?* Cd. 103; Th. 137, 6; Gen. 2269. Nô feásceafte findan meahton æt dam ædelinge *the poor could not prevail with the prince,* Beo. Th. 4735; B. 2373: Exon. 13 a; Th. 23, 13; Cri. 368.

feá-sceaftig; *adj. Poor, destitute;* pauper, destitūtus, mĭser :—Feásceaftig ferp *poor soul,* Exon. 81 b; Th. 307, 19; Seef. 26.

feasten, es; *n. A fastness, fortress;* mūnīmentum :—Hí on dam feastene wæron *they were in the fastness,* Chr. 877; Erl. 79, 23. v. fæsten II.

feastlíce; *adv. Firmly, constantly, stoutly;* firmĭter, constanter :—Hí feastlíce fêngon *they stoutly engaged,* Chr. 1004; Erl. 139, 32 : 1008; Erl. 141, 17. v. fæstlíce.

FEÁWA, feá; *pl. nom. acc.* feáwe, feáwa, feá; *gen.* feáwera, feáwra, feára; *dat.* feáwum, feáum, feám; *adj.* FEW; pauci :—Feáwa dara manna mihte beón eardfæste *few of the men could abide in their dwellings* [*lit. could be earth-fast* or *settled*], Ors. 5, 4; Bos. 105, 10: Deut. 4, 27: Mt. Bos. 9, 37: Lk. Bos. 10, 2. Hit puhte him feáwa daga *it seemed to him a few* [*of*] *days,* Gen. 29, 20. Feáwe [Spl. feáwa] gewordene hí syndon *pauci facti sunt,* Ps. Lamb. 106, 39. Wesan dagas his feáwe [feáwa, Spl. 108, 7] *fiant dies ejus pauci,* 108, 8. Dâ dâ hig wæron on gerîme [MS. gehrime] feáwa odde scortum, feáwoste and eardbegendan odde inlænde his *when they were few* or *short in number,* [*yea*] *very few and inhabitants of it* [*Canaan*], Ps. Lamb. 104, 12. Hira feáwa on weg cômon *few of them came in the way,* Chr. 918; Erl. 104, 9: Deut. 28, 62. Inne on dæm fæstenne sæton feáwa cirlisce men *a few countrymen sat within the fastness,* Chr. 893; Erl. 88, 33. Feáwa synt de done weg findon *pauci sunt qui invĕniunt viam,* Mt. Bos. 7, 14: Lk. Bos. 13, 23. Feáwa synt gecorene *pauci sunt electi,* Mt. Bos. 20, 16: 22, 14. Drihten, gedô dæt heora menigo sý læsse donne ûre feáwena nû is, and tostencte hí geond eorpan libbende of dis lande *Dōmĭne, a paucis de terra dīvĭde eos in vīta eōrum,* Ps. Th. 16, 13. Ic dê of Caldêa ceastre alædde, feáwera [MS. feowera] manna *I led thee, one of a few, from the Chaldeans' city,* Cd. 100; Th. 132, 30; Gen. 2201. Eustatius ætbærst mid feáwum mannum *Eustace escaped with a few men,* Chr. 1048; Erl. 178, 4. Efter feáwum dagum *after a few days,* 1070; Erl. 206, 2. Be dissum feáwum forpspellum *by these few intimations,* Exon. 84 a; Th. 316, 11; Môd. 47. Ic dê feáwe dagas mînra mættra môde secge *paucĭtātem diērum meôrum enuntia mihi,* Ps. Th. 101, 21. Feáwa fixa *paucos pisciculos,* Mt. Bos. 15, 34: Mk. Bos. 8, 7. Feáwa untrume he gehælde *paucos infirmos cūrāvit,* Mk. Bos. 6, 5. Dû wære getrýwe ofer feáwa *super pauca fuisti fĭdēlis,* Mt. Bos. 25, 23. He bip witnod feáwum wîtum *vāpŭlābit paucis plāgis,* Lk. Bos. 12, 48. [*Wyc. Chauc. R. Glouc.* fewe : *Laym.* feue, feupe : *Orm.* fæwe : *Plat.* fege, vöge : *O. Sax.* fáh : *O. Frs.* fê : *O. H. Ger.* fôh : *Goth.* faus, faws :

Dan. faa : *Swed.* få : *Icel.* fár : *Lat.* paucus, paulus : *Grk.* παῦρος *few ;* παύω *I make to cease.*]

feáwera *of a few,* Cd. 100 ; Th. 132, 30 ; *gen. pl. of* feáwa.

feáwnes, feánes, -ness, e ; *f.* FEWNESS ; paucĭtas :—Đa feáwnesse oððe gehwǽdnesse dagena mínra cýþ me *paucĭtātem diērum meōrum nuntĭa mihi,* Ps. Lamb. 101, 24.

FEAX, fex, es ; *n. Hair of the head, the locks ;* cæsăries, cŏma, căpillus :—Nimeþ ðæt feax to *the hair holdeth on,* Med. ex Quadr. 4, 11 ; Lchdm. i. 344, 20 : L. M. 1, 87 ; Lchdm. ii. 156, 7. Ne feax ne fel *neither hair nor skin,* Exon. 74 a ; Th. 278, 1 ; Jul. 591 : Cd. 195 ; Th. 243, 18 ; Dan. 438. Feax *cæsăries,* Ælfc. Gr. 12 ; Som. 15, 53. Licgaþ æfter lande loccas todrifene, fex on foldan *throughout the land lie my driven locks, hair upon the ground,* Andr. Kmbl. 2853 ; An. 1429. God tofylleþ feaxes scadan, ðe hér on scyldum swǽrum eódon *Deus conquassābit vertĭcem căpilli perambulan/tium in delic/tis suis,* Ps. Th. 67, 21 : 68, 4. Bôcstafa brego bregdeþ feónd be ðam feaxe *the prince of letters shall draw the fiend by his hair,* Salm. Kmbl. 201 ; Sal. 100 : Beo. Th. 3298 ; B. 1647. Wið feallendum feaxe *for falling hair,* Med. ex Quadr. 4, 11 ; Lchdm. i. 344, 18. Mid hyre heáfdes feaxe *căpillis căpĭtis sui,* Lk. Bos. 7, 38. Swát ǽdrum sprong forþ under fexe *blood strang forth from the veins under his hair,* Beo. Th. 5926 ; B. 2967. Æled lǽtaþ on ðæs feóndes feax *they shall let fire upon the fiend's hair,* Salm. Kmbl. 261 ; Sal. 130 : Judth. 12 ; Thw. 25, 27 ; Jud. 281. He hæfde blæc feax *he had black hair,* Bd. 2, 16 ; S. 519, 34. [*Laym.* uæx : *O. Sax.* fahs, *n* : *O. Frs.* fax : *M. H. Ger.* vahs, *m* : *O. H. Ger.* fahs, *n.* cæsăries, cŏma : *Icel.* fax, *n. a mane.*] DER. blanden-feax, blonden-, gamol-, un-, up-, won-, wunden-.

feax-cláþ, es ; *m. A head-clo/th, hair-band, fillet ;* fascia crīnālis, Cot. 93.

feaxe ; *adj. Having hair ;* cŏmātus. DER. ge-feaxe.

feax-eacas, -eacon? *Hair hanging down the forehead, forelocks ;* antiæ frontis, sive a fronte dependentes, Cot. 6, Som. Ben. Lye.

feaxede, fexede ; *adj. Having long hair, long-haired ;* cŏmātus :—Sume men cweðaþ ðæt cométa síe feaxede [fexede, Th. 162, 9. col. 2, 3 ; 163, 10] steorra, forðǽm ðǽr stent lang leóma of, hwílum on áne healfe, hwílum on ælce healfe *some men say that a comet is a long-haired star, because there stands a long ray from it, sometimes on one side, sometimes on each side,* Chr. 891 ; Th. 162, 9-14, col. 1. DER. ge-feaxode, -fexode, síd-fexede.

feax-fang, es ; *m. A taking hold by the hair ;* cŏmæ prehensio :—Gif feax-fang geweorþ *if there be a taking hold of the hair,* L. Ethb. 33 ; Th. i. 12, 3 ; Wilk. 5, 1.

feax-feallung, e ; *f. Falling off or loss of the hair, the mange ;* crīnium amissio, alōpĕcia = ἀλωπεκία :—Feaxfeallung *alōpĕcia,* Ælfc. Gl. 11 ; Som. 57, 56 ; Wrt. Voc. 19, 58.

feax-gerǽdian ; *p.* ode ; *pp.* od [gerǽdian *to make ready*] *To dress or trim the hair ;* crīnes compōnĕre, Som. Ben. Lye.

feax-hár ; *adj. Hoary-haired ;* cŏmam cānam hăbens :—Ic wæs feaxhár *I was hoary-haired,* Exon. 126 b ; Th. 487, 13 ; Rä. 73, 1.

feax-nǽdel, e ; *f. A hair-needle, curling-iron, crīsping-pin ;* călămistrum, ăcus crīnĭbus intorquendis sive crispandis adhĭbĭta :—Feaxnǽdel *călămistrum,* Ælfc. Gl. 1 ; Som. 55, 101 ; Wrt. Voc. 17, 4.

feax-net, -nett, es ; *n. A hair-net, net-work cap for confining the hair ;* rētĭcŭlum căpillis continendis, rīcŭla :—Feaxnet *rētĭcŭlum,* Ælfc. Gl. 4 ; Som. 55, 89 ; Wrt. Voc. 16, 59 : rīgŭla [= rīcŭla, Car. Ains.], Som. 55, 96 ; Wrt. Voc. 16, 66.

feax-preón, es ; *m. A hair-pin ;* discrīmĭnāle :—Uplegene *vel* feaxpreónas *discrīmĭnālia,* Ælfc. Gl. 4 ; Som. 55, 99 ; Wrt. Voc. 17, 2.

feax-sceacga, an ; *m. A bush of hair ;* cæsăries, crīnium fascĭcŭlus, Som. Ben. Lye.

feax-sceacged ; *part. Having hair, hairy ;* cŏmātus, Cot. 54.

feber-ádl, e ; *f. A fever-disease, fever ;* febris :—Forleórt ða of feber-ádlum *dīmīsit eam febris,* Mt. Kmbl. Lind. 8, 15. v. fefer-ádl.

febrig ; *adj. Feverish ;* febrĭcŭlōsus :—Gif he sý febrig *if he be feverish,* Herb. 1, 28 ; Lchdm. i. 78, 26.

Februarius, i ; *m. Lat. February ;* nōmen mensis :—Sígeþ Februarius *February approaches,* Menol. Fox 35 ; Men. 18. v. Sol-mōnaþ.

fec, es ; *n. A space, portion of time ;* spătium, tempŏris intervallum :—Æfter litlum fece *after a little time,* Chr. 1015 ; Erl. 152, 4. v. fæc.

FECCAN, feccean, fæccan ; *p.* feahte, fehte ; *pp.* feaht, feht *To* FETCH, *bring to, draw ;* addūcĕre, tollĕre, afferre, haurīre :—Đæt he sceolde hine feccan *that he should fetch him,* Bd. 4, 1 ; S. 564, 43 : Chr. 1017 ; Erl. 161, 10 : Gen. 27, 42, 45 : Ex. 2, 5. Com án wíf wæter feccan *vēnit mŭlier haurīre ăquam,* Jn. Bos. 4, 7, 15. He his dôhter lét feccean *he caused his daughter to be fetched,* Chr. 1121 ; Erl. 248, 35. Ic fecce wæter *afferam pauxillum ăquæ,* Gen. 18, 4. Híg fecceaþ ðíne sáwle fram ðé *they will fetch away thy soul from thee,* Lk. Bos. 12, 20. Đás menn ðé feccaþ *these men fetch thee,* Num. 22, 20. Gif preóst crisman ne fecce [fæcce MS. B.] *if a priest fetch not the chrism,* L. E. G. 3 ; Th. i. 168, 11. Se ðe ys uppan hys húse, ne gá he nyðer ðæt he ǽnig þing on his húse fecce *qui in tecto, non descendat tollĕre*

---

alĭquid de dŏmo sua, Mt. Bos. 24, 17 : L. Edg. C. 67 ; Th. ii. 258, 20. Đæt gé ðisne eówerne brôður feccon *that ye fetch this your brother,* Gen. 42, 34. [*Laym.* fæchen : *Orm.* fecchenn : *O. Frs.* faka *to prepare, make ready.*] DER. a-feccan, ge-.

fecele *a torch,* Som. Ben. Lye. v. fæcele, þæcele.

fecgan ; *p.* feah *To seize ;* răpĕre. DER. æt-fecgan, ge-.

FÉDAN ; *part.* fédende ; he fédeþ, fét, fétt ; *p.* ic, he fédde, ðú féddest, *pl.* féddon ; *pp.* féded, fédd. **I.** *to* FEED, *nourish, support, sustain, bring up, educate ;* pascĕre, cĭbāre, nutrīre, enutrīre, sustentāre, edūcāre :—Mægen mon sceal mid mete fédan *a man must feed strength with meat,* Exon. 90 b ; Th. 340. 22 ; Gn. Ex. 115. Wá eácniendum and fédendum on ðám dagum *væ autem prægnantĭbus, et nutrientĭbus in illis diēbus,* Mt. Bos. 24, 19 ; Lk. Bos. 21, 23. Đú us fédest teára hláfe *cĭbābis nos pāne lacrўmārum,* Ps. Th. 79, 5. Se deópa seáþ dreórge fédeþ *the deep pit feedeth the dreary,* Exon. 30 b ; Th. 94, 25 ; Cri. 1545 : 36 b ; Th. 118, 26 ; Gú. 245. He ðé fédeþ *ipse te enutriet,* Ps. Th. 54, 22. Eówer heofonlíca fæder híg fét *păter vester cœlestis pascit illa,* Mt. Bos. 6, 26. Se milda Metod fét eall ðætte grôweþ wæstmas on weorolde *the merciful Creator nourishes all fruits which grow in the world,* Bt. Met. Fox 29, 139 ; Met. 29, 70. He fétt ða ðe þurh dǽdbóte him to búgaþ *he feeds those who turn to him by repentance,* Homl. Th. ii. 396, 29. He me well fétt *me bĕne pascit,* Coll. Monast. Th. 22, 33 : 30, 27. Mægeþ and mæcgas fédaþ hine fægre *lasses and lads feed him kindly,* Exon. 113 a ; Th. 434, 9 ; Rä. 51, 8. God, ðú ðe me féddest fram cildháde óþ ðisne dæg *Deus, qui pascit me ab adolescentia mea in præsentem diem,* Gen. 48, 15. Mec seó friþe mæg fédde *the kind woman fed me,* Exon. 103 a ; Th. 391, 23 ; Rä. 10, 9. He fédde híg *sustentāvit eos,* Gen. 47, 17. He fédde me *edūcāvit me,* Ps. Spl. 22, 2. We ðé féddon *pāvimus te,* Mt. Bos. 25, 37. Féd freólíce feora wôcre *feed freely the living progeny,* Cd. 67 ; Th. 81, 8 ; Gen. 1342. Gif he nát hwá hine cwicne féde *if he knows not who may feed him living,* Exon. 90 b ; Th. 340, 21 ; Gn. Ex. 114. Đú bist féded on wélum his *pascĕris in dīvĭtiis ejus,* Ps. Spl. 36, 3 : Ps. Th. 130, 4. Fédd beón *pastus esse, pasci,* R. Conc. 10. **II.** *to bring forth, produce ;* gignĕre, prodūcĕre :—Wæstmas fédan *to bring forth fruits,* Cd. 46 ; Th. 59, 8 ; Gen. 960. Cucra wuhta, ðara ðe lyft and flód lǽdaþ and fédaþ *of living things, which air and flood train and bring forth,* 65 ; Th. 78, 25 ; Gen. 1298. Ides eaforan fédde *a female brought forth offspring,* 50 ; Th. 64, 23 ; Gen. 1054. Đá wearþ eafora féded *then was an heir brought forth,* 58 ; Th. 70, 27 ; Gen. 1159 : 82 ; Th. 103, 3 ; Gen. 1712. [*Wyc. Chauc.* fede : *Piers P.* feden : *Laym.* feden, ueden : *Orm.* fedenn : *Scot.* fede : *Plat.* voden, vöden, föden, füden : *O. Sax.* fôdjan, fuodjan : *Frs.* fieden : *O. Frs.* foda, feda : *Dut.* voeden : *Ger.* füttern : *M. H. Ger.* vuoten, vüeten : *O. H. Ger.* fuotjan : *Goth.* fodyan : *Dan.* föde : *Swed.* föda : *Icel.* fœða : *Lat.* pascĕre : *Grk.* πατέομαι *to eat* : *Sansk.* pitu, *m. nourishing food.*] DER. a-fédan, ge-.

fédels, es ; *m. A fatling ;* altĭlis :—Fédels *altĭle,* Ælfc. Gl. 22 ; Som. 59, 95 ; Wrt. Voc. 33, 51 : altĭlis, 114 ; Som. 80, 7 ; Wrt. Voc. 60, 43.

feder *a father,* Chr. 1052 ; Th. 319, 17 ; Hy. 8, 8 ; Hy. Grn. ii. 290, 8 : 8, 43 ; Hy. Grn. ii. 291, 43. v. fæder.

federa, fedra. an ; *m. An uncle, a father's brother ;* patruus :—Se wæs Ælfríces sunu Ædwines federan *he was the son of Ælfric, Edwin's uncle,* Chr. 634 ; Erl. 25, 25 : 737 ; Erl. 47, 24. Edwines fedran suna *Edwin's uncle's son,* Chr. 643 ; Erl. 27, 19. v. fædera.

fédesl, es ; *m? e ; f? A feeder, provider ;* obsōnātor :—Cyninges fédesl xx scillinga forgelde *let the king's feeder be paid for with twenty shillings,* L. Ethb. 12 ; Th. i. 6, 8.

féding, e ; *f. A feeding ;* pastio :—Seó féding ðara sceápa *the feeding of the sheep,* Past. 5, 2 ; Hat. MS. 10 b, 11. v. fédan *to feed.*

fédnes, -ness, e ; *f. Nourishment ;* nutrĭmentum :—On lustfullnysse ðær biþ synne fédnes *in delectātiōne fit peccāti nutrĭmentum,* Bd. 1, 27 ; S. 490, 25.

FEFER, fefor, es ; *m. A* FEVER ; febris :—Se fefer hine forlét *relīquit eum febris,* Jn. Bos. 4, 52. Gif him fefer derige *if fever vex him,* Herb. 46, 2 ; Lchdm. i. 148, 19. Se fefor *the fever,* Mt. Bos. 8, 15. Ǽr hym ðæs feferes wéne *before he expects the fever,* Herb. 2, 12 ; Lchdm. i. 84, 7. Wið fefre *for fever,* L. M. 1, 62 ; Lchdm. ii. 134. 14, 27. Wið ðone cólan fefor *against cold fever,* Herb. 138, 2 ; Lchdm. i. 256, 10. Đa feforas beóþ fram ándyde *the fevers will be forced away,* 143, 4 ; Lchdm. i. 266, 13. On mycelum feferum *magnis febrĭbus,* Lk. Bos. 4, 38. Wið ða stíðustan feferas, genim ðas sylfan wyrte and gedrige hý *for the strongest fevers, take this same herb and dry it,* Herb. 20, 3 ; Lchdm. i. 114, 16 : 38, 2 ; Lchdm. i. 138, 3. Ælces dæges fefer *an every day or quotidian fever,* L. M. 1, 62 ; Lchdm. ii. 134, 24. Þriddan dæges fefer *a tertian fever,* 1, 62 ; Lchdm. ii. 134, 21. Feórþan dæges fefer *a quartan fever,* Herb. 2, 12 ; Lchdm. i. 84, 5. [*Piers P.* feveres, *pl* : *Chauc.* fevere : *Plat.* fever, *n* : *Ger.* fieber, *n* : *M. H. Ger.* vieber, *n* : *O. H. Ger.* fiebar, *n* : *Dan.* feber, *m. f.* : *Swed.* feber, *m* : *Lat.* febris, *f.*]

fefer-ádl, fefor-ádl, e ; *f.* [ádl *a disease*] *Fever-disease, fever ;* febris :—Heó wæs swenced mid hǽto and mid bryne feferádle *she had been afflicted with the heat and burning of a fever,* Bd. 5, 4 ; S. 617, 28. Wið fefer-

ädle *for fever disease*, L. M. 1, 62; Lchdm. ii. 134, 13. Sleá đē Drihten mid feforädle and mid cíle *percŭtiat te Dŏmĭnus febri et frĭgŏre*, Deut. 28, 22.

fefer-fuge, an; *f. The herb feverfew*; febrĭfŭgia :—Feferfuge *febrĭfŭgia*, Ælfc. Gl. 40; Som. 63, 89; Wrt. Voc. 30, 39: Herb. 36; Lchdm. i. 134, 15. Genim feferfugean blóstman *take blossoms of feverfew*, Lchdm. i. 374, 3.

fefer-seóc; *adj. Fever-sick, feverish*; febrĭcĭtans, Cot. 88.

fefor *a fever*, Mt. Bos. 8, 15. v. fefer.

fefor-ädl *fever-disease, fever*, Deut. 28, 22. v. fefer-ädl.

FÉGAN; *p.* de; *pp.* ed *To join, bind, unite, fix*; jungĕre, pangĕre :— Heó fēgeþ mec on fæsten *she binds me in a fastness*, Exon. 107 a; Th. 407, 22; Rä. 26, 9. Freóndscipe fēgþ *it unites friendship*, Somn. 128; Lchdm. iii. 206, 4. Hió me on nearo fégde *she fixed me in a strait*, Exon. 124 b; Th. 479, 12; Rä. 62, 6. [*Laym.* fiede *wrote: Orm.* fe3est *joinest*; fe33ed, *pp. composed: Plat.* fögen: *O. Sax.* fógian: *Frs.* fuwgien: *O. Frs.* foga: *Dut.* voegen: *Ger.* fügen: *M. H. Ger.* vüegen: *O. H. Ger.* fuogͣan, fuogan: *Dan.* föie: *Swed.* foga: *Lat.* pācĭscor *to make a contract: Grk.* πήγνυμι *to join, fasten: Sansk.* paś *to bind.*] DER. ge-fégan, up-fégean.

feger, fegr *fair*; pulcher, Solil. præf. v. fæger.

fegere *fairly, beautifully*, Hy. 8, 43; Hy. Grn. ii. 291, 43. v. fægere.

féging, e; *f. A conjunction*; conjunctio :—Geþeóđnes ođđe féging is conjunctio *a joining is a conjunction*, Ælfc. Gr. 5; Som. 3, 47, MS. D.

féhan, đú féhst, he féhþ *to take, seize*; captāre, Bt. 35, 5; Fox 164, 16: Exon. 107 b; Th. 410, 1; Rä. 28, 9. v. féhþ, fón.

FEL, felo, fæle; *adj.* FELL, *cruel, savage*; crūdēlis, sævus. [*Wyc.* fel, felli *crafty: Piers P.* fell *fierce: Chauc.* felle *strong, fierce: Laym.* felle, *pl. cruel: Scot.* fell *keen, hot, acute: O. Frs.* fal: *Dut. Kil.* fel *violent: O. Fr.* fel *cruel, wicked: Ital.* fello *wicked: Ir.* feal *bad, naughty, evil.*] DER. æl-fæle, eal-felo, wæl-fel.

FEL, FELL, es; *n. A* FELL, *skin, hide*; pellis, cŏrium, cŭtis :—Fel *pellis*, Wrt. Voc. 65, 11: 86, 37: 283, 33. Næs hyre feax ne fel fýre gemæled *neither her hair nor skin was marked by the fire*, Exon. 74 a; Th. 278, 1; Jul. 591. Fell *pellis*, Wrt. Voc. 71, 18. Felles ne rēcceþ *he cares not for my skin*, Exon. 127 a; Th. 488, 12; Rä. 76, 5. Đæt celf híg bærndon bútan đære wícstówe mid felle and mid flǽsce *vĭtŭlum cum pelle et carnĭbus crēmans extra castra*, Lev. 8, 17. Hie blód and fel þēgon *they ate the blood and skin*, Andr. Kmbl. 46; An. 23: Ors. 1, 1; Bos. 20, 37. Đæs cealfes flǽsc and fell and gór đū bærnst úte búton fyrdwícon *carnes vĭtŭli et cŏrium et fĭmum combūres fŏris extra castra*, Ex. 29, 14. Fell hongedon on seles wæge *the skins hung on the wall of the room*, Exon. 104 a; Th. 394, 15; Rä. 14, 3. Đæt gafol biþ on deófra fellum *the tribute is in skins of animals*, Ors. 1, 1; Bos. 20, 33: Boutr. Scrd. 20, 29: Gen. 27, 16. Se byrdesta sceall gyldan fíftyne mearþes fell *the richest must pay fifteen skins of the marten*, Ors. 1, 1; Bos. 20, 36. Sió wæs orþoncum gegyrwed dracan fellum *it was cunningly prepared with dragon's skins*, Exon. Th. 4183; B. 2088. [*Wyc. Piers P.* fel: *Chauc. Orm.* fell: *O. Sax.* fel, n: *Frs. O.Frs.* fel, n: *Dut.* vel, n: *Ger.* fell, n: *M. H. Ger.* vël, n: *O.H.Ger.* fel, n: *Goth.* fill, n: *Icel.* fell, n: *Lat.* pellis, *f. a skin, hide: Grk.* πέλλα, *f. a hide, leather.*]

FELA, fæla, feala, feola; *adj. indecl.* **I.** *with gen. Many, much*; multum, multa :—Nis nū fela folca *there is not now much people*; multum pŏpŭlōrum, Exon. 81 a; Th. 304, 8; Fä. 67. Náh ic fela goldes *I have not much gold*; multum auri, Exon. 119 b; Th. 458, 14; Hy. 4, 100. Fela sceal gebídan leófes and láþes *much shall abide of loved and loathed*, Beo. Th. 2125; B. 1060. Fela meoringa *many obstacles*; multa impĕdimentōrum, Cd. 145; Th. 181, 16; Exod. 62. Fela is đæra þinga *many a one is there of the things*, Bt. 41, 3; Fox 250, 10. Fela swylces *much of the same*, Coll. Monast. Th. 24, 13. **II.** *many things, much, very*; multa, multum, in prīmis, cum maxĭme :—Fela đū didest *multa fēcisti*, Ps. Spl. 39, 7: Ps. Spl. C. 31, 13. Hie fela wiston *they knew many things*; multa, Cd. 143; Th. 179, 16; Exod. 29. Fela ic hæbbe geþolod to dæg *multa passa sum hŏdie*, Mt. Bos. 27, 19. Fela fricgende *inquiring much*, Beo. Th. 2187; B. 2106. Hū fela *how many*; quam multa, Exon. 25 a; Th. 72, 27; Cri. 1179. He ongan hí fela lǽran *cœpit illos dŏcēre multa*, Mk. Bos. 6, 34. **III.** *so many … as*; tot … quot :—Ic ne mæg swá fela [gefón], swá fela swá ic mæg gesyllan *non possum tot cŭpĕre, quot possum vendĕre*, Coll. Monast. Th. 23, 27. [*Wyc.* fele, feel: *Piers P. Chauc.* fele: *Laym.* fele, feole, vele, uæle: *Orm.* fele: *Scot.* feil, fiel: *Plat.* veel: *O. Sax.* filu, filo: *Frs.* foll, full: *O. Frs.* fel, ful: *Dut.* veel: *Ger.* viel: *M.H.Ger.* vil: *O.H.Ger.* filo, filu: *Goth.* filu: *Icel.* fjöl-, used only as a prefix, *much: Lat.* plus: *Grk.* πολύς: *Sansk.* puru, pulu *much, many.*] DER. eal-fela, efen-, em-.

fela-fǽcne; *adj. Very crafty*; multĭdŏlōsus :—Wineleás mon genimeþ him wulfas to geféran felafǽcne deór *a friendless man takes wolves for his comrades very crafty animals*, Exon. 91 b; Th. 342, 26; Gn. Ex. 148.

fela-feald; *adj. Manifold*; multiplex :—Dómas đíne synd neowelnys micellu ođđe felafeald *jūdĭcia tua sunt abyssus multa*, Ps. Spl. 35, 6.

fela-frécne; *adj. Very wild* or *savage*; valde fĕrox :—Úr biþ fela-

frēcne deór *a wild bull is a very savage beast*, Runic pm. 2; Kmbl. 339, 9; Hick. Thes. i. 135, 3.

fela-geómor; *adj. Very sad*; valde tristis :—Gewát him se góda, felageómor *the good* [*king*] *departed, very sad*, Beo. Th. 5892; B. 2950.

fela-geong; *adj. Very young*; valde jŭvĕnĭlis :—He sægde felageongum *he said to the very young* [*man*], Exon. 80 b; Th. 303, 15; Fä. 53.

fela-geonge; *adj. Having travelled much*; valde peregrĭnātus :—Wilt đū frīcgan felageongne ymb forþgesceaft *wilt thou ask one who has travelled much about the creation?* Exon. 92 b; Th. 346, 23; Sch. 3.

fela-hrór; *adj. Very strenuous*; valde strēnuus :—Him Scyld gewát felahrór *Scyld departed very strenuous*, Beo. Th. 53; B. 27.

fela-leóf; *adj. Much-beloved*; valde cārus :—Sceal ic mínes felaleófan fǽhþu dreógan *I must endure enmities for my much-loved* [*friend*], Exon. 115 a; Th. 443, 6; Kl. 26.

fela-meahtig; *adj. Much mighty*; valde pŏtens :—Felameahtig God *the much mighty God*, Exon. 90 a; Th. 338, 10; Gn. Ex. 76. Bletsien đec fiscas and fuglas, felameahtigne *may fishes and birds bless thee, much mighty!* 55 a; Th. 194, 17; Az. 140: Th. 195, 14; Az. 156.

fela-módig; *adj. Very daring*; fortissĭmus :—Men from đæm holmclife hafelan bǽron felamódigra *the men bore from the shore the heads of the very bold*, Beo. Th. 3278; B. 1637.

felan; *p.* fæl, *pl.* fǽlon; *pp.* folen *To stick, adhere*; hærēre :—Đæt ic in ne fele *ut non inhæream*, Ps. Surt. 68, 15. v. feolan.

FÉLAN; *p.* de; *pp.* ed; *v. a. gen. To* FEEL, *perceive, touch*; sentĭre, tangĕre :—Heó féleþ mínes gemótes *she perceives my meeting*, Exon. 107 a; Th. 407, 23; Rä. 26, 9. Hí đæs félaþ *they feel it*, Exon. 103 a; Th. 389, 16; Rä. 7, 8. [*Wyc.* felen, feele: *Chauc.* fele: *Plat.* fölen: *O.Sax.* gi-fólian: *Frs.* fielen: *O. Frs.* féla: *Dut.* voelen: *Ger.* fühlen: *M.H.Ger.* vüelen: *O.H.Ger.* fuoljan, fuolén: *Dan.* föle.] DER. ge-félan.

fela-sinnig; *adj. Very sinful*; valde facĭnŏrōsus :—Đær đū findan miht felasinnigne secg *where thou mayest find the very sinful man*, Beo. Th. 2762; B. 1379.

fela-specol; *adj. Speaking much, loquacious*; magnĭloquus, lŏquax :—Mǽden felaspecol *a loquacious maiden*, Obs. Lun. § 7; Lchdm. iii. 186, 26. Tostencþ Drihten tungan đa felaspecolan *disperdat Dŏmĭnus linguam magnĭlŏquam*, Ps. Spl. 11, 3.

fela-specolnys, -nyss, e; *f. Talkativeness, loquacity*; lŏquācĭtas, Scint. 54.

fela-wlonc; *adj. Very stately*; valde magnĭfĭcus :—Mec brýd triedeþ, felawlonc, fótum *the bride treads me, very proud, with her feet*, Exon. 103 b; Th. 393, 28; Rä. 13, 7.

fel-cyrf, e; *f? [fel skin*, cyrf *a cutting off*] *The foreskin*; præpūtium, Cot. 217.

FELD, feald; *gen.* es; *dat.* a, e; *m. A* FIELD, *pasture, plain, an open country*; campus, campestria :—Se ǽđela feld wrídaþ under wolcnum *the noble field flourishes under the skies*, Exon. 56 a; Th. 199, 16; Ph. 26. Feld *campus*, Wrt. Voc. 80, 48. Weaxaþ hrađe feldes blóstman *the flowers of the field quickly grow*, Bt. Met. Fox 6, 19; Met. 6, 10. On felda đam đe deórmóde Dīran hēton *in the field which the brave men call Dura*, Cd. 180; Th. 226, 13; Dan. 170: Byrht. Th. 138, 56; By. 241. He sette fóretācn his on felda Taneos *pŏsuit prŏdĭgia sua in campo Taneos*, Ps. Spl. 77, 48. On đam felde *upon the plain*, Salm. Kmbl. 427; Sal. 214. Hie gesóhton Sennera feld *they sought the plains of Shinar*, Cd. 80; Th. 100, 23; Gen. 1668: 205; Th. 253, 27; Dan. 602. Híg fundon ánne feld *invēnērunt campum*, Gen. 11, 2. Habbaþ feldas eác fægere blisse *gaudēbunt campi*, Ps. Th. 95, 12: Ps. Lamb. 103, 8. On Moabes feldum *in campestrĭbus Moab*, Deut. 34, 8. On fealda *in campo*, Ps. Spl. 77, 15. Fealdas đíne beóþ gefylled of genihtsumnysse *campi tui replēbuntur ubertāte*, 64, 12. [*Piers P.* felde: *Wyc.* feld, feeld: *Chauc. R. Glouc.* feld: *Laym.* feld, ueld, feold, uald: *Orm.* feld: *O. Sax.* feld, m: *Frs.* fjild: *O. Frs.* feld, field: *Dut.* veld, n: *Ger.* feld, n: *M. H. Ger.* velt, n: *O.H.Ger.* feld, n: *Dan.* fælled, m. f: *Swed.* fält, n: *Icel.* feld, f.*] DER. here-feld, sun-, wæl-, wudu-.

feld-beó; *f. A field-bee, locust*; ǣpis campestris, attăcus = ἀττακός :—Feld-beó *atticus* [= *attăcus*], Wrt. Voc. 281, 38.

feld-ciric, e; *f.* -circe, an; *f. A field-church, country church*; campestris ecclēsia :—Feldcirice gríþbryce is, đǽr legerstów ne sig, mid þrittigum scillingum *the ' grith-bryce' of a field-church, where there is no burial-place, is thirty shillings*, L. C. E. 3; Th. i. 360, 21. Æt feldcircan *for a field-church*, L. Eth. ix. 5; Th. i. 342, 3.

felde *felled*, Exon. 109 b; Th. 419, 11; Rä. 38, 4; *p. of* fellan.

felde-fare, an; *f? A* FIELD-FARE? turdus pĭlāris :—Clodhamer *vel* feldefare *a field-fare*; scorellus? [turdus pĭlāris? Lin.], Wrt. Voc. 63, 27.

feld-elfen, e; *f. A wood fairy* or *nymph*; hāmādryas = ἁμαδρυάς :— Feld-elfen *moides?* Ælfc. Gl. 113; Som. 79, 109; Wrt. Voc. 60, 16.

feld-gangende, -gongende; *part. Field-going, moving over a plain*; campum perăgrans :—Feldgangende feoh *pĕcus campum peragrans*, Soul Kmbl. 161; Seel. 81: Salm. Kmbl. 45; Sal. 23. Feldgongende feoh *cattle traversing the field*, Exon. 99 a; Th. 371, 25; Seel. 81, note: Salm. Kmbl. 309; Sal. 154.

feld-hryðer, es; *n.* *A field ox* or *heifer;* campestris bos sive vĭtŭlus, Chart. ad calc. C. R. Ben.

feld-hûs, es; *n.* *A field-house, tent;* tentōrium, tabernācŭlum :—Feldhûsa mæst *greatest of tents,* Cd. 146; Th. 183, 3; Exod. 85. Bræddon æfter beorgum flotan feldhûsum *the sailors spread [themselves] amongst the hills with their tents,* 148; Th. 186, 3; Exod. 133: Cd. 154; Th. 191, 31; Exod. 223.

feld-land, es; *n.* *Field-land, a plain;* plānĭties. It is opposed to dûn-land *hilly land :*—Faraþ to Amorrēa dûne and to ôðrum feld-landum and dûn-landum and to unhêheran landum *vēnite ad montem Amorrhæōrum et ad cētĕra campestria atque montāna et hŭmĭliōra lŏca,* Deut. 1, 7: 11, 30.

feldlîc; *adj. Fieldlike, country, rural;* campester :—Feldlîc *campester,* Ælfc. Gr. 9, 18; Som. 10, 4. On feldlîcre stôwe *in lŏco campestri,* Lk. Bos. 6, 17.. On feldlîcum wunungum *in campestrĭbus habĭtācŭlis,* Jos. 10, 40.

feld-mædere, an; *f.* [mædere, mæddere *madder] Field-madder, rosemary;* rosmărīnum :—Feldmædere *rosmărīnum,* Glos. Brux. Recd. 42, 34; Wrt. Voc. 68, 49.

feld-minte, an; *f. Field* or *wild mint;* silvestris menta, mentastrum :—Feldminte *mentarium?* [= *mentastrum*], Glos. Brux. Recd. 43, 3; Wrt. Voc. 69, 18.

feld-more, an; *f:* -moru, e; *f.* [more *a root] A parsnip, carrot;* pastĭnăca :—Feldmore *parsnip,* L. M. 3, 14; Lchdm. ii. 316, 21. Feldmore [MS. -mora] *pastĭnăca,* Ælfc. Gl. 42; Som. 64, 32; Wrt. Voc. 31, 42. Nim feldmoran sǽd *take seed of parsnip,* L. M. 3, 12; Lchdm. ii. 314, 19: iii. 72, 3. Wyrtdrenc of feldmoran sele drincan *give to drink a herb-drink of parsnip,* L. M. 1, 48; Lchdm. ii. 122, 15. Dô on eala feldmoran *put parsnip in ale,* 1, 66; Lchdm. ii. 142, 5 : 3, 32 ; Lchdm. ii. 326, 17 : iii. 22, 18. Herba pastĭnăca silvātica, ðæt is feldmoru *the herb pastĭnāca silvātica, that is parsnip,* Herb. cont. 82, 1; Lchdm. i. 32, 25. Feldmoru biþ cenned on sandigum stôwum and on beorgum *parsnip is produced on sandy places and on hills,* Herb. 82, 1; Lchdm. i. 186, 3 : L. M. 2, 53; Lchdm. ii. 274, 26. Feldmore niðeweard *the nether part of parsnip,* L. M. 1, 40; Lchdm. ii. 104, 14.

feld-oxa, an; *f. A field* or *pasture ox;* pascuālis bos :—Feldoxan *pascuāles bŏves,* Hymn. in Dedic. Eccles.

feld-rude, an; *f. Wild rue;* silvestris rŭta, Ben. Lye : Lchdm. Glos. vol. iii. p. 325.

feld-swam, -swamm, es; *m. A field mushroom, toadstool;* fungus, Cot. 87.

feld-swop *bradigaco?* Cot. 25, Lye. Feld-uuop *bradigabo?* Glos. Epnl. Recd. 154. 72.

feld-wêsten, es; *n. A field waste* or *desert;* campestris solĭtūdo :—Begeondan Iordane on ðam feldwêstene wið ða reádan sǽ *trans Iordanem in solitūdĭne campestri contra măre rubrum,* Deut. 1, 1.

feld-wurma *the plant wild marjoram.* v. felt-wurma.

feld-wyrt, e; *f. Field-wort, gentian;* gentiāna :—Feldwyrt *gentiāna,* Wrt. Voc. 68, 7. Herba gentiāna, ðæt ys feldwyrt *the herb gentiāna, that is, field-wort,* Herb. cont. 17, 1; Lchdm. i. 12, 16. Ðeós wyrt ðe man gentiānam, and ôðrum naman feldwyrt nemneþ, heó biþ cenned on dûnum *this herb, which is called gentian, and by another name field-wort, is produced on downs,* Herb. 17, 1; Lchdm. i. 110, 2.

fele-ferþ? [sele = fela *many?] A kind of worm under blocks having many feet,* Som ; vermĭcŭla quædam multĭpĕda, Lye :—Feleferþ *centumpellio,* forte *centupĕda,* Ælfc. Gl. 17; Som. 58, 86; Wrt. Voc. 22, 4.

fêle-leás; *adj.* [fêlan *to feel] Devoid of feeling;* insensĭlis :—Biþ his lîf scæcen and he fêleleás *his life is departed and he devoid of feeling,* Exon. 87 b; Th. 329, 26; Vy. 40.

FELG, e; *f.* felge, an; *f.* A FELLY. *part of the circumference of a wheel;* canthus = κανθός, absis rŏtæ :—Ælces spācan biþ ôðer ende fæst on ðære nafe, ôðer on ðære felge *one end of every spoke is fixed in the nave, the other in the felly,* Bt. 39, 7; Fox 222, 3, 7, 10. Ða felga hangiaþ on ðam spācan *the fellies depend on the spokes,* 222, 13, 19, 21, 27. Neár ðám felgum *nearer to the fellies,* 222, 11. Felge [MS. felga] *canthus,* Ælfc. Gl. 2; Som. 55, 48; Wrt. Voc. 16, 21. Ðæt hweól hwerfþ ymbûton, and sió nafa, nêhst ðære eaxe, micle fæstlîcor and orsorglîcor ðonne ða felgan dôn *the wheel turns round, and the nave, being nearest to the axle-tree, goes much more firmly and more securely than the fellies do,* Bt. 39, 7; Fox 220, 30. [*Wyc.* felijs, felys *fellies :* Plat. falge, falge, f: Dut. velg, f : Ger. felge, f : M.H.Ger. vëlge, f : O. H. Ger. felga, f : Dan. fælge, m. f.]

felgan, ic felge, ðu filgst, filhst, he filgþ, filhþ, *pl.* felgaþ; *p.* fealg, fealh, *pl.* fulgon; *pp.* folgen *To stick to, betake oneself to, go* or *come under, below* or *beneath anything, to go into, enter a place, to undergo;* inhærēre, sŭbīre, inīre, intrāre :—Oþ he on fleáme fealh *until he betook himself to flight,* Ors. 4, 8; Bos. 89, 42. Hý ymb ða geatu feohtende wêron oþ hý ðærinne fulgon *they were fighting about the gates until they entered therein,* Chr. 755; Th. 87, 3, col. 1. Siððan inne fealh Grendles môdor *when Grendel's mother came in,* Beo. Th. 2567; B. 1281. He searonîþas fealh Eormenrîces *he underwent the guileful*

enmity of Ermanric, 2405; B. 1200. [*O. Sax.* bi-felhan *trūdĕre, mandāre, condĕre : Frs.* be-feljen : *O. Frs.* bifella : *Dut.* be-velen : *Ger.* be-fehlen *mandāre : M. H. Ger.* be-vëlhen *condĕre, mandāre : O. H. Ger.* fëlahan, felhan *condĕre : Goth.* filhan *to hide, bury : Icel.* fela : *Lat.* se-pĕlire *to hide, bury.*] DER. æt-felgan, be-, bi-, ge-, wið-. v. felan, feolan.

feligean; *p.* de; *pp.* ed *To follow;* sĕqui :—Uton gán and feligean fremdum godum *eāmus et sĕquāmur deos aliēnos,* Deut. 13, 2. v. fylgean.

fell, es; *n. A fell, skin;* pellis :—Fell *pellis,* Ælfc. Gr. 9, 28; Som. 11, 56: Wrt. Voc. 71, 18. Cealfes fell *vĭtŭli cŏrium,* Ex. 29, 14. v. fel *a skin.*

fell, es; *m. Ruin, death;* lapsus, ruīna :—Ðêh ðe fell curen synnigra cynn *though the race of sinners chose death,* Andr. Kmbl. 3217; An. 1611. - v. fyll.

fell; *adj. Fell, cruel, severe;* crūdēlis, Som. Ben. Lye. v. fel; *adj.*

fellan, fyllan; ic felle, ðu felest, felst, he feleþ, felþ, *pl.* fellaþ; *p.* felde, *pl.* feldon; *pp.* felled; *v. trans. To cause to fall, to fell, cut* or *throw down, strip off, destroy;* cædĕre, sternĕre, projicĕre, abjicĕre, dejicĕre, destruĕre :—Gefered ðær hit felde *borne where it was thrown down,* Exon. 109 b; Th. 419, 11; Rä. 38, 4. DER. a-fellan, be-. v. fyllan, feallan.

fellen; *adj.* [fel *skin] Made of skins;* pellĭceus :—Fellen gyrdel wæs ymbe his lendenu *ĕrat zōna pellĭcea circa lumbos ejus,* Mk. Bos. 1, 6. God worhte Adame and his wîfe fellene reáf and gescridde hí *fēcit Deus Adam et uxōri ejus tunĭcas pellĭceas et induit eos,* Gen. 3, 21. Fellen hæt *a hat made of skin, a felt hat;* gălērus *vel* pīleus, Ælfc. Gl. 18; Som. 58, 111; Wrt. Voc. 22, 26.

felle-wærc, es; *n. The falling sickness, epilepsy;* epĭlepsia = ἐπιληψία :—Ðæt deáh wið fellewærce *it is good for epilepsy,* L. M. 2, 1; Lchdm. ii. 178, 8. v. fylle-wærc.

fel-nys, -nyss, e; *f. Cruelty, fierceness;* crūdēlĭtas, Som. Ben. Lye.

fêlnyss, e; *f.* [fêlan *to feel] Feeling;* sensus :—Gærs and treówa lybbaþ bûtan fêlnysse ... nŷtenu lybbaþ and habbaþ fêlnysse bûtan gesceáde *grass and trees live without feeling ... beasts live and have feeling without reason,* Homl. Th. i. 302, 15, 16. DER. ge-fêlniss.

felo; *adj. Fell, baleful;* pern`iciosus. DER. eal-felo. v. fæle, fel; *adj.*

felofor, fealvor, es; *m. A species of water-fowl, the sultana-hen;* porphyrio = πορφυρίων :—Felofor *porphýrio,* Glos. Epnl. Recd. 161, 36.

felsan *to recompense;* expiāre, Som. Ben. Lye.

FELT, es; *m?* FELT; pannus *vel* lāna coactilis, impīlia, Som. Ben. Lye :—Felt *centrum?* vel *filtrum?* Ælfc. Gl. 21; Som. 59, 59; Wrt. Voc. 23, 20. [*Plat.* filt, *m : Dut.* vilt, *n : Ger.* filz *m. n. carded wool, felt : M. H. Ger.* vilz, *m. felt : O. H. Ger.* filz, *m : Dan.* filt, *m. f : Swed.* filt, *m.*]

felp *falls,* Bt. Met. Fox 5, 30; Met. 5, 15 ; *3rd pers. pres. of* feallan.

fel-tûn, es; *m. An enclosed place, garden, privy, dunghill;* secessus, latrīna, sterquilīnium :—Se wîsdom and ôðre cræftas licgaþ forsewene swâ swâ meox under feltûne *wisdom and other virtues lie despised like dirt on a dunghill,* Bt. 36, 1; Fox 172, 11. In feltûn *in secessu,* Mt. Kmbl. Lind. 15, 17: Mk. Skt. Lind. 7, 19. In feltûne oððe mixen *in sterquilīnium,* Lk. Skt. Lind. Rush. 14, 35.

felt-wurma, an; *m.* [felt = feald?] *The plant wild marjoram;* orīgănum, Som. Ben. Lye : Lchdm. Glos. vol. iii. p. 349, col. 2, 32.

felt-wyrt, e; *f. The plant mullein;* verbascum thapsus, Lin :—Ðeós wyrt, ðe man verbascum, and ôðrum naman feltwyrt nemneþ, biþ cenned on sandigum stôwum and on myxenum *this plant, which is named verbascum, and by another name mullein, is produced in sandy places and on dunghills,* Herb. 73, 1; Lchdm. i. 174, 19–21. Feltwyrt *avadonia?* Wrt. Voc. 79, 5.

fêmne, an; *f. A virgin, young woman;* virgo :—Fêmne sceal hire freónd gesêcan *the virgin shall seek her friend,* Menol. Fox 548; Gn. C. 44. v. fæmne.

FEN, fenn, fæn, fænn, es; *n. m. A* FEN, *marsh, mud, dirt;* pălus, lŭtum, līmus, sordes :—Ic fûlre eom ðonne ðis fen swearte *I am fouler than this swart fen,* Exon. 110 b; Th. 423, 33; Rä. 41, 31. Fenn *lŭtum,* Ælfc. Gr. 13; Som. 16, 6 : līmus, lŭtum, Ælfc. Gl. 57; Som. 67, 61; Wrt. Voc. 37, 48. Þyrs sceal on fenne gewunian *the spectre shall dwell in the fen,* Menol. Fox 545; Gn. C. 42 : Beo. Th. 2595; B. 1295. Se ðe môras heóld, fen and fæsten *who held the moors, the fen and fastness,* Beo. Th. 208; B. 104. Hió wyrcþ ðæt fenn ðe man hāteþ Meotedisc *it forms the fen which is called Mæotis,* Ors. 1, 1; Bos. 15, 19. He underfêhþ ðæt fenn ðara þweándra *he receives the dirt of the washers,* Past. 16, 5; Hat. MS. 21 b, 20. Is Élig ðæt land eall mid fenne and mid wæter ymbseald *est Elge pălūdĭbus circumdāta vel ăquis,* Bd. 4, 19; S. 590, 4. Is ðæt êglond fenne biworpen *the island is surrounded with a fen,* Exon. 100 b; Th. 380, 9; Rä. 1, 5. Fennas and môras *fens and moors,* Bt. 18, 1; Fox 62, 14. On ðám fennum *in pălūdĭbus,* Bd. 4, 19; S. 590, 5. Eall oþ ða fennas norþ *as far north as the fens,* Chr. 905; Erl. 98, 21: 1010; Erl. 143, 27. [*Piers P.* fen : *Wyc.* fen, fenne : *Laym.* fenne, uenne, *dat;* fenes, *pl: Scot.* fen : *Plat.* fenne : *Frs.* finne : *O. Frs.* fenne, fene : *Dut.* veen, *n : Kil.* ven, venne : *Ger.* fenne, *n : O. H. Ger.* fenna, fenni, *f : Goth.* fani, *n. mud, dirt : Icel.* fen, *n. a fen, quagmire.*]

fen-cerse, an; *f. Fen-cress, water-cress*; nasturtium officināle, Lin :—Wyl fencersan *boil water-cress*, L. M. 1, 8; Lchdm. ii. 52, 15: 1, 61; Lchdm. ii. 132, 5.

fencg = fēng *took*; *p. of* fôn, q. v.

fen-fearn, fen-fern, es; *n. The fen or water-fern, flowering fern, the herb christopher, osmund-royal*; osmunda rēgālis, Lin. salvia?—Fenfearn *salvia*, Ælfc. Gl. 42; Som. 64, 8; Wrt. Voc. 31, 19. v. fearn.

fen-fixas; *pl. m. Fen-fishes*; pălustres pisces, Som. Ben. Lye. v. fisc.

fen-freoðo; *indecl. f. Fen-asylum*; āsȳlum in pălūde :—He in fenfreoðo feorh alegde *he laid down his life in his fen-asylum*, Beo. Th. 1706; B. 851.

fen-fugelas; *pl. m. Fen-birds, Fen-fowl*; pălustres āves, Som. Ben. Lye. v. fugel.

feng, es; *m.* [fôn *to take*].   **I.** *a grasp, span, hug, embrace*; amplexus, captus :—Ic fâra feng feore gedîgde *from the grasp of foes I with life escaped*, Beo. Th. 1160; B. 578. Fȳres feng *the grasp of fire*, Salm. Kmbl. 707; Sal. 353.   **II.** *what is taken, booty*; captum, præda :—Hî feng woldon fôn *they would take the booty*, Chr. 1016; Th. 280, 30, col. 2: 33, col. 1. DER. an-feng, and-, fore-, ofer-, on-, to-, under-. v. fang.

fêng, *pl.* fêngon *took*, Beo. Th. 5970; B. 2989: Salm. Kmbl. 866; Sal. 432; *p. of* fôn.

fengel, es; *m. A prince*; princeps :—Wîsa fengel geatolîc gengde *the wise prince stately went*, Beo. Th. 2805; B. 1400. Snottra fengel *the sagacious prince*, Beo. Th. 2954; B. 1475: 4318; B. 2156. Hringa fengel *prince of rings*, 4680; B. 2345.

fen-gelád, es; *n. Fen-path*; pălustris via, pălus :—Hie warigeaþ frêcne fengelād *they inhabit the dangerous fen-path*, Beo. Th. 2722; B. 1359.

feng-net, -nett, es; *n. A net for catching*; retiacŭlum :—Feallaþ firenfulle on heora fengnettum *cădent in retiacŭlo ejus peccātōres*, Ps. Th. 140, 12.

fen-hlip, -hleoþ, es; *n.* [hliþ *a declivity, slope*] *A fen-slope, bank of a fen*; păluster clîvus, pălūdis rīpa :—Scolde Grendel fleón under fenhleoþu *Grendel must flee under the fen-slopes*, Beo. Th. 1645; B. 820.

fen-hôp, es; *n. A fen-heap or mound?* pălūdis agger?—He meahte fleón on fen-hôpu *he might flee to the fen-mounds*, Beo. Th. 1532; B. 764.

fénix, es; *m.* **I.** *the fabulous bird phœnix* = φοῖνιξ :—Fênix, swâ hätte ân fugel on Arabiscre þeóde, se leofaþ fîf hund geára, and æfter deápe eft arîst ge-edcucod, and se fugel getâcnaþ ûrne ærîst on ðam endenêhstan dæge *phœnix, so a bird in Arabia is called, which lives five hundred years, and after death rises again re-quickened, and the bird betokens our resurrection at the last day*, Ælfc. Gr. 9, 64; Som. 11, 56-58. Se fugel se is fênix hâten *the bird which is called phœnix*, Exon. 57 a; Th. 203, 19; Ph. 86. Fênix byrneþ *phœnix burns*, 59 a; Th. 213, 2; Ph. 218: 60 b; Th. 221, 26; Ph. 340.   **II.** *a genus of palms, the date tree or date palm*; phœnix dactylifĕra :—Ðær he heánne beám wunaþ ðone hâtaþ men fênix, of ðæs fugles noman *there it inhabits a lofty tree, which men call phœnix, from the bird's name*, Exon. 58 a; Th. 209, 21; Ph. 174.

fen-land, es; *n. Fen-land, marshy land*; pălustris terra :—Hî ealle Egypta awêston, bûtan ðæm fenlandum *they laid waste all Egypt, except the fen-lands*, Ors. 1, 10; Bos. 32, 26. He þurh ða fenland reów *he rowed through the fen-lands*, Guthl. 9; Gdwin. 50, 13.

fen-lîc; *adj. Fenlike, marshy, fenny*; păluster :—Fenlîc păluster, Ælfc. Gr. 9, 18; Som. 10, 4. Of ðam fenlîcum adelan *from the fenlike mud*, Homl. Th. ii. 472, 7. Betwyx ða fenlîcan gewrido ðæs wîdgillan wêstenes he âna ongan eardian *he began to dwell alone among the fenny thickets of the wide wilderness*, Guthl. 3; Gdwin. 22, 9.

fen-minte, an; *f. Fen-mint, water-mint*; silvestris menta, Lin :—Fenminte *fen-mint*, L. M. 1, 3; Lchdm. ii. 40, 8.

fenn *a fen, marsh, mud, dirt*, Past. 16, 5; Hat. MS. 21 b, 20: Ps. Spl. 17, 44. v. fen.

fennig, fenneg; *adj.* FENNY, *marshy, muddy, dirty*; pălustris, ulīgĭnōsus, lŭtōsus :—Fennig æcer ulīgĭnōsus ăger, Ælfc. Gl. 57; Som. 67, 70; Wrt. Voc. 37, 56. Gif sió hond biþ fennegu *if the hand is dirty*, Past. 13, 1; Hat. MS. 16 b, 8.

fenol *the herb fennel*; fênĭcŭlum, Wrt. Voc. 79, 8. v. finol.

fen-ȳce, an; *f.* [ȳce *a frog*] *A fen-frog*; pălūdis rāna :—Me is fenȳce fôre hreþre *a fen-frog is more rapid than I in its course*, Exon. 111 a; Th. 426, 9; Rä. 41, 71.

feó *for or with cattle or money*, Cd. 126; Th. 161, 2; Gen. 2659; Beo. Th. 2765; B. 1380; *dat. and instr. of* feoh.

feóde, *pl.* feódon *hated*, Ps. Th. 118, 163; *p. of* feón, feógan.

FEÓGAN, feógean, fiógan, feón, fión; *part.* feógende; ic feóge, he feógeþ, feóþ, *pl.* feógaþ, feógeaþ; *p.* feóde, *pl.* feódon, feódun, feódan *To hate, persecute*; ōdisse, ōdio häbĕre, infestāre :—Uton we firene feógan *let us hate crimes*, Exon. 98 a; Th. 366, 16; Reb. 13. He hî alȳsde of feógendra folmum *lībĕrāvit eos de mănu ōdientium*, Ps. Th. 105, 10. Ic uurihte wegas ealle feóge *omnem viam inīquam ōdio hăbui*, Ps. Th. 118, 128: 138, 19. Ða wēregan neát nales feógaþ frȳnd hiera *the brute animals hate not their friends*, Elen. Kmbl. 719; El. 360. Ðe me earwunga

ealle feógeaþ *qui ōdērunt me grātis*, Ps. Th. 68, 4: 73, 22. Ic feóde fācnes wyrcend *fäcientes prævärīcātiōnes ōdīvi*, Ps. Th. 100, 3: 118, 113. Hî Dryhtnes æ feódon *they hated the Lord's law*, Exon. 66 a; Th. 243, 21; Jul. 14: Elen. Kmbl. 711; El. 356. Ðe feódun sybbe *qui ōdērunt pācem*, Ps. Spl. C. 119, 6. Hî Godes tempel feódan *they hated God's temple*, Exon. 18 a; Th. 44, 27; Cri. 709. Ða ðe hine feódan *qui ōdērunt eum*, Ps. Th. 67, 1: 82, 2: 85, 16: 104, 21. Feógeaþ [fiógaþ MS. T.] yfel ōdīte mălum, Ps. Spl. C. 96, 10. [O. H. Ger. fiên: Goth. fiyan, fian: Icel. fjá *to hate*.]

feó-gȳtsung, e; *f. Money-desire or greed, avarice*; pĕcūniæ cŭpīdo, avāritia :—Ðæt he sceolde his treówe for feógȳtsunge and lufan forleósan *that he should lose his truth for desire and love of money*, Bd. 2, 12; S. 514, 40.

FEOH, fioh; *gen.* feós; *dat.* feó; *n.*   **I.** *cattle, living animals*; pĕcus, jūmenta :—Gif ðé become ōðres monnes giĕmeleás feoh [G and H] on hand *if the stray cattle of another man come to thy hand*, L. Alf. 42; Th. i. 54, 9. Feoh bûtan gewitte *the cattle without understanding*, Salm. Kmbl. 46; Sal. 23. Wiht seó ðæt feoh fēdeþ *a thing which feeds the cattle*, Exon. 109 a; Th. 416, 21; Rä. 35, 2. Ic sealde him gangende feoh *I gave him live stock [walking cattle]*, Cd. 129; Th. 164, 23; Gen. 2719.   **II.** *cattle being used in early times as a medium of exchange, hence Money, value, price, hire, stipend,* FEE, *reward*; pĕcūnia, merces :—Næbbe gé feoh on eówrum bígyrdlum *nōlīte possīdēre pĕcūniam in zōnis*, Mt. Bos. 10, 9. Se ðe his feoh to unrihtum wæstmsceatte ne syleþ *qui pĕcūniam suam nōn dĕdit ad ūsūram*, Ps. Th. 14, 6. Ðæt he him sealde wið feoh ðæt scræf *ut det illi spēluncam pĕcūnia*, Gen. 23, 9. Ic ðé ða fæhþe feó leánige *I will recompense thee for the strife with money*, Beo. Th. 2765; B. 1380.   **III.** *as property chiefly consisted of cattle, hence Goods, property, riches, wealth*; bŏna, dĭvĭtiæ, ŏpes :—His feoh onfôn fremde handa *dīripiant aliēni omnes dīvitias ejus*, Ps. Th. 108, 11. Ne wilniaþ nânes ōðres feós *wish for no other riches*, Bt. 14, 2; Fox 44, 22. We ðé feoh syllaþ *we will give thee wealth*, Cd. 130; Th. 165, 2; Gen. 2725: Ors. 2, 4; Bos. 43, 22. Þe Anglo-Saxon Rune ᚠ = f, the name of which letter in Anglo-Saxon is feoh *money, wealth*,—hence this Rune not only stands for the letter *f*, but for feoh *money*, as,—ᚠ [= feoh] byþ frófur fira gehwylcum *money is a consolation to every man*, Runic pm. 1; Kmbl. 339, 1; Hick. Thes. i. 135, 1. ᚠ [= feoh] on foldan *wealth on earth*, Exon. 19 b; Th. 50, 28; Cri. 808: Elen. Grm. 1270. [*Piers P.* fee: *Chauc.* fee: *Laym.* feoh, feo, *n.: Orm.* fe, fehh: *Plat.* vee, veih, *n. cattle: O. Sax.* fê, fio; *Hel.* fehu, *n.* pĕcus, ŏpes: *O. Frs.* fia, fya, *n.: Dut.* vee, *n.: Kil.* veech, vee pĕcus: *Ger.* vieh, *n.: M. H. Ger.* vihe, *n.: O. H. Ger.* fihu, *n.: Goth.* faihu, *n. cattle, goods: Dan.* fæ, *n.: Swed.* fä, *n.: Icel.* fé, *n. cattle, goods: Lat.* pĕcus, *n.: Lith.* pekus cattle: *Sansk.* paśu, *m. cattle.* 'The importance of cattle in a simple state of society early caused an intimate connection between the notion of cattle, and of money or wealth. Thus we have *Lat.* pĕcus *cattle*; pĕcūnia *money*; and *Goth.* faihu *cattle, possessions*, is identical with *O. H. Ger.* fihu, fehu; *Ger.* vieh *cattle*; *Icel.* fé *cattle, money*; *A. Sax.* feoh *cattle, riches, money, price, reward*,' Wgwd.] DER. cwic-feoh, hæðen-, woruld-.

FEOHAN, feón; *part.* feónde; *p.* feah, *pl.* fægon; *pp.* fegen *To rejoice, be glad, exult*; gaudēre, lætāri, exultāre :—Se feónde [MS. feond] gespearn fleótende hreáw *the exulting [raven] perched on the floating corpses*, Cd. 72; Th. 87, 11; Gen. 1447. [O. Sax. gi-fehôn *to make to rejoice: O. H. Ger.* gi-fëhan, gi-vëhan gaudēre.] DER. ge-feohan, -feáh.

feoh-bôt, fioh-bôt, e; *f. A pecuniary recompence*; nummāria compensātio :—Feohbôt arîseþ *a pecuniary recompence shall arise*, L. Eth. vi. 51; Th. i. 328, 4. Ðæt hî môston ðære fiohbôte [ðæra feohbôta MS. H.] onfôn *that they might receive the pecuniary recompence*, L. Alf. 49; Th. i. 58, 8.

feoh-ern, es; *n. A money-place, treasury*; gazophylacium = γαζοφυλάκιον, Som. Ben. Lye.

feoh-fang, es; *m. Fee-taking, taking a bribe*; pĕcūniæ acceptio :—For feohfange *for bribery*, L. C. S. 15; Th. i. 384, 8.

feoh-gafol, es; *n. Usury, a duty, tax*; ūsūra, Som. Ben. Lye.

feoh-georn; *adj. Desirous of money, avaricious, covetous*; avārus, Som. Ben. Lye.

feoh-gesteald, es; *n. Possession of riches*; dīvĭtiārum possessio :—Ne þorfton ða þegnas feohgestealda [MS. -gestealde] wēnan *the followers needed not expect possession of riches*, Exon. 75 b; Th. 283, 5; Jul. 685.

feoh-gestreón, es; *n. Treasure, riches*; thēsaurus = θησαυρός, dīvĭtiæ :—Næbbe ic feohgestreón *I have no riches*, Andr. Kmbl. 602; An. 301: Exon. 66 a; Th. 245, 10; Jul. 42. Elþeódig hafaþ mec bereáfod feohgestreóna *a stranger has bereaved me of my treasures*, Elen. Kmbl. 1818; El. 911: Salm. Kmbl. 64; Sal. 32: Exon. 67 a; Th. 248, 27; Jul. 102.

feoh-gîfre; *adj.* [gîfre *greedy*] *Greedy of money, avaricious, covetous*; pĕcūniæ ăvĭdus, ăvārus :—Wita sceal ne tó feohgîfre *the sagacious must not be too greedy of money*, Exon. 77 b; Th. 290, 21; Wand. 68.

feoh-gift, -gyft, e; *f. A money-gift, precious gift*; pĕcūniæ dōnum,

vel ˜largĭtio, prĕtĭōsum dōnum :—Fromum feohgiftum *with bounteous money-gifts,* Beo. Th. 41; B. 21. Nó he ꝺære feohgyfte scamigan þorfte *he needed not feel shame at the precious gift,* 2055; B. 1025. Æt feohgyftum *with money-gifts,* 2182; B. 1089.

feoh-gítsere, es; *m. A miser; pĕcūniæ ăvārus :*—Eálá! hwæt se forma feohgítsere wære on worulde alas! *that the first miser should have been in the world,* Bt. Met. Fox 8, 110; Met. 8, 55. Ꝺæm feohgítsere *to the miser,* Bt. 7, 4; Fox 22, 26.

feoh-gyrnes, -ness, e; *f. Money-desire, avarice;* avārĭtia, L. Ath. Lye.

feoh-gýtsung *desire of money, avarice.* v. feó-gýtsung.

feoh-hof, es; *n. A treasury;* ærārium, Som. Ben. Lye.

feoh-hord, es; *m. A money-hoard;* ærārium, Cot. 212.

feoh-hús, es; *n. A treasure-house;* ærārium, Ælfc. Gl. 108; Som. 78, 104; Wrt. Voc. 58, 19.

feoh-lǽnung, e; *f. Money-lending, mortgage;* fenĕrātio :—Feohlǽnung būtan borge *hypothēca* [= ὑποθήκη], Ælfc. Gl. 14; Som. 58, 14; Wrt. Voc. 14, 9.

feoh-leás, *adj. Moneyless, priceless; pĕcūniæ ĭnops, sine prĕtio :*—Ꝺa ꝺe feohleáse wæron him scipu begëton *they who were moneyless got themselves ships,* Chr. 897; Erl. 94, 27. Ꝺæt wæs feohleás gefeoht *that was a priceless fight,* Beo. Th. 4873; B. 2441.

feoh-leásnes, -ness, e; *f. Poverty; pĕcūniæ ĭnŏpia,* paupertas, Som. Ben. Lye.

feoh-sceat, -sceatt, es; *n. Money-tribute, wages;* trĭbūtum, merces :—Nó ic wiꝺ feohsceattum ofer folc bere Drihtnes dōmas *I bear not the Lord's decrees among nations for wages,* Cd. 212; Th. 262, 14; Dan. 744.

feoh-spillung, -spilling, e; *f. Money-wasting, profusion; pĕcūniārum* effūsio *vel* profūsio :—Man ꝺǽr ne gespǽdde būtan manmyrringe and feohspillinge *man gained naught there except loss of men and waste of money,* Chr. 1096; Erl. 233, 30.

feoh-strang, *adj. Money-strong, possessing cattle or money; pĕcuārius, pĕcūniōsus :*—Feohstrang man *pĕcuārius,* Ælfc. Gl. 58; Som. 67, 112; Wrt. Voc. 38, 35. Feohstrang *pĕcūniōsus,* 88; Som. 74, 71; Wrt. Voc. 50, 51.

feoht, es; *n. A* FIGHT, *battle;* pugna, prœlium :—Wæs he þencende ꝺæt he ꝺæt feoht forlēte *he was thinking that he would give up the fight,* Bd. 3, 14; S. 539, 39. God tǽceþ handa mīne to feohte *Deus dŏcet mănus meas ad prœlium,* Ps. Spl. 143, 1. [*Laym.* fæht, faht: *Scot.* fecht, facht: *O. Sax.* fehta, *f: Frs.* fjuecht: *O. Frs.* fiucht: *Dut.* ge-vecht, *n: Ger.* ge-fecht, *n: M. H. Ger.* vëhte, *f: O. H. Ger.* fehta, *f.*] DER. ge-feoht, inge-, ofer-, ūtge-.

FEOHTAN; *part.* feohtende; ic feohte, ꝺú feohtest, he feohteþ, fiht, *pl.* feohtaþ; *p.* ic, he feaht, ꝺú fuhte, *pl.* fuhton; *pp.* fohten *To* FIGHT, *contend, make war, combat, struggle;* prœliāri, pugnāre, bellāre, contendĕre, decertāre, collīdĕre :—Mec mīn freá feohtan hāteþ *my lord commands me to fight,* Exon. 102 b; Th. 389, 10; Rä. 7, 5: 104 b; Th. 398, 2; Rä. 17, 1. Gyf hwylc cyning wyle faran and feohtan agēn ōꝺerne cyning *quis rex itūrus committĕre bellum adversus ălium rēgem,* Lk. Bos. 14, 31. Ealle on ꝺone cining feohtende wæron *all were fighting against the king,* Chr. 755; Erl. 49, 35: 994; Erl. 133, 11. Ic feohte *prœlior,* Ælfc. Gr. 25; Som. 27, 7. Feohteþ se feónd *the fiend fights,* Salm. Kmbl. 995; Sal. 499: L. Eth. vii. 15; Th. i. 332, 14: L. C. S. 60; Th. i. 408, 12. Drihten fiht for eów *Dŏmĭnus pugnābit pro vōbis,* Ex. 14, 14: Wrt. Voc. 78, 1. Monige synd, ꝺe to me feohtaþ *multi qui bellant me,* Ps. Th. 55, 3: 56, 1. Cūþwulf feaht wiꝺ Bretwalas *Cuthwulf fought against the Brito-Welsh,* Chr. 571; Erl. 18, 12: 661; Erl. 35, 9: 871; Erl. 75, 19. Ꝺa litlingas fuhton on hire innoþe *collīdēbantur in ūtĕro ejus parvŭli,* Gen. 25, 22. Stuf and Wihtgár fuhton [fuhtun, Erl. 14, 22] wiꝺ Bryttas *Stuf and Wihtgar fought against the Britons,* Chr. 514; Erl. 15, 23. Wítodlíce mīne þegnas fuhton *ministri mei utique decertārent,* Jn. Bos. 18, 36. Be ꝺon ꝺe mon on cynges healle feohte *in case a man fight in the king's hall,* L. Alf. pol. 7; Th. i. 66, 7: 39; Th. i. 88, 2. Ꝺeáh him feohtan on firas monige *although many men fight against it,* Runic pm. 26; Kmbl. 344, 27; Hick. Thes. i. 135, 52. [*Piers P.* fighten : *Laym.* fæhten, fahten : *Orm.* fihhtenn : *Scot.* fecht : *O. Sax.* fehtan : *Frs.* fjuechten : *O. Frs.* fiuchta : *Dut.* vechten : *Ger.* fechten : *M. H. Ger.* vehten : *O. H. Ger.* fehtan : *Dan.* fegte, fægte : *Swed.* fäkta.] DER. a-feohtan, æt-, be-, bi-, ge-, ofer-, on-, wiꝺ-.

feohte, an; *f. A fight, combat;* pugna :—Wearþ him seó feohte tó grim *the fight was too severe for them,* Exon. 84 a; Th. 317, 16; Môd. 66. Nó ic gefrægn heardran feohtan *I have not heard of a harder fight,* Beo. Th. 1157; B. 576: Exon. 102 b; Th. 388, 7; Rä. 6, 4: Andr. Kmbl. 2045; An. 1025. We ꝺæt ellenweorc feohtan fremedon *we have achieved that valourous deed by fighting,* Beo. Th. 1922; B. 959.

feohtere, es; *m. A fighter, warrior;* pugnātor, bellātor, Ben. Lye.

feoht-lác, es; *n. A fighting, fight;* pugna :—Gif ciricgriþ abrocen beó, bētan man georne, sí hit þurh feohtlác, sí hit þurh reáflác *if church-peace be broken, be it through fighting, be it through robbery, let amends be strictly made,* L. Eth. ix. 4; Th. i. 340, 20: L. C. E. 3; Th. i. 360, 11: L. C. S. 48; Th. i. 402, 28.

feoht-wíte *a fine for fighting.* v. fyht-wíte.

feól *fell,* Beo. Th. 1549; B. 772; *p. of* feallan.

FEÓL, e; *f. A* FILE; līma :—Ic eom láf fýres and feóle *I am the leaving of fire and file,* Exon. 126 a; Th. 484, 7; Rä. 70, 4. Mín heáfod is homere geþuren, sworfen feóle *my head is beaten with a hammer, rubbed with a file,* 129 b; Th. 497, 18; Rä. 87, 2. [*Prompt. Parv.* file : *Dut.* vijl, *f: Ger.* feile, *f: M. H. Ger.* vîle, *f: O. H. Ger.* fîhala, fîla, *f: Dan.* fiil, *m. f: Swed.* fil, *m: Icel.* þél, *f. a file.*]

feola *many,* Bd. 5, 19; S. 637, 15. v. fela.

feolan, fiolan, felan; *p.* fæl, *pl.* fǽlon, fēlon; *pp.* folen, feolen. **I.** *to cleave, stick, adhere;* adhærēre :—Ꝺæt ic in ne fele *ut non inhæream,* Ps. Surt. 68, 15. **II.** *to reach, come, pass;* procēdĕre, pervēnīre :—Ne meahton hí ofer mere feolan *they could not pass over the sea,* Exon. 106 a; Th. 404, 10; Rä. 23, 5. DER. æt-feolan, be-, bi-, ge-, geond-.

feóld, *pl.* feóldon *folded up,* Ælfc. Gr. 24; Som. 25, 50: Exon. 107 a; Th. 408, 4; Rä. 27, 7; *p. of* fealdan.

feól-heard, *adj. File-hard, hard like a file;* instar līmæ dūrus :—Hí lēton of folman feólhearde speru *they let the file-hard spears from their hands,* Byrht. Th. 134, 63; By. 108.

feó-lif? [feó = feoh?] *Munificence, bounty;* munĭfĭcentia, D. Som. Ben. Lye.

feóll *fell,* Beo. Th. 5830; B. 2919; *p. of* feallan.

feóllon *fell,* Beo. Th. 2089; B. 1042; *p. pl. of* feallan.

feolo *many,* Cd. 222; Th. 290, 26; Sat. 421. v. fela.

feolu-fór, e; *f? A field-fare;* turdus pĭlāris?—Feolufór *torax?* Wrt. Voc. 289, 17. v. feala-fór.

feon, feonn, es; *m. A fen;* pălus :—Geond ꝺa feonnas *about the fens,* Chr. 1010; Erl. 143, 29: 656; Erl. 31, 10, 26. v. fen.

feón, he feóþ; *p.* feóde, *pl.* feódon *To hate;* ōdisse :—He feóþ sáwle his *ōdit ănĭmam suam,* Ps. Spl. C. 10, 6: Cd. 43; Th. 56, 13; Gen. 911: Exon. 31 a; Th. 97, 31; Cri. 1599. Ic unrihta gehwylc feóde *inĭquĭtātem ōdio hăbui,* Ps. Th. 118, 163. Hie ꝺē feódon *they hated thee,* Elen. Kmbl. 711; El. 356. v. feógan.

feón *to rejoice, be glad.* v. feohan, ge-feohn.

feónd, fiónd, fýnd, fiénd, es; *pl. nom. acc.* feóndas, fýnd, feónd; *gen.* feónda; *dat.* feóndum; *m.* [feógan, feón *to hate*] *A* FIEND, *enemy, foe, the devil;* ōsor, inĭmīcus, hostis, diabŏlus = διάβολος :—Seó ydelnes is ꝺære sáwle feónd *idleness is the soul's enemy,* L. E. I. 3; Th. ii. 404, 9. Ēhteþ feónd sáwle mīne *persĕquātur inĭmicus anĭmam meam,* Ps. Spl. 7, 5. Se feónd his diórlingas duguþum stēpte *the fiend decked his favourites with honours,* Bt. Met. Fox 15, 14; Met. 15, 7. Se fýnd, 1455; B. 725: 1500; B. 748. Feónd *hostis vel ōsor,* Wrt. Voc. 86, 45. Se feónd mid his gefērum eallum feóllon of heofnum *the devil with all his company fell from heaven,* Cd. 16; Th. 20, 10; Gen. 306: Salm. Kmbl. 140; Sal. 69: 995; Sal. 499. Nã fægnian fýnd mín ofer me *non gaudēbit inĭmīcus meus sŭper me,* Ps. Spl. 40, 12. Stearcheort onfand feóndes fótlǽst *the stout of heart found the foe's foot-trace,* Beo. Th. 4567; B. 2289. Gif ꝺū gemēte ꝺīnes feóndes oxan ōꝺe assan, lǽd hine to him *si occurrĕris bŏvi inĭmici tui aut asĭno erranti, reduc ad eum,* Ex. 23, 4: Lk. Bos. 10, 19. Se ꝺæm feónde ætwand *he escaped from the fiend,* Beo. Th. 289; B. 143: Bt. Met. Fox 25, 31; Met. 25, 16. Ꝺū feónd oferswīꝺest *thou shalt overcome thy foe,* Elen. Kmbl. 186; El. 93: Cd. 144; Th. 179, 21; Exod. 32. Ꝺū fiónd geflǽmdest *thou didst put the enemy [the devil] to flight,* Hy. 8, 25; Hy. Grn. ii. 290, 25. Genāmon me ꝺǽr strange feóndas *strong enemies took me there,* Rood Kmbl. 60; Kr. 30: 65; Kr. 33. Fýnd syndon eówere *they are your enemies,* Judth. 11; Thw. 24, 18; Jud. 195: 12; Thw. 26, 10; Jud. 320. Eówre fýnd feallaþ befóran eów *cădent inĭmīci vestri in conspectu vestro,* Lev. 26, 8: 16; Deut. 32, 31. Ꝺíne feónd fǽcne forwurdan *inĭmici tui sonāvērunt,* Ps. Th. 82, 2: 91, 8. Hý fæder ageaf on feónda geweald *her father delivered her up into her foes' power,* Exon. 68 a; Th. 252, 7; Jul. 159: Elen. Kmbl. 135; El. 68. Ic agilde wrace mīnum feóndum *reddam ultiōnem hostĭbus meis,* Deut. 32, 41, 43: Jos. 10, 25. Ealle ic mihte feóndas gefyllan *I might have felled all his foes,* Rood Kmbl. 75; Kr. 38. Ꝺū swutole mihtest tocnāwan ꝺíne frínd and ꝺíne fýnd [fiénd Cot.] *thou mightest clearly distinguish thy friends and thy foes,* Bt. 20; Fox 72. 21. Lufiaþ eówre fýnd *dilĭgĭte inĭmicos vestros,* Mt. Bos. 5, 44: Lk. Bos. 6, 27, 35. Hió ofer heora feónd fæste getrymede *confirmāvit eum sŭper inĭmicos ejus,* Ps. Th. 104, 20: 107, 12. Ne murnþ naūꝺer ne friénd ne fiénd *he regards neither friend nor foe,* Bt. 37, 1; Fox 186, 8. Wæs wera eꝺelland geondsended feóndum *the people's native land was overspread with enemies,* Cd. 92; Th. 118, 22; Gen. 1969. [*Piers P.* fend: *Wyc.* fend, feend: *Chauc.* feend: *Laym.* feond, ueond, *m: Orm.* feend: *Plat.* fijend, fijnd, *m: O. Sax.* fiond, fiund, fiunt, fiand: *Frs.* fynne: *O. Frs.* fiand, fiund, *m: Dut.* vijand, *m: Ger.* feind, *m: M. H. Ger.* vîant, vîent, vînt, *m: O. H. Ger.* fiant, fîent, *m: Goth.* fiyands, *m: Dan. Swed.* fiende, *m: Icel.* fjándi, *m.*] DER. eald-feónd, þeód-: ge-fýnd.

feónd-ǽt, es; *m. Eating of the sacrifice to an idol;* diabŏlĭca mandŭcātio :—Hí ꝺæs feondǽtes Finees awerede *Phinehas restrained them from eating of the sacrifice to an idol,* Ps. Th. 105, 24, notes, p. 445.

feónd-gráp, e; f. A hostile grasp; hostīlis arreptio :—Ðæt ic ānunga eówra leóda willan geworhte, oððe on wæl crunge, feóndgrápum fæst that I alone would work your people's will, or bow in death, fast in hostile grasps, Beo. Th. 1276; B. 636.

feónd-gyld, es; n. Devil-worship, sacrifice to devils, idolatry, an idol; diǎbǒli cultus, diabǒlicum sacrificium, idōlatria, idōlum :—Ðā he on ðam folce feóndgyld gebræc when he destroyed idolatry amongst the people, Ps. Th. 105, 24.

feóndlíc; adj. Fiendlike, hostile; hostīlis, hostīcus :—Feóndlíc hostīcus vel hostīlis, Ælfc. Gl. 84; Som. 73, 95; Wrt. Voc. 49, 3.

feóndlíce; adv. Hostilely; hostīlĭter :—Hyre þurh yrre ageaf andsware fæder feóndlíce her father in anger gave answer hostilely, Exon. 67 b; Th. 249, 27; Jul. 118.

feónd-rǽden, e; f. [rǣden a condition] Fiend-condition, enmity; inǐmici conditio, inǐmicĭtia :—Ic sette feóndrǽdene betweox ðé and ðam wífe inǐmicĭtias pōnam inter te et mǔliĕrem, Gen. 3, 15.

feónd-rǽs, es; m. A fiendish violence; hostīlis impĕtus :—Ic feóndrǽs gefremede, fǣhþe geworhte I committed fiendish violence, wrought enmity, Cd. 42; Th. 55, 26; Gen. 900.

feónd-sceaða, -scaða, an; m. A fiend-enemy, dire enemy, robber; hostis nǒcivus, latro :—Slóh ðone feóndsceaðan fágum mēce she [Judith] slew the dire enemy [Holofernes] with a blood-stained sword, Judth. 10; Thw. 23, 4; Jud. 104. Me to grunde teáh fáh feóndscaða a hostile foe drew me to the ground, Beo. Th. 1112; B. 554. Ic sceal forstolen hreddan, flýman feóndsceaðan I shall rescue the stolen, make the robber flee, Exon. 104 a; Th. 396, 5; Rä. 15, 19.

feónd-scipe, -scype, es; m. Fiendship, enmity; inǐmicĭtia, hostīlĭtas :—Ðæt ys se feóndscipe that is the enmity, Beo. Th. 5991; B. 2999; Exon. 95 a; Th. 354, 60; Reim. 68. For feóndscipe ðæs gemynegodan cyninges propter inǐmicĭtias mĕmǒrāti rēgis, Bd. 4, 13; S. 581, 42: Cd. 128; Th. 163, 1; Gen. 2691: Ps. Th. 105, 30. He Rǣdwaldes feónd-scipe fleáh he fled from the enmity of Rædwald, Bd. 3, 18; S. 545, 40, col. 2: Cd. 29; Th. 38, 21; Gen. 610: Exon. 122 a; Th. 468, 5; Phar. 3: Elen. Kmbl. 711; El. 356. Hí feóndscype rǣrdon they raised enmity, Exon. 66 a; Th. 243, 22; Jul. 14: Exon. 14 b; Th. 30, 28; Cri. 486. Fleónde Rǣdwaldes feóndscypas inǐmicĭtias Redualdi fŭgiens, Bd. 3, 18; S. 545, 38, col. 1.

feónd-seóc; adj. Fiend-sick, demoniac; dæmǒniǎcus :—Ðætte seó ylce eorþe mihte to hæle feóndseócra manna and ōðra untrumnyssa ut ipsa terra ad ǎbigendos ex obsessis corpǒribus dæmǒnes grǎtiæ salutāris hǎbēret effectum, Bd. 3, 11; S. 535, 35.

feónd-seócnes, -ness, e; f. Fiend-sickness, demonology; dæmǒniǎcus morbus, Som. Ben. Lye.

feóndulf? [feónd a fiend, ulf = wulf a wolf?] A fiend, enemy, rascal, scoundrel; furcīfer :—Feóndulf furcīfer, furca dignus, Glos. Prudent. Recd. 146, 82.

feóng, e; f. Hatred; ǒdium, Bd. 3, 11; S. 535, note 20. v. feóung.

feor; adj. Perverse, depraved; prǎvus :—Mid feorum lífe by a perverse life, Bd. 5, 13; S. 633, note 33. v. þweor.

FEOR, feorr, fíor; comp. fyrr, fyr, fíer; sup. fyrrest; adv.   I. FAR, at a distance; prǒcul, longe :—Ðā wǣron ðás wundru feor and wíde gemǣrsode and gecýðed quibus pătĕfactis ac diffāmātis longe lātēque mīrācŭlis, Bd. 3, 10; S. 535, 2 : 3, 16; S. 542, 16. Hyra heorte is feor [feorr, Mt. Bos. 15, 8] fram me cor eōrum longe est a me, Mk. Bos. 7, 6 : Bt. Met. Fox 24, 4 ; Met. 24, 2. Ðá gyt ðá he wæs feor his fæder, he hyne geseah when he was yet far from his father, he saw him, Lk. Bos. 15, 20. Nōht feor úrum mynstre non longe a monastērio nostro, Bd. 5, 4 ; S. 617, 5 : Cd. 50 ; Th. 63, 28 ; Gen. 1039. Feor and neáh far and near, Exon. 13 b ; Th. 24, 25 ; Cri. 390 : Cd. 143 ; Th. 177, 27 ; Exod. 1 : Beo. Th. 2447 ; B. 1221 : Andr. Kmbl. 1276 ; An. 638. We witan heonan nōht feor ōðer eálond nǒvimus insŭlam ǎliam esse non prǒcul a nostra, Bd. 1, 1 ; S. 474, 15 : Beo. Th. 3615 ; B. 1805. Feor ðú dydest cúþan míne fram me longe fēcisti nōtos meos a me, Ps. Lamb. 87, 9. Hit feor on ōðre wísan wæs it was far otherwise ; longe ǎlĭter ērat, Bd. 3, 14 ; S. 539, 44.   II. beyond, moreover ; ultra, porro :—Ge feor hafaþ fǣhþe gestǣled and moreover she hath set up a deadly feud, Beo. Th. 2684 ; B. 1340. [Piers P. Chauc. fer : R. Glouc. Wyc. fer, ferr : Laym. feor, fer, ueor, feorre : Orm. feorr : Plat. feere, fere afar : O. Sax. fer : Frs. fier : O. Frs. fír, fer : Dut. ver, verre : Ger. fern : M. H. Ger. vërre : O. H. Ger. fer : Goth. fairra : Dan. fiern : Swed. fjerran : Icel. fjarri far off : Lat. porro : Grk. πόρρω : Sansk. pra forth, away.] DER. un-feor.

feor, feorr ; comp. m. fyrra, firra ; f. n. fyrre, firre ; adj. Far, distant, remote ; longinquus, remōtus :—Feorres folclondes of a far country, Exon. 115 b ; Th. 444, 14 ; Kl. 47. Hér is gefered ofer feorne weg æðelinga sum innan ceastre here a noble is come from a long way off into the city, Andr. Kmbl. 2348 ; An. 1175 : 382 ; An. 191 : 504 ; An. 252.

feora of souls or beings, Exon. 38 a ; Th. 126, 7 ; Gú. 367 : Cd. 161 ; Th. 202, 7 ; Exod. 384 ; gen. pl. of feorh.

feoran ; p. feorude To remove afar off ; elongāre :—Ic feorude elongāvi, Ps. Spl. C. 54, 7. v. feorran.

---

feor-búend, es ; m. One dwelling far off ; prǒcul habǐtātor :—Nú ge feorbúend, mínne gehýraþ ánfealdne geþoht now ye far-dwellers, hear my simple thought, Beo. Th. 514 ; B. 254.

feor-cumen ; part. Come from afar ; perĕgrīnus, perĕger ventus :—Feorcumen [MS. feorcuman] man a far-come man, a foreigner, L. In. 20 ; Th. i. 114, 15, note 30, MS. B.

feor-cund, feorr-cund ; adj. Come from afar ; perĕgrīnus :—Gif feorcund mon, oððe fremde, bútan wege geond wudu gonge, and ne hríeme ne horn bláwe, for þeóf he biþ to prōfianne, oððe to sleánne oððe to aliésanne if a far-come man, or a stranger, journey through a wood out of the highway, and neither shout nor blow his horn, he is to be held for a thief, either to be slain or redeemed, L. In. 20 ; Th. i. 114, 15–116, 2.

feor-cýþ, -cýþþ, e ; f. A far country ; remōta terra :—Feorcýþþe beóþ sēlran gesōhte far countries are better [when] sought, Beo. Th. 3681, note ; B. 1838.

feord an army, force, expedition, Chr. 1066 ; Erl. 203, 11 : 1140 ; Erl. 265, 8. v. fyrd.

feordian ; p. ode ; pp. od To be at war ; bellum gĕrĕre :—Hí feordodan wið Ætlan Húna cininge they were at war with Ætla king of the Huns, Chr. 443 ; Erl. 11, 35. v. fyrdian.

feording military service, Chr. 675 ; Erl. 38, 2, note 6. v. fyrding.

feore to, for or with life, Exon. 39 a ; Th. 128, 32 ; Beo. Th. 1161 ; B. 578 ; dat. and inst. of feorh.

feores of life, Exon. 30 b ; Th. 95, 32 ; Cri. 1566 ; gen. of feorh.

feorg life, soul, spirit, Exon. 82 b ; Th. 311, 19 ; Seef. 94 : 104 a ; Th. 394, 14 ; Rä. 14, 3. v. feorh.

feorg-bold, es ; n. The dwelling of life, the body ; ǎnīmæ dǒmus, corpus :—Hrǽw cōlode, fæger feorgbold the corpse grew cold, the fair dwelling of life, Rood Kmbl. 145 ; Kr. 73.

feorg-bona, an ; m. A life-destroyer ; vītæ interfector :—He him feorgbona weorþeþ he becomes a life-destroyer to him, Exon. 97 a ; Th. 362, 24 ; Wal. 41. v. feorh-bana.

feorg-gedál, es ; n. Life-separation, death ; vītæ divortium, mors :—Siððan líc and leomu and ðes lífes gǣst asundrien somwist hyra þurh feorg-gedál when body and limbs and this life's spirit sunder their fellowship through death, Exon. 50 a ; Th. 172, 29 ; Gú. 1151. v. feorh-gedál.

FEORH, feorg, fiorh, ferh, fyorh ; gen. feores ; dat. inst. feore ; pl. nom. acc. feorh ; gen. feora ; dat. inst. feorum ; n. m.   I. life, soul, spirit ; vīta, ǎnīma :—Nǣniges mannes feorh to lore wearþ no man's life was lost, Bd. 4, 21 ; S. 590, 23 : Beo. Th. 2425 ; B. 1210 : Ps. Th. 106, 4. Nō wæs feorh æðelinges flǣsce bewunden the prince's soul was not surrounded with flesh, Beo. Th. 4839 ; B. 2424 : Exon. 103 a ; Th. 391, 9 ; Rä. 10, 2. Ðonne him ðæt feorg losaþ when his life perishes, 82 b ; Th. 311, 19 ; Seef. 94. Ne biþ him feores wēn there will be no hope of his life, L. M. 2, 51 ; Lchdm. ii. 264, 19 : Bd. 5, 3 ; S. 616, 8 : Bt. 14, 3 ; Fox 46, 27 : Exon. 115 b ; Th. 445, 4 ; Dóm. 2 : Cd. 162 ; Th. 203, 15 ; Exod. 404. Feores aþolian to endure life, Exon. 27 a ; Th. 81, 7 ; Cri. 1320. Feores berǣdan to deprive of life, Andr. Kmbl. 266 ; An. 133. Feores getwǣfan to separate from life, Beo. Th. 2871 ; B. 1433. Feores geunnan to grant life, L. Eth. ix. 1 ; Th. i. 340, 8 : L. C. E. 2 ; Th. i. 358, 26 : Andr. Kmbl. 358 ; An. 179. Feores ongildan to give up or sacrifice one's life, Andr. Kmbl. 2204 ; An. 1103. Feores onsæcan to make an attempt against one's life, Beo. Th. 3889 ; B. 1942. Feores onsēcan to bereave of life, Exon. 75 b ; Th. 283, 13 ; Jul. 679. Feores orwéna hopeless of life, Beo. Th. 329, 27 ; Vy. 40 : Andr. Kmbl. 2216 ; An. 1109. Feores rēcan to care for life, Byrht. Th. 139, 27 ; By. 260. Feores scyldig guilty of life, liable in one's life, L. Alf. pol. 4 ; Th. i. 64, 1 : L. Ath. i. 4, 6 ; Th. i. 202, 3, 12 : v. § 1, 4 ; Th. i. 230, 6 : L. Eth. iii. 16 ; Th. i. 298, 14 : v. 30 ; Th. i. 312, 6 : vi. 37 ; Th. i. 324, 17 : L. C. S. 58 ; Th. i. 408, 4. Feores þolian to forfeit life, L. C. S. 78 ; Th. i. 420, 10. Feores unnan to grant life, Exon. 68 b ; Th. 254, 3 ; Jul. 191. Feores unwyrðe unworthy of life, 30 b ; Th. 95, 27 ; Cri. 1563. Feores wyrðe worthy of life, L. Ath. iv. 4 ; Th. i. 224, 3. Ðæt man forgá þýfþe be his feore that a man forego theft by his life, L. Ath. i. 20 ; Th. i. 210, 3 : Exon. 105 b ; Th. 401, 28 ; Rä. 21, 18 : Beo. Th. 3690 ; B. 1843 : Ps. Th. 54, 24. Beorh ðínum feore salva ǎnīmam tuam, Gen. 19, 17 : Cd. 89 ; Th. 110, 14 ; Gen. 1838 : Beo. Th. 2590 ; B. 1293 ; Byrht. Th. 137, 31 ; By. 194 : Elen. Kmbl. 268 ; El. 134 : Andr. Kmbl. 3075 ; An. 1540. Á to feore for evermore, Exon. 32 b ; Th. 102, 25 ; Cri. 1678. Ǽfre to feore, Ps. Th. 118, 165 : Exon. 111 a ; Th. 425, 33 ; Rä. 41, 65. Áwa to feore, Ps. Th. 51, 8. Lange to feore, Ps. Th. 132, 4. Syððan to feore in æternum, 54, 22 : 101, 25 : 106, 8. To wídan feore for ever, Cd. 170 ; Th. 213, 5 ; Exod. 547 : Exon. 11 a ; Th. 15, 3 ; Cri. 230 : Beo. Th. 1871 ; B. 933 : Andr. Kmbl. 211 ; An. 106 : Elen. Kmbl. 421 ; El. 211 : Ps. Th. 71, 17. Hæbbe his feorh let him have his life, L. In. 5 ; Th. i. 104, 14 : L. Ath. v. § 1, 4 ; Th. i. 230, 7 : L. Edg. ii. 7 ; Th. i. 268, 24 : L. C. S. 26 ; Th. i. 392, 3 : Ors. 2, 5 ; Bos. 48, 23 : Chr. 937 ; Erl. 114, 2 ; Æðelst. 36. Ymb cyninges feorh sierwian to plot against the king's life, L. Alf. pol. 4 ; Th. i. 62, 15. Ðú ðín feorh hafast thou

hast thy life, Beo. Th. 3703 ; B. 1849 : Cd. 116 ; Th. 151, 17 ; Gen. 2510 : Andr. Kmbl. 1908 ; An. 956 : Exon. 47 b ; Th. 164, 10 ; Gú. 1009. Ðær he eardaþ ealne wídan feorh *where he shall dwell for evermore*, 14 a ; Th. 27, 31 ; Cri. 439. He mín feorg freoðaþ *he will protect my life*, 36 a ; Th. 116, 28 ; Gú. 214 : Apstls. Kmbl. 116 ; Áp. 58. He sylfes feore beágas bohte *he has bought rings with his own life*, Beo. Th. 6019 ; B. 3013 : Exon. 106 b ; Th. 406, 9 ; Rä. 24, 14. Hí bǽdan hiora feorum fóddurgeafe *pĕtĕrent escas anĭmābus suis*, Ps. Th. 77, 20 : Cd. 184 ; Th. 229, 32 ; Dan. 226 : Beo. Th. 147 ; B. 73. Freónda feorum *with the lives of friends*, Beo. Th. 2616 ; B. 1306. **II.** *a living being, person* ; hŏmo, persōna :—Ða yldestan Chus and Cham hátene wǽron, fulfreólíce feorh, frumbearn Chames *the eldest were called Cush and Canaan, most liberal beings*, Ham's *firstborn*, Cd. 79 ; Th. 97, 25 ; Gen. 1618. Feónda feorh feóllon þicce *the bodies of the foes fell thickly*, 95 ; Th. 124, 19 ; Gen. 2065. Feora fǽsl *offspring of the living*, 67 ; Th. 80, 17 ; Gen. 1330 : 67 ; Th. 81, 9 ; Gen. 1342 : 161 ; Th. 200, 23 ; Exod. 361 : 161 ; Th. 202, 7 ; Exod. 384. Ðæt is sárlíc ðæt swá fæger feorh sceolan ágan þýstra ealdor *it is grievous that the prince of darkness should own such beautiful beings*, Bd. 2, 1 ; S. 501, 15. [O. Sax. ferah, ferh, *n. life, soul* : Ger. ferch, *n. vīta, sanguis* : M. H. Ger. vërch, *n. life* : O. H. Ger. fërah, ferh, *n. ănĭma, vīta* : Goth. fairhwus *world* : Icel. fjör, *n. life*.] DER. geógoþ-feorh, geóguþ-, wíde-.

feorh-ádl, e ; *f. A mortal disease, fatal sickness* ; fatālis morbus :—Biþ his feorhádl getenge *his fatal sickness is near*, L. M. 3, 22 ; Lchdm. ii. 320, 20. Heródes lǽfde fíf suna, þrý he hét acwellan on his feorhádle, ǽrðan ðe he gewíte *Herod left five sons, three he commanded to be slain in his last illness, ere he departed*, Homl. Th. i. 478, 13.

feorh-bana, -bona, feorg-bona, an ; *m. A life-destroyer, murderer* ; vítæ interfector, hŏmĭcīda :—Ðú Abele wurde tó feorhbanan *thou hast been for a life-destroyer to Abel*, Cd. 48 ; Th. 62, 26 ; Gen. 1020. Hí gesáwon feorhbanan fuglas slítan *they saw birds tearing the murderers*, 96 ; Th. 125, 32 ; Gen. 2088. He ne meahte on ðam feorhbonan fǽhþe gebétan *he might not avenge the feud on the murderer*, Beo. Th. 4921 ; B. 2465.

feorh-bealo, -bealu ; *gen.* -bealowes, -bealuwes ; *n. Life-bale, mortal affliction, deadly evil* ; vítæ mǎlum, lētāle mǎlum :—Gúþdeáþ fornam, feorhbealo frécne, fyra gehwylcne leóda mínra *war-death, a cruel life-bale, has taken every man of my people*, Beo. Th. 4492 ; B. 2250. Ic me ðæt feorhbealo feor aswápe *I sweep that deadly evil far from me*, Exon. 106 b ; Th. 405, 20 ; Rä. 24, 5 : Beo. Th. 314 ; B. 156. Ðǽr wæs hondscíó, feorhbealu fǽgum *there was [his] glove, deadly evil to the fated*, 4160 ; B. 2077 : 5067 ; B. 2537.

feorh-ben, -benn, e ; *f.* [ben *a wound*] *A life-wound, mortal wound* ; lētāle vulnus :—Feorhbennum seóc *sick with mortal wounds*, Beo. Th. 5473 ; B. 2740.

feorh-berende ; *part. Life-bearing, living* ; vītam fĕrens, vīvens :—Heó wile gesécan æghwylcne feorhberendra *it will seek each of those bearing life*, Exon. 110 a ; Th. 420, 19 ; Rä. 40, 6 : Cd. 92 ; Th. 117, 17 ; Gen. 1955.

feorh-bold *the dwelling of life, the body*. v. feorg-bold.

feorh-bona *a life-destroyer, murderer*, Beo. Th. 4921 ; B. 2465. v. feorh-bana.

feorh-cwalu, ferh-cwalu, e ; *f. Life-slaughter, death* ; vítæ cædes, mors :—Æfter feorhcwale *after death*, Exon. 97 b ; Th. 364, 27 ; Wal. 77. He sóhte hú he sárlícast, þurh ða wyrrestan wítu, meahte feorhcwale findan *he sought how he could invent a death most painfully, through the worst torments*, 74 a ; Th. 276, 28 ; Jul. 573.

feorh-cwealm, es ; *m. A mortal pang, death, slaughter* ; mors, cædes :—Ne þearft ðú ðé ondrǽdan deáþes brógan, feorhcwealm nú giet *thou needest not dread the pain of death, the mortal pang as yet*, Cd. 50 ; Th. 63, 26 ; Gen. 1038. Deáh him feónda hlóþ feorhcwealm bude *though the band of fiends threatened death to him*, Exon. 46 a ; Th. 157, 6 ; Gú. 887. Mín sceal golden wurþan feorhcwealm *my slaughter shall be requited*, Cd. 55 ; Th. 67, 19 ; Gen. 1103.

feorh-cyn, -cynn, es ; *n. Living kind* ; vīventium gĕnus :—Bealocwealm hafaþ fela feorhcynna forþ onsended *pernicious death has sent forth many living kinds*, Beo. Th. 4524 ; B. 2266 : Exon. 89 a ; Th. 334, 10 ; Gn. Ex. 14.

feorh-dæg, es ; *pl. nom. acc.* -dagas ; *gen.* -daga ; *dat.* -dagum ; *m. A life-day* ; vítæ dies :—Ðæt Ismael feorhdaga on woruldríce worn gebíde *that Ishmael may abide many life-days in the world*, Cd. 107 ; Th. 142, 8 ; Gen. 2358.

feorh-dolh, -dolg, es ; *n. A life-wound, deadly wound* ; lētāle vulnus :—Geseóþ nú ða feorhdolg ðe gefremedon ǽr on mínum folmum *see now the deadly wounds which they ere inflicted on my palms*, Exon. 29 a ; Th. 89, 10 ; Cri. 1455.

feorh-eácen ; *part. Endued with life, living* ; vītā auctus, vīvens :—Feorheáceno cynn inc hýraþ eall *all races endued with life shall obey you two*, Cd. 10 ; Th. 13, 17 ; Gen. 204.

feorh-gebeorh ; *gen.* -gebeorges ; *n. Life's security, refuge* ; vítæ servātio, refūgium :—He gelǽdde ofer lagustreámas máþmhorda mǽst on

feorhgebeorh *he led the greatest of store-houses over the water-streams for refuge*, Cd. 161 ; Th. 201, 8 ; Exod. 369.

feorh-gedál, feorg-gedál, es ; *n. Life-separation, death* ; vítæ divortium, mors :—Sceal feorhgedál æfter wyrþan *death must afterwards take place*, Andr. Kmbl. 362 ; An. 181 : 2854 ; An. 1429 : Exon. 50 a ; Th. 174, 5 ; Gú. 1173.

feorh-gener, es ; *n. Life-safety, salvation of life* ; vītæ servātio :—Búton se cyningc him feorhgeneres unne *unless the king grant him salvation of life*, L. Edg. ii. 7 ; Th. i. 268, 25.

feorh-geníþla, an ; *m. A life-enemy, deadly foe* ; qui vítæ insĭdiātur, lētālis hostis :—He brægd feorhgeníþlan, ðæt heó on flet gebeáh *he dragged the deadly foe, that she bowed on the place*, Beo. Th. 3084 ; B. 1540 : 5859 ; B. 2933.

feorh-gifa, -giefa, an ; *m. Giver of life* ; vítæ dător :—Me onsende sigedryhten mín, folca feorhgiefa, gǽst háligne *my glorious Lord, Giver of life to people, sent a holy spirit to me*, Exon. 50 b ; Th. 176, 20 ; Gú. 1213. Geségon on heáhsetle heofones waldend, folca feorhgiefan *they saw on his throne heaven's Ruler, Giver of life to nations*, 15 b ; Th. 35, 10 ; Cri. 556.

feorh-gifu, -giefu, e ; *f. The gift of life* ; vítæ dōnum :—Secgas feorhgiefe gefégon *men rejoiced in the gift of life*, Exon. 94 a ; Th. 353, 1 ; Reim. 6.

feorh-góma, an ; *m.* [góma *the gums, jaws*] *Fatal* or *deadly jaws* ; fatāles fauces :—Se deópa seáþ mid wíta fela, frécnum feorhgómum, folcum scendeþ *the deep pit [hell] afflicts people with many torments, with rugged fatal jaws*, Exon. 30 b ; Th. 94, 32 ; Cri. 1549.

feorh-hord, es ; *n. Life's treasure, the soul, spirit* ; vítæ thēsaurus, ănĭma :—Líf biþ on síþe, fǽges feorhhord *life is on its journey, the spirit of the fated*, Exon. 59 a ; Th. 213, 7 ; Ph. 221. Hád wereþ feorhhord feóndum *armour defends the soul from foes*, Wald. 100 ; Vald. 2, 22 : Exon. 49 b ; Th. 170, 26 ; Gú. 1117 : Andr. Kmbl. 2365 ; An. 1184.

feorh-hús, es ; *n. Life's house, spirit's house, the body* ; vítæ vel ănĭmæ dŏmus, corpus :—Gár oft þurhwód fǽges feorhhús *the dart often pierced the body of the fated*, Byrht. Th. 140, 32 ; By. 297.

feorh-hyrde, es ; *m. Life-guardian* or *protector* ; vítæ custos vel protector :—He hine 'bæd ðæt he him feorhhyrde wǽre *he prayed that he would be his life-protector*, Bd. 2, 12 ; S. 513, 5 : Hy. 9, 8 ; Hy. Grn. ii. 291, 8.

feorh-lást, es ; *m. A life-step, step taken to preserve one's life, flight* ; vítæ vestigium, gressus vítæ servandæ causā lātus, fŭga :—He onweg ðanon on nicera mere, fǽge and geflýmed, feorhlástas bær *he bore his life-steps away thence to the monsters' mere, death-doomed and put to flight*, Beo. Th. 1697 ; B. 846.

feorh-leán, es ; *n. Life's reward* or *gift* ; vítæ præmium :—Woldon hie ðæt feorhleán fácne gyldan *they would requite life's gift with fraud*, Cd. 149 ; Th. 187, 12 ; Exod. 150.

feorh-lege, es ; *m.* [lege = leg, lagu *law*] *Life-law, fate, death* ; vítæ lex, fātum, mors :—Ðæt on ðone hálgan handa sendan to feorhlege fæderas usse *that our fathers lay their hands on the holy one unto death*, Elen. Kmbl. 913 ; El. 458. Ic on máþma hord mínne bebohte feorhlege *I have bought my fate for treasures' hoard*, Beo. Th. 5592 ; B. 2800.

feorh-líf, es ; *n. Life* ; vīta :—On ðínre gesihþe ne biþ sóþfæst ǽnig, ðe on ðisse foldan feorhlíf bereþ *non justĭfĭcābĭtur in conspectu tuo omnis vivens*, Ps. Th. 142, 2.

feorh-loca, an ; *m. Life's inclosure, the breast* ; ănĭmæ claustrum, pectus :—Eom ic, in mínum feorhlocan, breóstum, inbryrded to ðam betran hám *I am, in my life's inclosure, in my breast, impelled to the better home*, Exon. 42 a ; Th. 141, 11 ; Gú. 625.

feorh-lyre, es ; *m. Loss of life* ; vítæ perdítio :—Gif feorhlyre wurþe *if there be loss of life*, L. E. B. 3 ; Th. ii. 240, 14.

feorh-ner, -nere, es ; *n. Life's preservation* or *salvation, a refuge, sustenance, nourishment, food* ; vítæ servātio, refūgium, ălĭmentum, cĭbus :—Monigfealde sind gód ðe us dǽleþ to feorhnere Fæder ælmihtig. *manifold are the goods which the Father almighty distributes to us for life's preservation*, Exon. 96 b ; Th. 359, 33 ; Pa. 72 : 16 b ; Th. 38, 21 ; Cri. 610. Ðe worhte weoroda Dryhten to feorhnere fira cynne *which the Lord of hosts wrought for salvation to the race of men*, Elen. Kmbl. 1792 ; El. 898 : Cd. 190 ; Th. 237, 18 ; Dan. 339. Hí nó ðonan lǽtaþ on gefeán faran to feorhnere *they will not let them go thence in joy to a refuge*, Exon. 31 a ; Th. 97, 28 ; Cri. 1597. Fuglas heora feorhnere on ðæs beámes blédum náme [= námon] *birds took their refuge on the tree's branches*, Cd. 200 ; Th. 248, 3 ; Dan. 507. Hwílum him to honda, hungre geþreátad, fleág fugla cyn, ðǽr hý feorhnere fundon *sometimes the race of birds, forced by hunger, flew to his hands, where they found sustenance*, Exon. 46 a ; Th. 157, 10 ; Gú. 889. Beóþ Godes streámas góde wætere fæste gefylde, ðanan feorhnere findaþ foldbúend *flūmen Dei replētum est ăqua, părasti cĭbum illōrum*, Ps. Th. 64, 10.

feorh-rǽd, es ; *m. Life-benefit, an action tending to the soul's benefit* ; id quod vítæ prodest, actio ad ănĭmæ sălūtem tendens :—Ðæt hie feorhrǽd fremedon *that they should do what would benefit their souls*, Andr. Kmbl. 3306 ; An. 1656.

feorh-scyldig; *adj. Life-guilty, liable in one's life;* vītæ reus, morte dignus :—Gif feorhscyldig man cyning gesôhte *if a man who had forfeited his life sought the king,* L. Eth. vii. 4 ; Th. i. 330, 10. Se ðe ofslehþ man binnan ciricwagum, he biþ feorhscyldig *he who slays a man within church-walls, he is liable in his life,* vii. 13, 15 ; Th. i. 332, 8, 14.

feorh-seóc; *adj. Life-sick, mortally wounded;* letāliter vulnĕrātus :— Scolde Grendel ðonan feorhseóc fleón *Grendel must flee thence mortally wounded,* Beo. Th. 1644; B. 820.

feorh-sweng, es ; *m. A life-blow, deadly blow;* lētālis ictus :—Hond feorhsweng ne ofteah *his hand withdrew not the deadly blow,* Beo. Th. 4972 ; B. 2489.

feorh-þearf, e ; *f. Distress of life, urgent need;* vītæ necessĭtas :— Drihten me hraðe gefultuma æt feorhþearfe *Dŏmĭne ad adjŭvandum me festīna,* Ps. Th. 69, 1.

feorh-wund, e ; *f. A life-wound, mortal wound;* lētāle vulnus :—He ðær feorhwunde hleát *he sank there with a mortal wound,* Beo. Th. 4760; B. 2385.

feorlen; *adj. Far off, distant, remote;* longinquus :—Se gingra sunu ferde wræclíce on feorlen ríce *adolescentior fĭlius pĕregre profectus est in rĕgiōnem longinquam,* Lk. Bos. 15, 13. v. fyrlen.

feor-lond, es ; *n. A far country, distant land;* remōta terra :—Feorlondum on *in distant lands,* Exon. 95 b ; Th. 356, 12 ; Pa. 10.

FEORM, fiorm, fyrm, e ; *f.*   I. *food, provision, goods, substance;* victus, substantia, bŏna :—Nô ðū ymb mīnes ne þearft līces feorme leng sorgian *thou needest not longer care about my body's food,* Beo. Th. 906; B. 451. Hí bærndon and awēston ðæs cynges feorme hāmas [MS. hames] *they burnt and laid waste the king's provision-homes* [or *farms*], Chr. 1087; Erl. 224, 13. Twegra daga feorme *provision for two days;* firmam duōrum diērum, Th. Diplm. A. D. 950; 501, 23; 504, 14 : Chr. 777; Erl. 55, 10. Gewât him mid cnósle, ofer Caldēa folc feran mid feorme, fæder Abrahames *the father of Abraham departed with his family, with his goods, to travel over the Chaldeans' nation,* Cd. 83; Th. 104, 6; Gen. 1731 : 126; Th. 161, 2; Gen. 2659. Gewiton him eástan æhta lædan, feoh and feorme *they departed from the east leading their possessions, cattle and substance,* Cd. 80 ; Th. 99, 22 ; Gen. 1650.   II. *an entertaining, entertainment, feast;* hospĭtālĭtas, convīvĭum, cœna :—Gif mon cierliscne monnan fliéman feorme teó *if a man accuse a churlish man of the entertaining of a fugitive,* L. In. 30 ; Th. i. 120, 16. Ân dæl bisceope and his hîrêde for feorme and onfangenysse gesta and cumena *ūna portio episcŏpo et fămĭliæ propter hospĭtālĭtātem atque susceptiōnem,* Bd. 1, 27 ; S. 489, 7. Ðætte ælþeódige bisceopas sỹn þoncfulle heora gæstlíþnesse and feorme *ut episcŏpi peregrīni contenti sint hospĭtālĭtā'tis mŭnĕre oblāto,* 4, 5 ; S. 573, 3. To ðære ēcan feorme *to the eternal feast,* Homl. Th. ii. 372, 5. He gegearwode mycele feorme *magnam cœnam fēcit,* Mk. Bos. 6, 21 : Lk. Bos. 14, 12, 16 : Homl. Th. ii. 370, 31 : 372, 1, 3.   III. *a place where provisions are kept, provision-quarters of an army;* victus stătio :—Se here eódan him to heora gearwan feorme ût þuruh Hamtûnscîre into Bearrucscîre to Readingon *the army went to their ready provision-quarters out through Hampshire into Berkshire to Reading,* Chr. 1006 ; Th. 256, 20–22, col. 1.   IV. *use, benefit, profit, enjoyment;* ūsus, fructus :—Ða swiðe lytle feorme [fiorme MS. Hat.] ðara bôca wiston, forðæmðe hi heora nān wuht ongietan ne meahton *they got very little benefit from the books, because they could not understand anything of them,* Past. pref ; Cot. MS. [Chauc. farme *meal* : Laym. feorme, veorme *feast.*]   DER. bēn-feorm, bend-, cyning-, éaster-, eástor-, gyt-, swíþ-, winter-: or-feorme.

feorma; *adj. First;* prīmus :—Ða feorman men *the first men,* Exon. 73 a; Th. 272, 15; Jul. 499. v. forma.

feormend-leás; *adj. Wanting a polisher;* pŏlītōre cărens :—Geseah he orcas stondan, fyrnmanna fatu, feormendleáse, ðær wæs helm monig eald and ômig *he saw bowls standing, vessels of men of yore, wanting a polisher, there was many a helmet, old and rusty,* Beo. Th. 5516, note; B. 2761. v. feormynd.

feormere, es ; *m. One who supplies with food, a purveyor,* FARMER ; obsōnātor :—Se ðe mā manna [MS. manne] inlæde ðonne he sceole, bûton ðæs stîwerdes leáfe and ðæra feormera, gylde his ingang *he who introduces more men than he should, without leave of the steward and of the purveyors, let him forfeit his admission,* Cod. Dipl. 942 ; Kmbl. iv. 278, 19–21.

feorm-fultum, es ; *m. Food-support, purveyance;* victus auxilium, commeātus, prōcūrātio :—Ðæt him nān man ne þearf to feormfultume nān þingc syllan, bûton he sylf wille *that no man need give him anything as purveyance, unless he himself will,* L. C. S. 70 ; Th. i. 412, 22.

feormian; *† art.* feormende; *p.* ode, ade; *pp.* od, *v.a.* [feorm *food*].   I. *to supply with food, feed, support, sustain, entertain, receive as a guest, cherish, benefit, profit;* victum suppĕdĭtāre, epŭlāre, suscĭpĕre, suscĭpĕre hospĭtio, fŏvēre, cūrāre, vălēre :—Ðæt ic [cyning] bebeóde eallum mīnan gerēfan ðæt hi on mīnan āgenan rihtlíce tilian, and me mid ðam feormian; and ðæt him nān man ne þearf to feormfultume nān þingc syllan, bûton he sylf wille *that I [the king] command all my reeves that they justly provide on my own, and feed [supply with food, maintain] me*

therewith; and that no man need give them anything as purveyance [*food-support*], unless he himself be willing, L. C. S. 70 ; Th. i. 412, 22. Feorma, mihtig Dryhten, mīnre sāwle *mighty Lord, sustain my soul,* Exon. 118 b ; Th. 454, 33 ; Hy. 4, 42. Âh he feormendra lyt lifgendra *he has few of entertainers living,* Exon. 87 b ; Th. 329, 7 ; Vy. 30. Ðæt se, ðe hine feormode, and se, ðe gefeormod wæs, sỹn hí begen bisceopes dôme scyldig *that he, who entertained him, and he, who was entertained, be both guilty to the bishop's doom,* Bd. 4, 5 ; S. 572, 44. Feorma mec hwæðre, ðeáh ðe ic fremede mā gylta *yet cherish me, though I have committed more crimes,* Exon. 118 a ; Th. 453, 36 ; Hy. 4, 25. Feorma ðū in ðīnum ferþe gôd *cherish thou good in thy soul,* Exon. 80 b ; Th. 303, 10 ; Fä. 51 : Ps. Th. 77, 69. Forðon hí ongeáton ðætte seó hālwende onsægedness to ēcre alỹsnesse swîþrade and feormade ge líchoman and sāwle *for they understood that the wholesome sacrifice availed and profited* [vălēret] *to the eternal redemption both of body and of soul,* Bd. 4, 22 ; Whel. 318, 25–27.   II. *to feed on, devour, consume;* vesci, comĕdĕre, consūmĕre :—Fealo líg feormaþ *the yellow flame consumes and burns up the Phœnix,* Exon. 59 a ; Th. 213, 1 ; Ph. 218.   III. *to cleanse,* FARM *or cleanse out;* mundāre, purgāre, expiāre :—He feormaþ his bernes flôre *he will cleanse the floor of his barn,* Lk. Bos. 3, 17 ; purgābit āream suam, Vulg. He feormaþ ælc ðara, ðe blǣda byrþ, ðæt hyt bere blǣda ðe swîðor *omnem, qui fert fructum, purgābit eum, ut fructum plus afférat,* Jn. Bos. 15, 2. Seofon dagas ðū feormast ðæt weofod, Ex. 29, 37; *seuen daies thow shalt clense the auter,* Wyc ; septem diēbus expiābis altāre, Vulg.   DER. a-feormian, ge-.

feorm-riht, es ; *n. Right in an estate;* in prædio jus, Heming, p. 50, Mann.

feormþ, e ; *f. A harbouring, an entertaining, a cleansing;* susceptio, hospĭtium, purgātio. v. fyrmþ.

feormung, e ; *f.*   I. *a harbouring, an entertaining;* susceptio, hospĭtium :—Þurh wreccena feormunge *by the harbouring of exiles,* L. Alf. pol. 4 ; Th. i. 62, 16.   II. *a cleansing, polishing;* purgātio, pŏlītio :—Gif sweordhwíta ôðres monnes wǣpn to feormunge onfô *if a sword-polisher receive ano'her man's weapon for polishing,* L. Alf. pol. 19 ; Th. i. 74, 9.   DER. a-feormung, niht-.

feormynd [= feormend], es ; *m.* [feormian III. *to cleanse*] *A cleanser, furbisher, polisher;* purgātor, pŏlītor :—Feormynd swefaþ, ða ðe beadogrímman býwan sceoldon *the polishers are dead, who should prepare the war-helmet,* Beo. Th. 4505, note ; B. 2256.

feornes, -nys, -ness, -nyss, e ; *f.* FARNESS, *distance;* longinquĭtas :—Gif mycel feornys síþfætes betwihligeþ *si longinquĭtas itĭnĕris magna interjācet,* Bd. 1, 27 ; S. 491, 39.

feorr; *adj. Far, distant;* longinquus :—Ðeáh him mon feorr land gehēte *though a distant land was promised him,* Past. 50 ; Hat. MS. : Andr. Recd. 850 ; An. 423. v. feor ; *adj. far.*

feorr; *adv. Far, at a distance;* prŏcul, longe :—Hyra heorte is feorr fram me *cor eōrum longe est a me,* Mt. Bos. 15, 8. Hí feorr ætstôdon *de longe stĕtĕrunt,* Ps. Spl. 37, 12. Seó sunne gǣþ eall swā feorr adūne on nihtlícre tíde under ðære eorþan swā heó on dæg bufan up astîhþ *the sun goes quite as far down under the earth in the night time as it rises above it in the day,* Bd. de nat. rerum ; Wrt. popl. science 2, 22 ; Lchdm. iii. 234, 20. v. feor ; *adv.*

feorran, feorrane, feorren ; *adv. Afar, far off, at a distance, from far;* a longe, prŏcul, longe, e longinquo :—Ðær wǣron manega wîf feorran *ĕrant ĭbi mŭlĭĕres multæ a longe,* Mt. Bos. 27, 55 : Mk. Bos. 5, 6. Folgiaþ feorran ðære hālgan earce *follow at a distance from the holy ark,* Jos. 3, 3. Swîðe feorran ymbûton *very far about,* Bt. 39, 5 ; Fox 218, 11. Ic eom hider feorran gefered *I have journeyed hither from far,* Cd. 25 ; Th. 32, 4 ; Gen. 498 : Beo. Th. 728 ; B. 361 : Andr. Kmbl. 48 ; An. 24 : Elen. Kmbl. 1982 ; El. 993 : Rood Kmbl. 114; Kr. 57 : Salm. Kmbl. 357 ; Sal. 178 : Exon. 103 a ; Th. 389, 15 ; Rä. 7, 8 : Boutr. Scrd. 17, 11. Feorran and neán *from far and near,* Beo. Th. 1683 ; B. 839 : Exon. 60 b ; Th. 220, 26 ; Ph. 326 : Cd. 52 ; Th. 64, 8 ; Gen. 1047. Petrus hym fyligde feorrane *Petrus sequēbātur eum a longe,* Mt. Bos. 26, 58. Feorren, Cd. 89 ; Th. 110, 10 ; Gen. 1836.

feorran; *p.* de ; *pp.* ed *To remove to a distance, withdraw;* remŏvēre, elongāre :—Ne wolde feorhbealo feorran *he would not withdraw the mortal bale,* Beo. Th. 314; B. 156.   DER. a-feorran, of-.

feorran-cund; *adj. Having a distant origin, coming from afar;* e longinquo ortus :—Sôna him seleþegn, síþes wērgum, feorrancundum forþ wîsade *forthwith the hall-thane guided him forth, weary from his journey, coming from afar,* Beo. Th. 3594, note ; B. 1795. v. feor-cund.

feorren; *adv. From far;* e longinquo :—Uncer twega feorren cumenra *of us two come from far,* Cd. 89 ; Th. 110, 10 ; Gen. 1836. v. feorran ; *adv.*

feorsian, fyrsian ; *p.* ode ; *pp.* od *To go beyond, remove;* ultĕrius procēdĕre, elongāre :—Ðū meaht feorsian *thou mayest go beyond,* Bt. Met. Fox 24, 52 ; Met. 24, 26.   DER. a-feorsian, -fyrsian, afor-feorsian.

feor-studu, e ; *f. A slanting post?* obstipum, Som. Ben. Lye :— Feorstuðu *obstipum?* Wrt. Voc. 290, 11.

feorþ, es ; *n. The soul, spirit, life;* ănĭma, vīta :—Feorþ biþ on sîþe

his soul shall be on its journey, Exon. 87 b; Th. 328, 32; Vy. 26. v. ferþ.

feórþa, feówerþa; seó, đæt feórþe, feówerþe; *adj. The* FOURTH; quartus :—Wæs geworden æfen and mergen se feórþa dæg *the evening and morning were the fourth day*, Gen. 1, 19. Seó feórþe eá ys geháten Euſrates *flūvius quartus ipse est Euphrātes*, 2, 14. Hér bốc Boëties onginþ seó feórþe *here begins the fourth book of Boethius*, Bt. 35, 6; Fox 170, 24: 40, 4; Fox 240, 9. Đæt feórþe cyn *the fourth tribe*, Cd. 158; Th. 197, 20; Exod. 310. Feórþan *a ruler of a fourth part, tetrarch*; tetrarcha, Lk. Bos. 3, 1. On đære feórþan mǽgþe *generātiōne quarta*, Gen. 15, 16. Com se Hǽlend embe đone feórþan hancrẽd to him *Iẽsus quarta vigĭlia noctis vẽnit ad eos*, Mt. Bos. 14, 25. Đa folctogan feórþan síđe æđeling lǽddon to đam carcerne *the leaders of the people led the noble to the dungeon the fourth time*, Andr. Kmbl. 2915; An. 1460.

feórþes fót *four-footed;* quadrúpes :—Feórþes fót neát *a four-footed beast;* bestia quadrŭpes, Som. Ben. Lye.

feórþling, es; *m:* feórþung, e; *f. in Anglo-Saxon;* but *m. in Northumb.* v. *last example. A fourth part of a thing*, FARTHING; quadrans :—Đes feórþling ođđe feórþa [MS. feórþan] dæl þinges *hic quadrans*, Ælfc. Gr. 9, 37; Som. 12, 35. Ǽr đu agylde đone ŷtemestan feórþling [MS. feórþlingc] *dōnec reddas nŏvissĭmum quadrantem*, Mt. Bos. 5, 26: Lk. Bos. 12, 59. Geseah he sume earme wudewan bringan twegen feórþlingas *vidit quandam vidŭam paupercŭlam mittentem æra mĭnŭta duo*, Lk. Bos. 21, 2: Mk. Bos. 12, 42. Twegen [MS. tuoge] stycas, đæt is feórþung penninges *duo mĭnŭta, quod est quadrans*, Mk. Skt. Lind. 12, 42. Feórþungas, *acc. pl.* Lk. Skt. Lind. Rush. 21, 2.

feórþ-ríce, es; *n. Dominion over a fourth part;* tetrarchia = τετραρχία, Som. Ben. Lye.

feórþung, e; *f: but in Northumb. m. A fourth part, a farthing*, Mk. Skt. Lind. Rush. 12, 42. v. feórþling.

feorting, e; *f. Crĕpĭtus ventris* :—Feorting *pēdātio*, Ælfc. Gl. 79; Som. 72, 64; Wrt. Voc. 46, 22.

feor-weg, es; *m. A far or long way;* via longinqua :—Mín bigengea ʒewát bryce on feorweg *incŏlātus meus prolongātus est*, Ps. Th. 119, 5: Exon. 36 a; Th. 117, 22; Gú. 228. Drihten asent þeóda ofer eów of feorwegum *addūcet Dŏmĭnus sŭper te gentem de longinquo*, Deut. 28, 49: Beo. Th. 73; B. 37: Ps. Th. 67, 26. On feorwegas *in distant ways*, Andr. Kmbl. 1855; An. 930: Exon. 87 b; Th. 329, 1.

feorwit-georn; *adj. Curious, inquisitive;* cūriōsus, Som. Ben. Lye. v. firwet-georn.

feorwit-geornes, -ness, e; *f. Curiosity;* cūriōsĭtas, Som. Ben. Lye. v. firwet-geornes.

feós *of cattle, money*, or *wealth*, Ors. 2, 4; Bos. 43, 15: Chr. 999; Erl. 134, 36: Bt. 14, 2; Fox 44, 22; *gen.* of feoh.

feostnode *confirmed*, Chr. 656; Erl. 32, 22: 963; Erl. 121, 32, = fæstnode; *p.* of fæstnian.

feoter, feotur; *gen.* feotre, feoture; *f. A fetter;* compes :—Mið feotrum [Rush. feoturum] *compēdĭbus*, Mk. Skt. Lind. 5, 4. v. feter.

feóþ *shall hate*, Cd. 43; Th. 56, 13; Gen. 911. v. feón.

feođer-scẽte *four-cornered, square;* quadrangŭlus, quadrātus, Som. Ben. Lye. v. feówer-scŷte.

feotod, feotud *called for, fetched;* arcessītus, Som. Ben. Lye, = fetod; *pp.* of fetian.

feóung, fióung, feóng, e; *f. Hatred, enmity;* ŏdium, inĭmĭcĭtia :—His unriht and his feóung wurþ đeáh swíđe open *invĕnĭret inĭquĭtātem suam et ŏdium*, Ps. Th. 35, 2. Hí me setton feóunge for mínre lufan *pŏsuĕrunt ŏdium pro dilectiōne mea*, 108, 4. Hí ealdum feóungum [feóngum MS. B.] hine ẽhton *vĕtĕrānis eum ŏdiis insĕquĕbantur*, Bd. 3, 11; S. 535, 20. v. feógan, feón *to hate*.

FEÓWER, feówere; *nom. acc; gen.* feówera, feówra; *dat.* feówerum: *Sometimes used indecl.* FOUR; quátuor :—Wurdon feówer cyninges þegnas ofslægene *four king's thanes were slain*, Chr. 896; Erl. 94, 4: Cd. 75; Th. 93, 16; Gen. 1546: Ælfc. T. 25, 19, 20. Feówer síđon *four times;* quáter, Ælfc. Gr. 38; Som. 40, 67. Felamódigra feówer scoldon geferian to đæm goldsele Grendles heáfod *four of those much daring ones must convey Grendel's head to the gold-hall*, Beo. Th. 3279; B. 1637. Hwæt beóþ đa feówere fǽges rápas *what are the four ropes of the doomed man?* Salm. Kmbl. 663; Sal. 331: 667; Sal. 333. Þrittig wæs and feówere feores onsóhte wígena cynnes *there were thirty-four of the race of men bereft of life*, Exon. 75 b; Th. 283, 12; Jul. 679. Feówra sum *one of four*, L. Wih. 19; Th. i. 40, 17: 21; Th. i. 40, 21. Of đisum feówer bócum *of these four books*, Ælfc. T. 27, 17. From feówerum foldan sceátum *from the four corners of the world*, Exon. 20 b; Th. 55, 5; Cri. 879: Menol. Fox 419; Men. 211. Embe feówer wucan *after four weeks*, 30; Men. 15: 313; Men. 158. Ic sette feówer bẽc *I composed four books*, Bd. 5, 24; S. 647, 37. Sylle feówer scẽp for án *restĭtuet quátuor ŏves pro ūna ŏve*, Ex. 22, 1: Jn. Bos. 19, 23. Seó hæfde feówere fẽt under wombe *it had four feet under its belly*, Exon. 109 b; Th. 418, 10; Rä. 37, 3. [*Wyc.* foure: *Laym.* feour, feouwer, feowere, feor, fower, four: *Orm.* fowwerr, fowwre: *Plat.* veer: *O. Sax.* fiwar, fiuwar, fior: *Frs.* fjouver: *O. Frs.* fiuwer, fiower, fior: *Dut. Ger.*

M. H. Ger. vier: *O. H. Ger.* fior: *Goth.* fidwor: *Dan.* fire: *Swed.* fyre: *Icel.* fjórir: *Lat.* quátuor: *Grk.* τέσσαρες; *Æolic* πίσυρες: *Wel.* pedwar: *Lith.* keturì: *Sansk.* ćatur, ćatvāras.]

feówera; *gen. pl.* of feówer *four:* = feáwera; *gen. pl.* of feáwa *a few.*

feówer-feald; *adj.* FOURFOLD; quadruplus :—Gif ic ænigne bereáfode, ic hit be feówerfealdum agyfe *si quid alĭquem defraudāvi, reddo quadruplum*, Lk. Bos. 19, 8.

feówer-fealdan *to make fourfold;* quadruplĭcāre, Som. Ben. Lye.

feówer-fẽte, fíówer-fẽte, tiér-fẽte, fiđer-fẽte, fyđer-fẽte, -fóte, -fóte; *adj. Four-footed;* quadrŭpes :—Se ælmihtiga God eallum mancinne forgeaf đa feówerfẽtan deór *the almighty God gave to all mankind the four-footed beasts*, Ælfc. T. 8, 26. Ælces cynnes feówerfẽtes feós án *one of each kind of four-footed cattle*, 2, 4; Bos. 43, 15. Hí sceoldon [MS. sceoldan] bringan feówerfẽtes twá hwíte *of four-footed [cattle] they must bring two white*, 2, 4; Bos. 43, 8. Eádbyrht bisceop, feówerfóttra nýtena đone tẽđan dǽl, to þearfum syllan wolde *bishop Eadbert would give the tenth part of his four-footed cattle to the poor*, Bd. 4, 29; S. 608, 17. v. flox-fóte, feówer-scŷte.

feówer-gild, es; *n. A fourfold payment or compensation;* quadruplex compensātio :—Ælc tihtbýsig man gilde feówergilde *let every man of bad repute pay with fourfold compensation*, L. Eth. iii. 3; Th. i. 294, 10.

feówer-scŷte, fyđer-scŷte, fiđer-scŷte, -scíte, feđer-scíte, -sciţte, -scette; *adj.* [sceát *a corner*] *Four-cornered, quadrangular, square;* quadrangŭlus, quadrātus :—Seó burh is feówerscŷte *the city is quadrangular*, Ors. 2, 4; Bos. 44, 21.

feówertene *fourteen*, Mt. Kmbl. Rush. 1, 17. v. feówertyne.

feówerteóđa, *m;* seó, đæt, feówerteóđe, *f. n; adj. The fourteenth;* quartus dĕcĭmus :—Se wæs feówerteóđa fram Agusto đam Cásere *who was the fourteenth from Augustus Cæsar*, Bd. 1, 4; S. 475, 27. Đæs feówerteóđan dæges *of the fourteenth day*, Ex. 12, 18. On đam feówerteóđan dæge *quarta decima die*, Lev. 23, 5: Jos. 5, 10. Healdaþ đæt óþ đone feówerteóđan dæg đæs mónþes *servābĭtur usque ad quartam dĕcĭmam diem mensis hujus*, Ex. 12, 6.

feówerþa; seó, đæt feówerþe; *adj. The fourth;* quartus :—Is feówerþe lyft *the fourth is air*, Bt. Met. Fox 20, 122; Met. 20, 61. v. feórþa.

feówerþa-fæder [MS. feówerþe-fæder]; *indecl. in sing. A great-great-grandfather;* ăbāvus, Ælfc. Gl. 91; Som. 75, 12; Wrt. Voc. 51, 57.

feówerþe-mốder, *indecl. in sing;* but *dat. sing.* -mẽder; *pl. nom. acc.* -mốdra; *gen.* -mốdra; *dat.* -mốdrum; *f. A great-great-grandmother;* ăbāvia, Ælfc. Gl. 91; Som. 75, 13; Wrt. Voc. 51, 58.

feówertig; *gen.* feówertiges; *dat.* feówertigum, feówertig; *adj.* FORTY; quadrāginta :—Ne ofsleah ic híg, gif đær beóþ feówertig *non percūtiam propter quadrāginta*, Gen. 18, 29. Æfter đæra feówertigra daga getele *after the number of forty days*, Num. 14, 34. On feówertigum geárum *quadrāginta annis*, 14, 34: Jn. Bos. 2, 20. Hie begeton feówertig bearna *they begat forty [of] children*, Cd. 223; Th. 294, 22; Sat. 475: 228; Th. 306, 21; Sat. 667. Israhéla bearn æton heofonlícne mete feówertig wintra *fīlii Israel comĕdĕrunt Manna; n. [μάννα; n; אֶת־הַמָּן] quadrāginta annis*, Ex. 16, 34: Gen. 32, 15: 50, 3. Feówertig [feówertigum MS. B.] scillingum gebẽte *let him make amends with forty shillings*, L. Alf. pol. 10; Th. i. 68, 11.

feówertigeđa, feówertigođa; *m. n. -tigođe, f. n; adj. Fortieth;* quadrāgēsĭmus :—Feówertigeđa *quadrāgēsĭmus*, C. R. Ben. 25. On đam feówertigóđan [MS. feówerteóđan] geáre *in the fortieth year;* quadrāgēsĭmo anno, Deut. 1, 3.

feówertig-feald; *adj. Fortyfold;* quadrāgēnārius, Ælfc. Gr. 49; Som. 50, 19.

feówertig-líc; *adj. Of* or *belonging to forty;* quadrāgēnārius :—He bebeád đæt feówertiglíce fæsten healden beón *jejūnium quadrāginta diērum observāri prǽcēpit*, Bd. 3, 8; S. 531, 10. Ealle tíd đæs feówertiglícan fæstenes *tōtum quadrāgēsĭmæ tempus*, 3, 23; S. 554, 31.

feówertyne; *adj.* FOURTEEN; quătuordĕcim :—Feówertyne cneóressa *genĕratiōnes quătuordĕcim*, Mt. Bos. 1, 17. Cốmon feówertyne Geáta gongan *fourteen Goths came marching*, Beo. Th. 3287; B. 1641: Andr. Kmbl. 3185; An. 1595. Oþ-đæt feówertyne niht ofer Eástron *until fourteen nights after Easter*, L. In. 55; Th. i. 138, 8, MS. B. Rachel acende feówertyne suna *Rachel bore fourteen sons*, Gen. 46, 22.

feówra *of four*, L. Wih. 19; Th. i. 40, 17, = feówera; *gen. pl.* of feówer.

feówrþa, *m;* seó, đæt feówrþe; *adj. The fourth;* quartus :—Feowrþe is fýr *the fourth is fire*, Bt. 33, 4; Fox 128, 30. v. feórþa.

feówrtig; *adj. Forty;* quadrāginta :—Ceorliscum men feówrtigum scillingum gebẽte *cŏlōni quadrāginta sŏlĭdis emendet*, L. Alf. pol. 10; Wilk. 37, 23. v. feówertig.

feowung, e; *f.* [feohan *to rejoice] A rejoicing, an enjoying, glorying;* gaudium, glória, Hpt. Gl. 433; Leo A. Sax. Gl. 95, 10.

feówurtig; *adj. Forty;* quadraginta :—Đá đá he fæste feówurtig dagá and feówurtig nihta *cum jejūnasset quadraginta diēbus et quadraginta noctĭbus*, Mt. Bos. 4, 2. v. feówertig.

fer, es; *n.* I. *a going, journey;* ĭter :—Wið fere *juxta ĭter*, Ps. Spl. M. 139, 6. II. *a vessel, ship;* nāvis :—Wæs se sunu

Lamehes of fere acumen *the son of Lamech was come from the vessel* [=ark], Cd. 75; Th. 93, 12; Gen. 1544. v. fær; n.

fér, es; *m. A fever;* febris :—Wið ælces dæges fére *for an every day's fever,* L. M. cont. 1, 62; Lchdm. ii. 12, 28. v. fǽr, fefer.

fér, es; *m. Fear, terror;* tĭmor :—Mid fére foldbúende se micla dæg meahtan Dryhtnes bihlǽmeþ *the great day of the mighty Lord shall strike earth's inhabitants with fear,* Exon. 20 b; Th. 54, 13; Cri. 868. v. fǽr; m.

fera, an; *m. A companion;* sócius, Som. Ben. Lye. v. ge-fera.

feran, to ferenne ; *part.* ferende; *p.* ferde, *pl.* ferdon; *pp.* fered [fer a *journey*] *To go, make a journey, set out, travel, march, sail;* ire, iter fǎcěre, proficisci, transíre, migrāre, nǎvǐgāre :—He hine to cyninge feran hét *he called him to go to the king,* Bd. 3, 23; S. 554, 39; Cd. 109; Th. 144, 32; Gen. 2398: Exon. 28 b; Th. 86, 31; Cri. 1416: Beo. Th. 53; B. 27: Andr. Kmbl. 347; An. 174: Elen. Kmbl. 429; El. 215: Ps. Th. 118, 3: Bt. Met. Fox 4, 35; Met. 4, 18: Judth. 9; Thw. 21, 10; Jud. 12: Byrht. Th. 132, 64; By. 41. Ðá hí swá mycelne síþfæt feran sceoldan *when they must go so great a journey,* Bd. 3, 15; S. 541, 30: 1, 23; S. 485, 38. He on morne feran wolde *he wished to set out in the morning,* Bd. 2, 6; S. 508, 7. Ic wegas dine þence to ferenne fótum mínum *I think to go thy ways with my feet,* Ps. Th. 118, 59. Folc ferende *travelling people,* Cd. 80; Th. 99, 28; Gen. 1653: Exon. 103 a; Th. 390, 12; Rä. 8, 9: Ps. Th. 125, 5. Ic fere geond foldan *I travel over the earth,* Exon. 101 a; Th. 381, 2; Rä. 2, 5: Th. 140, 12. Ðú mid mildse mínre ferest *thou goest with my grace,* Andr. Kmbl. 3345; An. 1676. Mon fereþ feor *a man goes far,* Exon. 91 a; Th. 342, 20; Gn. Ex. 146; Salm. Kmbl. 614; Sal. 306: Menol. Fox 327; Men. 165. Ác fereþ gelóme ofer ganotes-bæþ *a ship* [lit. *oak] often saileth over the sea* [lit. *sea-fowl's bath*], Runic pm. 25; Kmbl. 344, 18; Hick. Thes. i. 135, 49. Ða ðe heonon feraþ *those who go hence,* Cd. 228; Th. 305, 29; Sat. 654: Exon. 102 a; Th. 385, 14; Rä. 4, 44. Ic ferde to foldan ufan from eþle *I went to earth from the realm above,* Cd. 224; Th. 295, 30; Sat. 495: Ps. Th. 142, 11. Mid Gode Noe ferde *Noe cum Deo ambŭlāvit,* Gen. 6, 9: Andr. Kmbl. 1323; An. 662: Exon. 42 b; Th. 143, 18; Gú. 663. Ferde his hlísa to Galilea ríce *prōcessit rūmor ejus in omnem rěgiōnem Gǎlǐlǽæ,* Mk. Bos. 1, 28: Homl. Th. ii. 358, 5. Sum sǽdere ferde to sáwenne his sǽd *a sower went to sow his seed,* ii. 88, 12: 90, 10. He ferde fram him and wæs fered on heofen *recessit ab eis et ferēbātur in cælum,* Lk. Bos. 24, 51. He eft hám ferde *he went home again,* Bd. 2, 9; S. 512, 5: 3, 11; S. 536, 9. Hilde of deáþe ferde to life *Hilda de morte transívit ad vītam,* Bd. 4, 23; S. 595, 32. He ferde ofer sǽ *he went over the sea,* Boutr. Scrd. 17, 7: 19, 2: Chr. 1140; Erl. 265, 39. God ferde forþ *ǎbiit Dǒmĭnus,* Gen. 18, 33. Ferde Constantius forþ on Breotone *Constantius died* [lit. *went forth*] *in Britain,* Bd. 1, 8; S. 479, 29. Hí ferdon to Róme *they went to Rome,* Chr. 737; Erl. 47, 22: Gen. 11, 31: Boutr. Scrd. 22, 18: Beo. Th. 3268; B. 1632. He hí lǽrde ðæt hí ferdon on ðæt geweorc ðæs Godes wordes *in ōpus eos verbi proficisci suādet,* Bd. 1, 23; S. 485, 39. Hí ferdon ongén ðone brýdguman *exiērunt obviam sponso,* Mt. Bos. 25, 1. Hí ofer sǽ ferdon *they went over the sea,* Chr. 1087; Erl. 226, 7, 12. Tíd is ðæt ðú fere *it is time that thou goest,* Exon. 51 b; Th. 179, 30; Gú. 1269: Andr. Kmbl. 448; An. 224. Ǽr gé furður feran *ere ye go further,* Beo. Th. 513; B. 254. DER. be-feran, for-, ge-, geond-, of-, ofer-, þurh-, to-. v. faran.

fer-bed, -bedd, es; *n. A bed for a journey;* ǐtǐněris lectus :—Ferbed *bajulna?* Ælfc. Gl. 66; Som. 69, 78; Wrt. Voc. 41, 32.

fér-blǽd, es; *m.* [fér-= fǽr- *sudden,* blǽd *a blast*] *A sudden* or *fearful blast;* repentīnus flātus :—Ic lǽran wille ðæt gé eówer hús gefæstnige, ðý-lǽs hit férblǽdum windas toweorpan *I will exhort that ye make your house firm, lest winds overthrow it with sudden blasts,* Exon. 75 a; Th. 281, 21; Jul. 649.

fercian; *p.* ode; *pp.* od *To bring, assist, help, support;* ferre, adjǔvāre, subvěnīre, sustentāre :—Hí fercodon ða scypo eft to Lundenne *they brought the ships again to London,* Chr. 1009; Th. 260, 31, col. 2. On ðisum life we ateoriaþ gif we us mid bigleofan ne ferciaþ *in this life we faint if we support not ourselves with food,* Homl. Th. i. 488, 33. DER. ge-fercian.

fér-clam; *gen.* -clammes; *m.* [fér-= fǽr- *sudden,* clam *what holds*] *A sudden seizing;* arreptio repentína angustiæ perícŭlōsæ, Grn. Exod. 119. v. oferclamme, clam, clom.

fercung, e; *f. A sustaining;* sustentātio, Som. Ben. Lye.

fercuþ; *adj. Frugal, thrifty;* frūgālis, frūgi, Cot. 203.

ferd *an army,* Chr. 1140; Erl. 265, 28. v. fyrd.

ferde, *pl.* ferdon *went,* Bd. 2, 9; S. 512, 5: Chr. 737; Erl. 47, 22; *p.* of feran.

ferd-faru, e; *f. A military expedition;* mīlĭtāris expědītio, expědītio contra hostes, Heming. p. 234, Lye. v. fyrd-faru.

ferd-mon, -monn, es; *pl. nom. acc.* -men; *m. A soldier;* mīles :—Ðæt feoh mon ðám ferdmonnum sellan sceolde *the money should be given to the soldiers,* Bt. 27, 4; Fox 100, 14. Cyning sceal hæbban ferdmen *a king must have soldiers,* 17; Fox 58, 33, MS. Cot. v. fyrd-man.

ferd-rinc, es; *m. A warrior, soldier;* bellātor, míles :—He fromne ferdrinc fere berode *he deprived the brave warrior of life,* Ps. C. 50, 22; Ps. Grn. ii. 277, 22. v. fyrd-rinc.

ferd-wíte *a fine for neglecting to pay the contribution to the army,* L. In. 51; Th. i. 134, 10, note 23, MS. B: Th. Diplm. A. D. 1044; 359, 3. v. fyrd-wíte.

ferd-wyrt, e; *f.* [= feld-wyrt?] *Field-wort? gentian?* gentiāna ?— Nim ferdwyrt *take gentian* (?), L. M. 1, 87; Lchdm. ii. 154, 15. v. feld-wyrt.

fere; *adj. Passable, able to go;* meābĭlis. DER. earfoþ-fere, eáþ-, ge-, un-, un-ge-.

fere *with life,* Ps. C. 50, 22; Ps. Grn. ii. 277, 22; *inst.* of ferh *life.*

fered *carried,* Lk. Bos. 24, 51; *pp.* of ferian.

fereld; *n. A way, going, step;* gressus :—Fulfrema stepas oððe paðas oððe fereldu míne on síþfætum dínum *perfĭce gressus meos in sēmĭtis tuis,* Ps. Lamb. 16, 5. v. færeld.

féren *fiery, burning;* igneus, ignītus, Som. Ben. Lye. v. fýren.

ferend, es; *m.* [*part.* of feran] *A traveller, messenger, sailor;* pere- grĭnātor, nuncius, nauta :—He hét gefetigan ferend snelle *he commanded swift messengers to be fetched,* Exon. 66 b; Th. 246, 12; Jul. 60. Him ða ferend on fæste wuniaþ *the sailors firmly rest on him,* 97 a; Th. 361, 25; Wal. 25.

fere-scæt, es; *m. Fare-scot, passage-money;* naulum, Cot. 138.

fere-soca, an; *m.* [ferh *a pig,* soca?=socc *a sock*] *A bag made of swine's skin;* sibæa :—Feresoca sibba, Wrt. Voc. 289, 1. v. Littleton, Glossārium Lātīno-barbārum *under* sibæa.

fergan; *p.* ede; *pp.* ed. **I.** *to carry, convey, bear;* portāre, vehěre, ferre :—We willaþ Hláford fergan to ðære beorhtan byrg *we will bear the Lord to the bright city,* Exon. 15 a; Th. 32, 26; Cri. 518: 104 b; Th. 397, 1; Rä. 16, 13. Bearn fergaþ and fédaþ fæder and módor *father and mother carry and lead the child,* 87 a; Th. 327, 21; Vy. 7. **II.** *to go;* íre :—Ic seah rǽplingas in ræced fergan *I saw captives going into a house,* Exon. 113 b; Th. 435, 2; Rä. 53, 1. v. ferian.

fer-grunden *ground to pieces, mangled,* Chr. 937; Erl. 114, 9, = for- grunden ; *pp.* of for-grindan.

ferh; *gen.* feres; *dat. inst.* fere; *n. m. Life;* vīta :—Ferh ellen wræc *power drove out life,* Beo. Th. 5406; B. 2706. He fromne ferdrinc fere berode *he deprived the brave warrior of life,* Ps. C. 50, 22; Ps. Grn. ii. 277, 22. Ealne wídan ferh *to all eternity,* Exon. 44 b; Th. 151, 3; Gú. 789. v. feorh.

ferh, es; *m. A pig;* porcus, Wrt. Voc. 286, 47. v. fearh.

ferh-cwæle? [=-cwalu?] *A murrain of hogs;* lues porcína, Som. Ben. Lye.

ferh-cwalu, e; *f. Life-destruction, slaughter;* interněcio, Cot. 114. v. feorh-cwalu.

ferht *fear, fright, dread;* pǎvor, tĭmor, Som. Ben. Lye. v. fyrhto.

ferht, es; *m. n. The mind;* mens :—He mæg rihtwísnesse findan on ferhte *he may find wisdom in his mind,* Bt. Met. Fox 22, 119; Met. 22, 60. v. ferhþ.

ferhþ, fyrhþ, ferþ, ferht, es; *m. n.* **I.** *the soul, spirit, mind;* animus, mens :—Ðín ferhþ bemearn *thy spirit mourned,* Cd. 106; Th. 139, 14; Gen. 2309: Elen. Kmbl. 347; El. 174: Salm. Kmbl. 358; Sal. 178. Ferhþes fóreþanc *forethought of mind,* Beo. Th. 2124; B. 1060. His geleáfa wearþ fæst on ferhþe *his faith became firm in his spirit,* Elen. Kmbl. 2071; El. 1037: Exon. 100 a; Th. 375, 2; Seel. 132: Cd. 40; Th. 53, 32; Gen. 870: Beo. Th. 1512; B. 754: Ps. Th. 85, 11. Ðæt he andsware ǽnige ne cunne findan on ferhþe *that he cannot find any answer in his mind,* Bt. Met. Fox 22, 103; Met. 22, 52: Beo. Th. 2337; B. 1166: Cd. 161; Th. 200, 11; Exod. 355: Elen. Kmbl. 2325; El. 1164. He wiste ferhþ guman *he knew the man's soul,* Cd. 134; Th. 169, 2; Gen. 2793. Ne lǽt ðú ðín ferhþ wesan sorgum asǽled *let not thy soul be bound with sorrows,* Cd. 100; Th. 132, 17; Gen. 2194. Noe læg ferhþe forstolen *Noah lay deprived of mind,* Cd. 76; Th. 95, 15; Gen. 1579: Ps. Th. 131, 2. Hí ferdon forþ ðonon, ferhþum fægne *they went forth thence, rejoicing in their minds,* Beo. Th. 3270; B. 1633: 6334; B. 3177. **II.** *life;* vīta :—Wídan ferhþ, *acc. for a long life, for ever,* Elen. Kmbl. 1598; El. 801. DER. collen-ferhþ, -ferþ, -fyrhþ: dreórig-, freórig-, gál-, gamol-, gleáw-, sár-, sárig-, stærced-, stearc-, sterced-, stíþ-, sweorcend-, swíþ-, swoncen-, swýþ-, wérig-, wíde-. v. feorh.

ferhþ-bana, an; *m. A life-destroyer, murderer;* vītæ destructor, inter- fector :—Fyrst ferhþbana *the first life-destroyer,* Cd. 162; Th. 203, 5; Exod. 399.

ferhþ-cearig; *adj. Anxious in soul;* animo sollicĭtus :—Sarra ongan, ferhþcearig, to were sínum mæþlan *Sarah, anxious in soul, began to speak to her consort,* Cd. 101; Th. 133, 28; Gen. 2217.

ferhþ-cleófa, an; *m. The mind's cave, breast;* mentis cŭbīle, pectus :— Eádig byþ se wer, se ðe him ege Drihtnes on ferhþcleófan, fæste gestandeþ *beātus vir, qui tĭmet Dǒmĭnum,* Ps. Th. 111, 1.

ferhþ-cófa, an; *m. The mind's cave, breast;* mentis cŭbĭle, pectus :—

On ferhþcófan *in his mind's cave* or *breast*, Cd. 123; Th. 157, 8; Gen. ⊕ 2603; Ps. Th. 108, 17.

ferhþ-frec; *adj. Bold in spirit;* anǐmōsus :—Ferhþfrecan Fin begeat sweordbealo *misery from the sword seized Fin the bold in spirit*, Beo. Th. 2296; B. 1146.

ferhþ-friðende *life-saving.* v. ferþ-friðende.

ferhþ-genîþla, an; *m. A life-enemy, deadly foe;* vǐtæ hostis, lētālis hostis :—Ic sweorde drep ferhþgenîþlan *I struck the deadly foe with my sword*, Beo. Th. 5754; B. 2881.

ferhþ-gewit *mental wit, understanding.* v. ferþ-gewit.

ferhþ-gleáw, fyrhþ-gleáw; *adj. Prudent in mind, sagacious;* anǐmo prūdens, săpiens :—Ðǣr hie Iudiþe fundon ferhþgleáwe *they found Judith there prudent in mind*, Judth. 10; Thw. 21, 29; Jud. 41. Þúsenda manna ferhþgleáwra *of a thousand sagacious men*, Elen. Kmbl. 653; El. 327.

ferhþ-grim *fierce of spirit.* v. ferþ-grim.

ferhþ-lîc *rational, just, equitable.* v. ferht-lîc.

ferhþ-loca, ferþ-loca, fyrhþ-loca, an; *m. Soul-inclosure, bosom, body;* mentis clausūra, pectus, corpus :—Ðæt ðîn nama, Crist, in ûrum ferhþlocan sî feste gestaðelod *that thy name, O Christ, be firmly established in our soul's inclosure*, Hy. 6, 5, 32; Hy. Grn. ii. p. 286, 5, 32. Ne willaþ eów andrǣdan fǣge ferhþlocan *dread ye not feeble bodies*, Cd. 156; Th. 194, 27; Exod. 267.

ferhþ-lufe *soul's love, mental love.* v. fyrhþ-lufe.

ferhþ-sefa, ferþ-sefa, firhþ-sefa, fyrhþ-sefa, an; *m. The mind's sense, intellect;* mens :—Cwēn gefeah on ferhþsefan *the queen rejoiced in her mind*, Elen. Kmbl. 1696; El. 850: 1787; El. 895.

ferhþ-wérig *soul-weary, sad.* v. ferþ-wérig, fyrhþ-wérig.

ferht-lîc; *adj. Rational, wise, just, equitable;* rationālis, săpiens, æquus :—Drihten ferhtlîc riht folcum dēmeþ *Dŏmĭnus jūdĭcābit pŏpŭlos in æquĭtāte*, Ps. Th. 95, 10.

ferian, ferigan, ferigean, fergan; *to* ferianne; *p.* ode, ede; *pp.* od, ed [fer = fær *a journey*]. **I.** *to carry, convey, bear, lead, conduct;* ferre, portāre, vehěre, dedūcěre, afferre :—Hēht wîgend ðæt hālige treó him befóran ferian *he commanded the warriors to carry the holy tree before him*, Elen. Kmbl. 215; El. 108: Cd. 67; Th. 80, 18; Gen. 1330. We ðē willaþ ferigan freólîce ofer fisces bæþ *we will gladly convey thee over the fish's bath [the sea]*, Andr. Kmbl. 585; An. 293. Hêt lífes brytta englas sîne ferigean leófne ofer lagufæsten *the giver of life commanded his angels to bear the dear one over the stronghold of the waves*, 1647; An. 825. To ferianne *ad portandum*, Gen. 46, 5. Ic ferige onbûtan *circumfěro*, Ælfc. Gr. 47; Som. 48, 33. Mec merehengest fereþ ofer flódas *the vessel conveys me over the floods*, Exon. 104 a; Th. 395, 13; Rä. 15, 7: 114 b; Th. 439, 16; Rä. 59, 4. Hî hine feriaþ ofer fisces bæþ *they bear it over the fish's bath [the sea]*, Runic pm. 16; Kmbl. 342, 17; Hick. Thes. i. 135, 31. Hwanon ferigeaþ gē fætte scyldas *whence bear ye your stout shields?* Beo. Th. 671; B. 333. Folc ðîn ðû feredest swā sceáp *deduxisti sîcut ŏves pŏpŭlum tuum*, Ps. Th. 76, 17. He ferode ðone to his mynstre mid ârwurþnysse *he bare it to his minster with honour*, Homl. Th. ii. 358, 7: Chr. 1009; Erl. 141, 23. Us ofer ârwélan æðeling ferede *a noble one conducted us over the realm of oars [the sea]*, Andr. Kmbl. 1706; An. 855. Hî ðone sanct ferodon to ðære byrig *they conveyed the saint to the city*, Homl. Th. ii. 518, 29. Ðē on folmum feredan *in mănĭbus portābunt te*, Ps. Th. 90, 12: 82, 3. Feriaþ mid eów of ðære eorþan wæstmum *afferte nōbis de fructĭbus terræ*, Num. 13, 21. He wæs fered on heofon *ferēbātur in cælum*, Lk. Bos. 24, 51. **II.** *to betake oneself to;* se gerĕre, versāri :—Ðû aclǣccræftum lange feredes *thou hast long betaken thyself to evil arts*, Andr. 2725; An. 1365. Hî on lîge feredon *they betook themselves to lying*, Ps. Th. 58, 12. **III.** *to go, depart;* vehi, îre :—Mid friþe ferian *to depart in peace*, Byrht. Th. 136, 68; By. 179. Ðonne God geond wéstena wîde feraþ *Deus, dum transgredĭĕris per desertum*, Ps. Th. 67, 8. [*Laym.* uerien: *Plat.* fören: *O. Sax.* fôrian: *Frs.* fieren: *O. Frs.* fera: *Ger.* führen: *M. H. Ger.* vüeren: *O. H. Ger.* fuorjan, fôrjan: *Goth.* faryan *to convey a ship, row: Dan.* fŏre: *Swed.* fôra: *Icel.* ferja *to transport, carry by sea.*] DER. a-ferian, æt-, ge-, of-, ôþ-, to-, wið-.

Feriatus, es; *m. A Spanish robber*, Ors. 5, 2; Bos. 102, 19. v. Uariatus.

feriend, ferigend, es; *m. [part. of* ferian *to bear, bring] A bringer, leader;* dux :—Flôdes ferigend [MS. B. feriend] *bringer of the flood*, Salm. Kmbl. 161; Sal. 80.

ferigan, ferigean *to carry, convey, bear*, Andr. Kmbl. 585; An. 293: 1647; An. 825: Ælfc. Gr. 47; Som. 48, 33: Beo. Th. 671; B. 333. v. ferian.

fering, e; *f. A going, travelling, journeying;* peregrînātio, ǐter :—On ðære feringe *in that journeying*, Exon. 87 a; Th. 326, 20; Wîd. 131. DER. forþ-fering.

fêringa *suddenly;* extemplo, imprōvîso, Prov. 3. v. fǽringa.

fêrlíc *sudden, unlooked for, horrible;* repentînus, horrendus, Som. Ben. Lye. v. fǽrlíc.

fêrlíce *suddenly,* Ps. Spl. T. 63, 4. v. fǽrlíce.

---

fern, es; *n. Fern;* fĭlix :—Fern [MS. B. fearn], Herb. 78; Lchdm. i. 180, 23. v. fearn.

fernes, -ness, e; *f. A going, passing;* gressus, transĭtus :—Ne ðær fernes is *non est transĭtus*, Ps. Th. 143, 18. DER. ofer-fernes.

ferran *to remove, take away.* DER. a-ferran. v. feorran.

fêrrece? [fēr = fȳr?] *A fire-pan;* bătillum, Cot. 161, Som. Fêrrece *vatilla*, Wrt. Voc. 287, 7.

ferren, ferlen; *adj. Far off, distant, remote;* longinquus :—On ferren [ferlen MS. Rl.] land *in regĭōnem longinquam*, Lk. Skt. Hat. 19, 12. v. feorlen, fyrlen.

fers, færs, fyrs, es; *n. A* VERSE, *sentence, title;* versus, carmen :—Periodos is clȳsing, oððe ge-endung ðæs ferses *a period is the conclusion, or ending of the sentence*, Ælfc. Gr. 50, 14; Som. 51, 18. Ic fersige oððe ic wyrce fers *versĭfĭcor*, 37; Som. 39, 3, MSS. C. D. Ongan he sôna singan ða fers *stătim ipse cœpit cantāre versus*, Bd. 4, 24; S. 597, 18.

FERSC; *adj.* FRESH, *pure, sweet;* dulcis :—Eufrates is mǣst eallra ferscra wætera, and is yrnende þurh middewearde Babilōnian burh *Euphrates is the greatest of all fresh waters [rivers], and runs through the middle of the city of Babylon*, Ors. 2, 4; Bos. 44, 10. Gyf se wǣta sealt byþ of ðære sǣ, hit byþ þurh ðære sunnan hǣtan to ferscum wæterum awend *if the moisture be salt from the sea, it is turned to fresh water through the heat of the sun*, Bd. de nat. rerum; Lchdm. iii. 278, 9–12; Wrt. popl. science 19, 3. [*Chauc.* freisshe: *Laym.* freche: *Plat.* frisk: *Frs.* fersck: *O. Frs.* fersk, fersch, farsch: *Dut.* versch: *Ger.* frisch: *M. H. Ger.* vrisch: *O. H. Ger.* frisc: *Dan.* frisk, fersk: *Swed.* frisk, färsk: *Icel.* frískr: *Wel.* ffres.]

fer-scipe, es; *m. Society, fellowship;* sŏcietas :—To healfum fô se cyng, to healfum se ferscipe *dĭmĭdium căpiat rex, dĭmĭdium sŏcĭetas*, L. Ath. v. 2; Wilk. 65, 19. DER. ge-ferscipe.

fer-scrifen; *part.* [= for-scrifen; *pp. of* for-scrîfan *to disregard, abandon] Disregarded, abandoned;* addictus :—Ferscrifen [MS. færscribæn] *addictus*, Glos. Epnl. Recd. 153, 53. Ferscrifen *addictus* [Lye *has* ferscrifen = ferscrifen? *abdictus*], Cot. 14.

fersian; *p.* ode; *pp.* od *To make verse;* versĭfĭcāre :—Ic fersige oððe ic wyrce fyrs *versĭfĭcor*, Ælfc. Gr. 37; Som. 39, 3.

ferþ, ferþþ; *gen.* -es; *dat.* -e; *m. n.* **I.** *the soul, spirit, mind;* anĭmus, mens :—Wæs ðære fǣmnan ferþ geblissad *the damsel's soul was rejoiced*, Exon. 69 b; Th. 259, 25; Jul. 287: 89 a; Th. 334, 21; Gn. Ex. 19. Hî gemêtton ferþþes frófre *they found comfort of soul*, 46 a; Th. 157, 21; Gû. 895. On ferþe fægn *rejoicing in mind*, Bt. Met. Fox 9, 73; Met. 9, 37: Andr. Kmbl. 2968; An. 1487. Gefeóþ gē on ferþþe *rejoice ye in spirit*, Exon. 14 b; Th. 30, 7; Cri. 476: 70 b; Th. 262, 5; Jul. 328. Ðînne ferþ, *acc. m. thy mind*, 88 b; Th. 333, 9; Gn. Ex. 1. Sum hafaþ fæstgongel ferþ *one has a constant soul*, 79 b; Th. 298, 4; Crä. 80: 81 b; Th. 307, 19; Seef. 26. Ferþum gleáw *sagacious in soul*, 128 a; Th. 493, 10; Rä. 81, 28. Ferþþum, 114 b; Th. 440, 15; Rä. 60, 3. **II.** *life;* vîta :—Lǣtaþ gāres ord ingedūfan in fǣges ferþ *let the javelin's point dig into the life of the doomed one*, Andr. Kmbl. 2665; An. 1334. DER. dreórig-ferþ, freórig-, sārig-, stearc-, swîþ-, wérig-, wîde-. v. ferhþ.

ferþ-friðende; *part.* [friðian *to protect*] *Life-saving;* vītam servans :—Forlēt ferþfriðende wellan on gesceap þeótan *he let his life-saving fountains be poured into a vessel*, Exon. 109 b; Th. 419, 25; Rä. 39, 3.

ferþ-gewit, -gewitt, es; *n. Mental wit, understanding;* mentis intellectus :—Ðeáh hî ferþgewit ǣnig ne cûðen *though they knew not any mental wit*, Exon. 25 a; Th. 73, 4; Cri. 1184.

ferþ-grim; *adj. Fierce of spirit;* anĭmo sævus :—Frēcne and ferþgrim *rugged and fierce of spirit*, Exon. 67 b; Th. 251, 6; Jul. 141: 96 b; Th. 360, 13; Wal. 5.

ferþ-loca, an; *m. The soul's inclosure, bosom;* mentis clausūra, pectus :—Hyre wæs Cristes lof in ferþlocan *praise of Christ was in her soul's inclosure*, Exon. 69 a; Th. 256, 19; Jul. 234: 76 b; Th. 287, 12; Wand. 13. v. ferhþ-loca.

ferþ-sefa, an; *m. [sefa the faculty of perceiving;* sensus] *The mind;* mens :—Fæstnian ferþsefan *to fix in the mind*, Exon. 92 b; Th. 347, 29; Sch. 20. v. ferhþ-sefa.

ferþþ *the soul, mind.* v. ferþ.

ferþþes, ferþþe *of a soul, to a soul*, Exon. 46 a; Th. 157, 21; Gû. 895: 14 b; Th. 30, 7; Cri. 476; *gen. and dat. of* ferþ.

ferþ-wérig; *adj. Soul-weary, sad;* mæstus :—Freórig and ferþwérig *trembling and soul-weary*, Exon. 49 b; Th. 171, 21; Gû. 1130: 20 a; Th. 52, 9; Cri. 831. v. fyrhþ-wérig.

ferwett-full; *adj.* [ferwett = fyrwet *curiosity] Curious, anxious;* sollĭcitus :—Ferwettfulle men *sollĭcĭti*, Lk. Skt. Rush. 12, 26.

fêsian, he fêseþ; *p.* ode; *pp.* od; *v. a. To drive away, put to flight;* fŭgāre, in fŭgam ăgĕre :—Ðæt oft on gefeohte ân fêseþ tyne *ut in pugna ûnus sæpe decem in fŭgam ĕgĕrit*, Lupi Serm. i. 14; Hick. Thes. ii. 103, 20. DER. to-fêsian. v. fȳsian.

feste; *adv. Fastly, firmly* :—Ic hæbbe genôg feste on gemynde *I have it firmly enough in my mind*, Bt. 36, 3; Fox 176, 24. v. fæste.

. **festen**, es; *n. A fastness, fortress;* mūnīmentum :—Hī manige festena and castelas abræcon *they demolished many fastnesses and castles,* Chr. 1094; Erl. 230, 35. v. fæsten II.

festen-mon, -monn, es; *m. A surety;* fīdējussor, Som. Ben. Lye. v. fēster-man.

fester *food, nourishment, foster-,* in the compounds fēster-bearn, -fæder, -man, -mōdor. v. fōster.

fēster-bearn, es; *n. A foster-child;* ǎlumnus :—Fēsterbearn *ǎlumni,* Martyrol. ad 22 Martii. v. fōster-bearn.

fēster-fæder, es; *m. A foster-father, nourisher;* altor, nutrītor :—Fēsterfæder *altor,* Wrt. Voc. 284, 72. Ætȳwde me mīn iú magister and fēsterfæder *appǎruit mǎgister quondam meus et nutrītor,* Bd. 5, 9; S. 622, 34. v. fōster-fæder.

fēster-man, es; *m. A foster-man, bondsman, security;* fīdējussor :—Ælc preóst finde him xii fēstermen *let every priest find for himself twelve bondsmen,* L. N. P. L. 2; Th. ii. 290, 15.

fēster-mōdor, -mōdur; *f. A foster-mother, nurse;* altrix, nutrix :—Fēstermōdor *altrix,* Wrt. Voc. 284, 73. Wīfmonna lǎreów and fēstermōdur *māter et nutrix fēmǐnārum,* Bd. 4, 6; S. 574, 17. v. fōster-mōdor.

festing-men, -menn *servants of the king entrusted to the keeping of the monasteries while going from place to place,* Th. Diplm. A. D. 823; 67, 2 : A. D. 828; 79, 30. v. fæsting-men.

festlíce; *adv. Firmly, vigorously;* firmĭter :—Hī on ða burh festlíce feohtende wǣron *they were vigorously fighting against the town,* Chr. 994; Erl. 133, 11. v. fæstlíce.

festnes, -ness, e; *f. A fastness, firmament;* firmāmentum :—Weorc handa his bodaþ festnes [MS. fesnesse] *ǒpĕra mǎnuum ejus annuntiat firmāmentum,* Ps. Spl. T. 18, 1. v. fæstnes.

festnian *to confirm;* confirmāre :—Ic Ceólrēd abbud ðas ūre selene mid Cristes rôde tācne trymme and festnie *I Ceolred abbot ratify and confirm this our gift with the sign of Christ's cross,* Th. Diplm. A. D. 852; 106, 10–12. DER. ge-festnian. v. fæstnian.

fēstrian; *p.* ode, ude; *pp.* od, ud *To foster, nourish;* nutrīre :—Fēstrud beón *nutrīri,* Scint. 81. v. fōstrian.

fet *fetches, brings,* Prov. Kmbl. 61; *3rd sing. pres.* of fetian.

fēt *to* or *for a foot, feet,* Ex. 21, 24 ; Ps. Lamb. 72, 2 : Mt. Bos. 18, 8 ; *dat. sing. and nom. acc. pl.* of fót.

fēt *feeds,* Mt. Bos. 6, 26, = fēdeþ; *3rd sing. pres.* of fēdan.

fetan; *p.* sæt, *pl.* sǣton; *pp.* feten *To make, travail, join;* fǎcĕre, procreāre, jungĕre. [*Goth.* fitan; *p.* fat, *pl.* fetum; *pp.* fitans *to travail in birth;* partūrīre.] v. fetian.

fête; *adj. Provided with feet, footed;* pĕdĭbus instructus. v. ān-fête, twý-, þrý-, feówer-.

FETEL; *gen.* feteles, fetles; *m. A girdle, belt;* cingǔlum, balteus :—Sweordum and fetelum *with swords and belts,* Bt. Met. Fox 25, 19; Met. 25, 10. Mid fetlum *with belts,* Bt. 37, 1; Fox 186, 5. [*Ger.* fessel, *f.: M. H. Ger.* vezzel, *m: O. H. Ger.* fazzil, fezzil, fezil, *m. balteus: Icel.* fetill, *m. a strap, belt.*]

fetel-hilt, es; *n. A belted hilt;* cǎpǔlus baltěo instructus :—He gefēng fetelhilt *he seized the belted hilt,* Beo. Th. 3130; B. 1563.

fetels, es; *m. A little vessel, bag;* vas, saccus :—Fōrwerede fetelsas *saccos vĕtĕres,* Jos. 9, 5. v. fætels.

FETER, fetor, e; *f. A* FETTER, *chain for the feet;* compes, pĕdĭca :—He fēdeþ swā on feterum *he feeds him thus in fetters,* Exon. 88 b; Th. 332, 20; Vy. 88 : Ps. Th. 78, 11. Ān sceal inbindan forstes fetre *one shall unbind fetters of frost,* Exon. 90 a; Th. 338, 9; Gn. Ex. 76. Ic mōdsefan mīnne sceolde feterum sǣlan *I must bind my thought in fetters,* 76 b; Th. 287, 29; Wand. 21: Salm. Kmbl. 141; Sal. 70. [*O. Sax.* feterôs, *pl. m : Ger.* fesser, *f.: M. H. Ger.* vëzzer, *f: O. H. Ger.* fëzzera : *Icel.* fjöturr, *m. a fetter of iron.*]

feterian *to fetter.* DER. ge-feterian.

feter-wrāsen *a chain, fetter.* v. fetor-wrāsen.

fēða, an; *m.*   I. *a band on foot, infantry, a host, troop, tribe, company;* phǎlanx pĕdestris, pĕdĭtes, lĕgio, ǎcies, tribus, cǎterva :—Eórod sceal getrume rīdan, fæste fēða stondan *a band of horse [= cavalry] shall ride in a body, a band of foot [= infantry] stand fast,* Exon. 90 a; Th. 337, 13; Gn. Ex. 64. Fēða [MS. fēðu] *lĕgio,* Ælfc. Gl. 7; Som. 56, 73; Wrt. Voc. 18, 25. Se earga fēða *ǎcies segnis Brittǒnum,* Bd. 1, 12; S. 481, 19, MSS. B, C. Fēða eal gesæt *the band all sat,* Beo. Th. 2853; B. 1424. Iudisc fēða *the tribe of Judah,* Cd. 158; Th. 197, 25; Exod. 312. Se fēða com up to earde *the company came up to their home,* 223; Th. 293, 19; Sat. 457. Ðǣr wæs ungemetlíc wæl geslagen Persa, and Alexandres næs nā mā ðonne hund-twelftig on ðam rǣde-here, and nigon on ðam fēðan *there was a very great slaughter made of the Persians, and no more than a hundred and twenty in Alexander's cavalry, and nine in the infantry,* Ors. 3, 9; Bos. 64, 20. He cwiþ to ðara synfulra sāwla fēðan *he shall say to the band of sinful souls,* Exon. 30 a; Th. 93, 1; Cri. 1519. Ic him on fēðan beföran wolde *I would [go] before him in the host,* Beo. Th. 4987; B. 2497: 5830; B. 2919: Cd. 220; Th. 284, 19; Sat. 324. Ðū here fýsest,

fēðan to gefeohte *thou leadest a host, a troop to battle,* Andr. Kmbl. 2377; An. 1190. Fôr fyrda mǣst, fēðan trymedan *the greatest of armies marched, the infantry were strong,* Elen. Kmbl. 70; El. 35. Fēðan sǣton *the bands sat,* Andr. Kmbl. 1182; An. 591. Ymb ðæt hēhsetl standaþ engla fēðan *hosts of angels stand around the throne,* Cd. 218; Th. 278, 13; Sat. 221: Beo. Th. 2659; B. 1327. Ðǣr wæs Persa x m ofslagen gehorsedra, and eahtatig m fēðena *there were slain ten thousand of the Persians' cavalry and eighty thousand of the infantry,* Ors. 3, 9; Bos. 65, 2 : 68, 9. Ne willaþ eów andrǣdan deáde fēðan *dread ye not dead bands,* Cd. 156; Th. 194, 26; Exod. 266. Hí bǣdon ðæt hí mōston ofer ðone ford faran, fēðan lǣdan *they gave orders to go over the ford, to lead the troops onward,* Byrht. Th. 134, 23; By. 88. Gerǣrud fēða *an arranged band;* ǎcies: getrimmed fēða *cǔneus:* gangende [MS. gangend] fēða *a moving band;* agmen, Ælfc. Gl. 7; Som. 56, 74, 79, 82; Wrt. Voc. 18, 26, 31. 34.   II. *a battle;* pugna :—He beald in gebēde bīdsteal gifeþ, fæste on fēðan *he bold in prayer maketh a stand, firmly in battle,* Exon. 71 a; Th. 265, 30; Jul. 389. DER. gum-fēða, here-.

fēðan; *p.* de; *pp.* ed *To lead;* dūcĕre :—Bearn fergaþ and fēðaþ fæder and mōdor *father and mother carry and lead the child,* Exon. 87 a; Th. 327, 21.

Fēðan-leag; *gen.* -leage; *f.* [*Flor.* Fethanleah: *Hunt.* Fedhalnea, Fedhanlea: *Matt. West.* Frithenleia] *Frethern, Gloucestershire?* — Hēr Ceáwlin and Cūþa fuhton wið Brettas in ðam stede ðe mon nemneþ Fēðanleag [Fēðanlea, Th. 35, 8, col. 1] *in this year* [A. D. 584] *Ceawlin and Cutha fought against the Britons at the place which is called Frethern,* Chr. 584; Th. 34, 9.

fēðe, es; *n. The power of going on foot, walking, going, motion, pace;* fǎcultas pĕdĭbus eundi, ambŭlātio, gressus, passus :—Ðæra hǣðenra anlícnyssa habbaþ fēt būtan fēðe *the idols of the heathen have feet without the power of going,* Homl. Th. i. 366, 27. An fēðe mihtigost *most powerful in walking,* Bt. 36, 5; Fox 180, 21. He nǎhte his fēðes geweald *he had no power of walking,* Homl. Th. i. 336, 9. Hit is nǣdrena gecynd ðæt heora fēðe biþ on heora ribbum *it is the nature of serpents that their power of going is in their ribs,* Ors. 4, 6; Bos. 84, 44. On fēðe lēf [MS. líf] *lame in walking,* Exon. 87 b; Th. 328, 16; Vy. 18. Sum sceal on fēðe gongan *one shall go on foot,* 87 b; Th. 328, 33; Vy. 27. Swift ic eom on fēðe *I am swift of pace,* Exon. 104 b; Th. 396, 10; Rä. 16, 2 : Beo. Th. 1944; B. 970. Habbaþ hringa gespong aíyrred me mīn fēðe *the clasping of rings has taken from me my power of going,* Cd. 19; Th. 24, 17; Gen. 379. He fēðe ne sparode *he spared not pace,* 117; Th. 153, 6; Gen. 2524.

fēðe-cempa, an; *m. A foot-soldier, champion;* pĕdester mīles :—Fēðecempa, nom. Beo. Th. 3092; B. 1544: 5698; B. 2853.

fēðe-gang, es; *m. A foot-journey;* pĕdester iter :—Ne mæg ic aldornere mīne swā feor heonon fēðegange gesēcan *I cannot seek my life's safety so far hence by a foot-journey,* Cd. 117; Th. 152, 1; Gen. 2513.

fēðe-georn; *adj. Desirous of going;* meandi cǔpĭdus :—Sió fēðegeorn fremman onginneþ *desirous of going it resolves to proceed,* Exon. 108 a; Th. 413, 21; Rä. 32, 9.

fēðe-gest, es; *m. A pedestrian guest;* pĕdester advēna :—Fēðegestas eódon in on ða ceastre *the pedestrian guests went into the city,* Elen. Kmbl. 1687; El. 845. Wæs gerýmed fēðegestum flet *the hall was cleared for the pedestrian guests,* Beo. Th. 3956; B. 1976.

fēðe-here, es; *m. A foot army, infantry;* pĕdestris exercĭtus, pĕdĭtātus :—On his fēðehere wǣron xxxii m *in his infantry were 32,000,* Ors. 3, 9; Bos. 64, 17.

fēðe-hwearf, es; *m. A company on foot, pedestrian multitude;* pĕdestris cǎterva :—On fēðehwearfum *amongst the pedestrian multitude,* Exon. 35 a; Th. 113, 24; Gū. 162.

fēðe-lāst, es; *m. A footstep, pace;* passus, gressus :—Hie fēðelāste forþ onettan *they hastened forth with pace,* Judth. 11; Thw. 23, 25; Jud. 139. Ferdon forþ ðonon fēðelāstum *they went forth thence with their footsteps,* Beo. Th. 3269; B. 1632.

fēðe-leás; *adj. Footless;* pĕdĭbus cǎrens :—Ðū scealt faran fēðeleás *thou shalt go footless,* Cd. 43; Th. 56, 6; Gen. 908 : Exon. 127 a; Th. 488, 7; Rä. 76, 3.

fēðe-man, -mann, es; *m. A footman* or *soldier;* pĕdestris mīles. pēdes, Som. Ben. Lye.

fēðe-mund; es; *f. A foot-hand;* mǎnus gressus. Used for the fore-feet of the badger :—Ic sceal fromlíce fēðemundum þurh steápne beorg strǣte wyrcan *I [a badger] shall strenuously work a road through a steep mountain with my fore-feet,* Exon. 104 b; Th. 397, 10; Rä. 16, 17.

FEÐER; *gen. dat. acc.* feðere; *pl. nom. acc.* feðera, feðra, feðre; *f.*   I. *a* FEATHER; penna, plūma :—Mid níre [=niwre] feðere *with a new feather,* Herb. 122, 1; Lchdm. i. 234, 13: L. M. 1, 39; Lchdm. ii. 102, 8. Gedô feðere on lee *put a feather in oil,* L. M. 1, 18 : Lchdm. ii. 62, 11. Swanes feðre, nom. pl. *swan's feathers,* Exon. 57 b; Th. 207, 6; Ph. 137. Wurp ða feðera wið æftan ðæt weofod *plūmas projĭciet prŏpe altāre,* Lev. 1, 16 : Cd. 72; Th. 88, 26; Gen. 1471. Se fénix ascæceþ feðre *the phœnix shakes its feathers,* Exon. 58 a;

Th. 207, 21; Ph. 145: 58 b; Th. 212, 5; Ph. 205. Feðrum bifongen *clad with feathers,* 61 a; Th. 224, 23; Ph. 380: Bt. Met. Fox 24, 10; Met. 24, 5. **II.** in the *pl.* sometimes used for *Wings;* ālæ, pennæ:—Mec wǣgun feðre on lifte *wings bore me in air,* Exon. 107 b; Th. 409, 20; Rä. 28, 4. Ic hæbbe swīðe swifte feðera, ðæt ic mæg flíógan ofer ðone heán hróf ðæs heofones *I have very swift wings, that I can fly over the high roof of heaven,* Bt. 36, 2; Fox 174, 4: Ps. Lamb. 54, 7: 138, 9. He gesihþ brimfuglas brǣdan feðra *he sees sea-fowls spread their wings,* Exon. 77 a; Th. 289, 13; Wand. 47. Cômon earnas on flyhte, feðerum hrēmige *eagles came in flight, exulting in their wings,* Andr. Kmbl. 1728; An. 866: Bt. Met. Fox 24, 17; Met. 24, 9. Fugel feðrum strong *a bird strong of wings,* Exon. 57 a; Th. 203, 18; Ph. 86: 57 b; Th. 206, 7; Ph. 123: 58 a; Th. 208, 29; Ph. 163: 60 b; Th 222, 11; Ph. 347. **III.** what is made of a feather, *A pen;* penna, cǎlǎmus:—Feðer *a pen;* penna, Wrt. Voc. 75, 16. Nim dîne feðere and wrît fiftig *take thy pen and write fifty,* Lk. Bos. 16, 6. [*Chauc.* feder: *Plat.* fedder: *O. Sax.* fethera, *f: Dut.* veder, veer, *f: Ger.* feder, *f: M. H. Ger.* vēdere, vēder, *f: O. H. Ger.* fedara, *f: Dan.* fjeder, *m. f: Swed.* fjäder, *m: Icel.* fjǫðr, *f: Lat.* penna, old forms pesna, petna, *f: Grk.* πτερόν, *n. a feather;* πέτομαι *to fly: Sansk.* pat *to fly.*] DER. halsre-feðer, hleow-, wríting-. v. fiðere.

feðer-, *four-,* used only in the compounds:—feðer-fóte, -sceátas, -scette, -scîte, -scitte. v. fiðer-, fyðer-.

feðeran, feðran *to provide with feathers* or *wings.* DER. ge-feðeran, -feðran.

feðer-bed, -bedd, es; *n. A feather-bed;* culcǐta:—Feðerbed culcǐtes [= culcǐta], Ælfc. Gl. 27; Som. 60, 102; Wrt. Voc. 25, 42.

feðer-berende; *part. Bearing feathers, feathered;* pennǐger, Cot. 150.

feðer-cræft, es; *m. The art of feather-embroidering;* plūmāria ars, Som. Ben. Lye.

feðere, feðre; *def.* se feðera, feðra; seó, ðæt feðere, feðre; *adj. Feathered;* pennis prædītus. DER. deáwig-feðere, haswig-, îsig-, salwig-, ūrig-.

feðer-fóte; *adj. Four-footed;* quadrǔpes:—Eádbyrht feðerfótra [MS. -fóta] neáta ðone tēðan dǣl to ðearfum syllan wolde *Eadbyrht would give the tenth part of four-footed cattle to the poor,* Bd. 4, 29; S. 608, 17, note, MS. B. v. feówer-fête, fiðer-fête, fyðer-fête, -fóte.

feðer-gearwe; *pl. f.* [gearwe *clothing*] *Feather-gear, the feathering of an arrow;* pennis vestitus:—Sceaft feðergearwum fūs *an arrow prompt with its feather-geâr,* Beo. Th. 6229; B. 3119.

feðer-geweorc, es; *n. Feather-embroidered work;* plūmārium ǒpus:—Feðergeweorc besiwed ǒpus plūmārium intextum, Cot. 145.

feðer-hama, -homa, an; *m. Feather-covering, feathers, plumage, wings;* plūmārum tegmen, plūma, pennæ, ālæ:—Geseó ic him his englas ymbe hweorfan mid feðerhaman *I see his angels encompass me with feathery wings,* Cd. 32; Th. 42, 6; Gen. 670. Eall bið geniwad, feorh and feðerhoma *all is renewed, its life and plumage,* Exon. 60 a; Th. 217, 14; Ph. 280. Ðæt he mid feðerhoman fleógan meahte *that he might fly with wings,* Cd. 22; Th. 27, 13; Gen. 417.

feðer-sceátas; *pl. m. Four corners* or *quarters;* quǎtuor plǎgæ:—Eall ðeós leóhte gesceaft feðersceátum full feohgestreóna *all this bright creation in its four quarters full of treasures,* Salm. Kmbl. 63; Sal. 32.

feðer-scette; *adj. Four-cornered;* quadrangǔlāris, in quǎtuor plǎgas porrectus:—Eall ðeós leóhte gesceaft feðerscette, full fyrngestreóna *all this bright creation, four-cornered, full of ancient treasures,* Salm. Kmbl. 63, MS. B; Sal. 32, note. v. feðer-scîte.

feðer-scîte, -scitte, -scette; *adj. Four-cornered, quadrangular;* quadrangǔlāris:—Feðerscîte tæfel *four-cornered tables;* tessera *vel* lepuscǔlæ, Ælfc. Gl. 61; Som. 68, 66; Wrt. Voc. 39, 49. Lytle feðerscitte flórstānas *little four-cornered floor-stones;* tessellæ, 61; Som. 68, 67; Wrt. Voc. 39, 50. v. feówer-scŷte, fiðer-scŷte, -scîte, fyðer-scŷte.

fêðe-spêdig; *adj. Speedy of foot;* lēvipēs:—Sum bið on londe snel, fēðespêdig *one is swift on land, speedy of foot,* Exon. 79 a; Th. 296, 18; Crä. 53.

fêðe-wíg, -wigg, es; *n? m? A foot-battle;* pědestris pugna:—Fēðe-wîges *of the foot-battle,* Beo. Th. 4717; B. 2364: Wald. 88; Vald. 2, 16.

feðm, es; *m. A bosom;* sĭnus:—On feðme heora *in sĭnu eōrum,* Ps. Spl. T. 78, 13. v. fæðm **II.**

feðra, feðre *feathers, wings,* Exon. 57 b; Th. 207, 6; Ph. 137: 58 b; Th. 212, 5; Ph. 205: 77 a; Th. 289, 13; Wand. 47; *nom. acc. pl. of* feðer.

feðrum *with feathers* or *wings,* Bt. Met. Fox 24, 10; Met. 24, 5: Exon. 60 b; Th. 222, 11; Ph. 347; *inst. pl. of* feðer.

fêðu *a band on foot, a host;* lēgio, Ælfc. Gl. 7; Som. 56, 73; Wrt. Voc. 18, 25. v. fêða.

fetian, fetigean, fetigan; he fetaþ, fet; *p.* fette; *pp.* fetod *To fetch, bring to, marry;* addǔcěre, applicāre, uxōrem dūcěre:—He hêht him fetigean to sprecan sîne *he bade to fetch his counsellors to him,* Cd. 126; Th. 161, 17; Gen. 2666. Fetigan, Judth. 10; Thw. 21, 26; Jud. 35.

He ôðer fetaþ *ăliam duxĕrit,* Mt. Bos. 19, 9. Ǣlc ydel fet unhǣlo *all idleness brings illness,* Prov. Kmbl. 61. Se forma fette wíf, and forþferde *primus, uxōre ducta, defunctus est,* Mt. Bos. 22, 25: Gen. 48, 10. Wæs to bûre Beówulf fetod *Beowulf was fetched to his bower,* Beo. Th. 2625; B. 1310. DER. ge-fetian, -fætian. v. feccan.

fetlum *with belts,* Bt. 37, 1; Fox 186, 5. v. fetel.

fetor, e; *f. A fetter;* compes:—Îsern fetor *forfex,* Cot. 86. Îsen fetor *bǎlus,* Cot. 23. v. feter.

fetor-wrǎsen, e; *f.* [wrǎsen *a chain*] *A fetter, chain;* cǎtēna, compes:—Hraðe siððan wearþ fetorwrǎsnum fæst *he was soon fast bound in fetters,* Andr. Kmbl. 2215; An. 1109.

fett; *adj. Fat;* pinguis:—He biþ anlícost fettum swînum *he is most like to fat swine,* Bt. 37, 4; Fox 192, 26. v. fætt.

fette *fetched, brought, married,* Gen. 48, 10: Mt. Bos. 22, 25; *p. of* fetian.

fettian; *p.* ode; *pp.* od [fitt *contention, strife, fight*] *To contend, strive, dispute;* certāre, contendēre, dispūtāre:—Saturnus and Saloman fettodon ymbe heora wîsdóm *Saturn and Salomon contended about their wisdom,* Salm. Kmbl. p. 178, 7.

feuer-fuge, an; *f. Feverfew;* febrīfūgia:—Feuerfuge *feverfew,* Lchdm. iii. 12, 25. v. fefer-fuge.

fex, es; *n. Hair of the head, the locks;* cæsăries:—Fex *cæsăries,* Ælfc. Gl. 69; Som. 70, 39; Wrt. Voc. 42, 47: 70, 32. v. feax.

fexede *having long hair, long-haired,* Chr. 891; Th. 162, 9, col. 2, 3; 163, 10. v. feaxede.

fic *deceit, fraud, guile.* DER. ge-fic.

FÍC, es; *m.* **I.** *a FIG, the fruit of the fig-tree;* fīcus: found at present only in the following compounds in the sense of a tree or fruit, etc.—fíc-æppel, -beám, -leáf, -treów. **II.** *a disease so called, the piles, hemorrhoids;* ficus:—Wið seóndum ômum, ðæt is fíc *for running erysipelas, that is the 'fig,'* L. M. cont. 1, 39; Lchdm. ii. 10, 7: L. M. 1, 39; Lchdm. ii. 102, 12. Lǣcedómas and drencas and sealfa wið fíce *medicines and drinks and salves for the 'fig,'* L. M. cont. 1, 57; Lchdm. ii. 12, 18. Gif se fíc [MS. uíc] weorþe on mannes setle geseten *if the 'fig' be settled on a man's fundament,* Lchdm. iii. 30, 16. Se blêdenda fíc *the bleeding 'fig,'* iii. 38, 8. Wið ðone blêdendne [MS. blêdende] fíc nim murran ða wyrt *for the bleeding 'fig' take the plant sweet-cicely,* iii. 8, 1. [*Plat.* fige, *f: Dut.* vijg, *f: Ger.* feige, *f: M. H. Ger.* vîge, *f: O. H. Ger.* fîga, *f: Lat.* ficus, *f.* and *m.*]

fíc-ádl, e; *f.* [fíc **II.** *the piles, hemorrhoids*] *The fig-disease;* fīcus morbus:—Wið fícádle drenc and beðing *a drink and fomentation for the fig-disease,* L. M. cont. 3, 48; Lchdm. ii. 302, 24: L. M. 3, 48; Lchdm. ii. 340, 1.

fíc-æppel, -appel, es; *m; pl. nom. acc.* -æppla; *n. A fig-apple* or *fruit, a fig;* fîcus, fîcum cārîca, Ælfc. Gl. 46; Som. 64, 125; Wrt. Voc. 32, 59. Ne hîg of þornum fícæppla ne gaderiaþ *neque de spînis colligunt fîcus,* Lk. Bos. 6, 44: Mt. Bos. 7, 16.

fíc-beám, es; *m.* [beám *a tree,* v. **I.**] *A fig-tree;* fîcus:—Fîcbeám *fîcus,* Ælfc. Gl. 46; Som. 64, 122; Wrt. Voc. 32, 56. Behealdaþ ðone fîcbeám *vĭdēte fîculneam,* Lk. Bos. 21, 29. Forwurdan heora wîngeardas and fîcbeámas *percussit vîneas eōrum et fîculneas eōrum,* Ps. Th. 104, 29.

fíc-leáf, es; *n. A fig-leaf;* fîci fôlium:—Hîg siwodon fícleáf and worhton him wǣdbréc *consuērunt fôlia fîcus et fēcērunt sibi pěrizōmǎta,* Gen. 3, 7.

ficol; *adj. FICKLE, crafty;* versîpellis, inconstans, Prov. 14.

fíc-treów, es; *n. A FIG-TREE;* fîcus:—Forscranc ðæt fîctreów *fîcus āruit,* Mk. Bos. 11, 21: Mt. Bos. 21, 20: Wrt. Voc. 80, 11. Ðæs fíctreówes of the *fig-tree,* Mk. Bos. 11, 13. Leornigeaþ bigspel be ðam fíctreówes *ab arbŏre fîci discĭte pǎrǎbŏlam,* Mt. Bos. 24, 32: Mk. Bos. 13, 28. Hî gesǎwon ðæt fíctreów forscruncen of ðam wyrtruman *vidērunt fîcum āridam factam a rādicĭbus,* 11, 20: Mt. Bos. 21, 19. He ofslôh wîngeardas heora and fíctreów heora *percussit vîneas eōrum et fîculneas eōrum,* Ps. Spl. 104, 31.

fíc-wyrm, es; *m. A FIG-WORM, a worm originating from the fig-disease;* vermis ex fîco morbo ŏriens:—Feallaþ ða fícwyrmas on ða beðinge *dēcĭdent fîci morbi vermes in balneo,* L. M. 3, 48; Lchdm. ii. 340, 4.

fíc-wyrt, e; *f. The herb FIG-WORT;* fîcāria herba, fîcus, Ælfc. Gl. 41; Som. 63, 119; Wrt. Voc. 31, 6.

fieder *a father,* Cant. Moys. Ex. 15, 2; Thw. 2þ, 2. v. fæder.

fiell, es; *m. A fall, ruin, destruction;* cāsus, lapsus, ruĭna:—He wirþ swîðe raðe on fielle *he very quickly falls,* Past. 39, 3; Hat. MS. 53 b, 17. v. fyll.

fiénd *a fiend:*—Murnþ nâuðer ne friénd ne fiénd *regardeth neither friend nor foe,* Bt. 37, 1; Fox 186, 8. v. feónd.

fiénd-wíc, es; *n. An enemy's dwelling, a camp;* hostium vîcus, castra:—Hî feóllon on middele fiéndwîce heora *cecĭdērunt in mĕdio castrōrum eōrum,* Ps. Spl. T. 77, 32.

fier; *adv.* [fier, *comp.* of feor, *adv. far*] *Farther;* longius, ultĕrius:—Ðeáh ðû nû fier [fyr MS. Bod.] sîe ðonne ðû wǣre *though thou art now

farther than thou wast, Bt. 5, 1 ; Fox 8, 33. We areccan ne mågon ðæt fædrencynn fier ôwihte *we cannot reckon the paternal kin any degree farther*, Exon. 11 b ; Th. 16, 5 ; Cri. 248. v. fyr, fyrr.

fiér *four*, in the compound fiér-fête. v. feówer.

fierd, e ; *f. An army, force, expedition* ; exercĭtus, expēdītio :—Of ðære fierde *from the army*, Chr. 823 ; Erl. 62, 18 : 876 ; Erl. 78, 9 : 885 ; Erl. 82, 23 : 919 ; Erl. 104, 26. Ær sió fierd gesamnod wære *ere the army was assembled*, Chr. 894 ; Erl. 90, 21. v. fyrd.

fierdian ; *p.* ede ; *pp.* ed *To march* ; proficisci :—Mid ðære scîre ðe mid him fierdedon *with the division which marched with him*, Chr. 894 ; Erl. 90, 33. v. fyrdian.

fierdleás ; *adj. Without a force or army, unprotected* ; exercĭtu cärens :—Hit ðonne fierdleás wæs *it was then without a force*, Chr. 894 ; Erl. 90, 13. v. fyrdleás.

fieren-full *wicked*, Bt. Met. Fox 15, 13 ; Met. 15, 7, note. v. firen-full.

fiér-fête ; *adj. Four-footed* ; quadrŭpes :—Sume fiérfête *some are four-footed*, Bt. Met. Fox 31, 21 ; Met. 31, 11. v. feówer-fête.

FIERSN, fyrsn, e ; *f. The heel* ; calx :—Ðû scealt fiersna sǽtan *thou [the serpent] shalt lie in wait for her [Eve's] heels*, Cd. 43 ; Th. 56, 17 ; Gen. 913. [*Ger.* ferse, *f* : *M.H.Ger.* vërsen, *f* : *O.H.Ger.* fērsana, fërsina, fērsna, *f* : *Goth.* fairzna, *f* : *Grk.* πτέρνα, *f. the heel* : *Sansk.* pãrshṇi, *m. f. the heel.*]

fierst, es ; *m. The ceiling of a chamber* ; lăquear :—Fierst *lăquear*, Glos. Epnl. Recd. 158, 66. v. fyrst II.

fierst, es ; *m. A space of time, time* ; tempŏris spătium, tempus :—Forgif ðû me fierst and ongiet *give me time and understanding*, Exon. 118 a ; Th. 453, 28 ; Hy. 4, 21. v. fyrst.

FIF FIVE ; quinque.   1. *generally indecl* :—Hyra fíf wǽron dysige, and fíf gleáwe *quinque ex eis ĕrant fătuæ, et quinque prūdentes*, Mt. Bos. 25, 2 : Lev. 26, 8. Cômon ða fíf cynegas *ascendērunt quinque rēges*, Jos. 10, 5, 16. Ðǽra fíf hláfa *quinque pānum*, Mt. Bos. 16, 9. Of fíf hláfum *from five loaves*, Andr. Kmbl. 1179 ; An. 590 : Jn. Bos. 6, 13. We nabbaþ hêr bûton fíf hláfas and twegen fixas *non hăbēmus hic nĭsi quinque pānes et duos pisces*, Mt. Bos. 14, 17 : Lk. Bos. 9, 13, 16 : Jn. Bos. 6, 9 : Gen. 14, 9 ; 47, 2. Wintra hæfde fíf and hundteontig *he had a hundred and five winters*, Cd. 56 ; Th. 69, 5 ; Gen. 1131 : 59 ; Th. 71, 29 ; Gen. 1178 : 85 ; Th. 106, 26 ; Gen. 1777. Fíf sîdon *quinquies*, Ælfc. Gr. 38 ; Som. 40, 67. Fíf wintra fæc *lustrum quinquennium*, Ælfc. Gl. 16 ; Som. 58, 70 ; Wrt. Voc. 21, 57.   2. *but nom. acc. pl.* fife ; gen. fifa ; dat. fifum *are sometimes found* :—Fife ciningas lâgon *five kings lay [dead]*, Chr. 937 ; Th. 204, 1, col. 2 ; 205, 1 ; Æðelst. 28. Burga fife wǽron under Norþmannum *five towns were under the Northmen*, Chr. 942 ; Th. 208, 39 ; Edm. 5. Git sceolon fife geár *adhuc quinque anni restant*, Gen. 45, 6. Ðǽr fife [gimmas] wǽron *there were five [gems]*, Rood Kmbl. 16 ; Kr. 8. Him togeánes fife fôron folc-cyningas *five kings of nations marched against them*, Cd. 93 ; Th. 119, 3 ; Gen. 1974. Beóþ fife on ânum hûse todǽlede *ĕrunt quinque in dŏmo ûna divîsi*, Lk. Bos. 12, 52. Wǽron fife eorla and idesa *there were five men and women*, Exon. 112 b ; Th. 432, 1 ; Rä. 47, 6. Wintra hæfde twâ hundteontig and fife *he had two hundred and five winters*, Cd. 83 ; Th. 104, 28 ; Gen. 1742. Án ðissa fifa *one of these five*, Bt. 33, 3 ; Fox 126, 14. Bûton fifum *except five*, Chr. 897 ; Erl. 95, 28. [*Laym.* fif, uiuen : *Plat.* five, fiwe : *O. Sax.* fif, vîf : *Frs.* fyf : *O. Frs.* fif : *Dut.* vijf : *Ger.* fünf : *M. H. Ger.* vunf, vünf : *O. H. Ger.* fimf, finf : *Goth.* fimf, fif : *Dan. Swed.* fem : *Icel.* fimm : *Corn.* pemp : *Lat.* quinque : *Grk.* πέντε : *Æolic* πέμπε : *Sansk.* pañćan.]

Fíf burhga or burga ; *pl. f. The Five towns*, viz. *Leicester, Lincoln, Nottingham, Stamford, and Derby* ; quinque civitâtes :—On fif burhga geþincþe *in the assembly of the Five towns*, L. Eth. iii. 1 ; Th. i. 292, 6. Ferde se æðeling ðanon in to fíf burgum [burhgum, Th. 276, 7, col. 2] *the noble went thence to the Five towns*, Chr. 1015 ; Th. 276, 7, col. 1 ; 277, 7 ; 1013 ; Th. 270, 17, col. 2.

fife *five*. v. fíf 2.

fíf-ecgede ; *adj. Five-edged, five-cornered* ; quinquangŭlus :—Fífecgede *quinquangŭlus*, Ælfc. Gr. 49 ; Som. 50, 61.

fifel, es ; *n? m? A sea-monster, monster, giant* ; monstrum mărînum, gigas :—Þurh fifela gefeald forþ onette *through the field of the monsters he hastened forth*, Wald. 76 ; Vald. 2, 10. [*Icel.* fífl, *m.* 1. *a fool, clown, boor.* 11. *a monster, giant.*]

fifel-cyn, -cynn, es ; *n. A monster-race* ; monstrōrum mărînōrum gĕnus :—Fifelcynnes eard *the monster-race's abode*, Beo. Th. 209 ; B. 104.

fifel-dór, es ; *n. Monster* or *terror-door, the river Eider*, the boundary between Holstein and Schleswig ; monstrōrum mărînōrum porta :—Bî fifeldôre *by the monster-door*, Exon. 85 a ; Th. 321, 8 ; Wíd. 43.

fifele ? *a buckle, button* ; fibŭla, Som. Ben. Lye. v. figel.

fifel-streám, es ; *m. The frightful* or *horrid stream, the ocean* ; ōceănus :—Nǽnigne merehengesta mâ ðonne ǽnne ferede on fifelstreám *he led not more than one of the sea-horses on the ocean*, Bt. Met. Fox 26, 51 ; Met. 26, 26.

fifel-wǽg, es ; *m. The terrific wave, the ocean* ; ōceănus :—Léton ofer

---

fifelwǽg scrîðan bronte brimþisan *they let the high ships go over the ocean*, Elen. Kmbl. 473 ; El. 237.

fif-feald ; *adj. Five-fold* ; quintuplex, quīnărius :—Fíffeald *quīnărius*, Ælfc. Gr. 49 ; Som. 50, 16.

fif-fealde, -falde, an ; *f. A butterfly* ; pāpĭlio, Som. Ben. Lye :—Fíffealde *pāpĭlio*, Wrt. Voc. 281, 40. Fífaldæ *pāpĭlio*, Glos. Epnl. Recd. 160, 78.

fif-flére ; *adj.* [flôr *a floor*] *Five-floored, five-storied* ; quinque tăbŭlātis constans :—Se arc wæs fífflére *the ark was five-floored*, Boutr. Scrd. 21, 6.

fif-hund, -hundred *five hundred* ; quingenti :—Fífhund *quingenti*, Ælfc. Gr. 49 ; Som. 49, 48. Fífhund sîðon *five hundred times* ; quingenties, 49 ; Som. 50, 32. Fífhund cempena ealdor *a chief of five hundred soldiers* ; cohors, Ælfc. Gl. 7 ; Som. 56, 61 ; Wrt. Voc. 18, 14. Fundon fifhund forþsnotterra *they found five hundred of eminently wise men*, Elen. Kmbl. 757 ; El. 379. Fífhundred *quingenti*, Num. 1, 46.

fif-leáf, es ; *n.* -leáfe, an ; *f. Fiveleaf, cinquefoil* ; potentilla reptans, quinquefôlium :—Fífleáfe, Ælfc. Gl. 43 ; Som. 64, 54 ; Wrt. Voc. 31, 64 : 68, 69 : 79, 33 : 286, 40 : Herb. 3 ; Lchdm. i. 86, 20. Fífleáfan seáw *juice of fiveleaf*, Herb. 3, 2 ; Lchdm. i. 86, 24. Genim fífleáfan wyrtwalan *take the root of fiveleaf*, Herb. 3, 3 ; Lchdm. i. 86, 28. Genim fífleáfan ða wyrt *take the herb fiveleaf*, Herb. 3, 5 ; Lchdm. i. 88, 3, 9, 11, 14, 17, 20.

fifta ; *m.* : seó, ðæt fifte ; *adj. The* FIFTH ; quintus :—Se fífta dæg *the fifth day*, Gen. 1, 23. Fífta wæs Eádwine, Norþan Hymbra cyning *the fifth was Edwin king of the Northumbrians*, Chr. 827 ; Erl. 64, 3. Hêr onginnþ seó fífte bôc Boëties *here begins the fifth book of Boëthius*, Bt. 40, 4 ; Fox 240, 9. Ær ðam fiftan geáre *before the fifth year*, Lev. 19, 25. Ðæt gê habbon wǽstmas, and syllaþ ðam cynge ðone fiftan dǽl *ut frŭges hăbēre possîtis, quintam partem rēgi dăbĭtis*, Gen. 47, 24, 26.

fifta fæder ; *m. The fifth father* ; ătăvus :—Felix, se pápa wæs his [Gregories] fífta fæder *Felix, the pope was his [Gregory's] fifth father, that is*—reckoning Gregory's father as the first generation, his fifth father would be his great-grandfather's grandfather, Homl. Th. ii. 118, 9.

fiftegða *the fifteenth*, Bd. 4, 26 ; S. 602, 21. v. fífteóða.

fif-tene *fifteen* ; quindĕcim :—Fíftena sum *one of fifteen*, Beo. Th. 420 ; B. 207 : Cd. 69 ; Th. 84, 14 ; Gen. 1397. v. fíf-tyne.

fifteogoða ; *adj. The fiftieth* ; quinquăgēsimus :—Se fifteogoða *quinquăgēsimus*, Ælfc. Gr. 49 ; Som. 50, 1. v. fiftigoða.

fifteóða, fiftéða, fíftegða, fýfteogeða ; seó, ðæt fífteóðe ; *adj. The* FIFTEENTH ; quintus dĕcimus :—Môna [MS. mone] se fífteóða *the fifteenth moon*, Lchdm. iii. 190, 29. Ðam fíftéóðan geáre *anno quinto dĕcimo*, Lk. Bos. 3, 1. Under ðam fíftéðan dæge Kalendarum Octobris *sub die quinta decima Kalendas Octobres*, Bd. 4, 17 ; S. 585, 20. Heó leórde ðý fíftéóðan dæge *transivit die quinta dĕcima*, 4, 23 ; S. 592, 39. Ðý fíftegðan geáre *in the fifteenth year*, 4, 26 ; S. 602, 21.

fiftig FIFTY ; quinquăginta :—Fíftig yntsena scolfres *quinquăginta siclos argenti*, Deut. 22, 29. Fíftig wintra *fifty winters*, Beo. Th. 5459 ; B. 2733. Fíftig wintru, 4424 ; B. 2209. Se wæs fiftiges fôtgemearces lang *he was fifty feet of measure long*, 6076 ; B. 3042.

fiftigfeald ; *adj. Fiftyfold, containing fifty* ; quinquăgēnārius :—Fíftigfeald *quinquăgēnārius*, Ælfc. Gr. 49 ; Som. 50, 19.

fiftigoða, fíftigeða, fifteogoða ; *m.* : seó, ðæt fíftigoðe ; *adj. The fiftieth* ; quinquăgēsimus :—Fíftigoða *quinquăgēsimus*, Gr. Dial. 2, 2. Fíftigeða, C. R. Ben. 25. Ðæt fíftigoðe [MS. fífteóðe] gêr biþ hálig *the fiftieth year shall be holy* ; sanctificābis annum quinquăgēsimum, Lev. 25, 10.

fif-tyne, -tene *fifteen* ; quindĕcim :—Fíftyne fæðma *fifteen [of] cubits*, Gen. 7, 20. Fíftyne suna *fifteen [of] sons*, Boutr. Scrd. 21, 32. He slôh fiftyne men *he slew fifteen men*, Beo. Th. 3169 ; B. 1582. He on wéstenne wîceard geceás fiftynu geár *he chose a dwelling in the wilderness fifteen years*, Exon. 46 b ; Th. 158, 13 ; Gú. 908. Fíftyno, *acc. n.* Cd. 57 ; Th. 70, 10 ; Gen. 1151.

fif-wintre ; *adj. Of* or *belonging to five years, five years old* ; quinquennis :—Fífwintre *quinquennis*, Ælfc. Gr. 49 ; Som. 50, 45.

figan *to be* or *become an enemy, be at enmity* ; inĭmīcāri, inĭmīcĭtias exercēre, Som. Ben. Lye. v. feógan.

figel ? fifele ? *A buckle, button* ; fibŭla, Cot. 85, Lye.

fihle, es ; *m? n? A cloth, rag* ; pannus :—Fihles reáðes *panni rŭdis*, Mt. Kmbl. Lind. 9, 16.

fiht *fights*, Ex. 14, 14 ; *3rd sing. pres.* of feohtan.

fihtung, e ; *f. A fighting* ; pugnātio, dīmĭcātio, Som. Ben. Lye.

fiht-wíte, es ; *n. A fine for fighting* ; pugnæ mulcta :—He áh fihtwíte *he has fines for fighting*, L. C. S. 15 ; Th. i. 384, 3, note 6, MS. B. v. fyht-wíte.

fild ; *adj. Of* or *pertaining to a level field, even, flat, level* ; campester :—Seó burh wæs getimbred on fildum lande *the city [Babylon] was built on level land*, Ors. 2, 4 ; Bos. 44, 20.

fild, es ; *m? n?* e ; *f? A milking, the quantity of milk drawn at one milking* ; lactis quantĭtas sĕmel mulcta :—Gif fild sý awyrd *if a milking be spoilt*, L. M. 1, 67 ; Lchdm. ii. 142, 14. DER. fild-cum'þ.

fild-cumb, es ; *m.* [cumb II. *a liquid measure*] *A milk-pail* ;

mulctrāle, mulctrum :—Gif meoluc sīe awyrd, bind tosomne wegbrǽdan and giþrifan and cersan, lege on ðone fildcumb, and ne sete ðæt fæt niðer on eorþan seofon nihtum *if milk be spoilt, bind together waybroad and cockle and cress, lay them on the milk-pail, and set not the vessel down on the earth for seven nights*, L. M. 3, 53; Lchdm. ii. 340, 23-25.

filgst, filhst, he filgþ, filhþ *stickest to, sticks to; 2nd and 3rd pers. pres. of* felgan.

filian; *p.* filide *To follow;* sēqui :—Fīf eówer filiaþ hira hundteontig *persēquentur quinque de vestris centum āliēnos*, Lev. 26, 8. He filide me *he followed me*, Deut. 1, 36. v. fylgean.

filiende; *part. Rubbing;* fricans, Cot. 90.

fill, e; *f. Fulness, satiety, gluttony;* sătietas, inglŭvies :—He þurh fille unriht gefremode *he did wrong through gluttony*, L. Pen. 16; Wilk. 95, 58. v. fyll.

fille, an; *f. The plant thyme;* serpyllum = ἔρπυλλον :—Fille *serpyllum*, Wrt. Voc. 79, 47; Lchdm. iii. 34, 30.

filled *filled*, = fylled; *pp. of* fyllan.

film, e; *m. A* FILM, *skin, husk;* cŭtīcŭla, Som. Ben. Lye. v. fylmen.

filma, an; *m. A cleft;* rīma, Cot. 180.

filstan *to help, aid, assist* :—Gif he nelle filstan *if he will not help*, L. N. P. L. 54; Th. ii. 298, 19. v. fylstan.

filþ *filth, impurity, rottenness,* Som. Ben. Lye. v. fylþ.

FIN, finn, es; *m. A* FIN; pinna :—Ne ete gē nāune fisc būton ða ðe habbaþ finnas and scilla *ye shall not eat any fish except those that have fins and scales*, Lev. 11, 9. [*Plat.* finne, *f: Dut.* vin, *f: M. H. Ger.* vinne, *f: Dan.* finne, *m. f: Swed.* fena, *f: Lat.* pinna, *f.*]

fin? *A heap, pile;* strues, Cot. 195, Lye. DER. wudu-fin.

fina, an; *m. A woodpecker;* pīcus :—Fina *picus,* Ælfc. Gl. 38; Som. 63, 26; Wrt. Voc. 29, 46: 77, 31: 281, 4: Glos. Brux. Recd. 36, 33; Wrt. Voc. 62, 33.

FINC, es; *m. A* FINCH; fringilla :—Finc *fringilla*, Glos. Brux. Recd. 36, 37; Wrt. Voc. 62, 37: Glos. Epnl. Recd. 156, 57. [*Plat.* fink, finke, *m: Dut.* vink, *m: Ger.* fink, finke, *m: M. H. Ger.* vinke, *m: O. H. Ger.* finco, fincho, *m: Dan.* finke, *m. f: Swed.* fink, *m: Wel.* pinc, *m.*] DER. gold-finc, rago-.

fincer, es; *m. A finger;* digitus :—Dô hider fincer ðīnne *infer digitum tuum huc*, Jn. Rush. War. 20, 27. v. finger.

Finchamstede, -stæde, es; *m.* FINCHAMPSTEAD, *Berkshire;* lŏci nōmen in agro Berkeriensi :—Ðises geáres tó ðan sumeran, innan Barrucscīre æt Finchamstæde, ān mere blôd weóll *in the summer of this year* [A.D. 1098], *at Finchampstead in Berkshire, a pool welled out blood*, Chr. 1098; Th. 364, 4.

FINDAN, to findanne; ic finde, ðu findest, findst, fintst, finst, he findeþ, fint, *pl.* findaþ; *p.* fand, fond, funde, *pl.* fundon; *pp.* funden; *v. trans. To* FIND, *invent, imagine, devise, contrive, order, dispose, arrange, determine;* invēnīre, dispōnēre, consŭlēre :—Hīg ne mihton nāne fundan *non invēnērunt*, Mt. Bos. 26, 60: Bd. 1, 15; S. 483, 39. Ne mihte earmsceapen āre findan *nor might the poor wretch find pity*, Andr. Kmbl. 2260; An. 1131: 1960; An. 982. Tó findanne *to find*, Ps. Th. 76, 16. Ic hine finde ferþ staðelian *I find him strengthening his spirit*, Exon. 71 a; Th. 264, 14; Jul. 364: 67 a; Th. 247, 20; Jul. 81. Ðǽr ðú wraðe findest *there thou shalt find help*, Elen. Kmbl. 168; El. 84: Andr. Kmbl. 2698; An. 1351. Findst ðú ðǽr fíf mǽgþa *thou findest there five generations*, Boutr. Scrd. 22, 19, 20. Finst ðú *thou findest*, Bt. 18, 3; Fox 66, 11. Se ðe forstolen flǽsc findeþ *he who finds stolen flesh*, L. In. 17; Th. i. 114, 2. Nimþ eall ðæt hió fint *she will seize all she finds*, Bt. Met. Fox 13, 68; Met. 13, 34. Ðǽr hí fulle dagas findaþ sôna *dies plēni invēnientur in eis*, Ps. Th. 72, 8: 64, 10. Se cyning to nytnysse fand his leódum *rex ūtĭlĭtāti suæ gentis cōnsŭluit*, Bd. 2, 16; S. 520, 3. Heó nô reste fand *she found no rest*, Cd. 72; Th. 87, 30; Gen. 1456: 94; Th. 123, 6; Gen. 2040. Ic grundhyrde fond *I found the ground-keeper*, Beo. Th. 4279; B. 2136: Exon. 49 b; Th. 171, 2; Gú. 1120. Ic funde *I found*, Beo. Th. 2977; B. 1486: Gen. 12, 20. Ðú fundest *thou foundest*, Ps. Th. 16, 3. Swā we ǽr fundon *as we before determined*, L. Alf. pol. 18; Th. i. 72, 10. Wolde ic ðæt ðú funde ða *I would that thou wouldst find them*, Elen. Kmbl. 2157; El. 1080: Cd. 72; Th. 87, 6; Gen. 1444. Se cyng hæfde funden, ðæt ... *the king had contrived, that* ..., Chr. 918; Erl. 104, 3. [*Piers P.* fynden : *Laym.* finde, finden, ifinde, uinden, uinden : *Orm.* findenn : *Plat.* finnen : *O. Sax.* findan : *Frs.* fynnen : *O. Frs.* finna : *Dut.* vinden : *Ger.* finden : *M. H. Ger.* vinden : *O. H. Ger.* findan : *Goth.* finþan : *Dan.* finde : *Swed.* finna : *Icel.* finna.] DER. a-findan, an-, ge-, ofer-, on-, to-.

findele, an; *f?* es; *n? An invention, a device;* adinventio, inventum, Som. Ben. Lye.

findig; *adj. Considerable, good, heavy;* pondĕrōsus :—Findig corn *heavy corn*, Lye. DER. ge-findig.

finel, es; *m. Fennel;* fēuīcŭlum :—Fineles *of fennel*, Herb. 97, 1; Lchdm. i. 210, 8, MS. B. v. finol.

FINGER; *gen.* fingeres, fingres; *dat.* fingre; *pl. nom. acc.* fingras; *gen.* fingra, fingrena; *m. A* FINGER; digitus :—Finger *digitus*, Wrt. Voc. 71, 26. Send Lazarum, ðæt he dyppe his fingeres lið on wætere, and

mīne tungan gecǽle *mitte Lazarum ut intingat extrēmum digĭti sui in aquam, ut refrīgĕret linguam meam*, Lk. Bos. 16, 24. Gif ic on Godes fingre deófla ūtadrīfe *si in digĭto Dei ejĭcio dæmōnia*, 11, 20. On ðæm lytlan fingre *in the little finger*, Bt. Met. Fox 20, 359; Met. 20, 180. Ne gelýfe ic, būton ic dô mīnne finger on ðæra nægla stede *nisi mittam digĭtum meum in lŏcum clāvōrum non crēdam*, Jn. Bos. 20, 25, 27: Lev. 4, 17. Wulfere mid his fingre gewrāt on Cristes mēl *Wulfhere wrote with his finger on Christ's cross*, Chr. 656; Erl. 32, 23. Nellaþ hīg ða mid heora fingre æt-hrīnan *digĭto suo nōlunt ea mŏvēre*, Mt. Bos. 23, 4: Lk. Bos. 11, 46. Fingras *digĭti*, Wrt. Voc. 64, 78: 283, 18. Rand sceal on scylde, fæst finger gebeorh *a boss shall be on the shield, the sure protection of fingers*, Menol. Fox 535; Gn. C. 38: Elen. Kmbl. 239; El. 120. Ic geseó heofonas ðīne, weorc ðīnra fingra [MS. fingrena] *vidēbo cœlos tuos, ŏpĕra digĭtōrum tuōrum*, Ps. Lamb. 8, 4. Sum mæg fingrum hearpan stirgan *one can awaken the harp with fingers*, Exon. 17 b; Th. 42, 6; Cri. 668: Beo. Th. 3015; B. 1505. [*Laym.* finger, fenger : *O. Sax.* fingar, *m: Frs.* finger : *O. Frs.* finger, fingr, *m: Dut.* vinger, *m: Ger. M. H. Ger.* finger, *m: O. H. Ger.* fingar, *m: Goth.* figgrs, *m: Dan.* finger, *m. f: Swed.* finger, *m. n: Icel.* fingr, *m.*] DER. eáre-finger, gold-, hring-, lǽce-, leáw-, middel-, scyte-.

finger-æppel, es; *m: nom. acc. pl.* -æppla, -appla; *n. A* FINGER-APPLE, *finger-fruit, a date;* dactÿlus :—Fingerappla *dactÿlos*, Mone B. 542. Fingerapplum *dactÿlis*, 3830.

finger-lic; *adj. Of* or *belonging to a finger* or *ring;* digĭtālis, annŭlāris, Wrt. Voc. 65, 2.

fini; *adj. Decayed, mouldy;* corruptus, mūcĭdus :—Finie hlāfas *mouldy loaves*, Jos. 9, 5. v. fynig.

finiht; *adj.* [fin *a fin*] *Having fins, finny;* pinnĭger :—Scilfixas finihte *finny shell fishes*, L. M. 2, 37; Lchdm. ii. 244, 25.

Finn, es; *m. Fin, the king of the North Frisians* :—Finn [MS. Fin] Fresna cynne *Fin of the race of the Frisians*, Scôp. Th. 55; Wīd. 27. Be Finnes eaferum *in Fres-wæle of Fin's offspring in Friesland*, Beo. Th. 2140; B. 1068. v. Finns buruh.

Finnas; *gen.* a; *pl. m.* I. the Finns generally, including Scride-finnas and Ter-finnas, are the inhabitants of the north and west coast from Halgoland [v. map in Ors. Bos.] to the White Sea, as defined by Ohthere in the following example :—Ne mētte Ohthere nān gebūn land, syððan he fram his āgnum hāme [Hālgoland, *q. v.*] fôr; ac him wæs ealne weg wēste land on ðæt steór-bord, būtan fisceran, and fugeleran, and huntan, and ðæt wǽron ealle Finnas *Ohthere had not met with any inhabited land, since he came from his own home* [Halgoland]; *but the land was uninhabited all the way on his right, save by fishermen, fowlers and hunters, and they were all Finns*, Ors. 1, 1; Bos. 20, 3-6. Ða Finnas and ða Beormas sprǽcon neáh ān geþeóde *the Finns and the Biarmians spoke nearly the same language*, 1, 1: Bos. 20, 14: 19, 29. II. *Finwood, between Gothland and Smöland, in the south of Sweden* :—Ðá Beówulf sǽ ôþbær, flôd æfter faroþe, on Finna land *then the sea bore Beowulf away, the flood along the shore, on the Fins' land*, Beo. Th. 1165; B. 580. Not *Finland*, but the *Fins' land;* for how could Beowulf, in his swimming-match with Breca, be borne by the sea to Finland? Thorpe thinks the following extract may, however, afford a solution of the difficulty,—'Their [the Fins'] name is probably still to be found in the district of Finved [Finwood], between Gothland and Smöland. This inconsiderable and now despised race has, therefore, anciently been far more widely spread, and reached along the Kullen [the chain of mountains separating Norway from Sweden] down to the Sound, and eastward over the present Finland,' *Petersen, Danmarks Historie i Hedenold* i. p. 36. Ic wæs mid Finnum *I was with the Fins*, Scôp. Th. 153; Wīd. 76. DER. Scride-finnas, Ter-.

finnas *fins*, Lev. 11, 9; *pl. nom. acc. of* fin.

Finns buruh = Finnes burh; *gen.* -burge; *f. Finnsburg* :—Swylce eal Finnes buruh [MS. Finns] fýrenu wǽre *as if all Fin's castle were on fire*, Fins. Th. 72; Fin. 36. *This Finnsburg is no doubt the same as the Finneshäm mentioned by Beowulf*,—Swylce hie æt Finnes hām findan meahton *such as they might find at Finnesham*, Beo. Th. 2316; B. 1156. v. Finn.

FINOL, finul, finel, fynel, fenol, es; *m:* finule, finugle, an; *f. The plant* FENNEL; fēnĭcŭlum :—Finol *fēnĭcŭlum*, Glos. Brux. Recd. 41, 28; Wrt. Voc. 67, 43: L. M. 2, 34; Lchdm. ii. 238, 29. Genim finoles wyrttruman *take roots of fennel*, 1, 37; Lchdm. ii. 90, 6: 2, 11; Lchdm. ii. 188, 19: 2, 16; Lchdm. ii. 194, 23. Of ðam finole *from the fennel*, 2, 14; Lchdm. ii. 190, 22. Seóþ on ðam ecede ðone finol *seethe the fennel in the vinegar*, 2, 16; Lchdm. ii. 194, 26. [*Ger.* fenchel, *m: M. H. Ger.* venchel, *m: O. H. Ger.* fenachal, fenihil: *Lat.* fēnĭcŭlum, *n.*]

finol-sǽd, es; *n. Fennel seed;* fēnĭcŭli sēmen :—Finolsǽd gnīd to duste *reduce fennel seed to dust*, Lchdm. iii. 28, 3.

finst *findest*, Bt. 18, 3; Fox 66, 11, = findest; *2nd sing. pres. of* findan.

finta, an; *m.* I. *a tail;* cauda :—Ðonne is se finta fægre gedǽled *then is the tail* [*of the phœnix*] *beautifully divided*, Exon. 60 a; Th. 218, 15; Ph. 295. II. *what follows, a sequel, the consequence*

of an action; consĕquentia :—Ðonne he ðæs fācnes fintan sceáwaþ *when he sees the consequence of treachery,* Exon. 83 b; Th. 315, 17; Môd. 32: Exon. 74 b; Th. 278, 31; Jul. 606.

fintst, he fint *findest, finds,* Bt. Met. Fox 13, 68; Met. 13, 34; *2nd and 3rd pers. pres. of* findan.

finugle, an; *f. Fennel;* fēnĭcŭlum :—Wyl on ealoþ finuglan *boil fennel in ale,* L. M. 1, 39; Lchdm. ii. 104, 1: 1, 66; Lchdm. ii. 142, 2. v. finol.

finul, es; *m.* finule, an; *f. Fennel;* fēnĭcŭlum :—Genim ðysse wyrte wyrttruman, ðe man fēnĭcŭlum, and ôðrum naman finul nemneþ *take roots of this herb, which is named* fēnĭcŭlum, *and by another name fennel,* Herb. 126, 1; Lchdm. i. 238, 1: 382, 1. Genim finules niðeweardes *take some of the netherward part of fennel,* L. M. 1, 60; Lchdm. ii. 130, 18. Finule *fennel,* Lchdm. iii. 34, 30. v. finol.

fióde *hated,* Bt. 39, 1; Fox 212, 5; *p. of* fiógan, fión.

fiógan, fión; *p.* fióde, *pl.* fiódon *To hate;* ōdisse :—Fiógaþ yfel *ōdīte mălum,* Ps. Spl. T. 96, 10. Ðæt is unriht æghwelcum men ðæt he ôðerne fióge *it is wicked in every man that he should hate another,* Bt. Met. Fox 27, 47; Met. 27, 24. v. feógan.

fioh; *gen.* fiós; *dat.* fió; *n. Cattle, property, a portion;* pĕcus, ŏpes, dos :—Gif ðē becume ôðres monnes giémeleás fioh on hand *if the stray cattle of another man come to thy hand,* L. Alf. 42; Th. i. 54, 9: L. Ethb. 81; Th. i. 24, 1. v. feoh.

fioh-bót, e; *f. A pecuniary recompence;* nummāria compensātio, L. Alf. 49; Th. i. 58, 8. v. feoh-bót.

fiolan; *p.* fæl, *pl.* fælon; *pp.* folen *To reach, proceed, come;* procēdĕre, pervēnīre :—Hit fiolan ne mæg eft æt his ēþle *it cannot come again to its own region,* Bt. Met. Fox 20, 308; Met. 20, 154. v. feolan.

fión; *p.* fióde, *pl.* fiódon *To hate;* ōdisse :—Ic fióde cyrcean awyrgedra *ōdīvi ecclēsiam malignantium,* Ps. Spl. T. 25, 5. Hit nære nō manna ryht, ðæt hiora ænig ôðerne fióde *it would not be right in men, that any of them should hate another,* Bt. 39, 1; Fox 212, 5. v. feógan.

fiónd *a fiend,* Hy. 8, 25; Hy. Grn. ii. 290, 25. v. feónd.

fiónd-geld, es; *n. Devil-worship,* Mt. Lind. Stv. 4, 24. v. feónd-gyld.

fior; *adv. Far, at a distance;* prŏcul, longe :—Hió biþ swiðe fior hire selfre beneoðan *she is very far beneath herself,* Bt. Met. Fox 20, 443; Met. 20, 222. v. feor.

fiorh; *gen.* fiores; *dat.* fiore; *n. Life, spirit;* vīta, ănĭma :—Būton hiora āgnum fiore *except their own life,* Bt. 39, 11; Fox 230, 1. v. feorh I.

fiorm *use, benefit, profit, enjoyment,* Past. pref; Hat. MS. v. feorm IV.

fiórþa, seó, ðæt fiórþe; *adj. The fourth;* quartus :—Seó [MS. þio] fiórþe bôc *the fourth book,* Bt. 40, 4; Fox 240, 9, note 14. v. feórþa.

fióung, e; *f. Hatred;* ōdium :—Mid unrihtre fióunge *with evil hatred,* Bt. 39, 1; Fox 210, 24. DER. unriht-fióung. v. feóung.

fiówer-fête; *adj. Four-footed;* quadrŭpes :—Sume biþ fiówerfête *some are four-footed,* Bt. 41, 6; Fox 254, 27. v. feówer-fête.

fîr, es; *n. Fire;* ignis :—Þurh ðæs fîres fnæst *through the fire's blast,* Exon. 74 a; Th. 277, 29; Jul. 588. v. fŷr.

FIRAS, fyras; *gen.* a; *dat.* um; *pl. m. Living beings, the chief of living beings, men, mankind;* hŏmĭnes, vĭri, gĕnus hūmānum :—Firas monige *many men,* Runic pm. 26; Kmbl. 344, 28; Hick. Thes. i. 135, 52. Me wītan ne þearf Waldend fira *the Ruler of men need not upbraid me,* Beo. Th. 5476; B. 2741: 182; B. 91: Andr. Kmbl. 581; An. 291: 1840; An. 1072: Elen. Kmbl. 2153; El. 1078: 2343; El. 1173. Biþ ânra gehwylc flæsce bifongen fira cynnes *every one of the race of men shall be invested with flesh,* Exon. 63 b; Th. 234, 5; Ph. 535: 73 a; Th. 273, 1; Jul. 509: 92 b; Th. 347, 18; Sch. 14. Fira bearn *children of men,* Cd. 21; Th. 26, 17; Gen. 408. Firum uncūþ *unknown to men,* Bt. Met. Fox 4, 78; Met. 4, 39. Teóde firum foldan freá Ælmihtig *terram custos hūmāni gĕnĕris omnĭpŏtens creāvit,* Bd. 4, 24; S. 597, 23. [*O. Sax.* firihôs, *pl. m. men, people, mankind: Icel.* firar, *pl. m. men, people.*]

fird, e; *f. A force, army, expedition;* exercĭtus, expĕdītio :—Ne mehte seó fird hie hindan offaran *the force could not overtake them,* Chr. 894; Erl. 93, 7: 895; Erl. 93, 22: 905; Erl. 98, 19. Fôr Eádweard cyng mid firde to Steanforda *king Edward went with an army to Stamford,* 922; Erl. 108, 17. v. fyrd.

fird-cræft, es; *m. A war design, an expedition;* expĕdītio :—Mid hiora firdcræfte *by their expedition,* Num. 22, 4.

firding, e; *f. An expedition, army;* expĕdītio, exercĭtus :—Swiðe nicel folc ðū hæfst on ðînre firdinge to ðam gefeohte *very much people thou hast in thine army for the battle,* Jud. 7, 2. v. fyrding.

fird-stemn, es; *m. An army-corps;* exercĭtus cohors :—Ðá se firdstemn fôr hâm, ðá fôr ôðer ūt *when the army-corps went home, then another went out,* Chr. 921; Th. 195, 19.

FIREN, fyren, es; *pl. nom. acc.* firene, firena; *f.*   **I.** *a wicked deed, sin, crime;* scĕlus, crīmen, peccātum :—Næs ðær gefremed firen æt giftum *there was no sin committed at the nuptials,* Hy. 10, 17; Hy. Grn. ii. 293, 17. Nū eft gewearþ flæsc firena leás *flesh is again become void of sins,* Exon. 9 b; Th. 8, 25; Cri. 123: Elen. Kmbl. 2625; El.

1314: Salm. Kmbl. 897; Sal. 448. Firina gehwylc *each sin,* Exon. 8 b; Th. 4, 21; Cri. 56. Lýsde of firenum *released from sins,* 25 b; Th. 74, 22; Cri. 1210: Elen. Kmbl. 1814; El. 909. Uton we firene feógan *let us hate crimes,* Exon. 98 a; Th. 366, 16; Reb. 13: Ps. Th. 58, 3. Firena fremman *to perpetrate crimes,* Cd. 1; Th. 2, 14; Gen. 19: Salm. Kmbl. 632; Sal. 315.   **II.** *tribulation, torment, suffering, pain;* tribŭlātio, tormentum, crŭciātus :—Mid firenum *with torments,* Exon. 29 a; Th. 88, 16; Cri. 1441: 41 b; Th. 139, 26; Gū. 599. Wæron ealle fægen in firnum *they were all glad in their sufferings,* Cd. 223; Th. 292, 3; Sat. 435. [*O. Sax.* firina, *f. a wicked deed, crime, sin: O. Frs.* firne, ferne, *f.: O. H. Ger.* firina, *f. crimen, scĕlus, făcĭnus: Goth.* fairina, *f. crimination: Icel.* firn, *n. pl. a shocking thing, abomination.*] DER. folc-firen, hell-.

firen-bealu; *gen.* -bealuwes; *n. A sinful evil;* peccātum scĕlestum :—On him Dryhten gesihþ firenbealu lāþlíc *in them the Lord shall see loathly sinful evil,* Exon. 26 b; Th. 78, 19; Cri. 1276.

firen-cræft, es; *m. A sinful craft, wickedness;* scĕlesta ars, nēquĭtia :—Hí Dryhtnes ǽ feódon þurh firencræft *they hated the Lord's law in their wickedness,* Exon. 66 a; Th. 243, 21; Jul. 14.

firen-dǽd, fyren-dǽd, -dêd, e; *f. A wicked or sinful deed, crime;* scĕlestum făcĭnus :—Ðæt hie firendǽda tô frece wurdon *that they were too audacious in wicked deeds,* Cd. 121; Th. 155, 29; Gen. 2580: Exon. 118 a; Th. 453, 35; Hy. 4, 25. Firendǽda, Ps. C. 50, 44; Ps. Grn. ii. 277, 44. Firendǽdum fáh *stained with sinful deeds,* Exon. 22 b; Th. 62, 13; Cri. 1001: 31 b; Th. 99, 31; Cri. 1633.

firen-earfeðe *a sinful woe.* v. fyren-earfeðe.

firen-fremmende; *part. Committing sins;* scĕlĕra committens :—Ðæt he for ælda lufan firenfremmendra fela þrówade *that he suffered much for love of men committing crimes,* Exon. 24 a; Th. 69, 9; Cri. 1118.

firen-full, fyren-full, -ful; *adj. Sinful;* făcĭnŏrōsus, scĕlestus :—Swá firenfulle heora aldorþægn unreordadon *thus the sinful addressed their principal chief,* Cd. 214; Th. 268, 34; Sat. 65. Gif ðū wylt ða firenfullan fyllan mid deáþe *if thou wilt fell the wicked with death,* Ps. Th. 138, 16. Firenfulra *of the wicked,* Exon. 40 b; Th. 135, 30; Gū. 532: Ps. Th. 81, 4: 124, 3.

firen-georn; *adj. Sinful;* peccandi prōnus :—Firengeorne men *sinful men,* Exon. 31 b; Th. 98, 12; Cri. 1606.

firenian, firnian, fyrenian, fyrnian; *p.* ede; *pp.* ed.   **I.** *to sin;* peccāre :—Firenaþ ðus ðæt flæschord *thus will the body sin,* Exon. 99 b; Th. 373, 3; Seel. 103. Ða ðe firnedon beóþ beofigende *they who sinned shall be trembling,* Cd. 227; Th. 303, 29; Sat. 621.   **II.** *to revile;* călumniāri :—Heó firenaþ mec wordum *she reviles me with words,* Exon. 105 b; Th. 402, 24; Rä. 21, 34. [*O. H. Ger.* firinôn scĕlĕrāre: Goth.* fairinon *to criminate.*]

firenlíc; *adj. Wicked;* mālĭtiōsus, mālignus :—Hió me wrāþra wearn worda sprǽcon, fæcne, firenlícu *they spoke to me a multitude of wrathful words, deceitful, wicked,* Ps. Th. 108, 2.

firenlíce *vehemently, rashly.* v. fyrenlíce.

firen-ligerian *to commit fornication;* fornĭcāri. v. fyren-ligerian.

firen-lust, fyren-lust, es; *m. Sinful lust, luxury, wantonness;* libido, luxūria :—Mid ðý ðá ongon firenlust weaxan *cæpit cum quĭbus luxūria crescĕre,* Bd. 1, 14; S. 482, 22: Past. 27; Cot. MS. Hí firenlusta frece ne wǽron *they were not desirous of luxuries,* Bt. Met. Fox 8, 29; Met. 8, 15. Þurh firenlustas *through sinful lusts,* Exon. 29 b; Th. 90, 32; Cri. 1483: 44 a; Th. 150, 8; Gū. 775.

firen-synnig; *adj. Sinful;* făcĭnŏrōsus, scĕlestus :—Firensynnig folc *sinful people,* Exon. 28 a; Th. 84, 25; Cri. 1379.

firen-þearf *great distress, dire need.* v. fyren-þearf.

firen-weorc, es; *n. A wicked work, crime;* scĕlestum ŏpus, scĕlus :—Hí firenweorc beraþ *they bear their wicked works,* Exon. 26 b; Th. 80, 1; Cri. 1301: 28 a; Th. 85, 30; Cri. 1399.

firen-wyrcende; *part. Evil-doing, committing sin;* mălum făciens, peccans :—Me of folmum afere firenwyrcendra *take me out of the hands of those committing evil,* Ps. Th. 70, 3. Ic fyrenwyrcende oft elnade *I often emulated evil-doing* [*men*], 72, 2.

firen-wyrhta *an evil-doer, sinner.* v. fyren-wyrhta.

firgen, fyrgen, es; *n. A mountain, mountain-woodland;* mons, saltus. [*Goth.* fairguni, *n. a mountain: Icel.* Fjörgyn, *f. Mother-earth.*] DER. firgen-beám, -bucca, -gát, -holt, -stream.

firgen-beám *a mountain-tree.* v. fyrgen-beám.

firgen-bucca *a mountain-buck.* v. firgin-bucca.

firgend-streám *a mountain-stream,* Andr. Kmbl. 3144; An. 1575. v. firgen-stream.

firgen-gát, firgin-gát, e; *pl. nom. acc.* -gǽt; *f. A mountain-goat, chamois;* montāna *vel* saltuensis capra, ĭbex :—Firgengát [MS. firing-gát] ibex, Ælfc. Gl. 20; Som. 59, 39; Wrt. Voc. 23, 2. Firgengát *mountain-goat,* Cot. 109: 116. Firgingǽt [MS. -gǽtt] ibĭces, Glos. Epnl. Recd. 158, 31.

firgen-holt *a mountain-wood.* v. fyrgen-holt.

firgen-streám, fyrgen-streám, firgend-streám, firigend-streám, es; *m,*

A mountain-stream, the ocean; montānum *vel* saltuense flūmen, oceānus;—Hió ðæt líc ætbær under firgenstreám *she bore the corpse away under the mountain-stream,* Beo. Th. 4263; B. 2128. Fugel on firgenstreám lôcaþ georne *the bird looks earnestly into the mountain-stream,* Exon. 57 a; Th. 204, 20; Ph. 100. Wæs ic firgenstreámum swíðe besuncen *I was deeply sunk in mountain-streams,* 103 b; Th. 392, 4; Rä. 11, 2. Ymb ealra land gehwilc flôwan firgenstreámas *mountain-streams* [*shall*] *flow over every land,* Menol. Fox 555; Gn. C. 47. Fleów firgendstreám *the mountain-stream flowed,* Andr. Kmbl. 3144; An. 1575. Ofer firigendstreám *over the ocean,* Andr. Kmbl. 779; An. 390.

firgin-bucca, an; *m. A mountain-buck, wood-buck;* montānus *vel* saltuensis cǎper :—Firginbucca ðæt ys wudubucca *a mountain-buck that is a wood-buck,* Med. ex Quadr. 5, 1; Lchdm. i. 348, 2. v. firgen-bucca.

firgin-gǽtt *mountain-goats,* Glos. Epnl. Recd. 158, 31. v. firgen-gát.

firhþ-sefa, an; *m. The mind;* mens :—On firhþsefan *in his mind,* Elen. Kmbl. 425; El. 213. v. ferhþ-sefa.

firige *let him make a fire,* L. Pen. 14; Wilk. 95, 30. v. fýrian.

firigend-streám *a mountain-stream, the ocean,* Andr. Kmbl. 779; An. 390. v. firgen-streám.

firing-gát *a mountain-goat,* Ælfc. Gl. 20; Som. 59, 39; Wrt. Voc. 23, 2. v. firgen-gát.

firmetan; *p.* firmette, *pl.* firmetton; *pp.* firmeted *To request, pray,* pětēre, rôgāre :—Rômane hí firmetton ðæt hí ðæt gewin forlēton *the Romans requested them that they would leave off the siege,* Ors. 4, 8; Bos. 89, 21.

firna *sins, crimes,* Cd. 216; Th. 274, 27; Sat. 160; *acc. pl. of* firen.

firne *crime,* Cd. 227; Th. 305, 3; Sat. 641; *dat. of* firen.

firnian *to sin,* Cd. 227; Th. 303, 29; Sat. 621. v. firenian.

firnum, fyrnum; *adv. [dat. or inst. pl. of* firen *a sin, crime] Fearfully, intensely;* formīdōlōse, immānīter :—Nǽre firnum ðæs deóp merestreám *the sea-stream would not be so fearfully deep,* Cd. 39; Th. 51, 26; Gen. 832.

firra; *m.* firre; *f. n. adj.* [*comp. of* feor, *adj. far*] *Farther;* ultěrior :—On ðære firran Ispānie *in the farther Spain,* Ors. 4, 11; Bos. 97, 26. v. fyrra.

firran *to remove, take away.* DER. a-firran. v. feorran.

fír-scofl *a fire-shovel;* bātillum, Som. Ben. Lye. v. fýr-scofl.

first, es; *m. A rafter, beam, perch;* tigillum, pertǐca :—First *paratica?* [*=pertica*], Wrt. Voc. 290, 3. v. fyrst.

first, es; *m. A space of time, time;* tempǒris spătium, tempus :—Ðá wæs first agán *then was the time expired,* Andr. Kmbl. 293; An. 147. Ôþ ðone first ðe hie wurdon swíðe meteleáse *until the time that they were very destitute of food,* Chr. 918; Erl. 104, 12 : Bt. 38, 1; Fox 194, 27. v. fyrst.

first *first,* Chr. 675; Erl. 39, 28. v. fyrst.

first-mearc *an interval of time;* intercǎpēdo, Som. Ben. Lye. v. frist-mearc.

firþriende *furthering;* promǒvens, M. A. 1, p. 223, Lye. v. fyrþran.

firwet *curiosity.* DER. firwet-georn, -geornes. v. fyrwet.

firwet-georn; *adj. Very inquisitive, curious;* cūriōsus :—Ða ðe firwet-georne weorþaþ *they who are very inquisitive,* Bt. 39, 3; Fox 216, 4 : Bt. Met. Fox 28, 151; Met. 28, 76. v. fyrwet-georn.

firwet-geornes, -ness, e; *f. Curiosity, anxiety;* sollicītūdo, Cot. 60. v. fyrwet-geornes.

FISC, es; *pl. nom. acc.* fiscas, fixas, fisceas; *gen.* fisca, fixa; *dat.* fiscum, fixum; *m. A FISH;* piscis :—Fisc *piscis,* Wrt. Voc. 65, 60 : 77, 57 : 281, 54. Fise sceal on wætere cynren cennan [MS. cynran cennen] *the fish shall propagate his kind in the water,* Menol. Fox 514; Gn. C. 27 : Salm. Kmbl. 841; Sal. 420. Híg brohton him dǽl gebrǽddes fisces, and beóbread *illi obtulērunt ei partem piscis assi, et făvum mellis,* Lk. Bos. 24, 42 : Mt. Bos. 7, 10 : Deut. 4, 18. We ðé willaþ ferigan freólíce ofer fisces bæþ *we will freely convey thee over the fish's bath,* Andr. Kmbl. 586; An. 293 : Exon. 116 b; Th. 447, 14; Dôm. 39. Nim ðone ǽrestan fisc *take the first fish,* Mt. Bos. 17, 27 : Jn. Bos. 21, 13. Bletsien ðec fiscas and fuglas *may fishes and birds bless thee,* Exon. 55 a; Th. 194, 16; Az. 140 : 97 b; Th. 364, 33; Wal. 80. Ða fixas, ðe wǽron on ðam flôde, wurdon deáde *pisces quī ērant in flūmǐne, mortui sunt,* Ex. 7, 21 : Ors. 5, 4; Bos. 105, 15. Earmra fisca *of poor fishes,* Salm. Kmbl. 164; Sal. 81 : Bt. Met. Fox 11, 133; Met. 11, 67. Híg betugon mycele menigeo fixa *conclūsērunt piscium multītūdǐnem cōpiōsam,* Lk. Bos. 5, 6 : Mt. Bos. 15, 34 : Mk. Bos. 6, 43 : 8, 7. Hí gefēngon þreó hund fixa missenlícra cynna *they caught three hundred fishes of diverse kinds,* Bd. 4, 13; S. 583, 1. Mid fiscum *with fishes,* Exon. 22 a; Th. 60, 10; Cri. 967 : 126 b; Th. 487, 19; Rä. 73, 4. He afēdde of fixum twǽm and of fíf hláfum fíf þūsendo *he fed five thousand from two fishes and from five loaves,* Andr. Kmbl. 1178; An. 589 : Mk. Bos. 6, 41. We nabbaþ hér, būton fíf hláfas and twegen fixas *non hăbēmus hic, nisi quinque pānes, et duos pisces,* Mt. Bos. 14, 17 : Lk. Bos. 9, 13 : Jn. Bos. 6, 9 : 21, 10 : Gen. 1, 26. Heora fisceas forwurdan *occīdit pisces eōrum,* Ps. Th. 104, 25. [*Wyc.* fische : *Chauc.* fissch, fissche : *Laym.* fisc, uisc, *m* : *Orm.* fisskess *fishes, pl* : *Plat.* fisk, *m* : *O. Sax.* fisc, visc, *m* : *Frs.* fisck : *O. Frs.* fisk : *Dut.* visch, *m* : *Ger.* fisch, *m* : *M. H. Ger.* visch, *m* :

O. H. Ger. fisc, *m* : *Goth.* fisks, *m* : *Dan.* fisk, *m. f* : *Swed.* fisk, *m* : *Icel.* fiskr, *m* : *Lat.* piscis, *m* : *Wel.* pysg, *m* : *Corn.* pesc, pysc, pisc, *m* : *Armor.* pesc : *Ir.* iasg, iasc, *m* : *Gael.* iasg, éisg, *m.*] DER. eá-fisc, horn-, hran-, hron-, mere-, sǽ-.

fiscaþ, es; *m. A fishing;* piscātus :—Ðǽr biþ swýðe mycel fiscaþ *there is very much fishing,* Ors. 1, 1; Bos. 22, 14. v. fiscoþ.

fisc-bryne *fish-brine;* piscium salsūgo :—Fiscbryne *liguamen?* vel gārum, Ælfc. Gl. 32; Som. 62, 13; Wrt. Voc. 27, 66.

fisc-cynn, -cinn, es; *n. The fish kind, kind of fishes;* piscium gěnus :—Is heofena ríce gelíc asendun nette on ða sǽ, and of ǽlcum fisccynne gadrigendum *sǐmǐle est regnum cœlōrum săgēnæ missæ in măre, et ex omni gěněre piscium congrěganti,* Mt. Bos. 13, 47. God gesceóp ðá ða micelan hwalas and eall libbende fisccinn on heora hiwum *then God created the great whales and every living kind of fishes after their kinds,* Gen. 1, 21 : Ælfc. T. 8, 25.

fisceran *=fiscerum with fishers,* Ors. 1, 1; Bos. 20, 5; *dat. pl. of* fiscere. v. fugeleran, drían.

fiscere, es; *m.* I. *A FISHER;* piscātor :—Ic eom fiscere *ěgo sum piscātor,* Coll. Monast. Th. 23, 1 : Wrt. Voc. 73, 40. Hí wǽron fisceras *ěrant piscātōres,* Mt. Bos. 4, 18 : Mk. Bos. 1, 16. Ðæra Terfinna land wæs eall wēste, būtan ðǽr huntan gewícodon, oððe fisceras, oððe fugeleras *the land of the Terfinns was all waste, save where the hunters, fishers or fowlers encamped,* Ors. 1, 1; Bos. 20, 9. Ða fisceras eódon, and wôxon heora nett *piscātōres descenděrant et lăvābant rētia,* Lk. Bos. 5, 2. Fiscerum [MS. fisceran] *with fishers,* Ors. 1, 1; Bos. 20, 5. II. *the bird king-fisher;* alcēdo :—Fiscere *rapariolus?* [*= rǐpārǐolus?*], Ælfc. Gl. 38; Som. 63, 44; Wrt. Voc. 29, 62.

fisc-hús, es; *n. A fishing-house;* piscīnāle, Ælfc. Gl. 108; Som. 78, 105; Wrt. Voc. 58, 20.

fiscian, fixian; *p.* ode; *pp.* od *To fish;* piscāri :—Ðonne gē fiscian willaþ *when ye wish to fish,* Bt. 32, 3; Fox 118, 12.

fisc-mere, es; *m. A fish-pond;* piscina, vīvārium, Som. Ben. Lye.

fisc-naþ, es; *m? A fishing;* piscātus :—On fiscnaþe *by fishing,* Bd. 4, 13; S. 582, 41. v. fisc-nôþ.

fisc-net, -nett, es; *n. A net of fishes, fishing net;* piscium rēte, piscā-tōrium rēte :—Hí tugon hyra fiscnett *trăhentes rēte piscium,* Jn. Bos. 21, 8. Hwý gē ne settan on sume dūne fiscnet eówru *why do ye not set your fishing nets on some hill?* Bt. Met. Fox 19, 21; Met. 19, 11.

fisc-noþ, -naþ, es; *m? A fishing;* piscātus :—Seó þeód ðone cræft ne cūðe ðæs fiscnoþes *the people knew not the art of fishing,* Bd. 4, 13; S. 582, 43.

fiscoþ, fiscaþ, fixoþ, es; *m? A fishing;* piscātus :—On fiscoþe, Ors. 1, 1; Bos. 19, 30 : on fixoþ *afysshynge* (Tyndale) Jn. Bos. 21, 3.

fisc-pól, es; *m? A fish-pool, fish-pond;* piscina, vīvārium :—Fiscpôl *vīvārium,* Ælfc. Gl. 98; Som. 76, 94; Wrt. Voc. 54, 38 : 80, 66 : piscina, Som. 76, 95; Wrt. Voc. 54, 39. On fiscpôle *in a fish-pool,* Lchdm. iii. 212, 15.

fisc-wēr, es; *m.* [wēr II. *a draught of fishes*] *A draught of fishes;* piscium captūra :—Lǽtaþ eówre nett on ðone fiscwēr *laxāte rētia vestra in captūram* [*piscium*], Lk. Bos. 5, 4.

fisc-wylle, -welle; *adj.* [cf. weallan *to swarm*] *Full of fish, abounding in fish;* piscǐbus abundans, piscōsus :—Ðæt eálond is fiscwylle *the island is abounding in fish,* Bd. 1, 1; S. 474, 41. Fiscwyllum wæterum *flūviis piscōsis,* 1, 1; S. 473, 15. Fiscwelle *bisarius?* [*= piscārius*], Wrt. Voc. 66, 8.

fisting, e; *f. Fesciculatio? forte* fistulātio, Som. 72, 65; Ælfc. Gl. 79; Wrt. Voc. 46, 23.

fit, fitt, es; *n? Strife, a fight, contest;* rixa, pugna, certāmen :—He slôh and fylde feónd on fitte *he struck and felled the enemy in fight,* Cd. 95; Th. 124, 33; Gen. 2072. v. fettian, fitung.

fit, fitt, e; *f. A song, poem;* cantilēna, carmen :—Ðá se Wīsdóm ðas fitte asungen hæfde *when Wisdom had sung this song,* Bt. 30, 1; Fox 106, 29. On fitte *in song, verse,* Bt. Met. Fox introduc. 17; Met. Einl. 9.

fiter-sticca, an; *m. A tent-nail;* clāvus tentōrii :—Fitersticca *clāvus tentōrii,* Ælfc. Gl. 110; Som. 79, 42; Wrt. Voc. 59, 14.

FIÐELE, an; *f. A fiddle,* Som. Ben. Lye. [*Piers P.* fithele : *Chauc.* fithul : *Laym.* fiðele : *Plat.* fidel, *f* : *Dut.* vedel, veel, *f* : *Ger.* fiedel, fidel, *f* : *M. H. Ger.* videle, videl, *f* : *O. H. Ger.* fidula, *f* : *Dan.* fiddel, *m. f* : *Icel.* fiðla, *f* : *M. Lat.* fidula, vidula : *Lat.* fīdes, *f. a string, guitar.*]

fiðelere, es; *m. A fiddler;* fiðicen :—Fiðelere *fiðicen,* Ælfc. Gr. 9, 12; Som. 9, 25; Wrt. Voc. 73, 61.

fiðelestre; *f.* [fiðele *a fiddle,* -estre *a female termination,* q. v.] *A female fiddler;* fiðicīna, Wrt. Voc. 73, 62.

fiðer- *four-* in the compounds fiðer-fēte, -scýte. v. fyðer-, feówer.

fiðer-berende; *part. Bearing wings, winged;* āliger, Cot. 9 : 170.

fiðere, es; *n. A wing :* āla : more often found in the *pl. nom. acc.* fiðera, fiðeru, fyðera, fyðeru, fýðru; *gen.* fiðera, fyðera, fyðerena; *dat. inst.* fiðerum, fiðrum, fýðerum; *n :* also the forms are sometimes found *pl. nom. acc.* fiðeras, fyðeras; *m. Wings;* ālæ, pennæ :—Gif his óðer fiðere forod biþ *if one of its wings* [lit. *one wing of it*] *is broken,*

U

Homl. Th. ii. 318, 29. Fiðera [Spl. fyðera : Lamb. fyðeras] beóþ culfran fægeres seolfres *pennæ cŏlumbæ sunt deargentātæ*, Ps. Th. 67, 13. Sindon ða fiðru hwît *the wings are white*, Exon. 60 a ; Th. 218, 20 ; Ph. 297. Bearn manna under wæfelse oððe on gescyldnesse ðînra fiðera [Spl. fyðera] hopiaþ *filii hŏmĭnum in tegmĭne ālārum tuāruṁ spērābunt*, Ps. Lamb. 35, 8 : 56, 2 : 60, 5. Gehȳd me under ðînra fiðera [Lamþ. fyðerena] sceade *sub umbra ālārum tuārum protĕge me*, Ps. Th. 16, 8. Under fiðerum [Th. fiðrum : Lamb. fyðerum] his ðû hopudest *sub pennis ejus spērābis*, Ps. Spl. 90, 4 : Lk. Bos. 13, 34. Nabbaþ hî æt fiðrum fultum *they have no help from wings*, Bt. Met. Fox 31, 15 ; Met. 31, 8. Fleáh ofer fiðera [Th. fiðeru : Lamb. fyðru] winda *vŏlāvit sŭper pennas ventōrum*, Ps. Spl. 17, 12 : Homl. Th. ii. 318, 27. Abred of ða fiðeru *take off the wings*, Lev. 1, 17 : Ps. Th. 54, 6 : 138, 7 : Salm. Kmbl. 528 ; Sal. 263. Se fôtum tredeþ fiðru [Spl. fyðeru : Lamb. fyðeras] winda *qui ambŭlat sŭper pennas ventōrum*, Ps. Th. 103, 4 : Bt. Met. Fox 24, 1 ; Met. 24, 1 : Exon. 65 a ; Th. 241, 7 ; Ph. 652 : 109 b ; Th. 418, 18 ; Rä. 37, 7. Ac ðær ic môste ðîn môd gefiðerigan mid ðam fiðerum, ðæt ðû mihtest mid me fliógan *but if I were allowed to furnish thy mind with wings, that thou mightest fly with me*, Bt. 36, 2 ; Fox 174, 6 : Ps. Th. 60, 3 : 62, 7 : 148, 10. Hî mid hyra fiðrum weardiaþ [MS. wearþ] *they protect with their wings*, Exon. 13 b ; Th. 25, 3 ; Cri. 395 : 55 a ; Th. 195, 23 ; Az. 160 : 60 b ; Th. 220, 7 ; Ph. 316 : 88 b ; Th. 332, 21 ; Vy. 88 : Elen. Kmbl. 1482 ; El. 743. Him fiðeras ne fultumaþ *wings support them not*, Bt. 41, 6 ; Fox 254, 26. v. feðer II.

fiðer-féte, -fôte ; *adj. Four-footed ;* quădrŭpes :—Ælcum fiðerfêtum neáte *for any four-footed beast*, Med. ex Quadr. 1, 3 ; Lchdm. i. 328, 13. Eallum fiðerfêtum nȳtenum *to all four-fóoted beasts*, 1, 3 ; Lchdm. i. 330, 4. Fiðerfôte fugel *a four-footed bird, griffin ;* griffus, gryps = γρύψ, Wrt. Voc. 78, 2. v. feówer-fête.

fiðerian, fiðerigan, fiðrian *to give wings to, provide with wings*. DER. ge-fiðerian.

fiðer-leás ; *adj. Wingless ;* ālis cărens :—Sum sceal of heán beáme fiðerleás feallan *one wingless shall fall from a high tree*, Exon. 87 b ; Th. 328, 23 ; Vy. 22.

fiðer-scȳte, -scîte ; *adj. Four-cornered, quadrangular, square ;* quadrangŭlus, quadrāta :—Fiðerscȳte setel *siliquastrum vel cathedra quadrāta*, Ælfc. Gl. 116 ; Som. 80, 66 ; Wrt. Voc. 61, 44. Seó cyrce wæs eal of fiðerscîtum marmstânum geworht *the church was built all of quadrangular marble stones*, Homl. Th. ii. 496, 35. v. feówer-scȳte.

fiðru *wings*, Exon. 60 a ; Th. 218, 20 ; Ph. 297 : 65 a ; Th. 241, 7 ; Ph. 652 ; *pl. nom. acc. of* fiðere.

fiðrum *to* or *with wings*, Bt. Met. Fox 31, 15 ; Met. 31, 8 : Elen. Kmbl. 1482 ; El. 743 ; *pl. dat. and inst. of* fiðere.

fittan ; *p.* te ; *pp.* ed *To sing ;* cantāre :—Nû ic sitte gēn ymb fisca cynn *now again I sing about* [*the*] *kind of fishes*, Exon. 96 b ; Th. 360, 5 ; Wal. 1. [*Dut.* vitten *to criticise*.]

fitung, fytung, e ; *f. A fighting, quarreling ;* rixa :—Ascŭnige man swîðe fracodlîce fitunga *let a man earnestly shun shameful fightings*, L. Eth. vi. 28 ; Th. i. 322, 14.

fîwan *to hate ;* ŏdio hăbēre, inĭmīcāri, Som. Ben. Lye. v. feógan, feón.

fixas *fishes*, Ex. 7, 21 : Mt. Bos. 14, 17 : Lk. Bos. 9, 13 ; *pl. nom. acc. of* fisc, *q. v.*

fixen, e ; *f. A she-fox*, VIXEN ; vulpes fēmĭna, Som. Ben. Lye.

fixen ; *adj.* [fox *a* fox] *Of* or *belonging to a fox ;* vulpīnus :—Fixen hȳd *a fox-skin*, Med. ex Quadr. 3, 15 ; Lchdm. i. 342, 11.

fixian ; *p.* ode ; *pp.* od [fisc = fix *a fish*] *To fish ;* piscāri :—Ic fixige *piscor*, Ælfc. Gr. 25 ; Som. 27, 11. For hwî ne fixast ðû on sǽ *cur non piscāris in mări?* Coll. Monast. Th. 24, 1. v. fiscian.

fixoþ, es ; *m* ? *A fishing ;* piscātus :—Ic wylle gān on fixoþ *vādo piscāri*, Jn. Bos. 21, 3. v. fiscoþ.

FLÁ, flaa ; *gen. dat. acc.* flân ; *pl. nom. acc.* flân ; *gen.* flâna ; *dat.* flânum ; *f.* [flae, *gen.* flaan = flân ; *f.*] *An arrow, a dart, javelin ;* săgitta, tēlum, jăcŭlum :—Flâ săgitta vel *tēlum*, Wrt. Voc. 84, 27 : Ælfc. Gr. 8 ; Som. 7, 60 : Ælfc. Gl. 52 ; Som. 66, 35 ; Wrt. Voc. 35, 24. Flaa *tēlum* vel ŏbĕliscus = ὀβελίσκος, 53 ; Som. 66, 63 ; Wrt. Voc. 35, 49. Wîdnyt *vel* flâ jăcŭlum vel *funda*, 18 ; Som. 58, 106 ; Wrt. Voc. 22, 21. Wearþ Alexander þurhscoten mid ânre flân underneoþan ôðer breóst *Alexander was shot through with an arrow underneath one breast*, Ors. 3, 9 ; Bos. 68, 27. He gedēþ his flân fȳrena *săgittas suas ardentĭbus effēcit*, Ps. Th. 7, 13 : 90, 6 : Deut. 32, 42. Flâna scûras *showers of arrows*, Elen. Kmbl. 234 ; El. 117 : Judth. 11 ; Thw. 24, 33 ; Jud. 221. Sî he mid stânum oftorfod oððe mid flânum ofscotod *lăpidĭbus opprĭmētur aut confŏdiētur jăcŭlis*, Ex. 19, 13 : Ps. Th. 10, 2. [*Chauc.* flo ; *pl.* flone : *Laym.* fla, flo : *Icel.* fleinn, *m. a dart.*] v. flân.

fiacea *flakes of snow ;* flocci nĭvis, Som. Ben. Lye.

fiacge, an ; *f. A poultice ;* cataplasma, Cot. 55.

flacor ; *adj. Flickering ;* vŏlĭtans :—Flacor flânþracu feorhhord onleác *the flickering arrow's force unlocked life's treasury*, Exon. 49 b ; Th. 170, 25 ; Gû. 1117. Ofer scildhreáðan sceótend sendaþ flacor flânge-weorc *warriors send flickering arrow-work over the shield's defence*, 17 b ; Th. 42, 21 ; Cri. 676.

---

flésc *flesh ;* căro, Ælfc. Gl. 69 ; Som. 70, 31 ; Wrt. Voc. 42, 39. v. flǽsc.

fléh *a flea ;* pūlex, Som. Ben. Lye. v. fleá.

fléh *a flea ;* pūlex, Som. Ben. Lye. v. fleá.

flém, es ; *m. Flight ;* fūga :—He deófla afyrseþ and on flǽme gebringeþ *he sends away devils and puts them to flight*, L. C. E. 4 ; Wilk. 128, 15. v. fleám.

fléman, fléman ; *p.* de ; *pp.* ed *To cause to flee, put to flight ;* fŭgāre. DER. ge-fléman, -fléman. v. flȳman.

flén *a lance ;* frămea, Ps. Spl. 16, 14. v. flân.

flére, an ; *f. An earlap ;* pinnŭla auris :—Flǽran = eár-læppan *pinnŭlæ aurium = auricŭlæ*, Ælfc. Gl. 71 ; Som. 70, 84 ; Wrt. Voc. 43, 16. v. eár-læppa.

FLÆSC, es ; *pl. nom. acc.* flǽsc ; *gen.* flǽsca, flǽscea ; *dat.* flǽscum ; *n :* flǽsc, es ; *n.* FLESH ; căro :—Se gást is hræd, and ðæt flǽsc ys untrum *spirĭtus promptus est, căro autem infirma*, Mt. Bos. 26, 41 : Mk. Bos. 14, 38. Ðæt Word wæs geworden flǽsc, and wunode on us *the Word became flesh, and dwelt in us*, Homl. Th. i. 40, 17 : Exon. 9 b ; Th. 8, 25 ; Cri. 123 : 16 b ; Th. 37, 23 ; Cri. 597. Sôþlîce mîn flǽsc is mete, and mîn blôd ys drinc *căro ěnim mea věre est cĭbus, et sanguis meus věre est pōtus*, Jn. Bos. 6, 55 : Lk. Bos. 3, 6 : Gen. 2, 23 : 6, 3 : Ps. Spl. 15, 9 : Ps. Lamb. 55, 4 : 77, 39. Ge-endung ealles flǽsces com ætfôran me *finis ūnĭversæ carnis věnit cōram me*, Gen. 6, 13, 19 : Jn. Bos. 1, 13. In flǽsce *in the flesh*, Bt. Met. Fox 20, 475 ; Met. 20, 238 : Apstls. Kmbl. 73 ; Ap. 37. Ryht ædelo biþ on ðam môde, næs on ðam flǽsce *true nobility is in the mind, not in the flesh*, Bt. 30, 2 ; Fox 110, 19. Beóþ twegen on ânum flǽsce *ěrunt duo in carne una*, Mt. Bos. 19, 5 : Mk. Bos. 10, 8. Þurh ðæt flǽsc *through the flesh*, Exon. 27 a : Th. 80, 12 ; Cri. 1306 : 13 b ; Th. 26, 17 ; Cri. 418. Flǽsce bifongen *invested with flesh*, 84 a ; Th. 316, 13 ; Môd. 48 : 98 a ; Th. 368, 33 ; Seel. 34. Genam he ân ribb of his sîdan and gefylde mid flǽsce *tŭlit ūnam de costis ejus et replēvit carnem pro ea*, Gen. 2, 21. Beóþ ða syngan flǽsc scandum þurhwaden *the sinful flesh shall be penetrated with scandals*, Exon. 26 b ; Th. 78, 31 ; Cri. 1282. Flǽsca gehwylc *omnis căro*, Ps. Th. 144, 21. He afédeþ flǽscea [MS. flǽcsea] æghwylc *qui dat escam omni carni*, 135, 26. [*Piers P.* flesshe : *Wyc.* fleisch, fleixh, flehs : *Laym.* flæsce, flas, flæs : *Orm.* flæsh : *Plat.* fleesk, fleesch, *n :* O. Sax. flêsk, fleisk, *n : Frs.* flæsck, flæsch : *O. Frs.* flask, flesk, *n : Dut.* vleesch, *n : Ger.* fleisch, *n : M. H. Ger.* vleisch, *n : O. H. Ger.* fleisc, *n : Dan.* flesk, *n.* bacon, pork : *Swed.* flåsk, *n.* pork, bacon : *Icel.* flesk, *n.* pork, ham, bacon.]

flésc-ǽt, es ; *m.* [ǽt *food*] *Flesh food ;* carneus victus, R. Ben. 36.

flésc-côfa, an ; *m.* [flǽsc *flesh*, côfa *a chamber*] *The flesh chamber, the body, flesh ;* căro :—Gefæstna mid ege ðînum flǽsccôfan mîne *confige timōre tuo carnes meas*, Ps. Lamb. 118, 120.

flésc-cwellere, es ; *m. A butcher, hangman ;* lănius, carnīfex, Som. Ben. Lye.

flésc-cȳping, e ; *f.* [cȳping II. *a market-place, markeṫ*] *A flesh-market, meat-market ;* măcellum :—Flǽsccȳping [MS. flæc-cyping] *mă-cellum*, Ælfc. Gl. 55 ; Som. 67, 14 ; Wrt. Voc. 37, 8.

flésceht ; *adj. Fleshy, fleshly ;* carneus, Som. Ben. Lye.

flésc-gebyrd, e ; *f. Flesh-birth, incarnation ;* incarnātio : — Flǽsc-gebyrde *incarnātiōnis*, Mone B. 499.

flésc-hama, -homa, an ; *m. Flesh-covering, the body, a carcase ;* carnis tegmen, corpus :—Læg mîn flǽschoma in foldan bigrafen *my body lay buried in earth*, Exon. 29 a ; Th. 89, 32 ; Cri. 1466 : 47 b ; Th. 163, 35 ; Gû. 1004. Bil eal þurhwôd fǽgne flǽschoman *the falchion passed all through her fated carcase*, Beo. Th. 3140 ; B. 1568 : Andr. Kmbl. 307 ; An. 154. Lǽgon on greóte fǽgra flǽschaman *the carcases of the slain lay on the sand*, 2171 ; An. 1087.

flésc-hamian *to become incarnate ;* carnem humānam induĕre. v. hama, ge-flǽschamod.

flésc-hord, es ; *n. The flesh-hoard, the body ;* carnis thesaurus, corpus :—Firenaþ ðus ðæt flǽschord *thus will the body sin*, Exon. 99 b ; Th. 373, 3 ; Seel. 103 ; Soul Kmbl. 203.

flésc-hús, es ; *n. A flesh-house ;* carnis officīna :—Flǽschûs *carnāle*, Ælfc. Gl. 108 ; Som. 78, 102 ; Wrt. Voc. 58, 17.

flésc-líc ; *adj. Fleshly, carnal ;* carnālis :—Unrihtlîc biþ ðæt se cristena mann flǽsclîce lustas gefremme *unlawful it is for the christian man to indulge in fleshly lusts*, Homl. Th. ii. 100, 18. Swâ swâ ða gôdan fæderas gewuniaþ heora flǽsclîce bearn þreágean *sicut bŏni patres carnālĭbus filiis sŏlent discĭplīnam tĕnēre*, Bd. 1, 27 ; S. 490, 16. Hwæt gôdes mǽgan we secgan on ða flǽsclîcan unþeáwas *what good shall we say of the fleshly vices?* Bt. 31, 1 ; Fox 110, 25 : Boutr. Scrd. 21, 43 : Past. 11, 4 ; Hat. MS. 15 a, 17.

flésc-lícnes, -ness, -nys, -nyss, e ; *f. Fleshliness, incarnation ;* incarnātio :—Se ðe wile smeágan ymbe ða gerȳnu Cristes flǽsclîcnysse *he who will inquire about the mystery of Christ's incarnation*, Homl. Th. ii. 278, 35 : 280, 22.

flésc-mangere, es ; *m. A fleshmonger, butcher ;* carnis vendĭtor, măcellārius, lănius, Cot. 57 : 125 : Cod. Dipl. 1291 ; A. D. 996 ; Kmbl. vi. 135, 17, 18.

flæsc-maðu, e; f. A fleshworm, maggot; vermis carnem infestans, Ælfc. Gl. 24; Som. 60, 19; Wrt. Voc. 24, 23.

flæsc-mete, es; pl. nom. acc. -mettas; m. FLESH-MEAT, flesh; carnĕus cĭbus, căro:—Hū wæs mancynne flǣscmete alỹfed æfter ðam flōde why was fleshmeat allowed to mankind after the flood? Boutr. Scrd. 21, 16. Mid flǣscmete with flesh-meat, L. C. S. 47; Th. i. 402, 24. Gē etaþ flǣscmettas eówre hreáwe mandūcābĭtis carnes vestras crūdas, Coll. Monast. Th. 29, 11: Ps. Lamb. 49, 13.

flǣscnes, -ness, e; f. Incarnation; incarnātio, Hem. 57. DER. ge-flǣscnes.

flǣsc-strǣt, e; f. A FLESH-STREET, meat-market; carnāle, carnis officīna, măcellum:—Flǣscstrǣt [MS. flæc-strǣt] măcellum, Ælfc. Gl. 55; Som. 67, 14; Wrt. Voc. 37, 8.

flǣsc-tawere, es; m. A flesh-tawer or tormentor, an executioner; lănio, carnĭfex:—Hyldere, oððe cwellere, oððe flǣsctawere [MS. flǣc-tawere] lănio, vel lănista, vel carnĭfex, vel măcellārius, Ælfc. Gl. 113; Som. 79, 120; Wrt. Voc. 60, 27.

flǣsc-wyrm, es; m. A FLESH-WORM, maggot; tĕrēdo, vermis carnem infestans:—Wið flǣscwyrmum against flesh-worms, L. M. 1, 51; Lchdm. ii. 124, 19.

flǣðe-camb [MS. -comb], flede-camb, es; m. A weaver's comb; pecten, pectĭca, Glos. Brux. Recd. 40, 15; Wrt. Voc. 66, 23.

flāh; adj. Insidious, artful, deceitful, fraudulent; subdŏlus, fraudŭlentus, infestus:—Đonne ðæt gecnāweþ flāh feónd gemāh when the deceitful impious fiend knows that, Exon. 97 a; Th. 362, 19; Wal. 39.

flān, es; m. f. [flān; gen. flānes; m flān; gen. e; f.] An arrow, a dart; săgitta, tēlum:—Þurh flānes flyht through the flight of an arrow, Byrht. Th. 133, 56; By. 71. Fram flāne fleógendre a săgitta vŏlante, Ps. Spl. 90, 6: Beo. Th. 4868; B. 2438. Đīne flāna synt afæstnode [MS. afæstnade] on me săgittæ tuæ infixæ sunt mihi, Ps. Th. 37, 2: 44, 7: Ps. Spl. 56, 6. Ic afæstnie mīne flāna on him săgittas meas complēbo in eis, Deut. Grn. 32, 23. v. flā.

flān-boga, an; m. An arrow-bow; arcus săgittis aptus:—Se ðe of flānbogan fyrenum sceóteþ who wickedly shoots from his arrow-bow, Beo. Th. 3492; B. 1744: 2870; B. 1433.

flān-geweorc, es; n. Arrow-work; jaculatōrius apparātus:—Flacor flāngeweorc flickering arrow-work, Exon. 17 b; Th. 42, 21; Cri. 676.

flān-hred; adj. arrow-swift; săgittārius expedītus, Grn. Reim. 72.

flāniht; adj. Belonging to darts; ad tēla pertĭnens, jăcŭlatōrius, jăcŭlātus, Cot. 112. v. flān.

flān-þræc, -þracu; gen. -þræce; pl. nom. geṇ. acc. -þraca; f. Arrows' force; săgittārum impĕtus:—Wið flānþræce, Exon. 71 a; Th. 265, 20: Jul. 384. Flānþracu, Exon. 49 b; Th. 170, 25; Gū. 1117.

flāt, pl. fliton strove, contended; p. of flītan.

FLAXE, an; f. A FLASK, bottle; flasca, flasco, lăgēna:—Flaxe flasca, Ælfc. Gl. 25; Som. 60, 65; Wrt. Voc. 25, 7. Twā treówene fatu wīnes fulle, ða syndon on folcisc flaxan gehātene duo lignea vāsa vīno plēna, quæ sunt vulgo flascōnes vŏcāta, Greg. Dial. 1, 9: 2, 13. Ic bicge hýda and fell, and wyrce of him flaxan ȩ̃o ẽmo cūtes et pelles, et făcio ex iis flascōnes, Coll. Monast. Th. 27, 37. [Plat. flaske, f: Dut. flesch, f: Ger. flasche, f: M. H. Ger. vlasche, vlesche, f: O. H. Ger. flasca, f: Dan. flaske, m. f: Swed. flaska, f: Icel. flaska, f: M. Lat. flasca, flasco, Du Cange.] DER. wæter-flaxe.

flax-fōte, flox-fōte, flohten-fōte; adj. Broad-footed, flat-footed, web-footed; palmĭpes:—Đa fugelas ðe on flōdum wuniaþ syndon flaxfōte, ðæt hī swimman mægen [MS. magon] the birds that dwell in waters are web-footed, that they may swim, Hexam. 8; Norm. 14, 15.

FLEÁ, an; m. I. a FLEA; pūlex:—Fleá pūlex, Wrt. Voc. 78, 68. Κόνυζα fleán acwelleþ fleabane kills fleas, Herb. 143; Lchdm. i. 266, 2. Gorst cwelþ ða fleán gorse killeth the fleas, 142; Lchdm. i. 264, 15. Wið fleán against fleas, 142; Lchdm. i. 264, 14. v. fleó. II. a speck, speck or disease in the eye; albūgo, -ĭnis, f. măcŭla:—Wið fleán and wið eágena sāre against white specks and against sore of eyes, Herb. 24; Lchdm. i. 120, 16. [Plat. flo, flö a flea: Dut. vloo, f. a flea: Ger. floh, m. a flea: M. H. Ger. vlôch, m. a flea: O. H. Ger. flôh, flôch, m. a flea: Icel. fló, f. a flea: Lat. pūlex, f. a flea.] DER. eág-fleá. v. fleah.

fleág flew, Exon. 46 a; Th. 157, 9; Gū. 889; p. of fleógan.

fleah a flea; pūlex, Glos. Epnl. Recd. 161, 42. v. fleá.

fleah, fleó, flió, flié, flíg; indecl. n: fleá, an; m. A white spot in the eye; albūgo:—Þurh ðone æpl ðæs eágan mon mæg geseón, gif him ðæt fleah on ne gǣþ, gif hine ðonne ðæt fleah mid ealle ofergǣþ, ðonne ne mæg he nôht geseón a man can see with the pupil of the eye, if the white speck does not spread over it, if the white speck spreads all over it, then he cannot see anything, Past. 11, 4; Hat. MS. 15 b, 4. Se hæfþ eallinga fleah on his mōdes eágum he has altogether a white speck in the eyes of his mind, 11, 4; Hat. MS. 15 b, 1.

fleáh flew, Ps. Spl. 17, 12; p. of fleógan.

fleáh fled, Ps. Lamb. 113, 3; p. of fleón.

fleám, flǣm, es; m. [fleón to flee] Flight; fŭga:—Đæt eówer fleám on wintra ne geweorþe ut non fiat fŭga vestra in hieme, Mt. Bos. 24, 20: Chr. 998; Erl. 135, 19. Wurdon feówer on fleáme folccyningas four kings of nations were in flight, Cd. 95; Th. 125, 4; Gen. 2074: Chr. 477; Erl. 12, 31: L. C. E. 4; Th. i. 360, 29: Jos. 7, 4. Nū sceal æðelingas gefricgean fleám eówerne now nobles shall hear of your flight, Beo. Th. 5771; B. 2889: Ps. Th. 141, 5: Ps. Spl. 88, 23. Fleám gewyrcan to take to flight, Byrht. Th. 134, 9; By. 81. Efne ic feor gewīte, fleáme dǣle ecce elongāvi fŭgiens, Ps. Th. 54, 7: Andr. Kmbl. 3087; An. 1546. Crist nolde ða þrówunge mid fleáme forbūgan Christ would not by flight avoid his passion, Homl. Th. i. 206, 6: Chr. 937; Erl. 114, 3; Æðelst. 37. [Laym. flæm, fleam, flem flight.]

fleáming a runaway, Grm. Gr. ii. 351, 11. v. flýming.

fleán; p. flôh, pl. flôgon; pp. flagen To flay, pull off the skin; excŏriāre, deglŭbĕre, Cot. 61. [Laym. flan, fiean to flay: Dut. Kil. vlaen vlaeghen: Swed. flå: Icel. flá.] DER. be-fleán.

fleard, es; n. Trifles; nūgæ:—Gif friþgeard sī on hwæs lande, abūton stān, oððe treów, oððe wille, oððe swilces ænige fleard if there be an inclosed space on any one's land, about a stone, or a tree, or a well, or any trifles of such kind, L. N. P. L. 54; Th. ii. 298, 17. Flearde fraude, Mone B. 1530. [Orm. flærd mockery: Scot. flird: Icel. flærð, f. deceit.] DER. ge-fleard.

fleardian; p. ode; pp. od To trifle, err; nūgāri, errāre:—Fleardian nūgāri, Off. Episc. 7: errāre, Scint. 31.

fleát floated, Beo. Th. 3822; B. 1909; p. of fleótan.

fleaðe, fleoðe, an; f. The water-lily; nymphæa alba, Lin:—Of fleaðan wyrte from the plant of the water-lily, L. M. 2, 51; Lchdm. ii. 264, 20.

fleá-wyrt, e; f. FLEA-WORT, flea-bane; pūlicāria, psyllium = ψύλλιον, cōnyza = κόνυζα:—Fleáwyrt parirus? Wrt. Voc. 287, 23.

FLEAX, flex, es; n. FLAX; līnum:—Of ðære eorþan cymeþ ðæt fleax flax comes from the earth, Past. 14, 6; Hat. MS. 18 b, 13. Fleax līnum, Wrt. Voc. 82, 6. Þurh ðæt fleax by the flax, Past. 14, 6; Hat. MS. 18 b, 14. Swīðe hwīt fleax very white flax; bissum [= byssus = βύσσος], Ælfc. Gl. 62; Som. 68, 94; Wrt. Voc. 40, 5. [Wyc. flax, flaxe, flex, flexe: Chauc. flex: Plat. flass, n: Frs. flæges: O. Frs. flax, n: Dut. vlas, n: Ger. flachs, m: M. H. Ger. vlahs, m: O. H. Ger. flahs, m: Lat. flectĕre, plectĕre: Grk. πλέκειν to plait, twine, twist, weave.]

fleaxen; adj. Flaxen; līneus, Som. Ben. Lye.

fléc flesh, Chr. 1137; Gib. 239, 27. v. flǣsc.

fled a dwelling, abode, Lchdm. iii. 54, 17. v. flet.

flēd, es; n. [flōd a flood] A flowing, flood; flūmen:—Eá in flēde the river in its flow, Cd. 12; Th. 15, 12; Gen. 232: Andr. Kmbl. 3006; An. 1506. cf. Grein, inflēde.

flēde; adj. Flooded, overflowed; tŭmĭdus:—Wæs seó eá to ðan flēde the river was so flooded, Ors. 2, 5; Bos. 48, 13. Seó eá flēde wæs the river was flooded, Ors. 2, 4; Bos. 44, 7. Tiber flēdu wearþ the Tiber was flooded, Ors. 4, 7; Bos. 87, 20. DER. ofer-flēde.

flēding, e; f. A flowing, an inundation; fluxus:—Se ele geswāc ðære flēdinge the oil ceased from the flowing, Homl. Th. ii. 180, 2.

flēge a fly; cŭlĭcem, Mt. Kmbl. Lind. 23, 24. v. fleóge.

flēgende flying; vŏlans, Bd. 1, 7, Lye, = fleógende; part. of fleógan.

flēma, an; m. A fugitive; profŭgus:—Đū flēma scealt wīdlast wrecan thou shalt go a fugitive into far exile, Cd. 48; Th. 62, 27; Gen. 1020: L. C. S. 13; Th. i. 382, 23: Obs. Lun. § 7; Lchdm. iii. 186, 23. v. flýma.

flēman; p. de; pp. ed To cause to flee, put to flight; fŭgāre. DER. ge-flēman. v. flǣman, flýman.

flene, an; f. What is made soft, batter:—Wyl ða flenan boil the batter, L. M. 1, 38; Lchdm. ii. 98, 11. v. flyne.

fleó a flea; pūlex, Ælfc. Gl. 23; Som. 60, 6; Wrt. Voc. 24, 10. v. fleá.

fleó; indecl. n. A white speck, disease of the eye; albūgo:—Đæs eágan wǣron mid fleó and mid dimnesse twelf mōnþ ofergān whose eyes had been for a twelvemonth overspread with the white speck and with dimness, Guthl. 22; Gdwin. 96, 14. v. fleah.

FLEÓGAN, flíogan, to fleógenne; part. fleógende; ic fleóge, ðū fleógest, he fleógeþ, pl. fleógaþ; p. ic, he fleág, fleáh, ðū fluge, pl. flugon; pp. flogen [fleóge a fly]. I. v. intrans. To FLY as with wings; vŏlāre:—Đæt he mid feðerhoman fleógan meahte that he might fly with wings, Cd. 22; Th. 27, 14; Gen. 417: Bt. Met. Fox 24, 3; Met. 24, 2. Ic hæbbe swīðe swifte feðera, ðæt ic mæg flíogan ofer ðone heán hróf ðæs heofones I have very swift wings, that I can fly over the high roof of heaven, Bt. 36, 2; Fox 174, 5. Hwā me sealde to fleógenne fiðeru swā culfran quis dăbit mihi pennas sīcut cŏlumbæ, et vŏlābo? Ps. Th. 54, 6. Geseah he ða wērian gāstas þurh ðæt fyr fleógende he saw the accursed spirits flying through the fire, Bd. 3, 19; S. 548, 34: Bt. Met. Fox 31, 22; Met. 31, 11. Gif ic mīne fiðeru gefó, fleóge ǣr leóhte si sumpsĕro pennas meas ante lūcem, Ps. Th. 138, 7. Se fugel fleógeþ the bird flies, Exon. 60 b; Th. 220, 18; Ph. 322: Beo. Th. 4539; B. 2273. Me of hrife fleógaþ hylde pílas shafts of battle fly from my belly, Exon. 105 a; Th. 399, 4; Rä. 18, 6. Fleág fugla cyn the race of birds flew, Exon. 46 a; Th. 157, 9; Gū. 889: 86 b; Th. 326, 12; Wid. 127. He fleáh ofer fyðru winda vŏlāvit super pennas ventōrum, Ps. Lamb. 17, 11: Cd. 72; Th. 87, 29; Gen. 1456. Đa englas twegen him on twā healfa

flugon *the two angels flew on both sides of him*, Bd. 3, 19; S. 548, 32 : Exon. 43 a; Th. 146, 14; Gú. 709. **II.** *v. intrans. To flee, flee from*; fúgěre, effúgěre :—Ðæt he nolde fleógan *that he would not flee*, Byrht. Th. 139, 56; By. 275. Fleógende *fúgiens*, Ps. Spl. 54, 7. Hí fleógaþ mid ðám feóndum *they flee with the fiends*, Exon. 116 a; Th. 446, 6; Dóm. 18. v. fleón **I.** [*Laym.* fleon : *Orm.* fleghenn : *Plat.* flegen : *Frs.* flega : *O. Frs.* fliaga : *Dut.* vliegen : *Ger.* fliegen : *M. H. Ger.* vliegen : *O. H. Ger.* fliugan, fleogan : *Dan.* flyve : *Swed.* flyga : *Icel.* fljúga.] DER. be-fleógan, forþ-, ge-, of-, óþ-, up-, ymb-.

FLEÓGE, an; *f.* A FLY; *musca* :—Fleóge *musca*, Wrt. Voc. 77, 53; 281, 33. For ðé ic gebidde and ðeós fleóge færþ fram ðé órábo Dómĭnum et recédet musca a Pharaóne, Ex. 8, 29. Ðæt ðær ne beóþ náne fleógan *ut non sint ibi muscæ*, 8, 22. Ic sende on ðé eall fleógena cynn *ěgo immittam in te omne gěnus muscárum*, 8, 21, 24. He adráf ða fleógan fram Pharaone *abstŭlit muscas a Pharaóne*, 8, 31 : Ps. Th. 89, 10. Hundes fleóge *a dog-fly*; cynomya = κυνόμυια, Ælfc. Gl. 21; Som. 59, 79; Wrt. Voc. 23, 37 : 23; Som. 59, 119; Wrt. Voc. 23, 73 : Ps. Spl. 104, 29. Hundes fleóge *ricinus*, Ælfc. Gl. 21; Som. 59, 80; Wrt. Voc. 23, 38. [*Laym.* fleʒen, fleie, *pl. flies* : *Plat.* flege, *f.* : *O. Sax.* fliuga, *f.* : *Dut.* vlieg, *f.* : *Ger.* fliege, *f.* : *M. H. Ger.* vliege, *f.* : *O. H. Ger.* fliuga, fleoga, fliega, *f.* : *Dan.* flue, *m.* : *Swed. Icel.* fluga, *f.*] DER. buttor-fleóge.

fleógende; *part. Flying, winged*; vŏlans, vŏlŭcer :—Fleógende *vŏlŭcer*, Ælfc. Gr. 9, 18; Som. 9, 66.

fleógendlíc; *adj. Flying, winged*; vŏlātĭlis :—Fleógendlíc *vŏlātĭlis*, Ælfc. Gr. 9, 28; Som. 11, 41.

fleóg-ryft, es; *n.* [fleóge *a fly*, ryft *a garment, veil, curtain*] *A fly-net, net for keeping off flies*; vēlámen ad muscas prŏhĭbendas, cōnōpeum = κωνωπεῖον :—Fleógryft *cōnōpeum*, Cot. 46. v. fleóh-net.

fleógynða, fleógenda, an; *m.* [fleógende, *part. of* fleógan *to fly*] *A flying creature, bird, fowl*; vŏlātile :—Ic oncneów ealle fleógyndan heofones *cognóvi omnia vŏlātĭlia cæli*, Ps. Spl. C. 49, 12; ic oncneów all ða fleógendan [MS. flégendan] heofenes *cognóvi omnia vŏlātĭlia cæli*, Ps. Surt. 49, 11 : Ps. Spl. C. 77, 31.

fleóh-cyn, -cynn, es; *m. A kind of flies*; muscárum gěnus :—Fleóh-cynnes feala fliugan on gemǽru *sciniphes in omnĭbus fīnĭbus eōrum*, Ps. Th. 104, 27.

fleóh-net, -nett, es; *n. A fly-net, net for keeping off flies*; cōnōpeum = κωνωπεῖον :—Fleóhnet *cōnōpeum*, Ælfc. Gl. 84; Som. 73, 92; Wrt. Voc. 48, 30. Fleóhnet *vel* micgnet *cōnōpeum*, 106; Som. 78, 42; Wrt. Voc. 57, 24. Ðér wæs eallgylden fleóhnet *there was an all-golden fly-net*, Judth. 10; Thw. 22, 3; Jud. 47. v. fleóg-ryft.

FLEÓN, flión, to fleónne, fliónne; *part.* fleónde, fliónde; ic fleó, ðú flíhst, flýhst, he flíhþ, flýhþ, *pl.* fleóþ, flióþ, flýþ; *p.* ic, he fleáh, ðú fluge, *pl.* flugon; *pp.* flogen. **I.** *v. trans. To* FLEE, *escape, avoid*; fúgěre, effúgěre, vītāre :—Ic heonon nelle fleón fótes trym *I will not flee hence a footstep*, Byrht. Th. 138, 68; By. 247; Andr. Kmbl. 3074; An. 1540. He sceal swíðe flión disse worulde wlite *he must quickly flee this world's splendour*, Bt. Met. Fox 7, 60; Met. 7, 30. Ðú tilast wædle to fliónne *thou toilest to avoid poverty*, Bt. 14, 2; Fox 44, 7. Fleónde *fúgiens*, Ps. Lamb. 54, 8 : Cd. 95; Th. 125, 17; Gen. 2080. Se wlite ðæs líchoman is swíðe fliónde *the beauty of the body is very fleeting*, Bt. 32, 2; Fox 116, 17. Ic fleó *fúgio*, Ælfc. Gr. 36; Som. 38, 20 : 28, 6; Som. 32, 47. He flíhþ ða wædle *he flees from poverty*, Bt. 33, 2; Fox 122, 33. He flýhþ yfla gehwilc *he flees every evil*, Exon. 62 b; Th. 229, 25; Ph. 460 : 81 a; Th. 305, 3; Fä. 82. Fleóþ his ansýne, ða ðe hine feódan *fúgiant a fácie ejus, qui ōdērunt eum*, Ps. Th. 67, 1 : 103, 17. Hí flýþ [Cott. flióþ] ðæt hí hatiaþ *they avoid what they hate*, Bt. 41, 5; Fox 252, 27. Sǽ geseah and heó fleáh *máre vīdit, et fúgit*, Ps. Lamb. 113, 3 : Bt. Met. Fox 1, 40; Met. 1, 20. Hwæt is ðé sǽ ðæt ðú fluge *quid est tibi máre quod fugisti?* Ps. Lamb. 113, 5. Ða hyrdas flugon *pastóres fúgerunt*, Mt. Bos. 8, 33 : Ps. Lamb. 30, 12 : Elen. Kmbl. 267; El. 134. Fleóþ on feorweg *flee far away*, Exon. 36 a; Th. 117, 22; Gú. 228. Ðæt ic mán fleó *that I flee evil*, Ps. Th. 93, 14. **II.** *to put to flight, rout, conquer*; fúgāre, vincěre :—Hundteóntig eówer fleóþ hira tyn þúsendu *your hundred shall put to flight their ten thousands*, Lev. 26, 8. **III.** *v. intrans. To fly as with wings*; vŏlāre :—Ic fleó *vŏlo*, Ælfc. Gr. 36; Som. 38, 16; Ps. Lamb. 54, 7. Culfran fleóþ him floccmǽlum *doves fly flockwise*, Homl. Th. i. 142, 9. v. fleógan **I.** [*Wyc.* fle : *R. Glouc.* fle : *Laym.* fleon : *Orm.* fleon, fleo : *Plat.* flugten : *O. Sax.* fliohan : *Frs.* flan : *O. Frs.* flia : *Dut.* vlieden : *Ger.* fliehen : *M. H. Ger.* vliehen : *O. H. Ger.* fliuhan : *Goth.* þliuhan : *Dan.* flye : *Swed.* fly : *Icel.* flýja.] DER. a-fleón, æt-, be-, for-, in-, ofer-, ongeán-, óþ-, þurh-, to-, up-, út-, úta-, út-óþ-.

fleós, es; *n. A fleece*; vellus :—Gilde ðæt fleós mid twám pæningum *let the fleece be paid for with two pence*, L. In. 69; Th. i. 146, 11, note 23, MS. B. In fleós *in vellus*, Ps. Surt. 71, 6. v. flýs.

FLEÓT, fliét, es; *m.* fleóte, an; *f.* **I.** *a place where vessels float, a bay, gulf, an arm of the sea, estuary, the mouth of a river, a river, stream*; hence the names of places, as *Northfleet, Southfleet, Kent*; and in London, *Fleetditch*; sĭnus, æstuārĭum, rīvus :—Se Abbod

Petrus wæs besenced on sumne sǽs fleót, se wæs háten Am-fleót *abbas Petrus demersus est in sĭnu máris, qui vŏcātur Amfleat*, Bd. 1, 33; S. 499, 6, note. Fleót *æstuārium*, Cot. 14. Ispánia land is eall mid fleóte ymbhæfd *the country of Spain is all encompassed with water*, Ors. 1, 1; Bos. 24, 3. Fleótas *æstuāria*, Glos. Epnl. Recd. 154, 46 : Wrt. Voc. 63, 69. **II.** *a raft, ship, vessel*; rătis, návis :—Ic gebycge bát on sǽwe, fleót on faroþe *I buy a boat on the sea, a vessel on the ocean*, Exon. 119 b; Th. 458, 13; Hy. 4, 100. [*Laym.* fleote *a fleet of ships* : *Plat.* fleet *a small river* : *O. Frs.* flet, *n. a river* : *Dut.* vliet, *m. a rivulet, brook* : *Ger.* fliesz, *m. n. fluentum* : *M. H. Ger.* vliez, *m. n. a rivulet* : *O. H. Ger.* fluz, *m. a river* : *Icel.* fljót, *n. a river*.]

fleótan; *part.* fleótende; ic fleóte, ðú flýtst, he flýt, *pl.* fleótaþ; *p.* fleát, *pl.* fluton; *pp.* floten [fleót *a stream*] *To* FLOAT, *swim*; fluctuāre, nātāre, nāvĭgāre :—Ðæt scip sceal fleótan mid ðý streáme *the ship must float with the stream*, Past. 58; Hat. MS. Nó he fram me flódðům feor fleótan meahte *he could not float far from me on the waves*, Beo. Th. 1089; B. 542. Se feónde [MS. feond] gespearn fleótende hreáw *the exulting [fowl] perched on the floating corpses*, Cd. 72; Th. 87, 12; Gen. 1447. Fleótendra ferþ nó ðær fela bringeþ cúþra cwidegiedda *the spirit of sea-farers brings there not many known songs*, Exon. 77 a; Th. 289, 26; Wand. 54. Ageót ele uppon wæter oððe on óðrum wætan, se ele flýt búfon *pour oil upon water or on another fluid, the oil will float above*, Homl. Th. ii. 564, 13. Oft scipu scríðende scrinde fleótaþ *illic náves pertransībunt*, Ps. Th. 103, 24. Fleát fámigheals forþ ofer ýðe *the foamy necked one floated forth over the wave*, Beo. Th. 3822; B. 1909. [*Piers P.* fleten : *Wyc. Chauc.* flete : *Orm.* fletenn : *Scot.* fleit, flete : *Plat.* fleten : *O. Sax.* fliotan : *O. Frs.* fliata : *Dut.* vlieten : *Ger.* fliessen : *M. H. Ger.* vliuzen : *O. H. Ger.* fliuzan, fleozan : *Dan.* flyde : *Swed.* flyta : *Icel.* fljóta : *Lat.* fluěre *to flow* : *Grk.* πλέιν *to navigate* : *Sansk.* plu *to float, swim*.] DER. a-fleótan.

fleóte, an; *f. A stream, river*; rīvus :—To ðære fleótan *to the stream*, Cod. Dipl. Apndx. 123; A.D. 774; Kmbl. iii. 381, 7. v. fleót **I.**

fleoðe, an; *f. The water-lily* :—Of fleoðan wyrte *of the plant of the water-lily*, L. M. 2, 51; Lchdm. ii. 266, 28. v. fleaðe.

fleótig; *adj. Swift, fleet, rapid*; cěler, vělox :—Swift wæs on fóre, fleótga [= fleótiga] on lyfte [MS. fleotgan lyfte] *it was swift in its course, rapid in the air*, Exon. 113 b; Th. 434, 22; Rä. 52, 4.

fleót-wyrt, e; *f. Floatwort, seaweed?* alga? L. M. 2, 52; Lchdm. ii. 268, 28.

fleów, *pl.* fleówon *flowed, issued*, Jn. Bos. 19, 34 : Ps. Lamb. 77, 20; *p. of* flówan.

fleówþ *flows*, Ex. 3, 17, = flēwþ; *3rd sing. pres. of* flówan.

flēre *having a floor, floored.* DER. fíf-flēre.

flēring, e; *f.* A FLOORING; contābŭlātio :—On ðære nyðemestan flēringe wæs heora gangpyt and heora myxen, on ðære óðre flēringe wæs ðæra nýtena fóda gelogod, on ðære [MS. ðone] þriddan flēringe [MS. flēringa] wæs seó forme wunung, and ðær wunodon ða wildeór and ða rēðan wurmas, on ðære feorþan flēringe [MS. flēringa] wæs ðæra tamra nýtena steall, on ðære fíftan flēringe wæs ðæra manna wunung mid wurþmynte gelogod *on the lowermost flooring [of the ark] was their privy and dunghill, on the second flooring the food of the cattle was placed, on the third flooring was the first dwelling, and there dwelt the wild beasts and fierce serpents, on the fourth flooring was the stall of the tame cattle, on the fifth flooring the dwelling of the men was placed with honour*, Boutr. Scrd. 21, 6-10 : Homl. Th. i. 536, 11, 13 : ii. 164, 5. Ðú macast þreó flēringa binnan ðam arce *tristēga fácies in arca*, Gen. 6, 16. DER. up-flēring.

flés, es; *n.* A FLEECE; vellus :—Be sceápes gonge mid his flése *of a sheep's going with its fleece*, L. In. 69; Th. i. 146, 9, note 20, MS. G. v. flýs.

flésc, es; *n. Flesh*; căro :—We hæfdon hláf and flésc genóh on Egipta lande *in terra Ægypti sedēbāmus sŭper ollas carnium et comědēbāmus pánem in sătŭrĭtāte*, Ex. 16, 3. v. flǽsc.

fleswian; *p.* ede; *pp.* ad *To mutter, whisper*; susurrāre :—Mid ðý he ða geswippre múþe lĭcettende ǽrend rehte [MS. wrehte] and leáse fleswede *when he then told a feigned message with his crafty mouth, and falsely whispered*; cum sĭmŭlātam lēgātiōnem ōre astúto volvěret, Bd. 2, 9; S. 511, 20.

FLET, flett, es; *n.* **I.** *the ground, floor of a house*; āręa :—Ne cume on bedde, ac licge on flette *let him not come into a bed, but lie on a floor*, L. P. M. 2; Th. ii. 286, 21. Heó on flet gecrong *she sank on the ground*, Beo. Th. 3141; B. 1568 : 3085; B. 1540. **II.** *a dwelling, habitation, house, cottage, hall*; hăbĭtātio, dŏmus, căsa, aula :—Gif ðæt flet geblódgad wyrþe *if the house be stained with blood*, L. H. E. 14; Th. i. 32, 14. Gif man mannan an óðres flette mánswara háteþ *if one man call another a perjurer in another's cottage*, 11; Th. i. 32, 4 : L. In. 39; Th. i. 86, 21. Him se ædela geaf giestlĭþnysse fægre on flette *the noble gave them a fair entertainment in his dwelling*, Cd. 112; Th. 147, 29; Gen. 2447. Beo. Th. 2054; B. 1025. Scilling agelde ðam ðe ðæt flet áge *let him pay a shilling to him who owns the dwelling*, L. H. E. 11, 12, 13; Th. i. 32, 6, 9, 12. Hí fǽrlíce flet ofgeáfon *they suddenly gave up the hall*, Exon. 77 a; Th. 290, 7; Wand. 61 :

Beo. Th. 3903; B. 1949: 4039; B. 2017. [*Laym.* ulette *floor : Scot.* flet, flett *a house : Plat.* flet *a bedroom in the upper floor of a peasant's house : O. Sax.* flet, fletti, *n. the floor of a house, deal, house, hall : O. Frs.* flet *a house : Ger. dial.* fletz *aula,* ărea *: M. H. Ger.* vletze, *n.* ārea *: Icel.* flet, *n. a set of rooms, house.*]

flét, e; *f. Cream, skimming, curds;* flos lactis, lactis crĕmor exemptus, coagŭlum :—Flét *flos lactis,* Cot. 37. Hwít sealt dó on reám oððe góde fléte *put white salt into cream or good skimmings,* L. M. 3, 10; Lchdm. ii. 314, 2. v. fléte.

fléte, fliéte, flýte, an; *f.:* flét, e; *f.* [fleótan *to float*] What floats on the surface, hence,—*Cream, skimming, curds;* flos lactis, lactis crĕmor exemptus, coagŭlum :—Genim cūmeoluc bútan wætere, læt weorþan to flétum, geþwer to buteran *take cow's milk without water, let it become cream, churn it to butter,* L. M. 1, 44; Lchdm. ii. 108, 22. Hafa clǽne flétan *have clean curds,* L. M. 1, 2; Lchdm. ii. 38, 19. Menge wið flétan, and nán óðer molcen þicge *let him mingle it with curds, and eat no other milk-food,* L. M. 2, 51; Lchdm. ii. 264, 26.

flet-gesteald, flett-gesteald, es; *n. Dwelling-place, household goods;* hăbĭtātio, dŏmestĭcæ ŏpes :—Lamech onféng fletgestealdum *Lamech succeeded to the dwelling-places,* Cd. 52; Th. 65, 31; Gen. 1074.

fleðe-camb, es; *m. A weaver's comb;* pecten, pectĭca, Ælfc. Gl. 110; Som. 79, 47; Wrt. Voc. 59, 18. v. flæðe-camb.

flet-mon *a sailor,* Som. Ben. Lye. v. flot-man.

flet-pæþ *a house-path, floor.* v. flett-pæþ.

flet-rest, e; *f. Domestic couch, sleeping quarters in the hall;* lectus domestĭcus :—Sum fletreste gebeág *one bowed to the domestic couch,* Beo. Th. 2487; B. 1241.

flet-sittend, es; *m. A court-resident;* in aula sĕdens :—Ðá wæs fletsittendum fægere gereorded *there was a feast fairly arranged to the court-residents,* Beo. Th. 3580; B. 1788. Ða ic Freáware fletsittende nemnan hýrde *whom I heard the court-residents call Freaware,* 4049; B. 2022. Ðær wæron boren æfter bencum orcas fulle fletsittendum *there were full jugs carried along the benches to the court-residents,* Judth. 10; Thw. 21, 15; Jud. 19: 21, 24; Jud. 33.

flett *the floor of a house, a dwelling, habitation;* sēdes, hăbĭtātio, Som. Ben. Lye. v. flet.

flett-gesteald, es; *n. Household goods, domestic wealth;* domestĭcæ ŏpes :—Geomor fæder flettgesteald freóndum dǽlde *Gomer distributed his father's domestic wealth to his friends,* Cd. 79; Th. 97, 11; Gen. 1611. v. flet-gesteald.

flett-pæþ, es; *pl. nom. acc.* -paðas; *m. A house-path, floor;* dŏmi sēmĭta, păvĭmentum :—Ðæt ðú flettpaðas míne trǽde *that thou hast trodden my house-paths,* Cd. 130; Th. 165, 10; Gen. 2729.

flet-werod, es; *n. Court-host, the court-retainers;* aulĭci :—Is mín fletwerod, wígheáp, gewanod *my court-host, the company in war, is diminished,* Beo. Th. 957; B. 476.

fléuwþ *flows,* Ps. Lamb. 57, 9, = flêwþ; *3rd sing. pres. of* flówan.

fléwsa, an; *m.* [flówan *to flow*] *A flowing, flux;* fluxus :—Wið innoþes fléwsan *for flux of inwards,* Herb. 53, 2; Lchdm. i. 156, 14: Med. ex Quadr. 6, 9; Lchdm. i. 352, 15. Wið wífes fléwsan *for flux of a woman,* Herb. 89, 2; Lchdm. i. 192, 12: 128; Lchdm. i. 240, 2: 178, 6; Lchdm. i. 312, 10. Ðý sylfan dæge hyt ðone fléwsan belúceþ *eòdem die fluxum comprĭmet,* 178, 6; Lchdm. i. 312, 16: 175, 3; Lchdm. i. 308, 1. Heó ða fléwsan gewríþ *it stops the flux,* 128; Lchdm. i. 240, 5.

fléwst, he fléwþ *flowest, flows,* Ex. 3, 8; *2nd and 3rd sing. pres. of* flówan.

flex, es; *n. Flax;* līnum :—Smeócende flex he ne adwæscþ *līnum fūmĭgans non extinguet,* Mt. Bos. 12, 20. Eall hira flex and hira bernas wǽron fordóne *linum et hordeum læsum est,* Ex. 9, 31. v. fleax.

flicce, es; *n? A flitch of bacon;* succĭdia, perna :—Flicce *perna,* Wrt. Voc. 86, 13 : 286, 51. [*Plat.* flikke, *m. a spot, piece : Ger.* fleck, *m. n;* flecke, *m. a rag, piece, spot, place : M. H. Ger.* vlëc, *m. a piece : O. H. Ger.* fleccho, *m. măcŭla : Dan.* flik, flikke, *m. f. a piece, rag : Swed.* flik, *m. a lap : Icel.* flik, *f. a rag;* flikki, *n. a flitch of bacon.*]

flicerian, flicorian; *p.* ode; *pp.* od [fleógan *to fly*] *To move the wings, flutter;* mōtāre ālas, vŏlĭtāre :—Ic flicerige *vŏlĭto,* Ælfc. Gr. 36; Som. 38, 16. Swá earn his briddas spænþ to flihte and ofer híg flicerаþ *sīcut ăquĭla prŏvŏcans ad vŏlandum pullos suos et sŭper eos vŏlĭtans,* Deut. 32, 11. Án blac þrostle flicorode ymbe his neb *a black thrush flickered about his face,* Homl. Th. ii. 156, 22. [*Dut.* flakkeren, flikkeren : *Ger.* flackern : *M. H. Ger.* vlackern : *O. H. Ger.* flokarôn.]

flié; *indecl. n. A white speck, disease of the eye;* albūgo :—Wið flié eágsealf *an eye-salve for the white speck,* L. M. 1, 2; Lchdm. ii. 32, 12, 17, 18, 20, 23, 26 : 3, 2; Lchdm. ii. 308, 9. Ægðer mæg adón flié of eágan *either can remove the white speck from the eye,* 3, 2; Lchdm. ii. 308, 26. v. fleah.

fliéman feorm, e; *f. The harbouring of a fugitive;* fŭgĭtivi susceptio, L. In. 30; Th. i. 120, 16. v. flýman fyrmþ.

fliés, es; *n. A fleece;* vellus :—Be sceápes gonge mid his fliése. Sceáp sceal gongan mid his fliése óþ midne sumor, oððe gilde ðæt fliés mid twám pæningum *of a sheep's going with its fleece. A sheep shall go*

with its fleece until midsummer, or let the fleece be paid for with two pence, L. In. 69; Th. i. 146, 9–11. v. flýs.

fliét, es; *m. A raft, ship, vessel;* rătis, nāvis :—Fliét *rătis,* Cot. 200. v. fleót II.

fliéte, an; *f. Cream, curds;* flos lactis, coagŭlum :—Fliéte *verbĕrātum :* geþworen [MS. geþrofen] fliéte *churned cream;* lactudiclum? Wrt. Voc. 290, 27, 28. Dó on ðæt fæt swá fela swá ðara fliétna ðǽron clifian mæge *put into the vessel as much of the curds as may cleave thereon,* L. M. 1, 2; Lchdm. ii. 38, 20. v. fléte.

flíg; *indecl. n. A white speck, disease of the eye;* albūgo, Wrt. Voc. 285, 2. v. fleah.

fligan; *p.* de; *pp.* ed *To put to flight;* fūgāre. DER. a-fligan.

flige-wíl, es; *m.* [flíge = flyge *vŏlātus;* wíl *a wile, deceit,* q. v.] *A flying wile, dart of Satan;* vŏlans astūtia, diabŏli sagitta :—Gefylled feóndes fligewílum, fācensearwum *filled with the fiend's [Satan's] flying darts, with treacherous wiles,* Exon. 83 b; Th. 315, 6; Mód. 27.

flíhst, he flíhþ *fleest, flees,* Bt. 33, 2; Fox 122, 33; *2nd and 3rd pres. sing. of* fleón.

fliht, es; *m. A flight;* vŏlātus :—Swá earn his briddas spænþ to flihte *sīcut ăquĭla prŏvŏcans ad vŏlandum pullos suos,* Deut. 32, 11: Exon. 13 b; Th. 25, 11; Cri. 399. v. flyht.

flíma, an; *m. A runaway, fugitive;* profūgus, Cot. 151. v. flýma.

flind, e; *f. Genetrix,* Cot. 98, Lye.

FLINT, es; *m.* FLINT, *a rock;* sĭlex, petra :—Flint *sĭlex,* Ælfc. Gl. 58; Som. 67, 94; Wrt. Voc. 38, 19 : 85, 21. Flinte ic eom heardra *I am harder than flint,* Exon. 111 b; Th. 426, 23; Rä. 41, 78. Ðæt ðú gesomnige flint unbrǽcne *that thou unite the unfragile flint,* Exon. 8 a; Th. 1, 11; Cri. 6: Salm. Kmbl. 202; Sal. 100. Flintum heardran *harder than flints,* Exon. 25 a; Th. 73, 13; Cri. 1189. Híg cômon to ðam flinte, and Moyses ætfóran him eallum slóh mid ðære girde tûwa ðone flint, and fleów sóna of ðam flinte wæter *they came to the rock, and Moses struck the rock twice with his rod before them all, and immediately water flowed from the rock,* Num. 20, 10, 11. [*M. H. Ger.* vlins, *m. sĭlex : Dan.* flint, *m. f : Swed.* flinta, *f.*]

flint-grǽg; *adj. Flint-grey;* cānus :—Ic sceal to staðe þýwan [MS. þyran] flintgrǽgne flód *I shall impel the flint-grey flood to the shore,* Exon. 101 b; Th. 383, 31; Rä. 4, 19.

flió; *indecl. n. A white speck, disease of the eye;* albūgo, Glos. Epnl. Recd. 153, 12. v. fleah.

fliógan *to fly;* vŏlāre :—Ic mæg fliógan ofer ðone heán hróf ðæs heofones *I can fly over the high roof of the heaven,* Bt. 36, 2; Fox 174, 5. v. fleógan.

flión *to flee;* fŭgēre :—He sceal flión ðisse worulde wlite *he must flee this world's splendour,* Bt. Met. Fox 7, 60; Met. 7, 30. v. fleón.

flís *a fleece;* vellus, Wrt. Voc. 66, 30 : 282, 13. v. flýs.

FLÍT, es; *n. Scandal, contention, strife;* scandālum, contentio :—Togeánes sunu mōdor ðíne ðú settest flít *adversus filium matris tuæ pōnĕbas scandālum,* Ps. Spl. T. 49, 21. [*Laym.* flit, *n. dispute : Scot.* flyte : *Plat.* flit, fliit, fliet, *m. diligence : O. Sax.* flit, *m. contention, contest : O. Frs.* flit *diligence : Dut.* vlijt, *f. diligence : Ger.* fleiss, *m : M. H. Ger.* vlîz, *m : O. H. Ger.* flîz, *m.*] DER. ge-flít, sund-flít.

flíta, an; *m.* [flítan *to contend*] *A fighter, striver, foe.* DER. ge-flíta, wið-, wiðer-.

flítan; *part.* flítende; ic flíte, ðú flítest, flítst, he flíteþ, flít, *pl.* flítaþ; *p.* flát, *pl.* fliton; *pp.* fliten *To strive, contend, dispute, rebel;* contendĕre, certāre, dispūtāre, jurgāre :—Ic flítan gefrægn on fyrndagum mōdgleáwe men, gewǽsan ymbe hyra wísdóm *I have learnt that in days of yore men wise of mood contended, struggled about their wisdom,* Salm. Kmbl. 359; Sal. 179. Ðam ðe wylle on dóme wið ðé flítan, and niman ðíne tunecan, læt him tó ðínne wǽfels *ei, qui vult tecum jūdĭcio contendĕre, et tūnĭcam tuam tollĕre, dimitte ei et pallium,* Mt. Bos. 5, 40. Flítende *contending,* Beo. Th. 1836; B. 916. Hwí flítst ðú wið ðínne nēxtan *quāre percŭtis proxĭmum tuum?* Ex. 2, 13. Flíteþ *strives,* Exon. 95 a; Th. 354, 17; Reim. 62. Ne flít he *non contendet,* Mt. Bos. 12, 19. Flát he wið ánne Israhēliscne man *jurgātus est cum vĭro Israhēlīta,* Lev. 24, 10: Bd. 4, 16; S. 584, note 31. Me þincþ nú ðæt ðín gecynd and ðín gewuna flíte swíðe swíðlíce wið ðæm dysige *methinks now that thy nature and thy habit contend very powerfully against error,* Bt. 36, 4; Fox 178, 28. [*Scot.* flyte; *p.* flet *to scold : M. H. Ger.* vlîzen : *O. H. Ger.* flîzan.] DER. ofer-flítan, óþ-, wiðer-.

flít-cræft, es; *m. The art of disputing, logic;* disceptandi ars, dialectĭca :—Flítcræft *dialectĭca,* Mone B. 3030.

flít-cræftlíc; *adj. Of or belonging to disputation, dialectical, logical;* dialectĭcus = διαλεκτικός :—Mid flítcræftlícum *dialectĭcis,* Mone B. 3147.

flítend, es; *m.* [flítende, *part. of* flítan *to strive*] *A wrangler, quarrelsome person;* certans, lĭtĭgans :—Flítend *certans,* Cot. 181. Flítend *lĭtĭgans,* Mone B. 2927.

flíter-cræft, es; *m. The art of disputing, logic;* dialectĭca, Som. Ben. Lye. v. flít-cræft.

flítere, es; *m. A brawler, wrangler, schismatic;* rābŭla, schismātĭcus

σχισματικός:—Flítere *răbŭla*, Cot. 208: Glos. Epnl. Recd. 161, 81. Flítera *schismaticŏrum*, Mone B. 2816.

flít-ful, -full; *adj. Contentious, dialectical;* contentiōsus, dialectĭcus = διαλεκτικός:—Flítfulles *dialectĭcæ*, Mone B. 3304. Flítfulra *dialectĭcōrum*, 3164. DER. ge-flítful.

flít-georn, -gern, es; *m. One desirous of contention, a quarreller;* lītigātor, vītilīgātor, rixātor:—Flítgern *lītĭgātor*, Prov. 25. DER. ge-flítgeorn.

flítlíce *contentiously, earnestly, eagerly;* certātim, stŭdiōse. DER. ge-flítlíce.

flít-mǽlum; *adv.* [mǽlum, *dat. pl. of* mǽl, *n.*] *By strife, strifewise, eagerly, earnestly;* certātim:—Flítmǽlum *certātim*, Mone B. 199. DER. ge-flítmǽlum.

FLÓC, es; *n. A sole, kind of flat fish;* plātessa, passer:— Flóc *plātessa*, Glos. Brux. Recd. 39, 67; Wrt. Voc. 65, 70: 281, 49. Flóc *pansor*? [=*passer*], Ælfc. Gl. 102; Som. 77, 80; Wrt. Voc. 56, 4. Fage and flóc *plātesias et plātessas*, Coll. Monast. Th. 24, 12, 13. [*Icel.* flóki, *m. a kind of halibut;* passer, sŏlea.]

flocan; *p.* ede; *pp.* ed *or* floccan *To clap, strike;* plaudĕre, complōdĕre:—Heó floceþ hyre folmum *she claps with her hands*, Exon. 105 b; Th. 402, 23; Rä. 21, 34.

FLOCC, es; *m.* A FLOCK, *band, company, division;* grex, cāterva, turma:—Gif Esau cymþ to ánum flocce and ðone ofslihþ, se óðer flocc byþ gehealden *si vēnĕrit Esau ad ūnam turmam et percussĕrit eam, ălia turma servābĭtur*, Gen. 32, 8. Mid ðam mánfullum flocce *with the ungodly company*, Ælfc. T. 34, 22: 35, 8. Him mon mid óðrum floccum sóhte *they were sought by other bands*, Chr. 894; Erl. 90, 14. Ic híg eft ongeán oferfare mid twǽm floccum [MS. floccon] *cum duābus turmis regrĕdior*, Gen. 32, 10. [*Wyc.* floc: *Chauc.* flok: *Laym.* floc *a host: Orm.* flocc: *Dan.* flok, *m. f:* Swed. flock, *m. a crowd: Icel.* flokkr, *m. a troop, band.*]

flocc-mǽlum; *adv.* [mǽlum, *dat. pl. of* mǽl, es; *n. a measure,* q.v.] *By flocks, flockwise, in companies;* grĕgātim, cātervātim:— Fleóþ him floccmǽlum *they fly by flocks,* Homl. Th. i. 142, 9: Num. 2. 34. Hí hý flocmǽlum slógon *they slew them in companies,* Ors. 2, 5; Bos. 46, 6. Hí ferdon ǽghweder flocmǽlum *they went everywhere in flocks,* Chr. 1011; Erl. 145, 25.

floc-rád, e; *f. A riding company, a troop;* turma:—Ðá fundon hie óðre flocráde, ðæt rád út wið Lygtúnes *then they raised another troop, which rode out towards Leighton,* Chr. 917; Erl. 102, 15. Fóron hie æfter ðæm wealda hlóþum and flocrádum *they went through the wood in bands and troops,* 894; Erl. 90, 13.

FLÓD, es; *m. n.* I. *a flowing of water, flow, flowing water, wave, tide,* FLOOD, *sea, running stream, river;* flūmen, fluctus, fluentum, æstus, accessus, flūvius:—Ðæt flód [*n.*] eóde of stówe ðære winsumnisse to wætrienne neorxna wang; ðæt flód [*n.*] ys ðanon tōdǽled on feówer eán *flūvius egrēdiēbātur de lŏco voluptātis ad irrigandum părădīsum; flūvius inde dīvĭdĭtur, in quătuor căpita,* Gen. 2, 10. Flód [*m. or n.*] vel yrnende eá *flūmen,* Ælfc. Gl. 97; Som. 76, 73; Wrt. Voc. 54, 17. Flód [*m. or n.*] *flūmen vel flūvius,* Wrt. Voc. 80, 57; Flód [*m. or n.*] accessus, Ælfc. Gl. 105; Som. 78, 35; Wrt. Voc. 57, 17. Hwenne ðæt flód [*n.*] byþ ealra hehst and ealra fullost *when the tide is highest and fullest of all,* Chr. 1031; Erl. 162, 5: 897; Erl. 96, 6. Se flód [*m.*] onsprang the *flood departed,* Andr. Kmbl. 3269; An. 1637. Com flówende flód [*m. or n.*] æfter ebban ... se flód [*m.*] út gewát *the flowing tide came after the ebb ... the tide receded,* Byrht. Th. 133, 45, 58; By. 65, 72. Cynn ða ðe flód [*m. or n.*] wecceþ inc hýraþ *races which the water bringeth forth shall obey you two,* Cd. 10; Th. 13, 18; Gen. 204: Beo. Th. 1095; B. 545: Andr. Kmbl. 3091; An. 1548: Exon. 106 a; Th. 404, 12; Rä. 23, 6. Flódes [*m. or n.*] ryne *flūmĭnis impĕtus,* Ps. Lamb. 45, 5. Ðæs sǽes flódes [*m. or n.*] weaxnes *an increasing of the sea's tide,* Bd. 5, 3; S. 616, 16. Hie on flódes [*m. or n.*] fædm ceólum lácaþ *they sail in ships on the bosom of the sea,* Andr. Kmbl. 503; An. 252: Beo. Th. 83; B. 42: Salm. Kmbl. 161; Sal. 80. On Iordanes flóde [*m. or n.*] *in Iordānis flūmĭne,* Mk. Bos. 1, 5. Se wuldorcyning gesette ýþum heora onrihtne ryne, rúmum flóde [*m. or n.*] *the king of glory appointed to the waves, to the spacious flood, its just course,* Cd. 8; Th. 10, 36; Gen. 167: Exon. 25 a; Th. 72, 8; Cri. 1169: Beo. Th. 3780; B. 1888: Andr. Kmbl. 530; An. 265. Cyning út gewát on fealene flód [*m.*] *the king departed on the dusky flood,* Chr. 937; Erl. 114, 2; Æðelst. 28: Beo. Th. 3904; B. 1950: Andr. Kmbl. 841; An. 421: Exon. 101 b; Th. 383, 31; Rä. 4, 19. Sió eá forþ mid micle flóde [*m. or n.*] út on ða sǽ flóweþ *the river flows forth out to the sea with a great flow,* Ors. 1, 1; Bos. 15, 20: Cd. 8; Th. 10, 15; Gen. 157: Andr. Kmbl. 1907; An. 956: Exon. 103 b; Th. 392, 3; Rä. 11, 2. Ðǽr cómon flód [*n.*] *vēnērunt flūmĭna,* Mt. Bos. 7, 27. Upahófon flód [*n.*] Driht, upahófon flódas [*m.*] stefne his, upahófon flód ýþe his *elēvāvērunt flūmĭna Dŏmĭne, elēvāvērunt flūmĭna vōcem suam, elēvāvērunt flūmĭna fluctus suos,* Ps. Spl. 92, 4, 5. Flódas [*m.*] feágaþ oððe hafetiaþ mid handa samod *flūmĭna plaudent mănu sĭmul,* Ps. Lamb. 97, 8. Fǽmige flódas [*m.*] *foamy floods,* Cd. 100; Th. 133, 19; Gen. 2213: Ps. Th. 68, 14: Exon. 125 b; Th. 482, 19; Rä. 67, 4. Flóda [*m. or n.*] begong *the floods' course,* Beo. Th. 2999; B. 1497: Ps. Th. 65, 11. Ða fugelas ðe on flódum [*m. or n.*] wuniaþ syndon flaxfóte *the birds which dwell in waters are web-footed,* Hexam. 8; Norm. 14, 14: Exon. 22 a; Th. 61, 5; Cri. 980. Ofer flód, *n.* [flódas, *m.* Lamb.] he gegearwode hine *sŭper flūmĭna prǣpărāvit eum,* Ps. Spl. 23, 2. Ðú adrygdest flód, *n.* [flódas, *m.* Spl.] *tu siccasti flŭvios,* Ps. Lamb. 73, 15. He gewende to blóde heora flódas [*m.*] *convertit in sanguĭnem flūmĭna eōrum,* 77, 44: Andr. Kmbl. 1811; An. 908. II. *the Flood, deluge;* dilŭvium:— Ýdode ðæt flód [*n.*] ofer eorþan *aquæ dilŭvii inundāvērunt sŭper terram,* Gen. 7, 10, 17: Mt. Bos. 24, 39: Lk. Bos. 17, 27: Boutr. Scrd. 21, 11, 13. Flód [*m. or n.*] ofslóh giganta cyn *the flood slew the race of giants,* Beo. Th. 3383; B. 1689: Cd. 69; Th. 83, 28; Gen. 1386. Ic gebringe flódes [*m. or n.*] wæteru ofer eorþan, ðæt ic ofsleá eall flǽsc *ĕgo addūcam aquas dilŭvii sŭper terram, ut interfĭciam omnem carnem,* Gen. 6, 17: 7, 6, 7: 9, 11. Noe lyfode þreóhund geára and fíftig geára æfter ðam flóde [*m. or n.*] *vixit Noe post dilŭvium trecentis quinquāgĭnta annis,* Gen. 9, 28: Mt. Bos. 24, 38: Boutr. Scrd. 21, 12, 13, 16, 18, 29: Cd. 75; Th. 93, 13; Gen. 1544. Ic wille mid flóde [*m. or n.*] folc acwellan *I will destroy the people with a flood,* 64; Th. 78, 20; Gen. 1296: Boutr. Scrd. 21, 21, 22. Flódas [*m.*] Noe oferlǽþ *Noah sailed over the floods,* Cd. 161; Th. 200, 25; Exod. 362. [*Laym.* flod, ulod, *n: Orm.* flod: *Plat.* flood, *f: O. Sax.* flôd, fluod, *m. f. n;* fluot, *f: Frs.* floede: *O. Frs.* floed, flod, *n: Dut.* vloed, *m: Ger.* fluth, *f: M. H. Ger.* vluot, *f. m: O. H. Ger.* flôt, fluot, *f;* flôz *fluxus: Goth.* flôdus, *f: Dan.* flod, *m. f:* Swed. flod, *m. a flood, river: Icel.* flóð, *n. inundation, deluge.*] DER. brim-flód, Cofer-, drenc-, geofon-, heáh-, lagu-, mere-, nép-, sǽ-, wæter-, will-.

flód-blác; *adj. Flood-pale, made pale by water, that is, by drowning;* per ăquam pallĭdus:—Flódblác here *the flood-pale host,* Cd. 167; Th. 209, 11; Exod. 497.

flóde, an; *f. A place where anything flows, a channel, sink, gutter;* cloāca, lăcūna, Cot. 44: 193, Som. Ben. Lye.

flód-egsa, an; *m. Flood-dread;* ăquārum terror:—Flódegsa becwom gástas geómre *flood-dread seized on their sad souls,* Cd. 166; Th. 206, 4; Exod. 446.

flód-líc; *adj.* FLOODLIKE; flŭviālis:—Flódlíc *flŭviālis,* Ælfc. Gr. 9, 28; Som. 11, 36.

flód-weard, e; *f. A flood-guard, sea-wall;* măris custōdia, măris mūrus:—Flódwearde slóh *he struck the sea-wall* [i. e. *the wall caused by dividing the Red Sea*], Cd. 167; Th. 209, 3; Exod. 493.

flód-weg, es; *m. Flood-wood, watery way, the sea;* mărīna via, măre:—Sǽmen fóron flódwege *the seamen went on the sea,* Cd. 147; Th. 184, 12; Exod. 106. Fór flódwegas *went the watery ways,* Exon. 109 b; Th. 418, 22; Rä. 37, 9: 82 a; Th. 309, 4; Seef. 52.

flód-wudu; *m. Flood-wood, a ship;* mărīnum lignum, nāvis:—Swá we ofer cald wæter ceólum líðan, geond sídne sǽ flódwudu fergen *as if we journey in vessels over the cold water, convey our ships through the wide sea,* Exon. 20 a; Th. 53, 21; Cri. 854.

flód-wylm; *m. Flood-boiling, raging flood;* ăquārum fluctus:— Flódwylm ne mæg manna ǽnigne gelettan *a raging flood may not hinder any man,* Andr. Kmbl. 1032; An. 516.

flód-ýþ, e; *f. A flood-wave;* măris unda:—Nó he fram me flódýþum feor fleótan meahte *he could not float far from me on the flood-waves,* Beo. Th. 1088; B. 542.

floga, an; *m.* [flogen, *pp.* of fleógan *to fly;* fleón *to flee*] *One who flies or flees, a fugitive;* fŭgitīvus. DER. án-floga, gúþ-, lyft-, uht-, wíd-.

flogen *flown; pp.* of fleógan.

flogen *fled, escaped; pp.* of fleón.

flogettan; *p.* te; *pp.* ed *To fluctuate;* fluctuāre, Scint. 77.

flóh, e; *f. That which is flown off, a fragment, piece;* fragmen, frustum:—Flóh stánes *a piece of stone;* glēba sĭlicis, Cot. 99.

flohten-fóte; *adj. Web-footed;* palmĭpes:—Ne sé flohtenfóte fugelas *let him not eat web-footed birds,* L. M. 1, 36; Lchdm. ii. 88, 9. v. flax-fóte.

flooc, es; *n. A sole;* plātessa, Glos. Epnl. Recd. 161, 31. v. flóc.

FLÓR; gen. flóre; **dat.** flóre, flóra; **acc.** flór, flóre; *f:* flór, es; *m. A* FLOOR; păvĭmentum, sŏlum, ārea:—Flór on húse *a floor in a house;* excussōrium, Ælfc. Gl. 29; Som. 61, 34; Wrt. Voc. 26, 33. Flór *păvĭmentum,* Wrt. Voc. 290, 10. Flór *păvĭmentum vel sŏlum,* Wrt. Voc. 81, 7. Breda þiling vel flór on to þerscenne *a joining of planks or a floor to thresh on,* Ælfc. Gl. 57; Som. 67, 73; Wrt. Voc. 37, 59. Scipes flór *a ship's floor, gangway;* fóri, Ælfc. Gl. 103; Som. 77, 116; Wrt. Voc. 56, 36. Is glisnaþ glæshluttur, flór forste geworht *ice glittereth transparent as glass, a floor caused by frost,* Runic pm. 11; Kmbl. 341, 18; Hick. Thes. i. 135, 22. Flór ættre weól *the floor* [of hell] *boiled with venom,* Cd. 220; Th. 284, 8; Sat. 318: 213; Th. 267, 17; Sat. 39. Swá swá ælces húses wah biþ fæst ǽgðer ge on ðære flóre, ge on ðæm hrófe, swá biþ ǽlc gód on Gode fæst, forðam þe he is ǽlces gódes ǽgðer ge hróf ge flór *as the wall of every house is fixed both to the floor and to the roof, so is every good fixed in God, for he is both the roof and*

the floor of every good, Bt. 36, 7; Fox 184, 11–14. Ætfealh mīn sáwul flóre [flóra, Spl.] *adhæsit păvĭmento anĭma mea*, Ps. Th. 118, 25. He gang æfter flóre *he went along the floor*, Beo. Th. 2636; B. 1316. Đū dæm wættere foldan to flóre gesettest *thou settest the earth for a floor to the water*, Bt. Met. Fox 20, 181; Met. 20, 91. On flóra *on the floor*, Cd. 215; Th. 271, 24; Sat. 110: Homl. Th. ii. 56, 33: 334, 35. He gefeóll on đa flór *he fell on the floor*, Bt. 1; Fox 4, 3: 33, 4; Fox 130, 4. He feól on đa flóre, Bt. Met. Fox 1, 161; Met. 1, 81: Judth. 10; Thw. 23, 8; Jud. 111. He feormaþ his bernes flóre *purgābit āream suam*, Lk. Bos. 3, 17. On fågne flór feónd treddode *the fiend trod on the variegated floor*, Beo. Th. 1454; B. 725. [*Orm.* flor: *Plat.* floor: *Dut.* vloer, *m.: Ger.* flur, *f. field: M. H. Ger.* vluor, *m. sĕges: O. H. Ger.* flūr *sĕges: Icel.* flór, *m. a floor, pavement: Wel.* llawr, *m. a floor.*] DER. bere-flór, helle-, þirsce-, þyrscel-, up-.

flór-stán, es; *m. A floor-stone, stone used for pavement;* tessĕra pavimento stĕrnendo designāta :—Lytle federscitte flórstánas *little four-cornered floor-stones;* tessellæ, Ælfc. Gl. 61; Som. 68, 67; Wrt. Voc. 39, 50.

flot, es; *n.* [floten, *pp.* of fleótan *to float*] *Water deep enough for sustaining a ship, the sea;* æqua sătis alta ad nāvem sustĭnendam, măre :—Ongan eorla mengu to flote fȳsan *the multitude of warriors began to hasten to the sea*, Elen. Kmbl. 451; El. 226: Andr. Kmbl. 3393; An. 1700. Wæron đa útlagas ealle on flote *the outlaws were all afloat* [lit. *on the sea*], Chr. 1070; Erl. 209, 24. We willaþ on flot feran *we will depart on the sea*, Byrht. Th. 132, 64; By. 41: Chr. 937; Erl. 114, 1; Æđelst. 35. [*Plat.* flot: *Dut.* vlot: *Ger.* floss: *M. H. Ger.* vlôz, *m. river, raft: Icel.* flot; á flot *on* or *afloat.*]

FLOTA, an; *m.* [floten, *pp.* of fleótan *to float*]. I. *a ship, vessel, fleet;* nāvis, classis :—Flota stille bād on sole *the vessel abode still in the mud*, Beo. Th. 608; B. 301: 426; B. 210. Næs se flota swā rang *no fleet was so insolent*, Chr. 975; Erl. 125, 26: 1006; Erl. 140, 6. Mid dæm flotan *with the fleet*, 904; Erl. 98, 12. Lǣt nū geferian flotan úserne to lande *let our ship now go to land*, Andr. Kmbl. 794; An. 397: Beo. Th. 594; B. 294. II. *a sailor, pirate;* nauta, pīrata :—Flota módgade *the sailor proudly moved*, Cd. 160; Th. 198, 32; Exod. 331. Brǣddon æfter beorgum flotan feldhūsum *the sailors spread themselves amongst the hills with their tents*, 148; Th. 186, 3; Exod. 133: 154; Th. 191, 31; Exod. 223. Đa flotan, wícinga fela *the pirates, vikings many*, Byrht. Th. 133, 25; By. 72. ´ [*Scot.* flote *a fleet: Dut.* vloot, *f. a fleet: Ger.* flotte, *f. a. fleet: Dan.* flaade, *m. f: Swed.* flotta, *f: Icel.* floti, *m. a fleet.*] DER. æg-flota, ge-, hærn-, sǣ-, scip-, wǣg-.

floten *floated, swam;* *pp.* of fleótan.

floterian, flotorian; *p.* ode; *pp.* od *To* FLUTTER, *be disquieted or troubled, be carried by the waves;* fluctuāre, fluctibus ferri :—Đín heorte floteraþ on gýtsunge *thy heart flutters or is disquieted with covetousness;* cor tuum fluctuat avāritia, Homl. Th. ii. 392, 28. Flotorode *fertur fluctĭbus*, Glos. Prudent. Recd. 150, 1. Flotorodon *prævŏlant*, 150, 10.

flot-herge, es; *m. A naval force;* nāvālis exercĭtus :—Hygelác cwom faran flotherge *Hygelac came faring with a naval force*, Beo. Th. 5822; B. 2915. v. here, herge *an army.*

flotian; *part.* flotigende; *p.* ode; *pp.* od [floten, *pp.* of fleótan *to float*] *To float;* fluitāre :—Beó ān scip flotigende swā nēh đan lande swā hit nýxt mǣge *let a ship be floating as near the land as it nearest can*, Chr. 1031; Erl. 162, 6.

flot-man, -mann, -mon, -monn, es; *m. A float-man, sailor, pirate;* nauta, pīrāta :—Wícing ođđe flotman *pīrāta*, Wrt. Voc. 73, 74. Flotmen *pĭrātæ*, Lupi Serm. i. 14; Hick. Thes. ii. 103, 19. Flotmanna *nautārum*, Mone B. 114. Flotmonna freá *chief of mariners* [*Noah*], Cd. 72; Th. 89, 3; Gen. 1475.

flot-scip, es; *n. A floating ship, light bark;* barca, cĕlox :—Flotscip barca, Ælfc. Gl. 103; Som. 77, 100; Wrt. Voc. 56, 22: Glos. Brux. Recd. 37, 18; Wrt. Voc. 63, 32. Flotscip *cĕlox*, Ælfc. Gl. 103; Som. 77, 114; Wrt. Voc. 56, 34.

flot-smere, es; *n.* [smeru *fat, grease*] *Floating fat, scum of a pot;* pinguédo ollæ sŭpernātans, Som. Ben. Lye.

flot-weg, es; *m. A sea-way, the sea;* mărĭna via, măre :—He sceolde faran on flotweg *he must journey on the sea*, Exon. 123 b; Th. 475, 1; Bo. 41.

FLÓWAN; *part.* flówende; *ic* flówe, đū flówest, flēwst, he flóweþ, flēwþ, *pl.* flówaþ; *p.* fleów, *pl.* fleówon; *pp.* flówen *To* FLOW, *issue;* fluĕre, fluctuāre, inundāre :—Đæt ealle eán eft flówan mǣgon *that all waters may flow again*, Boutr. Scrd. 21, 16. Flówan mót ȳþ ofer eall lond *the wave may flow over all the land*, Salm. Kmbl. 644; Sal. 321: Ps. Th. 77, 21: 104, 36: Menol. Fox 555; Gn. C. 47. Com flówende flód *the flood came flowing*, Byrht. Th. 133, 44; By. 65. Ic flówe *fluo*, Ælfc. Gr. 28, 5; Som. 32, 4. Lagu flóweþ ofer foldan *water shall flow over the earth*, Exon. 115 b; Th. 445, 2; Dóm. 1: Bt. Met. Fox 5, 28; Met. 5, 14: Ps. Th. 67, 2: 68, 1: 103, 10: 147, 7. On đæt land đe flēwþ meolece and hunie *in terram quæ fluit lacte et melle*, Ex. 3, 8: Num. 13, 28: 14, 8: 16, 14: Ps. Spl. 57, 8: Bd. de nat. rerum; Wrt. popl. science 15, 19; Lchdm. iii. 268, 16. Lybbendes wætres flód flówaþ of his innoþe *flūmĭna de ventre ejus fluent ăquæ vīvæ*, Jn. Bos. 7, 38: Ps. Lamb. 147, 18. Sǣstreámas flówaþ *sea-streams flow*, Ps. Th.

92, 5. Fleów blód út and wæter *exīvit sanguis et ăqua*, Jn. Bos. 19, 34. Fleów firgend-streám *the mountain-torrent flowed*, Andr. Kmbl. 3144; An. 1575. He slóh stán and fleówon wæteru, and burnan fleówon ođđe ȳþgodon *percussit petram et fluxērunt ăquæ, et torrentes inundāvērunt*, Ps. Lamb. 77, 20: 104, 41. Đeáh đe wealan flówen *dīvĭtiæ si affluant*, Ps. Th. 61, 11. [*Chauc.* flowen: *Orm.* flowenn: *Plat.* floien, flojen: *Dut.* vloeien: *M. H. Ger.* vlæjen, vlæen: *O. H. Ger.* flawjan, flewēn: *Icel.* flóa *to flood: Lat.* flu-ĕre: *Grk.* πλώ-ω *to swim: Sansk.* plu *to float, swim.*] DER. a-flówan, æt-, be-, forþ-, geond-, of-, ofer-, to-, to-be-, under-.

flówednys, -nyss *a flowing, flux, torrent.* DER. ofer-flówednys, to-.

flównys, -nyss, e; *f. A flowing, flux, torrent;* fluxus, torrens :—Đæt wíf wæs þrówiende blódes flównysse *mŭlier fluxum pătiēbātur sanguĭnis*, Bd. 1, 27; S. 494, 5. Burnan ođđe flównyssa unrihtwísnyssa gēdrēfdun me *torrentes inĭquĭtātis conturbāvērunt me*, Ps. Lamb. 17, 5. DER. ofer-flównys.

flox-fóte; *adj. Web-footed;* palmĭpes, Hexam. 8; Norm. 14, 15, note x. v. flax-fóte.

fluge *fleddest;* fugisti, Ps. Lamb. 113, 5; *2nd pers. sing. p.* of fleón.

flugol; *adj.* [fleógan *to fly;* fleón *to flee*] *Apt to fly* or *flee, flying swiftly, swift;* fŭgax :—Flugol *fŭgax*, Ælfc. Gr. 9, 60; Som. 13, 43.

flugon *flew*, Bd. 3, 19; S. 548, 32; *p. pl.* of fleógan.

flugon *fled, escaped*, Cd. 166; Th. 206, 15; Exod. 452; *p. pl.* of fleón.

flustrian; *p.* ode; *pp.* od *To plait, weave;* plectĕre :—Flustriende *plectens*, Cot. 176, Som. Ben. Lye.

fluton *floated, swam;* *p. pl.* of fleótan.

flýcþ *flees*, Chr. 473; Ing. 16, note o, = flýhþ; *3rd pers. pres.* of fleón.

flyge, es; *m.* [fleógan *to fly*] *A flying, flight;* vŏlātus :—Se fugel flyges cunnode *the bird made trial of his flying*, Exon. 17 a; Th. 40, 28; Cri. 645. Wiđ flyge gāres *against an arrow's flight*, 79 a; Th. 297, 11; Crä. 66. Ic sceal on flyge earda neósan *I shall in flight visit lands*, Cd. 215; Th. 271, 28; Sat. 112. [*Ger.* flug, *m: M. H. Ger.* vluc, *m: O. H. Ger.* flug, *m: Icel.* flug, *n;* flugr, *m. vŏlātus.*] DER. a-flyge.

flyge-reów; *adj.* [reów *wild, fierce, cruel*] *Wild-flying, wild in flight;* vŏlātu fĕrus :—Flygereówe þurh nihta genipu neósan cwômon, hwæđere ... *the wild-flying [evil spirits] came in the darkness of night to find out, whether ...*, Exon. 37 b; Th. 123, 10; Gū. 320.

flyge-wíl *a flying wile, cunning trick.* v. flige-wíl.

flyht, fliht, es; *m.* [fleógan *to fly*] *A flight;* vŏlātus :—Wæs đæs fugles flyht dyrne and dēgol *the bird's flight was hidden and secret*, Exon. 17 a; Th. 40, 15; Cri. 639. On flyhte *in flight*, Elen. Kmbl. 1485; El. 744: Cd. 215; Th. 271, 29; Sat. 112. Se đe nafaþ fugles flyht *who has not the flight of a bird*, Salm. Kmbl. 451; Sal. 226: Exon. Th. 41, 12; Cri. 654. Earnas feredon sáwle flyhte on lyfte *eagles conveyed the soul in flight through the sky*, Andr. Kmbl. 1732; An. 868: Nicod. 26; Thw. 14, 36. [*Laym.* fliht, fluht, flut: *Orm.* flihht: *Scot.* flicht: *Plat.* flugt, *f: O. Sax.* fluht, *f: Frs.* flechte: *O. Frs.* flecht, *f: Dut.* vlugt, *f: Ger.* flucht, *f: M. H. Ger.* vluht, *f: O. H. Ger.* fluht, *f: Dan.* flugt, *m. f: Swed.* flykt, *m.*]

flyht-cláþ, es; *m. A joining, binding* or *tying together;* commissūra, conjunctūra, ligātūra, Som. Ben. Lye.

flýhþ, đū flýhst *flees, thou fleest*, Exon. 81 a; Th. 305, 3; Fä. 82; *3rd* and *2nd pers. pres.* of fleón.

flyht-hwæt; *adj. Flight-prompt;* in vŏlātu strēnuus :—Weras mundum mearciaþ on marmstáne frætwe flyhthwates *men design with hands in marble stone the plumage of the prompt in flight* [*phœnix*], Exon. 60 b; Th. 221, 15; Ph. 335. Se fēnix ascæceþ feđre, flyhthwate *the phœnix shakes its feathers, prompt for flight*, 58 a; Th. 207, 21; Ph. 145.

flýma, flēma, an; *m. One who flees, a runaway, an exile, outlaw, a man who had fled for any offence, and whose flight was equivalent to a conviction;* profūgus, fŭgĭtīvus, exul :—Đū bist flýma geond ealle eorþan *profūgus ĕris sŭper terram*, Gen. 4, 12: 4, 16. He monigra geára tíde flýma wæs *multo annōrum tempŏre profūgus văgābātur*, Bd. 2, 12; S. 513, 3: Ps. Th. 77, 37. Beó he syddan flýma *let him be henceforth a fugitive*, L. Ath. i. 2; Th. i. 200, 10: i. 20; Th. i. 210, 13, 14. DER. here-flýma.

flýman; *p.* de; *pp.* ed *To cause to flee, put to flight, rout, banish;* fŭgāre :—Ic sceal flýman feóndsceadan *I shall cause the hostile-spoiler to flee*, Exon. 104 a; Th. 396, 5; Rä. 15, 19. Hí mec sóna flýmaþ *they soon put me to flight*, 105 a; Th. 398, 12; Rä. 17, 6. Hie God flýmde *God routed them*, Th. 127, 24; Gen. 2115. DER. a-flýman, ge-, út-, úta- [-flǣman, -flēman]. v. fleón.

flýman fyrmþ, flíeman feorm, e; *f. A fugitive's food* or *support, the offence of harbouring a fugitive, the penalty for such an offence;* fŭgitīvi suscepto :—Đis syndon đa gerihta đe se cyning áh ofer ealle men on Wes-sexan; đæt is ... and flýmena fyrmþe *these are the rights which the king possesses over all men in Wessex; that is ... and [the penalty] for harbouring a fugitive*, L. C. S. 12; Th. i. 382, 14: Th. i. 382, 21. Gif mon cierliscne monnan flíeman feorme teó *if a man accuse a churlish man of harbouring a fugitive*, L. In. 30; Th. i. 120, 16.

flýming, es; *m. A fugitive, runaway, exile;* profūgus, fŭgĭtīvus, exul, Som. Ben. Lye. v. fleáming, flýma.

flyne, flene, an; _f. What is made soft, batter_; fluĭdum quid :—Gewyrce to flynan micelne citel fulne _work a large kettle full into a batter_, L. M. I, 38; Lchdm. ii. 98, 6. Geót ða flynan on _pour the batter on_, 1, 38; Lchdm. ii. 98, 10.

FLÝS, flís, flíés, flés, fleós, es; _n. A fleece, wool_; vellus, lānūgo :—Ðis flýs _hoc vellus_, Ælfc. Gr. 9, 32; Som. 12, 12. Gilde ðæt flýs mid twām pæningum _let the fleece be paid for with two pence_, L. In. 69; Th. i. 146, 11, MS. H. Mid his flýse _with its fleece_, L. In. 69; Th. i. 146, 9, 10, MSS. B. H. He nyðerastíhþ swā swā rēn on flýs _descendet sĭcut plŭvia in vellus_, Ps. Lamb. 71, 6 : Ps. Th. 147, 5. Of flýsum mĭnra sceápa wǽron gehlyde þearfena sídan _the sides of the poor were clothed with the fleeces of my sheep_, Job Thw. 165, 2. Wulle flýsum _with fleeces of wool_, Exon. 109 a; Th. 417, 12; Rä. 36, 3. Flýs _lānūgo_, Cot. 122. [_Piers P._ flus : _Plat._ fliis _vellus_ : _Dut._ vlies, _n_ : _Ger._ vlies, fliesz, _n_ : _M. H. Ger._ vlies, _n_.]

flýte, an; _f. Cream_; flos lactis :—Dó flýtan to _add cream_, L. M. I, 34 : Lchdm. ii. 80, 23. _v._ flēt.

flýte, es; _m?_ [fleótan _to float_] _What floats, hence,—A boat, punt_; pontōnium :—Flýte _pontōnium_, Ælfc. Gl. 103; Som. 77, 103; Wrt. Voc. 56, 25 : 63, 35.

flýþ, es; _m. Flight_; vŏlātus :—Forgeaf ðām fugelum flyþ geond ðas lyft _he gave to the birds flight through this air_, Hexam. 8; Norm. 14, 10. _v._ flyht.

flýþ _flee, flee from, avoid_, Bt. 41, 5; Fox 252, 27; _pres. pl._ of fleón.

flýtst, he flýt _floatest, floats_, Homl. Th. ii. 564, 13; _2nd and 3rd pers. pres._ of fleótan.

fnæd, es; _pl. nom. acc._ fnadu, fnado; _gen._ fnada; _dat._ fnadum; _n. A hem, edge, fringe_; fimbria :—Fnæd _fimbria_, Wrt. Voc. 81, 66. Ân wíf æt-hrán hys reáfes fnæd _mŭlier tĕtĭgit fimbriam vestĭmenti ejus_, Mt. Bos. 9, 20 : Bd. 1, 27; S. 494, 6, MS. B : Ps. Th. 132, 3. Híg mǽrsiaþ heora reáfa fnadu _magnĭficant fimbrias_, Mt. Bos. 23, 5. Fnado _vel_ læppan _fimbriæ_ [MS. _timbria_], Ælfc. Gl. 64; Som. 68, 128; Wrt. Voc. 40, 33. On fnadum gyldenum _in fimbriis aureis_, Ps. Lamb. 44, 14.

fnæs, es; _pl. nom. acc._ fnasu; _gen._ fnasa; _dat._ fnasum; _n. A fringe_; fimbria :—Mid gyldnum fnasum _in fimbriis aureis_, Ps. Th. 44, 15. _v._ fæs, fnæd.

FNÆST, es; _m. A puff, blast, breath_; flātus, anhēlĭtus :—Ure fnæst ateoraþ _our breath faileth_, Hexam. 4; Norm. 8, 18. Þurh ðæs fíres fnæst _through the fire's blast_, Exon. 74 a; Th. 277, 29; Jul. 588. Hyt bringþ forþ ðone [MS. ðane] fnæst _it will bring forth the breath_, Lchdm. iii. 100, 13 : 116, 24. Fnæstas [MS. fnæstiaþ] swíðe beóþ fortogene _the breathings are very hard drawn_, L. M. 2, 36; Lchdm. ii. 242, 7. [_O. H. Ger._ fnastōn _anhēlāre_ : _Dan._ fnyse _to puff_ : _Swed._ fnysa _to snort_ : _Icel._ fnasa _to sneeze_ : _Grk._ πνέω _I blast, puff._]

fnæstiaþ, L. M. 2, 36; Lchdm. ii. 242, 7,=fnæstas? _pl._ of fnæst.

fneósung, e; _f. A sneezing_; sternūtātio, sternūtāmentum :—Snytingc _vel_ fneósung _sternūtātio vel sternūtāmentum_, Ælfc. Gl. 79; Som. 72, 62. [_Wyc._ fnesynge, fnesing : _Icel._ fnasan, fnōsun _a sneezing._]

fnésan _to sneeze._ [_Icel._ fnœsa _to sneeze._] DER. ge-fnésan.

fnora, an; _m. A sneezing, sneeze_; sternūtātio, Wrt. Voc. 289, 4.

fó _I take_; _1st sing. pres. indic._ of fón. Ne ne fó he _he may not take_, L. Ælf. C. 30; Th. ii. 354, 2; _3rd sing. pres. subj._ of fón.

foca, an; _m. A cake baked on the hearth_; pānis sub cĭnĕre pistus :—Wirc focan _fac subcĭnĕrĭcios pānes_, Gen. 18, 6.

FÓDA, an; _m. FOOD, nourishment_; ălĭmentum :—On ðære ōðre flēringe wæs ðæta nýtena fóda gelogod _on the second flooring_ [_of the ark_] _the food of the cattle was placed_, Boutr. Scrd. 21, 8. Fóda fýres, holt _food of fire, wood_, Scint. 12. Būton ðam gōdspellícan fōdan _without the evangelical food_, Homl. Th. ii. 396, 31. [_Orm._ fode : _Plat._ fōde, vōde : _Goth._ fōdeins. _f_ : _Dan._ føde, _m. f_ : _Swed._ föda, _f_ : _Icel._ fæði, _n_.]

fódder, fóddor, fóddur, fóder, fódor; _gen._ fódres; _dat._ fódre; _n._   **I.** FODDER, _dry food for cattle, hay, corn, provender, food generally_; jūmenti pābŭlum, fœnum, ĕdūlium, pābŭlum, esca, victus :—Ða ungesceádwísan neát ne wilniaþ nānes ōðres feós to eácan ðam fódre _the irrational cattle desire no other wealth in addition to the fodder_, Bt. 14, 2; Fox 44, 23. Wolde syllan his assan fóddur _ut dăret jūmento pābŭlum_, Gen. 42, 27. Fódder neátum _fœnum jumentis_, Ps. Th. 103, 13. We fódder horsum ūrum habbaþ _pābŭla ĕquis nostris hăbēmus_, Coll. Monast. Th. 31, 29. Fóddur, Ps. Th. 77, 20; [mettas, Ps. Spl. 77, 21] _ut pĕtĕrent escas anĭmābus suis_. Fóddor, Exon. 96 a; Th. 357, 28; Pa. 36. Fódor, Runic pm. 25; Kmbl. 344, 17; Hick. Thes. i. 135, 49. Brúceþ fódres _has an enjoyment of food_, Runic pm. 28; Kmbl. 345, 6; Hick. Thes. i. 135, 55. Gif ðam ðe ðæs beþurfe fýr and fóddor _let him give fire and food to him who needs it_, L. Pen. 15; Th. ii. 282, 26.   **II.** _a case from which anything is fed, a case, cover, sheath_; thēca = θήκη :—Fódder _thēca_, Ælfc. Gl. 53; Som. 66, 68; Wrt. Voc. 35, 54. _v._ boge-fódder. [_Laym._ fodder, uodder _fodder, meat_ : _Plat._ foder, voder, voer : _Dut._ voeder, voer, _n. fodder, provender_ : _Ger._ futter, _n_ : _M. H. Ger._ vuoter, _n_ : _O. H. Ger._ fuotar, _n_ : _Goth._ fōdr, _n. a sheath_ : _Dan. Swed._ foder, foer, _n_ : _Icel._ fóðr, _n. pābŭlum._] _v._ fóðer.

fódder-brytta, an; _m. A fodder-distributor, fodderer, herdsman_;

---

pābŭlātor :—Horshyrde _vel_ fódderbrytta _pābŭlātor_, Ælfc. Gl. 9; Som. 56, 122; Wrt. Voc. 19, 6.

fóddor-þegu, fóddur-þegu, fódor-þegu, e; _f._ [þegu _a taking, receiving_] _A taking or receiving food, food_; cĭbi acceptio, cĭbus :—Ðæt hie tobrugdon, blōdigum ceaflum, fíra flǽschoman him to fóddorþege _that they tore asunder, with bloody jaws, the bodies of men for their food_, Andr. Kmbl. 320; An. 160. Lēton him ða betweonum tān wísian hwylcne hira ǽrest ōðrum sceolde to fóddurþege feores ongildan _they let the lot decide between them which of them first should give up to the rest his life for food_, 2203; An. 1103. Ðǽr hí mētaþ fódorþege gefeán [MS. gefeon] _where they find the joy of taking food_, Exon. 59 b; Th. 215, 4; Ph. 248.

fóddur-wéla, an; _m. Abundance of food_; cĭbi ōōpia :—Fere fóddurwēlan folescipe dreógeþ [_a ship_] _performs the bringing_ [i. e. _a ship brings_, Grn.] _abundance of food to people_, Exon. 108 b; Th. 415, 12; Rä. 33, 10.

fódnôþ, es; _m? Food, nourishment_; ălĭmentum, Som. Ben. Lye.

fódrere, es; _m. A fodderer, forager_; pābŭlātor :—Þunor ofslōh xxiv heora fódrera _thunder killed twenty-four of their foragers_, Ors. 4, 1; Bos. 78, 1.

fóg, es; _n. A joining, joint_; conjunctio, commissūra, Som. Ben. Lye. DER. ge-fóg, stán-ge-.

fóge _fitly, aptly, comprehensibly._ DER. un-ge-fóge.

fógere, es; _m. A suiter, wooer_; prŏcus :—Fógere [MS. foghere] _prŏco_, Mone B. 4287. _v._ wōgere.

fóh _take_ :—Fóh to me _take from me_; accĭpe a me, Cd. 228; Th. 308, 2; Sat. 686; _impert._ of fón.

fóh _comprehensible, measurable, moderate._ DER. un-ge-fóh.

fóhlíc _comprehensible, measurable, moderate._ DER. un-ge-fóhlíc.

fóhlíce _comprehensibly, measurably, moderately._ DER. un-ge-fóhlíce.

fohten _fought, contended_; _pp._ of feohtan.

FOLA, an; _m. A FOAL, colt_; pullus, poledrus :—Cicen oððe brid oððe fola _pullus_, Wrt. Voc. 77, 37. Fola _poledrus_, Ælfc. Gl. 20; Som. 59, 50; Wrt. Voc. 23, 11. Hí gemētton ðone folan úte _invēnērunt pullum fŏris_, Mk. Bos. 11, 4, 5 : Mt. Bos. 21, 2, 5. [_Piers P._ fole : _Plat._ falen, vale : _Frs._ fole : _O. Frs._ folla, _m_ : _Dut._ volen, veulen, _n_ : _Ger._ fohle, _m_ ; füllen, _n_ : _M. H. Ger._ vole, vol, _m_ ; vüli, vüln, _n_ : _O. H. Ger._ folo, _m._ pullus, poledrus; fuli, _n._ pullus, pultrinus : _Goth._ fula, _m_ : _Dan._ fole, _m. f_ ; føl, _n_ : _Swed._ föl, _n_ : _Icel._ foli, _m_ : _Lat._ pullus, _m. a young animal_ : _Grk._ πῶλος, _m. f. a foal._]

FOLC, es; _n._ [Folc being a neuter noun, and a monosyllable, has the _nom. and acc. pl._ the same as the _nom. and acc. sing_ : _it is a collective noun_ in English, _and has not the plural form_ folks _but by a modern corruption_] _The_ FOLK, _people, common people, multitude, a people, tribe, family_; pōpŭlus, gens, nātio, vulgus, plebs, cīves, hŏmĭnes, exercĭtus, multĭtūdo :—Twā folc beóþ tōdǽled, and ðæt folc oferswíþ ðæt ōðer folc _two nations shall be divided, and the one folk shall overcome the other folk_, Gen. 25, 23. Ðæt folc wæs Zachariam geanbĭdigende ĕrat plebs expectans Zacharĭam, Lk. Bos. 1, 21. Micel folc mid hym _cum eo turba multa_, Mt. Bos. 26, 47. Hie awerede ðæt folc _the people defended it_, Chr. 921; Erl. 106, 10, 33. Gif folces man syngaþ _if a man of the people sin_, Lev. 4, 27. Ðæs folces hlísa _the people's praise_, Bt. 30, 1; Fox 108, 16. He slōh folces Denigea fýftyne men _he slew of the Danes' folk fifteen men_, Beo. Th. 3168; B. 1582. Folces hyrde _the people's shepherd_, Beo. Th. 1224; B. 610 : 3668; B. 1832 : 5282; B. 2644. Eallum folce to friþe _to the peace of all the people_, L. Edg. S. 15; Th. i. 278, 7. Eádmund cyning cýþ eallum folce _Edmund king makes known to all people_, L. Edm. S; Th. i. 246, 17. Se ðe sý folce ungetrýwe _he who may be untrue to the people_, L. C. S. 25; Th. i. 390, 17. Man swencte ðæt earme folc _one harassed the poor people_, Chr. 999; Erl. 135, 32. Se eorl earfoþlíce gestylde ðæt folc _the earl hardly stilled the people_, Chr. 1052; Erl. 187, 4, 3. Þurh úre folc _throughout our folk_, L. In. prm; Th. i. 102, 9. Beó se þeóf útlah wið eall folc _let the thief be an outlaw to all people_, L. C. S. 30; Th. i. 394, 24. He gesōhte Sūþ-Dena folc _he sought the people of the South-Danes_, Beo. Th. 931; B. 463 : 1049; B. 522 : 1390; B. 693 : 2362; B. 1179. Folce gestēpte sunu Ohtheres _with people he supported Ohthere's son_, Beo. Th. 4776; B. 2393. Ða folc fǽhþe towehton _the people excited enmity_, 5888; B. 2948 : 2849; B. 1422. Freáwine folca _friend of peoples_, 864; B. 430 : 4038; B. 2017 : 4849; B. 2429. Folcum gefrǽge _famed among nations_, 109; B. 55 : 530; B. 262 : 3715; B. 1855. Mec wolcna strengu ofer folc byreþ _the clouds' strength bears me over people_, Exon. 103 a; Th. 390, 5; Rä. 8, 6. Folgad folcum _followed by peoples_, Cd. 226; Th. 300, 4; Sat. 559. [_Laym._ folc, uolc : _Orm._ follc : _O. Sax._ folk, folc, _n_ : _Frs._ folck : _O. Frs._ folk, _n_ : _Dut._ volk, _n_ : _M. H. Ger._ volc, _m_ : _O. H. Ger._ folc, folch, folk, _n_ : _Dan. Swed._ folk, _n_ : _Icel._ fólk, _n_.] DER. dryht-folc, here-, mægen-, sige-, sūþ-, wíd-.

folc-ágende, _part. Folk-owning_; pŏpŭlum possĭdens :—Bealg hine swíðe folcágende _the folk-owning_ [_man_] _was much irritated_, Exon. 68 a; Th. 253, 26; Jul. 186 : Beo. Th. 6218; B. 3113. Nis se foldan sceat mongum gefēre folcágendra _the tract of earth is not easy of access to many folk-owning_ [_men_], Exon. 56 a; Th. 198, 4; Ph. 5.

folc-bealo; _gen._ -bealowes; _n. Folk-torment, torment by many, a great torment;_ ingens mălum _vel_ crŭciātus:—Petrus and Paulus þrŏwedon on Rôme folcbealo þreálíc _Peter and Paul suffered grievous torment by the people at Rome,_ Menol. Fox 248; Men. 125.

folc-bearn, es; _n. A folk-child, a child of man;_ pŏpŭli filius, hŏmĭnis filius:—Swilc biþ mægburh menigo ðínre, folcbearnum frome _such shall be the family of thy people, excellent in children,_ Cd. 100; Th. 132, 16; Gen. 2194. þurh ðē eorþ-búende ealle onfóþ, folcbearn, freoðo and freóndscipe _through thee all dwellers upon earth, the children of men, shall receive peace and friendship,_ 84; Th. 105, 28; Gen. 1760.

folc-beorn _a popular man._ v. folc-biorn.

folc-biorn, es; _m. A popular man;_ pŏpŭlāris vir:—Folc-biorn, Beo. Th. 4444; B. 2221.

folc-cú; _f. The folk's cow, a cow of the herd;_ pŏpŭli vacca:—Under folc-cúm [MS. folcum] _inter vaccas_ pŏpŭlōrum, Ps. Th. 67, 27; among the kien of puplis, Wyc. 67, 31. v. cú.

folc-cúþ; _adj. Known to the people, folk-known, well-known, public, celebrated;_ pŏpŭlis nōtus, publĭcus, cĕleber:—Wæs his freádrihtnes folc-cúþ nama Agamemnon _his lord's celebrated name was Agamemnon,_ Bt. Met. Fox 26, 18; Met. 26, 9. Folc-cúþne ræd _a discourse known to nations,_ Bt. Met. Fox introduc. 18; Met. Einl. 9. Be folc-cúþum strætum _by the public roads,_ Bd. 2, 16; S. 520, 4.

folc-cwén, e; _f. Folk's queen, queen of the people;_ pŏpŭli rēgīna:—Eóde freólícu folc-cwén to fúran sittan _the noble queen of the people went to sit by her lord,_ Beo. Th. 1286; B. 641.

folc-cyning, es; _m. Folk's king, king of nations, king of the people;_ pŏpŭli rex:—Nealles folc-cyning fyrdgesteallum gylpan þorfte _the people's king needed not to boast of his comrades in arms,_ Beo. Th. 5738; B. 2873: 5460; B. 2733. Folc-cyninge _for the king of nations,_ Cd. 131; Th. 166, 25; Gen. 2753. Fífe folc-cyningas _five kings of nations,_ 93; Th. 119, 4; Gen. 1974: 95; Th. 125, 5; Gen. 2074. cf. O. Sax. folk-kuning.

folc-dryht, -driht, e; _f._ [dryht, driht _a multitude_] _A multitude of people, an assemblage;_ pŏpŭli multĭtūdo, cŏmĭtātus:—Folcdryht wera bifóran _before the assemblage of men,_ Exon. 23 b; Th. 66, 5; Cri. 1067. Folcdriht, Cd. 64; Th. 76, 24; Gen. 1262.

folce-firen, e; _f. A folk-crime, public crime;_ pŏpŭli scĕlus:—Wǽrlogona sint folcefirena hefige _the public crimes of the faithless are heavy,_ Cd. 109; Th. 145, 23; Gen. 2410.

folc-getrum, es; _n. A host of people;_ exercĭtus:—Mid heora folcgetrume _with their band of people,_ Cd. 95; Th. 123, 18; Gen. 2046, note. v. folc-getrum.

folc-egsa, an; _m. Folk-terror;_ publĭcus terror, formīdo:—Ðú towurpe fæsten his for folcegsan _pŏsuisti munītiōnes ejus in formīdĭnem,_ Ps. Th. 88, 33.

folc-firen _a folk-crime._ v. folce-firen.

folc-freá, an; _m. Folk's lord, lord of a nation;_ pŏpŭli dŏmĭnus:—Hie ðæt cúþ dydon heora folcfreán _they made that known to their nation's lord,_ Cd. 89; Th. 111, 7; Gen. 1852.

folc-frig, folc-frý; _adj. Folk-free;_ liber ăpud plēbem:—Beó he syððan folcfrig _be he afterwards folk-free,_ L. C. S. 45; Th. i. 402, 17. Se sié folcfrý _let him be folk-free,_ L. Wih. 8; Th. i. 38, 15. cf. _Grm. RA._ 349.

folc-gefeoht, es; _n. Folk-battle, a great battle, pitched battle;_ publĭca pugna, plēnum prælium:—Ða Sciððie noldon hine gesēcan to folcgefeohte _the Scythians would not attack him in a pitched battle,_ Ors. 2, 5; Bos. 46, 5. Wurdon ix folcgefeoht gefohten _nine great battles were fought,_ Chr. 871; Erl. 77, 7: 887; Erl. 87, 9. On þrim folcgefeohtum _in three pitched battles,_ Ors. 3, 9; Bos. 66, 11. cf. _Icel._ fólk-orrusta.

folc-gemót, -mót, folces gemót, es; _n. A folk-meeting;_ pŏpŭli consessus. The folc-gemót was _a general assembly of the people_ of a town, city or shire, and was held annually on the first of May, but it could be convened on extraordinary occasions by ringing the moot-bell,—' Cum ăliquid vēro inŏpīnātum, vel dŭbium, vel mălum contra, regnum, vel contra cŏrōnam dŏmĭni rēgis, forte in ballīvis suis sŭbito emersērit, dēbent, stătim pulsātis campānis quod Anglĭce vŏcant môtbel convŏcāre omnes et ūnĭversos, quod Anglĭce dīcunt folcmōte, i. e. vŏcātio et congrĕgātio pŏpŭlōrum, et gentium omnium, quia ĭbi omnes convēnīre dēbent ... Stătūtum est quod dēbent pŏpŭli omnes, et gentes ūnĭversæ sĭngŭlis annis, sĕmel in anno scīlicet convēnīre, scīlicet in căpĭte kal. Maii,' Th. Anglo-Saxon Laws, vol. i. 613, note a. The folc-gemót was forbidden to be held on Sundays:—On folcgemóte _at the folk-moot,_ L. Alf. pol. 34; Th. i. 82, 12, 13: L. Ath. i. 2; Th. i. 200, 8: iv. 1; Th. i. 220, 23. On folcgemóte [-móte, L.], L. Ath. i. 12; Th. i. 206, 11. On folces gemóte, L. Alf. pol. 22; Th. i. 76, 5. Gif he folcgemót [folces gemót, MS. H.] mid wǽpnes bryde arǽre _if he disturb the folk-moot by drawing his weapon,_ L. Alf. pol. 38; Th. i. 86, 16. Sunnan dæges we forbeódaþ ǽlc folcgemót, búton hit for mycelre neódþearfe sí _we forbid every Sunday folk-moot, unless it be for great necessity,_ L. C. E. 15; Th. i. 368, 16: L. N. P. L. 55; Th. ii. 298, 22. Sunnan dæges freóls healde man georne, and folcgemóta on ðam hálgan dæge geswíce man georne _let Sunday's festival be diligently kept, and folk-moots be carefully abstained_

from on that holy day, L. Eth. v. 13; Th. i. 308, 11: vi. 22; Th. i. 320, 12: L. Edg. C. 19; Th. ii. 248, 14. v. folc-mót, folc-land. v. Stubbs' Const. Hist. folk-moot.

folc-geréfa, an; _m. A folk-reeve, a people's governor;_ pŏpŭli præpŏsĭtus:—Folcgeréfa _actionātor,_ Ælfc. Gl. 5; Som. 56, 25; Wrt. Voc. 17, 30. v. Du Cange, sub voce Actionator.

folc-geriht, es; _n. Folk-right;_ publĭcum jus:—Feola syndon folcgerihtu _there are many folk-rights,_ L. R. S. 21; Th. i. 440, 25. v. folc-riht.

folc-gesetness, e; _f. A decree or ordinance of the people;_ plēbiscītum, Som. Ben. Lye.

folc-gesíþas; _gen._ -gesíþa; _m. The nobles of a country;_ păres, nōbĭles, gentis cŏmites, pŏpŭlāres:—Syndon deáde folcgesíþas _the nobles of the country are dead,_ Cd. 98; Th. 128, 29; Gen. 2134: Bt. Met. Fox 1, 140; Met. 1, 70. Wið ðám nēhstum folcgesíþum _with the nearest rulers of the people,_ Cd. 193; Th. 241, 29; Dan. 412.

folc-gestǽlla, an; _m. An adherent, follower;_ gentis cŏmes:—Cræft folcgestǽlna _a force of adherents,_ Cd. 15; Th. 18, 10; Gen. 271. v. folc-gestealla.

folc-gestealla, -gestǽlla, an; _m. A noble companion;_ gentis cŏmes, pŏpŭlāris:—Mid swilcum mæg man fón folcgesteallan _with such, one may obtain adherents,_ Cd. 15; Th. 19, 6; Gen. 287.

folc-gestreón, es; _n. A public treasure;_ pŏpŭli dīvĭtiæ:—Ða leóde leng ne woldon Elamitarna aldor swíðan folcgestreónum _those nations would no longer strengthen the Elamites' prince with the public treasures,_ Cd. 93; Th. 119, 17; Gen. 1981.

folc-getǽl, es; _n. A number of people;_ pŏpŭli nŭmĕrus:—On folcgetæl fíftig cista _in the number of people [were] fifty bands,_ Cd. 154; Th. 192, 9; Exod. 229.

folc-geþrang, es; _n. Folk-throng, a crowd;_ pŏpŭli căterva:—þurh ðæt folcgeþrang _through the crowd,_ Ors. 3, 9; Bos. 68, 30.

folc-getrum, folce-getrum, es; _n. Folk-host;_ exercĭtus:—Folcgetrume gefaren hæfdon _they had come with a host,_ Cd. 93; Th. 119, 29; Gen. 1987. DER. getrum.

folc-gewinn, es; _n. Folk's war, battle;_ bellum:—Wæs monig Gota gelysted folcgewinnes _many a Goth was desirous of battle,_ Bt. Met. Fox 1, 19; Met. 1, 10.

folcisc; _adj. Folkish, common, vulgar, popular;_ rustĭcus, plēbĕius:—Gif man folciscne mæsse-preóst mid tĭhtlan belecge _if a man charge a secular mass-priest with an accusation,_ L. Eth. ix. 21; Th. i. 344, 19: L. C. E. 5; Th. i. 362, 16. Folcisce men _common men,_ Bt. 30, 1; Fox 108, 23: 35, 6; Fox 168, 24. Ðæt hí folciscra gemóta geswícan _that they abstain from popular meetings,_ L. Eth. vi. 44; Th. i. 326, 21.

folc-léasung, e; _f. Public lying, slander;_ publĭcum mendācium:—Gif mon folclǽsunge gewyrce _si quis publĭcum mendācium confingat,_ L. Alf. pol. 28; Wilk. 41, 19. v. folc-leásung.

folc-lagu, e; _f. Folk or public law;_ publĭca lex:—Gif hwá folclage wirde _if any one corrupt the law of the people,_ L. N. P. L. 46; Th. ii. 296, 22. Folclaga wyrsedon _the laws of the people were corrupted,_ Lupi Serm. i. 5; Hick. Thes. ii. 100, 19.

folc-land, -lond, es; _n._ [folc _folk,_ land _land_]. I. _the land of the folk or people._ It was the property of the community. It might be occupied in common, or possessed in severalty; and, in the latter case, it was probably parcelled out to individuals in the folc-gemót, _q. v._ or court of the district, and the grant sanctioned by the freemen who were there present. While it continued to be folc-land, it could not be alienated in perpetuity; and, therefore, on the expiration of the term for which it had been granted, it reverted to the community, and was again distributed by the same authority. Spelman describes folc-land as ' terra pŏpŭlāris, quæ jūre commūni possidētur—sīne scripto,' Gloss. Folcland. In another place he distinguishes it accurately from bóc-land: ' Prædia Saxōnes duplĭci tĭtŭlo possīdēbant; vel scripti auctōrĭtate, quod bóc-land vocābant, vel pŏpŭli testĭmōnio, quod folc-land dīxēre,' Id. Bocland:—Eác we cwǽdon hwæs se wyrðe wǽre ðe óðrum ryhtes wyrnde, áðor oððe on bóc-lande oððe on folc-lande, and ðæt he him geándagode of ðam folc-lande, hwonne he him riht worhte befóran ðam geréfan. Gif he ðonne nán riht næfde ne on bóc-lande ne on folc-lande, ðæt se wǽre ðe rihtes wyrnde scyldig xxx scillinga wið ðone cyning; and æt óðrum cyrre, eác swá: æt þriddan cyrre, cyninges oferhýrnesse, ðæt is cxx scillinga, búton he ǽr geswíce _also we have ordained of what he were worthy who denied justice to another, either in book-land or in folk-land, and that he should give him a term respecting the folk-land, when he should do him justice before the reeve. But if he had no right either to the book-land or to the folk-land, that he who denied the right should be liable in 30 shillings to the king; and for the second offence, the like: for the third offence, the king's penalty, that is, 120 shillings, unless he previously desist,_ L. Ed. 2; Th. i. 160, 10-17. All lands, whether bóc-land or folc-land, were subject to the Trĭnóda Necessĭtas. Under this denomination are comprised three distinct imposts, to which all landed possessions, not excepting those of the church, were subject, viz:—[a] Brycg-bót _for keeping the bridges, and highways in repair._ [b] Burh-bót _for keeping the burghs, or fortresses, in_

an efficient state of defence. [c] Fyrd *a contribution for maintaining the military and naval force of the kingdom* :—Gif hwá Burh-bóte, oððe Brycg-bóte, oððe Fyrd-fare forsitte; gebéte mid hund-twelftigum scillinga ðam cyninge on Engla lage, and on Dena lage, swá hit ǽr stód *if any one neglect Burh-bót, or Brycg-bót, or Fyrd-fare; let him make amends with one hundred and twenty shillings to the king by English law, and by Danish law, as it formerly stood*, L. C. S. 66; Th. i. 410, 8–10. Þegenes lagu is, ðæt he sý his bóc-rihtes wyrðe, and ðæt he þreó þinc of his lande dó, fyrd-færeld, and burh-bóte, and brycg-geweorc [MS. bryc-] *thane's law is, that he be worthy to make his will, and that he perform three things for his land, military service, repairs of fortresses, and of bridges*, L. R. S. 1; Th. i. 432, 1–3. II. Folk-land was subject to many burthens and exactions from which book-land was exempt. The possessors of folk-land were bound to assist in the reparation of royal vills, and in other public works. They were liable to have travellers and others quartered on them for subsistence. They were required to give hospitality to kings and great men in their progress through the country, to furnish them with carriages and relays of horses, and to extend the same assistance to their messengers, followers, and servants, and even to the persons who had charge of their hawks, horses, and hounds. Such at least are the burthens from which lands are liberated when converted by charter into book-land. 2. Folk-land might be held by freemen of all ranks and conditions. It is a mistake to imagine with Lambarde, Spelman, and a host of antiquaries, that it was possessed by the common people only. Still less is Blackstone to be credited, when, trusting to Somner, he tells us it was land held in villenage by people in a state of downright servitude, belonging, both they and their children and effects, to the lord of the soil, like the rest of the cattle or stock upon the land. [Blackstone, ii. 92.]—A deed published by Lye, exposes the error of these representations. [*Anglo-Saxon Dict.*, App. ii. 2.] Alfred, a nobleman of the highest rank, possessed of great estates in book-land, beseeches King Alfred, in his will, to continue his folk-land to his son, Æthelwald; and if that favour cannot be obtained, he bequeaths, in lieu of it, to his son, who appears to have been illegitimate, ten hides of book-land at one place, or seven at another. From this document it follows, first, that folk-land was held by persons of rank; secondly, that an estate of folk-land was of such value, that seven, or even ten hides of book-land were not considered as more than equivalent to it; and, lastly, that it was a life-estate, not devisable by will, but in the opinion of the testator, at the disposal of the king, when by his own death it was vacated. 3. It appears also from this document, that the same person might hold estates both in book-land and in folk-land; that is, he might possess an estate of inheritance of which he had the complete disposal, unless in so far as it was limited by settlement; and with it he might possess an estate for life, revertible to the public after his decease. In the latter times of the Anglo-Saxon government it is probable there were few persons of condition who had not estates of both descriptions. Every one was desirous to have grants of folk-land, and to convert as much of it as possible into book-land. Money was given and favour exhausted for that purpose. 4. In many Saxon wills we find petitions similar to that of Alfred; but in none of them is the character of the land, which could not be disposed of without consent of the king, described with the same precision. In some wills, the testator bequeaths his land as he pleases, without asking leave of any one [Somner's *Gavelkind*, 88, 211; Hickes, *Pref.* xxxii; *Diss. Epist.* 29, 54, 55, 59; Madox, *Formul.* 395]; in others he earnestly beseeches the king that his will may stand, and then declares his intentions with respect to the distribution of his property [Lambarde, *Kent*, 540; Hickes, *Diss. Epist.* 54; Gale, i. 457; Lye's *Append.* ii. 1, 5; Heming, 40];—and in one instance he makes an absolute bequest of the greater part of his lands, but solicits the king's consent to the disposal of a small part of his estate [Hickes, *Diss. Epist.* 62.] There can be no doubt that book-land was devisable by will, unless where its descent had been determined by settlement; and a presumption, therefore, arises, that where the consent of the king was necessary, the land devised was not book-land, but folk-land. If this inference be admitted, the case of Alfred will not be a solitary instance, but common to many of the principal Saxon nobility. 5. That folk-lands were assignable to the thegns, or military servants of the state, as the stipend or reward for their services, is clearly indicated in the celebrated letter of Bede to Archbishop Ecgbert [Smith's *Bede*, 305–312]. In that letter, which throws so much light on the internal state of Northumberland, the venerable author complains of the improvident grants to monasteries, which had impoverished the government, and left no lands for the soldiers and retainers of the secular authorities, on whom the defence of the country must necessarily depend. He laments the mistaken prodigality, and expresses his fears that there will be soon a deficiency of military men to repel invasion, no place being left where they can obtain possessions to maintain them suitably to their condition. It is evident from these complaints, that the lands so lavishly bestowed on the church had been formerly the property of the public, and at the disposal of the government. If they had been book-lands, it

could have made no difference to the state whether they belonged to the church or to individuals, since in both cases they were beyond its control, and in both cases were subject to the usual obligations of military service. But if they formed part of the folk-land, or property of the public, it is easy to conceive how their conversion into book-land must have weakened the state, by lessening the fund out of which its military servants were to be provided. 6. A charter of the eighth century conveys to the see of Rochester certain lands on the Medway, as they had been formerly possessed by the chiefs and companions of the Kentish kings. [*Text. Roffens.* 72, edit. Hearne; Kemble, *Cod. Dipl.* No. cxi.] In this instance folk-land, which had been appropriated to the military service of the state, appears to have been converted into book-land, and given to the church, L. Th. ii. Glossary, Folc-land: Sandys' *Gavel.* 97. v. Stubbs' Const. Hist. folk-land. v. fyrd, scip-fyrd, bóc-land.

folc-lár, e; *f. Popular instruction, a sermon*; pōpŭlāris institūtio *vel* instructio, hŏmīlia, sermo, Cot. 143, Som. Ben. Lye. v. lár.

folc-leásung, e; *f. Folk-leasing, public lying, slander*; publĭcum mendācium, călumnia :—Be folcleásunge gewyrhtum. Gif mon folcleásunge gewyrce, mid nánum leóhtran þinge gebéte ðonne him mon aceorfe ða tungan of *of those committing slander. If a man commit slander, let him make amends with no lighter thing than that his tongue be cut out*, L. Alf. pol. 32; Th. i. 80, 19–82, 1.

folc-líc; *adj. Folklike, common*; pōpŭlāris, commūnis :—Folclíc lár hŏmīlia [MS. ŏmīlia = ὁμιλία], Ælfc. Gl. 35; Som. 62, 75; Wrt. Voc. 28, 53. He sǽde ðæt he folclíc man wǽre *rustĭcum se fuisse respondit*, Bd. 4, 22; S. 591, 6: Nar. 18, 4.

folc-lond *folk-land*; pōpŭli terra, Exon. 115 b; Th. 444, 14; Kl. 47. v. folc-land.

folc-mægen, es; *n. People's force*; pōpŭli rōbur :—Ðá ðær folc-mægen fór *then there marched a people's force*, Cd. 160; Th. 199, 31; Exod. 347.

folc-mǽgþ, e; *f. A nation-tribe, tribe*; nātio, tribus :—Folc-mǽgþa *of nation-tribes*, Cd. 64; Th. 77, 18; Gen. 1277.

folc-mǽlum *in bands*, Chr. 1011; Erl. 145, 5, = floc-mǽlum. v. floccmǽlum.

folc-mǽre; *nom. pl. n.* folc-mǽro; *adj. Folk-known or popular*; cĕlĕber, pōpŭlōsus :—Ofer folc-mǽro land *over celebrated lands*, Cd. 86; Th. 108, 5; Gen. 1801.

folc-mót, es; *n. A popular assembly*; pōpŭli consessus :—On folcmóte *at the folk-moot*, L. Ath. i. 12; Th. i. 206, 11, note 25. v. folc-gemót.

folc-néd, e; *f. A people's need*; pōpŭli necessĭtas :—Him wísode wolcen unlytel daga ǽghwylce, swá hit Drihten hét; and him ealle niht, óðer beácen, fýres leóma, folcnéde heóld *a large cloud directed them every day, as the Lord commanded it; and to them all night, another sign, a pillar of fire, supplied the people's need*, Ps. Th. 77, 16.

folc-rǽd, -rǽd, es; *m. A public benefit, that which serves for the good of the people*; publĭcum bĕnĕfĭcium :—Dryhten gumena folcrǽd fremede *the Lord of men did public benefits*, Andr. Kmbl. 1243; An. 622. He folcrǽd fremede *he accomplished public benefit*, Beo. Th. 6004, note; B. 3006.

folc-rǽden, -rǽdenn, e; *f. A nation's law*; plēbiscītum :—Sum mæg folcrǽdenne gehycgan *one may deliberate a nation's law*, Exon. 79 a; Th. 295, 32; Crä. 42.

folc-riht, -ryht, es; *n. Folkright, common law, public right, the understood compact by which every freeman enjoys his rights as a freeman*; publĭcum jus, commūne = τὸ κοινόν :—Arǽre up Godes riht; and heonanforþ lǽte manna gehwylcne, ge earmne ge eádigne, folcrihtes wyrðe, and him man rihte dómas déme *let God's right be exalted; and henceforth let every man, both poor and rich, be worthy of folk-right, and let a man have right dooms judged to him*, L. C. S. 1; Th. i. 376, 10: L. Ed. 11; Th. i. 164, 20: L. Edg. ii. 1; Th. i. 266, 4: L. Eth. vi. 8; Th. i. 316, 28. Hit he becwæþ mid fullan folcrihte *he bequeathed it with full folkright*, L. O. 13; Th. i. 184, 1: 2; Th. i. 178, 13. To folcryhte *to folk-right*, L. Ath. i. 2; Th. i. 200, 7: i. 8; Th. i. 204, 7: i. 23; Th. i. 212, 1. He him forgeaf wícstede wéligne, folcrihta gehwylc, swá his fæder áhte *he had given him the wealthy dwelling place, every public right, as his father had possessed*, Beo. Th. 5209; B. 2608. Gesealde wǽpna geweald ofercom mid ðý feónda folcriht *he gave him power of weapons with which he overcame the folkright [liberty] of enemies*, Cd. 143; Th. 179, 1; Exod. 22.

folc-riht, -ryht; *adj. According to folk-right, lawful*; secundum publĭcum jus, lēgālis :—Síe he wyrðe folcryhtre [-rihtre MS. G.] bóte *let him be worthy of lawful compensation*, L. Alf. 13; Th. i. 46, 25.

folc-sæl, es; *pl. nom. acc.* -salo; *n. A folk-building*; pōpŭlāris ædes :— Ic folcsalo bærne *I burn public structures*, Exon. 101 a; Th. 381, 3; Rä. 2, 5.

folc-scearu, -sceru, -scaru, e; *f. A division of the people, nation, multitude*; nātio, provincia :—Ðæt hie hine onsundne gebrohten of ðære folcsceare *that they should bring him uninjured from that tribe of people*, Cd. 90; Th. 112, 17; Gen. 1872: 114; Th. 149, 20; Gen. 2477. Ðú úsic woldest on ðisse folcsceare besyrwan *thou wouldest deceive us among*

this nation, 127; Th. 162, 12; Gen. 2680: 136; Th. 171, 16; Gen. 2829: Andr. Kmbl. 1368; An. 684: Elen. Kmbl. 1933; El. 968. Geond ða folcsceare *among the nation-host*, Cd. 85; Th. 106, 34; Gen. 1781. On ðisse folcscere *in this country*, Elen. Kmbl. 804; El. 402. Búton folcscare *except the host of people*, Beo. Th. 146; B. 73.

folc-sceaða, an; *m. People's tyrant, villain*; pŏpŭli tyrannus:—Ðæs weorudes ða wyrrestan fâ folcsceaðan feówertyne gewiton in forwyrd sceacan *of the host the worst, hateful villains, fourteen departed into destruction*, Andr. Kmbl. 3184; An. 1595.

folc-scipe, es; *m. People*; nātio, pŏpŭlus:—Fere fôddurwĕlan folcscipe dreógeþ [*a ship*] brings [lit. *performs the bearing of*] *abundance of food to people*, Exon. 108 b; Th. 415, 13; Rä. 33, 10. [*O. Sax.* folkskepi.]

folc-slite, es; *m. A folk-slit, sedition*; sēdĭtio:—Folcslite *vel* æswîcung, sacu, ceást sēdĭtio, Ælfc. Gl. 15; Som. 58, 38; Wrt. Voc. 21, 30.

folc-stede, -styde, es; *m. Folk* or *dwelling-place*; pŏpŭli lŏcus, habĭtācŭlum:—Folcstede gumena *the dwelling-place of men*, Andr. Kmbl. 40; An. 20. On folcstede *in the folk-place*, Chr. 937; Erl. 114, 7; Æðelst. 41: Exon. 102 b; Th. 388, 21; Rä. 6, 11. On ðam folcstede *in the folk-place*, Judth. 12; Thw. 26, 10; Jud. 320: Andr. Kmbl. 357; An. 179. Ic gehêt ðē folcstede *I promised thee a dwelling-place*, Cd. 100; Th. 132, 31; Gen. 2201. Folcstede frætwian *to decorate the dwelling-place*, Beo. Th. 152; B. 76. Se ðe gegân dorste folcstede fâra *he who durst go into the folk-place of the hostile*, Beo. Th. 2930; B. 1463. Ðær folcstede fægre wǣron *where the dwelling-places were fair*, Cd. 91; Th. 116, 8; Gen. 1933. Fram ðam folcstyde *from the folk-place*, Cd. 93; Th. 120, 25; Gen. 2000.

folc-stów, e; *f. A public place, country place*; pŭblicus *vel* rustĭcus lŏcus:—He ferde ge þurh mynsterstôwe ge þurh folcstôwe *discurrĕre per urbāna et rustĭca lŏca sŏlēbat*, Bd. 3, 5; S. 526, 27.

folc-sweót, es; *m.* [sweót, *m. a band*] *A multitude of people, multitude*; pŏpŭli multĭtūdo, catĕrva:—Folcsweóta mǣst *greatest of multitudes*, Cd. 171; Th. 215, 2; Exod. 577.

folc-talu, e; *f. Folk-reckoning, genealogy*; pŏpŭli enŭmĕrātio, genealŏgia:—On folctale *in the genealogy*, Cd. 161; Th. 201, 29; Exod. 379.

folc-toga, an; *m. A popular leader, commander* or *leader of the people*; pŏpŭli dux, princeps:—Frome folctogan *pious leaders*, Andr. Kmbl. 15; An. 8. Ferdon folctogan *the nation's chieftains came*, Beo. Th. 1682; B. 839. Fyllan folctogan *to fell the people's chieftains*, Judth. 11; Thw. 24, 17; Jud. 194. [*O. Sax.* folk-togo.]

folc-truma, an; *m.* [truma *a band, troop*] *A host of people, people*; pŏpŭli cohors, pŏpŭlus:—Cweðe eall folctruma, sŷ ðæt, sŷ ðæt oððe beó hit swâ *dîcet omnis pŏpŭlus, fiat, fiat*, Ps. Lamb. 105, 48. Folctruman andettaþ ðē *pŏpŭli confĭtēbuntur tĭbi*, 44, 18. Drihten dēmþ folctruman *Dŏminus jŭdĭcat pŏpŭlos*, 7, 9; 9, 9: 46, 4.

folcú [folc *people*, cú *a cow*] *A cow of the herd*:—Under folcúm *inter vaccas pŏpŭlōrum*, Ps. Th. 67, 27. Folcûm, for folc-cûm, from folcú, like wildeór, wyrtruma, for wild-deór, wyrt-truma, etc. v. folc-cú.

folc-wĕlig, -wĕleg; *adj. Rich in people, populous*; pŏpŭlo dîves, abundans:—Folcwĕlega *populous*, Cot. 153.

folc-weras; *gen.* -wera; *pl. m. Men of the people, people*; pŏpŭlares, pŏpŭlus:—Hâtaþ Fîson folcweras *people call it Pison*, Cd. 12; Th. 14, 21; Gen. 222: 89; Th. 110, 30; Gen. 1846. [*O. Sax.* folk-werós.]

folc-wíga, an; *m. A warrior*; bellātor:—Folcwígan wicge wegaþ *warriors on horseback bear me*, Exon. 104 a; Th. 395, 26; Rä. 15, 13.

folc-wita, an; *m. A senator*; pŭblicus consĭliārius:—Sum biþ folcwita *one is a senator*, Exon. 79 b; Th. 297, 33; Crä. 77.

fold-ærn, es; *n.* [folde *the earth*, ærn *a place*] *An earth-place, a cave, sepulchre*; terrēnus lŏcus, sepulcrum:—Foldærne fæst *fast in the earth-house*=sepulcrum, Exon. 18 b; Th. 45, 36; Cri. 730: 47 b; Th. 163, 36; Gú. 1004.

fold-bold, es; *n.* [folde *the earth*, bold *a dwelling*] *The land-dwelling, royal palace*; terrestris dŏmus, rēgia aula, arx:—Ne feól fæger foldbold *the fair earthly dwelling fell not*, Beo. Th. 1550; B. 773.

fold-búend, -búende; *noun from pres. part.* v. búend, *pl. m. Earth-dwellers, earth's inhabitants, inhabitants of a land* or *country*; terrĭcŏlæ:—Ðanan feorhnere findaþ fold-búend *thence earth's inhabitants find nourishment*, Ps. Th. 64, 10: Beo. Th. 4541; B. 2274. Ðone Grendel nemdon foldbúende *whom earth's inhabitants named Grendel*, Beo. Th. 2714; B. 1355: Elen. Kmbl. 2026; El. 1014: Exon. 25 a; Th. 72, 25; Cri. 1178: 121 a; Th. 465, 9; Hö. 101. Hŷ ongytan mihton ðæt wæs fôremǣrost foldbúendum receda *they might perceive what was the grandest of houses to earth's inhabitants*, Beo. Th. 624; B. 309: Bt. Met. Fox 8, 8; Met. 8, 4: Exon. 53 a; Th. 186, 24; Az. 24. Deáþ rícsade ofer foldbúend *death ruled over earth's inhabitants*, Exon. 45 b; Th. 154, 17; Gú. 844. Mid fére fold-búende se micla dæg meahtan Dryhtnes bihlǣmeþ *the great day of the mighty Lord shall strike earth's inhabitants with fear*, Exon. 20 b; Th. 54, 14; Cri. 868. Ðæt eorþwaran ealle hæfden foldbúende fruman gelícne *that all mortals, inhabitants of the earth, had a like beginning*, Bt. Met. Fox 17, 3; Met. 17, 2. Ðone fugel hâtaþ foldbúende Filistina fruman uasa mortis *the inhabitants*

of the land, the princes of the Philistines, call the bird vāsa mortis, Salm. Kmbl. 560; Sal. 279. Ic hæbbe me on hrycge ðæt ǣr hâdas wreáh foldbúendra *I have on my back what ere covered the persons of dwellers on earth*, Exon. 101 a; Th. 381, 18; Rä. 2, 13: 32 b; Th. 106, 2; Gú. 35.

FOLDE, an; *f.* I. *the earth, dry land*; tellus, terra:—He gesêceþ fægre land ðonne ðeós folde *he shall seek a fairer land than this earth*, Cd. 218; Th. 277, 32; Sat. 213: 84; Th. 106, 3; Gen. 1765: 100; Th. 133, 2; Gen. 2204: Exon. 73 a; Th. 272, 14; Jul. 499: 120 a; Th. 460, 21; Hö. 20: Bt. Met. Fox 11, 86; Met. 11, 43: 20, 118; Met. 20, 59. Folde wæs ðâ gyt græs ungrêne *the earth was as yet not green with grass*, Cd. 6; Th. 7, 35; Gen. 116: 12; Th. 14, 7; Gen. 215: Exon. 43 b; Th. 146, 26; Gú. 715. Stôd bewrigen folde mid flôde *the dry land stood covered with water*, Cd. 8; Th. 10, 15; Gen. 157. Geblissad mid ðâm fægrestum foldan stencum *made blissful by the sweetest odours of earth*, Exon. 56 a; Th. 198, 11; Ph. 8: Cd. 161; Th. 201, 9; Exod. 369. Foldan bearm or fæðm *the bosom of the earth*, Beo. Th. 2278; B. 1137: 2790; B. 1393: Exon. 93 b; Th. 351, 4; Sch. 75: 125 b; Th. 482, 20; Rä. 67, 4. Foldan sceát *a region* or *tract of the earth*, Exon. 9 a; Th. 5, 21; Cri. 72: 20 b; Th. 55, 6; Cri. 879: 116 a; Th. 445, 20; Dôm. 10: Bt. Met. Fox 4, 103; Met. 4, 52: Cd. 75; Th. 92, 26; Gen. 1534: 199; Th. 247, 25; Dan. 502: 213; Th. 265, 6; Sat. 3: Beo. Th. 193; B. 96. On ðisse foldan *on this earth*, Salm. Kmbl. 953; Sal. 476: Cd. 121; Th. 155, 24; Gen. 2577: Exon. 19 b; Th. 50, 28; Cri. 808: Beo. Th. 2396; B. 1196: Menol. Fox 283; Men. 143: Rood Kmbl. 261; Kr. 132. Teóde firum foldan Freá ælmihtig *fĭliis hŏmĭnum terram omnĭpŏtens creāvit*, Bd. 4, 24; S. 597, 24: Cd. 8; Th. 10, 9; Gen. 154: Exon. 12 b; Th. 20, 22; Cri. 321. II. *a land, country, district, region, territory*; rēgio, tractus, plăga, terrĭtōrium:—Wæs wera gŭþhergum êðelland geondsended, folde feóndum *the people's native land was overspread with hostile bands, their country with enemies*, Cd. 92; Th. 118, 22; Gen. 1969: Exon. 56 a; Th. 199, 21; Ph. 29. Unlytel dæl sîdre foldan geondsended wæs bryne *no small part of the wide land was overspread with burning*, Cd. 119; Th. 154, 5; Gen. 2551. Nyste hine on ðære foldan fira ænig *none of the men in the land knew him*, Salm. Kmbl. 547; Sal. 273: Menol. Fox 29; Men. 15. Ðæt land gesêc ðe ic ðē ŷwan wille, brâde foldan *seek the land which I will show thee, a spacious country*, Cd. 83; Th. 105, 12; Gen. 1752: Exon. 123 b; Th. 474, 27; Bo. 37: Salm. Kmbl. 431; Sal. 216. Ðú eart hyht ealra ðe feor on sǣ foldum wuniaþ *thou art the hope of all who dwell in lands far in the sea* [i.e. *in islands*], Ps. Th. 64, 6. III. *the ground, soil*; hūmus, sŏlum:—He gefeóll to foldan *he fell to the ground*, Judth. 12; Thw. 25, 27; Jud. 281: Andr. Kmbl. 1474; An. 738: Exon. 29 a; Th. 88, 34; Cri. 1450: Elen. Kmbl. 1970; El. 987. Him heortan blôd foldan gesêceþ *his heart's blood seeks the ground*, Salm. Kmbl. 316; Sal. 157: Exon. 103 b; Th. 393, 17; Rä. 13, 1. Foldan begræfen *buried in the ground*, Elen. Kmbl. 1944; El. 974: Exon. 63 a; Th. 231, 17; Ph. 490: Ps. Th. 142, 4. IV. *earth, clay*; terræ lĭmus, lūtum:—Cd ðone ǣrestan ælda cynnes of ðære clǣnestan foldan geworhte *God made the first of the race of men from the purest earth*, Exon. 44 b; Th. 151, 14; Gú. 795. [*Laym.* folde: *O. Sax.* folda, *f: Icel.* fold, *f. a field, earth*.]

fold-græf, es; *n. An earth-grave*; sepulcrum:—He ahóf of foldgræfe he raised [*it*] *from an earthly grave*, Elen. Kmbl. 1686; El. 845. Of foldgrafum *from the earth-graves*, Exon. 23 a; Th. 63, 27; Cri. 1026.

fold-grǣg; *adj.* [grǣg *grey*] *Earth-grey, earth-coloured*; instar terræ cānus:—Eá of dúne sceal foldgrǣg fêran *earth-coloured water shall proceed from a hill*, Menol. Fox 521; Gn. C. 31.

fold-hrêrende; *part. touching, moving on, the earth*; terram tangens vel peragrans:—Deóra foldhrêrendra *of earth-enlivening beasts*, Exon. 95 b; Th. 356, 2; Pa. 5. cf. mold-hrêrende.

fold-ræst, e; *f. Earth-rest*; sepulcrālis requies:—Weorþeþ foldrǣs'e æt ende *shall be at the end of their earth-rest*, Exon. 23 a; Th. 63, 34; Cri. 1029.

fold-wæstm, es; *m. Earth-fruit*; quidquid terra gignit:—Fægrum foldwæstmum *with fair fruits of earth*, Exon. 65 a; Th. 241, 10; Ph. 654.

fold-weg, es; *m.* I. *earth-way*; terrestris via:—On foldwege *on the earth-way*, Cd. 95; Th. 123, 24; Gen. 2050: 116; Th. 151, 17; Gen. 2510: 139; Th. 174, 4; Gen. 2873: Beo. Th. 3271; B. 1633. Foldwegas, Beo. Th. 1736; B. 866: Exon. 96 a; Th. 358, 25; Pa. 51. II. *the earth in general*; terra:—On ðissum foldwege *on this earth*, Exon. 30 a; Th. 93, 22; Cri. 1530. On foldwege *on the earth*, Andr. Kmbl. 412; An. 206. Cwicra ǽngum on foldwege *to any living on earth*, Exon. 51 a; Th. 177, 8; Gú. 1224.

fold-wéla, an; *m. Earth-wealth*; terrestres ŏpes:—Foldwéla fealleþ *earthly wealth decays*, Exon. 95 a; Th. 354, 59; Reim. 68.

fold-wong, es; *m. Earth-plain*; terræ campus:—On foldwong *on earth's plain*, Exon. 22 a; Th. 60, 25; Cri. 975.

folgaþ, es; *m.* I. *a train, retinue*; id quod sĕquĭtur, cŏmĭtātus:—Him wæs lâþ to amyrrene his âgenne folgaþ *he was loath to injure his own retinue*, Chr. 1048; Erl. 178, 12. II. *service of*

a follower; cŏmĭtis servĭtus, ministĕrium :—Hwæt is betere đonne đæs cyninges folgaþ *what is better than the king's service?* Bt. 29, 1; Fox 102, 6. Heó fægerne folgaþ hæfdon uppe mid englum *they had a fair service above with angels,* Cd. 220; Th. 284, 30; Sat. 329. Ic gewát folgaþ sécan *I departed to seek my service,* Exon. 115 a; Th. 442, 8; Kl. 9. Âhte ic fela wintra folgaþ tilne, holdne hláford *I had for many years a good service, a kind lord,* 100 b; Th. 379, 25; Deór. 38. v. folgoþ. v. Stubbs' Const. Hist. comitatus.

folgen *stuck to, went into; pp. of* felgan.

folgere, es; *m.* I. *a* FOLLOWER, *attendant, disciple;* assecla, pĕdĭsĕquus, aßectātor :—Folgere *assecla,* Ælfc. Gl. 113; Som. 79, 131; Wrt. Voc. 60, 35. Hwæt wille we sprecan be đam cyninge, and be his folgerum *what shall we say about the king, and about his followers?* Bt. 29, 1; Fox 104, 10. Đý þriddan dæge þeóda Wealdend arás, and he feówertig daga folgeras síne rúnum arétte *on the third day the Ruler of nations arose, and for forty days he comforted his followers [=disciples] with words,* Hy. 10, 35; Hy. Grn. ii. 293, 35. II. *one of a class of freemen who has no dwelling of his own, but is the follower or retainer of another, for whom he performs certain agricultural services;* folgārius, ūnus ex lībērōrum ordĭne qui ălĭcūjus clientēlæ *vel* servĭtio sese addĭcit, fămŭlus qui fŏco proprio cāret, aut sub stīpendio et servĭtii ălĭcūjus præstātĭōne possĭdet :—Folgere gebýreþ, đæt he on twelf mónþum ii æceras geearnige, ŏđerne gesáwene and ŏđerne unsáwene; sǽdige sylf đane, and his mete, and scóung, and glófung him gebýreþ: gyf he máre geearnian mæg [MS. mæig], him biþ sylfum fremu *folgārio compĕtit, ut in duodĕcim mensĭbus duas acras hăbeat, ūnam sēmĭnātam, ăliam non; sed ĭdem sēmĭnet eam, et victum suum, et calciamenta dēbet hăbēre, et cĭrotĕcas [=chirothĕcas] : si plus deservit, ipsi commŏdum ĕrit,* L. R. S. 10; Th. i. 438, 4–7: L. C. S. 20; Th. i. 386, 23. DER. æfter-folgere.

folgian; *p.* ode, ade, ede; *pp.* od, ad, ed; *v. trans. dat. and acc.* I. *to* FOLLOW, *go behind, run after, pursue;* sĕqui, insĕqui :—Míne sceáp gehýraþ míne stefne, and hig folgiaþ me *ŏves meæ vōcem meam audiunt, et sĕquuntur me,* Jn. Bos. 10, 27. He folgode feorhgeníþlan *he pursued his deadly foes,* Beo. Th. 5858; B. 2933. Þegn folgade *a thane went behind it,* Exon. 109 b; Th. 419, 8; Rä. 38, 2: 129 a; Th. 495, 4; Rä. 84, 2. We sóþfæstes swađe folgodon *we followed the true one's track,* Andr. Kmbl. 1346; An. 673. Đæt mínre spræce spéd folgie *that success follow my word,* Ps. Th. 55, 4. Gif ceorl acwyle be libbendum wífe and bearne, riht is đæt hit đæt bearn médder folgie *if a husband die, his wife and child yet living, it is right that the child follow the mother,* L. H. E. 6; Th. i. 30, 4. Đæt đære spræce spéd folgode *that success would follow thy speech,* Cd. 109; Th. 144, 4; Gen. 2384. II. *to follow as a servant, attendant or disciple;* cŏmĭtāri, adhærēre alicui, servīre, subdĭtus esse :—Cwǽdon hí đæt him nænig mæg leófra nære đonne hira hláford, and hí næfre his banan folgian noldon *they said that no kinsman was dearer to them than their lord, and they would never follow [=serve] his murderer,* Chr. 755; Erl. 50, 20. Folgian líchoman luste *to follow [=serve] the body's lust,* R. Ben. 4. Ne mæg nán þeów twám hláfordum þeówian: he ánum folgaþ and ŏđerne forhogaþ *nēmo servus pŏtest duŏbus dŏminis servīre: ūni adhærēbit et altĕrum contemnet,* Lk. Bos. 16, 13. He forlǽteþ láre đíne and mánþeáwum mínum folgaþ *he shall desert thy doctrine and follow my evil customs,* Elen. Kmbl. 1857; El. 930. Him folgiaþ in đam gladan hám gǽstas gecorene *chosen spirits follow [=serve] him [Christ] in that glad home,* Exon. 64 b; Th. 237, 16; Ph. 591. He folgode ánum burhsittendum men đæs ríces *adhæsit ūni cīvium rĕgĭōnis illīus,* Lk. Bos. 15, 15: Homl. Th. ii. 500, 10. Dô đæt mid đæs ealdormonnes gewitnesse đe he ǽr in his scíre folgode *let him do it with the knowledge of the alderman whom he before followed in his shire,* L. Alf. pol. 37; Th. i. 86, 4, 7: L. Ath. i. 8; Th. i. 204, 5: i. 22; Th. i. 210, 21: iv. 1; Th. i. 220, 21. We lǽraþ, đæt ǽnig preóst ne underfó ŏđres scólere, búton đæs leáfe đe he ǽr folgode *we enjoin, that no priest receive another's scholar, without leave of him whom he previously followed,* L. Edg. C. 10; Th. ii. 246, 15. Đeáh hie hira beággyfan banan folgedon *though they followed [=served] their ring-giver's murderer,* Beo. Th. 2209; B. 1102. Đæt ǽlc folgie swylcum hláforde swylcum he wille *that each follow [=serve] such lord as he will,* L. Ath. iv. 1; Th. i. 222, 1. Wæs on eorþan éce Drihten feówertig daga folgad folcum, ǽr he to heofonríce astáh *on earth the Lord eternal was followed [=attended] by people for forty days, ere he ascended into heaven,* Cd. 226; Th. 300, 4; Sat. 559. DER. æfter-folgian, ge-. v. fylgean.

folgoþ, folgaþ, es; *m.* [folgoþ = folgaþ; *3rd sing. pres. of* folgian *to follow.*] I. *that which follows,—A train, retinue;* id quod sĕquitur, cŏmĭtātus :—Á to his folgoþe and to his þenunge đa æđelestan men cómon the noblest men always came to his retinue and to his service,* Bd. 3, 14; S. 540, 11. On Swegenes eorles folgoþe *among the train of earl Sweyn,* Chr. 1048; Erl. 178, 16. II. *service of a follower,—A service, office, official dignity;* cŏmĭtis servĭtus, ministērium, officĭum, præpŏsĭtūra :—Se biscop amanige đa oferhýrnesse æt đam gerēfan đe hit on his folgoþe sý *let the bishop exact the penalty for contempt from the reeve in whose service it may be,* L. Ath. i. 26; Th. i. 214, 3. He folgode Iuliane, and he on đam folgoþe ealle fúlnysse forbeáh, lybbende swá swá

munuc *he followed Julian, and in that service he avoided all foulness, living as a monk,* Homl. Th. ii. 500, 12. On đý eahtateóđan geáre đe Óswold arcebisceop to folgoþe féng *in the eighteenth year [from that] in which archbishop Oswald took office,* Cod. Dipl. 620; A. D. 978; Kmbl. iii. 168, 23. Beó se geréfa búton his folgoþe *let the reeve be without [=deprived of] his official dignity,* L. Ath. v. § 11; Th. i. 240, 19. Ualentinianus wæs Iulianuses cempena ealdorman : he him bebeád đæt he forléte đone his cristendóm ođđe his folgoþ; đá wæs him leófre đæt he forléte his folgoþ đonne đone cristendóm *Valentinian was chief of Julian's soldiers: he [Julian] commanded him to give up christianity or his office; then it was dearer to him to give up his office than christianity,* Ors. 6, 33; Bos. 129, 16–19. Habbaþ folgoþa cyst mid Cyninge *they [the angels] have the choicest of services with their King,* Exon. 13 b; Th. 24, 26; Cri. 390. III. *condition of life;* condĭtio vītæ :—Óđer biþ unlǽde, ŏđer biþ eádig... hwædres biþ hira folgoþ betra *one is miserable, the other is fortunate... of which of them is the condition better?* Salm. Kmbl. 740; Sal. 369. DER. under-folgoþ.

FOLM; *gen. dat.* folme; *acc.* folm, folme; *pl. nom. acc.* folme, folma; *f.:* folme, an; *f. The palm of the hand, the hand;* palma, mănus :—Folm mec mæg bifón *the hand may grasp me,* Exon. 111 a; Th. 425, 6; Rä. 41, 52 : Ps. Th. 79, 15. Of sceađan folme *from the hand of the foe,* Andr. Kmbl. 2268; An. 1135. Ne hafaþ hió fót ne folm *it has not foot nor hand,* Exon. 110 a; Th. 420, 27; Rä. 40, 10. Heó genam cúþe folme *she took the well-known hand,* Beo. Th. 2610; B. 1303: Salm. Kmbl. 339; Sal. 169 : Ps. Th. 128, 5. Mægþ scearpne mēce of sceađe abræd swíđran folme *the woman [Judith] drew the sharp sword from its sheath with her right hand,* Judth. 10; Thw. 22, 26; Jud. 80: Beo. Th. 1500; B. 748. For đám næglum đe đæs Nergendes fét þurhwódon and his folme *for the nails which pierced the Saviour's feet and his hands,* Elen. Kmbl. 2130; El. 1066: Exon. 108 b; Th. 415, 3; Rä. 33, 5. Hæfde unlifigendes gefeormod fét and folma *he had devoured the feet and hands of the lifeless,* Beo. Th. 1494; B. 745. Náh geweald fóta ne folma *he shall not have the power of feet nor of hands,* Exon. 107 b; Th. 410, 12; Rä. 28, 15. Me of folmum afere firenwyrcendra *take me out of the hands of those committing sin,* Ps. Th. 70, 3 : Beo. Th. 319; B. 158. Geseóþ đa feorhdolg đe gefremedon on mínum folmum and on fótum *see the deadly wounds which they inflicted on my palms and in my feet,* Exon. 29 a; Th. 89, 12; Cri. 1456. On đone eádgan andwlitan helfúse men hondum slógun, folmum areahtum, and fýstum eác *wicked men struck on the blessed visage with their hands, with outstretched palms, and fists also,* Exon. 24 a; Th. 69, 23; Cri. 1125. Ic đe wreó and scylde folmum mínum *I will cover and shield thee with my hands,* Cd. 99; Th. 131, 4; Gen. 2171: Exon. 28 b; Th. 87, 9; Cri. 1422: Beo. Th. 1449; B. 722: Judth. 10; Thw. 23, 1; Jud. 99: Andr. Kmbl. 1044; An. 522: Elen. Kmbl. 2150; El. 1076: Ps. Th. 68, 5. [O. Sax. folmôs, *m. pl. the hands:* O. H. Ger. folma, *f.* palma : Swed. famla *to grope:* Dan. famle *to grope:* Icel. fálma *to grope about:* Lat. palma, *f.:* Grk. παλάμη, *f. the palm of the hand.*] DER. beadu-folm, gearo-, min-.

folme, an; *f.* [folm *the palm of the hand*] *The hand;* mănus :—Worhte his folme foldan drige *his hand made the dry land,* Ps. Th. 94, 5. Forlét drenga sum daroþ fleógan of folman *one of the warriors let fly a dart from his hand,* Byrht. Th. 136, 12; By. 150. Đa ísenan næglas, đe wæron adrifene þurh Cristes folman *the iron nails, which were driven through Christ's palms,* Homl. Th. ii. 306, 16. v. folm.

fon *a fann,* Lk. Skt. Rush. 3, 17. v. fann.

FÓN, to fónne; ic fó, đú féhst, he féhþ, *pl.* fóþ; *p.* ic, he féng, đú fénge, *pl.* féngon; *impert.* fóh, *pl.* fóþ; *subj. pres.* fó, *pl.* fón; *p.* fénge, *pl.* féngen; *pp.* fangen, fongen; *v. trans. To grasp, catch, seize, to seize with hostile intention, take, undertake, accept, receive;* mănu comprehendĕre, captāre, căpĕre, accipĕre :—Ne sceolde fón bispell *should not take a fable,* Bt. 35, 5; Fox 166, 20. Mæg man fón folcgesteallan *one may take his adherents,* Cd. 15; Th. 19, 6; Gen. 287. On ōđer weorc to fónne *to take to other work,* Bt. 39, 4; Fox 218, 4: Chr. 1009; Erl. 142, 28. Heó him to-geánes féng *she grasped at him,* Beo. Th. 3089; B. 1542. Se đe mec féhþ ongeán *he who is hostile towards me,* Exon. 107 b; Th. 410, 1; Rä. 28, 9: Beo. Th. 3515; B. 1755. We fóþ nú on đa axunga đǽr we hí ǽr forléton *we will now take up the questions where we before left them,* Boutr. Scrd. 18, 44. Féngon Ædelwulfes twegen suna to ríce *Æthelwulf's two sons took to the kingdom,* Chr. 855; Erl. 70, 17. Ne preóst ne fó to woruldspræcum *let not a priest take to worldly conversations,* L. Ælf. C. 30; Th. ii. 354, 2. Đú féhst on uncúþe *thou takest to the unknown,* Bt. 35, 5; Fox 164, 16. Hér beóþ fangene seólas and hronas *here are caught seals and dolphins,* Bd. 1, 1; S. 473, 16. Hí feng woldon fón *they would take the booty,* Chr. 1016; Erl. 156, 28, 12. Đá féng Ælfred to đam ríce *then Ælfred took to the kingdom,* Chr. 871; Erl. 76, 3: Jud. 13, 1. Fóh to me *take from me;* accipe a me, Cd. 228; Th. 308, 2; Sat. 686. Fóþ him on *accipĭte eum,* Bd. 5, 13? Lye. Ǽlas fongene beóþ *anguillæ căpiuntur,* Bd. 4, 19; S. 590, 5. [Piers P. fangen, fongen: Chauc. fonge: Laym. fon, ifon: Orm. fon: O. Sax. fáhan, fangan: Frs. fean, fangen: O. Frs. fa: Dut. vangen, vaan: Ger. fangen, fahen: M. H. Ger. vāhen: O. H. Ger.

fāhan: *Goth.* fahan: *Dan.* faa, faae: *Swed.* få, fånga: *Icel.* fá, fanga: *Lat.* pangĕre *to fasten: Grk.* πήγνυμι *to fasten: Sansk.* paś *to bind.*] DER. a-fón, æt-, an-, be-, bi-, for-, fór-, fóre-, ge-, ofer-, on-, þurh-, to-, under-, úta-, wið-, ymb-, ymbe-.

fond *found,* Cd. 119; Th. 154, 1; Gen. 2549; *p. of* findan.

fongen *taken:*—Ǽlas fongene beóþ *anguillæ căpiuntur,* Bd. 4, 19; S. 590, 5; *pp. of* fón.

FONT, es; *m.* A FONT, *fountain,* Som. Ben. Lye. [*Lat.* fons; *gen.* fontis, *m.*] v. font-wæter.

font-bæþ, es; *n.* A *font-bath, baptism;* baptismus, Som. Ben. Lye.

font-wæter, es; *n. Font, fountain* or *spring water;* fontāna ǎqua:— Wyrc drenc font-wæter *make a font-water drink,* L. M. 3, 62; Lchdm. ii. 350, 6. v. fant-wæter.

foor, es; *m.* A *pig, hog;* porcaster:—Foor *porcaster,* Ælfc. Gl. 19; Som. 59, 28; Wrt. Voc. 22, 69: Glos. Epnl. Recd. 161, 39. v. fór.

FOR, *prep. dat. acc. and inst.* **I.** *with the dative;* cum dătīvo. **1.** FOR, *on account of, because of, with, by;* pro, propter, per:—Nys ðeós untrumnys nā for deáþe, ac for Godes wuldre *infirmĭtas hæc non est ad mortem, sed pro glōria Dei,* Jn. Bos. 11, 4. Ðæt he ðone dǽl Willferþe for Gode gesealde *to brūcanne ut hanc [partem] Vilfrido, ūtendam pro Dōmĭno offerret,* Bd. 4, 16; S. 584, 11. Eardas rūme Meotud arǽrde for moncynne *the Creator established spacious lands for mankind,* Exon. 89 a; Th. 334, 15; Gn. Ex. 16. Aguldon me yfelu for gōdum *retribuēbant mihi măla pro bŏnis,* Ps. Spl. 34, 14. He wearþ sārig for his synnum *he was sorry for his sins,* Exon. 117 a; Th. 450, 15; Dōm. 88. Ne dyde ic for fācne, ne for feóndscipe, ne for wihte *I did it not for fraud, nor for enmity, nor for aught,* Cd. 128; Th. 162, 34; Gen. 2691. Ðe for ðám lārum com *that came by reason of those wiles,* Cd. 29; Th. 37, 32; Gen. 598. Moyses wearþ gebýsgad for heora yfelum *vexātus est Moyses propter eos,* Ps. Th. 105, 25. Ðæt hí dydon for ðǽm þingum *they did it for these reasons,* Bt. 35, 4; Fox 162, 21. Ūre gást biþ swiðe wíde farende for his gecynde, nalles for his willan *our spirit is very widely wandering, by reason of its nature, not by reason of its will,* Bt. 34, 11; Fox 152, 4, 5. For hwilcum þingum *quas ob res,* Ælfc. Gr. 44; Som. 46, 15. Se wæs in ðam fire for Freán meahtum *he was in the fire by the Lord's power,* Exon. 54 a; Th. 189, 26; Az. 65. For dæge oððe for twám *per ūnum aut duos dies,* Ex. 21, 29. **2.** *according to;* pro, sěcundum, juxta:—Eall sió lufu biþ for gecynde, nallas for willan *omne illud dēsĭdĕrium juxta nātūram est, non juxta vŏluntātem suam,* Bt. 34, 11; Fox 152, 14, 15. Ic gelýfe to ðé, ðæt ðú me, for ðínum mægenspēdum, nǽfre wille ānforlǽtan *I believe in thee, that thou, according to thy great power, never wilt desert me,* Andr. Kmbl. 2572; An. 1287. For ðam, for ðan, for ðon, for ðam ðe, for ðan ðe, for ðon ðe *for that, for that which, for this reason that, because, for that cause, therefore.* **II.** *with the accusative;* cum accūsātīvo. *For, instead of;* pro, lŏco, vĭce:—Archelāus ríxode on Iudēa þeóde for ðæne Hērōdem [= Ἡρώδης] *Archelāus [= Ἀρχέλαος] regnāvit in Jūdæa pro Hērōde,* Mt. Bos. 2, 22. Eáge for eáge, and toþ for tóþ *ŏcŭlum pro ŏcŭlo, et dentem pro dente,* Mt. Kmbl. Hat. 5, 38. Nafast ðú for āwiht ealle þeóda *pro nihil hăbēbis omnes gentes,* Ps. Th. 58, 8. Hǽfdon heora Hláford for ðone hēhstan God *they held their Lord for the most high God,* Bt. Met. Fox 26, 88; Met. 26, 44. **III.** *with the instrumental;* cum āblātīvo. *For, on account of, because of, through;* pro, propter, per:— We sunna fela didon for ūre disige *we committed many sins through our foolishness,* Hy. 7, 107; Hy. Grn. ii. 289, 107. Hine feor forwræc Metod for ðý māne *the Creator banished him far for that crime,* Beo. Th. 220; B. 110. Acol for ðý egesan *trembling for the terror,* Andr. Kmbl. 2533; An. 1268. Hæleþ wurdon acle arāsad for ðý rǽse *the men were seized with fear on account of its force,* Exon. 74 a; Th. 277, 27; Jul. 587. Ne murn ðú for ðí mēce *mourn not for the sword,* Wald. 43; Vald. 1, 24. For ðý, for ðí, for ðý ðe, for ðí ðe *for that, therefore, wherefore, because;* proptĕrea, quia. [*Piers P. Chauc.* for: *Laym.* for, uor: *Orm.* for: *Plat.* för, vör: *O. Frs.* fori, fóre, for: *Dut.* voor: *Ger.* für: *M. H. Ger.* vür, vüre: *O. H. Ger.* fora, furi: *Goth.* faur, faura: *Dan.* for: *Swed.* för: *Icel.* fyrir: *Lat.* pro.]

for- is used in composition in Anglo-Saxon exactly as the English *for:* it often deteriorates, or gives an opposite sense, or gives strength to the words before which it is placed; in which case it may be compared with Gothic *fra-,* Dutch and German *ver-* [different from the Dutch *voor,* and German *vor*]. Forbeódan *to forbid;* fordēman *to condemn;* forcúþ *perverse, corrupt;* fordón *to destroy, to do for.*—Sometimes fór denotes an increase of the signification of the word before which it is placed, and is then generally to be in English *very;* valde, as fór-eáðe *very easily,* Homl. Th. ii. 138, 35: fór-oft *very often,* Bd. de nat. rerum; Wrt. popl. science 11, 8; Lchdm. iii. 256, 16. For- and fór-, or fóre- are often confounded, though they are very different in meaning; as forseón [*Flem.* versien] *to overlook, despise;* fór- or fóreseón [*Flem.* veursien] *to foresee.*—If a word, having for, fór or fóre prefixed, cannot be found under for-, fór- or fóre-, it must be sought under the simple term, and the sense of the preposition added; thus, fór- or fóre-sendan is from sendan *to send,* and fór-, fóre *before, to send before, etc.* [On the vowel in fór, fóre, see remark in the preface.]

FŌR; fōre; *prep. dat. acc. Before, fore;* ante, cōram, in conspectu, præsente *vel* audiente ǎlĭquo, præ, priusquam. **I.** *dat:*—Fór Gode and fór [fóre Cott.] mannum *cōram Deo et hŏmĭnĭbus,* Bd. 5, 20; S. 641, 37. He fór eaxlum gestód Deniga freán *he stood before the shoulders of the lord of the Danes,* Beo. Th. 722; B. 358. Fór horde *before the hoard,* Beo. Th. 5555; B. 2781. Ic hefde dreám micelne fór Meotode *I had great joy before the Creator,* Cd. 214; Th. 269, 34; Sat. 83. We fór Dryhtene iu dreámas hefdon *we formerly had joys before the Lord,* 214; Th. 267, 26; Sat. 44. He gehālgode fór heremægene wín of wætere and wendau hēt *he hallowed before the multitude wine from water and bade it change,* Andr. Kmbl. 1172; An. 586. Geónge þúhton men for his eágum *they seemed young men before his eyes,* Cd. 111; Th. 146, 28; Gen. 2429. Wlytig heaw fór bearnum manna *spĕciōsus forma præ fĭliis hŏmĭnum,* Ps. Spl. 44, 3. **II.** *acc:*—Ne dear forþgán fór ðé *I dare not come forth before thee,* Cd. 40; Th. 54, 2; Gen. 871. He his módor fór ealle menn geweorþode *he esteemed his mother before all mankind,* Rood Knibl. 184; Kr. 93. Fór ðæt folc cōram pŏpŭlo, Ps. Th. 67, 8. [*Wyc.* for-fore-, as for-goer *a fore-goer:* *Plat.* vor: *O. Sax.* for, far, fur, furi: *Dut.* voor: *Ger.* vor: *M. H. Ger.* vor, vore: *O. H. Ger.* fora, furi: *Goth.* faur, faura: *Dan.* for: *Swed.* för: *Icel.* fyrir: *Lat.* præ: *Grk.* πρό *before: Sansk.* pra- *before.*]

fór, e; *f.* [fór, *p. of* faran *to go*] A *going, setting out, journey, course, way, approach;* ĭtio, profectio, ĭter, cursus, sēmĭta, accessus:—Fór wæs ðý beorhtre *the course was the brighter,* Exon. 105 a; Th. 400, 11; Rä. 20, 8. Me is fenýce fóre hreþre *a fen-frog is more rapid than I in its course,* 111 a; Th. 426, 10; Rä. 41, 71. He hine oftéah ðære fóre *subtraxit se illi profectiōni,* Bd. 5, 9; S. 623, 23: Ps. Th. 104, 33. He ðyder on ðære fóre wæs *he was on the journey thither,* Guthl. 16; Gdwin. 68, 1: Exon. 112 b; Th. 430, 19; Rä. 44, 11: 120 a; Th. 461, 9; Hö. 33. He sóna ongann fýsan to fóre *he soon began to hasten for the way,* Cd. 138; Th. 173, 12; Gen. 2860. Ne can ic Abeles ór ne fóre *I know not Abel's coming nor going,* 48; Th. 61, 33; Gen. 1006. Ðú scealt ða fóre geferan *thou shalt go the journey,* Andr. Kmbl. 431; An. 216: 673; An. 337: Exon. 40 b; Th. 136, 8; Gú. 538. Ðú ongeáte fóre míne *intellexisti sēmĭtam meam,* Ps. Th. 138, 2. Hí wendon heora fóre to Cantwarbyrig *they went their way to Canterbury,* Chr. 1009; Erl. 142, 17: 1004; Erl. 139, 24. Ðara láreówa fóre heaðoradon *doctōrum arcēbant accessum,* Bd. 4, 27; S. 604, 29. DER. forþ-fór, sǽ-.

fór, foor, es; *m.* A *pig, hog;* porcaster:—Fór *porcaster,* Wrt. Voc. 286, 48.

fór *went,* Gen. 31, 31; *p. of* faran.

fóra, L. C. S. 33; Th. i. 396, 17, note 51 *has this reading for* fór, *or* fóre *before;* ante, *q. v.* under for-, *or* fóre.

forad; *part. adj. Broken, weakened, void;* fractus, lăbĕfactus:—Gif se earm biþ forad būfan elmbogan *if the arm be broken above the elbow,* L. Alf. pol. 54; Th. i. 94, 24: 62, 63; Th. i. 96, 14, 17. Gif ða earm-scancan beóþ forad *if the arm-bones be both broken,* 55; Th. i. 94, 26. Beó ðæt ordāl forad *let the ordeal be void,* L. Ath. i. 23; Th. i. 212, 9: iv. 7; Th. i. 228, 1. v. forod.

fóra-gleáwlíce *providently, carefully, prudently;* prōvĭde, R. Ben. interl. 3. v. fóre-gleáwlíce.

for-aldod *antiquated,* Solil. 11, = for-ealdod; *pp. of* for-ealdian.

fóran; *prep. Before;* ante:—Fóran Andreas mæssan *before Andrew's mass-day,* Chr. 1010; Erl. 144, 13. ¶ Fóran ongeán *opposite;* contra:— Fóran ongeán eów *contra vos,* Mt. Bos. 21, 2. Fóran ongén Galileam *contra Galilæam,* Lk. Bos. 8, 26. Fóran ongeán ða burh *ex adverso contra urbem,* Jos. 8, 5. Fóran-to *before,* Chr. 920; Erl. 104, 31. v. fóran-to. DER. æt-fóran, be-, bi-, on-, to-, wið-.

fóran; *adv. In front, before;* ante, antequam, prius:—Wonnum hyrstum fóran gefrætwed *adorned in front with dark trappings,* Exon. 113 b; Th. 436, 2; Rä. 54, 8: Chr. 894; Erl. 93, 11. Is se fugel fæger fóran *the bird is fair before,* Exon. 60 a; Th. 418, 10; Ph. 292. DER. be-fóran, bi-, on-.

fór ān, *only;* tantum, tantummŏdo:—Gelýf fór ān μόνον πίστευε, *tantummŏdo crēde,* Mk. Bos. 5, 36. Fór ān ic beó hāl, gyf ic hys reáfes æthríne *si tĕtĭgĕro tantum vestīmentum ejus, salva ĕro,* Mt. Bos. 9, 21. Fór ān eówre yrfe sceal beón hēr *ōves tantum vestræ et armenta remāneant,* Ex. 10, 24.

fóran-bodig, es; *n. The forebody, chest;* pectus:—Fóran-bodig *vel* breóst-bedern [MS. breost-beden] *thōrax* = θώραξ [MS. tōrax], Ælfc. Gl. 73; Som. 71, 26; Wrt. Voc. 44, 12.

fóran-dæg, es; *m. Before day* or *dawn;* antelūcănum tempus, Som. Ben. Lye.

fóran-heáfod, es; *n. The forehead;* antērior pars căpĭtis, frons:—On fóran-heáfde *on the forehead,* Homl. Th. ii. 266, 13: Nar. 15, 13.

fóran-niht; es; *f. The fore-night, early part of the night, dusk of the evening;* antērior pars noctis, crĕpuscŭlum:—Lǽd hine ūt of ðam hūse on fórannihte *lead him out of the house in the dusk,* Herb. 8, 2; Lchdm. i. 98, 18: fram foran-nihte *per noctem,* Nar. 35, 9.

fóran-onsettende; *part.* [*part. of* fóran-onsettan] *Closing in;* præclūdens, Bd. 5, 1; S. 613, 31, note. v. fóre-settan.

fôran-to; *prep. Before;* ante :—Fôran-to Eástron *before Easter,* Chr. 921; Erl. 104, 37.` Fôran-to middum sumera *before midsummer,* 920; Erl. 104, 31 : fôran-to uhtes antelúcānum tempus, Nar. 15, 31. v. to-fôran.

fôr-arn *ran before,* Jn. Bos. 20, 4; *p. of* fôr-yrnan.

fôra-sceáwian; *p.* ode; *pp.* od *To foresee, forethink, consider;* prævĭdēre, præcōgitāre, consĭdĕrāri :—Fôrasceáwod beón *consĭdĕrāri,* R. Ben. interl. 64. v. fôre-sceáwian.

fôra-sceáwung, e; *f. Foresight, forethought, consideration* :—Fôrasceáwung *consĭdĕrātio,* R. Ben. interl. 34. v. fôre-sceáwung.

fôr-áþ, es; *m. A fore-oath, an oath first taken;* præjūrāmentum, antejūrāmentum :—Ofgá his spræce mid fôráþe *let him begin his suit with a fore-oath,* L. O. D. 6; Th. i. 354, 31. v. fôre-áþ.

for-bæran *to forbear* :—Hwá mæg forbæran *who can forbear?* Bt. 36, 1; Fox 172, 13. v. for-beran.

for-bærnan, -bearnan, to -bærnenne; *part.* -bærnende; *p.* -bærnde, *pl.* -bærndon; *pp.* -bærned, -bærnd; *v. trans. To burn up, consume;* ūrĕre, combūrĕre :—Nerón hêt forbærnan ealle Rôme burh *Nero commanded to burn up all the city of Rome,* Bt. 16, 4; Fox 58, 3 : Cd. 138; Th. 173, 8; Gen. 2858 : Exon. 30 b; Th. 94, 21; Cri. 1543 : Beo. Th. 4258; B. 2126. Isaac bær wudu to forbærnenne ða offrunge *Isaac bare wood to burn the offering,* Homl. Th. ii. 60, 26 : Mt. Bos. 13, 30. Swá swá lêg forbærnende muntas *sĭcut flamma combūrens montes,* Ps. Spl. 82, 13. Ic forswæle oðde forbærne *ūro,* Ælfc. Gr. 28, 4; Som. 31, 11. Man hine forbærneþ *one burns him,* Ors. 1, 1; Bos. 22, 44. Ða ceafu he forbærnþ on unadwæscendlícum fýre *pāleas combūret igni inextinguĭbĭli,* Mt. Bos. 3, 12 : Bt. 15; Fox 48, 22 : 33, 4; Fox 130, 12. Hí hine forbærnaþ *they burn him,* Ors. 1, 1; Bos. 22, 26. Líg forbærnde ða árleásan *flamma combussit peccātōres,* Ps. Lamb. 105, 18 : Boutr. Scrd. 22, 40 : Chr. 685; Erl. 40. 20. Hí ǽr Mul forbærndon *they had formerly burnt Mul,* Chr. 694; Erl. 43, 21 : 894; Erl. 91, 25 : 1001; Erl. 136, 31 : 1055; Erl. 190, 4. Nim astíne sticcan . . . forbærn ðone ôðerne ende *take a stick . . . burn the one end,* Bd. de nat. rerum; Wrt. popl. science 17, 15; Lchdm. iii. 274, 4. Ðæt seó sunne mid hyre hǽtan middaneardes wæstmas forbærne *that the sun with her heat burn up the fruits of the earth,* Wrt. popl. science 9, 6; Lchdm. iii. 250, 17. Ðæt he werod forbærnde *that it [the pillar of fire] would burn up the host,* Cd. 148; Th. 185, 16; Exod. 123. Hwí ðeós þyrne ne sí forbærned *quare non combūrātur rūbus,* Ex. 3, 3 : Chr. 687; Erl. 42, 1 : Cd. 146; Th. 182, 3; Exod. 70 : Exon. 22 b; Th. 62, 26; Cri. 1007. Beó se forbærnd *combūrētur,* Jos. 7, 15. cf. *Ger.* verbrennen.

for-bærnednes, -ness, -nyss, e; *f. A burning up;* ustio :—Wið forbærnednysse [-nesse MS. B.] *for a burning,* Herb. cont. 168, 2; Lchdm. i. 62, 19 : Herb. 168, 2; Lchdm. i. 298, 10.

for-bærst, *pl.* -burston *burst asunder,* Beo. Th. 5354; B. 2680: Bt. 18, 4; Fox 68, 6; *p. of* for-berstan.

for-barn *burnt,* Beo. Th. 3236; B. 1616; *p. of* for-beornan.

for-beád *forbade,* Cd. 30; Th. 40, 11; Gen. 637; *1st and 3rd sing. p. of* for-beódan.

for-beáh *avoided,* Byrht. Th. 141, 21; By. 325; *p. of* for-búgan.

for-bearan *to forbear,* Scint. 11. v. for-beran.

for-bearn *burnt,* Boutr. Scrd. 22, 33; *p. of* for-beornan.

for-bearnan, *p.* de; *impert. pl.* -bearnaþ; *pp.* ed *To burn up, consume by fire;* combūrĕre :—Lædaþ hig forþ and forbearnaþ hig *prodūcĭte eam ut combūrātur,* Gen. 38, 24. Hí forbearndon Beorn ealdorman *they consumed Beorn alderman,* Chr. 779; Erl. 55, 36 : 1052; Erl. 185, 4. v. for-bærnan.

for-bêgan; *p.* de; *pp.* ed *To bow down, bend down, humble, abase, destroy;* deprĭmĕre, hūmĭliāre, ēmĭnuĕre :—Ðæt ge gúpfreáu gylp forbêgan *that ye may humble the warrior's pride,* Andr. Kmbl. 2668; An. 1335 : 3141; An. 1573 : Cd. 223; Th. 294, 8; Sat. 468. v. for-býgan.

for-beódan, -biódan, to -beódanne; *part.* -beódende; *p.* ic, he -beád, ðú -bude, *pl.* -budon; *pp.* -boden [*Ger.* ver-bieten] *To FORBID, prohibit, restrain, suppress;* prohĭbēre, vĕtāre, interdīcĕre :—Nelle gê hig forbeódan cuman to me *nōlīte eos prohĭbēre ad me vĕnīre,* Mt. Bos. 19, 14; L. C. S. 77; Th. i. 418, 24. To forbeódanne *to forbid,* L. Alf. 49; Th. i. 56, 1. Ðisne we gemêton forbeódende ðæt man ðam Cásere gafol ne sealde *hunc invēnimus prohĭbentem trĭbūta dāre Cæsāri,* Lk. Bos. 23, 2. Ic forbeóde *prohibeo* : ic forbeád *prohĭbui* : forboden *prohĭbĭtum,* Ælfc. Gr. 26, 2; Som. 28, 34, 35. Ic forbeóde *vĕto,* Ælfc. Gr. 24; Som. 25, 49. Búton ðú forgange ðæt ic ðé forbeóde *unless thou forgo that which I forbid thee,* Homl. Th. i. 14, 8 : Chr. 675; Erl. 38, 22. Fram eallum wege yfelum ic forbeád fêt míne *ab omni via mǎla prohĭbui pĕdes meos,* Ps. Spl. 118, 101. Ðone hire forbeád Drihten *which the Lord forbade her,* Cd. 30; Th. 40, 11, 29; Gen. 637, 646 : Gen. 3, 1: Mt. Bos. 3, 14. We him forbudon *prohĭbuimus eum,* Mk. Bos. 9, 38; Lk. Bos. 9, 49. Ne forbeód him ná ðíne tunecan *tŭnicam nōli prohĭbēre,* 6, 29: Num. 11, 28. Lætaþ ða lytlingas to me cuman, and ne forbeóde gê him *suffer the little ones to come unto me, and forbid them not,* Mk. Bos. 10, 14: Lk. Bos. 18, 16. Sunnan daga cýpinga forbeóde man georne *let Sunday marketings be strictly forbidden,* L. Eth. ix. 17; Th. i.

♄

344, 7. Hit forboden wæs *it was forbidden,* iii. 8; Th. i. 296, 13 : Chr. 1048; Erl. 177, 21. Ðú Adame sealdest wæstme ða inc wǽron fæste forbodene *thou gavest to Adam the fruits which were strictly forbidden to you two,* Cd. 42; Th. 55, 16; Gen. 895.

for-beódendlíc; *adj. Forbidding-like, dissuasive;* prohĭbĭtōrius, dehortātōrius :—Sume synd *dehortātīva,* ðæt synd forbeódendlíce oðde mistihtendlíce *some are dehortātīva, which are dissuasive,* Ælfc. Gr. 38; Som. 40, 8.

for-beornan, -byrnan; *p.* -bearn, -barn, -born, *pl.* -burnon; *pp.* -bornen, -burnen; *v. n. To burn up, be destroyed by fire, be consumed;* combūri, ignĭbus consūmi :—On ðære Sodomitiscra gewítnunge forbearn seó eorþe *in the punishment of the Sodomites the earth was burnt,* Boutr. Scrd. 22, 33. Forbarn broden mǽl *the drawn brand was burnt,* Beo. Th. 3236; B. 1616 : 3338; B. 1667. Hit gelamp, ðæt se ylca tún forbarn [forborn, col. 2], and seó cyrice *evēnit, vīcum eundem, et ipsum pārĭter ecclēsiam ignĭbus consūmi,* Bd. 3, 17; S. 544, 27, col. 1: Chr. 816; Erl. 62, 7. Forburnon xv túnas *fifteen towns burned,* Ors. 6, 1; Bos. 115, 37. He geseah, ðæt seó þyrne barn and næs forburnen *vĭdēbat, quod rūbus ardēret et non combūrērētur,* Ex. 3, 2: Bd. 3, 17; S. 544, 20, col. 1. Wǽron ða bende [MS. benne] forburnene *the bands were burnt,* Cd. 195; Th. 243, 12; Dan. 435.

for-beran; *p.* -bær, *pl.* -bǽron; *pp.* -boren [for *for*; beran *to bear*] *To FORBEAR, abstain, refrain, restrain, bear with, endure, suffer;* abstĭnēre, sustĭnēre, comprĭmēre, reprĭmēre, tŏlĕrāre, pāti, ferre :—Ðæt he ðone breóstwylm forberan ne mihte *that he might not restrain the fervour of his breast,* Beo. Th. 3759; B. 1877. Hí firenlustas forberaþ in breóstum *they restrain sinful lusts in their breasts,* Exon. 44 b; Th. 150, 9; Gú. 776. Seó æftere cneóris ealle gemete is to forberanne *sĕcunda gĕnĕrātio a se omni mŏdo dēbet abstĭnēre,* Bd. 1, 27; S. 491, 9. Ic forbær ðé *sustĭnui te,* Ps. Spl. 24, 22. Yfelu forberan ne sceal *mǎla tŏlĕrāre non dēbet,* Past. 21, 5; Hat. MS. 31 b, 2. Hú lange forbere ic eów *quousque pătiar vos?* Mt. Bos. 17, 17. Ðonne him mon yfel dô, he hit sceal geþyldelíce forberan *when one does him evil, he shall patiently endure it,* Glostr. Frag. 112, 18 : Mk. Bos. 14, 4. [cf. *Goth.* frabairan *to endure.*]

fôr-beran, fôre-beran; *p.* -bær; *pp.* -boren [for *fôre before*; beran *to bear*] *To fore-bear, to bear or carry before, to prefer;* præferre :—Ðæt ic forbær rúme regulas and rêðe môd geongra monna *that I preferred the lax rules and rough minds of young men,* Exon. 39 b; Th. 131, 22; Gú. 459. Ðætte nǽnig bisceop hine ôðrum forbere *ut nullus episcŏpōrum se præfĕrat altĕri,* Bd. 4, 5; S. 573, 10.

for-berstan, he -birsteþ; *p.* -bærst, *pl.* -burston; *pp.* -borsten *To break, burst asunder, fail;* contĕri, dirumpi, exstingui :—Wên nǽfre forbirsteþ *hope never fails,* Exon. 64 a; Th. 236, 2; Ph. 568. Heora bogan forberstaþ *arcus eōrum contĕrātur,* Ps. Th. 36, 14. Forbærst sweord Beówulfes *Beowulf's sword burst asunder,* Beo. Th. 5354; B. 2680: Bt. 18, 4; Fox 68, 6. Ðæt him forberste se sweora *that his neck break,* L. Eth. iii. 4; Th. i. 294, 16; Prov. Kmbl. 19. Wæs him beót forborsten *their threat failed,* Cd. 4; Th. 5, 11; Gen. 70.

fôr-bétan *to make full amends for anyone or anything;* compensāre pro ǎlĭquo, Som. Ben. Lye. v. fôre-bétan.

for-bígan, -bígean; *p.* de; *pp.* ed *To bow down, bend down, humble, abase, depreciate, avoid, pass by;* hūmĭliāre, prætĕrīre :—Bælc forbígde he humbled *he humbled their pride,* Cd. 4; Th. 4, 15; Gen. 54: Th. 5, 12; Gen. 70: Exon. 85 b; Th. 321, 19; Wíd. 48: Wald. 47; Vald. 1, 26. Litlingas nellaþ forbígean (cf. forbúgan) me *parvŭli nōlunt prætĕrīre me,* Coll. Monast. Th. 29, 3. v. for-bêgan.

for-bígels, es; *m. An arch, a vault, an arched roof;* arcus, fornix, cǎmĕra = καμάρα :—Forbígels *arcus,* Ælfc. Gl. 29; Som. 61, 32; Wrt. Voc. 26, 31. v. bígels.

for-bindan; *p.* -band, *pl.* -bundon; *subj. pres.* -binde, *pl.* -binden; *pp.* -bunden *To bind or tie up;* allĭgāre :—Ne forbinden gê ná ðæm þyrstendum oxum ðone múþ *ye may not tie up the mouth of the thirsting oxen,* Past. 16, 5; Hat. MS. 21 b, 7.

for-biódan *to forbid* :—He wel meahte ðæt unriht him eáðe forbiódan *he might well easily forbid that injustice to him,* Bt. Met. Fox 9, 108; Met. 9, 54. v. for-beódan.

for-birsteþ *fails,* Exon. 64 a; Th. 236, 2; Ph. 568; *3rd sing. pres. of* for-berstan.

for-bláwan; *p.* -bleów, *pl.* -bleówon; *pp.* -bláwen *To blow away, inflate;* inflāre :—Com án wind, ond forbleów hie út on sǽ *there came a wind, and blew them out on to the sea,* Ors. 5, 4; Bos. 105, 19. Gif mon síe forbláwen *if a man be inflated,* L. M. 2, 34; Lchdm. ii. 240, 4.

for-blindian; *p.* ode, ade; *pp.* od, ad *To blind;* obcæcāre :—Wæs forblindad *ērat obcæcātum,* Mk. Skt. Rush. 6, 52. v. blendan *to blind.*

fôr-bóc, e; *f.* [fôr *a journey,* bóc *a book*] *A journey-book, itinerary;* itĭnĕrārium :—Fôrbóc [MS. fôrebóc], síþbóc *itĭnĕrārium,* Mone B. 1994.

for-bod, es; *n. A forbidding, prohibition, countermand;* prohĭbĭtio :—Ðæt hit ðara manna forbod wǽre *that it was forbidden by those men* [lit. *that it was the forbidding of those men*], L. Alf. pol. 41; Th. i. 88, 19. On Godes forbode *with God's prohibition,* L. N. P. L. 61; Th. ii. 300, 12.

fôr-boda, an; *m. A foreboder, forerunner, messenger*; prænuntius:— Gôdes fôrboda *God's messenger,* L. N. P. L. 2; Th. ii. 290, 6.

for-boden *forbidden,* L. Eth. iii. 8; Th. i. 296, 13; *pp. of* for-beódan.

for-bogen *avoided,* App. Lit. Scint. Lye; *pp. of* for-búgan.

for-boren *forborne, restrained, endured,* Bt. 38, 4; Fox 204, 18: L. M. 1, 45; Lchdm. ii. 114, 8; *pp. of* for-beran.

for-born *burnt,* Chr. 816; Erl. 62, 7; *p. of* for-beornan.

for-borsten *bursted, failed,* Cd. 4; Th. 5, 11; Gen. 70; *pp. of* for-berstan.

for-brecan; *part.* -brecende; *đû* -brecest, -bricst, -brycst, he -breceþ, -bricþ; *p.* -bræc, *pl.* -bræcon; *pp.* -brocen *To break, break in two, bruise, crush, violate*; frangĕre, confringĕre, contĕrĕre, commĭnuĕre, viŏlāre:— Wolde heofona helm helle weallas forbrecan *heaven's chieftain would break down hell's walls,* Exon. 120 a; Th. 461, 13; Hö. 35. Stefn Drihtnes forbrecendes cederbeám, and forbricþ Drihten cederbeám đæs holtes *vox Dŏmĭni confringentis cedros, et confringet Dŏmĭnus cedros Lĭbăni,* Ps. Spl. 28, 5. Đû forbrycst đone earm đæs synfullan *thou shalt break the arm of the sinful,* Ps. Th. 9, 35. Ic sumra fêt forbræc bealosearwum *I have broken the feet of some by wicked snares,* Exon. 72 b; Th. 270, 30; Jul. 473. He helle dûru forbræc *he brake hell's door,* Cd. 223; Th. 294, 8; Sat. 468: Ps. Spl. 106, 16. Forbræcon Rômâne heora áþas *the Romans broke their oaths,* Ors. 3, 8; Bos. 63, 31: Cd. 37; Th. 49, 27; Gen. 798. Forbrec odđe tobryt earm đæs synfullan *contĕre brachium peccātōris,* Ps. Lamb. second 9, 15. Ne forbrece [MS. forbræce] gê nân bân on him *os non commĭnuĕtis,ex eo,* Jn. Bos. 19, 36. Đæt man forbræce hyra sceancan *ut frangĕrentur eōrum crūra,* 19, 31. Hie gebod Godes forbrocen hæfdon *they had broken God's command,* Cd. 33; Th. 43, 30; Gen. 698.

for-bredan; *p.* -bræd, *pl.* -brudon; *pp.* -broden *To transform*; transformāre:—Sceolde beornas forbredan *should transform men,* Bt. Met. Fox 26, 149; Met. 26, 75: Bt. 38, 1; Fox 194, 31. DER. bredan.

for-bregdan; *p.* -brægd, *pl.* -brugdon; *pp.* -brogden *To cover*; obdū-cĕre:—Ic mist-helme contrīvit brægd leóman *I covered the light of their eyes with a mantle of mist,* Exon. 72 b; Th. 270, 25; Jul. 470.

for-brict *crushed,* L. E. I. 2; Th. ii. 404, 5,=for-britt; *pp. of* forbritan.

for-bricþ *breaks,* Ps. Spl. 28, 5; *3rd sing. pres. of* for-brecan.

for-brittan; *p.* -britte; *pp.* -britted, -britt *To break in pieces, smash, bruise*; confringĕre, contĕrĕre:—God forbriteþ têþ heora on mûþe heora *Deus contĕret dentes eōrum in ōre ipsōrum,* Ps. Spl. 57, 6. Hû he forbritte ealle his bîgengan *quŏmŏdo contrīvĕrit omnes cultōres ejus,* Deut. 4, 3. Beóþ ælce uncysta forbritte [MS. forbricte] *all vices shall be crushed,* L. E. I. 2; Th. ii. 404, 5. v. for-bryttan.

for-brocen *broken,* Cd. 33; Th. 43, 30; Gen. 698; *pp. of* for-brecan.

for-brycst *breakest* or *shalt break,* Ps. Th. 9, 35; *2nd sing. pres. of* for-brecan.

for-brytednys, -nyss, e; *f. Bruisedness, sorrow*; contrītio:—Forbrytednys and ungesælignys [synd] on wegum heora *contrītio et infēlĭcĭtas* [*sunt*] *in viis eōrum,* Ps. Spl. 13, 7.

for-bryttan, -brittan; he -bryteþ, -brytt; *p.* -brytte; *pp.* -bryted, -bryt *To break in pieces, smash, bruise, crush*; confringĕre, contĕrĕre, conquassāre:—Tocwysed hreód he ne forbrytt *arundĭnem quassātam non confringet,* Mt. Bos. 12, 20. Moises forbrytte đæt celf eall to duste *Moyses vĭtŭlum contrīvit usque ad pulvĕrem,* Ex. 32, 20. Forbryt đû earm synfulles *contĕre brachium peccātōris,* Ps. Spl. second 9, 18. Đæt đû sî forbryt *dōnec contĕrāris,* Deut. 28, 24. Ælc đe fylþ ofer đone stân, byþ forbryt *omnis, qui cecĭdĕrit sŭper illum lăpĭdem, conquassābĭtur,* Lk. Bos. 20, 18.

for-budon *forbade,* Mk. Bos. 9, 38; *p. pl. of* for-beódan.

for-búgan; *part.* -búgende; *p.* -beáh, *pl.* -bugon; *impert.* -búh, *pl.* -búgaþ; *pp.* -bogen; *v. trans. To bend from, pass by, decline, avoid, shun, eschew*; recēdĕre, praetĕrīre, declīnāre, evītāre, devītāre:—He mæg forbúgan đa þegnunga *he can decline the ministrations,* Past. 7, 2; Hat. MS. 12 a, 14: Wald. 25; Vald. 1, 15. Hû man sēlost mæg synna forbúgan *how a man may best avoid sin,* Ælfc. T. 15, 2: Homl. Th. i. 82, 26: 206, 6: Num. 22, 26. Se wer wæs forbúgende yfel *ĕrat vir recēdens a mălo,* Job Thw. 164, 3. Næs đæt nâ se Godrîc đe đa gûþe forbeáh *this was not the Godric who had fled from the war,* Byrht. Th. 141, 21; By. 325. Đâ he đæt geseah, he hine forbeáh *viso illo, praetĕrivit;* Lk. Bos. 10, 31, 32: Num. 22, 23. Forbûh *devīta,* Scint. 88. Forbúgaþ unrihtwîsnysse *eschew unrighteousness,* Homl. Th. i. 28, 21: 180, 13. Æghwylc cristen mân unriht hæmed georne forbúge *let every christian man carefully eschew unlawful concubinage,* L. Eth. v. 10; Th. i. 306, 26: vi. 11; Th. i. 318, 11. Forbogen beón *evītāri,* App. Lit. Scint. Lye. (*Orm.* forrbuȝhenn *to avoid, refuse.*)

for-búgennys, -nyss, e; *f. An avoiding, eschewing, a declining*; declīnātio, Som. Ben. Lye.

for-burnen *burnt,* Ex. 3, 2; *pp. of* for-beornan.

for-burnon *burnt,* Ors. 6, 1; Bos. 115, 37; *p. pl. of* for-beornan.

for-bŷgan, -bîgan, -bîgean, -bêgan; *p.* de; *pp.* ed *To bow down, bend down, abase, humble, destroy*; deprimĕre, humiliāre, imminuĕre:—He

---

hellwarena heáp forbŷgde *he humbled the multitude of hell's inmates,* Exon. 18 b; Th. 46, 3; Cri. 731: Exon. 120 a; Th. 461, 13; Hö. 35: v. bŷgan.

for-byrd, e; *f. A forbearing, an abstaining from*; abstĭnentia:—Đæt nân forbyrd nære æt geligere betwuh nânre sibbe *that there should be no abstaining from concubinage between any kindred,* Ors. 1, 2; Bos. 27, 15.

for-byrdian, -byrdigan; *p.* ode; *pp.* od *To forbear, wait for*; sustĭnēre:—Sâwla ûre forbyrdigaþ Driht *ănima nostra sustĭnet Dŏmĭnum,* Ps. Spl. 32, 20.

for-byrnan *to burn up*:—Hig forbyrnaþ *they burn up,* Jn. Bos. 15, 6. v. for-beornan.

FORCA, an; *m. A* FORK; furca:—Litel forca *furcilla,* Ælfc. Gl. 66; Wrt. Voc. 41, 37. [*Laym.* forken, furken, *pl. the gallows: Plat.* furke, forke, fork, *f: Dut.* vork, *f: M. H. Ger.* furke, *f: Icel.* forkr, *m: Lat.* furca, *f: Wel.* ffwrch, *m;* fforch, *f: Armor.* forc'h, *f.*]

for-ceorfan; *part.* -ceorfende; ic -ceorfe, đû -ceorfest, -cirfst, -cyrfst, he -ceorfeþ, -cyrfþ, *pl.* -ceorfaþ; *p.* ic, he -cearf, đû -curfe, *pl.* -curfon; *pp.* -corfen *To cut* or *çarve out, cut down, cut off* or *away, cut through, divide*; excīdĕre, concīdĕre, succīdĕre, incīdĕre, intercīdĕre:—Đî-læs đe se Hláford hâte us mid deáþes æxe forceorfan *lest the Lord command to cut us down with the axe of death,* Homl. Th. ii. 408, 28. Forceorfende *intercīdens,* Ps. Lamb. 28, 7. Ic forceorfe *succido, incido,* Ælfc. Gr. 28, 4; Som. 31, 34. Đû forcirfst heora horsa hôhsina *ĕquos eōrum subnervābis,* Jos. 11, 6. Đû forcyrfst hit *thou wilt cut it down,* Homl. Th. ii. 408, 8. Drihten se rihtwîsa forheáweþ odđe forcyrfþ hnollas synfulra *Dŏmĭnus justus concīdet cervīces peccātōrum,* Ps. Lamb. 128, 4. Đæt heó healfne forcearf đone sweoran him *so that she half cut through his neck,* Judth. 10; Thw. 23, 5; Jud. 105. Rômâne Leóne đæm pâpan his tungan forcurfon *the Romans cut out the tongue of Pope Leo,* Chr. 797; Erl. 58, 13: Ors. 4, 6; Bos. 86, 33. Forceorf hine, hwî ofþricþ he đæt land *succīde illam, ut quid ĕtiam terram occŭpat?* Lk. Bos. 13, 7: Homl. Th. ii. 408, 4. Ælc treów, đe gôdne wæstm ne bringþ, byþ forcorfen *omnis arbor, quae non făcit fructum bŏnum, excīdētur,* Mt. Bos. 3, 10: Homl. Th. ii. 406, 32. Đæt we ne beón forcorfene *that we may not be cut down,* 408, 25.

for-ceówan; *p.* -ceáw, *pl.* -cuwon; *pp.* -cowen *To chew off, bite off*; corrōdĕre:—Forceáw he his âgene tungan *he bit off his own tongue,* Bt. 16, 2; Fox 52, 24.

for-cerran *to avoid.* v. for-cyrran.

for-cinnan, ic -cinne, đû -cinnest, he -cinneþ, *pl.* -cinnaþ; *p.* ic, he -can, đû -cunne, *pl.* -cunnon; *pp.* -cunnen [for, cinnan gĕnĕrāre] *To repudiate*; rejĭcĕre:—Hine forcinnaþ đa cyrican ge tûnas *the churches as well as houses shall repudiate him,* Salm. Kmbl. 215; Sal. 107.

for-cirfst *cuttest* or *shalt cut,* Jos. 11, 6; *2nd sing. pres. of* forceorfan.

for-clingan; *p.* -clang, *pl.* -clungon; *pp.* -clungen *To shrink up*; marcescĕre:—Wæron sume on forclungenum treówe ahangene *some were hung up on a shrunken tree,* Nath. 8. [*Orm.* forrclungenn *withered.*]

for-clŷsan; he -clŷseþ, -clŷst; *p.* de; *pp.* ed [clŷsan *to close, shut*] *To close* or *shut up*; occlūdĕre:—Đis sceal to đâm eárum [MS. đan earen] đe wind odđe wæter forclŷst *this shall [do] for the ears which wind or water closes up,* Lchdm. iii. 92, 24.

for-cneów, es; *n. A progeny, race*; progěnies, Lye.

for-cnídan; *p.* ic -cnâd, đû -cnîde, -cnyde, *pl.* -cnidon; *pp.* -cniden *To beat* or *break into pieces, dash* or *throw down*; commĭnuĕre, contĕrĕre, collīdĕre:—Ic gewanie odđe forcníde hig swâ swâ dust *commĭnuam eos ut pulvĕrem,* Ps. Spl. 17, 44. Ealle trumnysse hláfes he forcnâd *omne firmāmentum pānis contrīvit,* 104, 15. Setl his on lande đû forcnyde *sēdem ejus in terra collīsisti,* 88, 43. v. for-gnîdan.

for-corfen *cut down,* Mt. Bos. 3, 10; *pp. of* for-ceorfan.

for-cuman; *p.* -com, -cwom, *pl.* -cômon, -cwômon; *pp.* -cumen, -cymen *To surpass, overcome, destroy, harass, wear out*; sŭpĕrāre, vexāre:—Hæfde đâ se snotra sunu Dauides forcumen and forcŷđed Caldēa eorl *then had the wise son of David overcome and surpassed in knowledge the earl of the Chaldeans,* Salm. Kmbl. 353; Sal. 176: Andr. Kmbl. 2651; An. 1327. Yrfe đîn eall forcôman *hærēdĭtātem tuam vexāvērunt,* Ps. Th. 93, 5. Bring us hælo líf, wērigum wîteþeówum, wôþe forcymenum *bring to us weary slaves, oppressed by weeping, a life of health,* Exon. 10 a; Th. 10, 13; Cri. 151. [*O. Sax.* far-kuman; *Ger.* ver-kommen *to overcome, destroy.*]

fôr-cuman; *p.* -com, -cwom, *pl.* -cômon, -cwômon; *pp.* -cumen [fôr *before;* cuman *to come*] *To* FORE-COME, *go before, prevent*; praevĕnīre:— Arîs, Drihten, fôrcum hî *exurge, Dŏmĭne, praevĕni eos,* Ps. Spl. 16, 14. Ic fôrcom on rîpunga *praevĕni in matūrĭtāte,* 118, 147. [*Ger.* vorkommen *to come before, occur.*]

for-curfon *cut out,* Chr. 797; Erl. 58, 13; *p. pl. of* for-ceorfan.

for-cúþ; *comp. m.* -cúþera, -cúþra; *sup. m.* -cúþesta, -cúþosta; *adj.* [cúþ *known, excellent*] *Perverse, bad, infamous, wicked*; perversus, mălus, nēquam:—Mânfull odđe forcúþ *nēquam,* Ælfc. Gr. 9, 78; Som. 14, 30. Se yfela, swâ he oftor on đære fandunge abrýþ, swâ he forcúþra biþ *the oftener the evil man sinks under temptation, the more wicked he will be,* Homl. Th. i. 268, 30. Wearþ he and ealle his geferan forcúþran *and*

wyrsan đonne ǽnig óđer gesceaft *he and all his companions became more wicked and worse than any other creature*, i. 10, 35. Hí habbaþ đæs mennisces đone betstan dǽl forloren, and đone forcūþestan [forcūþeran MS. Bod.] gehealden *they have lost the best part of humanity, and kept the worst* [*worse*], Bt. 37, 3; Fox 192, 4. Oft đa eallra forcūþestan men cumaþ to đam ānwealde and to đam weorþscipe *the most wicked men of all often come to power and dignity*, 16, 3; Fox 54, 21. Hwæđer he wolde đám forcūþestum mannum folgian *would it follow the most wicked men?* 16, 3; Fox 54, 10, 27. Đa Sodomitiscan menn wǽron đa forcūþostan *hŏmĭnes Sŏdŏmītæ pessĭmi ĕrant*, Gen. 13, 13. [*Goth.* frakunþs *despised.*] DER. unforcūþ.

for-cūþlíce; *adv. Perversely, across*; perverse, transverse :— Đǽra cynega swuran forcūþlíce trǽdon *colla rēgum pĕdĭbus calcārent*, Jos. 10, 24.

for-cweđan; *p.* -cwæþ, *pl.* -cwǽdon; *pp.* -cweden *To rebuke, censure, revile, refuse, reject*; incrĕpāre, maledīcĕre, recūsāre, rejĭcĕre :— Ne sceal hine mon cildgeong ne forcweđan *one must not while a young child rebuke him*, Exon. 89 b; Th. 336, 14; Gn. Ex. 49. Đa fortrūwodan forsióþ óđre menn and eác forcweđaþ [MS. forcueđaþ] *the presumptuous despise and also revile other men*, Past. 32, 1; Hat. MS. 39 b, 27. Se wisa Catulus forcwæþ Nonium đone rícan *the wise Catulus censured Nonius the rich*, Bt. 27, 1; Fox 94, 32. Drihten forcwæþ swelce ælmessan *the Lord rejected such alms*, Past. 45, 4; Hat. MS. 65 a, 26.

for-cwolstan; *p.* te; *pp.* ed *To swallow down*; haurīre :— Fífleáfan seáwes þrý bollan fulle lytle sceal forcwolstan *he shall swallow down three little bowls of the juice of cinque-foil*, L. M. 1, 4; Lchdm. ii. 48, 18.

for-cwom, *pl.* -cwōmon *came upon*; supervēnit, supervēnĕrunt :— Egsa me and fyrhtu ealne forcwōmon *timor et trĕmor vēnērunt sŭper me.* Ps. Th. 54, 5. v. for-com, -cōmon; *p.* of for-cuman.

for-cwysan; *p.* de; *pp.* ed *To shake violently*; conquassāre :— He forcwysde heáfda on eorþan manigra *he shook violently the heads of many in the earth*, Ps. Spl. 109, 7.

for-cymen *overcome, harassed, worn out*, Exon. 10 a; Th. 10, 13; Cri. 151; *pp.* of for-cuman.

for-cyrfst, he -cyrfþ *cuttest down, he cuts down*, Homl. Th. ii. 408, 8: Ps. Lamb. 128, 4; *2nd and 3rd sing. pres.* of for-ceorfan.

for-cyrran; *p.* de; *pp.* ed *To turn again, subvert, avoid*; pervertĕre, subvertĕre, evītāre :— Būton deáþ hí ne mágon forcyrran *except they cannot avoid death*, Bt. 41, 2; Fox 246, 8.

for-cýđan; *p.* de; *pp.* ed *To surpass or excel in knowledge*; scientia excellĕre *vel* superāre :— Hæfde se snotra sunu Davides forcumen and forcýđed Caldēa eorl *the wise son of David had overcome and surpassed in knowledge the leader of the Chaldeans*, Salm. Kmbl. 353; Sal. 176: 411; Sal. 206.

FORD; *gen.* fordes; *dat.* forde, forda; *m. A* FORD; vădum :— Ford vădum, Ælfc. Gl. 97; Som. 76, 66; Wrt. Voc. 54, 10: 80, 51. Hie flugon ofer Temese būton ǽlcum forda *they fled over the Thames without any ford*, Chr. 894; Erl. 90, 28. Neáh đam forda, đe man hǽt Welinga ford *near the ford which is called Wallingford*, Ors. 5, 12; Bos. 110, 20. Æt đam forda [Th. forde] *at the ford*, Byrht. Th. 134, 8; By. 81. Đa Walas adrifon sumre eá ford ealne mid scearpum pílum greátum *the Welsh staked the ford of a river all with great sharp piles*, Chr. Erl. 5, 9, 12. Ofer đone ford *trans vădum*, Ælfc. Gr. 47; Som. 47, 38: Byrht. Th. 134, 22; By. 88: Beo. Th. 1140; B. 568. He oferfōr đone ford *transīvit vădum*, Gen. 32, 22. He mihte fordas oferrídan, đonne he to hwylcere eá cōme *he might ride over the fords, when he came to any river*, Bd. 3, 14; S. 540, 17. [*Laym.* uord, ford : *Scot.* firth, frith *a bay*: *O. Frs.* forda : *Dut. Kil.* voord *vădum* : *Ger.* furt, *f*: *M. H. Ger.* vurt, *m*: *O. H. Ger.* furt, *n*: *Dan.* fjord, *m. f. a bay, gulf*: *Swed.* fjärd, *m. a bay*: *Icel.* fjörðr, *m*: *Grk.* πόρος, *m. a ford, ferry.*]

for-dǽdla *a destroyer.* v. mān-fordǽdla.

for-dǽlan; *p.* de; *pp.* ed *To deal out, expend*; dispensāre, erŏgāre :— Seó fordǽlde on lǽcas eall đæt heó áhte *quæ in mĕdĭcos erogāverat omnem substantiam suam*, Lk. Bos. 8, 43. [*Goth.* fradailjan *to give away*: *Dut.* ver-deelen *to divide, distribute*: *Ger.* ver-theilen *to distribute.*]

for-deáþ *destroys, does for*, Wanl. Catal. 112, 65, col. 2, = for-dēþ; *3rd sing. pres.* of for-dón.

for-dēman, to for-dēmanne; *p.* de; *pp.* ed *To condemn, damn*; dijudĭcāre, damnāre, condemnāre :— Đæt hig hine gesealdon đám ealdron to dōme, and to đæs dēman ānwalde to fordēmanne *ut tradērent illum princĭpātui, et potestāti præsĭdis*, Lk. Bos. 20, 20. On middele sóþlice godas he fordēmþ *in mĕdio autem deos dijūdĭcat*, Ps. Spl. 81, 1. Đá geseah Iudas đe hyne belǽwde, đæt he fordēmed wæs, đá ongan he hreówsian *tunc vĭdens Iudas, qui eum tradĭdit, quod damnātus esset, pænĭtentia ductus*, Mt. Bos. 27, 3. Nellen gē dēman, đæt gē ne sýn fordēmede *judge not, that ye be not condemned*, 7, 1. Đæt man cristene men, for ealles tō lytlum, to deáþe ne fordēme *that christian men, for all too little, be not condemned to death*, L. Eth. v. 3; Th. i. 304, 17. [*O. Sax.* fardōmjan : *O. H. Ger.* firtuoman : *Dut.* verdoemen *to condemn.*]

for-dēmednes, -ness, e; *f. Condemnation, proscription*; condemnātio, proscriptio :— Þurh tyn winter full Godes cyricena bærnesse, and unscead-

---

diendra fordēmednesse, and slege hāligra martyra unblinnendlíce dōn wæs *per dĕcem annos, incendiis ecclēsiārum, proscriptiōnĭbus innŏcentum, cædĭbus martўrum incessābĭlĭter acta est*, Bd. 1, 6; S. 476, 25.

for-demman; *part.* -demmende; *p.* de; *pp.* ed *To shut or dam up*; obtūrāre :— Swá swá nædran deáfe, and fordemmende eáran heora *sĭcut aspĭdis surdæ, et obtūrantis aures suas*, Ps. Spl. T. 57, 4. [*Goth.* faurdammjan *to stop up*: *Ger.* verdammen *to embank, dam up.*]

for-dēn *done for, destroyed, defiled*, Exon. 25 b; Th. 74, 15; Cri. 1207; *pp.* of for-dōn.

for-dēþ *does for, destroys*, L. Edg. S. 14; Th. i. 278, 1; *3rd pres. sing.* of for-dōn.

for-dettan *to shut up*; obtūrāre, Prov. 21. v. for-dyttan.

for-dician; *p.* ode; *pp.* od *To obstruct, shut, or fence off with a ditch*; fossā obstruĕre, Som. Ben. Lye.

for-dilgian, -diligian; *p.* ode, ade; *pp.* od, ad *To blot out, destroy*; dēlēre, obnūblāre, oblĭtĕrāre :— He wolde ealle his þeóde fram đam gingrum óþ đa yldran fordōn and fordilgian *he would do for and blot out all his nation from the younger to the elder*, Bd. 3, 24; S. 556, 13: 5, 21; S. 643, 26. He đá óđer werod đære [MS. đara] mānfullan þeóde fornam and fordilgade *sic cētĕras nefandæ milītiæ cōpias delēvit*, S. 504, 7: 5, 13; S. 633, 34. Đæt hí óþ forwyrd ǽghwǽr fordiligade ne wǽron *ne usque ad internĕciōnem usquequaque delērentur*, Bd. 1, 16; S. 484, 17. [*Orm.* forrdillȝenn : *Dut.* ver-delgen : *Ger.* ver-tilgen *to extirpate, destroy.*]

for-dimmian; *p.* ode; *pp.* od *To make very dim, darken, obscure*; obnūbĭlāre, obfuscāre, obscūrāre, R. Conc. 1.

for-dōn; *p.* for-dónne; he -dēþ; *p.* ic, he -dyde, đū -dydest, *pl.* -dydon; *subj.* -dō, *pl.* -dōn; *p.* -dyde, đū -dyde, *pl.* -dyden; *pp.* -dōn, -dēn. **I.** *to do for, destroy, kill*; perdĕre, destruĕre, dēlēre, contĕrĕre, interfĭcĕre, occīdĕre :— Ondrǽdaþ đone, đe mæg sāwle and líchaman fordōn on helle *tĭmēte eum, qui pŏtest et ănĭmam et corpus perdĕre in gehennam*, Mt. Bos. 10, 28 : Mk. Bos. 3, 6 : Gen. 18, 23 : Chr. 1013; Erl. 149, 2, 24 : L. Ath. iv. 1; Th. i. 220, 23. He wolde ealle his þeóde fordōn and fordilgian *tōtam ejus gentem dēlēre et extermĭnāre decrēverat*, Bd. 3, 24; S. 556, 13 : Deut. 9, 19. He wolde Aaron fordōn *vŏluit Aaron contĕrĕre*, Deut. 9, 20. Đæt he mǽge fordōn đa unsceđđendan *ut interfĭciat innŏcentem*, Ps. Th. 9, 28. He sēcþ hine to fordōnne *quærit perdĕre eum*, Ps. Th. 36, 32. Ic fordō hig *ego disperdam eos*, Gen. 6, 13. Đe đæue scyldigan rihtlíce fordēþ *who lawfully does for the guilty*, L. Edg. S. 14; Th. i. 278, 1. Be đam wífmen đe hire bearn fordēþ *de mulĭĕre quæ infantem suum occīdit*, L. Ecg. P. cont. ii. 2; Th. ii. 180, 3. Se bisceop towearp and fordyde đa wigbed *pontĭfex ipse polluit ac destruxit eas āras*, Bd. 2, 13; S. 517, 18 : Chr. 986; Erl. 130, 11 : 1075; Erl. 214, 15 : Deut. 9, 4. Đú fordydest ǽlcne man *perdĭdisti omnem*, Ps. Lamb. 72, 27. Se here fordydon eall đæt he oferferde *the army destroyed all that it passed over*, Chr. 1016; Erl. 157, 12. Hí fordydon me *consummāvērunt me*, Ps. Lamb. 118, 87. Đæt ic hig fordō *ut contĕram eum*, Deut. 9, 14. Đæt he fordō *ut perdat*, Jn. Bos. 10, 10 : Bt. Met. Fox 20, 260; Met. 20, 130. Đæt we hig fordōn *ut perdāmus illos*, Gen. 19, 13. Đý-lǽs hí fordōn óđra gesceafta *lest they destroy other creatures*, Bt. 39, 13; Fox 234, 9. Đæt he eów ne fordyde *ne dēlēret vos*, Deut. 9, 25. Đý-lǽs hí óđra fordydon ǽdela gesceafta *lest they should destroy other noble creatures*, Bt. Met. Fox 29, 91; Met. 29, 45. Hú oft ic hæbbe fordōn đa Egiptiscan *quotiens contrīverim Ægyptios*, Ex. 10, 2. **II.** *to seduce, defile, corrupt*; sedūcĕre, scĕlĕrāre :— Đeáh heó dearnenga fordōn wurde mid ligenum *though she* [*Eve*] *was secretly seduced with lies*, Cd. 30; Th. 39, 22; Gen. 629. Deáþfirenum forden *defiled by deadly sins*, Exon. 25 b; Th. 74, 15; Cri. 1207. On đa firenum fordōne sorgum wlítaþ *on which the defiled by sins shall sorrowfully look*, Exon. 24 a; Th. 68, 16; Cri. 1104. Đǽr wæs cirm micel, fordēnera gedræg *there was a great noise, a tumult of the defiled*, Andr. Kmbl. 85; An. 43. Seóđeþ swearta lēg synne on fordōnum *the swart flame of sin shall seethe on the corrupted*, Exon. 28 b; Th. 62, 2; Cri. 995. [*O. Sax.* fardōn : *Dut.* ver-doen *to destroy, kill*: *Ger.* ver-thun *to waste.*] Used by Shakespeare.

for-drencan; *p.* -drencte; *pp.* -drenced, -drenct *To make drunk, inebriate, intoxicate*; madefācĕre, inebriāre :— Uton fordrencan úrne fæder mid wíne *let us make our father drunk with wine*, Gen. 19, 32, 33. Nis đæs mannes fæsten náht, đe hine sylfne on forhæfednysse dagum fordrencþ *the man's fasting is naught who inebriates himself on days of abstinence*, Homl. Th. ii. 608, 24. Đás men sindon mid muste fordrencte *these men are drunken with new wine*, i. 314, 22, 23.

for-drífan; *p.* -dráf, *pl.* -drifon; *pp.* -drifen *To drive away, force, compel, drive out, eject, banish*; pellĕre, prŏpellĕre, compellĕre, cōgĕre, expellĕre :— Sumne sceal hreóh fordrífan *the tempest shall drive one away*, Exon. 87 a; Th. 328, 10; Vy. 15. Hine se streám fordráf *the stream drove him*, Ors. 2, 4; Bos. 44, 3: Judth. 12; Thw. 25, 25; Jud. 277: Andr. Kmbl. 538; An. 269. Norþhymbra fordrifon heora cining Alhrēd of Eoferwíc *the Northumbrians drove their king Alhred from York*, Chr. 774; Erl. 53, 33 : 954; Erl. 119, 6. Fordríf hí *expelle eos*, Ps. Th. 5, 11. Sió wunode on đam íglande đe se cyning on fordrifen wearþ *she dwelt in the island on which the king was driven*, Bt. 38, 1; Fox 194, 21. Hió geseah đone fordrifenan cyning *she saw the driven king*, 194, 23.

Lufiaþ fordrifene, fordamđe gē sylfe wǽron fordrifene and ūtancymene on Egipta lande *vos ǎmǎte pěrěgrīnos, qui et ipsi fuistis advěnæ in terra Ægypti*, Deut. 10, 19, 18. [*Laym.* men al for-dreuen: *O. Sax.* fordrīđan: *Dut.* ver-drijven: *Ger.* ver-treiben *to drive away, banish.*]

for-drincan; *p.* -dranc, *pl.* -druncon; *pp.* -druncen *To make drunk, inebriate*; madefǎcěre, ebriāre:—Gedrēfde hī syndon and astyrede syndon swā swā fordruncen [MS. fordruncon] man *turbāti sunt et mǒti sunt sīcut ebrius*, Ps. Lamb. 106, 27. Abigail forswīgode· đæt dysig hiere fordruncnan hlāfordes *Abigail concealed the folly of her drunken lord*, Past. 40, 4; Hat. MS. 55 a, 13. [*Laym.* for-drunkene cnihtes.]

for-drugian, -druwian; *p.* ode; *pp.* od *To dry up, parch, wither*; arescěre, siccāri:—He forheardaþ and fordrugaþ indǔret et arescat, Ps. Lamb. 89, 6. Hió wǽre fordrugod to duste *it would be dried to dust*, Bt. Met. Fox 20, 207; Met. 20, 104. [*Dut.* ver-droogen: *Ger.* vertrocknen *to dry up.*]

for-druncen, -druncn *drunken*, Past. 40, 4; Hat. MS. 55 a, 13; *pp. of* for-drincan.

for-druwian; *p.* ode; *pp.* od *To dry up, wither*; arescěre:—He byþ aworpen ūt swā twīg, and fordruwaþ *mittětur fǒras sīcut palmes, et arescet*, Jn. Bos. 15, 6. [*A. R.* vor-druwede, *pp. pl.*] v. for-drugian.

for-dwilman; *p.* de; *pp.* ed *To confound*; confunděre:—Đa mistas fordwilmaþ đa sōþan gesiehþe *the mists confound the true sight*, Bt. 5, 3; Fox 14, 17.

for-dwīnan, he -dwīneþ, -dwīnþ; *p.* -dwān, *pl.* -dwinon; *pp.* -dwinen *To dwindle away, vanish*; evānescěre:—Fordwīneþ heó sōna *it soon will dwindle away*, Herb. 2, 2; Lchdm. i. 80, 17. Mannes ege hrædlīce fordwīnþ *awe of man quickly vanishes*, Homl. Th. i. 592, 12. Se sceocca fordwān of his gesihþe *Satan vanished from his sight*, ii. 504, 4. [*Chauc.* hondes for-dwīned: *Dut.* ver-dwijnen *to vanish.*]

for-dyde, *pl.* -dydon *did for, destroyed*, Deut. 9, 1: Ps. Lamb. 118, 87: for-dyde, *pl.* -dyden *should do for, destroy*, Deut. 9, 25: Bt. Met. Fox 29, 91; Met. 29, 45; *p. indic. and p. subj. of* for-dōn.

for-dyttan; *part.* -dyttende; *p.* -dytte; *pp.* -dytted, -dytt, -dyt *To close or shut entirely up, stop up*; oppilāre, clauděre, obstruěre:—Swā swā nǽddran deáfre, and fordyttendre hire eáran *sīcut aspǐdis surdæ, et obtūrantis aures suas*, Ps. Lamb. 57, 5. Ǽlc unrihtwīsnes fordyt mūþ hire *omnis inǐqũitas oppilābit os suum*, 106, 42. Is fordyt mūþ sprecendra unrihte þing *obstructum est os lǒquentium inǐqua*, 62, 12. Đa wilspringas đǽre miclan niwelnisse wurdon fordytte *clausi sunt fontes abyssi*, Gen. 8, 2. [*Laym.* for-dut, *pres. sing. indic.*]

fore = for, *q. v*; *prep. dat. acc.* I. *for, on account of, for the sake of*; pro, propter, per; *with the dative*; cum dǎtīvo:—Ne syndon to lufianne đa wīsan fore stōwum, ac for gōdum wīsum stōwe syndon to lufianne *non pro lǒcis res, sed pro bŏnis rēbus lǒca amanda sunt*, Bd. 1, 27; S. 489, 41. Fore miltsum *for his mercies*, Exon. 46 b; Th. 159, 25; Gū. 932. He lāþ biþ ǽghwǽr fore his wonsceaftum *he is everywhere unwelcome on account of his misfortunes*, 87 b; Th. 329, 10; Vy. 32. He fore his mondryhtne mōdsorge wæg *he bare mental sorrow for his master*, Exon. 48 a; Th. 165, 5; Gū. 1024. Nō mearn fore fǽhþe and fyrene *he mourned not on account of his enmity and crime*, Beo. Th. 273; B. 136. Gē scofene wurdon fore oferhygdum in ēce fýr *ye were thrust into eternal fire on account of pride*, Exon. 41 b; Th. 140, 6; Gū. 606. II. = for, *q. v. for, on account of, for the sake of*; pro, propter, per; *with the accusative*; cum accusātīvo:—Gehālgode fore hine Damiānum *consecrǎvit pro eo Damiānum*, Bd. 3, 20; S. 550, 33. III. sometimes fore is separated from its case, v. III. *in* fōre:—Đæt he hine fore gebǽde *that he might pray for him*, Bd. 5, 5; S. 618, 2. He ahongen wæs fore moncynnes mānforwyrhtum *he was hanged for the evil deeds of mankind*, Exon. 24 a; Th. 67, 27; Cri. 1095. Se þegn fore fæder dǽdum swefeþ *the thane sleeps for his father's deeds*, Beo. Th. 4125; B. 2059.

fōre = fōr; *prep. dat. acc.* I. *before*; cōram, ante, in conspectu, præsente *vel* audiente ǎliquo, ante; *with the dative*; cum dǎtīvo:—Se ār Godes ǎnne wīsfæstne wer gehālgode fōre đam heremægene *the messenger of God consecrated a wise man before the host*, Andr. Kmbl. 3299; An. 1652. Fela gē fōre monnum mīđaþ *ye conceal much before men*, Exon. 39 a; Th. 130, 10; Gū. 436. Hý fōre leódum leóhte blīcaþ *they shall shine brightly before the people*, 26 a; Th. 76, 13; Cri. 1239. Gehealdne sind sāwle wiđ synnum fōre sigedēman *souls have been preserved from sins before the judge triumphant*, Exon. 23 b; Th. 65, 28; Cri. 1061. Fōre Waldende *before the Lord*, 23 b; Th. 66, 12; Cri. 1070. Fōre onsýne ēcan Dryhtnes standaþ stīþferhþe *the stout-hearted stand before the face of the eternal Lord*, Andr. Kmbl. 1441; An. 721. Fōre eágum *before the eyes*, Exon. 27 a; Th. 81, 15; Cri. 1324. II. *before*; ante, *with the accusative*; cum accusātīvo:—Sendon hira bēne fōre bearn Godes *they sent their petition before the Son of God*, Andr. Kmbl. 2056; An. 1030. Ne sceal ic mīne onsýn fōre eówere mengu mīđan *I shall not conceal my countenance before your multitude*, Exon. 43 a; Th. 144, 17; Gū. 679. Fōre þreó niht *before three nights*, Andr. Kmbl. 369; An. 185. III. sometimes fōre follows its case or is separated from it:—On đone Drihten đe đes hāligdōm is fōre hālig *by the Lord*

before whom this relic is holy, L. O. 1, 2; Th. i. 178, 3, 12. Đes ār me fōre stondeþ *this messenger stands before me*, Exon. 69 b; Th. 259, 5; Jul. 277. Cumaþ him fōre *come before him*, Ps. Th. 94, 6. Scīneþ đē leóht fōre *the light shines before thee*, Cd. 29; Th. 38, 30; Gen. 614. Him wēpan fōre *plǒrēmus cǒram eo*, Ps. Th. 94, 6.

fōre; *adv. Before, aforetime, formerly*; antea, ōlim, quondam:—He on Ægypta lande worhte fōre wundur mǽre *he aforetime did great wonders in the land of Egypt*, Ps. Th. 77, 14.

fōre; *gen. dat. acc. of* fōr *a going, journey, course, approach*, Exon. 111 a; Th. 426, 10; Rä. 41, 71: Bd. 5, 9; S. 623, 23: 4, 27; S. 604, 29. v. fōr, e; *f*.

fōre- *before*, used in composition as the English *fore-*.

fōre-ætýwian; *p.* ede; *pp.* ed *To fore-show, to go before and show the way*; præmonstrāre, Som. Ben. Lye. v. æt-eówian, -ýwan.

for-ealdian, -ealdigean, -ealldian; *p.* ode; *pp.* od [for-, eald *old*] *To grow or wax old, become old*; senescěre, veterascěre, inveterascěre:—Wyrceaþ seódas, đa đe ne forealdigeaþ *fǎcǐte vōbis saccǔlos, qui non veterascunt*, Lk. Bos. 12, 33. Bearn elelendisce forealdodon *fīlii aliēni inveterāti sunt*, Ps. Lamb. 17, 46. Forealldodon đa gewritu *the writings waxed old*, Bt. 18, 3; Fox 64, 37. Ne forealdige đeós hand ǽfre *nunquam inveterascat hæc mānus*, Bd. 3, 6; S. 528, 24. Ǽlc ānweald biþ sōna forealdod *every power soon becomes old*, Bt. 17; Fox 60, 10: 39, 8; Fox 224, 11. Đe forealdode wǽron *who were grown old*, Homl. Th. ii. 500, 4. [*Ger.* ver-alten *to grow old.*] DER. ealdian, eald.

fōre-astreccan; *p.* -astreahte, -astrehte; *pp.* -astreaht, -astreht *To lay or stretch out before*; prōsterněre:—Đæt he fōreastrehte hig on wēstene *ut prǒsternēret eos in deserto*, Ps. Spl. T. 105, 25. [*Ger.* vor-strecken *to stretch forth.*]

fōre-aþ; *p.* -aþ, es; *m. A fore-oath, an oath first taken*; antejūrāmentum, præjūrāmentum, præjūrātio:—So called because it was that by which every accuser or plaintiff commenced his accusation or suit against the accused or defendant. To this the defendant opposed his own fōre-aþ, thereby pleading not guilty to the charge. The oaths both of plaintiff and defendant were supported by consacramentals, respecting the number of which see L. H. 66, § 8; Th. i. 569: v. also Āþ II, III. If the fōre-aþ of the accuser failed, the charge was quashed and the accused set at liberty:—Ofgā ǽlc man his tīhtlan mid fōreaþe *let every man begin his charge with a fore-oath*, L. Ath. i. 23; Th. i. 212, 5. Agife đone fōreaþ on feówer ciricum *let him make his fore-oath in four churches*, L. Alf. pol. 33; Th. i. 82, 7. Ofgā his spræce mid fōraþe *let him begin his suit with a fore-oath*, L. O. D. 6; Th. i. 354, 31: L. Ath. iv. 2; Th. i. 222, 16. Ofgā man ānfealde lāde mid ānfealdan fōraþe and þrýfealde lāde mid þrýfealdan fōraþe *one may proceed to a simple exculpation with a simple fore-oath and to a threefold exculpation with a threefold fore-oath* L. C. S. 22; Th. i. 388, 15; cf. Schmid, Ges. der Angelsachsen, foraþ.

fōr-eáđe; *adv. Very easily*; perfǎcile:—God mæg fōreáđe unc ǽt fōresceáwian *God can very easily provide food for us two*, Homl. Th. ii. 138, 35.

fōre-beácen, -beácn, es; *n. A fore-token, prodigy, wonder*; prodǐgium, portentum, ostentum:—Ic eom swā fōrebeácen folce manegum *tamquam prodǐgium factus sum multis*, Ps. Th. 70, 6. Fōrebeácna prodǐgiōrum 104, 23. He sigetācen sende manegum, fōrebeácn feala folce Ægipta *mīsit signa et prodǐgia in mědio Ægypti*, 134, 9: Ps. Lamb. 77, 43: Mt. Bos. 24, 24. Sōþlīce leáse cristas and leáse wītegan arīsaþ, and wyrcþ, fōrebeácnu *exsurgent ěnim pseudochristi, et pseudoprophētæ, et dǎbunt signa et portenta*, Mk. Bos. 13, 22: Deut. 13, 1. Būton gē tācna and fōrebeácna geseón, ne gelýfe gē *except ye see signs and wonders, ye will not believe*, Jn. Bos. 4, 48: Nar. 50, 21: -beácno, Blickl. Hom. 117, 30.

fōre-beón *to be before or over, to preside*; præesse, Scint. 32, 58. v. fōre-eom, fōre-wesan.

fōre-beran; *part.* -berende; *p.* -bær, *pl.* -bǽron; *pp.* -boren *To prefer*; præferre:—He sundorlīf and munuclīf wæs fōreberende eallum đām weólum and ārum đæs eorþlican rīces *ěrat vītam prīvātam et mǒnachǐcam cunctis regni dīvǐtiis et hǒnōrǐbus præferens*, Bd. 4, 11; S. 579, 8. v. fōre-beran.

fōre-bétan; *p.* -bētte; *pp.* -bēted [fōre *before, full, entire*; bētan *to make amends*] *To make full amends to or for anyone or anything*; compensāre prō ǎliquo:—Lādige mid his māgan, đe fǽhþe mōton mid-beran, odđe fōrebétan *let him clear himself with his kinsmen, who must bear the feud with him, or make full amends for it*, L. Eth. ix. 23; Th. i. 344, 27: L. C. E. 5; Th. i. 362, 23. Gif he nyte hwā him fōrebēte *if he know not who shall make full amends for him*, L. Ed. 9; Th. i. 164, 12: L. Ath. i. 8; Th. i. 204, 8.

fōre-birig; *dat. s. of* fōre-burh *a vestibule*, Ex. 29, 32.

fōre-bodian; *p.* ode; *pp.* od *To* FOREBODE, *announce, declare*; annuntiāre, prōnuntiāre:—Mūþ mīn fōrebodaþ rihtwīsnysse đīne *os meum annuntiābit justǐtiam tuam*, Ps. Spl. 70, 16. Fōrebodaþ tunge [MS. tunga] mīn spræca đīne *prōnuntiābit lingua mea elǒquium tuum*, 118, 172.

fōre-breóst, es; *n. The fore-breast, breast, chest*; præcordia, thōrax = θώραξ:—Fōrebreóst *præcordia*, Ælfc. Gl. 73; Som. 71, 23; Wrt. Voc. 44, 9.

X

fôre-burh; *gen.* -burge; *dat.* -byrig, -birig; *f.*   I. *a fore-court,
entrance-court, vestibule*; vestībŭlum :—Hig etaþ ða hlāfas on ðæs geteldes
fôrebirig *comēdent pānes in tabernācŭli testĭmōnii vestĭbŭlo*, Ex. 29,
32.   II. *a wall before a fortification*; mūrus ante
mūrum, dictum ex eo quod pro mūnītione sit [*Du Cange*] :—Fôreburh
promūrāle, Ælfc. Gl. 55; Som. 66, 118; Wrt. Voc. 36, 38.

fôre-býsen, e; *f.* [fôre, býsen *an example, model*] *A fore-model, an,
example*; exemplum :—Arcebisceop sceal hālgian and getryman mid gōdan
mynegunga and fôrebýsene *an archbishop shall hallow and strengthen
them with good admonitions and example*, Chr. 694; Th. 67, 43.

fôre-ceorfan; *p.* -cearf, *pl.* -curfon; *pp.* -corfen [fôre *fore*, ceorfan *to
cut*] *To cut off the front*; præcīdĕre :—Ic fôreceorfe *præcīdo*, Ælfc. Gr.
28, 4; Som. 31, 35.

fôre-ceorfend, es; *m.* [fôre-ceorfende, *part. of* fôre-ceorfan] *A fore-
cutter, front tooth*; præcīsor, Wrt. Voc. 282, 73.

fore-costian, -costigan; *p.* ode; *pp.* od [=for-costian] *To profane,
pollute*; profānāre :—Gyf rihtwīsnys mīn hī forecostigaþ *si justītias
meas profānāvĕrint*, Ps. Spl. C. 88, 31.

fôre-cuman; *part.* -cumende; ic -cume, ðū -cumest, -cymest, -cymst,
he -cumeþ, -cymeþ, -cymþ, -cimþ, *pl.* -cumaþ; *p.* -com, -cwom, *pl.* -cōmon,
-cwōmon; *pp.* -cumen *To come forth, come before, prevent*; prævĕnīre :—
Ðæt ðū sí fôrecumende Drihtnes onsȳne in andetnesse *quo præoccŭpando
fāciem Dŏmĭni in confessione*, Bd. 4, 25; S. 599, 42. God fôrecymeþ
me *Deus prævĕniet me*, Ps. Spl. 58, 10. Fôrecymþ *prævĕniet*, 67, 34.
Ic fôrecom oððe ic fôrhradode on rīpunga oððe on rīpnysse *prævĕni in
matūrĭtāte*, Ps. Lamb. 118, 147. Ðū fôrecôme hine on bletsunge swētnysse *prævĕnisti eum in benedictiōnĭbus dulcēdĭnis*, Ps. Spl. 20, 3. Fôre-
cômon eágan mīne to ðē on dægrēd *prævēnērunt ŏcŭli mei ad te dilūcŭlo*,
Ps. Spl. 118, 148 : 17, 21. [*Goth.* faura-qîman.]   v. fôr-cuman.

fôre-cweðan; *p.* -cwæþ, *pl.* -cwǣdon; *pp.* -cweden *To foresay, predict*;
prædĭcĕre, propŏnĕre :—Hēt he him sillabas and word fôrecwæðan *addĭdit
et syllābas ac verba dīcenda illi propŏnĕre*, Bd. 5, 2; S. 615, 13. Ealle
ðās þing swā se bisceop fôrecwæþ, of endebyrdnysse gelumpon and
gefyllede wǣron *quæ cuncta ut prædīxĕrat antistes, ex ordĭne complēta
sunt*, 3, 15; S. 541, 37. Swā swā we on ðysse ǣrran bēc feáwum
wordum fôrecwǣdon *ut præcēdente libro paucis dixĭmus*, 4, 1; S. 563,
18. Sume men eác swylce sægdon, ðæt heó, þurh witedōmes gāst,
ða ādle fôrecwǣde [MS. -cwede], ðe heó on forþferde *sunt ĕtiam qui
dīcant, quia per prophētīæ spīrĭtum, pestĭlentiam qua ipsa esset mŏrĭtūra,
prædixĕrat*, 4, 19; S. 588, 15. Swā hit fôrecweden wæs *ut prædictum
ĕrat*, 3, 15; S. 542, 3. [*Goth.* faura-qiþan.]   DER. cweðan.

fôre-cwide, es; *m. A foretelling, prophecy*; prædictio, Som. Ben. Lye.

fôre-cymeþ, -cymeþ *prævĕniet*, Ps. Spl. 58, 10 : 67, 34.   v. fôre-cuman.

fôre-cynn; *pl. n. Ancestors, predecessors, progenitors*; antecessōres,
prædecessōres, progĕnĭtōres, Som. Ben. Lye.

fôre-cynren, es; *n. A progeny*; prôgĕnies, Cot. 154.

fored; *part. Broken, fractured*; fractus :—Gif monnes ceácan mon
fôrslihþ ðæt hie beón fôrede *if a man smite another's cheeks that they be
broken*, L. Alf. pol. 50; Th. i. 94, 15, note 34. Se fôreda fôt [MS. foot]
the fractured foot, Past. 11, 2; Hat. MS. 15 a, 4.   v. forod.

fôre-dûru, e; *f.* : -dŷr, e; *n. A fore-door, porch, an entry, hall*;
vestĭbŭlum, propýlæum = προπύλαιον :—Fôredýre *vestĭbŭla*, Cot. 190.

fôre-eom [fôre *before*, eom *am*] *I am before* or *over, I preside*;
præsum :—Ic begīme oððe ic fôre-eom *præsum*, Ælfc. Gr. 32; Som. 36,
32.   v. fôre-wesan.

fôre-fæder, fôre-fæderas FOREFATHER, FOREFATHERS; mājōres.   v. forþ-
fæderas.

fore-feng, -fong, es; *m.* [=for-feng] *A seizing, rescuing*; apprehensio :—Be forstolenes monnes forefonge *of seizing a stolen man*, L. In.
53; Th. i. 134, 15. Be forefonge [forefenge MSS. B, G, H.], 72; Th.
i. 148, 5. Be forstolenes ceápes forefonge *of the rescuing of stolen
property*, 75; Th. i. 150, 4.   v. for-fang.

fôre-fôn, ic -fô; *p.* -fêng, *pl.* -fêngon; *pp.* -fangen *To take before,
anticipate*; antĭcĭpāre :—Fôrefêngon wæccan eágan mīne *antĭcĭpāvērunt
vigĭlias ŏcŭli mei*, Ps. Spl. C. T. 76, 4. Raðe fôrefô us mildheortnysse
ðīne *cĭto antĭcĭpent nos mĭsĕricordiæ tuæ*, Ps. Spl. C. 78, 8.

fôre-gân; *p.* -eóde; *pp.* -gân *To go before, precede*; præcēdĕre :—
Mildheortnys and sôþfæstnys fôregâð ansýne ðīne *misĕricordia et vērĭtas
præcēdent fāciem tuam*, Ps. Spl. C. 88, 15. Oðer fôre-eóde ða sunnan
ūna *sōlem præcēdēbat*, Bd. 5, 23; S. 645, 24.   v. fôre-gangan, fôr-gân.

fôre-gangan; *part.* -gangende; *p.* -geóng, -gêng, *pl.* -geóngon,
-gêngon; *pp.* -gangen *To go before, precede*; præcēdĕre :—Hí wǣron
fôregangende in ðone lēg *they were going before into the flame*, Bd. 3, 19;
S. 548, 31. Hine sôþfæstnes fôregangeþ *justītia ante eum ambŭlābit*, Ps.
Th. 84, 12. Hwæt ðǣr fôregange oððe hwæt ðǣr æfterfylige we ne
cunnon *quid autem præcessĕrit quidve sĕquātur ignōrāmus*, Bd. 2, 13;
S. 516, 22. [*Ger.* vor-gehen *to precede*.]   v. fôre-gân, fôr-gangan.

fôre-gehât, es; *n. A fore-promise, vow*; prōmissio :—Ðæt fôregehât
forgifenysse *that we habbaþ fram Gode prōmissio remissiōnis, quam
hăbēmus a Deo*, Bd. Whelc. 341, 27. On ðīnum fôregehātum *in pro-
missiōnĭbus tuis*, 341, 26.

fôre-genga, an; *m.*   I. *a fore-goer, fore-runner, predecessor ;
predecessor* :—Ðætte swā ǣðele fôregenga swylcne yrfeweard hæfde *that
so noble a predecessor should have such an heir*, Bd. 3, 6; S. 528, 33 :
3, 9; S. 533, 12 : 4, 30; S. 609, 6. Laurentius bii his fôregengan
bebyrged wæs *Lawrence was buried beside his predecessor*, Bd. 2, 7;
S. 509, 6. Ða fôregengan, yldran usse *those ancestors, our parents*, Exon.
62 a; Th. 228, 13; Ph. 437. On hiora fôregengena dagum *in diēbus
antĭquis*, Ps. Th. 43, 2.   II. *a fore-runner*; prodrŏmus = πρόδρομος :—
Hæfde fôregenga fýrene loccas *their fore-runner had fiery locks*, Cd.
148; Th. 185, 9; Exod. 120. Ðone fôregengan Fæder ælmihtig gesette
the almighty Father had placed that fore-runner, Exon. 40 b; Th. 134,
7; Gū. 504. [*Dut.* voor-ganger : *Ger.* vor-gänger *a predecessor*.]

fôre-genge, an; *f. A fore-goer, female servant*; ancilla :—Hyre fôre-
genge [MS. fôregenga] blâc-hleór ides *her servant, the pale-faced woman*,
Judth. 11; Thw. 23, 18; Jud. 127.

fôre-gesettan; *part.* -gesettende; *p.* -gesette; *pp.* -gesett, -geset *To
place before*; præpŏnĕre :—Fôregesettendum ðām swýðe hālgan gôd-
spellum *præpŏsĭtis sacrosanctis evangĕliis*, Bd. 4, 17; S. 585, 27. [*Goth.*
faura-gasatjan *to present*.]   v. fôre-settan.

fôre-gewîtnys, -nyss, e; *f. False witness*; falsum testĭmōnium :—Ðæt
heora ǣnig on fôre-gewîtnysse sý *quod eōrum ălĭquis in falso testĭmōnium
sit*, L. Ath. i. 10; Wilk. 58, 22; Lambd. 49, 12, = wôhre gewîtnesse,
Th. i. 204, 23; *dat. sing. f.* of wôh and gewîtnes.

fôre-gilpan; *p.* -gealp, *pl.* -gulpon; *pp.* -golpen *To boast greatly*;
valde jactāre :—Ðæt he wǣre cumen to ðām gôdan tīdum ðe Rômāne eft
fôregulpon *that he was come to the good times of which the Romans
afterwards boasted greatly*, Ors. cont. 4, 7; Bos. 12, 13.

fôre-gisel; *gen.* -gîsles; *m.* [gîsel *a hostage*] *A foremost hostage,
principal* or *eminent hostage*; præstans *vel* electus obses :—Salde se here
him fôregîslas and micle âþas *the army gave him eminent hostages with
great oaths*, Chr. 878; Erl. 80, 16 : 877; Erl. 79, 24. Norþhymbre- and
Eást-Engle hæfdon Ælfrêde cyninge âþas geseald, and Eást-Engle fôregîsla
vi *the Northumbrians and East-Angles had given oaths to king Alfred,
and the East-Angles six principal hostages*, Chr. 894; Erl. 90, 4.

fôre-gleáw; *adj. Very prudent*; provĭdus, præ aliis săpiens :—Fôre-
gleáw *provĭdus*, R. Ben. 64 : Homl. Th. ii. 152, 2. Fôregleáwe ealde
ûþwitan *very prudent ancient philosophers*, Menol. Fox 328; Men. 165.

fôre-gleáwlíce; *adv. Providently, prudently*; provĭde, R. Ben.
interl. 3.

fôre-gleáwnes, -ness, e; *f. Providence, prudence, carefulness*; provĭ-
dentia, Som. Ben. Lye.

fôre-gulpon *boasted greatly*, Ors. cont. 4, 7; Bos. 12, 13; *p. pl.* of
fôre-gilpan.

fôre-heáfod *the forehead*; frons, Som. Ben. Lye.   v. fôr-heáfod.

fôre-mǣre; *def.* se fôre-mǣra; *sup.* -mǣrost, -mǣrest; *adj. Fore-great,
very honourable, illustrious, eminent, famous, celebrated*; præclārus, illustris,
excellens, fāmōsus, celeberrĭmus :—Ic nǣfre ne geseah ne gehýrde nǣnne
wîsne mon ðe mā wolde bión wrecca, and, earm, and ælþiódig, and
forsewen, ðonne wêlig, and weorþ, and rîce, and fôremǣre on his āgnum
earde *I never saw nor heard of any wise man who would rather be an
exile, and miserable, and foreign, and despised, than wealthy, and
honourable, and powerful, and eminent in his own country*, Bt. 39, 2;
Fox 212, 17. Is mîn land nū fôremǣre, and me swýðe unbleó *hærēdĭtas
mea præclāra est mihi*, Ps. Th. 15, 6. Hwǣr is nū se fôremǣra and se
arǣda Rômwâra heretoga *where is now the illustrious and the prudent
consul of the Romans?* Bt. 19; Fox 70, 6. Hæfde gefohten fôremǣrne
blǣd Iudith *Judith had gained illustrious honour*, Judth. 11; Thw. 23,
15; Jud. 122. Fôremǣre Simon and Iudas symble wǣron Drihtne dŷre
the celebrated Simon and Jude were always dear to the Lord, Menol.
Fox 378; Men. 190. Hū he fôremǣrost seó *how he may be most
illustrious*, Bt. 3, 2; Fox 122, 34 : 18, 3; Fox 64, 35. Se wer se fôre-
mǣresta *the most eminent man*, Bd. 5, 20; S. 641, note 37.

fôre-mǣrlíc; *adj. Eminent*; præclārus :—Hū weorþlíc and hū fôre-
mǣrlíc *how honourable and how eminent*, Bt. 33, 1; Fox 120, 34.

fôre-mǣrnes, fôr-mǣrnes, -ness, e; *f. Greatness, eminence, renown,
glory*; clārĭtas :—Weorþscipe and fôremǣrnes *dignity and renown*, Bt.
34, 6; Fox 142, 7 : 33, 1; Fox 122, 12.

fôre-manian; *p.* ode; *pp.* od *To fore-warn*; præmŏnēre :—He fôre-
manod wæs *præmŏnĭtus fŭĕrat*, Bd. 5, 10; S. 623, 39.

fôre-meahtig, fôre-mihtig; *adj. Prepotent, most mighty*; præpŏtens :—
Ða fôremeahtige folces rǣswan *the prepotent chieftains of the folk*, Cd. 80;
Th. 100, 24; Gen. 1669. Ðǣr he ealdordôm onfêhþ, fôremihtig ofer
fugla cynn *where it [the phœnix] receives supremacy, most mighty over
the race of birds*, Exon. 58 a; Th. 208, 21; Ph. 159 : Cd. 208; Th.
257, 33; Dan. 667.

fôre-mearcod; *part. Fore-noted*; prænŏtātus, Cot. 157.

fôre-mihtig; *adj. Prepotent, most mighty*; præpŏtens, Cd. 208; Th.
257, 33; Dan. 667.   v. fôre-meahtig.

fôre-mihtiglíce, -mihtlíce; *adv. Most mightily*; strēnue, Cot. 202.

fôre-múnt, es; *m. A fore-mount, promontory*; promontōrium, Cot.
149.

fōrene? *before;* ante, cĭtius, Lye:—Gif hine hwā fōrene [MS. A. of the 12th century has fōra] forstande *if any one stand up for him,* L. C. S. 33; Th. i. 396, 17; Wilk. 139, 22, 23. v. Schmid, s. v. forstandan.

fōre-rím, es; *m. A prologue, preface;* prolŏgus:—Onginneþ fōrerím *incipit prolŏgus,* Mt. Kmbl. Præf. p. 1, 1.

fōre-rynel, fōr-rynel, es; *m.* [fōre, fōr *before;* rynel, es; *m. a runner*] *A fore-runner;* præcursor:—Iohannes his fōrerynel wæs on lífe ge on deáþe *John was his fore-runner both in life and in death,* Ælfc. T. 24, 20: Bt. 36, 1; Fox 170, 28, MS. Cot. v. fōr-rynel.

fōre-sacan; *p.* fore-sōc [=for-sacan] *to forbid;* prohĭbēre:—Foresōc oðŏe forbeád *prohĭbēbat,* Mt. Lind. Kmbl. 3, 14.

fōre-sǽde *foretold, predicted,* Mt. Bos. 24, 25; *p. of* fōre-secgan.

fōre-sǽgde *foretold, told,* Bd. 3, 15; S. 541, 16: biseno foresægde *parabolam proposuit,* Mt. Kmbl. 13, 24; *p. of* fōre-secgan.

fōre-sǽge *should provide;* provīdēret, Bd. 4, 1; S. 565, 8; *3rd sing. imperf. subj. of* fōre-seón.

fōre-sændan *to send before,* Ælfc. Gr. 28, 4; Som. 31, 41, MS. D. v. fōre-sendan.

fōre-sáwe *foresawest;* prævīdisti, Ps. Th. 138, 2; *2nd sing. p. of* fōre-seón.

fōre-sceáwere, es; *m. A foreshewer, foreseer;* prævīsor, Consid. ætātum lunæ in mŏdo gĕnitis, Lye.

fōre-sceáwian, fōre-sceáwigan, fōr-sceáwian; *p.* ode; *pp.* od *To foreshew, foresee, provide;* præ-ostendĕre, pōnĕre in conspectu, prævidĕre, providĕre:—Ic fōresceáwige *prævideo,* Ælfc. Gr. 26, 5; Som. 29, 3. God fōresceáwaþ him sylf ða offrunge *Deus providēbit sĭbi victĭmam,* Gen. 22, 8. He him fōresceáwode sumne heretogan *he provided them a leader,* Jud. 6, 8. Đæt he fōresceáwode hū he hig gecígde *ut vidēret quid vŏcāret ea,* Gen. 2, 19. Ic wisce ðæt hig fōresceáwodon hira ende *ūtinam nŏvissima providērent,* Deut. 32, 29. Hū hit gebýreþ to fōresceáwigenne *quōmŏdo oporteat providĕre,* L. Ecg. P. cont. i. 1; Th. ii. 170, 3. DER. sceáwian.

fōre-sceáwung, fōr-sceáwung, e; *f. A* FORESHEWING, *foreseeing, foresight, providence;* providentia:—Beó ðē ān fōresceáwung *let there be one providence to thee,* Basil. admn. 3; Norm. 38, 17. Fōresceáwung Godes *God's providence,* Bt. 39, 4; Fox 216, 30: 39, 5; Fox 218, 21. Com hit mid Godes fōresceáwunge and bletsunge *it came with God's providence and blessing,* Homl. Th. i. 92, 22: Hexam. 8; Norm. 14, 15. On ðara þinga fōresceáwunge *in rērum providentia,* Bd. 4, 10; S. 578, 7. þurh godcundan fōresceáwunga *through divine providence,* Bt. 39, 13; Fox 224, 6: fōresceáuung *prudentia,* Rtl. 108, 25.

fōre-scyttels, es; *m.* [fōre, scyttels *a bolt, bar*] *A fore-bolt, bar;* repāgŭlum:—Đæt ǽnig elda meahte swā fæstlíce fōrescyttelsas ō inhebban *that any one should ever raise up such firm bars,* Exon. 12 a; Th. 20, 4; Cri. 312.

fōre-secgan; *p.* -sǽgde, -sǽde; *pp.* -sǽgd, -sǽd *To* FORE-SAY, *foretell, predict, announce;* præfāri, prædicĕre, prædĭcāre, pronuntiāre, annuntiāre:—Ic fōresecge oðŏe bodige *prædico,* Ælfc. Gr. 47; Som. 48, 40. Đæt se bisceop Aidan ðām scypfarendum ðone storm towardne fōresægde *ut episcŏpus Aidan nautis tempestātem fŭtūram prædixĕrit,* Bd. 3, 15; S. 541, 16: Ps. Th. 118, 172: 147, 8. Gerīses to fōresægcane godspell *oportet prædĭcari evangelium,* Mk. Skt. Lind. 13, 10. Iosue cwæþ ðā to ðām fōresǽdan ǽrendracum *Joshua then spoke to the aforesaid messengers,* Jos. 6, 22. [*Ger.* vor-sagen *to recite to a person.*]

fōre-sendan; *ic* -sende; *p.* -sended *To send before;* præmittĕre:—Ic fōresende *præmitto,* Ælfc. Gr. 28, 4; Som. 31, 41.

fōre-seón, to -seónne; *p,* ic, he -seah, ðū -sáwe, *pl.* -sáwon; *pp.* -sewen *To see before,* FORESEE, *provide;* prævidĕre, providĕre:—Swylce eác he heora andlyfene is to þenceanne and to fōreseónne *de eōrum quŏque stĭpendio cōgĭtandum atque providendum est,* Bd. 1, 27; S. 489, 21. Đū ealle míne wegas wel fōresáwe *omnes vias meas prævidisti,* Ps. Th. 138, 2. He fōreseah Godes cyricum and mynstrum micle frēcnesse towearde *he foresaw much peril awaiting God's churches and monasteries,* Bd. 3, 19; S. 549, 46: 3, 15; S. 542, 4. Đæt he him on his biscopscíre gerisene stōwe fōresǽge and sealde, on ðære ðe he mid his geferum wunian mihte *ut in diœcēsi sua providēret et dāret ei lŏcum, in quo cum suis apte dēgĕre pŏtuisset,* 4, 1; S. 565, 8. [*Ger.* vor-sehen *to foresee, provide.*]

fōre-seónd, es; *m. One who foresees, a provider;* provīsor:—Lícode ðam árfæstan fōreseónde úre hǽlo *plăcuit pio provĭsōri sălūtis nostræ,* Bd. 4, 23; S. 595, 13.

fōre-seónes, -ness, -nys, -nyss; *f. A foreseeing, foresight, providence;* provīsio, providentia:—Heó ða cúþestan andsware ðære upplícan fōreseónesse onfēng *accēpit ipsa certissĭmum sŭpernæ provisiōnis responsum,* Bd. 4, 7; S. 575, 1. Mid ða godcundan fōreseónesse *divīna provisiōne,* 5, 6; S. 619, 21. Mid ða árfæstan fōreseónysse úres alýsendes *pia redemptōris nostri provisiōne,* 4, 9; S. 576, 26.

fōre-setnes, -ness, -nys, -nyss, e; *f.* I. *a thing proposed, proposition, purpose, intention;* propŏsĭtio, propŏsĭtum:—Wæs seó cwēn lustfulliende ðære gōdan fōresetnesse and willan ðæs iungan *the queen rejoiced at the young man's good purpose and will,* Bd. 5, 19; S. 637, 32: 5, 20; S. 642, 17. Hēredodon hí his gemynd and his fōresetnesse

laudāvērunt ejus prŏpŏsĭtum, 5, 19; S. 637, 26: 4, 23; S. 593, 15. Ic ontýne on sealmlofe ingehygdnessa oðŏe fōresetnysse míne *apĕriam in psaltērio propŏsĭtiōnem meam,* Ps. Lamb. 48, 5. Ic sprece fōresetnyssa fram frymþe *lŏquar propŏsĭtiōnes ab initio,* 77, 2. II. *that which is placed before, a preposition;* præpŏsĭtio:—Præpŏsĭtio mæg beón gecweden on Englisc fōresetnyss *præpŏsĭtio may be called in English a fore-setting,* Ælfc. Gr. 47; Som. 47, 10: 5; Som. 3, 52.

fōre-settan; *p.* -sette, *pl.* -setton; *pp.* -seted, -sett *To set before, propose, shut, close in;* præpōnĕre, propōnĕre, præclūdĕre:—Hí ða ylcan Eald-Seaxan næfdon ágenne cyning, ac ealdormen wǽron heora þeóde fōresette *non hăbent rēgem iidem antīqui Saxōnes, sed satrăpas suæ genti præpŏsĭtos,* Bd. 5, 10; S. 624, 23. He fōresette on his mōde ðæt he wolde cuman to Rōme *propŏsuit ănĭmo vĕnīre Rōmăm,* 5, 19; S. 637, 23. Hí nā fōresetton ðē on gesihþe his *non propŏsuērunt te in conspectu suo,* Ps. Spl. 85, 13: 53, 3. Gemĕtton [MS. gemettan] we us storme fōresette *invĕnimus nos tempestāte præclūsos,* Bd. 5, 1; S. 613, 31.

fōre-settendlíc; *adj.* Set before, prepositive; præpŏsĭtīvus, Som. Ben. Lye.

fōre-seuwenes, -ness, e; *f.* [=for-sewennes] *A despising, contempt, dishonour;* · contemptus, dedĕcus:—On mínre unwurþuesse and fōreseuwenesse *on account of my unworthiness and dishonour,* Bt. 5, 1; Fox 10, 23.

fōre-singend, es; *m. A fore-singer, one who pitches tunes, a precentor;* præcentor. Ælfc. Gl. 33; Som. 62, 37; Wrt. Voc. 28, 19.

fōre-sittan, *part.* -sittende; *p.* -sæt, *pl.* -sǽton; *pp.* -seten *To sit before or in front, to preside;* præsidēre:—Wæs fōresittende se Arcebiscop þeodórus *the Archbishop Theodore was presiding,* Bd. 4, 5; S. 571, 25. Fōresittendum Theodóre *præsidente Theodōro,* 4, 17; S. 585, 24. [*Ger.* vor-sitzen *to preside.*]

fōre-smeagan-smeágean *to premeditate;* præmĕdĭtāri:—Ne scyle gē on eówrum heortum fōresmeágean, hū gē andswarion *pōnĭte in cordĭbus vestris non præmĕdĭtāri, quemadmŏdum respondĕatis,* Lk. Bos. 21, 14. Foresmeagan *scrutari, investigare,* Hpt. Gl. DER. smeágan.

fōre-smeáung, e; *f. Premeditation;* præmĕdĭtātio, Som. Ben. Lye.

fōre-snotor; *adj. Highly sagacious;* prudentissimus:—Fōresnotre men *highly sagacious men,* Beo. Th. 6305; B. 3163.

fōre-spæc, e; *f. A speaking for or together, an assenting, agreement;* astipŭlātio:—Đæt eall gelǽst sý ðæt on úre forespæce stænt *that all be fulfilled which stands in our agreement,* L. Ath. v. § 3; Th. i. 232, 8. v. fore-spræc.

fōre-spræce, e; *f. A fore-speech, preface;* præfātio:—Fōrespæc *præfātio,* Ælfc. Gl. 90; Som. 74, 126; Wrt. Voc. 51, 39. v. fōre-spræc.

fore-speca, an; *m.* [=for, speca *a speaker*] *One who speaks for another, a sponsor, an advocate, a patron;* prolŏcūtor, advŏcātus :—Fore-speca [=for-speca] *causīdĭcus, advŏcātus,* Ælfc. Gl. 48; Som. 65, 67; Wrt. Voc. 34, 2: Th. Diplm. A. D. 997; 539, 33; 540, 15.

fore-specen; *part. Fore-spoken, aforesaid;* præfātus, prædictus :—Dō ðæt [MS. dæs] leán to ðám fōrespecenan gōdum *add that reward to the aforesaid goods,* Bt. 37, 2; Fox 190, 2. v. fōre-sprecen.

fore-spræc, -spæc, e; *f.* [=for, spræc *a speech*] *A speaking for, a defence, an assenting, agreement;* defensio, excūsātio, astipŭlātio:—Ic secge ðæt sió forespræc ne dýge, nāuðer ne ðam scyldigan, ne ðam ðe him foreþingaþ *I say that the defence does no good, neither to the guilty, nor to him who pleads for him,* Bt. 38, 7; Fox 210, 6.

fōre-spræc, fōre-spæc, e; *f.* [fōre- *fore-,* spræc *a speech*] *A fore-speech, preface, introduction, a speaking before for another, a fore-promise;* præfātio, præ-sponsio:—Ðis is seó fōrespræc hū S. Gregorius ðas bōc gedihte, ðe man Pastoralem nemnaþ *this is the preface how St. Gregory made this book which people call Pastoral,* Past. pref; Cot. MS. Beóþ ða ungewittigan- cild gehealdene on ðam fulluhte þurh fōrespræce ðæs godfæder *unknowing children are saved in baptism by the fore-promise of the godfather,* Bd. Whelc. 180, 44.

fore-spreca, -spræca; *m.* [=for-speca] *One who speaks for another, an advocate;* prolŏcūtor, advŏcātus :—Đæt he beó mín freónd and forespreca, and ðære [MS. ðara] hálgan stōwe freónd and forespræca *that he be my friend and advocate, and the friend and advocate of the holy place,* Th. Diplm. A. D. 972; 524, 34–525, 1. He gebond feónda foresprecan *he bound the advocate of fiends* [*the devil*], Exon. 18 b; Th. 46, 6; Cri. 733. Cleopedon feónda foresprecan *the advocates of the fiends cried out,* 34 a; Th. 118, 7; Gū. 236. [*Ger.* für-sprecher, *m. an advocate.*]

fōre-sprecen, -specen, fōr-sprecen; *part.* FORE-SPOKEN, *aforesaid, forementioned;* præfātus, prædictus :—Se fōresprecena here *the fore-mentioned army,* Chr. 896; Erl. 93, 34. Se fōresprecena Godes man *præfātus clēricus,* Bd. 1, 7; S. 477, 5. He on ðæt fōresprecene mynster gedón and geþeóded wæs *he had been put in and joined to the aforesaid monastery,* 5, 19; S. 637. 29.

fōre-stæppan *to siep* or *go before, precede,* Ælfc. Gr. 28, 4; Som. 31, 30: Ps. Lamb. 88, 15. v. fōre-steppan.

fōre-stæppend, es; *m.* [fōre-stæppende; *part. of* fōre-stæppan] *A stepper* or *goer before;* præcessor:—Se ðe fōrestæppend ys *qui præcessor est,* Lk. Bos. 22, 26.

fóre-stæppung, e; _f._ _A stepping before, preventing, anticipation;_ præventio, anticipátio, Som. Ben. Lye.

fóre-stæpþ _steps before, precedes,_ Homl. Th. ii. 82, 18; _pres._ of fóre-stapan.

fóre-standan; _p._ -stód, _pl._ -stódon, _pp._ -standen _To stand before, to excel;_ præstáre:—Fórestandan _præstáre,_ Cot. 149.

fóre-standende; _part. Standing before;_ præstans:—Biscop oððe fóre-standende _antistes,_ Ælfc. Gr. 9, 26; Som. 11, 9.

fóre-stapan; _p._ -stapþ; _p._ ic, he -stóp, ðú -stópe, _pl._ -stópon; _impert._ -stape, -stæpe, _pl._ -stapaþ; _pp._ -stapen _To step before, prevent, come_ or _go before, precede;_ prægrédi, prævéníre, præíre, præcédére:—Forðan ðú fórestópe hine on blætsungum _quóniam prævénisti eum in benedictiónibus,_ Ps. Lamb. 20, 4. Arís eálá Drihten, fórestæpe oððe fórhrada hine _exsurge Dómine, præveni eum,_ 16, 13. Ða ðe fórestópon hine þreádon, ðæt he súwode _qui præibant, incrépabant eum, ut táceret,_ Lk. Bos. 18, 39. Fýr ætfóran him fórestæpþ _ignis ante ipsum præcédet,_ Ps. Lamb. 96, 3: Homl. Th. ii. 82, 18. Paulus fórestóp Stephanum _Paul preceded Stephen,_ Homl. Th. ii. 82, 22.

fóre-steall, es; _m._ [fóre _before,_ steall from stellan _to leap_] _A leaping before, forestalling, rescue;_ assultus, interceptio:—Ða Iudéiscan ealdras geornlíce smeádon hú hí Hǽlend Crist acwellan mihton, ondrédon him swá-ðeáh ðæs folces fóresteall _the Jewish elders earnestly deliberated how they might slay Jesus Christ, but they dreaded a rescue by the people,_ Homl. Th. ii. 242, 14. v. fóre-steall.

fóre-steóra, an; _m. A fore-steerer, man at the prow of a ship;_ próréta, próræ conductor, Cot. 149.

fóre-steppan, -stæppan, ic -steppe, -stæppe, he -stepþ, _pl._ -steppaþ, -stæppaþ; _p._ -stepede = -stepte? _pp._ -steped = -stept? _To step_ or _come before, to prevent, go before, precede;_ prægrédi, prævéníre, anticípáre, præcédére:—Mín God fórscýt [MS. forscytte] oððe fórestepþ me _Deus meus præveniet me,_ Ps. Lamb. 58, 11. Fórhradien oððe fóresteppen [MS. forhradian oððe foresteppan] us ðíne mildheortnessa _antícípent nos misericórdiæ tuæ,_ 78, 8. Ic fórestæppe _præcédo,_ Ælfc. Gr. 28, 4; Som. 31, 30. Mildheortnys and sóþfæstnys fóresteppaþ [Lamb. fórestæppaþ] ansýne ðíne _misericordia et véritas præcédent fáciem tuam,_ Ps. Spl. 88, 15.

fóre-stígan; _p._ -stág, -stáh, _pl._ -stigon; _pp._ -stigen _To go before, to excel;_ excellére:—Ic fórestíge _excelleo,_ Ælfc. Gr. 26, 2; Som. 28, 45, MS. C. DER. stígan.

fóre-stihtod, -stihtud; _part._ [stihtian _to dispose, order_] _Fore-appointed_ or _ordained, determined;_ prædestínátus, definítus:—Fórestihtod, fóre-stihtud _prædestínátus,_ Scint. de Prædest. Æfter ðam ðe fórestihtod wæs _secundum quod definítum est,_ Lk. Bos. 22, 22.

fóre-stihtung, e; _f. A fore-appointment;_ prædestínátio, dispensátio:—Mid fórestihtunge ðære godcundan árfæstnesse _by the dispensation of the divine mercy,_ Bd. 4, 29; S. 607, note 42: Homl. Th. ii. 364, 29.

fóre-stóp, ðú -stópe, _pl._ -stópon _stepped before, prevented, went before, preceded,_ Ps. Lamb. 20, 4: Lk. Bos. 18, 39: Homl. Th. ii. 82, 22; _p._ of fóre-stapan.

fóre-swerian; _p._ ic, he -swór, ðú -swóre, _pl._ -swóron; _pp._ -sworen _To_ FORESWEAR, _declare before;_ antejúráre:—Ðæt land, ðe ic fóreswór heora fæderum _terram, pro qua_ [ante-] _jurávi patríbus eórum,_ Num. 14, 23. Ðæt land, ðe ðú hira fæderum fóreswóre _terram, pro qua_ [ante-] _jurasti patríbus eórum,_ 11, 12.

fóre-tácen, -tácn, es; _n. A_ FORE-TOKEN, _presage, sign, wonder;_ præságium, prodígium:—Fóretácn ecra góda _a fore-token of eternal blessings,_ Bt. 40, 2; Fox 236, 21: Ps. Spl. 77, 48: 70, 8. Ðæt bíþ fóretácna mǽst _that shall be the greatest of fore-tokens,_ Exon. 21 a; Th. 55, 34; Cri. 893. He sette on him word tácna heora and fóretácna _pósuit in eis verba signórum suórum et prodigiórum,_ Ps. Spl. 104, 25. He sende fóretácna _emísit prodígia,_ 134, 9.

fóre-tácnian; _p._ ode; _pp._ od _To foreshow;_ præmonstráre:—Ðætte seó sáwl prówiende wæs, ðætte se líchoma fóretácnode _quod anima passa sit cáro præmonstrábat,_ Bd. 3, 19; S. 549, 17.

fóre-teohung, -teohhung, e; _f. Predestination;_ prædestínátio:—Sió godcunde fóreteohhung is ánfeald and unawendendlíc _the divine predestination is simple and unchangeable,_ Bt. 39, 6; Fox 220, 16. Be ðære fóreteohunga Godes _concerning the predestination of God,_ Bt. titl. xxxix; Fox xviii. 16. v. fóre-tiohung.

fóre-teón; _p._ -teóde; _pp._ -teód _To pre-dispose, pre-ordain;_ prædispónére, præordináre:—Swá ðé bearn weorþaþ geboren syððan, ða ylcan ic ǽr fóreteóde _ecce nátio filiórum tuórum quibus dispósui,_ Ps. Th. 72, 12. Swá monige swá fóreteóde wǽron to ǽcum lífe _quotquot erant præordináti ad vitam æternam,_ Bd. 2, 14; S. 517, 36.

fóre-teþ; _pl. m. The fore-teeth;_ præcisóres, Ælfc. Gl. 71; Som. 70, 101; Wrt. Voc. 43, 30. v. tóþ _a tooth._

fóre-þanc, es; _m. Forethought, consideration;_ consíderátio:—Bíþ andgit ǽghwǽr sélest, ferhþes fóreþanc _understanding is everywhere best, forethought of mind,_ Beo. Th. 2124; B. 1060. Náhton fóreþances wísdómes gewitt _they had no sense of wisdom's foresight,_ Elen. Kmbl. 712; El. 356. Ða hát-heortan hie mid náne fóreþance nyllaþ gestillan

the furious will not calm themselves with any consideration, Past. 40, 6; Cot. MS. v. fóre-þonc.

fóre-þanclíce; _adv. Considerately, prudently;_ consíderáte, províde, Past. 15, 5, Lye.

fóre-þancolnes, -ness, e; _f. Forethought, prudence;_ prúdentia:—Seó smeáung míṇre heortan wile sprecan fóreþancolnesse _meditátio cordis mei lóquetur prúdentiam,_ Ps. Th. 48, 3.

fóre-þancul, -þoncol, for-þoncol; _adj. Forethinking, provident, prudent;_ próvidus, prúdens:—Se fóreþancula wer _the provident man,_ Past. 41, 5; Hat. MS. 57 b, 16.

fore-þencan; _p._ -þohte, _pl._ -þohton; _pp._ -þoht [=for-þencan] _To distrust, despair;_ diffídére, desperáre:—Ðý-læs ˌhe hine for ðære wynsuman wyrde fortrúwige, oððe for ðære réðan foreþonce _lest he on account of pleasant fortune should be arrogant, or on account of the affliction should despair,_ Bt. 40, 3; Fox 238, 18.

fóre-þencan, -þencean; _p._ -þohte, _pl._ -þohton; _pp._ -þoht _To_ FORETHINK, _consider beforehand;_ præcógitáre, præmédítári:—Se láreów sceal mid geornfullíce ingehygde fóreþencean _the teacher must consider beforehand with careful meditation,_ Past. 15, 5; Hat. MS. 20 a, 1.

fóre-þingere, es; _m._ [fóre = for, þingere _a pleader_] _One who pleads for another, an intercessor;_ intercessor:—Sceolon ða æðelan Godes þeówas beón his folces foreþingeras _the noble servants of God should be the intercessors of his people,_ Homl. Th. ii. 224, 11.

fóre-þingian, for-þingian; _p._ ode; _pp._ od [fóre = for, þingian _to plead_] _To plead for anyone, intercede, defend;_ intercédére, defendére:—Ic secge ðæt sió foresprǽc ne dýge, náuðer ne ðam scyldigan, ne ðam ðe him foreþingaþ _I say that the defence does no good, neither to the guilty, nor to him who pleads for him,_ Bt. 38, 7; Fox 210, 7. Foreþinga for synnum mínum _intercéde pro peccátis meis,_ Wanl. Catal. 293, 28, col. 2: 294, 25, col. 1. Ne cweðe ic ná ðæt ðæt yfel síe ðæt mon helpe ðæs unscyldigan, and him foreþingie _I do not say that it is wrong that a man should help the innocent, and defend him,_ Bt. 38, 7; Fox 210, 4: L. Alf. pol. 21; Th. i. 76, 3: 24; Th. i. 78, 10.

fóre-þingiend, es; _m. One who pleads for another, an intercessor;_ intercessor:—Us Drihten sealde ðé foreþingiend _nobis Dóminus dédit te intercessórem,_ Wanl. Catal. 294, 34, col. 1.

fóre-þingrǽden, e; _f. A pleading for anyone, intercession;_ intercessio:—Þurh foreþingrǽdena háligra martíra ðínra _per intercessiones sanctórum martýrum tuórum,_ Wanl. Catal. 294, 16, col. 1. Ic gyrne fultum ðínre foreþingrǽdene ðú háligoste mægden and þrówystre _implóro auxílium tuæ interventiónis sanctíssima virgo et martyr,_ 294, 6, col. 2.

fóre-þingung, e; _f. A pleading for anyone, intercession;_ intercessio:—Se Hǽlend hét gehwilcne óðerne aþweán fram fúlum synnum mid fóre-þingunge _the Saviour commanded each to wash the other from foul sins by intercession,_ Homl. Th. ii. 242, 33. Þurh foreþingunga ealra háligra ðínra gehýr me _per intercessiones ómnium sanctórum tuórum exaudi me,_ Wanl. Catal. 294, 20, col. 2. Mid gódum foreþingungum _with good ntercessions,_ Bd. 4, 3; S. 568, 21, note, MS. Ca.; Rtl. 49, 34.

fóre-þonc, -þanc, es; _m. Fore-thought, providence;_ providentia:—Ananias, Azarias and Misahel þurh fóreþoncas fýr gebýgdon _Hananiah, Azariah and Mishael escaped the fire through providences,_ i. e. _through their trust in the provisions of God,_ Dei provísiónibus ignem superárunt, Exon. 55 b; Th. 197, 16; Az. 191: 118 a; Th. 454, 22; Hy. 4, 37. Se fóreþonc is sió godcunde gesceádwísnes, sió ðe eall fórewát _providence is the divine intelligence, which foreknows all,_ Bt. 39, 5; Fox 218, 26. Se godcunda fóreþonc heaðeraþ ealle gesceafta _the divine providence restrains all creatures,_ 39, 5; Fox 218, 30: 39, 5; Fox 220, 1, 2: 39, 6; Fox 220, 11. Be ðam godcundan fóreþonc stýreþ ðone ródor and ða tunglu _with respect to divine providence,—the divine providence rules the sky and the stars,_ Bt. 39, 8; Fox 224, 3–7.

fóre-þoncol; _adj. Sagacious, prudent;_ próvidus, prúdens:—Ðæt fóreþoncle men _quod sagacious men said,_ Exon. 25 a; Th. 73, 19; Cri. 1192. v. fóre-þancul.

fóre-þýstrian; _p._ ede; _pp._ ed _To darken;_ obscúráre:—He sende þýstru and foreþýstrede _misit tenebras et obscúrávit,_ Ps. Spl. 104, 26. v. for-þeóstrian.

fóre-tíge, es; _m._ [tíge from tígan _to bind_] _A fore-binding place, market;_ fórum:—Heó ys gelíc sittendum cnapum [MS. cnapun] on fóretíge símilis est pueris sedentíbus in fóro, Mt. Bos. 11, 16.

fóre-timbrigende; _part. Building before, shutting up;_ præclúdens, Bd. 5, 1; S. 613, 31, note.

fóre-tiohung, -tiohhung, -teohung, -teohhung, e; _f. A fore-appointing, predestination;_ prædestínátio:—Ær hit wæs Godes fóretiohung _before it was God's predestination,_ Bt. 39, 6; Fox 220, 11: 39, 4; Fox 216, 31. Be ðære Godes fóretiohunge _concerning the predestination of God,_ 40, 5; Fox 240, 13. Sió godcunde fóretiohhung _the divine predestination,_ 40, 6; Fox 242, 9. DER. tiohhian _to determine._

fóre-týnd; _part. p. Foreclosed;_ præclúsus:—Gemettan we us ǽghwanan gelíce storme fóresette and fóretýnde _invénimus nos pári tempestáte præclúsos,_ Bd. 5, 1; S. 613, 31.

fôre-wæs *was before* or *over*, Bd. 5, 18; S. 635, 35; *p. of* fôre-wesan.

fôre-ward, e; *f. An agreement, compact, treaty*; pactum, fœdus :—His brôðer griþ and fôrewarde eall æftercwæþ *his brother renounced all peace and agreement*, Chr. 1094; Erl. 229, 30, 31. Bûton he ða fôrewarda geheólde *unless he kept the agreements*, Erl. 229, 32 : Cod. Dipl. 732; A.D. 1016-1020; Kmbl. iv. 10, 16. v. fôre-weard, e; *f.*

fôre-ward; *adj. Forward, fore, former, early*; prônus, antêrior, prior :—On fôrewardre ðyssere bêc ys awriten be me *in the fore part of this book it is written of me*, Ps. Th. 39, 8. v. fôre-weard; *adj.*

fôre-warde, an; *f. An agreement*; pactum :—Seó fôrewarde ǽr wæs gewroht *the agreement was formerly made*, Chr. 1094; Erl. 229, 34. v. fôre-weard, e; *f.*

fôre-weall, es; *m. A fore-wall, bulwark*; propugnāculum :—Syndon ða fôrewealles gestêpte oþ wolcna hróf *the fore-walls are raised to the clouds' roof* [*the water-walls in the Red Sea*], Cd. 158; Th. 196, 25; Exod. 297.

fôre-weard, -ward, fôr-word, -werd, e; *f.*: fôre-warde, an; *f. A* FORE-WARD, *precaution, contract, agreement, compact, treaty, provision*; præcautio, pactum, fœdus :—Wurdon ða fôrewearda full worhte *the contracts were completed*, Chr. 1109; Erl. 242, 22. To ðān ylcan fôreweardum [MS. foreweardan] *with the same provisions*, Cod. Dipl. 731; A.D. 1013-1020; Kmbl. iv. 10, 6. Fôreweard *exordium*, Rtl. 69, 17. DER. weard, e; *f.* [*Dut.* voor-waarde, *f. condition, terms, pre-contract.*]

fôre-weard, es; *m. A forewarder, scout*; antecursor, explôrātor :—Siðdan Scipia geahsode ðæt ða fôreweardas wǽron feor ðam fæstenne gesette, he ða dýgellīce gelǽdde his fyrde betuh ðām weardum *when Scipio learned that the scouts [forewarders] were set far from the fastness, he then secretly led his army between the warders*, Ors. 4, 10; Bos. 95, 12. v. weard; *m.*

fôre-weard, fôr-weard, -werd, -ward; *adj.* FORWARD, *fore, former, early*; prônus, antêrior, prior :—Lǽteþ fôreweard hleór on strangne stān *he shall let his cheek [fall] forward on a strong stone*, Salm. Kmbl. 228; Sal. 113. In fôreweardum Danieles dagum *in the early days of Daniel*, Chr. 709; Erl. 42, 30. On fôreweard Eásterfæsten *in the fore [part of the] Easter-fast*; incipiente Quadragēsima, Bd. 5, 2; S. 614, 37. Fôreweard feng ðara [MS. ðære] lippena togædere *the fore-grasp of the lips together*; rostrum, Ælfc. Gl. 71; Som. 70, 95; Wrt. Voc. 43, 26. Fôreweard fôt *the fore [part of the] foot, the sole of the foot*; planta, Ælfc. Gl. 75; Som. 71, 95; Wrt. Voc. 45, 3. Ða sylfan tiid [= tīd] folc habbaþ fôreweard geár *at the same time people have the fore [part of the] year*, Menol. Fox 12; Men. 6. Fôrewearde heáfod *the forehead*; frons, Wrt. Voc. 70, 28. We sceolon mearcian úre fôreweardan heáfod mid Cristes rôde tācne *we should mark our foreheads with the sign of Christ's cross*, Homl. Th. ii. 266, 11. Fôreweard lencten *the early spring*; ver nôvum, Ælfc. Gl. 95; Som. 76, 12; Wrt. Voc. 53, 26. Hit wæs fôreweard middæg *it was the fore [part of] midday*; hōra secunda diei, Bd. 4, 32; S. 612, 5. Wæs fôreweard niht *it was the early [part of] night*; prīma hōra noctis, Bd. 2, 12; S. 513, 19. On fôrewearde niht *in the early [part of] night*; primo tempore noctis, Bd. 5, 13; Whelc. 412, 15. Fôreweard nôsu *the fore-nose, extremity of the nose*; pīrŭla [*q. v. in* Du Cange], Ælfc. Gl. 71; Som. 70, 90; Wrt. Voc. 43, 21. On ðæs cyninges rīce fôreweardum *in the fore [part of the] reign of the king*; cujus regni princĭpio, Bd. 5, 2; S. 614, 24: 5, 23; S. 646, 3. Be ðisses bisceopes lífes stealle fôreweardum *of the early state of this bishop's life*; de cujus pontifĭcis stātu vītæ ad priôra repĕdantes, Bd. 5, 19; S. 637, 2. Drihten ðē gesett on fôreweard and nâ on æftewweard *constĭtuet te Dŏmĭnus in căput et non in caudam*, Deut. 28, 13. Ðū gesetst me on heáfod oððe on fôrewearde þeóda *constĭtues me in căput gentium*, Ps. Lamb. 17, 44. [*Dut.* voor-waarts; *adv. forward.*]

fôre-werd; *adj. Forward, fore, former, early*; prônus, antêrior, prior, prīmus :—On fôrewerdne morgen ic drífe sceáp mīne to heora lease *in primo māne mīno ŏves meas ad pascua*, Coll. Monast. Th. 20, 11. Fôrewerd swira *căpĭtium*, Wrt. Voc. 282, 42. Fôrewerd nāsù *pīrŭla*, 282, 65. On fôrewerd ðære bôc oððe on heáfde bêc awriten is be me *in căpĭte libri scriptum est de me*, Ps. Lamb. 39, 9. v. fôre-weard; *adj.*

fôre-wesan; *p.* ic, he -wæs, ðú -wære, *pl.* -wǽron [fôre *before*, wesan *to be*] *To be before, to preside*; præesse :—Ðyssum tídum fôrewæs Norþan Hymbra rīce se strangesta cyning *his tempŏribus regno Nordanhymbrôrum præfuit rex fortissĭmus*, Bd. 1, 34; S. 499, 18 : 5, 18; S. 635, 35. v. wesan *to be*.

fôre-wís; *adj. Forewise, foreknowing*; præscius, Cot. 149.

fôre-witan, fôr-witan; ic, he -wât, ðū -wâst, *pl.* -witon; *p.* -wiste, *pl.* -wiston; *pp.* -witen *To foreknow*; præscire :—He eall fôrewāt hū hit geweorþan sceal *he foreknows all how it shall come to pass*, Bt. 39, 5; Fox 218, 27.

fôre-wítigian; *p.* ode, ade; *pp.* od, ad *To foresay, prophesy*; prænuntiāre :—Se mycla hunger, ðe wæs fôrewítegad on Actibus Apostŏlôrum *the great famine, which was foretold in the Acts of the Apostles*, Chr. 47; Erl. 7, 24.

fôre-wítig-wittig; *adj. Foreknowing*; præscius :—Fôrewitig towerdra þinga *præscius fŭtūri*, Ælfc. Gr. 41; Som. 44, 12; Hpt. Gl.

fôre-witol; *adj.* [witol *knowing*] *Foreknowing*; præscius, Lye.

fôre-witung, e; *f. A foreknowing, foretelling, presage*; præsāgium, Som. Ben. Lye.; Hpt. Gl.

fore-wrêgan; *p.* de; *pp.* ed *To accuse strongly*; valde accūsāre :—He būtan leahtrum wæs clǽne gemêted ðara þinga ðe hine mon forewrêgde *he was found without crimes clean of the things of which he was accused*; absque crīmine accūsātus fuisse inventus est, Bd. 5, 19; S. 639, 30.

fore-wrítan; *p.* -wrât, *pl.* -writon; *pp.* -writen *To proscribe, banish*; proscrībĕre, Som. Ben. Lye.

fore-writennes, -ness, e; *f. Proscription, banishment, exile*; proscriptio, Som. Ben. Lye.

fore-wyrcan; *p.* -worhte; *pp.* -worht *To work for, do anything for anyone*; făcĕre alĭquid pro alĭquo :—Se man ðane ôðerne æt rihte gebrenge, oððe riht forewyrce *let the man bring the other to justice, or do justice for him*, L. H. E. 15; Th. i. 34, 2.

fôre-wyrd, e; *f.* [fôre, wyrd *an event*] *A deed done before*; antefactum, Som. Ben. Lye.

for-fang, -feng, fore-feng, -fong, es; *m.* **I.** *a seizing* or *rescuing of stolen* or *lost property*; apprehensio :—Be forstolenes mannes forfenge *of seizing a stolen man*, L. In. 53; Th. i. 134, 15, note 32. Be forstolenes ceápes forfenge *of the rescuing of stolen property*, 75; Th. i. 150, 4, note 7. **II.** *the reward for rescuing such property*; merces, quæ bŏnōrum surreptōrum restĭtūtōri dătur :—Forfang ofer eall fíftyne peninga *the reward for rescuing stolen property shall be everywhere fifteen pence*, L. Ff; Th. i. 224, 21. Embe forfang, witan habbaþ gerǽdd, ðæt man ofer eall Engle-land gelícne dôm healde; ðæt is æt men fiftene peninga, and æt horse eal [MS. heal] swā ... Hwīlon stôd, ðæt man æt ǽlcon þeófstolenan orfe ... and be his forfange sylle, ðæt is, æt ǽlcon scill. penig, sý ðæs cynnes orf ðe hit sý, gyf hit man æt þeófes handa ahret; gyf hit ðonne elles on hýdelse funden sý, ðonne mæg ðæt forfangfeoh leóhtre beón *concerning the reward for rescuing stolen property, the counsellors have determined, that one shall hold like judgment all over England; that is for a man fifteen pence, and for a horse as much ... Formerly it stood, that for all stolen cattle ... and on its rescue one should pay, that is, for every shilling a penny, be the cattle of whatever kind it may, if one rescues it from the hands of the thief; but if otherwise it be found in a hiding-place, then the reward for rescuing may be less*, Th. i. 224, 24-226, 5.

for-fangen *forfeited*, L. Alf. pol. 2; Th. i. 62, note 9; *Seized*, Cd. 205; Th. 254, 19; Dan. 614; *pp. of* for-fôn.

forfang-feoh; *gen.* -feós; *n. The reward for rescuing stolen cattle* or *lost property*; merces, quæ bŏnōrum surreptōrum restĭtūtōri dătur :—Gyf hit ðonne elles on hýdelse funden sý, ðonne mæg ðæt forfangfeoh leóhtre beón *if otherwise it be found in a hiding-place, the reward for rescuing it may be less*, L. Ff; Th. i. 226, 5.

for-faran; *p.* -fôr, *pl.* -fôron; *pp.* -faren [for-, faran *to go*]. **I.** *to go* or *pass away, perish*; perīre :—Seó scipfyrd [MS. scipfyrde] ælmǽst earmlíce forfôr *almost all the ship-force perished miserably*, Chr. 1091; Erl. 227, 35. Hī mǽst ealle forfôron *they almost all perished*, 910; Erl. 101, 8, 33 : 1096; Erl. 233, 22. **II.** *to cause to pass away, cause to perish, to destroy*; perdĕre :—Forfare hý man mid ealle *let a man totally destroy them*, L. E. G. 11; Th. i. 174, 2 : L. C. S. 4; Th. i. 378, 9. Ðæt man ða sāwla ne forfare ðe Crist mid his āgenum lífe gebohte *that a man cause not the souls to perish which Christ bought with his own life*, L. C. S. 3; Th. i. 378, 2. Wæs swíðe feala manna forfaren *very many men were destroyed*, Chr. 1025; Erl. 163, 10. Mycel orfes wæs ðæs geáres forfaren *much cattle was destroyed this year*, 1041; Erl. 169, 8. Wearþ micel his heres forfaren *many of his army were destroyed*, 1067; Erl. 204, 9. Fordoes ł forfæras *perdiderit*, Mt. Kmb. Lind. 10, 39.

fôr-faran; *p.* -fôr, *pl.* -fôron; *pp.* -faren [fôr *before*, faran *to go*] *To go before, get in front of*; præīre :—Fôrfôron him ðone mūþan fôran on úter mere *they got in front of them before the mouth [of the river] in the outer sea*, Chr. 897; Erl. 95, 21. [*O. Sax.* furfaran *to precede*.]

for-fêhþ *surprises*, Exon. 20 b; Th. 54, 25; Cri. 874; *3rd sing. pres. of* for-fôn.

for-feng *a seizing of stolen property*, L. In. 75; Th. i. 150, 4, note 7, MS. H. v. for-fang.

for-feran; *p.* de; *pp.* ed [for-, feran *to go*] *To go* or *pass away, perish*; perīre :—Fôrneáh ælc tilþ on mersclande forferde *very nearly all the tilth in the marsh-land perished*, Chr. 1098; Erl. 235, 13.

for-fleón; *p.* -fleáh, *pl.* -flugon; *subj. pres.* -fleó, *pl.* -fleón; *pp.* -flogen [for-, fleón *to flee*] *To flee away from, escape*; fŭgĕre, effugĕre :—Ic forfleó mīne hlǽfdian *a fácie dŏmĭnæ meæ ĕgo fŭgio*, Gen. 16, 8. Ðæt gē ðās towerdan þing forfleón *that ye escape those future things*, Lk. Bos. 21, 36.

for-fôn; ic -fô, ðú -fêhst, he -fêhþ, *pl.* -fôþ; *p.* ic, he -fêng, ðū -fênge, *pl.* -fêngon; *pp.* -fangen, -fongen [for-, fôn *to take*]. **I.** *to be deprived of anything, forfeit*; ǎlĭquo prīvāri, amittĕre :—Næbbe his āgne forfongen [hæbbe his āgen forfangen MS. H.] *let him not have forfeited his own [let him have forfeited his own*, MS. H.], L. Alf. pol. 2; Th. i. 62, 6. **II.** *to take violently* or *by surprise, clutch, arrest, seize*;

vehementer căpĕre, imprŏvīso adventu căpĕre, prehendĕre, apprehendĕre, deprehendĕre:—Swā þeóf sorgleáse hæleþ semninga forfêhþ slǽpe gebundne *as a thief suddenly surprises careless mortals bound in sleep*, Exon. 20 b; Th. 54, 25; Cri. 874. Ealle deáþrǽs forfêng *the death-rush clutched them all*, Andr. Kmbl. 1990; An. 997. Ǽr đū đa miclan meaht mín forfênge *ere thou didst arrest my great power*, Exon. 73 a; Th. 273, 26; Jul. 522. Forfóh đone frǽtgan, and fæste geheald *seize the proud one [the devil], and firmly hold [him]*, Exon. 69 b; Th. 259, 18; Jul. 284. For đam gylpe gumena drihten forfangen wearþ, and on fleám gewāt *for that boast the lord of men [Nebuchadnezzar] was seized [with madness], and in flight departed*, Cd. 205; Th. 254, 19; Dan. 614. [*O. Sax.* farfahan : *Ger.* verfangen.]

for-fôr, *pl.* -fôron *passed away, perished* :—Seó scipfyrd earmlîce forfôr *the ship-force miserably perished*, Chr. 1091; Erl. 227, 35 : 910; Erl. 101, 8 ; *p.* of for-faran.

fôr-fôr, *pl.* -fôron *went before, got in front of* :—Fôrfôron *went before*, Chr. 897; Erl. 95, 21 ; *p.* of fôr-faran.

for-fylden [fylden = fealden, *pp.* of fealdan *to fold up*] *Filled up, stopped, opposed*; obstructus, Cot. 148.

for-gǽgan; *p.* de; *pp.* ed *To transgress, prevaricate*; transgrĕdi, praetĕrîre, praevāricāre :—Ic forgǽge praetĕreo, Ælfc. Gr. 30, 5; Som. 35, 2. Hí Godes bebod forgǽgdon *they transgressed God's command*, Homl. Th. i. 112, 14. Đæt he Godes beboda ne forgǽge *that he transgress not God's commandments*, i. 604, 20. Ic geseah ǽslîtendras ođđe đa forgǽgendan *vidi praevāricantes*, Ps. Lamb. 118, 158.

for-gǽgednys, -nyss, e; *f. A transgression, prevarication, stubbornness*; transgressio, praevāricātio, perversitas :—Hí wǽron deádlîce for đære forgǽgednysse *they became mortal through the transgression*, Boutr. Scrd. 20, 29. Cain wiste his fæder forgǽgednysse *Cain knew his father's transgression*, 20, 40. Þurh Adames forgǽgednysse *through Adam's transgression*, Homl. Th. ii. 268, 31. We sceolon úre forgǽgednysse geandettan *we ought to confess our transgressions*, ii. 98, 25. Đæt gĕ ne beón scildige scamlîcre forgǽgednysse *ne sîtis praevāricātiōnis rei*, Jos. 6, 18.

for-gæt, *pl.* -gǽton *forgot*, Ps. Lamb. 77, 11 : 118, 61, = for-geat, *pl.* -geáton ; *p.* of for-gitan.

for-gán, to -gánne ; he -gǽþ ; *p.* -eóde, *pl.* -eódon ; *pp.* -gán *To* FOR-GO, *abstain from, pass over, neglect*; abstînêre, transcendĕre, praetĕrîre :—Đæt he smeáge hwæt him sý to dônne and to forgánne *that he meditate what is for him to do and what to forgo*, L. C. S. 85; Th. i. 424, 6. We lǽraþ, đæt man freólsdagum and fæstendagum forgá áþas and ordéla *we enjoin, that a man on feast-days and fast-days forgo oaths and ordeals*, L. Edg. C. 24; Th. ii. 248, 28 : 25; Th. ii. 250, 1. He forgǽþ đæs húses dúru *transcendet ostium dŏmus*, Ex. 12, 23. Se đe đis forgǽþ [MS. forgæiþ], his sáwul losaþ *he who neglects this, his soul shall perish*, Homl. Th. i. 92, 2 : pricle ne forgǽs *iota non praeteribit*, Mt. Kmbl. Lind. 5, 18.

fôr-gán, fôre-gán ; he -gǽþ ; *p.* -eóde, *pl.* -eódon ; *pp.* -gán *To go before, precede, stand out, project*; praecêdĕre, prôdîre :—Fôrgǽþ swá swá of fætnysse unrihtwîsnys heora *prôdit quasi ex adîpe inîquîtas eôrum*, Ps. Spl. 72, 7. [*Dut.* voor-gaan : *Ger.* vor-gehen *to go before.*]

fôr-gangan, fôre-gangan ; *p.* -geóng, -gêng, *pl.* -geóngon, -gêngon ; *pp.* -gangen *To go before, precede* ; praeire, praecêdĕre :—Mildheortnes and sôþfæstnes fôrgangaþ đînne andwlitan *misericordia et vēritas praeibunt ante făciem tuam*, Ps. Th. 88, 13. v. fôr-gán.

for-geaf, đū -geáfe, *pl.* -geáfon *forgave, gave, gavest*, Cd. 30; Th. 40, 20; Gen. 642 : Gen. 3, 12 ; *p.* of for-gifan.

for-geald *paid for, repaid*, Job Thw. 168, 17 ; *p.* of for-gildan.

for-geat, đū -geáte, *pl.* -geáton *forgot, hast forgotten*, Gen. 24, 67 : Ps. Lamb. 41, 10 : Jud. 3, 7 ; *p.* of for-gitan : for-geáte *should forget*, Ors. 6, 3 ; Bos. 118, 4 ; *p. subj.* of for-gitan.

for-gedón ; *p.* -gedyde, *pl.* -gedydon ; *pp.* -gedón *To do for, destroy* ; perdĕre :—Ǽr Rômaburh abrocen wǽre and forgedón *ere the city Rome was broken into and done for*, Bd. 1, 11 ; S. 480, 10, note. v. for-dón.

for-gef = for-geaf, *the perf. also for* for-gif, *the impert. of* for-gifan *to give, forgive*, Andr. Kmbl. 971; An. 486 : Ps. C. 50, 45 ; Ps. Grn. ii. 277, 45 : 50, 63 ; Ps. Grn. ii. 278, 63 : 50, 139 ; Ps. Grn. ii. 280, 139 : 50, 154 ; Ps. Grn. ii. 280, 154.

for-gefenes, -ness, e; *f. Forgiveness*, Ps. C. 50, 37 ; Ps. Grn. ii. 277, 37. v. for-gifnes.

for-geldan *to pay for, repay, return, give, render* ; reddĕre, retribuĕre :—Ic forgelde heom *retribuam eis*, Ps. Lamb. 40, 11. Twentig scillinga forgelde *let him pay twenty shillings*, L. Ethb. 22 ; Th. i. 8, 6 : 7; Th. i. 4, 9 : 12 ; Th. i. 6, 8 : 26 ; Th. i. 8, 12, 13 : 32 ; Th. i. 12, 2. Hine man forgelde *let a man pay for him*, L. H. E. 4 ; Th. i. 28, 7 : 11 ; Th. i. 32, 7. Đa mágas healfne leód forgelden *let his kindred pay half the fine [for slaying a man]*, L. Ethb. 23 ; Th. i. 8, 8. v. for-gildan.

for-géman *to neglect*, Prov. 19. v. for-gýman.

for-gémeleásian ; *p.* ode; *pp.* od *To neglect* ; neglîgĕre :—Swylc geréfa swylc đis forgémeleásige *quîlibet praefectus qui hoc neglîgit*, L. Ath. iv. 1; Wilk. 62, 38. v. for-gýmeleásian.

fôr-gesettenys, -nyss, e; *f. A proposition* ; propŏsitio :—Ic atýne on

saltere fôrgesettenysse mîne *ăpĕriam in psaltĕrio propŏsitiōnem meam*, Ps. Spl. 48, 4. v. fôre-setnes.

for-get *forgets*, Bt. 3, 2 ; Fox 6, 9, = for-git, -giteþ ; *3rd pres. sing. of* for-gitan : for-getst *forgettest*, Ps. Lamb. 43, 24, = for-gitst ; *2nd pres. sing. of* for-gitan.

for-geton *forgot*, Deut. 32, 18 : Mt. Bos. 16, 5, = for-geáton ; *p. pl.* of for-gitan.

for-giefan, *pp.* -giefen *To give, forgive, bestow, give up* ; đáre, dêdĕre, remittĕre, dimittĕre, Exon. 93 a ; Th. 348, 25 ; Sch. 33 : 28 a ; Th. 85, 33 ; Cri. 1400 : 49 a ; Th. 170, 4 ; Gú. 1106 : 39 a ; Th. 130, 2 ; Gú. 432. v. for-gifan.

for-gieldan *to pay for, repay, requite* ; reddĕre :—Đæt he hine scolde forgieldan *that he should pay for it*, Past. 63 ; Hat. MS. We đe nú willaþ womma gehwylces leán forgieldan *we will now pay thee retribution for every crime*, Exon. 41 a ; Th. 137, 16 ; Gú. 560 : 117 a ; Th. 450, 1 ; Dôm. 81. Forgield me đín líf *give me thy life*, 29 b ; Th. 90, 20; Cri. 1477. Forgielde he hine *let him pay for him*, L. In. 35, 36 ; Th. i. 124, 9, 18 : 9 ; Th. i. 108, 5 : 11 ; Th. i. 110, 4 : 31 ; Th. i. 122, 6. v. for-gildan.

for-giemeleásian; *p.* ode; *pp.* od *To neglect* ; neglîgĕre :—Gif hwá adulfe pytt, and forgiémeleásode đæt he hine betýnde *if anyone dug a pit, and neglected to inclose it*, Past. 63 ; Hat. MS. v. for-gýmeleásian.

for-gietan *to forget* ; oblīvisci :—Hý sceolon forgietan đære gesceafte *they shall forget the world*, Exon. 92 a ; Th. 345, 4 ; Gn. Ex. 183. v. for-gitan.

for-gifan, -gyfan, -giefan ; *p.* ic, he -geaf, đū -geáfe, *pl.* -geáfon ; *pp.* -gifen. **I.** *to give, grant, supply, permit, give up, leave off* ; đáre, dônáre, praebĕre, indulgĕre, dêdĕre, relinquĕre :—Đæt wíf đæt đū me forgeáfe *mûlier, quam dêdisti mihi*, Gen. 3, 12. Manegum blindum he gesihþe forgeaf *caecis multis dônāvit visum*, Lk. Bos. 7, 21. He forgeaf wíd-brádne wêlan *he gave wide-spread bliss*, Cd. 30; Th. 40, 20; Gen. 642. Siđđan đis gedón wæs, gesceóp God Adam, and him sáwle forgeaf *after this was done, God created Adam, and gave him a soul*, Ælfc. T. 4, 25-5, 1. Đisum men ic forgife hors *huic hŏmĭni do ĕquum*, Ælfc. Gr. 7; Som. 6, 21. Ne biþ đæt forgifen đætte alýfed biþ *non indulgêtur quod licet*, Bd. 1, 27 ; S. 496, 1. He him his bearn forgeaf *he gave up his child to him*, Cd. 141; Th. 177, 4 ; Gen. 2924. Hlyst ýst forgeaf *the storm left off being heard [hearing]*, Andr. Kmbl. 3171; An. 1588. **II.** *to* FORGIVE, *remit* ; remittĕre, dimittĕre, condônāre :—Eádige beóþ đa, đe him beóþ heora unrihtwísnesse forgifene *beāti, quorum remissae sunt inīquĭtātes*, Ps. Th. 31, 1. Forgifaþ, gif gĕ hwæt agén ǽnigne habbaþ *dimittĭte, si quid hăbĕtis adversus ăliquem*, Mk. Bos. 11, 25. Fæder, forgif him *Păter, dimitte illis*, Lk. Bos. 23, 34. He forgifþ hit *he will forgive it*, Cd. 30; Th. 41, 25 ; Gen. 662. [*Dut.* ver-geven : *Ger.* ver-geben *to forgive, pardon.*]

for-gifenlîc, -gifendlîc, -gyfendlîc, -gyfenlîc ; *comp. m.* ra ; *f. n.* re ; *sup.* ost ; *adj.* **I.** *giving, dative, or giving [case]* ; dātīvus :—Dātīvus is forgifendlîc *dative is giving* : Mid đam casu biþ geswutelod ǽlces þinges gifu *the gift of everything is declared by this case*. Đisum menn ic forgife hors *huic hŏmĭni do ĕquum*, Ælfc. Gr. 7; Som. 6, 19. **II.** *forgiving, pardonable, bearable* ; remissus, tolerābilis :—Ic eów secge, đæt Sodom-warum, on đam dæge, biþ forgifenlîcre đonne đære ceastre *dîco vôbis, quia Sŏdŏmis, in die illa, remissius êrit quam illi cîvîtāti*, Lk. Bos. 10, 12.

for-gifnes, -gyfnes, -ness, -nyss, -gifenes, -gyfenes, -gyfennes, -gifeniss, -gifenys, -gefenes, -ness, e; *f.* FORGIVENESS, *remission, indulgence, permission* ; remissio, vĕnia, indulgentia :—Sý on đære bôte forgifnes [forgyfnes MS. A.] *let there be a remission in the compensation*, L. Edg. ii. 1 ; Th. i. 266, 5 : L. Edg. S. 1 ; Th. i. 272, 9 : 9 ; Th. i. 276, 3. Dô him his synna forgifenisse *grant him forgiveness of his sins*, Chr. 1086 ; Erl. 222, 39. Đæt he đa gýmeleáste to forgyfenesse [forgyfnysse MS. F.] lǽte *that he grant forgiveness of the neglect*, L. Edg. S. 1 ; Th. i. 270, 17. His forgifnesse gumum to helpe dǽleþ dôgra gehwam Dryhten weoroda *the Lord of hosts dealeth his forgiveness each day for help to men*, Exon. 14 a ; Th. 27, 7 ; Cri. 427. Se nǽfþ on êcnysse forgyfenesse *non hăbêbit remissiōnem in æternum*, Mk. Bos. 3, 29. On hyra synna forgyfenesse *in remissiōnem peccātôrum eôrum*, Lk. Bos. 1, 77 : 3, 3. On synna forgyfennesse *in remissiōnem peccātôrum*, Mt. Bos. 26, 28. Đæt fiftigođe gêr biþ hálig, and forgifenisse gêr *sanctîfîcābis annum quinquāgēsĭmum, et vŏcābis remissiōnem*, Lev. 25, 10. Mín unrihtwísnysse is mǽre đonne ic forgifenysse wyrđe sý *mâjor est iniqûîtas mea, quam ut vĕniam mĕrear*, Gen. 4, 13. Đis ic cwêđe æfter forgifenysse nalæs æfter bebode *hoc autem dîco sĕcundum indulgentiam, non sĕcundum impĕrium*, Bd. 1, 27 ; S. 495, 45. To forgefenesse gáste mínum *for forgiveness to my soul*, Ps. C. 50, 37 ; Ps. Grn. ii. 277, 37. [*Dut.* ver-giffenis, *f. pardon, forgiveness.*]

for-gifung, e; *f. A giving, gift, donation* ; dônātio :—Forgifung dônātio, Ælfc. Gl. 13 ; Som. 57, 115 ; Wrt. Voc. 20, 52.

for-gildan, -gyldan, -gieldan, -geldan ; he -gildeþ, -gilt ; *p.* ic, he -geald, đū -gulde, *pl.* -guldon ; *subj. pres.* -gilde, *pl.* -gilden ; *p.* -gulde, *pl.* -gulden ; *pp.* -golden *To pay for, make good, repay, requite, recompense,*

reward; reddĕre, exsolvĕre, compensāre, retrĭbuĕre :—Him wile éce Ælmihtig forgildan *the eternal Almighty will repay them,* Exon. 62 b; Th. 230, 17; Ph. 473. He him dǽre lisse leán forgildeþ *he will pay him a reward for that affection,* Exon. 14 a; Th. 27, 22; Cri. 434. Eall he hit forgilt *he will recompense it all,* Bt. 42; Fox 258, 28. Swá hwæt swá man dǽr of forstæl, ic hit forgeald *whatsoever has been stolen therefrom, I have repaid it;* quidquid furto pĕrībant, a me exigēbas, Gen. 31, 39 : Job Thw. 168, 17 : Beo. Th. 3087; B. 1541 : 5929; B. 2968 : Cd. 158; Th. 197, 31; Exod. 315 : 226; Th. 301, 8; Sat. 578. Ða forguldon yfele for gódum *retrĭbuērunt mǎla pro bŏnis,* Ps. Spl. 37, 21 : Chr. 1039; Erl. 167, 20. Forgilde hine be his were *let him pay for him according to his value,* L. In. 11; Th. i. 110, 4, note 14, MS. H: 9; Th. i. 108, 5, note 14, MS. H: L. Ath. i. 3; Th. i. 200, 1, 15: L. Edg. ii. 4; Th. i. 266, 25; Andr. Kmbl. 774; An. 387. Forgildan hý hine be his were *let them pay for him according to his value,* L. Ath. i. 1; Th. i. 198, 24. Ðæt hine man forgulde *that a man should pay for him.* L. Ath. v. § 6, 3; Th. i. 234, 11 : Ps. Th. 65, 13. Gif ðú gód dést, hit biþ ðé mid góde forgolden; gif ðú ðonne yfel dést, hit biþ ðé mid yfele forgolden *if thou doest good, it shall be repaid thee with good; but if thou doest evil, it shall be repaid thee with evil,* Gen. 4, 7: Cd. 35; Th. 47, 6; Gen. 756 : Beo. Th. 5679; B. 2843: Judth. 11; Thw. 24, 31; Jud. 217 : Menol. Fox 302; Men. 152. Him wǽron eft forgoldene feówertyne þúsend sceápa *fourteen thousand sheep were repaid him,* Job Thw. 168, 19. [*Dut.* ver-gelden: *Ger.* ver-gelten *to reward, recompense.*]

for-gíman *to neglect,* Ex. 9, 21. v. for-gýman.

for-gímeleásian; *p.* ode; *pp.* od *To neglect entirely;* omnīno neglĭgĕre, neglĭgĕre :—Gif gé forgímeleásiaþ Drihtnes· bebod eówres Godes *if ye neglect the command of the Lord your God,* Deut. 8, 19. v. for-gýmeleásian.

for-gitan, -gytan, -gietan; ic -gite, ðú -gitest, -gitst, he -giteþ, -gitt, -git, *pl.* -gitaþ; *p.* ic, he -geat, -gæt, ðú -geáte, *pl.* -geáton, -gǽton, -géton, *impert.* -git, *pl.* -gitaþ; *subj. pres.* -gite, *pl.* -giton; *p.* -geáte, *pl.* -geáten; *pp.* -giten; *v. trans. gen. acc.* [for-, gitan *to get*] *To* FORGET, *neglect;* oblivisci, neglĭgĕre :—Hú lange wilt ðú, Drihten, mín forgitan *quousque, Dŏmĭne, oblivīscĕris me?* Ps. Th. 12, 1 : 118, 109. Ic forgite *obliviscor,* Ælfc. Gr. 29; Som. 33, 54. Ic forgite [MS. forgeite] *negligo,* 28, 5; Som. 31, 50. Hú lange, eálá Drihten, forgitst ðú me *usquequo, Dŏmĭne, oblivīscĕris me?* Ps. Lamb. 12, 1 : Ps. Th. 41, 11. Ðæt man forgitt ða ǽrran geár *that the former years shall be forgotten,* Gen. 41, 30. Ne he ne forgit his wedd *neque oblivīscētur pacti,* Deut. 4, 31 : Ps. Th. 9, 32 : Bt. Met. Fox 3, 11; Met. 3, 6. Sýn gecyrrede to helle ealle þeóda ða ðe forgitaþ God *convertantur in infernum omnes gentes qui oblivīscuntur Deum,* Ps. Lamb. 9, 18. Ic forgeat to etanne mínne hláf *oblitus sum comĕdĕre pǎnem meum,* 101, 5 : 118, 153, 176. Ǽ díne ic ne forgæt *lēgem tuam non sum oblītus,* Ps. Lamb. 118, 61, 109, 141. For hwí forgeáte ðú mín *quāre oblītus es mei?* 41, 10. Nǽfre náuht he ne forgeat *he has never forgotten anything,* Bt. 42; Fox 258, 1 : Bd. 3, 2; S. 525, 13 : Gen. 24, 67 : Ps. Spl. 9, 12. Ne we ne forgeáton ðé *nec oblīti sŭmus te,* Ps. Lamb. 43, 18. Gé forgéton Drihten *oblītus es Dŏmĭni,* Deut. 32, 18. Hig his hálgan ǽ forgeáton *they forgot his holy law,* Jud. 3, 7 : Ps. Lamb. 105, 21 : 118, 139 : Cd. 227; Th. 305, 6; Sat. 642. Hig forgéton his welldǽda *obliti sunt benefactōrum,* Ps. Lamb. 77, 11. Hig forgéton ðæt hig hláfas námon *obliti sunt pānes accĭpĕre,* Mt. Bos. 16, 5 : Cd. 149; Th. 186, 25; Exod. 144. Ne forgit ðú þearfena *ne oblivīscāris paupĕrum,* Ps. Lamb. second 9, 12 : 44, 11 : Ps. Th. 73, 18, 22. Gemunaþ and ne forgitaþ, hú swíðe gé gremedon Drihten *mĕmento et non oblivīscāris, quōmŏdo ad īrācundiam provŏcāvĕris Dŏmĭnum,* Deut. 9, 7. Oþ-ðæt he forgite ða þing, ðe ðú him dydest *dōnec oblivīscātur eōrum, quæ fēcisti in eum,* Gen. 27, 45. ᐧ Ðæt gé nǽfre ne forgiton Drihtnes wedd *ne quando oblivīscāris pacti Dŏmĭni,* Deut. 4, 23 : 6, 12. Ðæt he hí de-læs forgeáte *that he should the less forget them,* Ors. 6, 3; Bos. 118, 4 : Cd. 40; Th. 52, 25; Gen. 849. Ðe ðú forgiten hafst *which thou hast forgotten,* Bt. 36, 2; Fox 174, 22 : Ps. Lamb. second 9, 11: Ps. Th. 77, 13. Manige licggaþ deáde, mid ealle forgitene *many lie dead, entirely forgotten,* Bt. 19; Fox 70, 13 : Bt. Met. Fox 10, 120; Met. 10, 60. Án ðé is forgeten *unum tibi deest,* Mk. Skt. Lind. 10, 21. [*Dut.* ver-geten: *Ger.* ver-gessen *to forget.*]

for-gitel *forgetful, forgetting.* v. for-gytel.

for-gitelnes, -ness, e; *f. Forgetfulness, a forgetting;* oblivio :—Ne forgitelnes byþ ðæs þearfan *non oblivio ĕrit paupĕris,* Ps. Lamb. 9, 19. v. for-gytelnes.

for-gitennes, -ness, e; *f. Forgetfulness, oblivion;* oblivio, Som. Ben. Lye.

for-glendrad; *part. p. Conglūtĭnātus, allectus* :—Gebíged oðde forglendrad oðde gelímod is to eorþan wambe úre *conglūtĭnātus est in terra venter noster,* Ps. Lamb. 43, 25.

for-glendran; *p.* ade, ede; *pp.* ad, ed [glendran *to devour*] *To eat greedily, devour voraciously;* lurcāri, devŏrāre :—Forglendrad *lurcātus,* Cot. 124. Ealle heora snytru beóþ yfele forglendred *omnis sǎpientia eōrum devŏrāta est,* Ps. Th. 106, 26 : Blickl. Hom. 99, 9. Forglendred

serviunculus? Wrt. Voc. 290, 49. Forglendrad *conglūtĭnātus?* = glūtītus devoured, vel glūtĭnātus *glued together,* Ps. Lamb. 43, 25.

for-gnád *rubbed together, broke,* Ps. Lamb. 104, 16; *p.* of for-gnídan.

for-gnagan; *p.* -gnóg, *pl.* -gnógon; *pp.* -gnagen [for-, gnagan *to gnaw*] *To gnaw* or *eat up;* corrōdĕre, comĕdĕre :—On eallum grówendum þingon hig forgnagaþ *omnia quæ nascuntur corrōdent, sive comĕdent,* Ex. 10, 5. Gærstapan forgnógon swá hwæt swá se hagol belǽfde *locusts gnawed up whatsoever the hail had left,* Homl. Th. ii. 194, 1.

for-gnídan, -gnýdan, -cnídan; he -gnít; *p.* ic, he -gnád, ðú -gnide, *pl.* -gnidon; ᐧ*pp.* -gniden [for-, gnídan *to rub*] *To rub together, dash* or *throw down, break;* contĕrĕre, allĭdĕre, elĭdĕre :—He forgnád oððe he tobrytte treów gemǽru heora *contrivit lignum fīnium eōrum,* Ps. Lamb. 104, 33, 16 : Ps. Spl. 106, 16. Grin forgniden is, and we alýsde synd *lǎqueus contrītus est, et nos lībĕrāti sŭmus,* Ps. Spl. 123, 7. Heorte forgnidene God nã beheóld *cor contrītum Deus non despĭcies,* Ps. Spl. 50, 18. He forgnít hine *allīdit illum,* Mk. Bos. 9, 18. Forðon ðú forgníde me *quia allīsisti me,* Ps. Spl. 101, 11. Drihten arǽreþ ealle forgnidene *Dŏmĭnus erĭgit omnes elīsos,* Ps. Spl. 144, 15.

for-gnidennys, -nyss, e; *f. Contrition, sorrow;* contrītio :—Tobrytednys oððe forgnidennys and ungesǽliguys [syndon] on wegum heora *contrītio et infēlicĭtas [sunt] in viis eōrum,* Ps. Lamb. 13, 3.

for-gnísednys, -nyss, e; *f. Bruisedness, sorrow, contrition;* contrītio, Som. Ben. Lye.

for-gnóg, *pl.* -gnógon *gnawed up,* Homl. Th. ii. 194, 1; *p.* of forgnagan.

for-gnýdan; *pp.* -gnyden *To dash* or *throw down;* elĭdĕre :—On eorþan forgnyden, fǽmende he tearflode *elīsus in terram, vŏlūtābātur spūmans,* Mk. Bos. 9, 20. v. for-gnídan.

for-golden *paid for, repaid,* Judth. 11; Thw. 24, 31; Jud. 217; *pp.* of for-gildan.

for-grand *crushed,* Beo. Th. 852; B. 424; *p.* of for-grindan.

for-gráp *grasped,* Beo. Th. 4695; B. 2353; *p.* of for-grípan.

for-grindan; *p.* -grand, *pl.* -grundon; *pp.* -grunden [for-, grindan *to grind*] *To grind thoroughly, grind to pieces, grind down, crush, pulverize, mangle, consume, destroy;* commŏlĕre, contĕrĕre, contundĕre, confringĕre, pulvĕrāre, lǎcĕrāre, demōliri :—Forgrindan *commŏlĕre,* Cot. 35. Ic forgrand.gramum *I fiercely (?) crushed [them],* Beo. Th. 852; B. 424. Ðǽr læg secg manig, gárum forgrunden *there lay many a warrior, ground to pieces by javelins,* Chr. 937; Th. 202, 21, col. 2; Ædelst. 18. Billum forgrunden *ground down with swords,* Andr. Kmbl. 826; An. 413. Biþ beorhtast nesta bǽle forgrunden *the brightest of nests is pulverized by the fire,* Exon. 59 a; Th. 213, 20; Ph. 227. Wundum forgrunden *mangled with wounds,* Chr. 937; Erl. 114, 9; Ædelst. 43. Glēdum forgrunden *consumed or destroyed by fire,* Beo. Th. 4659; B. 2335: 5347; B. 2677.

for-grípan; *p.* -gráp, *pl.* -gripon; *subj. pres.* -grípe, *pl.* -grípen; *pp.* -gripen [for-, grípan *to grasp*] *To grasp, snatch away, seize, assail, overwhelm;* corrĭpĕre, comprehendĕre, apprehendĕre, vi affĕrre, obruĕre :—Ádle forgripen *languōre correptus,* Bd. 5, 7; S. 620, 40, note. He þohte forgrípan gumcynne *he resolved to overwhelm mankind,* Cd. 64; Th. 77, 14; Gen. 1275. Ðonne fýr ǽpplede gold gífre forgrípeþ *when fire greedily grasps appled gold,* Exon. 63 a; Th. 232, 15; Ph. 507: Ps. Th. 58, 12. He æt gúþe forgráp Grendeles mǽgum *he in conflict grasped Grendel's kinsmen,* Beo. Th. 4695; B. 2353. Æbylignes yrres ðínes hí forgrípe *indignātio īræ tuæ apprehendat eos,* Ps. Th. 68, 25. Ðonne we hine forgrípen *when we seize him,* Ps. Th. 70, 10: 138, 9. Deáh gé minne flǽschoman fýres wylme forgrípen *though ye assail my body with fire's heat,* Exon. 38 a; Th. 124, 31; Gú. 346. [*O. Sax.* fargrípan *to seize for destruction:* Ger. ver-greifen *to take away.*]

fór-grípan; *p.* -gráp, *pl.* -gripon; *subj. pres.* -grípe, *pl.* -grípen; *pp.* -gripen *To take before, carry off prematurely, pre-occupy;* præripĕre, præ-occūpāre :—Wæs heó mid deáþe fórgripen *illa morte prærepta est,* Bd. 3, 8; S. 532, 27 : 3, 29; S. 561, 17. Dý-læs hit sí mid deáþe fórgripen *ne morte præ-occūpētur,* 1, 27; S. 492, 30, note. [*Ger.* vorgreifen *to anticipate, forestall.*]

for-grówan; *p.* -greów, *pl.* -greówon; *pp.* -grówen [for-, grówan *to grow*] *To grow up, grow into;* increscĕre :—Se ǽr in dæge wæs dýre, scrídeþ nú deóp feor, brondhord geblówen, breóstum in forgrówen *copper was dear in [that] day, now it circulates wide and far, an ardent treasure flourishing, grown up in the hearts,* Exon. 94 b; Th. 354, 16; Reim. 46.

for-gulde *should pay for* or *repay,* Ps. Th. 65, 13; *p. subj.* of forgildan. For-guldon *paid for,* Ps. Spl. 37, 21; *p. pl.* of for-gildan.

for-gyfan; *pp.* -gyfen *To give, forgive, supply;* dăre, ministrāre, remittĕre, dimittĕre, Lk. Bos. 7, 48 : Mt. Bos. 6, 12 : 18, 21: Mk. Bos. 2, 7 : Lk. Bos. 6, 37 : Bd. 1, 25; S. 486, 29 : Exon. 28 a; Th. 85, 9; Cri. 1388. v. for-gifan.

for-gyfendlíc, -gyfenlíc; *adj. Forgiving, pardonable, tolerable;* remissus :—Tyro and Sydone byþ forgyfendlícre [MS. forgyfendlícur] on dómes dæg, ðonne eów *it shall be more pardonable for Tyre and Sidon in the day of judgment, than for you;* Tyro et Sidoni remissius ĕrit in die jūdicii quam vōbis, Mt. Bos. 11, 22. Sodomwara lande byþ forgyfenlícre

on dómes dæg, ðonne ðé *terræ Sŏdŏmŏrum remissius ĕrit in die jŭdĭcii, quam tĭbi*, Mt. Bos. 11, 24: Lk. Bos. 10, 14. v. for-gifenlīc.

for-gyfenes, -gyfennes, -gyfnes, -ness, -nyss *forgiveness, remission*, Mt. Bos. 26, 28: Lk. Bos. 3, 3: L. Edg. ii. 1; Th. i. 266, 5, MS. A: L. Edg. S. 1; Th. i. 270, 17, MS. F. v. for-gifnes.

for-gyldan; ic -gylde, ðú -gylst; *subj. pres.* -gylde, *pl.* -gylden; the other inflections as in for-gildan *To pay for, repay, requite, recompense, reward*:—Hwî nolde God him forgyldan his bearn be twîfealdum *why would not God repay him his children twofold?* Job Thw. 168, 23: L. Ath. v. § 8, 8; Th. i. 238, 10. Hêht forgyldan *commanded to pay for*, Beo. Th. 2112; B. 1054: Fins. Th. 79; Fin. 39: Lk. Bos. 10, 35: Ps. Th. 88, 29: Ps. Lamb. 141, 8: L. Ethb. 4; Th. i. 4, 3: L. In. 9; Th. i. 108, 5, note 14, MS. B: 11; Th. i. 110, 4, note 14, MS. B: L. Ath. i. 1; Th. i. 198, 17: i. 2; Th. i. 200, 11: L. Edm. S. 1; Th. i. 248, 4: Ps. Th. 141, 9: Beo. Th. 1916; B. 956: L. Ath. i. 6; Th. i. 202, 16: Byrht. Th. 132, 47; By. 32.

for-gyltan *to become guilty, to commit;* committĕre, Scint. Ben. Lye. [*Orm.* forgilltenn : *A. R.* vorgulte *p. p.*] v. gyltan.

for-gŷman, -gîman ; *p.* de ; *pp.* ed [for, gŷman *to take care*] *To neglect, pass by, transgress;* negligĕre, prætĕrīre, transgrĕdi :—He ða forþgesceaft forgyteþ and forgŷmeþ *he forgets and neglects the future state*, Beo. Th. 3506; B. 1751. Hwî forgŷmaþ ðîne leorningcnihtas úre yldrena lage *quāre discĭpŭli tui transgrĕdiuntur tradĭtiōnem sēniōrum?* Mt. Bos. 15, 2. Hwî forgŷme gē Godes bebod for eówre lage *quāre vos transgrĕdimĭni mandātum Dei propter tradĭtiōnem vestram?* 15, 3. Se ðe Drihtnes word forgîmde, he forlêt his men and nŷtenu úte *qui neglexit sermōnem Dŏmĭni, dimīsit servos suos et jŭmenta in agris*, Ex. 9, 21. Ic næfre ðîn bebod ne forgŷmde *nunquam mandātum tuum prætĕrīvi*, Lk. Bos. 15, 29. Hie þegnscipe Godes forgŷmdon *they neglected the service of God*, Cd. 18; Th. 21, 20; Gen. 327. Forgŷmdon hig ðæt *illi neglexĕrunt*, Mt. Bos. 22, 5. Ne forgŷm ðú ðînes Drihtnes steóre *be not heedless of thy Lord's correction*, Homl. Th. ii. 328, 21. [*O. Sax.* fargûmôn *to neglect.*]

for-gŷmednes, -ness, e ; *f. Neglect;* neglĭgentia, Som. Ben. Lye.

for-gŷmeleásian, -gîmeleásian, -giémeleásian, -gémeleásian ; *p.* ode ; *pp.* od [for-, gŷmeleásian *to neglect*] *To neglect entirely;* omnīno neglĭgĕre :—Forgŷmeleásian *neglĭgĕre*, Scint. 81: Fulg. 18. Gif he forgŷmeleásaþ his hláfordes gafol *if he neglect his lord's tribute*, L. Edg. S. 1; Th. i. 270, 15. Swylc gerêfa swylc ðis forgŷmeleásie *such reeve as may neglect this*, L. Ath. iv. 1; Th. i. 222, 2. Forgŷmeleásod beón *neglectus esse, neglĭgi*, R. Ben. 36.

forgŷmeleásnes, -ness, e ; *f. Carelessness, neglect;* neglĭgentia, Som. Ben. Lye.

fŏr-gyrd, es ; *m. A fore-girdle, martingale;* antela, cingŭlum illud quod ante pectus ĕqui tendītur, Som. Ben. Lye. v. forþ-gyrd.

for-gytan; ic -gyte, ðú -gytest, -gytst, he -gyteþ, -gyt, *pl.* -gytaþ; *impert.* -gyt, *pl.* -gytaþ; *subj.* -gyte, *pl.* -gytan; *pp.* -gyten *To forget;* oblīvisci :—Nylle ðú forgytan ealle edleánunga oððe edleán his *nōli oblīvisci omnes retrĭbūtiōnes ejus*, Ps. Lamb. 102, 2: Ps. Th. 118, 93: Ps. Lamb. 118, 16, 83, 93: 136, 5: Ps. Th. 43, 25: Beo. Th. 3506; B. 1751: Ps. Lamb. 76, 10: 43, 21: 49, 22: 73, 19, 23: Ps. Th. 136, 5: Ps. Lamb. 77, 7: 58, 12. The other forms as in for-gitan.

for-gytel, -gytol, -gyttol ; *adj. Forgetful, forgetting;* oblĭviōsus :—He næs forgytel [forgyttol, Homl. Th. ii. 118, 19] *he was not forgetful*, Nat. S. Greg. Els. 5, 11. Forgytele we ne synt ðé *nec oblīti sŭmus te*, Ps. Lamb. 43, 18. He nis forgytol clypunge þearfena *non est oblītus clāmōrem paupĕrum*, 9, 13.

for-gytelnes, -gitelnes, -ness, -nyss e ; *f. Forgetfulness, forgetting, oblivion;* oblivio :—On lande forgytelnysse *in terra oblīviōnis*, Ps. Lamb. 87, 13. Forgytelnesse geseald *ic eom oblīviōni dātus sum*, 30, 13. Forgytelnesse sŷ geseald seó swîðre mîn *oblīviōni dētur dextĕra mea*, 136, 5.

for-habban; *part.* -hæbbende ; *p.* -hæfde, *pl.* -hæfdon; *impert.* -hafa, *pl.* -habbaþ; *pp.* -hæfed, -hæfd; *v. trans. To hold in, restrain, retain, abstain, refrain;* tĕnēre, contĭnēre, cŏhĭbēre, prŏhĭbēre, abstĭnēre :—Ne meahte wæfre mód forhabban in hreðre *he might not retain his wavering courage in his heart*, Beo. Th. 2306; B. 1151: 5211; B. 2609. He ðær sum fæc on forhæbbendum lîfe lifede *ălĭquandiu contĭnentissĭmam gessit vitam*, Bd. 5, 11; S. 626, 16. Ðæt mynster ôþ gyt to dæge Englisce menn ðær on ælþeódignysse hî forhabbaþ *quod vĭdēlicet mŏnastērium usque hŏdie ab Anglis tĕnētur incŏlis*, 4, 4 ; S. 571, 17. Forbeód oððe forhafa oððe bewere tungan ðîne fram yfle *prŏhĭbe linguam tuam a mălo*, Ps. Lamb. 33, 14. Hit forhæfed gewearþ ðætte hie sædon swefn cyninge *it was denied them that they should say the dream to the king*, Cd. 179; Th. 225, 1; Dan. 147. Hyra eágan wæron forhæfde *ŏcŭli illōrum tĕnēbantur*, Lk. Bos. 24, 16.

for-hæfedesta; *m. sup. Most continent;* contĭnentissĭmus :—Se hálgesta wer and se forhæfedesta *vir sanctissĭmus et contĭnentissĭmus*, Bd. 4, 3; S. 569, 41; *sup.* of for-hæfed, *pp.* of for-habban.

for-hæfednes, -hæfdnes, -ness, -nys, -nyss, e ; *f. Restraint, continence, abstinence;* contĭnentia, abstĭnentia :—Forhæfednyss [MS. -hefednyss] *abstĭnentia*, Ælfc. Gr. 43 ; Som. 45, 7. He hæfde swŷðe mycle georn-

nysse sibbe and sôþre lufan and forhæfdnesse and eádmódnysse *stŭdium vĭdēlicet pācis et cārĭtātis, continentiæ et hŭmĭlĭtātis*, Bd. 3, 17; S. 545, 7. Ða fægerestan bŷsne his gingrum forlêt, ðæt he wæs micelre forhæfdnysse and forwyrnednesse lîfes *sălūberrĭmum abstĭnentiæ vel contĭnentiæ clēricis exemplum relĭquit*, 3, 5 ; S. 526, 21. On forhæfednysse and on eádmódnysse *in continence and in humility*, 4, 3; S. 569, 1, 37. Lifde se man his lîf on mycelre forhæfdnesse *the man lived his life in great continence*, 4, 25; S. 599, 28. Ðæt is wundor ðæt ðú swá rêðe forhæfednesse and swá hearde habban wylt *mirum quod tam austēram tĕnēre continentiam vĕlis*, 5, 12; S. 631, 33.

for-hæl, -hæle, -hælon ; *p. indic. subj. indic. pl.* of for-helan *to conceal*, Glostr. Frag. 4, 20.

for-hĕlde, es ; *m?* [for, hælde, *p.* of hælan *to heal*] *An offence;* offensa, Cot. 148, Lye.

for-hátan; *p.* -hêt, -hêht ; *pp.* -háten [for, hátan *to call*] *To renounce, forswear;* renuntiāre, ejurāre :—Bûton he hit forháten hæbbe *unless he have forsworn it*, L. Ælf. P. 47; Th. ii. 384, 30.

for-hátena, an ; *m.* [hátan *to call* or *name*] *An ill-named, or a reprobate person;* fāmōsus, perdĭtus :—Ðá se forhátena spræc *then spake the reprobate one*, Cd. 29; Th. 38, 20; Gen. 609.

fŏr-heáfod, es ; *n. The fore part of the head*, FOREHEAD, *skull;* ancĭput? calvārium :—Fŏrheáfod *ancĭput?* Ælfc. Gl. 69; Som. 70, 34; Wrt. Voc. 42, 42. Fŏrheáfod *vel* heáfodpanne *calvārium*, 69; Som. 70, 33; Wrt. Voc. 42, 41.

for-healdan *to withhold, keep back, disregard;* detĭnēre, neglĭgĕre, contemnĕre :—Hæfdon hŷ forhealden helm Scylfinga *they had disregarded the helm of the Scylfings* [*had deserted him*], Beo. Th. 4751; B. 2381: Bt. 29, 1; Fox 102, 17. [*Ger.* ver-halten *to reserve, withhold, conceal.*]

for-healden *polluted;* incestus, Cot. 105.

fŏr-heard; *adj. Very hard;* prædūrus :—Wulfmær forlêt fŏrheardne gár faran eft ongeán *Wulfmær let the piercing dart fly back again*, Byrht. Th. 136, 24; By. 156.

for-heardian; *p.* ode ; *pp.* od *To harden, become hard;* indūrāre :—He forheardaþ and fordrugaþ *indūret et arescat*, Ps. Lamb. 89, 6. [*Dut.* ver-harden *to harden:* Ger. ver-härten *to grow hard, to harden.*]

for-heáwan; *p.* -heów ; *pp.* -heáwen *To hew* or *cut down, cut in pieces, slaughter;* concīdĕre, occīdĕre :—Hŷ forheówan Heaðobearda þrym *they slaughtered the host of Heathobeards*, Scóp. Th. 99; Wîd. 49: Byrht. Th. 135, 9; By. 115. [*Ger.* ver-hauen *to cut down.*]

for-helan, he -hilþ; *p.* -hæl, *pl.* -hælon; *subj. p.* -hæle, *pl.* -hælen; *pp.* -holen *To cover over, hide, conceal;* celāre, occultāre, abscondĕre :—Ðe hit forhelan þenceþ *who seeks to conceal it*, Exon. 91 a; Th. 340, 25; Gn. Ex. 116. Hú mæg ic forhelan Abrahame, ðe ic dón wille *num celāre potĕro Abraham, quæ factūrus sum?* Gen. 18, 17. Forhele ic incrum Hĕrran hearmes swá fela *I will conceal from your Lord so much calumny*, Cd. 27; Th. 36, 29; Gen. 579. Gif he hit forhilþ *if he hide it*, Lev. 5, 1. Ne biþ ðær wiht forholen *there shall be naught concealed*, Exon. 23 b; Th. 65, 14; Cri. 1054. Ðæt he ðæs hálgan hæse forhæle his hláforde *that he should conceal the saint's command from his Lord*, Glostr. Frag. 4, 20. Ðæt mîne cræftas and ánweald ne wurden forgitene and forholene *that my talents and power should not be forgotten and concealed*, Bt. 17; Fox 60, 9. [*Dut.* ver-helen : *Ger.* ver-hehlen *to conceal.*]

for-hergian, -heregian, to -hergianne; *part.* -hergiende, -hergende; *p.* ode, ade, ede ; *pp.* od, ad, ed *To lay waste, destroy, ravage, devastate, plunder;* vastāre, devastāre, depŏpŭlāre :—Ne wile he ealle ða rîcu forsleán and forheregian *will he not slay and destroy all the kingdoms?* Bt. 16, 1; Fox 50, 3. Mid ðŷ se ylca cyning gedyrstelîce here lædde to forhergianne Pehta mǽgþe *idem rex, cum tĕmĕre exercĭtum ad vastandam Pictōrum prōvinciam duxisset*, Bd. 4, 26; S. 602, 16. Forhergiende *depŏpŭlans*, 1, 15; S. 483, 44. Forhergende, 4, 7; S. 574, 30. Ceadwala eft forhergode Cent *Ceadwalla again ravaged Kent*, Chr. 687; Erl. 43, 2: 1000; Erl. 137, 2. Ecgfrið Norþan-Hymbra cyning sende wered and fyrd on Hibernia Scotta eálonde, and hî ða unscæððendan þeóde, and symble Angelcynne ða holdestan earmlîce forhergodon *Ecgfrid rex Nordanhymbrŏrum misso Hĭberniam exercĭtu vastāvit mĭsĕre gentem innoxiam et nātiōni Anglōrum amīcissĭmam*, Bd. 4, 26; S. 602, 7: Ceadwalla and Mul Cent and Wieht forhergedon *Ceadwalla and Mul ravaged Kent and Wight*, Chr. 686; Erl. 40, 25. Fêng to rîce Honorius, twám geárum ǽr Rôma burh abrocen and forhergad wǽre *Honorius succeeded to the sovereignty, two years before the city Rome was broken into and devastated*, Bd. 1, 11; S. 480, 10. Seó hreównes ðæs oft cwedenan wôles feor and wîde eall wæs forheregod and fornumen *tempestas sæpe dictæ clādis lāte cuncta depŏpŭlans*, 4, 7; S. 574, 30, MS. B. Hî forhergode wǽron *they were plundered*, Chr. 1013; Erl. 149, 19. [*Ger.* ver-heeren *to destroy, lay waste.*]

for-hergung, -heriung, e ; *f. A molesting, devastation, annoyance, trouble;* vastātio, infestātio :—Mid forhergunge gebysmerad *disgraced by pillage*, Ors. 2, 4; Bos. 45, 1: Cot. 108.

for-hicgan, -higan; *p.* ede, de ; *pp.* ed *To neglect, reject, despise, condemn;* despĭcĕre, spernĕre :—Se wæs middangeard forhicgende *he was despising the world;* cum esset contemptu mundi insignis, Bd. 5, 9; S. 623, 25. Se ðe

me forhigþ *qui spernit me*, Jn. Bos. 12, 48. We forhicgaþ on arīsendum on us *spernēmus insurgentes in nōbis*, Ps. Spl. 43, 7. Driht nā forhigede and ne forseah bēne þearfena *Dŏmĭnus non sprēvit neque despexit deprecātiōnem paupĕris*, 21, 23. Nā he forhigde bēne heora *non sprēvit prĕcem eōrum*, 101, 18. v. for-hycgan.

for-hilþ hides, Lev. 5, 1; *3rd sing. pres. of* for-helan.

for-hogednes, -hogodnes, -hogydnys, -ness, e; *f.*: for-hogung, e; *f.* Contempt, disdain; contemptus :—Fatu on forhogednysse hæfde *vāsa despectui hăbĭta*, Bd. 3, 22; S. 552, 15. Gefylled we synd forhogodnesse *replēti sŭmus despectiōne*, Ps. Spl. M. C. 122, 4.

for-hogian; *p.* ede, ode; *pp.* ed, od [hogian *to be anxious*] To neglect, despise, accuse; neglĭgĕre, spernĕre :—Hwylc wracu him forhogiende æfter fyligde *quæ illos spernentes ultĭo sĕcŭtā sit*, Bd. 2, 2; S. 502, 4. Ealle middaneardlīce þing swā swā ælfremede forhogigende *despising all earthly things as entirely foreign ones*, Nat. S. Greg. Els. 35, 4. He forhogaþ, ðæt he hīre uncre lāre *mŏnĭta nostra audīre contemnit*, Deut. 21, 20. Driht nā forhogode and ne forseah bēne þearfena *Dŏmĭnus non sprēvit neque despexit deprecātiōnem paupĕris*, Ps. Spl. C. 21, 23. Forhogedun Drihtnes bebod *contempsistis impĕrium Dŏmĭni*, Deut. 9, 23. Ða Sundor-hālgan forhogodon ðæs Hælendes geþeaht *Pharĭsæi consĭlium Dei sprēvērunt*, Lk. Bos. 7, 30. We forhogien on arīsendum on us *spernēmus insurgentes in nōbis*, Ps. Spl. T. 43, 7. Forhogedre āre heora anddetnesse *contempta revĕrentia suæ professiōnis*, Bd. 4, 25; S. 601, 15. Gif he ðonne eów forhogige, sī ðonne he fram eów forhogod *sin autem vos sprēvĕrit, et ipse spernătur a vobis*, 2, 2; S. 503, 12, 13.

for-hogung contempt, Ps. Spl. 118, 22. v. for-hogednes.

for-hogydnys contempt, Cambr. MS. Ps. 118, 22. v. for-hogednes.

for-holen concealed, hidden, Exon. 23 b; Th. 65, 14; Cri. 1054: Lk. Skt. Lind. 8, 17; *pp. of* for-helan.

forhŏrwade *was dirty*; obsorduit, Hymn.

fŏr-hradian, -hradigan; *p.* ode; *pp.* od *To hasten before, anticipate, prevent*; prævĕnīre, præoccŭpāre :—Utan fōrhradian his ansýne on andetnesse *præoccŭpēmus făciem ejus in confessiōne*, Ps. Lamb. 94, 2. Se sylfa dēaþ ðære ādle yldinge fōrhradaþ *death itself prevents the tarrying of the disease*, Homl. Th. ii. 124, 12. Fōrhradode Godes mildheortnys us *God's mercy prevented us*, ii. 84, 13. Ðonne hie fōrhradigaþ ðone tīman gōdes weorces *when they anticipate the time of a good work*, Past. 39, 3.

fōr-hraðe; *adv. Very quickly, soon*; cĭto, confestim :—Æfter ðam ðæs fōrhraðe *very soon after that*, Chr. 921; Erl. 107, 6, 24. v. fōr-raðe.

for-hrēred; *part. Annulled, made void*; cassātus :—Forhrēred *cassāta*, Ælfc. Gl. 49; Som. 65, 99; Wrt. Voc. 34, 28. v. hrēran.

forhswebung, e; *f. A storm*; prŏcella, Ps. Spl. T. 106, 25.

FORHT; *adj.* I. *fearful, timid, affrighted*; tĭmĭdus, păvĭdus, territus, trĕpĭdus :—Ne beó ðū on sefan tō forht *be not thou too fearful in mind*, Andr. Kmbl. 196; An. 98: Beo. Th. 1512; B. 754. Næs he forht *he was not afraid*, 5927; B. 2967: Andr. Kmbl. 2172; An. 1087: Rood Kmbl. 41; Kr. 21. Heó com forht *trĕmens vēnit*, Lk. Bos. 8, 47. To hwī synt gē forhte *quid tĭmĭdi estis?* Mt. Bos. 8, 26: Mk. Bos. 4, 40. We beóþ forhte on ferþþe *we are fearful in soul*, Exon. 70 b; Th. 262, 5; Jul. 328: Ps. Th. 64, 8: Bd. 5, 19; S. 640, 33. He sent on eów forhte heortan *dăbit tĭbi cor păvĭdum*, Deut. 28, 65. Nō ðý forhtra *be ye not the more fearful*, Exon. 35 b; Th. 114, 14; Gū. 172. II. *terrible, dreadful, formidable*; terrĭbilis, formĭdŏlōsus :—Ne wile forht wesan brōðor ōðrum *a brother will not be formidable to another*, Exon. 112 b; Th. 430, 20; Rā. 44, 11. On ða forhtan tíd *in that dreadful time*, Hy. 10, 56; Hy. Grn. ii. 294, 56. [O. Sax. foraht, forht, furht: O. H. Gĕr. forht tĭmĭdus, tĭmens: Goth. faurhts.] DER. an-forht, ge-, un-.

forht-full; *adj. Fearful*; formĭdŏlōsus, Coll. Monast. Th. 22, 21.

forhtian, forhtigan, forhtigean, forhtgean; *to* forhtianne; *part.* forhtiende, forhtigende; *p.* ode, ede, ode; *pp.* od, ed [forht *affrighted*, and the terminations -an, -anne, -gan]. I. v. *intrans. To be afraid or frightened, tremble*; păvēre, trĕmēre, trĕpĭdāre, formĭdāre :—Ongan he forhtian, and sārgian *cœpit păvēre, et tædēre*, Mk. Bos. 14, 33: Boutr. Scrd, 21, 22. Ongunnon hī forhtigan *they began to be afraid*, Bd. 1, 23; S. 485, 30. Forhtigean, Ps. Th. 13, 7. To heora mōde gelæddum ðære forhtiendan tíde *reducto ad mentem trĕmendo illo tempŏre*, Bd. 4, 3; S. 569, 25. Flugon forhtigende *trembling they fled*, Cd. 166; Th. 206, 15; Exod. 452; Bd. 4, 7; S. 575, 8. Ic forhtige *formīdo*, Ælfc. Gr. 36; Som. 38, 50. Hie forhtiaþ *they will be afraid*, Rood Kmbl. 227; Kr. 115: Ps. Th. 67, 9. Ðǽr hig forhtodon mid ege *illic trĕpĭdāvērunt tĭmōre*, Ps. Lamb. 52, 6. He bæd ðæt ne forhtedon nā *he bade that they should not be afraid*, Byrht. Th. 132, 25; By. 21. Ne sý eówer heorte gedrēfed, ne ne forhtige gē *non turbētur cor vestrum, neque formīdet*, Jn. Bos. 14, 27. Ðæt ōðre forhtian *that others may fear*, Homl. Th. ii. 300, 15. II. v. *trans. To fear, be frightened at, dread*; tĭmēre :—Ic ne forhtige wiht I *fear nothing*, Ps. Th. 61, 2: 54, 2. Ne forhtast ðū on dæge flān on lyfte *non tĭmēbis a săgitta vŏlante in die*, 90, 6. Ðe Drihten forhtaþ *qui tŭmet Dŏmĭnum*, 127, 5: 60, 4. Ða ðē on feore forhtigaþ, ða me on fægere geseóþ *qui tŭment te, vĭdēbunt me*, 118, 74. Ne nān þing ne forhtgeaþ *fear nothing*, Deut. 1, 20. DER. a-forhtian, on-.

forhtiendlíc, forhtigendlíc; *adj. Timorous, fearful*; metĭcŭlōsus, Cot. 129.

forht-líc; *adj. Timid, fearful, trembling*; trĕpĭdus, ĭerrĭbĭlis :—Him forhtlíce fǽrspel bodedon *they fearful announced to them the sudden news*, Judth. 12; Thw. 25, 5; Jud. 244. Fleóþ forhtlíce þunres brōgan *they, being afraid, shall flee the terror of [thy] thunder;* a vōce tŏnĭtrui tui formīdābunt, Ps. Th. 103, 8. On ða forhtlíce sorgum wlītaþ *on which, they, frightened, look sorrowfully*, Exon. 24 a; Th. 68, 15; Cri. 1104. [O. Sax. forhtlík *terrible*.]

forht-líce; *adv. Fearfully, tremblingly*; trĕpĭde :—Æghwylc wille feores forhtlíce aþolian *every one will fearfully endure life*, Exon. 27 a; Th. 81, 7; Cri. 1320: R. Ben. interl. 5.

forht-mōd; *adj. Mind-frighted, timid, pusillanimous*; trĕpĭdus anĭmo, păvĭdus :—He forhtmōd wāfode *he was hesitating, being frightened in mind*, Ælfc. T. 35, 23. Ic sceal eaforan mīne forhtmōd fergan *I, being timid, must convey my children*, Exon. 104 b; Th. 397, 1; Rā. 16, 13.

forhtnys, fyrhtnes, -ness, e; *f. Fear, amazement, terror, dread*; tĭmor :—Ðā aforhtode Isaac micelre forhtnisse *expāvit Isaac stupŏre vehĕmenti*, Gen. 27, 33.

forhtra *more fearful* :—Ne beóþ gē ðý forhtran *be ye not the more fearful*, Cd. 156; Th. 194, 11.

forhtudon = forhtodon *trĕpĭdāvērunt*, Ps. Spl. 13, 9; *p. of* forhtian *to fear, tremble*.

forhtung, e; *f.* [forht, ung] *Fear*; păvor :—Būton blācunge and forhtunge *without paleness and fear*, Homl. Th. i. 72, 28: ii. 560, 15. On forhtunge *in păvōre*, Ps. Lamb. 30, 23.

for-hwǣga, -hwāga; *adv. At least*; saltem :—Forhwǣga on fíf mílum oððe on syx mílum fram ðæm feó *at least within five or six miles from the property*, Ors. 1, 1; Bos. 22, 35. Forhwǣga on ānre míle fram ðæm tūne *at least within one mile from the town*, 1, 1; Bos. 22, 30.

for-hwām *wherefore, why*. v. hwā *who: interrog*.

for-hwerfan *To transform, pervert*; transformāre, pervertĕre :—Cnihtas wurdon ealle forhwerfde to sumum dióre *the men were all transformed to some beast*, Bt. Met. Fox 26, 172; Met. 26, 86: Bt. 38, 1; Fox 196, 2. Eówra sāwla mā forhwerfdon ðonne hie gerihton *they have perverted more of your souls than they have directed*, L. Alf. 49; Th. i. 56, 18. v. forhwyrfan.

for-hwí, -hwig *For why, wherefore*; quāre, cur, Ps. Th. 113, 5; Nicod. 4; Thw. 2, 19.

for-hwon *why*; quāre, Bd. 2, 6; S. 508, 14: 2, 12; S. 513, 37.

for-hwyrfan, -hwerfan; *part.* -hwyrfende; *p.* -hwyrfde; *pp.* -hwyrfed, -hwyrfd. I. *to change for or from, transform, transfer, remove*; avertĕre, transformāre :—He forhwyrfþ eów of ðam lande *he will remove you from the land*, Deut. 28, 63. Sī se man awirged, ðe forhwyrfe his freóndes landgemǽro *maledictus hŏmo, qui transfert termĭnos proxĭmi sui*, Deut. 27, 17. II. *to turn aside, pervert, deprave*; subvertĕre, pervertĕre, deprāvāre :—Ðisne we gemētton forhwyrfende ūre þeóde *huuc invĕnīmus subvertentem gentem nostram*, Lk. Bos. 23, 2. Swylce he ðis folc forhwyrfde *as if he perverted this people*, 23, 14. Ðā forhwyrfed wæs *when it was perverted*, Exon. 8 a; Th. 3, 11; Cri. 34. Mid forhwyrfedum forhwyrfed ðū bist *cum perverso pervertĕris*, Ps. Spl. T. 17, 28. Hwyrf ðē wið ða forhwyrfdan *cum perverso pervertĕris*, Ps. Th. 17, 25.

for-hycgan *To despise, reject*; despicĕre, contemnĕre, spernĕre :—Ðe forhycgeaþ God *who despise God*, Ps. Th. 52, 6. Ðæt ic ne forhycge I *reject it not*, Exon. 63 b; Th. 235, 4; Ph. 552.

for-hýdan *To hide*; abscondĕre :—Forhýddan meinwitgyrene *abscondērunt mĭhi lăqueōs*, Ps. Th. 139, 5.

for-hygde-líc; *adj. Despisable*; contemptĭbilis :—Forhygdelíc oððe forsewen *contemptus*, Ps. Lamb. 118, 141.

for-hylman; *p.* de; *pp.* ed *To cover over, conceal*; obdueĕre, occŭlĕre :—Ne dorste forhylman Hælendes bebod *he dared not conceal the Saviour's command*, Andr. Kmbl. 1469; An. 736.

for-hýnan; *p.* -hýnde; *pp.* -hýned, -hýnd [hýnan *to humble, put down*] *To cast down, humble, oppress, waste*; hŭmĭliāre, opprĭmĕre, vastāre :—Ðone forhýndon and þearfan gerihtlæcaþ *hŭmĭlem et paupĕrem justĭfĭcāte*, Ps. Lamb. 81, 3. Forhýned *cast down*, Ors. 3, 7; Bos. 62, 10. Wǽron Pene forhýnde *the Carthaginians were cast down*, Ors. 4, 10; Bos. 95, 30. Mid ðam bryne Rōme burh wæs swīðe forhýned *the city Rome was brought very low by that burning*, Ors. 6, 1; Bos. 115, 41.

for-hyrdan; *p.* de; *pp.* ed; *v. trans. To harden against, to harden*; obdūrāre :—Nǽfre gē heortan geþanc deorce forhyrden *nolĭte obdūrāre corda vestra*, Ps. Th. 94, 8.

for-lǽcan; *p.* -léc, -leólc; *pp.* -lácen *To seduce, betray, deceive*; sedūcĕre, decipĕre :—Ðū leóda feala forleólce and forlǽrdest *thou hast deceived and seduced many people*, Andr. Kmbl. 2727; An. 1366. Forléc hie mid ligenum *he seduced her with lies*, Cd. 30; Th. 40, 30; Gen. 647. Hie seó wyrd forleólc *fate deceived them*, Andr. Kmbl. 1227; An. 614. He wearþ on feónda geweald forlácen *he was betrayed into the foes' power*, Beo. Th. 1811; B. 903.

for-lǽdan; *p.* -lǽdde; *pp.* -lǽded, -lǽdd, -lǽd *To mislead, lead astray, seduce*; sedūcĕre :—Forlǽdan and forlǽran *to mislead and pervert*, Cd. 23

Th. 29, 18; Gen. 452 : 32; Th. 43, 17; Gen. 692. Ic bepǽce oððe forlǽde *sedûco*, Ælfc. Gr. 47; Som. 48, 53. He ðæs folces ðone mǽstan dǽl mid ealle forlǽdde *he wholly misled the greatest part of the people*, Ors. 1, 12; Bos. 35, 41. Hie forlǽddon swǽse gesîþas *they misled their dear associates*, Beo. Th. 4084; B. 2039. Forlǽdd be ðám lygenum *misled by lies*, Cd. 28; Th. 37, 31; Gen. 598. Ðeáh heó wurde forlǽd mid lygenum *though she was misled with lies*, 30; Th. 39, 23; Gen. 630. Past. 58; Hat. MS. Men synt forlǽdde *men are misled*, Cd. 33; Th. 45, 18; Gen. 728. [*O. Sax.* farlêdean : *Dut.* ver-leiden : *Ger.* ver-leiten *to mislead, seduce* : *Laym.* forledeþ *leads astray*.]

for-lǽge *neglected, disgraced* :—Ðý-læs seó mynegung [MS. mynugung] forlǽge *lest the giving notice should be neglected*, L. Ath. v. § 7; Th. i. 234, 29; *subj.* of forlicgan. v. licgan.

for-lǽran ; *to* -lǽranne; *p.* -lǽrde; *pp.* -lǽred *To misteach, deceive, seduce, corrupt, pervert*; declpôre, sedûcêre, corrumpêre :—Forlǽdan and forlǽran *to mislead and pervert*, Cd. 23; Th. 29, 18; Gen. 452 : 32; Th. 43, 17; Gen. 692. Handweorc Godes to forlǽranne *to deceive God's handywork*, 33; Th. 44, 3; Gen. 703. Ðú leóda feala forleólce and forlǽrdest *thou hast deceived and seduced many people*, Andr. Kmbl. 2727; An. 1366. Hie seó wyrd forlǽrde *fate mistaught them*, 1227; An. 614 : Elen. Kmbl. 415; El. 208. Ðe hig forlǽrdon *who deceived them*, Num. 31, 16. Ðú me forlǽred hæfst *thou hast seduced me*, Cd. 38; Th. 50, 34; Gen. 818 : Ex. 14, 11. [*Dut.* ver-leeren *to unteach*.]

for-lǽtan ; ic -lǽte, ðú -lǽtest, -lǽtst, he -lǽteþ, -lǽteþ, *pl.* -lǽtaþ; *p.* -lét, -leórt, -leót, -léton; *pp.* -lǽten [for, lǽtan]. **I.** *to let go, permit, suffer*; permittêre :—Sum eorþlíc ǽ forlǽtaþ *some earthly law permits*, Bd. 1, 27; S. 491, 2. **II.** *to relinquish, forsake, omit, neglect*; relinquêre, omittêre, prætêrîre :—Forlǽt se man fæder and môder, and geþeót hine to his wîfe *the man shall leave father and mother, and join himself to his wife*, Gen. 2, 24. [*Dut.* ver-laten : *Ger.* ver-lassen *to leave, quit, abandon, forsake*.]

for-lǽtennys, -lǽtnys, -nyss, -ness, e; *f. A leaving, remission, desolation, loss*; intermissio, remissio, desôlátio, perdítio :—Þeóstru ne synd nán þing bûton leóhtes forlǽtennyss *darkness is nothing but the departure of light*, Boutr. Scrd. 20, 46. On synna forlǽtnysse bæþe *lavacro peccátôrum remissiônis*, Bd. 2, 14; S. 518, 10. On synna forlǽtnesse *in remissiônem peccátôrum*, 5, 6; S. 620, 3. On forlǽtnysse *in desôlátiônem*, Ps. Spl. 72, 19. On forlǽtennysse *in perditiône*, 87, 12. Forlǽtnes gôda *loss of goods*, Lchdm. iii. 172, 2.

for-leás *lost*, Beo. Th. 5715; B. 2861; *p.* of for-leósan.

for-leó *seduced, deceived*, Cd. 30; Th. 40, 30; Gen. 647; *p.* of for-lácan.

for-legen *fornicated, committed fornication*, Gen. 38, 24; *pp.* of for-licgan. [*Orm.* forrlezenn.]

for-legenes, -legnes, -ness, -nys, -nyss, e; *f. Fornication*; fornicátio :—Bûton forlegenysse þingum *excepta fornícátiônis causa*, Mt. Bos. 5, 32. He swylce unálýfeddre forlegnesse and egeslícre wæs besmiten *fornícátiône pollûtus est tâli*, Bd. 2, 5; S. 506, 39.

for-legere, es; *m. A fornicator*; fornícátor, Som. Ben. Lye. v. for-liger, es; *m.*

for-legis, -legiss, e; *f. A fornicatress, harlot*; mêretrix :—Ðú hæfst forlegisse andwlitan *frons mêretrícis facta est tîbi*, Past. 52, 2; Hat. MS. Cwæþ Crist be Marian ðære forlegisse *Christ spoke of Mary the harlot*, Past. 52, 9; Hat. MS.

for-legystre, an; *f. A harlot*; mêretrix, Som. Ben. Lye. v. for-legis.

for-leógan ; *p.* -leág, *pl.* -lugon; *pp.* -logen [leógan *to lie*] *To lie greatly, belie*; valde mentíri, ementíri :—Hí mid leásum gewitum forleógan woldon *they would lie with false witnesses*, Homl. Th. ii. 248, 16. Leáse gewitan hine forlugon *false witnesses belied him*, Homl. Th. i. 44, 28. Mænige synd forsworene and swýðe forlogene *permulti sunt perjúri et mendáces*, Lupi Serm. 1, 12; Hick. Thes. ii. 102, 41.

for-leólc *seduced, deceived*, Andr. Kmbl. 1227; An. 614; *p.* of for-lácan.

for-leósan, he -lýst; *p.* ic, he -leás, ðú -lure, *pl.* -luron; *subj. pres.* -leóse, *pl.* -leósen; *p.* -lure, *pl.* -luran, -luren; *pp.* -loren *To lose, let go, destroy*; amittêre, perdêre, destruêre :—He wolde forleósan lîca gehwilc *he would destroy each body*, Cd. 64; Th. 77, 26; Gen. 1281. His treówe for feógýtsunge forleósan *fidem suam amôre pêcûniæ perdêre*, Bd. 2, 12; S. 514, 40. Ic forleóse *amitto*, Ælfc. Gr. 28, 4; Som. 31, 41. Gif he forlýst án of ðám *si perdídêrit ûnam ex illis*, Lk. Bos. 15, 4. Ic forleás *perdídêram*, Lk. Bos. 15, 9. Ðú forleóse láþra gehwylcne *mayest thou destroy every one of my enemies*, Ps. Th. 142, 12. Ðam ðe ǽr his elne forleás *to him who had before lost his courage*, Beo. Th. 5715; B. 2861. Ðú nǽne myrhþe ne forlure, ðá ðá ðú hie forlure *thou didst lose no pleasure, when thou didst lose them*, Bt. 7, 1; Fox 16, 18. Ðý-læs ic mîn gehát forleóse *ne fidem mei promissi prævárícer*, Bd. 4, 22; S. 592, 2. Hí sylfe þurh ðæt forluran *they ruined themselves through that*, 3, 1; S. 523, 23. Gê eówra yldrena hwetstán forluron *ye have lost the whetstone of your elders*, Ors. 4, 13; Bos. 100, 24. Ðæt he forlure ða gestrión *that he would lose the treasures*, Past. 7, 1; Hat. MS. 12 a, 5. Ðú forloren hæfst ða woruldsǽlþa *thou hast*

lost the worldly prosperity, Bt. 7, 1; Fox 16, 7. [*Dut.* ver-liezen : *Ger.* ver-lieren *to lose*.]

for-lét *left*, Cd. 70; Th. 84, 29; Gen. 1405; *p.* of for-lǽtan.

for-létenes, -létnes, -ness, e; *f. A leaving, leaving off, end*; intermissio, reliquiæ :—Synd forlétnesse manna gesibsumum *sunt relíquiæ hômíni pacífîco*, Ps. Spl. T. 36, 39 : R. Ben. interl. 15. v. for-lǽtennys.

for-licgan, -licggan, -ligan; *p.* -læg, *pl.* -lǽgon; *pp.* -legen [licgan *to lie*] *To lie in a forbidden manner, fornicate, commit fornication*; fornícári, adultêrâre :—Ðá forlæg heó hý sôna *then she soon committed fornication*, Ors. 3, 6; Bos. 58, 6 : 4, 4; Bos. 80, 21. Ðæt nán wíf heó ne forlicge *that no woman commit fornication*, L. C. S. 54; Th. i. 406, 4, 7 : 51; Th. i. 404, 22 : L. E. G. 3; Th. i. 168, 5 : 4; Th. i. 168, 19 : L. N. P. L. 63; Th. ii. 300, 20. Gif beweddodu fǽmne hie forlicgge *if a betrothed woman commit fornication*, L. Alf. pol. 18; Th. i. 72, 11. Sceolan þeófas and forlegene lifes ne wênan *thieves and fornicators shall not hope for life*, Exon. 31 b; Th. 98, 21; Cri. 1611 : L. Alf. pol. 10; Th. i. 68, 8. Forligende *fornícans*, Obs. Lun. § 4; Lchdm. iii. 186, 2.

for-liden *part.* [for-, liden, *pp.* of líðan *to sail*] *Shipwrecked*; naufrâgus :—Gemildsa me, nacodum, forlidenum *pity me, naked, shipwrecked*, Apol. Th. 11, 19 : 14, 1, 9 : 15, 11 : 21, 7, 13, 14, 15, 20 : 22, 1, 22 : 24, 16 : 25, 9.

for-lidennes, -ness, e; *f. Shipwreck*; naufrâgium :—Hwǽr gefôre ðú forlidennesse *where hast thou suffered shipwreck?* Apol. Th. 21, 19.

for-lignes, -lignes, -ness, -nys, -nyss, e; *f. Fornication, adultery*; fornícátio :—Ne wæs acenned of unrihthǽmede ne þurh dyrne forligenysse *non de adultêrio vel fornícátiône nâtus fuêrat*, Bd. 1, 27; S. 495, 21. Ymb hiora hetelícan forlignessa ic hit eall forlǽte *I pass over all about their hateful adulteries*, Ors. 1, 8; Bos. 31, 38. v. for-legenes.

for-liger, -ligr, es; *pl. nom. acc.* -ligeru, -ligru, -ligra; *n. Fornication, adultery*; fornícátio, adultêrium :—For forligere *ob fornícátiônem*, Mt. Bos. 19, 9 : Jn. Bos. 8, 41 : Homl. Th. ii. 322, 28 : L. Edm. S. 4; Th. i. 246, 5. Se óðer heáfodleahter is gecweden forliger *the second chief sin is called fornication*, Homl. Th. ii. 220, 3. Innan of manna heortan cumaþ forligeru *ab intus de corde hômínum procêdunt fornícátiônes*, Mk. Bos. 7, 21. Forligru *fornícátiônes*, Mt. Bos. 15, 19. Ǽnig cristen man ne ǽnige forligru ne begange *let not any christian man commit fornication*, L. C. E. 7; Th. i. 364, 24. Ascúnige man swíðe fûle forligra *let a man earnestly shun foul fornications*, L. Eth. vi. 28; Th. i. 322, 15.

for-liger, -ligr, -lír, es; *m. A fornicator, adulterer*; fornícátor, ádulter :—Ðæt Abraham nǽre forliger [MS. -ligr] geteald *ut Abraham non compútátus ádulter esset*, Boutr. Scrd. 22, 21. v. hor-cwén *an adulteress*. Forligr *adulter*, Wrt. Voc. 86, 68. He is forlír *he is an adulterer*, Homl. Th. ii. 208, 17. God fordêmþ ða dyrnan forlíras *God condemns secret adulterers*, ii. 324, 7.

for-liger; *adj. Adulterous*; ádulter :—Yfel cneórys and forliger [μοιχαλίς *adulterous*] sêcþ tácn *genêrátio mâla et adultêra signum quærit*, Mt. Bos. 12, 39.

forliger-bed, -bedd, es; *n. A bed of fornication*; fornícátiônis lectus :—On forligerbeddum *in beds of fornication*, Homl. Th. i. 604, 30.

for-liggang, es; *n? Lúpânar, prostíbulum*, Cot. 194.

for-ligr, es; *m. A fornicator*, Boutr. Scrd. 22, 21. v. for-liger, es; *m.*

for-ligr, es; *n. Fornication*, Mt. Bos. 15, 19. v. for-ligenes; *f.*

for-ligrian ; *p.* ode; *pp.* od [for-liger *a fornicator*] *To fornicate*; fornícári :—Ðú forspildest ealle ða ðe forligriaþ fram ðé *perdídisti omnes qui fornícantur abs te*, Ps. Spl. 72, 26.

for-lír *a fornicator*, Homl. Th. ii. 208, 17 : 324, 7. v. for-liger, es; *m.*

for-liðednes, -ness, e; *f.* [líðan *to sail*] *Shipwreck*; naufrâgium, Som. Ben. Lye.

for-logen *lied greatly*, Lupi Serm. 1, 12; Hick. Thes. ii. 102, 41; *pp.* of for-leógan *to lie*.

for-lor, es; *m. Destruction, perdition, loss*; perditio :—Hæleða forlor *men's perdition*, Cd. 33; Th. 45, 4; Gen. 721. Ic ofslóg ðis folc and to forlore gedyde *I slew and destroyed this people*, Past. 37, 2; Hat. MS. 49 b, 23 : Andr. Kmbl. 2846; An. 1425. Mid hæleða forlore *with men's perdition*, Cd. 35; Th. 47, 8; Gen. 757. Ðéh ðe he hý mid micle forlore ðæs folces begeáte *though he took it with great loss of the people*, Ors. 3, 9; Bos. 67, 28. [*O. Sax.* farlor.]

for-loren *forlorn, lost*, Bd. 2, 5; S. 507, 41; *pp.* of for-leósan.

for-lorenes, -ness, e; *f.* FORLORNNESS, *destruction*; perditio :—Ic geseó me stôwe gegearwode beón ǽccre forlorenesse *mihi lôcum despício æternæ perditiônis esse præpárátum*, Bd. 5, 14; S. 634, 29. On lyre oððe on forlorenesse *in perditiône*, Ps. Lamb. 87, 12.

for-lure *hast lost, didst lose*, Exon. 28 a; Th. 85, 30; Cri. 1399; *2nd sing. p.* of for-leósan : for-lure *would lose*, Chr. 81; Erl. 8, 4 : Past. 7, 1; Hat. MS. 12 a, 5; *p. subj.* of for-leósan.

for-luron *lost, have lost*, Ors. 4, 13; Bos. 100, 24; *p. pl.* of for-leósan.

fór-lustlíce ; *adv. Very willingly, gladly*; líbentissime :—Ic wille fór-lustlíce, for ðínum lufum *I will gladly* [*do so*], *for love of thee*, Bt. 22, 2; Fox 78, 12. [Cf. beon forrlisst *to be very desirous*, Orm.]

for-lýst *loses*, Mk. Bos. 9, 41; *3rd sing. pres. of* for-leósan.

FORMA; *m* : forme; *f. n : def. adj. The first, earliest*; prīmus :—Se forma ys Simon *the first is Simon*, Mt. Bos. 10, 2 : 22, 25 : Bt. 15; Fox 48, 22 : Cd. 143; Th. 179, 2; Exod. 22 : Exon. 18 b; Th. 45, 16; Cri. 720 : Beo. Th. 1437; B. 716 : Menol. Fox 17; Men. 9 : Bt. Met. Fox 8, 109; Met. 8, 55. Hū gesǽlig seó forme eld was ðises middangeardes *how happy was the first age of this world*, Bt. 15; Fox 48, 2 : Bt. Met. Fox 8, 7; Met. 8, 4; Boutr. Scrd. 21, 8. Ðis wæs ðæt forme tácn *this was the first miracle*, Jn. Bos. 2, 11. On ðone forman dæg *on the first day*, Boutr. Scrd. 19, 4 : Bd. de nat. rerum; Wrt. popl. science 4, 12; Lchdm. iii. 238, 15 : Cd. 48; Th. 61, 17; Gen. 998: Byrht. Th. 133, 68; By. 77. Forman sīðe *for the first time*, Beo. Th. 4562; B. 2286 : Exon. 84 b; Th. 319, 3; Wíd. 5 : Cd. 17; Th. 21, 4; Gen. 319. Gebletsode Metod monna cynnes ða forman twā *the Lord blessed the first two of mankind*, Cd. 10; Th. 12, 31; Gen. 194. On forman *at first*, Blickl. Homl. 127, 20. [*Wyc.* forme *in* forme-fadris : *Chauc.* forme : *Laym.* uorme, forme : *Orm.* forrme : *O. Sax.* formo : *O. Frs.* forma : *Goth.* fruma *the first* : *Icel.* frum- in compounds, *the first.*]

fōr-mǽl, fōr-mál, e; *f.* [fōr = fōre, mǽl *a speech, discourse*] *An agreement, a treaty*; fœdus, pactum :—Wið ðam ðe he eall ðæt lǽste ðæt uncer fōrmǽl wæs *on condition that he fulfil all that was our agreement*, L. O. 1; Th. i. 178, 8. Æfter ðam fōrmálum [MS. -málan] *according to the treaties*, L. Eth. ii. 1; Th. i. 284, 11.

fōr-mǽrnes, -ness, e; *f. Brightness, glory, renown*; clārĭtas :—Fōrmǽrnes and genyht *renown and abundance*, Bt. 34, 6; Fox 140, 23, note 8. v. fōre-mǽrnes.

fōr-maneg, -moni; *adj. Very many*; permultus :—Heora fōrmanega oft féngon to ánwealde *very many of them often undertook the government*, Jud. Thw. 161, 26.

for-meltan, -myltan; *p.* -mealt, *pl.* -multon; *pp.* -molten; *v. intrans. To melt away, become liquid, liquefy*; liquescĕre, liquĕfĭĕri :—Hēt wǽpen eall formeltan *he commanded the weapons all to melt away*, Andr. Kmbl. 2294; An. 1148. Formealt oðđe hnesce geworden is eorþe *liquĕfacta est terra*, Ps. Lamb. 74, 4 : Ex. 16, 21. Ealle ða scipu formultan *all the ships were consumed*, Ors. 5, 4; Bos. 105, 14. [*Dut.* ver-smelten *to melt, dissolve* : *Ger.* ver-schmelzen *to melt away.*]

for-mengan; *p.* de; *pp. To join together, mingle*; conjungĕre, Past. 21, 1? Lye. [*Dut. Ger.* ver-mengen *to mix, mingle, confuse.*] v. mengan.

formesta; *m* : formeste; *f. n : def. adj.* [*sup. of* forma *the first*] *Foremost, first, bĕst, most valiant*; prīmus, strēnuissĭmus :—Wæs he se wer se formesta *ĕrat vir ipse strēnuissĭmus*, Bd. 5, 20; S. 641, 37. v. fyrmest.

fōr-mete, es; *m.* [fōr *a journey*, mete *food*] *Fare-meat, provision for a journey*; cĭbus in itĭnĕre sūmendus, Gr. Dial. 2, 13 : Deut. 15, 14.

for-molsnian; *p.* ode, ede; *pp.* od, ed [molsnian *to corrupt*] *To putrefy, corrupt, make rotten, decay*; putrĕfăcĕre, tabĕfăcĕre, măcĕrāre :—To duste formolsnod *decayed to dust*, Wanl. Catal. 20, 4; Homl. Th. i. 218, 25. Se ylca God, ðe ealle þing of náhte geworhte, mæg arǽran ða formolsnedan líchaman of ðam duste *the same God, that wrought all things from naught, can raise up the decayed corpses from the dust*, Homl. Th. ii. 608, 6.

fōr-moni; *adj. Very many*; permultus :—Fōrmoni man *many a man*, Byrht. Th. 138, 52; By. 239. v. fōr-maneg.

for-myltan *to melt* :—Ic formylte *lĭquor*, Ælfc. Gr. 29; Som. 33, 44. v. for-meltan.

for-myrþrian; *p.* ode; *pp.* od *To kill, murder, destroy utterly*; occīdĕre, enĕcāre, perdĕre :—Gif wif hire cild formyrþrige innan hire si *mŭlier infantem suum intra se. perdĭdĕrit*, L. M. I. P. 10; Th. ii. 268, 5.

FORN, e; *f? A trout?* turnus :—Forn *turnus?* Ælfc. Gl. 102; Som. 77, 72; Wrt. Voc. 55, 76. [*Ger.* fohre, fore, forelle, *f. a trout* : *Ger. Swiss dial.* forne : *M. H. Ger.* vorhen, *f* : *O. H. Ger.* forahana, forhana *trutta* : *Dut.* voorn, *f*; vóren, *m. a roach.*]

fōrn, fōrne; *adv. Before; cōram* :—Gesæt Benedictus fōrn ongeán ðam Riggon *Benedict sat opposite to Riggo*, Homl. Th. ii. 168, 15. Óþ-ðæt he eft cume hyre fōrne geán *until he again comes opposite to it*, Bd. de nat. rerum; Wrt. popl. science 8, 13; Lchdm. iii. 248, 17. v. fōran; *prep.*

for-nam, *pl.* -námon *took away, destroyed, consumed*, Beo. Th. 2415; B. 1205 : Ps. Th. 77, 53; *p. of* for-niman.

forne; *prep. acc. For*; pro, propter :—Gif hwa hine forne forstande *if anyone will stand up for him*, L. Eth. i. 4; Th. i. 284, 3, note 8. v. for; *prep.* v. forene.

fōrne; *adv. Before, sooner*; prius, cĭtius :—Se óðer leorningcniht fōram Petrus fōrne *ille ălius discĭpŭlus prǣcŭcurrit cĭtius Petro*, Jn. Bos. 20, 4. v. fōran; *adv.* [*O. Sax.* forana.]

fōr-neáh, fōr-neán; *adv. Very nearly, nigh, nearly, almost, about*; prŏpe, fēre, pǽne, paulo mĭnus, circiter :—Fōrneáh *fēre*, Ælfc. Gr. 33; Som. 37, 50. Fōrneáh oðđe hwæt-hwega hí fordydon me on eorþan *paulo mĭnus consummāvērunt me in terram*, Ps. Lamb. 118, 87 : 93, 17.

Seó upastíhþ fōrneán óþ ðone mōnan *it extends upwards very nearly to the moon*, Bd. de nat. rerum; Wrt. popl. science 17, 4; Lchdm. iii. 272, 18. Fōrneán *fēre*, Ælfc. Gr. 38; Som. 41, 45. Mīne fōrneán astyrode synt fēt *mei pǽne mōti sunt pĕdes*, Ps. Lamb. 72, 2. Fōrneán þreó þūsend *circĭter tria millia*, Ælfc. Gr. 47; Som. 47, 42, 43.

fōr-nefe, an; *f. A nephew's daughter*; proneptis, Som. Ben. Lye. v. nefe.

Fornētes folm, e; *f. Fornet's palm*; Fornēti palma :—Wyl on eówe meolce Fornētes folm *boil Fornet's palm in ewe's milk*, L. M. 1, 70; Lchdm. ii. 144, 22. Nim Fornētes folm *take Fornet's palm*, 1, 71; Lchdm. ii. 146, 4. The *Icel.* has Fornjótr; *gen.* Fornjóts, the name of an eóten, es; *m. a giant.* Fornjótr's three sons had control over *air, fire*, and *wind*. In the Gl. Cleop. folm is glossed *mănus, the hand* or *palm.* As this refers to the palm only, it leaves us in difficulty what variety is intended by Fornet's palm. It must, however, be one of the chief species, as Fornjótr was a chief god of the heathen Icelanders.

for-niman, -nyman; *p.* -nam, -nom, *pl.* -námon, -nōmon; *pp.* -numen; *v. trans. To take away, deform, plunder, destroy, ransack, waste, consume, devour*; răpĕre, perdĕre, extermĭnāre, vastāre, consūmĕre, dĕvŏrāre :—Ðú hí eáðe miht forniman *thou mayest easily consume them*, Ps. Th. 72, 16 : 118, 36. Eów in beorge bǽl fornimeþ *fire shall consume you upon the hill*, Elen. Kmbl. 1153; El. 578. Se ðe fornimþ þearfan on dýgelnysse *qui dĕvŏrat paupĕrem in abscondĭto*, Cant. Abac. Lamb. fol. 190 b, 14. Hig fornymaþ hyra ansýna *extermĭnant făcĭes suas*, Mt. Bos. 6, 16. Hine wyrd fornam *fate took him away*, Beo. Th. 2415; B. 1205 : 2877; B. 1436 : 4245; B. 2119. Líg eall fornam *the flame consumed all*, Cd. 119; Th. 153, 34; Gen. 2548: Andr. Kmbl. 1988; An. 996: 3061; An. 1533. Swylt ealle fornom secga hlōþe *death destroyed all the band of men*, Exon. 75 b; Th. 283, 5; Jul. 675 : 59 b; Th. 216, 15; Ph. 268. Se Brytta þeóde fornom *qui gentem vastāvit Brittōnum*, Bd. 1, 34; S. 499, 20. Him írenne ecga fornámon *iron edges had taken them away from him*, Beo. Th. 5649; B. 2828. Fōrneáh hí fornámon me on lande *paulo mĭnus consummāvērunt me in terra*, Ps. Spl. C. 118, 87. Fornōmon [MS. -noman] *have consumed*, Exon. 78 a; Th. 292, 14; Wand. 99. Wylt ðū we secgaþ ðæt fŷr cume of heofone, and fornime hig *vis dīcĭmus ut ignis descendat de cǽlo, et consūmat illos?* Lk. Bos. 9, 54. Ðæs mannes wlite wyrþeþ eall fornumen mid onsīgendre ylde *the beauty of man becomes thoroughly destroyed by approaching old age*, Basil admn. 8; Norm. 50, 20. Swā swā sceáp from wulfum and wildeórum beóþ fornumene, swā ða earman ceasterwaran toslitene and fornumene wǽron fram heora feóndum *sīcut agni a fĕris, ĭta mĭsĕri cīves discerpuntur ab hostĭbus*, Bd. 1, 12; S. 481, 26, 27: Homl. Th. ii. 416, 12.

for-nýden; *p.* -nýdde; *pp.* -nýded *To force greatly, compel*; cōgĕre :—Wydewan syndon wīde fornýdde on unriht to ceorle *vidŭǽ crebro injuste ad nuptias trăhuntur*, Lupi Serm. i. 5; Hick. Thes. ii. 100, 25.

for-nyman *to take away, deform, disfigure*, Mt. Bos. 6, 16. v. forniman.

forod, forad, fored, forud; *adj. part.* [v. nacod *naked*] *Broken, fractured, violated*; fractus, violātus :—Wæs him gylp forod *their vaunt was broken*, Cd. 4; Th. 5, 10; Gen. 69. Ðá wearþ hire mid ánum wyrpe án ribb forod *then with one throw one of its ribs was broken*, Ors. 4, 6; Bos. 84, 41. Gif se earm biþ forod *if the arm be broken*, L. Alf. pol. 54; Th. i. 94, 24, note 57. Gif monnes ceácan mon forslihþ, ðæt hie beóþ forode *if a man smite another's cheeks, so that they be broken*, L. Alf. pol. 50; Th. i. 94, 15 : Ps. Th. 30, 12. Foredum sceancum *with broken legs*, H. R. 101, 21.

fōr-oft; *adv. Very often*; persǽpe :—Se deófol sǽwþ fōroft mánfullīce geþohtas into ðæs mannes heortan *the devil very often sows evil thoughts in the heart of man*, Boutr. Scrd. 20, 16. Swā swā we sylfe fōroft gesáwon *as we ourselves have very often seen*, Bd. de nat. rerum; Wrt. popl. science 12, 9; Lchdm. iii. 260, 2 : Wrt. popl. science 11, 8; Lchdm. iii. 256, 16.

fōron *went*, Ps. Spl. 65, 11; *pl. p. of* faran *to go*.

for-pǽran; *p.* de; *pp.* ed *To turn away, pervert, ruin, destroy*; pervertĕre, perdĕ :—He ðæs óðres sáwle forpǽrþ þurh his yfelum tihtingum *he perverts the other's soul by his evil instigations*, Homl. Th. ii. 226, 31 : 208, 20. Hie forpǽraþ ðæm edleáne *merĭtum pervertunt*, Past. 39, 3; Hat. MS. 53, 8. Wolde we sylfe ne forpǽraþ *if we do not destroy ourselves*, Homl. Th. i. 216, 9 : ii. 50, 5. Adam us forpǽrde þurh ánes æpples þigene *Adam ruined us by the eating of an apple*, Homl. Th. ii. 330, 32. Ðæt he ðone man forpǽre *that he may destroy the man*, Boutr. Scrd. 20, 20.

for-pyndan; *p.* de; *pp.* ed *To turn away*; rĕmŏvēre, reprĭmĕre :—Ðæt Éuan scyld is eal forpynded *the sin of Eve is all turned away*, Exon. 9 a; Th. 7, 7; Cri. 97. [*Icel.* pynda prĕmĕre, vexāre.] v. pynding.

fōr-rád *rode before* :—Fōrrád sió fierd hie fōran *the force rode before them*, Chr. 894; Th. 166, 7; *p. of* fōr-rídan, q. v.

fōr-radian *to hasten before, prevent*, Nat. S. Greg. Els. 23, 4 : 24, 6. v. fōr-hradian.

for-rǽdan; *p.* -rǽdde; *pp.* -rǽded; *or p.* -reord, -rēd; *pp.* -rǽden, *v. a. to give counsel against, to condemn, plot against, deprive by*

treachery, wrong; condemnāre, insĭdias părāre:—We beódaþ ðæt man Cristene men for ealles tô lytlum to deáþe ne forrǽde *we command that Christian men be not for altogether too little condemned to death,* L. C. S. 2 ; Th. i. 376, 19. Eádweard man forrǽdde and syððan acwealde *they plotted against Edward and afterwards murdered him,* Lupi Serm. i. 9 ; Hick. Thes. ii. 102, 10. Ðæt man his hláford of lífe forrǽde *that a man deprive his lord of life,* Lupi Serm. i. 9 ; Hick. Thes. ii. 102, 7. [Cf. *Icel.* ráða af dögum *to kill.*] Gif man gehádodne man forrǽde æt feó oððe æt feore *if any one wrong a man in holy orders as to money or as to life,* L. C. S. 40 ; Th. i. 400, 5 : L. E. G. 12 ; Th. i. 174, 6. [*Ger.* ver-rathen *to betray.*]

fŏr-raðe ; *adv. Very quickly;* cito :—Hí Godes bebod tobrǽcon forraðe *they broke the commandment of God very quickly,* Ælfc. T. 5, 6 : Gen. 20, 7.

for-rîdan ; *p.* -rád, *pl.* -ridon ; *pp.* -riden *To ride before, intercept;* præequĭtāre, intercĭpĕre :—Forrád sió fierd hie fôran *the force rode before them,* Chr. 894 ; Erl. 90, 25. Ða men hie fôran forrîdan mehton bútan geweorce *the men they might intercept outside the work,* 894 ; Erl. 93, 11. [*Laym. p. p.* forriden : *Ger.* vor-reiten *to ride before.*]

fŏr-rĭdel, es ; *m. A fore-rider, outrider, harbinger ;* præcursor :—Cyning Totilla sende his forrídel cýðan his tocyme ðam hálgan were *king Totila sent his harbinger to announce his coming to the holy man,* Homl. Th. ii. 168, 10. [*A. R.* vorrideles : *Ger.* vor-reiter *a fore-rider.*]

for-rotian ; *p.* ode, ade, ede ; *pp.* od, ad, ed [for-, rotian *to rot*] *To become wholly rotten, to rot, putrefy;* computrescĕre :—Ða fixas acwelaþ and ða wæteru forrotiaþ *pisces mŏrientur et computrescent ăquæ,* Ex. 7, 18. Hit forrotode *computruit,* 16, 20. Gemolsnad flǽsc *vel* forrotad *corrupted flesh;* tābes, Ælfc. Gl. 12 ; Som. 57, 74 ; Wrt. Voc. 20, 16. Ðæt sió rêþnes ðæs wînes ða forrotedan wunde clǽnsige *that the harshness of the wine may cleanse the corrupted wound,* Past. 17, 10 ; Hat. MS. 25 a, 9. [*A. R.* vorrotien : *Dut. Ger.* ver-rotten *to rot, putrefy, mortify.*]

for-rotodnys, -rotednys, -nyss, e ; *f. Rottenness, corruption ;* putrēdo, pus :—Mín flǽsc is ymbscrýd mid forrotodnysse *my flesh is covered with corruption,* Job Thw. 167, 36 : Prov. 12 : Homl. Th. ii. 282, 11. Deós forrotednyss *hoc pus,* Ælfc. Gr. 8 ; Som. 7, 35.

fŏr-rynel, fôre-rynel, es ; *m. A forerunner ;* præcursor :—Is se fôrrynel fǽger and sciéne *the forerunner* [*morning star*] *is fair and shining,* Bt. Met. Fox 29, 49 ; Met. 29, 25. Iohannes wæs Cristes forrynel *John was Christ's forerunner,* Homl. Th. i. 484, 34 : 356, 21 : Bt. 36, 1 ; Fox 170, 28. Ðæs mǽran fôrryneles *of the great forerunner,* Homl. Th. i. 364, 6.

for-sacan ; *p.* -sôc, *pl.* -sôcon ; *pp.* -sacen *To declare an opposition, oppose, object to, refuse, give up, forsake ;* detrectāre, recūsāre, desĕrĕre :—Gange ân mynet ofer ealne ðæs cynges ânweald, and ðone nân man ne forsace *let one money pass throughout the king's dominion, and that let no man refuse,* L. Edg. ii. 8 ; Th. i. 270, 1. Forsôc ðæne triumphan *refused the triumph,* Ors. 2, 4 ; Bos. 42, 43. He ðæt wæs eall forsacende *he was giving up all that,* 1, 12 ; Bos. 36, 16. v. sacan.

for-sêcan *to punish,* Exon. 38 a ; Th. 125, 2 ; Gú. 348. v. for-sêcan.

for-sêde, *p.* -sêde *accused,* Homl. Th. i. 50, 14, 16 ; *p.* of for-secgan.

for-sæt, *pl.* -sæton *delayed, deferred, obstructed,* Cd. 138 ; Th. 173, 10 ; Gen. 2859 : 114 ; Th. 150, 10 ; Gen. 2480 ; *p.* of for-sittan.

for-sáwon *rejected, despised,* Elen. Kmbl. 2633 ; El. 1318 ; *p. pl.* of for-seón.

for-scáden *scattered,* Exon. 39 b ; Th. 131, 1 ; Gú. 449 ; *pp.* of for-scádan. v. for-sceádan.

for-scæncednys, -nyss, e ; *f.* [for-, screcednes *supplantātio*] *A supplanting, deceit ;* supplantātio, fraus :—Man miclode ofor me hleóhrǽsc-nesse oððe forscæncednysse *hŏmo magnĭfĭcāvit sŭper me supplantātiōnem,* Ps. Lamb. 40, 10.

for-scapung, -sceapung, e ; *f. A bad action, fault, crime;* perversa actio, scĕlus :—Hí sǽdon ðæt hió wǽre for Fetontis forscapunge *they said that it was for the fault of Phaëton,* Ors. 1, 7 ; Bos. 30, 35. On mislícre forsceapunge *by various misdeeds,* 1, 11 ; Bos. 35, 2.

for-sceádan, -scádan ; *p.* -sceód, *pl.* -sceódon ; *pp.* -sceáden, -scáden [sceádan *to separate*] *To scatter, disperse;* dispergĕre :—Ðæt ða giemmas wǽren forsceádne [forsceadene, Cot.] æfter ðǽm strǽtum *that the gems were scattered along the streets,* Past. 18, 4 ; Hat. MS. 26 b, 25. Gê sind forscádene *ye are scattered,* Exon. 39 b ; Th. 131, 1 ; Gú. 449.

for-sceáf *cast down,* Cd. 153 ; Th. 190, 25 ; Exod. 204 ; *p.* of for-scúfan.

for-sceamian, -scamian, -scamigan ; *p.* ode ; *pp.* od [sceamian *to be ashamed*] *To be greatly ashamed ;* erŭbescĕre :—Forsceamian *erŭbescĕre,* Scint. 8. Hie forscamige *let it shame them,* Past. 21, 1 ; Hat. MS. 29 a, 26. [*Orm.* forrshamedd *much ashamed.*]

for-sceap, es ; *n.* [from sceapen *formed, created ; pp.* of sceppan *to create*] What is for- or mis-shapen *a fault, crime ;* mălefactum :—Me nǽdre to forsceape scyhte *the serpent incited me to crime,* Cd. 42 ; Th. 55, 22 ; Gen. 898.

fôr-sceáwian ; *p.* ode ; *pp.* od *To foreshew, foresee;* præ-ostendĕre, pŏrēre *in conspectu, providēre* :—Ic fôrsceáwode Driht on gesihþe mínre

symble *providēbam Dŏmĭnum in conspectu meo semper,* Ps. Spl. 15, 8. [*Ger.* vor-schauen *to foresee.*] v. fôre-sceáwian.

fôr-sceáwudlíce ; *adv. Providently, carefully, prudently ;* prōvide, Procem. R. Conc.

for-sceáwung, e ; *f. Providence ;* prŏvĭdentia :—Þurh Godes fôrsceáwunge *by the providence of God,* Homl. Th. i. 234, 21. v. fôre-sceáwung.

for-scending, e ; *f.* [scendan *to confound*] *Confusion ;* confúsio :—Mid forscendinge *præ confūsiōne,* Lk. Skt. Rush. 21, 25.

for-sceóppan ; *p.* -scóp, *pl.* -scópon ; *pp.* -sceápen *To re-create, transform, deform ;* transformāre :—Sume, hí sǽdon, ðæt hió [Circe] sceolde forsceóppan to león *some, they said, she* [*Circe*] *should transform to a lioness,* Bt. 38, 1 ; Fox 194, 33. v. for-sceppan.

for-sceorfan ; *p.* -scearf, *pl.* -scurfon ; *pp.* -scorfen [sceorfan *to gnaw, bite*] *To gnaw or eat off;* arrōdĕre :—Gærstapan ælc wuht forscurfon, ðæs ðe on ðam lande wæs grówendes *locusts ate off everything that was growing in the land,* Ors. 5, 4 ; Bos. 105, 17, notes, p. 24, 7, MS. L.

fôr-sceótan, he -scýt, *pl.* -sceótaþ ; *p.* -sceát, *pl.* -scuton ; *pp.* -scoten *To shoot before, anticipate, come before, prevent ;* anticĭpāre, prævĕnīre :—Ða ungesǽligan menn ne mágon gebidan hwonne he [deáþ] him to cume, ac fôrsceótaþ hine fôran *unhappy men cannot wait till he* [*death*] *comes to them, but anticipate him beforehand,* Bt. 39, 1 ; Fox 212, 3. Fôrscýt ðæt hwílendlíce wíte ða écan geniðerunge *the transient punishment will prevent eternal damnation,* Homl. Th. i. 576, 2. Mín God fôrscýt [MS. forscytte] oððe fôrestepþ me *Deus meus prævĕniet me,* Ps. Lamb. 58, 11. [*Ger.* vor-schiessen.]

for-sceppan, -sceóppan ; *p.* -sceóp, *pl.* -sceópon ; *pp.* -scepen *To transform ;* transformāre :—Heó alle forsceóp Drihten to deóflum *the Lord transformed them all to devils,* Cd. 16 ; Th. 20, 14 ; Gen. 308. Scinnan forscepene [*their*] *beauty transformed,* Cd. 214 ; Th. 269, 12 ; Sat. 72.

fôr-scip, es ; *n. The forepart of a ship, the prow ;* prōra :—Ancersetl [MS. anfer-] *vel* fôrscip *prōra,* Ælfc. Gl. 83 ; Som. 73, 73 ; Wrt. Voc. 48, 12.

for-scranc *shrank up, dried up, withered,* Gen. 32, 25 : Mt. Bos. 21, 19 : Mk. Bos. 4, 6 ; *p.* of for-scrincan.

for-scrang *shrank up, dried up,* Ps. Spl. 128, 5, = for-scranc ; *p.* of for-scrincan.

for-screncan, -scræncan ; *p.* -screncte, -scræncte ; *pp.* -scrænct, -screnct [screncan *to trip up*] *To supplant, overcome, oppress, cast down;* supplantāre, opprimĕre, elīdĕre :—Ða ðe leahtras forscrencaþ belimpaþ to Godes ríce *those who overcome sins belong to God's kingdom,* Homl. Th. i. 198, 23. Forscrænc hine *supplanta eum,* Ps. Lamb. 16, 13. Ðú forscrǽnctest onarísende on me *supplantasti insurgentes in me,* 17, 40. Forscrenct *elísa vel dejecta,* Ælfc. Gl. 78 ; Som. 72, 36 ; Wrt. Voc. 45, 68. Crist arǽrþ ða forscrenctan *Christ raises the oppressed,* Homl. Th. ii. 414, 23.

for-screncend, es ; *m.* [*part.* of forscrencan] *A supplanter ;* supplantātor :—Iacob is gecweden, forscrencend *Jacob is interpreted, a supplanter,* Homl. Th. i. 198, 21.

for-scrífan ; *p.* -scráf, *pl.* -scrifon ; *pp.* -scrifen [scrífan *to judge*].   **I.** *to condemn, proscribe ;* condemnāre, proscrībĕre :—He ðæt scyldige werud forscrifen hefde *he had proscribed the guilty host,* Cd. 213 ; Th. 267, 5 ; Sat. 33. Grendel fífelcynnes eard weardode hwíle, siððan him Scyppend forscrifen hæfde *Grendel inhabited a while the monster-race's abode, after the Creator had proscribed him,* Beo. Th. 213 ; B. 106.   **II.** *to write or cut into, cut down ;* incīdĕre, succīdĕre :—Awríteþ he on his wǽpne wællnota heáp, bealwe bócstafas bill forscrífeþ *he writes upon his weapon a heap of fatal marks, baleful letters he cuts into the bill,* Salm. Kmbl. 323–326, note ; Sal. 161, 162. Forscríf hine *succíde illam,* Lk. Skt. Hat. 13, 7, 9. [*Ger.* ver-schreiben *to prescribe.*]

fôr-scríhan ; *p.* -scráh, *pl.* -scrigon ; *pp.* -scrigen [scríhan *dīcāre*] *To abdicate, resign, give up ;* abdĭcāre :—Forscráh *abdĭcāvit,* Cot. 205.

for-scrincan, he -scrincþ ; *p.* -scranc, *pl.* -scruncon ; *pp.* -scruncen [for-, scrincan *to shrink*] *To shrink up, dry up, dwindle away, wither ;* emarcescĕre, exarescĕre, arefĭeri, arescĕre :—He forscrincþ *arescit,* Mk. Bos. 9, 18. Æt-hrán he his sine on his þeó and heó ðǽrrihte forscranc *tĕtĭgit nervum fĕmŏris ejus, et stătim emarcuit,* Gen. 32, 25. Sǽd forscranc *sēmen exāruit,* Mk. Bos. 4, 6 : Lk. Bos. 8, 6. Sóna forscranc ðæt fíctreów *arefacta est contĭnuo fĭculnea,* Mt. Bos. 21, 19. Hig forscruncon *āruērunt,* Mt. Bos. 13, 6. Mín hýd is forscruncen *my skin is shrunk up,* Job Thw. 167, 37. Hí gesáwon ðæt fíctreów forscruncen of ðam wyrtruman *vĭdērunt fīcum arĭdam factam a radīcibus,* Mk. Bos. 11, 20. On ðám porticon læg mycel menigeo forscruncenra *in his portĭcibus jăcēbat multĭtūdo magna arĭdōrum,* Jn. Bos. 5, 3.

for-scrufon *ate off,* Ors. 5, 4 ; Bos. 105, 17, = for-scurfon ; *p. pl.* of for-sceorfan.

for-scruncen *shrank up, dried up, withered,* Job Thw. 167, 37 : Mk. Bos. 11, 20 ; *pp.* of for-scrincan.

for-scruncon *dried up,* Mt. Bos. 13, 6 ; *p. pl.* of for-scrincan.

for-scúfan ; *p.* -sceáf, *pl.* -scufon ; *pp.* -scofen *To cast down ;* amŏvēre, dispellĕre :—Wlance forsceáf mihtig engel *a mighty angel cast down their pride,* Cd. 153 ; Th. 190, 25 ; Exod. 204. v. scúfan.

for-scúnian, -scúnigean; *p.* ode; *pp.* od [scúnian *to shun*] *To blush, feel shame;* erŭbescĕre, Scint. 4.

for-scurfon *gnawed* or *ate off*, Ors. 5, 4; Bos. 105, 17, notes, p. 24, 7, MS. L; *p. pl.* of for-sceorfan.

for-scyldigian, -scyldegian, -scyldgian; *p.* ode; *pp.* od [scyldigian *accūsāre*] *To make guilty, to criminate, condemn;* reum fácĕre, damnáre:—Hreówlíce gefǽrþ se ðe hine sylfne forþ forscyldigaþ *he fares roughly who constantly criminates himself,* L. Pen. 12; Th. ii. 280, 28. Forscyldegod *scělěrātus* vel *facǐnŏrōsus,* Wrt. Voc. 86, 65. Wurdon hí deádlíce and forscyldegode þurh ágenne cyre *they became mortal and guilty through their own choice,* Homl. Th. i. 112, 16. He wæs forscyldgod *he was guilty,* i. 12, 21. Ne slihþ se déma ðone forscyldgodan sceaðan, ac he hæt his underþeóddan hine belífian *the judge slays not the condemned robber, but he commands his subordinates to deprive him of life,* ii. 36, 9. [Cf. *Ger.* ver-schulden *to be guilty.*]

for-scyppan *to transform.* v. for-sceóppan.

fôr-scýt *shoots before, prevents* or *will prevent,* Homl. Th. i. 576, 2; *pres.* of fôr-sceótan.

fôr-scyttan; *p.* -scytte, *pl.* -scytton; *pp.* -scytted *To shoot before, prevent;* prævĕníre:—Hí heófodon folces synna, and heora wrace on him sylfum fôrscytton *they bewailed the people's sins, and prevented their punishment on themselves,* Homl. Th. i. 540, 31. Ðæt ða sceortan wítu ðises geswincfullan lífes fôrscytten [MS. forscyttan] ða toweardan, ðe nǽfre ateoriaþ *that the short punishments of this painful life may prevent those to come, which will never fail,* Homl. Th. ii. 328, 34. DER. scyttan.

for-seah, ðú -seáge *despised, thou despisedst,* Exon. 40b; Th. 134, 23; Gú. 512: Ps. Spl. 88, 37; *p.* of for-seón.

for-seárian; *p.* ode; *pp.* od [seárian *to sear*] *To dry up, wither;* arēre, arescĕre:—Ic forseárige *āreo,* Ælfc. Gr. 26, 2; Som. 28, 44. Se líchama ᵹewyrþeþ *to* duste and forseáraþ *the body turns to dust and withers,* Basil admn. 8; Norm. 50, 17: Homl. Th. ii. 92, 3. Adruwode oððe forseárode swá swá blýwnys oððe croccsceard mægen mín *āruit tamquam testa virtus mea,* Ps. Lamb. 21, 16. Mín hýd forseárode *my skin withered,* Job Thw. 167, 37. Ðonne hit forealdod biþ and forseárod *when it is grown old and withered,* Bt. 39, 8; Fox 224, 11.

for-sécan, -sécan; *p.* -sóhte, *pl.* -sóhton; *pp.* -sóht *To afflict, punish;* pœna afflicĕre:—Ðeáh ðe gé hine sárum forsécen *though ye sorely afflict it,* Exon. 38a; Th. 125, 2; Gú. 348. Sárum forsóht *afflicted with sorrows,* Elen. Kmbl. 1862; El. 933. DER. sécan.

for-secgan; *p.* -sægde, -sǽde; *pp.* -sægd, -sǽd *To for-say, mis-say, pretend, deny, say against, accuse;* prædícĕre, diffámáre, něgáre, accūsáre:—Se ðe óðerne mid wó forsecgan wille *he who shall accuse another wrongfully,* L. C. S. 16; Th. i. 384, 20: L. Edg. ii. 4; Th. i. 266, 22. Se óðerne to deáþe forsegþ *he traduces another to death,* Homl. Th. ii. 208, 19. Be ðon ðe mon óðerne forsecgaþ *in case any one accuse another,* L. Edg. ii. 4, titl; Th. i. 266, 21. Swá hwá swá óðerne forsǽde *whosoever accused another,* Homl. Th. i. 50, 16. Ða leásan gewitan hine forsádon *the false witnesses accused him,* i. 50, 14.

for-ségon *despised, rejected, renounced,* Elen. Kmbl. 778; El. 389; *p. pl.* of for-seón.

for-sendan; *p.* -sende; *pp.* -sended *To send away, send into banishment, banish;* dimittĕre, relēgáre, deportáre:—Sume on wræcsíþ forsende *some he sent away into banishment,* Ors. 3, 7; Bos. 60, 39. He hine siððan forsende *he afterwards banished him,* 3, 7; Bos. 59, 26. He wearþ snúde forsended *he was quickly banished,* Beo. Th. 1812; B. 904. [*Ger.* ver-senden *to send away.*]

fôr-sendan *to send before.* v. fôre-sendan.

for-seón, -sión; ic -seó, ðú -sihst, -sixst, he -sihþ, -syhþ, *pl.* -seóþ; *p.* -ic, he -seah, ðú -sáwe, -seáge, *pl.* -sáwon, -ségon; *impert.* -seoh; *subj.* he -seó; *pp.* -sewen *To overlook, despise, contemn, scorn, be ashamed of, neglect, reject, renounce;* despícĕre, temnĕre, contemnĕre, spernĕre, erŭbescĕre, neglĭgĕre, posthăbĕre, rejícĕre:—We á sculon ídle lustas forseón *we should ever despise idle lusts,* Exon. 19a; Th. 47, 18; Cri. 757: Boutr. Scrd. 21, 43. Ôþ-ðæt ðú meahte ǽlc eorþlíc þing forsión *until thou mayest look down upon every earthly thing,* Bt. Met. Fox 24, 14; Met. 24, 7. Ic forseó *temno,* Ælfc. Gr. 28, 4; Som. 31, 17. Ic fracuþe forseó feóndas míne *ego vĭdēbo inĭmīcos meos,* Ps. Th. 117, 7. Ic forseó *posthăbeo,* Ælfc. Gr. 47; Som. 48, 31. Ðú forsihst [-sixst, Lamb.] on gerecum on gedréfednysse *despicis in opportunitātĭbus in tribulātiōne,* Ps. Spl. second 9, 1. He forsihþ ðás eorþlícan gód *he despises these earthly goods,* Bt. 12; Fox 36, 25: Gen. 16, 5. Se ðe me and míne spæca forsyhþ, ðone mannes Sunu forsyhþ *qui me erubŭerit et meos sermōnes, hunc Fīlius hŏmĭnis erubescet,* Lk. Bos. 9, 26: Mk. Bos. 8, 38. Gif gé míne ǽ and míne dómas forseóþ *si sprevĕrĭtis lēges meas et jūdĭcia mea,* Lev. 26, 15. Gúþlác mán eall forseah *Guthlac despised all sin,* Exon. 34a; Th. 108, 4; Gú. 67: 40b; Th. 134, 23; Gú. 512. Ðú forseáge Cristum ðínne *despexisti Christum tuum,* Ps. Spl. 88, 37. Hie mána gehwylc forseón *they rejected every sin,* Elen. Kmbl. 2633; El. 1318. Forsáwon hyra séllan *they despised their superior,* Exon. 84a; Th. 317, 5; Mód. 61. Gé blindnesse bóte forségon *ye renounced the remedy of blindness,* Elen. Kmbl. 778; El. 389. Ne forseoh ǽfre, ðæt ðú sylfa ǽr, mid ðínum

---

handum hér geworhtest *ŏpĕra manuum tuārum ne despĭcias,* Ps. Th. 137, 8: 54, 1: Ps. Lamb. 26, 9. Gif preóst óðerne forseó oððe gebismirige *if a priest despise or insult another,* L. N. P. L. 29; Th. ii. 294, 17. Wæs mǽrþa fruma tó swíðe forsewen *the source of marvels was too greatly despised,* Chr. 975; Erl. 126, 16; Edg. 42. Bióþ forsewene heora láreówas *their teachers are despised,* Bt. Met. Fox 13, 74; Met. 13, 37. Forhygdelíc oððe forsewen *contemptus,* Ps. Lamb. 118, 141. [*Orm.* forrseon *to despise:* . *Ger.* ver-sehen *to see wrong.*]

for-seónnes, -ness, e; *f.* A *looking down upon, contempt;* despectio, contemptus, Som. Ben. Lye. v. for-sewennes.

for-seten *obstructed,* Ors. 4, 6; Bos. 84, 13; *pp.* of for-sittan.

for-settan; *p.* -sette, *pl.* -setton; *pp.* -seted, -sett *To obstruct;* obstruĕre:—Hí ðone heofonlícan weg forsetton *they obstructed the heavenly way,* Bd. 3, 19; S. 548, 4. [*Ger.* versetzen *to misplace, obstruct.*]

fôr-settan; *p.* -sette, *pl.* -setton; *pp.* -seted, -sett *To set before; pro-*pŏněre:—Gif ic ne fôrsette ðé Hierusalem *si non propŏsuěro Hierŭsălem,* Ps. Th. 136, 6. Hig ne fôrsetton God tofôran ansýne heora *non propŏsuērunt Deum ante conspectum suum,* Ps. Lamb. 53, 5. [*Ger.* vor-setzen *to set before.*]

fôr-settednys, -nyss, e; *f.* [fôrseted, *pp.* of fôrsettan; -nyss] A *proposition;* propŏsĭtio:—Ic sprece fôrsettednyssa of frymþe *lŏquar propŏsĭtiōnes ab inĭtio,* Ps. Spl. 77, 2. v. fôre-setnes.

for-sewen *despised,* Ps. Lamb. 118, 141; *pp.* of for-seón.

for-sewenlíce; *comp.* -lícor; *adv. Contemptibly, ignominiously;* contemptĭbĭlĭter, turpĭter:—Swá he forsewenlícor biþ gewítnod for Godes naman, swá his wuldor biþ máre fôr Gode *the more ignominiously he is tortured for the name of God, the greater shall his glory be before God,* Homl. Th. i. 486, 23.

for-sewennes, fore-seuwenes, -ness, -nyss, e; *f.* A *looking down upon, contempt;* contemptus, despectio:—Gefylled we synd forsewennysse *replēti sŭmus despectiōne,* Ps. Spl. 122, 4, 5. For his forsewenness *out of contempt for him,* Ors. 4, 4; Bos. 81, 13. Forsewennyss *contemptus,* Ælfc. Gr. 28, 4; Som. 31, 17.

for-sewestre, an; *f. She who despises;* contemptrix, Som. Ben. Lye.

for-sihst, -sihþ *despisest, despiseth,* Ps. Spl. second 9, 1: Gen. 16, 5; 2nd *and* 3rd *sing. pres.* of for-seón.

for-singian *to sin greatly,* L. Pen. 12; Wilk. 95, 9. v. for-syngian.

for-sión *to despise,* Past. 32, 1; Hat. MS. 39 b, 27. v. forseón.

for-síþ, es; *m.* A *going away, departure, death;* exĭtium, ŏbĭtus, mors:—Sóna æfter his forsíþe wæs ealra witena gemót on Oxna forda *soon after his death there was a meeting of all the counsellors at Oxford,* Chr. 1036; Erl. 164, 12. v. forþ-síþ.

for-síðian; *p.* ode; *pp.* od [síðian *to journey*] *To perish; ĭter fătāle* iníre:—Hæfde ða forsíðod sunu Ecgþeówes *Ecgtheow's son had then perished,* Beo. Th. 3104, note; B. 1550.

for-sittan; he -siteþ; *p.* -sæt, *pl.* -sǽton; *pp.* -seten *To mis-sit, to be absent from, neglect, delay, defer, diminish, obstruct, besiege;* abesse a, neglĭgĕre, supersĕdēre, desĕrĕre, præstruĕre, obsĭdēre:—Be ðon ðe gemót forsitte *of him who is absent from the council,* L. Ath. i. 20; Th. i. 208, 25, 26. Be ðon ðe man fyrde forsitte *in case a man neglect the army,* L. In. 51; Th. i. 134, 7, 8. Ne forsæt he ðý síðe *he delayed not the journey,* Cd. 138; Th. 173, 10; Gen. 2859. Ne he tíd forsæt *he deferred not the time,* Exon. 37b; Th. 122, 26; Gú. 311. Ðæt eágena bearhtm forsiteþ and forsworceþ *the twinkling of the eyes diminishes and darkens,* Beo. Th. 3538; B. 1767. Hí hæfdon ðone weg forsiten *they had blockaded the way,* Ors. 4, 6; Bos. 84, 13. Fearras forsǽton me *tauri obsēdērunt me,* Ps. Spl. 21, 11: Cd. 114; Th. 150, 10; Gen. 2489.

for-sixt *despisest,* Ps. Lamb. second 9, 1, = for-sihst; 2nd *sing. pres.* of for-seón.

for-slǽgen *slain,* Chr. 882; Erl. 82, 13; *pp.* of for-sleán.

for-slǽhþ *breaks,* L. Ethb. 50; Th. i.16, 1; 3rd *sing. pres.* of for-sleán.

for-slagen *slain,* Ors. 3, 7; Bos. 62, 10; *pp.* of for-sleán.

fôr-slāwian; *p.* ode; *pp.* od [slāwian *to be slow*] *To be slow, unwilling;* pĭgēre:—Ic wát, ðæt ðú náht né forslāwodest *I know that thou wouldest not be unwilling,* Bt. 10; Fox 28, 15.

for-sleán; he -slæhþ, -slyhþ, -slihþ; *p.* -slóh, *pl.* -slógon; *pp.* -slegen, -slægen, -slagen [sleán *to strike*] *To strike with violence, smite, break, slay, kill, destroy;* vehementer fĕríre, percŭtĕre, frangĕre, occīdĕre, interfícĕre:—Se ðe cinbán forslæhþ mid xx scillingum forgelde *let him who breaks the chin-bone pay for it with twenty shillings,* L. Ethb. 50; Th. i. 16, 1. Gif monnes ceácan mon forslihþ [forslyhþ, H] ðæt hie beóþ forode, gebéte mid xv scillinga *if one smite a man's cheeks, that they be broken, let him make amends with fifteen shillings,* L. Alf. pol. 50; Th. i. 94, 14. He ealle ða rícostan forsleán hét *he commanded [them] to slay all the most powerful,* Ors. 3, 7; Bos. 60, 38. Ercol hí swíðe forslóh and fordyde *Hercules grievously slew and destroyed them,* Ors. 1, 10; Bos. 33, 34. Forslegen Sodoma folc *the slaughtered people of Sodom,* Cd. 94; Th. 122, 5; Gen. 2022. Hí forslegene wurdon *they were slain,* Ors. 1, 13; Bos. 37, 5. Ða men wæron forslægene *the men were slain,* Chr. 882; Erl. 82, 13. He hí forslagen hæfde *he had slain them,* Bt. 16, 2; Fox 54, 2: Ors. 3, 7; Bos. 62, 10. [*Ger.* verschlagen.]

for-slegen *slain, slaughtered*, Cd. 94; Th. 122, 5; Gen. 2022; *pp. of* for-sleán.

for-sliet, es; *m.* [sliet = slíte *a slit*] *Slaughter, massacre*; internēcio, Cot. 108.

for-slihþ *smites*, L. Alf. pol. 50; Th. i. 94, 14; *3rd sing. pres. of* for-sleán.

for-slítan; *p.* -slát, *pl.* -sliton; *pp.* -sliten [slítan *to tear*] *To tear with the teeth, to devour*; mordícus lacĕrāre, comĕdĕre :—Lēt [wyrm] hiora wyrta wæstme forslítan *he let [the worm] devour the fruit of their plants*, Ps. Th. 77, 46. [*O. Sax.* farslítan *to tear up, consume.*]

for-slóh *slew*, Ors. 1, 10; Bos. 33, 34; *p. of* for-sleán.

for-slyhþ *smites*, L. Alf. pol. 50; Th. i. 94, 14, MS. H; *3rd sing. pres. of* for-slean.

for-smorian; *p.* ode; *pp.* od; *v. trans.* *To smother, choke, suffocate, stifle*; suffŏcāre :—Hí synd mid heora lífes lustum forsmorode ... woruldcara and wélan forsmoríaþ *dæs módes þrotan they are choked with the pleasures of their life ... worldly cares and riches choke the throat of the mind*, Homl. Th. ii. 92, 8–11. On úrum gástlícum fulluhte biþ se deófol forsmorod fram us *in our spiritual baptism the devil is stifled by us*, ii. 200, 19.

for-sóc, *pl.* -sócon *refused*, Chr. 1070; Erl. 208, 4; *p. of* for-sacan.

for-sogen *sucked* or *drawn out*, L. M1 2, 7; Lchdm. ii. 186, 17; *pp. of* for-súgan.

fór-sorged; *part.* [fór, sorgian *to sorrow*] *Made very sad, grieved, sorrowful*; tristātus, triste factus, Som. Ben. Lye.

for-sóþ; *adv.* FORSOOTH, *truly, certainly*; certe :—Wite ðu forsóþ *know thou assuredly*, Bt. 14, 3; Fox 46, 16. Ic forsóþ wát *vērum nōvi*, Bd. 3, 13; S. 538, 33. Saga him forsóþ *dic ergo illi*, Bd. 5, 9; S. 622, 37.

for-spanan, he -spaneþ, -spenþ; *p.* -spon, -speón, *pl.* -spónon, -speónon; *pp.* -spanen, -sponen; *v. trans.* *To entice, seduce*; illicĕre, sedúcĕre :—Gehwá se ðe óðerne to leahtrum forspenþ is manslaga *every one who entices another to sins is a manslayer*, Homl. Th. ii. 226, 30. Hine his hyge forspeón, dæt he ne wolde Drihtnes word wurþian *his mind seduced him, that he would not revere the Lord's word*, Cd. 18; Th. 22, 34; Gen. 350. Forspanen beón *seductum esse, sedúci*, Prov. 30, Lye. [*O. Sax.* for-far-spanan *to entice.*]

for-spanc, -spanc *an enticement, allurement.* v. for-spanincg.

for-spanend; es; *m.* *A seducer*; seductor, Som. Ben. Lye.

for-spanincg, -spannincg, e; *f.* *An enticement, allurement*; illĕcebra, Scint. 21, Lye.

for-speca, fore-speca, -spreca, -spræca, an; *m.* *One who speaks for another, a defender, advocate*; advŏcātus, patrōnus :—Forspeca vel mundbora *advŏcātus, patrōnus*, vel *interpellātor*, Ælfc. Gl. 106; Som. 78, 62; Wrt. Voc. 57, 42. Slaga sceal his forspecan on hand syllan, and se forspeca mágum *the slayer shall give pledge to his advocate, and the advocate to the kinsmen*, L. Edm. S. 7; Th. i. 250, 14, 15, 16. Ðe hire forsprecan [-specan MS. B.] synd *who are her advocates*, L. Edm. B. 1; Th. i. 254, 5.

for-specan; *p.* -spæc, *pl.* -spæcon; *pp.* -specen [for-, specan, sprecan *to speak*] *To speak in vain, speak negatively, deny*; frustra dícĕre, nĕgāre :—Hæbbe he dæt eall forspecen *let him have spoken that all in vain*, L. C. S. 27; Th. i. 392, 6. Ne sý forspecen ne forswígod *let it not be denied nor concealed*, L. Ath. v. § 8, 9; Th. i. 238, 15.

fór-spédian; *p.* ode; *pp.* od *To speed forward, to prosper*; prospĕrāre :—Eálá ðu Driht gehæl me, eálá ðu Driht wel to fórspédienne *O Dŏmine salvum me fac, O Dōmine bĕne prospĕrāre*, Ps. Spl. T. 117, 24. v. spédan.

for-spendan; *p.* de; *pp.* ed [for-, spendan *to spend*] *To spend utterly, to consume*; consūmĕre :—Swíðost ealle hys spéda hý forspendaþ *they squander almost all his property*, Ors. 1, 1; Bos. 22, 45.

for-spennen, e; *f.* *An enticement*; lēnŏcĭnium :—Forspennene lēnŏcĭnia, Mone B. 671. v. for-spenning.

for-spennend, es; *m.* *A whoremonger*; léno, Ælfc. Gr. 9, 3; Som. 8, 49 : Mone B. 3130. v. for-spanend.

for-spennestre, -spennystre, an; *f.* *A bawd*; léna, Ælfc. Gr. 9, 3; Som. 8, 49.

for-spenning, e; *f.* *An enticement, allurement*; illĕcebra, lēnŏcĭnium :—Forspenningce *illĕcebras*, Mone B. 4614. Mid forspenningce lēnŏcĭnio, 3098. Forspenningce lēnŏcĭnia, 6013 : 6274.

for-spenþ *entices*, Homl. Th. ii. 226, 30; *3rd sing. pres. of* for-spanan.

for-speón *seduced*, Cd. 18; Th. 22, 34; Gen. 350; *p. of* for-spanan.

for-spild, es; *m.* *Destruction*; perdítio :—On forspild *into destruction*, Past. 40, 5; Cott. MS.

for-spildan; *p.* de; *pp.* ed [spild *destruction*] *To bring to naught, destroy*; perdĕre :—Sum sceal on geóguþe, mid Godes meahtum, his earfoþsíþ forspildan *one shall in youth, with God's power, bring to naught his hard lot*, Exon. 88 a; Th. 330, 31; Vy. 59.

for-spillan, -spyllan; *p.* de; *pp.* ed [spillan *to spill, spoil, destroy*] *To spill, lose, waste, destroy, disperse*; perdĕre, disperdĕre, dissĭpāre :—Darfus wolde hine sylfne forspillan *Darius would destroy himself*, Ors. 3, 9; Bos. 65, 40. Alýfþ reste-dagum wel to dónne, hwæðer ðe yfele?

sáwla gehælan, hwæðer ðe forspillan *lĭcet sabbătis benefăcĕre, an mŭle? anĭmam salvam făcĕre, an perdĕre?* Mk. Bos. 3, 4. Se ðe wyle hys sáwle hále gedón, he hig forspilþ; and se ðe wyle hig for me forspyllan, se hig fint *qui vŏluĕrit anĭmam suam salvam făcĕre, perdet eam; qui autem perdĭdĕrit anĭmam suam propter me, invĕniet eam*, Mt. Bos. 16, 25. Ðú forspildest ealle ða ðe forligríaþ fram ðē *perdĭdisti omnes qui fornĭcantur abs te*, Ps. Spl. 72, 26. He his gód forspilde *dissĭpasset bŏna ipsius*, Lk. Bos. 16, 1 : 15, 13. Ne forspil ðú sáwle míne *ne perdas anĭmam meam*, Ps. Spl. 26, 9. Dæt he fordó oððe forspille of lande gemynd heora *ut perdat de terra mĕmōriam eōrum*, Ps. Lamb. 33, 17. [*Dut.* ver-spillen *to spend, waste.*]

for-spillednes, -nys, -ness, -nyss, e; *f.* [forspilled, *pp. of* forspillan *to spill*; -nes, -ness] *A spilling, waste, perdition, destruction*; perdĭtio :—Forhwí wæs ðisse sealfe forspilledness geworden *ut quid perdĭtio ista unguenti facta est?* Mk. Bos. 14, 4. Ne forwearþ hyra nán, búton forspillednysse bearn *nēmo ex eis pĕriit, nĭsi fílius perdĭtiōnis*, Jn. Bos. 17, 12. Se weg is swíðe rúm ðe to forspilledness gelæt *spatiōsa via est, quæ dŭcit ad perdĭtiōnem*, Mt. Bos. 7, 13.

for-spreca *one who speaks for another, an advocate*, L. Edm. B. 1; Th. i. 254, 5. v. for-speca.

fór-sprecen; *part.* *Fore-spoken, fore-mentioned*; præfātus :—Todælde se fórsprecena here on twá *the fore-mentioned army divided into two*, Chr. 885; Erl. 83, 22. v. fóre-sprecen.

for-spyllan *to lose* :—Wyle forspyllan *will lose*, Mt. Bos. 16, 25. v. for-spillan.

for-spyrcan; *p.* te; *pp.* ed [spearca *a spark*] *To dry out, empty*; exarescĕre, arēre :—Forspyrcende synd míne mearhcófan *ossa mea aru-ērunt*, Ps. Th. 101, 3.

FORST, es; *m.* FROST; gĕlu :—Se hearda forst *the hard frost*, Exon. 56 b; Th. 201, 19; Ph. 58 : 111 a; Th. 425, 11; Rä. 41, 54. Forst gĕlu, Ælfc. Gl. 94; Som. 75, 101; Wrt. Voc. 52, 51 : 76, 30 : Ps. Th. 148, 8. Hwílum hára scóc forst of feaxe *sometimes the hoar frost shook from my hair*, Exon. 130 a; Th. 498, 27; Rä. 88, 8. Án sceal inbindan forstes fetur *one shall unbind frost's fetters*, 90 a; Th. 338, 9; Gn. Ex. 76 : Beo. Th. 3222; B. 1609 : Salm. Kmbl. 708; Sal. 353. Forste gefeterad *fettered with frost*, Menol. Fox 407; Men. 205 : Homl. Th. i. 84, 15. Forstas and snáwas *frosts and snows*, Cd. 192; Th. 239, 31; Dan. 378. [*Chauc.* froste: *Orm.* frosst: *O. Sax.* frost, *m*: *Frs.* froast: *O. Frs.* frost, forst: *Dut.* vorst, *f*: *Ger.* frost, *m*: *M. H. Ger.* vrost, *m*: *O. H. Ger.* frost, *m*: *Goth.* frius, *n*: *Dan.* frost, *m. f*: *Swed.* frost, *m*: *Icel.* frost, *n.*] DER. rím-forst.

for-stæl, *pl.* -stælon *stole*, Gen. 27, 36: Mt. Bos. 28, 13. *p. of* for-stelan.

fór-stæpþ *steps before, goes before*, Ps. Spl. 96, 3; *pres. of* fór-stapan.

for-stal *an assault, fine for an assault*, L. C. S. 12; Th. i. 382, 14. v. fór-steal.

for-stalian; *p.* ede; *pp.* ed [stalian *to steal*] *To steal away*; aufŭgĕre :—Gif wítepeów hine forstalie *if a penal slave steal himself away*, L. In. 24; Th. i. 118, 6. Gif he hine forstalede *if he should have stolen himself away*, L. Ath. v. § 6, 3; Th. i. 234, 7.

for-standan, -stondan; he -stent; *p.* -stód, *pl.* -stódon; *pp.* -standen; *v. trans.* I. *to stand up for, to defend, aid, help, benefit, avail*; defendĕre, prodesse :—Gif hine nelle forstandan *if he will not stand up for him*, L. In. 62; Th. i. 142, 6. Twelfhyndes mannes áþ forstent vi ceorla áþ *a twelve hundred man's oath stands for the oath of six churls*, L. O. 13; Th. i. 182, 19. Dæt his gewitnes eft næht ne forstande *that his witness avail again nothing*, L. Ath. i. 10; Th. i. 204, 24. Gif hine hwá forstande *if any one stand up for him*, L. Ath. i. 1; Th. i. 198, 25. Gif hine hwá fórene forstande *if any one defend him*, v. § 1, 4; Th. i. 230, 4 : v. § 8, 2; Th. i. 236, 12: L. Eth. i. 4; Th. i. 284, 2 : L. C. S. 33; Th. i. 396, 17. He mihte hord forstandan *he might defend the treasure*, Beo. Th. 5903; B. 2955. Forstond ðu mec *protect thou me*, Exon. 118 b; Th. 455, 31; Hy. 4, 58. Hwá forstandeþ hie, gif ðú hie ne scyldest *who shall defend it, if thou dost not shield it*, Blickl. Homl. 225, 18. Hwæt forstód ðám betestum mannum—oððe hwæt forstent hit *what did it help the best men—or what does it profit?* Bt. 18, 4; Fox 68, 7, 9. Ne forstent dæt þweál náuht *the washing profits nothing*, Past. 54; Hat. MS. II. *to understand*; intelligĕre :—Uneáðe ic mæg forstandan ðíne acsunga *I can scarcely understand thy questions*, Bt. 5, 3; Fox 12, 15. Selfe forstódon his word onwended *they themselves understood his words [to be] perverted*, Cd. 37; Th. 48, 2; Gen. 769. v. under-standan. [Like *Dut.* ver-staan: *Ger.* ver-stehen *to understand.*]

fór-standan, -stondan; *p.* -stód, *pl.* -stódon; *pp.* -standen *To stand before* or *against, withstand, oppose, hinder*; resistĕre, impĕdīre :—Ne meahte seó weáláf wíge fórstandan *the miserable remnant could not withstand in battle*, Bt. Met. Fox 1, 44; Met. 1, 22. Ne mágon gé him ða wíc fórstondan *to him ye may not hinder the dwelling's*, Exon. 42 b; Th. 144, 7; Gú. 674. Ic him dæt fórstonde *I hinder them from that*, Exon. 105 a; Th. 398, 15; Rä. 17, 8. Godes engel fórstód ðone weg *stĕtit angĕlus Dŏmini in via*, Num. 22, 22. v. wiðstandan *to withstand*.

fór-stapan; he -stæpþ; *p.* -stóp, *pl.* -stópon; *pp.* -stapen *To st.p* or

go before, precede; præcēdĕre:—Fŷr ætfōran him fôrstæpþ [Lamb. fôre-stæpþ] *ignis ante ipsum prœcēdet*, Ps. Spl. 96, 3. v. fôre-stapan.

fôr-steal, -steall, -stal, fôre-steall, es; *m*. [fôr, fôre *before*; steal *from* stellan *to leap, spring*; therefore, at least originally, *an assault, consisting in one man springing or placing himself before another, so as to obstruct his progress*, Thorpe's Glos. to A. Sax. Laws]. I. *an assault*; assultus sŭper ălĭquem in via rēgia factus, viæ obstructio:—Gif hwā fôr-steal oððon openne wiðercwyde ongeán lahriht Cristes oððe cyninges gewyrce *if any one commit an assault or open opposition against the law of Christ or of the king*, L. Eth. v. 31; Th. i. 312, 8: vi. 38; Th. i. 324, 21. In L. H. 80, § 2; Th. i. 586, 2, it is said,—' Si in via rēgia fiat assultus sŭper ălĭquem, fôrestel est.' II. *the fine for an assault*; mulcta pro assultu:—Ðis syndon ða gerihta ðe se cyning āh ofer ealle men on West-Sexan [MS. Wes-Sexan], ðæt is ... fôrsteal *these are the rights which the king enjoys over all men in Wessex, that is ... the fine for assault*, L. C. S. 12; Th. i. 382, 14, note 27, MS. G. Switelige ic hēr hwæt se eáca is ðe ic ðærto ge-unnen hæbbe ... ðæt syndan fôr-steallas *I here declare what the augmentation is which I have thereto granted ... that is the fines for assaults*, Th. Diplm. A.D. 1035; 333, 32: A.D. 1066; 411, 32. See also Schmid Glos. forsteal.

for-stelan, he -steleþ, -stelþ, -stylþ, *pl*. -stelaþ; *p*. -stæl, *pl*. -stǽlon; *pp*. -stolen *To steal with violence, rob, deprive*; fūrāri, surrĭpĕre, prīvāre:—Sēcende forstelan sāwla *quærens fūrāri ănĭmas*, Ps. Lamb. fol. 142, 8. Gif ceorl ceáp forstelþ [-stylþ MS. B; -steleþ MS. H.] *if a churl steal property*, L. In. 57; Th. i. 138, 15: L. Alf. 15; Th. i. 48, 5, MS. H. Gif hwā befæst his feoh to hyrdnysse and hit man forstylþ ðam, ðe hit underfēhþ, gif man ðone þeóf finde, gilde be twīfealdon *si quis commendāvĕrit amīco pĕcūniam in custōdiam et do, qui suscĕpĕrat, furto ablāta fŭĕrit, si invĕnĭtur fur, duplum reddet*, Ex. 22, 7. Ðǽr þeófas hit delfað and forstelaþ *ubi fūres effŏdiunt et fūrantur*, Mt. Bos. 6, 19, 20. Ǽr he ætbræd me mīne frumcennedan and nū ōðre sīþe forstæl mīne bletsunga *prīmŏgĕnĭta mea ante tŭlit et nunc sĕcundo surrĭpuit benedictiōnem meam*, Gen. 27, 36. Secgeaþ, ðæt hys leorningcnihtas forstǽlon hyne dīcĭte, *quia discĭpŭli fūrāti sunt eum*, Mt. Bos. 28, 13. Gif frigman mannan forstele *if a freeman steal a man*, L. H. E. 5; Th. i. 28, 10: 7; Th. i. 30, 7: L. In. 46; Th. i. 130, 12. Gif hine man forstǽle *if any one should steal him*, L. Ath. v. § 6, § 3; Th. i. 234, 4: L. Alf. 15; Th. i. 48, 5. Iacob niste, ðæt Rachel hæfde ða andlícnyssa forstolen *Iacob ignōrābat, quod Rachel fūrāta esset idōla*, Gen. 31, 32: Exon. 92 a; Th. 345, 18; Gn. Ex. 190. Ferhþe forstolen *deprived of life*, Cd. 76; Th. 95, 15; Gen. 1579. Gif mon forstolenne ceáp befēhþ *if a man attach stolen cattle*, L. In. 47; Th. i. 132, 4: 75; Th. i. 150, 5. Be forstolenes ceápes forefonge *of the rescuing of stolen property*, 75; Th. i. 150, 4. Be forstolenum flǽsce *of stolen flesh*, 17; Th. i. 114, 1.

for-stent *stands for, avails, profits*, L. O. 13; Th. i. 182, 19: Bt. 18, 4; Fox 68, 9; *3rd sing. pres.* of for-standan. v. standan.

forst-líc; *adj. Frost-like, frozen*; glăciālis:—Forstlíc glăciālis, Ælfc. Gl. 94; Som. 75, 104; Wrt. Voc. 52, 54.

for-stód, *pl*. -stódon *stood for, availed, profited, understood*, Bt. 18, 4; Fox 68, 7: Cd. 37; Th. 48, 2; Gen. 769; *p*. of for-standan.

fôr-stód, *pl*. -stódon *stood before or against, withstood*, Num. 22, 22; *p*. of fôr-standan.

for-stolen *stolen*, Gen. 31, 32; *pp*. of for-stelan.

for-stondan *to stand up for, defend, protect*, Exon. 118 b; Th. 455, 31; Hy. 4, 58. v. for-standan.

fôr-stondan *to stand before or against, oppose, hinder*, Exon. 42 b; Th. 144, 7; Gū. 674: 105 a; Th. 398, 15; Rä. 17, 8. v. fôr-standan.

fôr-strang; *adj. Very strong*; prævălĭdus:—Fôrstrangne oft wíf hine wríþ [*though*] *very strong, a woman often binds him*, Exon. 113 a; Th. 434, 2; Rä. 51, 4.

for-stylþ *steals*, Ex. 22, 7; *3rd sing. pres.* of for-stelan.

for-styntan *to break, knock, blunt*; contundĕre, Cot. 48: 177. DER. stintan.

for-sūgan; *p*. -seág, *pl*. -sugon; *pp*. -sogen [sūgan *to suck*] *To suck or draw out*; exsūgĕre:—Wið forsogenum magan oððe aþundenum *for a drawn out or puffed up stomach*, L. M. 2, 7; Lchdm. ii. 186, 17.

for-sūwian, -sūgian; *p*. ode, ade; *pp*. od, ad; *v. trans. To pass over in silence, keep silent*; silentio prætĕrīre, tăcēre, reticēre:—We wyllaþ sume forsūwian *we will pass some in silence*, Homl. Th. ii. 138, 26. We woldon iówra Rōmāna bismara beón forsūgiende *we would pass in silence over the shames of you Romans*, Ors. 3, 8; Bos. 63, 23. Gif hí unriht sprǣcaþ, oððe riht forsūwiaþ *if they speak the wrong, or keep silent the right*, Job Thw. 166, 14: Homl. Th. i. 56, 18. Ic secge ðæt ic ǽr forsūwode *I say that which I before kept silent*, Boutr. Scrd. 18, 27. Iob Godes hērunge ne forsūwade *Job kept not God's praise silent*, Job Thw. 166, 16. Hwí wæs ðæra engla syn forsūgod on ðære bēc Genesis *why was the angels' sin passed over in silence in the book of Genesis?* Boutr. Scrd. 17, 19. Ǽlc cræft biþ forsūgod, gif he biþ būtan wísdôme *every craft is passed over in silence, if it be without wisdom*, Bt. 17; Fox 60, 10, MS. Cot. v. for-swígian.

for-swǽlan; *p*. de; *pp*. ed *To burn, burn up, consume, scorch*; ūrĕre, exūrĕre, combūrĕre, concrĕmāre, exæstuāre:—Ic forswǽle oððe forbærne ūro, Ælfc. Gr. 28, 4; Som. 31, 11. Hí wendon to Wealinga forda, and ðæt eall forswǽldon *they turned to Wallingford and burnt it all*, Chr. 1006; Th. 256, 26, col. 1. Fŷr forswǽlþ wudu, swā swā líget forswǽlende dūna *ignis combūrit silvam, sícut flamma combūrens montes*, Ps. Lamb. 82, 15. Ðā hit [sǽd] upeóde, seó sunne hit forswǽlde *when it* [*the seed*] *grew up, the sun scorched* [*burnt up*] *it*, Mk. Bos. 4, 6, quando exortus est sol, exæstuāvit [ἐκαυματίσθη], Vulg. Onleóht breóst and dīnre lufe forswǽl *illūmĭna pectŏra tuŏque ămōre concrĕma*, Hymn. Surt. 36, 12. Hí wurdon mid swæflenum fŷre forswǽlede *they were burnt up with sulphurous fire*, Boutr. Scrd. 22, 32: Homl. Th. ii. 496, 27. We sind mid līgum forswǽlede *we are scorched up with flames*, Homl. Th. ii. 494, 20. [*Laym. p.* forswǽlde, *pp.* forswæled.]

for-swāpan; *p*. -sweóp; *pp*. -swāpen *To sweep away*; verrĕre, protrūdĕre:—Hie wyrd forsweóp *fate has swept them away*, Beo. Th. 959; B. 477. Hafaþ us God forswāpen on ðās sweartan mistas *God has swept us into these dark mists*, Cd. 21; Th. 25, 9; Gen. 391. Ealle wyrd forsweóp [MS. forsweof] mīne māgas *fate has swept away all my kinsmen*, Beo. Th. 5621; B. 2814. [*Cf. O. Sax.* forswīpan *to sweep away.*]

for-swealh, -swealg *swallowed up, devoured*, Ex. 7, 12: Beo. Th. 2249; B. 1122; *p.* of for-swelgan.

for-sweal *died away*, Cot. 65: 190; *p.* of for-sweltan.

for-swelan; *p*. -swæl, *pl*. -swǽlon; *pp*. -swolen [swelan *to burn*] *To burn up, kindle*; combūri:—Hit fǽringa fŷre byrneþ, forswǽlþ under sunnan *it suddenly burns with fire, kindles under the sun*, Exon. 63 b; Th. 233, 29; Ph. 532.

for-swelgan, -sweolgan, he -swelgeþ, -swilgeþ, -swelhþ, *pl*. -swelgaþ; *p*. ic, he -swealh, -swealg, ðū -swulge, *pl*. -swulgon; *subj. pres.* -swelge, *pl*. -swelgen; *p*. -swulge, *pl*. -swulgen; *pp*. -swolgen, -swelgen [swelgan *to swallow*] *To swallow up, devour, absorb*; devŏrāre, deglutīre, absorbēre:—Baru sond willaþ rēn forswelgan *the bare sand will swallow up the rain*, Bt. Met. Fox 7, 27; Met. 7, 14: Exon. 35 a; Th. 113, 30; Gū. 164. Wēn is ðæt hí us wyllen forsweolgan *forsĭtan deglūtissent nos*, Ps. Th. 123, 2. Ic forswelge absorbeo, Ælfc. Gr. 26, 2; Som. 28, 51. Hit eorþe forswelgeþ *the earth swallows it up*, Ps. Th. 57, 6. Forswilgeþ *devours*, Exon. 113 a; Th. 433, 22; Rä. 50, 11. He forswelhþ hig *absorbet eos*, Ps. Lamb. 57, 10. Ða ðe wudewena hūs forswelgaþ *qui devŏrant dōmos vĭduārum*, Mk. Bos. 12, 40: Ps. Spl. 13, 8: Exon. 22 b; Th. 62, 4; Cri. 996. Aarones gird forswealh ealle heora girda *devŏrāvit virga Aaron virgas eōrum*, Ex. 7, 12: Cd. 119; Th. 154, 17; Gen. 2557: Ps. Th. 77, 50. Seó eorþe forswealh Dathan and Abiron *Dathan atque Abiron terra absorbuit*, Deut. 11, 6: Ps. Spl. 105, 17. Grendel leófes mannes líc forswealg *Grendel devoured the beloved man's body*, Beo. Th. 4167; B. 2080: Andr. Kmbl. 3179; An. 1592. Ðe ðū forswulge *which thou hast swallowed up*, Cd. 43; Th. 57, 34; Gen. 938. We forswulgon hine *devŏrāvĭmus eum*, Ps. Spl. 34, 28: Ps. Lamb. 123, 3. Ne me forswelge deóp *lest the deep swallow me up*, Ps. Th. 68, 15. Wǽnunga wæteru forswulgen us *forsĭtan ăqua absorbuisset nos*, Ps. Lamb. 123, 4. Eall wísdom heora forswolgen is *omnis săpientia eōrum devŏrāta est*, 106, 27. Syndon hí æt stāne forswolgene *absorpti sunt juxta petram*, Ps. Th. 140, 8. Heó beóþ forswelgene *they shall be swallowed up*, 57, 8. [*Ger.* ver-schwelgen *to waste in excess.*]

for-swelhþ *swallows up*, Ps. Lamb. 57, 10; *3rd sing. pres.* of for-swelgan.

for-sweltan, he -swilt; *p*. -swealt, *pl*. -swulton; *pp*. -swolten *To die away, perish*; permŏri:—Manig wíf forswilt for hire bearne *many a woman dies because of her child*, Bt. 31, 1; Fox 112, 11, note 17. Forswealt *disparuit*, Cot. 65: 190.

for-sweóf, Beo. Th. 5621, note, = for-sweóp *swept away*; *p.* of for-swāpan.

for-sweógian; *p*. ode; *pp*. od *To pass over in silence, keep silent*; silentio prætĕrīre:—We ne durron forsweógian ... gif we hit forsweógiaþ *we dare not keep silent ... if we keep it silent*, L. Ælf. P. 1; Th. ii. 364, 11, 13. v. for-swígian.

for-sweolgan *to swallow up, devour*, Ps. Th. 123, 2. v. for-swelgan.

for-sweóp *swept away*, Beo. Th. 959; B. 477; *p.* of for-swāpan.

for-sweorcan, he -sworceþ; *p*. -swearc, *pl*. -swurcon; *pp*. -sworcen [sweorcan *to dim*] *To be very dark, to darken, obscure*; cālīgāre, obscūrāre:—Eágena bearhtm forsiteþ and forsworceþ *the brightness of the eyes diminishes and darkens*, Bt. 19, 8; B. 1767. Seó sunne biþ forsworcen *sol obscūrābĭtur*, Mt. Bos. 24, 29. On forsworcenan *in obscūro*, Prov. 7.

for-swerian; *p*. -swôr, *pl*. -swôron; *pp*. -sworen *To* FORSWEAR, *to swear falsely, perjure*; ejūrāre, pējĕrāre:—He sigewǽpnum forsworen hæfde *he had forsworn martial weapons*, Beo. Th. 1613; B. 804. Ic forswerige pējĕro, Ælfc. Gl. 84; Som. 73, 98; Wrt. Voc. 49, 6. Ne forswere ðū *non perjūrābis*, Mt. Bos. 5, 33. Gyf gehādod man forswerige oððe forlicge, gebête ðæt be ðæm ðe seó dǽd sý *if a man in orders swear falsely or fornicate, let him make amends for it according as the deed may be*, L. E. G. 3; Th. i. 168, 5. Gif hwylc lǽwede man hine forswerige, fæste iv geár *if any layman perjure himself, let him fast four years,*

L. Ecg. P. ii. 24; Th. ii. 192, 6, 14. Forsworen *perjúrus*, Wrt. Voc. 86, 69: Gen. 24, 8. We ne beóþ forsworene *ērimus mundi ab hoc jūrāmento*, Jos. 2, 20. He hine forsworenne and trýwleásne clypode *he called him forsworn and faithless*, Chr. 1094; Erl. 229, 32. Đa forswerenan mid forsworenum forwurþaþ *perjurers shall perish with perjurers*, Homl. Th. i. 132, 24. [*Ger.* sich ver-schwören *to conspire*.]

for-swígian, -sweógian, -swúgian, -súwian, -súgian, -sýgian, to -swígianne, -swígienne; *p.* ode, ade, ede; *pp.* od, ad, ed.   **I.** *v. trans.* *To pass over in silence, keep silent, conceal;* silentio prætěrīre :—Betwih đás þing nis to forswígianne, hwylc heofonlíc wundor and mægen ætýwed wæs, đa his bân gefunden and gemēted wæron *inter quæ nequaquam silentio prætereundum reor, quid virtūtis ac mīrācŭli cælestis fŭěrit ostensum, cum ossa ejus inventa sunt*, Bd. 3, 11; S. 535, 9. Nis us đonne se hlísa to forswígienne *nec silentio prætereunda opinio*, 2, 1; S. 501, 1. Forswíged yrfe-bóc [MS. -bec] *suppressum testāmentum*, Ælfc. Gl. 13; Som. 57, 104; Wrt. Voc. 20, 43.   **II.** *v. intrans. To be silent;* reticěre :—He rícum mannum nő for áre ne for ege næfre forswígian wolde *nunquam dīvītibus hŏnōris sive tĭmōris grātia reticēbat*, Bd. 3, 5; S. 527, 10. [*Ger.* ver-schweigen *to pass over in silence*.]

for-swilgeþ *swallows up, devours*, Exon. 113 a; Th. 433, 22; Rä. 50, 11; *3rd sing. pres. of* for-swelgan.

for-swilt *dies*, Bt. 31, 1; Fox 112, 11, note 17; *3rd sing. pres. of* for-sweltan.

fŏr-swíþ; *adj. Very strong, very great;* prævălĭdus :—Is đín meaht fŏrswíþ *is thy power very great?* Exon. 92 b; Th. 348, 11; Sch. 26.

for-swíđan; *he* -swíþ; *p.* ede; *pp.* ed *To overcome;* reprĭmēre :—Se đas orsorgnesse đe he hēr hæfþ ne forswíþ mid đære gesceádwísnesse his ingeþonces *he does not overcome the prosperity he has here with prudence of mind*, Past. 50, 1; Hat. MS. Seó him sára gehwylc symle forswíđede *which constantly overcame each of his pains*, Exon. 46 b; Th. 160, 5; Gú. 939. Forsuíđa *confundere*, Rtl. 50, 13; *præcedere*, 32, 21.

fŏr-swíđe; *adv. Very strongly, very much, vehemently, utterly;* valde, vehĕmenter :—Hí wurdon gehergode and gehýnde fŏrswíđe eahtatýne geár *afflicti sunt et vehĕmenter oppressi per annos dĕcem et octo*, Jud. 10, 8 : Ps. Th. 84, 8. Næfde se here Angelcyn ealles fŏrswíđe gebrócod *the army had not utterly broken up the English race*, Chr. 897; Erl. 94, 29.

for-swolgen *swallowed up, devoured*, Ps. Lamb. 106, 27; *pp. of* for-swelgan.

for-sworcen *darkened, obscured*, Mt. Bos. 24, 29; *pp. of* for-sweorcan.

for-sworceþ *darkens*, Beo. Th. 3538; B. 1767; *3rd sing. pres. of* for-sweorcan.

for-sworen *forsworn, perjured*, Gen. 24, 8; *pp. of* for-swerian.

for-sworennys, -nyss, e; *f.* [forsworen, *pp. of* forswerian *to forswear;* -nys, -nyss] *False swearing, perjury;* pejěrātio, perjūrium :—Cýpmannum gedafenaþ đæt hí sóþfæstnysse healdon, and lofian heora þing búton láþre forswerennysse *it is fitting to merchants that they hold truth, and praise their things without hateful perjury*, Homl. Th. ii. 328, 9.

for-swúgian; *p.* ode; *pp.* od *To pass over in silence;* silentio prætěrīre :—Ælc ánweald biþ forswúgod, gif he biþ bútan wísdôme *every power is passed over in silence, if it be without wisdom*, Bt. 17; Fox 60, 10. *v.* for-swígian.

for-swulge *hast swallowed up* or *devoured*, Cd. 43; Th. 57, 34; Gen. 938; *2nd sing. p. of* for-swelgan.

for-swulgen *would have swallowed up* or *devoured*, Ps. Lamb. 123, 4; *subj. p. pl. of* for-swelgan.

for-swulgon *swallowed up, devoured*, Ps. Spl. 34, 28; *p. pl. of* for-swelgan.

for-sýgian; *p.* ode, ede; *pp.* od, ed *To pass over in silence, conceal;* silentio prætěrīre :—Hú wéne we hú monegra māran bismra hý forsýgedon *can we think how many greater reproaches they concealed?* Ors. 4, 4; Bos. 80, 27. *v.* for-swígian.

for-syhþ *despises*, Lk. Bos. 9, 26; *3rd sing. pres. of* for-seón.

fŏr-syngian, -singian; *p.* ode, ade; *pp.* od, ad [syngian *to sin*] *To sin greatly;* multum peccāre :—Ne wurþ ænig man on worlde swá swíđe fŏrsyngad, đe he wiđ Gode gebētan ne mæge *no man in the world is so very sinful, that he may not make atonement to God*, L. Pen. 12; Th. ii. 282, 1. [Cf. *Ger.* sich versündigen *to sin against*.]

fŏr-tácen [= fóre-tácen] *a fore-token;* portentum, Ælfc. Gl. 5; Som. 56, 12.

for-teáh *misled, seduced*, Exon. 11 b; Th. 17, 14; Cri. 270; *p. of* for-teón. *v.* for-teón.

for-tendan; *p.* -tende [= -tendede], *pl.* -tendon; *pp.* -tended [for-, tendan *to burn*] *To burn off* or *away, sear;* inūrěre :—Đǽm mæden-cildum [MS. -cildan], đa wíf fortendon đæt swýđre breóst fóran, đæt hí weaxan ne sceolde, đæt hí hæfden đý strengran scyte; forđon hí mon hēt on Creácisc Amázanas, đæt is on Englisc fortende *from the female children, the women burnt off the right breast so far that it should not grow, that they might have stronger shot; therefore, they are called in Greek Amazons, that is in English seared*, Ors. 1, 10; Bos. 33, 10-13. *The Latin of Ors. is,*—fēmĭnas stŭdiōse nutriunt, inustis infantium dexteriōrĭbus mamillis, ne sagittārum jactus impĕdīrentur, unde Amázŏnes dictæ, Ors. Hav. Lib. I. Cap. xv, p. 65, 3-4. [Amazons = 'Αμάζονες,

-όνων, *pl. f.* ά- *without*, μαζός *a breast*, or ά-, άμ- *intensive, and* άζω *to dry, parch,* or *sear*.]

Fortende, a; *pl. f.* [*pp. of* fortendan *to burn off* or *away, sear*] *The seared ones, Amazons;* Amāzŏnes, Ors. 1, 10; Bos. 33, 13.

for-teón, -tión; *impert.* -teó, -teoh, *pl.* -teóþ; *subj.* -teó, *pl.* -teón [for-, teón *to draw, lead*] *To mislead, seduce;* sedūcěre. *v.* teón, tión.

forþ; *adv.* [faran *to go*] FORTH, *thence, hence, forwards, onwards, henceforth, further, still;* inde, hinc, prorsum, porro, dehinc, deinceps, tāmen :—Abraham eóde forþ *Abraham went forth*, Gen. 18, 16 : Num. 22, 35 : Jud. 16, 30. Alædaþ míne bân forþ mid eów *efferte ossa mea hinc vōbiscum*, Ex. 13, 19 : Beo. Th. 1229; B. 612 : Cd. 111; Th. 147, 12; Gen. 2438 : Exon. 21 b; Th. 57, 20; Cri. 921 : Elén. Kmbl. 2207; El. 1105. Forþ on leóht gelæded *brought forth into light;* prolātum in lūcem, Bd. 4, 19; S. 588, 37. Teáh heora óđer forþ fægere bôc *one of them drew forth a beautiful book*, Bd. 5, 13; S. 632, 36; 633, 5. Gewát se dæg forþ *the day was going forth*, Lk. Bos. 9. 12. Hí ne mihton đanon fleón, ne forþ ne underbæc *they could not flee thence, neither forwards nor backwards*, Jos. 8, 20 : Cd. 118; Th. 153, 8; Gen. 2535. Cynríc rícsode forþ xxvi wintra *Cynric reigned on for twenty-six years*, Chr. 534; Erl. 14, 33. Swá forþ swá he mihte *as far as he could*, Bd. 3, 17; S. 545, 16 : 5, 21; S. 643, 5. Heald forþ tela niwe sibbe *hold well henceforth our new kinship*, Beo. Th. 1901; B. 948 : Cd. 22; Th. 28, 17; Gen. 437. Gif đú forþ his willan gehýrsum beón wylt *si deinceps voluntāti ejus obsecundāre vŏlŭeris*, Bd. 2, 12; S. 515, 27. He lēt đæt forþ on his bôsme awunian *he let it still remain in his bosom*, Bd. 3, 2; S. 525, 13 : Cd. 17; Th. 21, 7; Gen. 320 : Exon. 11 a; Th. 13, 31; Cri. 211. and swá forþ *and so forth, and so on*, Ælfc. Gr. 25; Som. 26, 59 : Homl. Th. ii. 198, 18 : Bd. de nat. rerum; Wrt. popl. science 8, 26; Lchdm. iii. 250, 7. On cnihthâde and swá forþ eallne đonne giógoþhâd *in childhood and then throughout youth*, Bt. 38, 5; Fox 206, 24. [*O. Sax.* forđ : *Frs.* fort, ford : *O. Frs.* forth, ford : *Dut.* voort : *Ger.* fort : *M. H. Ger.* vort.] *v.* forþon = furþ-um, *dat. of an old adj.* forþ, furþum-líc.

forþ; *prep. Out of, forth;* e, ex : used in composition, Som. Ben. Lye.

for-đa; *adv. For that cause, therefore;* proptěrea :—Forđa bletsode đe God on ècnysse *proptěrea benedixit te Deus in æternum*, Ps. Spl. 44, 3. *v.* for-đam; *adv.*

forþ-acígan; *p.* de; *pp.* ed *To call forth;* provocāre :—He monige forþacígde *he called forth many*, Bd. 5, 14; S. 635, 6.

forþ-agán; *part. Gone forth, passed;* prætěritus, peractus :—Tíma ys forþagán *hōra prætěriit*, Mt. Bos. 14, 15 : Mk. Bos. 6, 35. Forþagāne đý wintre *peracta hiěme*, Bd. 4, 28; S. 606, 22.

for-đam, for-đæm, for-đan, for-đon, for-đam-đe, for-đæm-đe, for-đan-đe, for-đon-đe; *conj.* [*for that which*] *For that, for that reason which, for, because;* nam, quia :—Eádige synd đa gástlícan þearfan, forđam hyra ys heofena ríce *blessed are the poor in spirit, for theirs is the kingdom of heaven*, Mt. Bos. 5, 3 : Ps. Spl. 24, 22 : Beo. Th. 301; B. 149 : Cd. 167; Th. 209, 30; Exod. 507 : Runic pm. 20; Kmbl. 343, 15; Hick. Thes. i. 135, 40. Swíđost he fór đyder for đæm horshwælum, forđæm hí habbaþ swýđe æđele bân on hyra tôþum *he went there chiefly for the walruses, because they have very good bone in their teeth*, Ors. 1, 1; Bos. 20, 16, 28 : Bt. Met. Fox 5, 76; Met. 5, 38. Me đæt gelærdon leóde míne đæt ic đē sôhte, forđan hie mægenes cræft mínne cúđon *my people counselled me that I should seek thee, because they knew my capacity of strength*, Beo. Th. 840; B. 418 : Ps. Spl. 6, 2 : Apstls. Kmbl. 93; Ap. 47 : Menol. Fox 42; Men. 21. Hí wíte þoliaþ forđon hie þegnscipe Godes forgýmdon *they suffer torment because they neglected the service of God*, Cd. 18; Th. 21, 19; Gen. 326 : Exon. 10 a; Th. 11, 11; Cri. 169 : Beo. Th. 4688; B. 2349 : Ps. Spl. 11, 1 : Bd. 4, 19; S. 587, 30. Eádige synd đa đe nú wēpaþ, forđamđe hí beóþ gefrêfrede *blessed are they who weep now, for they shall be comforted*, Mt. Bos. 5, 4, 5, 6, 7, 8, 9, 10, 12 : Cd. 184; Th. 230, 1; Dan. 226 : Bt. Met. Fox 20, 73; Met. 20, 37. Næfþ đys word [willan] nænne imperatívum, forđande se willa sceall beón æfre frig *this verb [to will] has no imperative, for the will must always be free*, Ælfc. Gr. 32; Som. 36, 11 : Homl. Th. ii. 290, 1, 3, 25. Forđonđe sió sunne đær gæþ neár on setl, đonne on ôđrum lande, đær syndon lýđran wedera đonne on Brittannia *because the sun in its setting goes nearer there than in any other land, there are milder weathers than in Britain*, Ors. 1, 1; Bos. 24, 20, 32 : Mt. Bos. 7, 13 : Ps. Spl. 1, 7 : Exon. 25 b; Th. 74, 7; Cri. 1203 : Beo. Th. 1010; B. 503.

for-đam, for-đæm, for-đan, for-đon; *adv. For that cause, consequently;* proptěrea, idcirco, ídeo :—Forđam ic secge eów *ídeo dīco vōbis*, Mt. Bos. 6, 25 : 12, 27, 31 : Cd. 5; Th. 6, 32; Gen. 97. Ne mốst đu wesan forđam ormôd *thou must not consequently be dejected*, Bt. Met. Fox 5, 58; Met. 5, 29. He arás of deáþe, and forđan synd đás wundru gefremode on him *ipse surrexit a mortuis, et ídeo virtūtes operantur in eo*, Mt. Bos. 14, 2 : Beo. Th. 1362; B. 679 : Cd. 217; Th. 276, 25; Sat. 194 : Andr. Kmbl. 915; An. 458 : Elen. Kmbl. 618; El. 309. Wæs he sóþfæstnysse wer, and he forđon eallum wæs leóf *he was a man of truth, and was consequently dear to all*, Bd. 3, 15; S. 541, 22 : Cd. 9; Th. 11, 9; Gen. 172 : Exon. 10 a; Th. 10, 7; Cri. 148 : Beo. Th. 6035;

B. 3021: Menol. Fox 382; Men. 192: Ps. Th. 54, 20: Salm. Kmbl. 921; Sal. 460.

forþ-aræsan; *p.* de; *pp.* ed *To rush forth;* prosilīre:—Ic forþaræse *prosilio,* Ælfc. Gr. 30, 3; Som. 34, 43. Forþaræsde of his bedde *prosīliit ex lecto suo,* Greg. Dial. 1, 2.

forþ-ascúfan; *p.* -sceáf, *pl.* -scufon; *pp.* -scofen *To shove forth, drive forward;* propellĕre, Exon. 129 b; Th. 498, 1; Rä. 87, 6.

forþ-asendan; *p.* -sende; *pp.* -sended, -send *To send forth;* emittĕre:—Binnan þrým dagum he mæg ðone migþan forþasendan *within three days he may send forth the urine,* Herb. 7, 3; Lchdm. i. 98, 8. Forþasend *emissus,* Greg. Dial. 1, 12.

forþ-asettan; *p.* -sette; *pp.* -seted *To set forth, appoint, make;* propōnĕre, pōnĕre, statuĕre:—Ic ðone frumbearn forþasette oðer eorþcyningas ealra heáhstne *ego primogĕnĭtum pōnam illum, excelsum præ rēgĭbus terræ,* Ps. Th. 88, 24.

forþ-asliden *passed* or *gone before, tumbled* or *fallen down;* prælapsus, prolapsus, Som. Ben. Lye. DER. a-slīdan.

forþ-ateón; *p.* -teáh, *pl.* -tugon; *pp.* -togen *To draw forth, bring forth, produce;* proferre, prodūcĕre, edūcĕre:—Forþateónde *prodūcens,* Ps. Lamb. 103, 14. Seó eorþe forþateáh grōwende wirte *protūlit terra herbam vĭrentem,* Gen. 1, 12. God ðā forþateáh of ðære moldan ælces cynnes treów *produxitque Dŏmĭnus Deus de hūmo omne lignum,* Gen. 2, 9. He forþateáh wæter of stāne *eduxit ăquam de petra,* Ps. Lamb. 77, 16. Forþ-atogen *progenitus,* Hpt. Gl.

forþ-atincg, e; *f.* *An exhorting, exhortation, encouraging;* exhortātio, Proœm. R. Concord.

forþ-aurnen; *part.* *Run forth, elapsed;* elapsus:—Nalæs micelre tīde forþaurnenre *non multo elapso tempŏre,* Bd. 4, 6; S. 573, 37.

forþ-bǽro; *f. indecl.* *A bringing forth, a production;* procreātio, productio:—Forþbǽro tīd *the time of production,* Cd. 6; Th. 8, 31; Gen. 132. Cf. onbǽru. Or is forþ-bǽro *adj. f.?* Cf. *O. H Ger.* frambari *inclytus,* Icel. frábærr *surpassing;* and forþ-genge for similar adjectival forms.

forþ-becuman, -bicuman; *p.* -com, -cwom, *pl.* -cómon, -cwómon; *pp.* -cumen *To come forth, proceed;* procēdĕre:—He gesyhþ fram hwylcum wyrttruman seó besmitenes forþbecom *vĭdet a qua rādĭce inquĭnātio illa processĕrit,* Bd. 1, 27; S. 497, 8: Ps. Th. 72, 6.

forþ-beran; *p.* -bereþ, -bireþ; *p.* -bær, *pl.* -bæron; *pp.* -boren *To bear* or *carry forth, bring forth, bring forward, produce;* proferre, perhĭbĕre:—Ðone æðelan Albanum seó wæstmberende Bryton forþbereþ *Albānum egregium fēcunda Britannia profert,* Bd. 1, 7; S. 476, 34. Ðætte ealle openlīce be heora dǽde þurh andetnesse forþbǽron *ut omnes pălam quæ gessĕrant confĭtendo proferrent,* 4, 27; S. 604, 23 : Blickl. Homl. 25, 2; 101, 30. Ðæt he gewitnesse forþbǽre be ðam leóhte *ut testĭmōnium perhĭbēret de lūmĭne,* Jn. Bos. 1, 8.

forþ-berstan; *p.* -bærst, *pl.* -burston; *pp.* -borsten *To burst* or *break forth;* erumpĕre, Som. Ben. Lye.

forþ-beseón; *p.* -beseah, ðū -besāwon; *pp.* -besewen *To look forth, look out;* prospĭcĕre:—He forþbeseah of heánnysse hālgan his *prospexit de excelso sancto suo,* Ps. Lamb. 101, 20.

forþ-bicuman; *p.* -bicwom, *pl.* -bicwómon; *pp.* -bicumen *To come forth· proveniŕe:*—Forþbicwom Godes þegna blǽd *the prosperity of God's servants came forth,* Exon. 18 a; Th. 44, 28; Cri. 709. v. forþ-becuman.

forþ-blǽstan; *p.* te; *pp.* ed [blǽst *a blast*] *To blast forth, puff out, burst out;* insufflāre, erumpĕre, Cot. 74.

forþ-blāwan; *p.* -bleów, *pl.* -bleówon; *pp.* -blāwen *To blow forth, belch out;* eructāre, Cot. 78.

forþ-boren; *part.* [*pp.* of forþ-beran] *Born forth, noble-born, high-born;* clāris parentĭbus ortus, nōbĭlis:—We lǽraþ ðæt ǽnig forþboren preóst ne forseó ðone læsborenan *we enjoin that no high-born priest despise the lower born,* L. Edg. C. 13; Th. ii. 246, 20.

forþ-brengan; *p.* -brohte; *pp.* -broht [forþ, brengan *to bring*] *To bring forth, produce, fulfil, accomplish;* proferre, prodūcĕre, dedūcĕre, efficĕre:—Wel forþbrengeþ hit *it brings forth well,* Bt. Met. Fox 29, 142; Met. 29, 71. Se Metod eallra gesceafta ealle forþbrengþ *the Creator of all things produces them all,* Bt. 39, 13; Fox 234, 19. Forþbrohte *proferret,* Bd. 4, 24; S. 596, 35. He forþbrohte swylce flód wæteru *deduxit tamquam flūmina ăquas,* Ps. Lamb. 77, 16.

forþ-bringan; *p.* -brang, *pl.* -brungon; *pp.* -brungen [forþ, bringan *to bring*] *To bring forth, produce, fulfil, accomplish;* proferre, prodūcĕre, efficĕre:—Gif he ðone āþ forþbringan ne mæg *if he cannot bring forth the oath,* L. Ath. iv. 6; Th. i. 224, 17. He ne mæg ðæt forþbringan *he cannot accomplish it,* Bt. 18, 3; Fox 64, 29. Yfel man yfel forþbringþ *mălus hŏmo profert mălum,* Lk. Bos. 6, 45: Mt. Bos. 12, 35. Ealle ða wæstmas ðe eorðe forþbringeþ *all the fruits that earth produces,* Blickl. Homl. 39, 17. Ðe swā manig ungelimp wæs forþbringende *which was bringing forth so many misfortunes,* Chr. 1086; Erl. 220, 23.

forþ-brohte *brought forth,* Ps. Lamb. 77, 16; *p.* of forþ-brengan.

forþ-bylding, e; *f.* *An instigation, incitement, emboldening;* incĭtātio:—Heora feónda forþbylding *the emboldening of their foes,* Chr. 999; Erl. 135, 38.

---

forþ-clipung, e; *f.* *A calling forth, provoking, an appeal;* provŏcātio, evŏcātio, Som. Ben. Lye.

forþ-clypian; *p.* ode; *pp.* od *To call forth, provoke;* provŏcāre:—Forþclypiende us betwynan *provŏcantes invĭcem,* Gal. 5, 26.

forþ-cuman; he ·cymeþ, -cymþ, *pl.* -cumaþ; *p.* -com, *pl.* -cómon; *subj. pres.* -cume, -cyme, *pl.* -cumen, -cymen; *pp.* -cumen, -cymen *To come forth* or *forward, proceed, succeed, arrive;* procēdĕre, pervĕnīre, advĕnīre:—Metod hēht leóht forþcuman *the Creator bade light to come forth,* Cd. 6; Th. 8, 11; Gen. 122. Ðonne forþcumaþ fyrenfulra þreát hīge onlīc *cum exŏrientur peccātōres sicut fēnum,* Ps. Th. 91, 6. Siððan hit forþcume *after it is come forth;* postquam nātus sit, L. M. I. P. 10; Th. ii. 268, 6. Ðæt ǽlc spræc hæbbe āndagan hwænne hit forþcume *that every suit have a term when it shall come forward,* L. Ed. 11; Th. i. 164, 21. Gif se āþ forþcume *if the oath succeed,* L. Eth. i. 1; Th. i. 280, 15; 282, 7. Ðæt he forþcume to ðǽm gesǽlþum *that he may arrive at the felicities,* Bt. Met. Fox 21, 16; Met. 21, 8. Ðonne ic forþcyme *when I come forth,* Exon. 125 a; Th. 480, 28; Rä. 64, 8. Wæs forþcumen geóc æfter gyrne *comfort was come forth after sorrow,* Andr. Kmbl. 3167; An. 1586. Forþcymene, *pp. pl. come forth,* Exon. 104 a; Th. 394, 28; Rä. 14, 10.

forþ-cyme, es; *m.* *A coming forth, egress;* egressus, effūsio:—On ðæra cilda forþcyme *in effūsiōne infantum,* Gen. 38, 28.

forþ-cyme *may come forth* or *forward,* Exon. 125 a; Th. 480, 28; Rä. 64, 8; *subj. pres.* of forþ-cuman.

forþ-cymen *come forth,* Exon. 104 a; Th. 394, 28; Rä. 14, 10; *pp.* of forþ-cuman.

forþ-cýðan; *p.* de; *pp.* ed *To declare, pronounce;* pronuntiāre, declārāre, Hymn. Lye.

forþ-dón; *p.* -dyde; *pp.* -dón *To put forth;* proferre:—Hēt he his tungan forþdón of his mūþe, and him eówian *linguam proferre ex ōre, ac sibi ostendĕre jussit,* Bd. 5, 2; S. 615, 6.

fŏr-þearle; *adv.* *Very much, greatly;* valde, vehĕmenter:—He behýdde his swiðran hand, ofsceamod *fŏrþearle he hid his right hand, greatly ashamed thereof,* Ælfc. T. 37, 13: Jud. 3, 8.

fŏr-þearlíce; *adv.* *Very severely, strictly;* districte, R. Ben. 2.

for-þencan; *p.* -þohte, *pl.* -þohton; *pp.* -þoht *To misthink, disdain, despise, distrust, despair;* dedignāri, diffĭdĕre:—Ðæt is nū git ðīnre unrihtwīsnesse ðæt ðū eart fullneáh forþoht; ac ic nolde ðæt ðū ðē forþohtest; forðam se se ðe hine forþencþ, se biþ ormód *it is still thy fault that thou art almost despaired; but I was unwilling that thou shouldest distrust thyself; for he who distrusts himself is without courage,* Bt. 8; Fox 24, 15–18. He lǽrde ðæt ða þearfan hý ne forþohton *he taught that they should not despise the poor,* Ps. Th. arg. 48. He fela worda spræc, forþoht þearle *he uttered many words, greatly despaired,* Bt. Met. Fox 1, 163; Met. 1, 82. [*Ger.* ver-denken *to think wrong, blame.*] v. fore-þencan.

for-þeón; *p.* -þeóde; *pp.* -þeód *To oppress;* opprĭmere, subīgĕre:—Scīrne scīman sceadu forþeóde *shadow oppressed the bright splendour,* Rood Kmbl. 108; Kr. 54. [*O. H. Ger.* fardúhian *opprimere.*]

for-þeóstrian; *p.* ode, ade; *pp.* od, ad *To darken, be dark;* obscūrāre:—He aserde þeóstru and forþeóstrade oððe swearc *mīsit tenebras et obscūrāvit,* Ps. Lamb. 104, 28. [*Ger.* ver-düstern *to darken.*] v. a-þýstrian.

forþ-fæderas; *gen.* a; *dat.* um; *pl. m.* Forefathers; mājōres:—Abrahames forþfæderas *Abraham's forefathers,* Ælfc. T. 7, 26. Forþfæderas *tritavi,* Hpt. Gl. 426. v. fórefæder.

forþ-faran; *p.* -fór, *pl.* -fóron; *pp.* -faren *To go forth, depart, die;* discēdĕre, abīre, defungi:—Ðætte hī ǽgðer ge forþfaraþ ge eftcumaþ *that they both depart and return,* Bt. 33, 4; Fox 128, 8. On ðam ilcan geáre he forþfór *in the same year he died,* Chr. 571; Erl. 19, 18. Forþfaren *defunctus,* Ælfc. Gr. 41; Som. 44, 31: Wrt. Voc. 85, 58. Ðā Hēródes wæs forþfaren *defuncto Hērōde,* Mt. Bos. 2, 19: Chr. 685; Erl. 41, 34: Homl. Th. ii. 158, 4. Synd forþfarene, ðe ðæs cildes sāwle sōhton *defuncti sunt, qui quærēbant anĭmam puĕri,* Mt. Bos. 2, 20. [*Laym.* forðfaren *pp. dead.*]

forþ-faru; e; *f.* *A going forth, departure, death;* ŏbĭtus, Som. Ben. Lye. [*Laym.* forðfare *departure, death.*]

forþ-feran; *p.* de; *pp.* ed. *To go forth, depart, die;* decēdĕre, defungi, mŏri, expīrāre:—He ðær forþferan sceolde *he should die there,* Bd. 3, 29; S. 561, 25: 4, 11; S. 579, 29, 42. Hī ðær cýddon hine forþferende *quem ibidem ŏbiisse narrāvĕrint,* 3, 29; S. 561, 4. Se Hǽlend asende his stefne and forþferde *Iesus emissa vōce magna expīrāvit,* Mk. Bos. 15, 37. Forþferde ðæt wīf *mŭlier defuncta est,* Mt. Bos. 22, 27: Lk. Bos. 16, 22: Bd. 3, 29; S. 561, 17: 4, 11; S. 579, 14; 580, 3: Chr. 101; Erl. 9, 10: 534; Erl. 14, 32: 544; Erl. 17, 5. Cūþrēd and Coenbryht on ānum geáre forþferdun *Cuthred and Cenbyrht died in one year,* Chr. 661; Erl. 34, 13. He forþfered wæs *defunctus est,* Bd. 2, 3; S. 505, 3. Hī wurdon fǽrlíce forþferede *they suddenly died,* Homl. Th. ii. 174, 15. Ðā mette he ðane man forþferedne þe ǽr untrum wæs *then he found the man dead that before was ill,* Blickl. Homl. 217, 18.

forþ-ferednes, -ness, e; *f.* *A going forth, departure, death;* ŏbĭtus,

Y

transmigrātio :—Ongeáton hí on ðon, ðæt heó to ðon ðider com, ðæt heó hire sǽde ða neáh-tíde hire forþferednesse *ex quo intellexēre quod ipsa ei tempus suæ transmigrātiōnis in proxĭmum nunciāre vēnisset*, Bd. 4, 9; S. 577, 34, MS. C.

forþ-fering, e; *f. A going forth, deceasing, dying*; defunctio, decessio, Scint.

forþ-fleógan; *p.* -fleáh, *pl.* -flugon; *pp.* -flogen *To fly forth*; evōlāre :—Hie lēton forþfleógan flāna scūras *they let fly forth showers of arrows*, Judth. 11; Thw. 24, 33; Jud. 221.

forþ-flówan; *p.* -fleów, *pl.* -fleówon; *pp.* -flówen *To flow forth*; effluĕre :—Genihtsum wæter forþflóweþ *plentiful water flows forth*, Bd. 5, 10; S. 625, 24.

forþ-fōr, e; *f.* [fōr *a going*] *A going forth, departure, death*; exĭtus, ŏbĭtus, mors :—Forðamðe him cūþ forþfōr toweard wǽre *eo quod certus sĭbi exĭtus esset*, Bd. 3, 19; S. 547, 16. Ðæt is gesægd ðæt he wǽre gewis his sylfes forþfōre, of ðam ðe we nū secgan hȳrdon *præscius sui ŏbĭtus exstĭtisse, ex his quæ narrāvĭmus, vĭdētur*, 4, 24; S. 599, 14 : 3, 19; S. 547, 17. He læg æt forþfōre *incĭpiĕbat mŏri*, Jn. Bos. 4, 47 : Bd. 4, 24; S. 598, 28, 37 : 5, 3; S. 616, 17. Be his forþfōre *de ŏbĭtu ejus*, 2, 3; S. 504, 13. Heora gemynde and forþfōre mid mæssesange mǽrsade syndon *their memory and decease are celebrated with mass-song*, 2, 3; S. 504, 41.

forþ-forlǽtan; *p.* -forlēt, *pl.* -forlēton; *pp.* -forlǣten *To let forth, send forth*; emittĕre :—Egeslícne cwide Weard ofer ðæt fǽge folc forþforlǽteþ *the Lord shall send forth a dreadful utterance over the fated people*, Exon. 30 a; Th. 92, 34; Cri. 1518.

forþ-forlǽtenes, -ness, e; *f. A free permission, license, fault*; derelictio :—On ðara mánfulra forþforlǽtenesse *on account of the license of the wicked*, Bt. 5, 1; Fox 10, 24.

forþ-framian, -fremian; *p.* ode; *pp.* od [fremian *to advance, avail*] *To grow up, ripen*; pubescĕre :—Forþframiende *pubescens*, Cot. 150.

forþ-fromung, e; *f.* [fromung *a going*] *A going forth, going away, departure*; profectio :—Geblissod is Egypt on forþfromunge heora *lætāta est Ægyptus in profectiōne eōrum*, Ps. Spl. C. 104, 36.

forþ-gān; *p.* -eóde, *pl.* -eódon; *pp.* -gān *To go forth, proceed, go or pass by*; exīre, procēdĕre, prætĕrīre, transīre :—Raulf wolde forþgān mid his folce *Ralph would go forth with his people*, Chr. 1075; Erl. 213, 18. Ða hwíle ðe ic forþgā *dōnec transeam*, Ex. 33, 22. Þūsend geár beforan eágan ðínum, swā swā dæg estra [= giestra] se forþgǽþ *mille anni ănte ŏcŭlos tuos tanquam dies hesterna quæ prætĕriit*, Ps. Spl. 89, 4. Ða þing ðe of ðam men forþgáþ, pa hine besmítaþ *quæ de hŏmĭne procēdunt illa sunt, quæ commūnicant hŏmĭnem*, Mk. Bos. 7, 15. Ða he forþeóde quo *transeunte cōram eo*, Ex. 34, 6. Ða ðe forþeódon *qui prætĕrībant*, Ps. Spl. C. 128, 7. Hý on heora dagum butu forþeódon *ambo processissent in diebus suis*, Lk. 1, 7.

forþ-gang, es; *m.*   **I.** [gang I. *a going*] *a going forth, progress, advance*; processus, progressus :—Ðæs cyninges ríce ge fōreweard ge forþgang *cūjus rēgis regni et princĭpia et processus*, Bd. 5, 23; S. 646, 3. Se hæfþ forþgang fōr Gode and fōr worulde *he shall have progress before God and before the world*, Ælfc. T. 1, 7.   **II.** [gang II. *latrīna*] *a passage, drain, privy*; meātus, secessus, latrīna :—Forþgang *meātus*, Ælfc. Gl. 71, 75; Wrt. Voc. 44, 57. Eall ðæt on ðone mūþ gǽþ, gǽþ on ða wambe, and byþ on forþgang asend *quod in os intrat, in ventrem vādit, et in secessum emittĭtur*, Mt. Bos. 15, 17 : Mk. Bos. 7, 19. [vorðgong *progress, A. R.*]

forþ-gangan, -gongan; -gangan; *p.* -géong, *pl.* -géongon; *pp.* -gangen, -gongen *To go forth, proceed, go before, precede*; procēdĕre, progrĕdi, præcēdĕre :—Hēt hyssa hwǽne forþgangan *he commanded each of the youths to go forth*, Byrht. Th. 131, 5; By. 3. Forþgangendre tíde *procēdente tempŏre*, Bd. 3, 19; S. 547, 30. Forðgeonga *prægredi*, Mk. Skt. Lind. 2, 23.

forþ-gebrengan; *p.* -gebrohte; *pp.* -gebroht *To bring forth or forward, make known*; edūcĕre, proferre :—Hí se hlísa ne mæg forþgebrengan *fame cannot bring them forward*, Bt. Met. Fox 10, 124; Met. 10, 62.

forþ-geclypian; *p.* ode; *pp.* od *To call forth, incite, provoke*; provōcāre, Scint.

forþ-gecȳgan; *p.* de; *pp.* ed *To call forth*; provōcāre :—He hí to gefeohte forþgecȳgde *he called them forth to battle*, Bd. 1, 16; S. 484, 20.

forþ-gefaran; *p.* -gefór, *pl.* -geföron; *pp.* -gefaren *To go by, go forth, pass*; transīre :—Nymne seó clænsunge tíd forþgefare *nĭsi purgātiōnis tempus transiĕrit*, Bd. 1, 27; S. 493, 39. Wulfríc forþgefaren wæs *Wulfric was departed [dead]*, Chr. 1061; Th. 329, 37 : 560; Erl. 17, 16 : Nar. 40, 9.

forþ-geferan; *p.* de; *pp.* ed *To go forth, depart, die*; decēdĕre, mŏri :—Ðara monige forþgeferdon on Drihten *many of whom died in the Lord*, Bd. 5, 11; S. 626, 34 : 2, 14; S. 518, 1.

forþ-gefremman; *p.* ede; *pp.* ed [gefremman *to effect, bring to pass*] *To move forwards, cause to advance*; promōvēre :—Hine God ofer ealle men forþgefremede *God advanced him above all men*, Beo. Th. 3440; B. 1718.

forþ-gelǽdan; *p.* de; *pp.* ed *To lead* or *bring forth, produce, conduct*; prodūcĕre, provĕhĕre :—He wolde manna rím forþgelǽdan *he would lead forth a number of men*, Cd. 222; Th. 289, 24; Sat. 402. Se forþgelǽdeþ on muntum híg *qui prodūcit in montĭbus fœnum*, Ps. Spl. 146, 9. Se ðe hine to heánnysse cyneríces forþgelǽdde *qui se ad regni ăpĭcem provĕhēret*, Bd. 2, 12; S. 514, 19: Blickl. Homl. 205, 32.

forþ-gelang; *adj. Dependent*; pendens, nixus :—On wīsum scrifte biþ swíðe forþgelang forsyngodes mannes nȳdhelp *on wise confession is greatly dependent the needful help to a sinful man*, L. Pen. 1; Th. ii. 278, 2 : 9; Th. ii. 280, 12.

forþ-geleoran; *p.* de; *pp.* ed *To pass forth, pass away, depart, die*; transīre, decēdĕre, mŏri :—Monige forþgeleordon on Drihten *many died in the Lord*, Bd. 5, 11; S. 626, 34, MS. T: 2, 14; S. 518, 1, MS. T. Nymne seó clǽnsunge tíd forþgeleore *nĭsi purgātiōnis tempus transiĕrit*, 1, 27; S. 493, 39, MSS. B. T. Ða ongeat he ðone mann, and him to gemynde com ðæt he his hrægle onfēng ðá he forþgeleored wæs *cognōvitque hŏmĭnem, et quia vestīmentum ejus mŏrientis accēpĕrit, ad mĕmōriam reduxit*, 3, 19; S. 549, 3 : Th. Chart. 138, 4.

forþ-genge; *adj. Progressive, increasing, effective*; pŏtens :—Hū mæg se leáfa beón forþgenge, gif seó lār [MS. lare] and ða lāreówas ateoriaþ *how can the faith be increasing if the doctrine and the teachers fail?* Ælfc. Gr. pref; Som. 1, 34. Ðæt hit þurh ðone fultum síe forþgenge *that it become effective through help*, Past. 14, 1; Hat. MS. 17 b, 2.

forþ-geong, es; *m. A going forth, progress, process*; processus :—On forþgeonge ðæs ǽrendgewrites *in processu epistŏlæ*, Bd. 1, 13; S. 481, 43. v. forþ-gang.

forþ-georn; *adj. Desirous to go forth, impetuous*; vehĕmens :—Swā dyde Æðeric, fús and forþgeorn *thus did Ætheric, eager and impetuous*, Byrht. Th. 139, 68; By. 281.

forþ-geótan; *p.* -geát, *pl.* -guton; *pp.* -goten *To pour forth*; profundĕre :—Ongeán ðam rǽse ðæs forþgotenan streámes *contra impĕtum flŭvii decurrentis*, Bd. 5, 10; S. 625, 7. He, forþgotenum teárum of inneweardre heortan, Drihtne his willan bebeád *profūsis ex imo pectŏre lacrȳmis, Dŏmĭno sua vōta commendābat*, 4, 28; S. 606, 42.

forþ-gesceaft, e; *f.*   **I.** *the created things, creation, world*; creātūra, res creātæ, mundus :—Fyrn forþgesceaft Fæder ealle bewāt *the Father guards all the ancient creation*, Exon. 128 a; Th. 492, 4; Rä. 81, 9 : 92 b; Th. 346, 24; Sch. 3.   **II.** *the future world, state, or condition*; stātus fūtūrus :—Is seó forþgesceaft dígol and dyrne *the future condition is dark and secret*, Menol. Fox 584; Gn. C. 61. He ða forþgesceaft forgyteþ and forgýmeþ *he forgets and neglects the future state*, Beo. Th. 3505; B. 1750: Exon. 80 b; Th. 303, 20; Fä. 56. Ðæt ic an forþgesceaft feran mōte *that I may come to a future state*, Ps. C. 50, 52; Ps. Grn. ii. 278, 52.

forþ-geseón; *p.* -geseah, *pl.* -gesāwon; *pp.* -gesewen *To see forth, onward, in front*; provĭdēre :—Hí forþgesāwon lífes lātþeów *they saw the guide of life in front*, Cd. 147; Th. 184, 7; Exod. 103.

forþ-gestapan; *p.* -gestōp, *pl.* -gestōpon; *pp.* -gestapen *To step forth*; progrĕdi :—He to forþgestōp dracan heáfde neáh *he had stept forth near to the dragon's head*, Beo. Th. 4568; B. 2289.

forþ-gestígan; *p.* -gestāh, *pl.* -gestigon; *pp.* -gestigen *To go forth or forwards, to advance, ascend*; prodīre, procēdĕre, ascendĕre :—Ðæt ǽnig forþgestígeþ *that any shall advance*, Exon. 78 b; Th. 294, 24; Crä. 20. Ðæt we eáðe māgon upcund ríce forþgestígan *that we may easily ascend to the realm on high*, 93 a; Th. 348, 28; Sch. 35.

forþ-gestrangian; *p.* ode, ade; *p.* od, ad *To make very strong, strengthen much*; confortāre :—Ofer me syndon, ða ðe me ēhton, forþgestrangad *confortāti sunt sŭper me qui me persequuntur*, Ps. Th. 68, 5.

forþ-gesȳne; *adj. Visible*; conspĭcuus :—Fela biþ on foldan forþgesȳnra geongra geofona *there are many early gifts ever visible on earth*, Exon. 78 a; Th. 293, 15; Crä. 1.

forþ-gewāt *went forth, passed*, Ps. Lamb. 89, 4; *p. of* forþ-gewítan.

forþ-gewendan; *p.* de; *pp.* ed *To go* or *turn out*; prodīre :—Ðæt ǽlc man ðe fere wǽre forþgewende *so that every man who was able to go should turn out*, Chr. 1016; Erl. 153, 31.

forþ-gewítan; *p.* -gewāt, *pl.* -gewiton; *pp.* -gewiten *To go forth, proceed, go by, pass, depart, die*; procēdĕre, transīre, prætĕrīre, decēdĕre, mŏri :—Swā swā brȳdguma forþgewítende of brȳdbūre his *tanquam sponsus procēdens de thălămo suo*, Ps. Spl. 18, 5. Oþ-ðæt forþgewíteþ unriht *dōnec transeat inĭquĭtas*, 56, 2. Swylce gysternlíc dæg ðe forþgewāt *tanquam dies hesterna quæ prætĕriit*, Ps. Lamb. 89, 4 : Bd. 4, 9; S. 577, 35. Forþgewít and ríce *procēde et regna*, Ps. Spl. 44, 5. Prætērĭtum tempus is forþgewiten tíd *prætĕrĭtum tempus is the past tense*, Ælfc. Gr. 20; Som. 23, 7, 10, 12, 13. Se forþgewitena tíma *the past tense*, Som. 23, 14. Ðone forþgewitenan tíman, Som. 23, 9.

forþ-gewitenes, -ness, e; *f. A going forth, departure*; profectio :—Blissade ðæt þeóstre folc on forþgewitenesse oððe fǽre heora *lætāta est Ægyptus in profectiōne eōrum*, Ps. Lamb. 104, 38.

forþ-gongan; *part.* -gongende; *p.* -géong, *pl.* -géongon; *pp.* -gongen *To go forth, proceed*; procēdĕre, præcēdĕre :—Forþgongende *going forth*, Exon. 14 a; Th. 27, 5; Cri. 426: Bd. 1, 8; S. 479, 20 : 1, 1;

S. 474, 24. Forþgongendre yldo *ævo præcēdente*, 4, 19 ; S. 587, 32. v. forþ-gangan.

forþ-gyrd, fōr-gyrd, es ; *m. A fore-girdle, martingale, the girdle which passes between the fore-legs of a horse from the nose-band to the girth ;* antela [ab ante et telon, quod est longum, compōnitur, Du Cange, sub voce], cingŭlum illud quod ante pectus ĕqui tendĭtur, crassius lōrum quo pectus, partim ad ornāmentum, partim ad firmandam sellam cingĭtur :—Forþgyrd *antela*, Ælfc. Gl. 20 ; Som. 59, 53 ; Wrt. Voc. 23, 14 : 84, 4.

forþ-heald, -heold ; *adj. Bent forward, inclined downwards, stooping ;* incurvus, prōnus, proclīvus :—Hwōn forþheald *paulŭlum incurvus*, Bd. 2, 16 ; S. 519, 33. He lang fæc forþheald licgende wæs *aliquandiu prōnus jăcens*, 4, 31 ; S. 610, 14. Forþheold *proclīvus*, Ælfc. Gr. 47 ; Som. 48, 39. Forðhald ł gebeged *inclinata*, Lk. Skt. Lind. 13, 11.

forþ-healdan ; *p.* -heóld, *pl.* -heóldon ; *pp.* -healden *To hold to, follow out, maintain ;* exsĕqui :—Mid ðy he ðæt langre tĭde forþheóld and dyde quod *dum multo tempŏre sēdŭlus exsĕquĕrētur*, Bd. 4, 25 ; S. 600, 24.

forþ-heold ; *adj. Stooping ;* proclīvus, Ælfc. Gr. 47 ; Som. 48, 39. v. forþ-heald.

forþ-here, -herge, es ; *m. The front or van of an army ;* frons exercĭtūs :—Hie getealdon on ðam forþherge fēðan twelfe *they numbered twelve bands in their van*, Cd. 154 ; Th. 192, 1 ; Exod. 225.

forþ-hreósan, he -hrýst ; *p.* -hreás, *pl.* -hruron ; *pp.* -hroren *To rush forth ;* proruĕre :—Forþhrýst *proruit*, Scint. 26.

for-ði, for-ðī-ðe ; *conj. For that, for, because, therefore ;* quia, quŏniam, ĭtāque :—Nā forðīðe heó of Moyse sý *non quia ex Moyse est*, Jn. Bos. 7, 22 : Ps. Lamb. 77, 22. Forðīðe he slóh stān *quŏniam percussit petram*, Ps. Lamb. 77, 20. v. for-ðý ; *conj.*

for-ði, for-ðī ðonne ; *adv. For that cause, consequently, wherefore ;* quamobrem, proptĕrea, quapropter, ĭdeo, idcirco :—Forhwī oððe forðī quamobrem, Ælfc. Gr. 38 ; Som. 40, 58. Forðī ðonne *qua propter :* forðī *ĭdeo, idcirco, proptĕrea*, 44 ; Som. 46, 17, 18. Forðī gehýrde Drihten *ĭdeo audīvit Dŏmĭnus*, Ps. Lamb. 77, 21 : Homl. Th. ii. 288, 22, 25. v. for-ðý ; *adv.*

forþian ; *p.* ode ; *pp.* od *To further, aid, assist, advance, perform ;* promŏvēre :—He ne muge hit forþian *he may not perform it*, Chr. 675 ; Erl. 38, 11 : 1052 ; Erl. 182, 2. Ðæt he Godes circan forþige *ut Dei ecclēsias promŏveat*, L. I. P. 2 ; Wilk. 147, 34. DER. ge-forþian.

for-ðig ; *conj. For, because ;* ĕnim, etĕnim, quia, quŏniam :—Forðig he āhte ægðer ge Engla land ge Normandige *for he owned both the land of the English as well as Normandy*, Chr. 1085 ; Erl. 218, 3–4. v. for-ðý ; *conj.*

for-ðig ; *adv. For that cause, consequently ;* proptĕrea :—Forðig ic eów sæde *proptĕrea dixi vōbis*, Jn. Bos. 6, 65. v. for-ðý ; *adv.*

for-þingian ; *p.* ode ; *pp.* od *To plead for anyone, intercede ;* intercēdĕre :—Būton se hlāford ðone wer forþingian wille *unless the lord will intercede for the man*, L. Alf. pol. 21 ; Wilk. 39, 34. v. fore-þingian.

for-þiófan *to thieve, steal ;* fūrāri :—Ðæt ðū ne forstele oððe ne forþiófe *ne fūrēris*, Mk. Skt. Lind. 10, 19. v. þeófan, þiófan.

forþ-lǣdan ; *p.* de ; *pp.* ed *To lead or bring forth, produce ;* prodūcĕre :—Se ðe forþlǣdeþ windas of goldhordum his *qui prodūcit ventos de thesauris suis*, Ps. Lamb. 134, 7. Freódrihten hine forþlǣdde to ðam hālgan hām *the Lord led him forth to the holy home*, Cd. 226 ; Th. 300, 18 ; Sat. 566.

forþ-lǣdnys, -nyss, e ; *f. A bringing forth, production ;* prolātio, productio :—On ðæs tuddres forþlǣdnysse *in prōlis prolātiōne*, Bd. 1, 27 ; S. 493, 21.

forþ-lǣstan ; *p.* -lǣste ; *pp.* -lǣsted *To follow out, accomplish, fulfil ;* ăgĕre, perăgĕre :—Ðæt for intingan ðæs godcundan eges ǣne sīþe for his scylde onbryrded ongan, swā he ǽr eft for intingan ðære godcundan lufan lustfulligende ðām ēcum mēdum fæstlīce forþlǣste *quod causa dīvini timōris sĕmel ob reātum compunctus cæpĕrat, jam causa dīvini āmōris delectātus præmiis indefessus ăgēbat*, Bd. 4, 25 ; S. 600, 24.

forþ-lǣtan ; *p.* -lēt, -lēton ; *pp.* -lǣten *To let forth, send forth, emit ;* emittĕre :—Swylce word he ðær forþlēt *such words he let forth there*, Nicod. 11 ; Thw. 6, 5 : Blickl. Homl. 133, 29.

forþ-leoran ; *part.* -leorende ; *p.* de ; *pp.* ed *To go forth, proceed ;* procēdĕre :—Wuldriende hāligne Gāst forþleorendne of Fæder and of Suna unasecgendlīce *glōrĭfĭcantes Spĭrĭtum sanctum, procēdentem ex Patre et Fīlio inenarrābĭlĭter*, Bd. 4, 17 ; S. 586, 13, note.

forþ-lífian ; *p.* -lāf, *pl.* -lifon ; *pp.* -lifen [lĭfan *to leave*] *To stand out, appear ;* promĭnēre :—Mid ðý me of sweoran forþlífaþ seó reádnes and bryne ðæs swyles *dum mihi de collo rŭbor tŭmōris, ardorque promĭneat*, Bd. 4, 19 ; S. 589, 30.

forþ-lócian ; *p.* ode, ade ; *pp.* od, ad *To look forth ;* prospĭcĕre :—Dryhten of heofene forþlōcade ofer bearn monna *Dŏmĭnus de cælo prospexit sŭper fīlios hŏmĭnum*, Ps. Surt. 52, 3 : Blickl. Homl. 217, 31 ; 219, 18.

forþ-lútan ; *p.* -leát, *pl.* -luton ; *pp.* -loten *To fall forwards, fall down ;* procĭdĕre :—He forþleát on his andwlitan *procĭdĕret in fāciem*, Bd. 4, 3 ; S. 569, 11. Forþloten *prōnus, proclīvus*, Scint. 6 : Prov. 29.

forþ-mǣre ; *adj. Very great ;* præclārus :—Gewīteþ on westrōdor

forþmǣre tungol faran *the very great star departs to go into the western sky*, Exon. 93 b ; Th. 350, 25 ; Sch. 69.

forþ-man *one very rich or wealthy ;* prædīves, Som. Ben. Lye.

for-þoht *despaired*, Bt. 8 ; Fox 24, 16 ; *pp.* of for-þencan.

for-þohte, ðū -þohtest *despaired, hast despaired*, Bt. 8 ; Fox 24, 17 ; *p.* of for-þencan.

for-þolian ; *p.* ode ; *pp.* od *To be deprived of, want ;* prīvāri, cărēre :—Wāt se ðe sceal his winedryhtnes lārcwidum longe forþolian *he knows who must long be deprived of his dear lord's lessons*, Exon. 77 a ; Th. 288, 29 ; Wand. 38.

for-ðon, for-ðon-ðe ; *conj. For that, for, because ;* quia, quŏniam :—Forðon ðū ofslōge ealle *quŏniam tu percussisti omnes*, Ps. Spl. 3, 7. Forðonðe wyste Drihten weg rihtwīsra *quŏniam nōvit Dŏmĭnus viam justōrum*, 1, 7. v. for-ðam ; *conj.*

for-ðon = for-ðam ; *adv. For that cause, consequently, therefore ;* proptĕrea, ĭdeo :—Forðon ne arīsaþ ða ārleáse on dōme *ĭdeo non resurgunt impii in jūdicio*, Ps. Spl. 1, 6.

forþ-on ; *adv.* [= forþ-an, forþ-um = furþ-um] *At first, indeed, also ;* primo, ĕtiam :—Nō forþon ānlēpe *no, not even [also] one*, Ps. Th. 13, 2. v. furþ-um.

fōr-þoncol ; *adj. Forethoughtful, prudent ;* prōvĭdus, prūdens :—Ðū ahýddest ðás from snottrum and fōrþonclum *abscondisti hæc a săpientĭbus et prūdentĭbus*, Mt. Kmbl. Rush. 11, 25. v. fōre-þancul.

forþ-onettan ; *p.* te ; *pp.* ed *To hasten forth ;* porro festīnāre :—Fæder on fultum forþonetteþ *the Father hastens forth to his aid*, Exon. 62 b ; Th. 229, 15 ; Ph. 455 : 108 a ; Th. 412, 9 ; Rä. 30, 11. He forþonette *he hastened forth*, Exon. 120 a ; Th. 461, 26 ; Hö. 41 : Wald. 77 ; Vald. 2, 10.

forþ-ongangan *to go forth, proceed ;* procēdĕre :—Hie gesāwon fyrd Faraonis forþongangan *they saw the host of Pharaoh go forth*, Cd. 149 ; Th. 187, 25 ; Exod. 156. v. forþ-gangan.

forþ-onloten ; *part.* [forþ *forth, forwards ;* onloten, *pp.* of onlūtan *to incline to, bow*] *Fallen forwards, prostrate ;* provŏlūtus, Gr. Dial. 1, 8.

forþ-onsendan ; *p.* de ; *pp.* ed *To send forth ;* emittĕre :—He in folc Godes forþonsendeþ of his brægdbogan biterne strǣl *he [the devil] sends forth, amongst God's people, the bitter arrow from his deceitful bow*, Exon. 19 a ; Th. 47, 33 ; Cri. 764. Hī nædran forþonsendon *they sent forth snakes*, Elen. Kmbl. 240 ; El. 120. Ðæt ðū forþonsende wæter *that thou send forth water*, Andr. Kmbl. 3011 ; An. 1508.

forðor *further, more*, Mt. Kmbl. Lind. 6, 25, 30 : Mk. Skt. Lind. 6, 51 : Lk. Skt. Lind. Rush. 22, 71. v. furðor.

forþ-rǣsan ; *p.* de ; *pp.* ed *To rush forth, spring forth, spring up, rise up ;* proruĕre, exsilīre, sălīre, exsurgĕre :—Biþ on him will forþrǣsendes wæteres on ēce līf *fiet in eo fons ăquæ sălientis in vītam æternam*, Jn. Bos. 4, 14. He ðā awearp his reáf, and forþrǣsde and to him com *qui projecto vestimento suo exsĭliens, vēnit ad eum*, Mk. Bos. 10, 50. Forþrǣsdon of ðǣm wītum *exsurrexērunt a supplĭciis*, Martyrol. ad 26, Mart.

for-þrǣstan ; *p.* te ; *pp.* ed *To entirely bruise, break ;* contĕrĕre, Ps. Spl. C. 45, 9 : 104, 15, 31 : 123, 7. Hpt. Gl. 425 ; 441. v. þrǣstan.

for-þriccan *to tread under, oppress*, Som. Ben. Lye. v. for-þryccan.

forþ-riccednes, -ness, e ; *f. A pressing, an oppression, distress, anxiety ;* pressūra :—Þeóda forþriccednes *pressūra gentium*, Lk. Bos. 21, 25.

forþ-riht ; *adj. Right forth, distinct, plain ;* hence, forþriht spræc *plain speech, prose ;* prōsa = prorsa, *i. e.* proversa, Som. Ben. Lye.

forþ-rihte ; *adv. Distinctly, plainly, manifestly ;* expresse, plāne, dīrecte, C. R. Ben. 29. Forþrihte *indeclinabĭliter*, Hpt. Gl. 406. [*Orm*, forrþrihht *straightway*.]

for-þringan ; *p.* -þrang, *pl.* -þrungon ; *pp.* -þrungen [þringan *to crowd, throng, rush upon*] *To snatch from any one, protect from any one ;* erĭpĕre ălĭcui, defendĕre ab ălĭquo :—Ðæt he ne meahte ða weálāfe wīge forþringan *þeódnes þegne that he might not by war protect the sad remnant from the king's thane*, Beo. Th. 2173 ; B. 1084. [*Orm*, forrþrungenn *oppressed : Ger*. verdrängen *to push away*.]

for-þryccan, -þrycan ; *p.* -þrycte ; *pp.* -þrycced, -þryct *To tread under, oppress greatly, suppress, overwhelm ;* opprĭmĕre, supprĭmĕre :—Ðære wambe flēwsan he forþryceþ *it suppresses the flux of the stomach*, Med. ex Quadr. 6, 9 ; Lchdm. i. 352, 17. Næs ǽnig ðara ðæt mec þreám forþrycte *there was not any of them that overwhelmed me with reproofs*, Exon. 73 a ; Th. 273, 22 ; Jul. 520. Þreám forþrycced *oppressed with afflictions*, 50 a ; Th. 174, 1 ; Gū. 1171 : Elen. Kmbl. 2551 ; El. 1277. Gesihst ðū nū ðæt ða rihtwīsan sint lāðe and forþrycte *seest thou now that the virtuous are hated and oppressed ?* Bt. 3, 4 ; Fox 6, 23.

for-þryct *oppressed*, Bt. 3, 4 ; Fox 6, 23 ; *pp.* of for-þryccan.

forþ-ryne, es ; *m. An onward course ;* procursus :—Heóldon forþryne eástreámas heora *river-streams held their onward course*, Cd. 12 ; Th. 14, 8 ; Gen. 215.

for-þrysmian ; *p.* ode, ede ; *pp.* od, ed [þrysmian *to suffocate*] *To suffocate, choke, strangle ;* suffōcāre :—Eornfullness ðisse worulde, and leásung ðissa woruldwelena forþrysmiaþ ðæt wurd *solicĭtūdo sæcŭli istĭus, et fallācia dīvitiārum suffōcat verbum*, Mt. Bos. 13, 22. Ða þornas hyt

forþrysmodon *spīnæ suffŏcāvērunt illud*, Lk. Bos. 8, 7. Ða synd for-þrysmede *qui suffŏcantur*, 8, 14.

forþ-scencan *to drink to;* propīnāre, Cot. 149.

forþ-scrīđan; *p.* -scrāþ, *pl.* -scridon; *pp.* -scriđen *To go forth, pass on, depart;* prōdīre, decēdēre :—Dagas forþscridon [MS. forþscridun] *days passed on*, Exon. 47 a; Th. 160, 12; Gū. 942. Đonne dōgor beóþ on moldwege mīn forþscriđen *then my day on earth will be departed*, 48 a; Th. 164, 16; Gū. 1012.

forþ-scype, es; *m. A going forth, growth;* profectus :—For his forþ-scype onstyred *mōtus ejus profectībus*, Bd. 1, 34; S. 499, 28, note.

forþ-sîþ, es; *m.* [sîþ *a journey*] *A going forth, departure, death;* progressus, ābĭtus, ŏbĭtus :—Forþsīþes georn *glad of departure*, Exon. 123 b; Th. 475, 2; Bo. 41: 124 b; Th. 479, 21; Rä. 63, 2. Æfter Ōswaldes forþsīþe *after Oswald's death*, Chr. 992; Erl. 130, 37; Hy. 7, 72; Hy. Grn. ii. 288, 72. Hređer innan born, afýsed on forþsîþ *his spirit burned within, bent on departure*, Exon. 46 b; Th. 158, 19; Gū. 911: 50 a; Th. 173, 2; Gū. 1154: 52 b; Th. 182, 34; Gū. 1320. He wæs đær ôþ Hērōdes forþsîþ *ērat ibi usque ad ŏbĭtum Hērōdis*, Mt. Bos. 2, 15.

forþ-sīđian; *p.* ode; *pp.* od [sîđian *to journey*] *To go forth, depart, die;* prōdīre, discēdēre, mŏri, Som. Ben. Lye.

forþ-snoter, -snotter; *adj.* [snoter *wise*] *Very wise;* săpientissĭmus :—Elene hēht gefetian on fultum forþsnoterne *Elene bade to fetch to her aid the very wise* [man], Elen. Kmbl. 2104; El. 1053. Forþsnoterne, 2320; El. 1161. Fundon fīfhund forþsnotterra *they found five hundred very wise* [men], 758; El. 379.

forþ-spell, es; *n.* [spell *a history*] *A speaking out, saying, intimation;* effātum, dictum :—Be đissum feáwum forþspellum *by these few intimations*, Exon. 84 a; Th. 316, 11; Mōd. 47.

forþ-spôwnes, -ness, e; *f.* [spôwan *to succeed*] *Great success,* hence *An advance, a growth, prosperity;* profectus :—To forþspôwnesse gedē-fenre heánesse *ad profectum dēbĭti culmĭnis*, Bd. 2, 4; S. 505, 17.

forþ-sprecan; *p.* -spræc, *pl.* -sprǣcon; *pp.* -sprecen *To speak forth, speak out;* prolŏqui :—Ic sceal forþsprecan gên ymbe Grendel *I shall speak forth again about Grendel*, Beo. Th. 4145; B. 2069.

forþ-stæppan; *part.* -stæppende *To step forth, proceed,* Homl. Th. ii. 90, 11. v. forþ-steppan.

forþ-stapan; *p.* -stôp, *pl.* -stôpon; *pp.* -stapen *To step or go forth, proceed, to go or pass by;* prōgrĕdi, prōdīre, procēdēre, prætērīre :—Forþ-stôp swylce of rysele heora unrihtwīsnes *prōdiit quăsi ex ădĭpe iniquĭtas eōrum*, Ps. Lamb. 72, 7. Đā he lyt-hwôn forþstôp *cum prŏcessisset paulŭlum*, Mk. Bos. 14, 35. Đa đe forþstôpon hine gremedon *qui prætĕrĭvērunt blasphēmābant eum*, 15, 29.

forþ-steallian; *p.* ode; *pp.* od *To come to pass;* posthac lŏcum hăbēre :—Sceal seó wyrd swā đeáh forþsteallian *that event shall yet come to pass*, Cd. 109; Th. 144, 15; Gen. 2390.

forþ-stefn, es; *m.* [stefn *a prow*] *A fore-prow, prow;* prōra :—Forþ-stefn scipes *prōra nāvis*, Lye.

forþ-steppan, -stæppan; *part.* -stæppende: *p.* -stepede = -stepte? *pp.* -steped = -stept? *To step or go forth, proceed;* prōgrĕdi, prōdīre, pro-cēdēre :—Of ansŷne đīnre dôm mīn forþsteppe *de vultu tuo judĭcium meum prōdeat*, Ps. Lamb. 16, 2. He is swā swā brŷdguma forþsteppende of brŷdbūre his *ipse est tamquam sponsus procēdens de thălămo suo*, Ps. Lamb. 18, 6: Homl. Th. ii. 90, 11. Đa þing đe forþsteppaþ [MS. forþ-stappaþ] of mīnum welerum *quæ procēdunt de lăbiis meis*, Ps. Lamb. 88, 35.

forþ-stôp, *pl.* -stôpon *went forth, proceeded, passed by*, Ps. Lamb. 72, 7: Mk. Bos. 14, 35: 15, 29; *p.* of forþ-stapan.

forþ-swebban, -swefian; *p.* -swefede; *pp.* -swefed *To prevail, profit;* proficēre :—Nāht forþswefaþ fŷnd *nĭhil prŏficiet inĭmīcus*, Ps. Spl. T. 88, 22.

forþ-tēge, forþ-tîge, -tŷge, es; *m. A fore-court, porch, entrance;* vesti-bŭlum, fŏris :—On đam forþtēge *in ipsis fŏrĭbus*, Prov. 8. Forþtŷge *vestibŭlum, atrium*, Hpt. Gl. 496; Leo A. Sax. Gl. 384, 56. v. fôre-tîge.

forþ-teón; *p.* -teáh, *pl.* -tugon; *pp.* -togen *To lead forth, make known, discover, betray, render up;* prōdĕre, Som. Ben. Lye.

forþ-tîhan; he -tîhþ; *p.* -tāh, *pl.* -tigon; *pp.* -tigen *To draw forth;* protrăhĕre, extrăhĕre :—Meaht forþtîhþ heofoncondelle *his might draweth forth heaven's candle*, Exon. 93 a; Th. 349, 29; Sch. 53. v. tîhan I.

forþ-tihting, e; *f.* [tihting *persuasion*] *An exhortation;* exhortātio, Epil. Reg. Concord.

forþum; *adv. Even, indeed;* quĭdem, saltem :—Nǽnig forþum wæs *none indeed was*, Exon. 46 a; Th. 157, 22; Gū. 895. v. furþum.

for-þunden; *part. p.* [þindan; *p.* þand; *pp.* þunden *to swell*] *Swollen up;* tŭmĭdus :—Gyf seó wund forþunden sŷ *if the wound is swollen up*, Herb. 90, 2; Lchdm. i. 198, 11.

forđung *an armament.* DER. scip-forđung. v. fyrdung.

forþ-weard, es; *m. A forward guard, pilot;* prōrēta :—Forþweard scipes *the pilot of the ship*, Cd. 71; Th. 86, 26; Gen. 1436.

forþ-weard, -werd; *adj.*   **I.** *in a forward direction, forward;* prōnus :—Forþweard *forward*, Exon. 106 a; Th. 403, 25; Rä. 22, 13:

126 b; Th. 487, 4; Rä. 72, 23. Â swā hit forþwerdre beón sceolde, swā wæs hit lætre *always as it should be more forward, so was it later*, Chr. 999; Erl. 134, 32.   **II.** *tending towards any one;* ălĭquem versus tendens :—Forþweard to đē *tending towards thee*, Ps. Cot. 50, 79; Ps. Grn. ii. 278, 79.   **III.** *everlasting, continual;* sempĭternus :—Ic forþweardne gefeán hæbbe *I have everlasting joy*, Exon. 64 a; Th. 236, 4; Ph. 569. Fremum forþweardum *with continual benefits*, Cd. 12; Th. 13, 29; Gen. 210.

forþ-weaxan; *p.* -weóx, *pl.* -weóxon; *pp.* -weaxen *To grow or break forth;* procrescĕre, prorumpĕre :—Forþweóx his feóndscipe *prorūpit ejus ŏdium*, Gr. Dial. 2, 27.

forþ-weg, es; *m. An onward course, a going forth, departure, journey;* progressus, profectio, ābĭtus, ŏbĭtus :—Fūs forþweges *desirous of departure*, Exon. 108 a; Th. 412, 20; Rä. 31, 3. Ferede in forþwege *borne on their journey hence*, 77 b; Th. 291, 12; Wand. 81: Rood Kmbl. 247; Kr. 125. He of ealdre gewāt on forþweg *he departed from life on his way forth*, Beo. Th. 5243; B. 2625: Cd. 148; Th. 185, 27; Exod. 129. On forþwegas *on their ways forth*, 160; Th. 200, 1; Exod. 350: 144; Th. 179, 22; Exod. 32.

forþ-werd [= -weard] *Forthward, those who are present;* præsens :—Đis gemet [*imperativus*] sprecþ forþwerd *this mood* [*imperative*] *speaketh to those present*, Ælfc. Gr. 21; Som. 23, 23. v. bebeódendlīc gemet.

forþ-wîf, es; *n. A married woman, mother,* hence *A matron;* matrōna, Wrt. Voc. 72, 78.

forþ-wîsian; *p.* ode, ade; *pp.* od, ad *To guide forth, direct;* dirĭgĕre :—Him seleþegn forþwîsade *the hall-thane guided him forth*, Beo. Th. 3595; B. 1795.

for-đý, for-đŷ-đe, for-đī, for-đī-đe, for-đig; *conj. For that, for, because, therefore;* nam, quia, ĭtăque :—Forđý đam cræftegan ne mæg nǽfre his cræft losigan *because to the skĭlful his skill can never be lost*, Bt. 19; Fox 70, 2. Nân mon forđý ne rît đe hine rîdan lyste *no man rides because he lists to ride*, Bt. 34, 7; Fox 144, 6, 12.

for-đŷ, for-đī, for-đig; *adv. For that cause, consequently;* proptērea, ĭdeo :—Forđŷ Moyses eów sealde ymbsnydenysse *proptĕrea Moyses dĕdit vōbis circumcisiōnem*, Jn. Bos. 7, 22: Bt. 19; Fox 70, 1: Bt. Met. Fox 20, 385; Met. 20, 193. [*Orm.* forrþl = *Laym.* for þl.]

for-þyldian; -þyldigan, -þyldegian, -þylgian; *p.* ode; *pp.* od *To sustain, bear, endure, suffer, be patient, wait patiently;* sustĭnēre, tolĕrāre, pāti :—For đē ic forþyldegode hosp *propter te sustĭnui opprobrium*, Ps. Spl. 68, 10: 54, 12: Homl. Th. ii. 174, 10. Hî forþyldegodon [Lamb. forþyldigodon] sāwle mīne *sustĭnuērunt anĭmam meam*, Ps. Spl. 55, 7. Ic forbær đē ođđe forþylgode đē *sustĭnui te*, 24, 22. Geþola ođđe forþyldiga Drihten *sustĭne Dŏmĭnum*, Ps. Lamb. 26, 14.

for-þylman, -þylmian; *p.* de, ode; *pp.* ed, od *To encompass, over-whelm, cover over, obscure;* invŏlvĕre, obvŏlvĕre, obscūrāre :—He his sylfes đær bān gebringeþ, đa ǽr brondes wylm on beorhstede forþylmde it [*the phœnix*] *brings its own bones there, which the fire's rage had before encompassed on the mound*, Exon. 60 a; Th. 217, 23; Ph. 284. Þeóstrum forþylmed *overwhelmed with darkness*, Elen. Kmbl. 1530; El. 767: Judth. 10; Thw. 23, 12; Jud. 118. Þeóstru ne beóþ for-þylmode ođđe forsworcene to đē *tĕnebræ non obscūrābuntur a te*, Ps. Lamb. 138, 12.

forþ-yppan; *p.* te; *pp.* ed *To make known, publish, declare;* pro-mulgāre, publĭcāre, prōdĕre, Cot. 150: Ps. Vos. 16, 3.

forþ-yrnan; *part.* -yrnende; *p.* -arn, *pl.* -urnon; *pp.* -urnen *To run forth or before, precede;* præcurrĕre :—Wæs, æfter forþyrnendre tîde, ymb fīfhund wintra and tū and hundnigontig fram Cristes hidercyme it was, *according to the time preceding, about five hundred and ninety-two years from Christ's coming hither*, Bd. 1, 23; S. 485, 18.

for-þyrrian; *p.* ode; *pp.* od [þyr *dry*] *To dry up;* perarescĕre :—Đæt đa sŷn forþyrrode *that they are dried up*, L. M. 2, 27; Lchdm. ii. 222, 5.

for-þýstrian *to darken.* v. for-þeóstrian.

for-tîhan; he -tîþ; *p.* -tāh, *pl.* -tigon; *pp.* -tigen *To draw against or over, cover over with anything, darken, obscure;* obdūcĕre :—Mid gedwol-miste fortîþ mōd *covers over the mind with the mist of error*, Bt. Met. Fox 22, 67; Met. 22, 34. DER. tîhan I. [*Germ.* vorziehen.]

for-tió *may cover over;* *subj. pres.* of for-tión.

for-tión; *impert.* -tió, -tióh, *pl.* -tióþ; *subj.* -tió, *pl.* -tión *To draw against or over, cover over, obscure;* obdūcĕre :—Đæt mōd mid đam gedwol-miste fortió *may cover over the mind with the mist of error*, Bt. 35, 1; Fox 156, 1. v. for-teón.

for-tîþ *covers over, obscures*, Bt. Met. Fox 22, 67; Met. 22, 34; *pres.* of for-tîhan.

for-togen; *part. Tugged or drawn together;* contractus :—Fortogen *turmĭnōsus* [= *tormĭnōsus*], Ælfc. Gl. 2; Som. 55, 35; Wrt. Voc. 16, 10.

for-togenes, -ness, e; *f. A tugging, drawing together, griping, cramp, convulsion;* contractio, convulsio; spasmus :—Wiđ fortogenesse innan *for inward griping or colic*, L. M. 2, 33; Lchdm. ii. 236, 32.

for-tredan, đū -tretst, -trydst, -trytst; *p.* -træd, *pl.* -trǣdon; *pp.* -treden *To tread upon, tread under foot;* conculcāre, calcāre :—Đæt đū cunne fortredan đas woruld *that thou mayest tread down this world*, Homl. Th.

ii. 392, 34. Ic fortrede *conculco*, Ælfc. Gr. 47 ; Som. 48, 43. Fortretst ðū ða woruldlícan styrunga *thou wilt tread down worldly commotions*, Homl. Th. ii. 392, 25. Ðū fortrydst leóna and dracena *thou shalt be a treader down of lions and of dragons*, Ps. Spl. 90, 13. Ðū fortrytst eorþan *conculcābis terram*, Cant. Abac. Lamb. fol. 190 a, 12. Wēnunga þeóstru fortredaþ me *forsitan tenebræ conculcābunt me*, Ps. Lamb. 138, 11. Wegferende ðæt sǽd fortrǣdon *the wayfarers trod the seed down*, Homl. Th. ii. 90, 15 : i. 544, 28. Būton ðæt hit sȳ fram mannum fortreden *nisi ut conculcētur ab hŏminibus*, Mt. Bos. 5, 13. Hierusalem biþ fram þeódum fortreden *Jerūsalem calcābitur a gentibus*, Lk. Bos. 21, 24. Seó fortredene heorte *the trodden down heart*, Homl. Th. ii. 90, 16. [*Chauc.* fortroden *trodden down* : *Ger.* ver-treten *to tread down.*]

for-treding, e ; *f. A treading down, crushing ;* conculcātio, contrītio, Som. Ben. Lye.

for-trūgadnes *over-confidence, precipitancy*, Ps. Spl. T. 51, 4. v. fortrūwodnes.

for-trūwian, -trūwigan ; *p.* ode, ude ; *pp.* od, ud *To be over-confident, rash, to presume ;* præsūmĕre, præcĭpĭtāre :—Ðū ðē fortrūwodest [MS. fortrūwudest] *for ðinre rihtwīsnesse thou wast over-confident on account of thy virtue*, Bt. 7, 3 ; Fox 22, 13. Ðȳ-lǽs he hine for ðære wynsuman wyrde fortrūwige *lest he through the pleasant fortune should be presumptuous*, 40, 3 ; Fox 238, 17. Ða fortrūwodan *the presumptuous*, Past. 32, 1 ; Hat. MS. 39 b, 25, 26. Ða fortrūwudan, 32, 1 ; Hat. MS. 40 a, 2, 12. Ðǽm fortrūwodum monnum *to presumptuous men*, 49, 5 ; Hat. MS.

for-trūwodnes, -trūgadnes, -ness, e ; *f. Over-confidence, precipitancy, presumption, arrogance ;* præcĭpĭtātio, præsumptio, arrŏgantia :— For eówerre fortrūwodnesse *for your presumption*, Past. 32, 1 ; Hat. MS. 40 a, 25. Ða fortrūwodnesse and ða ānwilnesse an Corinctheum Paulus ongeat swīðe wiðerweardne wið hine *the presumption and obstinacy of the Corinthians Paul saw [to be] greatly opposed tō himself*, 32, 1 ; Hat. MS. 40 a, 16. Ðū lufedest ealle word fortrūgadnesse *dīlexisti omnia verba præcĭpĭtātiōnis*, Ps. Spl. T. 51, 4.

for-trūwung, e ; *f. Over-confidence, presumption ;* præcĭpĭtātio :—On ðære fortrūwunga and on ðam gilpe *by presumption and by arrogance*, Bt. 3, 1 ; Fox 6, 4.

for-trydst, -trytst *treadest down*, Ps. Spl. 90, 13 : Cant. Abac. Lamb. fol. 190 a, 12 ; *2nd sing. pres.* of for-tredan.

for-tyhtan ; *p.* te ; *pp.* ed *To draw away, lead astray, seduce ;* sedūcĕre :—Se ealda feónd forlǽrde lygesearwum, leóde fortyhte *the old fiend mistaught with lying snares, led astray the people*, Elen. Kmbl. 416 ; El. 208.

for-tyllan ; *p.* de ; *pp.* ed *To draw off from the object, seduce ;* sedūcĕre :—Ðonan us se swearta gǽst forteáh and fortylde *whence the dark spirit drew away and seduced us*, Exon. 11 b ; Th. 17, 14 ; Cri. 270. v. tillan.

fōr-tymbrian ; *p.* ode, ede ; *pp.* od, ed *To build before* or *in front of, stop up, obstruct ;* obstruĕre :—Fōrtymbred is mūþ sprecendra unrihtu *obstructum est os lŏquentium inīqua*, Ps. Spl. C. 62, 10.

for-tȳnan ; *p.* de ; *pp.* ed *To shut in, stop, hinder ;* interclūdĕre :—Hī mid gelomlícum oncunningum tiledon ðæt hī hine me heofonlícan weg fōrsetton and fortȳndon *qui crebris accūsātiōnibus ūter illi cœleste interclūdĕre contendēbant*, Bd. 3, 19 ; S. 548, 4.

forud ; *part. Broken, fractured, worn out, decayed ;* fractus, contrītus :— Se foruda fōt and sió forude hond *the fractured foot and the fractured hand*, Past. 11, 2 ; Cot. MS. On ðisum þrīm stelum stynt se cynestól, and gif ān biþ forud, he fylþ adūn sōna *the throne stands on these three pillars, and if one is decayed, it soon falls down*, Ælfc. T. 41, 6. v. forod.

for-ūton ; *conj. Without, besides, except ;* sĭne, nĭsi :—Se fīr forbearnde ealle ðe minstre, forūton feáwe bēc *the fire burnt all the monastery except a few books*, Chr. 1122 ; Erl. 249, 8. v. būtan ; *conj.*

for-wærnan ; *p.* de ; *pp.* ed *To deny, refuse ;* recūsāre :—Gif he byrigan forwærne *if he refuse to give a pledge*, L. H. E. 9 ; Th. i. 30, 15. v. for-wyrnan.

for-wandian, -wandigan ; *p.* ode ; *pp.* od [wandian *to fear*]. I. *v. trans. To reverence, have in honour ;* vĕrēri, revĕrēri :—Mīnne sunu hig forwandiaþ *revĕrēbuntur fīlium meum*, Mk. Bos. 12, 6 : Lk. Bos. 20, 13. II. *v. intrans. To be afraid, be confounded, hesitate ;* confundi, cunctāri :—Nellaþ forwandian ðæt hī ne syllon sōþfæstnysse wið sceattum *they are not afraid to betray truth for money*, Homl. Th. ii. 244, 23. Hig forwandiaþ ðæt hig ne dōn mīnum suna swā *they will be afraid to do so to my son*, Mt. Bos. 21, 37. Forwandiaþ ðæt hie mid ðǽm kycglum hiera worda ongeán hiera ierre worpigen *they hesitate to hurl the darts of their words against their anger*, Past. 40, 5 ; Hat. MS. 55 b, 4. He forwandode ðæt he swā ne dyde *he hesitated to do so*, 49, 5 ; Hat. MS. Gescamian and forwandian, ðe ðe sēcaþ sáwle mīne *let them be ashamed and confounded that seek after my soul*, Ps. Spl. T. 69, 2 : Ps. Spl. 39, 19. Nā hī forwandian ofer me *non confundantur sŭper me*, 68, 9.

for-wandung, e ; *f. Shyness, shame, dishonour ;* revĕrentia, ignō-

minia :—Ðū wāst forwandunga mīne *tu scis revĕrentiam meam*, Ps. Spl. 68, 23.

fōr-ward *a fore-ward, precaution*, Chart. ad calc. C. R. Ben. Lye. v. fōre-weard, e ; *f.*

for-warþ *perished*, Cd. 213 ; Jun. 92, 2, = for-wearþ ; *p.* of for-weorþan.

for-weallen ; *part. Thoroughly boiled ;* excoctus, percoctus, Som. Ben. Lye ; *pp.* of for-weallan. v. weallan.

fōr-weard ; *adj. Forward, fore ;* antĕrior :—Is se fugel fæger fōrweard hiwe *the bird is fair of hue in front [forward]*, Exon. 60 a ; Th. 218, 8 ; Ph. 291. Fōrweard heáfod *the forehead ;* frons [obcăput, Wrt. Voc. 64, 26]. Hig beóþ on fōrwearde and gē on æfteweard *ipse ĕrit in căput et tu ĕris in caudam*, Deut. 28, 44. v. fōre-weard ; *adj.*

fōr-weard ; *adv. Onwards, continually, always ;* semper :—Gif hie wolden lāre Godes fōrweard fremman *if they would always perform God's precepts*, Cd. 37 ; Th. 49, 6 ; Gen. 788.

for-wearþ *perished*, Cd. 121 ; Th. 156, 14 ; Gen. 2588 ; *1st and 3rd sing. p.* of for-weorþan.

for-weaxan ; *p.* -weóx, *pl.* -weóxon ; *pp.* -weaxen, -wexen *To overgrow, grow immoderately, swell ;* excrescĕre, turgescĕre :—Ðȳ-læs hie to ðæm forweóxen ðæt hie forseáreden *lest they should grow so much that they should wither away*, Past. 40, 3 ; Hat. MS. 54 b, 17. Wið ðon ðe man on wambe forweaxen sȳ *in case that a man be overgrown in the belly*, Herb. 2, 4 ; Lchdm. i. 80, 22. Forwexen *overgrown*, 40, 1 ; Lchdm. i. 140, 16 : 53, 1 ; Lchdm. i. 156, 9 : 69, 1 ; Lchdm. i. 172, 7. [*Ger.* ver-wachsen *to overgrow.*]

for-weddad *for-weddad ;* *pp.* [wed *a pledge*] *Pledged ;* oppignĕrātus :—Forweddod [MS. for-weddad] feoh *pledged property ;* fīdūcia, Ælfc. Gl. 14 ; Som. 58, 13 ; Wrt. Voc. 21, 8.

for-wegan ; *p.* -wæg, *pl.* -wǣgon ; *pp.* -wegen *To kill ;* interficere :— Ðæt he on foldan lag forwegen mid his wǣpne *that he lay slain on the field with his weapon*, Byrht. Th. 138, 30 ; By. 228.

fōr-wel ; *adv. Very well, very ;* valde :—Him nǣfre seó gītsung fōrwel ne lícode *covetousness never very well pleased him*, Bt. titl. xvii ; Fox xii. 24 : Bt. 17 ; Fox 58, 24. Ólǽcþ ðes middangeard fōrwel menige *this world flatters very many*, Homl. Th. i. 490, 14 : ii. 158, 30 : Ps. Th. 131, 6. Wurdon geworhte wundra fōrwel fela *very many wonders were wrought*, Homl. Th. ii. 152, 28 : 292, 34. Fōrwel oft *very often ;* multŏtiens, Ælfc. Gr. 49 ; Som. 50, 35.

for-wēnan ; *p.* de ; *pp.* ed *To overween, think too highly of ;* nĭmium æstĭmāre :—Forwēned *insŏlens*, Cot. 186. v. wēnan.

for-weoren = for-woren ; *part. p.* [for-, woren, *pp.* of forweosan, v. weosan] *Tottering, decayed ;* marcĭdus, decrēpĭtus :—Eorþgráp hafaþ waldendwyrhtan, forweorene [MS. forweorone], geleorene *earth's grasp* [i. e. *the grave*] *holdeth its mighty workmen, decayed, departed*, Exon. 124 a ; Th. 476, 14 ; Ruin. 7. Forworen *decrēpĭtus*, Hpt. Gl. 456 ; Leo A. Sax. Gl. 84, 60.

for-weornan ; *p.* de ; *pp.* ed *To refuse ;* recūsāre :—He forweornde swīðe *he refused vehemently*, Chr. 1046 ; Erl. 174, 16. Ne forweorn ðū me *refuse thou not me*, Hy. 3, 54 ; Hy. Grn. ii. 282, 54. v. for-wyrnan.

for-weornian ; *p.* ode ; *pp.* ed *To dry up, wither away, fade, grow old, rot, decay ;* marcescĕre, sĕnescĕre, tābescĕre :—Eal forweornast, lāmes gelícnes *thou art all rotting, image of clay !* Exon. 98 a ; Th. 368, 8 ; Seel. 18. Ðonne forweornaþ he and adeádaþ *then it decays and dies*, Homl. Th. i. 168, 31. Hȳ forweorniaþ *they wither away*, Salm. Kmbl. 629 ; Sal. 314. Ðæt gē hrædlíce forweornion *that ye may speedily fade*, Homl. Th. i. 64, 15.

for-weorþan ; *p.* ic, he -wearp, ðū -wurpe, *pl.* -wurpon ; *subj. p.* -wurpe, *pl.* -wurpen ; *pp.* -worpen *To cast, cast away, reject ;* jăcĕre, projĭcĕre, repellĕre :—Se feónd hogode on ðæt micle morþ men forweorpan *the foe thought to cast men into that great perdition*, Cd. 32 ; Th. 43, 16 ; Gen. 691. Ðū forwurpe mīn word *tu projēcisti sermōnes meos*, Ps. Th. 49, 18. Mæg secgan se ðe wyle sōþ sprecan ðæt he gūþgewǣdu forwurpe *he who will speak the truth can say that he cast away his armour [war-garments]*, Beo. Th. 5736 ; B. 2872. Hwī forwurpe ðū me oððe forhwī ūtaþȳgdest ðū me *quāre repŭlisti me ?* Ps. Lamb. 42, 2. [*Goth.* frawairpan : *Orm.* forrwerrpenn : *O. Sax.* farwerpan : *Ger.* ver-werfen *to reject.*] DER. weorpan.

for-weorþan, -wurþan ; ic -weorþe, ðū -weorþest, -wyrst, he -weorþeþ, -wyrþ, *pl.* -weorþaþ, -wyrþaþ ; *p.* ic, he -wearþ, ðū -wurde, *pl.* -wurdon ; *pp.* -worden *To become nothing, to be undone, to perish, die ;* ad nihĭlum devĕnīre, pĕrīre, intĕrīre, defĭcĕre :—Swā sceal ælce sáwl forweorþan æfter ðam unrihthǣmede, būton se mon hweorfe to góde *so shall every soul perish after unlawful lust, unless the man turn to good*, Bt. 31, 2 ; Fox 112, 27 : 34, 9 ; Fox 148, 12. Sceolon hig ealle samod forweorþan *pĕrībunt sĭmul ?* Gen. 18, 24 : Ps. Th. 118, 176. Ðū forwyrst *pĕrībis*, Ex. 9, 15. Óþ-ðæt ðiós eorþe eall forweorþeþ *until this earth shall all perish*, Bt. Met. Fox 11, 170 ; Met. 11, 85. Sīþfæt ārleásra forwyrþ oððe losaþ *iter impiōrum pĕrībit*, Ps. Lamb. 1, 6. Hī forweorþaþ *pĕrībunt*, Ps. Spl. 79, 17 : Ps. Th. 63, 5 : 67, 2 : 72, 22. Hig forwyrþaþ oððe losiaþ *ipsi pĕrībunt*, Ps. Lamb. 101, 27. Seó mænegeo forwearþ *the multitude perished*, Cd. 121 ; Th. 156, 14 ; Gen. 2588 : 213 ;

Th. 266, 13; Sat. 21: Chr. 655; Erl. 28, 1. Ealle nýtenu neáh for-wurdon *nearly all the cattle died*, Ors. 1, 7; Bos. 30, 31: Chr. 593; Erl. 18, 33. Ðý-læs ðû forweorþe *lest thou perish*, Cd. 116; Th. 151, 3; Gen. 2503. Hî forweorþan *ad nihĭlum devĕnient*, Ps. Th. 57, 6. Ðâ wênunga ic forwurde on eáþmôdnesse mînre *tunc forte pĕrissem in hŭmĭlĭtâte mea*, Ps. Lamb. 118, 92. Ðæt hî forwordene weorþen syððan, on worulda woruld and to wîdan feore *ut intĕreant in sĕcûlum sĕcŭli*, Ps. Th. 91, 6. v. for-wurþan, wurþan.

for-weorþenes, -ness, e; *f. A coming to nothing, perishing, ruin*; intĕrĭtus:—Ðis wæs swîðe gedeorfsum geár hêr on lande and þurh orfcwealm and wæstma forweorþenesse *this was a very grievous year in the land, both through murrain of cattle and perishing of fruits*, Chr. 1103; Erl. 239, 3. v. for-wordenes.

fôr-weorþfullíc; *adj. Very worthy, very excellent*; præclârus:—Fôr-weorþfullíc wêla *very excellent wealth*, Bt. 29, 1; Fox 102, 14.

for-weosnian *to pine, fade* or *wither away*; tâbescĕre, languescĕre, marcescĕre, Som. Ben. Lye. v. for-wisnian.

fôr-werd, e; *f. A fore-ward, precaution, contract, agreement*; præcautio, pactum:—Hêr swutelaþ ymb ða fôrwerda ðe Wulfríc and se arcebisceop geworhton *here is made known concerning the agreements which Wulfric and the archbishop made*, Cod. Dipl. 738; A.D. 1023; Kmbl. iv. 25, 29. v. fôre-weard, e; *f.*

for-werednys, -nyss, e; *f. Old age*; sĕnium:—On ylde and for-werednysse *in sĕnectam et sĕnium*, Ps. Spl. 70, 19.

for-wernan; *p.* de; *pp.* ed *To refuse*; recûsâre:—Se arcebisceop him ânrædlíce forwernde *the archbishop constantly refused him*, Chr. 1048; Erl. 177, 24. Hî forwerndon heom ægðer ge upganges ge wæteres *they refused them both landing and water*, 1046; Erl. 171, 5. v. for-wyrnan.

fôr-wernedlíce; *adv. Against one's will, very grievously, hardly*; ægre, anguste, Som. Ben. Lye.

fôr-werod, -wered; *part. p.* [werian *to wear*] *Worn out, very old*; attrîtus, vĕtus:—Seó endlyfte tîd biþ seó fôrwerode ealdnyss *the eleventh hour is very late* or *very great oldness*, Homl. Th. ii. 76, 22. On fôr-werodre ealdnysse *in very old age*, 76, 26. Næs his reáf hôrig ne tosigen, ne his scôs fôrwerode *his raiment was not dirty nor threadbare, nor his shoes worn out*, i. 456, 21: ii. 94, 11. Næron eówre reáf fôrwerode *non sunt attrîta vestimenta vestra*, Deut. 29, 5. Fôrwerede fetelsas *saccos vĕtĕres*, Jos. 9, 5. [Laym. uorwerien *to spend*.]

for-weryþ *shall destroy*; destruet, Ps. Spl. 51, 5, = for-werþþ [Ps. Lamb. towyrþþ *destruet*, 51, 7] for-weorþeþ; *3rd sing. pres. of* forweorþan.

for-wexen *overgrown*, Herb. 69, 1; Lchdm. i. 172, 7, = for-weaxen *pp. of* for-weaxan.

for-wiernan, -wirnan; *p.* de; *pp.* ed *To hinder, prevent, keep from, withhold*; arcêre, rĕtinêre:—Ðæt ða Deniscan him ne mehton ðæs rîpes forwiernan *that the Danish might not hinder them from the harvest* Chr. 896; Erl. 94, 7. Ðæt mann forwierne his sweorde blôdes, ðæt hwâ forwirne his lâre ðæt he mid ðære ne ofsleá ðæs flæsces lustas *keeping one's sword from blood is withholding one's instruction, and not slaying with it the lusts of the flesh*, Past. 49; Hat. MS. v. for-wyrnan.

for-wird, e; *f. Loss, destruction, ruin, perdition*; perdĭtio, intĕrĭtio:—Hira forwirde dæg ys gehende *juxta est dies perdĭtiônis*, Deut. 32, 35. He generode hî of forwirdum heora *erĭpuit eos de intĕrĭtiônĭbus eôrum*, Ps. Spl. 106, 20. v. for-wyrd.

for-wisnian; *p.* ode, ade; *pp.* od, ad *To wither* or *wizen away, dry up, decay*; marcescĕre, arescĕre, tâbescĕre, putrescĕre:—Wyrt forwisnaþ, weorþeþ to duste *herba indûret, et arescat*, Ps. Th. 89, 6: 101, 23. Ðæt biþ forwisnad wraðe sôna, ær hit afohten foldan losige *quod priusquam evellâtur, arescit*, 128, 4. To hwan drehtest ðû me eal forwisnad *wherefore didst thou torture me all decayed?* Soul Kmbl. 36; Seel. 18.

fôr-witan; *p.* -wiste, *pl.* -wiston; *subj. pres.* -wite; *pp.* -witen *To foreknow, know beforehand*; præscîre:—Ðæs ðe ðû fôrwite hwâm ðû gemiltsige *that thou mayest know beforehand whom thou pitiest*, Apol. Th. 11, 21. v. fôre-witan.

fôr-witolnes, -ness, e; *f. Foreknowledge, diligence, industry*; præscientia, industria, R. Ben. interl. 27.

fôr-wlencean; *p.* -wlencte; *pp.* -wlenced [wlenco *pride*] *To exalt, fill with pride, make very proud*; exaltâre, arrŏgantia implêre:—Ðonne hine ne mâgon ða wêlan fôrwlencean *when the riches are not able to make him proud*, Past. 26; Hat. MS. 35 b, 2. Forwlencte *proud*, Blickl. Homl. 199, 14.

fôr-word, es; *n. A fore-word, stipulation, agreement*; præcautio, pactum:—Ðæt hire frýnd ða fôrword habban *that her friends have the stipulations*, L. Edm. B. 7; Th. i. 256, 2. Ðis synd ða fôrword ðe Æðelrêd cyng and ealle his witan wið ðone here gedôn habbaþ *these are the agreements which king Æthelred and all his counsellors have made with the army*, L. Eth. ii. prm; Th. i. 284, 6. cf. fôre-weard, e; *f.*

for-worden *perished*, Ps. Th. 91, 6; *pp. of* for-weorþan.

for-wordenes, -weorþenes, -ness, e; *f.* [*pp.* forworden *perished*] *A coming to nothing, perishing, ruin*; intĕrĭtus:—Ðis wæs swîðe gedyrfsum geár hêr on lande þurh wæstma forwordenessa *this was a very grievous year in the land through the perishing of fruits*, Chr. 1105; Erl. 240, 15.

for-wordenlíc *damnable*; damnabĭlis, Som. Ben. Lye.

fôr-worht *obstructed*, Chr. 901; Erl. 96, 31; *pp. of* fôr-wyrcan.

for-worhta, an; *m.* [*pp. of* for-wyrcan] *A misdoer, malefactor*; scĕlestus, mălefactor:—Ða forworhtan, ða ðe firnedon, beóþ beofigende *the malefactors, they who sinned, shall be trembling*, Cd. 227; Th. 303, 28; Sat. 620.

for-worhte *did wrong, sinned, ruined, convicted, condemned, forfeited*, Cd. 40; Th. 53, 6; Gen. 857: Exon. 21 b; Th. 57, 20; Cri. 921, = *p. of* for-wyrcan.

for-wrecan; *p.* -wræc, *pl.* -wrǣcon; *pp.* -wrecen [wrecan *to drive*] *To drive out, banish, expel*; expellĕre, propellĕre, fûgâre:—Ðý-læs hit ýþa þrym forwrecan meahte *lest the force of the waves might drive it out*, Beo. Th. 3843; B. 1919. He hine feor forwræc *he banished him far*, 219; B. 109. Hý forwrǣcon wîcinga cynn *they expelled the race of the vikings*, Scôp Th. 95; Wîd. 47. Eart ðû âna forwrecen on Hierusalem *tu sôlus peregrînus es in Jerusalem?* Lk. Bos. 24, 18.

for-wrêgan, fore-wrêgan; *p.* de; *pp.* ed [wrêgan *to accuse*] *To accuse strongly*; vehementer accûsâre:—Brihtríc forwrêgde Wulfnôþ to ðam cyninge *Brihtric accused Wulfnoth to the king*, Chr. 1009; Erl. 141, 29. Ða Wǣlisce men forwrêgdon ða eorlas *the Welshmen accused the earls*, 1048; Erl. 178, 24. He wæs oft to ðam cyninge forwrêged *he had often been accused to the king*, 952; Erl. 118, 27: 1068; Erl. 206, 33. Se wearþ wið hine forwrêged *hic diffâmâtus est ăpud illum*, Lk. Bos. 16, 1.

for-wrîtan; *p.* -wrât, *pl.* -writon; *pp.* -writen [wrîtan *to cut, carve, engrave, write*] *To cut asunder*; dissecâre:—He forwrât wyrm on middan *he cut the worm asunder in the middle*, Beo. Th. 5403; B. 2705.

for-wrîðan; *p.* -wrâþ, *pl.* -wridon; *pp.* -wriden *To bind up, stanch*; obligâre, supprimĕre:—Gif ðû ne mæge blôd-dolh forwrîðan *if thou canst not stanch a blood-running wound*, L. M. 3, 52; Lchdm. ii. 340, 19.

for-wûndian; *p.* ode, ede; *pp.* od, ed *To wound badly, ulcerate*; grăvĭter vulnĕrâre:—Gif mon ôðrum ða geweald uppe on ðam sweoran forwûndie [-wûndige MS. H.] *if a man wound the tendons on another's neck*, L. Alf. pol. 77; Th. i. 100, 11. Eall ic wæs mid strǣlum forwûndod *I was all wounded with arrows*, Rood Kmbl. 124; Kr. 62: Cd. 216; Th. 273, 4; Sat. 131. Se læg on his dûra swýðe forwûndod *qui jăcêbat ad jânuam ejus ulcĕrĭbus plênus*, Lk. Bos. 16, 20. Forwûnded mid wommum *wounded with sins*, Rood Kmbl. 27; Kr. 14. Ða men wǣron forwûndode *the men were badly wounded*, Chr. 882; Erl. 83, 11: 897; Erl. 96, 13. [*Ger.* ver-wunden *to wound*.]

for-wurdon *perished*, Ors. 1, 7; Bos. 30, 31; *p. pl. of* for-weorþan.

for-wurþan *to perish*; pĕrîre:—Ðæt eall Egipta land môt forwurþan *quod pĕrierit Ægyptus*, Ex. 10, 7: Mt. Bos. 8, 25; Hy. 7, 112; Hy. Grn. ii. 289, 112. v. for-weorþan.

for-wyrcan, -wyrcean; *p.* -worhte, -wyrhte; *pp.* -worht, -wyrht [for-, wyrcan *to work, do*]. **I.** *to miswork, do wrong, sin*; măle ăgĕre, delinquĕre, peccâre:—Ðæt ðâm forworhtum mannum beó ðe mâra ege for fŷre gesomnunge *that to the wrong doing men there may be the more fear for our assemblage*, L. Ath. v. § 8, 3; Th. i. 236, 16. He wiste forworhte, ða he ǣr wlite sealde *he knew [they had] done wrong whom he had before gifted with beauty*, Cd. 40; Th. 53, 6; Gen. 857. Iudas hine sylfne ahêng, and rihtlíce gewráþ ða forwyrhtan þrotan, seó ðe belǣwde Drihten *Judas hanged himself, and justly bound the sinful throat, which had betrayed the Lord*, Homl. Th. ii. 250, 15. **II.** *to do for, destroy, ruin, convict, condemn*; perdĕre, destruĕre, labefactâre, condemnâre:—Ða Perse ondrêdon ðæt man ða brycge forwyrcean wolde *the Persians dreaded that they would destroy the bridge*, Ors, 2, 5; Bos. 46, 8. Gif hwâ hine sylfne forwyrce on mænigfealdum synnum *si quis seipsum multĭfâriis peccâtis labefactâvĕrit*, L. M. I. P. 44; Th. ii. 276, 28: L. E. G. 4; Th. i. 168, 22. He biþ egeslíc to geseónne ðâm ðǣr mid firenum cumaþ forþe forworhte *he shall be dreadful to see to those who come ever done for with crimes*, Exon. 21 b; Th. 57, 20; Cri. 921. Wâ me forworhtum *woe to me ruined!* 75 a; Th. 280, 20; Jul. 632. Se ðe þýfþe oft forworht wǣre openlíce *he who has often been convicted openly of theft*, L. Ath. v. § 1, 4; Th. i. 228, 25. Ðe forworht wǣre *who has been condemned*, L. E. G. 10; Th. i. 172, 16. Ne dýde man ǣfre on Sunnan dæges freólse ǣnigne forwyrhtne [forworhtne MS. B.] man *let not a man ever put any condemned man to death on the festival of Sunday*, L. C. S. 45; Th. i. 402, 10: L. E. G. 9; Th. i. 172, 14. **III.** *to forfeit*; amittĕre:—Ðæt man sceolde ge-earnian ða wununga on heofenan rîce, ðe se deófol forwyrhte mid môdignysse *that man should merit the dwellings in the kingdom of heaven, which the devil had forfeited through his pride*, Homl. Th. i. 12, 28. Gif hwâ freót forwyrce *if any one forfeit his freedom*, L. Ed. 9; Th. i. 164, 10: L. Edg. ii. 2; Th. i. 266, 13: L. In. 5; Th. i. 104, 15. Ic forworht hæbbe hyldo ðîne *I have forfeited thy favour*, Cd. 48; Th. 62, 33; Gen. 1024: Blickl. Homl. 25, 1: L. Alf. pol. 42; Th. i. 90, 20: L. Eth. vii. 16; Th. i. 332, 16. [*Ger.* verwirken *to forfeit*.]

fôr-wyrcan, -wyrcean; *p.* -worhte; *pp.* -worht [fôr *before*, wyrcan *to work, do*] *To work* or *place before, obstruct, barricade*; oppônĕre, obstruĕre:—Se cing gehâwode hwǣr man mihte ða eá fôrwyrcan [fôr-

wyrcean, col. 2] *the king observed where the river might be obstructed,* Chr. 896; Th. 173, 36, col. 1. He hæfde ealle ða geatu fôrworht into him *he had barricaded all the entrances against him,* Chr. 901; Erl. 96, 31. Synt ðissa heldôra wegas fôrworhte *the ways of these hell-doors are obstructed,* Cd. 19; Th. 24, 21; Gen. 381.

for-wyrd, -wird, e; f. [wyrd *fortune;* for-weorþan *to perish*] Loss, damage, destruction, perdition, ruin, death; detrimentum, intĕritus, intĕritio, perditio, pernicies, internĕcio:—Hēr is geswutelod ûre forwyrd *here is made manifest our destruction,* Judth. 12; Thw. 25, 30; Jud. 285. He alŷsde ðîn lîf of forwyrde *qui redĭmit de intĕrĭtu vîtam tuam,* Ps. Th. 102, 4: 106, 19: Ps. Lamb. 9, 16: Boutr. Scrd. 17, 23: 20, 16. Hwæt fremaþ ænegum menn, ðeáh he ealne middaneard gestrŷne, gyf he hys sâwle forwyrd þolaþ *quid prodest hŏmĭni, si mundum ūnĭversum lucrētur, anĭmæ vēro suæ detrĭmentum pătĭātur?* Mt. Bos. 16, 26: Lk. Bos. 9, 25. Ðâ sió wērge sceólu hreósan sceolde in wîta forwyrd *when the wretched crew must fall into the ruin of punishment,* Elen. Kmbl. 1526; El. 765: Frag. Kmbl. 16; Leás. 10: Andr. Kmbl. 3234; An. 1620. Ðæt hî ôþ for-wyrd æghwær fordîligade ne wæron *ne usque ad internĕciōnem usquequaque delērentur,* Bd. 1, 16; S. 484, 17. Of forwyrdum heora *de intĕrĭtiōnibus eōrum,* Ps. Lamb. 106, 20. 2. for-wyrd, es; *n.* is *neuter* in the following examples:—Ðîn andbîdaþ ðæt ēce forwyrd *the eternal perdition awaits thee,* Homl. Th. i. 598, 9. God forlēt hî to ðam ēcan forwyrde *God will abandon them to the eternal perdition,* i. 112, 23.

for-wyrht, es; *n. A sin, crime;* peccātum. DER. mân-forwyrht.

for-wyrhta, an; *m.* [for *for,* wyrhta *a workman*] *One who does anything for another, an agent, vicegerent;* institor, procūrātor:—Ðe nænne forwyrhtan næfde *who had no agent,* L. Ath. v. 2; Th. i. 230, 20. Se ðe swâ geþogenne forwyrhtan næfde, swôre for sylfne *he who had not such a prosperous vicegerent, swore for himself,* L. R. 4; Th. i. 192, 5.

for-wyrhte *destroyed, forfeited,* Homl. Th. i. 12, 28; *p.* of for-wyrcan.

for-wyrnan, -weornan, -wiernan, -wirnan, -wernan; *p.* de; *pp.* ed *To prohibit, deny, refuse, restrain, prevent, hinder;* prohibēre, recūsāre, denĕgāre, renuĕre:—Him ðær se geonga cyning ðæs oferfæreldes forwyrnan myhte *where the young king might prevent his going over,* Ors. 2, 4; Bos. 45, 9. Se ilca forwyrnþ ðære [MS. ðæræ] sæ ðæt heó ne môt ðone þeorscwold oferstæppan ðære eorþan *the same restrains the sea that it may not overstep the threshold of the earth,* Bt. 21; Fox 74, 25. Me ðæs forwyrnde Waldend heofona *the Lord of heaven hath denied it me,* Cd. 101; Th. 134, 3; Gen. 2219; Exon. 34b; Th. 111, 31; Gū. 135. Ic ne forwyrnde woroldrædenne *he refused not worldly converse,* Beo. Th. 2288; B. 1142. Forwyrnde beón afrêfrod sâwle mîn *renuit consōlāri anĭma mea,* Ps. Spl. 76, 3. Þearfum forwyrndon, ðæt hî under eówrum þæce môsten ingebûgan *ye prohibited the needy, that they might enter under your roof,* Exon. 30a; Th. 92, 4; Cri. 1504. Ðæt ðû me ne forwyrne *that thou deny me not,* Beo. Th. 862; B. 429. Ðŷ-læs eów weges forwyrnen to wuldres byrig *lest they prohibit you the way to glory's city,* Exon. 75b; Th. 282, 18; Jul. 665. Me hwîlum biþ forwyrned willan mînes *sometimes I am denied my will,* 72a; Th. 268, 32; Jul. 441. [O. Sax. far-wernian *to refuse: Laym. pp.* furwurnen.]

for-wyrnednes, -ness, e; *f. A restraining, continence, forbidding;* continentia:—He wæs micelre forhæfdnysse and forwyrnednesse lîfes *he was of great abstinence and continence of life,* Bd. 3, 5; S. 526, 21.

for-wyrpnes, -ness, e; *f. A rejection;* abjectio:—Ic eom forwyrpnes oððe aworpennys folces *ēgo sum abjectio plēbis,* Ps. Lamb. 21, 7.

for-wyrst, he -wyrþ *shalt perish, perishes,* Ex. 9, 15: Ps. Lamb. 1, 6; *2nd and 3rd sing. pres. and fut. of* for-weorþan.

for-wyrþaþ *perish,* Ps. Lamb. 101, 27, = for-weorþaþ; *pl. pres. of* for-weorþan.

for-yldan; *p.* -ylde; *pp.* -ylded *To put off, defer;* differre, sŭpersĕdēre:—Ne mæg mon foryldan ðone deóran sîþ *no one may put off the severe journey,* Salm. Kmbl. 721; Sal. 360. Ðe he to medmicelre tîde forylde dôn [MS. doan] *quam ad brĕve tempus făcĕre sŭpersēdit,* Bd. 5, 13; S. 633, 23: Blickl. Homl. 213, 24; 95, 25

for-yrman; *p.* ode; *pp.* ed [yrman *to afflict*] *To afflict greatly, harass;* vehementer affligĕre:—Hî hî ealle foryrmdon *they harassed them all,* Bd. 1, 12; S. 480, 36.

fôr-yrnan; *p.* -arn, *pl.* -urnon; *pp.* -urnen *To run before;* præcurrĕre:—Se ôðer leorningcniht fôrarn Petrus *ille ălius discĭpŭlus præcŭcurrit Petro,* Jn. Bos. 20, 4. [Ger. vor-rennen *to run before.*]

fôr-yrnere *a fore-runner;* præcursor, Som. Ben. Lye. v. fôr-rynel.

FÓSTER, fôstor, fôstur; *gen.* fôstres; *n. A* FOSTERING, *nourishing, rearing, feeding, food, nourishment, provisions;* edŭcātio, nutrīcium, pastio, alimentum, victus:—Ic gegaderige in to ðê of deórcynne and of fugelcynne gemacan, ðæt hî eft to fôstre beón *I will gather in to thee mates of beast-kind and of bird-kind, that they afterwards may be for food,* Homl. Th. i. 20, 35. Be fundenes cildes fôstre. To fundenes cildes fôstre ðŷ forman geáre geselle vi scillinga, ðŷ æfterran twelf, ðŷ þriddan xxx; siððan, ðæs hŷ wlite *of the fostering of a foundling* [lit. *of a found child*]. *Let six shillings be paid for the fostering of a foundling for the first year, twelve for the second, thirty for the third; afterwards, according to its appearance,* L. In. 26; Th. i. 118, 17–20: 38; Th. i.

126, 5. Mon sceal sellan, to fôstre, x fata hunies, ccc hlâfa, etc. *one shall give, as provisions, ten vats of honey, three hundred loaves, etc.* L. In. 70; Th. i. 146, 16. He gecŷðde hwæðer he mænde ðe ðæs môdes fôster ðe ðæs lîchoman *he made known whether he meant the feeding of the mind or of the body;* pastiōnem cordis an corpŏris suāderet, apĕruit, Past. 18, 6; Hat. MS. 27a, 21. [*Laym.* uoster *a foster-child: Plat.* voedster: *Dut.* voedster, *f. a nurse: Dan. Swed.* foster, *n. embryo, child: Icel.* fóstr, *n. the fostering of a child.*] v. fôda *food.*

fôster-bearn, fēster-bearn, es; *n. A* FOSTER-BEARN *or child;* ălumnus, Cot. 9.

fôster-brôðor; *m. A* FOSTER-BROTHER; collactāneus:—Fôsterbrôðor ălumnus, Wrt. Voc. 284, 74.

fôster-cild, es; *n. A* FOSTER-CHILD; ălumnus, Wrt. Voc. 72, 39.

fôster-fæder, fēster-fæder, es; *m. A* FOSTER-FATHER, *nourisher, bringer up;* altor, nutrîtor, Wrt. Voc. 72, 37. Fôsterfæder *ălumnus,* Ælfc. Gl. 86; Som. 74, 36; Wrt. Voc. 50, 18. [*Orm.* fossterfaderr.]

fôster-land, fôstor-land, es; *n. Land assigned for the procuring of provisions;* fundus cībāriis emendis assignātus:—He gean [MS. geun] ðæs landes æt Wihtrîces hamme ðâm Godes þeówum, to fôsterlande *he gives the land at Wittersham to God's servants, as foster-land,* Th. Diplm. A. D. 1032; 329, 27. Se cyning ðæt land geaf into Cristes cyrcean ðân hîrēde to fôsterlande *the king gave the land to Christ-church as foster-land for the convent,* Th. Diplm. A. D. 1052; 368, 17.

fôster-leán, fôstor-leán, es; *n. Foster-loan, remuneration for rearing a foster-child;* edŭcātiōnis præmium, nutrīcii merces:—Is to witanne hwâm ðæt fôsterleán gebŷrige *it is to be known to whom the remuneration for fostering belongs,* L. Edm. B. 2; Th. i. 254, 8.

fôster-ling *a* FOSTERLING, *foster-child,* Som. Ben. Lye. v. fôstor-ling.

fôster-man *a foster-man, bondsman, security.* v. fēster-man.

fôster-môdor, -môder, fôstor-môdor, fēster-môdor, -môdur, fæster-môdor; *f. A* FOSTER-MOTHER, *nurse;* altrix, nutrix:—Hwæðer hit oncneówe his fôstermôdor *whether it knew its foster-mother,* Bt. 3, 1; Fox 4, 30. Fôstermôder *altrix* vel *nutrix,* Wrt. Voc. 72, 38. Ic gean mînre fôstermēder ðæs landes æt Westune *I give to my mother the land at Weston,* Th. Diplm. 560, 25.

fôster-nôþ, fôstor-nôþ, es; *m? A pasturage, pasture;* pascua:—On stôwe fôsternôþes me he gestaðelode *in lŏco pascuæ me collŏcāvit,* Ps. Spl. T. 22, 1.

fôster-sweostor; *f. A* FOSTER-SISTER; collactānea, Som. Ben. Lye.

fôstor *a fostering, nourishing, food, nourishment,* Som. Ben. Lye. v. fôster.

fôstor-land, es; *n. Land assigned for the procuring of provisions:—* Ðæt ylce land hî gefreódon Godes þeówan to brŷce into fôstorlande *they freed the same land for the use of God's servants as foster-land,* Th. Diplm. A. D. 963–975; 297, 33. v. fôster-land.

fôstor-leán, es; *n. Remuneration for fostering;* nutrîcii merces:—Hæfde Nergend fôstorleán fæmnan forgolden, ēce to ealdre *the Saviour had repaid the virgin the reward for fostering, in eternal life,* Menol. Fox 301; Men. 152. v. fôster-leán.

fôstor-ling, es; *m. A fosterling, foster-child;* ălumnus, verna, vernŭla:—Fôstorling *vernŭla,* Ælfc. Gl. 8; Som. 56, 103; Wrt. Voc. 18, 53. Inberdling vel fôstorling *verna* vel *vernăculus,* 86; Som. 74, 34; Wrt. Voc. 50, 17. [*Laym.* fosterling.]

fôstor-môdor; *f. A foster-mother;* altrix:—Ðæs mædenes fôstor-môdor into ðam bûre eóde *the maiden's foster-mother went into the chamber,* Apol. Th. 2, 7, 11, 12, 15, 19, 23: Nar. 40, 7. v. fôster-môdor.

fôstor-nôþ, es; *m? A pasture;* pascua:—Sceáp fôstornôþes his *ŏves pascuæ ejus,* Wanl. Catal. 223, 37, col. 2: 291, 23, col. 1. v. fôster-nôþ.

fôstraþ, es; *m. Food, victuals;* esca, cibus:—Met oððe fôstraþ wæs, Mt. Kmbl. Lind. 3, 4. Hlâf oððe fôstraþ *pānem,* Jn. Lind. War. 6, 31. Fôstraþ *manna,* Jn. Lind. War. 6, 49. Fôstraþas *epimēnia* = ἐπιμήνια *provisions for a month, a month's rations,* Som. Ben. Lye.

fôstre, an; *f. A fosterer, nurse;* altrix, nutrix. DER. cild-fôstre.

fôstrian; *p.* ode; *pp.* od *To* FOSTER, *nourish;* ălĕre, nutrīre, Som. Ben. Lye. v. fēstrian. [*Orm.* fosstrenn *to nourish: Laym.* fostrien.]

fôstur, es; *n. A fostering, feeding, food, nourishment;* edŭcātio, pastio, nutrīcium:—Fôstur feormian *to give food, to foster, cherish,* Ps. Th. 77, 69. v. fôster.

FÓT; *nom. acc. gen.* fôtes; *dat.* fēt, fôte; *pl. nom. acc.* fēt, fôtas; *gen.* fôta; *dat. inst.* fôtum; *m.* I. *a* FOOT; pēs, *gen.* pĕdis; *m:*—Gyf ðîn swîðre oððe ðîn fôt ðe swîcaþ *si mănus tua, vel pēs tuus scandălizat te,* Mt. Bos. 18, 8. Ne cume me fôt ofermôdignysse *ne vĕniat mihi pēs superbiæ,* Ps. Spl. 35, 12. Swâ his fôt gestôp *where his foot stepped,* Andr. Kmbl. 3163; An. 1584. Nâmen ðâ ðæt fôtspure ðe wæs undernæðen his fôte *then* [*they*] *took the footstool, that was underneath his foot,* Chr. 1070; Erl. 209, 8. Ðæt ic heonon nelle fleón fôtes trym *I will not flee hence a footstep,* Byrht. Th. 138, 68; By. 247. On ânum fêt *on one foot,* Exon. 108b; Th. 415, 5; Rä. 33, 6. On fôte *in the foot,* 108b; Th. 414, 8; Rä. 32, 17. Mid fôte *pĕde,* Ps. Th. 65, 5. Sylle fôt wið fêt *reddat pĕdem pro pĕde,* Ex. 21, 24: Ps. Spl. 90, 12: Lk. Bos. 4, 11. Standende wæron fêt ûre on cæfertûnum ðînum *stantes ĕrant pĕdes nostri in atriis tuis,* Ps.

Spl. 121, 2: Cd. 19; Th. 24, 18; Gen. 379. Sindon fealwe fôtas *the feet are yellow*, Exon. 60 a; Th. 219, 22; Ph. 311: Ps. Th. 121, 2: 131, 7. Ge-eádmēdaþ oððe gebiddaþ fôtsceamol his fôta *adōrāte scabellum pĕdum ejus*, Ps. Lamb. 98, 5: Exon. 107 b; Th. 410, 12; Rä. 28, 15. Ðe-læs hig mid hyra fôtum hig fortredon *ne forte conculcent eas pĕdibus suis*, Mt. Bos. 7, 6. Hæfde gefeormod fêt and folma *he had devoured feet and hands*, Beo. Th. 1494; B. 745. **II.** *the foot*; pēs, gen. pĕdis:—*The foot of a man, a measure of length, was divided into twelve equal parts or inches.* v. ynce, es; *m. inch; and an inch is three barley-corns in length.* In Anglo-Saxon times, the people and their rulers were satisfied with the simplest weights and measures, thus a yard was three feet, of twelve inches each foot, while an inch was in length three barley-corns. In our day, the legislature passed an act so late as July 30, 1855. It is styled, *An Act for legalising and preserving the restored standards of weights and measures.* This Act includes the weights of George the Fourth, 1824, in which the pounds avoirdupois is fixed by a standard weight, kept in the office of the Exchequer, and one equal seven-thousandth part of such pound avoirdupois shall be a grain. Thus our measures and weights are so recently fixed by standards. v. fôt-gemet, eln, ynce, met-geard, geard, gyrd. Nigon fôta, and ix scæfta munda, and ix bere-corna *nine feet, and nine half feet, and nine barley-corns or three inches*, L. Ath. iv. 5; Th. i. 224, 9. [Wyc. Piers P. Chauc. foot: Laym. Orm. fôt: Plat. voot, *m*: O. Sax. fôt, fuot, *m*: Frs. foet: O. Frs. fot, *m*: Dut. ٠oet, *m*: Ger. fusz, *m*: M. H. Ger. vuoz, *m*: O. H. Ger. fuoz, *m*: Goth. fotus, *m*: Dan. fod, *m. f*: Swed. fot, *m*: Icel. fótr, *m*: Lat. pēs, gen. pĕd-is, *m*: Grk. πούς, gen. ποδός, *m*: Pers. پا pa; *pl.* پایان payan: Lith. pádas *sole of the foot*: Sansk. पद् pad, पाद् pād, पाद् pāda, *m.* from पद् pad *to go*.]

fôt-ádl, e; *f. A foot-disease, the gout*; podagra:—Wæs Mellitus mid fôtádle swîðe gehefigad *ĕrat Mellitus podagra grăvātus*, Bd. 2, 7; S. 509, 12. Wið fôtádle *against gout*, Lchdm. i. 376, 1.

fôt-bred, es; *n. A foot-board, stirrup*; tăbella in qua pĕdes requiescunt, astrăba [*q. v.* in Du Cange]:—Fôtbred [MS. fôtbret] *astrăba*, Ælfc. Gl. 3; Som. 55, 67; Wrt. Voc. 16, 40.

fôt-cops, -cosp, es; *m. A fetter, shackle for the feet*; pĕdica, compes:—Fôtcops *compes* vel *cippus*, Wrt. Voc. 86, 31. Hig ge-eádmēttan on fôtcopsum fêt his *humiliăvērunt in compĕdibus pĕdes ejus*, Ps. Lamb. 104, 18. Hine ne mihte nân man mid fôtcopsum gehæftan *no man could confine him with fetters*, Homl. Th. ii. 378, 27 : Mk. 5, 4. To gewrîðenne cyningas heora on fôtcopsum *ad alligandos rēges eōrum in compĕdibus*, Ps. Spl. 149, 8.

fôt-cosp, es; *m. A fetter*; compes:—Hî ge-eádmētton on fôtcospum fêt his *humiliăvērunt in compĕdibus pĕdes ejus*, Ps. Spl. C. 104, 17. v. fôt-cops.

fôt-côðu, e; *f. A foot-disease, the gout*; podagra, Hpt. Gl. 471, 472; Leo A. Sax. Gl. 24, 28.

fôt-cypsed; *part. Fettered*; compĕdītus, Som. Ben. Lye. DER. gefôtcypsed.

fôte; *adj. Provided with feet, footed*; pĕdātus. DER. feðer-fôte, fiðer-, flax-, flohten-, flox-, fyðer-. v. fête.

fôt-ece, es; *m. Foot-ache, the gout*; pĕdis dŏlor, podagra = πόδαγρα:—Wið fôtece *for foot-ache*, L. M. 1, 27; Lchdm. ii. 68, 12, 19, 20, 23.

fôt-gemearc, es; *n. A foot-mark, length of a foot*; ūnius pĕdis longĭtūdo:—Se lêgdraca wæs fîftiges fôtgemearces lang *the fire-dragon was fifty feet of measure long*, Beo. Th. 6077; B. 3042.

fôt-gemet, es; *n. A foot-measure, foot-band, fetter*; pĕdis mensūra, compes:—Hî ge-eádmētton on fôtgemetum fêt his *humiliăvērunt in compĕdibus pĕdes ejus*, Ps. Spl. T. 104, 17.

fôt-gewǽde, es; *n. Foot-clothing*; pĕdum indūmentum, R. Ben. 55.

fôþ *take; pl. impert.* of fôn, *q. v*:—On fôþ hine *accĭpĭte eum*, Bd. 5, 13; S. 633, 14.

FÔÐER, fôður, es; *n.* **I.** *food, food for cattle, fodder*; ălĭmentum, jūmenti pābŭlum:—Fôðres ne gîtsaþ *it is not desirous of food*, Exon. 114 b; Th. 440, 1; Rä. 59, 11. Twentig pŭnd-wǽga fôðres *twenty pounds weight of fodder*, L. In. 70; Th. i. 146, 20. Se ceorl, se ðe hæfþ ôðres oxan ahŷrod, gif he hæbbe ealle on fôðre to agifanne, agife ealle. Gif he næbbe, agife healf on fôðre, healf on ôðrum ceápe *the ceorl, who has hired another's oxen, if he have to pay all in fodder, let him give it all. If he have not, let him pay half in fodder, and half in other goods*, 60; Th. i. 140, 8–11. **II.** *that which food is carried,—a basket*; cophĭnus = κόφινος:—Genômon ceawlas *vel* fôður tŭlērunt cophĭnos, Mt. Lind. Stv. 14, 20. **III.** *that in which food for cattle is carried,—a cart or cart-load, about 19 or 20 cwt.* a heavy weight, as we now use the word for a FOTHER of lead, that is 19½ cwt ; vêhes, plaustrum, nunc massa *vel* vŏlūmen plumbi:—He scolde gife sixtiga fôðra wuda, and twælf fôður græfan, and sex fôðra gearda *he should give sixty loads of wood, and twelve loads of gravel, and six loads of faggots*, Chr. 852; Erl. 67, 37: Cod. Dipl. 508; A. D. 963; Kmbl. ii. 398, 20. [*Laym.* iii. 22 uoðere, foðer *a load*: O. Sax. fôðer,

uoðer vêhes: Dut. voeder, *n. a cart-load*: Ger. fuder, *n. a cart-load, tun*: M. H. Ger. vuoder, *n. a cart-load, tun*: O. H. Ger. fuotar, *n. thēca, plaustrum*.] v. fôdder.

fôþorn, es; *m.* [fôn *to grasp, catch*; þorn *a thorn*] *A fothorn, surgeon's instrument*; tĕnācŭlum:—Wið ðam niðeran tôþece, slît mid ðê fôþorne ôþ-ðæt hie blêden *for the nether tooth-ache, slit [the gums] with the fothorn till they bleed*, L. M. 1, 6; Lchdm. ii. 52, 8.

fôt-lǽst, -lâst, es; *m. A foot-step, foot-trace*; vestīgium pĕdis, trāmes:—Se wyrm onfand feóndes fôtlâst *the worm found the foe's foot-trace*, Beo. Th. 4567; B. 2289. Fôtlǽstas [MS. fôtlæst] ðîne ne beóþ oncnâwen *vestīgia tua non cognoscentur*, Ps. Spl. 76, 19: Blickl. Homl. 203, 36.

fôt-mǽl, es; *n. A foot-mark* or *print, foot-space*; signum *vel* mensūra pĕdis:—Ic wille nǽfre ðê myntan ne furh ne fôtmǽl *I will never appoint for thee neither furrow nor foot-mark*, L. O. 13; Th. i. 184, 7. He nǽfde ðâ ealles landes bûton seofon fôtmǽl *he had not then but seven feet of all his land*, Chr. 1086; Erl. 221, 2. Ðæt he nolde fleógan fôtmǽl landes *that he would not flee a foot-space of land*, Byrht. Th. 139, 57; By. 275. On twentigum fôtmǽlum feor *twenty feet deep*, Elen. Kmbl. 1658; El. 831: Nar. 35, 2; 36, 12.

fôt-mǽlum; *adv. By footsteps, step by step, by degrees*; pĕdĕtentim, grădātim, R. Conc. 5: Cot. 95. v. mǽl, es; *n.* **III.**

fôt-ráp, es; *m. A rope of a ship which fastens the sail*; prōpes:—Fôtráp prōpes, Ælfc. Gl. 84; Som. 73, 87; Wrt. Voc. 48, 25.

fôt-sceamel, -sceamol, -scamel, -scamul, es; *m. A footstool*; pĕdum scăbellum, subpĕdāneum:—Ôþ-ðæt ic asette ðîne fŷnd to fôtsceamele ðînra fôta *dōnec pōnam inĭmĭcos tuos scăbellum pĕdum tuōrum*, Lk. Bos. 20, 43: Ps. Lamb. 109, 1. Ge-eádmēdaþ fôtsceamol his fôta *adōrāte scăbellum pĕdum ejus*, Ps. Lamb. 98, 5: Mt. Bos. 22, 44: Mk. Bos. 12, 36. Under ðînum fôtscamele *under thy footstool*, Homl. Th. i. 314, 32. Seó eorþe ys Godes fôtscamul *terra scabellum est pĕdum Dei*, Mt. Bos. 5, 35. Fôtscamul *scabellum vel subpĕdāneum*, Ælfc. Gl. 66; Som. 69, 79; Wrt. Voc. 41, 33. [O. Sax. fôt-skamel: Germ. fuss-schemel.]

fôt-sîþ-gerif, es; *n. A taking away* or *stoppage of a foot-path*; līmes, Ælfc. Gl. 3; Som. 55, 72; Wrt. Voc. 16, 45.

fôt-sîþ-sticcel, es; *m. A cloak, mantle*; chlămys, ȳdis, *f.* = χλάμύs, ύδos, *f*:—Hacele *vel* fôtsîþsticcel *chlămys*, Ælfc. Gl. 65; Som. 69, 40; Wrt. Voc. 40, 67.

fôt-spor, es; *n. A foot-track, foot-trace*; pĕdis vestīgium:—On ðæt fôtspor *on the foot-track*, Lchdm. iii. 286, 3.

fôt-spure, es; *n. A foot-support, foot-rest*; pĕdum fultūra:—Hî clumben upp to ðe hâlge rôde, nâmen ðâ ðe kynehelm of ûre Drihtnes heáfod, eall of smeáte golde, nâmen ðâ ðet fôtspure ðe wæs underneaðen his fôte, ðæt wæs eall of reáð golde *they climbed up to the holy cross, and took the crown, all of beaten gold, from our Lord's head, and took the foot-rest which was underneath his foot, which was all of red gold*, Chr. 1070; Erl. 209, 6–8.

fôt-stân, es; *m. A foot-stone, base, pedestal*; băsis = βάσιs, fultūra:—Fôtstân *fultūra*, Ælfc. Gl. 116; Som. 80, 72; Wrt. Voc. 61, 49.

fôt-swǽp; *gen.* -swǽdes; *pl. nom. acc.* -swaðu; *n*: fôt-swaðu, e; *f. A foot-trace, foot-print*; pĕdis vestīgium:—Ðæt ne sȳn astyrode oððe awende sîþstapla oððe wegas oððe fôtswaðu mîne *ut non mŏveantur vestīgia mea*, Ps. Lamb. 16, 5. Eall ðæt rȳmet, ðe eówer fôtswaðu on bestæpþ, ic eów forgife *omnem lŏcum, quem calcāvĕrit vestīgium pĕdis vestri, vōbis trādam*, Jos. 1, 3. Ðîne fôtswaða nǽron oncnâwene *vestīgia tua non cognoscentur*, Ps. Lamb. 76, 20.

fôt-swile, -swyle, es; *m. A foot-swelling*; pĕdis tŭmor:—Wið fôtswylum *for foot-swellings*, Med. ex Quadr. 4, 3; Lchdm. i. 342, 18. Ðes drænc is gôd wið fôtswilum *this drink is good for foot-swellings*, Lchdm. iii. 50, 12.

fôt-þweál, es; *n. A washing of the feet*; pĕdum lōtio:—Fôtþweál *pedĭlāvium*, Ælfc. Gl. 56; Som. 67, 27; Wrt. Voc. 37, 17. Fôtþweáles fæt *a vessel for washing the feet in*; pellŭviæ, 26; Som. 60, 88; Wrt. Voc. 25, 28.

fôt-wærc, es; *n. A pain in the foot*; pĕdis dŏlor:—Wið fôtwærce [MS. fôtwræce] *for a pain in the foot*, Med. ex Quadr. 3, 15; Lchdm. i. 342, 10.

fôt-welm, -wylm, es; *m*: fôt-wolma, an; *m. The sole of the foot*; pĕdis planta:—Fôtwelm *planta*, Ælfc. Gl. 75; Som. 71, 94; Wrt. Voc. 45, 2. Fôtwylm *planta*, Wrt. Voc. 71, 62. Mid ðære côðe he wæs ofset fram ðam hnolle ufan ôþ his fôtwylmas neoðan *with which disease he was afflicted from the crown above to the soles of his feet below*, Homl. Th. ii. 480, 12 : 508, 20. He hæfde ðæs brôðor fôtwolman on handa *plantam fratris tĕnēbat mănu*, Gen. 25, 25. Ðæt ðû næbbe nân þing hâles fram ðam fôtwolmum ôþ ðone hneccan *sanāri non possis a planta pĕdis usque ad vertĭcem tuum*, Deut. 28, 35.

fôwer *four*:—Cnut hit todǽlde on fôwer *Cnut divided it into four*, Chr. 1017; Erl. 785, 19, col. 1. v. feówer.

fox, es; *m. A* FOX; vulpes:—Fox *vulpes*, Ælfc. Gl. 19; Som. 59, 27; Wrt. Voc. 22, 68. Secgaþ ðam foxe *dīcĭte vulpi illi*, Lk. Bos. 13, 32. Foxas habbaþ holu *vulpes fŏveas hăbent*, 9, 58. Foxes dǽlas *vulpis partes*, Ps. Th. 62, 8. [*Laym.* fox, uox: *Orm.* fox: *Plat. Dut.* vos, *m*

Ger. fuchs, m: M. H. Ger. vuhs, m; vohe, f: O. H. Ger. fuhs, m; foha, f: Goth. fauho,]

foxes cláte, an; f. Fox's clote, burdock; arctium lappa, Lin:—Wið hundes dolge, foxes cláte, etc. for wound by a hound, burdock, etc. L. M. 1, 69; Lchdm. ii. 144, 11.

foxes clífe, an; f. The greater burdock; arctium lappa, Lin:—Genim clifwyrt, sume men hátaþ foxes clife, sume eáwyrt take burdock, some men call it fox's cliver or the greater burdock, some riverwort, L. M. 1, 15; Lchdm. ii. 58, 3: Lchdm. iii. 74, 10.

foxes fót, es; m. Fox's foot, bur reed, a water plant; sparganum simplex, xiphion = ξιφίον:—Genim ðysse wyrte wyrttruman, ðe man xiphion, and óðrum naman foxes fót, nemneþ take a root of this plant, which is named xiphion, and by another name fox's foot, Herb. 47, 1; Lchdm. i. 150, 16.

foxes glófa, an; m. [foxes glófa MS. B.] Foxglove; digitális purpúrea, Lin:—Wið óman genim ðysse wyrte leáf ðe man στρύχνος μανικός, and óðrum naman foxes glófa [MS. foxes clófa] nemneþ for inflammatory sores, take leaves of this wort, which is named sólánum insánum or Sodómeum, and by another name foxglove, Herb. 144; Lchdm. i. 266, 18. Mr. Cockayne says, in note b on this passage,—'Strychnos manikos is Sólánum insánum or Sodómeum fairly drawn, MS. V. fol. 60 a, not an English plant, and certainly not foxglove. The leechdoms here recorded seem derived from what Dioskorides says of the στρύχνος κηπαῖος: namely, τὰ φύλλα καταπλασσόμενα ἁρμόζει πρὸς ἐρυσιπέλατα καὶ ἕρπητας; and so on of κεφαλαλγία and στόμαχος καυσούμενος and ὠταλγία [iv. 71].' v. clif-wyrt foxglove.

fra from, fro, Chr. 656; Erl. 31, 10: 963; Erl. 123, 2. v. fram.

fraced abominable, Ælfc. T. 34, 25. v. fracoþ; adj.

fraced-líce; comp. -lícor; adv. Shamefully, disgracefully; turpiter:—Hwæt is fracedlícor quid est turpius? Ælfc. Gr. 48; Som. 49, 15. v. fracoþ-líce.

fraceþ an insult, Exon. 66 b; Th. 246, 34; Jul. 71. v. fracoþ, es; n.

fracod vile, abominable, useless, Coll. Monast. Th. 18, 11: Beo. Th. 3155; B. 1575. v. fracoþ; adj.

fracod-líc shameful, L. Eth. vi. 28; Th. i. 322, 14. v. fracoþ-líc.

fracod-líce; adv. Shamefully; turpiter:—Hí wyllaþ fracodlíce him betwynan sacian they will shamefully quarrel among themselves, Homl. Th. ii. 292, 35. v. fracoþ-líce.

fracoþ, fracuþ, fracod, fraced; adj. Vile, filthy, unseemly, hateful, abominable, worthless, useless; turpis, detestábilis, indecórus:—Is úser líf fracoþ and gefrǽge our life is vile and infamous, Cd. 189; Th. 235, 10; Dan. 304: Salm. Kmbl. 67; Sal. 34: Exon. 10 b; Th. 12, 33; Cri. 195. Ne wæs ðæt [MS. ðær] húru fracoðes gealga that was indeed no vile [man's] gibbet, Rood Kmbl. 20; Kr. 10. We bióþ folcum fracoðe we shall be hateful to the people, Andr. Kmbl. 817; An. 409. Fracoðest vilest, Salm. Kmbl. 702; Sal. 350. Wæs úre líf fracuþ and gefrǽge our life has been vile and infamous, Exon. 53 a; Th. 186, 23; Az. 24. Hí fracuðe and earme wǽron they were worthless and worthless, Bd. 3, 21; S. 551, 26. Hwæt rece we hwæt we sprecan, búton hit riht spræc sý, and beháfe, næs ídel, oððe fracod quid cúrámus quid lóquámur, nisi recta locútio sit, et útilis, non ánilis, aut turpis? Coll. Monast. Th. 18, 11. Næs seó ecg fracod hilde ince the edge was not useless to the warrior, Beo. Th. 3155; B. 1575. On ðam fracodan gilte in fácinóre, Jos. 7, 15. On his fracedum dǽdum in his abominable deeds, Ælfc. T. 34, 25.

fracoþ, fraceþ, es; pl. nom. acc. fracoðu, fracoðu; n. [fracoþ vile] An insult, contumely; turpitúdo, contúmélia:—Búton fracoðum without insults, Ps. Th. 54, 22. Me ða fraceðu sind mǽste weorce these insults are the greatest trouble to me, Exon. 66 b; Th. 246, 34; Jul. 71: 73 b; Th. 274, 31; Jul. 541. Fracoþ abominatio, Lk. Skt. Lind. 16, 15.

fracoðe, fracuðe; adv. Shamefully; turpiter:—He mæg úre fýnd gedón fracoðe to náhte he can shamefully destroy our enemies, Ps. Th. 59, 11: 88, 28. Ic fracuðe forseó feóndas míne I shamefully despise my enemies, 117, 7: 62, 8.

fracoþ-líc, fracuþ-líc, fraceþ-líc, fracod-líc; adj. Heinous, ignominious, shameful; turpis:—Ðam folctogan fracuþlíc þúhte it seemed heinous to the chieftain, Exon. 69 a; Th. 256, 2; Jul. 225. Fracodlíce fitunga shameful fightings, L. Eth. vi. 28; Th. i. 322, 14. Ðæt wíte ðæs fracoþ-lícostan [fraceþlécestan MS. Hat.] deáþes he geceás he chose the punishment of the most ignominious death, Past. 3, 1; Cot. MS.

fracoþ-líce, fracuþ-líce, fracod-líce, fraceþ-líce; adv. Shamefully, disgracefully, wickedly; turpiter:—Biþ us swíðe fracoþlíce [fracuþlíce MS. Cot.] óðer fót unscód one of our feet is very disgracefully unshod, Past. 5, 2; Hat. MS. 11 a, 17. Ic fracoþlíce feóndrǽs gefremede I wickedly committed the fiendish violence, Cd. 42; Th. 55, 25; Gen. 899.

fracoþ-nes, -ness, e; f. Vileness, obscenity; turpitúdo, obscénítas, Cot. 143.

fracu, e; f. Wickedness, impudence; protervítas. DER. neód-fracu, scyld-.

fracuþ vile, Exon. 53 a; Th. 186, 23; Az. 24. v. fracoþ; adj.

fracuðe; adv. Shamefully, Ps. Th. 62, 8: 117, 7. v. fracoðe.

fracuþ-líc heinous, Exon. 69 a; Th. 256, 2; Jul. 225. v. fracoþ-líc.

fracuþ-líce disgracefully, Past. 5, 2; Cot. MS. v. fracoþ-líce.

frǽ- before, in a greater degree, very, exceedingly; præ-: found in the compounds frǽ-beorht, -fǽtt, -mǽre, -micel, -ófestlíce. v. freá-.

frǽ-beorht exceedingly bright; præclárus, Lye. v. freá-beorht.

fræc; adj. Voracious, greedy; gúlósus:—Fræc [MS. fræt] gúlósa, Mone B. 3533. v. frec.

frǽcednys, -nyss, e; f. Danger, peril; perículum:—Saca mid frǽcednysse hit getácnaþ it betokens disputes with peril, Somn. 122; Lchdm. iii. 204, 33. v. frécednes.

frǽcenes, frǽcnes, -ness, -nyss, e; f. Danger, peril; perículum:—On frǽcenesse heora stealles in perículum sui státus, Bd. 4, 25; S. 601, 17. Bútan frǽcnesse without danger, Herb. 30, 4; Lchdm. i. 126, 24, MS. B. Bútan frǽcnysse, 63, 2; Lchdm. i. 166, 7, MSS. B. H. v. frécennes.

frǽcenful; adj. Dangerous, perilous; perículósus:—Móna se þreótteóða frǽcenful ys to angennene þing the thirteenth moon is perilous for beginning things, Obs. Lun. § 13; Lchdm. iii. 190, 11: 15; Lchdm. iii. 190, 30: 17; Lchdm. iii. 192, 14. v. frécenful.

frǽc-genga, an; m. A fugitive, apostate; profúgus, apostáta, Som. Ben. Lye.

frǽclíce; adv. Greedily; ávide:—Frǽclíce bát ávide momordit, Gr. Dial. 1, 4.

frǽc-máse, an; f. The nun bird, titmouse; párus cærúleus:—Frǽcmáse sigatula? Glos. Brux. Recd. 36, 38; Wrt. Voc. 62, 38. v. frec-máse.

frǽcne; adj. Grievous, dire, dangerous; dírus, perículósus:—Awend ðín ansýne fram mínum frǽcnum firenum turn thy face from my grievous sins, Ps. Ben. 50, 10; Ps. Grn. ii. 149, 10. Ðæt hí ne þorftan in swá frǽcne síþfæt feran ne tam perículósam peregrinátiónem adíre debérent, Bd. 1, 23; S. 485, 37. v. frécne; adj.

frǽcne; adv. Fiercely, severely, hardly; dúre, atrócíter, audacter:—Abrahames cwén spræc frǽcne on fǽmnan Abraham's wife spoke severely against the damsel, Cd. 103; Th. 136, 22; Gen. 2262: Ps. Th. 64, 3: 90, 12. Ðonne hit ðé frǽcnost þynce when it seems worst to thee, Prov. Kmbl. 75. v. frécne; adv.

frǽcnes, -ness, -nyss danger, Herb. 30, 4; Lchdm. i. 126, 24, MS. B: 63, 2; Lchdm. i. 166, 7, MSS. B. H. Blickl. Homl. 109, 7. v. frécennes.

frǽ-fǽtt; adj. Very fat; præpinguis, Cot. 177.

frǽfele; adj. Saucy; audax, prócax, Som. Ben. Lye. [Scot. frewall frivolous: Plat. wrevel, wrewel, m. obstinacy, impudence: O. Frs. frevelhed boldness: Dut. wrevel, m. stubbornness, contumacy: Ger. frevel bold, frivolous; frevel, m. boldness, crime, insolence, impudence: M. H. Ger. vrevel, vrävel bold, impudent; vrevele, vrevel, f. m. boldness, impudence: O. H. Ger. frafali contúmax, protervus; fravali, f. temérítas, protervitas: Lat. frívólus empty, trifling, worthless, frivolous.]

frǽfellíce; adv. Saucily; procáciter, Som. Ben. Lye.

frǽfelnes, -ness, e; f. Sauciness, faction; procácitas, factio, Cot. 213.

frǽg, ðú frǽge, pl. frǽgon asked, hast asked, inquired; p. of fricgan.

frǽge, frége known, famous. DER. ge-frǽge, -frége; adj.

frǽge, frége an inquiring, knowing, hearsay. DER. ge-frǽge, -frége, es; n.

frǽgin asked, Bd. 2, 1; S. 501, 9: 4, 5; S. 572, 21, = frægn; p. of frignan.

frǽgn asked; interrógávit, Bd. 2, 12; S. 513, 37, 38; p. of frignan.

frǽ-mǽre, -mére; adj. Very great, famous, excellent; egrégius, eximius, Cot. 77. v. freá-mǽre.

frǽmde strange, foreign, L. Wih. 28; Th. i. 42, 23: Somn. 79; Lchdm. iii. 202, 20. v. fremede.

frǽ-micel; adj. Very great, famous; præ-magnus, eximius, Cot. 178.

frǽm-sum; adj. Kind; benignus:—Gedó frǽmsume frófre ðíne make thy comfort kind, Ps. C. 50, 130; Ps. Grn. ii. 279, 130. v. frem-sum.

frǽng asked, Bd. 3, 14; S. 541, 3, = frægn; p. of frignan.

frǽ-ófestlíce; adv. Very hastily, very quickly; præprópére, Cot. 178.

frǽt; adj. Obstinate, proud; perversus, superbus:—Háteþ ðæt ðú, on ðis frǽte folc, onsende wæter he commandeth that thou send water upon this obstinate people, Andr. Kmbl. 3010; An. 1508: Exon. 28 a; Th. 84, 15; Cri. 1374. Frǽtre þeóde to the proud people, Andr. Kmbl. 1141; An. 571.

frǽt, ðú frǽte, pl. frǽton devoured, devouredst, Beo. Th. 3167; B. 1581: Ps. Th. 34, 23; p. of fretan.

frǽtewe, frǽtewa ornaments, Bd. 1, 29; S. 498, 10, note. v. frætwe.

frǽtewung, e; f. An ornament; ornámentum:—Heofonas and eorþe and eall heora frǽtewung cæli et terra et omnis ornátus eórum, Gen. 2, 1. v. frætwung.

frǽtig; def. se frætga; adj. Proud, perverse, wicked; superbus, perversus:—Forsóh ðone frǽtgan seize the proud one [the devil], Exon. 69 b; Th. 259, 18; Jul. 284.

frǽt-læppa, an; m. Dew-lap; pálear:—Frǽtlæppa runia vel páleáre, Ælfc. Gl. 99; Som. 76, 123; Wrt. Voc. 54, 63.

frǽttewian, frætwian, fretwian, frætwan; p. ode, ede; pp. od, ed To adorn, deck, embroider, trim; ornáre:—Ða burh timbrum and gyfum eác frǽttewodon and weorþodon urbem ædificiis ac donáriis adornárunt, Bd. 3, 19; S. 547, 24. Ðe ðone sele frætweþ who adorns the hall

Exon. 117 a; Th. 450, 24; Dôm. 92. Ic wylle frætwian mec *I will prepare myself*, Exon. 119 a; Th. 456, 23; Hy. 4, 71. Hî odde hî sylfe frætwiaþ *aut seipsas adornent*, Bd. 4, 25; S. 601, 17. Sáwle frætwaþ hálgum gehygdum *they adorn their souls with holy meditations*, Exon. 44 b; Th. 150, 14; Gú. 778. Ða de geolo godwebb geatwum frætwaþ *those who embroider the yellow godly garment with ornaments*, Exon. 109 a; Th. 417, 26; Rä. 36, 10. De mec frætwede *who adorned me*, 124 b; Th. 479, 15; Rä. 62, 8. Folcstede frætwan *to deck a dwelling-place*, Beo. Th. 152, note; B. 76. Brídels frætwan *to deck the bridle*, Elen. Kmbl. 2396; El. 1199. Hyrstum frætwed *adorned with ornaments*, Exon. 104 a; Th. 395, 22; Rä. 15, 11: 107 b; Th. 411, 1; Rä. 29, 6: 108 b; Th. 414, 15; Rä. 32, 20. [*Chauc.* fret *wrought*: *O. Sax.* fratahôn *to adorn, ornament, decorate: Goth.* us-fratwyan *to make ready, to outfit.*] DER. ge-frætewian, -frætwian, ymb-.

frætwe, frætewe, frætuwe, frætwa, frætewa; *pl. f. Ornaments, adornments, decorations, treasures*; ornámenta, ornátus, res pretiōsæ:—Holtes frætwe *the decorations of the wood*, Exon. 57 a; Th. 202, 22; Ph. 73. Ða wæstmas, foldan frætwe *the fruits, the treasures of the earth*, 59 b; Th. 215, 22; Ph. 257. Wangas grēne, foldan frætuwe *green fields, the ornaments of the earth*, Menol. Fox 411; Menol. 207. Ic dara frætwa þanc secge *I say thanks for these ornaments*, Beo. Th. 5580; B. 2794. Frætwa hyrde *the guardian of the treasures*, 6258; B. 3133. Máþma fela frætwa *many treasures, ornaments*, 74; B. 37. Dám frætwum *to these precious things*, 4332; B. 2163. He dám frætwum fēng *he received the ornaments*, 5970; B. 2989. On frætewum *in his garnishments, viz. armour*, 1928; B. 962. Secgas bæron beorhte frætwe *the warriors bare bright arms*, 434, note; B. 214. He frætwe geheóld, bill and byrnan *he held the armour, the falchion and coat of mail*, 5233; B. 2620. Frætwe and fætgold *ornaments and plated gold*, 3846, note; B. 1921. Hafa wunden gold, feoh and frætwa *have the twisted gold, the wealth and ornaments*, Cd. 98; Th. 128, 21; Gen. 2130: 136; Th. 171, 17; Gen. 2829: Exon. 51 b; Th. 179, 3; Gú. 1256: Beo. Th. 1797; B. 896. Cyricean frætewa *ornámenta ecclēsiæ*, Bd. 1, 29; S. 498, 10, note. Frætwum gefyrdred *furthered by the treasures*, Beo. Th. 5561; B. 2784: 4114; B. 2054. [*O. Sax.* fratahi, *f? ornaments.*] DER. gold-frætwe.

frætwednes, fretwednes, frætwædnys, -ness, -nyss, e; *f. An adorning, ornament, a trifle*; ornátio, ornámentum, crēpundia:—He sende cyricean frætwednesse *misit ornámenta ecclēsiæ*, Bd. 1, 29; S. 498, 10. On eorþlícre frætwædnysse [fretwednesse MS. Ca.] *in earthly adorning*, 3, 22; S. 552, 20: Blickl. Homl. 195, 11; 127, 3; 207, 25. Frætwednessa *crepundia*, Cot. 56. DER. hrægel-gefrætwodnes.

frætwung, frætewung, e; *f. An adorning, adornment, ornament*; ornátus, ornámentum:—He micele swidor lufode dære heortan clænnysse donne dæra stána frætwunge *he much more loved cleanness of heart than the adornment of stones*, Homl. Th. i. 508, 22. On disum getelde wæron fornean unasecgendlíce frætwunga *in this tabernacle were almost unspeakable ornaments*, ii. 210, 11. DER. world-frætwung.

fragendlíc; *adj.* [= framigendlíc, *q. v.*] *Beneficial*; salubris, salútáris:—Fragendlíc lǣcedôm *a beneficial medicine*, Herb. 159; Lchdm. i. 288, 2, MS. B.

FRAM, from; *prep. dat.*　　I. FROM; a, ab:—Ic adilige done mannan fram dære eorþan ansíne, fram dam men óþ da nýtenu, fram dam slincendum óþ da fugelas *delēbo hŏminem a fácie terræ, ab hŏmine usque ad anĭmantia, a reptĭli usque ad volucres cœli*, Gen. 6, 7. Gewítaþ fram me *discēdĭte a me*, Ps. Th. 6, 7: Ps. Spl. 30, 15: Mt. Bos. 1, 17, 21, 22.　　II. *with verbs of speaking, Concerning, about, of; cum verbis lŏquendi*, de:—Dæt he fram Sigemunde secgan hýrde ellendædum *that he, concerning Sigemund, had heard tell of valiant deeds*, Beo. Th. 1754; B. 875. Nó ic fram de swylcra searunída secgan hýrde *never have I heard speak about thee of such hostile snares*, Beo. Th. 1167; B. 581.　　III. fram *is sometimes placed after its case:*—He hine forwræc mancynne fram *he banished him from mankind*, Beo. Th. 221; B. 110. [*Chauc.* fra: *Laym.* fram, from: *Orm.* fra: *O. Sax.* fram, vram: *O. H. Ger.* fram: *Goth.* fram: *Dan.* fra: *Swed.* fram *forward*, forth; *frán* from: *Icel.* fram *forward*; *frá* from.]

fram; *adj. Valiant, stout, firm*; strēnuus:—Geong and fram *young and valiant*, Bd. 4, 15; S. 583, 25. He wæs fram to Godes compe *he was stout for God's battle*, Andr. Kmbl. 467; An. 234. v. from.

fram-acyrran; *p.* de; *pp.* ed *To turn from or away, take from*; avertēre, auferre:—Framacyr yrre dín fram us *averte iram tuam a nóbis*, Ps. Spl. 84, 4. Se brýdguma him biþ framacyrred *auferētur ab eis sponsus*, Mk. Bos. 2, 20.

fram-adôn, he -adēþ; *p.* -adyde; *pp.* -adôn *To do or take from or away, cut off*; auferre, abscídēre:—Sóna heó done fefer framadēþ *it will soon take away the fever*, Herb. 12, 5; Lchdm. i. 104, 15. Mildheortnesse his he framadeþ *misericordiam suam abscídet*, Ps. Lamb. 76, 9.

fram-adrífan, -adrýfan; *p.* -adráf, *pl.* -adrifon *To drive from or away, expel*; expellēre:—Gif gē me framadrýfaþ *si me expellĭtis*, Coll. Monast. Th. 29, 23.

fram-ahyldan; *p.* de; *pp.* ed *To turn from or away*; declínāre:—

Hió him framahyldeþ *it will turn from them*, Med. ex Quadr. 1, 2; Lchdm. i. 328, 10.

fram-anýdan; *p.* -anýdde; *pp.* -anýded, -anýdd *To force from or away, drive away*; repellēre:—Ða feforas beóþ framanýdde *the fevers will be forced away*, Herb. 143, 4; Lchdm. i. 266, 13.

fram-a-teón; *p.* -ateáh, *pl.* -atugon; *pp.* -atogen *To draw away from*; abstrǎhēre, extrǎhēre:—Framatuge *extraxisti*, Ps. Vos. 21, 8. Framatogen *detractus, ablātus*, Cot. 69. v. teón I.

fram-ateran; *p.* -atær, *pl.* -atǣrou; *pp.* -atoren *To tear from or asunder, to tear in pieces*; diripēre:—Ic framatere *dirĭpio*, Ælfc. Gr. 28, 3; Som. 30, 64.

framaþ *does good, avails*, Herb. 146, 2; Lchdm. i. 270, 4, = fremaþ; *3rd sing. pres.* of fremian.

fram-atíhan; he -atíhþ; *p.* -atáh, *pl.* -atigon; *pp.* -atigen *To draw away from*; abstrǎhēre:—Donne he framatíhþ hine *dum adtrǎhit* [*abstrǎhet*, Ps. Surt. 9, 30] *eum*, Ps. Spl. second 9, 11. v. tíhan I.

fram-awendan; *p.* de; *pp.* ed *To turn from or away*; avertēre, Scint. 53.

fram-aweorpan, -wurpan; ic -aweorpe, -awurpe; *p.* -awearp, *pl.* -awurpon; *pp.* -aworpen *To cast from, throw away*; abjicēre:—Ic framawurpe *abjicio*, Ælfc. Gr. 28, 6; Som. 32, 39. DER. weorpan.

fram-bringan; *p.* -brang, *pl.* -brungon; *pp.* -brungen *To bring from or away*; dedúcēre:—Gyf he done him eádelíce frambringan ne mǣge *if he cannot easily bring it away from him*, Herb. 158, 2; Lchdm. i. 284, 24.

fram-búgan; *p.* -beáh, *pl.* -bugon; *pp.* -bogen *To turn from or away, leave*; deflectēre, declínāre:—Dæt him da frambugon [MS. frambugan], de hí betst getreówodon *that those left them, whom they most trusted*, Ors. 2, 5; Bos. 47, 44.

fram-fleón; *p.* -fleáh, *pl.* -flugon; *subj. p.* -fluge, *pl.* -flugen; *pp.* -flogen *To flee from*; aufūgēre:—Dæt hí him framflugen *that they should flee from them*, Ors. 1, 7; Bos. 30, 10.

fram-gewítan, from-gewítan; *p.* -gewát, *pl.* -gewiton; *pp.* -gewiten *To go away from, depart from*; discēdēre:—Hie him framgewítaþ *they depart from him*, Bt. 8; Fox 26, 10.

framian *to avail, profit*; vălēre, prōdesse, R. Ben. 64, 72: R. Conc. 7. v. fremian.

framigendlíc; *adj.* [framigende, *part.* of tremian and líc] *Profitable, beneficial*; salubris, salútáris:—Dæt sylfe is framigendlíc lǣcedôm ongeán ealle ǣttru *the same is a beneficial medicine against all poisons*, Herb. 159; Lchdm. i. 288, 2.

framlíce; *adv. Strongly, firmly, stoutly*; fortĭter, strēnue:—Des Cāsere framlíce rehte da cynewísan *this Cæsar firmly ruled the kingdom*, Bd. 1, 5; S. 476, 7: 4, 10; S. 578, 6. Benedictus done síþfæt framlíce to Rôme geferde *Benedict stoutly went his journey to Rome*, Bd. 5, 19; S. 637, 45. v. from-líce.

fram-scipe, es; *m. A fellowship, association, fraternity*; collēgium:—Framscipe muneca *collēgium monachōrum*, Bd. 3, 5; S. 526, 18, note, MSS. Ca. O.

fram-síþ *a going from or away, departure*, Som. Ben. Lye. v. from-síþ.

fram-sídian; *p.* ode; *pp.* od *To go from or away, depart*; abscēdēre, Som. Ben. Lye.

fram-standan; *p.* -stód, *pl.* -stódon; *pp.* -standen *To stand away from, stand aloof*; abstáre, Som. Ben. Lye.

fram-swengan; *p.* de; *pp.* ed *To shake from or away, shake off*; excútēre:—Framswengde *excussit*, Cot. 179.

fram-weard; *adj. Turned from or away, averse, froward, perverse*; aversus, perversus, Som. Ben. Lye. v. from-weard.

fram-wísum; *adv. Wisely*; săpienter:—Ættrene beóþ gegalene framwísum *venefĭci incantantis săpienter*, Ps. Spl. 57, 5.

fran *asked, inquired*; *p.* of frinan.

franca, an; *m. A javelin, lance*; lancea, frǎmea, hasta:—He lēt his francan wadan þurh dæs hysses hals *he let his javelin go through the youth's neck*, Byrht. Th. 135, 59; By. 140. He done forman man mid his francan ofsceát *he shot the foremost man with his javelin*, 134, 1; By. 77. Francan wǣron hlúde *the javelins were loud*, Cd. 93; Th. 119, 20; Gen. 1982 [*Icel.* frakka]. v. Grm. Gesch. D. S. p. 359.

Francan; gen. Francena, Francna; *dat. Francum; pl. m.: France*; gen. Franca; *pl. m. The Franks*; Franci:—Hēr Ald-Seaxe and Francan gefuhton *in this year* [A. D. 780] *the Old Saxons and the Franks fought*, Chr. 780; Erl. 54, 3; 881; Erl. 82, 5. Of Francena cyningcynne *de gente Francōrum rēgia*, Bd. 1, 25; S. 486, 32. Francena cyning *Francōrum rex*, 3, 19; S. 550, 2. Wid Francena ríce *against the kingdom of the Franks*, 4, 1; S. 565, 1. Cyrdon hí to Pipne Francna cyninge *diverṭērunt ad Pippinum dúcem Francōrum*, 5, 10; S. 624, 2: Chr. 855; Erl. 68, 29; 885; Erl. 82, 34. Ymb ii geár dæs de he of Francum ,còm, he gefór *two years after he came from the Franks, he died*, Chr. 855; Th. 126, 2, col. 2, 3: 890; Erl. 86, 32. Franca cyng *king of the Franks*, Chr. 1070; Th. 347, 7: 1077; Th. 351, 14. DER. Eást-Francan. v. Grm. Gesch. D. S. cap. xx.

Franc-land, Fronc-land, Frang-land, es; *n. Frank-land, the country*

of the Franks; Francōrum terra, Francia:—Nāmon [MS. noman] hī him wealhstōdas of Franclande mid *accēpērunt de gente Francōrum interprĕtes,* Bd. 1, 25; S. 486, 24. On ðam mynstre ðe on Franclande wæs getimbred *in monastērio quod in regiōne Francōrum constru*ʋum *est,* 3, 8; S. 531, 13. On Francland [Froncland, Th. 150, 23, col. 1; Frangland, 151, 23, col. 2, 3] *into Frank-land,* Chr. 882; Th. 150, 23, col. 2, 3.

Franc-rice, es; *n. The kingdom of the Franks;* Francōrum regnum :—He hæfde ǽrendo sum to Breotone cyningum of Francrīce *he had an errand to the kings of Britain against the kingdom of the Franks,* Bd. 4, 1; S. 565, 1, MS. B: Chr. 1060; Erl. 193, 32.

FRĀSIAN, freásian; *p.* ade; *pp.* ad *To ask, inquire, tempt;* interrŏgāre, conquīrĕre, sciscitāri, tentāre :—Frāsiaþ [MS. frasias] *conquīrĭtis,* Mk. Skt. Lind. 9, 16. Wæs mǽst Babilōn burga, ōþ-ðæt Baldazar, þurh gylp, grome Godes freásade [MS. frea sæde] *Babylon was greatest of cities, until Belshazzar, through vain glory, fiercely tempted God,* Cd. 209; Th. 259, 22; Dan. 695. [O. Sax. frēsôn *to try, tempt:* M.H. Ger. vreisen *to endanger:* O.H. Ger. freisôn *periclĭtāri:* Goth. fraisan *to try, tempt;* fraistubni, *f. temptation:* Dan. friste *to try, tempt:* Swed. fresta: Icel. freista.] DER. ge-frāsian.

frāsung, e; *f. An asking, inquiring, tempting, temptation;* interrŏgātio, tentātio :—Hý to Gūþlāces gáste gelǽddun frāsunga fela *they brought many temptations to Guthlac's spirit,* Exon. 35 a; Th. 113, 19; Gū. 160. Mid frāsung *interrŏgātiōne,* Mt. Kmbl. Præf. p. 19, 9.

FREÁ [= freaha], freó; *gen.* frean; *m. A lord, master, the Lord;* dŏminus :—Freá sceáwode fyrngeweorc *the lord beheld the ancient work,* Beo. Th. 4560; B. 2285. Freá Ælmihtig *the Lord Almighty,* Cd. 1; Th. 1, 9; Gen. 5: 101; Th. 134, 24; Gen. 2229. Freá moncynnes *Lord of mankind,* Bt. Met. Fox 17, 17; Met. 17, 9. Swá neáh wæs sigora Freán þūsend aurnen *so nearly a thousand [winters] of the Lord of victories had elapsed,* Chr. 973; Erl. 124, 23; Edg. 15. Habbaþ we to dæm mǽran ǽrende Deniga freán *we have an errand to the famous lord of the Danes,* Beo. Th. 547; B. 271. Ðis is hold weorod freán Scyldinga *this is a band attached to the lord of the Scyldings,* 587; B. 291. Wígheafolan bær freán on fultum *he bore the helmet to bring aid to his lord,* 5316, note; B. 2662. To hire freán sittan *to sit by her lord,* 1287; B. 641. Ic Freán þanc secge, ēcum Dryhtne *I say thanks to the Lord, the eternal Ruler,* 5581; B. 2794. He ðone wísan wordum hnægde freán Ingwina *he addressed with words the wise lord of the Ingwines,* 2642; B. 1319. Gūþ nimeþ freán eówerne *war shall take away your lord,* 5068; B. 2537. Ðonne we geferian freán ūserne, leófne mannan *when we bear our lord, the dear man,* 6206; B. 3107. [O. Sax. fráho, frôho, frôio, frô, *m:* O.H. Ger. frô, *m. dŏminus: Goth.* frauya, *m. lord:* Icel. Freyr, *m. name of the god Freyr.*] DER. ágend-freá, aldor-, folc-, gūþ-, heáh-, líf-, mān-, sin-.

freá-, frǽ- *before, in a greater degree, very, exceedingly;* præ-: found in the compounds freá-beorht, -bodian, -dréman, -fætt, -gleáw, -hræd, -mǽre, -micel, -ōfestlíce, -reccere.

freá-beorht, -briht, frǽ-beorht; *adj. Exceedingly bright, glorious;* præclārus, clarissimus :—Eálá freábeorht folces [MS. folkes] scippend *O! glorious creator of people,* Hy. 2, 1; Hy. Grn. ii. 281, 1. Eálá freábrihta folces Scyppend, Ps. Lamb. fol. 183 b, 15. Blickl. Homl. 229, 28.

freá-bodian; *p.* de; *pp.* od *To proclaim, declare;* pronuntiāre :—Freábodaþ odðe mǽrsaþ tunge mín spǽce ðíne *pronuntiābit lingua mea elŏquium tuum,* Ps. Lamb. 118, 172.

freá-dréman; *p.* de; *pp.* de *To rejoice exceedingly, shout for joy;* jubilāre :—Fægniaþ odðe freádrēmaþ Gode on stefne wynsumnesse odðe blisse *jubilāte Deo in vōce exsultātiōnis,* Ps. Lamb. 46, 2: 97, 4.

freá-drihten, freah-drihten; *gen.* -drihtnes; *m. A lord, master, the Lord;* dŏminus :—Wæs his freádrihtnes folc-cūþ nama Agamemnon *his lord's celebrated name was Agamemnon,* Bt. Met. Fox 26, 17; Met. 26, 9. Abraham, ðín freádrihten *Abraham, thy lord,* Cd. 130; Th. 165, 9; Gen. 2729. Freádrihten mín *O my Lord,* 42; Th. 54, 29; Gen. 884. He wolde freahdrihtnes feorh ealgian *he would defend his lord's life,* Beo. Th. 1596, note; B. 796.

freá-fætt *very fat.* v. frǽ-fætt.

freá-gleáw; *adj. Very prudent;* prudentissimus :—Hie ðǽr fundon freágleáwe ædele cnihtas *they found there very prudent noble youths,* Cd. 176; Th. 221, 15; Dan. 88.

freah-drihten *a lord, master,* Beo. Th. 1596, note; B. 796. v. freá-drihten.

freá-hræd; *adj. Very quick, speedy, swift;* prŏpĕrus, expĕdītus, Som. Ben. Lye.

freá-mǽre, frǽ-mǽre; *adj. Very renowned;* celeberrimus :—Firum freámǽrne eard weardian *to inhabit a country very renowned to men,* Exon. 95 b; Th. 356, 11; Pa. 10.

freá-micel *very great, famous.* v. frǽ-micel.

freá-ōfestlíce *very hastily, very quickly.* v. frǽ-ōfestlíce.

freá-reccere, es; *m. A chief ruler, prince;* princeps :—Freárecceras odðe ealdras ēhton me būton ge-earnungum *princĭpes persĕcūti sunt me grātis,* Ps. Lamb. 118, 161.

freás, *pl.* fruron *froze; p. of* freósan.

---

freatewung, e; *f. An adorning, adornment, ornament;* ornātus, ornāmentum, Som. Ben. Lye. v. frætwung.

freáum *to chieftains,* Exon. 94 b; Th. 353, 53; Reim. 32; *dat. pl. of* freá.

freá-wine, es; *m. A dear* or *beloved lord;* dŏminus cārus :—Syððan freáwine folca swealt *when the beloved lord of people perished,* Beo. Th. 4703; B. 2357: 4849; B. 2429. He of hornbogan his freáwine flāne geswencte *he laid low his dear lord with an arrow from his horned bow,* 4867; B. 2438. Cf. Grm. D. M. 82, 192.

freá-wrāsen, e; *f. A noble* or *royal chain, a diadem;* nōbilis torquis, diadēma = διάδημα :—Se hwīta helm hafelan werede, since geweorþad, befongen freáwrāsnum *the bright helmet guarded his head, ornamented with treasure, encircled with noble chains,* Beo. Th. 2906; B. 1451.

FREC, fræc; *adj. Desirous, greedy, gluttonous, audacious, bold;* avĭdus, gŭlōsus, audax, temĕrārius :—Gífere *vel* frec ambro [q. v. in Du Cange], Ælfc. Gl. 88; Som. 74, 83; Wrt. Voc. 50, 63. Frec ambro, Wrt. Voc. 86, 50. Hí firenlusta frece ne wǽron *they were not desirous of luxuries,* Bt. Met. Fox 8, 30; Met. 8, 15. Ðæt hie firendǽda tó frece wurdon *that they were too audacious in wicked deeds,* Cd. 121; Th. 155, 30; Gen. 2580. [Dut. vrec, *m. a miser:* Ger. frech *rash, impertinent:* M.H. Ger. vrēch: O.H. Ger. frēh, frēch *avārus, cupĭdus, arrŏgans:* Goth. friks in faihu-friks *desirous for money, avaricious:* Dan. frāk: Swed. fräck: Icel. frekr *greedy, voracious.*] DER. ferhþ-frec, gūþ-.

freca, an; *m.* [frec *bold*] *A bold man, warrior, hero;* bellātor, hēros = ἥρως :—Gefēng fetelhilt freca Scyldinga *the Scylding's warrior seized the belted hilt,* Beo. Th. 3131; B. 1563; Andr. Kmbl. 2328; An. 1165. Moyses bebeád frecan arīsan *Moses bade the bold arise,* Cd. 154; Th. 191, 20; Exod. 217. DER. hild-freca, scyld-, sweord-, wíg-.

frēcednes, -ness, -nyss, frǽcednys, -nyss, e; *f. Danger, peril, hazard;* perīculum, discrīmen :—Ne ða tobeótiendan frēcednesse ðam eágan mennisc hand gehǽlan mihte *human hand could not save the eye from the threatening danger,* Bd. 4, 32; S. 611, 23. Ahred fram frēcednysse *saved from peril,* Homl. Th. ii. 304, 30. Forðam he geþristade ðæt he hine sylfne on geweald sealde swylcere frēcednysse *quod se ille discrīmini dāre præsumpsisset,* Bd. 1, 7; S. 477, 16. Frēcednysse helle gemētton me *perīcula inferni invēnērunt me,* Ps. Lamb. 114, 3. He ferde fram eallum frēcednyssum ðises lǽnan lifes *he went from all the perils of this frail life,* Homl. Th. ii. 516, 2. v. frēcennes.

frēcelsod; *part. Put in danger, endangered;* perīclītātus :—Frēcelsod qui perīclītātus est, Cot. 151.

frēcen; *gen.* frēcnes; *n. Peril, danger;* perīculum, discrīmen :—Frēcnes ne wēnaþ *they think not of peril,* Exon. 96 b; Th. 361, 16; Wal. 20. Ðǽr is ealra frēcna mǽste *there is the greatest of all perils,* Cd. 24; Th. 31, 21; Gen. 488.

frēcendlíc; *adj. Dangerous;* perīculōsus :—Hú frēcendlíc ðæt dysig is *how dangerous the error is!* Bt. 32, 3; Fox 118, 6. Ða habbaþ sum yfel frēcendlícre ðonne ǽnig wíte síe on ðisse worulde *they have an evil more dangerous than any punishment in this world is,* 38, 3; Fox 200, 27. v. frēcenlíc.

frēcenful, frǽcenful, -full; *adj. Harmful, dangerous, perilous;* perīcŭlōsus :—Se þunor byþ frēcenfull [MS. P. frēcenful] for ðæs fyres sceótungum *thunder is harmful from the shootings of the fire,* Bd. de nat. rerum; Lchdm. iii. 280, 14; Wrt. popl. science 19, 27. Of frēcenfulre forliðennysse *perīculōso naufrāgio,* Mone B. 685, 686.

frēcenlíc, frǽcenlíc; *adj. Dangerous, perilous;* perīcŭlōsus :—Ðæt ðære tíde blódlæswu wǽre frēcenlíc *quia perīculōsa sit illius tempŏris phlebŏtŏmia,* Bd. 5, 3; S. 616, 16. Ðæt is hefig dysig, and frēcenlíc fira gehwilcum *that is a grievous folly, and dangerous to every man,* Bt. Met. Fox 19, 3; Met. 19, 2: Bt. 14, 1; Fox 42, 13.

frēcenlíce; *adv. Dangerously;* perīcŭlōse :—Scipio frēcenlíce gewundod wearþ *Scipio was dangerously wounded,* Ors. 4, 8; Bos. 89, 40: Lchdm. iii. 156, 26.

frēcennes, frǽcenes, frēcednes, frēcenis, frēcnes, -nis, -ness, -niss, -nyss, e; *f. Danger, peril, hazard, mischief, harm;* perīculum, discrīmen, mālum :—Betwuh ða frēcennesse stōwe *inter perīculōsa lŏca,* Cot. 111. For ege māran frēcennesse *mētu majŏris perīculi,* Bd. 4, 32; Whelc. 365, 18. Būton mycelre frēcennesse *without much peril,* Ps. Th. 9, 26: Bd. 3, 19; S. 548, 33. Frēcennyssa helle gemētton me *perīcula inferni invēnērunt me,* Ps. Spl. 114, 3. He ongon ða frēcenisse onweg adrīfan *cœpit perīculum abigĕre,* Bd. 2, 7; S. 509, 25. Ðǽr seó frēcnis mǽst wæs *where the danger was greatest,* 2, 7; S. 509, 24. To swylcre frēcnesse *discrīmini,* Bd. 1, 7; S. 477, 16, MS. B: Herb. 30, 4; Lchdm. i. 126, 24. Būton frēcnysse *without harm,* 63, 2; Lchdm. i. 166, 7. He fóreseah micle frēcennesse *he foresaw much peril,* Bd. 3, 19; S. 549, 46. Mid frēcnysse deáþes *mortis perīculo,* 1, 27; S. 493, 26. He oferwon frēcnessa fela *he overcame many perils,* Exon. 35 a; Th. 113, 3; Gū. 152. Mænige ætberstaþ frēcnyssa *multi evādunt perīcula,* Coll. Monast. Th. 25, 1. Se hālga wer in ða ǽrestan ældu gelufade frēcnessa fela *the holy man in his early age loved much mischief,* Exon. 34 a; Th. 108, 31; Gū. 81.

freceo *a glutton;* lurco, Cot. 120. v. frec.

frecgenga? *apostacy;* apostăsia = ἀποστασία, Cot. 16. Lye.

frec-mǽse, frǽc-mǽse, an; *f. The nun bird, titmouse;* pārus cærūleus :— Frecmǽse *sigitula?* Wrt. Voc. 281, 9.

FRÉCNE, frǽcne; *adj. Horrible, savage, audacious, wicked, daring, dangerous, perilous;* dīrus, asper, austērus, atrox, audax, perīcūlōsus :— Đǽr đû findest frécne feohtan *there thou wilt find a savage contest,* Andr. Kmbl. 2699; An. 1352. Đæt biþ frécne wûnd *that is a perilous wound,* Exon. 19 a; Th. 48, 12; Cri. 770. He āna genéđde frécne dǽde *he alone ventured on the daring deed,* Beo. Th. 1782; B. 889. Be đǽre frécnan côđe *of the dangerous disorder,* L. M. 2, 33; Lchdm. ii. 236, 12. He sceal fleón đone frécnan wlite đises middaneardes *he should avoid the dangerous splendour of this earth,* Bt. 12; Fox 36, 20. On đa frécnan tîd *tempŏre discrīmĭnis,* Bd. 1, 8; S. 479, 21. Frécne þûhton egle eálāda *the fearful sea-ways seemed terrible,* Andr. Kmbl. 880; An. 440. Hwonne him Freá frécenra sîþa reste ageáfe *when the Lord should give him rest from his perilous journeyings,* Cd. 71; Th. 86, 8; Gen. 1427: Ps. Th. 143, 8. To frécnum þingum *for daring things,* Lchdm. iii. 158, 16. Đæt he him afirre frécne geþohtas *that he should banish from him wicked thoughts,* Cd. 219; Th. 282, 10; Sat. 284. He frécnu gestreón funde *he found dangerous wealth,* Bt. Met. Fox 8, 115; Met. 8, 58. Đe đa frécnan deórwurþnessa funde *who found the dangerous treasures,* Bt. 15; Fox 48, 24. [*O. Sax.* frôkan *wild, bold, impudent.*] DER. fela-frécne: ge-frécnod.

frécne, frǽcne; *adv. Horribly, savagely, fiercely, severely, insolently, boldly, dangerously;* atrocīter, dūre, audacter, perīcūlōse :— Se wrāđa boda fylgde him frécne *the dire messenger boldly followed him,* Cd. 32; Th. 43, 9; Gen. 688: Beo. Th. 1923; B. 959: 3386; B. 1691. Híe hit frécne genéđdon *they severely oppressed it,* Cd. 170; Th. 214, 17; Exod. 570: Exon. 105 b; Th. 401, 23; Rä. 21, 16: Ps. Th. 67, 2: 103, 33: 104, 25. Đæt him hit frécne ne meahte sceđđan *that it might not dangerously wound him,* Beo. Th. 2069; B. 1032: Ps. Th. 114, 3.

frécnen-sprǽc, e; *f. An audacious or hostile speech;* audax *vel* hostīlis sermo :— Gyf Frysna hwylc frécnensprǽce đæs morđorhetes myndgiend wǽre *if any of the Frisians, by audacious speech, should call to mind* [lit. *should be a rememberer of*] *this deadly feud,* Beo. Th. 2213, note; B. 1104.

frecnes? *glis,* Cot. 96, Som. Lye : also *clammy earth;* argilla, Som. Ben. Frecnis *glus,* Glos. Epnl. Recd. 157, 25.

frécnes, -nis, -ness, -nyss *danger, peril,* Bd. 2, 7; S. 509, 24 : 3, 19; S. 549, 46: Coll. Monast. Th. 25, 1. v. frécennes.

frécne-stîg, e; *f. A dangerous way or path, steep place, precipice;* præcĭpĭtium, Som. Ben. Lye.

frédan; *p.* de; *pp.* ed [frôd *wise, prudent*] *To feel, perceive, know, be sensible of;* sentīre. DER. ge-frédan.

fréfergende = fréfrigende *comforting; part.* of fréfrian, Cd. 220; Th. 284, 7; Sat. 318.

fréfran; *p.* ede; *pp.* ed *To comfort, console;* consōlāri :— Ic findan meahte đone đe mec freóndleásne fréfran wolde *I might find one who would comfort me friendless,* Exon. 76 b; Th. 288, 9; Wand. 28: Andr. Kmbl. 733; An. 367. Hwîlum ic fréfre đa ic ǽr winne on *sometimes I comfort those whom ere I war against,* Exon. 102 b; Th. 389, 13; Rä. 7, 7: 27 b; Th. 82, 19; Cri. 1341. Hî earme fréfraþ *they comfort the poor,* 33 b; Th. 106, 29; Gū. 48. Đû me fréfredest *tu me consōlātus es,* Ps. Th. 85, 17 : Blickl. Homl. 135, 23. Cwæþ he đæt gewunalîce word đara fréfrendra *dixit sōlĭto consōlantium sermōne,* Bd. 5, 5; S. 681, 9. Fréfrede *consōlāti,* Ps. Spl. 125, 1. DER. ge-fréfran. v. fréfrian.

fréfrend, es; *m. A comforter, consoler;* consōlātor :— Mêđra fréfrend *comforter of the weak,* Exon. 62 a; Th. 227, 13; Ph. 422. Fréfrend ic sôhte, findan ic ne mihte *consōlantem me quæsīvi, et non invēni,* Ps. Th. 68, 21 : 31, 8 : Blickl. Homl. 135, 33 : 131, 23. v. fréfriend.

fréfrian; *p.* ode, ade; *pp.* od *To comfort, console;* consōlāri :— Đæt hig woldon hî fréfrian *ut consōlārentur eas,* Jn. Bos. 11, 19. Hwænne fréfrast đû me *quando consōlābĕris me?* Ps. Spl. 118, 82. Đæt he fréfrige me *ut consōlētur me,* 118, 76. Đû fréfrodest me *tu consōlātus es me,* 85, 16 : 118, 50 : Ps. Th. 118, 82. Fréfra đîne mæcgas on môde *comfort thy young men in mind,* Andr. Kmbl. 842; An. 421. He hêran ne wolde Fæder fréfergendum* [= fréfrigendum] *he would not obey the comforting Father,* Cd. 220; Th. 284, 7; Sat. 318. [*Laym.* uroefrien; *p.* freuerede: *Orm.* froffrenn, frofrenn : *O. Sax.* frôbrean : *O. H. Ger.* flôbarjan, fluobarēn.] DER. a-fréfrian, ge-. v. frôfor.

fréfriend, es; *m. A comforter, the Comforter, the Paraclete;* consōlātor, paraclētus :— Ne cymþ se fréfriend to eów *Paraclētus non vĕniet ad vos,* Jn. Bos. 16, 7 : 14, 16 : Ps. Th. 134, 14.

fréfrung, e; *f. A comforting, comfort, consolation;* consōlātio :— He nolde nāne fréfrunge underfôn *nōluit consōlātiōnem accĭpĕre,* Gen. 37, 35.

fregn *asked, inquired,* Andr. Kmbl. 2327; An. 1165, = frægn; *p.* of frignan.

fregnan *to inquire,* Mt. Kmbl. Lind. 21, 24 : Mk. 11, 29. v. frignan.

fremde *did, effected,* Cd. 181; Th. 227, 11; Dan. 185, = fremede; *p.* of fremman.

fremde *foreign, strange,* Beo. Th. 3387; B. 1691. v. fremede.

fremdian; *p.* ode; *pp.* od *To alienate, estrange;* aliēnāre, R. Ben. 4.

fremdnys, -nyss, e; *f. Strangeness, the condition of a foreigner;* peregrīnĭtas, Som. Ben. Lye.

freme; *adj. Good, strenuous, bold;* bŏnus, strēnuus :— Fremu folces cwén *the folk's bold queen,* Beo. Th. 3868; B. 1932. v. fram; from; *adj.*

freme, an; *f. Advantage, profit, benefit, good;* commŏdum, quæstus, emŏlūmentum, bŏnum :— Hýþ *vel* freme *commŏdum, quæstus,* Ælfc. Gl. 81; Som. 73, 25; Wrt. Voc. 47, 30. Đæs we māgon fremena gewinnan *of what we may gain of advantages,* Cd. 22; Th. 28, 18; Gen. 437. Đû us unfreóndlîce fremena þancast *thou thankest us unkindly for our benefits,* Cd. 128; Th. 162, 31; Gen. 2689: 89; Th. 110, 24; Gen. 1843 : 135; Th. 170, 27; Gen. 2819. Gesǽton land unspédigran fremena gehwilcre *they inhabited a land more barren of every good,* 46; Th. 59, 13; Gen. 963. v. fremu.

freme *do, effect, perform,* Ps. Th. 68, 17; *impert.* of fremman.

FREMEDE, fremde, frempe, frǽmde; *adj. Strange, foreign, estranged from, devoid of;* aliēnus, peregrīnus, aliēnātus, aversus, remōtus, expers :— He biþ fremede Freán ælmihtigum *he shall be estranged from almighty God,* Salm. Kmbl. 67; Sal. 34. Đonne beó we fremde fram eallum đǽm gôdum *then should we be cut off from all those good things,* St. And. 8, 10. Feorcund mon ođđe fremde *a far-coming or a strange man,* L. In. 20; Th. i. 114, 15 : L. Edg. ii. 7; Th. i. 268, 21 : L. C. S. 25; Th. i. 390, 24 : Ps. Spl. C. T. 68, 11. Me biþ se éđel fremde *the land is strange to me,* Exon. 105 a; Th. 398, 6; Rä. 17, 3 : Cd. 5; Th. 7, 13; Gen. 105 : Beo. Th. 3387; B. 1691: Ps. Th. 136, 4. Đe đara gefeána sceal fremde weorþan *who shall be devoid of those joys,* Andr. Kmbl. 1780; An. 892 : Hy. 6, 30; Hy. Grn. ii. 286, 30. On fremdes fæđm *into the embrace of a strange* [*man*], Cd. 92; Th. 118, 26; Gen. 1971. Fremdre meówlan *of a strange damsel,* Exon. 80 b; Th. 302, 20; Fä. 39 : Bt. Met. Fox 3, 21; Met. 3, 11. On fremedum *in aliēno,* Lk. Bos. 16, 12. Ne lǽne đînum brêđer nān þing to hîre, ac fremdum menn *non fænērāběris fratri tuo ad ūsūram pĕcūniam, sed aliēno,* Deut. 23, 20. On lande fremdre *in terra aliēna,* Ps. Spl. 136, 5. Ne đû fremedne god gebiddest *neque adōrābis deum aliēnum,* Ps. Th. 80, 9. Wilt đû fremdne monnan grétan *wilt thou address a strange man?* Exon. 92 b; Th. 346, 20; Sch. 1. Him folcweras fremde wǽron *the people were strange to him,* Cd. 89; Th. 110, 31; Gen. 1846. Folca fremdra *of strange people,* Ps. Th. 104, 39. Of fremedum *ab aliēnis,* Mt. Bos. 17, 25, 26. Đæt đa þing đîne āgene sîen, đa đe heora āgene gecynd đé gedydon fremde *that those things can be thine own, which their own natures have made foreign to thee,* Bt. 14, 1; Fox 40, 32. Gif đû fremdu godu bigongest *if thou wilt worship strange gods,* Exon. 67 b; Th. 250, 2; Jul. 121. On đa fremdan þîstro *into the strange darkness,* Bt. 3, 2; Fox 6, 10. [*Piers P.* fremmed *strange : Chauc.* fremde, fremed *foreign, strange : Orm.* fremmde *strange, not of kin : Scot.* fremyt, fremmyt : *Plat.* fromd, frömd : *O. Sax.* fremiđi, fremethi, fremit : *Frs.* freamd : *O. Frs.* framd, fremed : *Dut.* vreemd : *Ger.* fremd : *M. H. Ger.* vremede, vremde : *O. H. Ger.* framadi, fremidi : *Goth.* framaþs : *Dan.* fremmed : *Swed.* främmande : *Icel.* framandi *a man of distinction, stranger.*] v. Grm. R. A. pp. 396 sqq. Schmid. s. v. fremde.

fremede, *pl.* fremedon *made, did, performed,* Elen. Kmbl. 942; El. 472 : Bd. 1, 8; S. 479, 26; *p.* of fremman.

fremednes, -ness, -nyss, e; *f.* [fremed, *pp.* of fremman *and* -ness, -nyss] *An accomplishment, fulfilment;* peractio :— Næþ đæt swefen nǽnige fremednesse gôdes ne yfeles *the dream has no accomplishment for good or evil,* Lchdm. iii. 154, 17. Nǽnige fremednysse *no fulfilment,* iii. 156, 1.

fremeþ *performs, practises,* Beo. Th. 3406; B. 1701; *3rd sing. pres.* of fremman.

fremfull; *adj.* [freme *good,* ful, full *full*] *Beneficent, profitable;* bĕnĕficus :— Đa đe ānweald ofer hig habbaþ synd fremfulle genemned *qui pŏtestātem hăbent sŭper eos bĕnĕfici vŏcantur,* Lk. Bos. 22, 25.

fremfullîce; *adv. Effectually, beneficially;* efficācĭter, R. Ben. interl. Prol.

fremfulnes, -ness, e; *f. Profitableness, utility;* utilĭtas, R. Ben. 53.

fremian, fremoian; *part.* fremiende; hit fremaþ; *p.* ode; *pp.* od [fremman] *To profit, do good, be good or expedient, avail;* prōfĭcĕre, prōdesse, expĕdīre, vălēre :— Ne mid seglinge ne mid rôwnesse ôwiht fremian *nĕque vēlo nĕque remĭgio quicquam prōfĭcĕre,* Bd. 5, 1; S. 613, 26. Biþ heó fremiende to his clǽnsunge ǽrit *in expiatiōnem ejus prŏficiens,* Lev. 1, 4. Hwæt fremaþ ǽnegum menn *quid prodest hŏmĭni?* Mt. Bos. 16, 26 : 15, 5. Gyf se wǽta byþ māre đonne đæt fýr, đonne fremaþ hit *if the moisture is more than the fire, then it does good,* Bd. de nat. rerum; Wrt. popl. science 19, 23 ; Lchdm. iii. 280, 9. Eów fremaþ đæt ic fare *expĕdit vōbis ut ĕgo vādam,* Jn. Bos. 16, 7 : Mt. Bos. 19, 10. Đæt hyt naht ne fremode *quia nihil prŏfĭcĕret,* Mt. Bos. 27, 24 : Mk. Bos. 5, 26. Đonne biþ gesýne, hwæt him his swefn fremion *tunc appārēbit, quid illi prōsint somnia tua,* Gen. 37, 20. [*Orm.* frame *profit : Swed.* främja *to forward, advance : Icel.* frama *to further.*]

FREMMAN, to fremmanne; ic fremme, đû fremest, he fremeþ, *pl.* fremmaþ; *p.* fremede, fremde, *pl.* fremedon; *impert.* freme, *pl.* fremmaþ;

subj. pres. fremme, *pl.* fremmen; *pp.* fremed.　　**I.** *to advance;* promōvēre :—Ðæt ic eáðe mæg ânra gehwylcne fremman and fyrðran freónda mînra *that I may easily advance and further every one of my friends,* Andr. Kmbl. 1867; An. 936: Beo. Th. 3669; B. 1832. Sume ic to gefîite fremede *some I have urged to strife,* Exon. 72 b; Th. 271, 18; Jul. 484.　　**II.** *to* FRAME, *make, do, effect, perform, commit;* făcĕre, patrāre, efficĕre, perfĭcĕre, perpetrāre :—Ðe ðone unrǽd ongan ǽrest fremman *who first began to frame that evil counsel,* Cd. 1; Th. 3, 4; Gen. 30: Andr. Kmbl. 133; An. 67: Beo. Th. 4991; B. 2499: Exon. 67 b; Th. 250, 27; Jul. 133. Sæcce to fremmanne *to make strife,* Exon. 129 b; Th. 496, 28; Rä. 85, 21. Ic gûþe fremme *I make war,* Exon. 105 b; Th. 402, 5; Rä. 21, 25. Ne fremest ðû riht wið me *thou doest not right towards me,* Cd. 102; Th. 135, 19; Gen. 2245: Exon. 54 b; Th. 191, 33; Az. 97. He sôþ fremeþ *he performs truth,* Exon. 81 a; Th. 304, 35; Fä. 80. Sume stale fremmaþ *quidam furtum perpetrant,* Bd. 1, 27; S. 490, 9: 491, 36: Exon. 44 b; Th. 150, 17; Gû. 780. Ic andsæc fremede *I made denial,* Elen. Kmbl. 942; El. 472: Exon. 17 a; Th. 40, 23; Cri. 643: Beo. Th. 6004; B. 3006: Andr. Kmbl. 1237; An. 619: Cd. 177; Th. 222, 18; Dan. 106. He fremede swâ and Freán hýrde *he did so and obeyed the Lord,* Cd. 73; Th. 90, 10; Gen. 1493: 130; Th. 165, 21; Gen. 2735. Ne ic firene fremde *I have not committed crimes,* Ps. Th. 58, 3: Cd. 181; Th. 227, 11; Dan. 185. Hî ða godcundan gerýno clǽnre heortan fremedon *they performed the divine mysteries with a clean heart,* Bd. 1, 8; S. 479, 26: Beo. Th. 6; B. 3: Elen. Kmbl. 1288; El. 646: Menol. Fox 254; Men. 128: Exon. 26 b; Th. 79, 16; Cri. 1291: Cd. 149; Th. 187, 5; Exod. 146. He help freme *do me help or give me help,* Ps. Th. 68, 17. Fremmaþ gē nû leóda þearfe *perform ye now the people's need,* Beo. Th. 5593; B. 2800. Ðæt ðû hospcwide ǽfre ne fremme wið Godes bearne *that thou never make contemptuous words against God's son,* Elen. Kmbl. 1046; El. 524: Andr. Kmbl. 2708; An. 1356. Fremme se ðe wille *let him perform [it] who will,* Beo. Th. 2011; B. 1003. Ǽr gē fremmen yfel *ere ye commit evil,* Cd. 113; Th. 149, 4; Gen. 2469. Nô hwæðre he ofer Offan eorlscype fremede *yet he could not effect supremacy over Offa,* Exon. 85 a; Th. 320, 31; Wíd. 37: Beo. Th. 4274; B. 2134. [*Laym.* fremmen, uremmen *to perform, frame: O. Sax.* fremmian, fremman *to perform, execute: O. Frs.* frema *to commit, effect: O. H. Ger.* ga-fremjan: *Dan.* fremme *to promote: Icel.* fremja *to further: Armor.* framma *to join.*] DER. ge-fremman : ǽ-fremmende, firen-, gôd-, gûþ-, heaðo-, mân-, nâht-, ryht-, till-, wôh-.

fremming, e; *f.* A *framing, an effect, efficacy;* fabrĭcātio, effectus, effĭcācia :—Fremming *effectus,* Ælfc. Gr. 11; Som. 15, 15: Homl. Th. i. 8, 7.

frem-sum, fræm-sum; *adj.* Kind, *benign, courteous;* benignus :—He þearfum and ellreordigum symble eáþmôd and fremsum and rûmmôd wæs *pauperibus et peregrinis semper humilis, benignus et largus fuit,* Bd. 3, 6; S. 528, 11: Ps. Spl. 68, 20: Ps. Th. 134, 3. Syleþ us fremsum gôd Drihten *Dŏminus dăbit benignĭtātem,* 84, 11.

fremsumlíce; *adv.* Kindly, *benignly;* benigne :—Ðâ wæs he fremsumlíce onfangen *cum benigne susceptus,* Bd. 3, 11; S. 536, 12: 1, 25; S. 487, 15.

fremsumnes, -ness, -nys, -nyss, e; *f.* [fremsum, -nes, -ness] Kindness, *benefit, benignity, liberality;* benĕfĭcium :—For fremsumnysse *pro benignĭtāte,* Bd. 1, 27; S. 493, 7: Ps. Spl. C. 84, 13. Ðû geáres hring mid gyfe bletsast, and ðíne fremsumnesse wylt folcum dǽlan benĕdíces cŏrōnæ anni benignĭtātis tuæ, Ps. Th. 64, 12. Ðe ðam godcundum fremsumnessum *de benĕfĭciis dīvīnis,* Bd. 4, 24; S. 598, 17.

frempe; *adj.* Strange, *foreign;* aliēnus, externus :—Ðæt ríce tweógende cyningas and frempe forluron and towurpon *regnum rēges dŭbii vel externi disperdĭdērunt,* Bd. 4, 26; S. 603, 17. Hî awurpon ða aldermenn ðæs frempan cyninges *they cast off the aldermen of the strange king,* 3, 24; S. 557, 45: Lk. Skt. Lind. 24, 18: Jn. 10, 5. v. fremede.

FREMU, e; *f.* Advantage, *profit, gain, benefit;* commŏdum, emŏlŭmentum, quæstus, fructus, benĕfĭcium, sălus :—Hwelc fremu is ðé ðæt, ðæt ðû wilnige ðissa gesǽlþa *what advantage is it to thee, that thou desirest these goods?* Bt. 14, 1; Fox 42, 8: 26, 3; Fox 94, 12. Ðe ðissum folce to freme stondaþ *which for this folk's prosperity stand,* Exon. 67 b; Th. 350, 7; Jul. 123; 54 a; Th. 191; Az. 81: Nar. 39, 18. Ðæt we sceoldon [MS. sceolde] fremena friclan, and us fremu sēcan *that we might desire benefits, and seek to us advantage,* Cd. 89; Th. 110, 25; Gen. 1843. Ne ðǽr freme mêteþ fira ǽnig *no man findeth profit there,* Exon. 68 b; Th. 255, 22; Jul. 218. Neorxna wang stôd, gifena gefylled, fremum forþweardum *paradise stood, filled with gifts, with continual benefits,* Cd. 12; Th. 13, 29; Gen. 210: Exon. 113 a; Th. 434, 10; Rä. 51, 8. DER. un-fremu. v. freme, an; *f.*

fremung, freomung, fromung, e; *f.* Advantage, *profit, good;* commŏdum, profectus, benĕfĭcium :—Ðæt gē gehycgen ymbe ða fremunge gôdra weorca *that ye meditate on the advantage of good works,* L. E. I. prm; Th. ii. 400, 32. For heora fremunge *for their good,* ii. 400, 36.

Frencisc; *def.* se Francisca; *adj.* Belonging to *France;* Francus :—þurh ðone Frenciscan ceorl Hugon *through the French churl Hugo,* Chr. 1003; Erl. 139, 1. Mid mycclum werode Frenciscra manna *with a great*

multitude of Frenchmen, Chr. 1052; Erl. 181, 30. Mid ðám Frenciscum mannum *with the Frenchmen,* Chr. 1052; Erl. 186, 6. Ða Frencisce menn *the Frenchmen,* Chr. 1052; Erl. 187, 7, 26. [*Laym.* frensc.]

Frencisca, an; *m.* A *Frenchman;* Francus :—Ægebertus, se Frencisca, wæs gehâdod *Ægebert, the Frenchman, was ordained,* Chr. 650; Th. 51, 2, col. 2.

frênd *friend* or *friends;* amīcus, amīcos :—Ðæt ðû swutole mihtest tocnâwan ðíne frênd and ðíne fýnd *that thou mightest clearly distinguish thy friends and thy foes,* Bt. 20; Fox 72, 20, MS. Cot. v. freónd.

FREÓ, frió, freoh, frioh, frig, frí, frý; *adj.* FREE, *having liberty or immunity, noble, glad, joyful;* līber, sui jūris, ingĕnuus, nōbĭlis, lætus :—Heó ðâ freó on hire fôta gangum blíðe hâm wæs hweorfende *ipsa lībero pĕdum incessu dŏmum læta reversa est,* Bd. 4, 10; S. 578, 32. Beó he freó *he shall be free,* L. Alf. 11; Th. i. 46, 3, MS. H: L. In. 3; Th. i. 104, 3, MS. B: Bt. 34, 8; Fox 144, 23. Hû wolde ðé lícian, gif hwylc swíðe ríce cyning wæsðe nǽnne freóne mon on eallon his ríce *how would it please thee, if some very powerful king had not any free man in all his realm?* 41, 2; Fox 24, 25, MS. Cot. Gif he mǽgburg hæbbe freó *if he have a free kindred,* L. In. 74; Th. i. 148, 19. Ðǽr freó, môton eard weardigan *where free, they might inhabit a country,* Andr. Kmbl. 1196; An. 598. Ðâ wearþ worn aféded freóra bearna *then was a number of noble children brought forth,* Cd. 79; Th. 99, 6; Gen. 1642: 131; Th. 166, 26; Gen. 2753. Lǽt me freó lǽdan, eft on êðel *let me lead them free, back into their country,* 98; Th. 128, 22; Gen. 2130: Bt. 41, 2; Fox 244, 30; MS. Cot. Ðæt hý ðý freóran hyge gefēngen *that they might receive the gladder spirit,* Exon. 30 a; Th. 92, 22; Cri. 1512. [*Chauc.* fre: *Laym.* freo: *Orm.* freo, fre: *Plat.* fri, frij: *O. Sax.* frí in frí-lík *free-born: Frs.* fry: *O. Frs.* fri: *Dut.* vrij: *Ger.* frei: *M. H. Ger.* vrî: *O. H. Ger.* frî: *Goth.* freis: *Dan. Swed.* fri: *Icel.* frí.] DER. mûþ-freó.

freó; *indecl. m.* A *lord, master;* dŏmĭnus :—Freó ðæt bihealdeþ *my master beholds that,* Exon. 105 a; Th. 399, 3; Rä. 18, 5. v. freá.

freó; *indecl. f.* A *woman;* mŭlier ingĕnua :—Oþ-ðæt he funde freó fægroste *until he found the fairest woman,* Cd. 23; Th. 29, 28; Gen. 457. [*O. Sax.* frî.] v. Grim. D. M. 279.

freó-bearn, es; *n.* One *free-born, a noble child;* prōles ingĕnua, fīlius nōbĭlis :—Freóbearn *vel* æðelborene cild *lībĕri,* Ælfc. Gl. 91; Som. 75, 23; Wrt. Voc. 51, 67. Freóbearn Godes *the noble son of God,* Exon. 17 a; Th. 40, 24; Cri. 643. Freóbearn wurdon alǽten líges gange *the noble children were delivered from the course of the flame,* Cd. 187; Th. 232, 19; Dan. 262.

freó-bearn-fæder; *m.* A *father of noble children;* nōbĭlium fīliōrum pater, Cd. 163; Th. 206, 1; Exod. 445.

freó-borh; *gen.* -borges; *m.* A *free surety, pledge, bondman;* fidejussus, L. Ed. C. 20; Wilk. 201, 53, col. 2. v. friþ-borh.

freó-brôðor; *m.* An *own brother;* germānus frāter :—Him frumbearnes riht freóbrôðor oþ-þah *his own brother took from him his firstborn's right,* Cd. 160; Th. 199, 14; Exod. 338.

freó-burh; *gen.* -burge; *f.* A *free city;* lībĕra arx :—He scolde gesêcean freóburh *he should seek the free city,* Beo. Th. 1390; B. 693.

freócenness; *danger, peril;* perīcŭlum, Som. Ben. Lye. v. frêcenness.

freód, e; *f.* Affection, *good-will, friendship, peace;* ămor, dilectio, amīcĭtia, pax, grātia :—Næs ðǽr mâra fyrst freóde to friclan *there was no more time to desire peace,* Beo. Th. 5105, note; B. 2556. Swâ ðû wið me freóde gecýddest *as thou hast manifested affection to me,* Andr. Kmbl. 780; An. 390. Freóde ne woldon healdan *they would not hold peace,* Beo. Th. 4946; B. 2476. Ic forworht hæbbe ðíne lufan and freóde *I have forfeited thy love and good-will,* Cd. 48; Th. 63, 2; Gen. 1026: Exon. 10 a; Th. 11, 5; Cri. 166: Beo. Th. 3418; B. 1707. Ðæt ðû wille syllan sǽmannum feoh wið freóde *that thou wilt give treasures to the seamen for their friendship,* Byrht. Th. 132, 60; By. 39.

freód *liberty, privilege,* Th. Diplm. A. D. 970; 243, 20. v. freót.

freóde, *pl.* freódon *freed,* Chr. 777; Erl. 55, 22: 963; Erl. 121, 30; *p.* of freógan, freón.

freó-dôm, frió-dôm, frý-dôm, es; *m.* FREEDOM, *liberty;* lībertas, emancĭpātio :—Ðæt is se freódôm, ðætte mon môt dôn ðæt he wile *that is freedom, that a man may do what he will,* Bt. 41, 2; Fox 246, 4, MS. Cot. Freódôm *emancĭpātio,* Ælfc. Gl. 112; Som. 79, 93; Wrt. Voc. 60, 24. Ðâm he geaf micle gife freódômes *to these he gave the great gift of freedom,* Bt. 41, 2; Fox 246, 1. Be ðam freódôme *concerning freedom,* 41, 2; Fox 246, 13. Nis nân gesceádwís gesceaft ðæt næbbe freódôm *there is no rational creature which has not freedom,* 40, 7; Fox 242, 17: 34, 8; Fox 144, 26. Freódôm onfêngon *lībertātem recēpērunt,* Bd. 3, 24; S. 557, 46: 4, 26; S. 602, 31.

freó-drihten, -dryhten, es; *m.* A *noble lord* or *master;* ingĕnuus *vel* nōbĭlis dŏmĭnus :—Onfôh ðissum fulle, freódrihten mín *accept this cup, my noble lord,* Beo. Th. 2343; B. 1169. Freódrihten hine forþlǽdde to ðam hâlgan hâm, heofna Ealdor *the noble Lord, the Prince of heaven, led him forth to the holy home,* Cd. 226; Th. 300, 17; Sat. 566: 225; Th. 299, 10; Sat. 547. Wâst ðû freódryhten, hû ðeós âdle scyle ende

gesettan *knowest thou, noble master, how this disease shall have an end?* Exon. 47 b; Th. 163, 16; Gú. 994.

frê-ófestlíce *very hastily, quickly, speedily;* præprŏpĕre, festinanter, expĕdíte, Som. Ben. Lye. v. frê-ófestlíce.

FREÓGAN, freón; ic freó, he freóþ, *pl.* freógaþ, freóþ; *p.* freóde, *pl.* freódon; *impert.* freó; *subj. pres.* freóge; *pp.* freód [freó *free*]. I. *to free, make free;* manumittĕre, lībĕráre :—Man sceal freógan ǽlcne þeówan *one shall free every slave;* revertĕtur hŏmo ad possessiōnem suam, Lev. 25, 10. Ic hit freó *I free it,* Chr. 963; Erl. 122, 2. He freóde ðæt mynster [MS. mynstre] *he freed the monastery,* 777; Erl. 55, 22. Hí hit freódon *they freed it,* 963; Erl. 121, 30. Freó hine on ðam seofoðan geáre *free him in the seventh year;* in septĭmo anno dimittes eum lībĕrum, Deut. 15, 12. Ðonne ðú hine freóge *when thou freest him;* quem lībertáte donávĕris, 15, 13. Ðæt he scolde freón his mynster [MS. mynstre] *that he would free his monastery,* Chr. 777; Erl. 55, 18. II. *to honour, like, love;* honōráre, dílĭgĕre, ămáre :—Ic ðec for sunu wylle freógan *I will love thee as a son,* Beo. Th. 1900; B. 948. Nǽnig óðerne freóþ swá him God bebeád *no one loves another as God commanded him,* Frag. Kmbl. 70; Leás. 37. Ðú ðín ágen mōst mennen ateón swá ðín mód freóþ *thou mayest treat thine own servant as thy mind liketh,* Cd. 103; Th. 136, 15; Gen. 2258. Ða gecorenan freógaþ folces Weard *the chosen shall love the Lord of mankind,* Exon. 32 a; Th. 100, 27; Cri. 1648; 114 a; Th. 436, 36; Rä. 55, 12. Freóþ hý fremde monnan *strange men love them,* 90 b; Th. 339, 32; Gn. Ex. 103. Fæder and mōdor freó ðú *love thou father and mother,* 80 a; Th. 300, 21; Fä. 9. Hit gedéfe biþ ðæt mon his winedryhten freóge *it is fitting that a man love his dear lord,* Beo. Th. 6334; B. 3177. [Laym. freoien, freoiȝen, ureoiȝen *to set free:* Plat. frijen *to free, woo:* O. Sax. friohan *to love:* O. Frs. friaia, fraia, fria *to free:* Dut. vrijen *to woo:* Ger. freien *to woo;* be-freien *to free:* M. H. Ger. vríen, vríȝen *to free:* Goth. friyon, frion *to love:* Dan. frie *to woo, deliver:* Swed. fria *to free, save, court:* Icel. frjá *to pet.*] DER. be-freón, ge-freógan, -freón.

freó-gyld *a free guild* or *society;* lībĕrum sodālitium. v. frý-gyld.

freoh; *adj. Free;* liber :—Ic neom freoh *non sum liber,* Coll. Monast. Th. 20, 7: Ps. Spl. 87, 4. Gif he freoh sý *if he be free,* L. Wg. 8; Th. i. 188, 3: L. Ath. i. 24; Th. i. 212, 14. He gewát freoh fram deáþes sárnysse *he departed free from the pain of death,* Homl. Th. i. 76, 13. v. freó.

freó-lác; es; *n. A free offering, oblation;* lībĕra oblátio :—Ðú onféhst onságednesse rihtwísnesse, freóláca and offrunga *acceptábis sacrifícium justitiæ, oblătiōnes et holocausta,* Ps. Lamb. 50, 21.

freó-lǽta, frig-lǽta, an; *m. One made free, a freedman;* libertus :—Freólǽta *libertus,* Ælfc. Gl. 8; Som. 56, 106; Wrt. Voc. 18, 55. Freó-lǽtan sunu *the son of a freedman;* libertínus, 8; Som. 56, 107; Wrt. Voc. 18, 56.

freólíc, freólĕc, frílíc; *adj. Free, noble, ingenuous, comely, goodly;* liber, ingĕnuus, egrĕgius, dĕcens :—Eádward, Engla hláford, freólíc wealdend *Edward, lord of the English, a noble ruler,* Chr. 1065; Erl. 196, 25; Edw. 6. Se eafora wæs Enoc háten, freólíc frumbearn *the offspring was called Enoch, a comely first-born,* Cd. 59; Th. 72, 19; Gen. 1189. Freólíc fyrdsceorp *a goodly war-vest,* Exon. 104 a; Th. 395, 25; Rä. 15, 13: Cd. 55; Th. 67, 29; Gen. 1108. Freólíc wíf *the noble woman,* Beo. Th. 1234; B. 615. Freólícu meówle *a goodly damsel,* Exon. 124 b; Th. 479, 2; Rä. 62, 1. Freólícu mǽg *a comely maiden,* Cd. 50; Th. 64, 21; Gen. 1053: 101; Th. 134, 18; Gen. 2226. Freólícum lībĕro, Mone B. 1341. Ðæt he brohte wíf tó háme, fæger and freólíc *that he should bring to his home a wife, fair and goodly,* Cd. 83; Th. 103, 22; Gen. 1722. Bearn freólícu tú *two comely children,* 82; Th. 102, 30; Gen. 1708. Mid his twegen suno, freólíco frumbearn *with his two sons, comely first-born,* Exon. 112 b; Th. 431, 31; Rä. 47, 4. Fǽmne freólícast *most noble damsel,* 9 a; Th. 5, 20; Cri. 72. [O. Sax. frílík.] DER. ful-freólíc.

freólíce, friólíce; *comp.* freólícor; *adv.* FREELY, *without hindrance, with impunity;* lībĕre, impúne :—Ðæt he mihte freólíce Gode þeówian *that he might freely serve God,* Bd. 3, 19; S. 547, 31: Ps. Spl. 93, 1: Cd. 67; Th. 81, 8; Gen. 1342: Andr. Kmbl. 585; An. 293. Seó sáwl færþ swíðe freólíce [friólíce Cott.] tó heofonum *the soul goes very freely to the heavens,* Bt. 18, 4; Fox 68, 14. Heó deófla bígengum freólíce þeówedon *dæmŏnicis cultĭbus impúne serviēbant,* Bd. 2, 5; S. 507, 38. Ðæt hí for gewillnunge ðara écra góda ðý freólícor winnen *pro appĕtítu æternōrum bŏnōrum lībĕrius labōráre,* 4, 25; S. 601, 7.

FREÓLS; es; *m. sometimes, but rarely, n.* I. *freedom, immunity, privilege;* lībertas, immúnĭtas, privilēgium :—Ic ðisne freóls on Róme gefæstnode *I confirmed this freedom at Rome,* Th. Diplm. A. D. 856; 116, 5. Gif man his mǽn an wiofode freóls gefe, se síe folcfrý *if any one give freedom to his man at the altar, let him be folk-free,* L. Wih. 8; Th. i. 38, 15: Cod. Dipl. 925; Kmbl. iv. 263, 27. Ic forgyfe ðisne freóls tó ðære hálgan stōwe æt Scireburnan *I give this immunity to the holy place at Sherborne,* Th. Diplm. A. D. 864; 125, 5. Se arcebisceop spæc to me ymbe Christes circean freóls; ðá lýfde ic him ðæt he mōste

niwan freóls settan; ðá cwæþ he ðæt he freólsas genóge hæfde; ðá nam ic ða freólsas *the archbishop spoke to me about the privilege of Christ's church; then I allowed him to institute a new privilege; then he said that he had privileges enough; then I took the privileges,* Cod. Dipl. 731; A. D. 1013–1020; Kmbl. iv. 9, 32, 35; 10, 1, 3. II. *a time of freedom, a holy day, feast, festival, the celebration of a festival;* festum, festi celebrátio :—Ðæt man sceal fæstan ælce Frigedæg, bútan hit freóls sý *that a man shall fast every Friday, unless it be a festival,* L. Eth. v. 17; Th. i. 308, 23: L. C. E. 16; Th. i. 368, 26. Tó ðam eáster-lícan freólse *to the paschal feast,* Lk. Bos. 2, 42: L. Eth. v. 14; Th. i. 308, 14, 16, 17: L. C. E. 16; Th. i. 368, 25. Gif mæsse-preóst folc miswyssige æt freólse and æt fæstene *if a mass-priest misdirect the people about a festival and about a fast,* L. E. G. 3; Th. i. 168, 8. On Sunnan dæges freólse *on the festival of Sunday,* L. E. G. 9; Th. i. 172, 14. Be mæsse-daga freólse *of the celebration of mass-days,* L. Alf. pol. 43; Th. i. 92, 1. Sunnan dæges freóls healde man georne *let a man diligently keep the festival of Sunday,* L. Eth. v. 13; Th. i. 308, 10: vi. 22; Th. i. 320, 11. Freólsa and fæstena healde man rihtlíce *let a man rightly keep festivals and fasts,* L. Eth. v. 12; Th. i. 308, 8: v. 15; Th. i. 308, 18: vi. 22; Th. i. 320, 10: L. C. E. 14; Th. i. 368, 10. [O. Frs. frihals, frihelse *freedom:* O. H. Ger. frihalsi lībertas: Goth. frei-hals, m.: Icel. frelsi, f. *freedom.*] DER. gál-freólsas, heáh-freóls, sunder-freóls.

freóls; *adj. Free;* liber :—Sý ðis land ǽlces þinges freóls *let this land be free of everything,* Cod. Dipl. 923; Kmbl. iv. 263, 5. v. freó.

freóls-ǽfen, es; *m. A festival-eve, vigil;* festi vigília :—Man mót, freólsǽfenum [MS. freólsǽfenan], faran betweonan Eferwíc and six míla gemete *one may travel, on festival-eves, between York and a distance of six miles,* L. N. P. L. 56; Th. ii. 298, 26.

freóls-bóc, e; *f. A charter of freedom;* lībertátis charta = χάρτης :—Ðis is seó freólsbóc to ðan mynstre æt Byrtúne, ðe Æðelréd cyng ǽfre écelíce gefreóde *this is the charter of freedom to the monastery at Burton, which king Æthelred for ever freed,* Th. Diplm. A. D. 1002; 548, 29.

freóls-brice, -bryce, es; *m.* [freóls *a feast, festival;* brice, bryce *a breaking, breach*] *A breach or violation of a festival;* festi violátio :—On freólsbricum [MS. freólsbricon] *in breaches of festivals,* L. Eth. vi. 28; Th. i. 322, 19. Freólsbrycas *breaches of festivals,* Wulfst. 109, 152.

freóls-dæg, es; *m. A feast-day, festival-day;* festus dies :—Geneálǽhte freólsdæg azimorum, þis is gecweden eástre *appropinquábat dies festus azymōrum, qui dícítur pascha,* Lk. Bos. 22, 1. On ðam freólsdæge *in die festo,* Mt. Bos. 26, 5. Gif hláford his þeówan freólsdæge nýde to weorce *if a lord oblige his servant to work on a feast-day,* L. E. G. 7; Th. i. 172, 2. Be freólsdagum and fæstenum *of festivals and fasts,* L. Edg. i. 5; Th. i. 264, 17: L. Eth. v. 18; Th. i. 308, 24: L. C. E. 17; Th. i. 370, 2. Freólsdæg *festívĭtas, solemnĭtas, vel celebrĭtas, vel cere-mōnia,* Ælfc. Gl. 56; Som. 67, 23; Wrt. Voc. 37, 13. DER. heáh-freólsdæg.

freóls-dóm, es; *m. Freedom, liberty;* lībertas :—Ciricean freólsdóm [MS. freólsdóme] gafola *to the church freedom from imposts,* L. Wih. 1; Th. i. 36, 15. v. freó-dóm.

freóls-geár, -gér, es; *n. A feast-year, jubilee;* annus jubílæus, Cot. 106.

freóls-gefa, an; *m. A freedom-giver;* manumissor :—Gif man his mæn freóls gefe, freólsgefa áge his erfe *if any one give freedom to his man, let the freedom-giver have his heritage,* L. Wih. 8; Th. i. 38, 16.

freólsian; *p.* ode; *pp.* od [freóls *a holy day*]; *v. trans. To keep holy day, to celebrate;* celebráre diem festum :—Sce. Eádweardes mæssedæg witan habbaþ gecoren, ðæt man freólsian sceal ofer eal Engla land *the witan have chosen, that St. Edward's mass-day should be celebrated over all England,* L. Eth. v. 16; Th. i. 308, 21: L. C. E. 17; Th. i. 370, 7. Wirc six dagas and freólsa ðone seofoðan *sex diēbus ŏpĕrábĕris, die septĭmo cessábis,* Ex. 34. 21. Freólsiaþ Drihtnes restedæg *sabbátĭzes sabbătum Dŏmĭno,* Lev. 25, 2. Beó ðú gemyndig ðæt ðú ðone restendæg freólsige *be thou mindful that thou keep holy the day of rest,* Homl. Th. ii. 198, 4: E. Eth. v. 14; Th. i. 308, 15. [Orm. freollsenn.] DER. ge-freólsian.

freóls-líce; *adv. Solemnly, freely;* sollennĭter, lībĕre :— Freólslíce sollennĭter, R. Concord. 8. In ðæm he freólslíce meahte lifian *in which he might freely live,* Bd. 3, 19; S. 547, note 30. v. freólíce.

freóls-man; *gen.* -mannes; *m. A freeman;* liber :—Ic wylle, ðæt ða ðe to mínre áre fón ðæt hí fédon twentig freólsmanna *I will, that those who succeed to my property feed twenty freemen,* Cod. Dipl. 694; Kmbl. iii. 295, 6. v. freó-man.

freóls-stów, e; *f. A festival-place;* lŏcus in quo festívĭtas consecrátæ diei celebrári solébat :—On freóls-stówum [MS. -stówan] *in festival-places,* L. C. S. 38; Th. i. 398, 17.

freóls-tíd, e; *f. A feast-tide;* festívum tempus :—Æt ðissere freólstíde *at this feast-tide,* Homl. Th. ii. 264, 17. Sce. Marian freólstída ealle weorþie man georne *let all St. Mary's feast-tides be strictly honoured,* L. Eth. v. 14; Th. i. 308, 13. Freólstídan and fæstentídan *at festival-tides and fast-tides,* L. C. S. 38; Th. i. 398, 17. DER. heáh-freólstíd.

freólsung, e; *f. A feasting, celebrating a feast;* sollennĭtas :—On middele freólsunga ðíne *in mĕdio sollennĭtátis tuæ,* Ps. Spl. 73, 5. Healde

mon ælces Sunnan dæges freólsunge *let a man keep every Sunday's festival*, L. C. E. 14; Th. i. 368, 11: L. Edg. i. 5; Th. i. 264, 18.

freom; *adj. Firm, strong, powerful*; firmus, strēnuus, fortis:—Ðā com Metod freom on fultum *then came the powerful Lord to his aid*, Cd. 134; Th. 169, 1; Gen. 2793; 143; Th. 178, 19; Exod. 14. Se wæs mā on cyriclícum þeódscypum gelǽred, ðonne he freom wǽre in weoroldþingum *măgis ecclesiasticis disciplīnis instĭtūtum, quam in sēcŭli rēbus strēnuum*, Bd. 4, 2; S. 566, 18. v. from.

freó-mǽg, -mág, es; *m. A relation, kinsman*; consanguĭneus, germānus:—Cain freómǽg ofslōh, brōðor sīnne *Cain slew his kinsman, his brother*, Cd. 47; Th. 60, 18; Gen. 983. Ðeáh ðū from scyle freómāgum feor gewítan *though thou shalt depart far from thy kindred*, 50; Th. 63, 28; Gen. 1039: 161; Th. 200, 12; Exod. 355. Freómǽgum feor *far from my kindred*, Exon. 76 b; Th. 287, 28; Wand. 21: 85 b; Th. 321, 28; Wíd. 53.

freó-man, frí-man, frig-man, -mann, es; *m. A freeman, free-born man*; līběræ conditiōnis hŏmo, vir ingĕnuus:—Ðæt ælc freóman getreówne borh hæbbe *that every freeman have a true surety*, L. Eth. i. 1; Th. i. 280, 7: L. C. S. 20; Th. i. 386, 19. Hwæt gifest ðū me freómanna to frófre *what givest thou me for men's comfort?* Cd. 99; Th. 131, 12; Gen. 2175.

freomian, *part.* freomigende *To profit, be good, avail*; prōdesse, vălēre:—Ðæt ðære ylcan stōwe myl wið fýre wæs freomigende *ut pulvis lŏci illius contra ignem văluĕrit*, Bd. 3, 10; S. 534, 16. v. fremian.

freomlíc; *adj. Profitable, advantageous*; ūtĭlis, commŏdus:—Nerón náht freomlíces ongan on ðære cynewísan *Nero began nothing profitable in the state*, Bd. 1, 3; S. 475, 20.

freomung, e; *f. Profit, advantage, good*; profectus:—In ða tíd his bisceophádes swā mycel gástlíc freomung ongon beón in Angelcynnes cyricum, swā nǽfre ǽr ðon beón mihte *tantum profectus spĭritālis tempŏre præsŭlātus illīus Anglōrum ecclēsiæ, quantum nunquam antea potuĕre, cæpērunt*, Bd. 5, 8; S. 621, 30. v. fremung.

freón; *p.* freóde; *pp.* freód *To free, love*; līběrāre, ămāre, Chr. 777; Erl. 55, 18. v. freógan.

freó-nama, -noma, an; *m. A surname*; cognōmen:—Ðæs fæder wæs háten Oeríc, wæs his freónama Oesc *cūjus păter Oeric, cognōmento Oisc*, Bd. 2, 5; S. 506, 33: 4, 2; S. 565, 39: 5, 19; S. 637, 39. Se pāpa hine nemde freónaman Clemens *the pope named him by surname Clement*, 5, 11; S. 626, 23.

freónd, friónd, es; *pl. nom. acc.* freóndas, frênd, frýnd, freónd; *gen.* freónda; *dat.* freóndum; *m.* [freónde *loving, part.* of freón, v. freógan, freón II. *to honour, like, love*] A FRIEND; ămīcus:—Se feónd and se freónd *the fiend and the friend*, Elen. Kmbl. 1904; El. 954: Exon. 43 a; Th. 144, 33; Gú. 687. Mānfulra and synfulra freónd *publicānōrum et peccătōrum amīcus*, Mt. Bos. 11, 19: Lk. Bos. 7, 34. He wæs Godes freónd *he was the friend of God*, Chr. 654; Erl. 29, 12: 656; Erl. 32, 28. Se hláford ne scrífþ freónde ne feónde *the lord regards not friend nor foe*, Bt. Met. Fox 25, 31; Met. 25, 16: Exon. 105 b; Th. 401, 23; Rä. 21, 16. Gif ðū áge freónd ǽnigne *if thou have any friend*, Cd. 116; Th. 150, 30; Gen. 2499: 135; Th. 170, 10; Gen. 2811: Beo. Th. 2774; B. 1385. Hwylc eówer hæfþ sumne freónd *quis vestrum habēbit amīcum?* Lk. Bos. 11, 5: Ps. Th. 90, 2. Me ðær freóndas gefrunon *friends discovered me there*, Rood Kmbl. 151; Kr. 76. Frýnd synd hie míne georne *they are my zealous friends*, Cd. 15; Th. 19, 7; Gen. 287: Exon. 115 b; Th. 443, 21; Kl. 33. Gê synd míne frýnd, gif gê dóþ ða þing, ðe ic eów bebeóde *vos amici mei estis, si fecērĭtis quæ ĕgo præcĭpio vōbis*, Jn. Bos. 15, 14: Ps. Spl. 37, 11: Ps. Th. 138, 15. Ðǽr mótan freónd sēman *there the friends must arbitrate*, L. Ethb. 65; Th. i. 18, 14: L. Eth. ix. 1; Th. i. 340, 7. He wæs freónda gefylled *he was deprived of his friends*, Chr. 937; Erl. 114, 7: Bt. 20; Fox 72, 14. Náh ic rícra feala freónda on foldan *I have not many powerful friends on earth*, Rood Kmbl. 261; Kr. 132: Apstls. Kmbl. 182; Ap. 91: Andr. Kmbl. 1868; An. 936: 2257; An. 1130. Hine his freóndum gecýðe *let notice of him be given to his friends*, L. Alf. pol. 42; Th. i. 90, 16. Ðæt inwitspell Abraham sægde freóndum sínum *Abram told that tale of woe to his friends*, Cd. 94; Th. 122, 11; Gen. 2025: 79; Th. 97, 12; Gen. 1611. Se hundrēdman sende hys frýnd to him *mīsit ad eum centŭrio amícos*, Lk. Bos. 7, 6: 15, 6, 9: Ps. Th. 87, 18. Heorot innan wæs freóndum afylled *Heorot within was filled with friends*, Beo. Th. 2040; B. 1018: 2256; B. 1126. [*Wyc.* frendesse *a female friend: Laym.* freond: *Orm.* freond, frend: *Scot.* frend *a relation: Plat.* frund, fründ, *m*: *O. Sax.* friund, *m. a friend, relation: Frs.* frjuen: *O. Frs.* friond, friund, *m*: *Dut.* vriend, vrind, *m*: *Ger.* freund, *m*: *M. H. Ger.* vriunt, *m*: *O. H. Ger.* friunt, friônt, friant, *m*: *Goth.* friyonds, *m. a friend*; friyondi, *f. a female friend: Dan.* frände, frænde, *m. f. a cousin, kinsman: Swed.* frände, *m. a relation: Icel.* frændi, *m. a kinsman.*] DER. weoruld-freónd.

freónd-heald; *adj.* [heald *inclined*] *Friend-inclined, friendly*; amīcābĭlis:—Cild biþ freóndheald *a child will be friendly*, Obs. Lun. § 17; Lchdm. iii. 192, 15.

freónd-lár, e; *f.* [lár *instruction*] *Friendly instruction*; fămĭliāris

instructio:—He hine on folce freóndlárum heóld *he maintained him among his people with friendly instructions*, Beo. Th. 4744; B. 2377.

freónd-laðu, e; *f. A friendly invitation*; invītātio fāmĭliāris:—Him wæs freóndlaðu bewægned *a friendly invitation was offered him*, Beo. Th. 2389; B. 1192.

freónd-leás; *adj.* FRIENDLESS; absque amícis:—Gif freóndleás man geswenced weorþe *if a friendless man be distressed*, L. C. S. 35; Th. i. 396, 22. Ic findan meahte ðone ðe mec freóndleásne frēfran wolde *I might find one who would comfort me friendless*, Exon. 76 b; Th. 288, 8; Wand. 28: L. Eth. ix. 22; Th. i. 344, 22: L. C. E. 5; Th. i. 362, 18. Be freóndleásan *of the friendless*, L. C. S. 35; Th. i. 396, 22, 26.

freónd-leást, e; *f. Want of friends, indigence*; amīcōrum inŏpia, indĭgentia:—Þurh freóndleáste *through want of friends*, L. C. S. 35; Th. i. 396, 23.

freóndlíc; *adj. Friend-like, friendly*; ămīcus, benignus:—Þurh ða freóndlícan englas *per ămicos angĕlos*, Bd. 5, 13; S. 633, 29.

freóndlíce; *adv. Like a friend, kindly*; ămīce, benigne:—We ðē freóndlíce wíc getǽhton *we kindly assigned to thee a dwelling-place*, Cd. 127; Th. 162, 25; Gen. 2686: 76; Th. 95, 16; Gen. 1579: Hat. MS. Freóndlícor *more kindly*, Beo. Th. 2058; B. 1027. DER. un-freóndlíce.

freónd-lufu, e; *f. Friendly love, friendship, love, intimacy*; amīcĭtia, cārĭtas, familiārĭtas:—Saga ðæt ðū síe sweostor mín, ðonne ðē leódweras fricgen, hwæt síe freóndlufu uncer twega *say that thou art my sister, when the men of the country ask thee what may be the intimacy of us two*, Cd. 89; Th. 110, 7; Gen. 1834.

freónd-mynd, e; *f. An amorous mind*; amātōria mens:—Ic me onēgan [MS. onagen] mæg ðæt me wrāþra sum, wǽpnes ecge, for freóndmynde, feore beneóte *I for myself may fear that some enemy, through amorous mind, may deprive me of life with a weapon's edge*, Cd. 89; Th. 109, 31; Gen. 1831.

freónd-rǽden, -rædden, -rædenn, e; *f. A friend-condition, friendship*; amīcĭtia:—Ðæt heó mínre ne gýme freóndrǽdenne *that she cares not for my friendship*, Exon. 66 b; Th. 246, 33; Jul. 71. Hig mihton náne freóndrǽdene wið hine habban *they would have no friendship with him*, Gen. 37, 4. Hie getreówlíce heora freóndrǽdenne healdaþ *they faithfully hold their friendship*, Bt. 21; Fox 74, 39: Exon. 67 a; Th. 249, 5; Jul. 107: Elen. Kmbl. 2413; El. 1208. Gif man wille fulle freóndrǽdene [freóndrǽddene MS. B.] habban *if a man will have full friendship*, L. E. G. 12; Th. i. 176, 2. God gefēgþ mid freóndrǽdenne folc togædere *God joins people together with friendship*, Bt. 21; Fox 74, 37.

freónd-scipe, -scype, es; *m.* FRIENDSHIP; amīcĭtia:—Is nū swā hit nō wǽre freóndscipe uncer *our friendship is now as it had not been*, Exon. 115 a; Th. 443, 4; Kl. 25. Þolige úre ealra freóndscipes, and ealles ðæs ðe he áge *let him forfeit the friendship of us all, and all that he has*, L. Ed. 8; Th. i. 164, 4: L. Ath. i. 26; Th. i. 214, 5. Be mínum freóndscipe *by my friendship*, i. prm; Th. i. 194, 5: L. Edg. S. 1; Th. i. 272, 5. Fram ðyssa muneca freóndscipe *by the friendship of these monks*, Bd. 3, 5; S. 526, 18. Man fullne freóndscipe gefæstnode *they confirmed full friendship*, Chr. 1014; Erl. 150, 14: 1016; Erl. 159, 3. Ðæt man frið and freóndscipe rihtlíce healde *that peace and friendship be lawfully observed*, L. Eth. v. 1; Th. i. 304, 10: vi. 8; Th. i. 316, 28. Git móston freóndscype fremman *ye might foster friendship*, Exon. 123 a; Th. 473, 21; Bo. 18. Se gefēhþ fela folca tosomne mid freóndscipe *he joins many people together with friendship*, Bt. Met. Fox 11, 179; Met. 11, 90. Freóndscipas niwe *new friendships*, Somn. 203; Lchdm. iii. 210, 2.

freónd-spéd, e; *f. An abundance of friends*; amīcōrum cōpia:—Ic ðam magorince sylle freóndspéd *I will give many friends to the youth*, Cd. 106; Th. 140, 19; Gen. 2330.

freónd-spédig; *adj. Rich in friends*; amīcōrum dīves:—Ðus mæg mihtig man, and freóndspédig, his dǽdbóte, mid freónda fultume, micelum gelíhtan *thus may a powerful man, and rich in friends, greatly lighten his penance, with the help of his friends*, L. P. M; Th. ii. 286, 13.

freó-noma, an; *m. A surname, noble name*; cognōmen:—Iob Sunu Waldendes freónoman cende *Job gave a noble name to the Lord's son*, Exon. 17 a; Th. 40, 9; Crí. 636. v. freó-nama.

freóra *of free*, Cd. 131; Th. 166, 26; Gen. 2753; *gen. pl.* of freó; *adj.*

freórig; *adj.* I. *freezing, chilled, frigid, frozen*; frígens, frígŏre rigens, frigĭdus, gĕlĭdus:—Ic wæs mundum freórig *my hands were chilled* [lit. *I was freezing in my hands*], Andr. Kmbl. 982; An. 491. Mec se wǽta wong, wundrum freórig, ǽrist cende *the humid field, wonderously frigid, first brought me forth*, Exon. 109 a; Th. 417, 8; Rä. 36, 1. Land wǽron freórig cealdum cýlegicelum *the lands were frozen with cold icicles*, Andr. Kmbl. 2520; An. 1261. II. *chilled with fear or sorrow, trembling, sad*; trĕmens, tristis:—He gefeóll freórig to foldan *he fell trembling to the ground*, Judth. 12; Thw. 25, 27; Jud. 281. Ongon hygegeómor, freórig and ferþwérig, fúsne grētan he, *sad in mind, trembling and weary of soul, resolved to greet the departing*

[*man*], Exon. 49 b; Th. 171, 21; Gú. 1130. Ferþloca freórig *a trembling body*, 76 b; Th. 288, 18; Wand. 33.

freórig-ferþ; *adj. Sad in soul;* tristis animo :—Cwom freórigferþ ðã seó fǽmne wæs he, *sad in soul, came to where the damsel was*, Exon. 52 b; Th. 182, 30; Gú. 1318.

freórig-mód; *adj. Sad in mind;* tristis animo :—He monge gehǽlde, ðe hine ádle gebundne gesóhtun, freórigmóde *he healed many, who, oppressed with malady, sad in mind, sought him*, Exon. 45 b; Th. 155, 14; Gú. 860.

freó-riht, es; *n. A free right, common right, right of a freeman;* libĕrōrum et ingenuōrum jus :—He ne beó syððan ǽniges freórihtes wyrðe *he shall not afterwards deserve any free right*, L. C. S. 20; Th. i. 386, 22.

FREÓSAN, hit freóseþ, frýsþ, frýst; *p.* freás, *pl.* fruron; *pp.* froren *To* FREEZE ; gĕlāre :—Forst sceal freósan *frost shall freeze*, Exon. 90 a; Th. 338, 1; Gn. Ex. 72. Men steorran mágon [MS. magan] geseón swá sutole swá on niht ðonne hit swíðe freóseþ *men may see the stars as plainly as at night when it freezes hard*, Homl. Blick. 93, 20. Hit frýst [frýsþ MS. D.] *gĕlat*, Ælfc. Gr. 22; Som. 24, 8. [*Wyc.* frees, freesede *froze : Plat.* fresen, freren : *Dut.* vriezen : *Ger.* frieren : *M. H. Ger.* vriusen : *O. H. Ger.* friusan, freosan : *Goth.* frius, *n. frost : Dan.* fryse : *Swed.* frysa : *Icel.* frjósa.] DER. ge-freósan : ofer-froren.

freót, freód, es; *m. Freedom, liberty, an enfranchisement, a setting a man free;* libertas, mănūmissio :—Þolie his freótes *let him forfeit his freedom*, L. E. G. 7; Th. i. 170, 17. We scylon todǽlan freót and þeówet *we ought to distinguish between freedom and slavery*, L. C. S. 69; Th. i. 412, 9 : L. Ed. 9; Th. i. 164, 10.

freót-gifa, an; *m. A giver of freedom, liberator, emancipator;* manumissor, Ælfc. Gl. 112; Som. 79, 91; Wrt. Voc. 59, 58.

freót-gifu, e; *f. The gift of freedom, emancipation, manumission,* manumissio :—Freótgifu [MS. freótgife] *manumissio*, Ælfc. Gl. 112; Som. 79, 92; Wrt. Voc. 60, 1.

freoða, an; *m. A protector, defender;* tūtor :—Ðú me, God, wǽre freoða *thou, O God, wast a protector to me;* refúgium meum es tu, Ps. Th. 70, 3.

freoðan; *p.* ede; *pp.* ed *To* FROTH; spūmāre, Som. DER. a-freoðan.

freoðian; *p.* ode, ade; *pp.* od, ad *To care for, maintain, cherish, protect, keep, observe;* consŭlĕre, sustentāre, fŏvēre, tuēri, observāre :—In eallum þingum ðǽre cirican eahtum and gódum he freoðode and fultemede *ecclēsiæ rēbus in omnibus consŭlĕre ac fŏvēre cūrāvit*, Bd. 2, 6; S. 508, 32. Ðæt mínes freán módwén freoðaþ *what my master's mind's thought will maintain*, Exon. 129 b; Th. 498, 3; Rä. 87, 7. God mín feorg freoðaþ *God will protect my life*, Exon. 36 a; Th. 116, 28; Gú. 214. Hie ælmihtig sigebearn Godes freoðode *the almighty victorious Son of God protected her*, Elen. Kmbl. 2292; El. 1147: Exon. 94 b; Th. 354, 3; Reim. 40 : 103 a; Th. 391, 14; Rä. 10, 5. Hine weoruda God freoðade on foldan *the God of hosts protected him on earth*, Exon. 38 a; Th. 126, 6; Gú. 367. Hí ðone heágan dæg healdaþ and freoðiaþ *they keep and observe the high day [Sunday]*, Hy. 9, 27; Hy. Grn. ii. 291, 27. DER. ge-freoðian. v. friðian.

freoðo, frioðo, friðo, fryðo, freðo; *indecl. f* : freoðu, friðu, e; *f. Peace, security, protection, a refuge;* pax, secūritas, tūtēla, asylum :—Seó [treów] ðé freoðo sceal in lífdagum weorþan *which [faith] shall be peace to thee in thy life's days*, Cd. 163; Th. 204, 21; Exod. 422. Wel biþ ðǽm ðe mót Drihten sécean, and to Fæder fæðmum freoðo wilnian *it shall be well to him who may seek the Lord, and desire peace in his Father's bosom*, Beo. Th. 379; B. 188: Exon. 121 a; Th. 465, 3; Hö. 98. Gif me freoðo Drihten an *if the Lord will grant me protection*, Cd. 89; Th. 110, 15; Gen 1838 : 183; Th. 229, 25; Dan. 222. Ic me freoðu to ðé wilnige *I desire peace from thee*, Ps. Th. 55, 8. Hí ðǽr lifgaþ á in freoðu Dryhtnes *they shall live there for ever in the Lord's peace*, Exon. 64 b; Th. 238, 1; Ph. 597. Þurh ðé eorþbúende ealle onfóþ freoðo and freóndscipe *through thee all dwellers upon earth shall receive peace and friendship*, Cd. 84; Th. 105, 28; Gen. 1760. Ic eów freoðo healde *I will hold you in protection*, Andr. Kmbl. 672; An. 336. Ne mihte earmsceapen findan freoðe *the poor wretch could not find protection*, 2261; An. 1132. Utan us to Fæder freoða wilnian *let us desire peace from our Father*, Exon. 19 a; Th. 48, 18; Cri. 773. [*O. Sax.* friðu : *O. H. Ger.* fridu.] DER. fenfreoðo. v. friþ.

freoðo-beácen, es; *n. A sign of peace, sign granting safety;* pācis signum, signum incolumĭtātem præbens :—Hine Waldend on tácen sette, freoðobeácen, ðý-læs hine feónda hwilc mid gúþ-þræce grétan dorste *the Lord set a token, a sign of peace, upon him [Cain], lest some enemy durst greet him with hostile force*, Cd. 50; Th. 64, 4; Gen. 1045.

freoðo-burh; *gen.* -burge; *f. A peaceful city, city of refuge, an asylum;* pācis arx, asylum :—He gesóhte freoðoburh *he sought the peaceful city*, Beo. Th. 1048; B. 522. v. friþ-burh.

freoðo-leás; *adj. Peaceless;* pāce cărens :—Swylc wæs ðæs folces freoðoleás tácen *such was the people's peaceless token*, Andr. Kmbl. 58; An. 29. v. friþ-leás.

freoðo-scealc, es; *m. A minister of peace;* pācis minister :—Swá se

---

engel, fǽle freoðoscealc, fǽmnan sægde *as the angel, the faithful minister of peace, said to the damsel*, Cd. 105; Th. 138, 33; Gen. 2301. Sprǽcon fǽle freoðoscealcas to Lothe *the faithful ministers of peace spake to Lot*, Cd. 115; Th. 150, 25; Gen. 2497.

freoðo-sibb *protecting peace.* v. friðu-sibb.

freoðo-spéd, friðo-spéd, e; *f. Abundance of peace, protecting power;* pācis cōpia, tutēlāris pŏtestas :—Enoch siððan ealdordóm ahóf, freoðospéd *Enoch then raised his sovereignty, his protecting power*, Cd. 60; Th. 73, 2; Gen. 1198.

freoðo-tácen *a token* or *sign of peace.* v. friðo-tácen.

freoðo-þeáw, es; *m. Peaceful behaviour* or *manner;* pacifĭci mōres :—Ðá wæs sibb on heofnum, freoðoþeáwas *then there was agreement in heaven, peaceful manners*, Cd. 4; Th. 5, 29; Gen. 79.

freoðo-wǽr, freoðu-wǽr, frioðo-wǽr, frioðu-wǽr, friðu-wǽr, e; *f. A covenant of peace, an agreement, compact;* pācis fœdus, pactum :—Wæs seó eorla gedriht ánes módes, fæstum fæðmum freoðowǽre heóld *the host of men was of one mind, held the covenant of peace in their firm breasts*, Cd. 158; Th. 197, 13; Exod. 306. Hí onféngon fulwihte and freoðuwǽre *they received baptism and the covenant of peace*, Andr. Kmbl. 3259; An. 1632. v. frioðo-wár, -waru.

freoðo-weard *a guardian of peace.* v. freoðu-weard.

freoðo-webba *a peace-weaver, an angel.* v. friðo-webba.

freoðo-webbe *a peace-weaver, woman.* v. freoðu-webbe.

freoðo-wong, es; *m. A peaceful plain;* pācis campus :—Freoðowong ðone ofereódon *they went over the peaceful plain*, Beo. Th. 5910; B. 2959.

freoðu *peace, security, protection,* Ps. Th. 55, 8 : Exon. 64 b; Th. 238, 1; Ph. 597. v. freoðo.

freoðu-wǽr *a covenant of peace*, Andr. Kmbl. 3259; An. 1632. v. freoðo-wǽr.

freoðu-weard, es; *m. A guardian of peace;* pācis custos :—Him wæs engel neáh fǽle freoðuweard *the angel was near him, a faithful guardian of peace*, Exon. 35 a; Th. 112, 15; Gú. 144.

freoðu-webbe, an; *f. A peace-weaver, woman;* pācis textrix, conciliatrix, mŭlier :—Ne biþ swylc cwénlíc þeáw, ðætte freoðuwebbe feores onsæce leófne mannan *such is no feminine usage, that a peace-weaver deprive a dear man of his life*, Beo. Th. 3888; B. 1942. Wídsíþ mid Ealhhilde, fǽlre freoðuwebban, hám gesóhte Eormanríces *Widsith with Ealhild, faithful peace-weaver, sought the home of Ermanric*, Exon. 84 b; Th. 319, 2; Wid. 6. v. Grm. And. u. El. 144.

freót-man, -mann, es; *m. A freedman;* libertus :—Hió hyre an ðara [MS. ðere] manna and ðæs yrfes, bútan ðám freótmannum [MS. -mannon] *she gives her the men and the stock, except the freedmen*, Cod. Dipl. 1290; A. D. 995; Kmbl. vi. 131, 10.

freó-wine, es; *m. A noble friend;* nōbĭlis *vel* princeps ămīcus :—Ðæt ðú me ne forwyrne, freówine folca *that thou deny me not, noble friend of people*, Beo. Th. 864, note; B. 430.

Fresan; *gen.* Fresena, Fresna; *pl. m. The Frisians;* Frisii, Fresōnes :—He mid Wilbrord ðone hálgan bisceop Fresena wæs wuniende *apud sanctissimum Fresōnum gentis archiepiscŏpum Vilbrordum mŏrābātur*, Bd. 3, 13; S. 538, 8: Beo. Th. 2191; B. 1093. Ðæt Swíþbyrht and Wilbrord biscopas wǽron Fresna þeóde gehálgode *that Swithbyrht and Wilbrord were consecrated bishops of the Frisians' nation*, Bd. 5, 11; S. 625, 28: Exon. 85 a; Th. 320, 11; Wíd. 27: Beo. Th. 5823; B. 2915. v. Frysa.

Fres-cyning, es; *m. A Frisian king;* Fresōnum rex :—Nalles he Frescyninge breóstweorþunge bringan móste *he could not bring the ornament to the Frisian king*, Beo. Th. 5000; B. 2503.

Fresisc; *adj. Of* or *belonging to Friesland, Frisian;* Frīsĭcus :—Nǽron hí náwðer ne on Fresisc gescæpene ne on Denisc *they were shapen neither as the Frisian nor as the Danish*, Chr. 897; Erl. 95, 15. Ðǽr wearþ ofslægen Lucumon, and ealra monna, Fresiscra and Engliscra, lxii *there was slain Lucumon, and of all the men, Frisian and English, sixty-two*, Chr. 897; Erl. 96, 4. v. Frysisc.

Fres-lond, es; *n. Friesland;* Frīsia :—Freslondum on Hreðles eafora swealt *Hrethel's offspring perished in the Frieslands*, Beo. Th. 4704; B. 2357. v. Frys-land.

FRETAN, ic frete, ðú fritest, fritst, he freteþ, friteþ, fritt, fryt, *pl.* fretaþ; *p.* ic, he fræt, ðú frǽte, *pl.* frǽton; *pp.* freten [for-, etan *to eat*?]. I. *to eat up, gnaw,* FRET, *devour, consume;* devŏrāre, consūmĕre, comĕdĕre :—Ða ðe wilniaþ fretan mín folc *qui devŏrant plēbem meam*, Ps. Th. 13, 9 : 26, 3 : Exon. 127 a; Th. 488, 11; Rä. 76, 5 : 87 b; Th. 329, 34 : Vy. 44 : Beo. Th. 6021; B. 3014 : 6220; B. 3114. Swá 'líg freteþ mórhǽþ *vĕlut flamma incendat montes*, Ps. Th. 82, 10. Friteþ wildne fugol *it eats the wild bird*, Salm. Kmbl. 596; Sal. 297 : 808 : Sal. 403. Deáþ misfédeþ oððe fritt híg *mors depascet eos*, Ps. Spl. T. 48, 14. Fýr fryt land mid his wæstme *ignis devŏrābit terram cum germĭne suo*, Deut. 32, 22. Gærstapan hit freteþ eall *locustæ devŏrābunt omnia*, Deut. 28, 38 : Ps. Th. 52, 5. He fræt fýftýne men *he devoured fifteen men*, Beo. Th. 3167; B. 1581: Exon. 112 b; Th. 432, 4; Rä. 48, 1. He fræt uncer wurþ *cŏmēdit prĕtium nostrum*, Gen. 31, 15 : Ps. Spl. 79, 14.

Fugelas hit fræton *võlucres comēdērunt illud*, Mk. Bos. 4, 4; frétun, Rush.: fréton, Mt. Lind. 13, 4: Gen. 37, 20. We hine fræton *.obsor-buimus eum*, Ps. Th. 34, 23: 104, 30. Wæron hie mid metelíéste gewægde, and hæfdon miclne dæl ðara horsa freten *they were distressed for want of food, and had eaten a great part of their horses*, Chr. 894; Erl. 92, 28. Swá hwylcne man swá hy gefóþ fretaþ hí hine *quoscunque capiunt comedunt*, Nar. 36, 4. Freotas *devorant*, Mk. Skt. Rush. 12, 40. II. *to break, burst;* frangĕre, rumpĕre:—Heó wære fræton *they brake their covenant*, Cd. 149; Th. 187, 7; Exod. 147. [*Piers P. Chauc.* frete: *Laym.* freten *to gnaw: Orm.* freteþþ *fretteth: Plat.* freten, vreten: *Dut.* vreten: *Ger.* fressen: *M. H. Ger.* vrëzzen: *O. H. Ger.* farëzzan, fire-zan, frezzan, frëzan: *Goth.* fra-ītan: *Dan.* fraadse: *Swed.* fräta, fråssa.]

fretere, es; *m. A glutton;* lurco, Som. Ben. Lye.

freðo; *indecl. f. Peace;* pax:—Gewít on freðo gangan, ût of earce go *forth in peace, out of the ark*, Cd. 73; Th. 89, 28; Gen. 1487. v. freoðo.

fretnes, -ness, e; *f. A devouring, ravening;* edācĭtas, vŏrācĭtas, Som. Ben. Lye.

fretol, frettol; *adj. Voracious, gluttonous;* ĕdax:—Frettol *ĕdax* vel *glutto*, Ælfc. Gl. 88; Som. 74, 81; Wrt. Voc. 50, 61.

frettan; *p.* te; *pp.* ed *To feed upon, eat up, consume;* depasci:—Hine [wíngeard] wilde deór wēstaþ and frettaþ *singŭlāris fĕrus depastus est eam* [*vineam*], Ps. Th. 79, 13. Hie ðæt corn forbærndon, and mid hira horsum fretton on ælcere efenēhþe *they burned the corn, and with their horses ate it up on every plain*, Chr. 894; Erl. 93, 12. Fretton *comĕderunt*, Mk. Skt. Lind. 4, 4.

fretwednes, fretwodnes, -ness, e; *f. An adorning, decoration;* ornātio, decŏrāmentum:—On eorþlicre fretwednesse *in earthly adorning*, Bd. 3, 22; S. 552, 20, note. Beóþ ðonne ûre hrægla fretwodnes on ðam ēcan fýre wítnode *then our decoration of garments will be punished in the eternal fire*, L. E. I. prm; Th. ii. 394, 11. v. frætwednes.

fretwian; *p.* ode; *pp.* od *To adorn;* ornāre, insignire:—Ic mærsige oððe fretwige *insignio*, Ælfc. Gr. 30; Som. 34, 60. v. frættewian.

fretwung *an adorning;* ornātio, Som. Ben. Lye. v. frætwung.

frí; *adj. Free, noble;* līber, ingĕnuus, nōbĭlis:—Fríes mannes wíf *the wife of a free man*, L. Ethb. 31; Th. i. 10, 6. Ic ðē on folcum frîne Drihten ēcne andete *I acknowledge thee amongst the people, a noble eternal Lord*, Ps. Th. 56, 11. v. freó; *adj.*

friá, an; *m. A lord, master;* dŏmĭnus:—Ðam ágenan frián *to the possessor*, L. Eth. iii. 4; Th. i. 294, 17. v. freá, ágen-frigea.

fría; *p.* ade; *pp.* ad I. *to love:*—Fríende was *complexus esset*, Mk. Skt. Lind. 9, 36. II. *to free:*—Ic fría *liberabo*, Rtl. 9, 40. We sie fríado *liberemur*, 7, 3. v. freógan.

frí-borh; *gen.* -borges; *m. A free surety, pledge, bondman;* fĭdejussio, L. Ed. C. 20; Wilk. 202, 11. v. freó-borh.

fric; *adj. Voracious:*—Fric ꝼ étere *vorax*, Mt. Lind. 11, 19. v. frec.

fricca, fryccea, an; *m. A crier, herald;* præco:—Hleówon hornboran, hreópon friccan *trumpeters sounded, heralds shouted*, Elen. Kmbl. 108; El. 54: 1097; El. 550. Hreópon friccan *heralds shouted*, Andr. Kmbl. 2314; An. 1158. Cristes fricca *Christ's crier*, Blickl. Homl. 163, 21. Sylle se friccea his stefne *let the crier give out his voice*, 163, 31.

fricgan, fricgean, fricggan; *part.* fricgende; ic fricge, ðú frigest, frigst, frihst, he frigeþ, frigþ, frihþ, *pl.* fricgaþ; *p.* ic, he fræg, ðú fræge, *pl.* frægon; *impert.* frige; *subj. pres.* fricge, *pl.* fricgen; *pp.* ge-frigen, -fregen, -frægen *To ask, inquire, question, find out, seek after, learn, get information of;* interrŏgāre, sciscĭtāri, pĕtĕre, condo accŷpĕre, compĕrire:—Wilt ðú fricgan felageongne ymb forþgesceaft *wilt thou ask one who has travelled much about the creation?* Exon. 92 b; Th. 346, 23; Sch. 3. Sceal bearna gehwylc leánes fricgan, ealles ðæs ðe we on eorþan ær geworhton [MS. geweorhtan], gódes oððe yfles *every child shall seek the reward of all that we ere did on earth, of good or evil*, Exon. 116 b; Th. 447, 18; Dóm. 41. Higelác ongan sínne geseldan fricgean *Higelac began to question his guest*, Beo. Th. 3974; B. 1985: Cd. 139; Th. 174, 33; Gen. 2887. Ðæs fricggan ongan folces aldor *the prince of the people began to inquire about it*, Elen. Kmbl. 313; El. 157: 1116; El. 560. Gomela Scylding, fela fricgende, feorran rehte *the aged Scylding, learning much, related* [*things*] *from* [*times*] *remote*, Beo. Th. 4218; B. 2106: Exon. 92 b; Th. 347, 17; Sch. 14. Fricge ic ðē, hwæðres biþ hira folgoþ betra *I ask thee, of which of them is the condition better?* Salm. Kmbl. 739; Sal. 369. Hí fricgaþ, hú... *they ask, how...*, Exon. 9 a; Th. 6, 30; Cri. 92. Frige mec fródum wordum *question me in prudent words*, Exon. 88 b; Th. 333, 8; Gn. Ex. 1. Frige hwæt ic hátte *find out what I am called*, Exon. 104 a; Th. 396, 6; Rä. 15, 19: 105 a; Th. 398, 20; Rä. 17, 10: 107 a; Th. 409, 9; Rä. 27, 26: 107 b; Th. 410, 13; Rä. 28, 15. Ðonne ðē leódweras fricgen *when the men of the country ask thee*, Cd. 89; Th. 110, 6; Gen. 1834. DER. ge-fricgan, un-fricgende. v. frignan.

frician; *p.* ode, ude; *pp.* od, ud *To dance;* saltāre:—Gē ne fricudun *non saltastis*, Mt. Bos. 11, 17.

friclan; *p.* ede; *pp.* ed; *with the gen. To desire, seek for;* appĕtĕre:—Ðæt we sceoldon [MS. sceolde] fremena friclan *that we might desire benefits*,

Cd. 89; Th. 110, 24; Gen. 1843. Næs ðær mára fyrst freóde to friclan *there was no time more to seek for friendship*, Beo. Th. 5105; B. 2556.

friclo; *indecl. f. An appetite;* appĕtītus:—Be ðære ofermiclan friclo, ðonne of ðære selfan cealdan ádle ðæs magan cymþ, ðæt sió ofermiclo friclo* and gīfernes arîst *of the excessive appetite, when from the same cold disease of the stomach it comes, that the excessive appetite and greediness arise*, L. M. 2, 16; Lchdm. ii. 196, 1, 2.

frico; *f. Usury;* usura, Mt. Lind. 25, 27. [Cf. *O. H. Ger.* frechí *avaritia.*]

frictrung, frictung; *f. Divination;* ariolatus, Gl. Mett. 10: Gl. Amplon. 45. v. frihtrung, freht.

fríd-hengest, es; *m. A stately horse:*—Hæfdon xi eóredmæcgas fríd-hengestas *the horsemen had eleven stately horses*, Exon. 106 a; Th. 404, 7; Rä. 23, 4.

friénd *friend:*—Ne murnþ náuðer ne friénd ne fiénd *he regards neither friend nor foe*, Bt. 37, 1; Fox 186, 7. v. freónd.

Friesa *a Frisian*, Chr. 897; Erl. 96, 2, 3. v. Frysa.

frig; *def.* se frigea; *adj. Free, noble;* liber, ingĕnuus, nōbĭlis:—Nelle ic gán ût ne beón frig *non egrĕdiar liber*, Ex. 21, 5. Gif hwá his ágenne geleód bebycgge, þeówne oððe frigne *if any one sell his own countryman, bond or free*, L. In. 11; Th. i. 110, 4: L. Wih. 14; Th. i. 40, 9: L. C. S. 20; Th. i. 388, 3. Gif God næfde on eallum his ríce náne frige sceaft *if God had not any free creature in all his kingdom*, Bt. 41, 2; Fox 244, 29. Gif beóþ frige *libĕri ĕritis*, Jn. Bos. 8, 33, 36: Bd. 3, 24; S. 557, 46. Gif se frigea ðý dæge wyrce *if a freeman work on that day*, L. In. 3; Th. i. 104, 5: 74; Th. i. 150, 1. Eal swá ælcan frigean men gebýreþ *sīcut omnis liber făcĕre dēbet*, L. R. S. 3; Th. i. 432, 23: L. In. 74; Th. i. 150, 3. v. freó.

frig, frigu? e; *f. Love, affection, favour;* ămor:—Sió weres friga wiht ne cûðe *she knew nothing of the love* [*affections*] *of man*, Exon. 13 b; Th. 26, 19; Cri. 419. Ðæt wæs geworden bûtan weres frigum *that was done without the favours of man*, 8 b; Th. 3, 17; Cri. 37.

Frig-dæg, Frige dæg, es; *m.* FRIDAY, *Friga's day, the day on which the heathens worshipped the goddess Friga*, or *Venus, the consort of Woden and protectress of matrimony;* dies Vĕnĕris:—Man singe ælc Frigdæge æt ælcum mynstre, ealle ða Godes þeówan, ân fiftig sealmas for ðone cyng *one shall sing every Friday, at every monastery, all servants of God fifty psalms for the king*, L. Ath. iv. 3; Th. i. 222, 18. Ælces Frige dæges fæsten *every Friday's fast*, L. Edg. i. 5; Th. i. 264, 23: L. C. E. 16; Th. i. 368, 25. Fæstan ælce Frige dæg *to fast every Friday*, L. Eth. v. 17; Th. i. 308, 23: vi. 24; Th. i. 320, 22. Ðis sceal on Frige dæg ofer twelftan dæg *this* [*Gospel*] *must be* [*read*] *on Friday after the twelfth day*, Rubc. Mt. Bos. 4, 12, 23; Notes. p, 574. For Friga v. Grm. D. M. p. 278; and for names of the days of the week in the several Teutonic dialects pp. 112-115.

frigea, an; *m. A lord, master;* dŏmĭnus:—Se ágena frigea *the possessor*, L. Eth. iii. 4; Th. i. 294, 18. DER. ágen-frigea. v. freá.

Frige æfen, es; *m. Thursday evening*, Homl. Th. i. 216, 21.

frigenes, frignes, -ness, -nyss, e; *f.* [frigen *asked, pp.* of fricgan *to ask;* ness, -ness] *An asking, inquiry, a question;* interrŏgātio, quæstio:—þurh his geornfulle frigenesse *repĕtīta interrŏgātiōne*, Bd. 5, 12; S. 631, 4. Wæs Édwine bealdra geworden on ðære frignesse *Edwin was become bolder on that inquiry*, Bd. 2, 12; S. 514, 10. Be monigum frignyssum ða ðe him nýdþearflíce gesewen wæron *de eis quæ necessāriæ vĭdēbantur quæstiōnĭbus*, 1, 27; S. 488, 33. DER. ge-frignys.

frigest, frigst, frihst, he frigeþ, frigþ, frihþ *inquirest, inquires; 2nd and 3rd pers. pres. pres.* of fricgan.

frig-lǽta, an; *m. One made free, a freedman;* libertus, Cot. 120. v. freó-lǽta.

frig-man, -mann, es; *m. A freeman;* hŏmo līber:—Gif frigman freólsdæge wyrce *if a freeman work on a festival-day*, L. C. S. 45; Th. i. 402, 12, note 28: 47; Th. i. 402, 21. Gif frigman frēum stelþ *if a freeman steal from a freeman*, L. Ethb. 9; Th. i. 6, 2. v. freó-man.

FRIGNAN; *part.* frignende, ic frigne, ðú frignest, he frigneþ, *pl.* frignaþ; *p.* ic, he frægn, frægen, frægin, fræng, fregen, fregn, ðú frugne, *pl.* frugnon; *impert.* frign, *pl.* frignaþ; *subj. pres.* frigne, *pl.* frignen; *pp.* frugnen *To ask, inquire;* interrŏgāre, sciscĭtāri:—Ic ðē frignan wille hwæt forlǽtest ðú me *I wish to ask thee why hast thou for-saken me*, Andr. Kmbl. 2824; An. 1414. He hine wæs frignende, for hwon he ðæt Godes eówde forlætan wolde *illum sciscĭtābātur, quāre grĕgem relinquĕret*, Bd. 2, 6; S. 508, 14: 2, 13; S. 515, 41. Ic fregno(a) *interrogabo*, Mt. Lind. 21, 24: Mk. 11, 29. Swá ðu hine wordum frignest *as thou askest him in words*, Elen. Kmbl. 1175; El. 589: Exon. 50 b; Th. 175, 27; Gū. 1201. Gif ðeós cwēn ûsic frigneþ ymb ðæt treó *if this queen asks us about the tree*, Elen. Kmbl. 1065; El. 534. Frægn gif him wære niht getǽse *he asked if he had had an easy night*, Beo. Th. 2643; B. 1319. Eft he frægn hwæt seó þeód nemned wære *rursus interrŏgāvit quod esset vocābŭlum illīus gentis*, Bd. 2, 1; S. 500, 16: 2, 12; S. 513, 37, 38. He frægen and axsode *interrogabat*, Nar. 17, 30. Frægin he hwylcum lande hí brohte wæron *interrŏgāvit de qua terra essent adlāti*, Bd. 2, 1; S. 501, 9: 4, 5; S. 572, 21. Ðá

Z

fræng hine his mæsse-preóst for hwon he weópe *quem dum presbyter suus quare lachrymāretur interrogasset*, Bd. 3, 14; S. 541, 3. Fregn freca óðerne *one warrior asked another*, Andr. Kmbl. 2327; An. 1165. Cýðeras unrehte ða ic nysse frugnon mec *testes iniqui quæ ignorābam interrogābant me*, Ps. Surt. 34, 11: 136, 3. Frign mec *interroga me*, Ps. Surt. 138, 23. Ðeáh hine rinca hwilc æfter frigne *though any man inquire about it*, Bt. Met. Fox 22, 91; Met. 22, 46. Gif he frugnen biþ *if he is asked*, 22, 104; Met. 22, 52: Invent. Crs. Recd. 1083; El. 542. [*Piers P.* fraynen: *Chauc.* freyne: *Laym.* fræine, fræinien: *Orm.* fragnenn: *O. Sax.* fregnan, fragon: *Frs.* freegjen: *O. Frs.* fregia: *Dut.* vragen: *Ger.* fragen: *M. H. Ger.* vragen: *O. H. Ger.* frāgen: *Goth.* fraihnan: *Swed.* frāga: *Icel.* fregna *to hear, ask*: *Lat.* prēc-or *I ask*: *Lith.* praszyti: *Sansk.* prach *to ask.*] DER. ge-frignan. v. frinan.

frignes, -ness, e; *f. Freeness, immunity*; lībertas, immūnītas, Chr. 796; Th. 102, note 1, 2.

frihtan *to fright, terrify*; terrēre, Som. Ben. Lye. v. fyrhtan.

frihtere, es; *m. A soothsayer, diviner*; hariŏlus:—Ða syndon gefeaxene swā frihteras *quasi divine*, Nar. 37, 2. The translator has read divīni for divīne.

frihþ *the soul, spirit, mind.* DER. stíþ-frihþ. v. ferhþ.

frihtrung, e; *f. Divination, sooth-saying*; hariŏlātio, Cot. 21. v. frictrung.

frílíc; *adj. Free, liberal*; līber, libĕrālis:—Frílíc gestreón *libĕrāle fænus*, Prov. 28. v. freólíc.

frí-man, -mann, es; *m. A freeman*; liber hŏmo:—Gif fríman edor-brecþe gedéþ, vi scillingum gebéte *if a freeman commit house-breaking, let him make amends with six shillings*, L. Ethb. 27; Th. i. 8, 15: 29, 31; Th. i. 10, 3, 6: L. Wih. 11; Th. i. 40, 1: L. N. P. L. 56; Th. ii. 298, 24. v. freó-man.

frimdig, frimdi, frymdi, firmdig; *adj. Inquisitive, asking, desirous*; inquisītīvus, desidĕrans, requīrens:—Man him sóna funde, ðæs ðe he frimdig wæs *one soon found for him, what he was desirous*, Ælfc. T. 36, 13. Swā gĕ frimdie wæron *sicut dicītis*, Ex. 12, 31. Hú mǽge gĕ ðæs frimdie beón *how can ye be asking for that?* Ex. 10, 10. Ðæt land ðe ðú me firmdig to wǽre ðæt ic ðé lénde *the land that thou wast desirous I should lease to thee*, Th. Chart. 162, 13.

frinan; *part.* frinnende; ic frine, ðú frinest, he frineþ, *pl.* frinaþ; *p.* ic, he fran, ðú frune, *pl.* frunon, frunnon; *impert.* frin, *pl.* frinaþ; *subj. pres.* frine, *pl.* frinen; *p.* frune, *pl.* frunen; *pp.* frunen *To ask, inquire, consult*; interrŏgāre, sciscĭtāri, consŭlĕre:—Se gesíþ ongan hine frinan, for hwon hine mon gebindan ne mihte *cœmes eum interrŏgāre cœpit quāre ligāri non posset*, Bd. 4, 22; S. 591, 24: Cd. 25; Th. 31, 34; Gen. 495: Beo. Th. 708; B. 351. Me sylfum frinnendum *mihimet sciscĭtanti*, Bd. 4, 19; S. 587, 26. Ne frine ic ðé for tǽle *I ask thee not for blame*, Andr. Kmbl. 1265; An. 633. Ic frine ðé *consŭlo te*, Ælfc. Gl. 86; Som. 74, 15; Wrt. Voc. 49, 38. Hwæt frinest ðú me *what askest thou of me?* Andr. Kmbl. 1257; An. 629. Frineþ he hwær se man síe *he will ask where the man is*, Rood Kmbl. 221; Kr. 112: Salm. Kmbl. 117; Sal. 58. Ða ic nyste hí frunon me *quæ ignorābam interrogābant me*, Ps. Spl. C. 34, 13: Ps. Th. 136, 3. Mid ðý hine frunnon his geferan, for hwon he ðis dyde *cum interrŏgārētur a suis, quāre hoc fācĕret*, Bd. 4, 3; S. 569, 16. Ne frine ðú æfter sǽlum *ask thou not after happiness*, Beo. Th. 2648; B. 1322. Frine me *interroga me*, Ps. Th. 138, 20. Ðæt heó hí frune hwæt hí sóhton *that she asked them what they sought*, Bd. 3, 8; S. 531, 39: Nar. 28, 22. v. be-frinan, ge-. v. frignan.

frínd *friends*, Bt. 20; Fox 72, 20, = frýnd; *pl.* of freónd.

frió; *adj. Free*; līber:—Frióra ǽghwilc fundie to ðæm écum góde *let every one of the free aspire to the eternal good*, Bt. Met. Fox 21, 3; Met. 21, 2. He gesceóp twá gesceádwísan gesceafta frió *he created two rational creatures free*, Bt. 41, 2; Fox 244, 30. v. freó.

frió-dóm, es; *m. Freedom, liberty*; lībertas:—Séce him hræðe fulne frió-dóm *let him quickly seek for himself full freedom*, Bt. Met. Fox 21, 15; Met. 21, 6. v. freó-dóm.

frioh; *adj. Free*; líber:—Beó he frioh *he shall be free*, L. Alf. 11; Th. i. 46, 3: L. In. 3; Th. i. 104, 3. v. freó.

frió-léta *a freedman*, Som. Ben. Lye. v. freó-lǽta.

friólíce *freely*, Bt. 18, 4; Fox 68, 14, note 4. v. freólíce.

friólsend, fríólsiend, es; *m. A deliverer, redeemer*; libĕrātor:—Drihten, fríólsend mín *Dŏmĭnus, libĕrātor meus*, Ps. Spl. T. 17, 1, 49. Fríólsiend mín libĕrātor meus, Ps. Spl. T. 69, 7. v. freóls.

friónd, es; *m. A friend*; amīcus:—Hine his mǽgum gebodie and his frióndum *let notice of him be given to his kinsmen and to his friends*, L. Alf. pol. 42; Th. i. 90, 9. v. freónd.

frioðo; *indecl. f. Peace, pardon*; pax, vĕnia:—He feóll to foldan, frioðo wilnode *he fell to the earth, implored pardon*, Andr. Recd. 1839; An. 920. v. freoðo.

frioðo-wǽr, frioðu-wǽr, e; *f. A covenant of peace*; pācis fœdus:—Hie getrúwedon fæste frioðuwǽre *they confirmed a firm covenant of peace*, Beo. Th. 2196; B. 1096. v. freoðo-wǽr, friðo-wǽr.

frioðo-waru, e; *f. Protection*; tutela:—He frioðo-wǽre bæd hláford

---

sinne *he prayed his lord for protection*, Beo. Th. 4554; B. 2282. [Cf. *O. Sax.* friðu-wara.]

Frisan; *pl. m. Frisians*; Frīsii:—Ðǽr wǽron Frisan mid *there were Frisians with them*, Chr. 885; Th. 154, 24, col. 1. v. Frysa.

frisca, an; *m. A bittern*; bútio, Som. Ben. Lye.

frist-mearc, e; *f.* [frist = first, fyrst *a space of time*] *An interval of time, intermission, respite*; intercăpēdo:—Fristmearc *intercăpēdo*, Glos. Epnl. Recd. 158, 19. v. fyrst-mearc.

friteþ, fritt *eats*, Salm. Kmbl. 596; Sal. 297: Ps. Spl. T. 48, 14; *3rd pers. pres.* of fretan.

FRIÞ, fryþ, es; *m. n. Peace, freedom from molestation, security guaranteed by law to those under special protection*, e. g. that of the Church, v. cyric-friþ. See Stubbs' Const. Hist. i. 180:—*It seems to have been used for the king's peace or protection in general, and to be the right of all within the pale of the law* [cf. *Icel.* fyrirgöra fé ok friði = *to be outlawed*]: *agreement, truce, league*; pax, tūtēla, refúgium:—Ðæt ðú wille niman friþ æt us: we willaþ eów friþes healdan *that thou wilt accept peace from us: we will keep peace with you*, Byrht. Th. 132, 56-65; By. 37-41. Ðis friþ, *n. this protection*, L. Alf. pol. 5; Th. i. 64, 9. Ðis is ðæt friþ, ðæt Ælfrēd cyning [cynincg MS.] and Gúþrum [Gyþrum MS.] cyning gecweden habbaþ *this is the peace, that king Alfred and king Guthrum have agreed upon*, L. A. G; Th. i. 152, 2: L. Ath. v. § 8, 9; Th. i. 238, 24. He nam friþ wið ðæt folc *he made peace with the people*, Ors. 5, 2; Bos. 102, 41. Friþes bót *a compensation or offering of peace, peace-offering, amends for a breach of the peace*, L. Edg. S. 14; Th. i. 278, 2: L. Eth. i. prm; Th. i. 280, 4: L. Eth. v. 26; Th. i. 310, 22: L. C. S. 8; Th. i. 380, 12, 13. Drihten is mín friþ *Dŏmĭnus est refúgium meum*, Ps. Th. 143, 2. Ðonne nam mon friþ and griþ wið hí, and nā-ðe-læs for eallum ðissum griþe and gafole, hí ferdon ǽghweder and heregodon úre earme folc *then they [Saxons] made truce and peace with them [Danes], nevertheless for all this peace and tribute, they went everywhere, and harried our miserable people*, Chr. 1011; Th. 266, 14-18, col. 1. Gif we aslaciaþ ðæs friþes *if we get neglectful of the peace*, L. Ath. v. § 8, 9; Th. i. 238, 21. To þearfe and to friþe *for the need and peace*, L. Edg. S. 2; Th. i. 272, 26. To gebeorge and to friþe eallum leódscipe *for security and peace to all the people*, L. Edg. S. 12; Th. i. 276, 21. Eallum folce to friþe *to the peace for all the people*, L. Edg. S. 15; Th. i. 278, 7. [*Piers P.* fryth *an inclosed wood*: *Laym.* frið *concord, amity*: *Orm.* friþþ *love, concord*: *Plat.* frede, free, *m*: *O. Sax.* friðu, *m*: *Frs.* freede, freed: *O. Frs.* fretho, frede, ferd, *m*: *Dut.* vrede, *m*: *Ger.* friede, *m*: *M. H. Ger.* vride, *m*: *O. H. Ger.* fridu, frido, *m*; frida, *f*: *Dan.* fred, *m. f*: *Swed.* frid, fred, *m*: *Icel.* friðr, *m*.] DER. cyric-friþ, un-, woruld-. For the difference in the meanings of friþ, *m. n*; friðo, friðu, *f*; griþ, *n*; and sib, *f*, v. griþ and sib.

friþ; *adj. Stately, beautiful*; splendĭdus, pulcher:—Seó friþe mæg *the stately woman*, Exon. 103 a; Th. 391, 22; Rä. 10, 9. [*Icel.* friðr *fair, beautiful, handsome.*]

friþ-áþ, es; *m. A peace-oath*; pācis jūrāmentum, Lye.

friþ-béna, an; *m.* [béna *a petitioner*] *A peace-petitioner, refugee*; pācis supplex:—Bútan hit friþbéna sý *unless it be a peace-petitioner*, L. Eth. v. 29; Th. i. 312, 1. Bútan friþbénan sindan *unless they are peace-petitioners*, vi. 36; Th. i. 324, 15.

friþ-borh; *gen.* -borges; *m. A peace or frank-pledge, peace-surety*; pācis fidejussio, L. Ed. C. 20; Th. i. 450, 24, 29; 451, 2, 4, 7: 21; Th. i. 451, 19, 20: 28; Th. i. 454, 18, 22. v. Stubbs' Const. Hist. i. 87.

friþ-bræc, -brec, e; *f. A peace-breaking, breach of the peace*; pācis violātio:—Gyf binnan byrig gedón biþ seó friþbræc *if the breach of the peace be committed within a city*, L. Eth. ii. 6; Th. i. 286, 30. Is ðæt friþbrec *that is a breach of the peace*, ii. 5; Th. i. 286, 26.

friþ-burh; *gen.* -burge; *dat.* -byrig; *f. A town with which one is at peace, one included in the 'friþ' or peace made between two parties*; pācis urbs:—Ðéh hit [the ship] gedriuen beó and hit ætfleó to hwilcre friþbyrig and ða menn útætberstan into ðære byrig ðonne habban ða men friþ *though it be driven and it escape to any town with which 'friþ' has been made, and the men get away into the town, then let the men have protection*, L. Eth. ii. 2; Th. i. 286, 1. v. Schmid, 204, note.

friþ-candel, e; *f. A peace-candle, the sun*; pācis lucerna, sol:—Folca friþcandel furðum eóde *the peace-candle [sun] of nations had just mounted*, Cd. 118; Th. 153, 15; Gen. 2539. DER. candel.

friþ-dóm, es; *m. Liberty, freedom*; libertas, Som. Ben. Lye.

friþe-leás; *adj. Peaceless*; sine pāce:—Hǽðene feóllon friðeleáse *the heathen fell without quarter being given them*, Elen. Kmbl. 253; El. 127. v. friþ-leás.

friþ-geár, es; *n. A year of peace or jubilee*; pācis annus, jūbĭlæus annus, Som. Ben. Lye.

friþ-geard, es; *m. An inclosed space, habitation of peace*; septum, pācis domicīlium:—Gif friþgeard sí on hwæs lande, abúton stán, oððe treów, oððe wille, oððe swilces ǽnige fleard *if there be an inclosed space on any one's land, about a stone, or a tree, or a well, or any trifles of such kind*, L. N. P. L. 54; Th. ii. 298, 16. Friþgeardum in *in the courts of*

peace [in heaven], Exon. 13 b; Th. 25, 12; Cri. 399. v. Th. L. Gl. s. v.

friþ-gedál, es; *n. A life or spirit-separation, death;* a páce divortium, ŏbitus :—He friþgedál fremman sceolde *he should effect separation from life,* Cd. 56; Th. 69, 27; Gen. 1142. v. ferþ *vita?*

friþ-gegilda, friþ-gegylda, an; *m.* [friþ-gild *a peace-guild*] *A member of a peace-guild;* congildo, sŏdális, sŏcius :—Ðis is seó gerǽdnis ðe ða biscopas and ða geréfan ðe to Lundenbyrig hýraþ gecweden habbaþ on úrum friþgegyldum, ǽgðer ge eorlisce ge ceorlisce *this is the ordinance that the bishops and reeves which belong to London have agreed on among the members of our peace-guilds, as well earlish as churlish,* L. Ath. v. prm; Th. i. 228, 6-9. v. ge-gilda.

friþ-georn; *adj. Peace-desirous, peaceable;* pācĭfĭcus :—Sibsume oððe friþgeorne *pacifici,* Mt. Kmbl. Lind. 5, 9.

friþ-gewrit, es; *n. Peace-writing, an article of peace;* pācis scriptum, artĭcŭlus pācis *vel* fœdĕris scripto consignāti :—Béte be ðam ðe ða friþgewritu sǽcgan *let him make amends according as the articles of peace say,* L. Ed. 8; Th. i. 164, 8.

friþ-gild, es; *n. A peace-guild, a society for the maintenance of peace and security;* fœderātōrum sodālicium. This name was given to certain guilds or clubs established during, or before, the reign of king Athelstan, for the repression of theft, the tracing of stolen cattle, and the indemnification of persons robbed, by means of a common fund raised by subscription of the members [gegildan]. The statutes of these guilds are contained in the JUDICIA CIVITATIS LUNDONIÆ, set forth, under royal authority, by the bishop and reeves of the city [v. Th. L. Gl.] :—Gif úre hláford us ǽnigne geþæncean mǽge to úrum friþgildum *if our lord should suggest to us any addition to our peace-guilds,* L. Ath. v. § 8, 9; Th. i. 238, 17. v. friþ-gegilda.

friþ-gisel, es; *m. A peace-pledge, peace-hostage;* obses pācis feriendæ causa dātus :—Ðæt man húru friþgíslas to heom lǽte *that at least peace-hostages be allowed them,* L. O. D. 9; Th. i. 356, 20.

friþ-hús, es; *n. A house of peace, refuge, an asylum;* pācis dŏmus, āsȳlum :—Friþhús *vel* generstede *asȳlum,* Ælfc. Gl. 110; Som. 79, 28; Wrt. Voc. 59, 2. [Cf. *O. Sax.* friðu-wîh.]

FRIÐIAN, freoðian; *p.* ode; *pp.* od; *v. a.* [friþ *peace*]. I. *to keep the peace,* 'friþ,' *towards, make peace, to protect, defend, keep;* pācĭfĭcāre protĕgĕre, tuēri :—Ðæt man eall friðige, ðæt se cyng friðian wille *that one shall keep the peace towards all that the king will,* L. Ath. i. 20; Th. i. 210, 2. Ǽlc ðæra landa, ðe ǽnigne friðige ðæra ðe Ængla land hergie *each of those lands which may keep the peace towards, afford protection to, any of those who ravage England,* L. Eth. ii. 1; Th. i. 284, 17. Man scolde fridian wiþ þonne man *peace should be made with the army,* Chr. 1004: Erl. 138, 22. Ðæt hie eall ðæt friðian woldon ðæt se cyng friðian wolde *that they would protect all that the king would protect,* Chr. 921; Erl. 108, 10, 11. Angunnon hergian ða ðe hý friðian sceoldan *they began to pillage those whom they ought to have protected,* Ors. 4, 1; Bos. 79, 1. Ne fúl náwar friðian ne feormian *that they shall not protect nor harbour a guilty one anywhere,* L. Ed. 7; Th. i. 162, 26. Ðæt hí Godes þeówas friðian and gridian *that they shall protect and defend God's servants,* L. E. B. 1; Th. ii. 240, 6. Hit friðaþ and fyrðraþ *it shelters and furthers,* Bt. 34, 10; Fox 148, 29. Ðæt ic friðian sceal *that I shall protect them,* Exon. 105 a; Th. 398, 14; Rä. 17, 7. Ealle Godes gerihta friðige man georne *one shall diligently keep all God's laws,* L. C. E. 14; Th. i. 368, 9, note 8. [*Piers P.* frythed *wooded; O. Sax.* friðon: *O. Frs.* frethia, frithia, ferdia: *Ger.* frieden tuēri: *M. H. Ger.* vriden: *O. H. Ger.* ga-fridôn pācāre, protĕgĕre: *Goth.* friþôn *to make peace: Dan.* frede: *Swed.* freda *to fence in, protect:- Icel.* friða *to pacify.*] DER. ge-friðian: ferþ-friðende.

friþ-land, es; *n. A land with which one is at peace, with which 'friþ' has been made;* pācis terra :—Hí ðone mǽstan hearm dydon ðe ǽfre here innon friþlande dôn sceolde *they did the greatest harm that ever an army could do in a land with which it was at peace,* Chr. 1097; Erl. 234, 22. [*Icel.* friðland *a friendly country,* v. Cle. and Vig. Dict.]

friþ-leás, friþe-leás, freoðo-leás; *adj. Peaceless, not included in a treaty of peace;* pācis expers :—Gif hwá ðæne friþleásan man healde *if any one keep a peaceless man,* L. C. S. 15; Th. i. 384, 5. [*Icel.* friðlauss, *outlawed.*]

friþlíc; *adj. Peaceable, gentle, mild;* pācĭfĭcus, clēmens, mītis :—Gerǽde man friþlíce steóra *let a man decree mild punishments,* L. Eth. vi. 10; Th. i. 318, 2: L. C. S. 2; Th. i. 376, 19.

friþ-líce; *adv. Peaceably, quietly;* pācĭfĭce, quiēte, Som. Ben. Lye.

friþ-mǽl, -mál, es; *n. An article of peace;* pācis pactio :—Ðis synd ða friþmál and ða fôrword *these are the articles of peace and the agreements,* L. Eth. ii. prm; Th. i. 284, 6.

friþ-man, fryþ-man, -mann, es; *m. One who is under special protection,* 'friþ':—Ǽlc ágenra friþmanna hiþ hæbbe *let each of those who are in our 'friþ' be unmolested,* L. Eth. ii. 3; Th. i. 286, 5, 7, 13.

friðo; *indecl. f. Peace;* pax :—On friðo Drihtnes *in the Lord's peace,* Cd. 57; Th. 70, 11; Gen. 1151. He benam his feónd friðo *he deprived his foe of peace,* Cd. 4; Th. 4, 21; Gen. 57. v. freoðo.

---

friðo-sibb *protecting peace.* v. friðu-sibb.

friðo-spéd, e; *f. Peaceful speed* or *prosperity;* pācis cōpia :—He friðospéde bæd [MS. friþo spebæd] gǽste sínum *he prayed for peaceful prosperity for his soul,* Exon. 114 b; Th. 440, 16; Rä. 60, 3. v. freoðo-spéd.

friðo-tácen, -tácn, es; *n. A peace-sign;* pācis signum :—Abraham sette friðotácn on his selfes sunu *Abraham set a sign of peace on his own son,* Cd. 107; Th. 142, 29; Gen. 2369. [*Icel.* friðar-tákn.]

friðo-wǽr, e; *f. A covenant of peace;* pācis pactum :—Ic manige geseah men ða ðe noldan heora friðowǽre fæste healdan *vidi non servantes pactum,* Ps. Th. 118, 158. v. freoðo-wǽr.

friðo-webba, an; *m. A peace-weaver, an angel;* pācis tector, angĕlus :—He up lôcade swá him se ár abeád, fǽle friðowebba *he looked up as the messenger commanded him, the faithful weaver of peace,* Elen. Kmbl. 175; El. 88. v. Grm. And. u. El. pp. 143-5.

friþ-scipe, es; *m. A state of peace;* pax :—To friþscipe *for peace,* L. R. S. 1; Th. i. 432, 5.

friþ-sócn, e; *f. A peace-refuge, an asylum;* asỹlum :—Ðæt he friþsócne geséce *that he may seek a refuge of peace,* L. C. E. 2; Th. i. 358, 25.

friþ-splot, -splott, es; *m?* [splot *a spot*] *A peace-spot* or *place;* pācis lōcus :—On friþsplottum *in peace-spots,* L. Edg. C. 16; Th. ii. 248, 5.

friþ-stól, fryþ-stól, es; *m. A peace-stool* or *seat, peace-place, asylum, sanctuary, refuge;* pācis sēdes *vel* lōcus, asȳlum, refūgium :—Se here com to his friþstóle [fryþstóle, Th. 256, 18, col. 2; 257, 18, col. 1] *the army came to its secure quarters,* Chr. 1006; Th. 256, 18, col. 1. Gif forworht man friþstól geséce *if a man who has forfeited his life seek a sanctuary,* L. Eth. vii. 16; Th. i. 332, 16. Ðú eart friþstól us, Drihten *Dōmĭne, refūgium factus es nōbis,* Ps. Th. 89, 1: 90, 9. Me is geworden Drihten to friþstóle *factus est mihi Dōmĭnus in refūgium,* 93, 21. [*Icel.* friðstóll.]

friþ-stów, e; *f. A peace-place, refuge, asylum;* pācis lōcus, refūgium, asȳlum :—Ðæt is seó án friþstów *this is the only refuge,* Bt. 34, 8; Fox 144, 29: Bt. Met. Fox 21, 31; Met. 21, 16. Gif he friþstówe geséce *if he seek an asylum,* L. Alf. 13; Th. i. 46, 25. v. Grm. R. A. 886 sqq.

friþ-sum; *adj. Peaceful, peace-making, pacific;* pācĭfĭcus :—Sibsume oððe friþsume *pācĭfĭci,* Mt. Kmbl. Rush. 5, 9. [Cf. *O. Sax.* friðu-samo; *adv. in peace: Icel.* friðsamr: *O. H. Ger.* friðu-samo.] DER. ge-friþsum.

friðu-sibb, e; *f. Protecting peace;* tūtēla pācis, tūtēla pācĭfĭca :—Cwén, friðusibb folca *the queen, the protecting peace of nations,* Beo. Th. 4038; B. 2017.

fritt eats, devours, Ps. Spl. T. 48, 14; *3rd sing. pres.* of fretan.

frocga a frog, Ps. Spl. 77, 50. v. frogga.

frocx? *A nightingale;* luscinia, luscicia? Cot. 121, Lye.

FRÓD; *def.* se fróda, seó, ðæt fróde; *comp. m.* fródra, *f. n.* fródre; *adj.* I. *wise, prudent, sage, skilful;* săpiens, prūdens, sciens, perītus :—Þing sceal gehégan fród wið fródne *the wise shall hold counsel with the wise,* Exon. 89 a; Th. 334, 20; Gn. Ex. 19: Menol. Fox 267; Men. 135: Beo. Th. 3693; B. 1844: Cd. 161; Th. 200, 11; Exod. 355: Elen. Kmbl. 685; El. 343. Se fróda *the sage* [*Isaiah*], Exon. 12 b; Th. 20, 32; Cri. 326. Heó hêht gefetigean fródne on ferhþe *she commanded [them] to fetch the prudent in mind,* Elen. Kmbl. 2325; El. 1164. Gemyne fróde fæder láre *remember [thy] father's wise lore,* Exon. 81 a; Th. 305, 26; Fä. 94. Þurh fród gewit *through wise mind,* Exon. 25 a; Th. 72, 26; Cri. 1178. Fródra and gódra gumena *of wise and good men,* Elen. Kmbl. 1270; El. 637. Fróde men *prudent men,* Salm. Kmbl. 849; Sal. 424. Frige mec fródum wordum *question me in prudent words,* Exon. 88 b; Th. 333, 8; Gn. Ex. 1. Hý beóþ ferþe ðý fródran *they will be the wiser in mind,* 107 a; Th. 408, 32; Rä. 27, 21. II. as wisdom and experience belong to old age, hence,—*Advanced in years, aged, old, ancient;* ætāte provectus, sĕnex, vĕtus, priscus :—Wintrum fród *advanced in years,* Cd. 107; Th. 141, 31; Gen. 2353: Exon. 58 a; Th. 208, 11; Ph. 154: Beo. Th. 5243; B. 2625: Andr. Kmbl. 1012; An. 506: Menol. Fox 133; Men. 66: Byrht. Th. 141, 4; By. 317. Fród cyn *the ancient race,* Cd. 143; Th. 179, 15; Exod. 29. Se fróda Constantínus *the aged Constantine,* Chr. 937; Th. 204, 18; Æðelst. 37: Beo. Th. 5848; B. 2928. Geárum fródne, *acc. advanced in years,* Exon. 126 b; Th. 485, 25; Rä. 72, 3. [*Plat.* frod, vrood: *O. Sax.* fród: *Frs.* froed: *O. Frs.* frod: *Dut.* vroed: *M. H. Ger.* vruot *healthy, brave: O. H. Ger.* fruot, frôt: *Goth.* froþs *prudent: Icel.* fróðr *learned.*] DER. geómor-fród, hige-, in-, un-.

fródian; *p.* ade; *pp.* ad *To be wise* or *prudent;* săpĕre :—[Ic] fródade [*I*] *was wise,* Exon. 94 b; Th. 353, 53; Reim. 32.

frófer *comfort, solace, consolation,* Hy. 9, 15; Hy. Grn. ii. 291, 15. v. frófor.

frófer-bóc, e; *f. A consolation-book;* consōlātiōnis lĭber :—Seó æftre fróferbóc Boētiuses *the second consolation-book of Boëthius,* Bt. 21; Fox 76, 2.

frófer-gást, es; *m. The consolation-ghost, the Holy Ghost;* consōlātiōnis Spīritus, Paraclētus :—Frófergást *paraclētus,* Wrt. Voc. 75, 47. v. frófor-gást.

fróferian, frófrian; *p.* ode; *pp.* od *To comfort;* consōlāri, Grm. Gr.

ii. 137, 11 : Som. Ben. Lye. Ðæt wíf nalde froefra *Rachel noluit con-solari*, Mt. Kmbl. Lind. 2, 18. v. frēfrian, frēfran.

frófernis, se; *f. Consolation*; consōlātio :—Gie babbaþ froefernise, *habetis consolationem*, Lk. Skt. Lind. 6, 24.

FRÓFOR, frófer, frófur; *gen.* frófre; *f.* : v. **II** ; but frófor and frófer are sometimes *m.* **I.** *comfort, solace, consolation, help, benefit, profit, refuge*; sōlāmen, sōlātium, consōlātio, auxĭlium, refŭgium :—Sió frófor *the comfort*, Bt. Met. Fox 21, 32; Met. 21, 16. Wæs frófor cumen *comfort was come*, Cd. 72; Th. 89, 4; Gen. 1475. Frófor eft gelamp sārigmódum *comfort afterwards came to the sad in mind*, Beo. Th. 5875; B. 2941. Sārge gē ne sōhton, ne him swǽslíc word frófre gē sprǽcon *the sorrowful ye sought not, nor a kindly word spoke ye to them*, Exon. 30 a; Th. 92, 21; Cri. 1512. In me frófre gǽst ge-eardode *in me the Spirit of comfort hath dwelt*, 10 b; Th. 13, 24; Cri. 207. Folce to frófre *for comfort to the people*, Beo. Th. 27; B. 14 : Menol. Fox 115; Men. 57. Hý symle frófre ðǽr fundon *they ever found comfort there*, Exon. 45 b; Th. 155, 15; Gū. 860: Andr. Kmbl. 190; An. 95. Him Dryhten forgeaf frófor and fultum *to them the Lord gave comfort and succour*, Beo. Th. 1400; B. 698. Frófra ðíne consōlātiōnes tuæ, Ps. Spl. 93, 19. Ðíne frófre, Ps. Th. 93, 18. Frófra Fæder *the Father of consolations*, Hy. 9, 8; Hy. Grn. ii. 291, 8. Hie fuhton ðē æfter frófre *they fought for help to thee*, Cd. 98; Th. 130, 3; Gen. 2154. Frófor mín *refūgium meum*, Ps. Spl. 17, 1 : 30, 4 : 58, 19. **II.** the following examples are *m.* :—Frófres ic ðē bidde *I ask thee for comfort*, Hy. 6, 1; Hy. Grn. ii. 286, 1. He geandbídode ðone frófer *he awaited the comfort*, Homl. Th..i. 136, 2. Nū behófige gē ðæs ðe swíðor ðæs bóclícan frófres *now need ye so much the more the comfort of books*, ii. 370, 18. Se mann ðe biþ dreórig, he behófaþ sumes frófres *the man who is sad needs some comfort*, ii. 370, 21. [*Laym.* froure, *dat.* frofre, frouere, froure : *Orm.* frofre, *acc* : *O. Sax.* frōbra, frófra, *f* : *O. H. Ger.* fluobara, *f.*] DER. hyge-frófor : frófer-bóc, -gást.

frófor-gást, frófer-gást, es; *m. The Spirit of comfort, the Holy Ghost, Paraclete*; consōlātiōnis Spīritus, Spīritus Sanctus, Paraclētus = Παρά-κλητος :—Se Hālga Gást is gehāten on Grēciscum gereorde Paraclitus, ðæt is, Fróforgāst, forðíðe he frēfraþ ða dreórian *the Holy Ghost is called in the Greek tongue Παράκλητος, that is Spirit of comfort, because he comforts the sad*, Homl. Th. i. 322, 21.

frófre gást, es; *m. The Spirit of consolation, the Holy Ghost, Paraclete*; consōlātiōnis Spīritus, Paraclētus :—Se Hālga Frófre Gást Paraclētus Spīritus Sanctus, Jn. Bos. 14, 26. v. frófor-gást.

frófrung, e; *f. Comfort, consolation*; consōlātio, Som. Ben. Lye. v. frēfrung.

frófur *comfort, consolation* :—Feoh byþ frófur fira gehwylcum *money is a consolation to every man*, Runic pm. 1; Kmbl. 339, 1; Hick. Thes. i. 135, 1 : 4; Kmbl. 340, 8; Hick. Thes. i. 135, 7. v. frófor.

FROGGA, froga, frocga, an ; *m. A* frog; rāna :—Frogga *rāna*, Ælfc. Gl. 24; Som. 60, 16; Wrt. Voc. 24, 20 : 78, 58. He asende on hig froggan [frocgan, Spl.] *misit in eos rānam*, Ps. Lamb. 77, 45. Acende eorþe heora ýcan oðde froggan [frogan, Spl.] *edidit terra eōrum rānas*, 104, 30. He afylde eal heora land mid froggum [MS. froggon] *he filled all their land with frogs*, Homl. Th. ii. 192, 20. [*Wyc.* froggis, *pl* : *Chauc.* frogges, *pl* : *R. Glouc.* frogge : *Plat.* pogge : *Dut.* vorsch, *m* : *Ger.* frosch, *m* : *M. H. Ger.* vrosch, *m* : *O. H. Ger.* frosc, *m* : *Dan.* frö *m. f* : *Swed.* frö, *n* : *Icel.* froskr, *m.*] v. frox.

froht; *adj. Timid*; Mk. Skt. Lind. 4, 40. v. forht.

frohtian; *p.* ade, *pp.* ad *To fear, to be in danger* :—From frohtendum, *a periclitantibus*, Mt. Kmbl. p. 15, 18. Frohtende *timidi*, Lind. 8, 26. Frohtade *timuit*, Rush. 14, 30. v. forhtian.

FROM, freom; *comp.* fromra; *sup.* fromest, frommast; *adj.* **I.** FIRM, *strong, stout, bold, strenuous*; fortis, strēnuus :—Ic eom on móde from *I am firm in mind*, Beo. Th. 5048; B. 2527 : Exon. 46 a ; Th. 156, 13; Gū. 874. Ic eom forþsíþes from *I am strenuous of departure*, 124 b; Th. 479, 21; Rä. 63, 2 : 126 b ; Th. 487, 6; Rä. 72, 24. Hý Gūþlāc in Godes willan fromne fundon *they found Guthlac firm in God's will*, 37 b ; Th. 123, 9; Gū. 320 : Ps. C. 50, 22 ; Ps. Grn. ii. 277, 22. Ðæt wǽron frome folctogan *those were bold leaders*, Andr. Kmbl. 15 ; An. 8 : Elen. Kmbl. 521; El. 261 : Ps. Th. 103, 5 : Bd. 5, 9; S. 622, 25. Wæs Bassa heora lätteów Ēdwines þeng cyninges se fromesta *vēnit illuc dūce Basso, milĭte rēgis Ēduini fortissĭmo*, 2, 20 ; S. 521, 42 : 3, 18; S. 546, 27, col. 2. Hió biþ frommast and swíðost *she is most strenuous and most strong*, Exon. 128 a; Th. 493, 1; Rä. 81, 23. **II.** *rich, abundant, excellent*; über, abundans, præstans :—Swilc biþ mǽgburg menigo ðínre, folcbearnum frome *such shall be the family of thy people, abundant in children*, Cd. 100; Th. 132, 16; Gen. 2194. Fromum feohgiftum *with rich money gifts*, Beo. Th. 41; B. 21. Fromra *præstantior*, Cot. 154. [*Orm.* frame *profit* : *Plat.* fram, fraam *pious* : *O. Sax.* from *virtuous* ; fruma, *f. benefit* : *Frs.* froem *useful* : *O. Frs.* fremo, from *beneficial* ; fruma *benefit* : *Dut.* vroom *virtuous, pious* : *M. H. Ger.* vrum, vrom *useful* : *O. H. Ger.* frum *efficax* ; fruma, *f. benefit* : *Dan. Swed.* from *pious, meek* : *Icel.* frómi *honest, guileless*.] DER. dǽd-from, hild-, orleg-, síþ-, un-.

from; *prep. dat. From*; a, ab :—From eásteweardan *from the eastward*, Bt. 18, 1 ; Fox 60, 31 : 16, 4; Fox 58, 11 : Exon. 25 a; Th. 73, 20 ; Cri. 1192 : Cd. 161; Th. 201, 26 ; Exod. 378 : Beo. Th. 3274; B. 1635. v. fram.

from; *adv. Forth*; fōras :—From ǽrest cwom *first came forth*, Beo. Th. 5106; B. 2556.

Fróm, e; *f.* FROME, *Somersetshire*; oppĭdi nōmen in agro Somer-setensi :—Hēr forþferde Eádrēd cining on Scē Clementes mæssedæg on Fróme *here king Eadred died on St. Clement's mass-day at Frome*, Chr. 955; Erl. 118, 6.

Fróm, e; *f.* FROME; flŭvii nōmen in agro Dorsetensi, Som. Ben. Lye. v. Fróm-múþa.

fromawǽlta; *pp.* -ed *To roll away* :—Stan fromawælted *lapidem revolutum*, Lk. Skt. Lind. 24, 2.

fromcerran; *p.* de, *pp.* ed *To turn from, avert* :—Fromcerr iorre ðín from us *averte iram tuam a nobis*, Rtl. 172, 35 ; 168, 17.

fromcumen; *to be rejected, reprobari*, Lk. Skt. Rush. 9, 22.

from-cyme, es; *m. A coming from, a race, progeny*; prōgēnies :—Fromcyme folde weorþeþ ðíne gefylled *the earth shall be filled with thy race*, Cd. 84; Th. 106, 2 ; Gen. 1765.

from-cyn, -cynn, es; *n.* **I.** *a from-kin, offspring, progeny, posterity*; prōgēnies, prōles :—Gif ðú wille habban holdne freónd ðínum fromcynne *if thou wilt have a faithful friend to thine offspring*, Cd. 106; Th. 139, 23 ; Gen. 2314. Ðæt ðú hyra fromcynn ýcan wolde *that thou wouldest increase their offspring*, Exon. 53 b; Th. 187, 19; Az. 33. Fyllaþ eówre fromcynne foldan sceátas *fill the regions of the earth with your offspring*, Cd. 75; Th. 92, 25 ; Gen. 1534 : 100; Th. 133, 1; Gen. 2204. **II.** *the race from which one springs, ancestry, origin*; gĕnus, ŏrĭgo :—Fród wæs mín fromcynn *my ancestry was ancient*, Exon. 127 b; Th. 490, 16 ; Rä. 80, 1 : Th. 491, 2 ; Rä. 80, 8. Nis ǽnig ðæs horsc, ðe ðín fromcyn mǽge, fira bearnum, sweotule gesēðan *there is not any so wise, who may manifestly declare thine origin to the children of men*, Exon. 11 a ; Th. 15, 26; Cri. 242.

from-doe :—Gisēne wērun swā fromdoe word ðäs *visa sunt sicut delera-mentum verba ista*, Lk. Skt. Rush. 24, 11.

from-faru, e; *f. An excess* :—Fromfarum *excessibus*, Rtl. 17, 15. v. faru.

from-fēran; *p.* de *To go out, from* :—Fromfoerde of ceastre *egre-diebatur de civitate*, Mk. Skt. Lind. 11, 9.

from[-gangan], -geonga, -gonga *To go away*; abire, Jn. Skt. Lind. Rush. 6, 67 ; Mt. Kmbl. Lind. 11, 7.

from-gebúga; *p.* -beáh, bēg *To turn from* :—Fromgebēg *declinavit*, Jn. Skt. Lind. 5, 15.

from-genimma *to take away*; diripere, Mt. Kmbl. Lind. 12, 29.

from-gewítan; *p.* -gewāt, *pl.* -gewiton ; *pp.* -gewiten *To go away, depart from*; discēdere :—Gif hit eallunga fromgewíte *if it should altogether depart*, Bt. 33, 4; Fox 130, 35. Ne syndon me fromge-witene *they have not departed from me*, Cd. 63 ; Th. 76, 11 ; Gen. 1255. v. fram-gewítan.

from-gibēgan; *p.* de *To turn from* :—Fromgibēgde, Jn. Skt. Rush. 5, 13.

from-hweorfan; *p.* -hwearf, *pl.* -hwurfon; *pp.* -hworfen *To turn from, go or depart from*; exīre, discēdere :—Freá hēt hie fromhweorfan neorxna wange *the Lord bade them depart from paradise*, Cd. 45 ; Th. 58, 9; Gen. 943 : 50; Th. 64, 9; Gen. 1047. Ðonne heó hwäm fromhweorfende *when they are departing from any one*, Bt. 7, 2 ; Fox 18, 16. Nǽfre ic fromhweorfe *I will never depart from* [*you*], Exon. 14 b ; Th. 30, 8; Cri. 476.

fromian; *p.* ode, ade; *pp.* od, ad *To profit, avail*; prōdesse, vălēre :—Ðætte seó hālwende onsægednes to ēcre alýsnesse swíþrade and fromade ge líchoman ge sáwle *quia sacrĭfĭcium salūtāre ad redemptiōnem vălēret et anĭmæ et corpŏris sempĭternam*, Bd. 4, 22 ; S. 592, 28. v. fremian.

from-lád, e; *f.* [from, lád *a way*] *A going from, departure, retreat* ; discessus, abĭtus :—Hwelc gromra wearþ feónda fromlád *what the fierce enemies' retreat had been*, Cd. 97; Th. 126, 20 ; Gen. 2098.

fromlíce, framlíce, *adv. Strongly, stoutly, boldly, strenuously, promptly, speedily*; audāciter, strēnue, prŏpēre :—Gāþ fromlíce ðæt gē gúþfreán gylp forbēgan *go boldly that ye may bow the warrior's pride!* Andr. Kmbl. 2666; An. 1334 : 2366; An. 1184: Judth. 10; Thw. 22, 1 ; Jud. 41. Ic sceal fromlíce fēðemundum þurh steápne beorg strǽte wyrcan *I shall strenuously work with my feet a road through a steep mountain*, Exon. 104 b; Th. 397, 9; Rä. 16, 17 : Cd. 95; Th. 123, 23 ; Gen. 2050: Bd. 5, 7 ; S. 620, 41. Fromlícor *more stoutly*, Exon. 111 a; Th. 425, 34 ; Rä. 41, 66. Fromlícast *most promptly*, 66 a ; Th. 245, 5 ; Jul. 40.

from-lócian; *p.* ode; *pp.* od *To look from* or *away, look back*; respicĕre :—Biþ hit swutol ðæt he biþ fromlóciende oferswíðed *it is manifest that he will be overcome on looking back*, Past. 51, 9 ; Hat. MS.

Fróm-múþa, Frómúþa, an; *m. The mouth of the river Frome in Dorsetshire, where the Frome discharges itself into Poole Bay*; Fromi ostium in agro Dorsetensi, ŭbi se in sĭnum illum ad quem *Poole* oppĭdum

assĭdet, Fromus exŏnĕrat :—Hér wende se here eft eástweard into Frŏm-múþan, and up eódon swā wīde swā hī woldon into Dorsǣton *here* [A. D. 998] *the army again went eastward into the mouth of the Frome, and they went up as far as they would into Dorsetshire,* Chr. 998; Erl. 134, 16. Cnut cyng com to Frŏmmúþan, and heregode ðā on Dorsǣtum, and on Wiltūnscīre, and on Sumersǣtum *king Cnut came to the mouth of the Frome, and then ravaged in Dorsetshire, and in Wiltshire, and in Somersetshire,* Chr. 1015; Th. 276, 12. To Frŏmúþan, Th. 277, 13.

fromnis, se ; *f. Strength, excellence :*—Ic geseah mīne gesǣlinesse and þa fromnisse mīnre iuguðe *ego respiciens felicitatem meam insigni numero juventutis,* Nar. 7, 22. v. from.

fromscipe, -scype, es ; *m. Exercise, a proceeding, progress;* exercĭtātio, profectus :—Geunrōtsod ic eom on bigonge oððe fromscipe mīnum contristātus sum in exercītātiōne mea, Ps. Spl. C. 54, 2. Wæs for his fromscype onstyred Ædon Sceotta cyning *mōtus ĕrat ejus profectĭbus Ædan rex Scottōrum,* Bd. 1, 34; S. 499, 28.

from-sīþ, es ; *m. A going from* or *away, departure;* discessus, abĭtus :—Fromsīþ freán *my lord's departure,* Exon. 115 b; Th. 443, 20; Kl. 33.

from-slit[t]nis, se ; *f. Desolation;* desolatio, Mk. Skt. Rush. and Lind. 13, 14.

from-swīcan; *p.* -swāc, *pl.* -swicon ; *pp.* -swicen *To withdraw, desert;* desciscĕre, desĕrĕre :—Ðeáh ðe he him fromswice *though he had withdrawn from them,* Cd. 46; Th. 58, 31; Gen. 954. Ða leóde him fromswicon *the nations deserted him,* Cd. 93; Th. 119, 18; Gen. 1981.

fromung, e ; *f. Profit, advantage, good;* profectus :—Micel fromung *much good,* Bd. 5, 8; S. 621, 30, note. v. freomung, fremung.

from-weard; *adj. From-ward, turned from* or *away, departing, about to depart;* aversus, abĭtūrus, moritūrus :—Ǽlc ðara ðe ðās woruldgesǣlþa hæfþ, he wāt ðæt hī [MS. he] him fromwearde beóþ *every one who possesses these worldly goods, knows that they will be departing from him,* Bt. 11, 2 ; Fox 34, 24. Ádl fǣgum fromweardum feorh óþ-þringeþ *disease will expel life from the fated, about to depart,* Exon. 82 b; Th. 310, 7; Seef. 71. [Laym. from-fram-ward.]

from-weardes; *adv. From-wards, in a direction away from :*—Gif hunta gebīte mannan, sleah þrȳ scearpan neáh fromweardes *if a hunting spider bite a man, strike three scarifications near, in a direction from [the bite],* L. M. 1, 68; Lchdm. ii. 142, 19.

from-wendan; *p. de To avert :*—Fromwoend *averte,* Rtl. 42, 13.

Fronc-land, -lond, es ; *m. Frank-land, the country of the Franks;* Francōrum terra :—On Froncland *into the land of the Franks,* Chr. 920; Erl. 104, 35. On Fronclond, 836; Erl. 64, 32 : 880; Erl. 82, 2. v. Franc-land.

frore, es ; *m. Frost, ice, icicle;* gĕlu, glăcies, stīria, Wald. 81; Vald. 2, 12. v. hilde-frore. [O. Nrs. freri, *pl.* rerar, *m. ice, frozen ground.*]

froren *frozen; pp.* of freósan.

frost, es ; *m. Frost, hoar-frost;* gĕlu, pruīna :—On frost *in pruīna,* Ps. Spl. C. T. 77, 52. v. forst.

frostig; *adj. Frosty;* gĕlidus, Som. Ben. Lye.

frŏuer, e ; *f. Comfort;* consŏlātio :—On ðisum geáre se ārwurþa muneca feder and frŏuer, Landfranc arcebisceop, gewāt of ðissum līfe *in this year* [A. D. 1089] *the venerable father and comfort of monks, archbishop Lanfranc, departed from this life,* Chr. 1089; Erl. 226, 14. v. frōfor.

frox, es ; *m. A frog;* rāna :—To ðē and to ðīnum folce and in to eallum ðīnum þeówum gāþ ða froxas *ad te et ad pŏpŭlum tuum et ad omnes servos tuos intrābunt rānæ,* Ex. 8, 4, 6, 9, 11, 13 : Ors. 1, 7; Bos. 29, 25. Ic sende froxas ofer ealle ðīne landgemǣro *I will send frogs over all thy borders,* Ex. 8, 2, 5, 8. Ðæt flōd awylþ eall froxum *ebulliet flŭvius rānas,* 8, 3, 12. v. frogga.

frugnen *asked,* Bt. Met. Fox 22, 104; Met. 22,´52 ; *pp.* of frignan.

frugnon *interrŏgābant,* Ps. Surt. 34, 11 ; *p. pl.* of frignan.

frum; *comp.* frumra ; *adj. Vigorous, strenuous, prompt, quick, rapid;* strēnuus :—Swift wæs on fōre, fuglum frumra *it was swift in its course, more rapid than birds,* Exon. 113 b; Th. 434, 21; Rä. 52, 4. v. from.

FRUM; *def.* se· fruma ; *adj. Original, primitive, first;* nātīvus, prīmĭtīvus, prīmus :—Frum, in composition, is used with the preceding meanings :—On ðære fruman gecynde *in the original nature,* Bt. 30, 2 ; Fox 110, 14. Ðone fruman sceaft geþencan *to remember the first creation,* Bt. 30, 2 ; Fox 110, 17, 21. Frūmes prīmæ, Rtl. 35, 13. Æt fruman *at first* [cf. æt ǽrestan], H. R. 103, 34. [Laym. frum *first*: Goth. fruma *the first* : Icel. frum- *the first* : Lat. prīmus *the first*.]

FRUMA, an ; *m.* [frum *primitive, first*]. **I.** *a beginning, commencement, origin ;* princĭpium, inĭtium, ŏrīgo, prīmordium, exordium :—Hī sendon ǣrendgewrit, wæs se fruma ðus awriten *mittunt epistŏlam, cūjus hoc princĭpium est,* Bd. 1, 13; S. 481, 41 : 4, 17; S. 585, 17 : Ps. Spl. 118, 160: Cd. 1; Th. 1, 10; Gen. 5 : Exon. 44 b; Th. 151, 15; Gū. 795: Beo. Th. 4608; B. 2309. Ðú eart ealles þinga fruma and ende *thou* [God] *art the beginning and end of all things,* Bt. 33, 4; Fox 132, 36: Bt. Met. Fox 20, 549; Met. 20, 275 : Andr. Recd. 1116; An. 556. On fruman wæs word *in princĭpio ĕrat verbum,* Jn. Bos. 1, 1 :

6, 64: Mt. Bos. 19, 4 : Bd. 1, 1 ; S. 474, 5 : 1, 27; S. 489, 13 : 4, 17: S. 586, 12 : Ps. Spl. C. 73, 2 : 76, 11 : 101, 26 : Boutr. Scrd. 17, 14 : Cd. 174; Th. 218, 7; Dan. 35 : Exon. 69 b; Th. 258, 33 ; Jul. 274; Bt. Met. Fox 17, 25 ; Met. 17, 13. Fram fruman gesceafte *ab inĭtio creātūræ,* Mk. Bos. 10, 6 : Chr. 655; Erl. 28, 2 : Bt. 33, 4 ; Fox 128, 7 : Exon. 25 a ; Th. 73, 20; Cri. 1192: Elen. Kmbl. 2282; El. 1142: Andr. Kmbl. 2969; An. 1487: Ps. Th. 92, 3 : 98, 4. Song he fe fruman moncynnes *cănĕbat de orīgĭne hūmāni gĕnĕris,* Bd. 4, 24; S. 598, 10 : 1, 15; S. 483, 21. Ealle men hæfdon gelīcne fruman *all men had a like beginning,* Bt. 30, 2 ; Fox 110, 8 : Cd. 64; Th. 77, 19; Gen. 1277. Of ðæs strȳnde monigra mǣgþa cyningcynn fruman lǣdde *de cūjus stirpe multārum provinciārum rēgum gĕnus orīgĭnem duxit,* Bd. 1, 15; S. 483, 31. Of ðam ða fruman aweallaþ Deorwentan streámes *de quo Deruentiōnis flŭvii primordia erumpunt,* 4, 29 ; S. 607, 10. Hie sealdon heora wæstma fruman *they should give their first-fruits,* Blickl. Homl. 41, 5. To ðǣm frummum *ad initia* Mt. Kmbl. p. 1, 5. **II.** *an originator, author, founder, inventor ;* auctor, inventor :—God is fruma eallra gesceafta *God is the author of all creatures,* Bt. Met. Fox 29, 161; Met. 29, 81. Sigores fruma *the Lord of triumph,* Exon. 12 a; Th. 19, 2 ; Cri. 294. Fyrnweorca Fruma *the Author of deeds of old,* 16 a ; Th. 36, 20; Cri. 579: Chr. 975; Erl 126, 15; Edg. 41: Elen. Kmbl. 1583; El. 793. Ealre synne fruma *the author of all sin,* Elen. Kmbl. 1540; El. 772: Salm. Kmbl. 887; Sal. 443. Tubal Cain sulhgeweorces fruma wæs *Tubal Cain was inventor of plough-work,* Cd. 52 ; Th. 66, 20; Gen. 1087. Hie leahtra fruman lārum ne hȳrdon *they obeyed not the doctrines of the author of crimes,* Elen. Kmbl. 1674; El. 839. Ðæt ðū onsægde synna fruman *that thou shouldest sacrifice to the author of crimes,* Exon. 71 a ; Th. 264, 10; Jul. 362. Gif hī [MS. he] ne þiówedon hiora fruman *if they served not their author,* Bt. 39, 13 ; Fox 234, 31 : Exon. 8 b; Th. 3, 31; Cri. 44. **III.** *a chief, prince, ruler, king;* prŏcer, princeps, rex :—Burgwarena fruma *chief of citizens,* Exon. 86 a ; Th. 324, 6 ; Wïd. 90. Filistina fruma *prince of the Philistines,* Salm. Kmbl. 555, 561; Sal. 277, 280. Herga fruma *ruler of hosts,* Exon. 20 a ; Th. 53, 4 ; Cri. 845. Ealles folces fruma *prince of all people,* 120 a ; Th. 461, 2 ; Hö. 29. Upengla fruma *prince of archangels,* Andr. Kmbl. 451; An. 226. Se fruma David *the king David,* Ps. C. 50, 20 ; Ps. Grn. ii. 277, 20. Melchisedec com fyrdrinca fruman grētan *Melchizedec came to greet the chief of warriors,* Cd. 97; Th. 127, 1; Gen. 2104: Ps. Th. 112, 7. Hie ahēngon herga Fruman *they hung up the Prince of hosts,* Elen. Kmbl. 419; El. 210. [Laym. frume *beginning* : Goth. frums, *m. beginning*.] DER. dǣd-fruma, eád-, gūþ-, hild-, land-, leód-, leóht-, līf-, ord-, þiód-, tīr-, wīg-.

frum-bearn, es ; *n. A firstborn;* primŏgĕnĭtus :—Frumbearn Godes *the firstborn of God,* Cd. 223; Th. 294, 13; Sat. 470: Exon. 48 a ; Th. 166, 17; Gū. 1044. Frumbearnes riht *the firstborn's right,* Cd. 160; Th. 199, 13; Exod. 338. Ic ðone [ðonne MS.] frumbearn forþasette *ego primŏgĕnĭtum pōnam illum,* Ps. Th. 88, 24.

frum-byrd, e ; *f. Birth, nativity :*—On mīnre frumbyrde dæixge *on the day of my birth,* Th. Chart. 369, 9.

frum-byrdling, es ; *m. Pūbe tĕnus,* Ælfc. Gl. 88; Som. 74, 70; Wrt. Voc. 50, 50. [Frumberdlinges *youths,* O. E. Homl. 2nd series, p. 41.]

frum-cend, e ; *f. Origin :*—Frūmes frūmcende (?) *prīmæ originis,* Rtl. 35, 13.

frum-cenned, -cend ; *def.* se -cenneda ; *part.* **I.** *first-begotten, firstborn;* primŏgĕnĭtus :—Ðæt wæs se frumcenneda *that was the firstborn,* Homl. Th. ii. 194, 9. He ofslōh ǽlc þing frumcendes on lande *percussit omne primogĕnĭtum in terra,* Ps. Lamb. 77, 51 : 104, 36. Ic frumcende gesette hine *ego primŏgĕnĭtum pōnam illum,* 88, 28. He ofslōh ǽlc frumcenned cyld *percussit omne primŏgĕnĭtum,* Ps. Spl. 77, 56. Óþ-ðæt heó cende hyre frumcennedan sunu *dōnec pĕpĕrit fīlium suum primogĕnĭtum,* Mt. Bos. 1, 25 : Lk. Bos. 2, 7. Ðe on ðæm lande frumcennede wǣron *who were firstborn in the land,* Ors. 1, 7; Bos. 30, 5. He ætbræd me mīne frumcennedan *primogĕnĭta mea tŭlit,* Gen. 27, 36. Frumcendo *primĭtiæ,* Rtl. 2, 27. **II.** *in grammar, primitive;* primĭtīvus :—Sume naman sind *primĭtīva,* ðæt sind frumcennede oððe fyrmyste *some nouns are primĭtīva, which are primitive or original,* Ælfc. Gr. 5 ; Som. 4, 7. Hī synd sume *primĭtīva,* ðæt synd frumcennede *some of them* [pronouns] *are primĭtīva, that is primitive,* 15 ; Som. 17, 32, 33. Frumcynned *primĭtīvus,* Hpt. Gl. 448.

frum-cneów, es ; *n. A first generation;* primĭtīva gĕnĕrātio :—Noe hæfde frumcneów gehwæs, fæder and mōder tuddorteóndra *Noah had the first generation of each of* [those] *producing offspring, father and mother,* Cd. 161; Th. 201, 12; Exod. 371. v. cneow II.

frum-cyn, -cynn, es ; *n.* **I.** *original kind, lineage, descent, origin;* prosāpia, ŏrīgo :—Ða ðe mǣgburge mǣst gefrunon, frumcyn feora *those who most understood kinship, the lineage of men,* Cd. 161; Th. 200, 23; Exod. 361. Ic eówer sceal frumcyn witan *I must know your origin,* Beo. Th. 509; B. 252. **II.** *a race, tribe;* gĕnus, gens :—Ðæt he ahredde frumcyn fira *that he saved the race of men,*

Exon. 8 a; Th. 3, 12; Cri. 35: Cd. 190; Th. 236, 6; Dan. 317. He slôh frumcynnes heora freán *he slew the princes of their race*, Ps. Th. 104, 31. He geceás Iudan him geswæs frumcynn *elēgit trĭbum Jūda*, 77, 67.

frum-cyrr, es; m. [cyrr *a turn, space of time*] *A first turn* or *time*; prīmæ vices:—Beó his weres scyldig æt frumcyrre *let him be liable in his fine [for slaying a man] for the first time*, L. Ath. i. 3; Th. i. 200, 21.

frum-gâr, es; m. I: frum-gâra, an; m. II. [frum *prīmus*; gâr *a spear*.]   I. *a chieftain, leader, prince, patriarch*; prīmĭpīlus, prŏcer, dux, princeps, patriarcha:—Geared se frumgâr wæs his freómágum leóf *Jared the patriarch was dear to his kindred*, Cd. 59; Th. 72, 7; Gen. 1183. Ne meahte he on ðam frumgâre feorh gehealdan *he could not keep life in the chieftain*, Beo. Th. 5704; B. 2856: Exon. 75 b; Th. 283, 24; Jul. 685. Gesamnedon herigeas folces frumgâras *the leaders of the people collected their bands*, Andr. Kmbl. 2137; An. 1070: Cd. 176; Th. 222, 7; Dan. 101: Judth. 11; Thw. 24, 18; Jud. 195. Of ðam frumgâran folc unrîm awôcon *from those patriarchs innumerable people sprang*, Cd. 124; Th. 158, 8; Gen. 2614.   II. se frumgâra Malalehel *the patriarch Mahalaleel*, Cd. 58; Th. 71, 11; Gen. 1169. Gif ðû ðam frumgâran brýde wyrnest *if thou deny to the patriarch his wife*, 126; Th. 161, 3; Gen. 2659. Ða frumgâran hâtene wæron Abraham and Aaron *the patriarchs were called Abram and Haran*, 82; Th. 102, 31; Gen. 1708. [Cf. *O. H. Germ.* prcper name Frumigêr.]

frum-gesceap, es; n. [frum *first*; gesceap *creation*] *The first creation*; prīma creātio, principium mundi:—Ðær biþ ôþýwed egsa mâra ðonne from frumgesceape gefrægen wurde *there shall be shown greater terror than had been heard of from the first-creation*, Exon. 20 a; Th. 52, 27; Cri. 840.

frum-gifu, e; f. *An original gift, privilege, prerogative*; prīmāria grātia, prærogātīva:—Frumgifu *vel* synder-wurþmynt *prærogātīva*, Ælfc. Gl. 99; Som. 76, 119; Wrt. Voc. 54, 61. Hpt. Gl. 457. [*Icel.* frum-gjöf *first gift*].

frum-gild, -gyld, es; n. *A first payment* or *compensation,—the first payment* or *instalment of the price* [wer] *at which every man was valued, according to his degree, to be paid to the kindred*, or *guild-brethren, of a slain person, as compensation for his murder*; prīma compensātio:—Gylde man ðæs weres ðæt frumgyld *let the first payment of the valuation be paid*, L. E. G. 12; Th. i. 174, 28; L. Edm. S. 7; Th. i. 250, 21.

frum-grîpa, an; m. *A first grasper, occupier*; prīmus captor, occŭpātor, Wulfst. par 4: Mann. Lye.

frum-heowung, e; f. *First formation* or *creation*; prīma formātio, Cot. 154.

frum-hrægl, es; n. *A first garment*; prīmus vestītus:—Hêt heora sceome þeccan Freá frumhrægle *the Lord bade them conceal their nakedness with the first garment*, Cd. 45; Th. 58, 8; Gen. 943.

frum-leóht, es; n. *First light, dawn*; prīma lux, aurōra:—To ðé ic wacige of frumleóhte *ad te de lūce vĭgĭlo*, Wanl. Catal. 47, 41.

frum-líc; adj. *Original*, Hpt. Gl. 433.   v. frymlíc.

frum-lyhtan; p. -lýhte *to dawn*:—Siððan hit frumlýhte *after it had dawned*, Blickl. Homl. 207, 35.

frum-meolc, -meoluc, e; f. *The first milk, nectar*; prīmum lac, nectar, Som. Ben. Lye.

frum-ræd, es; m. *The first* or *primary ordinance*; prīmum consĭlium:—Ðæra biscopa frumræd *the primary ordinance of bishops*, L. Eth. vi. 1; Th. i. 314, 4.

frum-ræden, e; f. *An original, previous ordinance, condition*:—Ða wæs first agân frumrædenne *then was expired the space of time previously fixed*, Andr. Kmbl. 294; An. 147.

frum-rîpa, an; m. [rîpa *a handful of corn*] *First-fruits*; prīmĭtiæ:—Ðîne teóðan sceattas, and ðîne frumrîpan gongendes and weaxendes, agyf ðû Gode *thy tithes, and thy first-fruits of moving and growing things, render thou to God*, L. Alf. 38; Th. i. 52, 31.

frum-sceaft, e; f.   I. *the first creation, the creation, beginning, origin, original state* or *condition*; prīma creātio, ŏrīgo, prīmĭtiva *vel* pristĭna condĭtio:—Sing me frumsceaft *canta princĭpium creatūrārum*, Bd. 4, 24; S. 597, 16. Moyses awrât ærest be frumsceafte *Moses wrote first of the creation*, Homl. Th. ii. 198, 15. Frumsceaft genesis, Jn. Skt. p. 1, 12. Gê mágon hwæt-hwego ongitan be eówrum frumsceafte, ðæt is God *ye can in some measure understand concerning your origin, that is God*, Bt. 26, 1; Fox 90, 4. Æt frumsceafte *at the beginning*, Exon. 99 a; Th. 371, 21; Seel. 79: Beo. Th. 89; B. 45: Andr. Kmbl. 1593; An. 798. He cûðe frumsceaft fira feorran reccan *he could relate the origin of men from [times] remote*, Beo. Th. 182; B. 91. Fýr clymmaþ on gecyndo, cunnaþ hwænne môte on his frumsceaft, eft to his êþle *fire climbeth in its nature, strives when it can towards its origin, back to its home*, Salm. Kmbl. 831; Sal. 415. He forlæt ærest lîfes frumsceaft *he first forsakes his original state of life*, Bt. Met. Fox 17, 48; Met. 17, 24.   II. *a created being, creature*; creātūra:—Hî hēredon lîfes Âgend, Fæder frumsceafta *they praised the Lord of life, the Father of all created beings*, Exon. 14 b; Th. 29, 33; Cri. 472: 84 a; Th. 317, 15; Môd. 66: Cd. 156; Th. 195, 9; Exod. 274.

frum-sceapen; part. *First formed* or *created*; prīmus formātus *vel* creātus:—Ðá ðá he geworhte Adam, ðone frumsceapenan mann *when he wrought Adam, the first created man*, Hexam. 14; Norm. 22, 14.

frum-sceat, -sceatt, es; m. [sceat *money, gain*] *First-fruits*; prīmĭtiæ:—He ofslôh frumsceateas ealles geswinces heora on geteldum Chames *percussit prīmĭtias omnis lăbōris eōrum in tăbernăcŭlis Cham*, Ps. Spl. 77, 56. He slôh frumsceattas oððe frumwæstmas ealles geswinces heora *percussit prīmĭtias omnis lăbōris eōrum*, Ps. Lamb. 104, 36.

frum-scepend-sceppend, es; m. *An author, originator, creator*:—Frumscepend *auctor*, Rtl. 16, 19; 123, 10.

frum-scyld, e; f. *Original sin*; princĭpālis *vel* căpĭtālis culpa:—Frumscylda gehwæs fæder and môdor *father and mother of every original sin*, Salm. Kmbl. 891; Sal. 445.

frum-setnes, se; f. *Authority*; auctoritas, Rtl. 123, 15.

frum-setnung, e; f. *Original formation*:—Middengeordes frumsetnung *constitutio mundi*, Jn. Skt. Rush. 17, 24.

frum-slæp, e; f. *First sleep*; prīmus somnus:—On frumslæpe *in the first sleep*, Ors. 2, 8; Bos. 51, 9: Cd. 177; Th. 222, 22; Dan. 108.

frum-spræc, e; f. *An original speech, a promise, covenant*; prædictum, promissum:—Fyl nû frumspræce *fulfil now thy promise*, Cd. 190; Th. 236, 24; Dan. 326: Exon. 53 b; Th. 188, 7; Az. 42.

frum-staðol, es; m. *An original station*; prīmĭtīva sēdes:—Ic mînum gewunade frumstaðole fæst *I dwelt fast in my original station*, Exon. 122 b; Th. 471, 18; Rä. 61, 3.

frum-stemn, es; m. *The fore-part of a ship, prow*; prōra, Glos. Brux. Recd. 37, 41; Wrt. Voc. 63, 55.

frum-stôl, es; m. *An original seat, mansion-house, a proper residence* or *station*; sēdes principālis:—Se frumstôl, ðe hie of adrifen wurdon *the original seat [paradise] from which they were driven*, Cd. 46; Th. 59, 14; Gen. 963. Habbaþ ða feówer frumstôl hiora, æghwilc hiora âgenne stede *the four [elements] have their proper station, each of them its own place*, Bt. Met. Fox 20, 126; Met. 20, 63. Ðæs fýres frumstôl *the fire's proper station*, 20, 250; Met. 20, 125. Healden ða mægas ðone frumstôl *let the kindred hold the paternal mansion*, L. In. 38; Th. i. 126, 6. [v. note in Schmid.]   In ðam frumstôle, ðe him Freá sette *in the first seat, which the Lord placed for them*, Exon. 93 a; Th. 349, 24; Sch. 51.

frum-talu, e; f. [talu *a tale, story*] *First words of witnesses, first accusation*; prīma testium dicta, prīma delāta:—We willaþ ðæt frumtalu fæste stande *we will that first words of witnesses stand fast*, L. N. P. L. 67; Th. ii. 302, 6.

frumþ, es; m: e; f. *A beginning*; princĭpium:—Ic frumþa God fóresceáwode *I saw the eternal God [lit. God of beginnings] face to face*, Elen. Kmbl. 689; El. 345.   v. frymþ.

frum-tíhtle, -týhtle, an; f. [frum *original, primitive, first*; tíhtle *an accusation, charge*] *A first accusation, first charge*; prīma accūsātio, prīma calumnia:—Ðæt he borh næbbe æt frumtýhtlan *that he have no surety at the first accusation*, L. C. S. 35; Th. i. 396, 24.

frum-wæstm, es; m. [wæstm *fruit*] *First-fruits*; prīmĭtiæ:—Frumwæstmas prīmĭtiæ, Ælfc. Gr. 13; Som. 16, 17. Heora frumwæstme fulle syndon *promptuāria eōrum plēna*, Ps. Th. 143, 16. He ofslôh frumwæstmas [-wæstme, Th.] ealles geswinca heora *percussit prīmĭtias omnis lăbōris eōrum*, Ps. Lamb. 77, 51: 104, 36.

frum-weorc, es; n. *An ancient work, the work of the creation*; ŏpus priscum, res in princĭpio creāta:—Woldon hie ædre gecýðan frumweorca fæder *they would at once proclaim the father of creation's works*, Andr. Kmbl. 1607; An. 805.

frum-wyrhta, an; m. *An author, creator*; auctor:—Léhtes frumwyrhte *lucis auctor*, Rtl. 37, 7.

frum-yldo, e; f. *The first age*; prīma ætas:—Frumyldo *prīmævus?* Cot. 3: Som. Ben. Lye.

frune *asked*, Bd. 3, 8; S. 531, 39; *p. s. subj. of* frinan: frunon, frunnon *asked*, Ps. Th. 136, 3: Bd. 4, 3; S. 569, 16; *p. pl. of* frinan.

frý; adj. *Free*; liber:—Betwyx deádum frý *inter mortuos liber*, Ps. Lamb. 87, 6. Gif hwylc swîðe rîce cyning næfde nænne frýne mon on eallon his rîce *if some very powerful king had not any free man in all his realm*, Bt. 41, 2; Fox 244, 25.   v. freó.

fryccea, an; m. *A crier, preacher, herald*; præco:—Se dumba fryccea *the dumb herald*, Past. 15, 3; Hat. MS. 19 a, 28.   v. fricca.

frý-dôm, es; m. *Freedom, liberty*; lībertas:—Se frýdôm *the freedom*, Bt. 41, 2; Fox 246, 4. Ða men habbaþ simle frýdôm *men have always freedom*, 40, 7; Fox 242, 25, 27, 28: 41, 2; Fox 244, 16, 21. v. freó-dôm.

frý-gyld, es; n. *A free guild* or *society*; līberum sodalĭtium *vel* collēgium, Som. Ben. Lye. v. friþ-gild.

fryhtendo; pres. part. *Trementes*, Rtl. 122, 16.   v. fyrhtian.

fryhtu, e; f. *Fright, terror*, Rtl. 59, 19.   v. fyrhtu.

frymdi; adj. *Inquisitive, asking, desirous, suppliant*; inquīsītīvus, requīrens, desīderans, supplex:—Ic eom frymdi to ðé *I am suppliant to thee*, Byrht. Th. 137, 1; By. 179   v. frimdig.

frymetling, e; f. [frum *original, first, primitive*] *A youngling, young*

cow; júvenca :—Cúhyrde gebýreþ ðæt he hæbbe ealdre cú meolc, vii niht syððan heó uige cealfod hæfþ, and frymetlinge býstinge xiv niht *it belongs to a cowherd that he have the milk of an old cow, seven nights after she has newly calved, and the biestings of a young cow fourteen nights*, L. R. S. 13; Th. i. 438, 19.

frymlíc; *adj.* [frym = frum *first*] *Primitive, first*; prīmĭtīvus :—Ongunnon hí ðæt apostolíce líf ðære frymlícan cyricean onhýrigean *cœpērunt apostŏlĭcam prīmĭtīvæ ecclēsiæ vītam imĭtāri*, Bd. 1, 26; Whelc. 78, 22. v. frymþelíc.

frymþ, e; *f. A harbouring, an entertainment*; susceptio, receptio :—Ælc mon mót onsacan frymþe *líf ðære frymlícan* every man may deny entertainment, L. In. 46; Th. i. 132, 1. v. fyrmþ.

frymþ, frumþ, es; *m:* e; *f.* [frum *original, first*] *A beginning, foundation, origin, first-fruits*; inĭtĭum, princĭpĭum, constĭtūtĭo, ŏrīgo, prīmĭtĭæ :—Næs his frymþ æfre *his origin never was*, Exon. 65 a; Th. 240, 12; Ph. 637. Ic sprece fóresetnyssa fram frymþe *lŏquar propŏsĭtĭōnes ab inĭtĭo*, Ps. Lamb. 77, 2: Ps. Spl. 101, 26: Mt. Bos. 19, 8: Lk. Bos. 1, 2. Sceal seó wyrd swā ðeáh forþsteallian, swā ic ðé æt frymþe gehét *that event shall yet come to pass, as I promised thee at the beginning*, Cd. 109; Th. 144, 16; Gen. 2390: 6; Th. 8, 30; Gen. 132: 174; Th. 218, 6; Dan. 35: Bt. Met. Fox 11, 75; Met. 11, 38: 13, 25; Met. 13, 13: Ps. Th. 70, 4: 104, 24. Of middangeardes frymþe *a constĭtūtĭone mundi*, Mt. Bos. 25, 34: Bd. de nat. rerum; Wrt. popl. science 13, 29; Lchdm. iii. 264, 10. Heó of ðære ylcan mægþe Eást-Engla líchoman frymþe lædde *de provincia eōrumdem Orientālium Anglōrum ipsa carnis orĭgĭnem duxĕrat*, Bd. 4, 19; S. 590, 8. Frymþas *prīmĭtĭæ*, Scint. Lye. Gefreoða úsic, frymþa Scyppend *protect us, Creator of beginnings!* Exon. 65 a; Th. 239, 32; Ph. 630: 44 b; Th. 151, 9; Gú. 792: Elen. Kmbl. 1002; El. 502. [*Orm.* frummþe.]

frymþelíc; *adj.* [frymþ *a beginning*] *Primitive, first*; prīmĭtīvus :—Ongunnon hí ðæt apostolíce líf ðære frymþelícan cyricean onhýrigean *cœpērunt apostŏlĭcam prīmĭtīvæ ecclēsiæ vītam imĭtāri*, Bd. 1, 26; S. 487, 32: 4, 23; S. 593, 41. On frymþelícum synne *originali peccato*, Rtl. 101, 20. Of ðam frymþlícan *from the original*, Blickl. Homl. 107, 5.

frymþ-yldo, e; *f. An early, original age*, Hpt. Gl. 462. Cf. frumyldo.

frýnd *friends*, Jn. Bos. 15, 14: Lk. Bos. 7, 6; *pl. nom. acc. of* freónd.

Frysa, Friesa, an; *pl. nom. acc.* Frysan, Frisan, Fresan; *gen.* Frysena, Frysna; *dat.* Frysum; *m. A Frisian;* Frīsius, Freso :—Se Frysa hine gewráþ *the Frisian bound him*, Homl. Th. ii. 358, 19, 22: Chr. 897; Th. 176, 32, 33, col. 2: 177, 32, 33. Sealde se ealdorman hine sumum Frysan of Lundene *the alderman sold him to a Frisian of London*, Homl. Th. ii. 358, 18. Be norþan-westan him syndon Frysan *to the north-west of them are the Frisians*, Ors. 1, 1; Bos. 18, 35: Bd. 5, 9; S. 622, 15: Chr. 886; Th. 154, 24, col. 2, 3; 155, 23, col. 1. He com on Frysena land *he came to the land of the Frisians*, Bd. 5, 9; S. 623, 27: 5, 10; S. 623, 35: 5, 11; S. 626, 18, 21: 5, 19; S. 639, 20. Gyf Frysna hwylc ðæs morðorhetes myndgiend wære *if any of the Frisians should be a rememberer of this deadly feud*, Beo. Th. 2212; B. 1104. Ic wæs mid Frysum *I was with the Frisians*, Exon. 85 b; Th. 322, 24; Wíd. 68: Beo. Th. 2418; B. 1207: 5816; B. 2912: Bd. 5, 11; S. 625, 42. He ge-eóde ða fyrran Frysan *he had overcome the farther Frisians*, Bd. 5, 10; S. 624, 3.

Frys-cyning *a Frisian king.* v. Fres-cyning.

Frysisc, Fresisc; *adj. Of or belonging to Friesland. Frisian;* Frīsĭcus :—Næron hie nāðor ne on Frysisc gesceapen ne on Denisc *they were shapen neither as the Frisian nor as the Danish*, Chr. 897; Th. 176, 2, col. 2; 177, 2. Ðær wearþ ofslegen Lucuman, and ealra manna, Frysiscra and Engliscra, lxii *there was slain Lucuman, and of all the men, Frisian and English, sixty-two*, 897; Th. 176, 34, col. 2; 177, 34.

Frys-land, Fres-lond, es; *n. Friesland;* Frīsia :—Be westan Eald-Seaxum is Ælfe múþa ðære eá and Frysland *to the west of the Old Saxons is the mouth of the river Elbe and Friesland*, Ors. 1, 1; Bos. 18, 36. Gewiton him wígend Frysland geseón *the warriors departed to see Friesland*, Beo. Th. 2277; B. 1126.

frýst, frýsþ *freezes*, Ælfc. Gr. 22; Som. 24, 8; *3rd sing. pres. of* freósan.

fryt *eats up, devours, consumes*, Deut. 32, 22; *3rd sing. pres. of* fretan.

fryþ, es; *n. m? Peace*; pax :—Seó láf [MS. lafe] wil ðone here fryþ nam *the remainder made peace with the army*, Chr. 867; Erl. 73, 16: 1036; Th. 294, 9, col. 2. Ðæt he ne beó nānes fryþes weorðe *that he be not worthy of any peace*, L. Eth. iii. 15; Th. i. 298, 12. v. friþ.

fryþ-gegylda *a member of a peace-guild*, L. Ath. v. prm; Wilk. 65, 5. v. friþ-gegilda.

fryþing *a furthering, furtherance*, L. E. I. 21; Th. ii. 414, 23, = fyrþring. v. fyrþrung.

fryþ-man, -mann. v. friþ-man.

fryðo; *indecl. f. Peace*; pax :—Brúcaþ mid gefeán fryðo *enjoy peace with delight*, Cd. 74; Th. 91, 16; Gen. 1513. v. freoðo.

fryþ-stól *an asylum, refuge*, Chr. 1006; Th. 256, 18, col. 2; 257, 18, col. 1. v. friþ-stól.

FUGEL, fugol, fugul; *gen.* fugeles, fugles; *m. A bird*, FOWL; ăvis, āles :—Ðes fugel *hæc* ăvis, Ælfc. Gr. 9, 28; Som. 11, 54: Lk. Bos. 13, 34: Cd. 72; Th. 88, 5; Gen. 1460: Exon. 17 a; Th. 40, 27; Cri. 645: Salm. Kmbl. 507; Sal. 254: Judth. 11; Thw. 24, 25; Jud. 207. Fugel *āles*, Ælfc. Gr. 10; Som. 14, 59. Ne wirce ge nāne andlícnissa ne nānes nýtenes ne fugeles *make no images of any beast or bird*, Deut. 4, 17. Wæs ðæs fugles flyht dyrne and dēgol *the bird's flight was hidden and secret*, Exon. 17 a; Th. 40, 15; Cri. 639: 57 b; Th. 206, 12; Ph. 125: Salm. Kmbl. 451; Sal. 226. Ic spearuwan swā some gelíce gewearþ, ānlícum fugele *factus sum sĭcut passer ūnĭçus*, Ps. Th. 101, 5: Exon. 108 a; Th. 413, 18; Rä. 32, 7. Fugle gelícost *most like to a bird*, Beo. Th. 442; B. 218. Ðone fugel hātaþ Filistina fruman uasa mortis *the princes of the Philistines call the bird* vāsa mortis, Salm. Kmbl. 559; Sal. 279: Exon. 17 a; Th. 40, 10; Cri. 636. Fugelas æton of ðam *ăves comēdērunt ex eo*, Gen. 40, 17, 19: Ps. Spl. 103, 13: Mk. Bos. 4, 4, 32: Lk. Bos. 9, 58: Exon. 61 a; Th. 222, 22; Ph. 352: Fins. Th. 9; Fin. 5: Ps. Th. 77, 27. Heofenan fuglas habbaþ nest *volucres cœli nīdos hăbent*, Mt. Bos. 8, 20: 13, 4: Cd. 200; Th. 248, 2; Dan. 507: Exon. 55 a; Th. 194, 16; Az. 140: Ps. Th. 104, 35. Ðæt hí gehíran óðerra fugela stemne *that they hear the sounds of other birds*, Bt. 25; Fox 88, 21: Gen. 7, 21. Hér wæs ðæt micle fugla wæl *in this year* [A.D. 671] *was the great destruction of birds*, Chr. 671; Erl. 34, 33. Ofer fugla cynn *over the race of birds*, Exon. 58 a; Th. 208, 22; Ph. 159: 60 b; Th. 221, 6, 16; Ph. 330, 335. Gif seó offrung biþ of fugelum *si de ăvĭbus oblātĭo fuĕrit*, Lev. 1, 14: Deut. 28, 26: Ps. Lamb. 78, 2. Hę spyraþ æfter fuglum *he seeks after birds*, Bt. 39, 1; Fox 210, 29: Exon. 126 b; Th. 487, 16; Rä. 73, 3: Judth. 12; Thw. 25, 37; Jud. 297: Ps. Th. 78, 2. Ða fugelas hne ne todælde *ăves non divīsit*, Gen. 15, 10: Ps. Spl. 8, 8: 49, 12: 77, 31: Bt. Met. Fox 13, 95; Met. 13, 48. Behealdaþ heofonan fuglas *respĭcĭte volātĭlia cœli*, Mt. Bos. 6, 26: Cd. 65; Th. 78, 26; Gen. 1299. [*Piers P.* fowel: *Chauc.* foule: *Wyc.* foulis *fowls*: *Laym.* foȝel, fuȝel, fowel: *Plat.* vagel, *m:* *O.Sax.* fugal: *Frs.* fugil, foeggel: *O.Frs.* fugel: *Dut.* Ger. M.H.Ger. vogel, *m:* *O.H.Ger.* fogal, fugal, *m:* *Goth.* fugls, *m:* *Dan.* fugl, *m.f:* *Swed.* fågel, *m:* *Icel.* fugl, fogl, *m*.] DER. brim-fugel, carl-, cwēn-, dop-, fen-, gūþ-, hen-, heofon-, here-, nē-, treó-, wudu-.

fugel-bana, -bona, an; *m. A bird-killer, fowler*; auceps :—Sum biþ fugelbona, hafeces cræftig *one is a fowler, skilful with the hawk*, Exon. 79 b; Th. 298, 5; Crä. 80.

fugel-cyn, fugol-cyn, -cynn, -cinn, es; *n.* FOWL-KIND; vŏlucrium gēnus :—Eallum nýtenum and eallum fugelcynne *cunctis anĭmantĭbus terræ omnique vŏlucri cœli*, Gen. 1, 30: 7, 8. Nim of fugelcinne seofen and seofen ægðres gecyndes *tolle de volātĭlĭbus septēna et septēna, mascŭlum et fēmĭnam*, Gen. 7, 3.

fugel-doppe, es; *m? A dipping-fowl, water-fowl*; mergŭlus, Ælfc. Gl. 36; Som. 62, 118; Wrt. Voc. 29, 14. v. dop-fugel.

fugeleran = fugelerum *with fowlers*, Ors. 1, 1; Bos. 20, 5; *dat. pl. of* fugelere.

fugelere, fuglere, es; *m. A* FOWLER; auceps :—Fugelere *auceps*, Wrt. Voc. 73, 45: Coll. Monast. Th. 25, 9. Ðær gewícodon fisceras oððe fugeleras *where fishers or fowlers encamped*, Ors. 1, 1; Bos. 20, 9. Fugelerum [MS. fugeleran] *with fowlers*, 1, 1; Bos. 20, 5.

fugeles leác, es; *n. Viumum?* Glos. Brux. Recd. 42, 30; Wrt. Voc. 68, 45.

fugeles wíse, fugeles wýse, an; *f. The plant larkspur*; delphīnium = δελφίνιον :—Fugeles wíse *delphin*, Cot. 211, Som. Ben. Lye. Fugeles wýse *delphinion*, Glos. Brux. Recd. 41, 69; Wrt. Voc. 68, 4.

fugel-hælsere, es; *m.* [hælsere *a diviner*] *A diviner by birds, soothsayer;* augur, Som. Ben. Lye. v. fugel-weohlere.

fugel-hwata, an; *m. A diviner by birds;* augur :—Fugelhwata *caragius*, Ælfc. Gl. 48; Som. 65, 69; Wrt. Voc. 34, 4. v. Du Cange *sub vōce* Caragus.

fugelian, fuglian; *p.* ode; *pp.* od *To fowl, catch birds;* aucŭpāri :—Ic fugelige *aucŭpor*, Ælfc. Gr. 25; Som. 27, 12, MS. D.

fugel-lím, es; *m. Bird-lime*; viscum, Cot. 194.

fugel-net, -nett, es; *n. A bird-net*; aucŭpātōrium rēte :—Fugelnet [MS. fugelint] *pendera* [= *panthēra* = πανθήρα], Wrt. Voc. 288, 77.

fugel-noþ, es; *m? Bird-catching, fowling*; aucŭpium :—On fugelnoþum *in fowlings*, Cod. Dipl. 715; A.D. 1006; Kmbl. iii. 350, 9.

fugeloþ *bird-catching, fowling.* v. fugoloþ.

fugel-timber, es; *n.* [timber *a frame, structure*] *A young bird;* avĭcŭla. pullus :—Biþ fæger fugeltimber *it is a fair young bird*, Exon. 59 a; Th. 214, 8; Ph. 236.

fugel-tras? *pl. m. Poles* or *forks for spreading nets;* āmĭtes, Cot. 13.

fugel-weohlere, es; *m.* [fugel *a bird*, weohlere = wiglere, wigelere *a soothsayer*] *A diviner by birds;* augur, auspex, Ælfc. Gl. 4; Som. 56, 4; Wrt. Voc. 17, 13.

fugel-wylle *abounding in birds.* v. fugol-wylle.

fuglere, es; *m. A fowler*; auceps, Wrt. Voc. 285, 15. v. fugelere.

fugles *of a bird* or *fowl*, Exon. 17 a; Th. 41, 11; Cri. 654; *gen. of* fugel.

fugles beán, e; *f. Vetch*, Gl. Mett. 919.

fuglian *to fowl*; aucŭpári:—Ic fuglige *aucŭpor*, Ælfc. Gr. 25; Som. 27, 12. v. fugelian.

fuglung, e; *f. Fowling, bird-catching*; aucŭpium, Wrt. Voc. 285, 19.

fugol, es; *m. A bird, fowl*; ăvis:—Friteþ wildne fugol *it eateth the wild bird*, Salm. Kmbl. 597; Sal. 298. Fugole gelícost *most like to a bird*, Andr. Kmbl. 994; An. 497. God gelædde ðære lyfte fugolas to Adame *Deus volătília cæli adduxit ad Adam*, Gen. 2, 19: Cd. 200; Th. 248, 14; Dan. 513. v. fugel.

fugol-cyn, -cynn, -cinn, es; *n. Fowl-kind*; vŏlucrium gĕnus:—Micel fugolcyn *much fowl-kind*, Ex. 16, 13. Fisccinn and fugolcinn *fish and fowl*, Ælfc. T. 8, 26. v. fugel-cyn.

fugoloþ, es; *m? Bird-catching, fowling*; aucŭpium:—Búton huntoþe and fugoloþe *besides hunting and fowling*, Homl. Th. ii. 576, 34. v. fugelnoþ.

fugol-wylle; *adj. Bird-springing, producing birds, abounding in birds*; ăvibus ăbundans:—Hit is fiscwylle and fugolwylle *it is abounding in fish and fowl*, Bd. 1, 1; S. 474, 41.

fugul, es; *m. A bird, fowl*; ăvis, vŏlucris:—Ne wæs ðæt ná fugul ána *it was not a bird only*, Exon. 109 b; Th. 418, 23; Rä. 37, 9. Heofones fugulas hit fræton *vŏlucres cæli comēdērunt illud*, Lk. Bos. 8, 5. Fugulum volătĭlibus, Ps. Spl. 78, 2. v. fugel.

fuhlas *birds, fowls*, Mt. Bos. 13, 32, = fuglas; *pl. nom. acc.* of fugel.

FŪHT; *adj. Moist, damp*; hŭmĭdus:—Deós wyrt biþ cenned on fúhtum and on wæteregum stówum *this herb is produced in damp and watery places*, Herb. 9, 1; Lchdm. i. 98, 25: 39, 1; Lchdm. i. 140, 5: 52, 1; Lchdm. i. 154, 26. [*Plat.* fucht: *Dut.* vocht, *n. moisture*; vochtig *damp, humid*: *Ger.* feucht: *M.H.Ger.* viuhte: *O.H.Ger.* fiuhti: *Dan.* fugtig: *Swed.* fukt, *m. moisture*; fuktig *moist*.]

fúhtiende; *part. Moist, damp*; hŭmĭdus, Som. Ben. Lye.

fuhton *fought*, Chr. 449; Erl. 12, 4; *p. pl.* of feohtan.

ful; *adj. Full, filled, complete, entire*; plēnus:—Ealra fúla ful *full of all foulness* [*impurities*], Elen. Kmbl. 1534; El. 769: 1875; El. 939: Cd. 166; Th. 206, 11; Exod. 450: Exon. 74 b; Th. 279, 12; Jul. 612: 78 b; Th. 294, 33; Crä. 24: 84 a; Th. 316, 4; Mód. 43. Æfþancum ful *filled with grudges*, Salm. Kmbl. 992; Sal. 497. Đá beád Swegen ful gyld and metsunge to his here ðone winter *Sweyn then commanded full tribute and provisions for his army during the winter*, Chr. 1013; Erl. 149, 24. v. full.

ful, *full*; *adv. Full, perfectly, very, well*; plēne, perfecte, valde:—Wyrd ne ful cúðe *he knew not well her destiny*, Exon. 66 a; Th. 244, 26; Jul. 33.

ful, full, es; *n.* **I.** *a cup*; pōcŭlum:—He ðæt ful geþah *he partook of the cup*, Beo. Th. 1261; B. 628. Him wæs ful boren *to him the cup was borne*, Beo. Th. 2388; B. 1192. Onfóh ðissum fulle *accept this cup*, Beo. Th. 2342; B. 1169. Full *the cup*, Exon. 106 b; Th. 406, 8; Rä. 24, 14. Drince þreó ful fulle nistig *let him drink three cups full fasting*, Herb. 3, 6; Lchdm. i. 88, 13. **II.** *what contains liquids, A collection of water, the sea, clouds*; receptācŭlum liquĭdi, măre, nûbes:—He ða frætwe wæg ofer ýða ful *he carried the ornament over the sea* [*lit. the cup of the waves*], Beo. Th. 2421; B. 1208. Ic wíde toþringe lagustreáma full *I widely disperse the clouds* [*lit. the collection of water-streams*], Exon. 102 a; Th. 385, 1; Rä. 4, 38. [*O. Sax.* ful, *n. a goblet*: *Icel.* full, *n. a goblet full of drink*.] DER. medo-ful, meodu-sele-.

ful-, full-, in composition, denotes the *fulness, completeness* or *perfection* of the meaning of the word with which it is joined. [Cf. *Goth.* fulla-.] v. full.

-ful, -full, e; *f. -ful*, as in bûc-ful *a bucketful*, hand-ful, -full *a handful*, q. v.

-ful, -full, the termination of many adjectives, as,—Bealo-ful, -full *baleful*: Car-ful, cear-ful *careful*: Ege-full *fearful*, etc.

FŪL; *adj.* FOUL, *dirty, impure, corrupt, rotten, stinking, guilty, convicted of a crime*; fœdus, immundus, sordĭdus, obscœnus, spurcus, pŭtĭdus, fœtĭdus, culpæ conscius, crīmĭne convictus:—Byrgen útan fæger, and innan fúl *a sepulchre fair without, and foul within*, Ps. Th. 13, 5. On ðínne fúlan múþ *in thy foul mouth*, 49, 17. In fúle wyllan *to the foul spring*, Cod. Dipl. 724; A.D. 1016; Kmbl. iii. 367, 13: 366, 31. Þurh fúle synne *through foul sin*, Exon. 29 b; Th. 90, 33; Cri. 1483. Ne náht fúles ne þicge *nec immundum quidquam comēdas*, Jud. 13, 4. Wið fúlne gálscipe *against foul lasciviousness*, L. C. E. 24; Th. i. 374, 9. Ascúnige man swíðe fúle forligra *let foul fornications be earnestly shunned*, L. Eth. vi. 28; Th. i. 322, 15. Swá fúle swá gǽt *as foul as goats*, Exon. 26 a; Th. 75, 34; Cri. 1231. Fúl wín *spurcum vīnum*, Ælfc. Gl. 32; Som. 61, 127; Wrt. Voc. 27, 54. Ic eom wyrslícre ðonne ðes wudu fúla *I am viler than this rotten wood*, Exon. 111 a; Th. 424, 33; Rä. 41, 48. Fúl fýr *of heora múþe* bláwende *de ôre ignem pŭtĭdum efflantes*, Bd. 5, 12; S. 628, 41: 5, 12; S. 630, 12. Ic fúlre eom ðonne ðis fen swearte, ðæt hér yfle adelan stinceþ *I am fouler than this black*

fen, that here smells badly of filth, Exon. 110 b; Th. 423, 32; Rä. 41, 31. Gif se mynetere fúl wurþe *if the minter be guilty*, L. Ath. i. 14; Th. i. 206, 20: v. § 1, 1; Th. i. 228, 14. Gif he ðonne fúl wurþe *if he then be convicted*, L. Eth. i. 1; Th. i. 280, 19: i. 2; Th. i. 282, 21: L. C. S. 30; Th. i. 394, 6. [*Piers P. Chauc. R. Glouc.* foul: *Laym.* ful, fule: *Orm.* fule: *Plat.* vuul, ful, fuul: *Frs.* fuwle, fule: *O. Frs.* ful: *Dut.* vuil: *Ger.* faul: *M.H.Ger.* vúl: *O.H.Ger.* fúl: *Goth.* fuls: *Dan.* fuul: *Swed.* ful: *Icel.* fúll: *Lat.* pŭter *foul, putrid*: *Lith.* pú-lei *putrid matter*: *Sansk.* pûti *putrid*; from the root pûy *to become foul* or *putrid*.]

FŪL, es; *n. Foulness, impurity, guilt, offence, fault*; illŭvies, impûrĭtas, culpa:—Fúl and wydel *illŭvies*, Cot. 105. Ealra fúla ful *full of all foulness* [*impurities*], Elen. Kmbl. 1534; El. 769. Ðár ǽnig þing fúles neáh ne cume *where nothing foul* [*of foulness*] *may come near*, L. Edg. C. 42; Th. ii. 252, 25. Se ðe ðæs fácnes and ðæs fúles gewita sý *he who is privy to the crime and the guilt*, L. Ath. v. § 1, 2; Th. i. 228, 22. Sleá man of ða hand ðe he ðæt fúl mid worhte *let the hand be struck off with which he wrought that offence*, i. 14; Th. i. 206, 21. v. fúl *foul*; *adj.*

fúl, es; *m. A convicted offender*; reus, qui scĕlĕris damnātus est:—Ðæt hý ne fúl náwár friðian ne feorman *nor that they anywhere protect or harbour a convicted offender*, L. Ed. 7; Th. i. 162, 25. Be ðon ðe fúl friðiaþ *concerning those who protect a convicted offender*, 8 titl; Th. i. 164, 1. v. fúl; *adj.*

fúl, e; *f.* fúle, an; *f. A foul, common* or *unconsecrated place, a highway where criminals were buried*; lŏcus profānus:—Sleá mon hine and on fúl lecge *let him be slain and be laid in a common place*, L. Eth. i. 4; Th. i. 284, 2. Hine man on fúlan lecge *let one lay him in a common place*, L. C. S. 33; Th. i. 396, 17. v. Th. L. Gl.

ful-æðele *full noble, very noble*. v. full-ædele.

ful-bealdlíce, -baldlíce; *adv. Full boldly, very boldly*; audācissĭme:—Ðe dínes siþes fulbealdlíce biddaþ *who full boldly pray for thy coming*, Ps. Th. 68, 7. He fulbaldlíce beornas lǽrde *he exhorted the warriors full boldly*, Byrht. Th. 140, 60; By. 311.

ful-beám; *gen.* fúlan beámes; *m. The black alder*; alnus nigra, rhamnus frangŭla:—Wyl on wætere fúlan beámes rinde *boil in water black alder rind*, L. M. 1, 32; Lchdm. ii. 78, 12.

ful-berstan; *p.* -bærst, *pl.* -burston; *pp.* -borsten; *v. intrans. To burst fully* or *thoroughly*; plēne rumpi, Off. Reg. 3.

ful-bétan, full-bétan; *p.* -bétte; *pp.* -béted *To make full amends, give satisfaction*; pĕnĭtus compensāre, sătisfăcĕre:—Ðæt he fulbéte *till he make full amends*, L. Pen. 12; Th. ii. 280, 29.

ful-blác; *adj.* [blác **I.** *bright, shining*] *Full bright, very bright*; præ-lūcĭdus:—On fulblácne beám *on the very bright tree*, Exon. 116 b; Th. 449, 4; Dóm. 66.

ful-blíðe *full glad, very joyful*. v. full-blíðe.

ful-boren; *part. Full-born, noble-born*; nōbĭlis nātu:—Mid eahta and feówertig fulborenra þegena *with eight and forty noble-born thanes*, L. Ath. iv. 7; Th. i. 228, 4.

ful-bót *full amends*; plēna compensātio, Som. Ben. Lye.

ful-brecan; *p.* -bræc, *pl.* -brǽcon; *pp.* -brocen *To break entirely, violate*; pĕnĭtus frangĕre, violāre:—Se ðe áðor fulbrece *he who violates either*, L. C. E. 2; Th. i. 358, 21.

ful-brice, -bryce, es; *m.* [ful *full*, brice *a breaking, breach*] *A full* or *entire breach of the peace*; plēna pācis violātio:—Gif fulbrice wyrþe *si plēna pācis violātio fĭeret*, L. E. B. 4, 6, 7; Th. ii. 240, 17, 23; 242, 3. Fulbryce, 5, 8; Th. ii. 240, 20; 242, 6.

ful-cáflíce *full quickly, very eagerly*. v. full-cáflíce.

ful-cléne; *adj. Full clean, very pure*; purissĭmus:—Ic ðíne gewitnesse wát fulclǽne *I know thy testimonies* [*are*] *very pure*, Ps. Th. 118, 14.

ful-cúþ, full-cúþ; *adj. Full known, well known, famous, public*; bĕne nōtus, insignis, publĭcus:—On fulcúþum gemynde *in famous memory*, Ælfc. T. 21, 1. Bí fulcúþum strǽtum *juxta publĭcos viārum transĭtus*, Bd. 2, 16; S. 520, 4, note, MS. T: Nar. 2, 15.

ful-cyrten; *adj. Very beautiful*; pulcherrĭmus:—Fulcyrtenu ceorles dôhtor *a churl's very beautiful daughter*, Exon. 106 b; Th. 407, 16; Rä. 26, 6.

ful-dón; *p.* -dyde, *pl.* -dydon; *pp.* -dón *To do fully, satisfy*; plēne ăgĕre, satisfăcĕre, R. Ben. 44.

ful-dysig *very foolish* or *ignorant*. v. full-dysig.

ful-dyslíce; *adv. Very foolishly*; stultissĭme:—Se Godes cunnaþ fuldyslíce *he tempteth God very foolishly*, Salm. Kmbl. 455; Sal. 228.

ful-earmlíce; *adv. Full miserably, very wretchedly*; miserrĭme:—Sum sceal fulearmlíce ealdre linnan *one shall full miserably lose his life*, Exon. 88 a; Th. 330, 20; Vy. 54.

ful-eáðe, full-eáðe; *adv. Full easily, very easily*; facillĭme:—Ða men ðe habbaþ unhále eágan ne mágon fuleáðe lócian ongeán ða sunnan *the men who have weak eyes cannot very easily look at the sun*, Bt. 38, 5; Fox 204, 27.

ful-endian *to end fully, complete*. v. full-endian.

ful-fealdan; *p.* -feóld, *pl.* -feóldon; *pp.* -fealden *To explain*; explĭcāre:—Ic fulfealde *explĭco*, Ælfc. Gr. 24; Som. 25, 52.

ful-fleón *to flee fully* or *completely, flee away.* v. full-fleón.

ful-fremedlíce, full-fremedlíce; *adv. Fully, completely, perfectly;* perfecte :—Ne mæg nán gesceaft fulfremedlíce understandan ymbe God *no creature can perfectly understand about God,* Homl. Th. i. 10, 2, 4.

ful-fremednys, full-fremednes, -ness, -nyss, e; *f. Fulfilment, perfection;* perfectio :—Hwǽr is ðínra dǽda fulfremednys *ubi est perfectio viārum tuārum?* Job Thw. 167, 16. Lifde he his líf on sóþfæstnysse and on fulfremednysse *duxit vītam in justītiæ perfectiōne,* Bd. 3, 27; S. 559, 29.

ful-fremman, full-fremman, to -fremmanne; he -fremeþ; *p.* -fremede; *pp.* -fremed *To fulfil, perfect, practise;* perfĭcĕre :—Ðínre unrihtgítsunga gewill to fulfremmanne *to fulfil the desire of thine evil covetousness,* Bt. 7, 5; Fox 24, 10. Ic fulfremme *perfício,* Ælfc. Gr. 28, 6; Som. 32, 37. He his mód went to ðám yflum and hí fulfremeþ *he turns his mind to the vices and practises them,* Bt. 35, 6; Fox 170, 20. Of múþe cildra and súcendra ðú fulfremedest lof *ex ōre infantium et lactentium perfēcisti laudem,* Ps. Lamb. 8, 3; Ps. Spl. 39, 9. Heáhsetl his [biþ] swá swá móna fulfremed on écnysse *thrōnus ejus [erit] sīcut lūna perfecta in æternum,* Ps. Spl. 88, 36. Beóþ fulfremede *estōte vos perfecti,* Mt. Bos. 5, 48. DER. un-fulfremed.

ful-freólíc; *adj. Very liberal;* pĕnĭtus lĭbĕrālis :—Ða yldestan Chus and Cham hátene wǽron, fulfreólíce feorh, frumbearn Chames *the eldest were called Cush and Canaan, most liberal beings, Ham's firstborn,* Cd. 79; Th. 97, 25; Gen. 1618.

ful-fyllan *to fulfil, accomplish.* v. full-fyllan.

ful-gán, full-gán; he -gǽþ; *p.* -eóde, *pl.* -eódon; *pp.* -gán; *with the dat. To fulfil, perform, carry out, follow, accomplish;* adimplĕre, perfĭcĕre, perăgĕre, obsĕqui, patrāre :—Ðæt hí mǽgen hiora wísdóme fulgán *that they can fulfil their wisdom,* Bt. 39, 2; Fox 212, 19. Ðe hiora willan fulgǽþ *which fulfils their will,* 39, 8; Fox 224, 18. Se ne hwyrfþ his mód æfter ídlum geþohtum, and him mid weorcum [ne] fulgǽþ *he turns not his mind after vain thoughts, and does [not] carry them out with works,* Ps. Th. 23, 4. He fulgǽþ his lustum and his plegan *he follows his lusts and his pleasure,* Homl. Th. i. 66, 11.

ful-gangan, -gongan, full-gangan; *p.* -geóng, *pl.* -geóngon; *pp.* -gangen *To fulfil, perfect, follow, accomplish, finish;* complēre, perfĭcĕre, obsĕqui, fīnīre :—God bǽdon ðæt hie his hearmsceare habban mósten fulgangan *they prayed God that they might have to fulfil his punishment,* Cd. 37; Th. 48, 27; Gen. 782. Gif we him fulgangan wyllaþ *if we will follow him,* Ors. 5, 1; Bos. 101, 15. Hit is riht ðæt ðú heora þeáwum fulgange *it is right that thou follow their manners,* Bt. 7, 2; Fox 18, 35.

ful-geare, -gearwe, -gere; *adv. Full well, very well, fully, thoroughly;* sătis bĕne, plēne, pĕnĭtus :—Ic nát fulgeare ymbe hwæt ðú gyt tweóst *I know not full well about what thou still doubtest,* Bt. 5, 3; Fox 12, 12 : Ps. Th. 117, 28. Hió ne fulgeare cúðon gesecggan be ðam sigebeácne *they could not fully tell about the victorious sign,* Elen. Kmbl. 334; El. 167. Ic fulgearwe wát ðæt he byþ wís and mildheort *I know full well that he is wise and merciful,* Ps. Th. 135, 1; Exon. 127 b; Th. 491, 1; Rä. 80, 7. Judas ne fulgere wiste be ðam sigebeáme *Judas did not thoroughly know about the victorious tree,* Elen. Kmbl. 1717; El. 860.

ful-gegán; *p.* -ge-eóde, -geóde, *pl.* -ge-eódon, -geódon; *pp.* -gegán; *with the dat. To fulfil, perform, carry out, follow;* complēre, perfĭcĕre, perăgĕre, obsĕqui :—Ða ðú lustgryrum eallum fulgeódest *when thou didst follow all horrid lusts,* Soul Kmbl. 47; Seel. 24. v. ful-gán.

ful-gehende; *prep. Full nigh, very near;* valde prŏpe :—Hine man byrigde ðam stýple fulgehende, on ðam súþ-portice *he was buried very near the steeple, in the south porch,* Chr. 1036; Erl. 165, 38; Ælf. Tod. 19.

ful-gemǽc; *adj. Very suitable;* aptissĭmus :—Ða ic me fulgemæcne monnan funde *when I found a man very suitable for me,* Exon. 115 a; Th. 442, 25; Kl. 18.

ful-genihtsum; *adj. Very abundant, quite sufficient;* sătis abundans, omnĭno amplus :—Fulgenihtsum is munuce *suffĭcit monacho,* R. Ben. 55.

ful-geódest *didst fulfil, didst follow,* Soul Kmbl. 47; Seel. 24; *2nd sing. p.* of ful-gegán.

ful-geómor; *adj. Full sad, very sad;* valde tristis :—Ic ðis giedd wrece bí me fulgeómorre *I recite this lay of myself very sad,* Exon. 115 a; Th. 441, 19; Kl. 1.

ful-georne, full-georne; *adv. Full earnestly, very diligently, full well;* diligentissĭme, optĭme :—He wiste fulgeorne ðæt God hine lufode *he knew full well that God loved him;* qui optĭme nōvĕrat Dŏmĭnum esse cum eo, Gen. 39, 3.

ful-gere *full well, fully, thoroughly,* Elen. Kmbl. 1717; El. 860. v. ful-geare.

ful-getreów *full true, very true.* v. full-getreów.

ful-gewépned *fully weaponed, fully armed.* v. full-gewépned.

ful-gleáwlíce *full wisely, very prudently.* v. full-gleáwlíce.

fulgon *entered,* Chr. 755; Erl. 50, 27; *p. pl.* of felgan.

ful-gongan *to fulfil, perfect;* perfĭcĕre :—Ðæt he wíslíce woruld fulgonge *that he wisely perfect the world,* Exon. 92 b; Th. 348, 3; Sch. 22. v. ful-gangan.

---

ful-hár; *adj. Full hoary, gray-haired;* cānus, albescens senectūte, Cot. 54.

ful-heálíce; *adv. Full highly, very highly;* altissĭme :—Hý singaþ fulheálíce hlúdan stefne *they sing full highly with loud voice,* Exon. 13 b; Th. 24, 23; Cri. 389.

ful-hearde *full strongly, very firmly* or *tightly.* v. full-hearde.

ful-hræðe *full quickly, immediately,* Bt. 22, 1; Rawl. 47, 7, note *f.* v. ful-raðe.

fulhtere, es; *m. A baptizer, baptist;* baptista :—To ðæm dæge Seint Iohannes ðæs fulhteres *on the day of Saint John the baptist,* L. Ath. i. prm; Th. i. 196, 19. v. fulluhtere.

fúlian; *p.* ode, ede; *pp.* od, ed; *v. n. To become foul, putrefy, rot, decay;* putrescĕre, computrescĕre, corrumpi :—Ðǽr is mid Eástum án mǽgþ, ðæt hí mágon cýle gewyrcan; and ð ðǽr licgaþ ða deádan men swá lange, and ne fúliaþ, ðæt hí wyrcaþ ðone cýle hine on *there is among the Esthonians a tribe that can produce cold; and, therefore, the dead men lie there so long, and decay not, because they bring the cold into them,* Ors. 1, 1; Bos. 23, 7. Míne wunda rote an and fúledon *computruērunt et deteriorāvērunt cicātrīces meæ,* Ps. Th. 37, 5; Ps. Surt. 37, 6. DER. a-fúlian.

fúlíce; *adv. Foully;* sordĭde, R. Ben. 82.

FULL, ful; *gen. m. n.* fulles, *f.* fulre : *def.* se fulla; seó, ðæt fulle : *comp. m.* fulra, *f. n.* fulre; *sup.* fullost; *adj.* FULL, *filled, complete, entire;* plēnus, sătiātus, confertus, intĕger :—Ðæt se weorþig full sǽte *that the street was* [lit. *sat*] *full,* Bd. 3, 6; S. 528, 18. Be-yrnþ se móna hwíltídum ðonne he full byþ on ðære sceade ufeweardre *the moon, when it is full, sometimes enters into the upper part of the shadow,* Bd. de nat. rerum; Wrt. popl. science 5, 14, 20; Lchdm. iii. 240, 22; 242, 1. He wæs full cyng ofer eall Engla land *he was complete king over all England,* Chr. 1036; Erl. 165, 10. Mildheortnysse Drihtnes full is eorþe *misericordia Dōmini plēna est terra,* Ps. Spl. 32, 5 : Exon. 8 b; Th. 4, 24; Cri. 57 : Cd. 18; Th. 21, 33; Gen. 333 : Beo. Th. 4816; B. 2412 : Ps. Th. 140, 1 : Salm. Kmbl. 63; Sal. 32. Ðes fulla mann *hic sătur,* Ælfc. Gr. 8; Som. 7, 26. Mín fulla freónd *my full friend,* Th. Diplm. A. D. 972; 524, 35. Se fulla móna *the full moon,* Bt. 39, 3; Fox 214, 29. Hwá is ðær ne wundrige fulles mónan *who is there that wonders not at the full moon?* Bt. Met. Fox 28, 81; Met. 28, 41. Hí gebrohton hie on fullum fleáme *they put them to full flight,* Chr. 917; Erl. 102, 18. On fullum mónan *at full moon,* Bd. de nat. rerum; Wrt. popl. science 15, 13; Lchdm. iii. 268, 10. He gewende súþweard mid fulre fyrde *he went southward with the entire army,* Chr. 1013; Erl. 148, 4 : 1014; Erl. 151, 4, 22 : 1022; Erl. 161, 35. Be fullan *abundanter,* Ps. Th. 30, 27 : Past. pref; Hat. MS. Man ða fulne [fulne, Erl. 150, 32], freóndscipe gefæstnode *they then confirmed full friendship,* Chr. 1014; Erl. 150, 14 : 1052; Erl. 187, 23 : 1013; Erl. 148, 19, 36 : Bt. Met. Fox 21, 15; Met. 21, 8. Sceolon ðone ryhtan dóm ǽnne geæfnan, egsan fulne *they shall suffer the one righteous doom, full of terror,* Exon. 28 a; Th. 84, 8; Cri. 1370. Hâteþ ðonne heáhcyning helle betýnan, fýres fulle *then the mighty king shall command [them] to close hell, full of fire,* Salm. Kmbl. 349; Sal. 174. He geseah unrihte eorþan fulle *he saw the earth filled with unrighteousness,* Cd. 64; Th. 78, 13; Gen. 1292. Moises hêt nyman ðæt gemetfæt full, and settan beføran Drihtne *Moses commanded [them] to take the measure full, and to set [it] before the Lord,* Ex. 16, 33. Gód gemet, and full hig syllaþ on eówerne bearm *mensūram bōnam, et confertam dābunt in sĭnum vestrum,* Lk. Bos. 6, 38. Beád ða Swegen full gild *Sweyn then commanded full tribute,* Chr. 1013; Erl. 149, 2. Gif hí fulle ne beóþ *si non fuĕrint satūrāti,* Ps. Th. 58, 15 : Ps. Spl. 143, 16. Hig fyldon twelf wylian fulle ðæra brytsena *they filled twelve baskets full of the fragments,* Jn. Bos. 6, 13. Ðǽr hig wǽron seofon dagas fulle *there were there seven full days,* Gen. 50, 10. Of ðære tíde, Paulinus, syx geár fulle, on ðære mǽgþe Godes word bodade and lǽrde *Paulīnus ex eo tempŏre sex annis continuis, verbum Dei in ea provincia prædĭcābat,* Bd. 2, 14; S. 517, 33. Hit is gecyndelíce ðæt ealle eorþlíce líchaman beóþ fulran on weaxendum mónan ðonne on wanigendum *it is natural that all earthly bodies are fuller at the increasing moon than at the waning,* Bd. de nat. rerum; Wrt. popl. science 15, 11; Lchdm. iii. 268, 8. Hwenne ðæt flód byþ fullost *when the tide is fullest,* Chr. 1031; Erl. 162, 6, 16. [*Chauc.* ful, full: *R. Glouc.* ful: *Laym.* ful, uul, uule, fulle, uulle: *Orm.* full: *Plat.* vull, full: *O. Sax.* ful, fol: *Frs.* fol: *O. Frs.* ful, fol: *Dut.* vol: *Ger.* voll: *M. H. Ger.* vol: *O. H. Ger.* foll, fol, foll: *Goth.* fulls: *Dan.* fuld: *Swed.* full: *Icel.* fullr: *Lat.* plēnus: *Grk.* πλήρης: *Lith.* pilnas: *Sansk.* pūrṇa *filled, full.*]

full; *adv. Fully, perfectly, entirely;* plēne, perfecte, omnĭno :—Þurh tyn winter full *for fully ten winters,* Bd. 1, 6; S. 476, 25. He sæt ðær tyn winter full *he remained there fully ten winters,* Bt. Met. Fox 26, 33; Met. 26, 17. v. ful; *adv.*

full, es; *n. A cup;* pōcŭlum :—Gedrinc his þreó full fulle *drink three cups full of it,* Herb. 1, 9; Lchdm. i. 74, 2. v. ful; *n.*

-full *-ful.* v. -ful, the termination of many adjectives.

ful-lǽst, -lést, -láste (?) es; *m. Help, aid, support;* auxilium, subsĭdium :—Is mægenwísa trum, fullésta mǽst, se ðas fare lǽdeþ *he is a firm army-*

leader, the greatest of supports, who leads this expedition, Cd. 170; Th. 213, 18; Exod. 554. Ðæt we hæfdon æt ðæm fýre leóht and fullǽste *that we might have light and help from the fire,* Nar. 13, 3. [*O. Sax.* fullêsti; *O. H. Ger.* folleist.]

ful-lǽstan, -lêstan; *p.* te; *pp.* ed To *help, aid, support;* opĭtŭlāri :— Ic ðē fullǽstu *I will support thee,* Beo. Th. 5330; B. 2668. ℞ [ôs] fullêsteþ [*the mind*] *gives aid,* Exon. 106 b; Th. 407, 1; Rä. 25, 8. Him men fullêstaþ *men aid them,* 119 a; Th. 457, 31; Hy. 4, 92. [*O. Sax.* fullêstian : *O. H. Ger.* folleistian.]

full-ǽðele; *adj. Full noble, very noble;* valde nōbĭlis :—Manege beóþ ǽgðer ge fullǽðele ge fullwêlige, and beóþ ðeáh fullunróte *many are both very noble and very wealthy, and yet are very unhappy,* Bt. 11, 1; Fox 32, 3.

Fullan-ham, -hom; *gen.* -hammes, -hommes; *m.* [*Asser* Fullonham: *Hunt.* Fulenham: *Sim. Dun.* Fulanham: *Brom.* Fullenham] FULHAM, *Middlesex;* lŏci nōmen in agro Middlesexiensi, ad rīpam Tămĕsis flūmĭnis :—Æt Fullanhæmme be Temese *at Fulham on the Thames,* Chr. 879; Th. 150, 3. On Fullanhomme *at Fulham,* 880; Th. 150, 12, col. 1.

full-bêtan; *p.* te; *pp.* ed To *make full satisfaction;* sătisfăcĕre :—Ic fullbête oððe behreówsige *sătisfăcio,* Ælfc. Gr. 37; Som. 39, 40. v. ful-bêtan.

full-blíðe; *adj. Full glad, very joyful;* lætissĭmus :—Ða Philistei fullblíðe wǽron *the Philistines were very joyful,* Jud. 16, 23.

full-cáflíce; *adv. Full quickly, very eagerly;* velocissĭme :—Se full-cáflíce brǽd of ðæm beorne blódigne gár *he very eagerly plucked the bloody dart from the chief,* Byrht. Th. 136, 19; By. 153.

full-cúþ; *adj. Full known, well known;* bĕne nōtus :—On cyninga bócum ys fullcúþ be ðám *in the books of the kings it is well known about them,* Jud. Thw. 161, 20.

full-dysig; *adj. Very foolish or ignorant;* perfecte stultus :—Fulldysig biþ se mann *the man is very foolish,* Hexam. 2; Norm. 4, 6.

full-eáðe; *adv. Very easily;* facillĭme :—Ne meht ðú fulleáðe cweðan ðæt ðú earm sê *thou canst not very easily say that thou art miserable,* Bt. 8; Fox 24, 22. v. ful-eáðe.

full-endian; *p.* ode; *pp.* od To *end fully, complete, finish;* complēre, finīre :—He bæd Cynebill ðæt he ða árfǽstan ongunnennesse fullendode *pĕtiit Cynibillum pia cœpta complēre,* Bd. 3, 23; S. 554, 39, note. [*Ger.* vollenden.]

full-eóde, *pl.* -eódon *went after, followed, aided,* Beo. Th. 6230, note; B. 3119: Cd. 98; Th. 130, 1; Gen. 2153; *p.* of full-gán.

fullere, es; *m. A* FULLER, *bleacher;* fullo :—His reáf wurdon glitiniende swá hwíte swá snáw, swá nán fullere ófer eorþan ne mæg swá hwíte gedón, Mk. Bos. 9, 3; *vestĭmenta ejus facta sunt splendentia et candĭda nĭmis vĕlut nix, quālia fullo non pŏtest sŭper terram candĭda făcĕre,* Vulg; his clothis ben maad schynynge and white ful moche as snow, and which maner clothis a fullere, *or walkere of cloth* may not make white on erthe, Wyc. Fulleras *fullōnes,* Ælfc. Gl. 9; Som. 57, 1; Wrt. Voc. 19, 12.

full-fleón, ic -fleó; *p.* -fleáh, *pl.* -flugon; *pp.* -flogen To *flee fully or completely, flee away;* perfŭgĕre :—Ic fullfleó *perfŭgio,* Ælfc. Gr. 28, 6; Som. 32, 49.

full-fremedlíce; *comp.* -lícor; *adv. Fully, completely, perfectly;* perfecte :—Nán man ne mæg fullfremedlíce secgan embe ðone sóþan God *no man is able to speak perfectly about the true God,* Hexam. 3; Norm. 4, 26. Ne eart ðú fullfremedlíce gefullod *non es perfecte baptĭzātus,* Bd. 5, 6; S. 620, 6 : 618, 38. Ǽrdon ðe he be ðám forþgewitenum gýmeleásnyssum his fullfremedlícor of ðære tíde geclǽnsade *priusquam prætĕrĭtas neglĭgentias perfectius ex tempŏre castĭgāret,* 3, 27; S. 559, 6. [*Orm.* fullfremeddlike.] v. ful-fremedlíce.

full-fremednes, -ness, -nyss, e; *f. Fulfilment, perfection;* perfectio :— Ðæt ic hæbbe manege men gelǽd to ðæm stæðe fullfremednesse on ðæm scipe mínes módes *that I have brought many men to the shore of perfection in the ship of my mind,* Past. 65; Hat. MS. Ðæt he fullfremednysse hæbbe *that it may have fulfilment,* Ælfc. Gr. 21; Som. 23, 27. DER. un-fullfremednes. [*Orm.* fullfremeddness.] v. ful-fremednys.

full-fremman, to -fremmenne; *p.* -fremede; *pp.* -fremed To *do fully, fulfil, finish, perfect, practise;* perfĭcĕre, perăgĕre, patrāre :—Syððan he ne mæg ðæne grundweall fullfremman *posteaquam fundāmentum non potuĕrit perfĭcĕre,* Lk. Bos. 14, 29. Hwæðer he hæbbe hine to fullfremmenne *si hăbeat ad perfĭciendum,* 14, 28. Ðæt ic fullfremme his weorc *ut perfĭciam ŏpus ejus,* Jn. Bos. 4, 34. Ðæt he hí eft fullfremme *that he practise them [the vices] again,* Bt. 35, 6; Fox 170, 18. Swá eówer heofonlíca fæder is fullfremed *sicut păter vester cœlestis perfectus est,* Mt. Bos. 5, 48 : Ælfc. Gr. 20; Som. 23, 12, 13. Ðeáh hí on manegum þingum síen fullfremede *though they are perfect in many things,* Past. 65; Hat. MS. [*Orm.* fullfremedd.] v. ful-fremman.

full-fyllan; *p.* -fylde; *pp.* -fylled To *fulfil, accomplish;* complēre :— Ic fullfylle *compleo,* Ælfc. Gr. 26; Som. 28, 29.

full-gán; he -gǽþ; *p.* -eóde, *pl.* -eódon; *pp.* -gán; *with the dat.* To *fulfil, perform, go after, follow, aid;* perfĭcĕre, perăgĕre, sĕqui, adjŭ-

vāre :—We ne móton fullgán úres Scippendes willan *we cannot perform our Maker's will,* Bt. 7, 5; Fox 24, 8. Se lyðra man fullgǽþ deófles willan *the wicked man fulfils the devil's will,* Homl. Th. i. 172, 18. Sceaft fláne fulleóde *the shaft went after the arrow,* Beo. Th. 6230, note; B. 3119. Hie me fulleódon *they well aided me,* Cd. 98; Th. 130, 1; Gen. 2153. v. ful-gán.

full-gangan; *p.* -geóng, *pl.* -geóngon; *pp.* -gangen; *with the dat.* To *fulfil, accomplish, finish;* perfĭcĕre, finīre :—Ðæt hí móstan ðam gewinne fullgangan *that they might finish the war,* Ors. 3, 1; Bos. 54, 21. v. ful-gangan.

full-georne; *adv. Full earnestly, very diligently;* diligentissĭme :—Ic míne earfeðu ealle fullgeorne fóre him sæcge *I tell all my troubles very diligently before him,* Ps. Th. 141, 2. v. ful-georne.

full-getreów; *adj. Full true, altogether true;* pĕnĭtus vērax :—We synd fullgetreówe *sŭmus pĕnĭtus vērāces,* Gen. 42, 31.

full-gewépned; *part. Fully weaponed, fully armed;* perfecte armātus :—Hí cómon onuppon ða munecas fullgewépnede *they came upon the monks fully armed,* Chr. 1083; Erl. 217, 11.

full-gleáwlíce; *adv. Full wisely, very prudently;* sapientissĭme, prudentissĭme :—Ic míne sáwle symble wylle fullgleáwlíce Gode underþeódan *I will always very prudently subject my soul to God,* Ps. Th. 61, 1 : 72, 13 : 106, 42.

full-hearde; *adv. Full strongly, very firmly or tightly;* firmissĭme, artissĭme :—He ðone fullhearde geband *he bound it very tightly,* Cd. 23; Th. 29, 3; Gen. 444.

fullian, fulligan, fulwian, to fullianne; *part.* fulligende; ic fullige, ðú fullast, he fullaþ, *pl.* fulliaþ; *p.* fullode, ede; *pp.* fullod, ed; *v. trans.* To FULL or *make white as a fuller* [fullere, q. v.], to *baptize;* albāre, candĭdum făcĕre, baptīzāre = βαπτίζειν. A word of doubtful origin. It is by some connected with the verb which appears in Gothic as weihan to *sanctify.* See fulluht. Ongunnon hí men lǽran and fullian *ipsi prædĭcāre et baptizāre cœpērunt,* Bd. 1, 26; S. 488, 4: 1, 27; S. 493, 25. Se ðe me sende to fullianne on wætere *qui misit me baptīzāre in ăquam,* Jn. Bos. 1, 33. Iohannes wæs on wéstene fulligende *fuit Joannes in deserto baptizans,* Mk. Bos. 1, 4. Ic fullige on wætere *ĕgo baptīzo in ăqua,* Jn. Bos. 1, 26. Hwí fullast ðú *quid baptĭzas?* 1, 25. Se ðe fullaþ on Hálgum Gáste *qui baptĭzat in Spīrĭtu Sancto,* 1, 33 : 3, 26 : L. C. E. 4; Th. i. 360, 30. Iohannes fullode ða ðe him to cómon *John baptized those who came to him,* Homl. Th. i. 352, 16 : Jn. Bos. 1, 28, 31 : 3, 22, 23 : 4, 2 : 10, 40. Lǽraþ ealle þeóda, and fulligeaþ hig *dŏcēte omnes gentes, baptizantes eos,* Mt. Bos. 28, 19. Ðæt he hine fullode *that he might baptize him,* 3, 13. Iohannes se Fulluhtere cwæþ, witodlíce ic eów fullige on wætere, to dǽdbóte; se ðe æfter me towerd ys...he eów fullaþ on Hálgum Gáste, Mt. Bos. 3, 11; *Joon Baptist saide, forsothe Y cristene* [=waische] *ȝou in water, in to penaunce; forsothe he that is to cumme after me...he shal baptise, or cristen ȝow in the Holy Goost,* Wyc: Joannes Baptista dixit, ĕgo quĭdem baptīzo vos in ăqua in pœnĭtentiam; qûi autem post me ventūrus est...ipse vos baptizābit in Spīrĭtu Sancto, Vulg. 'In Anturs of Arther, end of 13th century, we find, st. xviii. lines 4, 5 :—*pp.* Fulled *whitened, baptized:* R. Glouc. A. D. 1297; 3 *p.* Follede; *pp.* y-fulled, fulled; *s.* fullynge: Piers P. 1362, *Wrt. small 8vo. London, Pickering,* 1842, pp. 244, 322, fullynge *baptizing, whitening:* 398, fullynge *baptizing.* After this, we do not find fulled, y-fulled, fullynge; yet in *A. Sax.* Mk. Bos. 9, 3, we have fullere: *Wyc.* 1389, fullere [*or walkere of cloth,* note]: *Tynd.* 1526 *and Eng. version* 1611, fuller. Baptism and Baptym with the verb Baptise is used by Wycliffe, and Baptyme and Baptyzyn by the compiler of the Promptorium. Wycliffe also uses the 1st person of the verb I waisch in Mt. 3, 11; and the two forms of the *pp.* waischen, waischun, in Mt. 3, 6, and Mk. 10, 38, 39. The form Bapteme seems to have been introduced into the language, through the French, by Robert Manning, called de Brunne, from Bourne, near Depyng in Lincolnshire, in his translation of Peter Langtoft's Chronicle, and to have been current, with slight variation in the orthography, till nearly the middle of the 16th century = 1550. Thus the forms Baptim and Baptime appear in the version of the N. T. by Tyndale in 1526, and Baptym, Baptyme in that by Cranmer in 1539. In the version made by Coverdale and other Protestant exiles at Geneva in 1559, in the Anglo-Rhemish version made by Cardinal Allen and other Romanists at Rheims in 1559, and in the authorized version of 1611, the word is written Baptisme. This last form is also found in *Piers P.* p. 398. Ormin only uses the verb *to dip,* once :—Unnderr waterr dippesst, H. 1551. In *Goth.* and in other divisions of the Teutonic as well as in the *Swed.* and *Dan.* divisions of the Scandinavian branch of the Gothic language, a noun and verb are used expressive of *dipping,* e. g. *Goth.* daupyan, daupeins : *O. H. Ger.* doufan, doufa : *Dut.* doopen, doop : *Ger.* taufen, taufe : *Swed.* döpa, döpelse : *Dan.* döbe, daab.' Orm. ii. 626, 627. Dyppan is also used in the Rushworth Gloss. v. fulwian. DER. ge-fullian, -fulwian : un-gefullod.

fullian; *p.* ode; *pp.* od To *fulfil, perfect;* exsĕqui :—Gif gē bebodu willaþ mín fullian *if ye will fulfil my commandments,* Cd. 106; Th. 139, 29; Gen. 2317. Ðonne sceal he ðæt mid mildheortum weorcum

fullian *then shall he perfect that with works of mercy*, Blickl. Hom. 37, 19. Fullade 213, 16. [O. Sax. fullôn: O. H. Ger. fullén.] DER. lustfullian, ge-lustfullian, wist-fullian: un-gefullod. v. fyllan.

fúl-líc; *adj. Foul, base; foedus, turpis*:—Gárclifan etan ǽrende fúllíc getácnaþ *to eat agrimony betokens a disagreeable message*, Lchdm. iii. 198, 25. Ansíne fúllíce habban *to have a dirty face*, iii. 204, 10, 26.

fullíce; *comp.* -lícor; *adv. Fully, perfectly, completely; plēne*, perfecte:—Se ðe Englisc fullíce ne cúðe *qui Anglōrum linguam perfecte non nōvĕrat*, Bd. 3, 3; S. 525, 39: 2, 3; S. 504, 32. Fullícor *plēnius*, 4, 25; S. 600, 10.

fúl-líce; *comp.* -lícor; *adv. Foully, shamefully; foede, sordīde*, turpiter:—Gif hwá fúllíce hine sylfne besmíte *si quis foede seipsum polluĕrit*, L. M. I. P. 40; Th. ii. 276, 7: C. R. Ben. 44: Scint. 24.

fulligan, to fulligenne *To baptize;* baptizāre:—Diāconus môt fulligan cild *a deacon may baptize children*, L. Ælf. C. 16; Th. ii. 348, 14. Gif cild biþ to fulligenne *if there be a child to baptize*, 29; Th. ii. 352, 30. v. fullian.

fulligenne *to baptize*, Th. L. ii. 352, 30. v. fulligan.

full-mannod, -monnad; *part. Full manned, well peopled;* vĭris instructus, pŏpŭlo frĕquens:—Ðæt he hæbbe his land fullmannod [Cot. fullmonnad] *that he have his land well peopled*, Bt. 17; Fox 58, 32.

full-neáh; *adj. Full nigh, very near;* valde propinquus:—Wæs se feónd fullneáh *the foe was very near*, Cd. 32; Th. 43, 10; Gen. 688.

full-neáh; *adv. Full nearly, very nearly, almost;* fĕre:—Ðú eart fullneáh forþoht *thou art almost despairing*, Bt. 8; Fox 24, 16: Chr. 897; Th. 175, 39, col. 1. v. ful-neáh; *adv.*

fullnes, -ness, -nyss, e; *f.* FULNESS; plēnĭtūdo, Som. Ben. Lye.

fúlnes, -ness, e; *f. Foulness, stench;* foetor:—Seó wundriende swétnes ðæs miclan swæcces sôna ealle ða fúllnessa ðæs þýstran ofnes on weg aflýmede *omnem mox foetōrem tenebrōsæ fornācis effŭgāvit admīrandi hūjus suāvĭtas ŏdōris*, Bd. 5, 12; S. 629, 21. v. fúlnes.

fulloc, es; *n. Baptism;* baptismus:—We willaþ ðæt fulloc fæste stande *we will that baptism stand fast*, L. N. P. L. 67; Th. ii. 302, 6. v. fulluht.

full-oft; *adv. Full oft, very often;* sæpissĭme:—We beóþ fulloft geneádode *we are very often compelled*, Greg. Dial. pref; Hat. MS. 1 a, 19. Fulloft fyrwit frineþ *curiosity inquires very often*, Salm. Kmbl. 116; Sal. 57. v. ful-oft.

ful-longe; *adv. Full long, very long;* diutissĭme:—Ða gyldnan geatu ðe fullonge ǽr bilocen stôdon *the golden gates which very long before stood locked*, Exon. 11 b; Th. 16, 12; Cri. 252.

full-recen; *adj. Full quick, very quick;* citissĭmus:—Ðú meahtest ðé fullrecen on ðæm rôdere ufan siððan weorþan *thou, very quick, mayest afterwards advance into the sky above*, Bt. Met. Fox 24, 33; Met. 24, 17.

full-sláw; *adj. Full slow, very slow;* persegnis, Off. Reg. 15.

full-sôþ *full sooth, most truly*, L. Ælf. C. 6; Lambd. 128, 29. v. ful-sôþ.

full-strong; *adj. Full strong, most rigid;* valde sĕvērus *vel* rĭgĭdus:—Wæs ðæt eall fullstrong *that was all most rigid*, Cd. 220; Th. 284, 16; Sat. 322. v. ful-strang.

fulluht, fulwiht, fullwiht, fulwuht, es; *n.* [v. Grimm And. u. El. pp. 136-7] *Baptism;* baptismus:—Hwæðer wæs Iohannes fulluht? Mt. Bos. 21, 25; *of whennes was the baptem of Joon?* Wyc; baptismus Joannis unde ĕrat? Vulg: Mk. Bos. 11, 30: Lk. Bos. 20, 4: Ælfc. Gr. 9, 1; Som. 8, 22. Ðæt fulluht us aþwehþ fram eallum synnum *baptism washes us from all sins*, Homl. Th. ii. 48, 29: 46, 24, 33: 48, 18, 20: i. 94, 2. Fram gyfe ðæs hálgan fulluhtes *a sacri baptismātis gratia*, Bd. 1, 27; S. 493, 10. Fulluhtes bæþ *the bath of baptism*, 2, 14; S. 518, 4. Mǽge gyt beón gefullod on ðam fulluhte, ðe ic beó gefullod *pŏtestis baptismo, quo ego baptizor, baptizāri?* Mk. Bos. 10, 38, 39: Mt. Bos. 3, 7: Lk. Bos. 7, 29: 12, 50: L. C. E. 22; Th. i. 374, 3: L. Ælf. C. 27; Th. ii. 352, 19: L. Ælf. P. 20; Th. ii. 370, 32: Chr. 601; Erl. 21, 11: 942; Erl. 116, 20: Ps. Th. arg. 22. Se yfela preóst ne mæg næfre Godes þēnunge gefílan, naðer ne ðæt fulluht, ne ða mæssan *the evil priest cannot ever defile God's ministry, nor baptism, nor the mass*, L. Ælf. P. 41; Th. ii. 382, 14: L. Alf. 49; Th. i. 58, 25: Homl. Th. i. 208, 11: 306, 1: 312, 21: ii. 48, 1, 3, 4, 5: Lk. Bos. 3, 3. [*Orm.* fulluhht: *Laym.* fulluht.]

fulluht-bæþ, fulwiht-bæþ, es; *n.* [full, wiht, e; *f:* bæþ, es; *n.*] *A bath or font of baptism;* baptismi fons, baptistērium = Βαπτιστήριον:—Ðæt geryne onfón fulluhtbæþes *to receive the sacrament of the baptismal font*, Bd. 1, 27; S. 492, 31. Fulluhtebæþes, 3, 3; S. 525, 30. Ðá onféng Eádwine cyning fulluhtebæþe *then king Edwin received the bath of baptism*, 2, 14; S. 517, 23: 1, 27; S. 491, 29.

fulluht-ere, fulwiht-ere, es; *m.* [ful, full; uht, wiht; ere; es; *m.*] *A baptizer, the Baptist;* baptista:—On ðam dagum com Iohannes se Fulluhtere *in diēbus illis vēnit Joannes Baptista*, Mt. Bos. 3, 1: 14, 2: Mk. Bos. 6, 14: Lk. Bos. 7, 20, 28, 33: Homl. Th. i. 356, 7: 352, 22: 478, 1, 30. Syle me on ánum disce Iohannes heáfod ðæs Fulluhteres *da mihi in disco cắput Joannis Baptistæ*, Mt. Bos. 14, 8: Mk. Bos. 6, 24: Homl. Th. i. 350, 31: 352, 23: 364, 6. Be Iohanne ðam Fulluhtere *de* Joanne Baptista, Mt. Bos. 17, 13: Homl. Th. i. 356, 19: 476, 27: 484, 22. Sume secgeaþ Iohannem ðone Fulluhtere *alii dīcunt Joannem Baptistam*, Mt. Bos. 16, 14: Mk. Bos. 8, 28.

fulluht-nama, an; *m. The baptismal or christian name;* nōmen tempŏre baptizandi impŏsĭtum:—Hér Godrum se norþerna cyning forþferde, ðæs fulluhtnama wæs Æðelstán *here* [A. D. 890] *Guthrum the Northern* [i. e. *Danish*] *king died, whose baptismal name was Æthelstan*, Chr. 890; Erl. 86, 27:

fulluht-stów *a baptism-place, baptistery.* v. fulwiht-stów.

fulluht-þeáw, es; *m. The rite of báptism;* baptismi mos:—Cyning onféng fulluhtþeáwum *the king received the rite of baptism*, Bt. Met. Fox 1, 65; Met. 1, 33.

fulluht-tíd *time of baptism, baptismal time.* v. fulwiht-tíd.

fullunga; *adv. Fully:*—Fullunga *peramplius*, Rtl. 21, 8: Jn. Skt. Lind. note in the margin.

full-unrôt; *adj. Full sad, very unhappy;* valde tristis:—Manege beóþ ǽgðer ge fullǽdele ge fullwēlige, and beóþ ðeáh fullunrôte *many are both very noble and very wealthy, and yet are very unhappy*, Bt. 11, 1; Fox 32, 3.

full-wélig; *adj. Full wealthy, very rich;* valde dīves, ditissĭmus:—Manege beóþ fullwēlige *many are very wealthy*, Bt. 11, 1; Fox 32, 3.

full-weorþlíce *full worthily, very honourably*, Chr. 1036; Th. 294, 21, col. 2. v. ful-wurþlíce.

fullwiht, es; *n. Baptism;* baptismus:—Hú hí hine bǽdan fullwihtes bæþes *how they had asked him for a font of baptism*, Ors. 6, 34; Bos. 130, 30: Bd. 2, 14; S. 518, note 10: Andr. Kmbl. 3279; An. 1642. Mid ðý fullwihte *with baptism*, Exon. 121 b; Th. 467, 9; Hö. 136. v. ful-uht.

full-wíte, es; *n. Full fine;* plēna mulcta:—Gylde fullwíte [fulwíte MS. B.] *let him pay full fine*, L. C. S. 49; Th. i. 404, 7, 9: L. In. 43; Th. i. 128, 18, note 48, MSS. B. H. v. ful-wíte.

fullwon, e; *f. Baptism?* baptismus?—Fullwona bearn *children of baptism, christians*, Cd. 92; Th. 117, 9; Gen. 1951. v. fulluht.

full-wyrcan; *p.* -worhte; *pp.* -worht *To do fully, commit, accomplish, complete;* perficĕre, complēre:—Se godcunda ánweald hí tostencte ǽr hí hit fullwyrcan môston *the divine power dispersed them before they could complete it*, Bt. 35, 4; Fox 162, 25. [*Orm.* fullwrohht *finished*.] v. ful-wyrcan.

ful-mannod *full manned, well peopled.* v. full-mannod.

ful-moneg; *adj. Full many, very many;* permultus:—To fulmonegum dæge men synt forlædde *men are seduced for full many a day*, Cd. 33; Th. 45, 17; Gen. 728.

ful-neáh *full nigh, very near.* v. full-neáh; *adj.*

ful-neáh, full-neáh, ful-nēh; *adv. Full nearly, very nearly, almost;* prŏpe, fĕre:—Steorran hie ætíewdon fulneáh [fulnēh, Th. 29, 12, col. 1] *healfe tíd ofer undern stars shewed themselves very nearly half an hour after nine o'clock* [a. m.], Chr. 540; Th. 28, 13; 29, 12, col. 2. Fulneáh [fullneáh, Th. 175, 39, col. 1] tú swá lange *very nearly twice as long*, Chr. 897; Th. 174, 42; 175, 39, col. 2. Se yfela willa unrihthǽmdes gedréfþ fulneáh ǽlces libbendes monnes môd *the evil desire of unlawful lust disquiets the mind of almost every living man*, Bt. 31, 2; Fox 112, 25: 4; Fox 8, 18: 11; Fox 30, 18: Bt. Met. Fox 18, 8; Met. 18, 4.

fúlnes, fúllnes, fýlnes, -ness, e; *f.* FOULNESS, *impurity, stench;* foedĭtas, sordes, foetor:—Fúlnes [fŷlnes, Exon. 98 a; Th. 368, 7] eorþan, eal forwisnad *foulness of earth, all decayed*, Soul Kmbl. 35; Seel. 18. Unarǽfnendlíce fúlnes wæs upp aweallende *foetor incompărăbĭlis ebulliens ĕrat* Bd. 5, 12; S. 628, 25.

ful-oft, full-oft; *adv. Full oft, very often;* sæpissĭme:—Hie ablændaþ fuloft wísra monna geþoht *they very often blind the thought of wise men*, L. Alf. 46; Th. i. 54, 18. Sió wyrd fuloft dereþ unscyldegum *fate very often injures the guiltless*, Bt. Met. Fox 4, 71; Met. 4, 36: Beo. Th. 964; B. 480: Exon. 81 b; Th. 307, 16; Seef. 24: Cd. 216; Th. 274, 11; Sat. 152: Salm. Kmbl. 695; Sal. 347.

ful-raðe, -ræðe, -hræðe; *adv. Full quickly, immediately;* cĭtissĭme:—Fulraðe [Cott. fulræðe] ðæs ic clipode *immediately thereupon I spoke*, Bt. 22, 1; Fox 76, 8. Fulraðe yrnende *running very quickly*, Ors. 1, 1; Bos. 17, 21.

ful-recen *full quick, very quick.* v. full-recen.

ful-ricene; *adv. Full quickly, very quickly, immediately;* citissĭme:—Gif he múntas hríneþ, hí fulricene reócaþ *if he touches the mountains, they immediately smoke*, Ps. Th. 103, 30.

ful-riht; *adj. Full right, most right or direct;* valde rectus, directissĭmus:—Ðú ne mihtest gyt fulrihtne weg arēdian *thou hast not yet been able to find the most direct way*, Bt. 22, 2; Fox 78, 8.

ful-rihte; *adv. Full rightly, very rightly;* rectissĭme, Solil. 5.

ful-sárlíce; *adv. Full sorely, very harshly or violently;* tristissĭme, acerbissĭme, gravissĭme:—Ðæt mín sylfes fôt fulsárlíce asliden wǽre *that my own foot had very violently slipped*, Ps. Th. 93, 17.

ful-scrid; *adj. Full quick, very swift;* velocissĭmus:—Is ðes bát fulscrid, fugole gelícost glídeþ on geofene *this boat is very quick, it glideth on the ocean most like to a bird*, Andr. Recd. 996; An. 496.

ful-séfte; *adj. Full soft, very soft*; valde mollis :—Ic geworhte fulséfte seld, ðæt hī sǣton on *I made a very soft seat, which they sat on*, Ps. Th. 88, 3.

ful-sláw *full slow, very slow.* v. full-sláw.

ful-sméðe; *adj. Full smooth, very smooth*; levissĭmus :—Ðe fulsméðe sprǣce habbaþ *who have very smooth speech*, Frag. Kmbl. 20; Leás. 12.

ful-sóþ, full-sóþ; *adv. Full sooth, very truly*; verissĭme :—Fulsóþ hŷ secgaþ *they say very truly*, L. Ælf. C. 6; Th. ii. 344, 22.

ful-stincende; *part. Foul-stinking*; fœde ölens :—Ðū fūlstincendiste hell, geopena ðīne gatu *thou most foul-stinking hell, open thy gates*, Nicod. 27; Thw. '16, 3.

ful-strang, -strong, full-strong; *adj. Full strong, very severe* or *overwhelming*; valde sevērus vel rigĭdus :—Wæs him eal fulstrang *it was all very severe to them*, Cd. 218; Th. 278, 23; Sat. 226. Is ðeós þrag fulstrong *this moment is very overwhelming*, Exon. 72 b; Th. 270, 13; Jul. 464.

ful-swíðe; *adv. Very much, very*; valde :—Wēne ic fulswíðe *I think very much*, Exon. 120 a; Th. 461, 4; Hö. 30.

fulteman, fultemian *to assist, help, support*; jŭvāre, auxĭliāri :—Sió womb sceal fulteman ðǣm hondum *the belly must support the hands*, Past. 34, 3; Hat. MS. 44 a, 21. For ðæm ānwalde ðe ānra gehwilc fultemaþ *through the power which each one supports*, Bt. Met. Fox 25, 42; Met. 25, 21. v. fultuman.

ful-þiclíce; *adv. Full thickly, very often, very frequently*; persæpe, frequentissĭme :—Heó sprǣc to Adame fulþiclíce *she spoke to Adam very frequently*, Cd. 33; Th. 44, 6; Gen. 705.

ful-þungen; *part. Full grown, high, lofty*; celsus, R. Ben. 73.

fultom, es; *m. Help, aid, support*; auxĭlium, adjūtōrium :—Ðæt he ðone hālgan heáp bidde friþes and fultomes *that he implore the holy troop for peace and support*, Apstls. Kmbl. 181; Ap. 91. To fultome *for aid*, Chr. 601; Erl. 20, 12. v. fultum.

fultomian; *part.* fultomiende *To help, aid*; auxĭliāri :—Sóna eft, Gode fultomiendum, he meahte geseón and sprecan *soon after, God helping, he could see and speak*, Chr. 797; Erl. 58, 15. v. fultuman.

ful-trum; *adj. Full strong, very firm*; valde firmus :—Sēcaþ gē Drihten, and gē beóþ fultrume *quærĭte Dŏmĭnum, et confirmāmĭni*, Ps. Th. 104, 4.

ful-trúwian; *p.* ode; *pp.* od *To trust fully in, confide in*; pēnĭtus confīdĕre :—Ic nāt, hwī gē fultrūwiaþ ðam hreósendan wēlan *I do not know, why ye confide in these perishable riches*, Bt. 26, 2; Fox 94, 7.

fultum, fultom, es; *m.* **I.** *help, aid, assistance, support, succour*; auxĭlium, adjūtōrium, adjūmentum :—Him wæs fultum neáh *support was nigh to him*, Exon. 35 a; Th. 113, 20; Gū. 160. Fultum mín *adjūtōrium meum*, Ps. Lamb. 7, 11. Bæd fultumes wǣrfæst hæleþ *the righteous man sought their aid*, Cd. 94; Th. 122, 12; Gen. 2025: Ors. 3, 7; Bos. 59, 38: 3, 7; Bos. 60, 32. Hie Mæcedoniam on fultume wǣron *they had helped the Macedonians*, 2, 5; Bos. 46, 16: 2, 5; Bos. 47, 14, 33: 3, 7; Bos. 59, 35. Syle us nū on earfoðum ædelne fultum *da nōbis auxĭlium de trībŭlātiōne*, Ps. Th. 59, 10: 83, 6: Ps. Lamb. 19, 3. Him Drihten forgeaf frófor and fultum *the Lord gave to them comfort and succour*, Beo. Th. 1400; B. 698: 3674; B. 1835: Salm. Kmbl. 882; Sal. 440: Bt. Met. Fox 31, 15; Met. 31, 8. Oðer ǣhte heóld fæder on fultum *the other kept cattle in aid of his father*, Cd. 47; Th. 59, 35; Gen. 974: 95; Th. 125, 1; Gen. 2072: Exon. 62 b; Th. 229, 14; Ph. 455: Ors. 2, 5; Bos. 47, 27: 3, 7; Bos. 58, 29. Mid godcunde fultume *by divine aid*, 1, 5; Bos. 28, 5. **II.** *a helper, an army, forces*; adjūtor, cōpiæ :—Fultum mín and alŷsend mín beó ðū *adjūtor meus et lībĕrātor meus es tu*, Ps. Spl. 69, 7: 70, 8: Ps. Lamb. 18, 15. He gegaderode ðone fultum ðe he ðā mihte *he gathered what forces he then could*, Ors. 1, 12; Bos. 36, 1: 2, 5; Bos. 46, 27. He mid his fultume næs *he was not with his army*, 2, 5; Bos. 48, 15, 22: 3, 7; Bos. 59, 18. DER. feorm-fultum, mann-.

fultuman, fultumian, fultomian, fulteman, fultemian; *p.* ode, ede; *pp.* od, ed *To help, assist, aid, support*; jŭvāre, adjŭvāre, auxĭliāri, făvēre :—Hí woldon me mā fultumian *me pŏtius jŭvāre vellent*, Bd. 2, 13; S. 516, 9: Th. 114, 114. Ic fultumige *auxĭlior*, Ælfc. Gr. 25; Som. 26, 61: făveo, 26, 5; Som. 28, 66. Me God fultumeþ *Deus adjŭvat me*, Ps. Th. 53, 4: 88, 18. We eów fultumiaþ *we will aid you*, Chr. Erl. 3, 12. Him nāuðer ne fēt ne fiðeras ne fultumaþ *neither feet nor wings support them*, Bt. 41, 6; Fox 254, 26. Me ðíne dómas dǣdum fultumiaþ *jūdĭcia tua adjŭvābunt me*, Ps. Th. 118, 175. Fultumode Beorhtríc Offan *Beorhtric assisted Offa*, Chr. 836; Erl. 64, 32. Ðet hí him fultumedon *that they would aid them*, 868; Erl. 73, 22. DER. gefultuman, -fultumian, to-, to-ge-.

fultumend, fultumiend, es; *m.* [fultumende, fultumiende, *part. of* fultuman, fultumian] *A helper, assistant, co-operator*; adjūtor, co-ŏpĕrātor :—Ðe his gefera wæs and fultumend ðæs godcundan wordes *qui cōmes itĭnĕris illi et co-ŏpĕrātor verbi*, Bd. 3, 30; S. 562, 12. Ðonne biþ eádig ðe him ǣror wæs Iacobes God geára fultumiend *beātus, cūjus Deus Iacob adjūtor ejus*, Ps. Th. 145, 4: 70, 3: Ps. Lamb. 70, 7: Bd. pref; S. 471, 22.

fultum-leás; *adj. Without help, helpless*; sĭne auxĭlio :—Ðæt hí tó raðe woldon fultumleáse beón æt hiora bearnteámum *that they would very soon be without help from their posterity*, Ors. 1, 14; Bos. 37, 18.

ful-unrót *full sad, very unhappy.* v. full-unrót.

ful-wacor; *adj. Full watchful, very watchful*; pervĭgil, vĭgĭlans, Off. Reg. 5.

ful-wǣrlíc; *adj. Full wary, very cautious* or *prudent*; valde circum-spectus vel cautus, prudentissĭmus :—Ys hit fulwǣrlíc *it is very prudent*, Gen. 41, 33.

ful-wélig *full wealthy, very rich.* v. full-wélig.

fulwere, es; *m. A baptist*; baptista, Menol. v. fulluhtere.

fulwian; *p.* ode, ade; *pp.* od, ad *To baptize*; baptīzāre :—Fulwiaþ folc under röderum *baptize the people under the firmament*, Exon. 14 b; Th. 30, 23; Cri. 484. Hwæt fulwastu *quid baptizas*, Jn. Sk. Rush. 1, 25. Fulwande, fulwende *baptizans*, Lind. and Rush. 3, 23. Fulwad beón *baptizāri*, Bd. 1, 27; S. 492, 28. Fulwod beón, 1, 27; S. 493, 2, note. [*Laym.* fulwen.] v. fullian.

ful-wíde; *adv. Full widely, all around, round about*; circumcirca :—Lóca fulwíde ofer londbūende *look all around over the land-dwellers*, Cd. 228; Th. 307, 23; Sat. 684: Exon. 115 b; Th. 444, 13; Kl. 46. Wǣlhreówes [Nerônes] gewéd wæs fulwíde cūþ *the madness of the cruel [Nero] was full widely known*, Bt. Met. Fox 9, 10; Met. 9, 5.

fulwiht, es; *n. Baptism*; baptismus :—Wæs mid ðŷ folce fulwiht hæfen *baptism was raised up among the people*, Andr. Kmbl. 3285; An. 1645. Fulwihtes bæþ *the bath of baptism*, Bd. 2, 5; S. 507, 17: Chr. 604; Erl. 20, 18: Cd. 225; Th. 299, 8; Sat. 546: Elen. Kmbl. 978; El. 490. Būtan fulwihte *without baptism*, L. In. 2; Th. i. 102, 20: Chr. 601; Erl. 20, 13: 661; Erl. 34, 16. Ceadwalla fór to Rôme, and fulwihte onféng from ðam pāpan *Ceadwalla went to Rome, and received baptism from the pope*, Chr. 688; Erl. 42, 6: 878; Erl. 80, 18: Exon. 99 b; Th. 372, 3; Seel. 86: Andr. Kmbl. 3258; An. 1632: Elen. Kmbl. 383; El. 192. Iohannes wæs bodiende dǣdbóte fulwiht *fuit Joannes prædicans baptismum pœnĭtentiæ*, Mk. Bos. 1, 4: Chr. 565; Erl. 19, 6: 606; Erl. 20, 26: 661; Erl. 34, 18: Andr. Kmbl. 3268; An. 1637. Þurh fulwihte *through baptism*, Elen. Kmbl. 344; El. 172. Fulwihta calica *baptismata calicum*, Mk. Skt. Lind. 7, 4. v. fulluht.

fulwiht-bæþ, es; *n.* [MS. ful-wihte; bæþ, es; *n.*] *A bath* or *font of baptism*; baptismi fons :—Mon ðæt cild brohte to ðam hālgan þweále fulwihtebæþes *they brought the child to the holy washing of the baptismal font*, Guthl. 2; Gdwin. 10, 18.

fulwiht-ere, es; *m. A baptizer, the Baptist*; baptista :—Ne arās māra Iohanne Fulwihtere *non surrexit mājor Joanne Baptista*, Mt. Bos. 11, 11. Iohannes Fulwihteres *Joannis Baptistæ*, 11, 12. v. fulluhtere.

fulwiht-fæder, es; *m. A baptizer* :—Sancte Iohannes, Cristes fulwihtfæder *St. John, Christ's baptizer*, Blickl. Homl. 205, 17.

fulwiht-hád, es; *m. A baptismal vow* :—Ðæt hie heora fulwihrhādas wel gehealdan *that they keep well their baptismal vows*, Blickl. Homl. 109, 26.

fulwihðe *baptism*, L. Wih. 6; Th. i. 38, 9. v. fulluht.

fulwiht-stow, e; *f. A baptism-place, baptistery*; baptismātis lŏcus, baptistērium :—Ne wǣron cyrican getimbrede, ne fulwihtstowe *churches were not built, nor baptism-places* [baptistēria], Bd. 2, 14; S. 518, 16.

fulwiht-tíd, e; *f. Time of baptism, baptismal time*; baptismātis tempus :—Fulwiht-tíd [MS. -tiid] éces Drihtnes to us cymeþ *the baptismal time of the eternal Lord comes to us*, Menol. Fox 22; Men. 11.

fulwiht-wer, es; *m. A baptist* :—Seó gebyrd Sancte Iohannes ðæs fulwiht-weres *the birthday of St. John the Baptist*, Blickl. Homl. 161, 6.

ful-wíte, full-wíte, es; *n. A full fine;* plēna mulcta :—Gielde he fulwíte [fullwíte MSS. B. H.] *let him pay the full fine*, L. In. 43: Th. i. 128, 18: 72; Th. i. 148, 8: L. C. S. 49; Th. i. 404, 9, note 18, MS. G.

fulwod *baptized*, Bd. 1, 27; S. 493, 2, note; *pp. of* fulwian.

ful-wrǣtlíce; *adv. Full wondrously, very wonderfully*; mirissĭme :—Ðæt me on gescyldrum scínan mótan fulwrǣtlíce wundne loccas *that on my shoulders curled locks may shine very wonderfully*, Exon. 111 b; Th. 428, 6; Rä. 41, 104.

fulwuht, es; *n. Baptism*; baptismus :—Hér Birīnus bisceop bodude West-Seaxum fulwuht *in this year* [A. D. 634] *bishop Birinus preached baptism to the West-Saxons*, Chr. 634; Erl. 24, 9. v. fulluht.

ful-wurþlíce, full-weorþlíce; *adv. Full worthily, very honourably*; dignissĭme :—Hine man byrigde fulwurþlíce [fullweorþlíce, Th. 294, 21, col. 2], swā he wyrðe wæs *they buried him very honourably, as he was worthy*, Chr. 1036; Th. 294, 22, col. 1.

ful-wyrcan, full-wyrcan; *p.* -worhte; *pp.* -worht *To do fully, accomplish, commit*; perfĭcĕre :—Gif hwā griþbryce fulwyrce *if any one commit a breach of the peace*, L. C. S. 62; Th. i. 408, 22.

ful-yrre; *adj. Full angry, very angry*; valde irātus :—He fulyrre wód *he rushed forth very angry*, Byrht. Th. 139, 13; By. 253.

funde, *pl.* fundon; *pp.* funden *Found*, Cd. 72; Th. 87, 6; Gen. 1444: 122; Th. 156, 27; Gen. 2595: 174; Th. 220, 5; Dan. 66; *p. and pp. of* findan.

fundian, ic fundige; *p.* ode, ade, ede; *pp.* od, ad, ed *To endeavour to*

find, tend to, aspire to, strive, go forward, hasten, intend, desire; nīti, tendēre, intendĕre, properāre :—Ic wylle fundian sylf to ðam sīþe *I will hasten myself to the journey,* Exon. 119 a; Th. 456, 24; Hy. 4, 71; 89 b; Th. 336, 21; Gn. Ex. 52. Fundigende of ðissere worulde *hastening from this world,* Homl. Th. ii. 360, 2. Ic fundige to ðē *I hasten to thee,* Exon. 118 b; Th. 454, 28; Hy. 4, 40. Hwider fundast ðū *whither art thou hastening?* Cd. 103; Th. 137, 5; Gen. 2269. He fundaþ to ðæm weorþscipe ðæs folgoþes *he aspires to the honour of rule,* Past. 8, 2; Hat. MS. 12 b, 25: 11, 3; Hat. MS. 15 a, 9. Hī to ðē hīonan fundiaþ *they tend hence to thee,* Bt. 33, 4; Fox 132, 25, 38. Gif twegen men fundiaþ to ānre stowe *if two men are going to the same place,* 36, 4; Fox 178, 10: Past. 18, 1; Hat. MS. 25 b, 6. Nū ðū mōst feran ðider ðū fundadest *now thou mayest go whither thou desiredst,* Exon. 32 b; Th. 102, 12; Cri. 1671. Fundode wrecca of geardum *the stranger hastened from the dwellings,* Beo. Th. 2279; B. 1137. Hwæðer ðū nū ongite forhwý ðæt fýr fundige up *dost thou understand why fire tends upwards?* Bt. 34, 11; Fox 150, 19. Frīora æghwilc fundie to ðæm ēcum gōde *let every one of the free aspire to the eternal good,* Bt. Met. Fox 21, 4; Met. 21, 2. Swā hie fundedon *as they desired,* Cd. 115; Th. 150, 17; Gen. 2493: Exon. 106 a; Th. 404, 11; Rä. 23, 6. [*Laym.* fondien *to seek, try: O. Sax.* fundōn *to strive: O. H. Ger.* fundjan, fundēn *subire.*] DER. tofundian.

fundung, e; *f. A going, departure;* abītus, decessus :—He nolde on his fundunge ofer sǽ hīred healdan *he would not hold a court on his departure over sea,* Chr. 1106; Erl. 241, 2.

furan; *sulcare, scribere,* Hpt. Gl. 465, 507. v. furh.

FURH; *nom. gen. acc; dat.* fyrh; *dat. pl.* furum; *f. A* FURROW; sulcus :—Furh *sulcus,* Ælfc. Gl. 1; Som. 55, 17; Wrt. Voc. 15, 17: 289, 80. Ne furh ne fōtmǽl *neither furrow nor foot-mark,* L. O. 13; Th. i. 184, 7. Andlang ðære furh *along the furrow,* Cod. Dipl. 554; A. D. 969; Kmbl. iii. 38, 34. Andlang weges to ðære gedrifenan fyrh; andlang fyrh *along the way to the driven furrow; along the furrow,* 1172; A. D. 955; Kmbl. v. 332, 22 : Cod. Dipl. Apndx. 441; A. D. 956; Kmbl. iii. 437, 21. On ða furh on the furrow, 356; Kmbl. iii. 409, 5 : 441; A. D. 956; Kmbl. iii. 437, 23. Ðám drīum furum *in the dry furrows,* Bt. 5, 2; Fox 10, 31. [*Wyc.* forewis, forowis *furrows: Piers P.* furwe: *Plat.* fore, fare, *f: Frs.* furch, furge: *O. Frs.* furch, *f: Dut.* vóre, *f: Ger.* furche, *f: M. H. Ger.* vurch, *f: O. H. Ger.* furh, furuh, *f: Dan.* fure, *m. f: Swed.* fåra, *f: Icel.* furask *to be furrowed.*]

furh-wudu; *m. Fir-wood, a fir-tree;* pīnus, Gl. C. fol. 48 d; Lchdm. iii. 327, 39, col. 1.

furlang, furlung, es; *n. A* FURLONG; stādium :—On ðæt lange furlang *to the long furlong,* Cod. Dipl. 578; A. D. 973; Kmbl. iii. 97, 32. Bethania ys gehende Hierusalem ofer fýftyne furlang *ĕrat Bethania juxta Ierosŏlymam quási stădiis quindĕcim,* Jn. Bos. 11, 18. Twentig furlanga *stădia vĭginti,* 6, 19: Lk. Bos. 24, 13. Se is þreóra furlunga brād *qui est latitūdinis circiter trium stădiōrum,* Bd. 1, 25; S. 486, 20.

furþ-an, furþ-on, furþ-um; *adv.* [furþ = forþ *forth,* furþan, furþon, furþum, forþum, *dat.*] *Also, too, even, indeed, further, at first;* etiam, quĭdem, prīmo :—Ic secge eów sōþlīce, ðæt furþon Salomon on eallum hys wuldre næs oferwrigen swā swā ān of ðyson *dico autem vobis, quoniam nec Salomon in omni gloria sua coopertus est sicut unum ex istis,* Mt. Bos. 6, 29. He wēneþ furþon ðæt he man ne sý *he even thinks that he is not man,* Blickl. Homl. 179, 5. Ic furþum ongan būgan *I first* [prīmo] *began to dwell,* Exon. 50 b; Th. 176, 21; Gū. 1213.

FURÐOR, furður; *adv.* FURTHER, *more, forwards;* ultĕrius, ultra, amplius, porro :—Ne gang ðū ānne stæpe furður *go not thou one step further,* Jos. 10, 12: Cd. 223; Th. 292, 24; Sat. 445. Siððan he ðone fintan furður cūðe *when he further knew the sequel,* Exon. 74 b; Th. 278, 32; Jul. 606: Cd. 21; Th. 26, 3; Gen. 401. Eóde se sæster hwætes to lv penega, and eác furður *the sester of wheat went to fifty-five pence, and even further,* Chr. 1167; Erl. 167, 22. Ðæt he á furður wǽre ðonne ōðre brōðor *that he was always more than the other brethren,* Past. 17, 6; Hat. MS. 23 b, 1. Ic wille furður gán *I will go forwards,* Byrht. Th. 139, 1; By. 247. Furður dôn *to prefer, esteem,* Past. 17, 7; Hat. MS. 23 b, 14. [*O. Sax.* furður *further: O. Frs.* furthor, furdor *further: Ger.* fürder *moreover: M. H. Ger.* vürder *further: O. H. Ger.* furdir *ultĕrius.*]

furðra, *m;* furðre, *f. n: comp. adj.* FURTHER, *greater, superior;* ultĕrior, mājor, prior :—Nys se þeówa furðra ðonne se hláford *non est servus mājor dŏmĭno suo,* Jn. Bos. 13, 16. Hwilc cræft ðē geþūht betwux ðás furðra wesan *quæ ars tĭbi vidĕtur inter istas prior esse?* Coll. Monast. Th. 30, 13.

furðrung *a furthering, promoting, forwarding,* Somn. 2 : 17, Lye. v. fyrðrung.

furþ-um; *adv.* [*dat.* of forþ?] *Also, even, indeed, at first;* prīmo, ĕtiam :—Ne furþum nǽnig nǽre on heofenum *nor was there any even in heaven,* Blickl. Homl. 117, 27. He furþum ongan *he also began,* Cd. 63; Th. 75, 11; Gen. 1238. Ic furþum ongan *I first began,* Exon. 50 b; Th. 176, 21; Gū. 1213. v. furþum-līc.

furþum-līc; *adj.* [furþ = forþ *forth, onwards;* furþum = forþum, *dat.*

to onwards, excessive? līc] *Luxurious, indulgent;* luxŭriōsus, mollis, ventricōsus :—Sarðanapālus [MS. -olus] se sīþmesta cyningc, wæs swīðe furþumlīc man *Sardanapalus the last king was a very luxurious man,* Ors. 1, 12; Bos. 35, 15.

furður; *adv. Further, more;* ultĕrius, ultra :—Ǽr gē on land furður feran *ere ye proceed further into the land,* Beo. Th. 513; B. 254: 1527; B. 761: Exon. 73 b; Th. 274, 30; Jul. 541: Cd. 94; Th. 121, 22; Gen. 2014: Andr. Kmbl. 2976; An. 1491. Ðæt ðē cyning engla gefrætwode furður micle ðonne eall gimma cynn *that the king of angels adorned thee much more than all the kinds of gems,* 3035; An. 1520. v. furðor.

furum *in furrows,* Bt. 5, 2; Fox 10, 31; *dat. pl.* of furh.

FÚS; *adj. Ready, prepared, prompt, quick, eager, hastening, prone, inclined, willing, ready for death, dying;* promptus, cĕler, părātus, prōnus, cŭpĭdus, properæ morti devōtus, mŏrĭbundus :—Se ðe stōd fús on faroþe *he who stood ready on the beach,* Andr. Kmbl. 509; An. 255 : Exon. 126 b; Th. 487, 7; Rä. 72, 24: Byrht. Th. 139, 68; By. 281. He ferde siððan swīðe fús to Rōme *he, being very quick, afterwards went to Rome,* Ælfc. T. 30, 8: Cd. 23; Th. 28, 28; Gen. 443: 147; Th. 184, 6; Exod. 103. Ic eom sīþes fús *I am ready for the journey,* Beo. Th. 2955; B. 1475: Elen. Kmbl. 2436; El. 1219: Exon. 58 b; Th. 212, 10; Ph. 208. Is him fús hyge *their mind is ready for death,* Andr. Kmbl. 3327; An. 1666. Ealle ða gemoniaþ mōdes fúsne *all these admonish the prompt of mind,* Exon. 82 a; Th. 309, 1; Seef. 50: Andr. Kmbl. 3307; An. 1656. Ðū me fúsne frignest *thou askest me dying,* Exon. 50 b; Th. 175, 27; Gū. 1201: 49 b; Th. 171, 22; Gū. 1130. Geseah ic ðæt fúse beácen wendan wǽdum and bleóm *I saw the hastening beacon change in hangings and colours,* Rood Kmbl. 42; Kr. 21. Gesáwon randwīgan segn ofer sweóton, fús on forþweg *the warriors saw the sign over the bands, hastening on its onward way,* Cd. 148; Th. 185, 27; Exod. 129. Wǽron æðelingas eft to leódum fúse to farenne *the nobles were ready to go again to their people,* Beo. Th. 3614; B. 1805: Cd. 151; Th. 190, 9; Exod. 196. Ic of fúsum rád *I rode from the ready* [men], Exon. 130 a; Th. 498, 28; Rä. 88, 8. [*Orm.* fus *eager: Laym.* fuse, *pl. prompt, ready: O. Sax.* fūs *inclined, ready: O. H. Ger.* funs *prōnus, promptus: Dan.* fuse *to rush forth: Icel.* fúss *willing, wishing for.*] DER. bealo-fús, ellor-, grund-, hell-, hin-, ūt-, wæl-.

fús, es; *n. A hastening, progress;* festīnātio, progressus :—Se ðe leófra manna fús feor wlātode *who beheld afar the dear men's progress,* Beo. Th. 3836; B. 1916.

fúse; *adv. Readily, promptly;* părāte, prompte, Th. Anlct.

fús-leóþ, es; *n. A parting-song, death-song, dirge;* mŏrientis cantus, fūnebris nēnia :—Ðær wæs ýþfynde innan burgum fúsleóþ galen *there was easy to be found within the dwellings the death-song sung,* Andr. Kmbl. 3097; An. 1551. Ðū scealt fúsleóþ galan *thou shalt sing the death-song,* Exon. 17 a; Th. 39, 17; Cri. 623: 52 b; Th. 183, 1; Gū. 1321.

fúslic; *adj. Ready, prepared;* părātus :—Him Onela forgeaf his gædelinges gūþgewǽdu, fyrdsearu fúslic *Onela gave him his companion's battle-garments, ready martial gear,* Beo. Th. 5229; B. 2618. He geseah beorhte randas, fyrdsearo fúslicu *he saw bright shields, a war-equipment ready,* 469; B. 232.

fúslice; *adv. Readily, promptly, gladly;* prompte, lĭbenter :—Ðæt hī fúslīce gehýrdon, ða ðe him gelǽrde wǽron *ut lĭbenter ea, quæ dĭcĕrentur, audīrent,* Bd. 4, 27; S. 604, note 17, MS. T.

fús-trendel; *focus,* Hpt. Gl. 439.

fýfteógeða *the fifteenth* :—Forþferde he ðý fýfteógeðan dæge Kalendarum Martiarum *qui defunctus die decima quinta Kalendārum Martiārum,* Bd. 4, 5; S. 571, 36. v. fífteóða.

fýftyne *fifteen;* quindĕcim :—Ofer fýftyne furlang *over fifteen furlongs,* Jn. Bos. 11, 18. v. fíftyne.

fyht *a fight, battle,* Som. Ben. Lye. v. feoht.

fyhte-horn; *m. A fighting or battle-horn;* pugnātōrium cornu :—Ealra fyrenfulra fyhtehornas ic bealdlíce gebrece snióme *omnia cornua peccātōrum confringam,* Ps. Th. 74, 9.

fyhtling, es; *m. A fightling, soldier;* præliātor, Gr. Dial. 2, 3.

fyht-wíte, fiht-wíte, es; *n. A fine for fighting;* pugnæ mulcta :—Ðæt fyht-wíte *the fine for fighting,* L. E. G. 13; Th. i. 174, 27.

fyl, es; *m. A fall, ruin, destruction;* cāsus, intĕrĭtus :—Hý ðam feore fyl gehěhton *they threatened destruction to his life,* Exon. 40 b; Th. 135, 7; Gū. 520: Byrht. Th. 133, 57; By. 71: 139, 35; By. 264. DER. hrā-fyl. v. fyll, es; *m.*

fýlan; *p. de; pp. ed To foul, defile;* inquīnāre, fœdāre, contămĭnāre. DER. a-fýlan, be-, ge-. v. fúlian.

fylc, es; *n. A company, troop, tribe, country, province;* agmen, caterva, trĭbus, provincia. [*Icel.* fylki, *n. a county or shire.*] DER. æl-fylc, bí-, ge-.

fylcian; *p.* ade *To arrange troops* :—Harald his liþ fylcade *Harold drew up his force,* Chr. 1066; Erl. 200, 33. [*Icel.* fylkja.]

fyld, es; *m. A fold, volume;* vŏlūmen, Som. Ben. Lye. Hpt. Gl. 494.

fylde, *pl.* fyldon *filled,* Andr. Kmbl. 1046; An. 523: Jn. Bos. 6, 13; *p.* of fyllan.

FYLGEAN, fylgan, fylgian, fyligean, fylian, filian, feligean; *p.* de; *pp.* ed; *v. trans. dat. acc. To follow, attend, follow or carry out; sĕqui, insĕqui, exsĕqui:—Ðæt hearma swā fela fylgean sceolde monna cynne *that so many ills must follow to mankind*, Cd. 33; Th. 44, 15; Gen. 709: L. Eth. ii. 9; Th. i. 288, 29. Ongon se wīsdóm his gewunan fylgan *wisdom began to follow his custom*, Bt. Met. Fox 7, 2; Met. 7, 1: Exon. 122 a; Th. 468, 6; Phar. 3: Judth. 10; Thw. 21, 24; Jud. 33. Ðe him fylgian wolde *who would follow him*, Hy. 10, 39; Hy. Grn. ii. 293, 39. He ne lēt him ænig ne fyligean *non admisit quemquam se sĕqui*, Mk. Bos. 5, 37: 8, 34. Ða he on his weorcum wæs geornlíce fyligende *which he was diligently carrying out in his works*, Bd. 3, 28; S. 560, 17. We wǣron þē fylgende *we were following thee*, St. And. 2, 20. Him fyliende *sĕquentes se*, Jn. Bos. 1, 38. Ic fylige *sĕquor*, Ælfc. Gr. 36; Som. 38, 24. Ðū gedwolan fylgest *thou followest error*, Exon. 68 b; Th. 254, 25; Jul. 202. Gūþmecga him fylgeþ *the warrior pursues him*, Salm. Kmbl. 186; Sal. 92. Ic fylgde gódnysse *sĕquebar bonĭtātem*, Ps. Spl. 37, 21: Bt. Met. Fox 26, 108; Met. 26, 54. Se wrāða boda fylgde him *the fell messenger followed him*, Cd. 32; Th. 43, 9; Gen. 688. Ðe he ǣr fyligde [fylgde, MS. B.] *whom he before followed*, L. Ed. 10; Th. i. 164, 16. Gē gedwolan fylgdon *ye followed error*, Elen. Kmbl. 742; El. 371: Exon. 29 a; Th. 88, 16; Cri. 1441. Twegen leorningcnihtas fyligdon ðam Hǣlende *duo discĭpŭli sĕcūti sunt Jēsum*, Jn. Bos. 1, 37. Fyle [fylge MS. C.] ðū ðam persĕquĕre eam, Ps. Spl. 33, 14. Ðæt we Godes lage fylgean [fylgian MS. B.] *that we follow God's law*, L. C. S. 85; Th. i. 424, 7. Ðæt hí georne heora bócum and gebēdum fylgean *that they strictly attend to their books and prayers*, L. Eth. vi. 41; Th. i. 326, 3. [Wyc. foleweden, *p. pl*: Piers P. folwe, folwen: Chauc. folwe: Laym. folien, foluen, fulien; Orm. follʒhenn: O. Sax. folgón: Frs. folgjen: O. Frs. folgia, fulgia, folia: Dut. volgen: Ger. folgen: M. H. Ger. volgen: O. H. Ger. folgén, folkēn: Dan. følge: Swed. följa: Icel. fylgja.] DER. æfter-fylian, æt-fylgian, be-filgan, ge-fylgan, under-fylgan. v. folgian.

fylgend, es; *m. One who follows or carries anything out, a performer*; exsĕcūtor:—Ðara þinga ðe he ōðre lǣrde to dónne, he sylfa wæs se wilsumesta fylgend *eōrum quæ agenda dŏcēbat ĕrat exsĕcūtor devōtissĭmus*, Bd. 5, 22; S. 644, 4, note, MSS. B. C.

fylgestre; *f. sectatrix.* Hpt. Gl. 435.

fylging, e; *f. A following*:—Miþ fylginge *sectando*, Rtl. 16, 23; 56, 5.

fylging, e; *f. That which follows, a harrow*; occa, Cot. 143.

fylian *to follow*:—Fyle ðū ðam *persequere eam*, Ps. Spl. 33, 14. Fylidon, Mt. Kmbl. C. C. 4, 22. v. fylgean.

fyligean *to follow, attend, follow or carry out*, Mk. Bos. 5, 37: 8, 34: Bd. 3, 28; S. 560, 17. v. fylgean.

fylignes, -ness; *f. A following, completing, executing*; successio, exĕcūtio:—Ðæt to gódra dǣda fylignessum he hī aweahte *ut eos ad opĕrum bŏnōrum exĕcūtiōnem excĭtāret*, Bd. 3, 5; S. 526, 33.

FYLL, fill, fyllu, fyllo, e; *f. The* FILL, *fulness, plenty*; plēnĭtūdo, satŭrĭtas:—Drinc nū ðíne fylle *drink now thy fill*, Ors. 2, 4; Bos. 45, 36. Gē etaþ to fylle *comēdētis in satŭrĭtāte*, Lev. 26, 5. Fylle gefrægnod *known by its plenteousness*, Beo. Th. 2670, note; B. 1333. Fylle gefǣgon *they rejoiced in the plenty*, 2032; B. 1014. Næs hie ðǣre fylle gefeán hæfdon *they had no joy of that plenty*, 1128; B. 562. Ic sylle heora hungrium hlāf to fylle *pauperes ejus satŭrābo pānĭbus*, Ps. Th. 131, 16. [Ger. fülle, f: M. H. Ger. volle, f. m: vülle, f: O. H. Ger. folla, follî, fullî, f: Goth. fullei, fullo, f: Dan. fylde, m. f: Swed. fylle, n: Icel. fylli, fyllr, f.] DER. wist-fyll.

FYLL, fyl, fell, fiell, es; *m.* **I.** *a* FALL, *ruin, destruction, death*; cāsus, intĕrĭtus:—Crist is ofermōdigra fyll *Christ is the fall of the high-minded*, Ors. 3, 2; Bos. 55, 6. Æfter his fylle *after his death*, 6, 5; Bos. 119, 22. Mîne innoþas on ðam fylle tolocene wǣron interānea essent ruendo convulsa, Bd. 5, 6; S. 619, 31. Se bisceop sárgode be ðam fylle and mîne forwyrde *episcŏpus de cāsu et intĕrĭtu meo dŏlēbat*, 5, 6; S. 619, 32. Æt fylle *at the fall*, L. M. 1, 4; Lchdm. ii. 48, 14, note. **II.** *a* FALL, *case, inflection in grammar*; cāsus, inflectio:—Cāsus, ðæt is fyll oðře gebígednis *case, that is a declining or inflection*, Ælfc. Gr. 14; Som. 17, 23. [Orm. fall: O. Sax. fal, m: Frs. O. Frs. fal, fel, m: Dut. val, m: Ger. fall, m: M. H. Ger. val, m: O. H. Ger. fal, m: Dan. fald, n: Swed. fall, n. lapsus, cāsus, Rask Hald. Egils.] DER. wæl-fyll.

FYLLAN; ic fylle, ðū fyllest, fylst, he fylleþ, fylþ, *pl.* fyllaþ; *p.* fylde, fyllde, *pl.* fyldon; *impert.* fyl, *pl.* fyllaþ; *pp.* fylled, fyld; *v. trans. To* FILL, *replenish, satisfy, cram, stuff, finish, complete, fulfil*; implēre, replēre, satŭrāre, farcīre, supplēre, complēre:—Ðæt sceolon fyllan firengeorne men *sinful men shall fill that*, Exon. 31 b; Th. 98, 11; Cri. 1606: 124 b; Th. 479, 16; Rä. 62, 8. Ðæt he fyrngewyrht fyllan sceolde *that he should finish his former deeds*, 47 a; Th. 160, 16; Gū. 944. Ic crammige oðře fylle *farcio*, Ælfc. Gr. 30, 2; Som. 34, 36. Ic fylle *suppleo*, 26, 1; Som. 28, 29. Ðū fyllest [fylst Spl.] *ealra wihta gehwam bletsunga tu imples omne anĭmal bĕnĕdictiōne*, Ps. Th. 144, 17. He heáhgetimbro fylleþ fýres egsan *he shall fill the high structures with*

fire's horror, Exon. 22 a; Th. 60, 25; Cri. 975. Se ðe fylþ on gódum gewilnunge ðíne *qui replet in bŏnis desĭdērium tuum*, Ps. Spl. 102, 5. Hí fyllaþ mid feore foldan gesceafte *they shall fill earth's creation with their spirit*, Exon. 22 a; Th. 59, 15; Cri. 953: Ps. Th. 64, 5. He fylde hig *satūrāvit eos*, Ps. Spl. 104, 38. He wuldres fylde beorhtne boldwēlan *he filled the bright dwelling of wealth with glory*, Andr. Kmbl. 1046; An. 523; Hy. 10, 19; Hy. Grn. ii. 293, 19. Moises sprǣc ðás word befóran Israēla folce and hig fyllde ōþ ende *lŏcūtus est Moyses audiente ūnĭverso cœtu Israel verba carmĭnis hūjus et ad finem usque complēvit*, Deut. 31, 30. Hig fyldon twelf wylian fulle *implēvērunt duodĕcim cophĭnos*, Jn. Bos. 6, 13. Fyl nū ða frumsprǣce *fulfil now the saying of old!* Exon. 53 b; Th. 188, 7; Az. 42: Cd. 190; Th. 236, 24; Dan. 326. Tudre fyllaþ eorþan ælgrēne *fill the all-green earth with progeny*, 10; Th. 13, 2; Gen. 196: 75; Th. 92, 24; Gen. 1533. Beóþ ðíne feldas fylde mid wǣstmum *campi tui replēbuntur ūbertāte*, Ps. Th. 64, 12. Ðonne heofon and hel hæleða bearnum fylde weorþeþ *when heaven and hell shall be filled with the sons of men*, Exon. 31 a; Th. 97, 20; Cri. 1593. [Wyc. fill, fille: Piers P. fillen: Chauc. filled, *pp*: Laym. fulle, iuullen: Orm. fillenn: Plat. vullen: O. Sax. fullian: Frs. folljen: O. Frs. fullia, folla, fella: Dut. vullen: Ger. füllen: M. H. Ger. vüllen: O. H. Ger. fulljan: Goth. fulljan: Dan. fylde: Swed. fylla: Icel. fylla.] DER. a-fyllan, be-, ge-, ongeán-, samod-: æ-fyllende.

FYLLAN = fellan; ic fylle, ðū fyllest, he fylleþ, *pl.* fyllaþ; *p.* fylde, *pl.* fyldon; *pp.* fylled; *v. trans. To fell, cut down, cast down, throw down, destroy*; prosternĕre, cædĕre, dejĭcĕre, destruĕre:—Ðá us man fyllan ongan ealle to eorþan *then they began to fell us all to the ground*, Rood Kmbl. 146; Kr. 73. Fyllan, Judth. 11; Thw. 24, 17; Jud. 194. Gif ðū wylt ða firenfullan fyllan mid deáþe *if thou wilt fell the wicked with death*, Ps. Th. 138, 16. Ic beámas fylle *I fell trees*, Exon. 101 a; Th. 381, 11; Rä. 2, 9. Se grimmesta hungor hí fylde *fāmes acerbissima eos prostrāvit*, Bd. 4, 13; S. 582, 29: Cd. 35; Th. 46, 20; Gen. 747. Ða synsceaðan Godes tempel fyldon *the sinful cast down God's temple*, Exon. 18 a; Th. 44, 27; Cri. 709. Fyll ða oferhydigan *cast down the proud*, Ps. Th. 73, 22. Hergas fyllaþ *cast down the idols*, Exon. 14 b; Th. 30, 27; Cri. 486. [Chauc. felle: Laym. fallen: O. Sax. fellian: Frs. fellen: O. Frs. falla, fella: Dut. vellen: Ger. fällen: M. H. Ger. falljan, fellen: Dan. fælde: Swed. fälla: Icel. fella.] DER. a-fyllan, be-, ge-, of-, to-.

fyllend, es; *m. A fulfiller, performer*; exsĕcūtor:—Ðara þinga ðe he ōðre lǣrde to dónne, he sylfa wæs se wilsumesta fyllend *eōrum quæ agenda dŏcēbat ĕrat exsĕcūtor devōtissĭmus*, Bd. 5, 22; S. 644, 4.

fylle-seóo; *adj. Falling sick, epileptic, lunatic*; ĕpĭlepticus = ἐπιληπτικός, lunātĭcus:—Ðÿ-læs cild sý fylleseóc *lest the child be epileptic*, Med. ex Quadr. 5, 12; Lchdm. i. 350, 12. He ys fylleseóc *lunātĭcus est*, Mt. Bos. 17, 15. Wið fylleseócum men *for an epileptic man*, Med. ex Quadr. 8, 9; Lchdm. i. 358, 21. Heð fylleseócum helpeþ *it helpeth the epileptic*, Herb. 143, 1; Lchdm. i. 266, 5.

fylle-seócnys, -nyss, e; *f. The falling sickness, epilepsy*; ĕpĭlepsia = ἐπιληψία:—Wið fylleseócnysse *for the falling sickness*, Herb. 61, 3; Lchdm. i. 164, 9.

fylle-wærc, felle-wærc, es; *n. The falling sickness, epilepsy*; ĕpĭlepsia = ἐπιληψία:—Of ðæs magan ádle cumaþ hramma and fyllewærc *from the disease of the stomach come cramps and epilepsy*, L. M. 2, 1; Lchdm. ii. 174, 25.

fyllnis, se; *f. Fulness, that which makes full or complete, a supplement*:—Fyllnis *plenitudo*, Mt. Kmbl. Lind. 9, 16; Rtl. 100, 13. Fyllnis *supplementum*, Mk. Skt. Lind. 2, 21. Fylnis *perfectio*, p. 1, 13.

fyllu, e; *f.: fyllo; indecl. f. Fulness*; plēnĭtūdo:—Anfēng fǣmne fyllo *the woman received fulness*, Exon. 112 a; Th. 429, 15; Rä. 43, 5. v. fyll, e; *f.*

fyllung, e; *f. A fulfilling, performing*; perfectio, Som. Ben. Lye.

fylmen, es; *n. A film, thin skin, prepuce*; præpūtium, omentum:—Gē emsnīdaþ ðæt flǣsc eówres fylmenes *circumcidĕtis carnem præpūtii vestri*, Gen. 17, 11. Se werhādes man, ðe ne byþ ymsnīden on ðam flǣsce hys fylmenes, his sāwul biþ adilegod of his folce *masculus, cūjus præpūtii cāro circumcīsa non fuĕrit, delēbĭtur ănĭma illa de pŏpŭlo suo*, 17, 14: Homl. Th. i. 94, 32. Fylmena *films; omenta vel membrānæ*, Ælfc. Gl. 31; Som. 61, 93; Wrt. Voc. 27, 23: Cot. 133. Fylmen *omentum*, 74; Som. 71, 61; Wrt. Voc. 44, 43. On ðam fylmene *in præpūtio*, Homl. Th. i. 94, 13. Feóllon swylce fylmena of his eágum *there fell as it were films from his eyes*, Homl. Th. i. 386, 31.

fýlnes, -ness, e; *f. Foulness*; fœdĭtas, fœtor, fūlīgo:—Eorþan fýlnes, eal forweornast *foulness of earth, thou art all rotting*, Exon. 98 a; Th. 368, 7: Cot. 83. v. fúlnes.

fylst, he fylþ *fillest, he fills*, Ps. Spl. 144, 17: 102, 5; *2nd and 3rd sing. pres.* of fyllan.

FYLST, e; *f. Help, assistance*; auxĭlium:—Mid Godes fylste *with God's help*, Bt. Met. Fox 23, 14; Met. 23, 7: Ors. 1, 12; Bos. 35, 20. [Laym. fulste, vulste *aid, help*: O. Frs. fulliste, folliste, folste, *aid*. [Cf. fullǣst.]

fylstan, filstan, ic fylste, he fylsteþ; *p.* [fylstede =] fylste, *pl.* fylston; *subj. pres.* fylste, *pl.* fylsten, fylston; *pp.* fylsted; *v. trans. dat.* [fylst e;

f. help] To help, give help, aid, protect; adjŭvāre, auxĭlĭāri, protĕgĕre:—Ongan him fylstan *began to give help to them*, Byrht. Th. 139, 37; By. 265. Hig bícnodon hyra geferan, ðæt hí him fylston *annŭerunt sŏciis, ut adjŭvārent eos* [*that they should give help to them*], Lk. Bos. 5, 7. Him fylste Drihten *the Lord helped him*, Cd. 124; Th. 159, 8; Gen. 2631. Pirrus him fylste *Pyrrhus helped him*, Ors. 3, 11; Bos. 75, 28. Hí him fylston wel *they helped him well*, Cd. 114; Th. 149, 34; Gen. 2484. Arĭson and fylston eów *surgant et vos prŏtĕgant*, Deut. 32, 38. [*Laym.* fulsten.] DER. ge-fylstan, to-: ge-fylsta. Cf. fullæstan.

fylþ *falls*, Mt. Bos. 21, 44; *3rd pers. pres. of* feallan.

FÝLþ, e; *f.* FILTH, *impuri'y, rottenness*; spurcĭtia, putrēdo:—Hig synt innan fulle ealre fýlþe *intus plēna sunt omni spurcĭtia*, Mt. Bos. 23, 27. Wið áne cwēnan fýlþe adreógaþ *cum ūna meretrīce spurcĭtiem ăgunt*, Lup. Serm. 1, 11; Hick. Thes. ii. 102, 27, 29: Scint. 9. [*O. Sax.* fūlitha, *f.*: *O. H. Ger.* fūlida, *f.*]

fyl-wērig; *adj. Slaughter-weary*; cæde defessus:—Ðú hine geseón móste fylwērigne *thou mightest have seen him slaughter-weary*, Beo. Th. 1929; B. 962.

fýnd *a fiend, an enemy*, Ps. Spl. 40, 12. v. feónd.

fýnd, *pl.* of feónd: Lev. 26, 8, 16: Bt. 20; Fox 72, 21: Mt. Bos. 5, 44: Lk. Bos. 6, 27, 35.

fynde; *adj. Able to be found.* DER. eáþ-fynde, éþ-, ýþ-.

fyne, es; *n? Moisture, mould*; ūlīgo:—Fyne *allugo* [= *ūlīgo*], Ælfc. Gl. 106; Som. 78, 47; Wrt. Voc. 57, 28.

fynegian; *p.* ode; *pp.* od [fynig *mouldy*] *To become mouldy* or *musty*; mūcescĕre:—Ðæt ðæt hálige hūsel sceole fynegian *that the holy housel should become mouldy*, L. Ælf. C. 36; Th. ii. 360, 7.

fynel, es; *m. Fennel*; fēnĭcŭlum, Ælfc. Gl. 39; Som. 63, 68; Wrt. Voc. 30, 20. v. finol.

fynig, fini; *adj. Mouldy, musty, damp*; mūcĭdus, ulīgĭnōsus:—Gyf ðæt hūsel byþ fynig *if the housel be mouldy*, L. Ælf. C. 36; Th. ii. 360, 8, 13. Fynig *alluginatus* [= *ulīgĭnōsus*], Ælfc. Gl. 106; Som. 78, 48; Wrt. Voc. 57, 29.

fyorh; *gen.* fyores; *dat.* fyore; *n. Life*; vīta:—Fíf and hundteontig on fyore lífde wintra *he passed a hundred and five years in life*, Cd. 59; Th. 72, 10; Gen. 1184. v. feorh I.

fyr, fyrr, fier; *adv.* [*comp.* of feor; *adv. far*, q. v.] *Farther*; ultĕrius, longius:—Ðeáh ðú fyr séo ðonne ðú wære *though thou art farther than thou wast*, Bt. 5, 1; Fox 8, 33, note 7, MS. Bod. Ær gē fyr heonan feran *ere ye proceed farther hence*, Beo. Th. 510; B. 252: 288; B. 143. Fyr faran *longius īre*, Lk. Bos. 24, 28. Fyr fleón *to flee farther*, Ors. 1, 12; Bos. 36, 4.

FÝR, fír, es; *n.* FIRE, *a fire, hearth*; ignis, fŏcus:—Būton he hæbbe swá scearp andget swá ðæt fýr *unless he have an understanding as sharp as the fire*, Bt. 39, 4; Fox 216, 28. Fýr *ignis*, Wrt. Voc. 284, 11: Mk. Bos. 9, 44, 46: Ex. 22, 6: Lev. 10, 2: Ps. Spl. 49, 4. Fýr *ignis vel fŏcus*, Wrt. Voc. 82, 51. Him beforan fóron fýr and wolcen *fire and cloud journeyed before him*, Cd. 146; Th. 183, 18; Exod. 93: 169; Th. 212, 9; Exod. 536: 192; Th. 239, 22; Dan. 374. Ðæs fýres gecynd is hát and drīe *the nature of fire is hot and dry*, Boutr. Scrd. 18, 22, 23. In fýres fæðm *into the fire's embrace*, Beo. Th. 372; B. 185. Fýres feng *grasp of the fire*, 3532; B. 1764. Lágon ða óðre fýnd on ðam fýre *the other fiends lay in the fire*, Cd. 17; Th. 21, 10; Gen. 322: 24; Th. 31, 19; Gen. 487: 117; Th. 152, 17; Gen. 2521. Sý hyt forcorfen, and on fýr aworpen *excĭdētur, et in ignem mittētur*, Mt. Bos. 7, 19: 17, 15: Mk. Bos. 9, 43: Lk. Bos. 3, 9: Jn. Bos. 15, 6. Ne onæle gē nán fýr on ðam dæge *non succendētis ignem per diem sabbăti*, Ex. 35, 3: 22, 6. Mid fýre *with fire*, Bt. 39, 4; Fox 216, 25. He sweartade fýre and ættre *he blackened with fire and venom*, Cd. 214; Th. 269, 26; Sat. 79: 220; Th. 284, 21: Sat. 325: Beo. Th. 5183; B. 2595. [*Wyc.* fyr, fire: *Piers P.* fir: *Chauc.* fire: *R. Glouc.* fyur: *Laym.* fur: *Orm.* fir: *Scot.* fyre: *Plat.* vür, vüer, füer, *n*: *O. Sax.* fiur, *n*: *Frs.* fĵœr: *O. Frs.* fior, fiur, *n*: *Dut.* vuur, *n*: *Ger.* feuer, *n*: *M. H. Ger.* viur, viuwer, viwer, *n*: *O. H. Ger.* fiur, *n*: *Dan.* fyr, *m. f*: *Swed.* fyr, *m. a lighthouse, beacon*: *Icel.* fúrr, *m. fire*: *Lat.* prūna, *f. a burning coal*: *Grk.* πῦρ, *n*.] DER. ád-fýr, æled-, bæl-, heáh-, heaðo-, helle-, líg-, wæl-, wælm-, wan-, won-.

fyran; *p.* fyrde *To go*; īre:—Ine fyrde to Sce. Petres *Ine went to St. Peter's*, Text Rof. 61, 15. v. feran.

fýran; *adj. Fiery*; ignītus:—God gelogode fýran swurd *God placed a fiery sword*, Boutr. Scrd. 20, 30. v. fýren.

fýran; *p.* de; *pp.* ed *To castrate*; castrāre:—Báras fýran *apros castrāre*, Obs. Lun. § 3; Lchdm. iii. 184, 19. DER. a-fýran.

fyras; *gen.* fyra; *pl. m. Men*; hŏmĭnes:—Freá sceáwode fyra fyrngeweorc *the lord beheld the ancient work of men*, Beo. Th. 4561; B. 2286: 4007; B. 2001. Ænig ne wæs fyra cynnes *there was not any of the race of men*, Exon. 47 a; Th. 161, 19; Gú. 961: 63 a; Th. 231, 20; Ph. 492: 92 a; Th. 345, 22; Gn. Ex. 194. v. firas.

fýr-bær; *adj. Igniferus*, Hpt. Gl. 509.

fýr-bæþ; *gen.* -bædes, -baðes; *n. A fire-bath*; igneum balneum:—On fýrbæðe *in the fire-bath*, Elen. Kmbl. 1895; El. 949. In fýrbaðe *in the*

fire-bath, Exon. 20 a; Th. 52, 10; Cri. 831: 22 b; Th. 61, 18; Cri. 986.

fýr-bend, es; *m. A fire-band*; vincŭlum igne dūrātum:—Dúru onarn fýrbendum fæst *the door fast with fire-bands yielded*, Beo. Th. 1448; B. 722.

fýr-bēta, an; *m.* [bētan II. *to light* or *make a fire, kindle*] *One who looks after the fire*; fŏcārius, Ælfc. Gl. 30; Som. 61, 74; Wrt. Voc. 27, 3.

fýr-bryne, es; *m. A fire burning*; incendium:—Wearþ ungemetlíc fýrbryne mid Rómānum *an immense fire happened among the Romans*, Ors. 4, 7; Bos. 87, 18.

fyrclian; *p.* ode; *pp.* od *To flash, flicker*; fulgēre:—Swilce se beám ongeán weardes wið ðæs steorran ward fyrcliende wære *as if the beam were flashing towards the star from an opposite direction*, Chr. 1106; Erl. 240, 34. v. flicerian.

fýr-clom; *gen.* -clommes; *m.* [clom *a band, bond*] *A fire-bond*; vincŭlum ignītum *vel* igne dūrātum:—Ðis is þeóstre [ðeostræ MS.] hám, þearle gebunden fæstum fýrclommum *this is a dark home, strongly bound with fast fire-bonds*, Cd. 213; Th. 267, 16; Sat. 39.

fýr-cruce *a fire-cruse* or *pot, kettle*; cūcŭma, Som. Ben. Lye. DER. cruce.

fýr-cyn, -cynn, es; *n. A kind of fire*; igneum gĕnus:—Mycel fýrcyn and mycel bryne *a great kind of fire and a great burning*, Ors. 6, 1; Bos. 115, 36.

FYRD, fyrdung, e; *f.* I. *an army, the military array of the whole country*; exercĭtus, expĕdĭtio. To take part in the *fyrd* was the general duty of every freeman, even of the mere churl, but as forming one branch of the *trinoda necessitas* it belonged especially to owners of land. 'Every owner of land was obliged to the *fyrd* or expeditio; the owner of bookland as liable to the *trinoda necessitas* alone; the occupier of folkland as subject to that as well as to many other obligations from which bookland was exempted.' Stubbs' Const. Hist. i. 190, q. v. By the simple appellation of fyrd *the land-force was to be understood*. The naval armament was denominated the scip-fyrd. v. folc-land I [c]:—Be ðon ðe gesíþcund man fyrde forsitte. Gif gesíþcund mon, landágende, forsitte fyrde, geselle cxx scillinga and þolie his landes; unlandágende lx scillinga; cierlisc xxx scillinga; to fyrd-wíte [MS. fierd-wíte] *in case a gesithcund man neglects the fyrd. If a gesithcund man owning land, neglect the fyrd, let him pay 120 shillings and forfeit his land; one not owning land, 60 shillings; a churlish man, 30 shillings; as a fine for neglecting the fyrd*, L. In. 51; Th. i. 134, 7-10. II. *an army*; agmen, exercĭtus:—Fyrd sceal wið fyrde sacan *army shall strive against army*, Menol. Fox 565; Gn. C. 52: Cd. 146; Th. 183, 8; Exod. 88. On Faraones fyrde *in Pharaoh's army*, Exon. 122 a; Th. 468, 3; Phar. 2. Claudius, se cásere, fyrde gelædde on Breotone *Claudius, the emperor, led an army into Britain*, Bd. 1, 3; S. 475, 11: Cd. 145; Th. 181, 17; Exod. 62. Gesomnade he his fyrd wið West-Seaxum *he assembled his army against the West Saxons*, Bd. 2, 9; S. 512, 2: Cd. 149; Th. 187, 24; Exod. 156. Fór fyrda mæst *the greatest of armies marched*, Elen. Kmbl. 69; El. 35. Hí heora fyrd gesomnedon *they assembled their armies*, Bd. 3, 14; S. 539, 36. III. *an expedition*; expĕdĭtio:—Ðæt ic of ðisse fyrde feran wille *that I will flee out of this expedition*, Byrht. Th. 138, 16; By. 221. Ðeáh ðú mid us ne fare on fyrd *though thou go not with us in the expedition*, Ps. Th. 43, 11. Onginnaþ ymb ða fyrde þencean *they begin to think about the expedition*, Cd. 21; Th. 26, 18; Gen. 408: 32; Th. 43, 11; Gen. 689: 92; Th. 118, 7; Gen. 1961. IV. *a camp*; castrum:—Fyrd *castrum*, Ælfc. Gl. 7; Som. 56, 76; Wrt. Voc. 18, 28. [*Laym.* ferde, uerde, *f. an army*: *Orm.* ferd *an army*: *Scot.* ferde *an army*; *mod.* O. *Sax.* fard, *f. an expedition*: *Frs.* feard: *O. Frs.* ferd, *f. an expedition*: *Ger.* fahrt, fart, *f. īter*: *M. H. Ger.* vart, *f*: *O. H. Ger.* fart, *f. īter*: *Dan.* fart, færd, *m. f. an expedition*: *Swed.* fart, *m. a passage*: *Icel.* ferð, *f. travel*.]

fyrd *a ford*, found in the compound Twý-fyrd *Twyford*. v. ford.

fyrd-cræft *an expedition*. v. fird-cræft.

fyrderung, e; *f. A preparation* or *provision for an expedition*; expĕdĭtiōnis appărātus, Som. Ben. Lye.

fyrd-esne, es; *m. A warlike youth* or *man, warrior*; bellĭcōsus jŭvĕnis, bellātor:—In ðam ylcan gefeohte, Osfriþ his óðer sunu, ær him gefeóll, se hwatesta fyrdesne *in quo bello, ante illum ūnus fīlius ējus Osfrid, jŭvĕnis bellĭcōsus, cĕcĭdit*, Bd. 2, 20; S. 521, 15.

fyrd-færeld, es; *n.* [fyrd *an army*; færeld *a journey*] *A military expedition* or *service*; mĭlĭtāris expĕdĭtio:—Būtan ðysum þrim þingum, ðæt is, fyrdfærelde, and brigcgewurce, and burhbóte *except these three things, that is, military service, bridge-work, and reparation of fortresses*, Cod. Dipl. 715; A. D. 1006; Kmbl. iii. 350, 10. Ðæt he þreó þing of his lande dó, fyrdfæreld, and burhbóte, and brycgeweorc *ut ita făciat pro terra sua, scĭlicet, expĕdĭtiōnem, burhbōtam, et brigbōtam*, L. R. S. I; Th. i. 432, 2.

fyrd-faru, ferd-faru; e; *f. A military expedition* or *service*; mĭlĭtāris expĕdĭtio:—Gif hwá burhbóte, oððe bricgbóte, oððe fyrdfare forsitte *if any one neglect reparation of fortresses, or reparation of bridges, or military service*, L. C. S. 66; Th. i. 410, 8.

fyrd-geatwe, -geatewe; *gen.* a; *pl. f.* [geatwe *arms, trappings*] *Warlike trappings* or *arms*; bellīcōsus appărātus:—Yr byþ fyrdgeatewa [fyrdgeacewa MS.] sum *a bow is a part of warlike arms*, Runic pm. 27; Hick. Thes. i. 135, 54.

fyrd-gemaca, an; *m.* [gemaca *a companion*] *A companion in war, fellow-soldier*; commīlĭto :—Tytus asende bodan to hys fyrdgemacan, ðe wæs genemned Uespasianus *Titus sent messengers to his fellow-soldier, who was named Vespasian*, Nathan. 5.

fyrd-gestealla, an; *m. A comrade in arms, martial comrade*; expĕdītiōnis bellĭcæ sŏcius, commīlĭto :—Nealles folc-cyning fyrdgesteallum gylpan þorfte *the people's king needed not to boast of his comrades in arms*, Beo. Th. 5739; B. 2873. Wurdon Sodomware leófum bedrorene fyrdgesteallum *the inhabitants of Sodom were deprived of their beloved martial comrades*, Cd. 93; Th. 120, 23; Gen. 1999.

fyrd-getrum, es; *n.* [getrum *a band*] *A martial band, company of soldiers*; agmen, cŏhors :—Fūs fyrdgetrum *the prompt martial band*, Cd. 147; Th. 184, 6; Exod. 103. Hēht his herecist healdan georne, fæst fyrdgetrum *he bade his warlike band, the firm company, bear them boldly*, 151; Th. 189, 1; Exod. 178.

fyrd-hom, es; *m.* [hom *a covering, garment*] *A war-covering*; bellĭca vestis, lōrīca :—Ðæt heó ðone fyrdhom þurhfōn ne mihte *that she might not pierce through the war-covering*, Beo. Th. 3012; B. 1504.

fyrd-hrægl, es; *n.* [hrægel, hrægl *a garment*] *A war-garb*; bellĭca vestis, lōrīca :—Helm oft gescær, fæges fyrdhrægl *it often slashed the helmet, the war-garb of the fated*, Beo. Th. 3058; B. 1527.

fyrd-hwæt; *adj. Bold in warfare, warlike, brave*; bellīcōsus :—Ðæt wæron mǣrc men ofer eorþan, and fyrdhwate *those were famous men throughout the earth, and bold in warfare*, Andr. Kmbl. 16; An. 8: Elen. Kmbl. 2356; El. 1179: Apstls. Kmbl. 23; Ap. 12: Beo. Th. 3286; B. 1641.

fyrdian, fierdian, feordian; *p.* ode, ede; *pp.* od, ed [fyrd *an army*] *To go with an army, march, be at war*; proficisci, bellum gĕrĕre :—Fyrdode him togeánes *he marched against him*, Chr. 835; Th. 117, 18, col. 1, 2: 894; Th. 166, 17, col. 2: 167, 16, col. 1. Hī fyrdedon wið Ætlan Hūna cyninge *they were at war with Attila, king of the Huns*, 443; Th. 18, 30, col. 1.

fyrding, firding, e; *f.*   I. *an army, army prepared for war*; exercĭtus, procinctus :—Fyrding [MS. fyrdingc] *procinctus*, Ælfc. Gl. 87; Som. 74, 42; Wrt. Voc. 50, 24 : 72, 71. Mid ormǣtre fyrdinge *with an immense army*, Homl. Th. ii. 66, 2 : 194, 13.   II. *an expedition*; expĕdītio :—Geswicon ðære fyrdinge *they withdrew from the expedition*, Chr. 1016; Erl. 153, 29. v. fyrdung.

fyrdinga; *adv. In companies* or *flocks, by bands* or *multitudes*; catervātim, Som. Ben. Lye.

fyrdleás, fierdleás; *adj. Without an army* or *force*; exercĭtu cărens :—Hit ðonne fyrdleás wæs *it was then without a force*, Chr. 894; Th. 164, 29, col. 2 : 165, 29, col. 1, 2.

fyrd-leóþ, es; *n. A war-song*; mīlĭtāre carmen :—Fyrdleóþ agōl wulf on walde *a wolf sang a war-song in the wood*, Elen. Kmbl. 54; El. 27: Cd. 171; Th. 215, 3; Exod. 577.

fyrdlíc; *adj. Military, martial*; mīlĭtāris :—Hire fær is wiðmeten fyrdlícum truman *her course is compared to a martial band*, Homl. Th. i. 444, 5 : Jos. 11, 10.

fyrd-man, ferd-mon; *pl.* -men; *m. A military man, a soldier*; miles :—He sceal hæbban fyrdmen *he must have soldiers*, Bt. 17; Fox 58, 33.

fȳr-draca, an; *m. A fire-dragon, fire-drake*; ignĭvŏmus drăco :—Frēcne fȳrdraca *a fell fire-dragon*, Beo. Th. 5371; B. 2689.

fyrd-rinc, ferd-rinc, es; *m. A man of arms, warrior, soldier*; bellātor, miles :—Frōd wæs se fyrdrinc *skilful was the man of arms*, Byrht. Th. 135, 58; By. 140. Fyrdrincas frome *bold soldiers*, Elen. Kmbl. 521; El. 261. Se cyn fyrdrinca fruman grētan *who came to greet the chief of warriors*, Cd. 97; Th. 127, 1; Gen. 2104.

fyrdringnes *an exalting, promoting, advancing* or *furthering*; exaltātio, promōtio, Som. Ben. Lye. v. fyrðringnes.

fyrd-sceorp; es; *n. A war-vest*; bellĭcus ornātus :—Hwīlum hongige on wage freólíc fyrdsceorp *sometimes I hang on the wall a goodly war-vest*, Exon. 104 a; Th. 395, 25; Rä. 15, 13.

fyrd-scip, es; *n. A ship of war*; bellĭca nāvis :—Gif hwâ fyrdscip awyrde *if any one injure a ship of war*, L. Eth. vi. 34; Th. i. 324, 5. Ðæt man fyrdscipa gearwige *that ships of war be made ready*, vi. 33; Th. i. 324, 4.

fyrd-searu, -searo; *gen.* -wes; *n. A war-equipment*; bellīcus appărātus :—Him Onela forgeaf fyrdsearu fūslíc *Onela gave him a ready war-equipment*, Beo. Th. 5229; B. 2618. Fyrdsearo, 469; B. 232.

fyrd-sócn, e; *f.* [sócn *the seeking*] *The seeking of the army, military service*; mīlĭtia :—Ðæt hit sȳ gefreoð ealra þeówdómes, būton fyrdsócne, and burhgeweorce and bryggeweorce *that it shall be freed from all services, except military service, castle-building, and bridge-work*, Th. Diplm. A. D. 1061; 389, 30.

fyrd-stemn *an army-corps.* v. fird-stemn.

fyrd-tiber, es; *n.* [tiber *a sacrifice*] *A military sacrifice*; mīlĭtāris hostia :—Fyrdtiber [MS. fyrdtimber] *hostia exercĭtūs*, Cot. 103.

fyrd-truma, an; *m. A martial band, an army*; exercĭtus :—Swā egeslíc swā fyrdtruma *as terrible as a martial band*, Homl. Th. i. 442, 34.

fyrdung, e; *f.*   I. *an army prepared for war, armament*; exercĭtus :—Beó man georne ymbe fyrdunga *let the armaments be diligently attended to*, L. Eth. v. 26; Th. i. 310, 24 : vi. 32; Th. i. 322, 32.   II. *an expedition*; expĕdītio :—On fyrdunge *in the expedition*, L. C. S. 79; Th. i. 420, 14.   III. *a camp*; castra :—Fyrdunga oððe fyrdwícu *castra*, Ps. Lamb. 26, 3. v. fyrd.

fyrd-wæn, es; *m. A military waggon*; essēdum, Th. Diplm. A. D. 1050-1073; 430, 2.

fyrd-weard, e; *f. An army-guard, a military watch*; mīlĭtāris custōdia :—Sǣweard and heáfodweard and fyrdweard *sea-guard and head-guard and army-guard*, L. R. S. 1; Th. i. 432, 5.

fyrd-werod, -weord, es; *m. An army-host, phalanx*; turma, phălanx =φάλαγξ, Cot. 140. Micel stefn fyrdweorodes getrymnesse *a great sound of the arraying of a host*, Blickl. Homl. 91, 35.

fyrd-wíc, es; *n. An army-station, a camp*; castra :—Ðis ys Godes fyrdwíc *castra Dei sunt hæc*, Gen. 32, 2 : Ælfc. Gl. 7; Som. 56, 77; Wrt. Voc. 18, 29. Fyrdunga oððe fyrdwícu *castra*, Ps. Lamb. 26, 3. Hī feóllon on middele fyrdwíca heora *cĕcĭdērunt in mĕdio castrōrum eōrum*, Ps. Spl. C. 77, 32. To ðam fyrdwícum *to the camps*, Judth. 11; Thw. 24, 33; Jud. 220.

fyrd-wísa, an; *m. A leader of an expedition*; expĕdītiōnis dux :—Sum biþ heretoga, fyrdwísa *from one is a general, a bold leader*, Exon. 79 b; Th. 297, 32; Crä. 77.

fyrd-wíse, an; *f. A military manner* :—Se mon se ne wære mid his wǣpnum æfter fyrdwíson gegered *qui non legĭtĭmis indutus insignibusque armis*, Nar. 9, 28.

fyrd-wíte, ferd-wíte, es; *n. A fine for neglecting the* fyrd, L. C. S. 12; Th. i. 382, 14 : 15; Th. i. 384, 3 : Th. Diplm. A. D. 1066; 411, 31.

fyrd-wyrðe; *adj. Famous in war*; bello clārus :—Gang æfter fiŏre fyrdwyrðe man *the man famous in war went along the floor*, Beo. Th. 2637; B. 1316.

fyren, e; *f. A sin, crime*; peccātum, crīmen :—Deorce fyrene *dark sins*, Ps. Th. 108, 14. He ðære mægþe fleáh fyrene *he avoided the crimes of the people*, Cd. 92; Th. 116, 24; Gen. 1941: Exon. 48 a; Th. 166, 18; Gú. 1044. v. firen.

fȳren, fȳran; *def.* se fȳrena, seó, ðæt fȳrene; *adj. Fiery, burning, flaming*; ignītus, igneus, flammeus :—Is ðín ágen spræc innan fȳren, sylf swīðe hát *ignītum elŏquium tuum vehementer*, Ps. Th. 118, 140. Sió fȳrene sunne *the fiery sun*, Bt. 39, 13; Fox 232, 27. Swylce eal Finns buruh fȳrenu wære *as if all Fin's castle were on fire*, Fins. Th. 73; Fin. 36. Ðæt fȳrene swurd *the fiery sword*, Boutr. Scrd. 20, 33. Under ðam fȳrenan hrófe *under the fiery roof*, Cd. 185; Th. 230, 27; Dan. 239. God hēt him fȳrenne beám befôran wīsian *God commanded a pillar of fire to point out the way before them*, Ps. Th. 104, 34. Fȳren swurd *flammeum glădium*, Gen. 3, 24. Fȳrene sweorde *with a fiery sword*, Cd. 45; Th. 58, 17; Gen. 947 : 76; Th. 95, 8; Gen. 1575. Fȳrnum clommum *with fiery fetters*, Andr. Kmbl. 2756; An. 1380: Exon. 18 b; Th. 46, 7; Cri. 733. [*Orm.* firen: *Laym.* furen.]

fȳren cylle, an; *f. A fiery torch*; ignea fax, Bd. 5, 23; S. 645, 29, note, MS. B. v. fȳren þecelle.

fyren-dǣd, e; *f. A wicked deed*; scĕlestum făcĭnus :—He is mild-heort, and manþwǣre hiora fyrendǣdum *ipse est mĭsĕrĭcors, et prŏpĭtius fit peccātis eorum*, Ps. Th. 77, 37 : Beo. Th. 2006; B. 1001: Cd. 191; Th. 237, 30; Dan. 345. v. firen-dǣd.

fyren-earfeðe, es; *n. A sinful woe*; scĕlestum mălum :—Heó nyste ðæt swā fela fyrenearfeða fylgean sceolde *she knew not that so many sinful woes must follow*, Cd. 33; Th. 44, 14; Gen. 709.

fyren-full; *adj. Sinful, unjust, wicked*; inĭquus :—Fyrenfulle men geworhton *wicked men have wrought*, Soul Kmbl. 179; Seel. 90. Used as a noun, *One who is sinful, a sinner*; peccātor :—Swā ða fyrenfullan frēcne forweorþaþ *sic pĕreant peccātōres a făcie Dei*, Ps. Th. 67, 2 : 54, 2 : 57, 9. v. firen-full.

fȳrenfull; *adj. Fiery*; ignītus :—Is fȳrenfull spæc ðín swiðlíce *est ignītum elŏquium tuum vehementer*, Ps. Lamb. 118, 140.

fyrenfulnes, -ness, e; *f. Luxury, riot*; luxūria, tŭmultus, Som. Ben. Lye.

fyrenian, fyrnian; *p.* ede; *pp.* ed *To sin, commit adultery*; peccāre, moechāri :—Fyrnaþ ðus ðæt flǣschord *thus will the body sin*, Soul Kmbl. 203; Seel. 103. Ne fyrena ðū *non moechāběris*, Lk. Bos. 18, 20. DER. ge-fyrnian. v. firenian.

fyrenlíce; *adv. Vehemently, rashly*; vehementer :—Ðæt ðū tô fyren-líce feohtan sōhtest *that thou soughtest to fight too rashly*, Wald. 35; Vald. 1, 20.

fyren-ligerian; *p.* ede; *pp.* ed *To commit fornication*; fornĭcāri :—Hī fyrenligeredon on begĭmmingum his *fornĭcāti sunt in adinventiōnĭbus suis*, Ps. Spl. 105, 36.

fyren-lust, es; *m. Luxury;* luxŭria :—Ne gēmdon hie nānes fyren-lustes *they cared not for any luxury,* Bt. 15; Fox 48, 7. v. firen-lust.

fyren-þearf, e; *f. Dire distress;* nĭmia mĭsēria :—Fyrenþearfe ongeat *he perceived the dire need,* Beo. Th. 28; B. 14.

fýren þecelle, an; *f. A fiery torch;* ignea fax, Bd. 5, 23; S. 645, 29. v. þecelle.

fyrenum; *adv.* [*dat. pl.* of fyren *a sin, crime*] *Sinfully, criminally;* mǎle, sceleste :—Bona of flāubogan fyrenum sceóteþ *the slayer wickedly shoots from his arrow-bow,* Beo. Th. 3493; B. 1744. Fyrenum ge-syngad *criminally perpetrated,* 4874; B. 2441.

fyren-wyrcende; *part. Evil-doing, committing sin;* mǎlum fǎciens, peccans :—Ic fyrenwyrcende oft elnade *I often emulated evil-doing* [*men*], Ps. Th. 72, 2. v. firen-wyrcende, firen.

fiven-wyrhta, an; *m. An evil-doer, sinner;* mǎli actor, peccātor :—Hū lange fyrenwyrhtan foldan wealdaþ *how long shall evil-doers rule the earth?* Ps. Th. 93, 3. Đæt ic on wrāþne seáþ mid fyrenwyrhtum feallan sceolde *that I should fall with sinners into the horrible pit,* 87, 4.

fyres *furze,* Wrt. Voc. 285, 48. v. fyrs.

fyrest; *adj. First, front;* primus :—Æt đām feówer tóþum fyrestum *for the four front teeth,* L. Ethb. 51; Th. i. 16, 2. v. fyrst; *adj.*

fýr-feaxe; *adj.* [feaxe *having hair*] *Fiery-haired;* ignĭcōmus :—Fýr-feaxe [MS. -feaxa] *ignĭcōmus,* Cot. 170.

fýr-gearwunge; *pl. f. Fire-preparation, fuel;* fōmes, focŭlāria, Cot. 83.

fýr-gebræc, es; *n. A fire-crash;* ignis frǎgor *vel* strĕpĭtus :—Đæt fýrgebræc *the fire-crash,* Cd. 119; Th. 154, 24; Gen. 2560.

fyrgen, es; *n. A mountain, mountain-woodland;* mons, saltus :—Flet [MS. fled] þor on fyrgen hæfde *Thor had a dwelling on the mountain,* Lchdm. iii. 54, 17. DER. fyrgen-beám, -holt, -streám. v. firgen.

fyrgen-beám, es; *m. A mountain-tree;* saltuensis arbor :—He fyrgen-beámas ofer hārne stān hleónian funde *he found mountain-trees leaning over the hoar rock,* Beo. Th. 2833; B. 1414.

fyrgen-holt, es; *n. A mountain-wood;* montāna silva :—On fyrgen-holt *into a mountain-wood,* Beo. Th. 2791; B. 1393.

fyrgen-streám, es; *m. A mountain-stream;* montānum flūmen :—Đær fyrgenstreám niđer gewíteþ *where the mountain-stream flows downward,* Beo. Th. 2723; B. 1359. v. firgen-streám.

fýr-gnást, es; *m. A fire-spark;* scintilla :—Flugon fýrgnāstas *fire-sparks flew,* Andr. Kmbl. 3090; An. 1548.

fyrh *to a furrow,* Cod. Dipl. 1172; A. D. 955; Kmbl. v. 332, 22; *dat. sing.* of furh.

fýr-hāt; *adj. Fire-hot;* ut ignis ardens :—Fýrhāt lufu *a fire-hot love,* Elen. Kmbl. 1871; El. 937.

fýr-heard; *adj. Fire-hard;* igne dūrātus :—Eoforlíc scionon fāh and fýrheard *boar's likenesses shone variegated and fire-hard,* Beo. Th. 615; B. 305.

fýr-hole; *f. Catasta,* Hpt. Gl. 310. 'Catastæ, genus tormenti, i. e. lecti ferrei, quibus impositi Martyres, ignis supponebatur.' Du Cange.

fyrht, firht, freht, es; *n? A divining, divination, augury;* auspicium, hariolātio, augŭrium :—Odđe on blóte odđe on fyrhte *either by sacrifice or by divination,* L. C. S. 5; Th. i. 378, 22. On firhte L. N. P. L. 48; Th. ii. 296, 28. Ǽristum odđe frumum frehtum *primis auspĭciis,* Rtl. 97, 16. v. frihtrung.

fyrht; *adj. Timid;* tĭmĭdus :—On his sóþfæstnesse swylce dēmeþ on folce fyrhte þearfan *in sua justĭtia jūdĭcābit paupĕres hujus pŏpŭli,* Ps. Th. 71, 4. DER. god-fyrht. v. forht.

fyrhtan; *p.* fyrhte; *pp.* fyrhted *To* FRIGHTEN, *terrify, tremble;* terrēre tremere :—Gif lígette and þunorrāde eorþan and lyfte brēgdon and fyrhton *si corusci ac tonitrua terras et aëra terrērent,* Bd. 4, 3; S. 569, 13. Đū dōest đa fyrhta *facis eam tremere,* Rtl. 102, 21. DER. a-fyrhtan.

fyrhþ, es; *m. n.* I. *the soul, spirit, mind;* ǎnĭmus, mens :—Biþ fyrhþ afrēfred *the spirit is comforted,* Andr. Kmbl. 1275; An. 638. Ic ne can đæt ic nāt findan on fyrhþe *I cannot find what I know not in my mind,* Elen. Kmbl. 1278; El. 641: 391; El. 196. II. *life;* vīta :—Đū God Dryhten wealdest wīdan fyrhþ *thou Lord God rulest for ever,* Elen. Kmbl. 1518; El. 761. DER. stærced-fyrhþ, wīde-. v. ferhþ.

fyrhþ-gleáw; *adj. Wise-minded, prudent;* prūdens, sǎpiens, Elen. Kmbl. 1758; El. 881. v. ferhþ-gleáw.

fyrhþ-loca, an; *m. The soul-inclosure, breast;* mentis clausūra, pectus :—Wæs Cristes lof on fyrhþlocan fæste bewunden *Christ's praise was steadfastly enclosed within his breast,* Andr. Kmbl. 115; An. 58: 3138; An. 1572. v. ferhþ-loca.

fyrhþ-lufe, an; *f. Love of the soul, mental love;* anĭmi ǎmor :—Ic to ǎnum đē stađolige fæste fyrhþlufan *I keep the steadfast love of my soul firmly fixed to thee only,* Andr. Kmbl. 165; An. 83.

fyrhþ-sefa, an; *m. The mind;* mens :—Gē fyrhþsefan mínne cunnon *ye know my mind,* Elen. Kmbl. 1066; El. 534. v. ferhþ-sefa.

fyrhþ-wērig; *adj. Soul-weary, sorrowful;* mæstus :—Seó cwēn ongan fricggan fyrhþwērige, ymb fyrngewritu *the queen began to ask them,*

sorrowful, concerning the old scriptures, Invent. Crs. Recd. 1119; El. 560. v. ferþ-wērig.

fyrhtnes, -ness, e; *f. Fear;* tĭmor :—Mid micelre fyrhtnesse *with great fear,* Ors. 6, 30; Bos. 126, 14; Mt. Kembl. Lind. 14, 26. v. forhtnys.

FYRHTO; *indecl. in sing.* fyrhtu, e; *f. Fear,* FRIGHT, *dread, terror, trembling;* tĭmor, pǎvor, formīdo, terror, trĕmor :—Us fyrhto gegrāp *fear seized us,* Nicod. 21; Thw. 10, 32: Cant. Moys. Ex. 15, 19; Thw. 30, 19. Fyrhto odđe bifung *trĕmor,* Ps. Lamb. 47, 7. Đeós firhtu [fyrhtu, MS. D.] *hæc formīdo,* Ælfc. Gr. 36; Som. 38, 50: Ps. Th. 54, 4. Egsa me and fyrhtu forcwōmon *tĭmor et trĕmor vēnērunt sŭper me,* Ps. Th. 54, 5. On mínre fyrhto *in pǎvōre meo,* 30, 25. Geblissiaþ him on fyrhto [fyrhtu, Lamb.] *exultāte ei in trĕmōre,* Ps. Spl. 2, 11. Hī mycle fyrhto onstyredon đām monnum đe hī sceáwodon and gesáwon *they stirred up much fear in the men who beheld and saw them,* Bd. 5, 23; S. 645, 23: Exon. 119 a; Th. 457, 21; Hy. 4, 87. Ne him Godes fyrhtu georne ondrǣdaþ *non tĭmuērunt Deum,* Ps. Th. 54, 20: 77, 53. [O. Sax. forhta, f: O. Frs. fruchta: Dut. Kil. vrucht, vurcht: Ger. furcht, f: M. H. Ger. vorhte, f: O. H. Ger. forhta, f: Goth. faurhtei, f: Dan. frygt, m. f: Swed. fruktan, f.]

fýr-hús, es; *n. A* FIRE-HOUSE, *furnace;* cǎmīnus = κάμινος, fornax :—Fýrhús camīnātum? Ælfc. Gl. 107; Som. 78, 92; Wrt. Voc. 58, 7. Fýrhúses hlýwing *caumenæ* (?) *refúgium,* R. Concord. 11.

fyrian; *p.* ode; *pp.* od *To make a furrow, to plough, till;* proscindĕre aratro, Scint. 32.

fýrian; *p.* ode; *pp.* od *To make a fire, give warmth, to cherish;* fŏcum præbēre :—Fēde þearfan and scrýde and hūsige and fýrige *let him feed the needy, and clothe, and house, and fire them,* L. Pen. 14; Th. ii. 282, 16.

fyrlen, feorlen; *adj. Far off, distant, remote;* longinquus, distans, remōtus :—Đeáh đe he fyrlen sý *though he be far off,* Homl. Th. ii. 444, 9. For đære fyrlenan heáhnysse *for its remote elevation,* Bd. de nat. rerum; Lchdm. iii. 232, 15, note 7. Sum ǽđelboren man ferde on fyrlen land *hŏmo quidam nōbĭlis abiit in regĭōnem longinquam,* Lk. Bos. 19, 12: Homl. Th. ii. 122, 14. To fyrlenum eardum *to distant lands,* Gen. 20, 13. Mid fulluhte aþwagen fram his fyrlenum dǽdum *with baptism washed from his former deeds,* H. R. 107, 14.

fyrlen, es; *n. Distance;* distantia :—For đam mycclan fyrlene *on account of the great distance,* Boutr. Scrd. 18, 43.

fýr-leóht, es; *n. A fire-light;* igneum lūmen :—He fýrleóht geseah *he saw a fire-light,* Beo. Th. 3037; B. 1516.

fýr-leóma, an; *m.* [leóma *a ray of light, beam*] *A fire-beam;* igneus splendor :—Fýrleóma stód geond đæt atole scræf *a fire-beam stood through that horrid den,* Cd. 216; Th. 272, 32; Sat. 128.

fýrlíce *suddenly,* Num. 16, 35. v. fǽrlíce.

fýr-loca, an; *m. A fire-bond;* igneum claustrum :—Eart tū in fýrlocan feste gebunden *thou art fast bound in fire-bonds,* Cd. 214; Th. 268, 20; Sat. 58.

fyrm, e; *f. A feast;* ĕpŭlæ :—Đa Philistei micele fyrme geworhton *the Philistines made a great feast,* Jud. 16, 25. v. feórm.

fýr-mǽl, es; *m. A fire-mark;* mǎcŭla igne inusta :—Fýrmǽlum fāg *variegated with marks of fire,* Andr. Kmbl. 2269; An. 1136.

fyrmest, formest; *def.* se fyrmesta, seó, đæt fyrmeste; *sup. adj.* FOREMOST, *first;* prīmus :—Se đe wyle betweox eów beón fyrmest, sý he eówer þeów *qui vŏluĕrit inter vos prīmus esse, ĕrit vester servus,* Mt. Bos. 20, 27: Mk. Bos. 9, 35: Boutr. Scrd. 21, 5. Se fyrmesta and se betesta *the foremost and the best;* præstantissĭmus, Cot. 153. Se fyrmesta is eásterne wind *the first is the east wind,* Bd. de nat. rerum; Wrt. popl. science 17, 22; Lchdm. iii. 274, 13. Đis ys đæt mǽste and đæt fyrmeste bebod *hoc est maxĭmum et prīmum mandātum,* Mt. Bos. 22, 38. Agynn fram đam ýtemestan óþ đone fyrmestan *begin from the last to the first,* 20, 8. Manega fyrmeste beóþ ýtemeste, and ýtemeste fyrmeste *multi ĕrunt prīmi novissĭmi, et novissĭmi prīmi,* 19, 30: Mk. Bos. 10, 31: Lk. Bos. 13, 30. Đa fyrmestan *prīmi,* Mt. Bos. 20, 16. Fyrmest manna *first of men;* summas, prīmas, Ælfc. Gr. 9, 25; Som. 10, 58, 59: Chr. 1086; Erl. 221, 39. Fyrmeste [MS. fyrmyste] naman *prīmĭtivan ōmĭna,* 5; Som. 4, 8.

fyrmest; *sup. adv. At first, most, very well, best;* prīmo, maxĭme, optime :—Hie feónda gefæt fyrmest gesǽgon *they first saw the enemies' march,* Elen. Kmbl. 136; El. 68: Cd. 158; Th. 197, 21; Exod. 310. Swā he fyrmest meahte *as much as ever he could,* Bd. 2, 6; S. 508, 32: Elen. Kmbl. 632; El. 316: Ps. Th. 72, 6: 106, 29: 121, 7. Swā forþ swa wē fyrmest leornian mágon *as far as ever we can learn,* Bd. 5, 21; S. 643, 5: L. C. S. 11; Th. i. 382, 6: L. Eth. vi. 40; Th. i. 324, 28.

fyrmþ, frymþ, e; *f.* I. [feormian I. *to feed, support, entertain*] *A receiving to food, harbouring, an entertainment;* receptio ad vĭctum, susceptio :—Đis syndon đa gerihta đe se cyning āh ofer ealle men; đæt is ... and flýmena fyrmþe *these are the rights which the king possesses over all men; that is ... and* [*the penalty for*] *the harbouring of fugitives,* L. C. S. 12; Th. i. 382, 14. Ǽlc mon mōt onsacan fyrmþe *every man*

A a

may deny entertainment, L. In. 46; Th. i. 132, 1, note 3, MSS.
B, H. II. [feormian III. *to cleanse*] *A cleansing, washing;*
ablūtio, baptisma = βάπτισμα :—Calica fyrmþa *cǎlicum baptismǎta*, Mk.
Bos. 7, 4 : Hpt. Gl. 420.

FYRN; *adj. Ancient, old;* antiquus, priscus :—Fyrn forþgesceaft *the
ancient creation*, Exon. 128 a; Th. 492, 4; Rä. 81, 9. [*O. Sax.* fern :
Ger. firn, firne : *M. H. Ger.* virne : *O. H. Ger.* firni : *Goth.* fairneis :
Swed. forn *only in compounds;* as, forn-ålder, *m. antiquity :* *Icel.* forn :
Lith. pernay *anni prioris.*]

fyrn; *adv. Formerly, long ago, of old;* ōlim, prīdem, antīquĭtus :—Hū
mæg ic ðæt findan ðæt swā fyrn gewearþ *how can I find that which
happened so long ago?* Elen. Kmbl. 1261; El. 632 : 1279; El. 641.
Ðæt he bibūgan mǣge ðone bitran drync ðone Eue fyrn Adame geaf
that he may escape the bitter drink which Eve of old gave to Adam, Exon.
45 b; Th. 154, 11; Gū. 841 : 47 a; Th. 160, 20; Gū. 946 : Cd. 128;
Th. 163, 11; Gen. 2696. [*O. Sax.* forn, furn : *O. H. Ger.* forn *prius,
ōlim.*] DER. ge-fyrn, un-,

fyrn-dagas; *gen.* a; *dat.* um; *pl. m. Days of old, ancient days;*
priscæ dies :—Ðis is se ilca God ðone on fyrndagum fæderas cūðon *this
is the same God whom your fathers knew in days of old*, Andr. Kmbl.
1503; An. 753 : 1951; An. 978 : Cd. 223; Th. 293, 31; Sat. 463.
Swā hine fyrndagum worhte wǣpna smiþ *as the armourer wrought it in
ancient days*, Beo. Th. 2907; B. 1451. [*Laym.* i furn daȝen : *O. Sax.*
an furndagun.]

fyrn-geár, es; *n. A former* or *by-gone year;* priscus *vel* prætĕrĭtus
annus :—Fyrngeárum frōd *old with by-gone years*, Exon. 59 a; Th. 213,
3; Ph. 219 : Menol. Fox 483; Gn. C. 12. [*Piers P.* fernyere.]

fyrn-geára; *adv.* [*gen. pl. of* -geár] *In by-gone years, of old time;*
ōlim, antīquĭtus, Ps. Th. 94, 9.

fyrn-geflit, es; *n. An ancient strife, old conflict;* vĕtus lis *vel* rixa :—
Þurh fyrngeflīt *through the old conflict*, Elen. Kmbl. 1804; El. 904. Hī
guldon hyra fyrngeflītu fāgum swyrdum *they requited their ancient strifes
with stained swords*, Judth. 12; Thw. 25, 17; Jud. 264.

fyrn-geflita, an; *m. An enemy of old;* antiquus inĭmicus :—Būtan his
fyrngeflītan *except to his enemy of old*, Exon. 96 a; Th. 357, 25;
Pa. 34.

fyrn-gemynd, es; *n. An ancient reminiscence;* antīqua mĕmŏria :—
Ða ðe fyrngemynd mid Iudēum gearwast cūðon *they who best knew the
old memories among the Jews*, Elen. Kmbl. 654; El. 327.

fyrn-gesceap, es; *n. A decree of old;* ōlim constitūtum :—Ne wāt
ǣnig hū ða wīsan sind wundorlīce, fæger fyrngesceap, ymb ðæs fugles
gebyrd *not any knows how the conditions are wondrous, the fair decree of
old, concerning the bird's birth*, Exon. 61 a; Th. 223, 15; Ph. 360.

fyrn-gesetu; *pl. n. Ancient seats, a former dwelling-place;* pristīnum
dŏmicīlium :—Oþ-ðæt fyrngesetu eft gesēceþ *till it again seeks its
ancient seats*, Exon. 59 b; Th. 216, 5; Ph. 263.

fyrn-gestreón, es; *n. An ancient treasure;* antiquus thesaurus :—
Full fyrngestreóna *full of ancient treasures*, Salm. Kmbl. 64; Sal. 32,
MS. B.

fyrn-geweorc, es; *n. An ancient work;* priscum *vel* jam diu perfectum
ŏpus :—Ǣr ðon endige frōd fyrngeweorc *before his wise ancient work
shall end*, Exon. 57 a; Th. 203, 14; Ph. 48 : 57 a; Th. 204, 9; Ph.
95 : Andr. Kmbl. 1473; An. 738. Freá sceáwode fyra fyrngeweorc
the lord beheld the ancient work of men, Beo. Th. 4561; B. 2286.

fyrn-gewinn, es; *n. An ancient war;* vĕtus pugna :—On ðæm wæs
ōr writen fyrngewinnes *on which was engraved the origin of the ancient
war*, Beo. Th. 3382; B. 1689.

fyrn-gewrit, -gewryt, es; *pl. nom. acc.* -gewritu, -gewrito; *n. An
ancient writing, old scripture;* vĕtus *vel* prisca scriptūra :—Ðȳ-læs
toworpen sīen frōd fyrngewritu *lest the wise old scriptures should be
overturned*, Elen. Kmbl. 861; El. 431. Ða ðe fyrngewritu sēlest cunnen
those who best know the ancient writings, 746; El. 373 : 1117; El. 560.
Þurh fyrngewrito *through ancient writings*, 309; El. 155. On eallum
ðām fyrngewrytum *in all the ancient writings*, Salm. Kmbl. 15; Sal. 8.

fyrn-gewyrht, es; *n. A former work;* ōlim factum :—Ðæt he fyrn-
gewyrht fyllan sceolde *that he should finish his former work*, Exon. 47 a;
Th. 160, 15; Gū. 944.

fyrn-gid, -gidd, es; *n. An old prophecy;* vĕtus prŏphētia :—Fyrngidda
frōd *prudent in old prophecies*, Elen. Kmbl. 1079; El. 542.

fyrnian, he fyrnaþ; *p.* ede; *pp.* ed *To revile;* calumniāri :—Fyrnaþ
ðus ðæt flǣschord *thus it [the soul] shall revile the flesh*, Soul Kmbl.
203; Seel. 103. v. firenian.

fyrn-man, -mann, es; *m. A man of yore;* qui ōlim vixit :—Geseah he
fyrnmanna fatu *he saw vessels of men of yore*, Beo. Th. 5515; B. 2761.

fyrn-sceaða, an; *m. An old enemy* or *fiend;* antiquus inĭmicus :—Fāh
fyrnsceaða *a hostile fiend*, Andr. Kmbl. 2691; An. 1348.

fyrn-streámas; *pl. m. Ancient streams, the ocean;* prisca fluenta,
oceănus :—Fyrnstreáma geflotan *to the ocean-floater*, Exon. 96 b; Th.
360, 17; Wal. 7.

fyrn-syn, -synn, e; *f. A sin of yore;* priscum peccātum :—Fyrnsynna
fruma *the author of sins of yore*, Exon. 70 b; Th. 263, 9; Jul. 347.

fyrnum; *adv. With horror, horribly, intensely;* horrĭbĭlĭter :—Ðonne
cymþ forst fyrnum cald *then cometh frost intensely cold*, Cd. 17; Th. 20,
28; Gen. 316 : 38; Th. 50, 16; Gen. 809. v. firnum.

fyrn-weorc, es; *n. An ancient work, the creation;* priscum ŏpus,
creātio :—Fyrnweorca Freá *Lord of creation*, Andr. Kmbl. 2819; An.
1412 : Exon. 16 a; Th. 36, 20; Cri. 579.

fyrn-wita, -wiota, -weota, an; *m. An ancient sage, old counsellor,
prophet;* antiquus săpiens :—Frōd fyrnwiota *a wise old counsellor*, Elen.
Kmbl. 875; El. 438. Dauid cyning, frōd fyrnweota *king David, the
prudent prophet*, 685; El. 343. Wæs frōdan fyrnwitan feorh ūþgenge
life was departed from the wise old counsellor, Beo. Th. 4252; B. 2123.
Frōde fyrnweotan *wise ancient sages*, Andr. Kmbl. 1567; An. 785.
Wæs se wītedōm þurh fyrnwitan sungen *the prophecy was sung by old
seers*, Elen. Kmbl. 2305; El. 1154.

fȳr-panne, an; *f.* [fȳr *fire*, panne *a pan*] *A fire-pan, chafing-dish,
pan for burning odoriferous herbs;* batillum, Ælfc. Gl. 26; Som. 60,
95; Wrt. Voc. 25, 35.

fyrr; *adv.* [*comp. of* feor; *adv. far*, q. v.] *Farther;* ultĕrius, longius :—
We usse gesihþ fyrr upp ahōfan *longius vīsum lĕvāvĭmus*, Bd. 5, 1;
S. 613, 32 : Bd. de nat. rerum; Wrt. popl. science 3, 11; Lchdm. iii.
236, 9 : Cd. 122; Th. 156, 23; Gen. 2593. v. fyr.

fyrra, firra, *m;* fyrre, firre, *f. n. adj.* [*comp. of* feor; *adv. far*, q. v.]
Farther; ultĕrior :—He ge-eóde ða fyrran Frysan *ħe had overcome the
farther Frisians*, Bd. 5, 10; S. 624, 3.

fyrrest; *adv.* [*sup. of* feor : *adv. far*, q. v.] *Farthest;* longissĭme :—Se
mōna wæs ðære sunnan fyrrest *the moon was farthest from the sun*, Ors.
6, 2; Bos. 117, 14 : Bt. 39, 7; Fox 222, 21.

fyrs, es; *n. A verse;* versus, Ælfc. Gr. 37; Som. 39, 3. v. fers.

FYRS, es; *m.* FURZE, *furze-bushes;* genista, rhamnus, ulex eurōpæus,
Lin :—Fyrs rhamnus, Wrt. Voc. 80, 21. Fyrses berian *arciotidas*
[= ἀρκευθίδες *juniper-berries*], Glos. Brux. Recd. 43, 15; Wrt. Voc. 69,
30. Ǣr-ðan undergǣton eówre þornas fyrs *priusquam intelligĕrent
spinæ rhamnus*, Ps. Lamb. 57, 10 : Lchdm. iii. 86, 17. Swā hwā swā
wille sāwan westmbǣre land, atió ǣrest of ða þornas, and ða fyrsas
*whosoever will sow fertile land, let him first draw out the thorns, and the
furze*, Bt. 23; Fox 78, 22 : Bt. Met. Fox 12, 6; Met. 12, 3. [*Wyc.*
firse, frijse *gorst, furze :* *Piers P.* firses, *pl.*]

fȳr-scofl, e; *f? A fire-shovel;* batilla, Cot. 24.

fyrsian; *p.* ode; *pp.* od *To put far, remove, separate;* elongāre :—Ða
ðe fyrsiaþ hig fram ðē losiaþ *qui elongant se a te pĕribunt*, Ps. Lamb. 72,
27. DER. a-fyrsian.

fȳr-smeortende; *part. Fire-smarting;* ignītus :—Gnǣttas cōmon
ofer ðæt land mid fȳrsmeortendum bītum *gnats came over the land with
fire-smarting bites*, Ors. 1, 7; Bos. 29, 30.

fyrsn, e; *f. The heel;* calx, calcāneum, Cot. 38. v. fiersn.

fȳr-spearca, an; *m. A fire-spark;* scintilla :—Būton īsene fȳrspearcan
nǣsi ferreas scintillas, Coll. Monast. Th. 31, 5.

FYRST, first, fierst, es; *m.* I. *the first entrance, a threshold,
door;* līmen, Cot. 118. II. *the first in height, the top, ridge,
the inward roof, ceiling of a chamber;* culmen, lǎquear :—Fyrst lǎquear,
Ælfc. Gl. 29; Som. 61, 43; Wrt. Voc. 26, 42 : 82, 15. [*Ger.* first,
m. f. a gable, summit : *M. H. Ger.* virst, *m :* *O. H. Ger.* first, *m. culmen,
pinna.*]

FYRST, first, fierst, es; *m. A space of time, time, respite, truce;*
spātium tempŏris, tempus constitūtum, intercăpēdo :—Næs hit lengra
fyrst *it was not a longer space of time*, Beo. Th. 269; B. 134 : 5104;
B. 2555. Ne wæs se fyrst micel *the respite was not great*, Exon. 37 a;
Th. 121, 32; Gū. 292. Æfter miclum fyrste *post multum tempŏris*,
Mt. Bos. 25, 19 : 26, 73 : Ex. 17, 4 : Boutr. Scrd. 18, 32 : 20, 19.
Hæfde nȳdfara nihtlangne fyrst *the fugitive had a night-long space*, Cd.
154; Th. 191, 2; Exod. 208 : Andr. Kmbl. 1668; An. 836 : 2620;
An. 1311. Fyrst næfdon ðæt hī æton *nec spātium mandūcandi hăbēbant*,
Mk. Bos. 6, 31 : Chr. 1004; Erl. 139, 22. Ðȳ fyrste *in the time*, Beo.
Th. 5139; B. 2573. [*Laym.* first, uirst, urist, feorst : *Orm.* fresst :
Plat. ferst, *f :* *O. Sax.* vrist, *f :* *O. Frs.* ferst, frist, *n :* *Dut.* Kil.
verste, verst, frist, virst *dīlātio :* *Ger.* frist, *f :* *M. H. Ger.* vrist, *f :
O. H. Ger.* frist, *f. mŏra, spātium :* *Dan.* frist, *m. f :* *Swed.* frist, *m.*]
DER. lang-fyrst.

FYRST, first, fyrest; *adj.* FIRST; primus :—Fyrst ferhþ-bana *the first
life-destroyer*, Cd. 162; Th. 203, 5; Exod. 399. [*Wyc. R. Glouc.*
firste : *Piers P.* furste, ferste : *Orm.* firrste : *Plat.* foorste *a prince :
O. Sax.* furisto *first :* *Frs.* foarste : *O. Frs.* ferost : *Dut.* vorst, *m. a
prince :* *Ger.* fürst, *m. a prince :* *M. H. Ger.* vürst *first :* *O. H. Ger.*
furisto : *Dan. Swed.* först, förste : *Icel.* fyrstr *first, foremost.*]

fyrst; *adv. At* FIRST; primo :—Se biscop com fyrst to Elíg *the bishop
came first to Ely*, Chr. 963; Erl. 121, 20 : 123, 2.

fyrstan [fyrst *a space of time, respite*] *To give respite;* indūcias
facere, Som. Ben. Lye.

fȳr-stán, es; *m. A fire-stone, flint;* pȳrītes = πυρίτης :—Fȳrstán pȳrītes
vel fŏcāris lǎpis, Ælfc. Gl. 58; Som. 67, 105; Wrt. Voc. 38, 29.

fyrst-gemearc, es; *n. An appointed time, space of time;* tempus con-

stĭtūtum, tempŏris spătium:—Ne biþ ðæs lengra swice sāwelgedāles ðonne seofon niht fyrstgemearces *there will be no longer evasion of the soul-separation than seven nights of time's space*, Exon. 47 b; Th. 164, 9; Gū. 1009: Andr. Kmbl. 1861; An. 933.

fyrstig; *adj*. [forst *frost*] *Frosty*; gĕlĭdus:—Ðæt se winter wǽre ceald and fyrstig *that the winter was cold and frosty*, Bd. 3, 19; S. 549, 27.

fyrst-mearc, frist-mearc, e; *f*. [mearc *a mark*] *Marked* or *appointed time, a space of time, interval*; tempus constĭtūtum, tempŏris spatium, intercăpēdo:—Sunne oncneów fyrstmearc his *the sun knew his appointed time*, Ps. Spl. T. 103, 20. Him eft-cymeþ æfter fyrstmearce feorh *life returns to it after a space of time*, Exon. 59 a; Th. 213, 11; Ph. 223: Andr. Recd. 269; An. 133: Elen. Kmbl. 2065; El. 1034. Ymb geára fyrstmearc *after a space of years; interjecto tempŏre aliquanto*, Bd. 3, 17; S. 543, 47: Cd. 202; Th. 251, 8; Dan. 560. Būtan fyrstmearce ǽnigre reste *sine ulla quiĕtis intercăpēdĭne*, Bd. 5, 12; S. 628, 3.

fȳr-sweart; *def*. se -swearta; *adj*. *Fire-swart, blackened with fire*; igne obscūrātus:—Fǽreþ æfter foldan [se] fȳrswearta lēg *the fire-swart flame shall pass along the earth*, Exon. 22 a; Th. 61, 14; Cri. 984.

fȳr-tang *fire-tongs*; forceps ignĭāria, Som. Ben. Lye.

fȳr-þolle? *An oven*; clĭbānus:—Ðū setst hig swā swā fȳrþolle fȳres pōnes eos ut clĭbānum ignis, Ps. Spl. T. 20, 9.

fyrþran, fyrþrian; *p*. ede; ode; *pp*. ed, od [furðor *further*] *To further, support, advance, promote*; provehere, promŏvēre:—Ðæt ic eáðe mæg ānra gehwylcne fremman and fyrþran freónda minra *that I may easily advance and further every one of my friends*, Andr. Kmbl. 1867; An. 936. Ðæt hī mǽgen hēnan ða yflan, and fyrþrian ða gódan *that they may humiliate the evil, and further the good*, Bt. 39, 2; Fox 212, 22. Friðaþ and fyrþraþ *protects and supports*, Bt. 34, 10; Fox 148, 29. Ealle Godes gerihto fyrþrie man georne *let every one zealously further all God's dues*, L. E. G. 5; Th. i. 168, 25, note 28, MS. B. DER. gefyrþran.

fyrþringnes, -ness, e; *f*. *A furthering, furtherance, promotion*; promōtio, L. I. P. 3; Th. ii. 306, 21.

fyrþrung, e; *f*. *A furthering, furtherance, promotion*; promōtio:—Ceápes fyrþrung *furtherance of trade*, Somn. 167; Lchdm. iii. 208, 6.

fȳr-tor, -torr, es; *m*. *A fire-tower, light-house*; phărus = φάρος, Cot. 93.

FYRWET, -wit, -wyt, es; *n*. *Curiosity*; cūriōsĭtas:—Hyne fyrwet bræc *curiosity urged him*, Beo. Th. 5562; B. 2784: 3975; B. 1985. Mec ðæs on worulde full oft fyrwit frineþ *my curiosity enquireth very often about this in the world*, Salm. Kmbl. 117; Sal. 58. Hine fyrwyt bræc *curiosity urged him*, Beo. Th. 470; B. 232. He his fyrwites ganges gylt forgeaf *he forgave him the guilt of his walk of curiosity*, Homl. Th. ii. 138, 24. þurh fyrwet *through curiosity*, Exon. 9 a; Th. 6, 30; Cri. 92. [*O. Sax*. firiwit, *m. n*: *O. H. Ger*. firiwizzî, *f*. cūriōsĭtas, portentum: *Icel*. fyrir-wissa, *f. a foreboding*.]

fyrwet-georn, firwet-georn; *adj*. *Curious, inquisitive*; cūriōsus:—Fela biþ fyrwetgeornra *there are many inquisitive*, Exon. 90 b; Th. 339, 31; Gn. Ex. 102.

fyrwet-geornnes, se; *f*. *Curiosity*:—For fyrwetgeornnesse ðæs wundres *for curiosity on account of the miracle*, Blickl. Homl. 69, 22.

fyrwit, -witt, -wyt; *adj*. *Curious, inquisitive*; cūriōsus:—Menn ða ða fyrwytte [fyrwite, MS. L.] beóþ *men who are inquisitive*, Bd. de nat. rerum; Wrt. popl. science 15, 9; Lchdm. iii. 268, 5.

fyrwit *curiosity*, Salm. Kmbl. 117; Sal. 58. v. fyrwet.

fyrwitnys, -nyss, e; *f*. *Curiosity*; cūriōsĭtas:—Hefigtyme leahter is ungefóh fyrwitnys *immoderate curiosity is a grave sin*, Homl. Th. ii. 374, 3. Ðæt he his fyrwitnysse fæderlīce miltsode *that he would paternally compassionate his curiosity*, ii. 138, 19.

fȳr-wylm, es; *m*. *A fire-boiling, raging flame*; flamma æstuans:—Wyrm cwom óðre sīþe, fȳrwylmum fāh *the dragon came a second time, coloured with raging flames*, Beo. Th. 5335; B. 2671.

fyrwyt *curiosity*, Beo. Th. 470; B. 232. v. fyrwet.

fȳryn, es; *n*. *A fire*; ignis:—On fȳrynes midlene *de mĕdio ignis*, Deut. 5, 24. v. fȳr.

FȲSAN; *p*. de; *pp*. ed [fūs *ready, prompt, quick*]. **I**. *v. intrans. To hasten*; festĭnāre:—He ongan fȳsan to fōre *he began to hasten for the way*, Cd. 138; Th. 173, 12: Gen. 2860: Elen. Kmbl. 451; El. 226. **II**. *v. reflex. To speed oneself, make haste, take oneself away, hasten away*; se festĭnāre, propĕrāre, se abrĭpĕre:—He ongan hine fȳsan and to flote gyrwan *he began speedily to prepare* [lit. *to speed himself and to prepare*] *for sailing*, Andr. Kmbl. 3392; An. 1700. Gǽst hine fȳseþ on ēcnegeard *the soul hasteneth to an eternal mansion*, Exon. 51 a; Th. 178, 7; Gū. 1240. He fȳsde hine *he hastened himself*, 120 a; Th. 461, 9; Hö. 33. **III**. *v. trans. To incite, stimulate, to send forth, drive away*; stĭmŭlāre, incĭtāre, accĕlĕrāre, emittĕre:—Ðū here fȳsest to gefeohte *thou excitest the host to a battle*, Andr. Kmbl. 2376; An. 1189. He fȳsþ ðē of getelde *emigrābit te de tabernācŭlo*, Ps. Lamb. 51, 7. He fȳsde forþ flāna genehe *he sent forth arrows abundantly*, Byrht. Th. 139, 44; By. 269. Fȳse hī man ūt of ðysan earde *let them be driven out of this country*, L. Eth. vi. 7; Th. i. 316, 22: L. C. S. 4; Th. i. 378, 8. [*Laym*. fusen, fuse, ifusen *to proceed*,

rush, drive: *O. Sax*. fūsian *to incline, strive*: *Icel*. fȳsa *to exhort*.] DER. a-fȳsan: ge-fȳsan.

fȳsian, fēsian *to send forth, to drive away*; relēgāre:—Ðonne fȳsie hī man of earde *let them then be driven from the country*, L. E. G. 11; Th. i. 174, 1. v. fȳsan.

FȲST, e; *f*. *A* FIST; pugnus:—Fȳst *pugnus*, Ælfc. Gl. 72; Som. 71, 3; Wrt. Voc. 43, 57. Gif men cīdaþ and hira ōðer hys nēxtan mid ȳste slicþ *si rixāti fŭĕrint vĭri et percussĕrit alter proxĭmum suum pugno*, Ex. 21, 18. On ðone eádgan andwlitan helfūse men hondum slōgun, folmum areahtum, and fȳstum eác *wicked men struck on the blessed visage with their hands, with outstretched palms, and with fists also*, Exon. 24 a; Th. 69, 24; Cri. 1125; Blickl. Homl. 23, 33; Mk. Bos. 14, 65. [*Piers P*. fust: *Chauc*. fest: *R. Glouc*. fustes, *pl*: *Laym*. uustes, fustes, *pl*. fists: *Plat*. fuust, fust, *f*: *Frs. O. Frs*. fest, *f*: *Dut*. vuist, *f*: *Ger*. faust, *f*: *M. H. Ger*. vûst, *f*: *O. H. Ger*. fûst, *f*: *Dan*. pust, *n. a blow*: *Swed*. pust, *m. a blow with the fist, box on the ear*: *Icel*. pústr, *m. a box on the ear*.]

fȳst-gebeát, es; *n*. *A blow with the fist*; pugni ictus, Past. 1, 3, 6? Lye.

fȳst-slægen; *part*. *Struck with the fist*; pugno cæsus:—Fȳstslægenu wæs exalapārētur, pugno cæsus erat, Cot. 79.

fyðer-, fiðer-, feðer- *four-*, found only in the compounds,—fyðer-dǽled, -fēte, -hiwe, -ling, -rīca, -rīce, -scȳte. v. feówer.

fyðera, fyðeru, fyðru, *pl. nom. acc*; *gen*. fyðera, fyðerena; *dat. inst*. fyðerum; *pl. nom. acc*. fyðeras; *m. Wings*; ālæ, pennæ:—Fyðera [Lamb. fyðeras] culfran ofersylfrede *pennæ cŏlumbæ deargentātæ*, Ps. Spl. 67, 14. Sunu manna on wǽfelse fyðera ðīnra hihtaþ *fīlii hŏmĭnum in tegmĭne ālārum tuārum spērābunt*, Ps. Spl. 35, 8: 56, 2: 60, 4: 62, 8. Under sceade fyðerena ðīnra gescyld me *sub umbra ālārum tuārum prŏtĕge me*, Ps. Lamb. 16, 8. Under his fyðerum ðū trūwast ðē ðū gehihtest *sub pennis ejus spērābis*, Ps. Lamb. 90, 4. Hwilc silþ me fyðera swā swā culfran *quis dābit mihi pennas sīcut cŏlumbæ?* Ps. Spl. 54, 6. Seó henn hyre cicenu under hyre fyðeru gegaderaþ *gallīna congrĕgat pullos suos sub ālas*, Mt. Bos. 23, 37. Ofer fyðeru [Lamb. fyðeras] winda *sŭper pennas ventōrum*, Ps. Spl. 103, 4. He fleáh ofer fyðru winda *vŏlāvit sŭper pennas ventōrum*, Ps. Lamb. 17, 11. v. fiðere, es; *n*: but generally *pl*.

fyðer-dǽled; *part*. *Divided into four, quartered*; quadripartĭtus, Leo. 151.

fyðered *having wings, winged*; ālātus, Som. Ben. Lye.

fyðer-fēte, -fóte; *adj*. *Four-footed*; quadrŭpes:—Fyðerfēte nȳten *a four-footed animal*, Med. ex Quadr. 1; Lchdm. i. 326, 11. Fyðerfēte quadrŭpes, Ælfc. Gr. 9, 26; Som. 11, 6. Ne on fyðerfótum ne on creópendum *neither among the four-footed nor the creeping*, Homl. Th. i. 486, 28. v. feówer-fēte.

fyðer-hiwe; *adj*. *Four-formed*; quadriformis, Leo. 151.

fyðerling, es; *m*. *The fourth part of a number* or *measure, a farthing*; quadrans, Som. Ben. Lye. v. feórþling.

fyðer-rīca, an; *m*. *A ruler over a fourth part, tetrarch*; tetrarches, tetrarcha, æ; *m*.=τετράρχης, ου; *m*:—Ða sind gecwedene tetrarche, ðæt sind, fyðerrīcan; fyðerrīca biþ se ðe hæfþ feórþan dǽl rīces *who are called tetrarchs, that is, rulers over a fourth; a tetrarch is he who has a fourth part of a kingdom*, Homl. Th. i. 478, 21.

fyðer-rīce, es; *n*. *A tetrarchy*; tetrarchia, Som. Ben. Lye.

fyðer-scȳte; *adj*. *Four-cornered, quadrangular*; quadrangŭlus:—Se arc wæs fyðerscȳte *the ark was quadrangular*, Boutr. Scrd. 21, 3. v. feówer-scȳte.

fȳtung, e; *f*. *A fighting, quarrelling*; rixa:—Ascūnige man swȳðe fracodlīce fȳtunga *turpes rixæ admŏdum evĭtentur*, L. Eth. vi. 28; Wilk. 122, 23. v. fītung, feohtan *to fight*.

fyxum *fishes*, Hexam. 11; Norm. 20, 5, = fixum, fiscum; *dat. pl. of* fisc.

G

WHEN *g* is the last radical letter of an Anglo-Saxon word, and follows a long vowel or an *r*, it is often changed into *h*, but then the *g* is resumed when followed by a vowel; as,—Beáh *a ring*; *gen*. es; *m*. beáges *of a ring*; *pl*. beágas *rings*; burh *a town*; *gen*. e; *f*. burge *of a town*; beorh *a hill*; *gen*. es; *m*. beorges *of a hill*; *pl*. beorgas *hills*. The same change takes place after a short vowel in wah *a wall*; *gen*. wages. In the conjugation of verbs, in some cases, *h* is found taking the place of *g*; thus from belgan *to be angry*, bilhst, bilhþ; from āgan *to own*, āhte. **2**. *g* is generally inserted between the vowels *-ie*, making -ige, -igende, *etc*. the first sing. *pres*. and part of verbs in -ian. Thus, from lufian *to love*, bletsian *to bless*, *etc*. are formed ic lufige *I love*, ic bletsige *I bless*, lufigende *loving*, bletsigende *blessing*. **3**. In later English the place of the earlier *g* is often taken by *y*, sometimes by *w*; as,—Geár *a year*, dæg *a day*, dagas *days*. *etc*: morg(en) *morrow*, sorg = *sorrow*, *etc*. **4**. The Anglo-Saxon Rune ᚷ not only stands for the letter *g*, but for gifu *a gift*; because gifu is the Anglo-Saxon name of this Rune. v. gifu **II**. and RŪN.

A a 2

gâ *go, come* :—Gâ hider neár *come hither near;* accéde huc, Gen. 27, 21; *impert. of* gân.

gaad *a goad,* Som. Ben. Lye. v. gâd, e; *f.*

gaar-leece *garlic,* Som. Ben. Lye. v. gâr-leác.

gaast, es; *m. A ghost, spirit:* spĭrĭtus :—Gaast is God spĭrĭtus est Deus, Jn. Lind. Skt. 4, 24. v. gâst.

gabban; *p.* ede; *pp.* ed *To scoff, mock, delude, jest;* hence, perhaps, GABBLE, GIBBERISH; derĭdĕre, lūdĕre, illūdĕre, Som. Ben. Lye. [*Prompt.* gabbin *mentiri* : Piers P. gabbe *to lie* : Chauc. *to chatter, lie* : Scot. gab *to mock, prate* : Icel. gabba *to mock.*]

gabbung, e; *f. A scoffing, mocking,* GIBING, *jesting;* derīsio, irrīsio, illūsio, Som. Ben. Lye. [*Prompt.* gabbinge *mendacium* : Piers P. gabbynge *lying* : Scot. gabbing *mockery, jeering.*]

gabere, es; *m. An enchanter, a charmer;* incantātor, Som. Ben. Lye. v. galere.

gabote, an; *f. A platter, small dish, dessert-dish;* paropsis = παροψίς, Wrt. Voc. 290, 22.

gabul-roid? *a line, rod, staff, compass;* rădius, circĭnus = κίρκινος, Som. Ben. Lye. v. gafol-rand.

GÂD, e; *f. A point of a weapon, spear* or *arrow-head, sting, prick,* GOAD; cŭspis, acūleus, stĭmŭlus :—Gâd *cuspis,* Wrt. Voc. 288, 23. Gâd *stĭmŭlus,* Wrt. Voc. 75; 1. Se yrþling nâ gâde hæfþ, būton of cræfte mĭnum *ărātor nec stĭmŭlum hăbet, nĭsi ex arte mea,* Coll. Monast. Th. 30, 31. Hafaþ gūþmecga gyrde lange, gyldene gâde *the warrior has a long rod, a golden goad,* Salm. Kmbl. 183 : Sal. 91. [*Goth.* gazds, *m. a prick, sting:* Swed. gadd, *m. a sting:* Icel. gaddr, *m. a goad, spike, sting.*] DER. gâd-ísen.

GÂD, gæd, es; *n? A lack, want, desire;* defectus, pēnūria, desīdĕrium, appĕtītus :—Ðæt ðâm gēngum þrym gâd ne wære wiste ne wæde *that there should be no lack of food or clothing to the three youths,* Cd. 176; Th. 222, 10; Dan. 102 : Elen. Kmbl. 1981; El. 992. Ne biþ ðé ǽnigra gâd wilna *there shall not be to thee a lack of any pleasures,* Beo. Th. 1903; B. 949. Ne wæs me in healle gâd *there was not a want to me in the hall,* Exon. 94 a; Th. 353, 20; Reim. 15. Ne wyrþ inc wilna gæd *there shall not be to you two a lack of pleasures,* Cd. 13; Th. 15, 21; Gen. 236. Nis him wilna gâd, ne meara, ne mâþma, gif he ðin beneah *there is not to him a desire for pleasures, nor horses, nor treasures, if he lacks thee,* Exon. 123 b; Th. 475, 6; Bo. 43. [*O. Sax.* gâdea, *f. a want,* in meti-gēdea *lack of food* : *Goth.* gaidw, *n. a want.*]

gada *a companion, an associate.* DER. ge-gada.

GADERIAN, gadorigean, gadrian, gadrigean, gæderian, gædrian; to gaderigenne, gadrienne, gadrigenne; ic gaderie, gaderige, gadrige, ðú gaderast, gadrast, he gaderaþ, gadraþ, *pl.* gaderiaþ, gadriaþ; *p.* gaderode; *pp.* gaderod *To* GATHER, *gather together, collect, store up;* lĕgere, collĭgĕre, congrĕgāre :—Næs nân heáfodman ðæt fyrde gaderian wolde *there was not a chief man who would gather together a force,* Chr. 1010; Erl. 144, 10. Ðâ án ongann folc gadorigean *then one began to gather the people,* Andr. Kmbl. 3111; An. 1558. Ic wolde eác gadrian sum gehwæde andgyt of ðære béc *I would also gather some little information from the book,* Bd. de nat. rerum; Lchdm. iii. 232, 2. Gadrigean, Andr. Kmbl. 1562; An. 782. Ðá ongan se æðeling Eádmund to gaderigenne [gadrigenne, Th. 276, 33, col. 2 : gadrienne, 277, 33, col. 1] fyrde *then the etheling Edmund began to gather a force,* Chr. 1016; Th. 276, 33, col. 1. Ic gaderige ðyder eall ðæt me gewexen ys *illuc congrĕgābo omnia, quæ nāta sunt mihi,* Lk. Bos. 12, 18. Ic gadrige [gaderie, MS. D.] lĕgo, Ælfc. Gr. 37; Som. 39, 22. Se ðe ne gaderaþ mid me, se hit tostret *qui non collĭgit mecum, dispergit,* Lk. Bos. 11, 23. Hý gaderiaþ feoh, and nyton hwâm hý hyt gadriaþ *they gather up wealth, and know not for whom they store it up,* Ps. Th. 38, 8 : Lk. Bos. 6, 44 : Mt. Bos. 6, 26. Ðæt folc gaderode mid micle menio ðæra fugela *the people gathered together a great number of the birds,* Num. 11, 32 : Chr. 1015; Th. 277, 16, col. 1 : Bd. de nat. rerum; Wrt. popl. science 1, 2; Lchdm. iii. 232, 4. Ic næbbe hwyder ic míne wæstmas gadrige *non hăbeo quo congrĕgem fructus meos,* Lk. Bos. 12, 17. [*Wyc.* gadre, gader, gedere : *Chauc.* gadred *gathered* : R. Glouc. gedere *gathered* : Laym. gadere, gaderen : Orm. gaddrenn : Scot. gadyr : Plat. gadern, gaddern : Frs. gearjen : O. Frs. gaduria, gaderia, gadria, garia : Dut. gaderen : Ger. gattern : M. H. Ger. gatern, getern : Icel. gadda coarctāre, Rask Hald.] DER. ge-gaderian.

gaderigendlíc, gadrigendlíc; *adj. Collective, congregative;* collectīvus, congrĕgātīvus, Som. Ben. Lye.

gaderscype, es; *m. Matrimonium,* Hpt. Gl. 438.

gader-tang, gæder-tang, gæder-teng; *adj. Continuous, connected with, united;* contĭnuus, assŏcius, consŏcius :—Biþ sum corn sædes gehealden symle on ðære sáule sóþfæstnysse, þenden gadertang wunaþ gâst on líce *some grain of the seed of truth will be always retained in the soul, while the spirit dwells in the body united to it,* Bt. Met. Fox 22, 77; Met. 22, 9 : Scint. 1.

gader-tangnys, gæder-tangnys, -nyss, e; *f. A continuation,* Scint. 12.

gader-tengan, gæder-tengan; *p.* de; *pp.* ed *To continue, join;* contĭnuāre, Som. Ben. Lye.

gaderung, e; *f. A* GATHERING, *congregation, joining, council, assembly,*

crowd; congrĕgātio :—Cyrce oððe geleáfful gaderung *a church* or *faithful gathering;* ecclēsia, Wrt. Voc. 80, 72. DER. ge-gaderung.

gadinca? *Mūtĭnus, fascĭnum obscēnum;* membrum vĭrĭle :—Gadinca vel hnoc mūtĭnus, Ælfc. Gl. 22; Som. 59, 83; Wrt. Voc. 23, 49.

gâd-ísen, es; *n. A gad-iron, goad;* acūleus, stĭmŭlus :—Sticel vel gâdísen acūleus, Ælfc. Gl. 1; Som. 55, 15; Wrt. Voc. 15, 15. Ic hæbbe sumne cnapan þýwende oxan mid gâdísene *hăbeo quendam puĕrum minantem bŏves mid stĭmŭlo,* Coll. Monast. Th. 19, 27.

gadorigean *to gather,* Andr. Kmbl. 3111; Au. 1558. v. gaderian.

gador-wist, e; *f. A dwelling together, companionship, intercourse;* contubernium, Ælfc. Gl. 116; Som. 80, 59; Wrt. Voc. 61, 42 : Cot. 43. DER. ge-gadorwist.

gadrian, gadrigean *to gather,* Bd. de nat. rerum; Lchdm. iii. 232, 2 : Andr. Kmbl. 1562; An. 782. v. gaderian.

gadrigendlíc *collective;* collectīvus, Som. Ben. Lye. v. gaderigendlíc.

gǽ *yea, yes,* Mt. Kmbl. Rush. 17, 25. v. gea.

géc, es; *m. A cuckoo, gawk;* cŭcūlus :—Gæces súre *cuckoo-sorrel, wood-sorrel;* acētōsa, acĭdŭla, Som. Ben. Lye. v. geác.

gǽd, es; *n. A being together, fellowship, union;* sŏcĭĕtas :—Nolde gæd geador in Godes ríce, eádiges engles and ðæs ofermódan *there would not [be] any fellowship in God's kingdom, of the blessed angel and the proud together,* Salm. Kmbl. 899; Sal. 449.

gǽd *a lack, want,* Col. 13; Th. 15, 21; Gen. 236. v. gâd, es; *n.*

gǽdeling, es; *m. A companion;* cŏmes :—His gædelinges gūþ-gewædu *his companion's battle-garments,* Beo. Th 5227; B. 2617 : Cd. 193; Th. 242, 20; Dan. 422. [*Piers P. Chauc. R. Glouc.* gadeling *an idle vagabond* : *Laym.* gadelinges, *pl. men of base degree* : *O. Sax.* gaduling, *m. a relation, kinsman* : *M. H. Ger.* geteling, *m. a relation, fellow* : *O. H. Ger.* gataling, *m. consanguĭneus, părens* : *Goth.* gadiligs, *m. a cousin, relation.*]

gǽdere; *adv. Together.* DER. æt-gædere, to-. v. geador.

gǽderian, gǽdrian *to gather,* Ps. Spl. 38, 10 : Exon. 58 b; Th. 211, 6; Ph. 193. v. gaderian.

gǽf *gave,* Bd. 3, 24; S. 557, 34, = geaf; *p. of* gifan.

gǽfe, e; *f. Grace;* Mid Godes gæfe *by God's grace,* Th. Chart. 459, 2. v. gifu.

gǽfel, es; *n. A gift, offering, tribute;* hostia, trĭbūtum, Lk. Skt. Rush. 2, 24 : Mt. Kmbl. Lind. 17, 25 : Mt. Kmbl. Rush. 9, 9. v. gafol.

gǽfel-geroefa, -gehréfa, -hroefa; *m. A publican,* Mt. Kmbl. Rush. 5, 46; 9, 11, 10.

gǽgl *wanton;* lascīvus, Lye. v. gagol.

gǽgl-bǽrnes, bǽrnes, -ness, e; *f. Wantonness, luxury, riot;* lascivia, Cot. 118.

gǽlǽþ, gǽleþ? *A cage to sell* or *punish bondmen in;* catasta, Som. Ben. Lye :—Gǽleþ *catasta,* Wrt. Voc. 288, 24.

GǼLAN; *p.* de; *pp.* ed.   I. *v. trans. to hinder, delay, impede, keep in suspense;* retardāre, mŏrāri, impĕdīre :—Hú lange gǽlst ðú úre líf *quousque anĭmam nostram tollis?* Jn. Bos. 10, 24. Swǽ mon oft lett fundiendne monnan, and his færelt gǽlþ, swǽ gǽlþ se líchoma ðæt mód *as a man hastening forward is often hindered, and his journey impeded, so the body impedes the mind,* Past. 256, 6; Hat. MS. 48 a, 16. Deáh hine singale gēmen gǽle *though perpetual care impede him,* Bt. Met. Fox 7, 101; Met. 7, 51. He men gǽleþ ǽlces gódes *he hinders men in respect to every good thing,* Blickl. Homl. 179, 11 : 191, 20.   II. *v. intrans. to hesitate, delay;* cunctāri :—Scealcas ne gǽldon *the servants delayed not,* Elen. Kmbl. 1381; El. 692 : 1999; El. 1001. DER. a-gǽlan.

gǽldan *to pay, depend, suspend;* pendĕre, dependĕre, suspendĕre, Som. Ben. Lye. v. geldan, gildan.

gǽle? *Saffron;* crŏcus :—Gǽle, geolo crŏcus, Wrt. Voc. 288, 47.

gǽleþ, ðú gǽlest *sings,* thou singest, Beo. Th. 4912; B. 2460; *3rd and 2nd pers. pres. of* galan.

gǽlnys, -nyss, e; *f. Wearisomeness, tediousness, loathing, disgust;* tædium :—Slǽþ sáwel mín for gǽlnysse *dormītāvit anĭma mea præ tædio,* Ps. Spl. 118, 28. v. gálnes.

gǽlsa, an; *m. Luxury, extravagance;* luxus, luxūria :—Lust oððe gǽlsa *luxus,* Ælfc. Gr. 11; Som. 15, 10. Lybbende on his gǽlsan *vivendo luxūriōse,* Lk. Bos. 15, 13. Þurh fulne folces gǽlsan *propter pŏpŭli luxum consummātum,* Lupi Serm. i. 21; Hick. Thes. ii. 105, 39. Ic him monigfealde módes gǽlsan ongeánbere *I present manifold mind's extravagances to him,* Exon. 71 a; Th. 264, 19; Jul. 366 : Homl. Th. i. 544, 28. Gǽlso *sollicitudo,* Mt. Kmbl. Lind. 13, 22. DER. hyge-gǽlsa.

gǽlþ, ðú gǽlst *sings,* thou singest; *3rd and 2nd pers. pres. of* galan.

gǽmnian; *p.* ode; *pp.* od *To play, game;* lūsĭtāre :—Ðæt man ungemetlíce gǽmnige *that a man immoderately play,* Homl. Th. ii. 590, 26. v. gamenian.

gǽngang; *adj. Pregnant?* prægnans? :—Gif hió gǽngang weorþeþ *if she becomes pregnant,* L. Ethb. 84; Th. i. 24, 7. v. Schmid, p. 9, note to c. 84.

gǽn-hwyrft, es; *m.* [gǽn = geán, ongeán *again*] *A turning again;* conversio :—On gecerringe oððe on gǽnhwyrfte Drihten gehæftnesse oððe hæftnunge Siones *in convertendo Dŏmĭnus captīvĭtātem Sion,* Ps. Lamb. 125, 1.

gǽn-ryne, es; *m. A running against, meeting*; occursus:—Arís on mínum gǽnryne *exsurge in occursum meum*, Ps. Lamb. 58, 6. v. geán-ryne.

Gǽnt *Ghent in Flanders*, Chr. 881; Th. 150, 13, col. 3. v. Gent.

gǽp; *adj. Cautious, shrewd, subtle*; sāgax, cautus, Ben. Lye. v. geap, II.

gǽr, geár, es; *n. A year*:—Ûre gǽr beóþ asmeáde *anni nostri meditābuntur*, Ps. Lamb. 89, 9. v. geár.

gǽrcian; *p.* ode; *pp.* od *To prepare*; pārāre:—Ðû gǽrcodest on ðínre swëtnysse ðam þearfan *parasti in dulcēdine tua paupĕri*, Ps. Lamb. 67, 11. Hî gǽrcodon flâna heora on cocere [MS. kokere] *parāvērunt sāgittas suas in pharetra*, 10, 3. v. gearcian.

gǽrcung, e; *f. A preparation, practice*; exercĭtātio:—Gedréfed oððe geunrôtsod ic eom on mínre gǽrcunge [MS. gǽrcuncge] *contristātus sum in exercĭtātiōne mea*, Ps. Lamb. 54, 3. v. gearcung.

gǽr-getal, es; *n.* [gǽr = geár *a year*; getæl, getel *a number*] *A tale of years, number of years*; annōrum sēries:—Hit cymþ æfter fiftigum wintra his gǽrgetales *it comes after fifty winters of his number of years*, L. M. 2, 59; Lchdm. ii. 284, 22.

GǼRS, gers, græs, es; *n.* GRASS, *a blade of grass, herb, hay*; grāmen, herba, fænum:—Gǽrs *vel* wyrt *herba*, Ælfc Gr. 4; Som. 3, 20: Jn. Bos. 6, 10. Híg and gǽrs *hay and grass*, Andr. Kmbl. 76; An. 38: Bt. Met. Fox 20, 196; Met. 20. 98. Gyf he mâran gǽrses beþyrfe *if he need more grass*, L. R. S. 4; Th. i. 434, 17. Seó eorþe wæstm beraþ, ǽrest gǽrs, syððan ear, syððan fulne hwǽte on ðam eare *terra fructĭficat, prīmum herbam, deinde spīcam, deinde plēnum frumentum in spīca*, Mk. Bos. 4, 28: Gen. I, 11: Num. 22, 4. Ðâ he hêt ða menegu ofer ðæt gǽrs hî sittan *cum jussisset turbam discumbĕre sŭper fænum*, Mt. Bos. 14, 19: Ps. Spl. 105, 20. Ofer gǽrsa cîpas *sŭper grāmĭna*, Deut. 32, 2. [*R. Brun.* gres: *Laym.* græs, gras: *Orm.* gresess *herbs*: *Scot.* gers, gyrs: *O. Sax.* gras, *n*: *Frs.* gerz: *O. Frs.* gers, gres, *n*: *Dut. Ger.* gras, *n*: *M. H. Ger. O. H. Ger.* gras, *n*: *Goth.* gras, *n*: *Dan.* græs, *n*: *Swed.* gräs, *n*: *Icel.* gras, *n*.]

gǽrsama, gersuma, an; *m. Treasure*; ōpes:—He lêt nyman of hire ealle ða betstan gǽrsaman *he caused all the best treasures to be taken from her*, Chr. 1035; Th. 292, 22, col. 2. Gif he ne sealde ðe mâre gersuman *if he had not given the greater treasures*, Chr. 1047; Erl. 177, 7. v. gǽr-sum.

gǽrs-bed, -bedd, es; *n. A grass-bed, grave*; sub cæspĭte lectus, sepulcrum:—Ðonne he gâst ofgifeþ, syððan hine (?) gǽrsbedd sceal wunian *when he gives up his spirit, then must he inhabit a grave*, Ps. Th. 102, 15.

gǽrs-cîþ, es; *n. A blade of grass*; grāmĭnis germen:—Gǽrstapan cômon and frǽton ealle ða gǽrscîpas *locusts came and ate up all the blades of grass*, Ors. 1, 7; Bos. 29, 42.

gǽrs-grêne *grass-green*; grāmĭneus, herbĭdus, vĭrĭdis, Som. Ben. Lye.

gǽrs-hoppa, an; *m. A grass-hopper, locust*; lŏcusta, cĭcāda:—He cwæþ and com gǽrshoppa *dixit et vēnit lŏcusta*, Ps. Lamb. 104, 34: 108, 23. Cwômón gǽrshoppan *grass-hoppers came*, Ps. Th. 104, 30: 77, 46. [*Orm.* gress hoppe *locusts*.]

gǽrs-molde *grass-land*. v. græs-molde.

gǽrs-stapa, gǽrstapa, an; *m. A* GRASS-STEPPER, *locust*; lŏcusta:—Gǽrstapa lŏcusta, Wrt. Voc. 78, 61. He sǽde and com gǽrstapa *dixit et vēnit lŏcusta*, Ps. Spl. 104, 32: 108, 22. He sealde geswinc heora gǽrstapan *dĕdit lābōres eōrum lŏcustæ*, Ps. Lamb. 77, 46. Gǽrstapan cômon and frǽton ealle ða gǽrscîpas *locusts came and ate up all the blades of grass*, Ors. 1, 7; Bos. 29, 42: Homl. Th. ii. 192. 35. Gǽrstapan hit fretaþ eall *lŏcustæ devŏrābunt omnia*, Deut. 28, 38: Num. 13, 33: Ex. 10, 12: Jud. 6, 5: Mt. Bos. 3, 4. Se byrnenda wind brohte gǽrstapan *ventus ūrens levāvit lŏcustas*, Ex. 10, 13, 19: 10, 4.

gǽrs-swŷn, es; *n. A pasturage swine*; herbāgii porcus:—He sceal syllan gǽrs-swŷn *dēbet dāre porcum herbāgii*, L. R. S. 2; Th. i. 432, 9.

gǽrst *green like grass*; herbeus, Som. Ben. Lye.

gǽrs-tún, es; *m. A grass-enclosure, a meadow*; prātum, pascuum:—hence GERSTON, now used in Surrey and Sussex, in the same sense:—Be ceorles gǽrstûne: gif ceorlas gǽrstûn hæbben gemǽnne, oððe ôðer gedǽl-land to tŷnanne *of a churl's meadow: if churls have a common meadow or other partible land to fence*, L. In. 42; Th. i. 128, 5. Prātum quod Saxŏnĭce Garstûn appellātur, Cod. Dipl. 350; A. D. 930; Kmbl. ii. 166, 6: Cod. Dipl. Apndx. 461; A. D. 956; Kmbl. iii. 449, 19.

gǽrs-tûn-díc, es; *m. A grass-meadow-dike*; vallum circa prātum ductum:—On gǽrstûndíc sûðeweardne *to the south of the grass-meadow-dike*, Cod. Dipl. Apndx. 441; A. D. 956; Kmbl. iii. 438, 4.

gǽrsum, gersum, es; *m. n. Treasure, riches*; thēsaurus, ōpes:—He lêt niman of hyre ealle ða betstan gǽrsuma *he caused all the best treasure to be taken from her*, Chr. 1035; Erl. 164, 23: 1090; Erl. 226, 25. Hî betêhtan ðǽr ealla ða gǽrsume *they deposited there all the treasures*, 1070; Erl. 209, 17, 27, 33. Hî nâmen manega gersumas *they took many treasures*, Chr. 1070; Erl. 209, 13. For his mycele gersuma *for his great treasures*, 1090; Erl. 226, 38. [*Laym.* gǽrsume *treasure*: *Scot.* gersome *a sum paid by a tenant to a landlord on the entry of a lease.* The word seems to have been introduced from the Scandinavian, cf. *Icel.* gör-semi, ger-semi *a costly thing, jewel*; and see Cl. and Vig. Dict. for etymology.]

gǽrs-wong *a field of grass, grassy plain.* v. græs-wong.

gǽrs-yrþ, e; *f. Grass-land, pasturage*; herbāgium:—To gǽrsyrþe *de herbāgio*, L. R. S. 4; Th. i. 434, 17. See Schmid, p. 374, note.

gǽruwe, an; *f. Yarrow*; millefŏlium:—Gǽruwe *millefŏlium*, Ælfc. Gl. 40; Som. 63, 82; Wrt. Voc. 30, 32. v. gearwe.

gǽsne, gesne, geásne, gǽsine; *adj. Barren, sterile, empty, wanting, void of, lifeless*; stĕrĭlis, inānis, ĕgēnus, destĭtūtus, expers, exănĭmis:—Ðæt we gǽstes wlite, on ðâs gǽsnan tíd, georne biþencen *that we earnestly consider, in this barren time, the spirit's beauty*, Exon. 20 a; Th. 53, 13; Cri. 850. Ðis geár wæs gǽsne on mǽstene *this year was barren in mast-fruit*, Chr. 1116; Th. 371, 16. Hirdas lǽgon gǽsne on greóte *the keepers lay lifeless on the sand*, Andr. Kmbl. 2169; An. 1086. v. Grm. Andr. Elen. p. 124, 1085: Graff. IV. 267. [*Piers P.* gesen: Halliw. Dict. geson *scarce.*]

gǽst, gest, gist, giest, gyst, es; *pl. nom. acc.* gastas; *m.* **I.** *a* GUEST; hospes, sŏcius:—Gǽst inne swǽf *the guest slept within*, Beo. Th. 3605; B. 1800. Biþ symle gǽst *will ever be a guest*, Exon. 84 c; Th. 318, 9; Mod. 80. Gârsecges gǽst *the ocean's guest*, 97 a; Th. 301, 33; Wal. 29. Ferende gǽst *a journeying guest*, 103 a; Th. 390, 12; Rä. 8, 9. Gǽst ne grêtte *he greeted not the guest*, Beo. Th. 3790; B. 1893. Gasta werode *with the multitude of guests*, Cd. 67; Th. 81, 16; Gen. 1346. Gif hine sǽ byreþ gǽsta [gasta?] fulne *if the sea shall bear it [the vessel] full of guests*, Exon. 101 b; Th. 384, 20; Rä. 4, 30. **II.** *a stranger, an enemy*; vir' alĭenĭgĕnus, hostis:—Wæs se grimma gǽst Grendel hâten, wonsǽlig wer *the grim enemy was called Grendel, the unblest man*, Beo. Th. 204; B. 102: 4158; B. 2073. Ða se gǽst ongan glêdum spîwan *then the fiend [the dragon] began to vomit fire*, 4613; B. 2312. Hwonne gǽst cume to dûrum mínum, him biþ deáþ witod *when a stranger comes to my doors, death is decreed to him*, Exon. 104 b; Th. 396, 26; Rä. 16, 10. [*Piers P.* gest: *Wyc.* geste: *Chauc.* gest: *Laym.* gesst: *O. Sax.* gast, *m*: *Plat. Dut. Ger. M. H. Ger. O. H. Ger.* gast. *m*: *Goth.* gasts, *m*: *Dan.* giest, *m. f*: *Swed.* gäst, *m*: *Icel.* gestr, *m*.] DER. beód-gǽst, brim-, níþ-, wæl-.

gǽst, es; *m. The soul, spirit, mind*; spīrĭtus, anĭmus:—Him wæs gǽst geseald *a spirit was given to him*, Cd. 201; Th. 249, 21; Dan. 533. Nyle he ǽngum ânum ealle gesyllan gǽstes snyttru *he will not give all wisdom of mind to any one man*, Exon. 17 b; Th. 43, 5; Cri. 684. Gûþlâc in gǽste bær heofoncundne hyht *Guthlac bare heavenly hope in his spirit*, Exon. 35 a; Th. 112, 10; Gú. 141. Deáh ðe him onwrige wuldres cyning wîsdômes gǽst *though the king of glory revealed to them the spirit of wisdom*, Exon. 73 a; Th. 273, 15; Jul. 516. v. gâst.

gǽst goest, walkest, Gen. 3, 14; *2nd pers. pres. of* gân.

gǽst-ǽrn, -ern *a guest-place, guest-chamber, an inn.* v. gest-ǽrn.

gǽstan; *p.* te; *pp.* ed [gǽst, gǽst *a spirit, ghost*] *To gast, frighten, afflict, torment*; terrēre, crŭciāre, afflĭgĕre:—Hî gǽston Godes cempan gâre and lîge *they afflicted God's champions with spear and flame*, Exon. 66 a; Th. 243. 27; Jul. 17. [*Wyc.* gaste *to make greatly afraid*: *Piers P.* gaste *to scare* [*birds*], Cf. *Goth.* us-gaisjan, and v. Dief. ii. pp. 397-8.]

gǽst-berend, es; *pl. nom. acc.* -berend; *m. A spirit-bearer, man*; is qui spīrĭtum *vel* ănĭmum fert, hŏmo:—Ðâs gǽstberend gîman nellaþ *these spirit-bearers will not heed*, Exon. 31 a; Th. 97, 33; Cri. 1600: 78 a; Th. 293, 17; Crä. 2. Ic gǽstberend cwelle compwǽpnum *I kill the living with battle-weapons*, 105 b; Th. 401. 8; Rä. 21, 8.

gǽst-cund; *adj. Spiritual*; spīritālis:—Seó lufu in monnes môde getimbreþ gǽstcunde gife *love builds up spiritual grace in man's mind*, Exon. 44 a; Th. 148, 11; Gú. 743.

gǽst-cwalu, e; *f. Torment of soul*; ănĭmæ tormentum:—Ðǽr eów is hâm sceapen, grim gǽstcwalu *there a home is made for you, bitter torment of soul*, Exon. 42 b; Th. 142, 28; Gú. 651.

gǽst-gedâl, es; *n. Separation of soul and body, death*; ănĭmæ et corpŏris divortium, mors:—Ne he sorge wǽg gǽstgedâles *he sorrowed not for his soul's separation*, Exon. 49 a; Th. 170, 14; Gú. 1111. v. gâst-gedâl.

gǽst-gehygd, es; *n. Thought of mind*; ănĭmi cōgĭtātio:—Him seó unforhte ageaf andsware, þurh gǽstgehygd, Iuliana *the fearless Juliana gave him answer through her mind's thought*, Exon. 67 b; Th. 251, 20; Jul. 148. v. gâst-gehygd.

gǽst-gemynd, es; *n. Thought of mind or spirit*; ănĭmi cōgĭtātio:—Ic him gǽstgemyndum wille wesan underþýded *I will be subjected to him in my spirit's thoughts*, Exon. 41 a; Th. 138, 11; Gú. 574.

gǽst-genîþla, an; *m. A persecutor or foe of souls, the devil*; ănĭmārum insectātor *vel* hostis, diabŏlus:—Hæfde engles hiw gǽstgenîþla, helle hæftling *the foe of souls, the captive of hell, had an angel's form*, Exon. 69 a; Th. 257, 11; Jul. 245.

gǽst-gerýne, es; *n. A ghostly or spiritual mystery, a mystery of the mind*; spīrĭtāle mystērium, ănĭmi mystērium:—In godcundum gǽstgerýnum *in divine spiritual mysteries*, Exon. 36 a; Th. 117, 5; Gú. 219: 49 a; Th. 168, 31; Gú. 1086. Bî ðon Salomon song, gidda snottor, gǽstgerýnum *of whom Solomon, wise in song, sang in spiritual mysteries*, Exon. 18 a; Th. 45, 3; Cri. 713: 14 a; Th. 28, 2; Cri. 440. v. gâst-gerýne.

gǽst-gewinn, es; *n. Torment of soul*; ănĭmæ tormentum :—In đam grimmestan gǽstgewinne *in the bitterest torment of soul*, Exon. 41 a; Th. 137, 19; Gū. 561.

gǽst-hālig; *adj. Spirit-holy, holy in spirit*; in spīrĭtu sanctus :—Wǽr is ætsomne Godes and monna, gǽst-hālig treów *there is a compact together of God and men, a spiritual holy covenant*, Exon. 16 a; Th. 36, 31; Cri. 584. He fond fūsne on forþsīþ freán unwemne, gǽst-hāligne *he found his blameless master bent on departure, holy in spirit*, 49 b; Th. 171, 5; Gū. 1122. Gǽst-hālge guman *men holy in spirit*, 95 b; Th. 356, 33; Pa. 21: 45 b; Th. 154, 19; Gū. 845. v. gāst-hālig.

gǽst-, gast-, gest-, gyst-hūs, es; *n. A guest-house, guest-chamber*; hospĭtium :—Gǽst-hus *hospĭtium*, Wrt. Voc. 86, 44. [*Orm.* gessthus : *Ger.* gasthaus *inn.*]

gǽst-hof *a guest-house*, v. gast-hof.

gǽstlic *hospitable, ready for guests.* v. gastlíc.

gǽstlíc; *adj. Ghostly, spiritual*; spĭrĭtālis :—Giofu gǽstlíc *spiritual grace*, Exon. 8 b; Th. 3, 26; Cri. 42: 18 a; Th. 44, 7; Cri. 699: 71 a; Th. 265, 26; Jul. 387. þurh gǽstlícu wundor *through spiritual miracles*, Exon. 34 b; Th. 111, 14; Gū. 126. Mid gǽstlícum wǽpnum *with spiritual weapons*, 35 a; Th. 114, 24; Gū. 148. v. gástlíc.

gǽstlíce; *adv. Spiritually*; spĭrĭtālĭter :—Deáh he gōdes hwæt onginne gǽstlíce *though he attempt aught of good spiritually*, Exon. 71 b; Th. 266, 15; Jul. 398. v. gástlíce.

gǽst-līđe *kind to guests, hospitable.* v. gist-līđe.

gǽst-līđnes, gest-līđnes, giest-līđnys, -nyss, e; *f. Hospitableness, hospitality, entertainment of guests*; hospĭtālĭtas :—We willaþ eów on gǽst-līđnesse onfón *we will receive you in hospitality*, Bd. 1, 25; S. 487, 15. Đætte ælþeódige bisceopas sȳn þoncfulle heora gǽstlīđnesse and feorme *ut episcŏpi peregrīni contenti sint hospĭtālĭtātis mūnĕre oblāto*, Bd. 4, 5; S. 573, 3.

gǽst-lufe, an; *f. Soul's love, spiritual love*; spīrĭtālis ămor :—For gǽstlufan *for spiritual love*, Exon. 55 b; Th. 196, 11; Az. 172. Mid gǽstlufan *with spiritual love*, 55 b; Th. 197, 11; Az. 188.

gǽst-mǽgen, v. gist-mǽgen.

gǽst-sele *a guest-hall.* v. gest-sele.

gǽst-sunu; *gen.* -suna; *m. A spiritual son, Christ* :—Godes gǽstsunu *God's spiritual Son*, Exon. 17 b; Th. 41, 23; Cri. 660: 20 b; Th. 53, 35; Cri. 861. v. gāst-sunu.

gǽt, es; *n. A gate* :—Æt đam gǽte *ad ostium*, Bd. 3, 11; S. 536, 17: Mt. Lind. Stv. 7, 13. v. geat.

gǽt *goats*, Exon. 26 a; Th. 75, 34; Cri. 1231; Rtl. 119, 16; *pl. nom. acc. of* gāt.

gǽtan; *p.* de, te; *pp.* ed *To grant, to confirm* :—Ic gǽte *I confirm*, Chr. 675; Th. 59, 30. v. geátan.

gǽten; *adj.* [gāt *a goat*] *Of or pertaining to goats*; caprīnus :—Gǽten smeoro *goat's grease*, Med. ex Quadr. 6, 15; Lchdm. i. 354, 8. Gǽten roc [MS. rooc] *a garment made of goat-skins*; mēlōtes = μηλωτή, Ælfc. Gl. 63; Som. 68, 117; Wrt. Voc. 40, 27.

gǽþ *goes* :—He gǽþ *he goes*, Beo. Th. 4075; B. 2034; *3rd pers. pres. of* gān.

GAF; *adj. Base, vile, lewd*; turpis, vīlis, lŏquax :—Hwǽr biþ his gaf spræc *where will be his wanton discourse?* Basil admn. 9; Norm. 50, 28. [*Scot.* gaff *to talk loudly and merrily* (?)]. DER. ge-gaf; and cf. gafetung.

gaf *gave*, Salm. Kmbl. 114, note; Sal. 56; *p. of* gifan.

gafel, es; *n. Tax, tribute*; vectīgal, trĭbūtum :—Đæt he mǽge cyninges gafel forþbringan *that he can bring forth the king's tribute*, L. Wg. 7; Th. i. 186, 14, note 17. Hí Godes gafel lǽston *they rendered God's tribute*, L. Eth. ix. 43; Th. i. 350, 8. Gafeles andfengend *numĕrārius, numŭlārius, vectīgālis, receptor*, Cot. 142. v. gafol.

gafelian; *p.* ode; *pp.* od *To rent*; condūcĕre :—Ic geann đárto twegra hída đe Eádríc gafelaþ *I give thereto two hides which Eadric rents*, Cod. Dipl. 699; A.D. 997; Kmbl. iii. 305, 6. DER. ge-gafelod.

gafellíc; *adj. Tributary*; trĭbūto sive fisco pertĭnens, Cot. 85.

gafeluc, es; *m. A spear, javelin*; hastīle :—Gafelucas *hastīlia*, Ælfc. Gl. 52; Som. 66, 54; Wrt. Voc. 35, 41. [*R. Brun.* gauelokes *javelins*: *M. H. Ger.* gabilôt, gabylôt, *n. a javelin* : *Icel.* gaflok, *n. spĭcŭli gĕnus*, Rask Hald : *Fr.* javelot, *m. a javelin* : *It.* giavelotto, *m*: *Wel.* gaflach, *m. a fork, bearded spear* : *Ir.* gabhla *a spear, lance* : *Gael.* gobhlach *forked* : *Armor.* gavlod, *m. a javelin.*]

gaffetung, gafetung, e; *f. A scoffing, mocking*; dērisio :—Of đisum leahtre beóþ acennede mōdes unstæđđignys and ȳdel gaffetung *of this sin are born unsteadiness of mind and idle scoffing*, Homl. Th. ii. 218, 33. He forlǽt derigendlíce gaffetunga *he forsakes injurious scoffings*, Homl. Th. i. 306, 2. Đa wēlegan on heora gebeórscipe begáþ derigendlíce gafetunge *the wealthy in their feasting practise pernicious scoffing*, i. 330, 33. v. gaf.

gaflas; *pl. m. Forks, props, spars of a building, a gallows*; furcæ, patĭbŭlum, Som. Ben. Lye. [*O.H. Ger.* gabala *furca*: and v. Dief. ii. 402.]

gafol, gafel, gaful, es; *n.* [gifan *to give*] *Tax, tribute, rent, interest*; vectīgal, trĭbūtum, census, ūsūra :—Hyra ár is mǽst on đǽm gafole, đe

đa Finnas him gyldaþ: đæt gafol biþ on deóra fellum, and on fugela feđerum *their revenue is chiefly in the tribute, which the Finns pay them: the tribute is in skins of beasts, and in feathers of birds*, Ors. 1, 1; Bos. 20, 32–34. To gafle gesettan *to let out for rent*, Chr. 1100; Erl. 236, 6. Gafol *ūsūra*, Ælfc. Gr. 43; Som. 45, 4. Ætȳwaþ me đæs gafoles mynyt *ostendĭte mihi numisma census*, Mt. Bos. 22, 19: L. Edg. S. 1; Th. i. 270, 19: Exon. 16 a; Th. 35, 16; Cri. 559. Cyninges gafoles bīgerdel *a king's tribute-purse*; saccus *vel* fiscus, Ælfc. Gl. 65; Som. 69, 35; Wrt. Voc. 40, 63. Hí đone fīftan dǽl ealra hiora eorþwæstma đǽm cyninge to gafole gesyllaþ *they give the fifth part of all their fruits of the earth to the king for tribute*, Ors. 1, 5; Bos. 28, 31: Byrht. Th. 133, 6; By. 46. Ic nǽme đæt mín ys mid đam gafole *ego recēpissem quod meum est cum ūsūra*, Mt. Bos. 25, 27. Se đe feoh his ne sealde to gafole *qui pĕcūniam suam non dĕdit ad ūsūra..,* Ps. Lamb. 14, 5. Đæt him leófre wǽre wiđ hine to feohtanne, đonne gafol to gyldenne *that they would rather fight against him, than pay him tribute*, Ors. 1, 10; Bos. 32, 24, 28 : L. Edg. S. 1; Th. i. 270, 16: L. O. D. 9; Th. i. 356, 18: Chr. 991; Erl. 130, 21: 994; Erl. 132, 31. Đa đæt gafol nāmon *qui didrachma accipiēbant*, Mt. Bos. 17, 24, 25: 22, 17: Lk. Bos. 20, 22: 23, 2. Gafol sellan *to give tribute*, Cd. 93; Th. 119, 12; Gen. 1978. Đæt gē đisne gārrǽs mid gafole forgyldon *that ye buy off this warfare with tribute*, Byrht. Th. 132, 47; By. 32. Freólsdóm gafola *freedom from imposts*, L. Wih. 1; Th. i. 36, 15. [*M. Lat.* gablum: *Fr.* gabelle: *It.* gabella: *Span.* gabela *tax.* A Celtic origin has been suggested for this word, v. Dief. ii. 400–1.] DER. bere-gafol, ealu-, feoh-, hunig-, land-, mete-, neád-, rǽde-.

Gafol-, Gaful-ford; *gen.* -fordes; *dat.* -forde, -forda; *m.* [gafol *tribute*, ford *a ford: the tributary ford*] *Camelford, Cornwall*; lŏci nōmen in agro Cornubiensi :—Hēr wæs Weala gefeoht and Defna æt Gafolforda [Gafulforda, Th. 110, 111, 17, col. 1] *here* [A.D. 823] *there was a battle of the Welsh and Devonians at Camelford*, Chr. 823; Th. 110, 17, col. 2; 111, 17, col. 2, 3.

gafol-bere, es; *m. Barley paid as rent* :—Threó pund gauolbæres, Th. Chart. 145, 2.

gafol-, gaful-gylda, -gilda, -gelda, an; *m.*   I. *a tribute-payer, tributary, debtor*; trĭbūti reddĭtor, dēbĭtor :—Rōmāne hȳ to gafol-gyldum gedydon *the Romans made them tributaries*, Ors. 3, 8; Bos. 63, 38: Bd. 2, 5; S. 5c6, 20. Beón hig ealle gesunde and þeówion đē and beón đíne gafolgildan *cunctus pŏpŭlus salvābĭtur et serviet tĭbi sub trĭbūto*, Deut. 20, 11. Twegen gafolgyldan wǽron sumum lǽnende *duo dēbĭtōres ĕrant cuidam fænĕrātōri*, Lk. Bos. 7, 41: 16, 5.   II. *a rent-payer, a renter of land as opposed to the owner*: qui censum annum pendit, conductor :—Wealh gafolgelda [gafolgylda MSS. B. H.] *a foreign* [i.e. of British race] *tenant*, L. In. 23; Th. i. 118, 3. Gif he on gafolgeldan [gafolgildan MS. H.] hūse gefeohte, cxx scillinga to wíte geselle *if he fight in a tenant's house, let him pay cxx shillings as fine*, 6; Th. i. 106, 7.

gafol-gyldere; *m. A tribute-payer, tributary*; trĭbūti reddĭtor :—Đa Indiscan willaþ beón eówere gafolgylderas, and mid ealre sibbe eów underþeódan *the Indians will be your tributaries, and with all peace submit to you*, Homl. Th. ii. 482, 31.

gafol-heord, e; *f.* [gafol *a tax*, heord *a herd, flock*] *A taxable stock* or *hive of bees*; grex ad censum :—Beóceorle gebȳreþ, gif he gafolheorde healt, đæt he sylle đonne lande gerǽd beo. Mid us is gerǽd đæt he sylle v sustras huniges to gafole *it behoves a keeper of bees, if he hold a taxable hive* [stock of bees], *that he then shall pay what shall be ordered in the country. With us it is ordered that he shall pay five sustras of honey for a tax*; ‘beóceorle [id est, āpum custōdi, pertinet, si gavelheorde, id est, grēgem ad censum tĕneat, ut inde reddat sīcut ibi mos [MS. moris] ĕrit. In quibusdam lŏcis est instĭtūtum, reddi V. [MS. VI] mellis ad censum,' L. R. S. 5; Th. i. 434, 36–436, 2.

gafol-hwitel, es; *m. A tribute-whittle* or *blanket, a legal tender instead of coin for the rent of a hide of land*; trĭbūtāria sǎga :—Gafol-hwitel sceal beón æt hīwisce vi pæninga weorþ *a tribute-whittle from a hide* [of land] *shall be worth six pence*, L. In. 44; Th. i. 130, 5. Cf. Grm. R. A. p. 378. Perhaps hīwisc in the above passage should be translated ‘family;' cf. Th. Chart. 144, 31.

gafolian *to rent.* v. gafelian.

gafol-land, es; *n. Tribute-land, land let for rent* or *services*; trĭbūtāria terra :—Būton đam ceorle đe on gafollande sit *except the churl who resides on tribute-land*, L. A. G. 2; Th. i. 154, 2. Cf. Th. Chart. p. 144–5. [*Scot.* gaffol-land *rented*, or *liable to taxation.*]

gafollíc *of* or *belonging to tribute, tributary.* v. gafellíc.

gafol-mǽd, e; *f. A meadow, the mowing of which was part of the gafol due from the churls on an estate* :—Healfne æcer gauolmǽde, Th. Chart. 145, 3.

gafol-penig, es; *m. A tribute-penny*; trĭbūtārius dēnārius :—He sceal syllan on Michaeles mæssedæg x gafolpenigas *he shall give on Michael's mass-day ten tribute-pennies*; đǽre dēbet in festo Sancti Michaelis x dēnārios de gablo, L. R. S. 4; Th. i. 434, 10.

gafol-, gaful-rǽden, -rǽdenn, e; *f.* [gafol *tribute*, -rǽden *state, condition*] *Tribute*; trĭbūtum :—On sumum landum gebȳreþ mǽre gafolrǽden *in quibusdam lŏcis plus gabli reddĭtur*, L. R. S. 5; Th. i. 436, 3.

gafol-rand? *A pair of compasses;* circīnus = κίρκινος, Cot. 54, Som. Ben. Lye. v. gabul-roid.

gafol-swân, es; *m. A tribute-swain, a swine-herd, paying a tribute or part of his stock, for permission to feed his pigs on the land;* porcārius ad censum :—Gafolswāne gebȳreþ, ðæt he sylle his slyht be ðam ðe on lande stent. On manegum landum stent, ðæt he sylle ælce geáre xv swȳn to sticunge, x ealde, and v gynge; hæbbe sylf ðæt he ofer ðæt arære *gafol-swāne, id est, ad censum porcārio, pertīnet, ut suam occīsiōnem det secundum quod in patria stătūtum est. In multis lŏcis stat, ut det singŭlis annis xv porcos ad occīsiōnem, x vĕtĕres, et v juvĕnes; ipse autem hăbeat superaugmentum,* L. R. S. 6; Th. i. 436, 11–14.

gafol-tíning, e; *f. Material for fencing due as gafol* :—XVI gyrda gauoltīninga, Th. Chart. 145, 8.

gafol-wydu, a; *m. Wood furnished as gafol* :—IIII foðera aclofenas gauolwyda, Th. Chart. 145, 6.

gafol-yrþ, e; *f. The cultivation of tribute-land;* tribūtāriæ terræ arātio :—His gafolyrþe [MS. gauolyrþe] iii æceras erige, and sáwe of his ágenum berne *de arātūra gabli sui arābit iii acras, et semīnābit de horreo suo,* L. R. S. 4; Th. i. 434, 18.

gaful, es; *n. Tax, tribute, rent;* vectīgal, trĭbūtum :—Gaful *vectīgal,* Ælfc. Gr. 9, 5; Som. 9, 2. Alȳfþ gaful to syllanne ðam Cāsere *licet dări trĭbūtum Cæsări?* Mk. Bos. 12, 14: Exon. 68 a; Th. 251, 27; Jul. 151. v. gafol.

Gaful-ford *Camelford, Cornwall,* Chr. 823; Th. 110, 111, 17, col. 1. v. Gafol-ford.

gaful-gylda, an; *m. A tribute-payer, tributary;* trĭbūti reddĭtor :—He hí to gafulgyldum gesette on Angelþeodde *he made them tributaries among the English,* Bd. 1, 34; S. 499, 24. v. gafol-gylda.

gaful-ræden, -rædenn, e; *f. A tax, tribute;* census, trĭbūtum :—Ða byre onguldon gafulrædenne *the children paid the tax,* Exon, 47 a; Th. 161, 16; Gú. 959; Th. 274, 7; Jul. 529: Andr. Kmbl. 591; An. 296. v. gafol-ræden.

gagátes; *indecl. m. The agate* or *jet, a precious stone;* găgātes = γάγάτης :—Hér biþ eác gemēted gagátes, se stán biþ blæc-gym *here is also found the agate, the stone is a black gem,* Bd. 1, 1; S. 473, 24. Sceaf gagátes dæl ðæs stānes on ðæt wín *shave off a part of the stone agate into the wine,* L. M. 2, 65; Lchdm. ii. 296, 11. Be ðam stāne ðe gagátes hätte, is sǽd ðæt he viii mægen hæbbe *of the stone which is called agate, it is said that it hath eight virtues,* 2, 66; Lchdm. iii. 296, 29.

gagel, es; *m?* gagelle, gagille, gagolle, an; *f. Gale, sweet gale;* myrica gale, Lin :—Genim gagel *take gale,* L. M. 1, 36; Lchdm. ii. 86, 10: iii. 22, 21. Nim þré leáf gageles *take three leaves of gale,* Lchdm. iii. 6, 17. Genim gagéllan ... dó of ða gagellan *take gale ... remove the gale,* L. M. 2, 51; Lchdm. ii. 264, 27: 2, 53; Lchdm. ii. 274, 10. Genim gagollan *take gale,* 3, 14; Lchdm. ii. 316, 15. [*Prompt.* gawl *myrtus: Scot.* gale, gaul *a myrtle: Dut.* gagel, *m. a wild myrtle: Ger.* gagel *a myrtle-bush.*]

gagel-croppan; *pl. m. [croppa the top of a flower or herb] Catkins of gale;* myricæ panĭcŭlæ :—Genim gagelcroppan *take catkins of gale,* L. M. 1, 36; Lchdm. ii. 86, 20.

gagol, gægl, geagl; *adj. Lascivious, wanton;* lascīvus :—Gagol *lascīva,* Ælfc. Gl. 106; Som. 78, 46; Wrt. Voc. 57, 27. [*M. H. Ger.* gogel *licentious.*] v. gál.

gagol-bærnes, gægl-bærnes, -bērnes, -ness, e; *f. Wantonness, luxury, riot;* lascīvia, luxūria, Cot. 118.

gagul-suillan *to gargle;* gargarīzāre, Som. Ben. Lye.

-gal, -gil, -gel, *as sin-gal perpetual, continual* : wíd-gal, wíd-gil, wíd-gel, *wide-spread,* March. 38; p. 27, 8. v. wíd-gil, wíd-gal.

GÁL, es; *n. Lust, wantonness, lightness, folly;* lascīvia, lĭbído, luxūria, lĕvĭtas :—Hie hyra gál beswác *their folly deceived them,* Cd. 18; Th. 21, 21; Gen. 327. Gódes oððe gáles *of good or evil,* Exon. 23 a; Th, 64, 9; Cri. 1035. [*Cf. Icel.* gáll, *m. a fit of gaiety.*]

gál; *adj. Light, pleasant, wanton, licentious, wicked;* lĕvis, lĭbīdĭnōsus, luxūriōsus, mălus :—Ðam unstæðdigan and ðam gálan, ðú miht secggan, ðæt he [MS. hi] biþ winde gelícra, ðonne gemetfæstum monnum *to the inconstant and the light [man], thou mayest say that he is more like the wind, than modest men,* Bt. 37, 4; Fox 192, 23, note 20, MS. Cott. Ðæt he gesáwe ungelíce béc him berende beón þurh ða gódan gástas oððe þurh ða gálan *ut cōdices diversos per bŏnos sīve mălos spĭrĭtus sĭbi vĭdērit offerri,* Bd. 5, 13; S. 633, 25. Gecunnian hwæðer he wǽre gód oððe gál *to try whether he were good or bad,* Gu. 17; Gdwn. 74, 6. [*Orm.* gal *wanton: O. Sax.* gēl *merry: Dut. Ger.* geil *lustful: M. H. Ger.* geil *licentious: O. H. Ger.* geil *lætus, elātus, fĕrox, lĭbĭdĭnōsus: Dan.* geil *wanton:* and cf. *Icel.* gáli *a wag.*] DER. ealo-gál, hyge-, medu-, rúm-, symbel-, wín-.

GALAN; *part.* galende, ic gale, ðú gǽlest, gǽlst, he gǽleþ, gǽlþ, *pl.* galaþ, þ. gól, *pl.* gólon; *pp.* galen *To sing, enchant, call;* cănĕre, incantāre, insŏnāre, clāmāre :—Seó ne gehérþ stemne galendra, and ātterwyrhtan galendes wíslíce *quæ non exaudiet vōcem incantantium, et venēfici incantantis săpienter,* Ps. Lamb. 57, 6. Sorh-leóþ gǽleþ *he sings a sad lay,* Beo. Th. 4912; B. 2460. Se wísdóm gól gyd *wisdom sung*

---

a lay, Bt. Met. Fox 7, 3; Met. 7, 2. Wíf fyrd-leóþ gólon [MS. galan] *the women sang a martial song,* Cd. 171; Th. 215, 3; Exod. 577. Ða ðe gehȳrdon gryreleóþ galan Godes andsacan *those who heard the adversary of God sing the horrid lay,* Beo. Th. 1576; B. 786. Ðá wæs sigeleóþ galen *then was the song of triumph sung,* Elen. Kmbl. 248; El. 124: Andr. Kmbl. 3097; An. 1551. [*Chauc.* gale: *Scot.* gale *to cry: O. Sax.* galan: *O. H. Ger.* galan *cănĕre: Dan.* gale *to crow: Swed.* gala *to crow: Icel.* gala *to crow, sing.*] DER. a-galan, be-, on-: nihte-gale. See Grm. D. M. pp. 987, 1173.

galder-cræftiga *one crafty* or *skilful in enchantments, an enchanter,* L. Alf. 30; Th. i. 52, 9; MS. H. v. galdor-cræftiga.

galdere, es; *m. An enchanter, a charmer, sorcerer, diviner, soothsayer;* incantātor, augur, haruspex, Som. Ben. Lye. DER. wyrm-galdere. [*Cf. O. H. Ger.* kalstarari *incantator.*] v. galan.

galdor, gealdor, es; *pl. nom. acc.* galdor, galdru; *gen.* galdra; *dat.* galdrum; *n.* [galan *to sing, enchant,* q. v.] *An incantation, divination, enchantment, a charm, magic, sorcery;* incantātio, cantio, carmen, fascĭnātio :—þurh heora galdor *per incantātiōnes,* Bd. 4, 27; S. 604, 9. Sing ðæt galdor *sing the charm,* Lchdm. iii. 38, 3. Galdre bewunden *encircled by enchantment,* Beo. Th. 6097; B. 3052. Ne sceal nán man mid galdre wyrte besingan *no man shall enchant a herb with magic,* Homl. Th. i. 476, 8. Galdra fela *many sorceries,* Bt. Met. Fox 26, 106; Met. 26, 53: Deut. 18, 11. Nis ðé ende feor, ðæs ðe ic on galdrum ongieten hæbbe *thy end is not far off, from what I have understood by [thy] divinations,* Exon. 50 a; Th. 174, 19; Gú. 1180. Ðás galdor mon mæg singan on wunde *a man may sing these charms over a wound,* L. M. 3, 63; Lchdm. ii. 352, 5. Hig worhton óðer swilc þing þurh hira drȳcræft and þurh Egiptisce galdru *fecērunt etiam ipsi per incantātiōnes Ægyptiacas et arcāna quædam simĭlĭter,* Ex. 7, 11. Galdrum cȳdan *to inform by divination,* Elen. Kmbl. 321; El. 161. [*Laym.* galdere, *dat. magic: Icel.* galdr, galdr, *m. a song, charm, spell, witchcraft, sorcery.*] DER. cear-galdor-, gealdor-cræftiga, an; *m. One crafty* or *skilful in enchantments, an enchanter;* incantātor :—Ða fǽmnan, ðe gewunniaþ [MS. gewunniah] onfón galdorcræftigan, ne lǽt ðú ða libban *the women, who are wont to receive enchanters, suffer thou not to live,* L. Alf. 30; Wilk. 31, 26. gealdor, heáh-galdor.

galdor-cræft, gealdor-cræft, es; *m. The art of enchanting, magic art, incantation;* incantandi ars, măgĭca ars, incantātio :—On galdorcræftum *per incantātiōnes,* L. M. I. P. 39; Th. ii. 274, 32. He Iudéa galdorcræftum wiðstód *he withstood the magic arts of the Jews,* Andr. Kmbl. 332; An. 166. Ða ðe galdorcræftas begangaþ *those that practise magical arts,* Blickl. Homl. 62, 23.

galdor-cwide, es; *m. A magic saying, song;* măgĭcus sermo, cantus, Exon. 113 a; Th. 432, 28; Rä. 49, 7.

galdor-galere, es; *m. An enchanter, soothsayer;* incantātor, Cot. 118: 193.

galdor-leóþ, es; *n. A magic song, an enchantment, charm, spell;* incantātio, carmen, incantāmentum, Cot. 188.

galdor-word, es; *n. A magic word, word of incantation;* cantātiōnis verbum :—Ic galdorwordum gól *I sang in magic words,* Exon. 94 b; Th. 353, 37; Reim. 24.

galdra *of enchantments, of sorceries,* Bt. Met. Fox 26, 106; Met. 26, 53; *gen. pl.* of galdor.

galdru *enchantments,* Ex. 7, 11; *pl. nom. acc.* of galdor.

galdrygea, an; *m. An enchanter;* incantātor, Cot. 108.

galere, es; *m. An enchanter;* incantātor :—Galere *incantātor,* Wrt. Voc. 74, 38. DER. galdor-, wyrm-galere.

gál-ferhþ; *adj. Mind-lustful, licentious;* lĭbīdĭnōsus, lascīvus :—Gewát ðá se deófulcunda gálferhþ his beddes neosan *then the devilish [man] went lustful in mind to seek his bed,* Judith. 10; Thw. 22, 14; Jud. 62.

gál-freólsas; *pl. m. Licentious festivals;* lascīva festa, Lupercālia, Som. Ben. Lye.

gálfull; *adj. Lustful, licentious, luxurious;* lĭbīdĭnōsus, luxŭriōsus, Scint. 21: 28: 58.

gálfullíce; *adv. Lustfully, luxuriously;* lĭbĭdĭnōse, luxŭriōse, Scint. 13.

GALGA, gealga, an; *m. A gallows, gibbet, cross;* arbor infēlix, patĭbŭlum, crux :—Galga *patĭbŭlum,* Ælfc. Gl. 15; Som. 58, 30; Wrt. Voc. 21, 24. He of galgan his gǽst onsend *he sent forth his soul from a gallows,* Exon. 70 a; Th. 261, 4; Jul. 310: 72 b; Jul. 482: Beo. Th. 4883; B. 2446. He his blód ageát on galgan *he shed his blood on the cross,* Cd. 225; Th. 299, 15; Sat. 550: Menol. Fox 170; Men. 86: Elen. Kmbl. 957; El. 480. On galgum *on the cross,* Cd. 224; Th. 297, 3; Sat. 511. [*Chauc. R. Brun.* galwes, *pl: Plat.* galge: *O. Sax.* galgo, *m: O. Frs.* galga, *m: Dut.* galg, *f: Ger.* galgen, *m: M. H. Ger.* galge, *m: O. H. Ger.* galgo, *m: Goth.* galga, *m. a cross: Dan.* galge, *m. f: Swed.* galge, *m.: Icel.* gálgi, *m.*] See Grm. R. A. pp. 682–4.

galga-tré, es; *n. A gallows-tree, cross* :—Ðín ródes galgatré *tuum crucis patibulum,* Rtl. 23, 36. On ródes galgatree *in crucis patibulo,* 124, 1. v. galg-treów. [*Havel.* galwetre: *Icel.* gálga-tré.]

galg-mód; *adj.* [galg = gealh *sad;* mód *mind*] *Sad in mind, gloomy;* tristis animo :—His módor, gífre and galg-mód, gegán wolde sorhfulne

síþ *his mother, greedy and gloomy, would go a sorrowful journey*, Beo. Th. 2558; B. 1277. v. gealg-mód.

galg-treów, gealg-treów, es; *n. A gallows-tree, cross*; crūcis lignum, crux :—He wolde sume on galgtreówum [MS. galgtreówu] *he would* [*hang*] *some on gallows-trees*, Beo. Th. 5873; B. 2940.

Galiléa *Galilee*:—Sǽ Galilǽs *mǽre Galilǽæ*, Mk. Skt. Lind. 1, 16. Galiles, Jn. Skt. Lind. 6, 1. Of Galiléam ðǽm lande, Blickl. Homl. 123, 21. Witga of Galiléum *a prophet from Galilee*, 71, 16.

Galiléisc, Galiléisc; *adj. Galilean*:—Pilatus acsode hwæðer he wǽre Galileisc man *Pilātus interrŏgāvit si hŏmo Galilǽus esset*, Lk. Bos. 23,.6: 22, 59: Mk. Bos. 14, 70: Jn. Bos. 7, 52. Of ðære Galileiscan Bethsaida *a Bethsaida Galilǽæ*, Jn. Bos. 12, 21. Wið ða Galileiscan sǽ *juxta mǎre Galilǽæ*, Mt. Bos. 4, 18: 15, 29: Mk. Bos. 1, 16. Wéne gé, wǽron ða Galileiscan synfulle tofóran eallum Galileiscum *pŭtātis quod hi Galilǽi prǽ omnĭbus Galilǽis peccātōres fuĕrint?* Lk. Bos. 13, 2. On Galileisce dǽlas *in partes Galilǽæ*, Mt. Bos. 2, 22. Hwæt bídaþ gé Galilésce guman on hwearfte *why abide ye Galilean men about?* Exon. 15 a; Th. 32, 11; Cri. 511: Blickl. Homl. 123, 20.

Galleas *Gauls, the French*, Bd. 5, 11; S. 626, 27. v. Gallias.

Gallia ríce *the kingdom of the Gauls, France*, Bd. 4, 1; S. 564, 16: 5, 8; S. 621, 39. v. Gallias.

Gallias, Gallie, Galleas; *gen.* Gallia; *pl. m. The Gauls, the Franks*; Galli, ōrum; Galliæ, ārum; *pl. m* :—Ðǽr wæs Gallia ofslagen twá-hund þúsenda *ducenta millia Gallōrum interfecta sunt*, Ors. 5, 8; Bos. 107, 33; Hav. 329, 8: 4, 7; Bos. 89, 7. Gefeaht wið Gallie *adversum Gallos conflixit*, 4, 7; Bos. 89, 8; Hav. 251, 2. Hú sceolan we dón mid Gallia and Brytta bisceopum *quŭliter dēbēmus cum Galliārum Brittaniārumque episcŏpis ǎgěre?* Bd. 1. 27; S. 492, 10. Biscop Gallia ríces *bishop of the kingdom of the Gauls* [Galliārum], Bd. 5, 8; S. 621, 39. Galleas nemnaþ Trajectum *the Gauls call it Utrecht*, Bd. 5, 11; S. 626, 27. Monige gewunedon sécan Francna mynstro and Gallia *multi Francōrum vel Galliārum Monastēria adīre sŏlēbant*, Bd. 3, 8; S. 531, 17. Adrianus se abbad ða dǽlas Gallia ríces geferde and gesóhte *Adrian the abbot went and visited the parts of the kingdom of the Gauls*; partes Galliārum [regni] adiisset, Bd. 4, 1; S. 564, 16. Gallia rice *the kingdom of the Gauls*, Bd. 5, 23; S. 645, 31.

gál-líc; *adj. Lustful* :—Ǽlc gállíc ontendnys wearþ adwæsced *every lustful fervour was extinguished*, Th. Homl. ii. 156, 35. [*O. Eng. Homl.* galiche dede, i. 149, 16.]

Gallie; *gen.* a; *pl. m. The Gauls*; Galli :—Gallie oferhergodon land *the Gauls overran the lands*, Ors. 3, 4; Bos. 56, 9: 4, 7; Bos. 89, 8. v. Gallias.

Gallisc; *adj. Gaulish, belonging to Gaul*; Gallĭcus :—Ðǽr gefeaht Mallius wið ánne Galliscne mann *there Manlius fought with a man of Gaul*, Ors. 3, 4; Bos. 56, 16.

galluc, galloc, gallac, es; *m. The plant comfrey*; symphўtum officĭnāle, Lin :—Ðeós wyrt, ðe man *confirmam*, and óðrum naman galluc nemneþ, biþ cenned on mórum and on feldum, and eác on mǽdum *this herb, which is called confirma, and by another name comfrey, is produced on moors and in fields, and also in meadows*, Herb. 60, 1; Lchdm. i. 162, 10–12. Galluces moran *roots of comfrey*, Lchdm. iii. 6, 10. Genime galluc gesodenne *take sodden comfrey*, L. M. 1, 27; Lchdm. ii. 68, 15: 1, 31; Lchdm ii. 74, 11: 3, 73; Lchdm. ii. 358, 23. Galluc *adriatica* vel *mālum terrǽ*, Ælfc. Gl. 39; Som. 63, 70; Wrt. Voc. 30, 22: 79, 17. Galloc *galla*, Glos. Brux. Recd. 41, 46; Wrt. Voc. 67, 61. Gallac *symphўtum*, 42, 14; Wrt. Voc. 68, 29.

Galmanhó, Galmahó? *An Anglo-Saxon abbey at York, afterwards St. Mary's*; abbátiæ nōmen ǎpud Eborācum :—On ðysum geáre forþferde Síward eorl on Eoforwíc, and his líc líþ binnan ðam mynstre æt Galmanhó [Galmahó, Th. 324, 10, col. 2], ðe he sylf ǽr getimbrade, Gode to lofe and eallum his hálgum *in this year* [A.D. 1055] *earl Siward died at York, and his body lies within the monastery of Galmanho, which he himself had before built, to the glory of God and all his saints*, Chr. 1055; Th. 324, 8–12, col. 1.

gál-mód; *adj. Light-minded, licentious*; libĭdĭnōsæ mentis, lascīvus :—Se galmóda *the licentious* [Hŏlofernes], Judth. 12; Thw. 25, 12; Jud. 256. [*O. Sax.* gēl-mód.]

gálnes, -ness, -nyss, e; *f. Lustfulness, lust, luxury, wantonness*; lascīvia, libīdo, luxūria, Cot. 150: Scint. 12: 21: 81. He cnihtlice gálnysse næs begangende *he was not addicted to boyish levity*, Guthl. 2; Gdwn. 12, 16. [*Orm.* galness.]

gál-scipe, es; *m.* [gál *lust*, -scipe -*ship*] *Luxury, lustfulness, lasciviousness, wantonness, lewdness*; luxūria, libīdo, lascīvia, petulantia, satўrĭăsis = σατυρίασις :—He begæþ unǽtas and oferdrincas and gálscipe *comessatiōnĭbus vǎcat et luxūriæ atque convīviis*, Deut. 20, 21. We lǽraþ, ðæt man wið fúlne gálscipe warnige symle *we instruct, that one always guard himself against foul lasciviousness*, L. C. E. 24; Th. i. 374, 9. For gálscipe *for wantonness*, Cd. 18; Th. 22, 15; Gen. 342. Synwrǽnnys vel gálscipe *satўriasis*, Ælfc. Gl. 11; Som. 57, 49; Wrt. Voc. 19, 51.

gálsere, es; *m. A lustful man*; libĭdĭnōsus, Off. Reg. 15.

gál-smerc; *adj.* [smercian *to smirk, smile*] *Light, laughing, giggling*;

pĕtŭlans :—Gyf se munuc ne biþ gálsmerc, and eáðe and hræde on hleahtre *si mŏnǎchus non sit pĕtŭlans, et fǎcĭlis et proclīvis ad ridendum*, R. Ben. 7.

galung, e; *f. Incantation*, Hpt. Gl. 519.

Galwalas, galwealas, *nom. acc*; *gen.* a; *dat.* um; *pl. m.* [wealh *foreign*; cf. Bryt-walas] *Gauls, Frenchmen, people of Gaul in a body*, and as the name of a people is often used where according to later usage the name of their country would be found, the word may be translated *Gaul, France*; Galli, Gallia :—Hér wæs Brihtwald gehálgod to ærcebiscope fram Godune Galwala biscop *in this year* [A. D. 693] *Brihtwald was consecrated archbishop by Godun bishop of the Gauls*, Chr. 693; Erl. 43, 17. He gewát into Galwalum *he went into Gaul*, Chr. Erl. 5, 5, 14. Hér Ægelbryht of Galwalum [Galwealum, Th. 50, 2, col. 2, 3] onféng Wesseaxna bisceopdóme *in this year* [A. D. 650] *Ægelbyrht of Gaul received the bishopric of the West Saxons*, Chr. 650; Th. 50, 2, col. 1: 660; Th. 54, 16. He fór in Galwalas *he went into Gaul*, 380; Erl. 11, 2. v. Gallias.

gál-wrǽne; *adj. Luxurious, lecherous*; luxŭriōsus, Som. Ben. Lye.

gamel, gamol; *adj. Old, aged*; sĕnex, vĕtustus :—Wolde beddes neósan gamela Scylding *the aged Scylding would visit his bed*, Beo. Th. 3588; B. 1792. Wæs gylden hilt gamelum rince gyfen *the golden hilt was given to the aged warrior*, 3359; B. 1677: Elen. Kmbl. 2491; El. 1247. Gamele ne mōston hǎre heaðorincas hilde onþeón *the aged hoary chieftains might not prosper in battle*, Cd. 154; Th. 193, 3; Exod. 240. Ǽr he on weg hwurfe, gamol, of geardum *ere he, old, departed on his way from his courts*, Beo. Th. 535; B. 265: 115; B. 58. v. gomel. [*Icel.* gamall.]

gamelíc; *adj. Theatralis, ridiculosus*, Hpt. Gl. 459, 508.

GAMEN, gomen, es; *n.* GAME, *joy, pleasure, mirth, sport, pastime*; jŏcus, oblectāmentum, gaudium, jūbilum, lætĭtia, lūdus :—Gamen eft astáh *pastime rose again*, Beo. Th. 2325; B. 1160. Wynsum gamen *a pleasant game*; sáles, Ælfc. Gl. 16; Som. 58, 67; Wrt. Voc. 21, 54. Næs ðæt hérlic dǽd, ðæt hine swelces gamenes gilpan lyste *that was not a glorious deed, that he should wish to boast of such sport*, Bt. Met. Fox 9, 37; Met. 9, 19. Him to gamene *for his sport*, 9, 17; Met. 9, 9: 9, 91; Met. 9, 46. Ic mæg swegles gamen gehýran on heofnum *I can hear the joy of the firmament in heaven*, Cd. 32; Th. 42, 18; Gen. 675. Bǽdon híg sume, ðæt Samson móste him macian sum gamen *præcĕpērunt ut vocarētur Samson et ante eos lūdēret*, Jud. 16, 25. Gamena lŭdōrum : gamene *jŏco*, Mone B. 2807, 2808. [*Piers P.* gamen *a play* : Laym. game *a play* : *Scot.* gamyn *a game, play* : *O. Sax.* gaman, *n* : *Frs.* gammen : *O.Frs.* game, gome, *f*: *M. H. Ger.* gamen, *m. n*: *O. H. Ger.* gaman, *gaudium, jŏcus, lūdus* : *Dan.* gammen, *m. f*: *Icel.* gaman, *n. game, sport, pleasure, amusement*.] DER. glig-gamen, heal-.

gamenian, gamnian, gæmnian; *p.* ode; *pp.* od [gamen *game*] *To joke, play*; jŏcŭlāri, jŏcāri :—Gregorius gamenode mid his wordum *Gregory played with his words*, Homl. Th. ii. 122, 4. [*Icel.* gamma *to amuse, divert*.]

gamenlíce; *adv. Sportingly, deceitfully*; jŏcōse, callĭde :—Hí gamenlíce rǽddon *they counselled deceitfully*, Jos. 9, 3.

gamenung, e; *f. A gaming, jesting, playing*; lūsus, jŏcus :—Hwǽr biþ his gaf sprǽc and ða ídelan gamenunga *where will be his wanton discourse, and the idle jestings?* Basil admn. 8; Norm. 50, 28.

gamen-waðu *a joyous path*. v. gomen-waðu.

gamen-wudu *pleasure-wood, glee-wood, a musical instrument, harp*. v. gomen-wudu.

gamian *to game, play, sport*, Som. Ben. Lye v. gamenian.

gaming, e; *f. A* GAMING, *playing, gesticulation*; lūsus, gannātūra, sive mimica, gestĭcŭlātio, Cot. 203.

gamnian; *part.* gamnigende; *p.* ode; *pp.* od *To play*; lūděre :—Wæs him geþúht, swilce he gamnigende sprǽce *vīsus est eis quǎsi lūdens lŏqui*, Gen. 19, 14. v. gamenian.

gamol *old, aged*, Beo. Th. 115; B. 58: 535; B. 265. v. gomel.

gamol-feax; *adj. With hoary locks, grey-haired*; cānus :—Gamolfeax hæleþ *a hoary-headed hero*, Chr. 975; Erl. 126, 20; Edg. 46: Beo. Th. 1220; B. 608. v.-gomel-feax.

gamol-ferhþ; *adj. Advanced in age, aged*; ætāte provectus :—Gamolferhþ goldes brytta *the aged dispenser of gold*, Cd. 138; Th. 173, 26; Gen. 2867.

gán *yawned*; hiāvit; *p.* of gínan.

GÁN, to gánne; ic gá, ðú gǽst, he gǽþ; *pl.* gáþ; *p.* ic he eóde, ðú eódest; *pl.* eódon; *imp.* gá, *pl.* gáþ; *pp.* gán; *v. n.* [the conjugation is formed from two roots, the past tense being from root i; cf. Gothic iddja]; *To go, come, walk, happen*; īre, grǎdi, evĕnīre :—Uton gán and feligean fremdum godum *cǎmus et sequǎmur deos aliēnos*, Deut. 13, 1. Gearo to gánne *ready to go*, Homl. Th. ii. 32, 7. Ðú gǽst on ðínum breóste *super pectus tuum grǎdiēris*, Gen. 3, 14. He on flet gǽþ *he walks in the court*, Beo. Th. 4075; B. 2034. Gǽþ á wyrd swá hió sceal *fate goes ever as it must*, Beo. Th. 915; B. 455. Hí gáþ þheр́an, Andr. Kmbl. 3328; An. 1667. Gif gé gáþ æfter fremdum godum *if ye go after strange gods*, Deut. 11, 28. He sǽde unc eall swá hit siððan á eóde [or a-eode?] *he told us all as it always afterwards happened*; audīvĭmus quidquid **postea**

rei prŏbāvit eventus, Gen. 41, 13. Eóde eall seó ceasterwaru togeánes ðam Hǣlende *tōta cīvitas exiit obviam 'Jesu*, Mt. Bos. 8, 34: Bd. 1, 7; S. 478, 12. Sume for hungre heora feóndum on hand ẃodon *some for hunger went into the hands of their foes*, 1, 15; S. 484, 5. Gā hider *come hither*, Gen. 27, 21. Gáþ eów into ðære cyrcan unforhtlīce *go into the church fearlessly*, Homl. Th. i. 508, 1. [*Wyc.* gon, goon, goo: *Piers P.* goon: *Chauc.* gon, goon: *R. Glouc.* goon: *Laym.* goon: *Orm.* gan: *Plat.* gan. gaan; gaen: *O. Sax.* gān: *Frs.* gean: *O.Frs.* gan: *Dut.* gaan: *Ger.* gehen, gehn: *M.H. Ger.* gān, gēn: *O.H. Ger.* gān: *Dan.* gaae: *Swed.* gå: *Zend.* gā, gē *to go*: *Sansk.* gā *to go*.] DER. a-gān, æfter-, be-, bi-, for-, fóre-, forþ-, ful-, ge-, in-, of-, ofer-, ōþ-, þurh-, to-, under-, up-, upp-, ūt-, wið-, ymb-. v. gangan.

gancgan *to go*, Ps. Th. 85, 10. v. gangan.

Gandis, Gandes; *indecl. f. The river Ganges*; Ganges = Γάγγης :— Ðǣr licgeþ se mūþa ūt on ðone gārsecg ðære eá, ðe man hāteþ Gandis *there the mouth of the river, which is called Ganges, opens out into the ocean*, Ors. 1, 1; Bos. 16, 13, 17. Gandes seó eá is eallra ferscra wætera mǣst, būtan Eufrate *the river Ganges is the greatest of all fresh waters, except the Euphrates*, 2, 4; Bos. 43, 45. Æt Gande ðære eá, Nar. 3, 22.

GANDRA, ganra, an; *m. A* GANDER; anser :—Gandra *anser*, m. Ælfc. Gr. 9, 18; Som. 9, 59. [*Eng.* gander, *m: Ger.* gänserich, *m: Ger. dial.* gandert: *M.H. Ger.* ganzer, ganze, *m: O.H. Ger.* ganzo, *m: Icel.* gassi, *m. a gander*.]

ganet, es; *m. A gannet, sea-fowl, water-fowl, swan*; fūlica, cygnus :— Ganet *cygnus*, Glos. Prudent. Recd. 144, 32. Ofer ganetes bæþ [MS. baþ] *over the sea-fowl's bath*, Chr. 975; Erl. 125, 21. Ganetes hleóðor *the gannet's cry*, Exon. 81 b; Th. 307, 8; Seef. 20. Cômon of gārsecge ganetas fleógan *sea-fowls came flying from the ocean*, Ps. Th. 104, 35. v. ganot.

GANG, geng, gong, gung, es; *m.* I. GANG, *going, journey, step, way, path, passage, course (of time)*; īter, grădus? gressus, incessus, ambŭlātio, sēmĭta :—Beswīcan gangas [MS. M. stepas] mīne *supplantāre gressus meos*, Ps. Spl. C. 139, 5. Mīnne gang *gressum meum*, Ps. Th. 139, 5. Ganges, Beo. Th. 1940; B. 968. Him tǣcean lifes weg and rihtne gang to heofonum *to teach them the way of life and the right path to heaven*, Blickl. Homl. 109, 18. Ðīne gangas *gressus tui*, Ps. Th. 67, 23. Fóta gangas *pedum gressus*, 72, 1. Mīne gangas *meæ sēmĭtæ*, 138, 2. On ðære eá gang *in the river's course*, Ors. 2, 4; Bos. 44, 13. Heó freó on hira fóta gangum blīðe hām wæs hweorfende *ipsa lībēro pĕdum incessu dŏmum læta reversa est*, Bd. 4, 10; S. 578, 33. Heora geára gang *anni eorum*, Ps. Th. 77, 32. Geára gongum *in the course of years*, Elen. Kmbl. 1292; El. 648. II. *a passage, drain, privy*; latrīna, secessus :— Gang *latrīna, secessus*, Ælfc. Gl. 108; Som. 78, 121; Wrt. Voc. 58, 33. Ðonne him to gange lyst *when he desires the privy*, Hexam. 20; Norm. 28, 23: L. Ælf. C. 3; Th. ii. 344, 6: Homl. Th. i. 290, 19. [*Orm.* gang *a journey*: *Prompt.* gong *latrina*: *Scot.* gang *a journey*: *O.Sax.* gang, *m: O.Frs.* gong, gung, *m: Dut. Ger.* gang, *m: M.H. Ger.* ganc, *m: O.H. Ger.* gang, *m: Goth.* gaggs, *m: Dan.* gang, *m. f: Swed.* gång, *m. time: Icel.* gangr, *m;* göng, *n. pl. a passage*.] DER. be-gang, -gong, bi-, eder-, embe-, fēðe-, forþ-, ge-, hin-, hlāf-, hūsel-, in-, on-, setl-, stal-, stepe-, to-, up-, ūt-, wæfer-, ymb-, ymbe-.

gang *go, come*, Cd. 228; Th. 308, 32; Sat. 701: Gen. 27, 26; *impert. of* gangan.

gang *went*, Beo. Th. 2595; B. 1295; *p. of* geongan.

GANGAN, gongan, gancgan; *part.* gangende, gongende; ic gange, gonge, ðú gangest, gongest, he gangeþ, gongeþ, *pl.* gangaþ, gongaþ; þ. geóng, gióng, giéng, gēng, *pl.* geóngon, gióngon, giéngon, gēngon; *imp.* gang, gong; *pp.* gangen, gongen *To go, walk, turn out*; īre, meáre, vādēre, ambŭlāre, ingrĕdi, tendēre, evēnīre :—Ic gange *ambŭlo*, Ælfc. Gr. 19; Som. 22, 41. Gáng hider *accēde*, Gen. 27, 26: Num. 11, 21. He heonon gangeþ [gangaþ MS.] *he goes from hence*, Andr. Kmbl. 1782; An. 893. He of worulde gangende wæs *he was going from the world*, Bd. 4, 24; S. 598, 30. He ealle ða tīd mihte ge sprecan ge gangan *tōto eo tempŏre et lŏqui et ingrĕdi pŏtuit*, Bd. 4, 24; S. 598, 30. He to healle geóng *he went to the hall*, Beo. Th. 1855, note; B. 925. He ofer willan gióng *he went against his will*, 4810, note; B. 2409. Heó gióng [gien MS.] to Adame *she went to Adam*, Cd. 29; Th. 39, 15; Gen. 626. Ic to ðam grunde gēnge *I would go to the abyss*, Cd. 39; Th. 51, 29; Gen. 834. Forþ gangan *to go forward, to continue* :—Gange se teám forþ *let the warranty go forward*, L. Ed. 1; Th. i. 158, 13: Exon. 14 a; Th. 27, 5; Cri. 426. Ic ongitan mihte hu ðis gewinn wolde gangan *I should be able to know how this labour would turn out*, Ps. Th. 72, 13: 88, 3. [*Piers P.* gange, gangen: *Orm.* ganngenn: *Scot.* gang: *O.Sax.* gangan: *O.Frs.* gunga: *M.H. Ger.* gangen: *O.H. Ger.* gangan: *Goth.* gaggan: *Swed.* gånga: *Icel.* ganga.] DER. a-gangan, -gongan, æt-, be-, bi-, fōr-, fóre-, forþ-, ful-, ge-, in-, of-, ofer-, on-, ongeán-, þurh-, to-, under-, up-, ūt-, wið-, ymb-, ymbe-.

gang-dagas, gong-dagas; *pl. m.* [*dæg a day*] *Perambulation days, the three days before Ascension day or Holy Thursday, Rogation days, when the boundaries of parishes and districts were traversed*; dies perambŭlātiōnes *vel* processiōnis, rŏgātiōnum dies :—Betweox gang-dagum and

middum sumera *betwixt Rogation days and Midsummer*, Chr. 913; Erl. 102, 3: 1063; Erl. 195, 7. Ofer gang-dagas *after Rogation days*, L. Ath. i. 13; Th. i. 206, 15. Ðys Gôdspel sceal to Gang-dagon *this Gospel must be on the Rogation days [Gang-days]*, Rubc. Mt. Bos. 7, 7–14, notes, p. 575. Ðis sceal to Gang-dagon dæge twegen dagas, *this [Gospel] must be on the two days of the Rogation days*, Rubc. Lk. Bos. 11, 5–13? notes, p. 578. [*Icel.* gangdagar.]

gangel *going*. v. gongel. [*Icel.* göngull *strolling*.]

gangel-wæfre *a ganging weaver, spider*, Som. Ben. Lye. v. gongel-wæfre.

gangere, es; *m. A ganger, footman*; pedester, Som. Ben. Lye.

gang-ern, es; *n.* [gang II. *a privy*, ern *a place*] *A privy*; latrīna :— Goldhordhûs, dīgle gangern *hypodrŏmum vel spondoromum* ? [= *spidromum*, q. v. in Du Cange], Ælfc. Gl. 107; Som. 78, 81; Wrt. Voc. 57, 57.

gange-wifre, -wæfre, geonge-wifre, gonge-wifre, gongel-wæfre, an; *f. A ganging weaver, spider*; viātĭca arānea :—Ðú gedēst ðæt he aswint on his môde, and wyrþ swā tedre swā swā gangewifran nett *thou causest that he dwindles away in his mind, and becomes as frail as a spider's web*, Ps. Th. 38, 12. Swindan ðū dydest swā swā gangewæfre [āttercoppan MS.] sāwle his *tabescĕre fēcisti sīcut arāneam anĭmam ejus*, Ps. Spl. 38, 15.

gang-feormere, es; *m. A jakes-farmer, privy-cleanser*; fĭmārius, cloācārius, Som. Ben. Lye.

gang-geteld, es; *n. A travelling-tent, tent, pavilion*; tentŏrium ambŭlātōrium, pāpilio :—Gang-geteld *pāpĭlio*, Ælfc. Gl. 110; Som. 79, 40; Wrt. Voc. 59, 12.

gang-here, es; *m. A foot-army, infantry*; pedester exercĭtus :—Pirrus him com to mid ðam mǣstan fultume, ægðer ge on ganghere, ge on rādhere *Pyrrhus came to them with the greatest force, both in infantry, and in cavalry*, Ors. 4, 1; Bos. 76, 40.

gang-pyt, -pytt, es; *m. A privy*; latrīna :—On ðære nyðemestan flēringe wæs heora gangpyt and heora myxen *on the lowermost flooring [of the ark] was their privy and their dunghill*, Boutr. Scrd. 21, 7. v. gang II.

gang-setl, es; *n. A privy*; latrīna, Som. Ben. Lye. v. gang II.

gang-tún, es; *m. A privy*; latrīna, Som. Ben. Lye. v. gang II.

gang-weg, es; *m. A gang-way, way, road*; via :—Anes wænes gangweg *a road for one vehicle*; actus, Ælfc. Gl. 56; Som. 67, 50; Wrt. Voc. 37, 38. Twegra wæna gangweg *a road for two vehicles*; via, 56; Som. 67, 51; Wrt. Voc. 37, 39.

gang-wuce, an; *f. Rogation week, the week of holy Thursday*; perambŭlātiōnis septimāna :—Ðis sceal on þunres dæg, innan ðære Gang-wucan *this [Gospel] must be on Thursday in the Rogation week*, Rubc. Mk. Bos. 16, 14–20, notes, p. 578. Ðys Gôdspel gebýraþ on Wôdnes dæg, on ðære Gang-wucan to ðam uigilian *this Gospel belongs to the vigil on Wednesday, in the Rogation week*, Rubc. Jn. Bos. 17, 1–10, notes, p. 580.

GĀNIAN; *p,* ode; *pp.* od *To* YAWN, *gape, open*; hiāre, oscĭtāre, apērīre :—Gāniende *oscĭtans*, Cot. 147. Ðeáh ðe me synfulra, inwitfulra, mūþas on gānian *though the mouths of the sinful [and] deceitful yawn upon me*, Ps. Th. 108, 1. [*Plat.* janen: *Dut.* geeuwen: *Ger.* gähnen: *M.H. Ger.* gēnen: *O.H. Ger.* geinôn, ginôn, ginēn, giēn: *Icel.* gína: *Lat.* hiāre: *Grk.* χαίνειν *to yawn*, χαπc.]

GANOT, ganet, es; *m. A gannet, sea-fowl, water-fowl, fen-duck*; ăvis mărīna, fūlix, fūlĭca :—Ganot *fūlix*, Wrt. Voc. 62, 7: 280, 13. Ðā wearþ adrǣfed deórmôd hæleþ, Ôslac of earde, ofer ýþa gewealc, ofer ganotes bæþ *then the brave man, Oslac, was driven away from the land, over the billows' roll, over the gannet's bath [the sea]*, Chr. 975; Erl. 126, 20; Edg. 46: Beo. Th. 3727; B. 1861. Ác fereþ gelôme ofer ganotes bæþ *a ship* [lit. *oak*] *often saileth over the gannet's bath [the sea]*, Runic pm. 25; Kmbl. 344, 19; Hick. Thes. i. 135, 49. [*Plat.* gante: *Dut.* gent, *m. a male goose, gander: O.H. Ger.* ganazo, ganzo, *m. anetus*.]

ganra; *m. A gander*; anser, Ælfc. Gl. 36; Som. 62, 121; Wrt. Voc. 29, 17: 77, 33. v. gandra.

gānung, e; *f. A yawning*; oscĭtātio, Ælfc. Gl. 78; Som. 72, 59; Wrt. Voc. 46, 18.

GĀR, es; *m. A dart, javelin, spear, shaft, arrow, weapon, arms*; jacŭlum, pīlum, hasta, hastæ cuspis, săgitta, tēlum, arma :—Se gār *the dart*, Beo. Th. 3697; B. 1846. Fleág giellende gār on grome þeóde *the yelling shaft flew on the fierce nation*, Exon. 86 b; Th. 326, 13; Wīd. 128. Lǣtaþ gāres ord in gedûfan in fǣges ferþ *let the javelin-point plunge into the life of the doomed one*, Andr. Kmbl. 2662; An. 1332: Cd. 75; Th. 92, 2; Gen. 1522. Sende se sǣrinc sūþerne gār *the sea-chief sent a southern dart*, Byrht. Th. 135, 47; By. 134: 138, 48; By. 237. Gâre wunde *wounded by a dart*, Beo. Th. 2154; B. 1075: Exon. 66 a; Th. 243, 28; Jul. 17. Hí gewurdon scearpe gāras *ipsi sunt jăcŭla*, Ps. Th. 54, 21: 90, 6. Gāra ordum *with javelin-points*, Andr. Kmbl. 64; An. 32: Cd. 94; Th. 121, 32; Gen. 2019. Hý togædre gāras hlǣndon *they had inclined their weapons together*, Exon. 66 b; Th. 246, 8; Jul. 63: Elen. Kmbl. 235; El. 118. Gārum gehyrsted *adorned with javelins*, Andr. Kmbl. 90; An. 45: 2287; An. 1145: Chr. 937; Erl. 112, 18; Æðelst. 18. [*Chauc.* gere, *pl: Laym.* gar, gare, gǣre *a dart, spear,*

weapon: Plat. gere *a wedge: Kil.* gheer *fuscīna cuspǐdǐbus horrens,*
quibus pisces cǎpiuntur: O.Sax. gēr, m.: *Ger. M.H.Ger. O.H.Ger.*
gēr, m. *hastīle, jǎcǔlum, tēlum: Icel.* geirr, m. *a spear.*] DER. æt-gár,
bon-, frum-, hyge-, tite-, wæl-.

gára, an; m. *A spear-man.* v. frum-gára *in* frum-gár.

gára, an; m. [gár *a dart, point*] *An angular point of land, a promon-*
tory, corner of land; ōra prōmǐnens, angǔlus:—Ispania land is þrýscýte
... ān ðæra gárena líþ súþ-west, ongeán ðæt ígland, ðe Gades hátte *the*
country of Spain is three-cornered ... one of the corners lies south-west,
opposite the island which is called Cadiz, Ors. 1, 1; Bos. 24, 5.

gár-beám, es; m. *The wood* or *handle of a javelin, a spear-shaft;* cus-
pǐdis hasta:—Gárbeámes feng *a spear-shaft's grasp,* Cd. 155; Th. 193, 14;
Exod. 246.

gár-berend, es; m. *A javelin-bearer, soldier;* hastǐfer, tēlǐfer:—Grame
gárberend *the incensed javelin-bearers,* Byrht. Th. 139, 30; By. 262.
Gárberendra x hund *ten hundred javelin-bearers,* Cd. 154; Th. 192, 13;
Exod. 231.

gár-cēne; adj. *Spear-bold, bold in arms;* hastā audax:—Offa wæs
gárcēne man *Offa was a man bold in arms,* Beo. Th. 3921; B. 1958.

gár-clife, an; f. *Agrimony;* agrǐmōnia eupǎtōria:—Genim ðas wyrte,
ðe man agrimoniam, and ōðrum naman gárclife nemneþ *take this herb,*
which is named agrimony, and by another name garclive, Herb. 32, 1;
Lchdm. i. 130, 3. Genim gárclifan *take garclive,* L. M. 2, 51; Lchdm.
ii. 266, 8. Gárclifan etan ǽrende fúllíc getácnaþ *to eat agrimony betokens*
a disagreeable message, Somn. 20; Lchdm. iii. 198, 24. v. agrimonia.

gár-cwealm, es; m. *Spear-slaughter;* nex tēlo patrāta, clādes:—
Se ðe eall gemǎn gárcwealm gumena *who all remembers the slaughter of*
men, Beo. Th. 4092; B. 2043.

Gár-Dene; gen. a; dat. um; pl. m. *The spear-Danes, Danes who*
fought with spears, armed or *warlike Danes;* hastǎti Dǎni:—We Gár-
Dena, in geárdagum, þeódcyninga þrym gefrunon *we have heard of the*
renown of the Gar-Danes' great kings in days of yore, Beo. Th. 1; B. 1.
He sǽcce ne wēneþ to Gár-Denum *he expects not warfare from the Gar-*
Danes, 1206; B. 601: 3717; B. 1856: 4982; B. 2494.

gare *yare, ready, finished;* paratus, effectus:—Wæs ðæt mynstre gare
the monastery was finished, Chr. 656; Erl. 30, 19. v. gearo.

gár-faru, e; f. *A martial expedition,* v. faru III; turma hastifera:—
Þúfas wundon ofer gárfare *the standards fluttered over the martial band,*
Cd. 160; Th. 199, 23; Exod. 342. Ne þearf him ondrǽdan deófla
strǽlas, gromra gárfare *he need not dread the shafts of devils, the armed*
band of the hostile, Exon. 98 a; Th. 49, 5; Cri. 781. [Or gárfaru
flight of spears, cf. hægelfaru.]

gár-getrum, es; n. *A troop armed with spears, javelins:*—Gárgetrum
ofer scild-hreádan sceótend sendaþ flacor flángeweorc *the spear-troop, the*
archers, send over the shields the quivering arrows, Exon. 17 b; Th. 42,
18; Cri. 674.

gár-gewinn, es; n. *Spear-war;* hastātōrum pugna:—Wǽron þearle
gelyste gárgewinnes *they were very desirous of the spear-war,* Judth. 12;
Thw. 26, 3; Jud. 308. Ne lǽt ðé ahweorfan grim gárgewinn *let not the*
fierce javelin-strife turn thee away, Andr. Kmbl. 1915; An. 960.

gár-heáp, es; m. *A spear-band, armed band;* hastǐfera turma:—Hæfdon him beácen arǽred in ðam gárheápe *they had a signal reared in the*
armed band, Cd. 160; Th. 198, 11; Exod. 321.

gár-holt, es; n. [holt *lignum*] *A javelin-shaft, javelin;* hastæ lignum,
hasta:—Ðæt ic ðé to geóce gárholt bere *that I may bear the javelin-shaft*
for thy succour, Beo. Th. 3673; B. 1834.

gár-leác, es; n. [gár *a spear,* leác *a leek:* from its tapering acute
leaves] GARLIC; allium:—Gárleác *allium,* Ælfc. Gl. 41; Som. 63, 111;
Wrt. Voc. 30, 59: 286, 6. Genim gárleáces þreó heáfdu *take three heads*
of garlic, L. M. 2, 32; Lchdm. ii. 234. 19. Gárleáces iii clufe *three*
cloves of garlic, 3, 62; Lchdm. ii. 350, 8. Nim gárleáces gódne dǽl
take a good deal of garlic, Lchdm. iii. 12, 15. Nim gárleác *take garlic,*
L. M. 1, 47; Lchdm. ii. 118, 12: 1, 58; Lchdm. ii. 128, 10: 1, 63;
Lchdm. ii. 138, 3: 2, 56; Lchdm. ii. 276, 15. Wið gárleác gemenged
mingled with garlic, L. M. 1, 31; Lchdm. ii. 72, 4. [*Icel.* geirlaukr.]

gár-mitting, -mittung, e; f. *A meeting of spears* or *javelins, a battle:*—
Ðæt hí beadoweorca beteran wurdon, on campstede, cumbolgehnástes,
gármittunge [gármittunge, Th. 207, 3, col. 2] *that they were the better*
[*the victors*] *in works of war, on the battle-field, at the conflict of*
banners, at the meeting of javelins, Chr. 937; Th. 207, 3, col. 1;
Æðelst. 50.

gár-níþ, es; m. *A spear-battle, spear-war;* hastātōrum pugna:—
Geríseþ gárníþ werum *spear-war is fitting for men.* Exon. 91 a; Th. 341,
19; Gn. Ex. 128.

gár-rǽs, es; m. *A rush of spears, battle, war, warfare;* hastārum
impětus, proelium:—Ðæt gé ðisne gárrǽs mid gafole forgyldon *that ye*
buy off this warfare with tribute, Byrht. Th. 132, 46; By. 32.

gár-secg, -sæcg, es; m. [gár *a spear,* secg *man*]. I. *a spear-man,*
the ocean; hŏmo jǎcǔlo armǎtus, oceǎnus. The myth of an armed
man,—a spear-man is employed by the Anglo-Saxons as a term to denote
the Ocean, and has some analogy to the personification of Neptune holding

his trident. Spears were placed in the hands of the images of heathen
gods, as mentioned by Justin.—Per ea adhuc tempǒra rēges hastas pro
diadēmǎte habēbant, quas Grǣci sceptra dixēre. Nam et ab orǐgǐne rērum,
pro diis immortālǐbus větēres hastas coluēre; ob cujus religiōnis memǒriam
adhuc deōrum simulacris hastæ adduntur, l. xliii: c. iii:—Úre yldran ealne
ðysne ymbhwyrft ðyses middangeardes, cwæþ Orosius, swá swá Oceanus
ymblígeþ útan, ðone man gársecg háteþ, on þreó todǽldon *our forefathers,*
said Orosius, divided into three parts, all the globe of this mid-earth,
which the ocean that we call Garsecg, surrounds, Ors. 1, 1; Bos. 15, 2-4.
Asia is befangen mid Oceanus—dǽm gársecge—súþan, and norþan, and
eástan *Asia is encompassed by the ocean—the garsecg—on the south, and*
north, and east, 1, 1; Bos. 15, 8. Be norþan ðǽm beorgum, andlang
ðæs gársecges, óþ ðone norþ-eást ende ðyses middangeardes, ðǽr Bore seó
eá scýt út on ðone gársecg *to the north of the mountains, along the*
ocean to the north-east end of this mid-earth, there the river Bore shoots
out into the ocean, Ors. 1, 1; Bos. 18, 5-7. Gársecges deóp *the ocean's*
deep, Cd. 157; Th. 195, 24; Exod. 281. Gársecges begang *the circuit*
of ocean, Andr. Kmbl. 1059; An. 530. II. *a sea;* mǎre:—And
norþ óþ ðone gársecg, ðe man Cwén-Sǽ hǽt *and north to the sea, which*
is called the White Sea, Ors. 1, 1; Bos. 18, 27. Fuglas cōmon of gársecge *āves ex māri vēnērunt,* Ps. Th. 104, 35. Út on gársæcge *out in*
the sea, 96, 1.

gár-þrǽc, e; f. *Attack of javelins, battle;* hastōrum impětus, pugna:—
Æt gárþrǽce *in the attack of javelins,* Elen. Kmbl. 2369; El. 1186.

gár-þrist; adj. *Spear-bold, daring with a spear;* hastā audax:—Gúþ-
heard, gárþrist *warlike, spear-bold,* Elen. Kmbl. 407; El. 204.

gár-torn, es; m. [torn *anger*] *Spear-anger, rage of darts;.* íra tēlis
manifestāta:—Hí gártorn geótaþ gífrum deóðe *they shall pour the rage of*
darts upon the greedy devil, Salm. Kmbl. 291; Sal. 145.

garuwe, an; f. *Yarrow;* millefǒlium, Herb. 90; Lchdm. i. 194, 4,
MS. B. v. gearwe.

garwan *ready, prepared,* Chr. 1006; Erl. 140, 17, = gearwan; dat.
def. of gearo, q. v.

gár-wíga, an; m. *A spear-fighter, warrior;* hastātus bellātor:—Byrne
ne meahte geongum gárwígan geóce gefremman *the corslet could not*
afford aid to the young warrior, Beo. Th. 5341; B. 2674: 5614;
B. 2811.

gár-wígend, es; m. *A spear-fighter, warrior;* hastātus bellator:—
He úsic gárwígend góde tealde *he accounted us warriors good,* Beo. Th.
5275; B. 2641.

gár-wudu; gen. -wuda; m. *Spear-wood, a javelin;* hastæ lignum,
hasta:—Hie to gúþe gárwudu rǽrdon *they raised the spear-wood to battle,*
Cd. 160; Th. 198, 20; Exod. 325.

gast *a guest;* hospes, Cot. 102. DER. gast-hof, -hús, -líc. v. gǽst.

GÁST, gǽst, es; m. I. *the breath;* hālǐtus, spīrāmen:—Ne ne
is gást on múþe heora *there is not breath in their mouth,* Ps. Spl. 134, 17.
Ðæt ic ofsleá eall flǽsc, on ðam ðe ys lífes gást *that I may slay all flesh,*
in which is the breath of life, Gen. 6, 17. Mid gáste múþes his *with the*
breath of his mouth, Ps. Lamb. 32, 6. Blǽde oððe gáste *spīrāmine,*
Hymn Surt. 43, 36. II. *the spirit, soul,* GHOST; spīrǐtus, anǐmus,
ǎnǐma:—Gást *spīrǐtus,* Wrt. Voc. 76, 31. Se gást is hrǽd *spīrǐtus*
promptus est, Mt. Bos. 26, 41: Gen. 45, 27: Num. 11, 25, 26: Soul
Kmbl. 17; Seel. 9. Nó man scyle his gástes lufan wið Gode dǽlan *a*
man ought not to divide his spirit's love with God, Cd. 173; Th. 217, 11;
Dan. 21: Andr. Kmbl. 310; An. 155: Salm. Kmbl. 131; Sal. 65.
Hwyder ic gange fram gáste ðínum *quo ībo a spīrǐtu tuo?* Ps. Spl. 138, 6:
Num. 11, 17, 25: Elen. Kmbl. 939; El. 471: Exon. 73 a; Th. 113, 18;
Gú. 159. Bidde ic weoroda God, ðæt ic gást mínne agifan móte *I pray*
[*thee*] *God of hosts, that I may give up my spirit,* Andr. Kmbl. 2831;
An. 1418; Salm. Kmbl. 110; Sal. 54: Menol. Fox 340; Men. 171:
Elen. Kmbl. 958; El. 480. Gástas hwurfon, sóhton engla éðel *souls*
departed, sought the home of angels, Andr. Kmbl. 1280; An. 640:
Exon. 100 a; Th. 375, 6; Seel. 134. Gásta weardas *the guardians of*
spirits, Cd. 2; Th. 3, 25; Gen. 41. Gásta helm *the protector of spirits,*
God, Cd. 86; Th. 107, 22; Gen. 1793. Arǎs Metodes þeów gástum
togeánes *the Lord's servant* [*Lot*] *arose towards the spirits* [*angels*], 111;
Th. 140, 30; Gen. 2430. Folc wearð afǽred, flódegsa becwom gástas
geómre *the folk was affrighted, the flood-dread seized on the sad souls,*
166; Th. 206, 5; Exod. 447. Se hálga Gást *the holy Ghost;* Spīrǐtus
sanctus, Mk. Bos. 13, 11: Lk. Bos. 1, 15, 35: 2, 25, 26: Jn. Bos. 20, 22:
Elen. Kmbl. 2287; El. 1145. Se unclǽna gást *the unclean spirit,* Mt.
Bos. 12, 43: Mk. Bos. 1, 23: 5, 13: Lk. Bos. 4, 36: Elen. Kmbl. 603;
El. 302. Se werega gást *the accursed spirit, the devil,* Cd. 216; Th.
272, 27; Sat. 126. Werige gástas *accursed spirits, devils, demons,* Cd.
227; Th. .304, 15; Sat. 630. [*Piers P.* goost: *Chauc.* gost, goost:
R. Brun. gaste: *Laym.* gæst, gast, gost: *Orm.* gast: *Scot.* gest *a ghost,*
spirit: Plat. geest, m.: *O. Sax.* gēst, gást, geist, m.: *Frs.* gæst: *O. Frs.*
gast, iest, m.: *Dut.* geest, m.: *Ger. M.H.Ger. O.H.Ger.* geist, m.:
Goth. gaisyan *to be frightened: Dan.* geist, m. f.: *Swed.* gast, m. *an evil*
spirit, ghost.] DER. ǽrend-gást, cear-, ellen-, ellor-, geósceaft-, heáh-,
helle-, wuldor-.

gást-berend a *spirit-bearer, soul-bearer, living person, man.* v. gǽst-berend.

gást-bona, an; m. *The soul-killer, the devil;* anǐmi destructor, diǎbǒlus :—Ðæt him gástbona geóce gefremede *that the spirit-slayer would afford them help,* Beo. Th. 356; B. 177.

gást-cófa, an; m. *The spirit's chamber, breast;* anǐmi cūbǐle, pectus :— Hǐ habbaþ in gástcófan grimme geþohtas *they have fierce thoughts in their breast,* Frag. Kmbl. 22; Leas. 13.

gást-cund *spiritual.* v. gǽst-cund.

gást-cwalu *torment of soul.* v. gǽst-cwalu.

gást-cyning, es; m. *A spirit-king, God;* spǐrǐtālis rex, Deus :—Siððan wit ǽrende gástcyninge agifen habbaþ *after we two have performed the errand to the king of spirits* [*God*], Cd. 139; Th. 174, 24; Gen. 2883.

gást-gedál, gǽst-gedál, es; n. *Separation of soul and body, death;* anǐmæ et corpǒris divortium, mors :—Ðá he ðas woruld þurh gástgedál ofgyfan sceolde *when he must give up this world through death,* Cd. 55; Th. 68, 33; Gen. 1127 : Exon. 45 a; Th. 153, 32; Gú. 834.

gást-gehygd, gǽst-gehygd, es; n. *Thought of mind or spirit;* anǐmi cōgǐtātio :—Ðæt ðú sylfa miht ongitan gleáwlíce gástgehygdum *that thou thyself mayest prudently understand it with the thoughts of thy spirit,* Andr. Kmbl. 1722; An. 863.

gást-gemynd *thought of mind or spirit.* v. gǽst-gemynd.

gást-geníþla *a persecutor or foe of souls, the devil.* v. gǽst-geníþla.

gást-gerýne, gǽst-gerýne, es; n. *A ghostly or spiritual mystery, a mystery of the mind;* spǐrǐtāle mystērium, ănǐmi mystērium :—Him ða æðelingas ondsweorodon gástgerýnum *the princes answered him in spiritual mysteries,* Andr. Kmbl. 1716; An. 860: Elen. Kmbl. 378; El. 189: 2294; An. 1148.

gást-gewinn *torment of soul.* v. gǽst-gewinn.

gást-hálig, gǽst-hálig; adj. *Spirit-holy, holy in mind;* anǐmi sanctus :— Witgan sungon, gast-hálige guman, be Godes bearne *prophets, men holy in spirit, sung of the son of God,* Elen. Kmbl. 1120; El. 562.

gast-hof, es; n. *A guest-house, guest-chamber;* hospǐtium :—In ðam gast-hofe *in the guest-house,* Exon. 19 b; Th. 21, 24; Cri. 821. [*Ger.* gasthof *inn.*]

gást-hús, es; n. *A guest-house, guest-chamber;* hospǐtium :—On heora gast-húsum is gramlíc inwit *nēquǐtia est in hospǐtiis eōrum,* Ps. Th. 54, 15. v. gǽst-hús.

gást-leás; adj. *Lifeless, dead;* exănǐmis, mortuus :—Gefærenne man brohton on bǽre, gingne, gástleásne *they brought a dead man on a bier, young, lifeless,* Elen. Kmbl. 1746; El. 875.

gastlíc; adj. *Hospitable, ready for guests;* hospǐtālis :—Neorxna wang stód gód and gastlíc *paradise stood good and ready for guests,* Cd. 11; Th. 13, 27; Gen. 209.

gástlíc, gǽstlíc; adj. *Ghostly, spiritual;* spǐrǐtālis :—Gástlíc hreám *a cry of spirits, ghostly cry,* Nicod. 27; Thw. 15, 5. Leoðolíc and gástlíc *thé bodily and the ghostly,* Andr. Kmbl. 3254; An. 1630. Gé gástlícne god-dreám forségon *ye despised spiritual joy divine,* Exon. 41 b; Th. 139, 32; Gú. 602. Ðæt he healde gástlíce lufe *that he hold spiritual love,* Frag. Kmbl. 74; Leás. 39. Ðæt gástlíce folc *pǒpǔlus spǐrǐtālis,* Bd. 1, 27; S. 496, 28. Eádige synd ða gástlícan þearfan, forðam hyra ys heofena ríce *beáti sunt paupěres spǐrǐtu, quoniam ipsōrum est regnum cælōrum,* Mt. Bos. 5, 3.

gástlíce, gǽstlíce; adv. *Spiritually;* spǐrǐtāliter :—Ðæt hálige húsel is gástlíce Cristes líchama *the holy housel is spiritually Christ's body,* Homl. Th. i. 34, 19. Ðæt húsel is Cristes líchama, ná líchamlíce, ac gástlíce *the housel is Christ's body, not bodily, but spiritually,* L. Ælf. C. 36; Th. ii. 360, 16 : Bd. de nat. rerum; Wrt. popl. science 19, 25; Lchdm. iii. 280, 11 : Cd. 220; Th. 283, 7; Sat. 301.

gást-lufe *soul's love, spiritual love.* v. gǽst-lufe.

gást-sunu, gǽst-sunu; gen. a; dat. a, u; acc. u; pl. nom. acc. a, o, u; gen. a, ena; dat. um; m. *A spiritual son, Christ;* spǐrǐtālis filius, Christus :—Ahangen wæs on Caluarie Godes gástsunu *the spiritual Son of God was hanged up on Calvary,* Elen. Kmbl. 1342; El. 673.

gat, es; pl. nom. acc. u, a, o; n. *A* GATE; porta :—Ðá se Hǽlend geneálǽhte ðære ceastre gate *when the Saviour approached the gate of the city,* Lk. Bos. 7, 12 : Exon. 12 b; Th. 20, 15; Cri. 318 : Ps. Spl. 117, 19 : Ps. Th. 126, 6. v. geat.

GÁT; nom. acc; gen. gáte, gǽte; dat. gǽt; pl. nom. acc. gǽt, gét; gen. gáta; dat. gátum; f. *A she-*GOAT; capra :—Ic blǽte swá gát *I bleat as a goat,* Exon. 106 b; Th. 406, 17; Rä. 25, 2. Gát *capra vel capella,* Wrt. Voc. 78, 33 : 287, 36 : 288, 16. Gǽte blód *goat's blood,* Med. ex Quadr. 6, 4; Lchdm. i. 352, 3. Gǽte flǽsc *goat's flesh,* L. M. 1, 31; Lchdm. ii. 72, 8. Gǽte horn *a goat's horn,* Med. ex Quadr. 6, 1; Lchdm. i. 350, 17. Gǽte meolc *goat's milk,* L. M. 1, 7; Lchdm. ii. 52, 13. Genim ðæt wæter ðe innan gát byþ *take the water which is inside a goat,* Med. ex Quadr. 6, 10; Lchdm. i. 352, 19. Geoffra me áne þriwintre gát *sūme mihi capram trīmam,* Gen. 15, 9 : Lev. 3, 12 : 4, 28 : 5. 6. Hý beofiaþ fóre Freán, swá fúle swá gǽt *they shall tremble before the Lord, as foul as goats,* Exon. 26 a; Th. 75, 34; Cri. 1231. He asyndrode twáhund gáta *sepǎrāvit capras ducentas,* Gen. 32, 14. Gáta

---

hús *a goat-house;* caprīle, Ælfc. Gl. 108; Som. 78, 112; Wrt. Voc. 58, 27. Gáta loc *an enclosure for goats,* Wrt. Voc. 288, 20. Gáta hierde *a goatherd,* 288, 21. Gif seó offrung beó of gátum *si oblātio est de capris,* Lev. 1, 10. Drihten toscǽt hí on twá, swá swá scéphyrde toscǽt scép fram gátum : gelogaþ he ða scép on his swíðran hand, and ða gǽt on his wynstran *the Lord will part them into two, as a shepherd parts sheep from goats : he will place the sheep on his right hand, and the goats on his left,* Homl. Th. ii. 106, 27–29. Buccan oððe gǽt geseón ferþrunge getácnaþ *to see bucks or goats betokens advancement,* Somn. 126; Lchdm. iii. 206, 2. Gif ðú gesihst manega gǽt, ýdel getácnaþ *if thou seest many goats it betokens vanity,* 273; Lchdm. iii. 214, 1. Wæterbuca *vel* gát *tippǔla* [= *an insect that runs swiftly over the water, the water-spider, water-spinner*] Ælfc. Gl. 23; Som. 60, 10; Wrt. Voc. 24, 14. [*Chauc.* gat : *Laym.* gat, got : *Orm.* gat : *Dut.* geit, *f :* *Ger.* geisz, *f :* *M. H. Ger. O. H. Ger.* geiz, *f :* *Goth.* gaits, *f :* *Dan.* ged, *m. f :* *Swed.* get, *f :* *Icel.* geit, *f :* *Lat.* hædus, *m. a young goat, kid :* *Wel.* gid, giten, *f. a she-goat, young goat.*] DER. firgen-gát.

gát-bucca, an; m. *A he-goat;* cǎper :—Gát-buccan hyrde *a keeper of a he-goat,* Ælfc. Gl. 20; Som. 59, 37; Wrt. Voc. 22, 78.

Gátes héued, es; n. [*Goat's head*] GATESHEAD, *near Newcastle, Durham;* oppǐdi nōmen juxta Nǒvum Castrum in agro Dunelmensi, *capræ cǎput significans,* Som. Ben. Lye : Bd. 3, 21; S. 125, note 37. v. Hrége-heáfod.

gáte-treów, es; n. *A cornel tree?* cornus sanguinea? Lin :—Genim bircean, elebeám, gátetreów, ǽlces treówes dǽl *take birch, olive-tree, cornel-tree, a part of each tree,* L. M. 1, 36; Lchdm. ii. 86, 8.

gáþ go, Deut. 11, 28 : Mt. Bos. 9, 13; *pl. pres. indic. and impert. of* gán.

gaderian *to gather,* Som. Ben. Lye. v. gaderian.

gát-hyrde, es; m. *A* GOAT-HERD; caprārius :—Be gát-hyrde *gát-hyrde gebýreþ his heorde meolc ofer Martinus mæssedæg, and ǽr ðam his dǽl hwǽges, and anticcen of geáres geógoþe, gif he his heorde wel begýmeþ de caprario : caprārio convēnit lac grěgis sui post festum Sancti Martini, et antea pars sua mesguii, et caprīcum annǐculum, si běne custōdiat grěgem suum,* L. R. S, 15; Th. i. 438, 26–29.

gauel *a tribute,* Ps. Spl. T. 54, 11. v. gafol.

gauel-sester; es; m. *A measure of rent ale;* sextārius vectǐgālis cerevisiæ, Som. Ben. Lye. v. gafol, sester.

ge; conj. *And, also;* et :—Ánra gehwylc, sóþfæst ge synnig, sēceþ Meotudes dóm *every one, just and sinful, shall seek the Creator's doom,* Exon. 63 b; Th. 233, 11; Ph. 523 : Bt. Met. Fox 26, 171; Met. 26, 86: Ps. Th. 66, 6. Ge ... ge *both ... and;* et ... et. He bebýt ge windum ge sǽ *et ventis et mǎri impěrat,* Lk. Bos. 8, 25 : Jn. Bos. 2, 15 : Bt. 41, 3; Fox 248, 28: Chr. 835; Erl. 64, 28: Bt. Met. Fox 9, 3; Met. 9, 2 : 20, 25, 26; Met. 20, 2; Andr. Kmbl. 1083; An. 542. Ge mid býsenum heofonlíces lífes ge eác mid monungum *et exemplis vitæ cælestis et monitis,* Bd. 4, 19; S. 588, 3 : 2, 12; S. 512, 30, 31. Ge ... and *both ... and,* Cd. 35; Th. 46, 30–33; Gen. 752, 753. Ge eác swá same *and in like manner,* Bt. Met. Fox 11, 19; Met. 11, 10. Ge swylce *and also,* Beo. Th. 4508; B. 2258. Æghwæðer ge ... ge *either ... or;* vel ... vel, Bd. 2, 12; S. 513, 14, 15. Æghwæðer ge on mete, ge on hrægl, ge on æghwilcum ðinge *both in meat, and in dress, and in every thing,* Blickl. Homl. 219, 29. Ægðer ge ... ge *both ... and,* Bt. 41, 2; Fox 246, 5. Ægðer ge on spræce, ge on þeáwum, ge on eallum sidum *both in speech, and in manners, and in all customs,* Bt. 18, 2; Fox 62, 29; 41, 5; Fox 254, 19–21. [*O. Sax.* ge, gi, ja *and.*]

ge-, *or* æg-, prefixed to pronouns. v. æg-.

ge-, a preposition, originally meaning *with,* but found only as a prefix. v. Schleicher, Die Deutsche Sprache, p. 224. In accordance with this meaning it often gives a collective sense to nouns to which it is prefixed, as, ge-bróðor *brothers;* ge-húsan *housefolk;* ge-magas *kinsmen;* ge-macan *mates;* ge-gylda *a member of a corporation or guild;* ge-wita *a witness, accomplice;* ge-fera *a companion, attendant;* gescý *shoes.* Ge- sometimes gives to a neuter verb an active signification, as winnan *to fight,* ge-winnan *to win by fighting :*—Wið God winnan *to fight* [*war*] *with God,* Cd. 18; Th. 22, 26; Gen. 346. Sige on him ge-wann *he gained* [*won*] *a victory over him,* Num. 21, 1. Rídan *to ride;* ge-rídan *to reach by riding, arrive at :*—Ic on' wicge ríde *I ride on a horse,* Exon. 127 a; Th. 489, 14; Rä. 78, 7. Ge-rád Æðelwold ðone hám æt Winburnan *postea invāsit Æthelwaldus villam ǎpud Winburnam,* Gib. 99, 37 : Chr. 901; Erl. 97, 11. On this power of ge-, Mr. Earle, in Chr. p. 321, remarks :—' A strong instance is ge-winnan [1090] = *to win;* which sense, now so intimately identified with this root, is not in the simple verb winnan, until compounded with ge-. Winnan *to toil, fight, contend;* ge-winnan *is to get by striving, fighting, contending,* i. e. *to win,*' A.D. 685; p. 40, 16 : p. 4, 25. Ge- often seems void of signification; as, ge-sǽlþ *bliss;* ge-líc *like;* ge-sūnd *sound, healthy.* In verbs it seems sometimes to be a mere augment, e. g. in the following :—Ðæt wíf genam ðá of ðæs treówes wæstme and geæt and sealde hire were : he æt ða *mūlier tǔlit de fructu illīus et comēdit dēditque viro suo, qui comēdit,* Gen. 3, 6. It often changes the signification from literal to figurative; as, healdan *to hold;* ge-healdan *to observe, preserve;* fyllan *to fill;* ge-fyllan

to fulfil; biddan *to bid, require*; ge-biddan *to pray.* In the Rushworth Gloss. the prefix is often gi-. [*Wyc. Piers P. Chauc.* y-: *Laym.* i-: *O. Sax.* gi-: *O. Frs.* ge-, gi-, ie-: *Dut. Ger.* ge-: *M. H. Ger.* ge-, gi-: *O. H. Ger.* ga-, ka-, gi-, ki-, ge-, ke-: *Goth.* ga-: *Dan. Swed.* ge-.]

gē *ye, you*; vos, ὑμεῖς; *gen.* eówer [iwer] *your, of you*; vestrum *vel* vestri, ὑμῶν; *dat.* eów [íów, iu, iuh, iuih, iwh] *to you*; vobis, ὑμῖν; *acc.* eów [íów, iu, iuh, iuih, iwh], eówic *you*; vos, ὑμᾶς; *pl. of pers. pron. 2nd pers.* ðū *thou* :—Ne ondrǽde gē *fear ye not*, Mt. Bos. 10, 28. Gē ðe on hūse standaþ *you who stand in the house*; tu qui stātis in dŏmo, Ps. Th. 133, 2. Gebíde gē on beorge *abide you on the mount*, Beo. Th. 5051; B. 2529. Hwylc eówer *quis vestrum?* Mt. Bos. 6, 27 Ān eówer *ūnus vestrum*, 26, 21. Ic sylle eów *dăbo vobis*, Ex. 6, 8. Ic secge eów *dīco vobis*, Mt. Bos. 6, 16: 7, 7. Gyf gē ða lufiaþ ðe eów lufiaþ *si dīlĭgĭtis eos qui vos dīlĭgunt*, Mt. Bos. 5, 46. On eów becymþ Godes rīce *pervĕnit in vos regnum Dei*, Mt. Bos. 12, 28. Eówic grētan hēt *bade to greet you*, Beo. Th. 182; B. 3095. Hwanon eágorstreám eówic brohte *whence hath the ocean-stream brought you?* Andr. Kmbl. 518; An. 259: 1764; An. 884. Sibb sý mid eówic *peace be with you*, Exon. 75 b; Th. 282, 25; Jul. 668. [*Wyc.* ʒee, ʒe: *Piers P.* ye: *Chauc. Orm.* ʒe: *O. Sax.* gi, ge: *O. Frs.* gi, i: *Ger.* ihr: *M. H. Ger.* ir: *O. H. Ger.* īr: *Goth.* yus: *Dan. Swed.* i: *Icel.* ér.]

GEÁ, *adv.* YEA; ĕtiam :—'Quod est, lingua Anglōrum, verbum adfirmandi et consentiendi,' Bd. 5, 2; S. 183, 35. Geá, Drihten, ðū wāst ðæt ic ðē· lufige, *yea, Lord, thou knowest that I love thee*, Jn. Bos. 21, 15, 16; ĕtiam, Dŏmĭne, Vulg. Cweþ [cwæþ MS.] nū geá *say now yea*, Bd. 5, 2; S. 615, 9. [*Wyc.* ʒea, ʒhe: *Piers P.* ye: *Chauc.* ya, ye, yhe: *Orm.* ʒa: *O. Sax.* gĭ: *Frs.* ja: *O. Frs.* ie, ge: *Dut. Ger.* ja: *M. H. Ger. O. H. Ger.* jā: *Goth.* ya, yai: *Dan. Swed.* ja, jo: *Icel.* já *yes, yea.*]

GEÁC, es; *m.* A *cuckoo, gawk*; cŭcūlus :—Geác *cŭcūlus*, Ælfc. Gl. 37; Som. 63, 16; Wrt. Voc. 29, 38: 63, 3: 281, 31. Geác monaþ geómra reorde, singeþ sumeres weard *the cuckoo exhorts with mournful voice, summer's warden sings*, Exon. 82 a; Th. 309, 6; Seef. 53. Siððan ðū gehýrde galan geómorne geác on bearwe *when thou hast heard the sad cuckoo sing in the grove*, 123 b; Th. 473, 30; Bo. 22. Geácas geár budon *cuckoos announced the [time of] year*, Exon. 43 b; Th. 146, 27; Gū. 716. ¶ Geáces sūre, an; *f. Cuckoo-sorrel, wood-sorrel*; oxălis acetōsella, Lin :—Geáces sūre vel þrīlēfe *trifōlium*, Ælfc. Gl. 39; Som. 63, 72; Wrt. Voc. 30, 24. Genim geáces sūran *take cuckoo-sorrel*, L. M. I, 2; Lchdm. ii. 38, 14: 1, 38; Lchdm. ii. 96, 22: 3, 48; Lchdm. ii. 340, 2: iii. 12, 30. [*Scot.* gowk: *Dut.* koekpek, *m*: *Ger.* kuckuk, kukuk, gauch, *m. a cuckoo, gawk, simpleton*: *M. H. Ger.* gouch. *m*: *O. H. Ger.* gouch, gauch, *m. cŭcūlus, stultus*: *Dan.* giøg, *m. f*: *Swed.* gök, *m*: *Icel.* gaukr, *m*: *Fr.* coucou, *m*: *It.* cuculo, *m*: *Span.* cuco, cuclillo, *m*: *Lat.* cŭcūlus, *m*: *Grk.* κόκκυξ, *m*: *Sansk.* kokila, *m*.] v. Grm. D. M. pp. 640 sqq.

ge-aclian; *p.* ode, ade; *pp.* od, ad *To frighten, excite*; terrēre, terrōre percellĕre :—Ðā ðæt folc gewearþ egesan geaclod *then was the people terrified with fear*, Andr. Kmbl. 1609; An. 805: Elen. Kmbl. 2255; El, 1129. Cyning wæs egsan geaclad *the king was excited with terror*, 113; El. 57: Exon. 69 b; Th. 258, 20; Jul. 268.

geácnod *increased*, Elen. Kmbl. 681; El. 341, = ge-eácnod; *pp. of* ge-eácnian.

geácnung *a conceiving*; conceptio, Som. Ben. Lye. v. ge-eácnung.

ge-acsian, -acsigan; *p.* ode, ade; *pp.* od, ad *To find out by asking, discover, learn, hear*; resciscĕre, discĕre, agnoscĕre, audīre :—Ic wolde geacsigan and gewitan hwæt be ðē ðon sceolde *I would find out and know what should be done about thee*, Bd. 5, 12; S. 630, 30. Gyf se dēma ðis geacsaþ *si hoc audītum fuĕrit a præsĭde*, Mt. Bos. 28, 14. Ðā se pápa ðæt geacsade *when the pope heard it*, Bd. 2, 17; S. 520, 15; S. 10; S. 625, 20. We geacsodan *agnōvimus*, Bd. pref; S. 472, 16. Gif hine mon geacsige *if he be discovered*, L. In. 39; Th. i. 126, 9, MS. B. v. ge-ascian.

geacsung *an asking, inquiry*; inquīsītio, Som. Ben. Lye. v. ge-ascung.

ge-ádlian; *p.* ode, ede; *pp.* od, ed [ádlian *to be sick, to languish*] *To be sick, to languish, become impotent*; languescĕre :—ðām porticon læg mycel menigeo geádledra *in his portĭcis jăcēbat multĭtūdo magna languentium*, Jn. Bos. 5, 3. Ðæt ūre mōd þurh wærscipe wacole beón, ðæt hí þurh orsorhnysse ne asleacion, ne þurh nytennysse geádlion *that our minds may be vigilant through heedfulness, that through security they· slacken not, nor through ignorance become impotent*, Homl. Th. i. 610, 17.

geador; *adv. Together, altogether*; ūna, simul :—Þenden gǽst and líc geador sídedan *while soul and body journeyed together*, Exon. 76 a; Th. 285, 15; Jul. 714: Bt. Met. Fox 13, 98; Met. 13, 49: Salm. Kmbl. 899; Sal. 449. Gecyre ic ætsomne S. R. geador *I turn at once S and R together*, Exon. 123 b; Th. 475, 16; Bo. 48. Geátmæcgum geador ætsomne *for the Gothic warriors altogether*, Beo. Th. 987; B. 491. DER. eal-geador, on-geador. v. eador.

ge-æbiligan; *p.* de; *pp.* ed *To make angry, offend*; irrītāri :—Ðone ðe he ǽr mid forsewennysse geæbiligde *whom he had before angered by*

negligence, Homl. Th. ii. 592, 16. Gif hí us geæbiligdon *if they have offended us*, ii. 100, 33.

ge-ǽfenian, -ǽfnian; *p.* ode, ede; *pp.* od, ed [ǽfen *evening*] *To draw towards evening, become evening*; vesperascĕre, advesperascĕre :—Geǽfnaþ me ve·sperasco, Ælfc. Gr. 35; Som. 38, 10. Geǽfenedan dæge *advesperascente die*, Prov. 7.

ge-æfenlǽcan *to imitate*, Ben. Lye. v. ge-efenlǽcan.

ge-ǽfēstian *to envy* :—Giæfīstiaþ *invidet*, Rtl. 122, 1. v. æfēstian.

ge-æfnan; *p.* de; *pp.* ed [æfnan *to perform, execute*]. **I.** *to perform, execute, perpetrate, accomplish, complete, make*; perfĭcĕre, patrāre, præstāre, făcĕre :—He nele lāþes wiht ǽngum geæfnan *he will not perpetrate aught of harm to any*, Exon. 96 a; Th. 357, 23; Pa. 33: 95 b; Th. 356, 28; Pa. 18. Se eádga wer mægen unsōfte elne geæfnde *the blessed man with difficulty strenuously exerted his power*, 49 a; Th. 168, 21; Gū. 1081. We ðæt geæfndon swā *we thus accomplished it*, Beo. Th. 1081; B. 538. Síe sió bǽr gearo ædre geæfned *let the bier be quickly made ready*, 6203; B. 3106: 2218; B. 1107. **II.** *to stir up, excite*; excitāre :—Ic nolde þurh gielpcwide ǽfre geæfnan æbylg Godes *I would not through vaunting speech ever excite the anger of God*, Exon. 50 b; Th. 176, 16; Gū. 1211. **III.** *to bear, suffer, endure*; sufferre, sustĭnēre :—Hí sceolon ðone ryhtan dōm ǽnne geæfnan *they shall suffer the one righteous doom*, Exon. 27 b; Th. 84, 7; Cri. 1370. Ic yrmþu geæfnde *I suffered miseries*, 28 b; Th. 87, 24; Cri. 1430. v. ge-efnan.

ge-æhtan, -æhtian; *p.* te, ode; *pp.* ed, od [æht *valuation, estimation*] *To value, prize*; æstimāre :—Wæs gifu Hróþgāres oft geæhted *the gift of Hrothgar was often prized*, Beo. Th. 3774; B. 1885. Gebēte swā hit mon geæhtie *let him make amends as it may be valued*, L. Alf. 26; Th. i. 50, 26, MS. H. v. ge-eahtian.

ge-æhtendlíc; *adj. Valuable, estimable*; æstimābĭlis, Som. Ben. Lye.

ge-æhtle, an; *f.* [æht *valuation, estimation*] *Estimation, consideration*; æstimātio, delĭberātio :—Hý, on wĭggetawum, wyrðe þinceaþ eorla geæhtlan, *they, in their war-equipments, appear of the estimation of earls*, Beo. Th. 743; B. 369. Grein and Heyne give geæhtla *persecutor*; cf. ēhtan; then eorla geæhtlan would mean *warriors.*

ge-æhtung, e; *f. Deliberation, counsel*; consĭlium :—Nā hí wel syððan his geæhtunge āhwǽr heóldan *non sustĭnuērunt consĭlium ejus*, Ps. Th. 105, 11.

ge-ælged; *part. Coloured, painted, tanned, sunburnt*; cŏlōrātus, sōle fuscātus, Som. Ben. Lye.

ge-æmtian, -æmettigan, -æmtogian; *p.* ode; *pp.* od [æmtian *to be at leisure*] *To be unoccupied, be at leisure, be void*; văcuum esse, văcāre :—Ðe hie selfe geæmettigian sceoldon *who ought to keep themselves unoccupied*, Past. 18, 4; Swt. 134, 5; Cot. MS.: Swt. 4, 3. Geæmtiaþ eów, and geseóþ ðæt ic eam God văcāte, et vĭdēte quŏniam ĕgo sum Deus, Ps. Lamb. 45, 11. He wæs geæmtogod *he was void*, Homl. Th. i. 290, 21.

ge-ændung, e; *f. An end, a finish*; consummātio :—On graman geændunge *in īra consummātiōnis*, Ps. Lamb. 58, 14. v. ge-endung.

ge-ænged; *part.* [ænge *narrow, troubled, anxious*] *Troubled, anxious*; anxius :—Ge-ængedu *anxia*, Cot. 18.

ge-ǽrendian, -ǽrndian; *p.* ode; *pp.* od [ǽrendian *to go on an errand*] *To go on an errand, to ask, tell, intercede*; mandātum deferre, nuntiāre, interpellāre :—Se ðe him mæge geǽrendian [ge-ērendian MS. B: geǽrndian MS. H.] *who can do his errands*, L. In. 33; Th. i. 122, 13. Ðæt he him sceolde Gaiuses miltse geǽrendian *that he might ask the mercy of Caius for them*, Ors. 6, 3; Bos. 117, 36. He geǽrendaþ [geǽrndaþ MSS. A. G.] to Gode sylfum ymbe ælce neóde ðe man beþearf *he intercedes to God himself about every need a man may have*, L. C. E. 22; Th. i. 372, 29. Him geǽrendode Blyþþryþ his cwēn, ðæt he him wunonesse stōwe sealde on sumum eálande bí Ríne *qui, interpellante Blithrydæ conjuge sua, dĕdit ei lŏcum mansiōnis in insūlā quādam Rheni*, Bd. 5, 11; S. 626, 13. [*O. Sax.* habda giārundid *had performed his business.*] v. ǽrendian.

ge-ǽrnan, he -ærneþ; *p.* de; *pp.* ed. **I.** *v. intrans. To run*; currĕre :—Ðā geærndon hí sume þrage and efthwurfon *then they ran for some time and returned*, Bd. 5, 6; S. 619, 9. **II.** *v. trans. To run for, to gain by running*; cursu certāre, palma cursu contendĕre :—He nimþ ðone læstan dǽl, se nýhst ðæm tūne ðæt feoh geærneþ *he takes the least part, who nearest the town, gains [by running] the property*, Ors. 1, 1; Bos. 22, 40. DER. ærnan, yrnan, irnan.

ge-ærnian; *p.* ode; *pp.* od *To earn, deserve*; mĕrēri, promĕrēri :—Hí geærnian mägen *illi promĕrēri pŏtĕrint*, L. Alf. pol. 39; Wilk. 44, 42. v. ge-earnian.

ge-ærwe; *adj.* [arg *wicked, depraved*] *Perverse, wicked*; prāvus :—Nā tocleofode me heorte geærwe *non adhǽsit mihi cor prāvum*, Ps. Spl. T. 100, 4.

ge-æswicod; *part. Offended, scandalized*; scandălīzātus, Som. Ben. Lye. DER. æ-swícian.

ge-ǽt *ate*, Gen. 3, 6; *p. of* ge-etan.

ge-ǽðed; *part.* [āþ *an oath, a swearing*] *Sworn*; jūrātus :—Swā

geǽðedra manna sýn twegen oððe þrý to gewitnysse *of such sworn men let there be two or three as witness*, L. Edg. S. 6 ; Th. i. 274, 18.

ge-ǽðele ; *adj. Congenial, in accordance with one's nature, race* [v. ǽðelo] ; congĕnĭtus :—Swá him geǽðele wæs from cneómǽgum *as was to them natural from their kindred*, Chr. 937 ; Erl. 112, 7 ; Æðelst. 7. v. on-ǽðele. cf. gecynde.

ge-ǽðelian ; *p.* ode ; *pp.* od ; *v. trans. To render celebrated, renowned, excellent, to ennoble, improve* ; nōbĭlĭtāre :—Ðú geǽðelodest ealle gesceafta *thou ennobledst all creatures*, Hy. 7, 64 ; Hy. Grn. ii. 288, 64. Ðú eart geǽðelod geond ealle world *thou art renowned throughout all the world*, 7, 26 ; Hy. Grn. ii. 287, 26. [*Laym.* i-æðelien *to honour.*]

ge-ǽtred, -ǽttred, -ǽttrad, -ǽttrud ; *part.* [ātor *poison, venom*] *Poisoned, envenomed, poisonous* ; infectus, toxĭcātus, vĕnēnātus :—Forwearþ micel heres for geǽtredum gescotum *many of the army died from poisoned arrows*, Ors. 3, 9 ; Bos. 68, 38. Geǽttred *infectus*, Cot. 104. Hæfde he twigecgede handseax geǽttred *hăbēbat sīcam bicĭpĭtem toxĭcātam*, Bd. 2, 9 ; S. 511, 15. Geǽttrad flaa *a poisoned arrow*, Ælfc. Gl. 53 ; Som. 66, 65 ; Wrt. Voc. 35, 51. Geǽttrude nýtenu *vĕnēnāta anĭmālia*, Scint. 7.

ge-ǽwnod ; *part.* [ǽwnian *to marry, wed*] *Married* ; nuptus :—Ruth wearþ geǽwnod Iessan ealdan fæder *Ruth was married to the grandfather of Jesse*, Ælfc. T. 12, 17.

geaf *gave* :—He nallas beágas geaf *he gave no rings*, Beo. Th. 3443 ; B. 1719 ; *p.* of gifan.

geafel, es ; *m? A fork* :—Hine ufan mid īsenum geaflum ðydon *from above pierced him with iron forks*, Homl. Th. i. 430, 5. [*Gaffle a dung-fork*, Halliwell : *Ger.* gabel : cf. *O. H. Ger.* isarngabala, *f.* tridens.] v. gaflas.

geafla ; *p.* ode, ade ; *pp.* od, ad *To glorify* :—Geafade hine *glorificavit eum*, Rtl. 78, 32.

geaflas ; *pl. m. The jaws* ; fauces :—Geaflas *fauces*, Cot. 91. Ðæt nebb lixeþ swá glæs oððe gim, geaflas scýne innan and útan *the beak [of the Phœnix] glitters like glass or gem, the jaws comely within and without*, Exon. 60 a ; Th. 219, 1 ; Ph. 300. Biþ ðæt heáfod tohliden, geaflas toginene *the head shall be split open, the jaws distended*, Exon. 99 b ; Th. 373, 17 ; Seel. 110. Ðam ða geaflas beóþ nǽdle scearpran *whose jaws are sharper than a needle*, 100 a ; Th. 373, 32 ; Seel. 118.

geafle ? *a lever* ; palanga, vectis, Som. Ben. Lye.

geafol-monung, e ; *f.* :—Sittende to geafol-monunge *sedens ad teloneum*, Mk. Skt. Rush. 2, 14.

ge-aforud ; *part.* [aforud *exalted*] *Lifted up* ; sublīmātus, Som. Ben. Lye.

geafu, e ; *f. A gift* ; dōnum :—Ic mót meorda hleótan, gingra geafena *I may obtain rewards, new gifts*, Exon. 48 a ; Th. 164, 21 ; Gú. 1015. v. gifu.

ge-ágen ; *adj. Own* :—His geágenes ðances *of his own accord*, Th. Chart. 159, 5. v. ágen.

ge-ágennud ; *part.* [ágen *own*] *Adopted* ; adoptīvus :—Geágennud bearn *an adopted child* ; fīlius adoptīvus. Som. Ben. Lye.

geagl, geahl, es ; *m.* [also *n. v.* the last example] *The jowl, jaw* ; mandĭbŭla, rictus, fauces :—Geagl *rictus* Prooem. R. Concord. On ðam geagle in ðe jowl, L. M. 1, 4 ; Lchdm. ii. 46, 8. To swillanne ðone geagl *to swill the jowl*, 1, 1 ; Lchdm. ii. 24, 10 : 1, 4 ; Lchdm. ii. 48, 15, 19. Biþ ðæt heáfod tohliden, geaglas toginene *the head shall be split open, the jaws distended*, Soul Kmbl. 215 ; Seel. 110 : 229 ; Seel. 118. Ðæt geagl to swillanne *to swill the jowl*, L. M. 1, 1 ; Lchdm. ii. 24, 12, 22, 26, 29.

geagl *light, frolicsome, lascivious*, Bd. 5, 6 ; Whelc. 390, 39, MS. C. v. gagol.

geaglisc, geglesc ; *adj. Light, frolicsome, lascivious* ; lĕvis, lascīvus :—Ic wæs mid geaglisce [geglescum MS. B : geagle MS. C.] móde oferswýðed *I was overcome with a frolicsome mood* ; lascīvo supĕrātus anĭmo, Bd. 5, 6 ; Whelc. 390, 39. v. gagol.

geagl-swile, es ; *m. A swelling of the jowl* ; faucium tŭmor :—Lǽcedóm wið geaglswile *a remedy for jowl-swelling*, L. M. 1, 4 ; Lchdm. ii. 46, 7. Wið geaglswile [MS. gealhswile] *for jowl-swelling*, 1, 4 ; Lchdm. ii. 44, 8.

geagn-cwide, es ; *m. A reply, answering again* ; responsum :—Grimme geagncwide *with grim response*, Elen. Kmbl. 1047 ; El. 525. v. gegn-cwide.

ge-ágnian, -áhnian ; to -ágnianne, -áhnianne ; *p.* ode, ade, ede ; *pp.* od, ad, ed *To own, possess, inherit, appropriate to one's self, claim as one's own* ; possīdēre, herēdĭtāre, vindĭcāre sibi :—Hwí sceal he him ánum geágnian ðæt him bám is forgifen *why should he appropriate to himself only that which is given to both?* Homl. Th. ii. 102, 29 : Ors. 5, 4 ; Bos. 104, 17 : Cd. 86 ; Th. 109, 27 ; Gen. 1829. Nán man hit náh to geáhnianne [geágnianne MS. A.] *no man ought to claim possession of it*, L. C. S. 24 ; Th. i. 390, 13. Ic geáhnige *possīdeo*, Ælfc. Gr. 26, 5 ; Som. 29, 5. He his gecorenan on ðisum middanearde geágnaþ *he owns his chosen in this world*, Homl. Th. ii. 72, 28. Ða geyrfweardiaþ oððe geáhniaþ land *ipsi herēdĭtābunt terram*, Ps. Lamb. 36, 9. Ðú geág-

nadest, Ps. Th. 79, 16. Parthe him ðæt ríce geáhnedon *the Parthians took the kingdom to themselves*, Ors. 5, 4 ; Bos. 104, 35. Oþ-ðæt se ágenfrigea him ðæt orf geáhnige *till the proprietor claims the cattle for his own*, L. Edg. S. 11 ; Th. i. 276, 16. Sceal monna gehwilc wesan geágnod me *every man shall be appropriated to me*, Cd. 106 ; Th. 140, 1 ; Gen. 2321. [*Goth.* ga-áiginōn : *Laym.* iahnien.]

ge-ágniendlíc, -ágnigendlíc ; *adj. Owning, possessive* ; possessīvus :—Genitivus is gestrýnendlíc oððe geágniendlíc *the genitive [case] is producing or possessive*, Ælfc. Gr. 7 ; Som. 6, 17. Sume synd geágnigendlíce, ða geswuteliaþ ða þing ðe beóþ geágnode *some are possessive, which make known the things which are owned*, 5 ; Som. 4, 55.

geagninga ; *adv. Clearly, truly, certainly* ; plāne, prorsus, certe :—Ðú scealt geagninga wisdóm onwreon *thou shalt truly display wisdom*, Elen. Kmbl. 1343 ; El. 673. v. gegninga.

geahl, es ; *m. The jowl, jaw* ; fauces :—God forbriteþ téþ, heora on múþe heora, tuxlas oððe geahlas leóna tobrycþ Drihten *Deus contĕret dentes eōrum in ōre ipsōrum, mŏlas leōnum confringet Dŏmĭnus*, Ps. Spl. 57, 6. v. geagl.

ge-áhnian *to own, possess, appropriate to one's self* :—Ic geáhnige *possĭdeo*, Ælfc. Gr. 26, 5 ; Som. 29, 5 : Ors. 5, 4 ; Bos. 104, 35 : L. Edg. S. 11 ; Th. i. 276, 16. v. ge-ágnian.

ge-áhnung, e ; *f. An appropriation, possession, owning* ; approprĭātio, possessio, Som. Ben. Lye.

ge-ahsian ; *p.* ode ; *pp.* od *To find out by asking, discover, learn, hear* ; fando accīpĕre, rescīscĕre, discĕre :—Ðá Latinus hyre wer geahsode *when Collatinus her husband heard it*, Ors. 2, 2 ; Bos. 41, 32 : 3, 11 ; Bos. 75, 26. We geahsodon ðæt úre gefēran sume to eów cómon *we have heard that some of our fellows have come to you*, L. Alf. 40 ; Th. i. 56, 14, MS. G : Ors. 3, 11 ; Bos. 74, 41. Gif hine mon geáhsige *if he be discovered*, L. In. 39 ; Th. i. 126, 10. Hæbbe ic geahsod, ðæt . . . *I have heard that . . .*, Beo. Th. 870 ; B. 433. v. ge-ascian.

geal, *pl.* gullon *yelled* ; *p.* of gellan.

geal-ádl, e ; *f.* [gealla *gall, bile*] *Gall-disease, the jaundice* ; ictĕrus = ἴκτερος, aurūgo :—Of gealádle cymeþ greát yfel . . . se líchoma ageolwaþ swá gód seoluc *from jaundice comes great evil . . . the body becomes yellow like good silk*, L. M. 1, 42 ; Lchdm. ii. 106, 19–22.

gealchattan *p.* te ; *pp.* ed *To ordain, frame, devise* ; concinnāre :—Tunge ðín gealchatte oððe gereónode fácnu *lingua tua concinnābat dŏlos*, Ps. Lamb. 49, 19.

geald *possibly, perhaps* ; forte, forsĭtan, Jos. 9, 8. v. weald ; *adv.*

geald *paid*, Beo. Th. 2099 ; B. 1047 ; *p.* of gildan.

gealdor, es ; *n. An incantation, a charm, lore* ; incantātio :—Be ðam gealdre *through that lore*, Exon. 83 a ; Th. 313, 26 ; Mód. 6. Sing ðis gealdor *sing this charm*, L. M. 3, 63 ; Lchdm. ii. 350, 28 : 3, 24 ; Lchdm. ii. 322, 6. v. galdor.

gealdor-cræft, es ; *m. The art of enchanting, incantation* ; incantātio :—On ǽniges cynnes gealdorcræftum *per alĭcŭjus gĕnĕris incantātiōnes*, L. Ecg. P. iv. 18 ; Th. ii. 208, 32. v. galdor-cræft.

gealdor-cræftiga, an ; *m. One crafty or skilful in enchantments, an enchanter* ; in arte incantandi perĭtus, incantātor :—Ða fǽmnan, ðe gewuniaþ onfón gealdorcræftigan ne lǽt ðú ða libban *the women, who are wont to receive enchanters, suffer thou them not to live*, L. Alf. 30 ; Th. i. 52, 9. v. galdor-cræftiga.

gealewe *yellow* ; flāvus, Som. Ben. Lye. v. geolo.

gealga, an ; *m. A gallows, gibbet, cross* ; patĭbŭlum, crux :—Fracoðes gealga *a malefactor's gibbet*, Rood Kmbl. 20 ; Kr. 10. Ðone óðerne he hét hón on gealgan *altĕrum suspendit in crŭcem*, Gen. 40, 22 : Deut. 21, 22 : Past. 3, 1 ; Swt. 33, 20 ; Hat. MS. 8 b, 7 : Apstls. Kmbl. 44 ; Ap. 22 : Rood Kmbl. 80 ; Kr. 40. v. galga.

ge-algian, -ealgian ; *p.* ode ; *pp.* od *To protect, defend* ; tuēri, defendĕre :—Hér stynt eorl, ðe wile gealgian ēðel ðysne *here stands an earl, who will defend this land*, Byrht. Th. 133, 18 ; By. 52. Ðæt hí, æt campe, wið láþra gehwæne, land gealgodon *that they, in conflict, should defend the land against every foe*, Chr. 937 ; Th. 203, 4, col. 2 ; Æðelst. 9. v. ealgian.

gealg-mód, galg-mód, gealh-mód ; *adj.* [gealg = gealh *sad* ; mód *mind*] *Sad in mind, gloomy, furious* ; tristis anĭmo, furiōsus :—Gealgmód guma *the furious man*, Exon. 73 b ; Th. 274, 10 ; Jul. 531 : 74 b ; Th. 278, 15 ; Jul. 598. Hie eágena gesihþ aguton gealgmóde gára ordum *they, furious, thrust out the eyesight with javelins' points*, Andr. Kmbl. 63 ; An. 32 : 1125 ; An. 563.

gealg-treów, es ; *n. A gallows-tree, cross* ; crux :—Dryhten þrowode on ðam gealgtreówe for guman synnum *the Lord suffered on the cross for the sins of man*, Rood Kmbl. 289 ; Kr. 146. v. galg-treów.

gealh ; *adj. Sad, angry* ; tristis :—Unrót *vel* gealh *tristis*, Ælfc. Gl. 88 ; Som. 74, 88 ; Wrt. Voc. 51, 1. Se ðe biþ ungeðyldig, and mid gealgum móde ceoraþ ongeán Gode *he who is impatient and passionately murmurs against God*, Homl. Th. i. 472, 8.

gealh-mód ; *adj. Sad in mind, gloomy* ; tristis anĭmo :—Grim and gealhmód *grim and gloomy*, Cd. 184 ; Th. 230, 8 ; Dan. 230. v. gealg-mód.

gealh-swile *a swelling of the jowl*, L. M. 1, 4; Lchdm. ii. 44, 8. v. geagl-swile.

GEALLA, ealla, an; *m.* I. GALL, *bile*; fel, bīlis:—Gealla *fel vel bīlis*, Ælfc. Gl. 76; Som. 71, 111; Wrt. Voc. 45, 17. Đe cymeþ of togotennysse đæs geallan *which cometh of effusion of the gall*, Herb. 141, 2; Lchdm. i. 262, 12, MS. O: 146, 2; Lchdm. i. 270, 4, MS. H. Hig sealdon hym wîn drincan mid geallan gemenged *dĕdĕrunt ei vinum bĭbĕre cum felle mistum*, Mt. Bos. 27, 34: Exon. 29 a; Th. 88, 13; Cri. 1439. Wið seóndum geallan *for straining out bile*, L. M. 3, 11; Lchdm. ii. 314, 7. II. *a gall, fretted place on the skin*; intertrīgo:—Wið horses geallan *for a horse's gall*, L. M. 1, 88; Lchdm. ii. 156, 21. Lācna đone geallan mid *cure the gall therewith*, 1, 88; Lchdm. ii. 156, 21. [*Orm.* galle: *O. Sax.* galla, *f*: *Dut.* gal, *f*: *Ger. M. H. Ger.* galle, *f*: *O. H. Ger.* galla, *f*: *Dan.* galde, *m. f*: *Swed.* galle, *m*: *Icel.* gall, *n*: *Lat.* fel, *n*: *Grk.* χολή, *f*; χόλος, *m*.]

gealled; *part. Galled, fretted*; intertrīgĭnōsus:—Gif hors geallede sîe *if a horse be galled*, L. M. 1, 88; Lchdm. ii. 156, 18.

geallig; *adj. Acris, tristis*, Hpt. Gl. 456.

gealp *boasted*, Beo. Th. 5160; B. 2583; *p. of* gilpan.

ge-an ic, he *I give, he gives*, Th. Diplm. 560, 24; 1st and 3rd pres. of ge-unnan.

geán; *prep. Against, over against, on the opposite side*; contra:—Mónaþ is đonne se móna gecyrþ niwe fram đære sunnan, ôþ-đæt he eft cume hyre fôrne geán *a month is when the moon returns new from the sun, until it [the moon] again comes opposite it [the sun]*, Bd. de nat. rerum; Wrt. popl. science 8, 13; Lchdm. iii. 248, 17, note 30. On đæm clife on đæm is geán bearwum *on the cliff which is over against the woods*, Blickl. Homl. 209, 35. [*Orm.* 3æn.] v. on-geán.

geána; *adv. Yet, still*; adhuc:—Get geána *adhuc*, Mt. Kmbl. Lind. 15, 16. v. gén.

ge-anbīdian; *part.* -anbīdiende, -anbīdigende; *p.* ode; *pp.* od [anbīdian *to abide*] *To abide, await, wait for, expect*; expectāre, sustĭnēre:—Đes man wæs ôþ Israhēla frófor geanbīdiende *hŏmo iste expectans consōlātiōnem Israel*, Lk. Bos. 2, 25. Đæt folc wæs Zachariam geanbīdigende *ĕrat plebs expectans Zachariam*, 1, 21. Hî þrŷ dagas me geanbīdiaþ *jam trĭduo sustĭnent me*, Mk. Bos. 8, 2. Geanbīda Drihten, werlîce dô đū, and sŷ gestrangod heorte đîn, and geanbīda Drihten *expecta Dŏmĭnum, virīlĭter ăge, et confortētur cor tuum, et sustĭne Dŏmĭnum*, Ps. Spl. 26, 20.

ge-anbyrdan, ge-onbyrdan; *p.* de; *pp.* ed *To strive against, resist*; repugnāre, resistĕre:—Gif he gewyrce đæt man hine afylle þurh đæt đe he ongeán riht geanbyrde *if he act so that he be killed because he strove against right*, L. C. S. 49; Th. i. 404, 13. v. anbyrdnys.

ge-ancsumian; *p.* ode; *pp.* od *To make anxious, vex*; anxiāre:—Wæs geancsumod mîn heorte *anxiārĕtur cor meum*, Ps. Lamb. 60, 3. v. ge-angsumian.

geán-cyme, es; *m. A coming against, meeting*; occursus:—Đæt đū yfele geáncymas ne ondrǽde *ut occursus mălos ne formīdes*, Herb. 111, 3; Lchdm. i. 224, 19.

geán-cyr, -cyrr, es; *m. A turning against, coming against, meeting*; occursus:—Fram heán heofone is ûtgang his, and geáncyr his ôþ to heáhnesse his *a summo cœlo est egressio ejus, et occursus ejus usque ad summum ejus*, Ps. Spl. 18, 7.

ge-ándagian; *p.* ode; *pp.* od; *v. a. To appoint a day or term*; diem dīcĕre:—Đæt he him geándagode of đam folclande *that he should give him a term respecting the folk-land*, L. Ed. 2; Th. i. 160, 12. v. ándagian.

ge-andettan, -ondettan; *p.* te; *pp.* ed *To confess*; confĭtēri:—Se seóca sceal geandettan đam sacerde *the sick must confess to the priest*, L. Ælf. C. 32; Th. ii. 354, 28: L. Alf. pol. 14; Th. i. 70, 15, note 38. Gif he hine geandette *if he confess himself*, L. Alf. pol. 5; Th. i. 64, 22: L. In. 71; Th. i. 148, 3, note 4. v. andettan.

ge-andswarian; *p.* ode; *pp.* od *To answer*; respondēre:—Đa ne geandswarode he hyre *qui non respondit ei verbum*, Mt. Bos. 15, 23. v. and-swarian.

ge-andwerdian; *p.* ode; *pp.* od [andweard *present*] *To present, bring before one*; præsentāre:—Đa hēt he đone biscop mid his preóstum samod geandwerdian *then commanded he to bring the bishop together with his priests before [him]*, Homl. Th. i. 416, 4. Geandweardod beôn *præsentātus esse, præsentāri*, R. Ben. 7. Giondweardad *præsentātus*, Rtl. 4, 28.

ge-andwyrdan, -andwerdan; *p.* -andwyrde; *pp.* -andwyrded, -andwyrd *To answer*; respondēre:—Ne mihton hig agēn đis him geandwyrdan *non pŏtĕrant ad hæc respondēre illi*, Lk. Bos. 14, 6: Bt. 41, 2; Fox. 244, 23. Geandwyrde [geandwerde MS. G.] he đam ôðrum swā hundrēde riht þence *let him answer to the other as shall seem right to the hundred*, L. C. S. 27; Th. i. 392, 6. Him wæs geandwyrd đus *he was answered thus*, Gen. 19, 21.

ge-āned; *part.* [ān *one*] *Made one, united*; adūnātus:—Oþ-đæt đe hî wǽron on ænne lēg geānede *usque ad in immensam adūnāti sunt flammam*, Bd. 3, 19; S. 548, 21. [Cf. *Ger.* vereint: *O. H. Ger.* gaeinôn *adunare*.]

geán-fær, es; *n. A going again, returning, return*; rĕdĭtus:—Him widcwæþ se cyng ælces geánfæres [MS. geánfares] to Engla lande *the king prohibited him from all return to England*, Chr. 1119; Erl. 247, 34.

ge-angsumian, -ancsumian, -anxsumian; *p.* ode; *pp.* od *To vex, make anxious or uneasy*; angēre, anxiāre:—Ic geangsumige *ango*, Ælfc. Gr. 28, 5; Som. 31, 56.

geán-hweorfan; *p.* -hwearf, *pl.* -hwurfon; *pp.* -hworfen *To turn again, return*; rĕdīre, Hpt. Gl. 409; Leo A. Sax. Gl. 229, 21.

geán-hworfennis, se; *f. A return*; obvia quæque, ad propria limina reversio, Hpt. Gl. 470.

geán-hwyrft *a turning again.* v. gǽn-hwyrft.

ge-ánlǽcan; *p.* -lǽhte; *pp.* -lǽht *To make one, join, unite*; unāre, unīre:—Ic geánlǽce [MS. -lace] *ūno, ūnio*, Ælfc. Gr. 37; Som. 39, 29. Þurh đæs Hālgan Gāstes tócyme wurdon ealle gereord geánlǽhte *through the advent of the Holy Ghost all languages became united*, Homl. Th. i. 318, 24. Geánlǽcan *adsciscere, miscere*, Hpt. Gl. 504.

ge-anlīcian; *p.* ode; *pp.* od [līc *like*] *To make like, liken*; assĭmĭlāre:—For hwam geanlīcie we heofena rîce *cui assĭmĭlābimus regnum Dei?* Mk. Bos. 4, 30.

ge-anmētan; *p.* -anmętte; *pp.* -anmēted, -anmētt *To encourage*; anĭmāre:—He him to fultume com, and hine swîđe geanmętte *he came to his help and greatly encouraged him*, Ors. 3, 10; Bos. 70, 45. Wæs Demetrius swîđe þearle geanmętt *Demetrius was very greatly encouraged*, 3, 11; Bos. 75, 25.

geánnis, se; *f. A meeting*; obviam itio, Hpt. Gl. 513.

geán-ryne, geán-ryne, es; *m. A running against, meeting*; occursus:—Arîs on geánryne mînne *exurge in oscursum meum*, Ps. Spl. 58, 5.

geán-þingian; *p.* ode, ade; *pp.* od, ad [þingian *to address, speak*] *To speak again, answer, reply*; respondēre:—Him brego engla geánþingade *the Lord of angels replied to him*, Cd. 48; Th. 62, 5; Gen. 1009.

geánunga; *adv. Directly*:—Geánunga foron đa sunnan *directly before the sun*, Bd. de nat. rerum; Wrt. popl. science 5, 29; Lchdm. iii. 242, 12, note. v. gegnunga.

ge-anwyrde; *adj. Known, manifest, confessed*; professus:—Ic eom geanwyrde monuc *professus sum monachus*, Coll. Monast. Th. 18, 23. He đæs geanwyrde wæs ætfôran eallum đām mannum *he confessed it before all the men*, Chr. 1055; Erl. 189, 5. v. note where the Latin is given, ipse ante cognovit ita esse.

ge-anxsumian; *p.* ode, ade; *pp.* od, ad *To make anxious, vex*; anxiāre:—Geanxsumad is ofer me gāst mîn *anxiātus est sŭper me spĭritus meus*, Ps. Lamb. 142, 4. v. ge-angsumian.

geap, gæp; *comp. m.* geappra, *f. n.* geappre; *adj.* I. *crooked, bent, curved*; curvus, pandus:—Geap *curvus*, Cot. 50. Geap stæf *a crooked letter*, Salm. Kmbl. 250; Sal. 124: 269; Sal. 134. Geapum, gebīgedum *pando*, Mone B. 90. II. *not straightforward, deceitful, crafty, cunning, shrewd, astute*; fallax, callīdus, astūtus:—Geap *callīdus*, Wrt. Voc. 49, 11. Seó næddre wæs geappre đonne ealle đa ôðre nŷtenu *serpens ĕrat callīdior cunctis animantĭbus terræ*, Gen. 3, 1. Cild geap *an astute child*, Obs. Lun. § 2; Lchdm. iii. 184, 14: § 9; Lchdm. iii. 188, 11. DER. hinder-geap. Grein writes geáp, in support of which may be noticed 3æp in the Ormulum. Layamon also has the word, and it occurs in Piers P.

geáp *took*, Exon. 106 b; Th. 405, 29; Rä. 24, 9; *p. of* geópan.

GEÁP; *adj. Open, spread out, extended, broad, roomy, spacious, wide*; pătens, pătŭlus, amplus, lātus:—Gim sceal on hringe standan, steáp and geáp *a gem shall stand in a ring, prominent and broad*, Menol. Fox 505; Gn. C. 23. Steáp and geáp *high and wide*, Salm. Kmbl. 827; Sal. 413. Reced hlifade, geáp and goldfáh *the mansion towered, spacious and golden-hued*, Beo. Th. 3604; B. 1800. Munt is hine ymbûtan, geáp gylden weal *a mountain is about him, a lofty golden wall*, Salm. Kmbl. 511; Sal. 256. Sum sceal on geápum galgan rîdan *one shall ride on the extended gallows*, Exon. 87 b; Th. 239, 12; Vy. 33. Under geápne hróf *under the spacious roof*, Beo. Th. 1677; B. 836. [Cf. *Icel.* gaupn *both hands held together* in the form of a bowl; geypna *to encompass*.] DER. horn-geáp, sǽ-.

geáp, geápu, e; *f.* [geáp *roomy, spacious*] *Expanse, room*; latĭtūdo, spātium:—Đās hofu dreórigiaþ, and đæs teáfor geápu *these courts are dreary, and its purple expanse [?]*, Exon. 124 a; Th. 477, 27; Ruin. 31.

geápan, geápian; *p.* te, ode; *pp.* ed, od *To* GAPE, *open*; pandēre, Cot. 158.

geápes; *adv.* [*gen. of* geáp *broad, spacious, roomy*] *In width, wide*; lāte:—Strûdende fŷr, steápes and geápes, forswealh eall eador *the ravaging fire swallowed all together, high and wide*, Cd. 119; Th. 154, 16; Gen. 2556. So Bouterwek takes it, but the word is rather a neuter genitive after 'eall;' cf. vv. 2548-9.

geáplic; *adj. Crafty, cunning, deceitful*; subdŏlus, callīdus:—Hî mid geaplīcre fare ferdon to Iosue *they went to Joshua with deceitful expedition*, Jos. 9, 6.

geáplīce; *adv. Deceitfully, boldly*; subdŏle, procācĭter, Prov. 21.

geap-neb; *adj.* [geap *crooked*; neb *the head, face, beak, nib*] *Crooked-*

nibbed, with a bent beak, arched; curvātus :—Standeþ me hēr on eaxelum Ælfheres láf, gód and geapneb *Ælfhere's legacy stands here on my shoulders, good and crooked-nibbed,* Wald. 94; Vald. 2, 19.

geap-scipe, es; *m. Craft, cunning, deceit, fraud;* astūtia, fraus :— Eall heora geapscipe wearþ ameldod Israhēla bearnum *all their deceit was made known to the children of Israel,* Jos. 9, 16. Þurh his geapscipe he begeat ðone castel *through his cunning he obtained the castle,* Chr. 1090; Erl. 226, 25.

geápung, e; *f. A heaping, heap, pile;* cŭmŭlus :—Fôþ him on, and on geápunga eówre niðerunge gelǽdaþ *accipite, et in cŭmŭlum damnātiōnis vestræ dūcĭte,* Bd. 5, 13; S. 633, 14, note 13, MS. B. v. heápung.

gear, *pl.* gurron *sounded, creaked; p.* of georran.

GEÁR, gēr, gǽr, es; *n. A* YEAR; annus :—Ôðer com geár *another year came,* Beo. Th. 2272; B. 1134. Ðis wæs feorþes geáres his rîces *this was in the fourth year of his reign,* Chr. 47; Th. 10, 13, col. 1. On geáre *in the year,* Menol. Fox 218; Men. 110. Ðrîwa on gére *thrice a year,* Thw. Exod. 23, 17. Hæfde me ēce geár ealle on mōde *annos æternos in mente hăbui,* Ps. Th. 76, 5 : Lk. Bos. 2, 36. Þreó and þritig geára *three and thirty years,* Cd. 224; Th. 296, 16; Sat. 503. Geárum frôd *old in years,* 109; Th. 143, 19; Gen. 2381. Men hâtaþ ðysne dæg geáres dæg, swylce ðes dæg fyrmest sý on geáres ymbryne *men call this day [new] year's day, as if this day were the first in the year's circuit,* Homl. Th. i. 98, 16. [*Wyc.* ʒeer, ʒer, ʒeers, ʒerys *years : Piers P.* yere : *Chauc.* yer, yere : *R. Brun.* ʒere : *Laym. Orm.* ʒer : *Plat.* jaar, jar, *n* : *O. Sax.* gēr, jâr, *n* : *Frs.* jier : *O. Frs.* ier, iar, ger, *n* : *Dut.* jaar, *n* : *Ger.* jahr, jar, *n* : *M. H. Ger.* jâr, *n* : *O. H. Ger.* jâr, *n* : *Goth.* yér, *n* : *Dan.* aar, *n* : *Swed.* âr, *n* : *Icel.* ár, *n* : *Bohem.* gar, *m. f. spring* : *Zend.* yâre, *n. year.*] DER. freóls-geár, fyrn-. v. Grm. D. M. p. 715.

geara, *adv.* [gearo? *ready*] *Utterly, altogether, well, enough, very much;* pĕnĭtus, prorsus, bĕne, sătis, valde :—He hēt geara forbærnan Rômâna burig he [*Nero*] *commanded utterly to burn up the city of the Romans,* Bt. Met. Fox 9, 18; Met. 9, 9. Ðú geara canst *tu bĕne nosti,* Bd. 1, 27; S. 439, 2 : Ps. Th. 75, 1 : 81, 5. Ðonne mon me geofe geara þúsende goldes and seolfres *super millia auri et argenti,* 118, 72.

geara, *gen. pl.* of geare, q. v. *furniture, gear for horses.*

geára; *adv.* [*gen. pl.* of geár *a year*] YORE, *formerly, of old, long since, once;* ōlim, antīquĭtus, quondam :—Se geára hider fram ðam eádigan Gregorie sended wæs *qui olim huc a beato Gregorio directus fuit,* Bd. 2, 3; S. 504, 44. Ic þeódenmádmas geára forgeáfe *I princely treasures gave of old,* Cd. 22; Th. 26, 21; Gen. 410. Ðú on geóguþfeore geára gecwǽde *thou in youthful life long since didst say,* Beo. Th. 5322; B. 2664 : Ps. Th. 73, 12 : 80, 10 : 104, 6 : 118, 152. Geára iû, Exon. 76 b; Th. 287, 30; Wand. 22 : 84 a; Th. 316, 31; Môd. 57 : Bt. Met. Fox 1, 1; Met. 1, 1. [*Laym.* ʒære, ʒare : *Chauc.* yore.] DER. ǽr-geára, fyrn-, geó-, iû-, un-.

gearcian, gærcian; *p.* ode; *pp.* od [gearo *ready*] *To prepare, make ready, procure, furnish, supply;* părāre, præpărāre, appărāre, exhĭbēre, præbēre :—Ic gearcige exhĭbeo, *præbeo,* Ælfc. Gr. 26, 2; Som. 28, 35, 36 : 47; Som. 48, 43. On lâfum ðînum ðú gearcast [MS. gearcost] andwlitan heora *in relĭquiis tuis præpărābis vultum eōrum,* Ps. Spl. 20, 12. On him gearcode fæt deáþes *in eo părāvit vāsa mortis,* 7, 14 : Gen. 19, 3. [*Piers P.* yarken *to make ready* : *R. Glouc.* yarkede, *p. prepared* : *Laym.* ʒarkien, ʒarekien, ʒearkien *to get ready* : *Orm.* ʒarrkenn *to prepare, make ready.*] DER. ge-gearcian.

gearcung, e; *f. A preparation, preparing;* præpărātio, appărātus :— Gearcunge heortan heora gehýrde eáre ðîn *præpărātiōnem cordis eōrum audīvit auris tua,* Ps. Spl. second 9, 20 : 32, 14. Gearcung appărātus, Ælfc. Gl. 87; Som. 74, 44; Wrt. Voc. 50, 26. [*Orm.* ʒarrking.]

gearcung-dæg, es; *m. A preparation-day, day before the Sabbath;* præpărātiōnis dies, parascēve = παρασκευή, dies azýmōrum :—On ðam forman gearcungdæge *prīma die azýmōrum,* Mt. Bos. 26, 17.

geár-cyning, es; *m. A year-king, consul;* consul, Cot. 48. v. consul.

geár-cyningdōm, es; *m. A year-kingdom, a consulate;* consŭlātus, Som. Ben. Lye.

GEARD, es; *m. An inclosure, inclosed place,* YARD, GARDEN, *court, dwelling, home, region, land;* septum, lŏcus septus, hortus, ārea, habĭtācŭlum, domĭcĭlium, rĕgio :—Se Godes cwide is weorþmynda geard *the word of God is the garden of worship,* Salm. Kmbl. 168; Sal. 83. On gearde deáþes sceade *in rĕgiōne umbræ mortis,* Mt. Bos. 4, 16. Ðæt ǽlc cóme to his ágenum gearde *that each should come to his own land,* Ors. 5, 14; Bos. 114, 18. On geard *at home,* Menol. Fox 215; Men. 109. In ēcne geard *into the eternal home,* Exon. 44 a; Th. 149, 17; Gú. 763 : 51 a; Th. 178, 8; Gú. 1241. Geard ymbtynde *sepem circumdedit,* Mt. Kmbl. Rush. 21, 33. Brâde synd on worulde grēne geardas *in the world are broad green regions,* Cd. 25; Th. 32, 30; Gen. 511. Ǽr he on weg hwurfe of geardum *ere he went away from his courts,* Beo. Th. 535; B. 265 : Exon. 64 a; Th. 236, 23; Ph. 578. In geardum *at home,* Exon. 16 b; Th. 13, 11; Cri. 201 : 50 b; Th. 175, 13; Gú. 1194 : 61 a; Th. 223, 5; Ph. 355 : Beo. Th. 25; B. 13. Wit forlēton on heofonrîce gōdlíce geardas *we two have lost in the heavenly kingdom goodly courts,* Cd. 35; Th. 46, 6; Gen. 740 : Beo. Th. 2272; B. 1134. On Fæder

geardas *in the dwellings of the Father,* Salm. Kmbl. 832; Sal. 415 : Exon. 105 b; Th. 401, 7; Rä. 21, 8. [*Wyc.* ʒerd *a field, garden : Piers P.* yerd *habitation : Chauc.* yerde : *O. Sax.* gard, *m* : *O. Frs.* garda, *m* : *Dut. Kil.* gærde, gærd *hortus : Ger.* garten, *m* : *M. H. Ger.* garte, *m* : *O. H. Ger.* garto, gart, *m. hortus, dŏmus : Goth.* gards, *m. house : Dan.* gaard, *m. f* : *Swed.* gård, *m* : *Icel.* garðr, *m* : *Lat.* hortus, *m* : *Grk.* χόρτος, *m. an inclosed place, feeding-place : Slav.* grad, gorod *a fence.*] DER. eador-geard, eard-, fæder-, friþ-, leód-, middan-, ort-, wîn-, wyrm-, wyrt-.

geard, e; *f. A staff, rod, stake, fagot;* băcŭlum, virga, pālus, fascis :— He scolde gifan [MS. gife] sex fôður gearda *he should give six loads of fagots,* Chr. 852; Erl. 67, 38. DER. cyne-geard. v. gyrd.

geár-dagas; *pl. m.* [geár, dæg] YORE-DAYS, *days of yore, days of years, time of life;* dies antîqui, annōrum dies :—In [on] geardagum *in days of yore,* Exon. 11 b; Th. 16, 11; Cri. 251 : 77 a; Th. 289, 6; Wand. 44 : Cd. 21; Th. 287, 16; Sat. 368 : Beo. Th. 2; B. 1 : 2712; B. 1354 : 4458; B. 2233. In geárdagan, Menol. Fox 231; Men. 117. Úre geárdagas *dies annōrum nostrōrum,* Ps. Th. 89, 10. Scyle gumena gehwylc on his geárdagum georne biþencan *every man should in the days of his years well consider,* Exon. 19 b; Th. 51, 26; Cri. 822 : 61 a; Th. 225, 4; Ph. 384 : Elen. Grm. 1267 : L. Eth. vii. 24 : Th. i. 334, 21. [*Icel.* í árdaga *in days of yore.* Cf. Gen. 47, 9, 'The days of the years of my pilgrimage are an hundred and thirty years.']

geár-dagum; *adv.* [*dat. pl.* of geárdæg, *nom. pl.* -dagas] *In days of yore, formerly;* ōlim, antiquĭtus :—Hie gesetton ðâ Sennar geárdagum *then they occupied Shinar in days of old,* Cd. 80; Th. 99, 36; Gen. 1657 : Exon. 16 a; Th. 35, 17; Cri. 559 : Andr. Kmbl. 3036; An. 1521 : Elen. Grm. 291 : 834.

geardlíc; *adj. Worldly, mundane;* mundiālis, mundānus, Som. Ben. Lye.

geare; *pl. f. Furniture,* GEAR *for horses;* appărātus :—Geara feng *the grasp of the gear, the bit;* harpax vel lŭpus, Ælfc. Gl. 3; Som. 55, 69; Wrt. Voc. 16, 42 : 105; Som. 78, 32; Wrt. Voc. 57, 14. v. gearwe; *pl. f.*

geare, gearwe, gearuwe, gearewe, gere; *adv.* [gearo? *ready*] *Entirely, clearly, certainly, well, very well, enough;* pĕnĭtus, plăne, certe, bĕne, valde, optĭme, sătis :—Ic wât geare *I well know,* Beo. Th. 5306; B. 2656 : Bt. Met. Fox 20, 188; Met. 20, 94. Ic cann swâ geare *I so well know,* Cd. 27; Th. 37, 1; Gen. 583. Nû gē geare cunnon *now ye well know,* Exon. 16 a; Th. 36, 9; Cri. 573. Hî wiston geare *certi sunt,* Lk. Bos. 20, 6. Swîðe geare, Ps. Th. 101, 5. Gearor, *comp.* Ors. 5, 14; Bos. 114, 11. [*O. Sax.* garo *quite, entirely : O. H. Ger.* garo, garawo *penitus, prorsus : Ger.* gar : *Icel.* görva, gerva *quite.*]

geáre; *adv. Formerly, of old;* ōlim :—Geáre ic ðæt ongeat *jam ōlim intellexĕram,* Bd. 2, 13; S. 516, 29. DER. geó-geáre. v. geára; *adv.*

gearewe; *adv. Entirely, well, very well;* pĕnĭtus, prorsus, bĕne, optĭme, Ps. Th. 55, 4, 11 : 68, 3 : 118, 118. v. geare; *adv.*

gearewe, an; *f. Yarrow;* millefŏlium, Glos. Brux. Recd. 41, 45; Wrt. Voc. 67, 60. [*O. Sax.* gare : *O. H. Ger.* garawa *millefolium : Ger.* schaf-garbe *common yarrow;* ʒarow, Wrt. Voc.] v. gearewe.

ge-arfoþ, es; *n. Trouble;* molestia :—He sceal geþolian manige gearfoðu *he shall suffer many troubles,* Bt. 31, 1; Fox 110, 26. DER. earfoþ, es; *n.*

ge-arfoðe; *adj. Difficult;* diffĭcĭlis, molestus :—Hû gearfoðe ðis is to gereccanne ! *how difficult this is to explain !* Bt. 39, 4; Fox 216, 33. DER. earfeðe; *adj.*

geár-gemearc, es; *n. A year's limit* or *space;* anni defīnītio *vel* spătium :—Siððan ic ongon on ðone ânseld búgan geárgemearces *after I had dwelt in the hermitage for a year's space,* Exon. 50 b; Th. 176, 24; Gú. 1215.

geár-geriht, es; *n. A yearly due;* annuum dēbĭtum :—Gif preóst geárgerihta unmynegode lǽte, gebēte ðæt *if a priest let the yearly dues pass unreminded, let him make amends for it,* L. N. P. L. 43; Th. ii. 296, 15.

geár-gerím, es; *n. A year-number, number of years, numbering by years;* annōrum nŭmĕrus :—Ymb þritig geárgerímes *after thirty, numbering by years,* i. e. *after thirty years,* Bt. Met. Fox 28, 59; Met. 28, 30. v. geár-rím.

geár-getal *a tale of years, number of years.* [Cf. *O. Sax.* gér-tal : *O. H. Ger.* jár-zala *a full year.*] v. gǽr-getal.

ge-árian; *p.* ode; *pp.* od. v. trans. *with the dat.* I. [ár I. *honour*] *To give honour, to honour;* honōrāre, honŏrĭfĭcāre :—Onsegdnis lofes geáraþ mec *sacrĭfĭcium laudis honŏrĭfĭcābit me,* Ps. Surt. 49, 23. Hý beóþ geárode and uppahefene *honōrāti et exaltāti fuĕrint,* Ps. Th. 36, 19. II. [ár II. *kindness, favour, mercy*] *To have mercy or compassion upon any one, be merciful to, pity, pardon;* propĭtium esse, misĕrēri, parcĕre :—Þolige he landes and lífes, búton him se cyning geárian wylle *let him forfeit land and life, unless the king will be merciful to him,* L. C. E. 2; Th. i. 358, 21 : L. C. L. 60; Th. i. 408, 15 : L. Eth. vii. 16; Th. i. 332, 18. Geára me, ēce Waldend ! *have compassion upon me, eternal Ruler !* Hy. 1, 2; Hy. Grn. ii. 280, 2. Ðæt se Dēma us geárige *that the Judge may have compassion on us,* Homl. Th. ii. 126, 13. Wæs Abrahame leófre ðæt he Godes hǽse gefylde, ðonne he his leófan bearne

geárode *it was dearèr to Abraham to fulfil God's command, than to have compassion on his beloved son,* Boutr. Scrd. 23, 5 : Ps. Th. arg. 34. III. [ár III. *property*] *To endow* :—Ðurh ðone tocyme we wǽron geweorðode and gewelgade and geárode *through that advent we were honoured and enriched and endowed,* Blickl. Homl. 105, 24.

geárlíc; adj. *Yearly, annual*; annuus :—Ðes geárlíca ymryne *this yearly course,* Homl. Th. ii. 98, 23. Ge ðæs libbendes yrfes, ge ðæs geárlíces westmes *both of live stock and of yearly fruit,* L. Ath. i. prim; Th. i. 194, 17. Geárlícne tíman *annuum tempus,* Hymn. Surt. 106, 33. Geárlíc wuldor *annuam glóriam,* 79, 34. Geárlíce tída gesette wǽron *the yearly seasons were fixed,* Bd. de nat. rerum; Wrt. popl. science 7, 25; Lchdm. iii. 246, 23.

geárlíce; adv. *Yearly, from year to year*; annuātim, Cot.

geár-mǽlum; adv. [mǽlum, *dat. pl.* of mǽl, es; n.] *Yearly*; quotannis :—Ríce geármǽlum weóx *the kingdom increased year by year,* Bt. Met. Fox 1, 10; Met. 1, 5.

GEARN, gern; es; n. YARN, *spun wool*; pensum, lāna nēta :—Gearn *pensum,* stāmen, lāna, Cot. 85. Unwunden gearn *unwound yarn, a ball* or *clew of yarn*; glōmus, Ælfc. Gl. 111; Som. 79, 67; Wrt. Voc. 59, 36. [*Dut.* garen, *n. thread, yarn* : *Ger.* M.H.Ger. O.H.Ger. garn, *n. filāmen* : *Dan. Swed.* garn, *n* : *Icel.* garn, *n.*] DER. nett-gern.

gearnfull; adj. *Anxious*; sollicītus :—Gearnfulle *sollicīti,* Lk. Skt. Lind. 12, 11. Gearnfull *austerus,* 19, 22. v. geornful.

ge-arnian; p. ode; pp. od [earnian *to earn*] *To earn, merit*; měrēri :—Sceal mon lofes [MS. leofes] gearnian *a man shall merit praise,* Exon. 91 a; Th. 342, 9; Gn. Ex. 140. v. ge-earnian.

ge-arnung, e; f. [earnung *an earning*] *Merit, reward*; měrītum :—Nǽnig efenlíc ðam, ǽr ne siððan, in worlde gewearþ, wífes gearnung *no woman's reward in the world was equal to that, before nor after,* Exon. 8 b; Th. 3, 23; Cri. 40. v. ge-earnung.

gearn-winde, gern-winde, es; m? [windan *to wind*] *A yarn-winder, reel*; rhombus = ῥόμβος :—Gearn-winde *conductum,* Wrt. Voc. 66, 19.

GEARO, gearu; *gen. m. n.* -wes, -owes; f. -re, -rwe; *def.* se gearwa; adj. YARE, *ready, prepared, equipped, complete*; promptus, pārātus, instructus, perfectus :— Gearo wyrde on gespræce *factus est lŏquēla promptus,* Bd. 5, 2; S. 615, 29. Gearo is mín heorte *pārātum est cor meum,* Ps. Th. 56, 9. Gearo ic eom *pārātus sum,* 118, 60 : Ps. Spl. 16, 13 : 107, 1. Wes tú gearo *pārātus esto,* Bd. 5, 19; S. 640, 44. He wæs gearo gúþe *he was ready for war,* Andr. Kmbl. 467; An. 234. Ic beó gearo sóna *I shall be ready at once,* Beo. Th. 3655; B. 1825 : 6202; B. 3106. Ða wæs gearo gyrnwræce Grendeles mŏdor *then was Grendel's mother ready with vengeance for wrongs,* 4242; B. 2118. Swā gearwe swā seó leó sícut leo pārātus, Ps. Th. 16, 11. Óþ-ðæt he Adam gearone funde *until he found Adam ready,* Cd. 23; Th. 29, 25; Gen. 455: Bt. Met. Fox 7, 67; Met. 7, 34. Gearwe, *acc. s. f.* Beo. Th. 2017; B. 1006: Exon. 45 b; Th. 155, 17; Gū. 861. Ðæt hý grim helle fýr gearo to wíte seóþ *that they shall see hell's grim fire ready for punishment,* 26 b; Th. 78, 7; Cri. 1270. Beornas gearwe on stefn stigon *the warriors ready [or equipped] stept on the prow,* Beo. Th. 428; B. 211: Ps. Th. 124, 5 : 141, 4. Ealle þing synt gearwe *omnia sunt pārāta,* Mt. Bos. 22, 4. Ða flotan stŏdon gearowe wícinga fela *the pirates stood ready, many Vikings,* Byrht. Th. 133, 59; By. 72 : 134, 47; By. 100. Searwum gearwe *equipped with arms,* Beo. Th. 3631; B. 1813. Geseah Metod geofonhúsa mǽst gearo hlífigean *the Creator saw the greatest of sea-houses arise complete,* Cd. 66; Th. 79, 35; Gen. 1321. Geofum biþ gearora *with gifts is more prepared,* Exon. 128 b; Th. 493, 15; Rä. 81, 31. [*Chauc.* yare : *R. Glouc.* ȝare : *Laym.* ȝaru, ȝaru : *O. Sax.* garu : *Ger.* gar *ready* : *M. H. Ger.* gar, gare : *O. H. Ger.* garo, garaw.] DER. ánwíg-gearo, eal-, un-.

gearo, gearu; adv. *Promptly, readily, entirely, altogether*; prompte, omnīno, prorsus :—Ðæt ic goldǽht gearo sceáwige *that I may promptly behold the gold-treasure,* Beo. Th. 5490; B. 2748. Gē ða fægran gesceaft gearo forsēgon *ye utterly despised the fair creation,* Exon. 41 b; Th. 139, 33; Gū. 602 : 17 b; Th. 7, 31; Cri. 109. Se mec gearo [or geáro; see next word] on bende legde *he who altogether laid me in bonds,* 105 b; Th. 402, 14; Rä. 21, 29. v. geare; adv.

geáro; adv. *Of yore, formerly, of old*; ōlim :—Be ðam wealle, ðe geáro Rōmāne Breotone eálond begyrdon *juxta mūrum, quo ōlim Rōmāni Brittaniam insŭlam præcinxēre,* Bd. 3, 22; S. 552, 30. v. geára.

gearo-brygd, e; f. [bregdan *to vibrate*] *A prompt vibration*; prompta pulsātio :—Ah he gleóbeámes gearobrygda list *he has skill in prompt vibrations of the harp,* Exon. 79 a; Th. 296, 13; Crä. 50.

gearod *clothed, endowed,* Bt. 14, 3; Fox 46, note 7, MS. Cott. = gearwod; *pp.* of gearwian.

gearo-folm; adj. [folm *a hand*] *Ready-handed*; promptus mănu :—He grápode gearofolm *he ready-handed grasped [me],* Beo. Th. 4176; B. 2085.

gearo-gongende *going quickly* or *swiftly.* v. gearu-gongende.

gearolíce; adv. *Readily, clearly*; prompte, plāne :—Ic ðæt gearolíce ongiten hæbbe *I have clearly understood that,* Elen. Kmbl. 575; El. 288:

Exon. 100 a; Th. 378, 2; Deór. 10. [*O. Sax.* garolíko : *O. H. Ger.* garalíhho.]

gearo-snotor, -snottor, gearu-snottor; adj. *Very wise*; valde săpiens :—Gidda gearosnotor *very wise in songs,* Elen. Kmbl. 835; El. 418. Giedda gearosnottor, Exon. 18 a; Th. 45, 2; Cri. 713.

gearo-þoncol; adj. *Very considerate* or *prudent*; valde considĕrātus vel providus :—Hí ðæt idese ageáfon gearoþoncolre *they gave it to the very prudent woman,* Judth. 12; Thw. 26, 23; Jud. 342.

gearowe *prepared, ready,* Jud. 4, 13 ; *dat. s. f.* of gearo.

gearo-wita, an; m. *Intellect, understanding*; intelligentia, intellectus :—Ðeáh we fela smeán, we habbaþ litellne gearowitan búton tweón *though we contemplate many things, we have little understanding free from doubt,* Bt. 41, 5; Fox 254, 10 : 39, 8; Fox 224, 4.

gearo-wyrdig, gearu-wyrdig; adj. *Ready in words, speaking with ease* or *fluency, eloquent*; verbis promptus, fācundus :—Se wítga song, gearowyrdig guma ðæt gyd awræc *the prophet sang, the eloquent man recited the lay,* Exon. 84 a; Th. 316, 19; Mōd. 51.

geár-rím, es; n. *A year-number, a year* [?], *number of years*; annōrum nŭmĕrus :—Seó tíd gegǽþ, geár-rínum, ðæt ða geongan leomu geloden weorþaþ *the time passes, in a number of years* [or *by years*], *that the young limbs be grown,* Exon. 87 a; Th. 327, 17; Vy. 5. [Cf. *O. Sax.* gēr-tal *a year.*]

geár-þenung, e; f. *A yearly service, annual service*; annuum ministĕrium :—Gif preóst misendebirde ciriclíce geárþenunga, dæges oððe nihtes, gebēte ðæt *if a priest misorder the annual services of the church, by day or by night, let him make amends for it,* L. N. P. L. 38; Th. ii. 296, 7.

geár-torht; adj. *Yearly bright, every year glorious*; quotannis splendidus :—Ðá him wæstmas brohte, geártorhte gife, grēne folde *when the green earth should bring fruits to him, yearly-bright gifts,* Cd. 76; Th. 94, 13; Gen. 1561.

gearu; adj. *Yare, ready, prepared*; promptus, pārātus, Beo. Th. 2223; B. 1109 : Cd. 178; Th. 223, 32; Dan. 128 : Ps. Th. 61, 2, 7: Andr. Kmbl. 2716; An. 1360: 3157; An. 1581: Jn. Bos. 7, 6: Ps. Th. 107, 1: Elen. Grm. 604. v. gearo; adj.

gearu-gongende; part. *Going quickly* or *swiftly*; expĕdīte incēdens :—Ic eom to ðon bleáþ, ðæt mec mæg gearugongende gríma abrēgan *I am so timid, that a phantom going swiftly may frighten me,* Exon. 110 b; Th. 423, 6; Rä. 41, 17.

gearu-snottor; adj. *Very wise*; valde săpiens :—Hie ænne betǽhtoń giddum gearusnottorne *they gave up one very skilled in songs,* Elen. Kmbl. 1168; El. 586. v. gearo-snotor.

gearuwe *prepared, ready,* Bd. 4, 2; S. 565, 34; *acc. pl.* of gearu. v. gearo; adj.

gearuwe, an; f. *Yarrow*; millefŏlium :—Seó reáde gearuwe *the red yarrow,* Lchdm. iii. 24, 2. v. gearwe.

gearuwe; adv. *Entirely, well, very well*; pĕnītus, prorsus, bĕne, optime, Ps. Th. 53, 2 : 61; 11 : 62, 1 : 70, 1 : 118, 21 : 138, 11 : 139, 12. v. geare; adv.

gearu-wyrdig; adj. *Ready in words, eloquent*; verbis promptus :—Sum biþ gearu-wyrdig *one is eloquent,* Exon. 78 b; Th. 295, 21; Crä. 36. v. gearo-wyrdig.

gearwa *prepared*; pārātus; *nom. m. def.* of gearo; adj.

gearwe; comp. gearwor; sup. gearwost, gearwast; adv. *Entirely, well, very well, enough*; pĕnītus, prorsus, bĕne, optime, sătis, Cd. 52; Th. 67, 10; Gen. 1098 : 107; Th. 141, 10; Gen. 2342 : Beo. Th. 528; B. 265: Exon. 48 a; Th. 164, 28; Gū. 1018 : Bd. 5, 6; S. 618, 30: Ps. Th. 142, 9. Gearwor, Andr. Kmbl. 1864; An. 934: Exon. 73 b; Th. 275, 27; Jul. 556: Beo. Th. 6141; Elen. Grm. 945. Gearwost, Beo. Th. 1435; B. 715. Gearwast, Elen. Grm. 329. v. geare.

gearwe *prepared*; pārāta :—Ealle míne þing synt gearwe *omnia pārāta sunt,* Mt. Bos. 22, 4; *nom. pl. n.* of gearo; adj.

gearwe, an; f. *Clothing, attire*; vestītus, hăbĭtus :—Ic on his gearwan geseó ðæt he is ǽrendsecge uncres Hearran *I see by his attire that he is the messenger of our Lord,* Cd. 30; Th. 41, 16; Gen. 657. v. gearwe; *pl. f.*

gearwe; *pl. f. Clothing, attire,* GEAR, *adornment, arms, armour*; vestītus, hăbĭtus, arma :—Enoch cwic gewāt mid Cyning engla of ðyssum lǽnan lífe, on ðám gearwum ðe his gást onfēng, ǽr hine to monnum mŏdor brohte *Enoch alive departed with the King of angels from this frail life, in the vestment which his soul received, ere his mother brought him amongst men,* Cd. 60; Th. 73, 29; Gen. 1212 : Menol. Fox 150; Men. 76. Óþ-ðæt on Gúþmyrce gearwe bǽron *till they bore their arms against the Æthiopians,* 145; Th. 181, 11; Exod. 92: 151; Th. 190, 3; Exod. 193. [*O. Sax.* garuwi, f : *O. H. Ger.* garawi, f.] DER. feðer-gearwe.

gearwe, gearuwe, gearewe, gæruwe, garuwe, an; f. YARROW; millefŏlium, achillæa millefŏlium, Lin :—Ðas wyrte man *millefŏlium* and on úre geþeóde gearwe nemneþ *this plant is named* millefŏlium *and in our language yarrow,* Herb. 90, 1; Lchdm. i. 194, 6: Wrt. Voc. 79, 23. Wylle gearwan on buteran *boil yarrow in butter,* L. M. 1, 60; Lchdm. ii. 130, 22 : 2, 56; Lchm. ii. 276, 19: 3, 30; Lchdm. ii. 324, 25. Wyl on

meolcum ða reádan gearwan *boil in milk the red yarrow*, L. M. 3, 65; Lchm. ii. 354, 9. v. gearewe.

ge-árweorþian, -árwurþian; *p.* ode, ede; *pp.* od, ed *To honour*; honorificāre:—Me swíðe geárweorþede syndon freónd ðíne *mihi nimis honorificāti sunt amīci tui*, Ps. Lamb. 138, 17.

gearwian, gerwian, geiwan, girwan, gierwan, gyrwan, gyrian, girian, gierian; *p.* ode, ade, ede; *pp.* od, ad, ed *To make ready, prepare, procure, supply, put on, clothe;* parāre, præparāre, præstāre, induĕre, vestīre:—Ðū gæst beföran Drihtnes ansýne, his wegas gearwian *præibis ante faciem Dŏmini, parāre vias ejus,* Lk. Bos. 1, 76: Exon. 58 b; Th. 210, 21; Ph. 189: 119 a: Th. 456, 27; Hy. 4, 73: Elen. Kmbl. 1997; El. 1000. Wísdōm oððe snytro gearwiende lytlingum *săpientiam præstans parvŭlis,* Ps. Spl. 18, 8. Óþ on ēcnysse ic gearwie sǽd ðín *usque in æternum præparābo sēmen tuum,* 88, 4. He lífes weg gǽstum gearwaþ *he prepares life's way for souls,* Exon. 34 a; Th. 108, 11; Gū. 71: 117 a; Th. 450, 21; Dōm. 91. Ic gearwode leóhtfæt cyninge mínum *părāvi lucernam Christo meo,* Ps. Spl. 131, 18. Ðū gearwodest wlite mínum mægn *præstĭtisti dĕcōri meo virtūtem,* 29, 8. Grinu hí gearwodon fótum mínum *laqueum părāvĕrunt pĕdĭbus meis,* Ps. Spl. 56, 8. Sumum wun-dorgiefe þurh goldsmiþe gearwad weorþeþ *to one a wondrous skill in goldsmith's art is provided,* Exon. 88 a; Th. 331, 25; Vy. 73. Gear-wian us togēnes grēne strǽte up to englum *let us prepare before ourselves a green path to the angels above,* Cd. 219; Th. 282, 15; Sat. 287. Hū gē eówic gearwige *quid induamini,* Mt. Kmbl. Rush. 6, 25: 27, 29. Ðæt selfe wæter ðegnunge gearwode beforan his fótum *the very water did reverence before his feet,* St. And. 22, 19. [*Piers P.* gare: *R. Brun.* ʒared, *pp.* prepared: *Laym.* ʒærwen *to make ready:* O. *Sax.* garuwian, gerwean, girwian *to make ready, prepare:* O. H. *Ger.* garawēn, garwēn, garawjan.] v. Grm. D. M. 984. DER. a-gearwian, ge-.

gearwung, e; *f. A making ready, preparation;* præparātio:—Of gearwunge eardunge his *de præparāto habitācŭlo suo,* Ps. Spl. T. 32, 14. Gearwunga dæg *parasceue,* Jn. Skt. Lind. 19, 31. DER. ge-gearwung.

ge-árwurþian; *p.* ode; *pp.* od *To honour;* honorificāre:—Ðæt hí sín geárwurþode fram mannum *ut honorificentur ab hŏmĭnĭbus,* Mt. Bos. 6, 2: Ps. Lamb. 36, 20. v. ge-árweorþian.

gearwutol; *adj. Austere:*—Gearwutol *austerus,* Lk. Skt. Lind. 19, 21, 22.

ge-ascian, -acsian, -ahsian, -axian; *p.* ode, ade; *pp.* od, ad [acsian *to ask*] *To find out by asking, learn, hĕar;* fando accĭpĕre, discĕre, audīre:—Geascode he ðone cyning on Meran tūne *he learnt [that] the king [was] at Merton,* Chr. 755; Erl. 48, 28. Ðá geascade se cyng ðæt ðæt hie ūt on hergaþ fōron *then the king heard that they were gone out to ravage,* 911; Erl. 100, 24. We geascodon ðæt úre geferan sume to eów cōmon *we have heard that some of our fellows have come to you,* L. Alf. 49; Th. i. 56, 14: Exon. 100 a; Th. 378, 24; Deór. 20. Habbaþ we geascad ðæt se Ælmihtiga worhte wer and wíf *we have heard that the Almighty created man and woman,* 61 b; Th. 225, 22; Ph. 393.

ge-ascung, e; *f.* [ascung *asking*] *An asking, inquiry;* interrogātio, inquīsītio:—Būton be gemynde and be geascunga *except by memory and by inquiry,* Bt. 42; Fox 256, 25.

ge-asmirian; *p.* ode, ede; *pp.* od, ed [smyrian, smirian *to smear*] *To smear, anoint;* ungĕre, inungĕre:—Bring clǽne ofenbacene hláfas mid ele geasmirede būtan beorman *pānes scĭlicet absque fermento conspersos ŏleo,* Lev. 2, 4.

geásne; *adj. c. gen. Deprived of, void of;* expers:—He sceal gōdra gum-cysta geásne hweorfan *he shall pass away, deprived of good blessings,* Exon. 71 a; Th. 265, 15; Jul. 381. Ða sind geásne gōda gehwylces *those are void of every good,* 68 b; Th. 255, 18: Jul. 216. v. gēsne, gǽsne.

ge-asyndrod; *part. Sundered, separated;* sequestrātus, R. Ben. interl. 43. v. a-syndran.

geat, *pl.* geáton *got;* *p.* of gitan.

GEAT, gat, es; *pl. nom. acc.* u, a, o; *n. A gate, door;* porta, ostium, jānua:—Ic eom sceápa geat *ego sum ostium ŏvium,* Jn. Bos. 10, 7, 9: 10, 1, 2. Gangaþ inn þurh ðæt nearwe geat, forðonðe ðæt geat is swýðe wíd *intrāte per angustam portam, quia lāta porta est,* Mt. Bos. 7, 13, 14. Ðǽr is geat gylden *there is the golden gate,* Cd. 227; Th. 305, 19; Sat. 649. Þurh ðæs wealles geat *through the gate of the wall,* Judth. 11; Thw. 23, 32; Jud. 151: Exon. 71 b; Th. 266, 21; Jul. 401. Ðá he genéaláehte ðære ceastre gate *cum appropinquāret portæ civitātis,* Lk. Bos. 7, 12. Heó ðæt geat ðæs mynstres ontýnde *illa apĕruit jānuam Monas-tērii,* Bd. 3, 11; S. 536, 18. Ða gyldnan geatu hát ontýnan *bid open the golden gates,* Exon. 1 b; Th. 16, 10; Cri. 251: 16 a; Th. 36, 15; Cri. 576. Opnyaþ me gatu rihtwísnysse *apĕrīte mihi portas justĭtiæ,* Ps. Spl. 117, 19: Exon. 12 b; Th. 20, 15; Cri. 318. On gaton *in portis,* Ps. Th. 126, 6. [*Piers P.* yates, *pl. gates;* gate *a way: Chauc.* yate *a gate;* gate *a street, way: Laym.* ʒat: *Orm.* ʒate *a gate;* gate *a way: Scot.* yet, yett *a gate:* O. *Sax.* gat, *n. a hole:* Frs. gat: O. *Frs.* gat, iet, *n. a hole:* Dut. gat, *n. a hole:* Ger. gasse. *f. a thoroughfare, narrow road:* M. H. *Ger.* gat, *n. a hole;* gazze, *f. a narrow road:* O. H. *Ger.* gaza, *f. vīcus, plătea:* Goth. gatwo, *f. plătea:* Dan. gat, *m. f. an aperture, opening:* Swed. gata, *f. a*

street, lane: Icel. gat, *n. a hole;* gata, *f. a way.*] DER. ben-geat, burh-, fæsten-, hord-, weall-.

Geát, es; *m. Geat,* Exon. 100 a; Th. 378, 13; Deór. 15. See Grimm D. M. 341-5.

geát *poured out,* Bd. 2, 6; S. 508, 9; *p.* of geótan.

GEÁTAN, gǽtan, gētan; *p.* de te; *pp.* ed *To grant, confirm, assent to;* concēdĕre, confirmāre, assentīri:—Ic geáte ðē *I grant to thee,* Chr. 656; Th. 53, 38: 675; Th. 59, 33. Ic Ædgár geáte and gife tó. dæi *I Edgar grant and give to-day,* 963; Th. 220, 33. Se æðeling him geátte *the ætheling granted it to them,* 1066; Th. 337, 30. Ealle hit geátton *all confirmed it,* 963; Th. 221, 25. [*Laym.* ʒetten *to grant:* Orm. ʒatenn *to grant, allow:* O. *Frs.* gēta, gáta *confirmāre: Icel.* játa, játta *to say 'yes,' assent.*] v. geá.

GEÁTAS, Iótas, Iútas, Eótenas [v. eóten, II.]; *gen.* a; *dat.* um; *pl. m.* I. *the Jutes, the ancient inhabitants of Jutland, who, with the Angles and Saxons, colonized Britain;* Jutæ, pŏpŭlus Chersŏnēsi Cymbrĭcæ, qui relicta patria ūna cum Saxŏnĭbus Anglisque Britanniam occupāvērunt. Though the Jutes are now regarded as Danes, they were, in the earliest times, distinguished as a separate people, and were probably the descendants of earlier Gothic settlers in Jutland, while the Danes = Dene, were an invading nation. Thus Hengest was a Jute, and Healfdene, his lord, a Dane. The Eótenas = Jötnar, were apparently a still earlier Finnish race, from whom the Gothic conquerors probably derived their trolls and giants. Both Jóti, *pl.* Jótar, and iötunn, *pl.* iötnar, are rendered in A. *Sax.* by eóten; *pl.* eótenas. From the Ynglinga-Saga, c. 5, we learn that before the time of Skiold, the seat of the Danish kings was in Reitgothland = Jutland, but Skiold transferred it to Lethra in Seeland, of which he was the founder:—Cōmon hí of þrim folcum ðǽm strangestan Germanie, ðæt [is,] of Seaxum, and of Angle, and of Geátum. Of Geáta fruman syndon Cantware, and Wihtsǽtan, ðæt is seó þeód ðe Wiht ðæt Eálond oneardaþ . . . And of Engle cōman East-Engle and Middel-Engle, and Myrce, and eall Norþhembra cynn, is ðæt land ðe Angulus is nemned betwyh Geátum and Seaxum *advēnērant autem de trĭbus Germānĭæ pŏpŭlis fortiōrĭbus, id est, Saxŏnĭbus, Anglis, Jutis. De Jutārum orīgĭne sunt Cantuārii et Victuārii, hoc est, ea gens, quæ Vectam tĕnet Insŭlam . . . De Anglis vēnēre Orientāles Angli, Mediterrānei Angli, Merci, [et] Nordanhymbrōrum prōgĕnies, id est, de illa patria quæ Angŭlus dīcĭtur inter provincias Jutārum et Saxŏnum,* Bd. 1, 15; S. 483, 20-26. II. *the GAUTS, the inhabitants of the south of Sweden,* which in ancient times comprehended nearly the whole of South-Sweden = A. *Sax.* Geát-land, *Icel.* Gautland *the land of the Gauts,* which must be distinguished from *Icel.* Gotar, and A. *Sax.* Gotland *the land of the Goths,* q. v; Gauti in Suecia = Γαυτοί, Procopius Bell. Goth. 2, 15:—We synt gumcynnes Geáta leóde *we are of the race of the Gauts' nation,* Beo. Th. 526; B. 260: 730; B. 362. Ic wæs mid Hrēþ-Gotum, mid Sweóm and mid Geátum, and mid Sūþ-Denum *I was with the Hreth-Goths, with the Swedes, and with the Gauts, and with the South-Danes,* Exon. 85 b; Th. 322, 4; Wid. 58: Beo. Th. 392; B. 195: 2347; B. 1171: 4391; B. 2192. Beó wid Geátas glæd *be cheerful towards the Gauts,* Beo. Th. 2350; B. 1173. DER. Gūþ-Geátas, Sǽ-, Weder-. See Grimm Geschichte d. D. S. pp. 512, 312.

ge-atelod; *part.* [ge, atol, atel *dire, terrible*] *Misshapen, deformed, hideous;* deformis, deformātus:—Geatelod *deformis,* Cot. 66: *deformā-tus,* 202.

geáþ, e; *f. Foolishness, lightmindedness, luxury, mockery;* stultĭtia, lascīvia, luxūria, ludibrium:—Ðú, on geáþe, hafast ofer witena dōm wísan gefongen *thou, in foolishness, hast taken thy course against wise men's judgment,* Exon. 67 a; Th. 248, 16; Jul. 96. Þeódum ýwaþ wísdōm weras, siððan geóguþe geáþ gǽst aflíhþ *men manifest wisdom to people, when the spirit puts to flight the lightmindedness of youth,* 40 a; Th. 132, 19; Gū. 475. Ðý-læs ðæt wundredan weras and idesa, and on geáþ gutan *lest men and women should wonder thereat, and pour it forth in mockery,* 50 b; Th. 176, 8; Gū. 1206. [Gĕác a cuckoo: *Icel.* gaúð, *f. a barking.*]

geatolic; *adj. Ready, prepared, equipped, stately;* părātus, instructus, ornātus:—Ðǽr wæs on eorle geatolíc gūþscrūd *there was on the man a prepared war-dress,* Elen. Kmbl. 515; El. 258: Beo. Th. 435; B. 215: 4314; B. 2154. Wísa fengel geatolíc gengde *the wise prince went stately,* 2806; B. 1401.

geat-torr, es; *m. A gate-tower;* portam hăbens turris:—Sind geat-torras berofen *the gate-towers are despoiled,* Exon. 124 a; Th. 476, 7; Ruin. 4.

geatwan; *p.* ede; *pp.* ed *To make ready, equip, adorn;* părāre, or-nāre:—Frætwed, geatwed *adorned, equipped,* Exon. 107 b; Th. 411, 1; Rä. 29, 6.

geatwe; *gen.* a; *dat.* um; *acc.* a; *pl. f. Arms, trappings, garments, ornaments;* armāmenta, vestimenta ornāmenta:—Twegen englas gescel-dode and gesperode and mid heora geatwum gegyrede, efne swá hie to campe fēran woldon *two angels with shields and spears and with their equipments, just as if they meant to go to battle,* Blickl. Homl. 221, 28. Freólíce in geatwum [MS. geotwum] *in trappings goodly,* Chr. 1066;

B b

Th. 334, 35, col. 1; Edw. 22. Geatwum *with ornaments*, Exon. 109 a; Th. 417, 26; Ra. 36, 10. Ic geondseah recedes geatwa *I looked over the ornaments of the house*, Beo. 6167; B. 3087. DER. eóred-geatwe, fyrd-, gryre-, gúþ-, here-, hilde-. v. ge-tawe.

geat-weard, es; *m. A gate-ward, door-keeper, porter*; ostiārius:—Ðæne se geatweard lǽt in *huic ostiārius apĕrit*, Jn. Bos. 10, 3. Geatweard *januārius*, Wrt. Voc. 81, 16.

ge-aurnen; *part.* [aurnen *run out, pp.* of a-yrnan] *Over-run, overtaken*; cursu apprehensus, Som. Ben. Lye.

ge-aworpen; *part.* [ge, and *pp.* of a-weorpan *to throw away*] *Cast or thrown away*; abjectus, Som. Ben. Lye.

ge-axian; *p.* ode; *pp.* od [acsian *to ask*] *To find out by asking, learn, hear*; exquīrĕre, resciscĕre, audīre:—Swā hwā swā ðæt geaxaþ, he hlihþ eác mid me *quicumque audiĕrit, corrīdēbit mihi*, Gen. 21, 6. Æfter ðǽre tíde ðe he geaxode fram ðǽm tungelwītegum *secundum tempus exquisiĕrat a magis*, Mt. Bos. 2, 16. Geaxodon ða cynegas *audiērunt rēges*, Jos. 5, 1: L. Alf. 49; Th. i. 56, 14, MS. H. Geaxode dómas *responsa*, Ælfc. Gl. 14; Som. 57, 131; Wrt. Voc. 20, 68. v. ge-ascian, ge-acsian.

ge-bacen; *part.* BAKED; coctus:—Gesoden, gebacen *coctus*, Ælfc. Gl. 31; Som. 61, 86; Wrt. Voc. 27, 16; 82, 71. DER. bacan; *p.* bóc, *pl.* bócon; *pp.* bacen *to bake*.

ge-bád *abode, dwelt, remained*, Jn. Bos. 8, 9; *p.* of ge-bídan.

ge-bæc, es; *n.* [bacan *to bake*] *Anything baked*; quod est tostum:—Ic geseah swefen, ðæt ic hæfde þrí windlas mid meluwe ofer mín heáfod, and on ðam ufemystan windle wǽre manegra cynna gebæc *ego vīdi somnium, quod tria canistra fărīnæ habērem super căput meum, et in ūno canistro, quod ĕrat excelsius, portāre me omnes cĭbos, qui fĭunt arte pistōria*, Gen. 40, 17.

ge-bæcu; *pl. n. Back parts, hinder parts*; postĕriōra:—Synd gebæcu hire hrycges on blācunge goldes *sunt postĕriōra dorsi ejus in pallōre auri*, Ps. Lamb. 67, 14. He slóh heora fýnd on gebæcum *percussit inĭmīci suos in postĕriōra*, 77, 66. v. bæc.

ge-bæd *prayed*, Ps. Th. 108, 3; *p.* of ge-biddan.

ge-bǽdan; *p.* -bǽdde; *pp.* -bǽded [bǽdan *to compel*] *To compel, constrain, force, impel, urge, oppress*; compellĕre, cōgĕre, persuādēre, impellĕre, urgēre, prĕmĕre:—Mid rihtre nýdþearfnysse gebǽded *justa necessĭtāte compulsus*, Bd. 2, 2; S. 502, 27. Mid nýde gebǽded *necessĭtāte cōgente*, 3, 24; S. 556, 7: Exon. 70 b; Th. 263, 2; Jul. 343: Bt. Met. Fox 6, 28; Met. 6, 14. Nīþa gebǽded *constrained by hatred*, Exon. 68 b; Th. 254, 27; Jul. 203. Mon sceal gebídan ðæs he gebǽdan ne mæg *a man ought to wait for what he cannot hasten* [*compel to come*], 90 b; Th. 340, 2; Gn. Ex. 105. Hie gecwǽdon ðæt ne hie to ðam gebéde he mihte gebǽdan *they said that he could not force them to that prayer*, Cd. 182; Th. 228, 15; Dan. 202. Strǽla storm strengum gebǽded, scóc ofer scyld-weall *a storm of shafts, impelled from strings, rushed over the shield-wall*, Beo. Th. 6226; B. 3117. Býsigum gebǽded *oppressed with labour*, 5153; B. 2580; 5644; B. 2826. [*Goth.* gabaidjan.]

ge-bǽlded; *part.* [ge-, *pp.* of bǽldan *to animate*] *Made bold, animated*; anĭmātus:—Wæs Laurentius mid ðæs apostoles swingum and trymenessum swíðe gebǽlded *apostŏli flagellis sĭmul et exhortātiōnibus anĭmātus ĕrat Laurentius*, Bd. 2, 6; Wilk. 124, 7.

ge-bǽndan; *p.* de; *pp.* ed [ge, and bænd a *band*] *To bind*; vincīre:—Ic hine gebǽndan hēt *I commanded* [*them*] *to bind him*, Salm. Kmbl. 551; Sal. 275.

ge-bǽr *bare, bore*, Gen. 39, 19; *p.* of ge-beran *to bear, bring forth*.

ge-bǽran; *p.* de; *pp.* ed [ge-, *and* bǽru *bearing, habit*] *To bear one's self, behave* or *conduct one's self*; se gerere:—Ne gefrægn ic ða mǽgþe sēl gebǽran *never have I heard of the tribe bearing themselves better*, Beo. Th. 2029; B. 1012: 5640; B. 2824: Fins. Th. 77; Fin. 38. Ne scule gē wið hine gebǽran swā swā wið feónd *ye must not behave to him as to an enemy*, Past. 46, 8; Swt. 356, 7; Hat. MS. 68 a, 14. We gebǽraþ swelce we hit nyten *we behave as though we know it not*, 28, 4; Swt. 194, 4; Hat. MS. 37 a, 25. Ðæt hí gebǽrdon wel *that they should bear themselves well*, Judth. 10; Thw. 21, 20; Jud. 27: Bd. 4, 25; S. 600, 32: Ps. Th. 113, 6. [*Laym.* i-bere: *O. Sax.* gi-bārian: *O. H. Ger.* ga-baran.]

ge-bǽrd *natural quality, nature*; indŏles, Som. Ben. Lye. v. ge-byrd, II.

gebǽrd-stán, es; *m. Calcisiva?* Ælfc. Gl. 58; Som. 67, 102; Wrt. Voc. 38, 27: *forte* gebǽrn-stán *vel* gebǽrned stán *calx vīva*, Som. 67, 102.

ge-bǽrmed; *part.* [ge, and *pp.* of byrman *to ferment with barm or leaven*] *Fermented, leavened*; fermentātus:—Gebǽrmed hláf *leavened bread*; pānis fermentātus, Som. Ben. Lye. v. ge-byrman.

ge-bǽrnan; *p.* -bǽrnde; *pp.* -bǽrned [ge, *and* bærnan *to burn*] *To burn*; ūrĕre:—Ne ðǽ sunne on dæge gebǽrne *per diem sol non ūret te*, Ps. Th. 120, 6.

gebǽrn-lím *quicklime*; calx vīva, Som. Ben. Lye.

gebǽr-scipe, es; *n. A feast*, Lk. Skt. Lind. 14, 13. v. gebeór-scipe.

ge-bǽru, *gen.* e; *acc.* e, u; *f.* ge-bǽro; *f.* indecl. Or ge-bǽre; *n.*; *pl.* u. See the cognate words at the end. [baero, bǽru a *bearing*] BEARING, *state, habit* or *disposition of body* or *mind, manner, conduct, behaviour, demeanour, manners in society, society*; gestus, hăbĭtus, mōres,

consortium, consuētūdo:—Biþ swā fæger fugles gebǽru *the bird's bearing* [*demeanour*] *is so pleasing*, Exon. 57 b; Th. 206, 12; Ph. 125. We on gewritu setton þeóda gebǽru *we have set in writing the conduct of the people*, Elen. Kmbl. 1314; El. 659. Gehýrde beornes gebǽro *she heard of the conduct of the man*, 1416; El. 710. Ðæt he sceáwode monna gebǽru *that he might behold men's behaviour*, Exon. 38 b; Th. 127, 17; Gú. 387: Ors. 4, 10; Bos. 92, 37. Swylce habban sceal blíðe gebǽro *shall such have a blithe demeanour?* Exon. 115 b; Th. 444, 8; Kl. 44: 115 a; Th. 442, 31; Kl. 21. On gebǽrum *ex hăbĭtu ejus*, Bd. 4, 22; S. 591, 33: Ps. Th. 34, 15. He swíðor lufade wífa gebǽra, ðonne wǽpnedmanna *he loved the society of women more than of men*, Ors. 1, 12; Bos. 35, 16. On ðæs wifes gebǽrum onfundon ðæs cyninges ðegnas ða unstilnesse *by the woman's cries* [?] *the king's thanes discovered the disturbance*, Chr. 755; Erl. 100, 2. Cf. *Laym.* wide me mihte iheren Brutten iberen, iii. 125. [*O. Sax.* gi-bāri, *n*: *O. H. Ger.* ga-bāri, *n.*]

ge-bǽtan; *p.* -bǽtte; *pp.* -bǽted [ge, *and* bǽtan *to bridle*] *To bit, bridle, curb*; frēnum ĕquo *vel* ăsino injĭcĕre, frēnāre:—Ðā wæs Hróþgāre hors gebǽted *then a horse was bitted for Hrothgar*, Beo. Th. 2803; B. 1399. He gebǽtte his āgen weorc *he curbed his own work*, Bt. Met. Fox 11, 152; Met. 11, 76. Hæþþ se Alwealda ealle gesceafta gebǽt mid his bridle *the Almighty has restrained all creatures with his bridle*, Bt. Met. Fox 11, 45; Met. 11, 23.

ge-bǽte, -bǽtel, es; *n.* [ge, *and* bǽte a *bit of a bridle*] *A bit of a bridle, a bridle, trappings*; lūpātum, cāmus, frēnum:—Ðæt gebǽtel of ateáh *he took the bridle off*, Bd. 3, 9; S. 533, 34. Mid ðǽm gebǽtum *with the trappings*, Bd. 3, 14: S. 540, 22.

ge-ban, -bann, -benn, es; *n.*   **I.** *a command, ordinance, decree, proclamation*; mandātum, stătūtum, decrētum:—Brād is ðín gebann *lātum est mandātum tuum*, Ps. Th. 118, 96. Ðīne ealle gebann *omnia mandāta tua*, 118, 86. Ðīnre ǽ geban *lēgis tuæ mandātum*, 58, 10: Elen. Grm. 556. Þurh hláfordes geban *by his lord's decree*, L. Edg. H. 7; Th. i. 260, 14. Gif preóst biscopes geban forbúge *if a priest decline* [*to obey*] *the bishop's edict*, L. N. P. 4; Th. ii. 290, 20.   **II.** ge-bann, -bonn, es; *n. the indiction*; indictio, edictum. The indiction is a cycle or revolution of 15 years, like the date of the year from the Birth of our Saviour. Indiction was introduced by Augustine, through the influence of Gregory the Great. It was used by the Roman emperors in the solemn *Edictum* or *Indictio*, relative to the taxes, and adopted by the Church to denote the cycle of 15 years. The number of the Indiction was thus easily ascertained, add 3 to the year of our Lord and divide by 15, and the remainder will be the year of Indiction. If there be no remainder the Indiction will be 15. Bede, in his *De Rătiōne Tempōrum*, says plainly,—Si vis scīre quŏta sit Indictio, sūme annos Dŏmĭni, et adjĭce tria, partīre per xv, et quod remansĕrit, ipsa est Indictio anni præsentis, *Cap. xiv.* Indiction is useful in ascertaining the exact year in a reign, *etc*:—Ðam mildestan cyninge Cantwara, Wihtrǽde, ríxigendum, ðē fíftan wintra his ríces, ðý niguþan gebanne, in ðære stōwe ðy hātte Berghāmstyde, ðǽr wæs gesamnad eádigra geþeahtendlíc ymcyme *in the reign of the most mild king of the Kentish-men, Wihtræd, in the fifth year of his reign, the ninth indiction, in the place which is called Berham, where was assembled a deliberative assembly of the great men*, L. Wih. pref; Th. i. 36, 4-7. Thus, Wihtrǽd began to reign A. D. 691; this gives A. D. 696 for the deliberative assembly; add 3 by rule, the sum, 699, divided by 15, leaves 9 remainder after the division, or the year of the Indiction as in the preceding example. Ríxiendum ussum Dryhtene ðæm Hǽlendan Crist. Æfter ðon ðe agān wæs ehta hund wintra and syx and hundnigontig efter his acennednesse, and ðý feówerteóðan gebonn-gēre; ðā, ðý gēre, gebeón [*p.* of gebannan] Æðelrēd ealderman alle Mercna weotan tosomne to Gleaweceastre *under the rule of our Lord Jesus Christ. When 896 winters were passed after his birth, and in the 14th indiction-year; then, in that year, alderman Æthelred assembled all the witan of the Mercians together at Gloucester*, Th. Diplm. A. D. 896; 139, 4-13. Thus, Æthelred assembled the witan at Gloucester in the year 896; 896 + 3 = 899; this after division by 15 leaves a remainder 14, or the year of Indiction, as stated in the foregoing example. Geban *edictum*, Ælfc. Gl. 87; Som. 74, 43; Wrt. Voc. 50, 25. [*O. Sax.* ban, *n. mandātum*: *O. Frs.* ban, bon, *n*: *Dut.* ban, *m*: *Ger.* bann, *m. edictum, interdictum, proscriptio*: *M. H. Ger.* ban, *m*: *O. H. Ger.* pan, *m. scĭtum, anathēma*: *Dan.* band, *m. f*: *Swed.* bann, *n*: *Icel.* bann, *n. interdictum, excommunĭcātio, prohĭbĭtio*.]

ge-band *bound*, Gen. 22, 9; *p.* of ge-bindan.

ge-bannan, -bonnan; *p.* -beónn, *pl.* -beónnon; *pp.* -bannen [ge, *and* bannan *to summon*].   **I.** *to command, order, proclaim*; jŭbēre, mandāre, edīcĕre:—Ðā ic gefrægn weorc gebannan manigre mǽgþe *then I heard* [*him*] *command the work to many a tribe*, Beo. Th. 149; B. 74.   **II.** *to summon, call together*; cĭtāre, convŏcāre:—Folc biþ gebonnen ealle to sprǽce *all people shall be summoned to judgment*, Exon. 117 b; Th. 451, 8; Dóm. 100. Ðā gebeón Æðelrēd ealderman alle Mercna weotan tosomne *then alderman Æthelred summoned all the 'witan' of the Mercians together*, Th. Diplm. 139, 11. [*Laym.* i-bannen *to summon.*]

ge-barn *burned*, Beo. Th. 5388; B. 2697; *p. of* ge-beornan.

ge-bāsnian; *p.* ade; *pp.* ad [ge, *and* bāsnian *to expect*] *To expect*; exspectāre:—Gebāsnade rīc Godes *expectābat regnum Dei*, Lk. Skt. Lind. 23, 51.

ge-bātad, -bātod; *part. Abated*; mitīgātus, Cot. 135.

ge-beácnian, -bēcnian, -bícnian; *p.* ode; *pp.* od [ge, *and* beácnian *to beckon*] *To point out, indicate, make signs*; indīcāre, nuntiāre, innuere:—Đā him gebeácnod wæs *then it was indicated to him*, Beo. Th. 283; B. 140. We woldon mid gebeácnian đa sōþfæstnesse *we would therewith point out the truth*, Bt. 35, 5; Fox 166, 16. Gebēcnadon feder his *innuebant patri ejus*, Lk. Skt. Lind. 1, 62. [*O. Sax.* gi-bōknian *to shew, indicate*: *O. H. Ger.* ga-bauhnjan *adnuere, figurare.*]

ge-beácnung, -bícnung, e; *f.* [ge, *and* beácnung *a beckoning*] *A presage, sign, a speaking by tropes or figures, predicament*; præsāgium, catēgŏria = κατηγορία:—Gebeacnunge *catēgŏriæ*, Cot. 57.

ge-beád *offered*, Chr. 755; Erl. 50, 5, 15; *p. of* ge-beódan.

ge-beág, -beáh *bowed*, Beo. Th. 2487; B. 1241: 3085; B. 1540: 5128; B. 2567; *p. of* ge-búgan.

ge-beágian, -bēgian; *p.* ode; *pp.* od *To crown*:—Mid lawere ge-beágod *crowned with laurel*, Blickl. Homl. 187, 28. Gebēgde, 203, 30.

ge-bealg, -bealh [ge, *and* bealg *was angry, p. of* belgan *to be angry*] *made angry, irritated, enraged*, Bt. 27, 1; Fox 94, 32; Lk. Bos. 15, 28.

ge-bearg, -bearh *secured, protected*, Beo. Th. 5134; B. 2570: 3101; B. 1548; *p. of* ge-beorgan.

gebeár-scipe *a feast*, Lk. Skt. Lind. 9, 14. v. gebeór-scipe.

ge-beát, es; *n. A beating, blow*:—Drihten worhte āne swipe of rāpum, and hî ealle mid gebeáte ūtacynde *the Lord made a scourge of ropes and hurried them all out with beating*, Homl. Th. i. 406, 8. [*Laym.* i-beat *beating, striking*: *M. H. Ger.* gebōz.] DER. fȳst-gebeát.

ge-beátan; *p.* -beót, *pl.* -beóton; *pp.* -beáten *To beat, strike*; tundĕre, fĕrīre:—Hređles eafora swealt, bille gebeáten *Hrethel's offspring perished, beaten by the falchion*, Beo. Th. 4707; B. 2359. Gebeáten fisc *mĭnūtal*, Ælf. Gl. 31; Som. 61, 98; Wrt. Voc. 27, 27. Gebeáten flǣsc *martisia* vel *baptitura*, 31; Som. 61, 99; Wrt. Voc. 27, 28.

ge-bēcan [ge, *and* bócian *to book* or *charter*] *to grant by book* or *charter, to charter*, Hem. p. 480.

ge-bēcnend, es; *m. A discoverer, discloser, informer*; index:—Ge-bēcnend mîn *index meus*, Ps. Surt. 72, 14. v. ge-beácnian.

ge-bēcnendlíce, -bēcniendlíce; *adv. Figuratively*; allēgŏrīce, Cot. 1.

ge-bēd, -bēdd; *gen.* es; *pl. nom. acc.* -bēd, -bēdu, -bēdo; *n.* [The other dialects seem to point to 'gebed:' *O. Sax.* gibed: *O. H. Ger.* gabet: *Ger.* gebet.] I. *a prayer, petition, supplication*; ōrātio, prĕces, supplicātio:—Gebēd mîn on bōsme mînum sȳ gecyrred *ōrātio mea in sinum meum convertētur*, Ps. Spl. 34, 16. Gehȳr mîn gebēd *exaudi orātiōnem meam*, Ps. Th. 54, 1. Đū mînes gebēdes bēne gehȳrdest *exaudīvisti vōcem orātiōnis meæ*, 114, 1: 129, 1. Beald in gebēde *bold in prayer*, Exon. 71 a; Th. 265, 28; Jul. 388. Wæs wacigende on Godes gebēde *ĕrat pernoctans in orātiōne Dei*, Lk. Bos. 6, 12. Hie to gebēde feóllon *they fell to prayer*, Cd. 37; Th. 48, 18; Gen. 777. Hȳ gebēdu sēcaþ *they seek prayers*, Exon. 44 b; Th. 150, 20; Gū. 781: Cd. 181; Th. 227, 24; Dan. 191. Đæt hî bēna and gebēdu sendan and geótan *qui prĕces fundant*, Bd. 1, 27; S. 492, 8. His gebēdo mihte gesēcan *ad deprecandum Dŏmĭnum advĕnīre dēbēret*, 3, 23; S. 554, 11. Mid đȳ he đá đæt gebēdd gefylde *cum orātiōnem complēret*, Bd. 5, 1; S. 614, 7. Wesan đîne eáran eác gehȳrende and beheldende on eall ge-bēdd esnes đînes *fiant aures tuæ intendentes in orātiōnem servi tui*, Ps. Th. 129, 2. II. *a religious service, an ordinance*; verbum legĭtĭmum, cærĭmōnia:—Gehealdaþ đis gebēd on ēcnysse *custōdi verbum istud legĭtimum in æternum*, Ex. 12, 24. DER. gebēd, *q. v. for cognates.*

gebed-clȳfa [ge, bed *a bed*, clȳfa, II. *a cave, den*] an; *m. A den*; spēlunca:—Swā swā leo on gebedclȳfan *quăsi leo in spēlunca*, Ps. Spl. C. second 9, 10: 103, 23. v. bed-clȳfa.

ge-bedda, -bedde [(?) cf. heals-gebedda, Beo. 63], an; *f. A bed-fellow, consort, wife*; consors tŏri, uxor:—His gebedde [MS. gebædda] wæs gecíged Elisabeth *his wife was named Elizabeth*, Wanl. Catal. 4, 13: Cd. 86; Th. 109, 25; Gen. 1828. Wolde wígfruma sēcan cwēn to gebeddan *the martial leader would seek the queen as bed-companion*, Beo. Th. 1334; B. 665: Runic pm. 29; Kmbl. 345, 16; Hick. Thes. i. 135, 58. Sægde Lameh leófum gebeddum unārlíc spel *Lamech told a wicked tale to his dear consorts*, Cd. 52; Th. 66, 29; Gen. 1091. Gebed wîf *uxor*, Mt. Kmbl. pp. 14, 16. [*O. Sax.* gi-beddio: *O. H. Ger.* ga-betti *or* -betta *a bed-fellow.*]

ge-bēd-dagas; *pl. m. Prayer-days*; Lĭtānia mājor: this *greater Litany* is for St. Mark's day, and the *Less Litany*, Lĭtānia mĭnor, is for gang-dagas *the Rogation days*:—In Letānia mājōre: đās dagas synd gehātene Letānĭæ, þæt sint, Gebēd-dagas *on the greater Litany: these days are called Lĭtānĭæ, that is, Prayer-days*, Homl. Th. i. 244, 11.

ge-bēded *compelled, driven*, Chr. 937; Erl. 112, 33, =ge-bǣded; *pp. of* ge-bǣdan.

ge-beden *demanded, intreated*, Lk. Bos. 1, 63; *pp. of* ge-biddan.

gebed-giht, e; *f. Bed-time*; contĭcĭnium:—Cwyltíd *vel* gebedgiht *contĭcĭnium*, Ælfc. Gl. 16; Som. 58, 63; Wrt. Voc. 21, 50.

ge-bēd-hūs, es; *n. A prayer-house, an oratory, house of prayer*; ōrātōrium, dŏmus orātiōnis:—Habbaþ đa wîc gebēd-hūs *the dwellings have a prayer-house*, Bd. 5, 2; S. 614, 33. Mîn hūs biþ genemned gebēd-hūs *dŏmus mea dŏmus orātiōnis vŏcābitur*, Mk. Bos. 11, 17. Godes cyrce is ūre gebēd-hūs *God's church is our prayer-house*, Homl. Th. ii. 584, 3. [*O. H. Ger.* gabethūs.]

ge-bēdian, bēdigan; *p.* ode; *pp.* od *To pray, pray to, worship*; ōrāre, adōrāre:—Đæt he wolde Rōme gesēcan, and him đǣr gebēdigan *that he would visit Rome, and worship there*, Bd. 5, 9; S. 622, 21, note, MS. T. DER. ge-bēd, ge-biddan.

ge-bēd-man, -mannes; *m. A prayer-man, one whose duty it is to pray, one of the clergy, worshipper*; ōrātor, adōrātor:—He sceal hæbban gebēd-men and fyrdmen and weorcmen *he must have prayer-men and soldiers and workmen*, Bt. 17; Fox 58, 33. Sōþe gebēd-men gebiddaþ fæder on gāste and on sōþfæstnesse *vēri adōrātōres adōrābunt Patrem in spĭrĭtu et vērĭtāte*, Jn. Bos. 4, 23.

ge-bēd-rǣden, -rǣddenn, -rēddenn, e; *f. The office of prayer, prayer*; precātiōnis offĭcium, prĕces:—Heó hî ealle eádmōdlíce heora gebēdrǣddenne bæd *se omnium prĕcibus humĭlĭter commendāvit*, Bd. 3, 8; S. 531, 34: R. Ben. 52. Hî beóþ on ealdra eorþlícra gebēdrǣdenne đe Cristene wǣron *they shall be in the prayers of all earthly folk who have been Christians*, Blickl. Homl. 45, 37. He nelle gehȳran đæs gîmeleásan mannes gebēdrǣdene *he will not listen to the prayers of the negligent man*, 57, 4.

gebed-scipe, es; *m. Bed-fellowship, cohabitation, marriage*; cohabĭtātio:—Þurh đone gebedscipe *through cohabitation*, Exon. 9 a; Th. 5, 29; Cri. 76: Cd. 57; Th. 70, 4; Gen. 1148: 100; Th. 133, 25; Gen. 2216.

ge-bēd-stōw, e; *f. A prayer-place, place where prayers have been offered, an oratory*; ōrātiōnis lŏcus, ōrātōrium:—In đære gebēdstōwe æfter đon monige mægen and hǣlo tācen gefremede wǣron *in cūjus lŏco orātiōnis innŭmĕræ virtūtes sanĭtātum noscuntur esse patrātæ*, Bd. 3, 2; S. 524, 28. He ne mæg lenge gewunian in gebēdstōwe *he may not longer remain in the place of prayer*, Exon. 71 a; Th. 265, 4; Jul. 376. On heora gebēdstōwe *in their place of prayer*, Blickl. Homl. 133, 19.

ge-bēgan; *p.* de; *pp.* ed; *v. trans. To cause to bow, bend, bow down, recline, press down, humble, crush*; flectĕre, incurvāre, humĭliāre, deprĭmĕre:—Gebēgdon sāwle mîne *incurvāvērunt anĭmam meam*, Ps. Surt. 56, 7: Lk. Skt. Lind. 9, 58. Se đe hine ahefeþ he biþ gebēged and se đe hine gebēges he ahæfen biþ *qui se exaltaverit humiliabitur et qui se humiliaverit exaltabitur*, Mt. Kmbl. Lind. 23, 12. Heó sceáf in đæt neowle genip, nearwe gebēged *thrust them into that deep darkness, closely pressed down*, Cd. 223; Th. 292, 26; Sat. 446. Burga fîfe wǣran under Norþmannum nȳde gebēgde on hǣđenra hæfteclommum lange þrage *five towns were under the Northmen by necessity bowed down in the bonds of the heathen for a long space*, Chr. 941; Th. 210, 7, col. 1; Edm. 9. DER. bēgan *to bow*, ge-bȳgan.

ge-bēgdnes, -bēgednes, -ness, e; *f. Crookedness*; aduncĭtas, oblīquĭtas, Som. Ben. Lye.

ge-bēgendlíc; *adj. Bending, flexible*; flexībĭlis, Som. Ben. Lye. v. ge-bȳgendlíc.

gebēldan; *p.* de:—Eđiluald hit [the book] ūta giđryde and gibēlde *Ethelwald made it firm on the outside and covered it*, Jn. Skt. p. 188, 3. See note 8, p. viii. Or is it the verb gebēldan [from bald] used in the sense of 'strengthen?' cf. note 7, on giđryde and the connection suggested with đryþ.

ge-belg, -belh, es; *m. Anger, offence*; īra, offensio:—Us is acumendlícere eówer gebelh, đonne đæs Ælmihtigan Godes grama *your displeasure is more tolerable to us than the anger of the Almighty God*, Homl. Th. i. 96, 6. Bd. de Sapientĭbus, Som. Ben. Lye. DER. belgan.

ge-belgan, he -bylgþ, -bilhþ; *p.* -bealg, -bealh, *pl.* -bulgon; *pp.* -bolgen. I. *v. reflex. acc.* [ge, *and* belgan *to irritate*] *To make one angry, irritate, enrage*; īra se tumefăcĕre, irrĭtāre, exaspĕrāre:—Se wîsa Catulus hine gebealg *the wise Catulus made himself angry*, Bt. 27, 1; Fox 94, 32. Đá gebealh he hine *tunc ille indignātus est*, Lk. Bos. 15, 28: 13, 14; Ors. 4, 4; Bos. 81, 12. Gebulgon đa tyne hî be Iacŏbe and Iohanne *dĕcem coepērunt indignāri de Jacobo et Joanne*, Mk. Bos. 10, 41. II. *trans. dat. To anger, incense*; irrĭtāre, exaspĕrāre:—Đæt he ēcean Dryhtne bitre gebulge *that he had bitterly incensed the eternal Lord*, Beo. Th. 4651; B. 2331. Đá wæs Herodes swȳđe gebolgen *tunc Hērōdes irātus est valde*, Mt. Bos. 2, 16: 26, 8: Cd. 4; Th. 4, 16; Gen. 54. Torne gebolgen *swollen with anger*, Beo. Th. 4794; B. 2401. Mid gebolgne hond *with wrathful hand*, Exon. 37 a; Th. 120, 19; Gū. 274. III. *intrans. To be angry*; indignāri, irasci:—Gebulgon wiđ đa twegen gebrōđru *indignāti sunt de duōbus fratrĭbus*, Mt. Bos. 20, 24.

ge-belimpan; *p.* -belamp, *pl.* -belumpon; *pp.* belumpen *To happen, occur, befall*; evĕnīre, accĭdĕre, contingĕre:—Hit gebȳraþ đæt hit ge-belimpe *oportet hæc fĭeri*, Mk. Bos. 13, 7. DER. be-limpan, II.

B b 2

ge-bén *a praying, prayer;* prĕces, Ben. Lye.　Hiora ĕcelícum giboene *eorum perpetua supplicatione,* Rtl. 73, 38 : 74, 12.　v. bén.

ge-bend, es ; *n. A band;* vinculum :—Gebend tungæs his *vinculum linguæ ejus,* Mk. Skt. Lind. 7, 35.

ge-bendan, -bændan ; *p.* -bende ; *pp.* -bended, -bend.　I. *to bend;* flectĕre, tendĕre :—He hornbogan hearde gebendaþ *confrēgit cornua arcuum,* Ps. Th. 75, 3.　He gebende his bogan *he bent his bow,* Homl. Th. i. 502, 15.　Of gebendum bogan *from a bended bow,* Guthl. 4 ; Gdwin. 28, 2.　II. *to bind, fetter;* vincīre :—Swā gebend he wæs wuniende, ōþ he his líf forlét *he remained so bound until he gave up his life,* Ors. 5, 2 ; Bos. 103, 1.　Hieremias se wítega wearþ oft gebend *Jeremiah the prophet was often in bonds,* Ælfc. T. 18, 23.　DER. bendan.

ge-bénlíc *prayer-like, nun-like;* vestālis, Som. Ben. Lye.

ge-benn, es ; *n. A command, edict,* Cot. 79.　v. ge-ban.

ge-bennian ; *p.* ode ; *pp.* od, ad *To wound;* vulnĕrāre :—Bille geben-nad *wounded with a sword,* Exon. 102 b ; Th. 388, 3 ; Rä. 6, 2.　DER. ben, benn *a wound.*

ge-bénsian *to pray* :—Gi-boensandum dínum *supplicibus tuis,* Rtl. 51, 29.　v. bénsian.

ge-beod, es ; *n. A prayer, supplication;* prĕces :— Dæghwamlíce Drihtne béna and gebeoda borene beón sceoldan *cotídie Domino prĕces offerri dēbērent,* Bd. 3, 14 ; S. 540, 6.　Gebeodo dína *deprecatio tua,* Lk. Skt. Lind. 1, 13 : Rtl. 14, 36.　v. ge-béd.

ge-beódan ; *p.* -beád, *pl.* -budon ; *pp.* -boden [ge-, beódan *to command*].　I. *to command, order, summon;* jūbēre, mandāre :—Hét gebeódan byre Wihstānes hæleða monegum boldāgendra, đæt hie bǽlwudu feorran feredon *Wihstan's son bade command many house-owning men, that they should convey pile-wood from afar,* Beo. Th. 6211 ; B. 3110 : Elen. Kmbl. 551 ; El. 276.　II. *to announce, proclaim;* annun-tiāre :—Hit beó seofon nihtum geboden ǽr *let it be announced seven days before,* L. Ath. i. 20 ; Th. i. 208, 27 ; Cd. 183 ; Th. 229, 27 ; Dan. 223.　III. *to offer, propose, give, grant;* offerre, præbēre :—Hiera se æđeling gehwelcum feoh and feorh gebeád *to each of them the noble offered money and life,* Chr. 755 ; Erl. 50, 5, 15.　Gebudon him Perse đæt hí hæfdon iii winter sibbe wið hí *the Persians proposed that they should have peace with them for three years,* Ors. 3, 1 ; Bos. 52, 27. [*O. Sax.* gibiodan : *O. H. Ger.* ga-biutan, -piotan : *Ger.* gebieten.]

ge-beón, -beónn *commanded, assembled,* Cod. Dipl. 1073 ; A. D. 896 ; Kmbl. v. 140, 8 : Th. Diplm. A. D. 896 ; 139, 11 ; *p.* of ge-bannan.

ge-beón *been,* Chr. 1096 ; Erl. 233, 3.　v. beón.

ge-beór, es ; *m. A guest;* hospes, convíva :—Đá đæt đa gebeóras ge-sáwon *quod cum convívæ conspícĕrent,* Bd. 3, 10 ; S. 534, 33.　Gebeór convíva, Ælfc. Gr. 7 ; Som. 6, 45 : Scint. 63 : Homl. Th. i. 484, 1 ; 528, 9.　DER. beór.

ge-beoran, to -beoranne [ge-, beoran *to bear*] *To bear, bring, offer;* ferre, proferre :—Đám đe se deáþ tobeótaþ, bútan ænigre yldinge is to gebeoranne *his quibus mors inmĭnet, sĭne ulla dīlātiōne profĕrenda est,* Bd. 1, 27 ; S. 493, 30.

ge-beorc, es ; *n? A barking;* latrātus :—Gemenged stemn is đe biþ bútan andgite, swylc swá is hryđera gehlów, and horsa hnægung, húnda gebeorc, treówa brastlung, et cætera *confused voice is what is without understanding, such as is the lowing of oxen, and the neighing of horses, the barking of dogs, the rustling of trees, etc,* Ælfc. Gr. 1 ; Som. 2, 34–36.

ge-beorg, -beorh -berg ; *gen.* -beorges, -beorhges ; *n.* [ge-, *and* beorg *a protection, refuge*] *A defence, protection, safety, refuge;* præsīdium, refūgium, tutāmen, tuítio :—Leófsunu ahóf bord to gebeorge *Leofsunu raised up his buckler for defence,* Byrht. Th. 138, 64 ; By. 245 : 135, 40 ; By. 131.　Britwalum to gebeorge *for the protection of the Brito-Welsh,* Chr. 189 ; Erl. 9, 26 : Bd. 1, 12 ; S. 480, 32.

ge-beorgan, to -beorganne ; *p.* ic, he -bearg, -bearh, đú -burge, *pl.* -burgon ; *pp.* -borgen [ge-, beorgan *to save*] *To save, protect, defend, secure, spare, preserve;* servāre, salvāre, tuēri, defendĕre, arcēre, parcēre :—Ne mæg nán man őđerne wyrian and him sylfum gebeorgan *no man may curse another and save himself,* Homl. Th. ii. 36, 3 : Gen. 19, 19, 20 : Boutr. Scrd. 22, 3.　Áge he þreóra nihta fierst him to gebeorganne *let him have a space of three days to save himself,* L. Alf. pol. 2 ; Th. i. 62, 2.　Đú him yfele dagas ealle gebeorgest *mitiges eum a diēbus mālis,* Th. 93, 12.　Scyldweall gebearg líf and líce *the shield-wall secured life and body,* Beo. Th. 5134 ; B. 2570.　Đæt gebearh feore *which protected his life,* 3101 ; B. 1548 : Cd. 197 ; Th. 246, 6 ; Dan. 475.　Gebeorh đé on đam munte *in monte salvum te fac,* Gen. 19, 17 ; Homl. Th. i. 416, 17.　Đæt hí him gebeorgen bogan and strǽle *ut fūgiant a fácie arcús,* Th. 59, 4.　Ne biþ us geborgen *we shall not be secure,* Homl. Th. i. 56, 18. [*O. Sax.* gi-bergan : *O. H. Ger.* ga-pergan.]

ge-beorglíc *safe, cautious, prudent, becoming,* L. Edg. ii. 1 ; Th. i. 266, 6, note 12, MS. G.　v. ge-beorhlíc.

ge-beorh ; *gen.* -beorges ; *m.* [ge-, *and* beorh *a hill, mountain*] *A mountain;* mons :—Gebeorh Godes *mons Dei,* Ps. Th. 67, 15.　[*Ger.* gebirge.]

ge-beorh ; *gen.* -beorges, -beorhges ; *n. A defence, protection, refuge;*

tuítio, refūgium :—Dryhten ys úre gebeorh *Deus noster refūgium est,* Ps. Th. 45, 1 : Ps. Spl. C. 9, 9 : 17, 1.　To gebeorhge đæs sǽs *for the sea's protection,* Bd. 1, 12 ; S. 481, 12.　Wolde he đám gebeorh gewarnian đe he heora láre onféng *vŏlens scīlicet tuītiōnem eis, quos et quórum doctrī-nam suscēperat, præstāre,* 2, 5 ; S. 506, 30, MS. B.　DER. ge-beorg.

ge-beorhlíc, -beorglíc ; *adj. Safe, cautious, prudent, becoming ;* tūtus, circumspectus, dĕcens :—Gebeorhlícre ys me faran to eá, mid scype mýnum, đænne faran mid manegum scypum, on huntunge hranes *tūtius est mihi íre ad amnem, cum náve mea, quam íre cum multis nāvibus, in venātiōnem bălænæ,* Coll. Monast. Th. 24, 21.　Gebeorhlíc *circumspectus,* R. Ben. 64.　Swilce hit fór Gode gebeorhlíc sý and fór weorulde aberend-líc *as it may be becoming before God and tolerable before the world,* L. Edg. ii. 1 ; Th. i. 266, 6 : L. C. S. 2 ; Th. i. 376, 14.

ge-beorhnys, -nyss, e ; *f. A refuge ;* refūgium :—On húse gebeorh-nysse *in dŏmum refúgii,* Ps. Spl. C. 30, 3.

gebeorh-stów, e ; *f. A place of refuge ;* refūgium :—Đú eart mín gebeorhstów on mínum earfođum *tu es mihi refūgium a pressūra,* Ps. Th. 31, 8.

ge-beorhtian ; *p.* ode ; *pp.* od [ge-, beorhtian *to shine, brighten*] *To make bright, brighten, glorify ;* clārĭfĭcāre :—Đú Fæder, gebeorhta me mid đé sylfum *clārĭfĭca me tu, Pater, ăpud temetipsum,* Jn. Bos. 17, 5. [*Goth.* ga-bairhtjan.]

ge-beornan ; *p.* -barn, *pl.* -burnon ; *pp.* -bornen, -burnen [ge-, beornan *to burn*].　I. *v. intrans. To burn, be on fire, be consumed ;* ardēre, combūri :—Sió hand gebarn mōdiges mannes *the hand of the bold man burned,* Beo. Th. 5388 ; B. 2697.　II. *v. trans. :—Seó eorþe wæs* to axsan geburnen *the earth was burnt to ashes,* Ors. 4, 2 ; Bos. 79, 19.

ge-beór-scipe, -scype, es ; *m.* [ge-, beór *beer,* -scipe -*ship*] BEER-SHIP, *convivial society, a drinking party, feast, an entertainment ;* pōtātio, com-pōtātio, coena, convivium :—Hig lufigeaþ đa fyrmestan setl on gebeór-scypum *ămant prīmos recūbĭtus in coenis,* Mt. Bos. 23, 6 : Jn. Bos. 12, 2 ; 21, 20.　Dyde mycelne gebeórscype *fécit convivium magnum,* Lk. Bos. 5, 29 : Gen. 21, 8 : 40, 20.　In gebeórscipe *in convivio,* Bd. 4, 24 ; S. 597, 4.　On gebeórscipe *at a feast,* L. In. 6 ; Th. i. 106, 11.

ge-beorþor ; *g.* -beorþres ; *n.* [ge-, beoþor *child-birth*] *A birth ;* nātus :—Þurh đa burþran we wæron gehælde, and þurh đæt gebeorþor we wurdon alýsde *through the issue we were saved, and through the birth we were redeemed,* Homl. Blickl. 105, 21.

ge-beót, es ; *n.* [ge-, beót *a threatening*].　I. *a threatening, threat, boast ;* comminātio, mĭnæ :—Alýs us, Drihten, fram his gebeóte and mihte *redeem us, Lord, from his threatening and might,* Homl. Th. i. 568, 22.　Swá fela þeóda wurdon todǽlede æt đære wundorlícan byrig đe đa entas woldon wircean mid gebeóte æfter Noes flóde, ǽr đan đe hí toferdon *so many [of] nations were divided at the wonderful city which the giants would build with boasting after the flood of Noah, before they parted,* Ælfc. T. 39, 10–12.　II. *a promise ;* promissum :—Ofer eald gebeót *contrary to the old promise,* Exon. 123 b ; Th. 475, 13 ; Bo. 47.　[*Laym.* ibeot.]　DER. word-gebeót.

ge-beótian ; *p.* ode, ede ; *pp.* od, ed [ge-, beotian, II. *to boast, vow, promise*] *To promise in a boastful manner, to vow ;* glōriōse pollicēri :—Gebeótode án þegena, đæt he mid sunde đa eá oferfaran woldon *one of the officers vowed that he by swimming would cross over the river,* Ors. 2, 4 ; Bos. 44, 2, 4.　Antigones and Perþica gebeótedan, đæt hý woldan him betweonum gefeohtan *Antigonus and Perdiccas vowed that they would fight with one another,* Ors. 3, 11 ; Bos. 72, 41.　Wit gebeótedon, đæt wit on gársecg út aldrum néđdon *we two vowed that we would venture our lives out on the ocean,* Beo. Th. 1076 ; B. 536 : 964 ; B. 480.

ge-beótung, e ; *f.* [ge-, beótung *a threatening*] *A threatening ;* com-minātio :—Gebeótung *fascinātio?* Cot. 90.

ge-beran ; he -bireþ, -byreþ, -byrþ ; *p.* -bær, *pl.* -bǽron ; *pp.* -boren [ge-, beran *to bear*] *To bear, bring forth ;* ferre, părĕre :—Ne mihton nánuht libbendes geberan *they could not bring forth anything alive,* Ors. 4, 1 ; Bos. 78, 22 : Exon. 10 b ; Th. 13, 19 ; Cri. 205.　Rachel gebær Beniamin *Rachel bare Benjamin,* Gen. 35, 19.　Him wíf sunu gebær *his wife bare a son to him,* Cd. 132 ; Th. 167, 31 ; Gen. 2774.　Đá wearþ Abrahame Ismael geboren *then Ishmael was born to Abraham,* 105 ; Th. 138, 26 ; Gen. 2297 : Andr. Kmbl. 1379 ; An. 690.

geberbed, *pp. Vermiculatus :—*Giberbedo sulfere *vermiculatas argento,* Rtl. 4, 5.　[Cf. *O. H. Ger.* furben, furbian *mundare, purgare.*]

ge-bered ; *part. Vexed, oppressed, crushed ;* vexātus, măcĕrātus, elī-sus :—Gebered beón măcĕrāri, Cot. 136.　Gebered wæs *vexābātur,* Mk. Skt. Lind. 5, 15, 18.　Gebered *vexāti,* Mt. Kmbl. Lind. 9, 36.　Ge-bered *elīsus,* Mk. Skt. Lind. 9, 20.　Beren gebered corn *tipsane* [= *ptĭsăna* = πτισάνη *barley, crushed and cleaned from the hulls*], Ælfc. Gl. 12 ; Som. 57, 86 ; Wrt. Voc. 20, 27.

ge-berg, es ; *n. A defence, refuge ;* refūgium :—Geworden is Dryht geberg þearfena *factus est Dŏminus refūgium pauperum,* Ps. Surt. 9, 10 : 58, 17 : 89, 1.　v. ge-beorg.

ge-berhtan, -byrhtan, -birhtan ; *p.* te ; *pp.* ed [ge-, berhtan *to shine*] *To make bright, brighten, enlighten ;* illūmināre, clārĭfĭcāre :—Đe wuhta

gehwæs wlite geberhteþ *which brightens the beauty of everything*, Bt. Met. Fox 21, 64; Met. 21, 32.

ge-berian; *p.* ede; *pp.* ed [ge-, berian *to happen*] *To happen*; evēnīre, accīdĕre:—Geberian compĕtĕre, C. R. Ben. 37. Geberede hit dæt Ercules com to him *it happened that Hercules came to him*, Bt. 16, 2; Fox 52, 34, note 10, MS. Cot: Bt. Met. Fox 25, 61; Met. 25, 31.

ge-bernan [ge-, bernan *to burn*] *To burn*; combūrere:—Geberneþ combūret, Lk. Skt. Lind. 3, 17.

ge-berst, es; *m?* *A bursting, eruption*; eruptio:—Wið ōmena geberste *against bursting of erysipelas*, L. M. 1, 39; Lchdm. ii. 100, 2.

ge-bésmed; *part. Bosomed, bent, crooked*; sīnuātus, Som. Ben. Lye. v. ge-bósmed.

ge-bétan, he -béteþ, *pl.* -bétaþ; *p.* bétte, *pl.* bétton; *pp.* -béted, -bétt; *v. trans.* [ge-, bétan *to amend*]. **I.** *to make better, improve, mend, amend, repair*; emendāre, repārāre:—Gimmas ne scearpnesse gebétaþ *gems do not improve sharpness*, Bt. 34, 8; Fox 144, 33. Ðæt hí gebétton *that they repaired*, Ors. 3, 1; Bos. 54, 15: Bt. 20: Fox 70, 35. Geboeton netta hiora *reficientes retia sua*, Mt. Kmbl. Lind. 4, 21. Geboeta *curare*, 4, 24. Giboeted wæs ðá fýr *accenso autem igni*, Lk. Skt. Rush. 22, 55. **II.** *to make strong, fortify, surround with a wall*; confirmāre, mūnīre, mūrāre:—Sceáwiaþ ðæt land hwæðer hit wæstmbære sí, and ða burga gebétte oððe bútan weallum *consīdĕrāte terram, quālis sit, hŭmus pinguis, et urbes quāles, mūrātæ an absque mūris*, Num. 13, 20. **III.** *to make amends, reparation*, 'bót' *for, repent*:—Ðonne sceolan we mid úre ánre sáule forgyldan and gebétan ealle ða ðing ðe we ǽr ofor his bebod gedydon *then must we with our soul alone make recompence and amends for all things that we have previously done against his command*, Blickl. Homl. 91, 16; 63, 34; 57, 27: Ors. 1, 1; Bos. 23, 5; H. R. 107, 4. Hea geboeton *pæniterent*, Lk. Skt. Lind. 10, 13. **IV.** *to obtain a remedy against, to get* 'bot' *from, avenge*:—Ðú wille cweðan ðæt ða welgan habban mid hwam hí mǽgen ðæt [*hunger, thirst, cold*] gebétan *you will say that the rich have wherewith they can remedy that*, Boeth. 26, 2; Fox 92, 37. Ne meahte on ðam feorh-bonan fǽhþe gebétan *could not avenge the feud on the murderer*, Beo. Th. 4922; B. 2465. [*Goth.* ga-bótjan: *O. Sax.* gi-bótean: *O. H. Ger.* ga-bōzian.]

ge-beterian, -betrian; *p.* ode; *pp.* od [ge-, beterian *to make better*, betera *better*]; *To better, make better*; meliōrāre, emendāre:—Ðe mid ðære láre gebeterode wǽron *who were bettered by that instruction*, Homl. Th. i. 406, 32. Ða scamfæstan beóþ oft mid gemetlícre láre gebetrode *the modest are often improved with moderate instruction*, Past. 31, 1; Swt. 205, 23; Hat. MS. 39 b, 5.

ge-beterung *an amending, bettering, making better*; emendātio, instauratio, Som. Ben. Lye.

ge-beðian; *p.* ode, ede; *pp.* od, ed; *v. trans.* [ge-, beðian *to bathe*] *To wash, bathe, foment, cherish, warm*; lăvāre, fŏvēre:—Mid ðam wætere ða eágan gebeða *bathe the eyes with the water*, Herb. 88; Lchdm. i. 192, 5. Wearþ his lǽcum geþúht ðæt hí on wlacum ele hine gebeðedon *it seemed good to his physicians that they should bathe him in lukewarm oil*, Homl. Th. i. 86, 23. Byþ langum ǽrdamðe heó eft gebeðod sý *it is long before it is again warmed*, Bd. de nat. rerum; Wrt. popl. science 9, 21, 22; Lchdm. iii. 252, 8, 10. Of ðam wíne sýn ða lyðu gebeðede *let the joints be bathed with the wine*, Herb. 89, 5; Lchdm. i. 192, 25.

ge-bétt *amended, reformed*, Bd. 1, 21; S. 485, 8: 1, 27; S. 492, 17; *pp. of* ge-bétan.

ge-bétung, -béttung, e; *f.* [gebétan *to better*] *A bettering, amending, repairing, renewing, restoring*; emendātio, instaurātio:—Be ciricena gebétunge *of the repairing of churches*, L. Edm. E. 5; Th. i. 246, 9. Be burga gebéttunge *of repairing of fortresses*, L. Ath. i. 13; Th. i. 206, 13.

ge-bicgan, -bicgean *to buy, purchase*, Exon. 90 a; Gu. Ex. 82: L. Edg. ii. 3; Th. i. 266, 18: L. Eth. ii. 1; Th. i. 284, 13. v. ge-bycgan.

ge-bícnian, -býcnian; *p.* ode, ede; *pp.* od, ed [ge-, bícnian *to beckon, nod*]. **I.** *to beckon, nod*; innuĕre:—Ic gebícnige [gebýcnige MS. D.] *innuo*, Ælfc. Gr. 28, 3; Som. 30, 48. **II.** *to point out, shew, indicate, betoken*; indĭcāre, signĭfĭcāre, portendĕre:—Ic gebícnige [gebýcnige MS. D.], Ælfc. Gr. 37; Som. 39, 40. Hí gebícniaþ sum þing niwes *they betoken something new*, Bd. de nat. rerum; Wrt. popl. science 16, 23; Lchdm. iii. 272, 7. Pirrus gebícnede eft hú him se sige gelícode *Pyrrhus afterwards shewed how the victory pleased him*, Ors. 4, 1; Bos. 77, 35. Gebýcna hit eal me *tell it all to me*, St. A. 44, 12. v. ge-beácnian.

ge-bícnigendlíc; *adj. Pointing out, shewing, indicative*; indĭcātīvus:—Gebícnigendlíc gemet *indĭcātīvus mŏdus*, Ælfc. Gr. 21; Som. 23, 18.

ge-bícnung; e; *f.* [ge-, bícnung *a sign*] *A presage, sign*; præsāgium:—Þurh heofenlícre gebícnunge *through a heavenly sign*. Hom. Th. ii. 306, 7. v. ge-beácnung.

ge-bídan, he -bídeþ, -bít; *p.* -bád, *pl.* -bidon; *pp.* -biden [ge-, bídan *to bide, abide*] *To abide, tarry, remain, await, look for, expect, meet with, experience, endure*; mănēre, remănēre, expectāre, consĕqui, sustinēre, tŏlĕrāre:—Ðæt feorhdaga on woruldríce worn gebíde *that he may abide many life-days in the world's realm*, Cd. 107; Th. 142, 10; Gen. 2359.

Gebídaþ hér *sustĭnēte hic*, Mt. Bos. 26, 38. Dreámleás gebád *he continued joyless*, Beo. Th. 3445; B. 1720. He gebád ðǽr sylf *remansit sōlus Jēsus*, Jn. Bos. 8, 9. Ne mæg feónd gebídan *foe may not await him*, Exon. 30 a; Th. 93, 23; Cri. 1520. Hig gebídon his *erant expectantes eum*, Lk. Bos. 8, 40. He ðæs frófre gebád *he from that* [*time*] *met with comfort*, Beo. Th. 14; B. 7: Exon. 41 b; Th. 140, 11; Gu. 608. Óðres ne gýmeþ to gebídanne yrfeweardes *he cares not to wait for another heir*, Beo. Th. 4895; B. 2452. Fela sceal gebídan leófes and láþes *much shall he experience of loved and hated*, 2125; B. 1060. [*Laym.* i-biden: *Goth.* ga-beidan *to abide, endure*: *O. Sax.* gi-bídan *to experience.*]

ge-biddan; *p.* -bæd, *pl.* -bǽdon; *pp.* -beden; *often followed by a reflexive dative* [ge-, biddan *to ask, pray*] *To pray, pray to, worship, adore*; ōrāre, adōrāre, cŏlĕre:—Uton gebiddan us *let us pray*, Homl. Blick. 139, 30. Ðonne we us gebiddaþ *when we pray*, Bt. 41, 2; Fox 246, 21. Ðonne gé eów gebiddon *cum ōrātis*, Mt. Bos. 6, 5. Ðonne ðú ðé gebidde *cum orāvēris*, 6, 6. Lǽr us us gebiddan *dŏce nos ōrāre*, Lk. Bos. 11, 1. For ðé gebitt *ōrābit pro te*, Gen. 20, 7. Ic him á gebæd *ego autem ōrābam*, Ps. Th. 108, 3. Ne ðú fremedne god gebiddest *neque adōrābis deum aliēnum*, 80, 9. Gebiddaþ him ðǽr to *adōrant eum*, Ex. 32, 8. Gebiddaþ on gesihþe his *adōrābunt in conspectu ejus*, Ps. Spl. 21, 28. Ic me to him gebidde *eum cŏlo*, Bd. 1, 7; S. 477, 34. Gebiddande *orans*, Mt. Kmbl. 26, 39.

ge-bierde; *adj. Inborn, natural*; innātus, natūrālis, Cot. 106. v. gebyrde.

ge-biesgian *to occupy, afflict, overcome*, Exon. 96 a; Th. 358, 2; Pa. 39. v. ge-býsgian.

ge-bígan, *p.* de; *pp.* ed; *v. trans.* [ge-, bígan *to bow, bend*] *To bow, bend, turn, inflect or decline a part of speech, twist, bow down, humble, bring under, subdue, crush*; flectĕre, inflectĕre, declīnāre, humiliāre:—He hí to fulluhte gebígde *he brought them to baptism*, H. R. 101, 26. Se sceal heán wesan niðer gebíged *he shall be low bowed down*, Exon. 84 a; Th. 316, 28; Mód. 55: Bd. 4, 10; S. 578, 28: Gen. 27, 29. Ealle naman beóþ gebígede on fíf declíuungum *omnia nōmĭna quinque declīnātiōnĭbus inflectuntur*, Ælfc. Gr. 7; Som. 6, 2: 14; Som. 16, 56: Exon. 24 a; Th. 69, 26; Cri. 1126: Ors. 3, 9; Bos. 64, 15: Ælfc. T. 30, 5. Ps. Th. 106, 15. v. ge-býgan.

ge-bígednys, -nyss, e; *f. A bending, inflection, declining, declension, case*; declīnātio, cāsus:—Gebígednys *cāsus*, Ælfc. Gr. 15; Som. 17, 30. Cāsus, ðæt is fyll oððe gebígedniss *a case, that is a fall or inflection*, Ælfc. Gr. 14; Som. 17, 23. Ða pronōmĭna ðe habbaþ vŏcātīvum, ða habbaþ six casus, and ða óðre ealle nabbaþ bútan fíf gebígednyssa *the pronouns which have a vocative have six cases, and all the other have but five cases*, Ælfc. Gr. 18; Som. 20, 55. Nemnigendlíc gebígednys *vel* nemnigendlíc cāsus *Nominative case*, Ælfc. Gr. 7; Som. 6, 16. Gestrýnendlíc, geágniendlíc *Genitive*, 6, 17: Forgifendlíc *Dative*, 6, 19: Wrēgendlíc *Accusative*, 6, 22: Clipigendlíc, oððe gecígendlíc *Vocative*, 6, 24, 25: Ætbredendlíc *Ablative and Instrumental*, 6, 27, *q. v.*

ge-bígendlíc; *adj. Bending, flexible, declined with cases*; flexibĭlis, cāsuālis:—Be ðám six gebígendlícum hiwum *de sex casuālĭbus formis*, Ælfc. Gr. 14; Som. 17, 19.

ge-bígeþ, -bigþ *buys*, L. Ethb. 77; Th. i. 22, 1: Mt. Bos. 13, 44, = ge-bygeþ; *pres. of* ge-bycgan.

ge-bihþ, e; *f.* [cf. byht *a dwelling, abode*] *An abode, habitation*; domĭcīlium:—On mislícum monna gebihþum *in the various abodes of men*, Exon. 45 b; Th. 154, 22; Gú. 846.

ge-bild; *adj. Bold, brave, confident*; audax, fortis, fīdens:—He mid gebildum móde hine ealne gedranc *he drank it all with a bold mind*, Homl. Th. i. 72, 25. v. gebyldan.

ge-bilegan *to make angry, to be angry*, Som. Ben. Lye. v. ge-belgan.

ge-bilod; *pp.* [bile *a bill, beak*] *Having a bill or beak*; rostrātus:—Ða fugelas, ðe be flǽsce lybbaþ, syndon clyferféte and scearpe gebilode *the birds which live by flesh are cloven-footed and sharp-billed*, Hexam. 9; Norm. 14, 19.

ge-bind, es; *n. A binding, fastening*; ligātūra, strictūra:—Ofer waðema gebind [or waðema-gebind, cf. ýþ-gebland] *over the watery band*, i. e. *the surface of the water*, Exon. 76 b; Th. 288, 1; Wand. 24: 77 a; Th. 289, 32; Wand. 57. Gebynd *strictura*, Ælfc. Gl. 11; Wrt. Voc. 19, 50. [Cf. *Goth.* ga-binda, -bindi *a band*.] v. ís-gebind.

ge-bindan; ic -binde, ðú -bintst, -binst, he -bint, *pl.* -bindaþ; *p.* ic, he -band, -bond, ðú -bunde, *pl.* -bundon; *pp.* -bunden [ge-, bindan *to bind, tie*]. **I.** *to bind, tie up*; ligāre, allĭgāre, vincīre, constringĕre:—Hine nán man ne mihte gebindan *neque quisquam pŏtĕrat eum lĭgāre*, Mk. Bos. 5, 3: 6, 17; Cd. 184; Th. 230, 6; Dan. 229: Salm. Kmbl. 556; Sal. 277. Sorg and slǽp earmne ánhogan oft gebindaþ *sorrow and sleep often bind a poor lone-dweller*, Exon. 77 a; Th. 288, 33; Wand. 40. Ðú mec fæste fetrum gebunde *thou didst bind me fast with fetters*, Exon. 72 a; Th. 268, 11; Jul. 433: 98 a; Th. 368, 28; Seel. 31. He geband ðá his sunu *cum alligasset fĭlium suum*, Gen. 22, 9: Homl. Th. ii. 414, 18: Cd. 23; Th. 29, 3; Gen. 444: Beo. Th. 845; B. 420. Ðære moldan sumne dǽl he gebond on his sceáte *a part of the mould he tied*

up in his clothing, Bd. 3, 10; S. 524, 23: Exon. 18 b; Th. 46, 5; Cri. 732. Hie handa gebundon *they bound the hands,* Andr. Kmbl. 96; An. 48: 2446; An. 1224. Ceácan heora gewríþ oððe gebind *maxillas eórum constringe,* Ps. Spl. 31, 12. Gif he hí ne gebunde *if he had not bound them,* Bt. 35, 2; Fox 158, 1, note, MS. Cot. Se wæs gebunden *qui ĕrat vinctus,* Mk. Bos. 15, 7: Bd. 1, 27; S. 497, 31, 32: Cd. 35; Th. 45, 30; Gen. 734: Exon. 13 a; Th. 23, 7; Cri. 365: Andr. Kmbl. 2792; An. 1398: Bt. Met. Fox 5, 78; Met. 5, 39: Judth. 10; Thw. 23, 11; Jud. 115: Beo. Th. 3490; B. 1743. Wæs his gewuna ðæt he him forgeáfe ænne gebundenne *sŏlēbat dimittĕre illis ūnum ex vinctis,* Mk. Bos. 15, 6: Bd. 1, 27; S. 497, 33: Chr. 796; Erl. 58, 12: Exon. 102 b; Th. 387, 20; Rä. 5, 8. He gehýrde heáh gnornunge ðæra ðe gebundene bitere wǽron *audivit gĕmĭtum vincŭlātōrum,* Ps. Th. 101, 18: Cd. 19; Th. 24, 18; Gen. 379: Andr. Kmbl. 1893; An. 949. **II.** *to deceive* [?]; fallĕre :—He hine on ðære wēnunge [wenunge Thorpe] geband *he deceived him in that hope,* Ors. 3, 7; Bos. 59, 25. [*Goth.* ga-bindan: *O. Sax.* gi-bindan.]

ge-bíraþ *becomes,* L. Edg. C. 64; Th. ii. 258, 8. v. ge-býrian.

ge-bird, e; f. *Birth, origin* :—Forðam sín ealle men ánra gebirda *because all men are of one origin,* L. Edg. C. 13; Th. ii. 246, 22. v. ge-byrd.

gebirg, es; n. *Taste* :—On gebirge *in gustu,* Rtl. 116, 5.

ge-birhtan, he -birht; p. -birhte; pp. -birhted, -birht *To make bright, brighten, illuminate* ; illūmĭnāre :—Ðe ealle þing gebirht *which brightens all things,* Bt. 34, 8; Fox 144, 37. Ealle steorran weorþaþ onlíhte and gebirhte of ðære sunnan *all stars are lighted and made bright by the sun,* 34, 5; Fox 140, 5. v. ge-berhtan.

ge-bírigan *to taste,* Mt. Kmbl. Hat. 27, 34. v. ge-býrgan.

ge-bisgian *to occupy, afflict, agitate,* Exon. 50 a; Th. 173, 34; Gú. 1170. v. ge-býsgian.

ge-bismerian, -bismrian, -bysmerian, -bysmrian; p. ode, ede; pp. od, ed [ge-, bismerian *to mock*] *To mock, laugh at, deride, provoke* ; illūdĕre, irrīdēre, derīdēre, exacerbāre :—Draca ðes ðe ðú hywodest to gebismrienne him *drăco iste quèm formasti ad illūdendum ei,* Ps. Lamb. 103, 26. Se ðe eardaþ on heofenum gebismeraþ oððe hyscþ hig *qui hăbĭtat in cœlis irrīdēbit eos,* 2, 4. Dú, Drihten, gebysmerast hí *tu, Domine, derīdēbis eos,* 58, 9. Hí heánne God gebysmredon [MS. gebysmredan] *exacer-bāvĕrunt Deum excelsum,* Ps. Th. 77, 56.

ge-bisnere, es; m. *An imitator* :—Gibisnere *imitator,* Rtl. 45, 14.

ge-bisnian *to inform, imitate* :—Gibisnendo *informanda,* Rtl. 103, 30. We gibisnia *imitemur,* 52, 3. Gebisened *imitandam,* Lk. Skt. p. 6, 20. v. gebysnian.

ge-bisnung *an example* ; exemplum, Som. Ben. Lye. v. ge-bysnung.

ge-bit, -bitt, es; n. [ge-, biten, pp. of bítan *to bite*] *A biting, biting together, grinding, gnashing* ; morsus, strīdor :—Ðǽr biþ wóp and tóþa gebitt *there shall be weeping and gnashing of teeth,* Homl. Th. 126, 20.

ge-bítan *to bite* :—Gebítes ł to-slítes *adlidit,* Mk. Skt. Lind. 9, 18.

ge-biterian ; p. ode; pp. od [ge-, biterian *to embitter*] *To make bitter* ; amarefacere :—Hí sealdon him gebiterod wín *dăbant ei myrrhātum vīnum* [*amarefactum vīnum, vīnum myrrha imbūtum*], Mk. Bos. 15, 23.

ge-bitt *prays or will pray* ; ŏrābit, Cot. 3, 7; 3rd pres. of ge-biddan.

ge-blǽd, -es; m. [ge-, blǽd **I.** *a blast, blowing*] *A blowing out in the skin, blister* ; vēsīca in cūte. DER. þorn-geblǽd, þystel-, wæter-, wyrm-, ýs-.

ge-blǽdfæst ; adj. [blǽd *fruit*] *Fruitful* ; fertĭlis :—Beorht and ge-blǽdfæst *bright and fruitful,* Cd. 5; Th. 6, 15; Gen. 89.

ge-bland, -blond, es; n. [ge-, bland *a mixture, confusion*] *A mixture, mingling, commotion* ; commixtio, turba :—Ofer æra gebland *over the mingling of the waves,* Chr. 937; Erl. 112, 26; Æðelst. 26. Árýþa geblond *commotion of the oar-waves,* Andr. Kmbl. 1063; An. 532. DER. ár-gebland, ear-, snǽw-, sund-, ýþ-. v. bland.

ge-blandan, -blondan; p. -bleónd, -blénd, pl. bleóndon, -bléndon; pp. -blanden, -blonden [ge-bland]. **I.** *to blend, mix, mingle* ; mis-cēre, turbāre :—Hí me geblēndon unswētne drync *they mixed for me an unsweet drink,* Exon. 29 a; Th. 88, 10; Cri. 1438; Andr. Kmbl. 65; An. 33. Wurman geblonden *mixed with scarlet,* Exon. 60 a; Th. 218, 14; Ph. 294. Hie him sealdon attor drincan ðæt mid myclen lybcræfte wæs geblanden *they gave them poison to drink mixed by powerful magic,* Blickl. Homl. 229, 12. [Cf. *O. Sax.* baluwes gi-blandan.] **II.** *to stain, colour, corrupt* ; infĭcĕre :—Geblende *infĕcit,* Cot. 112. Wæs seó hǽwene lyft heolfre geblanden *the azure air was corrupted with gore,* Cd. 166; Th. 208, 1; Exod. 476.

ge-blann *ceased,* Mk. Skt. Lind. 6, 51; p. of ge-blinnan.

ge-bláwan ; p. -bleów, pl. -bleówon; pp. -bláwen [ge-, bláwan *to blow*] *To blow* ; flāre, sufflāre :—Gebleów *sufflāvit,* Jn. Skt. Lind. 20, 22.

ge-blecte [?] *destroyed* ; extermĭnāvit, Ps. Spl. C. 79, 14.

ge-bledsian ; p. ode; pp. od [ge-, bledsian *to bless*] *To bless* ; benedĭ-cĕre :—Gebledsod wearþ engla ēðel *the dwelling of the angels was blessed,* Andr. Kmbl. 1048; An. 524: 1079; An. 540: 1873; An. 939: 3434; An. 1721.

ge-blégenad ; part. [ge-, blégen *a blain, blister*] *Blistered* ; ulcĕrātus, Som. Ben. Lye.

ge-blénd, pl. -bléndon *mixed,* Exon. 29 a; Th. 88, 10; Cri. 1438; p. of ge-blandan.

ᵗ ge-blendan ; p. -blende; pp. -blended, -blend [ge, blendan *to blind*] *To blind, make blind* ; cœcāre :—Gē habbaþ eówre heortan geblende ye *have your hearts blinded,* Mk. Bos. 8, 17. [*Goth.* ga-blindjan.]

ge-bleód, -blíód; part. [ge-, bleoh, bleó *a colour, hue, complexion*] *Coloured, of different colours, variegated, gifted with beauty, beautiful in countenance* ; colōrātus, versĭcŏlor, spĕcie præditus, aspectu formĭtus :—Ða wyrta greówon, mid menigfealdum blóstmum mislíce gebleóde *the plants grew, diversely coloured with manifold blossoms,* Hexam. 6; Norm. 10, 36. Oþýweþ Cristes onsýn, on sefan swēte sínum folce, gebleód wundrum *Christ's countenance shall appear, sweet in mind to his people, wondrously gifted with beauty,* Exon. 21 a; Th. 56, 32; Cri. 909.

ge-bleoh, -bleó; gen. -bleós; n. [ge-, bleoh *a colour*] *A colour* ; color :—Mid swá wlitigum blóstmum hí oferstígaþ ealle eorþlíce gebleoh *with such beautiful blossoms they excel all earthly colours,* Homl. Th. ii. 464, 9.

ge-bleów *blew,* Jn. Skt. Lind. 20, 22; p. of ge-bláwan.

ge-bletsian, -bledsian; p. ode, ade; pp. od [ge-, bletsian *to bless*] *To bless, consecrate* ; benedĭcēre, consecrāre :—Ic ðē gebletsige *benedīcam tibi,* Gen. 12, 2, 3: 17, 16. Ic wát, ðæt se biþ gebletsod, ðe ðú gebletsast *nōvi ĕnim, quod benedictus sit, cui benedīxĕris,* Num. 22, 6. Gebletsode Romulus mid his bróðor blóde ðone weall *Romulus blessed [consecrated] the wall [of Rome] with his brother's blood,* Ors. 2, 2; Bos. 41, 5. God gebletsode ðone seofeðan dæg and hine gehálgode *Deus benedixit diei septĭmo et sanctĭfĭcāvit illum,* Gen. 2, 3: 5, 2: 24, 1. Dú gebletsadest bearn Israhēla *benedixit dŏmui Israel,* Ps. Th. 113, 21. Miltsa us mihtig Drihten, and us on móde eác gebletsa nú *Deus misereātur nostri, et benedicat nōbis,* 66, 1. Ðæt ǽnig preóst ne forlǽte ða circan ðe he to gebletsod wæs *that no priest forsake the church to which he was consecrated,* L. Edg. C. 8; Th. ii. 246, 8. Sý gebletsod se ðe com on Drihtenes naman *benedictus qui vēnit in nōmĭne Domĭni,* Mt. Bos. 21, 9: 23, 39. Ðú gebletsad eart *thou art blessed,* Cd. 192; Th. 241, 18; Dan. 406: 83; Th. 105, 13; Gen. 1752.

ge-blinnan ; p. -blann, pl. -blunnon; pp. blunnen [ge-, blinnan *to cease*] *To cease, desist* ; cessāre, desistĕre :—Geblann ðæt wind *the wind ceased,* Mk. Skt. Lind. 6, 51.

ge-blíód ; part. *Coloured, variegated* ; colōrātus, variegātus :—Gebliód reáf *vestis variegāta,* Prov. 31. v. ge-bleód.

ge-blissian ; part. -blissiende; p. ode, ade; pp. od, ad [ge-, blissian *to rejoice*]. **I.** *v. intrans. To rejoice, be glad* ; lætāri, gaudēre :—Ðē gebýrede gewistfullian and geblissian *epŭlāri et gaudēre oportēbat,* Lk. Bos. 15, 32 : Jn. Bos. 5, 35. Geblissiaþ on Drihtne *lætāmĭni in Dŏmĭno,* Ps. Spl. 31, 14: Mt. Bos. 5, 12. **II.** *v. trans. To make to rejoice, gladden, fill with bliss, bless* ; lætifĭcāre, benedĭcēre :— Rihtwīsnyssa Drihtnes rihte synt, geblissiende heortan *justĭtiæ Dŏmĭni rectæ sunt, lætĭfĭcantes corda,* Ps. Lamb. 18, 9. Ðú geblissast hine *lætĭfĭcābis eum,* 20, 7. Pater Noster hálige geblissaþ *the Pater Noster gladdens the holy,* Salm. Kmbl. 80; Sal. 40: Ps. Spl. 45, 4. Frófra dīne geblissodon sáwle mīne *consōlātiōnes tuæ lætĭfĭcāvērunt anĭmam meam,* 93, 19. Ðú ðisne middangeard milde geblissa *do thou kindly bless this mid-earth,* Exon. 11 b; Th. 16, 7; Cri. 249. Iudas wæs miclum geblissod *Judas was greatly rejoiced,* Elen. Kmbl. 1749; El. 876: 2249; El. 1126. Ðá wæs Gūþláces gǽst geblissad *then was Guthlac's spirit gladdened,* Exon. 43 a; Th. 145, 14; Gú. 694: 56 a; Th 198, 9; Ph. 7. Eálá! heofoncund Þrýnes, bráde geblissad geond brytenwongas *oh! heavenly Trinity, widely blessed over the spacious world!* 13 a; Th. 24, 5; Cri. 380. [*Laym.* i-blissed.]

ge-blissung, e; f. *A rejoicing, joyousness, hilarity* ; hĭlārĭtas, Procem. R. Conc.

ge-blódegian, -blódgian; p. ode, ade; pp. od, ad [ge-, blódegian *to make bloody*] *To make bloody, cover with blood* ; cruentāre :—He ge-blódegod wearþ sáwuldrióre *he was made bloody with life-gore,* Beo. Th. 5378; B. 2692. Swilce ðǽr lǽge on ðam disce ánes fingres liþ eal geblódgod *as if there lay in the dish the joint of a finger all covered with blood,* Homl. Th. ii. 272, 27; Wanl. Catal. 43, 16. Gif ðæt flet geblódgad wyrþe *if the dwelling be covered with blood,* L. H. E. 14; Th. i. 32, 14.

ge-blond *a mixture,* Andr. Kmbl. 1063; An. 532. v. ge-bland.

ge-blondan ; pp. -blonden *To blend, mix, mingle* ; miscēre :—Áttre geblonden *mixed with venom,* Cd. 216; Th. 272, 34; Sat. 129. v. ge-blandan.

ge-blót, es; n. [ge-, blót *a sacrifice*] *A sacrifice* ; sacrifícium :—Bútan geblóte *without sacrifice,* Ors. 5, 2; Bos. 102, 14. Hí swylc geblót and swylc morþ dónde wǽron *they made such sacrifices and such murders,* 1, 8; Bos. 31, 8.

ge-blówan ; p. -bleów, pl. -bleówon; pp. -blówen [ge-, b'ówan *to blow*] *To blow, flourish, bloom, blossom* ; flōrēre, efflōrēre :—Wyrt geblóweþ *herba flōreat,* Ps. Th. 89, 6. Ðæt gē on his wícum wel geblówan *in atriis dŏmus Dei nostri flōrēbunt,* 91, 12. Se æðela feld wrídaþ under

wolcnum, wynnum geblówen *the noble field flourishes under the skies, blooming with delights*, Exon. 56 a; Th. 199, 18; Ph. 27: 56 b; Th. 200, 27; Ph: 47. Geseh ge geblówene bearwas, blǽdum gehrodene *he saw blooming groves, adorned with blossoms*, Andr. Kmbl. 2894; An. 1450: Exon. 51 a; Th. 178, 25; Gú. 1249. He geseah geblówen treǫw wæstm-berende *he saw a full-blown tree bearing fruit*, Blickl. Homl. 245, 8.

ge-bócian; *p.* ode; *pp.* od [ge-, bócian *to give by charter*]. **I**. *to give* or *grant by book* or *charter, to charter*; libro *vel* charta dōnāre :—Ðis is seó bóc, ðe Æðelstán cing gebócode Friþestáne bisceope *this is the charter which king Athelstan chartered to bishop Frithestan*, Th. Diplm. A. D. 938; 187, 19: 966; 218, 12. Gebócode Æðelwulf [MS. Aðelwulf] cing teóðan dǽl his landes, ofer ealle his rīce, Gode to lofe *king Æthelwulf chartered the tenth part of his land over all his kingdom for the glory of God*, Chr. 856; Th. 124, 22, col. 3: Text. Rof. 115, 22. **II**. *to furnish with books*; libris instruĕre :—Gē preóstas sculon beón gebócode *ye priests shall be furnished with books*, L. Ælf. P. 44; Th. ii. 382, 36.

ge-bod, es; *n.* [ge-, bod *a command*] *A command, order, mandate*; jussum, mandātum :—Is ðæt Þeódnes gebod *it is God's command*, Exon. 56 b; Th. 202, 12; Ph. 68: Menol. Fox 457; Men. 230. Be ðæs cyninges gebode *by the king's command*, Bt. 39, 13; Fox 234, 13. Gif preóst ofer arcediācones gebod mǽssige *if a priest celebrate mass against the archdeacon's command*, L. N. P. L. 7; Th. ii. 290, 25: Chr. 901; Erl. 98, 3. Ðú gebod Godes lǽstes *thou hast performed God's mandate*, Cd. 27; Th. 36, 14; Gen. 571: 33; Th. 43, 29; Gen. 698: Ps. Th. 118, 87. Hí woldon onwendan eall ða gebodu *they would change all the orders*, Ors. 6, 10; Bos. 120, 33. [*O. Sax.* gi-bod: *O. H. Ger.* ga-pot: *Ger.* gebot.]

ge-boden *announced*, L. Ath. i. 20; Th. i. 208, 27; *pp.* of ge-beódan.

ge-bodian; *p.* ode, ade; *pp.* od, ad [ge-, bodian *to tell*] *To tell, make known, announce, proclaim*; nuntiāre, annuntiāre :—Se ðæt láþspell æt hám gebodode *who made known the sad story at home*, Ors. 2, 4; Bos. 43, 37: Hy. 10, 13; Hy. Grn. ii. 293, 13. Ðæt ðǽr nán to láfe ne wearþ ðæt hit to Róme gebodade *so that there was none left to tell it at Rome*, Ors. 4, 11; Bos. 97, 30: Exon. 10 b; Th. 13, 14; Cri. 202. [*Laym.* i-boded.]

gebod-scipe, es; *m.* [gebod *a command*] *A commandment*; mandātum :—Gif hie brecaþ his gebodscipe *if they break his commandment*, Cd. 22; Th. 28, 3; Gen. 430. [*O. Sax.* gi-bodskepi, *n.*]

ge-bogen *submitted*, Chr. 1013; Erl. 148, 2, 21; *pp.* of ge-búgan.

ge-bógian; *p.* ode; *pp.* od [ge-, bógian *to inhabit*] *To inhabit*; incŏlĕre :—Hí gebógodon eástdǽl middaneardes *they inhabited the east part of the earth*, Boutr. Scrd. 21, 30, 31, 32. v. ge-búgian.

ge-boht *bought*, Ælfc. Gl. 86; Som. 74, 33; Wrt. Voc. 50, 16; *pp.* of ge-bycgan: ge-bohte, *pl.* -bohton *bought, redeemed*, Gen. 39, 1: L. C. E. 18; Th. i. 370, 28: Chr. 1016; Erl. 159, 23; *p.* of ge-bycgan.

ge-bolged; *part. Caused to swell, made angry*; tumĭdus, indignātus, Som. Ben. Lye.

ge-bolgen *offended, angry*, Mt. Bos. 2, 16; *pp.* of ge-belgan.

ge-bolstrod; *part.* [ge-, bolster *a bolster*] *Guarded, environed, defended, supported* or *bolstered up*; stipātus, Som. Ben. Lye.

ge-bond *bound, tied up*, Bd. 3, 10; S. 543, 23; *p.* of ge-bindan.

ge-boned; *part. Polished, burnished*; pŏlītus :—He hæfþ ðiderynn gedón ii mycele gebonede róda, and ii mycele Cristes béc gebonede, and iii gebonede scrín, and ii gebened altare *he has placed therein two large burnished crosses, and two large Christ's books [= Gospels] polished, and three burnished shrines, and one burnished altar*, Th. Diplm. A. D. 1050-1073; 429, 11-18. Ic gean Sĉe Eádmunde twegea gebonedra horna *I give to St. Edmund two polished horns*, Th. Diplm. A. D. 1046; 564, 12. [*Swed.* bona *to polish with wax, to rub*; *Dan.* bone *to cleanse, make clean, to burnish, polish*.]

ge-bonn, es; *n. The indiction*; indictio, Th. Diplm. A. D. 896; 139, 10: Cod. Diplm. 1073; Kmbl. v. 140, 8. v. ge-ban **II**.

ge-bonnan; *pp.* bonnen *To summon, call together* :—Folc biþ gebonnen *mankind shall be summoned*, Exon. 117 b; Th. 451, 8; Dóm. 100. v. ge-bannan.

ge-bonn-gér, es; *n.* [gebonn *indiction*; gér, geár *a year*] *The indiction-year*; indictiōnis annus, Cod. Dipl. 1073; A. D. 896; Kmbl. v. 140, 8: Th. Diplm. A. D. 896; 139, 10. v. ge-ban **II**.

ge-boren *born*, Chr. 381; Erl. 10, 2; *pp.* of ge-beran.

ge-borga *a protector, guardian*; tūtor, DER. lind-geborga.

ge-borgen *defended, safe, secure*, Homl. Th. i. 56, 18; *pp.* of ge-beorgan.

ge-borhfæstan; *p.* te; *pp.* ed [ge-, borhfæstan *to fasten by pledge* or *surety*] *To determine* or *fasten by a surety*; intertiāre [q. v. in Du Cange], ǽpud sequestrum depōnĕre, Cot. 107.

ge-borsnung, e; *f. Corruption*; corruptio :—Ne ðú ne selst hálige ðinne geseón geborsnunga *nec dăbis sanctum tuum vĭdēre corruptiōnem*, Ps. Spl. 15, 10. v. ge-brosnung.

ge-bósmed; *part.* [ge-, bósum, bósm *the bosom*; sĭnus] *Bosomed,*

bent, crooked; sĭnuātus :—Gebósmed segelbósmas sinuāta carbăsa, Cot. 185.

ge-bótad; *part. Bettered, mended*; resartus :—Ðá him gebótad wæs *when he was better*, Chr. 1093; Erl. 228, 30. v. ge-bétan.

ge-brǽc; *p.* -brǽc, *pl.* -brǽcon *broke, didst break*, Bd. 3, 2; S. 525, 2: Ps. Th. 73, 13; *p.* of ge-brecan.

ge-brǽc, es; *n.* [ge-, bræc *a breaking*] *A breaking, crashing, noise*; fractio, frăgor, strĕpĭtus :—Ðá wearþ borda gebrǽc *then there was a crashing of shields*, Byrht. Th. 140, 28; By. 295: Beo. Th. 4510; B. 2259. [*O. Sax.* gi-brak: *O. H. Ger.* ka-preh *fragor*.] v. ge-brec.

ge-brǽceo; *indecl. n. A cough*; tussis :—Wið gebrǽceo *for cough*, Herb. 124, 2; Lchdm. i. 236, 15: 126, 1; Lchdm. i. 236, 24. Heó gebrǽceo útatyhþ *it draweth out cough*, 124, 1; Lchdm. i. 236, 12.

ge-brǽcseóc, -brǽcsióc; *adj.* [ge-, brǽcseóc *epileptic, lunatic*] *Epileptic, lunatic*; epilepticus = ἐπιληπτικός, lunāticus :—Gebrǽcsióce *epileptici, comĭtiāles*, Cot. 46.

ge-brǽcseócnes, -ness, e; *f.* [ge-, brǽcseócnes *epilepsy*] *The falling sickness, epilepsy*; morbus comĭtiālis, epilepsia, Som. Ben. Lye.

ge-brǽd *drew, brandished*, Beo. Th. 5118; B. 2562; *p.* of ge-bredan.

ge-brǽdan; *to* -brǽdenne; *p.* de; *pp.* ed [ge-, brǽdan *to make broad*] *To make broad, broaden, extend, spread*; dīlātāre, ampliāre, extendĕre, expandĕre, sternere :—Merestreám ne dear ofer eorþan sceát eard gebrǽdan *the sea-stream dares not extend its province over the region of the earth*, Bt. Met. Fox 11, 132; Met. 11, 66. Ðæt mód wilnaþ to gebrǽdenne his ǽgen lof *the mind desires to extend its own praise*, Past. 65, 4; Swt. 463, 36; Hat. MS. Bt. 18, 2; Fox 64, 15. He his cyricean wundorlícum weorcum gebrǽdde *ecclesiam suam mirīficis ampliāvit opĕribus*, Bd. 5, 20; S. 641, 40. Ealle ða telgan ðú æt sǽstreámas sealte gebrǽddest *extendisti palmĭtes ejus usque ad măre*, Ps. Th. 79, 11. Ðreátas gebrǽdon wēdo hiora *turba straverunt vestimenta sua*, Mt. Kmbl. Lind. 21, 8. Miþ stáne gebrǽded *lapide stratus*, Jn. Skt. Lind. 19, 13.

ge-brǽdan; *p.* -brǽdde; *pp.* -brǽded, -brǽdd, -brǽd [ge-, brǽdan *to roast*] *To roast, broil*; torrēre, assāre :—Eton ealle ðæt flǽsc on fȳre gebrǽded *ĕdent carnes assas igni*, Ex. 12, 8. Ne ne eton gé of ðam nán þing hreówes, ne mid wætere gesoden, ac sig hit eall on fȳre gebrǽd *non comĕdētis ex eo crŭdum quid, nec coctum ăqua, sed tantum assum igni*, 12, 9. Hig brohton him ðǽl gebrǽddes fisces *illi obtŭlērunt ei partem piscis assi*, Lk. Bos. 24, 42. Genime ðysse ylcan wyrte wyrttruman gebrǽde on hátan axan *let him take roots of this same herb roasted on hot ashes*, Herb. 60, 3; Lchdm. i. 162, 17.

ge-brǽgd *drew*, Beo. Th. 3133; B. 1564; *p.* of ge-bregdan.

ge-brǽgd, es; *m.* [ge-, brǽgd *deceit*] *Deceit, fraud*; fraus, fallācia :—Gebrǽgdas oððe leásunga ðæra wlenca *fallācia divĭtiārum*, Mt. Kmbl. Lind. 13, 22. [*Cf. Icel.* bragð *a trick*.]

ge-brǽgdnys, -nyss, e; *f. Craft, deceit*; astus, Cot. 18.

ge-breadian; *p.* ode; *pp.* od, ad [= ge-bredian] *To restore the flesh* or *body* :—Ðonne [Fēnix] þurh briddes hād gebreadad weorþeþ *eft of ascan then [the Phœnix] through youth's state is restored again from ashes*, Exon. 61 a; Th. 224, 8; Ph. 372.

ge-brec, -brǽc, es; *n.* [ge-, brec *a breaking, crash*] *A breaking, crashing, clamour, noise*; fractio, frăgor, strĕpĭtus :—Se dæg biþ dæg gebreces *the day will be a day of clamour*, Past. 35, 5; Swt. 245, 5: Hat. MS. 46 a, 17. He gehýrde ðæt gebrec ðara storma *audīto frăgōre procellārum*, Bd. 5, 1; S. 614, 3. Gebrecu feraþ ofer dreohtum [MS. dreontum] *the crashes go over multitudes*, Exon. 102 a; Th. 385, 14; Rä. 4, 44. Se biþ gebreca hlúdast *that is loudest of crashes*, 102 a; Th. 385, 6; Rä. 4, 40.

ge-brecan, he -breceþ, -bryceþ; *p.* -bræc, ðú -brǽce, *pl.* -brǽcon; *pp.* -brocen; *v. trans.* [ge-, brecan *to break*] *To break, bruise, crush, destroy, shatter, waste*; frangĕre, confringĕre, contrībŭlāre, contĕrĕre, conquassāre, attĕrĕre :—Ealra fyrenfulra fyhtehornas ic bealdlíce gebrece snióme *omnia cornua peccātōrum confringam*, Ps. Th. 74, 9. Heáfod he gebreceþ hæleða mæniges *conquassābit căpĭta multa*, 109, 7. Se snáw gebryceþ burga geatu *the snow destroys the gates of towns*, Salm. Kmbl. 613; Sal. 306. Ðú gebrǽce ðæt dracan heáfod deópe wætere *tu contrībŭlasti căpĭta drăcōnum sŭper ăquas*, Ps. Th. 73, 13. He him on fæðm gebrǽc *he crushed them into his grasp, i. e. subdued them*, Cd. 4; Th. 4, 32; Gen. 62: 97; Th. 127, 15; Gen. 2111: Bd. 3, 2; S. 525, 2. He ða mǽgþe mid grimme wæle and herge gebrǽc *provinciam illam sæva cæde ac depopŭlătiōne attrīvit*, 4, 15; S. 583, 26, MS. C. Se þuma gebrocen wæs *the thumb was broken*, 5, 6; S. 619, 24: Andr. Kmbl. 2944; An. 1475. [*Goth.* ga-brikan: *O. H. Ger.* ga-brechan.]

gebrec-drenc, es; *m. A drink for epilepsy*; epilepticus pōtus, arteriaca? Cot. 14. v. ge-brǽcseóc.

ge-bredan; *p.* -brǽd, *pl.* -brudon; *pp.* -broden [ge-, bredan *to draw*] **I**. *to draw, unsheath, brandish*; stringĕre, evagīnāre, vibrāre :—He sweord gebrǽd *he drew his sword*, Beo. Th. 5118; B. 2562. Sweord gebrudon ða synfullan *glădium evagīnāvērunt peccātōres*, Ps. Spl. 36, 14. Gif hwá his wǽpn gebrede *if any one draw his weapon*, L. Alf. pol. 7; Th. 66, 9. Ic ðȳ wǽpne gebrǽd *I brandished the weapon*, Beo. Th. 3333; B. 1664. Cyning wælseaxe gebrǽd *the king brandished his deadly knife*, 5400;

B. 2703. **II** *to draw breath, take breath, inspire*; inspīrāre :—Đeáh he late meahte oreþe gebredan *though he could slowly take breath*, Exon. 49 b; Th. 172, 4; Gú. 1138. **III.** *to weave, plait*; nectĕre, plectĕre :—Spyrte biþ of rixum gebroden *a basket is plaited of rushes*, Homl. Th. ii. 402, 8. Herebyrne hondum gebroden *a martial corslet woven with hands*, Beo. Th. 2891; B. 1443. **IV.** *to feign, pretend*; simūlāre :—Gebræd he hine seócne *he feigned himself sick*, Chr. 1003; Erl. 139, 9.

ge-bredian; *p.* ode; *pp.* od, ad *To restore the flesh or body*; pulpōsum reddere :—Him folgiaþ fuglas scýne, beorhte gebredade, blissum hrēmige *beautiful birds follow him, brightly restored, blissfully exulting*, Exon. 64 b; Th. 237, 18; Ph. 592. v. ge-breadian.

ge-brēgan; *p.* de; *pp.* ed [ge-, brēgan *to give fear*] *To frighten, terrify*; terrēre, perterrēre :—Wæs his mód mid đam beótungum gebrēged *his mind was frightened by the threats*, Bd. 2, 12; S. 513, 14. Ic wæs mid đysse ongrislícan wæfersýne gebrēged *I was terrified by this horrible sight*, 5, 12; S. 628, 9. We sind gebrēgede *we are terrified*, Homl. Th. i. 578, 27.

ge-bregd, -brægd, es; *m. Craft, cunning*; astūtia :—Dryhten dæleþ sumum tæfle cræft, bleóbordes gebregd *the Lord allots to one skill at the table, cunning at the coloured board*, Exon. 88 a; Th. 331, 20; Vy. 71.

ge-bregd, es; *n.* [ge-, bregdan *to move to and fro*] *A moving to and fro, agitation, tossing*; vibrātio, agītātio, jactātio :—Nis đær on đam londe wedra gebregd hreóh under heofonum, ne se hearda forst *there is not in that land tossing of tempests rough under heaven, nor the hard frost*, Exon. 56 b; Th. 201, 17; Ph. 57.

ge-bregdan; *p.* -brægd, *pl.* -brugdon; *pp.* -brogden [ge-, bregdan *to vibrate, draw*]. **I.** *to draw, unsheath*; stringĕre, exīmĕre :—He hringmǽl gebrægd *he drew the ringed sword*, Beo. Th. 3133; B. 1564. He gebrægd his sweord *exēmit gladium suum*, Mt. Kmbl. Rush. 26, 51. **II.** *to feign, pretend* :—Se đe đa gebregdnan dómas dēmde *he who hath judged false judgments*, Blickl. Homl. 99, 32. [v. brægden.] v. gebredan.

ge-brēgdnes, -ness, e; *f. Fear, dread*; tĭmor, terror, Som. Ben. Lye.

gebregd-stafas; *pl. m.* [gebregd *craft, cunning*; stafas, *pl. of* stæf *a letter*] *Literary arts*; artes litĕrārum :—Ic íglanda eallra hæbbe bóca onbýrged þurh gebregdstafas *I have tasted the books of all islands through literary arts*, Salm. Kmbl. 4; Sal. 2.

ge-brehtnian; *p.* ade, ode; *pp.* ad, od *To become bright* :—Đætte he gebrehtnige *se clarificari*, Jn. Skt. p. 6, 17. Gibrehtnad [gebereht-nad, Lind.] is *clarificatus est*, Jn. Skt. Rush. 13, 31.

ge-brehtnis, se; *f. Brightness* :—Gebrehtnis *clarificatio*, Jn. Skt. p. 6, 15.

ge-brēman; *p.* de; *pp.* ed [ge-, brēman *to celebrate*] *To celebrate, make famous, honour*; celebrāre, honōrāre :—He wolde gebrēman đa Iudēiscan *he would honour the Jews*, Som. Lye.

ge-brengan; *p.* -brohte, *pl.* -brohton; *pp.* -broht; *v. trans.* [ge-, brengan *to bring*] *To bring, lead, produce, bear*; ferre, dūcĕre, prodū-cĕre :—He wēnþ đæt đone mon ǽr mǽge gebrengan on fǽrwyrde *that he thinks may bring the man earlier to a terrible fate*, Past. 62; Swt. 457, 11; Hat. MS: Salm. Kmbl. MS. A. 176; Sal. 87: 296; Sal. 147. Gif đú gebrengest *if thou bringest*, Salm. Kmbl. MS. A. 178; Sal. 88. Iudith gebrohte heáfod on đam fætelse *Judith put the head into the sack*, Judth. 11; Thw. 23, 17; Jud. 125. Đú us to eádmēdum gebrohtest *thou broughtest us to humility*; nos humiliasti, Ps. Th. 89, 17. Hý hit gebrohton burgum in innan *they brought it within the towns*, Exon. 75 b; Th. 284, 2; Jul. 691: 40 b; Th. 135, 24; Gú. 529. On þeówote gebroht *brought into slavery*, Ors. 3, 9; Bos. 66, 20. Đǽr wæs gebroht wín *there was wine brought*, Chr. 1012; Th. 269, 21, col. 1. [*O. Sax.* gi-brengean.]

ge-brengnis, -niss, e; *f. Food, support*; victus, Mk. Skt. Lind. 12, 44.

ge-brice, -bryce, es; *m.* [ge-, brice *a breaking*] *A breaking, breach*; confractio :—Gyf nā Moyses gecoren his stóde on gebrice [Lamb. gebryce] on gesihþe his *si non Moyses electus ejus stetisset in confractiōne in conspectu ejus*, Ps. Spl. 105, 22.

ge-bridlian, -bridligan; *p.* ode; *pp.* od [ge-, bridlian *to bridle*] *To bridle, restrain*; frēnāre :—He đa gesceafta nū gebridlod [MS. gebridlode] hæfþ *he has now bridled the creatures*, Bt. 21; Fox 74, 32. Đæt hí hira mód gebridligen *that they bridle their mind*, Past. 33, 1; Swt. 215, 7; Hat. MS. 41 a, 8.

ge-brihtan; *p.* te; *pp.* ed [ge-, brihtan *to brighten*] *To brighten, make beautiful*; illumināre, pulchrum reddĕre :—Gebrihted *beautiful*, Menol. Fox 272; Men. 137.

gebringan, he -bringeþ, -brincþ; *p.* -brang, -brong; *pp.* brungen [ge-, bringan *to bring*] *To bring, lead, adduce, produce, bear*; ferre, dūcĕre, addūcĕre, prodūcĕre, offerre :—He mæg đone lǽdan gást fleónde gebringan *he may bring the evil spirit to flight*, Salm. Kmbl. 176; Sal. 87: Bt. 32, 1; Fox 114, 4. Storm oft holm gebringeþ *the sea often brings a storm*, Exon. 89 b; Th. 336, 19; Gn. Ex. 51. Đe hine gebrincþ to đære byrig *which brings him to the city*, Homl. Th. i. 164, 9: 198, 20. Đa hine on yrre gebringaþ *they bring him to anger*; in ira provŏcant, Ps. Th. 65, 6.

Đæt he hine on orwēnnysse gebringe *that he may bring him to despair*, Boutr. Scrd. 20, 17: Homl. Th. i. 8, 13: Rood Kmbl. 275; Kr. 139. Đæt we đone gebringen [MS. gebringan] on ádfære *that we bring him on the way to the pile*, Beo. Th. 6010; B. 3009: Homl. Th. i. 164, 11.

ge-britnod; *part.* [ge-, brytnian *to dispense*] *Bestowed*; impensus, Som. Ben. Lye.

ge-brittan *to exhibit, give, to crumble, break small*; exhībēre, impen-dĕre, friāre, Som. Ben. Lye.

ge-broc, es; *n.* [ge-brocen, *pp. of* ge-brecan *to break*] *A breaking, broken piece, fragment*; fractio, fragmentum :—Sum biþ mid đæs innoþes gebrocum gemenged *some is mingled with fragments of the inwards*, L. M. 2, 56; Lchdm. ii. 276, 26. Đara gebroca *fragmentorum*, Mt. Kmbl. Rush. 14, 20: 15, 37. [*Goth.* ga-bruka *a fragment*.]

ge-bróc, es; *n.* [ge-, bróc *affliction*] *Affliction, sorrow*; dŏlor :—Đēh eów lytles hwæt swelcra gebróca on becume *though only a little of such sorrows comes upon you*, Ors. 3, 7; Bos. 62, 26.

ge-brocen *broken*, Bd. 5, 6; S. 619, 24; *pp. of* ge-brecan.

ge-brocen *enjoyed*, Exon. 38 b; Th. 127, 29; Gú. 392; *pp. of* ge-brúcan.

ge-brócod, -brócad, -bróced, -brócud [or -brocod?]; *part. p.* [ge-, brócod; *pp. of* brócian *to oppress, vex*] *Afflicted, broken up, injured*; afflictus, confractus :—Gif se synfulla biþ gebrócod *if the sinful be afflicted*, Homl. Th. i. 472, 3: 474, 19. Næfde se here Angelcyn ealles forswíđe gebrócod *the army had not all too much afflicted the English race*, Chr. 897; Erl. 94, 30. Sume gebrócode wæron *some were injured*, 978; Erl. 127, 12: Homl. Th. i. 476, 19. Đa óđre gebrócade aweg cómon *the others came away afflicted*, Ors. 4, 1; Bos. 78, 1. Hie wæron gebrócede *they were afflicted*, Chr. 897; Erl. 94, 30. We ealle on hǽđenum folce gebrócude wæron *we were all afflicted by the heathen folk*, Cod. Dipl. 314; A. D. 880–885; Kmbl. ii. 113, 16. [Cf. *O. H. Ger.* ga-brochōn *confringere*.]

ge-brocseóc; *adj. Lunatic, frantic*; phreneticus :—Sum gebrocseóc man *phreneticus quidam*, Bd. 4, 3; S. 570, 10. v. ge-brǽcseóc.

ge-broden *drawn, unsheathed*; *pp. of* ge-bredan.

ge-brogne, an; *n. A bush* :—Gistígeþ swoelce gibrogne *ascendet sicut virgultum*, Rtl. 19, 33.

ge-broht *brought*, Ors. 3, 9; Bos. 66, 20; *pp. of* ge-brengan.

ge-broiden *entwined*, Chr. 1104; Erl. 239, 19. v. ge-bredan III. *to weave*.

ge-brosnod, -brosnad; *part. p.* [ge-, brosnod, *pp. of* brosnian *to corrupt*] *Corrupted, decayed*; corruptus :—Gebrosnad is hús under hrófe *the house is decayed under the roof*, Cri. 8 a; Th. 2, 3; Cri. 13: 9 a; Th. 6, 15; Cri. 84. Rotudon and gebrosnode synd dolhswaðo míne *putruērunt et corruptæ sunt cicātrīces meæ*, Ps. Spl. 37, 5. Đa gebros-nodan bán *the corrupted bones*, Hy. 7, 88; Hy. Grn. ii. 289, 88.

ge-brosnodlíc; *adj. Corrupted* :—Đeós world is gebrosnodlíc *this world is corrupted*, Blickl. Homl. 115, 3.

ge-brosnung, -borsnung, e; *f.* [ge-, brosnung *corruption*] *A decaying, corruption*; corruptio :—Hí hire líchoman gemetton swá ungewemmedne and swá gesundne, swá swá heó wæs fram gebrosnunge lícumlícre willnunge clǽne and unwemme *intĕmĕrātum corpus invenēre, ut a corruptiōne concŭpiscentiæ carnālis ĕrat immūne*, Bd. 3, 8; S. 532, 36: 3, 19; S. 550, 15.

ge-brot, es; *n.* [ge-, brot *a fragment*] *A fragment*; fragmentum :—Of đam gebrote hý námon seofon wilian fulle *de fragmentis tūlērunt septem sportas plēnas*, Mt. Bos. 16, 37. Man nam đa gebrotu đe đár belifon, twelf cýpan fulle *sublātum est quod superfuit illis, fragmentōrum cophīni duodĕcim*, Lk. Bos. 9, 17.

ge-brot, es; *m. A barn-keeper*; granatārius, frumenti præfectus, N. Som. Ben. Lye.

ge-bróđor, -bróđer, -bróđra, -bróđru, -bróđro *brethren, used as the pl. of* bróđor, bróđer *for brothers collectively*; fratres conjuncti :—Begen đa gebróđor *both the brethren*, Andr. Kmbl. 2053; An. 1029: Ps. Th. 98, 6. Ic seah vi gebróđor *I saw six brethren*, Exon. 104 a; Th. 394, 12; Rä. 14, 2: 98 a; Th. 366, 12; Reb. 11. Đa gebróđer begen ætsamne *the brothers both together*, Chr. 937; Th. 206, 17, col. 1; Æđelst. 57. Wyt sind gebróđra *we two are brethren*; nos duo fratres sūmus, Gen. 13, 8. Gé synt ealle gebróđru *omnes vos fratres estis*, Mt. Bos. 23, 8: Mk. Bos. 10, 29. Twegen ǽwe gebróđro *duo germāni fratres*, Bd. 1, 27; S. 490, 28. Be đám gebróđrum twǽm *by the two brethren*, Beo. Th. 2387; B. 1191: Andr. Kmbl. 2027; An. 1016. [*Laym.* i-broðeren: *O. Sax.* gi-broðar: *O. H. Ger.* ga-pruoder: *Ger.* gebrüder.] v. bróđor.

ge-bróđorscipe, es; *m. Brothership, brotherhood, fraternity*; frater-nitas :—Đyllícne gebróđorscipe hý heóldon [MS. hĕaldan] him betweonum *such brotherhood they had among them*, Ors. 3, 11; Bos. 76, 6.

ge-brotu *fragments*, Lk. Bos. 9, 17; *pl. nom. acc. of* ge-brot.

ge-browen *brewed*, Ors. 1, 1; Bos. 22, 17: Homl. Th. i. 352, 7; *pp. of* breówan.

ge-brúcan; *p.* -breác, *pl.* -brucon; *pp.* -brocen [ge-, brúcan *to use, enjoy*] *To enjoy, eat*; perfrui, edere, manducare :—Hí đæs blǽdes gebrocen hæfdon *they had enjoyed the success*, Exon. 38 b; Th. 127, 29; Gú. 393.

Midðȳ sacerdhâd gebrêce *cum sacerdotio fungeretur*, Lk. Skt. Lind. 1, 8. Ðætte hia gebrêcon *manducarent*, Jn. Skt. Lind. 18, 28 : 6, 58.

ge-brudon *drew, unsheathed*, Ps. Spl. 36, 14; *p. pl.* of ge-bredan.

ge-bryce *a breaking, breach*, Ps. Lamb. 105, 23. v. ge-brice.

ge-bryceþ *breaks, destroys*, Salm. Kmbl. 613; Sal. 306; *3rd sing. pres.* of ge-brecan.

ge-brýcgan *to use :*—Gibrýcgende *utenda*, Rtl. 97, 33. v. brýcian.

ge-brýcsian; *p.* ade, ode; *pp.* ad, od *To use, enjoy :*—Gebrýcsiaþ *utuntur*, Rtl. 118, 39. Gebrýcsade *functus est*, 195, 1. v. brýcian, brícsian.

ge-bryddan; *p.* de; *pp.* ed *To frighten, terrify;* terrēre :—Gif ðú mec gebringest, ðæt ic sí gebrydded þurh ðæs cantices cwide Cristes lïnan *if thou wilt bring me, that I may be frightened through the word of the canticle of Christ's discipline*, Salm. Kmbl. 32; Sal. 16. v. broddetan.

ge-bryidan; *p.* de; *pp.* id [ge-, bryidan *to take*] *To take;* tollĕre, sūmĕre :—Ðonne mon hæfþ his æhte gebryid *when a man has taken* [Th. *discovered*] *his property*, L. O. 2; Th. i. 178, 11.

ge-brýsed; *part. p.* [ge-, brýsed, *pp.* of brýsan *to bruise*] *Bruised;* contrîtus :—Ðæt he his preósta ænne of horse fallende and gebrýsedne gelíce gebiddende and bletsigende fram deáþe gecyrde *ut clēricum suum cadendo contrîtum, æque ōrando ac benedīcendo a morte revocāvĕrit*, Bd. 5, 6; S. 618, 24.

ge-brýsednes, -ness, e; *f. A bruising;* contūsio, Som. Ben. Lye.

ge-brytan; *p.* te; *pp.* ed [ge-, brytan *to break*] *To break up, destroy;* confringĕre, extermĭnāre :—Gebrytte hine eofor of wuda *extermĭnāvit eam ăper de sylva*, Ps. Spl. C. 79, 14. Gebryted wið ecede *broken up with vinegar*, Med. ex Quadr. 5, 1; Lchdm. i. 348, 3.

ge-búan; *p.* -búde, *pl.* -búdon; *pp.* -búen, -bún [ge-, búan *to dwell*]. I. *intrans. To dwell, abide;* habĭtāre, versāri alĭquo lŏco :—Hí gebúdon betweoh Capadotiam and Pontum *they abode between Cappadocia and Pontus*, Ors. 1, 10; Bos. 32, 36. II. *v. a. acc. To inhabit, occupy;* inhabĭtāre, incŏlĕre :—Hū hit [ðæt hús] Hring-Dene gebún hæfdon *how the Ring-Danes had occupied it* [*the house*], Beo. Th. 235; B. 117. Ne sceal ðes wong gebúen weorþan *nor shall this field be occupied*, Exon. 37 a; Th. 120, 24; Gú. 276; Blickl. Homl. 121, 33.

ge-budon *proposed*, Ors. 3, 1; Bos. 52, 27; *p. pl.* of ge-beódan.

ge-búdon *abode*, Ors. 1, 10; Bos. 32, 36; *p. pl.* of ge-búan.

ge-búgan; *p.* ic, he -beág, -beáh, ðú -buge, *pl.* -bugon; *impert.* -búh, *pl.* -búgaþ; *pp.* -bogen [ge-, búgan *tô bow*]. I. *v. intrans. To bow* or *bow down oneself, bend, submit, turn, turn away, revolt;* se flectĕre vel inclīnāre, curvāre, declīnāre, transfūgĕre :—He cwæþ ðæt he wolde to fulluhte gebúgan *he said that he would submit to baptism*, Homl. Th. ii. 26, 10 : Boutr. Scrd. 22, 43 : Bt. Met. Fox 25, 128; Met. 25, 64. Heó on flet gebeáh *she bowed to the floor*, Beo. Th. 3085; B. 1540 : 5953; B. 2980. Se wyrm gebeáh snúde tosomne *the worm quickly bent together*, 5128; B. 2567. Hí gebugon to Iosue and to Israhéla bearnum *transfūgĕrit ad Iosue et ad filios Israel*, Jos. 10, 4. Ne ðú ne gebúh fram dære æ on ða swiðran healfe ne on ða wynstran *ne deolines ab lēge ad dextēram vel ad sinistram*, 1, 7. Ðæt ðú to sæmran gebuge *that thou should bow to worse*, Exon. 71 a; Th. 264, 9; Jul. 361. Eall folc him to gebogen wæs *all people submitted to him*, Chr. 1013; Erl. 148, 2, 21 : L. Edm. S. 4; Th. i. 250, 1. Ðe ær fram him gebogene wæron *who had formerly turned from them*, Ors. 2, 5; Bos. 45, 44. II. *v. trans. acc. To bow to, turn towards;* inclīnāre ad :—Sum fletreste gebeáh *one bowed to his domestic couch*, Beo. Th. 2487; B. 1241. Monig snellíc særinc selereste gebeáh *many a keen seaman bowed to his hall-couch*, 1385; B. 690. DER. in-gebúgan.

ge-búgian, -bógian; *p.* ode; *pp.* od; *v. trans.* [ge-, búgian II. *to inhabit, occupy*] *To inhabit, occupy;* inhabĭtāre, incŏlĕre :—Hý hit ne mágon ealle gebúgian *they cannot inhabit it all*, Bt. 18, 1; Fox 62, 10.

ge-búh *turn from*, Ors. 1, 7; *impert.* of ge-búgan.

ge-bún *inhabited*, Ors. 1, 1, § 13; Bos. 20, 2, 3, 7; *pp.* of ge-búan, q.v.

ge-bunden *bound*, Mk. Bos. 15, 7; *pp.* of ge-bindan.

gebundennes, -ness, e; *f.* [ge-bunden, *pp.* of ge-bindan *to bind*] *A binding, an obligation;* oblĭgātio :—Gibundennises *ligandi*, Rtl. 59, 11. Ða abúgendan on gebundennesse oððe to bændum *declīnantes in oblĭgātiōnes*, Ps. Lamb. 124, 5.

GEBÚR, es; *m. A dweller, husbandman, farmer, countryman*, BOOR; incŏla, agricŏla, cŏlōnus :—Gif he on gebúres húse gefeohte *if he fight in a boor's house*, L. In. 6; Th. i. 106, 8. Gebúres gerihte *rights of the boor*, Th. i. 434, 3. See the section to which this heading belongs for an account of the relation of the 'gebur' to his lord. [Cf. *Icel.* búi [in compounds] and bónde [v. Cl. and Vig. Dicty. s. v.], and see Kemble's Saxons in England, i. 131 : *Plat.* buur, *m*; in earlier time *a neighbour, a citizen;* now *a farmer, a peasant: Dut. Frs.* boer, *m : Ger.* bauer, *m:* in *Silesia* gebaur, *m*. The *Old Franc.* and *Al.* writers designate by puarre, buara *an inhabitant*, and by gibura, giburo *a peasant, a farmer*. From the *A.-S.* búan *to dwell, inhabit*.] DER. neáh-gebúr.

gebúr-gerihta; *pl. n. A boor's* or *farmer's rights* or *dues;* cŏlōni consuetūdines :—Gebúrgerihta sýn mislíce, gehwâr hý sýn hefige, gehwâr

eác medeme *geburi consuetudines inveniuntur multimode, et ubi sunt onerose et ubi sunt leviores aut medie*, Th. i. 434, 4.

ge-burh-scipe, es; *m. A township;* municipium, municipatus :—On ðam ylcan geburhscipe [MS. B. gebúrscipe] *in the same township*, L. Ed. 1; Th. i. 158, 21. v. burh-scipe.

ge-burnen *burnt*, Ors. 4, 2; Bos. 79, 19; *pp.* of beornan.

ge-búr-scipe, es; *m.* [ge-búr *a dweller;* scipe *state, condition*] *A neighbourhood, an association of the dwellers in a certain district acknowledged by the state;* colonia, vicinia, consociatio :—On ðam ylcan gebúrscipe *in the same neighbourhood*, L. Ed. 1; Th. i. 158, 21 [MS. B].

ge-býa; *p.* -býde *To dwell :*—Gibýaþ mið ðæm *habitabit cum eis*, Rtl. 71, 3. Gebýde *habitavit*, Mt. Kmbl. Lind. 1, 23; 4, 13. Gibýe *posside*, Rtl. 165, 20. v. gebúgian.

ge-bycgan, -bicgan, -bicgean; ic -bycge, -bicge, ðú -bygest, -bigest, he -bygeþ, -bigeþ, -bigþ, *pl.* -bycgaþ, -bicgaþ; *p.* -bohte, *pl.* bohton; *pp.* -boht *To buy, procure, purchase, redeem;* emere, redimere :—Hí meahton hefonríce gebycgan [MS. gebycggan] *they could buy the kingdom of heaven*, Past. 59, 2; Swt. 449, 15; Hat. MS. Cyning sceal mid ceápe cwêne gebicgan *a king shall buy a queen with goods*, Exon. 90 a; Th. 338, 22; Gn. Ex. 82. [For this use of the verb see Grimm R. A. pp. 421 sqq. where similar phrases in other dialects are given.] Ðæt hý môston friþ gebicgean *that they might buy peace*, L. Eth. ii. 1; Th. i. 284, 13. Ic gebycge bât *I buy a boat*, Exon. 119 a; Th. 458, 11; Hy. 4, 99. Ðæt hí man beágum gebycge *that one may buy her with bracelets*, Menol. Fox 551; Gn. C. 45 : L. H. E. 16; Th. i. 34, 3 : L. C. S. 15; Th. i. 384, 11. Bútan he hine æt ðam cynge gebicge *unless he buys it of the king*, L. Edg. ii. 3; Th. i. 266, 18. Gif mon hwelcne ceáp gebygeþ *if a man buy any kind of cattle*, L. In. 56; Th. i. 138, 10. Gif man mægþ gebigeþ *if a man buy a maiden*, L. Eth. 77; Th. i. 22, 1. Se man gebigþ ðone æcer *homo emit agrum illum*, Mt. Bos. 13, 44. Hine gebohte Putifar *emit eum Putiphar*, Gen. 39, 1 : Cd. 149; Th. 187, 15; Exod. 151 : Beo. Th. 1951; B. 973 : 4955; B. 2481. God us ðeópum ceápe gebohte *Deus redemit nos alto pretio*, L. C. E. 18; Th. i. 370, 28 : Exon. 29 a; Th. 89, 27; Cri. 1463 : 98 a; Th. 368, 25; Seel. 30. Ðú blôde gebohtest bearn Israêla *thou hast redeemed the children of Israel with thy blood*, Hy. 8, 26; Hy. Grn. ii. 290, 26. Lundenwaru him friþ gebohton *the Londoners bought themselves peace*, Chr. 1016; Erl. 159, 23. Nænig usic mið leáne gebohte *nemo nos conduxit*, Mt. Kmbl. Rush. 20, 7. Geboht þeówa *emptitius*, Ælfc. Gl. 86; Som. 74, 33; Wrt. Voc. 50, 16 : Gen. 17, 12.

ge-bycnian *to beckon, shew, indicate*, St. A. 44, 12 : Evan. Nic. 4, 13 : Ælfc. Gr. 28, 3; Som. 30, 48, MS. D : 37; Som. 39, 40, MS. D. v. gebícnian.

gebýdan *to abide, wait*. v. gebídan.

ge-býgan, -bígan, -býgean, -bígean, -bêgan; *p.* de; *pp.* ed; *v. trans. To bow, bend, turn, inflect or decline a part of speech, recline, twist, bow down, humble, abase, bring under, subdue, crush;* flectere, incurvare, inflectere, declinare, reclinare, torquere, humiliare, confringere :—Gebígdum cneówum *flexis genibus*, Bd. 4, 10; S. 578, 28. Hý gebýgdon sáwle mîne *incurvaverunt animam meam*, Ps. Spl. 56, 8 : Gen. 27, 29. Ðá hí hwæsne beág ymb mín heafod gebýgdon *then they twisted a sharp crown around my head*, Exon. 29 a; Th. 88, 25; Cri. 1445. Hý ealle to him gebígde *he brought them all under him*, Ors. 3, 9; Bos. 64, 15 : 5, 3; Bos. 104, 11. Ðæt hig ealle leóda sceoldan gebígan to geleáfan *that they should subdue all nations to the faith*, Ælfc. T. Lisle 30, 5. Íserne steng gebígeþ *vectes ferreos confringit*, Ps. Th. 106, 15 : 72, 17; 143, 18. v. býgan.

ge-býgean, -bígean; *v. trans. To bow, bend, turn, bow down, subdue, crush :*—Ðú miht leon and dracan liste gebýgean *conculcabis leonem et draconem*, Ps. Th. 90, 13. Gebígean to synnum *adigere ad peccata*, Alb. resp. 68 [Lye]. v. ge-býgan.

ge-býgednys, -nyss, e; *f. A bending, declining, declension, case.* v. gebígednys.

ge-býgel; *adj. Subject, submissive, obedient;* subjectus :—Gebýgle to dônne *to make obedient*, Chr. 1091; Th. 358, 38 : 1105; Th. 367, 22.

ge-býgendlíc; *adj. Bending, flexible, declined with cases.* v. gebígendlíc.

ge-býgeþ *buys*, L. In. 56; Th. i. 138, 10; *pres.* of ge-bycgan.

ge-byld, e; *f.* [byld *boldness*] *Boldness, courage;* audācia :—Calep hig gestilde and cwæþ mid gebylde *Caleb quieted them and said with courage*, Num. 13, 31 : Jos. 4, 9.

gebyld; *adj. Bold, courageous;* audax :—Gebyld swiðe ðurh God, Jud. 4, 14.

ge-bylded, -bælded, -byld; *part.* [ge-, byldan *to make bold*] *Emboldened, encouraged, animated;* corrŏbŏrātus, anĭmātus :—Wæs Laurentius mid ðæs apostoles swingum and trymnessum swíðe gebylded *apostŏli flagellis sĭmul et exhortatiōnibus anĭmātus ĕrat Laurentius*, Bd. 2, 6; S. 508, 22. He wið mongum stód ealdfeónda elne gebylded *he stood against many of the old fiends, emboldened with courage*, Exon. 39 b; Th. 130, 31; Gú. 446. Se Barac, gebyld swiðe þurh God, feaht him togeánes *Barak, much encouraged by God, fought against them*, Jud. 4, 14. Hý wæron gebylde

they were encouraged, Ors. 4, 1; Bos. 77, 25. We us bletsiaþ gebylde ðurh God *we bless ourselves emboldened by God*, H. R. 105, 17.

ge-bylgan; *p.* de; *pp.* ed *To cause to swell, to make angry :*—Leóhtlíce gebylged *leviter indignata*, Bd. 4, 9; S. 577, 24. v. ge-belgan.

ge-bylged *made angry; pp. of* ge-bylgan.

ge-byrd; *gen. dat.* -byrde; *acc.* -byrde, -byrd; *pl. nom. gen. acc.* a; *dat.* um; *f.:* ge -byrdo; *indecl. in* s; *f.:* found in both *s.* and *pl.* without any apparent difference of meaning. **I.** *birth, origin, beginning, parentage, family, lineage;* nativitas, origo, stirps, genus :—Bearnes þurh gebyrde *through the birth of a child*, Exon. 9 a; Th. 5, 28; Cri. 76. þurh bearnes gebyrd *through child-birth*, 8 b; Th. 3, 18; Cri. 38. On dæg gebyrde *die natalis*, Mt. Kmbl. Rush. 14, 6. Wîtgan cýðdon Cristes gebyrd *prophets announced Christ's birth*, 8 b; Th. 5, 5; Cri. 65. Bearnes gebyrda *the infant's birth*, 18 b; Th. 45, 24; Cri. 724: L. Edg. C. 13; Th. ii. 246, 22. Cennan bearn mid gebyrdum *to bring forth children by birth*, Exon. 89 a; Th. 334, 32; Gn. Ex. 25. Wæs he líchomlícre gebyrde æðeles cynnes *erat carnis origine nobilis*, Bd. 2, 7; S. 509, 15. Of ðære cynelícan gebyrdo *de stirpe regiâ*, 5, 7; S. 621, 8, note 8. Be ðam gebyrdum *concerning parentage*, Bt. 30, 1; Fox 108, 19. **II.** *nature* [*what a man is* natu *by birth, or to what he is* natus *born*], *quality, state, condition, lot, fate;* natura, qualitas, conditio, sors, fatum :—God âna wât ymb ðæs fugles gebyrd *God alone knows concerning the bird's nature*, Exon. 61 a; Th. 223, 16; Ph. 360. Ic cann engla gebyrdo *I know the nature of the angels*, Cd. 27; Th. 37, 2; Gen. 583. Æghwilc gylt be hys gebyrdum *every one pays according to his condition*, Ors. 1, 1; Bos. 20, 35. Nâh seó môdor geweald bearnes blǽdes, ac sceal on gebyrd faran an æfter ânum *the mother hath not power over her child's happiness, but according to his fate* [*what he is born to*] *one shall go after another*, Salm. Kmbl. 770; Sal. 384. Hie on gebyrd hruron gâre wunde *they fell according to their fate, wounded by the spear*, Beo. Th. 2153; B. 1074. Or in the last two instances may 'gebyrd' be referred to 'gebyrian' *to happen?* [*O. Sax.* gi-burd, *f. nativitas, genus:* Ger. geburt, *f.:* Goth. ga-baurþs, *f.*] DER. eág-gebyrd, eorl-, sib-, weoruld-. v. beran.

ge-byrd; *part. p.* [beard a beard] *Bearded;* barbâtus :—Gebyrd barbâtus, Ælfc. Gr. 43; Som. 45, 11. Gebyrdne hine he gesihþ *he sees himself bearded*, Lchdm. iii. 200, 4.

ge-byrd; *part. p. Burdened :*—Gebyrde sindun *onerati estis*, Mt. Kmbl. Rush. 11, 28.

ge-byrd-dæg, es; *m. A birth-day;* natalis dies :—On Herodes gebyrddæge *die natalis Herodis*, Mt. Bos. 14, 6.

ge-byrde, -bierde; *adj. Inborn, innate, natural;* innatus, ingenitus, naturalis :—Ne him nis gebyrde ðæt hî ðé folgien *it is not natural to them that they should follow thee*, Bt. 14, 1; Fox 40, 34. Him gebyrde is ðæt he gencwidas gleáwe hæbbe *to him it is natural that he should have prudent replies*, Elen. Kmbl. 1183; El. 593.

ge-byrdelíce; *adv. Suitably, orderly :*—Ymbsittaþ ða burg swíðe gebyrdelíce *ordinabis adversus eam obsidionem*, Past. 21, 5; Swt. 160, 19.

ge-byrdo *birth, nature, condition.* v. ge-byrd.

ge-byrd-tíd, e; *f. Birth-tide, time of birth;* natale tempus :—Se dæg com Herodes gebyrdtíde *dies accidit Herodis natalis*, Mk. Bos. 6, 21: Gen. 40, 20. Fram gebyrdtíde brêmes cyninges *from the birth-tide of the glorious king*, Chr. 973; Th. 224, 36; Edg. 12.

ge-byrd-wiglære, es; *m. A birth-diviner;* ex natalibus divinator, astrologus, Ælfc. Gl. 4; Wrt. Voc. 17, 14.

ge-byrd-wítega, an; *m. A birth-prophet, an astrologer;* ex natalibus propheta, astrologus, mathematicus, Ælfc. Gl. 112; Wrt. Voc. 60, 12.

ge-byre, es; *m. The time at which anything happens, a favourable time, an opportunity;* occasio, opportunitas :—Hwonne him eft gebyre weorþe, hâm cymeþ *when there shall again be an opportunity to him he will come home*, Exon. 90 b; Th. 340, 3; Gn. Ex. 105. [*O. H. Ger.* gaburi, *f. eventus, casus.*] v. byre, ge-byrian.

ge-byredlíc; *adj. Suitable, fitting, due;* debitus, congruus :—Herenissa gibyredlíce *laudes debitas*, Rtl. 165, 22. Gibyredlícre wordunge *congruo honore*, 78, 10; 8, 23.

ge-byredlíce; *adv. Conveniently;* convenienter, Rtl. 16, 31.

ge-byrelíc beón :—Ne sint gebyrelíco Iudea to Samaritaniscum *non coutuntur Iudæi Samaritanis*, Jn. Skt. Lind. 4, 9.

ge-byreþ *bears, produces*, L. Ethb. 78; Th. i. 22, 4. v. ge-beran.

ge-byreþ *happens, becomes, behoves.* v. ge-byrian.

ge-byrgan; *p.* de; *pp.* ed *To bury;* sepelire :—Wæs on helle gebyrged *sepultus est in inferno*, Lk. Bos. 16, 22. v. byrgan.

ge-byrgan; *p.* de; *pp.* ed *To taste;* gustare :—Nô he fôddor þigeþ, nemne mele-deáwes dǽl gebyrge *it touches not food, except that of honeydew it tastes a portion*, Exon. 59 b; Th. 215, 30; Ph. 261: Cd. 24; Th. 31, 10; Gen. 483. v. byrgan.

ge-byrhtan; *p.* te; *pp.* ed *To make bright, brighten;* illûmināre, clârifîcāre :—Ys his nama fôr him neóde gebyrhted *præclārum nômen eôrum côram ipso*, Ps. Th. 71, 14. v. ge-berhtan.

gebyrhte *declared.*

ge-byrian, -býrigan, -bírian; *3rd sing.* eþ; *p.* ede; *pp.* ed; *3rd sing.*

---

aþ; *p.* ode; *pp.* od. [The cognate words point to a short vowel.] **I.** *v. intrans. To happen, to fall out, to pertain to, belong to;* evenire, accidere, contingere, pertinere ad :—Ðonne hit gebýrigan mæg *when it may happen*, Bt. Met. Fox 4, 22; Met. 4, 11. Syle me mínne dǽl mínre ǽhte, ðe me to gebýreþ *da mihi portionem substantiæ quæ me contingit*, Lk. Bos. 15, 12. Hit nis nâuht ðæt mon cwiþ ðæt ǽnig þing weás gebýrige *it is naught* [*nothing*] *that men say that anything happens by chance*, Bt. 40, 5; Fox 240, 28: Ps. Th. 4, 5. Ðâs ðing gebýrigeaþ ǽryst *oportet primum hæc fieri*, Lk. Bos. 21, 9. Men cwǽdon gió ðonne him hwæt unwénunga gebýrede, ðæt ðæt wǽre weás gebýred *men said formerly, when anything happened to them unexpectedly, that it happened by chance*, Bt. 40, 6; Fox 242, 4: 16, 2; Fox 54, 3. Gebýrode, Ex. 14, 28. And feng to ealle ðam landum ðe ðǽr-to gebýredon *and took to all the lands which thereto belonged*, Chr. 910; Erl. 101, 6. **II.** *v. impers. It pertains to, it is fitting* or *suitable, it becomes, it behoves;* pertinet ad, convenit, oportet, decet :—Swâ gebýreþ ǽlcum Cristnum men *as it becómeth every Christian man*, Ps. Th. 39, Arg. Swâ ðǽr-to gebýrige *as may thereto be becoming*, L. Eth. vi. 22; Th. i. 320, 11: L. Ath. v. 1, 4; Th. i. 230, 3. Ne gebýraþ hit swâ *non ita convenit*, Gen. 48, 18. Him ne gebýraþ to ðám sceápum *non pertinet ad eum de ovibus*, Jn. Bos. 10, 13. Him gebýrede to ðám þearfum *de egenis pertinebat ad eum*, 12, 6. Hine man byrigde swâ him wel gebýrede *they buried him as well became him*, Chr. 1036; Th. 294, 22. On ealle þeóda gebýraþ beón ðæt gôdspel gebodod *in omnes gentes oportet prædicari evangelium*, Mk. Bos. 13, 10. [*Orm.* 3rd pres. birrþ *it becomes*, 3rd *p.* birrde: *Havl. p.* birde: *R. Brun.* burd: *Gaw. gloss.* burde: *O. Sax.* giburian *accidere, evenire, contingere:* Ger. gebühren: *O. H. Ger.* gaburjan *pertinere, contingere:* O. Nrs. byrja *incipere, inchoare, decere.*] v. býrian.

ge-byrigednes, -ness, e; *f. A burial;* sepultûra :—Æfter monigum geárum his gebyrigednesse *post multos sepultûræ annos*, Bd. 4, 32; Whelc. 365, 31.

ge-byrman; *p.* de; *pp.* ed *To ferment with* BARM, *to leaven;* fermentare :—Bryden wæs ongunnen ðætte Adame Eue gebyrmde *the drink was prepared which Eve fermented for Adam*, Exon. 47 a; Th. 161, 6; Gû. 954. Þrymme gebyrmed *fermented with greatness*, 84 a; Th. 316, 2; Môd. 42. Ne beó nân beorma on eówrum húsum; swâ hwilc man swâ ytt gebyrmed, forwyrþ *non erit fermentum in domibus vestris; quicumque comederit fermentatum, peribit*, Ex. 12, 15: 12, 19. v. beorma.

ge-byrmed BARMED, *fermented, leavened;* fermentatus, Ex. 12, 15, 19. v. ge-byrman.

ge-byrnod; *part. p.* [byrne *a coat of mail*] *Furnished with a coat of mail;* lôrîcātus :—Gebyrnod *lôrîcātus*, Ælfc. Gr. 43; Som. 45, 12. [*Laym.* i-burned.]

ge-byr-tíd, e; *f. Birth-tide;* natale tempus, Chr. 1087; Th. 353, 34. v. ge-byrd-tíd.

ge-býsgian [or - bysgian?], -bîsgian, -býsigan, -biesgian; *p.* ode, ade; *pp.* od, ad [ge, býsgian *occupare, affligere, tribulare*] *To occupy, busy, afflict, trouble, vex, oppress, overcome, agitate, weaken, destroy;* occupare, affligere, turbare, vexare, opprimere, conripere, conficere :—He mid gýmeleáste húru us gebýsgaþ *saltem negligentia nos occuparet*, Bd. Whelc. 310, 20. Ðonne hî hî gebýsgiaþ mid woruldlícum hordum *when they busy themselves with worldly treasures*, Homl. Th. i. 524, 14. Ic eom lég býsig, fýre gebýsgad *I am a busy flame, with fire occupied*, Exon. 108 a; Th. 412, 21; Râ. 31, 3. Môde gebýsgad *in mind afflicted*, Exon. 87 b; Th. 328, 20; Vy. 20: 47 b; Th. 162, 34; Gû. 985. Is môdigra mægen miclum gebýsgod *the strength of the valiant is much troubled*, Andr. Kmbl. 790; An. 395. Moyses wearþ gebýsgad *for heora yfelum vexatus est Moyses propter eos*, Ps. Th. 105, 25: 76, 6. Wintrum gebýsgad *oppressed with years*, Exon. 58 a; Th. 208, 28; Ph. 162: 62 a; Th. 227, 25; Ph. 428. Ádle gebýsgad *with disease oppressed*, 49 a; Th. 170, 10; Gû. 1109. Slǽpe gebiesgad *with sleep overcome*, Exon. 96 a; Th. 358, 2; Pa. 39. Ne ðǽr wæter fealleþ, lyfte gebýsgad *water falls not there, agitated in air*, Exon. 56 b; Th. 201, 26; Ph. 62. Wearþ môdgeþanc miclum gebýsgad, þurh ðæs þeódnes word, ombehtþegne *the mind of the disciple was greatly agitated through his lord's words*, 50 a; Th. 173, 34; Gû. 1170. Sceaða biþ gebýsigod, swíðe gestilled *the fiend shall be destroyed, made very still*, Salm. Kmbl. 234; Sal. 116.

ge-býsigan *to occupy, afflict, overcome*, Salm. Kmbl. 234; Sal. 116. v. ge-býsgian.

ge-bysmerian *to deride*, Ps. Lamb. 58, 9. v. ge-bismerian.

ge-bysmrian *to mock, deride, provoke*, Ps. Th. 77, 56. v. ge-bismerian.

ge-býsnian [or -bysnian; cf. Goth. busns]; *p.* ode; *pp.* od *To give or set an example;* exemplum dare :—Se' man biþ hérigendlíc, ðe óðrum gebýsnaþ *the man is praiseworthy who sets an example to others*, Homl. Th. ii. 406, 17. v. býsnian.

ge-býsnung, e; *f.* [býsnung *an example*] *An example;* exemplum :—He sealde sóþe gebýsnunge *he gave true example*, Ælfc. T. Lisle 38, 3. Mâ manna beóþ gecyrrede þurh his gebýsnunge to Godes hêrunge *more* [*of*] *men will be turned through his example to the praise of God*, HomL

Th. i. 494, 23. Ne dô ge nâ be his gebýsnùngum *do ye not according to his examples*, Homl. Th. ii. 48, 35.

ge-býtlian [*or rather* -bytlian, cf. botl]; *p.* ode; *pp.* od [býtlian *to build*] *To build*; ædíficāre :—Eal Godes geladung is ofer ðam stáne gebýtlod *all God's church is built on that stone*, Homl. Th. i. 368, 18.

ge-bytlu; *indecl. f. A building* :—Man bytlode áne gebytlu, and ða wyrhtan worhton ða gebytlu on ðam Sæternes-dæge, and wæs ðá forneán geendod *they were building a building, and the workmen were making the building on the Saturday, and it was then very nearly finished*, Homl. Th. ii. 580, 32; 172, 23; 580, 21. He gýmþ grædelíce his gafoles, his gebytlu *he attends greedily to his rent, his buildings*, i. 66, 11; 68, 2. He eów sylþ micle burga and ða sélustan gebytlu *he will give you great cities and the best buildings*, Deut. 6, 10. v. botl.

ge-bytlung, e; *f.* [bytlung *a building*] *A building*; ædifícium :—Ic inc ealle ða gebytlunge gewisslíce tǽhte *I shewed you two plainly all the building*, Homl. Th. ii. 172, 27; 16.

ge-cǽlan; *p.* de; *pp.* ed; *v. trans. To cool*; refrigerare :—Send Lazarum, ðæt he dyppe his fingeres liþ on wætere, and míne tungan gecéle *mitte Lazarum, ut intingat extremum digiti sui in aquam, ut refrigeret linguam meam*, Lk. Bos. 16, 24.

ge-cǽlcian; *p.* ode; *pp.* od *To whiten*; dealbare :—Gecǽlcad *dealbatus*, Mt. Kmbl. Lind. 23, 27.

ge-cǽnenis, gecǽnes *a calling, vocation*. v. gecigednes. [Cf. gecænnan?]

ge-cǽnnan *to declare, clear, prove*; advocare, purgare, manifestare :—Hine gecænne ðæt he ðane banan begeten ne mihte *let him prove that he could not obtain the slayer*, L. H. E. 2; Th. i. 28, 2: 4; Th. i. 28, 8. DER. cennan *to declare*, q. v. and cf. *Goth.* ga-kannjan *to make known*.

ge-cafstrian; *p.* ode; *pp.* od [cæfester *a halter*] *To bridle, restrain*; frænare, restringere :—Swelce sió geþyld hæbbe ðæt môd gecafstrod *as if patience has restrained the mind*, Past. 33, 4; Swt. 218, 22; Cot. MS. 42 a.

ge-camp, -comp, es; *m.* [camp *a contest, war*] *Warfare, a contest, battle*; mīlitia, certāmen, pugna :—Gecampes feld *certāmĭnis campus*, Greg. Dial. 2, 3. On gecampe *in warfare*, Byrht. Th. 136, 18; By. 153. Iosue com mid gecampe to him mid eallum his here *vēnit Iosue et omnis exercĭtus cum eo adversus illos*, Jos. 11, 7. In gecomp *in agonia*, Lk. Skt. Lind. 22, 44.

ge-campian, -compian; *p.* ode; *pp.* od *To fight* :—He wolde gecompian wiþ ðone awerigdan gást *he wished to fight with the accursed spirit*, Blickl. Homl. 29, 17.

ge-canc, es; *n.* [?] *A mock, gibe*; ludibrium, vituperium, Som : Hpt. Gl. 441, 510. [Cf. *Icel.* kank, *n.*; kank-yrði *gibes*; kankast *to jeer, gibe*; cank *to talk of anything, to cackle*, Halliwell : *Scot.* cangle *to quarrel*.]

ge-ceápian; *p.* ode; *pp.* od [ceápian *to bargain*] *To buy, purchase, trade*; ĕmĕre, negotiari :—He sæde, ðæt man nâne burh ne mihte ýþ mid feó geceápian *he said that no city could be more easily bought with money*, Ors. 5, 7; Bos. 106, 16. Geoweorþa geceápode mid his feó æt ðam consule *Jugurtha bribed the consul with his money*, 5, 7; Bos. 106, 10, 12. Ðone mândrinc geceápaþ *he buys the deadly drink*, Exon. 106 b; Th. 406, 7; Rä. 24, 13. Gif he hit næbbe beforan gódum weotum geceápod *if he have not bought it before good witnesses*, L. In. 25; Th. i. 118, 14: L. Ethb. 77; Th. i. 22, 1: Gen. 43, 21. Hú feolu ĕghwelc geceápad wêre *quantum quisque negotiatus ĕsset*, Lk. Skt. Rush. 19, 15.

ge-cearfan, -ceorfan; *p.* earf *To kill, cut off* or *up*; interficere, decollare :—Gie soecas mec gecearfa *quaeritis me interficere*, Jn. Skt. Lind. 8, 37; 40. Ðone ic gecearf *quem ego decollavi*, Mk. Skt. Lind. 6, 16.

ge-ceás *chose*, Bd. 1, 6; S. 476, 17; *p.* of ge-ceósan.

ge-cêgan *to call, to call upon*, Ps. Spl. 48, 11: 49, 1. v. ge-cígan.

ge-cêgung, -cígung, e; *f. A calling*; invocatio :—Giceigingcum ũsum *invocationibus nostris*, Rtl. 97, 37.

ge-cêlan; *p.* de; *pp.* ed. **I.** *v. trans. To make cold, to cool, allay*; refrigerare :—Ðæt man ne mæge wæterseóces þurst gecêlan *that any one might not allay the thirst of a watersick (dropsical) man*. **II.** *v. intrans. To become cold, to be refreshed*; refrigerari :—Forlæt me ðæt ic gecêle ǽrðam ðe ic gang *remitte mihi ut refrigerer priusquam abeam*, Ps. Spl. 38, 18. v. cêlan, calan.

gecele *an icicle*. v. gicel.

ge-celf; *adj. Great with calf* :—Ðæt ic hæbbe hnesce litlingas, and gecelfe cý mid me *that I have tender children and incalving cows with me*, Gen. 33, 13; quod parvŭlos hăbeam tĕnĕros, et boves fetas mecum, Vulg. Gen. 33, 13.

ge-cêlnes, -nys, -nyss, e; *f. Coolness*; refrigerium :—For wegferendra gecêlnysse *ob refrigerium viantium*, Bd. 2, 16; S. 520, 6. v. cêl-nes.

ge-cenenis, se; *f. A delight*, Som.

ge-cennan; *p.* de; *pp.* ed. **I.** *to beget, bring forth, produce* :—Gicende *edidit*, Rtl. 108, 29. From forleigere ne aru we gecenned *ex fornicatione non sumus nati*, Jn. Skt. Lind. 8, 41. [Cf. *O. H. Ger.* ki-chennan, *gignere*.] **II.** *to clear, declare, prove*; purgare, advocare, manifestare :—Gif he gecenne *if he prove*, L. Eth. ii. 8; Th. i. 288, 17.

Ic ðê êcne God ǽnne gecenne *I confess thee the only everlasting God*, Grn. Hy. 10, 4. DER. cennan. v. gecænnan.

ge-cennice, an [?]; *f. Genetrix*, Rtl. 68, 39.

ge-ceolan; *p.* de; *pp.* ed; *v. trans. To make cold, to cool*; refrigerare, Lk. Skt. Lind. 16, 24. v. gecêlan, calan.

ge-ceósan; to geceósanne, geceósenne; ic -ceóse, ðú -ceósest, -cýst, -císt, he -ceóseþ, -cýsþ, -cýst, *pl.* -ceósaþ; *p.* -ceás, *pl.* -curon; *pp.* -coren *To elect, choose, decide, prove, approve*; eligere, præeligere, seligere, asciscere, petere, nancisci :—Nû monna gehwylc geceósan môt swá helle hiénþu swá heofones mǽrþu *now every man may choose either hell's humiliations or heaven's glories*, Exon. 16 b; Th. 37, 9; Cri. 590. He wolde geceósan *he would choose*, Bd. 4, 11; S. 579, 9 : Salm. Kmbl. 780; Sal. 389. Swá ðê leófre biþ to geceósanne *ut tibi placeat eligere*, Elen. Kmbl. 1210; El. 607. To geceósenne *to choose*, Beo. Th. 3706; B. 1851. Gif ðú ða swiðran healfe gecíst *si tu dextĕram elĕgĕris*, Gen. 13, 9. Eall ðæt folc heom ðæt gecuron *all the people approved for themselves of that plan*, St. And. 36, 14. He hyht geceóseþ *he chooseth hope*, Frag. Kmbl. 77; Leas. 40 : Exon. 79 b; Th. 298, 21; Crä. 88 : Ps. Th. 64, 4 : Exon. 61 a; Th. 225, 1; Ph. 382. Ðonne hine man to gewitnysse gecýsþ *when he is chosen as witness*, L. Edg. S. 6; Th. i. 274, 15. Hý wíc geceósaþ *they choose a habitation*, Exon. 97 a; Th. 362, 16; Wal. 37 : 95 a; Th. 354, 36; Reim. 56 : Ps. Th. 136, 7. Se geceás Maximianum to fultume his ríces *he chose Maximianus to the help of his kingdom*, Bd. 1, 6; S. 476, 17 : Ex. 18, 25. Cain geceás wíc *Cain chose a dwelling*, Cd. 50; Th. 64, 17 : Gen. 1051 : 91; Th. 115, 29 ; Gen. 1927 : 129; Th. 164, 3; Gen. 2709 : Beo. Th. 2407; B. 1201 : 4930; B. 2469 : 5270; B. 2638 : Exon. 45 b; Th. 154, 34; Gú. 852 : 46 b; Th. 158, 12; Gú. 907 : Elen. Kmbl. 2076; El. 1039 : 2330; El. 1166 : Apstls. Kmbl. 38; Ap. 19 : Ps. Th. 77, 67 : 131, 14 : Byrht. Th. 135, 5; By. 113. Gecuron híg ða gódan on hyra fatu *elegerunt bonos in vasa*, Mt. Bos. 13, 48 : Gen. 6, 2 : Ors. 1, 14; Bos. 37, 26 : Ps. Th. 105, 27. Ðê wíc geceós on ðissum lande *choose thee a habitation in this land*, Cd. 130; Th. 164, 30; Gen. 2722 : Beo. Th. 3523; B. 1759 : Exon. 80 b; Th. 303, 3; Fä. 47. Ðeáh hí gecure bútan cræftum cyninga dysegast *though the most foolish of kings chose them without skill*, Bt. Met. Fox 15, 21; Met. 15, 11. Se foresprecena wer for hine in bisceop-háde wæs gecoren *the aforesaid man was chosen into bishophood for him*, Bd. 4, 23; S. 594, 29 : 4, 1; S. 564, 12. Ðætte eallra heora dôme gecoren wære *ut universorum judicio probaretur*, Bd. 4, 24; S. 597, 31. Ðá Abraham gewât Drihtne gecoren *then Abraham, the chosen of the Lord, departed*, Cd. 86; Th. 109, 5; Gen. 1818 : 179; Th. 225, 7; Dan. 150 : 212; Th. 261, 35; Dan. 736 : Andr. Kmbl. 647; An. 324 : Exon. 108 a; Th. 413, 23; Rä. 32, 10. He wiste ðone láreów gecorenne *he knew the teacher chosen*, Exon. 47 b; Th. 162, 18; Gú. 977. Witodlíce manega synt geladode, and feáwa gecorene *multi enim sunt vocati, pauci vero electi*, Mt. Bos. 22, 14 : Ælfc. Gl. 7; Som. 56, 64. Torhte twelfe wǽron, Dryhtne gecorene *bright were the twelve, chosen unto the Lord*, Apstls. Kmbl. 10; Ap. 5 : Elen. Kmbl. 2115; El. 1059 : Cd. 83; Th. 104, 12; Gen. 1734 : 176; Th. 221, 23; Dan. 92 : Hy. 7, 53; Hy. Grn. ii. 288, 53 : Ps. Th. 131, 5 : Exon. 25 b; Th. 75, 19; Cri. 1224 : 15 a; Th. 31, 18; Cri. 497 : 12 b; Th. 21, 7; Cri. 331 : 64 b; Th. 237, 21; Ph. 593 : 63 b; Th. 234, 16; Ph. 541 : 74 b; Th. 279, 13; Jul. 613 : 66 a; Th. 243, 26; Jul. 16 : 74 b; Th. 278, 29; Jul. 605 : 33 a; Th. 163, 29; Th. 149, 29; Gú. 769. He hæfde cempan gecorone *he had chosen champions*, Beo. Th. 417; B. 206. Simon sacan ongon wið ða gecorenan Cristes þegnas *Simon began to strive against the chosen ministers of Christ*, Exon. 70 a; Th. 260, 18; Jul, 299 : 31 b; Th. 100, 1; Cri. 1635 : Ps. Th. 104, 38 : 107, 5 : Hy. 9, 42; Hy. Grn. ii. 292, 42. Ic mínum gecorenum cûðe gesette *deposui testamentum electis meis*, Ps. Th. 88, 3 : 105, 5 : 131, 18 : Exon. 61 b; Th. 225, 12; Ph. 388. [*Goth.* ga-kiusan *to test, approve* : *O. H. Ger.* gi-chiosan *discernere, probare, approbare, eligere.*] v. ceósan.

ge-ceówan; *p.* -ceáw, *pl.* -cuwon; *pp.* -cowen [ceówan *to chew*] *To chew*; rūmĭnare :—Sume dweorgedwostlan geceówaþ *some chew pennyroyal*, L. M. 2, 32; Lchdm. ii. 236, 11. Lege dweorgedwostlan gecowene on ðone nafolan *lay chewed pennyroyal on the navel*, 2, 30; Lchdm. ii. 228, 20.

ge-cêpan; *p.* -cêpte; *pp.* -cêpt *To buy*; ĕmĕre :—Hí ðæt ríce hæfdon dióre gecêpte *they had dearly bought that kingdom*, Bt. Met. Fox 26, 37; Met. 26, 19. v. ge-cýpan.

ge-cerran; *p.* de; *pp.* ed *To turn, return* :—Ic gecyrre on mín hús *revertar in domum meam*, Mt. Bos. 12, 44. Gecerreþ ðæt folc *commovet populum*, Lk. Skt. Lind. 23, 5. Gecerre hine *let him turn*, Bt. 35, 1; Fox 156, 10. From wind gecerred *a vento motus*, Lk. Skt. Lind. 7, 24. v. cerran.

ge-cerring, e; *f. A turning, conversion*; conversio :—On gecerringce oððe on gænhwyrfte *in convertendo*, Ps. Lamb. 125, 1.

ge-cíaþ *call*, Ps. Lamb. 19, 8, = ge-cígaþ, *pres. pl.* of ge-cígan.

ge-cíd, es; *m. n? Strife*; lis :—Gecíd *lis*, Rtl. 162, 28.

ge-cídan; *p.* -cídde, *pl.* -cíddon, -cídon; *pp.* -cíded, -cídd *To chide, quarrel, strive*; litigare, rixari :—Gecídon oððe getugon Iudéas bituih

litigabant Judæi adinvicem, Jn. Skt. Lind. 6, 52. Gif on gebeórscipe hie gecíden *if they quarrel in a feast*, L. In. 6; Th. i. 106, 11.

ge-cígan, -cígean, -cýgan, -cégan; *p.* -cígde, -cýgde, -cégde; *pp.* -cíged, -cýged, -cýgd, -céged [ge, cígan *to call*]; *v. trans.* To call, name, call upon, invoke, call forth, provoke, incite; vocare, nominare, invocare, provocare, incitare :—Ne com ic rihtwíse to gecígeanne, ac ða synnfullan *non veni vocare justos, sed peccatores*, Mt. Bos. 9, 13. Ðú gecígst his naman Ysmaél *vocabis nomen ejus Ismael*, Gen. 16, 11. Him Dryhten gecýgþ *the Lord calls him*, Exon. 62 b; Th. 229, 13; Ph. 454. Drihten gecégde eorþan *Dominus vocavit terram*, Ps. Spl. 49, 1. Hí gecégdon naman heora *vocaverunt nomina sua*, Ps. Spl. 48, 11. Se wæs gecíged Godwine *he was called Godwine*, Chr. 984; Erl. 130, 3; Ælfc. Gr. 22; Som. 24, 4; Bd. 1, 7; S. 477, 31: 4, 19; S. 588, 30. Hí gewunedon to gebédum gecígde beón *they were accustomed to be called to prayers*, 4, 23; S. 595, 41. On ðam þeódlande ðe is gecýged Élige *in regione quæ vocatur Elge*, Bd. 4, 19; S. 588, 1: 4, 23; S. 593, 20, 35. Seó is gecýged Solente *quod vocatur Solvente*, 4, 16; S. 585, 2. Ðú, Drihten, [eart] wynsum eallum gecýgendum ðé *tu, Domine, [es] suavis omnibus invocantibus te*, Ps. Spl. 85, 4. On dagum mínum ic gecýge hine *in diebus meis invocabo eum*, Ps. Lamb. 114, 2. He gecýgde me *invocavit me*, Ps. Spl. 88, 26. Hine hí gecýgdon *eum provocaverunt*, Ps. Spl. 77, 4. Ða to yrre beóþ gecígde *they shall be provoked to anger*, Ps. Th. 7, 7. Folc gecýgde naman ðínne *populus incitavit nomen tuum*, Ps. Spl. 73, 19.

ge-cígednes, -cýgednes, -ness, e; *f.* A calling; vŏcātio :—Óþ ðone dæg his gecígednesse of middangearde *usque ad diem suæ vŏcātiōnis*, Bd. 5, 12; S. 631, 34. Gecígednes *vocatio, vocabulum, nomen*, Hpt. Gl. 441, 466.

ge-cígendlíc; *adj.* [cígan *to call, invoke*] Calling, addressing; vocativus :—Vocativus is clipigendlíc oððe gecígendlíc *vocative is calling or invoking*, Ælfc. Gr. 7; Som. 6, 25. v. clipigendlíc.

ge-cígnes, se; *f.* A calling, entreaty :—Ofer mínre gecígnesse ðú gesettest ealle ðíne apostolas to mínre byrgenne *without my entreaty thou hast appointed all the apostles to be present at my burial*, Blickl. Homl. 143, 29.

ge-cind, es; *n.* also, e; *f.* A kind, nature, sort; generatio, genus, conditio :—And of fugelcinne seofon, and seofon ægþres gecindes *et de volatilibus caeli septena, et septena cujuslibet generationis*, Gen. 7, 3. Fram gecinde *a generatione*, Ps. Spl. T. second 9, 7. v. ge-cynd.

ge-ciist *choosest*, Gen. 13, 9; *2nd sing. pres.* of ge-ceósan.

ge-cláded, *part.* Clothed, clad; vestitus :—Hí geségon hine gecláded oððe gegerelad *vident illum vestitum*, Mk. Skt. Lind. 5, 15.

ge-clǽman; *p.* de; *pp.* ed To smear; linere :—Geclǽm ealle ða seamas mid tyrwan, *smear all the seams with tar*, Homl. Th. i. 20, 33. v. O. Engl. Homl. i. 225, 17, i-clem.

ge-clǽne; *adj.* Clean, pure :—Giclǽno heart innwardo *pura cordis intima*, Rtl. 163, 1.

ge-clǽnsian, -clǽnsigan, -clǽsnian, -clánsian; *p.* ode, ede; *pp.* od, ed [clǽnsian *to cleanse*] To cleanse, purify; mundāre, purgāre :—Gyf ðú wylt, ðú miht me geclǽnsian *si vis, potes me mundāre*, Mt. Bos. 8, 2; Mk. Bos. 1, 40: Elen. Kmbl. 1352; El. 678. Saul ne meahte his wambe geclǽnsian *Saul could not purify his stomach*, Past. 28, 6; Swt. 197, 24; Hat. MS. 38 a, 9. Geclǽnsa oððe afeorma me *munda me*, Ps. Lamb. 50, 4. Ic beó geclǽnsod *mundābor*, 50, 9: Mt. Bos. 8, 3: Mk. Bos. 1, 40, 41: Bt. 38, 4; Fox 202, 29. Geclǽnsedra *castigātior*, Bd. 4, 31; S. 611, 1.

ge-clǽnsung, e; *f.* A cleansing, purifying; purifĭcātio :—Æfter Iudéa geclǽnsunge *secundum purificātiōnem Judæōrum*, Jn. Bos. 2, 6.

ge-clǽsnian; *p.* ode; *pp.* od To cleanse, purify; mundāre, purgāre :—Saul ne meahte his wambe geclǽsnian *Saul could not purify his stomach*, Past. 28, 6; Swt. 196, 24; Cot. MS. Óðer dǽl sceal beón geclǽsnod *the other part shall be cleansed*, Bt. 38, 4; Fox 202, 29, MS. Cot. v. ge-clǽnsian.

ge-clánsian; *p.* ode; *pp.* od To cleanse :—Geltas geclánsa, ða ðe ic gefremede *cleanse the sins which I have committed*, Ps. C. 50, 39; Ps. Grn. ii. 227, 39: 50, 112, 127; Ps. Grn. ii. 279, 112, 127. v. ge-clǽnsian.

ge-cleofian; *p.* ode, ede; *pp.* od, ed [clifian, cleofian *to cleave, adhere*] To cleave, adhere, stick; adhærēre :—Geþeódde oððe gecleofede on flóre sáwle mín *adhæsit pavimento anima mea*, Ps. Lamb. 118, 25.

ge-clibs, -cleps, -clebs, -clysp a clamour, outcry; clamor :—Ne wend ðú ðe on ðæs folces geclysp *turn thou not thyself to the people's cry*, L. Alf. 41; Th. i. 54, 7. [Cf. clypian.]

ge-cliht; *part.* Collectus :—Hand gecliht [or hand-gecliht?] *manus collecta* vel *contracta, pugnus*, Som. [Cf. Scot. cleik *to seize as by a hook :* A. R. clahte [*p. tense*] *seized;* clech *unguis :* Mod. Engl. clutch.]

ge-clungen *dried up, shrivelled;* contractus, *pp.* of geclingan :—Hý beóþ cealde geclungne *they are shrivelled with cold*, Salm. Kmbl. 609; Sal. 304: Exon. 59 a; Th. 213, 17; Ph. 226.

ge-clútod; *adj.* [clút *a patch*] CLOUTED, *patched, nailed;* consutus, clavatus :—Geclútode bytta *patched bottles* [A. V. *wine bottles old, and*

rent, and bound up], Jos. 9, 5. Gesceód mid geclúdedum scón *shod with clouted shoes*, Dial. 1, 4.

ge-clypian, -clipian; *p.* ode, ede; *pp.* od, ed [clypian, clipian *to call*] To call, call upon, invoke; vŏcāre, invŏcāre :—He his naman geclipode *invŏcāvit nōmen ejus*, Gen. 12, 8. Manega synt geclypede *multi sunt vŏcāti*, Mt. Bos. 20, 16. [*Still retained in* y-clept.]

ge-cnǽwe; *adj.* Knowing, conscious, aware, acknowledging; cognoscens, conscius :—Se synfulla stód feorran, gecnǽwe his misdǽda *the sinful stood afar off, conscious of his misdeeds*, Homl. Th. ii. 428, 27. Se cwellere bæd forgifenysse, gecnǽwe his mánes *the murderer prayed for forgiveness, acknowledging his crime*, 510, 20. We sind gecnǽwe ðæt . . . *we are aware that* . . . , 378, 9. Híg ealle wǽron ðæs gecnǽwe *omnes testimonium illi dabant*, Lk. Bos. 4, 22.

ge-cnáwan; ic -cnáwe, ðú -cnáwest, -cnǽwst, he -cnáweþ, -cnǽwþ, *pl.* -cnáwaþ; *p.* -cneów, *pl.* -cneówon; *pp.* -cnáwen To know, perceive, understand, recognise; noscere, agnoscere, sentire, cognoscere :—Ne meahton [meahtan MS.] ða ðæs fugles flyht gecnáwan *they might not know the bird's flight*, Exon. 17 a; Th. 41, 12; Cri. 654: Bt. Met. Fox 12, 46; Met. 12, 23: Beo. Th. 4101; B. 2047. Ðonne ðæt gecnáwen fláh feónd gemáh *when the deceitful impious fiend knows that*, Exon. 97 a; Th. 362, 17; Wal. 38. Heonon-forþ ge hyne gecnáwaþ *henceforth ye shall know him*, Jn. Bos. 14, 7. He ðæt gecneów *he knew that*, Exon. 46 b; Th. 159, 22; Gú. 930: Mk, Bos. 14, 69. Ða he ða lác gecneów *qui agnitis muneribus*, Gen. 38, 26. Ðæt ðú gecnáwe ðæt ðis is sóþ *that thou may know that this is true*, Exon. 70 b; Th. 263, 27; Jul. 356. Hí hine gecneówon *cognoverunt eum*, Mk. Bos. 6, 54. Gif mín fæder me handlaþ and me gecnǽwþ *if my father handleth me and knows me*, Gen. 27, 12. Ic ðæt gecneów *I perceived that*, Exon. 72 a; Th. 269, 1; Jul. 443. Ge mágon sóþ gecnáwan *ye may know the truth*, Andr. Kmbl. 3115; An. 1560: 3032; An. 1519: Elen. Kmbl. 1413; El. 708. Ðæt geðeóde ðe we ealle gecnáwan mægen *the language that we can all understand*, Past. Swt. 6, 8. Ic hafu gecnáwen ðæt ðú Hǽlend eart middangeardes *I have perceived that thou art the Saviour of the world*, Elen. Kmbl. 1613; El. 808. Ðú miht ða sóðan gesǽlþa gecnáwan *thou mayest recognise the true goods*, Bt. 23; Fox 78, 32; 80, 2.

ge-cnedan; *p.* -cnæd, *pl.* -cnǽdon; *pp.* -cneden To mix, mingle, spread, knead; depsere :—Gecned nú hrǽdlíce þrí sestras smedeman depse nunc tres mensuras similaginis, Gen. 18, 6. Gecned hine mid meocle *knead it with milk*, Th. An. 119, 5. Óððæt sie gecnoeden all *donec fermentaretur totum*, Lk. Skt. Lind. 13, 21. Gecneden sealf cataplasma, Cot. 209.

ge-cneord; *adj.* Diligent, intent; intentus, sollers :—Wæs he on willsumnesse háligra gebéda gecneord and geornfull *erat orātiōnum devotiōni sollertissime intentus*, Bd. 4, 28; S. 606, 34.

ge-cneordlǽcan to study, be diligent, Hpt. Gl. 412, 432. v. cneordlǽcan.

ge-cneordlíc; *adj.* Diligent :—Swilce hí swuncon on wíngeardes biggenge mid gecneordlícere teolunge *as if they had laboured in the cultivation of the vineyard with diligent tilling*, Homl. Th. ii. 74, 33.

ge-cneordlíce; *adv.* Diligently; studiose :—Ða ðe woldon woruld-wisdom gecneordlíce leornian *those who wished diligently to learn philosophy*, Homl. Th. i. 60, 27.

ge-cneordnys, -nyss, e; *f.* [cneordnys *diligence*] Diligence, study, an invention; dīligentia, stūdium, adinventio :—Gecneordnysse stūdium, Greg. Dial. 9, 2. Gremedon hine on gecneordnyssum his *irritāvērunt eum in adinventiōnibus suis*, Ps. Spl. 105, 28.

ge-cneórednis, se; *f.* Genealogy; genealogia, Hpt. Gl. 552.

ge-cneów *knew, perceived*, Gen. 38, 26: Elen. Kmbl. 2278; El. 1140; *p.* of ge-cnáwan.

ge-cneówian; *p.* ode; *pp.* od [cneówian *to kneel*] To bend the knee, kneel; genuflectĕre :—He on díglum stówum gecneówige gelóme *let him frequently kneel in secret places*, L. Pen. 16; Th. ii. 282, 30.

ge-cnocian to beat, pound, Herb. 64; Lchdm. i. 168, 6, MS. B. v. ge-cnucian.

ge-cnoden *given, dedicated*, Bt. Met. Fox 1, 63; Met. 1, 32. v. cnódan.

ge-cnucian, -cnocian; *p.* ode, ede, ude; *pp.* od, ed, ud [cnucian *to beat*] To beat, pound; tundĕre, pertundĕre :—Gecnuca hý mid swínenum góre *pound it with swine dung*, Herb. 9, 3; Lchdm. i. 100, 11. Mid gecnucedum [MS. gecnucedon] ŏe *oleo tūso*, Ex. 29, 40. Genim ða wyrte gecnucude [gecnocode MS. B.] *take the herb pounded*, Herb. 64; Lchdm. i. 168, 6.

ge-cnyce, es; *n.* A bond; nexus :—Gicnyccum *nexibus*, Rtl. 59, 13; 66, 25. v. gecnyttan.

ge-cnyrdlǽcan to study. v. cneordlǽcan.

ge-cnyssan, -cnysan; *p.* ede, de; *pp.* ed [cnyssan *to press, trouble*] To press, trouble, strike, beat, overcome; premĕre, tribulāre, pulsāre, icĕre :—Unsóþfæstne wer yfel gecnysseþ *vǐrum injustum mǎla cǎpient*, Ps. Th. 139, 11. Gecnyssed *ictus*, Ælfc. Gr. 43; Som. 44, 55. Wurdon Rómáne gecnysede *the Romans were overcome*, Ors. 3, 11; Bos. 71, 19.

ge-cnyttan, -cnyhtan; *p.* -cnytte; *pp.* -cnytted, -cnytt, -cnyt [cnyttan *to tie*] To tie or fasten to, to annex; adnectĕre, alligāre :—Gecnyttan

adnectĕre, Cot. 4. Bende gicnyhtest *vinculo nexius ti*, Rtl. 108, 21. Betere him ys đæt án cwyrnstán sí to hys swyran gecnytt *expĕdit ei ut suspendātur möla asĭnāria in collo ejus*, Mt. Bos. 18, 6. Gecnyt, Mk. Bos. 9, 42 : Lk. Bos. 17, 2. Gicnyht, Rtl. 109, 41; Jn. Skt. Lind. 11, 44. [*Laym.* i-cnutten; *p. pl. knotted*.]

ge-cœlan; *p.* de; *pp.* ed; *v. trans. To cool, refresh, revive*; refrigerare :—Forlétaþ me đæt ic sie gecœled ærđon ic gewíte *remitte mihi ut refrigerer prius quam abeam*, Ps. Surt. 38, 14. v. cælan, calan.

ge-cope; *adj. Fit, proper*; congruus, opportūnus :—We sculon geleornian đæt we gecope tíd [MS. tiid] arédigen *we must learn to arrange a proper time*, Past. 38, 5; Swt. 277, 1; Hat. MS. 51 b, 8. Hwæt him gecopust sié *what is most fit for them*, 13, 2; Swt. 77, 26; Hat. MS. 17 a, 1; Swt. 275, 18.

ge-coplíce; *adv Fitly, well, readily*; apte, congrue :—Ic geó hwílum gecoplíce funde *I formerly readily invented*, Bt. 2 ; Fox 4, 9.

ge-copsende; *part.* [cops *a fetter*] *Fettered*; compēdītus :—Đæt he gehérde geomrunga gecopsendra ođđe gefótcypstra *ut audíret gěmǐtus compĕdītōrum*, Ps. Lamb. 101, 21.

ge-coren; *pp.* of geceósan *Chosen, choice, fit, good, beloved, dear* :—Mín gecorena *dilectus meus*, Mt. Bos. 12, 18. Đone gicoren *Christum*, Rtl. 4, 36; 82, 36. Đe gecorena *Messias*, Jn. Skt. Lind. 4, 25. Gecoren is to ríc godes *aptus est regno dei*, Lk. Skt. Lind. 9, 62. Đú gecorene *optime*, Lk. Skt. Lind. 1, 3; 8, 15. Sanctus Iohannes eallum Godes hálgum is gecorenra *St. John is more beloved than all God's saints*, Blickl. Homl. 167, 26. Đa gecorenistan dune *the goodliest mountain*, Deut. 3, 25.

ge-corenes, -corennes, -ness, -nys, -nyss, e ; *f.* [corenes *an election*] *An election, choice, choiceness, goodness* ; electio, electus, probĭtas :—Seó gecorennys stent on Godes fóresceáwunge *the election stands in God's providence*, Homl. Th. ii. 524, 25. Ne ic on heora gecorenesse becume æfre *non comminābor cum electis eōrum*, Ps. Th. 140, 6. Đe gelýfedre yldo wǽron ođđe on gecorenesse heora þeáwa máran and beteran wǽron *quæ vel ætāte provectæ vel probĭtāte ĕrant mōrum insigniōres*, Bd. 3, 8; S. 531, 33 : Mk. Skt. p. 2, 1.

ge-corenlíc; *adj. Choice, elegant*; elĕgans, Cot. 74.

ge-corenlíce; *adv. Choicely, elegantly*; elĕganter, Cot. 77.

ge-corenscipe, es ; *m. Election, excellence*; electio, excellentia :—Gecorenscip *electio*, Mt. Kmbl. p. 12, 11 : Rtl. 2, 27. Gicorenscipe *excellentia*, Rtl. 54, 21.

ge-corónian; *p.* ode; *pp.* od *To crown* :—Đú us gecorónadest *coronasti nos*, Ps. Th. 5, 13.

ge-cosped; *part. p.* [cosp *a fetter*] *Fettered* ; compēdītus :—Drihten tolýsþ gecospede ođđe đa gefótcypstan *Dŏmĭnus soluit compēdītos*, Ps. Lamb. 145, 8.

ge-cost; *adj.* [cost *tried*] *Tried, proved, chosen*; probātus :—Til mon, tiles and tomes meares, cúþes and gecostes *a good man has care for a good and tame horse. known and tried*, Exon. 91 a; Th. 342, 14; Gn. Ex. 143. Heápe gecoste *with a chosen company*, Elen. Kmbl. 538; El. 269. Swyrd ecgum gecoste *swords tried in their edges*, Judth. 111; Thw. 24, 39; Jud. 231. Đa đe seolfres beóþ since gecoste *qui probāta sunt argento*, Ps. Th. 67, 27. Đæt sind đa gecostan cempan *these are the proved champions*, Exon. 33 b; Th. 107, 21; Gú. 62. [Cf. *Goth.* ga-kusts; *p. trial, test: O. H. Ger.* gi-costôt *proved*.] v. gecostian.

ge-costian, -costnian; *p.* ode; *pp.* od. [costian *to tempt*] *To tempt, try, prove* ; tentāre, probāre :—He gecostaþ wildeóra worn *it tryeth the multitude of beasts*, Salm. Kmbl. 610; Sal. 304. Ne eart đú clǽne gecostad *thou art not thoroughly proved*, Exon. 41 a; Th. 136, 36; Gú. 552 : 40 b; Th. 134, 13; Gú. 507. [*O. Sax.* gi-kostôn : *O. H. Ger.* gi-costôt *proved, tried*.]

ge-costnes, -ness, e ; *f.* [costnes *a temptation*] *A temptation, trial, proving* ; probātio :—Se wæs of dæghwamlícre gecostnesse đæs mynstres becom to áncerlífe *qui de monastěrii probātiōne ad heremitcam pervēněrat vitam*, Bd. 3, 19; S. 549, 42.

ge-costnian; *p.* ode; *pp.* od *To try*; tentare :—Gecostna me *tenta me*, Ps. Lamb. 25, 2. He wæs fram Satane gecostnod *tentabatur a Satane*, Mk. Bos. 1, 13.

ge-costung, e ; *f. Tribulation, trial*; tribulatio, Mk. Skt. Lind. 13, 24.

ge-cræftan; *p.* -cræfte; *pp.* -cræfted, cræft [cræftan *to build*; cræft *art*] *To contrive, build*; molíri, machināri :—Ic gecræfte, đæt se cempa ongon Waldend wundian *I contrived that the soldier did wound the Lord*, Exon. 70 a; Th. 259, 30; Jul. 290. Đæt Godes tempel wæs wundorlíce gecræft *the temple of God was wonderfully contrived*, Homl. Th. ii. 574, 29.

ge-cræftgian; *p.* ade; *pp.* ad [cræft I. *power, strength*] *To strengthen, make powerful*; firmare, roborare :—Đa rícu of nánes mannes mihtum gecræftgade ne wurdon *the kingdoms were not strengthened by the powers of any man*, Ors. 2, 1; Bos. 39, 2.

ge-cráwan *to crow* :—Hona gesang ł gecráwæ *gallus cantavit*, Mt. Kmbl. Lind. 26, 74.

ge-crincan; *p.* -cranc, *pl.* -cruncon; *pp.* -cruncen *To yield, fall*; occumbere, ruere :—He under rande gecranc *he fell beneath his shield*, Beo. Th. 2423; B. 1209: Byrht. Th. 139, 7; By. 250: 141, 19; By. 324. v. crincan.

ge-cringan; *p.* -crang, -crong, *pl.* crungon ; *pp.* crungen *To sink, fall, die*; occumbere, mori :—Heó on flet gecrong *on the ground she sank*, Beo. Th. 3141; B. 1568: 5003; B. 2505: 2679; B. 1337: Apstls. Kmbl. 120; Ap. 60 : Exon. 124 b; Th. 477, 30; Ruin 32. Gárulf gecrang *Garulf fell*, Fins. Th. 63; Fin. 31 : Exon. 77 b; Th. 291, 9; Wand 79. Stíđmód gecrang *firm of mind he died*, Apstls. Kmbl. 144; Ap. 72. v. cringan.

ge-cristnian; *p.* ode, ade; *pp.* od, ad [cristnian *to christianize*] *To christianize, catechize* ; catechízāre :—He đone cyning gecristnade, and hine eft æfter fæce mid fulluhtbæþe aþwógh mid his þeóde *cum rex ipse catechizātus, fonte baptism, cum sua gente abluĕrĕtur*, Bd. 3, 7; S. 329, 13. Syđđan he gecristnad wæs *cum catechizārētur*, 2, 14; S. 517, 27 : Blickl. Homl. 211, 29 : 213, 15 : 215, 22. Ne mót gefullod inne mid đam gecristnedan etan *non licet baptizato cum catecumeno comedere*, Th. Lg. ii. 144, 25.

ge-croced; *adj. Croceus, coccineus*, Hpt. Gl. 528.

gecrod, es; *n. A crowd*; turba. v. hlóþ-gecrod, lind-: creódan.

ge-cuman, -cyme; *p.* -com, *pl.* -cômon; *pp.* -cumen *To come, go*; venire, ire :—Seueriana gecom to đæra hálgena byrgenum *Severiana came to the graves of the saints*, Homl. Th. ii. 312, 27. Gecum to mínum þeówan Saulum *go to my servant Saul*, Homl. Th. i. 386, 19. Of nánum óđrum gecumen *come from none other*, Ælfc. T. 2, 26. Æfter meh gecyme *post me venire*, Mt. Kmbl. Lind. 16, 24; 17, 10 : Jn. Skt. Lind. 5, 40; 7, 27. [*Goth.* ga-kwiman: *O. H. Ger.* ka-queman.]

ge-cundelíc; *adj. Natural*; natūrālis :—Gê wénaþ đæt gê nán gecundelíce gód ne gesælþa in eów selfum nabbaþ *ye think ye have no natural good or happiness within yourselves*, Bt. 14, 2; Fox 44, 16. v. ge-cyndelíc.

ge-cunnan; *p.* -cúđe *To know* :—Huu alle bispello gie gecunnas ł giecunna gie mágon [Rush. gicunniga] *quomodo omnes parabolas cognoscetis*, Mk. Skt. Lind. 4, 13. Ic đe gecúđe ǽr đan đe ic đe gesceópe *I knew thee ere I created thee*, Ælf. Test; Swt. Rdr. 70, 433. [*Goth.* ga-kunnan *to know*.]

ge-cunnian; *p.* ode, ade; *pp.* od, ad *To try, enquire, experience*; probare, explorare, experiri :—Đæt hi móstan gecunnian hwylc heora swiftost hors hæfde *that they should try which of them had the swiftest horse*, Bd. 5, 6; S. 618, 42 : Nar. 25, 29. Đe đone wígend aweccan dorste ođđe gecunnian, hú *who dared to awake the warrior or to enquire how...*, Judth. 12; Thw. 25, 14; Jud. 259. Ic hæbbe gecunnad cearselda fela *I have experienced many places of sorrow*, Exon. 81 b; Th. 306, 9; Seef. 5. v. cunnian.

gecure, gecuron *chose*; gecoren *chosen*. v. geceosan.

ge-cúþ, *known*. v. gecunnan.

ge-cwæþ, đú -cwǽde, *pl.* -cwǽdon *Said, spoke, pronounced*, Cd. 202 ; Th. 251, 10; Dan. 561 : Beo. Th. 5322; B. 2664 : Chr. 1014; Erl. 150, 16; *p.* of ge-cweđan.

ge-cweccan :—Gecwecton đegnas his đa croppas *vellebant discipuli ejus spicas*, Lk. Skt. Lind. 7, 1.

ge-cwed, -cwid, -cwyde *a word, command*. v. cwide.

ge-cweden *spoken, called, ordained*, Chr. 456; Th. 22, 5, col. 2, 3 : L. Ath. v. § 12, 1; Th. i. 240, 32; *pp.* of ge-cweđan.

ge-cwednis, se ; *f. Vocabulum, nomen*, Hpt. Gl. 441.

ge-cwed-rǽden, ne; *f. An agreement*, Ors. 5, 12; Bos. 111, 23.

ge-cwellan *to kill* :—Đa suno gecuoellas hia *filii morte adficient eos*, Mk. Skt. Lind. 13, 12. Đætte hia woere gecuelledo *ut interficerentur*, Lk. Skt. Lind. 23, 32. [*O. H. Ger.* ge-quelit *cruciatus*.]

ge-cwelman *to destroy*. v. ge-cwylman.

ge-cwelmbǽran *to be tortured*; extorqueri, cruciari, Hpt. Gl. 470.

ge-cwéman; *p.* de; *pp.* ed [cwéman *to please*] *To please, satisfy, propitiate*; plăcēre, satisfăcēre :—He ne mihte đám folcum mid gifum gecwéman *he had not power to satisfy the people with rewards*, Ors. 3, 7; Bos. 60, 45. Pilatus wolde đam folce gecwémam *Pilātus völens pŏpŭlo satisfăcēre*, Mk. Bos. 15, 15. Gif đú godum ussum gecwémest *if thou wilt propitiate our gods*, Exon. 68 a; Th. 252, 27; Jul. 169. Đe him dǽdum gecwémde *who pleased him by deeds*, 46 b; Th. 159, 6; Gú. 922. Sume gecwémdon englum *some have given pleasure to angels*, Homl. Th. ii. 286, 12. God, đú đe mid hreównisse gicuoemes ł gicómed biþ *Deus qui pænitentia placaris*, Rtl. 8, 33. [*Laym.* i-quemen *to please*.]

ge-cwémdun *pleased*, Exon. 21 a; Th. 57, 14; Cri. 918, = gecwémdon; *p. pl.* of gecwéman.

ge-cwéme; *adj.* [cwéme *pleasant, pleasing*] *Pleasant, pleasing, grateful, acceptable, fit*; jŏcundus, grātus, plăcǐtus, complăcǐtus :—Noe wæs Gode gecwéme and gife ætfóran him gemétte *Noe invĕnit grātiam coram Domino*, Gen. 6, 8. Seó wæs Criste gecwéme *she was acceptable to Christ*, Exon. 69 b; Th. 258, 2; Jul. 259 : Elen. Kmbl. 2097; El. 1050. Gecwéme sý him spræc mín *jŏcundum sit ei elŏquium meum*, Ps. Spl. 103, 35. Forđam hyt wæs swá gecwéme befóran đé *quoniam sic fuit plăcǐtum ante te*, Mt. Bos. 11, 26; Jn. Bos. 8, 29. Đǽr is bráde land in heofonríce Criste gecwémra *there is a spacious land in heaven's kingdom of the grateful to Christ*, Cd. 218; Th. 278, 5; Sat. 217. Gecwémre *complăcǐtior*, Ps. Spl. 76, 7. Swá him gecwémast

wæs *as was most pleasing to him*, H. R. 103, 6. [Laym, A. R. i-queme
pleasing. Cf. *O. H. Ger.* biquáme: *Ger.* bequem.]

ge-cwémedlíc; *adj. Well pleased*; beneplăcĭtus :—Gecwémedlíc is
Drihtne *beneplăcĭtum est Dŏmĭno*, Ps. Lamb. 146, 11. Ne ne on
glywcum weres gecwémedlíce oððe welgecwéme biþ him *nec in tĭbiis
vĭri beneplăcĭtum ĕrit ei*, 146, 10. v. ge-cwémlíc.

ge-cwémednes, -ness, -nys, -nyss, e; *f. Satisfaction, pleasure, content-
ment*; beneplăcĭtum :—Gode to gecwémednesse *to the pleasure of God*,
L. Ælf. C. 33; Th. ii. 376, 38. Gode to gecwemednysse *to God's con-
tentment*, Homl. Th. i. 180, 10. v. ge-cwémnes.

ge-cwéming, e; *f. A pleasing*; beneplăcĭtum :—On gecwéminge
ðínre *in beneplăcĭto tuo*, Ps. Spl. 88, 17.

ge-cwémlíc; *adj. Agreeable, well pleased*; placĭtus, placatus, complacatus, congruus, beneplăcĭtus :—Gecwémlíc *congruus*, R. Ben. interl.
43. Gecwémlíc is Drihtne on his folce *beneplăcĭtum est Dŏmĭno pŏpŭlo
suo*, Ps. Lamb. 149, 4. In tíde gicuoemlícum *in tempore placĭto*, Rtl.
19, 7; 18, 29. Gicuoemlíce *placatus*, 43, 17; 35, 43. Gicuoemlíc
complacatus, 69, 11. Gicuæmlíc *supplex*, 166, 5.

ge-cwémlíce; *adv. Agreeably, acceptably* :—Hú fela wítegan under
ðære ǽ Gode gecwémlíce drohtnodon *how many prophets under the
old law passed their life acceptably to God*, Homl. Th. ii. 78, 33;
576, 4.

ge-cwémnes, -nys, -ness, -nyss, e; *f. A pleasing, satisfaction, appeasing*; plăcātio, beneplăcĭtum :—He ne selþ Gode gecwémnesse his *non
dăbit Deo plăcātiōnem suam*, Ps. Lamb. 48, 8. On gecwémnesse folces
ðínes *in beneplăcĭto pŏpŭli tui*, 105, 4. Tíma gecwémnysse *tempus bene-
plăcĭti*, Ps. Spl. 68, 16. Martha wæs geornful ðæt heó ðon Hǽlende to
gecwémnesse ðegnode *Martha was desirous to minister to the Saviour
to his satisfaction*, Blickl. Homl. 67, 29. Gicuoemnise hæbbendo *suffi-
centiam habentes*, Rtl. 13, 15.

ge-cwémsum; *adj. Illibatus*, Hpt. Gl. 520.

ge-cweðan; he -cweðeþ, -cwyþ; *p.* ic, he -cwæþ, ðú -cwǽde, *pl.*
-cwǽdon; *pp.* -cweden *To say, speak, call, pronounce, agree, resolve,
order*; dícĕre, lŏqui, profári, pronunciáre, pangĕre, stătuĕre :—Se næfre
nǽnig word gecweðan mihte *qui ne ūnum quidem sermōnum unquam
profári pŏtĕrat*, Bd. 5, 2; S. 614, 43. He ðæt word gecwæþ *he spake
the word*, Elen. Kmbl. 687; El. 344: 878; El. 440: Andr. Kmbl. 1791;
An. 898: 2600; An. 1301. Ðe Drihten wið eów gecwæþ *quod pĕpĭgit
vobiscum Dŏmĭnus*, Deut. 9, 9. Hí ǽfre ǽlcne Deniscne cyng útlah of
Engla lande gecwǽdon *they pronounced every Danish king an outlaw
from England for ever*, Chr. 1014; Erl. 150, 34. On ðære stówe ðe is
gecweden Creacan ford *in the place which is called Crayford*, Chr. 456;
Th. 22, 5, col. 2, 3: H. R. 105, 9. Éce Drihten gecwyþ *the Lord
eternal shall speak*, Cd. 227; Th. 304, 9; Sat. 627. Ðú gecwǽde ðæt
ðú ne alǽte dóm gedreósan *thou saidst that thou wouldst not let thy
greatness sink*, Beo. Th. 5322; B. 2664. Swá seó stefn gecwæþ *thus
spake the voice*, Cd. 202; Th. 251, 10; Dan. 561: 203; Th. 252, 22;
Dan. 582. Iulianus se cásere gecwæþ to gefeohte *the emperor Julian
gave order for a battle*, Homl. Th. ii. 502, 4. Swá hit gecweden wæs
as it was agreed, L. Ath. v. § 12, 1; Th. i. 240, 32: L. A. G. prm;
Th. i. 152, 4. Ða deófolgildan gecwǽdon ðæt hí woldon ðone apostol
to heora hǽðenscipe geneádian *the idolaters agreed to force the apostle
to their idolatry*, Homl. Th. i. 70, 24; H. R. 101, 20. [Laym. i-queðen:
Goth. ga-kwithan *to agree*: *O. Sax.* gi-queðan *to speak, declare*:
O. H. Ger. gi-quedan *dicere*.]

ge-cwician, -cwycian; *p.* ode, ude; *pp.* od, ud [cwician *to quicken*]
To quicken, create; vivĭficāre, creāre :—Dó me æfter ðínum wordum
wel gecwician *vivĭfica me secundum verbum tuum*, Ps. Th. 118, 25.
Heortan clǽne gecwica in me God *cor mundum crea in me Deus*, Ps.
Surt. 50, 12. Ðæt ðú me on rihtes rǽd gecwycige *in æquĭtāte tua vivĭfica me*, Ps. Th. 118, 40. He bebeád and gecwicode synd *ipse man-
dāvit et creāta sunt*, Ps. Spl. C. 32, 9: 101, 19. Hí biþ gecwicude
creăbuntur, Ps. Spl. C. 103, 31. [*Goth.* ga-kwiujan *to quicken, make
alive*: *O. H. Ger.* ki-chuuichan.]

ge-cwide, v. cwide, p. 180, col. 2. [Cf. *O. H. Ger.* ka-qhuit, ke-
chuiti, *f. sententia.*]

ge-cwid-rédden, -cwid-ráden, -cwyd-ráden, -cwed-ráden, -rádenn, e;
f. An agreement, a contract, statute, conspiration; ratio, pactorum, conventio, conspiratio :—He oferbræc heora gecwidrǽdene *he broke through
their agreement*, Ors. 3, 6; Bos. 57, 40. Geweordene gecwydrǽdene
conventione facta, Mt. Bos. 20, 1. Gecwidrǽdden *conspiratio*, Ælfc. Gl. 49;
Som. 65, 87; Wrt. Voc. 34, 19. Ðæt wæs seó gecwydrǽden *that was the
agreement*, Ors. 5, 12; Bos. 111, 26.

ge-cwis *a conspiracy, consent*; conspiratio, Cot. 46: Hpt. Gl. 519.
[*Goth.* ga-kwiss consent.]

ge-cwyd-rǽden *agreement*, Ors. 5, 12; Bos. 111, 21, 26: Mt. Bos.
20, 2. v. ge-cwidrǽden.

ge-cwylman; *p.* de; *pp.* ed [cwelman, cwylman *to torment*] *To
afflict, torment, punish, destroy, kill*; pūnīre, trucidāre, mortĭficāre :—Ná
ðæt án me, ac eác swylce míne geféran mid ánum slege he mæg gecwyl-
man *non sōlum me, sed etiam meos sŏcios ūno ictu pŏtĕrat mortĭficāre*,

Coll. Monast. Th. 24, 33. Ðæt hí gecwylmen rihte heortan *ut trucident rectos corde*, Ps. Spl. C. 36, 15. Ðæt he byþ gecwylmed *ut pūnĭē-
tur*, Ps. Lamb. 36, 13. Mid ormǽtre angsumnysse gecwylmed *afflicted
with excessive pain*, Homl. Th. i. 88, 6.

ge-cwylmful; *adj. Pernicious*; perniciosus, Hpt. Gl. 428.

ge-cwyþ *speaks*, Cd. 227; Th. 304, 9; Sat. 627; *3rd sing. pres. of*
ge-cweðan.

ge-cýgan *to call, call upon, invoke, provoke, incite*, Exon. 62 b; Th.
229, 13; Ph. 454: Ps. Spl. 73, 19: 77, 64: 85, 4. v. ge-cígan.

ge-cygd *strife, contention, debate*; jurgium, Bd. 1, 14; S. 482, 26. v.
gecíd.

ge-cýgednes, -ness, e; *f. A calling*; vŏcātio :—On ðam dæge ðe
geneálǽhte hyre gecýgednesse of ðyssum life *imminente die suæ vŏcātiōnis*, Bd. 3, 8; S. 531, 31. v. ge-cígednes.

ge-cyn, -cýnn, es; *n. Nature*; natura :—Ðæt is of untrumnisse ðæs
gecynnes *ex infirmitate naturæ est*, Bd. 1, 27; S. 494, 15.

ge-cynd, ge-cind, *acc.* ge-cýnd, ge-cynde; *f. also* ge-cynd, ge-cynde,
nom. acc.; *gen.* -cyndes; *dat.* -cynde; *pl. nom. acc.* -cyndu, -cyndo, -cynd;
gen. -cynda; *dat.* -cyndum; *n.*   I. *nature, kind, manner, condition,
gender*; natura, indoles, ingenium, proprietas, modus, qualitas, conditio,
genus :—For his ágenre gecynde *from its own nature*, Bt. 13; Fox 38,
7. On swíðe lytlon hæfþ seó gecynd genóg *with very little nature has
enough*, Bt. 14, 1; Fox 42, 10. Is sió þridde gecynd betere *the third
nature is better*, Bt. Met. Fox 20, 373; Met. 20, 187. On ða beteran
gecynd *into the better nature*, Andr. Kmbl. 1176; An. 588. Hú his
gecynde biþ *what its nature [sex] is*, Exon. 61 a; Th. 223, 8; Ph. 356.
Wǽstma gecyndu *kinds of fruits*, 33 a; Th. 104, 30; Gú. 15. Cristes
gecyndo *the natures of Christ*, Salm. Kmbl. 819; Sal. 409. On feówer
gecynd *in four kinds*, 996; Sal. 499. Æfter gecynde *de genere*, Ælfc.
Gr. 6; Som. 5, 27. II. *generation, nakedness*; generatio, natales,
partes, genitales, verenda :—Ðurh clǽne gecynd *by pure generation*,
Hy. 9, 11; Hy. Grn. ii. 291, 11: 9, 52; Hy. Grn. ii. 292, 52. Be-
heledon heora fǽderes gecynd *operuerunt verenda patris sui*, Gen. 9,
23. III. *offspring*; proles :—Hyra gecynda on weorold bringaþ
prolem reddunt, Nar. 35, 26. [Cf. *O. Sax.* kind: *O. H. Ger.* kint:
Ger. kind.]

ge-cynd-bóc, e; *f. Genesis* :—Seó bóc ys geháten Genesis ðæt ys
gecyndbóc *the book is called Genesis, that is the book of generation*, Thw.
Hept. p. 2, 33.

ge-cynde; *adj.* [cynde *natural*] *Natural, innate, inborn, genial*;
natūrālis, innātus, ingĕnitus, ingĕnuus :—Gif se weorþscipe ðam wélan
gecynde wǽre *if dignity were natural to wealth*, Bt. 27, 3; Fox 98,
25. Swá him gecynde wæs *as was natural to him*, Beo. Th. 5386; B.
2690: Bt. 36, 4; Fox 178, 12. Gecynde riht *jus natūrāle*, Ælfc. Gl.
12; Som. 57, 90; Wrt. Voc. 20, 31. Gefrǽgn ic hebréos in Hierusalem cyningdóm habban, swá him gecynde wæs *I have heard that the
Hebrews had kingly sway in Jerusalem, as was natural to them*, Cd.
173; Th. 216, 8; Dan. 3. Þurh gecynde cræft *through natural
virtue*, Chr. 975; Erl. 126, 9; Edg. 35. Céne men gecynde ríce *bold
men [have] inborn sway*, Exon. 89 b; Th. 337, 3; Gn. Ex. 59. Hæfdan
him gecynde cyningas twegen *they had two kings of their own race*, Bt.
Met. Fox 1, 11; Met. 1, 6.

ge-cyndelíc; *adj.* [cyndelíc *natural*] *Natural, according to nature*;
natūrālis :—Hit is gecyndelíc ðæt ealle eorþlíce líchaman beóþ fulran on
weaxendum mónan, ðonne on wanigendum *it is natural that all earthly
bodies are fuller at the increasing moon than at the waning*, Bd. de nat.
rerum; Wrt. popl. science 15, 11; Lchdm. iii. 268, 7. Gecyndelíce
dohtor *filia natūrālis*, Bd. 3, 8; S. 531, 21. Gecyndelíces gódes *of
natural good*, Bt. 27, 3; Fox 100, 4. Hí nán gecyndelíc gód on him
selfum nabbaþ *they have no natural good in themselves*, Bt. 27, 3; Fox
98, 30: 27, 4; Fox 100, 18. Ne forléton hí nó ðæt gecyndelíce gód
they would not lose the natural good, 27, 3; Fox 100, 6.

ge-cyndelíce; *adv. Naturally*; natūrālĭter :—Ealle gesceafta gecyndelíce fundiaþ to cumanne to góde *all creatures naturally desire to
come to good*, Bt. 35, 4; Fox 160, 15.

gecynde-sprǽc, e; *f. A natural speech, an idiom*; proprietas
linguæ, idioma, Ælfc. Gl. 101; Som. 77, 41.

ge-cynd-lim, es; *n. A birth-limb, womb*; vulva :—Gecyndlim ontýnende
vulvam aperiens, Lk. Bos. 2, 23: Hpt. Gl. 441.

ge-cyndnys, -nyss, e; *f. A nation*; nātio :—Gecyndnys bearna dínra
ic ascunode *nātiōnem filiōrum tuōrum reprobāvi*, Ps. Spl. 72, 15.

ge-cýpan, -cépan; *p.* -cýpte; *pp.* -cýpt [cýpan *to sell*] *To buy, purchase*; ĕmĕre :—Wyrsan wígfrecan gecýpan *to buy a worse warrior*, Beo.
Th. 4986; B. 2496. Ðæt ic ðé gecýpte *which I bought for thee*, Exon.
29 b; Th. 90, 11; Cri. 1472.

ge-cýpe; *adj. For sale* :—Ðær wǽron gecýpe hryðeru *there were oxen
for sale*, Homl. Th. i. 402, 17.

ge-cypsed; *part. p. Fettered*; compĕdītus :—Ingá on gesyhþe ðíne
geómrunga gecypsedra *introeat in conspectu tuo gĕmĭtus compĕdītōrum*,
Ps. Spl. 78, 11. Driht tolýseþ gecypsede *Dŏminus solvit compĕdĭtos*,
Ps. Spl. 145, 6.

ge-cyrnlad; *adj. Having kernels :*—Gecyrnlade appla *pomegranates,* Hpt. Gl. 496.

ge-cyrran; *p.* de; *pp.* ed. I. *to turn, convert;* vertere, convertere :—We sceolan ða wundor gecyrran on sóðfæstnesse geleáfan *we must apply those wonders to the belief in the truth,* Blickl. Homl. 17, 10. Ic gecyrre feónd mīnne *converto inimicum meum,* Ps. Spl. 9, 3. Manega israhela bearna he gecyrþ to drihtne *multos filiorum israel convertet ad dominum,* Lk. Bos. 1, 16. Gif hē ðæt Cristene folc mid lufan ne mehton gecyrron *if they could not by love convert Christian people,* Blickl. Homl. 45, 22. Ðīne heortan to rǣde gecyr *turn thy heart to counsel,* Blickl. Homl. 113, 27: Ps. Th. 114, 7; 84, 5. Heora líf he hæfþ to gefeán gecyrred *their life he hath turned to joy,* Blickl. Homl. 85, 24; 57, 30; 59, 13. II. *to turn [one's self], go, return;* verti, reverti, ire :—Ic wille ðæt he libbe and to Gode gecyrran *I will that he live and turn to God,* Blickl. Homl. 97, 34; 101, 15. Gecyrraþ to me ðonne gecyrre ic to eów. He ðonne gecyrde to us *turn to me then will I turn to you. He turned to us then,* Blickl. Homl. 103, 1. Ðū ne gecyr from ðīnre ðeówene *turn not from thy servant,* 89, 12: Ps. Th. 58, 14: Andr. Kmbl. 2158; An. 1080. Hī symle sculon ðone ylcan ryne eft gecyrran *they ever must go again the same course,* Bt. Met. Fox 11, 74; Met. 11, 37. Ðā gecyrdon ða twā and hund-seofontig *reversi sunt septuaginta duo,* Lk. Bos. 10, 17. Hwænne he sý fram gyftum gecyrred *quando revertatur a nuptiis,* Lk. Bos. 12, 36.

ge-cyrred-nes, -ness, e; *f. A turning, conversion :*—Æfter his gecyrrednysse, Gregorius þēnode þearfum *after his conversion Gregory ministered to the poor,* Homl. Th. ii. 118, 35. v. acyrrednes.

ge-cyrring, e; *f. Converting, changing;* conversio, C. R. Ben. 62: Ps. Spl. T. 9, 3.

ge-cyspyd *fettered,* Ps. Spl. 78, 11. v. cyspan.

ge-cyssan; *p.* -cyste; *pp.* -cyssed [cyssan *to kiss*] *To kiss;* osculāri :—Gecyste cyning þegn betstan *the king kissed the best of thanes,* Beo. Th. 3744; B. 1870. Gecyste foet his *osculabatur pedes ejus,* Lk. Skt. Lind. 7, 38.

ge-cýð, -cýðð, e; *f. A country, native country;* patria, natale solum :—On hiora āgenre gecýþþe *in their own country,* Bt. 27, 3; Fox 100, 1. v. cýð.

ge-cýðan; *p.* -cýðde, -cýdde; *pp.* -cýðed, -cýð. I. *to make known, tell, relate, proclaim, announce, inform ;* nuntiare, annuntiare, referre, effari, monere :—Ða andsware gecýðan *to make known the answer,* Beo. Th. 714; B. 354: 4638; B. 2324: Ps. Spl. 101, 24. Gecýð *make known,* Exon. 50 a; Th. 173. 4; Gū. 1155. Sōþ gecýðan *to tell the truth,* Elen. Kmbl. 1173; El. 588. Se ðæt orleg-weorc ðam ebriscan eorle gecýðde *who announced that fatal work to the Hebrew leader,* Cd. 94; Th. 122, 4; Gen. 2021: Andr. Kmbl. 1568; An. 785: 1718; An. 861. Swā hie gecýðde wǣron *as they were informed,* Cd. 195; Th. 243, 9; Dan. 433. Him wæs gecýðed *nuntiatum est illi,* Lk. Bos. 8, 20. Ða wearþ hit Constantine gecýð *it was told to Constantine,* H. R. 3, 11. II. *to declare, reveal, manifest, shew, perform, confirm, testify, prove;* declarare, revelare, edocere, manifestare, monstrare, perhibere, testari, probare :—Ðæt wille ic gecýðan, ðæt ða rícu of nānes mannes mihtum swā gecræftgade ne wurdon *that will I declare, that the kingdoms were not strengthened by the powers of man,* Ors. 2, 1; Bos. 39, 1. God wolde gecýðan hwylcre geearnunge se hālga wer wǣre *Deus qualis meriti vir fuerit demonstrare voluit,* Bd. 1, 33; S. 499, 8; H. R. 15, 31. Se inlíca dēma mannum gecýðde *internus arbiter edocuit,* 3, 15; S. 541, 19. He gecýðeþ ðē wisdómes gife *he will shew thee the gift of wisdom,* Elen. Kmbl. 187; El. 595. Swā ðū hyldo wið me gecýðdest *as thou hast manifested grace to me,* Andr. Kmbl. 780; An. 390. Ðæt ðíne leóde gecýðdon *that thy people shewed,* Salm. Kmbl. 654; Sal. 326. Wundor wæs gecýðed *the miracle was manifested,* Cd. 208; Th. 257, 6; Dan. 653: 212; Th. 263, 11: Dan. 760. Gecýðan mid āþe *to prove* or *declare on oath,* L. In. 16; Th. i. 112, 7: 17; Th. i. 114, 2: L. Ed. 1; Th. i. 160, 5. Tree of wæstm his gecýðed biþ *arbor fructu suo cognoscitur,* Lk. Skt. Lind. 6, 44. III. *to make celebrated, renowned, famed;* notum facere, inclytum reddere :—Cyning cystum gecýðed *the king for virtues famed,* Beo. Th. 1850; B. 923: 530; B. 262: Exon. 41 a; Th. 137, 3; Gū. 553. [*O. Sax.* gi-kūðian: *O. H. Ger.* ga-chundan.] v. cýðan, cūð.

ge-cýðelíc; *adj. Manifest, made known;* manifestatus, Alb. resp. 10. v. cýðlíc.

ge-cýðig; *adj. Knowing, cognizant :*—Gicýðig cognitor, Rtl. 41, 23. [Cf. *Ger.* kundig *acquainted with.*]

ge-cýðnes, -ness, -nys, -nyss, e; *f. Testimony, testament, manifestation;* testimonium, testamentum :—Manega sǣdon leáse gecýðnysse *multi testimonium falsum dicebant,* Mk. Bos. 14, 56. Ðes calic is niwe gecýðnes on mínum blóde *hic est calix novum testamentum in sanguine meo,* Lk. Bos. 22, 20. Ps. Spl. 49, 6, 17. Drihten, ðíne gecýðnessa sindon swíðe geleáflíce *Lord, thy testimonies are very faithful,* Homl. Th. ii. 42, 14. Seó ealde gecýðnis *the Old Testament,* Thw. Hept. p. 2, 14. Nū neálǣceþ ǣgðer ge ðín onwrigennes ge uncer gecýðnes *now approaches both the discovery of thee [as false] and the manifestation of us two [as true],* Blickl. Homl. 187, 23. v. cýðnes.

ged, gedd, es; *n. A song, proverb, poem,* Bt. Met. Fox 2, 10; Met. 2, 5. Gedd *proverbium,* Jn. Skt. Lind. 10, 6; 16, 25. v. gid.

ge-dæftan; *p.* -dæfte; *pp.* dæft *To put in order, make ready, prepare :*—Ða ðe mid ðām [treowum] Cristes weig gedæfton *those who with the [trees] prepared Christ's way,* Homl. Th. i. 212, 34. He eów betǣcþ mycele healle gedæfte *ipse vobis ostendet cenaculum magnum stratum,* Lk. Bos. 22, 12: Mk. Bos. 14, 15. v. dæftan.

ge-dæfte; *adj. Mild, gentle, meek :*—Ðīn cyning cymþ to ðē, gedæfte *rex tuus venit tibi, mansuetus,* Mt. Bos. 21, 5. [Cf. *Orm.* daffte *humble, quiet.*] The later sense of 'daft' *foolish, stupid,* may be compared with the slang sense of 'soft.'

ge-dæftlíce, -dæftelíce, -deftlíce; *adv. Fitly, seasonably;* opportūne, commōde :—Ic ðē beóde ðæt ðū stande on ðissum wordum, and hie lǣre ǣgðer ge gedæftlíce ge ungedæftlíce *I charge thee to abide by these words, and teach them both seasonably and unseasonably,* Past. 15, 6; Swt. 96, 15; Hat. MS. 20 a, 21. Gedæftelíce *seasonably,* 15, 6; Swt. 96, 17; Hat. MS. 20 a, 22.

ge-dǣlan; *p.* de; *pp.* ed *To divide, part, impart, separate, distribute, share, partake :*—Seoððan se líchoma and se gāst gedǣlde beóþ *after the body and the spirit shall be separated,* Blickl. Homl. 111, 30. Ic gedǣle bā Sicimam et convallem, ða ǣr samod wǣron *dividam Sichimam et convallem,* Ps. Th. 59, 5. Hine gedǣlaþ *dividet eum,* Mt. Kmbl. Rush. 24, 51. He sceole wiþ ðæm líchomon hine gedǣlon *he must separate himself from the body,* Blickl. Homl. 97, 21. He hine wiþ ðas world gedǣleþ *he separates himself from the world,* 125, 11; 21, 26: Exon. 10 b; Th. 102, 6; Cri. 1668: Beo. Th. 4836; B. 2422: Exon. Th. 115, 32; Gū. 198. Ne mæg mín líchoma wiþ deáþ ge-dǣlan *my body cannot separate [itself] from [i. e. avoid] death,* Exon. Th. 124, 25; Gū. 343; 146, 19; Gū. 712. Gedǣle woeron ł todǣldon woedo mīno *partiti sunt vestimenta mea,* Jn. Skt. Lind. 19, 24. Gif he ǣr nele ðone sēlestan dǣl Gode gedǣlan *if he will not before give the best part to God,* Blickl. Homl. 195, 7. Ðæt we gedǣlan ðone teóþan dǣl *that we distribute the tenth part,* 39, 19. Gedǣled ðearfendum mannum *given to the poor,* 69, 8; 75, 23; Beo. Th. 143; B. 71: Exon. Th. 371, 19; Seel. 78: Past. 63; Swt. 459, 12. Sceolde he worc ðæs gewinnes gedǣlan *he should get pain on account of that struggle,* Cd. Th. 19, 24; Gen. 296. [*Goth.* ga-dailjan : *O. Sax.* gi-dēlian : *O. H. Ger.* ki-teilan *to divide, impart, distribute.*]

ge-dǣledlíce; *adv. Apart, separately :* separatim, Cot. 201.

ge-dǣman *to obstruct, dam;* obstruere, Serm. Creat.

ge-dǣrsted; *part.* [dǣrst *leaven*] *Leavened, fermented;* fermentatus :—Gedǣrsted is all *fermentatum est totum,* Mt. Kmbl. Lind. 13, 33. Oþ-ðæt sié gedǣrsted oððe gecnoeden all *donec fermentaretur totum,* Lk. Skt. Lind. 13, 21.

ge-dafen; *part.* [dafen *becoming*] *Becoming, fit, suitable ;* dēcens, congruus, convēniens :—Gif ðē gedafen þince *if it seem becoming to thee,* Exon. 67 a; Th. 247, 32; Jul. 87. This points to a verb 'gedafan,' corresponding to the Gothic ' gadaban ;' *convenire, decere.* [Cf. gedafenian.]

ge-dafenian, -dafnian, -dæfnia; *p.* ode; *pp.* od *To be becoming* or *fit, to behove ;* decere, convēnīre : chiefly used impersonally, *it behoves, it is becoming* or *fit, ought :*—Ic axige hwæðer hit mihte gedafnian Abrahame *I will ask whether it was becoming to Abraham,* Boutr. Scrd. 21, 47. Lāreówum gedafenaþ ðæt hí mid wísdómes sealte geleáffulra manna mód sylton *it befits teachers that they salt the minds of believing men with the salt of wisdom,* Homl. Th. ii. 536, 16: L. E. I. 24; Th. ii. 420, 32. Me gedæfnaþ *me oportet,* Jn. Skt. Lind. 9, 4. Ðē gedæfineþ *te oportet,* 3, 7. Ðē gedafenaþ *te dēcet,* Ps. Th. 64, 1: 92, 7: Ælfc. Gr. 33; Som. 37, 20: Andr. Kmbl. 633; An. 317. Me gedafenaþ ōðrum ceastrum Godes ríce bodian *aliis civitātibus oportet me evangelizāre regnum Dei,* Lk. Bos. 4, 43: Ælfc. Gr. 33; Som. 37, 20. Gedafenode *dēcuit,* 33; Som. 37, 21: Bd. 4, 11; S. 579, 11. Hit gedafnode ðæt se Ælmihtiga ǣrest ðæt hwílendlíce leóht geworhte *it was becoming that the Almighty first created the temporary light,* Boutr. Scrd. 19, 4: 21, 39. Gedæfnad is ūs *decet nos,* Mt. Kmbl. Lind. 3, 15.

ge-dafenigendlíce; *adv. Consequently;* consequenter, Scint. 11.

ge-dafenlíc, -dæfenlíc; *adj.* [ge-dafen *becoming*] *Becoming, fit, decent, convenient, agreeable;* dēcens, congruus, convēniens, hābilis :—Ðæt is gedafenlíc ðæt ðū Dryhtnes word on hyge healde *it is fit that thou shouldst keep in mind the word of the Lord,* Elen. Kmbl. 2333; El. 1168: Bt. Met. Fox 31, 42; Met. 31, 21: Bd. 4, 23; S. 594, 43. Hit gedafenlíc is ðæt his reáf ne beó hórig *it is becoming that his vestment be not dirty,* L. Ælf. C. 22; Th. ii. 350, 20. Gedafenlíc þeódnes [MS. seodnys] *hābilis conjunctio,* Ælfc. Gl. 99; Som. 76, 118; Wrt. Voc. 54, 60. Us dæg endebyrdnysse mid gedafenlícre cymþ *nōbis dies ordine congruo vēnit,* Hymn. Surt. 38, 3. Nis nā gedafenlíc ðæt ðes man āna beó *it is not fitting that this man be alone,* Homl. Th. i. 14, 17. Uæs gedæfenlíc [gidæfenlic, Rush.] *oportebat,* Jn. Skt. Lind. 4, 4.

ge-dafenlíce; *adv. Fitly, properly, justly;* dēcenter, convenienter,

juste :—God gewræc swíðe gedafenlíce on ðam árleásan men his árleáse geþoht *God very justly avenged his wicked thought on this wicked man*, Ors. 6, 31; Bos. 128, 33.

ge-dafenlícnes, -nys, -ness, -nyss, e ; *f.* *Decency, convenience, an opportunity ;* dĕcentia, convĕnientia, opportūnitas :—Eton mid gedafenlícnysse *juxta convĕnientiam comēdāmus*, Bd. Whelc. 228, 43. On gedafenlícnessum *in opportūnitātĭbus*, Ps. Lamb. 9, 10 : second 9, 1.

ge-daflíc ; *adj. Convenient, fitting ;* conveniens, congruus, Hpt. Gl. 415.

ge-dafniendlíc ; *adj. Suitable,* Hpt. Gl. 433, 497.

ge-dǽl, es ; *n.* *A division, separation, parting, distribution ;* dīvīsio, sepărātio, dīvortium, distribūtio :—Ðĕ is gedǽl witod líces and sáwle *a separation of body and soul is decreed to thee,* Cd. 43 ; Th. 57, 19 ; Gen. 930 : Beo. Th. 6128 ; B. 3068. Ic uncres gedǽles onbád earfoþlíce *I awaited our parting in sorrow,* Soul Kmbl. 74 ; Seel. 37 : Bd. 1, 15 ; S. 483, 37. Se hæfde heortan unhneáweste hringa gedǽles *he had the most liberal heart in the distribution of rings,* Scóp Th. 148 ; Wíd. 73. Æfter ðæs líchoman gedǽle and ðære sáwle *after the separation of the body and soul,* Bt. 18, 4 ; Fox 68, 12. Ðú ondrǽtst ðé on ðam gedǽle *thou fearest to distribute,* Homl. Th. ii. 104, 25. Se todǽlde sǽ reáde on gedǽl *qui dīvīsit măre rubrum in dīvīsiōnes,* Ps. Spl. 135, 13. [Cf. *O. Eng. Homl.* elmes i-dal *almsgiving.*] DER. deáþ-, ealdor-, feorh-, friþ-, gást-, híw-, líf-, nýd-, sáwel-, ðeóden-, woruld-gedǽl.

ge-dǽl-land, -dæl-land, es ; *n.* *Partible land, land belonging to several proprietors ;* sepărābilis terra :—Gif ceorlas gærstún hæbben gemǽnne, oððe gedǽlland to týnanne *if churls have a common meadow or partible land to fence,* L. In. 42 ; Th. i. 128, 6. v. note. Híd gedǽllandes, Kmbl. Cod. Dipl. iii. 6, 11.

geddian ; *p.* ode, *pp.* od *To sing ;* cantare :—Ðá ongan he geddian *then began he to sing,* Bt. 31, 2 ; Fox 112, note 25. Se scóp geddode *the poet sang,* 35, 5 ; Fox 166, 8. v. giddian.

geddung, giddung, e ; *f.* *A similitude, parable, riddle ;* similitudo, parabola :—In geddungum *in parabolis,* Lk. Skt. Lind. 8, 10. Geddung *parabola,* 18, 9 ; 19, 11. Geddung ł onlícnis *similitudo,* 13, 6. v. gidding.

ge-deágod *dyed, coloured,* DER. twi-gedeágod. v. deágian.

ge-deápian ; *p.* ade, ode ; *pp.* ad, od *To deepen, become deep* [?] :—Gideóþadon niólnisso *preruperunt abyssi,* Rtl. 81, 24. [Cf. *Goth.* ga-diupjan *to deepen, dig deeply.*]

ge-deáðian ; *p.* ode ; *pp.* od *To kill ;* mortificare :—Gedeáða ðú *mortifica,* Rtl. 48, 14. v. ge-dǽðan.

ge-deccan ; *imp.* -dec. [deccan *to cover*] *To cover ;* tĕgĕre :—Gedec ánne cláþ ðǽr mid *cover a cloth therewith,* Herb. 78, 2 ; Lchdm. i. 182, 3. Gedeced mid wyrtum *covered with spices,* Homl. Th. ii. 260, 35. v. Leo 607. 39. v. ge-þeccan.

GE-DÉFE, -doefe ; *comp.* -ra ; *superl.* -est, -ust ; *adj. Becoming, fit, proper, seemly, convenient, agreeable, decent, quiet, mild, meek, gentle, kind, benevolent ;* congruus, convĕniens, dĕcens, opportūnus, hŏnestus, quiĕtus, mansuĕtus, bĕnignus :—Swá hit gedéfe wæs *as it was fit,* Beo. Th. 3345 ; B. 1670 : Ps. Th. 60, 6 : 117, 13. Ne biþ ðæt gedéfe deáþ *that is not a seemly death,* Exon. 91 a ; Th. 340, 26 ; Gn. Ex. 117. Beóþ gē gedoefe *estote vos perfecta,* Mt. Kmbl. Rush. 5, 48. Noe wæs dómfæst and gedéfe *Noah was just and meek,* Cd. 64 ; Th. 78, 2 ; Gen. 1287 : Exon. 41 a ; Th. 136, 34 ; Gú. 551 : Beo. Th. 2458 ; B. 1227. Gedéfe is ðín milde mód *bĕnigna est mĭsĕrĭcordia tua,* Ps. Th. 68, 16. Gedéfe sacerd *sacerdos quiĕtus,* Nar. 37, 25. Eart ðú on lifigenda lande se gedéfa dǽl *tu es portio mea in terra vīventium,* 141, 5. On tíde gedéfre *in tempŏre opportūno,* Ps. Spl. C. 144, 16 : Bd. 4, 1 ; S. 564, 3. Doþ gedéfne dóm *with fitting judgment,* Exon. 41 b ; Th. 138, 26 ; Gú. 582 : Bd. 4, 1 ; S. 564, 4. Dó gedéfe mid me, Drihten, tácen *fac mecum, Dŏmĭne, signum in bŏno,* Ps. Th. 85, 16. Ða synd líðe and gedéfe *they are meek and gentle,* Homl. Th. i. 550, 20. Sýn hí adílgad of gedéfra eác ðæra lifigendra leófra bócum *deleantur de libro vīventium,* Ps. Th. 68, 29. Wuna mid us ðæt ðú us gedéfra gedó *stop with us to improve us,* St. And. 24, 8. Deórust and gedéfust *dearest and fittest,* 102, 16. Ealra dēmena ðam gedéfestan *to the most benevolent of all judges,* Exon. 93 a ; Th. 350, 4 ; Sch. 58. [*Goth.* ga-dóbs *fitting.*] DER. lǽr-gedéfe.

ge-défe ; *adv. Becomingly, decently ;* dĕcenter :—Ic eom on ðínum dómum gedéfe glæd *jŭdīcia tua jūcunda,* Ps. Th. 118, 39 : 124, 4.

ge-défelíc ; *adj. Fit, becoming, decent, honest ;* honestus :—Ðǽr syndon gedéfelíce menn *sunt ibi homines honesti,* Nar. 37, 32.

ge-défelíce ; *adv. Becomingly, fitly, decently, properly ;* dĕcenter, opportūne :—Sóna ðæs ðe gehálgod wæs, ða dyde mon his líchoman in, and on ðære cyrican norþ-portice gedéfelíce wæs bebyriged *mox vēro ut dedĭcāta est, intro inlātum, et in portĭcu illius aquĭlōnālis dēcenter sepultum est,* Bd. 2, 3 ; S. 504, 34. He symle gedéfelíce æftercwæþ *he always repeated [them] properly,* 5, 2 ; S. 615, 15.

ge-defen ; *part. Fit, proper, due ;* dēbĭtus :—Gedefen *dēbĭtus,* Cot. 61 : Th. An. 101, 10. To forþspównesse gedefenre heánesse *ad profectum dēbĭti culmĭnis,* Bd. 2, 4 ; S. 505, 17. v. gedafen.

ge-defenlíc ; *adj. Fit, proper, due ;* dēbĭtus :—Mid gedefenlícre ege *dēbĭto cum tĭmōre,* Bd. 4, 3 ; S. 569, 28. v. gedafenlíc.

ge-défnes, -ness, e ; *f. Quietness, mildness, gentleness ;* mansuētūdo :—Oferbecymþ gedéfnes sŭpervĕnit mansuĕtūdo, Ps. Lamb. 89, 10.

ge-deftlíce ; *adv. Fitly, moderately ;* dĕcenter :—Gif ðú wile hǽl beón, drinc ðé gedeftlíce *if thou wilt be healthy, drink in moderation,* Prov. Kmbl. 61. v. ge-dæftlíce.

ge-dégan, ge-dégean *to pass through, escape ;* pertransíre :—Oft úre sáwl swýðe frécne hlimman gedégde hlúdes wæteres ; wēne ic forðon ðæt heó wel mǽge ðæt swýðre mægen sáwel usser wæteres wēnan ðæs wel gedégean *torrentem pertransivit anima nostra ; forsitan pertransisset anima nostra aquam intolĕrābĭlem,* Ps. Th. 123, 4. Gif he wille sylf Godes dómas gedégan *if he himself wish to be uncondemned,* Blickl. Homl. 43, 12. v. gedígan.

ge-dégled *hidden ;* abscondĭtus, Lk. Skt. Lind. 12, 2. v. ge-díglian.

ge-delf, es ; *n. A delving, the act of digging, a trench ;* fossio, fossa :—Mid gedelfe *by digging,* Ors. 2, 4 ; Bos. 44, 12. He lēt delfon an mycel gedelf *he had a great trench dug,* Cod. Dipl. Kmbl. iv. 58, 5.

ge-delfan ; *p.* -dealf, *pl.* -dulfon ; *pp.* dolfen *To dig, delve ;* fodere, effodere :—Wæs ðǽr sum hláw ðone men gedulfon *there was a mound which men had dug,* Guthl. 4 ; Gdwin. 26, 6. Ðé wearþ helle seáþ niðer gedolfen *the pit of hell was dug beneath for thee,* Exon. 71 b ; Th. 267, 30 ; Jul. 423.

ge-déman ; *p.* de ; *pp.* ed *To deem, judge, determine, ordain, decree, doom, condemn ;* jūdĭcāre, decernĕre, sancīre, condemnāre :—He wile gedéman dǽda gehwylce *he will judge each deed,* Exon. 15 b ; Th. 33, 13 ; Cri. 525. Ðæt he æghwelcne on riht gedémeþ *that he judge every one righteously,* L. Alf. 49 ; Th. i. 56, 30 : Ps. Th. 57, 10. He gedémde úrne Drihten to deáþe *he condemned our Lord to death,* Ors. 6, 3 ; Bos. 117, 42. Gedémdon [MS. gedémden] *sanxērunt,* Mone B. 1940. Se ðe undóm gedéme *he who shall doom unjust doom,* L. C. S. 15 ; Th. i. 384, 7. Swá gedémed *is as is ordained,* Th. 207, 26 ; Ph. 147. He gedémed hæfde ðæt Ceólwulf æfter him cyning wǽre successōrem fore Ceoluulfum decrēvisset, Bd. 5, 23 ; S. 646, 1 : Cd. 186 ; Th. 231, 11 ; Dan. 245. Fýnd syndon eówere gedémed to deáþe *your enemies are condemned to death,* Judth. 11 ; Thw. 24, 19 ; Jud. 196. [*Goth.* ga-dōmjan.]

ge-deóful-geld *idolatry.* v. deófolgeld.

ge-deorf, es ; *n. Labour, trouble, tribulation ;* lăbor, trĭbŭlātio :—Micel gedeorf ys hit *magnus lăbor est,* Coll. Monast. Th. 20, 5, 7. Byþ mycel gedeorf *ĕrit trĭbŭlātio magna,* Mt. Bos. 24, 21. Hæfst ðú ǽnig gedeorf *hăbestu ălĭquem lăbōrem?* Coll. Monast. Th. 20, 9. For his micclum gedeorfum *for his great labours,* Homl. Th. ii. 522, 3 : 82, 33.

ge-deorfan ; *p.* -dearf, *pl.* -durfon ; *pp.* -dorfen *To labour :*—Micel ic gedeorfe *multum laboro,* Coll. Monast. Th. 20, 25. In Ors. 4, 6 ; Bos. 86, 3, Heora scipa gedurfon L and C *perhaps we should read gedufon sank,* cf. 85, 38, gedeáf [gedráf], *and* Ors. 1, 7 ; Bos. 30, 24, Ða gedufon hí ealle and adruncon. [*A. R.* i-dorven ; *pp. grieved, injured.*]

ge-deorfleás ; *adj.* This word in Glos. Prudent. Recd. 151, 73 is explained *nil prosperum.* The natural meaning would be *without labour, trouble,* which hardly agrees with that given above. Leo 230, 38, to connect the two, suggests the meaning *without effort,* so *without result, success.*

ge-deorfnys, -nyss, e ; *f. Trouble, tribulation ;* trĭbŭlātio :—God is úre fultum on gedeorfnyssum oððe on gedréfednyssum *Deus est noster adjūtor in trĭbŭlātiōnĭbus,* Ps. Lamb. 45, 2.

ge-deorfsum ; *adj. Troublesome, grievous ;* mŏlestus, grăvis :—Ðis wæs swíðe gedeorfsum geár *this was a very grievous year,* Chr. 1103 ; Erl. 239, 1.

ge-derian ; *p.* ode, ede ; *pp.* od, ed *To injure, hurt ;* lædĕre :—Ðyssum wordum ðá gecwedenum, hine sóna se wind onwearp fram ðære byrig, and dráf ðæt fýr on ða ðe hit ǽr onbærndon, swá ðæt hí sume mid ðam fýre gederede wǽron *quo dicto, stătim mŭtāti ab urbe venti, in eos qui accendĕrant flammārum incendia retorsērunt, ita ut ălĭquot læsi,* Bd. 3, 16 ; S. 543, 7–12, col. 1.

ge-dícian ; *p.* ode ; *pp.* od. *To make a dike* or *mound ;* vallum facere :—Eardædon Bryttas binnan ðam díce, ðe we gemynegodon ðæt Severus hēt þwyrs ofer ðæt eálond gedícian *habitabant Brittones intra vallum, quod Severum trans insulam fecisse commemoravimus,* Bd. 1, 11 ; S. 480. v. dícian.

ge-dieglan *To hide, cover ;* velare :—He wolde ðara scamfæstna giemelieste mid liðelícum wordum gedieglan *he would cover* [velare] *the negligence of the modest with gentle words,* Past. 31, 2 ; Swt. 207, 23 ; Hat. MS. 39 b, 23. v. ge-díglan.

ge-diernan ; *p.* de ; *pp.* ed *To conceal ;* cēlāre :—Se ðe þiéfþe gedierne, forgielde ðone þeóf be his were *let him who conceals the theft pay for the thief according to his value,* L. In. 36 ; Th. i. 124, 17. v. ge-dyrnan.

ge-dígan, -dýgan, -dégan, ic -díge, ðú -dígest, he -dígeþ, *pl.* -dígaþ ; *p.* de ; *pp.* ed *To endure, carry through, tolerate, overcome, escape ;* ēti, perpĕti, perferre, superāre, evadere :—Swá mæg unfǽge gedígan

weán so *an undoomed* [*man*] *may escape calamity*, Beo. Th. 4572; B. 2291. Ðū aldre gedīgest *thou escapest with life*, 1327; B. 661. He gedīgeþ *he escapes*, 606; B. 300. He feore gedīgde *he escaped with life*, 1161; B. 578. Feore gedýged *escaped with life*, Exon. 39 a; Th. 128, 21; Gū. 407. Ðæt wīf ne gedīgþ hyre feore *the woman will not escape with her life*, Nar. 50, 10. Ðara monna hit ælc gedīgde *hominibus idem morsus non usque ad interitum nocebant*, Nar. 16, 11. Sume hit ne gedýgdan mid ðam life *some did not escape with life*, Chr. 978; Erl. 127, 12. v. dýgan, gedēgan.

ge-dígl[i]ian, -dēglan, -dýglan; *p.* ode, ede; *pp.* od, ed, ad *To hide, conceal, cover;* abscondere, operire :—Gedeigla *abscondere*, Mt. Kmbl. Lind. 5, 14. Gedeigeldes *abscondisti*, 11, 25. Gedēglod *opertum*, 10, 26. Gidēglad [delgad] *abscondita*, Rtl. 25, 7. Helme gedýgled *concealed by a covering*, Hy. 11, 13. [Cf. *O. H. Ger.* tougilian *to hide*.]

ge-díhligean *to hide, make private, detach, separate;* velare. secernere, separare :—Eádgār, mid rýmette gedīhligean hēt ða mynstra *Edgar commanded the monasteries to be made private* or *detached*, Th. Diplm. A.D. 963–975; 231, 4. v. ge-díglan.

ge-diht, es; *n. A composition* :—Fela fægere godspel we forlǽtaþ on ðisum gedihte *many excellent gospels we omit in this composition*, Homl. Th. ii. 520, 1. [Cf. *Ger.* gedicht.]

ge-dihtan; *p.* -dihte; *pp.* -dihted, -diht. I. *to put in order, dispose, compose, arrange, conspire;* disponere, componere, conspirare :—Nū sindon twā bēc gesette on endebyrdnisse to Salomones bōcum, swilce he hīg gedihte *now two books are set in order after Solomon's books, as if he composed them*, Ælfc. T; Swt. A. S. Rdr. 69, 402. Bēda ðe ðas bōc gedihte *Bede who composed this book*, Swt. A. S. Rdr. 102, 224. Ðā gedihton ða Iudeas *jam conspiraverant Judæi*, Jn. Bos. 9, 22. Gediht *digestus, ordinatus*, Hpt. Gl. 409. II. *to order, direct, appoint;* dirigere, dictare :—Hīg dydon swā, swā him gedihte Iosue *they did as Joshua directed them*, Josh. 6, 23. Ðis gewrit wæs to ānum menn gediht *this writing was directed to a particular man*, Ælfc. T; Swt. A. S. Rdr. 56, 1. [*Laym.* so idæde idihte.] v. dihtan.

ge-dihtnung *a disposing.* v. dihtnung.

ge-dilgian; *p.* ede, ode; *pp.* ed, od *To blot out* :—Gidilge *dele*. Rtl. 168, 19; 19, 1.

ge-dirnan; *p.* de; *pp.* ed *To conceal, keep secret;* cēlāre :—Se ðe forstolen flǽsc findeþ and gedirneþ *he who finds stolen flesh and keeps it secret*, L. In. 17; Th. i. 114, 2, note 1. v. ge-dyrnan.

ge-dofung, e; *f. Dotage;* deliramentum, Hpt. Gl. 416.

ge-dolgian; *p.* ode; *pp.* od *To wound;* vulnerāre :—Deópe gedolgod *deeply wounded*, Exon. 113 b; Th. 435. 25; Rä. 54, 6.

ge-dōn; ic -dō, ðū -dēst, he -dēþ, *pl.* -dōþ; *p.* -dyde, *pl.* -dydon; *pp.* -dēn, -dōn *To do, make, put, cause, effect, reach a place;* facere :—Ic sceal cunnan hwæt ðū gedōn wille *I shall know what thou wilt do*, Andr. Kmbl. 684; An. 343. Ðū ne miht ǽnne locc gedōn hwītne *non potes unum capillum album facere*, Mt. Bos. 5, 36. Gedō dē hālne *salvum te fac*, Lk. Bos. 23, 37: 8, 48. Ðæt gefeoht wæs gedōn mid micelre geornfulnesse *the battle was fought* [*done*] *with much earnestness*, Ors. 3, 9; Bos. 64, 45. Ðæt hit gedōn wǽre *that it was done*, Andr. Kmbl. 1530; An. 766. Swā fela wundra swā we gehýrdon gedōne *quanta audivimus facta*, Lk. Bos. 4, 23. Ðæt he us ðæt cūþ gedō *that he make that known to us*, Blick. Homl. 139, 31. Hie gedōþ ðæt ǽgðer biþ ofer froren *they cause each to be frozen over*, Ors. 1, 1; Bos. 23, 9: Past. Swt. 7, 8: Ps. Th. 82, 12. Ðone eádigan Matheum he gedyde gangan *he caused the blessed Matthew to go*, St. And. 14, 13. We syndon niwe to ðissum geleáfan gedōn *we are newly turned to this faith*, 24, 9. Streównesse him under gedōn *to put litter under him*, Blickl. Homl. 227, 12. On cweartern gedōn *to put in prison*, Jn. Bos. 3, 24. Fōron ōð ðæt hie gedydon æt Sæferne *they went until they reached the Severn*, Chr. 894; Erl. 92, 14; 93, 5: 895; Erl. 94, 2, 15. Fōron ðæt hie gedydon innan Sæferne mūðan *they went so as to get within the mouth of the Severn*, Chr. 918; Erl. 102, 24. [*O. Sax.* gi-dōn.] DER. dōn.

ge-drǽfan; *p.* de; *pp.* ed *To drive, push, urge, trouble;* pellere, urgere, perturbare :—Wōd-þrag gedrǽfþ sefan ingehygd *lust urges the thoughts of mind*, Bt. Met. Fox 25, 83; Met. 25, 42: 18, 5; Met. 18, 3. v. drǽfan, gedrīfan.

ge-drǽfnes, ness, e; *f. A disturbance;* perturbatio, Bt. Met. Fox 22, 121; Met. 22, 61.

ge-drǽg, ge-dreag, es; *n. A dragging, band, multitude, tumult;* tractus, turma, tumultus :—He wolde sēcan deófla gedrǽg *he would seek the band of devils*, Beo. Th. 1516; B. 756. Eác ðon breost-ceare sin-sorgna gedreag sý æt him *even when care of breast, multitude of constant sorrows be at him*, Exon. 115 b; Th. 444, 10; Kl. 45. Ðǽr wæs fordēnera gedrǽg *there was a tumult of undone men*, Andr. Kmbl. 85; An. 43. Ðǽr wæs wīde gehýred earmlīc ylda gedrǽg *then was widely heard the wretched tumult of mortals*, 3108; An. 1557.

ge-drāf *drove, was wrecked*, Ors. Cot. MS. 4, 6; Bos. Notes, p. 20, col. 2, § 10. v. ge-drífan.

ge-dræg *multitude, tumult*, Exon. 22 b; Th. 62, 11; Cri. 1000: 103 a; Th. 389, 19; Rä. 7, 10. v. gedrǽg.

ge-dreccan; *p.* -drehte; -*pp.* -dreht, -dreaht *To vex, afflict, torment, oppress;* vexare, affligere, tribulare, opprimere :—He hæfþ on slǽpe ðýn wýf gedreht *he hath vexed thy wife in her sleep*, Nicod. 6; Thw. 3, 15. Beornas, gretaþ hýgegeómre hreówum gedreahte *men sad in mind with griefs afflicted shall wail*, Exon. 22 b; Th. 61, 34; Cri. 994. Hī scondum gedreahte *they shamefully tormented*, Exon. 26 b; Th. 79, 32; Cri. 1299: 30 a; Th. 92, 15; Cri. 1509. For meteleáste gedrehte *for want of food oppressed*, Andr. Kmbl. 78; An. 39. Of unclǽnum gāstum gedrehte *vexati a spiritibus immundis*, Lk. Bos. 6, 18: 7, 6.

ge-dreccednys, se; *f. Tribulation, affliction* :—Ðonne beóþ swilce gedreccednyssa swilce nǽron ǽr *then shall be such tribulations as were not before*, Homl. Th. i. 4, 1. Līchamlíc gedreccednys *bodily affliction*, 454, 26.

ge-drecte *oppressed.* v. gedreccan.

ge-drēfan; *p.* de; *pp.* ed *To disturb, trouble, vex, offend;* turbare, conturbare, confundere, scandalizare :—Hwī gedrēfe gyt me *quare* [*vos duo*] *conturbatis me*, Ps. Th. 41, 5. Se Hǽlend gedrēfde hyne sylfne *Jesus turbavit seipsum*, Jn. Bos. 11, 33: Lk. Bos. 24, 37. Ðū gedrēfest deópe wǽlas *tu conturbas profundos vortices*, Ps. Th. 64, 7. Ðū gedrēfst grūnd sǽs *tu confundas profundum maris*, Ps. Spl. 64, 7. Beóþ gedrēfde þeóda *turbabuntur gentes*, Ps. Spl. 64, 8. Swā hwā swā gedrēfþ ǽnne of ðyssum lytlingum *whosoever shall offend one of these little ones*, Mk. Bos. 9, 42. [*O. Sax.* ge-drōbian.] v. drēfan.

ge-drēfedlíc; *adj. Troublesome;* turbulentus, Ors. 1, 7; Bos. 30, 4.

gedrēfednes, -drōfednes, se; *f. Trouble, disturbance, confusion, vexation, tribulation, offence, scandal;* perturbatio, conturbatio, confusio, tribulatio, scandalum :—Būtan gedrēfednesse ðe menn þrōwiaþ *a conturbatione hominum*, Ps. Th. 30, 22. For gedrēfednesse sǽs swēges and ýða *præ confusione sonitus maris et fluctuum*, Lk. Bos. 21, 25: Mt. Bos. 13, 21: Lk. Bos. 17, 1.

ge-drēfnis, niss, e; *f. Disturbance, confusion;* perturbatio :—To ǽrēcte ðisse gedrēfnisse storm Sǽberhtes deáþ *auxit procellam hujusce perturbationis mors Sabercti*, Bd. 2, 5; S. 507, 6; Hpt. Gl. 463. v. gedrēfednes; ge-drēfnes.

ge-dreht, *oppressed, afflicted.* v. gedreccan.

ge-drēme, -drýme; *adj. Melodious, harmonious, joyous;* canōrus, consōnus, lætus :—Beóþ on heora hūsum blíðe gedrēme *lætābuntur in cubilibus suis*, Ps. Th. 149, 5. Hī ealle samod mid gedrēmum sange Godes wuldor hleoðrodon *they all together celebrated God's glory with melodious song*, Homl. Th. i. 38, 7. On gedrēmum lofsangum *in harmonious hymns*, 600, 9.

ge-drencan; *p.* -drencte; *pp.* -drenced *To drench, drown;* submergere, demergere :—Se wǽg gedrencte [-drecte MS.] dugoþ Egypta *the wave drowned the army of the Egyptians*, Cd. 167; Th. 209, 16; Exod. 500. Deáþe gedrenced *drenched with death*, 144; Th. 179, 25; Exod. 34. Ðū [bist] to helle gedrencged *tu ad infernum demergeris*, Lk. Skt. Lind. 10, 15.

ge-dreog, es; *n. A rubbing* :—Swīnes rysl his scōn to gedreoge *swine's fat for rubbing his shoes*, Homl. Th. ii. 144, 29.

ge-dreóg, es; *n. A retiring, modesty;* modestia, R. Ben. 8.

ge-dreógan; *p.* -dreág, -dreáh, *pl.* -drugon; *pp.* -drogen *To perform, finish, bear, suffer;* perficere, tolerare, pati :—Gedrogen hæfde *had finished*, Beo. Th. 5446; B. 2726. Wīf gedrōg *mulier patiebatur*, Mt. Kmbl. Lind. 9, 20. v. dreógan.

ge-dreóh; *adj. Sober* :—We lǽraþ ðæt man, æt ciric-wæccan, swíðe gedreóh sī *we teach that man, at the church wakes, be very sober*, L. Edg. 28; Th. ii. 250, 12.

ge-dreóhlíce; *adv. Discreetly, modestly, cautiously;* patienter, modeste, prudenter, L. C. S. 76; Th. i. 418, 6.

ge-dreósan; *p.* -dreás, *pl.* -druron; *pp.* -droren; *v. intrans. To fall together, disappear, fail;* cadere, corruere, deficere, Beo. Th. 3513; B. 1754: 5325; B. 2666: Ps. Th. 101, 9: Exon. 77 a; Th. 288, 25; Wand. 36. [*Goth.* gadriusan.]

ge-drep, es; *n. A stroke;* ictus :—Þurh daroþa gedrep *through the stroke of darts*, Andr. Kmbl. 2886; An. 1446.

ge-drettan; *p.* -drette; *pp.* -drett *To consume;* consūmere :—Beóþ gedrette eác gescende *confundantur et deficiant*, Ps. Th. 70, 12. [*Or does gedrette = gedrehte ?*]

ge-drif, e; *f. A fever;* febris, Mk. Skt. Rush. 1, 31. v. drif.

ge-dríf, -drif [?], es; *n. What is driven, stubble;* stipula :—Gesete hī swā swā gedríf ætforan ansýne windes *pone illos sicut stipulam ante faciem venti*, Ps. Spl. T. 82, 12. [Cf. *Icel.* drif *driven snow.*]

ge-dríf, es; *n. A driving, movement* :—Ðæs lyftes gedríf, ðæs wæteres gedríf *the regions of air and water*, Salm. Kmbl. 186, 22. [Cf. *Icel.* drífa *a fall of snow.*]

ge-drífan; *p.* -dráf, *pl.* -drifon; *pp.* -drifen *To drive, go adrift, be driven, cast away* or *lost;* agere, agi, ventis jactari, naufragāre :—Ðēh scyp gedrifen [MS. gedriuen] beó *though a ship be driven*, L. Eth. ii. 2; Th. i. 286, 1. Rōmāne oferhlæstan heora scipa ðæt heora gedráf [gedeaf *Laud.*] cc and xxx, and lxx wearþ to lāfe, and uneáðe genered *the Romans overloaded their ships, so that 230 of them were lost, and 70*

C c

were left, and with difficulty saved, Ors. 4, 6; Th. 400, 20. Ðæt scip gedrifen wæs *naviculo jactabatur*, Mt. Kmbl. Lind. 14, 24.

ge-driht, -dryht, e; *f. A host, company;* turma, cohors :—Wæs seó eorla gedriht ánes módes *the host of men was of one mind*, Cd. 158; Th. 197, 10; Exod. 304: Exon. 22 b; Th. 63, 3; Cri. 1014.

ge-drihþ, e; *f. Forbearance, sobriety;* patientia, sobrietas, L. T. P. 9; Th. ii. 314, 34.

ge-drinc, -drync, es; *n. A drinking;* compotatio, convivium :—We lǽraþ ðæt man ænig gedrinc, and ænig unnit ðǽr ne dreóge *we teach that man suffer not there any drinking nor any vanity*, L. Edg. 28; Th. ii. 250, 12: Exon. 88 a; Th. 330, 27; Vy. 57: Ors. 1, 1; Bos. 22, 25.

ge-drincan; *p.* -dranc, *pl.* -druncon; *pp.* -druncen *To drink;* bibere :—Grúndleás gítsung gilpes and ǽhta gedrinceþ to dryggum dreósendne wélan *the bottomless avarice of glory and possessions drinks to the dregs perishable wealth*, Bt. Met. Fox 7, 31; Met. 7, 16. Ðæt wín is gedruncen *bibitur vinum*, Ælf. Gr. 19; Som. 22, 47: Bd. 5, 5; S. 618, 13: Gen. 27, 25.

ge-dripan *to drip.* v. gedrypan.

ge-dróf; *adj. Dirty, muddy;* turbidus, lŭtōsus :—On ðæm gedrófum wætere *in the muddy water*, Past. 54, 1; Swt. 421, 8; Hat. MS.

ge-drófednys *trouble*, Scint. 50. v. ge-dréfednys.

ge-drófenlíc; *adj. Troublous :*—Ðeós world is gedrófenlíc *this world is troublous*, Blickl. Homl. 115, 3.

ge-drugian; *p.* ode, ade; *pp.* od, ad *To become dry, wither;* arescere :—Ficbeám gedrugade *ficus aruit*, Mk. Skt. Lind. 11, 21; 4, 6: Ps. Th. 68, 22. Gedrugad wæs *arefacta est*, Mt. Kmbl. Lind. 21, 19. v. drugian.

ge-druncen *drunk*, Bd. 5, 5; S. 618, 13; *pp.* of ge-drincan.

ge-druncnian; *p.* ode, ade; *pp.* od, ad *To sink, drown :*—Gedruncnadon *mergerentur*, Lk. Skt. Lind. 5, 7.

ge-drygan; *p.* de; *pp.* ed *To dry :*—Gédrygde his foet *extersit pedes ejus*, Jn. Skt. Lind. 11, 2. Gidrygedo *abstersa*, Rtl. 98, 24.

ge-dryht, -driht, e; *f. A host, company, band of retainers :*—Engla gedryht *a company of angels*, Exon. 22 b; Th. 63, 3; Cri. 1014: 60 b; Th. 222, 13; Ph. 348. Ðǽr cyninges giefe brúcaþ eádigra gedryht *there the band of the blessed enjoy the king's grace*, Exon. 32 a; Th. 101, 26; Cri. 1664. Ðínra secga gedryht *the band of thy men*, Beo. Th. 3349; B. 1672. v. dryht.

ge-dryhta, an; *m. A comrade;* commilito, Grm. ii. 736, 40.

ge-dryhtu; *pl. n. Elementa, sidera, fortunæ*, Hpt. Gl. 462. [Cf. droht?]

ge-drýme; *adj. Melodious, joyous;* lætus :—Drihta gedrýmost *most joyous of multitudes*, Cd. 146; Th. 182, 21; Exod. 79: Hpt. Gl. 513, 519. v. ge-dréme.

ge-drync *drinking*, Ors. 1, 1; Bos. 22, 25. v. ge-drinc.

ge-drypan; *p.* -drypte; *impert.* -dryp, -drype; *pp.* -dryped *To drop;* stillāre :—Beolonan seáw on eáre gedryp *drop juice of henbane on the ear*, L. M. 1, 3; Lchdm. ii. 40, 14. Gedrype on *drop* [*it*] *on*, 1, 3; Lchdm. ii. 40, 7.

ge-drysnan; *p.* ade, ede; *pp.* ad, ed *To put out, quench, extinguish, vanish;* extinguĕre, evanescĕre :—Ðæt fýr ne biþ gedrysned *ignis non extinguitur*, Mk. Skt. Lind. 9, 44, 48. He gedrysnade from égum hiora *ipse evanuit ex oculis eorum*, Lk. Skt. Lind. 24, 31.

ge-dúfan, he -dýfþ; *p.* -deáf, we -dufon; *pp.* -dofen; *v. intrans. To plunge, to duck, sink, dive, be drowned;* mergi :—Heó gedúfan sceolun in ðone deópan wælm *they must dive into the deep fire*, Cd. 213; Th. 266, 30; Sat. 30: Exon. 41 a; Th. 137, 6; Gú. 555. Gedeáf *sank*, Ors. 4, 6; Bos. 85, 38. Ðæt ðæt sweord gedeáf *so that the sword dived*, Beo. Th. 5394; B. 2700: Cd. 228; Th. 306, 27; Sat. 670. Ðá gedufon hí ealle and adruncon *then they all sank and were drowned*, Ors. 1, 7; Bos. 30, 24. He wearþ gedofen *coepit mergi*, Mt. Bos. 14, 13.

ge-dugan; *p.* -deáh *To thrive*, Shrn. 13, 1.

ge-dwǽlan; *p.* -dwǽlde *To seduce, lead astray :*—Ðæt is hefig dysig, ðæt ða earman men mid ealle gedwǽleþ of ðæm rihtan wege *that is a grievous folly that altogether seduces the miserable men from the right way*, Bt. Met. Fox 19, 6; Met. 19, 3. [*Or* gedwæleþ = gedweleþ *from* gedwellan.]

ge-dwǽs; *adj. Foolish, dull, stupid :* — Gedréfede syndon, hearde onhrérede men anlícast, hú druncen hwylce gedwǽs spyrige *turbati sunt et moti sunt ut ebrius*, Ps. Th. 106, 26. v. dwǽs.

ge-dwelian, -dweligan.   I. *to deceive, lead astray :*—Ðæt his me nán man gedweligan mæg *that no man can seduce me from it*, Bt. 23, 3; Fox 126, 18. Ne weorðe ic ðínra dóma gedweled ǽfre *judicia tua non sum oblitus*, Ps. Th. 118, 30.   II. *to err :*—Ic gedwelede swá ðæt dysige scép *erravi sicut ovis*, Ps. Th. 118, 176. v. dwelian *and* ge-dwellan.

ge-dwellan;   I. *to deceive, lead astray*, Bt. 23, 3; Fox 126, 18, note 6. Dysge and gedwealde *foolish and led astray*, Exon. 24 b; Th. 69, 29; Cri. 1128.   II. *to err :*—Gedwellas *erratis*, Mt. Kmbl. Lind. 22, 29. v. dwellan *and* ge-dwelian.

ge-dweola, -dweolda, an; *m. Error, heresy;* error, hærĕsis :—Se ge-

dweola wæs on ðam Nyceniscan sinoþe geniðerad *the error was put down in the Nicene synod*, Bd. 1, 8; S. 479, 36. Gé gedweolan lifdon *ye lived in error*, Invent. Crs. Recd. 623; El. 311. Se Arrianisca gedweolda *Arriāna hærĕsis*, Bd. 1, 8; S. 479, 27. v. ge-dwola.

ge-dwild, -dwyld, es; *n. Error, heresy;* error, hærĕsis :—On ðám tídum árás Pelaïes gedwild geond middangeard *in those times the heresy of Pelagius arose throughout the world*, Chr. 380; Erl. 11, 6. On gedwilde *into error*, Cd. 1; Th. 2, 22; Gen. 23. Ðú scealt ðrówian ðínra dǽda gedwild *thou shalt expiate the error of thy deeds*, 43; Th. 57, 2; Gen. 922. Dyrnra gedwilda *of dark errors*, Exon. 71 a; Th. 264, 22; Jul. 368. Deorcum gedwildum *by dark errors*, 72 b; Th. 270, 4; Jul. 460.

ge-dwimere, -dwomere; *m. A juggler, sorcerer;* nebulo, Hpt. Gl. 514, 515.

ge-dwimor, -dwimer, -dwymer, es; *n. An illusion, delusion, apparition, phantom;* error, fallācia, phantasma = φάντασμα, phantāsia = φαντασία :—Gedwimor *phantasma* vel *phantāsia*, Ælfc. Gl. 78; Som. 72, 54; Wrt. Voc. 46, 14: 77, 7. Hí wéndon ðæt hit sum gedwimor wǽre *they thought that it was an apparition*, Homl. Th. ii. 388, 24: Jud. 15, 19. Hine drehton nihtlíce gedwimor *nightly phantoms tormented him*, Homl. Th. i. 86, 18. Swylcra gedwimera *of such illusions*, L. C. S. 5; Th. i. 378, 22. On manegum mislícum gedwimerum *with many various delusions*, L. Edg. C. 16; Th. ii. 248, 7.

ge-dwimorlíce; *adv. Illusorily, fantastically*, Homl. Th. ii. 140, 16.

ge-dwínan; *p.* -dwán, *pl.* -dwinon; *pp.* -dwinen *To dwindle or vanish away, disappear;* evanescere, disparere :—Ðæt hálige sǽd gedwán and gewát *the holy seed was wasted away and departed*, Blickl. Homl. 55, 29. His drýcræftas gedwinon *his magic vanished*, Shrn. 135, 1.

ge-dwola, -dweola, an; *m.*   I. *error, madness, heresy;* error, errātum, vesānia, hærĕsis :—Se mennisca gedwola *human error*, Bt. 33, 2; Fox 122, 22. Se Arrianisca gedwola *Arriāna hærĕsis*, Bd. 1, 8; S. 479, 33: Bt. Met. Fox 1, 81; Met. 1, 41. Oþ ða tíde ðæs Arrianiscan gedwolan *usque ad tempŏra Arriānæ vesāniæ*, Bd. 1, 8; S. 479, 18. Gé gedwolan fylgdon *ye followed error*, Elen. Kmbl. 742; El. 371: Bt. Met. Fox 26, 108; Met. 26, 54. Ðæt ða beóþ oñ gedwolan gelædde *ut in errōrem indūcantur*, Mt. Bos. 24, 24: Gen. 21, 14: 37, 15: Bt. Met. Fox 26, 78; Met. 26, 39. Þurh deópne gedwolan *through deep error*, Andr. Kmbl. 1221; An. 611: Exon. 70 a; Th. 260, 22; Jul. 301. Gedwolena rím *a number of errors*, 71 a; Th. 264, 23; Jul. 368. For mínum gedwolum *pro meis errātĭbus*, Bd. 4, 25; S. 601, 3.   II. *a heretic;* hærĕticus :—Begeat se gedwola ðæs cáseres fultum to his gedwylde *the heretic got the emperor's support to his heresy*, Homl. Th. i. 290, 11, 17, 28. Done ealdan gedwolan *the old deceiver*, Blickl. Homl. 7, 12.

ge-dwol-cræft, es; *m. A deceptive art, deception :*—Mid heora gedwolcræftum *with their deceptions*, Blickl. Homl. 61, 25. Ða ðe gedwolcræftas begangaþ *those who practise divination*, 63, 14.

ge-dwolen [*pp.* of *strong verb* ge-dwelan. v. dwelan]; *adj. Erroneous, wrong, perverse :*—Dǽdum gedwolene *in deeds perverse*, Cd. 91; Th. 116, 14; Gen. 1936: Exon. 66 a; Th. 243, 19; Jul. 13: 103 b; Th. 393, 8; Rä. 12, 7. [Cf. *O. H. Ger.* gi-tiuolin *sopitus*.]

ge-dwol-godas; *pl. m. False gods, idols;* falsi dei, Idōla : — To gedwolgoda weorþunge *idolōrum cultui*, Lupi Serm. i. 4; Hick. Thes. ii. 100, 3. Ne dear man gewanian on hǽðenum ænig ðæra þinga ðe gedwolgodum [MS. -an] broht biþ *ne ausus est quispiam e pāgānis eōrum quidquam commĭnuĕre quæ deōrum simulacris allāta fuĕrant*, i. 4; Hick. Thes. ii. 100, 6, 11.

ge-dwolian; *p.* ede; *pp.* ed *To err :*—Ic gedwolede swá swá sceáp ðæt forwearþ *I have erred as the sheep that perished*, Blickl. Homl. 87, 30. Gé swíðe gidwoligas *vos multum erratis*, Mk. Skt. Rush. 12, 27: Mt. Kmbl. Lind. 18, 12.

ge-dwol-man, gedwol-mon, es; *m. An erring man, a heretic, impostor;* hæreticus : — Arrius hátte án gedwolman *there was a heretic called Arius*, Homl. Th. i. 290, 3, 5, 25: 110, 6.

gedwol-mist, es; *m. Mist of error;* errōris nĕbŭla :—Mid ðam gedwolmiste *with the mist of error*, Bt. 35, 1; Fox 156, 1: Bt. Met. Fox 22, 65; Met. 22, 33.

ge-dwolsum; *adj. Erroneous;* errōneus :—Hit biþ swíðe gedwolsum *it is very erroneous*, Ælf. Pref. Gen. 4, 10.

ge-dwol-þing *an erroneous thing, deceit, imposture.*

ge-dwomer, es; *n. Necromancy*, Hpt. Gl. 515.

ge-dwyld, es; *n. Error, heresy;* error, hærĕsis :—Ðæt æftere gedwyld *novissĭmus error*, Mt. Bos. 27, 64. Ic wille him dón edleán heora gedwyldes *I will give them a reward for their error*, Boutr. Scrd. 22, 37. Forwearþ ðes gedwola mid his gedwylde *this heretic perished with his heresy*, Homl. Th. i. 290, 29: ii. 506, 27: Boutr. Scrd. 18, 30. Ðæt he mid his hálgan láre middaneardlíc gedwyld adwæscte *that he might extinguish worldly error by his holy doctrine*, Homl. Th. ii. 90, 13: Deut. 4, 19. v. ge-dwild.

ge-dwymer, es; *n. An illusion;* error :—Swylcra gedwymera *of such illusions*, L. C. S. 5; Th. i. 378, 22, note 66. v. ge-dwimor.

ge-dwymorlíc; *adj. Illusive;* phantasticus, Dial. 2, 10.

ge-dýgan; *p.* de; *pp.* ed *To escape:*—Hwæðer mæge wunde gedýgan *which may escape from wound*, Beo. Th. 5056; B. 2531: 5091, note; B. 2549. Gedýgdon *escaped*, Exon. 55 b; Th. 197, 17; Az. 191. Gedýged, 39 a; Th. 128, 21; Gū. 407. v. ge-dígan.

ge-dyn, es; *m. A din, noise;* frāgor, clangor:—Se dæg biþ dæg gedynes ofer ealle [MS. ealla] truma ceastra *the day will be a day of din over all strong cities*, Past. 35, 5; Swt. 245, 6; Hat. MS. 46 a, 17. Gedyne micle *with a great din*, Exon. 102 a; Th. 385, 16; Rä. 4, 45.

ge-dyngan; *p.* ede; *pp.* ed *To dung, manure;* stercōrāre :—Hit ðonne mid ðam gedynged wearþ *then it was thus manured*, Ors. 1, 3; Bos. 27, 23.

ge-dyppan, -deppan *to dip, baptize:*—Đā gedeped [wæs] baptizatus, Mt. Kmbl. Rush. 3, 16.

ge-dýran; *p.* de; *pp.* ed *To glorify, endear;* glorifīcāre :—Dreámum gedýrde *endeared by joys*, Exon. 32 a; Th. 100, 21; Cri. 1645.

gedýre, es; *n.* [or -dyre, y *from* u; cf. *Goth.* daur] *A door-post;* postis ad fores :—On ægðrum gedýre *in utro poste*, Ex. 12, 23. On ægðer gedýre *on each door-post*, Ex. 12, 7. Hí mearcodou mid blóde on heora gedýrum Tau, ðæt is, róde tācen *they marked on their door-posts Tau, that is, the sign of the cross*, Homl. Th. ii. 266, 8: 264, 1. v. ofer-gedýre.

ge-dyrfsum; *adj. Afflictive;* calamitosus, Lye.

ge-dyrnan, -diernan, -dirnan; *p.* de; *pp.* ed *To conceal, hide, keep secret;* cēlāre, occultāre :—Se ðe forstolen flǣsc findeþ and gedyrneþ *he who finds stolen flesh and keeps it secret*, L. In. 17; Th. i. 114, 2. Se ðe ða þýfþe gedyrne, forgylde ðone þeóf be his were *let him who conceals the theft pay for the thief according to his value*, 36; Th. i. 124, 17, note 40, MS. B. Đonne hit gedyrned weorþeþ *when it is hidden*, Exon. 91 a; Th. 340, 27; Gn. Ex. 117.

ge-dýrsian; *p.* ode; *pp.* od *To glorify;* glorifīcāre :—Dóme gedýrsod, Judth. 12; Thw. 25, 40; Jud. 300.

ge-dyrst, e; *f. Tribulation;* tribulatio? [Th] :—Ic ðē hālsie deópe in gedyrstum, ðæt ðū us gemiltsie *I beseech thee deeply in tribulations, that thou us pity*, Exon. 121 a; Th. 465, 22; Hö. 108. [*O. H. Ger.* gaturst, *f. audacia.*]

ge-dyrste-líce; *adv. Boldly, daringly, rashly;* temere, audaciter, Bd. 4, 26; S. 602, 16. v. dyrste-líce.

ge-dyrstig; *adj. Bold;* audax, protervus, Exon. 72 a; Th. 268, 12; Jul. 431: Past. 32, 1; Swt. 209, 15; Hat. MS. 40 a, 8: Guthl. 20; Gdwn. 84, 20. v. un-gedyrstig, dyrstig.

ge-dyrstigan; *p.* ede; *pp.* ed *To dare, presume;* audēre, præsumēre:— Ðe gedyrstigedon ðæt hí Eástran heóldan būtan heora rihtre tíde *qui Pascha non suo tempŏre observāre præsumērent*, Bd. 5, 21; S. 642, 40.

ge-dyrstig-nes, -ness, e; *f. Boldness;* audacia, Past. 13, 2; Swt. 79, 17; Hat. MS. 17 a, 16: Nar. 19, 11. v. dyrstignes.

ge-dyrst-lǣcan; *p.* -lǣhte; *pp.* -lǣht *To dare;* audere :—He ne gedyrstlǣcþ ðæt he furðon orðige oððe sprece *he dare not even breathe or speak*, Homl. Th. i. 456, 9: Ælfc. Gr. 41; Som. 43, 29. v. dyrst-lǣcan.

ge-dysig; *adj. Foolish.* v. dysig.

gee *yea, yes.* v. gea.

ge-eácnian, ic -eácnige, ðū -eácnigast, he -eácnaþ, *pl.* -eácniaþ; *p.* ode; *pp.* od *To increase, conceive, become pregnant;* augēri, concipĕre, augēre :—Ic hine bletsige and geeácnige *benedīcam ei et augēbo eum*, Gen. 17, 20. Efnenū geeácnode unrihtwísnesse *ecce partūri injustĭtiam*, Ps. Lamb. 7, 15. Hí geeácnodon unrihtwísnysse *augēbant injustĭtiam*, Jud. 4, 1: Elen. Grm. 342. Elizabeþ his wíf geeácnode *Elizabeth his wife conceived*, Lk. Bos. 1, 24. Đū on innoðe geeácnast *thou shalt conceive in thy womb*, 1, 31. In synnum geeácnod wæs he *was conceived in sins*, Ps. C. 50, 61; Ps. Grn. ii. 278, 61. Der. to-geeácnian. v. eácnian.

ge-eácnung, e; *f. A conceiving, conception;* conceptio, conceptus :— Đæt he bodige hire geeácnunge *to proclaim her* [Maria] *conception*, Blickl. Homl. 143, 24. Ic gemenigfilde ðíne yrmþa and ðíne geeácnunga *multiplicabo ærumnas tuas et conceptus tuos*, Gen. 3, 16. v. eácnung.

ge-eádgian; *p.* ode, ade; *pp.* od, ad *To bless:*—Gieadgade hine *beatifīcavit illum*, Rtl. 88, 26.

ge-eádmédan, -eáþmédan, he -eádmédeþ; *p.* -medde, -métte; *pp.* -méded, -mét; *v. a. To humble, humiliate, subdue, submit one's self, humble one's self, deign, condescend, adore, worship;* humiliare, dignari, condescendere, adorare :—Se gehnysta gāst and geeádméded ingeþancum *the bruised heart and humbled by reflections*, Ps. C. 50, 128; Ps. Gen. ii. 279, 128. Ic geeádméded eom *humiliatus sum*, Ps. Th. 141, 6. Hí hí geeádmétte *he humiliated* [*subdued*] *them*, Jud. 11, 33. Se ðe hyne sylfne geeaþmēt *qui se humiliaverit*, Mt. Bos. 23, 12: 18, 4. Hine to him geeaþmédde *he submitted himself to him*, 8, 2: Bd. 5, 3; S. 616, 9. We cōmon us him to ge-eádmédenne *venimus adorare eum*, Mt. Bos. 2, 2. Geeámédun ðē ealle mægþa *may all nations adore thee*, Gen. 27, 29: Ex. 11, 8; Mt. Bos. 20, 20. v. ge-eáþmédan, eádmédan.

ge-eádmódian, -eáþmódian *to humiliate, deign:*—Se ðe ne wyle geeádmódian ingangan *qui non vult humiliatus ingredi*, Bd. 5, 14; S. 634, 19. Đæt he ge-eádmódige *ut ipse dignetur*, 2, 2; S. 502, 19. v. eád-módan.

---

ge-eádmódlíce; *adv. Humbly;* humiliter, Bd. 2, 2; S. 503, 11. v. eádmódlíce.

ge-eæd-leǽnian, ic -eædleǽnige *to repay, reward*, Ps. Spl. T. 17, 22. v. ed-leǽnian.

ge-eærfoðod *troubled.* v. eærfoðian.

ge-eahtian, -ehtian, -æhtian; *p.* ode; *pp.* od *To estimate, value;* æstimāre :—Gebēte swā hit mon geeahtige *let him make amends as it may be valued*, L. Alf. 26; Th. i. 50, 26: L. Alf. pol. 32; Th. i. 82, 2.

ge-ealdian; *p.* ode; *pp.* od, ad *To grow old;* senescere :—Geealdad biþ *is become old*, Exon. 62 a; Th. 227; 23; Ph. 427. v. ealdian.

ge-ealgian *to defend*, R. Ben. 69, Lye. v. ge-algian.

ge-eán; *adj. Yeaning;* enītens, pariens :—Đū wāst ðæt ic hæbbe hnesce litlingas, and ge-eáne eówa mid me *thou knowest that I have tender infants and yeaning sheep with me*, Gen. 33, 13; tu scis [MS. nosti = novisti], quod parvŭlos hăbeam tĕnĕros et oves fētas mecum, Vulg. Gen. 33, 13. v. gecelf. Der. eánian [?].

ge-eardian; *p.* ode; *pp.* od *To dwell;* inhabitāre :—In me gǣst geeardode *the spirit dwelt in me*, Exon. 11 a; Th. 13, 25; Cri. 208: Ps. Lamb. 26, 4.

ge-earfoþ, es; *n. Trouble;* trībūlātio :—He sceal geþolian manige geearfoðu [MS. gearfoðu] *he shall suffer many troubles*, Bt. 31, 1; Fox 110, 26.

ge-earnian, -igan; *p.* ode; *pp.* od *To earn, deserve, enjoy;* mereri, promereri, frui :—Ic ge-earnige *mereor*, ðū ge-earnast *mereris*, he geearnaþ *meretur*, ic ge-earnode *merui vel meritus*, Ælfc. Gr. 27; Som. 29, 64, 65: 33; Som. 36, 49. Đæt heó ðý ēþ meahte ðæt ēce ríce in heofonum geearnian *quo facilius perpetuam in cælis patriam posset mereri*, Bd. 4, 23; S. 593, 12. Đæt se man sceolde ða myrhþe geearnian *that man should enjoy the pleasure* [gaudium], Hexam. 17; Norm. 24, 23. Hie ne mágon geearnigan ðæt gē heora wundrigen *they cannot deserve that ye should admire them*, Bt. 13; Fox 40, 8. He geearnode *meruit*, Bd. 4, 23; S. 593, 6. He hí hæfþ geearnod mid his hearpunga *he has earned her by his harping*, Bt. 35, 6; Fox 170, 7.

ge-earnung, e; *f. Earning, desert, merit;* meritum :—For heora lífes geearnunge *for their life's earning* [*desert*]; præ merito virtutum, Bd. 3, 8; S. 531, 23. Nu ic ongite ðæt sió sóþe gesǣlþ stent on gódra monna geearnunga *now I understand that true happiness stands on the merit of good men*, Bt. 39, 2; Fox 212, 12. Be geearnunge *de merito*, Ps. Lamb. 7, 5. Geearnunga *merita*, Cot. 129. Būtan geearnungum *sine merito, immerito, gratis*, Ps. Lamb. 34, 7; 68, 5: 108, 3: 118, 161: 119, 7. Der. earnung.

ge-eáþmédan *to humiliate, submit one's self, condescend, vouchsafe, deign*, Mt. Bos. 8, 2: Bd. 5, 3; S. 616, 9. v. ge-eádmédan.

ge-eáþmódian *to humiliate, condescend, vouchsafe, deign :*—Drihten wæs geeáþmódad to onwreónne *dominus revelare dignatus est*, Bd. 4, 23; S. 595, 35. v. ge-eádmódian.

ge-eáwan; *p.* de; *pp.* ed; *v. trans. To shew, manifest, bestow;* ostendere, manifestare, præbere :—Geeáude him alle rícas middangeardes *ostendit ei omnia regna munda*, Mt. Kmbl. 4, 8. Him wæs wunden gold ēstum geeáwed *on him was twisted gold kindly bestowed*, Beo. Th. 2392; B. 1194: Exon. 60 b; Th. 221, 14; Ph. 334: 66 b; Th. 246, 29; Jul. 69: Bt. 39, 8; Fox 224, 12: Elen. Grm. 102: Elen. Kmbl. 1570; El. 787. Der. eáwan, ýwan.

ge-ebbian; *p.* ode, ade; *pp.* od, ad *To ebb;* recedere, refluere :—Đā ðæt wæter wæs geebbod fram ðām scipum *when the water had ebbed from the ships*, Chr. 897; Th. 176, 26, col. 2. v. ebbian.

ge-ebolsian, -eofulsian; *p.* ode, ade; *pp.* od, ad *To blaspheme*, Mk. Skt. Lind. and Rush. 15, 29; Mt. Kmbl. Lind. 27, 39.

ge-ēcan *to add, increase :*—His sylfes synna geēceþ *increases his own sins*, Blickl. Homl. 97, 9; 37, 17; 121, 32. v. ge-ícan.

ge-edbyrdan; *p.* de; *pp.* ed *To cause to be born again, to regenerate;* facere ut aliquis renascatur, regenerare :—Đonne he unc hafaþ geedbyrded ōðre síðe *when he hath caused us two to be born again a second time*, Exon. 99 b; Th. 372, 30; Seel. 100.

ge-edcégan; *p.* de; *pp.* ed *To recall;* revŏcāre :—Ne geedcēg ðū me on midlunge mínra daga *ne revŏces me in dimīdio diērum meōrum*, Ps. Lamb. 101, 25.

ge-edcenned *regenerated;* regeneratus, Jn. Bos. 3, 5.

ge-edcucian, -cwician; *p.* ode, ede; *pp.* od, ed *To requicken, revive;* revīviscĕre :—Ic geedcucige *revivisco*, Ælfc. Gr. 35; Som. 38, 9. Đes mín sunu wæs deád, and he geedcucode *hic fīlius meus mortuus ĕrat, et revixit*, Lk. Bos. 15, 24, 32: Homl. Th. ii. 26, 27: 28, 5. His cealdan limu geedcucodon *his cold limbs requickened*, i. 534. 35. He wearþ ðā geedcucod æfter lytlum fyrste *he then after a little space revived*, ii. 504, 28, 8. Geedcuced *redivīvus*, Ælfc. Gl. 35; Som. 62, 91; Wrt. Voc. 28, 68. His gāst wearþ geedcwicod *revixit spirĭtus ējus*, Gen. 45, 27. Geedcwycode *brought to life again*, Nicod. Thw. p. 18, 15.

ge-edhiwod; *part. p. Conformatus*, Som.

ge-edhyrt; *adj. Recreatus*, Gl. Prud. 201.

ge-edlǣcan; *p.* -lǣhte; *pp.* -lǣht *To repeat :*—Đonne mōt he geornlíce warnian, ðæt he eft ðām yfelum dǣdum ne geedlǣce *then must he*

diligently take heed that he do not afterwards repeat those evil deeds, Homl. Th. ii. 602, 24. Geedlǽcend, geedlǽht, *reciprocus,* Hpt. Gl. 450, 460, 481, 484.

ge-edlǽsian; *p.* ode; *pp.* od *To restore;* restituere:—Ðú ðe geedlǽsast *qui restituis,* Ps. Lamb. 15, 5.

ge-edleánénd, es; *m. A rewarder,* Som.

ge-edlian *to renew,* Som.

ge-edniwian, -edneowian; *p.* ode, ade; *pp.* od, ad *To restore, renew, renovate, change;* restituere, renŏvāre, innŏvāre:—Helias geedniwaþ ealle þing *Elias restĭtuet omnia,* Mt. Bos. 17, 11: Mk. Bos. 9, 12. Geedniwod eald hrægel *renovāta antīqua vestis,* Ælfc. Gl. 63; Som. 68, 105. Se mōna biþ þreottyne sīðon geedniwod [MS. geedniwad] *the moon is thirteen times changed* [*renewed*], Lchdm. iii. 248, 24. Biþ geedniwad moncyn *mankind shall be renewed,* Exon. 23 a; Th. 64, 20; Cri. 1040: Ps. Th. 103, 28. Se firdstemn hie geedneowade *the army-corps renovated it,* Chr. 921; Erl. 107, 33. Gást riht geedneowa on innoþum mīnum *spīrĭtum rectum innŏva in viscĕrĭbus meis,* Ps. Lamb. 50, 12. Se man ðe æfter dǽdbōte his mānfullan dǽda geedniwaþ *the man who after repentance renews his sinful deeds,* Homl. Th. ii. 602, 25.

ge-edstaðelian, *p.* ode; *pp.* od *To restore;* instaurare, suscitare:—Ða hǽr beóþ ealle geedstaðelode *the hairs shall be all restored,* Homl. Th. ii. 542, 35: i. 62, 11, 12. Se cyng fērde and ða burh gæedstaðelede *the king went and restored the town,* Chr. 1092; Erl. 228, 15: Th. Apol. 27, 5: Hpt. Gl. 456.

ge-edstaðelung, e; *f. A renewing;* repărātio, C. R. Ben. 48.

ge-edstaðolian. v. ge-edstaðelian.

ge-eddrawen; *part. p. Twisted again* or *back;* retortus, Som.

ge-edwistian; *p.* ode; *pp.* od *To feed, support:*—He geedwistode me *educavit me,* Ps. Lamb. 22, 2.

ge-edwyrpan; *p.* te; *pp.* ed *To recover, revive;* revīviscĕre:—Ðá æt nýhstan onfēng he gáste and wearþ geedwyrped *tandem recepto spīrĭtu revixit,* Bd. 4, 22; S. 590, 36.

ge-efenlǽcan; *p.* -lǽhte; *pp.* -lǽht, -lǽced; *v. trans. To be like, equal, to imitate;* æquāre, assimĭlāri, imĭtāri:—Nellen ge eornostlīce him ge-efenlǽcan *nolite ergo assimĭlāri eis,* Mt. Bos. 6, 8. Hwylc biþ geefenlǽced drihtne *quis æquālĭtur Domino,* Ps. Spl. 88, 7: Wanl. catal. 5, 1. Ongann Augustinus mid his munecum to geefenlǽcenne ðæra apostola líf *Augustine with his monks began to imitate the life of the apostles,* Homl. Th. ii. 128, 32. Ðæt hí ðám flǽsclícum geefenlǽcon *that they imitate the fleshly,* 82, 15. v. efenlǽcan.

ge-efenlǽcestre, an; *f. A female imitator,* Scint. 13, Lye.

ge-efenlǽcung; e; *f. Imitation:*—To geefenlǽcunge ðæra eádigra apostola *in imitation of the blessed apostles,* Homl. Th. ii. 148, 23.

ge-efenlíc; *adj. Equal,* Bd. 4, 29; S. 608, 3, note, MS. Ca. See next word.

ge-efenlícad; *part. p. Made equal;* æquātus:—Ðæt he swá geefenlícad wǽre mid ða gife his þingeres *quătēnus æquātus grātia suo intercessōri,* Bd. 4, 29; S. 608, 3.

ge-efesian, -efsian; *p.* ode; *pp.* od *To cut in the form of eaves, to round, shear, clip, crop;* tondēre:—Ne he næs geefesod ne bescoren *he was not clipped nor shorn,* Homl. Th. ii. 298, 20. Ic næs nǽfre geefsod ne nǽfre bescoren *ferrum nunquam ascendit super caput meum,* Jud. 16, 17. DER. efesian.

ge-efnan; *p.* ede; *pp.* ed *To do, perform, carry out, sustain:*—Eft geblōweþ and geefneþ swá óþ ðæt æfen cymeþ *it blows again, and does so until even comes,* Ps. Th. 86, 6. Hió geefnede swá *she did so,* Elen. Kmbl. 2028; El. 1015. Hwá gedēþ ǽfre, ðæt he ðæt geefne *quis sustinebit?* Ps. Th. 129, 3. Ealdor geefnan *to spend* [*one's*] *life,* Salm. Kmbl. 711; Sal. 355. v. efnan; ge-æfnan.

ge-efn[i]an; *p.* ade, [e]de; *pp.* ed *To make even, liken, compare:*—Byrgennum ða ilco geefnade *monumentis eos compărans,* Mt. Kmbl. p. 19, 12. Giefndes *coequasti,* Rtl. 57, 13. Geefuad *æquatus,* Bd. 4, 29; S. 608, 3, note. Geefned biþ *assimilabitur,* Mt. Kmbl. 7, 24. [*O. H. Ger.* ge-ebanôn *explanare, æquare.*]

ge-efstan; *p.* -efste; *impert.* -efst; *pp.* -efsted, -efst *To hasten, make haste, be quick;* festināre, accelĕrāre:—Geéfst oððe hrada ðæt ðú alýse me *accelĕra ut eruas me,* Ps. Lamb. 30, 2. DER. efstan.

ge-egesian; *p.* ode; *pp.* od *To frighten;* terrēre:—Hí wurdon geegesode *they were frightened,* Ors. 5, 3; Bos. 104, 5. v. ge-egsian.

ge-eggian; *p.* ede *To egg on, urge, excite:*—Ða biscobas geeggedon ðone ðreát *Pontifices concitaverunt turbam,* Mk. Skt. Lind. 15, 11.

ge-eglan, -eglian; *p.* de, ede, ode; *pp.* ed *To trouble, injure;* molestāre:—Hyra líce ne wæs ōwiht geegled *their bodies were not injured aught,* Cd. 191; Th. 237, 27; Dan. 344: Shrn. 99, 9: 154, 4.

ge-egsian, -egesian; *p.* ode; *pp.* od *To frighten;* terrēre:—He hý mid his wordum geegsode *he frightened them with his words,* Ors. 2, 3; Bos. 42, 13: Jud. 7, 22. Geegsod *frightened,* 4, 17.

ge-ehtian; *p.* ode; *pp.* od *To estimate, value;* æstimāre:—Ðæt hie mon nã undeórran weorþe mōste lésan ðonne hie mon be ðam wǽre geehtige *which must not be redeemed at any cheaper rate than it is estimated according to his value,* L. Alf. pol. 32; Th. i. 82, 2, note 8. v. geeahtian.

ge-elnian; *p.* ode; *pp.* od *To strive with zeal after another;* zēlāre:—Ic geelnode ofer ða unrihtwísan *zēlāvi super inīquos,* Ps. Spl. T. 72, 3.

ge-embehtan; *p.* ade *To minister;* ministrare:—Geembehta ministrare, Lk. Skt. Lind. 10, 40. He geembihtæs *ministrat,* Mt. Kmbl. p. 15, 15. Ðætte he geembehtade *ut ministraret,* Mk. Skt. Lind. 10, 45: 15, 41.

ge-emnettan, -emnittan, -emnyttan; *p.* te; *pp.* ed *To make even* or *level, compare;* æquāre, exæquāre:—Deáþ geemnet ða rīcan and ða heánan *death levels the rich and the poor,* Bt. 19; Fox 68, 34. Gif we úre unþeáwas geemnettaþ be his hǽsum *if we level our vices by his commands,* Homl. Th. ii. 316, 1. Heó hí sylfe to hwelpum geemnette *she compared herself to the whelps,* 114, 10. Geemnittan *exæquāre,* Scint. 9. Ðæt heó ðone dæg and ða niht geemnytte *that it might make even the day and the night,* Bd. de nat. rerum; Lchdm. iii. 238, 24. Geemnettan *quadrare, congruere,* Hpt. Gl. 506.

ge-emnian; *p.* ode; *pp.* od *To make even, match;* adæquare, Som. [Cf. ge-efnian.]

ge-enogd; *part. p. Anxious, careful,* Som. [Cf. ange, enge.]

ge-endadung, e; *f. Finishing, consummation:*—Giendadunge *consummatu,* Rtl. 105, 28.

ge-ende, es; *m. An end,* Som.

ge-endebredian; *p.* ade; *pp.* ad *To set in order,* Rtl. 69, 4: 109, 4.

ge-endebrednian; *p.* ade; *pp.* ad *To set in order;* ordinare:—Ðætte hia geendebrednadon *ordinare,* Lk. Skt. Lind. 1, 1. Geendebrednege *ordinare,* Mt. Kmbl. p. 7, 2.

ge-endebyrdan; *p.* -byrde; *pp.* -byrded, -byrd *To set in order, arrange, dispose;* ordināre, dispōnēre:—Manega þohton ðæra þinga race geendebyrdan *multi cōnāti sunt ordĭnāre narrātiōnem rērum,* Lk. Bos. 1, 1. Heó ðæt sóna mid reogollíce lífe gesette and geendebyrde *she soon settled and ordered it with regular life,* Bd. 4, 23; S. 593, 28. Rihte Godes dōme geendebyrded wæs æfter synne ðæs ǽrestan mannes *est digno Dei jŭdĭcio þost culpam ordīnātum,* 1, 27; S. 494, 13. Gif heora mōd wǽre geendebyrd *if their minds were ordered,* Bt. 21; Fox 76, 1: Bt. Met. Fox 11, 199; Met, 11, 100.

ge-endian, -endigan, to -endianne; *p.* ode, ade; *pp,* od, ad. **I.** *v. trans. To end, finish, complete, accomplish;* finīre, consummāre, perficĕre:—Ðes man agan timbrian, and ne mihte hit geendian *hic hŏmo cœpit ædĭfĭcāre, et non pŏtuit consummāre,* Lk. Bos. 14, 30. Ǽr heó hit geendigan mōste *ere she might end it,* Bd. 3, 8; S. 532, 28. Se cyning mid árleásre cwale ofslegen wæs, and ðæt ylce geweorc his æfterfyligende Oswalde forlēt to geendianne *rex ipse impio nece occīsus, ŏpus ĭdem successōri suo Osualdo perfĭciendum relīquit,* 2, 14; S. 517, 33. Ic geendige *finio,* Ælfc. Gr. 30, 5; Som. 34, 57. Man ðæt geendaþ on ǽfynne *man ends it in the evening,* Ps. Th. 103, 22. Oþoniél geendode his dagas *mortuus est Othoniel,* Jud. 3, 11: Chr. 189; Erl. 9, 27. Hyt ys geendod *consummātum est,* Jn. Bos. 19, 30: Mk. Bos. 13, 4. Ðe nó geendad weorþeþ *which shall not be ended,* Exon. 32 a; Th. 100, 12; Cri. 1640: 63 a; Th. 232, 1; Ph. 500. **II.** *to come to an end:*—Ðá geendode se gebeorscipe *then the feast came to an end,* Th. Apol. 18, 8. Siððan Eádgár geendode *since Edgar died,* Swt. A. S. Rdr. 106, 44: 68, 365. Geendiaþ ealle on *ans they all end in -ans,* Ælfc. Gr. Som. 43, 46.

ge-endung, -ændung, e; *f. An end, finish, death;* finis, consummātio, mors:—Geendung ealles flǽsces *finis ūnīversæ carnis,* Gen. 6, 13. Ðonne cymþ seó geendung *tunc vēniet consummātio,* Mt. Bos. 24, 14. Óþ ðisre worulde geendunge *until the end of this world,* Boutr. Scrd. 17, 18: 20, 20; Homl. Th. ii. 74, 10. On geendunga *in consummātiōne,* Ps. Spl. 58, 14. Æfter geendunge ðæra ealdra manna *after the death of the old men,* Jud. Thw. 153, 20: Homl. Th. ii. 122, 18.

ge-engd, -enged; *past p. Anxious, sad.* v. ange.

ge-eofot, es; *n. A debt;* dēbĭtum:—Gif mon on folces gemōte geeofot uppe *if a man declare a debt at a folk-moot,* L. Alf. pol. 22; Th. i. 76, 6, MS. H. v. eofot.

ge-eorsian; *p.* ode; *pp.* od *To be angry;* irasci:—Wæs geeorsod on hát-heortnesse Drihten on folce his *irātus est fŭrōre Dŏmĭnus in pŏpŭlo suo,* Ps. Lamb. 105, 40. v. ge-yrsian.

ge-eówan *to shew, discover;* ostendere:—He hit eft gehýt and eft geeówþ it [*the divine providence*] *again hides it and again discovers it,* Bt. 39, 8; Fox 224, 12. v. ge-eáwan, eówian.

ge-ǽrendian *to go on an errand,* L. In. 33; Th. i. 122, 13, note 37, MS. B. v. ge-ǽrendian.

ge-erfeweardian; *p.* ade *To inherit:*—Gierfeueardade *hereditavit,* Rtl. 45, 35: 84, 37.

ge-erian; *p.* ede, ode, ade; *pp.* ed, od, ad *To ear, plough;* arare:—Geerod [geered MS. C; geerad MS. D.] *aratus,* Ælfc. Gr. 19; Som. 22, 45. Ðæt land is geerod [geered MS. C.] *aratur terra,* 19; Som. 22, 46: Heming. p. 134.

gees *geese,* L. In. 70; Th. i. 146, 18, = gés; *pl. of* gós.

ge-etan; *p.* ic, he ǽt, ðú ge-ǽte, *pl.* ge-ǽton; *pp.* ge-eten *To eat together, to eat, to consume;* comedere, edere:—Elnung húses ðínes geet mec [me æt, Bos.] Jn. Skt. Lind. 2, 17. Ðæt híg ǽton: ðá híg

geeten hæfdon, hig wunedon ðær *ut ederunt : cum comedissent, manserunt ibi*, Gen. 31, 54. Gif ðu ðæs treówes wæstm geetst *if thou eatest the fruit of this tree*, Homl. Th. i. 14, 2.

ge-éðan ; *p.* de; *pp.* ed [éðe *easy*] *To make easy* or *light, alleviate* ; lĕvāre :—Ðæt ðu hygesorge heortan mínre geéðe *that thou alleviate the sorrow of my heart*, Exon. 50 a ; Th. 174, 17 ; Gú. 1179.

ge-epcucigan *to revive.* v. ge-edcucian.

gef, *pl.* gēfon *Gave* :—Ge him hleoþ gēfon *ye gave them shelter*, Exon. 27 b ; Th. 83, 11 ; Cri. 1354 ; *p. of* gifan.

gef *if*, Bt. 36, 4 ; Fox 178, 27. v. gif.

ge-fã [= ge-faa, ân ; *m.* [fãh *a foe*] *A foe, an enemy* ; inimicus, adversarius :—Gif se man [MS. mon] his gefān wite *if the man know his foe*, L. Alf. pol. 42 ; Th. i. 90, 2, 4, 14. Ðã mētte hine his eald-gefāna sum, and hine ofstang *then one of his old foes met him, and stabbed* [*killed*] *him*, Ors. 3, 7 ; Bos. 62, 22. To bismere his gefān [MS. gefaan = gefān = gefāum = gefāhum] *in mockery to his foes*, Homl. Th. i. 226, 28. v. fāh, fã.

ge-fadian ; *p.* ode, ade, ede; *pp.* od, ad, ed *To set in order, dispose, arrange, regulate* ; ordināre, dispōnĕre :—Se ðe awent of Ledene on Englisc sceal gefadian hit swā ðæt ðæt Englisc hæbbe his ágene wisan *he that translates from Latin into English must arrange it so that the English have its own manner*, Thw. Hept. p. 4, 9. Se Fæder gefadaþ ealle þing *the Father disposes all things*, Homl. Th. ii. 606, 3. He gefadode wið ða burhware *he arranged with the townsfolk*, Chr. 1052; Erl. 184, 21: Homl. Th. i. 278, 19. Hí ða gebytlunge gefadedon *they arranged the building*, ii. 172, 30. Gefadige [gefadie MS. B.] man ða steóre swā hit fōr Gode sý gebeórhlíc and fōr worulde aberendlíc *let the correction be regulated so that it be becoming before God and tolerable before the world*, L. C. S. 2 ; Th. i. 376, 13. Gefadad *disposed*, Th. Diplm. A. D. 972 ; 522, 12.

ge-fadung, e ; *f. A disposing, arranging* ; dispōsītio :—He nære nã ælmihtig, gyf him ǽnig gefadung earfoðe wære *he would not be almighty if any arranging were difficult to him*, Bd. de nat. rerum ; Wrt. popl. science 19, 6; Lchdm. iii. 278, 14.

ge-fæd, es ; *n?* *Order, decorum* ; decōrum :—Mid gefæde *with decorum*, L. Edg. C. 4 ; Th. i. 244, 15.

ge-fæd ; *adj.* [ge-fadian *to set in order*] *Orderly* ; dispōsītus :—Ðæt preósta gehwilc to sinoþe hæbbe gefædne man to cnihte *that every priest at the synod have an orderly man for servant*, L. Edg. C. 4 ; Th. ii. 244, 14.

ge-fædera, an ; *m. A godfather* ; compater :—Mauricius wæs his gefædera *Mauricius was his godfather*, Homl. Th. ii. 122, 24. [*O. H. Ger.* geuatero *compater* : *Ger.* gevatter.] v. cumpæder.

ge-fæderan, *pl.* v. suhtor-gefæderan.

ge-fædere, ge-federe, an ; *f. A godmother* ; commater, susceptrix :—Ǽfre ne geweorþe, ðæt Cristen man gewífige on his gefæderan *let it never be that a Christian man marry with his godmother*, L. Eth. vi. 12 ; Th. i. 318, 17: L. C. E. 7 ; Th. i. 364, 22. [*O. H. Ger.* gi-uatara ; *Ger.* gevatterin.]

ge-fædlíce ; *adv. Orderly, quietly* ; quiēte, Glos. Prudent. Recd. 145, 78.

ge-fædred ; *part. Fathered*, Ors. 3, 7 ; Bos. 60, 19. v. ge-fædrian.

ge-fædrian ; *p* ede ; *pp.* ed *To* FATHER, *to adopt* or *to ascribe to any one as a son* or *daughter* ; adoptare, patri filium vel filiam ascribere :—Ða þrý gebróþra næron nã Philippuse gemēdred, ac wæron gefædred *the three were not brothers of Philip by their mother* [*mothered*], *but they were by their father* [*fathered*], Ors. 3, 7 ; Bos. 60, 19.

ge-fægen, -fagen ; *adj. Glad, rejoiced* ; lætus :—Ic bió swíðe gefægen *I shall be very glad*, Bt. 40, 5 ; Fox 240, 25, MS. Cot. Hie ðæs gefægene wērun *they were rejoiced thereat*, Chr. 855 ; Erl. 68, 31: 878 ; Erl. 80, 11.

ge-fægerian ; *p.* ode ; *pp.* od *To adorn*, ornare, Som.

ge-fægnian ; -fagnian, -fagenian ; *p.* ode ; *pp.* od *To rejoice, be glad, exult* ; gaudēre, exultāre :—Ic geblissige and ic gefægnige on ðe *lætābor et exultābo in te*, Ps. Lamb. 9, 3. Geblissiaþ, and gefægniaþ on ðām dagum *gaudēte in illa die et exultāte*, Lk. Bos. 6, 23. Blissian and gefægnian þeóda *lætentur et exultent gentes*, Ps. Spl. 66, 4.

ge-fægnung, e ; *f. Exultation* ; exultātio :—Ðon gefylled is tunge úre gefægnunge *tunc repleta est lingua nostra exultātiōne*, Ps. Spl. 125, 2 : 104, 41 : 44, 17. v. fægning.

ge-fǽgon *rejoiced.* v. gefeón.

ge-fǽlan, -fællan ; *p.* de; *pp.* ed *To overturn, overthrow, throw down* ; prosternere, Ps. Vos. 105, 25 ; Lk. Skt. Lind. 20, 18. v. a-fælan.

ge-fǽllnis, -fǽlnis, se; *f. A fall*, Lk. Skt. Lind. 2, 34 ; *transmigration*, Mt. Kmbl. Lind. 1, 12.

ge-fǽlsian ; *p.* ode, ade ; *pp.* od, ad *To cleanse, purify, expiate* ; lustrāre, pūrificāre, expiāre :—Hé wolde gefælsian foldan mǽgþe *he would purify the race of earth*, Exon. 10 a ; Th. 9, 33 ; Cri. 144: 12 b ; Th. 20, 19 ; Cri. 320. Heorot is gefælsod *Heorot is purified*, Beo. Th. 2357 ; B. 1176: 3245 ; B. 1620: Apstls. Kmbl. 132 ; Ap. 66. Fýre gefælsad *purified with fire*, Exon. 127 b ; Th. 490, 21 ; Rä. 80, 5.

ge-fǽr, es ; *n. A going, journey, course, march, expedition* ; profectio, expēditio :—Ðisses fugles gefær *this bird's course*, Exon. 62 a ; Th. 227, 20 ; Ph. 426. On gefære *in profectiōne*, Ps. Spl. 104, 36. Ðæs ðe hie feónda gefær fyrmest gesǽgon *after they first saw the enemies' march*, Elen. Kmbl. 135 ; El. 68.

ge-fǽran [= ge-fēran] ; *p.* de ; *pp.* ed *To lead, bring* :—Ic eów hebbe hãm gefærde alle *I have brought you all home*, Cd. Th. 270, 18 ; Sat. 92. [Cf. *O. Sax.* gi-fōrian *to bring.*]

ge-fǽrnys, se; *f. A transmigration*, Som.

ge-fǽrrēden, ge-fǽrscipe. v. gefērrǽden, gefērscipe.

ge-fǽstan ; *p.* -fæste; *pp.* -fæsted *To place* ; locare :—Monn gefæste ða *homo locavit eam*, Mk. Skt. Lind. 12, 1. v. fæstan.

ge-fǽstan ; *p.* -fæste *To fast* :—Gefæsta *jejunare*, Lk. Skt. Lind. 5, 34 : Mt. Kmbl. Lind. 4, 2 ; 6, 16.

ge-fǽsten, es ; *n. A fast* ; jejunium, Rtl. 16, 41.

ge-fǽstnian ; *p.* ode, ade ; *pp.* od, ad *To fix, fasten, secure, confirm, betroth* ; figere, firmare, confirmare, infigere, despondere :—Iulius him mid gewritum gefæstnod *Julius secured it to him by writings*, Ors. 5, 13 ; Bos. 112, 31. Gefæstnade *secured*, Bd. 1, 5 ; S. 476, 10. Gefæstnode, 4, 28 ; S. 605, 24. Gefæstnode synd þeóda *infixæ sunt gentes*, Ps. Spl. 9, 15. Gifæstnad *desponsata*, Lk. Skt. Rush. 1, 27.

ge-fǽstnung, e ; *f. A fastening, securing, defence* ; munimen, Rtl. 37, 15.

ge-fǽtan ; *p.* -te *To pack up* ; convasare :—Ðæt gold hí gefætaþ on ða myran *the gold they pack on the mares*, Nar. 35, 12. v. fæt.

ge-fǽtian *to fetch, send for*, Cd. Th. 297, 22 ; Sat. 521. v. gefetian.

ge-fǽtnian ; *p.* ode ; *pp.* od *To fatten, anoint* ; impinguare, unguere :—Ðu amæstest oððe ðu gefætnodest on ele heáfod mín *impinguasti in eleo caput meum*, Ps. Lamb. 22, 5. v. fætnian.

ge-fǽttian ; *p.* ode ; *pp.* od *To fatten, anoint* ; impinguare, pinguefieri ; Ps. Vos. 19, 3. Gefætted *incrassatum*, Mt. Kmbl. Rush. 13, 15. v. ge-fætnian.

ge-fagen ; *adj. Glad, joyful* ; lætus :—Gefagen biþ, gif hit ǽfre to cuman mæg *it will be joyful if it ever may come thereto*, Bt. 25 ; Fox 88, 29. v. ge-fægen.

ge-fagnian, -fagenian ; *p.* ode ; *pp.* od *To rejoice, be glad, exult* ; gaudēre, exultāre :—Manega on his acennednysse gefagniaþ *multi in nativitāte ejus gaudēbunt*, Lk. Bos. 1, 14. Gefagnode ðæt cild on hyre innoþe *exultāvit infans in utero ejus*, 1, 41. Ic blissie and ic gefagenie on ðe *lætābor et exultābo in te*, Ps. Spl. T. 9, 2. v. ge-fægnian.

ge-fãh, ge-fãhmon *an enemy.* v. fãh, fãhman.

ge-fana, an ; *m. A standard*, Som.

ge-fandod, -fondad ; *past. p.* Beo. Th. 4900; B. 2454: 4592; B. 2301. [*Laym.* i-fonded.] v. fandian.

ge-fangennes, se ; *f. A taking, laying hold of, apprehension*, Som.

ge-fara, an ; *m. A companion* ; sŏcius :—Ic eom fyrdrinces gefara *I am a soldier's companion*, Exon. 127 a ; Th. 489, 3 ; Rä 78, 2. Hi heora wǽpen hwyrfdon wið heora gefaran *in sŏcios arma vertēre incipiunt*, Bd. 1, 15 ; S. 483, 5. v. ge-fēra.

ge-faran ; *p.* fōr ; *pl.* -fōron, -fōran ; *pp.* faren. I. *intrans. To go, proceed, reach by going, arrive* ; ire, proficisci, meare :—[He] walde gefara *voluit exire*, Lk. Skt. Lind. 1, 43. Swā feor swā man on ánum dæge gefaran mæg *as far as one can journey in a day*, Thw. Num. 11, 31. Eall under hróf gefōr *all came under the roof*, Gen. 1360. Oþ ðæt drihtweras gefōr ðær is botlwela bethlem hâten *until the men arrived where is a village called Bethel*, Cd. Th. 107, 33 ; Gen. 1798. II. *to depart, die* :—His fæder gefærþ *his father dies*, Blickl. Homl. 131, 25. Bearn hraðe gefaraþ [*their*] *children soon die*, Boeth. 11, 1 ; Fox 32, 10. Ne wēne ic ðæt ǽnig wære ðe ðæt atellan mihte, ðæt on ðam gefeohte gefōr *I do not suppose that anybody could reckon* [*the number*] *that died in that battle*, Ors. 3, 11 ; Bos. 75, 9. Gefōr Æðerēd cyning *king Ethelred died*, Chr. 871 ; Erl. 76, 1. Hý æt nýhstan ne ahsedan hwæt ðæra gefarenra wære *at last they did not ask how many there were dead*, Ors. 4, 4 ; Bos. 80, 12. III. *to proceed, get on, fare* :—Hú se mânscaða gefaran wolde *how the wicked spoiler meant to proceed*, Beo. Th. 1481 ; B. 738. Eustatius cýdde hú hí gefaren hæfdon *Eustace told how they had fared*, Chr. 1048; Erl. 178, 6. We nyton hwæt Moises gefaren hafþ *we know not what has become of Moses*, Exod. Thw. 32, 1, 23. IV. *v. trans. To get by going, experience, occupy, reach, obtain, go against* :—Hú mæg ic hit on ðrim dagum gefaran *how can I perform the journey in three days*, Blickl. Homl. 231, 23 : 235, 35. Hie wræcstōwe gefōran *they had reached the place of exile*, Cd. Th. 6, 20 ; Gen. 91. Ic wisce ðæt ic eft forlidennesse gefare *I wish that I may again suffer shipwreck*, Th. Apol. 12, 10 : 21, 19. Ðænne gefærþ he sige on ǽghwylcum gefeohte *then shall he obtain victory in every battle*, H. R. 17, 10. Twegen æðelingas gefōran ðæt lond *two princes occupied that land*, Ors. 1, 10 ; Bos. 32, 35. Philippus gefōr heora burh Philip took their town, 3, 7 ; Bos. 60, 6. Ne dorste he genēðan ðæt he hie mid firde gefōre *he dare not venture to attack them with an army*, 1, 10 ; Bos. 33, 31. Cf. gerídan. [*O. Sax.* gifaran *takes an accusative.*]

gefe a gift, Bd. 2, 13; S. 516, 6: Mt. Kmbl. Lind. 23, 18, 19. v. gifu.

ge-feá, an; m. Joy, gladness, glory, favour; gaudium:—Ðes mín gefeá is gefylled this my joy is fulfilled, Jn. Bos. 3, 29. Mid gefeán with joy; gaudio, 3, 29. Bodan cýþdon sóþne gefeán messengers announced real joy, Exon. 14 a; Th. 28, 23; Cri. 451. Se biþ gefeána fægrast that shall be the fairest of joys, 32 b; Th. 102, 1; Cri. 1666: 15, 11. On gefean with joy, Ps. Spl. 20, 6.

ge-feagan, -feán. v. ge-feohan, -feón.

ge-feaht, es; n. A battle; prælium:—Ðǽr nán hefilíc gefeaht ne wearþ there was no hard battle there, Chr. 868; Erl. 73, 26. Mycclum gefeahtum in great battles, 755; Erl. 49, 26. v. ge-feoht.

ge-feald, es; n. A fold, inclosure, field; septum, áger:—Þurh fífela gefeald forþonette he hastened forth through the field of the monsters, Wald. 76; Vald. 2, 10.

ge-fealdan; p. -feóld, pl. -feóldon; pp. -fealden To fold up, wrap; plicáre, involvĕre:—Ne læg hyt ná mid línwǽdum, ac onsundron gefealden on ánre stówe non cum linteamīnībus pŏsĭtum, sed sepărātim invŏlūtum in ūnum lŏcum, Jn. Bos. 20, 7. Miððý gefeáld ðæt bóc cum plicuisset librum, Lk. Skt. Lind. 4, 20.

ge-feálic; adj. Pleasant, joyous, delightful; lætus:—Ðǽr is éðellond fæger and gefeálic there is a country fair and joyous, Exon. 42 a; Th. 141, 18; Gú. 628: 44 b; Th. 151, 18; Gú. 797.

ge-feallan; p. -feól, -feóll, pl. -feóllon; pp. feallen To fall; cadere, decidere:—Ic gefealle be gewyrhtum fram feóndum mínum decidam merito ab inimicis meis, Ps. Spl. 7, 4. Ðǽr Pharaon gefeól, on ðam Reádan Sǽ et excussit Pharaonem in Mari Rubro, Ps. Th. 135, 15. He eorþan gefeóll he fell to earth, Beo. Th. 5661; B. 2834: 4207; B. 2100. Me fela ðínra edwíta on gefeóllon opprobria exprobantium tibi ceciderunt super me, Ps. Th. 68, 9. Ðá gefeól hire mód on his lufe then she fell in love with him, Th. Apol. 17, 18: 1, 13. Sóðlíce ðín dôhtor gefeól on swégcræft, ac heó næfþ hine ná wel geleornod thy daughter indeed has attempted [?] music, but she has not learnt it well, 16, 23. v. feallan.

ge-fearh-sugu, e; f. [fearh a farrow] A farrowing sow; prægnans sus, forda:—Gefearhsugu forda, Wrt. Voc. 286, 49.

ge-fearrian; p. ade; pp. ad To remove to a distance, go away; avellere, discedere, abscedere:—He gefearrad wæs from him ipse avulsus est ab eis, Lk. Skt. Lind. 22, 41. Gifearria abscedat, Rtl. 98, 22; discedat, 120, 31. v. feorran, afyrran.

ge-feastian; p. ode, ade; pp. od, ad To entrust, commit; commendare:—Gefeastadon commendaverunt, Lk. Skt. Lind. 12, 48. v. gefæstan.

ge-feaxe; adj. [feax hair] Having hair; cōmātus:—Wǽron men æðelíce gefeaxe the men had beautiful hair [lit. the men were beautifully haired], Bd. 2, 1; S. 501, 8.

ge-feaxen; adj. Having hair, haired; cōmātus:—Ða syndon gefeaxene swá frihteras they have hair as soothsayers have, Nar. 37, 1. v. gefeaxode.

ge-feaxode, -fexode; adj. Having hair, haired; cōmātus:—Ða wǽron hwítes líchaman and fægres andwlitan men, and æðelíce gefeaxode [gefexode, Homl. Th. ii. 120, 19] they were men of white complexion and fair countenance, and having noble hair, Nat. S. Greg. Els. 12, 1. v. feaxede.

ge-feccan, -feccean; p. -feahte, -fehte; pp. -feaht, -feht To fetch, bring to; addūcĕre:—He mæg ða sáwle gefeccan under foldan it can fetch back the soul under the earth, Salm. Kmbl. 139; Sal. 69. He him hêt to wífe gefeccean Cleopatran he commanded [them] to bring Cleopatra to him for a wife, Ors. 5, 13; Bos. 112, 44: Blickl. Homl. 187, 15.

ge-fecgan; p. -feah To seize; arrĭpĕre:—He wolde ðæs beornes beágas gefecgan he would seize the chieftains gems, Byrht. Th. 136, 33; By. 160.

ge-fédan; ðú -fédst; p. -fédde; pp. -féded, -fédd, -féd To feed, nourish; pascĕre, enutrīre:—Ðú gefédst me enutries me, Ps. Lamb. 30, 4. Ic eom geféd pascor, Ælfc. Gr. 33; Som. 36, 44. MS. D.

ge-federe, an; f. A godmother; susceptrix, L. C. E. 7; Th. i. 365, note 18. v. ge-fædere.

ge-fég, -feig, es; n. A joining, juncture; commissura, junctura, Cot. 43: Ælfc. Gl. 62; Som. 68, 82; Wrt. Voc. 39, 65: Compago, 70; Som. 70, 57; Wrt. Voc. 42, 65. Gefeig formula, Lye. Gefég borda a joining of boards, Ælfc. Gl. 62; Som. 68, 82. Mennisce handa hit ne mihton towurpan, for ðam fæstum gefége ðæs feóndlícan temples human hands could not overthrow it because of the fast joining of the devilish temple, Homl. Th. ii. 510, 14. [Ger. gefüge.]

ge-fégan, -fégean; p. de; pp. ed; v. trans. To join, unite, compact, compose; jungĕre, conjungĕre, compingĕre, compōnĕre:—Con he sídne ræced fæste gefégan he can firmly compact the spacious dwelling, Exon. 79 a; Th. 296, 8; Crä. 48: 79 a; Th. 297, 10; Crä. 66. Ic ða ged ne mæg gefégean I cannot compose the songs, Bt. Met. Fox 2, 11; Met. 2, 6. Ic gefége compōno, Ælfc. Gr. 28, 3; Som. 30, 57. Conjunctio gefégþ togædere ǽgðer ge naman ge word a conjunction joins together both nouns and verbs, 5; Som. 3, 48, 51: Bt. 21; Fox 74, 37. Se gefébþ fela folca tosomne he joins many people together, Bt. Met. Fox

11, 177; Met. 11, 89. Gefég ðǽs bricas join these fragments, Homl. Th. i. 62, 7. Ne weorþaþ hí nǽfre tosomne geféged they are never united together, Bt. 16, 63; Fox 56, 7: Bt. Met. Fox 20, 231; Met. 20, 116: 20, 241; Met. 20, 121. Gifoega sociare, conciliare, Rtl. 104, 12: 74, 18.

ge-fége; adj. Fit, adapted; aptus, Grm. i. 735, 5. [Ger. gefüge flexible.] v. ungefége.

ge-fégednes, se; f. Figure, shape, a joining, Som.

ge-fegian to rejoice. v. gefeón.

ge-fégineg, -fégung, e; f. A joining, composing, conjunction; compositio, conjunctio:—Seó geþeódnys oððe gefégineg is conjunctio the joining is a conjunction, Ælfc. Gr. 5; Som. 3, 47. v. ge-þeódnes.

ge-fégniss, e; f. Companionship; societas, Rtl. 109, 25: 106, 4.

geféhst catchest; capis, Coll. Monast. Th. 23, 7.

geféhþ seizes, Bt. 39, 1; Fox 212, 1. v. ge-fón.

ge-félan; p. de; pp. ed To feel, perceive; sentīre:—Ðæt hit man gefélan mihte that it might be felt, Ors. 1, 7; Bos. 30, 4: Exon. 24 b; Th. 69, 33; Cri. 1130: 25 a; Th. 72, 28; Cri. 1179. Geféleþ fácnes cræftig ðæt him ða férend on fæste wuniaþ the skilled in guile feels that the voyagers firmly rest on him, 97 a; Th. 361, 23; Wal. 24. Gefélde ic me beótiende and wyrpende me mĕlius hăbĕre sentīrem, Bd. 5, 6; S. 620, 12. Gefélde he his líchoman healfne dǽl mid ða ádle geslægene beón sensit dimĭdiam corpŏris sui partem languōre depressam, 4, 31; S. 610, 15: 3, 2; S. 525, 15: 3, 9; S. 534, 11. He ðæs wítes worn gefélde he felt the force of the torment, Cd. 214; Th. 269, 23; Sat. 77.

ge-felgan; p. -fealh, pl. -fulgon; pp. -folgen To stick to; inhærēre:—He ðære godspellícan láre georne gefealh he earnestly stuck to the gospel lore, Bd. 3, 22; S. 552, 43. v. felgan.

ge-fellan; p. -felde; pp. -felled To fill, fulfil:—Se gefelde xx daga he had fulfilled twenty days, St. And. 4, 23. v. gefyllan.

ge-fellan; p. -felde; pp. -feld To cause to fall, fell, kill:—Hie gefélde wurdon fram Alexandre they were killed by Alexander, Nar. 38, 11. v. gefyllan.

ge-félniss, e; f. A feeling, perception, sense; sensus:—Bútan ǽnigre gefélnisse without any feeling, Bd. 4, 11; S. 580, 2. DER. félnyss.

ge-felsode expiated. v. gefælsian.

ge-fend, es; m. A giver:—Gefend largitor, Rtl. 108, 16. v. gifend.

ge-feng, es; n. A taking, capture, captivity; captura, captivitas:—On gefeng in capturam, Lk. Skt. Lind. 5, 4. On gefeng fiscana in captura piscium, 5, 9. Gefeng captivitas, Rtl. 83, 3. v. feng.

gefeó take, Coll. Monast. Th. 21, 31, = gefó; pres. of gefón, q. v.

ge-feógan to hate. v. ge-fía.

ge-feohan to rejoice:—Gefeoh nú on ferþe rejoice now in mind, Hy. 11, 1; Hy. Grn. ii. 294, 1. v. gefeón.

ge-feoht, -fioht, -feht, es; n. A fight, battle, contest, war, preparation for war; prælium, pugna, congressio, bellum, procinctus:—Ðæt ungemetlíce mycle gefeoht the very great battle, Ors. 1, 9; Bos. 32, 1: Homl. Th. ii. 538, 14: Chr. 603; Erl. 20, 15: 868; Erl. 72, 28. Gefeoht congressio, Ælfc. Gl. 14; Som. 57, 125; Wrt. Voc. 20, 62. On dæge gefeohtes in die belli, Ps. Lamb. 139, 8. Ðú here fýsest to gefeohte thou incitest a host to battle, Andr. Kmbl. 2377; An. 1190: 2393; An. 1198: Elen. Kmbl. 2365; El. 1184. To gefeohte in procinctu, Ælfc. Gl. 101; Som. 77, 35; Wrt. Voc. 55, 40. Gé gehýraþ gefeoht and sace ye shall hear of battle and strife, Homl. Th. ii. 538, 2, 13: Bt. 15; Fox 48, 15. Ðonne gé geseóþ gefeoht and twýrǽdnessa cum audiĕritis præwhen lia et sedĭtiōnes, Lk. Bos. 21, 9: Mt. Bos. 24, 6: Ps. Lamb. 139, 3. Ðonne gé gehýraþ gefeohtu and gefeohta hlísan, ne ondrǽde gé eów cum audiĕritis bella et opiniōnes bellōrum, ne timuĕritis, Mk. Bos. 13, 7: Mt. Bos. 24, 6. Miclum gefeohtum in great battles, Chr. 755; Erl. 48, 25: L. In. 6; Th. i. 106, 1, note 1. Gefehto and woeno gefehtana prælia et opiniones præliorum, Mt. Kmbl. Lind. 24, 6. [Laym. i-fiht.]

ge-feohtan; p. -feaht, pl. -fuhton; pp. -fohten. I. to fight; pugnare:—And gif he ðonne wið hine gefeohtan ne mæg and if he may not fight against him, Lk. Bos. 14, 32. Ðe teáh mine fingras to gefeohtanne qui docet digitos meos ad bellum, Ps. Th. 143, 1. He wel gefeaht he fought well, Ors. 5, 13; Bos. 112, 34. Margareta wiþ ðone deófol gefæht Margaret fought with the devil, Nar. 39, 28. Gif hwá gefeohte on cyninges huse, sie [sy MSS. B. H.] he scyldig ealles his ierfes [yrfes MSS. B. H.] if any one fight in the king's house, let him be liable in all his property, L. In. 6; Th. i. 106, 6. Ðeáh hit sié on middum felda gefohten though it be fought on mid-field, L. In. 6; Th. i. 106, 10: Judth. 11; Thw. 23, 15; Jud. 122. II. to obtain by fighting; pugnando acquirere:—Ðæt he ne meahte wiht gefeohtan that he could not gain aught by fighting [lit. to fight], Beo. Th. 2171; B. 1083. Dôm gefeohtan to gain glory by fighting, Bryht. Th. 135, 37; By. 129. Hæfde ða gefohten foremǽrne blǽd Judith Judith had gained exceeding great glory, Judth. 11; Thw. 23, 15; Jud. 122. [Cf. Ger. erfechten.] v. feohtan.

gefeoht-dæg, es; m. A fight-day, day of battle; dies belli:—On gefeohtdæge, Ps. Th. 139, 7.

ge-feolan; p. -fæl, pl. -fǽlon; pp. -folen, -feolen To stick to, persist;

insistere:—Ðæt he ðám hálwendum ongynnessum georne gefeole *ut cœptis salutaribus insisteret*, Bd. 5, 19; S. 637, 11. v. feolan.

ge-feón, -feohan, -feagan, -feagian; ic -feó, ðú -fehst, he -fehþ, -fiþ, -feaþ, *pl.* -feóþ; *p.* -feah, -feh, *pl.* -fægon; *pp.* -fegen [The Northern Gospels have weak forms] *To be glad, rejoice, exult;* lætari, delectari, gaudere, exultare :—Ic gefeó *gaudeo*, Jn. Skt. Lind. 11, 15. Gefeaþ *gaudebit*, 16, 20, 22. Manige on his gebyrd gefeóþ *many shall rejoice at his birth*, Blickl. Homl. 165, 10. Míne weleras gefeóþ *gaudebunt labia mea*, Ps. Th. 70, 21. Gefeah blíðe-mód ðæs ðe . . . *glad of mind rejoiced that* . . . , Cd. 72; Th. 88, 21; Gen. 1468. Bona weorces gefeah *the destroyer rejoiced at the work*, Exon. Th. 464, 17; Hö. 88: Elen. Kmbl. 220; El. 110. Secg weorce gefeh *the warrior in the work rejoiced*, Beo. Th. 3143; B. 1569: 3253; B. 1624. Fylle gefǽgon *they rejoiced at the plenty*, Beo. Th. 2032; B. 1014. Leóhte gefǽgun *they rejoiced in the light*, Exon. Th. 31, 32; Cri. 504. Gefeade *exaltavit* [misread by the translator *exultavit*], Jn. Skt. Lind. 3, 14. Gefeade *exultavit*, 8, 56. Gefeoh *rejoice*, Hy. 11, 1; Hy. Grn. ii. 294, 1. Gefeóþ mid me *rejoice with me*, Blickl. Homl. 191, 22. Gefeaþ *gaudete*, Mt. Kmbl. Lind. 5, 12. Eal rihtgelýfed folc sceal gefeón on ðone his tocyme *all right-believing folk ought to rejoice at his advent*, Blickl. Homl. 167, 14. Ðonne mótan we in ðære engellícan blisse gefeón *then may we in angelic bliss rejoice*, 83, 3. Gefeage *exultare*, Jn. Skt. Lind. 5, 35: 3, 14. Gifeaga *gaudere*, Rtl. 34, 3. Gifeagia *gaudere*, 69, 30. Gefeónde for Paules eáðmódnesse *rejoicing on account of Paul's humility*, Blickl. Homl. 141, 4. He wæs gefeónde myclum gefeán *he was rejoicing with great joy*, 233, 2. Hio wǽron gefeónde mycle gefeán, 249, 16. Gefeándo woeron *gavisi sunt*, Mk. Skt. Lind. 14, 11. Gefagen wǽron *gavisi sunt*, Mt. Kmbl. Lind. 2, 10.

ge-feormian; *p.* ode; *pp.* od. *v. a.* **I.** *to entertain, harbour, receive as a guest, feed, cherish, support;* suscipere, hospitio suscipere, epulare, fovere, curare :—Sanctus Albanus for ðam cuman, ðe he gefeormode [MS. gefeormade] gegyrede hine *Saint Alban arrayed himself for the stranger whom he entertained*, Bd. 1, 7; S. 477, 9. Ðæt se, ðe hine feormode, and se, ðe gefeormod wæs, sýn hí begen bisceopes dóme scyldig *that he who entertained him, and he who was entertained, be both liable to excommunication;* susceptor et is qui susceptus est excommunicationi subjacebit, 4, 5; S. 573, 1. Búton ðæs bisceopes leáfe, ðe hí on his scíre gefeormode [MS. gefeormaðe] sín *without the bishop's leave, in whose diocese they may be entertained*, 4, 5; S. 573, 5. We ðe gefeormedon *we entertained thee*, Cd. 127; Th. 162, 24; Gen. 2686. Ðonne mon motnan betýhþ ðæt he ceáp forstele, oððe forstolenne gefeormie *when a man charges another that he steal cattle, or harbour the stolen*, L. In. 46; Th. i. 130, 13. Geóca mihtig Dryhten mínre sáwle, gefreoða hyre and gefeorma hý *save my soul, O mighty Lord, protect it and cherish it*, Exon. 118 b; Th. 456, 3; Hy. 4, 61. **II.** *to feed on, devour;* vesci, comedere :—Hie ða behlidenan him to lífnere gefeormedon *they feed on the dead* [mortuos] *to* [*save*] *their lives*, Andr. Kmbl. 2181; An. 1092. Grendel unlifigendes gefeormod fét and folma *Grendel devoured the feet and hands of the lifeless*, Beo. Th. 1493; B. 744. **III.** *to cleanse, farm* or *cleanse out, Provncl;* mundare :—Ðæt hí ða bán woldon upádón, and onþweán and gefeormian *that they would take up the bones to wash and cleanse*, Bd. 4, 19; S. 589, 11. Hát gefeormian mín blód *bid* [*them*] *wipe away my blood*, Blickl. Homl. 183, 26. v. feormian.

ge-fér, es; *n.* *A company, society;* cŏmĭtātus :—Eart ðú úres geféres ðe úre wiðerwinna *noster es an adversāriōrum* [?], Jos. 5, 13. Wéndon ðæt he on heora feére wǽre *existĭmantes illum esse in cŏmĭtātu*, Lk. Bos. 2, 44.

ge-féra, an; *m.* *A companion, comrade, associate, fellow, colleague, fellow-disciple, man, servant;* sŏcius, contŭbernālis, cŏmes, condiscĭpulus, vir, puer :—Geféra *contŭbernālis* vel *sŏcius*, Ælfc. Gl. 116; Som. 80, 63; Wrt. Voc. 61, 41: Ælfc. Gr. 5; Som. 5, 20. Geféran áþ *a companion's oath*, L. O. 6; Th. i. 180, 17. Ðæt wíf ðæt ðú me forgeáfe to geféran *mŭlier quam dēdisti mihi sŏciam*, Gen. 3, 12: Exon. 76 b; Th. 288, 13; Wand. 30. He geseh swǽsne geféran *he saw his dear comrade*, Andr. Kmbl. 2018; An. 1011 : 2040; An. 1022. Ædele geféran Philippus and Iacob feorh agéfau for Meotudes lufan *the noble companions Philip and James gave their lives for the love of God*, Menol. Fox 158; Men. 80: Gen. 14, 10: Chr. 755; Erl. 50, 25. Bæd se gesíþ hine ðæt he eóde in to ánum his geférena *rogātus est ab eodem cŏmĭte intrāre ad unum de pŭeris ejus*, Bd. 5, 5; S. 617, 36 : 1, 7; S. 476, 29. Cwæþ Thomas to hys geférum *dixit Thomas ad condiscĭpŭlos*, Jn. Bos. 11, 16: Bd. 2, 3; S. 504, 29 : 3, 21; S. 551, 9. Ceós ðé geféran and feoht ongén Amalech *chŭse thee vĭros et pugna contra Amalec*, Ex. 17, 9. Wordes geféra *a verb's companion, an adverb;* adverbium, Ælfc. Gr. 5; Som. 3, 34. Gefoera *condiscipulus*, Jn. Skt. Lind. 11, 16. [*Laym: A. R.* i-vere.]

ge-féran; *p.* -férde; *pp.* -féred. **I.** *v. intrans.* *To go, travel, go on, behave, fare, get on, come, get to a place* :—He geférde óð ðæt he Adam funde *he journeyed until he found Adam*, Cd. 23; Th. 29, 20; Gen. 453. Frécne geférdon *daringly they behaved*, Beo. Th. 3386;

B. 1691. Ðá ðis cúþ wæs hú ða óðre geférdon *when this was known how the others had fared*, Chr. 1009; Erl. 142, 8: Cd. 214; Th. 268, 29; Sat. 62. Ne mæg ðær unwitfull ǽnig geféran *no deceitful man can get there*, Cd. 45; Th. 58, 19; Gen. 948. Ic eom hider feorran geféred *I have come hither from far*, 25; Th. 32, 4; Gen. 498. **II.** *v. trans.* *To perform a journey, reach* or *get by going, obtain, attain, experience, suffer* :—Ðú scealt ða fóre geféran *thou shalt perform that journey*, Andr. Kmbl. 431; An. 215; 388; An. 194. Se hit mæg hrædlícor geféran *who can perform the journey more speedily*, Blickl. Homl. 231, 24, 25. Ðe ðæt upplíce ríce geférdon *who reached the realm on high*, Homl. Th. i. 542, 26: Chr. 988; Erl. 131, 10: Beo. Th. 6119; B. 3063. Ðæs siges ðe hie geféred hæfdon *for the victory that they had obtained*, Blickl. Homl. 203, 33. Ðá férdon ða Pyhtas and geférdon ðis land norðanweard *then the Picts went and got the north part of this land*, Chr. Erl. 3, 13. Hafast ðú geféred ðæt ðé weras ehtigaþ *thou hast attained* [*this*] *that men will esteem thee*, Beo. Th. 2446; B. 1221. Hí ðær geférdon máran hearm ðonne hí ǽfre wéndon *they there suffered greater hurt than they ever expected*, Chr. 994; Erl. 132, 21: Andr. Kmbl. 2801; An. 1403.

ge-fercian; *p.* ode; *pp.* od *To support, sustain;* sustentāre :—Úre hwílendlíce líf biþ mid mettum gefercod *our transitory life is sustained by meats*, Homl. Th. ii. 462, 20.

ge-fére; *adj.* *Easy of access;* făcĭlis accessu :—Nis se foldan sceát mongum gefére *the tract of earth is not easy of access to many*, Exon. 55 b; Th. 198, 3; Ph. 4. [Cf. *O. H. Ger.* kifuari *apta*, Grff. iii. 600.] v. fére.

ge-ferian, -fergan; *p.* ode, ede; *pp.* od, ed *To carry, convey, bear, lead, conduct;* ferre, vehĕre, dūcĕre :—Feówer scoldon geferian to ðæm goldsele Grendles heáfod *four must convey Grendel's head to the goldhall*, Beo. Th. 3281; B. 1638: Andr. Kmbl. 793; An. 397. He geferode hine mid mycclum wurþscipe to Scæftes byrig *he conveyed it with great honour to Shaftesbury*, Chr. 980; Erl. 129, 33. Ðæt he úsic geferge in Fæder ríce *that he convey us into his Father's kingdom*, Exon. 12 b; Th. 22, 1; Cri. 345. Ðonne we geferian freán úserne ðær he longe sceal on ðæs Waldendes wære geþolian *then we bear our lord to where he shall long endure in the All-powerful's care*, Beo. Th. 6205; B. 3107. Ðæt hie út geferedon dýre máþmas *that they might convey out the precious treasures*, 6252; B. 3130. Godes gást wæs geferod ofer wæteru *spīritus Dei fĕrēbātur sŭper ăquas*, Gen. 1, 2: Boutr. Scrd. 19, 2: Nicod. 31; Thw. 18, 10. Feorran gefered *conveyed from afar*, Salm. Kmbl. 357; Sal. 178: Andr. Kmbl. 529; An. 265: Elen. Kmbl. 1982; El. 993. Se arc wæs geferud ofer ða wæteru *arca fĕrēbātur sŭper aquas*, Gen. 7, 18.

gefér-lécan; *p.* -léhte; *pp.* -léht *To keep company* or *fellowship, accompany, associate;* assŏciāre :—Ic geférlǽce associo? Ælfc. Gr. 30, 5; Som. 34, 52. He hí geférlǽcþ on ánnysse his geladunge *he associates them in the unity of his church*, Homl. Th. i. 496, 24. He biþ gemǽnscipe ðære hálgan geladunge geférlǽht *he is associated in the communion of the holy church*, i. 494, 19. Ðær beóþ geférlǽhte on ánre súsle, ða ðe on lífe on mándǽdum geþeódde wǽron *there shall be associated in one torment those who in life were united in evil deeds*, Homl. Th. i. 132, 20 : 414, 34.

ge-fér-rǽden, -réden, -rédin, -rædenn, e; *f.* **I.** *companionship, fellowship, congregation, church;* societas, comitatus, ecclesia, synagoga :—Hwá wolde on ðære geférrǽddene [MS. B. geférrǽdene] beón ðe he wǽre *who would be in that fellowship that he was*, L. Ed. 4; Th. i. 162, 5 : Ors. 5, 12; Bos. 111, 23. He hæfde on his geférrǽdene cratu and rídende men *habuit in comitatu currus et equites*, Gen. 50, 9. Smerede ðé God ðín mid ele blysse for geférrédinum ðínum *unxit te Deus tuus, oleo lætitiæ præ consortibus tuis*, Ps. Spl. C. 44, 9. Gyf he híg ne gehýrþ, sæge hyt geférrǽdene *quod si non audierit eos: dic ecclesiæ*, Mt. Bos. 18, 17 : Jn. Bos. 9, 22. **II.** *familiarity, friendship;* familiaritas, amicitia :—Ðæs cyninges geférrǽden mæg nǽnigne mon gedón weligne *the king's familiarity can make no man wealthy*, Bt. 29, 3; Fox 102, 2. v. ge-fér-scipe.

ge-fér-rǽdnes, -ness, e; *f. Society;* societas, Lye.

ge-fér-scipe, -scype, es; *m.* *Society, fellowship, brotherhood;* sŏcĭĕtas, cŏmĭtātus, clērus :—To healfum fó se cyng, to healfum se geférscipe *let the king take half, half the fellowship*, L. Ath. v. § 1, 1; Th. i. 228, 18. Þolige ægðer ge geférscipes ge freóndscipes *let him forfeit both their society and friendship*, L. Eth. ix. 27; Th. i. 346, 11: L. C. E. 5; Th. i. 362, 32: L. N. P. L. 45; Th. ii. 296, 19. Of geférscipe ðæs bisceopes Deosdedit de *clēro* Deusdedit *episcŏpi*, Bd. 3, 29; S. 561, 12 : 4, 1; S. 564, 18 : 5, 6; S. 618, 28 : 5, 19; S. 639, 3: L. E. B. 12; Th. ii. 242, 18. For lufan ðínre and geférscype *for thy love and fellowship*, Exon. 51 a; Th. 177, 24; Gú. 1232: Nicod. 11; Thw. 6, 3. Wið ðone geférscipe *with the fellowship*, L. Ath. v. § 1, 1; Th. i. 228, 20. Se cræftga geférscipas fæste gesamnaþ *the artificer firmly unites societies*, Bt. Met. Fox 11, 185; Met. 11, 93. Of hiora gefoerscipe *de eorum societate*, Rtl. 75, 28.

ge-férscipian *to unite, accompany* :—Gifoerscipia *unitare*, Rtl. 110, 18. Gifoerscipeþ *comitentur*, 93, 13.

ge-festnian ; *p.* ode; *pp.* od *To fasten, make fast, confirm, shut up, imprison ;* firmāre, confirmāre, inclūdĕre :—He đæt mid āþe gefestnode *he confirmed that with oath,* Chr. 1091; Erl. 228, 4. Se cyng genam Roger eorl his mæg, and gefestnode hine *the king took earl Roger his kinsman and imprisoned him,* 1075; Erl. 214, 5. Đe be swylcre gewittnesse gefestnod is *which is confirmed by such witness,* Th. Diplm. A. D. 856; 117, 18. v. ge-fæstnian.

ge-fetelsod ; *adj.* [fetel *a girdle, belt*] *Polished, trimmed, ornamented ;* perpŏlītus, adornātus :—Twā sweord gefetelsode *two swords trimmed ; duos glădios optĭme adornātos,* Text. Roff. 110, 15.

ge-feterian, -fetrian ; *p.* ode, adĕ; *pp.* od ad *To fetter, bind ;* compĕdīre, vincīre :—He đa strangan mæg streámas gefeterian *he can fetter the strong streams,* Ps. Th. 65, 5. He gefeteraþ fæges monnes handa *he fetters the hands of the doomed man,* Salm. Kmbl. 317; Sal. 158. He gefeterode fēt and honda bearne sīnum *he fettered the feet and hands of his child,* Cd. 140; Th. 175, 27; Gen. 2902. Đa wæron gefeterade fæste togædre *who were fettered fast together,* Exon. 113 b; Th. 435, 7; Rä. 53, 4.

ge-fēđe ; *adj. Lying at the feet,* Gl. Prud. 1046. *Contentus, conscriptus,* Hpt. Gl. 499.

ge-feđeran, -feđran ; *p.* ede; *pp.* ed *To feather, give wings to ;* ālas addĕre :—Ic sceal ǽrest điu mōd gefeđeran *I shall first give wings to thy mind,* Bt. 36, 1; Fox 172, 31, MS. Cot. Gefeđran, Bt. Met. Fox 24, 8; Met. 24, 4. v. ge-fiđerian.

ge-fetian, -fetigan, -fetigean ; *p.* -fetode, -fetede, -fette ; *pp.* -fetod *To fetch, bring ;* addūcĕre, accīre, afferre :—Elene hēht gefetian on fultum forþsnoterne hæleđa gerǽdum *Elene bade [them] fetch to her aid the very wise in the councils of men,* Elen. Kmbl. 2103; El. 1053: Beo. Th. 4387; B. 2190. Gefetigan, Exon. 66 b; Th. 246, 11; Jul. 60. Hēt heó sōna hire þīnenne gān and đa cyste hire to gefetigean *stătim jussit ire ministram et capsellam addūcĕre,* Bd. 3, 11; S. 536, 27 : Elen. Kmbl. 2319; El. 1161. Swā strang đæt ǽs him gefetede *so strong that it got prey for itself,* Chr. 975; Erl. 125, 29. He of helle hūþe gefette sāwla manega *he from hell fetched spoils, many souls,* Hy. 10, 30; Hy. Grn. ii. 293, 30: Gen. 24, 11. Đa men of Lundenbyrig gefetodon đa scipu *the men of London brought away the ships,* Chr. 896; Erl. 94, 17. Hý gefetton Escolāfius đone scīnlācan *they fetched Æsculapius the magician,* Ors. 3, 10; Bos. 70, 30. Hwænne me Dryhtnes rōd gefetige *when the Lord's cross shall fetch me,* Rood. Kmbl. 274; Kr. 138. Gefetod *accītus,* Cot. 7. Gefotad *accersitus,* Mk. Skt. Lind. 15, 44.

ge-fetrian *; p.* ode, ade, ede; *pp.* od, ad, ed *To fetter, bind ;* compĕdīre, vincīre :—Đone he gefetrade fȳrnum teágum *whom he fettered with fiery shackles,* Exon. 96 a; Th. 359, 9; Pa. 60. Drihten đa gefetredan alȳseþ *Dŏmĭnus solvit compĕditos,* Ps. Th. 145, 7. v. ge-feterian.

ge-fettan. v. gefetian.

ge-fette, *pl.* -fetton *Fetched, brought,* Gen. 24, 11 : Ors. 3, 10 ; Bos. 70, 30; *p.* of ge-fetian.

ge-fexode *having hair, haired,* Homl. Th. ii. 120, 19. v. ge-feaxode.

ge-fīa, -fiáge *to hate :*—Gefiáge *odisse,* Jn. Skt. Lind. 7, 7. Gefīeþ *odit,* 3, 20 : 12, 20. Gefīweþ *odiet,* Lk. Skt. Lind. 16, 13. Gefīadon *oderant,* 19, 14. v. gefeógan.

ge-fīc, es ; *n. Fraud, deceit ;* fraus :—Mid fǽcne gefice *with fraudulent deceit,* Elen. Kmbl. 1150; El. 577.

ge-fiht *a fight, battle,* Chr. 1128; Erl. 257, 1. v. ge-feoht.

ge-filce. v. gefylce.

ge-filde, es ; *n. A field, plain ;* campus :—Be norþan Capadocia is đæt gefilde đe man hǽt Temeseras *to the north of Cappadocia is the plain which is called Themiscyra,* Ors. 1, 1; Bos. 17, 7.

ge-fillan ; *p.* -filde; *pp.* -fild *To fulfil, finish, complete ;* implēre, complēre :—Đū gefilst Godes hǽse and his bebodu *implēbis impĕrium Dei et præcepta ejus,* Ex. 18, 23. God gefilde on đone seofeđan dæg his weorc *complēvit Deus die septĭmo ŏpus suum,* Gen. 2, 2 : Deut. 31, 24. Gefild *fulfilled,* Chr. 605; Erl. 21, 27. v. ge-fyllan.

ge-findan *; p.* -fand, -fond, *pl.* -fundon; *pp.* -funden *To find ;* invĕnīre :—His bān gefundon and gemēted wǽron *ossa ejus inventa sunt,* Bd. 3, 11; S. 535, 10: Chr. 963; Erl. 121, 36.

ge-findig ; *adj. Finding, receiving, capable ;* capax :—Numol ođđe gefindig *capax,* Ælfc. Gr. 9, 60; Som. 13, 42.

ge-finegod *; part. p.* [fynegian *to become mouldy*] *Mouldy ;* mūcĭdus :—Đe nū sind gefinegode *which are now mouldy,* Jos. 9, 12.

ge-fioht, es ; *n. A battle ;* prælium :—Aulixes to đam gefiohte fōr *Ulysses went to the battle,* Bt. 38, 1; Fox 194, 6. v. ge-feoht.

ge-firenian, -firnian ; *p.* ode; *pp.* od *To sin ;* peccāre :—We gefirenodon mid fæderum ūrum *peccāvimus cum patribus nostris,* Ps. Spl. C. 105, 6. Ic gefirnode *I sinned,* St. And. 10, 19 : Mt. Kmbl. Rush. 27, 4. v. gefyrenian.

ge-firn ; *adv. Long ago,* Th. Apol. 19, 25. v. gefyrn.

ge-firnian. v. ge-firenian.

ge-fiđerhamod ; *part. p. Provided with a covering of feathers :*—He wæs egeslīce gefiđerhamod *he was frightfully feather-clad,* Homl. Th. i. 466, 27. [Cf. Thorpe's North. Myth. i. 52.]

ge-fiđerian, -fiđerigan, -fiđrian, -fyđerian ; *p.* ode, ade; *pp.* od, ad *To give wings to, provide with wings ;* ālas addĕre, pennis instruĕre :—Ic sceal ǽrest đīn mōd gefiđerian *I must first give wings to thy mind,* Bt. 36, 1; Fox 172, 31. Gefiđerigan, 36, 2; Fox 174, 6. Gefiđrade [MS. gefriđade] fugelas *vŏlātĭlia pennāta,* Ps. Th. 77, 27.

ge-flǽman ; *p.* de; *pp.* ed *To cause to flee, put to flight :*—Đū fiónd geflǽmdest *thou didst put to flight the enemy,* Hy. 8, 25; Hy. Grn. ii. 290, 25. v. ge-flȳman.

ge-flǽschamod ; *part. p. Incarnate ;* incarnātus :—Se wearþ geflǽschamod *who was incarnate,* Homl. Th. ii. 596, 32 : i. 40, 24 : 284, 22.

ge-flǽscnes, -ness, e ; *f. Incarnation ;* incarnātio :—Ǽr Cristes geflǽscnesse *before Christ's incarnation,* Chr. Erl. 4, 22.

ge-fleard, es ; *n. A trifling, nonsense, madness :* — Gefleard *delīramentum,* Hpt. Gl. 416.

ge-flēman ; *p.* de; *pp.* ed *To cause to flee, to rout :*—Hæfde đā Drihten seolf feónd geflēmed *then the Lord himself had routed the foe,* Cd. 223; Th. 293, 30; Sat. 463: Chr. 938; Th. 204, 9, col. 1; Ædelst. 32. v. ge-flȳman.

ge-flēme ; *adj. Fugitive ;* fugitivus, Rtl. 147, 15.

ge-fleógan ; *p.* -fleág, -fleáh, *pl.* -flugon; *pp.* -flogen *To fly, fly over ;* volare, transvolare :—He hēht his heáhbodan hider gefleógan *he commanded his archangel to fly hither,* Exon. 12 a; Th. 19, 4; Cri. 295. Ne mæg ǽnig đone mearcstede fugol gefleógan *nor may any bird fly over the boundary place,* Salm. Kmbl. 435; Sal. 218.

ge-fleón, -fleóhan ; *p.* -fleáh, *pl.* -flugon *To flee, escape :*—Gefleá *fugere,* Mt. Kmbl. Lind. 3, 7. Se to ānre đara burga gefiiéhþ *who to one of those cities escapes,* Past. 21, 7 ; Swt. 167, 20; Hatt. MS. Geflēg *fugit,* Rtl. 147, 15. Alle geflugon *omnes fugerunt,* Mt. Kmbl. Lind. 26, 56. Đætte gifiēga *ut fugiant,* Rtl. 118, 31. Ǽr he on đa wēstenu middangeardes gefiuge *antequam in desertas orbis terrarum abiret solitudines,* Nar. 6, 6.

ge-fleów *overflowed,* Ors. 1, 3; Bos. 27, 28; *p.* of ge-flōwan.

ge-fliēman ; *p.* de; *pp.* ed *To cause to flee, to drive away ;* fugare, Past. 61, 2; Hat. MS. v. ge-flȳman.

ge-flit *a fan to clean corn ;* vannus, Cot. 33.

ge-flīt, -flȳt, es ; *n. Contention, strife, contest, dispute, discussion ;* contentio, lis, certāmen, concertātio, rixa :—Agoten is geflīt ofer ealderas *effūsa est contentio sŭper princĭpes,* Ps. Lamb. 107, 40 : Bd. 1, 1; S. 473, 30. Đis geflīt *hæc lis,* Ælfc. Gr. 9, 29; Som. 11, 62. Sume ic to geflȳte fremede *I have urged some to strife,* Exon. 72 b; Th. 271, 18; Jul. 484; Bd. 5, 6: S. 619, 4. On geflīt *in contest,* Beo. Th. 1734; B. 865. We on geflītum sǽton *we sat in discussion,* Salm. Kmbl. 862; Sal. 430: H. R. 9, 3. Uton towurpan hwætlīcor đās geflītu *dissolvāmus citius has contentiones,* Coll. Monast. Th. 31, 23: Elen. Kmbl. 884; El. 443: 1905; El. 954. Heó gehȳrde martyra geflītu *she heard of the struggle of martyrs,* Nar. 40, 13. To geflītes *emulously, eagerly,* Apol. Th. 10, 5.

ge-flīta. v. fyrn-geflīta.

ge-flītan, -flȳtan ; *p.* -flāt, *pl.* -fliton ; *pp.* -fliten *To strive, fight, dispute ;* contendĕre, certāre :—Cynewulf and Offa gefliton ymb Benesingtūn *Cynewulf and Offa fought at Benson,* Chr. 777; Th. 93, 11, col. 1. Ne geflīttes *non contendet,* Mt. Kmbl. Lind. 9, 34. Geflīton *disputaverant,* Mt. Skt. Lind. 9, 34. Geflītan [-flītta, Lind.] *contendere,* Mt. Kmbl. Rush. 5, 40.

ge-flītful ; *adj. Contentious ;* contentiōsus :—Geflītful *contentiōsus,* Ælfc. Gl. 85; Som. 74, 10; Wrt. Voc. 49, 33 : 74, 31: Hpt. Gl. 502.

ge-flītfullíc ; *adj. Contentious ;* contentiōsus :—Wæs geflītfullíc senoþ æt Cealchýþe *there was a contentious synod at Chalk,* Chr. 785; Erl. 56, 7.

ge-flītgeorn ; *adj. Contentious ;* contentiōsus, R. Ben. 71.

ge-flītlíce ; *adv. Contentiously, emulously ;* certātim :—Đæt ge wēpned ge wīfmen geflītlíce dydon *quod vĭri et fēmĭnæ certātim făcĕre consuērunt,* Bd. 5, 7 ; S. 621, 15.

ge-flītmǽlum ; *adv. Contentiously, emulously ;* certātim, R. Ben. interl. 72.

ge-flota, an ; *m. A floater, swimmer :*—Fyrnstreáma geflotan *to the ocean's floater [the whale],* Exon. 96 b; Th. 360, 17; Wal. 7. v. flota.

ge-flōwan ; *p.* -fleów, *pl.* -fleówon; *pp.* -flōwen *To overflow ;* inundāre :—Swā hit đære eá flōd ǽr gefleówn *as the flowing of the river formerly flowed over it,* Ors. 1, 3; Bos. 27, 28.

ge-flȳman, -flǽman, -flēman ; *p.* de; *pp.* ed *To cause to flee, put to flight, drive away, banish ;* fugare, in fugam vertere, expellere :—His ēhtendas ealle geflȳme *odientes eum in fugam convertam,* Ps. Th. 88, 20: Ors. 1, 10; Bos. 32, 25. Feónd wæs geflȳmed *the fiend was put to flight,* Exon. 34 b; Th. 110, 13; Gū. 107: Cd. 187; Th. 232, 17; Dan. 261. v. flȳman.

ge-flȳt ; *n. Contention, strife, schism ;* contentio, lis, schisma :—Geflȳt *schisma,* Ælfc. Gr. 9, 1; Som. 8, 23. v. ge-flīt.

ge-flȳtan ; *p.* -flāt, *pl.* -flytor; *pp.* -flyten *To strive, fight ;* contendĕre, certāre :—Cynewulf and Offa geflyton ymb Benesingtūn *Cynewulf and Offa fought at Benson,* Chr. 777; Erl. 55, 1. v. ge-flītan.

ge-fnæd, es; *n. A hem:*—Gif ic huru his reáfes gefnædu hreppe *if I only touch the hems of his garment*, Homl. Th. ii. 394, 10. v. fnæd.

ge-fnésan *to sneeze;* sternūtāre :—Ðæt he gelóme gefnése *that he often sneezes*, L. M. 2, 59; Lchdm. ii. 282, 27.

ge-fóg, es; *n. A joining, joint:*—Ðæt ðú gesomnige síde weallas fæste gefóge *that thou unite the spacious walls with a fast juncture*, Exon. 8 a; Th. 1, 10; Cri. 6. From eallum heora gefógum *from all their joints*, Blickl. Homl. 101, 4. [Cf. *Ger*. gefüge.] v. fóg.

ge-folc *people, a troop.* v. folc.

ge-fole; *adj. Having a foal, milch :*—Ðrítig gefolra olfend-myrena mid heora coltum *thirty milch camels [camelos fætas] with their colts*, Gen. 32, 15.

ge-fón, ic ge-fó; ðú ge-féhst; he ge-féhþ, *pl.* ge-fóþ; *imp.* ge-fóh; *p.* ge-féng, *pl.* ge-féngon; *pp.* ge-fangen *To take, seize, catch ;* capere :—Ic sylle cync swá hwæt swá ic gefó *ego do regi quicquid capio*, Coll. Monast. Th. 22, 27. He geféhþ ðæt ðæt he æfter spyreþ *he seizes that which he tracks*, Bt. 39, 1; Fox 212, 1. Ðú byst men gefónde *homines eris capiens*, Lk. Bos. 5, 10. Ðonne ðú híg gefangen hæbbe *quando tu illos cepisti*, Gen. 44, 4. Hú geféhst ðú fixas? *quomodo capis pisces?* Coll. Monast. Th. 23, 7.

ge-fór *died*, Ors. 6, 3; Bos. 126, 40; *p. of* ge-faran.

ge-forht *timid.* v. forht.

ge-forþian; *p.* -forþode; *pp.* forþod *To carry out, perform, accomplish, further, promote :*—His feónd ne mihten ná geforþian heora fare *his enemies could not carry out their expedition*, Chr. 1085; Erl. 218, 14. He hæfde geforþod ðæt he his freán gehét *he had performed what he promised his lord*, Byrht. Th. 140, 16; By. 289; Hy. 9, 24; Hy. Grn. ii. 291, 24. He ðæt mynster wel geforþode ða hwíle ðe he ðær wæs *he advanced the monastery while he was there*, Chr. 1045; Erl. 171, 17. [*Laym.* i-forded.] v. forþian.

ge-forwearþan *to perish.* v. forweorþan.

ge-fórword; *part. Agreed upon, covenanted, bargained ;* compactus :—Gif hit swá gefórword biþ *if it be so agreed*, L. Edm. B. 4; Th. i. 254, 14: L. Eth. ii. 4; Th. i. 286, 19.

ge-fótcypsed, -cypst; *part.* [cops *a fetter*] *Bound with fetters ;* compēditus :—Infare on ðínre gesihþe geómrung gefótcypsedra *introeat in conspectu tuo gĕmitus compedītōrum*, Ps. Lamb. 78, 11: Ps. Spl. 101, 21. Drihten tolýsþ gecospede oðde ða gefótcypstan *Dŏmĭnus solvit compĕdĭtos*, Ps. Lamb. 145, 8.

ge-fræge, -frêge, es; *n. An inquiring, a knowing, knowledge, information, hearsay;* percontātio, cognĭtĭo, audĭtio :—Míne gefræge *in my knowledge, as I have heard, as I am informed*, Beo. Th. 1557; B. 776: 1679; B. 837: Cd. 58; Th. 71, 20; Gen. 1173: 161; Th. 201, 7; Exod. 368: Chr. 975; Erl. 126, 10; Edg. 36.

ge-fræge, -frêge; *adj. Known, renowned, celebrated, remarkable, noted, famous, notorious, infamous ;* nŏtus, mănĭfestus, cĕlĕber, fāmōsus :—Hæbbe ic gefrugnen ðætte is eástdælum on ædelast londa, firum gefræge *I have heard tell that in eastern parts there is a land most noble, renowned among men*, Exon. 55 b; Th. 197, 22; Ph. 3 : 44 b; Th. 151, 8; Gú. 792. Ic eom folcum gefræge *I am noted among people*, 130 b; Th. 500, 7; Rä. 89, 3: Beo. Th. 109; B. 55. Wæs úre líf fracuþ and gefræge *our life was vile and infamous*, Exon. 53 a; Th. 186, 23; Az. 24: Cd. 189; Th. 235, 10; Dan. 304. Hæleðum gefrægost *most famous among men*, 162; Th. 202, 27; Dan. 394. [*O. Sax.* gi-frági : *Icel.* frægr.]

ge-frægen, -fregen [*part. p. of* gefragan [?]; cf. gefragian] *Heard of, known* :—Egsa mára, ðonne from frumgesceape gefrægen wurde æfre on eorðan *greater terror than was ever heard of on earth since the creation*, Exon. 20 a; Th. 52, 28; Cri. 840. Ðara ðe ic ofer foldan gefrægen hæbbe *of those that I have heard of on earth*, Exon. 85 a; Th. 319, 25; Víd. 17: Beo. Th. 2397; B. 1196; Andr. Kmbl. 1374; An. 687: 2122; An. 1062. Gefregen, Exon. 53 b; Th. 188, 14; Az. 45. [Cf. *Icel.* freginn.] v. gefragian.

ge-frægnan, -fraignan, -fregnan, -frægnian; *p.* -frægn, -fraign, -frægnade, *pl.* -frugnon *To ask, inquire :*—Gifrægna *interrogare*, Jn. Skt. Lind. Gifregna, Rush. 21, 12. Gefraigne, Mk. Skt. Lind. 12, 34. Gefraign *interrogavit*, Lind. Gifrægn, Rush. 8, 5; 9, 16. Gefrægnade *interrogavit*, Lind. 15, 2. Gefraignade *sciscitabatur*, Mt. Kmbl. Lind. 2, 4. Gefrugnun *interrogaverunt*, 17, 10; Jn. Skt. Lind. 5, 12. Gefrugnon *interrogarent*, Jn. Skt. Lind. 1, 19. Gefraignaþ *interrogate*, Jn. Skt. Lind. 9, 21. Gefraignes *interrogate*, Mt. Kmbl. Lind. 10, 11: 2, 8. v. gefrignan.

ge-frægnian; *p.* ode; *pp.* od *To make famous :* — Gefrægnod, Beo. Th. 2670. [*Thorpe* gefréfrod.]

ge-fræpi[g]an; *p.* ede. **I.** *to accuse :*—Gefræpgedon *accusarent*, Mt. Kmbl. Lind. 12, 10. **II.** *to reverence :*—Gefræppegedon *reverebuntur*, Mk. Skt. Lind. 12, 6.

ge-frætewian, -frætwian, -frætwian; *p.* ode, ade, ede; *pp.* od, ad, ed *To adorn, deck, trim;* ornāre, redimīre :—Ic gefrætwige *orno*, Ælfc. Gr. 24; Som. 25, 41. Ic gefrætwige *redimio*, 30; Som. 34, 58. Ðé Cyning engla gefrætwode *the King of angels adorned thee*, Andr. Kmbl. 3034;

An. 1520. He gefrætwade foldan sceátas *he adorned earth's regions*, Beo. Th. 192; B. 96. He æfter fæce mid óðrum gástlícum mægenum gefrætewod ætýwde *postmŏdum cætĕris virtūtĭbus ornātus appāruit*, Bd. 3, 5; S. 527, 44: 3, 11; S. 535, 32. Ðær is geat gylden, gimmum gefrætewod *there is a golden gate decked with gems*, Cd. 227; Th. 305, 20; Sat. 649: 220; Th. 283, 21; Sat. 308. Fiðrum gefrætwad *adorned with wings*, Elen. Kmbl. 1482; El. 743: Exon. 59 a; Th. 214, 14; Ph. 239. Fægre gefrætwed *neatly adorned*, 59 b; Th. 217, 2; Ph. 274: 64 a; Th. 237, 4; Ph. 585.

ge-frætwodnes *an ornament.* v. frætwednes, hrægel-gefrætwodnes.

ge-fragian; *p.* ade *To learn by asking :*—Gefragade *exquisierat*, Mt. Kmbl. Lind. 2, 16.

ge-frásian; *p.* ade; *pp.* ad *To ask, inquire;* interrŏgāre, sciscĭtāri :—He gefrásade þegnas his *interrŏgābat discĭpŭlos suos*, Mt. Kmbl. Lind. 16, 13. Geascade oððe gefrásade *sciscĭtābātur*, Mt. Kmbl. Lind. 2, 4.

ge-freán *to free;* liberare, Ps. Spl. C. 43, 29.

ge-frécnod; *part.* [frécne *savage, wicked*] *Savage, evil, wicked, corrupted ;* atrox, scĕlestus :—Móde gefrécnod *corrupted in mind*, Cd. 181; Th. 227, 10; Dan. 184.

ge-frédan, ic -fréde, ðú -frédest, he -frédeþ, frêt, *pl.* -frédaþ; *p.* -frédde; *pp.* -fréded *To feel, perceive, know, be sensible of ;* sentīre :—Sió gefrédnes hine mæg gegrápian, and gefrédan ðæt hit líchoma biþ, ða hió ne mæg gefrédan hwæðer he biþ ðe blac ðe hwít *the feeling may touch it, and feel that it is a body, but cannot feel whether it be black or white*, Bt. 41, 4; Fox 252, 10, 11. Ðeáh ðe we hit gefrédan ne mágon *though we cannot perceive it*, Boutr. Scrd. 18, 44. Ic gefréde *sentio*, Ælfc. Gr. 30; Som. 34, 39 : 37; Som. 39, 8. Se líchama awent eorþan and anbídaþ æristes, and on ðam fyrste nán þing ne gefrét *the body turns to earth and awaits the resurrection, and in that space feels nothing*, Homl. Th. ii. 232, 25. Stánas ne gefrédaþ *stones have not sense*, i. 302, 14, 18. Heó on hire gefrédde ðæt heó of ðam wíte gehæled wæs *sensit corpŏre quia sanāta esset a plāga*, Mk. Bos. 5, 29. He gefrédde his deáþes neálæcunge *he was sensible of his death's approach*, Homl. Th. i. 88, 8 : 574, 16. Hí swurdes ecge ne gefréddon *they felt not the sword's edge*, 544, 22. Ðæt he gefréde *that he has sense*, 302, 21.

ge-frédendlíc; *adj. Sensible, perceptible;* sensĭbĭlis :—Stemn is geslagen lyft, gefrédendlíc on hlyste *the voice is struck air, perceptible to the hearing*, Ælfc. Gr. 1; Som. 2, 29.

ge-frédmélum; *adv. Sensim, paulatim*, Hpt. Gl. 482.

ge-frédnes, -ness, e; *f. A feeling, sense, perception;* sensus :—Gesiht, and gehérnes, and gefrédnes ongitaþ ðone líchoman ðæs monnes *sight, and hearing, and feeling perceive the body of the man*, Bt. 41, 4; Fox 252, 7, 10.

ge-fréfran; *p.* ede; *pp.* ed *To comfort, console ;* consolari :—Ðæt híg hira fæder gefréfredon *ut lenirent dolorem patris*, Gen. 37, 35. Heó nolde beón gefréfred *noluit consolari*, Mt. Bos. 2, 18. Gefroefred, Mt. Kmbl. Lind. 5, 5. v. fréfran.

ge-fréfrian; *p.* ode; *pp.* od *To comfort, console;* consolari :—Ic gefréfrige *consolor*, Ælfc. Gr. 25; Som. 26, 64. Nú ys ðes gefréfrod *nunc hic consolatur*, Lk. Bos. 16. 25. v. fréfrian.

ge-frége, es; *n. A knowing, knowledge, hearsay;* cognĭtio, audĭtio :—Míne gefrége *in my knowledge, as I have heard*, Andr. Kmbl. 3251; An. 1628: Apstls. Kmbl. 50; Ap. 25. v. ge-fræge, es; *n.*

ge-frége; *adj. Known, celebrated, famous;* nŏtus, cĕlĕber, fāmōsus :—Læt ðe on gemyndum hú ðæt manegum wearþ fira gefrége *keep in thy mind how that was known among many men*, Andr. Kmbl. 1921; An. 963 : 2240; An. 1121. v. ge-fræge; *adj.*

ge-fremednes, -ness, e; *f. An accomplishment, fulfilment, effect;* perfectio, effectus :—He hrade ða gefremednesse ðære árfestan béne wæs fylgende *mox effectum piæ postulātĭōnis consĕcūtus est*, Bd. 1, 4; S. 475, 31.

ge-freman; *p.* ode; *pp.* od; *v. a. To finish, effect, bring to pass, accomplish, commit;* efficĕre, perficĕre, patrāre, committĕre :—Se gefremode fét [MS. fót] míne swá swá heortes *qui perfécit pĕdes meos tanquam cervōrum*, Ps. Spl. 17, 35. Ðe he gefremode *quod patrārat*, Gen. 2, 2: Jos. 7, 17. Ic ne gemune nánra his synna ðe he gefremode *I will remember none of his sins which he has committed*, Homl. Th. ii. 602, 19. Forðan synd ðás wundru gefremode on him *ideo virtūtes ŏpĕrantur in eo*, Mt. Bos. 14, 2. Árleásnes ða scilde on me gefremode *impiety perpetrated that guilt against me*, Th. Apol. 2, 19.

ge-fremman; *p.* -fremede; *pp.* -fremed *To promote, perfect, perform, commit :*—Hie mihtan æghwæt gefremman *they could accomplish anything*, Blickl. Homl. 137, 1. / Ðæt weorc to gefremmenne *to perform that work*, Homl. Th. ii. 122, 10. Ic hæla gefremme *sanitates perficio*, Lk. Bos. 13, 32. Ðás ongunnenan ðing ðurh Godes fultum gefremmaþ *perform the things begun with God's help*, Homl. Th. ii. 148, 4. Swá hwæt swá he on mycclum gyltum gefremede *whatsoever he hath committed in great sins*, Blickl. Homl. 107, 14 : 189, 22. Seó stihtung wæs gefremed *the arrangement was completed*, 81, 29. Hine mihtig God ofer ealle men forþ gefremede *him mighty God advanced above all men*, Beo. Th. 3440; B. 1718. Ðæt hire mægen on untrumnesse gefremed and getrymed wære *ut virtus ejus in infirmitate perficeretur*, Bd. 4, 23; S. 595, 16. Ðæt gefremede mán *the perpetrated crime*, Th. Apol. 2, 5. v. fremman.

ge-fremniss, e; *f. Effect*; effectus. Rtl. 16, 41: 41, 11.

ge-fremðian *to curse*; anathematizare, Mk. Skt. Lind. 14, 71.

ge-freógan, -freón; *p.* -freóde; *pp.* -freód *To free, make free* :—Đonne mót hine se hlaford gefreógan *then must the lord free him*, L. In. 74; Th. i. 148, 18: L. Ælfc. C. 20; Th. i. 48, 25: Ps. Th. 93, 1. Gefreóde *freed*, Exon. 16 a; Th. 37, 4; Cri. 588. Gefreó us wiþ yfela *free us from evils*, Hy. 6, 31; Hy. Grn. ii. 286, 31. Gefreouad *liberatus*, Lk. Skt. Lind. 1, 74. v. freógan.

ge-freólsian; *p.* ode; *pp.* od *To liberate, deliver, set free* :—He wolde Adam gefreólsian *he would deliver Adam*, Blickl. Homl. 29, 20, 35. Ic ðé gefreólsige of ealre frécennesse *I will deliver thee from all danger*, 231, 3. Úre Drihten us gefreólsode *our Lord delivered us*, 83, 25. Đurh Cristes sige ealle hálige wæron gefreólsode *through Christ's victory all holy men were set free*, 31, 35.

ge-freódian; *p.* ode; *pp.* od *To protect, guard, free, keep* :—We wæron gefreoðode feónda gafoles *we were freed from devils' tribute*, Blickl. Homl. **105, 23.** Se ðe his ánum her feore gefreoðade *he who here protected only his life*, Exon. 39 a; Th. 128, 32; Gú. 413. Gefreoða hyre *protect it [the soul]*, Exon. 118 b; Th. 456, 3; Hy. Grn. ii. 284, 61. Gefreóde and gefreoðade folc *freed and protected the people*, Exon. 16 a; Th. 37, 4; Cri. 588. Gefreoðode, Andr. Kmbl. 2083; An. 1043. He lýfde ðæt friþ wiþ hý gefreoðad wære *he allowed that peace should be kept towards them*, Exon. 38 b; Th. 127, 7; Gú. 382. Đæt lond Gode gefreoðode *he kept that land for God*, 34 b; Th. 111, 7; Gú. 123. v. gefriðian.

gefrett *consumed*; devorávit, Lk. Skt. Lind. 15, 30. v. fretan.

ge-fricgan, -fricgean; *p.* -fræg, *pl.* -frægon; *pp.* -frigen *To learn by asking or by inquiry, hear of* :—Syððan hie gefricgeaþ freán úserne ealdorleásne *when they learn that our lord is lifeless*, Beo. Th. 5996; B. 3002. Gif ic ðæt gefricge *if I learn that*, 3656; B. 1826. Syððan æðelingas feorran gefricgean fleám eówerne *after nobles from afar shall hear of your flight*, 5770; B. 2889. Đæt ðæt folca fela gefrigen habbaþ *that which many peoples have heard of*, Cd. 190; Th. 236, 31; Dan. 329: Bt. Met. Fox 9, 54; Met. 9, 27. Đa ðe snyttrocræft ðurh fyrngewritu gefrigen hæfden *they who had learned wisdom through ancient writings*, Elen. Kmbl. 310; El. 155. We feor and neáh gefrigen habbaþ Moyses dómas hæleðum secgan *we far and near have heard that Moses gave laws to men*, Cd. 143; Th. 177, 28; Exod. 1.

ge-frige, es; *n. Inquiry, knowledge resulting from inquiry* :—Gefreogum gleáwe *men wise from the knowledge obtained by their inquiries*, Exon. 56 a; Th. 199, 22; Ph. 29.

ge-frigian *to embrace*, Mk. Skt. Lind. 10, 16.

ge-frignan, -fringan; *p.* -frægn, -fregn, *pl.* -frugnon; *pp.* -frugnen. **I.** *to ask*; interrogare :—Đa Euan gefrægn ælmihtig God *then almighty God asked Eve*, Cd. 42; Th. 54, 34; Gen. 887. **II.** *to learn by asking, hear of* :—Đá gefrægn Higeláces ðegn Grendles dæda *when Hygelac's thane heard of Grendel's deeds*, Beo. Th. 390; B. 194: 1155; B. 595. Eác we ðæt gefrugnon *also we have heard that*, Exon. 12 a; Th. 19, 15; Cri. 301: 100 a; Th. 378, 11; Deór. 14: Elen. Kmbl. 343; El. 172. Swá guman gefrungon *as men have heard*, Beo. Th. 1337; B. 666. Hæbbe ic gefrugnen *I have heard*, Exon. 55 b; Th. 197, 18; Ph. 1. Đá ic néðan gefrægn hæleþ to hilde *then I heard that heroes went daringly to war*, Cd. 95; Th. 124, 9; Gen. 2060: 92; Th. 118, 4; Gen. 1960: Beo. Th. 148; B. 74: 4961; B. 2484. Gefregn, Cd. 224; Th. 298, 1; Sat. 56. Gefregen, 218; Th. 278, 21; Sat. 225. Ne gefrægen ic ða mægðe sél gebæran *never have I heard of the tribe bearing itself better*, Beo. Th. 2026; B. 1011. [O. Sax. gifregnan.] v. ge-frægnan.

ge-frignys, -nyss, e; *f. Inquiry, questioning* :—Đis syndon andsware to geðeahtunge and to gefrignysse Sct. Augustinus *responsiones ad consulta Augustini*, Bd. 1, 27; S. 497, 44.

ge-frinan, ic -frine, ðú -frinst, he -frinþ, *pl.* -frinaþ; *p.* -fran, *pl.* -frunon; *pp.* -frunen *To learn by asking, find out, hear of* :—Đá gefran Ioseph ðæt Archelaus rixode on Iudea lande *then Joseph learned that Archelaus reigned in Judea*, Homl. Th. i. 88, 19. We ðeódcyninga ðrym gefrunon *we have heard of the glory of the great kings*, Beo. Th. 4; B. 2: Andr. Kmbl. 1; An. 1: Cd. 184; Th. 230, 19; Dan. 235. Me ðær dryhtnes ðegnas gefrunon *the Lord's servants found me there*, Rood Kmbl. 151; Kr. 76. Hie hæfdon gefrunen *they had learned*, Beo. Th. 1392; B. 694: 4797; L. 2403. v. ge-frignan.

ge-friólíc; *adj. Free*; liber, Rtl. 32, 9.

ge-friðian; *p.* ode; *pp.* od *To guard, protect, defend, deliver* :—He hie gefriðode *he protected her*, Judth. 9; Thw. 21, 3; Jud. 5: Bt. 39, 10; Fox 228, 11. Đæt hys yrþ sí gefriðod *that its produce be protected*, Th. An. 118, 20. He me gefriðode *eripuit me*, Ps. Th. 33, 4. Alýs me and gefriða me *libera me et eripe me*, 7, 1. Gefriðie *protegat*, 19, 1: Exod. 19, 4. v. ge-freoðian.

ge-froefred *comforted*; consolatus, Mt. Kmbl. Lind. 5, 5, = ge-fréfred; *pp.* of ge-fréfran.

ge-frohtian *to be afraid*; expavescere, Mk. Skt. Lind. 16, 6. v. forhtian.

ge-froren *frozen*. v. freósan.

ge-frunon *asked, understood*. v. gefrinan.

ge-frygnys *a question*. 'v. gefrignys.

ge-frýnd *friends* :—On ðam dæge wurdun Herodes and Pilatus gefrýnd *facti sunt amici Herodes et Pilatus in ipsa die*, Lk. Bos. 23, 12. v. freónd.

ge-fryþsum; *adj. Safe, fortified*; salvus, múnítus :—On stówe [MS. stówum] gefryþsumre *in lócum múnítum*, Ps. Spl. 70, 3. v. friþsum.

Gefðas, Gifðas, *pl. The Gepidæ* :—Mid Gefðum ic wæs *I was among the Gefths*, Exon. 85 b; Th. 322, 8; Víd. 60. Gifðum, Beo. Th. 4981; B. 2494. v. Grm. Gesch. D. S. 324.

ge-fullan *to fill* :—Đú gefullest me of blisse mid andwlitan ðínum *adimplébis me lætítia cum vultu tuo*, Ps. Spl. 15, 11. v. ge-fyllan.

ge-fullæstan; *p.* -læste; *pp.* -læst *To help, give aid, assist*; auxiliári :—Weoruda God gefullæste, ðæt seó cwén begeat willan in worulde *the Lord of Hosts gave aid, that the queen obtained her will in this world*, Elen. Kmbl. 2299; El. 1151.

ge-fullfremman *to perfect*. v. fulfremman.

ge-fullian; *p.* ode; *pp.* od *To become full, perfect* :—Gé geseóþ nú tódæge mínra gewinna wæstm gefullian *ye see now to-day the fruit of my toils come to perfection*, Blickl. Homl. 191, 23.

ge-fullian; *p.* ode; *pp.* od *To baptize*; baptizáre :—He gefullode ðone sunu *he baptized the son*, Homl. Th. i. 352, 20. Gyt beóþ gefullode ðam fulluhte, ðe ic beó gefullod *baptismo, quo ego baptizor, baptizari*, Mk. Bos. 10, 39. Gefullod, Mt. Bos. 3, 14, 16: Mk. Bos. 1, 9: 10. 38, 39: 16, 16: Lk. 3, 21. v. fullian.

ge-fultuma, an; *m. A helper*; adjútor :—Driht gefultuma mín and alýsend mín *Dómĭne adjútor meus et redemptor meus*, Ps. Spl. 18, 16.

ge-fultuman, -fultumian, -fultmian; *p.* ode, ede; *pp.* od, ed *To help, assist, help to, supply* :—Đæt hie sceoldan Martine gefultmian *that they should help St. Martin*, Blickl. Homl. 221, 31. Gefultumian *subministrare, concurrere, suppeditare*, Hpt. Gl. 446. Of ðem æfse ðe me God forgef and míne friónd to gefultemedan *of the inheritance that God gave me and my friends helped me to*, Th. An. 127, 21: 24. Búton him seó sóþe hreów gefultmige *unless true penitence succour them*, Blickl. Homl. 101, 8: 159, 34. Nymðe me drihten gefultumede *unless the Lord had helped me*, Ps. Th. 93, 16. Gefultuma me *adjuva me*, 69, 6. Đú gefultuma úrum misdædum *impietatibus nostris tu propitiaberis*, 64, 3. He wæs godcundlíce gefultumad *divinitus adjutus*, Bd. 4, 24; S. 596, 41.

ge-fultumend, es; *m. A helper* :—Đú eart mín alýsend, and mín gefultumend *liberator meus, adjutor meus*, Ps. Th. 17, 2.

ge-fulwian, -fulgwian; *p.* ode, ade; *pp.* od, ad *To baptize* :—Gefulwia *baptizari*, Mt. Kmbl. Lind. 3, 14. Gefulwas *baptizabit*, Mk. Skt. Lind. 1, 8. Se ðe gefulguas *qui baptizat*, Jn. Skt. Lind. 1, 33. Hine man gefulwade *he was baptized*, Blickl. Homl. 219, 1. Gefulguade *baptizabat*, Jn. Skt. Lind. 3, 22. Gefulwad, Blickl. Homl. 213, 14: Elen. Kmbl. 2085; El. 1044. Gifulgwado *baptizati*, Rtl. 26, 9.

ge-funden *found*, Bd. 3, 11; S. 535, 10; *pp.* of ge-findan.

ge-fýlan; *p.* ede; *pp.* ed; *v. a. To foul, defile, pollute*; inquinare, foedare, contaminare :—Đæt hí willaþ mid gegaf-spræcum Godes hús gefýlan *so that they will with idle speeches defile God's house*, L. Ælfc. C. 35; Th. ii. 356, note 2, line 22. Đæt man mid flæsc-mete hine sylfne gefýle *that any one should defile himself with flesh-meat*, L. C. S. 47; Th. i. 402, 24, note 57.

ge-fylce, -filce, es; *n. A collection of people, army, troop, division* :—Đa Wylisce menn gewinn up ahófon and syððan heora gefylce weóx hí hí on má todældon *the Welshmen raised war . . . and after their number had increased they separated into more [bands]*, Chr. 1094; Erl. 230, 36. Hí férdon mid miclum gefilce *they marched with a great army*, Thw. Hept. 162, 38. Send ðærto gefylce *send troops against it*, Past. 21, 5; Swt. 161, 6; Hatt. MS. Hie wærun on twæm gefylcum *they were in two divisions*, Chr. 871; Erl. 74, 16, 30: Nar. 19, 22. v. fylc.

ge-fylced *collected as an army*. v. fylcian.

ge-fylgan; *p.* -fylgde; *pp.* -fylged *To follow, attend upon, reach by following* :—Đæt him gefylgan ne mæg drýmendra gedryht *so that the flock of rejoicing ones cannot follow him*, Exon. 60 b; Th. 222, 12; Ph. 347. Gif ge ðisum léase leng gefylgaþ *if ye pursue this falsehood longer*, Elen. Kmbl. 1149; El. 576. Đa ilco gefylgdon him *illi secuti sunt eum*, Mt. Kmbl. Lind. 4, 20. Gefylgend wæs ł gefylgede *sequebantur*, Jn. Skt. Lind. 18, 15. Gifylge *assequi*, Rtl. 4, 20. Đætte erestes gefe we gifylga *ut resurrectionis gratiam consequamur*, 23, 40.

ge-fyllan; *p.* -fylde; *pp.* -fylled *To fell, cut down, cast down, destroy, deprive of*; cædere, destruere :—Đá wolde he ðæt gyld gefyllan *then he determined to cast down the idol*, Blickl. Homl. 221, 21, 32: Beo. Th. 5303; B. 2655. He gefylde ðone ealdan feónd *he cast down the old fiend*, Blickl. Homl. 87, 19: 221, 2, 4, 33. Freónda gefylled *deprived of friends*, Chr. 937; Erl. 114. 7; Æðelst. 41. Seó nædre gefylled wæs *the serpent was destroyed*, Ors. 4, 6; Bos. 84, 45.

ge-fyllan; *p.* ede, de; *pp.* ed; *v. a. To fill, fulfil, make a total, complete, finish, accomplish, satisfy*; implere, saturare :—Đus unc gedafenaþ ealle rihtwisnisse gefyllan, Mt. 3, 15. We sceolon ðone geleáfan mid gódum dædum gefyllan *we must complete the belief with good deeds*,

Blickl. Homl. 23, 10. Hî ne mágon ealle ðîne bletsunge gefyllan *they do not complete the sum of all thy blessings*, 157, 20. Ealle stôwa he gefylleþ *he fills all places*, 23, 20. Mîne geornnesse mid gôde ðû gefyldest *thou didst satisfy my longing with good*, 89, 5. He him gehêt his æriste swâ he mid sôðe gefylde *he promised them his resurrection as he truly performed*, 17, 4. Hî heofon-hláfe hálige gefylde *pane cæli saturavit eos*, Ps. Th. 104, 35. Dû geịyldest foldan and rodoras wuldres ðînes *thou hast filled earth and skies with thy glory*, Exon. 13 b; Th. 25, 29; Cri. 408. Oð ðæt ðû gefylle ðîne ðegnunge *until thou fulfil thy business*, Blickl. Homl. 233, 28, 12: Guthl. 5; Gdwn. 40, 25. On hire wæs gefylled ðætte on Cantica Canticorum wæs gesungen *in her was fulfilled what was sung in the Song of Songs*, Blickl. Homl. 11, 15: 13, 26. Gefylde, 15, 8. Æfter ðon ðe ða mæssan wæron gefyllede *after the masses were finished*, 207, 29: Lk. Bos. 4, 13. Ðæt hûs wæs gefylled of ðære sealfe swæces *domus impleta est ex odore ungenti*, Jn. Bos. 12, 3. Gifena gefylled fremum forðweardum *filled with gifts with continual benefits*, Cd. 11; Th. 13, 28; Gen. 209. Gefylled *consumtus, finitus*, Hpt. Gl. 457. Wel gefylde *bene pastos*, Th. An. 20, 31.

ge-fyllednes, -ness, -nys, -nyss, e; *f. A fulness, satiety, completion, finishing, end*; plēnitūdo, sătūritas, consummātio:—Astyrod biþ sæ and gefyllednys hyre *commŏveātur, măre et plēnītūdo ejus*, Ps. Spl. 95, 11: 97, 7. Cherubin is gereht gefyllednyss ingehydes *cherubin is interpreted the fulness of the mind*, Boutr. Scrd. 20, 33. On graman gefyllednysse *in īra consummātiōnis*, Ps. Spl. C. 58, 15. Of his gefyllednesse we ealle onféngon *de plēnītūdĭne ejus nos omnes accēpĭmus*, Jn. Bos. 1, 16. He asende gefyllednysse on sáwlum heora *mīsit sătūrĭtātem in anĭmas eōrum*, Ps. Spl. 105, 15. Oð ðissere worulde gefyllednysse *until the end of the world*, Homl. Th. i. 600, 18.

ge-fyllendlíc; *adj. Filling*; explētīvus, complētīvus:—Sume syndon gehâtene *explētīvæ* oððe *complētīvæ*, ðæt synd gefyllendlîce *some are called* explētīvæ *or* complētīvæ, *that is filling*, Ælfc. Gr. 44; Som. 45, 57.

ge-fylnes, -ness, e; *f. Fulness, fulfilment, performance, completion*:— On gefylnesse Godes beboda *in the performance of God's commands*, Blickl. Homl. 29, 9. For gefyllnesse ðæs heofonlîcan eðles *for the perfection of the heavenly country*, 81, 29. Ðe hie swâ mycle gefylnesse hæfdon *of which they had so great fulness*, 135, 24. Gifylnisse *plenitudinis*, Rtl. 83, 18.

ge-fylst *help*. v. fylst.

ge-fylsta, an; *m. A helper, an assistant*; adjūtor:—God mîn gefylsta is *Deus meus adjūtor est*, Ps. Spl. 17, 2: 27, 9. He him to gefylstan gesette *he appointed him his assistant*, Homl. Th. ii. 120, 13: Job Thw. 166, 39.

ge-fylstan; ic -fylste; *subj. pres.* -fylste; *p.* [-fylstede], -fylste, *pl.* -fylston; *pp.* -fylsted *To help, give help*; adjuvare:—Ðæt heó him gefylste *that she might assist them*, Ors. 3, 11; Bos. 73, 45. God gefylsteþ me *Deus adjuvat me*, Ps. Spl. 53, 4. Driht, to gefylstane me efste *Domine, ad adjuvandum me festina*, 69, 1. DER fylstan.

ge-fýnd *foes, enemies*:—Hîg wæron ær gefýnd him betwynan *antea inimici erant adinvicem*, Lk. Bos. 23, 12. v. feond.

ge-fyndig; *adj. Capable*; capax, Ælf. gr. 9, 60. v. gefindig.

ge-fyrenian, -fyrnian; *p.* ode, ede; *pp.* od, ed *To sin*; peccāre:—Ic gefyrenode *I have sinned*, Blickl. Homl. 235, 32, 34. We gefyrnedan mid ûrum fæderum *peccāvimus cum patrĭbus nostris*, Ps. Th. 105, 6. v. ge-firenian.

ge-fyrht, ge-fyrhted; *part. p. Terrified, affrighted*:—Ðá wæs se dêma swyðe gedrêfed and gefyrhted *then was the judge very much troubled and frightened*, Bd. 1, 7; S. 478, 44. Hie wæron to ðæs swyðe gefyrhte *they were so greatly terrified*, Blickl. Homl. 221, 34. [Cf. fyrhtan, gefyrhtian.]

ge-fyrhtian; *p.* ade; *pp.* ad *To frighten*:—Wîfo sume gefyrhtadon ûsig *mulieres quædam terruerunt nos*, Lk. Skt. Lind. 24, 22. Miþ fyrhto gefyrhtad *timore exterriti*, Mk. Skt. Lind. 9, 6.

ge-fyrhto; *p. Fear, doubt*:—Be ðære cennendre gefyrhtum ðæs bearnes weorðe ongyten wære *by the mother's fears the child's worth might be understood*, Blickl. Homl. 163, 27.

ge-fyrn; *adv.* [fyrn *formerly*] *Formerly, long ago, of old, of yore*; olim, prīdem:—Hû ne wæron ðás gefyrn forþgewitene *were not these long ago departed?* Bt. 19; Fox 70, 9. Ðû mid Fæder ðînne gefyrn wære efenwesende *thou with thy father of old was co-existent*, Exon. 12 b; Th. 22, 10; Cri. 349: 12 a; Th. 19, 16; Cri. 301. Gefyrn hî dydun dædbôte on hêran and on axan *olim cĭlĭcio et cĭnĕre pænĭtentiam egissent*, Mt. Bos. 11, 21: Lk. Bos. 10, 13: Ælfc. Gr. 38; Som. 39, 57. Gefyrn *prīdem*, 38; Som. 39, 56. Gefyrn ær *formerly*, Bt. 33, 3; Fox 126, 30: 37, 1; Fox 186, 25: Chr. 892; Erl. 89, 1.

ge-fyrþran; *p.* ede; *pp.* ed *To further, advance, promote, improve*; promovere, prosperare:—Heora sîþfæt wæs fram Drihtne sylfum gefyrþred [MS. gefyrþrad], *their journey was furthered by the Lord himself*, Bd. 4, 19; S. 588, 34. Wæs efstíðes georn, frætwum gefyrþred was *desirous of return, furthered by the treasures*, Beo. Th. 5561; B. 2784. Ânræd oretta elne gefyrþred *the steadfast champion advanced with*

valour, Andr. Kmbl. 1966; An. 985. Ic ðe gefyrþrede *I improved thee*, Bt. 8; Fox 24, 29. DER. fyrþran.

ge-fýsan; *p.* -fýsde; *pp.* -fýsed *To make ready, cause to hasten*:— Werod wæs gefýsed *the band was made ready*, Cd. 154; Th. 191, 28; Exod. 221. Gefýsed to fæder rîce *ready to depart to his father's kingdom*, Exon. 14 b; Th. 30, 5; Cri. 475. Winde gefýsed *hurried on by the wind*, Beo. Th. 440; B. 217. Secgas wæron sîðes gefýsde *the men were ready for the journey* [cf. sîðes fûs, B. 1475],'Elen. Kmbl. 520; El. 260. v. fýsan.

ge-fystlian; *pp.* -lad *To beat with the fists, buffet*; pugnis impetere, Scint. 2.

ge-fyðerian; *p.* ode, ade, ede; *pp.* od, ad, ed *To feather, give wings to, provide with wings*; ālas addĕre, pennis instruĕre :—Gefyðerad flaa *săgitta* vel *spĭcŭlum*, Ælfc. Gl. 53; Som. 66, 64; Wrt. Voc. 35, 50. Fugelas gefyðerede *vŏlātĭlia pennāta*, Ps. Spl. 77, 31. v. ge-fiðerian.

ge-gada, an; *A fellow-traveller, a companion, associate*; comes, complex, conspirans, Ælfc. Gl. 86; Som. 74, 27, 28. He feóll ðá adûn and ealle his gegadan into helle wîte *he fell down then and all his companions into hell torment*, Swt. A. S. Rdr. 59, 93, 87. Afeóll se deófoll mid his gegadum *the devil fell with his companions*, Hexam. 10; Norm. 16, 18. v. gædeling.

ge-gaderian; *p.* ode; *pp.* od *To gather, unite*; colligere, conjungere :— Se fela folca fæste gegadraþ *he unites many people*, Bt. Met. Fox 11, 180; Met. 11, 90. Gegaderade *conjuncti*, Ps. Th. 67, 24: Chr. 973; Th. 224, 32. v. gadorian, ge-gæderian.

ge-gaderscype, -gæderscype, es; *m. A joining, union, matrimony*; jugalitas, Hpt. Gl. 411, 416.

ge-gaderung, e; *f. A gathering, congregation, assembly, crowd*; congregatio, turba :—Se Hælend beáh fram ðære gegaderunge *Iesus declinavit a turba*, Jn. Bos. 5, 13: Ps. Spl. 39, 14; Ælfc. Gl. 87; Som. 74, 47. Gegaderung lîchoman *copula carnis*, Bd. 1, 27; S. 495, 30. Gegaderung *congregatio*, Th. An. 30, 7. Rihtwîsra manna gegaderung is gecweden heofenan rîce *a gathering of righteous men is called the kingdom of heaven*, Homl. Th. ii. 72, 25. v. gaderung.

ge-gador-wist, e; *f. An assembly for feasting*; contubernium, Ælfc. Gl. 93; Som. 75, 87. v. gador-wist.

ge-gæde *a collection, congregation*; congregatio, R. Ben. interl. 2. v. gæd.

ge-gæderian, -gaderian; *p.* ode, ade; *pp.* od, ad *To gather, join*; colligere, congregare :—Searwum gegædraþ bân gebrosnad *he gathers skilfully the perished bones*, Exon. 59 b; Th. 216, 17; Ph. 269. Beóþ gegædrad gæst and bân-sele and body shall be joined, 117 b; Th. 451, 11; Dôm. 102. Wyt beóþ gegæderode *we two shall be gathered*, 100 a; Th. 376, 23; Seel. 159. Ðam biþ gæst gegæderad Godes bearn *God's child will be a guest associated with him*, 84 b; Th. 318, 9; Môd. 80. v. gæderian, gegaderian.

ge-gælen, -galen, *enchanted*, Ps. Spl. 57, 5. v. galan.

ge-gæncg, es; *m. A society, meeting, an assembly*; cœtus :—Ðe wæs on ðam gegæncge ðær man Crist bænde *who was in the company where Christ was bound*, Ælf. ep. 1st, 50; Th. ii. 386, 23.

ge-gærwan *to prepare*. v. gegerwan.

ge-gaf; *adj. Base, wanton, lewd*:—He wæs gegaf spræce *he was wanton in talk*, Homl. Th. i. 534, 2. [Or gegaf-spræce; *adj.* (?).]

ge-gafelian; *p.* ode; *pp.* od *To impose a fine, proscribe*, Hpt. Gl. 517.

ge-gafelod *confiscated*; infiscatus, Cot. 108, 194. v. gegafelian.

ge-gaf-spréc, e; *f. Idle, wanton, scoffing speech* :—Ðá wæs seó tunge teartlícor gewîtnod *for his gegafspræce then was the tongue more sharply punished for his wanton speech*, Homl. Th. i. 330, 34. Men willaþ bysmorlîce plegian and mid gegafspræcum Godes hûs gefýlan *men will play shamefully and defile God's house with wanton speeches*, L. Ælfc. C. 35; Th. ii. 357 note, 3. v. gaf.

ge-gán; *p.* -eóde, -íóde; *pp.* -gán. **I.** *to go, go or pass over, come to pass, happen*; ire, præterire, evenire:—Heó mihte gegán ofer eall ðis eálond *vellet totam perambulare insulam*, Bd. 2, 16; S. 520, 2. Se ðe gryre-sîþas gegán dorste *who durst go ways of terror*, Beo. Th. 2929; B. 1462. Swâ geostran-dæg gegán wære *sicut dies hesterna quæ præteriit*, Ps. Th. 89, 4. Hû ðæt gegeóde, ðæt . . . *how that came to pass, that* . . ., Exon. 14 a; Th. 28, 7; Cri. 443. Eall ðás wundor geeódon in ussera tîda tîman *all these wonders happened in the period of our times*, 43 b; Th. 147, 11; Gú. 725. **II.** *to occupy, overcome, overrun, subdue*; occupare, vincere, subigere:—Ðæt ðû hám on us [hus MS.] gegán wille *that thou wilt occupy a home with us*, Exon. 36 b; Th. 118, 21; Gú. 243. Eádmund cyning Myrce geeóde *king Edmund subdued Mercia*, Chr. 942; Th. 208, 33; Edm. 2: Bd. 1, 2; S. 475, 4: 2, 5; S. 506, 20: Ors. 3, 7; Bos. 58, 39: 3, 9; Bos. 65, 44. Ne geeódon ûre foregengan nâ ðas eorðan mid sweorda ecgum *non enim in gladio suo possidebunt terram*, Ps. Th. 43, 4. Seo burh wæs gegán *civitas capta erat*, Jos. 8, 21. **III.** *to observe, practise, exercise, effect, accomplish*; observare, exercere, perficere, efficere:— Gif gê ðæt tâcen gegáþ *if ye observe that sign*, Cd. 106; Th. 140, 8; Gen. 2324. Ðæt se hálga þeów elne geeóde *which the holy minister*

zealously practised, Exon. 34 b ; Th. 111, 19 ; Gú. 129 : Ps. Th. 118, 40. Hie elne geeódon *they effected by strength*, Beo. Th. 5826 ; B. 2917. **IV.** used with an adjective [cf. such an expression as ' to go lame'] :—He wæs wérig gegán *fatigatus ex itinere*, Jn. Bos. 4, 6. **ge-gang** *an event, a fate.* v. gegong.

ge-gangan, -gongan ; *pp.* -gangen, -gongen. **I.** *to go, happen, take place, befal, to fall to one's share, to come in ;* ire, evenire, accidere :—Ne mágon hî ofer gemǽre máre gegangan *non transgredientur terminum*, Ps. Th. 103, 9. Ful oft ðæt gegongeþ *full oft it happens*, Exon. 87 a ; Th. 327, 9 ; Vy. 1 : 117 a ; Th. 451, 3 ; Dóm. 98. Ðá wæs gongen gumum unfródum, ðæt . . . *then it had befallen the youthful man, that* . . . , Beo. Th. 5634 ; B. 2821. Ealles ðæs andlyfenes ðe him gegonge *of all the livelihood which comes in to them*, Bd. 1, 27 ; S. 489, 6. **II.** *to exercise, effect, accomplish ;* exercere, perficere, efficere :—Ic díne bebodu bealde gegange *exercebor in mandatis tuis*, Ps. Th. 118, 78. He hæfde elne gegongen, ðæt . . . *he had effected by his valour, that* . . . , Beo. Th. 1791 ; B. 893. **III.** *to go against with hostile intention, to pass over, overcome, subdue, conquer, obtain, acquire ;* aggredi, transgredi, superare, subigere, obtinere, adipisci, possidere :—Gif fríman edor gegangeþ *if a freeman pass over a fence*, L. Ethb. 29 ; Th. i. 10, 3. Hí þohton Italia ealle gegongan *they thought to conquer all Italy*, Bt. Met. Fox 1, 24 ; Met. 1, 12. Ic mid elne sceal gold gegangan *I shall with valour obtain the gold*, Beo. Th. 5065 ; B. 2036 : 6162 ; B. 3085 : Ps. Th. 78, 12. v. gán.

ge-geafian ; *p.* ede, ode ; *pp.* ed, od *To bestow gifts upon :*—Ic hine mid deórweorðum gyfum gegeafede *dignis eum muneribus honoravi*, Nar. 8, 16. Gigeafiga *præstolari* [=*præstare?*], Rtl. 20, 15. v. gegifod.

ge-gealt = ge-healt, Deut. 7. 12. v. gehealdan.

ge-gearcian ; *p.* ode ; *pp.* od *To prepare :*—Ðá hét se cyngc scipa gegearcian and him æfter faran, ac hit wæs lang ǽr ðam þe ða scipa gegearcode wǽron *then the king bade prepare ships and go after him, but it was long before the ships were ready*, Th. Ap. 7, 16–7 : Homl. Th. ii. 84, 16. v. gearcian.

ge-gearcung-dæg, es ; *m. Preparation-day ;* parasceve = παρασκευή :—Hit wæs eástra gegearcung-dæg *erat parasceve Paschæ ἦν παρασκευὴ τοῦ πάσχα*, Jn. Bos. 19, 14, 31. v. gearcung.

ge-gearnian, Blickl. Homl. 35, 36. v. ge-earnian.

ge-gearwian, -gearwigean ; *p.* ode, ede ; *pp.* od, ad *To prepare, make ready, provide with, endue :*—Ða láreowas sceolan Drihtnes weg gegearwian to heora módum *the teachers ought to prepare the Lord's way for their minds*, Blickl. Homl. 81, 7. Gegearwigean, Cd. 23 ; Th. 29, 30 ; Gen. 458. Ða áne ðe mid clǽnum geleáfan hie to ðæm. gegearwiaþ *those only who with pure belief prepare themselves for it*, Blickl. Homl. 185, 10. Gegearwode he ðæm éce forwyrde *he prepared for them eternal perdition*, 159, 19 : 233, 33. Gegearewadest, Ps. Th. 64, 10. Gegearwiga we *paremus*, Mk. Skt. Lind. 14, 12. Ða wearþ werod gegearewod to campe *then was the band made ready for battle*, Judth. 11 ; Thw. 24, 21 ; Jud. 199. Ðæt his líf ðæm his naman wæs gelíce gegearwod *his life was ordered like to his name*, Blickl. Homl. 167, 32. Gáste gegearwod *endued with spirit*, Cd. 10 ; Th. 12, 17 ; Gen. 187 : Elen. Kmbl. 1774 ; El. 889. v. gearwian.

ge-gearwung, e ; *f. A preparation ;* præparatio :—Gegearwung setles ðínes *præparatio sedis tuæ*, Ps. Spl. 88, 14. v. gearwung.

ge-gearwungness, e ; *f. A preparation ;* præparatio :—Gearcunga oððe gegearwungnessa heortan gehýrde *præparatio cordis audivit*, Ps. Lamb. second 9, 17.

ge-gegnian ; *p.* ode, ade ; *pp.* od, ad *To meet ;* obviare, Rtl. 45, 23.

ge-géman ; *p.* de ; *pp.* ed *To heal, cure, amend, treat* [as *a patient*] :—Ðæt hea gegéme all unhǽlo *ut curarent omnem languorem*, Mt. Kmbl. Lind. 10, 1 : Mk. Skt. Lind. 3, 2. Gegémde t gehǽlde *curavit*, 6, 5. Gegéma *corrigere*, Mt. Kmbl. p. 1, 9. Gegémed, L. Ædelb. 62 ; Th. i. 18, 8. [See the note, and also Schmid, p. 8, note.]

ge-geótan ; *p.* -geát ; *pp.* -goten *To found, cast :*—He of golde gegeát and geworhte *he cast and wrought them of gold*, Nar. 19, 29. Ða gelícnessa wǽron gegotene *the images were cast*, 32.

ge-gerela, -gyrela, -girla, an ; *m. Clothing, apparel, habit, garment, robe ;* amiculum, stola :—Hwǽr agylte he ǽfre on his gegerelan *where trespassed he ever in his clothing?* Blickl. Homl. 169, 1. His gegirla hine geswutelaþ *his garment betrays him*, Th. Ap. 14, 3 : 12, 8. Bringaþ raðe ðæne sélestan gegyrelan, Lk. Bos. 15, 22 : Mk. Bos. 12, 38.

ge-gerelad, -gerlad ; *part. Clothed ;* indutus :—Gegerlad is Drihten mid strǽncþe *indutus est Dominus fortitudinem*, Ps. Lamb. 92, 1. Gegerelad *vestitus*, Mk. Skt. Lind. 1, 6.

ge-gerwan, -gærwan, -girwan, -gierwan, -gyrwan ; *p.* -gerede ; *pp.* -gered, -gerwed *To prepare, make ready, clothe, array, adorn, furnish :*—Ne hýrde ic cymlícor ceól gegyrwan hilde wǽpnum *I never heard of furnishing a comelier vessel with weapons of war*, Beo. Th. 76 ; 13, 38. Ðér ðú scealt ád gegærwan *there shalt thou prepare a pile*, Cd. 138 ; Th. 173, 3 ; Gen. 2855. Ic his sacerdas mid hǽlu gegyrwe *sacerdotes ejus induam salutare*, Ps. Th. 131, 17. Heó alegde hire hrægl ðe heó mid

gegyred wæs and hie gegyrede mid ðon sélestan hrægle *she laid aside the garment that she was clothed with, and arrayed herself with the finest garment*, Blickl. Homl. 139, 6, 7 : 89, 35 : 103, 3. Ðǽr weofod inne wlitelíce geworhtan and gegyredon *therein they wrought and adorned an altar beautifully*, 205, 6 : Beo. Th. 6265 ; B. 3137. Gegyre ðú hiné *clothe him*, Blickl. Homl. 37, 21. Mid heora geatwum gegyrede *equipped*, 221, 29 : Nar. 4, 13. Golde gegyrwed *adorned with gold*, Beo. Th. 1110 ; B. 553. Ymb frætwum útan gegyrede *circumornatæ*, Ps. 143, 15. Sió wæs orðoncum eall gegyrwed *this was all cunningly prepared with devilish arts*, Beo. Th. 4181 ; B. 2087. Heardum tóþum and miclum hit wæs gegyred *duris munitum dentibus*, Nar. 21, 1.

ge-gifod ; *part. Enriched with gifts :*—Se cyng him wel gegifod hæfde on golde and on seolfre *the king had bestowed many gifts of gold and silver on him*, Chr. 1001 ; Erl. 136, 17. v. gegeafian.

ge-gild, ge-gyld, es ; *n. A guild, society, or club ;* societas, fraternitas :—We for his lufon ðis gegyld gegaderodon *for love of him we have gathered this guild*, Th. Diplm. 608, 7. v. gild.

ge-gilda, -gylda, an ; *m. A person who belongs to a guild, club,* or *corporation, a guild-brother, a companion, fellow* [v. Kmbl. Sax. Eng. i. 262, 259] ; congildo, socius, sodalis :—Gieldan ða gegildan healfne *let his guild-brethren pay half*, L. Alf. pol. 27 ; Th. i. 78, 24 : 28 ; Th. i. 80, 3 : L. In. 16 ; Th. i. 112, 8 : 21 ; Th. i. 116, 6 : L. Ath. v. § 8, 6 ; Th. i. 236, 36 : Hick. Thes. ii. Dis. Epist. pp. 20–21. v. gild ; and Schmid, s. v.

ge-gild-heall, e ; *f. A guild-hall :*—Orc hæfþ gegyfen ðæ gegyld-healle ðam gyldscipe to ágenne *Orc hath given the guild-hall for the guild to own*, Kmbl. Cod. Dipl. iv. 277, 21.

ge-giwian ; *p.* ade, ode ; *pp.* ad, od *To demand, ask ;* postulare, petere :—Swǽ hwæt ðú gegiuas *quidquid petieris*, Mk. Skt. Lind. 6. 23. Gegiwade *postulans*, Lk. Skt. Lind. 1, 63.

ge-gladian ; *p.* ode ; *pp.* od *To make glad, gladden, appease ;* lætifícáre, exhilárare, plácáre :—Flódes ryne gegladaþ burg Godes *flúminis impétus lætíficat civitátem Dei*, Ps. Lamb. 45, 5 ; Homl. Th. i. 288, 8. Cúþbertus hit mid cossum gegladode *Cuthbert gladdened it with kisses*, ii. 134, 21. Ðæt he gegladie anséne on ele *ut exhiláret fáciem in óleo*, 103, 15. Ðæt he ðé mid his lácum gegladige *that he appease thee with his gifts*, Gen. 32, 20. Gegladan *mitigare, repropitiare*, Hpt. Gl. 515.

ge-gléded ; *part.* [gléd *a burning coal*] *Kindled ;* accensus :—Wæs gegléded fýr on Iacobe *ignis accensus est in Iacob*, Ps. Th. 77, 23.

ge-glendrian ; *p.* ade, ode ; *pp.* ad, od *To precipitate :*—Ðætte hia geglendradon hine *ut præcipitarent eum*, Lk. Skt. Lind. 4, 29.

ge-glengan, -glencan, -glæncan, -glencgan, -glengcan ; *p.* -glengde, -glencde ; *pp.* -glenged, -glencged, -glengd, -glend *To adorn, embellish, set in order, compose ;* ornáre, cómére, componére :—Gé preóstas sculon eówerne hád healdan árwurþlíce, and mid gódum þeáwum symle geglencan *ye priests should religiously observe your order, and always adorn it with good habits*, L. Ælf. P. 5 ; Th. ii. 366, 2. Ic geglenge *cómo*, Ælfc. Gr. 28, 4 ; Som. 31, 13. Ic smicere geglencge *orno*, Ælfc. Gl. 99 ; Som. 76, 116 ; Wrt. Voc. 54, 58. Nerón hine mid ælces cynnes gimmum geglengde *Nero adorned himself with gems of every kind*, Bt. 28 ; Fox 100, 27 : Bt. Met. Fox 15, 7 ; Met. 15, 4. Ðæt he æfter medmiclum fæce in sceópgereorde mid ða mǽstan swétnesse and inþrydnesse geglencde, and in Englisc gereorde wel gehwǽr forþbrohte *hoc ipse post pusillum verbis poëticis maxima suávitáte et compunctióne compósitis, in sua, id est, Anglórum lingua proferret*, Bd. 4, 24 ; S. 596, 35. Ðæt hit wǽre geglenged mid gódum stánum and gódum gifum, *quod bónis lápidibus et dónis ornátum esset*, Lk. Bos. 21, 5 : Elen. Kmbl. 179 ; El. 90. Geglenged *discrimínátus*, Ælfc. Gl. 61 ; Som. 68, 48 ; Wrt. Voc. 39, 32. Godes geladung is geglencged mid deórwurþre frætewunge *God's church is adorned with precious ornament*, Homl. Th. ii. 586, 17. Heó wæs geglengd þurh Godes wundra *it was embellished by the miracles of God*, Th. Diplm. A. D. 970 ; 241, 6. Ða bióþ sweordum and fetelum swíðe geglende *who are greatly adorned with swords and belts*, Bt. Met. Fox 25, 20 ; Met. 25, 10.

ge-glengendlíc ; *adj. Splendid, brilliant ;* pomposus, delicatus, Hpt. Gl. 435.

geglesc *light, frolicsome, lascivious*, Bd. 5, 6 ; Whelc. 390, 39, MS. B. v. geaglisc.

ge-glídan ; *p.* -glád, *pl.* -glidon ; *pp.* -gliden *To glide, fall ;* labi :—Ðá he sceolde into geglídan Nergendes níþ *when he must fall into the Saviour's hate*, Cd. 221 ; Th. 288, 6 ; Sat. 376. v. glídan.

gegn, geagn, geán, gén ; *adv. Again ;* contra :—Brego geán þingade *the Lord spoke again*, Cd. 48 ; Th. 62, 5 ; Gen. 1009.

gegn-cwide, es ; *m. A reply, answering again ;* responsum :—Ðínra gegncwida [MS. -cwiða] *of thy replies*, Beo. Th. 739 ; B. 367.

Gegnes-burh *Gainsborough, Lincolnshire*, Chr. 1013.

ge-gnídan ; *p.* -gnád, *pl.* -gnidon ; *pp.* -gniden *To rub, rub together, comminute ;* fricare, defricare, fricando comminuere, planare, levigare :—Nim ðas ylcan wyrte dryg he ðonne and gegníd to duste *take this same wort, then dry it, and rub it to dust*, Herb. 90, 10 ; Lchdm. i.

196, 12. Genim ðas wyrte on wætre gegnidene *take this wort rubbed in water*, Herb. 84, 1; Lchdm. i. 188, 1. Ic gegnîde *plano vel levigo*, Ælfc. Gl. 36; Som. 62, 8. v. gnîdan.

gegninga, -nunga; *adv. Plainly, wholly, altogether, certainly, directly;* omnino :—Ðær ðú gegninga gûðe findest *there wilt thou straightway find war*, Andr. Kmbl. 2697; An. ·1351. Ðæt hit gegnunga from Gode côme *that it came directly from God*, Cd. 32; Th. 42, 35; Gen. 683: Exon. 44 b; Th. 150, 27; Gú. 785.

gegn-pæþ, es; *m. A path along which one goes to oppose another*, Exon. 104 b; Th. 397, 27; Rä. 16, 26.

gegn-slege, es; *m. A striking back again, exchange of blows, battle*, Andr. Kmbl. 2711; An. 1358.

gegnum; *adv. Forward;* obviam :—For hwam ne môton we ðonne gegnum gangan *why then may we not go forward?* Salm. Kmbl. 705; Sal. 352. Eódon ðá gegnum ðanonne *they thence went on forward*, Judth. 11; Thw. 23, 21; Jud. 132: Beo. Th. 633; B. 314: 2813; B. 1404. [Cf. Icel. gegnum *through*.]

ge-gnysan *to dash against*, Ps. 136, 12. v. forgnidan.

ge-gôded. v. gegôdian.

ge-gôdian; *p.* ode; *pp.* od *To bestow goods upon, enrich* :—Ða mynstru he genihtsumlîce to dæghwomlîcum bigleofan gegôdode *he abundantly enriched those minsters for daily subsistence*, Homl. Th. ii. 118, 30; H. R. 105, 6: Chr. 1086; Erl. 220, 39. Ðonne ðú Hierusalem gegôdie *in die Hierusalem*, Th. 136, 7. Apollonius ðe ðurh us gegôdod is *Apollonius who is enriched by us*, Th. Ap. 18, 20. Ða sîn gegôded *utuntur*, Hpt. Gl. 447, 494. Gegôded *fretus*, 503; *acquisitus, adeptus*, 513. v. gôdian.

ge-gogud *relying on;* fretus, R. Conc. v. ge-gôded [?].

ge-golden; *part. Paid, performed;* præstitus, L. In. 71.

ge-gong, -gang *fate, a falling out, an accident;* fatum, Cot. 48.

ge-gongan *to go over, conquer*, Bt. Met. Fox 1, 24; Met. 1, 12. v. gegangan.

ge-goten *poured out, molten, melted*, Kmbl. Sal. and Sat. 61; Sat. 31. v. ge-geótan.

ge-græppian; *p.* ade; *pp.* ad *To seize*, Mt. Kmbl. Lind. 14, 31.

ge-grâpian; *p.* ode; *pp.* od *To grope, touch;* palpâre :—Sió gefrêdnes hine mæg gegrâpian *the feeling may touch it*, Bt. 41, 4; Fox 252, 10 : Ps. Th. 113, 15 : 134, 18. Hand hî habbaþ and hîg ne gegrâpiaþ *mănus hábent et non palpâbunt*, Ps. Lamb. second 113, 7. Ða he hyne ggrápod hæfde *palpâto eo*, Gen. 27, 22.

ge-gremian, -gremman; *p.* ode, ede; *pp.* od, ed *To irritate, provoke, excite, incense, inflame;* exaspěrâre, provôcâre, exâcerbâre :—Ðe in eorre [hine] gegremmaþ *qui in îra* [*eum*] *provôcant*, Ps. Surt. 67, 7. Hwæt hit swîður gehierste and gegremige *what more scorches and excites it?* Past. 21, 6; Swt. 165, 2; Hat. MS. 32 a, 15. Gegremod wearþ se gûþrinc *the chief was incensed*, Byrht. Th. 135, 54; By. 138. Hî wæron gûþe gegremede *they were made fierce by battle*, Judth. 12; Thw. 26, 2; Jud. 306: Cd. 4; Th. 4, 29; Gen. 61.

ge-grétan; he -grét, *pl.* -grétaþ; *p.* -grétte, *pl.* -grétton; *pp.* -gréted *To approach, come to, address, greet, welcome;* adire, alloqui, salutare :—Wîf sceal eodor æþelinga [MS. e] ǽrest gegrétan *the wife shall the nobles' chief first greet*, Exon. 90 a; Th. 339, 7; Gn. Ex. 90. Holdne gegrétte meaglum wordum *he addressed his faithful friend in powerful words*, Beo. Th. 3964; B. 1980. Hie ðá gegrétte *he then addressed them*, Andr. Kmbl. 507; An. 254. Ðæt we mágon úre frýnd geseón and úre siblingas gegrétan *that we may see our friends and greet our kinsmen*, Homl. Th. ii. 526, 33. Man tǽleþ and mid yfle gegréteþ ða ðe riht lufiaþ *men blame and insult those that love right*, Swt. A. S. Rdr. 110, 164. v. grétan.

ge-gréwþ *grows*, Bt. 34, 10; Fox 148, 27; *3rd sing. pres. of* ge-grówan.

ge-grin *a snare*, Ps. Spl. T. 24, 16. v. grin.

ge-grinan; *p.* ode; *pp.* od *To ensnare;* illaqueare, Prov. 6.

ge-grind, es; *n. A grinding or rubbing together, a noise, whizzing, clashing, commotion;* collisio, contritio, frágor :—Grîmhelma gegrind *the crashing of helmets*, Cd. 160; Th. 198, 29; Exod. 330: 95; Th. 124, 15; Gen. 2063. Geótende gegrind grund eall forswealg *the abyss swallowed up the pouring commotion*, Andr. Kmbl. 3178; An. 1592.

ge-grindan; *p.* -grand, *pl.* -grundon; *pp.* -grunden *To grind together, sharpen, grind to powder;* commolere, perîricare :—Gegrindaes *comminuet*, Lk. Skt. Lind. 20, 18. Gegrunden [MS. gegrunde] *commolitus*, Ælfc. Gl. 36; Wrt. Voc. 28, 78. Gegrundene gâras *the sharpened arrows*, Byrht. Th. 134, 64; By. 109. DER. grindan.

ge-grip *a gripe, seizing*, Ps. gripa.

ge-grípan; *p.* -gráp, *pl.* -gripon; *pp.* -gripen *To gripe, grasp, seize;* capere, rapere, prehendere, apprehendere, comprehendere, arripere, coripere, eripere :—Máran ðonne ðú in hreðre mǽge môde gegrîpan *too great for thee to comprehend in thy breast with thy mind*, Exon. 92 b; Th. 348, 10; Sch. 26: Bt. Met. Fox 10, 138; Met. 10, 69. Feónd sáwle mîne gegrîpeþ *inimicus animam meam comprehendat*, Ps. Spl. 7, 5 : Salm. Kmbl. 226; Sal. 112. Us fyrhto gegráp *fear seized us*, Nicod. 21;

Thw. 10, 33: Cd. 140: Th. 175, 32; Gen. 2904: Cant. Moys. Surt. 188, 15: Nar. 44, 13. Ða gegripon ða unclǽnan gâstas ǽnne of ðám mannum *then the unclean spirits seized one of the men*, Bd. 3, 19; S. 548, 47 : Ps. Spl. 39, 16: Cant. Moys. Ex. 15, 17. Gegrîp wǽpn and scyld *apprehende arma et scutum*, Ps. Spl. 34, 2. Êhtaþ gê and gegrîpaþ hine *persequimini et comprehendite eum*, Ps. Spl. 70, 12. Ðî læs áhwænne gegrîpe swâ swâ leó sáwle mîne *ne quando rapiat ut leo animam meam*, Ps. Spl. 7, 2. Ða wæs he fram deófle gegripen *then he was seized by a devil*, Bd. 3, 11; S. 536, 13: Ps. Spl. 17, 31. On tintregum gegripene *tormentis comprehensos*, Mt. Bos. 4, 24. Geneálǽcende he hî upahôf, hyre handa gegripenre *accedens elevavit eam, apprehensa manu ejus*, Mk. Bos. 1, 31. Hî wurdon gegripene fram môderlicum breóstum *they were snatched from their mothers' breasts*, Homl. Th. i. 84, 8. v. grîpan.

ge-gripennis, -niss, e; *f. A taking, seizing, snare;* correptio, captio :—Gegripennis ðone ðe he behýdde togegrîpe hine *captio quam abscondit apprehendat eum*, Ps. Spl. T. 34, 9.

ge-griþian; *p.* ode, ede; *pp.* od, ed. I. *v. intrans. To make peace;* pácîficâre :—Ealle Eást-Centingas gegriþedan wið hî *all the East Kentians made peace with them*, Chr. 1009; Th. 261, 20, col. 2. II. *v. trans. To protect;* tuêri :—Syndon cyrcan wáce gegriþode *churches are weakly protected*, L. I. P. 25; Th. ii. 340, 11.

ge-grówan; *p.* -greów, *pl.* -greówon; *pp.* -grówen *To grow;* succrescere :—Ne gegréwþ hit ðǽr *it will not grow there*, Bt. 34, 10; Fox 148, 27. v. grówan.

ge-grunded *grounded, founded*.

ge-grundon *ground*. v. ge-grindan.

ge-grundweallian; *p.* ode; *pp.* od *To found;* fundâre :—He ofer sǽs gegrundweallode hine *ipse super mǎria fundâvit eum*, Ps. Spl. 23, 2.

ge-grynd, es; *n. A plot of ground* :—Aðelwold gesealde twâ gegrynd *Æthelwold gave two plots of ground*, Thorpe Chart. 231, 22.

ge-gryndan; *p.* de; *pp.* ed *To found*, Mt. Kmbl. Lind. 7, 25 [MS. gewrynded].

ge-gyddian; *p.* ode; *pp.* od *To sing;* cantâre :—Ic ðás word gegyddode *I sang these words*, Nicod. 27; Thw. 15, 40. v. giddian.

ge-gyfan *to bestow*. v. gifan.

ge-gyld, es; *n. A guild, society or club*. v. ge-gild.

ge-gyld; *adj. Golden, gilded;* deaurâtus :—On gyrlan gegyldum *in vestîtu deaurâto*, Ps. Spl. 44, 11. Gyldena, *vel* gegylde fatu *gilded vessels*, Ælfc. Gl. 67; Som. 69, 97; Wrt. Voc. 41, 48. v. gylden.

ge-gylda, an; *m. A member of a guild, club, or corporation, a companion, fellow*. v. ge-gilda.

ge-gyldan; *p.* -geald *To yield, pay, give, requite;* reddere, tribuere, retribuere :—Him God wolde æfter ðrowinga ðonc gegyldan *to him God would, after sufferings, requite favour*, Exon. 39 b; Th. 130, 23; Gú. 442. v. gildan.

ge-gyld-scipe, es; *m. A guild-ship, society;* sodalitas, L. Ath. v. § 8, 6; Th. i. 236, 35. v. gild-scipe.

ge-gyltan; *p.* -gylte; *pp.* -gylt *To become guilty, to offend, sin;* peccâre :—Ðeáh ðe he self gegyltan ne meahte *although he himself could not sin*, Past. 49; Swt. 385; 17; Hat. MS. Ðeáh ðe hwâ gegylte *though any one become guilty*, Ors. 1, 12; Bos. 36, 44.

ge-gymmod; *part. Gemmed, set with gems;* gemmâtus :—Gegymmod gemmâtus, Ælfc. Gr. 43; Som. 45, 16.

ge-gyrdan; *p.* -gyrde; *pp.* -gyrded, -gyrd *To gird;* præcingěre :—Eaxle gegyrde *girded shoulders*, Exon. 106 b; Th. 486, 14; Rä. 72, 14.

ge-gyrela, -gyryla *a garment*. v. gegerela.

ge-gyrian; *p.* ode; *pp.* od, wed; *v. a. To clothe, put on, adorn, endow;* vestire :—Ðú gegyrydist, Ps. Spl. C. 103, 2. Ðone lîchoman gegyredon *clothed the body*, Bd. 4, 30; S. 609, 21. Gegyrewod *endowed*, Bt. 14, 3; Fox 46, 14. v. gegerwian.

ge-gyrnan; *p.* de; *pp.* ed [gyrnan *to yearn*] *To desire, seek;* desîděrâre, pětěre :—Ic friþ wille æt Gode gegyrnan *I will desire peace from God*, Exon. 36 a; Th. 117, 24; Gú. 229. Ðonne ðæt gegyrnaþ ða ðe him Godes egsa hleónaþ ofer heáfdum *when they, over whose heads the fear of God impendeth, desire that*, 33 b; Th. 106, 18; Gú. 43.

ge-gyrnendlic; *adj. Desirable;* desiderabilis, Ps. Spl. T. 18, 11.

ge-gyrwan. v. gegerwan.

ge-habban; ðú -hæfst. -hafast, *pl.* -habbaþ; *p.* -hæfde; *pp.* -hæfed, -hæfd *To hold, be* [*ill*]; habere, tenere :—Gehafa gebyld on me *patientiam habe in me*, Mt. Bos. 18, 26: Exon. 105 a; Th. 398, 19; Rä. 17, 10. Ðara synna gê gihabbaþ *quorum peccata retinueritis*, Jn. Skt. Lind. 20, 23: Past. 51, 9; Swt. 401, 32; Hat. MS. Æfter ðisum wordum wearþ gemót gehæfd *after these words a meeting was held*, Homl. Th. ii. 148, 1. Ðær ðær wǽron gehæfde hâte baþu *where hot baths were kept*, i. 86, 21. Mîn cneów is yfele gehæfd *my knee is diseased*, 134, 33 : 150, 7.

ge-haccod *hacked, cut*. v. haccan.

ge-hâda, an; *m. One of the same state or order;* qui ejusdem stătus vel ordînis est :—Mid twám his gehâdan *with two of his fellow ecclesiastics*, L. Eth. ix. 19, 20; Th. i. 344, 14, 16: L. C. E. 5; Th. i. 362, 12, 15.

ge-hâdian; *p.* -hâdode; *pp.* -hâdod *To ordain, consecrate;* consecrare :—Hér Vitalianus se pápa gehâdode Theodorus to arcebiscop *in*

this year pope Vitalianus consecrated Theodore archbishop, Chr. 668 ; Erl. 35, 27 : 1070 ; Erl. 208, 2. Hēr Paulinus wæs gehádod Norþhymbrum to biscepe *in this year Paulinus was consecrated bishop of Northumbria*, 625 ; Erl. 22, 11. Mauricius hine gehádian hēt *Mauricius ordered that he should be ordained*, Homl. Th. ii. 122, 32 : Bd. 3, 7 ; S. 530, 30. v. hádian.

ge-hádod, -háded ; *def.* se ge-hádoda ; *part. In holy orders*; ordinátus :—Nū, gē habbaþ gehīred be gehádodum mannum *now ye have heard concerning men in orders*, L. Ælf. P. 41 ; Th. ii. 382, 16 ; Wilk. 169, 23. Se gehádoda *one in holy orders*; ordinātus, 42 ; Th. ii. 382, 23 ; Wilk. 169, 34. Be gehádedum mannum *concerning men in holy orders*; de ordinatis, Th. ii. 364, 7 ; Wilk. 161, 1. He ælces mannes gehádodes and læwedes yrfenuma beón wolde *he wanted to be the heir of every man, cleric and lay*, Chr. 1100 ; Erl. 236, 7.

ge-hæft ; *adj.* [-hæft ; *pp.* of ge-hæftan] *Bound, captive*; captus :—Ōþ ðære gehæftan wylne *to the captive slave*, Ex. 12, 29. Nyle he gehæfte nā forhycgan *vinctos suos non sprevit*, Ps. Th. 68, 34. Ða gehæftan *vinctos*, 67, 7. Gehæftum *captivis*, Lk. Bos. 4, 18.

ge-hæftan, he -hæft ; *p.* -hæftede, -hæfte ; *pp.* -hæfted, -hæft *To take, take captive, cast into prison, detain, bind* ; captare, captivare, vincire :—Swā hwæt swā hīg gehæftaþ, *quicquid ceperint*, Th. An. 23, 11. Hī gehæftaþ on sāwle rihtwīses *captabunt in animam justi*, Ps. 93, 21. Abraham geseah ānne ramm be ðǽm hornum gehæft *Abraham saw a ram caught* [*captus*] *by his horns*, Gen. 22, 13. On ēcnesse gehæft *for ever binds*, Bt. 19 ; Fox 70, 18. Mid ðӯ me God hafaþ gehæfted on ðam healse *with which God hath fastened me by the neck*, Cd. 19 ; Th. 24, 29 ; Gen. 385 : Judth. 10 ; Thw. 23, 11 ; Jud. 116. He hæfde ænne ðeófman gehæftne *habebat vinctum*, Mt. Bos. 27, 16. Handa synt gehæfte *my hands are manacled*, Cd. 19 ; Th. 24, 19 ; Gen. 380 : Exon. 16 a ; Th. 35, 22 ; Cri. 562. Hīg mycelum ege gehæfte wǽron *timore magno tenebantur*, Lk. Bos. 8, 37. Drihten hīg gehӯrde ðæt hīg gehæfton wiþ hine, Josh. 11, 20 [?].

ge-hæftednes, -ness, e ; *f. A captivity*; captīvitas :—Gecyr Drihten gehæftednesse ūre *converte Dŏmine captīvitātem nostram*, Ps. Lamb. 125, 4.

ge-hæftfæst ; *adj. Captive* ; captivus, Hpt. Gl. 434.

ge-hæftnan, -hæftnian ; *p.* ede, ade ; *pp.* ed, ad *To take, lay hold of, take captive* ; comprehendĕre, captivāre :—Ðū me gehæftnedest [gehæftnadest, Exon. 98 a ; Th. 368, 29] *thou didst take me captive*, Soul Kmbl. 63 ; Seel. 32. Sӯ ēhtende oððe ēhte feónd mīne sāwle and gehæftnige hī oððe gegrīpe hī *persĕquātur inĭmicus anĭmam meam et comprehendat*, Ps. Lamb. 7, 6. Ða ðe ǽr gehæftnede wǽron *who before were held captive*, Blickl. Homl. 87, 7 : 89, 29.

ge-hæftnys, -nyss, e ; *f. Captivity* ; captīvitas :—Ðonne awent oððe acyrreþ God gehæftnysse oððe hæftnōde folces his *cum convertit Deus captīvitātem plĕbis suæ*, Ps. Lamb. 52, 7. v. ge-hæftednes.

ge-hæft-world, e ; *f. A captive world* ; Ðeós gehæftworld, Blickl. Homl. 9, 4.

ge-hǽgan ; *pp.* -hǽged *To surround as with a hedge* :—Folc wæs gehǽged *the people was hemmed in*, Cd. 151 ; Th. 188, 17 ; Exod. 169. [Cf. *Icel.* hegna *to hedge, fence* (?) ; and see Grein, gehǽgan.]

ge-hǽge, es ; *n. Land hedged in, a paddock, garden* ; hortus, pratum, Mone B. 618 : Hpt. Gl. 419, 439.

ge-hǽlan ; *p.* -hǽlde ; *pp.* -hǽled *To heal, cure, save* ; sanare, salvare :—Untrume gehǽlan *to heal the sick*, Lk. Bos. 9, 2. He gehǽlde manega folc *he saved much people*, Gen. 50, 20. Ðæt gē him sǽra gehwylc hondum gehǽlde *that ye should heal with hands each of his sores*, Exon. 42 b ; Th. 144, 12 ; Gū. 677.

ge-hǽld *a keeping, regarding* ; observatio, Bd. 4, 23 ; S. 594, 16. v. ge-heald.

ge-hǽled ; *comp.* gehǽledra, gehǽldra, gehǽldre ; *adj. Safe, secure, good* ; tutus, Bd. 2, 2 ; S. 503, 39.

ge-hǽman ; *p.* de ; *pp.* ed *To lie with, cohabit, commit fornication* ; concumbĕre :—Gif he mid gehǽme *if he lie with her*, L. Alf. pol. 11 ; Th. i. 68, 16.

ge-hǽnan *to accuse, condemn*, Jn. Skt. Lind. 8, 6 ; 8, 10. v. gehénan.

ge-hǽnan ; *p.* de ; *pp.* ed *To stone* :—Ic gemētte ðær Archelaus gehǽnedne *I found there Archelaus stoned*, St. And. 44, 18. v. hǽnan.

ge-hǽp ; *adj. Fit* :—On stōwe gehǽppre *in loco apto*, Th. An. 21, 13.

ge-hǽre ; *adj. Hairy* :—Wǽron hie swā gehǽre swā wildeór *pilosus in modum ferarum*, Nar. 22, 5.

ge-hǽt ; *part. Made warm, heated* ; călĕfactus :—Ðæt sӯ gehǽt *let it be heated*, Herb. 23, 2 ; Lchdm. i. 120, 8.

ge-hǽtan *to promise* ; promittere, Bt. 20 ; Fox 70, 33. v. ge-hátan.

ge-hafa *have*, Mt. 18, 26 ; *imp.* of ge-habban.

ge-hafen *raised up, fermented*, Ælfc. Gl. 66 ; Wrt. Voc. 41, 15. v. ge-hebban.

ge-hagian ; *p.* ode ; *pp.* od ; *v impers. To please* :—Swā hwylc swā ðæt sió ðæt hine to ðan gehagige ðæt he ða ōðoro lond begeotan wille *whoever it be that is ready to take the other lands*, Kmbl. Cod. Dipl. ii. 120, 24, v. onhagian.

ge-hál ; *ad. Entire, whole, healthy* ; intĕger, sānus :—Gemētte he ðæt

---

fæt swā gehál, ðæt ðǽr nán cīnu on næs gesewen *he found the vessel so whole that no chink was visible in it*, Homl. Th. ii. 154, 22 : 166, 11 : Bt. 34, 12 ; Fox 152, 27. On gehálum þingum *in health*, Homl. Th. ii. 352, 22.

ge-haldan ; *pp.* -halden *To keep, preserve, hold* ; servāre, recondĕre, tēnēre :—On ðam heó wilnode gehaldan ða ārwurþan bān hire fæderan *in quo desīdērābat hŏnōranda patrui sui ossa recondĕre*, Bd. 3, 11 ; S. 535, 16. Mid ðӯ hine ðā nǽnig man ne gehaldan ne gebindan mihte *cum a nullo vel tēnēri vel ligāri pŏtuisset*, 3, 11 ; S. 536, 16. Ðǽr hī nū gehaldene syndon *in qua nunc servantur*, 3, 11 ; S. 535, 11 : 3, 6 ; S. 528, 29. v. ge-healdan.

ge-halding, e ; *f. A holding, keeping* ; custōdia :—On gehaldinge sprǽca ðīne *in custōdiendo sermōnes tuos*, Ps. Spl. C. 118, 9.

ge-hálgegend, es ; *m. One who hallows* ; dicator, Hymn. Surt. 64, 19.

ge-hálgian ; *p.* ode, ade ; *pp.* od, ad *To consecrate, dedicate, initiate, ordain, hallow, make holy, sanctify* ; consecrāre, dedĭcāre, sacrāre, initĭāre, ordĭnāre, sanctĭficāre :—Hēt se pāpa hine to bisceope gehálgian *the pope commanded to consecrate him bishop*, Bd. 3, 7 ; S. 529, 9 : 3, 24 ; S. 556, 19. Ðæt hīg woldon hīg sylfe gehálgian *ut sanctĭfĭcārent seipsos*, Jn. Bos. 11, 55. Siððan ðū gehálgast hira handa *postquam initĭāvēris mănus eōrum*, Ex. 29, 9, 35. Ðū gehálgast ðæt gehálgode anribb and ðone bōh *sanctĭfĭcābis et pectuscŭlum consecrātum et armum*, 29, 27, 36. He gehálgode wīn of wætere *he hallowed wine from water*, Andr. Kmbl. 1171 ; An. 586 : 3298 ; An. 1652. Wælhreów Criste gehálgode offrunge *tӯrannus Christo sacrāvit victĭmam*, Hymn. Surt. 52, 11. Gif preóst on treowenan calice hūsl gehálgige *if a priest consecrate housel in a wooden chalice*, L. N. P. L. 14 ; Th. ii. 292, 20. Ðis hūs ðē gehálgod ys *hæc dŏmus tibi dedĭcāta est*, Hymn. Surt. 141, 18 : L. Ælf. C. 25 ; Th. ii. 352, 13. Sӯ ðīn nama gehálgod *hallowed be thy name*, Homl. Th. ii. 596, 5 : Hy. 6, 3 ; Hy. Grn. ii. 286, 3 : 7, 18 ; Hy. Grn. ii. 287, 18. He wæs gehálgod fram Scottum *ordĭnātus a Scottis*, Bd. 3, 24 ; S. 557, 22. On gehálgodre cirican *in a consecrated church*, L. Edg. C. 30 ; Th. ii. 250, 19.

ge-hálgung, e ; *f. A consecration, sanctification, sanctuary* ; consecrātio, sanctĭfĭcātio, sanctuārium :—He ingelǽdde hine in munt gehálgunge hys *induxit eos in montem sanctĭfĭcātiōnis suæ*, Ps. Surt. 77, 54 : 131, 8. On gehálgunge hys *in sanctĭfĭcātiōne ejus*, Ps. Spl. C. T. 95, 6.

ge-hálsian ; *p.* ode ; *pp.* od *To adjure, exorcise* :—Ic gihálsige *adjuro*, Rtl. 113, 24. Gihálsad *adjuratus*, 120, 35. Gihálsia *exorcizare*, 119, 7. Ic gihǽlsiga *exorcizo*, 120, 21.

ge-hámettan ; *p.* te ; *pp.* ed *To appoint a home, domicile* ; dŏmum assignāre :—Ðæt hī hine to folcryhte gehámetten *that they domicile him to folk-right*, L. Ath. i. 2 ; Th. i. 200, 7.

ge-hámian ; *p.* ode, ade ; *pp.* od, ad *To make [one's self] familiar with* (?) :—Aldred hine gihámadi mið ðæm ðriim dǽlum *Aldred made himself familiar with the three parts*, Jn. Skt. 188, 7. [See p. ix. note 1.]

ge-hát, es ; *n. A promise, vow* ; promissum, votum :—Gemunde heofonweardes gehát *he remembered the promise of heaven's guardian*, Cd. 86 ; Th. 107, 28 ; Gen. 1796. Ðe ðam geháte getrūwode *he trusted to the promise*, 33 ; Th. 44, 9 ; Gen. 706. Ðæt ic mīn gehát hēr agylde *ut reddam vota mea*, Ps. Th. 60, 6. Gehát gehēt *votum vovit*, 131, 2 : Bd. 3, 27 ; S. 559, 8. [O. H. Ger. ki-heiz. v. Grm. R. A. p. 893.] DER. ge-hátan.

ge-hata *a hater, an enemy* ; inimicus, Cot. 74.

ge-hátan, -hǽtan, he -hát, -hǽt ; *p.* -hēt, *pl.* -hēton, -hēht, *pl.* -hēhton ; *pp.* -háten. *I. to call, name* ; vocare, nominare :—Swā ðū gehâten eart *as thou art called*, Exon. 8 b ; Th. 4, 26 ; Cri. 58. Crist wæs on ðӯ eahteoþan dæg Hǽlend gehâten *Christ was on the eighth day named Jesus*, Menol. Fox 7 ; Men. 4. Is gehâten Saturnus *is called Saturn*, Bt. Met. Fox 28, 48 ; Met. 28, 24. Ðæt land ðe ys gehâten Euilaþ *omnem terram Hevilath*, Gen. 2, 11 : Jud. 4, 2, 6. **II.** *to call, command, promise, vow, threaten* ; vocare, accessere, jubere, spondere, promittere, vovere :—Fōre waldende gǽþ bī noman gehâtene *they shall go before the Lord, called for by name*, Exon. 23 b ; Th. 66, 16 ; Cri. 1072. Him ðæt eall gehǽt his *his security commands all that to him*, Bt. Met. Fox 25, 104 ; Met. 25, 52. Him sibbe gehâteþ *he shall promise peace to them*, Exon. 27 b ; Th. 82, 16 ; Cri. 1339. Ic ðē gehâte *I vow to thee*, Cd. 98 ; Th. 129, 5 ; Gen. 2139. Gehâtaþ Drihtne *vovete Domino*, Ps. Th. 75, 8. Ðeáh ðe gē me deáþ gehâten *though ye have threatened death to me*, Exon. 36 a ; Th. 116, 23 ; Gū. 211 : 40 b ; Th. 135, 7 ; Gū. 520. v. hátan.

ge-haðerian ; *p.* ode ; *pp.* od *To restrain* ; cohĭbēre :—Wambe sǽr gehaðeraþ *it restraineth sore of stomach*, Med. ex Quadr. 2, 2 ; Lchdm. i. 334, 8. Ðā ðæt ðā geseah se ðe hine gebohte, ðæt he mid bendum ne mihte gehaðerod beón *cumque vidisset qui emērat, vincŭlis eum non pŏtuisse cohĭbēri*, Bd. 4, 22 ; S. 592, 9. Ic am gehaðrad *coarctor*, Lk. Skt. Lind. 12, 50. v. ge-heaðerian.

ge-háþyrt ; *adj. Irritated, angry* :—Ðā wearþ se hálga wer geháþyrt *the holy man was irritated*, Homl. Th. ii. 176, 18.

ge-háþyrtan ; *p.* te ; *v. reflex. To become angry* :—Se Godes wiðersaca hine ðā geháþyrte *the adversary of God then became angry*, Homl. Th. i. 450, 9.

ge-hátian; *p.* ode, ude; *pp.* od, ud *To become* or *be hot;* concălescĕre :—Gehátude heorte mín on in me *concăluit cor meum intra me*, Ps. Spl. 38. 4.

ge-hát-land, es; *n. Land of promise* :—Be inngonge ðæs gehátlondes *about the entrance of the promised land*, Bd. 4, 24; S. 598, 12.

ge-háwian; *p.* ode, ade; *pp.* od, ad *To look at, view, observe, examine, survey, inspect;* intuēri, aspicĕre, circumspicĕre :—Se cing geháwode [geháwade, col. 1] hwær man mihte ða eá forwyrcean *the king observed where the river might be obstructed,* Chr. 896; Th. 172, 35, col. 2; 173, 35 : Shrn. 178, 7 : 179, 21.

ge-heáf; *adj.* [heáh *high*] *Lifted up, exalted;* exaltātus :—Wæs Bryten swýðe geheáf *Britain was very much exalted*, Bd. 1, 6; S. 476, 27, MS. B. [*A. R.* i-heied.]

ge-heald, -hæld, es; *m.* [?] *n.* [?] I. *a holding, keeping, guard, observing;* observantia :—He sende him stafas and gewrit be gehealde rihtra Eástrana *he sent him a letter and epistle about the holding of right Easters,* Bd. 5, 21; S. 643, 8. Habbaþ gē gehæld *habetis custodiam*, Mt. Kmbl. Lind. 27, 65 : Rtl. 123, 31 : Shrn. 36, 30. II. *a keeper, guardian, protection;* custos, tūtēla :—Willem eorl sceolde beón [MS. ben] his geheald *earl William was to be his guardian,* Chr. 1070; Th. 347, 7. Ælfgár eorl gesóhte Griffines geheald on Norþwealan *earl Ælfgar sought Griffith's protection in North Wales,* 1055; Th. 325, 20. He beó ðærto geheald and mund under me *let him be thereto guardian and patron under me,* Thorpe Chart. 391, 17. v. ge-hyld.

ge-heald; *adj.* v. ge-hyldra.

ge-healdan, -haldan, to -healdenne; ic -healde, ðú -healdest, -hiltst, he -healdeþ, -healt, -helt, -hylt, *pl.* -healdaþ; *p.* -heóld, -hióld, ðú -heólde, *pl.* -heóldon, -hióldon; *impert.* -heald, *pl.* -healdaþ; *subj. pres.* -healde, *pl.* -healden; *p.* -heólde, *pl.* -heólden; *pp.* -healden. I. *to keep, hold, observe, keep in, retain, reserve, preserve, save, defend, protect;* custodire, servāre, observāre, continēre, reservāre, ſalvāre, defendĕre :—Ðæt ic díne word mihte wel gehealdan *ut custodiam verbum tuum,* Ps. Th. 118, 101: Andr. Kmbl. 426; An. 213. Se ðe him God syleþ gumena ríce to gehealdenne *to whom God gives an empire over men to hold,* Scóp Th. 269; Wíd. 134. Ic gehealde wegas míne *custōdiam vias meas,* Ps. Lamb. 38, 2. Gif ðú híg gehiltst *si custōdiĕris ea,* Deut. 7, 12 : Ex. 34, 6. Drihten gehealdeþ dóme ða lytlan *custōdiens parvŭlos Dŏmĭnus,* Ps. Th. 114, 6. Se stranga gewǽpnod his cǽfertún geheald *fortis armātus custōdit atrium suum,* Lk. Bos. 11, 21 : Ps. Lamb. 120, 5. God hine gehelt æghwonan *God preserves him everywhere,* Bt. 12; Fox 36, 37. Drihten gehylt ðē fram ælcum yfele *Dŏmĭnus custōdit te ab omni mălo,* Ps. Lamb. 120, 7. Ic ðē fordig geheóld *ídeo custōdivi te,* Gen. 20, 6. Ðú eágan míne wið teárum geheólde *thou hast kept mine eyes from tears,* Ps. Th. 114, 8. Hí ðæt word geheóldon betwux *verbum contĭnuērunt ăpud se,* Mk. Bos. 9, 10. Hie sibbe innan bordes gehióldon *they preserved peace at home,* Past. pref; Swt. 3, 7; Hat. MS. Geheald ðú, mín folc, míne fǽte ǽ *attendĭte, pŏpŭle meus, lēgem meam,* Ps. Th. 77, 1. Ðec ǽ wið firenum geheald *preserve thyself ever from sins,* Exon. 81 a; Th. 305, 27; Fä. 94. Fæder alwalda mid ārstafum eówic gehealde *may the all-ruling Father hold you with honour,* Beo. Th. 640; B. 317. Ðæt he cóme and ða burh geheólde *that he would come and defend the city,* Jos. 10, 6. Ðæt sǽd sí gehealden ofer ealre eorþan brádnisse *ut salvētur sēmen super făciem ūnivērsæ terræ,* Gen. 7, 3 : Jos. 2, 13 : Mt. Bos. 9, 17. Gehealdne, *pp. pl.* Exon. 23 b; Th. 65, 26; Cri. 1060. Mid gehealdan *to satisfy,* Bt. 13; Fox 38, 34. Wel gehealden *well contented, satisfied,* Bt. 18, 3; Fox 64, 27: Basil admn. 9; Norm. 52, 22. II. *to hold, occupy, possess;* tĕnēre, possidēre :—On eówrum geþylde gē gehealdaþ eówre sáwla *in pătientia vestra possĭdēbĭtis ănĭmas vestras,* Lk. Bos. 21, 19. He frætwe geheóld fela missera *he held the armour many years,* Beo. Th. 5253; B. 2620.

ge-heald-dagas; *pl. m. Kalends* :—Gehealddagas vel hálige dagas *kalendæ,* Ælfc. Gl. 96; Som. 76, 26; Wrt. Voc. 53, 35.

ge-healden; *part. p. Satisfied* :—Beó gehealden on ðínum gecynde ðonne hæfst ðú genóh *be satisfied in thy kind, then hast thou enough,* Kmbl. Sal. 264, 21. v. gehealdan.

ge-healdnys, -nyss, e; *f. A keeping;* custōdia :—On gehealdnysse ðara *in custōdiendis illis,* Ps. Lamb. 18, 12.

ge-healdsum; *adj. Keeping, sparing, frugal;* parcus :—Ðæt he síe gehealdsum on ðæm ðe he healdan scyle oðða dǽlan *that he is frugal in what he ought to keep or give away,* Past. 20, 2; Swt. 149, 18; Hat. MS. 29 b, 9.

ge-healdsumnys, -nyss, e; *f. A keeping, observance, preservation, abstinence;* custōdia, observātio, conservātio, abstĭnentia :—We rǽdaþ on bócum, ðæt ðeós gehealdsumnys wurde árǽred on ðone tíman ðe gelamp on ánre byrig ðe Uigenna ís gecweden micel eorþstyrung *we read in books, that this observance was established at the time when a great earthquake happened in a city which is called Vienna,* Homl. Th. i. 244, 15. Ðæt he wǽre on gehealdsumnysse ðæs bebodes his Scyppende underþeód *that he was subject to his Creator in the keeping of the commandment,* Boutr. Scrd. 17, 29. For gehealdsumnysse sóþre eádmódnysse beóþ forwel oft Godes gecorenan geswencte *for preservation of true humility God's chosen are very often afflicted,* Homl. Th. i. 474, 10. Mid ðære gehealdsumnysse *with abstinence,* i. 318, 8.

ge-heálgian; *p.* ode; *pp.* od *To consecrate, hallow;* consecrāre, sacrāre :—Theodór bisceop on Hrófes ceastre Quichelm to bisceope geheálgode *Theodōrus in civĭtāte Hrofi Cuichelmum consecrāvit episcŏpum,* Bd. 4, 13; S. 581, 8. Ðǽr se bisceop towearp and fordyde ða wigbed ðe he sylf ǽr geheálgode *ubi pontĭfex polluit et destruxit eas quas ipse sacrāvĕrat āras,* 2, 13; S. 517, 18. v. ge-hálgian.

ge-healt *keeps, guards, protects,* Lk. Bos. 11, 21: Ps. Lamb. 120, 5; *3rd sing. pres.* of ge-healdan.

ge-healtsumnys *captivity.*

ge-heáne *servire,* Rtl. 42, 40. v. gehýnan.

ge-heápod; *part. Heaped* or *piled up;* coacervātus :—Gód gemet, and full, and geheápod, and oferflówende híg syllaþ on eówerne bearm *mensūram bŏnam, et confertam, et coagĭtātam, et sŭpereffluentem dăbunt in sĭnum vestrum,* Lk. Bos. 6, 38 : Blickl. Homl. 175, 17. v. ge-hýpan.

ge-heáðorian, -heaðerian, -heaðrian; *p.* ode, ade; *pp.* od, ad *To restrain, control, compress;* cohĭbēre, coartāre, coangustāre :—Hafaþ geheaðorad heofona Wealdend ealle gesceafta *the Ruler of the heavens has controlled all creatures,* Bt. 13, 11; Met. 13, 6: Bt. 21; Fox 74, 9: 25; Fox 88, 5. Ðæt se secg wǽre hergum geheaðerod *that the man should be restrained with harryings,* Beo. Th. 6136; B. 3072. He eft semninga swíge gewyrþeþ, in nédcleofan nearwe geheaðrod it [*the wind*] *again suddenly becomes silent, narrowly compressed in its close bed,* Elen. Kmbl. 2550; El. 1276.

ge-heáw, es; *n. A striking together, a gnashing, grinding;* concussio, stridor :—Tóþa geheáw *a gnashing of teeth,* Cd. 221; Th. 285, 18; Sat. 339.

ge-heáwan; *p.* -heów; *pp.* -heáwen *To hew, cut, cut in pieces;* dolare, cædere, concidere :—Wicg hornum geheáweþ *heweth the war-horse with his horns,* Salm. Kmbl. 313; Sal. 156: Beo. Th. 1368; B. 682: Judth. 10; Thw. 22, 33; Jud. 90: 12; Thw. 25, 36; Jud. 295: Bd. 4. 19; S. 588, 27. Ðæt wæs geheáwen of carre *quod erat excisum de petra,* Mk. Skt. Lind. 15, 46. DER. heáwan.

ge-hebban; *p.* -hóf; *pp.* -hafen *To heave up, raise up, ferment;* elevare, fermentare :—Gehafen hláf *fermentatus panis,* Ælfc. Gl. 66; Wrt. Voc. 41, 15. Gehebbes ða ilco *levabit eam,* Mt. Kmbl. Lind. 12, 11. Gehefen biþ *exaltabitur,* Lk. Skt. Lind. 14, 11. v. hebban.

ge-hédan; *p.* de; *pp.* ed. I. *to hide, conceal;* condĕre, abscondĕre :—Is ðæt fýr on stánum gehéded *fire is hidden in stones,* Bt. Met. Fox 20, 302; Met. 20, 151. II. *to acquire, obtain, seize;* obtinēre, deprehendĕre :—Ær he gehéde ðæt he ǽr æfter spyrede *until he seizes that which he before sought after,* Bt. Met. Fox 27, 29; Met. 27, 15. Forðonðe he ne úðe ðæt ǽnig óðer man ǽfre mǽrþa má gehédde under heofenum ðonne he sylfa *because he would not grant that any other man had ever obtained more glories under heaven than himself,* Beo. Th. 1014; B. 505. v. ge-hýdan.

ge-héed; *adj.* [= ge-heáf] *Exalted;* exaltātus :—Wæs Bryten gehéed *Britain was exalted,* Bd. 1, 6; S. 476, 27.

ge-hefigian, -hefegian, -hefgian; *p.* ode; *pp.* od, ad; *v. trans. To make heavy* or *sad, to load, burden, weigh down, increase the weight of, aggravate;* gravare, contristare, vexare, deprimere, aggravare :—He handa gehefegaþ *he makes the hands heavy,* Salm. Kmbl. 319; Sal. 159. Ðonne biþ gehefgad haswig-feðra, gomol, geárum fród *then the variegated-feathered* [*phœnix*] *becomes sad, old, advanced in years,* Exon. 58 a; Th. 208, 9; Ph. 153. Ðé-læs eówer heortan gehefegode sýn on oferfylle *ne forte graventur corda vestra in crapula,* Lk. Bos. 21, 34. Swá swá hefig byrden mín unriht synt gehefegode ofer me *sicut onus grave iniquitates meæ gravatæ sunt super me,* Ps. Th. 37, 4. Wæs mid swá mycelre untrumnesse his líchoman gehefigad *tanta erat corporis infirmitate depressus,* Bd. 4, 23; S. 594, 26: Lk. Bos. 9, 32 : Num. 11, 17. Heora synn ys swíðe gehefegod *peccatum eorum aggravatum est,* Gen. 18, 20.

ge-hégan; *p.* -hégde, -héde *To do, perform, effect, hold* :—Ðing gehégan *to have a meeting,* Beo. Th. 855; B. 425: Andr. Kmbl. 1859; An. 932: Exon. 89 a; Th. 334, 19; Gn. Ex. 18. Seonoþ gehégan *to hold a synod,* Exon. 89 b; Th. 231, 23; Ph. 493: 116 a; Th. 445, 17; Dóm. 9. Hie ðing gehégdon *they held a meeting,* Andr. Kmbl. 314; An. 157: 2100; An. 1051: 2991; An. 1498. [*See* heyja *in Cl. and Vig. Icel. Dict; Grimm writes* gehegan = *sepire,* And. u. El. 101.]

ge-helan; he -hele, -hileþ; *p.* -hæl, *pl.* -hǽlon; *pp.* -holen *To conceal, hide, cover up;* cēlāre, occŭlere, tĕgĕre :—Se ðe dearnenga bearn gestriéneþ and gehileþ [gehelep MSS. B. H.] *he who secretly begets a child and conceals it,* L. In. 27; Th. i. 120, 2. Ic ðé háte ðæt ðú hí gehele and gehealde, oþ-ðæt ic wite hwæt God wylle *te silentio tĕgĕre vŏlo, dōnec sciam quid vĕlit Deus,* Bd. 5, 19; S. 640, 37. Woldon hí and wéndon ðæt hí ðǽr mihton dígle and geholene beón fram andsýne ðæs unholdan cyninges *occŭlendos se a făci rēgis victōris crēdĭdissent,* 4, 16; S. 584, 25.

ge-hélan; *p.* de; *pp.* ed *To heal, save;* sānāre, salvum făcĕre :—Gehél me of eallum æhtendum *salvum me fac ex omnĭbus persēquentĭbus,* Ps. Lamb. 7, 2. v. ge-hǽlan.

ge-helian; _p._ ede; _pp._ ed _To conceal, hide, cover over_; cēlāre, claudēre:—Se pitt wæs geheled mid ānum stāne _os ejus grandi lăpĭde claudēbātur_, Gen. 29, 2.

ge-helmian; _p._ ode, ede; _pp._ od, ed _To cover with a helmet, crown_; gāleāre, cŏrōnāre:—Ðū gehelmodest us _cŏrōnasti nos_, Ps. Spl. 5, 15. Of wuldre and weorþmynt ðū gehelmedest hine _de glōria et hŏnōre cŏrōnasti eum_, Ps. Spl. T. 8, 6. Gehelmod _gāleātus_, Ælfc. Gr. 43; Som. 45, 11. [_Laym._ i-helmed: _O. H. Ger._ gehelmot.]

ge-helpan; _p._ -healp, -heolp, _pl._ -hulpon; _pp._ -holpen; _gen. dat. To assist, preserve, to be sufficient_; adjuvare, subvenire, suppetere. **I.** _cum gen._:—Ðonne hie māgon ðīn gehelpan _when they can help thee_, Bt. 14, 1; Fox 42, 10. Ðū gehelpest ðysses menniscan cynnes _thou shalt help this human race_, Blickl. Homl. 9, 8. Ðū mīn hæfst geholpen _thou hast assisted me_, Bt. 41, 4; Fox 250, 18. **II.** _cum dat_:—Him ðā Ioseph gehealp _then Joseph helped them_, Ors. 1, 5; Bos. 28, 6. Ðæt wīf, ðe eówrum life gehealp _the woman who preserved your life_, Jos. 6, 22. He wolde gehelpan ðearfum _he wished to help needy people_, Swt. A. S. Rdr. 102, 226. v. helpan.

ge-helt _preserves_, Bt. 12; Fox 36, 37; _3rd sing. pres._ of ge-healdan.

ge-hēn; _adj. Fallen, low_:—Ða gehēno _kadŭca_, Rtl. 189, 31. v. heán.

ge-hēnan; _p._ de; _pp._ ed _To humble, accuse, condemn, despise_; humiliare, accusare, condemnare, spernere:—Gehēned ic eom _humiliatus sum_, Ps. Vossii, 37, 8. Hine gehēnan [MS. gehena] _illum accusare_, Lk. Skt. Lind. 23, 2. He gehēned wæs _he was condemned_, Cd. 217; Th. 276, 18; Sat. 190. Gehēneþ mec _spernit me_, Lk. Skt. Lind. 10, 16. v. hēnan.

ge-hendan; _p._ de; _pp._ ed _To hold_; tĕnēre:—Me ðīn seó swīðre ðǽr gehendeþ _tenēbit me dextĕra tua_, Ps. Th. 138, 8.

ge-hende; _adj. Neighbouring, next_; vicinus:—On gehende tūnas _in proximos vicos_, Mk. Bos. 1, 38: 6, 36. Ðā fērdon hī to gehendre byrig _then they went to a neighbouring city_, Homl. Th. i. 456, 5. Ðæt hȳ ðǽr gehendaste wǽron on gehwylc land ðanon to winnanne _that they there should be most handy for waging war thence on every land_, Ors. 3, 7; Bos. 61, 5.

ge-hende; _adv. Near, at hand_; prope:—Sumor is gehende _æstas est prope_, Lk. Bos. 21, 30. Godes rīce is gehende _Dei regnum est prope_, 21, 31: Gen. 19, 20: Exod. 2, 12: Deut. 31, 14. Hī wǽron swā gehende ðet ǽgðer heora on ōðer hāwede _they were so near that each of them looked on the other_, Chr. 1003; Erl. 139, 8. Ða mynstra gehendor ðam wæterscipe timbrian _to build the monasteries nearer to the water_, Homl. Th. ii. 160, 32: i. 106, 19.

ge-hende; _prep. dat. Nigh, near_: juxta:—Me gehende _juxta me_, Gen. 45, 10: 12, 11. He wæs gehende ðam scype _he was near the ship_, Jn. Bos. 6, 19. He læg ðeódne gehende _he lay by his prince_, Byrht. Th. 140, 27; By. 294: Ælfc. Gr. 47. Som. 47, 34.

ge-hendnys, -nyss, e; _f. Nearness, proximity, vicinity_; proxīmĭtas, vicīnĭtas:—Gehendnys _vicīnĭtas_, Glos. Prudent. Recd. 139, 47. Ða geswuteliaþ gehendnysse _they express vicinity_, Ælfc. Gr. 5; Som. 4, 50. On gehendnysse his mynstres _in the neighbourhood of his monastery_, Homl. Th. ii. 174, 5.

ge-hentan; _p._ te; _pp._ ed _To take, seize_; căpĕre, prehendēre:—Hió abīt hæleða gehwilcne ðe hió gehentan mæg _she devours every man whom she can seize_, Bt. Met. Fox 13, 64; Met. 13, 32. Eall ðæt hie gehentan mehton _all that they could seize_, Chron. 905; Erl. 98, 17.

ge-heofegian; _p._ ode, ede; _pp._ od, ed; _v. trans. To make heavy, load, weigh down_; gravare, Mt. Kmbl. Hat. 26, 43. v. ge-hefigian.

ge-heold, es; _m? A keeping, observing_; custōdia, observātio:—Hī sōþfæstnysse and ārfæstnesse and clænnesse, and ōðra gāstlīcra mægena geheold, and swyðost sibbe and Godes lufan geornlīce lǽrde _justītiæ, pietūtis et castīmōniæ, cætērārumque virtūtum, sed maxīme pācis et cārĭtātis custōdiam dŏcuit_, Bd. 4, 23; S. 593, 40. On geheoldum [MS. geheoldan] unrihta Eástrena _in the keeping of unright Easters_, 5, 24; S. 646, 39. v. geh=ald.

ge-heóld, ðū -heólde; _pl._ -heóldon _kept, observed_, Gen. 20, 6: Ps. Th. 114, 8: Andr. Kmbl. 691; An. 346; _p._ of ge-healdan: ge-heólde, _pl._ -heólden _would save_, Jos. 10, 6; _p. subj._ of ge-healdan.

ge-heolp _preserved_, Jos. 6, 22; _p._ of ge-helpan.

ge-heóran; _p._ de; _pp._ ed _To hear_; audire:—Geheór nū _hear now_, Bt. 35, 5; Fox 116, 21. Ne geheórþ _hears not_, Bt. 18, 2; Fox 64, 3. Ne geheórdon _heard not_, 18, 2; Fox 64, 12. v. gehȳran, hȳran.

ge-heordnes, -ness, -nys, -nyss, e; _f. A keeping, guard, watch_; custōdia:—On geheordnesse ðara edieán manige [is] _in custōdiendis illis retrĭbūtio multa_ [_est_], Ps. Spl. T. 18, 12. Gesete Driht geheordnysse mūþes mīnes _pōne Dŏmĭne custōdiam ōri meo_, Ps. Spl. 140, 3. v. ge-hyrdnes.

ge-heordung, e; _f. A keeping, guard, watch_; custōdia:—Ic sette mūþe mīnum geheordunga _pŏsui ōri meo custōdiam_, Ps. Spl. T. 38, 2.

ge-heort; _comp._ ra; _adj. Hearty, animated, courageous_; anĭmæquus:—On geheortum hyge _in a courageous soul_, Exon. 81 a; Th. 305, 14; Fä. 86. Beó geheortra _anĭmæquior esto_, Mk. Bos. 10, 49.

ge-heowian; _p._ ode, ade; _pp._ od, ad _To form_; formāre:—Dracan ðū ðysne geheowadest _drāco iste, quem formasti_, Ps. Th. 103, 25: Blickl. Homl. 87, 32: 31, 16. v. ge-hiwian.

ge-heowung. v. gehiwung.

geher _an ear of corn_, Mk. Skt. Rush. 4, 28. v. ear.

ge-hēran; _p._ de; _pp._ ed _To hear_; audīre:—Ic ne sceal ǽfre gehēran ðære byrhtestan bēman stefne _I shall never hear the brightest trumpet's sound_, Cd. 216; Th. 275, 14; Sat. 171: 220; Th. 284, 27; Sat. 328. Ic gehēre helle scealcas grundas mǽnan _I hear hell's ministers bemoaning the gulfs_, 216; Th. 273, 7; Sat. 133. We gehērdon wuldres swēg _we heard the sound of glory_, 218; Th. 279, 13; Sat. 237. Gehēr ān spell _hear a discourse_, Bt. 37, 1; Fox 186, 1: 35, 5; Fox 166, 21, note 24. Ðā sió stefn gewearþ gehēred of heofenum _then the voice was heard out of heaven_, Andr. Kmbl. 335; An. 168. v. ge-hȳran.

ge-hercnian; _p._ ode; _pp._ od _To hear_:—Gehercnadon _audientes_, Mt. Kmbl. Lind. 22, 22.

ge-hergian; _p._ ode, ade; _pp._ od, ad _To ravage, plunder, afflict, harrow, take captive_; vastāre, spŏliāre, afflīgĕre, captīvum dūcĕre:—He on ðam fyrste helle gehergode _he harrowed hell in that space of time_, Homl. Th. ii. 608, 1. Ðe hie gehergod hæfdon _which they had plundered_, Chr. 895; Erl. 93, 19. Gehergad _ravaged_, Ors. 3, 11; Bos. 72, 22. Ðæt ūre wīf and ūre cild wurdon gehergode _ut uxōres ac libĕri nostri dūcantur captīvi_, Num. 14, 3: Jud. 10, 8: Gen. 31, 26: Shrn. 96, 12.

ge-hērian [_or_ -herian; _cf. Goth._ hazjan]; _p._ ode, ede; _pp._ od, ed [hērian _to praise_] _To praise, honour, glorify_; laudāre, hŏnōrāre, celebrāre:—Unlǽde biþ se ne can Crist gehērian _wretched is he who cannot honour Christ_, Salm. Kmbl. 48; Sal. 24. On Gode byþ gehērod mīn sāwl _in Dŏmĭno laudābĭtur anima mea_, Ps. Th. 33, 2. Ðeáh he seó ānum gehēred _though it be praised in one_, Bt. 30, 1; Fox 108, 14; Blickl. Homl. 71, 16. On Gode we beóþ gehērode _in Dŏmĭno laudābĭmur_, Ps. Lamb. 43, 9. He wæs gehiered _he was praised_, Blickl. Homl. 165, 1.

ge-hēring, e; _f. A hearing, hearsay, tidings_; audītio:—Fram gehēringe yfelre he ne ondrǽt _ab audītiōne măla non tĭmēbit_, Ps. Lamb. 111, 7.

gehēr-nes, -ness, e; _f. Hearing_; auditus:—In gehērnesse _audiendo_, Bd. 4, 24; S. 598, 6. Dryhten ic gehērde gehērnisse [gehírnesse, Ps. Trin. Camb. fol. 244, 7] ðīne _Dŏmĭne audīvi auditum tuum vocem tuam_, Cant. Abac. Surt. 189, 2: Jn. Skt. Rush. 12, 38. v. ge-hȳrnes.

ge-hēt _promised_, v. ge-hātan.

Gehhol, Gehhel, es; _n. Yule, Christmas_, L. Alf. pol. 5; Th. i. 64, 23: 43; Th. i. 92, 3. v. geól.

ge-hicgan, -hicggan, -hicgean, -higgan _to study, search out._ v. ge-hycgan.

ge-hídan; _p._ de; _pp._ ed _To hide, conceal_; condēre, abscondēre:—Ðe ic hafa on stānfate gehided _which I have hidden in a stone chest_, Wald. 63; Vald. 2, 3. v. ge-hȳdan.

ge-hiénan _to humble._ v. ge-hȳnan.

ge-hiéran. v. ge-hȳran.

ge-hierstan _to fry._ v. ge-hyrstan.

ge-hiérsum; _adj. Obedient_; obēdiens:—Hie him alle gehiérsume dydon _they made all obedient to him_, Chr. 853; Erl. 68, 11. v. ge-hȳrsum.

ge-hiérsumian _to make obedient_, Chr. 853; Th. 122, 22, col. 1. v. ge-hȳrsumian.

ge-higd, e; _f_: es; _n. Thought, meditation_; cōgĭtātio:—Sende mihtig God his milde gehigd _misit Deus misĕricordiam suam_, Ps. Th. 56, 4. Heortan gehigdum _in the heart's thoughts_, Elen. Kmbl. 2445; El. 1224. v. ge-hygd.

ge-hihtan, -hyhtan; _p._ -hihte; _pp._ -hihted. **I.** _to hope, trust_; spērāre:—Betere is gehihtan on Drihtne ðonne gehihtan on ealdrum _bŏnum est spērāre in Dŏmĭno quam spērāre in princĭpĭbus_, Ps. Lamb. 117, 9. On hys naman ðeóda gehyhtaþ _in nōmĭne ejus gentes spērābunt_, Mt. Bos. 12, 21. **II.** _to rejoice_; exultāre:—Muntas gehihtaþ swā swā rammas _montes exultasti sīcut arietes_, Ps. Spl. 113, 6.

ge-hild, es; _n. A secret place_:—On gehildum _in abditis_, Ps. Spl. T. 16, 13.

ge-hilep _conceals_, L. In. 27; Th. i. 120, 2; _3rd sing. pres._ of ge-helan.

ge-hilt, es; _n. A hilt, handle_; căpŭlus:—He gegrāp sweord be gehiltum _he seized the sword by the hilt_, Cd. 140; Th. 176, 1; Gen. 2905. [_O. H. Ger._ gehilze.]

ge-hiltst _keepest_, Ex. 34, 6; _2nd sing. pres._ of ge-healdan.

ge-hínan _to oppress_, Ex. 5, 9: L. Alf. 35; Th. i. 52, 23, note 64. v. ge-hȳnan.

ge-hindred, -hindrad, -hyndred; _part. Hindered_; impĕdītus:—Biþ eall se here swīðe gehindred [gehindrad, 252, 33, col. 1; gehyndred, col. 2] _all the army will be greatly hindered_, Chr. 1003; Th. 253, 32.

ge-hióld, _pl._ -hióldon _kept, preserved_, Past. pref; Swt. 3, 7; Hat. MS; _p._ of ge-healdan.

ge-hióran; _p._ de; _pp._ ed _To hear_; audīre:—Ða [MS. ðe] eáran ongitaþ ðæt hī gehióraþ _the ears perceive that which they hear_, Bt. 41, 4; Fox 252, 8. v. ge-hȳran.

ge-hiowian; *p.* ade; *pp.* ad *To form, fashion;* formāre :—Đū ge-hiowades mec *formasti me,* Ps. Surt. 138, 5 : 103, 26. v. ge-hiwian.

ge-híran; *p.* -hírde; *pp.* -híred *To hear;* exaudire :—Gehír, God! mín gebed *exaudi, Deus! ortiōnem meam.* Đys is gebed, and nā hǽs this is a prayer, and not a command, Ælfc. Gr. 33; Som. 37, 52. v. ge-hýran, hýran.

ge-hírness, e; *f. Hearing;* auditus :—Ic gehíre gehírnesse dīne *audivi auditum tuum* [*vocem tuam*], Ps. Trin. Camb. fol. 244, 7. v. ge-hírnes.

ge-hírsumnes, se; *f. Obedience* :—For his gehírsumnisse đe he hæfde to Gode *for his obedience to God,* Swt. A. S. Rdr. 62, 181.

ge-hiscan *to hate;* abominari :—Đæne wer gehiscþ drihten *virum abominabitur dominus,* Ps. Lamb. 5, 8.

ge-hiwad; *p. part. Coloured;* purpuratus, Lk. Skt. p. 9, 2. [*A. R.* i-heouwed.]

ge-hiwian, -hywian, -heowian, -hiowian; *p.* ode, ade, ede; *pp.* od, ad, ed. **I.** *to form, fashion, make, transform, transfigure;* formāre, plasmāre, fingĕre, figūrāre, transfigūrāre :—Đū đe gehiwast sārnesse on bebode *qui fingis lābōrem in præcepto,* Ps. Lamb. 93, 20. Sió godcunde fōreteohhung eall þing gehiwaþ *the divine predestination fashions everything,* Bt. 39, 6; Fox 220, 17. Đū gehiwadest me *formasti me,* Ps. Th. 138, 3. Handa ne đíne geworhton and gehiwedan *mănus tuæ fēcērunt me et plasmāvērunt me,* 118, 73. He wæs gehiwod befōran him transfigūrātus est ante eos, Mt. Bos. 17, 2. Seó heáfodstōw gescrepelíce gehiwad ætýwde to đam gemete hyre heáfdes *lŏcus cāpĭtis ad mensūram cāpĭtis illíus aptissĭme figūrātus appāruit,* Bd. 4, 19; S. 590, 2. **II.** *to seem, appear, pretend;* simŭlāre :—Đeáh đe he hit swā gehiwige *though he may so pretend,* Homl. Th. i. 6, 18. Seó gehiwode anlícnys getidode đām toslitenum mannum hwílendlíc líf *the apparent likeness imparted to the torn men transitory life,* ii. 240, 17. Gehiwed *dissimŭlatus,* Hpt. Gl. 517. Ne lufa đū đīnne brođor mid gehiwodre heortan *do not love thy brother with a dissembling heart,* Basil admn. 5; Norm. 46, 4.

ge-híwian, -hiewian; *p.* ode; *pp.* od *To marry* :—Forđæm hit is awriten đæt hit síe betere đæt mon gehiewige đonne he birne, forđæm bútan synne he mæg gehíwian *for it is written that it is better to marry than to burn, because a man may marry without sin,* Past. 51, 9; Swt. 401, 33; Hat. MS.

ge-hiwung, -hywung, -heowung, e; *f. A form, fashion, shape, position, predicament;* figmentum, cătēgŏria :—He oncneów gehywunge úre *ipse cognōvit figmentum nostrum,* Ps. Spl. C. 102, 13. Gehiwunge *cătēgŏriæ,* Cot. 57. Drihten, đū wást míne geheowunga *Lord, thou knowest my fashioning,* Blickl. Homl. 89, 15.

ge-hladan; *p.* -hlód, -hleód, *pl.* -hlódon; *pp.* -hladen, -hlæden. **I.** *to load, burden, freight, heap up;* onĕrāre, impŏnĕre, congĕrĕre, cŭmŭlāre :—Đe he on foldan on his gǽste gehlód *which he on earth loaded on his soul,* Exon. 23 a; Th. 64, 10; Cri. 1035. He sǽbat gehleód *he loaded the sea-boat,* Beo. Th. 1795, note; B. 895, note. Hí gehlódon werum and wífum wǽghengestas *they loaded the ocean-stallions with men and women,* Elen. Kmbl. 467; El. 234: Cd. 174; Th. 220, 2; Dan. 65. Biþ seó mōdor wistum gehladen *the mother is laden with provisions,* Exon. 128 a; Th. 492, 16; Rä. 81, 16. Đa wǽron ofætes gehlædene *which were laden with fruit,* Cd. 23; Th. 30, 4; Gen. 461. **II.** *to draw* [*water*]; haurire :—To gehladanne *haurire,* Jn. Skt. Lind. 4, 15.

ge-hlǽg, es; *n. Scorn, ridicule* :—Hí gehlǽges tilgaþ *they strive after scorn,* Exon. 116 a; Th. 446, 1; Dóm. 15. [Cf. *Icel.* hlǽgi *ridicule,* and hlíhan.]

ge-hlǽnian *to make lean, thin.* v. lǽnian.

ge-hlǽstan; *p.* -hlǽste; *pp.* -hlǽsted, -hlǽst *To load, adorn* :—Mid đý hí þæt scyp gehlæsted hæfdon *when they had freighted the ship,* Bd. 5, 9; S. 623, 17: Exon. 52 a; Th. 182, 8; Gú. 1307. Đa eádigan mægþ beágum gehlæste *the blessed maid adorned with rings,* Judth. 10; Thw. 21, 30; Jud. 36.

ge-hlađen *invited.* v. ge-lađian.

ge-hleápan; *p.* -hleóp, *pl.* -hleópon; *pp.* -hleápen *To leap, dance;* salire, saltare :—Meotud gehleápeþ heá dúne *the Creator shall leap the high downs,* Exon. 18 a; Th. 45, 10; Cri. 717. He gehleóp đone eoh *he leaped upon the horse,* Byrht. Th. 137, 20; By. 189.

ge-hleód *loaded,* Beo. Th. 1795, note; B. 895, note; *p.* of ge-hladan.

ge-hleodu *vaults,* Exon. 21 a; Th. 56, 23; Cri. 905; *pl. nom. acc.* of ge-hlid.

ge-hleótan; *p.* -hleát, *pl.* -hluton; *pp.* -hloten *To share* or *appoint by lot, to get, receive;* sortiri, nancisci :—He đæs weorc gehleát *he got pain for this,* Cd. 131; Th. 166, 10; Gen. 2745: Ps. Th. 105, 24. Se eádiga Matheus gehleát to Marmadonia *St. Matthew was allotted to Mermedonia,* Blickl. Homl. 229, 6. Gehluton [MS. gehlutan] *they obtained,* Ps. Th. 113, 2. Gehloten, Exon. 95 a; Th. 355, 18; Reim. 79. Hit wæs gehloten to Iosepes bearna lande *it was allotted to the land of the children of Joseph,* Jos. 24, 32. Ic wæs gehloten mid ānum wífe in ānes ceorles đeówdōme *I was allotted with a woman to the service of a certain man,* Shrn. 38, 13. [*Laym.* i-leoten *to fall to one's lot.*] v. hleótan.

ge-hleóþ; *adj. Harmonious;* consonus :—Đæt hí đysne letanían and

---

antefn gehleóþre stæfne sungan *quia hanc litaniam consona voce modularentur,* Bd. 1, 25; S. 487, 24.

ge-hleów *a lowing.* v. gehlów.

ge-hleow; *adj. Sheltered, warm* :—Ond đā on gehliúran dene and on wearmran we gewícodon *in apriciore valle sedem castrorum inveni,* Nar. 23, 4. [Cf. *Icel.* hlýr *warm.*] v. unhleow.

ge-hlếđa, an; *m.* [hlōþ] *A companion, comrade;* sŏcius :—Wulf sang ahóf, holtes gehlếđa *the wolf uplifted his song, the companion of the forest,* Elen. Kmbl. 225; El. 113. Se đe ǽr bær wulfes gehlếđan *who ere bore the wolf's companion,* Exon. 130 b; Th. 499, 30; Rä. 88, 23. DER. wil-gehlếđa.

ge-hlid, es; *pl. nom. acc.* -hlidu, -hleodu; *n. A lid, covering, roof, an inclosure, a vault;* tectum, clausūra, septum :—Ic cann ealle heáh-heofona gehlidu *I know all the roofs of the high heavens,* Cd. 27; Th. 37, 3; Gen. 584: Exon. 15 a; Th. 32, 25; Cri. 518. Đonne bearn Godes þurh heofona gehleodu ōþýweþ *when the son of God shall appear through heaven's vaults,* 21 a; Th. 56, 23; Cri. 905.

ge-hlidad; *part.* [ge-hlid *a lid*] *Lidded, covered with a lid;* opercŭlo tectus :—Seó wæs gerisenlíce gehlidad mid gelíce stāne *opercŭlo sĭmĭlis lăpĭdis aptissĭme tectum,* Bd. 4, 19; S. 588, 32.

ge-hlihan; *p. pl.* gehlogun *to deride.* v. hlihan.

ge-hlioran *to pass over.* v. leoran.

ge-hliþ, es; *pl. nom. acc.* -hliđo; *n. A lid, covering, roof;* tectum :—Sceolde he sēcan helle gehliđo *he must seek the roofs of hell* [or *gates of hell* : cf. *Icel.* hlið *a gate*], Cd. 36; Th. 47, 21; Gen. 764. v. ge-hlid.

ge-hlód, *pl.* -hlódon *loaded,* Exon. 23 a; Th. 64, 10; Cri. 1035: Elen. Kmbl. 467; El. 234; *p.* of ge-hladan.

ge-hlot, es; *n. A lot;* sors :—Đæt gehlot *sors,* Jos. 7, 14, 17.

ge-hloten *appointed by lot.* v. ge-hleótan.

gehlot-land, es; *n. Land appointed by lot, an inheritance;* terra sorte assignāta, possessio :—Híg hine bebirgdon on his gehlotland *sepĕliĕrunt eum in fīnĭbus possessiōnis suæ,* Jos. 24, 30.

ge-hlów, -hleów *a lowing of beasts;* mugitus :—Hryđera gehlów *lowing of oxen,* Ælfc. Gr. 1; Som. 2, 35.

ge-hluttrad; *part.* [hluttran *to purify*] *Purified, made clear;* defæcātus :—Gehluttrad wín *defæcātum vīnum,* Ælfc. Gl. 32; Som. 62, 6; Wrt. Voc. 27, 60.

ge-hlýd; *part. Covered;* tectus :— Of flýsum mínra sceápa wǽron gehlýde þearfena sídan *the sides of the poor were covered with the fleeces of my sheep,* Job Thw. 165, 2. v. ge-hlywan.

ge-hlýd, -hlýde, es; *n. A cry, clamour, noise, tumult, murmuring;* clāmor, tumultus, murmur :—Mycel gehlýd wæs on đære menigeo be him *murmur multum ĕrat in turba de eo,* Jn. Bos. 7, 12: Mt. Bos. 27, 24: Homl. Th. ii. 336, 18. Gehlýde mín to đē becume *clāmor meus ad te pervĕniat,* Ps. Th. 101, 1. He geseah mycel gehlýd *vĭdet tumultum multum,* Mk. Bos. 5, 38: Bd. 5, 12; S. 628, 30: Homl. Th. ii. 252, 17: 546, 16: Basil admn. 2; Norm. 34, 15. Mid ānþræcum gehlýde *with a horrible clamour,* Homl. Th. ii. 508, 17.

ge-hlyn, es; *n. A noise, din;* clangor :—Đā wæs on healle wælslihta gehlyn *then was in the hall the din of slaughters,* Fíns Th. 57; Fin. 28.

ge-hlyst, es; *n. Hearing;* auditus, R. Ben. 67. DER. hlyst.

ge-hlystan; *p.* -hlyste; *pp.* -hlysted. **I.** *to listen, hear;* auscultare, audire :—Gehlyste me *audiat me,* Mk. Bos. 7, 16. Beornas gehlyston *men listened,* Byrht. Th. 134, 31; By. 92. **II.** *to obey;* obedire :—On hlyste eáran gehlyste me *in auditu auris obediunt mihi,* Ps. Spl. 17, 46. DER. hlystan.

ge-hlystfull; *adj. Exorable, gracious;* audire volens, deprecabilis, Ps. Lamb. 89, 13. DER. hlyst.

ge-hlyta, an; *m. A companion;* consors :—Fōr gehlytum đīnum *præ consortĭbus tuis,* Ps. Spl. 44, 9.

ge-hlytto *fellowship;* consortium, Rtl. 38, 43.

ge-hlyttrod; *part. Purified, pure;* mērācus :—Gehlyttrod wín *mērācum vīnum,* Ælfc. Gl. 32; Som. 62, 7; Wrt. Voc. 27, 61. v. ge-hluttrad.

ge-hlywan; *p.* de; *pp.* ed *To cover, shelter* :— Of flýsum mínra sceápa wǽron gehlywde đearfena sídan *the sides of the needy were covered with the fleeces of my sheep,* Homl. Th. ii. 448, 18. v. hleow.

ge-hnád, es; *n. A conflict, fight;* immanitas, Chr. 937; Erl. 114, 15. v. ge-hnǽst.

ge-hnǽcan; *p.* te; *pp.* ed *To check, restrain, bruise, destroy;* reprimĕre, contĕrĕre, allidĕre :—Heó gehnǽceþ đa anginnu *it checketh the beginnings,* Herb. 148, 1; Lchdm. i. 272, 15: 163, 6; Lchdm. i. 292, 19. Đū me ahófe and gehnǽctest eft *elĕvans allīsisti me,* Ps. Th. 101, 8.

ge-hnǽgan, -hnǽgean, -hnégan; *p.* -hnǽgde, -hnǽde; *pp.* -hnǽged, -hnǽgd; *v. trans. To bend down, humble, cast down, subdue;* declīnāre, hŭmĭliāre, dejĭcĕre, subĭgĕre :—Đū miht oferhydige eáđe mid wunde heáne gehnǽgean *tu hŭmĭliasti sīcut vulnĕrātum sŭperbum,* Ps. Th. 88, 9. Đū hí mid fýre fācnes gehnégest *in ignem dejĭcies eos,* 139, 10. He fyrenfulle wið eorþan niđer ealle gehnégeþ *hŭmĭliat peccātōres usque ad terram,* 146, 6. Hie on wætere wicg gehnǽgaþ *they cast down the horse in the water,* Salm. Kmbl. 312; Sal. 155. Đū goda ussa gilp gehnǽgdest *thou humbledst the glory of our gods,* Andr. Kmbl. 2640; An. 1321:

Ps. Th. 118, 71. He gehnǽgde helle·gást *he subdued the spirit of hell*, Beo. Th. 2552; B. 1274: Andr. Kmbl. 2383; An. 1193. Mín Drihten ðē gehnǽde in helle *my Lord hath trodden thee down in hell*, Blickl. Homl. 241, 5. Hyne Hetware hilde gehnǽgdon *the Hetwaras subdued him in war*, Beo. Th. 5825; B. 2916. Ðæt gē wiðerfeohtend gehnǽgan *that ye may subdue your adversary*, Andr. Kmbl. 2368; An. 1185. Blǽd is gehnǽged *glory is humbled*, Exon. 82 b; Th. 311, 7; Seef. 88: Ps. Th. 142, 3. Wǽron ða mǽgþe mid hefigran þeówdóme gehnǽgde *provincia grāviōre servitio subacta*, Bd. 4, 15; S. 583, 30.

ge-hnǽst, -hnást, es; *n. A conflict, slaughter*; conflictus, prœlium:— Æfter ðæm gehnǽste *after the battle*, Cd. 94; Th. 121, 24; Gen. 2015: Chr. 937; Erl. 114, 15, note 9. DER. cumbol-, hóp-, wolcen-. v. hnítan.

ge-hnégan *to humble, cast down*, Ps. Th. 139, 10: 146, 6. v. ge-hnǽgan.

ge-hnesctun, -hnescod *softened*. v. hnescian.

ge-hnígan; *p.* -hnáh, -hnág, *pl.* -hnigon; *pp.* -hnigen *To bow, bow the head*; inclinare, inclinare se:—Heán sceal gehnígan *the humble shall bow*, Exon. 91 a; Th. 340, 28; Gn. Ex 118. v. hnígan.

ge-hnyscan *to crush*; conterere, Mt. Kmbl. Rush. 21, 44. [Cf. hnesc.]

ge-hnyst; *part. p. Contrite*:—Se gehnysta gást *the contrite spirit*, Ps. C. 50, 127; Ps. Grn. ii. 279, 127. [Cf. hnossian *and* cnyssan (?).]

ge-hoered *heard*. v. ge-hýran.

ge-hoferod; *part. Hump-backed*; gibbĕrōsus:—Ðe wǽron gehoferode *who were hump-backed*, Homl. Th. ii. 586, 23.

ge-hogde, -hogode. v. ge-hycgan.

ge-hola, an; *m. A protector*:—Ðam ðe him lyt hafaþ leófra geholena *to him who has for himself few dear protectors*, Exon. 76 b; Th. 288, 15; Wand. 31.

ge-holen *hidden*, Bd. 4, 16; S. 584, 25; *pp. of* ge-helan.

ge-hón, -hongian; *pp.* -hongen, -hoen *To hang, hang with*:—Ðætte he gehongiga *that he hang*, Mt. Kmbl. Lind. 18, 6. He sē gehoen *crucifīgaʹur*, 26, 2. Wudu biþ blédum gehongen *the wood will be hung with fruits*, Exon. 56 a; Th. 200, 9; Ph. 38: 566; Th. 202, 18; Ph. 71.

ge-honge; *adj. Having an inclination to*:—Teala gehonge *inclined to good*, Exon. 94 b; Th. 354, 8; Reim. 42.

ge-hopp *a little bag*; folliculus, Cot. 87.

ge-horian; *pp. ad To spit*:—Gehorogæ *conspuere*, Mk. Skt. Lind. 14, 65. Gehoræd biþ *conspuetur*, Lk. Skt. Lind. 18, 32. v. horu.

ge-hornian; *p.* ade *To insult* [?]:—Mið sceofmum miclum gehornadon *contumeliis affecerunt*, Mk. Skt. Lind. 12, 4. v. gehornung.

ge-hornung, e; *f. Sadness, grief*, Som.

ge-horsian; *p.* ode, ade, ude; *pp.* od, ad, ud *To horse, to set or mount on a horse, to supply with a horse*; equitem facere, equo instruere *vel* imponere: *as yet found only as pp.*:—Here gehorsude wurdon *the army was horsed* [*mounted*], Chr. 867; Th. 130, 28, col. 3: Gehorsade, 130, 28, col. 2: 131, 28, col. 1, 2: Gehorsude, 130, 27, col. 1. Ælfréd æfter ðam gehorsudan [gehorsudan, col. 1; -sedum, 147, 3, col. 1; sedun, col. 2] here mid fyrde rád óþ Exancester *Alfred with his force rode after the mounted army to Exeter*, Chr. 877; Th. 146, 1, col. 3. Ða Denan wurdon gehorsode *the Danes were horsed* [*mounted*], Chr. 1010; Th. 264, 2, col. 2. DER. horsian.

ge-horsod [*pp. of* ge-horsian] *Horsed, mounted*; equo imposĭtus *vel* instructus:—Ðá com him ðǽr ongeán twá hund þúsenda gehorsodes [MS. gehorsades] folces *then came against him* [*Alexander*] *two hundred thousand horsemen* [*horsed folk, cavalry*], Ors. 3. 9; Bos. 67, 43. v. ge-horsian.

ge-hradian; *p.* ode; *pp.* od *To hasten*; accelerare:—Sóna wól ealra monna gehradode *continuo omnium lues scelerum adcelerāvit* Bd. 1, 14; S. 482, 23: 4, 19; S. 588, 33. v. ge-radod.

ge-hrǽcan *to set in order, direct*; dirĭgĕre:—Weorc handa ussera gehrǽce *ŏpus mănuum nostrārum dirĭge*, Ps. Lamb. 89, 17. v. ge-reccan

ge-hrǽdnys, -nyss; *f. What passes swiftly, swiftness, fewness*; paucitas, Ps. Spl. 101, 24.

ge-hrán *touched*, Exon. 47 b; Th. 163, 28; Gú. 1000; *p. sing. of* ge-hrínan.

ge-hreás *rushed*. v. ge-hreósan.

ge-hrec, es; *n. Government, management*:—Mid mycele gehrece *sedulo moderamine*, Bd. 3, 7; Whelc. 179, 8. v. ge-rec.

ge-hréfan; *p.; pp.* ed [hróf *a roof*] *To roof, cover*; tĕgĕre:—Gehréf hit eall *roof it all*, Homl. Th. i. 20, 32. Holme gehréfed *covered with water*, Exon. 101 a; Th. 381, 12; Rä. 2, 10.

ge-hrehte *corrected*; correxi, Bd. 5, 24; S. 648, 25. v. ge-rehte.

ge-hréman; *p.* de *To cry, implore*:—Gihrēmaþ and woepaþ gē *plorabitis et flēbitis vos*, Jn. Skt. Rush. 16, 20. Gihrēme we *implōramus*, Rtl. 37, 3.

ge-hremmed; *part. Hindered*; impĕdītus:—Gehremmed beón *impĕdiri*, R. Ben. 52.

ge-hreónis, se; *f. Repentance*, Rtl. 102, 45.

ge-hreósan; *p.* -hreás, *pl.* -hruron; *pp.* -hroren *To rush, fall, glide away, to fail*; ruere, cadere, labi, deficere:—Hrófas sind gehrorene *the*

roofs *are fallen*, Exon. 124 a; Th. 476, 5; Ruin. 3. Ða cōmon hí to sumre ceastre gehrorenre *venerunt ad civitātulam quandam desolatam*, Bd. 4, 19; S. 588, 29. Ic ðus gehroren eom ond aweg gewiten *I* [*Babylon*] *am thus fallen and passed away*, Ors. 2, 4; Bos. 44, 35. Móna niðer gehreúseþ *the moon shall fall down*, Exon. 21 b; Th. 58, 22; Cri. 939. Swíðe oft se micla anweald ðara yfelena gehríst swíðe fǽrlíce *very often the great power of the wicked falls very suddenly*, Bt. 38, 2; Fox 198, 8. Gehreósaþ *labuntur*, Exon. 95 a; Th. 354, 34; Reim. 55. DER. hreósan.

ge-hreóðan *to adorn*. v. ge-hroden.

ge-hreów, es; *n. A lamenting*; lamentatio:—Ðǽr biþ gehreów and hlúd wóp *there shall be lamenting and loud weeping*, Exon. 22 b; Th. 62, 9; Cri. 999. DER. hreów.

ge-hreówan; *p.* -hreáw, *pl.* -hruwon; *pp.* -hrowen *To rue, repent, grieve, pity*; pœnitere, dolere, miserere:—Mec his bysgu gehreáw *his affliction grieved me*, Exon. 43 a; Th. 144, 31; Gú. 686. Generally *impers.* hit-hreóweþ, -hrýwþ; *p.* hit-hreáw *It rues, it repents, it grieves, it pities*; pœnitet, dolet, miseret:—Him ðæt gehreówan mæg *that may rue them*, Cd. 225; Th. 298, 29; Sat. 540. Mec æt heortan gehreáw *I repented at heart* [lit. *it repented me at heart*], Exon. 29 b; Th. 91, 18; Cri. 1494: Cd. 221; Th. 288, 2; Sat. 374. DER. hreówan.

ge-hrepod [*pp. of* ge-hrepian *to touch*] *touched*; tactus:—He wæs gehrepod mid heortan sárnisse wiðinnan *tactus dolore cordis intrinsecus*, Gen. 6, 6. Gehrepod *tactus*, Ælfc. Gr. 43; Som. 44, 56.

ge-hréran; *p.* de *To move*:—Mægen heofunas bióþ gehroered *virtutes cælorum commovebuntur*, Mt. Kmbl. Lind. 24, 29.

ge-hrespan *to tear*:—Hý him sylfum gehrespaþ *diripiēbant sibi*, Ps. Th. 43, 12.

ge-hrifan; *p.* ede; *pp.* ed [hrif *the womb*] *To bring forth*; părĕre:—Gecende sárnessa and gehrifede oððe acende unrihtwísnesse *concēpit dolōrem et pĕpĕrit inĭquitātem*, Ps. Lamb. 7, 15.

ge-hrínan, -rínan; he -hríueþ, -hrínþ; *p.* -hrán, *pl.* -hrinon; *pp.* -hrinen *To touch, take hold of, seize, affect*; tangĕre, contingĕre, răpĕre, affectāre:—Ne ofer ðæt syððan him ówiht gehrínan dorste *neque unquam exinde eum audēret contingĕre*, Bd. 3, 12; S. 537, 14, MS. B: 3, 17; S. 544, 28. Ða mǽran tungl áuðer óðres rene ā ne gehríneþ *these splendid stars never touch each other's course*, Bt. Met. Fox 29, 20; Met. 29, 10. Hí gehrínþ hér sumu wracu *some punishment affects them here*, Past. 55; Swt. 429, 19; Hat. MS. Me sár gehrán *pain hath touched me*, Exon. 47 b; Th 163, 28; Gú. 1000. Heó sóna wæs gehrinen and genumen of middanearde *rapta confestim de mundo*, Bd. 4, 19; S. 589, 5: 4, 8; S. 575, 30. Hia gehrínadon ł gehrínad hæfde *tetigerunt*, Mt. Kmbl. Lind. 14, 36.

ge-hrinenes, -ness, e; *f. A touch*; tactus:—Mid ðý gehrinenesse ðæra [MS. ðære] ilcena gegyrlena *tactu indūmentōrum eōrumdem*, Bd. 4, 19; S. 589, 32.

ge-hríst *falls*. v. hreósan.

ge-hroden [*pp. of* ge-hreóðan *to adorn*] *adorned*; ornatus:—Biþ seó mōdor hordum gehroden *the mother is adorned with treasures*, Exon. 128 a; Th. 492, 17; Rä. 81, 17. Eoforlíc gehroden golde *a boar's likeness adorned with gold*, Beo. Th. 614; B. 304. Grēne stondaþ gehroden hyhtlíce beorhtast bearwa *the brightest of groves stands green, gloriously adorned*, Exon. 57 a; Th. 203, 4; Ph. 79. Ðec gemētte, meahtum gehrodene *he found thee adorned with virtues*, 12 b; Th. 21, 6; Cri. 330: Judth. 10; Thw. 21, 27; Jud. 37. Geseh he bearwas blǽdum gehrodene *he saw groves adorned with blossoms*, Andr. Kmbl. 2896; An. 1451: Exon. 97 b; Th. 364, 21; Wal. 74.

ge-hror, es; *n. A fall, ruin, death*:—Ðonne ðæt gelumpe ðæt hí of middangearde genumene wǽron ðý ylcan gehrore ðe hí óðre gesáwon *cum eas eodem quo cæteros exterminio raptari e mundo contingeret*, Bd. 4, 7; S. 574, 38. v. gehreósan, *and cf. Icel.* hrör *cadaver*.

ge-hroren *fallen*, Exon. 124 a; Th. 476, 5; Ruin. 3; *pp. of* ge-hreósan.

ge-hrorenes, -ness, e; *f. Affliction, ruin*; ærumna:—Gecerrod oððe gewend ic eom on gehrorenesse oððe yrmþum mínum *conversus sum in ærumna mea*, Ps. Lamb. 31, 4.

ge-hruron, -hroren *rushed down, destroyed, was desolate*. v. gehreósan.

ge-hruxl *a noise, disturbance*; tumultus, Dial. 2, 10.

ge-hrýne, es; *n. A mystery, sacrament*; mystērium:—Ðǽr Godes nama gelóme gecýged biþ, and ðæt [MS. ða] hálige gehrýne on mæssesange geoffrod, nis nǽnig tweó ðæt ðǽr biþ Godes engla andweardnes *where God's name is frequently invoked, and the holy mystery offered in the mass service, there is no doubt that the presence of God's angels is there*, L. E. I. 10; Th. ii. 408, 24. v. ge-rýne.

gehþ *a station*, Ex. MS. Conb. p. 233. v. giht.

gehðo, gehðu, geohðu, geoðu, giohðu, giðu, e; *f. Care, anxiety*; cura, solicitudo:—Gomol on gehðo eówic grētan hēt *the aged* [*prince*] *in sadness commanded to greet you*, Beo. Th. 6181; B. 3095. Gehðo

mǽnan *to bemoan misery*, Andr. Kmbl. 3095; An. 1550. Iudas cwæþ dæt he dæt on gehðu gespræce *Judas said that he spoke that in trouble*, Elen. Kmbl. 1331; El. 667. Ne meahte he ða gehðu bebúgan *he could not avoid the sorrow*, 1215; El. 609. Ic sceal gehðu mǽnan *I must lament my cares*, Exon. 71 b; Th. 266, 1; Jul. 391. Oft mec gehða gemanode *often sorrow hath admonished me*, 50 a; Th. 174, 22; Gú. 1181. Sceal se gǽst cuman gehðum hrēmig *the ghost shall come moaning with anxiety*, 98 a; Th. 367, 18; Seel. 9: 9 a; Th. 6, 27; Cri. 90: Elen. Kmbl. 643; El. 322: 1059; El. 531. Geohðo mǽnaþ *they lament their grief*, Andr. Kmbl. 3329; An. 1667. Ic þurh geohða sceal dǽda fremman *I must do deeds with sorrow*, Andr. Kmbl. 132; An. 66. Sceal se gǽst cuman geohðum hrēmig *the spirit shall come sadly lamenting*, Soul Kmbl. 18; Seel. 9. He ðǽr ána sæt geoðum geómor *he sat there alone sad with sorrows*, Andr. Kmbl. 2015; An. 1010. Gomel on giohðe gold sceáwode *the aged [man] beheld the gold in sorrow*, Beo. Th. 5578; B. 2793. Giohðo mǽnde he *bewailed his afflictions*, 4527; B. 2267. Geómrian on gihða *to mourn in spirit*, Salm. Kmbl. 701; Sal. 350. Éðelléase ðysne gyst-sele gihðum healdeþ *the homeless held in memory this guest-hall*, Cd. 169; Th. 212, 5; Exod. 534. v. Grm. And. u. El. p. 97.

ge-hú; *adv.* *In any manner* :—He is gecweden hláf ðurh getácnunge and lamb and leó and gehú elles *he is called bread typically and lamb and lion and in any other way*, Homl. Th. ii. 268, 17. Ðeáh de heó sý ge-býged gehú *though it be bent anyhow*, Hexam. 6; Norm. 10, 30.

ge-hugod; *part. p. Minded, disposed* :—Boda bitre gehugod *the messenger bitter of purpose*, Cd. 33; Th. 45, 11; Gen. 725.

ge-huntian; *p.* ode; *pp.* od *To hunt* :—Hí gehuntiaþ *venantur*, Nar. 38, 6.

ge-húsan; *pl. m. Housefolk, those of the household;* dōmestici :—Mannes fýnd, hys gehúsan *inimici hōminis, dōmestici ejus*, Mt. Bos. 10, 36.

ge-húsed; *part. Housed, having a house;* dōmum hábens :—Gehúsed snægl *a housed* or *shelled snail;* testūdo, Ælfc. Gl. 23; Som. 60, 1; Wrt. Voc. 24, 5.

ge-húslian; *p.* ode; *pp.* od *To give the eucharist, housel* :—He hēt ðǽr hine gehúslian *he commanded them to give him the eucharist*, Homl. Th. ii. 186, 29. Se hálga sacerd Iustinus him eallum gemæssode and gehúslode *the holy priest Justin said mass to them all and houseled them*, i. 430, 29. Gehúslod beón *communicari*, R. Conc. 5.

ge-hússcype, es; *m. A house, household, family, race;* dōmus :—Gehússcype Israhel bletsiaþ Driht *dōmus Israhel benedīcite Dōmino*, Ps. Spl. C. 134, 19.

ge-hwá; *m.* -hwæt; *n. g.* -hwæs; *pron. Every one, whoever, who;* quisque, quis. This word is often found with a genitive :—Forðí sceal gehwá on his Drihtne wuldrian *therefore shall every man glory in his Lord*, Homl. Th. ii. 526, 12. Hwæt gehwá náme *quis quid tolleret*, Mk. Bos. 15, 24. Fæder-æðelo gehwæs *the ancestry of each*, Cd. 161; Th. 200, 24; Exod. 361. Ðonne fēran sceal ánra gehwæs sáwl of líce *when the soul of each one shall go from the body*, Exon. 54 b; Th. 191, 24; Az. 93: 64 b; Th. 238, 3; Ph. 598. Ðec sóþfæstra gehwæs sáwle and gástas lōfiaþ *the souls and spirits of all the just praise thee*, Cd. 192; Th. 240, 31; Dan. 395. He ðeóda gehwam hefonríce forgeaf *he to every people gave heaven's kingdom*, Th. 40, 19; Gen. 641. Ic leófra gehwone lǽran wille *I will teach each dear one*, Exon. 19 b; Th. 51, 14; Cri. 816. Háteþ cuman to gemóte moncynnes gehwone *bids come to the meeting every man*, 23 a; Th. 63, 30; Cri. 1027. Ðæt fýr nimeþ ðurh foldan gehwæt *the fire shall seize everything on earth*, 22 b; Th. 62, 18; Cri. 1003. [O. Sax. gi-hwe *quisque*.]

ge-hwǽde; *adj. Little, moderate, scanty* :—Hí wǽron gehwǽde acwealde *they were killed while little*, Homl. Th. i. 84, 21: ii. 162, 2: Gen. 19, 20. Úre gehwǽda wæstm *our little fruit*, Homl. Th. 526, 22. Seó gehwǽde oferflówendnys *the slight superfluity*, i. 332, 14: Mt. Bos. 6, 30: Bd. de nat. rerum; Wrt. popl. science 1, 1; Lchdm. iii. 232, 1.

ge-hwǽdnes, -hwēdnes, es; *f. Sparingness, paucity, fewness, subtilty;* parcitas, paucitas :—Gehwǽdnis *humilitas, mediocritas*, Hpt. Gl. 403, 467. Gehwǽdnysse dagena mínra gecýþ me *paucitatem dierum meorum nuntia mihi*, Ps. Spl. 101, 24.

ge-hwǽmlíc; *adj. Each, every* :—Dæge gehwǽmlíce *cotidie*, Lk. Skt. Lind. 9, 23.

ge-hwǽr, -hwár; *adv. On every side, everywhere;* undique, ubique :—Se symle leofaþ gehwǽr on uurím gódum *qui innumeris semper vivit ubique bonis*, Bd. 2, 1; S. 500, 23. His gebyrd and goodnys sind gehwǽr cúþe *his birth and goodness are known everywhere*, Homl. Th. i. 2, 16. Nemnaþ men dǽue mónaþ gehwǽr Iulius *men name that month everywhere July*, Chr. 975; Erl. 124, 33; Edg 25: Elen. Kmbl. 2364; El. 1183. Wel wíde gehwǽr *everywhere far and wide*, Menol. Fox 118; Men. 59. Ðeáh ðú heaðorǽsa gehwǽr dohte *though thou hast in martial exploits everywhere succeeded*, Beo. Th. 1057; B. 526: Elen. Kmbl. 1092; El. 548. Gehwár hí syn hefige gehwár eác medeme *in some places they are heavy, in others moderate*, Th. Ll. i. 434, 4. [Laym. i-hwær, i-war: A. R. i-hwar.]

ge-hwǽðer; *pron. Both, each, either;* uterque, promiscuus :—Wæs

gehwæðer óðrum láþ *each was hateful to the other*, Beo. Th. 1633; B. 814. Gehwæðer incer *either of you two*, 1173; B. 584. He biþ him self gehwæðer fæder and sunu *it is to itself both father and son*, Exon. 61 a; Th. 224, 12; Ph. 374. Se willa bēga gehwæðres ge . . . ge . . . *her will in both respects both . . . and . . .* , Elen. Kmbl. 1925; El. 964: Beo. Th. 2091; B. 1043. Ðǽr wearþ monig mon ofslægen on gehwæðre hond *there was many a man slain on each side*, Chr. 853; Erl. 68, 19: 871; Erl. 74, 12.

ge-hwæðere; *adv. Yet, however* :—Weorðeþ heó ðeáh oft niða bearnum to helpe and to hǽle gehwæðere *it becomes oft however help and safety nevertheless to the children of men*, Runic pm. 10; Kmbl. 341, 12. v. hwæðere.

ge-hwæðeres; *adv. Anywhere, on every side, every way;* undique :—Wæs gehwæðeres waa· *there was woe on every side*, Bt. Met. Fox 1, 50; Met. 1, 25. v. ge-hwæðer.

ge-hwanon; *adv. From all sides* :—Fela ðearfan gehwanon cumene *many needy come from all sides*, Swt. A. S. Rdr. 97, 78.

ge-hwearf, -hwyrf, es; *n. A change, exchange;* commūtātio, permūtātio :—Gehwearf *commūtātio*, Ælfc. Gl. 81; Som. 73, 26; Wrt. Voc. 47, 31.

ge-hwearf *returned.* v. ge-hweorfan.

ge-hweled; *part. Inflamed;* inflammātus :—Ðæt ðǽrinne gehweled biþ *which is inflamed therein*, Past. 38, 3; Swt. 273, 22; Hat. MS. 51 a, 12: Swt. 275, 5.

ge-hweorf; *adj.* I. *versed, practised, clever;* versutus :—Sum biþ ðegn gehweorf on meoduhealle *one is a thane familiar in the meadhall*, Exon. 79 a; Th. 297, 15; Crä. 68. v. hwearf. II. *converted* :—Nymðe gē gewerfe beón *nisi conversi fueritis*, Mt. Kmbl. Rush. 18, 3. [Cf. *Goth.* ga-hwairbs.]

ge-hweorfan; *p.* -hwearf, *pl.* -hwurfon; *pp.* -hworfen. I. *act.* *To turn;* convertere :—Manige sindon ðe ðú gehweorfest to heofonleóhte *there are many whom thou shalt turn to the light of heaven*, Andr. Kmbl. 1947; An. 976. Gehweorf úre hæftnēd *converte captivitatem nostram*, Ps. Th. 125, 4. Gehweorf us, mægena God *Domine Deus virtutum, converte nos*, 79, 4. Gehweorf nú ðíne ansýne *turn now thine eye*, 79, 14. II. *intrans. To turn, go away, depart, die, pass as property, fall as a lot;* verti, abire, redire, excidere :—Ymb ofn útan alet gehwearf *the fire turned round about the oven*, Cd. 186; Th. 232, 3; Dan. 254. Mán eft gehwearf ðǽr *their sin turned again thither*, Andr. Kmbl. 1388; An. 694: Lk. Bos. 8, 55: 17, 7: 24, 52. Siððan to reste gehwearf *after he had gone to rest*, Cd. 177; Th. 222, 23; Dan. 109. Ǽr ic of ðysum lífe gehweorfe *ere I depart from this life*, Hy. 3, 53; Hy. Grn. ii. 284, 53. Hit on ǽht gehwearf Denigea freán *it passed into the possession of the Danes' lord*, Beo. Th. 3363; B. 1679: 2424; B. 1210: 4422; B. 2208. Ðá se tán gehwearf ofer ǽnne ealdgesíþa *then the lot fell on one of the old comrades*, Andr. Kmbl. 2208; An. 1105. v. hweorfan.

ge-hwerfnes *a conversion.* v. ge-hwyrfednes.

ge-hwettan; *p.* te; *pp.* ed *To whet, excite;* excitāre :—He gehwette and tihte ðæra Iudéiscra manna heortan *he whetted and instigated the hearts of the Jews*, Homl. Th. i. 26, 31.

ge-hwider; *adv. Whithersoever, anywhere, everywhere;* alicubi :—Ðonon eóde gehwyder ymb *inde circumquaque exire consueverat*, Bd. 3, 17; S. 543, 26: Bt. Met. Fox 25, 26; Met. 25, 13.

ge-hwylc, -hwelc, -hwilc; *pron. Each, every one, all, whoever, whatever;* quisque, unusquisque :—Gē gehwilce uncóðe gehǽldon *ye healed every disease*, Homl. Th. i. 64, 23. Of gehwilcum burgum *from every city*, 86, 29. Nú smeádon gehwilce men *now some men have enquired*, ii. 268, 7. Dǽda gehwylcra *of all deeds*, Elen. Kmbl. 2563; El. 1283. Háteþ arísan folc ánra gehwylc *bids each folk arise*, Exon. 23 a; Th. 63, 28; Cri. 1026. Ðæt he wiste hú mycel gehwylc gemangode *ut sciret quantum quisque negotiatus esset*, Lk. Bos. 19, 15. Sió gesceádwísnes sceal on gehwelcum waldan *reason shall rule in each one*, Bt. Met. Fox 20, 394; Met. 20, 197. Ongan ánra gehwylc cweðan *cœperunt singuli dicere*, Mt. Bos. 26, 22: Deut. 24, 16. Lifigendra gehwylc *every one living*, Cd. 219; Th. 282, 12; Sat. 285. And hiera se æðeling gehwelcum feoh and feorh gebeád *and the atheling offered each of them money and life*, Chr. 755; Erl. 50, 5. He beheóld heora ánra gehwilcne *he observed each one of them*, Th. Ap. 12, 24.

ge-hwyrf, es; *n. Exchange* :—Be gehwyrfe *of exchange*, L. Ath. i. 10; Th. i. 204, 16, 21, note 23, 31. v. ge-hwearf.

ge-hwyrfan, -hwerfan, -hwirfan, -hwiersan; *p.* de; *pp.* ed *To change, turn, convert;* mutare, convertere :—Hyra woruld wæs gehwyrfed *their world [life] was changed*, Cd. 17; Th. 21, 3; Gen. 318. Flód gehwerfde ða ceastre *a flood overturned the city*, Shrn. 77, 12. Hwylc ðonne gēna gehwyrfed byþ *quoadusque justitia convertatur in judicium*, Ps. Th. 93, 14. Hí gehwyrfde synd *conversi sunt*, Ps. Spl. 77, 46: Exon. 10 b; Th. 12, 20; Cri. 188. Mín drihten, ðú de gehwyrfest ealle sáule *my Lord, thou who convertest all souls*, Blickl. Homl. 249, 14. Manige Israhela bearna he gehwyrfþ to heora drihtne *many of the children of Israel he shall turn to their Lord*, 165, 13. Ic ðé bidde for dínum naman ðæt ðú gehwyrfe on me ealle eáþmódnesse dínra beboda *I beseech*

thee for thy name that thou devolve on me all submission to thy commands, 147, 11. Paulinus gehwerfde Édwine Norþymbra cyning to fulwihte *Paulinus converted Edwin king of Northumbria to christianity*, Chr. 601; Erl. 20, 12. Hér wæs Paulus gehwierfed *in this year Paul was converted*, 34; Erl. 6, 14: 30; Erl. 6, 9. His word bióþ gehwirfdo to unnyttre oferspræce *his words will be perverted to useless loquacity*, Past. 21; Swt. 164, 18; Cot. MS. Hí wurdon gehwyrfede to deórwurðum gimmum *they were turned into precious stones*, Homl. Th. i. 64, 5: Th. An. 28, 35. On heáf gehwyrfede *turned to mourning*, Blickl. Homl. 195, 17: 233, 5. Ic wæs gehwyrfed on mínne líchoman *I was restored to my body*, 155, 25.

ge-hwyrfednes, -hwyrfenes, -ness, e; f. *A conversion, change*; conversio:—Ðara geleáfan and gehwyrfednesse *quórum fídei et conversióni*, Bd. 1, 26; S. 488, 13. In ða tíd heora gehwyrfenesse *tempŏre suæ conversiónis*, 4, 5; S. 572, 39.

ge-hwyrftnian *to tear* (?):—His æfterfolgeras hit siððan totugon and totǽron ðam gelícost ðonne seó leó bringaþ *his hungregum hwelpum hwæt to etanne hý ðonne gecýðaþ on ðam ǽte hwylc heora mǽst mæg gehwyrftnian *his successors afterwards rent and tore it most like to when the lion brings its hungry whelps something to eat, then they show in that food which of them can tear it most*, Ors. 3, 11; Bos. 71, 39, note.

ge-hycgan, -hicgan; p. -hogde, -hogede, -hogode; pp. -hogod [see March, § 222] *To think, conceive, consider, devise, reflect, be mindful, think about, care, intend, resolve* :—Ne mæg ic ðeáh gehycgan hwý him on hige ðorfte á ðý sǽl wesan *I cannot, however, conceive why it need be the better in mind for them*, Bt. Met. Fox 15, 17; Met. 15, 9. Sceal gehycgan hæleða ǽghwilc ðæt he ne abǽlige bearn wealdendes *every man must be mindful that he offend not the son of the powerful*, Cd. 217; Th. 276, 25; Sat. 195: 219; Th. 282, 7; Sat. 283. Ðú gehycgan meaht ðæt gé willaþ ða on wuda sécan *you may consider that you will seek them in the wood*, Bt. Met. Fox 19, 34; Met. 19, 17. Sum in mǽðle mæg folcrǽdenne gehycgan *one in council can devise a nation's law*, Exon. 79 a; Th. 295, 33; Crä. 42: Cd. 203; Th. 252, 29; Dan. 586. Gehyge on ðínum breóstum ðæt ðú inc bǽm meaht wíte bewarigan *reflect in thy breast that thou from you both mayest ward off punishment*, Cd. 27 : Th. 35, 29; Gen. 562. Fela gé fore monnum míðaþ ðæs ðe gé in móde gehycgaþ *much ye before men conceal of what ye in mind devise*, Exon. 39 a; Th. 130, 11; Gú. 436. Hú ðú yfle gehogdes *how thou didst devise evilly*, 28 a; Th. 85, 29; Cri, 1398. Ðá ðú gehogodest sæcce sécean *when thou didst resolve to seek conflict*, Beo. Th. 3981; B. 1988: Cd. 209; Th. 259, 5; Dan. 687; Andr. Kmbl. 857; An. 429. Hæfde on án gehogod ðæt he gedǽde swá hine drihten hét *his purpose had continually been to do as the Lord commanded him*, Cd. 140; Th. 175, 9; Gen. 2892. Ðæt hió ðæs niwan taman náuht ne gehicgge *that she care nothing about the new tameness*, Bt. Met. Fox 13, 52; Met. 13, 26. On drihten helpe gehogedan *speravit in domino*, Ps. Th. 113, 18: Exon. 33 a; Th. 105, 5; Gú. 18. [Goth. ga-hugjan : O. Sax. gi-huggian.]

ge-hýd, e; f: es; n. *A thought*; cŏgītātio :—In sefan gehýdum *in the mind's thoughts*, Cd. 212; Th. 261, 27; Dan. 732. DER. mis-gehýd. v. ge-hygd.

ge-hýd; part. p. *Exalted*; exaltātus, Hpt. Gl. 440. v. geheád.

ge-hýd; part. p. *Provided with a skin*, Nar. 50, 5.

ge-hýdan, -hídan, -hédan; he -hýdeþ, -hýt, pl. -hýdaþ; p. -hýdde; pp. -hýded, -hýdd. **I.** *to hide, conceal*; condĕre, abscondĕre :—He hit gehýt and gehelt *it hides and preserves it*, Bt. 39, 8; Fox 224, 11: 39, 13; Fox 234, 19. Sumne dreórighleór in eorþscræfe eorl gehýdde *a man sad of countenance has hidden one in an earth-grave*, Exon. 77 b; Th. 291, 19; Wand. 84: Beo. Th. 4463; B. 2235. Hí wiston ðæt hine gehýdan hæleþ Iudéa *they knew that the men of Judea had hidden him*, Exon. 119 b; Th. 460, 6; Hö. 13. Læg mín flǽschoma niþre gehýded, in byrgenne *my body lay hidden beneath, in the sepulchre*, 29 a; Th. 89, 34; Cri. 1467: Elen. Kmbl. 2182; El. 1092. Heofona ríce is gelíc gehýddum goldhorde on ðam æcere símíle est regnum cælórum thésauro abscondíto in agro, Mt. Bos. 13, 44. Fint he ðǽr ða ryhtwísnesse gehýdde mid ðæs líchoman hæfignesse *he will there find the wisdom concealed by the heaviness of the body*, Bt. 35, 1; Fox 156, 11. Sticiaþ gehýdde beorhte cræftas *bright virtues lie hid*, 4; Fox 8, 15: 32, 3; Fox 118, 23. **II.** *to watch, guard, heed*; observāre :—Ðæt heó gehýden hǽlan [MS. hælun] míne calcāneum meum observābunt, Ps. Th. 55, 6. **III.** *to bring into safety, make firm, fasten*; alligāre :—Hý ehýdaþ heáhstefn scipu to ðam unlonde oncyrrápum *they fasten the high-prow'd ships to the false land with anchor-ropes*, Exon. 96 b; Th. 361, 1; Wal. 19. v. hédan and hýdan.

ge-hýdnes, se; f. *Comfort, security* (?) :— Ðýlæs hie gedwelle sió gehýdnes and ða getǽsu ðe hie on ðǽm wege habbaþ *lest the comfort and pleasures that they have on the way seduce them*, Past. 50, 1; Swt. 387, 13; Hat. MS. See the note on this passage, Swt. 491-2. Or is the word connected with gehýdan? cf. gehýdan **III**. and the *subsidia itineris* of the original Latin.

ge-hygd, -higd, -hýd, e; f: es; n. *Thought, cogitation, meditation, deliberation, consultation*; cŏgītātio, mĕdītātio, consĭlium :—Sceal on

leóht cuman heortan gehygd *his heart's thought shall come into light*, Exon. 23 a; Th. 64, 17; Cri. 1039: 77 b; Th. 290, 28; Wand. 72. On mínre gehygde heortan ealre *in tóto corde meo*, Ps. Th. 137, 1: 118, 58: 54, 20. Þurh deóp gehygd *through deep thought*, Exon. 72 a; Th. 268, 13; Jul. 431: Cd. 221; Th. 285, 28; Sat. 344. Sete on Drihten ðín sóþ gehygd *jacta in Deum cōgĭtātum tuum*, Ps. Th. 54, 22. Ne biþ ðǽr wiht forholen monna gehygda *there shall be naught of men's cogitations concealed*, Exon. 23 b; Th. 65, 15; Cri. 1055. On sefan gehygdum *in the mind's thoughts*, 39 b; Th. 130, 27; Gú. 444: 81 a; Th. 305, 14; Fä. 88. Eálá ðæt we nú mágon geseón on ussum sáwlum synna wunde, mid líchoman leahtra gehygdu eágum *alas that we now may see in our souls wounds of sin, with the body's eyes wicked cogitations!* 27 a; Th. 80, 32; Cri. 1315. Ðú ána canst ealra gehygdo *thou alone knowest the thoughts of all men*, Andr. Kmbl. 136; An. 68: 399; An. 200. Hí sáwle frætwaþ hálgum gehygdum *they adorn their souls with holy meditations*, Exon. 44 b; Th. 150, 15; Gú. 779: 62 b; Th. 229, 22; Ph. 459. Landágende men ic lǽrde ðæt hie heora gafol mid gehygdum aguldon *I taught landowners to pay their taxes carefully*, Blickl. Homl. 185, 22. [Goth. ga-hugds; f: O. Sax. gi-hugd; f.] DER. breóst-, gást-, in-, inn-, mód-gehygd.

ge-hyht, es; m. *A hope, comfort, refuge*; refúgium :—Drihten trumnes mín and gehyht mín *Dŏminus firmāmentum meum et refúgium meum*, Ps. Spl. T. 17, 1.

ge-hyhtan; p. te *To hope, trust* :—We sceolan gehyhtan on godes ða gehálgodan cyricean *we must trust in God's holy church*, Blickl. Homl. 111, 8. On his naman ðeóda gehyhtaþ *in nomine ejus gentes sperabunt*, Mt. Bos. 12, 21. On hine gehyhtton *trusted in him*, Blickl. Homl. 103, 12: 159, 18. Ðæt on ðínum upstige geblissian and gehyhton ealle ðíne gecorenan *that in thy ascension all thine elect may rejoice and trust*, 87, 25. v. ge-hihtan.

ge-hyhtlíc; adj. *Seasonable, fit, commodious*; opportunus, R. Ben. 53. v. hihtlíc.

ge-hylced; part. p. *Divaricatus*, Gl. Prud. 758.

ge-hyld, es; n. *Regard, observation, keeping, concealing*; observantia, custodia :—In gehylde rihtra Eástrana *in the keeping of right Easter*, Bd. 2, 4; S. 505, 25. Ic wæs on ðínum gehylde begangen *in observatiónibus tuis exercebor*, Ps. Th. 76, 10. [Him] hálige heápas on gehyld bebeád *commended to his protection the holy bands*, Cd. 161; Th. 202, 3; Ex. 382. Lǽdan on gehyld Godes *to lead into God's protection*, Andr. Kmbl. 2091; An. 1047: 234; An. 117. Háligra gehyld *the preservation of the holy ones*, Exon. 55 b; Th. 196, 4; Az. 169. He is manna gehyld *he is the protection of men*, Beo. Th. 6104. On heofona gehyld *into the protection [?] of the heavens*, Exon. 15 b; Th. 34, 20; Cri. 545. Thorpe translates *into heaven's vault*, Grein has *recessus, arcanum?* Or could the word have the sense of *space*, cf. Ger. gehalt, gehaltig? Cf. *also* geheald *subst. and adj. and* gehild.

ge-hyldan; p. -hylde; pp. hylded *To keep, hold, forbear*; custodire, conservare, differe :—Gehylde *forbore*; distulit, Ps. Spl. 77, 25.

ge-hyldan *to bend, incline* :—To gehyldanne *declinare*, Ps. Lamb. 16, 11.

ge-hyldig; adj. *Patient*; patiens, Ps. Spl. 7, 12.

ge-hyldness, e; f. *Keeping, observance* :—On heora gehyldnesse *in custodiendis illis*, Ps. Th. 18, 10.

ge-hyldra; m. e; f. n; compar. of geheald (?) *Safer* :—Ðǽm gehyldrum wegum *tuta itinera*, Nar. 6, 3. Ðohtan ðæt him wíslícre and gehyldre wǽre *they thought that it would be wiser and safer for them*, Bd. 1, 23; S. 485, 31. On gehældran stowe *in tutiore loco*, Bd. 2, 2; S. 503, 39.

ge-hylmd, -hylmed; adj. *Galeatus*, Cot. 97. *Frondosus*, 89.

ge-hylt *keeps*, Ps. Lamb. 120, 7; 3rd sing. pres. of ge-healdan.

ge-hýnan, -hénan, -hínan; p. de; pp. ed *To humble, oppress, waste, destroy*; humiliare, opprimere, damnare :—Uton gehýnan hit *opprimamus eum*, Ex. 1, 10. Eágan ofermódra ðú gehýnyst *oculos superborum humiliabis*, Ps. Spl. C. 17, 29. Gehýnyþ *humiliat*, Ps. Spl. C. M. 74, 7. Híg gehýndon eos *oppresserunt*, Ex. 1, 11. Gehýned *damnatus*, C. R. Ben. 58. Gehéned, Ps. Vos. 37, 8. v. ge-hínan, hýnan.

ge-hyndred; part. *Hindered*; impédítus :—Biþ eall se here swýðe gehyndred *all the army will be greatly hindered*, Chr. 1003; Th. 252, 33, col. 2. v. ge-hindred.

ge-hyngran; p. -hyngerde *To be hungry* :—Mec gehyncgerde *esurivi*, Mt. Kmbl. Lind. 25, 42. Ic gehwyncgerde *esurivi*, 25, 35. Hine gehyngerde *esuriit*, 12, 3. Gihyńcrede *esuriit*, Mk. Skt. Rush. 11, 12. Eádgo ða ðe nú gehyncres *beati qui nunc esuritis*, Lk. Skt. Lind. 6, 21. Gehyngrede hundas *hungry dogs*, Shrn. 145, 3.

ge-hýpan; p. de; pp. ed *To heap* :—Ðonne hit gehýpþ yfel ofer yfele *when it heaps evil upon evil*, Homl. Th. i. 410, 21.

ge-hýran, -híran, -héran; to -hýranne, -hýrenne; part. -hýrende; ic -hýre, ðú -hýrest, -hýrst, he -hýreþ, -hýrþ, pl. -hýraþ; p. ic, he -hýrde, ðú -hýrdest, pl. -hýrdon; impert. -hýr, pl. -hýre, -hýraþ; subj. pres. -hýre, pl. -hýron; p. -hýrde, pl. -hýrden; pp. -hýred. **I.** v. trans. *To hear, give ear to*; audire, exaudire :—Forðamðe gé ne mágon gehýran míne

spǽce *quia non pŏtestis audīre sermonem meum*, Jn. Bos. 8, 43 : Bd. 3, 5 ; S. 527, 22, 35. To eallum ðe ðis ylce stǽr becyme úres cynnes to rǽdanne oððe gehýranne *omnes ad quos hœc eadem histŏria pervēnīre potĕrit nostrœ natiōnis lĕgentes sīve audientes*, 5, 24 ; S. 649, 6. Ic ðæt gehýre, ðæt ðis is hold weorod *I hear that this is a friendly band*, Beo. Th. 585 ; B. 290: Exon. 72 b ; Th. 270, 6 ; Jul. 461. Gehýrest ðú uncerne earne hwelp *hearest thou our active whelp?* 101 a ; Th. 380, 30; Rä. 1, 16. Georne gehýreþ heofoncyninga hýhst hæleða dǽde *the highest of heaven's kings will earnestly hear men's deeds*, 117 b ; Th. 451, 22 ; Dóm. 107 : 19 b ; Th. 50, 9 ; Cri. 797. Ðænne hí ðæt word gehýraþ *qui cum audiĕrint verbum*, Mk. Bos. 4, 16, 18, 20. Ic gehýrde hine ðíue dǽd and word lofian *I heard him praise thy deed and words*, Cd. 25 ; Th. 32, 23 ; Gen. 507 : 26 ; Th. 33, 23 ; Gen. 524. Ðú gehýrdest me *exaudisti me*, Ps. Spl. 118, 26 : Ps. Th. 114, 1, 2. We ðis nǽfre gehýrdon hæleðum cýðan *we have never heard this declared to men*, Elen. Kmbl. 1317 ; El. 660 : 727 ; El. 364 : Apstls. Kmbl. 125 ; Ap. 63. Gáþ and cýðaþ Iohanne ða þing ðe gé gehýrdon and gesáwon *euntes renunciāte Ioanne quœ audistis et vĭdistis*, Mt. Bos. 11, 4 : Lk. Bos. 7, 22 : Jn. Bos. 14, 24. Gehýr me Drihten God mín *exaudi me Dŏmĭne Deus meus*, Ps. Spl. 12, 3 : 68, 17 : 142, 7. Gehýre gé ðæs sáwendan bigspell *vos audīte parăbŏlam sēminantis*, Mt. Bos. 13, 18. Gehýraþ me *audīte me*, Ps. Th. 65, 14. Ǽr he dómdæges dyn gehýre *before he shall hear doomsday's din*, Salm. Kmbl. 546 ; Sal. 272 : Exon. 13 a ; Th. 22, 31 ; Cri. 360. Wearþ Stephanes bén gehýred *Stephen's prayer was heard*, Homl. Th. i. 52, 32, 33. **II. v. intrans.** *To hear*; audīre :—Gehýran mæg ic rúme *I can hear from far*, Cd. 32 ; Th. 42, 14 ; Gen. 673. Se ðe hæbbe eáran to gehýrenne, gehýre *qui hăbet aures audiendi, audiat*, Mt. Bos. 13, 9. Geworden ic eom swá swá man ná gehýrende *factus sum sīcut hŏmo non audiens*, Ps. Spl. 37, 15 : Mt. Bos. 13, 13. Ic gehýre *audio*; ðú gehýrst *audis*; he gehýrþ *audit*, Ælfc. Gr. 30; Som. 33, 57, 58. Deáfe gehýrdon *the deaf heard*, Andr. Kmbl. 1154; An. 577. Ðé-læs híg mid eárum gehýron *nequando auribus audiant*, Mt. Bos. 13, 15 : Mk. Bos. 4, 12. **III.** *to obey*; obēdīre :—Hie Drihtne gehýrdon *they obeyed the Lord*, Cd. 196 ; Th. 245, 2 ; Dan. 456 : Exon. 62 a ; Th. 228, 26 ; Ph. 444 : Ps. Th. 17, 42.

ge-hýran ; *p.* de ; *pp.* ed *To hire*; conducere, locare :—Ðæs híredes ealdor gehýrde wyrhtan *the chief of the household hired workmen*, Homl. Th. ii. 74, 7. Behíring *vel* gehýred feóh *locatio*, Ælfc. Gl. 13 ; Som. 57, 123 ; Wrt. Voc. 20, 60. v. be-híring.

ge-hyrdan ; *p.* de ; *pp.* ed ; *v. trans. To harden, to strengthen*; durare, indurare, Exon. 88 a ; Th. 331, 26 ; Vy. 74. v. hyrdan.

ge-hyrde. v. ge-hyrwan.

ge-hyrdnes, -ness, e ; *f. A keeping, guard, watch*; custōdia :—Sete gehyrdnessa múþe mínum *pōne custōdiam ōri meo*, Ps. Lamb. 140, 3.

ge-hyrned ; *part. Horned*; cornūtus :—Gehyrned *cornūtus*, Ælfc. Gr. 43 ; Som. 45, 17 : Ex. 34, 29, 30. Byþ he ymlíce gehyrned *he is equally horned*, Bd. de nat. rerum ; Wrt. popl. science 15, 2 ; Lchdm. iii. 266, 22.

ge-hýrnes, se ; *f. A hearing, report*; auditus :—Of gehýrnysse gé gehýraþ, and gé ne ongytaþ *auditis, et non intelligetis*, Mt. Bos. 13, 14 : Blickl. Homl. 55, 31. DER. hýrnes.

ge-hyrst, e ; *f. An ornament*; ornāmentum :—Man reliquias rēran onginneþ, háliga gehyrste *man begins to elevate relics, holy ornaments*, Menol. Fox 146 ; Men. 74. Gehyrsto *phalerœ*, Lye.

ge-hýrst *hearest*, Ælfc. Gr. 30 ; Som. 33, 57, 58 ; *2nd sing. pres. of* ge-hýran.

ge-hyrstan ; *p.* -hyrste ; *pp.* -hyrsted, -hyrst *To adorn, ornament, decorate*; adornāre, ornāre, decŏrāre :—He gehyrsteþ wél *he adorns the metal work*, Exon. 88 a ; Th. 331, 27 ; Vy. 74. Golde gehyrsted *adorned with gold*, Elen. Kmbl. 662 ; El. 331 : Andr. Kmbl. 90 ; An. 45. Ða bióþ mid fetlum gehyrste *who are adorned with belts*, Bt. 37, 1 ; Fox 186, 6.

ge-hyrstan, -hierstan ; *p.* -hyrste ; *pp.* -hyrsted, -hyrst *To fry, roast*; frīgĕre :—Hí cócas gehyrstan *cooks roasted them*, Ps. Th. 101, 3. Gehyrsted síe *frigētur*, Cot. 87. Gehyrst hláf *frixius pānis*, Ælfc. Gl. 66 ; Som. 69, 69 ; Wrt. Voc. 41, 23. Ðæt ðas sídan ðe gehirsted is *eat this side that is roasted*, Shrn. 116, 6. [*O. H. Ger. giharstit frixus.*]

ge-hyrstan ; *p.* te *To murmur* :—Gehyrston *murmurabant*, Lk. Skt. Lind. 15, 2.

ge-hýrsum, -hiérsum ; *adj. Obedient, obliging, ready to serve*; obēdiens, officiōsus :—Wæs Abraham Gode gehýrsum *Abraham was obedient to God*, Boutr. Scrd. 23, 4 : Homl. Th. ii. 162, 26 : Mt. Bos. 6, 24. Éstful *vel* gehýrsum *officiōsus*, Ælfc. Gl. 115 ; Som. 80, 54 ; Wrt. Voc. 61, 32. Hí woldon him beón gehýrsume *they would be obedient to him*, Chr. 1083 ; Erl. 217, 6. [*O. H. Ger. and Ger. gehōrsam.*]

ge-hýrsumian, -hiérsumian ; *p.* ode, ade ; *pp.* od, ad. **I.** *to obey, be obedient to*; obēdīre, pārēre :—Ic gehýrsumige obēdio, Ælfc. Gr. 30, 5 ; Som. 34, 56 : *pāreo*, 26, 2 ; Som. 28, 48. Ðe heora lustum gehýrsumiaþ *who obey their lusts*, Homl. Th. ii. 82, 13. **II.** *to make obedient, bring into subjection*; subjicĕre :—Ðæt he him Norþ-Wealas gehýrsumode [gehiérsumade, col. 1] *that he might make the North Welsh*

obedient to him, Chr. 853 ; Th. 122, 22, col. 2. [*O. H. Ger. gihôrsamón to obey.*]

ge-hýrsumlíce ; *adv. Obediently*; obēdienter, Som. Ben. Lye.

ge-hýrsumnys, -nyss, e ; *f. Obedience, subjection*; obēdientia :—God wolde fandian Abrahames gehýrsumnysse *tentāvit Deus Abraham*, Gen. 22, 1 : Boutr. Scrd. 19, 26 : Chr. 1091 ; Erl. 228, 3.

ge-hyrtan ; *p.* -hyrte ; *pp.* -hyrted, -hyrt [hyrtan *to hearten, encourage*; heorte *the heart*] *To encourage, animate, refresh*; confortāre, animāre, refrigerāre :—Beó ðú húru gehyrt, and hicg þegenlíce *be thou only encouraged, and strive nobly*, Jos. 1, 18. Ðæt ðínre wylne sunu sý gehyrt *that the son of thy slave may be refreshed*; ut refrigeretur filius ancillœ tuæ, Ex. 23, 12. Drihten us gehyrte *the Lord encouraged us*, Homl. Th. ii. 538, 12. Mín werod gehyrted wæs *my army was encouraged*, Nar. 8, 17. Gehyrtan *refocillāre, confortāre*, Hpt. Gl. 478. Se læg dæg and niht geswógen. He wearþ ðá gehyrt *he lay day and night senseless. He then revived*, Homl. Th. ii. 356, 27.

ge-hýrþ *hears*, Ælfc. Gr. 30 ; Som. 33, 58 ; *3rd sing. pres. of* ge-hýran.

ge-hyrwan ; *p.* de ; *pp.* ed *To make game of, despise, disparage, traduce, vex, oppress*; cavillāri, contemnĕre, detrăhĕre :—Elene ne wolde ðæs wilgifan word gehyrwan *Elene would not despise the dear prince's word*, Elen. Kmbl. 442 ; El. 221 : Exon. 39 b ; Th. 131, 27 ; Gú. 462. He gehyrweþ fuloft hálge láre *he very often traduces holy lore*, 117 a ; Th. 449, 12 ; Dóm. 70. Hý ðæs láreowes word ne gehyrwdon *they despised not the teacher's words*, 14 b ; Th. 29, 8 ; Cri. 459. Beóþ ða gehyrwede *they are despised*, Ps. 52, 6 ; Ps. Grn. ii. 150, 6. Seó langung hine swíðe gehyrde and ðreáde *that longing much oppressed and afflicted him*, Blickl. Homl. 113, 14. Hí wurdon gehergode and gehyrde *they were wasted and oppressed*, Jud. 10, 8. [*O. H. Ger. harwjan exasperare.*]

ge-hyscan ; *p.* te *To mock, deride* :—Úre fýnd gehyscton us *inimici nostri subsannauerunt nos*, Ps. Lamb. 79, 7. Gehiscþ *abominabitur*, 5, 8.

ge-hyspan ; *p.* de, te *To deride, mock, scoff*; insultāre, exprobrāre, Hpt. Gl. 441. Se god ðe on heofonum ys híg gehyspþ *qui habitat in cœlis irridebit eos*, Ps. Th. 2, 4.

ge-hyspendlíc ; *adj. Despicable, abominable* :—Hí syndon gehyspendlíc geworden *sunt abominabiles facti*, Ps. Lamb. 13, 1.

ge-hýt *hides*, Bt. 39, 8 ; Fox 224, 11 ; *3rd sing. pres. of* ge-hýdan.

ge-hýðegod ; *part. p.* :—Gehyðegode *expedita*, Gl. Prud. 229.

ge-hýðelíc ; *adj. Favourable, seasonable*; opportunus, Ps. Spl. 31, 7 ; Hpt. Gl. 470.

ge-hýþnes, se ; *f. Opportunity.*

ge-hýwan ; *p.* de ; *pp.* ed ; *v. trans. To shew*; ostendere :—Ðú gehýwdest ðam eorle bán Iosephes *thou shewest the man the bones of Joseph*, Elen. Kmbl. 1570 ; El. 787. v. geýwan.

ge-hywian ; *p.* ode ; *pp.* od. **I.** *to form, fashion*; fingĕre :—Se ðe gehywode synderlíce heortan heora *qui finxit singillātim corda eōrum*, Ps. Lamb. 32, 15. **II.** *to seem, pretend*; sĭmulāre :—Ðeáh ðe hit swá gehywod wǽre *though it seemed so*, Job Thw. 166, 6. Mid gehywedan móde *with feigned mind*, Th. Ap. 3, 2. v. ge-hiwian.

ge-hywung *a form, fashion, shape*, Ps. Spl. C. 102, 13. v. ge-hiwung.

ge-ícan, -íecan, -ýcan, -iécan ; *p.* -ícte, -íhton ; *pp.* -íced, -íct *To eke, increase, add, enlarge*; augere, extendere :—Heó ongan his mǽg-burge geícean sunum and dohtrum *she began his kindred to increase with sons and daughters*, Cd. 56 ; Th. 69, 8 ; Gen. 1132. Eall geíceaþ *increase all things*, 74 ; Th. 91, 18 ; Gen. 1514. Ofer eall ðæt geícte *adjecit hoc supra omnia*, Lk. Bos. 3, 20. Æðelinga rím feorum geícte *he increased the number of men with lives*, 58 ; Th. 70, 33 ; Gen. 1162. Bizantium wæs fram Constantino geíced *Byzantium was enlarged by Constantine*, Ors. 3, 7 ; Bos. 61, 10 : Th. Diplm. A.D. 864 ; 125, 19. v. écan.

ge-ícendlíc ; *adj. Added to, adjective*; adjectivus :—Geícendlíc nama *a noun adjective*, Som.

ge-íchte, -íhton *added*; *p.* of ge-ícan.

ge-ídlian ; *p.* ade *To make* or *become vain, empty* :—Giídladest *vacuasti*, Rtl. 33, 3. Giídlege *vanescat*, 98, 24.

ge-íermed, -irmed ; *p.* de ; *pp.* ed ; *v. trans. To afflict*; Past. 28, 1 ; Swt. 188, 16.

ge-iéwan, ýwan ; *p.* de ; *pp.* ed ; *v. trans. To shew*; ostendere :—He ðæt beácen geseah ðæt him on heofonum ǽr geiéwed wearþ *he saw the beacon which to him before in heaven was shewn*, Elen. Grm. 102. v. ýwan, eáwan.

ge-íhtnyss, e ; *f. An addition, epact*, Lye.

ge-illerocaþ *surfeited*; crapulatus, Ps. Spl. C. 77, 71.

ge-incfullian ; *p.* ade ; *pp.* ad *To offend, scandalize* :—We ðonne ðyles geincfulligæ hiæ *ut autem non scandalizemus eos*, Mt. Kmbl. Rush. 17, 27. Se ðe ne biþ in me geincfullad *qui non fuerit scandalizatus in me*, 11, 6 : 15, 12.

ge-inlagian ; *p.* ode ; *pp.* od [ge, inlagian] *To inlaw, to restore to the protection of the law*; inlagare, intra legum protectionem accipere :—Man geinlagode Swegen eorl *Earl Sweyn was inlawed*, Chr. 1050 ; Erl. 176, 6. Willem se cyng Eádgár geinlagode and ealle his men *William the king inlawed Edgar and all his men*, 1074 ; Erl. 212, 5.

ge-innian; *pp.* -innod *To bring in, include, to fill, supply, charge;* præstare, includere :—Wolde God geinnian ðone lyre *God would supply the loss,* Homl. Th. i. 12, 24 : 180, 18 : L. In. 62 ; Th. i. 142, 4 : Th. Apol. 23, 7. Súsle geinnod *with sulphur filled,* Cd. 2 ; Th. 3, 28 ; Gen. 42. He hæfþ geinnod ðæt ær geútod wæs *he has included what before was excluded,* Cod. Ex. p. 1.

ge-inseglian, -insegelian ; *p.* ode ; *pp.* od, ud *To seal, to impress with a seal ;* signare, obsignare :—Hú næron ðás geinseglude on mínum goldhordum ? *whether these thingis ben seelid in myn tresouris ?* Wyc ; nonne hæc signata in thesauris meis ? Deut. 32, 34. Annas and Caiphas ðæt loc geinseglodon *Annas et Caiphas illud claustrum obsignarunt,* Nicod. 14 ; Thw. 7, 2. Lá hú ne ðás þingc geinseglode on goldhordum mínum *nonne hæc signata in thesauris meis,* Cant. Moys. Isrl. Lamb. 194 a, 34 : Th. Apol. 20, 10 : 21, 2.

ge-irgan ; *p.* de ; *pp.* ed *To make cowardly, terrify,* Jos. 2, 9. v. ge-yrgan.

ge-irman ; *p.* de ; *pp.* ed *To afflict ;* afflígĕre :—Ðæt hie elles ne síen geirmed *that they be not altogether afflicted,* Past. 28, 1 ; Swt. 189, 16 ; Hat. MS. 36 b, 5. v. ge-yrman.

ge-iukod ; *part. p. Yoked* :—Geiukodan oxan *junctis bobus,* Th. An. 19, 19.

ge-lác, es ; *n.* [lácan *to move* as e.g. *the waves do, to sport, play*] *Motion, commotion, tumultuous assembly, play* :—Sealtýða gelác *the tossing of the salt waves,* Exon. 82 a ; Th. 308, 5 ; Seef. 35 : 115 a ; Th. 442, 3 ; Kl. 7 : Ps. Th. 118, 136 : Bt. Met. Fox 20, 345 ; Met. 20, 173 : 26, 57 ; Met. 26, 29. Sweorda gelác *the play of swords,* i.e. *battle,* Beo. Th. 2084 ; B. 1040 : 2340 ; B. 1168. Gelác engla and deófla *hosts of angels and devils,* Exon. 21 a ; Th. 56, 5 ; Cri. 896. Ðurh heard ge'ác *through hard fortune,* Andr. Kmbl. 2185 ; An. 1094. v. bord-, lind-, lyft-, scín-gelác.

ge-lácan ; *p.* -léc *To play a trick on, delude :*—On hý geléc ðæt hý mid him wunnon *he deluded them into making war with him,* Ors. 3, 7 ; Bos. 60, 2. [Cf. *Icel.* leika á *to play a trick on.*]

ge-lácian, ic, he -lácige ; *p.* ode ; *pp.* od [lác *a gift*] *To give, bestow, present one with a thing ;* munerare, munerare aliquem aliqua re :— Gelácige mid eádigum gifum *donis beatis munerabit.* Mid ēcum dō, mid hálgum dínum, wuldre beón gelácod *eterna fac, cum sanctis tuis, gloria munerari,* Te Deum, 21 ; Lamb. 195 b, 21.

ge-lácnian, -lácnigan ; *p.* ode ; *pp.* od *To heal, cure ;* sānāre, mĕdēri :— Gif hine mon gelácnian mæge *if he can be healed,* L. Alf. pol. 69 ; Th. i. 98, 8. His sáwle wunda dædbētende gelácnian *to heal the wounds of his soul by doing penance,* Homl. Th. i. 124, 14. Gelácnigan, Exon. 27 a ; Th. 80, 19 ; Cri. 1309. Ic gelácnige *mĕdeor,* Ælfc. Gr. 27 ; Som. 29, 56. Gelácna ðú hý *heal thou them,* Hy. 1, 5 ; Hy. Grn. ii. 280, 5. He wæs gelácnod *he was cured,* Ors. 3, 7 ; Bos. 61, 44. Mon geseah hine laman gelácnian *people saw him healing the lame,* Blickl. Homl. 177, 16. Hine gelácnode *curam ejus egit,* Lk. Skt. 10, 34, note.

ge-lád, es ; *n. A way, path, road, course ;* via, trāmes :—Ofersór he uncúþ gelád *he traversed an unknown way,* Cd. 145 ; Th. 181, 9 ; Exod. 58 : 158 ; Th. 197, 27 ; Exod. 313. Ofer deóp gelád *over the deep way,* i.e. *ocean,* Andr. Kmbl. 380 ; An. 190 : Exon. 51 b ; Th. 179, 23 ; Gú. 1266. v. fen-gelád. See Kmbl. Cod. Dipl. iii. xxvi.

ge-ládian ; *p.* ode ; *pp.* od *To clear, vindicate, excuse ;* purgare, exculpare, excusare :—Geládige hine *let him clear himself,* L. C. S. 44 ; Th. i. 402, 5 : 29 ; Th. i. 392, 16. Ðonne biþ he self geládod wiþ hine selfne *then shall he himself be acquitted towards himself,* Past. 21 ; Swt. 151, 18 ; Hat. MS.

ge-læccan, -læccean ; he -læcþ ; *p.* he -læhte, *pl.* -læhton ; *pp.* -læht *To take, catch, seize, apprehend, comprehend ;* capere, arripere, comprehendere :—Ðæt híg woldon hine gelæccean and to cyninge dôn, Jn. Bos. 6, 15. Híg gelæhton hys hand, Gen. 19, 16 : Mk. Bos. 9, 18. Ða Englisce men gelæhton of ðám mannon má . . . *the English men captured of those men more* . . . , Chr. 1087 ; Erl. 225, 26. Hwæt gelæhtest ðú *quid cepisti,* Th. An. 22, 5. Germanus gelæhte ðone pistol æt Gregories ærendracan and hine totær *Germanus took the letter from Gregory's messenger and tore it to pieces,* Homl. Th. ii. 122, 29. Hét sóna gelæccan Stranguilionem *he bade seize Stranguilio at once,* Th. Apol. 25, 25. Ðis þing ic gelæhte *I have comprehended this thing ;* hanc rem apprehendi, Ælfc. Gr. 7 ; Som. 6, 24.

ge-lædan, -lēdan ; *part.* -lædende ; he -lædeþ, -lǽdt, -lǽt, *pl.* -lǽdaþ ; *p.* ic, he -lǽdde, ðú -lǽddest, *pl.* -lǽddon ; *impert.* -læd, *pl.* -lǽdaþ ; *subj. pres.* -lǽde, *pl.* -lǽden ; *pp.* -lǽded, -lǽdd, -lǽd *To lead, conduct, bear, bring, derive, bring out, bring forth, produce, bring up ;* dūcĕre, dedūcĕre, ägĕre, indūcĕre, deferre, perferre, derīvare, edūcĕre, prodūcĕre, edūcāre :— He wile folc gelædan in dreáma dreám *he will lead the people into joy of joys,* Exon. 16 a ; Th. 36, 21 ; Cri. 579 : 73 b ; Th. 274, 13 ; Jul. 532. Gelædende híg nítenum prodúcens fænum jumentis, Ps. Spl. 103, 15. Ic gelæde derivo, Ælfc. Gl. 61 ; Som. 68, 46 ; Wrt. Voc. 39, 30. Me engel to ealle gelædeþ spówende spēd *an angel will bring to me all prosperous success,* Exon. 36 a ; Th. 117, 15 ; Gú. 224 : 33 b ; Th. 107, 9 ; Gú. 56. Ðe to lífe gelædt *quæ dūcit ad vitam,* Mt. Bos. 7, 14. Ðe to forspilled-

nesse gelæt *quæ dūcit ad perdītiōnem,* 7, 13. Ða ðe feorran ðider feorh gelædaþ *they who lead their life thither from afar,* Andr. Kmbl. 564 ; An. 282. Ðú gelæddest me *deduxisti me.* Ps. Spl. 60, 3 : Ps. Th. 114, 8. Moyses fyrde gelædde *Moses led the march,* Cd. 145 ; Th. 181, 17 ; Exod. 62 : 162 ; Th. 203, 2 ; Exod. 397. He gelædde me *edūcāvit me,* Ps. Spl. C. 22, 2. Ðæt gē on fāra folc feorh gelæddon *that ye would lead your life among a hostile people,* Andr. Kmbl. 860 ; An. 430. Gelæd me on rihtwísnesse ðínre *deduc me in justītia tua,* Ps. Lamb. 5, 9 : 138, 23. Ne gelæd ðú us on costnunge *ne nos indūcas in tentātiōnem,* Mt. Bos. 6, 13. Ðæt ðú gelæde hláf of eorþan *ut edūcas pānem de terra,* Ps. Spl. 103, 16. His líchoma wæs to Turnum gelæded *corpus Turōnis delūtum,* Bd. 4, 18 ; S. 587, 9, 12. He wæs gelædd óþ ða þriddan heofonan *he was led to the third heaven,* Bd. de nat. rerum ; Wrt. popl. science 2, 4 ; Lchdm. iii. 232, 26. He wæs fram Háligum Gástum gelæd on sumum wēstene *ăgēbātur a spīritu in desertum,* Lk. Bos. 4, 1 : Chr. 693 ; Erl. 43, 19.

ge-lædenlíc ; *adj. What is easily led* or *beaten out, malleable ;* ductilis :—On býman gelædenlícum *in tubis ductilibus,* Ps. Spl. M. 97, 6.

ge-léafa, an ; *m. Belief, faith ;* fides :—He wolde ðone Cristes geléafan gerihtan *he would set right the faith of Christ,* Chr. 680 ; Erl. 41, 14. v. ge-leáfa.

ge-léafa, an ; *m. Leave, permission ;* permissio :—Be ðæs cynges geléafan *by the king's leave,* Chr. 1043 ; Erl. 170, 1.

ge-léafan *to believe.* v. ge-lýfan.

ge-léafan ; *p.* de ; *pp.* ed *To leave ;* derelinquĕre :—Ðē geléafed is se þearfa *tibi derelictus est pauper,* Ps. Lamb. second 9, 14. Ðæt geléafed wæs *quod superfuit,* Mt. Kmbl. Lind. 15, 37.

ge-læht, *pl.* ge-læhte ; *pp. Taken ;* captus, comprehensus :— Híg beóþ gelæhte *comprehenduntur,* Ps. Lamb. second 9, 2 ; *pp.* of ge-læccan.

ge-læmed ; *part. Lamed ;* claudus factus :—Gif eaxle gelæmed weorþeþ *if a shoulder be lamed,* L. Ethb. 38 ; Th. i. 14, 2.

ge-længed, -længd ; *part. Lengthened, drawn out :*—Eardberengnes mín afeorrad oððe gelængd is *incolatus meus prolongatus est,* Ps. Lamb. 119, 5. v. langian.

ge-lǽr ; *adj. Void, empty ;* vacuus, Som. [*Laym.* i-lær.]

ge-léran ; ic -lære, ðú -lærest, -lærst, he -læreþ, -lérþ, *pl.* -læraþ ; *p.* -lærde ; *pp.* -lǽed, -lǽrd *To teach, educate, instruct, advise, persuade, induce ;* dōcēre, erūdīre, persuādēre :—We ðē mágon eáðe sēlre gelǽran *we may easily teach thee better,* Andr. Kmbl. 2706 ; An. 1355 : Beo. Th. 562 ; B. 278. Se gelærde peohtas to fullwihte *he brought the Picts by his teaching to baptism,* Shrn. 89, 33. Gif he ða cwēne gespannan and gelǽran mihte ðæt heó brúcan wolde his gesynscipe *si regīnæ posset persuādēre ejus ūti connūbio,* Bd. 4, 19 ; S. 587, 30. Næfre ðú gelǽrest, ðæt ic dumbum and deáfum deófolgieldum gaful onhāte *never shalt thou induce me, that I promise tribute to dumb and deaf idols,* Exon. 67 b ; Th. 251, 22 ; Jul. 149. Ðæt gebrōcode flǽsc gelǽrþ ðæt upahæfene mōd *the afflicted flesh teaches the proud mind,* Past. 36, 7 ; Swt. 257, 14 ; Hat. MS. 48 a, 22. Hí á sibbe gelǽraþ *they shall ever teach peace,* Exon. 89 a ; Th. 334, 23 ; Gn. Ex. 20. He gelǽrde ealle Crēcas ðæt hý Alexandre wiðsócon *he persuaded all the Greeks to strive against Alexander,* Ors. 3, 9 ; Bos. 64, 6 : Cd. 222 ; Th. 290, 10 ; Sat. 413 : Th. Apol. 10, 18. Ðú us gelǽrdest ðæt we Hēlende hēran ne sceoldon *thou persuadest us that we should not obey the Saviour,* 214 ; Th. 268, 10 ; Sat. 53. Me gelǽr *dōce me,* Ps. Th. 118, 68. Gelǽred *doctus,* Ælfc. Gr. 8 ; Som. 7, 41 : 39 ; Som. 42, 47, 56. Ic eom gelǽred *dōceor ;* ðú eart gelǽrd *dōcēris ;* he is gelǽrd *dōcētur,* 27 ; Som. 29, 21. Beóþ gelǽrede gē ðe dēmaþ eorþan *erūdīmīni qui judīcātis terram,* Ps. Spl. 2, 10.

ge-lǽmed ; *part. p. Learned ;* doctus :—Albinus wæs betst gelǽred *Albinus was most learned,* Bd. Pref ; S. 471, 23. He is gleáwest úre gelǽred *he is the most skilfully instructed of us,* H. R. 11, 9. Mid gelǽredre handa he swang ðone top *with skilful hand he whipped the top,* Th. Apol. 13, 13.

ge-lǽrednes, -ness, -nys, -nyss, e ; *f. Learning, knowledge, skill ;* erūditio, pĕrītia :—Wæs Cúþberhte swá mycel getýdnes and gelǽrednes to sprecanne *Cudbercto tanta ĕrat dīcendi pĕrītia,* Bd. 4, 27 ; S 604, 19. Ðá se cyning his gelǽrednesse geseah *cujus erūdītiōnem vidēns rex,* 3, 7 ; S. 529, 46. On gelǽrednysse *in erūdītiōne,* 3, 21 ; S. 551, 13.

ge-lǽstan ; to -lǽstenne ; he -lǽsteþ, -lǽst ; *p.* -lǽste ; *pp.* -lǽsted, -lǽst.   **I.** *to do, perform, accomplish, fulfil, discharge, execute, pay ;* făcĕre, perfĭcĕre, patrāre, præstāre, persolvĕre :—Ic náuht ne tweóge ðæt ðú hit mæge gelǽstan *I doubt not that thou canst perform it,* Bt. 36, 3 ; Fox 174, 31 : Elen. Kmbl. 2329 ; El. 1166. Ic ða wǽre sóþe gelǽste *I will truly execute the compact,* Cd. 106 ; Th. 139, 11 ; Gen. 2308. Gif we sóþ and riht symle gelǽstaþ *if we always perform truth and right,* Hy. 7, 75 ; Hy. Grn. ii. 288, 75. Beót eal wið ðē he sóþe gelǽste *he truly fulfilled all his promise to thee,* Beo. Th. 1053 ; B. 524 : Byrht. Th. 132, 13 ; By. 15. Ðe ær Godes hyldo gelǽston *who ere executed God's pleasure,* Cd. 17 ; Th. 21, 9 ; Gen. 321 : Chr. 878 ; Erl. 81, 16 : Ors. 4, 9 ; Bos. 91, 17. Hwænne man ðæt gelǽste *when it shall be fulfilled,* L. Edg. H. 7 ; Th. i. 260, 13 : L. In. 4 ; Th. i. 104, 10 : L. E. G. 6 ; Th. i. 170, 4. He hæfde wordbeót leófum gelǽsted *he had performed*

his promise to the beloved, Cd. 132; Th. 167, 7; Gen. 2762: 109; Th. 144, 25; Gen. 2395. Ðæt gafol wæs gelǽst the tribute was paid, Chr. 1012; Erl. 146, 10: 1007; Erl. 141, 13. II. to accompany, follow, attend, serve; cŏmĭtāri, sĕqui, persĕqui :—He wolde gelǽstan freán to gefeohte he would accompany his lord to the fight, Byrht. Th. 132, 5; By. 11. Mec mín gewit gelǽsteþ my intellect attends me, Exon. 38 a; Th. 125, 1; Gú. 347. Swá lange swá me líf gelǽst as long as life attends me, L. Edg. S. 12; Th. i. 276, 19: 16; Th. i. 278, 12. Ðæt hý him æt ðám gewinnum gelǽston that they would serve him in the wars, Ors. 4, 9; Bos. 91, 30. Ðæt hine ðonne wíg cume leóde gelǽsten that the people serve him when war comes, Beo. Th. 47; B. 24. III. v. intrans. To continue, remain, last, endure; mănēre, dŭrāre :—Ne mæg hús on munte lange gelǽstan a house cannot long remain on a mountain, Bt. Met. Fox 7, 37; Met. 7, 19. Ðæt eówre blǽda gelǽston ut fructus vester măneat, Jn. Bos. 15, 16.

ge-lǽswian; p. ode; pp. od [lǽswian to feed] To feed :—Gilesua pasce, Jn. Skt. Lind. 21, 17. Ic eom gelǽswod pastus sum, Ælfc. Gr. 33; Som. 36, 44.

ge-lǽt leads, Mt. Bos. 7, 13; 3rd sing. pres. of ge-lǽdan.

ge-lǽtan, -lǽtan; p. -leórt; pp. -lǽten To allow, make over to any one :—Eádgár æðeling wearþ belandod of ðám ðe se eorl him ǽror to handa gelǽten hæfde Edgar Atheling was deprived of those lands which the earl had before made over to him, Chr. 1091; Erl. 227, 24. Ðú gelétas permittas, Rtl. 59, 5. Ne geleórt ǽnigne monno to fylganne non admisit quemquam sequi, Mk. Skt. Rush. 5, 37. Ðú gileórtest concessisti, Rtl. 76, 36.

ge-lǽte, es; pl. -lǽtu; n. [lǽtan to let go, leave] A going out, ending, meeting; exitus, occursus :—To wega gelǽtum to the meetings of ways, Mt. 22, 9. Twegra wega gelǽtu meetings of two ways, Cot. 110. Æt ðæra wæga gelǽte, Gen. 38, 21. v. weggelǽte.

ge-lafian; p. ode, ede; pp. od, ed To wash, lave, refresh; refĭcĕre :—He winedryhten his wætere gelafede he laved his liege lord with water, Beo. Th. 5438; B. 2722.

ge-lagian; p. ode; pp. od To establish by law, constitute, decree; lēge sancīre :—Ðe Eádgár cyningc gelagode which king Edgar decreed, L. Eth. ix. 7; Th. i. 342, 13. Hú hit gelagod wæs how it was constituted, L. Ælf. P. 41; Th. ii. 382, 17. Ðe gelagod is to gedwolgoda weorðunge that is appointed for the worship of false gods, Swt. Rdr. 105, 27.

ge-lagu; n. (?) A collection of water :—Ofer holmǫ gelagu over ocean's flood, Exon. 82 a; Th. 309, 28; Seef. 64. v. lagu.

ge-landa. v. ge-londa.

ge-landian; p. ode; pp. od. I. to land, arrive; accedere ad terram, Som. [Cf. ge-lendan.] II. to enrich with lands or possessions; terris locupletare :—Ðe gelandod sý who has lands, L. Lund. 11. Opposed to be-landian. v. be-landian.

ge-lang, -long; adj. Along (in the phrase along of), belonging, depending, consequent :—Æt ðé is úre lýf gelang our life is along of thee (thou hast saved our lives, A. V.), Gen. 47, 25. Seó gescyldnys is æt úrum Fæder gelang protection comes from our Father, Homl. Th. i. 252, 4: Ps. Th. 61, 1: Beo. Th. 2757; B. 1376. Nis me wiht æt eów leófes gelong I am not dependent upon you for anything dear, Exon. 37 a; Th. 121, 5; Gú. 284: 115 b; Th. 444, 11; Kl. 45. Ðæt wæs swíðor on ðam gelang that was rather owing to this reason, Ors. 4, 10; Bos. 94, 35. Gif hit on preóste gelang sý if it be along of the priest, L. M. I. P. 42; Th. ii. 276, 15: Bd. 3, 10; S. 534, 37. On heofonum sind láre gelonge instruction comes from heaven, Exon. 36 a; Th. 117, 12; Gú. 223. Frægn se Scipio hine on hwý hit gelang wǽre Scipio asked him to what it was owing, Ors. 5, 3; Bos. 103, 42. Ðær is help gelong help comes from there, Exon. 75 a: Th. 281, 13; Jul. 645: 83 a; Th. 313, 8; Seef. 121. [Laym. ilong: O. Sax. gilang.]

ge-langian, -langjan; p. ode; pp. od; v. trans. [ge, lang·an to long for] To call for, send for, deliver, liberate; convocare, arcessere, accersire, liberare :—Ðú gelangast to ðé ðíne leófostan frýnd thou shalt call to thee thy most beloved friends, Jos. 2, 18. Gelangode to him ða bróðru convocavit ad se fratres, Greg. Dial. 2, 3. He hét gelangian ðone hálgan láreów he ordered the holy teacher to be sent for, Homl. Th. ii. 308, 5. He gelangode him to his swustur he sent for his sister, i. 86, 30. He bæd ðæt him man sumne mæsse-preóst gelangode he asked them to send for a priest, ii. 26, 9. Ic gelangige arcesso [MS. accerso], Ælfc. Gr. 28, 1; Som. 30, 35. Wearþ ðá eft gelangod se geleáffulla apostol of ðam íglande so was the faithful apostle liberated from that island, Ælfc. T. Grn. 16, 28.

ge-lǽst, es; n. [v. ge-lǽstan] Duty, due; officium :—To ǽlcum ðara gelǽste to each of those duties, L. Ædelst. 3; Th. i. 230, 23: 232, 5. Gelǽst votum, Ps. 64, 2, Blickl. Gl. [Cf. fullǽst, and O. Sax. gilêsti an act, deed.]

ge-lǽstfull; adj. Helpful, officious :—Ðæt ǽlc man wǽre óðrum gelǽstfull that every man should be helpful to other, L. Ædelst. 5, 4; Th. i. 237, 11.

ge-laþ; adj. Hostile :—Gelaþe the foes, Cd. 153; Th. 190, 28, note;

Exod. 206, v. láðe, 207, 3; Exod. 461; and cf. ge-fýnd. [Owl and Night. ilað.]

ge-laðian; p. ode, ade, ede; pp. od, ad, ed To invite, bid, call, summon, assemble, congregate; invītāre, vŏcāre, arcessĕre, ciere, congrĕgāre :—Mágon we Ioseph to us geladian can we invite Joseph [to come] to us, Nicod. 20; Thw. 10, 3: Bd. 4, 1; S. 563, 34. Ic gelangige oððe geladige cieo, Ælfc. Gr. 37; Som. 39, 26: 30, 5; Som. 34, 52. Sum man worhte mycele feorme, and manega geladode hŏmo quidam fēcit cœnam magnam, et vocavit multos, Lk. Bos. 14, 16: Chr. 449; Erl. 13, 2. He to Bethania his þegna gedryht geladade he assembled his band of disciples in Bethany, Exon. 14 b; Th. 29, 5; Cri. 458. Geladede se gesíþ hine to his hâme the earl invited him to his home, Bd. 3, 22; S. 553, 29. Ðonne ðú byst to gyftum geladod cum invītātus fŭeris ad nuptias, Lk. Bos. 14, 8. Ða ðe geladode wǽron, ne synt wyrðe qui invītāti ĕrant, non fŭerunt digni, Mt. Bos. 22, 8: Jn. Bos. 2, 2. Wǽron ealle ða wíf beforan Rômāna witan geladode all the women were summoned before the Roman senators, Ors. 3, 6; Bos. 58, 21.

ge-laðung, e; f. A congregation, assembly, church; congrĕgātio, convŏcātio, ecclēsia :—Geladung convŏcātio, Ælfc. Gl. 30; Som. 61, 51; Wrt. Voc. 26, 50. On middele geladunge ic hērige ðé in mĕdio ecclēsiæ laudābo te, Ps. Spl. 21, 21. On Godes geladunge in God's church, Homl. Th. i. 412, 1, 21: 502, 6. Ic gelýfe on ða hálgan geladunge I believe in the holy church, ii. 596, 21: 598, 11. On geladunga háligra in ecclēsia sanctōrum, Ps. Spl. 88, 6. On gesamningum oððe on geladungum ic bletsige ðé in ecclēsiis bĕnĕdīcam te, Ps. Lamb. 25, 12.

ge-laured of or belonging to laurels; laureus, Som.

geld, es; n. A payment, society, worship, service, Ælfc. Gl. 35; Som. 62, 76: Cot. 76: Prov. 22. v. gild.

geldan, ic gelde, ðú geltst, gelst, he gelt, pl. geldaþ; p. geald, pl. guldon; pp. golden To pay, restore, render, make an offering, serve, worship :—Geld ðæt ðú áht to geldanne redde quod debes, Mt. Kmbl. Lind. 18, 28: Bt. 41, 3; Fox 248, 22, note 27: L. Wih. 12; Th. i. 40, 4, 6: L. H. E. 10; Th. i. 32, 2. v. gildan.

gelde; adj. That has yeaned, brought forth; effeta, Cot. 75.

gelden golden. v. gylden.

ge-leáf leave, license. v. leáf.

ge-leáfa, an; m. [leáfa belief] Belief, faith, confidence, trust; fĭdes, fĭdūcia :—Se rihta geleáfa us tǽcþ, ðæt we sceolon gelýfan on ðone Hálgan Gást the right faith teaches us that we should believe in the Holy Ghost, Homl. Th. i. 280, 22: Elen. Kmbl. 2070; El. 1036. Geleáfa fĭdes, Ælfc. Gr. 12; Som. 15. 54. Dæges ór onwóc leóhtes geleáfan the dayspring of bright belief awoke, Apstls. Kmbl. 131; Ap. 66: Elen. Kmbl. 1928; El. 966. On rihtum geleáfan in right faith, Bt. 6; Fox 14, 31. Hí monige hrædlíce fram deófolgyldum to Cristes geleáfan gecyrdon multos in brĕvi ab idŏlătria ad fĭdem convertĕrent Christi, Bd. 5, 10; S. 624, 9: Chr. 565; Erl. 17, 21. Ðú ðone geleáfan hæfst thou hast the belief, Bt. 5, 3; Fox 12, 11. Nú we wyllaþ secgan eów ðone geleáfan ðe on ðam crēdan stent we will now declare to you the faith which stands in the creed, Homl. Th. i. 274, 23: 292, 9, 10: 294, 8. Habbaþ geleáfan habēte fĭdūciam, Mt. Bos. 14, 27. Ic hæbbe me fæstne geleáfan up to ðam ælmihtegan Gode I have firm trust in the Almighty God above, Cd. 26; Th. 34, 26; Gen. 543: 205; Th. 256, 19; Dan. 643: Andr. Kmbl. 670; An. 335. Eom ic leóhte geleáfan fægre gefylled I am fairly filled with bright belief, Exon. 42 a; Th. 141, 8; Gú. 624: 62 b; Th. 230, 28; Ph. 479: 75 a; Th. 281, 28; Jul. 653. [O. Sax. gi-lôbo: O. H. Ger. ki-lauba: Ger. glaube: and cf. Goth. ga-laubeins.]

ge-leáfful, -full; adj. Full of belief, believing, faithful, holy; fĭdēlis, crēdŭlus :—Heó wundrade hú he swá geleáfful, on swá lytlum fæce, and swá uncýðig, ǽfre wurde gleáwnysse þurhgoten she wondered how he, so full of belief, in so short a space, and so ignorant, could ever be saturated with prudence, Elen. Kmbl. 1916; El. 960. Getreówe, oððe geleáfful fĭdēlis, Wrt. Voc. 74, 27. Cyrce, oððe geleáfful gaderung a church or faithful gathering; ecclēsia, 80, 72. Wyrd gescreáf ðæt he, swá geleáfful, weorþan sceolde Criste gecwēme fortune ordained that he, so full of faith, should become accepted of Christ, Elen. Kmbl. 2093; El. 1048. Ne geleáffulle gecwēme·synd·on cýðnesse his nec habĭti sunt in testamento ejus, Ps. Spl. 77, 41. On geleáffullum bócum in holy books, Ælfc. T. 13, 22. Ealle þing synd ðam geleáffullum acumendlíce omnia sunt possĭbĭlia crēdenti, Boutr. Scrd. 20, 26. Ofer geleáffulle eorþbúgende super fĭdēles terræ, Ps. Th. 100, 6. Ða beorhtan steorran getácniaþ ða geleáffullan on Godes geladunge the bright stars betoken the faithful in God's church, Bd. de nat. rerum; Wrt. popl. science 4, 4; Lchdm. iii. 238, 4.

ge-leáffulnes, -ness, -nys, -nyss, e; f. Faithfulness, belief, trust; fĭdēlĭtas, crēdŭlĭtas :—Geleáffulnys crēdŭlĭtas, Ælfc. Gr. 9, 25; Som. 10, 64. We sceolan andettan ða sóðan geleáffulnesse on úrne Drihten we must confess the true belief in our Lord, Blickl. Homl. 111, 6.

ge-leáfhlystend, es; m. A catechumen; catechumenus, Hpt. Gl. 457, 458.

ge-leáfleás; adj. Unbelieving :—Ðone geleáfleásne ent the unbelieving giant, Swt. Rdr. 66, 323.

ge-leáfleást, -leáflýst, e; f. Want of faith, unbelief, infidelity, unfaith-

fulness; infĭdēlĭtas, incrēdŭlĭtas:—For hyra geleáfleáste *on account of their unbelief,* Basil admn. 4; Norm. 42, 1. Drihten Hǽlend þreáde mid wordum ðæra Iudeiscra þwyrnysse and geleáfleáste *the Lord reproved with words the perversity and unbelief of the Jews,* Homl. Th. ii. 110, 4. Nū sind adwæscede ealle geleáflýstu *now all infidelities are extinguished,* i. 226, 2 : Deut. 1, 40.

ge-leáflíc; *adj. To be believed, credible, faithful;* crēdĭbĭlis :—Nis hit nā geleáflíc ðæt se wurm Euan bepǽhte, and se deófol spræc þurh ða næddran *it is not to be believed that the serpent deceived Eve, but the devil spoke through the serpent,* Boutr. Scrd. 19, 40. Ðíne gecýdnyssa sindon swíðe geleáflíce *thy testimonies are very faithful,* Homl. Th. ii. 43, 15. Ðíne gecýdnyssa [MS. -kýdnyssa] geleáflíce gewordene synt swíðe *testĭmōnia tua crēdĭbĭlia facta sunt nimis,* Ps. Lamb. 92, 5.

ge-leáfnes-word, es; *n. A pass-word,* Beo. Th. 496.

ge-leáfsum; *adj. Faithful, credible, credulous;* fĭdēlis, crēdĭbĭlis :—Ðín gewitnes is weorcum geleáfsum *testĭmōnia tua crēdĭbĭlia facta sunt,* Ps. Th. 92, 6. Wǽron forþgongende ða cristenan men and ða geleáfsuman *the christian men and the faithful went forth,* Bd. 1, 8; S. 479, 20. Seó ætýwnys heofonlíces wundres geopnode hū ārwyrþlíce hí wǽron to onfónne eallum geleáfsumum *mirăcŭli cælestis ostensio, quam revĕrenter eæ suscĭpiendæ a cunctis fĭdēlĭbus essent, patĕfēcit,* 3, 11; S. 535, 34, note: 5, 24; S. 646, 32.

ge-leáh; *p. of* ge-leógan.

ge-leahtrian; *p.* ode, ade; *pp.* od, ad *To accuse, complain of, rebuke;* crĭmĭnāri, accūsāre :—He wæs geleahtrad from Gode *he was rebuked by God,* Past. 46, 6; Swt. 355, 1; Hat. MS. 67 b, 14.

ge-leánian; *p.* ode; *pp.* od *To reward, repay, recompense;* reddĕre, tribuĕre, rependĕre :—Ne mágon we geleánian him mid lāþes wihte *we may not reward him with aught of hostility,* Cd. 21; Th. 25, 15; Gen. 394. Him ðæt geleánaþ lífes Waldend *the Lord of life will repay him that,* Exon. 117 a; Th. 450, 9; Dóm. 85. Biþ hiora yfel geleánod be heora gewyrhtum *their wickedness is recompensed according to their deserts,* Bt. 38, 3; Fox 202, 4.

ge-leás; *adj. False;* falsus :—Ne underfó geleáse gewitnysse *non suscĭpies vōcem mendācii,* Ex. 23, 1.

ge-leást, e; *f. Carelessness, negligence;* incuria, Som.

ge-leaðian; *p.* ade; *pp.* ad *To invite:* invītāre :—Hengest and Horsa, from Wyrtgeorne geleaðade Bretta kyninge, gesóhton Bretene *Hengest and Horsa, invited by Vortigern, king of the Britons, sought Britain,* Chr. 449; Erl. 12, 1. v. ge-laðian.

ge-leccan; *part.* -leccende; ic -lecce, ðū -lecest, -lecst, he -leceþ, -lecþ, *pl.* -leccaþ; *p.* -lehte; *pp.* -leht *To moisten, wet;* hūmectāre, rĭgāre :—Geleccende muntas ofer ðām uferum his *rĭgans montes de sŭpĕriōrĭbus suis,* Ps. Spl. 103, 14. Mid mínum teárum strecednysse míne oððe míne beddinge ic beþweá oððe ic gelecce *lacrĭmis meis strātum meum rĭgābo,* Ps. Lamb. 6, 7. Sió mildheortnes ðæs lāreówes geþwǽnþ and gelecþ ða breóst ðæs gehiérendes *the kindness of the teacher softens and moistens the breast of the hearer,* Past. 18, 5; Swt. 137, 8; Hat. MS. 27 a, 12. For ðam sýpe heó biþ geleht *by the moistening it never was wet,* Bt. 33, 4; Fox 130, 6. Ðā sóna mínne ðurst gelehte *I then at once slaked my thirst,* Nar. 12, 11.

ge-lecgan; *p.* -legde; *pp.* -leged, -legd, -léd *To lay;* pōnĕre :—Hí ðec gelegdon on lāþne bend *they laid on thee the loathsome band,* Cd. 225; Th. 298, 26; Sat. 539. Hwǽr he geléd wǽre *ubi pōnĕrētur,* Mk. Bos. 15, 47. He wæs unscyldig ðæs ðe him geléd wæs *he was guiltless of that which was laid to him,* Chr. 1053; Erl. 187, 21.

ge-lécnian, -leicnian *to cure,* Mt. Kmbl. Lind. 12, 10, 22. v. lācnian.

ge-lédan; *p.* -lédde; *pp.* -léded, -lédd *To lead;* dūcĕre :—Ðe ic hebbe to helle hām gelédde *which I have led home to hell,* Cd. 215; Th. 270, 11; Sat. 88. v. ge-lǽdan.

ge-lédd; *part. p. Malleable, ductile;* ductilis :—On býman geléddon *in tubis ductilĭbus,* Ps. Spl. T. 97, 6.

ge-léfan *to allow, permit,* Mt. Kmbl. Lind. 12, 10 : Mk. Skt. Lind. 11, 16. v. ge-lýfan.

ge-léfan; *p.* de; *pp.* ed *To believe, confide, trust;* crēdĕre, confīdĕre :—Gif gē willaþ mínre mihte geléfan *if ye will believe my power,* Cd. 219; Th. 280, 6; Sat. 251. Geléfst ðū ðæt seó wyrd wealde disse worulde *dost thou believe that fortune governs this world?* Bt. 5, 3; Fox 12, 1. v. ge-lýfan.

ge-léfed; *part.* [léf *infirm, weak*] *Corrupted, injured;* putrĭdus :—Se milte wyrþ geléfed *the milt becomes corrupted,* L. M. 2, 36; Lchdm. ii. 244, 10. Hér sindon ðurh synnleáfa sāre geléfede to manege *here through impunity in sin too many are injured,* Swt. Rdr. 110, 174. v. ge-lýfed.

ge-léfenscipe, es; *m. Permission, excuse;* excusatio, Jn. Skt. Lind. 15, 22.

ge-leht *wet,* Bt. 33, 4; Fox 130, 6; *pp. of* ge-leccan.

ge-lend; *part. p. Provided with land:*—Gyf he wel gelend biþ *si bonam terram habeat,* L. R. S. 5; Th. i. 436, 5. [Cf. belendan, gelandian.]

---

ge-lend, e; *f. Fat, lard;* adeps, axungia, Ælfc. Gl. 73; Som. 71, 35. v. gelynd.

gelenda, an; *m. A man of landed property, a rich man;* dives, Som : Hpt. Gl. 480.

ge-lendan, he -lent; *p.* -lende; *pp.* -lended, -lend *To approach, come, arrive, go, proceed;* applĭcāre, accēdĕre, procēdĕre :—Ic gelende mid scipe *applĭco,* Ælfc. Gr. 24; Som. 25, 53. Ðæt scip gelent mid ðý streáme *the ship goes with the current,* Past. 58; Swt. 445, 13; Hat. MS. Conon gelende to Ahtene *Conon came to Athens,* Ors. 3, 1; Bos. 54, 12 : Chr. 886; Erl. 85, 10. He wæs on hergaþ gelend on ðæt ilce ríce *he had arrived on a plundering expedition in the same kingdom,* 894; Erl. 92, 3. Heo on Norþhumbrelond gelændon mid æscum *they came to Northumbria with their boats,* Th. An. 120, 17 : Shrn. 191, 15.

ge-lendan; *p.* de *To endow with land:*—Ða seofon mynstru he gelende mid his ǽgenum *those seven monasteries he endowed with his own lands,* Homl. Th. ii. 118, 29. v. ge-lend, ge-lendian, be-lendan.

ge-léned; *part. p. Lent :*—Geléned feoh *res credita,* Ælfc. Gl. 14; Som. 58, 2 : Wrt. Voc. 20, 70. v. lǽnan.

ge-lengan; *p.* de; *pp.* ed *To prolong, lengthen;* prolongāre, protēlāre :—Heora unriht gelengdon *prolongāvērunt inĭquĭtātem suam,* Ps. Th. 128, 2. Eówre dagas sín gelengede *protēlentur dics vestræ,* Deut. 5, 33 : Homl. Th. ii. 576, 26.

ge-lenge; *adj. Belonging, related;* pertinens, pertingens :—Ða ðe ðurh geleáfan us gelenge beóþ *those who through belief are related to us,* Homl. Th. ii. 314, 14. Yrfeweard líce gelenge *an heir of my body,* Beo. Th. 5457; B. 2732. Leahtrum gelenge *attached to vices,* Exon. 71 a; Th. 264, 28; Jul. 371. v. ge-lang.

ge-lent *goes,* Past. 58; Swt. 445, 113; Hat. MS; *3rd sing. pres. of* ge-lendan.

ge-leód, es; *m. One of a nation, a fellow-countryman, compatriot;* conterraneus, compatriota :—Gif hwá his ágenne geleód bebycgge *if any one sell his own countryman,* L. In. 11; Th. i. 110, 3.

ge-leódan; *p.* leád, *pl.* -ludon; *pp.* -loden *To spring, grow, descend;* crescere, germinare :—From ðām gumrincum folc geludon *nations grew from these patriarchs,* Cd. 75; Th. 93, 28; Gen. 1553. Oþðæt ða geongan leomu geloden weorþaþ *till the young limbs be grown,* Exon. 87 a; Th. 327, 20; Vy. 6 : Elen. Kmbl. 2451; El. 1227 : Runic pm. 18; Kmbl. 343, 1; Hick. Thes. i. 135. DER. leódan.

ge-leofian; *p.* ode, ade; *pp.* od, ad *To live;* vīvĕre :—Ne geleofaþ man nāht miriges, ða hwíle ðe mon deáþ ondrǽt *there is no mirth in life when there is dread of death,* Prov. Kmbl. 16. Gyf swā biþ geleofad *si sic vīvĭtur,* Cant. Ezech. Lamb. fol. 185 a, 16. v. ge-lifian.

ge-leófst *believest,* Bt. 5, 3; Fox 14, 10, = ge-lýfst; *2nd sing. pres. of* ge-lýfan.

ge-leógan; *p.* -leáh, *pl.* -lugon; *pp.* -logen *To lie, belie, deceive;* mentīre, fallēre :—Be ðām ðe hiora gewitnessa befóran biscope geleógaþ *of those who belie their testimonies before a bishop,* L. In. 13; Th. i. 110, 10, MS. B. Him seó wēn geleáh *hope deceived him,* Beo. Th. 4636; B. 2323 : Andr. Kmbl. 2150; An. 1076. Gelugon hý him *they deceived themselves,* Exon. 118 b; Th. 455, 27; Hy. 4, 56.

ge-leómod, -leómad; *part.* [leóma *a ray of light*] *Rayed, furnished with rays;* rădiātus :—Cométæ synd geleómode [MSS. R. P. L. geleómode] *comets are furnished with rays,* Bd. de nat. rerum; Wrt. popl. science 16, 20; Lchdm. iii. 272, 4.

ge-leoran; *p.* de; *pp.* ed *To go, depart, emigrate, die;* ire, migrāre, emigrāre, deficēre :—Mec geleoran lǽt *let me depart,* Exon. 116 b : Th. 455, 3; Hy. 4, 44 : Bd. 4, 23; S. 596, 11. Ic nā geleore *non emigrābo,* Ps. Spl. C. 61, 6. Seó rǽdelse, and ðæt geþeaht úrra feónda geleorde [MS. geleorode], ðā hí hit endian sceoldon *inĭmici defēcērunt frāmeæ in finem,* Ps. Th. 9, 6. Ðonne heora hwylc of weorulde geleored wæs *cum quis eórum de sæcŭlo fuisset evŏcātus,* Bd. 4, 23; S. 595, 41, note. Sægde Hilde of weorulde geleoran *nunciavit Hild migrasse de sæcŭlo,* 596, 11. Ne gelioraþ non prætēribit, Mt. Kmbl. Lind. 24, 34. Dóhter mín geliored is *filia mea defuncta est,* 9, 18.

ge-leorednes, -ness, -nys, -nyss, e; *f. A going, removing, transmigration;* transĭtus, transmigrātio :—Fram Dauide óþ Babilōnis geleorednysse, and fram Babilōnis geleorednesse óþ Crist *a David usque ad transmigrātiōnem Babylōnis, a transmigrātiōne Babylōnis usque ad Christum,* Mt. Bos. 1, 17. v. ge-leornes.

ge-leoren; *part. Gone away, departed;* defunctus :—Eorþgráp hafaþ waldendwyrhtan, forweorene [MS. forweorone], geleorene *earth's grasp* [i. e. *the grave*] *holds its powerful workmen, decayed, departed,* Exon. 124 a; Th. 476, 14; Ruin. 7.

ge-leorendlíc, -liorendlíc; *adj. Transitory;* transiens, Rtl. 28, 1.

ge-leornes, -ness, e; *f. A going, removing, departure, death;* transĭtus, transmigrātio :—Wæs gemēted ðætte hire geleornes wæs in ða ilcan tíd ðe hire þurh ða gesihþe ætýwed wæs *inventum est eadem hōra transĭtum ejus illis ostensum esse per visiōnem,* Bd. 4, 23; S. 596, 22. Ongeáton hí on ðon, ðæt heó to ðon ðider com, ðæt heó hire sǽde ða neáhtíde hire geleornesse *ex quo intellexēre quod ipsa ei tempus suæ transmigrātiōnis in proximum nunciāre venisset,* 4, 9; S. 577, 34. In

geliornisse *in transmigratione*, Mt. Kmbl. Lind. I, 11. To geliornisse herodes *ad obitum Herodis*, 2, 15. In dálum geliornesse *in partes Galileæ*, 2, 22. This gloss is to be explained by the old interpretation of the Hebrew name, according to which *Galilea = transmigratio.*

ge-leornian; *p.* ode, ede; *pp.* od, ed *To learn, inquire*; discĕre, disquírĕre :—Swá swá heó æt gelǽredum wǽpnedmonnum geleornian mihte *prout a doctis vīris discĕre pŏtĕrat*, Bd. 4, 23 ; S. 593, 28 : 4, 18 ; S. 587, 1. He nǽfre ǽnig leóþ geleornode *nil carmĭnum alĭquando dĭdĭcĕrat*, 4, 24 ; S. 597, 4 : Ps. Th. 118, 7. Hú hí ðás þing geleornodon *quomŏdo hæc dĕdĭcissent*, Bd. 4, 23 ; S. 596, 20. Geleornedon his byrelas him betweónum, hú hý him mihton ðæt líf oþþringan *his cupbearers inquired among themselves how they might take away his life*, Ors. 3, 9 ; Bos. 69, 9.

GE-LES, -lis, es ; *n. Reading, study, learning*; studium, lectura :— Gelis *studium*, Nar. 1, 20. On gelesum háligra gewrita gelǽred *in studiis scripturarum institutus*, Bd. 5, 20 ; S. 641, 33. Betweoh geleoso ðære gŏdcundan leornunge *inter studia divīnæ lectionis* Bd. 3, 13 ; S. 538, 29. [Cf. *O. Sax.* lesan : *Icel.* lesa : *O. H. Ger.* lesan, ga-lesan *to read.*]

ge-lésan; *p.* de; *pp.* ed *To redeem, save, spare :*— Gilésdes usig *redemisti nos*, Rtl. 29, 19. Ic gilése scíp míno *ego parcam oves meas*, 10, 3. Giléseno redemti, 24, 38.

ge-lésniss, e ; *f. Redemption*, Rtl. 12, 33.

ge-leswian *to feed*; pascere, Jn. Skt. Lind. 21, 17.

ge-lét *an ending, a meeting.* v. ge-lǽte.

ge-leðran; *p.* ede; *pp.* ed *To lather*; saponem illinere, sapone bullas excitare :—Ðæt heó sý eall geleðred *so that it may be all lathered*, Lchdm. iii. 2, 3. v. lyðran.

ge-lettan; ðú -letest; *p.* -lette; *pp.* -lett, -let ; *v. a. To hinder, delay, let, stop*; retardare, impedire :—Hí hine mágon gelettan *they may delay it*, Bt. 41, 2 ; Fox 246, 9. Hine seó eá lange gelette ðæs oferfæreldes *the river long hindered him from passing over*, Ors. 2, 4 ; Bos. 43, 45. Ðú geletest láþ werod *thou shalt stop the hostile force*, Elen. Kmbl. 187 ; El. 94. To hraðe hine gelette lidmanna sum ðá he ðæs eorles earm amyrde *too soon one of the seamen hindered him when he disabled the earl's arm*, Byrht. Th. 136, 40 ; By. 164. Ne lǽt ðec síðes getwǽfan láde gelettan lifgendne monn *let not living man divert thee from the course, hinder thee from the way*, Exon. 123 b ; Th. 474, 3 ; Bo. 24 : 37 b ; Th. 123, 29 ; Gú. 330. Ac hit wæs ðá ðurh Eádríc ealdorman gelet swá hit ðá ǽfre wæs *but matters were hindered by alderman Eadric as they always were then*, Chr. 1009 ; Erl. 143, 1. He wearþ gelet *he was hindered*, 1075 ; Erl. 213, 17. v. lettan.

gelew ; *adj. Yellow, bay*; flávus :—On horse gelewum sittan hýnþe getácnaþ *to sit on a bay horse betokens humiliation*, Lchdm. iii. 202, 29. v. geolo.

ge-léwan; *p.* de; *pp.* ed *To betray, deceive, weaken, injure*; prodere :— Geléwend *prodens*, Lye. Gif hit byþ deád oððe geléwed *if it is dead or hurt*, Exod. 22, 10, 14. (*Or does* geléwed *here =* geléfed? cf. alēuaþ and geuntrumaþ, Homl. Th. i. 4, 22 ; *and* Swt. Rdr. 110, 174, note) [*Goth.* ga-léwjan *to betray.*]

ge-líc [-líce?], es ; *n. Likeness, similitude*; sĭmĭlĭtūdo :—Næfdon hí máre monnum gelíces ðonne ingeþonc *they had no more likeness to men than the mind*, Bt. Met. Fox 26, 186 ; Met. 26, 93. [Cf. *Goth.* ga-leiki.]

ge-líc ; *comp. m.* -lícra ; *f. n.* -lícre ; *superl.* -lícost, -lícast, -lícust ; *adj. Like, alike, similar, equal*; sĭmĭlis, æqualis :—Næs se wæstm gelíc *the fruit was not alike*, Cd. 23 ; Th. 30, 13 ; Gen. 466 : Bt. 38, 6 ; Fox 208, 17 : Exon. 89 a ; Th. 334, 21 ; Gn. Ex. 19. Heofena ríce is geworden gelíc senepes corne *sĭmĭle est regnum cælōrum grāno sĭnāpis*, Mt. Bos. 13, 31, 33 : 22, 2 : Lk. Bos. 13, 18, 19, 20, 21 : Ps. Spl. 48, 12, 21. Ealle men hæfdon gelícne fruman *all men had a like beginning*, Bt. 30, 2 ; Fox 110, 7 : Andr. Kmbl. 988 ; An. 494. Ic ðé mæg andreccan sprǽce gelíce [MS. gelícne] *I can relate to thee a similar tale*, Bt. Met. Fox 26, 4 ; Met. 26, 2. Ic ǽnig ne métte wið ðé gelíc *I have not met any like unto thee*, Exon. 73 b ; Th. 275, 13 ; Jul. 549. Ealle hí beóþ gelíce acennede *they are all born alike*, Bt. 30, 2 ; Fox 110, 9 : Beo. Th. 4334 ; B. 2164. Wirc ðé twá stǽnene tabulan ðám óðrum gelíce *præcīde tibi duas tăbūlas lăpĭdeas instar priōrum*, Ex. 34, 1 : Ps. Th. 65, 5. Se líchoma wæs slǽpendum men gelícra ðonne deádum *the body was more like a sleeping than a dead man*, Bd. 4, 19 ; S. 589, 16 : Ps. Th. 88, 5. Gelícre *sĭmĭlior*, Ælfc. Gr. 5 ; Som. 5, 5. Slǽp biþ deáþe gelícost *sleep is most like death*, Salm. Kmbl. 624 ; Sal. 311 : Bt. Met. Fox 25, 36 ; Met. 25, 18 : 26, 176 ; Met. 26, 88. Rêce hí gelícast ricene geteoriaþ *sīcut defĭcit fūmus, defĭciant*, Ps. Th. 67, 2 : 102, 5. Ís byþ gimmum gelícust *ice is most like gems*, Runic pm. 11 ; Hick. Thes. i. 135, 21 ; Kmbl. 341, 17. Didimus ðæt ys Gelýcost on ure geðeóde Didimus, *that is in our language twin*, Jn. 20, 24 : 21, 2. [*Chauc.* ilik : *Goth.* ga-leiks : *O. Sax.* gi-lík : *O. Icel.* glíkr : *O. H. Ger.* ge-lich : *Ger.* gleich.]

ge-líca, an ; *m.* also ge-líce, an ; *f. An equal*; æqualis, par, æqualitas :— Nán man nis his gelíca on eorþan *non sit ei similis in terra*, Job. Thw. 164, 17. Micel is ðæt ongin ðínre gelícan *great is the attempt for thy equal* [cf. *Ger.* für Deinesgleichen ; *colloquial English* for the like of you], Exon. 67 b ; Th. 250, 16 ; Jul. 128. Nán þing nis ðín gelíca *no*

thing is thine equal, Bt. Met. Fox 20, 74 ; Met. 20, 37 : Homl. Th. ii. 576, 22. [*Laym.* (his) iliche : *O. H. Ger.* (min) gilicho.]

ge-lícan *to liken, imitate* :—To gelícanne *ad imitandum*, Rtl. 22, 36. Gelíced biþ *assimilabitur*, Mt. Kmbl. Lind. 7, 24. [*Goth.* ga-leikon : *O. H. Ger.* ki-lihhan : *Ger.* gleichen.]

ge-lícbisnung, e ; *f. Imitation*; imitatio, Rtl. 76, 1.

ge-liccettan; *p.* te; *pp.* ed *To flatter, dissemble*; assentari, simulare, Som.

gelíce ; *adv. Likewise, also, as*; pariter, Ps. Spl. 67, 7 : Mt. Bos. 27, 44. Gelíce swá swá heó bebeád *likewise as she commanded*, Bd. 4, 19 ; S. 588, 19 : Blickl. Hom. 17, 4. He dyde swá gelíce *fecit similiter*, Mt. Bos. 20, 5. Elpendes hýd wyle drincan wǽtan gelíce án spinge deþ *an elephant's hide will imbibe water as a sponge doth*, Ors. 5, 7 ; Bos. 107, 11. Ðǽm biscopum ðe hér on worlde syndon swýðe gelíce gegange ðǽm biscope ðe Paulus geseah *it shall happen to those bishops that are in this world as it did to the bishop that St. Paul saw*, Blickl. Homl. 45, 4 : 59, 4. Nis ðæt nó be eallum dēmum gelíce to secgenne *that is not to be said of all judges alike*, 63, 16. Ne wǽron ðás ealle gelíce lange *these were not all alike long*, 119, 3. His líf ðǽm his naman wæs gelíce gegearwod *his life was ordered in accordance with his name*, 167, 32. Gelíce sê lêg hie cwylmde gelíce ða Cristenan him mid heora wǽpnum hýndon *they were killed alike by the lightning and laid low by the weapons of the Christians*, 203, 16 : Nar. 14, 10. Ðon gelícost ðe ðǽr sum mon gestóde *just as if a man had stood there*, Blickl. Homl. 203, 35. Emne ðon gelícost ðe he ne cúðe *just as if he didn't know*, Cd. 92 ; Th. 116, 28 ; Gen. 1943. Efne ðǽm gelícost swylce *just as if*, Blickl. Homl. 221, 14.

ge-líogan, -lícgean; *p.* -læg, *pl.* -lǽgon ; *pp.* -legen. I. *to lie, lie near, together*; jacere, adjacere, conjacere :—Mægen-stán him on middan geligeþ *a huge stone lies in the middle of it*, Bt. Met. Fox 5, 32 ; Met. 5, 16. Stedewangas strǽte gelícgaþ *fixed plains lie near the road*, Andr. Kmbl. 668 ; An. 334. On ðǽm gelæg *in quo jacebat*, Lk. Skt. Lind. 5, 25. Ðá heó ðǽr on gelegen wæs *when she had lain down there*, Ors. 5, 13 ; Bos. 113, 23. II. *to lie down, fail, cease, loiter, delay*; deficere, cessare :—Windblond gelæg *the wind-storm ceased*, Beo. Th. 6284 ; B. 3146. Ne mihte se niþ betwux him twám gelícgean *the strife between the two could not be appeased*, Ors. 3, 11 ; Bos. 75, 36.

ge-líc-gemaca, an ; *m. An equal*; compar, Ælfc. Gr. 9, 51.

ge-líchamod, -homod ; *part. p. Incarnate*; :—Drihten wæs gelíchomod *the Lord became incarnate*, Blickl. Homl. 33, 15.

ge-lícian ; *p.* ode; *pp.* od; *with dat.* I. *to please, delight*; placere, acquiescere, delectare :—Ic gelícige *placebo*, Ps. Th. 114, 8. Gelícaþ [gelícige, Lamb. 14 ; Spl. 18] ðé Dryhten *complaceat tibi Domine*, Ps. Surt. 39, 14. Ðæt ðé gelíciaþ *ut te complaceant*, Ps. Spl. 18, 15. On ðé ic gelícode *in te complacui*, Mk. Bos. 1, 11. II. *impers. it pleases*; placet :—Me gelícaþ *placet mihi*, Ælfc. Gr. 33 ; Som. 37. 17. v. lícian.

ge-líclíc ; *adj. Likely, fit*; aptus :—Swíþor ðonne hit gelíclíc síe *more strongly than is proper*, L. M. 2, 16 : Lchdm. ii. 194, 14 : Hpt. Gl. 506.

ge-líclíce ; *adv. Equally* :—Gelíclíc *æqualiter*, Jn. Skt. p. 4, 10.

ge-lícnes, -ness, e ; *f.* I. *a likeness, image, resemblance*; similitudo, imago :—Uton wircean man to andlícnisse, and to úre gelícnisse *faciamus hominem ad imaginem et similitudinem nostram*, Gen. 1, 26. Ælc man hæfþ þreó þing on him sylfum untodǽledlíce and togædere wyrcende, swá swá God cwæþ, ðáða he ǽrest mann gesceóp. He cwæþ, 'Uton gewyrcean mannan to úre gelícnysse.' And he worhte ðá Adam to his anlícnysse. On hwilcum dǽle hæfþ se man Godes anlícnysse on him? On ðære sáwle, ná on ðam líchaman *every man has three things in himself indivisible and working together, as God said when he first created man. He said, 'Let us make man in our own likeness.' And he then made Adam in his own likeness. In which part has man the likeness of God in him? In the soul, not in the body*, Homl. Th. i. 288, 11–17. He worhte of seolfre ǽnne heáhne stýpel on stánweorces gelícnysse *he wrought a high tower of silver in the form of stone-work*, H. R. 99, 23. Uton gewyrcan mannan to úre anlícnysse and to úre gelícnysse *faciamus hominem ad imaginem nostram et similitudinem nostram*, Hexam. 11 ; Norm. 18, 15. II. *a parable, proverb*; parabola, proverbium :—Arecce us gelícnisse ðas *edissere nobis parabolam istam*, Mt. Kmbl. Rush. 15, 15. Gé secgaþ me ðas gelícnesse, Eálá lǽce, gehǽl ðé sylfne *diceris mihi hanc similitudinem* [*proverbium*], *Medice, cura teipsum*, Lk. Bos. 4. 23. [*O. H. Ger.* gelíhnessi *parabola* : *Ger.* gleichniss.]

ge-lícung, e ; *f. A liking.* v. lícung.

ge-líden *sailed*, Exon. 20 b ; Th. 53, 30 ; Cri. 858 ; *pp. of* ge-líðan.

ge-liese *care, learning.* v. ge-les.

ge-lífan, -líefan ; *p.* de; *pp.* ed *To believe, trust*; crēdĕre, confīdĕre :— Gif hie willen gelíefan ðætte Godes ríce hiera síe *if they will believe that God's kingdom is theirs*, Past. 36, 5 ; Swt. 253, 9 ; Hat. MS. 47 b, 8. Se ðe him to ðam hálgan helpe gelíeþ, he ðǽr gearo findeþ *he who trusteth himself to the holy one for help, he findeth it there readily*, Wald. 111 ; Vald. 2, 27. Abram gelífde Gode *crēdĭdit Abram Deo*, Gen. 15, 6. Ðæt hie gelíefon on ðínne naman *that they may believe on thy name*, Blickl. Homl. 247, 25. v. ge-lýfan.

ge-lifedlíce *lawfully.* v. ge-lýfedlíce.

ge-líffæstan; *p.* -líffæste; *pp.* -líffæsted, -líffæst *To make alive, quicken;* **vivifĭcāre**:—God geworhte ǽnne mannan, and hine gelíffæste, and he wearþ ðá mann gesceapen on sáwle and on líchaman *God made one man, and made him alive, and he then became man with soul and body*, Homl. Th. i. 12, 29. Se súnu gelíffæst ðá de he wyle *filius quos vult vivificat*, Jn. Bos. 5, 21. He wolde swá synfulle sáwle gelíffæstan *he would quicken so sinful a soul*, Homl. Th. i. 496, 15: ii. 206, 17. Mid gesceádwísre sáwle gelíffæst *quickened by a rational soul*, 270, 20.

ge-lífian; *p.* ode; *pp.* od *To live* [cf. *Ger.* erleben]:—Gif he hit gelifode *if he had lived*, Chr. 1093; Erl. 229, 8. v. ge-leofian.

ge-lígenod; *part. p. Convicted of lying:*—Se apostol Paulus ne biþ gelígenod *the apostle Paul is not shewn to be false*, Homl. Th. i. 54, 1.

ge-liger, es; *n. A lying with, fornication, adultery;* concubĭtus, conjūgium, fornĭcatio, adultĕrium:—He sǽde ðæt his nama wǽre spiritus fornicationis ðæt is dernes geligeres gást *he said that his name was spiritus fornicationis, that is, spirit of fornication*, Shrn. 52, 27: 130, 14. To geligere *concubĭtu*, Ors. 1, 2; Bos. 27, 13. Æt geligere de conjūgio, Bos. 27, 15. Geligre *fornicatiōni*, Bos. 27, 9. [*Goth.* ga-ligri. Cf. forliger.]

ge-ligernes, ness, e; *f. Fornication, adultery;* fornĭcātio, libīdo:—For hyre geligernesse *for her lustfulness*, Ors. 1, 2; Bos. 27, 11.

ge-líhtan; *p.* -líhte *To lighten, mitigate, assuage;* alleviare:—Mid ánre mæssan man mæg alýsan xii daga fæsten and mid x mæssan man mæg gelíhtan iiii monða fæsten and mid xxx mæssan man mæg gelíhtan xii monða fæsten *with one mass a man may redeem a xii days' fast, and with x masses a man may lighten a iiii months' fast; and with xxx masses a man may lighten a xii months' fast*, L. Pen. 19; Th. ii. 286, 6–9: 14. Ðonne hie willaþ him selfum ðæt yfel ðæt hie ðurhtugon to swíðe gelíhtan *when they wish to make too light of the evil they have done*, Past. 21; Swt. 159, 20; Hat. MS. Ic mínne ðurst geléhte [?] *I assuaged my thirst* [or gelehte from geleccan], Nar, 12, 11. [*A.R.* i-lihted *alleviated*: *O.H.Ger.* gi-líhten *lenire*.]

ge-líhtan; *p.* -líhte *To alight, approach, come:*—Gelíht of his horse *desiliit ab equo suo*, Gr. Dial. 1, 2. Ðá gelíhte se cuma *then the stranger alighted*, Homl. Th. ii. 134, 34. He gelíhte to ðæm hearge *propiabat ad fanum*, Bd. 2, 13; S. 517, 11. Segde ðætte sealfa god wolde helwarum hám gelíhtan *said that God himself would come home to the dwellers in hell*, Cd. 222; Th. 291, 16; Sat. 431.

ge-líhtan; *p.* -líhte *To shine, grow light;* lucere, lucescere:—Ðæt he gelíhte allum *ut luceat omnibus*, Mt. Kmbl. Rush. 5, 15.⋆ Gelihted *lucescit*, Lind. 28, 1. v. gelýhtan.

ge-líman; *pp.* ed *To glue or join together, connect;* conglutinare:—Gelímp ða friénd togædere *joins the friends together*, Bt. 24, 3; Fox 84, 1. Gelímed fæste tosomne *joined fast together*, Bt. 35, 2; Fox 156, 35. Gelímod *conglutinatus*, Ps. Lamb. 43, 25.

ge-limp, es; *n. An event, accident, a chance;* accĭdens, cāsus:—Ðara in gelimpe lífe weóldon *of those who in chance possessed life*, Exon. 36 b; Th. 118, 13; Gú. 239. Is ǽnig óðer on eallum ðám gelimpum búton godes yrre ofer ðas deóde swutol and gesýne *is there anything else that happened, and visible in these events but God's anger over this people?* Swt. A.S. Rdr. 109, 137. Ðá forhtede ðe biscop for ðam fǽrlíce gelimpe *then the bishop was afraid on account of that dangerous case*, Th. An. 121, 5: Th. Ap. 1, 12. Ðú woldest witan his naman and his gelimp *you wanted to know his name and what had befallen him*, 16, 4: 15, 20, 26.

ge-limpan; he -limpeþ, -limpþ; *p.* -lamp, -lomp, *pl.* -lumpon; *subj. p.* -lumpe, *pl.* -lumpen; *pp.* -lumpen *To happen, occur, befall, come to pass, take place;* accĭdere, evĕnire, contingĕre:—Ðæt gelimpan sceal ðætte lagu flóweþ ofer foldan *it shall happen that water shall flow over the earth*, Exon. 115 b; Th. 445, 1; Dóm. 1: 117 b; Th. 452, 5; Dóm. 116. Hit eft gelimpeþ ðæt se líchoma lǽne gedreóseþ *it afterwards befalls that the body miserably sinks*, Beo. Th. 3511; B. 1753. Gyf hyt gelimpþ ðæt he hyt fint *si contĭgĕret ut invĕniat eam*, Mt. Bos. 18, 13. Ðá gelamp hit *then it happened*, Gen. 40, 1: Homl. Th. ii. 120, 14. Frófor eft gelamp sárigmódum *comfort afterwards came to the sad of mood*, Beo. Th. 5875; B. 2941. Ðá sió tíd gelomp *when the time came*, Bt. Met. Fox 26, 34; Met. 26, 17: Bt. 18, 4; Fox 66, 27. Ealle ðas ungesǽlþa us gelumpon þurh unrǽdas *all these calamities befell us through evil counsels*, Chr. 1011; Erl. 145, 1. Gif ðé ðæt gelimpe *if that befall thee*, Elen. Kmbl. 879; El. 441: Beo. Th. 1862; B. 929. Geseón hwæt us gelumpe *vidēre quid nōbis accĭdĕret*, Bd. 5, 1; S. 614, 3: Exon. 35 a; Th. 113, 32; Gú. 165. Gregorius Gode þancode ðæt Angelcynne swá gelumpen wæs, swá swá he sylf geornlíce gewilnode *Gregory thanked God that it had so happened to the English nation, as he himself had earnestly desired*, Homl. Th. ii. 130, 28: Beo. Th. 1653; B. 824.

ge-limpfull; *adj. Fit, suitable:*—Ðæt he gedó disne weig gelimpfulran *that he make this way better*, Shrn. 163, 25.

ge-limplíc; *adj. Fit, seasonable, suitable, meet, ordered by fate, fatal;* compĕtens, congruus, opportūnus, fātālis:—On gelimplícre tíde *in tempŏre opportūno*, Ps. Spl. 144, 16: Bd. 4, 24; S. 597, 10. Swá hwǽr swá he gelimplíce stówe findan mihte *wheresoever he could find a suitable place*, 3, 19; S. 547, 5: 5, 3; S. 616, 25.

ge-limplíce; *comp.* -lícor; *adv. Fitly, seasonably, opportunely;* op-

portūne:—Ðæt hí oncnáwen hú gelimplíce úre God ða ánwaldas and ða rícu sette *that they might know how seasonably our God settled the empires and the kingdoms*, Ors. 2, 1; Bos. 40, 7. Gelimplíce he us lǽrde hú we us gebiddan sceoldan *fortunately he hath taught us how we ought to pray*, Blickl. Homl. 19, 35. Gelimplícor *opportūnius*, Bd. 3, 29; S. 561, 29.

ge-limpwíse, an; *f. An event;* eventus, quod evenit, Hpt. Gl. 457.

ge-lióma, an; *m. A light;* lumen, Mone B. 174.

ge-lioran *to pass over.* v. ge-leoran.

ge-liornes *a going, death.* v. ge-leornes.

ge-lírde *emigrated.* v. ge-leoran.

ge-lis *study, learning.* v. ge-les.

ge-lísian *to slip, slide:*—Be ðæm is awriten se ðe nylle onscúnian his lytlan scylda ðæt he wille gelísian to máran *it is written that he who will not shun his little sins will glide into greater*, Past. 57, 2; Swt. 437, 20; Hat. MS. v. note.

ge-lisþelícnis, se; *f. Opportunity;* opportunitas, Ps. Spl. T. 9, 9.

ge-líðan; *p.* -láþ, *pl.* -lidon; *pp.* -liden, -liden *To go, move, sail, advance, proceed, come;* íre, meáre, advĕhi, profĭcisci, vĕnire:—Mænig tungul mǽran ymbhwyrft hafaþ on heofonum, sume hwíle eft lǽsse gelíðaþ, ða ðe lácaþ ymb eaxe ende *many a star has a greater circuit in the heavens; sometimes again, they move in a less, that sport about the end of the axis*, Bt. Met. Fox 28, 43; Met. 28, 22. Ǽr ðon we to lande geliden hæfdon *ere that we had sailed to land*, Exon. 20 b; Th. 53, 30; Cri. 858: Elen. Kmbl. 498; El. 249. Ðæs ðe lencten geliden hæfde werum *after spring had come to men*, Menol. Fox 57; Men. 28.

ge-líðan; *p.* -wǽhte; *pp.* -wǽht *To soften, calm, appease;* lēnire:—Ic gelíðewǽce lēnio, Ælfc. Gr. 30, 5; Som. 34, 56. His afyrhte mód swíðe fægerlíce mid his frófre he gelíðewǽhte *he gently appeased his troubled mind with his comfort*, Ælfc. T. 37, 24.

ge-líðian, -líðegian; *p.* ode; *pp.* od *To soothe, soften, mitigate, relieve, appease;* lēnire, mītĭgāre, plācāre:—Styrunge ýþa hire ðú gelíðegast [gelíðegost MS.] *mōtum fluctuum ejus tu mītigas*, Ps. Lamb. 88, 10. Gáte cýse niwe ongelegd ðæt sár gelíðegaþ *a new goat's cheese laid on relieveth the sore*, Med. ex Quadr. 6, 7; Lchdm. i. 352, 9. Ðú gelíðegodest ealne ðínne graman *mītigasti omnem īram tuam*, Ps. Lamb. 84, 4. Drihtnes yrre wearþ gelíðegod ongén ðæt folc *plācātus est Dŏmĭnus adversus pŏpŭlum suum*, Ex. 32, 14. His ðurst wæs gelíþad *his thirst was appeased*, Shrn. 130, 5. Forðæm is swíðe micel néddearf ðæt mon mid micelre gemetgunge swelcra scylda ðreáunga gelíðige *therefore it is very necessary that the chiding of such sins be tempered with great moderation*, Past. 21; Swt. 159, 3; Hat. MS.

ge-litlian; *p.* ode; *pp.* od *To diminish, lessen;* mĭnŏrāre:—Nýtenu heora he ne gelitlode oððe he ne gewanode *jūmenta eōrum non mĭnŏrāvit*, Ps. Lamb. 106, 38. Ic beóde mínum erfeweardum ðæt heo nǽfre ðis feoh gelitlian *I enjoin my heirs that they never diminish this money*, Th. Chart. 168, 22. v. ge-lytlian.

gellan, gillan, giellan, gyllan; *part.* gellende, gillende, giellende, gyllende; ic gelle, gille, gielle, gylle, ðú gilst, gielst, gylst; he gilleþ, gilþ, gielþ, gylleþ, gylþ, *pl.* gellaþ, gillaþ, giellaþ, gyllaþ; *p.* geal, *pl.* gullon; *pp.* gollen *To yell, sing, chirp;* stridere, sonare:—Gellende *yelling*, Exon. 94 b; Th. 353, 40; Reim. 25. Ic seah searo giellende *I saw a yelling machine*, 108 b; Th. 415, 1; Rä. 33, 4. Gyllende gryre *with yelling horror*, Cd. 167; Th. 208, 26; Exod. 489. Ic gielle swá hafoc *I yell as a hawk*, Exon. 106 b; Th. 406, 19; Rä. 25, 3. Gilleþ geómorlíce *he yelleth sadly*, Salm. Kmbl. 535; Sal. 267. Gylleþ grǽghama *the cricket chirps*, Fins. Th. 10; Fin. 6. Gielleþ ánfloga *the lone-flier yells*, Exon. 82 a; Th. 309, 25; Seef. 62. Hí gullon *they sung*, Andr. Kmbl. 253; An. 127. [*Plat.* gillen *to shriek: Frs.* galljen: *Dut.* galmen *to sound: Ger.* gellen, gällen *to sound, from* gal, gall *a sound: O.H.Ger.* galm: *Icel.* gella.] DER. bi-gellan.

gellet, es; *n? A large vessel or cup, basin;* alveus, pōcŭlum mājus:—Gescearfa ðás wyrto tosomne, dó on gellet *scrape these herbs together, put them into a basin*, L.M. 3, 48; Lchdm. ii. 340. 3.

GELM, gilm, es; *m. A* YELM, *handful;* manĭpŭlus:—Genim grēne mintan, gelm take green mint, a handful, L.M. 1, 48; Lchdm. ii. 120, 22: iii. 74, 18.

gelo; *adj. Saffron, yellow;* crocus, Som.

ge-loccian *to stroke gently;* demulcere, Som. [*O.H.Ger.* gi-locchon *mulcere*.]

ge-lócian; *p.* ode; *pp.* od *To look, behold, see;* respĭcĕre, aspĭcĕre:—Driht of heofonum on eorþan gelócaþ *Dŏmĭnus de cælo in terram aspexit*, Ps. Spl. 101, 20. Eágan his ofer þeóda gelóciaþ *ŏcŭli ejus sŭper gentes respĭciunt*, 65, 6. Gelóca on cýðnysse ðíne *respĭce in testāmentum tuum*, Ps. Spl. C. 73, 20.

ge-loda; *pl. Joints of the back:*—Geloda *vel* gelyndu *spondilia*, Ælfc. Gl. 74; Som. 71, 51; Wrt. Voc. 44, 34.

ge-loda, an; *m. A brother;* frater:—Gebroðru *vel* gelodan *fratres*, Ælfc. Gl. 92; Som. 75, 42; Wrt. Voc. 52, 3.

gelodr, e; *f. A part of the body about the chest, the backbone or spine?* pars corporis circa thoracem *vel* spinam?—Se maga biþ neáh ðære heor-

tan and ðære gelodre *the stomach is near the heart and the spine,* L. M. 2, 1; Lchdm. ii. 176, 3.

ge-lod-wyrt, e; *f. Silverweed;* potentilla anserina:—Gelodwyrt *heptaphyllon,* Recd. 42, 75; Wrt. Voc. 68, 10: Lchdm. ii. 78, 1: 98, 16.

ge-logian; *p.* ode; *pp.* od *To place, lodge, dispose, regulate;* ponere, disponere, reponere, collocare:—God gelogode ðone man *Deus posuit hominem,* Gen. 2, 8: Homl. Th. i. 12, 33. Ða geleáfullan folc híg sylfe gelogiaþ and heora líf for Gode *the faithful folk dispose themselves and their life for God,* Ælfc. T. Lisle 28, 13. Gelogaþ his ágen líf *regulates his own life,* Tract. de Spir. Septif: Homl. Th. i. 168, 11. Godes ðeów se ðe hád underféhþ sceal beón on wíson gelogod ðe God tǽhte *the servant of God who takes orders must be disposed in the manner that God has taught,* ii. 48, 31: i. 286, 13. Ðæt mynster he gelogode mid wellybbendum mannum *that monastery he filled with men of good life,* 506, 15. Ðá ðwóh man ða hálgan bán and gelogodon hí up *then the holy bones were washed and laid up,* Swt. Rdr. 100, 158. Hí gelogodon sče Ælfeáges hálgan líchaman on norþhealfe weofodes *they placed S. Ælfeg's holy body on the north side of the altar,* Chr. 1023; Erl. 163, 33. He begeat má castelas and ðǽr inne his ríderas gelogode *he got more castles and lodged his knights therein,* 1090; Erl. 226, 30. Geloga híg on ðære sélostan stówe *in the best of the land make them to dwell,* Gen. 47, 6. Ðás lamb ðe ðú gelogast on sundron *these lambs which thou hast set by themselves,* 21, 9. Ðone wudu gelogode *laid the wood in order,* 22, 9. He wæs gelogod to his folcum *he was gathered to his people,* Deut. 32, 50.

ge-logod; *part. p.* Arranged; appositus:—For ðære gelícnisse his gelogodan sprǽce *for the likeness of his disposed speech* or *style,* Ælfc. T. Lisle 17, 12.

GE-LÓMAN; *pl. m. Household stuff, furniture, utensils, tools;* supellex, instrumenta:—Ísern-gelóman *ferramenta ruralia,* Bd. 4, 28; S. 605, 32: Shrn. 146, 15. Ða mén hwílum ða íren-gelóman liccodan *milites nunc ferramenta lambendo,* Nar. 9, 19. v. andlóman.

ge-lóme; *adv. Often, frequently, continually, repeatedly;* sæpe, frequenter, contínuo, crebro:—Fregn gelóme freca óðerne *one warrior often asked the other,* Andr. Kmbl. 2327; An. 1165: Beo. Th. 1122; B. 559: Ps. Th. 54, 13: 62, 4. Ðonne hí gelóme sáwon swíðlíce rénas *when they frequently saw severe showers,* Boutr. Scrd. 21, 22: 17, 11. Wæs he se mon ǽfest on his dǽdum and gelóme on hálgum gebedum *erat relígiósis actibus, crebris précibus,* Bd. 4, 11; S. 579, 6. Oft and gelóme *very frequently,* Bt. Met. Fox 30, 10, 14; Met. 30, 5, 7: Chr. 887; Erl. 86, 11: 959; Erl. 119, 25. Oft gelóme *full oft, very often,* Cd. 75; Th. 93, 2; Gen. 1539. [*O. H. Ger.* ki-lómo *frequenter.*]

ge-lómed; *part. p. Having rays;* radiatus. v. ge-leómed.

ge-lómelíc *frequent,* Bd. 2, 7; S. 509, 32. v. ge-lómlíc.

ge-lómlǽcan; *p.* -lǽhte; *pp.* -lǽht *To frequent, to use often;* frequentare:—Gelómlǽcende word *frequentative verb,* Ælfc. Gr. 36; Som. 38, 14. Mid gelómlǽcendum hryrum *with frequent destructions,* Homl. Th. i. 578, 34: ii. 350, 19.

ge-lómlǽcing, -lómlǽcung, e; *f. Frequency, a frequenting, a common resort;* frequentatio, Ælfc. Gr. 36; Som. 38, 15.

ge-lómlǽcnys, -lómlícnes, ness, e; *f. A frequented* or *public place;* locus condensus, Ps. Spl. 117, 26.

ge-lómlíc, -lómelíc; *adj. Frequent, repeated;* frequens, creber:—Mid gelómlícra wundra wyrcnysse *virtútum frequentium operatióne,* Bd. 3, 13; S. 538, 39. Mid gelómlícum oncunningum *by frequent accusations,* 3, 19; S. 548, 3. Mid his gelómlícum bedum *crebris orátiónibus,* 2, 7; S. 509, 32.

ge-lómlíce; *comp.* -lícor; *superl.* -lícost; *adv. Often, frequently, repeatedly;* sǽpe, frequenter, crebro:—Gelómlíce *sǽpe,* Ælfc. Gr. 38; Som. 39, 52. Hwí fæste we and ða Sundor-hálgan gelómlíce *quare nos et Pharisǽi jejúnámus frequenter?* Mt. Bos. 9, 14: Bd. 3, 22; S. 552, 9: 3, 23; S. 554, 11. Búton hí hyra handa gelómlíce þweán *nisi crebro láverint mánus,* Mk. Bos. 7, 3: Bd. 3, 13; S. 538, 8: Hymn. Surt. 116, 14. Gelómlícor *oftener;* sǽpius, Ælfc. T. 22, 22: Ælfc. Gr. 38; Som. 39, 53. Gelómlícost *most frequently;* sǽpissime, Ors. 4, 4; Bos. 81, 3: Ælfc. Gr. 38; Som. 39, 53.

ge-lómlícian; *p.* ode; *pp.* od *To become frequent:*—Manig yfel we geaxiaþ hér on lífe gelómlícian and wæstmian *many an evil we learn has become frequent in this life and flourishes,* Blickl. Homl. 109, 2.

ge-lomp *happened,* Bt. 18, 4; Fox 66, 27; *p.* of ge-limpan.

ge-londa, an; *m. A fellow-countryman;* compatriota:—Be ðám monnum ðe hiora gelondan bebycgaþ *of those men who sell their countrymen,* L. In. 11; Th. i. 110, 1. Cf. ge-leód. [*O. H. Ger.* gi-lante *patriota.*]

ge-long. v. ge-lang.

ge-lósian; *p.* ode, ade; *pp.* od, ad *To lose, perish:*—We biðn gelósoad *perimus,* Mt. Kmbl. Lind. 8, 25. Gelósiga *perdet,* 16, 25. Ðæt gelósade *quod perierat,* 18, 11. [*Laym.* i-losed.]

ge-lostr *a gathering to form matter, imposthume;* suppuratio, Som.

ge-loten *dæg* oððe *ofernón latter part of the day;* suprema, Ælfc. Gl. 95; Som. 75; Wrt. Voc. 53, 14. v. lútan.

gelp, es; *m. Glory, vain-glory, pride;* glória, vána glória:—Ne

---

gýtsung, ne ídel gelp him on ne rícsode *neither avarice nor vain-glory reigned in him,* Bd. 3, 17; S. 545, 9. Gif he unnýtne gelp ágan wille *if he will possess unprofitable glory,* Bt. Met. Fox 10, 3; Met. 10, 2. v. gilp.

gelpan *to boast;* glóriári:—Gif hwá ðæs gelpþ *if any one boast of it,* Bt. 30, 1; Fox 108, 19, MS. Bod. v. gilpan.

gelp-scaða, an; *m. A boastful foe:*—Ðone gelpscaðan ríces berǽdan *to deprive that boastful foe of his power,* Bt. Met. Fox 9, 99; Met. 9, 49. v. gielp-sceaða.

gelsa. v. gǽlsa.

gelt, es; *m. A sin, crime, fault, debt;* delictum, débítum:—Geltas geclánsa ða ðe ic gefremede *cleanse the sins which I have committed,* Ps. C. 50, 39; Ps. Grn. ii. 277, 39. Gelt *débítum,* Prov. 24. v. gylt.

ge-lúcan; *p.* -leác, *pl.* -lucon; *pp.* -locen *To shut, lock, fasten, weave;* claudère, nectère:—Ðē gelúcaþ ríce heofona *quia claudítis regnum cælorum,* Mt. Kmbl. Rush. 23, 13. He geseah segn eallgylden, hond-wundra mǽst, gelocen leóðo-cræftum [*or* leoðo-cræftum?] *he saw an allgolden ensign, greatest of hand-wonders, woven by arts of song [by magic],* Beo. Th. 5531; B. 2769. [Cf. hand-locen.]

ge-ludon *descended.* v. geleódan.

ge-lufian; *p.* ode, ade; *pp.* od, ad *To love, esteem;* ámáre, dilígère:—Ne sceal se Dryhtnes þeów máre gelufian eorþan ǽhtwelan *nor shall the Lord's servant love more of earth's riches,* Exon. 38 a; Th. 125, 23; Gú. 358: 119 b; Th. 458, 26; Hy. 4, 106. Se hálga wer, in ða ǽrestan ældu, gelufade frécnessa fela *the holy man, in his first age, loved much mischief,* 34 a; Th. 108, 30; Gú. 80: 39 b; Th. 130, 25; Gú. 443: 43 a; Th. 144, 23; Gú. 682. Ic eom gelufod *ámor,* Ælfc. Gr. 25; Som. 26, 1, 6, 9, 12, 16. Ðú eart mín gelufoda sunu *tu es fílius meus dilectus,* Mk. Bos. 1, 11. Hí wǽron gelufode *ámáti sunt,* Ælfc. Gr. 25; Som. 26, 8, 11, 13, 16.

ge-luggian *to pull, lug;* vellere, Som.

ge-lugon *deceived,* Exon. 118 b; Th. 455, 27; Hy. 4, 56; *p. pl.* of ge-leógan.

ge-lumpe, *pl.* -lumpen *would happen,* Bd. 5, 1; S. 614, 3: Exon. 35 a; Th. 113, 32; Gú. 165; *subj. p.* of ge-limpan: ge-lumpen *happened,* Homl. Th. ii. 130, 28; *pp.* of ge-limpan: ge-lumpon *befell,* Chr. 1011; Erl. 145, 1; *p. pl.* of ge-limpan.

ge-lustfullian; *p.* ode; *pp.* od. I. *v. intrans. To be delighted, be pleased, rejoice;* delectári, lætári:—Hí gelustfulliaþ on mycelnysse sybbe *delectábuntur in multítúdíne pácis,* Ps. Spl. 36, 11. Gelustfulla on Drihtne *delectáre in Dómíno,* 36, 4. For ðysum gelustfullod is heorte mín *propter hoc lætátum est cor meum,* 15, 9. Ðe gelustfullaþ on yfelum lustum *that delights in evil pleasures,* Homl. Th. i. 496, 13. II. *v. trans. To delight, please;* delectáre, júváre:—Me gelustfullaþ *júvat me,* Ælfc. Gr. 33; Som. 37, 12. Gelustfullodon ðē dóhtra cyninga *delectáverunt te filiæ regum,* Ps. Spl. 44, 10. Ða welan gelustfulliaþ *riches afford pleasure,* Homl. Th. ii. 88, 20: 130, 9.

ge-lustfullíce; *comp.* -lícor; *adv. Willingly, earnestly, studiously;* stúdióse:—Nǽnig ðínra þegna neódlícor [MS. -lucor] ne gelustfullícor hine sylfne underþeódde to úra goda bigange ðonne ic *nullus tuórum stúdiósius quam égo cultúræ deórum nostrórum se subdídit,* Bd. 2, 13; S. 516, 5.

ge-lustfulling, e; *f. That which delights* or *pleases;* oblectamentum, Scint. 81.

ge-lustfulnys, -nyss, e; *f. Delight, pleasure;* delectátio:—Gelustfulnyssa [synd] on swíðran ðíne óþ on ende *delectátiónes [sunt] in dextéra tua usque in fínem,* Ps. Spl. 15, 11.

ge-lútan; *p.* -leát *To bow:*—Se bisceop eádmódlíce to ðam Godes were geleát *the bishop humbly bowed to the man of God,* Guthl. 17; Gdwin. 72, 17.

ge-lútian; *p.* ode; *pp.* od *To lie hid;* látère:—Ðæt ic gelútian ne mæg on ðyssum sídan sele *that I may not lie hid in this wide hall,* Cd. 216; Th. 273, 2; Sat. 130.

ge-lýcost *a twin;* gemellus:—Didymus, ðæt is gelýcost, Jn. 20, 24: 21, 2.

ge-lýfan, -lífan, -léfan; to -lýfanne, -lýfenne; *part.* -lýfende; ic -lýfe, ðú -lýfest, -lýfst, he -lýfeþ, -lýfþ, *pl.* -lýfaþ; *p.* ic, he -lýfde, ðú -lýfdest, *pl.* -lýfdon; *impert.* -lýf, *pl.* -lýfaþ; *pp.* -lýfed *To believe, confide, trust, hope;* crédère, confídère, speráre:—We sceolon on hine gelýfan *we should believe in him,* Homl. Th. i. 274, 27: 280, 22: 290, 31. To gelýfanne [-lýfenne, col. 1] to ðan leófan Gode *to trust in the beloved God,* Chr. 1036; Th. 294, 10, col. 2. Of ðyssum lytlingum on me gelýfendum *ex his pusillis crédentíbus in me,* Mk. Bos. 9, 42. Se Hǽlend wiste hwæt ða gelýfendan wǽron *sciebat Jesus qui essent credentes,* Jn. Bos. 6, 64. I do not believe that . . . , Bt. 5, 3; Fox 12, 4: Exon. 82 a; Th. 309, 33; Seef. 66. Gif ðú sóþne God lufast and gelýfest *if thou lovest and believest the true God,* 66 b; Th. 245, 21; Jul. 48: Cd. 203; Th. 252, 14; Dan. 578. Gelýfst ðú ðyses *crédis hoc?* Mk. Bos. 11, 26. He his Hláfordes hyldo gelýfeþ *he believes his Lord's kindness,* Exon. 120 b; Th. 463, 9; Hö. 67: 81 b; Th. 307, 21; Seef. 27. He gelýfþ on God *confídit in Deo,* Mt. Bos. 27, 43: Jn. Bos. 11, 25. Ðe on me gelýfaþ *qui in me crédunt,* Mt. Bos. 18, 6. Ic ðín bebod gelýfde *mandátis tuis crédidi,*

Ps. Th. 118, 66: Bt. 38, 1; Fox 194, 14. Đū mīnum wordum ne ge-lȳfdest *non crēdidisti verbis meis,* Lk. Bos. 1, 20: Jn. Bos. 1, 50. Hī nō gelȳfdon ðæt he God wǣre *they believed not that he was God,* Andr. Kmbl. 1123; An. 562: Elen. Kmbl. 1034; El. 518. Aarones hûs on Dryhten leófne gelȳfdan *dŏmus Aaron spērāvit in Dŏmīno,* Ps. Th. 113, 19. Gelȳf me crēde mihi, Jn. Bos. 4, 21. Gelȳfe gyt, ðæt ic inc mæg gehǣlan *crēdītis quia hoc possum făcĕre vōbis?* Mt. Bos. 9, 28. Gelȳfaþ for ðām weorcum *propter ŏpĕra ipsa crēdĭte,* Jn. Bos. 14, 11. Ne bepǣce nān man hine sylfne, swā ðæt he secge oððe gelȳfe ðæt þrȳ Godas syndon *let no man deceive himself, so as to say or believe that there are three Gods,* Homl. Th. i. 284, 16. Ðæt gē gelȳfon, ðæt se Hǣlend ys Crist *ut crēdātis, quia Jesus est Christus,* Jn. Bos. 20, 31: Ex. 4, 5. Ne gelȳfe ic me nū ðæs leóhtes furðor *I have no longer now any hope for myself of that light,* Cd. 21; Th. 26, 3; Gen. 401. [*Goth.* ga-laubjan: *O. Sax.* gi-lōbian: *O. H. Ger.* gi-louban: *Ger.* glauben.]

ge-lȳfan; *p.* de; *pp.* ed *To make dear* [leóf] :— Dryhtne gelȳfde *endeared to the Lord* [*faithful to the Lord,* Th.], Exon. 32 a; Th. 100, 22; Cri. 1645.

ge-lȳfan; *p.* de; *pp.* ed *To allow, permit;* concēdĕre, permittĕre :— Wæs him seó rōw gelȳfed þurh lytel fæc *repose was allowed them for a little time,* Exon. 35 b; Th. 115, 5; Gū. 185.

ge-lȳfed; *part. p.* [*pp.* of ge-lȳfan *to believe*] *One who believed, faithful;* religiosus, fidus, fidelis :— His [Constantīnes] mōdor wæs cristen, Elena gehâten, swīðe gelȳfed mann, and þearle eáwfæst *his* [*Constantīne's*] *mother was a christian, called Helena, a very faithful person, and very pious,* Homl. Th. ii. 306, 3: i. 60, 13. Com se ârwurþa Swīþhun to sumum gelȳfedan smiþe on swefne *the venerable Swithun came to a certain religious* [lit. *faithful*] *artisan in a dream,* Glostr. Frag. 2, 5. Wæs sum cyning gelȳfed swīðe on God *there was a king firmly believing on God,* Swt. Rdr. 95, 2: H. R. 101, 13. Hie wurdan hraðe gelȳfde *they immediately believed,* Blickl. Homl. 155, 5. Ealle ðing synd gelȳfedum mihtlīce *omnia possibilia credenti,* Mk. Bos. 9, 23.

ge-lȳfed; *part. p. Weakened, advanced* [*in age*] :— Ðara ðe gelȳfedre yldo *earum quæ ætate provectæ,* Bd. 3, 8; S. 531, 33 : 4, 24; S. 597, 3.

ge-lȳfedlic; *adj.* [ge-lȳfan *to allow*] *Allowable, permissible;* lícitus, permissus :— Nis hit nāht gelȳfedlíc *it is not allowable,* L. E. I. 39; Th. ii. 436, 35.

ge-lȳfedlíce; *adv. Faithfully, confidently;* confidenter :— Xersis swīðe gelȳfedlíce his þegene gehȳrde *Xerxes very confidently listened to his general,* Ors. 2, 5; Bos. 48, 9: 3, 1; Bos. 53, 15.

ge-lȳhtan; *p.* -lȳhte; *pp.* -lȳhted, -lȳht *To illumine, give light to* :— He blynde gelȳhte *he enlightened the blind,* St. And. 44, 34: Nic. 34; Thw. 20, 2. [*Goth.* ga-liuhtjan: *O. Sax.* gi-liuhtian.] v. līhtan.

ge-lymp *an accident.* v. ge-limp.

ge-lymplicnys, se; *f. Opportunity, occasion;* opportunitas, Ps. Spl. C. 9, 9.

ge-lynd, -lend, e; *f.* [lynd *fat*] *Grease, fat, fatness;* adeps, pinguedo :— Ys sāwl mīn swētes gefylled, swā seó fætte gelynd fægeres smeoruwes *sicut adipe et pinguedine repleatur animea mea,* Ps. Th. 62, 5. Gelynde *ex adipe,* 72, 6. Bringon gelynde *offerent adipem,* Lev. 3, 10. Nim león gelynde *take lion's fat,* Med. ex Quadr. 10, 2; Lchdm. i. 364, 24: 10, 4; Lchdm. i. 366, 4. DER. lynd.

ge-lyndu; *n. pl. Joints of the backbone* :— Geloda *vel* gelyndu *spondilia* [*Gk.* σπόνδυλος], Ælfc. Gl. 74; Som. 71, 51; Wrt. Voc. 44, 34.

ge-lȳsan; *pp.* ed *To redeem, loosen, dissolve, break* :— Eall his líchama wæs gelȳsed *all his body was broken,* Blickl. Homl. 241, 30. [Cf. to-lȳsan *and* ge-lēsan.]

ge-lȳsednes *redemption.* v. alȳsednys.

ge-lystan; *p.* -lyste; *pp.* -lysted, -lyst; *v. impers. with acc. of pers., gen.* of *thing; To please, cause a desire for anything* :— Ðegnas ðearle gelyste gârgewinnes *the thanes were very eager for the struggle,* Judth. 12; Thw. 26, 3; Jud. 307: Exon. 97 a; Th. 361, 22; Wal. 23. Gūðe gelysted *desirous of war,* Bt. Met. Fox 1, 18; Met. 1, 9. [*O. Sax.* gi-lustean: *O. H. Ger.* gi-lusten (*with the same government*): *Ger.* gelüsten.)

ge-lytfullíce; *adv. Prosperously;* prospere, Ps. Spl. C. 44, 5.

ge-lyðen; *part. p. Travelled* :— Se ylca Nathan wæs swā gelyðen ðæt he hæfde gefaren fram ǣlcum lande to ōðrum *this Nathan was so travelled that he had gone from every land to the other,* St. And. 26, 13. v. ge-líðan.

ge-lytlian, -litlian; *p.* ode, ade; *pp.* od, ad *To diminish, lessen, humble;* minuĕre, hŭmiliâre :— Ǣghwilc ælmesriht ǣlc man gelytlaþ oððe forhealdeþ *every almsright every man lessens or withholds,* Swt. Rdr. 106, 59. Ealle hire wæstmbǣro he gelytlade *he lessened all her* [*the earth's*] *fruitfulness,* Ors. 2, 1; Bos. 38, 8. Mīn líf gelytlad is *hŭmiliâvit vītam meam,* Ps. Th. 142, 3.

ge-maad *mad.* v. ge-mǣd.

ge-maca, an; *m.* and *f. A mate, an equal, companion;* par, socius :— Gemaca *hic et hæc par,* Ælfc. Gr. 9; Som. 9, 50. Of eallum nȳtenum ealles flǣsces twegen gemacan *of all beasts two of the same kind, male and female,* Gen. 6, 19. [*Laym.* i-maken: *O. Sax.* ge-mako: *O. H. Ger.* ka-mahho *socius.*] DER. fyrd-, heáfod-gemaca. [Cf. ge-mæcca.]

ge-macian; *p.* ode; *pp.* od *To make, cause* :— Hī heora lufigendne

gemaciaþ weligne ēcelíce *they make the lover of them rich eternally,* Homl. Th. ii. 88, 29. Ðone ðe he ǣr martyr gemacode *whom he had before made a martyr,* 82, 24. Hī ðæra cinga sehte gemacedon *they made peace between the two kings,* Chr. 1091; Erl. 228, 2. Ðæt land-folc gemacodon ðæt he nāht ne dyde *the folk of the country prevented him from doing anything,* 1075; Erl. 213, 20: Exod. 5, 21. He lēt castelas gemakian *he had castles built,* Chr. 1097; Erl. 234, 8. Eác is mōdsorg gode gemacod *also grief of mind is caused to God,* Cd. 35; Th. 47, 3; Gen. 755.

ge-mæc; *adj. Equal, like, well-matched, suited* :— Hī wíf habbaþ him gemæc *they are well-matched in marriage,* Bt. 11, 1; Fox 32, 4. Ge-mæcca ł gelíco *æquales* [*or v.* ge-mæcca?], Lk. Skt. Lind. 20, 36. Ic me ful gemæcne monnan funde *I found a man fully equal to me,* Exon. 115 a; Th. 442, 25; Kl. 18. [Cf. Grff. ii. 632.]

ge-mæcca, -mæccea, an; *m.* and *f. A companion, mate, consort, husband* or *wife* :— Twegen turturan gemæccan *a pair of turtle doves,* Blickl. Homl. 23, 27. Ne eart ðū ðon leófre nǣngum lifigendra menn to gemæccan ðonne se swearta hrefn *thou art not any dearer to any living man as mate than the swart raven,* Exon. 99 a; Th. 370, 6; Seel. 53. Boga sceal strǣle sceal mon to gemæccan *a bow must have an arrow, a man must to his mate,* Exon. 91 b; Th. 343, 10; Gn. Ex. 155. Gemæcca *conjunx,* Ælfc. Gr. 28; Som. 31, 54. Gif wíf wiþ ōðres gemæccan hǣmþ *si mulier cum alterius conjuge adulteraverit,* L. Ecg. P. iii. 10; Th. ii. 186, 7. Be Euan his gemæccan *by Eve his wife,* Gen. 4, 1: 28, 1: Homl. Th. ii. 498, 26. He onfēng hys gemæccean *accepit conjugem suam,* Mt. Bos. 1, 24. [*O. H. Ger.* gi-mahha *conjux.*] Cf. ge-maca.

ge-mæclíc; *adj. Relating to a wife, conjugal;* conjugalis, Scint. 58.

ge-mæcnes, -ness, e; *f. A companionship, mixture;* commixtio :— On ðæs líchoman gemæcnesse biþ willa *in carnis commixtiōne voluptas est,* Bd. 1, 27; S. 493, 20, MS. B.

ge-mæcscipe, es; *m. Fellowship, connection, cohabitation;* consortium, conjūgium, concūbitus :— þurh gemæcscipe *through cohabitation,* Exon. 10 b; Th. 13, 7; Cri. 199.

ge-mǣd; *adj.* [cf. *O. Sax.* ge-mēd *foolish:* *O. H. Ger.* ka-meit *stultus:* or ge-mǣd? v. Leo 29] *Troubled in mind, mad;* amens, Cot. 10, 169.

ge-mǣdan; *p.* de; *pp.* ed *To madden, make foolish* :— Swā gemǣde mōde bestolene dǣde gedwolene *so foolish bereft of mind erring in deed,* Exon. 103 b; Th. 393, 6; Rä. 12, 6. Gemǣded *vecors,* Lye. [Cf. *Laym.* Of witten heo weoren amadde (*later MS.* awed).] v. ge-mǣd.

ge-mǣdla, an; *m. Talk* :— Wiþ wíf-gemǣdlan geberge on neaht nestig rædices moran ðȳ dæge ne mæg ðē se gemǣdla sceððan *against a woman's chatter; taste at night fasting a root of radish, that day the chatter cannot harm thee,* L. M. 3, 57; Lchdm. ii. 342, 11. v. ge-maðel.

ge-mǣg, es; *m. A kinsman* :— Wit synt gemǣgas *we two are kinsmen,* Cd. 91; Th. 114, 14; Gen. 1904. v. mǣg.

ge-mægened; *part. p. Established, confirmed, strengthened;* confirmatus :— Gemǣgenad and gestrongad beón *to be confirmed and strengthened,* Bd. 4, 16; S. 584, 4.

ge-mægfæst; *adj. Gluttonous;* cibi deditus, Lye.

ge-mægnan. v. ge-mengan.

ge-mægþ, e; *f. Power, greatness;* pŏtentia :— Me nǣfre seó gemægþ ðisses eorþlícan anwealdes fōrwel ne lícode *the greatness of this earthly power never too well pleased me,* Bt. 17; Fox 58, 23.

ge-mǣgþ, e; *f. A family, tribe;* fămília, tribus :— Twā gemǣgþa *two families,* Ors. 3, 5; Bos. 57, 33.

ge-mǣhþ, e; *f. Greediness* :— Ic wolde witan hwæðer ðīn ealde gȳtsung and seó gemǣhþ eallunga of ðīnum mōde astȳfcod wǣre *I wanted to know whether thine old covetousness and greediness were altogether eradicated from thy mind,* Shrn. 184, 2. v. ge-māh.

ge-mǣl; *adj. Marked, stained* :— Earh ǣttre gemǣl *the arrow stained with poison,* Andr. Kmbl. 2663; An. 1333.

ge-mǣlan; *p.* de; *pp.* ed *To mark, stain* :— Scó hâlge stōd ungewemde wlite næs hyre feax ne fel fȳre gemǣled *the saint stood with spotless aspect, neither her hair nor skin was marked by the fire,* Exon. 74 a; Th. 278, 2; Jul. 591.

ge-mǣlan; *p.* de; *pp.* ed *To speak, harangue* :— Adam gemǣlde and to Euan spræc *Adam spoke and to Eve said,* Cd. 37; Th. 49, 10; Gen. 790. Offa gemǣlde *Offa spake,* Byrht. Th. 138, 34; By. 230: 53; By. 244.

gēmæn. v. gēmen.

ge-mǣnan; *p.* de; *pp.* ed [ge-mǣne *communis*]. **I.** *to* MEAN, *to signify;* sibi velle, significare :— Hwæt gemǣnþ ðæs lamb *quid sibi volunt agnæ istæ?* Gen. 21, 29. Ic wēne ðæt ðū nyte hwæt ðis gemǣne *I expect that thou wilt not know what this means,* Btwk. Scrd. 18, 26. Hwæt gemǣnaþ ðā ðreó ūtfaru? *Ðæt getācnaþ . . . what do the three outgoings mean? They indicate . . . ,* 21, 40. **II.** *to communicate, announce, pronounce, utter;* communicare, pronuntiare :— Hwīlum ic glidan reorde mūþe gemǣne *sometimes in a kite's voice I utter with my mouth,* Exon. 105 b; Th. 406, 24; Rä. 25, 6. **III.** *to give expression to one's feelings,* as, *of pain, to* MOAN, *to groan;* ingemiscere, plangere, Mk. Skt. Lind. 8, 12: Lk. Skt. Lind. 23, 27. **IV.** *to commune with oneself about anything, to consider;* colloqui, considerare :—

Se fæder hit gemænde stille *pater rem tacitus considerabat*, Gen. 37, 11. **V.** [mǽne *vilis, scelestus*] *to make common, contaminate, defile, violate;* communicare, coinquinare, violare:—Ðæt ðær ǽnig mon wordum ne worcum wǽre ne brǽce, ne þurh inwit-searo ǽfre gemǽnden *that there not any man by words or works should break the compact, nor through guileful art should ever violate it*, Beo. Th. 2207, note; B. 1101. [*Goth.* ga-mainjan *communicare alicui;* κοινῶν *vel* κοινωνεῖν τινί τι, *etiam, coinquinare vel communicare aliquid;* κοινῶν τι: *O. Sax.* gi-mēnian *to make known: O. H. Ger.* gi-meinen *dicere, monstrare, judicare.*] v. mǽnan.

ge-mǽncgan, -mǽngan; *p.* -mǽnced *To mix.* v. ge-mengan.

ge-mǽne; *adj. Common, general, mutual, in common;* communis:—Reord wæs ðá gieta eorþ-búendum án gemǽne *there was yet one common language to the dwellers upon earth*, Cd. 79; Th. 98, 27; Gen. 1636. Sib sceal gemǽne englum and ældum á forþ heonan wesan *a common peace shall be to angels and men henceforth for ever*, Exon. 16 a; Th. 36, 25; Cri. 581. Hwæt ys ðé and us gemǽne *what is common to thee and us?* Mt. Bos. 8, 29. Ne beó ðé nán þing gemǽne ongén ðisne rihtwísan *ne quid tibi sit commune adversus hunc justum*, 27, 19: Nicod. 6; Thw. 3, 11. Se ðe oferhogie ðæt he Godes bodan hlyste, hæbbe him gemǽne ðæt wið God sylfne *he who scorns to listen to God's preacher, let him have that between him and God himself*, L. C. E. 26; Th. i. 374, 27: Kmbl. Cod. Dipl. iii. 22, 27. Ðæt hí sceoldon habban sunu him gemǽne *that they should have a son common to them [between them]*, Cd. 13, 3: Cd. 100; Th. 133, 26; Gen. 2216. Gemǽne win *communis labor*, Bd. 2, 2; S. 502, 9. Gemǽne læs *compascuus ager*, Ælfc. Gl. 96; Wrt. Voc. 53, 54. Him eallum wǽron eall gemǽne *erant eis omnia communia*, Bd. 1, 27; S. 489, 15: Jos. 8, 2. Unc sceal worn fela mádma gemǽne *to us two shall be a great many common treasures*, Beo. Th. 3572; B. 1784. Ðá wæs synn and sacu Sweóna and Geáta, wróht gemǽne *then was sin and strife of Swedes and Goths, mutual dissension*, Beo. Th. 4938; B. 2473. Ðæt sceal Geáta leódum and Gár-Denum sib gemǽnum *so that there shall be peace to the Goths' people and to the Gar-Danes in common*, 3718; B. 1857. Hand gemǽne *a joined hand [in conflict];* manus conserta, 4281; B. 2137. [*Laym.* i-mæne: *O. Sax.* gi-mêni *communis, generalis, solitus: O. Frs.* ge-mêne: *O. H. Ger.* ga-meini: *Goth.* ga-mains *communis;* κοινός, συγκοινωνός.]

ge-mǽne-líc; *adj. Common, general;* communis, generalis:—Swá swá man gerǽde for gemǽnelícre neóde *so that the common need may be consulted for*, L. Eth. vi. 32; Th. i. 324, 1. Hí arísaþ on ðam gemǽnelícum dóme *they shall arise at the judgment of all*, Homl. Th. i. 84, 22, 24. Mid ða getýdnesse ge cyriclícra gewrita ge eac gemǽnelíca *cum eruditione litterarum vel ecclesiasticarum vel generalium*, Bd. 5, 23; S. 645, 15. Gemǽnelíce naman *appellative* or *common nouns;* appellativa nomina, Ælfc. Gr. 9, 3; Som. 8, 31.

ge-mǽnelíc nama, an; *m. A common noun;* appellativum nomen, Ælfc. Gr. 9; Som. 8, 31. v. ge-mǽnelíc.

ge-mǽne-líce; *adv. Commonly, in common, generally, mutually, in turn, one amongst another;* communiter, generaliter, invicem:—Ðæt hý ðæt feoh mihton him eallum gemǽnelíce to nytte gedón *that they might apply that wealth to the use of all in common*, Ors. 2, 4; Bt. 39, 13; Fox 234, 28. Iohannes ðá beád ðreóra daga fæsten gemǽnelíce *John then ordered a general fast of three days*, Homl. Th. i. 70, 8. Þurh hí sende gemǽnelíce ða þing eall ða ðe to cyrican bigange and þénunge nýdþearflíco wǽron *misit per eos generaliter universa quæ ad cultum erant ac ministerium ecclesiæ necessaria*, Bd. 1, 29; S. 498, 8. Ðæt gé lufion eów gemǽnelíce, swá ic eów lufode *ut diligatis invicem, sicut dilexi vos*, Jn. Bos. 15, 12, 17.

ge-mǽnigfealdian; *p.* ode; *pp.* od *To multiply:*—Gemænigfealdige ðis mihtig Dryhten ofer eów ealle *adjiciat Dominus super vos*, Ps. Th. 113, 22.

ge-mǽnigfyldan; *p.* ode; *pp.* od *To multiply, enlarge;* multiplicare:—Ðú gemænigfyldest sunú manna, Ps. Spl. 11, 9: 17, 16. Gemǽnigfylde beón, Ex. 1, 7.

ge-mǽn-nes, -ness, e; *f.* [ge-mǽne *communis*] *A communion, fellowship, connection;* communio, consortium, admixtio:—Hí sealdon hí ðǽr on ðara fǽmnena gemǽnnesse *they gave her up there to the society of the women*, Shrn. 127, 11. Ne ic ǽfre mid mannum mán-fremmendum ǽr gemǽnnesse micle hæbbe *cum hominibus operantibus iniquitatem non comminabor* [*Vulg.* communicabo, Ps. Surt. conbinabor], Ps. Th. 140, 6: R. Ben. proœm. Gemencgnyss [MS. B. gemǽnnes] wífes *admixtio conjugis*, Bd. 1, 27; S. 495, 18. Ðurh flǽsces gemǽnnysse *per carnis contubernium*, Hymn. Surt. 31, 32. [Hence the Kentish word *mennys* a large common.]

ge-mǽnnung, e; *f. Communion, fellowship;* communio, contubernium, Som.

ge-mǽn-scipe, es; *m. Communion, fellowship;* communio:—Ic gemǽnscipe getreówe ðínra háligra *I believe in the communion of thy saints*, Hy. 10, 52; Hy. Grn. ii. 294, 52: Wanl. Catal. 49, 16.

ge-mǽn-suman, -mǽn-sumian; *p.* ode, ade; *pp.* od, ad [ge-mǽne *communis*] *To do* or *have anything in common with another, to communicate to* or *share with another, to marry;* communicare, nubere:—Wylladon us ða þing gemǽnsuman [MS. gemǽnsumian] *ea nobis com-*

municare desiderastis, Bd. 1, 25; S. 487, 14. Gemǽnsumad *nuptus*, Mk. Skt. Lind. 12, 25. [*O. H. Ger.* ga-meinsamôn *communicare, participare.*] v. mǽn-sumian.

ge-mǽnsumnys, -nyss, e; *f. A communion, a participation*, also *the Sacrament of the Holy Communion;* communio:—Ne syndon hí for ðysse wísan to bescyrianne gemǽnsumnysse Cristes líchoman and blódes *non pro hac re sacri corporis ac sanguinis Domini communione privandi sunt*, Bd. 1, 27; S. 491, 27. Ðam gerýne onfón ðǽre hálgan gemǽnsumnysse *sacræ communionis sacramentum vel mysterium percipere*, Bd. 1, 27; S. 492, 35: 1, 27; S. 494, 23.

ge-mǽn-sumung, e; *f. A communion;* communio, R. Ben. 38.

ge-mǽran *to fix limits, determine:*—Gimǽrende *determinans*, Rtl. 164, 38.

ge-mǽran; *p.* de; *pp.* ed [mǽre] *To celebrate, divulge, spread abroad:*—Ðá ðeós gesyhþ wæs gemǽred *qua divulgata visione*, Bd. 4, 25; S. 601, 25: 3, 10; S. 535, note 2. Gemǽred wæs word ðis mið Iudeum *divulgatum est verbum istud apud Judæos*, Mt. Kmbl. Rush. 28, 15. Hiæ gemǽrdon hine *illi diffamaverunt eum*, 9, 31.

ge-mǽran; *p.* de; *pp.* ed [mǽra] *To enlarge:*—He merce gemǽrde wiþ Myrgingum *he enlarged his marches towards the Myrgings* [or *gemǽrde from gemǽran to determine?*], Exon. 85 a; Th. 321, 6; Víd. 42.

ge-mǽre, es; *pl. nom.* a, o, u; *n. An end, boundary, termination, limit;* finis:—Gemǽro *limes*, Ælfc. Gr. 9; Som. 11, 16. Gemǽre ðú settest *terminum posuisti*, Ps. Spl. 103, 10. Ne mágon hí ofer gemǽre gegangan *terminum non transgredientur*, Ps. Th. 103, 10. On Hwicna gemǽre and West-Sexna *in confinio Huicciorum et occidentalium Saxonum*, Bd. 2, 2; S. 502, 7: 5, 23; S. 646, 25: Exon. 93 a; Th. 349, 28: Sch. 53. Gemǽro eorðan *terminos terræ*, Ps. Spl. 2, 8. Oþ gemǽru *usque ad terminos*, 71, 8. Ðis sind ðæs londes gemǽra *these are the land's boundaries*, Kmbl. Cod. Dipl. iii. 78, 20. He ða gemǽro his rynes gefylde *metas sui cursus implevit*, Bd. 3, 20; S. 550, 25. Eall eorðan gemǽru *omnes fines terræ*, Ps. Th. 66, 6: 73, 16. Mycel sǽ and on gemǽrum wíd *mare magnum et spatiosum*, 103, 24. On gemǽru *in finibus eorum*, 104, 27: Bt. Met. Fox 29, 17; Met. 29, 9: Th. Apol. 9, 14. Cýð ðis folc ðæt hig ne gán ofer ða gemǽro *tell this people not to cross the bounds*, Exod. 19, 21, 12. v. Kmbl. Cod. Dipl. iii. viii sqq.

ge-mǽrsian, ic -mǽrsige; *p.* ode; *pp.* od *To magnify, glorify, celebrate;* magnificare, glorificare, celebrare:—Ðínne naman ic gemǽrsige *magnificabo nomen tuum*, Gen. 12, 2. Ðú Sunnan dæg sylf hálgodest and gemǽrsodest hine manegum to helpe *thou thyself didst sanctify Sunday and didst glorify it for help to many*, Hy. 9, 26; Hy. Grn. ii. 291, 26. On ðam dæge gemǽrsode se mihtiga Drihten Iosue ðone ǽðelan ætfóran Israhéla folce *in die illo magnificavit Dóminus Josue coram omni Israel*, Jos. 4, 14. Is ðín nama miltsum gemǽrsod *thy name is magnified with mercies*, Andr. Kmbl. 1087; An. 544: Hy. 7, 44; Hy. Grn. ii. 288, 44. He wæs fram eallum gemǽrsod *ipse magnificabatur ab omnibus*, Lk. Bos. 4, 15. Ic beó gemǽrsod on Pharaone *glorificabor in Pharaóne*, Ex. 14, 17. He wæs gemǽrsod ofer ealle óðre cyningas *he was celebrated above all other kings*, Ors. 4, 1; Bos. 76, 41.

ge-mǽrsung, -mǽrsung, e; *f. Magnificence;* magnificentia:—Ðæt hí cúðe wyrcan wuldor gemǽrsunge ríces ðínes *ut nótam faciant glóriam magnificentiæ regni tui*, Ps. Spl. 144, 12. Gimǽrsung *celebritas*, Rtl. 48, 20.

ge-mǽssian; *p.* ode; *pp.* od *To say mass to:*—Iustinus him eallum gemǽssode *Justin said mass to them all*, Homl. Th. i. 430, 29.

ge-mǽst; *part. p. Fat, fattened;* altilis. v. ge-mǽstan.

ge-mǽstan; *pp.* -mǽsted, -mǽst *To fatten;* saginare, pinguefacere, impinguare:—Híg wǽron gemǽste *erant impinguati*, Deut. 32, 15. Gemǽstra fugela *of fatted fowls*, Homl. Th. ii. 576, 34: Bd. Whelc. 378, 19. v. amǽstan, mǽstan.

ge-mǽtan; *p.* -mǽtte; *pp.* -mǽted; *v. impers. acc. To dream;* somniare, somnium videre:—Hwæt hine gemǽtte *what he had dreamed*, Cd. 178; Th. 223, 20; Dan. 112: Rood. Kmbl. 3; Kr. 2. Swá his man-drihten gemǽted wearþ *as his lord had dreamed*, Cd. 179; Th. 225, 21; Dan. 157. v. mǽtan.

ge-mǽte; *adj. Moderate, meet, fit;* modicus, aptus, Mod. Conf. 1; C. R. Ben. 55. [*O. H. Ger.* ge-mázer: *Laym.* i-mete.] v. mǽte.

ge-mǽt-fæstan; *p.* -fæste; *pp.* -fæsted, -fæst [gemet *a measure,* fæst *fast*] *To compare;* comparare, Ps. Lamb. 48, 21. v. ge-met-festan.

ge-mǽtgan; *p.* ede; *pp.* ed; *v. trans.* [mǽte *moderate*] *To make moderate, to limit, diminish;* moderare, moderari, minuere:—Ful oft hit eác ðæs deófles dugoþe gemǽtgeþ *full oft it also limits the devil's power*, Salm. Kmbl. 800; Sal. 399.

ge-mǽdian, -mǽdegian, -mǽdrian, -mēdrian; *p.* ode; *pp.* od *To honour, bestow something with honour upon one;* honōrāre, benigne conferre:—Búton he hwæne furðor gemǽdrian [gemǽdian, MS. B.] *unless he will more amply honour any one*, L. C. S. 12; Th. i. 382, 15; Th. i. 384, 4. For ðære micclan mǽrþe ðe he hine gemǽdegode *for the great glory which he honourably bestowed upon him*, Ælfc. T. 4, 11.

ge-mǽt-líc; *adj. Moderate;* modicus. v. un-ge-mǽt-líc.

ge-mágas; *pl. m. Kinsmen, relations;* consanguinei:—Wit synt gemágas *we two are kinsmen,* Cd. 91; Th. 114, 14; Gen. 1904. God hí gesceóp to gemágum *God created them as relations,* Bd. 24, 3; Fox 82, 31. v. mæg.

ge-máglic; *adj. Importunate, pertinacious:*—Mid gemáglícum wôpum *with importunate weeping,* Homl. Th. ii. 126, 1. v. ge-máhlíc.

ge-máglíce; *adv. Urgently, importunately:* — He tiht ǽlcne swíðe gemáglíce to gebedum *he exhorts everybody very urgently to prayers,* Homl. Th. i. 158, 13. v. ge-máhlíce.

ge-mágnys, se; *f. Perseverance, importunity, petulance:* — Sôðlíce gemáguys is ðam sôðan Dēman gecwême *truly importunity is pleasing to the true judge,* Homl. Th. ii. 126, 2. Asolcennys acenþ gemágnysse *slothfulness gives birth to petulance,* 220, 26.

ge-máh; *adj. Shameless, obstinate, stubborn, impious, wicked, importunate;* prŏcax, pervĭcax, pertĭnax, imprŏbus, importūnus:—Gemáh *prŏcax* vel *pervĭcax,* Ælfc. Gl. 88; Som. 74, 84; Wrt. Voc. 50, 64: 86, 52. Fláh feónd gemáh *the deceitful impious fiend,* Exon. 97 a; Th. 362, 19; Wal. 39: 64 b; Th. 237, 24; Ph. 595. Gemáh *importūnus,* Ælfc. Gl. 101; Som. 77, 45; Wrt. Voc. 55, 50.

ge-máh *made water;* minxit, Med. ex Quadr. 9, 13; Lchdm. i. 364, 1; *p.* of ge-mígan.

ge-máhlíc; *adj. Shameless, wanton, greedy;* prŏcax, ǎvĭdus:—Ðæt hit gemáhlíc wǽre and unrihtlíc *that it was greedy and unjust,* Ors. 1, 10; Bos. 32, 20. v. ge-máglíc.

ge-máhlíce; *adv. Importunately, peremptorily, boldly, pertinaciously:*—Se cyng hēt swýðe gemáhlíce ofer eall ðis land beódan *the king very peremptorily ordered it to be proclaimed over all this land,* Chr. 1095; Erl. 232, 22. Án blac ðrostle flicorode ymbe his neb swá gemáhlíce *a black throstle flitted about his face so boldly,* Homl. Th. ii. 156, 23: Gr. Dial. 1, 8. v. ge-máglíce.

ge-máhlícnes, e; *f. Importunity, perverseness, dishonesty;* importunitas:—Se forhwierfeda gewuna gemáhnesse *the perverse habit of wantonness,* Past. 13, 2; Swt. 79, 19; Hat. MS

ge-máhnes, -nys, -ness, -nyss, e; *f. Shamelessness, stubbornness;* prŏcācĭtas, pervĭcācĭa:—Gemáhnes *prŏcācĭtas,* Wrt. Voc. 86, 53. Gemáhnys *prŏcācĭtas* vel *pervĭcācĭa,* Ælfc. Gl. 88; Som. 74, 85; Wrt. Voc. 50, 65. v. ge-mágnys.

ge-máleca *importunate;* importunus, Cot. 2.

ge-málíce; *adv. Importunately;* importune, Cot. 189.

ge-mal-mægen *an assembly.* v. al-mægen.

ge-man *the hollow of the hand, sole of the foot;* vola, Cot. 198.

ge-man, ic, he *I remember, he remembers,* Beo. Th. 5259; B. 2633; Jn. Bos. 16, 21; *pres.* of ge-munan.

gēman; *p.* de; *pp.* ed *To care for, regard, heed, cure;* cūrāre:—Ne gēmdon hie nánes fyrenlustes *they cared not for any luxury,* Blt. 15; Fox 48, 7: Bd. 2, 6; S. 508, 39. Nǽnig mon ne sceal lufian ne ne gēman *his gesibbes gif he hine ǽrost agælde Godes ðeówðômes no man shall love or care about his relatives if he first have devoted himself to God's service,* Blickl. Homl. 23, 17: 67, 30. Hí nystan ne ne gēmdon *they neither knew nor cared,* 99, 30. Ic cymo and gēmo hine *ego veniam et curabo eum,* Mt. Kmbl. Lind. 8, 7: Lk. Skt. Lind. 10, 9. Nallaþ gie gēma *nolite sollicliti esse,* 12, 11. Ne gēmes ðū *non curas,* Mk. Skt. Lind. 12, 14. Gēmende *sollicliti,* Mt. Kmbl. Lind. 6, 25. v. gýman.

ge-mána, an; *m.* [ge-mǽne *communis*] *Companionship, society, fellowship, familiarity, marriage, intercourse, commerce, conjunction;* communio, societas, consortium, contubernium, commercium, concubitus:—Gifeon we on ðone gemánan Godes and manna and on ðone gemánan ðæs brýdguman and ðære brýde *let us rejoice in the union of God and men and in the union of the bridegroom and the bride,* Blickl. Homl. 11, 5. Ðonne he wæs mid his ágnum cynne ðonne he wæs on ðare ryhtwísera gemánan *he was then with his own kin when he was in the company of the righteous,* Bt. 5, 1; Fox 10, 12. Engla gemána *the society of angels,* Exon. 42 a; Th. 142, 10; Gū. 642: Ps. Th. 56, 4: Bd. 4, 23; S. 596, 13. Ðysse fǽmnan gemánan bæd *hujus virginis consortium petebat,* 2, 9; S. 510, 23, 26: Exon. 67 b; Th. 250, 14; Jul. 127: Jn. Skt. p. 1, 3: Rtl. 109, 31. Hrēman ne þorfte mǽcan gemánan *he needed not to exult in the falchion's intercourse,* Chr. 937; Th. 204, 24; Æðelst. 40. Wið dam ðe ðú mínes gemánan brúce *ut fruaris concubitu meo,* Gen. 38, 16: Med. ex Quadr. 5, 11; Lchdm. i. 350, 10. [*Goth.* ga-mainei: *O. H. Ger.* gameinî *f.*]

ge-mane, -mone; *adj. Having a mane:*—Ðara hæfda beóþ gemona swá leóna hǽfdo *their heads have manes like lions' heads,* Nar. 35, 29. [*Cf. O. H. Ger.* mana: *Icel.* mön *a mane.*]

ge-mang, -mong, es; *n.* 　　I. *a mingling together, mixture, crowd, throng, company, multitude, an assemblage, a congregation;* commixtio, turba, cœtus, sŏcietas:—Ic bebeóde wundor geweorþan on wera gemange *I command a miracle to be done in the midst of men,* Andr. Kmbl. 1460; An. 730. God mihtig stôd godum on gemange *Deus stĕtit in synăgôga deōrum,* Ps. Th. 81, 1. In heora gemange *in their congregation,* L. Wih. 23; Th. i. 42, 6: Nicod. 6; Thw. 6, 8. Gáras sendon in heardra gemang *they sent their darts into the throng of the brave,* Judth. 11; Thw.

24, 36; Jud. 225. On clǽnra gemang *in the company of the pure,* Elen. Kmbl. 191; El. 96: 216; El. 108: 236; El. 118. 　　II. *an assembly for legal or other business:*—Ne miltsa ðū þearfan on gemange *pauperis non misĕrēbĕris in jūdĭcio,* Ex. 23, 3. Ne mæg ic āna eówre gemang acuman *non văleo sōlus nĕgōtia vĕstra sustĭnēre,* Deut. 1, 12: Shrn. 40, 30.

ge-mang; *prep.* [ge-mang *a mixture*] AMONG; inter, in medio. 　　I. *dat:*—Ðeós sprǽc com ūt gemang brôþrum *exiit sermo iste inter fratres,* Jn. Bos. 21, 23. Arís gemang him *surge in medium,* Mk. Bos. 3, 3. Gemang ðām *interim,* Gen. 43, 1. Gemang ðām arás micel murcnung *interea ortum est murmur,* Num. 11, 1. 　　II. *acc:*—Ic eów sende swá sceáp gemang wulfas *ego mitto vos sicut oves in medio luporum,* Mt. Bos. 10, 16. DER. a-mang, on-.

ge-mangcennyss, e; *f. A mingling, confection;* confectio, debilitatio, Hpt. Gl. 450: Mone B. 1846.

ge-mangian; *p.* ode; *pp.* od *To traffic, trade;* nĕgōtiāri:—Ðæt he wiste hú mycel gehwilc gemangode *ut scīret quantum quisque nĕgōtiātus esset,* Lk. Bos. 19, 15. Hwæt forstent ǽnegum men, ðeáh he gemangige ðæt he ealne ðisne middangeard áge, gif he his sáule forspildeþ *what profits it any man, though he trade so as to obtain all this world, if he destroy his soul?* Past. 44, 10; Swt. 332, 9; Cot. MS.

ge-mangnys, se; *f. A mingling, confection;* commixtio. Som.

ge-manian, -monian, -monigan; *p.* ode, ade; *pp.* od, ad *To admonish, exhort, prompt, remind, remember;* admonere, hortari, suggerere, in memoriam rei reducere, recordari:—Seó sáwl ðurh ðæt gemynd gemanaþ *the soul through the memory reminds,* Homl. Th. i. 288, 28. Oft mec geómor sefa gemanode *oft my sad spirit has admonished me,* Exon. 50 a; Th. 174, 22; Gū. 1181. Se áuwealda hæfþ ealle his gesceafta mid his bridle getogene and gemanode *the Ruler has with his bridle restrained and admonished all his creatures,* Bt. 21; Fox 74, 7: Bt. Met. Fox 11, 47; Met. 11, 24. Gemanad *admonished,* Exon. 102 a; Th. 386, 23; Rä. 4, 66: Exon. 88 b; Th. 333, 19; Gn. Ex. 6: Cd. 49; Th. 63, 9; Gen. 1029. v. manian.

ge-mánna, an; *m. Fellowship,* Wanl. Catal. 23, 47. v. ge-mána.

ge-mannian; *p.* ode; *pp.* od *To man, supply with men, garrison;* viris vel mīlĭtibus instruĕre:—He hēt ða burg gemannian *he commanded to man the city,* Chr. 923; Erl. 110, 2, 5: 924; Erl. 110, 13.

ge-martyrian, -martirian, -martrian; *p.* ode, ade, ede; *pp.* od, ad, ed *To martyr;* martýrem făcĕre:—He hine gemartirode *he martyred him,* Homl. Th. ii. 478, 21. Hí Petrus and Paulus gemartredan *they martyred Peter and Paul,* Ors. 6, 5; Bos. 119, 21. He wæs for sôpfæstnysse gemartyrod *he was martyred for truth,* Homl. Th. i. 484, 33: Boutr. Scrd. 18, 8, 10. Wæs heáfde beslegen and gemartyrad se mon *decollātus est mīles,* Bd. 1, 7; S. 478, 39. Ðus wearþ gemartirod se mǽra apostol *thus was martyred the great apostle,* Homl. Th. ii. 300, 24: 478, 22: 496, 22.

ge-maðel, es; *n. Speech, conversation, talking, harangue;* sermo, ōrātio, sermōcĭnātio:—Úre heofenlíca Hláford nolde ðæra deófla gemaðeles ná máre habban *our heavenly Lord would not have any more of the devil's harangue,* Nicod. 29; Thw. 16, 39.

ge-mearc, es; *n. A boundary, limit;* lŏcus designātus:—Gewát him se æðeling to ðæs gemearces ðe him Metod tǽhte *the man departed to the limit which the Lord had shewn him,* Cd. 139; Th. 174, 28; Gen. 2885. DER. fôt-gemearc, fyrst-, geár-, míl-, þing-, word-.

ge-mearcan; to -mearcenne; *p.* ede; *pp.* ed *To mark, observe, keep;* observāre:—Getácna me ðær sēlast sý sáwle mínre to gemearcenne Meotudes willan *signify to me where it be best for my soul to observe the Creator's will,* Exon. 118 a; Th. 453, 7; Hy. 4, 11.

ge-mearcian; *p.* ode, ade; *pp.* od, ad *To mark, point out, describe, assign, appoint, determine;* nōtāre, signāre, designāre, assignāre, constituĕre, decernēre:—He gemet ne con gemearcian his mūðe móde síne *he cannot set bounds to his mouth with his mind,* Exon. 87 b; Th. 330, 18; Vy. 53. Ic wolde gesecgan hú Crēca gewinn, ðe of Lacedemonia ðære byrig ǽrest onstæled wæs, and, mid spellcwydum gemearcian *I wished to tell how the war of the Greeks was first raised from the city of the Lacedæmonians, and, in the language of history, to describe it,* Ors. 3, 1; Bos. 54, 34. Ðú him mete sylest, mǽla gehwylce, and ðæs tídlíce tíd gemearcast *tu das escam illis in tempŏre opportūno,* Ps. Th. 144, 16. Symle he twelf síþum tída gemearcaþ dæges and nihtes *it ever marks the hours of day and night twelve times,* Exon. 58 a; Th. 207, 24; Ph. 146. Se Hǽlend gemearcode ôðre twá and hundseofentig *designāvit Dŏmĭnus et alios septuaginta duos,* Lk. Bos. 10, 1: Bd. 3, 9; S. 534, 2. Hæfde hire wácran hige Metod gemearcod *to her the Creator had appointed a weaker mind,* Cd. 28; Th. 37, 17; Gen. 591: 38; Th. 50, 25; Gen. 814. Getácnod oðde gemearcod is ofer us leóht andwlitan ðínes *signātum est sŭper nos lūmen vŭltus tui,* Ps. Lamb. 4, 7. He is wuldre gemearcad *it is marked with glory,* Exon. 60 b; Th. 220, 11; Ph. 318. Hí hæfdon ǽlce scire on West-Sexum stíðe gemearcod mid bryne and mid hergunge *they had severely marked every shire of Wessex with burning and harrying,* Chr. 1006; Erl. 141, 2. Gemearca hú hý ǽr stódon *mark how they stood before,* Lchdm. i. 398, 5. v. ge-mercian.

ge-mearcod; *part. Marked;* signatus:—On ða gemearcodan lindan

on the marked linden or *lime tree*, Cod. Dipl. 1317 ; A.D. 1033 ; Kmbl. vi. 182, 2 : 1102 ; A.D. 931; Kmbl. v. 195, 14.

ge-mearcund. v. ge-mercung.

ge-meare *an end*, Ps. Lamb. 58, 14. v. ge-mǽre.

ge-mearr, es ; *n. A hindrance, error :*—Ðonne se Godes ðiów on ðæt gemearr ðære woruldsorga beféhþ *when the servant of God accepts the hindrance of worldly cares*, Past. 51, 7 ; Swt. 401, 20 ; Hat. MS. Ða gemearr ðe man drífþ on mislícum gewiglungum *the erroneous practices which are carried on with various spells*, L. Can. Edg. 16 ; Th. ii. 248, 4. Gemear *nugæ, errores*, Gl. Prud. 662. [Cf. *Goth.* ga-marzeins *a stumbling-block.*] v. myrran.

ge-mearr ; *adj. Wicked, fraudulent :*—Gif hwá gemearra manna wǽre *if there were any wicked man*, L. Edw. 1 ; Th. i. 160, note 2. v. ge-mearr.

ge-mec ; *adj. Equal, suited, matched :*—Oððe wíf habbaþ him gemæc oððe him gemece nabbaþ *either they are well-matched in marriage or have not wives suited to them*, Bt. 11, 1 ; Fox 32, 5. v. ge-mæc.

ge-mecca, an ; *m. and f. A consort, an equal :*—Ic Oswulf aldormonn ond Beorndryþ mín gemecca *I Oswulf alderman and Beornthryth my wife*, Th. Dipl. 459, 3 : 469, 30. Gemecca *conjunx*, Ælfc. Gl. 3 ; Wrt. Voc. 72, 9. Clippende to heora gemeccum *clamantes coæqualibus*, Mt. Kmbl. Rush. 11, 16. v. ge-mæcca.

ge-méd *mad.* v. ge-mǽd.

ge-méde, es ; *n. That which pleases, satisfies, due observance :*—Maga gemédu *the due observances of kinsmen*, Beo. Th. 499 ; B. 247. [*O. Sax.* gimôdi :—Ðemu manne te gimôdea *to satisfy the man :* *O. H. Ger.* gi-muati.] v. ge-méde ; *adj.*

ge-méde ; *superl.* -médost ; *adj. Agreeable, pleasing ;* acceptus, grátus :—Swá him gemédost wæs *as was most agreeable to them*, Andr. Kmbl. 1188 ; An. 594. Geméde *agreeable*, Bt. 11, 1 ; Bt. Fox 32, note 1. Gímoedo ł wala middangeardes *prospera mundi*, Rtl. 50, 6. [*O. H. Ger.* gi-muati : cf. *O. Sax.* gi-môdi, *n.*] DER. un-geméde.

ge-medemian ; *p.* ode ; *pp.* od [medeme] *To deign, deem worthy, honour, vouchsafe, moderate, humiliate, humble :*—Ic gemedemige ðé to ðam ðinge *dignor te illa re*, Ælfc. Gr. 41; Som. 44, 5. Ðætte hía mildelíce mið woere hire gisomnia ðú gimeodomiga *ut eam propitius cum viro suo copulare digneris*, Rtl. 108, 42 : 36. Ic ðancige mínum Gode ðe me gemedemode to his hálgum *I thank my God that has deemed me worthy to be among his saints*, Homl. Th. i. 424, 15. Ðú eart on écnesse gemedemod *thou art honoured for ever*, Blickl. Homl. 147, 12. Godes sunu gemedemode hine sylfne ðæt he wolde beón acenned of Marian *God's Son condescended to be born of Mary*, Homl. Th. 32, 7 : Blickl. Homl. 39, 17 : Nicod. 20 ; Thw. 10, 9. Crist sylf gemedemode ðæt he wolde gebígan his hálige heáfod to his ðeówan handum *Christ himself deigned to bow his head to his servant's hands*, Homl. Th. i. 40, 25. He wæs gemedomad on róde beón ahangen *he suffered the humiliation of being hung on the cross*, L. E. I. 21 ; Th. ii. 416, 28 : Blickl. Homl. 179, 9 : 139, 26. Gemedemud *temperatus*, Scint. 12.

ge-medemlíce, -meodomlíce ; *adv. Worthily ;* digne, Rtl. 18, 33 : dignanter, 34, 18.

ge-méder ; *f. A godmother ;* commater, Som.

ge-medmicel ; *adj. Small, mean, weak :*—Gimetomicla *infirma*, Rtl. 50, 11.

ge-médred ; *part. Mothered, of the same mother ;* uterinus, Ors. 3, 7 ; Bos. 60, 19. v. ge-médrian.

ge-médrian ; *p.* ede, ode ; *pp.* ed, yd *To* MOTHER, *to adopt* or *to have as a son* or *daughter ;* adoptare, habere sibi filium *vel* filiam :—Ða þrý gebróðra nǽron ná Philippuse gemédred *the three were not brothers of Philip by their mother [mothered]*, Ors. 3, 7 ; Bos. 60, 19. Geseah hys gemédrydan bróðor Beniamin *vidit Benjamin fratrem suum uterinum*, Gen. 43, 29.

ge-médryd ; *def.* se ge-médryda ; *part. p. Mothered, of the same mother*, Gen. 43, 29 : 44, 20. v. ge-médrian.

ge-meldian ; *p.* ode, ade ; *pp.* od, ad *To announce ;* nuntiare, ad-nuntiare :—Blód-gyte weorðeþ mongum gemeldad *bloodshed shall be announced to many*, Exon. 116 b ; Th. 448, 20 ; Dóm. 37 : Ps. Th. 61, 11.

géme-leás ; *adj. Negligent ;* negligens, C. R. Ben. 54. v. gýme-leás.

géme-leáslíce ; *adv. Negligently ;* negligenter :—For hwon sǽdest ðú Ecgbyrhte swá gémeleáslíce and swá wlætlíce ða þing ðe ic ðé bebeád him to secganne *quare tam negligenter ac tĕpide dixisti Ecgbercto quæ tibi dícenda præcépi?* Bd. 5, 9 ; S. 623, 9. Ða ðe unwærlíce and gémeleáslíce Gode hýraþ *those who heedlessly and carelessly serve God*, Blickl. Homl. 63, 22. v. gýme-leáslíce.

géme-leásniss, e ; *f. Negligence ;* negligentia, Rtl. 178, 11. v. gýme-leásness.

géme-lést, e ; *f. Negligence, carelessness ;* negligentia, incúria :—Þurh ðíne ágene gémeléste *through thine own negligence*, Bt. 5, 1 ; Fox 10, 2. Þurh heora gémelést *through their carelessness*, Chr. 1070 ; Erl. 209, 34. v. gýme-leást.

ge-meltan, -myltan ; *p.* -mealt, *pl.* -multon ; *pp.* -molten *To melt, digest :*—Beorgas gemeltaþ *the hills shall melt*, Exon. 22 a ; Th. 61, 2 ; Cri. 978. Gif his mete gemyltan nelle *if his meat will not digest*, Herb. i. 90, 9 ; Lchdm. i. 196, 6 : 1, 19 ; Lchdm. 76, 15. Ðæt sweord eal gemealt íse gelícost *the sword all melted just like ice*, Beo. Th. 3220 ; B. 1608 : 3235 ; B. 1615. Ne gemealt him se módsefa *his courage did not fail*, 5249 ; B. 2628. On hyre bryne gemultan ealle ða anlícnessa togædere *in its burning all the statues melted together*, Ors. 5, 2 ; Bos. 101, 21. Eorðe is gemolten *liquefacta est terra*, Ps. Th. 74, 3. Me wearþ gemolten mód on hreðre *defectio animo tenuit me*, 118, 53.

ge-men ; *nom. pl : gen.* -manna *Men :*—Wǽron ðærin gemanna hund twelftig ðúsenda *there were therein a hundred and twenty thousand men*, Salm. and Sat. Kmbl. 186, 1.

gémen ; *gen.* gémenne ; *f. Care ;* cúra :—Ǽlc mon'mæg witan hú hefig sorg men beóþ seó gémen his bearna *every one may know how heavy a trouble to a man is the care of his children*, Bt. 31, 1 ; Fox 112, 17 : 12 ; Fox 36, 38. Be ðære hæfegan gémenne bearna *concerning the heavy care of children*, 31, 1 ; Fox 112, 19. Mid micle gémænne and gewinne *cum magna cura ac labore*, Bd. 2, 7 : S. 509, 11. v. gýmen.

ge-mencgan *to mingle*, Ælfc. Gr. 28, 6 ; Som. 32, 33. v. ge-mengan.

ge-mencgednys, -nyss *a mingling together*, Bd. 1, 27 ; S. 495, 29. v. ge-mengednys.

ge-mend *a memorial.* v. ge-mynd.

gémend, es ; *m. A keeper ;* custos, Mt. Kmbl. p. 20, 4.

ge-mendful, -full ; *adj.* [ge-mend = ge-mynd *the mind, memory*] *Of good memory, mindful ;* mĕmor :—Cild biþ gemendful *a child will be of good memory*, Lchdm. iii. 186, 24.

ge-ménelíc ; *adj.* [ge-méne = ge-mǽne *common*] *Common ;* commúnis :—For geménelícre neóde *for the common need*, L. C. S. 10 ; Th. i. 382, 2, MS. A. v. ge-mǽnelíc.

ge-ménelíce ; *adv. In common, commonly ;* commúníter :—We mynegiaþ eów ealle geménelíce *we admonish you all in common*, Wanl. Catal. 111, 25, col. 2. v. ge-mǽnelíce.

ge-mengan, -mencgan ; *p.* de ; *pp.* ed *To mingle, commingle, mix, blend, confuse, unite, join, combine ;* miscére, commiscére, confundére, consóciáre, inficére :—Ðæt he wísdóm mǽge wið ofermetta gemengan *that he may mingle wisdom with sensuality*, Bt. Met. Fox 7, 16 ; Met. 7, 8. Ic gemencge *confundo*, Ælfc. Gr. 28, 6 ; Som. 32, 33. Ic gemenge *conficio*, Ælfc. Gl. 36 ; Som. 62, 99 ; Wrt. Voc. 28, 76. Ðú hí on ðisse worulde gemengest *thou unitest them in this world*, Bt. 33, 4 ; Fox 132, 24. He gemengeþ ðæt fýr wið ðam cíle *he mingles the fire with the cold*, 39, 13 ; Fox 234, 11 : Bt. Met. Fox 11, 182 ; Met. 11, 91. Ic me to' ðam plegan gemengde *lüdentibus me miscui*, Bd. 5, 6 ; S. 619, 11. Ðæt'we hit gemengen to ðam ǽrran *that we mix it with the preceding*, Bt. 34, 5 ; Fox 140, 13. Eorþe wearþ eall mid blóde máne gemenged *infecta est terra in sanguínibus eórum*, Ps. Th. 105, 28. Ðæt wæter and seó eorþe wǽron gemengede óþ ðone þriddan dæg *the water and the earth were commingled unto the third day*, Hexam. 4 ; Norm. 8, 15. Ðǽr gemengde beóþ onhǽlo gelác engla and deófla *there shall be mingled the whole assemblage of angels and of devils*, Exon. 21 a ; Th. 56, 4 ; Cri. 895 : Bd. 5, 23 ; S. 646, 4. Se ryhtwísa Déma se ðe hine on úrne geférscipe ðurh flǽsces gecynd gemengde *the righteous Judge who joined himself to our fellowship through fleshly nature*, Past. 21 ; Swt. 167, 23 ; Hat. MS.

ge-menged, -mencged ; *part. p. Mixed, mingled, confused ;* mixtus, commistus, confusus :—God sende rénscúr mid swefle gemenged *God sent a shower of rain mingled with brimstone*, Gen. 19, 24. Gemencged *mixtus*, Ps. Spl. 74, 7. Gemencged hund and wulf *commistus canis et lupus*, Wrt. Voc. 77, 79. Gemenged stemn is, ðe biþ bútan andgite, swylc swá is hryþera gehlów, hunda gebeorc, treówa brastlung *confused voice is what is without understanding, such as lowing of oxen, barking of dogs, rustling of trees, etc*, Ælfc. Gr. 1 ; Som. 2, 34, 3.

ge-mengednys, -mengdnys, -mencgednys, -mencgdnys, -mencgnys, -nyss, e ; *f. A mingling together, mixing, mixture, connection ;* commixtio, admixtio :—Seó gemengdnys ðæs flǽsces *carnis commixtio*, Bd. 1, 27 ; S. 495, 31. Se willa má waldeþ on ðam weorce ðære gemengdnysse *vŏluntas dómĭnátur in ŏpĕre commixtĭónis*, 1, 27 ; S. 495, 38. On ðæs líchoman gemengednysse biþ willa *in carnis commixtĭóne vŏluptas est*, 1, 27 ; S. 493, 20 : 1, 27 ; S. 495, 39. Æfter his wífes gemengednysse *post admixtĭónem conjúgis*, 1, 27 ; S. 496, 17. Hwæðere on ðam wordum is sweotol ðæt he wónysse nemde nalæs ða gemencgdnysse ðæs gesinscypes, ac ðonne sylfan willan ðære gemencgednysse *in quibus tămen verbis non admixtĭónem conjúgium iniquitátem nómĭnat, sed ipsam vidĕlĭcet vŏluptátem admixtĭónis*, 1, 27 ; S. 495, 28, 29. Seó alýfede gemencgnyss ipsa lícita admixtio, 1, 27 ; S. 495, 18. Æfter gemencgnysse ágenes wífes *post admixtĭónem propríæ conjúgis*, 1, 27 ; S. 495, 15. Bútan womme oððe gemencgednysse ðwyrlíces weorces *without blemish or admixture of perverse work*, Homl. Th. i. 544, 17. Ðære sǽ gemengednyssa *the minglings of the sea*, 610, 11 : 608, 20. [Cf. Lk. 21, 25.]

ge-mengung, e ; *f. A mixing, confusing ;* mixtura, Cot. 35.

ge-menigfealdan, -menigfildan ; *p.* de [menig *many*, feald *a fold, plait*] *To multiply, increase, extend ;* multiplicare, Ex. 32, 13 : Gen. 9, 27 : 32, 12.

gémenis, gémnis, se; *f. Care;* cura, Mt. Kmbl. Lind. *and* Rush. 22, 16.

ge-meodniss, e; *f. Worthiness, dignity;* dignitas, Rtl. 192, 37.

ge-meotu *boundaries, limits,* Andr. Kmbl. 907; An. 454, = ge-metu. v. ge-met.

ge-mercian; *p.* ode; *pp.* od *To mark out;* signāre :—Man hæfde ða buruh mid stacum gemercod *the city was marked out with stakes,* Ors. 5, 5; Bos. 105, 28. Gemercadon ðone stán *signantes lapidem,* Mt. Kmbl. Lind. 27, 66. Ðæt gemercad wǽre all ymb-hyrft *ut describeretur universus orbis,* Lk. Skt. Lind. 2, 1. v. ge-mearcian.

ge-mercung, e; *f. A description;* descriptio, Lk. Skt. Lind. 2, 2.

ge-mére, es; *n. A boundary, end;* finis :—Fram gemērum eorþan a *finibus terræ,* Ps. Spl. 60, 2. v. ge-mǽre.

ge-merran *to mar, spoil,* Lk. Skt. Lind. 13, 7. v. ge-myrran.

ge-mérsian, Mt. Kmbl. Lind. 9, 31; 28, 15. v. ge-mǽrsian.

ge-met, es; *nom. acc. pl.* -u, -a; *n.* **I.** *a measure, space, distance;* mensura, spatium, intervallum :—Gefylle gē ðæt gemet eówra fædera *vos implete mensuram patrum vestrorum,* Mt. Bos. 23, 32. On ðam ylcan gemete ðe gē metaþ *qua mensura mensi fueritis,* Mt. Bos. 7, 2 : Mk. Bos. 4, 24 : Lk. Bos. 6, 38 : Cd. 80; Th. 101, 4; Gen. 1677. Betweonan Eferwíc and six míla gemete *between York and a distance of six miles,* L. N. P. L. 56; Th. ii. 298, 27. **II.** *that by which anything is measured, a measure;* mensura, modius, satum :—Gemeta and gewihta rihte man georne *let measures and weights be carefully rectified,* L. C. S. 9; Th. i. 380, 24. Hæbbe ǽlc man rihte gemetu *modius æqualis et verus erit tibi,* Deut. 25, 15 : Lev. 6, 20 : 19, 36. On þrim gemetum melwes *in farinæ satis tribus,* Mt. Bos. 13, 33 : Lk. Bos. 13, 21. **III.** *measure, capacity, ability, power, etc;* mensura, facultas, potestas, vis :—Ne sceal se Dryhtnes þeów in his mód-sefan māre gelufian eorþan æhtwelan, ðonne his ānes gemet, ðæt he his líchoman lāde hæbbe *the Lord's servant shall not in his mind love more of earth's riches than his own measure, that he may have support for his body,* Exon. 38 a; Th. 125, 25; Gū. 359. Nis ðæt monnes gemet *it is not man's ability,* 92 b; Th. 348, 12; Sch. 27. Næs ðā monna gemet, ne mægen engla, ðæt eów mihte helpan *there was then no power of men, no angel's might, that could help you,* Cd. 224; Th. 295, 22; Sat. 490. Ofer mín gemet *above my power,* Beo. Th. 5750; B. 2879: 5059; B. 2533: Ps. Th. 59, 11: 107, 12. **IV.** *a fit or proper measure, and so metaph. measure, proportion, moderation, bounds, limit, boundary, means, way, manner;* mensura, modus, finis, terminus, limes, ratio :—Ðy læs he of gemete hweorfe *lest he turn from moderation,* Exon. 78 b; Th. 294, 35; Crä. 25: 83 a; Th. 312, 18; Seef. 111. He gemet ne con gemearcian his mūþe mód síne *he cannot set bounds to his mouth by his understanding,* 88 a; Th. 330, 17; Vy. 52. Gytsung gemet nāt *avarice knows no bounds,* Scint. 25. Ðās miclan gemetu middan-geardes *these great boundaries of middle-earth,* Exon. 20 a; Th. 52, 1; Cri. 827: Andr. Kmbl. 617; An. 309. Eal ic hit arǽfnede ðæt ic eów ǽteówe hwylcum gemete gē sceolan arǽfnan *I suffered it all to shew you how you ought to suffer,* Blickl. Homl. 237, 12. Ealle gemete *omni modo,* Bd. 1, 27; S. 491, 9. Ðysses gemetes *hujusmodi,* 2, 1; S. 500, 18 : 4, 9; S. 577, 7 : 4, 19; S. 589, 18. On ðam gemete *quemadmodum,* Ps. 36, 2, 21 : 32, 22. **V.** *a rule, order, law;* norma, regula, lex :—Fram ðām he ðæt gemet leornode regollíces þeódscipes *a quibus normam disciplinæ regularis didicerat,* Bd. 3, 23; S. 554, 35. Gemetu *normulæ,* Cot. 138 : Exon. 93 a; Th. 349, 14; Sch. 46. Ðínes mūþes gemet *lex oris tui,* Ps. Th. 118, 72. **VI.** 1. *a mood, the inflection of a verb expressing the mode or manner of action or being, abstracted from time—tense* tíd *q. v. and person* hād **IV.** *q. v. : such as, indicative* gebícnigendlíc, *q. v. : imperative* bebeódendlíc, *q. v. : subjunctive* under-þeódendlíc, *q. v. : infinitive* unge-endigendlíc, *q. v. :* modus :—Modus is gemet oððe ðare sprǽce wíse *a mood is mode [manner] or the manner [wise] of speaking,* Ælfc. Gr. 21; Som. 23, 17. 2. *a poetical measure, metre;* metrum :—And ðām wordum sóna monig word in ðæt ylce gemet Gode wyrðes songes to geþeódde *et eis mox plura in eundem modum verba Deo digni carminis adjunxit,* Bd. 4, 24; S. 597, 26. [*O. Sax.* gi-met : *O. H. Ger.* ki-mez.] DER. eln-gemet, un-. v. metan.

ge-met; *adj.* [ge-met **IV.** *a fit or proper measure*] *Fit, meet, proper;* aptus, congruus, conveniens :—Wearþ him hýrra hyge ðonne gemet wǽre *he had a loftier soul than were meet,* Cd. 198; Th. 247, 5; Dan. 492 : 186; Th. 231, 21; Dan. 250 : Andr. Kmbl. 2358; An. 1180. Swā him gemet þince *as to him may seem fit,* Beo. Th. 1379; B. 687 : 6107; B. 3057. Ðæt hit gemet wǽre *that it were fit,* Ps. Th. 143, 4 : Bt. Met. Fox 29, 86; Met. 29, 43. DER. un-ge-met.

ge-mét, es; *n. A meeting, assembly;* conventus :—Hí hæfdon ǽlce dæge heora witena gemét *they had their meeting of counsellors every day,* Jud. Thw. 161, 31. v. ge-mót.

ge-meta *measures,* L. C. S. 9; Th. i. 380, 24. v. ge-met.

ge-metan; *p.* -mæt *and* -mette, *pl.* -mǽton; *pp.* -meten; *v. trans.* **I.** *to measure, measure back or again;* metiri, remetiri :—On ðam ylcan gemete ðe gē metaþ, eów byþ gemeten *qua mensura mensi fueritis, remetietur vobis,* Mt. Bos. 7, 2 : Mk. Bos. 4, 24 : Lk. Bos. 6, 38. God

ðū ðe heofen mid honda gemettest and eorðan on ðínre fyst betýndest *God thou who has meted heaven with thy hand and enclosed the earth in thy fist* [cf. Isaiah 40, 12], St. And. 47, 2. **II.** *to measure by traversing* or *going over;* metiri transeundo :—And his cwēn mid him medo-stíg gemæt *and his queen with him measured the mead-way* [*way to the mead-hall*], Beo. Th. 1852; B. 924. v. metan.

ge-metan; *p.* -mette; *pp.* -mett, -met *To paint;* pingere, depingere :—Swylce hí gemette wǽron *as if they were painted,* Chr. 1104; Th. 367, 1 : Lchdm. iii. 206, 18 : Prov. 7. Gē sind gelíce gemettum ofergeweorcum *ye are like painted sepulchres,* Homl. Th. ii. 404, 17. v. metan *to paint.*

ge-métan; he -mēteþ, -mētt, -mēt; *p.* -mētte, *pl.* -mētton; *pp.* -mēted, -mētod, -mētt, -mēt *To find, find out, discover, come upon, meet with;* invēnīre, compērīre :—Ic gemēte *invēnio,* Ælfc. Gr. 30, 4; Som. 34, 49 : 37; Som. 39, 6. He holtes hleó heáh gemēteþ *he finds the wood's lofty shelter,* Exon. 62 a; Th. 227, 27; Ph. 429: Ps. Th. 54, 24 : 87, 12. Gemoetaþ *invenerit,* Lk. Skt. Lind. 12, 43. Ealc ðæra, ðe me gemētt, me ofslyþ *omnis qui invēnerit me, occidet me,* Gen. 4, 14. Se ðe gemēt hys sáwle, se forspilþ híg *qui invēnit animam suam, perdet illam,* Mt. Bos. 10, 39 : 24, 46 : Lk. Bos. 12, 37, 38, 43. Gē gemētaþ án cild hræglum bewunden, and on binne alēd *invēniētis infantem pannis involūtum, et pŏsītum in præsēpio,* 2, 12 : Mt. Bos. 11, 29 : Mk. Bos. 11, 2. Ðæs bisceopes líf ic gemētte biscope wyrðe beón *vitam episcŏpi episcŏpo dignam esse compēri,* Bd. 5, 6; S. 618, 30. Ðū gemēttes Meotod alwihta *thou hast met the Lord of all things,* Cd. 228; Th. 308, 23; Sat. 697. He gemētte stapul ǽrenne *he found a brazen pillar,* Andr. Kmbl. 2123; An. 1063 : 481; An. 241. Geswinc and angnys gemētton me *tribūlātio et angustia invēnērunt me,* Ps. Spl. 118, 143 : 75, 5. Gemēte gē hine *invēnies eum,* Deut. 4, 29. Gif ic gemēte fíftig rihtwísra wera *si invēnēro quinquaginta justos,* Gen. 18, 26, 28. Gif hwā þeóf gemēte *if any one find a thief,* L. C. S. 29; Th. i. 392, 14; L. In. 49; Th. i. 132, 12. Ðæt we ðíne onsýne milde gemēten *that we may find thy countenance mild,* Exon. 76 a; Th. 286, 13; Jul. 731. Swā hwylce swā gē gemēton *quoscumque invēnēritis,* Mt. Bos. 22, 9. Hí hæfdon neowne gefeán gemēted *they had met with new joy,* Eleu. Kmbl. 1738; El. 871 : 2447; El. 1225. He is gemēt *inventus est,* Lk. Bos. 15, 24, 32. Gif ðær beóþ gemētte feówertig rihtwísra *sin quadraginta ibi inventi fuērint,* Gen. 18, 29 : 2, 12. Gif we gemēte sín on moldwege oððe feor oððe neáh fundne weorðen *if we are met on earth's way or far or near are found,* Exon. 70 b; Th. 262, 17; Jul. 334. Gif hwilc mon sí gemētod on ðínum ðæm egeslícan dóme *if any man be found at thy awful judgment,* St. And. 47, 8.

ge-mete; *adv. Fitly, meetly, in a proper manner;* apte, congruenter, convenienter, Exon. 40 a; Th. 132, 13; Gū. 472 : Bt. Met. Fox 13, 36; Met. 13, 18. DER. un-gemete.

ge-meted = ge-mett *painted,* Som. 143? v. ge-metan.

ge-mētednes, -ness, e; *f. An invention, a discovery;* inventio, adinventio :—Syle heom æfter nearoþancnysse oððe māne gemētednessa oððe heora afundennysse *da illis secundum nequitiam adinventiōnum ipsōrum,* Ps. Lamb. 27, 4.

ge-metegian; *p.* ode; *pp.* od *To measure, moderate,* Ps. Spl. 38, 7. v. ge-metgian.

ge-meten; *part. Measured, measured back* or *again;* remensus, Mt. Bos. 7, 2. v. ge-metan.

ge-meteng *a meeting.* v. ge-mēting.

ge-met-fæst; *adj. Moderate, modest;* moderatus, modestus :—Ne hie ðam geþyldegum and ðam gemetfæstum simble ne wuniaþ *neither do they always dwell with the patient and moderate,* Bt. 11, 1; Fox 34, 3. Sió is swíðe gemetfæst *she is very modest,* 10; Fox 28, 20. Man gemetfæst *vir modestus,* Bd. 1, 16; S. 484, 18 : 4, 28; S. 606, 33 : Exon. 48 b; Th. 168, 19; Gū. 1080 : 95 b; Th. 357, 19; Pa. 31.

ge-met-fæstlíce; *adv. Modestly;* modeste :—He swā gemetfæstlíce hine sylfne beheóld *ita se modeste gerebat,* Bd. 5, 19; S. 637, 4.

ge-met-fæstnys, -nyss, e; *f. Moderation, modesty;* moderatio, moderamen, modestia :—Mycelre monþwǽrnysse and ǽrfæstnysse and gemetfæstnysse mon *summæ mansuetudinis et pietatis ac moderaminis vir,* Bd. 3, 3; S. 525, 32 : 3, 14; S. 540, 13. Petrus tihte geleáffulle wíf to eádmódnesse and gemetfæstnysse *Peter exhorted faithful women to humility and modesty,* Homl. Th. i. 98, 3. Gimetfæstnisse *modestiam,* Rtl. 13, 33.

gemet-fæt, es; *nom. acc. pl.* -fatu; *n. A measuring-vessel, a measure;* metatorium vas, mensura quǽvis definita :—Án gemetfæt full, ðe híg Gomor hēton, Ex. 16, 16, 33.

ge-met-festan; *p.* -feste; *pp.* -fested, -fest *To compare;* comparare :—Gemetfest *comparatus,* Ps. Spl. T. 48, 21.

ge-metgian, -metegian, -metigian; *p.* ode; *pp.* od. **I.** *v. trans. To measure, moderate, temper, regulate, order, govern, restrain;* mensurare, temperare, moderare, regere :—Heora wíte biþ gemetegod ǽlcum be his geearnungum *their punishment shall be measured to every one by his deserts,* Homl. Th. i. 294, 6. Efne gemetegode ðū settest dagas míne *ecce mensurabiles posuisti dies meos,* Ps. Spl. 38, 7. Hine se'lfne of dūne

ǽstīgende he cūðe gemetgian his hiéremonnum *se auditoribus condescendendo noverat temperare*, Past. 16, 2; Swt. 101, 15; Hat. MS. 21 a, 2: 35, 1; Swt. 237, 23; Hat. MS. 45 a, 4. Á sceal ðæt wiðerwearde ðæt ŏðer wiðerwearde gemetgian *ever must the contrary moderate the other contrary*, Bt. 21; Fox 74, 19: 40, 3; Fox 238, 25: Bt. Met. Fox 11, 107; Met. 11, 54. Gif ðū ne gemetgodest cēle and hǽto *if thou didst not moderate cold and heat*, Bt. Met. Fox 20, 224; Met. 20, 112: Salm. Kmbl. 879; Sal. 439. Beorhte steorran mōna gemetgaþ *the moon tempers the bright stars*, Bt. Met. Fox 4, 17; Met. 4, 9. Se gemetgaþ ðone bridel *he regulates the bridle*, Bt. 36, 2; Fox 174, 18. God gemetgaþ ealla gesceafta *God regulates all creatures*, Bt. 39, 13; Fox 234, 9: Bt. Met. Fox 13, 10; Met. 13, 5: 24, 78; Met. 24, 39. II. *to measure in the mind, to deliberate, meditate on; deliberare, meditari* :— Ic on ðīnum bebodum mōte gemetgian rǽd *meditabor in mandatis tuis*, Ps. Th. 118, 47. III. *v. intrans. To become moderate, to moderate one's self; moderari, temperari* :—Him gemetgaþ eall éðles leóma *to them shall all the bright fire of their home moderate itself*, Elen. Kmbl. 2584; El. 1293. v. metgian.

ge-metgung, e; f. *Moderation, temperance, a fit or proper measure, a direction, a regulation; moderatio, temperantia, modus, moderamen* :— Wisdōm is se hēhsta cræft, and se hæfþ on him feówer ōðre cræftas, ðara is ān wǽrscipe, ōðer gemetgung, þridde is ellen, feórþe rihtwīsnes *wisdom is the highest virtue, and it has in it four other virtues, of which one is prudence, another temperance, the third is fortitude, and the fourth justice*, Bt. 27, 2; Fox 96, 34, note. Ealla gesceafta onfōþ æt Gode endebyrdnesse, and andwlitan, and gemetgunge *all creatures receive from God order, and form, and measure*, Bt. 39, 5; Fox 218, 15, 20, 33. Mid ðam gemetgunge ðæs gesceádes gefrætewod *moderamine discretionis ornatus*, Bd. 3, 5; S. 527, 42. Swylce monige gemetgunge ðara rihtgelȳfedra gehǽlde ðære Rōmaniscan cyricean Angel-cynnes cyricum mid his lāre brohte *perplura Catholicæ observationis moderamina ecclesiis Anglorum sua doctrina contulit*, 3, 28; S. 560, 37. Hī būton gemetgunge ðæt wīn drincende wǽron *they drank the wine without moderation*, Ors. 2, 4; Bos. 45, 19. v. metgung.

ge-mēdgian; p. ode, ade; pp. od, ad [mēdig *wearied*] *To weary, fatigue, impair*; fātīgāre :—Wæs Gūþlāce mægen gemēðgad *Guthlac's strength was impaired*, Exon. 47 a; Th. 160, 27; Gū. 950.

ge-mēdrian; p. ode; pp. od *To honour*; hŏnōrāre :— Būton he hwǽne furðor gemēðrian wylle *unless he will more amply honour any one*, L. C. S. 15; Th. i. 384, 4, MS. A. v. ge-mǽdian.

ge-mēting, e; f. *A meeting, an assembly, association, a society*; conventus, conventio, conventĭculum, congrĕgātio :—Is undyrne uncer gemēting *our meeting is not secret*, Beo. Th. 4006; B. 2001. Gemētingc *conventus vel conventio*, Wrt. Voc. 72, 75. Ðū bewruge me fram gemētinge awyrgedra *protexisti mea conventu mālignantium*, Ps. Spl. 63, 2: Ps. Th. 105, 16. On gemētingum *in congrĕgātiōne*, 110, 1. Ne ic ne gederige gemētinga heora *non congrĕgābo conventĭcŭla eōrum*, Ps. Spl. 15, 4. To gemoetingum *conciliis*, Mk. Skt. Lind. 13, 9.

ge-metlǽcan; p. -lǽhte; pp. -lǽht *To moderate* :—We hit eft gemetlǽcaþ *we afterwards moderate it*, Past. 16, 2; Swt. 101, 12; Hat. MS.

ge-met-līc; adj. *Moderate, temperate, measurable, fit*; moderatus, temperatus, mensurabilis, aptus :—Hæle wīsfæst and gemetlīc *a man wise and moderate*, Exon. 81 a; Th. 305, 12; Fä. 87. Him gemetlīc seó may *be suitable for him*, Bt. 14, 2; Fox 44, 21: 40, 3; Fox 238, 21; Ps. Lamb. 38, 6. [O. H. Ger. ki-mezlih *mediocris*.]

ge-met-līce; adv. *Moderately, fitly*; moderate, modeste, apte :—To ðon gemetlīce *adeo moderate*, Bd. 4, 24; S. 598, 26. Gemetlīcost *most fitly*, Bt. Met. Fox 8, 32; Met. 8, 16. [O. H. Ger. ki-mezliho *commode*.]

ge-met-līcung, e; f. *Due measure, moderation*; moderatio, Som.

ge-mētnes, -ness, e; f. *A finding, discovery*; inventio :— Se dæg heora þrōwunga ge heora līchoman gemētnesse mid ǽrwurþe weorþunge on ðām stōwum mǽrsode syndon *dies passiōnis vel inventiōnis eōrum congrua illis in lōcis vĕnĕrātiōne celebrātur*, Bd. 5, 10; S. 625, 18. v. ge-mētednes.

ge-metsian; p. ode; pp. od *To furnish with provisions* :—Ðæt scip ðe Swegen eorl hæfde him silfum ǽr gegearcod and gemetsod *the ship that Earl Sweyn had before prepared and provisioned for himself*, Chr. 1052; Erl. 181, 14. v. metsian.

ge-mett *measure, manner*, Bd. 4, 9; S. 577, 7. v. ge-met.

ge-mettan; pl. m. *Eaters, partakers*; comestōres :—Ða gemettan ne mōston ðæs lambes bān scǽnan *the partakers might not break the bones of the lamb*, Homl. Th. ii. 282, 7. Ðām gemettum *to the partakers*, 282, 2.

ge-mette *painted*, Chr. 1104; Th. 367, 1. v. ge-metan.

ge-metu *measures, boundaries, laws*, Deut. 25, 15; Andr. Kmbl. 617; An. 309: Exon. 93 a; Th. 349, 14; Sch. 46. v. ge-met.

ge-miclian, -myclian; p. ode, ade; pp. od, ad *To enlarge, magnify, extol, glorify* :—Se Mǽda rīce swīðe gemiclade *who greatly enlarged the kingdom of the Medes*, Ors. 1, 12; Bos. 35, 28: Ps. Th. 147, 3. Se ðe reorda gehwæs ryne gemiclaþ *he who enlargeth the course of every speech*,

Exon. 8 b; Th. 4, 4; Cri. 47. Swīðe gemiclade se drihten miltheortnisse his *magnificavit dominus misericordiam suam*, Lk. Skt. Lind. 1, 58. Gemycla mīne sáuwle *magnify my soul*, Blickl. Homl. 159, 2. Gemycclige mīn sául Drihten *my soul magnify the Lord*, 13, 5. Gemicliaþ hine *glorificate eum*, Ps. Spl. 21, 22. Ðū gemiclast me *honorificabis me*, 49, 16.

ge-miclung, e; f. [mycel *much, great*] *Greatness, magnificence, glory*; magnificentia, Ps. Spl. 144, 5: 70, 21.

ge-midlian, -middlian; p. ode; pp. od [middel *middle*] *To divide, separate in the middle*; dimidiare :—Fācenfulle nā gemidliaþ dagas heora *dolosi non dimidiabunt dies suos*, Ps. Spl. C. 54, 27.

ge-midlian; p. ode; pp. od [medl *a bridle*] *To bridle, restrain* :— Gif hwa nyle gemidlian his tungan *if a man will not bridle his tongue*, Past. 38, 8; Swt. 281, 3; Hat. MS: 38, 1; Swt. 271, 13; Hat. MS.

ge-midlige *a bridle*, Lye. v. midl.

ge-mieltan *to melt, digest* :—Suā suā sió wamb gemielt ðone mete suā gemielt ðæt mōd mid ðære gescádwīsnesse his geþeahtes his sorga *as the belly digests food so does the mind digest its sorrows with wise reflection*, Past. 36, 8; Swt. 259, 6; Hat. MS. v. ge-myltan.

ge-mīgan; p. -māh, pl. -migon; pp. -migen *To water, pass water*; mingere :—Gif hwā ne mǽge gemīgan *if one cannot pass water*, Herb. 7, 3; Lchdm. i. 98, 5: 12, 1; Lchdm. i. 102. 19: 80, 1; Lchdm. i. 182, 12. Ðǽr se hund gemáh *where the hound watered*, Med. ex Quadr. 9, 13; Lchdm. i. 364, 1.

ge-milcian; p. ode, ade; pp. od, ad *To give milk, suckle*; lactare, Lk. Skt. Lind. 23, 29.

ge-mildscad; part. p. *Mixed with honey*; mulsus :—Gemildscad wæter melicraton, i. e. *mellis mistura, sc. cum aqua: hydromeli*. Gemildscad wīn *mulsum*, i. e. *mellis mistura cum vino*, Cot. 137; Lye. v. milisc.

ge-mildsian; p. ode; pp. od *To shew mercy, to pity*; mĭsĕrēri :— Nemne God me earmum and unwyrðum gemildsian wylle *unless God will shew mercy to me wretched and unworthy*, Bd. 3, 13; S. 538, 35. v. ge-miltsian.

ge-mildsiend, -miltsiend, es; m. *A pitier*; mĭsĕrātor :—Ðū Driht God gemildsiend *tu Dŏmĭne Deus mĭsĕrātor*, Ps. Spl. 85, 14. Ðū gōda cyngc and earmra gemiltsigend *thou good king and pitier of the poor*, Th. Apol. 18, 11.

ge-miltan; p. -milte; pp. -milted *To melt, soften, subdue*; liquefăcĕre, emollīre :—Woldon āninga ellenrōfes mōd gemiltan *they would entirely subdue the bold man's mood*, Andr. Kmbl. 2785; An. 1395. v. ge-myltan.

ge-miltsian, -mildsian, -milsian; p. ode; pp. od. I. *to shew mercy, have compassion, to pity, pardon*; mĭsĕrēri, propĭtiāri :—Ic gemiltsige ðysse menegu *mĭsĕreor sŭper turbam*, Mk. Bos. 8, 2: Ælfc. Gr. 27; Som. 29, 56. Árleásnyssum ūrum ðū gemiltsast *impĭetātibus nostris tu propĭtiāběris*, Ps. Spl. 64, 3: 24, 12. Gemiltsode se Hǽlend him *mĭsertus eōrum Jēsus*, Mt. Bos. 20, 34. Gemiltsa mīn *mĭsĕrēre mei Deus*, mĭsĕrēre *mei*, Ps. Spl. 56, 1: 50, 1: Ps. Th. 118, 132. Ðæt ðū gemiltsige me *that thou pardon me*, Hy. 3, 49; Hy. Grn. ii. 282, 49. Ðæt ðū us gemiltsie *that thou pity us*, Exon. 121 b; Th. 465, 24; Hö. 109. Gimildsa *propitiare*, Rtl. 89, 40. Ðætte he gimilsage *miserere*, 40, 19. II. *to make mild, make kind, soften*; propitium reddĕre, mītīgāre :—Ðæt Pater Noster Metod gemiltsaþ *the Pater Noster makes mild the Lord*, Salm. Kmbl. 81; Sal. 41.

ge-miltsiend. v. ge-mildsiend.

ge-miltsung, e; f. *Favour, mercy, pardon*; propĭtiātio :—Forðonðe mid ðē gemiltsung is *quia ăpud te propĭtiātio est*, Ps. Spl. 129, 4.

ge-mimor; adj. *Existing in the memory or mind [?], known; notus* :— Leden him wæs swā cūþ and swā gemimor swā Englisc ðæt him gecyndelīc wæs *linguam Latinam non minus quam Anglorum, quæ sibi naturalis est, noverit*, Bd. 5, 20; S. 641, 35. v. Grm. D. M. 352-3.

ge-mimorlīce; adv. *By heart*; memoriter, R. Ben. Inter. 13.

ge-mincged *mixed*. v. ge-mengan.

ge-mind, es; n. *A remembrance, memorial*; mĕmōriāle :—Ðū Driht on ěcnysse þurhwunast, and gemind ðīn on cynrine and cynrine *tu Dŏmĭne in æternum permănes, et mĕmŏriāle tuum in generātiōne et generātiōnem*, Ps. Spl. C. 101, 13. [Goth. ga-minþi *remembrance*.] v. ge-mynd.

ge-mindblīðe [blīðe *cheerful*] *A grateful remembrance, a memorial*; memoriale, Ps. Spl. T. 101, 13.

ge-mindig; adj. *Mindful*; mĕmor :—Gemindig biþ on worulde gecȳdnysse his *mĕmor ĕrit in sæcŭlum testāmenti sui*, Ps. Spl. 110, 5: 8, 5. Gif he sī gemindig mīnum [?] naman and ðīnes *if he be mindful of my name and thine*, Nar. 47, 9. v. ge-myndig.

ge-mindiglīcnys, -nyss; e; f. *A remembrance, memorial*; mĕmŏriāle :— Ðū Driht on ěcnysse þurhwunast, and gemindiglīcnys ðīn on cynrine and cynrine *tu Dŏmĭne in æternum permănes, et mĕmŏriāle tuum in generātiōnem et generātiōnem*, Ps. Spl. 101, 13.

ge-mittan; p. -mitte; pp. -mitted *To find, meet*; invĕnīre, obviam hăbēre :—On hwan mæg se iunga, on gōdne weg, rihtan ne rædran rǽd gemittan *in quo corrĭgit Jūnior viam suam?* Ps. Th. 118, 9. Gif ðū

E e

ðyslíce þegn gemittest *if thou meetest such a man*, Exon. 84 a; Th. 316, 8; Môd. 45. Hý gemittaþ mearclonde neáh heá hlincas *they meet lofty hills near the border-land*, 101 b; Th. 384, 5; Rä. 4, 23: 117 b; Th. 451, 15; Dôm. 104. Hine gemitte ân man *invēnit eum vir*, Gen. 37, 15: Cd. 103; Th. 137, 2; Gen. 2267. Efne we ðás eall on Eufraten sæcgean gehýrdon, syððan gemittan fôrwel manegu, on wudu-feldum *ecce audivimus ea in Euphrata, invēnimus ea in campis silvæ*, Ps. Th. 131, 6: Cd. 80; Th. 101, 24; Gen. 1687. Hie æt burhgeate beorn gemitton *they found the chief at the town-gate*, 111; Th. 146, 23; Gen. 2426. Gif gê gemitton Esau mînne brôður *si obvium hăbuĕris fratrem meum Esau*, Gen. 32, 17.

ge-mitting, -mittung, e; *f.* A *meeting, an assembly*; congressus:—Heora gemitting wæs æt Trefia ðære eá *their meeting was at the river Trebia*, Ors. 4, 8; Bos. 90, 2: 5, 7; Bos. 106, 20, 43. Ǽt heora gemittinge *in their meeting*, 4, 6; Bos. 85, 26. Wega gemittung *a meeting of ways*; compĭtum, Ælfc. Gl. 100; Som. 77, 5; Wrt. Voc. 55, 8.

gemme *a* GEM; gemma :—Sweor-gemme *a neck-gem* or *-lace*; monile, Cot. 170.

gêmnis, se; *f.* Cărĕ, *anxiety*; cura:—Ne is ðê gêmnise *non est tibi curæ*, Lk. Skt. Lind. 10, 40: 34: Mt. Kmbl. Lind 9, 12. Gêmnisse *sollicitudo*, 13, 22.

ge-môd; *adj.* [môd *mind*] *Of one mind, agreed*; concors:—Ðîne freónd næfst ðé swá gemôde swá swá ðu woldest *thou hast not thy friends in such agreement with thee as thou wouldest*, Shrn. 182, 5. Wæs ðu gemôd ðînum ðæm weðerwearde *esto consentiens adversario tuo*, Mt. Kmbl. Rush. 5, 25. Gemôde *conjurati*, Cot. 36. [Cf. gemêde.]

ge-môdod; *part.* [môd *the mind*] *Minded, disposed*; prônus, proclîvis:—Sume beoþ þwyrlíce gemôdode *some are perversely minded*, Homl. Th. i. 524, 18.

ge-môdsumian; *p.* ode; *pp.* od *To agree*; concordāre :—We geþiédaþ and gemôdsumiaþ to ðæra yfelena freóudscipe *we associate and agree in the friendship of the wicked*, Past. 46, 6; Swt. 355, 7; Hat. MS. 67 b, 18. [O. H. Ger. ki-môtsamôn *consacrare*.]

ge-môdsumnes, -ness, e; *f.* Agreement, concord; concordia :—He cýðde ðæt he nolde habban nâne gemôdsumnesse wið ða yfelan *he proclaimed that he would have no concord with the wicked*, Past. 46, 5; Swt. 353, 4; Hat. MS. 67 a, 21. [Cf. O. H. Ger. ki-môtsam *commodus*.]

ge-molsnian; *p.* ode, ade; *pp.* od, ad *To corrupt, decay, wither*; putrefacere, tabefacere, macerare, marcescere :—He ðær on moldan gemolsnaþ *he shall there rot in the earth*, Blickl. Homl. 109, 32. Mîne herewîc syndon gebrosnode and gemolsnode *my dwellings are decayed and perished*, 113, 26. Gemolsnad flǽsc *tabes*, Ælfc. Gl. 12; Wrt. Voc. 20, 16: Solil. 2. Swá gemolsnad wyrt *as a withered herb*, Ps. Th. 89, 6. v. molsnian.

ge-molten *molten, melted.* v. ge-meltan.

ge-mon ic, he *I remember, he remembers*, Exon. 74 b; Th. 280, 5; Jul. 624; Beo. Th. 3407; B. 1701. v. ge-munan.

ge-monan *to remember* :—Gemona *recordare*, Lk. Skt. Lind. 16, 25. Seó leó gemonþ [= geman] ðæs wildan gewunan hire eldrena [MS. eldrana] *the lioness remembers the wild manner of her parents*, Bt. 25; Fox 88, 12. v. ge-munan.

ge-mone. v. ge-mane.

ge-mong, es; *n.* A *mixture, crowd, throng, company*; commixtio, turba, cætus :—Ðær is sib bûtan nîþe hâlgum on gemonge *there is amity without envy among the holy*, Exon. 32 a; Th. 101, 19; Cri. 1661 : 59 b; Th. 216, 9; Ph. 265. On gemonge *in the throng*, Beo. Th. 3290; B. 1643. On clǽnra gemong *in the company of the pure*, Exon. 71 b; Th. 267, 24; Jul. 420 : Judth. 11; Thw. 24, 17; Jud. 193 : 12; Thw. 26, 1; Jud. 304. Wyrta gemong *aromata*, Lk. Skt. Lind. 23, 56. Ðæt gemong *mixtura*, Jn. Skt. Lind. 19, 39. v. ge-mang.

ge-mong *among.* v. ge-mang.

ge-monian, -monigan; *p.* ode, ade; *pp.* od, ad *To admonish, exhort, remind* :—Ealle ða gemoniaþ môdes fûsne fêran to sîþe *all these admonish the prompt of mind to go on a journey*, Exon. 82 a; Th. 308, 25; Seef. 50 : 88 b; Th. 333, 19; Gn. Ex. 6 : 52 a; Th. 182, 22; Gú. 1314 : Cd. 49; Th. 63, 9; Gen. 1029. v. ge-manian.

ge-monige *may remind*, Cd. 49; Th. 63, 9; Gen. 1029. v. ge-monian.

ge-monigfealdian; *p.* ode *To increase, multiply*; amplificare :—Ðætte gemonigfaldade † gewôxe *quod abundabat*, Mk. Skt. Lind. 12, 44. Gimonigfalda *multiplica*, Rtl. 8, 90. Gemonigfealdode *multiplied*, Blickl. Homl. 107, 25 : Bd. 5, 20; S. 641, 40. v. ge-mænigfealdian.

ge-monnad *manned, supplied with men.* v. ge-mannian.

ge-môt, es; *n.* A *meeting, coming together*, MOOT, *assembly, council*; conventus, congregatio, concursus :—Gârmitting gumena gemôt wǽpengewrixl *the meeting of spears, concourse of men, exchange of weapons*, Chr. 937; Erl. 114, 16; Ædelst. 50 : Exon. 72 a; Th. 268, 3; Jul. 426. Gif he leng bide lâðran gemôtes *if he should longer await a more hostile meeting*, 36 a; Th. 116, 15; Gú. 207 : Byrht. Th. 140, 40; By. 301. Híg hæfdon mycel gemôt *they held a great council*, Mt. Bos. 26, 4 : 26, 59 : 28, 12. Se gedwola cwæþ gemôt ongeán ðone bisceop *the heretic*

proclaimed a council against the bishop, Homl. Th. i. 290, 12. Ðú me oft aweredest wyrigra gemôtes *protexisti me a conventu malignantium*, Ps. Th. 63, 2: Andr. Kmbl. 2120; An. 1061: Exon. 34 a; Th. 109, 31; Gú. 98. Ðǽr monig beoþ on gemôt lǽded fore onsýne ēces dêman *there many a one shall be brought to the assembly before the face of the eternal Judge*, 19 b; Th. 50, 5; Cri. 795 : 21 b; Th. 58, 30; Cri. 943 : 23 a; Th. 63, 29; Cri. 1027. On gemôt cuman *to come to the assembly*, Elen. Kmbl. 558; El. 279. Gif hwá gemôt forsitte *if any one fail to attend the 'gemot*,' L. Athelst. 20; Th. i. 208, 26. Hwî biþ elles ǽlce dæge swelc seófung and swelce geflîtu and gemôt and dômas *why else is every day such sorrow and such contentions and assemblies and judgments*, Bt. 26, 2; Fox 92, 16. ¶ Witena gemôt *an assembly of the wise* [sapientum conventus, Bd. 3, 5; S. 527, 23]; *the supreme council of the Anglo-Saxon nation or parliament*. Mr. Kemble, in his 'Saxons in England,' vol. ii. page 203, A. D. 1849, says—'The proper [Anglo-] Saxon name for these assemblies was Witena gemôt, literally *the meeting of the witan* [or *the wise* or *experienced*]; but we also find,—Micel gemôt *the great meeting*; Sinoþlíc gemôt *the synodal meeting*; Seonoþ *the synod*. The Latin names are Concilium, Conventus, Synôdus, Synôdǎle conciliābǔlum, and the like. Although synôdus and seonoþ might more properly be confined to ecclesiastical conventions, the Saxons do not appear to have made any distinction; probably because ecclesiastical and secular regulations were made by the same body, and at the same time. . . . It is very probable that the . . . system of separate houses for the clergy and laity prevailed . . ., and that merely ecclesiastical affairs were decided by the king and clergy alone. It is probable that even in strictly ecclesiastical synods, the king had a presidency at least, as head of the church in his dominions, Cod. Dipl. 116; A. D. 767; Kmbl. i. 142, 143. There are some acts [of the Witena Gemôt], in which the signatures are those of clergymen only, others in which the clerical signatures are followed and, as it were, confirmed by those of the laity; and in one remarkable case of this kind, the king signs at the head of each list, as if he had in fact affixed his mark successively in the two houses, as president of each.' See above, Cod. Dipl. 116. Se cyng hæfde ðǽr [MS. ðæs] on morgen witena gemôt *on the morrow the king* [Edward] *had there a meeting of the wise*, Chr. 1052; Erl. 181, 9. Wæs ðá witena gemôt *then there was a meeting of the wise*, 1052; Erl. 184, 35. Hæfde Eádwerde cing witena gemôt on Lunden *king Edward had a meeting of the wise in London*, 1050; Erl. 176, 9. See also Stubbs' Const. Hist. i. cap. vi. Bisceopa gemôt *a meeting of bishops*, Bd. 1, 14: S. 482, 35. Be gemôtum *of moots*. And sêce man hundred-gemôt swá hit ǽr geset wæs; hæbbe man þriwa on geáre burh-gemôt; and tûwa, scir-gemôt, and ðǽr beó on ðære scire bisceop and se ealdorman, and ðǽr ǽgðer tǽcan ge Godes riht ge woruld-riht *and let the hundred-moot be attended as it was before fixed; and thrice in the year let a city-moot be held; and twice a shire-moot; and let there be present the bishop of the shire and the alderman, and there each expound both God's law* [right] *and the world's law*, L. Edg. ii. 5; Th. i. 268, 1–5. Ðás gemôt *these moots*, ii. 7; Th. i. 268, 15. See Schmid A. S. Gesetz. 595–6. DER. burh-gemôt, folc-, halle-, hundred-, scir-.

gemôt-ærn, -ern, es; *n.* [gemôt; ærn, ern *a place*] A *meeting-place, senate-house, hall*; conveniendi locus, aula :—Ahleópon ðá ealle, and hine mid heora metseaxum ofsticedon on heora gemôtærne [MS. gemôterne] *then* [the consuls and the senate] *all jumped up, and stabbed him* [Julius Cæsar] *with their daggers in their senate-house*, Ors. 5, 12; Bos. 112, 25. Gemôtern *in pretorio*, Mt. Kmbl. Lind. 27, 27.

ge-môtod *discussed*, Th. Chart. 172, 10. v. môtian.

gemôt-stede, es; *m.* A *meeting-place*; convĕniendi lŏcus :—On gemôtstede manna and engla *in the meeting-place of men and angels*, Soul. Kmbl. 296; Seel. 152.

gemôt-stôw, e; *f.* [gemôt, stôw *a place*] A *meeting-place, council*; conveniendi locus, concilium :—Gemôtstôw *vel* ceorla samnung *a meeting-place* or *a meeting of freemen*; compita, Ælfc. Gl. 55; Som. 66, 110; Wrt. Voc. 36, 32. Ic ne sæt mid gemôtstôwe ydelnyssa *non sedi cum concilio vanitatis*, Ps. Spl. T. 25, 4.

ge-mun; *adj. Mindful, having a recollection* :—Swá gemune menn wǽron ǽlces brôces *men had such a recollection of every trouble*, Ors. 1, 10; Bos. 34, 2. v. ge-myne.

ge-munan; ic, he -man, -mon, *pl.* -munon; *also* ic -mune, he -monþ, *pl.* -munaþ; *p.* -munde; *pp.* -munen [*a verb whose present tense is the past tense of a lost strong verb, cf. Lat.* memini]; *with gen. and acc.* To *remember, bear in mind, consider*; recordari, memorari, meminisse, meditari :—Gemunan his hâlegan cýðnesse *memorari testamenti sui sancti*, Lk. Bos. 1, 72. Gif he ne wile mid inneweardre heortan gemunan and geþencean *if he will not with sincere heart bear in mind and consider*, Blickl. Homl. 55, 11. Hie nellaþ gemunan ðone dæg heora forþfôre *they will not remember the day of their departure*, 61, 4. Ne geman heó ðære hefinysse *non meminit pressuræ*, Jn. Bos. 16, 21. Gif he ðæt eal gemon *if he remembers that all*, Beo. Th. 2375; B. 1185. Ic ðê ðæs leán geman *I will remember a reward for thee for it*, 2445; B. 1220. Ic gemune ðê *recordor tui*, Ælfc. Gr. 41; Som. 44, 2. Ic gemuna *meditabor*,

Ps. Spl. 62, 7. Seó leó gemonþ ðæs wildan gewunan hire eldrana *the lioness remembers the wild manner of her parents*, Bt. 25; Fox 88, 12. Hie ðæt eall gemunan and ðurh ðæt leóht gemanode beóþ *they remember all that and are admonished by the light*, Blickl. Homl. 129, 21: Bt. 16, 1; Fox 48, 30. Hie gemunaþ ða mycclan eáðmódnesse *they recollect his great humility*, Blickl. Homl. 129, 10. Ðonne gē gemunaþ Drihten eówerne God *when ye remember the Lord your God*, Deut. 4, 29. Ða gemunde God sunu Lameches *then God remembered Lamech's son*, Cd. 71; Th. 84, 33; Gen. 1407: 121; Th. 156, 8; Gen. 2585. Hig gemundon his worda *recordati sunt verborum ejus*, Lk. Bos. 24, 8. Gemundon weardas wíg-leóþ *the watchmen remembered the war-song*, 154; Th. 191, 26; Exod. 220. Gemun ðín mann-weorod *memento congregationis tui*, Ps. Th. 73, 2. Gemune ðú manigra bearna ðe on Edom synt *memento filiorum Edom*, 136, 7: 118, 49: Ps. Spl. 24, 6. Gemunaþ mínre spréce *mementote sermonis mei*, Jn. Bos. 15, 20. Gemunaþ ðæt gē silfe wéron þeówe on Egipta lande *remember that ye yourselves were slaves in Egypt*, Deut. 5, 15; Exon. Th. 281, 4; Jul. 641. Gemunon we úre dæghwamlícan synna *let us be mindful of our daily sins*, Blickl. Homl. 25, 14: Cd. 217; Th. 277, 11; Sat. 202. Gif hí ða geearnunga ealle gemundon *if they had remembered all the benefits*, Byrht. Th. 137, 35. Ne biþ gemunen *non memoretur*, Ps. Spl. 82, 4. v. munan.

ge-mund *meditation*; meditatio, Som.

ge-mundbyrdan; *p.* de; *pp.* ed [*mundbyrd protection*] *To protect, defend, patronize*; protēgĕre, tuēri :—Ða ic fór God wille gemundbyrdan *whom I will protect before God*, Cd. 113; Th. 149, 11; Gen. 2473. Ðæt he hine gemundbyrde *that he would protect him*, Bt. 35, 6; Fox 168, 21.

ge-mundian *to protect* :—Mildheortnys ána gemundaþ us on ðam micelum dóme *mercy alone will protect us at the great doom*, Homl. Th. ii. 102, 5. Gemunde ðisne heáp *protect this assembly*, H. R. 103, 31.

gēmung, e; *f. A marriage*; nuptiæ :—Ðe worhte gēmunge sunu his *qui fecit nubtias filio suo*, Mt. Kmbl. Rush. 22, 2 : 3 : 25, 10. Se ðe worhte gīmungo bearne his *qui fecit nuptias filio suo*, Rtl. 107, 15. Gīmungana *nuptiarum*, 108, 19 : 109, 23. [Cf. [?] *O. H. Ger.* gauma *epulæ*; and farmum ł gereordum *nuptias*, Mt. Kmbl. p. 19, 4.] v. gýmung.

gēmungian *to marry* :—Gimungia *nubat*, Rtl. 109, 35.

gēmunglic; *adj. Belonging to a marriage, nuptial*; nuptialis :—Hrægl gēmunglíc *vestis nubtialis*, Mt. Kmbl. Rush. 22, 12 : 11. Gīmungalíc *nuptialis*, Rtl. 108, 1.

ge-myltan, -miltan, -mieltan; *pp.* ed *To cause to melt, soften* :—Gold ðæt biþ ðurh ofnes fýr gemylted *gold that is melted by the fire of the furnace*, Elen. Kmbl. 2621; El. 1312. Gemyltyd is eorðe *liquefacta est terra*, Ps. Spl. C. 74, 3. Woldon ellenrófes mód gemiltan *they wished to subdue the bold man's courage*, Andr. Kmbl. 2785; An. 1395. v. ge-mieltan.

ge-mynan; *p.* de *To remember, remind* :—Dryhten gemynest ðú ðæt se forlærd cwæþ *sir, dost thou remember that that deceiver said?* Mt. Kmbl. Rush. 27, 63. Dú nú gemyndest ða word ðe ic ðē sæde *thou now rememberest the words that I said to thee*, Bt. 35, 2; Fox 156, 21. Ðæt he mec bí noman mínum gemyne *that he remember me by name*, Exon. 76 a; Th. 215, 28; Jul. 721. Gie gemynan *reminiscamini*, Jn. Skt. Lind. 16, 4. Gemyne ðú ðæt ðú ðisne ele send on ða sǽ *tu memento ut hoc oleum mittas in mare*, Bd. 3, 15; S. 541, 33. Gemyne ðē sylfne hú mycel yfel ðē gelamp *remember how great an evil befell thee*, Blickl. Homl. 31, 12. Gemyne ðis *remember this*, 113, 23, 24 : 225, 21 : Exon. 81 a; Th. 305, 25; Fä. 93 : Beo. Th. 1322; B. 659. God gemyne ðú Eádfriþ *O God, remember Eadfrith*, Mk. Skt. p. 1, 4. Gemynas gie *mementote*, Jn. Skt. Lind. 15, 20. v. ge-munan.

ge-mynd, es; *n.* e; *f. Mind, memory, memorial, memento, remembrance, commemoration* :—He fæste on gemynde hæfde *he had fast in mind*; memoriter retinuit, Bd. 4, 24; S. 597, 26. Gecerre hine to his gemynde *let him have recourse to his memory*, Bt. 35, 1; Fox 156, 10. Ðæs mannes sáwl hæfþ on hire þreó þing, ðæt is gemynd and andgit and willa. Ðurh ðæt gemynd se man geþencþ ða þing ðe he gehýrde oððe geseah oððe geleornode *man's soul has in it three things, that is memory and understanding and will. By the memory a man recollects the things that he has heard or seen or learned*, Homl. Th. i. 288, 18–21 : 28. Tubal Cain ðurh módes gemynd sulh-geweorces fruma wæs *Tubal Cain was the originator of plough-work by thought of mind*, Cd. 52; Th. 66, 16; Gen. 1085 : Exon. 17 b; Th. 41, 33; Cri. 665 : Bt. Met. Fox 22, 115; Met. 22, 58. Ðǽr se wísdóm á wunaþ on gemyndum *there wisdom ever dwells in mind*, 7, 79; Met. 7, 39. Me hæfþ ðeós gnornung ðære gemynde benumen *this grief has deprived me of the recollection*, Bt. 5, 3; Fox 12, 20. We witon swíþe lytel ðæs ðe ǽr us wæs búton be gemynde and be geacsunge *we know very little of that which was before us except by memory and by inquiry*, 42; Fox 256, 25. Heora gemynd is forgiten *the memory of them is forgotten*, Swt. A. S. Rdr. 57, 13. Ic wilnode ðæm monnum to léfanne ðe æfter me wǽren mín gemynd on gódum weorcum *I desired to leave to the men that should be after me my memory in good works*, Bt. 17; Fox 60, 16; Blickl. Homl. 197, 5. Ðín gemynd

memoriale tuum, Ps. Th. 101, 10 : Blickl. Homl. 171, 32. Ðis wæs gedón on mín gemynd *this was done in remembrance of me*, 69, 20. Ðæs hálgan biscopes gemynd *the commemoration of the holy bishop*, Shrn. 78, 23 : 86, 29 : 105, 30. Mannum to ēcre gemynde *for a perpetual remembrance to men*, 127, 22; 189, 15. Ðis to gemyndum habban *to have this as a memento*, 113, 34 : Beo. Th. 5600; B. 2804. Ne cwæþ he ðæt ná forðon ðe him wǽre ǽnig gemynd ðearfendra manna *he did not say that because he minded about the needy*, Blickl. Homl. 69, 10 : 61, 25 : 83, 16. Swá ic ðín gemynd rihte begange *sic memor fui tui*, Ps. Th. 62, 6 : 108, 16. Us is mid mycelre gemynde to geþencenne *we must bear well in mind*, Blickl. Homl. 29, 2. Gimynd *commemoratio*, Rtl. 62, 21. In gemyndum to habbanne *to be had in mind*, Nar. 4, 9 : 2, 8. [*Goth.* gamunds; *f. remembrance*: *O. H. Ger.* gi-munt; *f.*]

ge-mynd-benimming, e; *f.* Lethargy, Lye.

ge-mynd-dæg, es; *m. A commemoration day, day of birth* or *of death* :—Ðære abbudissan gemynd-dæg *cujus natalis*, Bd. 3, 8; S. 532, 39. Ðý dæge ðe his gemynd-dæg wǽre *die depositionis ejus*, Bd. 4, 30; S. 608, 35 : Th. Chart. 496, 4.

ge-mynd-drepen, e; *f. A mind stroke, a swoon, delirium*; mentis percussio :—On gemynd-drepen *in his mind's swoon*, Cd. 76; Th. 94, 34; Gen. 1571. Grn. has,—On gemynd drepen; *pp.* of drepan. DER. drepen.

ge-myndelíc; *adj. Belonging to memory, memorable*; mēmŏriālis, mēmŏrābilis :—Gemyndelíc *mēmŏriālis*, Ælfc. Gr. 9, 28; Som. 11, 35. Ðyssum tídum wæs sum gemyndelíc wundor, and ealdum wundrum gelíc on Breotone geworden *his tempŏribus mīrācŭlum mēmŏrābĭle, et antiquōrum sĭmĭle in Britannia factum est*, Bd. 5, 12; S. 627, 4 : 3, 16; S. 542, 14.

ge-myndelíce; *adv. By memory, without book*; mēmŏrĭter, sĭne libro :—Léraþ ðisne cantic Israēla bearn, ðæt híg hine gemyndelíce singon, and sí me to tácne ðis leóþ gemang Israēla folce *cantĭcum istud dŏcēte fīlios Israel, ut mēmŏrĭter tĕneant et ore decantent, et sit mihi carmen istud pro testĭmōnio inter fīlios Israel*, Deut. 31, 19.

ge-myndig, -mindig; *adj. Mindful, remembering* :—Wæs he gemyndig his bebodes *ipsi mēmŏr præcepti ejus*, Bd. 4, 25; S. 600, 14 : Ps. Spl. 118, 52. Wæs heó þearle gemyndig, hú heó ðone atolan eáðost mihte ealdre benǽman *she was very mindful how she might easiest deprive the fell one of life*, Judth. 10; Thw. 22, 23; Jud. 74 : Ps. Th. 73, 21 : 82, 4. Hý nǽron gemyndige manigfealdnesse mildheortnesse ðínre *non fuērunt mēmŏres multĭtūdĭnis mĭsĕrĭcordiæ tuæ*, Ps. Lamb. 105, 7. Beóþ hyra geóca gemyndge *they are mindful of their safety*, Exon. 33 b; Th. 107, 18; Gū. 60 : 39 a; Th. 129, 7; Gū. 417. Gemyndigra monna *of mindful men*, 34 b; Th. 111, 11; Gū. 125.

ge-myndigian; *p.* ode, ade; *pp.* od, ad *To remember, be mindful of, call to mind* :—Gemyndiga cýðnisse *memorari testamenti*, Lk. Skt. Lind. 1, 72. Ic gemyndige ða mǽran Raab and Babilonis *memor ero Rahab et Babylonis*, Ps. Th. 86, 2. Ðæt ðú ne gemyndgast æfter mandreáme ne gewittes wást bútan wildeóra ðeáw *that thou shalt not understand after the manner of the joy of man, nor know aught but the manner of wild beasts*, Cd. 203; Th. 251, 29; Dan. 571. Cwoen súðerne gemyndgade *reginam austri commemoram*, Mt. Kmbl. p. 16, 19. Ic God gemyndgade *memtor fui Dei*, Ps. Th. 76, 3 : 135, 24 : 142, 5. Gemyndgade mínes *memineris mei*, Mt. Kmbl. p. 4, 9. Gemyndgad biþ *memoratur*, p. 16, 15 : Lk. Skt. Lind. 1, 54. [*O. H. Ger.* gi-muntigōn *to remember*.]

ge-myndleás; *adj. Senseless, witless*; amens :—Sum gemyndleás wíf *a witless woman*, Homl. Th. ii. 188, 14. Gemyndleás *demens*, Ælfc. Gr. 47; Som. 48, 38.

ge-mynd-stów, e; *f. A monument* :—Gemyndstówa *monumenta*, Mt. Bos. 23, 29.

ge-myne; *adj. Mindful* :—Gif ðú ðǽr gemyne bist *si ibi recordatus fueris*, Mt. Kmbl. Rush. 5, 23.

ge-mynegian; *p.* ode; *pp.* od *To call to mind, remember, mention, admonish* :—He eall ða he in gehérnesse geleornian mihte mid hine gemynegode *ipse cuncta quæ audiendo discere poterat rememorando secum*, Bd. 4, 24; S. 598, 6. We gemynegodon *commemoravimus*, 1. 11; S. 480, 18. Ne gemynega ðú me mínra firena ðe ic geong dyde *delicta juventulis meæ ne memineris*, Ps. Ben. 24, 6. Ða wearþ he on swefne gemynegod *then was he admonished in a dream*, Homl. Th. i. 88, 22. Gemyngad *admonitus*, Mt. Kmbl. Rush. 2, 22 : Mt. Bos. 14, 8. Seó gemynegode cyninges dóhter *memorata regis filia*, Bd. 3, 24; S. 557, 3. v. mynegian.

ge-myntan; *p.* -mynte; *pp.* -mynted, -mynt *To determine, resolve*; stätuĕre, decernĕre :—Gregorius gemunde hwæt he gefyrn Angel-cynne gemynte *Gregory remembered what he of old had determined for the English race*, Homl. Th. ii. 126, 25. He befran hwam ða gebytlu gemynte wǽron. Him wæs gesǽd ðæt hí wǽron gemynte ánum sutere *he asked for whom those buildings were intended. He was told that they were meant for a shoemaker*, 354, 35. Hæfdon hie gemynted to ðam *they had resolved thereon*, Cd. 153; Th. 190, 10; Exod. 197. Ic hæfde gemynt ðē to árwurþienne on ǽhtum and on feó *decrēvĕram quĭdem

E e 2

magnĭfĭce hŏnōrāre te, Num. 24, 11 : Gen. 18, 33 : Bd. 3, 9 ; S. 534, 3 : Homl. Th. ii. 548, 31.

ge-myrran ; *p.* de ; *pp.* ed *To hinder, obstruct, force, trouble* ; impedire, turbare, obstruere :—Mōde gemyrde *disturbed in mind*, Andr. Kmbl. 1491 ; An. 747 : Ps. Th. 62, 9 : Exon. 71 b ; Th. 267, 8 ; Jul. 412. v. myrran.

ge-mȳþ ; *pl. n. The mouth of a river* ; ostium fluminis :—Æt đâm gemȳđum Tyne streámes *juxta ostium Tini fluminis*, Bd. 5, 6 ; S. 618, 28 : Cod. Dipl. Kmbl. iii. 48, 26. [O. H. Ger. ge-mundi *ostia*.]

GĒN, gién ; *adv. Again, moreover, besides, at length, yet, hitherto* ; iterum, denuo, adhuc, insuper, denique :—Đær he gēn ligeþ *there he still lies*, Exon. 18 b ; Th. 46, 9 ; Cri. 734. Swā he nū gēn dēþ *as he still does*, Beo. Th. 5711 ; B. 2589 : Exon. 29 a ; Th. 89, 17 ; Cri. 1458. Bidon ealle đær tyn niht đá gēn *all waited there yet ten nights*, 15 b ; Th. 34, 15 ; Cri. 542. Đā gién wæs yrre God *God was yet angry*, Cd. 131 ; Th. 166, 1 ; Gen. 2741. Wæs Iustus đā gēn lifigende *Iustus adhuc superstes*, Bd. 2, 7 ; S. 509, 10. Đæs gēn to tācne is *of that further is as proof*, 6 ; S. 508, 42. Ic sceal forđ sprecan gēn ymb Grendel *I shall go on to speak further about Grendel*, Beo. Th. 4146 ; B. 2070 : Exon. 96 b ; Th. 360, 5 ; Wal. 1 : Elen. Kmbl. 2434 ; El. 1218. Gién đē sunu weordeþ *yet there shall be a son to thee*, Cd. 100 ; Th. 132, 19 ; Gen. 2195. Gēn ic đē feores unnan wille *yet will I grant thee life*, Exon. 68 b ; Th. 254, 3 ; Jul. 191. Đā gēn Abrahame eówde heáhcyning *again the high king appeared to Abraham*, Cd. 98 ; Th. 130, 23 ; Gen. 2164. Đā gién seó fæmne spræc *then again spoke the woman*, Exon. 71 b ; Th. 267, 19 ; Jul. 417. Geornor đonne he gēn dyde *more eagerly than yet he had done*, 67 a ; Th. 249, 12 ; Jul. 110. Gēn strengre is *it is yet harder*, 10 b ; Th. 12, 28 ; Cri. 192 : 95 b ; Th. 357, 14 ; Pa. 28 : 97 a ; Th. 363, 8 ; Wal. 50.

gēn, gegn [?] ; *adj. Direct, short, near [of a road]* :—Đe đa gēnran wegas cūdan đara sīđfato *qui brevitates itinerum noverant*, Nar. 6, 7. [O. E. gein, v. Stratmann : *North E.* and *Scot.* gane, 'the ganest way :' *Icel.* gegn, ' hinn gegnsta vegr.']

gēna ; *adv. Yet, still, further* :—Đafodest đū gēna đæt me þeówmennen drehte *thou hast still permitted the slave-woman to vex me*, Cd. 102 ; Th. 135, 21 ; Gen. 2246. Næbbe ic synne wiþ hie gefremed gēna *I have not committed sin against her yet*, 125 ; Th. 160, 17 ; Gen. 2651. Nū gēna *still at the present time*, Exon. 34 b : Th. 111, 13 ; Gū. 126. Ic eom gēna swētran *I am yet sweeter*, 111 a ; Th. 425, 19 ; Rä. 41, 58. Ic wille đē ánre nū gēna bēne biddan *I will of thee one more boon require*, Andr. Kmbl. 950 ; An. 475. Mycel is nū gēna lād ofer lagustreám *great is now still our voyage over the lake-stream*, 844 ; An. 422. Cwico wæs đā gēna *was still living*, Beo. Th. 6178 ; B. 3093. v. gēn, geóna.

ge-nacian ; *p.* ode, ede, *pl.* odon, edon ; *pp.* od, ed *To make naked* or *bare* ; nudare, nudum facere :—Menigo genacedon đæt hūs *turba nudaverunt tectum*, Mk. Skt. Lind. 2, 4. DER. nacian.

ge-nacodian ; *p.* ode, ade ; *pp.* od, ad *To make bare, naked, to strip* ; nudare : — He hine middangeardes þingum ongyrede and genacodade [genacode] *he unclothed and stripped himself of worldly things*, Bd. 4, 3 ; S. 567, 24. DER. nacodian, nacod.

ge-næfd ; *part. p. Not had* :—Đonne sint hie đē pleólícran gehæfd đonne genæfd *then are they more dangerous to thee had than not had*, Bt. 14, 1 ; Fox 42, 22.

ge-nægan, -nēgan ; *p.* de ; *pp.* ed ; *c. acc. pers : gen. inst. rei To approach one with anything, address, approach, assail, assault* ; adire aliquem aliqua re, appellare, instare alicui, urgere, tribulare :—Hio sió cwēn ongan wordum genēgan *the queen began to address them with words*, Elen. Kmbl. 769 ; El. 385. Þeóf đe eorlas ungearwe yfles genægeþ *the thief who assaults with evil unprepared men*, Exon. 20 b ; Th. 54, 28 ; Cri. 875. Đā hyne gesōhton Heađoscylfingas, nīđa genægdon [MS. gehnægdan] *when the martial Scylfings him sought [and] assailed [him] in the wars*, Beo. Th. 4418 ; B. 2206. Nearwum genæged nȳd-costingum *assailed with painful troubles*, Exon. 49 b ; Th. 171, 13 ; Cri. 1126.

ge-næged [= gehnæged] ; *part. p. Subdued, humbled* ; subactus, Mt. Kmbl. Rush. 23, 12.

ge-nægled ; *part. p. Nailed* :—Genæglad on rōde *nailed on the cross*, Mt. Kmbl. Lind. 27, 22, 26, 31 : Exon. 90 b ; Th. 339, 14 ; Gn. Ex. 94. Genæglod, Homl. Th. i. 82, 25.

ge-næs, -nǣson *saved.* v. ge-nesan.

ge-nǣstan ; *p.* te *To contend* :—Se đe wiþ mægenđisan mînre genǣsteþ *he that contends against my main force*, Exon. 107 b ; Th. 410, 3 ; Rä. 28, 10. [Cf. ge-nǣtan.]

ge-nǣtan ; *pp.* -nǣt *To afflict, trouble* :—Đa underđiéddan mon sceal lǣran đæt hie elles ne sién genǣt ne geirmed *illos ne subjectio conterat*, Past. 28, 1 ; Swt. 189, 16 ; Hat. MS. Đonne genǣt he hine *humiliabit eum*, Ps. Th. 9, 30. [Goth. ga-naitjan *to maltreat*.]

ge-nág or -nag [?] *incumbens* [Grn.], *urgens* [Ettm.], Exon. 95 a ; Th. 354, 38, 40 ; Reim. 57, 58.

ge-namian ; *p.* ode ; *pp.* od [nama *a name*] *To name, call, appoint* ; appellare, vocare :—And Adam đā genamode ealle nȳtenu heora namum *and Adam then named all cattle by their names* ; appellavitque omne

jumentum nominibus suis, Gen. 2, 20. Hī wurdon genamode to đam ylcan gewinne đe heora fæderas on wǣron *they were nominated to the same warfare in which their fathers were*, Homl. Th. ii. 500, 4 : i. 88, 3. Būtan đære mægđe Leui đe næs genamod đǣr to *besides the tribe of Levi that was not named amongst them*, Swt. Rdr. 63, 224 : Homl. Th. i. 282, 20. DER. namian, nama. v. ge-nomian.

ge-namne = ge-numne [?]. v. ge-niman.

ge-nāþ *darkened* ; *p.* of ge-nīpan.

ge-nāþan ; *p.* -neóþ, *pl.* -neópon ; *pp.* -nāpen *To overwhelm* ; incumbere, obrepere, supervenire :—Se đe feóndum geneóþ *who overwhelmed the foes*, Cd. 166 ; Th. 207, 32 ; Exod. 475. v. nāþan.

gēn-cyme, es ; *m. A meeting* ; conventus, Ps. Spl. T. 63, 2.

gende = gengde, Beo. Th. 2806 ; B. 1401. Grein however compares *Icel.* gana *to rush.*

ge-neádian, -nēdian ; *p.* ode ; *pp.* od *To compel* : — Nolde swā-đeáh nænne to cristendôme geneádian *he would not however compel any one to christianity*, Homl. Th. i. 130, 14 : i. 70, 25. Næs Iohannes niid ēhtnysse geneádod đæt he Criste wiđsôce *John was not compelled by persecution to deny Christ*, i. 484, 31 : 88, 1. Geneádige *urgent*, Ps. Lamb. 68, 16. We bióþ genēdode *we are forced*, Past. 53 ; Swt. 417, 30 ; Hat. MS.

ge-neah, es ; *n. f.* [?] *Sufficiency, abundance :* — Mid geneahe *abundantly*, Vercel. Kmbl. ii. 81, 68 ; Leás. 36. [Cf. *Goth.* ga-nauha *sufficiency : O. H. Ger.* gi-nogi, Grff. ii. 1008.]

ge-neah *it is sufficient* ; sufficit, Exon. 93 a ; Th. 348, 29 ; Sch. 35. v. ge-nugan.

ge-neahhe, -neahe, -nehhe, -nehe ; *adv. Enough, sufficiently, abundantly, frequently, very much, earnestly, instantly* ; satis, sufficienter, frequenter, valde, sedulo, instanter :—Đara đe geneahhe noman scyppendes hergan willaþ *of those who sufficiently will praise the creator's name*, Exon. 8 b ; Th. 4, 5 ; Cri. 48 : Elen. Kmbl. 2313 ; El. 1158 : Beo. Th. 1570 ; B. 783. Nū ic his geneahhe neósan wille *now I will frequently visit him*, Exon. 43 a ; Th. 145, 7 ; Gū. 691 : 100 b ; Th. 379, 13 ; Deór. 32 : 77 a ; Th. 289, 31 ; Wand. 56. He wyscte geneahhe, đæt . . . he *wished earnestly, that . . .*, 100 b ; Th. 378, 33 ; Deór. 25 : Ps. Th. 62, 8 : 63, 1 : 65, 1 : 87, 3 : 114, 4 : 137, 7 : 149, 1. Swíđe genehhe *very frequently*, Hy. 3, 42 ; Hy. Grn. ii. 282, 42 ; L. E. I. 10 ; Th. ii. 408, 25. Geneahe *sufficiently*, Cd. 137 ; Th. 172, 12 ; Gen. 2843. Genehe *abundantly*, Byrht. Th. 139, 45 ; By. 269. Đær genehost brægd eorl Beówulfes ealde láfe *then very frequently drew a warrior of Beowulf's an ancient relic* [i. e. *very many of Beowulf's warriors, etc.*], Beo. Th. 1593 ; B. 794. DER. swíđ-geneahhe.

ge-neahhie, -neahhige, -nehhige ; *adv. Enough, sufficiently, abundantly, frequently, very much, earnestly, instantly* ; satis, sufficienter, frequenter, valde, sedulo, instanter, Ps. Th. 55, 7 : 67, 4 : 118, 25 : 65, 3 : 70, 5 : 85, 3. DER. swíđ-geneahhige.

ge-neáhsen ; *adj. Near* :—Hwīlum mōna sunnan sînes leóhtes bereáfaþ đonne hit gebyrigan mæg đæt swā geneáhsne weorđaþ *sometimes the moon deprives the sun of its light when it happens that they get so near*, Bt. Met. Fox 4, 23 ; Met. 4, 12.

ge-neálǣcan, -lǣcean ; *p.* -lǣhte ; *pp.* -lǣht *To approach, draw near, adhere* [wiþ *dat.* and *acc.*] :—Ne dorstan hie đære stówe geneálǣcan *they durst not approach the place*, Blickl. Homl. 199, 26. Hī ne dorston hine geneálǣcan *they durst not approach him*, 243, 13. Geneálǣcean, 77, 11 : Shrn. 76, 29. Nū geneálǣceþ mínum gebedum đæt ic bidde on đînre gesîhþe *appropiet oratio mea in conspectu tuo*, Ps. Spl. 118, 169. Geneálǣcþ *adhǣret*, Ps. Spl. C. 93, 20. He him geneálǣhte *he drew near to him*, Blickl. Homl. 15, 24 : 67, 2. Geneáhlǣhte *adhǣsit*, Ps. Spl. C. 101, 6. Me geneálǣhton me *appropinquaverunt*, Ps. Spl. 37, 11. Hī geneálǣhton *acceleraverunt*, Ps. Lamb. 15, 4. Folce geneálǣcendum *populo appropinquanti*, Ps. Spl. 148, 14.

ge-neálǣcing, e ; *f. An approach* :—Toforan đære geneálǣcincge đæs fefores *before the access of the fever*, Herb. 160 ; Lchdm. i. 288, 11.

ge-neán *to draw near, cleave, adhere* :—Gineá đū đóast *inherere facias*, Rtl. 34, 28. Đes cwom ł geneó *hic accessit*, Mt. Kmbl. Lind. 27, 58. v. ge-nēhwian.

ge-near, -ner *a refuge, protection* ; refugium :—Genear [gener, Lamb.] mîn eart đū *refugium meum es tu*, Ps. Spl. 90, 2. v. ge-ner.

ge-nearwian ; *p.* ode, ade ; *pp.* od, ad, ot *To narrow, straiten, constrain, confine, oppress, afflict* :—Hwīlum mec mîn freá fæste genearwaþ *sometimes my master fast confines me*, Exon. 101 b ; Th. 382, 24 ; Rä. 4, 1. Swā hit is genearwed *so is it narrowed*, Bt. 18, 1 ; Fox 62, 24. Fæste genearwad *fast confined*, Exon. 126 a ; Th. 484, 8 ; Rä. 70, 4. Mid eofer-spreótum hearde genearwod *hard pressed with boar-spears*, Beo. Th. 2881 ; B. 1438. Mid weres egsan hearde genearwod *with the fear of man sorely oppressed*, Cd. 43 ; Th. 56, 32 ; Gen. 921 : 123 ; Th. 157, 9 ; Gen. 2603. Genearwad biþ heorte mín *anxiaretur cor meum*, Ps. Spl. 60, 2. v. ge-nyrwian.

ge-neát, es ; *m. A companion, associate, vassal :* — Big-standaþ me strange geneátas đa ne willaþ me æt đam stríđe geswícan *strong companions stand by me who will not fail me at the strife*, Cd. 15 ; Th. 18, 36 ;

Gen. 284. Geneát *inquilinus*, Cot. 108: *parasitus*, 152. Byrhtwold wæs eald geneát [*or* eald-geneát, *q. v.*] Be cyninges geneáte *of a king's* ' geneat,' L. In. 19; Th. i. 114, 9: Chr. 897; Erl. 96, 3. Be ðon ðe monnes geneát stalige *in case a man's* ' *geneát* ' *steal*, L. In. 22; Th. i. 116, 9. [*Icel.* nautr: *O. H. Ger.* ganôz, Grff. ii. 1125: *Ger.* genoss.] v. Stubbs' Const. Hist. i. 149; Kemble's 'Saxons in England,' i. c. vii; Schmid A. S. Ger. s.v. DER. beód-, heorþ-geneát.

ge-neát-land, es; *n. Land granted for services or rent :—*Ægðer ge of ðegnes inlande ge of geneát-lande *both from a thane's inland and from* ' geneát-land,' L. Eádg. I, I; Th. i. 262, 8. v. in-land.

ge-neát-man, -mann, es; *m.* [v. ge-neát] *A tenant, one holding land on payment of rent,* ' gafol':—Gif geneátmanna hwilc forgýmeleásaþ his hláfordes gafol *if any* ' *geneat-man* ' *neglect the tribute due to his lord,* L. Eádg. Suppl; Th. i. 270, 16.

ge-neát-riht, es; *n. The conditions regulating the tenure of the* ' geneát-land :'—Geneát-riht is mistlic ðam ðe on lande stænt. On sumon he sceal land-gafol syllan . . . *villani rectum est varium et multiplex secundum quod in terra statutum est. In quibusdam terris debet dare land-gablum* . . ., LL. Th. i. 115, note.

ge-neát-scólu, e; *f. A band of companions :—*Ða ðegnas seó geneát-scólu, Exon. 75 b; Th. 283, 22; Jul. 684.

ge-nec *a light ship, a frigate*; liburnica, Cot. 120. v. naca.

ge-nédan, -niedan, -nýdan; *p.* de; *pp.* ed *To compel, force, urge :—* Ðú tunglu genédest ðæt hí ðé to hérað *thou compellest the stars to obey thee*, Bt. Met. Fox 4, 9; Met. 4, 5: 4, 30; Met. 4, 15. Seðe ðec genédes *quicunque te angariaberit*, Mt. Kmbl. Lind. 5, 41. Sihhem geniédde ðæt mæden *Sichem forced the maiden*, Past. 53, 5; Swt. 415, 22; Hat. MS. Genéddon Simon *angariaverunt Simonem*, Mk. Skt. Lind. 15, 21. Ealle Asiam hý genýddon ðæt hí him gafol guldon *they compelled all Asia to pay them tribute*, Ors. 1, 10; Bos. 32, 28. He næs nó genéded *he was not compelled*, Blickl. Homl. 29, 15. Ðæt Bryttas mid ðý mærran hungre genédde ða elreordian adrifan *ut Brittones fame famosa coacti barbaros pepulerint*, Bd. 1, 14; S. 482, 12.

ge-nédedlíc; *adj. Compulsory, forced*; coactus :— He geleornade ðæt Cristes þeówdóm sceolde beón wilsumlíc, nalæs genédedlíc *didicĕrat servĭtium Christi voluntārium, non coactitium esse debēre*, Bd. 1, 26; S. 488, 18.

ge-nefa, an; *m. A nephew*; nepos :— Caius his [Agustuses] genefa nolde gebiddan to ðam ælmihtigum Gode *Caius his [Augustus's] nephew would not worship the almighty God*, Ors. 6, 1; Bos. 116, 18.

ge-négan; *p.* de; *pp.* ed *To approach one with anything, to address*, Elen. Kmbl. 769: El. 385. v. ge-nægan.

ge-neh; *adv. Enough, sufficiently, abundantly :—*Ðonne sceolon we geneh geþencean emb úre sáula ðearfa *then ought we to consider very much about our souls' needs*, Blickl. Homl. 101, 32. v. ge-neahhe.

ge-nehhe, -nehe *enough, frequently*, L. E. I. 10; Th. ii. 408, 25. v. ge-neahhe.

ge-nehige, -nehge; *adv. Enough, very much, frequently :—*Hie genehge mid gebedum séceaþ *seek it frequently with prayers*, Blickl. Homl. 207, 3. v. ge-neahhie.

ge-nehlíce; *adv. Sufficiently, abundantly, frequently :—*Gé sceolon myngian eówre hýremen ðæt hig hyra gebedu genehlíce begán *ye shall admonish your parishioners that they sufficiently cultivate their prayers*, L. E. I. 29; Th. ii. 424, 39.

ge-néhlíce; *adv. Near :—*Ðæt reáf ðe he genéhlíce on him hæfde *the garment that he wore next his skin*, Guthl. 16; Gdwin. 68, 17.

ge-néhwian; *p.* ode, ade; *pp.* od, ad *To approach, draw near, adhere :—*Monn genéhwas wífe his *homo adhærebit uxori suæ*, Mt. Kmbl. Lind. 19, 5. Ánum genéhwaþ *uni adhærebit*, Lk. Skt. Lind. 16, 13. Genéhwade ánum *adhæsit uni*, 15, 15. [Cf. ge-neálǽcan.]

ge-nemnan; *p.* -nemde; *pp.* -nemned, -nemnod *To name*; nominare :— On ðære ceastre, ðe is genemned Nazareth *in civitate, quæ vocatur Nazareth*, Mt. Bos. 2, 23: 5, 19: Mk. Skt. Lind. 15, 7: Cd. 6; Th. 8, 27; Gen. 130: 217; Th. 277, 14: Sat. 205: 217; Th. 287, 13; Sat. 366. Ða genemde ðæra scypmanna án Scs. Martynus *then one of the sailors named St. Martin*, Shrn. 147, 8. Hí beóþ Godas genemnede [Cot. genemde] *they are named gods*, Bt. 37, 4; Fox 192, 9. Hí Angle genemnode wǽron *they were named Angles*, Homl. Th. ii. 120, 29.

ge-neósian; *p.* ode; *pp.* od [neósian *to visit*] *To visit, come to*; visĭtāre, adīre :— Beheald holdlíce, hú ðú hraðe wylle geneósian niða bearna ealra þeóda *intende ad visĭtandas omnes gentes*, Ps. Th. 58, 5. Hí ne mihton hine for ðære manegu geneósian *non potĕrant adīre eum præ turba*, Lk. Bos. 8, 19. Ðú geneósast hine *visĭtas eum*, Ps. Spl. 8, 5. Se gesǽliga his ealdcýþþe eft geneósaþ *the blessed [bird] again visits its old country*, Exon. 61 a; Th. 222, 20; Ph. 351. Forðam ðe he geneósode, and his folces alýsednesse dyde *quia visitāvit, et fecit redemptiōnem plebis suæ*, Lk. Bos. 1, 68, 78. Us mid hǽlo hér geneósa *visĭta nos in salutāri tua*, Ps. Th. 105, 4. Ðæt ic geneósige temple his *ut visĭtem templum ejus*, Ps. Spl. 26, 8.

ge-neósung, e; *f. A visiting, visitation*; visitatio :—Forðam ðe ðú ne ᴍɴᴄɴᴄówe ða tíde ðínre geneósunge *eo quod non cognoveris tempus*

visitationis tuæ, Lk. Bos. 19, 44: Scint. 21: Greg. Dial. 2, 35. v. neósung.

ge-neoðerian *to condemn.* v. ge-niðerian.

ge-ner, -near, es; *n. A refuge*; refugium, asylum, sanctuarium :— Ðú eart gener mín *tu es refugium meum*, Ps. Spl. 31, 9: Ps. Lamb. 90, 2. Hí óðer gener næfdon *they had not another refuge*, Ors. 1, 12; Bos. 36, 10. Ongin ðé generes wilnian *desire a refuge for thyself*, Exon. 36 b; Th. 119, 28; Gú. 261. v. ner, feorh-gener.

ge-nerenes, -ness, e; *f. A taking away, deliverance*; ereptio :—For generenesse heora freónda, ðara ðe of weorulde leordan *pro ereptiōne suōrum qui de sæcŭlo migrāvĕrant*, Bd. 4, 22; S. 592, 26. Ginerenis *ereptio*, Rtl. 30, 5.

ge-nerian, -nergan, -nerigan; *p.* ede, ode; *pp.* ed, od *To save, deliver, take away, set free, preserve, defend*; servare, redimere, liberare, eripere, salvum facere, defendere :—Se mec wile wiþ ðǽm níðum genergan *he will protect me against that malice*, Exon. 36 a; Th. 116, 24; Gú. 212. We mágon feorh generigan *we may save life*, Cd. 117; Th. 152, 22; Gen. 2524. Ic hine generige *eripiam eum*, Ps. Th. 90, 16. He generaþ hig *eripiet eos*, Ps. Spl. 33, 7. Oswio his ðeóde generede *Osuiu suam gentem liberavit*, Bd. 3, 24; S. 557, 14. Abraham Loth generede *Abraham saved Lot*, Cd. 121; Th. 156, 12; Gen. 2587. Ðú hí generedest *liberavisti eos*, Ps. 105, 8: Exon. 98 a; Th. 369, 28; Seel. 48. He híg generode of Egipta lande *he delivered them out of the land of the Egyptians*, Ex. 18, 9. Alýs me and genere *eripe me et libera me*, Ps. Th. 143, 8: 139, 1. Ðæt ðú generige oððe alýse me *ut eruas me*, Ps. Lamb. 39, 14: Ps. Th. 88, 41. Generigende *eripiens*, Ps. Spl. 34, 11. Genered *liberatus*, Bd. 4, 31; S. 610, 24. Genered *saved*, Beo. Th. 1658; B. 827. Hí sind fram graman generode *they are saved from wrath*, Homl. Th. ii. 120, 35. [Cf. ge-nesan.]

ge-nerwde *vexed.* v. ge-nyrwian.

ge-nesan; *p.* -næs, *pl.* -nǽson; *pp.* -nesen *To be saved, preserved, escape from :—*Se biþ hál and geneseþ on écnesse *he shall be safe and shall be preserved to eternity*, Blickl. Homl. 171, 26. Hróf ána genǽs ealles ánsúnd *the roof alone was saved wholly sound*, Beo. Th. 2003; B. 999. Se ða sæcce genæs *who had come safely from the conflict*, 3959; B. 1977: 4844; B. 2426: Cd. 94; Th. 121, 33; Gen. 2019. Ða ðe ða frécennesse and yrmðo genǽson *those who had survived the danger and misery*, Blickl. Homl. 203, 20: Ors. 4, 8; Bos. 90, 8: Fins. Th. 95; Fin. 47. Hý ðurh miltse meotudes genǽson *they have been saved through the Lord's mercy*, Exon. 26 a; Th. 77, 12; Cri. 1255. He níða gehwane genesen hæfde *he had survived every struggle*, Beo. Th. 4786; B. 2397. Ðæt híg mihton ða frécnesse genesan *that they might escape the danger*, Shrn. 38, 2. [*Goth.* ga-nisan *to be saved*: *O. Sax.* gi-nesan: *O. H. Ger.* ge-nesan: *Ger.* ge-nesen *to get well.*]

Génesburuh *Gainsborough.* v. Gegnesburh.

ge-nésta, an; *m. A neighbour*; proximus :— Mið ðǽm ginéstum sínum *apud proximos suos*, Rtl. 84, 37.

ge-néðan; *p.* de; *pp.* ed *To venture, attempt, strive :—*Ne dorste he genéðan ðæt . . . *he durst not venture to* . . ., Ors. 1, 10; Bos. 33, 30. Nú ðú Andreas scealt genéðan in gramra gripe *now shalt thou Andrew venture into the grasp of foes*, Andr. Kmbl. 1900; An. 952: 2702: An. 1353. Sió sunne uncúðne weg nihtes genéðeþ *the sun ventures on an unknown way by night*, Bt. Met. Fox 13, 117; Met. 13, 59: Exon. 100 a; Th. 374, 1; Seel. 119. He genéðde under ánne elpend *he ventured under an elephant*, Ors. 4, 1; Bos. 77, 20: 8; 90, 8. He ána geneðde frécne dǽde *he alone ventured on the daring deed*, Beo. Th. 1781; B. 889: 3317; B. 1656. Ðæt ic ealdre genéðde *that I should venture my life*, 4273; B. 2133: Apstls. Kmbl. 34; Ap. 17: 100; Ap. 50. Hie hit frécne genéðdon under wætera hrófas *they boldly ventured it under the waters' roofs*, Cd. 170; Th. 214, 17; Exod. 570: Beo. Th. 1923; 959. v. néðan.

geng *a privy*; latrina, Cot. 123. v. gang.

geng; *adj. Young*; jŭvĕnis :—Ðǽm gengum þrým *to the three young men*, Cd. 176; Th. 222, 9; Dan. 102. v. geong.

gengan; *p.* de, *pl.* don; *pp.* ed *To go, pass*; ire, meare, currere, ferri, converti :—Forhwí gengdest ðú on bæcling *quare conversus es retrorsum*, Ps. Th. 113, 5. He feára sum beforan gengde wong sceáwian *he with a few went before to view the plain*, Beo. Th. 2829; B. 1412. Him oft betwuh gnornword gengdon *words of sadness passed oft between them*, Cd. 37; Th. 47, 27; Gen. 767. Beornas cómon wicgum gengan *the men came riding on horses*, Andr. Kmbl. 2192; An. 1097. v. gán, gangan.

gengdon *passed*, Cd. 37; Th. 47, 27; Gen. 767; *p.* of gengan.

genge; *f. A GANG, flock, company*; grex :—Ðæt wæs Hereward and his genge *that was Hereward and his followers*, Chr. 1070; Erl. 207, 29. [*Laym. Orm.* genge *a host, retinue.*]

génge ic *I would go*, Cd. 39; Th. 51, 29; Gen. 834; *p. subj.* of gangan.

genge; *adj. Going, current, prevalent, valid :—*Ðeáh ðe ðæs cyninges béne mid hine swíðode and genge wǽre *preces regis illius multum valere apud eum*, Bd. 3, 12; S. 537, 19. Ðæt his sóþ fore us genge weorðe *that his truth be current before us*, Exon. 43 b; Th. 147, 35; Gú. 737.

Â ðín dóm sý gód and genge *ever be thy judgment good and valid*, 54 b; Th. 192, 20; Az. 109. Gód biþ genge and wiþ God lenge *good prevails and lasts before God*, 91 a; Th. 341, 4; Gn. Ex. 121. [*O. H. Ger.* gengi *usual: Ger.* gäng.]

ge-nídde, Ps. Vos. 58, 14: ge-níded *compelled;* coactus, Cot. 59: 106. v. ge-nédan.

ge-niédde *compelled, forced.* v. ge-nédan.

ge-nierede, -wod *vexed.* v. ge-nyrwian.

ge-niht, -nyht, es; *n: e; f. Abundance, fulness, sufficiency;* abundantia, ubertas:—Wēnst ðú ðæt se ânweald and ðæt geniht seó to forseónne *thinkest thou that power and abundance are to be despised?* Bt. 33, 1; Fox 120, 22, 24, 26. Hý beóþ oferdrencte on ðære genihte ðínes húses *inebriābuntur ab ūbertāte dŏmus tuæ,* Ps. Th. 35, 8. To genihte *in abundantia,* Ps. Th. 77, 25, 27: 84, 6: Menol. Fox 364; Men. 183. Ðú sealdest me wilna geniht *thou gavest me the fulness of my desires,* Soul Kmbl. 285; Seel. 146: Cd. 90; Th. 113, 21; Gen. 1890: Ps. Th. 4, 8. [*O. H. Ger.* ge-nuht *copia, abundantia.*]

ge-nihtlíce; *adv. abundantly;* abunde, Cot. 6.

ge-nihtsum, -nyhtsum; *adj.* I. *abundant, abounding, copious, rich, plentiful, fruitful;* abundans, uber, cōpiōsus, affluus, profluus:— Genihtsum *uber,* Ælfc. Gr. 9, 18; Som. 10, 7. Genihtsum wæter forþflōweþ *plentiful water flows forth,* Bd. 5, 10; S. 625, 24: Ps. Th. 85, 4: 143, 17. On ylde genihtsumre *in sĕnecta ūbĕri,* Ps. Spl. 91, 14. Ðæt hí wǽron genihtsume *ut essent profluí,* Hymn. Surt. 94, 5. Hladungum genihtsumum *haustĭbus affluis,* 58, 12. II. *satisfied;* sătiābĭlis:— Se ðe ǽr ne wæs nípes genihtsum *who ere was not satisfied with slaughter,* Cd. 93; Th. 120, 15; Gen. 1995. [*O. H. Ger.* ge-nuhtsam *abundans.*]

ge-nihtsumian, -nyhtsumian; *part.* -nihtsumigende; *p.* ode; *pp.* od *To abound, suffice;* abundāre, sufficĕre:—Hí synfulle and genihtsumigende on worulde, hí begeáton *welan ipsi peccātōres et abundantes in sæculo, obtinuērunt divĭtias,* Ps. Spl. 72, 12: 127, 3. Ic genihtsumige *abundo,* Ælfc. Gr. 38; Som. 41, 10. Se ungesǽliga gýtsere wile mâre habban ðonne him genihtsumaþ *the unhappy miser wishes to have more than suffices him,* Homl. Th. i. 64, 34. Ânes engles geearnung ne genihtsumode to alýsednysse ealles mancynnes *the merit of an angel was not sufficient for the redemption of all mankind,* Boutr. Scrd. 17, 37.

ge-nihtsumlíce, -nyhtsumlíce; *comp.* -lícor; *adv. Abundantly, plentifully, copiously, sufficiently;* abundanter, abunde, ūbertim, sufficienter:— He agylt genihtsumlíce ðám wyrcendum ofermódignysse *retrĭbuet abundanter făcientibus sŭperbiam,* Ps. Spl. 30, 30: Bd. 5, 19; S. 637, 48. Genihtsumlíce *abunde,* Ælfc. Gr. 38; Som. 41, 10. Ðær genihtsumlíce is sǽd *ubi ūbertim indīcātum est,* Bd. 1, 27; S. 494, 36: 4, 28; S. 605, 12. Genihtsumlícor *abundantius,* 3, 27; S. 559, 7.

ge-nihtsumnes, -nyhtsumnes, -ness, -nys, -nyss, -nis, -niss, e; *f. Abundance, plenty, copiousness, sufficiency;* abundantia, ubertas, cōpia:— Genihtsumnys *abundantia* vel *cōpia,* Wrt. Voc. 83, 40. Genihtsun nys *ubertas,* Ælfc. Gr. 9, 18; Som. 10, 7. Gemynd genihtsumnesse wynsumnesse ðíure hí bylcettaþ *mĕmōriam abundantiæ suāvĭtātis tuæ eructābunt,* Ps. Lamb. 144, 7. On genihtsumnysse mínre *in abundantia mea,* 29, 7: 77, 25. Híg beóþ gedrencte for genihtsumnisse húses ðínes *inebriābuntur ab ūbertāte dŏmus tuæ,* 35, 9. Ðǽre eorþan wæstmbǽrnysse and genihtsumnysse we nellaþ habban us to lífes brícum, ac to oferflówednyssum *the fruitfulness and abundance of the earth we will not have for the uses of life, but as superfluities,* Homl. Th. ii. 540, 10: 64, 35.

ge-niman, -nyman, -nioman; he -nimeþ, -nimþ; *p.* -nam, -nom, *pl.* -námon, -nómon; *imp.* -nim, *pl.* -nimaþ; *subj. p.* -náme, *pl.* -námen; *pp.* -numen *To take, take up, take away, assume, receive, accept, obtain, comprehend, enter into;* sūmĕre, tollĕre, auferre, assūmĕre, accĭpĕre, nancisci, comprehendĕre, inīre:—Forlǽt mec englas geniman on ðínne neáwest *let angels take me into thy presence,* Exon. 118 b; Th. 455, 13; Hy. 4, 49. Ðæt hí woldon his bán geniman *ut tollĕrent ossa illius,* Bd. 4, 30; S. 608, 28. He genimeþ hraðe ðære rôsan wlite *it taketh away the beauty of the rose,* Bt. Met. Fox 6, 24; Met. 6, 12: Cd. 80; Th. 23; Gen. 1209. Wintres dæg sigelbeorhtne genimþ hærfest *winter's day takes away the sun-bright autumn,* Menol. Fox 404; Men. 203. Hú lange dēmaþ ge unrihtwísnysse, and ansýne synfulra genimaþ *usquequo jūdĭcātis inĭquĭtātem, et făcies peccātōrum sūmĭtis?* Ps. Spl. 81, 2. Heó genam cúðe folme *she took the well known hand,* Beo. Th. 2609; B. 1302: 4850: B. 2429. He his folc genam swá fǽle sceáp *abstŭlit sĭcut oves pŏpŭlum suum,* Ps. Th. 77, 52, 69. Ðe ic to swá myclum cyninge genom *quod cum tanto rēge inii,* Bd. 2, 12; S. 513, 25. He feówer túnas genom *he took four towns,* Chr. 571; Erl. 18, 13: 584; Erl. 18, 24. On ðam ilcan ðú eard genáme *in quo hăbĭtas in idipsum,* Ps. Th. 73, 3: 72, 19. Genámon me ðǽr strange feóndas *strong enemies took me there,* Rood. Kmbl. 60; Kr. 30: 120; Kr. 60: Cd. 210; Th. 260, 10; Dan. 707. Þýstro ðæt ne genámon *tenebræ eam non comprehendĕrunt,* Jn. Bos. 1, 5. Hí genómon unlytel *they took not a little,* Chr. 921; Erl. 106, 14. Ðú ðé ânne genim to gesprecan *take thou one to thee for counsellor,* Exon. 80 a; Th. 301, 25; Fä. 24: Cd. 67; Th. 80, 27; Gen. 1335. Genimaþ eów árlíce lác *tollĭte hostias,* Ps. Th. 95, 8. Búton hwá þurh flánes flyht fyl genáme *unless any one through an*

arrow's flight obtained his fall, Byrht. Th. 133, 57; By. 71. Hēt se kásere ðæt he genáme on ðam biscope ealle godes béc *the emperor ordered him to take from the bishop all God's books,* Shrn. 123, 24. Ân byþ genumen *ūnus assūmētur,* Mt. Bos. 24, 40, 41: Gen. 2, 23. Geniman friþ *to make peace,* Chr. 865; Erl. 71, 12: Ors. 5, 7; Bos. 106, 21.

ge-nioman *to take, receive, obtain;* sūmĕre, nancisci:—Ðǽr gé to genihte geniomaþ wæstme *where ye shall obtain fruits in abundance,* Ps. Th. 67, 16. v. ge-niman.

ge-nip, es; *pl. nom. acc.* -nipu; *n. A mist, cloud, darkness, obscurity;* nēbŭla, cālīgo, nūbes, tĕnebræ:—Mist *vel* genip *nĕbŭla,* Ælfc. Gl. 94; Som. 75, 111; Wrt. Voc. 52, 61. Wearþ genip, and ofersceadede híg *facta est nūbes, et obumbrāvit eos,* Lk. Bos. 9, 34. Ðæt genip stód æt ðæs geteldes dura *the cloud stood at the door of the tabernacle,* Ex. 33, 10: Cd. 8; Th. 9, 9; Gen. 139. Moises eóde to ðam genipe *Moyses accessit ad cālīgĭnem,* Ex. 20, 21. Com stefen of ðam genipe *vox facta est de nūbe,* Lk. Bos. 9, 35. On ðæt genip *in nūbem,* 9, 34. In ðæt neowle genip *into the deep darkness,* Cd. 223; Th. 292, 25; Sat. 445: 217; Th. 275, 31; Sat. 180: Exon. 93 b; Th. 351, 12; Sch. 79. Ofer flóda genipu *over the mists of floods,* Beo. Th. 5608; B. 2808: 2724; B. 1360. Ðú ðe gesetst genipu upastínesse ðínne oððe ðínne upstíge *qui pōnis nūbem ascensum tuum,* Ps. Lamb. 103, 3: Ps. Spl. 77, 27. Sweart wolcen and genip *atra nubes,* Nar. 23, 23. [Cf. *Ger.* nebel: *Icel.* nifl.]

ge-nípan; *p.* -náp, *pl.* -nipon; *pp.* -nipen. I. *to darken, become dark;* cālīgāre, obnūbĭlāri:—Hú seó þrag gewát, genáp under niht-helm, swá heó nó wǽre *how the time has passed, has darkened under the veil of night, as if it had not been,* Exon. 77 b; Th. 292, 8; Wand. 96. II. *to rise as a cloud, to creep up* or *come suddenly upon one;* obrēpĕre, sŭpervĕnīre alĭcui:—Him ongén genáp atol ýþa gewealc *the terrible rolling of the waves rose as a cloud against them* [*came suddenly upon them*], Cd. 166; Th. 206, 20; Exod. 454.

ge-nirwed *vexed.* v. ge-nyrwian.

ge-niðerian, -niðrian, -neðerian, -nyðerian; *p.* ode, ade; *pp.* od, ad *To put down, bring low, subdue, humiliate, condemn:*—Nelle gé genyðerian and gé ne beóþ genyðerude *nolite condemnare et non condemnabimini,* Lk. Bos. 6, 37. Ne ic ðech geniðro *nec ego te condemnabo,* Jn. Skt. Lind. 8, 11. Eágan ofermódra ðú genyðeræst *oculos superborum humiliabis,* Ps. Spl. 17, 29. Útan cumene men eów genyðriaþ *strangers shall bring you low,* Deut. 28, 43. Ðú genyðerodest *tu humiliasti,* Ps. Spl. 88, 11. He ðurh his ðrowunga deófles ríce geneðerode *he through his passion put down the devil's kingdom,* Blickl. Homl. 7, 13. Alle geniðradon hine *omnes condemnaverunt eum,* Mk. Skt. Lind. 14, 64. On Godes dóme geniðerod *condemned at God's judgment,* Homl. Th. i. 60, 33. Geniðrad *damnatus,* Mt. Kmbl. Lind. 27, 3: Mk. Skt. Lind. 16, 16. Se ðe hyne upahefþ se byþ genyðerud *qui se exaltaverit humiliabitur,* Mt. Bos. 23, 12. Simon ne aríseþ nǽfre forðon ðe he is sóðlíce deád and on écum wítum genyðerod *Simon will never arise for he is really dead and sunk in eternal punishments,* Blickl. Homl. 189, 20; Judth. 10; Thw. 23, 9; Jud. 113. Ðurh Cristes sige ealle hâlige wǽron gefreólsode; swá ðonne beóþ ða synfullan genyðerade mid heora ordfruman swá he genyðerad wearþ *through Christ's victory all holy people were set free; so then the sinful shall be subdued with their chief as he was subdued,* Blickl. Homl. 33, 1: Chr. 1075; Erl. 214, 17.

ge-niðerung, -nyðerung, e; *f. Condemnation, humiliation, laying low:*—Ða ýttran ðeóstru is seó swearte niht ðære écan geniðerunge *the outer darkness is the black night of eternal condemnation,* Homl. Th. i. 530, 23. Ðæt he onfó ðære écan genyðerunga *that he receive the everlasting condemnation,* Blickl. Homl. 61, 32. For deófles genyðerunge *for the casting down of the devil,* 67, 3.

ge-níðla, an; *m. An enemy:*—Nǽfre ðú gelǽrest ðæt ic dumbum and deáfum deófolgieldum gǽste geníðlum gafol onhâte *never shalt thou induce me to promise tribute to dumb and deaf idols, foes to the spirit,* Exon. 68 a; Th. 251, 26; Jul. 151. DER. eald-, feorh-, gǽst-, láþ-, mán-, sweord-, torn-geníðla.

ge-níðle, an; *f.* [*or a, an; m?*] *Enmity, hate, fierceness:*—Fram hungres geníðlan *from the fierceness of hunger,* Elen. Kmbl. 1398; El. 701: 1216; El. 610. Ic onféng feonda geníðlan *I received the hate of foes,* Exon. 29 a; Th. 88, 15; Cri. 1440.

ge-niwian; *p.* ode; *pp.* od, ad *To renew, make new, change;* renovare, innovare:—Gǽst rihtne geniwa *spiritum rectum innova,* Ps. Spl. 50, 11. Biþ geniwod *renovabitur,* 102, 5. On sumum geáre byþ se mōna twelf síðon geniwod fram ðære hâlgan Eáster-tíde óþ eft Eástron; and on sumum geáre he biþ þreóttyne síðon geedniwad *in some years the moon is twelve times changed* [*renewed*] *from the holy Easter time till Easter again; and in some years it is thirteen times changed* [*renewed*], Lchdm. iii. 248, 22. Heáf wæs geniwad *the wail was renewed,* Cd. 144; Th. 179, 28; Exod. 35: Exon. 15 b; Th. 33, 22; Cri. 529: 60 a; Th. 217, 13; Ph. 279: Andr. Kmbl. 2020; An. 1012. v. niwian.

ge-niwung, e; *f. A renewing, recovering;* renovatio, Som.

gén-lád, e; *f. An arm of the sea, into which a river discharges itself;* brachium oceani, Som. v. lád.

gennelung, e; _f. Greatness_; magnificentia, Ps. Spl. 67, 37. v. ge-miclung [?]

ge-nóg, -nôh; _adj._ ENOUGH, _sufficient, abundant_; satis, sufficiens, abundans :—He hæfþ on his ágenum genôh _he has of his own enough_, Bt. 24, 4; Fox 86, 8. Ðǽr wæs genôg drinc sôna gearu _there was soon drink enough ready_, Andr. Kmbl. 3067; An. 1536. Hwæt druge ðú dugeða genôhra _what madest thou of the abundant blessings_, Cd. 42; Th. 55, 3; Gen. 888. Hí mágon geseón on him selfum synne genôge _they may see in themselves sins enough_, Exon. 26 a; Th. 77, 32; Cri. 1265. Ðú hæfst ælces gôdes genôh _thou shalt have abundance of every good thing_, Deut. 28, 11: Exon. 93 b; Th. 352, 8; Sch. 94: Cd. 29; Th. 39, 4; Gen. 619. [_Orm._ Laym. inoh: _Plat._ nog, genog: _O. Sax._ ginôg: _O. Frs._ enoch, anog, noch: _Dut._ genoeg: _Ger._ genug: _M. H. Ger._ genuoc, gnuoc: _O. H. Ger._ ginôg: _Goth._ ganôhs: _Dan._ nok: _O. Nrs._ gnogr.]

GE-NÓG, -nôh; _adv. Sufficiently, abundantly_, ENOUGH; satis, abunde :— Genôg sweotol hit is _it is sufficiently manifest_, Bt. 36, 3; Fox 176, 27. Genôg riht ðú segst _rightly enough thou sayest_, Bt. 33, 1; Fox 120, 17. Ðæt híg habbon líf and habbon genôh _ut vitam habeant et abundantius habeant_, Jn. Bos. 10, 10. Cwædon ðæt we fundon sumne swíðe micelne mere in ðæm wǽre fersc wæter and swête genôg _dixerunt ingens nos stagnum dulcissime aque inventuros_, Nar. 11, 27.

ge-nógan _to multiply_; multiplicare, Lye. [_O. H. Ger._ gi-nuogan.]

ge-nôh; _adj. Sufficient, abundant;_ abundans. v. ge-nóg.

ge-nôh _sufficiently_, Bt. 13; Fox 38, 22. v. ge-nóg; _adv._

ge-nom, pl. -nômon _took_ :—Weard genom _the guardian took_, Exon. 11 a; Th. 14, 22; Cri. 223: Chr. 921; Erl. 106, 14; _p. of_ ge-niman.

ge-nomian, -namian; _p._ ode; _pp._ od _To name, point out_; nominare, indicere, Exon. 24 a; Th. 68, 10; Cri. 1101.

ge-notian; _p._ ode; _pp._ od, ud _To use, consume_ :—Hie hæfdon hiora mete genotudne _they had consumed their provisions_, Chr. 894; Erl. 90, 31. [_Cf._ ge-nyttian.]

Gent, Gænt, Gend _Ghent, in Flanders;_ Gandavum, Chr. 880; Erl. 83, 2.

ge-nugan; _pres._ hit -neah [_Goth._ ganah] _To suffice, to be sufficient, not to be wanting;_ sufficere :—Gif us on ferðe geneah _if in our soul we be not wanting [if it is sufficient to us in our soul]_, Exon. 93 a; Th. 348, 29; Sch. 35: 90 a; Th. 337, 26; Gn. Ex. 70. Nǽnig mennisc tunge ne geneah ðæs acendan engles godcund mǽgen to gesecgenne _no human tongue is sufficient to tell the divine virtue of that begotten messenger_, Blickl. Homl. 165, 5. v. be-nugan, nugan.

ge-numen _taken_, Mt. Bos. 24, 40, 41; _pp. of_ ge-niman.

ge-nycled _knuckled, crooked;_ obuncus, Som.

ge-nýdan, -nêdan, -niédan, he -nýt; _p._ de; _pp._ ed _To compel, force, press;_ cogere, compellere, expellere :—Alexander ðæt folc to him genýdde _Alexander forced the people to him_, Ors. 3, 9; Bos. 65, 18, 19, 20. Genýddon, Mk. Bos. 15, 21. Genýt, Mt. Bos. 5, 41. Gást hine on wêsten genýdde _spiritus expulit eum in deserto_, Mk. Bos. 1, 12. Wǽron genýdde _were forced_, Ors. 3, 6; Bos. 58, 21. v. ge-nêdan.

ge-nýd-magas; _pl. m. Near relations_ :—Gif twegen genýdmagas _if two near relations_, L. E. and G. 4; Th. i. 168, 19, MS. B. v. nýdmaga.

ge-nýh; _adj. Near_ :—Gif twegen genýhe magas [genýhe-magas, Th. cf. neáh-mæg] _if two near kinsmen_, L. E. and G. 4; Th. i. 168, 19.

ge-nyht, es; _n_ : e; _f._ [_O. H. Ger._ ganuht,] _An abundance, plenty, sufficiency, fulfilment;_ abundantia, ubertas :—Ðeáh mon nú anweald and genyht to twæm þingum nemne _though any one call power and abundance two things_, Bt. 33, 1; Fox 120, 20. Ðætte genyht wǽre gesælða _that sufficiency was happiness_, 35, 3; Fox 158, 13. v. ge-niht.

ge-nyht-ful, -full; _adj. Plentiful;_ profusus, prodigus, Lye.

ge-nyhtlíce; _adv. Abundantly;_ abunde, Cot. 6.

ge-nyhtsum; _adj. Plentiful, abundant;_ abundans, uber, copiosus :— Feoh genyhtsum sældun ðǽm kempum _they gave much money to the soldiers_, Mt. Kmbl. Rush. 28, 12. v. ge-nihtsum.

ge-nyhtsumian, -nihtsumian; _p._ ode; _pp._ od _To suffice, abound;_ abundare :—Gemǽru and ðene genyhtsumiaþ hwǽte _convalles abundabunt frumento_, Ps. Surt. 64, 14. Genyhtsumegende _abundantes_, Ps. Surt. 72, 12. v. ge-nihtsumian.

ge-nyhtsumlíce; _adv. Abundantly, plentifully;_ abunde, abundanter :— Ða genyhtsumlíce dóeþ oferhygd _qui abundanter faciunt superbiam_, Ps. Surt. 30, 24. v. ge-nihtsumlíce.

ge-nyhtsum-nes, -ness, -nis, -niss, e; _f. An abundance, plenty;_ abundantia :—In mínre genyhtsumnisse _in mea abundantia_, Ps. Surt. 29, 7: 64, 12. v. ge-nihtsumnes.

ge-nyman _to take;_ assúmëre :—Ðú genymest gecýðnysse mîne þurh múþ ðínne _tu assûmis testâmentum meum per os tuum_, Ps. Spl. 49, 17. v. ge-niman.

ge-nyrwian, -nyrwan; _p._ ede, ode; _pp._ ed, od _To make narrow, compress, oppress_ :—Ic genyrwige _co-arto_, Ælfc. Gr. 47; Som. 48, 56. Ðíne fýnd ðé genyrwaþ _inimici tui coangustabunt te_, Lk. Bos. 19, 43. Ne genyrwe ofer me pyt múþ his _neque urgeat super me puteus os suum_, Ps. Spl. 68, 19. Genyrwyd [C], geniered [T] is ofer me gást mín _anxiatus est super me spiritus meus_, 142, 4. Swá genyrwod _so narrowed_, Btwk. Scrd. 21, 5. Hearde genyrwad _hardly constrained_, Exon. 13 a; Th. 23, 6; Cri. 364.

ge-nýt _compels._ v. ge-nýdan.

ge-nýðerian, -nyðerian; _p._ ode; _pp._ od, ad, ud _To humble, condemn_, Ps. Spl. 17, 29: Lk. 6, 37. v. ge-niðerian.

ge-nýðerung _humiliation, condemnation._ v. ge-niðerung.

ge-nyttian; _p._ ode; _pp._ od _To use, enjoy_ :—He hæfde eorþ-scrafa ende genyttod _he had enjoyed the last of his earth-dens_, Beo. Th. 6085; B. 3046. [_Cf._ ge-notian.]

GEÓ, gió; _adv. Formerly, of old, before_; quandam, olim, pridem :— Ða lióþ ðe ic, wrecca, geó lustbǽrlíce song, ic sceal nú heófiende singan _the lays which I, an exile, formerly with delight sung, I shall now mourning sing_, Bt. 2; Fox 4, 7: Bt. Met. Fox 10, 68; Met. 10, 34. Ðú wið Criste geó wunne _thou of old didst strive against Christ_, Exon. 71 b; Th. 267, 25; Jul. 420: 19 b; Th. 51, 11; Cri. 814: Cd. 106; Th. 139, 12; Gen. 2308: Menol. Fox 34; Men. 17. Wæs ðis eálond geó gewurþad mid ædelestum ceastrum _this island was formerly adorned with the noblest cities_, Bd. 1, 1; S. 473, 25. Geþenc se snottra fengel hwæt wit geó sprǽcon _do thou, sagacious prince, bear in mind what we have before spoken_, Beo. Th. 2957; B. 1476. Geó ǽr _long before_, Bd. 4, 19; S. 589, 17. Geó dagum _in days of old, formerly_, 4, 27; S. 605, note 2. Geó geára _formerly_, Bt. 31, 1; Fox 112, 15. Geó hwílum _in times of old, formerly_, 2; Fox 4, 9. [_Goth._ ju: _O. Sax._ giu: _O. H. Ger._ giu.]

geoc, gioc, geoht, gôc, ioc, es; _n_ : _pl._ geocu. I. _a_ YOKE; jugum :— Nimaþ mín geoc ofer eów _tollite jugum meum super vos_, Mt. Bos. 11, 29. Mín geoc ys wynsum _jugum meum suave est_, 11, 30. We weorpan fram us geoc heora _projiciamus a nobis jugum ipsorum_, Ps. Spl. 2, 3. Utan aweorpan heora geocu of us _projiciamus a nobis juga ipsorum_, Ps. Th. 2, 3. II. _a yoke of oxen_; boum jugum, boves jugo juncti :—Se ceorl hæfþ ôðres geoht [geoc: B. oxan] ahýrod _the ceorl has hired another's yoke_, L. In. 60; Th. i. 140, 8. Be hýr-geohte [hyr-geoce: B. hýr-oxan] _of a hired yoke_, 60; Th. i. 140, 7. III. _conjux_ :—Gebede † geoc _conjugem_, Mt. Kmbl. Lind. 1, 20. [_Goth._ juk: _O. H. Ger._ joh: _M. H. Ger._ joch.]

geóc, gióc, eóc, e; _f. Safety, help, aid, succour, comfort, consolation_; salus, auxilium, subsidium, consôlatio :—Mec geóc cyme _safety shall come to me_, Exon. 102 b; Th. 388, 9; Rä. 6, 5: Andr. Kmbl. 3618; An. 1587. Geóce gefégon _they rejoiced in the aid_, Exon. 43 b; Th. 146, 16; Gú. 710. Ne miht ðú me ofer ðisne dæg ænige helpe ne geóce gefremman _non mihi aliquid utilitatis aut salutis potes ultra conferre_, Bd. 5, 13; S. 632, 30. Nú we cunnon hyhtan ðæt we heofones leóht uppe mid englum ágan môton, gástum to geóce _now we can hope that we may possess the light of heaven above with the angels, for the comfort of our spirits_, Frag. Kmbl. 88; Leás. 46: Elen. Kmbl. 2491; El. 1247. Gnyrna to geóce _for a consolation of sorrows_, 2275; El. 1139. Se hálga his God geóce bæd _the holy one prayed to his God for aid_, Andr. Kmbl. 2060; An. 1032: 2132; An. 1569. Ðæt him gástbona geóce gefremede _that the spirit-slayer would afford them succour_, Beo. Th. 357; B. 177: 5342; B. 2674: Cd. 77; Th. 95, 31; Gen. 1587: 184; Th. 230, 14: Dan. 233. Beóþ hyra geóca gemyndge _they are mindful of their safety_, Exon. 33 b; Th. 107, 18; Gú. 60.

geocboga, an; _m. A yoke._ v. geoc.

geócend, es; _m. A preserver, Saviour;_ servator, salvator :—Wís biþ se ðe con ongytan ðone geócend _he is wise who can understand the preserver_, Exon. 54 a; Th. 191, 14; Az. 88. Gǽsta geócend _Saviour of souls_, 10 b; Th. 13, 5; Cri. 198: 49 a; Th. 170, 3; Gú. 1106: Andr. Kmbl. 1095; An. 548: 1801; An. 903: Elen. Kmbl. 1360; El. 682: 2151; El. 1077.

geócian; _p._ ode; _pp._ od; _gen. dat. To preserve, save;_ servare, salvare. I. _with the gen_ :—Geóca úser _preserve us_, Cd. 188; Th. 234, 14; Dan. 292. Geóca mínes gǽstes _save my soul_, Exon. 118 b; Th. 455, 5; Hy. 4, 45. II. _with the dat_ :—Geóca us _preserve us_, Exon. 53 a; Th. 185, 23; Az. 12. Geóca mínre sáwle _save my soul_, 118 b; Th. 455, 34; Hy. 4, 59.

geócor [or geocor? cf. geocsa]; _adj. Strong, fierce, harsh, dire, sad_ :— Geócor sefa, geómrende hyge _sad spirit, mourning mind_, Exon. 164, 33; Gú. 1021: 49 a; Th. 170, 13; Gú. 1111. On ða geócran tíd _in that grievous time_, 47 a; Th. 160, 26; Gú. 949. Hý sceolon forgietan ðara geócran gesceafte habban him gomen _they shall forget the harsh fate and have pleasure_, 92 a; Th. 345, 4; Gn. Ex. 183. Wiste his fingra geweald on grames grápum ðæt he wæs geócor he [Grendel] _knew that his fingers' power was in the gripe of the fierce one, so that he was sad_, Beo. Th. 1535. v. B. 765 for a different reading. Geócrostne síþ _a very sad journey_, Cd. 205; Th. 254, 25; Dan. 617. [_Cf. Goth._ juka _strife, anger._] v. Grm. And. u. El. 119.

geócre; _adv. Harshly, roughly_ :—Ða Babilone weard yrre andswarode corlum onmǽlde grimme ðǽm gingum and geócre oncwæþ _then the lord of Babylon angrily answered to the men, announced fiercely to the youths, and harshly spoke_, Cd. 183; Th. 229, 3; Dan. 211.

geocsa, an; _m. A sobbing;_ singultus :—Ðiós siccetung ðes geocsa _this sighing, this sobbing_, Bt. Met. Fox 2, 9; Met. 2, 5.

geoc-stecca, -sticca, an; _m. A bolt of a door, a bar;_ obex, Cot. 145.

geocsung, e; *f. Sobbing;* singultus, Ælfc. Gl. 99; Wrt. Voc. 54, 64.

geofa *a giver.* v. gifa.

geofan; *p.* geaf, *pl.* geáfon; *pp.* gifen *To give;* dare :—Nymþ ðú me ræd geofe *unless thou mayest give me counsel*, Ps. Th. 58, 1 : 118, 72. v. gifan.

geofen *the ocean*, Exon. 89 b; Th. 336, 20; Gn. Ex. 52. Geofenes *of the ocean*, Beo. Th. 729; B. 362. v. geofon.

ge-offrian; *p.* ode; *pp.* od *To offer, sacrifice* :—He hêt hine his leófan sunu geoffrian Gode to láce *he bade him offer his dear son as a sacrifice to God*, Btwk. Scrd. 23, 3. Abel geoffrode ða sélostan lác Gode *Abel offered the best sacrifices to God*, 18, 5 : 22, 9; Gen. 8, 20. Ðæt hí be hreówsunge Gode geoffrodon *that they should sacrifice to God by penitence*, Homl. Th. i. 68, 17. Geoffrod *sacrificed*, Lev. 4, 15.

geofian, *p.* ode; *pp.* od *To give, to endow;* dare, donare :—He mæg me geofian mid góda gehwilcum *he can endow me with every good*, Cd. 26; Th. 34, 31; Gen. 546. DER. geofu. v. gifian.

geofon, geofen, gifen, gyfen, es; *n. The sea, ocean;* mare, oceanus :—On geofones streám *on the ocean's wave*, Andr. Kmbl. 1704; An. 854 : Exon. 57 b; Th. 205, 25; Ph. 118. Geofon geótende *a gushing ocean*, Andr. Kmbl. 3014; An. 1510. [*O. Sax.* gebano.] v. Grm. D. M. 219.

geofon-flód, es; *m.* [geofon *a sea, ocean;* flód *a flood*] *A sea or ocean flood;* maris fluctus :— Dryhtnes bibod geofonflóda gehwylc georne bihealdeþ *each ocean flood strictly observes the Lord's command*, Exon. 54 b; Th. 193, 21; Az. 125.

geofon-hús, es; *n. A sea-house, vessel* :—Geofonhúsa mæst *greatest of sea-houses*, Cd. 66; Th. 79, 34; Gen. 1321.

geofon-ýþ, e; *f. A sea-wave, billow;* maris unda, Beo. Th. 1035; B. 515.

geofu, e; *gen. pl.* -a, -ena, -ona; *f. A gift, grace;* donus, gratia :— Beó geofena gemyndig *be mindful of gifts*, Beo. Th. 2351; B. 1173. Ðæt he dryhtnes mót geofona neótan *that he may partake of the Lord's gifts*, Exon. 61 a; Th. 225, 5; Ph. 384. Ne biddan we úrne Drihten ðyssa eorðlícra geofa *let us not ask our Lord for these earthly gifts*, Blickl. Homl. 21, 11. He hí mid missenlícum geofum gewelgode *ille eam* [*ecclesiam*] *diversis donis ditavit*, Bd. 1, 33; S. 499, 1 : Exon. 18 a; Th. 43, 10; Cri. 686 : 128 b; Th. 493, 15; Rä. 81, 31. Geofu wæs mid Gúþlác *grace was with Guthlac*, 40 a; Th. 134, 1; Gú. 501. v. gifu.

geó-geára; *adv. Of old;* olim, antiquitus, Ps. Th. 42, 3. v. geó, geára; *adv.*

geó-geáre; *adv. Of old;* olim, antiquitus :— Swá swá we geógeáre hýrdon *so as we of old have heard*, Ps. Th. 47, 7. v. geó, geáre.

geógelere, es : *m. A juggler;* præstigiator, Som. Geógulere *magus, haruspex, hariolus*, Hpt. Gl. 500, 502, 510. [*O. H. Ger.* gougulari : *Icel.* kuklari : *Ger.* gaukler.] v. Grff. iv. 134 : Grm. D. M. 990.

geógoþ-feorh *youthful life, youth*, Beo. Th. 1078; B. 537. v. geóguþ-feorh.

geógoþ-hád *youth*, Cd. 74; Th. 91, 4; Gen. 1507. v. geóguþ-hád.

geógoþ-lust, es; *m. Youthful pleasure, lust* :—Se líchoma geunlustaþ ða geógoþlustas to fremmenne *the body loathes to perform those youthful lusts*, Blickl. Homl. 59, 9.

geóguþ, geógeþ, giógoþ, geógaþ, gígoþ, iúguþ, e; *f.*   I. *YOUTH, the state of being young;* juventus, juvenilis ætas *vel* status :—Úre cnihthád is swylce undern-tíd, on ðam astíhþ úre geógoþ swá swá sunne dêþ ymbe ðære ðriddan tíde *our boyhood is as it were the third hour in which arises our youth as the sun does about the third hour*, Homl. Th. ii. 76, 15 : Elen. Kmbl. 2528; El. 1265. Of mínre geóguþe *a juventute mea*, Mk. Bos. 10, 20 : Blickl. Homl. 211, 26. Ðær is geógoþ búton ylde *there is youth without age*, 65, 17 : Exon. 32 a; Th. 101, 6; Cri. 1654. On geóguþe *in youth*, 34 a; Th. 108, 19; Gú. 75 : Ps. Th. 70, 4. Hie on geógoþe bu wlitebeorht wæron on woruld cenned *they both in youth beautiful were born into the world*, Cd. 10; Th. 12, 18; Gen. 187 : Ps. Th. 118, 141. On geógoþe *in youth*, Beo. Th. 4843; B. 2426. From gígoþe mínum *a juventute mea*, Mk. Skt. Lind. 10, 20 : Lk. Skt. Lind. 18, 21. Se férde on his iúguþe fram his freóndum *he went in his youth from his friends*, Swt. A. S. Rdr. 95, 3.   II. *the youth, young persons;* juventus, juvenes :—Eall sió gióguþ ðe nú is on Angelcynne *all the youth now in England*, Past. Pref; Swt. 7, 10; Hat. MS. Ðá wearþ iafeðe geógoþ afêded *then to Japhet was a youthful offspring born*, Cd. 78; Th. 96, 34; Gen. 1604. Óððæt seó geógoþ geweóx *until the youth grew up*, Beo. Th. 133; B. 66. Hyre byre Hrêdríc and Hróðmund and hæleða bearn giógoþ ætgædere *her sons Hrethric and Hrothmund and children of warriors, the youth together*, 2384; B. 1189: Cd. 176; Th. 220, 34; Dan. 81. Helpe gefremman gumena geógoþe *to give help to the young men of the people*, Andr. Kmbl. 3228; An. 1617. Duguþe and geógoþe *to old and young*, 304; An. 152 : Beo. Th. 323; B. 160. Heora geóguþ *juvenes eorum*, Ps. Th. 77, 64. Ic geseah míne gesælinesse and mín wuldor and ða fromnisse mínre iúguþe *ego respiciens felicitatem meam insigni numero juventutis*, Nar. 7, 22. [*O. Sax.* juguð : *O. H. Ger.* jugund : *Ger.* jugend.]

geóguþ-cnósl, es; *n.* [geóguþ *youth;* cnósl *progeny, a family*] *A*

youthful family, young progeny; novella família, libĕri :—Ic bíde ðær mid geóguþcnósle *I abide there with my young progeny*, Exon. 104 b; Th. 396, 25; Rä. 16, 10.

geóguþ-feorh, geógoþ-feorh; *gen.* -feores; *dat.* -feore; *n.* [geóguþ *youth,* feorh *life*] *Youthful life, youth;* juventus :—Sumum ðæt gegongeþ on geóguþfeore, ðæt se endestæf weálíc weorþeþ *it happens to one in youthful life that the end is miserable*, Exon. 87 a; Th. 328, 1; Vy. 10 : Beo. Th. 5321; B. 2664. On geógoþfeore *in youthful life*, 1078; B. 537.

geóguþ-hád, geógoþ-hád, es; *m. The state of youth, youth;* júventútis státus, júventus :—Ðú hafast geóguþhádes blæd *thou hast youth's prosperity*, Exon. 68 a; Th. 252, 25; Jul. 168 : Elen. Kmbl. 2531; El. 1267. Ðú me lærdest of geóguþháde *dŏcuisti me a jŭventúte*, Ps. Th. 70, 16. On geógoþháde *in youth*, Cd. 74; Th. 91, 4; Gen. 1507 : Blickl. Homl. 59, 5 : 211, 22.

geóguþ-hádnes, -ness, e; *f. The state of youth, youth;* ădŏlescentia :—On ða ærestan tíd mínre geóguþhádnesse *cum prímævo ădŏlescentiæ tempŏre*, Bd. 5, 6; S. 618, 36.

geóguþ-líc; *adj. Youthlike, youthful;* jŭvĕnílis :—Ic ne wæs mín mód íullfremedlíce bewerigende ðam geóguþlícum unalýfednessum *non animum perfecte a jŭvĕnílibus cohĭbens inlĕcebris*, Bd. 5, 6; S. 618, 39.

geóguþ-myru, we; *f. Youthful joy?* Exon. 109 b; Th. 419, 23; Rä. 39, 2.

Geóhhel-, geóhhel-dæg, es; *m. Yule-day, a day at Yule-tide* :—On ðone forman dæig on geáre ðæt is on ðone ærestan geóheldæig eall cristen folc wordiaþ cristes acennednesse *on the first day of the year, that is, on the first day of Yule all christian folk honour Christ's birth*, Shrn. 29, 26. On ðone eahteþan geóhheldæig biþ ðæs mónþes fruma ðe mon nemneþ ianuarius *on the eighth day of Yule is the beginning of the month that is called January*, 47, 13.

Geóhol, Geóhhol, es; *n. Yule, Christmas* :—Ðý twelftan dæge ofer geóhol *on the twelfth day after Yule*, Bd. 4, 19; S. 588, 8 : L. Alf. pol. 5; Th. i. 64, 23, note. v. geól.

geoht, es; *n. A yoke*, L. In. 60; Th. i. 140, 8 : 60; Th. i. 140, 7. v. geoc.

geohðu. v. gehðu.

GEÓL, giúl, iúl, geóhol, es; *n. YULE, Christmas;* festum nativitatis Domini :—On geól *at Christmas*, L. Alf. pol. 5; Th. i. 64, 23, note : Menol. Fox 59, note a. Ðý twelftan dæge ofer geóhol *Epiphaniæ*, Bd. 4, 19; S. 588, 8. Feówertig daga ær eástran and feówertig daga ær Cristes acennisse ðæt is ær geólum *fourty days before Easter and fourty days before Christ's birth, that is, before Christmas*, Shrn. 82, 11. [*Dan.* juul: *Swed.* jul, *m.* : *O. Nrs.* jól, *n. pl. festum jolense, festum natalitiorum Christi, festum quodvis, convivium.*] For this and the next word v. Grm. Gesch. D. S. c. vi, and Cl. and Vig. Icel. Dict. jól.

Geóla, Iúla, an; *m.* [geól *Yule*] *The YULE or Christmas month, that is, December* :—Se ærra geóla *the ere*, or *former yule, December.* Se æftera gèola *the after yule, January.* Se mónaþ is nemned on Leden *Decembris*, and on úre geþeóde se ærra geóla, forðan ða mónþas twegen syndon nemde ánum naman, óðer se ærra geóla [*December*], óðer se æftera mensis [*Januarius*] *hic vocatur Latine* December, *nostra vero lingua prior Geola, quoniam duo sunt menses qui uno nomine gaudent, alter Geola prior* [December], *alter posterior* [January], Hick. Thes. i. 212, 56; Shrn. 153, 23-6. [*Goth.* jiuleis, *m.*]

geolca, gioleca, geoloca, an; *m. A YOLK;* ovi vitellus :—Sceáwa nú on ánum æge, hú ðæt hwíte ne biþ gemenged to ðam geolcan, and biþ hwæðere án æg *look now on an egg, how the white is not mingled with the yolk, and yet it is one egg*, Homl. Th. i. 40, 28. On æge biþ gioleca on middan *in an egg the yolk is in the middle*, Bt. Met. Fox 20, 339; Met. 20, 170. Genim geolocan *take the yolk*, L. M. 1, 2; Lchdm. ii. 38, 7.

geold, es; *n. Charge, impost* :—Ne gafle ne geold *neither tax nor charge*, Chr. 675; Erl. 38, 1. Strange geoldes *heavy imposts*, 1124; Erl. 253, 21. v. gield.

ge-óleccan; *v. a. To allure;* blandiri :—Ðá hí ðé mæst geóleccan *when they most allure thee*, Bt. 7, 2; Fox 18, 1.

geole-wearte *a nightingale;* luscinus, Ælfc. Gl. 38; Som. 63, 37; Wrt. Voc. 29, 55.

geolhstor, geolstor, es; *m? Matter, corruption, poison, venom;* sanies :—Hire geolhstor út fleów *the matter flowed out from her*, Bd. 4, 19; S. 589, note 3. Geolster *virus, tabum*, Hpt. Gl. 517, 490.

geolna, an; *m. A kind of Egyptian stork;* ibis, Ælfc. Gl. 38; Som. 63, 30; Wrt. Voc. 29, 49.

geolo, geolu; *gen. m. n.* geolwes; *dat.* geolwum; *def.* se geolwa; *adj.* YELLOW; flavus :—Geolo godwebb *the yellow silk*, Exon. 109 a; Th. 417, 25; Rä. 36, 10. Geolwe linde [*acc. f.*] *yellow shield*, 5213; B. 2610. Him beóþ ða eágan geolwe *his eyes will be yellow*, L. M. 3, 62; Lchdm. ii. 348, 12. Geolo *flavus, fuscus*, Hpt. Gl. 510. Mid geolewere fáhnisse *crocea qualitate*, 419.

geolo-ádl, e; *f. The jaundice*, Lye.

geolo-blác; *adj. Pale yellow*, Lye.

geoloca, an; *m. A yolk;* ovi vitellus :—Genim hænne æges geolocan *take the yolk of a hen's egg*, L. M. 1, 2; Lchdm. ii. 38, 7. v. geolca.

geolo-hwît; *adj. Yellow-white;* mellinus, color stramineus, Lye.

geolo-rand, es; *m. A yellow disk, shield,* Beo. Th. 880; B. 438: Elen. Kmbl. 235; El. 118. v. Grm. A. u. E. 145.

geolo-reáð; *adj. Yellow-red;* croceus, Lye.

geolstrig; *adj. Poisonous;* virulentus, Hpt. Gl. 450, 453. Geolstru? Som.

geolwian; *p.* ode; *pp.* od *To become yellow;* flavescere. DER. a-geolwian.

geó-man, gió-man, -mann, es; *m. A man of old;* qui olim vixit. v. gió-man.

geómeleáslíce *carelessly.* v. gýmeleáslíce.

geómen *care.* v. gýmen.

geómerian *to groan, mourn, murmur,* Boutr. Scrd. 20, 43: Homl. Th. i. 142, 17. v. geómrian.

geómer-mód *sad of mind, sorrowful,* Cd. 40; Th. 53, 9; Gen. 858. v. geómor-mód.

geómerung *a groaning, moaning, lamentation,* Ps. Spl. 6, 6: Homl. Th. i. 142, 18: ii. 86, 16. v. geómrung.

geómian *to take care of.* v. gýman.

geómor, geómur, giómor; *adj. Sad, sorrowful, mournful, murmuring, miserable, wretched;* tristis, mæstus, quĕrŭlus, mĭser :—Hé wæs geómor sefa hís mínd was sad, Elen. Kmbl. 1251; El. 627: Beo. Th. 98; B. 49. He ðǽr ána sæt, geoðum geómor *he sat there alone, sad of mind,* Andr. Kmbl. 2015; An. 1010. Ic of grundum to ðé geómur cleopode *de profundis clāmāvi ad te,* Ps. Th. 129, 1. Ðæt wæs geómuru ides *that was a mournful woman,* Beo. Th. 2155; B. 1075. Ðeós geómre lyft *this murmuring air,* Cd. 163; Th. 205, 4; Exod. 430. Dust ne mæg andsware ǽnige gehátan geómrum gǽste *the dust cannot give any answer to the sad spirit,* Soul Kmbl. 211; Seel. 108: Apstls. Kmbl. 178; Ap. 89. Siððan ðú gehýrde galan geómorne geác on bearwe *when thou hast heard the sad cuckoo sing in the grove,* Exon. 123 b; Th. 473, 29; Bo. 22. He wæg hyge geómurne *he bare a mournful spirit,* 52 a; Th. 182, 15; Gú. 1310. In ðas geómran woruld *in this sad world,* 57 b; Th. 207, 10; Ph. 139: 63 a; Th. 232, 35; Ph. 517. Geómran stefne *with mournful voice,* Andr. Kmbl. 122; An. 61: 2254; An. 1128. Geómre gástas *sad spirits,* Cd. 4; Th. 5, 9; Gen. 69: 166; Th. 206, 5; Exod. 447. Geómrum to geóce *for salvation to the sad,* Exon. 9 b; Th. 8, 27; Cri. 124. [Laym. ʒeomere *doleful, miserable:* O. Sax. jámar *depressed, sad, sorrowful:* Dut. jammer, *n.* misery: Ger. jammer, *m.* misery: M. H. Ger. jâmer, âmer, *m.* pity: O. H. Ger. jâmar, âmar, *m.* mĭsĕria.] DER. fela-geómor, hyge-, mód-, síþ-, wine-.

geómor-fród; *adj.* [geómor *sad,* fród *old*] *Old with sadness;* mĭsĕre ætate provectus :—Ic eom geómorfród *I am old with sadness,* Cd. 101; Th. 134, 14; Gen. 2224.

geómor-gid, -gidd, -gyd, es; *n. A mournful song, dirge, lamentation;* lūgubris cantus, nēnia, lāmentātio :—Wæs geómorgidd wrecen *a mournful song was sung,* Andr. Kmbl. 3094; An. 1550. Geómorgyd, Beo. Th. 6291; B. 3150.

geómor-líc; *adj. Sad, sorrowful;* mæstus, flēbilis :—Biþ geómorlíc gomelum eorle to gebídanne, ðæt his byre ríde giong on galgan *it is sad for an aged man to experience that his child hang young on the gallows,* Beo. Th. 4879; B. 2444: Ors. 4, 5; Bos. 81, 31. [O. Sax. jámarlīk: O. H. Ger. jámarlîh: Ger. jämmerlîch.]

geómor-líce; *adv. Sadly;* · lūgubre :—He gilleþ geómorlíce *he yelleth sadly,* Salm. Kmbl. 535; Sal. 267.

geómor-mód, geómer-mód, giómor-mód; *adj. Sad of mind, sorrowful;* mæstus anĭmo :—Ongan geómormód to Gode cleopian *he sad of mind began to cry to God,* Andr. Kmbl. 2795; An. 1400: Beo. Th. 4094; B. 2044: Gen. 27, 34. Hie engel Drihtnes gemitte geómormóde *an angel of the Lord met her sad in mood,* Cd. 103; Th. 137, 3; Gen. 2268. Heó wǽron geómormóde *they were sorrowful,* Elen. Kmbl. 1107; El. 555: 825; El. 413. Gewitan him gangan, geómermóde *they retired, sad of mind,* Cd. 40; Th. 53, 9; Gen. 858. [O. Sax. jámar-mód.]

geómrian, geómerian, geómran; *part.* geómrigende, geómerigende, geómrende; *p.* ode; *pp.* od [geómor *sad,* sorrowful] *To be sad, to sigh, groan, murmur, mourn, sorrow, lament, bewail;* gĕmĕre, murmŭrāre, ingĕmĕre, ingĕmiscĕre, lūgĕre, quĕri :—Se ðe á wile geómrian on gihða *who for ever will mourn in spirit,* Salm. Kmbl. 701; Sal. 350. Béna geómrigende we asendaþ *prĕces gĕmentes fundīmus,* Hymn. Surt. 21, 13. Gáþ geómriende weras wíf samod *men and women together go sorrowing,* Andr. Kmbl. 3328; An. 1667: Bd. 1, 27; S. 497, 35: Gen. 42, 38: Mk. Bos. 5, 38: 8, 12. Geómerigende *mourning,* Boutr. Scrd. 20, 42. Gé, geómrende, gehðum mǽnaþ ye, *murmuring, grieve in spirit,* Exon. 9 a: Th. 6, 26; Cri. 90: 48 a; Th. 164, 34; Gú. 1021. Ic geómrige *gĕmo,* Ælfc. Gr. 28, 3; Som. 30, 58. Hí murcniaþ oððe geómriaþ *murmŭrābunt,* Ps. Spl. 58, 17. Hí geómeriaþ *they murmur,* Homl. Th. i. 142, 17. Ides geómrode giddum *the lady bewailed in songs,* Beo. Th. 2240; B. 1118. On ðone heofon behealdende, geómrode *sus-pĭciens in cœlum,* Mk. Bos. 7, 34: Jn. Bos. 11, 33, 38. Ne geómra ðú *be not thou sad,* Cd. 100; Th. 132, 25; Gen. 2198.

geómrung, geómerung, e; *f. A groaning, moaning, lamentation;* gĕmĭtus, lāmentum :—Brytta geómrung *gĕmĭtus Brittanŏrum,* Bd. 1, 13;

S. 481, 42. Fram geómrunga heortan mínre *a gĕmĭtu cordis mei,* Ps. Spl. 37, 8. On geómerunga mínre *in gĕmĭtu meo,* 6, 6: Bd. 5, 6; S. 619, 14. Hí getácniaþ háligra manna geómerunge *they betoken the groaning of holy men,* Homl. Th. i. 142, 18. Deáþes geómerunga me beeódon *the moanings of death surrounded me,* ii. 86, 16. On geómrungum *in gĕmĭtĭbus,* Ps. Spl. 30, 12. On geómrunga *in lamentation,* Blickl. Homl. 89, 14. For ðære geómrunga ðæs óðres deáðes *for sorrow at the other's death,* 113, 11.

geómur *sad, sorrowful,* Ps. Th. 129, 1. v. geómor.

geóna; *adv. Hitherto, yet;* adhuc :—Hwædd geóna me gwona is *quod adhuc mihi deest?* Mt. Kmbl. Lind. 19, 20. Geóna hlifigende *adhuc vivens,* 27, 63. Ða geóna [geone, Lind.] *athuc,* Jn. Skt. Rush. 11, 30: Mk. Skt. Rush. 5, 35. Ne ða geóna *nondum,* Jn. Skt. Rush. 7, 39: 8, 57. v. géna.

ge-onbyrdan; *p.* de; *pp.* ed *To bear against, strive against, resist :—* Gif he on ðone geonbyrde ðe hine slóg *if he strove against him who slew him,* L. In. 76; Th. i. 150, 18: L. E. G. 6; Th. i. 170, 13: L. Eth. v. 31; Th. i. 312, 11. v. ge-anbyrdan.

geonc *young,* Bt. 8; Rawl. 15, 13, note m. v. geong.

geond, giond; *prep. acc. Through, throughout, over, as far as, among, in, after, beyond;* per, trans, inter, post, ultra; kată :—He gǽd geond drige stówa *ambulat per loca arida,* Mt. Bos. 12, 43: 14, 35. Ða eóde geond Hiericho *tum perambulabat Jericho,* Lk. Bos. 19, 1. Beóþ mycele eorþan styrunga geond stówa *terræ motus magni erunt per loca,* 21, 11. Geond eorþan *throughout the earth,* Beo. Th. 538; B. 266: Cd. 227; Th. 305, 10; Sat. 644. Geond gehwilce wera *viritim,* Ælfc. Gr. 38; Som. 41, 5. Hí ealle beweópon Aarones forðsíþ geond ðrítig daga *they all mourned Aaron's death during thirty days,* Num. 20, 29. Ðé we þanciaþ geond ungeendode worulde *we will thank thee to all eternity,* Homl. Th. i. 76, 7. Geond to dæg *usque hodie,* Bd. 1, 1; S. 474, 28. Ðǽr se hálga stenc wunaþ geond wynlond *there a holy fragrance rests over the pleasant land,* Exon. 57 a; Th. 203, 10; Ph. 82. Geond sídne grund *over the wide abyss,* Cd. 6; Th. 8, 35; Gen. 134. Ðú geond holt wunast *thou shalt dwell among the groves,* Cd. 203; Th. 252, 6; Dan. 574. Geond ða þeóda *among the people,* Andr. Kmbl. 49; An. 25. Módes snyttru seów and sette geond sefan monna *he sowed and set the wisdom of mind in the minds of men,* Exon. 17 b; Th. 41, 30; Cri. 663. Mán wridode geond beorna breóst *wickedness blossomed in the breast of men,* Andr. Kmbl. 1535; An. 769. Geond feówertig daga *post quadra-ginta dies,* Num. 13, 22. Fæder folca gehwæs us féran hét geond ginne grund *the father of every nation bids us depart beyond the abysmal deep,* Andr. Kmbl. 661; An. 331. Sittaþ yfele men giond eorþrícu *wicked men sit in earthly kingdoms,* Bt. Met. Fox 4, 74; Met. 4, 37. Giond ðas wídan worulde *through this wide world,* 11, 89; Met. 11, 45. [Laym. ʒond per.] v. geondan, be-geondan.

geond; *adv. Yond, yonder, thither, beyond;* illuc :—Hider and geond *hither and thither,* Lye. Hyder geond *yonder,* Mt. Bos. 26, 36. [Chauc. yond: Goth. jaind *there.* Cf. Orm. ʒond *in* o ʒond half.]

geondan; *prep. acc. Beyond;* trans :—Ða sóne com Willelm eorl fram geondan sǽ *then earl William soon came from beyond sea,* Chr. 1052; Erl. 181, 29: 1048; Erl. 177, 28. v. geond, be-geondan.

geond-brǽdan; *p.* -brǽdde; *pp.* -brǽded [geond *over,* brǽdan *to spread*] *To overspread;* supersternĕre :—Hit geondbrǽded wearþ beddum and bolstrum *it was overspread with beds and bolsters,* Beo. Th. 2483; B. 1229.

ge-ondbyrde *strove against, resisted,* L. C. S. 49; Th. i. 404, 13, note 30. v. ge-onbyrdan.

ge-ondettan; *p.* te; *pp.* ed *To confess;* confĭtēri :— Ðe geondettaþ *that confess,* Blickl. Homl. 57, 27. Gif he hit geondette *if he confess it,* L. In. 71; Th. i. 148, 3. v. ge-andettan.

geond-faran; *part.* -farende; *p.* -fór, *pl.* -fóron; *pp.* -faren [geond *through,* faran *to go*] *To go through, pervade;* perambŭlāre, pervāgāri :— He langre tíde ealle heora mǽgþe mid gewēde wæs geonfarende *multo tempŏre totas eorum provincias debacchando pervăgātus,* Bd. 2, 20; S. 521, 27. Fram mangunge geondfarendre on þýstrum *a negotio perambŭlante in tenebris,* Ps. Lamb. 90, 6. Wæter wynsumu bearo ealne geondfaraþ *pleasant waters pervade all the grove,* Exon. 56 b; Th. 202, 10; Ph. 67.

geond-felan, -feolan; *p.* -fæl, *pl.* -fǽlon; *pp.* -folen [cf. (?) Goth. filhan: Icel. fela *to hide;* hence *to give into one's keeping;* so geondfolen fýre *utterly given up to fire.* Or may folen be taken from the literal meaning and so geondfolen compare with the preceding participle geinnod? The meaning of the verb in any case seems to be] *To fill throughout;* mplere, Cd. 2; Th. 3, 29; Gen. 43.

geond-féran; *p.* -férde; *pp.* -féred [geond *through,* féran *to go*] *To go through, traverse;* pertransīre, peragrāre :—Ne móstan ðé geondféran foldbúende *earth's inhabitants may not traverse thee,* Exon. 121 a; Th. 465, 11; Hö. 101. Gewunede he swýðost ða stówe geondféran, and in ðám túnum godcunde lǽre bodian, ða ðe in heágum mórum and in hréðum feor gesette wǽron *solēbat autem ea maxĭme lŏca peragrāre, illis prædĭcāre in vicŭlis, qui in arduis aspĕrisque montĭbus procul posĭti,*

Bd. 4, 27; S. 604, 26. Ic geondférde fela fremdra londa *I traversed many foreign lands*, Exon. 85 b; Th. 321, 22; Wíd. 50: 84 b; Th. 318, 23; Wíd. 3.

geond-flówan; *p.* -fleów, *pl.* -fleówon; *pp.* -flówen *To flow through*; pertransfluere :—Nales ðú geondflówan foldbüende *thou flowest not through earth's inhabitants*, Exon. 121 a; Th. 465, 16; Hö. 105. v. flówan.

geond-folen *filled throughout*, Cd. 2; Th. 3, 29; Gen. 43. v. geond-felan.

geond-geótan; *p.* -geát, *pl.* -guton; *pp.* -goten *To pour, pour out*; perfundére :—Ic geondgeóte *perfundo*, Ælfc. Gr. 28, 6; Som. 32, 33. Heó mid wópe and mid teárum wæs swýðe geondgoten *flētuque ac lacrȳmis perfūsa*, Bd. 4, 23; S. 596, 10.

geond-hweorfan; *p.* -hwearf; *pp.* -hworfen *To turn or pass through, go about, traverse*; pertransire, peragrare, perlustrare :— Ðonne maga gemynd mód geondhweorfeþ *when remembrance of friends passes through his mind*, Exon. 77 a; Th. 289, 21; Wand. 51. Hwílum cwēn flet eall geondhwearf *at times the queen went about all the hall*, Beo. Th. 4039; B. 2017. Ðonan ic ealne geondhwearf ēðel Gotena *thence I traversed all the country of the Goths*, Exon. 86 b; Th. 325, 9; Wíd. 109. Land eal geondhwearf *he travelled over all the land*, Salm. Kmbl. 372; Sal. 185. DER. hweorfan.

geond-hyrdan; *p.* de; *pp.* ed *To harden thoroughly*, Salm. Kmbl. 150, 28.

geond-innan; *prep. acc. Throughout*; per :—Geond woruld innan *throughout the world*, Exon. 14 b; Th. 29, 28; Cri. 469. Geond Bryten innan *throughout Britain*, 45 b; Th. 155, 5; Gū. 855 : 95 b; Th. 355, 43; Pa. 4.

geond-lácan; *p.* -léc; *pp.* -lácen *To go through or over, flow over*; pertransíre, transfluére :—Ðætte ðæt tírfæste lond geondláce laguflóda wynn *that the joy of water-floods sports over the glorious land*, Exon. 56 b; Th. 202, 15; Ph. 70.

geond-leccan; *part.* -leccende; *p.* -lehte; *pp.* -leht *To wet through, moisten, water*; rigáre :—Geondleccende muntas of heora uferum dǽlum *rigans montes de supēriōribus suis*, Ps. Lamb. 103, 13.

geond-líhtan; *p.* -líhte; *pp.* -líhted; *v. a.* [lýhtan, líhtan *to shine*] *To enlighten*; illūmināre :—Sunne endemes ne mæg ealle [gesceafta] geondlíhtan innan and ūtan *the sun cannot equally enlighten all [creatures] within and without*, Bt. Met. Fox 30, 24; Met. 30, 12.

geond-mengan; *p.* de; *pp.* ed [mengan *to mingle*] *To mingle, confuse*; perturbáre :— Mec ðæs full oft fyrwit frineþ, mód geondmengeþ *about this my curiosity full oft enquireth, it confuses my mind*, Salm. Kmbl. 119, MS. B; Sal. 59.

geond-sáwan; *p.* -seów, *pl.* -seówon; *pp.* -sáwen *To sow, scatter, spread abroad*; serere, spargere, disseminare :—Deáw-driás winde geondsáwen *the dew-fall is scattered by the wind*, Cd. 188; Th. 233, 19; Dan. 278. DER. sáwan.

geond-sceáwian; *p.* ode; *pp.* od [sceáwian *to look*] *To look at, survey*; perlustráre ocŭlis :— Georne geondsceáwaþ *earnestly surveys*, Exon. 77 a; Th. 289, 23; Wand. 52. Geondsceáwode he ða þing ðe to ðære stówe belumpon *he looked about at the things which appertained to the place*, Guthl. 3; Gdwin. 22, 17.

geond-scínan, -scán; *p.* -scínen *To shine upon, illuminate*; collustrare, illuminare :—Hit seó ēce ne mót geondscínan sunne *the eternal sun cannot shine on it*, Bt. Met. Fox 5, 88; Met. 5, 44 : Salm. Kmbl. 678; Sal. 339: Bt. 41, 1; Fox 244, 9. Sió sunne hine geondscínþ *the sun shines upon him*, Bt. 34, 5; Fox 140, 8.

geond-scínþ *shines upon*, Bt. 34, 5; Fox 140, 8; *3rd pres. of* geond-scínan.

geond-sécan; *p.* -sóhte, *pl.* -sóhton; *pp.* -sóht *To search thoroughly, pervade*; pervestigare :—Se gifra gæst grundas geondséceþ *the greedy guest shall pervade earth*, Exon. 22 a; Th. 60, 22; Cri. 973. His intinga wæs geondsóhte *his business was thoroughly searched*, Bd. 5, 19; S. 639, 28. DER. sécan.

geond-sendan; *p.* -sende; *pp.* -sended *To overspread*; perfundere :— Wæs gúþ-hergum wera ēðel-land wíde geondsended *the people's native-land was widely overspread with hostile bands*, Cd. 92; Th. 118, 21; Gen. 1968; Th. 154, 6; Gen. 2551.

geond-seón; *p.* -seáh *To see beyond or through*; perspicere, in conspectu habere, Beo. Th. 6166; B. 3087.

geond-smeágan; *p.* -smeáde; *pp.* -smeád *To search through, examine, discuss*; perscrūtári, discŭtĕre :—Ðæt we geondsmeáge ða dígolnysse úre heortan *that we search through the secrets of our heart*, Bd. 4, 3; Whelc. 266, 43, MSS. B. C.

geond-spǽtan; *p.* -spǽtte; *pp.* -spǽt *To spit or squirt through, syringe through, to squirt water as through a syringe or pipe*; siphonĭbus áquam exprimĕre :—Ðú hie ælce dæge mid pípan geondspǽt *do thou syringe through it every day with a tube*, L. M. 2, 22; Lchdm. ii. 208, 26.

geond-sprengan; *p.* de; *pp.* ed *To sprinkle over*; perspergere, perfudere :—Se awyrgeda gást ðæs ylcan preóstes heortan and geþanc mid his searwes ǽttre geondsprengde [-spregde, MS.] *the accursed spirit sprinkled over with the poison of his deceit the heart and mind of the same priest*, Guthl. 7; Gdwin. 44, 13. Me fugles wyn geond [-sprengde]

spēd-dropum *the bird's delight [feather] sprinkled me over with copious drops*, Rä. 27, 8.

geond-spreót *sprouted through or over, pervaded*; pergerminavit, pervasit, Exon. 8 b; Th. 3, 27; Cri. 42. v. spreótan.

geond-, gend-springan *percrebrescere, multiplicari*, Hpt. Gl. 473.

geond-stredan; *p.* -stredde; *pp.* -streded, -stred *To scatter about, sprinkle*; spargĕre :—Ic geondstrede *spargo*, Ælfc. Gr. 28, 4; Som. 31, 37. Geondstred *scattered over*, Homl. Th. ii. 536, 18.

geond-styrian; *p.* ede; *pp.* ed [geond, styrian *to move, stir*] *To move or stir violently, to agitate*; per omnes partes commovere, agitare :— Geondstyred *agitated*, Bt. Met. Fox 6, 29; Met. 6, 15.

ge-ond-swarian; *p.* ode; *pp.* od *To answer*; respondere, Lk. Skt. Lind. 10, 28. v. and-swarian.

geond-þencan; *p.* -þohte; *pp.* -þoht [þencan *to think*] *To think over, consider, contemplate*; animo lustráre, contemplári :—Ðonne ic eorla líf eal geondþence *when I consider all the chieftains' life*, Exon. 77 a; Th. 290, 5; Wand. 60. Se ðis deorce líf deópe geondþenceþ *he profoundly contemplates this dark life*, 77 b; Th. 291, 29; Wand. 89.

ge-ond-weardan, -wardan; *p.* de *To answer*, Blickl. Homl. 21, 21 : Mt. Kmbl. Lind. 3, 15 : 8, 8. v. ge-and-wyrdan.

ge-ond-weardian *to present*, Blickl. Homl. 181, 2 : Rtl. 4, 28, 30. v. ge-and-werdian.

geond-wlítan; *p.* -wlát, *pl.* -wliton; *pp.* -wliten. I. *v. trans. To look through, see through, look over*; perspicĕre, ocŭlis lustráre :—He selfa mæg sǽ geondwlítan *he can himself look through the sea*, Cd. 213; Th. 265, 18; Sat. 9: Beo. Th. 5335; B. 2771. Sunne woruld geondwlíteþ *the sun looks over the world*, Exon. 59 a; Th. 212, 16; Ph. 211. Ðæt ic ingehygd eal geondwlíte *that I can see through all his inward thoughts*, 71 b; Th. 266, 17; Jul. 399. II. *v. intrans. To look about, look around*; circumspectáre :—Sioh sylfa ðé geond ðas sídan gesceaft geondwlítan *see thyself look around this wide creation*, Exon. 8 b; Th. 4, 30; Cri. 60.

geond-yrnan; *p.* -arn, *pl.* -urnon; *pp.* -urnen *To run about*; discurrēre :—Ic geondyrne *discurro*, Ælfc. Gr. 47; Som. 48, 51.

ge-onet; *part. p. Hastened*; festinatus, Lye.

ge-ónētan [?] *To make useless* :—Giónétaþ ł gemerras *occupat*, Lk. Skt. Lind. 13, 7. Geónēt *occupatus*, Lye. [Cf. (?) Icel. ū-nýta *to make useless, destroy*.]

ge-onfenge; *adj. Taken* :—Ân geonfenge biþ *una assumetur*, Lk. Skt. Lind. 17, 35. v. onfenge.

geong, es; *m. A course, passage, journey*; cursus, meátus, iter :—Ongunnon him on úhtan ædelcunde mægþ gierwan to geonge *the noble women resolved ere dawn to prepare for a journey*, Exon. 119 b; Th. 459, 19, note; Hö. 2. Geong *iter*, Lk. Skt. Lind. 2, 44 : 8, 1.

geong *sighs*; gemitus :—Hér is Brytta geong [gnornung, B.] and geómerung *gemitus Brittanorum*, Bd. 1, 13; S. 481, 42.

GEONG, giong, geng, ging, giung, iung, gung; *def.* se geonga, seó, ðæt geonge; *comp.* geongra, gingra, gyngra; *superl.* gingest, gingst; *adj.* YOUNG, *youthful, new, recent, fresh*; jŭvĕnis, adolescens, nŏvellus, rĕcens :— Ðeáh ðe he geong sý folces hyrde *although he be a young shepherd of his folk*, Beo. Th. 3667; B. 1831: Rood Kmbl. 77; Kr. 39. Mǽden, oððe geong wífman *puella*, Wrt. Voc. 73, 5. Se geonga mann *adolescens*, Mt. Bos. 19, 22: Lk. Bos. 7, 14 : Ors. 2, 4; Bos. 45, 12 : Chr. 871; Erl. 75, 23. Ymb ðæs geongan feorh *about the young man's life*, Andr. Kmbl. 2236; An. 1119. On swá geongum feore *in so young a life*, Beo. Th. 3690; B. 1843. Me eáden wearþ, geongre *it was granted to me young*, Exon. 10 b; Th. 13, 11; Cri. 201. Ic ðé geongne gelǽrde *I taught thee young*, Bt. 8; Fox 24, 27 : Andr. Kmbl. 1101; An. 551: 2222; An. 1112. Cýse geongne onfón gestreón getácnaþ *to accept new cheese betokens gain*, Lchdm. iii. 200, 29. Ðǽr geong wiste wíc weardian *where he knew the young [woman] to be abiding*, Exon. 67 a; Th. 248, 6; Jul. 91. Ðæt he feorh geong eft onfón móte *that it may again receive a new spirit*, 62 a; Th. 228, 4; Ph. 433: 58 b; Th. 211, 3; Ph. 192. Sint geþreáde geonge gúþprincas *my young warriors are rebuked*, Andr. Kmbl. 783; An. 392: 1715; An. 860: 3060; An. 1533. Ða geongan leomu *the young limbs*, Exon. 87 a; Th. 327, 18; Vy. 5. Geongra gyfena *of recent gifts*, 65 a; Th. 239, 20; Ph. 624: 78 a; Th. 293, 16; Crä. 2. Geongum and ealdum *to young and old*, Beo. Th. 144; B. 72. He hēht hine geonge twegen men mid síþian *he bade two young men accompany him*, Cd. 138; Th. 173, 27; Gen. 2867 : Beo. Th. 4040; B. 2018. Geongra ic wæs, witendlíce ic ealdode *jūnior fui, etenim sēnui*, Ps. Spl. C. 36, 26. Gingra bróðor *a younger brother*, Exon. 130 a; Th. 499, 2, note; Rä. 88, 9. Seó gingre *the younger*, Cd. 123; Th. 158, 5; Gen. 2612. Ic gyngra wæs *jūnior fui*, Ps. Spl. 36, 26. Gingran bróðor *younger brothers*, Exon. 129 a; Th. 496, 10; Rä. 85, 12. Ioseph gingst wæs hys gebróðra *Joseph was the youngest of his brethren*, Ors. 1, 5; Bos. 28, 7. Se gingsta ys mid úrum fæder mĭnĭmus *cum patre nostro est*, Gen. 42, 13, 32. Fram ðam yldestan óþ ðone gingestan *a mājōre usque ad mĭnĭmum*, 42, 12. [Wyc. 30ng: Chauc. yong: Laym. ʒunge, ʒenge, ʒeonge: Orm. ʒung, ʒunng: Plat. jung, junk: O. Sax. jung: Dut. jong: Frs. jong: O. Frs. jung, jong:

Ger. jung: M. H. Ger. junc: O. H. Ger. jung: Goth. yuggs: Dan. Swed. ung: Icel. ungr: Lat. jŭvĕnis: Sansk. yuvan young.] DER. cild-geong, cniht-, ed-, fela-, heaðo-, magu-.

geóng went, Beo. Th. 1855, note; B. 925; p. of gangan.

geongan, ic geonge, ðū geongest, he geongeþ; p. gang, pl. gungon To go; ire :—He com to sele geongan he came to go [= he came or went] to the hall, Andr. Kmbl. 2624; An. 1313. Wutun geonga eamus, Mk. Skt. Lind. 14, 42: 12, 3, Geongende ambulans, 16, 12: Jn. Skt. Lind. 1, 36. Ic giungo, geongo, geonga vado, 13, 36, 33: 16, 7. Ic geonge I go, Exon. 106 a; Th. 403, 4; Rä. 22, 4. Heó to fenne gang she went to the fen, Beo. Th. 2595; B. 1295: 2022; B. 1009: 2636; B. 1316. Wyt on godes hūse gungan [gangan, MS.] in domo Dei ambulavimus, Ps. Th. 54, 13. Geonge for ðē care intret in conspectu tuo gemitus, Ps. Th. 78, 11. Nū ðū lungre geong hord sceáwian now go thou quickly and view the treasure, Beo. Th. 5480; B. 2743. Geong vade, Jn. Skt. Lind. 8, 11. v. gân, gangan.

geongerdôm subjection, Cd. 14; Th. 18, 3; Gen. 267. v. geongordôm.

geonge-wifre, an; f. A ganging-weaver, spider; viātica arānea :—Wǽron ānlícast ūre winter geongewifran, ðonne hió geornast biþ, ðæt heó afǽre fleógan on nette our years [lit. winters] were most like to a spider when it is most eager to terrify flies into its net; anni nostri sīcut arānea meditābuntur, Ps. Th. 89, 10. v. gange-wifre.

geong-líc; adj. Youthful, young; jŭvĕnilis :— Geonglíc jŭvĕnilis, Ælfc. Gr. 9, 28; Som. 11, 39. On geonglícum geárum in his young years, Homl. Th. ii. 118, 23. [O. H. Ger. junglich.]

geong-licnys, -nyss, e; f. Youth; jŭventus, Scint. 32.

geong-ling, es; m. A youngling, youth; jŭvĕnis :—Ðæt hí tǽcon sum gerād heora geonglingum that they teach some prudence to their younglings, Ælfc. Gr. pref; Som. 1, 30. [O. H. Ger. jungeling: Ger. jüngling.]

geongor-dôm, geonger-dôm, es; m. Youngership, minority, subjection, obedience, service, vassalage; jŭvĕnilis status, obsĕquium, obēdientia, ministĕrium :—Hwý sceal ic būgan him swilces geongordômes why shall I submit to him in such vassālage? Cd. 15; Th. 18, 34; Gen. 283. Unc wearþ God yrre forðon wit him noldon hnīgan mid heáfdum þurh geongordôme God was angry with us two because we two would not bow to him with our heads in subjection, 35; Th. 46, 12; Gen. 743: 30; Th. 41, 26; Gen. 662. Ðæt he Gode wolde geongerdôme þeówian that he would serve God in subjection, 14; Th. 18, 3; Gen. 267. [O. Sax. jungar-dôm.]

geongor-scipe youngership, service. v. giongor-scipe.

geongra, giongra, gingra, gyngra, giungra, an; m. A junior, disciple, vassal, subject, follower, attendant, servant; jūnior, adŏlescentŭlus, discĭpŭlus, assecla, sectātor, mĭnister :—Geongra ic eom adŏlescentŭlus sum ego, Ps. Spl. 118, 141. Ne wille ic leng his geongra wurþan I will no longer be his vassal, Cd. 15; Th. 19, 15; Gen. 291: 15; Th. 18, 23; Gen. 277. Þurh ǽnne ðara apostola geongrena through one of the followers of the apostles, Ors. 6, 11; Bos. 121, 8. He wolde Drihtnes geongran beswícan he would deceive the subjects of the Lord, Cd. 23; Th. 29, 15; Gen. 450.

geongre a female servant, maid-servant. v. gingre.

geónian; part. geóniende; p. ode; pp. od To yawn; hiāre :—Fore openre wunde and geóniendre pro aperto et hiante vulnĕre, Bd. 4, 19; S. 589, 19. Hí todydon heora mūþ ongeán me swā swā leó ðonne he geónaþ aperuerunt in me os suum, sicut leo rapiens, Ps. Th. 21, 11. v. gýnian.

geonlíc [= geonglíc?]; adj. Youthful : — For geonlíces mægdenes plegan for a young maiden's play, Shrn. 123, 7.

geonre; adv. There, yonder; illuc, Som. [Cf. Goth. jainar there.]

geónung, e; f. A YAWNING, braying, chattering; oscitatio, barritus, Cot. 95.

geópan, ic geópe, ðū gýpst, he gýpþ, pl. geópaþ; p. geáp, pl. gupon; pp. gopen To take up, take to oneself, receive; accĭpĕre :—Oþ-ðæt ic spǽte eal-felo āttor, ðæt ic ǽr geáp until I spit the very baleful venom which I took up before, Exon. 106 b; Th. 405, 29; Rä. 24, 9. [Cf. Scot. gowpen to lift or lade out with the hands: Icel. gaupn: O. H. Ger. coufan both hands held together in the form of a bowl.]

ge-openian; -openigean; p. ode; pp. od, ad. I. trans. To open, manifest, shew, reveal :—He bæd him engla weard geopenigean uncūðe wyrd he prayed the guardian of angels to reveal to him the unknown fate, Elen. Kmbl. 2201; El. 1102. Se anweald geopenaþ his yfel and gedēþ hit sweotol power reveals his evil and makes it plain, Bt. 16, 3; Fox 56, 20: Salm. Kmbl. 266, 2. He his goðcundnysse mihta mid ðam tācne geopenode he revealed the powers of his divinity with that miraclê, Homl. Th. ii. 54, 31: Gen. 18, 20. He heofonan ríces infær geopenode he opened an entrance to the kingdom of heaven, Homl. Th. ii. 128, 24: 260, 11: i. 78, 27. Geopena ongeán me lífes geat open to me the gate of life, 76, 3. Ðæt he geopenige that he shew, Past. 21; Swt. 159, 24; Hat. MS. God hine onwrýhþ ðeáh ðe wit hine ne geopenian God will reveal it though we two do not make it manifest, Blickl. Homl. 187, 17. Geopenod opened, 9, 8. II. intrans. To open :—Ðā geopenode seó sǽ togeánes Moysen the sea opened before Moses, Swt. A. S. Rdr. 63, 228.

ge-orettan, -oretan, -orrettan; p. te; pp. ted To disturb, confound; perturbare, confundere :—Ealle beóþ georette eác gescende omnes confundantur et conturbentur, Ps. Th. 82, 13. Georetan confundere, conturbare, Gl. Prud. 735. Georrettan infamare, Cot. 111.

georman-leaf, es; n. Mallow [?] L. Med. 1, 27; Lchdm. ii. 68, 12: 33; 80, 9.

GEORN; comp. m. geornra; f. n. geornre; sup. geornast; adj. Desirous, eager, anxious, ardent, zealous, studious, intent, careful, diligent; cupĭdus, appĕtens, sollĭcitus, studiōsus, intentus, dīlĭgens :—Cyning biþ anwealdes georn a king is desirous of power, Exon. 89 b; Th. 337, 4; Gn. Ex. 59. Georn wísdômes desirous of wisdom, 81 a; Th. 305, 15; Fä. 88. Forðam ðe ǽgðer ðæra folca wæs ðæs gefeohtes georn because the people on both sides were eager for the fight, Ors. 3, 8; Bos. 63, 35. Dǽda georn zealous in deeds, Cd. 188; Th. 233, 27; Dan. 282. Teónum georn anxious for mischiefs, 27; Th. 36, 34; Gen. 581. Azarias, dǽdum georn, Dryhten herede Azariah, ardent in deeds, praised the Lord, Exon. 53 a; Th. 185, 5; Az. 3. Ic beó lāreów georn I am a diligent instructor, 71 b; Th. 267, 3; Jul. 409. Mǽrþa georne eager for glory, Cd. 80; Th. 101, 5; Gen. 1677. Micle hý wǽron geornran ðæt hí him fram flugen they were much more eager that they should go from them, Ors. 1, 7; Bos. 30, 9. Geornast most eager, Ps. Th. 89, 10. [Piers P. yerne eagerly: Chauc. yerne brisk, quick: R. Brun. 3erne earnestly: Laym. 3eorne, 3eornen earnestly, eagerly: Orm. 3eorne, 3eornne, 3erne, 3errne, willingly, earnestly: O. Sax. gern desirous: Frs. jearn: O. Frs. ierne gerne willingly: Dut. gaarne willingly: Ger. gerne, gern willingly: M. H. Ger. gërne, gërn desirous: O. H. Ger. gern, gerni intentus, cupĭdus, stŭdiōsus, prōnus: Goth. gairns yearning for: Dan. gjerne gladly: Swed. gerna fain, willingly: Icel. gjarn eager, willing.] DER. ælmes-georn, clǽn-, dôm-, firen-, firwet-, fyrwet-, gilp-, glig-, ídel-, lof-, slāp-, weorþ-.

ge-orn rose; exortus est, surrexit, 38, 2; S. 605, 40; p. of ge-yrnan.

geornan, giornan, giornian; p. de, ade, ede; pp. ed To desire, beg; desīdĕrāre :—Gē geornaþ ðæt gē woldon eówerne naman tobrǽdan geond ealle [eallne, MS.] eorþan ye desire that ye should spread your name over all the earth, Bt. 18, 2; Fox 64, 4. Se cyng and his witan georndon friþes the king and his witan desired peace, Chr. 1011; Erl. 144, 21. To geornanne mendicare, Lk. Skt. Lind. 16, 3. Giornade, giornede, giornde begged, Mk. Skt. Lind. 10, 46: 1, 40: Jn. Skt. Lind. 9, 8. v. gyrnan.

georne, giorne, gyrne; comp. geornor; superl. geornost, geornast; adv. Eagerly, earnestly, diligently, carefully, zealously, willingly, readily, gladly, well; cŭpĭdè, enixe, dīlĭgenter, studiōse, prompte, lĭbenter, bĕne :—Ðæt fýr georne asēceþ innan and ūtan eorþan sceátas the fire shall eagerly seek within and without the tracts of earth, Exon. 22 b; Th. 62, 20; Cri. 1004: Cd. 29; Th. 38, 15; Gen. 606. Ic him georne unrihtes andsæc fremede I earnestly made denial to their injustice, Elen. Kmbl. 940; El. 471: 1197; El. 600: Cd. 103; Th. 137, 4; Gen. 2268: 137; Th. 172, 19; Gen. 2846. He sôhte georne æfter grunde he sought diligently along the ground, Beo. Th. 4577; B. 2294: Exon. 44 b; Th. 150, 11; Gū, 777: 57 a; Th. 204, 4; Ph. 92. He befran hí georne hwænne se steorra him æteówde diligenter didicit ab eis tempus stellæ, quæ appāruit eis, Mt. Bos. 2, 7: Ps. Th. 76, 6: 131, 5. Hæleþ hinfūse hýrdon to georne wrāðum wǽrlogan the death-devoted men too readily listened to the furious pledge-breaker, Andr. Kmbl. 1224; An. 612: Exon. 34 a; Th. 109, 24; Gū. 95. Ongan Dryhtnes ǽ georne cýðan he began gladly to proclaim the Lord's law, Elen. Kmbl. 398; El. 199: Cd. 32; Th. 42, 26; Gen. 679. Hit gôdode georne it prospered well, Chr. 959; Erl. 119, 13, 16: Bt. Met. Fox 20, 61; Met. 20, 31: 21, 39; Met. 21, 20. Geornor we woldon iówra Rômāna bismora beón forsū-giende we would more willingly be silent about the shame of you Romans, Ors. 3, 8; Bos. 63, 22: 3, 1; Bos. 53, 14. Swā he geornost mæge as he best may, Bt. Met. Fox 27, 58; Met. 27, 29. Geornast most diligently, Exon. 37 b; Th. 123, 25; Gu. 328.

geornes, geornys, gyrnes, gyrnys, -ness, -nyss, e; f. Earnestness, diligence, industry, care, endeavour; industria, stŭdium :—Mid ðysses cyninges geornesse hujus industria rēgis, Bd. 3, 6; S. 528, 30. He hæfde swýðe mycle geornysse sibbe stŭdium vĭdēlĭcet pācis hăbuit, Bd. 3, 17; S. 545, 7: 3, 28; S. 560, 31. Mīne geornnesse mid gôde ðu gefyldest thou didst satisfy my longing with good, Blickl. Homl. 89, 4.

georneste; adj. Earnest, serious :— Georneste seria, Cot. 195. v. eorneste.

georn-ful, -full; comp. -fulra; adj. Full of desire, eager, solicitous, anxious, strenuous, zealous, intent, diligent; sollĭcitus, stŭdiōsus, anxius, sēdŭlus, intentus, dīlĭgens :—On orde stôd Eádweard, gearo and geornful Edward stood in the array, ready and eager, Byrht. Th. 139, 54; By. 274. Geornfull ðū eart sollĭcita es, Lk. Bos. 10, 41. Wæs he on willsumnesse hāligra gebeda gecneord and geornfull erat orātiōnum devōtiõni sollertissĭme intentus, 4, 28; S. 606, 34. Ðæt he swā geornfulle gýmenne dyde him ða hǽla ūre þeóde tam sēdŭlam erga sālūtem nostræ gentis cūram gessĕrit, 2, 1; S. 501, 3: Hymn. Surt. 49, 21. Geornfulle men diligent men, Bt. 32, 3; Fox 118, 10. Se is yfla gehwæs geornfulra ðonne ic who is more zealous than I for every evil, Exon. 70 b; Th. 261, 33; Jul. 324. He wiste ðæt hý woldon georn-

fulran beón ꝺære wrace, ꝺonne óꝺre men *he knew that they would be more eager for revenge than others*, Ors. 2, 5; Bos. 47, 3.

geornful-líce; *comp.* -lícor; *adv.* [geornful *eager*] *Anxiously, diligently, earnestly*; stŭdiōse, dīligenter, sēdŭlo:—He húsulfatu and leóhtfatu geornfullíce gegearwode *vasa sancta et lumĭnāria stŭdiōsissĭme parāvit*, Bd. 5, 20; S. 642, 4. Swá he geornfullícor ꝺæs ēcan lífes gewilnode *he the more earnestly desired the eternal life*, Homl. Th. ii. 120, 8.

geornful-nes, giornful-nes, -nys, -ness, -nyss, e; *f. Eagerness, diligence, earnestness, zeal, fervour, devotion*; sollertia, dīligentia, industria, fervor, devōtio:—Sió geornfulnes [giornfulnes, MS. Hat.] eorþlícra þinga ablent ꝺæs mōdes eágan mid ꝺære costunga *the eagerness for earthly things blinds the eyes of the mind with temptation*, Past. 18, 2; Swt. 128, 15; Cot. MS. Deós geornfulnyss *hæc dīligentia*, Ælfc. Gr. 43; Som. 45, 6. He geornlíce gýmde ꝺæt he to lufan and to geornfulnesse awehte gódra dǣda *ad dilectiōnem vero et sollertiam bŏnæ actiōnis excitāre curābat*, Bd. 4, 24; S. 598, 19: 5, 13; S. 632, 8. Ða he ꝺa se cyning his gelǣrednysse and his geornfulnysse geseah *cujus erudītiōnem atque industriam videns rex*, 3, 7; S. 529, 46. Mid mycelre geornfulnesse devōtiōne magna, 3, 30; S. 562, 3: L. Edg. i. 5; Th. i. 264, 22. Ꝺone pipor ꝺa nǣddran healdaþ on heora geornfulnysse *piper quod serpentes servant sua industria*, Nar. 34, 22.

geornlíc; *adj. Desirable* :—Hit biþ geornlíc ꝺæt ... *it is desirable that* ..., Ors. 4, 13; Bos. 100, 28.

geornlíce; *comp.* -lícor; *superl.* -lícost; *adv. Earnestly, diligently, zealously, strenuously, carefully, willingly*; dīligenter, stŭdiōse, obnixe, sollícite, lībenter:—Faraþ and axiaþ geornlíce be ꝺam cilde *íte, et interrŏgāte dīligenter de puĕro*, Mt. Bos. 2, 8: Bd. 3, 11; S. 535, 28: 3, 19; S. 547, 14, 15: 4, 9; S. 576, 21: 5, 14; S. 634, 30. Ongan geornlíce on sefan sēcean weg to wuldre *she began earnestly in her mind to seek the way to glory*, Elen. Kmbl. 2293; El. 1148: Salm. Kmbl. 169; Sal. 84. He geornlíce on gebede hleóprede *obnixius orātiōni incumbĕret*, Bd. 4, 3; S. 569, 11: 3, 28; S. 560, 17. Hí bǣdon hyne geornlíce *rŏgābant eum sollícite*, Lk. Bos. 7, 4. Geornlíce Cyriacus on Caluarie hleór onhylde *Cyriacus willingly bent down his cheeks on Calvary*, Elen. Kmbl. 2192; El. 1097. Ꝺæt he wolde Paulinus ꝺone bisceop geornlícor gehýran be ꝺam Gode sprecende ꝺe he bodade *quia vellet ipsum Paulinum dīligentius audīre de Deo quem prædĭcābat, verbum fäciendum*, Bd. 2, 13; S. 516, 26, 30: 4, 9; S. 576, 34. Ꝺæt he geornlícost God weorþige *that he most zealously worship God*, Exon. 14 a; Th. 27, 19; Cri. 433.

geornung, gyrning, e; *f. A yearning, desire, diligence* :—Ic hæbbe geheórd seo kyninges Æꝺelrēdes geornunge *I have heard king Ethelred's desire*, Chr. 675; Erl. 37, 21. Geornung *industria*, Lye.

georran, girran, gyrran; ic georre, gyrre, ꝺu gyrst, he gyrþ, *pl.* georraþ; *p.* gear, *pl.* gurron; *pp.* gorren *To chatter, sound, creak*; sonare, stridere, garrīre:—Ic gyrre *garrio*, Ælfc. Gr. 36; Som. 38, 29. Strengas gurron *the ropes creaked*, Andr. Kmbl. 748; An. 374. [Cf. *Laym.* ȝurren þa stanes 28358 : garryng Morr. and Skt. Spec. 241, 163.]

ge-orsod *enraged*, Ps. Lamb. 105, 37. v. geyrsian.

georst *heath*. v. gorst.

georstan-dæg *yesterday*. v. gyrstan-dæg.

ge-ortréwan; *p.* de; *pp.* ed [trēwan *to trust*] *To despair*; dēspērāre:—Ꝺa þreó ꝺē ne lǣtaþ geortrēwan be ꝺam ēcan lífe *these three suffer thee not to despair of the everlasting life*, Bt. 10; Fox 30, 9. v. geortrúwian.

ge-ortrúwian, -trýwian; *p.* ode; *pp.* od [*or without*, treówian, trúwian *to trust*] *To distrust, despair*; diffīdĕre, dēspērāre:—Ꝺa ꝺē ne lǣtaþ geortrúwian þe ꝺis andweardan lífe *they suffer thee not to despair of this present life*, Bt. 10; Fox 30, 7. Se man lōcaþ underbæc, and geortrúwaþ Godes mildheortnysse *the man looks behind who despairs of God's mercy*, Homl. Th. i. 252, 10. Ꝺæt úre nán be his nēxtan ne geortrúwige *that none of us despair of his neighbour*, ii. 82, 27. Nis ꝺæt to geortrýwianne *nec diffīdendum est*, Bd. 4, 19; S. 587, 32. Ꝺæt ꝺú ne geortrýwe nánes gōdes on nánre wiꝺerweardnesse *that thou despair not of any good in any adversity*, Bt. 6; Fox 14, 35.

ge-orwénan; *p.* de; *pp.* ed [wēn *hope*] *To despair, to be out of hope*; dēspērāre:—Georwened *dēspērātus*, Ælfc. Gr. 47; Som. 48, 38. Ꝺæt he ꝺý earmlícor georwēnedre hǣlo hēr nú forwurde *quo misĕrābilius ipse dēspērāta sălūte pĕrīret*, Bd. 5, 14; S. 635, 3.

ge-orwyrþed *disgraced*; traductus, Cot. 171. v. onwurꝺe.

geó-sceaft, e; *f. That which has been determined of old, fate*:—Weras wyrd ne cúꝺon geósceaft grimme [MS. grimne] *men knew not their destiny, their grim fate*, Beo. Th. 2472; B. 1234. [Cf. frumsceaft, gesceaft.]

geó-sceaft-gást, es; *m. A fatal, dire spirit* [?] *or ancient spirit* [?]:—Ꝺanon wóc fela geósceaftgásta wæs ꝺæra Grendel sum *thence arose many dire spirits, Grendel was one of them*, Beo. Th. 2536; B. 1266.

geosterlíc; *adj. Of yesterday*; hesternus. v. gysternlíc.

geostra, giestra [estra, Ps. Spl. 89, 4] gystra, gyrsta; *adj. Of yesterday*; hesternus:—Geostran dæg *dies hesterna*, Ps. Th. 89, 4. Gioster doeg *heri*, Jn. Skt. Lind. 4, 52. Giestron *yesterday*, Exon. 111 a; Th. 424, 24; Rä. 41, 44. Gystran niht *yesternight*, Beo. Th. 2672; B. 1334.

Gyrstan dæg *heri*, Jn. Bos. 4, 52: Th. An. 22, 1. [*Laym.* ȝerstendæi (o, u): *Goth.* gistra dagis *to-morrow*, with which meaning the *Icel.* i gör occurs, v. Cl. and Vig. Dict. gær : *O. H. Ger.* gestre, gesteren *heri*; gestren *hesternum*: *Ger.* gestern : *Lat.* heri, hesternus.]

geot *yet*, Bt. 5, 3. v. gyt.

GEÓTAN; ic geóte, ꝺú gýtst, he gýt, *pl.* geótaþ; *p.* geát, gēt, *pl.* guton; *pp.* goten; *v. a.* **I.** *to pour, pour out, shed*; fundere, effundere, profundere :—Teáras geótan *to shed tears*, Exon. 10 b; Th. 11, 19; Cri. 173. Geát teáras *fundebat lachrymas*, Bd. 2, 6; S. 508, 9. He gēt ꝺæt blód uppan ꝺæt weofod *fudit sanguinem super altare*, Lev. 8, 24: Ex. 24, 6. Swá man gute wæter *as one would pour water*, Ps. Th. 78, 3. Ꝺý lǣs weras and idesa on geáþ gutan *lest men and women should pour it forth in mockery*, Exon. 50 b; Th. 176, 8; Gú. 1207. Ofer hleór goten *poured over the cheek*, Elen. Kmbl. 2264; El. 1133. **II.** *to flow, stream*; profluere, *v. n* :—He hâte lēt teáras geótan *he let hot tears flow*, Exon. 48 a; Th. 165, 16; Gú. 1029. Geofon geótende *the flowing sea*, Andr. Kmbl. 785; An. 393: 3014; An. 1510; Ps. Th. 17, 4. Mid geótendan here *with an overwhelming army*, Chr. 1052; Erl. 184, 17. **III.** *to found, cast*:—Gold and seolfur ꝺe hér geótaþ menn *gold and silver that men here found*, Ps. Th. 134, 15. Híg guton him hǣꝺenne god *they have made them a molten image*, Deut. 9, 12. [Cf. *Orm.* Moyses shollde ȝetenn himm a neddre : *Laym.* ȝeoten *to pour*: *Goth.* giutan: *O. Sax.* giotan: *Dan.* gyde: *Swed.* giuta *to cast*: *O. H. Ger.* giozan: *Ger.* giessen.] DER. a-geótan, be-, ge-, ofer-, on-, þurh-, to-.

geótende *arteries, veins*; arteriæ, Cot. 8.

geótere, es; *m. A pourer, melter, founder*; fūsor, flātor:—Se geótere *the founder*, Ors. 1, 12; Bos. 36, 27, 35. DER. ār-geótere.

geótton *confirmed*, Chr. 656; Th. 53, 32; *for* geátton. v. geátan.

Geoweorþa *Jugurtha*, Ors. 5, 7.

ge-oweꝺan *to subdue*; subjugare :—He bæd his twǣm sunum ꝺæt hí ꝺæs ríces ꝺriddan dǣl geoweꝺan sculdon *he ordered his two sons to subdue the third part of the kingdom*, Som. ge-ꝺeówan [?]

geoxa, geoxung *a sobbing, hiccup*, Cot. 109. v. geocsa.

gep *sly, cunning*, Scint. 3, 24, 65. v. geap.

ge-palmtwíged; *def.* se -twígeda, seó, ꝺæt -twígede; *part.* [palm-twíg *a palm-twig*] *Palm-twigged, adorned with palm-twigs*; palmæ rāmis ornātus:—Se gepalmtwígeda Pater Noster *the palm-twigged Pater Noster*, Salm. Kmbl. 23; Sal. 12. Ꝺæt gepalmtwígede Pater Noster, 77; Sal. 39.

ge-pilod *heaped* or *piled up*, Ex. 16, 14.

ge-píned; *part. p. Punished*:—Ꝺætte hia wēre gepíned *puniri*, Lk. Skt. p. 9, 4.

ge-plægde *danced*, Mt. Kmbl. Lind. 14, 6. v. plægan.

ge-plantod *part.* [plantian *to plant*] *Planted*; plantātus:—Sum man hæfde ān fíctreów geplantod on his wíngearde *arbŏrem fíci hăbēbat quidam plantātam in vinea sua*, Lk. Bos. 13, 6.

ge-portian *p.* ode; *pp.* od *To beat, pound*; contundĕre:—Geporta ꝺa wyrta tosomne *pound the herbs together*, Lchdm. iii. 4, 10. v. portian.

ge-pós, es; *n. The* POSE, *a cold in the head, catarrh*; grăvēdo:—Wiꝺ gepósu *for colds in the head*, Herb. 1, 1; Lchdm. i. 148, 12. Wiꝺ gepósum *for poses*, L. M. 1, 10; Lchdm. ii. 54, 17.

ge-price *a point* or *comma*; comma, Som.

ge-punian; *p.* ode, ude *To pound, beat, bray*; contundĕre:—Gepuna eall tosomne *pound all together*, Herb. 101, 3; Lchdm. i. 216, 13. Genim ꝺas ylcan wyrte gepunude [gepunode, MS. B.] *take this same herb pounded*, 129, 3; Lchdm. i. 240, 15: 75, 1; Lchdm. i. 176, 20.

ge-pyndan; *p.* -pynde; *pp.* -pynded, -pynd *To pound, impound, shut up*; circumclūdĕre:—Nellaþ hie gehæftan and gepyndan hiora mód *they will not restrain and shut up their mind*, Past. 39, 1; Swt. 283, 13; Hat. MS. 52 b, 26. Ꝺæt wæter biþ gepynd *the water is shut up*, 38, 6; Swt. 277, 6; Hat. MS. 51 b, 13.

gēr, es; *n.* **I.** *a year*; annus:—Hærfest biþ hreꝺeádegost, hæledum bringeþ gēres wæstmas *autumn is most joyous, [it] bringeth the fruits of the year to men*, Menol. Fox 477; Gn. C. 9. Wintras oꝺꝺe gēr *winters* or *years*, Glos. Prudent. Recd. 139, 23. **II.** *the Anglo-Saxon Rune* ᚷ = g, the name of which letter in Anglo-Saxon is gēr *a year*, hence. this Rune not only stands for the letter *g*, but for gēr *a year*. as,—ᚷ [gēr] byþ gumena hiht, ꝺonne God lǣteþ hrusan syllan beorhte blǣda beornum and þearfum *the year is the hope of men, when God letteth the earth give her bright fruits to rich and poor*, Runic pm. 12; Kmbl. 341, 20; Hick. Thes. i. 135. v. geár *winter*, II.

ge-rád. v. ge-rídan.

ge-rád, es; *n. Consideration, account, condition, reason, wisdom, prudence, manner*; ratio, conditio:—Ꝺá he ꝺæt gerád sette *cum coepisset rationem ponere*, Mt. Bos. 18, 24. Se hláford dyhte hym gerád *dominus posuit rationem cum eis*, 25, 19. Ꝺám ealdum gedafenaþ ꝺæt hí tǣcon sum gerád heora geonglingum *ad senes spectat juvenes prudentia erudire*, Ælfc. Gr. pref; Som. 1, 33. On ꝺæt gerád ꝺet he gesylle ælce geáre *on the condition that he give every year*, Th. Chart. 147, 31: Chr. 945;

Erl. 116, 31. To đam geráde đe . . . *on the condition that* . . . , Th. Chart. 168, 13. On đa ylcan geråd *under the same conditions*, Ps. Th. 9, argument 3. Crist awende úre stuntnysse to geráde *Christ turned our folly to wisdom*, Homl. Th. i. 208, 19. ¶ On đæt geråd *for that reason*, Ors. 1, 12; Bos. 36, 4. On đa geråd *on the condition* or *account*, Bt. 7, 3; Fox 22, 7: Chr. Erl. 3, 15: 1093; Erl. 229, 25.

ge-ráð; *adj. Considered, instructed, learned, skilful, expert, prudent, suited, conditioned;* consultus, consideratus, instructus, peritus, prudens, elegans, concinnus :—Gif ic đé geráde geméte *if I find thee instructed* [*skilful*], Bt. 5, 1; Fox 10, 16. Hí wurdon geráde wígcræfta *they became skilful in the arts of war*, Ors. 1, 2; Bos. 26, 29. Sió is swíđe wel geråd and swíđe gemetfæst *she is very prudent and very modest*, Bt. 10; Fox 28, 20: Beo. Th. 1751; B. 873. Ic him rúmne weg and gerádne tæhte *I might shew him a spacious and direct road*, Guthl. prol; Gdwn. 6, 3. On geráde spræce *into prose*, Bd. 5, 24; S. 648, 22. Geråd beón wiþ his wyrd *to be suited to his fortune*, Bt. 11, 1; Fox 32, 11. ¶ Đus geråd, swá geråd *such, of such sort*, Jn. Bos. 8, 5: Deut. 4, 32 : Basil admn. 2; Norm. 36, 30: Guthl. 3; Gdwn. 22, 2: Bt. 39, 11; Fox 230, 16. Hú geráð *of what kind*, Guthl. 17; Gdwn. 72, 2. [*Laym.* i-rad: *Goth.* ga-raids.] DER. un-ge-råd.

ge-rádegian; *p.* ode; *pp.* od *To reckon with :*—Anlíc đam cyninge đe hys đéowas gerádegode *adsimilatum regi qui voluit rationem ponere cum servis suis*, Mt. Bos. 18, 23. [Cf. ge-rádian.]

ge-rádian; *p.* ode; *pp.* od *To arrange, reason, argue;* disponere, rationem conferre, supputare cum aliquo :—Wiđerwearda gesceafta wæron gegaderode and gerádode *contrary creatures were united and arranged*, Bt. 35, 2; Fox 156, 36. v. ge-rædan.

ge-rádnes, -ness, e; *f. An agreement, a conspiracy;* conjuratio, Cot. 209.

ge-radod; *part. p. Quick;* citatus, Obs. Lun. 26; Lchdm. iii. 196, 7. v. ge-hradian.

ge-rádscipe, es; *m.* [geråd *consideration*, scipe *condition*] *Prudence;* prudentia :—He áwuht nafaþ on his módsefan rihtwísnesses ne gerádscipes *he has not aught in his mind of wisdom or prudence*, Bt. Met. Fox 22, 96; Met. 22, 48.

ge-ræc, es; *m? Opportunity;* opportunitas :—In geræcum *in opportunitatibus*, Ps. Spl. 9, 9.

ge-ræcan, -ræcean; *p.* -ræhte; *pp.* -ræht *To reach, obtain, seize, get, lay hold on, attain, reproach, present, offer :*—Sió fird hie geræcan ne mehte *the* [*English*] *force could not reach them*, Chr. 895; Erl. 93, 22: 894; Erl. 90, 11: Cd. 216; Th. 275, 10; Sat. 169. Geræcean, Blickl. Homl. 207, 22. Ne đú đé æfre ne læt wlenca geræcan *never do thou let pride lay hold on thee*, Bt. Met. Fox 5, 61; Met. 5, 31. Đæs landes máre geræcan *to obtain more of the land*, Chr. 921; Erl. 106, 21. Sige geræcan *to get the victory*, Ors. 3, 1; Bos. 53, 30: 9; 68, 11, 12. Andlífne geræcan *to get* [*one's*] *living*, Cd. 43; Th. 57, 26; Gen. 934. Of eágum teáras geræcan *to draw tears from the eyes*, L. Edg. C. iv; Th. ii. 288, 5. Đæt he þence đone sélestan hwet-stán on to geræcanne *that he think of applying the best whetstone*, Ors. 4, 13; Bos. 100, 30. To freán hond geræcan *to present to the lord's hand*, Exon. 90 b; Th. 339, 10; Gn. Ex. 92. Siđđan ic đurh hylles hróf geræce *when I reach through the hill's summit*, 104 b; Th. 397, 30; Rä. 16, 27. Đe geræcaþ wæpen *whom weapons reach*, 102 a; Th. 386, 7; Rä. 4, 58. Đú me geræhtest mid handa *extendisti manum tuam*, Ps. Th. 137, 7. Hyne Wulf wæpne geræhte *Wolf reached him with his weapon*, Beo. Th. 5923; B. 2965: 1117; B. 556: Byrht. Th. 135, 63; By. 142: 136, 29; By. 158. He đa burh geræhte *he took the town*, Ors. 2, 4; Bos. 44, 14. He hælu geræhte écan lífes *he obtained the salvation of eternal life*, Exon. 35 a; Th. 112, 12; Gú. 142. Đa scipo alle geræhton *seized all the ships*, Chr. 885; Erl. 82, 29: Cd. 119; Th. 154, 13; Gen. 2555. Hí đæt ríce geræht hæfdon *they had got that kingdom*, Bt. Met. Fox 26, 36; Met. 26, 18. Æfter đæm đe đa wíf hí swá scandlíce geræht hæfdon *after the women had so reproachfully addressed them*, Ors. 1, 12; Bos. 36, 12.

ge-ræd *elegans*, Cot. 80.

ge-ræd *advised;* *p.* of ge-rædan; *p.* -reórd.

ge-rædan; *p.* -reórd, -réd, -ræd *To give counsel, advise, bring about by advice;* consilium dare :—Đe him đone teónan geråd *who brought that injury upon them by his counsel*, Cd. 37; Th. 48, 12; Gen. 774: 37; Th. 49, 25; Gen. 797. [Cf. *O. Sax.* Siu bad, that he iru helpa gerêdi.] v. rædan; *p.* -reórd.

ge-rædan; *p.* de; *pp.* ed, -rædd, -ræd. **I.** *to arrange, dispose, direct, advise, determine, ordain, consult for, provide for;* decernere, statuere, edicere, consulere, providere :—Gerece and geråd đa rihtwísan *diriges justum*, Ps. Th. 7, 10: 24, 4. Gerædes *dispensas*, Rtl. 71, 11. Đæne ræd gerædde Síric arcebiscop *that counsel advised archbishop Sigeric*, Chr. 991; Th. 238, 28: 1052; Th. 320, 13, col. 1. Gyf đú đæt gerædest *if thou decidest on that*, Byrht. Th. 132, 54; By. 36: Exon. 92 a; Th. 344, 24; Gn. Ex. 178. Đa witan geræddan *the counsellors ordained*, L. E. G. 4; Th. i. 168, 15. Heó hire feax gerædde *crines composuit*, Bd. 3, 9; S. 534, 13. [Cf. *Icel.* greiða hár *to dress the hair.*] Biđon giræded *disponentur*, Rtl. 86, 24. Ic đone friþ gerædd hæbbe *I have ordained the peace*, L. Ath. v. § 11; Th. i. 240, 14. Geráð *ordained,*

§ 10; Th. i. 240, 2 : L. Eth. vi. 32; Th. i. 324, 1. [*Goth.* ga-raidjan *to enjoin : Icel.* greiða *to arrange.*] **II.** *to read;* legere :—Sý geråd *sit lectus*, C. R. Ben. 22. Hit is geråd on gewyrdelícum racum *it is read in historical narratives*, Homl. Th. i. 58, 9. Đonne geræde gé đás word beforan him đæt híg gehíron *then read these words before them that they may hear*, Deut. 31, 11. [Cf. ge-rádian.]

ge-ræde, es; *n:* ge-rædu, e; *f? A housing, harness, trappings, equipage;* phaleræ, apparatus :—Đa here-geata medemra þegna syndon hors and his geræda *the heriots of the medial thanes are a horse and his trappings*, L. C. S. 72; Th. i. 414, 12, MS. G: Bd. 3, 14; S. 540, 22, MS. B. Folc féreþ herega gerædum *the nation marches with martial equipage*, Cd. 209; Th. 259, 29; Dan. 699: Elen. Kmbl. 2105; El. 1054: 2213; El. 1108. v. ge-réde, ge-ræþle.

ge-ræde; *adj. Ready, swift, prompt, easy, plain, simple;* paratus, celer, promptus, expeditus, planus, simplex :—He gedyde míne fét swá geræde swá heorotum *qui perfecit pedes meos* [*celeres*] *tanquam cervi*, Ps. Th. 17, 32. Ge meterfers, ge geræðre spræce *et versibus heroicis, et simplici oratione*, Bd. 4, 28; S. 605, 13 : 5, 18; S. 636, 6; Bd. 5, 24; S. 648, 27. [*Icel.* greiðr *ready, free:* cf. *North. E.* gradely.] v. ræde, ge-ræd.

ge-ræden, ne; *f. A proposal, purpose, condition;* propositum, Rtl. 92, 36. On đa gerædene *on the condition*, Th. Chart. 484, 29.

ge-rædend, es; *m. A disposer;* dispositor, Rtl. 108, 16.

ge-ræding, es; *m. A decree;* consultum, Cot. 59; Lye.

ge-rædnes, -rædnis, -rédnys, -ness, e; *f. An ordinance, a decree, purpose, an intention, a resolution, condition;* consultum :—Đis is seó gerædnys đe Eádgár cyng gerædde *this is the ordinance that king Eadgar ordained*, L. Edg. i. pref; Th. i. 262, 2 : L. E. G. pref; Th. i. 166, 5 : L. Ath. v. pref; Th. i. 228, 6 : L. Eth. vi. 1, 2, 3; Th. i. 314, 2, 12, 19 : Cod. Dipl. ii. 150, 33 : Th. Chart. 168, 27. In đas gerédnisse *on this condition*, 104, 20. [Cf. *Goth.* ga-raideins *an ordinance.*]

ge-rædod; *part. p. Furnished with trappings, harnessed :*—Án gerædod hors *a harnessed horse*, Ælfc. T. Lisle 36, 12 : Th. Chart. 501, 5. v. ge-ræde; *subst.*

ge-ræf; *adj. Fixed;* fixus :—Gif mon folc-leásunge gewyrce and hió on hine geræf weorðe *if a man commit folk-leasing and it be fixed upon him*, L. Ælf. 32; Th. i. 80, 21, note.

ge-ræft *torn, distracted;* discerptus, Bt. 37, 1; Fox 186, 21.

ge-ræpan *to bind*, Bt. Met. Fox 13, 15; Met. 13, 8 : 25, 73, 96; 25, 37, 48. v. ræpan.

ge-ræsan; *p.* de; *pp.* ed [ræsan *to rush*] *To rush;* irruêre :—Đe wiđ swá miclum mægne geræsde *who rushed against so great a power*, Cd. 97; Th. 126, 15; Gen. 2095: Beo. Th. 5671; B. 2839. Hí geræsdon *they rushed*, Chr. Erl. 5, 7 : Shrn. 130, 22, 23.

ge-ræstan *to rest, sit;* quiescere :—Geræstun mid þone Hæland *discumbebant cum Jesu*, Mt. Kmbl. Lind. 9, 10 : Mk. Skt. Lind. 2, 15 : Jn. Skt. Lind. 21, 20. v. ræstan.

ge-ræswa, an; *m.* [ræswa *a chief*] *A chief, prince;* dux, princeps :—Cymeþ engla geræswa *the prince of angels cometh*, Salm. Kmbl. 223; Sal. 111.

ge-ræþle, an; *n. A harness, trappings;* phaleræ :—Hors and his geræþlan *a horse and his trappings*, L. C. S. 72; Th. i. 414, 12. v. ge-ræde.

ge-ræwen, -ræwud *set in rows, plaited, embroidered;* segmentatus :—Geræwen hrægel *segmentata vestis*, Ælfc. Gl. 63; Som. 68; Wrt. Voc. 40, 10.

ge-rafende, -rawende RIFTING, *cleaving;* infindens, Cot. 181.

Geransingas; *gen.* a; *pl. The Gergesenes :*—In lond Geransinga *in regionem Gerasenorum*, Mt. Kmbl. Rush. 8, 28.

ge-rár *a roaring, howling;* boatus, ululatus, Shrn. 50, 10. v. ge-rísan.

ge-rás. v. ge-rísan.

GERD, e; *f. A yard, rod, reed, twig, young shoot;* virga, arundo, Mt. Kmbl. Lind. 11, 7 : 12, 20. Sex fođur gerda *six fothers of faggots*, Th. Chart. 104, 27.

gerdel *a girdle*, Prov. 31. v. gyrdel.

gere; *adv. Entirely, well, very well;* penitus, bene, optime, Cd. 158; Th. 196, 14; Exod. 291. v. geare; *adv.*

ge-reápan *to bind.* v. ge-ræpan.

ge-reáfa, an; *m. A reeve, judge, count;* præfectus, judex, comes :—Ic bebeóde eallum mínum gereáfum *I command all my reeves*, L. Ath. i. prm: Th. i. 194, 14. v. ge-réfa.

ge-reáfian; *p.* ode; *pp.* od *To rob, steal, spoil :*—Gereófage *diripere*, Mk. Skt. Lind. 3, 27. Secgaþ đæt his đegnas gereáfodan his líc on us and forstélan *say that his disciples robbed his body from us and stole it away*, Blickl. Homl. 177, 29. Gereáfydon *diripiebant*, Ps. Spl. C. 43, 12. Đone deórwyrþan gym đone đe deófol wolde gereáfian *the precious jewel that the devil would steal*, Shrn. 155, 21.

ge-reahte, -reaht *related, explained, denoted, directed, ruled, reproved*, Exon. 34 b; Th. 110, 12; Gú. 106: Bt. Met. Fox 11, 197; Met. 11, 99; *p.* and *pp.* of ge-reccan.

ge-rec, es; *n. Rule, government, management, order, direction, explanation;* regimen, moderamen, ratio, directio, expositio :—On đara óđra mægþa gerece awunode *in illarum provinciarum regimine permansit*, Bd. 4, 12; S. 581, 28: 4, 23; S. 593, 26. Đone bisceophád mid mycele

gerece heóld and rihte *episcopatum sedulo moderamine gessit*, 3, 7; S. 530, 35 : Bt. 21; Fox 74, 29 : Bt. Met. Fox 22, 2; Met. 22, 1. Be efen-nihte æfter Anatolius gerece *concerning even-night [the equinox] after the explanation of Anatolius*, Bd. 5, 23 ; S. 648, 19, note.

ge-rec, es ; *n. A tumult :*—Gerec *tumultus*, Mt. Kmbl. Lind. 27, 24. [Cf. (?) *O. H. Ger.* ungareh *tumultus*.]

ge-reca, an ; *m. A governor, ruler, prefect;* præfectus :—Heáh gereca *summus præfectus*, Nat. S. Greg. Els. 21, 1.

ge-reccan, -recan, -reccean ; ic -recce, ðú -reccest, -recest, he -receþ, -recþ; *imp.* -rece ; *p.* -reahte, -rehte; *pp.* -reaht, -reht ; *v. trans.* I. *to put forth, shew, relate, express, denote, explain, interpret, translate;* exponere, demonstrare, narrare, referre, disserere, exprimere, interpretari, reddere :—Ic gereccan mæg *I can shew*, Bt. Met. Fox 25, 74; Met. 25, 37. Ic eów mæg gerecan [MS. Cot. gereccan] *I can shew you*, Bt. 11, 2; Fox 34, 7. Gê ðæt cunnon gereccan *ye know how to relate that*, Elen. Kmbl. 1294; El. 649: Homl. Th. ii. 118, 3. Nemn nú gif ðú hit gereccean mæge *declare it now if thou art able to shew it*, Blickl. Homl. 181, 14. Aristoteles hit gerehte *Aristotle has explained it*, Bt. 40, 6 ; Fox 242, 2. Wordum gereccan *to express in words*, 20; Fox 70, 28. Ðæt is gereht Crist *quod est interpretatum Christus*, Jn. Bos. 1, 38, 41, 42 : Exon. 9 b; Th. 9, 12 ; Cri. 133. Emanuhél, ðæt ys gereht on úre geþeóde, God mid us *Emanuel, which is, translated into our speech, God with us*, Mt. Bos. 1, 23 : Mk. Bos. 5, 41 : 15, 22. Gereccean þancas *referre gratias*, Procm. R. Conc. II. *to set forth, extend, direct, order, rule, control, reprove, correct, subdue, reduce to subjection;* exponere, extendere, dirigere, regere, corripere, corrigere, subigere, sub imperium redigere :—Sý on ðínre gesihþe mínes sylfes gebed gereht swá rícels byþ *dirigatur oratio mea sicut incensum in conspectu tuo*, Ps. Th. 140, 2. Sæd heora on worulda biþ gereht *semen eorum in seculum dirigetur*, Ps. Spl. 101, 29. He hie gereceþ to eallum gódum *he will direct them to all good*, Blickl. Homl. 79, 33. Ne biþ se ofer eorþan gereaht *non dirigetur super terram*, Ps. Th. 139, 11. Gerece on gesihþe ðíne weg mínne *dirige in conspectu tuo viam meam*, Ps. Spl. 5, 9: 24, 5; 39, 3 : Ps. Th. 118, 133. Hú Gúþlác his in Godes willan mód gerehte *how Guthlac directed his mind to God's will*, Exon. 34 a; Th. 108, 3 ; Gú. 67. Oþ-ðæt ðæs gewinnes God ende gereahte *until God directed an end of the strife*, 34 b; Th. 110, 12; Gú. 106. Ða witan gerehton Eádgife ðæt heó sceolde hire fæder hand gecléansian *the witan directed Eadgifu to clear her father's hand*, Chart. Th. 201, 33 : 70, 31. Míne fét to heofenum gereahte *my feet [shall be] directed to heaven*, Blickl. Homl. 191, 7. Gif hiora mód-sefa meahte weorþan staðol-fæst, gereaht þurh ða strongan meaht *if their mind might become stable, ruled by strong might*, Bt. Met. Fox 11, 197 ; Met. 11, 99. Me sóþfæst symble gerecce and mildheorte móde þreáge *corripiet me justus in misericordia et increpabit me*, Ps. Th. 140, 7 : Exon. 66 b; Th. 247, 4; Jul. 73. To gereccanne ðone gedwolan *ad corrigendum errorem*, Bd. 3, 30 ; S. 562, 9. In anwald gerehton *they reduced to subjection*, Bt. 1 ; Fox 2, 5. Mon gerehte ðæt yrfe cinge *the property was confiscated to the king*, Th. Chart. 173, 1. v. reccan.

ge-recce-líc; *adj. Stretched out, extended, strict, firm, steadfast;* extensus, strictus, firmus, Som.

ge-recednys, -recednes, -recenes, -nyss, e ; *f. A narration, history, report, an interpretation, a direction, correction;* narratio, relatio, historia, interpretatio, directio, correctio :—Ðæt gódspell æfter Matheus gerecednysse *the gospel according to the narration of Matthew*, Mt. Bos. titl : Mk. Bos. titl : Greg. Dial. 2, 15 : Th. Apol. 1, 1. To mægwlite andgytes and gástlícra gerecenessa ic to ætýcte *ad formam sensus et interpretationis eorum superadjeci*, Bd. 5, 23 ; S. 647, 35. Ðú gearwodest gerecednyssa, dóm and rihtwísnysse on Iacobe ðú dydest *tu parasti directiones, judicium et justitiam in Iacob tu fecisti*, Ps. Spl. 98, 4 : 96, 2.

ge-recenian; *p.* ode; *pp.* od *To explain;* exponere, explanare, interpretari :—Rún biþ gerecenod *a mystery shall be explained*, Cd. 169 ; Th. 211, 12 ; Exod. 525.

ge-rec-líce; *adv. In a direct course, directly, extensively, strictly, firmly;* extenso cursu vel modo, directe, stricte, firme :—Gereclíce rihte flóweþ *flows in a direct course straight along*, Bt. Met. Fox 5, 27; Met. 5, 14: 24, 16; Met. 24, 8: Bt. 35, 4; Fox 162, 1.

ge-réde, an ; *n. Harness, trappings;* phaleræ :—Hors and his geredan *a horse and his trappings*, L. C. S. 72; Th. i. 414, 12, note 39. [Cf. ge-ráde.]

GE-RÉFA, ge-reáfa, groefa, an ; *m. A prefect, steward, fiscal officer of the shire* or *county, judge, reeve* or *sheriff, count;* præpösitus, villicus, júdex, præfectus, cömes :—Fóreset vel geréfa *præpösitus*, Ælfc. Gl. 87 ; Som. 74, 37 ; Wrt. Voc. 50, 19. Cwæþ se geréfa *ait villicus*, Lk. Bos. 16, 3. Gif man biscopes esne tihte oððe cyninges, cænne hine on geréfan hand, oððe hine geréfa cléansie, oððe selle to swinganne *if any one accuse a bishop's servant* or *that of the king, he shall clear himself before the judge, either the judge shall clear him* or *give him up to be scourged*, L. Wih. 22 ; Th. i. 42, 4. Ðæs cynges geréfa *the king's reeve*, L. Eth. i. 4; Th. i. 282, 31: L. C. S. 33; Th. i. 396, 14. Gif hit se geréfa ne amanige mid rihte *if the reeve do not lawfully exact it [the fine]*, L. Ed. 5 ; Th. i. 162, 12. Ðæt ælc geréfa náme ðæt wedd on his ágenre scire, ðæt

hí ealle ðæt friþ healdan woldan *that each reeve should take a pledge in his own shire, that they would all hold the peace*, L. Ath. v. § 10; Th. i. 240, 1. Ðæt ælc geréfa fylste óðrum to úre ealra friþe *that every reeve may help another for the common peace of us all*, v. § 8, 4; Th. i. 286, 27. Ðæs landrícan and ðæs biscopes geréfa *the landlord's and the bishop's reeve*, L. Eth. ix. 8; Th. i. 342, 16: L. C. E. 8; Th. i. 366, 7. Iosep, se æðela geréfa, of Arimathia *Ioseph ab Arimathæa, nöbìlis decürio*, Mk. Bos. 15, 43. Sum wæs æhtwelig æðeles cynnes, ríce geréfa *there was a wealthy man of noble race, a powerful count*, Exon. 66 a; Th. 243, 31; Jul. 19. Se geréfa hét Iulianan *the count commanded Juliana*, 73 b; Th. 274, 9; Jul. 530. Geréfa mín *my steward*, Cd. 100 ; Th. 131, 25; Gen. 2181. Ealdorman oððe geréfa *cömes*, Wrt. Voc. 72, 61. Geréfa *consul*, Ælfc. Gl. 6; Som. 56, 49; Wrt. Voc. 18, 4. Nán man ne hwyrfe nánes yrfes bútan ðæs geréfan gewitnesse *let no man exchange any property without the witness of the reeve*, L. Ath. i. 10; Th. i. 204, 17, 18. On ælces geréfan manunge *in every reeve's district*, iv. 1; Th. i. 222, 9. Iohanna, Chuzan wíf, Herodes geréfan *Ioana, uxor Chusae, procurätöris Heródis*, Lk. Bos. 8, 3. He cwæþ to his geréfan *præcêpit dispensätöris dömus suæ dìcens*, Gen. 43, 16. Gecýðe cyninges geréfan *let them declare it to the king's reeve*, L. Alf. pol. 34; Th. i. 82, 17: 22; Th. i. 76, 5. Swá hie geþingian mægen wið cyning and his geréfan *as they can agree with the king and his reeve*, L. In. 73; Th. i. 148, 12. Gif man ðone geréfan teó *if any accuse the reeve*, L. C. S. 8; Th. i. 380, 19. On Lindcolene ceastre geréfan *pertingens ad præfectum Lindocolinæ civitatis*, Bd. 2, 16 ; S. 519, 20: Shrn. 120, 12: 123, 24. He sende his geréfan *misit præfectum suum*, 4, 1; S. 564, 42. Míne ealdormen and míne geréfan *my aldermen and my reeves*, L. Ath. i. prm; Th. i. 194, 10. Ic wille, ðæt bisceop and ða geréfan hit beódan *I will that the bishop and the reeves command it*, i. prm; Th. i. 194, 10. Ic wille, ðæt míne geréfan gedón, ðæt man agife ða ciricsceattas and sáwlsceattas *I will that my reeves cause that a man shall give the church-scots and the soul-scots*, i. prm ; Th. i. 196, 8. Ic Æðelstán cyningc cýðe [MS. cýð] ðám geréfan to hwilcere birig *I, Æthelstan king, make known to the reeves at each town*, i. prm ; Th. i. 194, 3. Eádwerd cyning být ðám geréfum eallum, ðæt ge déman swá rihte dómas swá ge rihtoste cunnon, and hit on ðære dómbéc stande *King Edward commands all the reeves, that ye pass the most righteous sentences you can, and as it stands in the doom book*, L. Ed. prm ; Th. i. 158, 3 : L. Eth. ix. 32; Th. i. 346, 29. Se sette geréfan geond eall ðæt ríce *qui constìtuat præpösìtos per cunctas regiones*, Gen. 41, 34. He hét sécan síne geréfan *he commanded to seek his officers*, Cd. 176 ; Th. 220, 31; Dan. 70. We ðær settan and geendebyrdedon úre geréfan *ordinarios proprætoresque nostros proposuimus*, Nar. 3, 25. From ðen groefa *a præside*, Mt. Kmbl. Lind. 28, 14 : Mk. Skt. Lind. 15, 5 : Jn. Skt. Lind. 19, 1, 4. See Stubbs' Const. Hist. and Schmid A. S. Gesetz. *s. v;* Kemble's Saxons in England, ii. c. 5; Grm. R. A. 752-4.

ge-réf-ærn, es ; *n. A court-house :*—Urbanus eode to his geréfærne *Urbanus went to his court-house*, Shrn. 106, 16.

ge-réf-land, es ; *n. Tributary land;* tributarium territorium, Cot. 106.

ge-réflang, es ; *m. A minister :*—Ða geréflanges of Cristes circean *the ministers of Christchurch*, Chart. Th. 317, 32.

ge-réf-mǽd, e ; *f.* '*The meadow which the reeve owned "ex officio," or over which, as common pasture, he exercised the right of superintendence,*' Cod. Dipl. Kmbl. iii. xxxiv.

ge-réf-scipe, es ; *m. Office of a geréfa :*—Ne heora nán geréfscipe ne drífe *let none of them practise any reeveship*, Homl. Th. ii. 94, 33.

ge-réf-scir or -scire, e ; *f. Stewardship;* villicatio :—Mín hláford míne geréfscire fram me nymþ *dominus meus aufert a me vilicatiönem*, Lk. Bos. 16, 3. Geréfscyre *præfectura*, Hpt. Gl. 438. v. scir, ge-sciran.

ge-regnian, -rénian ; *p.* ode; *pp.* od, ad *To put, dispose, adorn :*—Geregnian, *inficere*, Cot. 112. Hwæðer him leófre wære ðe he hý ealle acwealde ðe hý libbende to bismre gerénian héte *whether he would rather that he should kill them all* or *should order them to be put to shame while living*, Ors. 3, 8; Bos. 63, 14. Lii hit oftræd and hie to loman gerénode ðæt hie mec ænigre note nytte beón ne meahton *it trode down and made them cripples so that they could be of no use to me;* calcatos inutiles fecit, Nar. 15, 26. Ðonne hangaþ ðær eác búfan ðæm lástum geregnod swíðe mycel leóhtfæt *moreover there hangs, placed above the footsteps, a great lamp*, Blickl. Homl. 127, 29. Ðæs geregnedan *concinnati*, Cot. 57. Ne ðæt ne beoþ on ðý fægerre ðæt mid elles hwam gerénod biþ ðeáh ða gerénu fægeru síen ðe hit mid gerénod biþ *nor will that be the fairer which is adorned with something else though the ornaments be fair with which it is adorned*, Bt. 74, 3 ; Fox 46, 14 : 27, 1; Fox 96, 1. Golde geregnad *adorned with gold*, Beo. Th. 1558; B. 777. Gerénod *adorned*, Byrht. Th. 136, 35 ; By. 161: Judth. 12; Thw. 26, 21; Jud. 339. Girínad *ornatum*, Lk. Skt. Rush. 21, 5. [Cf. *Goth.* garaginon.]

ge-regnong *a making up;* confectio, Cot. 44.

ge-rehtan *made straight, set up;* erectus, Lk. Skt. Lind. 13, 13.

ge-rehte, -reht *related, explained, interpreted, directed*, Bt. 40, 6 ; Fox 242, 2 : Ps. Th. 140, 2 : Jn. Bos. 1, 38, 41, 42 ; *p. and pp.* of ge-reccan.

gerela, gierela, an ; *m. Apparel :*—Gif ðú wénst ðætte wundorlíc[e] gerela hwelc weorðmynd síe *if you suppose that that wonderful apparel is any*

honour, Bt. 14, 1; Fox 42, 18. Wynna gierelan gielplíces *the pleasures of pompous apparel*, Exon. 35 a; Th. 112, 3; Gú. 138: 38 b; 127, 22; Gú. 390. v. ge-gerela.

ge-rén, es; *n. An ornament* :—Ðeáh ða gerénu fægru síen ðe hit mid gerénod biþ *though the ornaments be fair with which it is adorned*, Bt. 14, 3; Fox 46, 15. Ða geréno *the ornaments*, Exon. 107 a; Th. 408, 20; Rä. 27, 15. Giríno ł glencas *ædificationes*, Mk. Skt. Rush. 13, 2.

ge-réne, es; *pl. nom. acc.* -u, -o, -a; *n. A mystery*; mysterium, Hy. 8, 11; Hy. Grn. ii. 290, 11. v. ge-rýne.

ge-rénian. v. ge-regnian.

gereófage. v. ge-reáfian.

ge-reohnung, e; *f. A making up*; confectio, Cot. 171. v. ge-regnong.

ge-reónian; *p.* ode; *pp.* od *To conspire, ordain, frame, devise*; conspirare, concinnare :—Ic gereúnige *conspiro*, Ælfc. Gr. 47; Som. 48, 42. Tunge ðín gereónode fácnu *lingua tua concinnabat dolos*, Ps. Lamb. 49, 19; thy tongue frameth deceit; *thi tunge ordeynde treccheries*, Wyc. Æfter manegum dagum gereónodon ða Iudeiscan hú hí done Godes cempan acwellan sceoldon *after many days the Jews conspired how they were to kill that champion of God*, Homl. Th. i. 388, 5.

ge-reónung, e; *f. A conspiracy, confederacy*; conjuratio :—Ne understenst ðú ðisra twegra manna gereónunge ongeán me *dost thou not understand the plot of these two men against me?* Homl. Th. i. 380, 7. Gereónung *fictio, mendacium*, Hpt. Gl. 459.

ge-reord, -reorde, es; *n. Language, speech, tongue, voice* :—Hí cunnon eall mennisc gereord *nationum linguis loquentes*, Nar. 37, 4: Bd. 1, 1; S. 474, 2: Hy. Grn. ii. 287, 19: 293, 43. Ðæt ys on úrum gereorde *that is in our language*, Thw. Hept. 155, 37: Swt. A. S. Rdr. 97, 55. To Norþhymbriscum gereorde *to the Northumbrian speech*, 58. Weorðlíce getýd ge on Ledenisc gereorde ge on Grecisc *Græcæ pariter et Latinæ linguæ peritissimus*, Bd. 4, 1; S. 563, 33: 2; 565, 28: Th. An. 18, 29. Ðá wǽron ða apostolas cweðende to him hwonon him ða wundorlícan gereordo cóman *then the apostles were saying to him whence came to him those wonderful speeches*, Blickl. Homl. 153, 9. Hwílum ic gereordum rincas laðige to wíne *sometimes with voices I invite men to wine*, Exon. 104 a; Th. 395, 31; Rä. 15, 16. v. reord.

ge-reord, -reorde, es; *n. A meal, refection, food* :—Sæt se Hǽlynd æt gereorde *discumbebat Iesus*, Mt. Bos. 26, 20. Hwǽr is mín gereord *ubi est refectio mea*, Mk. Bos. 14, 14. Ǽr his gereorde *ante prandium*, Lk. Bos. 11, 38: Gen. 19, 3. Be ðam liflícum gereorde *concerning the vital refection*, Homl. Th. ii. 262, 24. Oððæt ðæt gereorde gefylled wæs *until the meal was finished*, Bd. 5, 4; S. 617, 26. Cyninga gereordo *regum convivia*, Cot. 93. Him beád reste and gereorda *offered them rest and refreshment*, Cd. 112; Th. 147, 17; Gen. 2441: Exon. 96 a; Th. 357, 29; Pa. 36: Mt. Kmbl. Lind. 26, 7. Heofonlícu gereordu *heavenly food*, Shrn. 30, 28: 64, 2. Giriord *cæna, alimentum prandium, cibus*, Rtl. 70, 37: 99, 11: 107, 19: 116, 5. Gehriord *epula*, 116, 34. To gereordum ł farmum *ad nuptias*, Mt. Kmbl. p. 19, 4.

ge-reordan, -reordian; *p.* ode; *pp.* ad, od *To give food to, feed, take food, satisfy, refresh, feast*; cibare, saturare, satiare, epulari :—Ic gereordige *prandeo*, Ælfc. Gr. 26; Som. 29, 8. Ic gereordige *vescor*, 29; Som. 33, 50. Ic gereordige *reficio*, ic eom gereordod *reficior*, 37; Som. 39, 2. He hine gereordode mid ðam papan *he dined with the pope*, Chr. 1022; Erl. 161, 34. Giriordade hine *cibavit illum*, Rtl. 46, 9. He gereordode hí *saturavit eos*, Ps. Spl. C. 80, 15. Crist gereorde fíf þúsenda wera *Christ fed five thousand men*, Shrn. 48, 30. Ðæt ge eów gereordian *that ye may refresh yourselves*, Gen. 18, 5. Ǽr mǽle hine gereordige *that one take refection before the time*, Homl. Th. ii. 590, 25. Giriordiga we *epulemur*, Rtl. 25, 17. Giriord *satia*, 146, 17. Unbindaþ hí and gereordigaþ *unbind her and give her to eat*, Homl. Th. i. 458, 19. We willaþ mid ðý hláfe gereorde beón *pane illo refici volumus*, Bd. 2, 5; S. 507, 22. Ðá wæs flet-sittendum fægere gereorded *then were the sitters in the hall nobly feasted*, Beo. Th. 3581; B. 1788. Húsle gereordod *refreshed with the eucharist*, Exon. 51 b; Th. 180, 4; Gú. 1274. Gereordod, Andr. Kmbl. 770; An. 385. Ða ilco biðon geriorded *saturabuntur*, Mt. Kmbl. Lind. 5, 6: 14, 20. Hia síe giriordado *reficiantur*, Rtl. 15, 5.

ge-reord-hús, es; *n. A dining-room*; refectorium, triclinium, Ælfc. Gl. 107; Som. 78, 74; Wrt. Voc. 57, 51.

ge-reordig-hús *a dining-room*; refectorium, Lye.

ge-reording, -ung, e; *f. A meal, refection*; prandium, refectio :—Gearca us gereorduinge *prepare us a meal*, Homl. Th. i. 60, 18. On gereorduncge *in prandio*, Th. An. 28, 9. On ánre gereordinge *in una refectione*, 34, 37. Ofer wæteru gereordunga *super aquam refectionis*, Ps. Spl. 22, 2. Giriording *a meal*, Lk. Skt. Rush. 12, 19, 37.

ge-reordnes, -nys, se; *f. A repast, dinner, fulness*; refectio, Ps. Spl. C. 22, 2: Bd. 4, 28; S. 606, 1.

ge-reósan; *p.* -reás, *pl.* -ruron; *pp.* -roren *To fall*; cadere, Ps. Spl. second 9, 12. v. ge-hreósan.

ge-resp *convicted*; convictus, L. Alf. pol. 28; Th. i. 80, 21.

ge-rest, es; *n. A resting-place, couch*; accubitus, Rtl. 4, 11.

ge-resta, an; *f. One who rests with another, consort* :—Seó wæs Eádwardes cynges geresta *she was king Edward's consort*, Chr. 1076; Erl.

214, 32. Heó Balan sealde Iacobe to gerestan *Bilham dedit Iacobo quacum concumberet*, Gen. 30, 4.

ge-restan; *p.* te; *pp.* ed *To rest, remain, rest [one's self]* :—Ðæt he hine gerestan meahte *ad quiescendum membra*, Bd. 2, 6; S. 508, 9. Forðon ic ǽfre ne mæg ðære mód-ceare mínre gerestan *for I can never rest from my mind's sorrow*, Exon. 115 b; Th. 444, 1; Kl. 40. Templ Háliges Gástes snytro on to gerestenne *a temple for the wisdom of the Holy Ghost to dwell in*, Blickl. Homl. 163, 15. Ic me gereste *quiesco*, Ælfc. Gr. 28; Som. 30, 30. Mín hige geresteþ nó *my mind resteth not*, Elen. Kmbl. 2164; El. 1083: Exon. 8 b; Th. 4, 16; Cri. 53. On ðone seofoðan dú gerestest *on the seventh thou didst rest*, Hy. 9, 23; Grn. ii. 291, 23. Gif ic on ðunwange gereste *si dedero requiem temporibus meis*, Ps. Th. 131, 4. Gerest ðé *requiesce*, Lk. Bos. 12, 19: Homl. Th. ii. 104, 20. Girestun [Rush.] gehræston [Lind.] *requieverunt*, Lk. Skt. 12, 19.

ge-restscipe, es; *m.* I. *rest, ease*; quies, ótium, Som. Ben. Lye. II. *a cohabitation*; concúbitus :—To hyre gerestscipe hire wer ne sceal gangan *ad ejus concúbitum vir suus accédere non débet*, Bd. 1, 27; S. 493, 32.

ge-rétan; *p.* -rétte; *pp.* -réted, -rét *To restore, refresh, set right*; recreáre, reficére :—Wæs heó semninga mid gástlícre gesyhþe geréted *súbito visióne spíritáli recreáta*, Bd. 4, 9; S. 577, 19: 5, 1; S. 613, 22. Dú me hæfst gerétne mid ðínre gesceadwísnesse *thou hast comforted me with thy reasoning*, Bt. 22, 1; Fox 76, 12.

ge-reþra, an; *m. A sailor, rower*; nauta :—Gereþra [MS. gereþru] *nauta*, Ælfc. Gl. 103; Wrt. Voc. 56, 15. v. reþra.

ge-reþru; *pl. n. Rudder, helm* [the steering was done by means of an oar] :—Ða men ða ðe beóþ winnende in sciplícum gewinne híg ðonne begáþ ǽrost ða gereþru in ðære hýþe *qui in nauali prœlio demicaturi sunt ante in portu inflectant gubernacula*, Shrn. 35, 8: 9. Gereþru *vel* scip-getawu *aplustre*, Ælfc. Gl. 103; Wrt. Voc. 56, 19. Gereþra *aplustra*, Gl. Mett. 15. On ánum báte bútan ǽlcum gereþrum *in a boat without any means of steering*, Chr. 891; Erl. 88, 6, see note on this passage. 'Gereþrum' can however hardly be a case of 'gereþra' *nauta*, as the singular number would be used with 'ǽlc;' it is rather a plural like 'geatwe' or 'frætwe.'

gér-hwamlíce; *adv. Yearly*; annuatim, Som.

gerian; *p.* ode; *pp.* od *To clothe*; vestíre :—Ðám ðe ðone líchoman Cúþberhtes geredon *quibus corpus Cudbercti vestierant*, Bd. 4, 31; S. 611, 5, MS. B. v. gyrian.

ge-ricsian; *p.* ode; *pp.* od *To rule, govern*; regere, dominari, gubernare, Rtl. 8, 7: 26, 43: 38, 41.

ge-rídan; *p.* -rád; *pp.* -riden *To ride, reach* or *obtain by riding, get into one's power, subject* :—Ðá he gerád to Ecgbryhtes stáne *then he rode to Brixton*, Chr. 878; Erl. 80, 8. Se ðe næs gerád *he who rode to the ness*, Beo. Th. 5789; B. 2898. Ðá gerád he ða burg æt Tameworþige *then he rode and took the town at Tamworth*, Chr. 922; Erl. 108, 24: 901; Erl. 96, 26. Se here geridon Wesseaxna lond and gesǽton micel ðæs folces ofer sǽ adrǽfdon and ðæs óðres ðone mǽstan dǽl hie geridon *the [Danish] army rode to Wessex and occupied it; much of the folk they drove over sea and most part of the rest they got into their power*, 878; Erl. 78, 29-32. He gerád eall Norþhymbra land him to gewealde *he got all Northumberland into his power*, 948; Erl. 117, 9. Se cing lét gerídan ealle ða land ðe his módor áhte him to handa *the king caused all the lands that his mother owned to be brought under his own control*, 1043; Erl. 168, 8.

ge-ríd-men *horsemen, knights*; equites, Cot. 212.

ge-rif, es; *n. A seizing, taking away, a catching—as of fish*, also *that which is caught*; raptura, captura :—An gerif fisca, oððe án snǽs fisca oððe óðra þinga *one taking of fish, or one spear of fish, or of other things*, una sorta, Mone A. 141; Recd. 37, 77; Wrt. Voc. 64, 9: Ælfc. Gl. 98; Wrt. Voc. 54, 40. DER. fót-síþ-gerif.

ge-rifled, -riflod; *part. p. Wrinkled*; rugatus, Som.

ge-rifod; *part. p. Wrinkled* :—On ealdlícum geárum biþ ðæs mannes neb gerifod *in the years of old age man's face is wrinkled*, Homl. Th. i. 614, 14.

ge-riht, es; *n. What is right, a right, due, last office of the church, direction*; rectum, jus, ratio, officium :—Gif hwá ǽnigra godcundra gerihto forwyrne *if any one refuse any divine dues*, L. E. G. 6; Th. i. 170, 7. Godes gerihto *God's dues*, 5; Th. i. 168, 25: Homl. Th. i. 74, 22: Swt. A. S. Rdr. 105, 39: L. Eth. 5, 11; Th. i. 306, 30: Shrn. 208, 28. Ðis syndon ða gerihta ðe se cyning áh ofer ealle men on Wessexan *these are the rights which the king has over all men in Wessex*, L. C. S. 12; Th. i. 382, 12. Cynescipes gerihta *rights of royalty*, L. Edg. S. 2; Th. i. 272, 27: Chr. 1085; Erl. 218, 28. Ealla ða gerihta ðe ðǽr of arísaþ *all the rights arising therefrom*, 1031; Erl. 162, 4: 1074; Erl. 212, 6. Geriht *ratio*, Mt. Kmbl. Rush. 23, 23, 24. Heó to cyrcean eóde and hire gerihtan underfeng *she went to the church and received her rites*, Chr. 1093; Erl. 229, 11: Homl. Th. ii. 142, 9. Fóron to gefeohte forþ on gerihte *marched straight on to battle*, Judth. 11; Thw. 24, 23; Jud. 202. Man ána gǽþ mid his andwlitan up on gerihte *man alone walks with his face erect*, Bt. Met. Fox 31, 34; Met. 31, 17. On geryhte ongeán ðæne

mūþan *in a direction opposite the mouth,* Ors. 1, 1; Bos. 24, 8. On ge-rihte fram đam scipe to đam ancre *right from the ship to the anchor,* Shrn. 175, 19: Cod. Dipl. ii. 172, 20. DER. ald-, cyric-, geár-, woruld-geriht.

ge-riht; *adj.* RIGHT, *direct;* directus:—Đweoru beóþ on gerihte *erunt prava in directa,* Lk. Bos. 3, 5. [*Goth.* ga-raihts.]

ge-rihtan, -ryhtan; *p.* -rihte; *pp.* -rihted, -riht To set right *or straight, to direct, correct;* dirīgĕre, corrīgĕre, emendāre:—He wolde đone Cristes geléafan gerihtan *he would set right the faith of Christ,* Chr. 680; Erl. 41, 14. Đa þing đe he unfullfremed gemētte, mid heora fultume he đa gerihte and bētte *ea quæ mĭnus perfecta repĕrit, his quoque juvantĭbus corrĭgēbat,* Bd. 4, 2; S. 566, 3. Gerihtaþ Drihtnes weg *dirĭgĭte viam Dŏmĭni,* Jn. Bos. 1, 23. Fram sumum ungetýddum gerihted *a quodam impĕrīto emendātum,* Bd. 5, 24; S. 648, 24. Mín mundbyrd is geriht to đǽre róde *my protection is directed to the cross,* Rood Kmbl. 259; Kr. 131. [*Goth.* garaihtjan.]

ge-riht-lǽcan; *p.* -lǣhte; *pp.* -lǣht To justify, correct, direct, rectify, reprove;* rectificare, corrigere, arguere:—Se Hǽlend wolde đa synfullan gerihtlǽcan *the Healer [Saviour] would correct the sinful,* Homl. Th. ii. 470, 14. Đæt hys weorc ne sýn gerihtlǣht *ut non arguantur opera ejus,* Jn. Bos. 3, 20: Ps. Lamb. 36, 24. He đǽrbinnan wunode gerihtlǣcende đæt folc mid lāre to geleáfan *he dwelt therein directing the people by teaching to belief,* Swt. A. S. Rdr. 98, 113. Menn be his lāre heora líf gerihtlǣton *men by his instruction rectified their lives,* Homl. Th. ii. 146, 8. Gif we beóþ fram úrum đwyrnyssum gerihtlǣhte *if we be corrected from our perversities,* 124, 35.

ge-rihtnes, -ness, e; *f.* A setting right, correction;* correctio:—Be heora gerihtnesse *de illōrum correctiōne,* Bd. 5, 22; S. 644, 45. He wæs firena forgifnes and gerihtnes hǽþenra þeóda *he was forgiveness of sins and the setting right of heathen peoples,* Blickl. Homl. 163, 23.

ge-rihtreccan *to direct:*—Đē to gerihtrecenne đæt đū gesyhst myd đínes mōdes eágan god *to direct thee to see God with thy mind's eye,* Shrn. 177, 25.

ge-riht-wísian; *p.* ode; *pp.* od; *v. a.* To justify;* justificare:—He wolde hine sylfne gerihtwísian *ille vŏlens justĭfĭcāre seipsum,* Lk. Bos. 10, 29. Đū eart se đe me gerihtwísast *thou art he who justifieth me,* Ps. Th. 4, 1. Đa đe he him to clypode, đa đe he gerihtwísode, and đa đe he gerihtwísode, đa he gemǣrsode *those whom he called unto him he justified, and those whom he justified he glorified,* Homl. Th. ii. 366, 2. Hí synt gerihtwísode *justĭfĭcāta sunt,* Ps. Th. 18, 8. Gerihtwisud *justificatus,* Mt. Bos. 11, 19.

ge-rím, es; *n.* A number, computation, calendar, diary;* nŭmĕrus, compŭtātio, ephēmĕris = ἐφημερίς:—Đæs næs nā gerím *cujus non ērat nŭmĕrus,* Ps. Spl. 104, 32. Feówer and twentig wintra gerímes *twenty four winters in number,* Chr. 1065; Erl. 196, 26, 40; Edw. 7, 21: Cod. 224; Th. 296, 15; Sat. 502. Ofer gerím sŭper nŭmĕrum,* Ps. Spl. 39, 8: 38, 6. Ic ne mæg gerím witan heardra heteþonca *I cannot know the number of cruel enmities,* Exon. 70 a; Th. 261, 13; Jul. 314: Hy. 3, 17; Hy. Grn. ii. 281, 17. Gerím *ephēmĕrĭdes, nŭmĕrus quotĭdiānus,* Ælfc. Gl. 82; Som. 73, 51; Wrt. Voc. 47, 55. On getal gerímes *by reckoning of numbers,* Salm. Kmbl. 184, 7. On geríme *by number,* 192, 10. DER. đógor-gerím, geár-, heáfod-, niht-, þúsend-, un-, winter-.

ge-ríman, to -rímenne; *p.* de; *pp.* ed [ríman *to number*] To number, reckon;* nŭmĕrāre:—He āna mǣge ealle geríman *he alone can number all,* Cd. 163; Th. 205, 22; Exod. 439: Exon. 121 b; Th. 466, 4; Hö. 116. Đonne mæg he eác swilce geríman đínne ofspring *sēmen quŏque tuum nŭmĕrāre pŏtest,* Gen. 13, 16: Ps. Th. 104, 30. To gerímenne *to reckon,* Ors. 2, 5; Bos. 46, 39. Sceáwa heofon, hyrste gerím *behold the heaven, number its ornaments,* Cd. 100; Th. 132, 7; Gen. 2189. Đæm feówer bearn, forþ gerímed, in worold wōcon *to him four children, numbered forth, were born into the world,* Beo. Th. 118; B. 59.

ge-rímcræft, es; *m.* Arithmetic, art of numbering:*—Đe sēlost cunnon on gerímcræfte *that are best acquainted with arithmetic,* Bd. de nat. rerum; Wrt. popl. science 11, 1; Lchdm. iii. 256, 7: Hexam. 4: Norm. 8, 5.

ge-rímtæl, es; *n.* A number, reckoning:—Bión ou đǣm gerímtæle mid mínum brōþor *to be of the number with my brother,* H. R. 13, 11. [Cf. rímgetæl.]

ge-rínan; *pp.* -rinen To touch, take hold of, grip;* tangĕre, contingĕre, arrīpĕre:—Ne ofer đæt syđđan hine ō gerínan dorste *neque unquam exinde eum audēret contingĕre,* Bd. 3, 12; S. 537, 14. Wæs he sóna geríman líchomlíce untrumnysse *confestim languōre corpŏris tactus est,* 4, 3; S. 568, 37. Wæs he semninga fram deofle gerínen *sŭbĭto a diăbŏlo arreptus,* 3, 11; S. 536, 13, MS. B. v. ge-hrínan.

ge-ríne, es; *pl. nom. acc.* -u, -o, -a; *n.* A mystery;* mysterium:—Eów is geseald to witanne Godes ríces gerínu *vobis datum est nosse mysteria regni Dei,* Mk. Bos. 4, 11. v. ge-rýne.

ge-rinelíc; *adj.* Prosperous,* Hpt. Gl. 466.

ge-rinnan; *p.* -ran; *pp.* -runnen To run, run together, congeal, join;* coagulare, coaguli:—Nis nā gerunnen togædere seó Godcundnys and seó menniscnys *the divinity and the humanity are not mingled together,* Homl. Th. ii. 8, 5. Gerunnen is swā swā meolc heorte heora *coagulatum est sicut lac cor eorum,* Ps. Lamb. 118, 70. Munt gerunnen, dúne fæt,

to hwý wēne gē muntas gerunnene *mons coagulatus, mons pinguis, ut quid suspicamini montes coagulatos,* Ps. Spl. 67, 16. Gerunnen blōd *viscum,* 78; Som. 72, 52; Wrt. Voc. 46, 12. [*Goth.* ga-rinnan *to run together:* O. H. Ger.* gi-rinnan *coagulare.*]

ge-ríno *buildings;* ædificatiōnes, Mk. Skt. Rush. 13, 2. v. ge-rēn.

ge-ríp, es; *n.* [ríp *harvest*] A reaping, harvest;* messis:—Đæt geríp is micel *the reaping is great,* Homl. Th. ii. 530, 16. Geríp *messis,* Ælfc. Gr. 9, 28; Som. 11, 56: Wrt. Voc. 74, 69: Gen. 8, 22. Biddaþ đæs gerípes hláford, đæt he asende wyrhtan to his gerípe *pray to the lord of the reaping, that he send workmen to his reaping,* Homl. Th. ii. 530, 20. On Godes gerípe *in God's reaping,* 530, 19. Hwā gemenigfylt đæt geríp of feáwum cornum *who multiplies the harvest from a few grains of corn,* i. 184, 31.

ge-rípan; *p.* -ráp, *pl.* -ripon; *pp.* -ripen To reap;* mĕtĕre:—Hie heora corn geripon *they reaped their corn,* Chr. 896; Th. 172, 32, col. 2. On đæt gerád đe he ǽlce geáre gerípe *on the condition that each year he reap,* Cod. Dipl. ii. 398, 21.

ge-rípian; *p.* ode, ede; *pp.* od, ed [rípian *to ripen*] To ripen, grow old;* mātūrāri, sĕnescĕre:—Nǣron hí gerípode to slege *they were not ripe for slaughter,* Homl. Th. i. 84, 5. On wintrum gerípod *ripe in years,* ii. 24, 23. Mín hláford gerípod ys *dŏmĭnus meus vĕtŭlus est,* Gen. 18, 12. Gerípod *mātūrus,* C.·R. Ben. 43. Gerípod *mātūrus,* C.·R. Ben. 43.

ge-rísan; *3rd sing. pres.* -ríseþ, -ríst, *pl.* -rísaþ; *p.* -rás, *pl.* -rison; *pp.* -risen To behove, become, befit, suit;* dĕcēre, convĕnīre: generally used impersonally:—Gold geríseþ on guman sweorde *gold is fitting on a man's sword,* Exon. 91 a; Th. 341, 14; Gn. Ex. 126. Đe geríseþ lofsang *te dĕcet hymnus,* Ps. Spl. 64, 1: 92, 7. Cyninge geríst rihtwísnys *righteousness becomes a king,* Homl. Th. ii. 318, 32: i. 418, 8. Đe him betst geríst *which suits him best,* Bt. 34, 10; Fox 148, 20: Menol. Fox 117; Men. 58. Wera gehwylcum wíslícu word gerísaþ *to every man wise words are fitting,* Exon. 91 b; Th. 343, 34; Gn. Ex. 166. Swā đam þeódne gerás *as was fitting to the master,* 49 a; Th. 168, 34; Gú. 1087. Đæt đæm weorce nānum men ne geríse bēt to fandienne, đonne đam wyrhtan đe hit worhte *that it became no man better to prove the work than the workman who made it,* Ors. 1, 12; Bos. 36, 37.

ge-rísan; *pp.* -risen To seize, take;* rapere:—Geríseþ *rapit,* Mt. Kmbl. Rush. 13, 19. Gerísaþ *rapiunt,* 11, 12. Sóna wæs gerisen and genumen of middanearde *rapta confestim de mundo,* Bd. 4, 19; S. 589, 5, note.

ge-risen, -risne [?], es; *n.* A seizing;* rapina:—Ne begitest đū nā đæt ríce on gerisne woruldlícra þinga *non in præda, nec in rapina regnum tibi dabitur,* Guthl. 19; Gdwin. 78, 5. v. ge-rísan *to seize.*

ge-risene, -risne, -rysne; *adj.* Fit, convenient, proper;* congruus, decens, conveniens:—He sealde his lǣreowum gerisene stōwe and eþel heora hāde *doctoribus suis locum sedis eorum gradui congruum donaret,* Bd. 4, 26; S. 488, 19. Æfter gerisenre áre swā myclum B' *juxta honorem tanto Pontifici congruum,* 5, 19; S. 636, 45. Đa gerisno digna, Lk. Skt. Lind. 12, 48. Đis þinceþ gerisne *this seems fitting,* Cd. 114; Th. 149, 17; Gen. 2476. Swā gerysne ne wæs *as was not seemly,* 76; Th. 94, 22; Gen. 1565: 9; Th. 11, 2; Gen. 169: Beo. Th. 5299; B. 2653. Hit is ealles gerisnost *it is most fitting,* Blickl. Homl. 205, 24.

ge-risene, -risne, -rysne, es; [seems to occúr only in *pl.*] *n.* What is fitting, decent:—Godes hūs sindon innan bestrýpte ælcra gerisna *God's houses are stripped within of everything seemly,* Swt. A. S. Rdr. 106, 43: Th. Chart. 511, 4. Đæt heora gerisna nǣre đæt hý swā heáne hý gehohtan đæt hý heora gelícan wurdan *that it was not fitting for them [the Romans] to think themselves so low as to be their [the Carthaginians'] equals,* Ors. 4, 6; Bos. 86, 27: Cd. 93; Th. 242, 17; Dan. 420. Ne fremest đū gerysnu and riht wiþ me *thou dost not do what is fitting and right towards me,* 102; Th. 135, 19; Gen. 2245: 111; Th. 146, 4; Gen. 2432. Gif he mōt đǽr rihtes and gerysena onbrúcan *if he can there enjoy what is right and fitting,* Runic pm. 23; Kmbl. 344, 6. Ryhtum gerisnum *right fittingly,* Exon. 80 b; Th. 302, 2; Fä. 30.

ge-risenlíc; *comp. m.* -lícra, *f. n.* -lícre; *adj.* Convenient, suitable, befitting;* convĕniens, aptus:—Ne þuhte hit me nǣuht gerisenlíc *I should not think it at all suitable,* Bt. 41, 2; Fox 244, 27. Ægđer đara is swíđe nyt weorc and gerisenlíc *either is a very useful and befitting work,* Prov. Kmbl. 60. On đǽm gerisnlícan hēhsetle *on that seemly throne,* Blickl. Homl. 9, 26. Gerisenlíc me to wosanne *oportet me esse,* Lk. Skt. Lind. 2, 49. Đē is gerisenlícre đæt đū sí mid rihte ofersteled, đonne đū oferstele ōđerne man mid wōge *it is more befitting thee to be overruled with right than to overrule another with wrong,* Prov. Kmbl. 8: Bd. 2, 13; S. 516, 23.

ge-risenlíce; *comp.* -lícor; *adv.* Becomingly, fitly;* apte:—Seó wæs gerisenlíce gehlidad mid gelíce stāne *opercŭlo sĭmĭlis lăpĭdis aptissĭme tectum,* Bd. 4, 19; S. 588, 32: 3, 17; S. 544, 4, col. 1. Gerisenlícor *aptius,* 2, 13; S. 517, 2: 3, 29; S. 561, 29.

ge-risennes, -risnes, se; *f.* Conveniency, agreeableness, congruity;* convenientia, Cot. 58.

ge-rislíc; *adj.* Convenient, Bd. 5, 19; S. 636, 34, note. v. ge-risenlíc.

ge-risnian *to agree, accord;* convenire, Cot. 38.

ge-ríxian *to rule;* regnare, Lk. Skt. Lind. 19, 14. v. ge-rícsian.

gerla, an; *m. Tribute:*—To sellanne ðone gerlo *dare tributum,* Lk. Skt. Rush. 20, 22.

gérlíc; *adj. Yearly;* annuus, Rtl. 49, 25: Shrn. 208, 28.

Germania, e [=æ]; *f. Germany.* The Germania of Alfred extended from the Don on the east to the Rhine and the German Ocean on the west; and from the Danube on the south to the White Sea on the north; it therefore embraced nearly the whole of Europe north of the Rhine and the Danube. Its great extent will be seen by the countries mentioned in the notes from 5 to 39, and in the text of Ors. Bos. pp. 35–40. See also Cluverii Introductionis in universam Geographiam, Libri vi. Amstelædami, 4to. 1729, Lib. iii. Cap. 1. De veteri Germania, pp. 183–186, and the map of Europe, p. 72. Also the very learned work, Cluverii Germania antiqua, Lugd. Batavorum, Elzevir, Fol. 1616: Lib. 1: Cap. xi. De magnitudine Germaniæ antiquæ, pp. 94–98, also Lib. iii. Cap. xxxviii. pp. 157–162, and the map, p. 3. Also Cellarii Geographia Antiqua, Cantab. 4to. 1703, pp. 309–313. Warnefried's Hist. Longob. Lib. i. Cap. 1:—Nū wille we ymb Europe land-gemǽre reccan, swā mycel swā we hit fyrmest witon.—Fram ðære eá Danais, west ōþ Rín ða eá, [seó wylþ of ðæm beorge ðe man Alpis hǽt, and yrnþ ðonne norþ-ryhte on ðæs gársecges earm, ðe ðæt land ūtanymbliþ, ðe man Bryttannia hǽt];—and eft sūþ ōþ Donua ða eá, [ðære ǽwylme is neáh ðære eá Rínes, and is siðdan eást yrnende wiþ norþan Crēca land ūt on ðone Wendel-Sǽ];—and norþ ōþ ðone gársecg, ðe man Cwēn-Sǽ hǽt: binnan ðǽm syndon manega þeóda; ac hit man hǽt eall, GERMANIA *now we will speak, as much as we know, about the boundaries of Europe.—From the river Don, westward to the river Rhine,* [*which springs from the Alps, and then runs right north into the arm of the ocean, that lies around the country called Britain*];—*and again south to the river Danube,* [*whose spring is near the river Rhine, and which afterwards runs east by the country north of Greece into the Mediterranean Sea*];—*and north to the ocean, which is called the White Sea: within these are many nations, but it is all called* GERMANIA, Ors. 1, 1; Bos. 18, 20–28. Cōmon hí of þrim folcum ðam strangestan Germanie ðæt of Seaxum, and of Angle, and of Geátum' *advenerunt de tribus Germaniæ populis fortioribus, id est Saxonibus, Anglis, Jutis,* Bd. 1, 15; S. 483, 20.

gern *yarn, spun wool.* DER. nett-gern. v. gearn.

gernan; *p.* de; *pp.* ed *To desire;* desīdērāre:—He ðæs biscophādes gernde *he desired episcopal ordination,* Chr. 1048; Erl. 177, 23. v. gyrnan.

gern-winde, es; *m? A yarn-winder, reel;* conductum [ăpud textōres], Wrt. Voc. 282, 2. v. gearn-winde.

ge-rora. v. ge-hror.

ge-rósod *rosy, belonging to roses;* rosaceus, Som.

ge-rostod *roasted;* assus, Som.

ge-rótsian [= geunrōtsian?] *to make sad;* contristare, Rtl. 56, 20.

ge-rówen *rowed.* v. rōwan.

gers, es; *n. Grass;* herba:—Se ðe forþatýhþ wyrtcynren oððe gers þeówdōmes manna *qui prodūcit herbam servītūti, hŏmĭnum,* Ps. Lamb. 146, 8: Mk. Skt. Lind. 4, 28. v. gærs.

GERST; GRIST, *pearled barley;* frumentum quodvis tritum, Lye.

gersum, es; *m. n:* gersuma, an; *m. Treasure;* thēsaurus, Chr. 1070; Erl. 209, 13: 1090; Erl. 226, 38: 1047; Erl. 177, 7. v. gærsum, gærsama; and see Grm. D. M. 840.

ge-rūm, es; *n.* [rūm *space*] *Room, space;* spătium:—Hí nāuðer ne gestillan ne mōton, ne eác swíðor styrian, ðonne he him ðæt gerūm his wealdleðeres toforlǽt *they neither can be still, nor yet move farther, than he allows to them the space of his rein,* Bt. 21; Fox 74, 8. Eódon on gerūm eorlas āglēwe *the men learned in law went apart,* Elen. Kmbl. 639; El. 320. Cyning healdeþ me on heáðore, hwīlum lǽteþ eft on gerūm sceacan *the king holds me in restraint, sometimes again lets me go at large,* Exon. 105 b; Th. 401, 20; Rä. 21, 14.

ge-rūma, an; *m.* [rūm *room*] *A room, place, space;* lŏcus, spătium:—Ic his bīdan ne dear rēðes on gerūman *I dare not await him fierce in my place,* Exon. 104 b; Th. 397, 7; Rä. 16, 16.

ge-rūme; *adj. Ample, roomy, expanded, made open;* amplus, spătiōsus, dīlātus, pătēfactus:—Is mín mōd gehǽled, hyge ymb heortan gerūme *my mind is healed, the thoughts around my heart expanded,* Cd. 35; Th. 47, 11; Gen. 759. Syndon ðíne willan rihte and gerūme *thy wishes are right and great,* 188; Th. 234, 12; Dan. 291. [*Ger.* geraum *spacious:* O. H. Ger. kirūmo *opportunus.*]

ge-rumpen *rough, wrinkled;* rugosus:—Gerumpenu nædre *cerastes, coluber,* Cot. 38.

ge-rūna, an; *m. A counsellor:*—Gerūna *sinmistes vel consecretalis,* Ælfc. Gl. 7; Som. 56, 66; Wrt. Voc. 18, 18. Gerūna *a secretis, vel principis consiliarius,* 113; Som. 79, 127; Wrt. Voc. 60, 32.

ge-runnen *run together, congealed, joined;* coagulatus, Ps. Lamb. 67, 16: 118, 70: Ælfc. Gl. 33; Som. 12, 17; Wrt. Voc. 28, 1: 78; Som. 72, 52; Wrt. Voc. 46, 12; *pp.* of ge-rinnan.

gerwan, gerwian, gerwigan; *p.* ede, ode; *pp.* ed, od *To make ready prepare, make, construct;* părāre, præpărāre, făcĕre, construĕre:—Ciricean getimbran, gerwan Godes tempel *to build a church, to construct a temple*

---

of God, Andr. Kmbl. 3266; An. 1636. Gerwigan wífe hūs wexinge getácnaþ *to prepare* [*one's*] *house for a wife betokens increase,* Som. 205; Lchdm. iii. 210, 3. v. gearwian.

ge-ryd, -rid; *adj. Prepared, ready, usual;* paratus:—Ðeáh se graf geryd sī *though the grave be prepared,* Lchdm. iii. 355, 2, col. 1; Shrn. 184, 20. Moīses dyde on geryde orcas *Moses put it into the usual basons,* Ex. 24, 6.

ge-ryht. v. ge-riht.

ge-ryhtan *to set right;* dirĭgĕre:—He wolde ðone Cristes geleáfan geryhtan *he would set right the faith of Christ,* Chr. 680; Erl. 40, 12. v. ge-rihtan.

ge-rýman; *p.* de; *pp.* ed [rýman *to make room*] *To extend, enlarge, make room, open, manifest, expand;* dīlātāre, lŏcum dāre, apĕrīre, expandĕre:—Ongyn ðē scip wyrcan, on ðam ðū monegum scealt reste gerýman *begin thou to make a ship, in which thou shalt make room for resting-places to many,* Cd. 65; Th. 78, 36; Gen. 1304. Ic gerýme ðíne gemǽro dīlātăvēro termĭnos tuos, Ex. 34, 24. He ōðrum gerýmeþ wyrmum to wiste *he clears the way for other worms' repast,* Exon. 100 a; Th. 374, 9; Seel. 123. Ic him lífes weg gerýmde *I opened the way of life to them,* Rood Kmbl. 175; Kr. 89: Elen. Kmbl. 2496; El. 1249. Ðū me gerýmdes dīlātasti mihi, Ps. Th. 4, 1. Octauianus gerýmde Rōmāna rīce *Octavianus extended the Roman empire,* Homl. Th. i. 32, 18. Ðæt hie him ōðer flet eal gerýmdon *that they would wholly open to him another dwelling,* Beo. Th. 2177; B. 1086. Se weg biþ us gerýmed *the way is open to us,* Boutr. Scrd. 20, 32: Andr. Kmbl. 3159; An. 1582: Bt. Met. Fox 1, 37; Met. 1, 19: Homl. Th. i. 564, 18: 28, 12. Se ðe his godcundnesse mid sōþum wísum gerýmeþ *who truly manifests his divinity,* Blickl. Homl. 179, 24. Gif him swā byþ gerýmed *if he has opportunity,* Basil admn. 9; Norm. 52, 28. On ðam rýmette ðe se cing hēt gerýmen into ealdan mynstre *in the space that the king ordered to cede to the old monastery,* Ch. Th. 231, 26.

ge-rýne, -ríne, -rēne, es; *pl. nom. acc.* -u, -o, -a; *n. A mystery, a sacrament;* mysterium:—Ðæt dēgol wæs Dryhtnes gerýne *that was a secret mystery of the Lord,* Exon. 8 b; Th. 3, 25; Cri. 41. Ðæt monnum nis cūþ gerýne *that mystery is not known to men,* 9 a; Th. 7, 2; Cri. 95. Dryhtnes gerýne *the mystery of the Lord,* 49 a; Th. 169, 14; Gū. 1094: Lk. Bos. 8, 10. Ðæt word ðæs heofonlícan gerýnes *the word of the heavenly mystery,* Blickl. Homl. 17, 9: 7. Eów is geseald to witanne heofena ríces gerýnu *vobis datum est nosse mysteria regni cælorum,* Mt. Bos. 13, 11. Ða gerýnu Cristes menniscnysse *the mysteries of Christ's humanity,* Homl. Pasc. Lisle 12, 17. Hit forhæfed gewearþ ðætte hie sædon swefn cyninge, wyrda gerýnu *it was denied that they should tell the dream to the king, the mysteries of the fates,* Cd. 179; Th. 225, 4; Dan. 149. Engel Drihtnes wrāt in wāge worda gerýnu *the angel of the Lord wrote on the wall mysteries of words,* 210; Th. 261, 9; Dan. 723. On ðē wrāt wuldres God gerýno *on thee the God of glory wrote* [*his*] *mysteries,* Andr. Kmbl. 3020; An. 1513. Ðæt hie ðæt hālige gerýne ārwurþlíce breman mǽgen *that they may reverently celebrate the holy mystery,* L. E. I. 4; Th. ii. 404, 27: Bd. 1, 27; S. 496, 23, 43: 497, 2, 5. [*Goth.* ga-rūni *counsel: O. Sax.* gi-rūni *mystery: O. H. Ger.* ki-rūni *mysterium, sacramentum.*] DER. gāst-gerýne, gæst-, word-. v. rýne, iūn, gerýno, ge-rýnu.

ge-rýnelíc; *adj. Mystical;* mysticus:—Gerýnelíco word sprecende *mystica verba loquens,* Bd. 2, 1; S. 500, 26. Of gerýnelícum gāste *ex mystico spiramine,* Hymn. Surt. 43, 36. Ðās gerýnelícan þing *hæc mystica,* 94, 17: Blickl. Homl. 165, 35.

ge-rýnelíce *mystically;* mystice, Cot. 131.

ge-rýno; *indecl. n. A mystery:*—Ðis Eástorlíce gerýno us æteóweþ ðæs ēcean lífes sweotole bysene *this Easter mystery* [*Christ's resurrection*] *shews us a clear example of the life eternal,* Blickl. Homl. 83, 7. v. ge-rýne.

ge-rýnu; *indecl. f. A mystery:*—Ðeós gerýnu is wedd *this mystery is a pledge,* Homl. Th. ii. 272, 6. Þurh gāstlícere gerýnu *through a spiritual mystery,* 268, 29: 260, 12: 262, 22: Bd. de nat. rerum; Wrt. popl. science 14, 1; Lchdm. iii. 264, 11. [*O. H. Ger.* gi-riuna, *f.*] v. ge-rýne.

ge-rypon *reaped,* Chr. 896; Th. 172, 33, col. 1; = ge-ripon; *p. pl. of* ge-rípan.

ge-rysene *fit.* v. ge-risene.

gēs *geese,* L. In. 70; Th. i. 146, 18, MS. H; *pl. nom. acc.* of gōs.

ge-saca; *m. An adversary;* adversarius:—Geþafedon ðæt his gesacan *concesserunt id adversarii,* Bd. 2, 2; S. 502, 24. On gesacum on his adversaries, Cd. 4; Th. 4, 25; Gen. 59: Beo. Th. 3551; B. 1773. Gesaca *æmulus,* Ælfc. Gl. 114; Som. 80, 17; Wrt. Voc. 60, 51. v. sacan.

ge-sacan ? *p.* -sōc, *pl.* -sōcon; *pp.* -sacen *To oppose, strive against;* adversari:—Gesacan sceal sāwl-berendra, niðða bearna, gearwe stōwe *shall strive against the place prepared for those having souls, for the children of men,* Beo. Th. 2012, note; B. 1004. v. sacan.

ge-sacu; *f. Contention, hostility;* contentio, hostilitas, Beo. Th. 3479; B. 1737. v. sacu.

ge-sadelod, -sadolod; *part.* [sadelian *to saddle*] *Saddled;* strātus:—Twā hors, ān gesadelod and ōðer ungesadelod *two horses, one saddled*

and the other unsaddled, L. C. S. 72; Th. i. 414, 17. Eahta hors, feówer gesadelode [gesadolode, MS. A.] and feówer ungesadelode *eight horses, four saddled and four unsaddled*, 72; Th. i. 414, 5, 10. DER. un-gesadelod.

ge-sadian; *p.* ode, ade; *pp.* od *To satisfy, fill*; saturare:—Ƀeóþ gesadode oððe gefyllede treówa feldes *saturabuntur ligna campi*, Ps. Lamb. 103, 16. Drihten ðé gesadade mid ðý sélestan hwǽtecynnes holde lynde *Dominus adipe frumenti satiat te*, Ps. Th. 147, 3. v. sadian.

ge-sæccan *to dispute, discuss*; disserere, Mt. Kmbl. p. 11, 2.

ge-sæcgan *to say, tell*, Ps. Th. 77, 8. v. ge-secgan.

ge-sǽd *said, told, proved*, Ors. 1, 8; Bos. 31, 33, 34; *pp. of* ge-secgan.

ge-sǽgan; *p.* de; *pp.* ed [sǽgan *to cause to sink*] *To lay low, cast down*; prosternĕre, incurvāre:—Hǽfdon ealfela Eótena cynnes sweordum gesǽged *they had laid low full many of the Jutes' race with their swords*, Beo. Th. 1772; B. 884: Judth. 12; Thw. 25, 36; Jud. 294. Ic eom gesǽged, *incurvātus sum*, Ps. Th. 37, 8.

ge-sǽgde, -sǽde, *pl.* -sǽgdon *said, told*, Beo. Th. 4321; B. 2157: Bd. 4, 18; S. 587, 2: 1, 12; S. 481, 3; *p. of* ge-secgan.

ge-sǽgdnis, e; *f. A mystery*; mysterium, Mt. Kmbl. Lind. 13, 11.

ge-sǽgen *a saying, telling, tradition*, Bd. pref; S. 472, 8, 20, 25, 30: 5, 23; S. 647, 17: Blickl. Homl. 55, 26. v. ge-segen, ge-sagun.

ge-sæhtlian; *p.* ode, ade; *pp.* od, ad [sæhtlian *to reconcile*] *To reconcile*; reconciliāre:—Wearþ Eádgár wið ðone cyng gesæhtlad *Edgar was reconciled with the king*, Chr. 1091; Erl. 228.

ge-sæhtniss. v. ge-sehtniss.

ge-sǽlan; *p.* de; *pp.* ed [sǽlan *to bind, tie*] *To bind, tie*; ligāre:—Ða folan hý gesǽlaþ *they tie the foals*, Nar. 35, 11. Ðæt is se ealda feónd ðone he gesǽlde *that is the ancient fiend whom he bound*, Exon. 96 a; Th. 359, 7; Pa. 59. He ligeþ synnum gesǽled *he lies bound with sins*, 18 b; Th. 46, 12; Cri. 736: Beo. Th. 5521; B. 2764: Cd. 37; Th. 47, 23; Gen. 765: 200; Th. 248, 30; Dan. 251. Ðonne geméte gyt ðǽr eoselan gesǽlede *then shall ye find there an ass tied*, Blickl. Homl. 69, 36: Mt. Kmbl. Rush. 21, 2.

ge-sǽlan; hit -sǽleþ, -sǽlþ; *p.* de; *pp.* ed [sǽl *an occasion*] *To happen, come to pass, befall*; accidĕre, evenīre:—Hú gesǽleþ ðæt *how doth that happen?* Salm. Kmbl. 698; Sal. 348: Andr. Kmbl. 1021; An. 511: 1029; An. 515. Gif hit ǽfre gesǽlþ, ðæt . . . *if it ever happen that* . . ., Bt. Met. Fox 13, 43; Met. 13, 22: Th. Ch. 472, 4: 100, 5. Me gesǽlde ðæt ic mid sweorde ofslóh niceras nigene *it befell me that I slew with my sword nine monsters*, Beo. Th. 1152; B. 574: 1784; B. 890: 2504; B. 1250. Ðeáh eów nú gesǽle, ðæt . . . *though it now happen to you that* . . ., Bt. Met. Fox 10, 47; Met. 10, 24. Uncúþ hú him æt ǽhtum gesǽle *it is unknown how it may befall him in the matter of property*, Prov. Kmbl. 20.

ge-sǽlan; *p.* de *To be successful, succeed*:—Ðam ðe eahtan wile sáwla gehwylcre ðǽr he gesǽlan mæg *to him who will persecute every soul if he can manage it*, Exon. 37 b; Th. 123, 6; Gú. 318.

ge-sǽlig; *adv. Happily*; fauste, Cot. 89.

ge-sǽli; *adj. Happy*; félix:—Hweðer micel feoh mǽge ǽnigne mon dón swá gesǽline, ðæt he nánes þinges máran ne þurfe *can much money make any man so happy that he may need nothing more?* Bt. 26, 1; Fox 90, 13. v. ge-sǽlig.

ge-sǽlig, es; *m. One who carries a standard*; signifer, Hpt. Gl. 495.

ge-sǽlig, -sǽli; *comp.* ra; *superl.* ost, ust; *adj.* [sǽlig *happy*] *Happy, prosperous, blessed, fortunate*; félix, beatus, fortūnātus:—Seth wæs gesǽlig *Seth was happy*, Cd. 56; Th. 69, 19; Gen. 1138: 130; Th. 165, 28; Gen. 2738: Bt. Met. Fox 23, 3; Met. 23, 2. Se gesǽliga hlísa *félix rūmor*, Bd. 4, 23; S. 594, 41: Exon. 61 a; Th. 222, 17; Ph. 350. Ðæt gesǽlige weorud *the blessed company*, 26 a; Th. 76, 33; Cri. 1249. Wǽron swíðe gesǽlige *they were very happy*, Cd. 1; Th. 2, 12; Gen. 18: 220; Th. 282, 33; Sat. 296. Hí fram gesǽlgum tídum gilpaþ *they boast of happy times*, Ors. 5, 2; Bos. 103, 11: Exon. 32 a; Th. 101, 1, 17; Cri. 1652, 1660. Mǽrþa gesǽligost *most blessed of glories*, Salm. Kmbl. 136; Sal. 67. Cild gesǽligust *a very prosperous child*, Lchdm. iii. 196, 21. Se gesǽlgosta *the happiest*, Bt. 26, 1; Fox 90, 10.

ge-sǽlig-líc, -sǽl-líc; *adj. Happy, fortunate*; félix, fortūnātus:—Ðam ðe líf forgeaf gesǽliglíc *to him who gave him a happy life*, Cd. 137; Th. 172, 14; Gen. 2844: Exon. 23 b; Th. 66, 29; Cri. 1079. v. ge-sǽlig.

ge-sǽlig-líce, -sǽli-líce, -sǽl-líce; *adv. Happily*; félicĭter:—Gesǽliglíce *félicĭter*, Scint. 1. Manige habbaþ genóg gesǽliglíce [gesǽllíce, MS. Cot.] gewífod *many have married happily enough*, Bt. 11, 1; Fox 32, 5. Gesǽliglíce *félicĭter*, Bd. 5, 19; S. 639, 27.

ge-sǽlignes, -nys, -ness, -nyss, e; *f. Happiness*; félicĭtas:—Ðǽr biþ engla dreám, sib and gesǽlignes *there is joy of angels, peace and happiness*, Exon. 32 b; Th. 102, 23; Cri. 1677. Gif ðú wille ðysses lífes gesǽlignysse mid us brúcan *si vis pĕrennis vītæ félicĭtāte perfrui*, Bd. 1, 7; S. 477, 35.

ge-sǽli-líce *happily*; félicĭter, Bd. 5, 19; S. 639, 27. v. ge-sǽlig-líce.

ge-sǽl-líc; *adj. Happy*; félix:—Gesǽllíc mon *a happy man*, Bt. Met. Fox 2, 34; Met. 2, 17. v. gesǽlig-líc.

gesǽl-líce *happily*, Bt. 11, 1; Fox 32, 5, MS. Cot. v. gesǽlig-líce.

ge-sæltan; *pp.* -sælted, -sælt *To salt*, Mt. Kmbl. Lind. 5, 13: Mk. Skt. Lind. 9, 49.

ge-sǽlþ, e; *f.* [sǽlþ *happiness*] *Happiness, felicity, prosperity, wealth, good, advantage*; félicĭtas, prospĕrĭtas, bŏnum:—Sió sóðe gesǽlþ *the true happiness*, Bt. 23; Fox 78, 30: 34, 2; Fox 134, 32: 34, 4; Fox 138, 21, 24. God is full ǽlcere gesǽlþe *God is full of all happiness*, 34, 3; Fox 136, 20. Sóþra gesǽlþa *of true felicities*, Bt. Met. Fox 21, 49; Met. 21, 25. To ðǽm gesǽlþum *to the felicities*, 21, 7, 17; Met. 21, 4, 9. He selþ ða gesǽlþa ðǽm gódum *he gives felicities to the good*, Bt. 39, 2; Fox 214, 2, 5: 34, 1; Fox 134, 7. Ðú miht ða sóðan gesǽlþa gecnáwan *thou mayest discover the true goods*, 23; Fox 78, 32.

ge-sǽlþ *happens*, Bt. Met. Fox 13, 43; Met. 13, 22; *3rd sing. pres. of* ge-sǽlan.

ge-séman. v. ge-sýman.

ge-sǽt, *pl.* -sǽton *sat, sat down*, Beo. Th. 5427; B. 2717: Elen. Kmbl. 1732; El. 868; *p. of* ge-sittan.

ge-sǽtnys. v. ge-setnes.

ge-sagian *to say, tell*; dicere:—Gesaga him *tell them*, Beo. Th. 781; B. 388: Bd. 1, 7; S. 477, 30. v. sagian.

ge-sagu *a narration*, Lk. Skt. Lind. 1, 1.

ge-sagun, e; *f. A narration*, Lk. Skt. Rush. 1, 1.

ge-sealde *sold*; tradidit, Cd. 226; Th. 301, 2; Sat. 575, = ge-sealde; *p. of* ge-sellan.

ge-saldniss, e; *f. A giving*:—Ic berhtwulf rex ðas míne gesaldnisse trymme *I, king Berhtwulf, confirm this my gift*, Cod. Dipl. Kmbl. ii. 5, 32.

ge-sam, in composition, denotes *together, with*; simul, con. v. sam.

ge-sam-híwan; *gen.* -híwena, *pl. m. Married persons*; conjugati, conjugia:—Unriht gewuna is arisen betwih gesamhíwum *prava in conjugatorum moribus consuetudo surrexit*, Bd. 1, 27; S. 493, 34, note: Bd. 4, 5; S. 573, 14, note. v. gesinhíwan.

ge-samnian, -somnian; *p.* ode, ade, ede; *pp.* od, ad, ed. **I.** *to gather, collect*; congrĕgāre, collĭgĕre:—Se áncenneda ealle gesamnaþ *the only begotten one shall gather all*, Soul Kmbl. 102; Seel. 51. Valentinianus gesamnode weorod *Valentinian gathered an army*, Chr. 380; Erl. 11, 4: Cd. 174; Th. 219, 9; Dan. 52. He hí of sídfolcum gesamnade *de regiōnĭbus congrĕgāvit eos*, Ps. Th. 106, 2. Gesamnedon síde hérigeas folces frumgáras *the leaders of the people collected their wide bands*, Andr. Kmbl. 2135; An. 1069: Ps. Th. 125, 6. Us gesamna of wídwegum *congrĕga nos de nātiōnĭbus*, 105, 36. Wæs eall-geador to ðam þingstede þeód gesamnod *the people was collected together in the public place*, Andr. Kmbl. 2198; An. 1100: Elen. Kmbl. 563; El. 282. Mycle mænigeo wǽron gesamnode to hym *congrĕgātæ sunt ad eum turbæ multæ*, Mt. Bos. 13, 2: 26, 3. **II.** *to unite, join*; consōciāre, jungere:—Geférscipas fæste gesamnaþ *firmly unites societies*, Bt. Met. Fox 11, 186; Met. 11, 93: Bt. 21; Fox 74, 38. Se gesamnade sáwle to líce *he united the soul to the body*, Bt. Met. Fox 17, 23; Met. 17, 12. **III.** *v. intrans. To collect, come together*; congrĕgāri, convenīre:—Hí gesamniaþ congrĕgāti sunt, Ps. Th. 103, 21. Gesamnadon weras *the men collected together*, Andr. Kmbl. 3270; An. 1638.

ge-samning *a synagogue*; synăgōga, Ps. Th. 85, 13. v. ge-samnung.

ge-samnung, -somnung, -samning, -somning, e; *f. A meeting, assembly, council, union, congregation, synagogue, church*; conventus, conventio, concilium, congrĕgātio, synăgōga, ecclēsia:—Gesamnung *conventus, conventio*, Ælfc. Gl. 87; Som. 74, 48; Wrt. Voc. 50, 30. Se wæs ðære gesamnunge ealdor *ipse princeps synăgōgæ ĕrat*, Lk. Bos. 8, 41. Fram gesamnunge mycelre *a conciīio multo*, Ps. Spl. C. 39, 14. Ealra heora eágan on ðære gesamnunge wǽron on hyne behealdende *omnium in synăgōga ŏcŭli ĕrant intendentes in eum*, Lk. Bos. 4, 20: 8, 49: Jn. Bos. 6, 59: 18, 20. He eóde on reste-dæge on ða gesamnunge æfter his gewunan *intrāvit sĕcundum consuetūdĭnem die sabbăti in synăgōgam*, Lk. Bos. 4, 16. He lǽrde híg on hyra gesamnungum *dŏcēbat eos in synăgōgis eōrum*, Mt. Bos. 13, 54: 23, 6: Mk. 1, 39: 12, 39: Lk. Bos. 4, 44: 11, 43: 20, 46. On gesamnunga in synăgōgas, Lk. Bos. 21, 12. On gesamnunga hāligra *in ecclēsia sanctōrum*, Ps. Lamb. 149, 1. Þurh ða gesamnunga we wǽron gefreoþode feónda gafoles *through that union we were freed from devils' tribute*, Blickl. Homl. 105, 22.

ge-samodlǽcan *to put together*; conlocare, Blickl. Gl. 112, 8.

ge-sanco; *pl. n. Suckers*:—Gesanco *exigia*, Wrt. Voc. 287, 35.

ge-sárgian; *p.* ode, ade; *pp.* od, ad [sárgian *to afflict*] *To afflict, trouble, damage*; affligere, tribulāre:—Biþ untreó gesárgad *the faithless shall be afflicted*, Exon. 22 a; Th. 59, 34; Cri. 962: 22 a; Th. 60, 18; Cri. 971. Ne sceal nán mon siócne monnan gesárgodne swencan *no one ought to afflict a sick troubled person*, Bt. 38, 7; Fox 210, 20. Wǽron hie gesárgode *they were damaged*, Chr. 897; Erl. 96, 8.

ge-sáwan; *pp.* -sáwen *To sow*; seminare, Mt. Kmbl. Lind. 13, 3. DER. sáwan.

ge-sáweled *part. p. Provided with a soul*; animatus, Mk. Skt. p. 1, 11.

ge-scád *distance, reason*, Exon. 94 a; Th. 353, 16: Reim. 13. v. gesceád.

ge-scádenlíce; *adv. Separately, distinctly*; separatim, Cot. 198.

ge-scádwís *reasonable, intelligent*. v. ge-sceádwís.

ge-scádwíslíce; *comp.* or; *adv. Wisely, prudently, clearly*; prudenter, Ors. 1, 10; Bos. 32, 20: 2, 1; Bos. 38, 29.

ge-scádwyrt, e: *f. Oxeye*, Lchdm. ii. 274, 18; see the glossary at the end of the volume, and also iii. 328.

ge-scæft, e; *f. Creation*; creātio:—On ða beorhtan gescæft *on the bright creation*, Cd. 216; Th. 273, 20; Sat. 139. v. ge-sceaft.

ge-scænan, -sceánan, -scénan; *p.* de; *pp.* ed *To diminish, break, bruise, shake, shatter*; contērĕre, confringĕre, conquassāre:—God heora tóþas gescǽneþ *Deus contĕret dentes eōrum*, Ps. Th. 57, 5: 67, 21. Dú dæs myclan dracan heáfod gescǽndest *tu confrēgisti cáput dracōnis magni*, 73, 14. Ða he sylfa oft gebræc and gescǽnde *quas ipse ăliquando contrĭvĕrat*, Bd. 5, 12; S. 631, 27. Gesceányþ heáfda *conquassabit capita*, Ps. Spl. C. 109, 7.

ge-scænctest *thou hast given drink*; potasti, Ps. Lamb. 59, 3.

ge-scæned, -scæned [?]; *part. p. Ornamented* [?]:—Sweord swíðe gescæned, Salm. Kmbl. 444; Sal. 222. Cf. on dæm scennum scíran goldes, Beo. Th. 3392; B. 1694. Grein compares with *O. H. Ger.* giskeinan, and translates *made bright*; Kemble, again, translates *sheathed*.

ge-scænednes, -scéningnes, se; *f. A dashing together, a breaking*; collisio, Cot. 59.

ge-scap. v. ge-sceap.

ge-scafan, -sceafan; *p.* -scóf; *pp.* -scafen *To shave, scrape, plane*; radere, complanare:—Wið innoðes flĕwsan gáte horn gesceafen [gescafen, MS. B.] *for flux of inwards a goat's horn shaven*, Med. ex Quadr. 6, 9; Lchdm. i. 352, 15: 4, 12; Lchdm. i. 344, 23. v. scafan.

ge-scaldwyrt, e; *f. Talumbus*, Wrt. Voc. 289, 40.

ge-scamian; *p.* ode; *pp.* od. **I.** *v. intrans. To be ashamed, to blush*; erŭbescĕre:—Sýn gecyrred underbæc and gescamian, ða de wyllaþ me yfelu *avertantur retrorsum et erŭbescant, qui vŏlunt mihi măla*, Ps. Spl. 69, 3. **II.** *v. trans. impers. To shame, cause* or *bring shame to*; pŭdēre:—Sceal gescamian ða unrihtwísan *it shall shame the wicked*; erŭbescant impii, Ps. Th. 30, 20. Gescamige hí *let it shame them*; erŭbescant, Ps. Spl. 82, 16. v. ge-sceamian.

ge-scapennys, -nyss, e; *f. A creation, creating, formation*; creātio, figmentum:—Se emnihtes dæg is feórþa dæg ðysse worulde gescapennysse *the day of the equinox is the fourth day of the creation of this world*, Bd. de nat. rerum; Wrt. popl. science 4, 14, 16; Lchdm. iii. 238, 18, 20. He sylf oncneów hiwunga oðđe gescapennysse úre *ipse cognōvit figmentum nostrum*, Ps. Lamb. 102, 14. v. ge-sceapennys.

ge-scapu *pudenda*. v. ge-sceap.

gescea *a sobbing*; singultum, Wrt. Voc. 289, 35.

ge-sceád, -scád, es; *n.* **I.** *separation, distinction, difference*:—Ðæt gesceád *separatio*, Lk. Skt. Lind. 12, 51. Gesceád *distinctio*, Mt. Kmbl. p. 3, 3: Mk. Skt. Rush. 4, 12. Eálá mid hú micle gesceáde God todælde betwih leóht and ðýstru *O quam grandi distantia divisit deus inter lucem et tenebras*, Bd. 5, 14; S. 634, 37. He sceal gebencan ðæt gedál and ðæt gesceád *he must consider the distinction and the difference*, L. de Cf. 1; Th. ii. 260, 13. Gescád, Exon. 94 a; Th. 353, 16; Reim. 13. **II.** *power of distinguishing, reason, discretion, discrimination, an account, a reckoning, argument*:—Gĕ habbaþ gesceád ǽgðer ge gódes ge yfeles *ye can distinguish between good and evil*, Homl. Th. i. 176, 24. Fordý sealde God mannum gesceád *therefore has God given reason to men*, 96, 13: 7: Bt. Met. Fox 20, 436; Met. 20, 218: 22, 88; Met. 22, 44. On gesceád witan *to understand*, Exon. 83 b; Th. 314, 3; Mód. 8. Gesceád witan, cunnan [with *gen*; cf. the same phrase in *O. Sax.* wissun thingo gisked; and the *Ger.* bescheid wissen] *to be able to distinguish between things, to understand them*, Homl. Th. 186, 4: Beo. Th. 582; B. 288. Gesceád *discretio*, Bd. 1, 27; S. 496, 35. Gesceád agyldan *to render an account*, Mt. Bos. 12, 36: Homl. Th. i. 96, 20: ii. 50, 1. Dæt he mid gesceáde hine betealde unsynninne *that he proved himself sinless with reasoning*, 226, 11: Chr. 1070; Erl. 208, 17. For hwylcum gesceáde *propter quam rationem, quapropter*, Ælfc. Gr. 44; Som. 46, 16. Myd gewyssum gesceáde *propter certam rationem*, Nicod. 3; Thw. 2, 6. [*O. Sax.* gi-skêd: *O. H. Ger.* gi-skeit *distinctio, discretio, distantia*.]

ge-sceádan, -scádan; *p.* -scéd, -sceód; *pp.* -sceáden [in the Northern Gospels weak forms occur] *To separate, distinguish, discern, decide*:—Wĕron gesceádad from *exceptis*, Mt. Kmbl. Lind. 14, 21. Ðú de gesceádest *qui separasti*, Rtl. 182, 31: 36, 27. Gisceád *distingue*, 36, 29. Wolde hilde gesceádan *would decide the war*, Cd. 167; Th. 209, 25; Exod. 504: Elen. Kmbl. 298; El. 149. Rodera rǽdend hit on riht gescéd *the ruler of the firmament decided it aright*, Beo. Th. 3115; B. 1555. He biþ on ðæt wynstre weorud gesceáden *he will be assigned at the separation to the band on the left hand*, Exon. 117 a; Th. 449, 23; Dôm. 75. [*Goth.* ga-skaidan *to separate*: *O. H. Ger.* gi-sceidan.]

ge-sceáden; *adj. Rational*:—Nán nýten næfde nán gesceádne sáwle *no beast had a rational soul*, Btwk. Scrd. 19, 35.

ge-sceádlíce; *adv. Reasonably, rationally*; rationabiliter:—Ful gesceádlíce ðú me andswarast and fulrihte *thou answerest me very rationally and rightly*, Shrn. 184, 17: 165, 21. Man sceal gesceádlíce tosceádan ylde and geóguþe *we must discreetly distinguish between age and youth*,

L. de Cf. 4; Th. ii. 262, 4. Gesceádlícor *more rationally*, Bt. 39, 2; Fox 214, 7.

ge-sceádwís; *adj. Reasonable, rational, discriminating, intelligent, prudent, cautious*; rationalis:—God gesceóp twá gesceádwísan gesceafta *God created two rational creatures*, Bt. 41, 2; Fox 244, 30: 42, 1; Fox 256, 9. Ælce dǽde sceal gesceádwís dēma wíslíce tosceádan hú heó gedón sí and hwǽr and hwænne *in each deed an intelligent judge must distinguish how it be done, and where and when*, L. de Cf; Th. ii. 260, 27: Past. 21; Swt. 151, 6: Bt. Met. Fox 15, 27; Met. 15, 14. Mid gesceádwísum mægne *with intelligent power*, 20, 16; Met. 20, 8.

ge-sceádwíslíc; *adj. Reasonable*; rationalis, R. Ben. Interl. 2.

ge-sceádwíslíce; *adv. Rationally, prudently, sagaciously, discreetly, distinctly*; rationabiliter:—Ðú de gesceádwíslíce heora welst *thou that rulest them rationally*, Bt. 33, 4; Fox 128, 6: 21; Fox 74, 20. Hý him ða gesceádwíslíce andwyrdon *they answered him discreetly*, Ors. 1, 10; Bos. 32, 20. Ic wille gesceádwíslícor gesecgan *apertissime expedite curabo*, 2, 1; Bos. 38, 29.

ge-sceádwísnes, ness, e; *f. Reason, discretion*; ratio:—Geléf ðínre ágenre gesceádwísnesse *believe thine own reason*, Shrn. 199, 12: Bt. 33, 4; Fox 132, 9: Past. 11, 2; Swt. 65, 21; Hat. MS. 14 b, 27: Bt. Met. Fox 20, 375: Met. 188: 393; Met. 20, 197.

ge-sceafan *to shave, plane*, Med. ex Quadr. 6, 9; Lchdm. i. 352, 15: 4, 12; Lchdm. i. 344, 23. v. ge-scafan.

ge-sceaft, -scæft, -sceft, e; *f.*: es; *n.* **I.** *the creation, a created being* or *thing, creature, an element*; creātio, creātūra, plasma, ĕlĕmentum:—Eall deós mǽre gesceaft *all this great creation*, Rood Kmbl. 24; Kr. 12: 162; Kr. 82: Salm. Kmbl. 60; Sal. 30. Gesceaft *plasma*, Ælfc. Gr. 9, 1; Som. 8, 22. Fram fruman gesceafte *ab initio creātūræ*, Mk. Bos. 10, 6: Cd. 9; Th. 11, 7; Gen. 171. On ðisse lǽnan gesceafte *in this perishable creation*, Salm. Kmbl. 653; Sal. 326: 737; Sal. 368. Þurh ða ilcan gesceaft *through the same creature*, Elen. Kmbl. 365; El. 183: 2061; El. 1032. Ða widerweardan gesceafta betwux him winnaþ *contrary creatures strive with each other*, Bt. 21; Fox 74, 13: Exon. 68 a; Th. 253, 21; Jul. 183. Ealle gesceafte forhte geweorþaþ *all creatures shall tremble*, Andr. Kmbl. 2997; An. 1501: Cd. 191; Th. 239, 11; Dan. 368: Bt. Met. Fox 11, 16; Met. 11, 8. Hí wuldriaþ ǽdelne ordfruman ealra gesceafta *they glorify the noble origin of all creatures*, 18; Th. 25, 18; Cri. 402: 21 b; Th. 57, 29; Cri. 926: Andr. Kmbl. 652; An. 326: Elen. Kmbl. 1785; El. 894: Bt. 21; Fox 72, 29. Eallum his gesceaftum *to all his creatures*, 21; Fox 74, 2, 21: Salm. Kmbl. 672; Sal. 335. He gemetgaþ ða feówer gesceafta *he regulates the four elements*, Bt. 39, 8; Fox 224, 8: 33, 4; Fox 128, 29: Boutr. Scrd. 18, 20: 30, 7. Ofer ealle gesceafte *over all creatures*, Exon. 28 a; Th. 84, 33; Cri. 1388: 43 b; Th. 147, 25; Gú. 732. Biþ ðæt gesceaft swíðe nearu gebuht *the creation will appear very narrow*, Homl. Th. ii. 186, 7. He awende ðæt gesceaft *he changed the creature*, ii. 72, 10: i. 276, 8, 10, 14, 15, 20. Ða gesceafta tácnedon ðæt he wæs sôþ god *created things shewed that he was very God*, Shrn. 67, 16. Bodigaþ ǽlce gesceafte *prædicate omni creaturæ*, Mk. Skt. Rush. 16, 15: Rtl. 97, 12. Giscæf[t] *sexus*, 51, 7. Dú ðe gimetgaþ gescæfta wrixla *qui temperas rerum vĭces*, 164, 12. **II.** *a decree, destiny, fate, condition*; destināta, sors, fātum, condītio:—Ðæt is eald gesceaft *that is the ancient fate*, Salm. Kmbl. 772; Sal. 385. Nǽni eft cymeþ hider, ðe mannum secge hwylc sý Meotodes gesceaft *no one returns hither who may reveal to men what is the condition of the Creator*, Menol. Fox 592; Gn. C. 65. In gesceaft Godes *by God's decree*, Exon. 93 b; Th. 351, 3; Sch. 74. He sǽgde him wereda gesceafte *he told him the fates of peoples*, Cd. 180; Th. 225, 27; Dan. 160. [*Goth.* ga-skafts *creation, creature*: *O. Sax.* gi-skefti *decree of fate*: *O. H. Ger.* ga-skaft *creatura, elementum, habitus, fatum*.] DER. ealdor-gesceaft, eorþ-, forþ-, hand-, heáh-, land-, líf-, mǽl-, metod-, woruld-.

ge-sceamian, -sceomian, -scamian, -scomian; *p.* ode; *pp.* od. **I.** *v. intrans. To blush, be ashamed, be confounded*; erubescĕre, confundi:—Gesceamian [MS. gesceaman] oðđe gescende sýn ða sécendan sáwle míne *confundantur quærentes ănĭmam meam*, Ps. Spl. 34, 4. Gesceomadon alle fióndas his *erubescbant omnes adversari ejus*, Lk. Skt. Lind. 13, 17: 9, 26. **II.** *v. trans. To shame, cause* or *bring shame to, confound*; pŭdēre, confundēre:—Nú mæg ðám Cristenan gescomian *now may the Christians blush*, Ors. 4, 12; Bos. 99, 12. Ne gesceamaþ hý *it shall not confound them*; non confundentur, Ps. Th. 36, 18: 30, 1. Gesceamige heom *erubescant*, Ps. Lamb. 6, 11: Ps. Th. 30, 19. Gisceomiga *confundas*, Rtl. 125, 15. [*Goth.* ga-skaman sik *to be ashamed*.]

ge-sceandnys, -nyss, e; *f. A confusion*; confūsio:—Dú wást gesceandnysse míne *tu scis confūsiōnem meam*, Ps. Spl. 68, 23: 131, 19. v. ge-scendnys.

ge-sceánon. v. ge-scénan.

ge-sceap, -scæp, -scep, es; *pl. nom. acc.* -sceapu, -sceapo; *gen.* -sceapa, -sceapena; *n.* **I.** *a creation, created being* or *thing, creature*; creātio, creātūra:—Song he be middangeardes gesceape *cănēbat de creātiōne mundi*, Bd. 4, 24; S. 598, 9. Þurh ðæt beorhte gesceap *through that bright creature*, Elen. Kmbl. 1576; El. 790. Ðisses

gisceppes *hujus creationis*, Rtl. 21, 10.   **II.** *a decree, fate, destiny, condition, nature, form, shape;* fātum, destĭnāta, condĭtio, nātūra, indŏles, forma, spĕcies:—Ðæt ic sceolde wið gesceape mīnum on· bonan willan būgan *that I must submit to a murderer's will against my nature,* Exon. 126 b; Th. 486, 2; Rä. 72, 6. Ðeós woruld gesceap dreógeþ *this world fulfils its destiny,* 122 b; Th. 469, 25; Hy. 11, 7: Beo. Th. 6160; B. 3084. Swā mīn gesceapu wǽron *such were my decrees,* Exon. 103 a; Th. 391, 19; Rä. 10, 7 : 110 a; Th. 421, 26; Rä. 40, 24 : Cd. 76; Th. 95, 4; Gen. 1573. Ðæt ðīn líchoma leóhtra wurde, ðīn gesceapu scēnran *that thy body would become brighter, thy form more beauteous,* 25 ; Th. 32, 14 ; Gen. 503. God gesceapo ferede ǽghwylcum on eorþan eormencynnes *God has borne his decrees to every one of the human race on earth,* Exon. 88 b; Th. 333, 1; Vy. 95. Sinewealt gesceap volūbĭle schēma, Ælfc. Gl. 100; Som. 77, 14; Wrt. Voc. 55, 18. Giscæp *habitus,* Rtl. 103, 32.   **III.** *the privy members;* vĕrenda, pŭdenda :—Sumne dǽl ðæs felles æt foreweardan his gesceape *part of the foreskin,* Homl. Th. i. 94, 1. His gesceapu maðan weóllon *his members swarmed with vermin,* 86, 10 : ii. 512, 4: Gen. 9, 22. Wið gicþan ðæra gesceapa *against itch of the verenda,* Herb. 94, 4; Lchdm. i. 204, 22 : 123, 1; Lchdm. i. 234, 19. Ðæra gesceapena *of the verenda,* 103, 1; Lchdm. i. 218, 7. [*O. Sax.* gi-skap *creature;* gi-skapu, *pl.* decrees of fate, v. Grm. D. M. 817.] DER. frum-gesceap, fyrn-, heáh-.

ge-sceapen; *part. p. Formed, created :* — Adam wearþ ðā mann, gesceapen on sáwle and on líchaman *Adam then became man, formed with soul and body,* Homl. Th. i. 12, 30. v. sceppan *to create.*

ge-sceapennys, -sceapenys, -scapennys, -nyss, e; *f. A creation, creating, formation;* creātio :—God geswác ðære niwan gesceapennysse *God ceased from the new creation,* Boutr. Scrd. 17, 17. On ðæs mannes gesceapennysse *in the creating of man,* 19, 7. Se man ðe deófle geefenlǽcþ, se biþ deófles bearn, nā þurh gecynd oððe þurh gesceapenysse, ac þurh ða geefenlǽcunge and yfele geearnunga *the man who imitates the devil is a child of the devil, not by nature nor by creation, but by that imitation and evil deserts,* Homl. Th. i. 260, 13, 15.

ge-sceap-hwíl, e; *f. The time appointed by fate for dying :*—To gescæphwíle *at the appointed time,* Beo. Th. 52; B. 26. v. Grm. D. M. 817.

ge-sceaplíce; *adv. Properly, fitly, well;* apte :— Seó heáfodstów gesceaplíce gehiwad to ðam gemete hyre heáfdes *locus capitis ad mensuram capitis illius aptissime figuratus,* Bd. 4, 19; S. 590, 1, note.

ge-scearfan *to cut off;* succĭdere, Lk. Skt. Lind. 13, 9.

ge-sceát *shot forward, darted,* Beo. Th. 4628; B. 2319; *p.* of ge-sceótan.

ge-sceátaþ *fall to, shall fall to or be allotted to,* Ex. 29, 28, = ge-sceótaþ; *pres. pl.* of ge-sceótan.

ge-sceaþan. v. ge-sceþþan.

ge-sceaþian; *p.* ode; *pp.* od *To injure, harm, scathe :*—Hū he on manna sáulum mǽst gesceaþian mǽge *how he can most injure the souls of men,* L. C. E. 26; Th. i. 374, 31. [Cf. ge-sceþþan.]

ge-sceáwian; *p.* ode; *pp.* od.   **I.** *act. To shew, manifest, exhibit;* exhibēre, monstrāre, manifestāre :—Āre ne wolde gesceáwian *would not shew reverence,* Cd. 76; Th. 95, 19; Gen. 1581. Wile ðonne gesceáwian wlitige and unclǽne *then will he manifest the fair and the foul,* 227; Th. 303, 7; Sat. 609. Eorle monegum āre gesceáwaþ *to many a man he shews honour,* Exon. 100 b; Th. 379, 15; Deor. 33. He him wolde ārlíc bisceop-setl gesceáwian *he would shew [provide for] him an honourable bishop's-seat,* Bd. 3, 7; S. 530, 2.   **II.** *act. To see, behold, view, look round upon ;* vidēre, perspicere, circumspicere :—Heó endestæf gesceáwiaþ *they shall behold their end,* Cd. 225; Th. 298, 31 : Sat. 541. Ðæt deáþ ne gesceáwige *qui non videbit mortem,* Ps. Th. 88, 41. Mínre heortan gehygd gesceáwa *view the thoughts of my mind,* 138, 20. Hord ys gesceáwod *the hoard has been seen,* Beo. Th. 6161; B. 3084. Ðæt ðū ða bisne sweotole gesceáwige *that thou mayest clearly view the example,* Bt. 22, 2; Fox 78, 15. Him eallum gesceáwodum *iis omnibus circumspectis,* Lk. Bos. 6, 10.   **III.** *intrans. To see, consider ;* vidēre, considerāre :—And he scearpe ne mǽge gesceáwian *non considerat?* Ps. Th. 93, 9.

ge-sceft; *f. The creation, a creature;* creātio, creātūra :—In ðære ēcan gescefte [MS. gesceft] *in the eternal creation,* Cd. 228; Th. 306, 15; Sat. 664. Ealra gescefta *of all creatures,* 226; Th. 301, 20; Sat. 584 : 217; Th. 277, 13; Sat. 203. v. ge-sceaft.

ge-sceldod; *part. p. Provided with a shield :*—Twegen englas gesceldode *two angels with shields,* Blickl. Homl. 221, 28.

ge-scénan; *p.* de; *pp.* ed *To break, bruise, wound;* contĕrĕre, vulnĕrāre :—Hí woldon ǽninga heafolan gescénan *they would at once wound the head,* Andr. Kmbl. 2286; An. 1144. Forðon he ǽren dōr gescéeneþ *quia contrivit portas æreas,* Ps. Th. 106, 15. v. ge-scǽnan.

ge-scendan, -scindan, -scyndan; *p.* de; *pp.* ed *To shame, put to shame, confound, corrupt :*—Drihten hýg gescent *Dominus subsannabit eos,* Ps. Th. 2, 4. Gescendes *corrumpit,* Lk. Skt. Lind. 12, 33. Ðæt ðū híg gescindest *that thou didst shame her,* Gen. 20, 6. He us gescende and úre weorc *he hath put us and our deeds to shame,* Blickl. Homl. 243, 11. Hwā biþ gescended ðæt me forðæm ne scamige *who is shamed and*

I am not ashamed? Past. 21, 6; Swt. 165, 5; Hat. MS. Ne gescend me *non confundas me,* Ps. Th. 118, 116. Ðæt ic ne wese gescended *ut non confundar,* 80 : 87, 15 : 126, 6. Beóþ gescende míne fýnd *confundantur inimici mei,* 69, 2 : 82, 13. Ne wylt ðū me gescyndan *noli me confundere,* 118, 31. Ne beó ic gescynded *non confundar,* 6.

ge-scendnys, -scyndnys, se; *f. A confounding;* confusio :—Gescendnys, Ps. Spl. 43, 17 : 34, 30. Ðone deófol ðe ða synfullan gelǽt to gescyndnysse. Babilonia seó Chaldeisca burh is gereht gescyndnys *the devil that leads the sinful to confusion. Babylon, the Chaldean city, is interpreted 'confusion,'* Homl. Th. ii. 66, 21.

ge-sceó *shoes;* calceamenta, Mt. Kmbl. Lind. 3, 11. v. ge-scý.

ge-sceód; *part. p. Provided with shoes, shod :*—Gesceóde [gescóed, Lind.] mid calcum *calciatis sandalis,* Mk. Bos. 6, 9: Homl. Th. ii. 264, 9.

ge-sceón; *p.* de *To happen, come upon;* accidere, contingere :—Him níþ godes gesceóde *God's enmity came upon him,* Cd. 206; Th. 255, 7; Dan. 620. [Cf. Cod. Ex. Th. 226, 4; Ph. 400.] Him bonena hand hearde gesceóde [Kmbl. gesceód] *the hand of slayers had been hard upon him,* Andr. Kmbl. 36; An. 18. Egyptum wearþ ðæs dægweorces deóp leán gesceód *to the Egyptians for that day's work a deep requital was given,* Cd. 167; Th. 209, 29; Ex. 506.

ge-sceorf, es; *n. Scurf, the fur of the mucous membrane;* mucus intestinorum, L. M. 2, 35; Lchdm. ii. 240, 23. v. sceorf.

ge-sceorpan; *p.* -scearp *To scrape, carve in pieces;* conscindere minutatim, Herb. 57, 1; Lchdm. i. 160, 4. v. sceorfan, screpan.

ge-sceortian; *p.* ade *To fall short, fail :*—Midðý ðæt wín gesceortade *vino deficiente,* Jn. Skt. Lind. 2, 3.

ge-sceot, -scot, es; *n.*   **I.** *the collection of weapons necessary for shooting, a weapon that is shot or hurled, an arrow, dart :*—Nim ðīn gesceot ðīnne cocur and ðīnne bogan *take thy weapons, thy quiver and thy bow,* Gen. 27, 3. Ðū of heofenum dōm mid gescote sendest *de cœlo judicium jaculatum est,* Ps. Th. 75, 6. Ðǽr forwearþ micel Alexandres heres for gæētredum gescotum *there much of Alexander's army perished by poisoned arrows,* Ors. 3, 9; Bos. 68, 38. [*O. H. Ger.* gascoz *jaculum: Ger.* geschoss.]   **II.** *an advance [of money], a contribution, tribute* [cf. *Ger.* vorschiessen]. v. corn-, Rôm-gesceot.   **III.** *a part of a building shut off from the rest* [v. Cl. and Vig. Icel. Dict. skot, III ; *and* cf. *Ger.* geschoss *story of a house*] :—Gesceot bæftan ðæm heáhweofode *propitiatorium,* vel *sanctum sanctorum,* vel *secretarium,* vel *pastoforum,* Ælfc. Gl. 109; Som. 79, 26; Wrt. Voc. 59, 1. v. selegesceot.

ge-sceót, es; *n.*   **I.** *shooting, hurling :*—Ge mid gesceótum [or gesceotum ? v. gesceot] ge mid stána torfungum *both with shootings and flingings of stones,* Ors. 3, 9; Bos. 68, 19.   **II.** *rapid movement as of anything shot :*—Ða wǽmna flugon mid swiftum gesceóte on heora fīnd *the weapons flew with swift movement on to their enemies,* Jud. c. 16; Thw. 162, 8. v. sceót.

ge-sceótan; he -scýt, -scītt, *pl.* -sceótaþ; *p.* -sceát, *pl.* -scuton; *subj.* ic, ðū, he -sceóte, *pl.* -sceóten; *pp.* -scoten.   **I.** *to shoot forward, to rush or dart forward with a quick motion, send forth, expend, pay, to fall to any one's share, be allotted to;* cum impetu movere *vel* ruere, expendere, cedere in partem alicujus :—Draca hord eft gesceát, dryhtsele dyrne *the dragon again darted to his hoard, his secret hall,* Beo. Th. 4628; B. 2319. Ðæt feoh ðe ic for hyre āre gescoten [MS. gesceoten] hæbbe *the money which I have paid for her honour,* Th. Diplm. 558, 19. Ðū nást hwām hit [wela] gescýt *thou knowest not to whom it [wealth] shall fall,* Homl. Th. ii. 104, 9. Hit gescītt to his dǽle *it shall fall to his share,* Ex. 29, 26 Hīg gesceótaþ [MS. gesceataþ] to Aarones dǽle and his suna ēcre lage fram Israhēla bearnum *cedent in partem Aaron et filiorum ejus jure perpetuo a filiis Israel,* 29, 28. Ðeáh sumum men gesceóte læsse dǽl *though a less part be allotted to one man,* Homl. Th. ii. 272, 2; Jos. 9, 7.   **II.** *to bring before or refer to any one;* referre ad aliquem :—We lǽraþ, ðæt nān sacu ðe betweóx preostan sī, ne beó gescoten to worldmanna sōme *we enjoin that no dispute that be between priests be referred to the adjustment of secular men,* L. Edg. C. 7; Th. ii. 246, 4.

ge-sceppan, -scippan, -scyppan; *p.* -sceóp, -scóp, *pl.* -sceópon, -scópon; *pp.* -scæpen, -sceapen, -scoepen, -sceapen *To form, create;* formare, disponere, creare :—Ic gescippe *creo,* Ælfc. Gr. 26; Som. 29, 16. God gescypþ symle edniwan *God creates ever anew,* Boutr. Scrd. 18, 18. Ðá gesceóp Adam naman his wīfe *then Adam made a name for his wife,* Gen. 3, 20 : Boutr. Scrd. 19, 32. Hēr ǽrest gesceóp ēce Drihten heofon and eorþan *here the Lord eternal first created heaven and earth,* Cd. 5; Th. 7, 26; Gen. 112 : 12; Th. 14, 16; Gen. 219. God ðas world gescóp *God created this world,* Exon. 17 b; Th. 41, 22; Cri. 659: Salm. Kmbl. 936; Sal. 467. Hwæt! ðū ēce God! ealra gesceafta wundorlíce wel gescéope *O! eternal God! thou hast made all creatures wonderfully well,* Bt. Met. Fox 20, 10; Met. 20, 5 : Exon. 117 b; Th. 452, 14; Hy. 4, 1. Ǽr ðæt ðec ic gesceópe *prius quam te formarem,* Rtl. 55, 4. Nǽron nāwðer ne on Fresisc gescæpene ne on Denisc *they were formed neither on a Frisian nor on a Danish model,* Chr. 897; Erl. 95, 15. Ðá ðá híg wǽron gesceopene *when they were created,*

Gen. 2, 4. Mon wæs to Godes anlícnesse ǽrest gesceapen *man was to God's image first created*, Cd. 75; Th. 92, 16; Gen. 1529. Gesceapene híg synt *creata sunt*, Ps. Lamb. 32, 9: Ps. Th. 148, 5. Gescype scylfan on scipes bósme *make shelves in the ship's bosom*, Cd. 65; Th. 79, 4; Gen. 1306. God wolde þurh his ágene handa hine gescyppan *God would form him with his own hands*, Boutr. Scrd. 19, 10. To gescippenne *in order to create*, 3.

ge-sceppend, -scyppend, es; *m. A creator:*—Fram ðǽm heáhsetle úre gescyppendes *from the throne of our Creator*, Blickl. Homl. 11, 29.

ge-sceran; *p.* -scer, -scær; *pp.* -scoren *To cut, cleave*; secare, dissecare:—He him on heáfde helm gescer *he clave the helmet on his head*, Beo. Th. 5939; B. 2973. Helm gescær *he cut the helmet*, 3057; B. 1526. v. sceran.

ge-scerian, -scyrian, -scyrigan; *p.* ede; *pp.* ed. **I.** *to bestow, appoint, provide, ordain, destine*; tribuere, providere, ordinare, destinare:—He sceolde his Drihtne þancian ðæs leánes, ðe he him on ðam leóhte gescerede *he should thank his Lord for the reward which he bestowed on him in that light*, Cd. 14; Th. 17, 11; Gen. 258. Ic biddan wille ðæt ðú me ne gescyrige mid scyldhetum *I will pray that thou appoint me not among the guilty ones*, Andr. Kmbl. 169; An. 85 Is se rǽd gescyred monna cynne *this counsel is ordained for mankind*, Cd. 22; Th. 27, 28; Gen. 424. **II.** *to number, reckon*; numerare:—Se me beág forgeaf, on ðam siexhund wæs, smǽtes goldes, gescyred sceatta *he gave me a bracelet, on which six hundred sceats of beaten gold were numbered*, Exon. 86 a; Th. 324, 9; Wíd. 92. Ic wéne ðæt ðǽr screoda wǽre gescyred ríme siexhundreda *I believe that there were six hundred chariots reckoned by number*, 122 a; Th. 468, 10; Phar. 5. [O. Sax. gi-skerian *to ordain, arrange*.]

ge-scerpan, -scirpan, -scierpan; *p.* te *To sharpen*; acuere:—Ic gescirpe mín swurd *I will sharpen my sword*, Deut. 32, 41. Ðære culfran bilwitnesse gescierpan [-scirpan, Hat. MS.] *to sharpen the simplicity of the dove*, Past. 35, 1; Swt. 236, 23; Cot. MS.

ge-scerpan, -scirpan, -scyrpan; *p.* te *To clothe, furnish, adorn, deck*; vestire, ornare:—Ðeáh Neron hine gescerpte wlitegum wǽdum *though Nero clothed himself in beautiful garments*, Bt. Met. Fox 15, 4; Met. 15, 2. Gescyrpte, Bt. 28; Fox 100, 26. Ele andwlitan gescyrpeþ *oleum faciem exhilarat*, Ps. Th. 103, 15. Ðá ðæt folc hine gesceah swá gescyrpedne *when the people saw him so furnished* [i. e. *with sword and spear, and riding on the king's horse*], Bd. 2, 13; S. 517, 10: 5, 19; S. 638, 9, 10. Fugla cynn fiðerum gescyrped *volucres pennatæ*, Ps. Th. 148, 10. Ða ðe gescirped sind *qui vestiuntur*, Mt. Kmbl. Lind. 11, 8.

ge-sceððan [with the same form in the infinitive are to be found, apparently, two verbs, one belonging to the strong, the other to the weak conjugation. Corresponding to the *Gothic* verb skaþjan, skôþ is sceððan, scód; [cf. sceppan, scóp.] The infinitive 'sceadan' also occurs. Corresponding to *Icel.* skeðja, skaddi is sceððan, sceðede. There is besides the weak verb 'sceadian,' which corresponds to *Icel.* skaða, skaðaði, or *O. H. Ger.* scadôn, scadota. With regard to the form 'scód' instead of 'skôþ,' see Grm. And. u. El. 93] *To injure, hurt, oppress, be an enemy to*; nocere, adversari:—Gisceðeþ *nocebit*, Rtl. 8, 29. Ða ðe hríppum usum gesceððaþ and gefrettaþ *quæ messibus nostris adversantur et comedunt*, 147, 7. Ðæt him bám gescód *that injured them both*, Exon. 45 b; Th. 154, 14; Gú. 842: 38 b; Th. 127, 35; Gú. 396: 61 b; Th. 226, 4; Ph. 400. He manegum gescód *it proved a foe to many*, Cd. 167; Th. 208, 25; Exod. 488: 198; Th. 247, 1; Dan. 490: 209; Th. 258, 20; Dan. 678. Him hettende oft gescódan *enemies oft oppressed them*, Exon. 62 a; Th. 228, 23; Ph. 442. Him gesceðe scyldignis *ei noceat reatus*, Rtl. 103, 15. Gáste gesceððan *to injure the soul*, Andr. Kmbl. 1834; An. 919: Beo. Th. 2899; B. 1447. Gesceððed *læsus*, Lye.

ge-sceððendlíc; *adj. Hurtful:*—Alle gesceððendlíca *omnia nociva*, Rtl. 118, 33.

ge-scierpan *to sharpen*. v. ge-scerpan.

ge-scild, es; *n. A refuge*; refugium, Ps. Spl. T. 70, 4.

ge-scildan, -scyldan; *p.* de; *pp.* ed *To shield, cover, protect, defend*; protegere, tueri:—Ic gescilde ðé mínre swýðran handan *I will cover thee with my right hand*, Ex. 33, 22. Ic ðé gescilde on drihtenes name *I will protect thee in the name of the Lord*, Shrn. 15, 19. Gif ðé man scotaþ to ðú gescylst ðé *if you are shot at you shield yourself*, Homl. Th. ii. 538, 10. Giscildes *protegis*, Rtl. 52, 14. Us gescyldeþ scyppend engla *the Creator of angels protects us*, Andr. Kmbl. 867; An. 434: Exon. 68 b; Th. 255, 14; Jul. 214. He us gesceldeþ wið eallum feóndum *he will shield us from all enemies*, Blickl. Homl. 51, 14. Siððan hie heofonríces weard wið ðone hearm gescylde *after heaven's guardian had protected them against that hurt*, Cd. 196; Th. 245, 6; Dan. 458: Shrn. 90, 7: Mt. Kmbl. p. 7, 9. Giscilde *protegat*, Rtl. 49, 34. Wiþ egesan yfeles feóndes míne sáwle gescyld *a timore inimici eripe animam meam*, Ps. Th. 63, 1. Gescildan wið *to protect against*, Exon. 40 b; Th. 135, 23; Gú. 528. Heó is gescyld *she is protected*, Ors. 2, 4; Bos. 45, 3. Gescylded *protected*, Exon. 58 b; Th. 210, 4; Ph. 180: Bd. 3, 23; S. 555, 35.

ge-scildend, -scyldend, es; *m. A protector:*—Mín gescyldend *protector meus*, Ps. Th. 27, 8: Andr. Kmbl. 2583; An. 1293.

ge-scildnes, -scyldnes, -scildness, e; *f. Protection, defence, shielding*; tuitio, tutamen, tutela, defensio:—Þurh his gescildnisse synd ða fýnd on ðínum handum oferwunnene *through his protection are the enemies overcome in thy hands*, Gen. 14, 20: Homl. Th. ii. 140, 27. For heora gescyldnysse *ob eorum defensionem*, Bd. 1, 12; S. 481, 4: 2, 5; S. 506, 30. Gescyldnysse *protectionem*, Ps. Spl. 17, 37. Giscildniss *protectio, tuitio, defensio*, Rtl. 17, 9: 62, 8: 145, 30.

ge-scínan; *p.* -scán; *pp.* -scinen *To shine, shine upon, illuminate*; fulgere, collustrare, illuminare:—Ne mæg heó ealle gesceafta gescínan, ne ða gesceafta ðe heó gescínan mæg, ne mæg hió ealle endemest gescínan *she cannot shine upon all creatures, nor those creatures which she may shine upon can she shine upon all equally*, Bt. 41, 1; Fox 244, 7-9: Bt. Met. Fox 30, 17; Met. 30, 9: 30, 22; Met. 30, 11. Wuldres gim grund gescíneþ *the gem of glory illuminates the ground*, Exon. 57 b; Th. 205, 26; Ph. 118. Swá sió sunne hine gescínþ *as the sun shines upon him*, Bt. 34, 5; Fox 140, 8. Gescíneþ *lucet*, Jn. Skt. Lind. 1, 5. Giscína *fulgere*, Rtl. 67, 10: 86, 34. Gisceán *innituit* [= *enituit*], 45, 16.

ge-scincio; *pl. n. The fat about the kidneys.* v. Lchdm. iii. 361.

ge-scindan. v. ge-scendan.

ge-scipian; *p.* ode; *pp.* od *To provide with ships:*—Se micla here wurdon gescipode *the great army got ships*, Chr. 893; Erl. 88, 23.

ge-scippan. v. ge-sceppan.

ge-sciran *to act as a steward*; vilicare, Lk. Skt. Lind. 16, 2.

ge-scirpan. v. ge-scerpan.

ge-scirpla, -scyrpla, an; *m. Clothing, clothes*; vestitus:—Wǽron hie on gescirplan scipférendum onlíce *they were in clothing like seafarers*, Andr. Kmbl. 499; An. 250. Hwǽr beóþ ðonne his ídlan gescyrplan *where shall his vain garments be then?* Blickl. Homl. 111, 35.

ge-scítt *shoots forward, falls to, is allotted to*, Ex. 29, 26; *pres. of* ge-sceótan.

ge-scód, -scóed. v. ge-sceód.

ge-scóe, Mk. Skt. Rush. 1, 7: Lk. Skt. Rush. 10, 4: Jn. Skt. Rush. 1, 27. v. ge-scý.

ge-scola, an; *m. A fellow-debtor*; condebitor, Cot. 208. [*M. H. Ger.* geschol.] v. sculan.

ge-scoman. v. ge-sceamian.

ge-scot. v. ge-sceot.

ge-scotfeoht, es; *n. A fight with arrows or darts:*—Eft gewurdon on gescotfeohta scearpe gáras *ipsi sunt jacula*, Ps. Th. 54, 21: 75, 3.

ge-scrǽpe, -screope, -scroepe; *adj. Convenient, meet, fit for, accommodated*; aptus:— Breoton is gescrǽpe on lǽswe sceápa and neáta *Brittannia est apta alendis pecoribus ac jumentis*, Bd. 1, 1; S. 473, 13, 22. Giscroepo *aptus*, Rtl. 117, 14: Bd. 5, 6; S. 618, 41. DER. un-gescrǽpe. v. ge-screpelíce, ge-scropenys, ge-screope.

ge-screádian; *p.* ode, ede; *pp.* od, ed *To cut off, trim, prune*; sarpere:—Gif se wíngeard ne biþ onriht gescreádod *if the vineyard be not rightly pruned*, Homl. Th. ii. 74, 14. Gescreáded wíngeard *sarpta vinea*, Ælfc. Gl. 99; Som. 76, 125; Wrt. Voc. 54, 65. DER. screádian.

ge-screncan; *p.* te *To cause to shrink, to destroy, supplant:*—Wéron gescrencde *aruerunt*, Mt. Kmbl. Lind. 13 6. Ðú gescrencttyst onarísende me *supplantasti insurgentes in me*, Ps. Spl. C. 17, 41.

ge-screncednes, -ness, e; *f. A supplanting, an overturning*; supplantatio, Ps. Spl. C. 40, 10.

ge-screnge; *adj. Withered, shrunken, dry*; aridus, Lk. Skt. Lind. 6, 8.

ge-screope; *adj. Fit for, apt*; aptus:—Fela óðera gescreopa and gesynto he oncneów heofonlíce him forgifen beón *alia commoda et prospera cælitus sibi fuisse data intellexit*, Bd. 4, 22; S. 592, 20: Bd. 4, 19; S. 589, 42, note. v. ge-scrǽpe.

ge-screpelíce; *adv. Aptly, conveniently, fitly*; apte:— Gescrepelíce gehíwad *aptissime figuratus*, Bd. 4, 19; S. 590, 1. v. ge-scrǽpe.

ge-scrif, es; *n. A judgment, command, ceremony*; censura, edictum, ceremonia, Cot. 59: 79: 56. [Cf. *O. H. Ger.* gi-scrip *scriptura, forma*.] v. ge-scrífan.

ge-scrífan; *p.* -scráf, -screáf; *pp.* -scrifen. **I.** *to judge, deem, assign, impose, appoint*; judicare, assignare, imponere, designare:—Se ðe him gescráf weán *who to him had assigned misery*, Cd. 148; Th. 186, 16; Exod. 139. Swá him wyrd gescráf *so fate assigned to him*, Beo. Th. 5142; B. 2574: Elen. Grm. 1047: Bt. Met. Fox 1, 58; Met. 1, 29. Hió me lytle læs láðe woldan ðisses eorþweges ende gescrífan *paulo minus consummaverunt me in terra*, Ps. Th. 118, 87. Siððan gé agifen habbaþ sceattas gescrifene *when you have given the appointed sum*, Andr. Kmbl. 593; An. 297. **II.** *to shrive, impose penance, censure*; pœnitentiam imponere, reprehendere:—Manna sáwla lǽce sceal geþencan, hú he mannum heora dǽda gescrífe and hí þeáh-hwæðere ne fordǽme *the physician of men's souls must consider how he shall shrive their deeds and yet not condemn them*, L. de. Cf. 1; Th. ii. 260, 14.

ge-scrincan; *p.* -scranc; *pp.* -scruncen *To shrink, dry up:*—Giscrinca hia *arrescunt*, Rtl. 125, 35. For ðæm cíele him gescruncan ealle ða ǽdra ðæt him mon ðæs lífes ne wénde [*cum in præfrigidum amnem descen-*

disset] obriguit, contractuque nervorum proximus morti fuit, Ors. 3, 9; Bos. 64, 38. Gescriungon aruerunt, Mt. Kmbl. Lind. 13, 6. Mengo giscrungenra multitudo aridorum, Jn. Skt. Rush. 5, 3. Ða gescruncenan marcida, Cot. 133.

ge-scroepe. v. ge-scræpe.

ge-scropelîce fitly, meetly, Som. v. ge-screpelice.

ge-scropenys, -nyss, e; f. An applying, a fitting, accommodation; accommodatio, Som. DER. un-ge-screpnes. v. ge-scræpe.

ge-scrýdan, -scrîdan; p. -scrýdde; pp. -scrýd, -scýrd To clothe; induere, vestire:—God gescrîdde hî God clothed them, Gen. 3, 21. Mann hnescum gyrlum gescrýdne; nú ða ðe synt hnescum gyrlum gescrýdde synt on cyninga húsum hominem mollibus vestitum? ecce qui mollibus vestiuntur in domibus regum sunt, Mt. Bos. 11, 8. Mid wlite gescýrd is gescýrd is driht strangnysse decore indutus est, indutus est dominus fortitudinem, Ps. Spl. 92, 1. Myrce gescýrded shrouded in darkness, Andr. Kmbl. 2628; An. 1315.

ge-scryfu ceremonies; ceremoniæ, Som. v. ge-scrif.

ge-sculdre, -sculdru; pl. n. The shoulders; humeri:—Gesculdre palæ, Ælfc. Gl. 74; Som. 71, 45; Wrt. Voc. 44, 28. Middel gesculdru interscapilium, 74; Som. 71, 46; Wrt. Voc. 44, 29. Mid his gesculdrum scapulis suis, Ps. Th. 90, 4.

ge-scŷ, es; n. A pair of shoes; calceamentum, tegmentum pedis, caliga:—Gif he [man] ðonne cwiþ 'Nelle ic hîg habban to wîfe,' gâ ðæt wîf to him and nyme his gescŷ of his fôtum beforan ðâm ealdrum and spæte on his nebb and nemne hine ælc man on Israéla folce 'unsceóda' if he [the man] then say 'I will not have her to wife,' let the woman go to him and take his shoes off his feet before the elders and spit in his face, and let every man amongst the people of the Israelites call him 'the unshod,' Deut. 25, 8–10. In Idumea lande ic aþenige gescŷ mîn in Idumæam extendam calceamentum meum, Ps. Spl. 59, 9; Ps. Th. 59, 7 has On Idumea mîn gescy sende. Gescŷ calceamentum, Ps. Spl. 107, 10: Ps. Th. 107, 8. Ðæs gescŷ neom ic wyrþe to berenne non sum dignus calceamenta portare, Mt. Bos. 3, 11: Lk. Bos. 15, 22. Hwæt sind gescŷ bûton deâdra nýtena hýda what are shoes but the hides of dead cattle, Homl. Th. ii. 280, 29. [Goth. ga-skohi: O. Sax. gi-skohi: O. H. Ger. gi-scuohi; n.] v. ge-scóe.

ge-scŷfan; pp. -scyfen To eject; ejicere, Mt. Kmbl. p. 16, 4. [Cf. scûfan.]

ge-scyftan; pp. -scyft To share, distribute:—Beó seó æht gescyft swýðe rihte let the property be very fairly distributed, L. C. S. 71; Th. i. 414, 1.

ge-scyld, es; n. Guilt, debt; reatus, debitum:—Gescyldum reatibus, Rtl. 79, 22. All gescyld universum debitum, Mt. Kmbl. Lind. 18, 34.

ge-scyldan. v. ge-scildan.

ge-scyldend. v. ge-scildend.

ge-scyldigian, -scyldigian; p. ode; pp. od To prove guilty, charge with guilt, deserve punishment in consequence of guilt:—Ðæt hîg hine gescyldegodon ut caperent eum in sermone, Lk. Bos. 20, 20. Þurh ðæt gescildgade wîte per debitam pænam, Bd. 3, 19: S. 548, 30. [Cf. O. H. Ger. scildigon: Ger. schuldigen.]

ge-scyldnes. v. ge-scildnes.

ge-scyldru; pl. n. The shoulders:—Me on gescyldrum on my shoulders, Exon. 111 b; Th. 428, 4; Rä. 41, 103: 125 b; Th. 483, 17; Rä. 69, 4.

ge-scyndan. v. ge-scendan.

ge-scyndan; p. de. I. to hasten:—Heofon-torht swegl gescyndeþ the heaven-bright sun hastens, Exon. 94 b; Th. 351, 2; Sch. 74. II. to cause to hasten, to drive:—Ða twegen drýmen wurdon gescynde of ðam earde the two wizards were driven from the land, Homl. Th. ii. 476, 8. [Cf. a-, ge-, fýsan.]

ge-scyndnys a confusion; confusio, Ps. Spl. 70, 14. v. ge-scendnys.

ge-scŷnian to fear; metuere, Rtl. 32, 9. [Cf. scûnian.]

ge-scyppan. v. ge-sceppan.

ge-scýrd. v. ge-scrýdan.

ge-scyrian to ordain, number, reckon, Cd. 22; Th. 27, 28; Gen. 424: Exon. 86 a; Th. 324, 9; Wid. 92: 122 a; Th. 468, 10; Phar. 5. v. ge-scerian.

ge-scyrigan to appoint, Andr. Kmbl. 169; An. 85. v. ge-scerian.

ge-scyrpan. v. ge-scerpan.

ge-scyrtan; p. -scyrte; pp. -scyrted, -scyrt; v. a. [scyrt short]. I. to shorten, contract, lessen; abbreviare, minuere:—Ða spell ic sceal ge-scyrtan I must shorten the stories, Ors. 1, 8; Bos. 31, 29. Ðú his dagena tîd gescyrtest minorasti dies temporis ejus, Ps. Th. 88, 38. Gif drihten ðâs dagas ne gescyrte . . . he gescyrte ða dagas nisi breviasset dies . . . breviavit dies, Mk. Bos. 13, 20. Bûton ða dagas gescyrte wæron . . . ða dagas beóþ gescyrte, Mt. Bos. 24, 22. Heáp wæs gescyrted the crowd was diminished, Elen. Kmbl. 282; El. 141. II. to become short, be lessened, fail:—Ðætte gescyrte deficere, Jn. Skt. p. 3, 12: Lind. 2, 3.

ge-scýt shoots forward, falls or is allotted to, Homl. Th. ii. 104, 9; pres. of ge-sceótan.

GESE, gise, gyse [ge + se = geá + sî]; adv. YES; immo, etiam:—Gise, lâ gese, yes, oh yes, Bt. 16, 4; Fox 58, 15. v. geá.

ge-sealdniss. v. ge-saldniss.

ge-seáw; adj. [seáw juice] Juicy; sûci plēnus:—Geseáwe pýsan juicy peas, L. M. 2, 43; Lchdm. ii. 254, 15.

ge-sêcan, -sêcean; to -sæcanne, -sêcenne; part. -sêcende, ic -sêce, ðú -sêcest, -sêcst, he -sêceþ, -sêcþ, pl. -sêcaþ; p. -sôhte, pl. -sôhton; pp. -sôht; v. a. I. to seek, inquire, ask for; quærere, requirere, inquirere:—Ne mæg ic aldornere mîne gesêcan I cannot seek my life's safety, Cd. 103; Th. 136, 30; Gen. 2514. Gif he gesêcean dear wîg if he dare seek war, Beo. Th. 1373; B. 684. Heó mynster gesôhte monasterium petiit, Bd. 4, 19; S. 588, 5. Hie ðæs cnihtes cwealm gesôhton they sought the young man's death, Andr. Kmbl. 2244; An. 1123: Ps. Th. 70, 22. Ðæt ealra witegena blôd sý gesôht fram ðysse cneórysse ut inquiratur sanguis omnium prophetarum a generatione ista, Lk. Bos. 11, 50. II. to seek, go to, approach, look for, visit, come to; adire, ire vel proficisci, aliquo vel ad aliquem, visitare, venire, pervenire aliquo:—Wile nú gesêcan sâwla nergend gæsta giefstôl now the saviour of souls will seek the spirits' throne of grace, Exon. 16 a; Th. 36, 4; Cri. 571: Bd. 1, 23; 23; S. 485, 33: 3, 23; S. 554, 11. Nænig heora þohte ðæt he scolde eft eardlufan æfre gesêcean not one of them thought that he should ever seek his loved home again, Beo. Th. 1389; B. 692. Land swîðe feor to gesêcanne the land is very far to seek, Andr. Kmbl. 847; An. 424: Beo. Th. 3848; B. 1922. Ðonne ic ðas ilcan ôðre sîþe wîc gesêce when I seek this same dwelling a second time, Cd. 109; Th. 144, 23; Gen. 2394. He ôðer lîf eft gesêceþ he shall seek another life hereafter, Cd. 218; Th. 277, 30; Sat. 212: Salm. Kmbl. 316; Sal. 157: Exon. 97 a; Th. 361, 34; Wal. 29. Nales Dryhtnes gemynd siððan gesêcaþ they shall not seek the Lord's remembrance afterwards, 30 b; Th. 94, 10; Cri. 1538. He gesôhte Sûþ-Dena folc he sought the people of the South-Danes, Beo. Th. 930; B. 463: Cd. 128; Th. 163, 13; Gen. 2697: Andr. Kmbl. 759; An. 380. Hie gesôhton Sennera feld they sought the plains of Shinar, Cd. 80; Th. 100, 22; Gen. 1668: 111; Th. 146, 20; Gen. 2425. Ðæt land gesêc ðe ic ðê ýwan wille seek the land which I will show to thee, 83; Th. 105, 9; Gen. 1750: Cot. 3. III. to seek with hostile intention, to persecute, afflict, invade; hostiliter aggredi, invadere, corripere:—Gif úre fýnd us mid gefeohte gesêcaþ if our enemies make war upon us, Ex. 1, 10. Eorringa gesêceþ bôcstafa brego the prince of letters shall angrily seek him, Salm. Kmbl. 198; Sal. 98: Beo. Th. 5024; B. 2515. Ðæt he ðone wíd-flogan weorode gesêcan that he should seek the dragon [wide-flier] with a host, 4682; B. 2346. Geáta leóde gesôhton Gûþscilfingas the people of the Goths sought the warlike Scylfings, 5845; B. 2926: 4414; B. 2204. Gesôht; pp. Exon. 47 b; Th. 163, 11; Gû. 992: 49 b; Th. 170, 27; Gû. 1118. Hî scyndan sârum gesôhte they hastened forth sought with wounds, Exon. 72 b; Th. 271, 30; Jul. 490: 46 b; Th. 159, 21; Gû. 930: 47 b; Th. 163, 33; Gû. 1003. IV. to seek, go to, visit; ire, proficisci:—Ðú scealt sîþe gesêcan ðær sylfætan eard weardigaþ thou shalt seek in a journey where the cannibals defend the land, Andr. Kmbl. 349; An. 175. We ðê willaþ ferigan freólîce to ðam lande ðær ðê lust myneþ to gesêcanne we will freely convey thee to the land which desire urges thee to seek, 589; An. 295. Ðær mîn hyht myneþ to gesêcenne there my hope thinketh to visit, Exon. 48 b; Th. 167, 18; Gû. 1062. Ðæt him to môde sorg gesôhte that to his mind should come care, 37 b; Th. 123, 19; Gû. 325. V. to appoint, dispose, beset; exigere, disponere:—Hæfdon æglæcan sæcce gesôhte the wretches had appointed hostilities, Andr. Kmbl. 2265; An. 1134. Ðæt he mid âþsware to Abrahame, and to Isaac, eác gesôhte quod disposuit ad Abraham, et juramenti sui ad Isaac, Ps. Th. 104, 9. Synne gesôhte beset with sin, Exon. 74 b; Th. 280, 4; Jul. 624. DER. sêcan.

ge-seccan = ge-sêcean [?] or ge-feccan [?]:—Ides sceal dyrne cræfte hire freónd gesêccan gif heó nelle on folce geþeón ðæt hî man beágum gebycge a woman must by secret art get herself a friend if she do not wish publicly to succeed in being bought with rings, Menol. Fox 548; Gn. C. 44.

ge-sêcednes, -ness, e; f. A search, an inquiry, appeal; inquisitio, Som.

ge-secgan, -sæcgan, -secgean; to -secganne, -secgenne; p. -sægde, -sæde, pl. -sægdon, -sædon; impert. -sege; pp. -sægd, -sæd To say, tell, relate, declare, prove; dicere, narrare, indicere:—Mec Drihten hêht gesecgan the Lord commanded me to say, Exon. 42 b; Th. 144, 10; Gû. 676: 102 b; Th. 387, 29; Rä. 5, 12. Nelle ic ðê gesecgan I will not tell thee, Exon. 88 b; Th. 333, 11; Gn. Ex. 2: Elen. Kmbl. 1966: El. 985. Ic ðê sceal Meotudes mægenspêd I shall relate to thee the Creator's power, Exon. 92 b; Th. 348, 7; Sch. 24. Him sceolde se yldra eall gesæcgan narrabunt eam filiis suis, Ps. Th. 77, 8. Ic wille mîne leahterfulle þeáwas gesecgean I will confess my wicked ways; vitiosos mores corrigere, Bd. 3, 13; S. 538, 32. To gesecganne to say, Exon. 109 b; Th. 419, 1; Rä. 37, 13: Cd. 202; Th. 250, 9; Dan. 544. To gesecgenne to say. Cd. 163; Th. 205, 17; Exod. 437. Gif he hit gesêgþ if he saith it, Exon. 27 a; Th. 80, 22; Cri. 1310. Andreas Þeódne þanc gesægde Andrew said thanks to his Lord, Andr. Kmbl. 768; An. 384: Beo. Th. 4321; B. 2157. He gesæde swefen cyninge he said the dream to the king, Cd. 180; Th. 226, 2; Dan. 165: B. 4, 18; S. 587, 2. Ða

gesǽgdon Rómāne Bryttum *then the Romans said to the Britons*, Bd. 1, 12; S. 481, 3. Gesege me *dicito mihi*, Bd. 2, 12; S. 514, 1. Ðæt ðú gesecge sweostor mínre *that thou mayest say to my sister*, Exon. 50 a; Th. 172, 31; Gú. 1152: Bd. 4, 3; S. 568, 27. Wæs gesǽd hwám ðæt sweord geworht wǽre *it was said for whom that sword was wrought*, Beo. Th. 3396; B. 1696. Ic sceall ealle forlǽtan ða ðe of Perseo and of Cathma gesǽde syndon *I must pass over all things that are said of Perseus and Cadmus*, Ors. 1, 8; Bos. 31, 33, 34. Ðæt is gesǽd *that is proved*, Bt. 34, 9; Fox 146, 25, 27. DER. secgan.

ge-secggan *to say, tell*; dicere, narrare:—Hio him ne meahton ge-secggan be ðam sigebeácne *they could not tell him about the victorious sign*, Elen. Kmbl. 335; El. 168. v. ge-secgan.

ge-séclod; *part. Taken sick, ill*; ægrōtus:—Warþ se cyng geséclod *the king was taken sick*, Chr. 1093; Erl. 228, 22. v. ge-síclian.

ge-sédian *to satisfy*; satiare, Ps. Th. 106, 8.

ge-séfte; *adj. Soft, mild*; mītis:—Wǽron hyra gongas smēðe and gesēfte *their ways were smooth and soft*, Exon. 43 a; Th. 146, 3; Gú. 704. Swá him ēðost biþ, sylfum gesēftost *as to them may be easiest, softest to themselves*, Elen. Kmbl. 2587; El. 1295.

ge-sege *say, tell*, Bd. 2, 12; S. 514, 1; *impert. of* ge-secgan.

ge-segen, -sægen, -segn, e; *f. A saying, telling, conversation, relation, tradition*; dictum, narratio, relatio, traditio:—Mid gesegenum unrím geleáffulra witena *by the sayings of innumerable faithful witnesses*, Bd. pref; S. 472, note 25: Nar. 2, 6. Þurh gesegene ðæs ārwurþan biscopes Cynebyrhtes *through the conversation of the reverend bishop Cynebyrht*, Bd. pref; S. 472, 21. Mid Isses gesǽgene [gesegnum, MS. B.] ðæs ārwurþan Abbudes *by the conversation of the reverend abbot Isi*, 472, 20. Þurh swíðe getreówra manna gesægene *through the telling of very true men*, 472, 30: Bd. 5, 12; S. 631, 5, 11: 5, 23; S. 647, 17. v. segen.

ge-seglian; *p.* ode, ede; *pp.* od, ed. I. *to sail*; vēlĭfĭcāri:—Ðyder he cwæþ, ðæt nán man ne mihte geseglian on ānum mónþe *thither he said that a man could not sail in a month*, Ors. 1, 1; Bos. 21, 19. II. *to furnish with sails*; vēlis instruĕre:—Se ðe nafaþ ge-segled scip *who hath not a ship furnished with sails*, Salm. Kmbl. 450; Sal. 225.

ge-segn *a saying, telling, conversation*, Bd. pref; S. 472, note 20. v. ge-segen.

ge-segnian, -sénian; *p.* ode, ade; *pp.* od, ad [segnian, sēnian *to sign*] *To mark with the sign of the cross, to sign, bless;* crūcis signo signāre, bĕnĕdĭcĕre:—Fæder mancynnes hie gesegnaþ *the Father of mankind shall bless them*, Cd. 221; Th. 286, 30; Sat. 360: Salm. Kmbl. 807; Sal. 403. He heó gesēnaþ mid his swiðran hond *he shall bless them with his right hand*, Cd. 227; Th. 303, 18; Sat. 615. Se bisceop me ge-bletsode and gesegnode *the bishop blessed me and signed me*, Bd. 5, 3; S. 616, 33. Gesénode, 5, 3; S. 616, 25. His wuduwan ic wordum bletsige and gesegnade *viđuam ejus bĕnĕdīcens bĕnĕdīcam*, Ps. 131, 16. Gesénige hine *let him sign himself*, L. E. I. 29; Th. ii. 426, 9, 16. Gif heó gesegnod biþ *if it hath been blessed*, Salm. Kmbl. 812; Sal. 405. Gesunde and gesēnade *safe and blessed*, Exon. 27 b; Th. 82, 22; Cri. 1342.

ge-sehtian; *p.* ode; *pp.* od [sehtian *to settle*] *To settle, reconcile*; rĕconcĭliāre:—Ða heáfodmen ða bróðra gesehtodan *the chief men reconciled the brothers*, Chr. 1101; Erl. 237, 26.

ge-sehtness, e; *f. Reconciliation*:—To sibbe and to gesehtnesse *for peace and reconciliation*, Cod. Dipl. Kmbl. iii. 129, 22.

ge-selda, an; *m. One of the same dwelling, a companion, comrade*; contūbernālis, sŏcius:—Ic eom cyninges geselda *I am a king's companion*, Exon. 127 a; Th. 489, 5; Rä. 78, 3. Higelāc ongan sínne geseldan fricgean *Hygelac began to question his comrade*, Beo. Th. 3972; B. 1984: Exon. 77 a; Th. 289, 24; Wand. 53

ge-sele, es; *m.* [sele] *A tabernacle*; tăbernācŭlum:—On gesele ðínum *in tăbernācŭlo tuo*, Ps. Spl. T. 14, 1.

ge-selenis, -niss, e; *f. A handing over, giving, tradition*:—Æfter gimett giselenisse Cristes *secundum mensuram donatiōnis Christi*, Rtl. 83, 1. Æfter geselenise *juxta traditiōnem*, Mk. Skt. Lind. 7, 5. v. selenis.

ge-sélig; *adj. Happy*; fēlix:—Gebed dôn gesēligran tíman getācnaþ *to be repeating prayers betokens a happier time*, Lchdm. iii. 208, 23. v. ge-sǽlig.

ge-sélignes. v. ge-sǽlignys.

ge-sellan, -syllan; *p.* -sealde, -salde; *pp.* -seald *To give, give up, betray, sell*; dare, tradere, vendere:—Oðrum gesellan *to give to others*, Beo. Th. 2063; B. 1029. Ic ðé geselle *I will give thee*, Cd. 228; Th. 307, 25; Sat. 685. Me ða blǽda Eue gesealde *Eve gave me the fruits*, 42; Th. 54, 27; Gen. 883: Exon. 100 b; Th. 379, 31; Deór. 41. Ðú me ge-sealdest sweord *thou gavest me a sword*, 120 b; Th. 463, 18; Hö. 72. Ðe feorh gesealdon *who gave up life*, Andr. Kmbl. 3231; An. 1618: 865; An. 433. Inc is feoh geseald *cattle is given to you*, Cd. 10; Th. 13, 14; Gen. 202: 74; Th. 91, 23; Gen. 1516. Iudas gesalde Drihten Hǽlend *Judas sold* [*tradidit*] *the Lord Saviour*, 226; Th. 301, 2; Sat. 575. Ælfnóþ and Wulfmǽr feorh gesealdon *Ælfnoth and Wulfmær gave up their lives*, Byrht. Th. 137, 11; By. 184. DER. sellan.

ge-sélþ, e; *f. Happiness*; fēlĭcĭtas:—Gesélþe tíman hit getácnaþ *it betokens a time of happiness*, Lchdm. iii. 202, 10: 204, 23. We gyt næfdon ða gesélþa *we had not yet the happiness*, Chr. 1009; Erl. 141, 25. v. ge-sǽlþ.

ge-sém, es; *n. Reconciliation, an agreement, a compromise*; recon-cĭliātio, comprōmissum:—Siððan āne neaht ofer ðæt gesēm bíe *postquam ūna nox supra comprōmissum prætĕriit*, L. H. E. 10; Wilk. 8, 49.

ge-séman; *p.* de; *pp.* ed *To compose, settle, make peace with, reconcile, satisfy*; compōnĕre, conciliāre, reconciliāre, satisfăcĕre:—Ðæt hy gesēman wolde *that he would make peace with them*, Ors. 3, 7; Bos. 60, 33. Ðæt hí scioldon Wynflǽde and Leófwine gesēman *that they should reconcile Wynflæd and Leofwine*, Th. Diplm. A. D. 995; 288, 31: Past. 46, 4; Swt. 349, 12; Hat. MS. 66 b, 13: Byrht. Th. 133, 35; By. 60. Ðæt me gesēme snoterra mon *that a wiser man shall reconcile me*, Salm. Kmbl. 501; Sal. 251. Ðæt he hý ymbe ðæt ríce gesēmde *that he would satisfy them about the kingdom*, Ors. 3, 7; Bos. 60, 23. Siððan sió sace gesēmed sió *after the suit is settled*, L. H. E. 10; Th. i. 30, 19: Ors. 1, 12; Bos. 35, 39. Hí gesēmede beón ne mihtan *they could not be recon-ciled*, Chr. 1094; Erl. 230, 1: Homl. Th. ii. 338, 1.

ge-sencan; *p.* -sencte; *pp.* -senct *To sink, drown*; submergĕre:—Hí gesencte [synt] on ðære [MS. ðere] reádan sǽ [MS. sea] *they are drowned in the Red sea*, Cant. Moys. Ex. 15, 4; Thw. 15, 4.

ge-sendan; *p.* -sende; *pp.* -sended, -send *To send*:—Miððý gesende stefne *emissa voce*, Mk. Skt. Lind. 15, 37. Ðá wæs gesended *then was sent*, Blickl. Homl. 9, 28: Mt. Kmbl. Lind. 5, 13: 15, 17. Gesend *missus*, Ps. Lamb. 33, 8.

ge-séne; *adj.* v. ge-sýne.

ge-séne; *adv. Clearly*; manifeste, Jn. Skt. Lind. 11, 14.

ge-sénelic; *adj. Visible*:—Se gesénelica líchama *the visible body*, Blickl. Homl. 21, 24.

ge-sénelíce; *adv. Visibly*; visibiliter, Rtl. 103, 30.

ge-sénian *to mark with the sign of the cross, to sign, bless*, Cd. 227; Th. 303, 18; Sat 615: Bd. 5, 3; S. 616, 25; L. E. I. 29; Th. ii. 426, 9, 16: Exon. 27 b; Th. 82, 22; Cri. 1342. v. ge-segnian.

ge-seón, -sión, ic -seó, ðú -sihst, he -syhþ; *p.* -seah, ðú -sáwe, *pl.* -sáwon, -sēgon; *imp.* -syh, -seoh; *subj. pres.* ic -sáwe; *pp.* -sawen *To see*; videre, conspicere. I. *used absolutely or with acc*:—Ic geseóm menn *video homines*, Mk. Skt. Lind. 8, 24. He hér gesihþ *he here seeth*, Apol. Th. 14, 26. Ða liðende land gesáwon *the voyagers saw land*, Beo. Th. 448; B. 221. Ðá heó Isaac geseah *when she saw Isaac*, Gen. 24, 64. Ða he beseah, ðá geseah he olfendas *when he looked about then he saw the camels*, Gen. 24, 63. Abraham beseah upp and geseah þrí weras *Abraham looked up and saw three men*, Gen. 18, 2. Hie ðone heora scyppend gesēgon *they saw their creator*, Blickl. Homl. 121, 28: Exon. 15 b; Th. 35, 7; Cri. 554. Manega rihtwíse gewilnudon ða þing to geseónne ðe gē geseóþ and híg ne gesáwon *multi justi cupiērunt videre quæ videtis et non vidērunt*, Mt. Bos. 13, 17. Hwí fērde gē geseón . . . hwí fērde gē to geseónne *quid existis videre*, Lk. Skt. 7, 24, 25, note. Ðæt hie geseónde ne geseón *ut videntes non videant*, 8, 10. Cum and geseoh *veni et vide*, Jn. Bos. 1, 46. Ðíne gangas wǽron gesewene *visi sunt gressus tui*, Ps. Th. 67, 23: Shrn. 97, 30. Him wæs gesewen ðæt . . . *it seemed to him that . . .*, 111, 27: Blickl. Homl. 195, 20. Hie wurdon gesewene *they appeared*, 173, 25. Ic ðé gesáwe *that I saw thee*, Wald. 21; Vald. 1, 13. II. *with predicative adj. or part*:—Ic geseó mínne Crist cígende me *I see my Christ calling me*, Blickl. Homl. 187, 23: 59, 2. Hie Drihten gesáwon upastígende *they saw the Lord ascending*, 121, 22: 123, 25. Gesyhþ wínsele wēstne *he sees the wine-hall deserted*, Beo. Th. 4901; B. 2455: Cd. 37; Th. 48, 30; Gen. 783: 64; Th. 78, 12; Gen. 1292. Geseah lífes weard drige stówe wíde æteówde *life's guardian saw the dry place widely displayed*, 8; Th. 10, 28; Gen. 163. III. *with acc. and infin*:—Ða ðe he gesyhþ to Gode higian *those that he sees hurry to God*, Blickl. Homl. 29, 22. Hie ðæt leóht geseóþ scínan *they see the light shine*, 129, 7: Cd. 5; Th. 7, 20; Gen. 108: 32; Th. 42, 4; Gen. 669. Gesēgun ða dumban gesceaft gefēlan *they saw the dumb creation feel*, Exon. 24 b; Th. 69, 30; Cri. 1128. IV. *with infin*:—Geseah weard beran beorhte randas *the warder saw bright shields borne*, Beo. Th. 463; B. 229: 2051; B. 1023. V. *followed by a clause*:—Hie geseóþ hú God ða stówe geweorðaþ *they see how God honours the place*, Blickl. Homl. 129, 25: 229, 22: 41, 28: Ps. Th. 73, 19. He gesáwe ðæt he wǽre getogen *he saw that he was pulled*, Blickl. Homl. 43, 26: 145, 8. Ic mæg geseón hwǽr he sylf siteþ *I can see where he himself sits*, Cd. 32; Th. 41, 34; Gen. 666. v. seón.

ge-seórod; *part. p. Leavened*:—Geseorid hláf *acrizimus panis*, Ælfc. Gl. 66; Som. 69, 62; Wrt. Voc. 41, 18.

ge-set, es; *n. A sitting, lying in wait, ambush*; insidiæ:—Giseto *insidias*, Rtl. 37, 19. [Cf. O. H. Ger. gisez obsidio.]

ge-sete, *pl.* -setu, -seotu; *n. A seat, habitation, house*; sedes, domi-cilium, habitatio:—Ofer eall beorht gesetu *over all bright habitations*, Exon. 117 b; Th. 452, 7; Dôm. 117: 121 b; Th. 466, 3; Hö. 115. Sun-beorht gesetu *dwellings bright with the sun*, 59 b; Th. 217, 10;

Ph. 278 : 62 a ; Th. 228, 10 ; Ph. 436. On sēllan gesetu *to better dwellings*, 51 a ; Th. 178, 10 ; Gū. 1242. Ofer burga gesetu *over the cities' dwellings*, 26 a ; Th. 76, 16 ; Cri. 1240. Ges-otu, Cd. 227 ; Th. 302, 20 ; Sat. 602. Ða cynelícan burh porres and his cynelícan geseto *ipsam urbem regiam pori domumque*, Nar. 4, 20. To heora gesetum *to their lairs*, Blickl. Homl. 199, 7. [Cf. *O. H. Ger.* gesaze *habitatio, sedes, domicilium : O. Sax.* hôh-gisetu.]

ge-setednes, -nys, -ness, -nyss, e ; *f. A constitution, law, ceremony, religion ;* constitūtio, lex, cērēmōnia, religio :—Hwæt ys ðeós gesetednys *quæ est ista religio?* Ex. 12, 26. Fram middaneardes gesetednesse *a constitūtiōne mundi*, Mt. Bos. 13, 35. Begýmaþ ðisse gesetednysse *observābitis cērēmōnias istas*, Ex. 12, 25. To gesetednisse *for a law*, Gen. 47, 26. v. ge-setnes.

ge-setenes, -ness, e ; *f. A constitution, an appointment ;* constitūtio :—Ða gesetenes he lǽt standan *he allows this appointment to stand*, Bt. 21 ; Fox 74, 30. v. ge-setnes.

ge-sēðan ; *p.* de ; *pp.* ed [sēðan *to affirm*] To state as true, declare, prove, show, affirm ;* effāri, testificāri, vērificāre, contestāri, probāre :—Nis ǽnig ðæs horsc, ðe ðín fromcyn mǽge fira bearnum sweotule gesēðan *none is so wise who may manifestly declare thy origin to the children of men*, Exon. 11 b ; Th. 15, 18 ; Cri. 1240. Ne mágon gé ða word gesēðan *ye cannot prove the words*, Elen. Kmbl. 1160 ; El. 582 : Bt. 7, 3 ; Fox 20, 7. Ic gesweotelige oððe gesēðe ðé God *testificābor tibi Deus*, Ps. Lamb. 49, 7. Gehýr folc mín and ic gesēðe ðé *audi pōpulus meus et contestābor te*, Ps. Spl. 80, 8. Hí gesēðaþ and sprecaþ unrihtwísnysse *effābuntur et lŏquentur iniquitātem*, 93, 4. Ðǽre gesyhþe sóþ wæs gecýðed and gesēðed *cujus vēritas prŏbāta est*, Bd. 4, 8 ; S. 576, 10 : Cd. 208 ; Th. 257, 7 ; Dan. 254. Gesēðde, *pp. pl. proved*, Ps. Th. 118, 160. Ða wurdon mid manegum tácnum gesēðde *which were proved by many miracles*, Homl. Th. ii. 130, 11.

ge-sēðung, e ; *f. Assertion, affirmation ;* assertio, affirmatio, Hpt. Gl. 455.

ge-setl, es ; *n. A seat, settle :—*Ða foerþmestu gisedla æt feormum *primos discubitos in cenis*, Mk. Skt. Rush. 12, 39. v. setl.

ge-setnes, -setenes, -setednes, -ness, -nis, -niss, -nys, -nyss, e ; *f. Position, foundation, tradition, an institution, constitution, composition, ordinance, decree, law ;* pŏsītio, sītus, fundātio, trādītio, institūtio, constitūtio, compŏsītio, lex, pactum :—Cūþ is gehwilcum snotterum mannum, ðæt seó ealde ǽ wæs eáðelícre ðonne Cristes gesetnys sý *it is known to every intelligent man that the old law was easier than the institute of Christ*, Homl. Th. i. 358, 28, 30. Wæs se cyning becumen on swá mycle lufan ðǽre Rōmāniscan cyricean gesetnysse and ðǽre Apostolícan *rex tenēbātur ămōre Rōmānæ et Ăpŏstŏlícæ institūtiōnis*, 571, 32 : 5, 20 ; S. 642, 13. Be gesetnysse Breotene *de sītu Brītanniæ*, I, 1 ; S. 473, 6 : Nar. 1, 5. Ǽr middaneardes gesetnysse *before the foundation of the world*, Homl. Th. ii. 364, 27 : Mk. Bos. 7, 5. Be Godes gesetnysse *by God's ordinance*, Bd. de nat. rerum ; Wrt. popl. science 11, 22 ; Lchdm. iii. 258, 7 : Ælfc. T. 17, 24. Ðú cwǽde ðæt ǽlc wuht his rihte gesetnesse fuleóde, būtan menn ánum *thou saidst that every creature fulfilled its right institution, except man alone*, Bt. 5, 3 ; Fox 12, 9 : Homl. Th. ii. 330, 35. Rǽdaþ sume men ða leásan gesetnysse *some men read the false composition*, Homl. Th. ii. 332, 22 : i. 358, 14. Israhél syngode and ða gesetnisse gewemde *peccāvit Israel et prævārīcātus est pactum meum*, Jos. 7, 11. Sint heora gesetnessa swíðe mislíca *their institutions are very various*, Bt. 18, 2 ; Fox 64, 22. Healdende hira yldrena gesetnessa *tenentes trādītiōnem sēniōrum*, Mk. Bos. 7, 3. Ða gesetnessa sigora Wealdend lǽt geond ðas mǽran gesceaft mearce healden *the Lord of victories permits these constitutions to keep their limits over this great creation*, Bt. Met. Fox 11, 141 ; Met. 11, 71. Be gesetnessum and gemētum sprǽccynna *de fīgūris mŏdisque lŏcūtiōnum*, Bd. 5, 24 : S. 648, 42. Be heofenes gesetenissum *de statu cæli*, Nar. 1, 16.

ge-setnian ; *p.* ode, ade *To lie in wait ;* insidiari :—Herodia gesetnade him *Herodias insidiabātur illi*, Mk. Skt. Lind. 6, 19.

ge-settan ; *p.* -sette ; *pp.* -seted, -sett, -sett *To set, put, fix, confirm, restore, appoint, decree, settle, possess, occupy, place together, compose, make, compare, expose, allay :—*Ða apostolas hie gesetton on ðæm fægran neorxna wange *the apostles placed her in the fair paradise*, Blickl. Homl. 143, 25 : Exon. 28 a ; Th. 85, 13 ; Cri. 1390 : Ps. Spl. 18, 5. Hie on God ǽnne heora hyht gesetton *they should put their trust in God alone*, Blickl. Homl. 185, 15. Naman gesettan *to give a name*, 197, 29. He wæs to bóclícre lāre gesett *he was put to book-learning*, Shrn. 12, 16. Ðǽr is dryhtnes folc geseted to symle *there is the Lord's folk set to the feast*, Rood Kmbl. 279 ; Kr. 141. Ðæt hí hine Gode gesettan *to present him to God*, Lk. Bos. 2, 22. Under anweald gesett *sub potestate constitutus*, 7, 8 : 3, 13. Hwonne he ðisse worlde ende gesettan wolde *when he meant to fix the end of this world*, Blickl. Homl. 119, 9 : 27, 24. Ǽnne of heora aldormannum to bisceope he him gesette *he appointed one of their chief men as their bishop*, 247, 31 : Chr. 604 ; Erl. 20, 21. He Isaace wíf gesette *he fixed upon a wife for Isaac*, Gen. 24, 11. Heora gewinn mid ðam swíðe gesettan *therewith greatly confirmed their hostility*, Ors. 5, 10 ; Bos. 109, 5, note. Wilt ðú on ðas tíd gesettan Israhéla folca ríce *si in tempore hoc restitues regnum Israel?* Blickl. Homl. 117,

11. Gesete *restitue*, Ps. Spl. 34, 20. Hí him gesetton ðæt hyra ān lātteów wǽre *they decreed for themselves that there should be one leader of them*, Ors. 2, 4 ; Bos. 42, 26 : Shrn. 112, 18 : Blickl. Homl. 193, 3 : 61, 27. Dóm gesettan *to judge*, Gen. 18, 25. Gesette ýðum heora onrihtne ryne *he appointed the waves their proper course*, Cd. 8 ; Th. 10, 34 ; Gen. 166. He gefór ða burg and hét hie gesettan ǽgðer ge mid Engliscum mannum ge mid Deniscum *he gained the town and ordered it to be occupied by both English and Danes*, Chr. 922 ; El. 108, 31 : 886 ; Erl. 84, 26 : Mt. Bos. 21, 33, 41 : Ors. 3, 5 ; Bos. 56, 35. Ealne norþdǽl ðysses eálondes genôman and gesetton *omnem aquilonalem insulæ partem capessunt*, Bd. 1, 12 ; S. 481, 18 : Blickl. Homl. 79, 26. Heora éðel on heofenum sceolde eft gebúen and geseted weorðan mid hálgum sáwlum *their home in heaven should again be inhabited and peopled with holy souls*, 121, 33. Seó landbúnes is swíðost cýpemonnum geseted *hæc colonia est maxime negotiātōrum*, Nar. 33, 15. Gesettaþ *possidebit*, Ps. Spl. C. 68, 42 : 78, 12 : 82, 11. Of láme ic ðé leoðe gesette *of loam I formed thee limbs*, Exon. 28 a ; Th. 84, 31 ; Cri. 1380 : 33 a ; Th. 105, 12 ; Gū. 22. Ðú gesettest sunnan and mónan *tu fecisti solem et lunam*, Ps. Th. 73, 16. Ic ðé gesette manegra þeóda fæder *a father of many nations have I made thee*, Gen. 17, 5 : Homl. Th. ii. 136, 23. Ðæt tempel towearp æfter feówer hund geárum ðæs ðe hit gesett wæs *he destroyed the temple four hundred years after it was built*, Swt. A. S. Rdr. 68, 374. Swá hwæt swá ic ðé gehét eal ic hit gesette *whatsoever I have promised thee I will do it all*, Blickl. Homl. 147, 8. Seó tunge ðe swá monig hálwende word on ðæs scyppendes lof gesette *illa lingua quæ tot salutaria verba in laudem conditoris composuerat*, Bd. 4, 24 ; S. 599, 11 : Bt. 2 ; Fox 4, 7. Ða bóc ic gesette *I composed the book*, Guthl. prol ; Gdwin. 2, 8 : Homl. Th. i. 70, 7 : Th. Apol. 28, 13. Hiora birhto ne biþ to gesettanne wiþ ðǽre sunnan leóht *their brightness is not to be compared with the sun's light*, Bt. Met. Fox 6, 13 ; Met. 6, 7. Ðæt ðis ǽfre gesett sprǽc wǽre *that this should be a suit finally settled*, Th. Ch. 203, 4. Ðone storm he gesette and gestilde *tempestatem sedaverit*, Bd. 5, 1 ; S. 613, 8. Ðæt he ðæt yrre gesette *to allay their anger*, Ors. 4, 11 ; Bos. 98, 2 : Beo. Th. 4062 ; B. 2029. Ða earman ceasterwaran wǽron to hungre gesette *the miserable citizens were exposed to famine*, Bd. 1, 12 ; S. 481, 28. Ðæt land sum hit is to gafole gesett *some of the land is let*, Cod. Dipl. Kmbl. iii. 450, 19, 12.

ge-settnys, -nyss, e ; *f. Constitution, statute ;* stătūtum :—Ða ðe ða reogollícan gesettnysse háligra fædera gelufedon and cúðon *qui cănŏnīca patrum stătūta et dīligĕrent et nossent*, Bd. 4, 5 ; S. 571, 40. v. ge-setnes.

ge-setu *seats*, Th. 76, 16. v. ge-sete.

ge-séuling *a servant ;* minister, Lye.

ge-séunes *the sea ;* æquor, Lye.

ge-sewenlíc ; *adj. Visible :—*Ðíne gesceafta gesewenlíce and eác ungesewenlíce *thy creatures visible and also invisible*, Bt. 33, 4 ; Fox 128, 5 : Bd. de nat. rerum ; Wrt. popl. science 1, 12 ; Lchdm. iii. 232, 14 : Bt. Met. Fox 20, 13 ; Met. 20, 7 : 253 ; Met. 20, 127.

ge-sewenlíce ; *adv. Visibly :—*Ðú miht sóþlíce and gesewenlíce ðíne mihte gecýðan on Marian *thou canst truly and visibly make thy power known on Mary*, Blickl. Homl. 157, 3.

ge-sib, -sibb, -syb ; *adj. Peaceable, near, related, familiar ;* pācífícus, cognātus, prŏpinquus, fămĭliāris :—Ne bearh nū for oft gesibb gesibbum ðý má ðe fremdum *too often now has a kinsman no more protected a kinsman than a stranger*, Swt. A. S. Rdr. 107, 75. Sylle swá gesibre handa swá fremdre *give to a relation or to a stranger*, Cod. Dipl. Kmbl. ii. 114, 7. Nǽnig men ne sceal lufian ne gēman his gesibbes gif . . . *no man shall love or be mindful of his relative if . . .*, Blickl. Homl. 23, 17. Gisibbe *cognatos*, Lk. Skt. Rush. 14, 12. Tǽlende dígellíce gesibne his *dētrāhentem sēcrēto proximo suo*, Ps. Spl. 100, 5. Ðe him gesibbe wǽron *who were related to him*, Job Thw. 167, 3. Gesibbe ǽrendracan cǣdūcēātōres vel pācífici*, Ælfc. Gl. 53 ; Som. 66, 79 ; Wrt. Voc. 36, 6. Hý habbaþ freónda ðý má swǽsra and gesibbra *they will have more friends dear and near*, Exon. 107 a ; Th. 408, 34 ; Rä. 27, 22 : 84 a ; Th. 317, 21 ; Mód. 69. Snotor mid gesibbum sécean wolde Cananea land *the sagacious would seek the Canaanites' land with his kinsfolk*, Cd. 83 ; Th. 104, 8 ; Gen. 1738 : 79 ; Th. 97, 13 ; Gen. 1612. Gesibbra ærfeweard *a nearer heir*, Th. Chart. 483, 16. Sweolcum swelce him ðonne gesibbast wǽre *to such as may then be nearest of kin to him*, 105, 29. [*O. H. Ger.* gisibbo *consanguineus*.]

ge-sibbian ; *p.* ode, ade, ede ; *pp.* od, ad, ed [sibbian *to pacify*] To make peaceful, pacify, appease, gladden ;* pācāre, pācíficāre, conciliāre, lætificāre :—Ic gesibbige *concílio*, Ælfc. Gl. 76 ; Som. 74, 18 ; Wrt. Voc. 50, 2. He gesibbade ða cyningas betwih and ða folc *pācātis altĕrŭtrum rēgibus ac pŏpŭlis*, Bd. 4, 21 ; S. 590, 22. Gesibbedan sáwle míne *lætíficāverunt ănĭmam meam*, Ps. Th. 93, 18. Ða he hæfde ðone híréd gesibbodne *when he had reconciled the household*, Blickl. Homl. 225, 10. [*Goth.* ga-sibjon *to reconcile : O. H. Ger.* ge-sippot *united*.]

ge-sibbsum ; *adj. Peaceful ;* pācātus :—Salomon is gecweden gesibbsum on Englisc *Salomon is in English 'peaceful,'* Swt. A. S. Rdr. 67, 353. Sint to manienne ða gesibbsuman *the peaceful are to be admonished*, Past. 46, 5 ; Swt. 351, 3 ; Hat. MS. 66 b, 27. v. ge-sibsum.

ge-sibbsumnys, -nyss, e; f. *Peacefulness*; pax :—For gesibbsumnysse *for peacefulness*, Lev. 7, 32. v. ge-sibsumnes.

ge-siblíce; *adv. Peaceably*; pācĭfĭce :—Fæste gebunden gesiblíce togædere *fast bound peaceably together*, Bt. Met. Fox 20, 135; Met. 20, 68.

ge-sibling, es; m. [sibling *a relation*] *A relation*; prŏpinquus :—Mǽg vel gesibling prŏpinquus, Ælfc. Gl. 92; Som. 75, 39; Wrt. Voc. 51, 81.

ge-sibness, e; f. *Relationship*; affinitas, Lye.

ge-sibsum, -sybsum, -sibbsum; *adj.* [sibsum *peaceable*] *Peaceable, peaceful, loving peace*; pacatus, pācĭfĭcus :—Se đe of Gode cymþ he biþ gódes willan and gesibsum *that which comes from God is of good will and peaceful*, Past. 46, 3; Swt. 349, 1; Hat. MS. 66 b, 5, 7. On óđre wísan sint to manigenne đa gesibsuman *the peaceful are to be admonished in one way*, 46, 1; Swt. 345, 6; Hat. MS. 65 b, 22 : 46, 5; Swt. 351, 3; Hat. MS. 67 a, 12 : 46, 7; Swt. 355, 9; Hat. MS. 67 b, 19 : 47, 1; Swt. 3, 7, 15; Hat. MS. 68 a, 18, 19.

ge-sibsumian; *p.* ode; *pp.* od *To make peaceable, reconcile* :—Đē to him gesibsuma *reconcile thyself to him*, Homl. Th. i. 54, 20: Mt. Bos. 5, 24.

ge-sibsumlíce, -sybsumlíce; *adv.* [sibsumlíce *peaceably*] *Peaceably, peacefully*; pācĭfĭce :—Đa fuglas gesibsumlíce faraþ *the birds fly peacefully*, Past. 46, 4; Swt. 349, 22; Hat. MS. 66 b, 22. Forđamđe me witedlíce gesybsumlíce hí sprǽcon *quŏniam mihi quidem pācĭfĭce lŏquebantur*, Ps. Spl. 34, 23: Nicod. 20; Thw. 10, 15.

ge-sibsumnes, -sibbsumnes, -ness, -nys, -nyss, e; f. [sibsumnes *peacefulness*] *Peacefulness, concord, reconciliation*; pax, concordĭa, rĕconcĭliātĭo :—We mǽgon gecnáwan on đara ungesceádwísra niétena gesibsumnesse, hú micel yfel sió gesceádwíslíce gecynd þurh đa ungesibsumnesse gefremeþ *we can understand from the peacefulness of irrational animals how great a sin the rational race of man commits in being quarrelsome*, Past. 46, 4; Swt. 349, 25; Hat. MS. 66 b, 24; Lev. 7, 32.

ge-sibsumung, e; f. *A making peace, conciliation*; consĭliātĭo, Ælfc. Gl. 86; Som. 74, 16; Wrt. Voc. 49, 39.

ge-sícan; *p.* te; *pp.* ed [sícan *to give suck*] *To wean*; ablactāre :—Swá swá gesíced ofer módor his *sícut ablactātus sŭper matre sua*, Ps. Spl. 130, 4.

ge-síclian, -sýclian; *p.* ode; *pp.* od [seóc *sick*] *To be taken sick or ill, to be infirm*; ægrŏtāre, infirmāri :—Đæt his fæder wǽre gesíclod *quod ægrŏtāret pater suus*, Gen. 48, 1: Chr. 1003; Erl. 139, 10. Sum undercyning wæs, đæs sunu wæs gesýclod on Capharnaum *ĕrat quidam rēgŭlus, cujus fīlius infirmābātur Capharnaum*, Jn. Bos. 4, 46. Đá wearþ his hors gesíclod *his horse became ill*, Swt. A. S. Rdr. 100, 169.

ge-sída. v. heort-gesída.

ge-síe *to be*; esse, Mt. Kmbl. Lind. 6, 31.

ge-siehþ *sight*, Bt. 5, 3; Fox 14, 18. v. ge-sihþ.

ge-siftan; *p.* -sifte; *pp.* -sifted, -sift *To sift* :—Gesyft [*or* gesyfl?] melu *fine meal*, Ex. 12, 34.

ge-sig; n. *Victory* :—Đæt gesig *victoria*, Rtl. 28, 3.

ge-sígan; *p.* -sáh, *pl.* -sigon; *pp.* -sigen [sígan *to sink*] *To sink, fall, set as the sun*; cadere, labi, occĭdĕre ut sol :—Ǽr heó [sunne] fullíce gesígan onginne *before it [the sun] begin fully to sink*, Herb. 19, 5; Lchdm. i. 112, 21. Đæt he ána scyle gesígan æt sæcce *that he alone should sink in conflict*, Beo. Th. 5311; B. 2059. Đonne me ylde tíd on gesíge *in tempore senectutis*, Ps. Th. 70, 8. Đá to đam wage geság *then to the wall he sank*, Exon. 51 a; Th. 178, 13; Gú. 1243.

ge-sigefæstan; *p.* -fæste; *pp.* -fæsted, -fæst [sige *victory*] *To make triumphant, crown*; corrōbŏrāre, cŏrōnāre :—He đē gesigefæste sóþre miltse *qui coronat te in mĭsĕrātĭone*, Ps. Th. 102, 5. Hí synne geswencton and gesigefæston *they outwearied sin and triumphed*, Exon. 55 b; Th. 197, 13; Az. 189. We gesigefæstan đíne bǽre *let us crown thy bier*, Blickl. Homl. 149, 19 : 151, 9. Đæt ic mid Criste gesigefæsted wǽre *ipse cum Cristo coronandus*, Bd. 2, 5; S. 508, 21. Twegen cynelíce cnihtas mid syndriglícre Godes gyfe wǽron gesigefæste đuó rēgii pueri fratres spĕciāli sunt Dei grātia cŏrōnāti, 4, 16; S. 584, 21. Siendon đinne đómas gesigefæste *thy decrees are triumphant*, Cd. 188; Th. 234, 8; Dan. 288: Exon. 53 a; Th. 185, 18; Az. 9: Shrn. 146, 11. Drihten gesigefæsted *the Lord triumphant*, Blickl. Homl. 67, 14.

ge-sigfæstnian; *p.* ode; *pp.* od *To triumph, crown* :—He gesigfæstnade *triumphans*, Mt. Kmbl. 13, 3. Gesigfæstnad *coronandus*, Jn. Skt. 8, 12.

ge-siht, -sihþ, -siehþ, -syhþ, -sihtþ, e; f. *Sight, power of seeing, vision, something seen, aspect, respect*; vísus, acies oculorum, vísio, aspectus, conspectus, respectus :—Se ord on here ođđe scearp gesihþ *acies*, Ælfc. Gr. 5; Som. 4, 14. Yfel gesihþ *oculus malus*, Mk. Bos. 7, 22. Bodian blindum gesihþe *prædicare cæcis visum*, Lk. Bos. 4, 18 : Homl. Th. i. 64, 22: Blickl. Homl. 155, 5. Đú wást đæt gesiht and gehérnes ongitaþ đone líchoman đæs monnes *thou knowest that sight and hearing perceive the body of a man*, Bt. 41, 4; Fox 252, 6. Eágena gesihþ *eye-sight*, Andr. Kmbl. 60; An. 30: Ps. Th. 93, 9. Forhwan woldest đú đínre gesihþe me wyrnan *quid avertis faciem tuam a me?* 87, 14. He wun-

drode æfter đære gesihþe *he wondered at the sight*, Blickl. Homl. 153, 36 : 215, 31. Forht ic wæs for đære fægran gesyhþe *terrified I was at the fair sight*, Rood Kmbl. 41; Kr. 21. Đæt he sume gesihtþe geseah *quod visionem vidisset*, Lk. Bos. 1, 22. Engla gesihþe *visionem angelorum*, 24, 23. Þurh nihtlíce gesihþ *in a vision of the night*, Shrn. 63, 16: Lchdm. iii. 204, 31. Đære uplícan sibbe gesiehþ *the sight of the peace above*, Past. 21; Swt. 161, 16; Hat. MS. On ealles đæs folces gesihþe *in the sight of all the people*, Homl. Th. i. 60, 25 : Blickl. Homl. 121, 17 : 201, 5. On đínre gesyhþe *in conspectu tuo*, Ps. Th. 55, 7 : 137, 1 : Cd. 49; Th. 63, 20; Gen. 1035 Of heora gesihþum *from their sight*, Jud. 16, 3. Bútan gesyhþe *ærfæstnesse sine respectu pietatis*, Bd. 4, 12; S. 580, 41.

ge-sincan; *p.* -sanc, -sonc, *pl.* -suncon; *pp.* -suncen *To sink*; delābi :—Him in gesonc flacor flánþracu *the flickering arrow's force sank into him*, Exon. 49 b; Th. 170, 22; Gú. 1115. Đá ne meahton hi on đæm wætere gesincan *then they could not sink in the water*, Shrn. 103, 19.

ge-síne; *adj. Void, destitute*; expers :—Módum tǽcan đæt we gesíne ne sýn godes þeódscipes *to teach our minds that we be not destitute of God's communion*, Cd. 169; Th. 211, 18; Exod. 528. v. gésne, gǽsne.

ge-síne. v. ge-sýne.

ge-singalian; *p.* ode, ade *To continue, perpetuate* : — Gesyngalade continui, Ps. Spl. C. 88, 49.

ge-singallícode *continually*; continuatim, V. Ps. 140, 7. v. singallíce.

ge-singan; *p.* -sang, *pl.* -sungon; *pp.* -sungen *To sing*; cănĕre :—Sceal mon leóþ gesingan *a man shall sing songs*, Exon. 91 a; Th. 342, 8; Gn. Ex. 140: Menol. Fox 140; Men. 70. David þurh Godes gást Gode to lofe gesang *David through God's spirit sang to the praise of God*, Swt. A. S. Rdr. 67, 332. Mæssan gesingan *to sing mass*, Blickl. Homl. 45, 31: 207, 5. Đætte on Cantica Canticorum wæs gesungen *what was sung in the Song of Songs*, 11, 15.

ge-singe [= ge-sinhíge (?) v. ge-sinîg], an; f. *A wife* :—Ne meaht đú habban mec đē to gesingan *thou mayest not have me for thy wife*, Exon. 66 b; Th. 245, 34; Jul. 54. [Cf. ge-sinhíwan.]

ge-singian; *p.* ode; *pp.* od *To sin*; peccāre :—We habbaþ swíđe gesingod *we have greatly sinned*, Hy. 7, 115; Hy. Grn. ii. 289, 115. v. ge-syngian.

ge-sinhíwan, -hígan; *pl. m. Married persons*; conjuges, conjugati, conjugia :—Unriht gewuna is arisen betwih gesinhíwum *prava in conjugatorum moribus consuetudo surrexit*, Bd. 1, 27; S. 493, 34. Gesinhíwan *conjuges vel conjugales*, Ælfc. Gl. 86; Som. 74, 25; Wrt. Voc. 50, 7. Đæt líf đara gesinhíwena oferstígþ đæt líf đæs mægþhádes *the life of the married surpasses the life of virginity*, Past. 52, 8; Swt. 409, 29; Hat. MS. Tu gesinhíwan sprǽcon ymbe hine ealle niht *two married people were talking about him all night*, Shrn. 90, 2. Ealla đara monna hús bútan đara gesinhígna *all men's houses except the two married people's*, 5. Đara háligra gesinhína tíd *the holy man and wife's tide*, 55, 31. Wit sýn swá swá gesinhína [?] *we be as married people*, 40, 20. For gesinhíwum *pro conjugiis*, Bd. 4, 5; S. 573, 14. v. sin-híwan.

ge-sinîg [= sin-hîg, -híw?], e; f. *Marriage*; connubium : — Fore hálgum gesinîge ǽ *pro sacra connubii lege*, Rtl. 108, 14.

ge-sinîgan *to marry*; nubere :—Gesiniîgaþ *nubunt*, Lk. Skt. Lind. Gisinnîgo, Rush. 20, 34. v. ge-sinîg.

ge-sinîgscipe, es; m. *Marriage*; connubium, Rtl. 108, 23. v. sin-hîgscipe.

ge-sinlíce; *adv. Curiously, strictly*; curiose, R. Ben. 58.

ge-sinscipe, es; m. *Marriage, wedlock, matrimony*; in *pl. Married people*; connubium, Bd. 4, 5; S. 573, 14 : 19; S. 587, 30 : Shrn. 60, 2. Se mægþhád is hírra đonne se gesinscipe *virginity is more exalted than marriage*, Past. 52, 8; Swt. 409, 24; Hat. MS. He wæs seofan geár on gesinscipe geseted ǽr his biscopdóme *he was married for seven years before he was a bishop*, Shrn. 110, 1. Eác is gesynscipum micel þearf *for those married also there is much need*, L. E. I. 42; Th. ii. 440, 2.

ge-sinsciplíc; *adj. Conjugal, matrimonial*; conjugalis, L. E. I. 43; Th. ii. 440, 7.

ge-sión *to see, behold*; videre :—Wénaþ đa dysgan đæt ælc mon síe blind swá hí sint; and đæt nán mon ne mǽge seón [gesión, note] đæt hí gesión ne mágon *the foolish think that every man is blind as they are; and that no man is able to see what they cannot see*, Bt. 38, 5; Fox 206, 21. v. ge-seón.

ge-siowed *sewed together.* v. ge-siwed.

ge-sirwan, -serwan, -syrwan; *p.* ede; *pp.* ed. **I.** *to plot, contrive, conspire, deliberate* :—Se se đe đa synne gesireþ *he who designs the sin*, Past. 56, 6; Swt. 435, 6; Hat. MS. Đonne ne gesirede hit nó đæt hit þurhtuge swelce synne *then would it not have designed to carry out such sin*, Swt. 435, 4. Ic gesyrede *I plotted*, Exon. 72 b; Th. 270, 20; Jul. 468. Đý ne wrícþ Dryhten nó gelíce đa gesiredan synne and fǽrlíce þurhtogenan fordǽm sió gesirede syn biþ ungelíc eallum óđrum synnum *so the Lord does not punish equally the deliberate sin and the suddenly perpetrated, for the deliberate sin is unlike all other sins*, Past. 56, 7; Swt. 435, 13; Hat. MS. **II.** *to furnish with arms, equip* :—Gesyrwed secg *an armed man*, Byrht. Th. 136, 30; By. 159. v. ge-syrwan.

ge-síþ, es; *m.* [cf. ge-féra] *A companion, fellow, companion* or *follower of a chief or king;* socius, comes :—Gif mon elþeódigne ofsleá gif he mægleás síe healf kynincg [áh] healf se gesíþ *if one slay a foreigner, if he be kinless, half the king [has], half the companion,* L. In. 23; Th. i. 116, 16. Gif gesíþcund mon þingaþ wið cyning for his inhíwan, nâh he nâne wíteræðenne, se gesíþ *if a 'gesithcund' man compound a suit with the king for his household, he, the 'gesith,' shall not have any fee,* 50; Th. i. 134, 5. Se gesíþ gelaðede ðone cyning to his hâme *rex, rogatus a comite,* Bd. 3, 22; S. 553, 29. Him se gesíþ fultumade and ealle ða neáhmenn *juvante cŏmĭte ac vicĭnis omnĭbus,* 4, 4; S. 571, 14. Wæs sum gesíþ on neáweste *erat cŏmes in proximo,* 4, 10; S. 578, 18. Ðá bæd se gesíþ hine, ðæt he eóde on his hûs *rŏgāvit cŏmes eum in dŏmum suam ingrĕdi,* 5, 4; S. 617, 10: 5, 5; S. 617, 40. Daniel deóra gesíþ *Daniel, the beasts' associate,* Cd. 208; Th. 251, 24; Dan. 662. Hyre wæs hâlig gǽst singal gesíþ *to her the Holy Spirit was a constant companion,* Exon. 69 a; Th. 257, 4; Jul. 242. To hâm his gesíþes *in dŏmo cŏmĭtis,* Bd. 3, 14; S. 539, 43. He on ðæs gesíþes hûs ineóde *dŏmum cŏmĭtis intrāvit,* 5, 4; S. 617, 16. Wæs se bisceop gelaðod sumes gesíþes cyricean to hâlgianne *episcopus vocātus est ad dedicandam Ecclesiam comĭtis,* 5, 5; S. 617, 34: Shrn. 69, 32: 70, 23: 122, 18. On gesíþes hâd *in the condition of a comrade,* Beo. Th. 2598; B. 1297. Fram ðam ylcan gesíþe *ab eōdem comĭte,* Bd. 5, 4; S. 617, 9. To his treówum gesíþe *to his faithful companion,* Exon. 51 b; Th. 179, 29; Gú. 1269. He hæfde him to gesíþþe sorge and longaþ *he had for his companion sorrow and longing,* 100 a; Th. 377, 13; Deór. 3. Swǽse gesíþas *his dear companions,* Beo. Th. 57; B. 29: 4086; B. 2040: 5029; B. 2518. Fróde gesíþas, ealde ægleáwe hit getealdon *wise fellows, elders skilled in law computed it,* Menol. Fox 36; Men. 18. Ða gesíþas *the comrades,* Salm. Kmbl. 693; Sal. 346. Mec gesíþas sendaþ æfter hondum *comrades send me from hand to hand,* Exon. 108 a; Th. 412, 24; Rä. 31, 5. Hûþe feredon seccas and gesíþþas *warriors and allies carried away the spoil,* Cd. 95; Th. 124, 23; Gen. 2067: Judth. 11; Thw. 24, 22; Jud. 201. Gesíþa ða sǽmestan *the worst of companions,* Exon. 86 b; Th. 326, 7; Wíd. 125. Nǽnig swǽsra gesíþa *no one of the dear companions,* Beo. Th. 3872; B. 1934. Æðele cempa mid gesíþum *the noble champion with his companions,* 2630; B. 1313: 3852; B. 1924: 5257; B. 2632: Exon. 14 b; Th. 30, 1; Cri. 473. Ðæt wæs Satane and his gesíþum mid gegearwad *that was prepared for Satan and his associates with him,* Th. 93, 7; Cri. 1522: 123 b; Th. 474, 21; Bo. 33: 89 b; Th. 337, 2; Gn. Ex. 58: Salm. Kmbl. 907; Sal. 453. Þeóda þrymfæste þegnum and gesíþþum *famous nations with vassals and allies,* Cd. 91; Th. 114, 23; Gen. 1908. [For the technical meaning of 'gesith' see Stubbs' Const. Hist. under 'comitatus' and 'gesith;' Kemble's Saxons in England, i. 168; and Schmid's A. S. Gesetz. 'gesíþ.' *Goth.* ga-sinþja: *O. Sax.* gi-síð: *O. H. Ger.* gi-sindo.]

ge-síþ, -síþþ, es; *n. Company, fellowship;* comitatus :—Sweotol is ðæt ðe sóþ metod on gesíþþe is *it is plain that the true Lord is with thee,* Cd. 135; Th. 170, 3; Gen. 2807: 109; Th. 145, 5; Gen. 2401. [*O. H. Ger.* gi-sindi; *n.* comitatus: *O. Sax.* ge-síþi; *n.*]

ge-síþcund; *adj. Of the rank of a 'gesith:'*—Gif gesíþcund mon landâgende forsitte fyrde, geselle cxx scillinga and þolie his landes *if a 'gesithcund' man, owning land, neglect the 'fyrd,' let him pay* 120 *shillings and forfeit his land,* L. In. 51; Th. i. 134, 8. Gif gesíþcund man fare, þonne môt he habban his geréfan mid him, and his smiþ and his cildféstran *if a 'gesithcund' man go away, then may he have his reeve with him, and his smith and his child's fosterer,* 63; Th. i. 144, 2: 45; Th. i. 130, 9: 54; Th. i. 136, 12: 68; Th. i. 146, 7: L. Wih. 5; Th. i. 38, 4.

ge-síþcundlíc, -síþlíc; *adj. Pertaining to a companion :*—Swâ swâ he wære gesíþcundlícre [MS. Ca. gesíþlícre, MS. B.] gegaderunga *quasi comes copulæ carnalis,* Bd. 2, 9; S. 511, 1, note.

gesíþ-mægen, -mægnes; *n. A multitude of companions;* comitum turba :—For gesíþmægen, Exon. 90 a; Th. 339, 4; Gn. Ex. 89.

ge-síþman, -mon; *gen.* -mannes, -monnes; *m. A 'gesith;'* comes :—Se gesíþmon [gesíþmon, MSS. B. H.] *the 'gesith,'* L. In. 30; Th. i. 122, 1. v. ge-síþ.

ge-síþscipe, es; *m. A fellowship, society;* societas :—Nam he twegen bisceopas of Britta þeóde on gesíþscipe ðære hâlgunge *adsumtis in societatem ordinationis duobus de Brittonum gente episcopis,* Bd. 3, 28; S. 560, 27. Sum swíþe eald wífman wæs in his gesíþscipe *a very old woman lived with him,* Shrn. 36, 9. [*O. Sax.* gi-síðskepi.]

ge-síþwíf, es; *n. A woman of the class to which the 'gesith,' 'comes' belongs :*—Sca anastasiam ðære hâlegan gesíþwífes seó wæs swíþe æðele for worulde *St. Anastasia's the holy lady; she was very noble with respect to this world,* Shrn. 30, 26. All ða gesíþwíf and ða æðelan fǽmnan *all the ladies and noble women,* 87, 21. [Cf. ge-síþman.]

ge-sittan; *p.* -sæt, *pl.* -sǽton; *pp.* -seten. **I.** *to sit, sit down, settle, lean, recline;* sĕdēre, consĭdēre, discumbĕre :—Ic gesitte *I sit,* Exon. 73 a; Th. 272, 6; Jul. 495. Hí gesittaþ him on gesundum þingum *they sit in sound condition,* 89 b; Th. 337, 1; Gn. Ex. 58. He wið earm

gesæt *he leaned on his arm,* Beo. Th. 1503; B. 749: Cd. 223; Th. 291, 18; Sat. 432. Ðá eóde he into ðæs Fariseiscan hûse, and gesæt *ingressus dŏmum Pharisæi discŭbuit,* Lk. Bos. 7, 36. Alexander æt Somnite gemǽre and Rômâna gesæt *Alexander posted himself on the boundary of the Samnites and the Romans,* Ors. 3, 7; Bos. 58, 28. Gesǽton searuþancle sundor to rûne *the wise of thought sat apart in council,* Andr. Kmbl. 2323; An. 1163: Elen. Kmbl. 1732; El. 868. Twegen iunge men gesǽton æt me *two young men sat by me,* Bd. 5, 13; S. 632, 35. Him cierde eall ðæt folc to, ðe on Mercna lande geseten wæs *all the people who were settled in the Mercians' land submitted to him,* Chr. 922; Erl. 108, 34. Hie hæfdon heora stemn gesetenne *they had sat out their time of service,* Chr. 894; Erl. 90, 31. **II.** *to occupy, possess, inhabit;* possĭdēre, hăbĭtāre :—Ðeáh ðe wyrigcwydole Godes ríce gesittan ne mâgon *quamvis maledĭci regnum Dei possĭdēre non possint,* Bd. 4, 26; S. 602, 11. Sume sécaþ and gesittaþ hâmas on heolstrum *some seek and occupy houses in caverns,* Exon. 33 b; Th. 107, 3; Gú. 53: Cd. 170; Th. 213, 34; Exod. 562. Paulinus gesæt ðæt biscepsetl on Hrófes ceastre *Paulinus occupied the bishop's see at Rochester,* Chr. 633; Erl. 24, 7: 890; Erl. 87, 27: Beo. Th. 1270; B. 633. Hí folca gewinn fremdra gesǽton *lăbōres pŏpŭlōrum possēdērunt,* Ps. Th. 104, 39; 77, 56: Cd. 46; Th. 59, 9; Gen. 961. Bûtan ôðrum manegum gesetenum íglandum *besides many other inhabited islands,* Ors. 1, 1; Bos. 16, 25. Us is aléfed heofena ríce to gesittenne *we are permitted to occupy heaven's kingdom,* Blickl. Homl. 137, 15: Ors. 6, 34; Bos. 130, 23.

ge-siwed, -siwod, -siwud, -siuwed; *part. Sewed, patched;* sutus, assutus, consutus :—Gediht oððe gesiwed hrægel *acupicta vel Phrygia vestis,* Ælfc. Gl. 63; Som. 68, 107; Wrt. Voc. 40, 18. Mid golde gesiwud bend *nimbus,* 64; Som. 69, 13; Wrt. Voc. 40, 47. v. siwian.

ge-slǽpan, -slépan, -slépian [in the Northern glosses of the Gospels the verb is weak] *to sleep :*—He geslépde *dormiebat,* Mt. Kmbl. Lind. 8, 24. Geslépedon alle and geslépdon *dormitaverunt omnes et dormierunt,* 25, 5.

ge-sleán; *p.* -slóg, -slóh, *pl.* -slógon; *pp.* -slagen, -slægen, -slegen *To strike, pitch [a tent], smite, slay, quell, forge, fight, obtain by fighting :*—Hí lâgon swylce hí wǽron deáþe geslegene *they lay as if they were stricken by death,* Judth. 10; Thw. 21, 23; Jud. 31. Se geslagena biþ mid deáþe gegripen *the man stricken [by disease] is seized by death,* Homl. Th. ii. 124, 12. Ðǽr he geslóh his geteld *he pitched his tent there,* Gen. 12, 8. Wulfheard aldorman micel wæl geslóg and sige nom *alderman Wulfhard made a great slaughter and got the victory,* Chr. 837; Erl. 66, 5: 845; Erl. 66, 24: 823; Erl. 62, 17: 867; Erl. 72, 15: Bd. 1, 16; S. 484, 23. He geslóg xxv dracena *he slew xxv dragons,* Salm. Kmbl. 417; Sal. 214. Geslóh ðín fæder fǽhþe mǽste *thy father quelled the greatest feud,* Beo. Th. 922; B. 459. Geslægene grindlas *forged bars,* Cd. 19; Th. 24, 26; Gen. 383. Of ðære tíde hwílum Bryttas hwílum Seaxena sige geslógan *ex eo tempore nunc cives nunc hostes vincebant,* Bd. 1, 16; S. 484, 22. Offa geslóg cyneríca mǽst *Offa won the greatest of kingdoms,* Exon. 85 a; Th. 320, 32; Víd. 38: Th. 321, 11; Víd. 44. Hûþe ðe ic æt hilde geslóh *spoil that I gained in war,* Cd. 98; Th. 129, 25; Gen. 2149: Chr. 937; Erl. 112, 4; Æthelst. 4: Beo. Th. 5985; B. 2996. Ða bóreas geslógon eal his londríce *dario superato acceptaque in conditiones omni ejus regione,* Nar. 3, 24. Ðá þohte ic hwæðer ic meahte ealne middangeard me onweald gesleán *cogitabam si devicto orbe terrarum,* 29, 2. Óþ ðæt up gewát líg and þurh lust geslóh *until the flame went up and at will smote,* Cd. 186; Th. 231, 19; Dan. 249.

ge-sleccan; *p.* -slæhte *To make slack, enfeeble, weaken :*—Súslum geslæhte *weakened by torments,* Exon. 10 a; Th. 10, 8; Cri. 149. [Cf. *O. Sax.* an siuni gislekit.]

ge-sléfed; *pp. Having sleeves;* manicatus, manuleātus :—Gesléfed manuleātus vel manicātus, Ælfc. Gl. 3; Som. 55, 74; Wrt. Voc. 16, 47. DER. sléfan.

ge-sleht; *n.* v. bil-gesleht, ge-slyht.

ge-slit, es; *n. A bite, tearing :*—Ðæra næddrena geslit wæs deádlíc *the bite of those serpents was deadly,* Homl. Th. ii. 238, 30. Þurh deóra geslit *by the tearing of beasts,* 544, 2.

ge-slítan; *p.* -slát; *pp.* -sliten *To tear, rend, break :*—Middý geslitten wéron ða bendo *ruptis vinculis,* Lk. Skt. Lind. 8, 29.

ge-slóh *struck.* v. ge-sleán.

ge-slyht, -sleht, -sliht, es; *n. Battle, fight, conflict :*—Swâ he níþa gehwane genesen hæfde slíþra geslyhta *so he had come safely out of every enmity, every fierce conflict,* Beo. Th. 4787; B. 2398. v. bil-gesleht.

ge-smeágan, -smeán; *p.* -smeáde; *pp.* -smeád *To search, examine, consider;* scrutari, cogitare :—Hia gesmeádon miþ him *illi cogitabant secum,* Mk. Skt. Lind. 11, 31. Gismeáþ wegas ûsra *scrutemur vias nostras,* Rtl. 20, 21. Gismeága excogitare, 170, 5. Gesmeád sprǽc *sermo commentitius,* Ælfc. Gl. 100; Som. 77, 21; Wrt. Voc. 55, 25.

ge-smeáh; *gen.* -smeáges [?]; *n. Intrigue :*—Ðǽr wearþ se cyng of France þurh gesmeáh gecyrred *there the king of France was turned back by intrigue,* Chr. 1094; Erl. 230, 23.

ge-smecgan; *p.* ede; *pp.* ed [smæccan *to taste*] *To taste;* gustāre :—Ic gesmecge *gusto,* Ælfc. Gl. 5; Som. 56, 33; Wrt. Voc. 17, 37.

ge-smédan; *p.* de; *pp.* ed; *v. a. To make smooth* or *even, to soothe, soften;* complanare :—Se ele gesmḗð ða wunda *the oil sooths the wounds,* Past. 17, 10; Swt. 125, 10; Hat. MS.

ge-smicerad [smicere *elegant*]; *part. p. Worked, neatly made;* fabrefactus, Cot. 88, 184.

ge-smirian *to anoint,* Ex. 29, 29. v. ge-smyrian.

ge-smiten; *part. p. Anointed, smeared, smutted;* litus, unctus, Som. [*Goth.* ga-smeitan.]

ge-smiðian; *p.* ede; *pp.* ed; *v. trans. To forge, to make as a smith does;* fabricare :—Bend agimmed and gesmiðed diadema, Ælfc. Gl. 64; Wrt. Voc. 40, 46. [*Goth.* ga-smiþon: *O. H. Ger.* gi-smidon *cudere.*]

ge-smyltan; *p.* te; *pp.* ed [smylt *serene*] *To appease, quiet;* plācāre :—He ðone aþundenan sǽ gesmylte *tumīda æquŏra plācāvit,* Bd. 5, 1; S. 614, 8.

ge-smyrian, -smirian; *p.* ode, ede; *pp.* od, ed [smyrian *to smear*] *To smear, anoint;* ungĕre :—Hī word hira wel gesmyredon, ele anlīcast *mollĭerunt sermōnes suos sŭper ŏleum,* Ps. Th. 54, 21. Forðon gesmiride mec *propter quod unxit me,* Lk. Skt. Lind. 4, 18. Ðætte gesmiredon hine *ut ungerent eum,* Mk. Skt. Lind. 16, 1. Ðæt hīg sín gesmirode on ðam and hira handa gehālgode *ut ungantur in ea et consecrentur mănus eōrum,* Ex. 29, 29. Ðæt nǽfre ne afūlaþ ðæt mid hire gesmered biþ *that never becomes foul that is anointed with it,* Blickl. Homl. 73, 23. Gesmearuad oele hālgum *unctus oleo sancto,* Rtl. 198, 31.

gêsne; *adj. Lacking, wanting, destitute, lifeless;* expers, egenus, destitutus, exanimis :—Læg se fūla leáp gêsne *the foul corpse lay lifeless,* Judth. 10; Thw. 23, 8; Jud. 112. He funde ða on bedde his goldgifan gǽstes gêsne. lífes belidenne *he then found his goldgiver void of spirit, deprived of life,* 12; Thw. 25, 26; Jud. 279. v. gǽsne.

ge-snid, es; *n. A killing, slaughter;* occisio :—Swā swā sceáp to gesnide *sicut oves occisionis,* Ps. Lamb. 43, 23.

ge-sníþan; *p.* -snāþ; *pp.* -sniden *To cut, cut off* :—Summ monn gesnāþ him ða eárelipprica *quidam amputavit illi auricula,* Mk. Skt. Lind. 14, 47. Gif ðū stǽnen weofod me wyrce ne tymbra ðū ðæt of gesnidenum stānum *if thou wilt make me an altar of stone, thou shalt not build it of hewn stone,* Ex. 20, 25.

ge-snídan [?] *to lie down* :—Ðætte gesniða [Rush. gesnide] gedydon alle *ut accumbere facerent omnes,* Mk. Skt. Lind. 6, 39.

ge-sníþung, e; *f. A cutting;* dolatio, Som.

ge-snot *snot.* v. snot.

ge-snyttro; *f. n.* [?] *Wisdom* :—Gūþlāc wæs ealra gesnyttra goldhord *Guthlac was a treasure of all wisdom,* Guthl. 20; Gdwn. 92, 17.

gesoc, es; *n? Suck;* suctus :—Ðæt Sarra sceolde lecgan cild to hyre breóste to gesoce *quod Sara lac'tāret fīlium,* Gen. 21, 7.

ge-sod, es; *n? A cooking, boiling;* coctio, coctūra :—Gesod *coctio,* Wrt. Voc. 82, 70.

ge-soden; *part.* [soden, *pp.* of seóðan *to seethe*] *Seethed, sodden, cooked, boiled;* coctus, elixus :—Gesoden, gebacen *coctus,* Ælfc. Gl. 31; Som. 61, 86; Wrt. Voc. 27, 16: 82, 71. Gesoden mæt on wætere *elixus cĭbus,* 31; Som. 61, 87; Wrt. Voc. 27, 17. Gesoden wîn *defrŭtum vīnum,* 32; Som. 62, 8; Wrt. Voc. 27, 62.

ge-soecan *to seek, follow,* Jn. Skt. Lind. 13, 37. v. ge-sēcan.

ge-sóm; *adj. Unanimous, united, peaceable;* concors, pācĭficus :—Wǽron gesóme ða ðe swegl būan *those that inhabit the firmament were unanimous,* Cd. 5; Th. 6, 1; Gen. 82. Wit wǽron gesóme *we two were united,* Exon. 129 b; Th. 496, 27; Rä. 85, 21: Gen. 45, 24. Gesóme and to ðam geþwǽre ðæt heora nān ne mæg ōðerne mid ǽalle fordón *in union and in such accord that none can entirely destroy another,* Shrn. 165, 33.

ge-somnian; *p.* ode; *pp.* od *To assemble, collect;* congregare, colligere :—He us to dæge wolde on ðisse tíde gesomnian *he wished to assemble us to-day at this time,* Blickl. Homl. 139, 31. Gesomna cūe mesa *collect cow's dung,* L. M. 1, 38; Lchdm. ii. 98, 5. v. ge-samnian.

ge-somning, e; *f. A congregation;* congrĕgātio :—Seó Godes circe, þurh gesomninga sōðes and ryhtes, beorhte blíceþ *the church of God, through congregations of truth and right, brightly gleameth,* Exon. 18 a; Th. 44, 9; Cri. 700. v. ge-samnung.

ge-somnung, e; *f. A congregation, synagogue, church;* congrĕgātio, sÿnăgōga, ecclĕsia :—He com into hyra gesomnunge *venit in sÿnăgōgam eōrum,* Mt. Bos. 12, 9. On gesomnunge ingongan *ecclĕsiam intrāre,* Bd. 1, 27; S. 495, 7. Justus reahte ða gesomnunge *Justus rĕgēbat ecclĕsiam,* 2, 7; S. 509, 10. Gesomnunga folca ymbtrymdon ðē sÿnăgōga pŏpŭlōrum circumdābit te,* Ps. Spl. 7, 7. Beférde se Hǽlend ealle Galileam, lǽrende on hyra gesomnungum *circumĭbat Iesus tōtam Gălilæam, dŏcens in sÿnăgōgis eōrum,* Mt. Bos. 4, 23: 6, 2, 5: 9, 35. v. ge-samnung.

ge-somodlǽcan. v. ge-samodlǽcan.

ge-sóð *a soother, flatterer;* parasitus, Cot. 152.

ge-sóþfæstian; *p.* ode, ade; *pp.* od, ad *To justify* :—Bærsynnig gesóþfæstadon god *publicani justificaverunt deum,* Lk. Skt. Lind. 7, 29. He

---

wolde gesóþfæstiga hine seolfne *ille volens justificare seipsum,* 10, 29. Gesóþfæstad is snytro *justificata est sapientia,* Mt. Kmbl. Lind. 11, 19: 12, 37.

ge-sóþian; ic -sóþige; *p.* ode; *pp.* od *To prove the truth of, bear witness;* probare, testari :—Gif man ðæt gesóþige *if that be proved,* L. E. G. 6; Th. i. 170, 13. Menigo of hláfe and líchoma his gesóþade *plurima de pane et carne sua testatur,* Jn. Skt. p. 5, 2.

ge-sotig; *adj. Dirty,* Gl. Prud. 579.

ge-spænning, e; *f. An incitement, a provocation;* incitamentum, Som.

gespan *the tamarisk tree;* myrica, Cot. 131.

ge-span, -spon, es; *n. A prompting, enticing, persuasion, seduction;* suggestio, illectatio, persuasio, seductio, Past. 53, 7; Swt. 417, 20; Hat. MS: Cd. 33; Th. 45, 2; Gen. 720.

ge-span, -spann, -spon, es; *n. A joining, fastening together;* nexus :—Wíra gespann *joining of wires,* Andr. Kmbl. 604; An. 303. Wíra gespon, Elen. Kmbl. 2267; El. 1135. He is on helle hæft mid hringa gesponne *he is in hell bound with the clasping of rings,* Cd. 35; Th. 47, 17; Gen. 762. Searo-rūna gespon *the web of mysteries,* Exon. 92 b; Th. 347, 20; Sch. 15. v. ge-spannan, eaxle-gespan.

ge-spanan; *p.* -spón, -speón, *pl.* -spónon, -speónon; *pp.* -spanen, -sponen; *v. trans. To allure, entice, incite, persuade, induce, draw;* allicere, illicere, incitare, persuadere, inducere :—Ðe hine to dæm unfriðe gespón *who had allured him to a violation of the peace,* Chr. 905; Th. 182, 7, col. 1. Gif he ða cwēne gespanan [gespannan, MS.] and gelǽran mihte, ðæt heó brūcan wolde his gesynscipes *si reginæ posset persuadere ejus uti connubio,* Bd. 4, 19; S. 587, 29. Gespeón *persuadebat,* 2, 15; S. 518, 26. Swýðost gesponen [gesponnen, MS.] to onfōnne Cristes geleáfan *maxime persuasus ad percipiendam Christi fidem,* 3, 21; S. 551, 5. Wæs hām geladad and gesponen [gesponnen, MS.] *was called and drawn home,* 4, 23; S. 593, 17.

ge-spang, -spong, es; *n. A clasp, binding* :—Me habbaþ hringa gespong síþes amyrred *the rings' clasps have hindered me from going,* Cd. 19; Th. 24, 14; Gen. 377.

ge-spannan, -sponnan; *p.* -speón; *pp.* -spannen *To join, span, clasp, fasten* :—Gūþweard grímhelm gespeón *the leader clasped his helm,* Cd. 151; Th. 188, 27; Exod. 174. Ðā hēht cāsere gesponnan fiówer wildo hors to scride *then the emperor ordered to harness four wild horses to a chariot,* Shrn. 71, 34. [*O. H. Ger.* gi-spannan *tendere, conjungere.*] v. spannan.

ge-sparian; *p.* ede *To spare* :—Ne gisperede *non pepercit,* Rtl. 22, 17.

ge-sparrian; *p.* ode, ade; *pp.* od, ad *To shut;* claudere :—Gesparrado dure ðín *clauso ostio tuo,* Mt. Kmbl. Lind. 6, 6.

ge-spearn. v. ge-speornan.

ge-speca, an; *m. A speaker.* DER. eár-gespeca. v. ge-spreca.

ge-spédan; *p.* -spédde; *pp.* -spéded, -spédd *To speed, prosper, succeed;* progredi, prosperare, succedere :—Ðæs ðe blódgyte, wæll-fyll weres, wæpnum gespédeþ *because that bloodshedding, slaughter of man, speedeth by means of weapons,* Cd. 75; Th. 92, 12; Gen. 1527. Ac hí nāht nā gespéddan *but they succeeded naught,* Chr. 1036; Th. 293, 23, col. 2. Ac man ðǽr ne gespédde *but they didn't succeed there,* 1096; Erl. 233, 29. Ealle þinge swā hwæt swā he dēþ beóþ gespédde *omnia quæcunque faciet prosperabuntur,* Ps. Lamb. 1, 3.

ge-spédiglíce, *adv. Prosperously, successfully;* prospēre :—Gesundfullíce oððe gespédiglíce forþstæpe and ríxa *prospĕre procēde et regna,* Ps. Lamb. 44, 5.

ge-spédsumian *to prosper, succeed;* prosperari, Hpt. Gl. 491.

ge-spelia, an; *m.* [spelian *to represent*] *A substitute, deputy, vicegerent;* vĭcārius :—Cristen cyning is Cristes gespelia geteald on cristenre þeóde *a christian king is accounted Christ's vicegerent among christian people,* L. Eth. ix. 2; Th. i. 340, 12. Wið Cristes gespelian *against Christ's vicegerent,* ix. 42; Th. i. 350, 3. He wæs Æþelstānes b' gespelian siððan he unfere wæs *he was bishop Athelstane's substitute after he was unable to move,* Chr. 1055; Erl. 191, 12.

ge-spellian *to speak, tell* :—Miððý gespelledon *dum fabularentur,* Lk. Skt. Lind. 24, 15. [*Goth.* ga-spillon.]

ge-speoftad; *part. p. Spit upon* :—Gespeoftad biþ *conspuetur,* Lk: Skt. Lind. 18, 32. [*Cf.* speowian.]

ge-speón. v. ge-spanan, ge-spannan.

ge-speornan, -spornan; *p.* -spearn, *pl.* -spurnon; *pp.* -spornen *To tread upon, to perch, spurn;* calcare :—Ðæt heó fótum ne meahte land gespornan *so that she might not perch on land with her feet,* Cd. 72; Th. 87, 33; Gen. 1458: 72; Th. 87, 11; Gen. 1447. Ðæt se hearn-flota sond-lond gespearn *so that the floater of the surge spurned the sandy land,* Exon. 52 a; Th. 182, 11; Gū. 1308.

ge-speów *prospered,* Judth. 11; Thw. 24, 7; Jud. 175; *p. of* ge-spówan.

ge-sperod; *part.* [spere *a spear*] *Armed with a spear;* hastātus :—Gesperod *hastātus,* Ælfc. Gr. 43; Som. 45, 13: Blickl. Homl. 221, 28.

ge-spillan; *p.* de *To destroy, dissipate* :—Ðēr ne hrust gespilles *ubi neque ærugo demolitur,* Mt. Kmbl. Lind. 6, 20. Gespilleþ *ferdiderit,*

Lk. Skt. Lind. 17, 33. Alle gespilde *omnes perdidit*, 29 : Rtl. 107, 29. Ðǽr wǽron manege mid micel unrihte gespilde *there were many very wrongfully destroyed*, Chr. 1124; Erl. 253, 16 : Mt. Kmbl. Lind. 6, 19. Erfwardniso gispilledo *hereditates dissipatas*, Rtl. 21, 1.

ge-spittan; *p.* ed *To spit upon*; conspuere :—Gispitted biþ *conspuetur*, Lk. Skt. Rush. 18, 32. Gispittendum on mec *conspuentibus in me*, Rtl. 19, 17.

ge-spon, es; *n. An enticing, persuasion, artifice*; illectatio, persuasio, seductio :—Deófles gespon *the devil's artifice*, Cd. 33; Th. 45, 2; Gen. 720. v. ge-span.

ge-spón. v. ge-span.

ge-spón *allured, incited*, Chr. 905; Th. 182, 7, col. 1. v. ge-spanan.

ge-spong. v. ge-spang.

ge-sponnen *persuaded, drawn*, Bd. 3, 21; S. 551, 5 : 4, 23; S. 593, 17. For ge-sponen; *pp.* of ge-spanan.

ge-spornan. v. ge-speornan.

ge-spówan; *p.* -speów, *pl.* -speówon; *pp.* -spówen [spówan *to succeed*] *To succeed, prosper*; succēdĕre, prospĕrāre :—Hú hyre æt beaduwe ge-speów *how she prospered in battle*, Judth. 11; Thw. 24, 7; Jud. 175 : Andr. Kmbl. 2688; An. 1346. Him æt ðære byrig ne gespeów *he did not succeed at the city*, Ors. 4, 5; Bos. 82, 8.

ge-sprǽc, es; *n. Speech, discourse, conversation, advice* :—Se cyning wæs on gesprǽce wynsum *erat rex affatu jucundus*, Bd. 3, 14; S. 540, 8. Ic wæs mid his gesprǽce wel gerēted *allocutione ejus refecti*, 5, 1; S. 613, 22. Gearo on gesprǽce *loquela promptus*, 2; S. 615, 29. Ðá hæfde he gesprǽce and geþeaht *habito consilio*, 2, 13; S. 515, 40. Com for ge-sprǽce Finano ðæs biscopes *pervenire propter conloquium Finani episcopi*, 3, 22; S. 552, 41. Wæs gemyndig ðæs apostoles gesprǽces *was mindful of what the apostle said*, Shrn. 39, 5. Gesprǽcu, gesprécu *oracula*, Cot. 143, Lye. [*O. H. Ger.* ge-sprāche; *n.*]

ge-sprǽce; *adj. Eloquent, affable*; eloquens, affabilis :—Næs ic nǽfre gesprǽce *non sum eloquens*, Ex. 4, 10. He wæs eallum gesprǽce *erat affabilis omnibus*, Bd. 4, 28; S. 606, 34. [*O. H. Ger.* ge-sprāche *facetus, affabilis, disertus, orator.*] v. sprǽce.

ge-sprǽcelíc; *adj. Loquelaris* :—Ðás synd gehátene loquelares, loquela is sprǽc and loquelares synd gesprǽcelíce forþan ðe ðás syx prepositiones ne beóþ ná hwár ána ac beóþ ǽfre to sumum óðrum worde gefēgede, Ælfc. Gr. 47; Som. 48, 49. [*O. H. Ger.* ki-sprachlich *rhetoricus, urbanus.*]

ge-sprǽdan; *p.* de; *pp.* ed *To spread out, extend*; extendere :—Gesprǽd hond ðín *extende manum tuam*, Mt. Kmbl. Lind. 12, 13. Ge-sprǽde hond *extendens manum*, 8, 3.

ge-sprec, es; *n. The power of speech* :—He him sealde monnes gesprec *he should give him human speech*, Shrn. 76, 23. Sealde he dumbum gesprec *he gave speech to the dumb*, Andr. Kmbl. 1153; An. 577.

ge-spreca, an; *m. One who talks with another, a counsellor* :—Abraham wæs Godes gespreca *Abraham was one who talked with God*, Homl. Th. ii. 190, 12. Ðú ðé ánne genim to gesprecan symle spella and lára rǽd-hycgende *always take as thy counsellor one sagacious in discourses and doctrines*, Exon. 80 a; Th. 301, 26; Fä. 25.

ge-sprecan, -specan; *p.* -sprǽc, *pl.* -sprǽcon; *pp.* -sprecen *To speak, speak with, agree*; sometimes takes an accusative of the person spoken to :—Gif ðú him wuht hearmes gesprǽce *if thou hast said to him aught injurious*, Cd. 30; Th. 41, 24; Gen. 661. Feala worda gespæc se engel *many words spake the angel*, 15; Th. 18, 11; Gen. 271. Adam gespræc *Adam spoke*, 27; Th. 36, 31; Gen. 580. Ðe git on ǽrdagum oft ge-sprǽcon *which ye two in former days oft agreed upon*, Exon. 123 a; Th. 476, 16; Bo. 15 : 123 b; Th. 475, 24; Bo. 52. Feówer ða strengestan þeóda hý him betweonum gesprǽcan *the four strongest peoples agreed with one another*, Ors. 3, 10; Bos. 69, 33 : 6, 10; Bos. 120, 32. Mid ðý ðe hie ðis gesprecen hæfdon *when they had said this*, Blickl. Homl. 143, 14 : 191, 23 : Elen. Kmbl. 2568; El. 1285. God hí gespræc ðá *God addressed them then*, Homl. Th. ii. 456, 26 : 156, 16. Ðá wæs ic gesprecende ðone man *then was I conversing with the man*, Shrn. 36, 19. Plato hæfde hine gesprecen *Plato had conversed with him*, Swt. A. S. Rdr. 70, 443.

ge-sprengan; *p.* de; *pp.* ed *To sprinkle*; conspergere :—Ðú nymst ánne hláf mid ele gesprengedne *tolles unum panem oleo conspersum*, Ex. 29, 23.

ge-spring, es; *n. A spring*; fons, scaturigo :—Ðær wæs on blóde brim weallende, atol ýða gespring [geswing, Th.] eal gemenged *there was the surge boiling with blood, the foul spring of waves all mingled*, Beo. Kmbl. 1689. v. spring.

ge-springan; *p.* -sprang, -sprong, *pl.* -sprungon; *pp.* -sprungen. **I.** *v. intrans. To spring, bound, arise, go out, go forth*; prosilire, exoriri, abire, procedere :—Swá ðæt blód gesprang *as the blood sprang*, Beo. Th. 3339; B. 1667. Sigemunde g-sprong æfter deáþ-dæge ðóm unlytel *to Sigemund sprang after his death-day no little glory*, 1773; B. 884 : Exon. 92 a; Th. 345, 27; Gn. Ex. 196: Mt. Kmbl. Lind. 4, 24; Mk. Skt. Lind. 1, 28. **II.** *v. trans. To get by going* [?], *to cause to spring*; eructare :—Wíd-gongel wíf word gespringeþ *a rambling woman*

gets words [= *a bad reputation*, or *reproofs?*] *by wandering*, Exon. 90 a; Th. 337, 15; Gn. Ex. 65. [Or has gespringan the same meaning as in the following?] Fēwor streámas neirxna wong gespranc *quattuor flumina paradisi instar eructans*, Mt. Kmbl. p. 8, 5. Gisprunt [?] word *eructavit verbum*. Jn. Skt. p. 187, 26.

ge-sprucg *discord, strife*; seditio, Som. 171; Lye.

ge-spryng *a spring*. v. ge-spring.

ge-spunnen *spun*; netus, Som.

ge-spyrian; *p.* ede; *pp.* ed *To track, search, seek* :—Gif man spor gespirige *if one trace a track*, L. Ǽdelst. 5, 8; Th. i. 236, 20. Loca nú hwæðer ðú wille ðæt wit gespyrigen æfter ǽnigre gesceádwísnesse further *look now whether you wish us two to seek further after any argument*, Bt. 35, 5; Fox 162, 30, note.

GEST, es; *m.* GUEST, *stranger*; hospes :—For feorme and onfangenysse gesta *propter hospitalĭtātem atque susceptiōnem hospĭtum*, Bd. 1, 27; S. 489, 8 : Exon. 106 a; Th. 404, 30; Rä. 23, 15. Gest hine clǽnsie sylfes áþe on wiofode *let a stranger clear himself with his own oath at the altar*, L. Wih. 20; Th. i. 40, 19. DER. fēðe-gest, inwit-, wil-. v. gæst.

gést *a ghost, spirit*. v. cear-gést, gást.

gestæf-lǽred; *part.* [stæf *a letter*; lǽred *learned*; *pp.* of lǽran] *Versed in letters, literate, learned, booklearned*; litĕrātus :—Cild biþ gestæflǽred *a child will be booklearned*, Lchdm. iii. 184, 3 : 192, 15 : 194, 12.

ge-stǽlan; *p.* de; *pp.* ed *To set up, put upon, impute, accuse*; statuere, imponere in, imputare, arguere, accusare :—Ge feor hafaþ fǽhþe gestǽled *and moreover* [she] *hath a deadly feud set up*, Beo. Th. 2685; B. 1340. Ne mǽg on me fácnes frum-bearn fyrene gestǽlan *may not deceit's firstborn* [*the devil*] *impute crime to me*, Exon. 48 a; Th. 166, 18; Gú. 1044. He us ne mǽg ǽnige synne gestǽlan *he cannot accuse us of any sin*, Cd. 21; Th. 25, 10; Gen. 391. Ðý læs on me mǽge ídel spellung oððe scondlíc leágung beón gestǽled *ne aut fabulæ aut turpi mendacio dignus efficiar*, Nar. 2, 21. v. stǽlan. [Cf. (?) ge-stal.]

ge-stǽllan *to stall, stable* :—Ðá hēt he on ðæs pápan ciericean gestǽllan his blancan and monig óðer neát *he ordered his horse and many other cattle to be stabled in the pope's church*, Shrn. 51, 22.

ge-stén, es; *n. A groaning* :—Mín geár wǽron on sícetunga and on gestǽne *anni mei in gemitibus*, Ps. Th. 30, 11. [Cf. *Ger.* stöhnen.]

ge-stǽnan; *p.* de; *pp.* ed *To stone* :—Stephanus for Godes geleáfan wæs gestǽned *Stephen was stoned for belief in God*, Homl. Th. ii. 82, 21. In ǽ Moises bebeád us ðuslíc gestǽna *in lege Moses mandavit nobis hujusmodi lapidare*, Jn. Skt. Lind. 8, 5. Forðætt ðætte hiora werc mec ge-stǽnas *propter quod eorum opus me lapidatis*, Rush. 10, 32. Ðæt folc all gestǽnaþ usig *plebs universa lapidabit nos*, Lk. Skt. Lind. 20, 6. Óðer gestǽndon *alium lapidaverunt*, Mt. Kmbl. Lind. 21, 35.

ge-stǽnce. v. ge-stence.

ge-stǽppan *to step, go*; ire, ingredi :—Ðǽr nǽnig fira ne mæg fótum gestǽppan *where no man may step with feet*, Salm. Kmbl. 420; Sal. 210: Bt. Met. Fox 20, 279; Met. 20, 140. v. ge-steppan.

gest-ærn, -ern, gyst-ern, es; *n. A guest-place, guest-chamber, an inn*; hospitālis aula, hospitium, diversōrium :—Eódon hí on sumes tūngerēfan gestærn *qui intrāvērunt hospitium cujusdam villīci*, Bd. 5, 10; S. 624, 20: Lk. Skt. Lind. Rush. 22, 11.

ge-stæddig; *adj.* [stædig *firm*] *Steadfast, firm*; stăbilis, firmus :—Se án dēma is gestæddig and beorht *the only judge is steadfast and bright*, Bt. 36, 2; Fox 174, 20: Bt. Met. Fox 24, 84; Met. 24, 42 : 29, 171; Met. 29, 87. Ealle gesceafta onfóþ æt ðam gestæddigan Gode, endebyrdnesse, and andwlitan, and gemetgunge *all creatures receive from the steadfast God order, and form, and measure*, Bt. 39, 5; Fox 218, 14.

ge-stæddignes, -stæddines, -ness, -nys, -nyss, e; *f. Gravity, consistency, steadiness, maturity*; grăvitas, constantia, mātūritas :—Wæs he mycelre gestæddignysse wer *multæ grăvitātis ac vērītātis vir*, Bd. 3, 15; S. 541, 21. On lífes gestæddignesse *in consistency of life*, Past. 13, 1: Swt. 77, 14; Hat. MS. 16 b, 18. Ongan se bisceop lustfullian gestæddinesse his geþohta *dēlectābātur antistes constantia ac mātūrītāte cōgĭtātiōnis*, 5, 19; S. 637, 47. v. ge-stæddines.

ge-stæddines *consistency*, Bd. 5, 19; S. 637, 47: Shrn. 168, 2 : 175, 28. v. ge-stæddignes.

ge-stal *an obstacle, objection*; objectio, Cot. 144, Lye.

ge-stala, an; *m. A thief*; fur :—Ðæt he ne gestala nǽre *that he was not a thief*, L. In. 25; Th. i. 118, 15.

ge-stalian; *p.* ode; *pp.* od *To steal*; fūrāri :—Gyf gehádod man gestalige *if a man in orders steal*, L. E. G. 3; Th. i. 168, 4, MS. B.

ge-stalu, e; *f.* [stalu *theft*] *Theft*; furtum :—Ylce gestale *for every theft*, L. Ath. iv. 3; Th. i. 222, 22. Oft gē in gestalum stondaþ *oft ye are engaged in thefts* [*or gestalum from gestala?*], Exon. 40 a; Th. 132, 31; Gú. 481.

ge-standan, -stondan; *p.* -stód, *pl.* -stódon; *pp.* -standen. **I.** *to stand, stand still, remain, last, exist, be*; stāre, mănēre, existĕre, esse :—Heó mihte Gode willsumra wífmonna láreów and fēster-módur gestandan

ipsa Deo dēvōtārum māter ac nutrix posset existĕre fēmīnārum, Bd. 4, 6; S. 574, 17: Ps. Th. 118, 114. Eádig byþ se wer, se ðe him ege Drihtnes, on ferhþcleofan, fæste gestandeþ *beatus vir, qui tīmet Dŏmĭnum*, 111, 1: 113, 20. He fôr eaxlum gestôd Deniga freán *he stood before the shoulders of the Danes' lord*, Beo. Th. 722; B. 358: 813; B. 404: Andr. Kmbl. 1414; An. 707. Æðelinga bearn ymbe gestôdon *sons of nobles stood around*, Beo. Th. 5188; B. 2597: Rood Kmbl. 126; Kr. 63. His fôtas ǽr fæste gestôdan *stĕtērunt pĕdes ejus*, Ps. Th. 131, 7: 93, 18. Wese ðín milde môd geswíðed, and me to frôfre fæste gestande *fiat nunc mĭsĕrĭcordia tua, ut consōlētur me*, 118, 76. Ahsige hû lange seó sibb gestôde *let him ask how long the peace lasted*, Ors. 4, 7; Bos. 88, 6: Bd. 4, 23: S. 594, 40. Ðæt gestôd lytle leng ðonne vii hund wintra *that lasted a little longer than seven hundred years*, Ors. 6, 1; Bos. 115, 28, 20. Ðá gestôd seó cweorn *the mill stopped*, Shrn. 145, 28. Hǽlend ðá gestôd *the Saviour then stood still*, Blickl. Homl. 15, 23: 219, 10. Æfter ðære bêne gestôden him mæssan *after the prayer they attended mass*, Homl. Th. ii. 272, 15. Hie on eallum heora lífe orleahtre gestôdan *they continued blameless in all their life*, Blickl. Homl. 163, 17, 4. Hie mon to his andweardnesse hêht gestandan *they were ordered to stand in his presence*, 173, 11. Siððan hyt gestanden beó *when it be stood*, Herb. 1, 4; Lchdm. i. 72, 8. **II.** *to stand against any one, oppose, oppress, attack, urge, seize*; insurgĕre, ingruĕre, uıgĕre, corrĭpĕre :—He á wile ealra feónda gehwone fæste gestandan *he ever will firmly stand against every foe*, Salm. Kmbl. 196; Sal. 97. Fordam me fremde oft fácne gestôdon *quŏniam ălieni insurrexērunt in me*, Ps. Th. 53, 3. Ne mæg hûs nâht lange standan on ðam heán mûnte, gif hit full ungemetlíc wind gestent *a house cannot long stand on the high mountain if a violent wind press on it*, Bt. 12; Fox 36, 16: 38, 1; Fox 194, 10. Bûton ðú gestande ðone unrihtwísan and him his unrihtwísnysse secge *unless thou oppose the unrighteous man and tell him his unrighteousness*, Homl. Th. ii. 340, 23: i. 6, 24. Ðá gestôd hine swá micel líchamlíc costung *then so great a temptation of the body assailed him*, ii. 156, 25: 122, 17: Guthl. 20; Gdwn. 80, 5. Wæs heó gestanden mid hefigre untrumysse líchoman *she was seized with a heavy illness*, Bd. 4, 23; S. 595, 16: 5, 13; S. 632, 17: Blickl. Homl. 227, 6. [*Goth.* ga-standan: *O. Sax.* gi-standan.]

ge-stapan; *p.* -stôp, *pl.* -stôpon; *pp.* -stapen *To step, go*; gradi, ire, ingredi :—Ðá gestôp he to ánes wealles býge *then he stepped to a bend of a wall*, Ors. 3, 9; Bos. 68, 22: Andr. Kmbl. 3163; An. 1584. DER. stapan.

ge-starian; *p.* ode; *pp.* od [*starian* to stare] *To stare*; rectis ŏcŭlis intuēri :—He gestarode ðǽr gestaðelad wæs æðelíc ingong *he gazed where a noble entrance was placed*, Exon. 12 a; Th. 19, 27; Cri. 307.

gestaðel-fæstan; *p.* -fæste; *pp.* -fæsted [staðel a foundation, fæstan to make fast] *To found, establish*; stăbĭlīre :—Ic gestaðelfæste *stăbĭlĭo*, Ælfc. Gr. 30, 5; Som. 34, 54.

ge-staðelian, -staðolian; *p.* ode, ade; *pp.* od, ad [staðelian to found, establish] *To found, establish, build, erect, place, settle, strengthen, confirm, fortify, repair, restore*; fundāre, stăbĭlīre, ædĭfĭcāre, collŏcāre, lŏcāre, confortāre, confirmāre, restaurāre :—Ðe Eádgar cyng hêt Aðelwold gestaðelian *which king Edgar commanded Æthelwold to found*, Chr. 975; Erl. 127, 7: Shrn. 138, 1. Ðæt hí woldan his bân on ðære ylcan stôwe bûfan eorþan mid gedafenlícre ârwurþnesse gesettan and gestaðolian *ut ossa illius in eodem quidem lŏco, sed supra pāvĭmentum dignæ vĕnĕrātiōnis grātia lŏcārent*, Bd. 4, 30; S. 608, 32. Ðú ná gestaðolast hí *non ædĭfĭcābis eos*, Ps. Lamb. 27, 5: Mt. Bos. 18, 15. Meotud him ðæt môd gestaðelaþ *the Creator strengthens his mind*, Exon. 83 a; Th. 312, 11: Seef. 108. He gestaðolaþ and gemetgaþ ealle gesceafta *he establishes and regulates all creatures*, Bt. 25; Fox 88, 4. Gestrangaþ hý and gestaðeliaþ staðolfæstne geþoht *they strengthen and confirm the steadfast thought*, Salm. Kmbl. 477; Sal. 239. Ic geseó mônan and steorran, ða ðú gestaðelodest *vĭdebo lunam et stellas, quæ tu fundasti*, Ps. Spl. 8, 4: Ps. Th. 89, 8. Se þe middangeard gestaðelode *he who established the earth*, Andr. Kmbl. 323; An. 162: Cd. 6; Th. 7, 32; Gen. 115: Bd. 3, 23; S. 555, 4: Chr. 920; Erl. 104, 33. Ðǽr me he gestaðelode *ibi me collŏcāvit*, Ps. Spl. 22, 1: Bd. 4, 4; S. 570, 42. Þe wuldres blǽd gestaðolade *who established the increase of glory*, Andr. Kmbl. 1071; Bos. 83 a; Th. 312, 3; Seef. 104. Ðe hit gestaðolod wæs *qua fundāta est*, Ex. 9, 18: Ps. Th. 121, 5. Ðǽr gestaðelad wæs æðelíc ingong *where a noble entrance was placed*, Exon. 12 a; Th. 19, 28; Cri. 307: 67 a; Th. 249, 6; Jul. 107. Ðú wǽre gestaðelod þurh me *thou wast confirmed through me*, Soul Kmbl. 90; Seel. 45. Hí ðǽr gestaðelode wǽron *they were settled there*, Bd. 4, 4; S. 571, 1: Ps. Th. 138, 20.

ge-staðolfæstnian *to make firm*; solidare, Rtl. 22, 5.

ge-staðolian *to found, establish, strengthen, confirm*, Bd. 4, 30; S. 608, 32: Ps. Lamb. 27, 5: Bt. 25; Fox 88, 4: Andr. Kmbl. 1071; An. 536: Soul Kmbl. 90; Seel. 45. v. ge-staðelian.

ge-staðolung, e; *f. Firmness, stability*; stăbĭlĭtas :—Ðú ðe staðelodest eorþan ofer gestaðolung his *qui fundasti terram sŭper stăbĭlĭtātem suam*, Ps. Spl. T. 103, 6.

ge-steal, -steall, es; *n. Constitution, frame* :— Eal ðis eorþan gesteal

all this earth's frame, Exon. 78 a; Th. 293, 2; Wand. 110. [Cf. *O. H. Ger.* gistelli: *Ger.* gestell.]

ge-stealla, an; *m. A companion*; socius. DER. eaxl-, folc-, fyrd-, hand-, lind-, nýd-, will-gestealla. v. steal, steallian.

ge-steald, es; *n. A settled place, a station, dwelling-place, an abode*; stătio, dŏmĭcilium :—He lífes gesteald sceáwode *he beheld life's dwelling-place*, Exon. 12 a; Th. 19, 22; Cri. 304. Ðæt he walde wídanferhþ ēcra gestealda *that he shall rule for ever the eternal abodes*, Elen. Kmbl. 1601; El. 802.

gestêd-hors, es; *n.* [stêda a steed] *A stallion*; ĕquus admissārius vel ēmissārius :—He ðone cyng bæd ðæt he him wǽpen sealde and gestêdhors rŏgāvit sibi rēgem arma dăre et ĕquum ēmissārum, Bd. 2, 14; S. 517, 5.

ge-stefnan; *p.* ed; *pp.* ed [stefnian *to institute*] *To institute, place, fix*; instĭtuĕre :—Freá engla hêht wesan wæter gemǽne, stôwe gestefnde *the lord of angels bade the waters to be common, and their places fixed*, Cd. 8; Th. 10, 21; Gen. 160.

ge-stelan *to steal*; furari, Jn. Skt. Lind. 10, 10.

ge-steno, es; *n. Odour, smell*; odor :—Svoetnisse gistencs *suavitātem odoris*, Rtl. 3, 20: 12, 15.

ge-stence, -stænce; *adj. Fragrant, odorous* :— He hafaþ hwítne wyrtruman and swýðe gestencne [-stæncne, MS. B.] *it has a white and very fragrant root*, Herb. 156, 1; Lchdm. i. 282, 19.

ge-stencniss, e; *f. Odour*; odor, Rtl. 3, 22.

ge-steóran; *p.* ed; *pp.* ed *To steer, direct, control, correct*; contĭnēre, corrĭgĕre :—Hîg wistan ðæt hîg ne mihton manegum gesteóran *they knew that they might not control many*, L. E. G. prm; Th. i. 116, 14. Ðú his ýþum môtum gesteóran *mōtum fluctuum ejus tu mītīgas*, Ps. Th. 88, 8: Bt. 16, 4; Fox 58, 15. v. ge-stýran.

ge-stêpan; *p.* -stêpte; *pp.* -stêpt [stêpan *to raise*]. **I.** *to set erect, raise*; ērĭgĕre :—Syndon ða foreweallas fægre gestêpte *the forewalls are fairly raised*, Cd. 158; Th. 196, 26; Exod. 297. **II.** *to assist, sustain, support, help*; sublĕvāre, sustentāre, fulcīre, auxĭliāri :—He gestêpte sunu Ohtheres *he supported Ohthere's son*, Beo. Th. 4766; B. 2393.

ge-steped *stepped, introduced*; initiatus, Cot. 108; *pp.* of ge-steppan. v. steppan.

ge-steppan, -stæppan; *p.* -stepede = -stepte? *pp.* -steped = -stept? *To step, go*; gradi, ire, incedere :—Ðǽr nænig fira ne mæg fôtum gestæppan *where no man may step with feet*, Salm. Kmbl. 420; Sal. 210: Bt. Met. Fox 20, 279; Met. 20, 140. For hwí geunrôtsod gesteppe ic oððe gá ic *quare contristatus incedo*, Ps. Lamb. 42, 10. Gistepe ue *gradiamur*, Rtl. 51, 9. Gesteped *initiatus*, Cot. 108.

gest-ern, es; *n. A guest-place, guest-chamber* :—Ðæt gestern *diversōrium*, Lk. Skt. Lind. Rush. 22, 11. v. gest-ærn.

gest-hús, es; *n. A guest-house, guest-chamber*; hospĭtium :—Gân we sēcan ûre gesthús *let us go and seek our hostel*, Th. Apol. 18, 16. In gest-hûsum *in hospitiis*, Ps. Surt. 54, 16. v. gæst-hús.

ge-stieian, -sticcian; *p.* ode; *pp.* od, ed [stician *to stick*] *To stick, pierce, transfix* :—Hêt mon me ðæt ic ðorne swile gesticode *jussērunt me incidĕre tŭmōrem illum*, Bd. 4, 19; S. 589, 1. Gebýreþ ðæt ðæt môd wierþ gesticced mid ðære scylde gielpes *it happens that the mind is pierced by the sin of boasting*, Past. 33, 2; Swt. 217, 6; Hat. MS. 41 b, 1.

ge-stiéran; *p.* ode; *pp.* ed *To correct*; corrĭgĕre :—He him nolde gestiéran *he would not correct him*, L. In. 50; Th. i. 134, 5. v. ge-stýran.

gestig; *adj. Strange* :—Huonne ðec we sêgon gestig *quando te vidimus hospitem*, Mt. Kmbl. Lind. 25, 38.

ge-stígan; *p.* -stág, -stáh, *pl.* -stigon; *pp.* -stigen *To mount, ascend, descend* :—He me wolde on gestígan *he would mount upon me* [the cross], Rood Kmbl. 68; Kr. 34. In êcne geard up gestígan *to mount up to the eternal abode*, Exon. 44 a; Th. 149, 18; Gú. 763. Ðonne gestíge ic ofer ðone *then will I ascend upon it*, Blickl. Homl. 183, 4. Ðǽtte gestíge *ut descendat*, Rtl. 98, 10. Of dûne gestígdes ðú *descendes*, Mt. Kmbl. Lind. 11, 23. Ðæt we to ðam hýhstan hrôfe gestígan *that we may mount to the highest roof*, Exon. 18 b; Th. 47, 3; Cri. 749. Ðá ic on holm gestáh *when I went on the main*, Beo. Th. 1269; B. 632: Cd. 69; Th. 82, 29; Gen. 1369. Mihtig god on hira ænne gestág *the mighty God mounted on to one of them* [trees], Exon. 25 a; Th. 72, 13; Cri. 1172. Siððan ðú gestígest steápe dúne *after thou dost mount the lofty hills*, Cd. 137; Th. 172, 32; Gen. 2853: 227; Th. 303, 14; Sal. 612. Beddreste gestáh *mounted the couch*, 102; Th. 135, 25; Gen. 2248. Rôd ðe ic ǽr gestág *the cross which I mounted before*, Exon. 29 b; Th. 91, 15; Cri. 1492. Ic ðis lond gestág *I have reached this land*, 37 a; Th. 120, 28; Gú. 278: 15 a; Th. 32, 18; Cri. 514. [*Goth.* ga-steigan *to ascend, descend*: *O. Sax.* gi-stígan *with acc. and with prepositions*.]

ge-stihtian, -stihtan, -stitian; *p.* ode, ade, ede; *pp.* od, ad, ed [stihtian *to dispose*] *To dispose, order, determine*; dispōnĕre, appōnĕre :—Sunu unrihtwísnesse ne geýcþ oððe ne gestihteþ derian hine *fīlius inĭquĭtātis non appōnet nŏcēre eum*, Ps. Lamb. 88, 23. Ic gestihtode *dispŏsui*, Ps. Vos. 72, 25. Gestihtade he and funde ðæt he wolde land-fyrde ðider gelǽdan *terrestri quĭdem ĭtĭnĕre illo vĕnīre dispōnēbat*, Bd. 3, 15; S. 541,

26. Mellitus and Justus gestihtedon ðæt heó ðǽr wolden ðǽre wísan ende gebídan *Mellitus ac Justus ibi rērum fīnem expectāre dispōnentes*, 2, 5; S. 507, 35. Ðæt cúþ is ðæt ðæt mid Drihtnes mihte gestihtad wæs *quod Dŏmĭni nūtu dispŏsĭtum esse constat*, 1, 14; S. 482, 41: Ors. 6, 21; Bos. 123, 31.

ge-stihtung, e; *f.* [stihtung *a disposing*] *A dispensing, disposing, providence;* dispŏsĭtio, prŏvĭdentia :—Fram Godes gestihtunge *by God's providence,* Ors. 2, 1; Bos. 39, 3.

ge-stillan, -styllan; *p.* de; *pp.* ed [stillan *to rest*]. **I.** *v. intrans.* *To rest, cease, be still, quiet, mute;* quiescĕre, sēdāri, sĭlēre, obmutescĕre, rĕtĭcēre :—Hí ne mốten ǽfre gestillan *they may not ever be still,* Bt. Met. Fox 11, 51; Met. 11, 26. Seó gecyndelíce hǽtu gestilleþ on ðé *the natural heat shall be quiet in thee,* Blickl. Homl. 7, 28. Se wuldor-maga worda gestilde *the illustrious man ceased from words,* Exon. 48 b; Th. 167, 29; Gú. 1067: Andr. Kmbl. 1064; An. 532. On Sæterdæg híg gestildon *sabbăto sĭlŭērunt,* Lk. Bos. 23, 56. Tantalus gestilde *Tantalus became quiet,* Bt. 35, 6; Fox 170, 2. Sȳwa, and gestil *tăce obmūtesce,* Mk. Bos. 4, 39. He bebeád ðǽm winde ðæt he gestilde *he commanded the wind to be still,* Blickl. Homl. 235, 8. Ic bebeóde ðé ðæt ðú fram ðisse ungeþwǽrnysse gestille *I command thee to cease from this troubling,* Guthl. 8; Gdwn. 48, 17. **II.** *v. trans.* *To restrain, still, stop, stay, calm, keep in;* compescĕre, cŏhĭbĕre, sēdāre, mītĭgāre, rĕtĭnēre :—Hilde calla hêht ða folctogan fyrde gestillan *the herald of war bade the folk-leaders make the army still,* Cd. 156; Th. 194, 2; Exod. 254. Ða hátheortan hie mid nǽne fōreþonce nyllaþ gestillan *the furious will not calm themselves with reflection,* Past. 40, 6; Swt. 297, 4; Hat. MS. 55 b, 7. Hwá gestilleþ ðæt wán shall still that? Exon. 101 b; Th. 384, 30; Rä. 4, 35. Hí ðone storm gestildon *tempestātem sēdārent,* Bd. 3, 15; S. 541, 17. Hæfde Metod regn gestilled *the Creator had stilled the rain,* Cd. 71; Th. 85, 18; Gen. 1416: Salm. Kmbl. 236; Sal. 117.

ge-stincan; *p.* -stanc, *pl.* -stuncon; *pp.* -stuncen *To perceive by the sense of smelling;* olfacere aliquid, odorare, odorari :—Nas-þeorlu oððe nόsa hí habbaþ, and híg ne gestincaþ *nostrils or noses they have, and they smell not,* Ps. Lamb. second 113, 6. Hí nόse habbaþ náwiht gestincaþ *they have a nose* [*and*] *smell naught,* Ps. Th. 134, 17. Sume mágon gehíran, sume gestincan *some can hear, some smell,* Bt. 41, 5; Fox 252, 24. Ðonne gé ða swétan stencas gestincaþ *when ye smell the sweet odours,* Blickl. Homl. 59, 3. Hí ðæs landes lyft gestuncon *they smelt the air of the land,* Bd. 1, 1; S. 474, 35. Hí ná gestingcaþ [= gestincaþ] *they smell not,* Ps. Spl. 113, 14.

ge-stióran; *p.* de; *pp.* ed *To correct, restrain, direct, guide;* corrĭgĕre :—Wénst ðú ðæt se anwald eáðe ne meahte Godes Ælmihtiges him his yfeles gestióran *thinkest thou that the power of Almighty God could not keep him from his evil,* Bt. Met. Fox 9, 104; Met. 9, 52. v. ge-stýran.

ge-stir, -stirian. v. ge-styr, -styrian.

ge-stíran; *p.* de; *pp.* ed *To correct, restrain;* corrĭgĕre, cŏhĭbĕre :—Forðǽm ðæt ða wítu gestírdon [gestírden, MS. Cot.] ôðrum ðæt hí swá dôn ne dorsten *in order that the punishments might restrain others from daring to do so,* Bt. 39, 11; Fox 230, 7. v. ge-stýran.

ge-stíðian; *p.* ode, ude; *pp.* od, ud *To become hard, strong;* indurare :—Gistíðĭa induratam, Rtl. 102, 41. Mǽgen on him weóx and gestíðode *his power waxed and was strengthened,* Guthl. 2; Gdwn. 12, 26. Ðá ðá he gestíðod wæs *when he was grown up,* Homl. Th. ii. 38, 3.

ge-stitian; *p.* ode; *pp.* od *To dispose, order;* dispōnĕre :—Ða he gestitode to Abrahame *quod dispŏsuit ad Abraham,* Ps. Spl. C. 104, 8: Ps. Spl. T. 102, 12. v. ge-stihtian.

gest-líð; *adj. Hospitable;* hospitalis, Som.

gest-líðnes, -ness, -nyss, e; *f. Hospitableness, hospitality;* hospĭtālĭtas :—Ðá se fōresprecena Godes man fela daga mid him wæs on gestlíðnesse *cum præfātus clērĭcus alĭquot dĭēbus ăpud eum hospĭtārētur,* Bd. 1, 7; S. 477, 6. On gestlíðnysse *in hospitality,* 1, 7; S. 476, 37: 477, 16: 1, 27; S. 489, 26. v. gæst-líðnes.

ge-stondan. v. ge-standan.

ge-stόp *stepped, went,* Ors. 3, 9; Bos. 68, 22; *p.* of ge-stapan.

gestor-dæge *yesterday;* heri, Jn. Skt. Rush. 4, 52.

gestran-dæg *yesterday;* hesterna dies, Ps. Vos. 89, 4. v. gyrstan-dæg.

ge-strangian, -strongian; *p.* ode, ade; *pp.* od, ad [strangian *to strengthen*] *To make strong, strengthen, confirm, establish;* rŏbŏrāre, corrŏbŏrāre, confortāre, confirmāre :—Ðá wolde he heora geleáfan ge-strangian and getrymman *then would he strengthen and confirm their belief,* Homl. Th. i. 152, 34. Ic gestrangige *confirmo,* Cod. Dipl. Kmbl. iii. 349, 26: 350, 34. Ic heortan mannes gestrangie *ego cor hŏmĭnis confirmo,* Th. Anal. 29, 1. Earm mín gestrangaþ hine *brăchium meum confortābit eum,* Ps. Spl. 88, 21: Salm. Kmbl. 477; Sal. 239. He gestrangode hí *illos confortāvĕrit,* Bd. 1, 23; S. 485, 16. Bebeód Iosue and gestranga hine *præcĭpe Iosue et corrŏbŏra eum,* Deut. 3, 28. Ne biþ gestrangod man *non confortēmur hŏmo,* Ps. Th. 138, 4, 15. Wes ðú gestrangad and ne ondrǽd ðú ðé *be thou strengthened and fear not,* Blickl. Homl. 231, 2: Lk. Bos. 1, 80. Israéla folc wǽron swýðe gestrangode *filii Israel sunt rŏbŏrāti nĭmis,* Ex. 1, 7: Ors. 6, 35; Bos. 131, 1.

ge-streágung, e; *f. Vegetatio,* Hpt. Gl. 440.

ge-streáwian, -streόwian; *p.* ode; *pp.* od *To strew* :—Swylc hit eall gestreáwod wǽre mid wynsume blόstmen and wyrtgemangum *as if it all were strewed with pleasant flowers and spices,* Shrn. 15, 31.

ge-streccan; *p.* -streahte, -strehte; *pp.* -streaht, -streht *To stretch, spread;* sternere :—Wel gestreht bed *a well spread bed,* Lchdm. iii. 208, 4. v. streccan

ge-stredd; *part. p. Sprinkled;* sale conditus, Lye.

ge-streht *spread;* *pp.* of ge-streccan.

ge-strengan; *pp.* ed *to strengthen;* confortare :—Se cnæht gestrenced wes *puer confortebatur,* Lk. Skt. Lind. 1, 80.

ge-streόn, -strión, es; *n. Gain, product, emolument, wealth, riches, treasure, usury, business;* merces, mercātus, quæstus, lucrum, ēmŏlūmentum, ŏpes, thesaurus, ūsūra, nĕgōtium :—Gestreόn *quæstus vel lucrum,* Ælfc. Gl. 114; Som. 80, 6; Wrt. Voc. 60, 42. Swunce mǽre se ðe unriht gestreόn on his handa stόde *he should toil more, in whose hands lay the unjust gain,* L. Eth. ii. 9; Th. i. 290, 5. Sunu gestreόnes wæstm innoðes *filii mercis fructus ventris,* Ps. Spl. 126, 4. Fǽderes gestreόnes *patrĭmōnii,* Mone B. 3568. Ic hit witodlíce mid gestreόne onfénge *cum ūsūris ūtĭque exegissem illam,* Lk. Bos. 19, 23. Fram gestreόne *a nĕgόtio,* Ps. Spl. 90, 6. Mathusal magum dǽlde ǽdelinga gestreόn *Mathuselah distributed the chieftains' treasure to his brethren,* Cd. 52; Th. 65, 24; Gen. 1071: Bt. Met. Fox 8, 115; Met. 8, 58. Gestreόne *mercātu,* Mone B. 2588. Hý beόþ rúmmόde ryhtra gestreόna *they are liberal of just gains,* Exon. 33 b; Th. 106, 31; Gú. 49: 105 b; Th. 402, 18; Rä. 21, 31: 107 b; Th. 410, 23: Rä. 29, 3. Ðæt he æfter him to eallum his gestreόnum fénge *that he should take all his riches after him,* Ors. 5, 13; Bos. 112, 32. Æfter filiende gestreόn *sĕcŭtūra ēmŏlŭmenta,* Mone B. 623. Gehlόdon him hordwearda gestreόn *they loaded on themselves the riches of the treasure-wards,* Cd. 174; Th. 220, 3; Dan. 65: 208; Th. 257, 31; Dan. 666: 209; Th. 260, 4; Dan. 704. Gestreόn *usura,* Blickl. Gloss. Fram gestreόne gangendum *a negotio perambulante,* id. Ic wylle heora cýpan hér luflícor ðonne ic gebicge ðǽr ðæt sum gestreόn me ic begyte *volo vendere hic carius quam emi illic ut aliquod lucrum mihi adquiram,* Th. Anal. 27, 21. [O.Sax. gi-striuni: O.H.Ger. ki-striuni *lucrum.*]

ge-streόnan; *p.* de; *pp.* ed *To gain, get, obtain, acquire;* lucrāri, acquīrĕre :—Heora Criste sáule gestreόnan *suas Christo anĭmas lucrāri,* Hymn. Surt. 73, 7. Ðǽr is cúpre líf ðonne we on eorþan mǽgen ǽfre gestreόnan *there is a life more glorious than we may ever obtain on earth,* Cd. 226; Th. 302, 11; Sat. 597. Ðæt he manige þeόde úrum Drihtne þurh his láre gestreόnde *so that he gained many a nation for our Lord by his teaching,* Blickl. Homl. 121, 10.

ge-streόnful; *adj. Full of riches, copious, expensive, precious, sumptuous;* sumptuōsus :—Gestreόnfulre *sumptuōsā,* Mone B. 3566. Gestreόnful *copiosus, fructuosus,* Hpt. Gl. 443, 452, 491. His ða leόfan and ða gestreόnfullan bearn *his beloved and precious children,* Blickl. Homl. 131, 27.

ge-stric, es; *m? Strife, mutiny, sedition;* sēdĭtio :—Gesihþ leόn wédan feόndes gestric getácnaþ *the sight of a mad lion betokens sedition of an enemy,* Lchdm. iii. 206, 33.

ge-strician; *p.* ede *To knit* :—Gestricedon netta hiora *reficiebant retia sua,* Mt. Kmbl. Lind. 4, 21. [Cf. Ger. stricken.]

ge-strínan, -striénan; *p.* de; *pp.* ed *To obtain, get, acquire, beget, procreate;* acquīrĕre, gignĕre, procreāre :—Ðæt gé me mid rihte gestrínan mágon *what ye may justly acquire for me,* L. Ath. i. prm; Th. i. 196, 16. Se ðe bearn gestriéneþ *he who begets a child,* L. In. 27; Th. i. 120, 2: L. Alf. pol. 8; Th. i. 66, 19. Ðe hit on fruman gestríndon *who first acquired it,* 41; Th. i. 88, 19. v. ge-strýnan.

ge-strión, es; *n. Gain, wealth;* merces, ōpes :—Gió-monna gestrión sealdon unwillum éðelweardas *the country's guardians unwillingly gave up the wealth of men of old,* Bt. Met. Fox 1, 46; Met. 1, 23. v. ge-streόn.

ge-strod, es; *n. Banishment;* proscriptio, Cot. 194.

ge-strod, es; *n. Plunder* [?], *wealth* :—Ðæt hí ðȳ éþ mǽgen heora unriht gewitt forþbringan hí sind mid gifum and mid gestreόnum [Cot. gestrodum] gefyrðrode *flagitiosum facinus ad efficiendum præmiis incitari,* Bt. 3, 4; Fox 6, note 7. [Cf. ge-strúdan.]

ge-strogdniss, e; *f. A sprinkling;* conspersio, Rtl. 25, 15.

ge-strongian *p.* ode; *pp.* ad *To strengthen;* corrŏbōrāre :—Ceadwala wæs gestrongad *Ceadwalla was strengthened,* Bd. 4, 16; S. 584, 4. v. ge-strangian.

ge-strúdan; *p.* -streád, *pl.* -strudon; *pp.* -stroden *To destroy, plunder;* rapere, spoliare :—Godes cwide helle gestrúdeþ *God's word destroyeth hell,* Salm. Kmbl. 148; Sal. 73. Feoh gestrúdaþ *they destroy the cattle,* Salm. Kmbl. 310; Sal. 154. Ða wígan gestrudon [MS. gestrudan] gestreόna *the warriors plundered the treasures,* Cd. 174; Th. 219, 27; Dan. 61. v. strúdan.

ge-strýnan, -streόnan, -strínan, -striénan; *p.* de; *pp.* ed [gestreόn *gain*] *To gain, get, obtain, acquire. beget, procreate;* lucrāri, acquīrĕre, gignĕre, procreāre :—Ðæs ðe ic mόste mínum leόde swylc gestrýnan

because I have been able to acquire such for my people, Beo. Th. 5589; B. 2798: L. Ath. i. prm; Th. i. 196, 18: Homl. Th. ii. 46, 14. Ic gestrýne *gigno*, Ælfc. Gr. 28, 3; Som. 30, 57. Nǽnig fira to fela gestrýneþ *no man gains too much*, Exon. 91 a; Th. 342, 17; Gn. Ex. 144: L. C. S. 85; Th. i. 424, 13. Ðæt hý mid rihte gestrýnaþ *what they lawfully acquire*, L. Edg. S. 2; Th. i. 274, 3: Exon. 61 b; Th. 225, 21; Ph. 392. Ðín pund gestrýnde tyn pund *mna tua decem mnas acquisivit*, Lk. Bos. 19, 16, 18: Mt. Bos. 25, 16, 17, 20: Ps. Spl. 77, 59. He worn gestrýnde suna and dóhtra *he begat several sons and daughters*, Cd. 62; Th. 74, 11; Gen. 1220: Mt. Bos. 1, 2–16. Ðeáh he ealne middaneard gestrýne *si mundum universum lucrétur*, 16, 26: Mk. Bos. 8, 36. Ic hæbbe gestrýned óðre twá *alia duo lucratus sum*, Mt. Bos. 25, 22.

ge-strýnedlíc, -strýnendlíc; *adj. Producing, genitive*; genitivus :— Gestrýnedlíc oððe geágniendlíc *genitive or possessive*, Ælfc. Gr. 7; Som. 6, 17.

ge-strynge, es; *m. A wrestler, champion*; athleta :—Gestrynga plegstów *a place of wrestlers, a theatre*; athletarum locus, Cot. 151. [Cf. strang.]

gest-sele, gyst-sele, es; *m. A guest-hall*; hospitalis aula :—Ðe gest-sele gyredon *who prepared the guest-hall*, Beo. Th. 1992; B. 994.

ge-stun, es; *n.* [stunian *to stun*] *A noise, stun, crash, whirlwind*; strepitus, fragor, turbo :—Ðæt gestun and se storm brecaþ bráde gesceaft *the stun and the storm shall break the broad creation*, Exon. 22 b; Th. 61, 27; Cri. 991. Of gestune *from the whirlwind*, 102 a; Th. 386, 3; Rä. 4, 56. Þurh gestun *per turbinem*, Cot. 157.

ge-stuncon *smelt*, Bd. 1, 1; S. 474, 35; *p. pl. of* ge-stincan.

ge-stund, es; *n. A noise, din* :—Hí swá ungemetlícum gestundum fóron ðæt him þúhte ðæt hit eall betweox heofone and eorþan hleóðrode ðám egeslícum stefnum *they came with such immoderate noises that it seemed to him that between heaven and earth it all resounded with their voices*, Guthl. 5; Gdwn. 36, 28.

ge-stungen; *part.* [stungen, *pp. of* stingan *to pierce*] *Pierced*; transfixus :—He wæs mid spere on his sýdan gestungen *he was pierced in his side with a spear*, L. E. I. 21; Th. ii. 416, 31.

ge-styllan; *p.* de; *pp.* ed *To still, calm*; sédáre :—Se eorl gestylde ðæt folc *the earl stilled the people*, Chr. 1052; Erl. 187, 3. v. ge-stillan II.

ge-styllan; *p.* de *To spring, move rapidly* :—Hwílum he to eorþan gestylde *at times he descended to earth*, Exon. 17 a; Th. 40, 34; Cri. 648. Cyning engla munt gestylleþ gehleápeþ heá dúne *the king of angels shall mount a hill, shall leap the high downs*, 18 a; Th. 45, 9; Cri. 716. [Cf. a-stellan.]

ge-styltan; *p.* te *To be astonished, to be silent from astonishment* :— Gestylton ꞇ gesuígdon alle *stupebant omnes*, Mt. Kmbl. Lind. 12, 23. Folc gestylte [gistylted wæs, Rush.] *populus stupefactus est*, Mk. Skt. Lind. 9, 15. Gestyldon aldro *stupuerunt parentes*, Lk. Skt. Lind. 8, 56. v. ge-stillan [?].

ge-styr, -stir, es; *n. Movement, action* :—Gestir *actio*, Rtl. 187, 15.

ge-stýran, -stíran, -steóran, -stióran, -stíeran; *p.* de; *pp.* ed [stýran *to steer, rule*] *To steer, direct, rule, correct, restrain, withhold*; regére, corrigére, cohíbére, retínére :—Meaht ðú Adame eft gestýran *thou mightest afterwards rule Adam*, Cd. 27; Th. 36, 8; Gen. 568: Ors. 3, 1; Bos. 52, 36. Hám cymeþ nefne him holm gestýreþ *he will come home unless the ocean restrains him*, Exon. 90 b; Th. 340, 5; Gn. Ex. 106. Gif him Scipio ne gestýrde *if Scipio had not withheld them*, Ors. 4, 9; Bos. 91, 18: Judth. 10; Thw. 22, 13; Jud. 60. Forstond ðú mec and gestýr him *protect thou me and correct them*, Exon. 118 b; Th. 455, 31; Hy. 4, 58.

ge-styreniss, e; *f. Trouble, tribulation*; tribulatio, Rtl. 40, 39.

ge-styrian, -stirian; *p.* de; *pp.* ed [ge, styrian *to move, stir*] *To move, remove, excite, agitate*; amovere, agitare :—Nælle ðú gestyrege hine *noli vexare illum*, Lk. Skt. Lind. 8, 49. Biþ gestyred hiora orsorgnes [MS. orsornesse] *their prosperity will be removed*, Bt. 38, 2; Fox 196, 23. Ðú wǽre stronge gestyred *thou wast strongly excited*, Exon. 98 a; Th. 369, 22; Seel. 45. Ða wearþ swíðe gestired se here ongeán ðone biscop forðan ðe he nolde heom nán feoh beháten *then was the [Danish] army very much excited against the bishop because he would not promise them any money*, Chr. 1012; Erl. 146, 12. Mægna ða ðe sint in heofnum gestyred bíþon *virtutes quæ sunt in cælis movebuntur*, Mk. Skt. Lind. 13, 25. Forhuon arogie gestyred *quid turbamini*, 5, 39: Mt. Kmbl. Lind. 24, 6: Jn. Skt. Lind. 12, 27. Dóhter mín from diwble is gestyred *filia mea a dæmonio vexatur*, Mt. Kmbl. Lind. 15, 22.

ge-sufel; *adj. A word of uncertain meaning, but descriptive of a certain kind of bread* :—Ælc gegilda gesylle ǽnne gesufelne hláf *let each gild-brother give a 'gesufel' loaf*, L. Æthelst. 5, 8; Th. i. 236, 36. Mon geselle cxx gesufra hláfa *let cxx 'gesufel' loaves be given*, Th. Ch. 460, 32: 469, 3. v. sufel.

ge-súgian; *p.* ode; *pp.* od *To be silent*; tácére :— Gif ðú gesúgian meahte *if thou mightest be silent*, Bt. 18, 4; Fox 68, 4, MS. Cot. Gesúgode he *he was silent*, 17; Fox 58, 21, MS. Cot. v. ge-swígian.

ge-suirfed *polished, filed*; politus, Som.

ge-sund; *adj. Sound, healthy, entire, unhurt, safe, favourable, prosperous*; sanus, integer, salvus, incolumis, prosper, felix :—Ðæs ðe hí hyne gesunde geseón móston *for that they might see him sound*, Beo. Th. 3260; B. 1628: Exon. 74 a; Th. 276, 19; Jul. 568: 42 b; Th. 144, 4; Gú. 673: 23 b; Th. 66, 21; Cri. 1075. Beó gesund *ave, salve* Beóþ gesunde *avete, salvete*, Ælfc. Gr. 33; Som. 37, 42, 43. He cwæþ 'Wel gesund hláford apolloni' *he said 'All hail, lord Apollonius,'* Th. Apol. 7, 21. Ða cwæþ he to ánum cnapan 'Swá ðú gesund sý sege me' *then said he to a boy 'So be thou in health, tell me*, 6, 19. Híg cómon gesunde to hýde *they came to port safe and sound*, Shrn. 147, 10. Hý beóþ ðý gesundran *they will be the healthier*, Exon. 107 a; Th. 408, 28; Rä. 27, 19. Ðæt ic ðé lǽte brúcan sinces gesunde *that I will let thee enjoy wealth uninjured*, Cd. 126; Th. 161. 14; Gen. 2665. Ðonne beón híg ealle gesunde *cunctus populus salvabitur*, Deut. 20, 11. On ðǽre stówe we gesunde mágon bídan *in that place we may abide safe*, Cd. 117; Th. 152, 20; Gen. 2523: Exon. 27 b; Th. 82, 21; Cri. 1342: Beo. Th. 641; B. 318. Eálá ðú, Dryhten mín, dó ús gesunde *fac, O Domine, bene prosperare*. Ps. Th. 117, 23: Elen. Grm. 996: 1005. [O. Sax. gi-sund : O. H. Ger. ge-sunt : Ger. ge-sund.] DER. sund.

ge-sund-ful, -full; *adj. Full or quite sound, prosperous, successful*; prosperus :—Gesundfull síþfæt dó us, God *prosperum iter faciet nobis Deus*, Ps. Spl. 67, 21. His swíðre hand is gesundfull óþ ðis *his right hand is sound to this day*, Swt. A. S. Rdr. 98, 85.

ge-sundfullian; *p.* ode; *pp.* od *To be made prosperous, to be successful*; prosperari :— Swá hwæt swá he deþ beóþ gesundfullode *quæcumque faciet prosperabuntur*, Ps. Lamb. 1, 3. Gesundfulla *prosperare*, Ps. Spl. C. 117, 24.

ge-sundfullíc; *adj. Safe, sound* :—Ne biþ ǽfre ówiht gesundfullíces in ðam deófle *there is never aught sound in the devil*, Shrn. 38, 35.

ge-sundfullíce; *superl.* -licost; *adv. Safely, securely, successfully, prosperously*; túte, prospère :—Hí to ðisum íglande gesundfullíce becómon *they came safely to this island*, Homl. Th. ii. 128, 16. Begým gesundfullíce *intende prospère*, Ps. Spl. 44, 5. Hió færþ gesundfullícost *it goes most securely*, Bt. 39, 7; Fox 222, 22.

ge-sundfulnes, -fullnes, -ness, -nys, -nyss, e; *f. Soundness, healthiness, prosperity*; sánitas corpóris, prospéritas :—On ðínre gesundfulnesse *in thy health*, Bt. 6; Fox 14, 35. Se oferdrenc fordeþ untwílíce ðæs mannes sáwle and his gesundfullnysse *over-drinking surely destroys a man's soul and his soundness*, Ælfc. T. 43, 16. Ne breác se árleása Herodes his cyneríces mid langsumere gesundfullnysse *the impious Herod did not enjoy his kingdom in long health*, Homl. Th. i. 84, 34.

ge-sundig; *adj. Prosperous, favourable*; prosperus, secundus :—Gesundige windas *secundi venti*, Bd. 5, 1; S. 614, 9. v. ge-sundlíce.

ge-sundlíce; *adv. Prosperously* :—Gesundlíce *prosperare* [= *prospere* ?], Ps. Spl. 117, 24. We ða niht on ðære wícstówe gesundlíce wícodon *we stopped safely in the camp that night*; quieta nox fuit usque ad lucem, Nar. 21, 30.

ge-sundrian; *p.* ode; *pp.* od [sundrian, syndrian *to sunder*] *To separate, divide, sunder*; separare, discedére, disjungére :—Gesundrode sigora Waldend leóht wið þeóstrum *the Lord of triumphs sundered light from darkness*, Cd. 6; Th. 8, 18; Gen. 126: 8; Th. 9, 13; Gen. 141. Gesundrod wæs lago wið lande *water was separated from land*, 8; Th. 10, 26; Gen. 162. Of sceádes ꞇ gesundras *definiens*, Mt. Kmbl. p. 12, 13. Ðú ðe gesundradest *qui destinasti*, Rtl. 56, 31.

ge-súpan; *p.* ode; *pp.* -seáp, *pl.* -supon; *pp.* -sopen *To sup, sip, suck up, absorb*; absorbére :—Wén is ðæt hí us woldan wætre gelíce sóna gesúpan *forsitan velut aqua absorbuissent nos*, Ps. Th. 123, 3.

ge-suppan *to taste* :—Hia ðæt gebirigdon ꞇ gesupedon *gustaturos*, Mk. Skt. p. 4, 3. v. suppan.

ge-súwian *to be silent*. v. ge-swígian.

ge-swác *ceased, rested from*, Mt. Bos. 14, 32: Gen. 2, 3; *p.* of geswícan.

ge-swǽlan; *p.* de; *pp.* ed, ud *To light, kindle*; inflammare, accendere :—Geswǽlud spoon [= spón, *q. v.*] *vel* tynder *kindled chips or tinder*; fomes, Ælfc. Gl. 60; Som. 68, 35; Wrt. Voc. 39, 21.

ge-swæncan; *p.* te; *pp.* ed *To afflict, oppress*; affligére, opprímére :— Ða he gelomlíce geswæncte *whom he repeatedly oppressed*, Chr. 1105; Erl. 240, 11. v. ge-swencan.

ge-swæpa, -swæpo; *pl. n. Sweepings*; peripsema, sordes, Cot. 149, 169. Geswápa *ruina vel rudera*, Ælfc. Gl. 17; Som. 58, 96; Wrt. Voc. 22, 12. v. æsce-geswáp.

ge-swǽre, es; *n. Heaviness, affliction* :—Gisuoere *afflictionem*, Rtl. 41, 37. [Cf. O. H. Ger. swâri : Ger. schwere *weight*.]

ge-swǽre; *adj. Heavy, oppressed, afflicted* :—He lǽrde ælcne man ðe geswǽre and ofercumen, and eft gefríþod byþ, ðæt he swá ylce Gode þancode *he taught every man that is oppressed and overcome, and afterwards is saved, that he in the same way should thank God*, Ps. Th. 47, argument. [O. H. Ger. ge-swar.] v. ge-swǽre, *subst*; and swǽr.

ge-swǽs; *adj. Dear, familiar, kind*; cárus, familiáris, blandus :—He geceás Iudan him, geswǽs frumcynn *elegit tribus Juda*, Ps. Th. 77, 67.

Iohannes mid geswǽsum wordum ðæt folc tihte *John exhorted the people with kind words*, Homl. Th. i. 70, 34.

ge-swǽslǽcan; *p.* -lǽhte; *pp.* -lǽht *To flatter;* blandíri :—Ic geswǽslǽce *blandior*, Ælfc. Gr. 31; Som. 35, 49.

ge-swǽsnys, se; *f. A sweet word, a compliment, an enticement, allurement, a dainty;* blanditia :—Geswǽsnyssa *blanditiæ,* Ælfc. Gr. 13; Som. 16, 17.

ge-swǽtan; *p.* te *To sweat :*—Heó ná ne geswǽtte *she did not sweat,* Shrn. 150, 2.

ge-swǽþian; *p.* ode; *pp.* od *To track out, investigate :*—Geswæþodes *investigasti,* Ps. Spl. T. 138, 2. v. swæþ, swaþu.

ge-swǽðrung, e; *f. A failing, a want;* deliquium :—Se mon geswógunga þrówaþ and módes geswæðrunga *the man suffers swoonings and failings of the mind,* L. M. 2, 21; Lchdm. ii. 206, 9. v. sweðerian.

ge-swáp. v. æsce-geswáp.

ge-sweccan; *p.* te; *pp.* ed [sweccan *to smell*] *To smell;* odórári :—Næsþyrlu hí habbaþ and ná gesweccaþ *nares häbent et non odórábunt,* Ps. Spl. M. 113, 14.

ge-swefian, ic -swefige; *p.* ode; *pp.* od [swefan *to sleep*] *To cause to sleep, cast asleep, lull, appease;* sópíre, sópóráre :—Ic geswefige *sópio,* Ælfc. Gr. 30, 5; Som. 34, 57, MS. D. God geswefode ðone Adam *God caused Adam to sleep,* Homl. Th. i. 14, 20. Drihten on róde mid deáþe wæs geswefod *the Lord was put to sleep by death on the cross,* ii. 260, 18 : i. 496, 12 : Boutr. Scrd. 19, 37. Ic eom geswefod *sópórátus sum,* Ps. Lamb. 3, 6.

ge-swége; *adj.* v. ungeswége.

ge-swégsumlíce; *adv. Harmoniously, with one voice :*—Dá sǽde eall se þeódscipe geswégsumlíce *then all the people agreed in saying,* Shrn. 36, 17.

ge-swel, -swell, es; *n.* [swellan *to swell*] *A swelling, tumour;* tǔmor :—Wið ǽlcum heardum swile odðe-geswelle *for every hard tumour or swelling,* L. M. 1, 31; Lchdm. ii. 70, 20 : Herb. 86, 1; Lchdm. i. 188, 20 : 90, 1; Lchdm. i. 194, 19 : 109, 3; Lchdm. i. 222, 14. Hyt ðæt geswel gelíðigaþ *it relieves the swelling,* 109, 3; Lchdm. i. 222, 14 : 76, 1; Lchdm. i. 178, 20 : iii. 8, 28. Wið geswell *for a swelling,* Herb. 90, 4; Lchdm. i. 194, 18. Wið ealle geswell *for all swellings,* 130, 1; Lchdm. i. 240, 18. Dá wolde se heofenlíca lǽce ðæt geswell heora heortan gelácnian *then would the heavenly leech cure the swelling of their heart,* Homl. Th. i. 338, 23. Mislíce geswel and blǽdran *divers boils and blisters,* ii. 192, 30.

ge-swelgan; *p.* -swealg, -swealh, *pl.* -swulgon; *pp.* -swolgen [swelgan *to swallow*] *To swallow, devour;* devóráre, deglǔtíre :—Da mægenþreátas meredeáþ geswealh *the sea-death swallowed those mighty bands,* Cd. 169; Th. 210, 9; Exod. 512.

ge-swelge, es; *n. An abyss, gulf;* vorago, barathrum, charybdis, Hpt. Gl. 421, 513.

ge-swelgend; es; *m. An abyss, chasm;* vorago, Hpt. Gl. 507.

ge-sweltan; *p.* -swealt, *pl.* -swulton; *pp.* -swolten [sweltan *to die*] *To die, perish;* móri :—Men gesweltaþ *hŏmines moriemini,* Ps. Spl. 81, 6. Geswolten, Bd. 5, 6; S. 619, 18.

ge-swenc, es; *n. Labour, trouble :*—Þurh ðæt geswenc to éce reste becom *through that suffering came to the eternal rest,* Nar. 40, 2. v. ge-swinc.

ge-swencan, -swæncan; *p.* -swencte; *pp.* -swenced, -swenct [swencan *to disturb, vex*] *To disturb, agitate, trouble, vex, fatigue, outweary, afflict, harass, oppress;* pulsáre, agitáre, tríbǔláre, vexáre, fátígáre, afflígěre, afficěre, oppriměre :—Herodes cyning wolde geswencan sume of ðære gelaðunge *Herod the king would afflict some of the church,* Homl. Th. ii. 380, 25 : Salm. Kmbl. 299 ; Sal. 149. Híg eów to deáþe geswencaþ *morte afficient ex vobis,* Lk. Bos. 21, 16 : 8, 45. Sarai híg ðá geswencte and heó sóna fleáh út to ðam wéstene *affligiente igǐtur eam Sarai fǔgam iniit,* Gen. 16, 6. Hí synne geswencton *they outwearied sin,* Exon. 55 b; Th. 197, 12; Az. 189: Chr. 1116; Erl. 245, 35. Útancumene and elþeódige ne geswenc ðu nó *vex thou not comers from without and strangers,* L. Alf. 33; Th. i. 52, 14. Ic geswenced sý tríbǔlor, Ps. Th. 101, 2, 4: Bd. 4, 9; S. 576, 27. Synnum geswenced *oppressed with sins,* Beo. Th. 1954; B. 975 : 2741; B. 1368: Andr. Kmbl. 788; An. 394. He wæs geswenced mid grimmum gefeohte *he was wearied with fierce fighting,* Chr. Erl. 5, 30. He biþ geswenct óþ geár seofone *he will be troubled for seven years,* Lchdm. iii. 188, 12: 192, 4: 204, 14. Hí wurdan geswencte *vexáti sunt,* Ps. Th. 106, 38: 43, 23: Ors. 1, 7; Bos. 30, 30.

ge-swencednes, -swincednes, -swenctnes, -nis, -nys, -ness, -niss, -nyss, e; *f.* [geswencan, *pp.* of geswencan *to disturb, trouble, afflict*] *Sorrow, affliction, tribulation;* afflictio, tríbǔlátio :—Hí fórecómon me on ðæge geswencednysse mínre *prævenérunt me in die afflictiónis meæ,* Ps. Spl. 17, 21 : Homl. Th. ii. 456, 11. Æfter ðære geswencednysse *post tríbǔlátiónem illam,* Mk. Bos. 13, 24 : Ps. Spl. 54, 2. For ðam hwílwendlícum geswenctnessum [MS. e] *for the temporal afflictions;* temporales adflictiones, Bd. 4, 9; S. 577, 12. Nán ðyssera geswencednyssa ne becom on ðam ende ðæs eardes ðe ðæt godes folc on eardode *none of these afflictions*

came into that part of the country in which the people of God dwelt, Homl. Th. ii. 192, 25.

ge-sweógian; *p.* ode; *pp.* od *To be silent;* tácére :—Gesweógode he áne hwíle *he was silent for some time,* Bt. 39, 2 ; Fox 212, 10. v. geswígian.

ge-sweopornes, -swiopernis, -ness, -niss, e; *f. Cunning, craftiness, hypocrisy;* astutia, Mk. Skt. Rush. 12, 15. v. ge-swipornis.

ge-sweorc, -sworc, es; *n.* [sweorcan *to darken*] *A cloud, mist, smoke;* nūbes, nēbǔla, cālígo :—Gif hēr wind cymþ gesweorc upfæreþ *if wind comes here a cloud ascends,* Cd. 38; Th. 50. 12; Gen. 807. Cining geseah deorc gesweorc *the king saw a dark cloud,* 5 ; Th. 7, 19; Gen. 108. [*O. Sax.* gi-swerk: *O. H. Ger.* gi-swerc.]

ge-sweorcan, he -swyrcþ; *p.* -swearc, *pl.* -swurcon; *pp.* -sworcen *To become dark, be darkened, saddened, angry :*—Ródor eal geswearc *the heavens all grew dark,* Elen. Kmbl. 1709; El. 856: Beo. Th. 3583; B. 1789: Cd. 166; Th. 207, 4; Exod. 461. Seó eorþe wæs gesworcen and aþýstrod under his fótum *cālígo sub pedibus ejus,* Ps. Th. 17, 9. Dá geswearc se Godes man semninga and ongan heardlíce and bitterlíce wēpan *then suddenly the man of God became sad and began to weep sorely and bitterly;* solutus est in lacrymis vir Dei, Bd. 4, 25 ; S. 600, 29 : Exon. 77 a ; Th. 290, 3; Wand. 59. Geswearc ðá sweor *the father-in-law then grew angry,* 67 a; Th. 247, 13; Jul. 78. Cf. asweorcan. [*O. Sax.* gi-swerkan *to become dark, literally and metaphorically as in English :* *O. H. Ger.* ge-sworcen; *part. p.* turbulentus, nubilus.]

ge-sweorcnes, -ness, e; *f. Cloudiness, gloom, horror, affliction;* obscūrǐtus, horror, afflictio :—Ne ðær nǽfre biþ biternes, ne gesweorcnesse stów gemēted *nor is bitterness ever there, nor a place found for gloom,* L. E. I. prm ; Th. ii. 400, 9.

ge-sweorf, es; *m. n.* [?] *Filings;* limatura. DER. ár-gesweorf. v. geswyrf.

ge-sweorfan; *p.* -swearf, *pl.* -swurfon; *pp.* -sworfen *To file or rub off, to polish off;* expolíre :—To ásworfenum óran, to gesworfenum óran *sub expolita,* Glos. Prudent. Recd. 142, 19. v. sweorfan.

ge-sweoru, -swiru, -swyru; *pl. n. Hills;* colles :—Wurdan gesweoru swá on seledreáme swá on sceápum beþ sceóne lambru *colles vělut agni ǒvium,* Ps. Th. 113, 6. Mid wynngráfe weaxaþ geswiru [MS. gespiru] *exultátióne colles accingentur,* 64, 13. Muntas and geswyru *montes et omnes colles,* 148, 9 : 71, 3 : 113, 4.

ge-sweostor, -sweostra, -sweostro, -swustra, -swystra *sisters;* sorores; *used as the pl. of* sweostor :—His twá dóhtor, swáse gesweostor *his two daughters, own sisters,* Exon. 112 b; Th. 431, 29; Rä. 47, 3. Dær wǽron twá cwéna ða wǽran gesweostra *there were two queens who were sisters,* Ors. 1, 10; Bos. 33. 36. Hwæðer mótan twegen æwe gebróðro twá gesweostro on gesinscipe onfón *si debeant duo germani fratres singulas sorores accipere,* Bd. 1, 27; S. 490, 28. Dara eádigra gesweostra gemynd *the commemoration of the blessed sisters,* Shrn. 69, 18. [*O. Sax.* gi-swester: *O. H. Ger.* gi-suester.] DER. will-gesweostor. v. sweostor.

ge-sweotulian, -sweotlian; *p.* ode, ade; *pp.* od, ad *To manifest;* mǎnifestáre :—Gesweotula nú ðín sylfes weorc *manifest now thine own work,* Exon. 8 a; Th. 1, 16; Cri. 9. Biþ meaht gesweotlad *her might is manifested,* 128 a; Th. 492, 20; Rä. 81, 18. v. ge-swutelian.

ge-swerian, ic -swerige, -swerge; *p.* -swór, -sweór, *pl.* -swóron; *pp.* -sworen [swerian *to swear*] *To swear, take an oath;* júráre :—Ic ðæt geswerige þurh sunu Meotudes *this I swear by the son of the Creator,* Elen. Kmbl. 1368; El. 686. Ic geswerge *I swear,* Exon. 67 a ; Th. 247, 17; Jul. 80. Swá ic geswór wið Drihten *sícut júrávit Domino,* Ps. Th. 131, 2. Dú geswóre Apollonio *thou didst swear to Apollonio,* Apol. Th. 23, 5. He befóran his fæder gesweór, ðæt he nǽfre ne wurde Rómána freónd *he swore before his father that he would never become a friend of the Romans,* Ors. 4, 8; Bos. 89, 25. Him betweonum geswóran *they took an oath among themselves,* 1, 11; Bos. 34, 37 : 1, 14; Bos. 37, 16.

ge-swétan; *p.* -swétte; *pp.* -swéted, -swét [swéte *sweet*] *To make sweet, sweeten, season;* condíre, indulcáre, indulcóráre :—Ic geswéte synna lustas *I sweeten the delights of sins,* Exon. 71 a ; Th. 264, 24 ; Jul. 369. His bodunga mid sóþre lufe symle geswétte *he ever sweetened his preachings with true love,* Homl. Th. ii. 148, 28. Ic genam ða reliquias and mid swótum wyrtum gesweótte *I took the relics and sweetened with sweet herbs,* Nar. 49, 8. Geswéted wín *sweetened wine;* defrucatum, Wrt. Voc. 290, 58. Geswét wín *mēlícrátum* = μελίκρατον, Ælfc. Gl. 32; Som. 61, 113 ; Wrt. Voc. 27, 42. Geswét eced *sweetened vinegar;* oximellum, 32 ; Som. 61, 115 ; Wrt. Voc. 27, 44. On geswéttum wætere *in sweetened water,* Herb. 103, 3; Lchdm. i. 218, 3 : 33, 2 ; Lchdm. i. 132, 13 : 111, 2; Lchdm. i. 224, 17.

ge-sweðerian, -sweðrian; *p.* ode; *pp.* od *To weaken, destroy :*—Geswederad wæs se swyle *fuga tumoris secuta est,* Bd. 5, 3; S. 616, 39. Ðonne beoþ mín sorg gesweðrad *my sorrow will be stilled,* Exon. 48; Th. 164, 17; Gú. 1013. v. sweðrian, ge-swiðrian, ge-swæðrung.

ge-swétléht; *part. p. Made sweet* [?] :—Onsægnessa geswétléhte *holocausta medullata,* Blickl. Gloss.

ge-swic, es; *n. An offence;* scandalum, Ps. Spl. T. 118, 165: 49, 21. v. æ-, be-swic.

ge-swícan; ic -swíce, ðú -swícest, -swícst, he -swíceþ, -swícþ, *pl.* -swícaþ; *p.* -swác, *pl.* -swicon; *pp.* -swicen *To leave off, desist, stop, cease, rest from, turn from, withdraw, relinquish, fail, deceive, betray;* intermittere, desistere, cessare, quiescere, requiescere, deserere, discedere, relinquere, deficere, fallere, prodere. **I.** *v. n :*—He nolde geswícan *he would not cease,* L. Ælfc. C. 3; Th. ii. 344, 5. Ne wolde ic fram dínum bebodum geswícan *a mandatis tuis non erravi,* Ps. Th. 118, 110. Ic gedó, ðæt hira gemynd geswícþ of eallum mannum *cessare faciam ex hominibus memoriam eorum.* Deut. 32, 26. Geswác se wind *cessavit ventus,* Mt. Bos. 14, 32: Lk. Bos. 5, 4: 11, 1: Gen. 8, 22. Ic geswíce oððe ic forlǽte oððe ic me reste *quiesco,* Ælfc. Gr. 28, 1; Som. 30, 28. Geswác æt sæcce Beówulfes sweord *Beowulf's sword failed in the conflict,* Beo. Th. 5355; B. 2681. Gesuícas *mentientes,* Mt. Kmbl. Lind. 5, 11. **II.** *with the genitive:*—Wíle heó ðæs síðes geswícan *it will desist from its course,* Salm. Kmbl. 647; Sal. 323. Gif he unrǽdes ne geswíceþ *if he desist not from mischief,* Exon. 107 b; Th. 410, 7; Rä. 28, 12. Bútan he ðæs yfeles geswíce *except he desist from evil,* Ps. Lamb fol. 183 b, 20. Hí ðæs gefeohtes geswicon *they stopped the fight,* Ors. 3, 1; Bos. 54, 29. Ðæs fixnoþes geswícan *to cease from fishing,* Homl. Th. ii. 516, 11. Gif ðú unrǽdes ne geswícest *if thou cease not from evil counsel,* Exon. 67 b; Th. 250, 1; Jul 120. Gif we ðæs unrihtes geswícaþ *if we cease from evil,* Elen. Kmbl. 1030; El. 516. Gerǽddon [gerǽdden, MS.] ða witan ðæt man ælces yfeles geswác *the witan decreed that men should cease from every kind of evil,* Chr. 1048; Erl. 178, 33: Ps. Th. 58, 4. Hí nǽfre heora yfeles geswicon *they never ceased from their evil,* Chr. 1001; Erl. 137, 20. He geswác hys weorces *he rested from his work,* Gen. 2, 3. Gé hellfirena sweartra geswícaþ *ye turn from black hell-crimes,* Exon. 98 a; Th. 366, 4; Reb. 7. Geswícaþ ðære synne *turn from that sin,* Cd. 113; Th. 149, 1; Gen. 2468. Geswíc ðisses setles *relinquish this seat,* Exon. 36 b; Th. 119, 3; Gú. 249. **III.** *with the dative:*—Hí ðære heregunge geswicon *they ceased the ravaging,* Chr. 994; Erl. 132, 32. Hí geswicon ðære fyrdinge *they withdrew from the expedition,* 1016; Erl. 153, 29. Ðæt hí woldon [woldan, MS.] Rómánum geswícan *that they would relinquish the Romans,* Ors. 5, 10; Bos. 108, 29. Ðæt ic dínum lárum geswíce *that I relinquish thy doctrines,* Andr. Kmbl. 2582; An. 1292. Wélandes geworc ne geswíceþ monna ænigum *Weland's work deceiveth not any [of] men,* Wald. 3; Vald. 1, 2. Seó ecg geswác þeódne *the edge failed its Lord,* Beo. Th. 3053; B. 1524. Earm biþ se him his frýnd geswícaþ *miserable is he whom his friends betray,* Exon. 89 a; Th. 335, 22; Gn. Ex. 37. Ne ǽnig iuih giswíca *nemo vos seducat,* Rtl. 13, 29. Hine manoden ðæt he ne geswice Godes word to bodigenne *admonished him not to cease preaching God's word,* Shrn. 13, 33.

ge-swicennes, -swicenes, -ness, -nys, -nyss, e; *f. A ceasing, cessation, abstaining, repentance;* cessátio, resïpiscentia:—Búton geswicennesse *without abstaining,* L. N. P. L. 63; Th. ii. 300, 22. Mid geswicennysse yfelra dǽda *with cessation from evil deeds,* Homl. Th. ii. 48, 27: Ælfc. T. 29, 18. Þurh geswicenysse yfeles *by cessation from evil,* Homl. Th. ii. 332, 3. Búton ælcere geswicenesse *sine ulla resïpiscentia,* L. M. I. P. 20; Th. ii. 270, 21.

ge-swicn, e; *f. A cleansing, clearance;* purgátio:—Náh he ða geswicne *he shall not have the clearance,* L. In. 15; Th. i. 112, 5, MSS. B. H. [Cf. *Goth.* swiknei *purity;* Icel. sykna *blamelessness.*]

ge-swicnan; *p,* ede; *pp.* ed *To cleanse, clear;* purgáre:—Geswicne se hine be cxx hída *let him clear himself with cxx hides,* L. In. 14; Th. i. 110, 16: 15; Th. i. 112, 3: 52; Th. i. 134, 12. [Cf. *Goth.* swikns *innocent, pure;* Icel. sykn *free from guilt, innocent.*]

ge-swicneful; *comp.* -fulra; *adj. Treacherous, deceitful, harmful:*—Sint hie ðé geswicnefulran *they are more harmful to thee,* Bt. 14, 1; Fox 42, 22.

ge-swícung, e; *f. A ceasing, an intermission;* cessatio, R. Conc. pref. Mon. Angl.

ge-swígian, -swúgian; *p.* ode; *pp.* od. **I.** *to be silent:*—Monig mon hæfþ ðone unþeáw, ðæt he ne can nyt sprecan ne ne can geswígian *many a man has the bad habit, that he can say nothing to the purpose, nor yet hold his peace,* Prov. Kmbl. 47. Gif ðú geswúgian mihtest *if thou couldst be silent,* Bt. 18, 4; Fox 68, 4. He gesuígde *obmutuit,* Mt. Kmbl. Lind. 22, 12. Gesuígdon alle *stupebant omnes,* 12, 23. Ðá for ðæs byscopes hálignysse geswígdon eall ða deófolgyld *then on account of the bishop's holiness all the idols were silent,* Shrn. 151, 31. **II.** *to pass over in silence, with the genitive:*—Nelle ic lofes ðínes geswígian *I will not pass over thy praise in silence,* Ps. Th. 108, 1. Sóþes geswúgedon *were silent about the truth,* Swt. A. S. Rdr. 111, 202. Eác ic wille geswígian Tontolis and Philopes ðara scondlicestena spella *nec mihi nunc enumerare opus est Tantali et Pelopis facta turpia, fabulas turpiores,* Ors. 1, 8; Bos. 31, 24. **III.** *to silence:*—Fugol biþ geswíged *the bird is hushed,* Exon. 58 a; Th. 207, 22; Ph. 145. [O. H. Ger. gi-suígan *to pass over in silence.*]

ge-swígung *silence,* Lye.

ge-swin, -swins [?], es; *n. Melody;* modulatio:—Geswin *melody,* Exon. 57 b; Th. 207, 5; Ph. 137.

ge-swinc, -swing, es; *n.* [swinc *labour, trouble*] *Labour, exercise, inconvenience, fatigue, trouble, affliction, tribulation, torment, temptation, banishment;* lábor, exercítátio, incommodum, afflictio, tríbúlátio, tentátio, exsilium:—Geswinc *lábor,* Ælfc. Gr. 9, 21; Som. 10, 27. Com ðis geswinc ofer us *venit super nos ista tríbúlátio,* Gen. 42, 21: Ps. Surt. 21, 12. On tíd geswinces *in tempore tríbúlátiónis,* 36, 39: 17, 19. Ðú eall þing birest búton geswince *thou bearest all things without labour,* Bt. Met. Fox 20, 553; Met. 20, 277: Chr. 1016; Erl. 155, 3. On geswince *in exercútátióne,* Ps. Spl. 54, 2. Sum heard geswinc habban sceoldon *they must have some hard torment,* Cd. 17; Th. 20, 30; Gen. 317: Chr. 1085; Erl. 218, 10. Eallra geswinca *of all labours,* Bt. Met. Fox 21, 20; Met. 21, 10: 21, 28; Met. 21, 14. On mínum geswincum *in tentátiónibus meis,* Lk. Bos. 22, 28: Homl. Th. ii. 82, 23. Gé eodon on hyra geswinc *in lábóres eórum introistis,* Jn. Bos. 4, 38. Geswinc *exsilium,* Cot. 73.

ge-swinc-dæg, es; *m. A labour-day, day of toil;* tribulationis dies, Exon. 81 b; Th. 306, 4; Seef. 2.

ge-swincednes, -nis, -ness, -niss, e; *f. Tribulation;* tríbúlátio:—On geswincednisse *in tríbúlátióne,* Ps. Spl. C. 9, 9. v. ge-swencednes.

geswinc-ful, -full; *adj. Full of labour, laborious, troublesome, wearisome;* lábóriósus, incommódus:—Hit biþ swíðe geswincful *it is very laborious,* Past. 60; Swt. 453, 10; Hat. MS: Lchdm. iii. 188, 19: 192, 2, 23. Ðis wæs geswincfull *this was troublesome,* Chr. 1097; Erl. 234, 24. Sint hí ðé geswincfulran *they are more troublesome to thee,* Bt. 14, 1; Fox 42, 22, MS. Cot.

geswincfulnys, -nyss, e; *f. Sorrow, affliction, tribulation;* tríbúlátio:—Of eallum geswincfulnyssum he gehǽlde hine *de omnibus tríbúlátiónibus ejus salvábit eum,* Ps. Lamb. 33, 7.

ge-swincg, es; *n. Labour, toil;* lábor:—Léton ealles þeódscipes geswincg ðus leohtlíce forwurðan *they let the toil of all the nation thus lightly perish,* Chr. 1009; Erl. 142, 12. v. ge-swinc.

ge-swing, es; *n. Labour;* lábor:—Geswing is beforan me *lábor est ante me,* Ps. Spl. 72, 16: 89, 11. v. ge-swinc.

ge-swing, es; *n. A vibration;* vibrátio, fluctuátio:—Ofer ýða geswing *over the vibration of the waves,* Andr. Kmbl. 703; An. 352: Beo. Th. 1700; B. 848: Exon. 95 b; Th. 356, 7; Pa. 8.

ge-swingan; *p.* -swang, *pl* -swungon; *pp.* -swungen *To scourge, beat;* flagellare, verberare:—Hia geswingas iuih *flagellabunt vos,* Mt. Kmbl. Lind. 10, 17. God geswang Farao ðone cining mid ðám mǽstum wítum *flagellavit Dominus Pharaonem regem plagis maximis,* Gen. 12, 17: Jn. Skt. Lind. 19, 1. Ic wæs ealne ðæg geswungen *fui flagellatus tota die,* Ps. Th. 72, 11: Andr. Kmbl. 2791; An. 1398. Gie biþon geswinged *vapulabitis,* Mk. Skt. Lind. 13, 9. Gesuungun ł gesuinged biþ *flagellabitur,* Lk. Skt. Lind. 18, 32. Hia geþurscon ł geswungdon [MS. gesumgdon] *cædebant,* Mt. Kmbl Lind. 21, 8.

ge-swins. v. ge-swin.

ge-swip, es; *n. A scourge, whip;* flagellum, Som. v. swip.

ge-swip; *adj. Cunning, crafty;* astutus:—Geswippre múþe *ore astuto,* Bd. 2, 9; S. 511, 19.

ge-swiporlíce; *adv. Cunningly;* astute, V. Ps. 82, 3.

ge-swiporness, -swiforness, -swioporness, e; *f. Craft, cunning, art:* versutia:—Ðæs deófles geswipornysse syndon swíðe unasecgendlíce *the devil's arts are quite indescribable,* Shrn. 38, 35. Ðæs ealdan feóndes geswifornis *the old enemy's cunning,* 37, 14. Se ðe wiste geswipernise [-swiopornisse, Rush.] hiora *qui sciens versutiam eorum,* Mk. Skt. Lind. 12, 15.

ge-swiria, an; *m. A sister's son;* sororis filius, Cot. 35.

ge-swiru; *pl. n. Hills;* colles, Ps. Th. 64, 13. v. ge-sweoru.

ge-swíðan, -swýðan; *p.* de; *pp.* ed *To make strong, confirm, comfort:*—Mín earm hine mid mycle mægene geswýðeþ *brachium meum confortabit eum,* Ps. Th. 88, 18. He twelf apostolas mid his gástes gife geswíðde *he strengthened twelve apostles with the gift of his spirit,* Cd. 226; Th. 300, 29; Sat. 572. Hæfde he ðá geswíðed sóþum cræftum werodes aldor *he had then strengthened with true powers the chief of the band,* 143; Th. 179, 17; Exod. 30: 188; Th. 234, 7; Dan 288: Andr. Kmbl. 1394; An. 697: 1402; An. 701: Salm. Kmbl. 91; Sal. 45: Ps. Th. 118, 76: Exon. 13 a; Th. 24, 16; Cri. 385. Geswýðede, Ps. Th. 118, 77.

ge-swíðrian; *p.* ode, ade; *pp.* od, ad *To weaken, destroy;* imminuere, debilitare, conficere:—Mægen wæs geswiðrod *the might was destroyed,* Elen. Kmbl. 1393; El. 698: 1833; El. 918: 2526; El. 1264: Judth. 12; Thw. 25, 18; Jud. 266. Ne mót innan geondscínan sunne for ðǽm sweartum mistum ǽr ðǽm hí geswiðrad weoþen *the sun cannot shine through from within for the black mists before they are dissipated,* Bt. Met. Fox 5, 90; Met. 5, 45. Ðæt helle fýr wæs siððan geswiðrad *that hell-fire was afterwards mitigated,* Ors. 2, 6; Bos. 50, 20. v. ge-swedrian.

ge-swógen; *part. p. Senseless, inanimate, swooned:*—Se læg geswógen betwux ðám ofslegenum *he lay in a swoon amongst the slain,* Homl. Th. ii. 356, 27: Swt. A. S. Rdr. 66, 324. v. ge-swówung.

G g

ge-swógung, -swówung, e ; _f. Swooning,_ Lchdm. ii. 176, 13 : 194, 3.

ge-sworc, es ; _n. A cloud, mist;_ nĕbŭla :—Gesworc swá swá ahsan he tostredeþ _nĕbŭlam sīcut cĭnĕrem spargit,_ Ps. Spl. C. 147, 5. v. ge-sweorc.

ge-sworfen _rubbed off, polished off,_ Glos. Prud. Recd. 142, 19. v. ge-sweorfan.

ge-swúgian. v. ge-swígian.

ge-swungen _scourged, beaten,_ Andr. Kmbl. 2791 ; An. 1398 ; _pp. of_ ge-swingan.

ge-swurdod [sweord, swurd _a sword_] _armed with a sword;_ glädio cinctus :—Geswurdod _glädiātus,_ Ælfc. Gr. 43 ; Som. 45, 13.

ge-swustra, -swustru _sisters,_ Mk. Bos. 10, 29 : Homl. Th. ii. 458, 29. To mínre mēder and mínum geswustrum _to my mother and my sisters,_ Nar. 3, 8. v. ge-sweostor.

ge-swutelian, -swuteligan, -swytelian, -sweotulian, -sweotlian ; _p._ ode, ade, ude ; _pp._ od, ad, ud [sweotol _manifest, clear, open_] _To declare, publish, make known, explain, prove, manifest, show, glorify;_ monstrāre, demonstrāre, publĭcāre, exprĭmĕre, manĭfestāre, signāre, explānāre, prŏbāre, clārĭfĭcāre :—Ic wolde mid ðære gebícnunge geswutelian ðæt ic eom ðære stówe hyrde _I would manifest by that sign that I am the guardian of the place,_ Homl. Th. i. 504, I : L. C. E. 22 ; Th. i. 372, 26 : Ps. Spl. 79, 2 ; Jn. Bos. 14, 22. He wolde God geswutelian _clārĭfĭcātūrus esset Deum,_ 21, 19. Ic geswutelige _exprĭmo,_ Ælfc. Gr. 28, 4 ; Som. 31, 16 : Jn. Bos. 14, 21. He inc geswutelaþ mycele healle gedæfte _ipse vobis demonstrābit cænācŭlum grande strātum,_ Mk. Bos. 14, 15. Geswutelaþ _prŏbat,_ Glos. Prudent. Recd. 139, 25. He him lífes weig geswutelode _he manifested to them the way of life,_ Homl. Th. ii. 118, 16 : Boutr. Scrd. 20, 28 : 22, 2. Moses geswutelude ða ǽ _cǣpit Moyses explānāre lēgem,_ Deut. 1, 5. Geswutelie mid gewitnysse _let him show by witness,_ L. Eth. ii. 9 ; Th. i. 290, 10. Nis nán þing dígle, ðæt ne sý geswutelod _non est occultum, quod non manĭfestētur,_ Lk. Bos. 8, 17. Is geswutelod _signātum est,_ Ps. Th. 4, 7. Nú ys mannes sunu geswutelod, and God ys geswutelod on him _nunc clārĭfĭcātus est fīlius hŏmĭnis, et Deus clārĭfĭcātus est in eo,_ Jn. Bos. 13, 31, 32.

ge-swutelung, e ; _f. A making clear, plain, a manifestation, declaration_ :—Ðæt sum tácn wǽre on heora líchaman to geswutelunge ðæt hí on God belýfdon _that there might be some token on their body as a manifestation that they believed on God,_ Homl Th. i. 92, 32 : Cod. Dipl. Kmbl. ii. 300, 9. Hér is siú geswitelung ðære geræðnisse ðe ðius geférræden geræd hæfþ _here is the declaration of the ordinance that this society has decided upon,_ Th. Chart. 610, 27.

ge-swyrf, es ; _m. Filings;_ limatum :— Genim ánre yntsan gewihte geswyrfes of seolfre _take the weight of one ounce of the filings of silver,_ Herb. 101, 3 ; Lchdm. i. 216, 12.

ge-swyrfan _to file off, to polish;_ elimare, Cot. 71. v. ge-sweorfan.

ge-swyru; _pl. n. Hills;_ colles, Ps. Th. 71, 3 : 113, 4 : 148, 9. v. ge-sweoru.

ge-swystra _sisters_ :—Geswystrena bearn _sisters' children,_ Bt. 35, 4 ; Fox 162, 10. To mínre mēder and geswystrum _to my mother and my sisters,_ Nar. 1, 12. v. ge-sweostor.

ge-swytelian; _p._ ode ; _pp._ od _To make known, manifest, show;_ manĭfestāre :—Ðæt hí ðe sóþeste geswytelie _that he make manifest what is most true,_ L. Ath. iv. 7 ; Th. i. 226, 30. v. ge-swutelian.

ge-syb _peaceable, related,_ Soul Kmbl. 107 ; Seel. 54. v. ge-sib.

ge-sybsum; _adj. Peaceable;_ pācĭfĭcus :—Eádige synd ða gesybsuman _beati pācĭfĭci,_ Mt. Bos. 5, 9. v. ge-sibsum.

ge-sybsumlíce _peaceably,_ Ps. Spl. 34, 23. v. ge-sibsumlíce.

ge-sýcan, -sícan ; _p._ -sýhte _To give suck to, to suckle_ :—Ða breóst ðe swylce gesíhton _the breasts that gave such suck,_ Homl. Th. i. 84, 16.

ge-sýclian _to be infirm,_ Jn. Bos. 4, 46. v. ge-síclian.

ge-syd, es ; _n. A place in which to wallow, mud_ :—Sol _vel_ gesyd _volutabrum,_ Ælfc. Gl. 56 ; Som. 67, 32 ; Wrt. Voc. 37, 22.

ge-syflan _to provide 'sufol,'_ q.v. Salm. Kmbl. 807 ; Sal. 403.

ge-syfled hláf _panis lacticinio et ovorum luteo maceratus,_ Lye. v. ge-sufel.

ge-syfl-melu; _n. Dough_ :—Ðæt folc nam gesyflmelu [gesyft melu, Thw.] ǽr ðam hit gebyrmed wǽre _the people took their dough before it was leavened,_ Exod. 12, 34.

ge-syft. v. ge-syfl-melu.

ge-syhð. v. ge-siht.

ge-sylhð _a plough;_ aratrum, Som.

ge-syllan; _p._ -sealde ; _pp._ -seald _To give, deliver, betray, sell, give up;_ dare, donare :—Mycel feoh to gesyllanne _to give much money,_ Bd. 4, 19 ; S. 587, 29 ; Ps. Th. 110, 4 : 104, 10 : 117, 18. Gesyllon ðone oxan and todǽlon ðæt wurþ _let them sell the ox and divide the price,_ Ex. 21, 35. v. ge-sellan, sellan.

ge-sylt _salted;_ sale conditus :—Gyf ðæt sealt awyrþ, on ðam ðe hit gesylt biþ _if the salt be insipid, with what shall it be salted?_ Mt. Bos. 5, 13 : Mk. Bos. 9, 49 ; _pp. of_ ge-syltan. v. syltan.

ge-sýlð _happiness._ v. ge-sǽlð.

ge-sýman, -sēman, -sǽman ; _p._ de ; _pp._ ed _To load_ :—Se cyning gesýmde gold and seolfor uppan olfendas _the king loaded gold and silver_

---

upon camels, Homl. Th. i. 458, 23. Ða wǽron gesýmed mid feó and mid hrægle _that were laden with money and raiment,_ Gen. 45, 23. Ealle ðe gesýmede synt _omnes qui onerati estis,_ Mt. Bos. 11, 28. Ðeáh ðe we gesǽmde beón mid ðære berdene ðæs deádlíces líues _licet mortalis vitæ pondere pressi,_ Th. Chart. 317, 3.

ge-syndlíc; _adj. Prosperous, healthy, happy;_ prosperus :— On ðám gesyndlícan þingum . . . and on ðám wiðerweardum þingum _in prosperous . . . and in adverse circumstances,_ Bd. 4, 23 ; S. 595, 21.

ge-syndrian _to separate_ :—Gesyndrod sí hæ fram beodes dǽlnimunge _let him be separated from sharing in the table,_ R. Ben. interl. 24, Lye. On ðære gesyndredan híde _in the separate hide,_ Cod. Dipl. Kmbl. iii. 4, 8.

ge-sýne, -sēne, -sëne ; _adj. Visible, seen, evident, plain_ :—Ne mihte ic hire bedyrnan mínes módes unrótnesse for ðan hit wæs on mínum and-wlitan gesýne _I could not hide from her the disquiet of my mind for it was evident in my face,_ Shrn. 41, 25 : Ors. 1, 7 ; Bos. 30, 28 : Blickl. Homl. 93, 35. Ða fótlástas wǽron swutole and gesýne _the footsteps were plain and visible,_ 203, 36 : Andr. Kmbl. 1129 ; An. 565 : Beo. Th. 2811 ; B. 1403 : 4622 ; B. 2316 : Elen. Kmbl. 527 ; El. 264. Ðær wæs gesýne his seó sóþe spéd _videbitur in majestate sua,_ Ps. Th. 101, 14. Wæs gesýne ðæt . . . _it was evident that . . ,_ Blickl. Homl. 207, 11 : Beo. Th. 2515 ; B. 1255 : Andr. Kmbl. 1051 ; An. 526 : 1097 ; An. 549 : Elen. Kmbl. 487 ; El. 244. On me syndon ða dolg gesiéne _the wounds are visible on me,_ Rood Kmbl. 92 ; Kr. 46. Gesēne, Cd. 135 ; Th. 170, 1 ; Gen. 2806 : 218 ; Th. 278, 30 ; Sat. 230 : Chr. 1121 ; Erl. 248, 39. Ðæt hia gesēne síe _ut videantur,_ Mt. Kmbl. Lind. 6, 5, 16.

ge-sýnelíce; _adv. Visibly_ :—Ðæt tácen gesýnelíce bær _bore that token visibly,_ Bd. 3, 19 ; S. 549, 15.

ge-syngian, -singian ; _p._ ode, ade ; _pp._ od, ad _To sin, perpetrate crime, commit adultery;_ peccāre, mæchāri :—Ðæt ǽlc ðæra ðe wíf gesyhþ and hyre gewilnaþ, eallunga ðæt se gesyngaþ on hys heortan _quia omnis, qui vĭdĕrit mŭlĭĕrem ad concŭpiscendum eam, jam mæchātus eam in corde suo,_ Mt. Bos. 5, 28. Nú is gesēne ðæt we gesyngodon _now it is seen that we have sinned,_ Cd. 218 ; Th. 278, 31 ; Sat. 230. Ðæt wæs feohleás gefeoht, fyrenum gesyngad _that was a priceless fight, criminally perpetrated,_ Beo. Th. 4874 ; B. 2441.

ge-synlíce; _adv. More frequently;_ sæpius, R. Ben. 56.

ge-synto; _indecl. in sing ; gen. pl._ -synta, -synto ; _dat. pl._ -syntum ; _f. Health, welfare, safety, prosperity, success, advantage, profit, benefit;_ sānĭtas, sospĭtas, sălus, prospĕrĭtas :—Hí ðære gefeán ðære willendan gesynto onfóþ _cŭpĭtæ sospĭtātis gaudia redībunt,_ Bd. 4, 3 ; S. 570, 22. For heora gesynto _for their health,_ 3, 15 ; S. 541, 29. Ðe on eallum þingum máron gesynto hæfdon _qui măgis prospĕrantur in omnĭbus,_ 2, 13 ; S. 516, 8. Geunne me mínra gesynta _grant me my health,_ Judth. 10 ; Thw. 22, 34 ; Jud. 90 : Exon. 37 a ; Th. 122, 9 ; Gú. 303. Fela óðera gescreopa and gesynto _many other advantages and benefits,_ Bd. 4, 22 ; S. 592, 21. He hēt hine leóde swǽse sēcean on gesyntum _he bade him seek his own people in safety,_ Beo. Th. 3742 ; B. 1869 : Ps. Th. 114, 5. Him wǽre mín gesynto leófre ðonne hiora seolfra hǽlo _magis pro mea salute mori paratos,_ Nar. 30, 17 : 8, 6.

ge-syrwan; _p._ -syrede ; _pp._ -syrwed.　　I. _to arm_ [v. searu _armour_] :—Eode ða gesyrwed secg to ðam eorle _then went an armed man to the earl,_ Byrht. Th. 136, 30 ; By. 159.　[_Or_ gesyrwed _wily,_ searu _a wile;_ cf. gelýfed _having belief._]　　II. _to plot, machinate_ [searu _artifice_] :—Wom-dǽda ðe [MS. ðy] ic gesyrede _the ill-deeds that I have devised cunningly,_ Exon. 72 b ; Th. 270, 20 ; Jul. 468.

get, geta. v. git, gita.

gēt _she-goats,_ Som. 126 ; Lchdm. iii. 206, 2 ; _acc. pl. of_ gát.

get _a gate._ v. geat.

gēt _poured out_ :—He gēt ðæt blód uppan ðæt weofod _fudit sanguinem super altare,_ Lev. 8, 24 ; _p. of_ geótan.

ge-tácnian; _p._ ode, ade, ude ; _pp._ od, ad, ud [tácen, tácn _a sign, token_].　　I. _to denote by a sign, signify, betoken, show, instruct;_ signāre, signĭfĭcāre, dēnŏtāre, insĭnuāre, monstrāre, instrŭĕre :—Ic getácnige _signĭfĭco,_ Ælfc. Gr. 37 ; Som. 39, 36. Wæter getácnaþ on ðyssere stówe mennisc ingehýd _water in this place betokens human knowledge,_ Homl. Th. ii. 280, 1 : Boutr. Scrd. 21, 42 : Lchdm. iii. 198, 6, 7. Ða alecgendlícan word getácniaþ dǽde _the deponent verbs signify action,_ Ælfc. Gr. 19 ; Som. 22, 56. Eua getácnode Godes geláðunge _Eve betokened God's church,_ Ælfc. T. 6, 11, 13 : 7, 1. Adam getácnude úrne Hǽlend Crist _Adam betokened our Saviour Christ,_ 6, 8. Ðú me sóþfæstnysse weg getácna _viam justĭfĭcātiōnum tuārum insĭnua mihi,_ Ps. Th. 118, 27. Him gedafenaþ ðæt hí cunnon hwæt heó gástlíce getácnige _it is fitting that they know what it betokens spiritually,_ Homl. Th. ii. 264, 27. Mid ðý getácnod, ðæt . . . _by that is signified that . . . ,_ Bt. Met. Fox 31, 35 ; Met. 31, 18 : Boutr. Scrd. 19, 27, 28.　　II. _to sign, mark, witness, seal;_ signāre, insignīre, obsignāre :—He getácnaþ ðæt God is seolfa _signavit quia Deus vēra est,_ Jn. Bos. 3, 33. Ðone God Fæder getácnode _hunc Păter signavit Deus,_ 6, 27. Is eall heáhmægen tíre getácnod _all the lofty power is marked with glory,_ Elen. Kmbl. 1504 ; El. 754. Godes þeówas getácnude beón sceoldan _clērĭcos insignīri deceret,_ Bd. 5, 21 ; S. 642, 42.

ge-tácniendlíc, -tácnigendlíc; *adj. Bearing a sign, significative, typical*; significātīvus :—Ðæt getácniendlíce [getácnigendlíce, Homl. Th. ii. 278, 14] lamb wæs geoffrod æt heora Eáster-tíde *the typical lamb was offered at their Easter-tide,* Homl. Pasc. Lisle 11, 18.

ge-tácnung, e; *f.* [tácnung *a sign*] *A sign, signification, token, type*; significātio :—Ðæt unscæððige lamb hæfde getácnunge Cristes þrówunge *the innocent lamb was a token of Christ's passion,* Homl. Th. ii. 264, 29: 266, 1 : 276, 4 : 278, 7 : Jud. 16; Thw. 161, 6. Sume þing sind gecwedene be Criste þurh getácnunge *some things are said of Christ typically,* Homl. Th. ii. 268, 13, 16.

ge-tǽcan, -tǽcean, -tǽcan; *p.* -tǽhte; *pp.* -tǽht [tǽcan *to teach*] *To teach, instruct, show, declare, assign*; dōcēre, instruēre, ostendere, assignāre, offerre :—Ic hit ðé wille getǽcan *I will teach it thee,* Bt. 34, 9; Fox 146, 13 : 36, 1; Fox 172, 28. He cwæþ ðæt he mihte óðerne getǽcan [getǽcnan, MS. T.] *ostendēre posse se dixit alium,* Bd. 4, 1; S. 564, 2. Getǽcean, Ps. Th. 105, 25. Ðe ic ðé getǽce *which I will show thee,* Cd. 137; Th. 173, 1; Gen. 2854. Ðû me róde ródera cining ryhte getǽhtest *thou hast rightly shown me the cross of heaven's king,* Elen. Kmbl. 2148; El. 1075. Ðæt hie us fersc wæter and swéte getǽhton *ut dulcem aquam demonstrarent,* Nar. 10, 20: Guthl. 3; Gdwn. 20, 24. Him Dryhten hlyt getǽhte *God assigned to them a lot,* Andr. Kmbl. 12; An. 6: Beo. Th. 4031; B. 2013: Cd. 136; Th. 171, 32; Gen. 2837. We ðé wíc getǽhton *we assigned to thee a dwelling-place,* 127; Th. 162, 27; Gen. 2687. Weg rihtwísnyssa ðínra getǽc me *viam justificātiōnum tuārum instrue me,* Ps. Spl. 118, 27. Ðæt ðû me getǽhte *that thou teach me,* Andr. Kmbl. 969; An. 485. Ðæt he riht getǽhte *that he should declare the truth,* Elen. Kmbl. 1199; El. 601.

ge-tæl, -tel, -teal, es; *pl. nom. acc.* -talu; *n.* **I.** *a number, series, reckoning, computation*; numerus, series, computus, computatio :—Ðæra etendra getæl wæs fíf þúsenda wera *manducantium fuit numerus quinque millia virorum,* Mt. Bos. 14, 21. Seó Abbudisse hét hine [Cædmon] lǽran ðæt getæl ðæs hálgan stæres and spelles *the Abbess commanded [them] to teach him [Cædmon] the series of the holy story and narrative;* Abbatissa jussit illum [Cædmonem] seriem sacræ histōriæ docēri, Bd. 4, 24; S. 598, 5 : Homl. Th. ii. 222, 3. Getel is *numerus,* Ælfc. Gr. 13; Som. 15, 56: Num. Pref. Ágene naman habbaþ ánfeald getel, and nabbaþ mænigfeald ; eác sunne and móna syndon ánfealdes geteles *proper names have a singular number and have not a plural ; the sun and moon are also of the singular number,* 13; Som. 16, 1. Sume naman synd óðres cynnes on ánfealdum getele, and óðres cynnes on mænigfealdum getele *some nouns are of one gender in the singular number, and of another gender in the plural number,* 13; Som. 16, 25, 26. On fulfremedra hálgena geteal *in the number of perfect saints,* Nat. S. Greg. Els. 9, 2. God geíce fela þúsenda to ðison getale *Deus addat ad tuum numerum multa millia,* Deut. 1, 11. Twelf pund be getale *twelve pounds by tale,* Chart. Th. 577, 19. **II.** *a company, race, tribe*; centuria, tribus :—Getalu *vel* heápas *vel* hundredu *centurias,* Ælfc. Gl. 96; Som. 76, 25; Wrt. Voc. 53, 34. All getalu oððe cynn *omnes tribus,* Mt. Kmbl. Rush. 24, 30. Hie gemitton getalum myclum *they met in many tribes,* Cd. 80; Th. 101, 27; Gen. 1688. **III.** *a book of reckoning, a register, catalogue*; laterculum, catalogus = κατάλογος :—Getæl *laterculum,* Cot. 119 : *catalogus,* 31, 37, 104. DER. bold-getæl, -getel, folc-, rím-, ripc-, tigol-, winter-.

ge-tǽlan, -télan; *p.* ede; *pp.* ed *To accuse, reprove*; accusare, exprobrare, calumniari, reprehendere :—Ne meaht ðû nó getǽlan ðíne wyrd *thou canst not accuse thy fortune,* Bt. 10; Fox 28, 1. Ic mǽge getǽlan *I may reprove,* 32, 3; Fox 118, 27. Word his getǽla *verbum ejus repræhendere,* Lk. Skt. Lind. 20, 26. Óðerne getǽleþ *alterum contemnet,* Mt. Kmbl. Lind. 6, 24. Nǽfre getǽldon gé ða unsuinnigo *numquam condemnassetis innocentes,* 12, 7. Ðætte hé getǽldon him *ut accusarent eum,* 10: Mk. Skt. Lind. 3, 2. Ða ðé getǽled aron *quæ tibi objiciuntur,* 14, 60. DER. tǽlan.

getǽl-fæst; *adj. Measurable*; mensūrābĭlis :—Efne gemetelíce oððe getǽlfæste oððe ametendlíce ðú asettest dagas míne *ecce mensūrābĭles pŏsuĭsti dies meos,* Ps. Lamb. 38, 6.

getǽl-ríme, es; *n.* [getæl *a number*] *Succession*; successio :—On getælríme *in succession,* Salm. Kmbl. 76; Sal. 38.

ge-tǽnge; *adj. Incident*; incĭdens :—Gif hwylcum men sý ðæs feórþan dæges fefer getǽnge *if to any man there be a quartan fever incident,* Herb. 2, 12; Lchdm. i. 84, 5, MS. B. v. ge-tenge.

ge-tǽsan; *p.* de; *p.* [tǽsan *to tease*] *To pluck, tease*; carpēre :—Nim wǽte wulle wel getǽsede *take wet wool well teased,* Herb. 178, 6; Lchdm. i. 312, 13.

ge-tǽse, es; *n. An advantage*; commodum :—Ac geþenc ðæt ðú hym forwyrndest ælcra getǽsa ðá git becgen on líchaman wǽron and ðú hæfdest ælc good and he hefde ælc yfel ne mót he ðé nú ðý máre dón to getǽsan ðe ðú ðá hym woldest *but remember that thou didst refuse him every advantage when ye were both in the body and thou hadst every good and he had every evil; he cannot now do more for thy advantage than thou wouldest then do for his,* Shrn. 202, 31–4. Hió an Æþe flede ealra ðera getǽsa ðet ðǽr binnan beóþ *she gives to Æthelfled all the desirable things that are there within,* Th. Chart. 538, 37. Getǽse *commodum,* Cot. 59, Lye.

ge-tǽse; *adj. Meet, convenient, suitable, mild, easy*; accommodus, placidus, lenis :—Gif him wǽre niht getǽse *if he had had an easy night,* Beo. Th. 2645 ; B. 1320. Swá hit getǽsost wæs *as was most fitting,* Bt. Met. Fox 20, 22; Met. 20, 11. [O. H. Ger. ki-zeso *dextrum.* v. Grff. v. 708–10.]

ge-tǽsnes, se; *f. An opportunity, a saving, placing*; commoditas, Cot. 55.

ge-tal; *adj. Quick, ready, active*; agilis, velox, expeditus :—Wǽron hyra tungan getale teónan gehwylcre and to yfele gehwám ungemet scearpe *their tongues were swift to every wrong and to every evil exceeding sharp*; lingua eorum machæra acuta, Ps. Th. 56, 5. [O. H. Ger. ge-zal *agilis, rapidus, alacer.*]

ge-talian; *p.* ode, ade; *pp.* od, ad, ed *To tell, number, reckon, consider* :—Getalede *reputans,* Lk. Skt. Lind. 11, 38. Hēras heáfdes alle getalad aron *capilli capitis omnes numerati sunt,* Mt. Kmbl. Lind. 10, 30. Ueras getaled suelce fífo þúsendo *viri numero quasi quinque milia,* Jn. Skt. Lind. 6, 10. Miþ unrehtwísum getaled wæs *cum iniquis reputatus est,* Mk. Skt. Lind. 15, 28. v. ge-tellan.

getal-scipe, es; *m. Number*; numerositas :—Getalscipes and tídes *numerositatis et temporis,* Mt. Kmbl. p. 12, 14.

ge-talu *tribes*; tribus, Mt. Kmbl. Rush. 24, 30; *pl. nom.* of ge-tæl, **II.**

getan; *p.* de, te; *pp.* ed *To* GET, *take, obtain*; adipisci, capere, assequi :—Cwæþ he on mergenne mēces ecgum getan wolde *said he in the morning would take them with the edges of the sword,* Beo. Th. 5872; B. 2940. DER. a-getan. v.-gitan.

gétan; *p.* de, te; *pp.* ed *To grant, to confirm, assent to* :—Geáfon and gétton *gave and granted,* Chr. 675; Th. 59, 20. Gétton hit ælle ða óþre *all the others assented to it,* 656; Th. 53, 27. v. geátan.

ge-tang *lying, prostrate*; prostratus, C. R. Ben. 34, Lye.

ge-tanned; *part. Tanned*; cortice mācĕrātus :—Getannede hýd *subacta cŏria, vel mēdĭcāta, vel confecta,* Ælfc. Gl. 17; Som. 58, 104; Wrt. Voc. 22, 19.

ge-targed; *part. Provided with a shield*; scutatus, Hpt. Gl. 459. v. targe.

ge-tawa; *pl. f. Instruments*; instrumenta :—Mannes getawa *instrumenta genitalia,* L. M. 1, 29; Lchdm. ii. 70, 7. Ðis syndon ða getawa *these are the instruments,* L. E. I. 2; Th. ii. 404, 3. [O. H. Ger. gizawa *suppellex, stipendium.*] DER. gúþ-getawe, wíg-. v. taw, e; *f.*

ge-tawian; *p.* ode, ade; *pp.* od, ad [tawian *to prepare*] *To prepare, reduce or bring to ; parāre, redūcĕre ad* :—Getawian to yrmþe *redūcĕre ad mĭsĕrĭam,* Nathan. 7; St. And. 34, 18. Hý se æðeling to ðam bismre getawade *the prince brought them to shame,* Ors. 3, 8; Bos. 63, 15. To bysmere beóþ itawode ðæs earman lond-leódæ *to shame are brought this miserable people,* Th. An. 121, 9. v. tawian.

ge-teág, -teáh *drew, led, gave,* Cd. 162; Th. 203, 22; Exod. 407: Bd. 5, 18; S. 636, 4 : Beo. Th. 2093; B. 1044. v. ge-teón.

ge-teágan, -tégan; *p.* -téde; *pp.* -teád *To make, prepare* :—Ðæt land mid to teágenne. Ðá ðæt land ðá geteád wæs *to prepare the land with. When then the land was prepared*; preparata terra, Bd. 4, 28; S. 605, 33. Ðone ilcan mete ðe he hí ǽror mid tame getéde *the same food with which before he had made them tame [the prose has ða ilcan mettas ðe hí ǽr tame mid gewenedon,* Fox 88, 18], Bt. Met. Fox 13, 87; Met. 13, 44. [Cf. ge-tawian.]

ge-teal -teall *a number,* Nat. S. Greg. Els. 9, 2 : Chr. 1014; Erl. 151, 16. v. ge-tæl, **I.**

ge-teald, es; *n. A tent, tabernacle*; tabernāculum :—God afærþ ðé of getealde ðínum *Deus emigrābit te de tabernāculo tuo,* Ps. Spl. 51, 5. v. ge-teald.

ge-teáma, -týma, an; *m. An advocate, avoucher, a warranter*; advŏcātus, qui rei emptæ fĭdem præstat :—Ic wille ðæt gehwilc man hæbbe his geteáman *I will that every man have his warranter,* L. Ed. 1; Th. i. 158, 9 : L. Eth. ii. 8; Th. i. 288, 16. v. teám, ge-téman.

ge-tǽcan *to show* :—Is þearf ðæt ic ðé hí selfe getǽce *it is necessary that I show thee itself,* Bt. 33, 1; Fox 120, 1. v. ge-tǽcan.

ge-téde. v. ge-teágan.

ge-téh *drew,* Nicod. 30; Thw. 17, 31. v. ge-teón.

ge-tehhod *determined, decreed,* Bt. 7, 3; Fox 20, 30 = ge-teohhod; *pp.* of ge-teohhian.

ge-tel *a number*; numerus :—Gemænigfylde hí synt ofer getele *multĭplicati sunt super numerum,* Ps. Lamb. 39, 6 : Ælfc. Gr. 13; Som. 15, 56. v. ge-tæl, **I.**

ge-télan. v. ge-tǽlen.

getel-cræft, es; *m. Arithmetic,* Hpt. Gl. 479.

ge-teld, -tæld, -teald, es; *n.* [teld *a tent*] *A tent, tabernacle, pavilion,* TILT, *cover*; tentōrium, tabernāculum :—Geteld *tentōrium vel tabernācŭlum,* Wrt. Voc. 85, 84 : *scēna vel tabernācŭlum,* Ælfc. Gl. 56; Som. 67, 25; Wrt. Voc. 37, 15. God æteówde Abrahame on ðam dene Mambre, ðǽr ðǽr he sæt on his geteldes ingange *appāruit Abraham in convalle Mambre, sĕdenti in ostio tabernācŭli sui,* Gen. 18, 1 : Ps. Spl. 26, 9. Hwylc eardaþ on getelde ðínum *quis habĭtābit in tabernācŭlo tuo?* 14, 1. Hí aslógan án geteld *tĕtendērunt tentōrium,* Bd. 3, 17; S. 543, 34. On

sunnan gesette getelda his *in sōle pŏsuit tabernācŭlum suum*, Ps. Spl. 18, 5.

ge-teldung, e; *f. A tent, tabernacle*; tabernācŭlum:—On sunnan gesette geteldunge his *in sōle pŏsuit tabernācŭlum suum*, Ps. Spl. T. 18, 5 : 26, 9.

geteld-wurþung, e; *f. A celebration of tents, the feast of tabernacles*; scēnŏpēgia = σκηνοπηγία:—Getimbra hālgung *vel* geteldwurþung *scēnŏpēgia*, Ælfc. Gl. 3; Som. 55, 77; Wrt. Voc. 16, 50.

ge-telged *coloured, dyed*; coloratus, Cot. 49, 81. v. tælg.

ge-tellan, ic -telle, ðū -telest, he -teleþ, *pl.* -tellaþ; *p.* -tealde, *pl.* -tealdon; *pp.* -teald, -teled *To tell, number, reckon, esteem, consider, ascribe, assign*; numerare, computare, reputare, comparare, dinumerare:—Ruben and Simeon beóþ mid me getealde *Ruben et Simeon reputabuntur mihi*, Gen. 48, 5: Ps. Spl. C. 43, 25: Ps. Th. 118, 119. Hit getealdon ealde ægleáwe *elders skilled in laws reckoned it*, Menol. Fox 34; Men. 17: Cd. 154; Th. 191, 33; Exod. 224. Hwylc can getellan *quis novit dinumerare*, Ps. Spl. 89, 13. Ðā getealdon hie ðæt ðǽr wæs eác syx hund manna acweald *then they reckoned that there were six hundred men slain*, Blickl. Homl. 203, 27. Seó bóc ðe ys genemned on Englisc getel for ðam ðe Israhēla bearn wǽron on ðǽre getealde *the book that is called in English Numbers because in it the children of Israel were numbered*, Num. Pref : Ps. Th. 89, 11: Andr. Kmbl. 1765; An. 885: Mt. Bos. 10, 30. Ðæt is geteald ðæs lǽssan mílgetæles ðe stadia hátte ccc and þreó twentig *it is, reckoned by the smaller measure of distance that are called stadia, three hundred and twenty-three*, Nar. 36, 16 : 34, 27. Se biþ geteald Godes feónd *he will be accounted God's enemy*, Homl. Th. i. 162, 22. Án eórod is geteald to six þúsendum *a legion is reckoned at six thousand*, ii. 378, 29 : i. 68, 35. Ðæt Mæcedonisce gewinn ðæt mon mæg to ðám mǽstan gewinnum getellan *the Macedonian war which may be reckoned amongst the greatest wars*, Ors. 4, 11; Bos. 98, 18. Ðonne biþ he geteald to ðǽre fȳrenan eá *then shall he be assigned to the fiery river*, Blickl. Homl. 43, 24. Ðæt hí hiora ágnum godum getealde wǽron *that they might be ascribed to their own gods*, Ors. 1, 5; Bos. 28, 27. Ðā ðis Constantine geteald wæs *when this was told to Constantine*, H. R. 5, 27. Geteled rímes *reckoned by number*, Cd. 67; Th. 80, 30; Gen. 1336: 107; Th. 141, 14; Gen. 2344: Elen. Kmbl. 4; El. 2. Geteled ríme, Cd. 64; Th. 76, 27; Gen. 1263: 161; Th. 201, 15; Exod. 372: Andr. Kmbl. 2070; An. 1037. Tyn hund geteled *ten hundred in number*, Cd. 154; Th. 192, 15; Exod. 232: Andr. Kmbl. 1329; An. 665: Ps. Th: 90, 7. v. ge-talian.

ge-tēman, -tȳman; *p.* de; *pp.* ed *To vouch to warranty*; vocare ad warrantum. "Vouching to warranty. A process by which a person, in whose possession lost or stolen property was found, was compelled to show from whom he bought or had it, which latter was, in like manner, obliged to declare how it came into his hands, and so on to a third holder, beyond whom, provided he could prove lawful possession, the tracing might not proceed. The person from whom the accused party had the property, and who came forth as his warranter, was called the 'getȳma' or 'geteáma,' and the process itself 'teám,'" LL. Th. Glos. v. L. H. E. 7; Th. i. 30, 8: L. In. 35; Th. i. 124, 10.

ge-temesed, -temsud; *part. Sifted*; cribratus :— Hláfas getemeseda *panes propositionis*, Mt. Kmbl. Lind. 12, 4: Lk. Skt. Lind. 6. 4. Nim getemsud melu *take sifted meal*, Lchdm. iii. 134, 20.

ge-temian; *p.* ede; *pp.* ed *To tame*; domare :— Ic gewylde oððe temige [getemige, MS. C.] *domo*, Ælfc. Gr. 36; Som. 38, 19. Ða getemedon *domitos*, Th. An. 26, 7, 13. Se getemeda assa hæfde getácnunge ðæs Iudēiscan folces, ðe wæs getemed under ðǽre ealdan æ *the tamed ass betokened the Jewish people that was tamed under the old law*, Homl. Th. i. 208, 20. v. temian.

ge-temprian; *p.* ode; *pp.* od *To temper, moderate, govern, cure*; temperare :—Seó sunne ðā eorþan getempraþ *the sun tempers the earth*, Bd. de nat. rerum; Wrt. popl. Scienc. 9, 3; Lchdm. iii. 250, 14. Getemprie seó bilewitnys ðæt fȳr ðæt hit to rēðe ne sȳ *let the meekness temper the fire that it be not too fierce*, Homl. Th. ii. 46, 8. Án is ðæt gehwá hine sylfne getemprige mid gemete on ǽte and on wǽte *one is, that every one govern himself with moderation in food and drink*, i. 360, 12. Mót se ðe wile mid sóþum lǽcecræfte his líchaman getemprian *he who will may cure his body with true leechcraft*, 474, 35.

ge-temsud *sifted*. v. ge-temesed.

ge-tengan; *p.* de; *pp.* ed [tengan *to hasten, rush upon*] *To hasten, join, devote one's self to*; injungēre, dēdēre :—Hine sylfne getengde in Godes þeówdōm *he devoted himself to God's service*, Elen. Kmbl. 400; El. 200. Ðā getengde se Aristodemus to ðam heáhgerēfa *then Aristodemus hastened to the prefect*, Homl. Th. i. 72, 18. He sóna getengde wiþ ðæs drȳs *he at once hastened towards the magician*, 374, 4. Se þegn ðā ðǽr to geteingde *the servant then hastened thither*, Shrn. 14, 27.

ge-tenge; *adj. Near to, close to, pressing upon, oppressing*; propinquus, incumbens, gravis, molestus :—Geseah gold glitnian grunde getenge *he saw gold glitter lying on the ground*, Beo. Th. 5510; B. 2758: Elen. Kmbl. 2226; El. 1114: 456; El. 228: Bt. Met. Fox 31, 14; Met. 31, 7. Cyningas on heáhsetlum hrófe getenge *kings high-raised* [lit. *close to the roof*] *on thrones*, 25, 10; Met. 25, 5: Cd. 38; Th. 50, 14; Gen.

808 : Runic pm. Kmbl. 343, 2; Rūn. 18. Hundas deórum getenge *dogs pressing upon the animals*, Homl. Th. ii. 514, 25 : Shrn. 37, 14. Swā fela gásta wǽron getenge ðam ānum men *so many spirits were oppressing that one man*, 378, 30. Heora þurst ðe him getenge wæs *their thirst that was oppressive to them*, Ors. 5, 8; Bos. 107, 28 : 6, 4; Bos. 119, 4 : Nar. 8, 24: Bt. 5, 1; Fox 10, 24 : 10: Fox 30, 5. Bróhþreá Cananēa wearþ cynne getenge hunger se hearda *terrible calamity came upon the race of the Canaanites the hard famine*, Cd. 86; Th. 108, 31; Gen. 1814: 149; Th. 187, 9; Exod.-148 : 206; Th. 255, 25; Dan. 629 : 229; Th. 309, 18; Sat. 711. [Cf. *O. Sax.* bi-tengi : and v. *O. H. Ger.* gi-zengi, Grff. v. 680.] v. ge-tengan; ge-tingan.

getenys, gytenes, se; *f. A procuring, attaining*, GETTING, *instruction, education*; adeptio, institutio, Lye. Getenis *historia*, Hpt. Gl. 459.

ge-teód *determined, decreed*, Bd. 3, 24; S. 556, 12; *pp. of* ge-teón.

ge-teóde *formed, decreed, assigned*, Cd. 182; Th. 228, 19; Dan. 204: Exon. 88 b; Th. 333, 17; Gn. Ex. 5; *p. of* ge-teón.

ge-teóh; *gen.* -teóges; *n. Matter, material; pl. instruments, implements, utensils* :—Se ðis leóht onwráh and ðæt torhte geteóh tillíce onwráh *who this light displayed and the bright matter* [the universe] *revealed*, Exon. 94 a; Th. 352, 32; Reim. 2. Sulh-geteógo *ploughing implements*, Th. An. 118, 12. [*O. H. Ger.* ge-ziug *materia, suppellex, instrumentum* : *Ger.* ge-zeug.]

ge-teohhian, -teohian, -tiohhian, -tihhian; *p.* ode, ade; *pp.* od, ad *To appoint, determine, decree, assign*; stātuēre, decernēre, assignāre :—Ðā heó Gode ānum geteohode þeówian *cum Deo sōli servire decrēvisset*, Bd. 4, 23; S. 593, 7. Wæs óðer in geteohhod mǽrum Geáte *another dwelling had been assigned to the renowned Goth*, Beo. Th. 2605; B. 1300. Geteohod, Bd. 5, 14; S. 634, 31, note. Ðē sind heardlícu wítu geteohhad *stern torments are determined for thee*, Exon. 69 b; Th. 258, 13; Jul. 264: Blickl. Homl. 25, 25. Ðe his sylfes sáwle hafaþ deáþe geteohhad *who hath assigned his own soul to death*, 183, 33. Eall ðæt yfel, ðæt hí him geteohod hæfdon *all the evil that they had determined against him*, Ps. Th. 9, argument : 14 : 16, 13. Ðæt hí toweorpen ðæt God geteohhad hæfþ to wyrcanne *to destroy what God had determined to do*, 10, 3.

ge-teolod; *part. Gained*; lucrīfactus :—Ðonne sceal gehwá him æteówian hwæt he mid ðam punde geteolod hæfþ *then shall every one show to him what he has gained with the pound*, Homl. Th. ii. 558, 10. v. ge-tilian.

ge-teón, ic -teó, ðū -tȳhst, he -tȳhþ, *pl.* -teóþ; *p.* -teáh, -teág, -tēh, *pl.* -tugon; *pp.* -togen. **I.** *to draw, lead, incite, excite, constrain, restrain, bring up, instruct, bring to an end, complete, draw or bind together, string a musical instrument*; trahere, ducere, perducere, stringere, evaginare, excitare, constringere, educare, instituere, ad finem perducere, complere, nervis aptare *vel* instruere :—Woldon hine geteón in orwēnnysse *would draw him into despair*, Exon. 41 a; Th. 136, 24; Gū. 546. Ðás wíf wuna getēþ *has mulieres consuetudo constringit*, Bd. 1, 7; S. 494, 11. Ðū getīhst his heáhnisse *consummabis summitatem ejus*, Gen. 6, 16. He Adam fram helle getēh *he drew Adam from hell*, Nicod. 30; Thw. 17, 31. He monige to rihtre weorþunge ðǽre Drihtenlícan Eástrana geteáh and gelǽdde *multos ad Catholicam Dominici Paschæ celebrationem perduxit*, Bd. 5, 18; S. 636, 4. Ðā hí hæfdon getogen eall Creáca folc to ðǽm gewinnum *when they had drawn all the people of Greece to the wars*, Ors. 1, 14; Bos. 37, 14, 35. He geteág ealde láfe *he drew an ancient relic* [i. e. *a sword*], Cd. 162; Th. 203, 22; Exod. 407. Getogene ðȳ wǽpne *evaginata sica*, Bd. 2, 9; S. 511, 21. Folc to mánum getogen *excitatum ad scelera vulgus*, 2, 5; S. 507, 42. Hæfþ ealle gesceafta getogen *he has restrained all creatures*, Bt. Met. Fox 11, 48; Met. 11, 24. Ða ðe wǽron on rím-cræfte rihte getogene *those who were rightly instructed in the art of numbers*, Chr. 975; Th. 226, 31; Edg. 27. Swá getogen mann *a man so well instructed*, Homl. Th. ii. 122, 13; Th. Ap. 17, 18. Þeós fyrd wæs getogen ðȳ feorþan geáre his ríces *hoc bellum quarto imperii sui anno complevit*, Bd. 1, 3; S. 475, 15. Wæs heó mid micle sáre getogen *illa erat multo dolore constricta*, 5, 3; S. 616, 22. Wamb getogen *alvus constricta*, Med. ex Quadr. 6, 11; Lchdm. i. 352, 24. Mid tyn strengum getogen hearpe *a harp strung with ten strings*, Ps. Th. 143, 10. Ða organa wǽron getogene *the organs were played*, Th. Ap. 25, 15. **II.** *to bring as an offering or gift, contribute, bestow, give*; conferre :—Onweald geteáh wicga and wǽpna *gave possession of war-horses and weapons*, Beo. Th. 2093; B. 1044: 4337; B. 2165. Nó ðū him wearne geteóh *do not give them a denial*, 738; B. 366.

ge-teón, -tión; *p.* -teóde; *pp.* -teód *To make, form, frame, appoint, determine, decree, ordain, assign*; făcēre, stătuēre, constĭtuēre, decernēre :—Ðe him to gode geteóde *which he had formed to himself for a god*, Cd. 182; Th. 228, 19; Dan. 204. He us æt frymþe geteóde líf *he assigned life to us at the beginning*, Exon. 88 b; Th. 333, 17; Gn. Ex. 5 : 90 a; Th. 337, 28; Gn. Ex. 71 : Andr. Kmbl. 28; An. 14. He hine gegyrede mid grame wyrgþu, swā he hine wǽdum wrǽstum geteóde *induit se mălĕdictĭōne sicut vestimento*, Ps. Th. 108, 18. Hū woruld wǽre wundrum geteód *how the world was wondrously framed*, Cd. 177; Th. 222, 28; Dan. 111. Se ðe geteód hæfde *qui decrēvĕrat*, Bd. 3, 24;

S. 556, 12: Blickl. Homl. 19, 35. Geteód to ðæm ecan wîtum *destined to eternal torments*, 37, 4 : 31, 22. Ðonne biþ ðam heard dóm geteód *a hard sentence will be the lot of that man*, 95, 36. Þurh hwelces monnes hond mîn ende wære getiód *by what man's hand my death was determined*; cujus mortem percussoris manu cavendam habeam, Nar. 31, 19 : Th. Ch. 483, 15.

ge-teorian, -teorigan, -teorigean ; *p.* ode, ude ; *pp.* od, ud, ad ; *v. intrans.* *To fail, faint, be weary, languish, cease, perish*; deficere, fatigari, languere, exterminari :—Geteoriaþ *deficiant*, Ps. Th. 67, 2 : 103, 27 : Ps. Spl. 17, 39. Ic geteorode *ego defeci*, Ps. Spl. 38, 14 : 54, 11. Ðâ se mete geteorude ðe hig of Egipta lande brohton *when the food was consumed that they brought from Egypt*, Gen. 43, 2 : 47, 15 : Jn. Bos. 2, 3. Hî geteorodon *defecerunt*, Ps. Spl. 72, 19. Ûre dagas ealle geteorodun *omnes dies nostri defecerunt*, Ps. Th. 89, 9. Ðê læs hig on wege geteorian *ne deficiant in via*, Mt. Bos. 15, 32. Me is heorte geteorad *defecit cor meum*, Ps. Th. 72, 21. He sent on eów geteorigende eágan and môdes gnornunge *he shall send on you failing eyes and sorrow of mind*, Deut. 28, 65. Geteorigende ateoraþ *deficientes deficient*, Ps. Spl. 36, 21. Be wege hî geteorigeaþ *deficient in via*, Mk. Bos. 8, 3. For swiðlícre hætan geteorud *wearied by the excessive heat*, Herb. 114, 1 ; Lchdm. i. 226, 23. Beóþ geteorode *exterminabuntur*, Ps. Spl. 36, 9. Sume sceufon sume tugon and swiðe swætton oð ðæt hig geteorode wæron *some shoved, some pulled and sweated exceedingly until they were exhausted*, Shrn. 154, 27.

ge-teorung, e ; *f. A failing, fainting, languishing, weariness*; defectio, languor, fatigatio :—Geteorung nam me for synnullum *defectio tenuit me præ peccatoribus*, Ps. Spl. 118, 53 : 141, 3.

ge-teóðian; *p.* ode ; *pp.* od *To tithe, give a tenth part* :—Ic ealle ða landâre ðe ic on Angla þeóde hæfde Gode into hâlgan stôwon geteóðode *I gave a tenth part of all my landed property to God for holy places*, Chart. Th. 116, 27. v. teóðian.

ge-ter, es ; *n. A tearing*; dilaceratio, Hpt. Gl. 499.

ge-teran *to tear* :—Getearende *discerpens*, Mk. Skt. Lind. 9, 26.

ge-tése. v. ge-tǽse.

ge-tete *pomp, show, ostentation, magnificence.* v. ge-tot.

ge-téung. v. ofer-bæc-getéung.

ge-þaca, an ; *m. A thatcher, coverer* ; tector :—Sceal ðis sâwel-hûs fæge flæschoma leomu lâmes geþacan wunian wælreste *this soul-house, the doomed flesh-covering, the limbs, coverers of the earth* [lying upon the earth], *must inhabit the mortal resting-place*, Exon. 47 b ; Th. 164, 1 ; Gû. 1005.

ge-þæf; *adj.* [geþafian *to agree, consent*] *Agreeing, content*; consentiens :—He his nó geþæf wæs *he was not a consenting party to it*, Cod. Dipl. 183 ; A. D. 803 ; Kmbl. i. 222, 35 : R. Ben. 7.

ge-þæht, e ; *f.*: es ; *n. Counsel, consultation*; consilium :—Ðæt he wolde mid his freóndum sprǽce and geþæht habban *that he would have a counsel and consultation with his friends*, Bd. 2, 13 ; S. 515, 37. Giþæht *consilium*, Rtl. 1, 9. v. ge-þeaht.

ge-þæslǽcan *to fit, to be fit, to become*; aptare, quadrare, congruere, R. Ben. interl. 2 : Hpt. Gl. 506 ; 523.

ge-þæslíc ; *adj.* [þæslíc *fit*] *Fit, proper*; dĕcens, opportúnus :—Geþæslíc [MS. geþæsliic] *dĕcens*, Ælfc. Gr. 14 ; Som. 16, 44. On tîman geþæslícum oððe on gedafenlícre tîde *in tempŏre opportúno*, Ps. Lamb. 31, 6.

ge-þafa, an ; *m.* [geþafian *to consent*] *A favourer, supporter, helper, assenter, consenter* ; fautor, adjûtor :—He biþ ryhtes geþeahtes geþafa *he is the supporter of good designs*, Past. 42, 1 ; Swt. 306, 14 ; Hat. MS. 58 a, 17 : Cd. 22 ; Th. 127, 8 ; Gen. 414. Hwî ne eart ðû his geþafa *why art thou not an assenter to this?* Bt. 26, 2 ; Fox 92, 13 : L. De Cf. 7 ; Th. ii. 262, 30. Ic eom geþafa *I am convinced, I am an assenter*, Bt. 35. 2 ; Fox 156, 13 : 36, 5 ; Fox 180, 16 : 38, 2 ; Fox 196, 16. Gif ðê mon for rihtre scylde brócie, geþola hit wel and beó wel geþafa *if thou art afflicted for a just cause, bear it well and assent to it readily*, Prov. Kmbl. 45. Ðâ næs Æðelm nâ fullíce geþafa *then Æthelm did not fully assent*, Th. Ch. 171, 4. We sceolon beón geþafan *we must necessarily be consenters*, Bt. 34, 12 ; Fox 154, 7.

ge-þafian, -þafigan, -þafigean ; *p.* ode, ude ; *pp.* od, ud [þafian *to permit, allow, consent*] *To favour, support, permit, allow, admit, assent, consent, agree, approve, obey, submit to*; fávěre, sustiněre, siněre, admittěre, permittěre, assentíre, consentíre, obêdíre, concêděre :—Ðû deáþe sweltest gif ðû geþafian nelt môdges gemânan *thou shalt perish by death if thou wilt not consent to the proud one's fellowship*, Exon. 67 b ; Th. 250, 12 ; Jul. 126 : 41 a ; Th. 138, 7 ; Gû. 572 : Judth. 10 ; Thw. 22, 12 ; Jud. 60 : L. Alf. pol. 6 ; Th. i. 66, 5. He nolde geþafigan ðæt man hys hûs underdulfe *non siněret perfŏdi dŏmum suam*, Mt. Bos. 24, 43. Geþafigean, Bd. 2, 2 ; S. 502, 14. Ic geþafige *consentio*, Ælfc. Gr. 30, 2 ; Som. 34, 39 : 37 ; Som. 39, 37 ; Ps. Th. 130, 3. He ne geþafode ðæt hig ænig þing sprǽcon *non siněbat ea lŏqui*, Lk. Bos. 4, 41 : 12, 39. Se eádega wer idese lârum geþafode *the blessed man assented to the woman's counsels*, Cd. 101 ; Th. 134, 31 ; Gen. 2233 : Bd. 3, 23 ; S. 555, 2. Nâ hî geþafudon geþeaht his non sustinuêrunt consilium ejus, Ps. Th. 105, 13. Ðînum mǽge mân ne geþafa *approve not wickedness in thy kinsman*, Exon. 80 a ; Th. 301, 12 ; Fä. 18. Ne gê in ne gâþ, ne gê ne geþafiaþ ðæt ôðre ingân *vos non intrâtis, nec introeuntes sinŭtis intrâre*, Mt. Bos.

23, 13. Ðâs hwîlwendlícan gedrefednyssa we sceolon mid gefeán for Cristes naman geþafian *but these transitory tribulations we ought to submit to with joy for Christ's name*, Homl. Th. i. 556, 10 : Prov. Kmbl. 9 : Past. 21, 1 ; Swt. 151, 15 ; Hat. MS. Beágmund geþafie and mid wrîte I, *Beagmund, approve and consign*, Th. Ch. 475, 16.

ge-þafsum; *adj. Agreeing*; consentiens :—Wæs ðû geþafsum *esto consentiens*, Mt. Kmbl. Lind. 5, 25.

ge-þafsumniss, e ; *f. Agreement, consent*, Mt. Kmbl. p. 14, 14.

ge-þafung, e ; *f. Permission, allowance, assent, consent*; permissio, assensus, consensus :—Mid Earnulfes geþafunge *with Arnulf's permission*, Chr. 887 ; Erl. 86, 3. Be his geþafunge gecyrde se apostol *by his permission the apostle returned*, Homl. Th. i. 60, 6 : Th. Ch. 526, 21. On hûse Godes we eodon mid geþafunge *in dŏmo Dei ambulâvimus cum consensu*, Ps. Spl. C. 54, 15 : Bd. 1, 27 ; S. 497, 25. Ðyssum wordum ôðer ðæs cyninges wita and ealdormann geþafunge sealde, and to ðǽre sprǽce feng *cujus suasiŏni verbisque prudentĭbus alius optĭmâtum tribŭens assensum, continuo subdĭdit*, Bd. 2, 13 ; S. 516, 12 : 4, 8 ; S. 576, note 5.

ge-þanc, -þonc, -þang, es ; generally *m.* but sometimes *n.* [þanc *will*] *Mind, will, opinion, thought*; mens, animus, cŏgitatio :—Þincþ on his geþance *thinks in his mind*, R. Ben. 65. Ðone fǽlan geþanc frine *interrŏga me*, Ps. Th. 138, 20. Se Hǽlend geseh hyra heortan geðancas *Iesus videns cogitationes cordis illorum*, Lk. Bos. 9, 47 : Ps. Th. 91, 4 : 93, 11 : 128, 3 : 139, 8 ; all *m.* but the following three are *n.* :—Sôþlíce ðæt geþanc eode on hig, hwylc hyra yldest wǽre *intravit autem cogitatio in eos, quis eorum major esset*, Lk. Bos. 9, 46 : Byrht. Th. 132, 9 ; By. 13. Geþancu and geþeahtu *thoughts and plans*, Lchdm. iii. 214, 23. Ðone ilcan geþanc ic ðê ǽr sǽde *the same thought I have told thee before*, Blickl. Homl. 179, 28. Geþanges *mentis*, Ps. Spl. 67, 29.

ge-þancian, -þoncian ; *p.* ode, ede ; *pp.* ôd, ed [þancian *to thank*] *To thank, give thanks, reward*; grâtias agěre :—Geþance ðê þeóda Waldend, ealra ðǽra wynna ðe ic on worulde gebâd *I thank thee, Lord of the nations, for all the delights which I have experienced on earth*, Byrht. Th. 136, 57 ; By. 173. He geþancode Gode his sande *he thanked God for what he had sent*, Homl. Th. ii. 136, 18. We sceolon geþancian Gode ðæt he wolde asendan his âncennedan Sunu *we ought to thank God that he was willing to send his only-begotten Son*, 23, 2. We him his geswinces geþancedon, of ûrum gemǽnum feó *we would reward him for his labour out of our common money*, L. Ath. v. § 7 ; Th. i. 234, 27. We giþoncia *gratulamur*, Rtl. 74, 7 : 31, 1.

ge-þanc-metian ; *p.* ode ; *pp.* od *To deliberate, consider* ; considĕrâre :—Geþancmeta ðíne môde, on hwilce healfe ðû wille hwyrft dôn *deliberate in thy mind on which side thou wilt depart*, Cd. 91 ; Th. 115, 9 ; Gen. 1917.

ge-þancol, -þancul, -þoncol ; *adj.* [þanc *the mind, thought*] *Mindful, thoughtful, considerate, suppliant*; mĕmor, cŏgĭtâbundus, supplex :—Ic wæs gemyndig mǽrra dôma ðínra geþancol, þeóden Dryhten *mĕmor fui judiciôrum tuôrum a sêcŭlo, Dŏmine*, Ps. Th. 118, 52 : Ps. C. 50, 6 ; Ps. Grn. ii. 276, 6. Swâ hleóðrode hâlig cempa, þeáwum geþancul *thus spake the holy champion, in all his ways thoughtful*, Andr. Kmbl. 923 ; An. 462. Giþoncolo *intenti*, Rtl. 16, 31. Giþoncle *supplices*, 4, 24.

ge-þancol, -þoncol ; *adj. Thankful, grateful* :—Giþoncolo wosaþ gie *grati estote*, Rtl. 13, 39.

ge-þang. v. ge-þanc.

ge-þang, es ; *n. Growth* :—Gyfe pund, ðanon him wæs geseald se fæt and geþang *a pound of grace, thence was given him the fat and growth*, Salm. Kmbl. 180, 12.

ge-þawenian ; *p.* ode, ede ; *pp.* od, ed *To moisten* ; humectâre :—Hió mid ðæm wætere weorþeþ [weorþaþ, MS.] geþawened *it is moistened with the water*, Bt. Met. Fox 20, 204 ; Met. 20, 102.

ge-þeáh *thrived*, L. R. 3 ; Th. i. 190, 18 ; *p.* of ge-þeón.

ge-þeád. v. ge-þeód.

ge-þeaht, -þæht, e ; *f.*: es ; *n.* I. *counsel, consultation, deliberation, advice, thought, a determination, resolution, device, plan, purpose*; consĭlium, cŏgĭtâtio :—Geþeaht Drihtnes on ecnysse wunaþ *consĭlium Dŏmini in æternum mănet*, Ps. Spl. 32, 11 : Ps. Th. 88, 6. Ðæt geþeaht *the counsel*, Ps. Th. 9, 6. Hî ðǽre geþeahte wǽron *they were of the resolution*, Cd. 182 ; Th. 228, 21 ; Dan. 205. Hî nyllaþ geþafan beón ôðerra monna geþeahtes *they will not be supporters of the plan of other men*, Past. 42, 1 ; Swt. 305, 15 ; Hat. MS. 58 a, 2. On ânre geþeahte [MS. geþeaht] eodan togædere *consilium fêcêrunt in ûnum*, Ps. Th. 70, 9. On geþeahte *in consilio*, Ps. Spl. 1, 1 : Ps. Th. 105, 32. Bûtan geþeahte *without advice*; inconsulte, Bd. 3, 1 ; S. 523, 31. Of hiera âgnum geþeahte *from their own determination*, Past. 42, 1 ; Swt. 305, 18 ; Hat. MS. 58 a, 4. Ealle geþeaht ðín he getrymþ *omne consĭlium tuum confirmet*, Ps. Spl. 19, 4. Þurh monnes geþeaht *through man's device*, Cd. 29 ; Th. 38, 12 ; Gen. 605 : Elen. Kmbl. 2117 ; El. 1060. Hî forhogodon ðæs Hǽlendes geþeaht *consilium Dei sprêvêrunt*, Lk. Bos. 7, 30 : Bd. 2, 13 ; S. 515, 32, 40. Hî ân geþeaht ealle ymbsǽtan *cogĭtâvêrunt consensum in ûnum*, Ps. Th. 82, 5. Ðæt he him geþeaht sealde *ut consĭlium sibi dăret*, Bd. 4, 25 ; S. 599, 38. Ðû [God] eal gôd [MS. good] ânes geþeahte ðínes geþohtest *thou* [God] *didst conceive all good*

by the counsel of thyself alone, Bt. Met. Fox 20, 78; Met. 20, 39: Bt. 33, 4; Fox 128, 20. Mid geþeahte ðínum *with thy counsel,* Bt. Met. Fox 20, 173; Met. 20, 87. Geþancu and geþeahtu *thoughts and plans,* Lchdm. iii. 214, 24. He wiðcwyþ geþeaht ealdrum *reprŏbat consĭlia princĭpum,* Ps. Spl. 32, 10: Ps. Th. 55, 5. **II.** *a council, an assembly;* concĭlium :—Geþeaht awyrgedra ofsǽton me *concĭlium mălignantium obsēdit me,* Ps. Spl. 21, 15. Ic ne sǽt mid geþeahte ýdelnyssa *non sēdi cum concĭlio vānĭtātis,* 25, 4. On ðam geþeahte *in the council,* Homl. Th. i. 46, 5. DER. rǽd-geþeaht.

ge-þeaht *covered,* Cd. 73; Th. 90, 8; Gen. 1492; *pp.* of ge-þeccan.

ge-þeahta, an; *m. A counsellor;* consiliarius :—Hæfst ðú ǽnigne wísne geþeahtan *habes aliquem sapientem consiliarium,* Coll. Monast. Th. 30, 5.

ge-þeahtend, es; *m. A counsellor;* consĭliārius :—Se geþeahtend andsweraþ *consĭliārius respondit,* Coll. Monast. Th. 30, 37: 31, 21.

ge-þeahtendlíc; *adj. Deliberative:*—Geþeahtendlíc ymcyme *a deliberative convention,* L. Wih. pref; Th. i. 36, 7.

ge-þeahtere, es; *m. A counsellor;* consĭliārĭus :—Se wæs geþeahtere ðæs apostolícan pápan *qui consĭliārius ĕrat ăpostŏlĭci pāpæ,* Bd. 5, 19; S. 638, 14. DER. þeahtere.

ge-þeahtian; *p.* ode; *pp.* od *To take counsel, consult;* consĭliāri :—Geniman sáwle míne hí geþeahtodon *accĭpĕre ănĭmam meam consĭliāti sunt,* Ps. Lamb. 30, 17.

ge-þeahting, -þeahtung, -þæhtung, e ; *f. Counsel, consultation, deliberation, agreement;* consilium, consultātio, consultum, consensus :—Ic Ælfréd cinge mid geþeahtunge Æþerédes ercebisceopes *I, king Alfred, with the counsel of archbishop Athered,* Th. Ch. 484, 11. Hú egesfullíc he is in geþeahtingum ofer monna bearn *quam terrĭbĭlis in consĭliis sŭper fĭlios hŏmĭnum!* Bd. 4, 25; S. 601, 36. Se geþeahtingum hafaþ in hondum heofon and eorþan *who by his counsels holdeth in his power heaven and earth,* Exon. 43 a; Th. 140, 31; Gú. 618. To geþeahtunge *ad consulta,* Bd. 1, 27; S. 497, 43. Mid geþeahtunge *cum consensu,* Ps. Th. 54, 13. Geþæhtung *consilium,* Mt. Kmbl. Lind. 12, 14: 22, 15.

ge-þearf, ic, he; *I have, or he has need or necessity.* v. ge-þurfan.

ge-þearfian; *p.* ode; *pp.* od *To impose necessity;* necessitatem imponere :—Ðá him swá geþearfod wæs *as necessity thus was imposed upon them,* Beo. Th. 2211; B. 1103. v. þearfian.

ge-þeccan *p.*-þeahte *To cover;* tegere :—Lago hæfde geþeahte ēðel *the water had covered the country,* Cd. 73; Th. 90, 8; Gen. 1492. DER. þeccan.

ge-þegan; *p.* ede *To consume:*—Þurste geþegede *consumed with thirst,* Exon. 30 a; Th. 92, 17; Cri. 1510. v. a-, of-þegan. *But cf. also* ge-þéwan.

ge-þegnian, -þénian; *p.* ode; *pp.* od [þegnian *to serve*] *To minister, serve;* ministrāre :—Ðú hæfst to þance geþénod ðínum hearran *thou hast served thy lord so as to please him,* Cd. 25; Th. 32, 20; Gen. 506.

ge-þencan, -þencean, ic -þence, ðú -þencest, -þencst, he -þenceþ, -þencþ; *pl.* -þencaþ, -þenceaþ; *p.* ic, he -þohte, ðú -þohtest, *pl.* -þohton; *pp.* -þoht. **I.** *to think, conceive, perceive, reflect upon, weigh ;* meditari, considerare, pensare :—Hwylc eówer mæg sóþlíce geþencan ðæt he geeácnige áne elne to hys anlícnesse *quis autem vestrum cogitans potest adjicere ad staturam suam cubitum unum,* Mt. Bos. 6, 27: Exon. 77 a; Th. 289, 34; Wand. 58: 100 a; Th. 378, 6: Deor. 12. Ðú meaht sweotole geþencean *thou mayest clearly perceive,* Bt. Met. Fox 5, 2; Met. 5, 1. To geþencanne *to think,* Exon. 112 a; Th. 429, 3; Rä. 42, 8. Ðú [God] eal gód [MS. good] ánes geþeahte ðínes geþohtest, and hí ðá worhtest *thou [God] didst conceive all good by the counsel of thyself alone, and then didst create it,* Bt. Met. Fox 20, 79; Met. 20, 40. Snyttro geþencaþ weras wísfæste *think prudence, oh ye wise men!* Elen. Kmbl. 626; El. 313. **II.** *to think about, remember, consider maturely, to take to heart;* recogitare, iterum cogitare, reminisci :—He sceal geþencan gǽstes þearfe *he shall think about the need of his soul,* Exon. 23 b; Th. 65, 20; Cri. 1057. Geþenceþ *thinketh,* 117 a; Th. 449, 27; Dóm. 77. Ic geþence *reminiscor,* Ælfc. Gr. 29; Som. 33, 54. Ic ánne ánlépne ne mæg geþencean *I cannot remember a single one,* Past. pref; Swt. 3, 18; Hat. MS. Ðæt he ne mæg ende geþencean *that he cannot consider his end,* Beo. Th. 3473; B. 1734. Gif he hit geþencan can *if he can consider it,* Salm. Kmbl. 814; Sal. 406: Exon. 115 b; Th. 445, 8; Dóm. 4. Hwæt! ðú lyt geþohtest *lo! thou didst consider little,* Soul Kmbl. 45; Seel. 23. **III.** *to think of, bear in mind, remember;* recordari, cogitare, memor esse :—Mæg geþencan, ðæt geond ðas woruld witig Dryhten wendeþ geneahhe *he may bear in mind that throughout this world the sagacious Lord alternates abundantly,* Exon. 100 b; Th. 379, 10; Deór. 31: 83 b; Th. 314, 5; Mód. 9. Ðe his synna geþenceþ *who bears in mind his sins,* 117 a; Th. 450, 6; Dóm. 83. Sóþfæste beót geþenceaþ *the righteous think of the promise [of God],* Ps. Th. 106, 41: 118, 74. Ic ealde dagas geþohte *cogitavi dies antiquos,* 76, 5. Geþenc se snottra fengel hwæt wit sprǽcon *let the sagacious prince bear in mind what we have spoken,* Beo. Th. 2952; B. 1474: Exon. 13 a; Th. 23, 18; Cri. 370. **IV.** *to excogitate, devise, invent, conceive;* excogitare, struere, invenire :—Ðú meaht rǽd geþencan *thou mayest devise counsel,* Cd. 27; Th. 35, 28; Gen. 561. Mid swilcum mæg man rǽd geþencean *with such one may devise counsel,* 15; Th. 19, 4; Gen.

286. He worn geþenceþ hinderhóca *he devises a number of stratagems,* Exon. 83 b; Th. 315, 19; Mód. 33. Se geréfa hét ða hálgan margaretan on karcerne betýnan óþ ðæt he geþohte hú he hire mægþhád forspilde *the prefect ordered the holy Margaret to be shut up in prison until he had devised how he might destroy her virginity,* Nar. 41, 17. He cwæþ ðæt he nán ryhtre geþencan ne meahte *he said he could conceive nothing more right,* Th. Ch. 171, 15. Hý grófon ǽghwylcne stán swá se cásere geþohte *they carved every stone as the emperor devised,* Shrn. 146, 17. **V.** *to resolve, intend, wish;* intendere, velle :—Uton geþencan Hælende héran *let us resolve to obey the Saviour,* Cd. 227; Th. 305, 9; Sat. 644. Se awyrgda geþohte ðæt he heofencyninge héran ne wolde *the accursed one resolved that he would not obey heaven's king,* 220; Th. 284, 4: Sat. 316: 217; Th. 276, 11; Sat. 187. Ðú geþohtest ðæt ðú ðíne mægþhád Meotude sealdes *thou didst resolve that thou wouldest give to the Lord thy maidenhood,* Exon. 12 a; Th. 18, 23; Cri. 288. DER. þencan.

ge-þenian; *p.* ede; *pp.* ed *To stretch out, extend :*—Geþenede *extendens,* Mt. Kmbl. Lind. 12, 49: 14, 31.

ge-þénsum; *adj. Obsequious, obliging, serviceable;* officiosus :—He wearþ geset cumena þén ðæt he mynsterlícum cumum geþénsum wǽre *he was appointed servant of guests that he might attend upon the monastic guests,* Homl. Th. ii. 136, 24. Gif hwilc sibling ðé biþ swá geþénsum swilce ðín ágen fót *if any kinsman be as serviceable to thee as thy own foot,* i. 516, 15.

ge-þeód; *part. p. Captive;* captivus :—Geþeódo *captivi,* Lk. Skt. Lind. 21, 24. Fore geþeádum *pro captivis,* Rtl. 177, 19. v. ge-þeón.

ge-þeódan, he -þeót; *p.* -þeódde; *pp.* -þeód *To join, connect, unite, associate, apply, adjust, translate;* jungĕre, adjungĕre, conjungĕre, cōpŭlāre, sŏciāre, aptāre :—Ic geþeóde *conjungo,* Ælfc. Gr. 47; Som. 48, 42. Forðam forlǽt se man fæder and móder and geþeót hine to his wífe *quamobrem relinquet hŏmo patrem suum et matrem et adhærēbit uxōri suæ,* Gen. 2, 24: Mt. Bos. 19, 5. Ðe hí hie oftost to geþeódaþ *to whom they most frequently join themselves,* Bt. 16, 3; Fox 56, 34. He ðám wordum sóna monig word to geþeódde *eis mox plūra verba adjunxit,* Bd. 4, 24; S. 597, 27. Ðæt us Gode ðú geþeóddest *ut nos Deo conjungĕres,* Hymn. Surt. 31, 29. Ðonne mihte he ðara ríme geþeódan beón *posset eōrum nŭmĕro sŏciāri,* Bd. 5, 13; S. 633, 36: Ps. Th. 61, 5. Mihte swýðe well beón to him geþeóded se cwide ðe Iacob se heáh-fæder cwæþ *cui mĕrĭto pŏtĕrat illud quod Patriarcha dīcēbat aptāri,* Bd. 1, 34; S. 499, 25. Ðá wæs geþeóded hefig gefeoht *conserto grāvi prœlio,* 2, 20; S. 521, 10: 4, 21; S. 590, 12. Ðæt bearn fæderlícum setle ys geþeód *quod partus pāternæ sēdi jungĭtur,* Hymn. Surt. 89, 32. Of hwylce cneórysse sculon cristene men mid heora mágum him betwih on gesinscipe geþeódde beón *usque ad quŏtam generātiōnem fĭdēles dēbeant cum propinquis sibi conjūgio cōpŭlāri?* Bd. 1, 27; S. 490, 35 : 2, 3; S. 504, 17. He hét ðisne regul of læden-gereorde on englisc geþeódan *he ordered to translate this rule from Latin into English,* Lchdm. iii. 440, 28. v. ge-þýdan.

ge-þeóde, es; *n. Language, speech, idiom, translation;* lingua:—Nis nán mennisc geþeóde *non sunt sermones,* Ps. Th. 18, 3. Ðǽr ðǽr hine nán man ne can ne he nǽnne mon ne furðum ðæt geþeóde ne can *where no man knows him ne he any man, nor does he know even the language,* Bt. 27, 3; Fox 98, 23. Ðæt ys gereht on úre geþeóde *quod est interpretatum,* Mt. Kmbl. 1, 23: Mk. Skt. 5, 41: 15, 22: Homl. Th. i. 194, 1: Past. pref; Swt. 5, 13; Hat. MS. Ða Finnas and ða Beormas sprǽcon neáh án geþeóde *the Finns and the Permians spoke nearly one language,* Ors. 1, 1; Bos. 20, 15. Hér sind fíf geþeóde Englisc and Brittisc and Wilsc and Scyttisc and Pyhtisc and Bóc Leden *there are five languages here, English, British, Scotch, Pictish, and Latin,* Chr. Erl. 3, 2. Hí mihton sóna sprecan on ǽghwelc ðara geþeóda ðe under heofonum *they could at once speak in every language under heaven,* Shrn. 85, 16 : Bt. 35, 4; Fox 162, 26. Ðæt hér ðý mára wísdóm on londe wǽre ðý we má geþeóda cúþon *that there might be the more wisdom in the land the more languages we knew,* Past. pref; Swt. 5, 25; Hat. MS. Hát todǽlan heora geþeóde *divide linguas eorum,* Ps. Th. 54, 8. Ic ðá geþeóde to micclan gesceáde telede *I reckoned then a translation to make much difference,* Lchdm. iii. 442, 4. [Cf. O. H. Ger. ge-diuti, Grff. v. 131.]

ge-þeóde. v. ingeþeóde.

ge-þeódendlíc; *adj. Conjunctive, joining;* copulativus :—Copulativæ ðæt synd geþeódendlíce *copulativæ, that is, joining together,* Ælfc. Gr. 44; Som. 45, 36.

ge-þeódnes, -ness, -nyss, e ; *f.* [ge, þeódnes, -nys *a joining*]. **I.** *a joining, juncture, joint;* junctio, junctura, compages:—Seó geþeódnes ðæs heáfdes tobrocen wæs *the joining of the head was broken,* Bd. 5, 6; S. 619, 24. He wæs býgendlíc on ðám geþeódnessum his liþa *he was flexible in the joints of his limbs,* 4, 30; S. 608, 38. Monigra monna mód to geþeódnesse ðæs heofonlícan lífes onbærnde wǽron *multorum animi appetitum sunt vitæ cælestis accensi,* 4, 24; S. 596, 37. **II.** *a conjunction:*—Conjunctio mæg beón gecweden geþeódnyss *conjunctio may be called 'geþeódnyss,'* Ælfc. Gr. 44, 2; Som. 45, 24 : 5, 26; Som. 3, 50. **III.** *conjugation:*—Conjugatio verborum is worda geþeódnyss . . . Conjunctio mæg beón gecweden geþeódnyss forðan ðe on ðære

beóþ manega word geþeódde on ánre declínunge, Ælfc. Gr. 24; Som. 24, 19-23. **IV.** *a translation:*—Ðeáh ða scearpþanclan witan ðisse engliscan geþeódnesse ne behófien *though the acute wise men need not this English translation*, Lchdm. iii. 440, 32.

ge-þeódræden, e; *f. Fellowship, society:*—Ðonne biþ ðé sélre ðæt ðú heora geþeódrædene forbuge *then it will be better for thee that thou avoid their society*, Homl. Th. i. 516, 17.

ge-þeódsumness, e; *f. Assent, consent, agreement*, Lk. Skt. p. 8, 1.

ge-þeófian; *p.* ode, ade: *pp.* od, ad *To steal, thieve;* furári:—Gif hwá on cirican hwæt geþeófige *if any one thieve aught in a church*, L. Alf. pol. 6; Th. i. 66, 2. Ðæt he hæbbe ǽr geþeófad *that he had before thieved*, L. In. 48; Th. i. 132, 8, MSS. B. H.

ge-þeón, ic -þeó, *pl.* -þeóþ; *p.* -þeáh, *pl.* -þugon; *pp.* -þogen *To grow, grow up, increase, thrive, flourish, prosper;* crescĕre, proficĕre, vĭgēre:—Lofdǽdum sceal man geþeón *a man shall flourish by praiseworthy deeds*, Beo. Th. 50; B. 25: 1825; B. 910: Homl. Th. i. 12, 26. Erigende ic geþeó *arando profício*, Ælfc. Gr. 24; Som. 25, 18. Ic strangie oððe geþeó *vigeo*, 26, 3; Som. 28, 47. Fela ríccra manna geþeóþ Gode *many rich men thrive to God*, Homl. Th. i. 130, 33: ii. 22, 15. Gif þegen geþeáh ðæt he þénode cynge *if a thane thrived so that he served the king*, L. R. 3; Th. i. 190, 18: 5, 6; Th. i. 192, 7, 9. Wæs his fæder ǽrest cyninges þegn and ðá æt néhstan geþeáh ðæt he wæs cininges þegna aldorman *his father was first a king's thane, and at last rose to be chief of the king's thanes*, Blickl. Homl. 211, 21. Ðe Gode geþugon þurh gehaltsumnysse his beboda *who thrive to God through observance of his commandments*, Homl. Th. ii. 280, 32: i. 444, 16. Geþeóh tela *thrive well!* Beo. Th. 2441; B. 1218: Exon. 122 a; Th. 469, 13; Hy. 11, 1. Ðæt ic ðé geþeó þinga gehwylce *that I may thrive to thee in everything*, 118 a; Th. 453, 9; Hy. 4, 12: L. Wg. 7, 10; Th. i. 188, 1, 8. Se ðe for wísdóme wende to Scottum ðæt he ælþeódig on láre geþuge *who for the sake of wisdom had gone to Scotland that in a foreign land he might increase in learning*, Homl. Th. ii. 148, 19. Ðá ðá he geþogen wæs *when he was grown up*, 38, 9: L. Ælf. P. 40; Th. ii. 380, 27. Se ðe swá geþogenne forwyrhtan næfde *he who had not so prosperous a vice-gerent*, L. R. 4; Th. i. 192, 5. Wæl geboren and yfele geþogen *degener*, Ælfc. Gr. 9, 18; Som. 10, 6. Geþogen [geþeógend, MS.] on mægne *mactus virtúte*, 41; Som. 44, 14.

ge-þeón, -þeówan; *p.* -þeóde, -þeówde; *pp.* -þeód *To tame, oppress;* dŏmāre, opprimĕre:—Se mec ána mæg ēcan meahtum geþeón *who alone can tame me by his eternal powers*, Exon. 111 b; Th. 427, 14; Rä. 41, 91. Me ðínes yrres egsa geþeówde *the terror of thine anger oppressed me*, Ps. Th. 87, 16. v. ge-þýwan *and* ge-þeód *captive*.

ge-þeón; *p.* -þeóde *To do, commit, perform;* perficere, patrare:—Ðæt we siððan forþ ða séllan þing symle móten geþeón *that henceforth we may ever do those better things*, Exon 13 a; Th. 23, 31; Cri. 377. v. þeón.

ge-þeót, es; *n. Howling:*—Wulfa geþeót *howling of wolves*, Guthl. 8; Gdwn. 48, 4.

ge-þeót *shall join;* 3rd pres. sing. of ge-þeódan.

ge-þeówan *to oppress*, Ps. Th. 87, 16. v. ge-þeón, -þýwan.

ge-þeówian; *p.* ode, ade; *pp.* od, ad *To make a slave, enslave;* servĭtúti subjĭcĕre, in servĭtútem redĭgĕre:—Ǽr hine mon geþeówode *before he was made a slave*, L. In. 48; Th. i. 132, 9. Gif hwelc man biþ niwan geþeówad *if any man be newly made a slave*, 48; Th. i. 132, 7: Th. Chart. 553, 9. Syndon cradolcild geþeówode *infantes e cúnábulis sunt mancĭpáti*, Lupi Serm. i. 5; Hick. Thes. ii. 100, 30; Swt. A. S. Rdr. 106, 50.

ge-þersc *a stripe, blow;* verber, Dial. 1, 2.

ge-þerscan; *p.* -þearsc, *pl.* -þurscon *To strike, beat, thrash:*—Geþearsca *cædere*, Mk. Skt. Lind. 14, 65. To geþearscanne, 15, 15. Geþurscon *cederunt*, 12, 3: Mt. Kmbl. Lind. 21, 35.

ge-þéwan; *p.* -þéwde; *pp.* -þéwed, -þéwd *To oppress;* opprimĕre:—He sárig folc, geþéwde þurste, blissade *he gladdened the sorrowful people, oppressed with thirst*, Ps. Th. 106, 32. v. ge-þýwan.

ge-þicfyldan; *p.* de *To make thick;* densare, Gl. Prud. 970.

ge-þicgan, -þicgean; *p.* -þah *To take, accept, receive;* sumere, accipere:—Waldon ða swángeréfan ða læswe forður gedrífan and ðone wudu geþicgan ðonne hit aldgeryhto wéron *the swainreeves wanted to push the pasturage and take the wood further than the old rights extended*, Th. Ch. 70, 22. And hiera se æþeling gehwelcum feoh and feorh gebeád and hiera nænig hit geþicgean nolde *the atheling offered every one of them money and life and none of them would accept it*, Chr. 755; Erl. 50, 6. Hit on mete oððe on drince geþicganne *to take it [poison] in meat or drink*, Ors. 3, 6; Bos. 58, 16. He ðæt ful geþeah æt Wealþeon *he took the cup from Waltheow*, Beo. Th. 1261; B. 628: 1241: B. 618: Cd. 42; Th. 54, 30; Gen. 885. Ðǽr ic beág geþah *there I received a bracelet*, Exon. 85 b; Th. 322, 19; Víd. 65: 84 b; Th. 318, 24; Víd. 3. Londryht geþah *he received the land-right*, 100 b; Th. 379, 29; Deór. 40: Cd. 161; Th. 200, 10; Exod. 354. Boitius se hæle hátte se ðone hlísan geþah *Boethius the man was named who got that fame*, Bt. Met. Fox 1, 106; Met. 1, 53. Geþǽgon medoful manig *they took many a mead-cup*, Beo. Th. 2033; B. 1014.

ge-þicgan; *pp.* -þiged *To take:*—Seoððan wæs mēce geþiged [Th. geþinged] *afterwards was the sword taken*, Beo. Th. 3881; B. 1938. v. þicgan *wk*.

ge-þiédan. v. ge-þeódan.

ge-þiéfian; *p.* ode, ede; *pp.* od, ed *To steal;* furári, L. In. 48; Th. i. 132, 8. v. ge-þeófian.

ge-þihan; *p.* -þáh, -þág, -þæh *To thrive, prosper, grow;* vigere, proficere, crescere:—Ælc ðæra ðe Gode geþihþ *every one that thrives to God*, Homl. Th. ii. 454, 29. Eádig biþ se ðe in his ēðle geþihþ *happy is he who thrives in his country*, Exon. 89 a; Th. 335, 21; Gn. Ex. 37. Alexandreas monna cynnes mæst geþáh *Alexandreas prospered most of the race of men*, Exon. 85 a; Th. 319, 23; Wid. 16: 40 b; Th. 134, 16; Gú. 508: Cd. 149; Th. 186, 24; Exod. 143. [Cf. ge-þeón.] DER. þihan.

ge-þincð. v. ge-þingþu.

ge-þind. v. ge-þynd.

ge-þinde; *pl. m. Rivals;* æmulatores, Hpt. Gl. 429. [Cf. þindan.]

ge-þing, es; *n.* **I.** *a council, an assembly;* concilium, concio:—Ðá se þeóden ongan geþinges wyrcan *then did the prince form a council*, Cd. 197; Th. 245, 25; Dan. 468. Hēt hie upastandan to Godes geþinge *he bade them arise to God's assembly*, Andr. Kmbl. 1588; An. 7951. **II.** *a compact, an agreement, a condition;* pactum:—Be diernum geþinge *concerning a private compact*, L. In. 52; Th. i. 134, 11, 12: 50; Th. i. 134, 1: L. Ath. v. § 11; Th. i. 240, 16: Th. Ch. 465, 12. Hig him geþingo budon ðæt hie him óðer flet eal gerýmdon *they offered him conditions that they would wholly yield to him another dwelling*, Beo. Th. 2175; Th. 1085. v. Grm. R. A. 600. **III.** *what is impending over one, what is awaiting one, what is certainly to be expected* or *hoped for, fate, destiny;* quod est imminens *vel* expectandum, fatum, sors:—Bád beadwa geþinges *he awaited the fate of the battle*, Beo. Th. 1423; B. 709: 802; B. 398. Wéndon hie þearlra geþinga þrǽge hnágran *they expected a worse period of severe fates*, Andr. Kmbl. 3194; An. 1600: 1512; An. 757.

ge-þingan; *p.* -þang, *pl.* -þungon; *pp.* -þungen *To thrive, grow, become excellent:*—Metode geþungon Abraham and Loth *Abraham and Lot throve to the Lord* [cf. ge-þeón], Cd. 82; Th. 103, 7; Gen. 1714: Bt. Met. Fox 1, 14; Met. 1, 7. Æghwæder heora wæs ælþeódig ðǽr and hwæðere for heora lífes geearnunge geþungon ðæt hí búta wǽron Abbudissan on ðam mynstre *quæ utraque cum esset peregrina, præ merito virtutum ejusdem monasterii est abbatissa constituta*, Bd. 3, 8; S. 531, 23. Wát ic ðæt ðú wǽre on woruldríce geþungen þrymlíce *I know that thou wert in this world exalted gloriously*, Soul Kmbl. 328; Seel. 168. v. ge-þungen.

ge-þingan; *pp.* ed *To determine, fix, destine:*—Gif him ðonne Hrēdríc to hofum Geáta geþingeþ [MS. -ed] he mæg ðǽr fela freónda findan *if then Hrethric determine to come to the Goths' courts he can find there many friends*, Beo. Th. 3678; B. 1857. Hafaþ him geþinged hider þeóden usser *our prince hath determined to come hither*, Exon. 115 b; Th. 445, 9; Dóm. 5. [Cf. ge-þingian, II.] Wiste hilde geþinged *he knew war was destined*, Beo. Th. 1299; B. 647: Menol. Fox 326; Men. 164: 14; Men. 7.

ge-þingelíc; *adj. Concerning a council*, Cot. 179.

ge-þingere, es; *m. An intercessor:*—We biddaþ ðætte fore us geþingere astonde *quesumus ut pro nobis intercessor existat*, Rtl. 44, 36.

ge-þingian; *p.* ode; *pp.* od, ad. **I.** [ge-þing, II.] *to make terms with a person for one's self* or *for another, to be reconciled, to come to an agreement, to reconcile, settle a dispute, intercede, mediate:*—Swá hie geþingian mǽgen wið cyning and his geréfan *according to the terms they can make with the king and his reeve*, L. In. 73; Th. i. 148, 11: 62; Th. i. 142, 3: Cod. Dipl. ii. 58, 26. Ðá geþingadun wið ðæm wyrhtum *conventione facta cum operariis*, Mt. Kmbl. Rush. 20, 2: Chr. 694; Erl. 42, 15: 628; Erl. 24, 4. Ðǽr genam Hettulf Honoriuses sweostor and siððon wið hine geþingode *there Ataulf took the sister of Honorius and afterwards made an agreement with him*, Ors. 6, 38; Bos. 133, 15. Bútan ðú ǽr wið hí geþingige *unless thou first be reconciled to them*, Exon. 68 b; Th. 254, 16; Jul. 198. Geþinge wið ðínum bróðer *reconciliare fratri tuo*, Mt. Kmbl. Rush. 5, 24. Swá beóþ þeóda geþwǽre ðonne hý geþingad habbaþ *so are peoples in concord when they have made a treaty*, Exon. 89 b; Th. 336, 29; Gn. Ex. 57. Goda bæd ðæt se kynincg him geþingode wið Eádgife his bóca edgift *Goda asked that the king would arrange for him with Eadgifu the restoration of his charters*, Th. Ch. 202, 32. He geþingode fǽhþa mǽste *he settled the greatest feud*, Exon. 16 b; Th. 39, 2; Cri. 616: Blickl. Homl. 9, 6. Ná ðé geþingodre *none the more settled*, L. In. 22; Th. i. 116, 12, MS. B. Gehwilces mannes dǽda hine gewrēgaþ oððe geþingiaþ *every man's deeds accuse him* or *reconcile him* [to God], Boutr. Scrd. 20, 38. Ðæt me seó hálge wið ðone hýhstan cyning geþingige *that the holy one intercede for me to the most high king*, Exon. 76 a; Th. 285, 20; Jul. 717. Giþingage *intercedat*, Rtl. 66, 13: intervenire, 60, 42. Ðæt hí to ðam mildheortan Hǽlende hire geþingodon *that they would intercede for her to the merciful Saviour*, Homl. Th. ii. 112, 22: 528, 14: Past. 10, 2: Swt. 63, 2, 10:

Hat. MS. Geþinga us *intercede for us*, Exon. 12 b; Th. 21, 29; Cri. 342. **II.** *to determine*:—Hafaþ nû geþingod to us þeóden mæra *the great prince hath determined [to come] to us*, Cd. 226; Th. 302, 12; Sat. 598. [Cf. ge-þingan.]

ge-þingio *a provision; apparatio*, Cot. 8, Lye.

geþing-sceat, es; *m. Ransom*:—He ne sealde Gode nænne geþingsceat wið his miltse *he gave God no ransom for his mercy*, Past. 45; Swt. 339, 10; Hat. MS.

ge-þingþu, -þingcþu, -þincþ, -þyncþ, e; *f.*   **I.** *honour, dignity, rank; honor, dignitas*;—He becom to ðære cynelícan geþincþe *he arrived at the royal dignity*, Homl. Th. i. 82, 1. Eal folc ðone eádigan Gregorius to ðære geþincþe geceás *all folk chose the blessed Gregory to that dignity*, ii. 122, 22. Hû micelre geþincþe sý ðæt hálige mæden Maria *of how great dignity is the holy maiden Mary*, 22, 21. Godes gecorenan scínaþ on heofonlícum wuldre ælc be his geþingcþum; nû is geleáflíc ðæt seó eádige cwén mid swá micclum wuldre and beorhtnysse óðre oferstíge, swá micclum swá hire geþincþu óðra hálgena unwiðmetenlíce sind *God's elect shine in heavenly glory each according to his rank; now it is credible that the blessed queen excels others with so much brightness and glory, as much as her rank is not comparable with that of other saints*, i. 446, 2–5; Jud. Thw. p. 161, 21: Swt. A. S. Rdr. 98, 93; Homl. Th. ii. 450, 2. Sum ðungen láreow wæs on Engla lande Albin geháten and hæfde micele geþincþa *there was a certain distinguished teacher in England named Albin and he had great honour*, Boutr. Scrd. 17, 6. Him to wæron witode geþingþo *to him were destined honours*, Cd. 23; Th. 30, 31; Gen. 475. Geþyncþum *honourably*, Exon. 41 b; Th. 138, 16; Gú. 577.   **II.** *a court, legal assembly*:—Ðæt grið ðæt se ealdormann on fíf-burhga geþincþe sylle and ðæt grið ðæt man sylleþ on burhgeþincþe bête man *for the 'grith' which the alderman in the assembly of the five-burghs may give and for the 'grith' that is given in a burgh-assembly, let 'bót' be made*, L. Eth. iii. 1; Th. i. 292, 6. [Cf. ge-þungen and ge-þing (?).]

ge-þingung, e; *f. Intercession*:—Giþingunge *intercessione*, Rtl. 71, 17: 124, 36.

ge-þinnian, -þinngian, -þynnian; *p.* ode; *pp.* od *To thin, lessen, diminish, dispel; attenuāre*:—Ic hie sceal ærest geþinnian [geþinngian, MS. Bod.] *I must first dispel them*, Bt. 5, 3; Fox 14, 19.

ge-þióde *speech.* v. ge-þeóde.

ge-þióstrian; *p.* ode; *pp.* od *To obscure; obscūrāre*:—Seó sunne oferlíht ealle óðre steorran, and geþióstraþ mid hire leóhte *the sun outshines all other stars, and obscures [them] with her light*, Bt. titl. ix; Fox xii. 2. Sunna biþ geþióstrod *sol contenebrabitur*, Mk. Skt. Lind. 13, 24.

ge-þíwan; *p.* de; *pp.* ed *To threaten, rebuke, oppress*:—Simon me mid his englum geþíwde *Simon threatened me with his angels*, Homl. Th. i. 378, 2. Óþ-ðæt hio óðer folc egsan geþíwdan *until they oppressed other people with fear*, Ps. Th. 104, 11. v. ge-þýwan.

ge-þofta, an; *m. A companion, comrade; sōdālis, contŭbernālis*:—Onbræd se his geþofta and lócade to him *expergefactus sōdālis respexit eum*, Bd. 3, 27; S. 559, 17. Ðe ær his geþofta wæs *who was formerly his companion*, Ors. 3, 7; Bos. 61, 18: 3, 11; Bos. 74, 45. Ðæt ðú sí gemyndig dínes getreówan geþoftan *tui mĕmor sis fĭdissimi sōdālis*, Bd. 4, 29; S. 607, 25. Gemétte he ðone his geþoftan slæpendne *invēnit sōdālem dormientem*, 3, 27; S. 559, 14. Hé gesomnode wered his geþoftena *he collected a band of his companions*, Guthl. 2; Gdwn. 14, 2: Shrn. 196, 20. Geþofta *cliens*, Wrt. Voc. 291, 33.

ge-þoftian; *p.* ode, ade, ede; *pp.* od, ad, ed *To associate, join, to enter into an agreement; assŏciāre, societātem inīre*:—Geþoftade he wið Ptholomeus *he joined with Ptolemy*, Ors. 3, 11; Bos. 74, 26. Seleucus and Demetrias him togædere geþoftedan *Seleucus and Demetrius joined together*, 3, 11; Bos. 75, 14.

ge-þoftræden, e; *f. Companionship, fellowship, converse; consortium*:—God to him genam geþoftrædene *God held converse with him*, Homl. Th. i. 90, 20.

ge-þoftscipe, es; *m. Companionship, society; consortium*:—Ðýlæs he síe innan asliten from ðæm geþoftscipe ðæs incundan dēman *lest he be inwardly cut off from the society of the internal judge*, Past. 46, 5; Swt. 351, 24; Hat. MS. 67 a, 16, 20: Swt. 353, 3.

ge-þogen *grown up*, Homl. Th. ii. 38, 9; *pp.* of ge-þeón.

ge-þoht, es; *m. n.* [ge-þoht, *pp.* of ge-þencan *to think*] THOUGHT, *thinking, mind, determination; cōgĭtātio, mens*:—Ðæt wæs þreálíc geþoht *that was a guilty thought*, Elen. Kmbl. 851; El. 426: Exon. 115 b; Th. 444, 6; Kl. 43. Forðonðe mannes geþoht mægen andetteþ *quia cōgĭtātio hŏmĭnis confĭtēbitur tibi*, Ps. Th. 75, 7: 32, 10. Manna cynnes [MS. kynnes] casere hafaþ acenned on ðé ða unablinnu ðæs yfelan geþohtes *the tempter of mankind [lit. of the race of men] hath begotten in thee the unrest of this evil thought*, Guth. 7; Gdwn. 46, 10: Bd. 1, 27: S. 496, 32: Exon. 73 b; Th. 275, 14; Jul. 550. Minne gehýraþ ânfealdne geþoht *hear my simple thought*, Beo. Th. 517; B. 256: 1225; B. 610: Salm. Kmbl. 478; Sal. 239. Hwíle mid geþohte some-

times *with thought*, Hy. 3, 45; Hy. Grn. ii. 282, 45: Exon. 77 b; Th. 291, 27; Wand. 88. Ðæt geþohtas sýn awrigene of manegum heortum *ut revēlen'ur ex multis cordibus cōgĭtātiōnes*, Lk. Bos. 2, 35: Ps. Th. 138, 2. Gé sind earmra geþohta *ye are of poor thoughts*, Andr. Kmbl. 1488; An. 745: Bd. 2, 12; S. 513, 31. On geþohtum *in cōgĭtātiōnibus*, Ps. Th. 138, 17. Ðæt he him afirre frécne geþohtas *that he banish from him wicked thoughts*, Cd. 219; Th. 282, 10; Sat. 284: 217; Th. 277, 18; Sat. 206. The following examples are neuter:—Þurh dyrne [*or*= dyrnne?] geþoht *through dark counsel*, Exon. 115 a; Th. 442, 13; Kl. 12: Ps. Th. 139, 2.

ge-þohte *thought*, Cd. 217; Th. 276, 11; Sat. 187; *p.* of ge-þencean.

ge-þolian, to -þolianne, -þolienne, -þoligenne; *p.* ode, ade, ede; *pp.* od, ad, ed [þolienne *to bear, suffer*].   **I.** *to bear, suffer, endure, sustain; sufferre, pāti, sustĭnēre*:—Hea geþolas *patiuntur*, Mt. Kmbl. Lind. 5, 10: Mk. Skt. Lind. 9, 12. Hie geþolian sceolon carmlíc wíte *they shall suffer miserable torment*, Cd. 227; Th 304, 26; Sat. 636: Elen. Kmbl. 2582; El. 1292. Ðæt Andrea þúhte þeódbealo þearlíc to geþolianne *that seemed to Andrew a general evil hard to bear*, Andr. Kmbl. 2274; An. 1138: Beo. Th. 2842; B. 1419: Exon. 48 a; Th. 166, 7; Gú. 1039. To geþolienne, Andr. Kmbl. 3375; An. 1691. To geþoligenne, 3317; An. 1661. We hênþo geþoliaþ *we shall suffer punishment*, Cd. 222; Th. 289, 18; Sat. 399: Exon. 70 b; Th. 262, 30; Jul. 340. He feala wíta geþolode *he endured a multitude of torments*, Andr. Kmbl. 2979; An. 1492: Beo. Th. 297; B. 147. Ic ðæt for worulde geþolade *I suffered that for the world*, Exon. 28 b; Th. 87, 13; Cri. 1424; An. 9: Th. 88, 21; Cri. 1443. Geþoledan, Ps. Th. 145, 6. Geþola Drihtnes willan *sustĭne Dŏminum*, 26, 16: Andr. Kmbl. 213; An. 107.   **II.** *to have patience, endure, wait, remain; perdūrāre, mănēre*:—Ðú scealt geþolian sume hwíle *thou must bear [with me] for some time*, Bt. 39, 4; Fox 218, 8. Gif he inne geþolian wille *if he will remain within*, L. Alf. pol. 42; Th. i. 90, 6: Beo. Th. 6210; B. 3109. Se ðe geþolias on ende *qui sustĭnuerit in finem*; Mk. Skt. Lind. 13, 13: 14, 34.   **III.** *with the gen. To suffer loss of, forfeit, lose; cārēre*:—Ic geþolian sceal þinga æghwylces *I must forfeit everything*, Cd. 219; Th. 281, 17; Sat. 273.

ge-þonc, es; *m. n. Thought, mind, understanding; cōgĭtātio, mens*:—Gleáw on geþonce *cunning in thought*, Judth. 9; Thw. 21, 11; Jud. 13. Þurh glædne geþonc *through benign thought*, Exon. 12 b; Th. 20, 10; Cri. 315. Ðæt ic him monigfealde ongeánbere grimra geþonca *that I present manifold dire thoughts to him*, 71 a; Th. 264, 21; Jul. 367: 31 a; Th. 97, 1; Cri. 1584. Ic onsende in breóstsefan bitre geþoncas *I send into his mind bitter thoughts*, 71 b; Th. 266, 29; Jul. 405. He us geþonc syleþ, missenlícu mód *he gives us understandings, various minds*, 89 a; Th. 334, 7; Gn. Ex. 13. Breóst innan weóll þeóstrum geþoncum *his breast boiled within with dark thoughts*, Beo. Th. 4653; B. 2332: Exon. 54 a; Th. 190, 4; Az. 68. v. ge-þanc.

ge-þracen, *part. p. Prepared, decked; ornatus*:—Geþracen hors *mannus vel brunnicus*, Ælfc. Gl. 5; Som. 56, 18; Wrt. Voc. 17, 22. [Cf. ge-þræc *apparatus*, Lye.]

ge-þræc, -þrec, es; *n. Press, crowd, crush, tumult*:—Ac wæs flód to deóp atol ýda geþræc *but too deep was the flood, the fierce press of the waves*, Exon. 106 a; Th. 404, 13; Rä. 23, 7: 101 a; Th. 381, 26; Rä. 3, 2. Þurh þreáta geþræcu[?], 109 a; Th. 417, 17; Rä. 36, 6. Beorna geþrec *press of men*, Elen. Kmbl. 228; El. 114: Ps. C. 50, 44; Ps. Grn. ii. 277, 44: Exon. 102 a; Th. 386, 13; Rä. 4, 61. Geþrec *clangor*, Cot. 59, Lye.

ge-þræc *apparatus, adjutorium*, Cot. 1, Lye.

ge-þræstan; *p.* -þræste; *pp.* -þræst *To twist, hurt, torment, afflict; contĕrĕre, afflĭgere*:—Gefeóll he semninga on his earm ufan, and ðone swýðe geþræste and gebræc *repente corruens brachium contrīvit*, Bd. 3, 2; S. 525, 2. Se hæleþ heortan geþræste *qui sĭnat contrītos corde*, Ps. Th. 146, 3. Weorpen hí swá geþræste mid hungre ðæt hi eton swýnen flæsc *may they be so tormented with hunger as to eat swine-flesh*, Ps. Th. 16, 14. On ðám dagum ðe ic geþræsted wæs *in die afflictionis meæ*, 17, 19. Godes engel hí geþræste *angelus Domini adflĭgens eos*, 34, 6.

ge-þræstian *adducere, præjudicare*, Hpt. Gl. 440.

ge-þræstnes, -ness, e; *f. Affliction, contrition; contrĭtio*:—On swá mycelre geþræstnesse *in tanta contritiōne*, Bd. 5, 12; S. 627, 27.

ge-þráfod *corrected, chastised.* v. þráfian.

ge-þrang, es; *n. A throng, crowd, tumult; turba, tumultus*:—On geþrang *in the throng*, Byrht. Th. 140, 36; By. 299. [Cf. O. H. Ger. ge-threngi: Ger. ge-dränge.]

ge-þráwan, -þráwan; *p.* -þreów, *pl.* -þreówon; *pp.* -þráwen, -þráwen *To twist; torquere*:—Ðæt geþráwene [geþráwene, MS. Cot.] twín *byssus torta*, Past. 14, 6; Swt. 87, 11; Hat. MS. 18 b, 15. Geþráwan *torquere*, Hpt. Gl. 435.

ge-þreán; *p.* -þreáde; *pp.* -þreád *To reprove, rebuke, afflict, vex, constrain, compel; corripere, increpare, arguere, cogere, afflĭgere, coartare, urgere, vexare*:—Se ðe him sylfum leofaþ rihtlíce he is ýdel geþreád *he who lives for himself is rightly reproved as idle*, Homl. Th. ii. 78, 5.

Huelc from iúh geþreáþ mec *quis ex vobis arguit me?* Jn. Skt. Lind. 8, 46 : 16, 18. He geþreáde ðæt wind *ille increpavit ventum,* Lk. Skt. Lind. 8, 24 : 9, 55. Geþreá hine *increpa illum,* 17, 3. Ne geþreá me *neque corripias me,* Ps. Surt. 37, 2. From giþreándum *ab increpantibus,* Rtl. 19, 15. Hú beó ic geþreád *quomodo coarctor,* Lk. 12, 50. Ic wæs geþreád ðæt ic ðé sóhte *I was compelled to seek thee,* Exon. 70 b ; Th. 263, 3 ; Jul. 344. Egsan geþreád *afflicted with terror,* 30 b ; Th. 95, 28 ; Cri. 1564 : 33 b ; Th. 106, 22 ; Gú. 45 : Cd. 90 ; Th. 112, 4 ; Gen. 1865 : 126 ; Th. 161, 21 ; Gen. 2668 : Andr. Kmbl. 781 ; An. 391. He náhte his líchoman geweald ac he wæs mid godcundum mægene geþreád *he had no power over his body, but was afflicted by the divine might,* Blickl. Homl. 223, 12.

ge-þreátian ; *p.* ode, ade ; *pp.* od, ad [þreátian *to urge, press*] To urge, oppress, threaten, rebuke, compel, restrain, afflict, torment ; urgĕre, cōgĕre, afflīgĕre, trībulāre :—Ne meaht ðú mec geþreátian ðé to gesingan *thou canst not compel me to be thy wife,* Exon. 66 b ; Th. 245, 33 ; Jul. 54. On yrre ðú folc geþreátast *in ira populos confringes,* Ps. Th. 55, 6 : Exon. 68 a ; Th. 253, 6 ; Jul. 176. Se snáw hý geþreátaþ *the snow oppresseth them,* Salm. Kmbl. 607 ; Sal. 303. Geþýd and geþreátod *rebuked and threatened,* Andr. Kmbl. 871 ; An. 436 : 2231 ; An. 1117 : Elen. Kmbl. 1387 ; El. 695. Hungre geþreátad *oppressed by hunger,* Exon. 46 a ; Th. 157, 8 ; Gú. 888. Ðæt geþreátade mód biþ suíðe raðe gehwierfed to fióunga *the rebuked mind will very soon be turned to hatred,* Past. 21 ; Swt. 167, 13 ; Hat. MS. Hie hine hæfdon geþreátodne mid fýrenum racentum ðæt he ne móste gecweðan ' Miltsa me God ' *they had restrained him with fiery chains from saying ' Have mercy on me, O God !'* Blickl. Homl. 43, 30 : 221, 15. Geþreátad *coactus,* Mt. Kmbl. p. 8, 1. Petrus ongan giþreátiga hine *Petrus coepit increpare eum,* Mk. Skt. Rush. 8, 32 : Mt. Kmbl. Lind. 8, 26 : 20, 31. Sóna geþreátade þegnas his *statim coegit discipulos suos,* Mk. Skt. Rush. 6, 45.

ge-þrec. v. ge-þræc.

ge-þréstan, Ps. Surt. 146, 3. v. ge-þræstan.

ge-þring, es ; *n.* [ge-þringan *to press*] A press, tumult, crowd, throng ; tŭmultus, turba :—Ofer wætera geþring *over the throng of waters,* Chr. 975 ; Erl. 126, 21 ; Edg. 47 : Andr. Kmbl. 736 ; An. 368 : Beo. Th. 4271 ; B. 2132. Wæs giþring *there was a throng,* Lk. Skt. Rush. 8, 42. [*O. Sax.* ge-þring.] Cf. ge-þrang.

ge-þringan ; *p.* -þrang, -þringde [North. Gospels], *pl.* -þrungon ; *pp.* -þrungen To press, oppress ; comprimere, contendere, opprimere. **I.** *v. intrans* :—Ceól up geþrang *the keel pressed up,* Beo. Th. 3829 ; B. 1912. Deáþ in geþrong *death pressed in,* Exon. 45 a ; Th. 153, 34 ; Gú. 835. Hú he þurh ðæt folc geþrang *how he pressed through the people,* Ors. 3, 9 ; Bos. 68, 30. Geþrincgas to ingeonganne *contendite intrare,* Lk. Skt. Lind. 13, 24. Hæfde ðá se æþeling in geþrungen *then had the noble one pressed in,* Andr. Kmbl. 1080 ; An. 992. Wæs ðá ende-dógor neáh geþrungen *the final day had come near,* Exon. 46 b ; Th. 158, 10 ; Gú. 906. Ðære tíde ys neáh geþrungen *it is close upon the time,* Judth. 12 ; Thw. 25, 31 ; Jud. 287 : Cd 116 ; Th. 151, 15 ; Gen. 2509. **II.** *v. trans* :—Woldon Rómwara ríce geþringan *they would oppress the power of the Romans,* Elen. Kmbl. 80 ; El. 40 Me firenlustas díne geþrungon *me thy sinful lusts oppressed,* Exon. 98 b ; Th. 369, 2 ; Seel. 35 : Bt. Met. Fox 1, 5 ; Met. 1, 3. Geþringdon hine *comprimebant eum,* Mk. Skt. Lind. 5, 24. From ðæm here wæs geþringed t geþrungen *a turba comprimebatur,* Lk. Skt. Lind. 8, 42. Calde geþrungen wæron míne fêt *pinched with cold were my feet,* Exon. 81 b ; Th. 306, 16 ; Seef. 8. Wombe geþrungne *a swollen belly,* 129 a ; Th. 485, 3 ; Rä. 84, 2.

ge-þristian ; *p.* ode, ade ; *pp.* od, ad [þríst, þriste *bold*] To dare, presume ; audēre, præsūmĕre :—Ic ne geþristige *ego non audeo,* Coll. Monast. Th. 25, 5. Forðam he geþristade *quod se præsumpsisset,* Bd. 1, 7 ; S. 477, 15.

ge-þristlǽcan ; *p.* -lǽhte, -lǽcte To dare, presume, excite ; provocare :—Ne geþristlǽcaþ hí ô ðæt hí mánswergen on his noman *they never dare to sware falsely in his name,* Shrn. 109, 17. Þurh Albinus swíðost ic geþristlǽhte ðæt ic dorste ðis weorc ongynnan *hortatu præcipue ipsius Albini ut hoc opus adgredi auderem provocatus sum,* Bd. pref ; S. 472, 11. Eádréd biddeþ ðæt nán man geþristlíce his cynelícan gefe gewonian *Eadred prays that no man will presume to diminish his royal gift,* Cod. Dipl. Kmbl. ii. 304, 26. We geþristlǽcton *provocavimus,* Cot. 154.

ge-þrowian, -þrowigan ; *p.* ode, ade ; *pp.* od, ad To suffer :—Feolo geþrowia *multa pati,* Lk. Skt. Lind. 9, 22. Gê ondspyrnise geþrowiges *vos scandalum patiemini,* Mt. Kmbl. Lind. 26, 31. On hwylcre þeóden engla geþrowode *on which the prince of angels suffered,* Elen. Kmbl. 1714 ; El. 859. Se cyle geþrowode wið ða hæto *the cold should suffer by the heat,* Bt. 33, 4 ; Fox 128, 33. Geþrowade, 1123 ; El. 563. Twegen mid him geþrowedon *two suffered with him,* 1706 ; El. 855. Sunu monnes geþrowend biþ *Filius hominis passurus est,* Mt. Kmbl. Lind. 17, 12. Ðú bist geþrouad *tu cruciaris,* Lk. Skt. Lind. 16, 25. He swá mycel for úre lufan geþrowode *he has suffered so much for love of us,* Blickl. Homl. 25, 3 : 91, 12. Geþrowade, Elen. Kmbl. 1035 ; El. 519.

Deáþ he geþrowode for us *he suffered death for us,* Blickl. Homl. 85, 2 : Cd. 228 ; Th. 306, 18 ; Sat. 666. He æt ðæm unlǽdum Iudéum manig bysmor geþrowade *he suffered many contumelies at the hands of the wicked Jews,* Blickl. Homl. 23, 31.

ge-þruen [= ge-þuren] ; *part. p.* Pressed together, compact :—Eorþe is hefigre ôðrum gesceaftum þicre geþruen *earth is heavier than the other elements, more closely compact,* Bt. Met. Fox 20, 267 ; Met. 20, 134. v. ge-þweran.

ge-þryccan, -þrycgan ; *p.* -þrýde To press, compress, bind a book [?], restrain, express ; premere, comprimere, exprimere, operire :—Oðer geþrýde t awrát *alius expressit,* Mt. Kmbl. p. 3, 6. Eþiluald hit úta giþrýde *Ethewold bound* [?] *it,* Jn. Skt. p. 188, 3. See note, p. viii. Ðone fiónd úserne geþrycg *hostem nostrum comprime,* Rtl. 180, 18.

ge-þrýde. v. ge-þryccan.

ge-þryle *an assembly, a meeting* ; frequentia :—For þæs folces geþryle *for the folk's assembly,* Hom. 8, Cal. Jan. p. 18, Lye.

ge-þryscan ; *p.* te ; *pp.* ed To press, oppress, press down, depress ; premere, deprimere :—Ðæt hine ne geþrysce nán wiðermódnes to ormódnesse *non aspera ad desperationem premant,* Past. 14, 3 ; Swt. 83, 18 ; Hat. MS. 17 b, 26. Ðonne sió þreáung biþ ungemetgad ðonne biþ ðæt mód ðæs agyltendan mid ormódnesse geþrysced *cumque increpatio immoderate accenditur, corda delinquentium in desperatione deprimuntur,* 21, 7 ; Swt. 165, 19 ; Hat. MS.

ge-þrypian ; *p.* ede ; *pp.* ed [þryþ *power, strength*] To strengthen, arm ; corroborare :—Deáþ nimeþ wíga wælgífre wæpnum geþryþed ealdor ánra gehwæs *death, the blood-greedy warrior, armed with weapons, takes the life of every one,* Exon. 62 b ; Th. 231, 9 ; Ph. 486.

ge-þúf *growing, luxuriant* ; luxurians, Cot. 123, 198. v. þúf, þúfian.

ge-þugon *throve,* Homl. Th. ii. 280, 32 ; *p. pl.* of ge-þeón.

ge-þuhtsum ; *adj.* Abundant :—Hit wæs ǽr ðær singal druwung and sóna æfter ðam com geþuhtsum rén on eorþan *there had been there before continual drought, and directly after that came abundant rain on the earth,* Shrn. 113, 20. [Cf. þyhtig, ge-þyht.]

ge-þuild *patience,* Lk. Skt. Lind. 18, 7. v. ge-þyld.

ge-þun, es ; *n.* A noise ; clangor :—Us þúhte for þam geþune ðæt sió eorþe eall cracode *it seemed to us from the noise that the earth all cracked,* Ps. Th. 45, 3.

ge-þungen ; *part. p.* Grown, thriven, advanced [*morally, mentally,* etc.], excellent, pious, noble, perfect :—Leomum geþungen *perfect in its limbs,* Exon. 64 a ; Th. 241, 1 ; Ph. 649. On geþungenum wæstme *in mature growth,* Homl. Th. ii. 76, 25. Geþungen *emeritus, provectus,* Ælfc. Gl. 82 ; Som. 73, 52 ; Wrt. Voc. 47, 56. Ðæt nænig þing ne gedafenade swá æþelum cyninge and swá geþungenum *quia nulla ratione conveniat tanto regi,* Bd. 2, 12 ; S. 514, 38 : 2, 1 ; 501, 34 : Homl. Th. ii. 122, 14 : 126, 28. Gódne wer and geþungenne to biscopháde *virum bonum et aptum episcopatu,* Bd. 3, 29 ; S. 561, 11 : 4, 23 ; S. 594, 6. Sum æfast mann and geþungen *veracem ac religiosum hominem,* 3, 19 ; S. 549, 24. Sum geþungen and gedéfe sacerd *sacerdos quietus,* Nar. 37, 25. Mód geþungen *mens sobria,* Ps. Stev. ii. 202, 7. Wæs he swíðe geþungen on his þeáwum *he was very excellent in his conduct,* Blickl. Homl. 217, 6 : Judth. 11 ; Thw. 23, 19 ; Jud. 129. Cwén móde geþungen *the queen excellent of mind,* Beo. Th. 1252 ; B. 624. Þegen geþungen *an illustrious minister,* Andr. Kmbl. 1055 ; An. 528 : Exon. 69 b ; Th. 258, 8 ; Jul. 262. Ic ða geþungnestan nemde *I have named the most distinguished,* Chr. 897 ; Erl. 95, 6 : 905 ; Erl. 98, 30. Hafa ðú me to ðan geþungennestan wífe *have me as the most excellent wife,* Shrn. 40, 17. [*O. Sax.* gi-þungan.]

ge-þungenlíce ; *adv.* Soberly ; sobrie, Ps. Stev. ii. 201, 21.

ge-þungennes, -ness, e ; *f.* Increase, growth, piety, excellence, gravity :—Ðæt wæs ðæt templ ðære geþungennesse and ealre clænnesse *that was the temple of piety and all purity,* Blickl. Homl. 5, 20 : Shrn. 40, 2 : 44, 9. Geþungennis *incrementum,* Mk. Skt. p. 2, 6. Geþungennes perfectio, Mone Gl. 365.

ge-þurfan, ic -þearf ; *p.* -þorfte To have need or necessity ; indigere, necesse habere :—Ðýlæs ðé geþearfe to óðres mannes æhtum *lest thou have need of another man's goods,* Prov. Kmbl. 73. v. þurfan.

ge-þwǽnan ; *p.* de ; *pp.* ed To moisten, wet, soften ; irrigare, emollire :—Gif þat wæter hí ne geþwænde *if the water moisten it not,* Bt. 33, 4 ; Fox 130, 7. Ða adrugodan heortan geþwǽnan mid ðǽm flówendan ýdon [ýðum, MS. Cot.] his láre *corda arentia doctrinæ fluentis irrigare,* Past. 10, 1 ; Swt. 61, 19 ; Hat. MS. 14 a, 15 : 18, 5 ; Swt. 137, 8 ; Hat. MS. 27 a, 12. His læcas tiloden and ðone swile mid sealfum and mid beþenum geþwænan woldon *curabant medici tumorem adpositis pigmentorum fomentis emollire,* Bd. 4, 32 ; S. 611, 20.

ge-þwǽran *mansuescere, respirari,* Gl. Prud. 644, 714.

ge-þwǽre, -þwére ; *adj.* United, agreeing, consonant, harmonious, accordant, concordant, mild, gentle, peaceful ; concors, congruus, consŏnus, mansuētus, pācificus, plācidus :—Geþwǽre *concors,* Ælfc. Gr. 9, 44 ; Som. 13, 4, MSS. C. D. Geþwǽre sang *harmŏnia,* Ælfc. Gl. 34 ; Som. 62, 59 ; Wrt. Voc. 28, 39. Ðú noldest on eallum þingum beón geþwǽre ðæs unrihtwísan cyninges willan *thou wouldest not in all things be conformable*

to the will of the unrighteous king, Bt. 27, 2; Fox 96, 16. Sum hafaþ mód and word monnum geþwǽre *one has mind and words agreeable to men*, Exon. 79 b; Th. 298, 15; Crä. 85. Þegnas syndon geþwǽre *the thanes are united*, Beo. Th. 2464; B. 1230: Exon. 9 b; Th. 8, 33; Cri. 127: 89 b; Th. 336, 29; Gn. Ex. 57. Wurdon ealle gereord geán-lǽhte and geþwǽre *all languages became united and concordant*, Homl. Th. i. 318, 24. Ða geþwǽran yrfweardiaþ eorþan *mansuēti hæreditabunt terram*, Ps. Spl. 36, 11. On geþwǽrum limum *in agreeing limbs*, Bt. 33, 4; Fox 130, 39. Ðonne hit ǽfre geþwǽrust sý ondrǽt ðē ðonne ungeþwǽrnisse *when things go most smoothly, then expect trouble*, Prov. Kmbl. 75.

ge-þwǽrian, -þwérian; *p.* ode, ede; *pp.* od.   **I.** *to cause to agree, to make accordant, mild :—*He geþwǽrede ða ðe óþ ðæt ungeþwǽre wǽron *he brought those to agree who until then had disagreed*, Bd. 3, 6; S. 528, 31. God gemetgaþ ealla gesceafta and geþwǽraþ ðá hé betwuh him wuniaþ *God regulates all creatures and makes them agree when they exist together*, Bt. 39, 13; Fox 234, 10: 8; Fox 224, 9, Cot. MS. Geþwēraþ [geþweraþ?] Bt. Met. Fox 29, 94; Met. 29, 47. Ðú geþwéras *tu mitigas*, Ps. Spl. T. 88, 10. Geþwiǽrodes *mitigasti*, 84, 3.   **II.** *to be or become in accord, to agree, consent, be agreeable;* consentire, concordare, congruere, convenire :—Se eorl nolde ná geþwǽrian ðære infare *the earl would not consent to the entrance*, Chr. 1048; Erl. 178, 11. Uton geþwǽrian mid ðam yrþlinge *conveniamus apud aratorem*, Coll. Monast. Th. 31, 27. Gif twegen of eów geþwǽriaþ be ǽlcum þinge *si duo ex vobis consenserint de omni re*, Mt. Bos. 18, 19. Ðes ne geþwǽrode hyra geþeahte *hic non consenserat consilio eorum*, Lk. Bos. 23, 51. Hí geþwǽredon *sibi concordant*, Bd. 2, 2; S. 502, 16. Ða þing ðe geþwǽredon ǽnnysse ðære cyriclícan sibbe *ea quæ unitati pacis ecclesiasticæ congruerent*, 4, 5; S. 571, 42. Ðú ðe wǽre geþwǽrigende ðam Hǽlende *thou that wast consenting to the Saviour*, Nicod. Thw. 6, 24.

geþwǽr-lǽcan, -lécan; *p.* -lǽhte; *pp.* -lǽht *To agree, assent to;* concordāre, assentīre :—He sǽde ðæt heora þeáwas ne mihton his dihte geþwǽrlǽcan *he said that their manners could not accord with his disposition*, Homl. Th. ii. 158, 7. Se ðe sóþlíce God lufaþ nele he wiðerian ongeán his bebodum ac hí geþwǽrlǽhþ *he that truly loves God will not resist his commands but comply with them*, 522, 19. Seó sǽ and se móna geþwǽrlǽcaþ *the sea and the moon agree*, Bd. de nat. rerum; Wrt. popl. science 15, 15; Lchdm. 268, 12. Ða hǽðengyldan ðisum cwide geþwǽrlǽhton *the idolaters assented to this proposal*, Homl. Th. i. 70, 34. Geþwǽrlécan *to agree*, Boutr. Scrd. 21, 1.

ge-þwǽrlíce, -þwǽrelíce; *adv. Harmoniously, in accord :—*Sió sunne and se móna rícsiaþ swíðe geþwǽrelíce *the sun and moon rule very harmoniously*, Bt. 39, 13; Fox 234, 6. Geþwǽrlíce *consonanter*, Bd. 4, 17; S. 585, 35.

ge-þwǽrnes, -ness, niss, -e; *f. Concord, agreement, mildness;* concordia, mansuetudo :—Mid fægerre geþwǽrnesse *pulchra concordia*, Bd. 4, 23; S. 596, 23. Sibb and geþwǽrnyss *pax et concordia*, Coll. Monast. Th. 31, 25; Blickl. Homl. 109, 16. He ðæt ríce heóld on gódre geþwǽrnesse and on micelre sibsumnesse *he held the kingdom in great peace and tranquillity*, Chr. 860; Erl. 70, 23: 827; Erl. 64, 8. Geþwǽrnysse *mansuetudinem*, Ps. Spl. 44, 6: Prov. Kmbl. 23.

ge-þweán; *p.* -þwóh, *pl.* -þwógon; *pp.* -þwagen, -þwegen, -þwogen, -þwǽn *To wash :—*Ongann geþuoá foet his *coepit rigare pedes ejus*, Lk. Skt. Lind. 7, 38. Ða geþuógon ðæt nett *lavabant retiam*, 5, 2. Búta oftor geþuógon hondo *nisi crebro lavarent manus*, Mk. Skt. Lind. 7, 3. Búton hí geþwegene beón *nisi baptizentur*, Mk. Boś. 7, 4. Se ðe geþwǽn is *qui lavatus est*, Jn. Skt. Lind. 13, 10. Hwí he geþwogen nǽre *quare non baptizatus esset*, Lk. Bos. 11, 38.

ge-þweor, es; *n. Curd, what is coagulated;* coagulum, Coll. Monast. Th. 28, 19. v. buter-geþweor.

ge-þweran; *p.* -þwær, *pl.* -þwǽron; *pp.* -þworen, -þuren *To stir, beat* or *mix together, to churn, make thick* [*as butter from cream*], poetically, *to forge;* cudere :—Genim cú meoluc bútan wǽtere lǽt weorþan to flétum geþwer to buteran *take cow's milk, without water, let it become cream, churn it to butter*, L. M. 1, 44; Lchdm. ii. 108, 22. Geþworen [Lye] geþrofen [Wrt.] fliéte *churned cream;* lactudiclum, Wrt. Voc. 290, 28. Heoru hamere geþuren *the sword forged with the hammer*, Beo. Th. 2575; B. 1285: Exon. 129 b; Th. 497, 16; Rä. 87, 1. Eorþe is hefigre óðrum gesceaftum þicre geþruen *earth is heavier than the other elements, more closely compacted*, Bt. Met. Fox 20, 267; Met. 20, 134. [Cf. þwiril *verberaturium;* O. H. Ger. ga-dweran *confundere, miscere*, Grff. v. 278.]

ge-þwére; *adj. United, agreeing;* concors, Ælfc. Gr. 9, 44; Som. 13, 4: Shrn. 182, 5. v. ge-þwǽre.

ge-þwerian; -þweorian; *p.* ode, ede; *pp.* od, ed *To mix, mingle :—*Geþwere mix, L. M. 2, 51; Lchdm. ii. 264, 25. Geþweorod sint þegnas togædere *the ministers are mingled together* [cf. v. 66], Bt. Met. Fox 20, 143; Met. 20, 72. [Cf. ge-þweran, ge-þwǽrian; *and O. H. Ger.* tuaron, Grff. v. 278.]

ge-þwin [-þwing? Grn: cf. *O. Sax.* ge-þwing: *O. H. Ger.* ge-dwing], es; *n. Torment;* tormentum. v. hell-geþwin.

ge-þwinglod; *part. p. Compressed, fastened up :—*Ða Ismaheli hæfdon geþwinglode loccas *the Ishmaelites had their hair fastened up* [?], Shrn. 38, 5.

ge-þwit, es; *n. What is cut* or *shaved off, shavings, cuttings, chips;* assulæ :—Heo of ðære ilcan styþe spónas þweoton ond sceafþan [ðæt geþwit, MS. B.] nómon ond in wæter sendon *they cut off chips from the very stud* [*prop*] *and threw the cuttings into the water*, Bd. 3, 17, MS. T; S. 544, 44, col. 2, note. DER. þwítan.

ge-þýan; *p.* de; *pp.* ed *To press;* premere :—Geþýþ hý *presses them*, Salm. Kmbl. 607; Sal. 303: Salm. Kmbl. p. 150, 34. v. þýan.

ge-þýdan; *p.* de; *pp.* ed *To join, associate;* sŏciāre :—Monige to ðære ánnesse hí geþýddan þurh geleáfan ðære hálgan Cristes cyrican *plūres ūnĭtāti se sanctæ Christi ecclēsiæ crēdendo sŏciare*, Bd. 1, 26; S. 488, 12. Saga hú ðú ðec geþýde on clǽnra gemong *say how thou associatest thyself in the company of the pure!* Exon. 71 b; Th. 267, 22; Jul. 419. Us is swíðe mycel nédþearf ðæt we us geþýdon to úrum hálgum gebedum *there is very great need for us to betake ourselves to our holy prayers*, Blickl. Homl. 133, 8. Wit sceoldan beón tosamne geþýdde *we had to be joined together*, Shrn. 39, 19. He wæs Gúþláce neáh geþýded *he was nearly associated to Guthlac*, 47 a; Th. 162, 6; Gú. 971. v. ge-þeódan.

ge-þýde; *adj. Good :—*Sum biþ árfæst and ælmes-georn þeáwum geþýde *one is pious and charitable, morally good*, Exon. 79 a; Th. 297, 14; Crä. 68. [Cf. Goth. þiuþ.]

ge-þýht; *adj. Good, advantageous*, Exon. 94 a; Th. 353, 25; Reim. 18. [Cf. þyhtig, ge-þuhtsum.]

ge-þýlan *succumbere, consentire*, Hpt. Gl. 482.

ge-þyld, e; *f. Patience, resignation;* patientia :—Ðú me eart fǽle geþyld *tu es patientia mea*, Ps. Th. 70, 4: Ps. Spl. 61, 5: Job Thw. 167, 16. Sum þafaþ in geþylde ðæt he sceal *one allows what he must with patience*, Exon. 79 a; Th. 297, 20; Crä. 71. On geþylde *in patientia*, Lk. Bos. 8, 15. Gehafa geþyld on me *patientiam hábe in me*, Mt. Bos. 18, 26: Exon. 79 b; Th. 298, 3; Crä. 79: Beo. Th. 2795; B. 1395. Mid geþylde *with patience*, L. In. 6; Th. i. 106, 12: Ps. Th. 91, 13. Eal ðú hit geþyldum gehealdest *thou supportest it all patiently*, Beo. Th. 3415; B. 1705. Forber oft ðæt ðú wrecan mǽge geþyld biþ middes eádes *often forbear when vengeance is in your power, patience is half happiness*, Prov. Kmbl. 25. [*O. Sax.* gi-þuld: *O. H. Ger.* ge-dult: *Ger.* ge-duld.]

ge-þyldelíc; *adj. Patient :—*Crist us onstealde geþyldelíce bysene *Christ has set us an example of patience*, Blickl. Homl. 75, 29.

ge-þyldelíce; *adv. Patiently, quietly;* patienter :— Drihten deófles costunga geþyldelíce abær *the Lord bore the temptation of the devil patiently*, Blick. Homl. 33, 28: Bd. 1, 7; S. 477, 46.

ge-þyldig; *adj. Patient, long-suffering, quiet;* patiens, longănĭmis :—Þeáwfæst and geþyldig *upright and patient*, Cd. 126; Th. 161, 8; Gen. 2662: 92; Th. 116, 26; Gen. 1942. Geþyldig and swýðe mildheort *patiens et multum mĭserĭcors*, Ps. Spl. 144, 8: Ps. Th. 85, 14: 102, 8. Geþyldige hí beóþ *patientes ĕrunt*, Ps. Spl. C. 91, 14. Ða geþyldigan *sustĭnentes*, Ps. Spl. 36, 9. [*O. H. Ger.* ge-dultig: *Ger.* ge-duldig.]

ge-þyldigean, -þyldgian, -þyldian; *p.* ode; *pp.* od *To be patient, to bear patiently, endure, to bear, endure, sustain;* sustinere, patientiam habere, tolerare, pati :—He ne mæg geþyldgian ðæt he ðæt forhele *he cannot bear to conceal it*, Past. 33, 2: Swt. 216, 6, 8. Geþyldega *patientiam habe*, Mt. Bos. 18, 29. Ðē ic geþyldgode ealne dæg *te sustinui tota die*, Ps. Spl. 24, 5. Geþyldigendum *patientibus*, 102, 6. Geþyldiendium, Ps. Spl. T. 145, 5. [*O. H. Ger.* ge-dultian: *Ger.* ge-dulden.]

ge-þyldiglíce. v. ge-þyldelíce.

ge-þyll, es; *n. A breeze, air :—*Giþyll scendende *aura corrumpens*, Rtl. 121, 28.

ge-þylmédan *to make patient, bring down :—*Hí geþylméde synt *ipsi obligati sunt*, Ps. Lamb. 19, 9.

ge-þylmód; *adj. Patient;* patiens, Lye.

ge-þylmódness, e; *f. Patience;* patientia, Lye.

ge-þýn = ge-þýan *to press :—*He mæg ealla gesceafta on ánes weaxæpples [MS. -æples] onlícnisse geþýn *he can press all creatures into the likeness of a wax apple*, Salm. Kmbl. p. 150, 34.

ge-þyncan; *p.* -þúhte; *pp.* -þúht *To seem, appear :—*Ðǽr him wlitebeorhte wongas geþúhton *where appeared to them plains beautifully bright*, Cd. 86; Th. 108, 11; Gen. 1804. Se ðe to-dæg is úre folgere geþúht *he who to-day seems our follower*, Homl. Th. ii. 80, 20. His loccas and his beard wǽron gylden geþúht *his hair and his beard seemed of gold*, Nar. 43, 14: Homl. Th. ii. 80, 12. Ðonne wǽre geþúht swilce . . . *then it would have seemed as if . . .*, i. 578, 3. Is me geþúht *it seems to me*, Exon. 47 b; Th. 163, 6; Gú. 989: 49 a; Th. 169, 18; Gú. 1096. v. þyncan.

ge-þyncþ. v. ge-þincþ.

ge-þynd, es; *n. A swelling :—*Wið geþind *against a swelling*, Herb. 46, 4; Lchdm. i. 150, 1. [Cf. to-þunden.]

ge-þynge, es; *n. Growth, increase, advancement, honour* :—Ðætte he hæbbe forþgeong and geþyngo *that he may have advancement and honour*, Jn. Skt. p. 188, 11. Giþynge *provectum*, Rtl. 50, 21. [Cf. ge-þungen.]

ge-þynnian ; *p.* ode; *pp.* od *To thin, lessen, diminish* ; attenuäre :—Geþynnode synt eágan mîne *attenuäti öcüli mei*, Cant. Ezech. Lamb. fol. 185 a, 14. v. ge-þinnian.

ge-þyrst ; *adj. Thirsty* :—Se geþyrsta mon meolcode ða hinde *the thirsty man milked the hind*, Shrn. 130, 4.

ge-þýwan, -þýan, -þîwan, -þêwan, -þeón, -þeówan ; *p.* -þýwde, -þýde ; *pp.* -þýd *To press, impel, urge, force, impress, rebuke, oppress* ; prěměre, trüděre, urgěre, compellěre, imprîměre, incrěpäre, oppriměre :—Se snâw geþýþ hý and geþreátaþ *the snow presses and afflicts them*, Salm. Kmbl. 607 ; Sal. 303. Ðû Reádne Sǽ ricene geþýwdest *incrěpävit Märe Rubrum*, Ps. Th. 105, 9. He Ægypti egesan geþýwde mid feala tâcna *pösuit in Ægypto signa sua*, 77, 43. Hí mec þingum geþýdan *they pressed me violently*, Exon. 123 a ; Th. 472, 10 ; Rä. 61, 14. Geþýd and geþreátod *rebuked and threatened*, Andr. Kmbl. 871 ; An. 436. Gesâwon hí swilce mannes fôtlǽsta fæstlîce on ðam stâne geþýde *they saw as it were a man's footsteps firmly impressed on the stone*, Homl. Th. i. 506, 12.

ge-þýwe ; *adj. Customary, usual* :—Him geþýwe wæs ðæt he oft ðǽr wunode [*other version has* his gewuna wæs] *sæpius ibidem diverti ac manere consueverat*, Bd. 3, 17 ; S. 543, 24. Swâ him geþýwe ne wæs *as was not usual with him*, Bd. 2. 332. v. ungeþeáwe.

ge-tídan, -týdan ; *p.* de; *pp.* ed [tîdan *to betide*] *To betide, happen* ; contingere :—Getîdeþ oft *it often happens*, Bt. 33, 2 ; Fox 124, 13. Ðé-læs ðe ðé on sumum þingum wyrs getîde *ne deterius tibi ǎlǐquid contingat*, Jn. Bos. 5, 14. Getýdde hit, ðæt . . . *it happened that . . . ,* Bt. 16, 2 ; Fox 52, 34.

ge-tígan ; *pp.* -tíged *To tie, bind* :—Forhwon fealleþ se snâw wæstmas getîgeþ *why does the snow fall, bind up the fruits?* Salm. Kmbl. 606 ; Sal. 302. Ðǽr stôd ân ramm getíged be ðam hornum *there stood a ram tied by the horns*, Homl. Th. ii. 62, 3 : ·i. 206, 10 : Lk. Bos. 19, 30 : Mt. Bos. 21, 2.

ge-tigþian *to grant*, Cd. 131 ; Th. 166, 23 ; Gen. 2752. v. ge-tídian.

ge-tihhian ; *p.* ode, ade ; *pp.* od, ad *To appoint, determine, assign* ; stätuěre, decerněre, assignäre :—Swâ he æt fruman getihhod hæfde as *he at the beginning had determined*, Bt. 39, 3 ; Fox 220, 26. Hafast ðé ânum eall getihhad land and leóde *thou hast brought all the land and people on thyself*, Andr. Kmbl. 2642 ; An. 1322. v. ge-teohhian, -tiohhian.

ge-tihtan ; *p.* te *To incite, urge, persuade* :—Ic getihte hundas mîne *instigo canes meos*, Coll. Monast. Th. 21, 15. Getiht *suasum*, Ælfc. Gr. 26 ; Som. 28, 53. Getiht *instigatus, præmonitus, compunctus*, Hpt. Gl. 420.

ge-tíhtlod, -tîhtled, -týhtlod, -týhtled ; *part.* [tîhtlian *to accuse*] *Accused* ; accüsätus :—Gif se getîhtloda man mâran werude beó ðonne twelfa sum *if the accused man be of a larger company than twelve*, L. Ath. i. 23 ; Th. i. 212, 8. Nân man ne tæce his getîhtledan man fram him, ǽr he hæbbe ryht geworhte *let no one dismiss his accused man from him before he has done what is right*, i. 22 ; Th. i. 210, 23.

ge-tilian, -tilgan ; *p.* ode ; *pp.* od. • I. *to strive after, to get by striving, to obtain, procure, acquire* ; acquîrěre :—Ne ic mâran getilige to haldænne *nor do I strive to have more*, Shrn. 183, 3. Se ðe hit deþ him seluan éce hellewîte ungesǽliglîce getilaþ *he who does it will miserably get for himself everlasting hell torment*, Th. Chart. 117, 24. Hæbbe ic þearfe ðæt ic dîne hyldo getilge *I have need that I acquire thy grace*, Exon. 118 a ; Th. 454, 8 ; Hy. 4, 29. [*Goth.* ga-tilon *to obtain*.] II. *to treat a patient* ; curäre :—Ic wât hû ðín man getilian sceal *I know how you must be treated*, Bt. 5, 3 ; Fox 12, 32.

ge-tillan ; *p.* de ; *pp.* ed *To touch, reach, attain* ; tangere, attingere :—Astrece dîne hand and getill ealle ða þing ðe he âh *extende manum tuam et tange cuncta quæ possidet*, Job Thw. 165, 15. Weras blôda and fâcenfulle nâ healfe getillaþ *viri sanguinei et dolosi non dimidiabunt*, Blickl. Gl.

ge-tilþ, e ; *f. Gain* ; mercimonia, Hpt. Gl. 439.

ge-timbernes, -ness, e ; *f. A building, edification* ; ædîficätio :—To gemynde and to getimbernesse ðara æfterfyligendra *ad měmöriam ædîficätiönemque sěquentium*, Bd. 4, 7 ; S. 574, 25. Gitimbernise *ædificatio*, Rtl. 82, 36 : 83, 13.

ge-timbran, -timbrian, -timbrigean ; *part.* -timbriende ; *p.* ode, ade, ede ; *pp.* od, ad, ed [timbrian *to build*]. I. *to make of wood, to build, to build up, construct* ; ædĭficäre, construěre, exstruěre :—Cirircean getimbran *to build a church*, Andr. Kmbl. 3265 ; An. 1635. He hêt getimbrian cyrican of treówe *he commanded a church of wood to be built*, Chr. 626 ; Erl. 23, 40 : Bd. 2, 3 ; S. 504, 23. Getimbrigean, Mt. Bos. 26, 61. Getimbriende Hierusalem Drihten *ædĭficans Hierusalem Dömĭnus*, Ps. Spl. 146, 2. Ic getimbre hûs *I will build a house*, Exon. 36 a ; Th. 117, 9 ; Gû. 221. Ic getimbrie, Mk. Bos. 14, 58. Ofer ðisne stân ic getimbrige mîne cyrcan *over this stone I will build my*

church, Homl. Th. ii. 390, 2, 10, 11, 12. Nâ ðû getimbrast hí *non ædîficäbis eos*, Ps. Spl. 27, 7. He getimbreþ eardwîc niwe *he builds a new dwelling-place*, Exon. 62 a ; Th. 227, 28 ; Ph. 430 : Salm. Kmbl. 150 ; Sal. 74. Wâ, ðæt ðes towyrpþ Godes templ, and on þrim dagum hyt eft getimbraþ *vah qui destruis templum Dei, et in trǐduo illud reædîficas*, Mt. Bos. 27, 40 : Mk. Bos. 15, 29. On ðam seáþe ufan se eádiga wer, Gûthlâc, him hûs getimbrode *over the cistern the blessed man, Guthlac, built himself a house*, Guthl. 4 ; Gdwin. 26, 9 : Gen. 4, 17 : Ex. 24, 4. Ðe Rôme burh getimbredon *who built Rome*, Ors. 2, 1 ; Bos. 38, 41. Naman mînne on ferhþlocan fæste getimbre *fast build up my name within their hearts*, Andr. Kmbl. 3339 ; An. 1673. Deáh ðe ðæt port beó trumlîce on ælce healfe getimbrod *though the gate be firmly constructed on every side*, Homl. Th. ii. 432, 3. Ðǽr getimbred wæs tempel Dryhtnes *where the temple of the Lord was built*, Andr. Kmbl. 1333 ; An. 667 : Ors. 2, 1 ; Bos. 39, 30. Beóþ byrig mid Iudéum eft getimbrade *ædĭficäbuntur cĭvĭtätes Jüdæ*, Ps. Th. 68, 36 : Bd. 1, 1 ; S. 473, 27. II. *to build up the mind, instruct, edify* ; instruěre :—Ic getimbrige ðé on wege *instruam te in via*, Ps. Spl. C. 31, 10.

ge-timbru, -timbro ; *pl. gen.* -timbra ; *n. An edifice, a building, structure* ; ædǐficium, structüra :—Gé geseóþ ealle ða fægernessa ðissa getimbra . . . ealle ðás getimbro beóþ toworpene *ye see all the beauties of these buildings . . . all these buildings shall be destroyed*, Blickl. Homl. 77, 34-6 : Mt. Kmbl. Lind. Rush. 24, 1. Ðæt sind ða getimbru ðe nô [MS. nü] tydriaþ *these are the structures which shall not decay*, Exon. 32 b ; Th. 103, 5 ; Cri. 1683 : 39 b ; Th. 131, 16 ; Gû. 456 : Bd. 3, 8 ; S. 532, 30. Hruran and feóllan cynelîco getimbro and ânlîpie *ruĕbant ædĭficia publica sǐmul et prīväta*, Bd. 1, 15 ; S. 483, 45 : 3, 8 ; S. 532, 32 : Cd. 15 ; Th. 18, 20 ; Gen. 276. Getimbra hâlgung *scenophegia* [= scĕnŏpēgia], Ælfc. Gl. 3 ; Som. 55, 77 ; Wrt. Voc. 16, 50. [O. H. Ger. gizimbri ; *n. materia, ædĭficium* : Ger. ge-zimmer ; *n. timber-work* : and cf. Goth. ga-timrjo ; *f. a building*.]

ge-timbrung, e ; *f.* I. *an edifice, a structure, building* ; ædĭficium, ædĭficätio, structüra :—Getimbrung *ædĭficium*, Ælfc. Gl. 81 ; Som. 73, 12 ; Wrt. Voc. 47, 19 : 86, 26. Hí geswicon ðære getimbrunge *they ceased from the building*, Homl. Th. i. 318, 21. Ðæt hí him ætýwdon ðæs temples getimbrunge *ut ostenděrent ei ædĭficätiönes templi*, Mt. Bos. 24, 1 : Homl. Th. ii. 390, 13. Lôca hwylce getimbrunga *aspǐce, quäles structüræ*, Mk. Bos. 13, 1. II. *a definition* ; definîtio :—Getimbrung *definîtio*, Cot. 69.

ge-tîmian, -týmian ; *p.* ode ; *pp.* od [tîma *time*] *To happen, befall* ; accĭděre :—Getîmian to happen, Jud. 5 ; Thw. 156, 8. Getîmode hit ðæt he becom to heora byrig ðe wæs Gaza gehâten *it befell that he came to their city which was called Gaza*, Homl. Th. i. 226, 24 : 318, 15. Him getîmode swîðe rihtlîce *it happened very justly to them*, 88, 29 : ii. 160, 14 : 304, 24. Getîmige ðé swâ swâ ðú gelýfdest *be it to thee as thou hast believed*, i. 126, 21.

ge-ting, -tincg, e ; *f. Condition, state* :—Missenlîcræ yldo and getincge men *homines conditionis diversæ et ætatis*, Bd. 1, 7 : S. 478, 6.

ge-tingan ; *p.* -tang *To press upon, throng* :—Corþer ôðrum getang *one troop pressed on the other*, Andr. Kmbl. 276 ; An. 138.

ge-tingcræft, es ; *m. Mechanics*, Hpt. Gl. 479.

ge-tinge, -tynge *eloquence* ; lepor, Lye.

ge-tinge, -tincge, -tynge ; *adj. Skilful with the tongue, eloquent* :—Getinge *disertus*, Ælfc. Gr. 47 ; Som. 48, 51. Getingce *lepida vel facunda*, Ælfc. Gl. 100 ; Som. 76, 129 ; Wrt. Voc. 55, 1. Gif se Hǽlend gecure æt fruman getinge lâreówas *if the Saviour had chosen at first eloquent teachers*, Homl. Th. i. 578, 1. Wer getinge *vir linguosus*, Ps. Spl. C. 139, 12. v. ge-tynge.

ge-tingelic, -tyngelîc ; *adj. Pleasant in speech, affable, eloquent* ; lepidus, affabilis :—Getyngelîc *rhetoricus*, Hpt. Gl. 485 : Cot. 179.

ge-tingelîce ; *adv. Eloquently* :—Ðæt cild getingelîce spræc *the child spoke eloquently*, Homl. Th. ii. 490, 32.

ge-tingness, -tyngness, e ; *f. Eloquence, ease of speech* ; facundia :—Dumbum he forgeaf getingnysse *to the dumb he gave eloquence*, Homl. Th. i. 26, 12. Of woruldlîcre getingnysse *from worldly eloquence*, 578, 4 : ii. 140, 30 : Swt. A. S. Rdr. 69, 403. Metcundlîc getyngnis *metrica facundia*, Hpt. Gl. 409.

ge-tióde *appointed, determined, decreed*, Bt. Met. Fox 11, 76 ; Met. 11, 38 : 13, 26 ; Met. 13, 13 ; *p.* of ge-tión.

ge-tiohhian ; *p.* ode ; *pp.* od *To appoint, determine, ordain* ; stätuěre, decerněre :—Ðú ðæm winterdagum wundrum sceorta tîda getiohhast *thou appointest wondrously short times to winter-days*, Bt. Met. Fox 4, 41 ; Met. 4, 21. Swâ him æt frymþe Fæder getiohhode *as the Father appointed to them at the beginning*, 29, 78 ; Met. 29, 38. Swâ he getiohhod habbe *as he has ordained*, Bt. 41, 2 ; Fox 244, 20. v. ge-teohhian.

ge-tión, ic -tió, *pl.* -tióþ ; *impert.* -tió, *pl.* -tióþ ; *subj. pres.* -tió, *pl.* -tión *To draw, to attract* ; trähěre, attrahere :—Hwæðer nû gimma wlite eówre eágan to him getió *does now the beauty of gems attract your eyes to them?* Bt. 13 ; Fox 40, 2 : 38, 1 ; Fox 196, 15.

ge-tión ; *p.* -tióde ; *pp.* -tiód *To appoint, determine, ordain* ; stätuěre,

decernĕre :—Swâ him æt frymþe Fæder getióde *as the Father appointed to it at the beginning,* Bt. Met. Fox 24, 28 ; Met. 24, 14 : 13, 26 ; Met. 13, 13 : 11, 76 ; Met. 11, 38. v. ge-teón.

ge-tiorian. v. ge-teorian.

ge-titelian ; *p.* ode ; *pp.* od *To entitle, ascribe* :—Twâ bêc for ðære gelícnisse his gelogodan sprǽce man getitelode him *two books from the likeness to his style are ascribed to him,* Swt. A. S. Rdr. 69, 404.

ge-tíþ *draws, constrains,* Bd. 1, 27 ; S. 494, 11. v. ge-teón.

ge-tíðian, -týðian, -tigðian ; *p.* ode ; *pp.* od *To grant, allow* :—Him nolde Alexander ðæs getíðian *Alexander would not grant him that,* Ors. 3, 9 ; Bos. 65, 7. Ðæs him getíðaþ Drihten Crist *the Lord Christ grants him that,* Homl. Th. i. 76, 22. Ðû bæde me and ic ðê ne getíðode *you asked me and I did not grant thee,* Swt. A. S. Rdr. 57, 16 : Th. i. 76, 22 ; S. 525, 30. Gif he eów ðises ne getíðode *if he has not granted you this,* Homl. Th. ii. 144, 17. Hý him ðære bêne getiðdedon *they granted him the request,* Ors. 2, 5 ; Bos. 47, 43 : Cd. 131 ; Th. 166, 23 ; Gen. 2752. Getíða me *grant me,* Hy. 3, 2 : 55. Ic wille ðæt gé ealle getíðe mîne worde *I will that ye all allow my words,* Chr. 656 ; Erl. 31, 3. Hí his bênum getíðdodon *they should grant his prayers,* Swt. A. S. Rdr. 96, 42. Him wearþ ðæs getíðod *that was granted him,* 44 : Beo. Th. 4558 ; B. 2284.

ge-toge, es ; *n. A tugging, contraction, cramp, convulsion, spasm* ; contractio, convulsio, spasmus :—Wið sina getoge *for spasm of sinews,* Med. ex Quadr. 6, 23 ; Lchdm. i. 356, 3. v. ge-teóhn.

ge-togen *drawn, incited, restrained, educated, brought to an end, drawn together, strung,* Ors. 1, 14 ; Bos. 37, 14 : Bd. 2, 5 ; S. 507, 42. v. ge-teón.

ge-togennes, -ness, e ; *f. Cramp, convulsion* ; contractio, convulsio, Som.

ge-toht, es ; *n. A warlike expedition, battle* ; expǽditio bellīca, pugna :—Æt getohte *at the battle,* Byrht. Th. 134, 54 ; By. 104. v. tohte.

ge-torfian ; *p.* ode ; *pp.* od *To stone* :—Hig wæron myd stânum getorfode *they were stoned with stones,* St. And. 36, 19. v. torfian, of-torfian.

ge-tot, es ; *n. Pomp, splendour* ; pompa :—Ídel-wuldor ðæt is gylp oððe getot *vain-glory, that is pride or pomp,* Homl. Th. ii. 220, 28. Riggon ðe mid ðam leáslícum getote inneode *Riggo who entered with the false pomp,* 168, 16. Getote *pompa,* R. Ben. 7, Lye.

ge-trahtian, -trahtnian ; *p.* ode ; *pp.* od *To treat, explain, expound, consider* ; tractāre, expōnĕre, considĕrāre :—Sume ðas race we habbaþ getrahtnod on óðre stówe *some of this narrative we have expounded in another place,* Homl. Th. ii. 264, 23. Ðâ cwæþ Pilatus Hû clypedon hig and hû byþ hit getrahtnod on Hebreisc *then said Pilate ' How did they call out and how is it explained in Hebrew,'* Nicod. 4 ; Thw. 2, 31. Getrahtad *interpretatum,* Jn. Skt. Lind. 1, 38, 41 : 9, 7. Huætd on woeg gie getrahtade *quid in via tractabatis,* Mk. Skt. Lind. 9, 33. Habbaþ word gearu wið ðam æglǽcan eall getrahtod *we have words ready all considered against the wretch,* Andr. Kmbl. 2718 ; An. 1361.

ge-tredan *to tread down* ; conculcare :—Ðý læs hia getrede ða ilco miþ fótum hiora *ne forte conculcent eas pedibus suis,* Mt. Kmbl. Lind. 7, 6. Getreden biþ *conculcetur,* 5, 13 : Lk. Skt. Lind. 8, 5.

ge-tregian, *to despise* ; despicere :—Þû ne getregedest mǽdenes innoþ *tu non despexisti virginis uterum,* Te Deum, Lye.

ge-treminc *a fort, fortress* ; munimentum, Prov. 12, Lye.

ge-tremman ; *p.* -tremede ; *pp.* -tremed *To strengthen, establish, confirm* ; firmāre, confirmāre :—Eall úre líf he getremede *he strengthened all our life,* Blickl. Homl. 9, 36. Hwá hine heálíce torhtne getremede tungolgimmum *who had established it bright on high with starry gems,* Exon. 24 b ; Th. 71, 5 ; Cri. 1151. Me gâste ðíne, God, getreme *strengthen me, O God, with thy spirit,* Ps. C. 50, 102 ; Ps. Grn. ii. 279, 102. Getremed *confirmed,* 50, 133 ; Ps. Grn. ii. 279, 133 : Blickl. Homl. 17, 6 : 119, 14. v. ge-trymman.

ge-treówan, -triówan, -triéwan ; *p.* de ; *pp.* ed. **I.** *to trust, believe, have confidence, hope* ; confidere, credere, sperare :—Ic gemǽnscipe getreówe ðínra hâligra *I believe the communion of thy saints,* Hy. Grn. ii. 294, 52, 55 ; Ps. Th. 118, 15. Ic on ðín word getreówe *in verbum tuum speravi,* 114 : 62, 1, 7 : 129, 5 : 124, 1 : 129, 6. Ic on ðínum wordum getreówde *I trusted in thy words,* 5. Ic ðínum wordum getreówde *in verbum tuum speravi,* 118, 74. Ðú in êcne god ðínne getreowdes *thou hast trusted in thy eternal God,* Exon. 72 a ; Th. 268, 21 ; Jul. 435. Gûþlác sette hyht in heofonas hǽlu getreówde *Guthlac put his hope in heaven, trusted in salvation,* 39 a ; Th. 128, 19 ; Gû. 406. **II.** *to make true or credible* :—Ðín gewitnes is weorcum geleáfsum and mid sóþe is swíðe getreówed *testimonia tua credibilia facta sunt nimis,* Ps. Th. 92, 6. **III.** *to persuade, suggest* :—We getrêwaþ him *nos suadebimus ei,* Mt. Kmbl. Lind. 28, 14 : 27, 20. Ðe hálig gâst gitrióweþ íowih alle ða ðe swá hwæt ic cweðo íow *spiritus sanctus suggeret vobis omnia quæcumque dixero vobis,* Jn. Skt. Rush. 14, 26. **IV.** *to make one's self out to be true, to clear one's self* :—Getríówe hine fâcnes se ðe hine fêde *let him who brings him up clear himself of treachery,* L. Alf. 17 ; Th. i. 72, 5. Getríéwe hine ðæs sleges

let him clear himself of the slaying, L. In. 34 ; Th. i. 122, 17. v. ge-treówian, ge-treówsian, ge-trúwan.

ge-treówe, -trýwe, -trúwe, -trêwe ; *def.* se -treówa ; *comp.* -treówra ; *superl.* -treówest ; *adj. True, trusty, faithful* ; fīdus, fīdēlis :—Getreówe oððe geleáfful *fidēlis,* Wrt. Voc. 74, 27 : Ps. Lamb. 144, 14. Ǽlc getreówa man *every true man,* L. C. S. 23 ; Th. i. 388, 9, note 12, MS. B. Mid fulre gewitnesse and getreówre *with full and true witness,* L. Ath. v. § 10 ; Th. i. 240, 9. Gif þegen hæbbe getreówne man *if a thane have a true man,* L. C. S. 23 ; Th. i. 388, 16, MS. B. Ic wille him syllan mîne gewitnesse weorþe and getreówe *servabo testāmentum meum fidēle ipsi,* Ps. Th. 88, 25 : 118, 111. Hwæðer gê getreówe synd *whether ye are true,* Gen. 42, 33. Hý habbaþ freónda ðý mâ tilra and getreówra *they will have the more of excellent and faithful friends,* Exon. 107 a ; Th. 409, 2 ; Rä. 27, 23. Beó getreówra *be more trusty,* Prov. Kmbl. 76. Ðe he, getreóweste, gelufade *whom, most faithful, he loved,* Exon. 43 a ; Th. 144, 21 ; Gú. 681. DER. un-getreówe.

ge-treówfæstnian *to be faithful, firm, strong* :—Ðu getreówfæstnig *valeas,* Mt. Kmbl. p. 4, 9.

ge-treówfull ; *adj. Faithful* ; fīdēlis :—Getreówfull *fidēlis,* Ælfc. Gr. 9, 28 ; Som. 11, 38. Gecýðnys getreówfull *testimōnium fidēle,* Ps. Spl. 18, 8. Ðú góda þeów and ðú getreówfulla *thou good servant and faithful,* Blickl. Homl. 63, 26.

ge-treówfullíce ; *adv. Faithfully, confidently* ; fīdūciālĭter :—Getreówfullíce ic dême on ðam *fidūciālĭter ăgam in eo,* Ps. Spl. 11, 6.

ge-treówian, -triówian ; *p.* ode, ede ; *pp.* od, ed. **I.** *to trust, confide, hope* :—Nelle gê on ealdurmenn getreówian *nolite confide in principibus,* Ps. Th. 145, 2. Ic on ðín sóþfæst word getreówige *I will trust to thy true word,* Ps. Th. 118, 80, 43, 48 : 130, 5. Ic me on mînne Drihten getreówige *ego in te sperabo Domine,* 54, 24 : 70, 13. **II.** *to make a treaty, be confederate* [v. ge-treówþ] :—Ða beorn getreówedon betwuh him and sieredon ymbe ðone fæder *the children were confederates and plotted against the father,* Bt. 31, 1 ; Fox 112, 13. Getreówod *fæderatus,* Cot. 85, Lye. **III.** *to clear one's self* :—Getríówie hine *let him clear himself,* L. Alf. 36 ; Th. i. 84, 15. v. ge-treówan, ge-trúwian.

ge-treówleás, -trýwleás ; *def.* se -leása ; *adj. Without faith, unfaithful, perfidious* ; perfĭdus :—Se getreówleása cyning *rex perfĭdus,* Bd. 3, 24 ; S. 556, 11.

ge-treówleásnes, -ness, -nys, -nyss, e ; *f. Infidelity, perfidy* ; perfĭdia :—Hí þrowedon heora getreówleásnesse *suæ perfĭdiæ pœnas luēbant,* Bd. 5, 23 ; S. 645, 34. For heora getreówleásnysse *for their perfidy,* 2, 2 ; S. 504, 9 : 1, 8 ; S. 479, 34.

ge-treówlíc ; *adj. Faithful* ; fīdēlis :—Getreówlícu oððe getrýwe ealle bebodu his word *fidēlia omnia mandāta ejus sunt,* Ps. Lamb. 110, 8. Us is swíðe uncúþ hwæt úre yrfeweardas getreówlíces dón willon æfter úrum lífe *it is quite unknown to us how faithfully our heirs will act after our life,* Blickl. Homl. 51, 36.

ge-treówlíce, -triówlíce, -tríwlíce, -trýwlíce ; *adv. Faithfully* ; fīdēlĭter :—Ðe him getreówlíce þeówdon *qui illi fīdēlĭter serviērunt,* Bd. 3, 13 ; S. 538, 36 : 3, 23 ; S. 554, 13 : Swt. A. S. Rdr. 107, 81 : Blickl. Homl. 185, 24, 28. Getríwlíce, Th. Ch. 202, 26.

ge-treówsian, -trýwsian ; *p.* ode ; *pp.* od *To justify one's self, clear one's self, prove one's self innocent* ; se justificāre, se purgāre :—Getreówsie hine fâcnes *let him prove himself innocent of the treachery,* L. Alf. pol. 17 ; Th. i. 72, 5, note 8, MS. H : 36 ; Th. i. 84, 15, note 36, MS. B.

ge-treówþ, -trýwþ, e ; *f. A covenant, treaty, pledge* ; fædus, pignus :—He gemunde ðara getreówþa *recordātus est fœdĕris,* Ex. 2, 24. v. ge-trýwþ.

ge-trêwe ; *adj. True, faithful* ; fīdus, fīdēlis, Cot. 85. v. ge-treówe.

ge-tricce ; *adj. Faithful* [?] :—Gif he biþ eáþhylde and ðære stówe getricce *si contentus fuerit consuetudine loci,* R. Ben. 61, Lye. v. ge-tryccan.

ge-triéwan, -triówan. v. ge-treówan.

ge-trifulian *to rub down* ; triturare :—Genim ða reádan netlan getrifula *take the red nettle, bruise it,* L. M. 1, 1 ; Lchdm. ii. 20, 15.

ge-trimmed ; *part.* [ge-trymman *to draw up*] *Drawn up* ; instructus :—Getrimmed fêða *cuneus,* Ælfc. Gl. 7 ; Som. 56, 79 ; Wrt. Voc. 18, 31.

ge-triówlíce ; *adv. Faithfully* ; fīdēlĭter, Prov. 10. v. ge-treówlíce.

ge-tríwe ; *def.* se -tríwa ; *adj. True, faithful* ; fīdus, fīdēlis :—Ǽlc getríwa man *every true man,* L. C. S. 23 ; Th. i. 388, 9, note 12, MS. A. v. ge-treówe.

ge-trucian *to fail* ; deficere :—Ðâ ðæt wîn getrukede *deficiente vino,* Jn. Skt. 2, 3, col. 2.

ge-trudend, es ; *m. A seizer* ; raptor, Cot. 170, Lye.

ge-trúgung, e ; *f. A certainty, defence, refuge* ; confidentia, Ps. Vos. 88, 18.

ge-trum, es ; *n. A knot, band, mass, company, company of soldiers* ; nodus, caterva, cohors, exercitus :—Getrum *nodus, inter militāria,* Ælfc.

Gl. 7; Som. 56, 81; Wrt. Voc. 18, 33. Fyrd sceal ætsomne, tírfæstra getrum *the army shall be assembled, a band of warriors,* Menol. Fox 523; Gn. C. 32. Under tungla getrumum *under the troops of stars,* Salm. Kmbl. 285; Sal. 142. He eft gewát getrume micle *he returned with a great company,* Andr. Kmbl. 1413; An. 707: Beo. Th. 1849;. B. 922: Exon. 90 a; Th. 337, 12; Gn. Ex. 63. DER. án-getrum, folc-, fyrd-, gár-.

ge-truma, an; *m. A company, troop of soldiers;* cohors, exercĭtus:—Wið ðara cyninga getruman *with the king's troop,* Chr. 871; Erl. 74, 19, 21. Ðeáh hí wyrcen getruman wið me *si consistant adversum me castra,* Ps. Th. 26, 4. v. ge-trum.

ge-trumian; *p.* ode, ade; *pp.* od, ad. **I.** *to grow strong, to recover, to gain strength;* convălescĕre:—Ðá he getrumad wæs *ut conváluit,* Bd. 4, 22; S. 592, 3. **II.** *to make strong, confirm;* confirmáre:—Ðone ðú getrumodest *quem confirmasti,* Ps. Spl. 79, 16, 18. Getrummade *exortans,* Lk. Skt. Lind. 3, 18. Getrumade *firmavit,* 9, 51.

ge-trúwa, an; *m. Confidence;* confidentia:—Ælcum getrúwan ic gyrne fultum ðínre foreþingrædene *omni confidentia implóro auxilium tuæ interventiónis,* Wanl. Catal. 294, 4, col. 2.

ge-trúwian; *p.* ode, ede; *pp.* ed. **I.** *to trust, hope;* confidere, sperare:—Ða þe on heora feó getrúwigeaþ *confidentes in pecuniis,* Mk. Bos. 10, 24. He getrúwade ðæt he hine beswícan mihte *he trusted that he could circumvent him,* Ors. 2, 4; Bos. 45, 10: 4, 1; Bos. 78, 44. Ðǽm he getrúwode ðæt hie his giongorscipe fulgán wolden *of whom he expected that they would perform his service,* Cd. 14; Th. 16, 25; Gen. 248. Ðú mínum wordum getrúwodest *thou hast trusted my words,* 29; Th. 38, 28; Gen. 613: 33; Th. 44, 9; Gen. 706: Beo. Th. 3071; B. 1533: 5074; B. 2540. Beorges getrúwode wíges and wealles *in his hill he trusted, in his war and his wall,* 4634; B. 2322. Ic on ðínum wordum getrúwade *in verba tua speravi,* Ps. Th. 118, 147: 51, 6. **II.** *to make a treaty;* sancire:—Hie getrúwedon on twá healfa fæste frioðuwǽre *they confirmed on both sides a fast compact of peace,* Beo. Th. 2194; B. 1095. v. ge-treówian.

ge-trúwung, e; *f. Confidence, trust:*—Getrúwung úre *assumptio nostra,* Ps. Spl. C. 88, 18.

ge-tryccan *to trust:*—Getryccaþ *confidite,* Jn. Skt. Lind. 16, 33.

ge-trym, es; *m. n? A support;* firmámentum:—Æðele getrym eorþan weardaþ, biþ se beorht ahafen ofer beorgas *érit firmámentum in terra, in summis montium,* Ps. Th. 71, 16.

ge-trymman, -trymian, -trymigan, -tremman; he -trymmeþ, -trymþ; *p.* -trymde, -trymede; *pp.* -trymed, -trymmed, -trymd. **I.** *to confirm, strengthen, encourage, establish, found, set in order, arrange, draw up;* firmáre, confirmáre, múníre, confortáre, hortári, fundáre, instruĕre:—Ic Wærferþ bisceop mid mínre ágenre handa ðas sylene getrimme and gefæstnie *I, bishop Wærferth, with my own hand confirm and ratify this donation,* Th. Ch. 169, 3. Ða ðe mágon getrymian [getrymigan, MS. Bod.] *which may encourage thee,* Bt. 36, 1; Fox 172, 27. Ic getrymme ofer ðé eágan míne *firmábo súper te óculos meos,* Ps. Lamb. 31, 8: Ps. Th. 74, 3. Getrymmeþ rihtwíse Drihten *confirmat justos Dŏmĭnus,* Ps. Spl. 36, 18. Ealle geþeaht ðín he getrymþ *omne consĭlium tuum confirmet,* 19, 4: Ps. Lamb. 36, 18. Ðú getrymdest ofer me hand ðíne *confirmasti súper me manum tuam,* Ps. Spl. 37, 2. Ðú me getrymedest *exhortátus es me,* Ps. Th. 70, 20: 79, 14, 16. He ða ymbhwyrft eorþan getrymede *firmávit orbem terræ,* 92, 2: 104, 20: 131, 11. He beforan ðam geate his folc getrymede *he drew up his army before the gate,* Ors. 4, 10; Bos. 92, 41. Getrym me *confirma me,* Ps. Spl. 50, 13. Ðín weorc on us getryme *confirma hoc quod opĕrátus es in nobis,* Ps. Th. 67, 26. Eall ðín geþeaht he getrymie *omne consĭlium tuum confirmet,* 19, 4. Byþ his heorte getrymed *confirmátum est cor ejus,* 111, 7: 116, 2. Hit wæs ofer ðam stán getrymed *fundáta ĕrat súper petram,* Lk. Bos. 6, 48. He hæfde ðæt folc getrymmed *he had drawn up the troops,* Byrht. Th. 132, 27; By. 22. Worde [MS. word] Drihtnes heofonas [MS. heofones] getrymde synd *verbo Dŏmĭnĭ cæli firmáti sunt,* Ps. Spl. 32, 6. **II.** *v. reflex. To grow strong, gain strength, recover;* convălescĕre:—Ecbyrht hine ðære ádle getrymede *Ecgberct ægrĭtudĭnis convăluit,* Bd. 3, 27; S. 559, 23.

ge-trymnes, -ness, e; *f. An exhortation, persuasion, a setting in order, an arraying;* hortátus:—Mid his getrymnesse *ejus hortátu,* Bd. 1, 33; S. 498, 35. Gitrymniso *ortamenta,* Rtl. 56, 4. Fyrdweorodes getrymnes *the arraying of a host,* Blickl. Homl. 91, 36.

ge-trýwan; *p.* de *To trust, hope:*—Ða ðe noldan on hine getrýwan *those who would not trust in him,* Blickl. Homl. 159, 11. Ic getrýwe in ðone torhtestan þrýnesse þrym *I believe in the most glorious virtue of the Trinity,* Exon. 42 a; Th. 140, 28; Gú. 617. Mín sáwel on ðé swíðe getrýweþ *in te confidit anima mea,* Ps. Th. 56, 1. Mægene getrýweþ *trusts in its strength,* Frag. Kmbl. 65; Leás. 34: Cd. 27; Th. 36, 10; Gen. 569. Getrýwde hweðre on Ælmihtiges Godes miht *he trusted however in the power of Almighty God,* Blickl. Homl. 217, 23. v. ge-treówan.

ge-trýwe; *def.* se -trýwa; *adj.* TRUE, *faithful;* fídus, fĭdélis:—Beó blíðe, ðú góda þeów and getrýwa; forðamðe ðú wǽre getrýwe ofer lytle þing, ic gesette ðé ofer mycle *euge, serve bŏne et fĭdélis; quia súper pauca fuisti fĭdélis, súper multa te constĭtuam,* Mt. Bos. 25, 21, 23: 24, 45: L. C. S. 23; Th. i. 388, 9. He wearþ Criste getrýwe *he became faithful unto Christ,* Elen. Kmbl. 2068; El. 1035: Beo. Th. 2461; B. 1228. He eallum mannum sǽde and bodode ðæt wuldor his getrýwan þeówes *omnibus fĭdélis sui fámuli glóriam prædícabat,* Bd. 3, 13; S. 539, 10. Gif þegen hæbbe getrýwne man *if a thane have a true man,* L. C. S. 23; Th. i. 388, 16. Twegen getrýwe men *two true men,* 30; Th. i. 392, 26: 394, 21. Ðam getrýwestan witan *to the most faithful senator,* Ors. 5, 4; Bos. 105, 7. v. ge-treówe.

ge-trýwian; *p.* ode. **I.** *to trust:*—Ic on ðínum wordum wel getrýwade *in verbum tuum superaperavi,* Ps. Th. 118, 1. **II.** *to clear one's self:*—Getrýwie hine ðæs sleges *let him clear himself of the slaying,* L. In. 34; Th. i. 122, 15, MS. B. v. ge-treówian.

ge-trýwleás; *adj. Perfidious;* perfĭdus, Greg. Dial. 2, 14. v. ge-treówleás.

ge-trýwlíce; *adv. Faithfully, confidently;* fĭdélíter, fidúciáliter:—Ðæt flǽsclícnysse úres Drihtnes Hǽlendes Cristes getrýwlíce he gelýfe *ut incarnátiónem Dómĭni nostri Iésu Christi fĭdélíter crédat,* Ps. Lamb. fol. 201 b, 29: 202 b, 42. Getrýwlíce oððe baldlíce ic dó on him *fidúciálíter ăgam in eo,* Ps. Lamb. 11, 6. v. ge-treówlíce.

ge-trýwsian; *p.* ode; *pp.* od *To justify one's self;* se justificáre:—Ðæt he hine ðæs getrýwsige *that he may justify himself thereof,* L. Ed. 6; Th. i. 162, 18. v. ge-treówsian.

ge-trýwþ, e; *f. A covenant, treaty, pledge, faith, fidelity;* fœdus, pignus:—Ofer ealle ða getrýwþa ðe he him geseald hæfde *against all the pledges which he had given him,* Chr. 1001; Erl. 136, 15: 1093; Erl. 229, 19. Lytle getrýwþa wǽron mid mannum *there has been little faith amongst men,* Swt. A. S. Rdr. 104, 8: 107, 74; 111, 220. v. ge-treówþ.

ge-tucian; *p.* ode; *pp.* od *To torment, vex, punish;* púníre:—Swilce he for his synnum swá getucod wǽre *as if he was so tormented for his sins,* Job Thw. 167, 14. v. tucian.

ge-tucian; *p.* ode; *pp.* od *To adorn, dress* [?]:—Ðǽr stent cwén ðé on ða swýðran hand mid golde getucode, and mid ǽlcere mislícre fægernysse gegyred *adstitit regina a dextris tuis in vestitu deaurato circumamicta varietate,* Ps. Th. 44, 11.

ge-twǽfan; *p.* de; *pp.* ed *To separate, divert, detain, hinder, deprive:*—Ne lǽt ðú ðec síðes getwǽfan láde gelettan lifgendne monn *do not thou let any living man divert thee from thy course, hinder thy journey,* Exon. 123 b; Th. 474, 2; Bo. 23: Beo. Th. 3820; B. 1908: 963; B. 479. Sóna biþ ðæt ðec ádl oððe ecg eafoþes getwǽfeþ *soon will it be that disease or sword will deprive thee of vigour,* 3531; B. 1763. Sumne Geáta leód feores getwǽfde *one the Goths' prince separated from life,* 2871; B. 1433. Gúþ wæs getwǽfed *the contest was parted,* 3320; B. 1658. Swelap sǽ-fiscas sundes getwǽfde *the sea-fishes shall burn cut off from the ocean,* Exon. 22 b; Th. 61, 20; Cri. 987. Him se mǽra mód getwǽfde *the great one took courage from them,* Cd. 4; Th. 4, 14; Gen. 53: 148; Th. 185, 8; Exod. 119 [?].

ge-twǽman, -twéman; *p.* de; *pp.* ed [twǽman *to separate*] *To cut off, separate, divide;* separáre, sejungĕre, dívídĕre:—Ic hine ne mihte ganges getwǽman *I could not cut him off from his course,* Beo. Th. 1940; B. 968: L. N. P. L. 65; Th. ii. 300, 28. Ðá man getwǽmde ðæt ǽr wæs gemǽne Criste and cynincge *then was separated what was before in common to Christ and the king,* L. Eth. ix. 38; Th. i. 348, 20: Wald. 88; Vald. 2, 16. Ne getwǽme nán mann ða ðe God gesomnode *quod Deus conjunxit, hŏmo non sepáret,* Mt. Bos. 19, 6. Beó ǽlc sacu getwǽmed *let every strife be appeased,* L. Eth. v. 19; Th. i. 308, 30. Getwǽman *to alienate,* Basil admn. 4; Norm. 40, 29, note p.

ge-twancg, es; *n. Fraud, deception;* colludium, fraus, deceptio, Hpt. Gl. 442.

ge-tweó; *gen.* -tweón; *m. Doubt, ambiguity:*—In gituíá *in ambiguitate,* Rtl. 105, 9.

ge-tweógan, -tweón; *p.* -tweóde; *pp.* -tweód; *v. pers. and impers. To doubt, hesitate;* dúbitáre:—Ne getweóge ic náwuht be godes ǽcnessa *I do not at all doubt about God's eternity,* Shrn. 195, 4. Nó him treów getweóde *his faith doubted not in him,* Exon. 37 b; Th. 122, 25; Gú. 311: 40 b; Th. 134, 27; Gú. 515. Getuíga *hæsitare,* Mk. Skt. Lind. 11, 23. Forhwon getwiódes tú *quare dubitasti,* Mt. Kmbl. Rush. 24, 31: Lind. 28, 17. Ðá gehreów him ðæt hyne ǽfre swá on his geþohte getweóde *then he repented that he had ever so doubted in his mind,* Shrn. 155, 19.

ge-tweónian; *p.* ode; *pp.* od; *v. impers. To seem doubtful to any one;* dúbium vídéri alicui:—Getweónode hí hwæðer . . . *it seemed doubtful to them whether . . . ,* Ors. 1, 14; Bos. 37, 28.

ge-twífealdad; *part. Doubled;* duplĭcátus:—Biþ ðæt ǽfengyfel getwífealdad *the evening refection will be doubled,* L. E. I. 38; Th. ii. 436, 30.

ge-twífyldan, -twýfyldan *to double:*—Seó eahteoðe præteritum getwýfylt ðæt æftre stæfgefég *the eighth preterite doubles the second syllable,* Ælfc. Gr. 28, 8; Som. 33, 1. Hí beóþ getwyfylde *they are doubled,* Homl. Th. ii. 372, 35.

ge-twin, es; *m. A twin*:—Geminus ðæt is on ûre geþeóde getwyn *geminus, that is in our language twin*, Shrn. 155, 30. Hî wǽron ge-twinnas *that were twins*, 92, 22 : 134, 23 · Salm. Kmbl. 729; Sal. 364 : 216; Sal. 107 [?]. [Cf. *O. H. Ger.* zwinal, ge-zuinele *geminus*.]

ge-twis; *adj. Germanus*, Hpt. Gl. 477. Getwise *fratres germani*, Gl. M. 392.

ge-twisa, an; *m. A twin*:—Twegen getwisan *two twins*, Gen. 38, 27 : Swt. A. S. Rdr. 62, 197. Getwisan *gemini*, Ælfc. Gr. 13; Som. 16, 13 : Bd. de nat. rerum; Wrt. popl. science 7, 5; Lchdm. iii. 244, 24.

ge-tŷan; *p.* de; *pp.* -tŷd *To instruct, teach, imbue*; instruere, imbuere, docere:—He Sanctus Martinus fulfremedlîce on Godes ǽ and on Godes þeówdôm getŷde and lǽrde *he perfectly instructed and taught St. Martin in God's law and service*, Blickl. Homl. 217, 5. Ðîn lâr getŷde me *disciplina tua ipsa edocuit me*, Ps. Th. 17, 34 : Bt. 8; Fox 24, 25 : Ors. 5, 13; Bos. 112, 33. Gregorius wæs fram cildhâde on bôclîcum lârum getŷd *Gregorius was from childhood instructed in book learning*, Homl. Th. ii. 118, 17 : Bd. 1, 27; S. 489, 10 : Guthl. 2; Gdwin. 18, 11 : Nar. 1, 14. On snytrum sŷn swŷðe getŷde *eruditos corde in sapientia*, Ps. Th. 89, 14 : Elen. Kmbl. 2034; El. 1018.

ge-tŷd; *part. p. Skilful, learned*; peritus:—Wæs he se getŷdesta sangere *cantator erat peritissimus*, Bd. 5, 20; S. 642, 11. v. ge-tŷan.

ge-tŷdan; *p.* -tŷdde [v. (?) ge-tŷd] *To make learned, skilled, to in-struct*:—Dysine and ungelǽredne ic ðe underféng and ðâ ðe getŷdde and gelǽrede *foolish and ignorant I received thee, and then made thee wise and taught thee*, Bt. 7, 3; Fox 20, 10. Ic þohte ealra swîðost ymb ðone abbud ðe me getŷdde *I thought most of all of the abbot that had instructed me*, Shrn. 46, 33. [*Or* ge-tŷdde = ge-tŷde ?]

ge-tŷdan; *p.* de; *pp.* ed *To happen*; contingĕre:—Getŷdde hit, ðæt . . . *it happened that* . . . , Bt. 16, 2; Fox 52, 34.

ge-tyddrian; *p.* ode; *pp.* od *To produce, bring forth*:—Swilce he swâ fela wînboga getyddrode *as if it so many vine-branches brought forth*, Homl. Th. ii. 74, 7.

ge-tŷdnes, -ness, e; *f. Learning, knowledge, skill*; erūditio, pĕrītia:—Wæs Cûþberhte swâ mycel getŷdnes and gelǽrednes to sprecanne *Cud-bercto tanta ĕrat dīcendi pĕrītia*, Bd. 4, 27; S. 604, 19. Mid ða getŷd-nesse ge cyriclîcra gewrita ge eác gemǽnelîcra *cum erūditiōne litĕrārum vel ecclēsiastĭcārum vel genĕrālium*, 5, 23; S. 645, 15.

ge-tyhtan; *p.* te; *pp.* ed *To educate, teach, instruct*; erūdīre, dŏcēre, instruĕre:—Ðe ðû hine getyhtest *quem tu erūdiĕris*, Ps. Th. 93, 12. Byþ his heorte getrymed and getyhted *confirmātum est cor ejus*, 111, 7.

ge-tŷhtlod, -tŷhtled; *part. p. Accused*; accūsātus :—Ðe oft getŷhtlod wǽron *who have often been accused*, L. Ath. i. 7; Th. i. 202, 25, note 48. Se getŷhtleda man *the accused man*, i. 23; Th. i. 212, 8, note 19. v. ge-tîhtlod.

ge-tŷma, an; *m. An advocate, avoucher, a warranter*; advŏcātus :—Be getŷmum. Ðæt ǽlc man wite his getŷman *of warranters. That every man know his warranter*, L. A. G. 4; Th. i. 154, 12, 13. v. ge-teáma.

ge-tymbrian; *p.* ode; *pp.* od *To build*; ædĭfĭcāre:—Ðæt sŷn getym-brod weallas *ædĭfĭcentur mūri*, Ps. Spl. 50, 19. v. ge-timbrian.

ge-tŷme; *n. A team, yoke*; jŭgum :—Ic bohte ân getŷme oxena *jŭga boum ēmi quinque*, Lk. Bos. 14, 19. Fŷf hund getŷmu oxena *quingenta jŭga boum*, Job. Thw. 164, 5; Homl. Th. ii. 372, 23.

ge-tŷnan; *p.* de; *pp.* ed *To shut up, hide*; opĕrīre, inclūdĕre :—Se Hǽlend me in ðam engan hâm oft getŷnde *the Saviour often shut me up in the narrow dwelling*, Elen. Kmbl. 1839; El. 921. Foldan getŷned *hidden in earth*, 1441; El. 722. Ǽgo hiora getŷndon *oculos suos claus-erunt*, Mt. Kmbl. Lind. 13, 15. Getŷned wæs ðe dura *clausa est janua*, 25, 10.

ge-tŷne, es; *n. A court*; atrium :—On his getŷnum ðe ymb Dryhtnes hûs deóre sindan *in atriis dŏmus Dŏmĭni*, Ps. Th. 115, 8. [Cf. tûn.]

ge-tynge; *adj. Talkative*:—Se getynga wer *vir linguosus*, Ps. Th. 139, 11. [Cf. *O. H. Ger.* ge-zungel *loquax, facundus*; gi-zungili *verbositas*.] v. ge-tinge.

ge-tyrfian. v. ge-torfian.

ge-tyrian; *p.* ode; *pp.* od *To grow weary*; fătīgāre :—Ðeáh ðû getyrige *if thou shouldest grow weary*, Bt. 40, 5; Fox 240, 23. v. ge-teorian.

ge-uferian; *p.* ode; *pp.* od *To exalt, elevate, delay, put off* :—Ðæt he mid ðæs wurþmyntes wuldre geuferod wǽre *to be exalted with the glory of that honour*, Homl. Th. ii. 122, 26. Ic geseah ârleásne geufer-odne *vidi impium elevatum*, Ps. Lamb. 36, 35. Wæs ðâ þurh his langsume fær ðæra cildra slege geuferod *the children's slaying was delayed by his long journey*, Homl. Th. i, 80, 28.

ge-unârian *To dishonour* :—Hî hys cyn geunâredon *they dishonoured his race*, Ors. 1, 5; Bos. 28, 31. Sŷn geunârode *may they be dishonoured*, Ps. Spl. 34, 4.

ge-unclǽnsian *to make unclean, to pollute*; fœdare :—Romulus hiora angin geunclǽnsode mid his brôðor slege *Romulus polluted their under-taking with his brother's murder*, Ors. 2, 2; Bos. 40, 30.

ge-ungewlitegian; *p.* ode; *pp.* od *To deprive of beauty*:—Ôð ðre hwîle gegiereþ mid ðâm winsumestum wlitum ôðre hwîle eft geunge-wlitegaþ *at one time adorns with the most delightful beauty, at another again deprives of beauty*, Shrn. 195, 11.

ge-unlustian *to loathe*:—Se lîchoma geunlustaþ ða geógoþlustas to fremmenne *the body loathes to do the pleasures of youth*, Blickl. Homl. 59, 8.

ge-unlybba, an; *m. Poison [particularly when used in witchcraft]*:—Ne lǽt ðû lybban ða ðe geunlybban wyrcon *thou shalt not suffer a witch to live*, Ex. 22, 18. v. unlybba, lyblâc.

ge-unnan; ic, he -an; ðû -unne, *pl.* -unnon; *p.* -ûðe, *pl.* -ûðon; *subj.* -unne, *pl.* -unnen; *p.* -ûðe, *pl.* -ûðen; *pp.* -unnen *To give, grant, allow, concede*; concedere, indulgere, permittere, largiri :—Gif he us geunnan wile, ðæt we hine grētan môton *if he will grant to us that we may greet him*, Beo. Th. 698; B. 346 : Chr. 1095; Erl. 231, 25. Se cyning nolde him his feores geunnan *the king would not grant him his life*, Bt. 29, 2; Fox 104, 22 : Andr. Kmbl. 358; An. 179 : L. C. E. 2; Th. i. 358, 26. Hér sit mîn mǽge ðe ic geann ǽgðer ge mînes landes ge mînes goldes ge ealles ðe ic âh æfter mînon dæge *here sits my kins-woman, to whom I give both my land and my gold and all that I own, after my day*, Th. Chart. 337, 30 : 560, 9, 11, 15. Ǽrðon me geunne ēce dryhten, ðæt *until to me shall grant the eternal Lord, that*, Salm. Kmbl. 499; Sal. 250. Me geûðe ylda waldend, ðæt *the Ruler of men granted me, that*, Beo. Th. 3326; B. 1661. Ðû geûðest his bearne his cynerīces *thou hast given his kingdom to his child*, Homl. Th. ii. 576, 14. Ðæt ðæt him gôde menn geûðon *that which good men have given them*, Swt. A. S. Rdr. 106, 56. Hû Cnut cyncg and Ælfgifu seó hlǽfdige ge-ûðan heora preóste ðæt he môste ateón ðæt land swâ him sylfan leófast wǽre *how king Cnut and the lady Ælfgifu granted their priest that he might dispose of the land as he liked best*, Th. Chart. 328, 20 : Homl. Th. ii. 152, 15. God him geunne ðæt . . . *God grant him that . . .*, Chr. 959; Erl. 121, 5. Ðæra þinga wurðe ðe se cyng him geunnen hæfde *worthy of those things that the king had granted him*, 1046; Erl. 173, 3. [*O. Sax.* gi-unnan; *p.* -onsta : *O. H. Ger.* gunnen; *p.* gi-onsta, *both with the same cases as the English verb* : *Ger.* gönnen.]

ge-un-ret *saddened*; *pp.* of ge-un-rētan.

ge-un-rētan; *p.* -rētte; *pp.* -rēted, -rēt *To make sorrowful, sadden, trouble*; contristare :—Ðâ se engel cwedende ' Ne beó ðû Maria geunrēted ' *then the angel said ' Be not sorrowful, Mary,'* Blickl. Homl. 139, 15. Hŷ wurdon geunrētt mid manncwealme *they were troubled with pestilence*, Ors. 3, 10; Bos. 70, 27. Ðâ wearþ se cyning geunrēt for ðam âðe and for ðam ðe him mid sǽton nolde ðeáh hî geunrētan *et contristatus est rex propter jusjurandum et propter simul discumbentes noluit eam contristare*, Mk. Bos. 6, 26 : Mt. Bos. 14, 9 : Lk. Bos. 18, 23.

ge-un-rōtsian, -un-rōtsigean; *p.* ode; *pp.* od.   **I.** *to make sor-rowful, to offend*; contristare, contribulare, scandalizare :—Ðæt we hî ne geunrōtsigeon *ut non scandalizemus eos*, Mt. Bos. 17, 27. Ne sŷ ûre nân geunrōtsod *let none of us be sad*, Blickl. Homl. 149, 19 : Mt. Kmbl. Rush. 14, 9. Geunrōtsade swîðe *contristati valde*, Lind. 26, 22 : Mk. Skt. Lind. 10, 22.   **II.** *to become troubled, discontented*:—Ðæt se man geunrōtsige ongeán God for ungelimpum ðises andwerdan lîfes so *that a man becomes discontented with God for the mishaps of this present life*, Homl. Th. ii. 220, 16. Gâst geunrōtsod *spiritus contribulatus*, Ps. Spl. T. 50, 18.

ge-unsōþian; *p.* ode; *pp.* od *To disprove, refute, prove false*; refel-lĕre :—Gif se ôðer ðæt geunsōþian mǽge ðæt him man onsecgan wolde *if the other can disprove that which any one would charge to him*, L. Edg. ii. 4; Th. i. 266, 24 : L. C. S. 16; Th. i. 384, 22.

ge-unstillian; *p.* ode; *pp.* od. *To disquiet, disturb*; inquiĕtāre :— Ðætte ðâ mynster ðe ðe Gode gehâlgode syndon nǽnigum bisceope alŷfed sî in ǽnigum þinge hî geunstillian *ut quæque monastĕria Deo consecrāta sunt, nulli episcŏpōrum līceat ea in alĭquo inquiĕtāre*, Bd. 4, 5; S. 572, 35.

ge-unþwǽrian, -unþwǽrigan; *p.* ode; *pp.* od *To disagree, differ*; dissentīre, discordāre :—Ic geunþwǽrige *dissentio*, Ælfc. Gr. 37; Som. 39, 9. Ðætte hie selfe ne geunþwǽrigen [geunþwǽrien, MS. Cot.] ðǽm wordum ðe hie lǽraþ, mid ðý ðæt hie ôðer don, ôðer hie lǽraþ *that they themselves differ not from the words they teach, by doing one thing and teaching another*, Past. 48, 4; Swt. 371, 12; Hat. MS.

ge-untreówsian, -untrýwsian; *p.* ode; *pp.* od *To be offended*; scan-dălĭzāri :—Ðeáh ðe hig ealle geuntreówsion on ðē, ic nǽfre geuntreówsige *si omnes scandălĭzāti fŭĕrint in te, ĕgo nunquam scandălĭzābor*, Mt. Bos. 26, 33. Ealle gē wurþaþ geuntreówsode on me *omnes vos scandălum pătiemĭni in me*, 26, 31 : 13, 21.

ge-untrumian; *p.* ode; *pp.* od.   **I.** *v. trans. To enfeeble, make weak or sick;* infirmāre, debilĭtāre :—Deófol geuntrumaþ ða hâlan *the devil enfeebles the healthy*, Homl. Th. i. 4, 22. Ðe God sylf ǽr geuntru-mode *whom God himself had before enfeebled*, i. 4, 27. Þurh ðæs dracan blǽd eal seó menigu micclum wearþ geuntrumod *all the multitude were greatly sickened by the dragon's breath*, ii. 294, 23 : 296, 9 : 516, 17. Ðe wǽron geuntrumode *qui infirmābantur*, Jn. Bos. 6, 2 : Ps. Spl. 17, 38.   **II.** *v. intrans. To be enfeebled, be sick;* infirmāri, ægrōtāre :—

Hí geuntrumiaþ *infirmābuntur*, Ps. Spl. 9, 3. Ða geuntrumade he mid
dære mettrynnnesse podagre *then he was ill with the gout*, Shrn. 100, 18.

ge-untrýwsian; *p.* ode; *pp.* od *To be offended;* scandǎlīzāri :—Hig
wæron geuntrýwsode on him *scandǎlīzābantur in eo*, Mt. Bos. 13, 57.
v. ge-untreówsian.

ge-unwendness, e; *f. Unchangeableness:*—Ðeós ungewendnes *hæc
immutatio*, Ps. Th. 76, 9.

ge-unwurdod *dishonoured.* v. unweorðian.

ge-upped; *part. Revealed:*—Ne mihte Scs Neotus behýdd beón ðá
ðá God hine geupped habben wolde *St. Neot could not be hid when God
would have him revealed*, Shrn. 12, 15. v. ge-yppan.

ge-ūðe; *p.* of ge-unnan.

ge-ūtian; *p.* ode; *pp.* od *To eject, banish, alienate;* ejīcĕre, expellĕre :—Se cyng hine geūtode of earde *the king banished him from the
country*, Chr. 1002; Erl. 137, 29. Wæs Óslác geūtod of Angelcynne
Oslac was banished from England, 975; Erl. 127, 8. He beád ðæt
náðer ne dære stówe bisceop ne nánes bisceopes æftergenga ðæt land
næfre of dære stówe geūtode *he ordered that neither the bishop of the
place nor any bishop's successor should ever alienate that land from the
place*, Cod. Dipl. Kmbl. iii. 112, 9 : iv. 72, 27, 32. Cwædon hí ðæt hit
betere wære ðæt ic ða preóstas of Cristes cyrcean geūtode *they said it
would be better that I should expel the priests from Christchurch*, iii.
349, 14. Ic nelle geþafian ðæt æni man geūtige án æker landes *nolo
permittere ut quis unum jugerum excludat*, iv. 202, 15. Geūtian *exiliare*,
Hpt. Gl. 517.

ge-ūtlagian; *p.* ode; *pp.* od *To outlaw;* proscrībĕre :—Man geūtla-
gode Ælfgár eorl *earl Ælfgar was outlawed*, Chr. 1055; Erl. 188, 27 :
1020; Erl. 161, 22.

ge-wácian; *p.* ode; *pp.* od *To grow weak* or *lose energy, to flinch;*
languescere, obtorpescere :—Gif hý ðær ne gewácodan [gewícadon, Laud]
if they had not there lost energy [*stopped*], Ors. 3, 4; Bos. 56, 11.
v. wácian, wícian.

ge-wacsan. v. ge-wascan.

ge-wadan; *p.* -wód; *pp.* -waden. **I.** *v. intrans. To wade, go;*
vadere, ire :—Sár gewód ymb ðæs beornes breóst *pain went around the
man's breast*, Andr. Kmbl. 2494; An. 1248. Ord in gewód *the point
entered*, Byrht Th. 136, 26; By. 157 : Exon. 47 b; Th. 163, 29; Gū.
1001. Wundenstefna gewaden hæfde ðæt ða liðende land gesáwon *the
ship had gone* [*so far*] *that the sailors saw land*, Beo. Th. 446; B.
220. **II.** *v. trans. To pervade, go through:*—Flód blód gewód
blood pervaded the flood, Cd. 166; Th. 207, 6; Exod. 462. Elen. Kmbl.
2378; El. 1190. v. wadan.

ge-wǣcan, -wǣcean; *part.* -wǣcende; *p.* -wǣcte, -wǣhte; *pp.* -wǣct,
-wǣht *To weaken, affect, trouble, vex, afflict, oppress;* afficĕre, affligĕre :—
Heó nele ða andweardan myrhþe gewǣcan mid nánre care dære toweardan
ungesǽlþe *it will not trouble the present joy with any care for the future
unhappiness*, Homl. Th. i. 408, 21. Beóton hig ðone, and mid teónum,
gewǣcende, hine forlēton ídelne *illi hunc cædentes, et afficientes con-
tŭmēlia, dīmiserunt inānem*, Lk. Bos. 20, 11. Hí mid deáþe hí gewǣceaþ
morte afficient eos, Mk. Bos. 13, 12 : Homl. Th. ii. 542, 17. Hig eall
ðæt rīce myd forspyllednysse gewǣhton *they destroyed all that kingdom*,
St. And. 32, 32. Mid fefore gewǣht *suffering from fever*, Homl. Th. ii.
516, 30. Gewǣht ic eom *afflictus sum*, Ps. Spl. 37, 8. Ðe mid ðý
hungre gewǣcte wǣron *who were oppressed with the hunger*, Bd. 4, 13;
S. 582, 31. Gelomp us ðæt we wurdon earfoþlíce mid þurste geswencte
and gewǣcte *accidit nobis siti laborare*, Nar. 7, 30. We on ðínum yrre
synt swíðe gewǣhte *in īra tua defēcimus*, Ps. Th. 89, 9 : Jud. 6, 2 :
Homl. Th. ii. 396, 28.

ge-wǣccan *to watch* :—Ne mǣhtes ðú án huíl gewǣccæ *non potuisti
una hora vigilare*, Mk. Skt. Lind. 14, 37. Gewaccas *vigilate*, 13, 35.
Ðætte we giuæcge *ut vigilemus*, Rtl. 124, 23.

ge-wǣcednyss, e; *f. Weakness:*—Him nán þing ne eglaþ ǣnigre
brosnunge oðde gewǣcednysse *nothing pains him of any corruption or
weakness*, Homl. Th. ii. 552, 29.

ge-wǣde, -wǣde, es; *n. A garment, clothing;* vestimentum :—Saga
hwæt ðis gewǣde [gewædu, MS.] sý *say what this vestment is*, Exon.
109 a; Th. 418, 5; Rä. 36, 14. He nywolnessa him to gewǣde woruhte
abyssus amictus ejus, Ps. Th. 103, 7. Mīne gewǣda *vestimentum meum*,
68, 11 : Homl. Th. ii. 148, 30. Wǣpen and gewǣdu *arms and clothing*,
Beo. Th. 589; B. 292. Gewǣde *vestimentum*, Mt. Kmbl. Lind. 3, 4.
Mið his gewǣdum *vestimentis ejus*, 27, 31. He onfēng cynegewǣdum
purpuram sumpsit, Bd. 1, 6; S. 476, 19. Ongon me gewǣdum þeccan
he began to r'eck me with weeds, Exon. 103 a; Th. 391, 13; Rä. 10, 4.
[*O. Sax.* gi-wādi : *O. H. Ger.* gi-wāti *vestimentum, vestis.*]

ge-wǣdian, -wēdian; *p.* ode; *pp.* od *To dress, clothe, equip* :—Gi-
woedes *induite*, Rtl. 13, 31. Gewǣdod *equipped*, Chr. 992; Erl. 131,
34. Gewǣdod *vestitus*, Mt. Kmbl. Lind. 11, 8. Woere gewoedad *vestie-
batur*, Lk. Skt. Lind. 12, 27.

ge-wǣdod; *part. Prepared, equipped;* appǎrātus, instructus :—Hí ðæt
scip genámon eall gewǣpnod and gewǣdod *they took the ship all armed
and equipped*, Chr. 992; Erl. 131, 34. v. ge-wǣdian.

ge-wǣg *bore, carried*, Bd. 3, 16; S. 542, 22; *p.* of ge-wegan.

ge-wǣgan; *p.* ede; *pp.* ed. **I.** *to affect, weigh down, oppress;*
afficĕre, deprimere, vexare :—Wíne gewǣged *affected by wine*, Exon. 84 a;
Th. 315, 34; Mōd. 41. Wōpe gewǣged *oppressed with weeping*, Bt.
Met. Fox 2, 5; Met. 2, 3. Mid meteliéste gewǣgde *oppressed with lack
of food*, Chr. 894; Erl. 92, 27. **II.** *to frustrate;* frustrari, irritum
facere :—Cūþ sceal geweorþan ðæt ic gewǣgan ne mæg *that which I may
not frustrate shall become manifest*, Exon. 117 b; Th. 452, 3; Dōm.
115. v. wǣgan, a-wǣgan.

ge-wǣge, es; *n. A weight, measure:*—Gewǣge *weight*, Herb. 1, 15;
Lchdm. i. 74, 21 : 16; Lchdm. i. 76, 1. Gewege, 2; Lchdm. i. 70, 15,
note. Gewǣge [giwege, Rush.] *mensura*, Mk. Skt. Lind. 4, 24. Gewoege
ꝥ gemet *mensura*, Lk. Skt. Lind. 6, 38. v. ge-wegan.

ge-wǣgnian; *p.* ode; *pp.* od *To frustrate, deceive, disappoint;* frustrari, Cot. 83.

ge-wǣlan *to vex, afflict :*—Hie wēron gewǣlde *erant vexati*, Mt. Kmbl.
Rush. 9, 36. v. wǣlan, be-wǣlan.

ge-wǣltan *to roll :*—Gewǣlteno *provolutus*, Mt. Kmbl. 17, 14. He
gewǣlte stán micel to duru ðæs byrgennes *advolvit saxum magnum ad
ostium monumenti*, 27, 60.

ge-wǣmnednes, se; *f. A corruption;* corruptio :—Ānes wordes ge-
wǣmnednys *a corruption of a word, a barbarism;* barbarismus, Som.
v. ge-wemmednes.

ge-wǣmnod *armed;* Ælfc. T. 36, 22, q. ge-wǣpnod. v. ge-wǣpnian.

ge-wǣnian; *p.* ede; *pp.* ed. **I.** *to accustom;* assuefacere :—
Folc to ælmessan gewǣnian *to accustom the people to alms*, L. Edg. C.
55; Th. 256, 9. **II.** *to wean;* ablactare, Gen. 21, 8. v. ge-
wenian.

ge-wǣpnian, -wēpnian; *p.* ode; *pp.* od *To arm, furnish with weapons;*
armāre :—Ic gewǣpnige *armo*, Ælfc. Gr. 24; Som. 25, 41 : 36; Som.
38, 36, 37. He mid rōdetácne his mūþ and ealne his líchaman gewǣp-
node *he armed his mouth and all his body with the sign of the cross*,
Homl. Th. i. 72, 23. Se stranga gewǣpnod his cǣfertūn gehealt *fortis
armātus custōdit atrium sum*, Lk. Bos. 11, 21 : Ælfc. Gr. 43; Som. 45,
15. Hí ðæt scip genámon eall gewǣpnod and gewǣdod *they took the
ship all armed and equipped*, Chr. 992; Erl. 131, 34.

ge-wǣr; *adj. Aware;* conscius :—Hí his gewǣr wurdon *they were
aware of him*, Chr. 1095; Erl. 231, 39.

ge-wǣrlǣcan; *p.* -lǣhte, -lēhte; *pp.* -lǣht, -lēht *To remind, admonish;* commonefacĕre :—Cain wiste his fæder forgǣgednysse, and
næs þurh ðæt gewǣrlēht *Cain knew his father's transgression, and was
not admonished by it*, Boutr. Scrd. 20, 40.

ge-wǣrlan; *p.* de *To go, pass :*—Éghuoelc on weg his giwǣrlde
quisque in viam suam declinavit, Rtl. 19, 39. v. wǣrlan.

ge-wǣsc *a washing up* or *overflow of water;* alluvio :—Wǣtera
gewǣsc *aquarum alluvio*, Ælfc. Gl. 100; Wrt. Voc. 55, 26. v. wǣter-
gewæsc.

ge-wǣtan, -wētan; *p.* -wǣtte; *pp.* -wǣted, -wǣtt *To wet, to make
wet :*—Onsend Ladzarus ðætte he gewǣte his ýtemestan finger on wǣttre
send Lazarus, that he may wet the tip of his finger in water, Past. 43,
1; Swt. 309, 6; Hat. MS. · Strengas gurron wǣdo gewǣtte *the ropes
creaked wet with the waters*, Andr. Kmbl. 749; An. 375 : Ps. Th. 104,
36.

ge-wǣterian, -wǣtrian; *p.* ode; *pp.* od *To water, irrigate;* adǎquāre,
irrigāre :—Ðæt mǣge and cunne ōðerra monna inngeþonc giendgeótan and
gewǣterian [gewǣtrian, MS. Cot.] *that he may be able and know how to
irrigate and water the minds of others*, Past. 18, 5; Swt. 137, 10; Hat.
MS. 27 a, 14. Ic betǣce hig ðam yrþlincge, wel gefylde and gewǣterode
adsigno eos arātōri, bĕne pastos et adǎquātos, Coll. Monast. Th. 20, 31.
Teóh ðú forþ rēnscūras gif ðú miht and gewǣtera ðíne æceras *bring
forth rain-showers, if thou canst, and water thy fields*, Homl. Th. ii.
104, 1.

ge-wald, es; *m. n. Power, mastery, sway :*—Ða Denescan áhton wælstówe gewald *the Danes had the mastery of the battle-place*, Chr. 833;
Th. 116, 7, col. 1 : Cd. 214; Th. 268, 15; Sat. 55. v. ge-weald.

ge-waldan *to have power over.* v. ge-wealdan.

ge-walden. v. ge-wealden.

gewald-leðer *a power-leather, a rein*, Bt. Met. Fox 24, 77; Met. 24,
39. v. geweald-leðer.

ge-wana, -wona, an; *m. A lack, want :*—Huæd me gwona is *quid
mihi deest*, Mt. Kmbl. Lind. 19, 20, v. wana.

ge-wand *turned*, Beo. Th. 2007; B. 1001; *p.* of ge-windan.

ge-wanian; -wonian; *p.* ode; *pp.* od. **I.** *to lessen, diminish :*—
Se lāreow ne sceal ða inneran giémenne gewanian for dære ūterran abis-
gunge *the teacher is not to diminish his care of inner things for outer
occupations*, Past. 18, 1; Swt. 127, 8; Hat. MS. His cynelícan gefe
gewonian *to diminish his royal gift*, Cod. Dipl. Kmbl. ii. 304, 27. Ðone
hryre ðe se feallenda deófol on engla werode gewanode *the loss that the
falling devil caused in the host of angels*, Homl. Th. i. 32, 23 : 214, 24.
He his godcundnesse nán wiht ne gewanode *he did not at all diminish*

his divinity, Blickl. Homl. 91, 9. Gewanude, Th. Chart. 203, 36. Gewonade, Exon. 44 a; Th. 148, 19; Gū. 747. Būton he his flǽsclīcan lustas gewanige *unless he diminish his fleshly lusts*, Homl. Th. i. 96, 3: Past. 48, 1; Swt. 127, 12; Hat. MS. Ne gē nǽn þing ne gewanion *ye shall not diminish ought*, Ex. 5, 8. Gewonige, Cod. Dipl. Kmbl. ii. 100, 27. Is mīn flet-werod gewanod *my band of retainers is lessened*, Beo. Th. 958; B. 477: Cd. 24; Th. 31, 6; Gen. 481: Gen. 8, 1. Đā wæs dǽm tunglum gewonad heora beorhtnes *then had the stars their brightness diminished*, Shrn. 64, 22. **II.** *to be wanting:*—Giwonia deesse, Rtl. 71, 37.

ge-waran; *gen.* -warena; *pl. m*; used as a termination to denote *inhabitants, dwellers; incolæ:*—Đa Rōmāniscan ceastergewaran noldon geþafian đæt Gregorius đa burh forlēte *the Roman citizens would not consent that Gregory should leave the city*, Homl. Th. ii. 122, 13. v. waran.

ge-wardod *seen;* visus:—Þat he sȳ gewardod fram him *ut videatur ab illo*, R. Ben. interl. 49.

ge-warenian; *p.* ode; *pp.* od To *warn, guard;* cavere:—Ǽlc gleáw mōd hit gewarenaþ *every prudent mind guards itself*, Bt. 7, 2; Fox 18, 24. v. warenian.

ge-warian to *protect;* protegere, Hpt. Gl. 489, 500.

ge-warnian; *p.* ode; *pp.* od To *warn:*—God on swefne hī gewarnode *God warned them in a dream*, Homl. Th. i. 78, 29. Đā gewarnode man hī đæt đǽr wæs fyrd æt Lundene *then they had notice that there was a force at London*, Chr. 1009; Erl. 143, 12. Đā wearþ Godwine gewarnod *then was earl Godwin warned*, 1052; Erl. 183, 2. Gebeorh gewarnian *tuitionem præstare*, Bd. 2, 5; S. 506, 30, note.

ge-wascan, -wacsan; *p.* -wōcs To *wash:*—Ic hine mid mīnen handen gewōchs *I washed him with my hands*, Cod. Dipl. Kmbl. iv. 261, 1.

ge-wāt *departed; p.* of ge-wītan.

ge-wealc, es; *n. A rolling, motion, an attack;* volutatio, impetus:—Ýđa gewealc *a rolling of waves*, Ap. Th. 11, 1: Cd. 166; Th. 206, 21; Exod. 455: Exon. 81 b; Th. 306, 11; Seef. 6: 82 a; Th. 308, 28; Seef. 46: Beo. Th. 932; B. 464: Chr. 975: Erl. 126, 19; Edg. 45: Andr. Kmbl. 517; An. 259. Togeánes đān he manega gewealc and gewinn hæfde *against which he had many a struggle and contest*, Chr. 1100; Erl. 237, 9. v. ge-wilcþ, ge-wylc.

ge-wealcan; *p.* -weólc; *pp.* -wealcen To *roll;* volvere, revolvere:—Fām biþ gewealcen *the foam is rolled*, Exon. 101 a; Th. 382, 1; Rä. 3, 4.

ge-weald, -wald, es; *m. n.* **I.** *power, strength, might, efficacy;* potestas:—Þurh geweald Godes *through the power of God*, Cd. 1; Th. 1, 21; Gen. 11. Geweald hafaþ *shall have power*, Exon. 32 a; Th. 100, 29; Cri. 1649. Wiste his fingra geweald *knew the power of his fingers*, Beo. Th. 1533; B. 764. Gif hit geweald āhte *if it possessed power*, Bt. Met. Fox 22, 72; Met. 22, 36. Gif mon ōþrum đa geweald forsleá uppe on đam sweoran *if a man rupture the powers [tendons] on another's neck*, L. Alf. pol. 77; Th. i. 100, 10. **II.** *power over any thing, empire, rule, dominion, mastery, sway, jurisdiction, government, protection, keeping, a bridle-bit;* potestas, facultas, imperium, ditio, arbitrium, jus, camus:—Đæt he nāge đæra geweald *that he has no power over them*, L. Alf. pol. 77; Th. i. 100, 12: Jud. Thw. p. 153, 9. Đonne he his geweald hafaþ *when he has power over it*, Cd. 30; Th. 40, 7; Gen. 635: Bt. Met. Fox 9, 126; Met. 9, 63. Gesealde wǽpna geweald *gave power over weapons*, Cd. 143; Th. 178, 31; Exod. 20. Āhte bega geweald, līfes and deáđes *he had power of both, of life and death*, Exon. 40 a; Th. 133, 24; Gū. 494: Beo. Th. 3459; B. 1727: Shrn. 150, 13. On geweald gehwearf worold-cyninga *it passed into the power of worldly kings*, Beo. Th. 3372; B. 1684: Andr. Kmbl. 2547; An. 1275. His gewealdes *of his own accord*, L. Alf. 13; Th. i. 46, 21. Đæt se Gode mōte in geweald cuman *that he may come into God's dominion*, Exon. 32 b; Th. 103, 27; Cri. 1694: Cd. 10; Th. 13, 14; Gen. 202. Werþeóda geweald *the sway of nations*, 161; Th. 202, 4; Exod. 383. Wīnærnes geweald *jurisdiction over the wine-hall*, Beo. Th. 1312; B. 654. Đū scealt wǽpned-men wesan on gewealde *thou shalt be in subjection to man*, Cd. 43; Th. 56, 30; Gen. 920. Đæt mīn sāwul to đē sīđian mōte on đīn geweald *that my soul may proceed to thee, into thy keeping*, Byrht. Th. 136, 66; By. 178. Ic đē lǽte habban đis land to gewealde *I will let you rule this land*, H. R. 101, 33. Ic hine sealde to đīnum gewealde *I have given him into thy power*, Num. 21, 34. Đæt is God đe ealle þing on his gewealdum hafaþ *that is God, that hath all things in its power*, Salm. Kmbl. 178, 11: Blickl. Homl. 63, 3. Siđđan ic đā me hæfde đās þing on his be gewealdum *quibus in potestatem redactis*, Nar. 1, 17. Under hāligra hyrda gewealdum *under the protection of holy guardians*, Exon. 38 b; Th. 127, 15; Gū. 386: Ps. Spl. 31, 12. [O. Sax. gi-wald; *f. potestas, facultas, imperium:* Ger. gewalt; *f:* M. H. Ger. gewalt; *f:* O. H. Ger. gawalt; *m. f.*] DER. hand-geweald, ǽht-, nȳd-. v. ge-wealdes.

ge-weald, -wald, es; *m. n. Pudenda, inguen:*—Neáh đam gewealde *prope inguinem*, Herb. 104, 2; Lchdm. i. 218, 23 : 5, 5; Lchdm. i. 94, 22, 24. Đæt geweald, Lchdm. ii. 388, 9. v. ge-weald *power*.

ge-wealdan; *p.* -weóld; *pp.* -wealden To *wield, rule, have power over, command, control, cause.* **I.** *with gen:*—Ic gewealde ealles middan-eardes *I rule all the world*, Homl. Th. ii. 308, 21. Gregorius đæs pāpan setles geweóld *Gregory ruled the papal see*, 132, 18. Būton đū eác ūre gewelde *except thou make thyself altogether a prince over us*, Num. 16, 13. Gif he abilhþe āhwām on unriht āhwār geweólde gebēte hit georne and gif him abulge ǽnig man swīđe forgife đæt *if he anywhere have wrongly been the cause of offence to any man, let him diligently make amends; and if any man have much offended him, let him forgive it*, L. Pen. 16; Th. ii. 284, 7. Wǽpna gewealdan *to wield weapons*, Beo. Th. 3022; B. 1509. Swā heó đæs unlǽdan eáđost mihte wel gewealdau *so she most easily might have complete power over the wretch*, Judth. 10 ; Thw. 23, 3; Jud. 103. **II.** *with acc:*—Se đe gewylt đa đe he gesceóp *he who rules those whom he created*, Homl. Th. ii. 72, 27: Th. Chart. 239, 37. Đe ealne middangeard geweóld *who ruled all the world*, Homl. Th. i. 80, 7. Hālig God geweóld wīgsigor *holy God controlled victory in battle*, Beo. Th. 3112; B. 1554. **III.** *with instr:*—Nū leng ne miht gewealdan dȳ weorce *now canst thou no longer control the work*, Andr. Kmbl. 2729; An. 1367: Exon. 50 b; Th. 175, 24; Gū. 1199. Cyning geweóld his gewitte *the king got command of his senses*, Beo. Th. 5399; B. 2703.

ge-wealden; *part. Subject, under the power or control of any one, inconsiderable, small:*—God gedēþ him gewealdene worolde dǽlas sīde rīce *God puts under his power parts of the world, spacious realms*, Beo. Th. 3468; B. 1732. Hond biþ gelǽred wīs and gewealden *the hand is instructed, wise and under control*, Exon. 79 a; Th. 296, 4; Cra. 46; 91 a; Th. 341, 7; Gn. Ex. 122. Meahtig dryhten scyreþ sumum gūþe blǽd gewealdenne wīgplegan *the mighty Lord assigns to one glory in war, battle under his control, i.e. successful*, 88 a; Th. 331, 16; Vy. (9. Drincan gewealden wīnes for eówres magan mettrymnesse *modico vino utere propter stomachum*, Past. 43, 9; Swt. 319, 6; Hat. MS. Đā næfdon hī nān wīn būton on ānum gewealdenum butruce *in uno parvissimo vasculo*, Lchdm. iii. 362, col. 1. Būton swīđe gewealdenum dǽle eásteweardes đæs folces *except a small part of the people of the east of England*, Chr. 894; Erl. 91, 11: Ors. 4, 9; Bos. 92, 1. He myd us [wyrcþ] swā swā myd sumum gewealnum tōlum *he works with us as with some insignificant tools [or tools under his control, over which he has complete command ?]*, Shrn. 179, 28. v. Lchdm. iii. 361, col. 1. [Cf. ge-wealden-mōd.]

ge-wealdende; *adj. Powerful, mighty;* potens, validus:— Mid his gewealdendre hand *with his mighty hand*, Ps. Th. 113, 8. v. wealdende.

ge-wealdendlíce; *adv. Powerfully, mightily;* potenter, valide, Ps. Th. 135, 16.

ge-wealden-mōd; *adj. Subdued in mind, having the mind under control, self-controlled:*—Sum gewealdenmōd þafaþ in geþylde đæt he đonne sceal one, *self-controlled, suffers in patience what then he must* [cf. Luke 21, 19], Exon. 79 a; Th. 297, 19; Cra. 70. v. ge-wealden.

ge-wealdes; *adv.* [ge-weald *power*] Of *one's power, of one's own accord, willingly*; sponte:—Gif man hine sylfne gewealdes ofslihþ *si quis sponte seipsum occiderit*, L. M. I. P. 13; Th. ii. 268, 15. Se đe his gewealdes monnan ofsleá *he who slays a man of his own accord*, L. Alf. 13; Th. i. 46, 21, 26. Eówres gewealdes *quod ex vobis est*, Past. 46, 7, 8; Swt. 355, 19, 20, 25; Hat. MS.

ge-weald-leđer, ge-wald-leđer, es; *n.* [ge-, weald-leđer *a directing-leather*] A *power-leather, a directing-leather, a rein:*—Đonne he đæt gewealdleđer forlǽt đara bridla *when he shall let go the rein of the bridles*, Bt. 21; Fox 74, 31: Bt. Met. Fox 11, 55; Met. 11, 28: 11, 149; Met. 11, 75: 24, 77; Met. 24, 39, 155; Met. 29, 78.

ge-weallan to *boil, be hot;* fervescere, fervere, Rtl. 101, 26: 105, 3.

ge-weallod, -wealled; *part.* [weall *a wall*] Walled, *surrounded with a wall, fortified;* mūrātus, mūnītus:—Đa strengestan weras wuniaþ on đam lande and micele burga đǽr sind and mǽrlīce geweallode *cultōres fortissimos habet et urbes grandes atque mūrātas*, Num. 13, 29. On ceastre gewealledre *in cīvitāte mūnīta*, Ps. Spl. 30, 27.

ge-weardian. v. ge-weardod.

ge-wearmian; *p.* ode; *pp.* od To *become warm;* calere, calescere, Ælfc. Gr. 26, 2, 36.

ge-wearnian; *p.* ode; *pp.* od To *guard against, avoid:*—Hwǽr him wǽre fultum to sēcanne to gewearnienne swā đǽre hergunge *ubi quærendum est præsidium ad evitandas tam feras inruptiones*, Bd. 1, 14; S. 482, 37.

ge-wearþ *was, became, happened*, Beo. Th. 6115; B. 3061: Exon. 11 b; Th. 13, 30; Cri. 210: Andr. Kmbl. 613; An. 307; *1st and 3rd sing. p.* of ge-weorþan.

ge-weaxan; *p.* -weóx; *pp.* -weaxen To *grow, grow up;* crescere:— Gūþ sceal geweaxan *war shall grow*, Exon. 90 a; Th. 338, 27; Gn. Ex. 85. Moises geweóx *Moises creverat*, Ex. 2, 11. Geweaxen *auctus*, Exon. 99 b; Th. 372, 22; Seel. 96: Gen. 38, 11. Đǽm landbūendum is beboden, đæt ealles đæs đe him on heora ceápe geweaxe, hig Gode đone teóđan dǽl agyfen *to farmers it is commanded that of all which increases to them of their catile, they give the tenth part to God*, L. E. I.

35 ; Th. ii. 432, 29. Gyf hit geweaxen man sý fæste 1 geár *if he be a grown man let him fast one year*, L. Ecg. P. iv. 52 ; Th. ii. 218, note 11, line 9.

ge-wéd, es ; *n. A raging, madness ; fûror insânus, răbies :*—Wælhreówes [Nerônes] gewéd wæs fulwîde cûþ *the madness of the cruel [Nero] was full widely known*, Bt. Met. Fox 9, 9 ; Met. 9, 5. He langre tîde ealle heora mægþe mid gewéde wæs geondfarende *multo tempôre tôtas eôrum provincias debacchando pervŭgâtus*, Bd. 2, 20 ; S. 521, 27.

ge-weddian *to weed ;* herbis noxiis purgare, Cot. 178, 188, Lye.

ge-weddian *to betroth :*—Gewoedded *desponsata*, Lk. Skt. Lind. 1, 27.

ge-wéded. v. ge-wǽdod.

ge-weder, -wider, -wyder, es ; *pl. nom. acc.* -wederu ; *n.* [weder *weather*] *Weather, the temperature of the air ;* tempestas, cæli tempĕries :—Se sceortigenda dæg hæfþ lîðran gewederu ðonne se langienda dæg *the shortening day hath milder weather than the lengthening day*, Bd. de nat. rerum ; Wrt. popl. science 9, 21 ; Lchdm. iii. 252, 9. Godes miht gefádaþ ealle gewederu *God's power ordereth all weathers*, 19, 4 ; Lchdm. iii. 278, 13.

ge-wefan *to weave ;* texere, Exon. 95 a ; Th. 355, 1 ; Reim. 70 [v. Grmm. D. M. p. 387] : 111 b ; Th. 427, 2 ; Rä. 41, 85 : Ælfc. Gl. 63 ; Som. 68, 100, 101 ; Wrt. Voc. 40, 11, 12.

ge-wef[e], -wife, es ; *n. A web ;* textura. The word gets the meaning *fate, fortune*, from the spinning, which is the occupation of the Fates. Cf. Wyrd gewæf, Exon. 95 a ; Th. 355, 1 ; Reim. 70. See Grmm. D. M. 387 :—Gewife *fatum, fortuna*, Cot. 88 ; Lye. Him Dryhten forgeaf wîgspéda gewiofu *the Lord gave him the webs of success in war*, i. e. *he was successful in war*, Beo. Th. 1398 ; B. 697.

ge-wegan ; *p.* -wæg, *pl.* -wǽgon ; *pp.* -wegen. **I.** *to bear, carry, move, go, proceed ;* vehere, ire, procedere :—He to ðære byrig gewæg mycelne aad *advexit illi urbi plurimam congeriem*, Bd. 3, 16 ; S. 542, 22. To ðǽm readorlîcum blîðe ic sý gewegen rîces coelnesse ad *ethera letus vehar regni refrigeria*, Wanl. Catal. 304, 49. He wið ðam wyrme gewegan sceolde *he must proceed against the worm [dragon]*, Beo. Th. 4792 ; B. 2400. [Cf. *Icel.* vega *to fight*.] **II.** *to weigh, measure :*—Gewihþ *weighs*, L. M. 2, 67 ; Lchdm. ii. 298, 16-25. Gewegen biþ remetietur, Mt. Kmbl. Lind. 7, 2 : Mk. Skt. Lind. 4, 24. [Cf. a-wegan.]

ge-wélan ; *pp.* ed *To bind together :*—Þurh ðas þeóde gewélede togædere *through this people banded together*, Swt. A. S. Rdr. 108, 131.

ge-weldan *to rule, restrain ;* regere, cohibere :—Ðæt he hit ðonne [ne, MS. Cot.] mæge to his willan gewealdan [geweldan, MS. Cot.] *so that he then cannot restrain it according to his will*, Past. 17, 8 ; Swt. 119, 17 ; Hat. MS. 24 a, 6. DER. wealdan.

ge-welgian, -welegian ; *p.* ode, ade ; *pp.* od, ad *To enrich, make wealthy, endow ;* dîtare, dôtare :—Ðû gemǽnifyldest gewelgian hine *multîplicasti lôcuplêtâre eum*, Ps. Spl. 64, 9. Mid hire gestreóne he gewelgode Rôme burh *he enriched Rome with its wealth*, Ors. 5, 13 ; Bos. 113, 36 : Bd. 1, 33 ; S. 499, 1. Ic gewelegode Abram *ego dîtâvi Abram*, Gen. 14, 23. Hî nalæs mid deófolcræfte, ac mid godcunde mægene gewelgade côman *illi non dæmonîca. sed divîna virtûte prædîti vēniebant*, Bd. 1, 25 ; S. 487, 2 : 4, 13 ; S. 582, 39. Ða ðe geára on sacerdhâde ǽdellîce gewelegode wǽron *quos ôlim sacerdôtii grădu non ignobilĭter potītos*, 3, 19 ; S. 548, 38.

ge-welhwǽr ; *adv. Everywhere ;* ûbîque :—Is wîde cûþ þeódum gewelhwǽr *it is well known to people everywhere*, Menol. Fox 61 ; Men. 30 : Swt. A. S. Rdr. 105, 33.

ge-welhwilc ; *adj. Every :*—On gewelhwilcum ende *on every side*, Swt. A. S. Rdr. 106, 68 : 108, 121.

ge-welt-leðer, es ; *n. A power-leather, a rein*, Bt. Met. Fox 29, 155 ; Met. 29, 78. v. ge-weald-leðer.

ge-wéman ; *p.* de ; *pp.* ed [ge-, wéman *to persuade, entice*] *To turn, incline, seduce ;* inclînâre, sedûcĕre :—Hî næfdon ðone lâreów ðe cûþe hî to sóþfæstnysse wege gewéman *they had not the teacher who could incline them to the way of truth*, Homl. Th. ii. 400, 30 : ii. 498, 18. Hine wolde se deófol fram Gode gewéman *the devil would seduce him from God*, ii. 448, 28 : 478, 34 : 542, 19. Seó costnung gewémþ ðone man to syngienne *the temptation seduces the man to sin*, Boutr. Scrat. 23, 9. Hî eów to óðrum Gode gewémaþ *they will seduce you to another God*, Homl. Th. ii. 494, 9. Ðæt we ne sceolon nâ geþafian ðæt deófol us gewéme fram Cristes bróðorrǽdene *we should not allow the devil to seduce us from the brotherhood of Christ*, i. 260, 11.

ge-wemman ; *p.* -wemde ; *pp.* -wemmed, -wemd *To stain, defile, pollute, profane, corrupt, vitiate, mar, injure ;* coinquĭnâre, turpâre, pollŭĕre, profânâre, corrumpĕre, vîtiâre, contāmĭnâre, vĭolâre :—Ne mihte heora wlite gewemman wylm ðæs wæfran lîges *the heat of the flickering flame might not corrupt their beauty*, Cd. 185 ; Th. 231, 1 ; Dan. 240. Ic gewemme *corrumpo*, Ælfc. Gr. 28, 6 ; Som. 32, 21. Ðyder þeóf ne genealǽcþ, ne moþþe ne gewemþ *quo fur non appropiat, neque tĭnea corrumpit*, Lk. Bos. 12, 33. Hî on ðam temple gewemmaþ ðone resteðæg *in templo sabbătum vĭolant*, Mt. Bos. 12, 5. Ic honda gewemde *I have polluted my hands*, Cd. 52 ; Th. 672 ; Gen. 1094. Ðû gewemdest his hâlignesse on eorþan *profânasti in terra sanctĭtâtem ejus*, Ps. Th.

88, 32 : Exon. 29 b ; Th. 91, 5 ; Cri. 1487. Ða ðîn fǽle hûs ealh hâligne gewemdan *coinquĭnâvērunt templum sanctum tuum*, Ps. Th. 78, 1. Næs him gewenmed wlite *his beauty was not injured*, Andr. Kmbl. 2940 ; An. 1473 : Cd. 4 ; Th. 5, 13 ; Gen. 71 : Bd. 2, 12 ; S. 513, 15 : Ps. Spl. 13, 2. He geseah sîde sǽlwongas widlum gewemde *he saw the wide fertile plains defiled with pollutions*, Cd. 64 ; Th. 78, 16 ; Gen. 1294.

ge-wemmednys, se ; *f. Defilement, pollution :*—Ælfremed fram lîchamlîcere gewemmednysse *exempt from bodily defilement*, Homl. Th. i. 76, 15 : 90, 2 : ii. 478, 10 : 552, 24 : Blickl. Homl. 75, 6. Gewemmednyssa *prævaricationes*, Ps. Spl. 100, 3.

ge-wemming, -wemmincg, e ; *f. A corruption, violation, profanation ;* corruptio :—Be reste daga gewemminge *with regard to the profanation of sabbaths*, Nicod. 10 ; Thw. 5, 22.

ge-wemmodlîce ; *adv. Corruptly, impurely :*—Gewæmmodlîce we sprecaþ *corrupte loquimur*, Coll. Monast. Th. 18, 8.

ge-wén, e ; *f. Hope ;* spes. v ge-wéne.

ge-wénan ; *p.* de ; *pp.* ed *To hope, expect, suppose, think, esteem :*—Ne þurfon hî to meotude miltse gewénan *they need expect no mercy from the Lord*, Exon. 27 b ; Th. 83, 35 ; Cri. 1366. Nellaþ gê gewénan welan *nolite sperare in iniquitatem*, Ps. Th. 61, 10. On ǽrmergen ic on ðé gewéne *in matutinis meditabor in te*, 62, 6. Ic on God mínne gewéne *spero in Deum meum*, 68, 3 : 51, 7. Se sóþfæsta bóte gewéneþ *justus sperabit*, 63, 9. On his milde môd gewénaþ *sperant super misericordia ejus*, 146, 12 : 144, 16. Ic me dyslîcre ǽr þrage ne gewénde *I before expected not such a time for myself*, Exon. 72 a ; Th. 269, 21 ; Jul. 453. Gewéned ic eom *æstimatus sum*. Ps. Spl. 87, 4 ; 43, 25. Ðás beóþ men gewénede *hi putantur homines fuisse*, Nar. 35, 33.

ge-wend, es ; *n. A spiral shell, snail-shell ;* coclea, Ælfc. Gl. 49 ; Som. 65, 81 ; Wrt. Voc. 34, 13. [Cf. ge-wind, windan.]

ge-wendan ; *p.* -wende ; *pp.* -wended, -wend. **I.** *v. trans. To turn, change, translate, incline, bring about :*—Gif hit eówer ǽnig mǽge gewendan ðæt . . . *if any of you can bring it about that . . .*, Cd. 22 ; Th. 27, 35 ; Gen. 428. He cwæþ ðætte ǽghwilc ungemyndig rihtwîsnesse hine hræðe sceolde eft gewendan in to sînum môdes gemyndo *he said that every one unmindful of righteousness should speedily turn again to his mind*, Bt. Met. Fox 22, 113 ; Met. 22, 57. Wîcg gewende *he turned his steed*, Beo. Th. 635 ; B. 315. Gewend *conversus*, Lk. Bos. 22, 32. Ðis folc eall to yfele gewend ys *this people is all inclined to evil*, Ex. 32, 22. Him ðæt heáfod wæs adûne gewended *his head was turned down*, Blickl. Homl. 173, 4. Ne biþ ðé nó lîf afyrred ac biþ gewenden [?] in ðæt betere *life is not taken from thee but changed to the better*, Shrn. 119, 29. Ðonne weorþeþ sunne sweart gewended *then shall the sun be turned black*, Exon. 21 b ; Th. 58, 14 ; Cri. 935. **II.** *v. intrans. To turn [one's self], change, go, return :*—Wâ biþ ðam ðe sceal frófre ne wénan wihte gewendan *woe to the man that must expect no comfort, who must change [his condition] in nothing [whose state is hopeless and unchangeable ?]*, Beo. Th. 374 ; B. 186. He gewendeþ on ða wyrsan hand *he turns to the worse side*, Salm. Kmbl. 997 ; Sal. 500. Hwîlum hie gewendaþ on wyrmes lîc *sometimes they turn into the body of a snake*, 305 ; Sal. 152. Siððan nǽfre to unrihtum ne gewendaþ *never afterwards do they turn to iniquity*, Blickl. Homl. 193, 24 : Elen. Kmbl. 1230 ; El. 617. Drusiana hâm gewende *Drusiana went home*, Homl. Th. i. 60, 20. Drihten gewende to heofenum *the Lord returned to heaven*, 74, 19. Gewendon ealle heom hâm *they all went home*, Chr. 1052 ; Erl. 183, 11, 6, 12, 15. Ða wæs se cyng gewend ofer Temese *then the king was gone over the Thames*, 1006 ; Erl. 140, 29 : 1052 ; Erl. 183, 18.

ge-wéne ; *adv. Perhaps ;* forte :—Gewoene *forte*, Mk. Skt. Rush. 14, 2.

ge-wenge, es ; *n. The cheek ;* maxilla :—And ðam ðe ðé slihþ on ðîn gewenge *et qui te percutit in maxillam*, Lk. Bos. 6, 29 ; and to him ðæt schal smyte thee on o cheke, Wyc. Án strǽl hyne gewundode on hys óðer gewenge *an arrow wounded him in one of his cheeks*, Shrn. 97, 14. Gewenge *maxilla*, Ælfc. Gl. 71 ; Som. 70, 80 ; Wrt. Voc. 43, 13. v. wenge.

ge-wenian ; *p.* ede ; *pp.* ed. **I.** *to accustom, to accustom any one to one's self ;* assuefacere :—Gewenede hine sylfne to heora syulîcum þeáwum *he accustomed himself to their sinful manners*, Ælfc. T. Lisle 34, 20 : Bt. Met. Fox 29, 11 ; Met. 29, 6. Heora lâreówas him biódan ða ilcan mettas ðe hî ǽr tame mid gewenedon *their teachers offer them the same meats which they before accustomed the tame with or with which they before accustomed them to be tame*, Bt. 25 ; Fox 88, 18 : L. Edg C. 55 ; Th. ii. 256, 9. **II.** *to wean, to separate ;* ablactare, a lacte depellere, depellere, seducere :—Ðæt cild wearþ gewened *puer ablactatus est*, Gen. 21, 8. Se deófol wolde hine fram Gode gewenian *the devil would wean him from God*, Job. Thw. 165, 11. [*O. H. Ger.* ge-wenian assuefacere.] v. wenian.

ge-weold. v. ge-wild.

ge-weorc, -worc, -werc, es ; *n.* [ge-, weorc *a work*]. **I.** *work ;* ŏpus, ŏpuscŭlus :—Eue wæs geweorc Godes *Eve was God's work*, Cd. 38 ; Th. 51, 6 ; Gen. 822. Exon. 9 b ; Th. 8, 4 ; Cri. 112. Ðæt ðam þeódne wæs sîþes sigehwîl, sylfes dǽdum, worlde geweorces *that was a victorious moment to the prince of his enterprise, by his own deeds, of his worldly*

work, Beo. Th. 5415; B. 2711. He geseah eald enta geweorc *he saw the antique work of giants*, Andr. Kmbl. 2988; An. 1497: 2155; An. 1079. On ðæt geweorc *in ŏpus*, Bd. 1, 23; S. 485, 40. Ne wāciaþ ðás geweorc *these works fail not*, Exon. 93 b; Th. 351, 26; Sch. 86. Mǽre wurdon his wundra geweorc *great were his wondrous works*, 45 b; Th. 155, 2; Gū. 854: 40 a; Th. 133, 35; Gū. 500. Of geweorcum ārwurþra fædera *ex ŏpuscŭlis venerābĭlium patrum*, Bd. 5, 24; S. 647, 33. II. *a fort, fortress*; *arx*:—He of ðam geweorce wæs winnende wið ðone here *he warred on the army from the fortress*, Chr. 878; Erl. 80, 5: 896; Erl. 94, 3, 21. He worhte him geweorc æt Middeltūne *he wrought him a fortress at Middleton*, 892; Erl. 89, 14: 894; Erl. 92, 4, 11. ðe æt hām æt ðæm geweorcum wǽron *who were at home in the fortresses*, 894; Erl. 92, 18. Hī worhton tū geweorc *they wrought two forts*, 896; Erl. 94, 11. Geweorc *arx, figmentum, māchĭna*, Scint. 62: Cot. 85: 128, Lye. [*Goth.* ga-waurki: *O. Sax.* gi-werk: *O. H. Ger.* gi-werk.] DER. ǽr-geweorc, eald-, flān-, fyrn-, gold-, gūþ-, hand-, heáh-, land-, nīþ-, sulh-.

ge-weorht, es; *n. Work, deed, merit, desert*; *ŏpus, făcĭnus, mĕrĭtum*:—Ðætte rinca gehwylc ŏðrum gulde edleán on riht be geweorhtum *that every man should render rightly to other a reward proportionable to his deserts*, Bt. Met. Fox 27, 53; Met. 27, 27. v. ge-wyrht.

ge-weorhta, an; *m. One working with another, accomplice*:— Gif mæsse-preóst þeófa gewita and geweorhta beó *if a mass-priest be an accessory and accomplice of thieves*, L. Eth. ix. 27; Th. i. 346, 9. v. ge-wyrhta.

ge-weorp, es; *n. A throwing, tossing, dashing, what is thrown up, a heap*; *jactus, jactātio, projectio*:—Ofer waroþa geweorp *over the dashing of the waves*, Andr. Kmbl. 611; An. 306. Ðær ðū geseó tord-wifel on eorþan up weorpan ymbfó hine mid twām handum mid his geweorpe *when you see a dung-beetle in the earth throwing up mould, catch it with both hands along with his casting up*, L. M. iii. 18; Lchdm. ii. 318, 17. v. winter-geweorp, ge-wyrp.

ge-weorpan, -worpan; *p.* -wearp, *pl.* -wurpon; *pp.* -worpen. I. *to throw, cast*; *jacere, projicere*:—Hī habbaþ ingang swā mycelre brǽdo, swā mon mæg mid liðeran geworpan *they have an entrance of so much breadth, as one can throw with a sling*, Bd. 4, 13; S. 583, 11. Drihten hī gewyrpþ mid grine *the Lord will cast a snare upon them*; pluet super peccatores laqueos, Ps. Th. 10, 7. Middý gewearp woedo his *projecto vestimento suo*, Mk. Skt. Lind. 10, 50. Gewurpon būta *ejecerunt extra*, 12, 8, 41. Honda gewurpon on hine *manus injecerunt in eum*, 14, 46. Swā gewundade wrāðe slǽpe, sýn ðonne geworpene on wīdne hlǽw *sicut vulnerati dormientes, projecti in monumentis*, Ps. Th. 87, 5. II. *to turn one's self away, go away, depart, pass by*; *averti, abire, transire*:— Winter sceal geweorpan, weder eft cuman, sumor hāt *winter shall pass by, fair weather again shall come, hot summer*, Exon. 90 a; Th. 338, 11; Gn. Ex. 77. DER. weorpan.

ge-weorp, es; *n. Value, worth, price*, Th. Chart. 159, 1. v. ge-wyrþe.

ge-weorþan, -wiorþan, -wurþan, -wyrþan; he -weorþeþ, -weorþ, *pl.* -weorþaþ; *p.* ic, he -wearþ, ðū -wurde, *pl.* -wurdon; *subj. pres.* -weorþe, *pl.* -weorþen; *p.* -wurde, *pl.* -wurden; *pp.* -worden. I. *to be, be made, become, happen*; *fĭeri*:—Hū māgon ðás þing ðus geweorþan *quomŏdo possunt hæc fĭeri?* Jn. Bos. 3, 9: Elen. Kmbl. 909; El. 456. Ne sēc ðū þurh hlytas hū ðe geweorþan scyle *seek not by lots how it is to happen to thee*, Prov. Kmbl. 32. Hū geweorþeþ ðæt *how happeneth that?* Salm. Kmbl. 684; Sal. 341: Andr. Kmbl. 2872; An. 1439. Gif feaxfang geweorþ *if there be a taking hold of the hair*, L. Ethb. 33; Th. i. 12, 3. Ealle gesceafte forhte geweorþaþ *all creatures shall tremble*, Andr. Kmbl. 2298; An. 1502. He gewyrþ micelre mǽgþe *he shall become a great nation*, Gen. 21, 18. Ðes sige gewearþ Punicum *this victory happened to the Carthaginians*, Ors. 4, 6; Bos. 85, 23. Ic his mōdor gewearþ *I have become his mother*, Exon. 11 a; Th. 13, 30; Cri. 210: 9 a; Th. 6, 33; Cri. 93. Ðū ðissum hysse hold gewurde *thou hast been gracious to this man*, Andr. Kmbl. 1100; An. 550. Sió fǽhþ gewearþ gewrecen wrāþlīce *the quarrel was wrothfully avenged*, Beo. Th. 6115; B. 3061: Exon. 33 b; Th. 107, 26; Gū. 64: Chr. 592; Erl. 19, 34. Gewurdon manige wundor on manegum landum *many wonders happened in many lands*, Ors. 5, 10; Bos. 108, 16 Ðæt me Meotud moncynnes milde geweorþe *that the Lord of mankind be merciful to me*, Exon. 75 b; Th. 282, 23; Jul. 667: 78 b; Th. 294, 19; Cra. 17. Deáh mīn bān and blód būtū geweorþen eorþan to eácan *though my bones and blood both become an increase to earth*, 38 a; Th. 125, 9; Gū. 351. Saga, hū ðæt gewurde *say how that happened*, Andr. Kmbl. 1115; An. 558: Exon. 11 a; Th. 15, 19; Cri. 238. Ðæt word wæs flǽsc geworden *verbum căro factum est*, Jn. Bos. 1, 14: Homl. Th. i. 40, 17: Cd. 219; Th. 282, 5: Sat. 282: 223; Th. 293, 10: Sat. 453. Wæs onlīce bī hig geworden [swā bī Zachariam] gewearþ and bī Elizabeþ his wīfe *it had happened with them as it happened with Zacharias and his wife Elizabeth*, Shrn. 36, 12. We gesēgon windas and wǽgas forhte gewordne *we saw winds and waves become fearful*, Andr. Kmbl. 913; An. 457. II. *v. impers. cum acc. To happen, come to pass, befall, come together, agree, be agreeable*; *contingĕre, evĕnire, convĕnire, plăcēre*:—Ne mihte hī betwih him geþwǽrian and geweorþan *they might*

---

not accord and agree among themselves, Bd. 4, 4; S. 571, 2: Cd. 81; Th. 101, 32; Gen. 169, 1. Hū gewearþ ðe ðæs *how doth this befall thee?* Andr. Kmbl. 613; An. 307: Jud. 16, 21. Me gewearþ *convēnior*, Ælfc. Gr. 37; Som. 39, 6. Hý gewearþ, ðæt hý woidan to Rōmānum friþes wilnian *they agreed that they would seek peace from the Romans*, Ors. 4, 6; Bos. 86, 17: 5, 10; Bos. 108, 29: 6, 30; Bos. 126, 24: Gen. 20, 13. Ðā hī nānre sibbe gewearþ *when they could not agree upon any terms of peace*, Ors. 4, 11; Bos. 97, 19. Deáh ðe Rōmāne hæfde geworden ðæt . . . *though the Romans had agreed that . . .*, 4, 12; Bos. 98, 43. Hū ðone cumbolwīgan hæfde geworden *how it had befallen the warrior*, Judth. 12; Thw. 25, 15; Jud. 260. III. *cum dat*:—Ðā gewearþ ðam hlāforde and ðam hýrigmannum wiþ ānum peninge *then the lord and the labourers agreed on a penny*, Th. An. 73, 29: 74, 21 [*or acc.*]. Gewearþ him and ðam folce on Lindesīge ānes ðæt hī hine horsian sceolde *it was agreed between him and the people of Lindsey that they should provide him with horses*, Chr. 1014; Erl. 151, 1: Thw. 161, 30. Wyn ðū ongeán ðone wuldres cyning and gewurþe ðe and him *fight against the king of glory and let there be an agreement between thee and him*, Nicod. 27; Thw. 15, 14. [Cf. *O. Sax.* thea gumon alle giwarth that . . . : *Goth.* ga-wairþi *peace?*]

ge-weorþian, -wurþian, -wyrþian; *p.* ode, ade, ude; *pp.* od, ad, ud. I. *to set a price on, value*:—Ðone ðe wæs ǽr geweorþod *quem appretiaverunt*, Mt. Kmbl. 27, 9, note. II. *to distinguish, honour, dignify, adorn, worship, adore, celebrate, praise*; *insignīre, hŏnōrāre, ornāre, instruĕre, mactāre, adōrāre, celebrāre*:—Ðū hine gewuldrast and geweorþast *glōria et hŏnōre cŏrōnasti eum*, Ps. Th. 8, 6. Ðe beorht Fæder geweorþaþ wuldorgifum *the bright Father dignifies thee with glorious gifts*, Andr. Kmbl. 1875; An. 940: Bt. 14, 3; Fox 46, 13. Me geweorþode wuldres Ealdor *the Prince of glory honoured me*, Rood Kmbl. 177; Kr. 90: 185; Kr. 94. He Abrahames cynn geweorþude *he honoured Abraham's race*, Ps. Th. 104, 6. Geweorþie wuldres Ealdor eall ðeós eorþe, ēcne Drihten *omnis terra adōret te, Deus*, 65, 3. Gē wēnaþ ðæt ǽnig mæg mid frǽmdum welum beón geweorþod *ye think that one can be made honourable by external riches*, Bt. 14, 3; Fox 46, 10, 11. Wæs ēþfynde Afrisc meówle, golde geweorþod *the African maid was easy to be found, adorned with gold*, Cd. 171; Th. 215, 9; Exod. 580: 174; Th. 218, 18; Dan. 41: Elen. Kmbl. 2384; El. 1193. Wuldre geweorþad *honoured with glory*, Exon. 63 b; Th. 235, 2; Ph. 551: Beo. Th. 2904; B. 1450. Wīde is geweorþod hāligra tīd *the time of the saints is widely celebrated*, Menol. Fox 237; Men. 120: 306; Men. 154.

ge-weoton *went, departed*, Bd. 2, 5; S. 507, 34; *p. pl.* of ge-wītan.

ge-wēpan; *p.* -weóp, *pl.* -weópon; *pp.* -wōpen *To weep, lament*; *flere*:—Petrus geweáp bitterlīce *Petrus flevit amare*, Lk. Skt. Lind. 22, 62. Giweópun alle *flebant omnes*, Rush. 8, 52. Gewōpen *fletum*, Ælfc. Gr. 26, 1; Som. 28, 28.

ge-werc, es; *n. A fort, fortress*; *arx*:—Hie ðær gewerc worhton *they there wrought a fortress*, Chr. 896; Erl. 94, 16. v. ge-weorc.

ge-werdan; *p.* de; *pp.* ed *To hurt, injure*; *lædere, nocere*:—Gif hwā on ceáse wīf gewerde *if any one in strife hurt a woman*, L. Alf. 18; Th. i. 48, 17: 26; Th. i. 50, 24. v. ge-wyrdan.

ge-weredlǽht, -werodlǽht *sweetened, made sweet*; *indulcoratus*, Scint. 64. v. werod.

ge-werged; *part. Accursed*:—Ðara gewergedra *maledicorum*, Mt. Kmbl. p. 1, 11.

ge-wērigan, -wērigan; *p.* ode, ade; *pp.* od, ad *To weary, fatigue*; *fatigāre*:—He gewērgad sæt *he sat wearied*, Beo. Th. 5697; B. 2852: Exon. 51 a; Th. 178, 12; Gū. 1243. Mauritanie wǽron mid ðam gewērgode *the Mauritanians were wearied by it*, Ors. 5, 7; Bos. 107, 7. Ðe on lengtenādle gewērigade wǽron *who were wearied with ague*, Bd. 4, 6; S. 574, 7.

ge-werian; *p.* ode, ede; *pp.* od, ed *To put on, cover, clothe*; *induĕre, vestīre*:—Giwoeria *to cover, conceal*, Rtl. 103, 3. Ðe he mid gewered wæs *quibus indūtum ĕrat*, Mk. S. 608, note 39, 41. Ge-wered mid wæstme *covered with fruit*, Cd. 23; Th. 30, 5; Gen. 462. In hwītum hræglum gewerede englas ne ŏþeówdun *angels appeared not clad in white robes*, Exon. 14 a; Th. 28, 16; Cri. 447: 15 b; Th. 35, 3; Cri. 552. [*Goth.* ga-wasjan.]

ge-werian; *p.* ede, ode; *pp.* ed, od. I. *to defend, protect, take care of, make [land] free from claims*; *defendĕre, procurare*:—Ic gewerige *defendo*, Ælfc. Gr. 28, 6; Som. 32, 29. Se ðe land gewerod hæbbe *he who has defended land*, L. C. S. 80; Th. i. 420, 19. Þǽr of is gewerod ān and tuenti hīde *twenty-one hides of it are held in undisputed possession*, Schmid. A. S. Ges. p. 614, col. 1. See also p. 677. Ðonne his ðæs londæs hundseofontig hīda is nū eall gewǽred and ðā hit æst mīn lāford mæ to lǽt ðā wæs hit ierselæás *hujus terræ sunt lxx hidæ, et est modo tota bene procurata, quæ quando dominus meus michi eam tradidit omni peccunia caruit*, Th. Chart. 162, 26. Gange [ðæt land] into ðære stōwe swā gewered swā hit stande mid mete and mid mannum and mid ælcum þingan *let the land go afterwards to that place so provided as it may then be, with meat and with men and with everything*, 519, 3: Cod. Dipl. Kmbl. ii. 300, 10. II. *to associate with for the cause of defence.*

to make a treaty with; assŏciāre defensiōnis causa, jungere fœdĕre :—Nalæs æfter micelre tîde ðæt hî geweredon wið him, and heora wǽpen hwyrfdon wið Bryttas heora' gefaran *non multo post juncto cum his fœdĕre, in sŏcios arma vertĕrit,* Bd. 1, 15 ; S. 483, 4, 35. v. werian.

ge-wesan *to be together, converse, discuss :*—Ic flîtan gefrægn môd-gleáwe men gewesan ymbe hyra wîsdôm *I have learnt that wise men had disputes and discussions about their wisdom,* Salm. Kmbl. 363 ; Sal. 181. Grein writes 'gewêsan ;' *p.* -weós,' and compares 'ymbweoson' in the Northumbrian Gospels. But this word is wrongly written by Bouterwek, it should be 'ymbweoson,' see Mk. Skt. p. 1. The Durham Ritual glosses ' conversatio ' by 'giwosa,' and this may throw light on the meaning of 'gewesan.' Both *Goth.* and *O. H. Ger.* have the word ' gawisan, gi-wesan,' in the sense *to remain, abide; restare.*

ge-wêsan ; *p.* de ; *pp.* ed *To soak :*—Mid ecede gewêsed *soaked with vinegar,* Herb. 116, 3 ; Lchdm. i. 228, 24. Gewêsan *infîcere, miscere, fucare,* Hpt. Gl. 524. v. wôs.

ge-wêstan *to lay waste; desolare :*—Gewoested biþ *desolabitur,* Mt. Kmbl. Lind. 12, 25.

ge-wêðnis, se ; *f. Mildness;* lenitas :—Giwoeðnis *lenitas,* Rtl. 105, 1. v. wêðe.

ge-wîcan ; *p.* -wâc, *pl.* -wicon ; *pp.* -wicen *To give way, fail, depart, retire;* cedere, deficere, recedere :—To hwŷ, Driht, gewic [gewîte, Sur.] ðû feor *ut quid, Domine, recessisti longe,* Ps. Spl. second 9, 1. Ne his mægenes [mæges ?] lâf gewâc æt wîge *his kinsman's legacy failed not in the contest,* Beo. Th. 5251 ; B. 2529 : 5148 ; B. 2577. v. wîcan.

ge-wîcian ; *p.* ode ; *pp.* od *To dwell, lodge, encamp;* hospitare, castra metari :—Hý landes hæfdon ðæt hý mihton on gewîcian *they had land on which they could encamp,* Ors. 2, 5 ; Bos. 46, 36. Ic on fægerum scûan fiðera ðînra gewîcie *in umbra alarum tuarum spero,* Ps. Th. 56, 1. Se wilda fugel hûs getimbreþ and gewîcaþ ðǽr *the wild bird builds a house and dwells there,* Exon. 58 b ; Th. 212, 1 ; Ph. 203. Ðonne gewîceaþ faroþ-lácende on ðam eálonde *then the seafarers camp on that island,* 96 b ; Th. 361, 13 ; Wal. 19. Ðâ gewîcode he neáh ânre eá *then he encamped near a river,* Ors. 4, 6 ; Bos. 84, 31 : Chr. 894 ; Erl. 90, 8 : Blickl. Homl. 79, 14. v. wîcian.

ge-wiðer, -widor, es ; *pl. nom. acc.* -wideru, -widera, -widru ; *n. Weather, the temperature of the air, a tempest;* tempestas, cæli temperies :—Hî monige dagas windes and gewidor abidon *opportûnos alĭquot dies ventos ex-pectârunt,* Bd. 5, 9 ; S. 623, 19. Se sceortiganda dæg hæfþ lîðran ge-wideru [gewidera, MS. R.] ðonne se langienda dæg *the shortening day hath milder weather than the lengthening day,* Bd. de nat. rerum ; Lchdm. iii. 252, 9, MS. L : Bt. Met. Fox 11, 121 ; Met. 11, 61. On ðæm dæge eall godes folc sceal god biddan ðæt he him forgefe smyltelîco gewidra and genihtsume wæstmas *on that day all God's folk are to pray God to give them fair weather and abundant harvests,* Shrn. 74, 11. Ðonne wind styreþ lâþ gewidru *when the wind stirs hateful tempests,* Beo. Th. 2754 ; B. 1375. [Cf. *O. Sax.* un-giwideri : *O. H. Ger.* gi-witri *temperies, tempestas: Ger.* ge-witter.] v. ge-weder.

ge-widlian, -widligan ; *p.* ede ; *pp.* ed *To defile, contaminate, make common:* coinquinare, contaminare, Mk. Skt. Lind. 7, 15. v. widl.

ge-wîdmǽrsian ; *p.* ode ; *pp.* od *To publish, spread abroad, divulge, celebrate;* divulgare :—Ofer ealle Iudêa munt-land wǽron ðâs word ge-wîdmǽrsode *super omnia montana Iudǽæ divulgabantur omnia verba hæc,* Lk. Bos. 1, 65 : Mt. Bos. 28, 15. Iosep nolde hî gewîdmǽrsian *Joseph nollet eam traducere,* 1, 19.

ge-wif, es ; *n. An affection of the eye, web :*—Wið ǽlces cynnes brôc on eágon wið gewif *for every sort of malady in the eyes, for web,* Lchdm. iii. 290, 3. v. Hall. Dict. pin-and-web.

ge-wife *fortune, destiny;* fatum, Cot. 88. v. ge-wef.

ge-wîfian ; *p.* ode, ade ; *pp.* od, ad [wîfian *to take a wife*] *To take a wife, marry;* uxôrem dûcĕre :—Gewîfodon *duxĕrunt uxôres,* Jud. 3, 6. Ðæt cristen man gewîfige *that a christian man marry,* L. Eth. vi. 11 ; Th. i. 318, 13, 18 : L. C. E. 7 ; Th. i. 364, 23. Manige habbaþ genôg gesǽlilîce gewîfod *many have married happily enough,* Bt. 11, 1 ; Fox 32, 5. Gewîfad, Bd. 4, 22 ; S. 591, 7.

ge-wîfsǽlig ; *adj. Fortunate;* fato *vel* fortuna felix, Cot. 88, 194, 196, Lye [Cf. ge-wef.]

ge-wiglung, e ; *f. Soothsaying, divination, spell :* — Ða gemearr ðe man drîfþ on mislîcum gewiglungum *the erroneous practices that are carried on with various spells,* L. Edg. C. 16 ; Th. ii. 248, 4. v. wiglian.

ge-wiht, -wyht, -wihte, es ; *n. Weight;* pondus :—Twegra pundra gewiht *two pounds' weight;* dupondius, Ælfc. Gl. 59 ; Som. 67, 114 ; Wrt. Voc. 38, 37 : Th. Chart. 522, 22 : Salm. Kmbl. p. 180, 5. Gange ân gemet and ân gewihte *let one measure and one weight pass,* L. Edg ii. 8 ; Th. i. 270, 2. Nû hæbbe we hit broht ongén be ðam ylcan gewihte *quam nunc eodem pondĕre reportâvimus,* Gen. 43, 21 : 23, 16 : Lev. 26, 26. False gewihta *false weights,* L. Eth. v. 24 ; Th. i. 310, 13 : vi. 28 ; Th. i 322, 14. Gemeta and gewihta rihte man georne *let measures and weights be carefully rectified,* vi. 32 ; Th. i. 322, 30 : L. C. S. 9 ; Th. i. 380, 24. Gê etaþ hlâf be gewihte *ye shall eat bread by weight,* Lev. 26, 26. Mid twǽm hundred mancosan goldes be gewihte and mid v. pundan

be gewihte seolfres *for two hundred mancuses of gold by weight and for five pounds by weight of silver,* Th. Chart. 557, 28. See Turner's Hist. Anglo-Sax. ii. Appendix ii. [*Ger.* gewicht.]

ge-wil, -will, -wile, -wyle, es ; *n. A will, wish, pleasure;* vŏluntas, arbitrium, vōtum :—Ne wend ðû ðê nô on ðæs folces unriht gewil *turn thou not thyself to the unjust wish of the people,* L. Alf. 41 ; Th. i. 54, 7 : Hy. 7, 78 ; Hy. Grn. ii. 288, 78. On yfelra manna gewill *according to the will of evil men,* Bt. 4 ; Fox 8, 19 : Exon. 13 a ; Th. 23, 2 ; Cri. 362 : Ors. 1, 10 ; Bos. 34, 1 : 1, 12 ; Bos. 36, 33. Hit næs ne his gewile [-wyle, MS. A.] *it was not his will,* L. C. S. 76 ; Th. i. 418, 11.

ge-wilcþ, e ; *f. Rolling, motion [of waves] :* — Gewilcþ ŷðe *motum fluctuum,* Ps. Spl. M. 88, 10.

ge-wilcuman ; *p.* ode ; *pp.* od *To welcome;* salutare :—Se câsere hig gewilcumode *the emperor welcomed them,* L. Ælf. P. 23 ; Th. ii. 372, 30.

ge-wild, -weold, es ; *n. Power, control :* — Æfter ðæm ðe Alexander hæfde ealle Indie him to gewilde gedôn *perdomita Alexander India,* Ors. 3, 9 ; Swt. 132, 9. Geweoldum sylfes willum *spontaneously, of his own accord,* Beo. Th. 4446 ; B. 2222. [Cf. ge-weald, ge-wealdes, ge-wylde.]

ge-wildan *to exercise power over, rule over,* Gen. 3, 16 : Ps. Spl. 105, 38. v. ge-wyldan.

ge-wile, es ; *n. A will;* vŏluntas, L. C. S. 76 ; Th. i. 418, 11. v. ge-wil.

ge-wilnung *a wish, appetite,* Bd. 4, 25 ; S. 601, 7. v. ge-wilnung.

ge-willsum ; *adj. Desirable;* desiderābilis :—Hî hæfdon eorþan ge-willsum *habuêrunt terram desidĕrābĭlem,* Ps. Spl. C. 105, 23.

ge-wilnian, -wilnigan, to -wilnienne ; *p.* ode ; *pp.* od [wilnian *to desire*] *To wish, desire, expect, seek, strive for;* cŭpĕre, concŭpiscĕre, desîdĕrāre, expetĕre, ambîre :—Reáflácum nylle gê gewilnian *răpinas nôlîte con-cŭpiscĕre,* Ps. Spl. 61, 10 : Ps. Spl. 118, 20. Gôdes þegenas sceolon to ðam êcan life ǽfre gewilnian *God's servants must ever strive after the life everlasting,* Boutr. Scrd. 21, 44. He ne sceal gewilnian ða woruldlîcan þinge *he must not desire the things of this world,* 22, 44. Ðæt sum sume swîðe ondryslîcu, and eác to gewilnienne secgende wæs *ut quidam multa et trĕmenda, et desîdĕranda narrāvĕrit,* Bd. 5, 12 ; S. 627, 3. Ic gewilnige [gewilnie, MS. D.] *cŭpio,* Ælfc. Gr. 35 ; Som. 38, 8 : 28, 1 ; Som. 30, 39. Ic gewilnige *ambio,* 30, 5 ; Som. 35, 8. Ælc ðæra ðe wîf ge-syhþ and hyre gewilnaþ *omnis, qui vidĕrit mŭliĕrem ad concupiscendum eam,* Mt. Bos. 5, 28. Ic nânes eorþlices gestreónes ne flǽsclices lustes ne gewilnige *I desire no earthly treasure nor fleshly pleasure,* Homl. Th. i. 458, 31 : 512, 13. Gif hwâ gewilnigeþ to gewitane *if any one desires to know,* Chr. 1086 ; Erl. 221, 10. Gewilnod *ambîtus,* Ælfc. Gr. 30, 5 ; Som. 35, 10.

ge-wilnigendlîc, -wilniendlîc, -wilnindlîc ; *adj. Desirable;* desidĕrā-bilis :— For nâht hî hæfdon eorþan gewilnigendlîce *pro nihĭlo hăbuêrunt terram desidĕrābĭlem,* Ps. Spl. 105, 23. Gewilniendlîc *desîdĕrābĭlis,* Prov. 21. Gewilnindlîc, Prov. 8.

ge-wilnung, -wilnunge. e ; *f. A wish, desire, longing, seeking, appe-tite, will, vow;* concŭpiscentia, desîdĕrium, ambîtus, appĕtîtus, affectus, vōtum :— Mid gewilnunge *ambîtus,* Ælfc. Gr. 30, 5 ; Som. 35, 10. Of ge-wilnunge ic gewilnode etan mid eów ðâs eástron *desîdĕrio desîdĕrāvi hoc ʒascha mandûcāre vobiscum,* Lk. Bos. 22, 15. Hŷ fêrdon on gewilnunge heortan *transiêrunt in affectu cordis,* Ps. Spl. 72, 7 : Homl. Th. i. 136, 9, 31. For gewilnunge ðara êcra gôda *pro appĕtîtu æternôrum bonôrum,* Bd. 4, 25 ; S. 601, 7. Ôðra gewilnunga *reliqua concŭpiscentiæ,* Mk. Bos. 4, 19 : Num. 11, 34. Mid eallum gewilnungum *with all desires,* Homl. Th. ii. 118, 25. Ðæt ic agylde gewilnunga of dæge to dæge *ut reddam vôta mea de die in diem,* Ps. Spl. 60, 8.

ge-win, -winn, es ; *n.* [winnan *to fight*]. **I.** *a battle, contest, war, strife, quarrel, hostility, tumult;* certāmen, pugna, bellum, tŭmul-tus :—On ðâ tîde Troiâna gewin wearþ *the Trojan war happened at that time,* Bt. Met. Fox 26, 24 ; Met. 26, 12. Sceolde he worc ðæs gewinnes gedǽlan *he must get pain on account of that struggle,* Cd. 15 ; Th. 19, 24 ; Gen. 296 : 17 ; Th. 21, 12 ; Gen. 323 : Bt. Met. Fox 25, 101 ; Met. 25, 51. On ðam gewinne *in the contest,* Bt. 37, 1 ; Fox 186, 31 : 38, 1 ; Fox 194, 8 : Rood Kmbl. 129 ; Kr. 65. Hie gewin drugon *they fought,* Beo. Th. 1601 ; B. 798 : 1758 ; B. 877. Heora gewinn mid ðam swîðe geiécton *their quarrel was thus much strengthened,* Ors. 5, 10 ; Bos. 109, 4 : 5, 13 ; Bos. 112, 43. He his môdsefan wið ðam fǽrhagan fæste trymede feónda gewinna *he firmly strengthened his mind against the peril of the fiends' hostilities,* Exon. 46 b ; Th. 159, 29 ; Gû. 934. **II.** *labour, toil, sorrow, agony;* lâbor, trîbŭlātio, ăgônia :—Ðis gewin *hic lâbor,* Bd. 2, 1 ; S. 500, 29. Gewinn and sâr *lâbor et dŏlor,* Ps. Th. 89, 11 : 72, 13. Wæs gewinnes endedôgor neáh geþrungen *the final day of his labour was near at hand,* Exon. 46 a ; Th. 158, 6 ; Gû. 904 : Ps. Th. 127, 2. Ðû scealt wunian in gewinne *thou shalt continue in toil,* Exon. 16 b ; Th. 39, 14 ; Cri. 622 : 32 a ; Th. 101, 10 ; Cri. 1656 : Bd. 1, 23 ; S. 485, 17. He wæs on gewinne *factus in ăgônia,* Lk. Bos. 22, 44. Þurh mycel gewinn *with much toil,* Guthl. 16 ; Gdwin. 68, 5. On ge-winnum *in lâbôribus,* Ps. Th. 106, 11 : 72, 4. **III.** *fruit of labours, gain, profit;* fructus lâbôrum, lucrum, quæstus :—Hî folca gewinn frem-dra gesǽton *lâbôres pŏpŭlôrum possêdêrunt,* Ps. Th. 104, 39 : 77, 46. Gif hwilc man leoht deþ on nînum cirican of his gewinne *if any man*

puts a light in my church [bought] out of his gain, Nar. 47, 6, 15. [O. Sax. ge-win *strife*: O. H. Ger. ga-win *labor, certamen, quæstus*: Ger. ge-win *gain*.]

ge-wind, es; *n.* [windan *to bend*] *A winding, circuitous ascent*, Ælfc. Gl 55; Som. 67, 6; Wrt. Voc. 37, 4.

gewin-dæg, es; *m. A labour* or *trouble-day, battle-day; laboris vel tribulationis dies, pugnæ dies*:—On gewindæge *in the day of trouble*, Ps. Th. 77, 42. Of gewindagum weorþan sccolde líf alýsed *her life should be released from days of trouble*, Exon. 74 b; Th. 279, 9; Jul. 611: Cd. 205; Th. 254, 24; Dan. 616. Đonne cumbulgebrec on gewinndagum weorþan scoldun *when there should be crashings of banners in days of battle*, Ps. C. 50, 12; Ps. Grn. ii. 227, 12. v. win-dæg.

ge-windan; *p.* -wand, *pl.* -wundon; *pp.* -wunden. **I.** *v. trans. To twist, weave, bend, wind*; torquēre, plectere, implicāre:—Đa þegnas gewundun đæt sigbēg of þornum *milites plectentes coronam de spinis*, Jn. Skt. Lind. 19, 2. Ne hafu ic in heáfde hwíte loccas, wræste gewundne *I have not white locks on my head, delicately wound*, Exon. 111 b; Th. 427, 30; Rä. 41, 99. **II.** *v. intrans. To go, turn, turn about, revolve, roll*; íre, se vertĕre, volvĕre:—He meahte wídre gewindan *he might more widely turn about*, Beo. Th. 1530; B. 763. Se aglæca on fleám gewand *the miserable being turned to flight*, 2007; B. 1001: Homl. Th. i. 290, 19. Se líg gewand on láđe men *the flame rolled on to the hostile men*, Cd. 186; Th. 231, 22; Dan. 251.

ge-winde; *adj.*:—Đá hit wæs wel gewinde on đa burh *when the wind was in the right quarter [for blowing the flames] on to the town*; *ventum opportunum*, Bd. 3, 16; S. 542, 25.

ge-windwian; *p.* ode; *pp.* od *To blow*:—Seó onbláwnes đære heofonlican onfædmnesse sý gewindwod on đé *let the inspiration of the heavenly embrace be blown into thee*, Blickl. Homl. 7, 27.

ge-winful, -full; *adj. Full of labour, laborious, troublesome*; labōriōsus:—Agustinus đysses gewinfullan geflítes ende gesette *Augustinus hunc lābōriōsi certāminis fīnem fēcit*, Bd. 2, 2; S. 502, 17.

ge-winfullíc, -winnfullíc; *adj. Laborious, toilsome*; labōriōsus:—Đæt hí ne þorftan in swá frǽcne síþfætt, and on swá gewinfullícne, and on swá uncúþe ællþeódignysse féran *ne tam periculōsam, tam incertam peregrīnātiōnem adīre debērent*, Bd. 1, 23; S. 485, 37. Đæt hí nó má ne mihton swá gewinnfullícum fyrdum swencte beón *non se ultra tam labōriōsis expedītiōnibus posse fatigāri*, 1, 12; S. 481, 4.

ge-winfullíce; *adv. Laboriously, with difficulty*; labōriōse:—Đæt eahta and twentig wintra gewinnfullíce he heóld *id per annos viginti octo labōriōsissĭme tĕnuit*, Bd. 3, 14; S. 539, 17.

ge-winna, an; *m. An enemy, adversary, a foe, rival*; hostis, inĭmīcus, æmŭlus:—Cwom semninga hæleþa gewinna *the foe of men suddenly came*, Exon. 69 a; Th. 257, 7; Jul. 243. Gesaca *vel* gewinna *æmŭlus*, Ælfc. Gl. 114; Som. 80, 17; Wrt. Voc. 60, 51. Læddon leóde láþne gewinnan to carcerne *the people led their hated foe unto the prison*, Andr. Kmbl. 2500; An. 1251: 2603; An. 1303. Beóþ đé hungor and þurst hearde gewinnan *hunger and thirst will be hard adversaries to thee*, Exon. 36 b; Th. 118, 28; Gú. 246. Heora gewinnan hí éhtan *insĕquĭtur hostis*, Bd. 1, 12; S. 481, 23: 1, 23; S. 483, 13. Đa ǽrran gewinnan *priores inimīci*, S. 1, 12; S. 480, 33.

ge-winnan; *p.* -wan, -won, -wann, *pl.* -wunnon; *pp.* -wunnen. **I.** *to make war, fight, contend*; pugnāre, bellum gerere:—He ána gewon *he fought alone*, Exon. 39 a; Th. 129, 15; Gú. 21: Bd. 3, 19; S. 548, **2.** Hú hie wiđ đæm drý geflíton and gewunnon *how they contended and strove against the sorcerer*, Blickl. Homl. 173, 3. **II.** *to obtain by fighting, to conquer, gain, win*; pugna consequi, obtinēre, subjugāre:—Hú he mihte Normandige of him gewinnan *how he might conquer [win] Normandy from him*, Chr. 1090; Erl. 226, 25. Ne mágon we đæt on aldre gewinnan *we cannot ever obtain that*, Cd. 421; Th. 26, 6; Gen. 402. Ænig ne mæg friþ gewinnan *no one may gain peace*, Exon. 22 b; Th. 62, 14; Cri. 1001. Đæs đe he heora sáulum to hǽle and to lǽde gew nnan mihte *provided that he could win their souls to salvation and counsel*, Blickl. Homl. 227, 4. He hit gewan mid wisdóme *he gained it by wisdom*, Th. Ap. 4, 19. Chananéus đá wann wiđ Israéla bearn and sige on him gewann *the Canaanite fought against the children of Israel and gained a victory over them*, Num. 21, 1. Đone cyning đe hie ǽr mid unrihte gewunnon hæfde *the king that had before unjustly conquered them*, Bt. 16, 2; Fox 52, 22. On ágenum hwílum mid earfeþum gewunnen *laboriously gained in their own time*, Swt. A. S. Rdr. 106, 55. Đá wæs Rómana ríce gewunnen *then the empire of the Romans was conquered*, Bt. Met. Fox 1, 34; Met. 1, 17. [O. Sax. ge-winnan: O. H. Ger. ga-winnan: Ger. ge-winnen *to gain, obtain*.]

gewin-stów, e; *f. A place to contend in, battle-place, wrestling-place*; certāminis lŏcus, pălæstra, Ælfc. Gl. 29; Som. 61, 49; Wrt. Voc. 26, 48.

ge-wintred, -wintrad; *part. Grown to full age, full-aged, aged*; adultus:—Óþ-đæt hit gewintred síe *until it be of age*, L. In. 38; Th. i. 126, 7. Middý đū bist gewintrad *cum senueris*, Jn. Skt. Lind. 21, 18. Đeáh he gewintred wǽre *though he was aged*, Ors. 6, 31; Bos. 128, 7. Đæs gewintredan monnes *of a full-aged man*, L. Alf. pol. 26; Th. i. 78, 18. [Cf. M. H. Ger. ge-jāret.] DER. un-gewintred.

gewin-woruld, e; *f. A world of toil*; tribŭlātiōnis plēnus mundus:—Hý scofene wurdon on gewinworuld *they were thrust into a world of toil*, Exon. 45 a; Th. 153, 21; Gú. 829.

ge-wíred; *part. p. Made of wire*:—Hyre ealdan gewíredan preón an vi. mancussum *her old brooch made of [gold or silver] wire, worth six mancuses*, Th. Chart. 537, 34. v. wír.

Gewis, Giwis, es; *m. Gewis, the great grandfather of Cerdic*:—Se Cerdic wæs Elesing, Elesa Esling, Esla Gewising, Gewis Wiging *Cerdic was the son of Elesa, Elesa the son of Esla, Esla the son of Gewis, Gewis the son of Wig*, Chr. 495; Erl. 2, 5: 597; Erl. 20, 7. Giwis, 552; Erl. 16, 19. According to Asser it was from this name that the term Gevissæ, applied by Bede to the West Saxons, was derived. 'Gewis, a quo Britone totam illam gentem Gegwis nominant,' see Grmm. Gesch. D. S. 458. For the use by Bede, see Bd. 3, 7—'Gens Occidentalium Saxonum qui antiquitus Gevissæ vocabantur ... primum Gevissorum gentem ingrediens,' where the translation has 'West Seaxna þeód ... Đá com he ǽrest upp on West Seaxum.' See also 4, 15, 16. Smith's note on the word is 'Gevissæ. Saxonicum est pro Occidentalium. Sic Visigothi, præposita tantum Saxonica expletiva Ge.' See Thorpe's Lappenberg i. 109, note.

ge-wis, -wiss; *adj. Certain, sure, knowing, foreknowing*; certus:—Gewis be heora gerihtnesse *certus de illorum correctione*, Bd. 5, 22; S. 644, 45. Đæt is gesægd đæt he wǽre gewis his sylfes forþfóre *qui præscius sui obitus exstitisse videtur*, 4, 24; S. 599, 14. Wite đæt ǽrest gewiss đæt đæt mód byþ đære sáwle ǽge *know first that as certain, that the mind is the soul's eye*, Shrn. 178, 2. Gewis is *constat*, Hpt. Gl. 419. Đa uþwitan đe sǽdon đæt nǽfre nán wiht gewisses nǽre búton twæónunga *the philosophers that said that there was no certainty without doubt*, Shrn. 174, 25. Swá litel gewis funden *found so little certain*, Bt. 41, 4; Fox 250, 20. Gewis andgit *intelligence*, 5; Fox 252, 20, 30. We syndon gewisse đínes lífes *we are acquainted with thy life*, Guthl. 5; Gdwin. 30, 18. He hí gewisse gedyde and gelǽrde be ingonge đæs écan ríces đe *ingressu regni æterni certos reddidit*, Bd. 4, 16; S. 584, 35. On gewissum tídum *at certain times*, R. Ben. interl. 48. Of gewissum intingan *of certain causes*, R. Ben. interl. 63. Myd gewyssum gesceáde *with certain reason, wherefore*; propter certam rationem, quapropter, Nicod. 3; Thw. 2, 6. [O. H. Ger. giwis: Ger. gewiss *certus*.]

ge-wíscan, etc. v. ge-wýscan, etc.

ge-wisfullíce; *adv. Knowingly, expertly*; scienter, Greg. pref. lib. 2, Dial.

ge-wísian; *p.* ode; *pp.* od *To direct, teach, shew*:—Bǽdon đæt him gewísade waldend se góda hú hie libban sceoldon *prayed the good Ruler to direct them how they were to live*, Cd. 40; Th. 52, 27; Gen. 850.

ge-wislíce, -wisslíce; *adv. Certainly, exactly, truly, especially, besides*; videlicet, scilicet, sane, utique, porro:—Gewisslíce *sane*, Ælfc. Gr. 38; Som. 41, 45. Gyf sóþlíce gewis.íce rihtwísnysse sprecaþ *si vere utique justitiam loquimini*, Ps. Spl. C. 57, 1. Gewislíce án þing is neád-behéfe *porro unum est necessarium*, Lk. Bos. 10, 42. Đú miht blissigan gewisslíce *thou mayest certainly rejoice*, Homl. Th. ii. 132, 1. Se wítegode be Criste swíđe gewislíce swilce he godspellere wǽre *he prophesied about Christ with great exactness, as if he had been an evangelist*, Swt. A. S. Rdr. 69, 414. Gewislíce ic hæbbe *certe habeo*, Coll. Monast. Th. 30, 7. Ic nát náht gewislíce hwæđer đæs feós swá micel is *I do not know for certain whether there is so much money*, Th. Chart. 490, 15. Seó lenctenlíce emniht is gewisslíce on duodecima kl. April *the spring equinox is certainly on the twelfth day before the kalends of April*, Bd. de nat. rerum; Wrt. popl. science 11, 1; Lchdm. iii. 256, 8. Ic cweđe nú gewislícor *I say now more exactly*, 8, 23; Lchdm. iii. 250, 4: Th. Ap. 15, 24. Đæs đe hie gewislícost gewitan meahton *to the best of their knowledge*, Beo. Th. 2704; B. 1350.

ge-wiss, -wisslíce. v. ge-wis, ge-wislíce.

ge-wissend, es; *m. A director, ruler*; præceptor, rector, Hymn. Lye.

ge-wissian; *p.* ode, ade; *pp.* od *To make* or *cause to know, to instruct, inform, direct, command, govern*; docere, edocere, regere, præcipere, dirigere:—Đæt he đone iungan cniht gewissian sceolde *that he should instruct the young boy*, Ælfc. T. Lisle, 34, 3. To đam lande đe ic đé gewissige *unto a land that I will shew thee*, Boutr. Scrd. 21, 42. On đam regole đe us gewissaþ be đære hálgan Eástertíde *in the rule that directs us about the holy Eastertide*, Lchdm. iii. 256, 10. Heó gewissaþ and gescylt and gelǽt *it directs and protects and guides*, Homl. Th. i. 52, 15. Se đe gewylt and gewissaþ Israhéla folc *qui regit populum Israhel*, 78, 16. Swá swá him Gregorius ǽr gewissode as *Gregory had before directed him*, ii. 130, 22: Swt. A. S. Rdr. 64, 241. Se wítega hine gewissode đæt he cúđe gelýfan *the prophet directed so that he was able to believe*, 70, 444. Đú gewissa đa sacerdas *tu præcipe sacerdotibus*, Jos. 3, 8. Ic gean đara vi. punda đe ic 'Eádmunde mínon brēđer gewissod hæbbe *I give the six pounds that I have indicated to my brother Edmund*, Th. Chart. 559, 6. Gif đú nelt beón gewissod *if thou wilt not be directed*, Ælfc. T. Lisle, 40, 12.

ge-wissung, e; *f. Direction, instruction, guidance*:—For fela gewissungum đe seó án bóc hæfþ toforan đám óđrum *for many directions which that one book has above the others*, Swt. A. S. Rdr. 65, 295.

ge-wistfullian; *p.* ode; *pp.* od *To feast*:—Gewistfullian *epulari*, Lk.

Bos. 15, 23. Et drinc and gewistfulla *eat, drink, and feast*, Homl. Th. ii. 104, 21. Gewistfullien *epulentur*, Blickl. Gl. Ðæt ic mid mínum fréondum gewistfullode *ut cum amicis meis epularer*, Lk. Bos. 15, 29.

ge-wistian *to feast :*—Et drinc and gewista *comede bibe epulare*, Lk. Bos. 12, 19.

gewist-lǽcan; *p.* -lǽhte; *pp.* -lǽht *To feast*; epulari :—Ða ongunnon hig gewistlǽccan *cœperunt epulari*, Lk. Bos. 15, 24.

ge-wísung, e; *f. Direction* :—Be Godes sylfes gewísunge *by the direction of God himself*, Jud. pref. Thw. 153, 6.

ge-wit, -witt, es; *n.* I. *wits, senses, [right] mind, mind, intellect* :—Wíndruncen gewit *a mind stupefied with wine*, Cd. 212; Th. 262, 32; Dan. 753. Ðenden mec mín gewit gelǽsteþ *whilst my intellect attends me*, Exon. 38 a; Th. 125, 1; Gú. 347. He eft onhwearf wódan gewittes *he recovered from madness*, Cd. 206; Th. 255, 22; Dan. 628. Seó gedréfednes ðæt mód ne mæg his gewittes beréáfian *trouble cannot rob the mind of its faculties*, Bt. 5, 3; Fox 12, 25. Nú bidde ic ðé ðæt ðú hí on gewitte gebringe *now I beseech thee bring her to her wits*, Homl. Th. i. 458, 11: Exon. 67 b; Th. 251, 12; Jul. 144: 74 b; Th. 278, 13; Jul. 597. Sió wyrd cymþ of ðam gewitte ðæs ælmihtigan Godes *fate comes from the mind of the almighty God*, Bt. 39, 5; Fox 220, 1: Exon. 120 b; Th. 463, 30; Hö. 78: 78 b; Th. 294, 10; Crä. 13: Andr. Kmbl. 631; An. 316: 1344; An. 672. Bútan gewitte *irrational*, Salm. Kmbl. 46; Sal. 23. Se Hǽlend wódum monnum gewitt forgeaf *the Saviour gave reason to the insane*, Homl. Th. i. 480, 14: H. R. 105, 3: Andr. Kmbl. 69; An. 35: Bt. Met. Fox 26, 200; Met. 26, 100. He him gewit forgeaf *he gave him intelligence*, Cd. 14; Th. 16, 29; Gen. 250: Exon. 25 a; Th. 72, 26; Cri. 1178. Ic wát ðæt ðæt lýf á byþ and ðæt gewit *I know that life and mind will always exist*, Shrn. 199, 30, 26. Gehǽlde gewitte *sanato sensu*, Bd. 4, 3; S. 570, 13. II. *knowledge, understanding, consciousness* :—To syllenne his folce hys hǽle gewit *ad dandam scientiam salutis plebi ejus*, Lk. Bos. 1, 77. Lǽran sceal mon geongne monnan . . . sylle him wist and wǽdo oþ ðæt hine mon on gewitte alǽde *a young man must be taught . . . give him food and clothing until he be brought to understanding*, Exon. 89 b; Th. 336, 13; Gn. Ex. 47. Hwá meahte me swelc gewit gifan gif hit God ne onsende *who could give me such understanding if God did not send it*, Cd. 32; Th. 42, 10; Gen. 671. Cyning geweóld his gewitte *the king recovered consciousness*, Beo. Th. 5399; B. 2703. [O. Sax. gi-wit: O. H. Ger. ge-wizzi.]

ge-wita, an; *m. One who is cognisant of anything, a witness, an accessory*; testis, conscius :—Gewita *testis*, Wrt. Voc. 76, 21. Ælmihtig drihten ðe is ealra þinga gewita *the Lord Almighty that is cognisant of all things*, Lchdm. iii. 436, 20. Ðisæs is Oda gewita *of this is Oda witness*, Th. Chart. 510, 5. God sylf his is gewita *God is his own witness*, Homl. Th. ii. 126, 9: i. 84, 4: Ps. Th. 88, 31. Ða leásan gewitan *the false witnesses*, Homl. Th. i. 50, 14, 29: Swt. A. S. Rdr. 72, 497. Geweotan, Th. Chart. 480. 16. We þissa wundra gewitan sindon *we are witnesses of these wonders*, Exon. 43 b; Th. 147, 10; Gú. 724. Gif heó clǽne sý and ðæs fácnes gewita nǽre *if she be innocent and were not an accessory to the crime*, L. Ath. v. § 1, 1; Th. i. 228, 17. Ðæt ðú sý wommes gewita *that thou art an accessory to the crime*, Exon. 80 a; Th. 301, 14; Fä. 19: Frag. Kmbl. 12; Leás. 7. Wildeóra gewita *one who has the same knowledge [wit] as the beasts* [Grein and Bouterwek write gewita = socius], Cd. 206; Th. 255, 14; Dan. 624. [O. Sax. ge-wito: O. H. Ger. ki-wizo *conscius*.]

ge-witan; *p.* -wiste *To understand, know*; scire :—Hí woldon gewitan hwæt ðæt wǽre *dignoscere quid esset*, Bd. 3, 8; S. 532, 7; 4, 18; S. 587, 1; Beo. Th. 2705; B. 1350. Giuta *scire*, Rtl. 5, 18. Gif hwá gewilnigeþ to gewitane hú gedón mann he wæs *if any one wants to know what sort of man he was*, Chr. 1086; Erl. 221, 10. Ðone woeg giwutun *viam scitis*, Jn. Skt. Rush. 14, 4. Gewiste *sciens*, Mt. Kmbl. Lind. 16, 8: Exon. 108 a; Th. 412, 14; Rä. 30, 14. Embihtmen giwistun *ministri sciebant*, Jn. Skt. Rush. 2, 9. Ðæt ne sé gewitten *quod non scietur*, Mt. Kmbl. Lind. 10, 26. Gá and gewite *go and get to know*, Ap. Th. 13, 24.

ge-wítan; ic -wíte, ðú -wítest, -wítst, he -wíteþ, -wít, *pl.* -wítaþ; *p.* ic, he -wát, ðú -wite, *pl.* -witon; *pp.* -witeu. I. [wítan, I. *to see*] *to see, behold*; videre, spectare :—Gewíte and beseoh wíngeard ðisne *vide et visita vineam istam*, Ps. Th. 79, 14. II. *to turn one's eyes in any direction with the intention of taking that direction, to set out towards, start, pass over, to go, depart, withdraw, go away, retreat, retire, die*; transire, discedere. [*a*] *with the infin. of a verb of motion* :—Gewíteþ on weg faran engel *the angel departeth away*, Salm. Kmbl. 1003; Sal. 503. Gewát fleógan mid lácum hire *flew off with her offerings*, Cd. 72; Th. 88, 27; Gen. 1471: 8; Th. 9, 1; Gen. 135: Andr. Kmbl. 2496; An. 1249: Beo. Th. 1710; B. 853. Geweotan, Andr. Kmbl. 1602; An. 802. Gewít ðú nú féran *go now*, Cd. 83; Th. 104, 36; Gen. 1746. Gewát him hám síðian *went off home*, Cd. 98; Th. 130, 17; Gen. 2161: Beo. Th. 3930; B. 1963. [*b*] *with other infinitives* :—Ic gewíte sécan gársecges grund *I go and seek the ocean's bottom*, Exon. 101 a; Th. 381, 24; Rä. 3, 1. Heó on síþ gewát wésten sécan *she on her journey went seeking the desert*, Cd. 103; Th 136, 29; Gen. 2265: 93; Th. 120, 24; Gen. 1999: Beo. Th. 230; B. 115: 3811; B. 1903. Him Noe gewát

eaforan lǽdan *Noah went leading his offspring*, Cd. 67; Th 82, 2; Gen. 1356: 96; Th. 126, 21; Gen. 2008. [*c*] *followed by a clause :*—Gewát ðæt he in temple gestód wuldres aldor *the prince of glory went so as to stop in the temple*, Andr. Kmbl. 1411; An. 707: Exon. 52 a; Th. 181, 31; Gú. 1301. [*d*] *with prep. or adv. or adj :*—Hí forþ gewítaþ for ðæs sumeres hǽton *they shall fade away for the summer's heat*, Blickl. Homl. 59, 4. He forþ gewát *he died*, Cd. 52; Th. 65, 19: Rood Kmbl. 262; Kr. 133: Beo. Th. 2962; B. 1479. Ðá gewát se dæg forþ *dies cœperat declinare*, Lk. Bos. 9, 12. Fyrst forþ gewát *the time went on*, Beo. Th. 425; B. 210: Cd. 47; Th. 59, 36; Gen. 974: Exon. 49 a; Th. 170, 6; Gú. 1107. Se to forþ gewát þurh ðone æþelan *it [the dart] reached and pierced the noble man*, Byrht. Th. 136, 13; B. 150. Gif we gewítaþ fram ðé *if we depart from thee*, Blickl. Homl. 233, 31: 21, 12: Exon. 36 b; Th. 119, 1; Gú. 248. Ne syndon me from gewitene *they have not departed from me*, Cd. 63; Th. 76, 11; Gen. 1255. Me lǽrdon Rómáne ðæt ic gewát heonon onweg *the Romans advised me to depart away hence*, Blickl. Homl. 191, 14. Hwyder gewiton ða welan *whither has the wealth gone?* 99, 24. Ðonne gewitan ða sáula niðer *then down went the souls*, 211, 4: Exon. 97 a; Th. 361, 32; Wal. 28. Gewít of ðam menn *depart from the man*, Homl. Th. i. 458, 5: Blickl. Homl. 139, 13. Ðá he of life gewát *when he departed this life*, Beo. Th. 4934; B. 2471. Ǽr ðam ðæt óðer of gewíteþ *before the other goes away*, Bt. Met. Fox 29, 22; Met. 29, 11. Gewát ofer wægholm *went o'er the ocean*, Beo. Th. 439; B. 217. On fleám gewát *fled*, Cd. 205; Th. 254, 20; Dan. 614. He nǽfre onweg ne gewát *he has never departed*, Blickl. Homl. 117, 1: Ors. 2, 4; Bos. 44, 36. Gewiten under waðeman *retired under ocean*, Exon. 57 a; Th. 144, 11: Ph. 97. In ðæt églond up gewítaþ *they go up into that island*, 96 b; Th. 361, 8; Wal. 16. Ðonon ne gewát *he departed not thence*, Blickl. Homl. 121, 31. Ðæt us ðás tída ídle ne gewítan *that these times do not pass away without profit for us*, 129, 36. Seó deorce niht won gewíteþ *the dark night passes away murky*, Exon. 57 a; Th. 204, 17; Ph. 99. [*e*] *used absolutely :*—Gyf ðes calic ne mǽge gewítan *si non potest hic calix transire*, Mt. Bos. 26, 42. Nacode we wǽron acennede and nacode we gewítaþ *naked we were born and naked we depart*, Homl. Th. i. 64, 28. Heofon and eorþe mæg gewítan mín word nǽfre ne gewítaþ *heaven and earth may pass away; my words shall never pass away*, Blickl. Homl. 245, 5: 91, 21: 57, 30: Elen. Kmbl. 2552; El. 1277. Gif ðú gewítest *if you depart*, 225, 17. Hí ðǽrrihte æfter ðam drence gewiton *they died directly after the drink*, Homl. Th. i. 72, 21: Cd. 62; Th. 75, 7; Gen. 1236. Ðæt leóht gewát *the light vanished*, Elen. Kmbl. 188; El. 94. Gif he gewíte ǽr ðonne hia *if he depart before she does*, Th. Chart. 465, 30. Ðæt wuldor ðysses middangeardes is sceort and gewítende *the glory of this world is short and transitory*, Blickl. Homl. 65, 15. Ðare gewítende ǽhte ðises middaneardes *labentibus hujus seculi possessionibus*, Th. Chart. 317, 6: Bd. 3, 22; S. 552, 20. Dagas sind gewitene *days are passed away*, Exon. 82 b; Th. 310, 26; Seef. 80.

ge-wítendlíc; *adj. Transitory*; transitorius :—Hwæt is ðiós gewítendlíce sibb *what is this transitory peace*, Past. 46, 5; Swt. 351, 24; Hat. MS. 67 a, 17. Mín mód forhogode ealle ðás gewítendlícan þing *my mind despised all these transitory things*, Greg. Dial. Hat. MS. fol. 1 b, 14. Ðis lǽnelíce líf and ðis gewítendlíce *this poor and transitory life*, Blickl. Homl. 73, 9. Yrfenuma to wítendlícum ǽhtum *heir to transitory possessions*, Homl. Th. i. 56, 13.

ge-wítendnes, se; *f. Departure* :—Sǽdon his gewítendnesse *dicebant excessum ejus*, Lk. Bos. 9, 31.

ge-wítennes, se; *f. Departure* :—Ðá ðære tíde neálǽhte his gewítenesse *propinquante hora sui decessus*, Bd. 4, 24; S. 598, 24. On ðone ylcan dæg byþ ðæs bisceopes gewytennys se wæs nemned scs Cassius *on the same day is the bishop's departure who was named St. Cassius*, Shrn. 97, 36.

ge-witfæst; *adj. Of sound mind* :—Nǽnig deófolseóc ðæt he eít wel gewitfæst nǽre *no possessed person that was not in his right mind again*, Guthl. 15; Gdwin. 66, 17.

ge-wiðerworded; *part. p. Opposed*; adversatus, Rtl. 114, 1.

ge-witig. v. ge-wittig.

ge-wítigan, -wítgian; *p.* ode; *pp.* od *To prophesy* :—Wel gewítgade Esaias *bene prophetavit Esaias*, Mt. Kmbl. Lind. 15, 7: 11, 13. Swá hit gewítgod wæs *as it was prophesied*, Blickl. Homl. 93, 29: 83, 28.

gewit-leás; *adj. Witless, foolish, mad*; insanus, amens, stultus :—Gewitleás *amens*, Ælfc. Gr. 47; Som. 48, 35. Wurde ðú ðæs gewitleás ðæt ðú waldende þonc ne wisses *thou wast so witless that thou wast not grateful to the Lord*, Exon. 29 b; Th. 90, 12; Cri. 1473: Bt. Met. Fox 19, 92; Met. 19, 46.

ge-wit-leást, -witt-leást, e; *f. Folly, madness, phrensy*; stultitia :—On ðínre gewitleáste *in thy folly*, Homl. Th. i. 424, 16: Ælfc. T. Lisle 32, 24. Wið ða ádle ðe grécas frenésis nemnaþ ðæt is on úre geþeóde gewitlést ðæs módes *for the disease which the Greeks call φρένησις, that is, in our language, witlessness of the mind*, Herb. 96, 4; Lchdm. i. 210, 1.

gewit-loca, an; *m. A container of intelligence, the mind*; intelligentiæ clausura, animus, mens, pectus, Bt. Met. Fox 12, 52; Met. 12, 26: Exon. 123 a; Th. 473 13; Bo. 14.

ge-witnes, -ness, e ; *f.*   **I.** *knowledge, cognisance, witness, testimony* :—Oððe ðeós gewitness weorðeþ on heágum *si est scientia in excelso*, Ps. Th. 72, 9. Búton Godes willan and búton his gewitnesse *without God's will and without his knowledge*, Bt. 39, 9 ; Fox 212, 33 : Gen. 31, 27, 31. Gif he stalie on gewitnesse ealles his hírédes *if he steal with the cognisance of all his household*, L. In. 7 ; Th. i. 106, 16 : L. C. S. 76 ; Th. i. 418, 12. Wundorlíc is ðín gewitnes *mirabilia testimonia tua*, Ps. Th. 118, 129, 24. He wearþ gemyndig his gewitnesse *memor erit testamenti sui*, 110, 4. Ne yfel gewitnes ne wrégde *nor had evil witness accused them*, Blickl. Homl. 163, 1. Be leásre gewitnesse *of false witness*, L. C. S. 37 ; Th. i. 398, 9 : L. Ath. i. 10 ; Th. i. 204, 22. On hyra gewitnesse *they being witnesses*, Gen. 23, 9. On Moyses bóca gewitnesse *by the testimony of the books of Moses*, Blickl. Homl. 153, 5. Ðæt is to gewitnesse ðæt hit him ne lícode *that is for a testimony that they did not like it*, Past. 21, 6 ; Swt. 165, 13 ; Hat. MS. In gewitnisse hiora *in testimonium eorum*, Mt. Kmbl. Lind. 10, 14. Iohannes cýþ gewitnesse be him *Iohannes testimonium perhibet de ipso*, Jn. Bos. 1, 15. At ðis gewitnesse wæs seo kining Offa *at this witnessing was king Offa*, Chr. 777 ; Erl. 55, 12.   **II.** *used of persons* :—Ic Æthelmær gewitnys *I Æthelmær am witness*, Cod. Dipl. Kmbl. iii. 351, 12–18 : iv. 206, 6–9. Wynflæd gelædde hyre gewitnesse ðæt wæs Sigeríc arcebiscop, etc. *Wynflæd brought her witnesses, they were archbishop Sigeric, etc.*, Th. Chart. 288, 3 : 539, 31. Here ealre ðe hér bé gewitnesse *of all those that here are witnesses*, Chr. 675 ; Erl. 39, 21. Ymb huæd we willnias gewitnesa *quid desideramus testes*, Mk. Skt. Lind. 14, 63. Forðam arison ongeán me leáse gewitnessa *quoniam insurrexerunt in me testes iniqui*, Ps. Th. 26, 14 : Hy. 7, 94 ; Hy. Grn. ii. 289, 94. Beforan gewitnessum *before witnesses*, L. In. 25 ; Th. i. 118, 13. [See Grm. R. A. pp. 608, 779.]

ge-witnian ; *p.* ode ; *pp.* od *To punish, chastise* :—Se ðe mihte hine sóna on helle gewítnian *he that could at once punish him in hell*, Blickl. Homl. 33, 30 : Homl. Th. ii. 124, 22. Ic gewítnige *punio*, Ælfc. Gr. 30 ; Som. 34, 57. Hwí wurdon ða synfullan mid wætere gewítnode ? On Noes dagum gewítnode God manna gálnysse mid wætere . . . *why were the sinful punished with water? In Noah's days God punished men's wantonness with water . . .*, Boutr. Scrd. 22, 30 : Gen. 20, 18. Se man wæs stranglíce gewítnad *the man was severely punished*, Shrn. 73, 13 : Beo. Th. 6138 ; B. 3073.

ge-wítnung, e ; *f.* *Punishment* :—On ðære Sodomitiscra gewítnunge forbearn seó eorþe *in the punishment of the people of Sodom the earth was burnt*, Boutr. Scrd. 22, 33.

ge-witodlíce *truly* ; *certe, sane*, Ps. Spl. T. 57, 1. v. witodlíce.

gewit-scipe, es ; *m.* *A testimony, witnessing* ; *testimonium*, Bd. 1, 27, resp. 6 ; S. 492, 5, 6. [O. Sax. ge-wit-skepi *witness* : O. H. Ger. gi-wizscaf *testimonium*.]

gewit-seóc ; *adj.* *Mind-sick, lunatic, demoniac* ; *energumenus*, Ælfc. Gl. 78 ; Som. 72, 35 ; Wrt. Voc. 45, 67 : 75, 52.

gewit-seócnes, -ness, e ; *f.* *Insanity* ; insanitas, Som.

ge-wittig, -witig ; *adj.* *Wise, knowing, sane, conscious* ; intelligens :—Heó ðærrihte wearþ gewittig *she straightway became sane*, Homl. Th. ii. 24, 12 : 142, 19. Ne forlæt ðé nán ðe gewityg byt *nor does any one forsake thee that is wise*, Shrn. 166, 28. Sum biþ gewittig æt wínþege beórhyrde gód *one is expert at feasting, a good keeper of beer*, Exon. 79 b ; Th. 297, 26 ; Crä. 74 : Beo. Th. 6179 ; B. 3094.

ge-wixlan. v. ge-wrixlian.

ge-wlacian ; *p.* ode ; *pp.* od *To make lukewarm* ; tepefacere :—Ic eom gewlacod *tepesio*, Ælfc. Gr. 37 ; Som. 39, 38.

ge-wlætan ; *pp.* -wlæht ; *pp.* -wlæted, -wlæt *To defile, debase* ; fœdare :—Gif ðú swá gewlætne mon métst *if thou shouldest meet a man so debased*, Bt. 37, 4 ; Fox 192, 12. DER. wlætan.]

ge-wleccan, -wlecian ; *pp.* -wleht, -wleced *To make lukewarm* :—Genim ðysse ylcan wyrte seáw gewlæht [gewleht, MS. H. B.] *take of this same herb the juice made lukewarm*, Herb. 19 ; Lchdm. i. 114, 2 : 80 ; Lchdm. i. 184, 1. Gewleced *made lukewarm*, L. M. 1, 3 ; Lchdm. ii. 40, 21, 29. [Cf. ge-wlacian, wleccan.]

ge-wlencan ; *pp.* ed *To make proud, rich, to exalt* :—Ic Æþelréd eldorman gewelegod and gewlenced mid sume dæle Mercna ríces *I Ethelred alderman enriched and exalted with a part of the Mercians' realm*, Th. Chart. 129, 26. Wírum gewlenced *adorned with wires*, Elen. Kmbl. 2525 ; El. 1264. [O. Sax. gi-wlenkid.]

ge-wlitegian ; *p.* ode ; *pp.* ad, od *To form, adorn, make beautiful* ; formare, decorare, exornare, speciosum *vel* pulchriorem reddere :—Giwlitga *decorare*, Rtl. 105, 28. He gewlitegaþ ealle gesceafta *he adorns all creatures*, Shrn. 198, 12 : Salm. Kmbl. 793 ; Sal. 396. Hand his gewlitegodon *manus ejus formaverunt*, Ps. Spl. 94, 5. Wel gewlitegod *formosus*, Wrt. Voc. 72, 15. Wuldre gewlitegad *with glory beautified*, Exon. 55 b ; Th. 197, 8 ; Az. 187 : 57 b ; Th. 205, 23 ; Ph. 117 : 108 a ; Th. 413, 7 ; Rä. 32, 2 : 128 b ; Th. 493, 22 ; Rä. 81, 35 ; Andr. Kmbl. 1337 ; An. 669.

ge-wló ; *adj.* *Adorned* ; ornatus :—Seó eorþe wæstmum gewló *the earth with fruits adorned*, Cd. 85 ; Th. 107, 14 ; Gen. 1789. v. wló.

ge-wonian. v. ge-wanian.

ge-wópen *wept, lamented*, Ælfc. Gr. 26, 1 ; Som. 28, 28 ; *pp.* of ge-wépan.

ge-worc, es ; *n.* *A work* ; factúra :—On geworce ðínum *in factúra tua*, Ps. Spl. 91, 4. v. ge-weorc.

ge-worpan *to throw, cast*, Bd. 4, 13 ; S. 583, 11. v. ge-weorpan.

ge-worpen *thrown, cast* ; projectus, Ps. Th. 87, 5 ; *pp.* of ge-weorpan.

ge-woruht = ge-worht *wrought* ; *pp.* of ge-wyrcan, Runic pm. 11 ; Kmbl. 341, 18.

ge-wosa, -wesa *a being together, conversation* ; conversatio :—Ærfæst giwosa we gifylga bisene *piæ conversationis sequan:ur exempla*, Rtl. 51, 1 : 32. 32 : 74, 35.

ge-wræstan *to writhe, twist, join* ; intorquere, Cot. 4.

ge-wrædan *to be wroth, savage* :—Beran to him gewrædan gesihþ *if he sees a bear savage at him*, Lchdm. iii. 212, 4.

ge-wrádian ; *p.* ede *To make angry* :—Ðá gewrádede hine Landfranc *then Lanfranc was angry*, Chr. 1070 ; Erl. 208, 5.

ge-wrecan ; *p.* -wræc, *pl.* -wræcon ; *pp.* -wrecen *To wreak, avenge, revenge, punish* ; ulcisci, vindicáre, púníre :—Gebeótode Cirus ðæt he his þegen gewrecan wolde *Cyrus threatened that he would avenge his officer*, Ors. 2, 4 ; Bos. 44, 4 : Cd. 64 ; Th. 77, 13 ; Gen. 1274. Ic heora unriht gewrece egsan gyrde *visítabo in virga iniquítátes eorum*, Ps. Th. 89, 29. Se gewrycþ mynne teónan on ðé *he will avenge on thee my wrong*, Shrn. 96, 16. God gewrecþ on ðæm were *God will take vengeance on the man*, Blickl. Homl. 185, 25. Ná ðú úre gyltas egsan gewræce *avertisti ab ira indignátiónis tuæ*, 84, 3 : 98, 9. Ic ðæt eall gewræc *I have avenged all that*, Beo. Th. 4015 ; B. 2005 : 215 ; B. 107. Ðæt mægwinas míne gewræcon *my kinsmen avenged that*, 4952 ; B. 2479 : Cd. 94 ; Th. 123, 1 ; Gen. 2038. Hine hafaþ his heofonlíca Fæder swíðe gewrecen *his heavenly Father has amply avenged him*, Chr. 979 ; Erl. 129, 14 : Ors. 1, 14 ; Bos. 37, 17. Seó his unsynnige cwalu wæs gewrecen *his undeserved death was avenged*, Shrn. 93, 13.

ge-wrégan ; *p.* -wrégde ; *pp.* -wréged, -wréht [wrégan *to accuse*].   **I.** *to accuse* ; accúsáre :—Ða þwyran hæðengyldan ðone apostol to ðam cyninge gewrégdon *the perverse idolaters accused the apostle to the king*, Homl. Th. i. 470, 6 : Gen. 37, 2. Ðæt hí hine gewrégdon *ut accúsárent illum*, Mk. Bos. 3, 2. Secgaþ wyrdwríteras ðæt Herodes wearþ gewréged to ðam Rómániscan cásere *historians say that Herod was accused to the Roman emperor*, Homl. Th. i. 80, 6. Gytsung is gewréht wið God *covetousness is accused before God*, 256, 22.   **II.** *to stir up, excite, impel* ; concítáre :—Gifen biþ gewréged *the sea is impelled*, Exon. 101 a ; Th. 381, 29 ; Rä. 3, 3.

ge-wreot. v. ge-writ.

ge-wreðian ; *p.* ede ; *pp.* ed *To support* :—Mid his crycce hine gewreðede *supported himself with his crutch* ; baculo innitens, Bd. 4, 31 ; S. 610, 18, note.

ge-wrid, es ; *n.* *A place where shrubs grow, thicket* :— Betwyx ða fenlícan gewrido ðæs wídgillan wéstenes *amongst the fenny thickets of the wide wilderness*, Guthl. 3 ; Gdwin. 2, 10. Betwux ða þiccan gewrido ðara bremela *amongst the dense thickets of brambles*, 5 ; Gdwin. 36, 12. Gewrid *glomulus*, Cot. 95 : *fruticetum*, 90, Lye. [Cf. wríðan.]

ge-wridian ; *p.* ode ; *pp.* od *To flourish* :—Unarímed mengeo on manigfealdum ceápum geweóx and gewridode *the innumerable multitude of all sorts of cattle grew and flourished*, Blickl. Homl. 199, 2.

ge-wrinclod ; *part. p.* *Wrinkled, crooked, winding* :—Ðe gewrincloda díc *the winding dike*, Cod. Dipl. Kmbl. iv. 34, 9.

ge-wring, es ; *n.* [ge-wringan *comprimere*, wringan *to wring*, torquere] *What one can wring or press out, drink, strong drink* ; potus, sicera = σίκερα :—Sicera ðæs cynnes [MS. kynnes] gewring bútan wíne and wætere *what one can press out of every kind, except wine and water*, Ælfc. Gl. 32 ; Som. 61, 120 ; Wrt. Voc. 27, 48.

ge-wringan ; *p.* -wrang, *pl.* -wrungon ; *pp.* -wrungen *To wring* ; comprimere, constringere :—Gewring ða wós of hyre leáfon *wring the juice from its leaves*, Th. An. 116, 22. Munt gewrungen *mons coagulatus*, Ps. Lamb. 67, 16. Gewrungan *wrung*, Herb. 72, 2 ; Lchdm. i. 174, 11.

ge-writ, es ; *n.* *Something written, writing, scripture, inscription, a writing, letter, treatise, writ, charter, book* :—Oþ ðone first ðe hie wel cunnen Englisc gewrit arædan *until such time as they can read English writing well*, Past. pref. Swt. 7, 13, 17. Ne rædde gé ðis gewrit *nec scripturam hanc legistis*, Mk. Bos. 12, 10. Ðæt gewrit swá be him cwæþ *the Scripture thus spake about him*, Blickl. Homl. 167, 15 : 123, 6. Mid ðon worde ðæs godcundan gewrites *with the word of divine Scripture*, 33, 20. Ðæs hálgan gewrites *of holy writ*, Homl. Th. i. 82, 13. Ðis gewrit *inscribtio*, Mk. Bos. 12, 16. Ðá héht he rædan ðæt gewrit *then he ordered to read the letter*, Blickl. Homl. 177, 4, 35. Awrítaþ eówre naman on gewrite ðonne asænde ic ða gewrita mínre dóhtor . . . se cyngc nam ða gewrita and geinsegolde hí *write your names in a letter, then I will send the letters to my daughter . . . The king took the letters and sealed them*, Th. Ap. 20, 6–10 : Chr. 627 ; Erl. 25, 11. Se pápa seonde his gewrite to Engla lande *the pope sent his bull to England*, 675 ; Erl. 37, 15. Mid ðæs cynges gewrite *with the king's writ*, 1048 ; Erl. 177, 19. Án oxe ne án cú ne án swín ðæt næs gesæt on his gewrite and ealle

ða gewrita wǽron gebroht to him syððan *there was not an ox nor a cow nor a swine that was not put in his book* [*Doomsday Book*], *and all the writings were brought to him afterwards*, 1085; Erl. 218, 37: Homl. Th. i. 30, 2. Ðis gewrit *this treatise*, Swt. A. S. Rdr. 56, 1. Ðeáh ðe gewrita oft nemnan ealle ða land Media *though books often call all those lands Media*, Ors. 1, 1; Bos. 16, 30. Ðæs gewrito secgaþ *as books say*, Exon. 60 a; Th. 220, 1; Ph. 313: Chr. 889; Erl. 124, 22; Edg. 14: 109 b; Th. 420, 9; Rä. 40, 1. Swá wîtgan us on gewritum cýðaþ *as sages tell us in books*, 56 a; Th. 199, 24; Ph. 30: Elen. Kmbl. 1651; El. 827. We rǽdaþ on hálgum gewritum *we read in holy writings*, Homl. Th. ii. 356, 19. On gewritum *in scripturis*, Ps. Th. 86, 5. Us gewritu secgaþ *the Scriptures tell us*, Cd. 55; Th. 68, 23; Gen. 1121: 79; Th. 98, 15; Gen. 1630: 119; Th. 154, 30; Gen. 2563: Elen. Kmbl. 1345; El. 674. Ða hálgan gewreotu *the holy Scriptures*, Blickl. Homl. 15, 8: 17, 21. On gewritu settan *to record in books*, Elen. Kmbl. 1305, 1313; El. 654, 658. Tuegen hleáperas Ælfréd cyning sende mid gewritum *king Alfred sent two couriers with letters*, Chr. 889; Erl. 86, 24. Ûre bisceopas to me gewreoto sende *our bishops sent me letters*, Blickl. Homl. 187, 4. Ic hæfde ǽr on ôðre wîsan awriten ymbe mîn yrfe and hæfde monegum mannum ða gewritu ôðfæst *I had previously written in another way about my inheritance and had en'rusted the writings to many men*, Chart. Th. 490, 29: 541, 22. DER. ǽrend-, erfe-, firn-, hand-, mæg-, ofer-, riht-, yrfe-gewrit.

ge-wrítan; *p.* -wrát; *pp.* -writen *To write, to give or bestow by writing, to write along with others;* conscribere:—He lett gewrîtan hû mycel landes his arceb's hæfdon *he had written how much land his archbishops had,* Chr. 1085; Erl. 218, 29: Th. Chart. 296, 10. Werfriþ bisceop and seó heórédden æt Weogerna ceastre syllaþ and gewritaþ æþelræde and æþelflæde heora hláfordum *bishop Werfrith and the society at Worcester give and convey by writing to their lords Ethelred and Ethelfled*, Cod. Dipl. Kmbl. ii. 150, 4. Æþréd aldorman and æþelflæd mercna hláfordes mid us hit gewriotan *Ethelred alderman and Ethelfled, lords of the Mercians, joined with us in writing this*, 151, 2: Chr. 656; Erl. 32, 20. Seo kyning gewrát *the king signed*, 23. Ðes writ wæs gewriton *this writing was written*, 33, 9. Hwæt is gewriten *quid scribtum est*, Lk. Bos. 10, 26: Ps. Spl. 39, 11. Gewriten yrfe *legatum*, Ælfc. Gl. 13; 57, 96; Wrt. Voc. 20, 37. Gewriten yrfe-weard *legatarius*, Lye.

ge-writere, es; *m. A writer:*—Gewrîteres *scribæ*, Ps. Spl. T. 44, 2. v. wrîtere.

ge-wriðan; *part.* -wríðende; *p.* -wráð, *pl.* -wridon; *pp.* -wriðen *To bind, restrain, tie, tie together;* coartare, alligare:—Lim gewríðan *to bind the limb*, Homl. Th. ii. 136, 2. Ða myhta to gewríðenne *potestatem ligandi*, Th. Chart. 334, 7. Engel gewríðende oððe geswencende hig oððe genyrwiende *angelus coartans eos*, Ps. Lamb. 34, 5. Se heora unrótnesse gewríðeþ *qui alligat contritiones eorum*, Ps. Th. 146, 3. Gewríð *alligat*, Ps. Spl. 146, 3. Seó godcundnys gewráð ðone ealdan deófol *the divinity bound the old devil*, Homl. Th. i. 216, 28: ii. 416, 3. Iudas hine sylfne ahêng mid grine and rihtlíce gewráð ða forwyrhtan þrotan *Judas hung himself with a noose and rightly bound that wicked throat*, 250, 15. He his wunda gewráð *he bound up his wounds*, 356, 28. Ðonne gewríð ðú hý *then bind it*, Th. An. 116, 13. Ânra gehwilc manna is gewriðen mid rápum his synna *every man is bound with the ropes of his sins*, Homl. Th. i. 208, 3: 456, 9: 462, 1.

ge-wrixl, -wrixle, es; *n. A change, interchange, vicissitude, turn, course:*—Hwylc gewrixl sylþ se mann for hys sáwle *quam dabit homo commutationem pro anima sua?* Mt. Bos. 16, 26: Mk. Bos. 8, 37. Cêpena þinga gewrixle *commercium*, Ælfc. Gl. 16; Som. 58, 53; Wrt. Voc. 21, 41. Ne wæs ðæt gewrixle til ðæt hie on bâ healfa bicgan scoldon freónda feorum *nor was the exchange good, that they on both sides must buy with the lives of friends*, Beo. Th. 2613; B. 1304. Nú hæfþ God swíðe gesceádwíslíce geset ðæt gewrixle eallum his gesceaftum *God hath very wisely appointed change to all his creatures*, Bt. 21; Fox 74, 21: Bt. Met. Fox 11, 111; Met. 11, 56: Shrn. 168, 11. On hys gewrixles endebyrdnesse *in ordine vicis suæ*, Lk. Bos. 1, 8. Benedictus hæfde Paulus gewrixle *Benedictus tenuit Pauli vices*, Gr. Dial. 2, 17, Lye: Blickl. Homl. 91, 24. [Cf. wæpen-gewrixle.]

ge-wrixl; *adj. Changing, vicarious;* vicarius, alternans, aptus, Hpt. Gl. 460, 476, 506.

ge-wrixlian, -wixlian; *p.* ede; *pp.* ed. **I.** *to change:*—Gewixla *mutare*, Mt. Kmbl. p. 2, 17. **II.** *to get by exchange, obtain:*—Hie. hæfdon gewrixled wîta unrîm *they had got punishments innumerable*, Cd. 18; Th. 22, 3; Gen. 335. **III.** *to give in exchange, grant:*—Swá sceal gewrixled ðám ðe ǽr wel heóldon meotudes willan *so shall be granted to those that before well kept the Creator's will*, Exon. 26 a; Th. 77, 23; Cri. 1261.

ge-wuldorbeágian; *p.* ode; *pp.* od *To crown:*—Se gewuldorbeágaþ ðê *qui coronat te*, Ps. Spl. 102, 4. Ðú gewuldorbeágodest hine *tu coronasti eum*, 8, 6. Stephanus is on Leden coronatus ðæt we cweðaþ on Englisc gewuldorbeágod *Stephen is in Latin 'coronatus,' which we express in English by crowned*, Homl. Th. i. 50, 12; 52, 20.

ge-wuldrian; *p.* ode; *pp.* od *To glorify:*—Ic hine gewuldrige *glori-*

ficabo eum, Ps. Th. 90, 16. Gewuldradon *glorificaverunt*, Mt. Kmbl. Lind. 9, 8. Hie gesáwon ðæt heó wæs gewuldrod *they saw that she was glorified*, Blickl. Homl. 139, 25. Ðú eart gewuldrad *mirificatus es*, Ps. Th. 138, 12. Ðú gewuldroda cyning *thou glorified king*, Blickl. Homl. 147, 35.

ge-wun; *adj. Accustomed, usual:*—Gewune drenceas *usual drinks*, Herb. 68; Lchdm. i. 172, 6. Gewune *assuetæ*, Mone Gl. 435. [*O. H. Ger.* gi-won *solitus, suetus, adsuetus*, Grff. i. 869.] v. ge-wun; *adj.*

ge-wuna, an; *m. A custom, wont, manner, use, rite;* consuetudo:—Næs ðín gewuna ðæt ðú bútan ðínum diácone geoffrodest *it was not thy wont to offer without thy deacon*, Homl. Th. i. 418, 1. Wæs his gewuna ðæt he sægde *referre erat solitus*, Bd. 4, 19; S. 588, 42. Ðǽr wæs gewuna ðæm folce ðæt . . . *the people there were accustomed to . . .*, Blickl. Homl. 209, 6. Swá hit gewuna is *ut adsolet*, Ors. 3, 3; Bos. 55, 20. [*Or do the two last belong to* ge-wuna, *adj.?* (cf. ge-wunelíc.)] Is nú geworden to full yfelum gewunan ðæt menn swíðor scamaþ nú for gôddǽdum ðonne for misdǽdum *it has now become the very bad custom for men to be more ashamed of good deeds than of bad ones*, Swt. A. S. Rdr. 109, 161. Mid ðon gewunon ðære heofogoston gewemmednesse *by the practices of the most grievous impurity*, Blickl. Homl. 75, 6. Heó gemonþ ðæs wildan gewunan hire eldrana *she remembers the wild manner of her parents*, Bt. 25; Fox 88, 12: Bt. Met. Fox 13, 53; Met. 13, 27. Gewuna *ritus*, Ælfc. Gr. 38; Som. 41, 44. Æfter gewunan *after the custom*, Lk. 1, 9: 2, 27, 42: Blickl. Homl. 207, 18: Chr. 1070; Erl. 208, 2. Æfter ûron gewunon *nostro more*, Coll. Monast. Th. 33, 13. Of gewunan *from custom*, R. Ben. interl. 7. Ofer míne gewunan *contrary to my custom*. Ælf. T. Lisle 43, 7. [*O. Sax.* gi-wono: *O. H. Ger.* gi-wona *consuetudo*.]

ge-wuna; *indecl. adj. Accustomed:*—Dydon eall swá hí ǽr gewuna wǽron *they did just as they were wont to before*, Chr. 1006; Erl. 140, 6. Suæ ðætte he gewuna wæs *sicut consueverat*, Mk. Skt. Lind. 10, 1. Gewuna wæs se groefa *consueverat præses*, Mt. Kmbl. Lind. 27, 15: Cd. 166; Th. 207, 27; Gen. 473. [*O. Sax.* gi-wono.] v. ge-wun, *wuna, subst.*

ge-wunden *wound*, Exon. 111 b; Th. 427, 30; Rä. 41, 99; *pp. of* ge-windan.

ge-wundian; *p.* ode; *pp.* ed, od *To wound:*—And eft he hym sende ôðerne þeów and hí ðone on heáfde gewundodon, Mk. 12, 4. Hí hine mid spere gewundedon *they wounded him with a spear*, Homl. Th. i. 216, 23. Se swíðe gewundod wæs *he was sore wounded*, Chr. 755; Erl. 50, 8. v. wundian.

ge-wundorlǽcan *to make wonderful;* mirificare, Ps. Spl. 16, 8.

ge-wunelíc, -wunolíc; *adj. Accustomed, wonted, usual, ordinary;* consuetus:—Ðam folce wæs gewunelíc ðæt . . . *it was usual with the people to . . .*, Jud. 7, 8. Ðæm eádberhte wæs gewunelíc ðæt he wunode on dýgolre stôwe *that Eadberht was in the habit of dwelling in a secret place*, Shrn. 82, 9: 88, 1. Eall ðæt wæs gewunelíc on ðisan lande *all that was usual in this land*, Chr. 1100; Erl. 236, 13: Blickl. Homl. 85, 29. Gewunelícre mildheortnyssa *solita clementia*, Hymn. Surt. 11, 25. On úre wísan us to spræcþ swá ðæt we þurh ða gewunelícan sprǽce ða þing oncnáwan ðe us uncúþe wǽron *speaks to us in our manner so that through the speech to which we are accustomed we may understand those things that were unknown to us*, Boutr. Scrd. 21, 2. [*O. H. Ger.* ge-wonelich *consuetus*: Ger. ge-wöhnlich.]

ge-wunelíce; *adv. According to custom, ordinarily, commonly;* rite:—Swíðe gewunelíce *very commonly*, Ælf. T. Lisle 17. Gewunelíce *rite*, Ælfc. Gr. 38; Som. 41, 44. Ðæt mynster ðe gewunelíce is Magigeo nemned *monasterium quod Muigeo consuete vocatur*, Bd. 4, 4; S. 571, 18. Heó oft gewunolíce cwǽde *solita sit dicere*, 4, 19; S. 589, 24.

ge-wunian; *p.* ode; *pp.* od. **I.** *to dwell, inhabit:*—Ne mágon ðǽr gewunian wídfêrende ne ðǽr elþeódige eardes brúcaþ *there may not dwell wide wandering men, nor there do strangers enjoy a home*, Andr. Kmbl. 557; An. 279: Cd. 220; Th. 284, 24; Sat. 326. Nǽfre gewurþe ðæt ðǽr on gewunige áwiht lifigendes *non sit qui inhabitet*, Ps. Th. 108, 7. Ðú in ðære stôwe stille gewunadest *in that place didst thou dwell quietly*, Exon. 121 a; Th. 465, 7; Hö. 100. Ic mînum gewunade frumstaþole fæst *I dwelt fast in my original station*, 122 b; Th. 471, 17; Rä. 61, 2. Siððan gást wîc gewunode in ðæs weres breóstum *since the spirit inhabited a dwelling in the man's breast*, Elen. Kmbl. 2073; El. 1038. Him on ðæt wêsten gewunode *dwelt in the wilderness*, Blickl. Homl. 199, 8. Him aspidas under welerum is gewunad fæste *venenum aspidum sub labiis eorum*, Ps. Th. 139, 3: Cd. 215; Th. 271, 9; Sat. 103. **II.** *to remain, stay, abide, continue:*—He leng on ðam lande gewunian ne mihte *he could not stop any longer in the country*, Blickl. Homl. 113, 11: Ap. Th. 7, 4. Hý ealdrihta ælces môsten wyrðe gewunigan *they should remain in the enjoyment of every ancient right*, Bt. Met. Fox 1, 73; Met. 1, 37: 2, 38; Met. 2, 19. Þurh ðínra dǽda spéd dagas hêr gewuniaþ *ordinatione tua perseverat dies*, Ps. Th. 118, 91. He on ðæs láreówes wære gewunade *he continued in the teacher's protection*, Exon. 37 b; Th. 123, 31; Gú. 331. **III.** *c. acc. To stop, live, associate with, continue in* or *with:*—Hie se leódfruma leng ne wolde gewunian *with them the prince no longer would abide*, Andr. Kmbl. 3320; An. 1636. Ne gewuna wyrsan [MS. wyrsa] *do not associate with an*

inferior, Exon. 80 a; Th. 301, 22; Fä. 23. Ðæt hine on ylde eft gewunigen wilgesîðas *that with him in his age remain in his loved comrades*, Beo. Th. 44; B. 22. Ðæt hý ðis lǽne lif long gewunien *that they continue long in this poor life*, Exon. 62 b; Th. 230, 33; Ph. 481. **IV.** *to be accustomed, wont :*—Se ârwyrþa bisceop gewunade oft secgan *reverentissimus antistes solet referre*, Bd. 3, 13; S. 538, 7: 4, 23; S. 594, 38: 24; S. 596, 31: 5, 2; S. 614, 26. Ða sǽde Sompeius ðæt Ioseph gewunode monige wundor to wyrcenne *Sompeius said that Joseph used to work many miracles*, Ors. 1, 5; Bos. 28, 12. Him gewunode ðæt he wæs geond ðæt wésten sundorgenga *was accustomed to go through the desert by itself*, Blickl. Homl. 199, 5. Swâ swâ he gewunode *sicut consueverat*, Mk. Bos. 10, 1. Ðes hálga wer wæs gewunod ðæt he wolde gân on niht to sǽ *this holy man was accustomed to go at night to the sea*, Homl. Th. ii. 138, 3. His môd to ðâm woruldsǽlþum gewunod wæs *his mind was accustomed to worldly prosperity*, Bt. 1; Fox 4, 2. [*O. Sax.* gi-wonon: *O. H. Ger.* gi-wonan *manere, solere, consuescere : Ger.* ge-wohnen *to be accustomed.*]

ge-wunsum; *adj. Pleasant :*—Swîðe gewunsum hit biþ ðæt mon wíf hæbbe and bearn *it is very pleasant to have wife and children*, Bt. 31, 1; Fox 112, 8. [Cf. wynsum.]

ge-wurde *wast, hast been*, Andr. Kmbl. 1100; An. 550; *2nd sing. p. of* ge-weorþan: ge-wurde *happened*, Andr. Kmbl. 1115; An. 558; *p. subj. of* ge-weorþan: ge-wurdon *happened*, Ors. 5, 10; Bos. 108, 16; *p. pl. of* ge-weorþan.

ge-wurms; *adj. Full of matter, suppurated*; purulentus, Cot. 185, Lye. v. wyrmsig, wyrms.

ge-wurþan; he -wurþ; *subj. pres.* -wurþe, *pl.* -wurþon. **I.** *to be, become;* fiëri :—Ne mæg nân þinc gewurþan bûtan godes willan *nothing can happen without God's will*, Th. Ap. 22, 7: 9, 5. Hit gewurþ him of mínum fæder, ðe on heofonum ys *fiet illis a patre meo, qui in cælis est*, Mt. Bos. 18, 19. Ic ðé hâte ðæt ðû hi gehele and gehealde ôþ-ðæt ic wite hwæt God wylle, hwæt be me gewurþe *quam te silentio tĕgĕre vŏlo, donec sciam quid de me fiĕri velit Deus*, Bd. 5, 19; S. 640, 38. Ðæt ðâs stânas to hláfe gewurðon *ut lăpĭdes isti pânes fiant*, Mt. Bos. 4, 3: 5, 18. **II.** *v. impers. cum acc. To happen, come to pass, come together, agree;* evĕníre, convĕníre :—Ne meahte hie gewurþan *they might not agree*, Cd. 81; Th. 101, 32; Gen. 1691. v. ge-weorþan.

ge-wurþian; *p.* ode, ade; *pp.* od, ad *To distinguish, honour, adorn, celebrate, praise;* insignîre, honôrâre, ornâre, celebrâre :—Ðæt gê gewurþien wuldres Aldor *that ye honour the chief of glory*, Cd. 156; Th. 195, 1; Exod. 270. On Dryhtnes naman se dæg is gewurþod *the day is celebrated in the Lord's name*, Hy. 9, 30; Hy. Grn. ii. 292, 30: 7, 59; Hy. Grn. ii. 288, 59. Hæfde he gewurþodne werodes aldor *he had honoured the prince of the multitude*, Cd. 143; Th. 179, 19; Exod. 31. Wæs ðis eálond gewurþad mid ðâm æðelestum ceastrum *insula ĕrat civĭtâtĭbus nobilissîmis insignîta*, Bd. 1, 1; S. 473, 26. v. ge-weorþian.

ge-wyder, es; *pl. nom. acc.* -wyderu, -wydera; *n. Weather, the temperature of the air;* tempestas, cæli tempĕries :—Bringþ sumor wearme gewyderu *summer brings warm weather*, Menol. Fox 177; Men. 90. Godes miht gefadaþ ealle gewydera *God's power ordereth all weathers*, Bd. de nat. rerum; Lchdm. iii. 278, 13, MS. R. Of untýdlícan gewyderum *from unseasonable weather*, Ors. 3, 3; Bos. 55, 20. v. ge-weder.

ge-wyht, es; *n. A weight;* pondus :—Gewyht *vel* pund *pondus*, Ælfc. Gl. 59; Som. 67, 113; Wrt. Voc. 38, 36. v. ge-wiht.

ge-wyld, -wild, es; *n. Power, dominion :*—Æfter ðam ðe Alexander hæfde ealle Inde him to gewyldon gedôn *perdomita Alexander India*, Ors. 3, 9; Bos. 67, 21. [Cf. ge-weald *in pl.*]

ge-wyldan, -wildan; he -wyld, -wild, -wylt; *p.* -wylde; *pp.* -wyld; *v. a. To exercise power over, to tame, subdue, conquer, temper, seize, take;* dominari, domare, subigere, prehendere, capere :—Hî gewildon heora *dominati sunt eorum*, Ps. Spl. 105, 38. He gewild ðé *ipse dominabitur tibi*, Gen. 3, 16. Dauid gewylde ðone wildan beran, and his ceaflas totær *David subdued the wild bear, and tore apart his jaws*, Ælfc. T. Lisle 13, 26: 14, 1. Hine nân man gewyldan ne mihte *nemo poterat eum domare*, Mk. Bos. 5, 4: Homl. Th. ii. 192, 25. Gewylt ealle þeóda *will subdue all the nations*, Deut. 31, 3. Heora flæsclícan gewilnunga gewyldaþ *they subdue their fleshly desires*, Homl. Th. i. 552, 24. Gewyld mid ðam ele ðe sý of lawer treówe gewrungan *temper with the oil which is wrung out of laurel*, Herb. 72, 2; Lchdm. i. 174, 11. Gewildaþ ða eorþan *subjicite terram*, Gen. 1, 28. Gewylde man hine *prehendat aliquis eum*, L. C. S. 25; Th. i. 390, 20: L. E. G. 4; Th. i. 168, 22. Seó burh wearþ gewyld *the city was taken*, Ælfc. T. Lisle 42, 20: Jud. 16, 7. Ðonne he hine hæfþ gewyldne *dum dominabitur pauperi*, Ps. Th. 9, 30. He hæfþ nû gewyld to mínum anwealde Scottas and Cumbras and eác swylce Bryttas *subditis nobis sceptris Scottorum, Cumbrorumque, ac Brittonum*, Th. Chart. 240, 3. Alexander hine [Poros] gewildne gedyde *Porus captus est*, Ors. 3, 9; Bos. 67, 35; Guthl. 12; Gdwin. 56, 23. Mid ele wel gewylde *well tempered with oil*, Herb. 12, 3; Lchdm. i. 104, 6. Ic me gedô allophilas ealle gewylde *mihi allophyli subditi sunt*, Ps. Th. 59, 7.

ge-wylde; *adj. Subject, under one's power or control, in one's possession :*—Him wæs gelîce gewylde his wynstre and his swîðre *utraque manu pro dextra utebatur*, Jud. 3, 15. Nis us nân lim swâ gewylde to ælcum weorce swâ us sind ûre fingras *we have no limb so at our disposal for every work as are the fingers*, Homl. Th. ii. 204, 7. Seó gewylde gleáwnes *consummata prudentia*, Nar. 2, 1. He hit eft gedyde unc swâ gewylde swâ hit ðâ wæs ðâ we hit him ôðfæstan *he should put it again as much under our control as it was when we entrusted it to him*, Th. Chart. 484, 30: Cod. Dipl. Kmbl. v. 120, 19. He ne funde nân mâre landes ðe ðiderynn gewylde wǽre ðonne twâ hîda landes *he found no more land belonging thereto than two hides*, Th. Chart. 429, 3. Swâ he swîðor syngaþ swâ he deófle gewyldra biþ *the more he sins the more he will be in the devil's power*, Homl. Th. i. 268, 24. v. un-gewylde.

ge-wyldor, es; *m. A ruler, governor;* rector, gubernator, Som.

ge-wyle, es; *n. A will;* vŏluntas, L. C. S. 76; Th. i. 418, 11, MS. A. v. ge-wil.

ge-wyllan; *pp.* ed *To boil :*—Gewyll *boil*, Herb. 12, 1; Lchdm. i. 102, 21. Wel gewyllede *well boiled*, 12, 3; Lchdm. i. 104, 6, MS. O. v. a-wyllan.

ge-wylwed *wallowed, rolled;* volutatus, Dial. 4, 2.

ge-wynsumian *to exult;* exultare, Rtl. 1, 17: 13, 37.

ge-wynsumlíc; *adj. Pleasant;* acceptus, desiderativus, Hpt. Gl. 412, 446.

ge-wyrcan, -wyrcean; *p.* -worhte, ðû -worhtest; *pp.* -worht. **I.** *to work, make, build, form, dispose, do, perform, celebrate, commit :*—Úre Drihten wolde mannan gewyrcan *our Lord would make man*, Hexam. 10; Norm. 16, 16: 11; Norm. 18, 14. Gewyrcean mycelne tor *to build a great tower*, Blickl. Homl. 187, 12: Beo. Th. 139; B. 69. Ðû miht wundor gewyrcean *tu facis mirabilia*, Ps. Th. 76, 11. Gif ic godes meahte willan gewyrcean *if I could do God's will*, Cd. 39; Th. 51, 31; Gen. 835. Ne meahte ic æt hilde mid Hruntinge wiht gewyrcean *I could not perform aught with Hrunting in fight*, Beo. Th. 3324; B. 1660. Ða noldon fleám gewyrcan *they would not fly*, Byrht. Th. 134, 9; By. 81. Hî woldon hyra Eástron gewyrcan *they would celebrate Easter*, Lk. Bos. 22, 7. God wille disse worlde ende gewyricean *God will put an end to this world*, Blickl. Homl. 109, 33. He nest gewyrceþ *it makes a nest*, Exon. 62 b; Th. 230, 9; Ph. 469. Hie gewyrcaþ ænne lîchoman *they form one body*, Bt. 34, 6; Fox 142, 16. Crist him to cwæþ 'Ic ðé geworhte' *Christ said to him 'I made thee,'* Blickl. Homl. 231, 28. Ðû eall geworhtest þing þearle gôd *thou didst make every thing exceeding good*, Bt. Met. Fox 20, 88; Met. 20, 44. For úres lifes dædum ðe we geworhtan *for our life's deeds that we have done*, Blickl. Homl. 63, 32. Geworhton me him to wæfersýne *made me a spectacle for themselves*, Rood Kmbl. 61; Kr. 31. Mycel yfel geworhtan *did much harm*, Chr. 993; Erl. 133, 3. Þeáh we æbylgþ wið hine oft gewyrcen *though we oft offend against him*, Elen. Kmbl. 1024; El. 513. Sió wund ðe him se eorþdraca ær geworhte *the wound that the dragon had before given him*, Beo. Th. 5418; B. 2712. Hie geweorc geworht hæfdon *they had made a fort*, Chr. 894; Erl. 90, 2. He hæfþ mon geworhtne *he hath made man*, Cd. 21; Th. 25, 18; Gen. 395. Synna ðe we wið Godes willan geworht habbaþ *the sins that we have done against God's will*, Blickl. Homl. 25, 15: 125, 4. Heora ciningas hæfdon sige geworht on heora feóndum *their kings had got victory over their foes*, 67, 9. Of glæse geworht *made of glass*, 127, 33. He nys swâ wel wið me geworht swâ he wæs *he is not so well disposed to me as he was*, Gen. 31, 5. **II.** *to get by working, gain, obtain, merit :*—Ic me mid Hruntinge dôm gewyrce *I with Hrunting will gain myself glory*, Beo. Th. 2986; B. 1491. Lof se gewyrceþ hafaþ heáhfæstne dôm *he gains praise, hath undying glory*, Exon. 97 a; Th. 327, 6; Vîd. 142. Se ðe gewyrceþ ðæt him wuldorcyning milde geweorþeþ *he who obtains that the king of glory becomes mild to him*, 63 b; Th. 234, 8; Ph. 536. Hû geworhte ic ðæt *how did I merit this?* Cd. 127; Th. 162, 3; Gen. 2675. **III.** *with gen.* [cf. wyrcan *with gen.*]:—For hwam nele mon him georne gewyrcan dryhtscipes *why will not man earnestly gain himself worship*, Salm. Kmbl. 774; Sal. 386.

ge-wyrd, e; *f. Event, fate, destiny, condition :*—Ðeós æþele gewyrd *this noble event* [*the crucifixion*], Elen. Kmbl. 1291; El. 647. Sume cwædon ðæt se steorra his gewyrd wǽre. Gewîte ðis gedwyld fram geleáffullum heortum ðæt ǽnig gewyrd sý bûton se ælmihtiga scyppend *some said that the star was his destiny. Let this error depart from believing hearts, that there is any destiny except the Almighty Creator*, Homl. Th. i. 110, 11. Fore giwyrd líchoman *pro conditione carnis*, Rtl. 66, 37. Gewyrd *vel* gecwide *conditio*, Ælfc. Gl. 13; Som. 57, 117; Wrt. Voc. 20, 54. Hit is of ðæra bisceopa gehlote and of heora âgenre gewyrde ðæt ðæt hý secgaþ *in potestate esse antistitis quid velit fingere*, Ors. 3, 9; Bos. 65, 34. Gewyrd *fatum, parca, fortuna*, Hpt. Gl. 529, 467. Binnan ðam wendun gewyrda and gewât Eádrǽd cyng *meanwhile matters changed and king Eadred died*, Th. Chart. 207, 22. [Cf. wyrd, ge-weorþan; *and see* ge-wyrde.]

ge-wyrdan, -werdan; *p.* de; *pp.* ed; *v. trans. To hurt, injure;* lædere, nocere :—Gif hwâ on ceáse wíf gewerde [gewyrde, MS. G.] *if any one in strife hurt a woman*, L. Alf. 18; Th. i. 48, 17. Gif hwâ gewerde [gewyrde, MS. G.] ôðres monnes wîngeard *if any one injure another*

man's vineyard, 26 : Th. i. 50, 24. Ne mæg ðær rēn ne snāw gewyrdan *neither rain nor snow can there injure,* Exon. 56 a; Th. 199, 1; Ph. 19. Hæfde hī hungor and þurst heard gewyrded *esurientes et sitientes,* Ps. Th. 106, 4.

ge-wyrde, -wyrd [?], es; *n. Speech, conversation, collection of words, sentence, rule* [?] :—Ðæt ic mæge sum rust on weg adrīfan of mīnre tungan ðæt ic mæge becuman to bræddran gewyrde *that I may clear some rust away from my tongue, so that I may attain to more copious speech,* Shrn. 35, 22. Wīsra gewyrdum *by the rules of wise men,* Menol. Fox 132; Men. 66. Gewyrd *verbositas,* Hpt. Gl. 439. [*Goth.* gawaurdi*: O. H. Ger.* ga-wurti *comma, brevis dictio,* Grff. i. 1023.] Cf. andwyrde; *and see* ge-wyrd.

ge-wyrdelíc; *adj. Historical, fortuitous* :—On gewyrdelícum racum *in historical narratives,* Homl. Th. i. 58, 9. Gewyrdelíc *historialis,* Hpt. Gl. 506; *fortuitus,* 410, 495. [Cf. *Ger.* geschichtlich *and* geschehen.]

ge-wyrdelíce; *adv. Accurately,* Swt. A. S. Rdr. 69, 414.

ge-wyrdlian; *p.* ede; *pp.* ed *To hurt, injure;* lædere, nocere, Bd. 3, 16; S. 543, 11, col. 2. v. wyrdan.

ge-wyrht, es; *n. Work, deed, merit, desert* :—Deág ðīn gewyrhtu *if thy deeds are good,* Exon. 80 a; Th. 300, 11 : Fä. 4. Ða heálícan gewyrhto Sancte Iohannes *the exalted deeds of St. John,* Blickl. Homl. 167, 5. Rǣcaþ ǽghwilcum men ágen gewyrhta *give to every man his deserts,* Hy. Grn. 7, 16. Be heora gewyrhtum *secundum opera eorum* Ps. Th. 27, 5 : 102, 10. Be gewirhton we þoliaþ ðās þing *deservedly do we suffer these things,* Gen. 42, 21. Ǽlcum men wrecan be his gewyrhtum *to punish every man according to his deeds,* Bt. 35, 6; Fox 168, 26. Būton gewyrhtum *undeservedly,* 22, 1; Fox 76, 15 : 38, 3; Fox 202, 3. Wæs him forgolden æfter his ágenum gewyrhtum *he was requited according to his own deeds,* Blickl. Homl. 45, 2. For heora gewyrhtum *for their deeds,* 125, 2 : Swt. A. S. Rdr. 108, 112. Mid gewyrhtum *deservedly,* Blickl. Homl. 89, 7. Seóþ ðonne on ēce gewyrht *they shall look then on an everlasting state* [*one whose character is determined by their deeds*], Exon. 116 b; Th. 448, 29; Dóm. 61. [*O. Sax.* gi-wurhti *deed : O. H. Ger.* ka-wuruht, Grff. i. 975.]

ge-wyrhta, an; *m. A worker, doer, fellow-worker, accomplice* :—Ǽlc ðe gewita oððe gewyrhta sī *every one who is cognisant or co-operating,* L. O. D. 6; Th. i. 354, 28. Þeófa gewita and geweorhta *an accessory and accomplice of thieves,* L. Eth. 9, 27; Th. i. 346, 9 : L. O. 3; Th. i. 180, 1. Nū gē mágon oncýððǽda wrecan on gewyrhtum *now may ye wreak on the doers* [*their*] *grievous deeds,* Andr. Kmbl. 2361; An. 1182. [Cf. *Goth.* ga-waurstwa *a fellow-worker.*]

ge-wyrman :—To gewyrmenne, Lchdm. i. 116, 1.

ge-wyrp, es; *n. A heap thrown up* [?] :—Andlang gewyrpes, Cod. Dipl. Kmbl. v. 78, 29. v. ge-weorp, sand-gewurp.

ge-wyrpan; *p.* -wyrpte; *pp.* -wyrped *To recover;* verti, recuperare :— Gif se seóca man eft gewurpþ *if the sick man recovers,* L. Ælfc. P. 47 : Th. ii. 384, 29. Godwine gesíclode and eft gewyrpte *Godwin fell sick and got better again,* Chr. 1052; Erl. 186, 13. He eft gewyrpte, and ðam orþe onfēng *he recovered again and got his breath,* Guthl. 20; Gdwin. 86, 17. He hyne gewyrpte, ðeáh ðe him wund hrine *he recovered, though the wound had touched him,* Beo. Th. 5944; B. 2976. He ða befrān on hwilcere tíde he gewyrpte *he then enquired at what hour he recovered,* Homl. Th. i. 128, 12. Sóna ðæt him bet wæs, and gewyrpte fram ðære untrumnysse *melius habere cœpit, et convalescens ab infirmitate,* Bd. 3, 13; S. 539, 7.

ge-wyrsmed, -wyrmsed; *part. p. Full of matter, suppurated* :—Gewyrsmed, *saporatus,* Wrt. Voc. 289, 20. v. wyrmsan, ge-wurms.

ge-wyrþan; *p.* ede; *pp.* ed *To estimate, value* :—Oðre ungesawene þing mon mót mid áþe gewyrþan and syððan be ðam gyldan *other unseen things may be estimated on oath, and then paid for accordingly,* L. O. D. 7; Th. i. 356, 7. Swā hit man gewyrþe *as it may be valued,* L. A. G. 3; Th. i. 154, 11.

ge-wyrþan; he -wyrþeþ, -wyrþ *To be, become, happen;* fieri :—Hū mihte ðæt gewyrþan *how might that happen?* Andr. Kmbl. 1145; An. 573. Cūþ ðæt gewyrþeþ *it shall be known,* Elen. Kmbl. 2381; El. 1192 : 2548; El. 1275. Hū gewyrþ ðis *quomodo fiet istud?* Lk. Bos. 1, 34. v. ge-weorþan.

ge-wyrðe, es; *n. Amount, content* :—Swā micel ðæt sý iii ægscylla gewyrðe *as much as three eggshells full,* Lchdm. iii. 14, 23. Ánes æges gewyrðe greátes sealtes *of rock salt the content of one egg,* 40, 10. [Cf. *Goth.* andwairþi *price, value.*]

ge-wyrþian; *p.* ode; *pp.* od *To distinguish, honour, dignify;* insignīre, hŏnōrāre :—Ðone sōþfæst cyning mid his sylfes miht gewyrþode *whom the just king honoured with his own power,* Cd. 143; Th. 178, 11; Exod. 10. Sigore gewyrþod *honoured with victory,* Andr. Kmbl. 232; An. 116. Ða ðe beóþ mid cræftum gewyrþode *who are dignified with virtues,* Bt. 30, 1; Fox 108, 25. v. ge-weorþian.

ge-wyrtian; *p.* ode; *pp.* od *To season with herbs, to spice, perfume* :— Gewyrtad mid hyra weldǣdum *perfumed with their good deeds,* Exon. 63 b; Th. 234, 20; Ph. 543. Gewyrtod wín [cf. *O. H. Ger.* der gewurzeto win] *factitium vinum,* Cot. 268, Lye. Sele him etan gewyrtodne

hen fugel give him to eat a fowl dressed with herbs, L. M 3, 12; Lchdm. ii. 314, 15.

ge-wyrtrumian *to root up, eradicate;* eradicare, Rtl. 65, 25.

ge-wyrtūn, es; *m. A garden* :—Ðær wæs fæger gewyrtūn *ubi erat hortus,* Jn. Skt. Lind. 18, 1.

ge-wýscan; *p.* te; *pp.* ed. **I.** *to wish, desire;* optare, desiderare :— Ic wolde gewýscan ðæt hī næfdon ða heardsǣlþa ðæt hī mihton yfel dōn *I would wish that they had not the unhappiness of being able to do evil,* Bt. 38, 2; Fox 198, 3. **II.** *to adopt* :—Him to gástlícum bearnum gewīscede *adopted as his spiritual children,* Homl. Th. i. 520, 31.

ge-wýscednys, se; *f. Adoption;* adoptio, R. Ben. interl. 2, Lye.

ge-wýscendlíc; *adj. Optative* :—Gewīscendlíc gemet *modus optativus,* Ælfc. Gr. 21; Som. 23, 28. Gewíscendlíce *optativa,* 38; Som. 40, 25.

ge-wýscendlíce; *adv. By adoption* :—God Fæder Ælmihtig hæfþ ǣnne Sunu gecyndelíce and menige gewíscendlíce *God, the Father Almighty has one Son naturally and many by adoption,* Homl. Th. i. 258, 26.

ge-wýscing; *f. Adoption,* R. Ben. 2, Lye.

ge-ýcan, -ýcean; *p.* te *To increase, add, eke* :—Se ðe ðisne freóls geýcean wille geýce God his gesynta *qui hanc libertatis dapsilitatem augere voluerit, augeat dominus ejus prosperitatem,* Cod. Dipl. Kmbl. iii. 138, 14. Swā swā sorge and ymbhogan geýceþ monnes mōd, swā geýcþ se cræft his āre *as sorrow and cares increase a man's mind so a craft increaseth his honour,* Prov. Kmbl. 59. Ða geýhte he sum bigspell *he added a parable,* Lk. Skt. 19, 11, MS. A. v. ge-ícan.

ge-yde *subdued, conquered,* Chr. 617; Erl. 23, 16. v. ge-gān.

ge-yflian; *p.* ode, ede; *pp.* od, ed. **I.** *to injure* :—Gif hine mon gefflige *if one injure him,* L. Alf. pol. 2; Th. i. 62, 3. Gif se cristena mann ðē geffelode *if the christian man hath done thee wrong,* Homl. Th. i. 54, 25. Næs heora neáta nān geyfled *jumenta eorum non sunt minorata,* Ps. Th. 106, 37. Mid frǽcedo geyfled *contumelia adfectos,* Mt. Kmbl. Lind. 22, 6. Hine geyflade he fell *sick,* Th. Chart. 272, 29. Him geyfelade and ðæt him stranglíce eglade *he fell sick and it afflicted him severely,* Chr. 1086; Erl. 220, 33. Lazarus wæs geyfled *Lazarus infirmabatur,* Jn. Skt. 11, 2.

ge-ylca; *prn. The same* :—Eall ðæ geylcan gerihta *all the same rights,* Th. Chart. 433, 36.

ge-ymnyttan. v. ge-emnettan.

ge-yppan; *p.* -ypte; *pp.* -ypped, -yped, -ypt *To open, reveal, declare, manifest, disclose* :—Ic geyppe *promo,* Ælfc. Gr. 28, 4; Som. 31, 12. Wit wēndon ðæt ðæt sand uncre swaðe geypte *we expected that the sand would discover our track,* Shrn. 42, 19. Se geypte hǣðenum dēman ðæt ðæs tiburtius wæs cristen *he disclosed to the heathen judge that this Tiburtius was a christian,* 116, 23. Him wæs on swefne geyped *it was revealed to him in a dream,* 112, 6. Hit þurh ǣnne þeówne mann geypped wearþ *it was discovered by a slave,* Ors. 3, 6; Bos. 58, 20 : Nicod. 17; Thw. 8, 25. Giypped sē *manifestetur,* Rtl. 13, 3 : 102, 43. Biþ geypped *sciatur,* Lk. Skt. Lind. 12, 2 : Andr. Kmbl. 2447; An. 1225 : Menol. Fox 311; Men. 159. Þurh hine wurdon manege geypte *through him were many discovered,* Chr. 1095; Erl. 232, 20.

ge-yrfian; *p.* ode; *pp.* od *To stock with cattle* :—Swā geirfað swā hit nū stent *so stocked as it now stands,* Th. Chart. 158, 10.

ge-yrfweardian *to inherit,* Ps. Lamb. 24, 14. v. yrfweardian.

ge-yrgan, -irgan; *p.* de; *pp.* ed *To make cowardly, terrify* :— Ealle synd geyrgede ðe eardiaþ on ðisum lande *all the inhabitants of the land do faint because of us,* Jos. 2, 24 : 8, 6 : Swt. A. S. Rdr. 108, 123. v. earg.

ge-yrman; *p.* de; *pp.* ed *To afflict, make miserable* :—Ðū mīne cūþe geyrmdest *thou didst afflict mine acquaintance,* Ps. Th. 87, 18.

ge-yrnan; *p.* -arn, -orn, *pl.* -urnon; *pp.* -urnen *To run, arise;* exoriri, surgere :—Ða georn ðær sóna upp genihtsumlíc yrnþ and wæstm *then an abundant crop and grain* [*fruit*] *soon rose* [*ran*] *up there,* Bd. 4, 28; S. 605, 40.

ge-yrsian; *p.* ode; *pp.* od. **I.** *to anger, make angry* :—Hý geyrsedon *irritaverunt,* Ps. Lamb. 105, 7. Irtacus ða wearþ swíðe geyrsod *Irtacus then became very angry,* Homl. Th. ii. 476, 34. **II.** *to be angry* :—He nele swā micclum swā we geearniaþ us geyrsian *he will not be angry with us so much as we deserve,* 126, 6. v. yrsian.

ge-ýwan, -eáwan, -eówan, -iéwan; ic -ýwe; ðū -ýwest, -ýwst; he -ýweþ, -ýwþ, *pl.* -ýwaþ; *p.* de; *pp.* ed; *v. trans. To shew, manifest, reveal;* ostendĕre, præbēre, manifestāre, monstrāre :—Þeóden engla his þegnum seolfne geýwde *the king of angels revealed himself to his disciples,* Elen. Kmbl. 974; El. 488. Me ðīn dôhtor hafaþ geýwed orwyrðu *thy daughter has shewn me indignity,* Exon. 66 b; Th. 246, 29; Jul. 69 : Elen. Kmbl. 1570; El. 787. DER. ýwan.

gi-; for most words beginning with this prefix see ge-.

giccan *to itch;* prurire :—Wið giccendre wombe *for an itching stomach,* [Cockayne prefers to translate the verb *to hiccup,* v. his Glossary], Lchdm. iii. 50, 13. Wið ōðrum giccendum blece *for other itching blotch,* 70, 27. [*Prompt. Parv.* ȝichin *prurire : A. R.* ȝicchen : *Chauc.* icche : *O. H. Ger.* iuchian *prurire, scalpere : Ger.* jucken *to itch.*]

giccig; *adj. Putrid;* putridus, purulentus, Hpt. Gl. 453.

GICEL, es; *m. An icicle* :—Íses gicel *stiria, stillicidia,* Ælfc. Gl. 16;

Som. 58, 68; Wrt. Voc. 21, 55. [*Icel.* jökull.]　DER. Cyle-, hilde-, hrím-, ís-gicel.

gicelig; *adj. Icy*; glacialis, Hpt. Gl. 454, 465.

gicel-stán, es; *m. A piece of ice, hailstone:*—He sent gicelstán *mittit chrystallum*, Blickl. Gl.

gicenes, se; *f. An itch*, or *burning in the skin*; prurigo, Cot. 156.

gicþa, gyhþa, an; *m. Itch, itching:* — Gicþa *pruritus*, Ælfc. Gl. 11; Som. 57, 62: Wrt. Voc. 20, 6: Past. 11; Swt. 70, 19; Cot. MS. Wiđ gicþan *against itch*, Herb. 21, 3; Lchdm. i. 116, 23: L. M. 2, 41; Lchdm. ii. 252, 19, 24: 2, 65; Lchdm. ii. 296, 6. [*Prompt. Parv.* ʒikthe *prurigo*: *O. H. Ger.* iuchido *prurigo, scabies.*] v. gihþa.

gicþa *hiccup*, Lchdm. ii. 4, 27.

gid, gidd, gied, giedd, gyd, gydd, ged, es; *n.*　　I. *a song, lay, poem*; cantus, cantilena, carmen, poema:—Gid oft wrecen *a song oft sung* [*recited*], Beo. Th. 2135; B. 1065. Gidda gemyndig *mindful of songs*, Beo. Th. 1741; B. 868. Đær wæs gidd and gleó *there was song and glee*, Beo. Th. 4216; B. 2105. Gliówordum gól gyd æt spelle *sung in metre a lay in his discourse*, Bt. Met. Fox 7, 4; Met. 7, 2. Geríseþ gleómen gied *a song is proper for a gleeman*, Exon. 91 b; Th. 344, 1; Gn. Ex. 167. Cúþ gyddum *known in lays* [*songs*], Beo. Th. 304; B. 151. Se wítga song and đæt gyd awræc *the prophet sang and recited the poem*, Exon. 84 a; Th. 316, 20; Mód. 51. Đæt ic đa ged ne mæg gefégean *that I cannot compose the poems* [*songs*], Bt. Met. Fox 2, 10; Met. 2, 5.　　II. as Old English or Saxon proverbs, riddles, and particular speeches were generally metrical, and their historians were bards, hence, *A speech, tale, sermon, proverb, riddle*; sermo, dictum, loquela, proverbium, ænigma:—Gyd æfter wræc *the speech afterwards recited*, Beo. Th. 4315; B. 2154. Mæg ic be me sylfum sóþ gied wrecan *of myself I can relate a true tale*, Exon. 81 b; Th. 306, 2; Seef. 1. On gewunon gyddes gehwyrfed *in consuetudinem proverbii versum*, Bd. 3, 12; S. 537, 27. On gydde *into a proverb*, 3, 12; 537, 30. Nú me đisses gieddes onsware ýwe *now shew me an answer of this riddle*, Exon. 114 a; Th. 437, 28; Rä. 56, 14. v. Grmm. D. M. 853.

giddian, gieddian, gyddan, giddigan; *p.* ode; *pp.* od *To sing, recite, speak:*—Ongan he giddian *he began to sing*, Bt. 31, 2; Fox 112, 24. Giddigan, 16, 4; Fox 56, 36: 21; Fox 72, 27. Se hiora cyning ongan đá singan and giddian *Tyrtæi ducis composito carmine et pro concione recitato*, Ors. 1, 14; Bos. 37, 29. Ongan đá gyddigan þurh gylp micel *began then to speak through great pride*, Cd. 205; Th. 253, 21; Dan. 599. Se wísdóm geoddode đus *wisdom recited this song*, Bt. 12; Fox 36, 6: Bt. Met. Fox 1, 168; Met. 1, 84. Wíga gyddode Beówulf mađelode *the warrior spake, Beowulf said*, Beo. Th. 1264: B. 630; Cd. 97; Th. 127, 6; Gen. 2106. Waldere gyddode wordum, Wald. 83; Vald. 2, 13. Đus fród guma in fyrndagum gieddade *thus sang a wise man in days of old*, Exon. 64 a; Th. 236, 8; Ph. 571. Gyddedon hæleþ in healle hwæt seó hand write *heroes in hall discussed what did the hand write*, Cd. 210; Th. 261, 18; Dan. 728.

gidding, giedding, e; *f. Song, saying, discourse:* — Iobes gieddinga *Job's songs*, Exon. 63 b; Th. 234, 32; Ph. 549. Mid gieddingum *with songs*, 92 b; Th. 347, 13; Sch. 12. To đyssere gereccednysse genám se apostol menigfealde gyddunga and gewitnyssa heáhfædera and wítegena *for this narrative the apostle took manifold sayings and testimonies of patriarchs and prophets*, Homl. Th. ii. 420, 11. Giddung *divinatio, cantus*, Hpt. Gl. 466. [*Chauc.* ʒedding.] v. gid.

gidig; *adj.* GIDDY; vertiginosus, Som.

gied, giedd, es; *n. A song, lay, riddle*, Exon. 91 b; Th. 344, 1: 114 a; Th. 437, 28; Rä. 56, 14: 18 a; Th. 45, 2; Cri. 713. v. gid.

giefa *a giver*. v. gifa.

giefan; *p.* geaf, *pl.* geáfon; *pp.* gifen *To give*; dare:—Ic eów meaht giefe *I will give you might*, Exon. 14 b; Th. 30, 11; Cri. 478. He us æt giefeþ *he giveth us food*, 16 b; Th. 38, 9; Cri. 604: 87 a; Th. 327, 23; Vy. 8. Đú us freódóm gief *do thou give us freedom*, Hy. 5, 10; Hy. Grn. ii. 286, 10. v. gifan.

giefernes, -ness, e; *f. Gluttony*; gula:—Gemidliaþ hiera giefernesse [gífernesse, MS. Cot.] *refrenant gulam*, Past. 46, 2; Swt. 345, 23; Hat. MS. 66 a, 9. v. gífernes.

gief-stól, es; *m. A gift-seat, throne of grace*; donorum thronus, gratiæ thronus:—Wile nú gesécan sáwla Nergend gæsta giefstól *now will the Saviour of souls seek the spirits' throne of grace*, Exon. 16 a; Th. 36, 6; Cri. 572: 77 a; Th. 289, 7; Wand. 44. v. gif-stól.

giefu, e; *gen. pl.* -end; *f. A gift, grace, favour*; donum, munus, gratia:—To giefe *as a gift*, or *freely, gratuitously*, Exon. 65 b; Th. 241, 19; Ph. 658: 96 b; Th. 359, 32; Pa. 71. God-bearn on grundum his giefe bryttaþ *the divine Child on earth his grace dispenseth*, 17 b; Th. 43, 2; Cri. 682. Us giefe sealde uppe mid englum *gave us favour above with angels*, 17 b; Th. 41, 24; Cri. 660: 32 a; Th. 101, 24; Cri. 1663. v. gifu.

gield, es; *n. A payment of money, recompense, substitute, offering, worship, service, a heathen deity:*—Sáwlum to gielde *for a recompense to their souls*, Exon. 23 b; Th. 66, 30; Cri. 1079. Wæs Abeles gield *was Abel's substitute*, Cd. 55; Th. 67, 32; Gen. 1109: 5; Th. 7, 5; Gen.

101: 47; Th. 60, 5; Gen. 977: Exon. 67 b; Th. 251, 17; Jul. 146: 58 a; Th. 253, 3; Jul. 174. v. gild.

gieldan, ic gielde, đú gieltst, gielst, he gieldeþ, gielt, *pl.* gieldaþ; *p.* geald, *pl.* guldon; *pp.* golden *To yield, pay, render, repay, requite:*—Sceoldon gombon gieldan *they must pay homage*, Cd. 93; Th. 119, 11; Gen. 1978. Werum gieldeþ gaful *pays tribute to men*, Exon. 108 b; Th. 415, 15; Rä. 33, 11: 34 a; Th. 109, 24; Gú. 95: 39 a; Th. 130, 9; Gú. 435. He đé mid wíte gieldeþ *he will requite thee with punishment*, 80 a; Th. 301, 15; Fä. 19: Bt. 41, 3; Fox 248, 22. v. gildan.

gieldra *older*, Th. Diplm. A. D. 901–909; 162, 18. v. ieldra.

giellan *to yell*, Exon. 106 b; Th. 406, 19; Rä. 25, 3: 82 a; Th. 309, 25; Seef. 62. v. gellan.

gielp. v. gilp.

gielpan *to glory, boast, vaunt*; gloriári, jactáre:—Đæt hý gielpan ne þorftan dædum *that they should not boast of deeds*, Exon. 36 a; Th. 116, 21; Gú. 210: 114 b; Th. 440, 4; Rä. 59, 12. v. gilpan.

gieman. v. gýman.

gien, giéna *again, still, yet*. v. gén. géna.

giéng *went*, Cd. 29; Th. 39, 15; Gen. 626; *p.* of gangan.

gierian; *p.* ede; *p.* ed *To clothe, deck, adorn*; induére, vestíre, ornáre:—Hæleþ gierede mec mid golde *a man adorned me with gold*, Exon. 107 a; Th. 408, 16; Rä. 27, 13. v. gearwian, gyrian.

gierstandæg *yesterday*. v. gyrstandæg.

gierwan; *p.* ede; *pp.* ed *To make ready, prepare, put on, clothe, adorn*; páráre, induére, vestíre:—Ongunnon him on uhtan ædelcunde mægþ gierwan to geonge *the noble women resolved to prepare for journey at dawn*, Exon. 119 b; Th. 459, 19; Hö. 2. Bearn fæder and módor gierwaþ *father and mother adorn the child*, Exon. 87 a; Th. 327, 23; Vy. 8. v. gearwian.

giest, es; *m. A guest:*—Mid giestum *with the guests*, Cd. 112; Th. 148, 11; Gen. 2455: 112; Th. 147, 15; Gen. 2440: Exon. 94 a; Th. 353, 11; Reim. 11. DER. gryre-giest, hilde-, ryne-, stæle-. v. gæst.

giest-líđnys, -nyss, e; *f. Hospitality, entertainment*; hospitálitas:—Him se ædela geaf giestlíđnysse *the noble* [*man*] *gave them entertainment*, Cd. 112; Th. 147, 28; Gen. 2446. v. gæst-líđnes.

giestron *yesterday*; hesterus:—Ic giestron wæs acenned *I was yesterday brought forth*, Exon. 111 a; Th. 426, 24; Rä. 41, 44. v. geostra.

giet, gieta *yet*. v. git, gita.

gietan. v. gitan.

gif, e; *f:* nom. acc. gif [as tíd, dæd] *A gift, grace*; donum, gratia:—Hú he his gif cýđde geond woruld *how he shewed his grace throughout the world*, Andr. Kmbl. 1150; An. 575.

gif, gief, gyf, gib; *conj. with indic. or subj. If, though, whether:*—For đý me þyncþ betre gif iów swæ þyncþ *therefore it seems to me better, if it seems so to you*, Past. pref; Swt. 7, 6. Gif hie brecaþ his gebodscipe *if they break his commandment*, Cd. 22; Th. 28, 3; Gen. 434. Gif ic ænegum þegne þeóden-mádmas forgeáfe *if to any follower I gave princely treasures*, 12; Th. 26, 19; Gen. 409. Gif đú him wuht hearmes gespræce he forgifþ hit đeáh *though thou didst speak to him aught of harm yet will he forgive it*, 30; Th. 41, 23; Gen. 661. Frægn gif him wære niht getǽse *asked whether the night had been pleasant to him*, Beo. Th. 2643; B. 1319. Đú wást gif hit is swá *we secgan hýrdon thou knowest if it is as we heard say*, 550; B. 272. [*Laym.* ʒif: *Orm.* ʒiff: *Piers P. Chauc.* ʒif, if: *O. Frs.* jef.]

gifa, gyfa, giefa, geofa, an; *m. A giver, bestower*; dator, largitor:—Me þincþ betere đæt hi forléte đa gyfe and folgyge đam gyfan *it seems to me better to leave the gift and follow the giver*, Shrn. 176, 19. Used in the following compounds:—ár-gifa, æt-, beáh-, beág-, blæd-, eád-, feorh-, gold-, hyht-, lác-, mádđum-, ræd-, sinc-, symbel-, wil-, will-. [*Laym.* rædʒive *counsellor*: *O. Sax.* mêđom-gíƀo: *O. H. Ger.* gebo dator.]

gifan, gyfan, giefan, geofan, giofan; ic gife; đú gifest, gifst; he gifeþ, gifþ, *pl.* gifaþ; *p.* geaf, gæf, gaf, gef, đú geáfe, gǽfe, *pl.* geáfon, géfon; *pp.* gifen, giefen, gyfen *To give*; dare, impertíre:—Hwá meahte me swelc gewit gifan *who could give to me such perception?* Cd. 32; Th. 42, 10; Gen. 672. Ic gife *impertíor*, Ælfc. Gr. 37; Som. 39, 13. Gife ic hit đé *I will give it thee*, Cd. 32; Th. 42, 26; Gen. 679. Us drincan gifest *potum dabis nobis*, Ps. Th. 79, 5. Hwæt gifst đú me *quid dabis mihi*, Gen. 15, 2. God gifeþ gleáw word godspellendum *dominus dabit verbum evangelizantibus*, Ps. Th. 67, 12: Hy. 7, 102; Hy. Grn. ii. 289, 102: Ælfc. Gr. 7; Som. 6, 22: Ps. Th. 68, 27. He nallas beágas geaf *he gave no rings*, Beo. Th. 3443; B. 1719. Gǽf wæstm his *dedit fructum suum*, Ps. Spl. T. 66, 5. Him scippend geaf [gaf, MS. A.] wuldorlícne wlite *the Creator gave it wondrous beauty*, Salm. Kmbl. 114; Sal. 56. Gé him hleoþ géfon *ye gave them shelter*, Exon. 27 b; Th. 83, 11; Cri. 1354. Weoruda waldend đé wist gife heofonlícne hláf *the Lord of hosts grant to thee food, heavenly bread*, Andr. Kmbl. 776; An. 388. On Moyses hand wearþ wíg gifen *into Moses' hand martial force was given*, Cd. 173; Th. 216, 11; Dan. 5. Đær wurdon đa áđas gesworene his dóhter đam Cásere to gifene *oaths were then sworn there to give* [*in marriage*] *his daughter to the emperor*, Chr. 1109; Erl. 242, 23. [*Laym.*

Orm. Chauc. Piers P. ȝiven : *O. Sax.* geban : *Goth.* gibən : *Icel.* gefa :
O. Frs. jeva : *O. H. Ger.* geban.] DER. a-, æt-, ed-, for-, of-gifan.

gifen *the sea,* Exon. 101 a ; Th. 381, 29 ; Rä. 3, 3. v. geofon.

gifende *giving, giving in marriage,* Cot. 216. v. gifan, gift.

gifer, es ; *m. A glutton :*—Gifer hätte se wyrm *the worm's name is glutton,* Exon. 99 b ; Th. 373, 31 ; Seel. 118. v. gifre.

giferlíce ; *adj. Greedily, eagerly ;* avide :—Ongan giferlíce ðæt gærs etan *virecta herbarum avidius carpere cœpit,* Bd. 3. 9 ; S. 533, 41. Giferlíce *pertinaciter,* Hpt. Gl. 424. [Cf. *Icel.* gífrliga *savagely :* Mod. *Icel. exorbitantly.*]

gifer-nes, -ness, e ; *f. Greediness, avarice, voracity, gluttony ;* aviditas, gula :—Giferuys biþ ðæt se man ǽr tíman hine gereordige oððe æt his mǽle to micel þicge mid oferflówendnysse ǽtes oððe wǽtes *greediness is a man's eating before the time, or taking too much at his meal with superfluity of meat or drink,* Homl. Th. ii. 218, 29. Him wæs metes micel lust ac ðeáh mid nānum ǽtum his gýfernysse gefyllan ne mihte *he had great craving for food but yet could he not with any viands satisfy his voracity,* i. 86, 6 : 168, 12. Ða niðtenu for ðære gewilnunge hiera gífernesse simle lócigeaþ to ðære eorþan *beasts because of their greedy desires ever look to the earth,* Past. 21 ; Swt. 154, 20 ; Cot. MS. Ðæt ríce ðæt ða ǽrestan men forworhtan þurh heora gífernesse *the kingdom that the first persons forfeited through their greediness,* Blickl. Homl. 25, 1 : Num. 11, 4 : Rt. 35, 6 ; Fox 170, 2. [*Orm.* gifernesse : *Laym.* ȝivernesse.]

gifeðe, es ; *n. What is granted by fate, lot, fortune, fate :*—Wæs ðæt gifeðe to swíð ðe ðone ðyder ontyhte *too strong was the fate that impelled him thither,* Beo. Th. 6163 ; B. 3085. On gifeðe *by chance,* Andr. Kmbl. 977 ; An. 489. v. Grmm. And. u. El. p. 108. [Cf. *Laym.* swulc ȝifueðe, 2nd MS. so moche god, v. 8118 : *Icel.* gipta *good luck.*] v. next word.

gifeðe, gyfeðe ; *adj. Given, granted* [*by fate*] ; datus, concessus :—Gief ðæt biþ him gifeðe *if that be granted him,* Cod. Dipl. Kmbl. iii. 50, 2 : Th. Chart. 470, 1 : 472, 1. Nó gifeðe wearþ Abrahame ðæt him yrfeweard wlítebeorht ides on worulde brohte *it was not granted to Abraham that the beautiful woman brought him an heir into the world,* Cd. 83 ; Th. 103, 31 ; Gen. 1726 : 101 ; Th. 134, 13 ; Gen. 2224 : Beo. Th. 5454 ; B. 2730. Gyfeðe, 1115 ; B. 555. Him ðæt gifeðe ne wæs *it was not granted him,* 3658 ; B. 2682. Hwæt him gúðweorca gifeðe wurde *what work of war should be assigned him,* Andr. Kmbl. 2134 ; An. 1068 : Beo. Th. 4976 ; B. 2491 : 604 ; B. 299. v. Grmm. D. M. 843. [*Laym.* ȝifeðe : *O. Sax.* gibiðu : *O. H. Ger.* gibedig.] v. ungifeðe, and preceding word.

gif-fæst ; *adj. Gifted with, capable of, fitted for ;* capax :—Sum biþ wôþbora giedda giffæst *one is a poet gifted with song,* Exon. 78 b ; Th. 295, 20 ; Crä. 36 : Cot. 57.

gif-heal, -heall, e ; *f. A gift-hall, hall in which gifts are distributed ;* aula in qua dominus dona distribuit :—Ymb ða gifhealle *around the gift-hall,* Beo. Th. 1680 ; B. 838.

gifian ; *p.* ode, ede ; *pp.* od, ed *To bestow gifts :*—Se cyng him cynelíce gifode *the king bestowed gifts upon him royally,* Chr. 994 ; Erl. 133, 32. Hió ða gifede mycele þinc ðam biscope *she gave great gifts to the bishop,* H. R. 17, 12.

gifl, giefl, gifel, gyfl, es ; *n. Food, meat, piece of food :*—Líc biþ wyrmes giefl *the body shall be the worm's food,* Exon. 100 a ; Th. 374, 15 ; Seel. 126. Dú wyrma gifl *thou food for worms,* 98 b ; Th. 368, 16 ; Seel. 22. Hí ðæt gyfl þegun *they ate that food,* 61 b ; Th. 226, 24 ; Ph. 410 : 45 a ; Th. 153, 8 ; Gú. 822. Húsle gereorded ðý æþelan gyfle *fed with the Eucharist, with the noble food,* 51 b : Th. 180, 5 ; Gú. 1275. Lytlum giefluın *with the little bits of meat,* 88 b ; Th. 332, 23 ; Vy. 89. v. æfen-gifl.

gifnes, -ness, e ; *f. A favour, grace ;* beneficium, gratia :—Ealle we beþurfon Godes gifnesse *we all have need of God's grace,* Hy. 7, 114, 110 ; Hy. Grn. ii. 289, 114, 110. DER. for-gifnes.

gifol, giful ; *adj. Generous, bountiful, liberal ;* largus :—He swá gifol is and swá rúmedlíce gifþ *he is so liberal and gives so abundantly,* Bt. 38. 3 ; Fox 202, 14.

gifre ; *adj. Useful, salutary :*—Niðum to nytte hæleþum gifre *of advantage to men, useful to warriors,* Exon. 107 b ; Th. 409, 12 ; Rä. 27, 28 : 113 a ; Th. 433, 6 ; Rä. 50, 3. v. ungifre, and cf. *Icel.* gæfr.

gifre ; *adj. Greedy, covetous, voracious, eager, desirous ;* avidus :—Gífre *gulosus,* Wrt. Voc. 86, 51. Gifere *vel* frec *ambro,* Ælfc. Gl. 88 ; Som. 74, 83 ; Wrt. Voc. 50, 63. Tantalus se cyning ðe ungemetlíce gífre wæs *Tantalus the king who was immoderately greedy,* Bt. 35, 6 ; Fox 170, 1. Ða farasei ða ðe gífre wæron *pharasæi qui erant avari,* Lk. Skt. 16, 14. Líg gold gífre forgrípeþ grædig swelgeþ *the flame voracious lays hold on gold, greedy devours it,* Exon. 63 a ; Th. 232, 15 ; Ph. 507 : 38 a ; Th. 124, 32 ; Gú. 346. Gífrum grápum *with greedy clutches,* 38 b ; Th. 126, 28 ; Gú. 378 : Andr. Kmbl. 2671 ; An. 1337. Gesyhst ðú nú ða sweartan helle grǽdige and gífre *seest thou now the black hell greedy and ravenous?* Cd. 37 ; Th. 49, 16 ; Gen. 793 : 213 ; Th. 267, 2 ; Sat. 82 : 217 ; Th. 276, 21 ; Sat. 192 : Exon. 82 a ; Th. 309, 24 ; Seef. 62. Se gífra gǽst *the greedy spirit,* 22 a ; Th. 60, 21 ; Cri. 973. Ic heora eom swíðe gífre *I am very desirous for them,* Bt. 22, 1 ; Fox

76, 20. Líg gǽsta gífrost *flame, most ravenous of spirits,* Beo. Th. 2250 ; B. 1123. Gífrost and grǽdgost *most rapacious and most greedy,* Exon. 128 a ; Th. 493, 2 ; Rä. 81, 24. [*Orm.* giferr : *Laym.* ȝifer : cf. *Icel.* gífr ; *n. pl. fiends.*]

gif-sceatt, es ; *m. A gift-treasure, present ;* donum pretiosum, munus :—Sǽliðende gifsceattas Geátum feredon *sea-voyagers bore gift-treasures for the Gauts,* Beo. Th. 761 ; B. 378.

gif-stól, gief-stól, es ; *m. A gift-seat, seat from which gifts are distributed, throne, throne of grace ;* donorum thronus, solium, gratiæ thronus :—Ðone gifstól grétan *to greet the throne,* Beo. Th. 338 ; B. 168. Brynewylmun mealt gifstól Geáta *the gift-chair of the Goths was consumed by flames of fire,* Beo. Th. 4643 ; B. 2327 : Exon. 16 a ; Th. 36, 6 ; Cri. 572. Sceal gifstól gegierwed stondan *a throne shall stand prepared,* Exon. 90 a ; Th. 337, 23 ; Gn. Ex. 69 : 77 a ; Th. 289, 7 ; Wand. 44.

gift, gyft, e ; *f.* I. *a gift ;* as a technical term, *the amount to be given by a suitor in consideration of receiving a woman to wife :*—Gif mon wíf gebycgge and sió gyft forþ ne cume *if a man buy a wife and the sum agreed upon be not forthcoming,* L. In. 31 ; Th. i. 122, 5. See the note. That matrimony in the olden times was a bargain may be seen by the words used in connection with it, e. g. gebycgan, in the passage above ; see also ge-ceápian, ceáp. For an account of such a bargaining see Njál Saga, c. 2. See also Th. i. 254–6, Cl. and Vig. Icel. Dict. mundr, and Grmm. R. A. pp. 419 sqq. II. in *pl. f.* and *n.* gifta, giftu *marriage ;* nuptiæ :—Giftu *nuptiæ,* Ælfc. Gr. 13 ; Som. 16, 22 : 28, 4 ; Som. 31, 20 : Mone Gl. 433 a. On ðam þriddan dæge wǽron ða gifta gewordene *die tertia nuptiæ factæ sunt,* Jn. Bos. 2, 1 : Mt. Bos. 22, 3. Críst wearþ to his gyftum gelaðod *Christ was invited to his marriage,* Homl. Th. i. 58, 10, 11 : Hy. 10, 17 ; Hy. Grn. ii. 293, 17. Æt ðǽm giftan sceal mæssepreóst beón *at the nuptials there shall be a mass-priest,* L. Edm. 13, 8 ; Th. i. 256, 6. Wífigende and gyfta syllende *nubentes et nubtum tradentes,* Mt. Bos. 24, 38 : Lk. Bos. 20, 34. Gifta dón hearm getácnaþ *to keep a wedding betokens harm,* Lchdm. iii. 208, 21 : L. Alf. 12 ; Th. i. 46, 17. [*Laym. Piers P.* ȝift gift : *O. Frs.* jeft : *O. H. Ger.* gift gift : *Goth.* fra-gifts espousal : *Icel.* gipt *a gift, wedding.*]

gift-búr, es ; *m. A wedding-chamber, bride-chamber :*—Swá swá brýdguma forþ gewítende of giftbúre his *tanquam sponsus procedens de thalamo suo,* Ps. Spl. T. 18, 5.

gifte, an ; *f. Dowry :*—Gilde be ðære giftan mǽþe *reddat pecuniam juxta modum dotis quam virgines accipere consueverunt,* Ex. 22, 17. [Cf. L. Alf. 29 ; Th. i. 52, 8 ; *and see* gift.]

giftelíc ; *adj. Belonging to a wedding ;* nuptialis, Cot. 139.

gift-hús, es ; *n. A wedding-house ;* nuptiarum domus :—Ðá wǽrun ða gyfthús mid sittyndum mannum gefyllede *impletæ sunt nubtiæ discumbentium,* Mt. Kmbl. 22, 10. v. gift-líc.

giftian ; *p.* ode ; *pp.* od *To give a woman in marriage :*—Ne wífiaþ hí ne ne gyftigeaþ *neque nubent neque nubentur,* Mk. Skt. 12, 25. Ne giftigeaþ hí ne wíf ne lǽdaþ *neque nubunt neque ducunt uxores,* Lk. Skt. 20, 35. [Cf. *Icel.* gipta *to give a woman in marriage ;* giptask *to marry :* O. H. Ger. gi-gift *venundatus, deditus.*]

gift-leóþ, es ; *n. A marriage-song ;* epithalamium, carmen nubentium, Ælfc. Gl. 82 ; Som. 73, 53 ; Wrt. Voc. 47, 57.

gift-líc ; *adj. Nuptial, belonging to a marriage ;* nuptialis :—Ðá geseah he ðær ænne man ðe næs mid gyftlícum reáfe gescrýd *vidit ibi hominem non vestitum veste nubtiali,* Mt. Kmbl. 22, 11, 12. Ðæt gyftlíce hús *the house where the marriage was,* Homl Th. ii. 70, 16. Giftlíc *sponsalia,* Hpt. Gl. 525 ; *nuptialis,* 491. Giftlíce *sponsalia,* Mone Gl. 354 a.

giftu. v. gift.

gifu, gyfu, giefu, giofu, geofu, gif, e ; *pl. nom. acc.* -a, -e ; *gen.* -a, -ena ; *f.* I. *a gift, grace, favour ;* donum, munus, beneficium, gratia, virtus, facultas :—Wæs gifu Hróþgáres oft gæ:hted *the gift of Hrothgar was often prized,* Beo. Th. 3773 ; B. 1884. Ðám he geaf micle gife freódómes *to these he gave the great gift of freedom,* Bt. 41, 2 ; Fox 246, 1. Ðæt hie ælmihtiges gife ánforléten *that they the Almighty's gift might lose,* Cd. 32 ; Th. 43, 19 ; Gen. 693. Ic ðam mago-rince mine sylle godcunde gife *I will give to the youth my divine grace,* 106 ; Th. 140, 17 ; Gen. 2329. We onféngon gife for gife *we have received grace for grace,* Jn. Bos. 1, 16. Heó gefylled wæs wísdomes gife *she was filled with the gift of wisdom,* Elen. Kmbl. 2285 ; El. 1144. Ðá him wæstmas brohte geártorhte gife gréne folde *when to him the verdant earth should bring fruits, yearly-bright gifts,* Cd. 76 ; Th. 94, 13 ; Gen. 1561. Sáulum on heofonum selest weorþlíca gifa *to souls in heaven thou wilt give worthy gifts,* Bt. Met. Fox 20, 453 ; Met. 20, 227. Næs hió to gnéþ gifa *she was not too sparing of gifts,* Beo. Th. 3864 ; B. 1930. Neorxna wang stód gifena gefylled *paradise stood filled with gifts,* Cd. 11 ; Th. 13, 28 ; Gen. 209 : Exon. 41 b ; Th. 138, 18 ; Gú. 578. Ðín mód trymeþ godcundum gifum *strengtheneth thy mind with divine gifts,* Cd. 135 ; Th. 170, 8 ; Gen. 2810. Brýdlíce gife *nuptialis dos,* Hpt. Gl. 515. Hláfordes gifu *impost due to the Lord,* L. Eth. 3, 3 ; Th. i. 292, 16 : L. C. S. 82 ; Th. i. 422, 1 : L. N. P. L. 67 ; Th. ii. 302, 7. See Thorpe's Glossary. To gifes *gratis,* Hpt. Gl. 478. Gá hire út to gife bútan feó

let her go out free without money, Ex. 21, 11: Num. 11, 5. To gife *gratis*, Gen. 29, 15. Gifum *gratis*, Ps. Spl. T. 34, 8.   II. *the Anglo-Saxon Rune* X = *g*, the name of which letter in Anglo-Saxon is gifu *a gift*,—hence, this Rune not only stands for the letter *g*, but for gifu *a gift*, as :— X [Gifu] gumena byþ gleng and herenys *a gift is the honour and praise of men*, Runic. pm. 7; Kmbl. 340, 23; Hick. Thes. i. 135. [*Orm.* gife: *Laym.* geve: *R. Brun.* give: *Kath.* gcoven, *pl:* Piers P. yeves: *O. Sax.* geba; *f. donum: O. Frs.* jeve: *O. H. Ger.* geba: *Goth.* giba : *Icel.* gjöf; *f. donum, munus.*] DER. beáh-gifu, brýd-, eád-, feorh-, freót-, frum-, hyht-, mǽddum-, morgen-, sinc-, sundor-, sundur-, sweord-, wóþ-, wuldor-, wundor-.

gifung, gyfung, e; *f. A giving, granting, assent, consent :*—Mid gyfunge ðære synne *peccati consensu*, Bd. 1, 27; S. 497, 11.

gigant, gygant, es; *m. A giant*; gigas :—Untydras onwócon, eotenas, swylce gigantas *unnatural progenies sprang forth, monsters, also giants*, Beo. Th. 226; B. 113. Swá swá gigant yrnþ on his weg *ut gigas ad currendam viam*, Ps. Th. 18, 6. Ne se gigant ne wyrþ ná gehǽled *nec gigas salvus erit*, 32, 14. He ðone gigant ofwearp *he struck down the giant*, Blickl. Homl. 31, 18. [*Lat.* gigas; *gen.* gigantis.]

gigant-mæcg, es; *m. Giant progeny*; filius gigantis, Cd. 64; Th. 76, 36; Gen. 1268.

gi-hrínian, Jn. Skt. p. 188, 4. v. ge-regnian.

gihsinga *exugia*, Cot. 73, Lye. v. Lchdm. i. lxx, note 6.

gihþa, an; *m. Itch, itching :*—Unaberéndlíc gyhþa ofereode ealne ðone líchaman *an unbearable itching overspread the whole body*, Homl. Th. i. 86, 12. [The word used in the passage of Josephus describing Herod's condition is κνησμός.] v. gicþa.

gihþig, giþig; *adj. Lymphaticus, vecors*, Hpt. Gl. 520.

gihþu. v. gehþo.

gild, geld, gield, gyld. es; *n.*   I. *a payment of money, a tribute, compensation, retribution, substitute;* solutio, tributum, compensatio, remuneratio, retributio :—Beád ðá Swegen full gild *then Sweyn commanded a full contribution*, Chr. 1013; Th. 273, 6. Ðis wæs swíðe hefigtýne geár þurh mænigfealde gylda *this was a very grievous year on account of manifold taxes*, Chr. 1096; Erl. 233, 25. Menn guldon him gyld *men paid him tribute*, 1066; Erl. 203. 8. On Abeles gyld *in compensation for Abel*, Cd. 55; Th. 67, 22; Gen. 1104: 153; Th. 190, 15; Exod. 199. On ðære sunnan gyld *in the sun's stead*, Exon. 24 a; Th. 68, 14; Cri. 1103. IX gylde forgylde *let him pay nine[-fold] for compensation*, L. Ethb. 4; Th. i. 4, 3.   II. *a* GUILD, *society, or club, to which payments were made for mutual protection and support, more extensive than our friendly societies;* societas, fraternitas. The members of the A.-Sax. guild were answerable for each other's conduct, and thus character was made of the very greatest importance. v. Kmbl. Sax. Eng. i. 251-253; Th. Chart. p. xvi; pp. 605-17 : Stubbs' Const. Hist. s. v.   III. *a payment to God, worship, service, sacrifice, offering;* cultus, sacrificium :— Ðú goda ussa gield forhogdest *thou hast despised the service of our gods*, Exon. 67 b; Th. 251, 17; Jul. 146. To ðam gielde *for that sacrifice*, Cd. 74; Th. 90, 26; Gen. 1501. His Waldende gilde onsægde *dedicated an offering to his Lord*, 137; Th. 172, 11; Gen. 2842: Bd. 2, 1; S. 501, note 12.   IV. *a heathen deity;* numen :—Gif ðú onsecgan nelt sóþum gieldum *if thou wilt not sacrifice to true deities*, Exon. 68 a; Th. 253, 3; Jul. 174.   V. *a visible object of worship, an idol;* idolum :— He sum gild bræc *he was destroying an idol*, Blickl. Homl. 223, 4: 221, 8, 20. Gyld of golde gumum arǽrde *reared up for the people an idol of gold*, Cd. 180; Th. 226, 22: Dan. 175: 182; Th. 228, 18; Dan. 204. [*O. Sax.* geld; *n. retributio, tributum, cultus divinus, sacrificium: O. Frs.* jeld; *n: O. H. Ger.* gelt; *n: Goth.* gild; *n. tributum, census, multa: Icel.* gildi, gjald; *n. tributum, pœna, prœmium, multa cœdis.*] DER. æfter-gild, [-geld, -gield, -gyld], án-, bryne-, deófol-, ed-, feónd-, friþ-, frum-, god-, hǽðen-, leód-, sceucc-, þeóf-, un-, wer-, wig-, wiðer-. v. Grmm. D. M. 34: R. A. 601, 649. [Cf. friþ-gild.]

gilda, gylda, an; *m. A member of a guild :*—Se gylda ðe óðerne misgrét *the guildbrother that insults another*, Th. Chart. 606, 22 : 609, 10. v. ge-gilda.

gildan, geldan, gieldan, gyldan, ic gilde, gielde, gylde, ðú giltst, gieltst, gyltst, gilst, he gildeþ. gilt, gielt, gylt, *pl.* gildaþ; *p.* geald, *pl.* guldon; *pp.* golden; *v. a; n. To yield, pay, restore, requite, give, render, make an offering, serve, worship;* reddere, solvere, tribuere, retribuere, rependere, restituere, servire, colere :—Gafol gyldan *to pay tribute*, Ors. 1, 10; Bos. 32, 24, 28 : Mt. Bos. 17, 24. Ic mín gehat Dryhtne gylde *vota mea Domino reddam*, Ps. Th. 115, 8 : 78, 13 : 93, 22. Se gylt ǽlcum be his gewyrhtum *he requites each according to his works*, Bt. 40, 7; Fox 244, 1 : Ors. 1, 1; Bos. 20, 35. Gilde ðæt ilce wíte ðæt se óðer sceolde gif he him ryhtes wyrnde *let him pay the like penalty that the other should if he had denied him justice*, L. Ath. i. 3; Th. i. 200, 18. Drihtne guldon gód *they paid good to the Lord*, Cd. 111; Th. 146, 9; Gen. 2419. Gilde he twífealdon *duplum restituet*, Ex. 22, 4, 7. Gilde ðone byrst *reddet damnum*, 22, 6, 11. Gild ðínum esne góðe dǽde *retribue servo tuo*, Ps. Th. 118, 17. Heaðo-rǽsas geald mearum and máððum *requited war-attacks with horses and treasures,*

Beo. Th. 2099; B. 1047. Ðæt ǽlc gulde óþrum edleán ǽlces weorces *that each should render to another recompense for every work*, Bt. 39, 1; Fox 212, 5 : Bt. Met. Fox 27, 51; Met. 27, 26. Mín sceal mid grimme gryre golden wurþan fyll and feorh-cwealm *my fall and murder shall be requited with grim horror*, Cd. 55; Th. 67, 18; Gen. 1102. Sceucc-gyldum swýðe guldan *servierunt sculptilibus eorum*, Ps. Th. 105, 26. Bebeád se cásere ðæt cristne men guldan deófolgeldum *the emperor ordered that christian men should worship idols*, Shrn. 88, 14, 22 : 74, 26. Deóflum geldan *to worship devils*, 110, 18. [*Laym. Orm.* ʒelden : cf. Shakspere's God ild you: *O. Sax.* geldan *reddere, retribuere, solvere, prœstare: O. Frs.* jelda: *O. H. Ger.* geltan *reddere, solvere, retribuere, sacrificare: Goth.* -gildan, fra-gildan; *p.* -gald, *pl.* -guldum; *pp.* -guldans *to repay, requite: Icel.* gjalda.] DER. a-gildan, an-, on-, for-, ge-, to-.

gildan *to gild.* v. gyldan.

gilden. v. gylden.

Gild-ford, Gyldford, Guldeford [Gild *a fraternity;* ford *a ford: Domesd.* Gilda ad vadum] GUILDFORD, *a town in Surrey, on the river Wey*, Lye.

gild-rǽden, gyld-rǽden, -rǽdenn, e; *f. The relation involved in membership of a guild :*—Gif he nele to bóte gebúgan þolige he ðære geférrǽdene and ǽlcere óðre gyldrǽdene *if he will not submit to make amends let him forfeit the fellowship and every other interest in the guild*, Th. Chart. 606, 31. Ðæt byþ rihtlice gecweden gyldrǽdene ðæt we ðus dón *that is very properly agreed upon as a part of guild-membership, that we do thus*, 607, 24.

gild-scipe, gyld-scipe, es; *m. A guild-ship, society;* sodalitas :—Án gildscipe is gegaderod on Wudeburg lande *a guild-ship is gathered at Woodbury land*, Th. Diplm. 608, 30 : 605, 8 : L. Edg. C. 9; Th. ii. 246, 12. v. gild.

gild-sester, es; *m. A measure belonging to a guild;* sextarius :— Sceóte ǽlc gegylda ǽnne gyldsester fulne clǽnes hwǽtes *let each guild-brother contribute one guild-measure full of clean wheat*, Th. Chart. 606, 7 : 611, 4.

gillan *to yell*, Salm. Kmbl. 535; Sal. 267. v. gellan.

Gillinga, Gillinga-hám GILLINGHAM, *in Dorsetshire, on the river Stour*, Chr. 1016; Erl. 156, 1, 18.

gillister, es; *n. Phlegm*, L. M. 1, 1; Lchdm. ii. 24, 18.

gillistre, an; *f. Phlegm, matter*, L. M. 1, 1; Lchdm. ii. 18, 17 : 72; 148, 6.

gilm, es; *m. A yelm, a handful of reaped corn, bundle, bottle;* manipulus :—Eówre gilmas stódon *your sheaves stood*, Gen. 37, 7. v. gelm.

gilp *powder, dust;* scobs, Cot. 181.

GILP, gelp, gielp, gylp, es; *m. Glory, ostentation, pride, boasting, arrogance, vain-glory, haughtiness;* gloria, ostentatio :— Se seofoþa heáfod-leahter is ídelwuldor ðæt is gylp *the seventh chief sin is vain-glory, that is pride*, Homl. Th. ii. 220, 27 : 218, 22. He nolde uán þing dón mid gylpe forðon ðe se gylp is án heáfod-leahter *he would do nothing in pride, for pride is a deadly sin*, i. 170, 24. Geþenc be ðam gebyrdum gif hwá ðæs gilpþ hú ídel and hú unnyt se gilp biþ *consider birth; if any one boast of that how vain and how useless is the boast*, Bt. 30, 1; Fox 108, 20 : Cd. 219; Th. 280, 12; Sat. 254 : 4; Th. 5, 10; Gen. 69 : Blickl. Homl. 243, 9. Gylpes ðú girnest *thou desirest glory*, Bt. 32, 1; Fox 114, 18. Hú Orosius spræc ymb Rómána gylp hú hí manega folc oferwunnan *how Orosius spoke of the glory of the Romans, how they overcame many peoples*, Ors. Nap. 12, 42. Is ðæt unnet gelp *that is useless glory*, Bt. Met. Fox 10, 34, 26; Met. 10, 17, 13. Nǽfre gielpes to georn *never too eager for fame*, Exon. 77 b; Th. 290, 12; Wand. 69. On ídel gylp *in vanitate sua*, Ps. Th. 51, 6. For ðínum ídlan gilpe *for thine idle boasting*, Blickl. Homl. 31, 14. [*Laym.* ʒælp, ʒelp : *Orm.* ʒellp : *O. Sax.* gelp : *O. H. Ger.* gelf *jactantia*, [*inanis*] *gloria.*]

gilpan, gielpan, gylpan, ic gilpe, gielpe, gylpe, ðú gilpst, gielpst, gylpst, he gilpþ. gielpþ, *pl.* gilpaþ, gielpaþ, gylpaþ; *p.* gealp, *pl.* gulpon; *pp.* golpen *To glory, boast, desire earnestly;* gloriari :—Gif ðú gylpan wille, gilp Godes *if thou wilt glory, glory in God*, Bt. 14, 1; Fox 40, 24. Nó ðæs gilpan þearf synfull sáwel *the sinful soul need not boast of this*, Exon. 116 b; Th. 449, 9; Dom. 68. Ðæt hine swelces gamenes gilpan lyste *that he liked to boast of such sport*, Bt. Met. Fox 9, 38; Met. 9, 19. Ðæt ðú wile gilpan *that thou wilt boast*, Salm. Kmbl. 409; Sal. 205. Ic wundrige forhwí hí gilpan swelces anwealdes *I wonder why they boast of such power*, Fox 104, 1. Gelpan ne þorfte *had no cause to boast*, Chr. 937; Erl. 114, 10; Æðelst. 44. Gylpan, Beo. Th. 4016; B. 2006: 5740; B. 2874. Ná ic ðæs gylpe *I boast not of that*, 1177; B. 586: 4116; B. 2055. Hú lange mánwyrhtan morðre gylpaþ *usque-quo peccatores gloriabuntur?* Ps. Th. 93, 3 : 73, 4. He geald ðæt him nówiht widstandan mihte *nihil resistere posse jactábat*, Bd. 3, 1; S. 524, 8. Hréþsigora ne gealp *he boasted not of glorious victories*, Beo. Th. 5160; B. 2583. Burga aldor gramlíce gealp *the ruler of towns angrily boasted*, Cd. 210; Th. 260, 23; Dan. 714. Swíðe gulpon *they exceedingly boasted*, 210; Th. 260, 20; Dan. 712. Sigore gulpon *they boasted of victory*, Cd. 94; Th. 121, 29; Gen. 2017. Firenum gulpon

they wickedly boasted, Exon. 36 b; Th. 118, 8; Gú. 236. Ðæt hí ne gulpan ðæs *that they may not boast of it*, Ps. Th. 74, 4. [Laym. ȝælpen, ȝelpen: *Orm.* ȝellpenn, ȝillpenn: *Chauc.* yelpe *to boast*.]

gilp-cwide, es; *m. A boastful speech* :—Ðam wífe ða word wel lícodon gilpcwide Geátes *well did those words please the woman, the boastful speech of the Gaut*, Beo. Th. 1284; B. 640: Exon. 50 b; Th. 176, 12; Gú. 1209. [*O. Sax.* gelp-quidi.]

gilpen ; *adj. Boastful* :—Ne mæg he geþyldgian ðæt he ðæt forhele ac wierþ ðonon gilpen *he cannot bear to conceal it, but becomes boastful on account of it*, Past. 33, 2 ; Swt. 216, 9 : Cot. MS. Wát ic ðæt wǽron Caldéas gúðe ðæs gilpne *I knew that the Chaldeans were so boastful in war*; Salm. Kmbl. 413 ; Sal. 207.

gilp-georn ; *adj. Desirous of glory* :—Se strangesta cyning and se gilpgeornesta *rex fortissimus et gloriæ cupidissimus*, Bd. 1, 34 ; S. 499, 19.

gilp-hlæden ; *part. p. Vaunt-laden* :—Cyninges þegn guma gilp-hlæden gidda gemyndig *a king's thane, a man filled with lofty themes, with memory rich in songs*, Beo. Th. 1740 ; B. 868.

gilplíc ; *adj. Ostentatious, pompous, proud, vain-glorious* :—Ðæt wǽre swíðe gilplíc dǽd gif Crist scute ðá adún *it would have been a very vainglorious act if Christ had thrown himself down then*, Homl. Th. i. 170, 21. Gierelan gielplíces *of pompous garb*, Exon. 35 a ; Th. 112, 3 ; Gú. 138 : 38 b ; Th. 127, 22 ; Gú. 390.

gilp-líce ; *adv. Proudly, vauntingly* ; arroganter, Cot. 1, Lye. [*O. H. Ger.* gelfligho *jactanter*.]

gilpna, an ; *m. A boaster* ; jactator :—Betra biþ se geþyldega wer ðonne se gilpna *melior est patiens arrogante*, Past. 33, 2 ; Swt. 216, 14 ; Cot. MS: 20 ; Swt. 148, 19.

gilp-plega, an ; *m. Play of which one may boast* [*war*] :—Gylpplegan gáres, Cd. 154 ; Th. 193, 2 ; Exod. 240.

gilp-sceaða, an ; *m. An arrogant, boasting criminal* :—Gielpsceaðan *boastful and wicked ones* [*the fallen angels*], Cd. 5 ; Th. 6, 29 ; Gen. 96. Ðone gelpscaðan *that proud and wicked man* [*Nero*], Bt. Met. Fox 9, 98 ; Met. 9, 49.

gilp-sprǽc, e ; *f. Boastful speech*, Beo. Th. 1966 ; B. 981.

gilp-word, es ; *n. A boastful word, a boast, vaunt* :—Hí him to gylpworde hæfdon ' ðæt him leófre wǽre ðæt hí hæfdon healtne cyning ðonne healt ríce ' *their boast was ' that they had rather have a halting king than a halting kingdom*,' Ors. 3, 1 ; Bos. 53, 26. Gylpword *boastful words*, Cd. 14 ; Th. 17, 23 ; Gen. 264 : Beo. Th. 1355 ; B. 675 : Byrht. Th. 139, 55 ; By. 274.

gilte, an ; *f. A* GILT, *a young sow* :—Gilte *suilla* vel *sucula*, Ælfc. Gl. 20 ; Som. 59, 34 ; Wrt. Voc. 22, 75. [ȝelte *scropha*, Wrt. Voc. 177, 7 : gilt *Hall. Dict: Icel.* gilta *a young sow* : *O. H. Ger.* galza, gelza *sucula*.]

GIM, gimm, gym, gymm ; *gen.* gimmes ; *m.* I. *a* GEM, *jewel* ; gemma :—Se stán bið blæc gym *the stone is a black gem*, Bd. 1, 1 ; S. 473, 24. Ðæt nebb líxeþ swá glæs oððe gim *the beak glitters like glass or gem*, Exon. 60 a ; Th. 218, 25 ; Ph. 300. Gim sceal on hringe standan steáp *the gem shall stand prominent in the ring*, Menol. Fox 504 ; Gn. C. 22 : Salm. Kmbl 570 ; Sal. 284. Gimmas líxton *jewels glittered*, Elen. Kmbl. 180 ; El. 90. Seó gesomnung ðara deórwyrþra gimma *the collection of the precious gems*, Blickl. Homl. 99, 28. Se ðe wæs gescríd mid golde and mid gimmum *he that was clad with gold and with gems*, Chr. 1086 ; Erl. 221, 3 : Cd. 227 ; Th. 305, 20 ; Sat. 649. Hí wurdon gehwyrfede to deórwurþum gimmum *they were turned to precious gems*, Homl. Th. i. 64, 5. II. *used metaphorically of the eye, the sun, stars, etc.* [cf. *Icel.* fagr-gim = *sun*] :—He his eágan ontýnde hálge heáfdes gimmas *he unclosed his eyes, the head's holy gems*, Exon. 51 b ; Th. 180, 7 ; Gú. 1276. Hluttor heofenes gim *the clear jewel of heaven, i.e. the sun*, 58 b ; Th. 210, 9 ; Ph. 183 : 63 a ; Th. 232, 33 ; Ph. 516 : Beo. Th. 4151 ; B. 2072 : Andr. Kmbl. 2538 ; An. 1270. Iunius on ðam gim astíhþ on heofenas up hyhst on geáre *June in which the gem* [*sun*] *rises in the heavens highest in the year*, Menol. Fox 216 ; Men. 109. Hálge gimmas heofontungol sunne and móna *holy gems, stars of heaven, sun and moon*, Exon. 18 a ; Th. 43, 22, 27 ; Cri. 692, 695. [*Laym.* ȝim : later MS. gim : *Icel.* [poetry] gim ; n : *O. H. Ger.* gimma ; *f.*]

gíman. v. gýman.

gimbǽre ; *adj. Gemmifer, bullifer*, Hpt. Gl. 417.

gim-cyn, gym-cyn, -cynn, es ; *n. A gem-kind, a precious stone, a gem* ; genus gemmarum, gemma :—Se forma feohgítsere gróf æfter gimcynnum *the first miser delved after precious stones*, Bt. Met. Fox 8, 114 ; Met. 8, 57 : 15, 8 ; Met. 15, 4. On ðære éðyltyrf niððas findaþ gold and gymcynn *in that country men find gold and gems*, Cd. 12 ; Th. 14, 29 ; Gen. 226 : Elen. Kmbl. 2046 ; El. 1024.

gíme-. v. gýme-.

gíming. v. gýmung.

gimmisc ; *adj. Jewelled* ; gemmeus :—Monige fatu gimmiscu *gemmea vasa*, Nar. 5, 13. [*O. H. Ger.* gimmisc *gemmarius*.]

gim-reced, es ; *m. n. A hall adorned with gems* :—Ne hí gimreced

setton searolíce *nor with art did they build palaces*, Bt. Met. Fox 8, 50 ; Met. 8, 25.

gim-rodor, es ; *m. A precious stone* ; draconites, dracontia, Cot. 63, Lye : Hpt. Gl. 431.

gim-stán, es ; *m. A gem, jewel, precious stone* :—Gimstán *gemma*, Wrt. Voc. 85, 23. Ðás gymstánas synd tocwýsede *these jewels are crushed*, Homl. Th. i. 62, 6, 13, 15, 21. Hí behwyrfdon heora áre on gymstánum *they turned their property into jewels*, 60, 28, 24. [*Laym.* ȝimston : *Icel.* gim-steinn.]

gim-wyrhta, an ; *m. A worker in gems, jeweller* :—Ðás gymwyrhtan secgaþ ðæt hí nǽfre swá deórwurþe gymstánas ne gemétton *the jewellers say that they never met with such precious jewels*, Homl. Th. i. 64, 9.

GIN, es ; *n. A gap, an opening, abyss* ; hiatus :—Gársecges gin *ocean's expanse*, Cd. 163 ; Th. 205, 3 ; Exod. 430. [*Icel.* gin *the · mouth of beasts*.]

gin ; *adj. Wide, spacious, ample* :—Beligeð úton ginne ríce *encompasseth ample realms*, Cd. 12 ; Th. 15, 7 ; Gen. 230 : 46 ; Th. 59, 2 ; Gen. 957. Eall ðes ginna grund *all this spacious earth*, Exon. 116 a ; Th. 445, 23 ; Dóm. 12 : 85 b ; Th. 321, 24 ; Vid. 51 : Beo. Th. 3106 ; B. 1551 : Judth. 9 ; Thw. 21, 1 ; Jud. 2. [Cf. *Icel.* ginn— ; and see Grmm. D. M. 297.]

gínan, ic gíne, ðú gínest, gínst, he gíneþ, gínþ, *pl.* gínaþ ; *p.* gán, *pl.* ginon ; *pp.* ginen *To yawn* ; hiare, Cot. 23. [*Icel.* gína ; *p.* gein *to yawn*.] Cf. ginian. DER be-gínan, to-.

gind. v. geond.

gin-fæst ; *adj. Very fast* or *lasting* ; firmissimus :—Onfón ginfæstum gifum *to receive very fast gifts*, Cd. 141 ; Th. 176, 28 ; Gen. 2919 ; Beo. Th. 2546 ; B. 1271 : 4370 ; B. 2182 : Exon. 68 a ; Th. 252, 24 ; Jul. 168 ; Bt. Met. Fox 20, 453 ; Met. 20, 227. [Grein renders by *amplus* ; see gin.]

ging ; *adj. Young* ; jǔvěnis :—Ic up ahóf eaforan gingne *I raised up a young offspring*, Elen. Kmbl. 706 ; El. 353 : 1746 ; El. 875. v. geong.

gingifer, gingiber, gingifere, an ; *f. Ginger* :—Gingifer *ginger*, L. M. 1, 14 ; Lchdm. ii. 56, 11 : 23 ; Lchdm. ii. 66, 3. Gingiber, Lchdm. iii. 92, 15. Ging¡fran broþ *broth of ginger*, L. M. 1, 18 ; Lchdm. ii. 62, 6. Genym gingiferan, *take ginger*, Lchdm. iii. 136, 17. [*Laym.* gingiuere. Cf. *French* gingembre : *O. French* gingibre : *Lat.* zingiber : *Gk.* ζιγγίβερις.]

gingra, an ; *m. A disciple, vassal, follower* ; discipulus, assecla :—He and his gingran awyrdaþ manna líchaman *he and his disciples injure men's bodies*, Homl. Th. i. 4, 24 : Cd. 217 ; Th. 276, 20 ; Sat. 191 : 224 ; Th. 298, 2 ; Sat. 526. His gingrum *to his disciples*, Bd. 3, 5 ; S. 526, 21. He his gingran sent *he sendeth his vassal*, Cd. 25 ; Th. 33, 5 ; Gen. 515 : 26 ; Th. 34, 32 ; Gen. 546. v. geongra.

gingre, an ; *f. A female servant, maid-servant* ; fámula :—Gingran sínre *to her maid-servant*, Judth. 11 ; Thw. 23, 21 ; Jud. 132.

ginian, geonian, gynian ; *p.* ode *To yawn, gape* :—Ic gynige hio, Ælfc. Gr. 24 ; Som. 25, 39. Gewíte scó sáwul út ne mæg se múþ clypian deáh ðe he gynige *if the soul depart the mouth cannot cry, though it gape*, Homl. Th. i. 160, 9. Mid gynigendum múþe *with gaping mouth*, ii. 176, 21 : 510. 33. Seó eorþe swá giniende bád *the earth remained gaping so*, Ors. 3, 3 ; Bos. 56, 3. [*Wick. p. pl.* ȝeneden : *O. H. Ger.* ginen, ginon hiare.] v. geonian.

ginnan. v. a-, an-, be-, on-, under-ginnan.

gínung, e ; *f. A yawning* ; hiatus, Cot. 23. [Cf. geonung.]

gin-, gynn-wísed ; *part. p. Well-directed, wise* :—Nǽnig monna wæs godes willan ðæs georn ne gynnwísed *no man was so eager for God's will nor so wise*, Exon. 45 a ; Th. 154, 8 ; Gú. 839.

gió ; *adv. Formerly, of old, before* ; quondam, olim, pridem :—Se wæs gió cyning *who was formerly king*, Bt. Met. Fox 26, 70 ; Met. 26, 35 : 28, 60 ; Met. 28, 30 : Bt. 16, 1 ; Fox 50, 7, Cot. MS: 38, 1 ; Fox 194, 3 : Elen. Kmbl. 871 ; El. 436 : Beo. Th. 5036 ; B. 2521. Æror gió *before*, Bt. Met. Fox 20, 490 ; Met. 20, 245. v. geó.

gioc, es ; *n. A yoke* ; jugum :—Ðæt swǽre gioc *the heavy yoke*, Bt. Met. Fox 10, 39 ; Met. 10, 20 : 9, 110 ; Met. 9, 55. v. geoc.

gióc. v. geóc.

giofan ; *p.* geaf, *pl.* geáfon ; *pp.* gifen *To give* ; dare :—Ne meahte se sunu Wonredes hond-slyht gifan [MS. giofan] *nor could the son of Wonred give a hand-stroke*, Beo. Th. 5937 ; B. 2972. v gifan.

giofolnes, se ; *f. Munificence, liberality* ; munificentia, Past. 44, 2 ; Swt. 321, 22 ; Hat. MS.

giofu, e ; *f. A gift, grace* ; donum, gratia :—Ðé cyning engla gefrætwode giofum *thee the king of angels adorned with gifts*, Andr. Kmbl. 3036 ; An. 1521. Ðæt wæs giofu gǽstlíc *that was a ghostly grace*, Exon. 8 b ; Th. 3, 26 ; Cri. 42. v. gifu.

giógoð, giógað *youth*. v. geóguð.

gioleca, an ; *m. A yolk* ; ovi vitellus, Bt. Met. Fox 20, 339 ; Met. 20, 170. v. geolca.

giolu. v. geolewe.

gió-man, -mann, es ; *m. A man of old* ; qui olim vixit :—Giómonna gestrión *the wealth of men of old*, Bt. Met. Fox 1, 46 ; Met. 1, 23. v. iú-man.

giómor; *adj. Sad, sorrowful*; mæstus :—Nū sceal ic wreccea giómor, singan sārcwidas *now shall I, a sad wretch, sing mournful songs*, Bt. Met. Fox 2, 6; Met. 2, 3. v. geómor.

giómor-mód; *adj. Sad of mind*; mæstus animo :—He, giómormód, giohðo mǣnde *he, sad of mind, bewailed his afflictions*, Beo. Th. 4526; B. 2267. v. geómor-mód.

giond; *prep. acc. Through, throughout, over, in*; per, in :—Waldeþ giond werþióda *he rules throughout nations*, Bt. Met. Fox 24, 70; Met. 24, 35 : 11, 126; Met. 11, 63 : 4, 74; Met. 4, 37 : 11, 89; Met. 11, 45. v. geond.

giong; *def.* se gionga; *adj. Young*; jŭvĕnis :—Se æðeling biþ giong in geardum *the noble* [*bird*] *is young in its dwelling*, Exon. 61 a; Th. 223, 5; Ph. 355 : Beo. Th. 4883; B. 2446. Se gionga cyning *the young king*, Ors. 2, 4; Bos. 45, 15. v. geong.

gióng *went*, Beo. Th. 4810, note; B. 2409; *p. of gangan*.

giongor-scipe, es; *m. Youngership, service*; juvĕnilis status, ministĕrium :—Ðæt hie his giongorscipe fyligan woldan *that they would follow his service*, Cd. 14; Th. 16, 26; Gen. 249. [O. Sax. jungar-skepi.]

giongra, an; *m. A vassal, follower, attendant*; assecla, sectātor :—Móton we hie us to giongrum habban *we may have them as our vassals*, Cd. 21; Th. 26, 16; Gen. 407. v. geongra.

giorne; *adv. Diligently*; dīligenter :—Gif ðū wilnige weorulddrihtnes heáne anwald ongitan giorne *if thou desirest diligently to behold the high power of the world's Lord*, Bt. Met. Fox 29, 5; Met. 29, 3. v. georne.

giornfulnes *earnestness*, Past. 18. 2; Hat. MS. 25 b, 21. v. geornfulnes.

giow, es; *m? A griffin*; gryps, gryphus :—Giow *gryphus*, Wrt. Voc. 62, 3. v. giw.

giowian. v. giwian.

Gipeswíc *Ipswich*, Chr. 993; Erl. 132, 4.

gipung, e; *f. Gaping*; os patulum, Gl. Prud. 991.

gird *a staff*, Ex. 4, 2. v. gyrd.

giren, girn *a snare*, Ps. Vos. 17, 6 : 24, 16 : 58, 6 : 65, 10. v. grin.

girian; *p.* ðū giredost' *To prepare*, Ps. Spl. 146, 8. v. gearwian.

girnan *to yearn, seek for, require*, Ex. 21, 22. v..gyrnan.

girran *to chatter*; garrīre. v. georran.

girwan; *p.* ede; *pp.* ed *To prepare*; pārāre :—Girwan up swǣsendo *to prepare a feast*, Judth. 9; Thw. 21, 7; Jud. 9. v. gearwian.

giscian *to sob, sigh*; singultire, Bt. 2; Fox 4, 9.

gise *yes*; immo, etiam :—Gise, lā gese *yes, O yes*, Bt. 16, 4; Fox 58, 15. v. gese.

gisel, gýsel; *gen.* gísles; *dat.* gísle; *m. A pledge, hostage*; obses :—Gýsel *obses*, Wrt. Voc. 72, 63 : Byrht. Th. 139, 36; By. 265. Būtan ānum Bryttiscum gísle *except one British hostage*, Chr. 755; Erl. 50, 8. Ecgferþ wæs to gísle geseald *Ecgfrid obses tenebatur*, Bd. 3, 24 : S. 556, 26. Ðū eádige Maria God ðē hafaþ to gísle on middangearde geseted *thou blessed Mary, God hath placed thee on earth as a surety*, Blickl. Homl. 9, 5. Hió genam ðone hene to gísle *she took the one as hostage*, Eleu. Kmbl. 1196; El. 600. He him āðas swór and gíslas salde *he swore oaths to them and gave hostages*, Chr. 874; Erl. 76, 28. Ða gyrnde he grides and gísla *then he required protection and hostages*, 1048; Erl. 180, 6. [Laym. ȝisles, pl: Icel. gísl: Dan. gidsel, gissel: Swed. gislan: Ger. geissel: O. H. Ger. kísal obses. v. Grm. R. A. 619.]

gíslian; *p.* ode, ade; *pp.* od *To give hostages or security*; obsides dare :—He gíslode and līne man ðeáhhwæðere ofslóh *he gave hostages and yet he was slain*, Chr. 1016; Erl. 154, 11. Man gíslade ða hwíle in to ðām scipum *hostages were sent to the ships during the time*, 994; Erl. 133, 29. Seó burhwaru gíslode *the town's people gave hostages*, 1013; Erl. 148, 8. Ða weasternan þægnas gíslodon *the western thanes gave hostages*, 17 : 1015; Erl. 153, 1. [Icel. gísla *to give as hostage*].

gi-sprunt. v. ge-springan.

GIST, gyst, es; *m.* YEAST, *barm, froth*; spuma cerevisiæ, Herb. 21, 6; Lchdm. i. 118, 10. Niwne gist *new yeast*, L. M. ii. 51, 1; Lchdm. ii. 266, 1. [Prompt. Parv. ȝeest spuma.]

gist, es; *m. A guest*:—Fundode gist of geardum *the guest hastened from the dwellings*, Beo. Th. 2280; B. 1138 : 3049; B. 1522 : Cd. 113; Th. 149, 9; Gen. 2472 : 115; Th. 150, 20; Gen. 2494. v. gæst.

gist *a storm*. v. yst.

gist-. v. gæst-, gest-.

gist-líðe; *adj. Kind to guests, hospitable*; hospes :—Būton cræft mín gistlíðe him beó *nisi ars mea hospita ei fuĕrit*, Coll. Monast. Th. 28, 11 : Shrn. 129, 26.

gist-mægen, es; *n. A force composed of guests*:—Ðær frome wǣron godes spellbodan hæfde gistmægen strengeo *there were bold messengers of God, the band of guests* [*the angels visiting Lot*] *had strength*, Cd. 115; Th 150, 20; Gen. 2494.

git, gyt; *nom. You two*, vos duo, σφῶϊ, σφώ; *gen.* incer *of you two*, vestrûm duorum, σφῶϊν, σφῷν; *dat.* inc *to you two*, vobis duobus, σφῶϊν, σφῷν; *acc.* inc incit *you two*, vos duos, σφῶϊ, σφώ; *personal pron. dual of ðū thou* :—Gif git ðæt fæsten fýre willaþ forstandan *if you two will protect that fastness from fire*, Cd. 117; Th. 152, 16; Gen. 2521. Git

me freóndscipe cýðaþ *you two will shew friendship to me*, 117; Th. 152, 3; Gen. 2514. Gýt nyton hwæt gyt biddaþ. Mǣge gyt drincan ðone calic ðe ic to drincenne hæbbe? Ða cwædon hig, Wyt māgon [vos duo] *nescitis quid* [vos duo] *petatis. Potestis* [vos duo] *bibere calicem quem ego bibiturus sum? Dicunt ei*, [nos duo] *possumus*, Mt. Bos. 20, 22. Hwæt.wylle gyt ðæt ic inc dó *quid vultis* [vos duo] *ut faciam vobis* [duobus]? 20, 32. Gelýfe gyt ðæt ic inc mæg gehǣlan [vos duo] *creditis, quia hoc possum facere vobis* [duobus]? 9, 28. Incer twega *of you two*; vestrûm duorum, Exon. 123 b; Th. 475, 14; Bo. 47. Ne gehwæðer incer *nor either of you two*, Beo. Th. 1173; B. 584. Sý incit *fiat vobis* [duobus], Mt. Bos. 9, 29. Restaþ incit hér *rest your two selves here*. Cd. 139; Th. 174, 19; Gen. 2880. Git Iohannis *thou and John*, Exon. 121 b; Th. 467, 7; Hö. 135. [Laym. ȝit : Orm. ȝitt : O. Sax. git; *dat. acc.* inc : Goth. *gen.* igkwara; *dat. acc.* igkwis : Icel. it; *gen.* ykkar; *dat. acc.* ykkr.]

git, giet, get, gyt; *adv. Still, yet* :—Hér mon mæg giet gesión hiora swæþ *their track may still be seen here*, Past. pref; Swt. 5, 15; Hat. MS. Be ðiosum git is swíðe ryhtlíce gecweden to ðæm wítegan *about which further is very rightly said to the prophet*, Swt. 162, 22; Cot. MS. And git hit is mâre and eác manigfealdre ðæt dereþ ðisse þeóde *and yet there are greater and more manifold things that hurt this people*, Swt. A. S. Rdr. 108, 106. Gyf heó gyt lyfaþ *if she yet lives*, Beo. Th. 1893; B. 944. Metod eallum weóld gumena cynnes swā he nū git dēþ *the Lord ruled all of the race of men as he yet does now*, 2121; B. 1058. He nyste ne ic ðā git *he did not know, nor I as yet*, Pref. Ælfc. Thw. 2, 2 : Gen. 8, 8 : Beo. Th. 1077; B. 536. Ðā gyt, Cd. 6; Th. 7, 35; Gen. 1160. Ðā giet, 63; Th. 75, 25; Gen. 1243. He abád ðā git ōðre seofon dagas *he waited then yet other seven days*, Gen. 8, 10. Abraham cwæþ ðā git *Abraham said further*, 18, 29. He sende to eallum ðām cynegum ðe cuce ðā git wǣron *he sent to all the kings that were still alive*, Jos. 11, 1 : Homl. Th. i. 72, 9. Ðā get ic furðor gefregen *I yet further learned*, Cd. 218; Th. 278, 21; Sat. 225. Ðā git hét, Chr. 921; Erl. 108, 3. Alwalda ðec góde forgylde swā he nū gyt dyde *may the Almighty repay thee with good as he has done until now*, Beo. Th. 1917; B. 956. Ā ic ðæt heóld nū giet *I have ever held that until now*, Exon. 120 b; Th. 463, 21; Hö. 73. Ic wille mid giddum get gecýðan hū *I will further make known in songs how* . . . , Bt. Met. Fox 13, 2; Met. 13, 1. Gif giet lǣst mína lâra *if even now he obey my counsels*, Cd. 29; Th. 39, 2; Gen. 618. Ne wæs ðā giet wiht geworden *there was as yet nothing made*, 5; Th. 7, 8; Gen. 103. Nǣfre git *never yet*, Beo. Th. 1171; B. 583. Swýðor gyt *yet more*, Judth. 11; Thw. 24, 11; Jud. 182.

gita, gieta, geta, gyta; *adv. Yet, still* :—Dóþ gieta swā *yet do they so*, Cd. 48; Th. 61, 7; Gen. 993. Gita *yet*, Bt. Met. Fox 23, 13; Met. 23, 7. Ne wearþ wæl mâre æfer gieta folces gefylled *never yet was greater slaughter of people made*, Chr. 937; Erl. 115, 15; Ædelst. 66 : Cd. 113; Th. 148, 34; Gen. 2466. Reord wæs ðā gieta eorþbüendum ân geméne *there was as yet one speech common to dwellers on earth*, 79; Th. 98, 25; Gen. 1635. Hiora nǣnig næs ðā gieta *any yet none of them existed*, Bt. Met. Fox 8, 24; Met. 8, 12. [O. Frs. jeta.]

GITAN, ic gite, gyte, giete, ðū gitst, he git, *pl.* gitaþ gytaþ, gietaþ; *p.* geat, *pl.* geáton; *pp.* giten *To GET, take, obtain*; adipisci, capere, assequi. Only found in the following compounds :—a-gitan, an-, and-, be-, bi-, for-, ofer-, on-, under- : and-git; andgit, -ful, -fullíce, -ol, -tácen : forgitol, ofergitol, -nes : ongitful, -nis : [O. Sax. -getan; bi-getan *invenire*, assequi, far-getan, for-getan *perdere e memoria. oblivisci* : O. Frs. jeta : for-jeta *oblivisci* : O. H. Ger. gezan *adipisci* : Goth. -gitan; *p.* -gat, *pl.* -gētum; *pp.* -gitans *adipisci* : O. Nrs. geta *adipisci, assequi, gignere, dare, præbere*.]

gipcorn, es; *n. Spurge laurel* :—Ðeós wyrt ðe man lactyridem and ōðrum naman gipcorn nemneþ *this plant which is called lacterida and by another name githcorn*, Herb. 113 : Lchdm. i. 226, 12 : L. M. ii. 65, 1; Lchdm. ii. 292, 9 : v. glossary. [Hall. Dict. gith *corn-cockle* : Palladius on Husbandrie gith *cockle*, x. 155.]

giprife, gitrife, an; *f. Cockle*; agrostemma githago :—Gyþrife, L. M. i. 38, 4; Lchdm. ii. 24, 27. Gíþrife, 5; Lchdm. ii. 92, 27. Genim gitrifan, 1, 5; Lchdm. ii. 18, 23.

gítsere, es; *m. An avaricious, a covetous person, miser* :—Ða ðe wǣron gítsaras *qui erant avari*, Lk. Skt. Lind. 16, 14. Se ungesǣliga gýtsere wile mâre habban ðonne him genihtsumaþ *the miserable covetous man wants to have more than suffices him*, Homl. Th. i. 64, 33, 35 : Bt. 16, 3; Fox 56, 16. He wæs se wyresta gítsere ðe he gesealde wið feó heofcones hláford *he was the worst covetous man because he sold for money the lord of heaven*, Blickl. Homl. 69, 13, 10. Gítseras ðe on mannum heora ǣhta on wóh nimaþ *covetous men who take their property from men wrongfully*, 61, 21. [A. R. ȝissare : M. H. Ger. gite-sære.]

gítsian; *p.* ode; *pp.* od *To covet, desire* :—Ða ðe ðæs welan gítsiaþ hí biþ symle wǣdlan on hyra móde *those who covet* [*worldly*] *wealth will ever be poor in their mind*, Prov. Kmbl. 50. Gýtsaþ *covets*, Beo. Th. 3502; B. 1749. Fóþres ne gítsaþ *it craves not food*, Exon. 114 b;

Th. 440, 1; Rä. 51, 11 : Bt. 26, 2 ; Fox 92, 17. Ðá ðú gítsiende blǽda náme *when thou coveting didst take the fruit*, Cd. 42 ; Th. 55, 7 ; Gen. 890. Ðú gítsigenda and ðú welega *thou covetous and wealthy man*, Blickl. Homl. 51, 1. Gýtsiendre heortan *insatiabili corde*, Ps. Spl. 100, 6. Mid gítsigendum eágum *with covetous eyes*, Homl. Th. i. 68, 26. Gýtsian *concupiscere*, Ps. Spl. 61, 10. [*A. R.* ȝiscen : *M. H. Ger.* gitsen.]

gítsung, e ; *f. Covetousness, avarice, cupidity, desire :*—Ða ðe ne sécaþ heora ágen gestreón þurh gýtsunge *those who do not seek their own gain through covetousness*, Homl. Th. ii. 74, 34. Se þrydda heáfodleahter is gýtsung *the third chief sin is avarice*, 218, 21 : 592, 6. Hí ongunnan gítsunge begán *concupierunt concupiscentias*, Ps. Th. 105, 12. Gítsung *avaritia*, 118, 36 : Mk. Skt. 7, 22. Þurh ða ungefyldan gítsunge woruldmonna *through the unsatisfied covetousness of worldly men*, Bt. 7, 3 ; Fox 20, 26. Grundleás gítsung gilpes and ǽhta *the boundless desire for glory and possessions*, Bt. Met. Fox 7, 29 ; Met. 7, 15 : Bt. 16, 3 ; Fox 56, 2. Nales he giémde þurh gítsunga lǽnes liíwelan *he cared not from covetousness for the frail wealth of this world*, Exon. 34 b ; Th. 111, 4 ; Gú. 121. Þurh his ágene gítsunga he ǽfre ðas leóde mid ungylde tyrwigende wæs *through his own avarice he was ever harassing this nation with bad taxes*, Chr. 1100 ; Erl. 236, 1 : 1086 ; Erl. 222, 24. From ðisse worulde gítsungum *from the desires of this world*, Blickl. Homl. 57, 23. [*Laym.* ȝitsung : *Orm.* ȝittsunng : *A. R.* ȝissung.]

giú. v. geó.

giuan. v. giwian.

Giúl *Yule, Christmas*. v. geól.

giung ; *def.* se giunga ; *adj. Young, youthful* ; *jŭvĕnis, adolescens :*—Wæs sum giung mon *ĕrat quidam adolescens*, Bd. 4, 32 ; S. 611, 17. Se giunga *the young man*, Cd. 224 ; Th. 297, 3 ; Sat. 511. Ic ðé giungne underféng *I took thee young*, Bt. 8 ; Fox 24, 23. v. geong.

giungra, an ; *m. A junior, disciple, follower* ; *discĭpulus, assecla :*—He ðæt ríce forlét and his giungrum bebeád *ipse relicto regno ac jŭvĕniōrĭbus commendáto*, Bd. 5, 7 ; S. 621, 10. v. geongra.

giw, giow, eow, es ; *m? A griffin, a four-footed bird* ; gryps = γρύψ, gríphus :—Giw *gríphus*, Wrt. Voc. 280, 5.

giwian, giowian, giwan ; *p.* ode ; *pp.* de *To ask* ; *petere, postulare :*—Wǽlde giwiga ɫ giuiade *postullasset*, Mt. Kmbl. Lind 17, 7. Giuiga *petere*, Rtl. 179, 34. Ic giuge wælle *petam*, Mk. Skt. Lind. 6, 24. Huu giues ðú *quomodo poscis*, Ju. Skt. Lind. 4, 9. Se ðe giuæþ *qui petit*, Mt. Kmbl. Lind. 7, 8. We giugaþ *poscimus*, Rtl. 52, 10. Giude *mendicans*, Mk. Skt. p. 4, 16. Ðæt hia giudon *ut peterent*, Mt. Kmbl. 27, 20. Giwig pete, Mk. Skt. Lind. [Rush. giowa] 6, 22. Giwas *petite*, Mt. Kmbl. Lind. 7, 7. Giuwende *petentes*, Mk. Skt. p. 4, 14. Giuendo *postulata*, 18. Giuiendum *petentibus*, Mt. Kmbl. Lind. 7, 11.

giwung, e ; *f. An asking, a petition* ; *postulatio*, petitio :—Fífo giunga *quinque petitionum*, Lk. Skt. p. 7, 2. Giwunges, Rtl. 39, 23.

glad. v. glæd.

glád *glided, slid*, Beo. Th. 4152 ; B. 2073 ; *p. of* glídan.

Glademúð *Gledmouth.* v. Cledemúð.

gladian ; *p.* ode. **I.** *to be glad* ; *exultare :*—Ða gladia worhtest *quos lætari fecisti*, Rtl. 94, 15. Ða ðe gedréfaþ me gladiaþ *qui tribulant me exultabunt*, Ps. Lamb. 12, 5. Abraham gladade ɫ glæd wæs *Abraham gavisus, est* ; *Wick.* Abraham gladide, Jn. Skt. Rush. 8, 56. Glada and blissa *be glad and rejoice*, Apol. Th. 7, 2. Ne gladige on ðæt cyning *let no king rejoice at that*, Lchdm. iii. 442, 35. **II.** *to make glad :*—Ic gladige *gratificor*, Ælfc. Gr. 37 ; Som. 39, 3. Drihten mid to gladienne *to make glad the Lord therewith*, Lev. 1, 3. Gladigan *demulcere*, Hpt. Gl. 476. [*Icel.* gleðja *to gladden* ; gleðjask *to become bright, glad.*]

glæd, es ; *n. Gladness, joy :*—Swá missenlíce meahtig dryhten eallum dǽleþ sumum earfeþa dǽl sumum geógoþe glæd *thus diversely does the mighty Lord allot to all, to one a share of troubles, to one the gladness of youth*, Exon. 88 a ; Th. 331, 14 ; Vy. 68. *Perhaps here the form given by Lye ǽr sun gó to glade*, v. Grm. D. M. 702-3. [*Cf. Icel.* gleði ; *f: Dan.* glæde *gladness, merriment :* and *A. R.* gledful.]

GLÆD ; *adj.* **I.** *shining, bright :*—Glæd mid golde *bright with gold*, Exon. 125 a ; Th. 480, 16 ; Rä. 64, 3. Wyrþ heó ungladu ðeáh heó ǽr gladu wǽre on to lócienne *it [the sea] becomes turbid though before it was bright to look at* [cf. glæshlutru on to seónne, 24], *and the Latin sordida visibus obstat*], Bt. 6 ; Fox 14, 26 : Bt. Met. Fox 5, 21 ; Met. 5, 11. Godes condelle glædum gimme *God's candle, the bright jewel [the sun]*, Exon. 57 a ; Th. 204, 3 ; Ph. 92 : 64 b ; Th. 237, 20 ; Ph. 593. Glæd seolfor *shining silver*, Cd. 129 ; Th. 164, 24 ; Gen. 2719. Óðer biþ golde glædra óðer biþ grundum sweartra *one is brighter than gold, the other darker than the depths*, Salm. Kmbl. 975 ; Sal. 488. Gimma gladost *brightest of jewels*, Exon. 60 a ; Th. 218, 3 ; Ph. 289. **II.** *glad, cheerful, joyous, bright :*—Ðá wærþ he swíðe glæd *then he was very glad*, Chr. 656 ; Erl. 30, 20. Glæd wæs *gavisus est*, Jn. Skt. Lind. 8, 56. Wosaþ glæd *exultate*, Lk. Skt. Lind. 6, 23. Glædman *hilaris*, Ælfc. Gl. 88 ; Som. 74, 87 ; Wrt. Voc. 50, 67. Ǽfre he biþ ánes módes and glæd þurhwunaþ *he is ever of one mind and continues cheerful*, Homl. Th. i. 456, 25 ; 72, 27. He wearþ glæd on

his ansýne *he was bright of face*, Guthl. 2 ; Gdwin. 12, 20. Wínes glæd *merry with wine*, Exon. 117 a ; Th. 449, 28 ; Dóm. 78. Glæd gumena weorud *a joyous band of men*, 32 a ; Th. 101, 5 ; Cri. 1654. Nolde gladu ǽfre syððan ætýwan, *she, joyous, would not ever afterwards appear*, Cd. 72 ; Th. 89, 14 ; Gen. 1480. Iacob byþ on glædum sǽlum *exultabit Jacob*, Ps. Th. 52, 8. Sefa wæs ðé glædra *her mind was the gladder*, Elen. Kmbl. 1909 ; El. 956. **III.** *pleasant, kind, mild, courteous :*—Glæd man *jucundus homo*, Ps. Th. 111, 5. Glade fǽmnan *virgines*, 148, 12. Glædman Hróþgár *courteous Hrothgar*, Beo. Th. 740 ; B. 367. Beó wið Geátas glæd geofena gemyndig *be kind to the Gauts, mindful of gifts*, 2350 ; B. 1173 : 1730 ; B. 863. Mín Drihten hine gedó glædne wiþ eówne *may my Lord make him kind towards you*, Gen. 43, 14. Ðæt we ðone Hǽlend hæbben us glædne *that we may have the Saviour propitious to us*, Th. Chart. 240, 26 : Exon. 12 b ; Th. 20, 10 ; Cri. 315. [*Icel.* glaðr *bright, glad : Dan.* glad *glad : O. H. Ger.* glat *limpidus, candidus : Ger.* glatt.]

glædene, an ; *Gladden.* v. Lchdm. ii. Glossary.

glædlíc ; *adj. Bright, pleasant, kind :*—Scíneþ ðé leóht glædlíc ongeán *the light shineth bright over against thee*, Cd. 20 ; Th. 38, 31 ; Gen. 615. Hú glædlíc biþ and gód swylce *quam bonum et quam jucundum*, Ps. 132, 1. Me gúþhere forgeaf glædlícne máþþum *Guthhere gave me a splendid jewel*, Exon. 85 b ; Th. 322, 31 ; Víd. 66.

glædlíce ; *adv. Gladly, pleasantly, kindly, cheerfully :*—He glædlíce fram heom eallum onfangen wæs *he was gladly received by them all*, Chr. 1014 ; Erl. 150, 17. He frǽfrode hig and spræc glædlíce *he comforted them and spake kindly* [unto them], Gen. 50, 21. He glædlíce all eorþlíc þing wæs oferhleápende *alacriter terrena quæque transiliens*, Bd. 2, 7 ; S. 509, 13. Nú ðú ðus rótlíce and ðus glædlíce to us sprecende eart *qui tam hilariter nobiscum loqueris*, 4, 24 ; S. 598, 38 : Cd. 109 ; Th. 143, 18 ; Gen. 2381.

glædman, Beo. Th. 740 ; B. 367. Thorpe and Kemble take this word as the oblique case of a noun = *gladness, pleasure* ; but see 'glæd.'

glæd-mód ; *adj. Glad-minded, cheerful, of good cheer, joyous, pleasant, kind, courteous :*—Glædmód wes ðú *animæquior esto*, Mk. Skt. Rush. 10, 49. Geát wæs glædmód *the Gaut was glad of mind*, Beo. Th. 3574 ; B. 1785 : Exon. 62 b ; Th. 229, 28 ; Ph. 462 : Andr. Kmbl. 2119 ; An. 1061. Guman glædmóde god wurðedon *the men with cheerful mind worshipped God*, Cd. 187 ; Th. 232, 14 ; Dan. 260. Gongaþ glædmóde go *with gladsome mind*, Exon. 16 a ; Th. 36, 14 ; Cri. 576. He biþ ðám gódum glædmód on gesihþe *he shall be to the good pleasant of countenance*, 21 a ; Th. 56, 36 ; Cri. 911. Glædmód *kind*, 48 a ; Th. 165, 27 ; Gú. 1035. [*O. Sax.* glad-mód.] v. glæd.

glædmódnes, se ; *f. Gladness, cheerfulness, joyfulness, kindness :*—Ac ðonne ðæt mennisce mód Godes glædmódnesse mid gódum weorcum ne geandsworaþ *sed cum largientem Deum humana mens boni operis responsione non sequitur*, Past. 50, 3 ; Swt. 391, 6.

glædnes, se ; *f. Gladness, joy, cheerfulness :*—Ongan se bisceop lustfullian glædnesse his dǽda *delectabatur antistes alacritate actionis*, Bd. 5, 19 ; S. 637, 47. Glædnisse miclo *gaudio magno*, Mt. Kmbl. 2, 10 : 13, 20 : 25, 21. Glædniso *lætitia*, Rtl. 57, 2.

glædscipe, es ; *m. Gladness, joy :*—Crist is mid ealles módes gledscype to herienne *Christ is to be praised with joy of all the mind*, Lchdm. iii. 436, 19. Glædscip mín *gaudium meum*, Jn. Skt. Rush. 3, 29. [*Laym.* gladscipe : *Orm.* gladdshipe : *A. R.* gledschipe.]

glædsted. v. glédstede.

glǽm, es ; *m. Brightness, splendour, radiance :*—Se æðela glǽm *the noble brightness [the sun]*, Exon. 51 b ; Th. 178, 31 ; Gú. 1252 : Th. 179, 18 ; Gú. 1263. Sunnan glǽm *the sun's radiance*, 59 b : Th. 215, 15 ; Ph. 253. Mín se swétesta sunnan scíma hwæt ðú glǽm hafast *my sweetest sunshine ah! thou hast radiant beauty*, 68 a ; Th. 252, 23 ; Jul. 167. Ðe ofóíþ glǽmes gréne folde *the green earth shall deny thee her beauty*, Cd. 48 ; Th. 62, 22 ; Gen. 1018. [*O. H. Ger.* gleimo *nitor.*]

glǽr, es ; *n. Amber* ; electrum, succinum, Ælfc. Gl. 51 ; Som. 66, 6 ; Wrt. Voc. 34, 66 : Wrt. Voc. 286, 68 : [*Cf. Icel.* gler *glass* ; and see Grm. Gesch. D. S. 499.]

GLÆS, es ; *n. Glass :*—Glæs *vitrum*. Ælfc. Gl. 51 ; Som. 66, 5 ; Wrt. Voc. 34, 65. Beorhtre ðonne glæs *brighter than glass*, Homl. Th. ii. 518, 10. Ðæt scíre glæs *the clear glass*, Exon. 26 b ; Th. 78, 33 ; Cri. 1283. Ðæt nebb lǽxeþ swá glæs oððe gim *the beak glitters like glass or gem*, 60 a ; Th. 218, 25 ; Ph. 300. Biþ ðonne se fléschoma ascýred swá glæs *then shall the body be as transparent as glass*, Blickl. Homl. 109, 36. Of glǽse geworht *made of glass*, 127, 33. Mid glase geworht *wrought with glass* ; comptos vitro parietes, Bt. 5, 1 ; Fox 10, 16. [*O. H. Ger.* glas, clas *vitrum, electrum : Icel.* gler.]

glǽsen ; *adj. Made of glass, glassy, grey* ; vitreus :—Glǽsen *vitreus*, Ælfc. Gr. 5 ; Som. 4, 60. Ðǽr is ahangen sum glǽsen fæt *there is hung a glass vessel*, Homl. Th. i. 510, 1 : ii. 158, 16 ; Blickl. Homl. 209, 4, 7. Hí toslógon his glæsenne calic *they broke his glass chalice*, Shrn. 114, 25. Sǽ glǽsen *mare vitreum*, Mt. Kmbl. p 10, 3. [*Piers P.* glasen : *Prompt. Parv.* glasyne : *O. H. Ger.* glesin.]

glæs-fæt, es; *n. A glass vessel, a glass :*—He sende him glæsfæt full wînes *misit ei calicem vini,* Bd. 5, 5; S. 618, 12. [*Laym.* glæsfat: *O. H. Ger.* glasfaz.]

glæs-hluttor; *adj. Clear as glass :*—Ða sǽ ðe ǽr wæs glæshlutru *the sea that before was clear as glass,* Bt. 6; Fox 14, 24: Bt. Met. Fox 5, 15; Met. 5, 8. Ís glisnaþ glæshluttur *ice glistens clear as glass,* Runic pm. Kmbl. 341, 16; Rún. 11.

Glæstinga-burh; *gen.* burge; *dat.* byrig; Glestinga-byrig, Glasting-byri; *f.* GLASTONBURY, *Somerset :*—He getymbrade ðæt menster æt Glæstingabyrig *he built the monastery at Glastonbury,* Chr. 688; Erl. 42, note.

glæterian *to glitter, shine;* splendescere, Hpt. Gl. 419.

glǽw. v. gleáw.

glappe, an; *f. Buckbean* [?], Lchdm. i. 398, 9: iii. 292, 7.

glas. v. glæs.

glauwnes. v. gleáwnes.

GLEÁM, es; *m. A joyous noise, jubilation, joy :*—Hæfdon gleám and dreám engla þreátas *the hosts of angels had joy and delight,* Cd. 1; Th. 2, 1; Gen. 12. [*Icel.* glaumr; *m. a merry noise, merriment, joy;* gleym-ask *to be merry.*]

GLEÁW; *adj. Clear-sighted, wise, skilful, sagacious, prudent, good;* sagax, prudens, astutus, sapiens, gnarus :—Gleáw *expertus,* i. e. *multum peritus,* Ælfc. Gl. 18; Som. 58, 121; Wrt. Voc. 22, 35. Gleáw *sagax vel gnarus,* Wrt. Voc. 76, 9. Gleáw þeów *servus prudens,* Mk. Skt. 24, 45: 25, 2, 4. Ic gehírde secgan ðæt ðú wǽre gleáw ðǽron *I heard say that thou wast skilled therein,* Gen. 41, 15. Sumne wîsne man and glǽwne *a discreet and wise man,* 33. Ða ongan he mid gleáwe móde þencean and smeágean *cœpitque sagaci animo conjicere,* Bd. 3, 10; S. 534, 20. Nis nǽnig swá gleáw *there is none so skilful,* Cd. 221; Th. 286, 10; Sat. 350: Exon. 11 a; Th. 14, 17; Cri. 220: 120 b; Th. 463, 27; Hö. 76: Andr. Kmbl. 2992; An. 1499. Sum biþ leóþa gleáw *one is skilled in songs,* Exon. 79 a; Th. 296, 16; Crä. 52: 79 b; Th. 298, 33; Crä. 94: Bt. Met. Fox 1, 103; Met. 1, 52. Æcraftig gleáw geþances *cunning in the law, wise of thought,* Cd. 212; Th. 262, 13; Dan. 743. Swá him se gleáwa bebeád Gregorius *as the wise Gregory commanded him,* Menol. Fox 198; Men. 100. þurh gleáwne geþanc *by skilful thought,* Cd. 52; Th. 66, 3; Gen. 1078: Ps. Th. 67, 12: Elen. Kmbl. 1185; El. 594. Ic andette ěcne Drihten ðæne goodan God forðan ic hine gleáwne wát *confitemini Domino quoniam bonus,* Ps. Th. 106, 1: 117, 1. Ioseph se ðe gingst wæs hys gebróðra and eác gleáwra ofer hí ealle *Joseph who was youngest of his brethren and wise beyond them all,* Ors. 1, 5; Bos. 28, 8. He wæs on ðám dagum gleáwast to wîge *he was in those days the most expert man in war,* 4, 1; Bos. 77, 8. On gecynde se gleáwesta man *vir natura sagacissimus,* Bd. 2, 9; S. 512, 13. Hwilc ðǽre geógoþe gleáwost wǽre *which of the youth were most skilful,* Cd. 176; Th. 221, 1; Dan. 81. [*Laym.* glæuest *most skilful: O. Sax.* glau: *Goth.* glaggwus *diligent: Icel.* glöggr: *Scot.* gleg *quick of perception: O. H. Ger.* glaw: *Ger.* glau.]

Gleáw-ceaster, Gleáwan-ceaster, Gléu-cester, Glǽw-cæster, Glǽw-cester, Glóu-cester, Glówe-ceaster; *gen. dat.* -ceastre GLOUCESTER, *a county town in the west of England :*—Æþelflæd líð binnan Gleáwceastre *Ethelfleda lies buried at Gloucester,* Chr. 918; Erl. 109, 7.

gleáwe; *adv. Wisely, prudently, well :*—Efne me God gleáwe fultumeþ *ecce Deus adjuvat me,* Ps. Th. 53, 4. Ðæt byþ secga gehwam snytru on frymðe, ðæt he Godes egesan gleáwe healde *initium sapientiæ timor Domini,* 110, 7: 142, 11. Gleáwast, 118, 99.

Gleáwe-cestre-scir *Gloucestershire,* Chr. 1122; Erl. 249, 15.

gleáw-ferhþ; *adj. Of a wise mind, sagacious :*—Gleáwferhþ hæleþ *a man wise of mind,* Cd. 57; Th. 70, 12; Gen. 1152: 112; Th. 147, 27; Gen. 2446.

gleáw-hycgende; *adj. Thinking wisely :*— Gif ðú onsecgan nelt gleáwhycgende *if thou, wisely considering, wilt not sacrifice,* Exon. 69 a; Th. 257, 24; Jul. 252.

gleáw-hýdig; *adj. Wise of thought, heedful, prudent, sagacious :*—Gleáwhýdig wíf *the woman wise of thought,* Judth. 11; Thw. 23, 30; Jud. 148: Elen. Kmbl. 1866; El. 935. Glæd man gleáwhýdig seteþ sóðne dóm þurh his sylfes word *jucundus homo disponet sermones suos in judicio,* Ps. Th. 111, 5.

gleáwlíc; *adj. Wise, wary, astute :*—On sprǽcum gleáwlíce *in loquelis astuti,* Coll. Monast. Th. 32, 29.

gleáwlíce; *adv. Prudently, wisely, clearly, well :*—Forþam ðe he gleáwlíce dyde *quia prudenter fecisset,* Lk Bos. 16, 8. Gleáwlíce *astute,* Blickl. Gloss. Gleoúlíce *clare,* Mk. Skt. Lind. 8, 25. Ða ðæra bæcistra ealdor gehírde hú glǽwlíce he ðæt swefen rehte *when the chief baker heard how well he explained the dream,* Gen. 40, 16: Exon. 9 b; Th. 9, 6; Cri. 130: 27 a; Th. 81, 24; Cri. 1328: Andr. Kmbl. 853; An. 427: Elen. Kmbl. 377; El. 189. Ic mîne sáwle wylle full gleáwlíce Gode underþeódan *nonne Deo subdita erit anima mea?* Ps. Th. 61. 1. Ic gewitnesse wîse ðîne ongeat gleáwlíce *initio cognovi de testimoniis tuis,* 118, 152: 106, 42: Andr. Kmbl. 1721: An. 863.

gleáw-mód; *adj. Of wise mind :*—Fród guma gleáwmod *a wise man*

---

†sagacious *in mind,* Exon. 64 a; Th. 236, 8; Ph. 571: 47 a; Th. 162, 13; Gú. 975: Andr. Kmbl. 3156; An. 1581: Cd. 193; Th. 243, 22; Dan. 440.

gleáwnes, glauwnes, se; *f. Prudence, skill, wisdom, ability, sagacity, acuteness :*—Gleáwnys *argutiæ,* Ælfc. Gl. 115; Som. 80, 48; Wrt. Voc. 61, 26. Gleáwnysse *prudentiam,* Ps. Spl. 48, 3: 104, 20. He hæfde ða gleáwnysse Godes beboda to healdenne and to lǽranne *industriam faciendi et docendi mandata cælestia,* Bd. 3, 17; S. 545, 9. Twegen geonge æðelingas mycelre glauwnesse men of Angelþeóde *duo juvenes magnæ indolis, de nobilibus Anglorum,* 3, 27; S. 558, 29. Wer well gelǽred and scearpre gleáwnysse *vir doctissimus atque excellentis ingenii,* 4, 23; S. 594, 35. Þeód is búton geþeahte and bútan glǽwnisse *the nation is void of counsel and of understanding,* Deut. 32, 28. Beheald ðas sunnan mid gleáwnysse *behold this sun intelligently,* Homl. Th. i. 284, 34. Seó orþonce gláunes *the ingenious skill,* Blickl. Homl. 99, 31. Mid gleáwnesse feónd oferfeohtaþ *with prudence they overcome the fiend,* Exon. 44 a; Th. 150, 6; Gú. 774: Elen. Kmbl. 1920; El. 962.

gleáwscipe, es; *m. Sagacity, wisdom :*—To rihtwîsra gleáwscype *ad prudentiam justorum,* Lk. Skt. 1, 17: 2, 47.

GLÉD, e; *f. Burning coal. live coal, gleed, ember, fire, flame;* pruna, carbo, flamna :—Gléd *pruna,* Ælfc. Gl. 30; Som. 61, 75; Wrt. Voc. 27, 5: 82, 53. Gléda fýres *carbones ignis,* Ps. Spl. 17, 14: Ps. Th. 17, 12. Swá rícels byþ ðonne hit gléda bærnaþ *sicut incensum,* 140, 2: 119, 4. Ða þegnas stódon æt ðám glédon *stabunt ministri ad prunas,* Jn. Skt. 18, 18: 21, 9. Gloedo *scintillæ,* Rtl. 86, 34. Me is leófre ðæt mînne líchaman gléd fǽdmie *I would rather that fire should embrace my body,* Beo. Th. 5298; B. 2652: 6220; B. 3114: Exon. 87 b; Th. 330. 4; Vy. 46: 108 a; Th. 412, 23; Rä. 31, 4. Goldfrætwe gléda forswelgaþ *flames shall devour the gold ornaments,* 22 b; Th. 62, 4; Cri. 996. Biþ eal ðes ginne grund gléda gefylled *all this spacious earth shall be filled with gleeds,* 116 a; Th. 445, 24; Dóm. 12: Elen. Kmbl. 2601; El. 1302. Glédum spíwan *to spit forth flames,* Beo. Th. 4614; B. 2312: 4659; B. 2335. [*O. Frs.* gled: *Icel.* glóð: *f. red-hot embers: O. H. Ger.* gluot *pruna: Ger.* gluth: *and cf. O. Sax.* glód-welo.]

gleddian; *p.* ode *To spatter :*—Gledda, Lchdm. iii. 292, 14.

gléd-egesa, an; *m. Terror caused by fire,* Beo. Th. 5293; B. 2650.

gléd-fæt, es; *n. A fire-vat, chafing-dish :*—Dó gléda an glédfæt *put live coals in a chafing dish,* L. M. 3, 62; Lchdm. ii. 346, 3.

gléd-stede, es; *m. A place for a fire, an altar :*—On ðam glédstyde *at the altar,* Cd. 86; Th. 108, 22; Gen. 1810. On ðæm glédstede gild onsægde *made an offering on the altar,* 137; Th. 172, 10; Gen. 2842.

glemm *a spot, blemish;* macula, Off. Reg. 15, Lye. [Cf. glam *a wound, sore,* Halliwell; *and see* heaðu-glem.]

glenc, glencg. v. glenge.

glendran *to devour, swallow;* devorare :—Monn glendrende † swelgande *homo vorax,* Mt. Kmbl. Rush. 11, 19. Olbendu glendrende *camelum glutienies,* 23, 24. v. for-glendran.

gleng, e; *f. An ornament, honour;* ornamentum, decus :—Gifu gumena byþ gleng *gift is an ornament of men,* Runic pm. Kmbl. 340, 24; Rún. 7. Alege nú ðíne glenga *now put off thine ornaments,* Ex. 33, 5, 6. Gebyrde hine gesihþ glæncge getácnaþ *if he sees himself bearded, it betokens honour,* Lchdm. iii. 200, 5.

glengan, glengcan; *p.* de; *pp.* ed; *v. a. To adorn, trim, deck, compose, set in order;* ornare :—Þeódnes cynegold sóþfæstra gehwone glengeþ *the prince's crown shall adorn each of the just,* Exon. 64 b; Th. 238, 19; Ph. 606. Glengdon heora leóhtfatu *ornaverunt lampades suas,* Mt. Skt. 25, 7: Exon. 94 a; Th. 353, 14; Reim. 12. Glenged *adorned,* 352. 30; Reim. 3. Glengede word *composita verba,* Lye.

glenge, es; *m. An ornament :*—Hwǽr beóþ ðonne ða glengeas and ða mycclan gegyrelan ðe he ðone líchoman ǽr mid frætwode *where shall then be the ornaments and the grand apparel with which he before decked his body?* Blickl. Homl. 111, 35. Glengas, 99, 24, 19: 115, 2. Gesih ðás glencas *vide has ædificationes,* Mk. Skt. Lind. 13, 2.

glengista [?] :—To ðon ðæt hwæt hwygo to ðære ongietenisse ðissa mínra þinga ðín gehts and glengista geþeóde *ut aliquid per novarum rerum cognitionem studio et ingenio possit accedere,* Nar. 1, 20.

glenglíc; *adj. Full of pomp;* pompa plenus, Cot. 154.

gleó-, glig-beám, es; *m. A glee-beam, harp;* musicum lignum, harpa :—Nis hearpan wyn, gomen gleóbeámes *there is no joy of harp, the mirth of the glee-beam,* Beo. Th. 4518; B. 2263. Sum mæg hearpan stirgan, gleóbeám grétan *one can awake the harp, touch the glee-beam,* Exon. 17 b; Th. 42, 9; Cri. 670. Gligbeám *tympanum,* Blickl. Gloss.

gleó-, glig-cræft, es; *m. Glee-craft, art of music, minstrelsy, playing;* ars musica, histrionia, mimica gesticulatio, Greg. Dial. 1, 9. [*Laym.* gleo-cræft.]

gleó-dreám, es; *m. Glee-joy, pleasure caused by music;* jubilum :—Nú se herewîsa hleahtor alegde, gamen and gleódreám *now the martial leader has ceased from laughter, sport and joy of music,* Beo. Th. 6034; B. 3021. [*Laym.* gleo-drem.]

gleó-gamen, -gomen, es; *n. Glee-pleasure, merriment, sport;* jocus, ludus jocularis. v. gleó, gamen.

gleó-hleóþriend *a glee-sounder, musician, minstrel.* v. gliw-hleóþriend.

gleó-mǽden *a glee-maiden.* v. gliew-mǽden.

gleó-man, glí-man, glii-man, gliig-man, glig-man, -mann, es; *m. A glee-man, musician, minstrel, jester, player, buffoon;* musicus, cantor, joculator, histrio, scurra, mimus, pantomimus :—Leóþ wæs asungen, gleómannes gyd *the lay was sung, the gleeman's song.* Beo. Th. 2324; B. 1160. Wera gehwylcum wíslíc word gerísaþ, gleómen gied *to every man wise words are fitting, song to the gleeman,* Exon. 91 b; Th. 344, 1; Gn. Ex. 167 : 87 a; Th. 326, 29; Wíd. 136. Gligman *mimus, jocista, scurra, pantomimus,* Ælfc. Gl. 61; Som. 68, 59, 60; Wrt. Voc. 39, 42, 43. Gligman *mimus vel scurra,* 73, 69 : *sophista, parasitus,* Hpt. Gl. 406, 483, 504 : *seductor,* Gl. Prud. Gif preóst glíman wurþe *if a priest become a gleeman,* L. N. P. L. 41; Th. ii. 296, 11. Monige welige menn fédaþ yfle gliigmen [giiimen, Cot. MS.] *nonnulli divites nutriunt histriones,* Past. 44, 6; Swt. 327, 7; Hat. MS. See Turner's History of the Anglo-Saxons, Bk. 7, c. 7.

gleomu, e; *f. Splendour* :—Gleoma gefrætwed *splendidly adorned,* Exon. 124 b; Th. 478, 1; Ruin. 34.

gleó-, gliw-stæf, es; *m. Joy* :—Gliwstafum *joyously,* Exon. 77 a; Th. 289, 22; Wand. 52. [Cf. *other compounds with* stæf, e. g. ǽr-, sár-stæf.]

gleow, gleó, gliw, glig, es; *n.* GLEE, *joy, music, musical accompaniment of a song, mirth, jesting, sport;* gaudium, musica, faceti&, mimus, ludibrium :—Đǽr wæs gidd and gleó *there was song and glee* [*music*], Beo. Th. 4216; B. 2105. And gegaderade, gleowe sungon, on dæra manna midle geornga, on tympanis, togenum streugum *conjuncti psallentibus, in medio juvenum tympanistriarum,* Ps. Th. 67, 24 : Cot. 84. v. Grm. D. M. 854. [*O. Nrs.* glý; *n. lætitia, gaudium.*] v. gliw, glig.

gleowian, gliowian, gliwian, glywian; *p.* ode; *pp.* od *To play on an instrument, sing, joke, jest, act the gleeman or buffoon;* fidicinare, jocari, scurrari, scurram agere :—Đá ongan se wísdóm gliowian *then wisdom began to sing,* Bt. 12; Fox 36, 6. Đæt ǽnig preóst ne gliwige *that no priest act the gleeman,* L. Edg. C. 58; Th. ii. 256, 16. He sumu þing ætgædere mid him sprecende and gleowiende wæs de dǽr ǽr inne wǽron *cum ibidem positi aliqua, una cum eis qui ibidem ante inerant, loquerentur ac jocarentur,* Bd. 4, 24; S. 598, 34. Mádena glywiendra *juvencularum tympanistriarum,* Ps. Th. 67, 27. [*Laym.* gleowien *to chant, play : A. R.* gleowede *was merry.*]

gleów-líce. v. gleáw-líce.

gleó-, glió-word, es; *n. A musical strain, a song,* Bt. Met. Fox 7, 3; Met. 7, 2.

gleow-stól, es; *m. A glee-stool, seat of joy; lætitiæ sedes vel sella* :—Đone gleowstól [MS. gleáw- *prudens, gnarus*] bródor mín ágnade *my brother possessed the seat of joy,* Exon. 130 a; Th. 499, 1; Rä. 88, 9.

glésan *to gloss, explain;* interpretari. v. next word.

glésing, glésincg, e; *f. A* GLOSSING, *interpretation, explanation;* glossa :—Đæt is glésincg donne mann glésþ da earfodan word mid eádran Lédene *that is glossing when one explains the difficult words with easier Latin,* Ælfc. Gr. 50; Som. 51, 43.

gléw. v. gleáw.

glid; *adj. Slippery, ready to glide;* lubricus, Ps. Spl. C. 34, 7.

glida, an; *m. A kite, glede* :—Glida *milvus,* Ælfc. Gl. 38; Som. 63, 29; Wrt. Voc. 29, 48 : 77, 14. Se de þurh reáflác gewilnaþ da þing de he mid his eágum widútan sceáwaþ se is glida ná culfre *he who by rapine desires the things that he sees with his eyes without, he is a kite, not a dove,* Homl. Th. i. 586, 6 : Exon. 106 b; Th. 406, 23; Rä. 25, 5. [*Icel.* gleða.]

GLÍDAN, he glídeþ, glít; *p.* glád, *pl.* glidon; *pp.* gliden *To* GLIDE, *slip, slide;* labi :—Sunne gewát to sete glídan *the sun went gliding to its setting,* Andr. Kmbl. 2610; An. 1306 : 2498; An. 1250 : Exon. 57 a; Th. 204, 24; Ph. 102 : Ps. C. 50, 145; Ps. Grn. ii. 280, 145. Deós bát glídeþ on geofene *this boat glideth over the ocean,* Andr. Kmbl. 995; An. 498 : Bt. Met. Fox 20, 340; Met. 20, 170 : 29, 54; Met. 29, 27. Seó sunne glít abútan *the sun glides round it,* Lchdm. iii. 258, 6. Heofenes gim glád ofer grundas *heaven's gem had glided over the earth,* Beo. Th. 4152; B. 2073 : Homl. Th. i. 78, 23 : Exon. 94 a; Th. 353, 15; Reim. 13 : Andr. Kmbl. 741; An. 371 : Chr. 937; Erl. 112, 15; Ædelst. 15 : Ors. 4, 6; Bos. 84, 37. Đá git glidon ofer gársecg *when ye glided over the ocean,* Beo. Th. 1034; B. 515. DER. a-glídan, be-, bi-, ge-, óþ-, to-.

glidder; *adj. Slippery;* lubricus.

gliddrian *to slip, totter;* nutare, Hpt. Gl. 503.

gliew-mǽden, es; *n. A glee-maiden, female musician;* tympanistria, Ps. Spl. T. 67, 27.

glig, gligg, es; *n. Glee, music, minstrelsy, jesting, sport;* gaudium, musica, faceti&, ludibrium :—Mid dæm glige [MS. Cot. dam gligge] *with the music,* Past. 26, 2; Swt. 183, 25; Hat. MS. 35 b; 8. Hí hæfdon him to glige his hálwende mynegunge *habebant inter se ludibrio salutarem ejus admonitionem,* Basil. admn. 9; Norm. 54, 20. v. g'eó, gliw.

glig-beám, es; *m. A glee-beam, timbrel, tabret;* tympanum, Ps. Spl. 80, 2 : 150, 4. v. gleó-beám.

glig-cræft. v. gleó-cræft.

glig-gamen, -gomen *glee-pleasure.* v. gleó-gamen.

glig-georn; *adj. Glee-loving, fond of sport;* gaudii cupidus, joci amans, Off. Episc. 3.

glí-man, glii-man, gliig-man, glig-man. v. gleó-man.

gliowian. v. gleowian.

glisian *to shine, glisten* :—Se glisigenda wibba *cicindela, the glow-worm,* Ælfc. Gl. 23; Som. 59, 123; Wrt. Voc. 23, 77. [*Laym.* cliseden *glittered : O. Frs.* glisa *splendere.*]

glisnian; *p.* ode; *pp.* od *To glisten, shine* :—Ís glisnaþ glæshluttur *ice glistens bright as glass,* Runic pm. Kmbl. 341, 16; Rún. 11. Se engel hæfde twegen beágas on hys handa da glysnodon swa rósan blósman *the angel had two rings on his hand, they shone like roses,* Shrn. 149, 29. [*Laym.* glissenede : *p. part. pl : Wick.* glisninge.]

glitenung; e; *f. A flash, gleam* :—Mid glitenungum *coruscationem,* Ps. Spl. 143, 8.

glitinian, glitenian; *p.* ode; *pp.* od *To glitter, glisten, shine* :—Geseah gold glitinian *he saw gold glisten,* Beo. Th. 5509; B. 2758. Heó glytenode on dæra engla mydle swá scýnende sunne *she glittered amid the angels as the shining sun,* Shrn. 149, 7. His reáf wurdon glitiniende *vestimenta ejus facta sunt splendentia,* Mk. Skt. 9, 3. [Cf. *Goth.* glitmunjan : *O. H. Ger.* glizinon.]

gliw; *n. Glee, joy, minstrelsy, mirth, jesting, drollery;* gaudium, musica, faceti&, mimus :—Đý læs de him con leóda worn, odde mid hondum con hearpan grétan, hafaþ him his gliwes giefe *unless he knows many songs, or with hands can greet the harp, has his gift of glee,* Exon. 91 b; Th. 344, 11; Gn. Ex. 172. Glæd wæs ic gliwum *glad was I in glee,* 94 a; Th. 352, 29; Reim. 3. Gumum to gliwe *for delight to men,* 57 b; Th. 207, 9; Ph. 139. Đæt geára iú gliwes cræfte mid gieddingum guman oft wrecan *what of yore, by art of minstrelsy, with their lays men oft related,* 92 b; Th. 347, 12; Sch. 11. Wynsum gliw *faceti&,* Ælfc. Gl. 115; Som. 80, 39; Wrt. Voc. 61, 19 : Cot. 132 : 214. v. gleow.

gliw-beám, es; *m. A glee-beam, timbrel, tabret;* tympanum, Ps. Spl. 149, 3. v. glig-beám, gleó-.

gliwere, es; *m. A jester, player, one who aims at pleasing with a view to gain, a flatterer;* parasitus, assentator, scurra, Hpt. Gl. 422 : Gl. Prud. 618.

gliw-hleóþriend, es; *m. A glee-sounder, musician, minstrel;* musicus, fidicen, Cot. 134. v. gleó-hleóþriend.

gliwian. v. gleowian.

gliwian; *p.* ede *To adorn* [?] :—Me gliwedon wrætlíc weorc smiþa, Exon. 107 a; Th. 408, 17; Rä. 27, 13.

gliw-stæf, es; *m.* v. gleó-stæf.

gloed. v. gléd.

gloed-scof *a fire-shovel, warming-pan,* Lye.

glof, es; *n. A cliff* :—Hafuc sceal on glofe wilde gewunian *the hawk shall dwell wild on the cliff,* Menol. Fox 494; Gn. C. 17. [Cf. *Icel.* gliufr; *n. an abrupt descent.*]

glóf, e; *a weak pl.* glófan *occurs; f. A* GLOVE; chirothéca = χειροθήκη :—Glóf hangode, sió [glóf] wæs gegyrwed dracan fellum *his glove hung, it was made with dragon's skins,* Beo. Th. 4177; B. 2085. Glóf *mantium?* Ælfc. Gl. 27; Som. 60, 118; Wrt. Voc. 25, 58. Wilfriþ cwæþ dæt he forléte his twá glófan on dam scipe *Wilfrid said that he had left his two gloves in the ship,* Guthl. 11; Gdwin. 54, 14, 9, title. He mid gyrde of dam húses hrófe da glófe gerǽhte *he reached the glove from the house-roof with a stick,* 22 : 56, 4. Earnian mid dam glófa him sylfum *deserviat per id cirotecas sibi,* L. R. S; Th. i. 438, 15. Foxes glófa *buglosse,* Wrt. Voc. 67, 24; Herb. 144; Lchdm. i. 266, 16. [*Laym.* gloven; *pl : Icel.* glófi; *m.*]

glófung, e; *f. A providing with gloves* :—Glófung him gebyreþ *he is to be provided with gloves,* L. R. S; Th. i. 438, 6.

glóf-wyrt, e; *f.* I. *lily of the valley;* convallāria mājālis, Lin :—Đeós wyrt de man *Apollinarem,* and ódrum naman glófwyrt nemneþ *this plant which is called* Apollīnāris, *and by another name glovewort,* Herb. 23, 1; Lchdm. i. 120, 3 : L. M. 1, 40; Lchdm. ii. 106, 7 : Wrt. Voc. 66, 62. II. *hound's tongue;* cynoglossum officīnāle, Lin :—Đeós wyrt de Engle glófwyrt, and ódrum naman hundes tunge hátaþ *this plant, which the English call glovewort, and by another name hound's tongue,* Herb. 42; Lchdm. i. 144, 3.

glóm, es; *m* [?] *Gloom, twilight, darkness* :—Glóm óðer *a second twilight, i. e. the twilight of evening, the first being that of morning* [?], Exon. 93 b; Th. 350, 30; Sch. 71. DER. ǽfen-, mist-, niht-glóm.

glómung, glómmung, e; *f. Twilight, gloaming;* crepusculum, Lye.

glówan *to glow like a coal of fire;* candere, Lye.

glydering, glyderung, e; *f. What glides away, a vision, an illusion;* visio, Cot. 84.

glywian *to play on an instrument;* part. glywiende, Ps. Spl. 67, 27. v. gleowian.

GNÆT, gnætt; *gen.* gnættes; *m.* GNAT; culex :—Gedrehnigeaþ done

I i

gnæt aweg *ye strain out the gnat*, Mt. Bos. 23, 24. Com hundes fleógan and gnættas *venit cœnomyia et cinipes*, Ps. Spl. 104, 29. Aaron slóh mid ðære girde on ða eorþan, and gnættas wǽron gewordene on mannum and on yrfe ; and ealle ðære eorþan dust wæs gewurden to gnættum ofer eall Egipta land *Aaron percussit pulverem terræ, et facti sunt sciniphes [gnats] in hominibus, et in jumentis ; omnis pulvis terræ versus est in sciniphes per totam terram Ægypti*, Ex. 8, 17, 16 : Ps. Th 104, 27. Gnæt̃tas cómon ofer eall· ðæt land *gnats came over all the land*, Or. 1, 7 ; Bos. 29, 29.

GNAGAN, ic gnage, ðú gnǽgest, gnægst, gnæhst, he gnægeþ, gnægþ, gnæhþ, *pl.* gnagaþ ; *p.* gnóh, *pl.* gnógon ; *pp.* gnagen, gnægen To GNAW, *bite ;* rodere :—Ic gnage *rodo*, Ælfc. Gr. 28, 4 ; Som. 31, 24. Ðæt gewrit beó geworpen músen to gnagene *illiusmodi litteraturæ membranula suricum morsibus corrodenda*, Chart. Th. 318, 29. [Gnagan = ge-nagan : *Icel.* gnaga, naga : *O. H. Ger.* nagan, gi-nagan.] DER. be-gnagan, for-.

gnāst, es ; *m. A* spark. [*O. E. Hom.* gnast : *Icel.* gneisti : *O. H. Ger.* gneisto.] DER. fýr-gnäst.

gneáð, gnēð ; *adj. Sparing, frugal, stingy, scanty, small ;* parcus :— Næs hió to gneáð *gifa she was not too sparing of gifts*, Beo. Th. 3864 ; B. 1930. He self lifde on gneáðum woroldlífe ân tunece wæs his gegerela and ðæt wæs hǽren and beren hláf wæs his gereorde *he himself lived a frugal life in the world, one tunic was his raiment, and barley bread was his food*, Shrn. 110, 4 : 77, 4. He ðám ðe on scearan máran wǽron on ðám mægnum eáðmódnesse and ·hýrsumnesse nôhte ðon læssa ne gnēðra wæs *eis quæ tonsura majores sunt virtutibus, humilitatis et obedientiæ, non mediocriter insignitus*, Bd. 5, 19 ; S. 637, 18. Of gnēðum, of lytlum *parcis*, Gl. Prud. 227. [Gnede *scanty, O. E. Misc. Morris.*]

gneáðlionis *frugality*, Hpt. Gl. 463.

gnēðelíce ; *adv. Sparingly, frugally ;* parce, Greg. Dial. 1, 7, Lye. [Cf. *A. R.* al þet mon wilneþ more þen heo mei gnedeliche leden hire lif bi, al his giscunge.]

gnēðen, gnēðn ; *adj. Moderate, temperate, modest, low ;* mediocris, modestus, Cot. 129, Lye.

gnēðenes, gnēðnes, se ; *f. Frugality, care ;* parcimonia, Cot. 81, 149, Lye.

GNĪDAN, ic gníde, ðú gníst, he gnít, *pl.* gnídaþ ; *p.* gnád, *pl.* gnidon ; *pp.* gniden To *rub, break, rub together, comminute ;* fricare, comminuere :— Hys leorningcnihtas ða eár mid hyra handum gnidon *his disciples rubbed the ears with their hands*, Lk. Bos. 6, 1. Gif ðú gang ofer his æcer brec ða eár and gníd *if thou go across his field pluck the ears and rub them*, Deut. 23, 25. Nim ǽnne sticcan and gníd to sumum þinge *take a stick and rub it against something*, Lchdm. iii. 274, 3. Gníd ða þungana and on ufan ðæt hēfd *rub the temples and the top of the head*, 292, 23. Gníd swíðe smale to duste *rub very small, to dust*, Herb. 1, 2 ; Lchdm. i. 70, 14. [*Dan.* gnide : *O. H. Ger.* gnítan *fricare*.] DER. for-gnídan, ge-.

gnidennys, -nyss, e ; *f. A* rubbing, contrition. v. for-gnidennys, Ps. Lamb. 13, 3.

gnidill *a pestle ;* pistillum, Som.

gnīding *a rubbing ;* frictio, Som.

gnīst, he gnít *rubbest, rubs ; 2nd and 3rd pers. pres. of* gnídan.

gnōh, *pl.* gnógon *gnawed, bit ; p. of* gnagan.

gnorn, es ; *m. Sorrow, sadness, affliction ;* mæstitia :— Ne biþ ðær ǽngum gódum gnorn ætýwed *no sorrow shall there be shewn to any good man*, Exon. 31 a ; Th. 96, 19 ; Cri. 1576. Gnorn þrowian *to suffer sadness*, Beo. Th. 5310 ; B. 2658.

gnorn ; *adj. Sorrowful, sad, dejected, complaining ;* mœstus :— Leónhwelpas grymetigaþ *gnorne catuli leonum rugientes*, Ps. Th. 103, 20. Flugon forhtigende gylp wearþ gnornra *they fled in terror, their boast became more sorrowful*, Cd. 166 ; Th. 206, 19 ; Ex. 454.

gnornan, gnornian ; *p.* ede, ode ; *pp.* ed, od *To grieve, mourn, be sad, bewail, lament ;* mœrere :—Ic gnornige *mereo*, Ælfc. Gr. 33 ; Som. 36, 49 : Ps. Th. 54, 2. Ic cúþlíce wât for hwon ðú gnornast *scio certissime quare mœres*, Bd. 2, 12 ; S. 513, 42. Gnornaþ *he grieves*, Exon. 82 b ; Th. 311, 14 ; Seef. 92 : 51 a ; Th. 178, 6 ; Gú. 1240. Gif hí fulle ne beóþ fela gnorniaþ *si non fuerint saturati, et murmurabant*, Th. 58, 15. Ðæt wíf gnornode *the woman mourned*, Cd. 37 ; Th. 48, 4 ; Gen. 770 : Beo. Th. 2239 ; B. 1117 : Elen. Kmbl. 2518 ; El. 1260. Swá gnornedon godes andsacan *thus lamented God's adversaries*, Cd. 219 ; Th. 282, 1 ; Sat. 280 : Exon. 38 b ; Th. 128, 7 ; Gú. 400. Ne scyle nân wís monn forhtigan ne gnornian *no wise man ought to fear or lament*, Bt. 40, 3 ; Fox 238, 8 : Cd. 219 ; Th. 281, 19 ; Sat. 274. Sceoldon wræcmæcgas ofgiefan gnornende grēne beorgas *the exiles, sorrowing, must give up the green hills*, Exon. 35 b ; Th. 116, 6 ; Gú. 203 : 42 b ; Th. 142, 29 ; Gú. 651. He fērde gnornigende *abiit mœrens*, Mk. Skt. 10, 22. Geómor and gnörngende *sad and sorrowing*, Blickl. Homl. 113, 29 : Cd. 39 ; Th. 52, 9 ; Gen. 841. Gnorniende cynn *a mourning race*, 216 ; Th. 273, 9 ; Sat. 134 : Ps. Th. 101, 4. Geonge for ðē gnornendra care ðara ðe on feterum fæste wǽran *intret in conspectu tuo gemitus compeditorum*, 78, 11. [*O. Sax.* gnornon.]

gnorn-cearig ; *adj. Sad, sorrowful*, Exon. 73 b ; Th. 274, 6 ; Jul. 529.

gnorn-hof, es ; *n. A house of grief, a prison*, Andr. Kmbl. 2016 ; An. 1010 : 2087 ; An. 1045.

gnorn-scendende ; *part. Hurrying away in sorrow*, Ps. Th. 89, 10.

gnorn-sorh, -sorg, e ; *f. Care, sorrow*, Exon. 52 a ; Th. 182, 13 ; Gú. 1309 : Elen. Kmbl. 1307 ; El. 655 : 1951 ; El. 977.

gnornung, e ; *f. Grief, lamentation, mourning ;* mœstitia :—Gnornung *meror*, Ælfc. Gr. 33 ; Som. 36, 51. Hēr is Brytta gnornung *gemitus Brittanorum*, Bd. 1, 13 ; S. 481, 42, note. Me hæfþ ðeós gnornung ðære gemynde benumen *this grief hath deprived me of the remembrance*, Bt. 5, 3 ; Fox 12, 20 : 7, 2 ; Fox 18, 10. Mid mycelre gnornunge ymbe ðæs cyninges slege *with great grief for the king's death*, Ors. 2, 4 ; Bos. 45, 24 : Chr. 975 ; Erl. 126, 13 ; Edg. 39. Seó árleáse helwarena stefn wæs gehýred and heora gnornung *the impious voice of the dwellers in hell was heard, and their lamentation*, Blickl. Homl. 87, 4 : 91, 30 : Cd. 220 ; Th. 285, 8 ; Sat. 334 : Exon. 40 b ; Th. 134, 29 ; Gú. 516. DER. heáh-gnornung.

gnorn-word, es ; *n. A word of sadness, mournful discourse* :—Him oft betuh gnornword gengdon *oft mournful words passed between them*, Cd. 37 ; Th. 47, 27 ; Gen. 767. [Cf. *O. Sax.* gorn-word.]

gnyran [?] *to creak ;* stridere :—Gnyrende *stridentes*, Lchdm. iii. 210, 12. See Skt. Etymol. Dict. gnarl.

gnyrn, es ; *m. n* [?] *Grief, sorrow, evil, wrong :*—Lác weorþade ðe hire brungen wæs gnyrna to geóce *the gift she honoured that was brought to her as a consolation of sorrows*, Elen. Kmbl. 2275 ; El. 1139. Þeóda waldend eallra gnyrna [MS. gnyrnra] leás *the ruler of nations, free from all evils*, 843 ; El. 422. Wlance drihtne guldon gód mid gnyrne *arrogant, they repaid good to the Lord with evil*, Cd. 111 ; Th. 146, 10 ; Gen. 2420. [Cf. gyrn.]

gnyrn-wracu, e ; *f. Revenge for injury* or *grief, enmity, hate*, Elen. Kmbl. 718 ; El. 359. [Cf. gyrn-wracu.]

GOD, es ; *m. God, the Deity, a god.* The following epithets occur :—dryhten, wealdend, nergend, hǽlend, sóþ, hálig, mihtig, ælmihtig, lifgende, ealwealda, heáhengla, heofona, heofonengla, heofonríces, gǽsta, mihta, mægena, weoruda, wuldres, sigores, sigora. Ân God ys gód, Mt. 19, 17. Nys nân man gód, búton God âna, Lk. 18, 19. Hú gód Israhēl God, Ps. Spl. 72, 1. Hēr is Godes lamb, Jn. 1, 29. Enoch fērde mid Gode, Gen. 5, 24. Ða leásan godas *false gods*, Blickl. Homl. 201, 30. Rachel forstæl hire fæder hǽðenan godas *Rachel furata est idola patris sui*, Gen. 31, 19. Hwí forstæle ðú me míne godas *cur furatus es deos meos*, 31, 30. Hǽðenan godas *heathen gods*, 31, 32. Hǽðenan godas *heathen gods*, 31, 33. Ne wirc ðú ðē agrafene godas *work not thou for thyself graven gods*, Ex. 20, 4. Drihten sylf ys Goda God, mǽre God, and mihtig, and egefull *the Lord himself is God of Gods, a great God, a mighty and a terrible*, Deut. 10, 17. Ne wyrc ðú ðē gyldne godas oððe seolfrene *make thou not to thyself golden or silver gods*, L. Alf. 10 ; Th. i. 44, 21 : Ex. 32, 31 : 23, 32 ; Jn. Skt. 10, 34, 35. Ða hǽðenan noldou beón gehealdene on feáwum godum. . . . Mânfullan men wǽron ða mǽrostan godas *the heathens would not be contented with few gods. . . . Guilty men were the mightiest gods*, Salm. Kmbl. p. 121, 40. [*Goth.* guþ ; *m* : *O. Sax. O. Frs.* god : *Icel.* guð ; *m. pl.* guðir *dii* : *O. H. Ger.* got ; *m* : *Ger.* gott.] v. Grm. D. M. pp. 12 sqq. and cf. god ; *n.*

god, es ; *n. A god* :—Hiora godu syndon drýcræfta láreówas *their gods are teachers of magical arts*, Ors. 1, 5 ; Bos. 28, 28. He wolde gesēcan helle godu *he would visit the gods of hell*, Bt. 35, 6 ; Fox 168, 13. Goddo [godo, Rush.] gie aron *dii estis*? Jn. Skt. Lind. 10, 34. God *deos*, Rush. 35. Godu, Ps. Th. 81, 6 : 94, 3. Syndon ealle hǽdene godu hilde deóful *omnes dii gentium dæmonia*, 95, 5, 4 : Exon. 74 b ; Th. 278, 16 ; Jul. 598. Gif ðú fremdu godu forþ bigongest *if thou dost continue to worship strange gods*, 67 b ; Th. 250, 2 ; Jul. 121. [*Goth.* guþa ; *n. pl* : *Icel.* goð ; *n. pl.*]

gód ; *adj.* GOOD ; bonus :—Þæs gódan gódnes biþ his ágen gód *the goodness of the good is his own good*, Bt. 37, 3 ; Fox 190, 14. Gód mann sóþlíce of gódum goldhorde bringþ gód forþ *bonus homo de bono thesauro profert bona*, Mt. Bos. 12, 35. Mæg ǽnig þing gódes beón of Nazareth *a Nazareth potest aliquid boni esse*? Jn. Bos. 1, 46. Crist, seðe ǽfre is gód ðeáh ðe we wâce sindon *Christ who is ever good, though we are weak*, Homl. Th. ii. 48, 20. Ðær wearþ Heáhmund bisceop ofslægen and fela gódra monna *there was bishop Heahmund slain and many good men*, Chr. 871 ; Erl. 74, 34. Þa men hie gefliémdon and hira gódne dǽl ofslógon *the men put them to flight and slew a good part of them*, 921 ; Erl. 106, 24 : 913 ; Erl. 102, 7. Genim giþcornes leáfa gode handfulle *take good handfuls of leaves of githcorn*, L. M. ii. 65, 1 ; Lchdm. ii. 292, 10. Me is on gómum gód and swēte ðín ágen word *quam dulcia faucibus meis eloquia tua*, Ps. Th. 118, 103. Gód is ðæt man Drihtne andette *bonum est confiteri domino*, 91, 1 : 132, 1. Cyning and cwēn sceolon geofum gód wesan *a king and queen shall be liberal*, Exon. 90 a ; Th. 338, 35 ; Gn. Ex. 84. Nis mon his gifena ðæs gód *there is no man so good in his qualities*, 82 a ; Th. 308, 15 ; Seef. 40. He is to freónde gód *he is good as a friend*, 67 a ; Th. 248, 28 ; Jul. 102. We ðær góde hwíle stódon *we stood there a good while*, Rood Kmbl. 140 ; Kr. 70. Him ðæt geleánaþ lífes waldend gódum dǽdum *the ruler of life will repay them that with benefits*, Exon. 117 a ; Th. 450, 13 ; Dóm. 87. Þurh góde dǽda Gode lícian *to please God by good deeds*, Blickl. Homl. 129,

34. Đâm đe gódes willan sýn *to those who are of goodwill*, 93, 10 : 37, 27. Gódes lífes bysene onstellan *to set an example of good life*, 81, 6. Wæs he swíđe æþelra gebyrda and gódra *he was of very noble and good birth*, 211, 19. Gôde sangeras *good singers*, 207, 31. [*Goth.* gôds, gôþs: *O. Sax. O. Frs.* gôd: *O. H. Ger.* guot: *Ger.* gut: *Icel.* gôđr.]

gôd, es; *n.* *Good, good thing, good deed, benefit, goodness, welfare* :— Æghwylc man sceal on worlde geearnian đæt him đæt gód môte to êcum mêdum gegangan, đæt him his freónd æfter gedêþ. Se getreówa man sceal syllan his gôd on đa tîd đe hine sylfne lyste his brúcan *each man must in this world deserve that the good that his friend does for him afterwards may conduce to eternal rewards. The true man must give his wealth at the time that it best pleases him to enjoy it himself*, Blickl. Homl. 101, 17. Hwæđer him yfel đe gôd under wunige *whether evil or good dwell in it* [*the heart*], Exon. 27 a ; Th. 82, 3 ; Cri. 1333. Gôd dôend *qui faciat bonum*, Ps. Th. 52, 4 : Gen. 2, 9 : Bt. 37, 3 ; Fox 192, 1. His gôd wæs swíđe gecýđed *his goodness was very famous*, Blickl. Homl. 217, 3 : Bt. Met. Fox 20, 57 ; Met. 20, 29. Đæt hêhste gôd *the supreme good*, 90 ; Met. 20, 46 : Bt. 32, 1 ; Fox 114, 5. Swâ hwæt swâ we to gôde dôþ *whatever good we do*, Blickl. Homl. 29, 8 : 215, 26 : Ors. 6, 8 ; Bos. 120, 12. On ôđres gôde beón gefeónde *to rejoice at another's good*, Blickl. Homl. 75, 20. Se đe gôd onginneþ *he who attempts good*, 21, 34. He Godes good on đære his đæde ongeat, *he perceived in that deed of his the goodness of God*, 215, 33. He mid gôde gyldan wille uncran eaferan *he will repay our offspring with good*, Beo. 2372 ; B. 1184. Alwalda đec gôde forgylde *may the Omnipotent reward thee with good*, 1916 ; B. 956. Him sylfum nænige gôde beón *to be of no good to themselves*, Blickl. Homl. 45, 16. For eallum đâm gôdum đe he me dyde *pro omnibus quæ retribuit mihi*, Ps. Th. 115, 3 : 102, 2 : Cd. 15 ; Th. 19, 14 ; Gen. 291 : Homl. Th. i. 76, 7 : Blickl. Homl. 29, 11. Bûton he mid ôđrum gôdum hit geêce *unless he add thereto other good deeds*, Blickl. Homl. 37, 25. Đa gaderige đyder mîne gôd *illuc congregabo bona mea*, Lk. Skt. 12, 18 : Gen. 24, 10 : Bd. 4, 24 ; S. 598, 4. He forsihþ eorþlícu gôd *he despises earthly goods*, Bt. Met. Fox 7, 84 ; Met. 7, 42.

Goda, an ; *m.* *A deity, god*; deus :—Ealra godena God *Deus deorum*, Ps. Th. 135, 2, 28. God godana *Deus deorum*, Rtl. 101, 10.

god-æppel *a quince apple*; cydonium, Cot. 34, 93.

god-bearn, es ; *n.* I. *a divine child, the Son of God*; divinus filius, Dei Filius :—Ahangen wæs Godbearn on galgan *God's Son was hanged on the cross*, Elen. Kmbl. 1434 ; El. 719. Gesêgon hî on heáhþu hláford stîgan Godbearn of grundum *they saw the Lord, the Son of God, ascend on high from earth*, Exon. 15 a ; Th. 31, 21 ; Cri. 499 : Andr. Kmbl. 1279 ; An. 640. II. *a god-child, a god-son*; filius lustricus, ex sacro fonte baptismi jam primum susceptus :—Godbearn to fela man forspilde *god-children, too many of them have been destroyed*, Swt. A. S. Rdr. 107, 94.

god-borh ; *gen.* -borges ; *m.* A word of uncertain meaning occurring only in L. Ælf. pol. 33 ; Th. i. 82, 4-8. q. v.

god-bôt, an ; *f.* *An atonement made to the church*, L. Æthel. 6, 51 ; Th. i. 328, 4.

godcund ; *adj.* *Of the nature of God, divine, religious, sacred* :—Seó godcunde æ *lex divina*, Bd. 1, 1 ; S. 474, 2. Wiotan ægđer ge godcundra hâda ge woruldcundra *wise men both of religious and secular orders*, Past. Pref ; Swt. 3, 3, 8 ; Hat. MS. Hêr sende Gregorius pâpa wel monige godcunde lâreówas *in this year that pope Gregory sent very many religious teachers*, Chr. 601 ; Erl. 20, 11. In godcundum mægne *in divine power*, Exon. 40 a ; Th. 134, 2 ; Gú. 501 : 17 a ; Th. 40, 13 ; Cri. 638. Godcunde bêc *sacred books*, Cd. 123 ; Th. 158, 4 ; Gen. 2612. [*O. Sax.* god-kund : *O. H. Ger.* gotchund : *Orm.* Laym. godcund.]

godcundlíc ; *adj.* *Divine* :—Bûton yldinge him becom seó godcundlíce wracu *without delay the divine vengeance came upon him*, Homl. Th. i. 86, 1. Đâ ongeat he đæt đær wæs godcundlíc mægen ondweard *then he perceived that there was divine power present*, Blickl. Homl. 217, 29.

godcundlíce ; *adv.* *Divinely, from heaven, by inspiration*; divinitus :— Godcundlíce *divinitus*, Ælfc. Gr. 38 ; Som. 42, 5. Đeáh he sê godcundlíce gesceádwís *though he be divinely rational*, Bt. 14, 2 ; Fox 44, 18 : Bd. 4, 3 ; S. 567, 10 : 4, 24 ; S. 596, 41.

godcundnys, se ; *f.* *Divine nature, Deity, Divinity, Godhead, divine service* :—Se God wunaþ on þrynnysse untodæledlíc and on ânnysse ânre Godcundnysse *the Deity exists in Trinity indivisible, and in unity of one Godhead*, Homl. Th. i. 276, 24. Seó hrepaþ swýđost ymbe Cristes godcundnysse *that* [*book*] *treats chiefly of Christ's divinity*, 70, 1. Đeós wyrt hæfþ mid hire sume wundorlíce godcundnesse *this plant has in it a certain wonderful divine quality*, Herb. 50, 1 ; Lchdm. i. 152, 24. Ond Wærferþ bisceop and se heóred habbaþ geseted đas godcundnysse *and bishop Werferth and the convent have established this divine office*, Chart. Th. 137, 28 : Cod. Dipl. Kmbl. v. 218, 32.

gôd-dǽd, e ; *f.* *A good deed, a benefit* :—Menn swîđor scamaþ nû for gôddædum đonne for misdædum *men are now more ashamed of good deeds than of misdeeds*, Swt. A. S. Rdr. 109, 161 : Exon. 26 b ; Th. 79,

7 ; Cri. 1287 : 65 b ; Th. 242, 6 ; Ph. 669. Ealra gôddǽda hî forgiten hæfdon *obliti sunt benefactorum ejus*, Ps. Th. 77, 13.

god-dôhtor ; *f.* A GODDAUGHTER :—Ic geann mînre goddôhtor đæt land æt Strættûne *I grant to my goddaughter the land at Stretton*, Chart. Th. 548, 5.

gôd-dônd, -dênd, es ; *m.* *One who does good, a benefactor*, Elen. Kmbl. 717 ; El. 359.

god-dreám, es ; *m.* *A heavenly joy*, Exon. 41 b ; Th. 139, 32 ; Gú. 602 : 51 b ; Th. 180, 1 ; Gú. 1273.

god-fæder ; *m.* I. *a* GODFATHER ; baptizati susceptor, patrinus :— Gif hwâ ôþres sleá godfæder *if any one slay another's godfather*, L. In. 76 ; Th. i. 150, 13. Hit wæs mînes godfæder gyfu *it was my god-father's gift*, Chart. Th. 545, 21. II. *God the Father, the Divine Father* ; Deus ille Pater, Divinus Pater :—Crist ys word and tunge God-Fæder ; *þurh hine synt ealle þincg geworht* Christ is the word and tongue of God the Father ; through Him are all things made, Ps. Th. 44, 2. Ic eom Crist . . . ic đe fullwie on mînne godfæder and on mec his efenêcne sunu and on đone hâlgan gâste *I am Christ . . . I baptize thee in the name of my heavenly Father and of me his co-eternal Son and of the Holy Ghost*, Shrn. 106, 13 : 118, 6. Đû sitest on đa swîþran hand đînum God-Fæder *thou sittest on the right hand of thy Divine Father*, Hy. 8, 31 ; Hy. Grn. ii. 290, 31. [*O. Sax.* god-fader *God the Father.*]

gôd-fremmende ; *part. pres.* used as a noun. *One doing good, acting bravely*, Beo. Th. 603 ; B. 299.

God-fyrht, -ferht, -friht ; *adj.* *God-fearing* :—To oft man gôdfyrhte leahtraþ *too often the god-fearing are reviled*, Swt. A. S. Rdr. 110, 163. Ic haue here godefrihte muneces *I have here godfearing monks*, Chr. 656 ; Erl. 32, 1. Đâ oungan Andreas grêtan godfyrhtne *then began Andrew to greet the godfearing man*, Andr. Kmbl. 2043 ; An. 1024 : 3030 ; An. 1518. Godferhte, Ps. C. 14 ; Grn. ii. 277, 14.

god-gespræce, es ; *n.* *An oracle* :—Wæs đis Godgespræce đysses gemetes *erat oraculum hujusmodi*, Bd. 2, 12 ; S. 513, 1.

god-gild, -gield, -geld, -gyld, gode-gild, es ; *n.* *An idol* :—He hêt wyrcan gyldeno godgeld and seolfrene . . . đâ abræc đæt mægden đæt gold and đæt seolfor of đæm godgeldum *he bade make golden idols and silver . . . then the maiden broke the gold and the silver off the idols*, Shrn. 106, 2-4 : 122, 9 : L. Alf. 32 ; Th. i. 52, 12. Đæt he gulde đæm hǽđnum godgyldum *that he should sacrifice to heathen idols*, Shrn. 101, 1 : Bd. 1, 7 ; S. 477, 13. He heora godgieldum eallum wiđsôc *he renounced all their idols*, Ors. 2, 5 ; Swt. 78, 4. Hî on Choreb cealf ongunnan him to godegylde georne wyrcean *fecerunt vitulum in Choreb, et adoraverunt sculptile*, Ps. Th. 105, 17. [*Cf.* deófol-gild.]

god-gildlíc ; *adj.* Phanaticus, Cot. 152, Lye.

god [gôd-?] -gim, es ; *m.* *A heavenly* [*an excellent?*] *gem*, Elen. Kmbl. 2225 ; El. 1114.

god-gyld. v. god-gild.

gôdian ; *p.* ode, ede ; *pp.* od, ed. I. *to be or become good, to improve, get better* :—Đonne gôdiaþ đæra lendena sâr and đæra þeóna swýđe hræđe *then the pains in the loins and thighs will very speedily get better*, Herb. 1, 28 ; Lchdm. i. 80, 1. On his dagum hit gôdode *in his days things improved*, Chr. 959 ; Erl. 119, 13 : Swt. A. S. Rdr. 105, 19. Gif his hreófla gôdigende wǽre *if his leprosy were getting better*, Homl. Th. i. 124, 27. Þurh đæt hit sceal on earde gôdian to âhte *by that means matters must somewhat improve in the land*, L. C. S. 11 ; Th. i. 382, 8. II. *to do good, make good, improve, endow, enrich* :— Mid eallum þingum gôdode *enriched* [*the place*] *with all things*, Lchdm. iii. 438, 10 : Chr. 963 ; Erl. 123, 28. Gyf ǽnig sý đe hit mid ǽnigan þingan geêcean wylle ođđe gôdian *si quis autem hanc nostram donationem largioribus amplificare muneribus studuerit*, Cod. Dipl. Kmbl. iv. 72, 29 : Lchdm. iii. 442, 14 : L. Pen. 14 ; Th. ii. 282, 9. Hig bǽdan đone bisceop đæt hig môstan đæt mynster gôdian *they asked the bishop that they might endow the monastery*, Cod. Dipl. Kmbl. iv. 290, 9 : L. Pen. 14 ; Th. ii. 282, 8.

goding ; *m.* *The son of God* :—Đe hǽlend seđe wæs goding *the Saviour who was the Son of God*, Lk. Skt. Rush. 4, 1.

gôd-leás ; *adj.* *Without good, miserable, wretched* :—Đis ungesǽlige geár and đæt gôdleáse *infaustus ille annus et omnibus bonis exosus*, Bd. 3, 1 ; S. 523, 33.

gôdlíc, -lec ; *adj.* *Goodly, good* :—Gôdlíc gumrinc *a goodly man*, Exon. 129 a ; Th. 495, 7 ; Rä. 84, 4. Gôdlíce geardas *goodly dwellings*, Cd. 35 ; Th. 46, 6 ; Gen. 740. Gôdlecran stôl *a goodlier throne*, 15 ; Th. 18, 31 ; Gen. 281. [*Laym.* godlich : *O. Sax.* gôdlîk : *O. Frs.* gôdlîk : *O. H. Ger.* guotlih.]

god-mægen, es ; *n.* *A divine power, divinity*; numen :—Ic bæd đa godmægen *orabam numina*, Nar. 24, 22. Hie ondrêdon đæt hie hiora godmægne sceoldon beón benumene *they feared that they should be deprived of their divinity*; de numinum suorum statu timentes, 28, 13.

god-môdor ; *f.* A GODMOTHER :—Æt đam fulwihte hyre onfêng sum godes þeów đære noma wæs rômâna . . . heó slêp æt đære godmôdor húse *a certain servant of God, whose name was Romana, was her sponsor at baptism . . . she slept at the godmother's house*, Shrn. 140, 24.

Godmundingahám *Goodmanham, between Pocklington and Beverley, a place a little to the east of York, beyond the river Derwent, where a famous Witena-gemót was convened by Edwin, king of Northumbria, in A.D. 625, to consider the propriety of receiving the Christian faith. The speeches were so much in favour of Christianity that the creed was at once received; these speeches are particularly worthy of notice,* Bd. 2, 13; S. 517, 17.

gódnes, -ness, e; *f. Goodness :*—Se hálga hí eft alýsde and lêt hí forþgán for his gódnysse *the holy man loosed them again, and let them proceed through his goodness,* Homl. Th. ii. 508, 22 : Ps. Th. 24, 8. Ðæs gódan gódnes biþ his ágen gód and his ágen edleán *the goodness of the good is his own good and his own reward,* Bt. 37, 3; Fox 190, 14: 33, 4; Fox 128, 15.

god-sǽd, es; *n. The fear of God, piety :*—Æþele cnihtas and ǽfæste ginge and góde in godsǽde *noble youths and pious, young and good in the fear of God,* Cd. 176; Th. 221, 19; Dan. 90.

godscipe, es; *m. Goodness :*—Godscipe *bonitas,* Rtl. 100, 11 : 12, 23.

god-scyld, e; *f. A sin against a god, impiety :*—Ic nýde sceal godscyld wrecan *I needs must avenge impiety,* Exon. 68 b; Th. 254, 29; Jul. 204.

god-scyldig; *adj. Guilty against God,* Exon. 45 a; Th. 153, 31; Gú. 834.

god-sibb, es; *m. A sponsor :*—Godsibbas and godbearn *sponsors and godchildren,* Swt. A. S. Rdr. 107, 94. [*Ayenb.* godzyb : *Piers P. Chauc.* gossib.]

gód-spédig; *adj. Rich in good,* Cd. 48; Th. 62, 4; Gen. 1009.

god-spell, es; *m. Gospel :*—Gódspel *evangelium, id est, bonum nuntium,* Ælfc. Gl. 8; Wrt. Voc. 75, 9. Hér ys godspellys angyn *initium euangelii,* Mk. Skt. 1, 1. Gelýfaþ ðam godspelle *credite euangelio,* 15. Matheus ongan godspell ǽrest wordum wrítan *Matthew began first to write the gospel in words,* Andr. Kmbl. 24; An. 12. [*Laym. Orm.* godspell : *Piers P.* godspel, gospel : *Chauc.* gospel : *O. Sax.* god-spell : *Icel.* guð-spjall : *O.H.Ger.* gotspel.]

gódspell-bóc, e; *f. A copy of the gospels :*—Saltere and pistolbóc and godspellbóc *a psalter, a copy of the epistles, and a copy of the gospels,* L. Ælf. C. 21; Th. ii. 350, 13. [*Orm.* goddspellboc : *Icel.* guðspjallabók.]

godspellere, es; *m. An evangelist :*—Iohannes se godspellere *John the evangelist,* Homl. Th. i. 58, 3, 27 : Chr. 84; Erl. 8, 6. [*A.R. Ayenb.* godspellere : *Chauc.* gospellere.]

godspellian; *p.* ode; *pp.* od *To declare the gospel; evangelize :*—Ic godspellige *evangelizo,* Ælfc. Gr. 24; Som. 25, 45. Godspellian [MS. A. godspel secgan] Salm. Kmbl. 132; Sal. 65. God gifeþ gleáw word godspellendum *Dominus dabit verbum evangelizantibus,* Ps. Th. 67, 12.

godspellíc; *adj. Evangelical :*—He fylgde ðæt weorc ðæt him gewunelíc wæs ðæt he godspellíce láre lǽrde *solitum sibi opus evangelizandi exsequens,* Bd. 3, 19; S. 547, 9: Homl. Th. ii. 586, 3. Mid ðysum wordum ða godspellícan gesetnysse ongan *with these words began the gospel narrative,* i. 70, 11, 18.

godspellisc; *adj. Evangelical :*—Ðæs godspellesca bodes *euangelicæ prædicationis,* Mk. Skt. p. 1, 11.

god-sprǽce, es; *n. An oracle :*—Wæs sum godsprǽce and heofonlíc onwrigenes *oraculum cæleste,* Bd. 2, 12; S. 512, 23. v. god-gesprǽce.

god-sprec, es; *n. An oracle :*—We neáh stódan ðǽm godsprecum *we stood near to the oracles,* Nar. 28, 32.

god-sunu, a; *m. A* GODSON *:*—Ða onféng he him and æt fulluhtbæþe nam æt ðæs B' handa him to godsuna *then he accepted him and took him from the font at the bishop's hand as his godson,* Bd. 3, 7; S. 529, 18. Hiora wæs óðer his godsunu óðer Æþerédes ealdormonnes *one of them was his godson, the other was alderman Ethelred's,* Chr. 894; Erl. 91, 29 : L. In. 76; Th. i. 150, 13.

god-þrym; *gen.* -þrymmes; *m. Divine majesty :*—He geseah ðone hǽlend silfne standan on his godþrimme *he saw the Saviour himself stand in his divine majesty,* Shrn. 32, 2. Melchisedech godþrym onwráh éces alwaldan *Melchisedec revealed the divine majesty of the eternal ruler of all,* Exon. 10 a; Th. 9, 24; Cri. 139.

god-web, gode-web, -webb; *gen.* -webbes; *n. A divine or very precious web, purple cloth, excellently woven material :*—Mid golde and mid godewæbbe gefrætewod *auro et purpura compositum,* Bd. 3, 11; S. 535, 32: Homl. Th. i. 62, 26. Godwéb mid golde gefágod *a purple garment variegated with gold,* Blickl. Homl. 113, 20. Weofod bewrigen mid baswe godwebbe *an altar covered with a purple pall,* 207, 17. Twegea bleó godwebb *fine cloth of two colours; ex duplici tinctura,* Past. 14, 6; Swt. 87, 9; Hat. MS. Heó bewand sce adrianes hand on godwebbe *she wrapped up St. Adrian's hand in fine linen,* Shrn. 59, 35. Gold and godweb iosepes gestreón *gold and purple, Joseph's treasures,* Cd. 171; Th. 215, 22; Exod. 587: Bt. Met. Fox 8, 49; Met. 8, 25. Geolo godwebb *yellow silk,* Exon. 109 a; Th. 417, 25; Rä. 36, 10. Godwebba cyst *choicest of textures* [*the veil of the temple*], Exon. 24 b; Th. 70, 8; Cri. 1135. [*O. Sax.* godu-webbi : *O. Frs.* god-wob : *Icel.* guð-vefr :

O. H. Ger. gota-, goto-, gotu-, cota-, coti-weppi *sericum, purpura, polymitum, byssus,* Grff. i. 646-8.]

god-webben; *adj. Purple :*—Ná mid golde ne mid godwebbenum hræglum *not with gold nor with purple raiment,* Blickl. Homl. 95, 19. [*O. H. Ger.* gotaweppin *carbaseus, hyacinthinus, purpureus, coccineus, sericus.*]

godweb-wyrhta, an; *m. A weaver of godweb :*—To ðam diólgum godwebwyrhtum *ad abditos seres,* Nar. 6, 15.

god-wrac, -wrec; *adj. Impious :*—Crist forlét mid him beón ðone godwracan þeóf *Christ let that impious thief* [*Judas*] *be with him,* Blickl. Homl. 75, 26. Ða æfêstgodon ðæt sume godwrece men *then certain wicked men were envious of that,* Shrn. 74, 28. Gangaþ út git godwrecan and gongaþ út git ródewyrðan *come out ye two wretches that deserve to be hanged,* 43, 8.

god-wrecnis, -niss, e; *f. Wickedness, impiety :*—Hefig mán is and godwrecnis ðæt mon hine menge mid his steópméder *cum noverca miscere crave est facinus,* Bd. 1, 27; S. 491, 10.

gold, es; *n. Gold :*—Ðæs landes gold ys golda sélost *the gold of that land is the best of all gold,* Gen. 2, 12 : Cd. 12; Th. 14, 29; Gen. 226. Abram wæs swíðe welig on golde *Abram was very rich in gold,* Gen. 13, 6. Cnihtas cúþ gedydon ðæt hie him ðæt gold to gode noldon habban *the youths made known that they would not have that gold* [*the golden image*] *as their god,* Cd. 182; Th. 228, 4; Dan. 197: 183; Th. 229, 9; Dan. 216. Reád gold *aurum obrizum,* Ælfc. Gl. 58; Som. 67, 110; Wrt. Voc. 38, 33. Ealle ðás goldsmiþas secgaþ ðæt hí nǽfre ǽr swá clǽne gold ne swá reád ne gesáwon *all these goldsmiths say that they never before saw such pure and such red gold,* Homl. Th. i. 64, 9. Eall mid reádum golde his cynestól geworhte he wrought his throne all with red gold, H. R. 101, 2. Hundtwelftig mancæs reádes goldes *a hundred and twenty mancuses of red gold,* Th. Chart. 232, 10 : 375, 28 : Bt. Met. Fox 19, 11; Met. 19, 6 : Cd. 109; Th. 145, 11; Gen. 2404. Wunden gold *twisted gold,* 91; Th. 116, 4; Gen. 1931 : Beo. Th. 2391; B. 1193. *Other epithets applied to gold are* æpled, beorht, fæted, fætt, hyrsted, scír, smǽte. Gearea gumum gold brittade *Jared dispensed gold to men,* Cd. 59 : Th. 72, 4; Gen. 1181. Goldes brytta *a dispenser of gold,* 137; Th. 173, 26; Gen. 2867. [*Goth.* gulth : *O. Sax. O. Frs. O. Ger.* gold : *Icel.* gull.] DER. cyne-, fæt-, heáfod-gold.

gold-ǽht, e; *f. A possession or treasure of gold,* Beo. Th. 5489; B. 2748.

gold-beorht; *adj. Bright with gold; auro splendens :*—Beorn monig goldbeorht scán *many a warrior shone bright with golden ornaments,* Exon. 124 b; Th. 477, 33; Ruin. 34.

gold-bleoh; *gen.* -bleós; *n. A golden colour; crisoletus, auricolor,* Ælfc. Gl. 49; Som. 65, 89; Wrt. Voc. 34, 21.

gold-blóma; *m. A golden mass :*—Se hálga Gást wunode on ðam gecorenan hordfæte . . . se goldblóma on ðas world becom and menniscne líchoman ouféng æt Sancta Marian *the Holy Ghost dwelt in the chosen treasury . . . the golden mass came into this world and received a human body from St. Mary,* Blickl. Homl. 105, 18. [*Or* blóma = *bloom, blossom;* cf. *Goth.* blóma : *Icel.* blómi ?] v. blóma.

gold-burh; *gen.* -burge; *f. A town where gold is distributed or which is ornamented with gold,* Andr. Kmbl. 3308; An. 1657: Cd. 119; Th. 154, 2 : Gen. 2549. v. Grm. A. u. E. xxxviii.

gold-fæt, es; *n. A golden vessel :*—Godes goldfatu *God's golden vessels,* Cd. 212; Th. 262, 36; Dan. 755. [*O. Sax.* gold-fat : *O. H. Ger.* golt-faz.]

gold-fæt [-fatu ?], es; *n. A thin plate of gold; bractea, lamina aurea :*—Stáne gelícast gladum gimme ðonne in goldfate sniþa orþoncum biseted weorþeþ *to a stone most like to a bright jewel when by the smiths' art it has been set in a bracelet,* Exon. 60 a; Th. 219, 7; Ph. 303.

gold-fáh; *adj. Variegated or adorned with gold :*—Hió becwiþ him hyre goldfágan treówenan cuppan *she bequeaths to him her wooden cup ornamented with gold,* Th. Chart. 536, 17 : Beo. Th. 621; B. 308: 5615; B. 2811. Goldfág scinon web æfter wagum *the hangings along the walls shone interwoven with gold,* 1993; B. 994. [*Laym.* gold-fah, -faȝe, -fawe.]

gold-fell, es; *n. Gold skin, gold leaf; bractea,* Cot. 24, Lye.

gold-finc, es; *m. A gold-finch; auricinctus, florentius,* Ælfc. Gl. 37; Som. 62, 126; Wrt. Voc. 29, 21 : 38; Som. 63, 36; Wrt. Voc. 29, 54.

gold-finger, es; *m. The ring-finger :*—Goldfinger *mei : us vel annularis,* Ælfc. Gl. 73; Som. 71, 21; Wrt. Voc. 44, 7 : L. Alf. pol. 59; Th. i. 96, 5 : L. Eth. 54; Th. i. 16, 12.

gold-frætwe; *pl. f. Gold ornaments,* Exon. 22 b; Th. 62, 3; Cri. 996.

gold-geweorc, es; *n. Gold-work, what is made of gold :*—Ðǽr wæs ðǽre sunnan anlícnys geworht of golde and heó wæs on gyldenum scryd and æt ðam wǽron gyldene hors . . . ða eode ðǽr egeslíc deóful út of ðam goldgeweorce and ðæt goldgeweorc eall todreás swá swá weax gemylt æt fýre *there was an image of the sun made of gold, and it was on a golden chariot, and there were golden horses to the chariot . . . then came there a horrible devil out of the goldwork, and the goldwork all fell away as wax melts at the fire,* Shrn. 156, 10-16.

gold-gifa, an; *m. A giver of gold, a liberal lord or chief :*—Funde ða

on bedde blácne licgan his goldgifan *he found then his lord lying pale on the bed*, Judih. 12; Thw. 25, 26; Jud. 279. Goldgyfan, Beo. Th. 5297; B. 2652. Cyninges ne cáseras ne goldgiefan *neither kings nor emperors nor lords*, Exon. 82 b; Th. 310, 31; Seef. 83. *See other compounds under* gifa.

gold-hama, an; *m. A gilded* or *golden coat of mail*, Elen. Kmbl. 1980; El. 992.

gold-hilted; *adj. Having a golden hilt*, Exon. 114 a; Th. 437, 27; Rä. 56, 14.

gold-hladen; *adj. Adorned with gold*, Fins. Th. 26; Fin. 13.

gold-hord, es; *m. n. A treasure, treasury*; thesaurus :— Nellen gē goldhordian eów goldhordas on eorþan . . . goldhordiaþ eów goldhordas on heofenan . . . ðær ðín goldhord is ðær is ðín heorte *nolite thesaurizare vobis thesauros in terra . . . thesaurizate vobis thesauros in cælo . . . ubi est thesaurus tuus, ibi est cor tuum*, Mt. Kmbl. 6, 19–21 : 2, 11 : 13, 44, 52 : Exon. 19 b; Th. 49, 18; Cri. 787. Goldhord dǽlan *to distribute treasure*, Cd. 173; Th. 216, 16; Dan. 2. Ðæt goldhord, ðæt yldum wæs lange behýded *the treasure that was long hidden from men*, Elen. Kmbl. 1578; El. 791. Goldhord *thesaurarium*, Ælfc. Gl. 109; Som. 79, 23; Wrt. Voc. 58, 63. He gesette ðone gársecg on his goldhorde *ponens in thesauris abyssos*, Ps. Th. 32, 6. Ðe forþlǽdeþ fægere windas of his goldhordum *qui producit ventos de thesauris suis*, Ps. Th. 134, 8.

gold-hord-hús, es; *n. A privy*; ypodromum, Ælfc. Gl. 107; Som. 78, 80; Wrt. Voc. 57, 57, see note.

gold-hordian; *p.* ode; *pp.* od *To hoard, lay up treasure*; thesaurizare. Mt. Kmbl. 6, 19, 20.

gold-hroden; *adj. Adorned with gold* :—Cwén goldhroden *the queen adorned with gold*, Beo. Th. 1232; B. 614 : 1285; B. 640 : 3900; B. 1948 : 4054; B. 2025 : Exon. 86 a; Th. 324, 29; Víd. 102.

gold-hwǽte; *adj. Greedy for gold*, Beo. Th. 6140; B. 3074.

gold-lǽfra, an; *m. Gold-leaf*; bractea, Cot. 207, Lye.

gold-mǽstling, -mǽslinc, es; *n. Brass, latten*; auricalcum, Ælfc. Gr. 8; Som. 7, 65; Wrt. Voc. 85, 8.

gold-máðum, es; *m. A precious thing made of gold, treasure*, Beo. Th. 4820; B. 2414.

gold-sele, es; *m. A hall in which gold is distributed*, or *one adorned with gold*, Beo. Th. 1434; B. 715 : 2510; B. 1253 : 3282; B. 1639 : 4172; B. 2083. [Cf. gold-burh.]

gold-siowod *auro satus, acupictus, segmentatus*, Cot. 178, Lye.

gold-smiþ; es; *m. A goldsmith, worker in gold*; aurifex :—Tubalcain wæs ǽgðer ge goldsmiþ ge ísensmiþ *Tubalcain was a worker both in gold and in iron*, Gen. 4, 22. Goldsmið *aurifex*, Coll. Monast. Th. 29, 35 : Homl. Th. i. 64, 8. Ðe Eádréd cyng gebócode Ælfsige his goldsmiþe *which king Edred gave by charter to his goldsmith Ælfsig*, Cod. Dipl. Kmbl. iii. 431, 24 : vi. 211, 7 : Bt. Met. Fox 10, 67; Met. 10, 34.

gold-smiþu, e; *f. The art of the goldsmith*, Exon. 88; Th. 331, 24; Vy. 73.

gold-spédig; *adj. Wealthy*, Exon. 66 a; Th. 245, 3; Jul. 39.

gold-þeóf, es; *m. One who steals gold*, L. Alf. pol. 9; Th. i. 68, 5.

gold-torht; *adj. Bright like gold*, Exon. 93 b; Th. 351, 11; Sch. 78.

gold-weard, es; *m. A guardian of gold* [*a dragon*], Beo. Th. 6154; B. 3081.

gold-wine, es; *m. A liberal and kindly prince*, Judith. 10; Thw. 21, 17; Jud. 22 : Beo. Th. 2346; B. 1171 : 2956; B. 1476 : 4829; B. 2419 : 5161; B. 2584 : Elen. Kmbl. 401; El. 201 : Exon. 77 a; Th. 288, 23; Wand. 35 : 76 b; Th. 287, 31; Wand. 22.

gold-wlanc; *adj. Splendidly adorned with gold*, Beo. Th. 3766; B. 1881 : Salm. Kmbl. 414; Sal. 207.

gold-wlencu, e; *f. A golden ornament* :—Ðonne ne gefultumiaþ ðære sáule ðara gimma frætwednes, ne ðara goldwlenca nán *then the adornment of the gems does not help the soul, nor any of the golden ornaments*, Blickl. Homl. 195, 11.

GÓMA, an; *m. The palate; in pl. the fauces* :—Góma *vel* hróf ðæs múþes *palatum vel uranon*, Ælfc. Gl. 71; Som. 70, 107; Wrt. Voc. 43, 35. Góma *palatum*, Wrt. Voc. 70, 53. Ðes góma *hæc faux*, Ælfc. Gr. 9, 71; Som. 14, 15. Ic eom on góman swétra ðonne ðú beóbreád blende mid hunige *sweeter am I on the palate than didst thou blend honeycomb with honey*, Exon. 111 a; Th. 425, 18; Rä. 41, 58 : 113 a; Th. 433, 11; Rä. 50, 6. Ic dó ðæt ðín tunge clifaþ to ðínum góman *linguam tuam adherescere faciam palato tuo*, Homl. Th. ii. 530, 28. Me syndan góman hǽse *raucæ factæ sunt fauces meæ*, Ps. Th. 68, 3 : Soul Kmbl. 216; Seel. 110. Me is on gómum gód and swéte ðín ágen word *quam dulcia faucibus meis eloquia tua*, Ps. Th. 118, 103 : 136, 5 : 149, 6. He ða grimman góman bihlemmeþ fæste togædre *he clashes fast together the fierce jaws*, Exon. 97 b; Th. 364, 26; Wal. 76. [*Prompt. Parv.* gome *gingiva* : *Icel.* gómr *the palate* : *O. H. Ger.* guomo, gaumo, giumo *guttur, faux, palatum* : *Ger.* gaum, gaumen *the palate*.]

gombe, an; *f. Tribute* :—Niéde sceoldon gombon gieldon and gafol sellan *needs must they pay tribute and tax*, Cd. 93; Th. 119, 11; Gen. 1978. Gomban gyldan, Beo. Th. 21; B. 11. [Cf. *O. Sax.* gambra, *used with* geldan.]

gomel, gomol, gamel, gamol; *adj. Advanced in age, aged, old, ancient*; ætāte provectus, senex, vetustus, vetus :—Se fugel weorþeþ gomel *the bird becomes old*, Exon. 59 b; Th. 215, 24; Ph. 258 : Beo. Th. 5578; B. 2793. Ahleóp se gomela *the aged* [*man*] *leapt up*, 2798; B. 1397 : 5695; B. 2851. Biþ geómorlíc gomelum eorle *it is sad for an aged man*, 4880; B. 2444. Gomele ymb gódne ongeador sprǽcon *the old spake together about the good* [*warrior*], 3194; B. 1595. Se on him gyrdeþ gomelra láfe *he girds on him the relic of the ancients*, 4079; B. 2036. Forbærst sweord Beówulfes, gomol and grǽgmǽl *Beowulf's sword burst asunder, ancient and grey-marked*, 5357; B. 2682. Se gomola eald úþwíta *the ancient old sage*, Exon. 81 a; Th. 304, 5; Fä. 65.

gomel-feax, gomol-feax, gamol-feax; *adj. Hoary-locked, grey-haired*; cānus :—Gomelfeax gnornaþ *the hoary-locked grieves*, Exon. 82 b; Th. 311, 14; Seef. 92.

gomel-ferhþ *aged*. v. gamol-ferhþ.

gomen, es; *n. Game, joy, mirth, sport*; jŏcus, jūbĭlium, lætĭtia, lūdus :—Nis ðær gomen in geardum *there is no mirth in the courts*, Beo. Th. 4909; B. 2459 : 4518; B. 2263 : 3554; B. 1775. v. gamen.

gomen-wǽdu, e; *f. A joyous path*; lætum iter :—Gewiton ealdgesíþas of gomenwáde *the old comrades departed from the joyous path*, Beo. Th. 1713; B. 854.

gomen-wudu; *gen.* -wuda; *m. Pleasure-wood, glee-wood, a musical instrument, harp*; lætĭtiæ lignum, harpa = ἄρπη :—Ðær wæs sang and swēg samod ætgædere, gomenwudu grēted *there were song and sound at once together, the glee-wood* [*was*] *touched*, Beo. Th. 2134; B. 1065. Ðær wæs gidd and gleó, hwílum he hearpan wynne, gomenwudu grētte *there was song and glee, at times he touched the joy of harp, the wood of mirth*, 4222; B. 2108.

gomol *old, ancient*, Beo. Th. 5357; B. 2682 : Exon. 81 a; Th. 304, 5; Fä. 65. v. gomel.

gomol-feax; *adj. Hoary-locked, grey-haired*; cānus :—Gomolfeax hælep *a hoary-locked hero*, Chr. 975; Th. 228, 27, col. 2, 3. v. gomel-feax.

Gomorringas; *pl. a The people of Gomorrha* :—Eorþe gomorringa *terra Gomorræorum*, Mt. Kmbl. Rush. 10, 15.

gond. v. geond.

gong. v. gang.

gongan. v. gangan.

gongel, *found in composition as in* fæst-gongel, wíd-. v. gangel-, gongel-wæfre.

gongel-wæfre, an; *f. A ganging weaver, a spider*; arānea viātica :—Wið gongelwæfran bíte *for the bite of a spider*, L. M. 2, 65; Lchdm. ii. 296, 17 : 2. 48; Lchdm. ii. 142. 23. v. gange-wifre.

gonge-wifre, an; *f. A ganging weaver, a spider*; arānea viātica :—Wið gongewifran bíte *against the bite of a spider* [*gangweaver*], L. M. 3, 35; Lchdm. ii. 328, 10. v. gange-wifre.

good. v. gód.

gop, es; *m. A captive, slave* [?]. Cf. geópan *and Icel.* hergopa; *f. one taken in war, a bondwoman. Or is the word connected with* geap *crafty?* —þurh gopes hond, Exon. 113 a; Th. 433, 5; Rä. 50, 3.

gor, es; *n. Dung, dirt*; fimus. lutum, coenum :—Ðæs cealfes flǽsc, and fell, and gor ðú bærnst úte bútan fyrdwícon *carnes vituli, et corium, et fimum combures foris extra castra*, Ex. 29, 14. Ðæs gores sunu, ðone we wifel nemnaþ *son of the dung, which we call* [*dung-*] *beetle*, Exon. 111 a; Th. 426, 11; Rä. 41, 72. Mid swínenum gore *with swine dung*, Herb. 9, 3; Lchdm. i. 100, 11. Feares gor *bull's dung*, Med. ex Quadr. 11, 10, 11, 12; Lchdm. i. 368, 5, 7, 9. Gor *sordem*, Wrt. Voc. 65, 34. [*Prompt. Parv.* gore *limus* : *Icel.* gor : *O. H. Ger.* gor *fimus*.]

gorst, gost, es; *m.* GORSE, *furze, bramble* :—Deós wyrt ðe man tribulus and óðrum naman gorst nemneþ *this plant, which is named tribulus, and by another name gorse*, Herb. 142, 1; Lchdm. i. 262, 16. Of gorstum *de tribolis*, Mt. Kmbl. Rush. 7, 16. Ne wínberian on gorste ne nimaþ *neque de rubo vindemiant uvam*, Lk. Skt. 6, 44. Ðá hēt ualerianus gebindan ðysne ypolitum on wildu hors ðæt hyne drǽgon on gorstas and on þornas *then bade Valerian to bind this Hypolitus on wild horses that they might drag him into the brambles and thorns*, Shrn. 117, 13. Iuniperi ðæt is gorst *juniper that is gorse*, L. M. i. 31, 3; Lchdm. ii. 72, 10. Gost *accidenetum*, Wrt. Voc. 33, 32. [*Prompt. Parv.* fyrrys *or* gorstys tre *ruscus*, p. 162, v. note.]

gorst-beám, es; *m. A bramble*; rubus :—Ofer ðone gorstbeám *super rubum*, Mk. Skt. 12, 26.

GÓS; *gen.* gés; *dat.* gés; *acc.* gós; *pl. nom. acc.* gés, gees; *gen.* gósa; *dat.* gósum; *f. A* GOOSE; anser :—Gós *auca*, Ælfc. Gl. 36; Som. 62, 119; Wrt. Voc. 29, 15 : 77, 32. Grǽg gós *a grey goose*, Cot. 99, Lye. Hwílum ic grǽde swá gós *sometimes I cry as a goose*, Exon. 106 b; Th. 406, 18; Rä. 25, 3. Gees [gés, MS. H.] *geese*, L. In. 70; Th. i. 146, 18. [*Icel.* gás : *O. H. Ger.* gans : *Lat.* anser : *Gk.* χήν.]

gós-fugol, es; *m. A goose*, Th. Chart. 471, 31.

gós-hafoc, es; *m. Goshawk*; aucarius, Ælfc. Gl. 36; Som. 62, 120; Wrt. Voc. 29, 16. [*Chauc.* gos-hauk : *Icel.* gás-haukr : *O. H. Ger.* gans-hapich.]

gost. v. gorst.

gôst, Shrn. 152, 35. v. gâst.

Gota, an; *m. A Goth*; Gothus; chiefly used in the *pl*; *nom. acc.* Gotan; *gen.* Gotena; *dat.* Gotum; *m. The Goths:* — Unrîm mânes se Gota fremede *the Goth perpetrated an excess of wickedness,* Bt. Met. Fox 1, 89; Met. 1, 45.   **I.** VISIGOTHS or *West Goths, under Alríca,* q. v. A. D. 382-410, etc:—Ða [MS. ðe] Gotan of Sciððiu mægþe, wið Rômâna ríce gewin upahôfon; and mið heora cyningum, Rædgota and Eallerîca [Alríca] wæron hâtne, Rômâne burig abrâecon *the Goths, from the country of Scythia, made war against the empire of the Romans; and with their kings, who were called Rhadgast and Alaric, sacked the Roman city* [A. D. 410], Bt. 1; Fox 2, 1. Seó hergung wæs þurh Alarîcum [*acc. Lat.*] Gotena cyning geworden *hæc inruptio per Alarîcum regem Gothorum facta est,* Bd. 1, 11; S. 480, 11. Ða Gotan coman of ðám hwatestan mannan Germania *the Goths came from the bravest men of Germany,* Ors. 1, 10; Bos. 34, 5, 11.   **II.** OSTROGOTHS, or *East Goths, under Ermanric, Þeódric,* q. v. A. D. 475-526, etc:—Gotan eástan of Sciððia sceldas lǽddon *Goths from the east led their army from Scythia,* Bt. Met. Fox 1, 2; Met. 1, 1. Hû Gotan gewunnon Rômâna ríce *how the Goths conquered the empire of the Romans,* Bt. titl. i; Fox x. 2. Eormanríc âhte wîde folc Gotena ríces *Ermanric possessed the wide nations of the kingdom of the Goths,* Exon. 100 b; Th. 378, 28; Deor. 23: 86 a: Th. 324, 3; Wid. 89: 86 b; Th. 325, 10; Wid. 109. Weóld Eormanríc Gotum *Ermanric ruled the Goths,* Exon. 85 a; Th. 319, 27; Wid. 18. [*Icel.* Goti, *pl.* Gotnar.] v. Grmm. Gesch. D. S. c. xviii.

Got-land GOTHLAND; Gothia, Ors. 1, 1; Bos. 22, 2.

gôt-woþe, an; *f. Goatweed;* ægopodium podagraria, L. M. i. 31, 7; Lchdm. ii. 74, 19: 38, 3; Lchdm. ii. 92, 7.

goung, e; *f. A sighing, sobbing, mourning;* gemitus:—On ðæs tuddres forþlǽdnysse biþ goung and sâr *in prolis prolatione gemitus,* Bd. 1, 27 resp. 8; S. 493, 21. [*Cf.*[?] *Gk.* γοάν *to sigh.*]

grad, es; *m.* [*Lat.* gradus] *A* GRADE, *step, order, degree, rank*; gradus, ordo:—Seofon stapas sindon ciriclîcra grada and hâligra hâda *seven are the degrees of ecclesiastical ranks,* L. E. B. 1; Th. ii. 240, 2. Blôd com uppon þám gradan and of þám gradan on þa flôre *blood came upon the steps and from the steps on the floor,* Chr. 1083; Erl. 217, 28. Æt sumum sǽle ætslâd se hâlga wer on ðám heálicum gradum æt ðam hâlgum weofode *on one occasion the holy man slipped on the tall steps at the holy altar,* Homl. Th. ii. 512, 11.

grǽd, es; *m. Greed, rapacity;* aviditas:—Fuglas hungrige grǽdum gîfre *birds hungry, greedily voracious,* Exon. 43 a; Th. 146, 15; Gú. 710. [*Goth.* grêdus: *Icel.* grâðr *hunger, greed.*]

grǽdan; *p. de To cry, call out;* clamare:—Ic grǽde swâ gôs *I cry like a goose,* Exon. 106 b; Th. 406, 18; Rä. 25, 3. Ðonne grǽt se lâreów swâ swâ kok on niht *prædicâtor clamat quasi gallus cantat in nocte,* Past. 63; Swt. 459, 32; Hat. MS. Hine mon sceal swîðe hlûde hâtan grǽdan oððe singan *he must be bidden to cry out or sing very loud,* L. M. 2, 5; Lchdm. ii. 182, 26. [*A. R. Piers P.* greden: *Laym.* grædde;]

grǽde, es; *m. Grass, a herb;* gramen:—Grǽde ulva, Ælfc. Gl. 42; Som. 64, 23; Wrt. Voc. 34, 33. Grǽdas gramina, Cot. 95, Lye.

grǽdig; *adj.* GREEDY, *covetous;* avidus:—Grǽdig vorator, Ælfc. Gl. 88; Som. 74, 82; Wrt. Voc. 50, 62. Ða fýnd heora grîpende wǽron swâ swâ grǽdig wulf *the devils were seizing them like the ravening wolf,* Blickl. Homl. 211, 1. Líg grǽdig swelgeþ londes frǽtwe *flame, greedy, swallows the land's treasures,* Exon. 63 a; Th. 232, 16; Ph. 507: Beo. Th. 242; B. 121: 3002; B. 1497. Sum to lyt hafaþ gôdes grǽdig *one hath too little, eager for goods,* Saln. Kmbl. 689; Sal. 344. Ðá getîmode swâ ðé þ ðam grǽdigan fisce ðe gesihþ ðæt ǽs and ne gesihþ ðone angel ðe on ðam ǽse sticaþ *then it befel as it does to the greedy fish that sees the bait but sees not the hook which sticks in the bait,* Homl. Th. i. 216, 10. Helle grǽdige and gîfre *hell greedy and ravenous,* Cd. 37; Th. 49, 16; Gen. 793: 217; Th. 276, 21; Sat. 192. León-hwelpas sécaþ ðæt him grǽdigum æt God gedême *catuli leonum ... quærant a Deo escam sibi,* Ps. Th. 103, 20. Gîfrost and grǽdgost *most rapacious and most greedy,* Exon. 128 a; Th. 493, 2; Rä. 81, 24. [*Goth.* grêdags: *O. Sax.* grâdag: *Icel.* grâðugr: *O. H. Ger.* grâtag.] DER. heoro-, hilde-, wǽl-grǽdig.

grǽdig-, grǽdi-, grǽde-lîce; *adv.* GREEDILY, *covetously;* avidè:—He gýmþ grǽdelîce his teolunge *he attends greedily to his gain,* Homl. Th. i. 66, 10. Ðás fugelas habbaþ feónda gelícnysse ðe gehwilce menn beswîcaþ and grǽdelîce grîpaþ to grimre helle *these birds are like the fiends, that deceive some men, and greedily snatch them to grim hell,* ii. 516, 10. Ðonne him hingraþ he yt grǽdilîce *when he is hungry he eats greedily,* Hexam. 20; Norm. 28, 21.

grǽdignes, se; *f.* GREEDINESS, *covetousness;* aviditas:—Grǽdinesse he lufode *covetousness he loved,* Chr. 1086; Erl. 222, 25. Eorþlícan grǽdignysse *greediness after earthly things,* Boutr. Scrd. 20, 11.

grǽf, graf, es; *n. A grave, trench:* — Æt openum grǽfe *at the open grave,* L. Æthelb. 22; Th. i. 8, 5: L. Eth. 5, 12; Th. i. 308, 4: 6, 20; Th. i. 320, 4; Exon. 82 b; Th. 311, 24; Seef. 97: 91 b; Th. 342, 29;

Gn. Ex. 149. Ic ongyte ðeáh ðæt ða worlde lustas ne sint eallunga awyrtwalode of ðínum môde ðeáh se graf geryd sí *I perceive however that worldly pleasures are not entirely rooted out of thy mind, though the trench be sufficient,* Shrn. 184, 20. [*O. Sax.* graf: *O. Frs.* greb: *O. H. Ger.* grab: *Ger.* grab; *n: Goth.* graba; *Icel.* gröf; *f.*] DER. eorþ-, fold-, mold-grǽf.

grǽf, es; *n. A graving instrument, a style:*—Grǽf graffium, Ælfc. Gl. 8; Wrt. Voc. 75, 17: graphium vel scriptorium, Ælfc. Gl. 80; Som. 72, 114; Wrt. Voc. 46, 71.

grǽfa, grǽfe [?], an;—Twælf fôður grǽfan, Chr. 852; Erl. 67, 38. Earle in his note on this word, p. 300, suggests a translation other than that given by previous editors. By them it has been translated ' coal,' he suggests 'gravel.' The word may be of Celtic origin, and so may be compared with Old French *grave,* of which *gravel* is a diminutive. Celtic forms are Bret. grouan *gravel:* Corn. grow *gravel, sand:* W. gro *pebbles.*

grǽfa, gréfa, an; *m. A pit, cave, hole:*—Grǽfe speluncam, Mt. Kmbl. Lind. 21, 13. See Cod. Dipl. Kmbl. iii. xxvii. [*Cf. Goth.* grôba; *f. a hole: Icel.* gröf; *f. a pit: O. H. Ger.* grôba; *f. fovea, scrobs, barathrum: Ger.* grube.]

grǽfe, grafere, es; *m. A graver, an engraver:*—Grǽfere sculptor vel celator, Ælfc. Gl. 81; Som. 72, 121; Wrt. Voc. 47, 4.

grǽf-hús, es; *n. A grave-house, house of the dead:*—Hell grim grǽf-hús *hell the grim house of the dead,* Cd. 228; Th. 309, 11; Sat. 708.

grǽf-seax, -sex, es; *n. A graving knife:*—Grǽfsex scalprum vel scalpellum vel cælum, Ælfc. Gl. 81; Som. 72, 125; Wrt. Voc. 47, 7.

grǽft, es; *m:* grǽft, e; *f.* [?] *Carving, graving, a carved* or *graven image:*—Grǽft sculptura, Ælfc. Gl. 81; Som. 72, 122; Wrt. Voc. 47, 5. Ealle ða ðe gebiddaþ grǽftas *omnes qui adorant sculptilia,* Ps. Lamb. 96, 7: Ps. Spl. C. 105, 19: Homl. Th. i. 464, 27. Írene grǽfta *ferrea sculptilia,* carpenta, Cot. 38, Lye. [*O. H. Ger.* graft, grefti; *f. cælatura, sculptura, sculptile.*]

grǽft-geweorc, es; *n. Carved* or *graven work, a graven image:*—Ne wirce ðú grǽftgeweorc *thou shalt not make any graven image,* Deut. 5, 8.

grǽg, grég; *adj. Grey:*—Grég glaucus, Ælfc. Gl. 79; Som. 72, 90; Wrt. Voc. 46, 47. Deorce grǽg elbus, Wrt. Voc. 46, 48. Grǽg hwǽte far, Ælfc. Gr. 9, 17; Som. 9, 52. Se grǽga mǽw *the grey mew,* Andr. Kmbl. 742; An. 371. Wulf se grǽga *the grey wolf,* Exon. 91 b; Th. 343, 3; Gn. Ex. 151: Chr. 937; Erl. 115, 13; Ædelst. 64. Sǽ grǽge glashluðre *the sea grey and clear as glass,* Bt. Met. Fox 5, 15; Met. 5, 8. Grǽgan sweorde *with a grey sword,* Cd. 138; Th. 173, 22; Gen. 2865: Beo. Th. 665; B. 330: 673; B. 334. [*Icel.* grâr: *O. Frs.* grê: *O. H. Ger.* grâw: *Ger.* grau.]

grǽg-, grǽ-gôs *a grey goose, wild goose:*—Grǽg-gôs canta, Wrt. Voc. 280, 15: 62, 11: Mone Gl. 314. [*Icel.* grâ-gâs.]

grég-hama, an; *m. A corslet, coat of mail:*—Gylleþ grǽghama *the corslet rattles,* Fins. Th. 10; Fin. 6. [*Cf.* grǽge syrcan, Beo. Th. 673; B. 334; *and* gullon gúþ searo, Andr. Kmbl. 253; An. 127. Grein takes the word as an adjective *= grey-coated, the grey-coated one,* i. e. *the wolf.* In support of this cf. scírham, and the passages given under ' grǽg,' in which that adjective is applied to the wolf.]

grég-hiwe, -hǽwe; *adj. Of a grey hue* or *colour,* Lye.

grég-mǽl; *adj. Of a grey colour,* Beo. Th. 5357; B. 2682. See under 'grǽg,' the passage in which that adjective is applied to weapons.

grǽp *a grip, furrow, ditch;* sulcus, Som.

grǽs, es; *n. Grass, plant;* grâmen:—On grêne grǽs *on the green grass,* Cd. 56; Th. 69, 17; Gen. 1137. Ða ðe of grǽses deáwe geworht wǽron *those that were made of the dew of grass,* Shrn. 66, 3. Sume hió twiccedan ða grasu mid hiora múþe *some of them pulled the grass with their mouth,* 41, 2: Past. 23, 1; Swt. 173, 20. v. gærs.

grǽs-hoppa, an; *m. A grass-hopper, locust:*—Grǽs-hoppa *locustæ,* Mt. Kmbl. Rush. 3, 4. Hý habbaþ fêt swylce grǽs-hoppan *pedes quasi locuste,* Nar. 35, 7. v. gærs-hoppa.

grǽs-molde, an; *f. Grassland, greensward;* campus graminibus viridis:—Beówulf grǽs-moldan træd *Beowulf trod the greensward* [grassy mould], Beo. Th. 3767; B. 1881.

grǽs-wang, -wong, es; *m. A grassy plain,* Exon. 57 a; Th. 203, 2; Ph. 78: 65 b; Th. 243, 5; Jul. 6.

grǽtan *to bewail.* v. grêtan.

grǽtta GRITS, *groats, bran;* farina crassior, furfur, Som. v. gryt.

graf. v. grǽf.

gráf, es; *m. n. A grove:*—Heó hæbbe ða wuduræddenne in ðæm wuda ðe ða ceorlas brúcaþ and éc ic hire lête to ðæt ceorla gráf *let her have right of pasturage in the wood which the 'ceorls' use, and besides I leave to her the 'ceorls'' grove,* Cod. Dipl. Kmbl. ii. 100, 14. Andlang ðære lytlan díc æt ðæs gráfes ende *along the little ditch at the end of the grove,* 249, 29. Forþ be ðam gráfe *along past the grove,* iii. 18, 31. Ðone gráf, 52, 23. Eác we wrîtaþ him ðone gráf ðǽrto. Ðis syndon ða gemǽru ðe to ðæm gráfe gebyriaþ *also we assign to him in addition the grove. These are the boundaries that belong to the grove,* 261, 5-7. [*Laym.* groue: *Prompt. Parv.* grove lucus.]

grafan, ic grafe, græfe, ðú græfest, græfst, he græfeþ, græfþ, *pl.* grafaþ; *p.* gróf, *pl.* grófon; *pp.* grafen. **I.** *to dig, delve, dig up;* fodere, effodere :—Ic be grunde græfe *I dig along the ground,* Exon. 106 a; Th. 403, 3: Rä. 22, 2. Ðæt fýr græfeþ grimlíce eorþan sceátas *the fire shall fiercely delve the tracts of earth,* Exon. 22 b; Th. 62, 19; Cri. 1004: 95 a; Th. 354, 55; Reim. 66. Se forma feohgîtsere gróf æfter golde *the first miser delved after gold,* Bt. Met. Fox 8, 113; Met. 8, 57: Exon. 109 a; Th. 416, 4; Rä. 34, 6: 130 a; Th. 498, 24; Rä. 88, 6. Ðæt ic grófe græf *that I may dig a grave,* Exon. 95 a; Th. 355, 3; Reim. 71. **II.** *to* GRAVE, *engrave, carve;* sculpere, cælare :—Ic grafe *sculpo,* Ælfc. Gr. 28, 4; Som. 31, 20. Ðonne hî wôhgodu worhtan and grófun *in sculptilibus suis emulati sunt eum,* Ps. Th. 77, 58. Ac hý grófon æghwylcne stán swá se cásere geþohte *they carved each stone as the emperor designed,* Shrn. 146, 16. [*Laym.* graven : *Prompt. Parv.* gravin *sculpere: O. Sax.* (bi-)graban : *Goth.* graban : *Icel.* grafa : *O.H.Ger.* graban.] DER. a-grafan, be-, bi-.

grafet, es; *n.* A *trench* [?] :—On ðæt lange grauet ðam lange grafette, Cod. Dipl. Kmbl. v. 193, 33: 195, 5, 7. Leo takes the word as a diminutive of ' gráf.'

gram, grom; *adj.* [grama *anger*] *Furious, fierce, wroth, angry, offended, incensed, hostile, troublesome* :—He swá grom wearþ on his môde *he became so incensed;* rex iratus, Ors. 2, 4; Swt. 72, 32: 6, 4; Swt. 260, 23. Drihten wæs ðam folce gram *the Lord was angry with the people,* Deut. 1, 37: Cd. 16; Th. 20, 2; Gen. 302. Wearþ se cyng swîðe gram wið ða burhware *the king was very angry with the citizens,* Chr. 1048; Erl. 178, 6. He wæs on his gáste gram *exacerbaverunt spiritum ejus,* Ps. Th. 105, 25. Ic eom nalæs grames môdes *non sum turbatus,* 118, 60. Of gramum folce *de populo barbaro,* 113, 1. Ðín ðæt grame yrre *thy fierce anger,* 68, 25: 84, 1: 108, 18. Seó eádge biseah ongeán gramum *the blessed maid looked on the fierce one* [*the devil*], Exon. 75 a; Th. 280, 12; Jul. 628: Cd. 27; Th. 36, 35; Gen. 582. Ða graman Gydena *the folcisce men hâtaþ Parcas the fierce goddesses whom common people call Parcæ.* Bt. 35, 6; Fox 168, 24. Grame gûþfrecan *fierce warriors,* Judth. 11; Thw. 24, 35; Jud. 224: Andr. Kmbl. 1833; An. 919: Ps. Th. 104, 30. Grame me forhogedon *my enemies despised me,* 118, 141: 104, 15: Judth. 12; Thw. 25, 2; Jud. 238. Grame manige fremde þeóda *many hostile and strange nations;* alienigenæ, Ps. Th. 82, 6: 118, 138: Exon. 126 b; Th. 485, 26; Rä. 72, 3. Ðær ða graman wunnon *where the fierce ones struggled,* Bèo. Th. 1559; B. 777. In gramra gripe *into the grasp of foes,* Andr. Kmbl. 433; An. 217: 1901; An. 953. Gromra, Cd. 114; Th. 150, 2; Gen. 2485. Deófla strǽlas gromra gârfare *the shafts of devils, the spears of fierce spirits,* Exon. 19 a; Th. 49, 5; Cri. 781. Ne beó ðú ælþeódegum gram *thou shalt not . . . oppress a stranger,* Ex. 23, 9. Ne beó ðú me gram *noli mihi molestus esse,* Lk. Skt. 11, 7: 18, 5. [*Laym.* gram : *Orm.* gramm : *O. Sax.* gram, *the gramo the devil* : *Icel.* gramr *wroth;* *pl.* gramir, gröm *fiends, demons;* see Grmm. D. M. 942–3: *O. H. Ger.* gram *iratus:* *Ger.* gram.]

GRAMA, an; *m. Anger, rage, fury, indignation, wrath, trouble;* ira, furor, molestia :—On graman ðínum *in ira tua,* Ps. Spl. 6, 1: 7, 6. Drihten wearþ yrre mid graman his folce *iratus est furore Dominus in populo suo,* Swt. A. S. Rdr. 73, 54–6: Gen. 19, 25. Ic ondrêd his graman and his yrre *I was afraid of his anger and hot displeasure,* Deut. 9, 19. Ðæne úre yldran for graman to deáþe gedêmdon *whom our elders for anger doomed to death,* H. R. 9, 23. Wel hî sind Dere gehátene forðan ðe hî sind fram graman generode *well are they named Dere* [= *de ira*], *for they are saved from wrath,* Homl. Th. ii. 120, 35: 124, 9. Se upplíca grama *the wrath of heaven,* 538, 28. Æppla gaderian graman getácnaþ *to gather apples betokens trouble,* Lchdm. iii. 212, 21. [*Laym.* grome, grame : *A. R.* grome *anger:* *Chauc.* grame : cf. *O.H.Ger.* grame; *f. exacerbatio:* *Ger.* gram; *m. grief.*]

gramatisc-cræft, es; *m. The art of grammar,* Bd. 4, 2; S. 565, 26.

gram-bǽre; *adj. Angry, passionate;* iracundus, Past. 40, 1; Swt. 289, 5; Hat. MS.

grame, grome; *adv. Fiercely, cruelly, hostilely,* Ps. Th. 57, 5: 68, 3: 93, 2: 123, 7. Grome, Cd. 64; Th. 76, 21; Gen. 1260: 184; Th. 230, 15; Dan. 233: Exon. 89 b; Th. 336, 21; Gn. Ex. 52.

gramfærnys, se; *f. Anger, fury* :—Ælc gramfærnys cymþ of deófle *omnis furor venit a diabolo,* L. Ecg. P. 4, 66; Th. ii. 226, 25.

gram-heort; *adj. Having a fierce, hostile heart* or *mind,* Beo. Th. 3368; B. 1682: Exon. 31 a; Th. 136, 14; Gú. 541: 102 b; Th. 387, 17; Rä. 5, 6. [*O. Sax.* gram-hert.]

gram-hycgende; *part. Having fierce, hostile thought* or *purpose,* Ps. Th. 68, 25.

gram-hydig, -hýdig; *adj. Fierce-minded, hostilely disposed* :—Gramhegdig, Ps. C. 50, 49; Ps. Grn. ii. 278, 49. Gromhýdig guma, Exon. 55 b; Th. 196, 6; Az. 170: 18 b; Th. 46, 8; Cri. 734: Beo. Th. 3502; B. 1749. Ðær næfre feóndes ne biþ gástes gramhýdiges gang *where never shall be fiend's or fierce spirit's walk,* Andr. Kmbl. 1696: Ps. Th. 73, 4. Gromhýdge me oft onginnaþ *injusti insurrexerunt in me,* 85, 13. Gromhýdge, Exon. 38 a; Th. 124, 31; Gú. 346: 116 a; Th. 445, 27; Dôm. 14. [*O. Sax.* gram-hugdig.]

gramlíc; *adj. Fierce, hostile, cruel* :—He hig betǽhte sumum gramlícan cininge Iabin gehâton *he gave them into the hands of a fierce king named Jabin,* Jud. 4, 2. Oñ heora gasthúsum is gramlíc inwit *nequitia in hospitiis eorum,* Ps. Th. 54, 15. [*Icel.* gramligr *vexatious.*]

gramlíce; *adv. Hostilely, evilly, fiercely* :—Gramlíce be Gode sprǽcan *male locuti sunt de Deo,* Ps. Th. 77, 20: 105, 12: Cd. 210; Th. 260, 23; Dan. 714.

gram-môd; *adj. Of fierce* or *cruel mind* :—Hine nǽnig man grammôdne ne funde *no one found him cruel,* Blickl. Homl. 223, 33.

gram-word, es; *n. A word or speech expressing anger, wrath, hate, evil* :—Ne gê wið gode ǽfre gramword sprecan *nolite loqui adversus deum iniquitatem,* Ps. Th. 74, 5.

grandor-, grondor-leás; *adj. Guileless* :—Geong grondorleás *young and guileless,* Exon. 69 b; Th. 258, 26; Jul. 271. [Cf. *Icel.* grandlauss, grandvarr *guileless.*]

gránian; *p.* ode; *pp.* od *To groan, lament, murmur* :—Gránude lamentatæ, Ps. Spl. C. 77, 69. Hî gránedan *murmuraverunt,* Ps. Th. 105, 20. [*Laym.* granien, gronie : *A. R.* gronen : *Prompt. Parv.* gronin *gemere:* cf. *O. H. Ger.* grînan *mutire:* *Ger.* greinen *to cry.*]

Grantabrycgscír *Cambridgeshire.*

Granta-ceaster GRANTCHESTER, *a village near Cambridge,* Bd. 4, 19; S. 588, 30.

Grantan-brycg, e; *f:* Grante brycg, e; *f:* Granta-brycg, e; *f.* [*Hunt.* Grantebrige : *Dunel.* Grantabric, Grantnebrige, Grantebryge : *Hovd.* Grauntebrigge] CAMBRIDGE, *the chief town in Cambridgeshire, and seat of the University;* Cantābrīgia, agri Cantabrigiensis oppidum primarium :—To Grantanbrycge *to Cambridge,* Chr. 875; Th. 144, 9, col. 2 : 145, 9, col. 2: 921; Th. 195, 29. To Grante brycge *to Cambridge,* Chr. 875; Th. 144, 9, col. 1, 3. Forbærndon Granta-bricge *they burned down Cambridge,* Chr. 1000; Th. 264, 5, col. 1: 264, 8, col. 2: 265, 7, col. 1.

gránung, e; *f.* GROANING, *lamentation;* gemitus :—Me ymbhringdon sár and sorga and gránung *circumdederunt me gemitus mortis,* Ps. Th. 17, 4. Mîn gránung ðe nis na forholen *gemitus meus a te non est absconditus,* 37, 9. Wununga on ðám ne ablinþ gránung *dwellings in which groaning ceases not,* Homl. Th. i. 68, 7: L. E. I; Th. ii. 400, 7.

gráp, e; *f. Grasp, clutch* :—Me fæste hæfde on grápe *fast had me in his grasp,* Beo. Th. 1114; B. 555: 881; B. 438. Hond earm and eaxle Grendles grápe *hand, arm, and shoulder, Grendel's grasp,* 1676; B. 836. On grápum *in the clutches,* 1534; B. 765: 3088; B. 1542: Andr. Kmbl. 2671; An. 1337: Exon. 38 b; Th. 126, 28: 47 a; Th. 162, 1. [*Icel.* greip; *f. the space between the thumb and the fingers, a grasp:* *O. H. Ger.* greifa; *f. bidens.*]

GRÁPIAN, grópian; *p.* ode; *pp.* od *To grope, touch, feel with the hands:* — Ic grópige *palpo,* Ælfc. Gr. 24; Som. 25, 42. Grápige, 36; Som. 38, 46. Handa hî habbaþ and hî nâ grápiaþ *manus habent et non palpabunt,* Ps. Spl. 113, 15. Se cuma his cneów grápode mid his hálwendum handum *the stranger felt his knee with his healing hands,* Homl. Th. ii. 134, 35. Hire wið healse heard grápode bánhringas bræc *the hard blade touched her neck, broke the bone-rings,* Beo. Th. 3137; B. 1566: 4176; B. 2085. On ðæt bánleáse brýd grápode hondum *touched with hands that boneless bride,* Exon. 112 b; Th. 431, 20; Rä. 46, 3. Hie wurdon sóna ablinde and grápodan mid heora handum on ða eorþan *they at once became blind and groped on the ground with their hands,* Blickl. Homl. 151, 6. Grápiaþ *palpate,* Lk. Skt. 24, 39. Þýstro swá þicce ðæt hig grápion *darkness that may be felt,* Ex. 10, 21. Ðæt ðú grápie on midne dæg swá se blinda dêþ on þístrum *thou shalt grope at noonday, as the blind gropeth in darkness,* Deut. 28, 29. Ðone líchoman he æteówde to grápigenne *he shewed the body to be touched,* Homl. Th. i. 230, 24. [*O. H. Ger.* greifon *palpare.*]

grápigendlíc; *adj. Tangible* :—His líchama wæs grápigendlíc . . . he æteówde hine grápigendlícne *his body was tangible . . . he shewed himself tangible,* Homl. Th. i. 230, 25, 26.

grasian *to graze* :—Oxan grasiende gesihþ *if he sees oxen grazing,* Lchdm. iii. 200, 9. [*Icel.* gresja *to graze.*]

grátan; *pl. Groats, the grain of oats without the husks* :—Nim átena grátan *take groats of oats,* Lchdm. iii. 292, 24. [Cf. *Icel.* grautr *porridge.*]

Greácas. v. Grécas.

GREÁDA, an; *m. A bosom;* sinus, gremium :—On Habrahames greádan *in sinum Abrahæ,* Lk. Skt. 16, 22, 23. Ða ðe beraþ on hira greádum ða á libbendan fatu *those who bear in their bosoms the ever-living vessels,* Past. 13, 1; Swt. 77, 6; Hat. MS. [*Ayenb.* greade: *Alis.* grede.]

GREÁT; *adj. Great, large, thick, coarse* :—Grǽt *grossus,* Ælfc. Gl. 89; Som. 74, 101; Wrt. Voc. 51, 14. Swá swá greát beám *like a great tree,* Bt. 38, 2; Fox 198, 9. Ædelword Æðelmǽres sunu ðæs grǽtan *Ethelward son of Ethelmer the great,* Chr. 1017; Erl. 161, 7. Tú hund greátes hláfes and þridde smales *two hundred great loaves and a third of small,* Th. Chart. 158, 25. God him send ufan greáte hagolstánas *God cast down upon them great hailstones,* Jos. 10, 11: Cd. 19; Th. 24, 27; Gen. 384. Ða wæron unmetlíce greáte heáhnisse *ingenti grossitudine*

atque altitudine, Nar. 4, 22. Wǽron hie swá greáte swá columnan ge eác sume grýttran *serpentes columnarum grossitudine aliquantulum proceriores*, 14, 15. Greáte swá stǽnene sweras micle *vastitudine columnarum*, 36, 12. Mid greátan sealte *with coarse salt*; cum sale marino, Herb. 37, 5; Lchdm. i. 138, 14. Mid scearpum pílum greátum *with sharp and large stakes*, Chr. Erl. 5, 10. [*Orm.* græt : *Laym.* græt, great : *Chauc.* gret, greet : *O. Sax.* grôt : *O. Frs.* grât : *O. H. Ger.* grôz : *Ger.* gross.]

Greátan leag, leá, e; *f. Probably Greatley, near Andover, Hants*; Greatanleagensis :—Ealle ðis wæs gesetted on ðam miclan synoð æt Greátanleage, on ðam wæs se ærcebisceop Wulfhelme, mid eallum ðǽm ǽdelum mannum, and wiotan [and Æðelstáne cyninge] *all this was established in the great synod at Greatley, in which was the archbishop Wulfhelm, with all the noblemen and witan* [and *King Athelstan*], L. Ath. i. 26; Th. i. 214, 7. To-écan ðám dómum ðe æt Greátanleá and æt Exanceastre gesette wǽron, and æt Þunresfelda *in addition to the dooms which were fixed at Greatley, and at Exeter, and at Thunresfeld*, v. pref; Th. i. 228, 9.

greáte wyrt, e; *f. Meadow saffron*; colchicum autumnale :—Deós wyrt ðe man hieribulbum and óðrum naman greáte wyrt nemneþ *this plant which is called* lepóβoλβos *and by another name great wort*, Herb. 22, 1; Lchdm. i. 118, 14 : L. M. ii. 52, 1; Lchdm. ii. 268, 22.

greátian; *p.* ode; *pp.* od *To* GREATEN, *to become great* or *large*; grandescere, grossescere :—On ðæs siwenígean eágum beóþ ða ǽpplas hále, ac ða brǽwas greátigaþ *in lippi oculis pupillæ sanæ sunt,* sed *palpebræ grossescunt*, Past. 11, 4; Swt. 69, 2; Hatt. MS. 15 a, 18. [*A. R.* greaten *to grow great* : *O. H. Ger.* grôzen *grossescere*.]

greátnes, se; *f.* GREATNESS; magnitudo, R. Ben. 55, Lye.

Grēc *Greek* :—Cwæþende in Grēc *saying in Greek*, Mt. Kmbl. Rush. 27, 46.

Grēcas, Greácas; *gen.* a; *dat.* um; *pl. m. The Greeks*; Græci :—Ðá gefélde he his líchoman healfne dǽl mid ða ádle geslægene beón, ðe Grécas nemnaþ paralysis, we cweðaþ lyft-ádl *then felt he that the half of his body was struck with the illness which the Greeks call paralysis, we call lift-ill*, Bd. 4, 31; S. 610, 16. Of Grēcum *from the Greek*, Ors. 5, 11; Bos. 109, 30. Ðá fôron hí on Greácas *then they went against the Greeks*, Ors. 5, 12; Bos. 110, 38. Greáca land *land of the Greeks*, 5, 11; Bos. 109, 28.

Grēcisc, Grēccisc; *adj. Greek, Grecian* :—Heora discipulas wǽron well gelǽrede ge on Grēcisc gereorde ge on Lēdennisc *eorum discipuli Latinam Græcamque linguam æque ut propriam in qua nati sunt norunt*, Bd. 4, 2; S. 565, 27 : 4, 1; S. 563, 33. Grēccisc, 5, 8; S. 622, 2. Grecus grēcisc of ðam grecisso and grecor ic leornige grēcisc *Grecus Greek of which grecisso and grecor I learn Greek*, Ælfc. Gr. 36; Som. 38, 32. On grēcisc *in Greek*, Jn. Skt. Lind. 21, 2. On indisc and on grēcisc sprecende *indice et græce loquentes*, Nar. 25, 16. Ða grēciscan onginnaþ hyra geár æt ðam sunnstede *the Greeks begin their year at the solstice*, Lchdm. iii. 246, 18. [*Laym.* grickisc : *O. H. Ger.* grecisc : *Ger.* griechisch.]

Grēc-land, es; *n. Greece* :—Dionisius gewende on ðam tíman fram Grēclande *Dionisius returned at that time from Greece*, Homl. Th. i. 558, 33. [*Laym.* griclond.]

grēdig. v. grǽdig.

Gregorius; *gen.* Gregories; *dat.* Gregorie; *acc.* Gregorium; *m. Gregory the Great, Pope* A. D. 590–604, *who sent Augustine and other missionaries to England in* 597 ;—Gregorius se hálga pápa, Engliscre þeóde apostol, wæs of æðelborenre mǽghe acenned. . . . Felix, se eáwfæsta pápa, wæs his fifta fæder. . . . Gregorius is Grēcisc nama [= Γρηγόριος *watchful, from* γρηγορέω *I watch*], se sweigþ on Lēdenum gereorde, *Uigilantius,* ðæt is on Englisc *Wacolre. Gregory the holy pope, the apostle of the English, was born of a noble family*. . . . *Felix, the pious pope, was his fifth father*. . . . *Gregorius is a Greek name which in the Latin tongue signifies Vigilantius, that is in English* Watchful, Homl. Th. ii. 116, 24; 118, 8, 12. Æt Gregories ǽrendracan *from Gregory's messenger*, Homl. Th. ii. 122, 29. Augustínus cýdde ðam eádigan Gregorie, ðæt Angelcynn cristendóm underfêng *Augustine announced to the blessed Gregory, that the English nation had received Christianity*, 130, 24. Ðæt ðæt folc Gregorium to pápan gecoren hæfde *that the people had chosen Gregory for pope*, 122, 31. Gregorius asende ǽrendracan to ðisum íglande. . . . Ðæra ǽrendracena naman synd,—Agustinus, Mellitus, Laurentius, Petrus, Iohannes, Iustus. Ðás láreówas asende se eádiga pápa Gregorius, mid manigum óðrum munecum, to Angelcynne. . . . Agustínus ðá mid his geférum ðæt synd gerehte feówertig ðe férdon be Gregories hǽse, óððæt hí becômon gesundfullíce to ðisum íglande *Gregory sent messengers to this island*. . . . *The names of these messengers are,—Augustinus, Mellitus, Laurentius, Petrus, Johannes, Justus. These teachers the blessed pope Gregory sent, with many other monks, to the English nation*. . . . *Augustine then with his companions, who are reckoned at forty men, journeyed by Gregory's command, till they came safely to this island*, Nat. S. Greg. Els. 28, 10–13; 28, 19–29, 6; 31, 15–32, 5.

gremettan *to rage, roar* :—Ic gremette *fremo*, Ælfc. Gr. 28; Som. 30, 60. [*O. H. Ger.* gremizon *fremere, rugire*.] v. grimetian.

gremetunc, gremetung, e; *f. A raging, roaring, murmuring*; fremitus, Prov. 19, Lye. v. grimetung.

gremian; *p.* ede; *pp.* ed *To provoke, irritate, exasperate, vex, revile* :—He ða óðre elpendas gremede *it irritated the other elephants*, Ors. 4, 1; Bos. 77, 23. Gremedon *exacerbaverunt*, Blickl. Gl. Hig me gremedon *they provoked me*, Lev. 26, 40 : Num. 11, 20 : Deut. 9, 7, 8. Ða ðe forþstôpon hine gremedon *prætereuntes blasphemabunt eum*, Mk. Skt. 15, 29 : Lk. Skt. 23, 39. [*Laym.* gromien, gramie *irritare* : *A. R.* gremeþ *irritat* : *Goth.* gramjan : *Icel.* gremja : *O. H. Ger.* gremian *irritare, objurgare* : *Ger.* grämen.]

Grēna-wíc, Grēne-wíc, es; *n.* GREENWICH, *near London*, Chr. 1013; Erl. 149, 4.

Grendel; *gen.* Grendles GRENDEL, *a monster destroyed by Béowulf* :—Grendel mǽre mearcstapa, se ðe môras heóld, fen and fæsten *Grendel the great traverser of the march, that ruled* [*held*] *the moors, the fen and fastness*, Beo. Th. 205–208; B. 102–104. [Grendel] reste genam þrítig þegna : gewát to hám mid ðære wælfylle [*Grendel*] *took thirty thanes in their rest : departed to his home with the slaughtered corpses*, 245–250; B. 122–125. Grendles môdor *Grendle's mother*, Beo. Th. 3078–3085; B. 1537–1540 : 3139–3141; B. 1567–1568.

GRĒNE; *adj. Green*; viridis :—Grēne *viridis*, Ælfc. Gl. 79; Som. 72, 80; Wrt. Voc. 46, 37. Wende man ðæt grēne to ðan weofode *let the green* [*side of the sods*] *be turned to the altar*, Lchdm. i. 398, 17. Grēne folde *the green earth*, Cd. 76; Th. 94, 14; Gen. 1561. Of grēnum áre geworht *wrought of green copper*, Blickl. Homl. 127, 7. On grēnum treówe *in viridi ligno*, Lk. Skt. 23, 31. Grēne eorþan *green earth*, Cd. 91; Th. 115, 18; Gen. 1921. Grēne bearwas *green groves*, 72; Th. 89, 13; Gen. 1480. Genim ðære ylcan wyrte leáf ðonne heó grēnost beó *take the leaves of the same plant when it is greenest*, Herb. 1, 4; Lchdm. i. 72, 7. [*O. Sax.* grôni : *O. Frs.* grêne : *Icel.* grænn : *O. H. Ger.* gruoni : *Ger.* grün.]

grēnian *to become green, to flourish*; virescere, Bt. Met. Fox 11, 114; Met. 11, 57. [*A. R.* greneþ; *pres. indic :* Ayenb. greni : *Prompt. Parv.* grenyn *vireo* : *Icel.* grôna : *O. H. Ger.* gruonan *virescere* : *Ger.* grünen.]

grēnnes, se; *f.* GREENNESS; viriditas, Bd. 3, 10; S. 534, 21.

grennian; *p.* ode *To grin, shew the teeth as an expression of pain, anger, etc* ; ringere :—Ic grennige *ringo*, Ælfc. Gr. 28; Som. 31, 63. He grennade and gristbitade *he grinned and gnashed his teeth*, Exon. 74 b; Th. 278, 11; Jul. 596. Grenniendum welerum hleahter forþbringan *ringentibus labiis risum proferre*, Scint. 55, Lye. [*Laym. A. R.* grennen : *Prompt. Parv.* grennyn *ringo* : *Icel.* grenja *to howl* : *O. H. Ger.* grennat *mutiet*.]

grennung, e; *f.* GRINNING; rictus, Som. [*A. R.* grennung : *Prompt. Parv.* grennynge *rictus*.]

greofa, greaua *a pot* ; olla, Cot. 146, 173, Lye.

greósan, ic greóse, ðú grýst, he grýst, *pl.* greósaþ; *p.* greás, *pl.* gruron; *pp.* groren *To frighten.* DER. be-greósan.

GREÓT, es; *n.* GRIT, *sand, dust, earth, gravel*; pulvis :—Hēt ðæt greót útawegan *he ordered the earth to be removed*, Homl. Th. i. 74, 24. Ðú scealt greót etan *dust shalt thou eat*, Cd. 43; Th. 59, 9; Gen. 909. Ic gewíte in greótes fæðm *I depart into dust's bosom*, Exon. 64 a; Th. 235, 13; Ph. 556 : Andr. Kmbl. 1587; An. 795 : Beo. Th. 6315; B. 3168. Of greóte *from the earth*, Exon. 59 b; Th. 216, 13; Ph. 267 : Andr. Kmbl. 3246; An. 1626. Sand is geblonden grund wið greóte *the sand is mixed together, the abyss with the strand*, 849; An. 425 : 475; An. 238 : 508; An. 254 : Exon. 52 a; Th. 182, 12; Gú. 1309. Hēr líþ úre ealdor on greóte *here lies our chief in the dust*, Byrht. Th. 140, 68; By. 315 : Andr. Kmbl. 2169; An. 1086 : Judth. 12; Thw. 26, 4; Jud. 308. Deáþ ðe hit sý greóte beþeaht líc mid láme *though with dust it be covered, the body with clay*, Exon. 117 a; Th. 451, 4; Dôm. 98 : Elen. Kmbl. 1666; El. 835. [*A. R.* greot : *Wick.* greet : *O. Sax.* griot ; *n.* sand, strand : *Icel.* grjót ; *n.* stones, rubble : *O. H. Ger.* grioz *glarea*, arena : *Ger.* gries *gravel, grit*.]

greótan, ic greóte, ðú grýtest, grýtst, he greóteþ, grýt, *pl.* greótaþ; *p.* greát, *pl.* gruton; *pp.* groten *To weep* ; flere, lacrimare :—Heó sceal oft greótan *she shall often weep*, Salm. Kmbl. 753; Sal. 376. Se ðe on sefan greóteþ *who weeps in spirit*, Beo. Th. 2689; B. 1342. [*O. Sax.* griotan *to weep*.]

greót-hord, es; *n.* [greót *grit, dust, earth* ; hord *hoard, treasure*] *An earthen treasure, i. e. the body* :—Greóthord gnornaþ gǽst hine fýseþ on ēcne geard *the body mourns, the spirit hastens to an eternal dwelling*, Exon. 51 a; Th. 178, 6; Gú. 1240.

grep *a furrow, burrow* [*Prompt. Parv.* gryppe or *a* gryppel *where watur rennythe away in a londe* : grip *a drain, ditch, trench*, Hall. Dict.] v. græp.

grétan, grǽtan; *p.* grēt, *pl.* grēton; *pp.* grēten, grǽten *To bewail, deplore, weep* ; plorare, deplorare, flere :—Lápsíþ grétan *to bewail the dire journey*, Cd. 145; Th. 180, 13; Exod. 44. Beornas grētaþ *men shall wail*, Exon. 22 b; Th. 61, 30; Cri. 992. Hú ða womsceaðan hyra eald-gestreón grēten *how the wicked doers shall bewail their works*

of old, Exon. 31 a; Th. 96, 10; Cri. 1572. [Goth. grétan: O. Nrs. gráta *plorare*.] DER. be-grétan.

grétan, he grét, *pl.* grétaþ; *p.* grétte, *pl.* grétton; *pp.* gréted. **I. to approach, come to, visit, touch, attack, treat** or **use in any way, know carnally**; appropinquare, adire, visitare, tangere, hostiliter aggredi, afficere, cognoscere :—Đē wyrmas gyt gífre grétaþ *the greedy worms yet come to thee*, Exon. 100 a; Th. 375, 14; Seel. 138. Đonne hine engel grétte *when the angel visited him*, 37 b; Th. 123, 25; Gú. 328. Nó he ðone gifstól grétan móste *he might not touch the throne* [*gift-seat*], Beo. Th. 339; B. 168. Sum mid hondum mæg hearpan grétan *one may touch the harp with hands*, Exon. 79 a; Th. 296, 11; Crä. 49. Siððan wæs eallum ðám óðrum swá mycel ege fram him, ðæt hí hine grétan ne dorstan *afterwards the others were in so much fear of him, that they durst not attack him*, Ors. 5, 2; Bos. 102, 3. On sceortne -as geendiaþ grécisce naman ac we ne grétaþ nú ða *Greek nouns end in short -as, but we shall not treat them now*, Ælfc. Gr. 9, 24; Som. 10, 57. Se dæl se ðæt flód ne grétte *the part that the water did not touch*, 1, 3; Bos. 27, 29. Gomen-wudu gréted wæs *the glee-wood was touched*, Beo. Th. 2134; B. 1065. Đæt he ne grétte goldweard ðone *that he should not assail that gold-ward* [*that dragon*], Beo. Th. 6154; B. 3081: Bd. 3, 11; S. 536, 41. Gif ðe ænig mid weán gréteþ *if any one entreat thee evil*, Cd. 83; Th. 105, 18; Gen. 1755. He ne grétte hí *non cognoscebat eam*, Mt. Bos. 1, 25. **II. to speak to, call upon, hail, greet, welcome, salute, take leave of, bid farewell to**; alloqui, invocare, ciere, salutare, lætari de, valedicere :—Gomol eówic grétan hét *the aged* [*prince*] *commanded to greet you*, Beo. Th. 6182; B. 3095: Past. Pref. Swt. 3, 1; Hat. MS. Ælfríc munuc grét Æðelwærd ealdorman *Ælfric the monk greets alderman Ethelward*, Pref. Thw. 1, 1. Đonne he on gaton gréteþ his grame feondas *cum loquetur inimicis suis in porta*, Ps. Th. 126, 6. Gif man mannan mid bismær wordum scandlíce gréte *if a man address another shamefully with abusive words*, L. H. E. 11; Th. i. 32, 5. Hý grétte blíðum wordum *he addressed her with kind words*, Exon. 68 a; Th. 252, 17; Jul. 164. His God grétte *addressed his God*, Andr. Kmbl. 2059; An. 1032. Ongunnon hine grétan *cœperunt salutare eum*, Mk. Bos. 15, 18. Cwén grétte guman on healle *the queen greeted the men in the hall*, Beo. Th. 1232; B. 614. Wulfas hilde grétton *the wolves hailed the battle*, Cd. 151; Th. 189, 8; Exod. 181. Wác ne grétton in ðæt rinc-getæl *the weak they welcomed not into that martial number*, Cd. 154; Th. 192, 18; Exod. 233. Hróþgar grétte Beówulf *Hrothgar took leave of Beowulf*, Beo. Th. 1308; B. 652. [Orm. gretenn: Laym. græten *to accost, greet*; *p.* grætte: O. Sax. grótian: N. Frs. groetjen: O. Frs. gréta: N. Dut. groeten: N. Ger. grüszen: M. H. Ger. grüezen: O. H. Ger. gruoʒan.] DER. ge-grétan.

gréting, e; *f.* **A greeting, salutation, present in acknowledgment of a favour done**; salutatio :—Hwæt seó gréting wære *qualis esset ista salutatio*, Lk. Bos. 1, 29. Đínre gréting stefn *vox salutationis tuæ*, 1, 44. Lufiaþ grétinga on strætum *diligitis salutationes in foro*, 11, 43. Pápa sende Eádwine grétinge *the pope sent to Edwin greeting*, Bd. 2, 10; S. 512, 20. Sendaþ mín heáfod án to grétinge and bringaþ mínre méder ðæt heó ðæt cysse *send my head only in greeting and bring it to my mother that she may kiss it*, Shrn. 139, 28. Đá brohte seó sée damiane medmicle grétinge gewritu secgaþ ðæt ðæt wære þreó ægero *then she brought St. Damian a slight acknowledgment; books say that it was three eggs*, 135, 17, 23.

gretta. v. gryt.

gríg-hund a greyhound, Cot. 173, Lye.

grillan; *p.* de To **provoke, offend** :—Hie willaþ grillan [griellan, Hat. MS.] óðre men *they like to provoke other men*, Past. 40, 4; Swt. 292, 19; MS. Cot. [A. R. gruellen *to make sad*: O. E. Homl. igruld, 2, 259, 30; and see other instances in Stratmann: cf. Icel. grellskapr *spite*: Ger. groll *rancour*.]

GRIM; *adj.* **Sharp, bitter, severe, fell, fierce, dire, savage, cruel, GRIM, horrible**; acer, immanis, sævus, crudelis, atrox, dirus :—He him æt his ende grim geweorþeþ and hine gelædeþ on éce forwyrd *he* [*the devil*] *will become cruel to him at his end, and will lead him into eternal perdition*, Blickl. Homl. 25, 13: Cd. 184; Th. 230, 8; Dan. 230. Đæt wæs grim cyning *that was a fierce king*, Exon. 100 b; Th. 378, 29; Deór. 23. Grim and grædig *savage and greedy*, Beo. Th. 242; B. 121. Mycel wól and grim *acerba pestis*, Bd. 1, 14; S. 482, 29. Se grimma wítedóm *dira præsagia*, 3, 14; S. 541, 9. Wæs se winter to ðæs grim ðæt manig man his feorh for cýle gesealde *the winter was so severe that many a man lost his life with the cold*, Blickl. Homl. 213, 31: Chr. 1005; Erl. 139, 37. Mid grimmum gefeohte *with severe fighting*, 5, 3; Byrht. Th. 133, 36; By. 61. On ðam grimmun dæge dómes ðæs miclan *on the terrible day of the great doom*, Exon. 25 b; Th. 74, 12; Cri. 1205. Đæt wæter wæs biterre and grimre to drincanne ðone ic æfre ænig óðer bergde *amariorem elleboro fluminis aquam gustavi*, Nar. 8, 29. Cýle ðone grimmestan *the most severe cold*, Blickl. Homl. 61, 35. Đeáh ðú wære wyrmcynna ðæt grimmeste *though thou hadst been of serpents the fiercest*, Soul Kmbl. 167; Seel. 83. [O. Sax. O. Frs. O. H. Ger. grim *acerbus, austerus, atrox, sævus, ferus*: Icel. grimmr: Ger. grimm.]

gríma, an; *m.* **I. a mask, visor, helmet** :—Gylden gríma a *golden helm*, Elen. Kmbl. 249; El. 125. Gríma *a mask*, Gl. Mett. 504. He míne sáwle swylce gehealde wið ehtendra egsan gríman *ut salvam faceret a persequentibus animam meam*, Ps. Th. 108, 30. [Icel. gríma *a sort of hood* or *cowl*.] See Grmm. D. M. 218-9. DER. beadu-, here-gríma. **II. a spectre; larva** :—Mec mæg gríma ábrégan *a spectre can terrify me*, Exon. 110 b; Th. 423, 7; Rä. 41, 17. v. eges-gríma in Appendix.

grimena, grimenæ *a caterpillar*; bruchus, Ps. Spl. T. 104, 32.

grimetan, grymetan, grimetian; *p.* ode, ede To **rage, roar, make a loud noise, grunt**; fremere, rugire, grunnire :—Synfull tóþum torn þolaþ teónum grimetaþ *peccator dentibus suis fremet*, Ps. 111, 9. Grimme grymetaþ *fiercely roars*, Exon. 128 a; Th. 491, 22; Rä. 81, 3. Leónhwelpas grymetigaþ *catuli leonum rugientes*, Ps. Th. 103, 20. Đá awédde he and grymetede *he went mad and cried aloud*, Th. Anal. 125, 8: Ps. Spl. 37, 8. Ecg grymetode *loud rang the blade* [*as it was drawn from the sheath*], Cd. 162; Th. 203, 24; Exon. 408. He gristbitade and grymetade *he gnashed his teeth and raged*, Exon. 74 b; Th. 278, 15; Jul. 598. Sume sceoldan bión eaforas and ðonne hí sceoldan hiora sár siófian ðonne grymetodan hí *some had to be boars and when they should lament their misfortune then they grunted*, Bt. 38, 1; Fox 194, 35. Grymetedon, Bt. Met. Fox 26, 163; Met. 26, 81. Forhwon grymetedon þeóda *quare fremuerunt gentes?* Ps. Spl. C. T. 2, 1. Grymetigan to **roar**, Bt. Met. Fox 13. 58; Met. 13, 29. Fíf manna sáwla hreówlíce gnorniende and grymetende *five men's souls miserably wailing and crying out*, Homl. Th. ii. 350, 28. Grimetende *rugientes*, Ps. Spl. 103, 22. Swíðe grymetende *cum ingenti murmure*, Nar. 14, 27. Brim grymetende *the roaring ocean*, Exon. 95 b; Th. 356, 6; Pa. 7. Swá grymetigende leó *as a roaring lion*, Guthl. 4; Gdwin. 26, 22. v. gremettan.

grimetung, grymetung, e; *f.* **Raging, roaring, grunting, loud noise**; murmur, fremitus, rugitus :—Swýnes grymetunge *swine's grunting*, Guthl. 8; Gdwin. 48, 3 : 46, 20. Leóna grymetunge *roaring of lions*, Shrn. 50, 9.

grim-helm, es; *m.* **A helmet with a visor**; galea larvata, Cd. 151; Th. 188, 27; Exod. 174 : 160; Th. 198, 29; Exod. 330 : Elen. Kmbl. 516; El. 258: Beo. Th. 674; B. 334. See Grmm. A. E. xxviii; and gríma.

gríming *witchcraft*; veneficium, Som.

grimlíc; *adj.* **Grim, fierce, cruel, sharp, severe, bloody**; atrox, dirus, cruentus, crudelis :—Đone grimlícan gársecg *the fierce ocean*, Homl. Th. i. 454, 15. Hit wyrþ ðonne egeslíc and grimlíc *things will then become awful and terrible*, Swt. A. S. Rdr. 104, 5. Đa Crétense hæfdon ðone grimlecan sige *cruentiorem victoriam Cretenses exercuerunt*, Ors. 1, 7; Swt. 42, 28. Se lægdraca grimlíc gryre *the fire-drake, that fierce horror*, Beo. Th. 6074; B. 3041. Đa gewin wæron grimlícran ðonne hý nú sýn *struggles were more bloodthirsty than they now are*; quod crudelius graviusque erat quam nunc est, Ors. 1, 2; Swt. 30, 23.

grimlíce; *adv.* **Fiercely, severely, cruelly** :—Đám mannum sceolan ða déman grimlíce stýran *those men must the judges severely restrain*, Blickl. Homl. 63, 15. Oft hí grimlíce Godes costodan *tentaverunt Deum*, Ps. Th. 77, 41. Spreceþ grimlíce *speaketh fiercely*, Soul Kmbl. 31; Seel. 16: Exon. 22 b; Th. 62, 19; Cri. 100, 4.

grimman, ic grimme, ðú grimst, he grimmeþ, grimþ, *pl.* grimmaþ; *p.* gram, grom, *pl.* grummon; *pp.* grummen. **I. to rage, roar, make a loud noise**; fremere :—Đú hie grimman meaht gehýran *thou mayest hear it* [*hell*] *rage*, Cd. 37; Th. 49, 17; Gen. 793. Hwæl-mere hlúde grimmeþ *the whale-mere* [*the sea*] *rages loudly*, Exon. 101 a; Th. 382, 3; Rä. 3, 5. [Cf. O. Sax. grimmid the grôto séo.] **II. to run with haste, hasten**; properare, currere, festinare :—Gúþmóde grummon *the warlike of mind hastened*, Beo. Th. 617; B. 306. [So Grein translates the verb, but may not the word be taken more nearly in the sense of the preceding passages 'loud and fierce was their shout?']

grimme; *adv.* **Grimly, fiercely** :—Hý him æfter ðæm grimme forguldon ðone wígcræft ðe hý æt him geleornodon *they afterwards gave him grim requital for the military skill they learnt from him*, Ors. 1, 2; Bos. 26, 30: Cd. 64; Th. 77, 15; Gen. 1275: 183; Th. 229, 2; Dan. 211: Beo. Th. 6017; B. 3012.

grimnes, se; *f.* **GRIMNESS, severity, fierceness, cruelty**; ferocitas, atrocitas :—Se deófol wile hit him mid grimnesse and mid yfele eall forgyldan *the devil will requite it all to him with cruelty and with evil*, Blickl. Homl. 55, 24. Hí sceoldan ðæm unriht-dóndum mid grimnesse stéran *they should restrain with severity all evil-doers*, 63, 12. On grimnesse *in exacerbatione*, Ps. Th. 94, 9. Cwædon to gúðláce mid grimnysse *fiercely they* [*evil spirits*] *spake to Guthlac*, Exon. 41 a; Th. 136, 33; Gú. 550. [Prompt. Parv. grymnesse *austeritas, rigor, horror, horribilitas*.]

grimsian; *p.* ode To **be fierce, cruel, to rage**; sævire :—Đá ðara treówleásra cyninga beboda wið cristenum monnum grimsedon *cum perfidorum principum mandata adversum Christianos sævirent*, Bd. 1, 7; S. 476, 36. He grimsigende forleás *sæviens disperderat*, 3, 1; S. 523, 29. Wól mid grimme wæle lange feor and wíde grimsigende *pestilentia acerba clade diutius longe lateque desæviens*, 27; S. 558, 15 : 4, 25; S. 601, 20.

grimsung, e; *f.* **Fierceness, roughness** :—Mid ungemetlícre grimsunge *multa asperitate*, Past. 17, 11; Swt. 125, 14; Hat. MS.

grin, gryn, e; *f*: es; *n. A snare, gin, noose*; laqueus:—Swā swā grin he becymþ on ealle *tanquam laqueus superveniet in omnes*, Lk. Skt. 21, 35: Ps. Th. 123, 7. Grines *laquei*, Ps. Lamb. 34, 7. Of grames huntan grine *de laqueo venantium*, Ps. Th. 123, 6: 90, 3. Geheald me wið ðare gryne *custodi me a laqueo*, 140, 11. On grine *in laqueum*, 68, 23. Gryne, 65, 10. Ic fô mid grine *laqueo*, Ælfc. Gr. 26; Som. 29, 17. Iudas fêrde and mid gryne hyne sylfne ahêng *Iudas wente awey and goyinge awey he hangide hym with a grane*, Wyc; laqueo se suspendit, Mt. Bos. 27, 5: Homl. Th. ii. 30, 22. Mid ðȳ ilcan grine *in laqueo isto*, Ps. Th. 9, 14. He rîneþ ofer ða synfullan grinu *pluet super peccatores laqueos*, Ps. Lamb. 10, 7: Ps. Th. 17, 5: 34, 9. Fôtum heó mînum grine gearwodon *laqueos paraverunt pedibus meis*, 56, 7: 141, 4. Mid grinum *laqueis*, Coll. Monast. Th. 25, 13. [*Ayenb.* gryn *snare.*]

GRINDAN, gryudan; *part.* grindende, ic grinde, grynde, ðû grintst, grinst, he grint, *pl.* grindaþ; *p.* ic, he grand, grond, ðû grunde, *pl.* grundon; *pp.* grunden *To* GRIND, *grind together, rub, rub together*; molere, commolere, terere, frendere, allidi, collidi:—Ic seah searo grindan wið greóte *I saw a machine grind against the dust*, Exon. 108 b; Th. 414, 30; Rä. 33, 4. Ic grynde *molo*, Ælfc. Gr. 28, 3; Som. 31, 3. Ic grinde *commolo*, Ælfc. Gl. 36; Wrt. Voc. 28, 77. Ðû grinst *thou grindest*, Homl. Th. i. 488, 25. Se hæruflota grond wið greóte *the floater of the surge [the ship] ground against the gravel*, Exon. 52 a; Th. 182, 12; Gû. 1309. Hî grundon ofer me mid tôðum heard *frenduerunt super me dentibus suis*, Ps. Spl. 34, 19: Andr. Kmbl. 746; An. 373. Twâ beóþ æt cwyrne grindende, ân byþ genumen, and ôðer byþ læfed *duæ molentes in mola, una assumetur, et una relinquetur*, Mt. Bos. 24, 41: L. Ethb. 11; Th. i. 6, 6. Sume ðara munecena cômon to grindanne *some of the nuns came to grind*, Th. Chart. 447, 1. DER. be-, for-, ge-grindan.

grindel, es; *m. A bar, bolt*; in *pl. lattice-work, hurdle*; crates:—Geslægene grindlas greáte *forged large gratings*, Cd. 19; Th. 24, 27; Gen. 384. Guest, English Rhythms, ii. 40, note 1, observes:—' As far as we can judge from the drawing which accompanies the description, the *grindel* was a kind of heavy iron grating, which rather encumbered the prisoner by its weight, than fixed him in its grasp.' [*O. H. Ger.* grintil *temo, repagulum, pessulum, obex, vectis*: cf. *Icel.* grind *a lattice-door*.]

grindere, es; *m. A grinder*; molitor, Som.

grind-tôðas *grinding teeth, the grinders*, Som.

gring, es; *n? Slaughter*; clades, Elen. Kmbl. 230; El. 115. v. gringan.

gringan, ic gringe, ðû gringest, gringst, he gringeþ, gringþ, *pl.* gringaþ; *p.* grang, *pl.* grungen; *pp.* grungen *To sink down, perish*; occumbere, prosterni:—On herefelda hæðene grungon *the heathen sank down upon the battle-field*, Elen. Kmbl. 252; El. 126. [Cf. cringan.]

gring-wracu, e; *f. Deadly punishment*, Exon. 69 b; Th. 258, 14; Jul. 265.

grinian, grynian; *p.* ode; *pp.* od [grin *a snare*] *To ensnare*; ligare, illaqueare. DER. be-, ge-grinian.

grînu; *adj. Avidius*, Ælfc. Gl. 79; Som. 72, 85; Wrt. Voc. 46, 42.

griósn *a pebble stone*; calculus, Prov. 20, Lye.

gripa, an; *m. A handful, a sheaf*; manipulus, pugillus:—Gripa *pugillus*, Hpt. Gl. 497. Genim ðysse ylcan wyrte gôdne gripan *take a good handful of this same plant*, Herb. 36, 4; Lchdm. i. 136, 4: 81, 5; Lchdm. i. 184, 18. Berende gripan heora *portantes manipulos suos*, Ps. Spl. 125, 8.

grîpan, ic grîpe, ðû grîpest, grîpst, he grîpeþ, grîpþ, *pl.* grîpaþ; *p.* grâp, *pl.* gripon; *pp.* gripen; *v. a. To* GRIPE, *grasp, seize, lay hold of, apprehend*; capere, rapere, prehendere, apprehendere:—Ic on Lothe gefrægn hæþne heremæcgas handum grîpan *I heard that the heathen leaders seized on Lot with their hands*, Cd. 114; Th. 149, 32; Gen. 2483: 219; Th. 281, 9; Sat. 269. Oþ ðæt ðê heortan grîpeþ âdl unlîðe *until severe disease gripeth thee at heart*, Cd. 43; Th. 57, 31; Gen. 936: Exon. 107 a; Th. 407, 19; Rä. 26, 7. Hwîlum flotan grîpaþ *sometimes they seize the sailor*, Salm. Kmbl. 304; Sal. 151. Grîpaþ lâre *apprehendite disciplinam*, Ps. Spl. 2, 12. Grâp on wrâðe *laid hands on his enemies*, Cd. 4; Th. 4, 30; Gen. 61: 69; Th. 83, 18; Gen. 1381: 95; Th. 125, 1; Gen. 2072: 119; Th. 153, 28; Gen. 2545: Beo. Th. 3006; B. 1501: Exon. 129 a; Th. 495, 8; Rä. 84, 4. Ðû ðe samod mid me swête gripe metas *qui simul mecum dulces capiebas cibos*, Ps. Spl. 54, 15: Cd. 42; Th. 55, 8; Gen. 891. Scearpe gâras gripon *the sharp arrows griped*, Cd. 95; Th. 124, 16; Gen. 2063. Swâ swâ leó hreáfiende ôðe gripende oðe gyrretynde and grymetende *sicut leo rapiens et rugiens*, Ps. Lamb. 21, 14: Blickl. Homl. 211, 1. [*Goth.* greipan: *O. Sax.* grîpan: *O. Frs. Icel.* grîpa. *O. H. Ger.* grîfan.] DER. be-, for-, ge-, to-ge-, ôþ-, wið-grîpan.

grîpe, es; *m. A gripe, vulture*; gryps, vultur. [*Laym.* grîpes, *pl*: *Icel.* grîpr: *O. H. Ger.* grif: *Prompt. Parv.* grype *vultur*, p. 212, note 4: Wrt. Voc. 252, 28 grype *vultur*: and see Nares' Glossary.]

gripe, es; *m. A gripe, grip, grasp, hold, clutch, seizure*; pugillus, prehensio, captus:—Se gripe ðære hand *pugillus*, Ælfc. Gl. 72; Som. 71, 1; Wrt. Voc. 43. Gripe *pugilla*, Recd. 38, 72; Wrt. Voc. 64, 75. Eorþ-grâp heard gripe hrusan *earth's grasp, the fast hold of the ground*, Exon.

124 a; Th. 476, 15; Ruin. 8. Gripe mêces oðe gâres fliht *the falchion's clutch or the javelin's flight*, Beo. Th. 3534; B. 1735: Andr. Kmbl. 373; An. 187: Exon. 67 b; Th. 250, 10; Jul. 125. Of gromra gripe *from the cruel ones' clutch*, Exon. 68 b; Th. 255, 16; Jul. 215: 71 b; Th. 265, 34; Jul. 391: Salm. Kmbl. 97; Sal. 48: Elen. Kmbl. 2601; El. 1302: Andr. Kmbl. 433; An. 217: 1901; An. 953. For mînum gripe *for my grasp*, Exon. 126 a; Th. 484, 11; Rä. 70, 6: Beo. Th. 2300; B. 1148. Staþole strengra ðonne ealra stâna gripe *stronger in position than the hold of all stones*, Salm. Kmbl. 154; Sal. 76. [*Laym.* gripen; *pl.* grasps: cf. *O. H. Ger.* grif: *Ger.* griff.] DER. fær-, mund-, nîð-, stân-, sweord-gripe.

gripennis, se; *f. Captivity*; captivitas, Som.

gripu, e; *f. A cauldron*:—Seó ærene gripu *the brazen cauldron*, Salm. Kmbl. 94; Sal. 46.

grîsan, ic grîse, ðû grîsest, grîst, he grîseþ, grîst, *pl.* grîsaþ; *p.* ic, he grâs, ðû grise, *pl.* grison; *pp.* grisen *To shudder, to be frightened*; horrere. [Me gries, A. R. 366, 7, note: gros, *p. King Horn.* 1314: his herte gros, *Man.* ed. Furn. 8532: him gros, Handl. Synne 7875.] DER. a-grîsan, grislîc, an-grislîc, -grisenlîc.

grislîc, gryslîc; *adj.* GRISLY, *horrible, dreadful, horrid*; horridus, horrendus, horribilis. [*Laym.* grislich: *Orm.* grisslis: *A.R.* grislich: *Ayenb.* grislich: *O. Frs.* gryslik: cf. *O. H. Ger.* grisenlich, Grff. iv. 301: *Ger.* grässlich.] This word seems to belong to 'grîsan' rather than to 'greósan,' so should be written with *i* rather than with *y*. The spelling in the Ormulum supports the short vowel. v. grîsan.

grist, es; *m.* [?] *Grist, corn for grinding*:—Grist *molitura*, Ælfc. Gl. 50; Som. 65, 107; Wrt. Voc. 34, 36. v. gyrst.

gristbâtian *to gnash the teeth*:—Gristbâtaþ mid his tôþum *fremet dentibus suis*, Ps. Th. 36, 12, note. [Gristbeatien, *Juliana*, 69, 17: *A. R.* gristbatede; and cf. *Laym.* gristbating.]

gristbâtung, e; *f. A grinding, gnashing*:—Gristbâtung tôþa *stridor dentium*, Mt. Kmbl. Rush. 8, 12. [*Laym. O. E. Homl.* grisbating.]

gristbitian; *p.* ode, ede *To gnash or grind the teeth*; frendere, stridere:—Ic cearcige oðe gristbitige *strideo vel strido*, Ælfc. Gr. 26; Som. 29, 7. Tôþum gristbitaþ [gristbittþ, Lind.] *stridet dentibus*, Mk. Skt. 9, 18. He grennade and gristbitade *he grinned and ground his teeth*, Exon. 74 b; Th. 278, 12; Jul. 596. Gristbitedon mid heora tôþum ongeán me *striderunt in me dentibus suis*, Ps. Th. 34, 16. He ongan mid his tôþum gristbitian *cæpit dentibus frendere*, Bd. 3, 11; S. 536, 14: Judth. 12; Thw. 25, 21; Jud. 271.

gristbitung, e; *f. A gnashing of the teeth*:—Tôþa gristbitung [gristbiottung, Lind.] *stridor dentium*, Mt. Kmbl. 8, 12: 13, 42, 50: Blickl. Homl. 185, 7: Cd. 220; Th. 285, 7; Sat. 334. Gristbiotung, Mt. Kmbl. Lind. 25, 30. Gristbittung, Lk. Skt. Lind. 13, 28.

gristel, gristl, es; *m. Gristle*; cartilago, Ælfc. Gl. 72; Som. 71, 8.

gristel-bân, es; *n. A gristle bone*; cartilageum os.

gristian *to grind, grate, gnash*, Hpt. Gl. 513.

gristlung, grystlung, e; *Gnashing, grinding*:—Tôþa grystlung *stridor dentium*, Lk. Skt. 13, 28.

gristra, an; *m. A baker of dough made from grist, a baker*; cerealis pistor, Ælfc. Gl. 50; Som. 65, 108; Wrt. Voc. 34, 37.

grið, es; *n.* I. *peace limited to place or time, truce, protection, security, safety*. [The word comes into use during the struggles with the Danes. *Icel.* grið (v. Cl. and Vig. Dict.) means first *home, domicile*, then in *pl. truce, peace, pardon*; friðr is the general word, grið the special, deriving its name from being limited in time or space (asylum):—Ælfsig ealdorman grið wið hî gesætte *alderman Leofsig made a truce with them*, Chr. 1002; Erl. 137, 25. Ðonne nam man grið and frið wið hî *then was truce and peace made with them*, 1011; Erl. 145, 3, 4. We willaþ wið ðam golde grið fæstnian *for the gold we will make a truce*, Byrht. Th. 132, 53; By. 35. Heó gesôhte Baldwines grið *she sought the protection of Baldwin*, Chr. 1037; Erl. 167, 3: 1048; Erl. 178, 34: 180, 17, 19. Ðâ gyrnde he grîðes and gîsla *then he required security and hostages*, 180, 6: 1095; Erl. 231, 25. Sette man him iv nihta grið *his safety was secured for four days*, 1046; Erl. 173, 4. Godes grið *protection belonging to the church*, Swt. A. S. Rdr. 107, 99. II. for the passages in which the word occurs as a technical term in the laws, see Thorpe, index to vol. i. of 'Ancient Laws and Institutes,' s. v. Schmid, p. 585, arranges the several ' griths ' under the following heads:—(1) Place; churches, private houses, the king's palace and precincts; (2) Time; fasts and festivals, coronation days, days of public gemots and courts, times when the fyrd is summoned; (3) Persons; clergy, widows, and nuns. On this word, Stubbs, i. 181, says—' The *grith* is a limited or localized peace, under the special guarantee of the individual, and differs little from the protection implied in the *mund* or personal guardianship which appears much earlier; although it may be regarded as another mark of territorial development. When the king becomes the lord, patron, and *mundborh* of his whole people, they pass from the ancient national peace of which he is the guardian into the closer personal or territorial relation of which he is the source. The peace is now the king's peace; ... the *frith* is enforced by the national officers, the *grith*

by the king's personal servants: the one is official, the other personal; the one the business of the country, the other that of the court. The special peace is further extended to places where the national peace is not fully provided for: the great highways ... are under the king's peace.' [A. R. Laym. griþ: Orm. griþþ.] DER. cyric-, hǽlnes-, hád-, hand-grið.

grið-brice, -bryce, es; m. [grið peace; brice, bryce a breach, violation] A breach of the peace; pacis infractio vel violatio:—Griðbrice infractio pacis, L. Th. ii. 531, 12. Bête man ðone griðbryce let a man make amends for a breach of the peace, L. Eth. ix. 4; Th. i. 340, 21: L. C. E. 3; Th. i. 360, 12.

gridian; p. ode, ede; pp. od, ed.   I. to make peace:—Lundene waru griðede wið ðone here the people of London made peace with the army, Chr. 1016; Erl. 159, 9. Griðode, 1046; Erl. 172, 6: 1070; Erl. 207, 19. Griðedon, 1068; Erl. 207, 2. Griðodon, 1087; Erl. 225, 15.   II. to protect, give 'grith:'—Hwílum heálíce hádas griðian mihton ða ðe ðæs beþorf once those of high rank could extend protection to those that needed it, L. Eth. 7, 3; Th. i. 330, 7. Godes þeówas griðedan protected God's servants, 24; Th. i. 334, 24: Swt. A. S. Rdr. 105, 37. Griðian and friðian, L. Eth. 6, 42; Th. i. 326, 16: L. C. E. 2; Th. i. 358, 11: 4; Th. i. 360, 28. [Laym. griðien.]

grið-lagu, e; f. Law concerning 'grith,' L. Eth. 7, 9; Th. i. 330, 22.

gridleás; adj. Without 'grith' or protection, unprotected, Swt. A. S. Rdr. 106, 41.

gritta grit, bran; furfur:—Ðás gritta hic furfur, Ælfc. Gr. 9, 22; Som. 10, 47. v. gryt.

groene green, Lk. Skt. Lind. Rush. 23, 31. v. grêne.

groetan to greet; groeting a greeting. v. grêtan, grêting.

gróf, pl. grófon carved, Bt. Met. Fox 8, 113; Met. 8, 57; p. of grafan.

grom. v. gram.

grópian. v. grápian.

grorn, es; m [?] Grief, sadness; luctus, mœror, Exon. 94 b; Th. 354, 22; Reim. 49.

grorne; adv. Sadly, mournfully, Exon. 25 b; Th. 74, 11; Cri. 1205.

grorn-hof, es; n. A house of sadness, of woe, Exon. 70 b; Th. 261, 32; Jul. 324.

grornian; p. ode To mourn, murmur:—Grornaþ eal middangeard all the earth shall mourn, Exon. 22 a; Th. 60, 18; Cri. 971. Grornadun murmurabant, Mt. Kmbl. Lind 20, 11.

grornung, e; f. Complaint, mourning:—Búta grornunge sine quærella, Lk. Skt. Lind. Rush. 1, 6.

grot, es; n. A particle, an atom; particula:—Nán grot rihtwísnesse no particle of wisdom, Bt. 35, 1; Fox 156, 6. Nán grot andgites no particle of sense, 41, 5; Fox 252, 22. Uneáþe ǽnig grot staþoles aðstód hardly any particle of foundation remained, Ors. 6, 1; Swt. 252, 23. [A. R. of al þe brode eorðe ne moste he habben a grot forte deien uppon, 260, 20: Havel. karf hem al to grotes, 472.]

GRÓWAN; part. grówende; ic grówe, ðú grówest, grêwst, he gróweþ, grêwþ, pl. grówaþ; p. greów, pl. greówon; pp. grówen To grow, increase, spring, sprout, spring up; crescere, frondere, virere, germinare, florere:—Lǽteþ hió ða blówan and grówan it lets these blow and grow, Exon. 109 a; Th. 417, 6; Rä. 35, 9: 90 a; Th. 338, 3; Gn. Ex. 73: Bd. 1, 27; S. 491, 5: Bt. Met. Fox 22, 84; Met. 22, 42: Salm. Kmbl. 969; Sal. 484. Spritte seó eorðe grówende gærs germinet terra herbam virentem, Gen. 1, 11: Ps. Spl. 64, 11: Cd. 5; Th. 6, 13; Gen. 88. Ic grówe frondeo, Ælfc. Gr. 26, 2; Som. 28, 42. Ic grówe vireo, 26, 2; Som. 28, 44: Mk. Bos. 4, 27. Leáf and gærs geond Bretene blóweþ and gróweþ leaves and grass blow and grow over Britain, Bt. Met. Fox 20, 198; Met. 20, 99: 29, 140; Met. 29, 70: Ps. Th. 91, 11: 146, 8: Exon. 91 b; Th. 343, 19; Gn. Ex. 159: Hy. 35; Hy. Grn. ii. 292, 35. Eall se dǽl ðæs treówes upweardes grêwþ all that part of the tree grows upwards, Bt. 34, 10; Fox 150, 2. Hí grówaþ geára gehwilce on lencten tíd they grow every year in spring time, Bt. Met. Fox 29, 133; Met. 29, 67: Ps. Th. 103, 12: 64, 11. Greów grew, Beo. Th. 3441: B. 1718. Ða greówon [MS. greowan] and blósmodon [MS. blosmodan] the lands grew and blossomed, Bd. 4, 13; S. 582, 35: Ps. Th. 106, 36, 37. Forhwí ælc sǽd grówe innon ða eorþan? why should every seed grow in the earth? Bt. 34, 10; Fox 148, 9. Hwæt druge ðú grówendra gifa? what madest thou of the growing gifts? Cd. 42; Th. 55, 6; Gen. 890. [O. Frs. grówa: Icel. gróa: O. H. Ger. grôen, grüen virescere.] DER. a-, for-, ge-grówan.

grównes, se; f. Growth:—Grównys hreódes viror calami, Bd. 3, 23; S. 554, 23. Ne com ðær nænig grównes up ne wæstmas ne furþan brordas nil omnino, non dico spicarum, sed ne herbæ quidem ex eo germinare contigit, 4, 28; S. 605, 34.

gruncan prurire, Gl. Prud. 595.

GRUND, es; m.   I. ground, bottom, foundation; fundus, fundamentum:—Grund fundamentum, Lk. Skt. Lind. 14, 29: 6, 48: Rtl. 82, 34. Ælc sǽ ðeáh heó deóp sý hæfþ grund on ðære eorþan every sea, though it be deep, hath its bottom in the earth, Lchdm. iii. 254, 20. Hordweard sóhte georne æfter grunde the keeper of the hoard sought eagerly along the floor [of the cave], Beo. Th. 4577; B. 2294: 5523;

B. 2765: 5510; B. 2758. Grunde getenge deep in the earth, i. e. lying, as it were, at the bottom of a hole, Elen. Kmbl. 2226; El. 1114. Me to grunde teáh he drew me to the bottom [of the sea], Beo. Th. 1111; B. 553: Cd. 39; Th. 51, 29; Gen. 834. Ufan to grunde from top to bottom, 228; Th. 309, 2; Sat. 703: 229; Th. 310, 15; Sat. 705: Salm. Kmbl. 61; Sal. 31. Sió gítsung ðe nænne grund hafaþ avarice which hath no bottom, Bt. Met. Fox 8, 92; Met. 8, 46. Mid fótum ne mæg grund geræcan cannot reach the bottom with his feet, Salm. Kmbl. 453; Sal. 227: Beo. Th. 2739; B. 1367: Exon. 97 a; Th. 361, 34; Wol. 29.   II. ground, earth, land, country, plain; terra, solum, campus:—Hie ðæt gild gebrǽcan and gefyldan eal óð grund they broke the idol to pieces and cast it all to the ground, Blickl. Homl. 221, 33. Eal ðes grund all this spacious earth, Exon. 116 a; Th. 445, 23; Dóm. 12: Cd. 5; Th. 7, 11; Gen. 104. Eall eorþan grund all the earth, 192; Th. 240, 5; Dan. 382. We men on grunde we men on the earth, Hy. Grn. ii. 292, 39; Hy. 9, 39. Neól ic fére and be grunde grǽfe prone I go and along the ground dig, Exon. 106 a; Th. 403, 3; Rä. 22, 2: 128 a; Th. 491, 23; Rä. 81, 3. Geond ealne yrmenne grund through all the earth, 14 b; Th. 30, 20; Cri. 481: 66 a; Th. 243, 14; Jul. 10: Cd. 6; Th. 8, 35; Gen. 134: 69; Th. 83, 31; Gen. 1388: Exon. 57 b; Th. 205, 26; Ph. 118. He grund gesóhte he fell to the ground, Byrht. Th. 140, 13; By. 287: Andr. Kmbl. 3199; An. 1602. Grund and sund earth and sea, 1494; An. 748. Geond grunda fela through many lands, Exon. 87 a; Th. 326, 30; Víd. 136. On grundum on earth, 17 b; Th. 43, 1; Crí. 682: 18 b; Th. 46, 28; Cri. 744. Of grundum, 18 a; Th. 44, 13; Cri. 702. Rúme grundas swilce eác rêðe streámas spacious plains and fierce streams, Judth. 12; Thw. 26, 30; Jud. 349. Grêne grundas, Andr. Kmbl. 1551; An. 777: Beo. Th. 2812; B. 1404: 4152; B. 2073: Chr. 937; Erl. 112, 15; Ædelst. 15.   III. a depth, sea, abyss, hell; profundum, abyssus:—On sǽs grund in profundum maris, Mt. Kmbl. 18, 6. On grund in abissum, Lk. Skt. 8, 31. Grund eall forswealg the abyss swallowed up all, Andr. Kmbl. 3179; An. 1592. Sǽs sídne grund the sea's spacious depth, Exon. 93 a; Th. 349, 2; Sch. 40: Menol. Fox 323; Men. 113: Andr. Kmbl. 786; An. 393: 849: An. 425: Beo. Th. 3106; B. 1551. Wese ic earmum gelíc ðe on sweartne grund syðdan astígaþ ero similis descendentibus in lacum, Ps. Th. 142, 7. Ic of grundum cleopode de profundis clamavi, 129, 1. Ofer deópnesse ealra grunda above the depth of all abysses, Blickl. Homl. 141, 9. Deorce grundas in abyssis, Ps. Th. 134, 6: Cd. 213; Th. 265, 19; Sat. 10. Of grunde brymmes de profundo pelagi, Rtl. 61, 33.   Of helle grunde from the depth of hell, Blickl. Homl. 21: Th. 4: 33, 19: 65, 14. On helle grunde in the depth of hell, Th. Chart. 309, 8. Hêt hine ðære sweartan helle grundes gýman bade him rule the black hell's abyss, Cd. 18; Th. 22, 25, 31; Gen. 346, 349. To grunde to hell, 219; Th. 281, 9; Sat. 269: 227; Th. 304, 21; Sat. 633. Gríp wið ðæs grundes stretch forth thy hands towards the abyss [hell], 228; Th. 308, 31; Sat. 701. Ðone deópan grund the deep abyss, Blickl. Homl. 103, 15. Hátne grund, Cd. 224; Th. 295. 13; Sat. 485. Grimne grund, Exon. 30 a; Th. 93, 16; Cri. 1527. Súsla grund, Elen. Kmbl. 1885; El. 944. Ðás grimman grundas these grim depths, Cd. 21; Th. 26, 15; Gen. 407: Cd. 219; Th. 280, 23; Sat. 260. On ðám grundum helle tintreges in profundis tartari, Bd. 5, 14; S. 634, 25: Salm. Kmbl. 976; Sal. 488. [O. Sax. O. Frs. grund: Icel. grunnr the bottom [of the sea, etc.]: O. H. Ger. grunt fundus, profundum: Ger. grund: cf. Goth. afgrundiþa abyss; grundu-waddjus a foundation.] DER. bryten-, sǽ-, wæter-grund; un-grund.

grund-bedd, es; n. The ground; solum, Exon. 128 a; Th. 493, 3; Rä. 81, 24.

grund-búende; pl. Inhabitants of the earth, Beo. Th. 2016; B. 1006: Salm. Kmbl. 578; Sal. 288.

grunde-hirde, es; m. A guard of the deep, Beo. Th. 4279; B. 2136.

grunde-swelge, -swilge, -swilige, -swylige, -swulie, -an; f. GROUNDSEL; senecio:—Ompre, grundeswelge, ontre dock, groundsel, radish, L. M. 1, 32; Lchdm. ii. 78, 25. Grundeswylige groundsel, Herb. 77, 1; Lchdm. i. 180, 5. Genim grundeswelgean take groundsel, L. M. 1, 22; Lchdm. ii. 64, 19: 1, 2; Lchdm. ii. 32, 5: 1, 51; Lchdm. ii. 124, 15.

grund-fús; adj. Ready for hell, hastening to hell:—Ðæt biþ feóndes bearn hafaþ grundfúsne gǽst that is a child of the devil, hath a spirit hastening hellwards, Exon. 84 a; Th. 316, 15; Mód. 49.

grundleás; adj. GROUNDLESS, bottomless, boundless, immense, unbounded, interminable, endless; fundo carens, profundissimus, immensus:—Se grundleás seáþ gǽsta gíemeþ the bottomless pit holds the spirits, Exon. 30 b; Th. 94, 26; Cri. 1546: Bt. 9, 4; Fox 22, 32; Grundleás gítsung boundless greed, Bt. Met. Fox 7, 29; Met. 7, 15: Exon. 97 a; Th. 362, 34; Wal. 46: Cd. 21; Th. 25, 7; Gen. 390. Wurdon grundleáse Geátes frige ðæt him seó sorglufu slǽp binom Geat's loves were boundless so that anxious love took from him sleep, Exon. 100 a; Th. 378, 12; Deór. 15.

grundleás-líc; adj. Bottomless, unbounded, boundless, immense:—Swá grundleáslícu costung such immense temptation, Past. 53, 6; Swt. 417, 10.

grundlinga, -lunga; *adv. From the very bottom or root, entirely, totally :*— Grundlunge oððe mid stybbe mid ealle *stirpatus :* grundlunga *funditus :* grundlinga oððe mid wyrttruman mid ealle *radicitus,* Ælfc. Gr. 38; Som. 42, 3, 4. Hí tobrǽcon ða burh grundlinga *they destroyed the city to its very foundations,* Homl. Th. ii. 66, 3; i. 72, 5. Grundlunge, ii. 164, 16.

grund-sceát, es; *m. A region of earth,* Exon. 8 b; Th. 3, 27; Cri. 42: 17 a; Th. 41, 2; Cri. 649.

grundsópa *ground soap;* saponaria officinalis :— Cartilago, Gl. C. Lchdm. iii. 329, col. 1.

grund-stán, es; *m. A foundation-stone :*— Grundstánas *cementum,* Ælfc. Gl. 116; Som. 80, 70; Wrt. Voc. 61, 47. [*Ger.* grund-stein.]

grund-wæg, es; *m. A foundation, the earth :*—He on grundwæge men of deáþe worde awehte *he [Christ] on this earth raised men from death by his word,* Andr. Kmbl. 1163; An. 582. [Cf. *Goth.* grunduwaddjus *foundation.*]

grund-wang, -wong, es; *m. The bottom, ground, floor, the earth :*—He ðone grundwong ongytan mihte *he could perceive the bottom [of the lake],* Beo. Th. 2996; B. 1496: 5533; B. 2770. Grundwong ofgyfan *to give up the earth, to die,* 5169; B. 2588.

grund-weall, es; *m. A foundation :*—Ðes grundweall *hoc fundamentum,* Ælfc. Gr. 8; Som. 7, 60. Ic lecge grundweall *fundo,* 37; Som. 39, 20. Se cræft is eallra bóclícra cræfta ordfruma and grundweall *that art is the beginning and foundation of all literary arts,* 50; Som. 51, 2: Wrt. Voc. 81, 6. Se grundweall ðara munta *fundamenta montium,* Ps. Th. 17, 7: Lk. Skt. 6, 48, 49: Homl. Th. ii. 588, 20: Chr. 654; Erl. 29, 11: Bt. Met. Fox 7, 67; Met. 7, 34. [*Orm.* grunndwall: cf. *Ger.* grundmauer.] v. grund-wæg.

grund-wela, an; *m. Earthly wealth :*—Him grundwelan ginne sealde hét ðam sinhíwum sǽs and eorþan tuddorteóndra teohha gehwilcre wæstmas fédan *he gave them ample riches of earth, bade for the man and wife each of sea's and land's productive tribes bring forth fruits,* Cd. 46: Th. 59, 1; Gen. 957.

grund-wyrgen, ne; *f. A wolf of the deep [Grendel's mother],* Beo. Th. 3041; B. 1518.

grunian; *p.* ode *To make a loud noise, grunt :*—Swýn grunaþ *sus grunnit,* Ælfc. Gr. 22; Som. 24, 10. [Cf. *O. H. Ger.* grun, grunni, Grff. iv. 328.]

grunung, e; *f. A crying out, roaring;* rugitus, barritus, mugitus, Hpt. Gl. 462, 508.

grut *vorago,* Hpt. Gl. 423, 507; grutte *abyssus,* 529.

grút; *indecl.* but also *dat.* grýt, Lchdm. iii. 28, 9; *f.* GROUT, *the wet residuary materials of malt liquor;* condimentum cerevisiæ :—Wyrc claːm of sǘrre rigenre grút oððe dǽge *work a paste of sour rye grout or of dough,* L. M. 3, 59; Lchdm. ii. 342, 17. Grút mealtes, i. 31, 7; Lchdm. ii. 74, 9. Genim ealde grút *take old grout,* i. 39, 2; Lchdm. ii. 100, 1: 28; Lchdm. ii. 68, 26: Lchdm. iii. 42, 28. [*Worte* siromellum, sed *growte* dicas agromellum, Wrt. Voc. 178, 3. Growtt *hoc idromellum,* 233, 33. Growte for ale *granomellum,* Prompt. Parv. 217, 3, where see note. *Mod. Engl.* grouts *grounds, dregs.*] Cf. next word; *also* cf. *Icel.* grautr; *m. porridge.*

grút; *pl. n. Fine meal :*—Grút *pollis,* Wrt. Voc. 290, 63: L. M. i. 61, 1; Lchdm. ii. 132, 15. VI ambra grúta *six measures of meal,* Th. Chart. 471, 13. [Cf. grytta, grot, greót, and the preceding word.]

grym. v. grim.

grymede *glyppus,* Ælfc. Gl. 76; Som. 71, 128; Wrt. Voc. 45, 31.

grymetan. v. grimetan.

grymetung v. grimetung.

gryn, es; *m. n* [?] *Lamentation, grief, affliction, evil :*—Fela ic láðes gebád grynna æt Grendel *much evil have I experienced, many a grief at Grendel's hands,* Beo. Th. 1864; B. 930. [Cf. *O. H. Ger.* grun; *m.* grunni; *f.* Grff. iv. 328; *and see* grunian, gyrn. *Or does* gryn = grin?]

gryndan; *pp.* ed.   I. *to found.* [*Ger.* gründen.] v. ge-gryndan.   II. *to come to the ground, to descend :*—Gryndende *descendens,* Cot. 68, Lye. v. agryndan.

grynde, es; *n. An abyss,* Cd. 220; Th. 285, 2; Sat. 331. v. æfgrynde (Appendix), un-grynde.

grynel, es; *m. Kernel;* toles, Mone Gl.

grynian *to ensnare.* v. grinian.

gryn-smiþ, es; *m. One causing grief, affliction, evil* [gryn, q. v.], Andr. Kmbl. 1833; An. 919.

gryre, es; *m. Horror, terror, dread, something horrible, dreadful :*—Óðrum on gryre wǽron to neósienne *aliis horrori erant visendum,* Bd. 4, 27; S. 604, 27. Him ðæs egesa stód gryre fram ðam gáste *terror was upon him therefore, horror from the spirit,* Cd. 201; Th. 249, 6; Dan. 526: Exon. 116 a; Th. 446, 12; Dóm. 21: 116 b; Th. 447, 22; Dóm. 43. Wæs se gryre læssa *the horror was less,* Beo. Th. 2569; B. 1282. Se légdraca grimlíc gryre *the firedrake, a fierce terror,* 6074; B. 3041: Cd. 195; Th. 243, 20; Dan. 439. Wið ðæs egesan gryre ːːgainst *the terror of that fear,* 197; Th. 245, 22; Dan. 467: 223; Th. ː93, 13; Sat. 454. Ðæt he in ðone grimman gryre gongan sceolde *that*

he should go into that fell and fearful place, Exon. 41 a; Th. 136, 18; Gú. 543. Hie wyrd forsweóp on Grendles gryre *fate has swept them off into the terrible power of Grendel,* Beo. Th. 960; B. 478: Cd. 143; Th. 178, 32; Exod. 20. Mid gryrum ecga *with the terrors of swords,* Beo. Th. 971; B. 483: 1187; B. 591. [*Laym.* grure: *A. R.* grure: *O. Sax.* gruri.] DER. fǽr-, helle-, hinsíð-, leód-, wæl-, wésten-, wíg-gryre.

gryre-bróga, an; *m. Terror, horror,* Exon. 20 a; Th. 53, 12; Cri. 849.

gryre-fæst; *adj. Terribly fast,* Elen. Kmbl. 1516; El. 760.

gryre-fáh; *adj. Terribly hostile or terrible in its variegated colouring,* Beo. Th. 5146; B. 2576.

gryre-gæst, es; *m. A dreadful guest,* Beo. Th. 5113; B. 2560.

gryre-geatwe; *pl. f. Terrible, warlike equipments,* Beo. Th. 653; B. 324.

gryre-hwíl, e; *f. A time of terror,* Andr. Kmbl. 935; An. 468.

gryre-leóþ, es; *n. A song of terror,* Beo. Th. 1576; B. 786: Byrht. Th. 140, 8; By. 285.

gryre-líc; *adj. Horrible, terrible,* Andr. Kmbl. 3101; An. 1553: Exon. 108 b; Th. 415, 27; Rä. 24, 3: Beo. Th. 2886; B. 1441: 4278; B. 2136.

gryre-síð, es; *m. A terrible way,* Beo. Th. 2928; B. 1462.

grystlung. v. gristlung.

gryt *grues,* Wrt. Voc. 287, 25.

grýto; *f. Greatness;* grossitudo :—Ungemetlícre grýto and micelnysse *vincens grossitudine,* Nar. 8, 22.

grytta *and* **gryttan;** *pl. f. Grits, groats, coarse meal :*—Ðás gritta *hic furfur,* Ælfc. Gr. 9, 22; Som. 10, 47. Gretta *furfures,* Wrt. Voc. 83, 21. Beren mela oððe grytta *barley meal or grits,* L. M. 2, 26; Lchdm. ii. 220, 8: 39; Lchdm. ii. 250, 2. Grytte, 18; Lchdm. ii. 200, 9. Of berenum gryttum *of barley grits,* 19; Lchdm. ii. 202, 7: 22; Lchdm. ii. 206, 19. Hwæte gryttan *apludes vel cantabra,* Ælfc. Gl. 50; Som. 65, 124; Wrt. Voc. 34, 53. [Cf. *A. R.* gruttene brede, 186, 11: *O. H. Ger.* gruzze *furfur:* Ger. grütze; *f.* grit, groats.]

gú-dǽd, e; *f. A deed done in the past,* Exon. 64 a; Th. 235, 12; Ph. 556. v. iú-dǽd.

guma, an; *m. A man;* vir, homo :—Grétte ðá guma óðerne *then one man took leave of another,* Beo. Th. 1309; B. 652. God ealle cann guman geþancas *Dominus novit cogitationes hominum,* Ps. Th. 93, 11. Wiste ferhþ guman *knew the man's mind,* Cd. 134; Th. 169, 2; Gen. 2793. Guman God wurþedon *the men worshipped God,* 187; Th. 232, 14; Dan. 260. Gumena aldor *ruler of men,* 89; Th. 111, 30; Gen. 1863. God gumena weard *God, the guardian of men,* 184; Th. 230, 22; Dan. 237. Gumena gehwylc *each man,* Exon. 19 b; Th. 51, 25; Cri. 821: 32 a: Th. 101, 5: Cri. 1654. Gumena bearn *the children of men,* Beo. Th. 1760; B. 878. Geared gumum gold brittade *Jared distributed gold to the people,* Cd. 59; Th. 72, 3; Gen. 1181. [*Laym.* gume, gome: *Piers P.* gome: *O. Sax.* gumo; *m. vir, homo: O. Frs.* goma: *O. H. Ger.* goma: *Goth.* guma: *O. Nrs.* gumi; *m. homo, vir, primipilus: Lat.* homo.] DER. brýd-, dryht-, þeód-guma.

gú-mann, es; *m. A man of old :*—Ðǽm gúmonnum *antiquis,* Mt. Kmbl. Rush. 5, 27. v. gió-man.

gum-cynn, es; *n. Mankind, men, a race, nation;* humanum genus, gens, natio :—He þohte forgrípan gumcynne *he resolved to destroy mankind,* Cd. 64; Th. 77, 14; Gen. 1275. Eom ic gumcynnes ánga ofer eorþan *amongst men on the earth I am unique,* Exon. 129 a; Th. 496, 11; Rä. 85, 12: Beo. Th. 5524; B. 2765. Swá hwylc mægþa ðone magan cende æfter gumcynnum *whatever matron brought forth this son amongst men,* Beo. Th. 1892; B. 944. We synt gumcynnes Geáta leóde *we are of the race of the Gauts' people,* 525; B. 260. [*O. Sax.* gumkunni.]

gum-cyst, e; *f. Manly virtue or excellence, munificence, liberality :*—Ðú ðé lǽr be ðon gumcyste ongit *learn from that, understand liberality,* Beo. Th. 3450; B. 1723. He siððan sceal gódra gumcysta geásne hweorfan *afterwards shall he pass away wanting in all noble virtues,* Exon. 71 a; Th. 265, 14; Jul. 381. Nú is þearf micel ðæt we gumcystum georne hýran *now is it very needful that we with virtuous zeal attend,* Andr. Kmbl. 3210; An. 1608. Abraham gumcystum gód golde and seolfre gesǽlig *Abraham, noble in his munificence, blessed with gold and silver,* Cd. 85; Th. 106, 10; Gen. 1769: 86; Th. 108, 23; Gen. 1810: Beo. Th. 2976; B. 1486. Gumcystum gód *brave [or munificent?],* 5079; B. 2543. *See the use of* cystum *under* cyst III.

gum-dreám, es; *m. The joys of men, this life :*—He gumdreám ofgeaf Godes leóht geceás *he gave up the joy of men, chose God's light,* Beo. Th. 4929; B. 2469.

gum-dryhten, es; *m. A lord of men;* virorum dominus, Beo. Th. 3289; B. 1642.

gum-féða, an; *m. A troop of men,* Beo. Th. 2807; B. 1401.

gum-man, -mann, es; *m. A famous man, a man;* vir clarus, homo, Beo. Th. 2061; B. 1028.

gum-ríce, es; *n. Power, rule over men, a kingdom, the earth :*—Nis ðé goda ǽnig on gumríce efne gelíc éce Drihten *non est similis tibi in*

diis, Domine, Ps. Th. 85, 7. On ðam gumríce *in that kingdom*, Elen. Kmbl. 2439; El. 1221. Gumríces weard *the king*, Cd. 180; Th. 226, 25; Dan. 176.

gum-rinc, es; *m. A man* :—Gódlíc gumrinc *a goodly man*, Exon. 129 a; Th. 495, 7; Rä. 84, 4. Dysiges folces gumrinca gyden *a goddess of the foolish people, of men*, Bt. Met. Fox 26, 105; Met. 26, 53: Cd. 75; Th. 93, 27; Gen. 1552.

gum-stól, es; *m. A throne*, Beo. Th. 3908; B. 1952.

gum-þegen, es; *m. A man*, Exon. 79 b; Th. 298, 11; Crä. 83.

gum-þeód, e; *f. A nation, people* :—Gumþeóda bearn *the children of men*, Cd. 12; Th. 15, 1; Gen. 226.

gund, es; *m. Matter, corruption*; pus, L. M. 1, 4: 2, 3; Lchdm. ii. 44, 23, 26. [*Prompt. Parv.* gownde of eye *ridda, allugo*. See note, and v. Hall. Dict. gound. *O. H. Ger.* gund, gunt *virus, pus, tabum, tabes*.] DER. heals-gund.

gung; *adj. Young, youthful*; jŭvĕnis, adolescens :— Ic eom gungre yldo *adolescentior sum*, Bd. 4, 25; S. 600, 3. Hí ofslógon ænne gungne Brytiscne man *they slew a young Briton*, Chr. 501; Erl. 15, 16. v. geong.

gungling *a youngling*.

GÚÞ, e; *f. [a poetical word] War, battle, fight*; bellum :—Gúþ nimeþ freán eówerne *war shall take away your lord*, Beo. Th. 5066; B. 2536: 4960; B. 2483: 3320; B. 1658: 2251; B. 1123. Sumne sceal gúþ abreótan *war shall crush one*, Exon. 87 a; Th. 328, 12; Vy. 16: 88 a; Th. 331, 15; Vy. 68. Bídan Grendles gúþe *to await Grendel's attack*, Beo. Th. 970; B. 483. Gúþe gefýsed *ready for battle*, 1265; B. 630: Byrht. Th. 137, 27; By. 192: 140, 30; By. 296: Andr. Kmbl. 467; An. 234. He gúþe ræs fremman sceolde *he had to perform a war-onslaught*, Beo. Th. 5245; B. 2626: 4712; B. 2356. Grimre gúþe *in fierce fight*, 1058; B. 527. Ðonne hie to gúþe gárwudu ræerdon *when to battle they reared the spearshaft*, Cd. 160; Th. 198, 19; Exod. 325: Beo. Th. 880; B. 438: 2948; B. 1472: Byrht. Th. 132, 8; By. 13: 134, 34; By. 94: Elen. Kmbl. 45; El. 23. Ðe ðé æsca tír æt gúþe forgeaf *who gave thee martial glory in fight*, Cd. 97; Th. 127, 11; Gen. 2109: Judth. 11; Thw. 23, 15; Jud. 123: Exon. 17 b; Th. 42, 17; Cri. 674: Beo. Th. 3074; B. 1535: Byrht. Th. 140, 9; By. 285; Chr. 937; Erl. 114, 10; Ædelst. 44: Andr. Kmbl. 2661; An. 1332. Æt ðære gúþe Gárulf gecrang *at the battle fell Garulf*, Fins. Th. 62; Fin. 31. Ðær ðú gúþe findest *there wilt thou find conflict*, Andr. Kmbl. 2698; An. 1351. Ær ðú gúþe fremme *before thou do battle*, 2708; An. 1356: Exon. 105 b; Th. 402, 5; Rä. 21, 25. Se ða gúþe genæs *he had come safe out of the battle*, Cd. 94; Th. 121, 33; Gen. 2019. Ðe ða gúþe forbeáh *who turned aside from the battle*, Byrht. Th. 141, 21; By. 315. Gúþe spówan *to thrive in battle*, Cd. 97; Th. 127, 23; Gen. 2115: Exon. 71 b; Th. 266, 4; Jul. 393: Salm. Kmbl. 249; Sal. 124. Ic geneþde fela gúþa *I dared many a conflict*, Beo. Th. 5017; B. 2512: 5080; B. 2543. Guma gúþum cúþ *a man distinguished in battles*, 4362; B. 2178: 3920; B. 1958. [*Icel.* gúðr *war* (only used in poetry) : *O. H. Ger.* gund, Grff. iv. 219.] See Grmm. D. M. 393.

gúþ-beorn, es; *m. A man of war, warrior*; vir bellicosus, bellator :—Gúþbeorna sum wicg gewende *one of the warriors turned his charger*, Beo. Th. 634; B. 314.

gúþ-bil, -bill, es; *n. A war-bill, a sword*, Beo. Th. 5162; B. 2584: 1610; B. 803.

gúþ-bord, es; *n. A warlike board, a shield*, Exon. 92 a; Th. 346, 11; Gn. Ex. 203: Cd. 128; Th. 163, 5; Gen. 2693. [*Icel.* gunn-borð *shield*.]

gúþ-byrne; *f. A coat of mail*, Beo. Th. 648; B. 321.

gúþ-cearu, e; *f. The care which is caused by battle*, Beo. Th. 2520; B. 1258.

gúþ-cræft, es; *m. Warlike power or skill*, Beo. Th. 254; B. 127.

gúþ-cwén, e; *f. A warrior queen*, Elen. Kmbl. 507; El. 254: 661; El. 331.

gúþ-cyning, es; *m. A warlike king*, Cd. 97; Th. 128, 8; Gen. 2123: Beo. Th. 401; B. 199: 3942; B. 1969: 4660; B. 2335: 5119; B. 2563: 5348; B. 2677.

gúþ-cyst, e; *f. Warlike excellence, bravery* :—Sunu simeones sweótum cómon þridde þeódmægen gúþcyste onþrang *the sons of Simeon came in troops, a third great force bravely pressed on* [*or* cyst = *troop, band, and* gúþcyste onþrang = *pressed on in phalanx*, cf. sweótum cómon], Cd. 160; Th. 199, 24; Exod. 343. [Cf. hilde-cyst.]

gúþ-deáþ, es; *m. Death got in fight*, Beo. Th. 4491; B. 2249.

gúþ-fana, -fona, an; *m. A military standard, ensign, banner*; signum vexillum :—Ðær wæs se gúþfana genumen ðe hí ræfen héton *there was the banner taken that they called the Raven*, Chr. 878; Erl. 81, 3. Ðæt heofonlíce tácn ðære hálgan róde is úre gúþfana wið ðone gramlícan deófol *the heavenly sign of the Holy Rood is our banner against the fierce devil*, H. R. 105, 16: 52. Ða gúþfonan *signa*, Ors. 6, 4; Swt. 260, 1. Ðær wæron vii hund gúþfanena genumen *there were seven hundred standards taken*, 4, 1; Bos. 77, 29: Th. Chart. 430, 1. Under gúþfanum *under the standards*, Judth. 11; Thw. 24, 32; Jud. 219. [*Icel.* gunn-fani :

O. H. Ger. gund-fano. Adopted in the French and Italian from the German, *O. Fr.* gun-fanon: *Ital.* gonfalone: hence *Mid. E.* gon-fanoun: *and* gun-faneur *a standard-bearer*, A. R. 300, 17.]

gúþ-flán; *m. f:* or gúþ-flá; *f. A war-dart*, Cd. 95; Th. 124, 15; Gen. 2063.

gúþ-floga, an; *m. One that flies to battle, a dragon*, Beo. Th. 5049; B. 2528.

gúþ-freá, an; *m. A warlike lord or prince*, Andr. Kmbl. 2667; An. 1335.

gúþ-frec; *adj. Bold in war* :—Gúþfrec guma *man bold in war*, Andr. Kmbl. 2235; An. 1119.

gúþ-freca, an; *m. A warrior*, Exon. 61 a; Th. 223, 1; Ph. 353 [or perhaps this passage should be put under the preceding word]. Grame gúþfrecan gáras sendon *fierce warriors hurled spears*, Judth. 11; Thw. 24, 35; Jud. 224. v. freca.

gúþ-fremmende; *part. pres. One doing battle or fighting*, Cd. 154; Th. 192, 14; Exod. 231: Beo. Th. 497; B. 246.

gúþ-fruma, an; *m. A warlike chief*, Beo. Th. 39; B. 20.

gúþ-fugel, es; *m. A bird of war, eagle*, Exon. 106 b; Th. 406, 22; Rä. 25, 5.

Gúþ-geátas; *pl. The warlike Gauts*, Beo. Th. 3080; B. 1538.

gúþ-geatwe; *pl. f. Warlike dress or equipments*, Beo. Th. 796; B. 395.

gúþ-geláca, an; *m. A companion, comrade in war, a warrior*, Elen. Kmbl. 86; El. 43.

gúþ-gemót, es; *n. A battle-meeting, battle, fight*, Cd. 95; Th. 124, 1; Gen. 2056: Exon. 104 b; Th. 397, 28; Rä. 16, 25.

gúþ-getawa; *pl. f. War-equipments*, Beo. Th. 5265; B. 2636.

gúþ-geþingu; *pl. n. The lot to be expected from impending war*, Andr. Kmbl. 2044; An. 1024: 2088; An. 1045. v. ge-þing III.

gúþ-gewæd, es; *n. A martial dress, war-weeds*, Beo. Th. 5228; B. 2617: 5453: B. 2730: 5694: B. 2851: 5735: B. 2871: 459; B. 227: 5240: B. 2623.

gúþ-geweorc, es; *n. A warlike work or deed*, Beo. Th. 1360; B. 678: 1967: B. 981: 3654; B. 1825.

gúþ-gewinn, es; *n. Battle, warlike contest*, Andr. Kmbl. 434; An. 217: Exon. 102 b; Th. 388, 10; Rä. 6, 5.

gúþ-hafoc, es; *m. A war-hawk, eagle* :—Earn grædigne gúþhafoc *the eagle, greedy war-hawk*, Chr. 937; Erl. 115, 13; Ædelst. 64. [Cf. *Icel.* gunnar-haukr.]

gúþ-heard; *adj. Stout in war*, Elen. Kmbl. 407; El. 204.

gúþ-helm, es; *m. A helm*, Beo. Th. 4967; B. 2487.

gúþ-here, es; *m. A martial band, an army*, Cd. 92; Th. 118, 18; Gen. 1967.

gúþ-horn, es; *m. A war-horn, trumpet*, Beo. Th. 2868; B. 1432.

gúþ-hréþ, es; *m. Glory in war*, Beo. Th. 1642; B. 819.

gúþ-hwæt; *adj. Active, vigorous in war, valiant*, Apstls. Kmbl. 113; Ap. 57. [*Icel.* gunn-hvatr.]

Gúþ-lác, es; *m. The hermit or saint of Crowland* [v. Crúland] *died at the age of 41, in* A.D. 714 :—Gúþlác se nama ys on Rómánisc, Belli munus *the name Guthlac is in Latin, Belli munus*, Guthl. 2; Gdwin. 10, 23. Se hálga Gúþlác ðás word gehýrde *the holy Guthlac heard these words*, 4; Gdwin. 30, 9. Onginne ic nú be ðam lífe ðæs eádigan weres, Gúþláces *I begin now concerning the life of the blessed man Guthlac*, Guthl. 4; Gdwin. 26, 2: Exon. 34 b; Th. 110, 15; 113, 17; 115, 29. Hæfde Gúþlác ðá on ylde six and twentig wintra ðá he æerest on ðam wéstene [Crúlande] gesæt *then Guthlac was six and twenty years of age when he first settled in the desert* [*Crowland*], Guthl. 3; Gdwin. 24, 3. Gúþlác æfter ðon fíftyne geár ðe he lædde his líf, ðá wolde God his þeów gelædan to ðære écan reste ðæs heofoncundan ríces *after Guthlac had led his life for fifteen years, then God would lead his servant to the eternal rest of his heavenly kingdom*, Guthl. 20; Gdwin. 78, 19-22. A.D. 714, Hér, forþferde Gúþlác se hálga here, A.D. 714, *the saint Guthlac died*, Chr. 714; Erl. 44, 5. On ðone ændleftan dæg ðæs mónðes biþ sce gúþláces geleornes ðæs anceran on brytene *on the eleventh day of the month is the departure of St. Guthlac the anchorite in Britain*, Shrn. 71, 2.

gúþ-leóþ, es; *n. A war-song*, Beo. Th. 3048; B. 1522.

gúþ-mæcga, an; *m. A warlike man*; bellicosus vir, Salm. Kmbl. 181; Sal. 90 [MS. A].

gúþ-maga, an; *m. A warlike man*; bellicosus vir, Salm. Kmbl. 181; Sal. 90 [MS. B].

gúþ-mód; *adj. Of warlike mind*, Beo. Th. 617; B. 306.

Gúþ-myrce; *pl. The Ethiopians*, Cd. 145; Th. 181, 10; Exod. 59. [Cf. Ælmyrca.]

gúþ-plega, an; *m. War-play, battle*, Byrht. Th. 133, 35; By. 61: Exon. 16 a; Th. 36, 8; Cri. 573: Apstls. Kmbl. 43; Ap. 22: Andr. Kmbl. 2737; An. 1371.

gúþ-ræs, es; *m. A warlike attack*, Andr. Kmbl. 3061; An. 1533: Beo. Th. 5974; B. 2991: 3159: B. 1577: 4844; B. 2426.

gúþ-reáf, es; *n. A warlike dress, armour*, Exon. 71 a; Th. 265, 26; Jul. 387.

gúþ-reów; *adj. Fierce in fight*, Beo. Th. 115; B. 58.

gúþ-rinc, es; *m. A man of war, warrior*, Beo. Th. 1681; B. 838: 3007; B. 1501: 3766; B. 1881: Byıht. Th. 135, 55; By. 138: Andr. Kmbl. 309; An. 155: 783; An. 392.

gúþ-róf; *adj. Famous in war*, Beo. Th. 1220; B. 608: Elen. Kmbl. 545; El. 273.

gúþ-sceaða, an; *m. One who harms by warlike attack*, Beo. Th. 4625; B. 2318.

gúþ-scear, es; *m. War-shearing, slaughter in battle*, Beo. Th. 2430; B. 1213. v. scear, inwit-scear.

gúþ-sceorp, es; *n. War-clothing*; vestitus *vel* ornatus bellicus, Judth. 12; Thw. 26, 15; Jud. 329.

gúþ-scrúd, es; *n. War-clothing*, Elen. Kmbl. 515; El. 258.

gúþ-searo; *n. Arms, armour*, Beo. Th. 435; B. 215: 661; B. 328: Andr. Kmbl. 253; An. 127.

gúþ-sele, es; *m. A war-hall, hall in which warriors sit*, Beo. Th. 890; B. 443.

gúþ-spell, es; *n. War-tidings*, Cd. 97; Th. 126, 18; Gen. 2097.

gúþ-sweord, es; *n. A sword*, Beo. Th. 4314; B. 2154.

gúþ-præc; *gen. -þræce; pl. nom. gen. acc. -þraca; f. War-force*; vis bellica :—Mid gúþþræce *with war-force*, Cd. 50; Th. 64, 6; Gen. 1046; 93; Th. 119, 2; Gen. 1973.

gúþ-preát, es; *m. A martial band*, Cd. 151; Th. 190, 2; Exod. 193.

gúþ-weard, es; *m. A war-guard, a king*, Cd. 151; Th. 188, 26; Exod. 174: Elen. Kmbl. 27; El. 14.

gúþ-weorc, es; *n. A warlike work* or *deed*, Andr. Kmbl. 2133; An. 1068.

gúþ-wérig; *adj. Weary with battle*, Beo. Th. 3176; B. 1586.

gúþ-wíga, an; *m. A warrior*, Beo. Th. 4230; B. 2112.

gúþ-wine, es; *m. A comrade, friend in war*, Beo. Th. 3624; B. 1810: 5463; B. 2735.

gúþ-wudu, a; *m. War-wood, a spear*, Fins. Th. 11; Fin. 6.

gycel-stán. v. gicel-stán.

gyd, gyddian. v. gid, giddian.

gyden, e; *f.* gydene, an; *f. A goddess;* dea :—Iuno wæs swíðe heálíc gyden *Juno was a very lofty goddess*, Salm. Kmbl. 121, 32. Sceolde bión gydene *was said to be a goddess*, Bt. 38, 1; Fox 194, 19: Bt. Met. Fox 26, 105; Met. 26, 53. Óþ he gemétte ða graman gydena *until he met the fierce goddesses*, 35, 6; Fox 168, 24. Seó hæfde gehátten heora gydenne Dianan *ðæt heó wolde hiere líf on fæmnháde alibban* she had promised their goddess Diana that she would live her life in virginity; virgo vestalis, Ors. 3, 6; Swt. 108, 17. [Cf. *Icel.* guðja: *O. H. Ger.* gutin, gutenna: *Ger.* göttin.]

gydenlíc; *adj. Nunlike, vestal;* vestalis, Cot. 179, Lye.

gyf. v. gif.

gyfa. v. gifa.

gyfan. v. gifan.

gyfen, es; *n. Ocean* :—Ne on gyfenes grund *not in ocean's bed* [*ground*], Beo. Th. 2792, note; B. 1394. v. geofon.

gyfl. v. gifl.

gyft. v. gift.

gyfu, e; *gen. pl.* -ena; *f. A gift, grace;* donum, gratia :—Gyfu gif hwylc is of me *donum quodcumque ex me*, Mk. Bos. 7, 11. Godes gyfu wæs on him *gratia Dei erat in illo*, Lk. Bos. 2, 40: Cd. 212; Th. 262, 5; Dan. 739. v. gifu.

gyfung. v. gifung.

gy-fylness, e; *f. Completion, end* :—Óþ ða gyfylnesse ðisse worlde *until the end of the world*, Blickl. Homl. 145, 16. v. ge-fylness.

gyhþa. v. gihþa.

gyhþu. v. gehþo.

gyld, gyldan. v. gild, gildan.

gylda. v. gilda.

gyldan; *p.* ede *To gild*, Chr. 1052; Th. 321, 25. [*Icel.* gylla : *O. H. Ger.* uber-guldete; *p.*]

gylden, gilden; *adj. Golden;* aureus :—Gylden wed *vel* feoh arra, Ælfc. Gl. 14; Som. 58, 11; Wrt. Voc. 21, 6. Gylden læfr bractea, 98; Som. 67, 111; Wrt. Voc. 38, 34. Gylden fel bractea, Cot. 27, Lye. Gyldena *vel* gegylde fatu crisendeta, Ælfc. Gl. 67; Som. 69, 97; Wrt. Voc. 41, 48. Ða stód ðær gyldenu onlícnes *then stood there a golden image*, Shrn. 88, 22. Ðær is geat gylden *there is a golden gate*, Cd. 227; Th. 305, 19; Sat. 649. On sumum gyldenum wecge *to a golden wedge*, Homl. Th. i. 60, 29. Under gyldenum beáge *under a golden crown*, Beo. Th. 2330; B. 1163. To ðam gyldnan gylde *to the idol of gold*, Cd. 182; Th. 228, 18; Dan. 204. Hring gyldenne *a golden ring*, Beo. Th. 5611; B. 2809. [*Laym.* gulden : *Orm.* gilden : *O. Sax.* guldin : *O. Frs.* gulden, golden, gelden : *Icel.* gullinn : *O. H. Ger.* guldin : *Ger.* gülden, gold.] DER. eal-gylden.

gylden-beáh, -beág, es; *m. A crown* :—Mid gehálgodon gildenbeáge *with the hallowed crown*, Lev. 8, 9.

gylden-feaxa; *adj. Having golden hair;* auricomus, Cot. 11, Lye. [Cf. ge-feaxe.]

gylding-wecg *a gold mine, a vein of gold;* aurifodina, Cot. 16, 167, Som.

gylian; *p.* ede *To yell, shout out* :—Styrmde and gylede *shouted and yelled*, Judth. 10; Thw. 21, 19; Jud. 25. v. gellan.

gyllan *yell, chirp*, Cd. 167; Th. 208, 26; Exod. 489: Fins. Th. 10; Fin. 6. v. gellan.

gylm. v. gilm.

gylp, and its compounds. v. gilp, etc.

GYLT, gilt, gelt, gielt, es; *m. Guilt, crime, sin, offence, fault, wrong, debt, fine, forfeiture* :—Gylt *facinus vel culpa*, Wrt. Voc. 86, 67. Adames gylt *Adam's guilt*, Blickl. Homl. 9, 5: 23, 5: Exon. 61 b; Th. 226, 19; Ph. 408. For ðam gylte ðe hig worhton ðæt gildene celf *for the sin of making the golden calf*, Ex. 32, 35; Deut. 9, 21. Eustatius hæfde gecýdd ðam cynge ðet hit sceolde beón máre gylt ðære burhwaru ðonne his *Eustace had told the king that it was more the citizens' fault than his*, Chr. 1048; Erl. 178, 9. Man geútlagode Ælfgár bútan ælcan gylte *Ælfgar was outlawed without any crime* [*being proved against him*], Chr. 1055; Erl. 188, 28: 189, 35. Æt ðam forman gylte ðære fiohbóte onfón *on the first offence to accept pecuniary compensation*, L. Alf. 49; Th. i. 58, 8: L. Alf. pol. 7; Th. i. 66, 12: L. In. 73; Th. i. 148, 11: L. Ath. 1, 11; Th. i. 206, 3: L. Edg. S. 2, 2; Th. i. 266, 13. Gif he ðær gylt gewyrce *if he there do wrong*, L. Ath. 1, 8; Th. i. 204, 8. Gylt ceápes *crime in business*, Lchdm. iii. 198, 10. Þurh forman gylt *through the first sin*, Cd. 48; Th. 61, 17; Gen. 998. Forgyf us úre gyltas *demitte nobis debita nostra*, Mt. Kmbl. 6, 12: Ps. Th. 84, 3. Gyltas *delicta*, Ps. Spl. 18, 13. Geltas, Ps. C. 50, 39; Ps. Grn. ii. 277, 39. Gieltas, Exon. 62 b; Th. 229, 26; Ph. 461. Forgifnesse ealra heora gylta *forgiveness of all their sins*, Blickl. Homl. 193, 24: Elen. Kmbl. 1631; El. 817. Gyltum forgiefene *given up to sins*, Exon. 39 a; Th. 130, 2; Gú. 432. Forgeaf him ðone gylt *debitum dimisit ei*, Mt. Kmbl. 18, 27, 32. Ealle ða gyltes ða belimpeþ to míne kinehelme *omnes forisfacturas que pertinent ad regiam coronam meam*, Th. Chart. 423, 3. [*Laym. A. R.* gult : *Orm.* gillt : *Ayenb.* gelt.] Cf. scyld.

gyltan; *p.* gylte; *pp.* gylt *To commit guilt* or *sin, to be guilty* :—Ðara gyltendra scylda *the sins of the guilty*, Past. 21; Swt. 167, 6; Hat. MS. [*Orm.* gilltenn : *Wicl.* gilten : *O. E. Homl.* gulte; *p.*] v. gylting, gyltend, a-, for-gyltan.

gylte GELT, *gelded;* castratus, Som.

gyltend, es; *m. A debtor, an offender;* debitor :—Gyltend *lapsus*, Rtl. 189, 25. Swá swá we forgyfaþ úrum gyltendum *sicut nos dimittimus debitoribus nostris*, Mt. Bos. 6, 12. v. gyltende.

gyltende. v. gyltan.

gyltig; *adj.* GUILTY, *liable, bound;* reus :—Swá hwylc swá swereþ on ðære offrunge ðe ofer ðæt weofud ys, se ys gyltig *quicumque juraverit in dono quod est super illud, debet*, Mt. Kmbl. 23, 18. [*A. R.* Heo is gulti of the bestes deaðe, 58, 17 : *Chauc.* gulty.]

gylting, e; *f. Sinning, sin* :—Gyiltincg *prævaricatio*, Rtl. 109, 41. Gultingum *delictis*, 66, 29. Gyltingum, 124, 42.

gyltlíc *wicked, sinful* :—Gé gehýrdon gyltlíce spræce *audistis blasphemiam*, Mt. Kmbl. 26, 65.

gym *a gem.* v. gim.

GÝMAN, géman, gíman, giéman; *p.* de *To care for, take care of, take heed to, heed, observe, regard, keep; cum gen. acc* :—Ic gýme mín wedd *I will keep my covenant*, Lev. 26, 42. Ic geornor gýme ymb ðæs gæstes forwyrd ðonne ðæs líchoman *I care more earnestly about the spirit's destruction than the body's*, Exon. 71 b; Th. 267, 12; Jul. 414. Ic ne gýme ðæs compes *I care not for the strife*, 105 b; Th. 402, 26; Rä. 21, 35 : Lev. 26, 43. Egesan ne gýmeþ *heeds not terror*, Beo. Th. 3519; B. 1757. Dryhten mín gýmþ *Deus curam habet mei*, Ps. Th. 39, 20. Oðres ne gýmeþ to gebídanne yrfeweardes *he cares not to wait for another heir*, Beo. Th. 4894; B. 2451. Se deópa seáð giémeþ gæsta ðe *deep pit keeps the spirits*, Exon. 30 b; Th. 94, 26; Cri. 1546. Se ðe ne giémeþ hwæðer his gæst síe earm ðe eádig *who heeds not whether his spirit be miserable or blessed*, Th. 95, 6; Cri. 1553. Swíðe geornlíce giémaþ ðæt hie ða eorþlícan heortan geléren *they take very diligent heed to instruct the wordly hearts*, Past. 21; Swt. 161, 15; Hat. MS. Gýmaþ, Ps. Th. 118, 122. Ðæt he ðone stán nime hláfes ne gýme *to take the stone and neglect the bread*, Elen. Kmbl. 1229; El. 616: Exon. 66 b; Th. 246, 32; Jul. 70. He ætes ne gímde *he did not care for food*, Swt. A. S. Rdr. 60, 110. Giémde, Exon. 34 b; Th. 111, 3; Gú. 121. Ðæt hig gímdon ðæs dæges and ðære nihte *to rule the day and the night*, Gen. 1, 18. Moises and Aaron and hira bærn gímdon ðæs temples *Moses and Aaron and their children took charge of the temple*, Num. 3, 38. Rihtes ne gýmdon *cared not for right*, Andr. Kmbl. 278; An. 139: Cd. 113; Th. 148, 20; Gen. 2459: Exon. 18 a; Th. 44, 22; Cri. 706. Hí gímdon hwæðer . . . *observabant si* . . . , Mk. Skt. 3, 2: Lk. Skt. 6, 7. Ne gím ðu drýcræfta *regard not the arts of wizards*, Lev. 19, 31, 26: Deut. 18, 10: Beo. Th. 3525; B. 1760. Gém *observe*, Bt. Met. Fox 29, 6; Met. 29, 3. Gýmaþ and warniaþ *intuemini et cavete*, Mt. Kmbl. 16, 6. Sceal ic nú æniges lustes gíman *shall I care now for any pleasure*, Gen.

18, 12. Ða ðe bet cunnon sceolon gýman ôðra manna *those who know better are to take care of other men*, Homl. Th. ii. 282, 1 : Ps. Th. 77, 10 : Exon. 31 a ; Th. 96, 5 ; Cri. 1569. Gif his ðé gĕman lyst *if you pleased to care about it*, Bt. Met. Fox 31, 2 ; Met. 31, 1. Gŷman ðæs grundes *to take charge of the abyss*, Cd. 18 ; Th. 22, 31, 25 ; Gen. 349, 346. [*Laym. A. R.* ȝemen : *Orm.* ȝemenn : *Piers P.* ȝeme : *Goth.* gaumjan : *O. Sax.* gômean : *Icel.* geyma : *Dan.* gjemme : *Swed.* gömma : *O. H. Ger.* goumon.] DER. for-, ge-, ofer-gýman. v. gĕman.

gŷme, an ; *f. Care :*—Hŷ ðæs wealles nâne gýman [giéman, Swt. 134, 21] ne dydan *they took no care of the wall*, Ors. 3, 9 ; Bos. 68, 24. [*Orm.* gom : *Laym. A. R.* ȝeme : *O. Sax.* gôma ; *f : Icel.* gaumr ; *m.* gaum ; *f. heed, attention : O. H. Ger.* gouma ; *f.* Grff. iv. 203.] Cf. gýmen.

gŷmeleás ; *adj. Careless, negligent, uncared for, wandering, stray ;* negligens :—Gýmeleás feoh [giémeleás fioh] *stray cattle*, L. Alf. 42 ; Th. i. 54, 9 : Ps. Th. 70, 10. Ða gîmeleasan men ðe heora lîf adrugon on ealre ídelnisse *careless men who passed their life in all frivolity*, Swt. A. S. Rdr. 56, 11. Gŷmeleáse *heedless*, Exon. 73 a ; Th. 271, 33 ; Jul. 491 : Blickl. Homl. 55, 30.

gŷmeleásian ; *p.* ede *To neglect, be careless, despise ;* negligere :—Monige gŷmeleásedon ðám gerýnum ðæs hâlgan geleáfan *aliqui, neglectis fidei sacramentis*, Bd. 4, 27 ; S. 604, 6. DER. a-, for-gýmeleásian.

gŷmeleáslíce, gémeleáslíce ; *adv. Carelessly ;* negligenter, R. Ben. 44, Lye.

gŷmeleásnys, se ; *f. Carelessness ;* negligentia :— Forþgewitenum gŷmeleásnyssum *præteritas negligentias*, Bd. 3, 27 ; S. 559, 5.

gŷmeleást, gîmelîst, gémelést, e ; *f. Carelessness, negligence, neglect ;* negligentia :—Hit gelamp þurh gŷmeleáste *evenit per culpam incuriæ*, Bd. 3, 17 ; S. 544, 27. For giémelêste *for negligence*, Past. 21 ; Swt. 165, 6. Gîmeleáste, Swt. A. S. Rdr. 68, 376. On heora âgenre gŷmeleáste *from their own carelessness*, Chr. 1016 ; Erl. 156, 11 : Bt. 5, 1 ; Fox 10, 2. Þurh preósta gŷmeleáste *through the negligence of priests*, Cod. Dipl. Kmbl. iii. 349, 6. Se Hælend ne forlêt to gŷmeleáste his gelufedan apostol *the Saviour did not leave his beloved apostle to neglect*, Homl. Th. i. 58, 33. [*Orm.* ȝemelæste.]

gŷmen, gĕmen ; *f. Care, heed, solicitude, diligence, superintendence, rule ;* cura :—Se rêða rên sumes ymbhogan ungémet gĕmen *the fierce rain of some anxiety, immoderate care* [cf. se rên ungemetlîces ymbhogan, Fox 36, 19], Bt. Met. Fox 7, 56 ; Met. 7, 28 : 101 ; Met. 7, 51. Ðînre gŷmenne ic wæs beboden *in te jactatus sum*, Ps. Th. 21, 8. Hêr onfêng Pilatus gŷmene ofer ða Iudêas *in this year Pilate received the government of Judæa*, Chr. 26 ; Erl. 7, 6 : to gŷmenne, Erl. 6, 7. Of his bisceoplícan gŷmenne *cura pastorali*, Cod. Dipl. Kmbl. iii. 348, 35. Se stæf getácnaþ gŷmene and hyrdrædene *the staff is a symbol of care and guardianship*, Homl. Th. ii. 280, 35. Man sceal healdan ðæt hálige húsl mid mycelre gŷmene *the holy eucharist must be kept with great care*, L. Ælf. C. 36 ; Th. ii. 360, 11. He swâ geornfulle gŷmenne dyde ymb ða hæla ûre þeóde *tam sedulam erga salutem nostræ gentis curam gesserit*, Bd. 2, 1 ; S. 501, 3. Weoruldsorge and gŷmenne forlætan *sæculi curas relinquere*, 4, 19 ; S. 587, 38. Gŷmene dô se Abbod *curam gerit abbas*, R. Ben. interl. 27, Lye. DER. be-, un-gŷmen.

gŷmend, es ; *m. A governor* ; gubernator, Scint. 32.

gŷmmian *jugulare, occidere, perfodere*, Hpt. Gl. 495.

gŷmung, e ; *f. A marriage, nuptial :* — To gŷmungum ðæs heofonlícum brýdguman eádig fæmne ineode *ad nuptias sponsi cælestis virgo beata intravit*, Bd. 3, 24 ; S. 557, 6. v. gĕmung.

GÝNAN *to* GAIN ; lucrari, Lye.

gynd *beyond.* v. geond.

gyngra *younger*, Ps. Spl. 36, 26 ; *comp. of* geong.

gyngra, an ; *m. A junior* ; adŏlescentŭlus :—Gyngra ic eom *adŏlescentŭlus sum ego*, Ps. Spl. M. 118, 141. v. geongra.

gynian. v. ginian.

gynnan. v. ginnan.

gypigend *yawning* ; hiulcus, Gl. Prud. 703.

GÝR *a fir tree* ; abies, Lchdm. iii. 328, col. 1. v. gyrtreów.

GYR, gyra ; *m.* gyru ; *f. Mud, fen, marsh :*—Gyr *lætamen*, Hpt. Gl. 516. On gyran torr [?], Cod. Dipl. Kmbl. iii. 412, 8. Gyrwe fenn *palus*, Ælfc. Gr. 9, 33 ; Som. 12, 29. Gyran, gyras *paludes*, Lye. [Cf. *O. Frs.* cere, gere *dirty water*.] v. gor, Gyrwas.

GYRD, gird, gerd, e ; *f. A staff, rod, twig*, as a measure of distance, *a yard*, as a measure of area, *the fourth part of a hide* ; virga, vìrgata :—Gyrd *virga*, Wrt. Voc. 80, 3. Ðín gyrd and ðín stæf *virga tua et baculus tuus*, Ps. Th. 22, 5. Ðû ðínes yrfes gyrde alŷsdest *liberasti virgam hæriditatis tuæ*, 73, 3. Hit ys gird *it is a rod*, Ex. 4, 2. Ber Aarones girde into ðam getelde *bear Aaron's rod into the tabernacle*, Num. 17, 10 : Mt. Kmbl. 10, 10 : Homl. Th. ii. 8, 11 : i. 62, 34. He gebletsode ða grênan gyrda *he blessed the green twigs*, 64, 1. Fiórþe half gird *three yards and a half*, Lchdm. iii. 362, col. 2. Landes sumne dǽl ðæt is ân gyrd *a certain portion of land, that is the fourth part of a hide*, Cod. Dipl. Kmbl. iii. 260, 32 : 263, 7. Ðis synd ðære ânre gyrde landgemǽro *these are the boundaries of the one rood*, ii. 208, 18 : L. In.

67 ; Th. i. 146, 1, 2 : L. R. S. 4 ; Th. i. 434, 24. Swâ swŷðe nearwelíce he hit lêtt ût aspyrian ðæt næs ân ælpig hîde ne ân gyrde landes ðæt næs gesæt on his gewrite *so very narrowly did he have things searched out that there was not a single hide nor a rood of land that was not put down in his book*, Chr. 1085 ; Erl. 218, 35. [*Orm.* ȝerrd : *A. R. Chauc.* Piers P. ȝerd : O. H. Ger. gardea, garda, gerta, kirta : Ger. gerte.]

gyrdan, girdan ; *p.* gyrde ; *pp.* gyrded *To* GIRD, *bind round* ; cingere :— Ðá ðú gingra wære ðú gyrdest ðé . . . ðonne ðú ealdast ôðer ðé gyrt *cum esses junior cingebas te . . . cum senueris alius te cinget*, Jn. Skt. 21, 18. Se ðe hine man gelome gyrt *qua semper præcingitur*, Ps. Th. 108, 19. He girde hine *he girded him*, Lev. 8, 7. Hine se hálga wer gyrde grǽgan sweorde *the holy man girded himself with a grey sword*, Cd. 138 ; Th. 173, 22 ; Gen. 2865 : Fins. Th. 27 ; Fin. 13. Gyrd nú ðín sweord ofer ðín þeóh *accingere gladium tuum circa femur*, Ps. Th. 44, 4 : Lk. Skt. 17, 8. Gyrded cempa *a belted warrior*, Beo. Th. 4162 ; B. 2078. [*Icel.* gyrða : *O. H. Ger.* gurten : *Ger.* gürten.] DER. be-, ge-, ymb-gyrdan.

gyrdel, es ; *m. A* GIRDLE, *belt, zone, purse* ; cingulum :—Gyrdel *zona vel zonarium vel brachile vel redimiculum*, Ælfc. Gl. 64 ; Som. 69, 28 ; Wrt. Voc. 40, 57. Gyrdel *cingulum vel zona vel cinctorium*, Wrt. Voc. 81, 47. Gyrdel *stropheum*, Hymn. Surt. 103, 33. Fellenne gyrdel *zonam pelliciam*, Mt. Kmbl. 3, 4 : Mk. Skt. 1, 6. We hâtaþ on lêden quinque zonas ðæt synd fíf gyrdlas *we call them in Latin quinque zonas, that is five girdles*, Lchdm. iii. 260, 20. Him bebeád ðæt hî ne nâmon feoh on heora gyrdlum *præcepit ne tollerent in zona æs*, Mk. Skt. 6, 8. [*Icel.* gyrðill ; *m. a girdle, purse : O. H. Ger.* gurtil ; *m. cingulum, cinctorium, strophium, balteum : Ger.* gürtel.] v. gyrdels.

gyrdel-bred, es ; *n. Pugillar*, Lye.

gyrdel-hring, es ; *m. Ligula*, Lye.

gyrdels, es ; *m. A girdle* :—Gyrdels *cingulum*, Recd. 40, 27 ; Wrt. Voc. 66, 35. Gyrdils *zonam*, Mt. Kmbl. Lind. Rush. 3, 4 : 10, 9 : Mk. Skt. Lind. Rush. 1, 6 : 6, 8 : Rtl. 79, 7. Gelíc gyrdelse *sicut zona*, Ps. Th. 108, 19 : Exon. 113 b ; Th. 436, 21 ; Rä. 55, 4 : 114 a ; Th. 436, 34 ; Rä. 55, 11. v. gyrdel.

gyrd-weg, es ; *m. A road with a fence on either side* [?], Cod. Dipl. Kmbl. iii. 412, 21.

gyrd-wíte, es ; *n. Punishment with a rod, the punishment that came upon the Egyptians through Moses' rod*, Cd. 143 ; Th. 178, 22 ; Exod. 15.

gyren = grin, Ps. Th. 118, 110.

gyrian, gyrigan ; *part.* gyrigende ; *p.* ede, *pl.* gyredon. **I.** *to prepare* ; preparare :—Gyrigende dûna *præparans montes*, Ps. Spl. 64, 7. Gyrede setl his *paravit sedem suam*, 102, 19. **II.** *to clothe* ; vestire, amicire :—Swylce eác ða gegyrelan ðone lîchoman Cuðbertes gyredon *sed et indumenta quibus corpus Cudbercti vestierant*, Bd. 4, 31 ; S. 611, 5. v. gearwian.

gyrla. v. gerela.

gyrman ; *p.* de *To cry out, roar :*—Ic gyrmde *rugiebam*, Ps. Lamb. 37, 8.

gyrn, es ; *n. Grief, affliction, trouble, evil, calamity, injury :*—Me biþ gyrn witod *grief will be appointed me*, Exon. 104 b ; Th. 396, 18 ; Rä. 16, 6. Gyrn æfter gomene *grief after joy*, Beo. Th. 3554 ; B. 1775. Alŷsed of leóðhete of gyrme *rescued from the popular hate, from calamity*, Andr. Kmbl. 2301 ; An. 1152 : 3168 ; An. 1587. He gilleþ geómorlíce and his gyrn sefaþ *mournfully he cries out, sighs forth his grief*, Salm. Kmbl. 536 ; Sal. 267. Gyrn þurh gástgedál *affliction through death*, Exon. 45 a ; Th. 153, 31 ; Gú. 834. Gyldaþ nú mid gyrne ðæt heó goda ussa meaht forhogde *requite now with evil her contempt of our gods' might*, 74 b ; Th. 279, 25 ; Jul. 619. Ðæs ða byre siððan gyrne onguldon *for that the children greviously paid*, 61 b ; Th. 226, 23. Wíta unrím grimra gyrna *torments numberless, grim troubles*, 68 a ; Th. 252, 34 ; Jul. 173 : 39 a ; Th. 129, 7 ; Gú. 417. [Cf. *O. Sax.* gornword.] v. gryn.

gyrnan, girnan ; *p.* de *To desire, beg, yearn :*—Ic ne me micles gyrne *I do not desire much for myself*, Exon. 37 a ; Th. 121, 20 ; Gú. 291. Glædmód gyrneþ ðæt he gôdra mæst dæda gefremme *joyous is eager to perform very many good deeds*, Bt. 7 ; Th. 229, 28 ; Fox. 492. Ðæt mæden hire deápes girnde *the maiden desired to die*, Apol. Th. 2, 24 : 3, 8. Ne gyrne gê ðæt eów man Láreówas nemne *vos nolite vocari Rabbi*, Mt. Kmbl. 23, 8. Gyrnende *orantes*, Mk. Skt. 11, 24. [*Laym.* ȝeornien, ȝernen, ȝirnen : *Orm.* ȝeornenn, ȝeonenn : *Piers P.* ȝerne : *Goth.* gairnjan : *O. Sax.* girnean, gernean : *Icel.* girna.] v. geornan.

gyrne ; *adv. Earnestly* ; enixe :— Hí gyrne cleopedon to Gode *they earnestly cried to God*, Chr. 1083 ; Erl. 217, 22. v. georne.

gyrnes, gyrnys, -ness, -nyss, e ; *f. Diligence, industry* ; industria :— þurh Osþryþe gyrnysse *per industriam Osthrydæ*, Bd. 3, 11 ; S. 535, 12. v. geornes.

gyrning. v. geornung.

gyrn-stæf, es ; *m. Affliction, trouble :*—Gleáw gyrn-stafa *skilled in afflicting*, Exon. 68 a ; Th. 257, 10 ; Jul. 245.

gyrn-wracu ; *f. Vengeance for trouble or injury :*—Geáro gyrnwræce *ready to revenge her grief*, Beo. Th. 4242 ; B. 2118 : 2281 ; B. 1138.

Mārum sārum gyldan gyrnwræce *with greater pains to revenge their trouble*, Exon. 39 a; Th. 128, 16; Gû. 405.

gyrran. v. georran.

gyrretynde *roaring*; rugiens, Ps. Lamb. 21, 11. v. gyrran.

gyrst *gnashing, grinding*; stridor, Hpt. Gl. 513. v. grist.

gyrst; *adj. Grinding, grating*; stridulus, Hpt. Gl. 513.

gyrstan-dæg, gestran-dæg, gysternlíc-dæg YESTERDAY; heri:—Gyrstan-dæg heri, Ælfc. Gr. 38; Som. 39, 57. Swā he wæs gyrstan-dæg and æran dæg *sicut erat heri et nudius tertius*, Gen. 31, 5.

gyr-treów, es; *n. A spruce fir*; abies, Ælfc. Gl. 46; Som. 64, 128; Wrt. Voc. 32, 62. v. gyr.

gyrwan; *p.* ede; *pp.* ed *To prepare, make ready, make, put on, clothe, adorn*; părāre, făcĕre, vestīre, ornāre:—Angan hine gyrwan *he began to prepare himself*, Cd. 23; Th. 28, 26; Gen. 442: Andr. Kmbl. 1590; An. 796. Ic hæbbe geweald micel to gyrwanne gódlecran stól on heofne *I have great power to form a better throne in heaven*, Cd. 15; Th. 18, 30; Gen. 281. Cyning mec gyrweþ since and seolfre *the king adorns me with treasure and silver*, Exon. 105 b; Th. 401, 10; Rä. 21, 9. Wer and wíf bearn mid bleóm gyrwaþ *man and wife adorn their child with colours*, 87 a; Th. 327, 14; Vy. 3. v. gearwian.

Gyrwas; *pl. The people of a district in which Peterborough was situated:*—Se wæs of Gyrwa mægðe *de provincia Gyrviorum*, Bd. 3, 20; S. 550, 22. Abbud ðæs mynstres ðe gecweden is Medeshâmstyde on Gyrwa[n] lande *Abbas monasterii quod dicitur Medeshamstedi in regione Gyrviorum*, 4, 6; S. 573, 41. v. gyr *a marsh*.

gyse *yes*:—Hig cwædon,—Eówer lāreów, ne gylt he gafol? Ðā cwæþ he, Gyse he dêþ *they said,—Your master, doth he not pay tribute? He said, Yes, he does*; dixerunt,—Magister vester, non solvit didrachma? Ait, Etiam, Vulg. Mt. Bos. 17, 25. v. gese.

gyst, es; *m. A guest*:—Ic wæs gyst mōdor cildum *factus sum hospes filiis matris meæ*, Ps. Th. 68, 8; Cd. 114; Th. 150, 1; Gen. 2485. DER. sele-gyst. v. gæst.

gyst-ern, es; *n. A guest-place, guest-chamber*:—To ðam gysterne *to the guest-chamber*, Judth. 10; Thw. 21, 29; Jud. 40. v. gæst-ærn.

gysternlíc-dæg *yesterday*:—Swylce gysternlíc dæg, ðe forþgewāt *tanquam dies hesterna quæ præteriit*, Ps. Lamb. 89, 4. v. gyrstan-dæg.

gyst-hús, es; *n. A guest-house, guest-chamber*; hospītium:—Hwār is mín gyst-hús *where is my guest-house*? Mk. Bos. 14, 14. v. gæst-hús.

gystigan *to lodge, to abide as a guest*; hospitari, Scint. 47.

gyst-sele, es; *m. A guest-hall*; hospītālis aula:—Éðelleáse ðysne gystsele gihþum healdaþ [MS. healdeþ] *the homeless hold this guest-hall in memory*, Cd. 169; Th. 212, 4; Exod. 534. v. gest-sele.

gyt *you two*; vos duo:—Gyt nyton hwæt gyt biddaþ [*vos duo*] *nescitis quid* [*vos duo*] *petatis*, Mt. Bos. 20, 22. v. git.

GYT, gyta. v. git, gita.

gytan. v. gitan, and its compounds.

gyte, es; *m. A pouring, shedding, inundation, flood*:—Beó his blódes gyte ofer úrum bearnum *his bloodshed be upon our children*, Homl. Th. ii. 252, 20. Gyte *inundatio*, Cot. 108, Lye. Ne mihton hí for ðam ormætan gyte heora fêt of ðære cytan astyrian *they could not move their feet from the cottage for the excessive flood*, Homl. Th. ii. 184, 6. Martyrdôm biþ gefremmed nâ on blódes gyte ânum *martyrdom is effected not by bloodshed only*, i. 544, 24; Mt. Kmbl. 23, 30. Mid teâra gytum *with sheddings of tears*, Blickl. Homl. 61, 20. [O. H. Ger. gussi *diluvium*; gussa *inundatio*; guz *fusio*, Grff. iv. 285: Ger. guss.]

gytenes. v. getenys.

gyte-sél, es; *m. Joy at the pouring out of wine*:—Ðā wæs Olofernus on gytesālum *then was Holofernes joyous in feasting*, Judth. 10; Thw. 21, 17: Jud. 22. [Cf. Ðā wæs on sālum sinces brytta, Beo. Th. 1218; B. 607 and 2345; B. 1171.]

gyte-streám, es; *m. A current, flowing stream*:—Ebbe *vel* gyte-streám *reuma*, Ælfc. Gl. 105; Som. 78, 38; Wrt. Voc. 57, 20: Recd. 37, 65; Wrt. Voc. 63, 78.

gyt-feorm [?], L. R. S. 21; Th. i. 440, 26.

gýtsere. v. gítsere.

gýtsian. v. gítsian.

gýtsung. v. gítsung.

H

IN Anglo-Saxon the letter *h* represents the guttural aspirate and the pure spirant. In later English the guttural *h* is generally represented by *gh*, e. g. leóht *light*, heáh *high*. Under certain circumstances *h* takes the place of *c* and *g*, see those letters. In some cases it is dropped, e. g. bleó for bleoh; seón, *p.* seah; nabban = ne habban. In the Northumbrian specimens the use of the initial *h*, especially in the combinations *hl, hn, hr*, is uncertain, e. g. eorta = heorta, haald = ald, hlíf = líf, lysta = hlysta, hnett = nett, nesc = hnesc, hræst = ræst, ræfn = hræfn. The name of the Runic letter was hægl *hail*:—Hægl byþ hwítust corna, Runic pm. 9; Kmbl. 341, 4; the forms accompanying the poem and given by Kemble are these, ᚻ, ᚼ, ᚺ.

ha ha; *interj. Ha ha!*—Ha ha and he he getācniaþ hlehter on léden and on Englisc *ha ha and he he denote laughter in Latin and in English*, Ælfc. Gr. 48; Som. 49, 17.

habban, to habbanne, hæbbene; *pres. part.* hæbbende; *pres. indic.* ic hæbbe, hafa, ðû hæfst, hafast, he hæfþ, hafaþ, *pl.* habbaþ, hæbbaþ; *p.* hæfde; *subj.* hæbbe, *pl.* hæbben, habban; *imper.* hafa, *pl.* habbaþ; *pp.* hæfed. **I.** *cum acc. To HAVE, possess, hold, keep*:—Swylce getrýwþa swā se cyng æt him habban wolde *such pledges as the king wished to have from him*, Chr. 1093; Erl. 229, 19. Búton se biscop hie mid him habban wille *unless the bishop want to have it with him*, Past. Pref. Swt. 9, 6. Ða læwedan willaþ habban ðone mônan be ðam ðe hí hine geseóþ and ða gelæredan hine healdaþ be ðisum foresædan gesceáde *laymen will have the moon according as they see it, and the learned hold of it according to the aforesaid distinction*, Lchdm. iii. 266, 10. Hé ða word nel on his heortan habban and healdan *he will not have and hold those words in his heart*, Blickl. Homl. 55, 8. Ðonne mâgon wê ûs God ælmihtigne mildne habban *then may we have God Almighty merciful to us*, 107, 17. Hât twelf weras nyman twelf stânas and habban forþ mid eów *bid twelve men take twelve stones and have them along with you*, Jos. 4, 3. Ða hét ic eald hrægl tóslîtan and habban wið ðæm fýre and sceldan mid *jussi ergo scissas vestes opponere ignibus*, Nar. 23, 30. Hwilce gerihtæ hé âhte tô habbanne *what dues he ought to have*, Chr. 1085; Erl. 218, 28: Cd. 15; Th. 18, 26; Gen. 279. Swā ða hālgan dydon ðe nâht ne gyrndon tô hæbbenne *as the saints did who did not desire to have anything*, Blickl. Homl. 53, 25. Se deáda byþ uneáðe ælcon men on neáweste tô hæbbenne *it will be a hard matter for any one to have the dead man in his neighbourhood*, 59, 15. Eall ðæt him wæs leófost tô āgenne and tô hæbbenne *all that he liked best to own and to have*, 111, 27. Ic hæbbe geweald micel *I have much power*, Cd. 15; Th. 18, 29; Gen. 280. Ic hafo, Beo. Th. 4307; B. 2510. Ic hafu, Exon. 48 a; Th. 166, 10; Gû. 1040. Ic hæbbe ðé tô secgenne sum þing *habeo tibi aliquid dicere*, Lk. Skt. 7, 40. Se hafaþ in hondum heofon and eorþan *who hath in his hands heaven and earth*, 42 a; Th. 140, 32; Gû. 619. Ðis leóht wê habbaþ wið nýtenu gemæne *this light we have in common with beasts*, Blickl. Homl. 21, 13. Wê habbaþ nédþearfe ðæt wê ongyton *we have need to perceive*, 23, 1. Ða his mære word habbaþ and healdaþ *qui facitis verbum ejus*, Ps. Th. 102, 19. Æfter ðisum hæfde se cyng mycel geþeaht *after this the king held a great council*, Chr. 1085; Erl. 218, 22: St. And. 32, 29: Chr. 1050; Erl. 176, 9. Hér hæfde se cyng his hîréd æt Gleáweceastre *in this year the king held his court at Gloucester*, 1094; Erl. 229, 27. Penda hæfde xxx wintra ríce and hé hæfde l wintra ðā ðā hé tô ríce féng *Penda reigned thirty years, and he was fifty years old when he came to the throne*, 626; Erl. 22, 14. Ðæt cilde hæfde læsse ðonne þrý mônðas ðæs þriddan geáres *the child was not quite two years and three months old*, Shrn. 104, 18: Cd. 55; Th. 68, 14: Gen. 1117. Iudas hæfde onlícnesse ðara manna ðe willaþ Godes cyricean yfelian *Judas was like those men that desire to do evil to God's church*, Blickl. Homl. 75, 23. Hæfde cista gehwilc gârberendra x hund *each troop contained a thousand warriors*, Cd. 154; Th. 192, 11; Exod. 230. Hé ongan ða cnyhtas tô āxienne for hwig ðæt folc ðone Hælend swā yfele hæfde. Hig cwædon Hig habbaþ andan tô hym *he asked the men why the people treated the Saviour so ill. They said, 'They bear malice to him,'* Nicod. 8; Thw. 4, 18. Hé sceal bión stræc wið ða ðe āgyltaþ ond for ryhtwîsnesse hé sceal habban andan to hira yfele *contra delinquentium vitia per zelum justitiæ erectus*, Past. 12; Swt. 75, 13. Óð ðet hé ðone castel hæfde *until he got the castle*, Chr. 1102; Erl. 238, 14. Hine se môdega mæg Higelâces hæfde be honda *the proud kinsman of Hygelac held him by the hand*, Beo. Th. 1632; B. 814. Æðelwulf his dôhtor hæfde him tô cuêne Ethelwulf had his daughter for his queen, Chr. 885; Erl. 84, 5. Heó hyt for Crystes andwlytan æfre hæfde *she ever considered it as Christ's countenance*, St. And. 38, 4. Eal þeódscype hine hæfde for fulne cyng *all the nation considered him as full king*, Chr. 1013; Erl. 148, 36: Bt. Met. Fox 26, 87; Met. 26, 44: Mt. Kmbl. 14, 5. Ða Seaxan hæfdun sige *the Saxons got the victory*, Chr. 885; Erl. 84, 8: 909; Erl. 101, 20. Hí hæfdon hine mid heom óþ ðet hí ofslôgon hine *they had him with them till they slew him*, 1046; Erl. 174, 20. Hí on gewunan hæfdon *they have been accustomed*, L. Eth. 9, 31; Th. i. 346, 28. Hine grame hæfdon tô hæfte *fierce men held him captive*, Ps. Th. 104, 15. Ða hæfdon monige unwîse menn him tô worde and tô leásungspelle *quidam ridiculam fabulam texuerunt*, Ors. 1, 7; Swt. 40, 7. Gif cniht wæpn brede gilde se hláford ân pund and hæbbe se hláford æt ðæt hé mæge *if a follower draw a weapon, let the lord pay one pound, and let the lord get from him what he can*, Th. Chart. 612, 25. Ðæt ærest is ðæt man tô ôðrum lǽðde hæbbe *the first kind* [*of murder*] *is for a man to bear enmity to another*, Blickl. Homl. 63, 36. Se ðe forhogaþ ðæt hé ǽnig gemynd hæbbe Drihtnes eáðmódnesse *he that neglects to have any recollection of the Lord's meekness*, 83, 16. Æghwilcum men biþ leófre swā hé hæbbe holdra freónda mâ *the more friends every man has the better he likes it*, 123, 1. Be ðam sacerde ðonne hé mæssaþ hwæt hé on him hæbbe *of the priest when he says mass what he is to have on*, L. Edg. C; Th. ii. 128, 19. Âwriten is ðæt ðîne englas

ðē on hondum habban *it is written that thine angels shall take thee in their hands*, 27, 14. Ða hwíle ðe wē ðæt líf on úrum gewealde habban *while we have the life in our power*, 101, 11. Uton geþencean hwylc handleán wē him forþ tō berenne habban *let us consider what recompense we have to produce for him*, 91, 14. Hafa ðē wunden gold *take for thyself the twisted gold*, Cd. 97; Th. 128, 18; Gen. 2128. Gif man frigne man æt hæbbendre handa gefó *if a freeman be taken with stolen goods upon him*, L. Wiht. 26; Th. i. 42, 15: L. Ath. 1, 1; Th. i. 198, 16: 4, pref. Th. i. 220, 11. Wē beóþ hæbbende ðæs ðe wē ǽr hopedon *we shall be in possession of that which before we hoped for*, Homl. Th. i. 250, 34. Is seó stōw on micelre ârwurþnysse hæfed *in magna veneratione habetur locus ille*, Bd. 3, 2; S. 524, 12. Mid ðý hē mid ðone gesíþ hæfed wæs *dum apud comitem teneretur*, 4, 22; S. 591, 32. Adam and Eva on bendum wǽron hæfde *Adam and Eve were held in bonds*, Blickl. Homl. 87, 26. **II.** *with partitive gen:—*Hæbbe ic his on handa *I have some of it in my hand*, Cd. 32; Th. 42, 23; Gen. 678. Se ðe ðara mihta hæbbe ârǽre cirican *he who has the means let him erect a church*, L. Pen. 14; Th. ii. 282, 5: L. E. I. 3; Th. ii. 404, 22. Hē ne mōste ðæs fyrstes habban ðe hē gewilnode *he might not have any of the respite that he desired*, Homl. Th. i. 414, 28. **III.** *with the gerundial infin. to express the future:—*Ðone calic ðe ic tō drincenne hæbbe *calicem quem ego bibiturus sum*, Mt. Kmbl. 20, 22 [cf. the formation of the future tense in the Romance languages]. **IV.** *with an uninflected participle:—*Ðū mē forlǽred hæfst *thou hast seduced me*, Cd. 38; Th. 50, 34; Gen. 818. Ðæs lífes ðe ðū hafast ofslegen *the life that thou hast slain*, Exon. 29 b; Th. 90, 25; Cri. 1479. For ðissum ælþeódigum ðe wē on ðissum carcerne betýned hæbbaþ *on account of this stranger whom we shut up in this prison*, Blickl. Homl. 245, 36. Gē habbaþ ūs gedōn láðe Pharaone *ye have made us hateful to Pharaoh*, Ex. 5, 21. **V.** *with an inflected participle, sometimes also with an uninflected participle as well:—*Ic mínes hláfordes hafa hyldo forworhte *I have forfeited my prince's favour*, Cd. 39; Th. 52, 1; Gen. 836. Ðū hæfst ðē wið dryhten dýrne geworhtne *thou hast made thyself dear to the Lord*, 25; Th. 32, 22; Gen. 507. Ðū hafast helle bereáfod and ðæs deáþes aldor gebundene *thou hast despoiled hell, and bound the prince of death*, Blickl. Homl. 87, 22. Ðín ǽgen geleáfa ðē hæfþ gehǽledne *thine own faith hath saved thee*, 15, 27: 85, 23. Ðás þing wē habbaþ be him gewritene *we have written these things about him*, Chr. 1086; Erl. 222, 40. Ðá cwæþ Iacob Bearnleásne gē habbaþ mē gedōnne *then said Jacob, Ye have made me childless*, Gen. 42, 36. Hie hine ofslægenne hæfdon *they had slain him*, Chr. 755; Erl. 50, 1: 867; Erl. 72, 9. [*Laym.* habben, han: *Orm.* habbenn, hafenn: *A. R.* habben: *Goth.* haban: *O. Sax.* hebbian: *O. Frs.* hebba, habba: *Icel.* hafa: *O. H. Ger.* haben.] DER. â-, æt-, be-, for-, ge-, of-, on-, wið-, wiðer-, ymb-habban: nabban: bord-, daroþ-, dreám-, eard-, lind-, rand-, searo-hæbbende.

haca, an; *m. A hook* [?], *bolt or bar of a door*; pessulus, Gl. Mett. 658. [*Icel.* haki: *Dan.* hage: *Swed.* hake *a hook*: *O. H. Ger.* hako, hakko *uncinus, furca: Ger.* haken *a hook, clasp:* and cf. *Icel.* haka *the chin.*] See Skeat's Dict. hake, hatch, hackle.

haccian; *p.* ode; *pp.* od *To hack;* concidere, secando comminuere, Lye. [*A. R.* hackede: *p: Chauc.* hakke: *O. Frs.* (to-)hakkia: *Dut.* hakken *to hew, chop: Dan.* hakke *to hack, hoe: Ger.* hacken *to chop, cleave.*] v. tó-haccian.

hacele, an; *f:* hæcla, an; *m* [?] *A cloak, mantle, upper garment, coat, cassock.* Lye gives the following meanings lacerna, subucula, capsula, mantilia, *pl :—*Hacele *clamis,* Ælfc. Gl. 65; Som. 69, 40; Wrt. Voc. 40, 67: 110; Som. 79, 51; Wrt. Voc. 59, 22: 284, 65. Ðá bewráh se ârleása geréfa his ansýna mid his hacelan *then the impious count covered his face with his cloak,* Nar. 42, 24. Ðá gegyrede heó hý mid hǽrenre tunecan and mid byrnan ðæt is mid lytelre hacelan *she dressed herself in a tunic of hair and in a byrnie, that is in a little cassock,* Shrn. 140, 30. Ðá sende him mon âne blace hacelan angeán *a black mantle* [sagum] *was sent to him,* Ors. 5, 10: Swt. 234, 22. Saulus heóld ealra ðæra stænendra hacelan *Saul held the garments of all those who were stoning* [Stephen], Homl. Th. ii. 82, 22: i. 48, 1. Hæcla *pallium,* Mt. Kmbl. Lind. 5, 40. [*Goth.* hakuls; *m. a cloak: O. Frs.* hexil [= hekil (?)]: *Icel.* hekla; *f. a kind of cowled or hooded frock:* hökull; *m. a priest's cope: O. H. Ger.* hachul *cuculla, casula.*] See Grmm. D. M. 873 ff. DER. mæsse-hacele. 'In the West of England the word hackle is specially used of the conical straw roofing that is put over bee-hives. Also, of the "straw covering of the apex of a rick," says Mr. Akerman, *Glossary of Wiltshire words,* v. Hackle.'—Earle's Chronicle, p. 338.

hacine *pusta,* Ælfc. Gl. 33; Som. 62, 21; Wrt. Voc. 28, 4.

hacod, es; *m. A pike*:—Hacod *lucius,* Ælfc. Gl. 102; Som. 77, 69; Wrt. Voc. 55, 72; 77, 73. Hacodas *lucios,* Coll. Monast. Th. 23, 33. [Haked *a large pike* (Cambridgeshire): *O. H. Ger.* hachit, hechit, hæcid *lucius, mugil: Ger.* hecht *a pike.*] v. haca.

HÁD, es; *m.* **I.** *person;* persona:—Ðū ne besceáwast nánes mannes hád *non respicis personam hominum,* Mt. Bos. 22, 16. Cyninges naman hæfde and wæs ðæs hádes well wyrþe *regis nomine ac persona dignissimus,* Bd. 3, 21; S. 550, 40, MS. B. Weorþian wē ða claþas his

---

hádes *let us honour the clothes of his person,* Blickl. Homl. 11, 9. Hē wæs on ánum háde twegra gecynda *he was of two natures in one person,* 33, 33. On þrým hádum efnespēdelícum *in tribus personis consubstantialibus,* Bd. 4, 17; S. 585, 38: Homl. Th. ii. 42, 26. Þrý hádas synd worda. Se forma hád is ðe sprecþ be him sylfum âna . . . Se óðer hád is ðe forma sprecþ tō . . . Se þridda hád is ðe ðam ðe se forma hád sprecþ tō ðam óðrum háde *there are three persons of verbs. The first person is he who speaks about himself alone . . . The second person is he whom the first speaks to . . . The third person is he about whom the first person speaks to the second person,* Ælfc. Gr. 22; Som. 23, 49–53. Hád ðæt is *persona,* 15; Som. 17, 30. **II.** *sex:—*Gewuldrad is se heánra hád *the humbler sex is glorified,* Exon. 9 a; Th. 7, 10; Cri. 99. Óðre monige ǽghwæðeres hádes *alii utriusque sexus,* Bd. 1, 7; S. 479, 12. Ælcere yldo and háde *omni ætati et sexui,* 1, 1; S. 473, 22. Ðæt hē ne forðon wíslíce háde ârede *ut ne sexui quidem miliebri parceret,* 2, 20; S. 521, 25. **III.** *degree, rank, order, condition:—*Hád *gradus,* Ælfc. Gr. 11; Som. 15, 17. Gehwylces hádes menn *men of every degree,* Blickl. Homl. 47, 34: L. Ecg. C. 32; Th. ii. 156, 19. Sundor ânra gehwilc herige in háde *let each one separately praise thee in their degree,* Cd. 192; Th. 239, 16; Dan. 371: 28; Dan. 377: Th. 240, 27; Dan. 393. Fore ælcum háde ciricelíca *pro omni gradu ecclesiastico,* Rtl. 175, 25: 193, 37. Wer on lǽwedum háde *vir in laico habitu,* Bd. 5, 13; S. 632, 7: 4, 11; S. 579, 19. Hē on lǽwedum háde beón sceolde *he had to lead the life of a layman,* Blickl. Homl. 213, 9. Heárra on háde *higher in rank,* L. Eth. 6, 52; Th. i. 328, 14. Þurh hálige hád gecýðed *made known by clerks,* Exon. 34 a; Th. 107, 27; Gū. 65. Seofon hádas syndon gesette on bócum tō Godes þénungum intō Godes circan *seven orders are appointed in books for God's ministries in God's church,* L. Ælfc. P. 34; Th. ii. 378, 1: L. Ælfc. C. 10; Th. ii. 346, 25. Monige sindon hádas under heofenum *many are the conditions under the heavens,* Exon. 33 a; Th. 104, 3; Gū. 2. Biscopes oððe óðera háda *episcopi vel reliquorum ordinum,* Bd. 2, 5; S. 506, 30. Wiotan ǽgðer ge godcundra háda ge woruldcundra *wise men both clerks and laymen,* Past. Pref. Swt. 3, 3. Bútan hálgum hádum *extra sacros ordines,* Bd. 1, 27; S. 489, 16. Mid myclum hádum biscopas and cyningas *those of high degree, as bishops and kings,* Blickl. Homl. 109, 23; Homl. Th. ii. 122, 27. Swá wē settaþ be eallum hádum ge ceorle ge eorle *so we ordain for all degrees both gentle and simple,* L. Alf. pol. 4; Th. i. 64, 3. Ðám ðe heora hádas mid clǽnnesse healdan *to those who keep their orders with purity,* Blickl. Homl. 43, 4. Gemǽnes hádes man *clericus,* L. Ecg. P. 2, 24; Th. ii. 192, 8: 16; Th. ii. 186, 31. Tō háde fōn *to take orders,* 4, 8; Th. ii. 206, 7. **IV.** *state, condition, kind, nature, form* [having the meaning which is preserved in the suffix -hood, -head] :—Leóht hafaþ hád háliges gástes *light hath the nature of the holy spirit,* Salm. Kmbl. 817; Sal. 408. Se heáþrym ðæs Godes hádes *the excellent glory of the Godhead,* Blickl. Homl. 131, 18. Onsýn yldran hádes *the aspect of an older state* [a more advanced age], Exon. 40 a; Th. 132, 12; Gū. 471. Wæs se súðduru hwæthwega háde mǽre *the south door was somewhat greater in form,* Blickl. Homl. 201, 15. On weres háde *in the form of a man,* Elen. Kmbl. 144; El. 72. Onwendan heora wuldor on ðæne wyrsan hád hǽðenstyrces *mutaverunt gloriam suam in similitudinem vituli,* Ps. Th. 105, 17. Hád oferhogedon hálgan lífes *they despised the state of a holy life,* Cd. 188; Th. 235, 2; Dan. 300. Fǽmnan hád *virginity,* Exon. 9 a; Th. 6, 31; Cri. 92: 14 a; Th. 28, 10; Cri. 444. Cildes hád, Exon. 65 a; Th. 240, 15; Ph. 639: 61 a; Th. 224, 7; Ph. 372. Þurh cnihtes hád onsýne wearþ *he became visible in the form of a youth,* Andr. Kmbl. 1824; An. 914. Hæleþa leófost on gesíþes hád *dearest of men as a comrade,* Beo. Th. 2598; B. 1297. Næs sinc-máððum sélra on sweordes hád *there was no better treasure among swords,* 4393; B. 2193. Þurh hǽstne hád *by violence,* Beo. Th. 2674; B. 1335: Exon. 8 b; Th. 4, 7; Cri. 49. Þurh monigne hád *in many a form,* 54 b; Th. 191, 34. Blis manigra háda cwicera cynna *the joy of many kinds of living creatures,* Menol. Fox 182; Men. 92: Exon. 33 a; Th. 105, 15; Gū. 23. [*Laym.* hád, hód: *Orm.* hád: *A. R.* hód: *Ayenb.* hód: *Goth.* haidus *manner, way: O. Sax.* hēd: *Icel.* heiðr *honour: O. H. Ger.* heit *persona, sexus, ordo, gradus.*]

-hád *a suffix forming abstract nouns,* e. g. bisceop-, cild-, man-, werhád, etc. In the oldest English it is found combined only with nouns, while in the later stages of the language, as in *O. Sax. O. Frs. O. H. Ger.* words are formed with it from adjectives. An early instance occurs in the Laud MS. of the Chronicle 'druncenhed,' 1070; Erl. 209, 35. In later English it takes two forms, -hode, -hede; in modern times, -hood, -head. [*O. Sax.* -hēd: *O. Frs.* -hēd, -hēde, -heid: *O. H. Ger.* -heit, -heiti: *Ger.* -heit: *Dan.* -hed.] v. hád.

hád-bót, e; *f. A recompence, compensation,* or *atonement for injury done to persons in holy orders,* or hád-bryce; sacri ordinis violati compensatio, L. E. B. 4; Th. ii. 240, 17: L. O. 12; Th. i. 182, 13.

hád-breca, an; *m. A violator of holy orders;* sacri ordinis violator: — Hád-brēcan *violators of holy orders,* L. C. S. 6; Th. i. 380, 2: Lupi Serm. i. 19; Hick. Thes. ii. 105, 3; Swt. A. S. Rdr, 110, 178.

hád-brice, -bryce, es; *m.* [hád II. *holy orders in the church;* brice *a violation, breach*] *An injury done to persons in holy orders, a violation of holy orders;* ordinis infractura, sacri ordinis violatio :—Gif hwá hádbryce gewyrce, gebête ðæt be hádes mǽðe *if any one do an injury to a person in holy orders, let him make amends for it according to the degree of the order,* L. C. S. 50; Th. i. 404, 16. On hádbricum [MS. hádbrican] *in breaches of holy orders,* L. Eth. vi. 28; Th. i. 322, 19 : v. 25; Th. i. 310, 18 : Swt. A. S. Rdr. 109, 148.

hádelíce ; *adv. Personaliter,* Hymn. Surt. 29, 13.

haderung [= hád-árung?] *Personarum acceptio,* Som.

hád-grip, es ; *n. Peace, security, or privilege of holy orders;* sacri ordinis pax, L. Eth. vii. 19; Th. i. 332, 25.

hádian ; *p.* ode; *pp.* od *To ordain* :—Tó ðan ðet hé hine hádian sceolde *in order that he might ordain him,* Chr. 1048; Erl. 177, 20. Léton hig hádian tó bisceopum *they got themselves ordained bishops,* 1053 Erl. 188, 14. Ealdorlícnys ðæt hé bisceopas hádian móste *ordinandi episcopos auctoritas,* Bd. 2, 8; S. 510, 5. Hine hádigean tó bysceope *in episcopatus consecrare gradum,* 3, 7; S. 529, 9, note. Sende hé hine tó hádiganne *misit eum ordinandum,* 3, 28; S. 560, 8. Hádigenne, L. Ælf. C. 17; Th. ii. 348, 26. Hér mon hádode Byrnstán bisceop tó Wintanceastre *in this year Byrnstan was ordained to the bishopric of Winchester,* Chr. 931; Erl. 110, 22. Ne hádige man ǽfre wudewan tó hrædlíce *never let a widow take the veil too hastily,* L. C. S. 74; Th. i. 416, 15. [*Laym.* hoded; *pp.* Orm. hædedd.]

hád-notu, e; *f. The employment, ministry, office belonging to holy orders* :—Búton hé forworhte ðæt hé ðære hádnote notian ne móste *unless he should do amiss so that he might not exercise the office which belongs to his orders,* L. R. 7; Th. i. 192, 16.

hádod ; *part. p. used as adj. Ordained, in orders, clerical as opposed to lay* :—Ða witan ge hádode ge lǽwede *the 'witan,' both clerical and lay,* Chr. 1014; Erl. 150, 4 : 1023; Erl. 162, 46 : L. Edm. S. pref : Th. i. 246, 20.

hádor, es; m. n [?] *The clear, serene sky;* serenum :—Under heofones hádor *under heaven's serene,* Beo. Th. 832; B. 416. [*Cf. O. H. Ger.* heiteri *serenum : Icel.* heið *the brightness of the sky.*] Cf. rodor, and see hádor ; *adj.*

hádor, hǽdor ; *adj. Clear* [applied both to light and to sound], *bright, serene* :—Hádor heofonleóma *the clear heaven-light,* Andr. Kmbl. 1675; An. 840 : 2918; An. 1458 : 178; An. 89 : Bt. Met. Fox 22, 47; Met. 22, 24. Scóp hwílum sang hádor on Heorote *at times the poet sang clear-voiced in Heorot,* Beo. Th. 998; B. 497. Seó sunne on hádrum heofone scíneþ *the sun shines in the clear sky,* Bt. 9; Fox 26, 15 : Bt. Met. Fox 28, 95; Met. 28, 48. Hǽdre heofontungol *the bright stars of the sky,* Exon. 18 a; Th. 43, 23; Cri. 693. Hádrum nihtum *in clear nights,* Bt. Met. Fox 20, 463; Met. 20, 232. Se ðe heofen þeceþ hádrum wolcnum *qui operit cælum nubibus,* Ps. Th. 146, 8. Singaþ hǽdrum stefnum *they sing with clear voices,* Elen. Kmbl. 1492; El. 748. [*O. Sax.* hêdor : *O. H. Ger.* heitar *clarus, splendidus, serenus, micans : Ger.* heiter : *Icel.* heið *bright (of the sky, stars).*]

hádre, hǽdre ; *adv. Clearly* [of light and of sound] :—Hádre scíneþ rodores candel *the lamp of the firmament* [the sun] *shines brightly,* Beo. Th. 3147; B. 1571. Hǽdre blícan, scínan *to shine brightly,* Exon. 57 b; Th. 205, 20; Ph. 115 : 120 b; Th. 462, 17; Hö. 53 : 51 b; Th. 179, 6; Gú. 1257. Swéga mǽste hǽdre *clearly with loudest melody,* 64 b; Th. 239, 10; Ph. 619 : 54 a; Th. 190, 26; Az. 79. Ðonne sió sunne sweotolost scíneþ of hefone *when from heaven shines the sun most clearly and brightly,* Bt. Met. Fox 6, 7; Met. 6, 4. [*O. Sax.* hêdro.]

hád-swépa *pronuba,* Ælfc. Gl. 93; Som. 75, 79; Wrt. Voc. 52, 29. v. next word.

hád-swápe, -swǽpe, an ; *f. A bridesmaid;* pronuba, paranymphus = παράνυμφος, Ælfc. Gl. 87; Som. 74, 56, 58; Wrt. Voc. 50, 38, 40 : 288, 80. [*Ettmüller compares* hád *in this word with Gothic,* heþjo *a chamber.*]

hádung, e ; *f. Ordination* :—On ðære smyrunge biþ lǽcedóm and ne biþ ná hádung *in the unction is healing and there is not ordination,* L. Ælfc. P. 48; Th. ii. 384, 33. Bisceopum gebyreþ ðæt hí ne beón tó feohgeorne æt hádunge *it is fitting for bishops not to be too eager after money at ordination,* L. I. P. 10; Th. ii. 316, 32. On ǽlcere hádunge se ðe gehádod biþ hé biþ gesmyrod mid gehálgodum ele *at every ordination he that is ordained is anointed with consecrated oil,* Homl. Th. ii. 14, 25 : 124, 2. Ðæt hé ne háding ne háleging ne dó *not to ordain nor consecrate,* Chr. 675; Erl. 38, 4. [*Orm.* hading.]

hæbbendlíc ; *adj. Habilis* :—Sume habbaþ sceortne i *amabilis* lufigendlíc, habilis hæbbendlíc, Ælfc. Gr. 9, 28; Som. 11. 41.

hæbbenga ; *adv. With constraint, constrainedly* [Somner gives this word and explains it by *cohibitio,* but it appears to be an adverb like eallenga, etc.]

hæbern. v. hæfern.

hæc ; *gen.* hæcce; *f. A hatch, grating, a gate made of latticework* [?] :—Of ðære ealdan hæcce, Th. Chart. 394, 15, 21 : 395, 10, 22, 28 : 396, 4, 5, 14. [*Prompt. Parv.* hec, hek, or hetche, or a dore *antica.* On this word the following note is given :—' " Antica, a gate, or a dore, or hatche *est antica domus ingressus ab anteriori,*" Ortus. "An heke *antica,*" Cath. Ang. "*Ostiolum* hek," Roy. MS. 17 c. xvii. f. 27. " Hatche of a dore *hecq,*" Pals. "*Guichét,* a wicket, or hatch of a doore," Cotg. Forby gives " hack, half-hack, a hatch, a door divided across." In the North, a heck-door is one partly latticed and partly panelled.' See also Skeat's Dict. *hatch.*] Cf. haca.

hæca *pessulus,* Som. v. haca.

hæcce, e ; *f. A crosier* :—Ðis mycel is gegolden of ðære cyricean W. cyninge . . . of ðam candelstæfe x pund and of ðære hæcce xxxiii marca *this much has been paid by the church* [of Worcester] *to king William . . . from the candlestick x pounds, and from the crosier xxxiii marks,* Th. Chart. 440, 4. Ðæt hæcce wæs eall of gold and of seolfre *the crosier was all of gold and silver,* Chr. 1070; Erl. 209, 9. Eall ðæt ðider com ðæt wæs ðone hæcce and sume scríne and sume róden *all that came there was the crosier and some shrines and some crucifixes,* 32. [Cf. haca.]

hæced. v. hacod.

hæcele. v. hacele.

hæcewol *exactor,* Ælfc. Gl. 8; Som. 56, 94; Wrt. Voc. 18, 44.

hæc-wer, es ; *m. A weir with a grate to take fish,* Cod. Dipl. Kmbl. iii. 450, 15, 22. [' A salmon-heck, a grate to take them in,' English Dialect Society, No. 30, p. 82. v. hæc.]

hæddern. v. heddern.

hædre ; *adv. Straitly, hardly, oppressively, anxiously;* arcte, anxie :—Hyge hædre [hearde, A.] wealleþ *my mind is agitated with anxiety,* Salm. Kmbl. 126; Sal. 62. [Míne sáwle] hædre gehogode hǽl *save* [my soul] *oppressed by anxious thoughts,* Exon. 118 b; Th. 456, 5; Hy. 4, 62.

hædre. v. hádre.

hæfd. v. heáfod.

hæfdling. v. efen-hæfdling, heáfodling.

hæfe, es ; *m. Leaven;* fermentum :—Warniaþ fram herodes hæfe *cavete a fermento herodis,* Mk. Skt. 8, 15. [Cf. *O. H. Ger.* hefo; *m. fæx : Ger.* hefen *yeast.*] v. ge-hafen.

Hæfeldan *the name of a Slavonic people* :—Wylte ðe man Hæfeldan hét, Ors. 1, 1; Bos. 18, 39 : 19, 18. In explanation of this double naming, Bosworth, p. 36 (translation), quotes 'Wilsos, Henetorum gentem ad *Havelam* trans Albim sedes habentem.' v. note 12.

hæfen, e ; *f. Having, property, possession* :—Be his ágenre hæfene *according to his own property,* Homl. Th. i. 582, 28 : 580, 22 : ii. 400, 2. [*Icel.* höfn; *f. a holding, possession : cf. O. H. Ger.* haba *possessio : Ger.* habe.]

hæfen, e ; *f :* hæfene, an ; *f. A haven, harbour, port* :—Of ǽiðre healfe ðare hæfene *from either side of the harbour,* Chr. 1031; Erl. 162, 5. Ic ann ða hæuene on Sandwíc *I grant the port of Sandwich,* Th. Chart. 317, 21. Ða hæfenan on Sandwíc *the port of Sandwich,* Chr. 1031; Erl. 162, 3 : 1090; Erl. 226, 26. [*Icel.* höfn; *f :* Dan. havn : Ger. hafen.]

hæfen-blǽte, es ; *m. A haven-bleater* [?], *a sea-gull;* bugium, Ælfc. Gl. 37; Som. 62, 128; Wrt. Voc. 29, 23.

hæfenleás ; *adj. Without property, poor, needy,* Ps. Lamb. 11, 5. v. hafenleás.

hæfenleást, e ; f. Poverty, penury :—Þurh hæfenleáste *through poverty,* Lchdm. iii. 442, 19 : Ps. Lamb. 43, 27. v. hafenleást.

hæfer, es ; *m. A he-goat, buck;* caper :—Hæfer *caper,* Wrt. Voc. 288, 17. Nim hæferes smera *take goat's grease,* Lchdm. iii. 14, 8. [*Icel.* hafr : *Lat.* caper.]

hæferbíte, es ; *m. Forceps,* Som.

hæferbléte, es ; *m. Bicoca,* Ælfc. Gl. 16; Som. 58, 54; Wrt. Voc. 21, 42 : 280, 28. [Cf. hammer-bleat *the snipe,* English Dialect Society, No. 20, p. 42.]

hæfern, es ; *m. A crab;* cancer, Wrt. Voc. 281, 63. Hæfern *concern = cancer* [?], 291, 31. v. wæter-hæfern.

hæft, es ; *m.* **I.** *one seized or taken, a captive* :—Hé licgan geseah hæftas in hylle *he saw captives lying in hell,* Cd. 229; Th. 309, 27; Sat. 717 : 217; Th. 277, 10; Sat. 202 : Exon. 10 a; Th. 10, 18; Cri. 154 : Andr. Kmbl. 2142; An. 1072. Wé ðé biddaþ ðæt ðú gehýre hæfta stefne *we beseech thee to hear the voice of the captives,* Exon. 13 a; Th. 22, 32; Cri. 360. Under hæftum *amid the captives,* Cd. 220; Th. 284, 9; Sat. 319. **II.** *one taken and enslaved, a slave, servant* :—Ða bebohtan bearn Iacobes Ioseph ðǽr hine grame hæfdon tó hæfte *in servum venumdatus est Ioseph,* Ps. Th. 104, 15. Hweorfon ða hǽðenan hæftas fram ðám hálgan cnihton *the heathen slaves went from the holy youths,* Cd. 187; Th. 232, 28; Dan. 267. Gearwe stódun hæftas heársume *ready stood the slaves obedient,* Exon. 43 a; Th. 145, 19; Gú. 697. [*Icel.* haftr *a prisoner, bondman : cf. Goth.* hafts *joined : O. Sax.* haft : *O. H. Ger.* haft *vinctus, captivus.*]

hæft, es ; *m.* **I.** *a bond, fetter;* vinculum :—Bútan hæftum *without bonds,* Salm. Kmbl. 823; Sal. 411 : Cd. 222; Th. 291, 8; Sat. 427. Tó hæftum geferian *to bring into bonds,* 216; Th. 274, 2; Sat. 148 : 215; Th. 270, 17; Sat. 92. Of hæftum lǽdan *to bring out of captivity,*

224; Th. 296, 20; Sat. 505: 225; Th. 299, 21; Sat. 553. **II.** *captivity, bondage, imprisonment, keeping;* captivitas, custodia:—Is ðes hæft tô ðan strang *this imprisonment is so severe,* Elen. Kmbl. 1403; El. 703: Cd. 171; Th. 215, 15; Exod. 583. Hê betæhte hine on ðam hæfte sixtyne cempum tô healdenne *he committed him to the keeping of sixteen soldiers to hold,* Homl. Th. ii. 380, 29. Hê of hæfte âhlôd folces unrîm *from captivity he drew forth people numberless,* Exon. 16 a; Th. 35, 34; Cri. 568: Andr. Kmbl. 2797; An. 1401: 2938; An. 1472. Him on hæft nimeþ *takes into bondage to him,* 11 b; Th. 16, 29; Cri. 260: 41 a; Th. 138, 1; Gû. 569: Cd. 189; Th. 235, 16; Dan. 307: Chr. 1036; Erl. 164, 31. In hæftum *in custodias,* Lk. Skt. Lind. 21, 12. [*Icel.* haft, hapt; *n. a bond, chain:* O.H. Ger. haft; *m: Ger.* haft; *m. clasp, rivet:* haft; *f. imprisonment.*]

hæft, hæfte, es; *n. A haft, handle;* manubrium:—Hæft and helfe *manubrium,* Ælfc. Gl. 52; Som. 66, 31; Wrt. Voc. 35, 20. Nim ðæt seax ðe ðæt hæfte sîe fealo hryðeres horn *take a knife, the handle of which is yellow ox-horn,* L. M. 2, 65; Lchdm. ii. 290, 22: 52; Lchdm. ii. 272, 21. Folc Ebrêa fuhton hæfte guldon hyra fyrngefiîtu fâgum sweordum *the Hebrew folk fought with the haft* [=sword, a part put for the whole, cf. ord, ecg?], *with stained swords repaid their quarrels of old,* Judth. 12; Thw. 25, 16; Jud. 263. [*Prompt. Parv.* heft *manubrium: Icel.* hepti; *n. a haft or kilt:* O.H. Ger. hefti *capulum, manubrium: Ger.* haft *haft, handle.*]

hæftan; *p.* hæfte; *pp.* hæfted, hæft *To seize, bind, arrest, make captive, imprison:*—Gif hê nite hwâ hine âborgie hæfton hine *if he knows not who will be his surety let them arrest him,* L. Ath. i. 20; Th. i. 210, 8. Seó stôw ðe ðû nû on hæft eart *the place in which you are now imprisoned,* Bt. 11, 1; Fox 32, 27. Hæft mid hringa gesponne *bound with the clasp of rings,* Cd. 25; Th. 47, 17; Gen. 762. Hringan hæfted *confined with rings,* Exon. 102 b; Th. 387, 8; Rä. 5, 2. Tô bodanne hæftedum *prædicare captivis,* Lk. Skt. Rush. 4, 18. [*Goth.* haftjan: O. Sax. heftian *to bind, fetter:* O.H. Ger. heftan: *Ger.* heften.] DER. be-, gehæftan. v. hæft.

hæfte-clomm, es; *m. Fetter, bond:*—On hæðenra hæfteclommum *in the fetters of heathen men,* Chr. 942; Erl. 116, 16.

hæfte-dôm, es; *m. Captivity, service,* Bt. Met. Fox 25, 129; Met. 25, 65.

hæften, e; *f. Captivity, custody:*—Ða betste of ðes eorles hîrêde innan ânan fæstene gewann and on hæftene gedyde *he took the best of the earl's household within a fortress and placed them in custody,* Chr. 1095; Erl. 231, 29.

hæft-encel, -incel, es; *m. A slave;* emptitius, Cot. 74, Lye.

hæfte-neód, e; *f. Custody, prison* [?] :—Ûre bân syndon tôworpene be helwarena hæfteneódum *dissipata sunt ossa nostra secus infernum,* Ps. Th. 140, 9. [Grein gives as the meaning of the word *studium captandi* vel *tribulandi;* but is not *infernum* here paraphrased as the 'prison of the dwellers in hell?']

hæfting, e; *f. A fastening:*—Belûcaþ ða ærenan gatu and ða hæftinga gehealdaþ ðæt wê ne beón gehæfte *close the brazen gates and keep the fastenings that we be not captured,* Nicod. 27; Thw. 15, 16. [Cf. *Ger.* heftung.]

hæftling, es; *m. A captive:*—Hæftling *captivus,* Ælfc. Gr. 28; Som. 32, 41. Ðâ âxode se ealdorman ðone hæftling hwæðer hê þurh drýcræft his bendas tôbræce *then the alderman asked the captive whether he broke his bonds by witchcraft,* Homl. Th. ii. 358, 10. Nabochodonosor hergode on Iudêiscre leóde and hî hæftlingas tô Babilone gelædde *Nebuchadnezzar warred on the Jewish people and led them captives to Babylon,* 58, 6: i. 108, 21: Gen. 31, 26.

hæft-mêce, es; *m. A hilted sword,* Beo. Th. 2918; B. 1457.

hæft-nêd, -niéd, -nýd, e; *f. Captivity, thraldom, custody:*—Israhêla folc on hæftnêde Babiloniscum cyninge þeówde *the people of Israel served the king of Babylon in captivity,* Homl. Th. ii. 84, 27. Lýsan of hæftnêde *to release from captivity,* Elen. Kmbl. 593; El. 297. On hæftnêde habban *to hold in captivity,* Blickl. Homl. 85, 23. On hæftnýde geledan *to lead into captivity,* Ps. Th. 14, argument: L. Ecg. C. 26; Th. ii. 152, 4. All Angelcyn ðæt bûton Deniscra monna hæftniéde wæs *all the English that were not held in subjection by the Danish men,* Chr. 886; Erl. 84, 28. On hæftnêd lædan, Blickl. Homl. 79, 22. Gehweorf ûre hæftnêd *converte captivitatem nostram,* Ps. Th. 125, 4. Se Drihten ðe ûs fram deófles hæftnêdum âlýsde *the Lord who redeemed us from the devil's thraldom,* Homl. Th. i. 546, 34. Twegen gerêfan on ðæra hæftnêdum wæs se apostol gehæfd *two counts in whose custody the apostle was held,* ii. 294, 21.

hæftnian; *p.* ede; *pp.* ed *To seize, capture:*—Hî hæftniaþ *captabunt,* Ps. Lamb. 93, 21. Hæftned lædde ða on hæftnêde lange lifdon *captivam duxit captivitatem,* Ps. Th. 67, 18.

hæft-noþ, -neþ, es; *m. Custody, keeping, imprisonment:*—On hæftnoþe biþ gehæfd *he will be imprisoned,* Lchdm. iii. 200, 34. On hæftneþe gebringan *to imprison,* Chr. 1095; Erl. 232, 21. Ðær hê on hæftneþe wæs *where he was imprisoned,* 1101; Erl. 237, 40.

hæftnung, e; *f. Captivity, fastening, confinement:*—Hê hine gewrâþ

gelomlîce ac hine ne mihte nânes cynnes hæftnung gehealdan *he often bound him, but no kind of fastening could hold him,* Homl. Th. ii. 358, 20. On hæftnunge *in captivity,* 86, 3 : Ps. Spl. 13, 11. Ær hê forðferde hê beád ðæt man sceolde unlêsan ealle ða menn ðe on hæftnunge wæron *ere he departed he ordered that all those men who were in confinement should be released,* Chr. 1086; Erl. 223, 39. Dôn on hæftnunge *to put into confinement, imprison,* 1087; Erl. 225, 36.

hæft-nýd. v. hæft-nêd.

hægel, hægl, es; *m.* **I.** *hail:*—Fýr, forst, hægel and gefeallen snâw *ignis, glacies, grando, nix,* Ps. Th. 148, 8. Hægl, Exon. 56 b; Th. 201, 22; Ph. 60. Cymeþ hægles scûr *a shower of hail cometh,* Cd. 38; Th. 50, 13; Gen. 808. Hæglas and snâwas *hails and snows,* Bt. 39, 13; Fox 234, 16. Heora wîngeardas wrâðe hægle nêde fornâmon *occidit in grandine vineas eorum,* Ps. Th. 79, 47. Sealde heora neát hæglum *tradidit grandini jumenta eorum,* 77, 48. **II.** the Anglo-Saxon rune ᚺ =*h,* the name of which letter is *hægl* :—ᚺ byþ hwîtust corna *hail is whitest of grains,* Hick. Thes. 135; Runic pm. 9; Kmbl. 341, 4. Hægelas twegen *two H's,* Exon. 112 a; Th. 429, 27; Rä. 43, 11. v. hagal.

hæghâl; *adj. Safe, uninjured;* incolumis:—Eftgiondwearda ûsig ârmorgenlîcum tîdum hæghâle *representa nos matutinis horis incolomes,* Rtl. 124, 15: 98, 39: 174, 37.

hægl-faru, e; *f. A hailstorm,* Exon. 78 a; Th. 292, 26; Wand. 105.

hægl-scûr, es; *m. A shower of hail, hailstorm,* Andr. Kmbl. 2515; An. 1259. v. hagal-scûr.

hæg-steald, hæge-, heh-, es; *m:* e; *f* [?] *One living in the lord's house, not having his own household, an unmarried person, a young person, bachelor, virgin;* mansionarius, cælebs, juvenis, virgo:—Hwæðer hê sig hægsteald ðe hæmedceorl *utrum cælebs sit an uxoratus,* L. Ecg. C. 1; Th. ii. 132, 28. Hegsteald *cælebs,* 14; Th. ii. 142, 13. Hægsteald môdige wîgend unforhte *youths courageous, warriors fearless,* Cd. 160; Th. 198, 24; Exod. 327. His hægstealdas *his young warriors,* Fins. Th. 81; Fin. 40. Hægstealdas and fæmnan *juvenes et virgines,* Ps. Th. 148, 12. Swilce geongum hægstealde *ut ephebo hircitallo,* Mone B. 3434. Hehstald *virgo,* Mt. Kmbl. Lind. 1, 23: Lk. Skt. Lind. 1, 27. Hehstaldo *virgines,* Rtl. 47, 36. Hehstaldun *virginibus,* Mt. Kmbl. Lind. 25, 1. Of heghstalde *de virgine,* Rtl. 126, 3. v. hago-steald.

hæg-steald; *adj. Unmarried, young:*—Hægstealdra, Cd. 89; Th. 111, 28; Gen. 1862; Beo. Th. 3782; B. 1889. See the preceding word.

hægsteald-hâd, es; *m. The unmarried state, bachelorhood, virginity:*—Hehstaldhâd *virginitas,* Rtl. 105, 19: Lk. Skt. Lind. 2, 36. Hægstealdhâd *cælibatus,* Mone B. 1419.

hægsteald-lîc; *adj. Virgin;* virginalis, Rtl. 66, 1.

hægsteald-man = hægsteald, *q.v.,* Cd. 151; Th. 190, 1; Exod. 192: Exon. 113 b; Th. 436, 18; Rä. 55, 3.

hægstealdnis, e; *f. Virginity:*—Hehstaltnisse *virginitatis,* Jn. Skt. p. 1, 3.

hægtesse, an; *f. A witch, hag, fury:*—Helle-rûne vel hægtesse *pythonissa,* Ælfc. Gl. 112; Som. 79, 102; Wrt. Voc. 60, 10. Hægtesse *Tissiphona,* 113; Som. 79, 115; Wrt. Voc. 60, 22. Gif hêr inne sý îsenes dæl hægtessan geweorc hit sceal gemyltan . . . gif hit wære ylfa gescot oðde hit wære hægtessan gescot nû ic wille ðîn helpan *if herein there be a bit of iron, a witch's work, it shall melt. . . . if it were an elf's shot or it were a witch's shot, now will I help thee,* Lchdm. iii. 54, 1–12. v. Grmm. D. M. 992.

hæg-þorn, es; *m. Hawthorn:*—Hægþorn *alba spina,* Ælfc. Gl. 48; Som. 65, 50; Wrt. Voc. 33, 46. Genim hægþornes leáf *take leaves of hawthorn,* Herb. 37, 6; Lchdm. i. 138, 17. Of ðam mappuldre tô ðam hægþorne *from the maple to the hawthorn,* Cod. Dipl. Kmbl. iii. 424, 3. [*Icel.* hagþorn: M.H. Ger. hagedorn.]

hæg-weard, hæcg-, es; *m. A hayward, the keeper of cattle in a common field, who prevented trespass on the cultivated ground,* L. R. S. 20; Th. i. 440, 11, 12. [*A.R.* heiward: *Prompt. Parv.* heyward *agellarius.* The following note is given on this word, p. 234:—'Bp. Kennett observes that there were two kinds of *agellarii,* the common herdward of a town or village, called *bubulcus,* who overlooked the common herd, and kept it within bounds; and the heyward of the lord of the manor, or religious house, who was regularly sworn at the court, took care of the tillage, paid the labourers, and looked after trespasses and encroachments: he was termed fields-man or tithing-man, and his wages in 1425 were a noble. "*Inclusarius* a heyewarde." "*Inclusorius* a pynner of beestes." "Haiward, haward *qui garde au commun tout le bestiail d'un bourgade.*"']

hæl, es; *n. Omen, auspice:*—Hæl sceáwedon *they observed the favourable omen* (for Beowulf's undertaking), Beo. Th. 414; B. 204. [*Icel.* heill; *n. omen, auspice:* O.H. Ger. heil *omen, auspiciu:n.*]

hæl, e; *f. Health, safety, salvation, happiness;* salus:—Seó hæl cymeþ symle fram Gode *salus a domino,* Ps. Th. 36, 38. Tô-dæg ðisse hîwrædene ys hæl geworden *this day is salvation come to this house,* Lk. Bos. 19, 9: Homl. Th. i. 582, 5. Cristes þênung is ûre hæl and folca âlýsednys

K k 2

Christ's service is our salvation and the redemption of peoples, ii. 586, 32. Him cymþ gód hǽl *good health will come to them,* Lchdm. i. 342, 9. Sý him hǽl *Osanna,* Mt. Kmbl. 21, 9. Hrædlíce heora hǽle brúcaþ *speedily they enjoy their health,* Homl. Th. i. 510, 8. Brúc ðisses beáges mid hǽle *use this collar with good fortune,* Beo. Th. 2438; B. 1217. Hêht hê Elenan hǽl ábeódan *he bade them greet Elene,* Elen. Kmbl. 2004; El 1003: Beo. Th. 1311; B. 653. Ðíne hǽle syle *salutare tuum da,* Ps. Th. 84, 6. [Laym. heal: O. Sax. hêli; f: Icel. heill; f. *good luck happiness:* O. H. Ger. heili; f. *salus.*] Cf. hǽl; n. and hǽlu.

hǽl; *adj. Hale, safe, whole, sound:*—Hǽle and trume *safe and sound,* Blickl. Homl. 171, 30. v. hál.

hǽla. v. hǽla.

hǽlan; p. de; pp. ed *To heal, make whole, cure, make safe, save; sanare, salvare:*—Ys hyt álýfed tô hǽlenne on restedagum *si licet sabbatis curare,* Mt. Bos. 12, 10. Earm heora ne hǽlþ hig *brachium eorum non salvabit eos,* Ps. Spl. 43, 4. Sweord mín ne hǽlþ mê *gladius meus non salvabit me,* 43, 8. Hǽl ús on heánessum *Hosanna in the highest,* Blickl. Homl. 72, 12: Jn. Skt. Rush. 12, 13. Hǽlaþ untrume *heal the sick,* Mt. Bos. 10, 8. Ic offrige míne lác Hǽlendum Criste *I will present my offerings to Jesus Christ,* Homl. Th. i. 416, 17. Hí hrǽdlíce hǽlde wǽron *sanavit eos,* Ps. Th. 106, 19. [Goth. hailjan: O. Sax. hêlean: O. Frs. hêla: O. H. Ger. heilan *sanare, curare, salvare:* Ger. heilen.]

hǽl-bǽre; *adj. Salutary,* Lye.

hǽle, es; *m. A man, brave man, hero* [a word occurring only in poetry]:—Frôd hǽle *the aged man,* Cd. 62; Th. 74, 14; Gen. 1217. Boitius se hǽle hâtte *that man was called Boethius,* Bt. Met. Fox 1, 105; Met. 1, 53: Cd. 74; Th. 90, 28: Gen. 1502: 112; Th. 147, 27; Gen. 2446: 121; Th. 156, 16; Gen. 2589: Andr. Kmbl. 287; An. 144. [Icel. (in poetry only), halr *a man.*]

hǽle, an; *f. Health, safety:*—On gode standeþ mín gearu hǽle *in Deo salutare meum,* Ps. Th. 61, 1.

hǽlend, hêlend, es; *m. A healer, Saviour, Jesus:*—Se Hǽlend ðe is genemned Crist *Iesus qui vocatur Christus,* Mt. Bos. 1, 16. Ðú nemst hys naman Hǽlend. Hê sôþlíce hys folc hál gedêþ fram hyra synnum *vocabis nomen ejus Iesum; ipse enim salvum faciet populum suum a peccatis eorum,* 1, 16. Iesus is on Léden Saluator and on Englisc Hǽlend *Jesus is in Latin Salvator and in English healer,* Homl. Th. ii. 214, 22: i. 94, 27: Shrn. 47, 28. Hǽlend Crist *Jesus Christ,* Homl. Th. i. 420, 32. Ðú eart sôþ hêlend *thou art the true Saviour,* Hy. Grn. 8, 16. [Laym. hǽlend (2nd MS. helare): Orm. hælennde: O. Sax. hêliand: O. H. Ger. heilant: Ger. heiland.]

hǽlendlíc; *adj. Healthy, salutary;* salvans, prosperus, Hpt. Gl. 442, 511. [O. H. Ger. heilantlih *salubris.*]

hǽleþ, heleþ, es; *m. A man, warrior, hero* [a word occurring only in poetry, but there frequently]:—Gleáwferhþ hæleþ *the man wise of mind,* Cd. 57; Th. 70, 12; Gen. 1152: 59; Th. 72, 6; Gen. 1182, 94; Th. 122, 13; Gen. 2026: Beo. Th. 383; B. 190: 668; B. 331. Hæleþas heardmôde *warriors stern-minded,* Cd. 15; Th. 19, 2; Gen. 285. Hæleþ hâtene wǽron Sem and Cham Iafeþ þridde *the heroes were named Shem and Ham, the third Japhet,* Cd. 75; Th. 93, 22; Gen. 1550. Hæleþa scyppend *creator of men,* Exon. 11 b; Th. 17, 7; Cri. 266: Cd. 98; Th. 129, 6; Gen. 2139: Andr. Kmbl. 41; An. 21. Hæleþa bearn *the children of men,* Cd. 35; Th. 46, 30; Gen. 752. Heleþa sceppend *creator of men,* Hy. Grn. 8, 34. [Laym. hæleþ, heleþ: O. Sax. helið: O. H. Ger. helid (appears first in 12th cent. v. Graff. iv. 544): Ger. held.]

hæleþ-helm, es; *m. A helm which makes the wearer invisible,* Cd. 23; Th. 29, 2; Gen. 444. [O. Sax. helið-helm: O. H. Ger. helot-, helanthelm *latibulum.*] v. Grm. D. M. 432, and cf. heoloþ-helm.

hǽletoþ, es; *m. Greeting, Hosanna,* Hpt. Gl. 467.

hǽlettung, e; *f. A greeting, salutation:*—Hǽlettungæ on gemôte *salutationes in foro,* Mt. Kmbl. Rush. 23, 7.

hælftre, e; *f. A halter;* Hǽlftre *capistrum,* Wrt. Voc. 84, 8. On hælftre *in camo,* Ps. Spl. C. 31, 12. Hǽlftra *chamos,* Coll. Monast. Th. 28, 1. [O. H. Ger. halftra *brachiale, capistrum:* Ger. halfter.]

hǽlig; *adj. Slippery, easily moved, fickle, inconstant;* levis:—Ðam ungestæþþegan and ðam hǽligan ðú miht seccgan ðæt hê biþ winde gelícra oððe unstillum fugelum *levis, atque inconstans studia permutat? nihil ab avibus differt,* Bt. 37, 4; Fox 192, 23. [Cf. Icel. hâll *slippery:* O. H. Ger. hâli *lubricus, caducus.*]

hǽling, e; *f. Healing:*—Ic nán yfel on hym næbbe gemét be hǽlinge *I have found no evil in him with regard to healing,* Nicod. 10; Thw. 5, 21. [Prompt. Parv. heelinge: O. H. Ger. heilunga *sanatio:* Ger. heilung.]

hǽlnes, se; *f.*   I. *haleness, salvation:*—Nú sint hǽlnesse dagas *now are the days of salvation,* Past. 36, 1; Swt. 246, 14.   II. *a sanctuary:*—On circan and on hǽlnessan *in churches and sanctuaries,* L. Eth. 7, 25; Th. i. 334, 26. v. hálignes.

hǽlnes-griþ, es; *n. Privilege of security belonging to a sanctuary,* L. Eth. 7, 19; Th. i. 332, 25.

hǽlo. v. hǽlu.

hǽlsend, es; *m. An augur,* Cot. 73, Lye.

hǽlsere, es; *m. A soothsayer, diviner;* aruspex, augur, extispex, Cot. 190: exorcista, Lye. v. hálsere.

hǽlsian *to foretell;* augurari, ariolari, auspicari, Cot. 14, 17, Lye. v. hálsian.

hǽlsung, e; *f. Divination, augury;* augurium, Cot. 11, Lye.

hǽlp, e; *f. Health, healing, cure:*—Ðám árist rihtwísnysse sunne and hǽlp is on hyre fiðerum *to them shall arise the sun of righteousness, and healing is on its wings,* Lchdm. iii. 236, 31. Ðes þegen bæd for his þeówan hǽlþe *this officer prayed for the health of his servant,* Homl. Th. i. 128, 1. For hǽlþe heora untrumra *for the healing of their sick,* ii. 396, 21. Úre líchamana hǽlþe wê áwendaþ *we pervert the health of our bodies,* 540, 9. Ealle ða wundra and hǽlþa áwrítan *to write down all the miracles and cures,* 28, 10. [O. H. Ger. heilida *sanitas, salus.*]

**hǽlu, hǽlo; indecl. f. Health, safety, salvation:*—Æt him is hǽlu mín *ab ipso salutare meum,* Ps. Th. 61, 1. Sý hǽlu úrum Gode ðe sitt ofer his þrymsetle *salvation be to our God that sitteth on his throne,* Homl. Th. i. 538, 18. Hǽlo, Exon. 13 b; Th. 26, 1; Cri. 411. Hǽlu bútan sáre *health without pain,* 32 a; Th. 101, 8; Cri. 1655. Tô hǽlo hýðe *to a haven of safety,* 20 b; Th. 53, 33; Cri. 860. For heora sáwla hǽlu *for the salvation of their souls,* Homl. Th. ii. 344, 1. Hǽlo, L. M. Th. i. 102, 7. Uton hǽlu sêcan *let us seek salvation,* Exon. 97 b; Th. 365, 11; Wal. 87. Drihten ús sealde hǽlu and éce álýsednysse *the Lord gave us salvation and eternal redemption,* Homl. Th. ii. 248, 25. Heó forstæl hire hǽlu *she stole her health,* 394, 12. Gif gie hǽlo beádas *si salutaveritis,* Mt. Kmbl. Lind. 5, 47. v. hǽl.

hǽlu-bearn, hǽlo-, es; *n. A child who brings salvation, the Saviour,* Exon. 16 a; Th. 37, 1; Cri. 586: 19 a; Th. 47, 12; Cri. 754.

hǽman; p. de; pp. ed *To lie with, have intercourse with, to marry;* concumbere, coire, nubere:—Wit wǽron swíðe unrôte geworden for ðý hǽmede ðe we wêndon ðæt wit hǽman sceoldon *we became very sad on account of the intercourse that we expected we should be obliged to have,* Shrn. 39, 21. Mid ðám hǽleþum hǽman wolden, Cd. 112; Th. 148, 18; Gen. 2458. Gif hwylc man wið óðres riht-ǽwe hǽmþ *if any man lie with the lawful wife of another,* L. Ecg. P. ii. 8; Th. ii. 184, 21. Hê hǽmþ unrihtlíce *he commits adultery,* Homl. Th. ii. 208, 16. Ðám mannum ðe deófol mid hǽmþ *for those women with whom the devil hath carnal commerce,* L. M. 3, 61; Lchdm. ii. 344, 8. Ne hǽmeþ ne hǽmde bióþ *neque nubent neque nubentur,* Mt. Kmbl. Rush. 22, 30: 19, 10. Gif hwilc carlman hǽmde wið wimman hire unþances *if any man lay with a woman against her will,* Chr. 1086; Erl. 222, 7: Num. 25, 1. Ne hǽm ðú unrihtlíce *commit not adultery,* Homl. Th. ii. 198, 7. Gif ǽnig man hǽme mid óðres wífe *if a man be found lying with a woman married to an husband,* Deut. 22, 22: L. Alf. pol. 10; Th. i. 98, 9.

hǽmed, es; *n. A lying with, sexual intercourse, marriage; coitus:*—Ða ðe rihtlíce healdaþ hyra ǽwe and for bearnes gestreóne hǽmed begáþ *those who rightly observe their marriage and for procreation of children have carnal intercourse,* Homl. Th. ii. 148, 22. Mægþhád biþ forloren on hǽmede *maidenhead is lost in intercourse,* ii. 10, 5: 220, 4. Be hǽmede de coitu, L. Ecg. C; Th. ii. 128, 26. On unrihton hǽmede *in adulterio,* Jn. Skt. 8, 4: Shrn. 132, 6. Ic wið brýde ne môt hǽmed habban *with a bride I may not 'have intercourse,* Exon. 105 b; Th. 402, 11; Rä. 21, 28. Hǽmed *connubium,* Mone Gl. 340. Hǽmeda *connubii convenientia,* 417. Hǽmeða *himeneas,* Ælfc. Gl. 9; Som. 56, 119; Wrt. Voc. 19, 2. Hǽmdo *nubtiæ,* Jn. Skt. Lind. Rush. 2, 1.

hǽmed-ceorl, es; *m. A married man:*—Hwæðer hê sig hægsteald ðe hǽmedceorl *utrum cælebs sit an uxoratus,* L. Ecg. C. 1; Th. ii. 132, 28.

hǽmed-gemána, an; *m. Matrimony, marriage;* matrimonium, Cot. 129, Lye.

**hǽmed-gifta, pl. f. Nuptials;* hymenæi, Cot. 102, Lye.

hǽmed-lác, es; *n. Sexual intercourse;* coitus, Exon. 112 a; Th. 429, 11; Rä. 43, 3.

hǽmed-scipe, es; *m. Marriage, matrimony;* connubium, Hpt. Gl. 482: lenocinium, seductio, 521.

hǽmed-þing, es; *n. Carnal intercourse, venery, matrimony:*—Sió lufu ðæs hǽmedþinges biþ for gecynde *the desire of intercourse is from nature,* Bt. 34, 11; Fox 152, 14: Blickl. Homl. 59, 16. Be hǽðenra manna hǽmedþincge *de gentilium hominum matrimonio,* L. Ecg. C; Th. ii. 128, 27. Gif hí him betwynan hǽmedþing fremmen *si inter se fornicationem commiserint,* 16; Th. ii. 144, 9. Be hǽmedþingum: eallum þyrrum líchomum hǽmedþing ne dugon *of venery: venery does not do for all dry constitutions,* L. M. 2, 27; Lchdm. ii. 222, 28: 36; Lchdm. ii. 244, 4.

hǽmed-wíf, es; *n. A married woman;* uxor, matrona, Cot. 136, Lye.

hǽmere, es; *m. One who lies with another;* concubinus, Lye.

hǽn, hen, henn, e; *f. A hen:*—Hǽn *gallina,* Recd. 36, 56; Wrt. Voc. 63, 10. Seó henn *gallina,* Mt. Kmbl. 23, 37: Lind. Rush. henne. Hænne æges geolocan *the yolk of a hen's egg,* L. M. 1, 2, 23; Lchdm

ii. 38, 6 : 3, 2 ; Lchdm. ii. 40, 10. [*Icel.* hæna : *O. H. Ger.* henna : *Ger.* henne.]

hǽnan ; *p.* de ; *pp.* ed *To stone* :—For hwylcum ðæra weorca wylle gē mē hǽnan . . . ne hǽne wē ðē for gódum weorce *propter quod eorum opus me lapidatis . . . de bono opere non lapidamus te*, Jn. Skt. 10, 32, 33 : 11, 8. Ðú ðe ða wítegan hǽnst *quæ prophetas lapidas*, Lk. Skt. 13, 34. Eall folc ús hǽnþ *plebs universa lapidabit nos*, 20, 6. Hǽne hine man mid stánum *let him be stoned with stones*, Lev. 20, 2. v. hán.

hænep, henep, es ; *m.* *Hemp* :—Henep, hænep, Herb. 27, 1 ; Lchdm. i. 124, 1, 3 : Lchdm. iii. 22, 31. [*Icel.* hampr : *O. H. Ger.* hanaf : *Ger.* hanf : *Lat.* cannabis : *Grk.* κάνναβις. 'Grimm and Kuhn both consider the Greek word borrowed from the East, and the Teutonic one from the Latin *cannabis*, which certainly made its way to them.' Curtius, i. 173.]

hǽn-fugul, hen-, es ; *m.* *A hen* :— Henfugel *gallina*, L. Ecg. C. 40 ; Th. ii. 164, 21. Gewurp tó sumum hen [hæn, MS. B.] fugule *throw it to a hen*, Herb. 4, 10 ; Lchdm. i. 92, 16. iiii hænfugulas *four hens*, Th. Chart. 509, 18. Ðǽr æfter swulten ða henne fugeles *after that the hens died*, Chr. 1130 ; Erl. 259, 25.

hænne-belle, an ; *f.* Henbane ; hyoscyamus, Lchdm. iii. 60, 7. Henne-belle, Herb. 5, 1 ; Lchdm. i. 94, 3, 6. Henne-belle *simphoniaca*, Ælfc. Gl. 40 ; Som. 63, 96 ; Wrt. Voc. 30, 42.

hænnewol ; *n. m.* Henbane, Lchdm. iii. Gloss.

hǽplíc ; *adj.* *Equal* ; compar, Cot. 35, Lye. v. ge-hǽp.

hæpse, an ; *f.* *A hasp, clasp, fastening* :—Hæpse *sera*, Wrt. Voc. 81, 20 : *clustella*, Hpt. Gl. 500. Sum slóh ða hæpsan *one struck the hasps* [*of the door*], Th. An. 124, 14. [*Prompt. Parv.* hespe of a dore *pessulum* : *Icel.* hespa *a hasp, fastening* : *Ger.* haspe.]

hæpsian ; *p.* ode ; *pp.* od *To hasp, fasten with a bolt* :—Ic scitte sum loc oððe hæpsige *sero*, Ælfc. Gr. 37 ; Som. 39, 21.

hǽr, hér, es ; *n.* *Hair, a hair* :—Hǽr *capillus*, Wrt. Gl. 70, 30 : *pilus*, Recd. 38, 21 ; Wrt. Voc. 64, 30. Hǽr *pili*, Ælfc. Gl. 70 ; Som. 70, 54 ; Wrt. Voc. 42, 62. Loccas *vel* unscoren hǽr *comæ*, 70, 56 ; Wrt. Voc. 42, 64. Gif hǽr tó þicce síe *if the hair be too thick*, L. M. i. 87, 3 ; Lchdm. ii. 156, 8. Ne sceal eów beón forloren án hǽr of eówrum heáfde *there shall not a hair of your head be lost*, Homl. Th. i. 236, 22. Ðú ne miht wyrcan án hǽr ðínes feaxes hwít oððe blacc *thou canst not make one hair of thy locks white or black*, 482, 19. His reáf wæs geworht of oluendes hǽre *his raiment was wrought of camel's hair*, ii. 38, 9. Ðæt íren ne cume on hǽre ne on nægle *that iron come not on hair, nor on nail*, L. Pen. 10 ; Th. ii. 280, 20. Ne losaþ ðæt heáfod ðonne ða hǽr beóþ ealle geedstaðelode *the head perishes not when the hairs are all restored*, Homl. Th. ii. 542, 35. Wið wiðerweard hǽr onweg tó áðónne *for contrarious hairs, to remove them*, Lchdm. i. 362, 8. Hǽras heáfdes *capilli capitis*, Mt. Kmbl. Lind. 10, 30. Hiora is mycle má ðonne ic mē hæbbe on heáfde nú hǽra feaxes *multiplicati sunt super capillos capitis mei*, Ps. Th. 68, 4. Mið hǽrum oððe fæx hire *capillis suis*, Jn. Skt. Lind. 11, 2. Se eádiga wæs blíðe on andwlitan mid hwítum hǽrum *the blessed man was cheerful in aspect, with white hair*, Homl. Th. ii. 186, 20. Mid olfendes hǽrum gescrýd *clothed with camel's hair*, i. 330, 2 : Mt. Kmbl. 3, 4. Ic beleás hérum ðám ðe ic hæfde *I lost the hairs that I had*, Exon. 107 a ; Th. 407, 36 ; Rä. 27, 5. [*O. Sax.* hār : *O. Frs.* hêr : *Icel.* hár : *O. H. Ger.* hār : *Ger.* haar.] For notices as to the importance attached to the hair in early times, see Grimm R. A. pp. 146, 240, 283, 339, 702 ; and see *feax* and its compounds. DER. hrycg-, tægl-hǽr.

hǽre, an ; *f.* *Hair-cloth, sack-cloth* ; cilicium, saccus :—Gefyrn hí dydon dǽdbóte on hǽran and on axan *olim in cilicio et cinere pœnitentiam egissent*, Mt. Kmbl. 11, 21. Mid hǽran gescrýd *clad in sackcloth*, Homl. Th. ii. 312, 27 : Ps. Spl. 34, 15. Se cyning dyde hǽran tó his líce *the king put sackcloth next to his skin*, Homl. Th. i. 568, 13. Ðú slite hǽran míne *concidisti saccum meum*, Ps. Spl. 29, 13. [*Laym.* ane ladliche here : *A. R.* here, heare, 'Iudit werede heare :' *Prompt. Parv.* hayre *cilicium*. *Cilicium, velamen factum de pilis caprarum* a heere. An haire *cilicium* : *Icel.* hæra ; *f* : *O. H. Ger.* hârra, hâra ; *f. cilicium, saccus*.]

hǽrean-fagol [?] *a hedge-hog* :—Stán gener hǽreanfagol *petra refugium herinaciis*, Ps. Spl. 103, 19. v. hatte-fagol.

hǽrelof. v. herelof.

hǽren ; *adj. Made of hair* ; cilicius :— Hē hine ða gegyrede mid hǽrenum hrægle swíðe heardum and unwinsumum *he clothed himself then with a garment of hair very hard and unpleasant*, Blickl. Homl. 221, 24. Wring þurh hǽrenne cláþ *wring through a hair cloth*, Lchdm. i. 382, 21. Reáf hǽren *vestimentum cilicium*, Ps. Lamb. 68, 12. [*Wick.* heeren : *M. H. Ger.* hærin : *Ger.* hären.]

hǽrenes. v. herenes.

hærfest, es ; *m. Harvest, autumn* :—Hǽrfest *autumnus*, Ælfc. Gl. 95 ; Som. 76, 9 ; Wrt. Voc. 53, 23. Autumnus is hærfeste, Lchdm. iii. 250, 11. Se hærfest welig on wæstmum *the autumn rich in fruits*, Bt. 14, 1 ; Fox 40, 27 : 21 ; Fox 74, 22 ; Bt. Met. Fox 29, 123 ; Met. 29, 62. Hǽrfest *æstatem*, Ps. Spl. 73, 18. Ðæt gewrixle ðara feówer týda ðæt is lencten and sumer and herfest and winter *the change of the four seasons*,

that is spring and summer and autumn and winter, Shrn 168, 12. Ðæs ilcan hærfestes *in the course of the same autumn*, Chr. 921 ; Erl. 107, 13. Foran tó hærfestes emnihte *ante æquinoctium autumnale*, L. Ecg. P. 11 ; Th. ii. 208, 2 : Th. Chart. 151, 11. On hærfæste *in autumno*, Coll. Monast. Th. 26, 5. Ðis wæs on hærfest *this was in autumn*, Chr. 918 ; Erl. 104, 16. [*Prompt. Parv.* herueste *autumpnus* : *Icel.* haust ; *n. autumn* : *O. H. Ger.* herbist ; *m. autumnus* : *Ger.* herbst *autumn*.]

hærfest-handful *a due belonging to the husbandmen on an estate* :— Eallum ǽhte-mannum gebyreþ hǽrfesthandful *omnibus ehtemannis jure competit manipulus Augusti*, L. R. S. 9 ; Th. i. 438, 1.

hærfestlíc ; *adj. Autumnal* :—Hærfestlíc dæg *autumnalis dies*, Ælfc. Gl. 95 ; Som. 76, 19 ; Wrt. Voc. 53, 29. On ðæs hærfestlícan emnihtes ryne *in the course of the autumnal equinox*, Lchdm. iii. 238, 28 : 252, 1.

hærfest-mónaþ ; *m. September*, Ælfc. Gr. 9, 18 ; Som. 9, 54. [*Cf. Robt. of Glouc.* þe nexte moneþ afturward, þat heruest moneþ ys, He let clepe aftur hym August ywys. *Icel.* haust-mánuðr : *O. H. Ger.* herbist-manoþ : *Ger.* herbist-monat *September*.]

hærfest-wǽta, an ; *m. Autumnal wet* ; humor æstatis, Ors. 3, 3 ; Swt. 102, 7.

hériht ; *adj. Hairy* ; crinitus, setosus, Cot. 186, Lye.

hǽring, es ; *m. A herring* :—Hwæt fēhst ðú on sǽ ? Hǽringcas *quid capis in mari ? Aleces*, Coll. Monast. Th. 24, 9. Ðes hǽring *hoc allec*, Ælfc. Gr. 9 ; Som. 14, 22. Hǽring *allec vel jairus vel taricius vel sardina*, Ælfc. Gl. 102 ; Som. 77, 80 ; Wrt. Voc. 56, 3. Hǽrinc *taricus vel allec*, Wrt. Voc. 77, 62. xxx þúsenda hǽryngys ǽlce eáre 30 thousand herrings every year, Cod. Dipl. Kmbl. iv. 172, 3. [*O. Frs.* hereng : *O. H. Ger.* harinc : *Ger.* häring.]

hǽring-tíma, an ; *m. Herring-season* :—Twegen hǽringc-tíman *two herring-seasons*, Th. Chart. 338, 34.

hǽrlíc. v. hérlíc.

hǽr-loccas ; *m. pl. Locks of hair, curls* ; cincinni, crines, Hpt. Gl. 526.

hærn, e ; *f. The tide, waves, sea* :—Hærn *æstus, flustrum*, Cot. 81, Lye. Hærn eft onwand *back went the waves*, Andr. Kmbl. 1062 ; An. 531. [*Icel.* hrönn *a wave.*]

hærn or hærne [?], es ; *m. n? The brain* :—It gǽde tó ðe hærnes *it went to the brains*, Chr. 1137 ; Erl. 262, 6. [*Prompt. Parv.* hernys or brayne *cerebrum* ; herne panne of þe hed *craneum* : *Icel.* hjarni ; *m* : *O. H. Ger.* hirni ; *n. cerebrum* : *Ger.* hirn ; *n.*]

hǽr-nǽdl, e ; *f. A hair-pin* ; calamistrum, Lye.

hærn-flota, an ; *m. A wave-floater, ship*, Exon. 52 a ; Th. 182, 9 ; Gú. 1307.

hǽr-sceard, es ; *n. Hare-lip* :—Wið hǽrscearde *for hare-lip*, L. M. 1, 13 ; Lchdm. ii. 56, 5. [*Cf. Frs.* haskerde *hare-lipped* : *Icel.* skarði *hare-lip* (a nickname) : *Ger.* hasenscharte *hare-lip*.]

hærpan. v. herpan.

hǽs, e ; *f. A command, hest, behest* :—Hǽs *jussio*, Ælfc. Gr. 9 ; Som. 8, 40. Gehír God mín gebed *exaudi Deus orationem meam*. On ðysum is gebed and ná hǽs *hear my prayer, O God. In this there is a prayer, not a command*, Ælfc. Gr. 33 ; Som. 37, 52 : Cd. 6 ; Th. 8, 14 ; Gen. 124. Be his hláfordes hǽse *by his lord's command*, Gen. 24, 10 : Ex. 18, 23 : Cd. 46 ; Th. 59, 18 ; Gen. 965 : 69 ; Th. 82, 31 ; Gen. 1370 : 85 ; Th. 106, 35 ; Gen. 1781. Bûton ǽnigre hǽse *abs quolibet jussu*, Ælfc. Gr. 47 ; Som. 47, 54. Under abbodes hǽsum *under the commands of an abbot*, Homl. Th. ii. 118, 29. [*Orm.* hæs : *Laym. A. R.* hest : cf. *Goth.* haiti.] DER. be-hǽs.

hǽsel *galerus*, Lye.

hæsel, es ; *m. The hazel* :—Hæsel *corilus*, Ælfc. Gl. 45 ; Som. 64, 95 ; Wrt. Voc. 32, 30. Hæsles ragu *the lichen of hazel*, L. M. i. 38, 8 ; Lchdm. ii. 96, 2 : L. M. 2, 52 ; Lchdm. ii. 270, 22. Hwít hæsel *wich hazel* ; ulmus montana : saginus, Ælfc. Gl. 45 ; Som. 64, 96 ; Wrt. Voc. 32, 31. [*Prompt. Parv.* hesyl *corulus, colurnus* : *Icel.* hasl ; *m* : *O. H. Ger.* hasal ; *m* : hasala ; *f. corylus, amygdalus* : *Ger.* hasel ; *f.*] For special virtue of the hazel see Grmm. D. M. 927, and cf. hæslen. Cf. also the *Icel.* hasla völl *to challenge to a duel on a field marked out by hazel-poles.*

hæsel-hnutu, e ; *f. A hazel-nut* :— Hæsl *vel* hæsel-hnutu *abellanæ*, Ælfc. Gl. 47 ; Som. 65, 43 ; Wrt. Voc. 33, 40. [*O. H. Ger.* hasal-nuz : *Ger.* hasel-nuss.]

hæsel-wrid, es ; *n. m* [?] *A hazel-thicket* :—Tó ðam miclan hæslwride *to the great hazel-thicket*, Cod. Dipl. Kmbl. ii. 250, 34. v. ge-wrid.

hæsel-wyrt, e ; *f. Asarabacca, asarum Europæum*, Lchdm. iii. 329, col. 2.

hǽsere, es ; *m. A commander, one who orders, commands, a master, lord* :—Hǽsere *præceptor*, Lk. Skt. Lind. 8, 24, 45 : 9, 49 : 17, 13 : 21, 7. Hǽsere *imperator*, Rtl. 192, 39.

hæslen ; *adj. Of hazel* :—Genim æt fruman hæslenne sticcan oððe ellenne wrít ðinne naman on ásleah þrý scearpan on gefylle mid ðý blóde ðone naman weorp ofer eaxle on yrnende wæter and stand ofer ðone man ða scearpan ásleá ðæt eall swíginde gedó *take, to begin with, a hazel or an elder stick, cut thy name thereon, cut three scores on the place, fill the*

name with the blood, throw it over thy shoulder into running water and stand over the man. Strike the scores, and do all that in silence, L. M. 1, 39; Lchdm. ii. 104, 6–11. Lǽt ꝥæt blōd on grēnne sticcan hæslenne weorp ꝺonne ofer weg ǣweg ꝺonne ne biþ nān yfel *let the blood run into a green spoon of hazel-wood, then throw it away over the road; then no harm will come of the bite*, 68; Lchdm. ii. 142, 19–21.

hǽst, hēst, e; *f. Violence, fury:*—Ic þurh hēst hrīno lāꝺgewinnum *I violently touch my foes*, Exon. 104 b; Th. 397, 31; Rä. 16, 28. Fǽre ne mōston wætres brōgan hǽste hrīnan *the terrors of the water might not with violence touch the vessel*, Cd. 69; Th. 84, 11; Gen. 1396. [Hǽste may also be taken either as *adj.* agreeing with brōgan (v. next word), or as an adverb.] Grein compares with *Goth.* haifsts.

hǽst, hǽste [?]; *adj. Violent, vehement, impetuous:*— Ðū Grendel cwealdest þurh hæstne hād heardum clammum *thou didst kill Grendel violently with hard grasps*, Beo. Th. 2674; B. 1335. Nǣfre ꝺū ꝺæs swīꝺlīc sār gegearwast þurh hæstne nīþ ꝺæt ꝺū mec onwende worda ꝺissa *never shalt thou, through vehement hate, pain so violent prepare as to turn me from these words*, Exon. 66 b; Th. 246, 3; Jul. 56. Ðæt sceal wrecan swefyl and sweart līg sāre and grimme hāt [Junius hāte] and hǽste hǽꝺnum folce *sulphur and swart flame, sorely and fiercely, hot and vehement shall avenge it on the heathen folk* [Junius' reading might be taken and hǽste would then be an adverb parallel with sāre *and* grimme: v. preceding word], Cd. 110; Th. 146, 2; Gen. 2416. [Cf. Grff. iv. 969, '*Si quis in curte episcopi armatus contra legem intraverit, quod alamanni* haistera hanti *dicunt:*' and for similar expressions, v. Grmm. R. A. 4.]

hǽste; *adv.* [?] See two preceding words.

Hǽstingas, Hestingas, Hæstinga ceaster *Hastings:*—And ꝺa hwīle com Willelm eorl upp æt Hestingan *and that time Earl William landed at Hastings*, Chr. 1066; Erl. 203, 3. Ðā fērde se cyng tō Hæstingan *then the king went to Hastings*, 1094; Erl. 229, 35. Hī heafdon ofergān Sūþseaxe and Hæstingas [Hæsting, l. 36] *they had overrun Sussex and Hastings*, 1011; Erl. 144, 27. Tō Hæstinga ceastre *at Hastings*, L. Ath. 1, 14; Th. i. 208, 2.

hǽstlīce; *adv. Violently, vehemently, fiercely*, Exon. 67 b; Th. 250, 33; Jul. 136. [Cf. *O. H. Ger.* heistigo biscoltan, Grff. iv. 1063.]

hǽswalwe *astur*, Som.

hǽt, hætt, es; *m. A hat, covering for the head*; pileus, mitra, tiara:—Fellen hæt *galerus* vel *pileus*, Ælfc. Gl. 18; Som. 58, 111; Wrt. Voc. 22, 26. Hæt *calamanca*, Wrt. Voc. 41, 8: *capitium*, 74, 57. Terrentius bær hæt on his heáfde, for ꝺon Rōmānē hæfdon gesett ꝺæt ꝺa ꝺe hæt beran mōston mōston ǣgþer habban ge feorh ge freóꝺom *Terentius pileatus, quod indultæ sibi libertatis insigne fuit*, Ors. 4, 10; Swt. 202, 25–29. [*Icel.* höttr *a hood, cowl: Dan.* hat.]

hǽtan; *p.* te; *pp.* ed *To heat, make hot:*—Ðæt fȳr ꝺe man ꝺæt ordāl mid hǽtan sceal *the fire with which the ordeal is to be heated*, L. Ath. 4, 7; Th. i. 226, 11: 14. Tō hǽtanne magan *to heat the stomach*, L. M. 2, 10; Lchdm. ii. 188, 16. Hit gelamp sume dæige ꝺæt ꝺæs swānes wīf hǽtte hire ofen and se king ꝺær big set *it happened one day that the herdsman's wife heated her oven, and the king sat by*, Shrn. 16, 15. Hǽt scenc fulne wīnes *heat a cup full of wine*, Lchdm. i. 370, 26: ii. 24, 25. [*Icel.* heita: *Ger.* heizen.]

hǽte, an; *f. Heat:*—Cīle and hǽte ne geswīcaþ *frigus et æstus non requiescunt*, Gen. 8, 22. Ðā ꝺa seó hǽte com ꝺā forscranc hit *when the heat came then it withered away*, Homl. Th. ii. 90, 30. On ꝺære hǽtan ꝺæs dæges *in the heat of the day*, Gen. 18, 1: Mt. Kmbl. 20, 12. For sunnan hǽtan *on account of the heat of the sun*, Herb. 100, 8; Lchdm. i. 214, 24: 114, 1; Lchdm. i. 226, 23. Wið eágena hǽtan *for heat of the eyes*, Lchdm. i. 352, 5. Eówre glēda nāne hǽtan mīnum līchaman ne gedōþ *your embers cause no heat to my body*, Homl. Th. i. 430, 12. Ðæt hellīce fȳr hæfþ unāsecgendlīce hǽtan and nān leóht *the fire of hell has heat unspeakable, but no light*, 532, 2. Ongan mid monegum hǽtum geswenced beón *multis cæpit æstibus affici*, Bd. 2, 12; S. 513, 31. Wið wunda hātum *for inflammations of wounds*, Herb. 2, 16; Lchdm. i. 84, 20. [Cf. *Icel.* heita *brewing*.] v. hǽtu.

hǽtera, hæteru, *pl. Garments:*—Hē hæfde ne hǽlþe ne hætera *he had neither health nor garments*, Homl. Th. i. 330, 14. Se hund tōtær his hæteru sticmǣlum *of his bæce the dog tore his garments to pieces off his back*, 374, 8. Sume hī cuwon heora hætera *some of them chewed their garments*, 404, 5. Gā hē ūt mid his hætron swyclon hē in com *let him go out with his garments such as he came in with*, Ex. 21, 4. [*Laym.* alle his hateren weoren totoren: *A. R.* hateren; *dat. pl:* Piers P. I have but one hatere: *Prompt. Parv.* hatyr, rent clothe *scrutum, pannucia: O. H. Ger.* hadarun; *dat. pl.* pannis, mastrugis: *Ger.* hader rag, clout.]

hǽþ, e; *f. A heath, waste, desert, uncultivated land:*—Hār hǽþ *the hoar heath*, Cd. 148; Th. 185, 5; Exod. 118. Bera sceal on hǽþe *the bear shall [live] on the heath*, Menol. Fox 518; Gn. C. 29. [*Goth.* haiþi: *Icel.* heiðr *a low barren heath or fell: Ger.* heide (12th cent: Grff. iv. 809).]

hǽþ, e; *f. Heath, heather:*—Hǽþ *marica* vel *brogus*, Ælfc. Gl. 46; Som. 65, 3; Wrt. Gl. 33, 3. Smeóce mid hǽþe *smoke with heath*,

---

Lchdm. i. 354, 24. v. Gloss. iii. 329, col. 2. [*Prompt. Parv.* hethe or lynge *bruarium: O. H. Ger.* heida *thymus, mirice: Ger.* heide, heidekraut.]

hǽþ-berige, an; *f. Heath-berry, bilberry;* vaccinium:—Hǽþbergean wīsan *heath-berry plants*, L. M. 3, 61; Lchdm. ii. 344, 10.

hǽþ-cole *Cassis*, *galea*, Cot. 32, 36, Lye.

hǽden. v. heden.

hǽꝺen; *adj.* HEATHEN, *pagan, gentile;* and *subst. a heathen:*—Twā folc ꝺæt is Iudēisc and hǽꝺen *two peoples, that is Jew and gentile*, Homl. Th. i. 206, 32. Ðes wæs hǽꝺen *hic erat samaritanus*, Lk. Skt. Rush. 17, 16. Gif ungefullod cild fǽrlīce biþ gebroht tō ꝺam mæssepreóste hē hit mōt fullian sōna ꝺæt hit ne swelte hǽꝺen *if an unbaptized child be brought to the mass-priest suddenly, he must baptize it at once, that it die not heathen*, L. Ælfc. 26; Th. ii. 352, 17: L. M. I. P. 42; Th. ii. 276, 15. Hēr sæt hǽꝺen here on Tenet *in this year a heathen [Danish] army sat in Thanet*, Chr. 865; Erl. 70, 31. Oꝺ ꝺone hǽꝺenan byrgels *up to the heathen tomb*, Cod. Dipl. Kmbl. ii. 250, 13. (The same phrase often occurs in the charters in the descriptions of boundaries.) Se hæfde wununge on hǽꝺenum byrgenum *he had his dwelling among the tombs*, Homl. Th. ii. 378, 26. Hēr hǽꝺne men ǽrest ofer winter sǽtun *in this year heathen [Danish] men first remained through the winter*, Chr. 855; Erl. 68, 23: 851; Erl. 66, 26. Bachsecg and Halfdene ꝺa hǽꝺenan cyningas *Bachsecg and Halfdene the heathen kings*, 871; Erl. 74, 17. Ða ealdan Rōmānī on hǽꝺenum dagum ongunnon ꝺæs geáres ymbryne on ꝺysum dæge *the old Romans, in heathen days, began the circuit of the year on this day*, Homl. Th. i. 98, 20. *Used substantively:*—Ðæt hē forgeáfe gōdne willan ꝺam seócan hǽꝺenan *that he would grant good will to the sick heathen*, ii. 24, 33. Sume ꝺa hǽꝺenan *some of the heathens*, i. 562, 28: 560, 8. Ða hǽꝺenan on Norþhymbrum hergodon *the heathens harried in Northumbria*, Chr. 794; Erl. 39, 19. Ðyssera hǽꝺenra fǽrlīcan deáþ *sudden death from these heathens*, Homl. Th. ii. 494, 31. Hǽꝺinra *gentium*, Lk. Skt. Lind. 21, 25. Hǽꝺenra þeównēd *thraldom under the heathen*, Cd. 189; Th. 235, 17; Dan. 307. Hē hī on heregeweald hǽꝺenum sealde *tradidit eos in manus gentium*, Ps. Th. 105, 30. Hie fērdon ongeán ꝺæm hǽꝺnum *they marched against the heathens*, Blickl. Homl. 203, 3. [Cf. *Goth.* haiþno; *f. a heathen, gentile woman: O. Sax.* hēðin: *O. Frs.* hēthen: *Icel.* heiðinn: *O. H. Ger.* heidan *ethnicus, gentilis, paganus, samaritanus: Ger.* heide *a heathen.*] v. Grmm. D. M. 1198.

hǽꝺena, an; *m. A heathen, gentile:*—Hǽꝺnana *gentium*, Lk. Skt. Rush. 21, 25. See preceding word.

hǽꝺen-cyning, es; *m. A heathen king:*—Herige hǽꝺencyninga *a band of heathen kings*, Cd. 174; Th. 219, 13; Dan. 54.

hǽꝺen-cynn, es; *n. A heathen race*, Cd. 119; Th. 153, 29; Gen. 2546.

hǽꝺen-dōm, es; *m. Heathendom, paganism:*—Hī gecwǣdon ꝺæt hī ǣnne God lufian woldon and ǣlcne hǽꝺendōm georne āweorpan *they agreed that they would love one God and zealously put away every kind of heathendom*, L. E. G. pref; Th. i. 166, 12. Wē lǽraþ ꝺæt preósta gehwilc cristendōm geornlīce ārǽre and ǣlcne hǽꝺendōm mid ealle ādwǣsce *we enjoin that every priest zealously promote Christianity, and totally extinguish every kind of paganism*, L. Edg. C. 16; Th. i. 248; Cd. 183; Th. 229, 23; Dan. 221. [*Orm.* hæþenndom 'and tatt [*the death of the soul with the body*] iss mikell hæþenndom to lefenn and to trowenn:' *Icel.* heiðin-dómr: *O. H. Ger.* heidan-tuom: *Ger.* heidenthum.]

hǽꝺen-feoh, gen. -feós; *n. A heathen sacrifice*, Exon. 66 b; Th. 245, 31; Jul. 53.

hǽꝺen-gild, -gield, -gyld, es; *n. Heathen worship, idolatry;* also *an idol:*—Ðis hǽꝺengyld deófles biggeng is *this idolatry is worship of the devil*, Homl. Th. i. 72, 4. Hǽꝺengield, Exon. 66 a; Th. 243, 23; Jul. 15. Tō ꝺam hǽꝺengilde bugon *they turned to the idol [Baal-peor]*, Num. 25, 2: 31, 16. Hē bæd hig georne ꝺæt hig būgan ne sceoldon fram Godes bigengum tō ꝺam bysmorfullum hǽꝺengilde *he prayed them earnestly not to turn from the worship of God to degrading idolatry*, Jos. 23, 7. Iulianus ꝺā ongann tō lufigenne hǽꝺengyld *Julian then began to love idolatry*, Homl. Th. i. 448, 30. Ealle ꝺa hǽꝺengyld ꝺe ꝺās Indiscan wurꝺiaþ *all the idols that these Indians worship*, 454, 14. Hǽꝺengield, Exon. 66 a; Th. 244, 4; Jul. 22. v. gild.

hǽꝺen-gilda, -gylda, an; *m. A heathen worshipper, heathen, an idolater:*—Hē is gehīwod tō cristenum men, and is earm hǽꝺengylda *he is in appearance a Christian, and is a miserable heathen*, Homl. Th. i. 102, 16. Se yldesta hǽꝺengylda *the chief idolater*, 72, 9. Se ofslōh ꝺæs hǽꝺengyldan sunu *which slew the idolater's son*, ii. 294, 19. Se ealdorman wolde ꝺa hǽꝺengildan forbǽrnan *the general then wanted to turn the idolaters*, 484, 8. v. gilda.

hǽꝺenisc; *adj. Heathenish, pagan:*—Heora biscepas sǽdon ꝺæt heora godas bǣdon ꝺæt him man worhte anfiteatra ꝺæt mon mehte ꝺone hǽꝺeniscan plegan ꝺǣrinne dōn *suasere pontifices, ut ludi scaenici diis expetentibus ederentur*, Ors. 3, 3; Swt. 102, 12. [*O. H. Ger.* heidanisc *gentilis: Ger.* heidnisch.]

hǽðen-mann, -monn, es; *m. A heathen :*—Hǽðinmonn *samaritanus,* Lk. Skt. Lind. 10, 33.

hǽðen-nes, se; *f. Heathenism, paganism;* gentilitas :—Ða ongunnon monige hǽðennysse þeáw forlǽtan *relicto gentilitatis ritu,* Bd. 1, 26; S. 488, 12. Hé tó hǽðennysse wæs gehwyrfed *ad apostasiam conversus est,* 3, 30; S. 561, 39. [Laym. hæðenesse: Chauc. 'as wel in Cristendom as in hethenesse,' Prol. 49 : Piers P. ' al was hethenesse some tyme Ingelond and Wales, 15, 435.]

hǽðen-scipe, es; *m. Heathenism, paganism :*—Wé forbeódaþ eornostlíce ǽlcne hǽðenscipe. Hǽðenscipe biþ ðæt man ídola weorðige ðæt is ðæt man weorðige · hǽðene godas and sunnan oððe mónan fýr oððe flód wæter-wyllas oððe stánas *we earnestly forbid all heathenism: heathenism is to worship idols, that is to worship heathen gods, and sun or moon, fire or water, springs or stones,* L. C. S. 5; Th. i. 378, 17, 20. Ðæt ys mycel hǽðenscype *id magnus est paganismus,* L. Ecg. P. 4, 20; Th. ii. 210, 19: L. N. P. L. 48; Th. ii. 296, 27: Chr. 634; Erl. 25, 31. Ða tungelwítegan ðe wǽron on hǽðenscipe wunigende hæfdon getácnunge ealles hǽðenes folces *the astrologers, who were yet heathens, betokened all heathen people,* Homl. Th. i. 106, 9 : 70, 25, 28. [Laym. hæðenescipe]

hǽðen-styrc, es; *m. A heathen stirk, calf used in heathen worship, the golden calf made by the Israelites :*—Hí on Choreb swylce cealf ongunnon him tó godegylde georne wyrcean; onwendan heora wuldor on ðæne wyrsan hád hǽðenstyrces hig etendes *fecerunt vitulum in Choreb, et adoraverunt sculptile; et mutaverunt gloriam suam in similitudinem vituli comedentis fænum,* Ps. Th. 105, 17.

Hǽðfeld *Hatfield in Hertfordshire :*—Hér gesæt Þeodorius ærcebiscop senoþ on Hǽðfelda *in this year archbishop Theodore presided over a synod at Hatfield,* Chr. 680; Erl. 40, 11.

hǽðiht; *adj. Heathy :*—In ða hǽðihtan lége *to the heathy lea,* Cod. Dipl. Kmbl. iii. 121, 21: 262, 22.

hǽð-stapa, an; *m. A heath-stepper, an animal which wanders over heaths or uncultivated country :*—Ðeáh ðe hǽðstapa hundum geswenced heorot holtwudu séce *although the heath-wanderer, the hart by the hounds wearied, seek that wood,* Beo. Th. 2740; B. 1368. Wulf hár hǽðstapa *the wolf, the grey wanderer of the heath,* Exon. 87 a; Th. 328, 6: Vy. 13.

Hǽðum, æt *Slesvig,* Ors. 1, 1; Bos. 21, 30, 39. [Cf. Ethelweard 'Anglia vetus sita est inter Saxones et Giotos, habens oppidum capitale, quod sermone Saxonico Sleswic nuncupatur, secundum vero Danos Haithaby.' *Icel.* Heiðabær.]

hǽðung [= hǽtung], e; *f. Heating, warming :*—Belimpþ seó hǽðung tó ðære hǽtan and seó onlíhting belimpþ tó ðære beorhtnysse *the heating belongs to the heat and the illumination to the brightness,* Homl. Th. i. 286, 3.

hǽting, e; *f. Calipeatum,* Wrt. Voc. 290, 43.

hǽtsan *to drive, urge, impel* [?] :—Hwílum mec mín freá hǽtst on enge *sometimes my lord drives me into a narrow place,* Exon. 101 b; Th. 383, 3; Rä. 4, 5.

hǽttian; *p.* ode *pp.* od *To take the hair and skin from a person's head :*—Ðonne dó man út his eágan and ceorfan of his nôse and eáran and uferan lippan oððe hine hǽttian *then let his eyes be put out and his nose and ears and upper lip be cut off; or let him have the hair and skin of his head pulled off,* L. C. S. 30; Th. i. 394, 14. [The Latin version here has 'aut corium capitis cum capillis (auferatur) quod Angli vocant behættie.' Another translation has 'vel decapilletur.'] Sume man hǽttode, Chr. 1036; Erl. 164, 39. In the note Earle quotes Florence of Worcester 'cute capitis abstracta.' Cf. Grmm. R. A. 703, where he quotes an explanation of the punishment by which the hair was dragged from a person's head, 'man windet im die haar mit einer kluppen oder knebel aus dem heupt.' He thinks the form hettian [hættian] has no sense, but may it not be connected with *hæt,* as it was just that part of the head which the hat covered that was affected? It was giving the victim the appearance of wearing a hat of a most ghastly kind.

hǽtu, hǽto; *indecl; f. Heat :*—Hǽtu *calor,* Ælfc. Gr. 4, 26. Þridde ágennys is seó hǽtu *the third property is the heat,* Homl. Th. ii. 606, 13, 18. Þýstro and hǽto *darkness and heat,* Cd. 21; Th. 25, 6; Gen. 389: Bt. Met. Fox 20, 146; Met. 20, 73. Hǽto *æstus,* Mt. Kmbl. Lind. 20, 12. Gif se líchoma hwǽr mid hefiglícre hǽto sý gebysgod *if the body be troubled anywhere with heavy inflammation,* Herb. 2, 6; Lchdm. i. 82, 8. Unácumendlíce hǽtu þrowiaþ and unásecgendlíce cýle *they suffer intolerable heat and unspeakable cold,* Homl. Th. i. 532, 1. [O. Frs. hête; f: O. H. Ger. heizi, heiz; f. æstus, fervor : O. Sax. hêt; n.] v. hǽte.

hǽtung. v. hǽðung.

hǽwen; *adj. Blue, azure, purple, discoloured :*—Hǽwen *glaucus,* Cot. 96: *jacinthina,* 185: *fulvus,* Lye. Ádó in ǽren fæt lǽt ðǽr in óð ðæt hit hǽwen sý *put into a brazen vessel, leave it therein until it be turned colour,* Lchdm. iii. 20, 18. Gyf ðæt húsl byþ fynig oððe hǽwen *if the housel be mouldy or discoloured,* L. Ælf. C. 36; Th. ii. 360, 9. Seó hǽwene lyft *the azure air,* Cd. 166; Th. 207, 33; Exod. 476. Genim ðás wyrte ðe grêcas brittanice and engle hǽwen hydele, Herb. 30; Lchdm. i. 126, 6. Hǽwene hnydele, iii. 24, 8. Ðeós wyrt hafaþ lange leáf and hǽwene *this plant hath long leaves and purple,* Herb. 133, 1; Lchdm. i. 248, 18: 150, 1; Lchdm. i. 274, 16. Seó heall wæs getymbred ynnan and útan myd grênum and myd hǽwenum and myd hwýtum *the hall was built within and without with green and with purple and with white,* Shrn. 156, 6. Hǽwen-grên *cæruleus,* Cot. 53, Lye. Hǽwendeáge *hyacinthinus,* Lye.

hafa and forms as from hafian. v. habban.

hafecere, es; *m. A hawker :*—Wé lǽraþ ðæt preóst ne beó hunta ne hafecere *we enjoin that a priest be not a hunter, nor a hawker,* L. Edg. C. 64; Th. ii. 258, 7.

hafela, hafala, heafela, heafola, an; *m. The head;* caput; κεφαλή :—Se hwíta helm hafelan werede *the bright helm guarded the head,* Beo. Th. 2901; B. 1448: 2658; B. 1327: 3564; B. 1780. Of ðæs hǽlendes heafelan *from the Saviour's head,* Exon. 15 a; Th. 31, 34; Cri. 505. Heafolan, Beo. Th. 5352; B. 2679. Hafalan, 896; B. 446.

hafe-leást, e; *f. Want of means, indigence :*—For hauelêste *from lack of means,* Chr. 675; Erl. 38, 12. v. hafen-leást.

hafen. v. hebban.

hafenian; *p.* ode; *pp.* od *To grasp, hold :*—Wǽpen hafenade heard be hiltum *he grasped the weapon hard by the hilt,* Beo. Th. 3151; B. 1573. Bord hafenode *he grasped his shield,* Byrht. Th. 132, 67; By. 42: 140, 57; By. 309. [O. H. Ger. hebinon, hefinon, Grff. iv. 737, 828.]

hafen-leás; *adj. Lacking means, poor, indigent;* inops :—Hafenleás *inops,* Wrt. Voc. 74, 20. Hê wæs swíðe welig wædlum and þearfum and symle him sylfum swíðe hafenleás *he was very wealthy for the poor and needy, and ever very indigent for himself,* Homl. Th. ii. 148, 34. Sum hafenleás man sceolde ágyldan healf pund ánum menn *a certain indigent man had to pay a man half a pound,* 176, 34. Se hafenleása 178, 6. Se ðe spéda hæþ and ða áspendan nele hafenleásum brêðer *he that hath riches and will not expend them for his brother who lacks,* 318, 11: 484, 33: 178, 19. v. hæfen-leás.

hafen-leást, e; *f. Lack of means, indigence;* inopia :—Wé ne sceolon ða wannspêdigan for heora hafenleáste forseón *we ought not to despise those who are without means for their indigence,* Homl. Th. i. 128, 23. Fela sind þearfan þurh hafenleáste and ná on heora gáste. Sind eác ôðre þearfan ná þurh hafenleáste ac on gáste *many are poor from want of wealth, and not in spirit. There are also other poor, not from want of wealth, but in spirit,* 550, 3–5, 11, 12, 17. Úre sáule hafenleáste *the indigence of our souls,* ii. 88, 26. Ða getímode swá micel hafenleást ðæt ða gebrôðra næfdon búton fíf hláfas tó heora ealra gereorde *then there befell so great a lack that the brethren had but five loaves for the refection of them all,* 170, 33. v. hæfen-leást.

hafetian *to clap* [as a bird with its wings, or a man with his hands], *applaud :*—Ic hafetige *plaudo,* Ælfc. Gr. 28; Som. 31, 28. Flôdas hafettaþ handum *flumina plaudent manu,* Ps. Spl. 97, 8. Ǽrðan ðe se hana hafitigende cráwe *before the cock clapping its wings crow,* Homl. Th. ii. 246, 4.

hafoc, hafuc, heafoc, es; *m. A hawk;* accipiter :—Heafuc *accipiter,* Wrt. Voc. 77, 15. Mid hafoce *accipitre,* Coll. Monast. Th. 25, 15, 17, 31, 37. Gôd hafoc *a good hawk,* Beo. Th. 4519; B. 2263. Sum sceal wildne fugol átemian heafoc *one shall tame the wild bird, the hawk,* Exon. 88 b; Th. 332, 16; Vy. 86. [Laym. havek: Icel. haukr: O. H. Ger. hapuh, habich: Ger. habicht.] DER. gôs-, gûþ-, mûs-, spearwealh-hafoc. The word is found in many names of places, see Cod. Dipl. Kmbl. vi. index.

hafoc-cynn, es; *n. The hawk species :*—Ne ete gê nán þing hafoccynnes ne earncynnes *eat nothing of the hawk-kind or the eagle-kind,* Lev. 11, 13.

hafoc-fugel, es; *m. A hawk :*—Ðeáh hafucfugel ábíte *etiamsi accipiter momorderit,* L. Ecg. C. 38; Th. ii. 162, 19.

hafoc-wyrt, e; *f. Hawk-weed* [?]; hieracium, L. M. 1, 14; Lchdm. ii. 56, 11.

hafud. v. heáfud.

hafud-æcer, es; *m* [?] :—Tióþa hafudæcer *decumanus,* Ælfc. Gl. 57; Som. 67, 78; Wrt. Voc. 38, 4.

hafud-land, es; *n. A headland, boundary :*—Hafudland *limites,* Ælfc. Gl. 57; Som. 67, 77; Wrt. Voc. 38, 3. ['Headland, the upper portion of a field, generally left unploughed for convenience of passage,' Cod. Dipl. Kmbl. iii. xxix. 'Adlands, those butts in a ploughed field which lie at right angles to the general direction of the others; the part close against the hedge. Salop,' Halliwell. So in Surrey, Engl. Dial. Soc. No. 12, p. 91. 'Headland, that is which is ploughed overthwart at the ends of the other lands,' No. 30, p. 82.]

haga, an; *m. A place fenced in, an enclosure, a haw, a dwelling in a town :*—Haga *sæpem,* Mk. Skt. Lind. 12, 1. Se haga binnan port ðe Ægelríc himsylfan getimbrod hæfde *the messuage within the town that Ægelric had built himself,* Cod. Dipl. Kmbl. iv. 86, 26: Th. Chart. 569, 2, 5: 514, 13: Cod. Dipl. ii. 150, 5, 11. Ðis syndon ðæs hagan gemǽru *these are the boundaries of the messuage* [in the previous part of the charter the gift is spoken of as *unam curtem*], iii. 240, 18. Ða hagan

ealle ðe hē be westan cyrcan hæfde *all the messuages that he had west of the church*, Th. Chart. 303, 10. Ænne hagan on porte *curtem unum in supradicta civitate*, Cod. Dipl. Kmbl. iv. 72, 27: iii. 213, 13. Quandam hospicii portionem in præfata civitate sitam, quæ patria lingua *haga* solet appellari, vi. 134, 24; cf. 135, 14, 25. Tō hagan þrungon *they pressed to the entrenchment*, Beo. Th. 5913; B. 2960: Beo. Th. 5777; B. 2892. [*Chauc.* hawe *yard: in Kentish dialect* haw *a yard, or enclosure: Icel.* hagi *a hedged field, a pasture*.] DER. bord-, cumbol-, fǽr-, swîn-, turf-, wîg-haga.

haga, an; m. *A haw, berry of the hawthorn; also used to signify any thing of no value* [?], [cf. Chaucer's 'not worth an hawe']:—Hagan *gignalia*, Ælfc. Gl. 47; Som. 65, 24; Wrt. Voc. 33, 24. Hagan *quisquilia*, 285, 31. [*Prompt. Parv.* hawe, frute *cinum, cornum, ramnum*.]

hagal, hagol, es; m. *Hail;* grando:—Hagol *grando*, Ælfc. Gl. 94; Som. 75, 100; Wrt. Voc. 52, 50: Homl. Th. ii. 192, 32. Hagol cymþ of ðám rêndropum ðonne hî beóþ gefrorene *hail comes of the raindrops when they are frozen*, Lchdm. iii. 278, 19. Rēn hagal and snáw hrusan leccaþ *rain, hail, and snow moisten the earth*, Bt. Met. Fox 29, 127; Met. 29, 64. Mid hagole *with hail*, Homl. Th. ii. 350, 8. Gesihþ hreósan hrîm and snáw hagle gemenged *sees rime and snow fall mingled with hail*, Exon. 77 a; Th. 289, 15; Wand. 48. [*Laym.* haȝel: *Icel.* hagl; *n. hail;* Hagall; *m. the name of the rune* h: *O. H. Ger.* hagal; *m: Ger.* hagcl.] v. hægel.

hagalian; p. ode *To hail:*—Hit hagalade stánum ofer ealle Rómane *saxea de nubibus grando descendens*, Ors. 3, 5; Swt. 104, 20. [*Icel.* hagla: *M. H. Ger.* hagelen.]

hagal-scûr, hagol-, es; m. *A shower of hail*, Ps. Spl. M. 104, 30: Menol. Fox 71; Men. 35. v. hægel-scûr.

haga-þorn, es; m. *Hawthorn:*—Of hagaþornum *de tribolis*, Mt. Kmbl. Lind. 7, 16. v. hæg-þorn.

hagian. v. on-hagian.

hagol-stán, es; m. *A hailstone:*—God him sende ufan greáte hagolstánas *God sent down upon them great hailstones*, Jos. 10, 11. Betwux ðám greátum hagolstánum *amid the great stones*, Homl. Th. i. 52, 18. [*Icel.* hagl-steinn: *M. H. Ger.* hagel-sten: *Ger.* hagel-stein.]

hago-spind, heago-, hecga-, es; m. n? *The cheek*:—Hagospind *genæ*, Wrt. Voc. 64, 41. Heagospind, 282, 56. Hecgaspind, Ælfc. Gl. 71; Som. 70, 78; Wrt. Voc. 43, 11. Heortes heagospind *a hart's cheek*, Lchdm. i. 336, 12. [Somner, Lye, and Wright print *swind* for *spind*, the form which occurs in the transcript by Junius; see note to passage quoted above from Lchdm. i. *Eágospind* occurs, Guthl. 20; Gdwin. 82, 4.] v. spind.

hago-steald, es; m. *One living in the lord's house, not having his own household, an unmarried person, a young person, young warrior:*—Hagosteald onwôc môdig from moldan *the young warrior* [*Christ*] *was roused exulting from earth*, Exon. 120 a; Th. 460, 23; Hö. 21. Heafoc weorþeþ tô hagostealdes honda geláered *the hawk becomes trained to the youth's hand*, 88 b; Th. 332, 28; Vy. 92. [*O. Sax.* haga-stald, -stold *a servant, young man: O. H. Ger.* haga-stalt, -stolt *mercenarius, cælebs: Ger.* hagestolz *old bachelor*.] v. Grmm. R. A. 484, *and* hæg-steald.

hago-steald, es; n. *Celibacy*, Exon. 105 b; Th. 402, 17; Rä. 21, 31.

hago-stealdmonn, es; m. = hago-steald, q. v. Exon. 104 a; Th. 395, 3; Rä. 15, 2.

Hagustaldes eá, eé, hám *Hexham*, Chr. 681: 685: 766: 780: 789: 806: Bd. 5, 23; S. 646, 30. [*Dun.* Hestaldesham, Hestaldeshige: *Ric.* Hestalasham: *Gerv.* Hestoldesham: *Kni.* Exseldesham.]

hagu-swind. v. hago-spind.

hal, es; n. *A secret place, a corner:*—Ðá gemétte hē hine hleonian on ðam hale his cyrcan wið ðam weofode *he found him leaning in the corner of his church against the altar*, Guthl. 20; Gdwin. 82, 22. On halum *in abditis*, Ps. Spl. 16, 13. [Cf. we beth honted from hale to hurne, Pol. Songs. Wrt. 150, 17. In one swiþe diȝele hale, O. and N. 2.] v. helan.

hál; adj. *Whole, hale, well, in good health, sound, safe, without fraud, honest;* often used in salutation:—Iosep áxode hwæðer hira fæder wære hál *Joseph asked whether their father were well*, Gen. 43, 27. Se man wæs sôna hál *statim sanus factus est*, Jn. Skt. 5, 9. Se biþ hál geworden *he shall be saved*, Blickl. Homl. 21, 36. Hē þurh ðæt sôna wearþ hál geworden *he was at once by that restored to health*, 223, 26. Gif hie mon gelácnian mæge ðæt hie hál síe *if it* [*the broken sinew*] *can be cured so that it be sound*, L. Alf. pol. 75; Th. i. 100, 4. Mannes sunu com sêcean and hál dôn ðæt forwearþ *venit filius hominis quærere et saluare quod perierat*, Lk. Skt. 19, 10. Gedô mē hálne *salvum me fac*, Mt. Kmbl. 14, 30: Mk. Skt. 5, 34. Hine ðǽm mannum hálne and gesundne âgeaf *restored him to the men safe and sound*, Blickl. Homl. 219, 21: 107, 17. Ðú mē behête hál and clǽne ðæt ðæt ðú mē sealdest *thou didst declare to me that what thou didst sell me was sound and clean*, L. O. 7; Th. i. 180, 22: 9; Th. i. 182, 4. Hē hyne hálne onfêng *he hath received him safe and sound*, Lk. Skt. 15, 27. Ic geaf hit on mînon hálan lîfe intô Cristes cyrcean *I gave it while of sound body to Christ's church*, Cod. Dipl. Kmbl. iv. 305, 12. Ðá betǽhte Ecgferþ on hálre tungan land and bôc Dunstáne *then Ecgferth in plain, unequivocal language delivered*

land and charter to Dunstan [cf. *Icel.* með heilum hug *sincerely*], Th. Chart. 208, 11. Hál wes ðú Iudêa cyning *Haue rex Iudæorum*, Mt. Kmbl. 27, 29: Lk. Skt. 1, 28: Andr. Kmbl. 1827; An. 916: Beo. Th. 818; B. 407. Hále wese gê *Havete*, Mt. Kmbl. 28, 9. Sŷ ðú hál leóf Iudéiscre leóde cyning *hail sir, king of the Jewish people*, Homl. Th. ii. 252, 28. Hál beó ðú *Have*, Mt. Kmbl. 26, 49. Beó ðú hál and sig gebletsod se ðe on Dryhtnes naman com *Osanna benedictus qui venit in nomine Domini*, Nicod. 4; Thw. 2, 32. [*Laym.* hal, hæl, hæil, hail, hol: Lauerd king wæs hæil [wassayl, later MS.], 14309: *Orm.* hal: *A. R.* hol: *Prompt. Parv.* hool: *Goth.* hails: *O. Sax. O. Frs.* hêl: *Icel.* heill: *O. H. Ger.* heil: *Ger.* heil.] v. ge-, un-hál.

halan [*or* hamlan] *afterbirth:*—Gáte geallan on wîne gedruncen wîfa halan him ofáðêþ *goat's gall, drunken in wine, removes women's afterbirth for them*, Lchdm. i. 356, 8. v. Gloss: Lchdm. ii.

hál-bǽre; adj. *Wholesome, salutary;* salutaris, Scint. 32, 78, Lye.

hald. v. heald.

hálettan; p. te *To salute, greet, hail:*—Sum man hine hálette and grētte and hine be his naman nemde *quidam eum salutans ac suo appellans nomine*, Bd. 4, 24; S. 597, 12: 2, 12; S. 514, 31: Blickl. Homl. 155, 20. Iohannes hálette on hie mycelre stefne *John greeted her with a loud voice*, 143, 15. Hie háletton on hie *they greeted her*, 139, 25.

hálettend, es; m. *The middle finger, the finger by which a sign of greeting is made:*—Hálettend midemesta finger *salutarius*, Wrt. Voc. 283, 21. Hæletend *salutaris*, Recd. 38, 72; Wrt. Voc. 64, 81.

hálettung, e; f. *Greeting, salutation:*—Æfter ðæs engles bletsunga and hálettunga swîgende þohte hwæt seó hálettung wǽre *after the angel's blessing and greeting she considered in silence what the greeting might be*, Blickl. Homl. 7, 16. Hálettunge, 3, 21. Hælettungæ *salutationes*, Mt. Kmbl. Rush. 23, 7.

half. v. healf.

hál-fæst; adj. *Salutary;* qui potest sanare, Lye.

hálga, an; m. *A saint:*—Biþ gesmyrod ealra hálgena hálga *the saint of all saints shall be anointed*, Homl. Th. ii. 14, 16. Ðæt wundor gelamp þurh ðæs hálgan mihte *that miracle happened through the saint's might*, 28, 28; Swt. A. S. Rdr. 102, 212. Fram ðam rihtwîsan Abel oþ ðam endenêxtan hálgan *from righteous Abel to the last saint*, Homl. Th. ii. 74, 5. Godes hálgan sind englas and men *God's saints are angels and men*, i. 538, 23: 574, 22: ii. 112, 31. Hálgena lîchaman ârison *the bodies of saints arose*, 258, 5. On ðone dæg æfter ealra hálgena mæssedæg *on the day after All Saints' day*, Chr. 1083; Erl. 217, 32. November se mônaþ onginþ on ealra hálgena mæssedæg *the month of November begins on All Saints' day*, Ælfc. Gr. 9; Som. 9, 55. [*Chauc.* halwe: *Mod. E.* in All Hallows: *Ger.* heilige *a saint*.] v. hálig.

hálgawaras; *pl. Holy people, saints:*—Gisungan hálgawaras *cantabant sancti*, Rtl. 47, 26. Hálgawara ðînra *sanctorum tuorum*, 62, 12. Hálgawæra, Mt. Kmbl. Lind. 27, 52. [Cf. hálig-waras.]

hálgian; p. ode; pp. od *To hallow, make holy, consecrate, sanctify:*—Hweðer hie ða ciriccean hálgian dorston on ôðre wîsan *whether they durst consecrate the church otherwise*, Blickl. Homl. 205, 21, 24. Ne miht ðú on ôðre wîsan bisceop hálgian búton ôðrum bisceopum *ordinare episcopum non aliter nisi sine episcopis potes*, Bd. 1, 27; S. 492, 3. Ðú scealt hálgian hîred ðinne *thou shalt hallow thy family*, Cd. 106; Th. 139, 15; Gen. 2310. Hér man hálgode Ælfêhg tô arcebiscope *in this year Ælfheah was consecrated archbishop*, Chr. 1006; Erl. 138, 2: 1050; Erl. 176, 22. Nis eów þearf ðæt gê ða ciriccan hálgian *there is no need for you to consecrate the church*, Blickl. Homl. 207, 1. Hweðer hie ða ciriccean hálgedon *whether they should consecrate the church*, 205, 11. Hálgig oððe hálga ðú *sanctifica*, Jn. Skt. Lind. 17, 17. Hálgiaþ eówer fæsten *sanctify ye a fast*, Blickl. Homl. 37, 32. Sŷ hálgad noma *hallowed be thy name*, Exon. 122 a; Th. 468, 19; Hy. 5, 2. [*Laym.* halȝien: *Orm.* hallȝhenn: *Prompt. Parv.* halwin *consecrare: O. Sax.* hêlagôn: *Icel.* helga: *O. H. Ger.* heilagôn: *Ger.* heiligen.]

Hálgo-land, es; n. *A district* [*fylki*] *of Norway, Hâlogaland:*—Ôhthere sǽde ðæt sió scir hátte Hálgoland ðe he on búde. Hē cwæþ ðæt nân mann ne búde be norþan him *Ohthere said that the district was called Halogaland that he lived in. He said that no one lived north of him*, Ors. 1, 1; Bos. 21, 16. See Aall's translation of the Heimskringla, p. 24, note.

hálgung, hálegung, e; f. *Hallowing, consecration, sanctification:*—Getimbra hálgung *scenophegia*, Ælfc. Gl. 13; Som. 55, 78; Wrt. Voc. 16, 50. Niuæs húses hálgung i cirica hálgung *encenia*, Jn. Skt. Lind. 10, 22. Geworden is Iudêa hálgung *facta est Iudæa sanctificatio*, Ps. Spl. 113, 2: 77, 59. Biscopes hálgung *episcopi ordinatio*, Bd. 1, 27; S. 492, 5. Ðeáh ealle circan habban hálgunge gelîce *though all churches have like consecration*, L. Eth. 9, 5; Th. i. 340, 27. Seðe ða hálgunge oððe ða lectionem ne mæg æfter þeáwe gefyllan *qui consecrationem vel lectionem non potest rite implere*, L. Ecg Ç. 31; Th. ii. 160, 16. Hē ne háding ne háleging ne dô *let him not ordain nor consecrate*, Cod. Dipl. Kmbl. v. 28, 34. [*Prompt. Parv.* halwynge of holy placys *consecracio, dedicacio: O. H. Ger.* heilagunga *sanctificatic: Ger.* heiligung.]

hálgung-ram; m. *A consecrated ram:*—For ðam hit ys hálgungram *for it is a ram of consecration*, Ex. 29, 22.

háli-. v. hálig-.

hálian ; *p. ode To become hale, whole, to heal, to get well* :—Lege tô ðam sáre hyt sceal berstan and hálian *lay to the sore; it shall burst and heal*, Herb. 148, 2 ; Lchdm. i. 272, 21. Hê ðá ongan trumian and háligan *ubi sanescere cœpit*, Bd. 4, 22 ; S. 591, 10. Ðonne hálaþ ðæt heáfod swýðe hraðe *the head will heal very quickly*, Herb. 1, 2 ; Lchdm. i. 70, 16 : 2, 6 ; Lchdm. i. 82, 10. [*O. H. Ger.* heilen *sanescere*.]

hálig ; *adj. Holy* ; sanctus, sacer :—Hálig *sanctus, almus*, Ælfc. Gr. 8 ; Som. 7, 41. Ðæt hálige gewrit *scriptura*, Jn. Skt. 17, 12. Se háliga frófre gást *paracletus sanctus spiritus*, 14, 26. Hálig sealt *holy salt*, L. M. 3, 62 ; Lchdm. ii. 346, 30 ; 344, 14. Háliges wæteres *some holy water*, 348, 2. Woroldlícra weorca on ðam hálgan dæge geswíce man georne *let people carefully abstain from worldly works on that holy day* [*Sunday*], L. Eth. 6, 22 ; Th. i. 320, 13. On ðone hálgan Ðunresdæg on holy Thursday, L. Alf. pol. 5 ; Th. i. 64, 24. Ða hálgan hádas *the clergy*, L. Edm. E. 1 ; Th. i. 244, 9. Hê spræc þurh hys hálegra wítegena múþ *locutus est per os sanctorum prophetarum ejus*, Lk. Skt. 1, 70. Ðám hálgum tídum *at those holy times*, L. C. S. 17 ; Th. i. 370, 9. Hálige béc *sacros libros*, L. Ecg. P. 3, 4 ; Th. ii. 196, 27. [*Laym.* hali, holy : *Orm.* haliʒ : *Wick.* hooli : *O. Sax.* hêlag : *O. Frs.* hêlich : *Icel.* heilagr : *O. H. Ger.* heilag : *Ger.* heilig.]

hálig-dæg, es ; *m. A holy day, Sunday* :—Be hálidæiges freólse *of the festival of Sunday*, L. C. S. 45 ; Th. i. 402, 8. On háligdagum *sabbatis*, Mk. Skt. Lind. 3, 2. [*A. R.* halidei : *Piers P.* halidai.]

hálig-dóm, es ; *m.* I. *holiness, sanctity* ; sanctimonia : — Háligdóm *sanctimonia*, Rtl. 100, 11. Mycel is se háligdóm and seó weorþung sancte Iohannes *great is the sanctity and worthiness of St. John*, Blickl. Homl. 167, 16. Búton ða heánesse ðæs háligdómes *nisi excellentia sanctitatis*, Past. 18, 3 ; Swt. 133, 14 : 57 ; Swt. 439, 23. II. *holy things, relics, holy work, a sacrament* :—Háligdóm *sacramentum*, Mk. Skt. p. 5, 11. On ðone Drihten ðe ðes háligdóm is fore hálig *by the Lord, before whom these relics áre holy*, L. O. 1 ; Th. i. 178, 3, 12. Wê sceolon on ðissum dagum fyligan úrum háligdóme út and inn *on these days we ought to follow our relics out and in*, Homl. Th. i. 246, 28. Ðæt hig bereáfodan æt háligdome and æt eallon þingan *they plundered the monastery of the relics and of every thing*, Chr. 1055 ; Erl. 188, 40. On ðam háligdóme swerian *to swear on the relics*, L. Eth. 3, 2 ; Th. i. 292, 14 : Th. Chart. 610, 31 : Chr. 1131 ; Erl. 260, 10. Ðý-læs ǽnig unclǽnsod dorste on swá micelne háligdóm fón ðære clǽnan ðegnenga ðæs sacerd hádes *ne aut non purgatus adire quisque sacra ministeria audeat*, Past. 7, 1 ; Swt. 51, 1. Tô háligdóm ðíre gesibsumnesse tô ásend *ad sacramentum tuæ reconciliationis admitte*, Lye. Þurh hálgum háligdóm Drihtnes líchaman and blôdes *per sacrosanctum sacramentum Domini corporis ac sanguinis*, Lye. Háligdóm and hálige béc handligan *reliquias et sacros libros manu tractare*, L. Ecg. P. 3, 4 ; Th. ii. 196, 27 : 12 ; Th. ii. 200, 7. Háligdóm and hádas and gehálgode Godes hús man sceal weorþian georne *holy things and holy orders and the hallowed houses of God must be zealously honoured*, L. Eth. 7, 28 ; Th. i. 336, 1 : 24 ; Th. i. 334, 23 : L. E. B. 1 ; Th. ii. 240, 9. Wê lǽraþ ðæt ealle ða þingc ðe weofode neáh beón, and tô cirican gebyrian, beón swíðe clǽnelíce and wurþlíce behworfene, and ðǽr ǽnig þingc fúles neáh ne cume ; ac gelogige man ðone háligdóm swíðe árwurþlíce *we enjoin, that all the things which are near the altar, and belong to the church, be very cleanly and worthily appointed, and where nothing foul may come near them ; but let the holy things be very reverently arranged*, L. Edg. C. 42 ; Th. ii. 252, 23–6. Þurh ealne ðane háligdóm ðe ic on Rôme for mê and for ealne þeódscype gesôhte *by all the relics that I sought out in Rome for myself and for all the nation*, Th. Chart. 117, 10. III. *a holy place, sanctuary* :—Ðínne háligdóm *sanctuarium tuum*, Ps. Lamb. 73, 7. Hê getimbrade his háligdóm *ædificavit sanctificium suum*, 77, 69. Tô ðæs háligdómes dura *to the door of the sanctuary*, Ex. 21, 6. Tôweard ðam háligdóme *toward the sanctuary*, Chr. 1083 ; Erl. 217, 20. Án is mid ðæs kynges háligdóme, ôðer is mid Leófríce eorle and ðæt þridde is mid ðam bisceop *one* [*of the writings*] *is in the king's sanctuary, a second is with earl Leofric, and the third is with the bishop*, Th. Chart. 372, 29 : 541, 25 : 571, 20. [*Laym.* halidom *relic* : *Orm.* halidom *holiness* ; *pl. sacred things* : *Icel.* helgir dómar *relics* ; helgidómr *a sanctuary* : *O. H. Ger.* heiligtuom *sacramentum, sanctuarium* : *Ger.* heiligthum *sacred thing, relic, sanctuary*.]

hálig-ern, es ; *n. A holy place, sanctuary* :—Háligern *sanctuarium*, Blickl. Gl. Háliern *sacellum*, Hpt. Gl. 482. On ðam hálierne *in the holy place*, Ex. 29, 30.

hálig-mônaþ, -mônþ, es ; *m. Holy month, September* : — On ðæm nigoþan mônþe on geáre biþ xxx daga se mônaþ hátte on léden septembris and on úre geþeóde háligmônaþ for ðon ðe úre yldran ðá ðá hí hǽðene wǽron on ðam mônþe hí guldon hiora deófolgeldum *in the ninth month in the year there are thirty days. The month is called in Latin September, and in our language holy month, because our ancestors, when they were heathen, sacrificed to their idols in that month*, Shrn. 124, 28–31 : 136, 27. Háligmônþ, Menol. Fox 325 ; Men. 164. [*Bede, De temporum ratione, c. 13, gives Halegmonath as the native equivalent of September, v. Grmm. Gesch. D. S. 56 sqq.*]

hálig-nes, -ness, e ; *f.* I. *holiness, sanctity* :—Hálygnyss *sanctitas*, Ælfc. Gr. 5 ; Som. 5, 22. Hálignys on hálignysse hys *sanctimonia in sanctificatione ejus*, Ps. Spl. 95, 6. On rihtwísnesse and on hálignesse *in righteousness and in holiness*, Blickl. Homl. 31, 36 : 155, 31. On hálignesse *in sanctitate*, Lk. Skt. 1, 75 : Ps. Th. 88, 32. II. *a holy thing, relic* :—Seó hálignis *the relic*, St. And. 42, 7. Ic háte ðé Veronix ðæt ðú ágif mé ða hálignysse ðe ðú myd ðé hæfst. Veronix him ðá swýðe wiðsóc and sǽde, ðæt heó náne hálignyssa myd hyre næfde *I command thee, Veronica, that thou give up to me the relic that thou hast with thee. Then Veronica vehemently refused and said that she had no relics with her*, 40, 31–4. III. *a holy place, sanctuary* :—Gecwǽdon ðæt hí hálignesse Godes gesettan *dixerunt, possideamus sanctuarium Dei*, Ps. Th. 82, 9. Hálignessa sindon tô gridleáse *sanctuaries are too unprotected*, Swt. A. S. Rdr. 106, 41. Inngongende and útgongende beforan Gode tô ðam hálignessum *quando ingreditur et egreditur sanctuarium in conspectu Domini*, Past. 15, 4 ; Swt. 93, 7. [*O. H. Ger.* heilagnissa *sanctificatio, sanctitas*.] Cf. hálig-dóm.

hálig-rift, -reft, -ryft, e ; *f. A holy garment, veil* :—Háligryft *theristrum*, Hpt. Gl. 525. Hió an hyre betsþ háliryft *she gives her best veil*, Th. Chart. 538, 7. Heó ðǽr háligryfte onféng *accepto velamine sanctimonialis habitus*, Bd. 4, 19 ; S. 587, 42 : Shrn. 94, 25 : Lchdm. iii. 430, 26. Sca hylda wæs xxxiii geára on lǽwedum háde and xxxiii geára under háligryfte *St. Hilda was for thirty-three years in the world and for thirty-three years in the cloister*, Shrn. 149, 5. Effigenia is ðæs Heofenlícan Cynges brýd and mid háligrefte gehálgod *Effigenia is the bride of the Heavenly King, and hallowed with the veil*, Homl. Th. ii. 476, 32. Matheús léde háligreft ofer hire heáfod *Matthew placed a veil on her head*, 478, 5.

hálig-wæcca, an ; *m. One who observes vigils* :—Beón eáðmóde and ælmysfulle and háligwæccan *ut humiles simus et eleemosynis largi et sanctarum vigiliarum studiosi*, L. Ecg. P. 4, 64 ; Th. ii. 224, 27.

hálig-wæter, es ; *n. Holy water* :—Sumne dǽl ðæs háligwæteres *de aqua benedicta*, Bd. 5, 4 ; S. 617, 19 : L. Ath. 4, 7 ; Th. i. 226, 24 : L. M. 1, 64 ; Lchdm. ii. 138, 28. Mid háligwætere *with holy water*, 62 ; Lchdm. ii. 136, 4. On háligwætre *in holy water*, 45, 1 ; Lchdm. ii. 110, 14.

hálig-waras, -ware ; *pl. Holy people, saints* :—Þerh múþe háligwara *per os sanctorum*, Lk. Skt. Lind. 1, 70. Hálgwara *sanctorum*, Rtl. 45, 1. [Cf. hálga-waras.]

halm, hals. v. healm, heals.

hálor *salvation* :—From hálor áhwyrfan, oncyrran *to turn, seduce from salvation*, Exon. 70 b ; Th. 262, 3 ; Jul. 327 : 71 a ; Th. 264, 6 ; Jul. 360 : 72 a ; Th. 268, 30 ; Jul. 440.

háls, e ; *f. Health, salvation* :—Ðæt hǽlubearn háls eft forgeaf *that saviour-child gave salvation again*, Exon. 16 a ; Th. 37, 3 ; Cri. 587. [*Icel.* heilsa *health*.] v. heáls-bóc.

hálsere, es ; *m. An exorcist* :—Hálsere *exorcista*, L. Ecg. C. 41 ; Th. ii. 166, 21 : Rtl. 194, 5. [*O. H. Ger.* heilisari *augur, aruspex*.] v. hálsian.

hálsian, heálsian [Ettmüller connects this verb in the sense *obsecrare* with *hals*, and writes halsian, healsian; the forms in which *ea* occurs seem to favour this writing, while reference to cognate dialects seems to point to *á*] *To beseech, entreat, implore, adjure, conjure, exorcise* :—Ic hálsige and bidde ðone gelǽredan ðæt hê ðæt ús ne wíte *I beseech and beg the learned not to blame us for it*, Guthl. prol ; Gdwin. 2, 10 : Blickl. Homl. 57, 33. Ic hálsige ðé þurh ðone lifiendan God *adjuro te per Deum vivum*, Mt. Kmbl. 26, 63 : Exon. 72 a ; Th. 269, 6 ; Jul. 446 : Blickl. Homl. 151, 22. Ic eów hálsige scucna englas ðæt gé leng ne beran *I adjure you, devils' angels, that ye bear him no longer*, 189, 7. Ic ðé hálsige for ðínre þeówene Sancta Marian *I entreat thee for the sake of thy servant Saint Mary*, 89, 17 : Exon. 73 b ; Th. 274, 26 ; Jul. 539 : Cod. 222 ; Th. 290, 28 ; Sat. 422. Ic ðé heálsige *I beseech thee*, Bt. 22, 2 ; Fox 78, 10. Ic heálsige *obsecro*, Past. 18, 6 ; Swt. 137, 17. Ic hálsigo ðec *exorcizo te*, Rtl. 100, 27 : 117, 34. Exorcista is on Englisc se ðe mid áþe hálsaþ ða áwyrgedan gástas ðe wyllaþ menn dreccan þurh ðæs Hǽlendes naman ðæt hý ða menn forlǽton *exorcista is in English he who with oath conjures the accursed spirits that will torment men, in the Saviour's name to leave those men*, L. Ælfc. C. 13 ; Th. ii. 348, 1. Hê ðone unlybban on Godes naman hálsode *he exorcised the poison*, Homl. Th. i. 72, 24. For ðam ðe he hálsode Israhéla bearn *for he had strictly sworn the children of Israel*, Ex. 13, 19. Hê hie heálsade *he entreated them*, Ors. 4, 6 ; Swt. 178, 14 : Beo. Th. 4270 ; B. 2132. Fæder and môdor hálsedon hí ðæt hí forlétan ðone cristes geleáfan *father and mother implored them to forsake the faith of Christ*, Shrn. 92, 13. Heálsa hine suá suá ðínne fæder *obsecra ut patrem*, Past. 25 ; Swt. 181, 2. On wigbedde tô hálsienne *in altari ad augurandum*, Cot. 17, Lye. [*Laym. A. R.* halsien : *Chauc.* halse : *O. H. Ger.* heilison *augurari* : cf. *Icel.* heilsa *to salute, greet*.] v. gehalsian [*where read ge-hálsian*] *and* hálsung.

hálsigend, es ; *m. An exorcist* :—Exorcista is hálsigend, L. Ælf. P. 34 ; Th. ii. 378, 6.

hálsigendlíc, hálsiendlíc; *adj. That may be entreated* :—Hálsiendlíc *deprecabilis*, Ps. Spl. M. 89, 15.

hálsigendlíce, hálsiendlíce; *adv. Importune*, Greg. Dial. 1, 2, Lye.

hálsung, heálsung, e; *f. Supplication, beseeching, entreaty, adjuration, exorcising, exorcism, augury, greeting* [?] :—Micel is seó hálsung ánd mǽre is seó hálgung ðe deófla áfyrsaþ *great is the exorcising and greater is the hallowing that drives away devils*, L. C. E; Th. i. 360, 28. Hálsung *exorcismus*, Mone Gl. 414. Mid wêpendre hálsunga hine bǽdon *with weeping supplication prayed him*, Blickl. Homl. 87, 8. Hê breác ealdre heálsunge *vetere usus augurio*, Bd. 1, 25; S. 486, 40. On hálsunge *in auspicium*, 2, 9; S. 510, 13. Mid eárum onfóh míne hálsunge *auribus percipe obsecrationem meam*, Ps. Th. 142, 1. Hálsunga dôþ *obsecrationes faciunt*, Lk. Skt. 5, 33. Se ðe hálsunga behealdaþ *quicunque exorcismos observat*, L. Ecg. C. 29, note; Th. ii. 154, 29. Hie [*the rich*] hǽfdon oforgedrync and dyslíce and unrǽdlíce hálsunga *they had excessive drinking and foolish and thoughtless greetings* [?], Blickl. Homl. 99, 21. On hálsungum *in obsecrationibus*, Lk. Skt. 2, 37. On hálsungum *precibus*, L. Ecg. C. 2; Th. ii. 136, 19. [*A. R.* halsung *supplication* : *O. H. Ger.* heilisunga *omen, auspicium* : cf. [?] *Icel.* heilsan *greeting*.]

hálsung-gebed, es; *n. Litany*, R. Ben. 9, Lye.

háls-wurþung, e; *f. A celebration because of safety*, Cd. 171; Th. 215, 11; Exod. 581. v. háls.

hál-wenda, an; *m. A saviour* :—Míne eágan habbaþ gesewen ðínne Hálwendan. Se hálwenda ðe hé embe sprǽc is úre Hǽlend Crist se ðe com tó gehǽlenne úre wunda ðæt sindon úre synna *mine eyes have seen thy Saviour* [*viderunt oculi mei salutare tuum*]. *The Saviour that he spoke about is Jesus Christ who came to heal our wounds, that is, our sins*, Homl. Th. i. 142, 32 : 136, 21. [Cf. Hǽlend.]

hál-wende; *adj. Conducive to health, salutary, healing, wholesome* :—Ðes hálwenda *hic saluber*, Ælfc. Gr. 9, 18; Som. 9, 64. Ðín word is hálwende *thy word is salutary*, Ps. Th. 118, 103. Hálwoende ðín *salutare tuum*, Lk. Skt. Lind. 2, 30. Se middangeard wæs mannum hálwende *the earth was healthful for men*, Blickl. Homl. 115, 8 : 209, 10. Ðisse sylfan wyrte sǽd on wíne gedruncen is hálwende ongeán áttres drync *the seed of this same plant is wholesome against a draught of poison*, Herb. 142, 6; Lchdm. i. 264, 13 : 157, 2; Lchdm. i. 284, 10. Hit is hálwende bôte *it is a healing remedy*, 374, 24. Wê mágon eów sellan hálwende geþeahte hwæt gé dôn mágon *possumus salubre vobis dare consilium quid agere valeatis*, Bd. 1, 1; S. 474, 14. Seó tunge ðe swá monig hálwende word gesette *illa lingua quæ tot salutaria verba composuerat*, 4, 24; S. 599, 11. Háte baþu ðe wǽron hálwende gecwedene ádligendum líchaman *hot baths that were said to be salutary for diseased bodies*, Homl. Th. i. 86, 21. Ða hálwendan men *the men who taught a saving faith, the disciples*, Blickl. Homl. 117, 8. Swá se lǽcedôm yldra byþ swá hé hálwendra byþ *the older the medicine is the more healing it is*, Herb. 130, 3; Lchdm. i. 242, 5.

hál-wendlíc; *adj. Salutary, healthful* :—Ðæs Hǽlendes tôcyme wæs hálwendlíc ǽgðer ge mannum ge englum *the Saviour's advent was salutary for both men and angels*, Homl. Th. i. 214, 22 : ii. 220, 20 : 564, 7. Him se bisceop hálwendlíce geþeaht forþbrohte *the bishop proposed to them salutary counsel*, Blickl. Homl. 205, 18.

hál-wendlíce; *adv. Salutarily* :—Hálwoendlíce *salubriter*, Rtl. 9, 29. Se ylca Hǽlend ðe nú hálwendlíce clypaþ on his godspelle *the same Saviour that now cries out salutarily in his gospel*, Homl. Th. i. 94, 9.

hál-wendnes, -ness, e; *f. Salubrity* :—Hibernia ge on brǽdo his stealles ge on hálwendnesse ge on smyltnysse lyfta is betere mycle ðonne Breotone land *Hibernia et latitudine sui status et salubritate ac serenitate aerum multum Brittaniæ præstat*, Bd. 1, 1; S. 474, 29.

ham, hom, es; *m. A covering, garment, shirt* :—Ham *camisa*, Wrt. Voc. 288, 48. [*Icel.* hamr *a skin*.] v. hama.

ham, hom, hamm, e; *f. The ham, the inner* or *hind part of the knee* :—Hamm *poples*, hamma *suffragines*, Ælfc. Gl. 75; Som. 71, 84, 83; Wrt. Voc. 44, 66, 65. Ham *poples*, 71, 50. Monegum men gescrincaþ his fét tó his homme . . . gebéde ða hamma *with many a man the feet shrink up to the ham . . . warm the hams*, L. M. 1, 26; Lchdm. ii. 68, 3–5. [*A. R.* mid hommen iuolden *with bent knees* : *Icel.* höm *the ham* or *haunch of a* : *O. H. Ger.* hamma *poples, suffrago*.]

ham, hom; *gen.* hammes; *m. A dwelling, fold*, or *enclosed possession.* 'It is so frequently coupled with words implying the presence of water as to render it probable that, like the *Friesic* hemmen, it denotes a piece of land surrounded with paling, wicker-work, etc., and so defended against the stream, which would otherwise wash it away.' Cod. Dipl. Kmbl. iii. xxvii, where see instances of the occurrence of the word in local names. It occurs as an independent word in the following passages :—Ðonne geúðe ic Ælfwine and Beortulfe ðes hammes be norþan ðære littlan díc, iii. 421, 15. Of ðam beorg tó Cwichemhamme; of ðam hamme, v. 157, 24. Ðonne up on æscméres hammas súþewearde; of ðan hammum, 338, 32. Ða hammas ða ðér mid rihte tôgebyriaþ, 383, 18.

hám, es; *m. Home, house, abode, dwelling, residence, habitation, house with land, estate, property*; *domus, domicilium, prædium, villa, mansio, possessio* :—Se hám is gefylled mid heofonlícum gástum *that abode* [*heaven*] *is filled with heavenly spirits*, Blickl. Homl. 25, 33 : 9, 7. Ðes atola hám *this horrid abode* [*hell*], Cd. 215; Th. 270, 26; Sat. 96. Tó cyninges háme *ad mansionem regiam*, L. R. S. 1; Th. i. 432, 7 : Shrn. 187, 7, 22. Ða gerád Æþelwald ðone hám æt Winburnan . . . and sæt binnan ðæm hám mid ðæm monnum ðe him tó gebugon and hæfde ealle ða geatu forworht *then Ethelwald rode and occupied the residence at Winborne and sat within with those men that had joined him, and he had blockaded all the entrances*, Chr. 901; Erl. 96, 26–30. Mínre yldstan dêhter ðæne hám æt Welewe and ðære gingestan ðone hám æt Welig *to my eldest daughter the vill at Wellow, and to the youngest the vill at Welig*, Th. Chart. 488, 29–33. Gif cyning æt mannes hám drincaþ *if the king drink at a man's house*, L. Eth. 3; Th. i. 4, 1 : L. H. E. 15; Th. i. 32, 17 : L. Alf. pol. 21; Th. i. 76, 1. Hǽlend com tó Lazares hám *Jesus had come to the home of Lazarus*, Blickl. Homl. 69, 21. Ða Noe ongan hám staðelian *then began Noah to establish his home*, Cd. 75; Th. 94, 4; Gen. 556. In hûs fadores mínes hámas meniga sint *in domo patris mei mansiones multæ sunt*, Jn. Skt. Lind. 14, 2 : 23. Nǽron ða welige hámas *there were not then splendid mansions*, Bt. 15; Fox 48, 4. Wæs forðon hæbbend monigra hámas *erat enim habens multas possessiones*, Mt. Kmbl. Lind. 19, 22. Hig cíptun ealle hire hámas *vendebant omnia prædia sua*, Gen. 47, 20. On hira hámon *in possessionibus suis*, 48, 6. Se cyng him wel gegifod hæfde on hámon and on golde and seolfre and forbærndon Tegntún and eác fela óðra gódra háma . . . and ðone hám æt Peonhó . . . and ðone hám æt Wealtham and óðra cotlífa fela *the king had given him many gifts of vills and of gold and silver. And they burned down Teignton and many other good vills too . . . , and the vill at Penhoc . . . , and the vill at Waltham, and many other hamlets*, Chr. 1001; Erl. 136, 16–32. Ðér hé rád betwih his hámum oððe túnum *equitantem inter civitates sive villas*, Bd. 2, 16; S. 520, 10. Abbud of Peortanea ðam hám *Abbas de Monasterio Peartanea*, S. 519, 28. Æt hám *domi*, Mk. Skt. 9, 33 : Lk. Skt. 9, 61. Ðú nére æt hám *you were not at home*, Cod. Dipl. Kmbl. iv. 26, 9. Hám, *acc. is used adverbially after verbs of motion* :—Ða hé hám com *cum venisset domum*, Mt. Kmbl. 9, 28. Hig cyrdon ealle hám *reversi sunt unusquisque in domum suam*, Jn. Skt. 7, 53. Ða se cing lýfde eallon Myrceon hám *the king allowed all the Mercians to go home*, Chr. 1049; Erl. 172, 37 : 1066; Erl. 200, 9. [*Goth.* haims; *f. a village* : *O. Sax.* hêm *a dwelling-place* : *Icel.* heimr *an abode, world, this world* : heim; *adv. home* : *O. H. Ger.* haim *domus, domicilium, patria* : haim; *adv.* Ger. heim.]

-hám, es; *m.* 'The Latin word which appears most nearly to translate it is *vicus*, and it seems to be identical in form with the Greek κώμη. In this sense it is the general assemblage of the dwellings in each particular district, to which the arable land and pasture of the community were appurtenant, the *home* of all the settlers in a separate and well defined locality, the collection of the houses of the freemen. Whenever we can assure ourselves that the vowel is long, we may be certain that the name implies such a village or community,' Cod. Dipl. Kmbl. iii. xxviii–ix. The distinction between -*ham* and -*hám* seems to have been lost before the Norman Conquest, as in the Chronicle one MS. has tô Buccingahamme, another tô Buccingahám, 918; Th. i. 190, col. 1, 2, l. 21. [*Icel.* -heimr, e. g. Álf-heimr *the abode of the elves* : *O. H. Ger.* -heim.]

hama, homa, an; *m. A covering.* [*Prompt. Parv.* hame *thyn skynne of an eye*, or *other like* : *K. Alis.* dragoun's hame (cf. *Icel.* hams *a snake's slough*) : *O. Sax. O. H. Ger.* hamo *in compounds* : and cf. *O. H. Ger.* hemidi *camisa, vestimentum*.] v. ham. DER. byrn-, cild-, feðer-, flǽsc-, gold-, grǽg-, heort-, líc-, wuldor-hama.

háma, an; *m. A cricket*; *cicada*, Wrt. Voc. 281, 48. [*O. H. Ger.* heimo *cicada, grillus* : *Ger.* heime, heimchen *cricket*.] v. Grmm. D. M. 1222.

hamacgaþ [?] :—Se ðe gelíþ raðe hé hamacgaþ *he who takes to his bed will quickly be up again*, Lchm. iii. 184, 21.

hám-bringan; *pp.* -broht *To bring a wife home, marry* :—Ne hí beóþ hámbroht ne geǽwnode *neque nubentur*, Mone Gl. 357. [Cf. *O. H. Ger.* heimbringa, Grff. 3, 201.]

hám-cúþ; *adj. Familiar* :—Ða hámcúþa stówa *familiaria loca*, Mt. Kmbl. p. 11, 1.

hám-cyme, es; *m. A coming home, return* :—Æfter twegra geára ymbryne æfter ðæs wælhreówan hámcyme *after two years had elapsed after the return of the cruel tyrant*, Homl. Th. i. 80, 31. [*Will.* homkome : *Icel.* heim-kváma, -koma *return home*.]

hamele, hamule, an; *f. An oar-loop*, but the word occurs only in a phrase, which may be borrowed from the Scandinavian. *Icel.* hamla *an oar-loop*, is used in the phrase, til hömlu = *per man* [v. Cl. and Vig. Dict.], and apparently with the same meaning we get Chr. 1039; Erl. 167, 15, 21 :—On his dagum man geald xvi scipan æt ælcere hamulan viii marc eall swá man ǽr dyde on Cnutes cynges dagum. Ða hí gerǽdden ðet man geald lxii scipon æt ælcere hamelan viii marc *in his days sixteen ships were paid, eight marks to each of the crew, just as before was done in king Cnut's days . . . Then they decided that sixty-two ships should be paid, to each man eight marks. William of Malmesbury says twenty marks were paid*

to the soldiers of each vessel, ii. 12. Florence of Worcester, Chr. 1040, says eight marks to each rower, and twelve to the steersman, ' octo marcas unicuique suæ classis remigi et xii unicuique gubernatori præcepit dependi.'

hamelian; *p. ode; pp. od To mutilate* :—Sume man hamelode *some were mutilated*, Chr. 1036; Erl. 164, 38. [*Chauc.* a foot is hameled of thi sorwe, Tr. and Cr. 2, 138 : hamling *the operation of cutting the balls out of the feet of dogs*, Hall. Dict. *where see also* hamel : *Icel.* hamla *to mutilate* :—Sumir vôru hamlaðir at höndum eða fôtum *some had their hands or feet cut off* : *O. H. Ger.* bi-hamalon *mutilare*, pe-hamaloter *mutilatus*, Grff. iv. 945.]

hamer, homer, hamor, es ; *m. A hammer* :—Hamor *porticulus*, Ælfc. Gr. 104 ; Som. 78, 13 ; Wrt. Voc. 56, 59. Cf. porticulus *a maylat*, 275, 1. ' Porticulus, malleus in manu portatus quo signum detur remigantibus,' Du Cange. Heoru hamere geþuren *the sword forged by the hammer*, Beo. Th. 2575 ; B. 1285. Carcernes dura hamera geweorc *the doors of the prison, the work of hammers*, Andr. Kmbl. 2155 ; An. 1079. Homra, Exon. 69 a ; Th. 256, 25 ; Jul. 237. Homera lâfe *with the sword*, 102 b ; Th. 388, 14 ; Rä. 6, 7 : Chr. 937 ; Erl. 112, 6. [*O. Sax.* hamur : *Icel.* hamarr : *O. H. Ger.* hamar : *Ger.* hammer.] v. Grnm. D. M. 165. DER. scip-hamor.

hamer-secg, homor-, es ; *m. Hammer-sedge*, L. M. i. 56, 2 ; Lchdm. ii. 126, 19.

hamer-wyrt, hamor-, e ; *f. Black hellebore*, Lchdm. iii. 330, col. 1 : ii. 390, col. 1.

hæmettan; *p. te To provide with a home, to house* :—Denewulf bisceop lýfde Beornulfe his mêge ðæt hê môste ða inberðan menn hâmettan tô Ebblesburnan nû hebbe ic hî hâmet *bishop Denewulf allowed Beornulf his kinsman to house the inborn people at Ebblesburn. I have now housed them*, Th. Chart. 152, 3–7. v. ge-hâmettan.

hâm-færeld, es ; *n. A going home* :—Ða Antigones ðæt ongeat ða forlêt hê ðæt setl ; ac Ymenis him wênde fram Antigones hâmfæreld micelra untreówþa *when Antigonus heard that he abandoned the siege : but Eumenes anticipated for himself great treachery from Antigonus' going home*, Ors. 3, 11 ; Bos. 73, 21. [Cf. *Icel.* heim-ferð, -för *a going home* : *O. H. Ger.* heim-fart.]

hâm-fæst ; *adj. Resident, dwelling at home* :—Hû mæg ðær ðonne ânes rîces monnes nama cuman ðonne ðær mon furðum ðære burge naman ne geheórþ ne ðære þeóde ðe hê on hâmfæst biþ *how can one great man's name come there, when the name of the town even and of the people among whom he dwells is not heard there*, Bt. 18, 2 ; Fox 64, 3 : L. Ed. 1 ; Th. i. 158, 22. Gif mon becume on his gefân and hê hine ær hâmfæstne ne wite *si quis superveniat in hostem suum, et eum antea residentem nesciat*, L. Alf. pol. 42 ; Th. i. 90, 15. [Cf. hâm-sittende.]

hâm-faru, e ; *f. Forcible entry into a man's house ; the same as* hâm-sôcn, q. v. [*Trev.* hamfare :—' Hamsokene oðer Hamfare *a rese imade in house, a fray made in an howse*,' ii. 95 : *Icel.* heim-för *an inroad*.]

hâm-hæn, -henn, e ; *f. A domestic fowl*, L. M. 2, 37 ; Lchdm. ii. 244, 25.

hâm-leás ; *adj. Homeless* :—Sceal hâmleás hweorfan *it must wander homeless*, Exon. 110 a ; Th. 420, 25 ; Rä. 40, 9.

hâm-scir, e ; *f. The office of an œdile* ; ædilitas, officium ædile, Cot. 71, Lye.

ham-scyld [?], L. Eth. 32 ; Th. i. 12, 1, where see note. Leo in his work on Anglo-Saxon Names quotes a passage from Richthofen in which *skeld* occurs in the sense of *fence* ; so that the crime referred to in the passage would be the breaking through the fence which surrounded the *ham*. v. the translation of Leo, p. 40, note 2.

hâm-sittende ; *part. Sitting, dwelling at home, resident* :—Wê beódaþ se mon se ðe his gefân hâmsittendne wite ðæt hê ne feohte ærðam ðe hê him ryhtes bidde *we command that the man who knows his foe to be dwelling at his home fight not before he demand justice*, L. Alf. pol. 42 ; Th. i. 90, 2 : Cd. 209 ; Th. 259, 6 ; Dan. 687 : Andr. Kmbl. 1372 ; An. 686 : Cd. 86 ; Th. 108, 33 ; Gen. 1815. [*O. Sax.* hêm-sittiandi.]

hâm-sôcn, e ; *f. Attack on a man's house ; also the fine paid for such a breach of the peace*. The following passage will illustrate the character of the offence :—' Hamsocna, quod domus invasionem Latine sonat, fit pluribus modis, extrinsecus vel intrinsecus accidenciis. Hamsocna est, si quis alium in sua vel alterius domo cum haraido assaliaverit vel persequatur, ut portam vel domum sagittet vel lapidet vel colpum ostensibilem undecunque faciat. Hamsocna est, vel hamfare, si quis premeditate ad domum eat, ubi hostem suum esse scit, et ibi eum invadat, si die vel nocte hoc faciat ; et qui aliquem in molinum vel ovile fugientem prosequitur, hamsocna judicatur. Si in curia vel domo, sedicione orta, bellum eciam subsequatur, et quivis alium fugientem in alium domum infuget, si ibi duo tecta sint, hamsocna reputetur,' L. H. 80, 10, 11 ; Th. i. 587, 14–25. Other passages in the earlier laws and charters are :—Wê cwêdon be hâmsôcnum seðe hit ofer ðis dô ðæt hê þolige ealles ðæs ðe âge and sí on cyninges dôme hwæðer hê lif âge *we have ordained respecting ' ham-socns' that he who shall commit it after this forfeit all that he owns, and that it be in the king's judgment whether he have his*

life, L. Edm. S. 6 ; Th. i. 250, 9 : L. Eth. 4, 4 ; Th. i. 301, 18. Ðis syndon ða gerihta ðe se cyning âh ofer ealle men on Wesseaxan ðæt is hâmsôcne *these are the rights which the king has over all men in Wessex that is [the fines for]* ' ham-socn,' L. C. S. 12 ; Th. i. 382, 13, see the note : 15 ; Th. i. 384, 6 : Th. Chart. 333, 32 : 359, 4 : 369, 14. Gif hwâ hâmsôcne gewyrce gebête ðæt mid fîf pundan ðam cyninge *if any one commit ' ham-socn,' let him pay a fine of five pounds to the king*, 63 ; Th. i. 408, 27. [*Scot.* hame-sucken *the crime of beating* or *assaulting a person within his own house* : *Icel.* heim-sôkn *an inroad* or *attack on one's home* : *O. Frs.* ham-, hem-sekenge *attack on one's house*.] v. sêcan, in its sense of *to seek with a hostile intent*.

hâm-steall, es ; *m. A homestead, residence* :—On his hâmstealle *at his homestead*, Cod. Dipl. Kmbl. iii. 255, 9. Ðane hâmstal ðet hê on set *the homestead at which he resides*, iv. 133, 8. [Homestall *a homestead*, Hall. Dict : *a mansion, seat in the country*, Bailey.]

hâm-stede, es ; *m. A homestead* :—Tô hâmstede *to the homestead*, Cod. Dipl. Kmbl. iii. 77, 7. v. p. xxxviii s. v. stede *for compounds in which the word occurs*. [*O. Frs.* heem-steed *domicile* : cf. *Icel.* heim-stöð *a homestead*.]

Hâm-tûn [*or* Ham-tûn ?] *Hampton, a common local name, used for both the present Northampton*, Chr. 917 ; Erl. 102, 12 ; *and Southampton*, Chr. 981 ; Erl. 129, 36 : *for other towns see the index to Cod. Dipl. Kmbl. vol. vi.*

Hâmtûn-scir, e ; *f. Hampshire*, Chr. 1001 ; Erl. 136, 5.

hamule. v. hamele.

hâm-weard ; *adv. Homeward, in the direction of home* ; domum versus, retro :—Ða heó hâmwerd wæs *when it was on its way home*, H. R. 103, 24. Ða hý hâmweard wêron *when they were on the way home*, Ors. 4, 6 ; Bos. 85, 38, Ægeas wearþ gelæht fram atelîcum deófle hâmwerd be wege ærðan hê tô hûse côme *Ægeas was seized by a horrible devil on the way home, before he came to his house*, Homl. Th. i. 598, 23. Æþelwulf ða him hâmweard fôr *Ethelwulf then journeyed homeward*, Chr. 855 ; Erl. 68, 29 : 885 ; Erl. 82, 30. Se esne hig hâmweard lædde tô hlâforde *the servant brought her home to his lord*, Gen. 24, 61.

hâm-weardes ; *adv. Homewards* :—Sió ôðeru fierd wæs hâmweardes *the other force was returning home*, Chr. 894 ; Erl. 91, 1. [*O. H. Ger.* heimwartes *domum versus* : *Ger.* heimwärts.]

hâm-weorþung ; *f. Honour or ornament to the house or home* :—Eofore forgeaf ângan dôhtor hâmweorþunge *he gave Eofor his only daughter, an ornament of his home*, Beo. Th. 5988 ; B. 2998.

hâm-weorud, es ; *n. The body of people connected with a ' ham ;'* vicani :—Ða com hê tô sumum hûse on æfentîd and eode on ðæt hûs ðær ðæt hâmweorud eall tô symble gesomnod wæs *pervenit ad vicum quendam vespere intravitque in domum in qua vicani cænantes epulabantur*, Bd. 3, 10 ; S. 534, 26.

hâm-wyrt, e ; *f. Home-wort* ; sempervivum tectorum, L. M. 3, 41 ; Lchdm. ii. 336, 4 : 1, 1 ; Lchdm. ii. 18, 19 : 1, 40 ; Lchdm. ii. 104, 14.

hana, an ; *m. A cock* :—Se hana creów *gallus cantavit*, Mk. Skt. 14, 68, 30, 72. [*Goth.* hana : *O. Sax.* hano : *Icel.* hani : *O. H. Ger.* hano : *Ger.* hano.]

han-crêd, -cræd, hon-, es ; *m. Cock-crowing, cock-crow, a division of the night* :—Hancrêd *conticinium* vel *gallicinium*, Ælfc. Gl. 94 ; Som. 75, 122 ; Wrt. Voc. 53, 4. Seó niht hæfþ seofan dælas . . . fîfta is gallicinium ðæt is hancrêd *the night has seven divisions . . . the fifth is gallicinium, that is, cock-crow*, Lchdm. iii. 244, 4. Hêr wæs se môna âþîstrod betwux hancrêd and dagunge *in this year the moon was eclipsed between cock-crow and dawn*, Chr. 795 ; Erl. 59, 25. On æfen ðe on midre nihte ðe on hancrêde ðe on morgen *sero, an media nocte, an galli cantu an mane*, Mk. Skt. 13, 35 ; Bd. 4, 23 ; S. 595, 27 : Homl. Th. i. 74, 21. Honcrêd, Exon. 99 a ; Th. 370, 32 ; Seel. 68. Ðone drenc on þreó þicge æt ðam þrim honcrêdum *let him take the drink at three times at the three cock-crowings*, L. M. 2, 65, 2 ; Lchdm. ii. 294, 5. Se cyning embe forman hancrêd ût gangende wæs *the king about the first cock-crowing was going out*, Lchdm. iii. 424, 34. Ða com se Hælend embe ðone feórþan hancrêd *quarta autem vigilia noctis venit Iesus*, Mt. Kmbl. 14, 25. [*O. Sax.* hano-krâd : *O. H. Ger.* hana-crât *gallicinium, galli cantus*.]

hand, hond, a ; *f.* HAND, *side, power, control* [cf. *mund*] ; used also of *the person from whom an action proceeds* :—Hand *manus*, Wrt. Voc. 64, 73. Middeweard hand *vola* vel *tenar* vel *ir*, Ælfc. Gl. 72 ; Som. 70, 130 ; Wrt. Voc. 43, 54. Dîn seó swýðre hand *dextera tua*, Ps. Th. 59, 5. Ðær unc hwîle wæs hand gemæne *there for a time we two had a hand to hand struggle* [cf. *Ger.* handgemein werden *to fight hand to hand*], Beo. Th. 4281 ; B. 2137. Sette Ephraim on his swiðran hand ðæt wæs on Israhêles wynstran hand and Manasses on his winstran hand ðæt wæs on Israhêles swiðran healfe *he placed Ephraim on his right hand, that was on Israel's left hand, and Manasseh on his left hand, that was on Israel's right hand*, Gen. 48, 13. Seó hælo his ðære swýðran handa *salus dextera ejus*, Ps. Th. 19, 6. Gif hê heáhre handa dyntes onfêhþ *if he receives a right* [?] *hand blow* [cf. *Icel.* hægri hönd *the right hand*, and see note on the passage for other translations. The analogy with the Icelandic, it may be observed, is not perfect, since the English does not (as in the

case of swíðre) use the comparative; so that the phrase may perhaps refer to the hand being raised for defence; or the reference may be to the upraised hand of the striker], L. Eth. 58; Th. i. 18, 1. God álýsde hí láðum of handa *quos redemit de manu inimici*, 106, 2. Tó onfónne æt bisceopes handa *to receive at the hand of the bishop*, L. C. E. 22; Th. i. 374, 3: Chr. 942; Erl. 116, 22. Æt Seaxena handa forwurðan *to perish at the hand of the Saxons*, 605; Erl. 21, 29. Mid brádre hand slógan *smote with open hand*, Blickl. Homl. 23, 32; Past. 41, 4; Swt. 303, 11. Mid ðære ylcan hand *with the same hand*, Lchdm. iii. 68, 15. Ða witan ðe ðá néh handa wǽron *the 'witan' that were near at hand*, Chr. 1100; Erl. 236, 19. Hand on handa *hand in hand*, Ap. Th. 19, 18. Ðeóf ðe æt hæbbendre handa gefangen sý *a thief who is taken with the stolen property upon him*, L. Ath. 1; Th. i. 198, 17. Siððan ic hond and rond hebban mihte *since I could lift hand and shield*, Beo. Th. 1316: B. 656: Andr. Kmbl. 18; An. 9. Ðǽr wæs micel wæl geslægen on gehwæðre hond *there was great slaughter made on either side*, Chr. 871; Erl. 74, 12: Byrht. Th. 135, 2; By. 112. On ǽgðera hand *on either hand*, L. Ath. 1, 23; Th. i. 212, 6. Wið ǽlce hand *on all sides, towards every one*, L. Ed. 10; Th. i. 164, 18. Ic wille ðæt hit gange on ða nýhstan hand mé *I will that it go to the next of kin to me*, Th. Chart. 491, 13: 481, 22. Ða witan gerehton ðæt heó sceolde hire fæder hand geclǽnsian be swá miclan feó *the 'witan' decided that she should clear her father in respect to so much money; sapientes decreverunt quod ego patrem meum purgare deberem, videlicet sacramento xxx librarum, easdem triginta libras patrem meum persolvisse*, 202, 1. Bútan contlandes béc and hé ða bóc unnendre handa hire tó lét *excepto libro de osterlande quem bona voluntate dimisit*, 37. Sí mé wuldres hyht hand ofer heáfod *may there be to me a hope of glory, hand over head*, i. e. *without difficulty* [hand-over-head *thoughtlessly extravagant; careless; at random; plenty*, Hall. Dict.], Lchdm. i. 390, 3, 5. Gif mon forstolenne man befó æt óðrum and síe sió hond óðcwolen sió hine sealde ðam men ðe hine mon ætbeféng *if a stolen man be attached in another's possession, and the hand [person] be dead that sold him to the man in whose possession he is attached*, L. In. 53; Th. i. 134, 17: 136, 2: 75; Th. i. 150, 5: L. Eth. 2, 8; Th. i. 288, 18, 20. His feoh onfón fremde handa *diripiant alieni omnes labores ejus*, Ps. Th. 108, 11. Handa díne *manus tuæ*, 118, 73. Se ðe ofer ðis fals wyrce þolige ðæra handa ðe hé ðæt fals mid worhte *he that after this makes counterfeit money, let him lose the hands with which he made the counterfeit*, L. C. S. 8; Th. i. 380, 17. Domiciana wearþ ácweald æt his witena handum *Domitian was killed by his senators*, Homl. Th. i. 60, 4. Gebindan handum and fótum *to bind hand and foot*, 570, 10. Be heora handum gebundne *bound by the hands*, Blickl. Homl. 209, 36. Sý ðeós gesetnys ðus hér geendod god helpe mínum handum *so let this composition here end, God help my hands*, Lchdm. iii. 280, 16. Ealle forgielden ðone wer gemǽnum hondum *let them all pay the wergild in common*, L. Alf. pol. 31; Th. i. 80, 17: L. E. G. 13; Th. i. 174, 21. Ða genam Sanctus Martinus hine be his handa *then St. Martin took him by the hand*, Blickl. Homl. 219, 19. Hit hyre on hand ágeaf *gave it into her hand*, Judth. 11; Thw. 23, 20; Jud. 130. Ðýlæs ðe eów on hand becume seó leáse gesetnys *lest the false canon come into your hands*, Homl. Th. i. 436, 30. Him ealle on hand eodan ða hǽðnan leóde *then all the heathen people submitted to them*, Blickl. Homl. 203, 23: Chr. 882; Erl. 82, 13. Gif hig on hand gáþ *if they submit*, Deut. 20, 11. Ealle ða burgware ne mehton hiene ǽnne geníéddan ðæt hé him an hand gán wolde *all the citizens could not force him, though a single man, to yield*, Ors. 3, 9; 134, 18. Ðæs wíte on eówre handa geeode *on that account punishment came upon you*, Ps. 57, 2. Hé ealle gesceafta on his handa hafaþ *he hath all creatures in his hand*, Blickl. Homl. 121, 15. Se ðe hie on handa hæfþ *who has it in his possession*, L. Eth. 2, 9; Th. i. 290, 17. Se hæfde his abbotríce s' Iohs of Angeli on hande *he held his abbacy of St. John of Angeli*, Chr. 1127; Erl. 255, 27, 34: 256, 2. Ealle hé hí oððe wið feó gesealde oððe on his ágenre hand heóld *all of them he either sold for money or kept in his own hands*, 1100; Erl. 236, 6, 9. Mann sette Ælfgár ðane eorldóm on handa ðe Harold ǽr áhte *the earldom that Harold had before was put into Alfgar's hands*, 1048; Erl. 180, 29. Se ðe ic hit nú on hand sette *he into whose hand I now put it*, L. O. 3; Th. i. 180, 3. Se ðe unriht gestreón on his handa stóde *he in whose hands was the unjust gain*, L. Eth. 2, 9; Th. i. 290, 5: Th. Chart. 369, 7. Þridde gewrit á mid ðam ðe ðæt land on hande stande *the third copy always with him in whose possession the land is*, Cod. Dipl. Kmbl. iv. 235, 31. Gyf neód on handa stande *if there be present need*, L. Edg. H. 2; Th. i. 258, 6. Biþ mannes sunu geseald on synfulra hand *the Son of man shall be given into the hands of sinful men*, Blickl. Homl. 73, 1. On hand syllan *to give a pledge* or *promise*:—Hé sealde him on hand mid Cristes béc ðæt hé wolde ðisne þeódscype swá wel haldan swá ǽnig kynge ætforan him betst dyde *he promised him on the Gospels that he would rule this people as well as the king who before him had ruled best*, Chr. 1066; Erl. 202, 31: 1064; Erl. 196, 1. Slaga sceal his forspecan on hand syllan *the slayer shall give pledge to his advocate*, L. Edm. S. 7; Th. i. 250, 14: L. Eth. 2, 8; Th. i. 288, 16. Gif hwá his hand on hand sylle *if any one deliver himself up*, L. Ed. 9; Th. i.

164, 10. Cyricean hyrde tó cristes handa *shepherd of the Church for Christ*, Blickl. Homl. 171, 7. Ðet land eall ábégdon Willelme tó handa *brought all the land in subjection to William*, Chr. 1073; Erl. 212, 1. Him becómon swá micele welan tó handa *so great wealth came into his hands*, Homl. Th. ii. 576, 30. Ic beóde ðé ðat ðú beríde ðás land ðam hælge tó hande *I enjoin thee that thou perequitate these lands into the possession of the saint*, Th. Chart. 369, 22. Swá Ælfríc hig mínre móder tó handa bewiste *as Alfric administered it on behalf of my mother* [cf. *Icel.* einum til handa], Cod. Dipl. Kmbl. iv. 222, 19: 226, 4. Sý ðæt forworht ðam cyningce tó handa *let it be forfeited to the king*, L. C. S. 13; Th. i. 382, 20. Tó Godes handa gefrætwod *equipped for God*, Homl. Th. í. 210, 32. Se cing lét gerídan ealle ða land ðe his módor áhte him tó handa *the king had all the lands that his mother owned brought into his own power*, Chr. 1043; Erl. 8: Cod. Dipl. Kmbl. iv. 222, 6. Drihten gewylt eów ealle þeóda tó handa *the Lord will reduce all nations to subjection to you*, Deut. 31, 3. Hí cwǽdon ðæt hí him ðet tó handa healdan scoldan *they said that they would hold it for him*, Chr. 887; Erl. 87, 3: 1036; Erl. 165, 6: L. I. P. 19; Th. ii. 326, 6. Drihten lét hí tó handa ðam hǽðenan leódscipe Madian *the Lord delivered them into the hand of Midian*, Jud. 6, 1: Chr. 1048; Erl. 180, 9. Hér leót Ceolréd Wulfréde tó hande ðet land of Sempigaham *in this year Ceolred let the land of Sempringham to Wulfred*, 852; Erl. 67, 33: 1091; Erl. 227, 7, 24: Anal. Th. 126, 14. Gif þeówwealh Engliscne monnan ofslihþ ðonne sceal se ðe hine áh weorpan hine tó honda hláforde *if a British slave kill an Englishman, then shall he who owns him give him up to the lord*, L. In. 74; Th. i. 148, 15: 56; Th. i. 138, 12: L. Alf. pol. 21; Th. i. 76, 1: 24; Th. i. 78, 10. Gá bisceope under hand *arbitrio episcopi se dedat*, L. Ecg. P. 4, 52, note; Th. ii. 218, 33. Hí wǽron geseald under sweordes hand *tradentur in manus gladii*, Ps. Th. 62, 8. Alle þinge ðe hí under honde habben *all things that they have in their possession*, Th. Chart. 582, 1, 19: Cod. Dipl. Kmbl. iv. 268, 32. [Cf. *Icel.* undir höndum einum *in one's power*.] Gelæddon under hand hæleþ hǽðenum déman *led the men in subjection to a heathen ruler*, Cd. 175; Th. 220, 14; Dan. 71. [*Goth.* handus: *O. Sax.* hand: *O. Frs.* hand, hond: *Icel.* hönd: *O. H. Ger.* hant: *Ger.* hand.] DER. mæg-, wǽpned-, wíf-hand.

hand [= and (?)] *also* :—Ymbe midne dæg and nóntíde eode se híredes ealdor út and dyde hand swá gelíce *exiit circa sextam, et nonam horam: et fecit similiter*, Anal. Th. 74, 4. Hí férdon swá tó Sandwíc and dydon hand ðæt sylfa *they went to Sandwich and did just the same*, Chr. 1052; Erl. 184, 5.

hand-bæftian, -beaftan, -beoftan; *p.* -bæftade, -beafte *To beat with the hands as an expression of grief*[?], *to lament* :—Ða ðe gemǽndon and hondbæftadon *quæ plangebant et lamantabantur*, Lk. Skt. Lind. 23, 27. Wé hondbeafton *lamentavimus*, 7, 32. [Cf. apon þair brestes fast þai beft, Met. Homl. xviii.] v. beaftan.

hand-bana, -bona, an; *m.* A *murderer, homicide, one who slays with his own hand* [αὐτόχειρ], Beo. Th. 925; B. 460: 2665; B. 1330: 4997; B. 2502. [*O. Sax.* hand-bano: *Icel.* hand-bani *the actual slayer*.]

hand-bell, e; *f.* A *hand-bell* :—Ðǽr nǽron ǽr búton vii upphangene bella and nú sind xiii upphangene and xii handbella *before there were but seven hung-up bells, and now there are thirteen hung-up bells and twelve hand-bells*, Th. Chart. 430, 6.

hand-bóc, e; *f.* A *hand-book, manual* :—Hand-bóc *manualis*, Wrt. Voc. 81, 46. Ða hálgan béc saltere and pistolbóc . . . sangbóc and handbóc *the holy books: psalter and epistle-book . . . book of canticles and manual*, L. Ælf. C. 21; Th. ii. 350, 14. [*Ger.* hand-buch.]

hand-bona. v. hand-bana.

hand-bréd, -bréd, e; *f.* A *hand's breadth* :—Handbréd *vel* span *palmus*, Ælfc. Gl. 72; Som. 70, 127; Wrt. Voc. 43, 52. [*Chauc.* an hande brede, hondbrede: *Prompt. Parv.* hañde brede *palmus*: cf. *Ger.* hand-breit, *adj.*]

hand-bred, es; *n.* The *palm of the hand*; palma :—Ðis handbred *hoc ir*, Ælfc. Gr. 8; Som. 7, 26. Handbred *palma*, Ælfc. Gl. 72; Som. 71, 2; Wrt. Voc. 43, 56. Hondbreoðo *palmas*, Mt. Kmbl. Lind. 26, 67. Sleánde mid handbredum *striking with the palms of their hands*, Homl. Th. ii. 248, 13. [*O. Frs.* hond-brede palma.]

hand-cláþ, es; *n.* A *hand-cloth, towel* :—Ic geseó Godes engel standende ætforan ðé mid handcláþe, and wípaþ ðíne swátigan limu *I see God's angel standing before thee with a handcloth, and he wipes thy sweaty limbs*, Homl. Th. i. 426, 30. [*Rel. Ant.* hand-cloð: *Icel.* hand-klæði *a hand-towel*.]

hand-cops, es; *m.* A *handcuff, manacle* :— Handcops *manice*, Wrt. Voc. 86, 33. Tó gewrídenne cyningas heora on fótcopsum and æðele heora on handcopsum ísynum *ad alligandos reges eorum in compedibus et nobiles eorum in manicis ferreis*, Ps. Spl. C. 149, 8.

hand-cræft, es; *m. Skill or power of the hand, handicraft* :—Ðes lama wǽdla búton handcræfte Godes beboda gefylde *this paralytic pauper without the use of his hands fulfilled God's commands*, Homl. Th. ii. 98, 17. Mid his handcræfte *with his manual skill* [*in tent-making*], i. 392, 16. Wé lǽraþ ðæt preósta gehwilc tó-eácan láre leornige handcræft

georne *we enjoin that every priest besides book-learning diligently learn a handicraft*, L. Edg. C. 11; Th. ii. 246, 17. Eác him gerísaþ hand-cræftas góde ꝺæt man on his híréde cræftas begange *good handicrafts are also befitting him, that crafts may be practised in his household*, L. I. P. 8; Th. ii. 314, 23. [*O. Sax.* hand-kraft *strength, power of hand.*]

hand-cræftig; *adj. Mechanicus*, Lye.

hand-cweorn, -cwyrn, e; *f. A hand-mill :*—Héton hine grindan æt hira handcwyrne *ordered him to grind at their mill*, Jud. 16, 21. [*Icel.* hand-kvern.]

hand-dǽd, e; *f. Handiwork*, Lye. [*O. H. Ger.* hant-tât *opus manuum.*]

hand-dǽda, an; *m. One who does a deed with his own hand :*—Ꝺonne wille ic ꝺæt eall seó mǽgþ sý unfáh bútan ꝺam handdǽdan *then I will that all the kindred be free from the feud except the actual doer of the deed*, L. Edm. S. 1; Th. i. 248, 6, 12 : L. Eth. 2, 5; Th. i. 286, 22. [Cf. hand-bana.]

hand-fæstan; *p.* -fæste *To pledge by giving the hand*, Lye. [*Orm.* hannd-fest *betrothed* : *Scot.* hand-fast *to betroth by joining hands* : *Icel.* hand-festa *to strike a bargain by shaking hands, to pledge, betroth.*] v. next word.

hand-fæstung, -fæstnung, e; *f. A giving of the hand by way of pledge or assurance :*—Handfestnung *mandatum*, Ælfc. Gl. 13; Som. 57, 110; Wrt. Voc. 20, 48. [*Scot.* hand-fasting, -fastnyng *marriage with the encumbrance of some canonical impediment, not yet bought off : Icel.* hand-festa, -festning, -festr *striking a bargain, the joining hands.* 'In the early Dan. and Swed. laws the stipulation to be given by the king at his coronation was called haand-fæstning.' Cf. *O. H. Ger.* hant-feste *emunitas, cautio, testamentum, privilegium.*]

hand-full, e; *f. A handful*; *manipulus :*—Nimaþ handfulle axan of ꝺam ofene *tollite plenas manus cineris de camino*, Ex. 9, 8. Nime áne handfulle *tollet pugillum plenum*, Lev. 2, 2. Nim micle handfulle secges *take a great handful of sedge*, L. M. 3, 67; Lchdm. ii. 354, 26. Genim micle twá handfulla *take two great handfuls*, 69; Lchdm. ii. 356, 12 : Herb. 81, 5; Lchdm. i. 184, 19. Berende handfulla hꝺra *portantes manipulos suos*, Ps. Lamb. 125, 6. [*Orm.* hanndfull : *A. R.* honful : *Icel.* hand-fyllr : *Ger.* hand-voll.]

hand-gang, -gong, es; *m. Laying on of hands :*—Handgang *manus impositio*, Ælfc. Gl. 112; Som. 79, 94; Wrt. Voc. 60, 3. [*Orm.* hannd-gang *laying on of hands* (used in connection with the Apostles, and with bishops at confirmation).]

hand-gecliht. v. ge-cliht.

hand-gemǽne. v. hand.

hand-gemót, es; *n. A hand-meeting, battle*, Beo. Th. 3056; B. 1526.

hand-gesceaft, e; *f. That which is formed by the hand, a creature*, Cd. 23; Th. 29, 24; Gen. 455.

hand-gesella, an; *m. A companion who is close to one's side, comrade*, Beo. Th. 2966; B. 1481.

hand-gestealla, an; *m. One whose place is close at one's hand, a comrade, an associate*, Beo. Th. 5186; B. 2596. [Cf. preceding word, *and* eaxl-gestealla.]

hand-geswing, es; *n. Stroke given by the hand :*—Ꝺær wæs heard handgeswing *there were hard blows dealt by the hand*, Elen. Kmbl. 229; El. 115.

hand-geweald, es; *n. Power :*—Hé hí on handgeweald hǽꝺenum sealde *tradidit eos in manus gentium*, Ps. Th. 105, 30.

hand-geweorc, es; *n. Work of the hand, handiwork :*—Ꝺæra hǽꝺenra anlícnyssa sind gyldene and sylfrene manna handgeweorc *the idols of the heathen are of gold and of silver, the work of men's hands*, Homl. Th. i. 366, 26 : Deut. 4, 28. His handgeweorc *the work of his hands* [*Adam and Eve*], Cd. 13; Th. 16, 11; Gen. 241 : Ps. Th. 18, 1. Gerece úre handgeweorc *opus manuum nostrarum dirige*, 89, 19. On his handgeweorc byþ gefangen se synfulla *in operibus manuum suarum comprehensus est peccator*, 9, 15. [*Hom: Rel. Ant.* hond-iwerc : *O. Sax.* hand-giwerk.]

hand-gewinn, es; *n. Labour of the hands, struggle, strife, fighting :*—Ꝺa munucas lifdon on hira ágenum handgewinne *the monks lived by the labour of their own hands*, Shrn. 37, 2. Be heora ágenum handgewinne lifigeaþ *proprio labore manuum vivant*, Bd. 4, 4; S. 571, 22 : 4, 28; S. 606, note 2. Hefig hondgewinn *a heavy struggle*, Exon. 73 b; Th. 273, 34; Jul. 526. Hé sceal fore hǽꝺenra handgewinne gást onsendan *he shall because of the heathens' warfare give up the ghost*, Andr. Kmbl. 372; An. 186.

hand-gewrit, es; *n. What is written by the hand, a deed, contract*; *chirographum :*—Handgewrit *cirographum*, Ælfc. Gl. 13; Som. 57, 113; Wrt. Voc. 20, 51. Hondgiwrit *chyrographum*, Rtl. 32, 39. Sum man wrát his handgewrit ꝺam áwyrgedan deófle *a certain man put his hand to a contract with the accursed devil*, Homl.

hand-gewriðen; *p. Hand-twisted*, Beo. Th. 3878; B. 1937.

hand-gift, e; *f. A wedding-gift*, Hy. 10, 18; Hy. Grn. ii. 293, 18. [Cf. *O. Sax.* hand-geba.] v. gift.

hand-gripe, es; *m. Grasp*, Beo. Th. 1934; B. 965. [*O. H. Ger.* hant-grif *pugillus* : *Ger.* hand-griff.]

hand-griþ, es; *n. Peace, protection, security*, L. E. G. 1; Th. i. 166, 21 : L. Eth. vi. 14; Th. i. 318, 24 : vii. 2; Th. i. 330, 5; L. C. E. 2; Th. i. 358, 20. v. Stubbs' Const. Hist. i. 182.

hand-hæbbende; *part. Having* [*stolen property*] *in one's hand* [cf. under *hand* the phrase *æt hæbbendre handa*] :—Sit handhabenda, sit non handhabenda *whether the thief be taken with the stolen property upon him or not*, L. Eth. iii. 6; Th. i. 218, 32.

hand-hamer, es; *m. A hand-hammer*; *malleus*, Cot. 135.

hand-hefe, es; *m. A burden :*—Ne gehrínaþ ꝺǽm hondhæfum *non tangitis sarcinas*, Lk. Skt. Lind. 11, 46.

hand-hrægl, es; *n. A cloth for the hands, towel, napkin*; *mantile*, Ælfc. Gl. 30; Som. 61, 70; Wrt. Voc. 26, 67.

hand-hrine, es; *m. A touch with the hand :*—Þurh handhrine Háliges Gástes *through a touch with the hand of the Holy Ghost*, Andr. Kmbl. 1999 : An. 1002.

hand-hwíl, e; *f. A moment :*—Nis ná eów tó gewitenne ꝺa tíd oꝺꝺe ꝺa handhwíle ꝺe mín Fæder gesette þurh his mihte *it is not for you to know the hour or the moment that my Father hath appointed through his might*, Homl. Th. i. 294, 26. [*Orm.* inn an hanndhwile *in a moment of time* : *A. R.* hondhwule : *Piers P.* handwhile.]

hand-hwyrft, es; *m. A turning of the hand, the time occupied by such a turning, a moment*, Lye. [Cf. hand-hwíl.]

handle, es; *n. A handle.* Cf. sulh-handla *stiba*, Ælfc. Gl. 1; Som. 55, 8; Wrt. Voc. 15, 8. [*Prompt. Parv.* handyl *manutentum* : *Jul.* hondlen; *dat. pl.*]

hand-leán, es; *n. A reward, recompence given by the hand, retribution :*—Uton wé geþencean hwylc handleán wé him forþ tó berenne habban *let us consider what recompence we have to offer him*, Blickl. Homl. 91, 13 : Cd. 143; Th. 178, 29; Exod. 12 : Beo. Th. 3087; B. 1541 : 4195; B. 2094. [*O. H. Ger.* hant-lôn *bravium.*]

handlian; *p.* ode; *pp.* od *To handle, feel :*—Gif mín fæder mé handlaþ si *attrectaverit me pater meus*, Gen. 27, 12. Hý ꝺa spǽce swá lange handleedon *they handled the suit so long*, Th. Chart. 302, 31. Hálige béc handligan *sacros libros manu tractare*, L. Ecg. P. iii. 4; Th. ii. 196, 28 : 12; Th. ii. 200, 7 : Lchdm. iii. 198, 23 : 204, 2; 208, 24. [*Laym.* hondlien : *Orm.* hanndlenn : *Icel.* höndla : *O. H. Ger.* hantalôn *tractare* : *Ger.* handeln.]

hand-lín, es; *n. A hand-cloth, napkin :*—Hand-lín *manualis*, Ælfc. Gl. 27; Som. 60, 117; Wrt. Voc. 25, 57. iiii subdiácones handlín *four sub-deacon's handcloths*, Th. Chart. 429, 23. [Cf. *Icel.* hand-lín *sleeves.*]

handlinga; *adv. With the hands :*—Nis be him geræd ꝺæt hé handling ǽnigne man ácwealde *it is not read of him that he killed any man with his own hands*, Homl. Th. i. 386, 1.

hand-locen; *pp. Fastened, woven by the hand*, Beo. Th. 649; B. 322.

handlung, e; *f. Touching, handling :*—Ꝺone ꝺe se eádiga Benedictus ná handlunge ac on beseónde fram his bendum álýsde *whom the blessed Benedict not by touching him, but by looking on him, had released from his bonds*, Homl. Th. ii. 182, 4.

hand-mægen, es; *n. Might, power of hand*, Cd. 14; Th. 16, 22; Gen. 247 : Andr. Kmbl. 1450 : An. 725. [*O. Sax.* hand-magan, -megin : *Icel.* hand-megin, -megn.]

hand-mitta, an; *m. The sixth part of an ounce*; *exagium*, Lye.

hand-nægl, es; *m. A finger-nail :*—Ꝺonne beóþ him ꝺa handnæglas wonne *then will his finger-nails be livid*, L. M. 3, 63; Lchdm. ii. 350, 22.

hand-plega, an; *m. Fighting :*—Heard handplega *hard fighting*, Cd. 160; Th. 198, 23; Exod. 327 : 95; Th. 124, 3; Gen. 2057 : Chr. 937; Erl. 112, 25. Hí nǽfre wyrsan handplegan on Angelcynne ne gemitton ꝺonne Ulfcytel him tóbrohte *they had never had more disastrous fighting in England than in their engagement with Ulfcytel*, Chr. 1004; Erl. 138, note 7. v. plega *for similar compounds.*

hand-preóst, es; *m. A chaplain*; *sacellanus*, Ælfc. Gl. 68; Som. 70, 13; Wrt. Voc. 42, 22. Stigand ꝺe was ꝺes cinges rǽdgifa and his hand-preóst *Stigand who was the king's counsellor and chaplain*, Chr. 1051; Erl. 182, 20.

hand-rǽs, es; *m. Onset, attack*, Beo. Th. 4150; B. 2072.

hand-róf; *adj. Distinguished for exploits accomplished by the hands* [*used of warriors*], Cd. 155; Th. 193, 15; Exod. 247.

hand-sceaft, e; *f. That which is formed by the hand, a creature*; *creatura*, Lye.

hand-sceát, es; *m. A napkin*; *manutergium, sudarium*, Lye.

hand-sció; *m. A glove*, Beo. Th. 4158; B. 2075. Grein considers this meaning to be inadmissible and translates *impetus manibus factus* ; but cf. 4177; B. 2085.

hand-scólu, -scálu, e; *f. A retinue :*—Mid his hondscóle *with his retinue*, Beo. Th. 3931; B. 1963. Handscále, 2638; B. 1317. [Cf. hand-gesella, geneát-scólu.]

hand-scyldig; *adj. Liable to the penalty of losing the hand :*—Se ꝺe gewundaþ man binnan ciricwagum se biþ handscyldig *he that wounds*

a man within church walls shall be liable to lose his hand, L. Eth. vii. 13; Th. i. 332, 9.

hand-seax, es; *n. A short sword, dagger :* — Hæfde hē twigecgede handseax *habebat sicam bicipitem*, Bd. 2, 9; S. 511, 15. Hæfdon handseax on heora handa *habentes in manibus vomeres*, 5, 13; S. 633, 16. Godes engel stôd mid handsexe *God's angel stood with a dagger*, Homl. Th. ii. 272, 17. Án handsecs on hundeahtotigan mancysan goldes *a dagger worth eighty mancuses of gold*, Th. Chart. 501, 3: 502, 16. Handsex, 527, 8. [*Laym.* hond-sæx: *Icel.* hand-sax *a short sword, dirk.*]

hand-selen, e; *f. A giving into the hand of another ;* mancipatio, Cot. 136, Lye. [Cf. *Icel.* hand-sal, -sala, -selja.]

hand-seten, e; *f. The setting of one's hand to a deed, etc., a signature, sign manual :* — Ðas trymeþ se forespecena kyng mid Cristes rôde tácne and his weotena hondsetena his geofa *thus the aforesaid king confirms his gifts with the sign of Christ's cross and the signature of his witan*, Cod. Dipl. Kmbl. ii. 304, 11: 14: 89, 11. Mê saldan heora hondsetene ðisse geræðnesse *they put their hands to this agreement*, 100, 29. Hēr is seó hondseten. Ego Óswald archiepiscopus, etc. *here are the signatures. I Oswald archbishop, etc.*, iii. 260, 13. Ælfrêd cing Ósulfe his hondsetene sealde *king Alfred gave his sign manual to Osulf*, ii. 133, 22.

hand-sliht, -slyht, es; *m. A slaying with the hand :* — Ne meahte hē ealdum eorle hondslyht giofan *he could not give a deadly blow to the old warrior*, Beo. Th. 5937; B. 2972: 5851; B. 2929. v. sliht *and its compounds.*

hand-smæll, es; *m. A slap with the hand :* — Sealdon him hondsmællas *dabant ei alapas*, Jn. Skt. Lind. Rush. 19, 3. v. smæll.

hand-spor, es; *n. A talon, claw*, Beo. Th. 1976; B. 986.

hand-stoc, es; *n. A handcuff, manacle ;* manica, Hpt. Gl. 525, 526.

hand-þegen, es; *m. An attendant, one of a retinue, servant :* — Ðá hē ðá ðyder fêrde ðá wǣron his handþegnas twegen *when he journeyed thither, two of his attendants were with him*, Guthl. 12; Gdwin. 62, 3. Willfriþ his preóst and his hond-þeng *Wilfrid his priest and attendant ;* clericus illius, Bd. 5, 19; S. 638, 27: Cd. 224; Th. 295, 12; Sat. 485. [Cf. hand-gesella, -preóst.]

hand-þweál, es; *n. A washing of the hands :* — Hândþweáles fæt *malluviæ*, Ælfc. Gl. 26; Som. 60, 87; Wrt. Voc. 25, 27.

hand-weorc, es; *n. Handiwork, work done by the hand :* — Handweorc Godes *the work of God's hand*, Cd. 167; Th. 209, 1; Exod. 492. Sinc hondweorc smiþa *treasure, the handiwork of artificers*, Exon. 105 b; Rä. 21, 7. Þurh ðæt handweorc *by manual labour*, L. E. I. 3; Th. ii. 404, 19.

hand-worht; *adj. Hand-wrought, made with hands :* — Ic tôwurpe ðis handworhte tempel *ego dissoluam templum hoc manu factum*, Mk. Skt. 14, 58. [*Goth.* handu-waurhts.] DER. un-handworht.

hand-wundor, es; *n. A wondrous thing wrought by hand*, Beo. Th. 5530: B. 2768.

hand-wyrm, es; *m. An insect supposed to produce disease in the hand :* — Handwyrm *surio vel briensis vel sirineus*, Ælfc. Gl. 24; Som. 60, 25; Wrt. Voc. 24, 28. Handwyrm *ureius*, Wrt. Voc. 288, 4. Hondwyrm, Exon. 111 b; Th. 427, 24; Rä. 41, 96: 125 b; Th. 482, 15; Rä. 67, 2. Við hondwyrmum, L. M. 1, 50; Lchdm. ii. 122, 21.

hand-wyrst, -wrist, e; *f. The wrist :* — Fædm betwux elboga and handwyrste *cubitum*, Ælfc. Gl. 70; Som. 70, 125; Wrt. Voc. 43, 51. [Halliwell gives *hand-wrists* as a Somersetshire word.]

hangian; *p.* ode; *pp.* od *To hang, be suspended, depend :* — Ic hongige *pendeo*, Ælfc. Gr. 26, 6; Som. 29, 11: Exon. 104 a; Th. 395, 21; Rä. 15, 11. Ðes hálga Hælend hangaþ unscyldig *this holy Jesus hangeth guiltless*, Homl. Th. ii. 256, 14; Beo. Th. 4886; B. 2447. Manega sind beboda mannum gesette ac hí ealle hangiaþ on ðisum twám wordum *many are the commandments appointed to men, but they all depend upon these two sentences*, Homl. Th. ii. 314, 21. Ðá ðá Crist hangode on rôde *for ûre ālýsednysse when Christ hung on the cross for our redemption*, 240, 22: Lk. Skt. 23, 39. Wíde sceós hangodan on hira fôtum and bogan hangodan on hiora eaxlum *wide shoes hung on their feet and bows hung on their shoulders*, Shrn. 38, 8. His loccas hangodon tô ðám ancleowum *his locks hung down to his ancles*, Homl. Th. i. 466, 25. Swá hálig wer hangian ne sceolde *so holy a man ought not to be hung*, 596, 30. Hangigende, 594, 5. Hangiende, ii. 260, 25. [*Laym.* hongien; *p.* hongede: *A. R.* hongede: *Wick.* hangide: *O. Sax.* hangôn: *O. Frs.* hangia: *Icel.* hanga: *O. H. Ger.* hangen; *p.* hangeta.]

hangra, an; *m.* '*A meadow or grassplot, usually by the side of a road ; the village green,*' Cod. Dipl. Kmbl. iii. xxix :— Of ðam hangran sûþ tô ðære strǣt *from the meadow south to the road*, 229, 27: V. 374, 29. Ealle ða hangran betweónan ðam wege and ðam ðe tô Stánleáge ligþ gebyriaþ ealle tô Fearneborgan *all the meadows between the road and that which goes to Stanley all belong to Farnborough*, iii. 409, 17. [*Anger in local names, e. g.* Shelfanger, Birchanger.]

hár; *adj. Hoar, hoary, grey, old ;* canus :— Hár hǣþ *the grey heath*, Cd. 148; Th. 185, 5; Exod. 118. Se hǣra wulf *the grey wolf*, Exon. 77 b; Th. 291, 15; Wand. 82. Hāres hyrste *the old warrior's arms*,

Beo. Th. 5968; B. 2988: 3360; B. 1678: Cd. 164; Th. 193, 4; Exod. 241: 151; Th. 189, 7; Exod. 181. On ðone hāran hæsel *to the grey [with lichens ?] hazel*, Cod. Dipl. Kmbl. iii. 279, 14. Æt ðære hāran apuldran *at the old apple-tree*, Chr. 1066; Erl. 202, 6. Of clife hārum *from the grey cliff*, Bt. Met. Fox 5, 25; Met. 5, 13. On brime hāran *on the grey sea*, Menol. Fox 423; Men. 213. Hē geseah sumne hārne stán *he saw a grey stone*, Blickl. Homl. 209, 32: Cod. Dipl. Kmbl. iii. 313, 26: Beo. Th. 1779; B. 887. Hārne middengeard *canescentem mundum*, Mt. Kmbl. p. i, 5. Hrím and forst hāre hildstapan *rime and frost, hoary warriors*, Andr. Kmbl. 2517; An. 1260. Hāre byrnan *grey byrnies* [cf. grǣge syrcan, Beo. Th. 673; B. 334], Judth. 12; Thw. 26, 15; Jud. 328. [*Chauc.* hoor: *Piers P.* hore: *Ayenb.* hore vrostes: *Alis.* hore al so a wolf: *Icel.* hárr.]

hara, an; *m. A hare :*—Hara *lepus*, Ælfc. Gl. 19; Som. 59, 21; Wrt. Voc. 22, 62. Se hara mid ðysse wyrte hyne sylfne gelácnaþ *the hare doctors itself with this plant*, Herb. 114, 1; Lchdm. i. 226, 22: Med. ex Quadr. 4; Lchdm. i. 342, 14, 16, 18. Haran man môt etan and hē biþ gôd wið lengtenádle and wið útsiht gesoden ou wætere and his geallan man mæg wið pipor mengan wið mūþsāre *leporem licet comedere, et bonus est contra dysenteriam et diarrhæum, in aqua elixus ; et fel ejus miscendum est cum pipere contra dolorem oris*, L. Ecg. C. 38; Th. ii. 162, 22. Genim haran wulle *take hare's fur*, L. M. 1, 65; Lchdm. ii. 354, 13. Ne onscûnode nǎn hara nǣnne hund *no hare was afraid of any hound*, Bt. 35, 6; Fox 168, 9. Ic gefeó hwílon haran *capio aliquando lepores*, Coll. Monast. Th. 21, 33. Hē sǣtte be ðám haran ðæt hí môsten freó faran *he decreed concerning hares, that they should go free*, Chr. 1086; Erl. 222, 30. [*Icel.* heri: *O. H. Ger.* haso: *Ger.* hase.]

haran hige *hare's foot ;* trifolium arvense :—Genim ðás wyrte ðe man *leporis pes* and ôðrum naman haran hige nemneþ *take this plant which is called leporis pes and by another name hare's foot*, Herb. 62; Lchdm. 164, 17.

haran-specel, -sprecel *viper's bugloss ;* echium vulgare, Lchdm. iii. 330.

haran-wyrt, hare- *harewort ;* lepidium latifolium, Lchdm. iii. 330. Harewirta [MS. winta] *colocasia*, Ælfc. Gl. 42; Som. 64, 9; Wrt. Voc. 31, 20.

hár-hune [*and* hár hune], an; *f. Horehound ;* marrubium vulgare :—Hárhune *marrubium vel prassium*, Ælfc. Gl. 43; Som. 64, 47; Wrt. Voc. 31, 67. Hárhune *marubium*, 79, 35. Rômáne marubium nemnaþ and eác angle hāre hune *the Romans name it marrubium, the English also call it horehound*, Herb. 46; Lchdm. i. 148, 14. Genim ða hāran hunan *take horehound*, L. M. 1, 45; Lchdm. ii. 110, 24. Genim hwíte hare hunan *take white horehound*, Lchdm. i. 374, 18.

harian, horian; *p.* ode *To cry ;* clamare :—Tô ðē ic horige *ad te clamabo*, Ps. Th. 27, 1, note. [*O. H. Ger.* haren *clamare*, Grff. iv. 978 sqq.]

hārian; *p.* ode *To grow grey :*—Ic hārige *caneo*, Ælfc. Gr. 26; Som. 28, 43. Ic sceolde wesan ceorl on hāriendum heáfde *I should have to be a husband when my head was growing grey*, Shrn. 39, 27.

Harold, Harald, es; *m.* I. *Harold, second son of Cnut :*—Hēr man geceás Harald ofer eall tô cinge and forsôc Harðacnut *in this year Harold was chosen everywhere king, and Hardacnut was renounced*, Chr. 1037; Erl. 166, 4. Hēr forþfêrde Harold cyng on Oxnaforda *in this year king Harold died at Oxford*, 1039; Erl. 167, 12. II. *Harold, son of earl Godwin :*—Hēr forþfêrde Eádward king and Harold eorl fêng tô ðam ríce and heóld hit xl wucena and ǣnne dæg *in this year departed king Edward and earl Harold came to the throne and held it forty weeks and one day*, 1066; Erl. 198, 1. Ðǣr wearþ ofslægen Harold kyng *there was king Harold slain*, 202, 10.

Harþacnut, Hardacnut, es; *m. Hardacnut, son of Cnut :*—On ðís ilcan geáre com Hardacnut cyng tô Sandwíc vii nihtum ǣr middan sumera. And hē wæs sôna underfangen ge fram Anglum ge fram Denum *in this same year king Hardacnut came to Sandwich seven days before midsummer. And he was at once received by both English and Danes*, Chr. 1039; Erl. 167, 17. Hēr forþferde Hardacnut cyng *in this year died king Hardacnut*, 1041; Erl. 167, 30.

hārung, e; *f. Greyness, hoariness, age :*—Ða meolchwítan hārunge *lacteam caniciem*, Ælfc. Gr. 50, 26; Som. 51, 64.

hárwelle; *adj. Hoary :*—Hárwelle *canescens*, Mt. Kmbl. p. i, 5.

hár-wenge; *adj. Hoary, grey-haired :*—Hē wearþ fǣrlíce geþuht cnapa and eft hárwenge *he suddenly appeared a youth, and again grey-haired*, Homl. Th. i. 376, 13. Hē hæfþ síde beardas hwôn hárwencge *he has a good deal of hair on his face, rather grey*, 456, 18.

hás; *adj. Hoarse :*—Hás *raucus*, Ælfc. Gr. 30; Som. 34, 38. Ic hæbbe sumne cnapan ðe nū hás ys *habeo quendam puerum qui modo raucus est*, Coll. Monast. Th. 19, 29. Mê syndan gôman hāse *raucæ factæ sunt fauces meæ*, Ps. Th. 68, 3. [*Piers P.* hos, hors: *Chauc.* hors: *O. and N.* hos: *Wick.* hoos, hors: *Icel.* háss: *O. H. Ger.* heis: *Ger.* heiser.]

há-sæta, an; *m. A rower:*—And gerædde man ða ðæt ða scipu gewendan eft ongeán tô Lundene and sceolde man setton ôðre eorlas and ôðre hásæton tô ðám scipum *it was decided that the ships should go back again to London, and other commanders and other rowers were to be appointed to the ships,* Chr. 1052; Erl. 183, 9. [Icel. há-seti (hár *a thole*) *a thole-sitter, an oarsman,* opposed to the captain or helmsman.]

hásian; *p.* ode; *pp.* od *To grow hoarse:*—Ic hásige raucio, Ælfc. Gr. 30; Som. 34, 38.

hás-ness, e; *f. Hoarseness:*—Hásnys raucedo, Ælfc. Gl. 10; Som. 57, 26; Wrt. Voc. 19, 32. Hásnyss raucedo, Ælfc. Gr. 9; Som. 8, 59. [*Prompt. Parv.* hoosnesse, hoorsnesse *raucitas, raucor.*]

hassuc, es; *m. Coarse grass, a place where such grass grows:*—On ðone hassuc, Cod. Dipl. Kmbl. iii. 223, 25. [v. *Prompt Parv.* p. 228, note 2, where a passage is quoted in which the phrase *usque ad tercium hassocum* occurs in the defining of a boundary. In Engl. Dial. Soc. No. 26, is the following:—'Hassock *or* Hassocks. A name sometimes assigned to *aira cæspitosa,* L. but more accurately regarded as a term indicating the large coarse tufts formed in meadows by this grass and some sedges, such as *Carex cæspitosa* and *C. paniculata.*' Cf. too No. 30:—'Hassocks. "Great tufts of rushes, etc., called in Suffolk *hassocks.*" No. 31. [Leicestershire]:—'Hassock a tuft of coarse rank grass; an ant-hill.']

hasu, heasu; *adj. Grey, ash-coloured, tawny;* cinereus, fulvo-cinereus:—Hê of earce forlêt haswe culufran *he let out of the ark a grey dove,* Cd. 72; Th. 87, 20; Gen. 1451. Hwílum ic onhyrge ðone haswan earn *sometimes I imitate the grey eagle,* Exon. 106 b; Th. 406, 21; Rä. 25, 4. Se haswa fugel, 57 b; Th. 206, 4; Ph. 121. Rêcas stígaþ haswe ofer hrôfum *grey smoke mounts on the roofs,* 101 a; Th. 381, 6; Rä. 2, 7. [Icel. höss *grey* (applied to the wolf and eagle as above): cf. Gen. and Ex. haswed, v. 1723. Grein quotes the following passage from Haupt's Zeitschrift, x. 346:—'*Hasu* wol ursprünglich wolfgrau, und adlergrau, jene gemischte Farbe von goldgelb und grau: bald überwiegt der Gedanke an das Goldgelbe (vgl. blond), bald das Grau der Mischung.']

hasu-fág; *adj. Grey-coloured:*—Hrægl is mín hasofág *my raiment is grey,* Exon. 103 b; Th. 392, 23; Rä. 12, 1.

hasu-páda, an; *m. One having a grey garment;* a term applied to the eagle, cf. hasu:—Ðane hasupádan, earn *the grey-coated one, the eagle,* Chr. 937; Erl. 115, 11, note.

haswig-feðera; *adj. Having grey plumage,* Exon. 58 a; Th. 208, 10; Ph. 153.

hát, es; *n. Heat:*—Hát biþ onæled *heat shall be kindled,* Exon. 116 a; Th. 445, 18; Dôm. 9: 116 b; Th. 447, 11; Dôm. 37. Hát and ceald *heat and cold,* Cd. 192; Th. 239, 29; Dan. 377; 216; Th. 273, 5; Sat. 132: Exon. 117 b; Th. 451, 20: Dôm. 106. Hát þrowian *to suffer heat,* Beo. Th. 5204; B. 2605. [O. Sax. hêt; n. cf. O. H. Ger. heiz, heizi; f. fervor, æstus.]

hát; adj. Hot, fervent, fervid, fierce [of pain, punishment, etc.]:—Wæs him seó Godes lufu tô ðæs hát and tô ðæs beorht on his heortan *the love of God was so fervent and bright in his heart,* Blickl. Homl. 225, 36. Hys gecynde is swíðe hát *its nature is very hot,* Herb. 158, 1; Lchdm. i. 284, 22: 124; Lchdm. i. 236, 11. Hungor se háta *fierce hunger,* Exon. 64 b; Th. 238, 32; Ph. 613. Wæs seó ádl hát *fierce was the disease,* 47 a; Th. 161, 1; Gú. 952: Homl. Th. i. 404, 6. Deós wyrt byþ cenned on hátum stôwum *this plant is produced in hot places,* Herb. 115, 1; Lchdm. i. 228, 6. Hê háte lêt teáras geótan *he let hot tears gush forth,* Exon. 48 a; Th. 165, 14; Gú. 1029. Swá háttra sumor swá mára þunor and líget on geáre *the hotter the summer the more thunder and lightning in the year,* Lchdm. iii. 280, 9. [*Orm.* hat: *Laym.* hat, hot: *A. R.* hot: *Chauc.* hot, hoot: *Prompt. Parv.* hoot: *O. Sax.* hêt: *Icel.* heitr: *O. H. Ger.* heiz: *Ger.* heiss: cf. *Goth.* heito; f. *a fever.*]

hát, es; *n. A promise, vow:*—Ic sendo hát fadores mínes *ego mitto promissum patris mei,* Lk. Skt. Lind. 24, 49. Hátes *promissionis,* Rtl. 14, 14. [*Orm.* hát: *Gen. and Ex.* hot: *Ps.* hates, hotes *vota: Icel.* heit; *n. a solemn promise, vow:* cf. *Goth.* haiti; *f. a command.*] v. ge-hát.

**Hátabaðan Bath:*—Æt Hátabaðum *at Bath,* Chr. 972; Erl. 125, 9. v. Baðan.

**HÁTAN; ic háte, ðú hátest, hætsþ, hê háteþ, hát, hæt, *pl.* hátaþ; þ. hêht, hêt, *pl.* hêhton, hêton; *pp.* háten. I. to bid, order, command, (a) with acc. and infin:*—Drihten hwæt hætst ðú mê dôn *Lord, what dost thou bid me do?* Past. 58; Swt. 443, 24. Drihten háteþ ða eorþan eft ágifan ðæt heó ǽr onfeng *the Lord shall bid the earth give up what it received before,* Blickl. Homl. 21, 30. Mid ðam gemete wê hátaþ ôðre men dôn sum þingc *with that mood* [*the imperative*] *we command other men to do something,* Ælfc. Gr. 21; Som. 23, 23. Hê hêht englas him tô cuman and hie cômon *he bade angels come to him and they came,* 181, 5: Andr. Kmbl. 729; An. 365. Ða hêt hê mê on ðysne síþ faran *then he bade me go on this journey,* Cd. 25; Th. 32, 7; Gen. 499.

Hie hine hêton ðæt ættor etan *they bade him eat the poison,* Blickl. Homl. 229, 17. Mid ðý ðe ðú mê háte of mínum líchoman gewítan *when thou shalt bid me depart from my body,* 139, 13. Hát mê cuman tô ðê *jube me venire ad. te,* Mt. Kmbl. 14, 28. (b) *with infin. only:*—Ælfrêd kyning háteþ grétan Wærferþ biscep and ðê cýðan háte *king Alfred bids greet bishop Werferth; and I would that it should be known to you,* Past. Pref; Swt. 3, 1–2. Ic Elfrêd dux hátu wrítan and cýðan an ðissum gewrite Elfrêde regi *I alderman Alfred order to be written and made known in this writing to king Alfred,* Chart. Th. 480, 13. Ðonne háteþ Sanctus Micael bláwan ða feówer bêman *then St. Michael will order the four trumpets to be blown,* Blickl. Homl. 95, 12. Hæt [Cot. hát] fealdan ðæt segl *gives order to furl the sail,* Bt. 41, 3; Fox 250, 14. Ða hêht hê Simon infeccan beforan hine *then he ordered that Simon should be brought in before him,* Blickl. Homl. 175, 1: Andr. Kmbl. 2459; An. 1231: Chart. Th. 137, 6. (c) *with a clause:*—Ic ðê háte ðæt ðú ðás gesyhþe secge mannum *I command thee to tell this vision to men,* Rood Kmbl. 187; Kr. 95. Hê hæt hine ðæt hê hine fealde swá swá bôc *he shall bid it fold itself as a book,* Ps. Th. 49, 5. Ðê háteþ heofona cyning ðæt ðú onsende *Heaven's king bids thee send,* Andr. Kmbl. 3008; An. 1507. Hêht ðæt hê cuôme tô him *he commanded that he should come to him,* Chart. Th. 47, 11. Hêt ðæt ðú ǽte *he bade that thou shouldst eat,* Cd. 25; Th. 32, 8; Gen. 500. (d) *without an object, or with acc. only:*—Gif ðú hǽtst ðonne mæg ic *if thou biddest, then I can,* Homl. Th. ii. 390, 31. Wê dydon swá ðú ús hête *we have done as thou didst command us,* i. 394, 21. Ða mon sceal swá micle má hátan ðonne biddan *those are to be so much the more commanded than entreated,* Past. 26; Swt. 181, 21. (e) *with a verb of motion omitted:*—Hêht ôðre dæge hie ealle þrý in beforan hine *he commanded that next day they should all three come in before him,* Blickl. Homl. 175, 18. Ða hêht hê him tô ealle his discipulos *he summoned to him all his disciples,* 225, 12: Cd. 127; Th. 161, 27; Gen. 2671: Elen. Kmbl. 305; El. 153. Hêt tôsomne síne leóde *summoned his people together,* Cd. 197; Th. 245, 26; Dan. 469. Maria hêht hý ôðre mid *Mary bade another accompany her,* Exon. 119 b; Th. 459, 35; Hö. 10. Ða wæs tô ðam dôme Daniel háten *then was Daniel summoned to the judgment,* Cd. 201; Th. 249, 19; Dan. 532. II. *to promise, vow:*—Gif ðú hætsþ hæðenfeoh *if thou dost vow heathen offerings,* Exon. 66 b; Th. 245, 31; Jul. 53. III. *to call, name, give a name to:*—Nolde hê nô ða rúmmôdnesse hátan mildheortnes ac ryhtwísnes *non hanc vocare misericordiam, sed justitiam maluit,* Past. 45, 1; Swt. 337, 2: Cd. 106; Th. 140, 13; Gen. 2327. Consul ðæt wê heretoha hátaþ *consul we call heretoha,* Bt. 1; Fox 2, 13. Ða deór hí hátaþ hránas *those deer they call rein-deer,* Ors. 1, 1; Bos. 20, 27: Cd. 80; Th. 99, 19; Gen. 1648. And tú hine hête ða flýman *and then you declared him a fugitive,* Chart. Th. 173, 6. God hêt ða fæstnisse heofenan *vocavit Deus firmamentum cælum,* Gen. 1, 8. Hê hêt his naman Adam *he called his name Adam,* 5, 2: Cd. 124; Th. 158, 7; Gen. 2613: Beo. Th. 5605; B. 2806. Rômâne hý tictatôres hêton *the Romans gave them the name of dictators,* Ors. 2, 4; Bos. 42, 28. Sum consul Boetius wæs háten *a certain consul whose name was Boethius,* Bt. 1; Fox 2, 13: Cd. 79; Th. 99, 13; Gen. 1645. Is ðæt deór pandher bí noman háten *that beast is called by the name of panther,* Exon. 95 b; Th. 356, 17; Pa. 13. Hí nemnaþ ða eá archoboleta ðæt is háten ðæt miccle wæter *archoboleta vocant quæ est aqua magna,* Nar. 35, 21. [*Laym.* haten, heht. In Chaucer this verb and the next are confounded, thus *highte* = *hátte* ; and *hight* is used for *háten.* *Goth.* haitan *to name, call, bid, command: O. Sax.* hêtan : *Icel.* heita *to call, name, promise, vow: O. Frs.* hêta : *O. H. Ger.* heizan, heizzan *nominare, appellare, jubere, præcipere: Ger.* heissen.]

**hátan; *pres.* and *p.* hátte, *pl.* hátton *To be called* or *named, have for a name:*—Cwæþ ðæt se hêhsta hátan sceolde Satan siððan *said that the highest should be called Satan afterwards,* Cd. 18; Th. 22, 22; Gen. 344. Ân eá of ðám hátte Fison *one river of them is called Pison,* Gen. 2, 11. Saga hwæt ic hátte *say what I am called,* Exon. 106 b; Th. 406, 13; Rä. 24, 16. Hú ne hátte hys môdor Maria *nonne mater ejus dicitur Maria?* Mt. Kmbl. 13, 55. Ðe swá hátte *that was thus called,* Cd. 180; Th. 226, 17: Dan. 1722: Bt. Met. Fox 1, 105; Met. 1, 53. On ðæm bôcum ðe hátton Apocalypsin *in the books called the Apocalypse,* Past. 58; Swt. 445, 35: Ors. 2, 4; Bos. 42, 34. [*Goth.* haitada *I am called: Icel.* heita, ek heiti: *O. H. Ger.* heizan, Grff. iv. 1077: *Ger.* heissen.]

háte; *adv. Hotly:*—Háte glôwende *hotly glowing,* Homl. Th. i. 424, 35: Cd. 19; Th. 24, 26; Gen. 383: 38; Th. 50, 18; Gen. 810: Judth. 10; Thw. 22, 36; Jud. 94. Swá hê hátost mæge *as hot as possible,* L. M. 1, 2; Lchdm. i. 34, 10: Exon. 59 a; Th. 212, 13; Ph. 209.

háten [?] *heated:*—Mid hátene ísene *with heated iron,* L. M. 2, 25; Lchdm. ii. 218, 24.

hát-heort, es; *n. Fury, anger, wrath;* iracundia:—Nú is gefylled ðæt mycelle hátheort and ðæt mycelle yrre ðyses ealdermannes *now is completed the great fury and the great wrath of this ruler,* Blickl. Homl. 151, 10.

hât-heort; *adj. Furious, angry, irascible, passionate, ardent*; furiosus, iracundus, fervens :—Gif hwylc man tô ðam hâtheort sig and strangmôd ðæt hê tô nânum worldrihte and sybbe fôn nelle wið ðæne ðe wið hine âgylt *si homo quis adeo furiosus et duro corde sit, ut nullum sæculare jus et pacem admittere velit cum eo qui in eum deliquerit*, L. Ecg. P. ii. 28; Th. ii. 194, 5. Ðes geréfa is swíðe hâtheort and hê ðé wile forleósan *this consul is very furious and will destroy thee*, Nar. 42, 4: Exon. 77 b; Th. 290, 16; Wand. 66. Ðonne ða hâtheortan hie mid nâne foreþonce nyllaþ gestillan *cum iracundi nulla consideratione se mitigant*, Past. 40, 5; Swt. 297, 3. Timotheus hê ongeat hâtheortran ðonne hê sceolde *ferventioris spiritus vidit esse Timotheum*, 3; Swt. 291, 22. Ðâ wæs heora sum réðra and hâtheortra ðonne ða ôðre *then was one of them fiercer and more furious than the others*, Blickl. Homl. 223, 6.

hât-heorte, an; *f. Anger, fury, rage :*—Ic ðé bletsige forðon ðú mé ne forléte út gangan mid mínre hâtheortan of ðisse ceastre *I bless thee that thou didst not let me go out of this city in my anger*, Blickl. Homl. 249, 15.

hâtheort-líce; *adv. Furiously, ardently, fervently :*—Ða ðe hê ǽr hâtheortlíce lufode *which he before ardently loved*, Blickl. Homl. 59, 9: 17. Hie wǽron tô ðon hâtheortlíce yrre ðæt hie woldan ðone cásere cwicenne forbærnan *they were so furiously angry that they wanted to burn the emperor alive*, 191, 11.

hâtheort-nes, -ness, e; *f. Wrath, anger, fury, rage, fervour, zeal :*—Ðeós hâtheortnys *hic furor*, Ælfc. Gr. 9, 21; Som. 10, 26. Ðis synt ða ídelnyssa ðisse worlde . . . hâtheortnys . . . *hæ sunt vanitates hujus mundi . . . furor . . .*, L. Ecg. P. i. 8; Th. ii. 174, 33. Sió hâtheortness ðæt môd gebringþ on ðæm weorce ðe hine ǽr nân willa tô ne spôn *mentem impellit furor, quo non trahit desiderium*, Past. 33, 1; Swt. 215, 8. Ðonne wyrþ ðæt môd beswungen mid ðam welme ðære hâtheortnesse *then is the mind scourged with the heat of anger*, Bt. 37, 1; Fox 186, 21. Hû gesceádwís se reccere sceal bión on his hâtheortnesse *quæ esse debet rectoris discretio fervoris*, Past. 21; Swt. 151, 6. Fýr ys onæled on mínre hâtheortnisse *a fire is kindled in mine anger*, Deut. 32, 22. Forlǽt yrre and hâtheortnesse *desine ab ira et derelinque furorem*, Ps. Th. 36, 8: Homl. Th. i. 360, 3.

hât-hirtan, -hiertan, -hyrtan; *p. te To make angry :*—Ðonne is micel þearf ðætte se, se ða hâtheortnesse ofercuman wille, ðæt hé hiene ongeán ne hâthirte *necesse est, ut hi, qui furentes conantur reprimere, nequaquam se in furore erigant*, Past. 40, 5; Swt. 296, 6.

haþoliþa, an; *m. The elbow joint :*—Læt him blôd of ðam hâlan haþoliþan *let him blood from the sound elbow*, L. M. 2, 51; Lchdm. ii. 264, 17. vide Glossary. s. v.

hât-hyge, es; *m. Anger, fury, wrath :*—Wé wǽron on ðínum hâthige hearde gedréfde *in furore tuo conturbati sumus*, Ps. Th. 89, 7. [Cf. hât-heort, -heorte.]

hatian, hatigean; *p. ode, ede; pp. od, ed To hate :* — Ne mæg middaneard eów hatian ac hê hataþ mé *non potest mundus odisse vos : me autem odit*, Jn. Bos. 7, 7. Ða ðe ðone rihtwîsan hatiaþ ða âgyltaþ *qui oderunt justum delinquent*, Ps. Th. 33, 21. Hie hatigaþ [hatigeaþ, Cot. MS.] hiera hiéramonna unþeáwas *they hate the vices of their subjects*, Past. 18; Swt. 137, 4. Ðôþ ðæm wel ðe eów ǽr hatedon *do well to those that formerly hated you*, 33; Swt. 222, 17. Hû ne hatige ic ða ealle, Dryhten, ða ðe ðé hatigaþ? Mid fulryhte hete ic hie hatode. Swa mon sceal Godes fiénd hatigean *do I not hate all those, O Lord, who hate thee? With a perfect hatred I hated them. So shall God's enemies be hated*, 46; Swt. 353, 5–8. Hê sceal rýperas and reáferas hatian and hýnan *he must hate and humiliate robbers and plunderers*, L. I. P. 2; Th. ii. 304, 19: Beo. Th. 4627; B. 2319. [*Goth.* hatan, hatjan : *O. Sax.* hatan, hatôn : *O. Frs.* hatia : *Icel.* hata : *O. H. Ger.* hazên, hazôn : *Ger.* hassen.]

hâtian; *p. ode; pp. od To become* or *get hot, to be hot :*—Hingrian þyrstan hâtian eall ðæt is of untrumnysse ðæs gecynnes *esurire, sitire, æstuare ex infirmitate naturæ est*, Bd. 1, 27; S. 494, 14. Nim ǽnne sticcan and gníd tô sumum þinge hit hâtaþ ðærrihte of ðam fýre ðe him on lútaþ *take a stick and rub it against something, it gets hot directly from the fire which lurks in it*, Lchdm. iii. 274, 4: Herb. 90, 13; Lchdm. ii. 198, 4. Hâtode heorte mín *concaluit cor meum*, Ps. Spl. C. 38, 4. Ôþ ðæt se clam hâtige *till the paste gets hot*, L. M. 3, 59; Lchdm. ii. 342, 19. Ðonne byþ heó sôna hâtigende *it will at once be getting hot*, Herb. 90, 8; Lchdm. i. 196, 4. [*O. H. Ger.* heizên *fervere*.]

hatigend, es; *m. One who hates, an enemy :*—Hatigend oððe feónd *osor*, Ælfc. Gr. 33; Som. 37, 1.

hatol. v. hetol.

hatte-fagol *a hedge-hog*, Ps. Spl. M. 103, 19.

hatung, e; *f. Hating, hate, hatred :*—Hatung Godes beboda *hate of God's commands*, Homl. Th. ii. 220, 6. Mid ðære réðan ehtnysse hatunge *with the hate of fierce persecution*, i. 84, 12. Ða unrihtwîsan ic hæfde on hatunge *iniquos odio habui*, Ælfc. Gr. 33; Som. 36, 61. Gê beóþ on hatunge eallum mannum *eritis odio omnibus*, Mt. Kmbl. 10, 22. On hatunga, Lk. Skt. 21, 17. Hê becom on hatunga his herges *he came*

to be hated by his army, Blickl. Homl. 193, 2. Bânu sume handlian hatunge getâcnaþ *to handle bones betokens hate*, Lchdm. iii. 208, 24. [*O. H. Ger.* hazunga *æmulatio*.]

hâtung, e; *f. A growing hot, heating :*—Wið wunda hâtunge *against heating of wounds*, Herb. 2, 16; Lchdm. i. 84, 20, note.

hât-wende; *adj. Burning, hot, torrid :*—Hâtwendne lyft *the torrid air*, Cd. 146; Th. 182, 12; Exod. 74.

hâwere, es; *m. An observer, a spectator :*—Ðýlæs hie síen tô ôðerra monna gefeohte holde hâweras, and dôn him selfe nâwuht *lest they be friendly spectators of other men's struggle, and themselves do nothing*; ne, si in hoc præsentis vitæ stadio ad certamen alienum devoti fautores, sed pigri spectatores assistant, Past. 34, 1; Swt. 229, 17. [*Laym.* hauwares, hæweres *spies*.]

hâwian; *p. ode; pp. od To view, look, observe, regard, survey, inspect :*—Ic hâwige bufan and ðú beneoþan *ego supra aspicio, tu infra*, Ælfc. Gr. 47; Som. 47, 49. Drihten lôcaþ of heofenum and hâwaþ hwæðer hê geseó ǽnigne ðæra ðe hine séce oððe hine ongite *Dominus de cælo prospexit ut videat si est intelligens aut requirens Deum*, Ps. Th. 13, 3. Nýtene gelíc ðe hâwaþ symle tô ðære eorþan *like a beast that ever looks to the ground*, Homl. Th. ii. 442, 8. Ælc man ðara ðe æágan heft ǽrest hâwaþ ðæs ðe hê geseón wolde oþ ðone first ðe hê hyþ gegehâwaþ *every man who has eyes first looks towards what he wants to see, until he has got it under his observation*, Shrn. 178, 6. Þreó þinc sint neódbehæfe ðâm eágan ælcere sáwle . . . ôðder ðæt heó hâwien ðes ðe heó geseón wolden þridde ðæt hî mágen geseón ðæt ðæt hî gehâwian *three things are necessary for the eyes of every soul . . . second that they look at what they want to see, third that they be able to see what they bring under their notice*, 179, 20. Gúþlâc eode sôna út and hâwode and hercnode *Guthlac went out at once and looked and listened*, Guthl. 6; Gdwin. 42, 15. Sôna swâ hî wǽron swâ gehende ðet ægðer on ôðer hâwede *as soon as they were so near as to be in sight of one another*, Chr. 1003; Erl. 139, 8. Hý mé hâwedon and mé beheóldon *ipsi consideraverunt et conspexerunt me*, Ps. Th. 21, 16. Drihten hâwa nû mildelíce on ðâs earman eorþan *Lord, look now mercifully on this miserable earth*, Bt. 4; Fox 8, 20. Hâwa ðæt se inra wind ðé ne tôwende *look that the inward wind do not cast thee down*, Homl. Th. ii. 392, 32. Hâwa hwæðer his ceaflas sîn tôswollene *notice whether his jowls be swollen*, Lchdm. iii. 140, 8. Hâwiaþ be gehwilcum *take notice in the case of each one*, Homl. Th. i. 332, 15. Nân mon ne scyle dôn his hond tô ðære sylg and hâwian underbæc *no man shall put his hand to the plough and look back*, Past. 51, 8; Swt. 403, 2. DER. be-, gehâwian.

hâwung, e; *f. Looking, observation :*—Ic eom gesceádwîsnes and ic eom ælcum manniscum môde on ðam stale ðe seó hâwung byþ ðâm eágum *I am Reason, and in every human mind I hold the same place that observation does in the eyes*, Shrn. 178, 10: 21.

hê; *m:* heó; *f:* hit; *n. He, she, it :*—Ðâ hê geffôr ðâ fêng his sunu tô ðam ríce *when he died his son came to the throne*, Chr. Erl. 2, 11. Him sprecendum hig cômon *eo loquente veniunt*, Mk. Skt. 5, 35. Hê hine miclum gewundode *he wounded him severely*, Chr. 755; Erl. 48, 34. Hê hiene him tô biscepsuna nam *he was godfather to him*, 853; Erl. 68, 14. Hê hire hand nam and heó sôna árâs *he took her hand and she at once arose*, Mk. Skt. 5, 41–2. Hê him þearle bebeád ðæt hí hyt nânum men ne sǽdon and hê hêt hire etan syllan *præcepit illis vehementer ut nemo id sciret et dixit dari illi manducare*, 43. Ðâ cuǽdon hie ðæt him nǽnig mæg leófra nǽre ðonne hiera hlâford and hie nǽfre his banan folgian noldon *then said they that no kinsman was dearer to them than their lord, and they would never follow his murderer*, Chr. 755; Erl. 50, 18–20. Ealle ðíne gebrôðru beóþ under hý þeówdôme *all thy brethren shall be servants to him*, Gen. 27, 37. Tô tâcne ðæt hê his gewald âhte *as a sign that he had had power over him*, Past. 28; Swt. 197, 22. Ða hæðenan hæfdon heora geweald *the heathen had power over them*, Jud. pref. l. 8. Gedrinc his þreó full fulle *drink of it three cups full*, Herb. 1, 9; Lchdm. i. 74, 1. Hæbbe ic his on handa *I have some of it in my hand*, Cd. 32; Th. 42, 23; Gen. 678. Eorðe and ealle hire gefyllednys and eal ymbhwyrft and ða ðe on ðam wuniaþ ealle hit syndon Godes ǽhta *earth and all its fulness, and all the globe and those who dwell on it, all are God's possessions*, Homl. Th. i. 172, 10. Etaþ ðisne hlâf hit is mín líchama *eat this bread, it is my body*, Homl. Th. ii. 266, 33. Ic hyt eom *ego sum*, Mt. Kmbl. 14, 27: 28. Hit ys âwriten, N. leofaþ se man be hlâfe ânum *scriptum est: Non in pane solo vivit homo*, 4, 4. Ða rînde hit *then it rained*, 7, 27. Hit ǽfenlǽcþ *advesperascit*, Lk. Skt. 24, 29. Hit gelamp *it happened*, Homl. Th. i. 70, 23. Hit wæs winter *hiemps erat*, Jn. Skt. 10,.21. Hit lícode Herode *it pleased Herod*, Mt. Kmbl. 14, 6. Ðonne hit tôcymþ ðæt hie hit sprecan sculon *when the time comes that they ought to speak*, Past. 46; Swt. 355, 10. Hit neálǽcþ ðam ende; and ðý hit is on worulde â swâ leng swa wyrse, and swâ hit sceal nýde for folces synnum fram dæge tô dæge ǽr Antecristes tôcyme yfelian swíðe; and hûru hit wyrþ ðonne egeslíc *it is drawing near the end; and therefore the longer it goes on the worse it is in the world, and so for the people's sins it needs must get very bad from day to day before*

Antichrist's coming; and especially then it will be awful, Swt. A. S. Rdr. 104, 1–5. Hwæt magon wē his nū dôn *what can we do now in the matter;* quid ergo faciemus, Past. 58; Swt. 443, 14. Sume hit ne gedȳgdan mid ðam life *some did not come out of it with life*, Chr. 978; Erl. 127, 12. Se arcebiscop ǣxode hȳrsumnesse mid āþswerunge at him and hē hit forsôc *the archbishop required obedience with an oath of him, and he refused it*, 1070; Erl. 208, 16 : 1039; Erl. 167, 19. Hī nāmon hit ðā on twā healfe Temese tô scipan weard *they took their way on both sides the Thames towards the ships*, 1009; Erl. 143, 11. Hū mæg ic hit on ðrim dagum gefaran? ac mā wēn is ðæt ðū onsende ðinne engel, se hit mæg hrædlícor gefaran ... ic hit ne mæg hrædlíce gefaran *how can I do it in three days? it is better to send thy angel who can do it more quickly ... I cannot do it quickly*, St. And. 4, 29–6, 2. Godes bearn nāmon him wíf *the sons of God took them wives*, Gen. 6, 2. Hie woldon ða men him tô mete dôn *they wanted to make the men food for themselves*, St. And. 4, 18. Sȳ ðæt ylfa ðe him síe *be the elf what it may*, L. M. 2, 65; Lchdm. ii. 290, 29. Beó him æt hām *let him be at home*, Deut. 24, 5 : Chr. 1009; Erl. 143, 14. Abraham stôd him under ðam treówe *Abraham stood under the tree*, Gen. 18, 8. Heó sæt hire feorran *she sat her down a good way off*, 21, 16. Heó eodon heom *they went*, Chr. 1006; Erl. 140, 17 : 21. Hí fleóþ him floccmǣlum *they fly in flocks*, Homl. Th. i. 142, 9. Ondrēd hē him *timuit*, Jn. Skt. 19, 8. Hæbbe hire ðæt heó hafaþ *let her have what she has*, Gen. 38, 23. Eác him wolde Eádríc his ealdre gelǣstan *Eadric for his part would follow his chief*, Byrht. Th. 132, 4; By. 11. Ðá bealh hē hine *indignatus est*, Lk. Skt. 15, 28. Ðá beþohte hē hine *then he bethought himself*, 17. Reste ðæt folc hit on ðam seofoþan dæge *let the people rest on the seventh day*, Ex. 16, 30. Hie æt Tharse ðære byrig hie gemētton *they met one another at the city of Tarsus*, Ors. 3, 9; Swt. 128, 2. Se eádiga Mathēus and se hāliga Andreas hie wǣron cyssende him betweónon *the blessed Matthew and the holy Andrew kissed one another*, St. And. 12, 19. Hī betwux him cwǣdon *inter se dicentes*, Mk. Skt. 1, 27. Hig grētton hig gesybsumum wordum *they greeted each other with words of peace*, Ex. 18, 7. Hī ðá hí gecyston *then they kissed each other*, Shrn. 89, 12. Hī micclum ege him ondrēdon and cwǣdon ælc tô ôðrum *timuerunt magno timore et dicebant ad alterutrum*, Mk. Skt. 4, 41 : Bt. Met. Fox 25, 21; Met. 25, 11. Sume hī cômon feorran *quidam ex eis de longe venerunt*, Mk. 8, 3. Nū sceal hē sylf faran *now must he himself come*, Cd. 27; Th. 35, 18; Gen. 556. Hire selfre suna *her own sons*, Beo. Th. 2234; B. 1115. Pilatus hymsylf āwrāt ealle ða þyng *Pilate himself wrote all the things*, Nicod. 34; Thw. 19, 33. On himselfum *in semetipso*, Past. 16, 2; Swt. 101, 1. Hū ne becȳpaþ hig twegen spearwan tô peninge *are not two sparrows sold for a penny*, Mt. Kmbl. 10, 29 : 5, 11. Hē dyde ðæt hí twelfe mid him wǣron *fecit ut essent duodecim cum illo*, Mk. Skt. 3, 14. Hī ealle þrȳ tôgædere grētton ðone cyngc *all three of them together saluted the king*, Th. Ap. 19, 22 : Homl. Th. ii. 384, 4. Gewiton hie feówer *they four departed*, Cd. 92; Th. 118, 12; Gen. 1964 : 191; Th. 238, 28; Dan. 361. Heora begra æhte *the property of both of them*, 90; Th. 113, 27; Gen. 1893. Him bām on breóstum *in the breasts of them both*, 10; Th. 12, 25; Gen. 190. Him eallum *to them all*, 156; Th. 194, 16; Exod. 261. Him twám hē wæs ætȳwed *duobus ex eis ostensus est*, Mk. Skt. 16, 12. Hē Ninus Soroastrem Bactriana cyning se cûðe manna ǣrest drȳcræftas hē hine oferwann and oflslôh [*Ninus*] *Zoroastrem Bactrianorum regem, eundemque magicæ artis repertorem, pugna oppressum interfecit*, Ors. 1, 2; Swt. 30, 10 : St. And. 4, 3, 6. Wæs hē se man in weoruldhâde geseted *in habitu sæculari constitutus*, Bd. 4, 24; S. 597, 3. Europa hió onginþ *Europa incipit*, Ors. 1, 1; Swt. 8, 14. Ða ðe his líf ðæs eádigan weres cûðon *those who were acquainted with the life of the blessed man*, Guthl. prol : Gdwin. 4, 26. Wē gesâwon Enac his cynryn *we saw the children of Anak*, Num. 13, 29, 33 : Deut. 1, 28. Nilus seó eá hire ǣwielme *the source of the river Nile*, Ors. 1, 1; Swt. 12, 19. Affrica and Asia hiera landgemircu onginnaþ of Alexandria *the boundaries of Africa and Asia begin from Alexandria*, 8, 28. Ðæt se hiera folgoþ hine ne ôðhebbe *istos ne locus superior extollat*, Past. 28; Swt. 189, 17. Ða ðe hiera mildheortlíce sellaþ *qui sua misericorditer tribuunt*, 44; Swt. 319, 16. Wē his syndon *we are his*, Ps. Th. 99, 2. Hyra ys heofonan ríce *ipsorum est regnum cælorum*, Mt. Kmbl. 5, 10. Hē biþ unscildig ðe hine slôh *then shall he that smote him be quit*, Ex. 21, 19. Dôþ síðfæt ðæs sēftne and rihtne ðe hē sylfa ástāh ofer sunnan up *iter facite ei, qui ascendit super occasum*, Ps. Th. 67, 4. Se wer ðe his tôhopa byþ tô swylcum Drihtne *vir cujus nomen Domini spes ejus*, 39. 4 : Elen. Kmbl. 324; El. 162. Mid mínum brôðer steffane ðe fiola gôddra dǣda siond be him āwritene *with my brother Stephen about whom many good deeds are written*, H. R. 13, 12 : Ps. Th. 145, 4. Ðām wítgum ðe god self þurht hí spec *the prophets by whom God himself spoke*, Shrn. 107, 11. Ǣlc nȳten biþ oððe hē oððe heó *every animal is either male or female*, Ælfc. Gr. 6; Som. 5, 35, 46. Woepen mon t hee and hiuu t wífmon *masculinum et feminam*, Mk. Skt. Lind. 10, 6. Hē t woepenmon *masculinum*, Lk. Skt. Lind. 2, 23. [In later English the Northern dialect is first found adopting the forms which in Modern English have replaced the oldest, and the innovation gradually

spread. Thus while the Northumbrian Metrical Psalter (before 1300) has þai, þair, þam in the plural, the declension in *Piers P.* is hij and þei, here, hem : and these forms with the exception of hij, are used by Wicklif and Chaucer. So with *she* for heó, which is still preserved in the Lancashire *hoo*. Amongst the cognate dialects the *O. Frs.* is that which agrees best with English. v. Hilfenstein, Comparative Grammar, p. 193.]

heá. v. heáh.

heaf, es; *n. Sea, water*, Beo. Th. 4947; B. 2477. [*Icel. Swed.* haf: *Dan.* hav *sea, ocean*.]

heáf, es; *m. Lamentation, mourning, weeping, wailing:*—Ðǣr is se ungeendoda heáf *there is the never-ending lamentation*, L. E. I; Th. ii. 394, 10 : 400, 7. Wôp and heáf micel *ploratus et ululatus multus*, Mt. Kmbl. Rush. 2, 18. Ðǣr biþ heáf *illic erit fletus*, 24, 51. Nis hēr nǣnig wôp ne nǣnig heáf gehȳred *there is no weeping nor wailing heard here*, Blickl. Homl. 85, 28 : 115, 15 : 219, 9 : Exon. 48 a; Th. 164, 32; Gú. 1020 : Ors. 4, 5; Bos. 81, 28. Ðū gehwyrfdest mínne heáf mē tô gefeán *convertisti planctum meum in gaudium mihi*, Ps. Th. 29, 11 : Blickl. Homl. 195, 17. v. heóf.

heáfan; *p.* heóf, hôf *To mourn, wail, lament:*—Ðæt wíf hôf hreówigmôd *the woman mourned repentant*, Cd. 37; Th. 48, 5; Gen. 771. Heófon gehygd *they lamented their purpose*, 221; Th. 285, 28; Sat. 344. v. heófan.

heáfd. v. heáfod.

heáfian. v. heófian.

heáflíc; *adj. Mournful, lamentable, grievous:*—Ðæt heáflíce gewrit *that mournful sentence*, Blickl. Homl. 123, 6.

heáfoc. v. hafoc.

heáfod; *gen.* heáfdes; *dat.* heáfde; *pl.* heáfdu [v. Ælfc. Gr. 15; Som. 18, 21–25] HEAD, *chief, source, 'the commencing point, or the highest point, of a stream, of a field, hill, etc.* In reference to running water, the head is exactly converse to the gemȳðe or mouths. In the Saxon charters the word is of frequent occurrence, and, as it seems, generally to denote rising grounds. It is hardly distinguishable from the compound heáh-heáfod, on-heáfod,' Cod. Dipl. Kmbl. iii. xxix :—Ðis forweard heáfod *hæc frons*, Ælfc. Gr. 9, 39; Som. 12, 60 : Wrt. Voc. 70, 28 : Homl. Th. ii. 266, 11. Æfteweard hæfod *occiput vel postea :* ofer healf heáfod *sinciput*, Ælfc. Gl. 69; Som. 70, 35, 36; Wrt. Voc. 42, 43, 44. Healf heáfod *hoc sinciput*, Ælfc. Gr. 9, 78; Som. 14, 24. Cûþ is ðæt se āwyrgda gāst is heáfod ealra unrihtwísra dǣda, swylce unrihtwíse syndon deófles leomo *it is known that the accursed spirit is the source of all unrighteous deeds, as also unrighteous men are members of the devil*, Blickl. Homl. 33, 7. Hine ðe wæs ǣrur heáfod tô ðam unrǣde *the man that had before been the author of that mischief*, Chr. 1087; Erl. 225, 10. Heáfod ealra heáhgesceafta *the chief of all exalted creatures*, Cd. 1; Th. i. 7; Gen. 4 : Hy. 7, 62; Hy. Grn. ii. 287, 62. Hē getimbrede ða burg Babylonie tô ðon ðæt heó wǣre heáfod ealra Asiria *Babyloniam urbem instauravit, caputque regni Assyrii ut esset instituit*, Ors. 2, 1; Swt. 60, 14. Stæfes heáfud *apicem*, Lk. Skt. Lind. 16, 17. Wið healfes heáfdes ece *for megrim*, L. M. I, 1; Lchdm. ii. 20, 14, 17, 21. Þolige hē heáfdes *let him lose his head*, L. Edg. S. 11; Th. i. 276, 13. His heáfdes segl *his head's sun* [*the eye*], Andr. Kmbl. 100; An. 50. His eágan hálge heáfdes gimmas *his eyes, his head's holy gems*, Exon. 51 b; Th. 180, 7; Gú. 1276. Hāt mē heáfde beceorfan *order my head to be cut off*, Blickl. Homl. 183, 16. Wið tôbrocenum heáfde *for a broken head*, L. M. I, 1; Lchdm. ii. 2, 10. On ðam heáfde foran *on the forehead*, 2, 64; Lchdm. ii. 288, 22 : 65; Lchdm. ii. 290, 23. His heáfod forweard mid ðære hálgan rôde tácne gewǣpnige *let him arm his head in front with the sign of the holy rood*, L. E. I. 29; Th. ii. 426, 8. Wē sceolon fyligan ûrum Heáfde and faran fram deófle tô Criste *we ought to follow our Head, and pass from the devil to Christ*, Homl. Th. ii. 282, 20. Ic ðē gesette eallum Israhélum tô heáfde *caput te constitui in tribubus Israel*, Past. 17, 4; Swt. 113, 10. Ðū settest ûs mænige men ofer heáfod *imposuisti homines super capita nostra*, Ps. Th. 65, 10. Hēr Offa hēt Æþelbryhte ðæt heáfod ofásleán *in this year Offa ordered Ethelbert's head to be struck off*, Chr. 792; Erl. 58, 2. Bûton hē healde iii niht hȳde and heáfod *unless we keep the hide and head three nights*, L. Eth. iii. 9; Th. i. 296, 18. Fare seó buruhwaru sylf tô and begyte ða banan cuce oððe deáde heora nȳhstan mâgas, heáfod wið heáfde *let the burghers themselves go and get the murderers, living or dead, or their nearest kinsmen, head for head*, ii. 6; Th. i. 286, 32. Æt ðam ôðran cyrre ne sȳ ðǣr nān ôðer bôt bûtan ðæt heáfod *the second time let there be no other reparation than the head*, i. 1, 2; Th. i. 282, 2, 23. Heáfdas feónda *capita inimicorum suorum*, Ps. Th. 67, 21. Hie heora heáfdu slôgan on ða wagas *they struck their heads against the walls*, Blickl. Homl. 151, 5. Hȳ habbaþ hunda heáfda *they have dogs' heads*, Nar. 34, 32. Ða heáfda wǣran ofácorfena *the heads were cut off*, Ors. 4, 1; Bos. 79, 7. Nim ðe leáces heáfda *take the heads of this leek*, Lchdm. i. 376, 3. Heáfdu, L. M. 2, 32; Lchdm. ii. 234, 20. Of Godes half and ealre hādode heáfde *on behalf of God and of all persons in orders*, Chr. 675; Erl. 37, 25 : 963; Erl. 123, 15. Swā swā heó on dæg dēþ bufan ûrum heáfdum *as by day it does above our heads*, Lchdm. iii. 234, 25. Ðone stān ðe

æt his heáfdum læg *the stone that lay at his head*, Past. 16, 3; Swt. 101, 16. Ðá cóman ðyder tu wild deór and heóldan ðone líchoman óðer æt ðǽm heáfdum óðer æt ðǽm fótum *then came thither two wild beasts and guarded the body, one at the head, the other at the feet*, Shrn. 83, 25 : Rood Kmbl. 126; Kr. 63. Heáfdan, Blickl. Homl. 145, 26. [*Laym.* heaved, hæfed : *Orm.* hæfedd : *A. R.* heaved : *Piers P. Chauc. Wick.* hed, heed. The cognate dialects seem to offer two forms, differing in the root vowel, each of which may be represented in the English. Thus heáfod may compare with *Goth.* haubiþ : *O. Sax.* hôbid : *O. H. Ger.* haupit, houbit; while hæfod, hafud may compare with *Icel.* höfuð; v. Cl. and Vig. Dict. s. v.]

heáfod-ǽdre, e; *f. The cephalick vein :*— Lǽt him blôd on ðam winestran earme of ðære heáfodǽdre *let blood from the cephalick vein in his left arm*, L. M. 2, 42; Lchdm. ii. 254, 7.

heáfod-bán, es; *n. Head-bone, skull :*— Monnes heáfodbán bærn tô ahsan *burn a man's skull to ashes*, L. M. 1, 53; Lchdm. ii. 126, 2. Wulfes heáfodbán bærn swíðe *burn a wolf's skull thoroughly*, 61; Lchdm. ii. 132, 3. [*Laym.* hæfd-, heued-bon *skull : Icel.* höfuð-bein.]

heáfod-beáh; *gen.* -beáges; *m. A head-ring, crown :*— Heáfodbeáh gyldenne *a golden crown*, Bt. 37, 2; Fox 188, 8.

heáfod-beorh; *gen.* -beorge; *f. A head-shelter, helmet*, Beo. Th. 2065; B. 1030.

heáfod-beorht; *adj. Having a bright, splendid head*, Exon. 105 a; Th. 400, 2; Rä. 20, 2.

heáfod-biscop, es; *m. A head-bishop, high priest :*—Abiathar ðæra Iudéiscra heáfodbiscop *Abiathar high priest of the Jews*, Homl. Th. ii. 420, 31.

heáfod-bolla, an; *m. A skull :*—Heáfodbollan stôwe *Golgotha*, Lye.

heáfod-bolster, es; *n. A pillow :*— Heáfdbolster *capitale*, Ælfc. Gl. 27; Som. 60, 104; Wrt. Voc. 25, 44. Under ðínum heáfodbolstre *under thy pillow*, L. M. 3, 58; Lchdm. ii. 342, 14.

heáfod-botl, es; *n. A chief dwelling, principal mansion :*—Dǽlon hí ðæt heáfodbotl him betweónan *let them share the chief dwelling between them*, Chart. Th. 529, 33 : 542, 10 : 597, 6. [*Icel.* höfuð-ból *a manor, domain.*]

heáfod-burh; *gen.* -burge; *f. Chief town, capital, metropolis :*—Forgeaf him wununge on Cantwarebyrig, seó wæs ealles his ríces heáfodburh *he gave him a dwelling in Canterbury, that was the chief town of all his kingdom*, Homl. Th. ii. 128, 31. Hí becómon æt nêxtan tô ánre heáfodbyrig Suanir geháten *they arrived at last at a chief town called Suanir*, 494, 2. Cartaina heora heáfodburh *Carthage their principal city*, Ors. 4, 6; Bos. 84, 29. [*Orm.* ʒerrsalæm wass hæfeddburrh off Issraeless riche : *Icel.* höfuð-borg *metropolis : O. H. Ger.* houpit-purch.]

heáfod-cláþ, es; *n. Head-cloth, head-dress :*— Heáfodcláþ *vel cappa capitulum vel capitularium*, Ælfc. Gl. 64; Som. 69, 14; Wrt. Voc. 40, 48. [*A. R.* hore heued-cloð sitte lowe, 424, 23.]

heáfod-cwide, es; *m.*   **I.** *a saying of especial importance :*—Ða iiii heáfodcwidas in Actibus Apostolorum ðus bebeódaþ *quattuor dicta præcipua in Actibus Apostolorum sic præcipiunt*, L. Ecg. C. 38; Th. ii. 162, 33.   **II.** *a chapter :*—Onginnaþ heáfudcuido *incipiunt capitulæ*, Rtl. 166, 17.

heáfod-cyrice, an; *f. A principal church, cathedral*, L. C. E. 3; Wilk. 127, 52. [*R. Glouc.* heued chirche of al Cristendom : *Icel.* höfuð-kirkja *high-church, cathedral.*]

heáfod-ece, es; *m. Head-ache :*—Wið heáfodece *for head-ache*, Lchdm. i. 4, 15 : Herb. 75, 6; Lchdm. i. 178, 15. [*A. R.* heavedeche.]

heáfodeht; *adj. Having a head* [of plants] :—Heáfdehtes porres *of a leek having a head to it*, L. M. 2, 30; Lchdm. ii. 230, 10.

heáfod-fæder; *m. A patriarch*, Lye.

heáfod-frætewnes, -ness, e; *f. A head-ornament*, Cot. 65, Lye.

heáfod-gemaca, -gemæcca, an; *m. An equal, a mate, fellow :*—Ða sylfan his heáfodgemacan hê forlêt *his very fellows he forsook*, Guthl. 2; Gdwin. 16, 16. Ic mæg sleán míne heáfodgemæccan [heáfudgemæccean, Cot. MS.] *I may beat my fellow-servants; cæperit percutere conservos suos*, Past. 17, 8; Swt. 121, 12. Feówra sum his heáfodgemacene *with three of his equals*, L. Wih. 19, 21; Th. i. 40, 17, 21. Mid heora heáfodgemacum *cum suis similibus*, Bd. 4, 22; S. 591, 8. [Cf. heáfod-mǽg.]

heáfod-gerím, es; *n. The chief number, majority; or number of heads*, i. e. *of men* [cf. the other compounds of gerím], Judth. 12; Thw. 26, 4; Jud. 309. v. next word.

heáfod-getel, es; *n. A principal, cardinal number :*— Cardinales numeros ðæt synd ða heáfodgetel, Ælfc. Gr. 49; Som. 49, 64.

heáfod-gewǽde, es; *n. A head-dress, veil :*—Ðæt beó ðê tô heáfodgewǽdon *let it be to thee for a veil*, Gen. 20, 16.

heáfod-gim; *m. f.* [?] *Jewel of the head, the eye*, Exon. 27 a; Th. 81, 19; Cri. 1331 : 89 b; Th. 336, 6; Gn. Ex. 44 : Andr. Kmbl. 62; An. 31.

heáfod-gold, es; *n. A crown :*—Ðú him sylst heáfodgold tô meþre *honore coronasti eum*, Ps. Th. 8, 6. [*Icel.* höfuð-gull *head-jewels.*]

heáfod-gylt, e; *f. A capital crime, deadly sin :*—Búton hine hwá mid heáfodgylte forwyrce ðæt hê weofuðþénunge ðanonforþ wyrðe sí *unless any one by deadly sin render himself unworthy thenceforth of the altar-service*, L. N. P. L. 2; Th. ii. 290, 8.

heáfod-hǽr, es; *n. A hair of the head :*—Heáfod-hǽr *capilli*, Ælfc. Gl. 70; Som. 70, 55; Wrt. Voc. 42, 63.

heáfod-hríéþo; *f. Head-roughness;* capitis *scabies*, L. M. 2, 30; Lchdm. ii 228, 13.

heáfod-land. v. hafud-land.

heáfod-leahter, es; *m. A capital offence, mortal sin :*— Ælc ðara manna ðe mid heáfodleahtre besmiten biþ *unusquisque eorum hominum, qui capitalibus criminibus polluti sunt*, L. M. I. P. 1; Th. ii. 266, 3. Se ðe ða heáfodleahtras wyrcþ and on ðám geendaþ hê môt forbyrnan on ðam êcum fýre *he who commits the deadly sins and dies in them shall burn in the everlasting fire*, Homl. Th. ii. 590, 17.

heáfod-leás; *adj. Headless :*— Heáfodleás bodig *truncus*, Ælfc. Gl. 73; Som. 71, 30; Wrt. Voc. 44, 16 : Exon. 104 a; Th. 395, 19; Rä. 15, 10.

heáfod-lencten-fæsten, es; *n. The chief Lent fast*, R. Concord.

heáfod-líc; *adj. Chief, capital :*—For heáfodlícum gyltum *pro capitalibus criminibus*, L. Ecg. C. 2; Th. ii. 134, 3. Ðæt wê ús healdan wið heáfodlícan leahtras *to keep ourselves from deadly sins*, Blickl. Homl. 37, 3.

heáfod-ling, es; *m. An equal, a fellow, mate :*— Heáfodlinges coǽquales, Mt. Kmbl. Lind. 11, 16. Heáfudlinges *conservos*, 24, 49. [*Laym.* has hevedling *chief, captain, like Ger.* häuptling.]

heáfod-mǽg, es; *m. A near relation, a relation in the first degree*, Cd. 60; Th. 73, 6; Gen. 1200 : 78; Th. 96, 36; Gen. 1605 : Beo. Th. 1180; B. 588 : 4308; B. 2151. v. next word; and cf. *Icel.* höfuð-niðjar, höfuðbarmsmenn *agnates :* v. also cneów-mǽgas, and see Grmm. R. A. pp. 468–70, for terms belonging to the body in their application to degrees of relationship.

heáfod-mága, an; *m. A near relation*, Andr. Kmbl. 1884; An. 944. v. preceding word.

heáfod-man, -mann, es; *m. A chief man, prince, captain, leader :*— Heáfodman *vel* þegn *primas*, Ælfc. Gl. 68; Som. 70, 5; Wrt. Voc. 42, 14 : Homl. Th. ii. 514, 14. Ðæt folc wearþ micclum ástyred, and ða heáfodmenn and ða bôceras *the people were much stirred up and the elders and the scribes*, i. 44, 30. Israhêla heáfodmen *heads of thousands in Israel*, Num. 1, 16 : 13, 3 : Jos. 23, 2. Ða heáfodmen *the lords* [*of the Philistines*], Jud. 16, 27 : Chr. 1069; Erl. 207, 15 : 1101; Erl. 237, 14, 25. Ðǽr on wǽron twægen heáfodmenn Cnut and Hácun eorl *in them were two leaders, Cnut and earl Hakon*, 1075; Erl. 214, 7. [*Laym.* hæfdmen, *pl : Orm.* hæfeddmann : *Icel.* höfuðs-maðr *a chief, leader : O. H. Ger.* haubitman *satrapa : Ger.* hauptmann *captain.*]

heáfod-mynster, es; *n. A chief minster, church*, L. Eth. ix. 5; Th. i. 340, 27 : L. C. E. 3; Th. i. 360, 17.

heáfod-panne, an; *f. A skull :*— Heáfodpanne *calvaria*, Wrt. Voc. 64, 24 : 282, 40. Forheáfod *vel* heáfodpanne *calvarium*, Ælfc. Gl. 69; Som. 70, 33; Wrt. Voc. 42, 41. Golgotha ðæt is heáfodpannan stôw *Golgotha quod est calvariæ locus*, Mt. Kmbl. 27, 33 : Jn. Skt. 19, 17. Heáfodpannena stôw, Mk. Skt. 15, 22. Hundes heáfodpanne *a dog's skull*, L. M. ex Quad. 13, 3; Lchdm. i. 370, 3 : L. M. 2, 55; Lchdm. ii. 342, 4.

heáfod-port, es; *m. A principal town*, Chr. 1086; Erl. 220, 21.

heáfod-ríce, es; *n. A chief kingdom, empire :*—Feówer heáfodrícu *quatuor regnorum principatus*, Ors. 2, 1; Swt. 58, 31.

heáfod-sár, es; *n. Pain in the head*, Herb. 4, 7; Lchdm. i. 90, 28.

heáfod-sién, -sýn; *f. The eye :*—Ðǽr him hrefn nimeþ heáfodsýne *there* [*on the gallows*] *shall the raven take from him his eye*, Exon. 87 b; Th. 329, 19; Vy. 36. Heáfodsiéna, Cd. 114; Th. 150, 11; Gen. 2490.

heáfod-slæge, es; *n. Head of a pillar* [?] ; *capital*, Cot. 50, Lye. [Cf. ofer-slege.]

heáfod-smæl *capitium*, Wrt. Voc. 288, 43.

heáfod-stede, es; *m. A chief place :*—Heora þeówas hie benôman heora heáfodstedes ðæt hie Capitoliam hêton *servi invaserunt Capitolium*, Ors. 2, 6; Swt. 86, 30. Hwílum wǽran heáfodstedas and heálíce hádas micelre mǽðe wyrðe *formerly the chief places and high ranks were entitled to much honour*, L. Eth. vii. 3; Th. i. 330, 6. [*O. Sax.* hôbid-stedi : *O. H. Ger.* houpit-stat *toparchia.*]

heáfod-stól, es; *m. A chief place, capital :*—Thêbana fæsten ðætte ǽr wæs ealra Crêca heáfodstól *the city of Thebes which before was the chief place of all Greece*, Ors. 3, 9; Swt. 124, 5 : 3, 11; Swt. 144, 19. [*Icel.* höfuð-stóll *a chief seat.*]

heáfod-stôw, e; *f. A place for the head :*—Seó heáfodstôw cræftiglíce geworht ætýwde *locus capitis fabrefactus apparuit*, Bd. 4, 19; S. 590, 1.

heáfod-swíma, an; *m. Swimming in the head, dizziness*, Cd. 76; Th. 94, 28; Gen. 1568. [*Icel.* höfuð-svími *dizziness in the head.*]

heáfod-sýn. v. heáfod-sién.

heáfod-þweál, es; *n. A washing of the head;* capitilavium, Ælfc. Gl. 56; Som. 67, 26; Wrt. Voc. 37, 16. [*O. H. Ger.* houbit-twehela *caput-lavium.*]

heáfod-wærc, es; *m. Pain in the head*, L. M. 1, 1; Lchdm. ii. 18,

5, 19. [*Prompt. Parv.* heedwerke, heedwarke *cephalia, cephalargia:* *Icel.* höfuð-verkr *head-ache.*]

heáfod-weard, es; *m. A chief guardian, chief officer:*—Cynnes heáfud-wærd *tribunus,* Jn. Skt. Lind. 18, 12. Ðæs herefolces heáfodweardas *the leaders of the army,* Judth. 12; Thw. 25, 3; Jud. 239.

heáfod-weard, e; *f. A guarding of the [lord's] head, attendance as a guard upon the king.* The word occurs in an enumeration of the services required of the thane and the 'geneat,' Th. i. 432, 8, 17. So in Beowulf it is said of Wiglaf that he 'healdeþ heáfodwearde,' keeps guard over the dead king, Beo. Th. 5811; B. 2909. [Cf. *Icel.* höfuð-vörðr *a body-guard.*]

heáfod-weard, e; *f. A chapter;* capitulum, Mt. Kmbl. p. 11, 17: 13, 13. [Cf. fore-weard.]

heáfod-wind, es; *m. A wind from one of the four chief points of the compass:*—Feówer heáfodwindas synd se fyrmesta is eásterne wind ... se óðer heáfodwind is súðerne ... se þridda heáfodwind hätte zephirus ... se feórþe heáfodwind hätte septemtrio, Lchdm. iii. 274, 12–23. [*Icel.* höfuð-vindr.]

heáfod-wísa, an; *m. A chief director, ruler,* Cd. 79; Th. 97, 28; Gen. 1619.

heáfod-wóp, e; *f. The voice,* Exon. 103 a; Th. 390, 17; Rä. 9, 3.

heáfod-wund, e; *f. A wound in the head,* L. Alf. pol. 44; Th. i. 90, 13, 14. [*O. Sax.* hóbid-wunda.]

heáfod-wylm, es; *m. Burning* or *heat in the head,* L. M. 1, 1; Lchdm. ii. 26, 2.

heáfod-wyrhta, an; *m. A chief workman,* Homl. Th. ii. 530, 7.

heafola. v. hafela.

heáf-sang; es; *m. An elegy,* Cot. 118, Lye.

heág. v. heáh.

heáge; *adv. High:*—Heáge flíhþ se earn *sublime volat aquila,* Ælfc. Gr. 41, 16. Beheald ðás sunnan hú heáge heó ástíhþ *behold this sun, how high it mounts,* Homl. Th. i. 286, 31.

heago-rún, e; *f. A mystery in which magic is involved, necromancy:*—Hú mambres ontýnde ða drýlícan béc his bréðer iamnes and him geopenude ða heagorúne ðæs deófelgildes his bróður *aperuit mambres libros magicos fratris sui iamnis et fecit nicromantiam et eduxit ab inferis idolum fratris sui,* Nar. 50, 14. [Cockayne has the following note:—'Heag hic pro *veneficus, magicus* sumendus; nostrum HAG.']

heago-spind. v. hago-spind.

HEÁH, héh; *adj.* HIGH, *tall, lofty, sublime, haughty:*—Heáh on bodige *statura sublimis,* Bd. 3, 14; S. 540, 7. Gyldenu onlícnes twelf elna heáh *a golden image twelve ells high,* Shrn. 88, 23. Se beám geweóx heáh *the tree grew high,* Cd. 202; Th. 251, 15; Dan. 564. . Hwæt elles getácnaþ se heá torr búton ðone heán foreþonc and ða gesceádwísnesse ðara gódena manna *what else does the high tower signify but the lofty forethought and the sagacity of good men,* Past. 56; Swt. 433, 24. Sió heá lár *lofty doctrine,* 63; Past. 459, 8. Seó heáge dún *the high mountain,* Homl. Th. ii. 384, 29. Heáh heofoncyning *heaven's high king,* Cd. 23; Th. 30, 7; Gen. 463. Hé on hrófe gestód heán landes *he on the summit stood of the high land,* 140; Th. 175, 21; Gen. 2898. Hie be hliðe heáre dúne eorþscræf fundon *they found a cavern by the side of a lofty hill,* 122; Th. 156, 26; Gen. 2594. Se deófol gesette hine uppan ðam scylfe ðæs heágan temples *the devil placed him upon the summit of the lofty temple,* Homl. Th. i. 166, 18. Seó eádignes ðæs heán heáhengles tíd *the blessedness of the festival of the great archangel,* Blickl. Homl. 197, 4, 24. From stæþe heáum *from the high shore,* Exon. 106 a; Th. 405, 6; Rä. 23, 19. Uppan ánre swíðe heáhre dúne *upon a very high mountain,* Homl. Th. i. 166, 23. Unriht on heán húse ácwædon *iniquitatem in excelso locuti sunt,* Ps. Th. 72, 6. On heágum *in excelso,* 9. Hóf ic míne eágan tó ðam heán beorge *levavi oculos meos in montes,* 120, 1. Fram ðam heágan cederbeáme *from the tall cedar,* Homl. Th. ii. 578, 4. Hát ðú mé ánne heáhne tor getimbrian *order a high tower to be built for me,* Blickl. Homl. 183, 3. Hé ásette míne fét on swíðe heánne stán, ðæt ys on swýðe heáh setl *statuit super petram pedes meos,* Ps. Th. 39, 2. Ofer heáne hróf heofenes ðisses *beyond the lofty roof of the sky,* Bt. Met. Fox 24, 5; Met. 24, 3. Ðone heán heofoñ *high heaven,* Cd. 35; Th. 45, 33; Gen. 736. Se ðe gebígde ðone heágan heofonlícan bigels *he who bowed the lofty vault of heaven,* Homl. Th. i. 170, 23: H. R. 103, 1. Ofer heáh wæter *over deep water,* Cd. 72; Th. 87, 19; Gen. 1451. Engel drihtnes lét his hand cuman in ðæt heá seld *the angel of the Lord brought his hand into that lofty hall,* 210; Th. 261, 7; Dan. 722. Wæron ófras heá streámas stronge *the shores were high, the streams strong,* Exon. 106 a; Th. 404, 14; Rä. 23, 7. Wé ceorfaþ heáh treówu on healle *altum silvæ lignum succidimus,* Past. 58, 6; Swt. 443, 36. Wesan heá mihte handa ðínre áhafen ofer hæleþas *may the excellent powers of thy hand be exalted over men,* Ps. Th. 88, 12. Heágum þrymmum *in excellent majesty,* Cd. 1; Th. 1, 16; Gen. 8. Hýd heáum ceólum *a haven for the tall ships,* Bt. Met. Fox 21, 22; Met. 21, 11. On heán muntum heortas wuniaþ *montes excelsi domus cervis,* Ps. Th. 103, 17. Ná geþafian ðæt se heárra derige ðam heánran, *not to permit the higher to hurt the lower,* L. I. P. 7; Th. ii. 314, 1. Stól

heáhran, heárran *a loftier throne,* Cd. 15; Th. 18, 16, 26; Gen. 274, 282. Hérra, Exon. 56 a; Th. 199, 20; Ph. 28. Tó hiéran háde to *a higher rank,* Past. pref. Swt. 7, 15: Chr. 897; Erl. 95, 14. Se mægþhád is hírra ðonne se gesinscipe *præeminere virginitatem conjugio,* Past. 52, 8; Swt. 409, 23. Wearþ him hýrra hyge *he had a haughtier mind,* Cd. 198; Th. 247, 2; Dan. 491. Hýrre ic eom heofone *higher am I than heaven,* Exon. 110 b; Th. 424, 12; Rä. 41, 38. Cwæþ ðæt his hergas hýrran wæron ðonne israéla éce drihten *said that his gods were superior to the everlasting lord of Israel,* Cd. 210; Th. 262, 26; Dan. 715. Ðéh ðe hí selfe wilnien ðæs heáhstan *etsi summa appetunt,* Past. 16, 4; Swt. 103, 16: Ps. Th. 112, 4. Se geworden is hwommona heágost *hic factus est in caput anguli,* 117, 21. Ðæs héhstan heofonríces, 90, 1. Seó is ealra dúna mæst and hígest *mons maximus et altissimus,* Nar. 37, 32. Se hýhsta ealra cyninga cyning *the most high king of all kings,* Exon. 32 b; Th. 103, 1; Cri. 1682. [*Goth.* hauhs: *O. Sax.* hôh: *O. Frs.* hách, hág: *Icel.* hár: *O. H. Ger.* hôh *altus, excelsus, celsus, excellens, sublimis: Ger.* hoch.]

heáh, heá; *adv. High:*—Bryne stígeþ heáh tó heofonum *the burning mounts aloft to heaven,* Exon. 63 a; Th. 233, 7; Ph. 521: Cd. 166 Th. 207, 15; Exod. 467: Ps. Th. 138, 6. Heáor *altius,* Bd. 3, 8; S. 532, 16. On ðam gim ástíhþ on heofenas up hýhst on geáre ... *in it* [*June*] *the sun mounts highest in the year,* Menol. Fox 218; Men. 110. v. heáge.

heáh-beorg, es; *m. A high mountain:*—Hé ðás heáhbeorgas healdeþ *swylce et altitudines montium ipse conspicit,* Ps. Th. 94, 4. [*Icel.* hábjarg *a high rock;* há-fjall *a high fell.*]

heáh-biscop, es; *m. An archbishop, chief bishop, pontiff:*—Birhtwald Bretone heáhbiscop *Birhtwald archbishop of Britain,* L. Wih. pref.; Th. i. 36, 8. Mid geþeahte Wulfhelmes mínes héhbisceopes *with the counsel of Wulfhelm my archbishop,* L. Ath. prm.; Th. i. 194, 13. Se heáhbiscop and se hálga Wilfriþ *Antistes eximius Vilfrid,* Bd. 5, 19; S. 636, 41. Heáhbiscop *pontifex,* 2, 3; S. 504, 44, note. Héhbiscop *pontifex,* Rtl. 72, 8: *archiepiscopus,* 194, 27.

heáh-boda, an; *m. An archangel:*—Héht sigores fruma his heáhbodan hider gefleógan *bade the triumphant Lord his archangel fly hither,* Exon. 12 a; Th. 19, 3; Cri. 295.

heáh-burh; *gen.* -burge; *f. A chief town, large town;* also *a town having an elevated situation:*—Ðær is Créca heáhburg *there is the chief town of the Greeks,* Bt. 1; Fox 2, 21: Beo. Th. 2258; B. 1127. Tó ðære heáhbyrig *to the chief town [Babylon],* Cd. 209; Th. 259, 30; Dan. 699. Se kásere geeode wel manega héhburh *the emperor conquered a good many of the principal towns,* Chr. Erl. 5, 13. Ic wát heáhburg hér áne neáh lytle ceastre *I know that near here is a town placed on high, a little city,* Cd. 117; Th. 152, 8; Gen. 2517.

heáh-bytlere, es; *m. A chief-builder, architect,* Lye.

heáh-cleófa, an; *m. A principal chamber:*—His brýdbúras and his heáhcleófan ealle wæron *eorcnanstánum unionibus et carbunculis ðæm* gimcynum swíðast gefrætwode *talami cubiliaque margaritis unionibusque et carbunculis nitebant,* Nar. 5, 2.

heáh-clif, es; *n. A high, lofty cliff:*—Beorgas gemeltaþ and heáhcleofu *the hills shall melt and the lofty cliffs,* Exon. 22 a; Th. 61, 3; Cri. 979. [Cf. heáh-beorg.]

heáh-cræft, es; *m. Excellent art* or *skill,* Exon. 109 a; Th. 417, 13; Rä. 36, 4.

heáh-cyning, es; *m. A chief, great king, God:*— Mid heáhcyning *with God,* Exon. 62 b; Th. 231, 3; Ph. 483. On ða swýðran healfe ðæs heáhcyninges *on the right hand of the great king [God],* Shrn. 118, 9: Cd. 6; Th. 8, 14; Gen. 124. Ðæt wæs hildesetl heáhcyninges *that was the war-seat [saddle] of the great king [Hrothgar],* Beo. Th. 2083; B. 1039.

heáh-deór, heá-, es; *n. A stag, deer:*—Swá swíðe hé lufode ða heádeór swilce hé wære heora fæder *he loved the stags as if he were their father,* Chr. 1086; Erl. 222, 29: Hexam. 9; Norm. 16, 3. [Cf. *Ger.* die hohe Jagd *the hunting of deer.*]

heáhdeór-hund, es; *m. A stag-hound, deer-hound, a dog for hunting great game:*—Twegen hafocas and ealle his heádórhundas *two hawks and all his deer-hounds,* Chart. Th. 501, 7. Twegen and twegen fédan ænne heádórhund *duo et duo pascant unum molossum,* L. R. S. 4; Th. i. 434, 20.

heáhdeór-hunta, an; *m. A stag-huntsman:*—Mínon heáhdeórhunton *to my stag-huntsman,* Chart. Th. 561, 24.

heáh-diácon, es; *m. An archdeacon*——Næs ná ðam ánum ðe Gode sylfum underþeódde syndon mid myclum hádum, biscopas and cyningas and mæssepreóstas and heáhdiáconas *not to those alone who are subject to God himself in high positions, as bishops and kings and archdeacons,* Blickl. Homl. 109, 24: Shrn. 17, 11.

heáh-ealdor, es; *m. A chief ruler:*—Hí cómon on ðæs heáhealdres hús *veniunt in domum arche-synagogi,* Mk. Skt. 5, 38.

heáh-ealdorman; *gen.* -mannes; *m. A chief alderman, ruler, patrician:*—Ætius mære man se wæs iú ér heáhealdorman *Ætius vir industris qui et patricius fuit,* Bd. 1, 13; S. 481, 40. Ðe hælend cwæþ tó ðæm

L l 2

hēhaldurmenn *ihesus ait archesynagogo*, Mk. Skt. Rush. 5, 36. Hēhaldormenn *patricius*, Rtl. 193, 5.

heáh-engel, es; *m. An archangel :*—Heáhencgel *archangelus*, Ælfc. Gl. 67; Som. 69, 102; Wrt. Voc. 41, 52. Micahel se heáhengel se wæs ealra engla ealderman *Michael the archangel who was the chief of all angels*, Blickl. Homl. 147, 2. Englas and heáhenglas *angels and archangels*, 103, 32: Homl. Th. i. 10, 13. [*Orm.* heh-enngell.]

heáh-fæder; *m. A patriarch; also the great Father, i. e. God:*—Heáhfæder *patriarcha*, Ælfc. Gl. 68; Som. 69, 118; Wrt. Voc. 41, 68. Hēhfæder *patriarcha*, Rtl. 195, 10. Cuoeþ lā hēhfæder *dixit abba pater*, Mk. Skt. Lind. 14, 36. Seó stondeþ on ða swýðran healfe ðæs heáhfæder *she stands on the right hand of the Father*, Shrn. 118, 9: Rood Kmbl. 266; Kr. 134. Ðeodosius se wæs ðære hǣðenre hēhfæder *Theodosius who was the patriarch of the heathens*, Nar. 40, 5. Be ðam heáhfædere Abrahame *concerning the patriarch Abraham*, Homl. Th. i. 46, 11. Jacob gestrýnde twelf suna, ða sind gehátene twelf heáhfæderas *Jacob begat twelve sons, who are called the twelve patriarchs*, ii. 190, 25 : i. 396, 9. [*Orm.* Godd heh-faderr *God the Father: O.H.Ger.* hōh-fater *patriarca*.]

heáh-fæst; *adj. Very fast, fixed :*—Hafaþ under heofonum heáhfæstne dōm *hath under heaven enduring glory*, Exon. 87 a; Th. 327, 8; Wid. 143.

heáh-fæsten, es; *n. A chief fortress, a city :*—Heáhfæsten *castrum*, Ælfc. Gl. 54; Som. 66, 109; Wrt. Voc. 36, 29. Hēhfæsten *polis* (πόλις), Rtl. 195, 14.

heáh-flód, es; *m. High tide* [as opposed to *neap tide*], *deep water :*—Heáhflód *malina :* nēpflód *ledona*, Ælfc. Gl. 105; Som. 78, 30, 29; Wrt. Voc. 37, 12, 11. Lēt fleógan hrefn ofer heáhflód *he let a raven fly over the deep water* [*of the deluge*], Cd. 71; Th. 87, 1; Gen. 1442. [Cf. *Icel.* hā-flæðr *a high flood-tide*.]

heáh-fore, e; *f. A heifer :*—Heáhfore *annicula vel vaccula :* fæt heáhfore *altilium*, Ælfc. Gl. 22; Som. 59, 85, 93; Wrt. Voc. 23, 44, 50. Heáhfru *altile*, Wrt. Voc. 287, 55. Eálond hwítre heáhfore *insula vitulæ albæ*, Bd. 4, 4; S. 570, 41. Gif hē hriðeru offrian wille bringe unwemme fear oððe heáfre *if he offer it of the herd, whether it be a male or female, he shall offer it without blemish*, Lev. 3, 1. Farra mīno and hēhfaro gislægno *tauri mei et altilia occisa*, Rtl. 107, 21. [hayfare *juvenca*, Wrt. Voc. 177, 7: *Prompt. Parv.* hekfere *juvenca.*]

heáh-freóls, es; *m. A high festival*, L. C. S. 48; Th. i. 404, 1.

heáhfreóls-dæg, es; *m. The day of a high festival*, L. Eth. vi. 25; Th. i. 320, 25.

heáhfreóls-tíd, e; *f. The time of a high festival*, L. Eth. vi. 22; Th. i. 320, 13.

heáh-geréfa, an; *m. A high reeve, reeve of high rank.* Kemble, Saxons in England, ii. 156, observes of this word, 'It is a name of very indefinite signification, though not of very rare occurrence. It is obvious that it really denotes only a reeve of high rank, I believe always a royal officer; but it is impossible to say whether the rank is personal or official; whether there existed an office called *heáhgeréfscipe* having certain duties; or whether the circumstance of the shire or other reeve being a nobleman in the king's confidence gave to him this exceptional title. I am inclined to believe that they are exceptional, and perhaps in some degree similar to the Missi of the Franks, officers dispatched under occasional commissions to perform functions of supervision, hold courts of appeal, and discharge other duties, as the necessity of the case demanded; but that they are not established officers found in all the districts of the kingdom, and forming a settled part of the machinery of government.' See also Stubbs' Const. Hist. i. 125, 343. Hēhgeréfa *proconsul*, Ælfc. Gl. 106; Som. 78, 58; Wrt. Voc. 57, 38. Befora undercyningum † hēhgeroefum *ante præsides*, Mk. Skt. Lind. 13, 9. Hēghgeroefa *comes*, Rtl. 193, 9. Cyninges heáhgeréfan gild iiii þúsend þrymsa *the 'wergild' of a king's high reeve four thousand 'thrymsas,'* L. Wg. 4; Th. i. 186, 8; Chr. 778; Erl. 55, 26: 779; Erl. 55, 36: 1001; Erl. 136, 6, 8, 23, 24: 1002; Erl. 137, 29.

heáh-gesamnung, e; *f. A chief assembly, synagogue :*—Ðā com sum of heáhgesamnungum *et venit quidam de archesynagogis*, Mk. Skt. 5, 22.

heáh-gesceaft, e; *f. An exalted creature :*—Hē is heáfod ealra heáhgesceafta *he is the head of all exalted creatures*, Cd. 1; Th. 1, 8; Gen. 4.

heáh-geþungen; *adj. Of high rank, distinguished :*—Ic lǣrde heáhgeþungene men ðæt hī ne āstigan on ofermēdu *I taught men of high rank not to be exalted in pride*, Blickl. Homl. 185, 13.

heáh-getimbrad; *adj. High-built*, Cd. 213; Th. 266, 29; Sat. 29. [*Icel.* hā-timbra *to build high.*]

heáh-getimbru, -getimbro; *pl. n. A lofty building, a place built on high*, Exon. 41 a; Th. 137, 9; Gū. 556: 22 a; Th. 60, 24; Cri. 974: 25 a; Th. 72, 34; Cri. 1182: Cd. 35; Th. 46, 5; Gen. 739. [*O.H. Ger.* hōh-gizimbri *pergama* (πέργαμα), *capitolia.*]

heáh-gnornung, e; *f. Deep grief, sorrow, mourning :*—Hē gehýrde heáhgnornunge ðæra ðe gebundene bitere wǣron *ut audiret gemitum vinculatorum*, Ps. Th. 101, 18.

heáh-god, es; *m. High God, the most High :*—Ic cleopige tō heáhgode *clamabo ad Deum altissimum*, Ps. Th. 56, 2.

heáh-grǣft; *adj. Carved in bas-relief :*—Heáhgrǣfte *anaglypha*, Cot. 7, Lye.

heáh-hád, es; *m. A high order, religious order :*—Heáhhádes men *men in holy orders*, L. I. P. 22; Th. ii. 334, 6.

heáh-heort; *adj. High-hearted, haughty, proud*, Cd. 202; Th. 250, 1; Dan. 540. [*Goth.* hauh-hairts *proud.*]

heáh-hlip, es; *n. A high hill*, Cd. 71; Th. 86, 31; Gen. 1439.

heáh-lǣce, es; *m. A physician of the greatest skill :*—Sc. cosmas and sc. damianus wǣron heáhlǣcas and hý lácnodon ǣghwylce untrumnesse monna *St. Cosmas and St. Damian were very excellent leeches, and cured every infirmity of men*, Shrn. 135, 13.

heáh-landríca, an; *m. Irenarcha :* εἰρηνάρχης, Lye.

heáh-láreów, es; *m. A chief teacher;* archimandrita, gymosophista, Lye.

heáh-líc, -líce. v. heá-líc, -líce.

heáh-lufe, an; *f. Deep love*, Beo. Th. 3912; B. 1954.

heáh-mæsse, an; *f. High mass*, L. E. I. 45; Th. ii. 440, 32, 34: 442, 3: Chr. 1125; Erl. 254, 2. [*Icel.* hā-messa: *Ger.* hoch-messe.]

heáh-miht, e; *f. Great, excellent power :*—On his heáhmihtum *in potestatibus ejus*, Ps. Th. 150, 2.

heáh-mód; *adj. Of high, lofty mind, noble, proud, haughty :*—Siððan hine sylfne heáhmód hefeþ on heánne beám *afterwards exultant raises itself on to a lofty tree*, Exon. 57 b; Th. 205, 13; Ph. 112. Se ðe hine sylfne āhefeþ heáhmódne se sceal heán wesan *he who exalts his proud self shall be abased*, 84 a; Th. 316, 25; Mód. 54. [*O.H.Ger.* hōh-mōti; Cf. *Ger.* hoch-müthig.]

heáh-módness, e; *f. Pride :*—Dryhten ongiet swíðe feorran ða heáhmódnesse *Deus alta a longe cognoscit*, Past. 41, 1; Swt. 301, 1.

heáh-nama, an; *A great, exalted name :*—Swā is gehálgod ðīn heáhnama *thus is thy great name hallowed*, Hy. 7, 18; Hy. Grn. ii. 287, 18.

heáh-, heán-, heá-nes, -ness, e; *f. Highness, height, highest point, elevation, loftiness, sublimity, excellence :*—Ðæs heánes wǣre ōð monnes swyran *its height was up to a man's neck*, Shrn. 81, 13. Sió heánes ðara munta *altitudo montium*, Past. 51, 5; Swt. 397, 36. Hū micel sió heánes is and hū soðlíc *quam sit vera excellentia*, 41, 1; Swt. 299, 4. Mægnes heánnes *excellentia virtutis*, Bd. 3, 13; S. 538, 38. Heánnise hiordes *celsitudo pastoris*, Rtl. 32, 21. Heánnisse ðínes mæht *sublimitatis tuæ potentia*, 97, 27. Þrittig fæðma on heáhnisse *thirty cubits in height*, Gen. 6, 15. Of eorþan heánesse ōð heofones heáhnesse *a summo terræ usque ad summum cæli*, Mk. Skt. 13, 27. On ðæs heáhnysse ufeweardre *on the very top of it* [*the stalk*], Herb. 173, 1; Lchdm. 302, 24. Wē ne mágon for ðære fyrlynan heáhnysse hī nǣfre geseón *we cannot ever see it* [*heaven*] *for its remote elevation*, Lchdm. iii. 232, 15. Hē hæfde swā mycele heánnesse on ðæt cynerīce *tantum in regno excellentiæ habuit*, Bd. 2, 16; S. 520, 8. For ðæs ríces heánesse him weóxon ofermetto *in tumorem superbiæ culmine potestatis excrevit*, Past. 17, 4; Swt. 113, 6, 20. Heó biþ áfeorrod swíðe feor from ðære sōðan heánesse *ab altitudine veræ celsitudinis elongatur*, 41, 2; Swt. 301, 20. On ðære heofonlícan heánnesse *in heaven on high*, Shrn. 82, 20: Exon. 65 a; Th. 239, 34; Ph. 631: Elen. Kmbl. 2247; El. 1125. Gode sý wuldor on heáhnesse *gloria in altissimis deo*, Lk. Skt. 2, 14. Ðīn mægen is swā mǣre swā ðæt ǣnig ne wāt ðā deópnesse drihtnes mihta ne ða heáhnisse heofena kyninges *thy power is so excellent that none knows the depth of the might of the lord nor the height of heaven's king*, Hy. 3, 35; Hy. Grn. ii. 282, 35. Āstīgend on heáhnisse *ascendens in altum*, Rtl. 83, 3. Ōsanna on heáhnessum *osanna in excelsis*, Mk. Skt. 11, 10: Ps. Th. 92, 5: Exon. 13 b; Th. 25, 34; Cri. 410: 10 a; Th. 10, 35; Cri. 162. Of heánessum *de alto*, Ps. Th. 143, 8.

heáh-rodor, es; *m. The lofty sky :*—Under heáhrodore *under the lofty sky*, Cd. 8; Th. 10, 3; Gen. 151.

heáh-sacerd, es; *m. A chief priest :*—Ða heáhsacerdas and ða bōceras *summi sacerdotes et scribæ*, Mk. Skt. 14, 1: 11, 27: 8, 31.

heáh-sǣ; *f. High, deep sea :*—Wealdend heofones and eorþan and heáhsǣ *ruler of heaven and of earth and of deep sea*, Bt. Met. Fox 11, 6; Met. 11, 3.

heáh-sǣl, e; *f. Great happiness :*—Mīnes mūþes mē mōdes willa on heáhsǣlum hraþe gebringe *voluntaria oris mei beneplacita fac*, Ps. Th. 118, 108.

heáh-sǣ-þeóf, es; *m. A chief pirate;* archi-pirata, Cot. 9, 171.

heáh-samnung, e; *f. A chief synagogue :*—Of hēhsomnungum *de arche-synagogis*, Mk. Skt. Lind. 5, 22. v. heáh-gesamnung.

heáh-sangere, es; *m. A chief singer, arch-chanter :*—Se ārwurþa wer Johannes S. Petres cyricean ðæs apostoles heáhsangere *vir venerabilis Johannes archicantator ecclesiæ S. Apostoli Petri*, Bd. 4, 18; S. 586, 23.

heáh-sceáwere, es; *m. A chief overlooker, overseer :*—Hēhsceáwere *pontifex*, Rtl. 21, 1.

heáh-sciremann, es; *m. A procurator :*—Hēhsciremenn *procuratores*, Rtl. 193, 11.

heáh-seld, es; *n. A throne :*—Ðonne wē tō hēhselde hnígan þencaþ *when we intend to bend to the throne*, Cd. 217; Th. 277, 21; Sat. 208:

221; Th. 287, 25; Sat. 372. Ymb ðæt hálge heáhseld godes *around the holy throne of God*, Exon. 64 b; Th. 239, 11; Ph. 619. Hēhselda wyn *the joy of thrones*, Cd. 213; Th. 267, 25; Sat. 43.

heáh-sele; *m. A high hall* :—Tó ðæm heáhsele *to the high hall*, Beo. Th. 1298; B. 647. [*Icel.* há-salr *a high hall.*]

heáh-setl, es; *n. A high seat, throne, seat of honour* [*at table*], *seat of justice* :—Ðín heáhsetl *thronum*, Ps. Th. 88, 26. Forðon hēhsedil godes is *quia thronus Dei est*, Mt. Kmbl. Lind. 5, 34. Ðonne crist siteþ on his cynestōle on heáhsetle *when Christ sitteth on his royal seat, on his throne*, Exon. 25 b; Th. 75, 7; Cri. 1218: Lchdm. iii. 426, 6. Se ríca man ðe sitt on his heáhsetle hraðe geswícþ hē his gebeórscipes gif ða þeówan geswícaþ ðæra teolunga *the great man that sits on his high seat will soon discontinue his feast if the servants discontinue the attendance*, Homl. Th. i. 272, 35. Ðá hē ðá sett on hēhsettle *sedente autem illo pro tribunali*, Mt. Kmbl. Rush. 27, 19. Fore ðæm hēhsedle *pro tribunali*, Jn. Skt. Lind. 19, 13. Be ðám unrihtwísum cyningum ða wē gesióþ sittan on ðám hēhstan heáhsetlum *concerning unjust kings whom we see sitting on the highest thrones*, Bt. 37, 1; Fox 186, 2. [*Laym.* hæh-setle *throne* : *O. H. Ger.* hōh-sedal *thronus, solium, triclinium* : cf. *Icel.* há-sæti *a high seat* (*at table*).]

heáh-stede, es; *m. A high place* :—Ðenden ðǽr wunaþ on heáhstede húsa sélest *whilst there in its lofty place the best of houses continues*, Beo. Th. 575; B. 285. [*Icel.* há-staðr *a high place.*]

heáh-stefn; *adj. Having a high stem* or *prow* :—Heáhstefn naca *the high-prowed boat*, Andr. Kmbl. 532; An. 266. Heáhstefn scipu *high-prowed ships*, Exon. 96 b; Th. 361, 2; Wal. 13.

heáh-strǽt, e; *f. High road* :—Swá in ða heáhstrǽt so *into the high road*, Cod. Dipl. Kmbl. iii. 167, 21. Tó ðære hēhstrǽte *to the high road*, 246, 20.

hēh-sunn [?]; *adj. Very sinful* :—Openlíce synnige ł hēhsunne *publicani*, Mk. Skt. Rush. 2, 15.

heáh-synn, e; *f. Mortal sin, crime, wickedness* :—Hēhsynn *crimen*, Rtl. 187, 25. Búta hēhsynne sint *sine crimine sunt*, Mt. Kmbl. Lind. 12, 5. Bebeorh ðé wið ða eahta heáhsynna *cave tibi ab octo capitalibus criminibus*, L. Ecg. C. pref; Th. ii. 132, 5. Hēhsynna *scelera*, Rtl. 5, 16. Hēhsynno *facinora*, 42, 15.

heáh-þearf, e; *f. Great need* :—Æt heáhþearfe *at my greatest need*, Ps. Th. 117, 16, 20, 27.

heáh-þegen; *m. A great, high* or *chief minister* or *servant* :—On ðam wǽron gecorene twelf heáhþegenas *in that were chosen twelve chief ministers* [*the twelve apostles*], Homl. Th. ii. 520, 24.

heáh-þegnung, e; *f. High service* :—Heáhþegnunga hāliges gástes *the high services of the holy Spirit*, Cd. 147; Th. 183, 23; Exod. 96.

heáh-þeód, e; *f. A great, chief people* :—Was sum æþela man on ðære hēhþeóde Myrcna ríce *there was a certain noble man in the great kingdom of Mercia*, Guthl. 1; Gdwin. 8, 2.

heáh-þrymness, e; *f. Great glory*, Hy. 7, 51; Grn. ii. 288, 51: ç, 43; Hy. Grn. ii. 292, 43.

heáhþu, hēhþo, hiéhþo; *generally indecl* ; *f. Height, high place, glory* :—Hē his áras of heáhþu hider onsendeþ *he will send his messengers hither from above*, Exon. 19 a; Th. 47, 24; Cri. 760: 19 b; Th. 49, 21; Cri. 789: 69 b; Th. 258, 10; Jul. 263. On hēhþo on high, Andr. Kmbl. 1745; An. 875: 1995; An. 1000. Of hēhþo *from above*, 2289; An. 1146. Of hiéhþa, Elen. Kmbl. 2171; El. 1087. Heofona heáhþu gereccan *to tell the glory of the heavens*, Exon. 116 a; Th. 446, 33; Dōm. 31. Heofona heáhþu gestígan *to mount to the heights of heaven*, 117 a; Th. 451, 2; Dōm. 97. Gesēgon hí on heáhþu hláford stígan of grundum *they saw the Lord ascend to heaven from earth*, 15 a; Th. 31, 19; Cri. 498. Heofonríces hēhþe, Cd. 17; Th. 21, 8; Gen. 323. In heáhþum *on high*, Exon. 13 b; Th. 26, 8; Cri. 414: 44 a; Th. 149, 27; Gū. 768. Of heáhþum *from on high*, 46 b; Th. 158, 17; Gū. 910. [*Goth.* hauhiþa *height, loftiness, exaltation* : *O. H. Ger.* hōhida ʌltitudo, culmen.]

heáh-þungen; *adj. Of high rank, distinguished, noble* :—Heáhþungen wer *the noble man* [*Moses*], Cd. 169: Th. 210, 18; Exod. 517. Hē befæste ðæt ríce heáhþungenum menn Harolde *he committed the kingdom to a noble man, to Harold*, Chr. 1065: Erl. 198, 11; Edw. 30. Ða kyningas and ða óðre heáhþungene men *kings and other men of high rank*, Ors. 1, 1; Swt. 20, 22. Móton wyt ðonne unc on heofonum heáhþungene beón *we two may then be exalted in heaven*, Soul Kmbl. 315; Seel. 161. v. heáh-geþungen.

heáh-tíd, e; *f. A high time, high day, festival, solemnity* :—Tó ǽghwilces apostoles heáhtíde fæste man and freólsige *at every apostle's festival let there be fasting and feasting*, L. Eth. v. 14; Th. i. 308, 15. Hēhtíde *solemnia*, Rtl. 8, 23: 9, 27. [*Icel.* há-tíð *a high day, festival.*]

heáh-timber, es; *n. A lofty building* :—Heáhtimbra gehwæs *of every lofty building*, Exon. 79 a; Th. 296, 2; Crä. 45. v. heáh-getimbru.

heáh-torras; *pl. m. Alpes*, Hpt. Gl. 454.

heáh-treów, e; *f. An excellent, noble compact*, Cd. 162; Th. 202, 14; Exod. 388.

heáh-weofod, es; *n. The high altar* :—Gesceot bæftan ðæm heáh-

weofode *propitiatorium* vel *sanctum sanctorum*, vel *secretarium*, vel *pastoforum*, Ælfc. Gl. 109; Som. 79, 27; Wrt. Voc. 59, 1.

heáh-weorc, es; *n. Lofty work* :—Æfter heáhweorce heofenes ðínes *secundum altitudinem cæli*, Ps. Th. 102, 11.

heáh-wita, an; *m. A chief councillor* :—Fērde se cyng him hám and ða ealdormenn and ða heáhwitan *the king went home and the aldermen and the chief 'witan,'* Chr. 1009; Erl. 142, 10. v. Kmbl. Saxons in England, ii. 209, 9.

heal, hal, es; *m. n.* [?] *A corner, an angle, a secret place* [?] :—Heal oððe hyrne *angulus*, Wrt. Voc. 80, 73. Ǽlc wag biþ gebiéged twiefeald on ðæm heale *duplex semper est in angulis paries*, Past. 35, 5; Swt. 245, 13. Ðá gemētte hē hine hleonian on ðam hale *his cyrcan wið ðam* weofode *he found him leaning in the corner of his church against the altar*, Guthl. 20; Gdwin. 82, 22. On halum *in abditis*, Ps. Spl. 16, 13. [Cf. we beth honted from hale to hurne, Pol. Songs Wrt. 150, 17. In one swiþe diȝele hale, O. and N. 2.]

heal. v. healh *and* heall.

heála, an; *m. Rupture, hydrocele* :—Gif hē hæfde heálan *si fuerit ponderosus*, Past. 11, 1; Swt. 65, 5. [Cf. *Icel.* haull; *m.* hernia : *O. H. Ger.* hola; *f.* [?] *hernia*, Grff. iv. 848.]

heal-ærn; es; *n. A house with a hall, palace*, Beo. Th. 156; B. 78.

heald, es; *n. Hold, guardianship, protection, rule* :—Hí gecuron Harold tó healdes ealles Engla landes *they chose Harold to rule over all England*, Chr. 1036; Erl. 164, 14. Wit synd ðisra landa hald and mund *we two will be a protection and a defence to these lands*, Cod. Dipl. Kmbl. iv. 73, 5. [*Orm.* hald *support* : *Icel.* hald ; *n. upholding, support, custody, keeping.*] v. ge-heald.

heald; *adj. Bent, inclined* :—Ðeáh hí síen ásigen tó yfele and ðider healde *though they are sunk to evil and thither inclined*, Bt. 24, 4; Fox 84, 29. Ealle bióþ of dúne healde wið ðære eorþan *all are bent down towards the earth*, 41, 6; Fox 254, 28. Ða men lágon áþænede on ðære eorþan mid of dúne healdum ondwleotan *the men lay stretched out on the ground with faces turned downwards*, Shrn. 81, 26. [*Icel.* hallr *leaning, sloping* : *O. H. Ger.* hald *clivus, obliquus, pronus.*]

healdan, haldan; *p.* heóld; *pp.* healden. I. *to* HOLD, *keep, grasp, retain, restrain, confine, contain* :—Hēht Petrus and Paulus on bendum healdan *ordered Peter and Paul to be kept in bonds*, Blickl. Homl. 189 17: Bt. Met. Fox 1, 141; Met. 1, 71. Gif se hláford wiste ðæt se oxa hnitol wǽre and hine healdan nolde *if the lord knew that the ox were wont to push with its horn, and would not keep it in*, L. Alf. 23; Th. i. 52, 12. Se wísa hilt his sprǽce and bítt tíman *the wise man restrains his speech and bides his time*, Past. 33, 4; Swt. 220, 14. Afene streám healt ðone norþende *the river Avon bounds the north side*, Cod. Dipl. Kmbl. iii. 466, 21. Jacob heóld ðone yldran bróðer Esau be ðam fēt *Jacob held the elder brother Esau by the foot*, Homl. Th. i. 110, 22 : Beo. Th. 1581; B. 788. Hē heóld his ǽhta him tó wlencum *he kept his possessions for his own glory*, Blickl. Homl. 53, 8. Judéi heóldon heora eáran *the Jews stopped their ears*, Homl. Th. i. 46, 33. Genim ðás ylcan wyrte and heald hý mid ðé *take this same plant and keep it with you*, Herb. 111, 3; Lchdm. i. 224, 22. Gif hē næbbe ǽhta ðonne healde hine man tó dōme *if he have no property, then let him be held to judgment*, L. Ed. 5; Th. i. 162, 21 : L. C. S. 43; Th. i. 402, 1. Se ðe ofer ðæm dæg hit healde ágyfe ðam bisceope ðæne penig and ðærtó xxx penega *he that keeps it* [*Peter's pence*] *beyond that day, let him pay the penny to the bishop and thirty pence besides*, L. C. E. 9; Th. i. 366, 16. Healde ðonne on his múþe of ðam ecede lange hwíle *let him hold some of the vinegar in his mouth a long while*, Herb. 181, 4; Lchdm. i. 318, 2. Hū nytt rehton wē nú and rímdon ða cǽga búton wē eác feáwum wordum ætiéwen hwæt hie healden *of what use were it to describe and enumerate the keys, unless in a few words we shew what they lock up*, Past. 23; Swt. 178, 12. Wæterfatu healdende ǽnlípige twýfealde gemetu oððe þrýfealde. Nis gecweden ðæt ða wæterfatu sume heóldon twýfealde gemetu, sume þrýfealde *waterpots holding singly two or three measures. It is not said that some of the waterpots held two, some three measures*, Homl. Th. ii 56, 21–5. II. *to hold, have, possess, occupy, inhabit* :—Hie leng ne mágon healdan heofonríce *they may not longer occupy the heavenly kingdom*, Cd. 35; Th. 45, 25; Gen. 732: 26; Th. 33, 34; Gen. 530. Fundon on sande hlínbed healdan ðone ðe him hringas geaf *they found him who had given them rings occupying a couch on the sand*, Beo. Th. 6060; B. 3034. Ðú ðe heofonhāmas healdest and wealdest *qui habitas in cælo*, Ps. Th. 122, 1. Hér Cynegils fēng tó ríce and heóld xxxi wintra *in this year Cynegils came to the throne and held it thirty-one years*, Chr. 611; Erl. 20, 34. Ðǽr heó ǽr mæste heóld worolde wynne *in whom before she had had her chief joy in this life*, Beo. Th. 2163; B. 1078: 6079; B. 3043. Úre ieldran ða ðe ðás stówa ǽr hióldon *our forefathers who occupied these places before*, Past. pref; Swt. 5, 14: Beo. Th. 2432; B. 1214. III. *to rule, govern* :—Hie sealdon ánum unwísum cyninges þegne Miercna ríce tó haldanne *they gave Mercia to a foolish king's thane to rule*, Chr. 874; Erl. 76, 28: Beo. Th. 3709; B. 1852. Gif hē hí rihtlícor healdan wolde ðonne hē ǽr dyde *if he* [*Ethelred*] *would rule them more righteously than he had done before*, Chr. 1014; Erl. 150, 7: 1083; Erl. 217, 5. Ðú

eorþbúende ealle healdest *gentes in terra dirigis*, Ps. Th. 66, 4. Heóld ðæt folc teala *he ruled that people well*, Cd. 62; Th. 74, 34; Gen. 1232: Beo. Th. 114; B. 57. Eác áh hláforda gehwylc ðæs for mycle þearfe ðæt hé his men rihtlíce healde *also every lord has very great need to rule his men with justice*, L. C. E. 20; Th. i. 372, 13.   **IV.** *to behave, conduct* [*one's self*]:—Hú se sacerd hine healdan sceal and se diácon *quomodo sacerdos et diaconus se gerere debeant*, L. Ecg. P. iii. pref. v; Th. ii. 194, 29. Nolde ða béc ágifan ǽr heó wyste hú getríwlíce hé hi [hine?] æt landum healdan wolde *she would not give up the charters before she knew with what faith he would conduct himself* [*or treat her?*] *as regarded the lands*, Chart. Th. 202, 27. Wé sceolan eall úre líf on eáðmódnesse healdan *we should lead all our life in humility*, Blickl. Homl. 13, 1. Heó hit heóld ǽr tó fæste wið hine *she had before dealt too hardly with him*, Chr. 1043; Erl. 168, 10. Gif hé hine heólde swá swá hé sceolde *if he conducted himself as he ought*, L. R. 7; Th. i. 192, 15. Ic lǽrde weras ðæt hie be him ánum getreówlíce hie heóldan *I taught husbands to act faithfully, having to do with their wives only*, Blickl. Homl. 185, 24.   **V.** *to guard, defend, keep, preserve, protect, maintain, sustain, regard, observe, take heed:*—Him behéton ðet hí woldon ðisne eard healdan *they promised him that they would defend this land*, Chr. 1012; Erl. 147, 10. Se ðe sceal healdan Israéla folc wið feóndum *qui custodit Israel*, Ps. Th. 120, 4. Ðá héht Neron healdan Simones líc þrý dagas *Nero ordered Simon's body to be kept three days*, Blickl. Homl. 189, 20. Hí ǽfre woldon fryþ and freóndscype in tó ðisan lande haldan *they would ever maintain peace and friendship towards this land*, Chr. 1066; Erl. 201, 37. Uton healdan unc ðæt wit ne wénan swá swá ðis folc wénþ *let us guard ourselves from thinking as this people thinks*, Bt. 40, 2; Fox 236, 28. Healdan ðone hálgan sunnan dæg *to keep the holy Sunday*, Lchdm. iii. 226, 2. Ðæt hé hý healdan wille swá wǽr his wíf sceal *that he will keep her as a man shall his wife*, L. Edm. B. 1; Th. i. 254, 6. Utan ǽnne cynehláford holdlíce healdan *let us loyally support one royal lord*, L. Eth. v. 35; Th. i. 312, 21: vi. 1; Th. i. 314, 11. His múþ hé sceal symble from yfelum wordum healdan *he shall ever keep his mouth from evil words*, L. E. I. 21; Th. ii. 416, 33. Clǽnnysse healdan *castitatem servare*, L. Ecg. P. iii. 5; Th. ii. 198, 2. Wé sceolan ða tén bebodu healdan *we ought to keep the ten commandments*, Blickl. Homl. 35, 11. Sceolde ic mínne bróðor healdon *am I my brother's keeper?* Gen. 4, 9. Ðǽre heorde ðe hí healdan sceoldan *to the flock that they should have kept*, Blickl. Homl. 45, 15. Hí ne dorstan nán gefeoht healdan wið Willelm cynge *they dared not have any battle with king William*, Chr. 1075; Erl. 214, 8. Oðer æt hám beón heora land tó healdanne oðer út faran tó winnanne *vicissim curam belli et domus custodiam sortiebantur*, Ors. 1, 10; Swt. 46, 17. Tó healdenne, Blickl. Homl. 11, 25. Se ðe hylt Israhél *qui custodit Israel*, Homl. Th. ii. 230, 7. Swá swá sealt hylt ǽlcne mete wið forrotodnysse *as salt preserves every meat from corruption*, 536, 19. Healdeþ meotudes ǽ *keeps the law of the Lord*, Exon. 62 b; Th. 229, 19; Ph. 457. Wið óðrum unþeáwum hí sylfe healdaþ *they keep themselves from other vices*, Homl. Th. ii. 550, 25. Ne ða Eástron swá healdaþ swá wé healdað *nec Pascha ita observant uti nos observamus*, L. Ecg. P. add. 5; Th. ii. 232, 18. Ðíne gebróðru healdaþ scép on Sichima *thy brethren are keeping sheep in Shechem*, Gen. 37, 13, 2. Ebréi healdaþ heora geáres annginn on lenctenlícre emnihte *the Hebrews keep the beginning of their year at the spring equinox*, Lchdm. iii. 246, 17. Ða gelǽredan hine healdaþ bé ðisum foresǽdan gesceáde *the learned consider it in accordance with the aforesaid distinction*, 266, 11. Ðú heólde míne líchaman wið ǽlce besmittennysse *thou hast kept my body from every defilement*, Homl. Th. i. 74, 30. Hine swá lange heóld óð ðæt man hire gryþ salde *she held the castle until they made terms with her*, Chr. 1076; Erl. 214, 18. Se cyng heóld his híred on Winccastre *the king held his court at Winchester*, 1085; Erl. 218, 39. Ðonne hí wǽron be eáston ðonne heóld man fyrde be westan *when the Danes were to the east then the 'fyrd' was assembled to the west*, 1009; Erl. 144, 5. Heó hyt swýðe deórwyrþlíce heóld *she held it very dearly*, St. And. 38, 3. Ða weardas heóldon ðæs cwearternes duru *the keepers kept the door of the prison*, Homl. Th. ii. 382, 4. Wé náðor ne heóldon ne láre ne lage Godes ne manna swá swá wé scoldon *we have not kept as we should the doctrine or law of God or men*, Swt. A. S. Rdr. 107, 80. Ðá heóldon ða Judéi on heálícum gewunan *the Jews then held it as a solemn custom*, Homl. Th. ii. 252, 8. Heald ðonne georne ðæt se mete sí gemylt *observe then carefully that the meat be digested*, L. M. 2, 69; Lchdm. ii. 284, 2. Heald ðæt hie ne hrínan eorþan ne wæstre *take care that they do not touch earth or water*, L. M. 3, 1; Lchdm. ii. 306, 7. Ásette gé ðone líchoman tó ðære byrgenne and hine ðǽr healdaþ swá ic eów bebeóde *put down the body in the tomb and keep it there as I shall bid you*, Blickl. Homl. 147, 32. Healden hie hie ðæt hie ne weorðen ealdormenn tó forlore hira hiéramonnum *caveat ne fiat subditis auctor ruinæ*, Past. 10, 2; Swt. 63, 16. Hit betere wǽre ðæt heora seht tógædere wurde ðonne hý ǽnige sace hym betweónan heóldan *it would be better for them to come to an agreement than to maintain a suit between them*, Chart. Th. 377, 4: Blickl. Homl. 109, 16.   **VI.** *to hold out, last, hold on, continue, hold with:*—Hé hét ðæt werod healdan feste wið feóndum *he bade that band stand fast against the foes*,

Byrht. Th. 134, 51; By. 102. Hé wel healdeþ stondeþ stíðlíce *it holds well, stoutly it stands*, Exon. 93 b; Th. 351, 27; Sch. 86. Feáwa óðre ðe mid ðam eorle gyt heóldan *a few others that still continued with the earl*, Chr. 1106; Erl. 241, 7. Ðá nolde seó burhwaru ábúgan ac heóldan mid fullan wíge ongeán *the citizens would not submit but held their ground against him by all warlike means*, Chr. 1013; Erl. 148, 12. Hig heóldon þurh ða brycge *they held on their way through the bridge*, 1052; Erl. 184, 23. Hí heóldon ofer sǽ tó Flandran *they took their way across the sea to Flanders*, 1075; Erl. 214, 9. [Cf. *halda* as a nautical term in Icelandic, Cl. & Vig. p. 233, col. 1.] [**Goth.** haldan *to hold, keep, keep sheep*: **O. Sax.** haldan: **O. Frs.** halda: **Icel.** halda: **O. H. Ger.** haltan *servare, custodire*: **Ger.** halten.] DER. an-, be-, for-, ge-, ofer-, tó-, ymb-healdan.

healdend, es; *m.* *One who holds, keeps, sustains, rules, a guardian, keeper, ruler:*—Hér liþ beheáfdod healdend úre *here lies our ruler beheaded*, Judth. 12; Thw. 25, 32; Jud. 290. Ic ðæs folces beó hyrde and healdend *I will be the people's shepherd and keeper*, Cd. 106; Th. 139, 25; Gen. 2315. Se hálga healdend and wealdend *the holy preserver and ruler*, Andr. Kmbl. 450; An. 225. Se healdend *the ruler*, Cd. 98; Th. 130, 17; Gen. 2161. From ðam healdende ðe mé hringas geaf *from the guardian who gave me rings*, Exon. 105 b; Th. 402, 1; Rä 21, 23. Mið haldendum *cum custodibus*, Mt. Kmbl. Lind. 27, 66. v. healdan.

heald-nes, -ness, e; *f.* *Holding, keeping, observance:*—Ealles mǽst ymb eástrena healdnyssa *maxime in Pascha observando*, Bd. 2, 4; S. 505, 7.

heálede; *adj.* *Ruptured, hydrocelous:*—Heálede *hirniosus*, Ælfc. Gl. 76; Som. 71, 126; Wrt. Voc. 45, 29. Heálede *ydropicus*, Wrt. Voc. 283, 62. Heálede *ponderosus*, Past. 11, 7; Swt. 73, 4, 9, 11: Herb. 78, 2; Lchdm. i. 182, 1: Lchdm. iii. 144, 26. [Cf. **Icel.** haula *ruptured:* **O. H. Ger.** holoht *ponderosus, cui humor viscerum in virilia labitur.*] v. heála.

healf, e; *f.*   **I.** *a half:*—Healfe ðý swétre *sweeter by half*, Bt. Met. Fox 12, 18; Met. 12, 9.   **II.** *side, part:*—Mid ðæm worde biþ gecýðed hwæþer healf hæfþ ðonne sige *with that phrase* [*asking permission to bury the dead*] *is declared which side has the victory*, Ors. 3, 1; Swt. 100, 9. Him be healfe stód cniht *by his side stood a youth*, Byrht. Th. 136, 16; By. 152. Fram ðære uferran healfe *from the upper part*, L. M. 1, 27; Lchdm. ii. 68, 14. On ðás healfe *hac:* on ða healfe *illac:* on ða swíðran healfe *dextrorsum:* on ða winstran healfe *sinistrorsum*, Ælfc. Gr. 38; Som. 40, 4, 6. Ðǽr stent lang leóma of hwílum on áne healfe hwílum on ælce healfe *there stands out from it a long light, sometimes on one side, sometimes on every side*, Chr. 891; Erl. 88, 20. On ǽgðere healfe *on either side*, 1014; Erl. 150, 15. Hí heregodon on heora healfe and Cnut on his healfe *they harried on their side and Cnut on his*, 1016; Erl. 154, 23: 1025; Erl. 163, 10. On twá healfe ðære eás *on both sides of the river*, 896; Erl. 94, 11. Gif ðú fǽrst tó ðære winstran hælfe ic healde ða swíðran healfe gif ðú ðonne ða swíðran healfe gecíst ic fare tó ðære winstran healfe *if thou wilt take the left hand then I will go to the right hand; or if thou depart to the right hand then I will go to the left*, Gen. 13, 9: 48, 13. [**Goth.** halba: **O. Sax.** halba: **O. Frs.** halve: **Icel.** hálfa: **O. H. Ger.** halb, halba, Grff. iv. 882-6: **Ger.** halbe.]

healf; *adj.* HALF:—Mé næs be healfan dǽle ðín mǽrþ gecýdd *thy greatness was not half told me*, Homl. Th. ii. 584, 23. Síe be healfum ðæm ðonne sió bót *let the fine then be half that*, L. Alf. pol 11; Th. i. 68, 18: 39; Th. i. 88, 2: L. M. 2, 65; Lchdm. ii. 292, 17. Gé ðær búgiaþ on ðam fíftan dǽle healfum londes and unlondes *ye there dwell in the half of the fifth part* [*in the tenth part*, cf. l. 25] *of land and not-land*, Bt. 18, 1; Fox 62, 23. Heó mid ðæm healfan dǽle beforan ðæm cyninge farende wæs swelce heó fleónde wǽre *with half the army she was going before the king as if she were fleeing*, Ors. 2, 4; Swt. 72, 7. Healfne sealde ðæm þearfan and mid healfum hine sylfne besweóp *he gave half* [*his cloak*] *to the poor man and wrapped himself up with half of it*, Blickl. Homl. 215, 7. Ðeáh ðú wylle healf míne ríce *licet demedium regni mei*, Mk. Skt. 6, 23: Lk. Skt. 19, 8. Habban hí ðone brýce healfne and healfne ða munecas *let them have half the usufruct, and the monk's half*, Chart. Th. 547, 19. Heó healfne forcearf ðone sweoran him *she half cut through his neck*, Judth. 10; Thw. 23, 4; Jud. 105. Sele ðonne ðæt healf tó drincanne *then give half of it to drink*, L. M. 2, 2; Lchdm. ii. 180, 23. Hie wǽron simle healfe æt hám healfe úte *always half of them were at home and half out*, Chr. 894; Erl. 90, 17: Ors. 2, 6; Swt. 86, 25. Ic wille ðæt man frigæ hǽalue míne men *I desire that half my men should be freed*, Chart. Th. 522, 5. Æfter óðer healf hund daga *after a hundred and fifty days*, Gen. 8, 3. Hé heóld ðæt ríce óðrum healfum læs ðe xxx wintra *he reigned twenty-eight years and a half*, Chr. 901; Erl. 96, 24. Hit biþ óðres healfes fótes gemet bufan ðæm heáfde *it is a foot and a half above the head*, Shrn. 69, 2. Se bát wæs geworht of þriddan healfre hýde *the boat was made of two and a half hides*, Chr. 891; Erl. 88, 9. Ic him sylle vii æcras feórþe hælfne on ánum stede and feórþe halfne an óðrum stede *I give him seven acres, three and a half in one place and three and a half in another*, Cod. Dipl. Kmbl. iii. 263, 12-15. Nán rén ne com ofer eorþan feórþan healfan

geáre *no rain came upon the earth for three years and a half*, Lchdm. iii. 276, 19. Ðæt wæs ehtoþe healf híd *that was seven hides and a half*, Chart. Th. 550, 12. Seofon and twentigoþan healfes fótes *twenty-six feet and a half long*, Lchdm. iii. 218, 4, 12, 16, 19. [*Goth.* halbs: *O. Sax.* half: *O. Frs.* half: *Icel.* hálfr: *O. H. Ger.* halb: *O. Frs.* has the same use of half *with the ordinals*, other, thredda, fiarda, etc., half; so *O. H. Ger.* has anðar halb, dritte halp, Grff. iv. 890: so *Ger. In Icel.* the ordinal is placed after hálfr, hálfr annarr, etc.]

healf-clǽmed; *adj. Half finished [of house built with mud]* :—Mín ðæt healfclǽmede hús *my half-finished mud-hut*, Shrn. 39, 20.

healf-clypigende; *adj. Semi-vowel* :— Healfclypigende *semivocales*, Ælfc. Gr. 2; Som. 2, 55, 56.

healf-cwic; *adj. Half alive, half dead* :—Halfcwic *semivivus; half dead*, Lk. Skt. Lind. 10, 30. Helfcuicne, Past. 17; Swt. 125, 8. Funde hiene ǽnne be wege licgan healfcucne *invenit in itinere solum relictum et extrema vitæ efflantem*, Ors. 3, 9; Swt. 128, 14. Sume healfcwice flugon on fæsten *some half-dead fled to the fastness*, Elen. Kmbl. 266; El. 133: Blickl. Homl. 203, 19.

healf-deád; *adj. Half dead, palsied on one side* :—Wið ðære healfdeádan ádle *for the half-dead disease* [hemiplegia], L. M. 2, 59; Lchdm. ii. 280, 1: L. M. 1, 79; Lchdm. ii. 152, 2.

healf-eald; *adj. Half grown, of middle age* :—Halfeald swín *half-grown swine*, L. M. 2, 37; Lchdm. ii. 246, 2.

healf-heáfod; *es; n. The fore part of the head*; sinciput, Ælfc. Gr. 9, 78; Som. 14, 24.

healf-hunding; *es; m. A creature having a dog's head* :— Healfhundingas *cenocephali*, Nar. 34, 30: 22, 15.

healf-hwít; *adj. Half white, whitish*; subalbus, Ælfc. Gl. 79; Som. 72, 73; Wrt. Voc. 46, 30.

healf-mann; *es; m. Half man* :—Halfmann *semivir*, Ælfc. Gr. 8; Som. 7, 23.

healf-penig-wurþ; *es; n. A halfpennyworth*, L. C. E. 12; Th. i. 366, 32.

healf-reád; *adj. Reddish* :—Healfreáde peran *crustumie vel volemis vel insana vel melimendrum*, Ælfc. Gl. 60; Som. 68, 40; Wrt. Voc. 39, 25.

healf-slǽpende; *adj. Half asleep* :—Ætýwde him gamalielus gást healfslǽpendum *the spirit of Gamaliel appeared to him when half asleep*, Shrn. 113, 5.

healf-soden; *adj. Half cooked* :—On healfsodenum mete *in semicocto cibo*, L. Ecg. C. 40; Th. ii. 166, 2: Med. ex Quadr. 7, 2; Lchdm. i. 356, 18.

healf-trendel; *es; n. A hemisphere* :—Healftryndel, *emisperia*, Ælfc. Gl. 49; Som. 65, 71; Wrt. Voc. 34, 6.

healfunga; *adv. By halves, partially, imperfectly* :— Ðe shundredes ealdor geneálǽhte ðam Hǽlende ná healfunga ac fulfremedlíce *this centurion did not approach the Saviour by halves, but fully*, Homl. Th. i. 126, 23. Hit is nyttre ðæt hit mon healfunga sprece *it is better that it should be said in part only*, Past. 31; Swt. 207, 7: 32; Swt. 209, 22. Gif wé healfunga and be sumum dǽle heora gódan weorc secgeaþ *si quædam illorum bona ex latere requiramus*, 211, 16.

healf-weard; *es; m. One who has a share of another's property or power* :—Hé sette hine on his húse to halfwearde ealra him his ǽhta anweald betǽhte *constituit eum dominum domus suæ, et principem omnis possessionis suæ*, Ps. Th. 104, 17.

healf-wudu; a; *m. Field-balm; calamintha nepeta*, L. M. 1, 47; Lchdm. ii. 118, 1.

heal-gamen; *es; m. Hall-mirth, song*, Beo. Th. 2136; B. 1066.

healh, halh [*in the declension the final* h *seems to be omitted before an inflection*]; *m.* A word of doubtful meaning. Kemble, Cod. Dipl. iii. xxix. translates it *hall*, probably originally a *stone building*. Leo, A. S. Names, p. 52, takes it to be the same word as *ealh*. Somner gives *healh-stán crusta, collyrida.* In form it agrees with Latin *calx.* The following are some of the passages in which the word occurs :—Se westra eásthealh, Cod. Dipl. iii. 19, 6. On ðone west halh, 18, 25. Oþ cyninges healh, i. 257, 33. On Scottes healh; of ðam heale, vi. 2, 2. In Streónes halh; of ðam hale, 214, 25. On Hengestes healh; of Hengestes heale, iii. 80, 20. In Titten halh, 52, 11. [The word seems to have the same force as *haga* in the same charter, as *æt Batenhale* and *æt Batanhagan* both occur.] Æt Wreodanhale, i. 166, 18. On Rischale; of Rischale, iii. 399, 18. On hwítan heal; of hwítan heale, iii. 444, 4-5. On ða halas, iii. 34, 13. On fearnhealas; of fearnhealan, iii. 81, 14-5. On cotan healas, v. 401, 34. Tó hǽþhalan; of hǽþhalan, iii. 77, 13. Streónes halh, Bd. 4, 23; S. 592, 37. On Streónes heale, Chr. 680; Erl. 40, 13. [Strenaeshalch quod interpretatur Sinus fari, Bd. 3, 25; S. 132, 7.]

healic; *m. A herring*; halec :— Healic óðer sǽfisc *herring or seafish*, Cod. Dipl. Kmbl. iii. 250, 26.

heá-líc; *adj. High, elevated, lofty, sublime, proud, chief, very great, noble, distinguished, deep, profound* :—Nán gereord nis swá heálíc swá Ebréisc *no language is so noble as Hebrew*, Homl. Th. ii. 86, 28. Abram ðæt is heálíc fæder *Abram, that is, great father*, i. 92, 13. Leóht swilce heálíc sunnbeám *a light like a splendid sunbeam*, Swt. A. S. Rdr. 100,

152. Swíðe heálíc nama *a name of great distinction*, Blickl. Homl. 167, 31: L. E. I. 40; Th. ii. 438, 11. Is án ðæra eahta winda aquilo geháten se blǽwþ heálíc and ceald *one of the eight winds is called aquilo; it blows high and cold*, Lchdm. iii. 276, 5. Heálíc on his weorcum *actione præcipuus*, Past. 12; Swt. 75, 8. Gebletsod ys Abram ðam heálícan Gode . . . and gebletsod ys se heálíca God *blessed be Abram of the most high God . . . and blessed be the most high God*, Gen. 14, 19, 20. Nis nán leahter swá heálíc ðæt man ne mǽge gebétan *there is no crime so deep that it may not be expiated*, Homl. Th. ii. 602, 20. Hé næs ácweald þurh ðam heálícan fylle *he was not killed by the fall from such a height*, 300, 20. Mid heálícum gedwylde *through profound error*, 506, 27. On heálícum gemóte *in a principal meeting*, Swt. A. S. Rdr. 67, 348. Ðæt lengtenfæsten mon sceal mid swíðe heálícre gýmene healdan *the fast of Lent ought to be kept with the very greatest care*, L. E. I. 37; Th. ii. 436, 5. Heálíc þingc ðú ðǽrmid ongitst *thereupon thou wilt observe a remarkable thing*, Herb. 57, 2; Lchdm. i. 160, 1. Swá heálícne dem his ágnes hryres *alta ruinæ suæ damna*, Past. 58, 2; Swt. 441, 26. Hafaþ heálíce stefne *hath an excellent voice*, Exon. 79 b; Th. 298, 31; Crä. 93. Heálíce bodan *archangels*, Homl. Th. i. 342, 26: L. Eth. vii. 2; Th. i. 330, 6. Gif hie hwæt suá heálícra yfela on him ongieten *if they perceive any very great evil in them; si qua valde sunt eorum prava*, Past. 28, 5; Swt. 197, 6. On heálícum muntum *on lofty hills*, Homl. Th. ii. 160, 29. Wé lǽraþ ðæt man wið heálíce synna scylde swýðe georne *we instruct people to guard very diligently against very great sins*, L. C. E. 23; Th. i. 374, 6. Heálíce gegaderunga *legitima conjugia*, L. Ecg. C. 28; Th. ii. 152, 35. Spræc heálíg word wið drihten sinne *spoke proud words against his lord*, Cd. 15; Th. 19, 21; Gen. 294. Ælc sáwul sý underþeód heálícrum anwealdum *let every soul be subjected to the higher powers*, Homl. Th. ii. 362, 17. Se is heálícost seðe ðone martyrdóm æfter Gode ástealde *he is most exalted who was the first martyr after God*, i. 50, 1. Ða recceras scoldon þencean ymb ðæt hélícuste and ða underþióddan scoldon dón ðæt unweorðlícre *a subditis inferiora gerenda sunt, a Rectoribus summa cogitanda*, Past. 18, 3; Swt. 131, 19.

hea-líce; *adv. Highly, on high, excellently* :— Is ðín mildheort mód áhafen heálíce *magnificatur misericordia tua*, Ps. Th. 107, 4: 137, 6. Heálíce ða Cyricean reccende *ecclesiam sublimiter regens*, Bd. 5, 19; S. 639, 12. Seó gódnys is of ðam Scyppende se ðe is heálíce gód *that goodness is from the Creator, who is supremely good*, Homl. Th. i. 238, 19. Se ðe on heofonum is heálíce sittende *who sitteth on high in heaven*, ii. 318, 3: 254, 27. Heálíce geweorþod *highly honoured*, Blickl. Homl. 125, 18. Ðus heálíce *in such a high degree*, 123, 2. Ðonne fremaþ hit heálíce *it will do very great good*, Herb. 4, 2; Lchdm. i. 90, 7. Hé wolde ðæt his lof ðe heálícor weóxe *he desired that his praise should grow the greater*, Blickl. Homl. 33, 30. Heálícost fremede *was beneficial in the highest degree*, Herb. 73, 3; Lchdm. i. 176, 10.

heá-lícness, e; *f. Loftiness, sublimity, greatness* :—Heálícnyss *sublimitas*, Hymn Surt. 74, 26. Seó heofenlíce heálícnyss wearþ geopenod *the greatness of heaven was revealed*, Homl. Th. i. 106, 31.

heall, e; *f. A hall, residence* :—Heall *aula*, Ælfc. Gl. 61, 107; Som. 78, 89; Wrt. Voc. 58, 4. Mycel and rúm heall *atrium*, 109; Som. 79, 21; Wrt. Voc. 58, 61. Seó heall ðæs Hálgan Gástes *the residence of the Holy Ghost*, Blickl. Homl. 163, 13. Heal, Beo. Th. 2307; B. 1151. On his ðære hálgan healle *in aula sancta ejus*, Ps. Th. 95, 8. Hé dreám gehýrde hlúdne in healle *loud merriment he heard in the hall*, Beo. Th. 178; B. 89: Cd. 210; Th. 261, 1; Dan. 719. Hie tó his healle ne tó his híréde eft wendan noldan *they would not return to his [Nero's] residence nor household*, Blickl. Homl. 173, 18. On cynges healle *in the king's hall*, L. Alf. pol. 7; Th. i. 66, 7, 8: L. R. 2; Th. i. 190, 17. Ða heofenlícan healle innférde *entered the heavenly hall*, Homl. Th. i. 52, 20. [*O. Sax.* halla: *Icel.* höll: *O. H. Ger.* halla *aula, palatium, templum*: *Ger.* halle.] DER. gif-, medo-heall.

heal-líc; *adj. Belonging to a hall* or *palace*; aulicus, palatinus, Cot. 194, Lye.

heall-reáf; *es; n. A piece of tapestry for a hall* :—Ælfwine ic geann ánes heallreáfes *I give to Alfwine a piece of tapestry*, Chart. Th. 530, 35.

heall-wahrift; *es; n. Tapestry for hanging on the wall of a hall* :— Ic geann mínum suna ánes heallwahriftes, Chart. Th. 530, 33.

HEALM, es; *n.* I. *haulm, straw, stem or stalk of grass, stalk of a plant* :—Healm *culmus*, Ælfc. Gl. 59; Som. 67, 127; Wrt. Voc. 38, 49. Healmes láf *stipulæ*, Som. 67, 129; Wrt. Voc. 38, 51. Gán and gadrion him sylfe ðæt healm *let them go and gather straw for themselves*, Ex. 5, 7. Swá windes healm *sicut stipulam ante faciem venti*, Ps. Th. 82, 10. Genim rigen healm and beren *take rye and barley straw*, L. M. 1, 72; Lchdm. ii. 148, 11. II. *a roof of straw* [?] :—Ciricsceat mon sceal ágifan tó ðam healme and tó ðam heorþe ðe se mon on biþ tó middum wintra *ciricsceattum debet reddere homo a culmine et mansione, ubi residens erit in Natali*, L. In. 61; Th. i. 140, 13. [*Prompt. Parv.* halm *stipula: Icel.* hálmr; *m. straw: O. H. Ger.* halm; *m. culmus, calamus, stipula, festuca: Ger.* halm: *Grk.* κάλαμος *a reed.*]

healm-streaw, es; *n. Straw, stubble* :—Healmstreaw *stipulam*, **Ps.** Spl. 82, 12.

healoc, es; *m. n.* [?] *A hollow, corner, bending:* — Hēr sint tácn áheardodre lifre ge on ðām læppum and healocum and filmenum *here are symptoms of a hardened liver both on the lobes and hollows and membranes,* L. M. R. 21; Lchdm. ii. 204, 5. [Cf. (?) *Prompt. Parv.* halke *angulus, latibulum: Chauc.* halke, *corner.*] v. holc, hylca.

heal-reced, es; *n. A palace:*—Hē healreced hātan wolde medoærn micel men gewyrcean *he would bid men make a palace, a great mead-house,* Beo. Th. 136; B. 68.

heals, hals, es; *m. The neck, the prow of a ship:*—Se hals *the neck,* Exon. 60 a; Th. 218, 22; Ph. 298. Gehæfted be ðam healse *fastened by the neck,* Cd. 19; Th. 24, 29; Gen. 385. Heals ealne ymbefēng he *clasped all the neck,* Beo. Th. 5376; B. 2691. Lēt his francan wadan þurh ðæs .hysses hals *he let his weapon pass through the man's neck,* Byrht. Th. 135, 60; By. 141. [*Orm.* halls: *Piers P. Chauc.* hals: *Prompt. Parv.* hals *collum, amplexatorium: Goth. O. Frs. O. Sax. O.H. Ger.* hals: *Icel.* háls *neck, part of the bow of a ship.*]

heals-beág, es; *m. A ring for the neck, necklace;* monile, collare, Beo. Th. 4350; B. 2172. [*O.H. Ger.* hals-pouc *torques.*]

heals-beorh; *gen.* -beorge; *f. A protection for the neck, gorget, hauberk,* Hpt. Gl. 521, 423. [*Icel.* háls-björg *a gorget: O.H. Ger.* hals-pirc, -perg *pectoria, lorica.*]

heáls-bóc, e; *f. A book which brings safety, an amulet, a phylactery,* Mt. Kmbl. 23, 5. [*Icel.* háls-bók *a book to swear upon.*] v. háls.

healsed, healsde, healscod *a cloth for the head:*—Healsed *caputium,* Cot. 170, Lye. In halsado *in sudario,* Lk. Skt. Lind. 19, 20. Mið halsodo *sudario,* Jn. Skt. Rush. [halscode, Lind.] 11, 44. Halsodu *sudarium* [hascode, Lind.] 20, 7.

healseta, an; *m.* Se ealdorman rād þurh sumne wudu ðā ræsde ān næddre of holum treowe æt ðam healsetan him on ðone bōsm and hyne tōslāt ðæt hē wæs sōna deád, Shrn. 144, 27.

heals-fæst; *adj. Stiff-necked. stubborn,* Cd. 102; Th. 135, 5; Gen. 2238.

heals-fang, es; *n.* A term occurring in the laws which Thorpe thus defines: 'The sum every man sentenced to the pillory would have had to pay to save him from that punishment had it been in use.' The word occurs in the following passages:—Gif ceorl būton wífes wísdōme deóflum gelde hē síe ealra his æhtan scyldig and healsfange *if a married man without his wife's knowledge sacrifice to idols let him be liable in all his possessions and his 'heals-fang,'* L. Wih. 12, 11, 14; Th. i. 40, 5, 2, 10. Gylde man cxx scill. tō healsfange æt twelfhyndum were. Healsfang gebyreþ bearnum brōðrum and fæderan ne gebyreþ nānum mǣge ðæt feoh būte ðam ðe sý binnan cneówe. Of ðam dæge ðe ðæt healsfang ágolden sý, . . . , L. E. G. 13; Th. i. 174, 23–7: L. Edm. S. 7; Th. i. 250, 20: L. Eth. vi. 51; Th. i. 328, 11: L. C. S. 37: Th. i. 398, 13: 45; Th. i. 402, 14: 61; Th. i. 408, 19: L. C. F. 14; Th. i. 428, 7: L. H. 11, 7, 10; Th. i. 521, 5, 10: 76, 6; Th. i. 582, 4. Schmid A. S. Gesetze, p. 609, suggests a different origin from that given by Thorpe, 'Es liegt nahe, an die Berechnung der Verwandtschaftsgrade nach den Gliedern des menschlichen Leiber zu denken, wo dann die nächsten Verwandten, die auf den Halsfang Anspruch haben, in den Hals zu stehen kommen könnten, und damit hängt vielleicht Zusammen, dass die Gradberechnungen nicht von dem gemeinschaftlichen Stammvater, sondern dessen Kindern beginnen, sodass die näherstehenden Verwandten als *binnan cneówe* befindlich bezeichnet werden konnten.' But while this explanation might suit the circumstances described in the passage given above, from Edmund's Laws, it would not be applicable in the earlier passage from Wihtræd's Laws. Schmid seems to refer the penalty, in its origin, too exclusively to cases of killing: 'Eine Geldbusse, die bei einer Tödtung in Verbindung mit dem Wergeld an die nächsten Verwandten des Getödteten gezahlt werden musste, die aber auch sonst zur Bestimmung der Grösse einer Busse genannt wird.' [Cf. *Icel.* háls-fang; *n. embracing:* háls-fengja *to embrace.*]

heals-gebedda, an; *f. A bedfellow, consort around whose neck the arms are thrown, one dearly loved,* Beo. Th. 126; B. 63. v. healsmægeþ.

heals-gund, es; *m. A swelling in the neck;* struma, L. M. 1, 4; Lchdm. ii. 44, 10, 13, 15, etc.

heálsian. v. hálsian.

heal-sittende; *pl. People sitting in a hall,* Beo. Th. 4035; B. 2015: 5728; B. 2868.

heals-mægeþ, e; *f. A virgin embraced and beloved,* Cd. 98; Th. 130, 6; Gen. 2155. v. heals-gebedda.

heals-mene, -myne, es; *m. A necklace, chain for the neck:* — Hē dyde gyldene healsmyne ymbe his swuran *he put a gold chain about his neck,* Gen. 41, 42. [*O. Sax.* hals-meni; *n: Icel.* háls-men; *n.*]

heals-ome, an; *f. A humour in the neck,* Lchdm. iii. 4, 26.

healsre-feðer, e; *f. The feathers of a pillow, down:*—Hnescre ic eom micle halsrefeðre *I am much softer than down,* Exon. 111 b; Th. 426, 28; Rä. 41, 80. [Cf. *O.H. Ger.* halsare *cervical.*]

heals-wiða, an; *m. A necklace:*—Me healswiðan hláford sealde *my lord has given me a chain for my neck,* Exon. 102 b; Th. 387, 13; Rä. 5, 4.

heals-wyrt, e; *f.* In Lchdm. ii. Gloss. are given the following plant-names:—I. *Campanula trachelium, Dan.* halsurt: *Ger.* halswurz, halskraut: *Du.* halskrind. II. *Hare's ear;* bupleurum tenuissimum. III. *Scilla autumnalis.* IV. *Symphytum album.*

HEALT; *adj.* HALT, *lame, limping:*—Healt *claudus,* Wrt. Voc. 75, 35. Gif hē healt weorþ *if he become lame,* L. Ethb. 65; Th. i. 18, 14. Hæfdon him tō lādteówe ǣnne wísne mon, þēh hē healt wǣre and him tō gielpworde hæfdon ðæt him leófre wǣre ðæt hie hæfdon healtne cyning ðonne healt ríce *they had as their leader a wise man though he was lame, and made it their boast that they had rather the king halted than the kingdom,* Ors. 3, 1; Swt. 96, 28–31: Mt. Kmbl. 18, 8. Him tō eodan blinde and healte *the blind and halt went to him,* Blickl. Homl. 71, 21: Nicod. 2; Thw. 1, 29: Elen. Kmbl. 2427; El. 1215: Andr. Kmbl. 1155; An. 578. [*Goth.* halts: *O. Sax. O. Frs.* halt: *Icel.* haltr: *O.H. Ger.* halz.]

heal-þegen, es; *m. A hall-thane, one who resides or is occupied in a hall,* Beo. Th. 287; B. 142: 1443; B. 719.

healtian; *p.* ode; *pp.* od *To halt, limp, be lame:* — Ic healtige *claudico,* Ælfc. Gr. 28; Som. 31, 27. Hī nū gyt heora ealdan gewunon healdaþ and fram rihtum stígum healtiaþ *ipsi adhuc inveterati et claudicantes a semitis suis,* Bd. 5, 22; S. 644, 19. Hý healtodan on heora wegum *claudicaverunt a semitis suis,* Ps. Th. 17, 43. Ne healtigeaþ leng *ut non claudicans quis erret,* Past. 11, 1; Swt. 65, 18.

heal-wudu, a; *m. The woodwork of a hall,* Beo. Th. 2639; B. 1317.

heamol, hamol [?]; *adj. Frugal;* frugi, Cot. 86, Lye. [Cf. (?) *O.H. Ger.* hamal *mutilus.*]

heán; *adj. Low, mean, abject, poor, humbled, humble:*—Hiora heorte wæs heán on gewinnum *humiliatum est in laboribus cor eorum,* Ps. Th. 106, 11. Ic heán gewearþ hē mē hraðe lýsde *humiliatus sum et liberavit me,* 114, 6. Nǣnig eft síðade heán hyhta leás *none returned cast down and hopeless,* Exon. 46 a; Th. 157, 25; Gū. 897. Ðā ðū heán and earm ǣrest cwōme *when abject and poor thou first didst come,* 39 a; Th. 129, 23; Gū. 425. Dēmaþ ðam rícan swā ðam heánan and ðam litlan swā ðam miclan *judge the high as the low, and the little as the great,* Deut. 1, 17: Homl. Th. i. 64, 30. Hū uncūþ biþ ǣghwylcum ǣnum men his lífes tíd ǣghweðer ge rícum ge heánum ge geongum ge ealdum *how unknown to every single man is the period of his life, both to the rich man and to the poor, to the young and to the old,* Blickl. Homl. 125, 8. Habbaþ mē gehnǣged heánne tō eorþan *humiliavit in terra vitam meam,* Ps. Th. 142, 3. Ǣgðer ge welige ge heáne *simul in unum dives et pauper,* 48, 2. Swā ríce swā heáne vel *divites vel pauperes,* Bd. 3, 5; S. 526, 30. Se scearpa deáþ ðe ne forlēt ne ríce menn ne heáne se hine genam *stern death who spares neither rich men nor poor, that seized him,* Chr. 1086; Erl. 220, 35. Hī hí sylfe lēton ǣgðer ge for heáne ge for unwræste *ultima propemodum desperatione tabuerunt,* Ors. 3, 1; Swt. 98, 22. Hī taliaþ ðē wyrsan for heánan gebyrdan ða ðe heora yldran on worolde ne wurdan welige *they account the worse, for their humble birth, those whose forefathers were not rich in a worldly point of view,* L. Eth. vii. 21; Th. i. 354, 2. Ne wandige hē ná for rícum ne for heánum *non vereri potentes neque humiles,* L. Ecg. P. i. 1; Th. ii. 172, 3. Heánra burhwered *vulgus* vel *plebs,* Ælfc. Gl. 8; Som. 56, 82; Wrt. Voc. 18, 37. Heánra man vel *ceorlic* ǣhta *peculium,* 13; Som. 57, 122; Wrt. Voc. 22, 59. Se heánra hád *the weaker sex,* Exon. 9 a; Th. 7, 10; Cri. 99. Ne se heárra derige ðam heánran *nor let the higher injure the lower,* L. I. P. 7; Th. ii. 314, 1. Ðeáh hit se lǣsta wǣre and se heánosta *though it were the least and the lowest,* Blickl. Homl. 169, 23. [*Laym.* hæne, hene: *Goth.* hauns: *O.H. Ger.* hōn *humilis, infamis.*]

heán; *p.* heáde; *pp.* heád *To raise, heighten, exalt, advance:*—Mid singalum bysenum ārfæstre wyrcnysse hē ongan heán and miclian *continuis piæ operationis exemplis provehere curavit,* Bd. 2, 4; S. 505, 19. Heáþ and hebbaþ *exalt and raise,* Exon. 93 a; Th. 349, 6; Sch. 42. [*Goth.* hauhjan *to exalt: O.H. Ger.* hōhjan *exaltare: Ger.* erhöhen *exalt, raise.*]

heáne; *adv. Ignominiously, shamefully, abjectly:*—Dū sylfa mē heáne gehnǣgdest *humiliasti me,* Ps. Th. 118, 71. Scyldigra scólu āscyred weorþeþ heáne from hálgum *the band of the guilty shall with shame be separated from the holy,* Exon. 31 b; Th. 98, 17; Cri. 1609: 75 b; Th. 283, 16; Jul. 681. Swā hē sýn fram ðínes handa heáne ādrifene *quidem ipsi de manu tua expulsi sunt,* Ps. Th. 87, 5. [In some of these passages the word may be a case of the adjective rather than an adverb.]

heá-nes. v. heáh-nes.

heán-líc; *adj. Ignominious, disgraceful, vile, poor:*—Tō heánlíc mē þinceþ ðæt gē mid ūrum sceattum tō scype gangon unbefohtene *too shameful methinks that ye with our treasures should go to your ships without a struggle,* Byrht. Th. 133, 25. Swíðe nearewe sent and swíðe heánlíce ða menniscan gesǣlþa *very scanty and very poor are human felicities;* anxia enim res est humanorum conditio bonorum, Bt. 11, 1; Fox 30, 26: Ors. 2, 5; Swt. 84, 12. [*O.H. Ger.* hōn-líh *infamis, fœdus, ridendus, dedecor, indecor.*]

heán-líce; *adj. Ignominiously, ingloriously, disgracefully, miserably,*

humbly :—Fauius heánlíce **hámweard** ōþfleáh *Fabius ignominiously fled homewards,* Ors. 3, 10; Swt. 140, 13. Ne lǣt swā heánlíce ðīn handgeweorc forwurþan *let not thine handiwork so miserably perish,* Hy. 7, 111; Hy. Grn. ii. 289, 111 : Exon. 8 a ; Th. 3, 4; Cri. 31 : 13 a ; Th. 23, 21; Cri. 372.

heán-mód ; *adj. Dejected, cast down, humiliated :*—Ic sceal sárigferþ heánmód hweorfan *with sorrowing spirit and with dejected mind must I go,* Exon. 52 b ; Th. 184, 32; Gú. 1353. Ic sceal feor ðonan heánmód hweorfan *I must go far thence with humiliated heart,* 71 a ; Th. 265, 32; Jul. 390.

heán-spédig ; *adj. Scantily, poorly endowed :*—Ðȳ lǣs hē forhycge heánspédigran *lest he despise the more scantily endowed,* Exon. 78 b ; Th. 295, 1; Crä. 26.

heáp, es ; *m.* [*generally, but* ðeós earme heáp *occurs,* Cd. 215; Th. 270, 9; Sat. 87.] *A* HEAP, *pile, great number, host, multitude, crowd, band, troop, body of people, assembly, company :*—Galað ðæt is gewitnesse heáp *Galand acervus testimonii interpretatur,* Past. 48, 2 ; Swt. 367, 5. Se hálga heáp hēhfædera and wītgena *the holy host of patriarchs and prophets,* Blickl. Homl. 81, 9. Fyrenfulra þreát heáp synnigra *peccatores,* Ps. Th. 91, 6. Þegna heáp *a troop of thanes,* Beo. Th. 805; B. 400. Be ðam gesǣligan heápe ðe mid ðam Hǣlende on ðisum lífe drohtnode of *the blessed company that lived with the Saviour in this life,* Homl. Th. ii. 520, 22. Of ðam yfelan heápe gehádodra manna be ðām ðe úre Drihten cwæþ 'multi dicunt mihi, etc.' *of that evil band of men to whom our Lord said,* '*many will say to me, etc.*' L. Ælfc. P. 40; Th. ii. 380, 36: Apstls. Kmbl. 17; Ap. 9. Sum sceal on heápe hæleþum cwēman *one shall in company give pleasure to men,* Exon. 88 a ; Th. 331, 32; Vy. 77. Gewíteþ mid ðȳ wuldre mære tungol faran on heápe *the great star departs accompanied with that glory,* 93 b ; Th. 350, 26; Sch. 69. Hwanon ferigeaþ gē heresceafta heáp *whence bear ye a heap of war shafts,* Beo. Th. 675; B. 335. Hengestes heáp *Hengest's band,* 2186; B. 1091. His ðone gecorenan heáp *electos suos,* Ps. Th. 104, 38 : L. Ælfc. P. 21; Th. ii. 372, 3. Getalu *vel* heápas *vel* hundredu *centurias,* Ælfc. Gl. 96; Som. 76, 25; Wrt. Voc. 53, 34. Hine ðā ða heápas frugnon hwæt hie wyrcean mihton ðæt hie Godes erre beflugon *when the multitudes asked him* [*John*] *what they could do to escape God's wrath,* Blickl. Homl. 169, 10: Cd. 161; Th. 202, 2; Exod. 382. Biscopan and gehálgodan heápan *for bishops and consecrated bodies,* L. Eth. vii. 24; Th. i. 334, 23. Heápum *in troops,* Cd. 81; Th. 101, 36; Gen. 1693: 189; Th. 235, 6; Dan. 302: Exon. 15 b ; Th. 34, 29; Cri. 549: Judth. 11; Thw. 23, 39; Jud. 163. [*O. Sax.* hóp : *O. Frs.* hâp : cf. *Icel.* hópr *a troop, flock : O. H. Ger.* houf *strues, acervus : Ger.* haufe.] DER. gár-, wíg-heáp.

heáp-mælum ; *adv. In heaps, by troops, bands, companies, flocks :*—Telle ðū and Aaron heápmælum *thou and Aaron shalt number them by their armies,* Num. 1, 3. Ne wæs ðā ylding tó ðon ðæt hī heápmælum cōman māran weorod of ðam þeódum ðe wē ǣr gemynegodon *non mora ergo confluentibus certatim in insulam gentium memoratarum catervis,* Bd. 1, 15 ; S. 483, 31. Ða dumban niétenu hie hie gadriaþ heápmælum and hie ætsomne fēdaþ *gregatim animalia bruta pascuntur,* Past. 46, 4; Swt. 349, 23. Hȳ him heápmælum sylfe on hand eodon *they flocked to surrender to him,* Ors. 4, 5; Bos. 83, 8. [*O. H. Ger.* huufmálum *catervatim.*]

heápung, e ; *f. A heaping, heap :*—Onfóþ hine and on ða heápunge eówre niðerunge gelǣdaþ *accipite et in cumulum damnationis vestræ ducite,* Bd. 5, 13 ; S. 633, 14.

hearch. v. **hearg.**

HEARD, hard ; *adj.* HARD, *harsh, austere, severe, rigorous, stern, stubborn, firm, hardy, brave :*—Hond and heard sweord *the hand and the hard blade,* Beo. Th. 5011; B. 2509. Ic wát ðæt ðū eart heard mann *scio quia homo durus es,* Mt. Kmbl. 25, 24. Heard is ðeós sprǣc *this is an hard saying* ; *durus est hic sermo,* Jn. Skt. 6, 60. Heó wæs ǣror ðam cynge hire suna swíðe heard *she had been before very hard to the king her son,* Chr. 1043; Erl. 168, 36 : Cd. 103; Th. 136, 20; Gen. 2261. Se mon se ðe nú dēmeþ ðǣm earmun būton mildheortnesse, ðonne biþ ðam eft heard dóm geteód *the man who now judges the poor without mercy, on him shall a hard sentence be then passed in requital,* Blickl. Homl. 95, 36 : Cd. 22; Th. 28, 7; Gen. 432. Him nǣnig gewin hēr on worlde tó lang ne tó heard þuhte *no labour here in the world seemed to him too long or too hard,* Blickl. Homl. 227, 3; Cd. 17; Th. 20, 30; Gen. 317. Hunger se hearda *severe famine,* 86; Th. 108, 32; Gen. 1815. Ðǣr wæs heard plega wælgāra wrixl *there was hard fighting exchange of deadly darts,* 93; Th. 120, 5; Gen. 1989: Elen. Kmbl. 229; El. 115. Hē wæs ānrǣd heard and hygeróf *he was resolute, hardy and noble-minded,* Andr. Kmbl. 465; An. 233: Beo. Th. 689; B. 342. Ðes hearda heáp *this stout band,* 868; B. 432. Wíges heard *bold in battle,* 1776; B. 886: Exon. 78 b ; Th. 295, 27; Crä. 39; Byrht. Th. 135, 38; By. 130: Andr. Kmbl. 1677; An. 841. Hē wæs heardes cynnes *he was of a brave race,* Byrht. Th. 139, 39; By. 266. Ðone deópan grund ðæs hātan lēges and ðæs heardan lēges *the deep abyss of hot and cruel flame,* Blick. Homl. 103, 15. Hine ðā gegyrede mid hærenum

hrægle swíðe heardum and unwinsumum *he clothed himself with raiment of hair very hard and unpleasant,* 221, 24. Ic hafu gecnāwen on heardum hyge ðæt ðū hǣlend eart middangeardes *I have acknowledged in my stubborn heart that thou art the saviour of the world,* Elen. Knbl. 1614; El. 809. Beóþ ðē hungor and þurst hearde gewinnan *hunger and thirst will be hard adversaries to thee,* Exon. 36 b ; Th. 118, 28; Gú. 246. Ða heardan heortan *the hard hearts,* Past. 21, 3; Swt. 154, 2. Ða heardan þrowunga ðe hē ādreág *the hard sufferings that he endured,* Blickl. Homl. 97, 15. Ic hine heardan clammum wrīðan þohte *I thought to bind him with hard bonds,* Beo. Th. 1931; B. 963. Mē þinceþ ðæt ðū wǣre ðām ungelǣredum mannum heardra ðonne hit riht wǣre *videtur mihi quia durior justo indoctis auditoribus fuisti,* Bd. 3, 5 ; S. 527, 32. Hige sceal ðē heardra ðē úre mægen lytlaþ *our courage shall be the stouter as our force lessens,* Byrht. Th. 140, 62 ; By. 312. Nō ic gefrægn heardran feohtan *I have never heard of a harder fight,* Beo. Th. 1157; B. 576. Nǣfre hē ǣr ne siððan heardran hæle fand *never before or since did he find a stouter warrior,* 1442; B. 719. Se líchoma ðonne on ðone heardestan stenc and on ðone fúlostan biþ gecyrred *the body then shall be turned to the strongest and foulest stench,* Blickl. Homl. 59, 12. Ða ðe gecwedene syndon ða heardestan men who [*the Scythians*] *are said to be very hardy men,* Ors. 1, 2 ; Swt. 30, 3. [*Goth.* hardus : *O. Sax.* hard : *O. Frs.* herd : *Icel.* harðr : *O. H. Ger.* hart, harti, hert, herti *durus, rigidus, asper, acer : Ger.* hart.]

heard-cwide, es ; *m. Harsh language, reproach, abuse, contumely :*—Ic geþolade hosp and heardcwide *I suffered scorn and contumely,* Exon. 29 a ; Th. 88, 22 ; Cri. 1444.

hearde ; *adv. Severely, very much, greatly, sorely :*—Ðā cwæþ se Hǣlend ðæt him hearde þyrste *then said Jesus that he was sore athirst,* Homl. Th. ii. 256, 31. Hearde ofsceamode *sorely ashamed,* 518, 31. Ðæs ðe wē wēnaþ and hearde ondrǣdaþ *according to what we expect and very much fear,* L. Ælfc. P. 40; Ll. ii. 380, 35. Hine ðæs heardost langode hwanne hē of ðisse worlde mōste *he very earnestly longed for the time when he might leave this world,* Blickl. Homl. 227, 1: Bt. 36, 2 ; Fox 174, 28.

heard-ecg ; *adj. Hard of edge :*—Ðā wæs on healle heardecg togen sweord *then in the hall was drawn the sword hard of edge,* Beo. Th. 2581; B. 1288: 2984; B. 1490: Elen. Kmbl. 1513; El. 758: Exon. 102 b; Th. 388, 15; Rä. 6, 8. v. *other compounds with ecg.*

heard-fyrde ; *adj. Difficult to carry :*—Ðǣr oninnan bær eorl hardfyrdne dǣl goldes *there within bore the earl a weighty portion of gold,* Beo. Th. 4483; B. 2245.

heard-heáwa, an ; *m. A chisel; scalprum,* Som.

heard-heort ; *adj. Hard-hearted, stiff-necked :*— Heardheort biþ se mann ðe nele þurh lufe ōðrum fremigan ðǣr ðǣr hē mæg *that man is hard of heart who will not from love benefit others when he can,* Homl. Th. i. 252, 19. Hwā is swā heardheort ðæt ne mæg wēpan swylces ungelimpes *who is so hard of heart that he cannot weep at such misfortunes,* Chr. 1086; Erl. 219, 40. Ðis folc is hardheort *thou art a stiff-necked people,* Ex. 33, 3, 5: Homl. Th. i. 108, 22 : ii. 258, 22. Gē sind ealra folca ungeleáfulluste and heardheorteste *ye are of all nations the most unbelieving and most stiff-necked,* Deut. 9, 6.

heard-heortness, e ; *f. Hard-heartedness :*—Hwæt is seó stǣnige eorþe būton heardheortnyss *what is the stony ground but hard-heartedness,* Homl. Th. ii. 90, 35. Þurh ðone wah seó heardheortnes ðara hiéremonna *per parietem duritia subditorum,* Past. 21, 3; Swt. 153, 24. Ic can eówre heardheortnisse *I know thy stiff neck,* Deut. 31, 27.

heard-hicgende ; *adj. Bold in purpose,* Beo. Th. 793; B. 394: 1602; B. 799.

heardian ; *p.* ode *To be or become hard, to harden :* — Ic heardige *dureo* and *duro,* Ælfc. Gr. 35 ; Som. 38, 6 : 37; Som. 39, 26: Herb. 1, 19; Lchdm. i. 76, 18 : 2, 11; Lchdm. i. 84, 4. Ðæt wyrmþ and heardaþ ðone magan *it warms and hardens the stomach,* L. M. 2, 10; Lchdm. ii. 188, 18. Ðonne onginþ sió heardian *then the liver begins to harden,* 19 ; Lchdm. ii. 200, 25.

hearding, es ; *m. A brave man, warrior, hero,* Elen. Kmbl. 50 ; El. 25 : 260; El. 130 : Runic pm. Kmbl. 344, 1 ; Rún. 22. [Cf. æðeling and v. Grmm. D. M. 316, 321.]

heard-líc ; *adj. Severe, fierce, hard, strict :*—Heardlíc eornost *severe seriousness,* L. I. P. 10 ; Th. ii. 318, 37 ; Andr. Kmbl. 3100 ; An. 1553: Exon. 116 b ; Th. 447, 10 ; Dóm 37. Heardlícu wítu *severe punishments,* 69 b; Th. 258, 11; Jul. 263.

heard-líce ; *adv. Hardly, sorely, harshly, sternly, bravely, stoutly :*— Heardlíce *duriter,* Ælfc. Gr. 38 ; Som. 41, 41. Se Godes man ongan heardlíce and bitterlíce wēpan *the man of God began to weep sorely and bitterly;* solutus est in lacrymis, Bd. 4, 25 ; S. 600, 29. Hē heardlíce gewon wið Æþelbald cyning *he struggled hard with king Ethelbald,* Chr. 741; Erl. 46, 30. Ðet landfolc hardlíce wiðstódon *the people of the country withstood them stoutly,* 1046; Erl. 171, 4. Hē sprᾱc heardlícor wið hig ðonne wið fremde men *he spoke more harshly to them than to strangers,* Gen. 42, 8. [*O. Sax.* hard-líko.]

heard-lícness, e ; *f. Hardness, severity, strictness :*—Sume hī sædon

ða heardlícnysse his lífes *some of them told the severity of his life*, Guthl. 17; Gdwin. 70, 15.

heard-mód, *adj. Of a hard, unyielding spirit, self-confident, stout-hearted, brave :*—Eádig biþ se man ðe symle biþ forhtigende and sóþlíce se heardmóda befylþ on yfel *blessed is the man that is ever fearing ; and verily the self-confident man shall fall into evil*, Homl. Th. i. 408, 30. Hæleþas heardmóde *heroes stouthearted*, Cd. 15; Th. 19, 2; Gen. 285. [O. H. Ger. hart-muat *obstinatus ;* hart-móti *constantia, obstinatio, duritia.* Cf. O. Sax. hard-módig : Icel. harð-móðigr.]

heard-módness, e; *f. Hardness of mind* or *heart :*—Stán is gesett ongeán ðone hláf forðan ðe heardmódnys is wiðerræde sóþre lufe *a stone is put in opposition to bread, because hardness of mind is contrary to true love*, Homl. Th. i. 252, 18.

heard-neb, -nebb; *adj. Having a hard beak* [epithet of the raven] :—Ðá cwæþ se hálga tó ðám heardnebbum *then said the saint to the ravens*, Homl. Th. ii. 144, 15. v. *other compounds* of neb.

heardness, e; *f. Hardness :*— For eówer heortan heardnesse *ad duritiam cordis vestri*, Mt. Kmbl. 19, 8 : Mk. Skt. 10, 5 : Ðú æteówdest ðínum folce heardnyssa *ostendisti populo tuo dura*, Ps. Lamb. 59, 5.

heardra, an; *m. The name of a fish :*—Heardra *mulus vel mugilis*, Ælfc. Gl. 102; Som. 77, 64; Wrt. Voc. 55, 68 : *mullus*, Wrt. Voc. 77, 63.

heard-ræd; *adj. Steadfast, firm*, Cd. 107; Th. 141, 21; Gen. 2348. [Cf. Icel. harð-ræði *hardiness*.]

heard-sælig; *adj. Having hard fortune, unfortunate, unhappy :*—Sum biþ wonspédig heardsælig hæle *one is indigent, an unfortunate man*, Exon. 78 b; Th. 295, 12; Crä. 32 : Bt. 31, 1; Fox 112, 20 : Exon. 115 a; Th. 442, 27; Kl. 19.

heard-sælness, e; *f. Misfortune, calamity :*—Ðá com eác seó ofer-mæte heardsælnes *then came also the excessive calamity*, Ors. 3, 5; Swt. 104, 17.

heard-sælþ, e; *f. A hard fate, ill fortune, misfortune, unhappiness, wickedness, misconduct :* — Gong inn and geseoh ða heardsælþa and ða sconde ðe ðás hér dóþ *ingredere et vide abominationes pessimas quas isti faciunt hic*, Past. 21, 3; Swt. 155, 8. Ic wolde gewýscan gif ic mihte ðæt hí næfdon ða heardsælþa ðæt hí mihton yfel dón *uti hoc infortunio cito careant, patrandi sceleris possibilitate deserti, vehementer exopto*, Bt. 38, 2; Fox 198, 4. Hit gebyrede þurh ða heardsælþa ðara wrítera ðæt hí for heora slæwþe and for gímeléste and for recceléste forléton unwriten ðara monna dæda ðe on hiora dagum foremæroste wæron *quam multos clarissimos suis temporibus viros scriptorum inops delevit oblivio*, ·8, 3; Fox 64, 33.

heard-wendlíce; *adv. Severely, strictly :* — Heardwendlíce [MS. B. heardlíce] *districtius*, Bd. 4, 25; S. 601, 40.

hearg-træf, es; *n. A heathen temple*, Beo. Th. 353; B. 175.

hearg-, herig-weard, es; *m. A guardian of a temple*, Andr. Kmbl. 2249; An. 1126.

hearh, hearch, herh, es; *m : pl.* hearga, *f. A temple, an idol :*—Se ylca hearh *quod fanum*, Bd. 2, 15; S. 518, 35. Sóna ðæs ðe hé gelíhte tó ðam hearge ðá sceát he mid his spere ðæt hit sticode fæste on ðam hearge *nec distulit ille, mox ut propiabat fanum, profanare illud, injecta in eo lancea quam tenebat*, 13; S. 517, 11. Siððan hé fór tó ðæm hearge ðe Egypti sædon ðæt hé wære Amones heora godes *inde ad templum Jovis Ammonis pergit*, Ors. 3, 9; Swt. 126, 23. Hé on ðam ylcan hearhge wigbed hæfde tó Cristes onsægdnyssa and óðer tó deófla onsægd-nysse *in eodem fano et altare haberet ad sacrificium Christi et arulam ad victimas dæmoniorum*, Bd. 2, 15; S. 518, 33. Hie onhnigon tó ðam herige *they bowed to the idol*, Cd. 181; Th. 227, 3; Dan. 181. Gif ænig man gelýfe on Moloches hearch *if any man believe on Moloch*, Lev. 20, 2. Hé hét his geféran tóworpon ealne hearh and ða getymbro and for-bærnan *jussit sociis destruere ac succendere fanum cum omnibus septis suis*, Bd. 2, 13; S. 517, 14. Ealle ða hearga(s?) [cf. Swt. 157, 7] *universa idola*, Past. 21, 3; Swt. 153, 22. Cwæþ ðæt his hergas hýrran wæron and mihtigran mannum tó friðe ðonne Israéla éce drihten *he said that his idols were greater and more mighty for the protection of men than the eternal Lord of the Israelites*, Cd. 210; Th. 260, 25; Dan. 715. On westhealfe Alexandres herga *aras Alexandri magni*, Ors. 1, 1; Swt. 8, 17. Ne ic ne clypige tó heora godum ne tó heargum ne gebidde mid míne múþe *nec memor ero nominum eorum per labra mea*, Ps. Th. 15, 4. Ða wuldriaþ in hergum heara *qui gloriantur in simulacris suis*, Ps. Stev. 96, 7. Ðá ongunnon hí ða heargas edniwian *cæperunt fana restaurare*, Bd. 3, 30; S. 561, 42 : 562, 15. Mid ðý hé sóhte hwá ða wigbed and ða heargas ðara deófolgylda mid heora hegum ðe hí ymbsette wæron ærest áídlian and tóweorpan scolde *cum quæreret quis aras et fana idólorum cum septis quibus erant circumdata primus profanare deberet*, 2, 13; S. 516, 39. Heora hergas tówearp *templa subvertit*, Ors. 3, 7; Swt. 114, 2 : Exon. 14 b; Th. 30, 28; Cri. 485. And geeáþmédaþ hira hearga *et adoraverint simulacra eorum*, Ex. 34, 15. Ne wirc gé eów hearga ne ágrafene godas . . . eówre hearga ic tóbrece *ye shall make you no idols nor graven image . . . I will cut down your images*, Lev. 26, 1, 30. [Icel. hörgr; *m.* ' a heathen place of worship, an altar of stone, erected

on high places, or a sacrificial cairn, built in open air, and without images,' Cl. and Vig. Dict : O. H. Ger. haruc, haruch, harug; *m. lucus, nemus, fanum, delubrum, ara.* The word perhaps occurs in the sense of *grove* in Exon. 54 b; Th. 192, 25; Az. 110. Grein so translates the word in this passage.]

HEARM, herm, es; *m.* HARM, *hurt, injury, evil, grief, affliction, pain, injurious speech, calumny, insult :*—Hýnþ *vel* lyre *vel* hearm *dispendium vel damnum vel detrimentum*, Ælfc. Gl. 81; Som. 73, 24; Wrt. Voc. 47, 29. Eác is hearm gode módsorg gemacod *pain also and heart-sorrow is caused to God*, Cd. 35; Th. 47, 2; Gen. 754. Nán hearm ne biþ ðeáh hit nó ne gewyrðe *there is no harm if it do not happen*, Bt. 41, 3; Fox 250, 4. Ic forhele ðæt mé hearmes swá fela Adam gespræc eargra worda *I will conceal that Adam spoke so much calumny, so many evil words to me*, Cd. 27; Th. 36, 30; Gen. 579 : 30; Th. 41, 24; Gen. 661 : Exon. 10 a; Th. 11, 15; Cri. 171. Hé onfunde Godes ierre on ðam hearme ðe his bearne æfter his dagum becom *in damnationem securituræ prolis ex eo iram judicis pertulit*, Past. 4, 1; Swt. 39, 4. Nó hé mid hearme gæst ne grétte *not with insult did he greet the guest*, Beo. Th. 3788; B. 1892. Husworde ongan herme hyspan *with words of contumely and insult began to revile him*, Andr. Kmbl. 1341; An. 671. Gif hwæs weorc forbyrnþ, hé hæfþ ðone hearm and biþ swá ðeáh ge-healden þurh fýr *if any one's work is consumed he has the loss, and yet shall be saved by fire*, Homl. Th. ii. 588, 30. Hí gefeordon máran hearm and yfel ðonne hí ǽfre wéndon ðæt heom ǽnig burhwaru gedón sceolde *they got more damage and hurt than they ever expected any citizens would cause them*, Chr. 994; Erl. 133, 13. Ǽr hí tó mycelne hearm gedydon *before they did too much harm*, 1004; Erl. 139, 20 : Cd. 196; Th. 245, 6; Dan. 458. Ealle synt uncre hearmas gewrecene *all our injuries are avenged*, 35; Th. 47, 12; Gen. 759. Nyste ðæt hearma swá fela fylgean sceolde monna cynne *knew not that so many ills to man-kind must follow*, 33; Th. 44, 13; Gen. 708 : Andr. Kmbl. 2889; An. 1447. Mé is ðæt hearma mǽst *that is greatest of griefs to me*, Byrht. Th. 138, 21; By. 233. [O. Sax. harm *pain, grief :* Icel. harmr *grief, sorrow, harm :* O. H. Ger. harm *calamitas, calumnia, contumelia, ærumna, injuria :* Ger. harm *grief, sorrow*.]

hearm = hreám [?] L. E. G. 6; Th. i. 170, 10, see note there and Schmid, p. 123.

hearm, herm; *adj. Causing harm* or *sorrow, grievous, injurious, evil, malicious :* — Herm bealowes gást *the malicious spirit of evil*, Cd. 228; Th. 307, 19; Sat. 682. Hé mé álýsde of hearmum worde *ipse liberavit me a verbo aspero*, Ps. Th. 90, 3. Ða inwit and fácen hycgeaþ on heortan þurh hearme geþoht *qui cogitaverunt malitias in corde*, 139, 2. Tugon longne síþ in hearmra hond *went a long journey into the power of evil ones*, Exon. 62 a; Th. 228, 20; Ph. 441. Ne hyld ðú míne heortan ðæt ic hearme word þuruh inwitstæf útforlǽte *ut non declines cor meum in verbum malum*, Ps. Th. 140, 5.

hearma, an; *m. A shrew-mouse* [?]; nebila, Ælfc. Gl. 19; Som. 59, 6; Wrt. Voc. 22, 50. [O. H. Ger. harmo *mygale*.]

hearm-cwalu, e; *f. Grievous destruction*, Exon. 31 b; Th. 98, 18; Cri. 1609.

hearm-cwedelian; *p.* ode *To speak ill of, calumniate :*—Ná hearm-cwedelodon mé ofermóde *non calumnientur me superbi*, Ps. Spl. 118, 122.

hearm-cwedan; *p.* -cwæþ *To revile, speak ill of :*—Mið ðý menn iuih harmcuedaþ *cum homines vos exprobraverint*, Lk. Skt. Lind. 6, 22. Hearmcuædon him *convitiabantur ei*, Mk. Skt. Lind. 15, 32.

hearm-cwedend, es; *m. A calumniator :*—Hé ða hermcweðend hýneþ *humiliabit calumniatorem*, Ps. Th. 71, 5.

hearm-cwide, es; *m. Injurious, abusive speech, calumny, blasphemy, a sentence pronouncing harm* or *sorrow :*—Heora hearran hearmcwyde *their lord's sentence*, Cd. 29; Th. 39, 12; Gen. 625. Judéa cynn wið godes bearne áhóf hearmcwide *the race of the Jews against God's son blasphemed*, Andr. Kmbl. 1121; An. 561 : 157; An 79. Áhrede mé hearmcwidum heánra manna *redime a calumniis hominum*, Ps. Th. 118, 134 : Exon. 24 a; Th. 69, 15; Cri. 1121. [O. Sax. harm-quidi : O. H. Ger. harm-qhuiti *calumnia*.]

hearm-cwidian, -cwiddian; *p.* ode *To revile, calumniate, speak ill of :*—Ongan hine hyspan and hearmcwiddigan [-cwidian, Cott.] *he began to revile and speak ill of him*, Bt. 18, 4; Fox 66, 33.

hearm-cwidol; *adj. Given to speak evil, calumnious :*—Gebiddaþ for hearmcwidele *orate pro calumniantibus*, Mt. Kmbl. 5, 44. Ða wǽron hí æfter æþelborennysse oferhýdige and hearmcwydole *in consequence of noble birth they were haughty and given to speak contemptuously of others*, Homl. Th. ii. 174, 8.

hearm-edwit, es; *n. Grievous reproach*, Ps. Th. 68, 21.

hearm-fullic; *adj. Harmful, hurtful :*—Swurdboran hine gewordene gesihþ hearmfullíc getácnaþ *to see one's self become a gladiator betokens something hurtful*, Lchdm. iii. 204, 26.

hearm-heortness, e; *f. Murmuring, grieving ;* murmuratio, Cot. 187, Lye.

hearmian; *p.* ode *To harm, hurt, injure :*—Gif preóst óðerne un-warnode lǽte ðæs ðe hé wite ðæt him hearmian wille *if a priest leave*

another unwarned of that which he knows will harm him, L. N. P. L. 33;
Th. ii. 294, 26: Lchdm. iii. 202, 33. Gif ðú híne forgitst hit hearmaþ
ðé sylfum and ná Gode *if thou forgettest him it harms thyself and not
God*, Homl. Th. i. 140, 31. Ðeáh ðe hit hearmige sumum *though it
may do harm to some*, H. R. 105, 36. [O. H. Ger. harmên *calumniari :
Ger.* härmen *to afflict, grieve :* cf. *Icel.* hermask *to be annoyed.*]

hearm-leóþ, es; *n. A sorrowful song, lamentation :*—Hearmleóþ galan
to sing a song of grief, Andr. Kmbl. 2256 ; An. 1129 : 2684 ; An. 1344.
Hearmleóþ ágôl earm and unlæd *wretched and miserable sang a mournful
song*, Exon. 74 b ; Th. 279, 18 ; Jul. 615.

hearm-líc ; *adj. Hurtful, injurious, painful, miserable, grievous :*—
Hearmlíc him wǽre ðæt hé wurþe ðá éce *it would have been hurtful for
him to become eternal then*, Hexam. 18 ; Norm. 26, 17. Ðæt wæs
hreówlíc and hearmlíc *that was sad and grievous*, Chr. 1057 ; Erl. 192,
21. [*O. Sax.* harm-lík.]

hearm-loca, an ; *m. An enclosed place where hurt or affliction is suffered, a prison :*—Wræcstôwe under hearmlocan gefôran *they reached
their place of exile in hell*, Cd. 5 ; Th. 6, 19 ; Gen. 91. Hé his maguþegne under hearmlocan hǽlo ábeád *he announced safety to his servant in
prison*, Andr. Kmbl. 189 ; An. 95 : 2058 ; An. 1031 : Elen. Kmbl. 1386 ;
El. 695.

hearm-plega, an ; *m. Strife*, Cd. 90 ; Th. 114, 2 ; Gen. 1898.

hearm-scearu, e ; *f. What is imposed as a punishment or penalty*
['was zur pein und qual auferlegt wird,' Grmm. R. A. 681] :—Wyrþ
him wíte gegearwod sum heard harmscearu *for them punishment will be
prepared, some severe penalty*, Cd. 22 ; Th. 28, 7 ; Gen. 432 : 37 ; Th.
48, 25 ; Gen. 781 : 38 ; Th. 51, 19 ; Gen. 829. [*O. Sax.* harm-skara :
O. Frs. herm-skere : *O. H. Ger.* harm-, haram-skara *plaga, percussio, afflictio, castigatio, contritio, dejectio, calamitas, supplicium, scantinea*,
Grff. vi. 529.]

hearm-sceaða, an ; *m. A grievous, pernicious spoiler*, Beo. Th. 1536 ;
B. 766.

hearm-slege, es ; *m. A grievous blow*, Exon. 28 b ; Th. 88, 4 ; Cri.
1435.

hearm-sprǽc, e ; *f. Slander ;* calumnia, Som.

hearm-sprǽcol ; *adj. Calumnious*, Som. v. hearm-cwidol.

hearm-sprǽcolness, e ; *f. Slandering, traducing*, Som.

hearm-stæf, es ; *m. Hurt, harm, sorrow, trouble, affliction :*—Wé nú
gehýraþ hwǽr ús hearmstafas onwôcan *we now hear whence troubles
arose for us*, Cd. 45 ; Th. 58, 1 ; Gen. 939. Ne môstun hý Gúþláces
gæste sceððan ... ac hý áhôfun hearmstafas *they might not injure
Guthlac's spirit ... but they raised up troubles*, Exon. 35 b ; Th. 115,
35 ; Gú. 200. [Cf. *other compounds of* stæf.]

hearm-tán, es ; *m. A twig of sorrow or evil*, Cd. 47 ; Th. 61, 4 ; Gen.
992.

hearpe, hærpe, an ; *f. A harp :*—Hearpe cithara, Wrt. Voc. 73, 56 :
Ps. Th. 56, 10. Psalm æfter hærpan sang *canticum :* ǽr hærpan sang
psalmus, Ælfc. Gl. 34 ; Som. 62, 57, 58 ; Wrt. Voc. 28, 37, 38. Ðǽr
wæs hearpan swég *there was the sound of the harp*, Beo. Th. 179 ; B.
89 : 4908 ; B. 2458 : 6039 ; B. 3023 : 4517 ; B. 2262 : 4221 ; B. 2107.
Se hearpan ǽrest handum sínum hlyn áwehte *he first awaked with his
hands the sound of the harp*, Cd. 52 ; Th. 66, 5 ; Gen. 1079. Ðonne
ðǽr wæs blisse intingan gedémed ðæt hí ealle sceoldan þurh endebyrdnesse be hearpan singan ðonne hé geseah ða hearpan him neáléécean
ðonne árás hé *cum esset lætitiæ causa ut omnes per ordinem cantare
deberent ille ubi adpropinquare sibi citharam cernebat surgebat*, Bd. 4,
24 ; S. 597, 6. Ic ðé on sealmfatum singe be hearpan *psallam tibi in
cithara*, Ps. Th. 70, 20 : Exon. 86 b ; Th. 325, 1 ; Víd. 105. Ne biþ
him tô hearpan hyge ... se ðe on lagu fundaþ *he has no mind to the harp
... who on the ocean puts forth*, 82 a ; Th. 308, 23 ; Seef. 44. Sum
sceal mid hearpan æt his hláfordes fôtum sittan feoh þicgan *one shall at
his lord's feet sit with the harp and receive treasure*, 88 a ; Th. 332, 4 ;
Vy. 80. Sum mid hondum mæg hearpan grétan *one with his hands can
touch the harp*, 79 a ; Th. 296, 11 ; Crä. 49 : 91 b ; Th. 344, 10 ; Gn.
Ex. 171 : 17 b ; Th. 42, 8 ; Cri. 669. [*Icel.* harpa : *O. H. Ger.* harfa
plectrum, chelys, psalterium, cythara : Ger. harfe.]

hearpe-, hearp-nægel, es ; *m. An instrument for striking the strings of
a harp :*—Hearpnægel *plectrum*, Ælfc. Gl. 71 ; Som. 70, 96 ; Wrt. Voc.
43, 27. Apollonius his hearpenægl genam *Apollonius took his harp-nail*,
Ap. Th. 17, 7.

hearpene, an ; *f. A nightingale ;* aëdon, Cot. 19, Lye.

hearpere, es ; *m. A harper :*—Hearpere citharedus, Ælfc. Gl. 114 ;
Som. 80, 8 ; Wrt. Voc. 60, 44 : citharista, 73, 55. Ân hearpere wæs on
ðære þeóde ðe Thracia hátte ... ðæs nama wæs Orfeus *there was a
harper in Thrace whose name was Orpheus*, Bt. 35, 6 ; Fox 166, 29 ;
Past. 23 ; Swt. 175, 7. [*Icel.* harpari : *O. H. Ger.* harfere *citharedus.*]

hearpestre, an ; *f. A female harper :*—Hearpestre citharista, Ælfc. Gl.
114 ; Som. 80, 9 ; Wrt. Voc. 60, 45.

hearpe-streng, es ; *m. A harp-string :*—Hé ða hearpestrengas mid
cræfte ástirian ongan *he began to move the strings of the harp skilfully*,
Ap. Th. 17, 8. [*Icel.* hörpu-strengr.]

hearpian ; *p. ode To play on the harp, to harp :*—Hé mihte hearpian
ðæt se wudu wagode *he could play on the harp so that the wood moved*,
Bt. 35, 6 ; Fox 166, 32 : Ap. Th. 16, 16. Fægere hé hearpaþ *pulcre
citharizat*, Ælfc. Gr. 38 ; Som. 41, 31. Ða hwile ðe he hearpode *whilst
he played on the harp*, Bt. 35, 6 ; Fox 170, 5. Stefen swǽ hearpara
hearpandra in hearpum sínum *vocem sicut cytharedorum cytharizantium
in cytharis suis*, Rtl. 47, 24.

hearp-sang, es ; *m. A song to the harp, a psalm :*—Hearpsang *psalmus*,
Ælfc. Gl. 34 : Som. 62, 56 ; Wrt. Voc. 28, 36.

hearp-slege, es ; *m. A striking, playing of the harp :*—On hearpan
and on hearpslege and on stefne sealmcwides *in cithara, in cithara et voce
psalmi*, Ps. Lamb. 97, 5. [*Icel.* hörpu-slagr *striking the harp.*]

hearp-swég, es ; *m. The sound of the harp :*—Sealmleóþ and hearp-
swég *psalterium et cythara*, Blickl. Gloss.

hearpung, e ; *f. Harping, playing on the harp :*—Hé hí hæfþ geearnod
mid his hearpunga *he hath deserved her by his harping*, Bt. 35, 6 ; Fox
170, 8.

hearra, herra, hierra, an ; *m. A lord.* The use of this word, which
occurs only in poetry, is noticeable. It occurs twenty-three times in that
part of the Genesis [vv. 235–851] for which Sievers claims an old Saxon
origin, and only four times elsewhere, Cd. 192 ; Th. 240, 28 ; Dan.
393 : Judth. 10 ; Thw. 22, 9 ; Jud. 56 : Byrht. Th. 137, 51 ; By. 204 :
Chr. 1065 ; Erl. 198, 13. [In the Heliand *herro* occurs frequently. *Icel.*
has *harri, herra : O. H. Ger. herro :* Grff. iv. 991.]

hearste-, hierste-panne, an ; *f. A frying-pan :*—Hé him tǽhte ðæt hé
him genáme áne íserne hearstepannan *tu sume tibi sartaginem ferream*,
Past. 21, 5 ; Swt. 161, 7 : 163, 22.

heart. v. heort.

hearwian *to cool ;* refrigerare, Lye.

heaðorian, heaðerian ; *p. ode To restrain :* — Se godcunda foreþonc
heaðeraþ ealle gesceafta *the divine providence restrains all creatures*, Bt.
39, 5 ; Fox 218, 31. Mid þearfednesse ge mid heora ungelǽrednesse
ðara láreówa fôre heaðoradon *paupertate aut rusticitate sua doctorum ar-
cebant accessum*, Bd. 4, 27 ; S. 604, 29. v. ge-heaðorian.

heaðu, heaðo *war ;* a word occurring only in compounds. The word
is found in proper names in Icelandic, e. g. Höð the name of a Valkyria,
Höðbroddr, Höðr the slayer of Baldr ; and in *O. H. Ger.* e. g. Hadu-praht,
v. Grmm. D. M. 204 : Cl. and Vig. Dict. höð. Cf. beadu, gúþ, hilde *and
their compounds.*

heaðu [= heáhþu ?] *indecl. f. The deep, the sea ;* altum :—Sceal hring-
naca ofer heaðu bringan lác and lustácen *over the deep shall the bark bring
gift and love token*, Beo. Th. 3729 ; B. 1862.

heaðu-byrne, an ; *f. A war-corslet*, Beo. Th. 3108 ; B. 1552.

heaðu-deór ; *adj. Brave, stout in war*, Beo. Th. 1380 ; B. 688 : 1548 ;
B. 772.

heaðu-fremmende ; *part. Doing battle, fighting*, Elen. Kmbl. 258 ;
El. 130.

heaðu-fýr, es ; *n. Fierce, hostile fire*, Beo. Th. 5037 ; B. 2522 : 5087 ;
2547.

heaðu-geong ; *adj. Young and active in battle* (?) [*Hickes reads* hearo],
Fins. Th. 3 ; Fin. 2.

heaðu-glemm, es ; *m. A wound got in fight*, Exon. 114 a ; Th. 438,
6 ; Rä. 57, 3. v. glemm.

heaðu-grim ; *adj. Very fierce, cruel with the cruelty of war :*—Hungur
heaðogrimne hearde *famne fierce and fell*, Ps. Th. 145, 6 : Beo. Th.
1100 ; B. 548 : 5375 ; B. 2691.

heaðu-helm, es ; *m. A war-helm, casque*, Beo. Kmbl. 6304 ; B. 3156.

heaðu-lác, es ; *n. Battle*, Beo. Th. 1172 ; B. 584 : 3952 ; B. 1974.

heaðu-lind, e ; *f. A linden war-shield*, Chr. 937 ; Erl. 112, 6 ;
Ædelst. 6.

heaðu-líðende ; *part. Sea-faring*, Beo. Th. 3600 ; B. 1798 : 5902 ;
B. 2955 : Andr. Kmbl. 851 ; An. 426.

heaðu-mǽre ; *adj. Illustrious in war*, Beo. Th. 5596 ; B. 2802.

heaðu-rǽs, es ; *m. A battle-rush, charge, onslaught*, Beo. Th. 1056 ;
B. 526 : 1119 ; B. 557 : 2099 ; B. 1047.

heaðu-reáf, es ; *n. War-dress, armour*, Beo. Th. 807 ; B. 401.

heaðu-rinc, es ; *m. A warrior*, Judth. 11 ; Thw. 24, 9 ; Jud. 179 :
Thw. 24, 29 ; Jud. 212 : Beo. Th. 745 ; B. 370 : 4923 ; B. 2466 : Cd.
154 ; Th. 193, 4 ; Exod. 241 : Bt. Met. Fox 9, 89 ; Met. 9, 45.

heaðu-rôf ; *adj. Famed for excellence in battle*, Beo. Th. 767 ; B.
381 ; 1732 ; B. 864 : 4388 ; B. 2191 : Exon. 59 a ; Th. 213, 21 ; Ph.
228 ; Menol. Fox 27 ; Men. 14.

heaðu-sceared ; *adj. In* Beo. Th. 5650 ; B. 2829 ; *according to
Thorpe the reading of the MS. is* scearede, *other editors read* scearde. *In
the former case may not the word be connected with* scear [q. v. *share in
ploughshare*] *used here of the blade of a sword*, heaðo-scear *a war-share,
blade? and* hearde heaðo-scearede = *with hard and deadly blades. If
scearde is taken, the Icel.* skarð *may be compared, and the word = notched,
hacked in battle.*

heaðu-seóc ; *adj. Wounded in fight*, Beo. Th. 5501 ; B. 2754.

heaðu-sigel, es ; *m. The sun* [the prefix seems to be used from seeing

the sun rise or set over the sea], Exon. 126 b; Th. 486, 17; Rä. 72, 16. [Cf. merecandel.]

heaðu-steáp; *adj.* *Standing out prominently in battle* [an epithet of the helmet], Beo. Th. 2494; B. 1245: 4312; B. 2153.

heaðu-swát, es; *m.* *War-sweat, blood shed in battle,* Beo. Th. 2924; B. 1460: 3216; B. 1606: 3340; 1668.

heaðu-sweng, es; *m.* *A blow given in fight,* Beo. Th. 5155; B. 2581.

heaðu-torht; *adj.* *Clear-sounding and of warlike import,* Beo. Th. 5109: B. 2553.

heaðu-wǽd, e; *f.* *Warlike weeds, dress,* Beo. Th. 78; B. 39.

heaðu-wælm, -welm ,-wylm, es; *m.* *Fierce, intense heat,* Cd. 17; Th. 21, 14; Gen. 324: 149; Th. 187, 8; Exod. 148: Beo. Th. 165; B. 82: 5630; B. 2819; Andr. Kmbl. 3082; An. 1544: Elen. Kmbl. 1154; El. 578: 2607; El. 1305.

heaðu-weorc, es; *n.* *A work of war, a fight,* Beo. Th. 5776; B. 2892.

heaðu-wérig; *adj.* *Weary from fighting,* Vald. 2, 17.

HEÁWAN; *p.* heów, *pl.* heówon; *pp.* heáwen To HEW, *cut, strike, smite* [with a sharp weapon]:—Gif mon ôðres wudu heáweþ unáliéfedne *if a man cut another's wood without leave,* L. Alf. pol. 12; Th. i. 70, 4. Mǽst ǽlc ôðerne æftan heáweþ mid scandlícum onscytum *almost all men calumniate* [lit. *strike from behind*] *each other with shameful attacks,* Swt. A. S. Rdr. 107, 84. Se seðe unwǽrlíce ðone wuda hiéwþ *is qui incaute ligna percutit,* Past. 21, 7; Swt. 167, 16. Wé heáwaþ ðone wudu *ligna succidimus,* 167, 6. Hé heów ôð ðæt hé on hilde gecranc *he smote with his sword until in fight he fell,* Byrht. Th. 141, 18; By. 324. Heów ðæt hors mid ðam spuran *he struck the horse with the spurs* [cf. Icel. höggva hest sporum], Elf. T. 36, 25. Ðá heówon hí ðone stán swá swyðe swá hí mihton *dolantes lapidem in quantum valebant,* Bd. 4, 11; S. 580, 5. Heówon hereflýman þearle mécum mylenscearpum *they smote sorely the flying with falchions sharp ground,* Chr. 937; Erl. 112, 23; Æðelst. 23: Byrht. Th. 137, 4; By. 181. Linde heówon *they hewed the linden shields,* Judth. 12; Thw. 26, 1; Jud. 304: Chr. 937; Erl. 112. 6; Æðelst. 6: Mt. Kmbl. 21. 8. [O.Sax. hawan, hauwan: O.Frs. hawa, howa: Icel. höggva: O.H.Ger. houwan: Ger. hauen.]

HEBBAN; *p.* hóf, *pl.* hófon; *pp.* hafen, hæfen To HEAVE, *lift up, raise:*—Ic míne handa tó ðé hebbe and þenige *expandi manus meas ad te,* Ps. Th. 87, 9. Tó ðé ic hæbbe mín mód *ad te levavi animam meam,* 24, 1: Hine sylfne hefeþ on heáhne beám *raises itself into a lofty tree,* Exon. 57 b; Th. 205, 13; Ph. 112: Ps. Th. 148, 14. Forðon hiora heáfod hebbaþ *propterea exaltabit caput,* 109, 8. Tó ðé ic míne eágan hóf *ad te levavi oculos meos,* 122, 1. Hófon hlúde stefne *raised a loud voice,* Cd. 170; Th. 214, 24; Exod. 574: Exon. 45 b; Th. 156, 8; Gú. 871. Hefe ðú ðíne handa *leva manum tuam,* Ps. Th. 73, 4. Hebbaþ upp eówre eágan *levate oculos vestros,* Jn. Skt. 4, 35. God bebeád his englum be ðé ðæt hí ðé healdon and on heora handum hebban *God has given his angels charge concerning thee, that they may preserve thee and lift thee up in their hands,* Homl. Th. i. 516, 30. Siððan ic hond and rond hebban mihte *since I could lift hand and shield,* Beo. Th. 1317; B. 656. Hé wæs upp hafen engla fæðmum *he was lifted up in angels' bosoms,* Exon. 17 a; Th. 41, 5; Cri. 651: 756; Th. 284, 7; Jul. 693. Wæs wóp hæfen *then was a cry raised,* Andr. Kmbl. 2311; An. 1157: Beo. Th. 6038; B. 3023. [*Goth.* hafjan: *O. Sax.* hebbian: *O. Frs.* heva: *Icel.* hefja: *O. H. Ger.* heffan, heuen *levare, extollere: Ger.* heben.]

hebbendlíc; *adj.* *Exalted;* exaltatus, Rtl. 181, 27.

hebel, hebeld, heben. v. hefel, hefeld, heofon.

Hebréisc; *adj.* *Hebrew:*—On Hebréisc specan *to speak in Hebrew,* Nicod. 4; Thw. 2, 28. v. Ebréisc.

HÉDAN; *p.* de To HEED, *take care, observe, attend, guard, take charge, take possession, receive:*—Lazarus ne móste ǽr on lífe hédan ðæra crumena his mýsan *before when alive Lazarus might not take the crumbs of his table,* Homl. Th. i. 330, 31. Wé hédaþ ðæra crumena ðæs hláfes and ða Judéiscan gnagaþ ða rinde *we take the crumbs of the bread and the Jews gnaw the crust,* ii. 114, 33. Ða Judéiscan ne hédaþ ná máre búton ðære stæflícan gereccednesse *the Jews pay attention to nothing but the literal narrative,* 116, 4. Ne hédde hé ðæs heafolan *he was not careful for his head,* Beo. Th. 5387; B. 2697. Bóte gesáwon héddon hereréafes *they saw their compensation, took possession of the war spoils,* Cd. 171; Th. 215, 14; Exod. 583. Héde seðe scire healde ðæt hé wite ā hwæt eald landrǽden sý *videat qui scyrum tenet, ut semper sciat que sit antiqua terrarum institutio,* L. R. S. 4; Th. i. 434, 32. Ðonne him forþsíþ gebyrige héde se hláford ðæs he lǽfe *when he dies let the lord take possession of what he leaves* [cf. 434, 27], 436, 9: L. In. 74; Th. i. 148, 19. Bisceopum gebyreþ ðæt hí hunda ne hafeca hédan tó swyðe *it is befitting for bishops not to care too much for hounds or hawks,* L. I. P. 10; Th. ii. 316, 30. Gif ðǽr nán man ne biþ ðe ðære heofoulícan bodunge hédan wille *if there be no man there that will heed the heavenly preaching,* Homl. Th. ii. 534, 16. [*O. Sax.* hôdian *to take care of, guard: O.H.Ger.* huoten *custodire, observare: Ger.* hüten.]

hed-cláþ, es; *m.* *A thick upper garment of coarse material, like a chasuble,* Med. ex Quadr. 4, 17; Lchdm. i. 346, 17. v. heden.

hédd-, hýdd-ern, es; *n. A storehouse:*—Hýddern *cellarium,* Wrt. Voc. 83, 5. Héddern *penu,* Ælfc. Gr. 11; Som. 15, 30: *poenum,* 13; Som. 16, 7. Besceáwiaþ ða hrefnas ðæt hig ne sáwaþ ne ne rípaþ nabbaþ hig héddern ne bern *considerate corbos quia non seminant neque metunt quibus non est cellarium neque horreum,* Lk. Skt. 12, 24. Swá swá mon héddern ontýnde ðara swétestena wyrta ðe on middangearde wǽron *quasi opobalsami cellaria esse viderentur aperta,* Bd. 3, 8; S. 532, 19. Drihten sent bletsunga ofer ðíne héddernu *the Lord send blessings upon thy storehouses,* Deut. 28, 8.

heden, es; *m. A hood, chasuble:*—Heden *casla,* Cot. 32, Lye. Sacerd ðonne hé mæssan singe ne hæbbe hé on heden ne cæppan *sacerdos cum missam cantat ne portet cucullum nec cappam,* L. Ecg. C. 9; Th. ii. 140, 9. Swá hwylc swá wile lectiones rǽdan ne biþ hé nýded tó ðon ðæt hé him ofdó his oferhacelan oððe heden ac gyf hé euangelium rǽde wyrpe him of heden oððe cæppan on hys gescyldro *quicunque lectiones legere velit, non necesse est ei cappam suam vel cucullum exuere; si autem evangelium legit, cucullum vel cappam super humeros dejiciat,* 20-24. Hǽden *mastruca,* Lye. [Cf. hede *dress,* Halliw. Dict: *Icel.* héðinn, a *jacket of fur or skin.*]

hefe, es; *m. Weight:*—Hú mihte hé gefrédan ǽniges hefes swǽrnysse ðá ðá hé ðone ferode ðe hine bær *how could he feel the heaviness of any weight when he carried one who bore him,* Homl. Th. i. 336, 26. Swilce hé búton hefe wǽre *as if he were without weight,* ii. 164, 35. On gemete and on hefe and on getale *in mensura et pondere et numero,* 586, 32. Hé micelne hefe gefrét æt hys heortan *he feels a great weight at his heart,* Lchdm. iii. 126, 10. Áwend hefas leahtra *evente moles criminum,* Hymn. Surt. 23, 7.

hefeld, hebeld, hefel, hebel, es; *m.* [?] *Thread for weaving:*—Hefeld *licium,* Ælfc. Gl. 110; Som. 79, 50; Wrt. Voc. 59, 21. [Hevel *fine twine,* Halliw. Dict: cf. *Icel.* hefill; *m.* *the clew-lines and bunt lines of a sail.*]

hefeld-gyrd, e; *f. A weaver's shuttle;* liciatorium, Cot. 120, Lye.

hefeldian, hefaldian *to fix the weft or woof:*—Ic hefaldige *ordior,* Ælfc. Gl. 111; Som. 79, 73; Wrt. Voc. 59, 42.

hefeld-, hefel-þrǽd, es; *m. A thread for weaving;* licium:—Gewríð tó ánum hefel [MSS. H. B. hefeld] þrǽde *bind it to a yarn thread,* Herb. 183; Lchdm. i. 320, 6. Ðá tóbræc hé ða rápas swá swá hefelþrǽdas *and he brake the withs as a thread of tow,* Jud. 16, 9. Hefelþrǽd *licium,* Cot. 193, Lye.

hefe-líc, *adj.* *Weighty, heavy, grievous, serious, grave, tedious, wearisome:* — Ðér nán hefelíc gefeoht ne wearþ *no serious fighting took place there,* Chr. 868; Erl. 72, 28. Se cyng lét beódan mycel gyld and hefelíc *the king had a great and grievous tax proclaimed,* 1083; Erl. 217, 34. Ðæs ilcan geáres wæs swíðe hefelíc geár *it was a very grievous year that same year,* 1085; Erl. 219, 18. For hefelícum gyltum *pro gravibus peccatis,* L. Ecg. P. i. 6; Th. ii. 174, 17. Nú bidde ic ða ðe hit cunnon and ðis rǽdon ðæt him hefelíc ne beó *now I beg that my explanation may not be tedious to those who know the subject and read this,* Lchdm. iii. 280, 10.

hefe-líce; *adv.* *Heavily, exceedingly, seriously, with difficulty:*—Hig hefelíce mid eárum gehýrdon *auribus graviter audierunt,* Mt. Kmbl. 13, 15. For ðære ilcan eádmódnesse hé ofermódgaþ innan micle ðý hefelícor *de hac ipsa humilitate graviter interius superbitur,* Past. 43, 3; Swt. 313, 3: 46, 5; Swt. 351, 6.

hefe-tíme; *adj.* *Troublesome, displeasing, tedious:*—Hit þuhte Moise swíðe hefetíme *Moses was displeased,* Num. 11, 10. v. hefig-tíme.

HEFIG, hefeg; *adj.* HEAVY, *weighty, oppressive, grievous, difficult, serious, grieved, important;* gravis, molestus:—Wæs torn were hefig æt heortan *in the man's heart was grievous anger,* Cd. 47; Th. 60, 11; Gen. 980. Suíðe hefig is *quam difficile est,* Mk. Skt. Lind. 10, 24. Hit swíðe hefegu scyld is *it is a very grievous crime,* L. E. I. 27; Th. ii. 422, 36. Bútan hefegum gefeohte *without heavy fighting,* sine ullo prælio, Bd. 1, 3; S. 475, 11. Áhófon hine of ðam hefian wíte *they lifted him off that heavy punishment,* Rood Kmbl. 121; Kr. 61. Heó is hefegou swæce *it is of unpleasant smell,* Herb. 151, 1; Lchdm. i. 276, 9: 143, 1; Lchdm. i. 264, 20. Wermód drincan sace hefige hit getácnaþ *to drink wormwood betokens a serious dispute,* Lchdm. iii. 198, 24: Herb. 132, 7; Lchdm. i. 248, 11. Tó hwon syndon gé ðyses weorces swá hefige *why are you so grieved at this work,* Blickl. Homl. 69, 15. Wurdon mé on yrre yfele and hefige *in ira molesti erant mihi,* Ps. Th. 54, 3. Hig bindaþ hefige byrðyna *alligant onera gravia,* Mt. Kmbl. 23, 4. Eorþe is hefige ôðrum gesceaftum *earth is heavier than the other elements,* Bt. Met. Fox 20, 265; Met. 20, 133. Wé mágon geþencean ðæt ðæt hefige is ðæt man mid synnum him sylfum geearnige edwit *we may consider, what is more important, that with sins a man may get disgrace for himself,* Blickl. Homl. 101, 24. Ða þing ðe synt hefegran ðære & quæ graviora sunt legis, Mt. Kmbl. 23, 23. Hí eów hefigran wísan budon tó healdanne ðonne wé him budon *they commanded you to keep a harder rule than we commanded them,* L. Ælf. 49; Th. i. 56, 15. Wið fótádle ðeáh ðe heó

hefegust sȳ *for gout, though it be very bad,* Herb. 132, 4; Lchdm. i. 246, 22. Mid ðon gewunon ðære heofogoston gewemmednesse synna *with the habit of the most grievous impurity of sins,* Blickl. Homl. 75, 6. [O.Sax. hebig: O.H.Ger. hebic, heuig *gravis, arduus, molestus.*]

hefige; *adv. Heavily, grievously, with difficulty, hardly :—*Ðæs wîte eft on eówre handa hefige geeóde *for that punishment came upon you heavily,* Ps. Th. 57, 2. Hefia *vix,* Lk. Skt. Lind. 9, 39. Forhwon âhênge ðû mec hefgor *why didst thou crucify me more painfully,* Exon. 29 b; Th. 91, 6; Gen. 1488. [O.H.Ger. heuigor *gravius.*]

hefigian; *p.* ode. I. *to make heavy, oppress, grieve, afflict, vex :—*Forðon sió byrðen ðære sconde hine diógollîce hefigaþ *quia gravit hunc in abditis pondus turpe,* Past. 11, 7; Swt. 73, 15. Ðone mete ðe hine hefigaþ on his breóstum *cibum, qui pectus deprimebat,* 54, 1; Swt. 419, 29. Ða ðe mê hefigiaþ *those who vex me,* Ps. Th. 37, 12. Wæs heó eft hefigod mid ðâm ærran sârum *erat prioribus aggravata doloribus,* Bd. 4, 19; S. 589, 5. Wolde mê hefigad beón mid sâre mînes sweoran *me dolore colli voluit gravari,* 589, 28. II. *to become heavy, to be aggravated* or *increased, to be burdened* or *oppressed :—*Hû sió byrðen wiexþ and hefigaþ *molem crescentis tentationis,* Past. 21, 5; Swt. 163, 12. Seó untrumnys dæghwamlîce weóx and hefegode *languor per dies ingravescebat,* Bd. 4, 3; S. 568, 38. Monigum monnum ðe heora eágan sârgedon and hefegodan *nonnulis oculos dolentibus,* 4, 19; S. 589, 35: Exon. 46 b; Th. 159, 20; Gû. 929: 47 b; Th. 163, 32; Gû. 1002. [*Laym.* heueȝe *to grow heavy, slumber : A. R.* heuegeþ *oppresses : Chauc.* hevieþ: *Prompt. Parv.* hevyyng *mestificio, gravo, aggravo, pondero.*]

hefig-lîc; *adj. Grievous, troublesome :*—Ne sig ðe hefilîc geþuht ðæt ðæt Sarra ðe sæde *let not that be grievous in thy sight which Sarah hath said,* Gen. 21, 12. Gif se lîchoma hwær mid hefiglîcre hæto sȳ gebysgod *if the body be anywhere troubled with inflammation,* Herb. 2, 6; Lchdm. i. 82, 8.

hefig-lîce; *adv. Heavily, grievously;* graviter :—Abraham undernam hefiglîce ðâs word *the thing was very grievous in Abraham's sight,* Gen. 21, 11. Hefiglîce *graviter,* Mt. Kmbl. Lind. 13, 15. Hefilîce, Lk. Skt. 11, 53. Ða weras mon sceal hefiglecor and stîðlecor læran and wîf leóhtlecor *illis [viri] graviora, istis [feminæ] injungenda sunt leviora,* Past. 24; Swt. 179, 16.

hefig-mód; *adj. Evil-minded, oppressive :*— Hefigmôde *molesti,* Ps. Spl. T. 54, 3.

hefig-ness, e; *f. Heaviness, slowness, weight, grief, affliction :*—Nân hæfignes ðæs lîchoman ne mæg eallunga âtîon of his môde ða rihtwîsnese *no heaviness of the body can altogether take away rectitude from his mind,* Bt. 35, 1; Fox 154, 29: 156, 12. Ne geman heó ðære hefinysse *non meminit pressuræ,* Jn. Skt. 16, 21. Yfelra ûserra hefignisse *malorum nostrorum pondere,* Rtl. 15, 30: Mt. Kmbl. Lind. 20, 12. Hefignise gebær *ægrotationes portavit,* 8, 17.

hefig-tîme, -tȳme; *adj. Grievous, wearisome, tedious, troublesome :*—Hefigtȳme leahter is ungefôh fyrwitnys *immoderate curiosity is a troublesome vice,* Homl. Th. ii. 374, 2. Gif hit is hefigtȳme on ðyssere worulde hit becymþ tô micelre mêde on ðære tôweardan *if it is productive of trouble in this world, it attains to a great reward in that which is to come,* i. 56, 4: Ælfc. Gen. Thw. p. 1, 6. Ne þince ðe tô hefitȳme tô gehȳrenne mîne spræce *do not let it seem too tedious to thee to hear my speech,* Basil admn. 7; Norm. 48, 12. Se hefigtîma cwide ðe se wîtega gecwæþ be sumum leódscipe *the grievous sentence that the prophet declared concerning a certain nation,* Swt. A. S. Rdr. 73, 543. Ða wudewan fram hefigtîmum heáfodece gehælde *healed the widow of a wearisome headache,* Homl. Th. i. 418, 22.

hefigtîmness, e; *f. Trouble, affliction, vexation :*— Ðone hê tealde him tô frȳnd ðe him sume hefigtȳmnysse on belædde *him he accounted his friend who brought some trouble upon him,* Homl. Th. ii. 546, 19. Hê is nû mid ylde ofsett, swylce mid gelomlæcendum hefigtȳmnyssum tô deáþe geþreád *it is now oppressed with age, as if wearied to death with frequent troubles,* i. 614, 21.

hefung, e; *f. Heaving, lifting up;* elevatio, speculatio, Lye.

HEG, hig, es; *n. Hay, grass;* fœnum :—Heg [Rush. hoeg] londes *fœnum agri,* Mt. Knbl. Lind. 6, 30. Ðã bebeád se hælend ðæt ðæt folc sæte ofer ðæt grêne hig *præcipit illis ut accumbere facerent omnes super viride fœnum,* Mk. Skt. 6, 39. Heig [Rush. heg] *fœnum,* Jn. Skt. Lind. 6, 10. Ðær nænig mann for wintres cȳle on sumera heg ne mâweþ *nemo propter hiemem fœna secet æstate,* Bd. 1, 1; S. 474, 32. Dô hig on ðîn bed *put hay on your bed,* Lchdm. iii. 178, 6. Wê gesâwon oft in cyrcean ægðer ge corn ge hig beón gehealdene *we have often seen both corn and hay kept in the church,* L. E. I. 8; Th. ii. 406, 30. [*Laym.* hey, heie : *Chauc.* hei, hai : *Goth.* hawi : *Icel.* hey : *O.H.Ger.* hewi, howe, hou *fœnum : Ger.* heu.]

HEGE, es; *m. A* HEDGE, *fence :*—Hege *sepes,* Wrt. Voc. 84, 56 : Ælfc. Gr. 9, 27; Som. 11, 24. Bebbanburh wæs ærost mid hegge betîned and ðæræfter mid wealle *Bamborough was first enclosed with a hedge and afterwards with a wall,* Chr. 547; Erl. 17, 9. Gâ geond ðâs wegas and hegas *exi in vias et sepes,* Lk. Skt. 14, 23. Ðû tôwurpe ealle hegas his *destruxisti omnes sepes ejus,* Ps. Spl. 88, 39. Gif hryðera hwelc sîe ðe

hegas brece *if there be any beast that breaks hedges,* L. In. 42; Th. i. 128, 12. Mid heora hegum ðe hî ymbsette wæron *cum septis quibus erant circumdata,* Bd. 2, 13; S. 516, 39 : Homl. Th. ii. 448, 22. From hegum a silvis, Rtl. 118, 35. [Hay, hey *in provincial words,* e. g. heybote, hayboot = hedgeboot *the right of getting wood for mending fences,* Engl. Dial. Soc. vols. iii. vi. Haies, hays *ridges of lands as district boundaries,* vol. iv : *Prompt. Parv.* hedge, hegge.] v. hæg- *and* haga.

hege-clife, an; *f. Hedge clivers;* galium aparine, L. M. 1, 9; Lchdm. ii. 54, 8.

hegegian *to hedge, fence,* L. R. S. 2; Th. i. 432, 16.

hege-ræwe, -rêwe, e; *f. A hedge-row :*—Ðanon on ða hegeræwe *thence to the hedge row,* Cod. Dipl. Kmbl. ii. 54, 11. Hegerêwe, iii. 48, 15.

hege-rife, an; *f. Heyriffe;* galium aparine, Lchdm. iii. Gloss. [*Prompt. Parv.* hayryf *rubea vel rubea minor, et major dicitur* madyr. v. note, p. 221. See English Plant-names, Engl. Dial. Soc. no. 26, p. 242 harif.]

heges-sugge *a hedge-sparrow,* Ælfc. Gl. 37; Som. 63, 5; Wrt. Voc. 29, 28. [*O. and N.* hei-sugge : *Flower and Leaf* hay-sogge; *Gloucestershire dialect* hay suck.]

heg-, hig-hûs, es; *n. A hay-house;* fœnile, Ælfc. Gl. 109; Som. 79, 20; Wrt. Voc. 58, 60.

heg-, hege-stów, e; *f. A place enclosed by a hedge* [?], Cod. Dipl. Kmbl. iii. 77, 27 : 213, 8, 9 : 263, 23, 26.

hêh. v. heáh.

hel [?] *a pretext :*—Mid yfelan helan earme men beswîcaþ *with evil pretexts defraud poor men,* L. I. P. 12; Th. ii. 320, 18. [Cf. *O.H.Ger.* hal *tegmen,* Grff. iv. 844.]

HEL, hell, helle, e; *f.* HELL, *the place of souls after death, Hades, the infernal regions, the place of the wicked after death :*—Helle *infernus,* Ælfc. Gl. 54; Som. 63, 103; Wrt. Voc. 36, 24: Ælfc. Gr. 8; Som. 11, 34. Satanas ðære helle ealdor cwæþ tô ðære helle . . . Seó hell swîðe grymme andswarode *Satan the ruler of Hell said to Hell . . . Hell answered very fiercely,* Nicod. 26; Thw. 13, 32, 40. In ðæt hâte hof ðam is hel nama *into that hot abode whose name is hell,* Cd. 217; Th. 276, 24; Sat. 193. Ðonne heofon and hel hæleþa bearnum fylde weorþeþ *when heaven and hell shall be filled with the children of men,* Exon. 31 a; Th. 97, 17; Cri. 1592. Hel nimeþ wærleásra weorud *hell shall take the host of the faithless,* 31 b; Th. 98, 26; Cri. 1613. Him hel onfêng *hell received him,* Beo. Th. 1709; B. 852. Helle gatu *portæ inferi,* Mt. Kmbl. 16, 18. Helle bearn *filium gehennæ,* 23, 15. Fȳr byrnþ ôð helle endas *a fire shall burn unto the lowest hell,* Deut. 32, 22. Oð helle *in infernum,* Mt. Kmbl. 11, 23. For ðam ða deádan ðe on helle beóþ ðîn ne gemunan ne ðê andetaþ swâ swâ wê dôþ *quoniam non est in morte qui memor sit tui : in inferno quis confitebitur tibi,* Ps. Th. 6, 4. On ðære sweartan helle *in the black hell,* Cd. 35; Th. 47, 16; Gen. 761. Hig intô helle cuce sîdodon *they went down alive into the pit,* Num. 16, 33. Ic fare tô mînum sunu tô helle *I will go down into the grave unto my son,* Gen. 37, 35. Uton nû brûcan ðisses undernmetes swâ ða sculon ðe hiora æfengife on helle gefecceán sculon *prandete tanquam apud inferos cænaturi,* Ors. 2, 5; Swt. 86, 2. Swâ ðæt fȳr on ðære helle seó is on ðam munte ðe Ætne hâtte *as the fire on the hell that is in mount Ætna,* Bt. 15; Fox 48, 20. Hire sâwle mon sceolde lædan tô helle *her soul was to be conducted to hell,* 35, 6; Fox 168, 5. [*Goth.* halja *Hades : O.Sax.* hel, hellia : *O.Frs.* hille : *Icel.* hel (*local and personal*): *O.H.Ger.* hella *gehenna, infernus, baratrum : Ger.* hölle.] v. Grm. D. M. 288–92 : 760–7. See compounds with helle.

hel- v. hell-.

HÊLA, hêla, an; *m. The* HEEL :—Hêla *calx,* Wrt. Voc. 283, 75. Hêl *calcaneum,* Jn. Skt. Lind. 13, 18. Genim haran hêlan [hælan MSS. H. B.] *take hare's heel* [lat. *talum*], Med. ex Quadr. 4, 17; Lchdm. i. 346 16. Heó gehȳden hælun mîne *ipsi calcaneum meum observabunt,* Ps. Th. 55, 6. Gif ðæt wîf mid ðâm hêlum stæpeþ *if the woman steps with the heels,* Lchdm. iii. 144, 14. [*O.Frs.* hêla, heila : *Icel.* hæll.]

helan; *p.* hæl, *pl.* hælon; *pp.* holen *To conceal, hide, cover :*—Gif ðû mê hylest ðîne heortan geþohtas *if thou dost conceal from me thy heart's thoughts,* Exon. 88 b; Th. 333, 12; Gn. Ex. 3. Ðonne eówaþ hê hî nalles ne hilþ *then it shews them and does not conceal them,* Bt. 27, 1; Fox 94, 26. Swâ hwâ swâ hilþ his gôdan weorc *si bona quæ agit occultat,* Past. 59, 4; Swt. 449, 29. Ða ðe hira gôd helaþ ðe hie dôþ *qui bona que faciunt abscondunt,* 23; Swt. 179, 9. Ic hæl mîne scylda *I concealed my sins,* Ps. Th. 31, 3 : L. E. I. 30; Th. ii. 426, 21. Ðû heora fyrene fæste hæle operuisti *omnia peccata eorum,* Ps. Th. 84, 2. Hê hit hæl swîðe fæste wið his brôðor *he concealed it very carefully from his brother,* Ors. 6, 33; Swt. 288, 14. Hê ðæt hæl ærest scê petre *he at first concealed that from St. Peter,* Shrn. 74, 20. Ealle ða ðe ðone gylt mid him wiston and mid him hælon *all those who were cognisant of that crime and joined with them in concealing it,* Ors. 4, 4; Bos. 80, 24. Hî hælon ðæt hî forhelan ne mihton *they hid what they could not keep hidden,* Lchdm. i. 392, 4. Ðû him fæste hel sôþan spræce *hide carefully from them true speech,* Cd. 89; Th. 110, 11; Gen. 1836. Nân ôðrum his þearfe ne hele *let no one conceal from another what it is needful for him to know,* L. I.

P. 10; Th. ii. 316, 20: Andr. Kmbl. 2329; An. 1166. Ða ðe willaþ helan ðæt hí tó góde dóþ *qui bona clam faciunt*, Past. 59; Swt. 447, 23. Nele hé ús nánwiht helan se ðe ús lǽt hyne sylfne cunnan *he will not conceal anything from us who lets us know himself*, Shrn. 202, 12. Ic ne mæg leng helan be ðam lífes treó *I cannot longer conceal concerning the tree of life*, Elen. Kmbl. 1408; El. 706. [*Chauc*. hele: *A.R*. i-holen, *part. p*: hele *to cover*, in the Surrey dialect: *O. Sax*. helan: *O. Frs*. hela: *O. H. Ger*. helan *celare, tegere*: *Ger*. hehlen.] DER. be-, for-helan.

held. v. hyld.

heldan. v. hyldan.

helde, an; *f. Allegiance, fealty* :— Hé ðǽr on ðæs cynges willelmes heldan tó cynge gesette *he placed Edgar there as king in allegiance to King William*, Chr. 1097; Erl. 234, 37. Heanríg ofer sǽ fór on ðæs cynges heldan *Henry went over sea as liege man of the king*, 1095; Erl. 231, 9. [Cf. un-helde; hyld, hyldo.]

helde, an; *f. Tansy*; tanacetum vulgare:—Helde *tanicetum*, Wrt. Voc. 79, 24: *tanaceta*, Ælfc. Gl. 40; Som. 63, 87; Wrt. Voc. 30, 33. Genim heldan *take tansy*, L. M. 1, 36; Lchdm. ii. 86, 20.

hele-. v. helle-.

helerung, e; *f. The turning of a balance*; trutinæ inclinatio, Cot. 136, Lye. v. helur-bled, heolorian, heolra.

helfe, es; *m. n.* [?] *Helve, handle* :—Hæft and helfe *manubrium*, Ælfc. Gl. 52; Som. 66, 31; Wrt. Voc. 35, 20. Sió æcs áwient of ðæm hielfe *ferrum lapsum de manubrio*, Past. 21, 7; Swt. 167, 1. Gaderode me hylfa tó ǽlcum ðara tóla ðe ic mid wircan cúðe *I gathered me handles for each of the tools that I could work with*, Shrn. 163, 6. [*Orm*. hellfe: *Prompt. Parv*. helve *manubrium*: *Wick*. helve: *O. H. Ger*. halap, halp, halbe, helbe *manubrium*. *Helve* is a word given as still belonging to the dialects of East Anglia.]

helfling, es; *m. A halfpenny*: — Ne becýpaþ hig fíf spearwan tó helflinge *are not five sparrows sold for two farthings*, Lk. Skt. 12, 6. [*O. H. Ger*. helbeling-*obolus*.]

helian; *p*. ode, ede *To hide, conceal, cover* : — Mín unriht ic ná ne helede wið ðé *injustitia mea non operui*, Ps. Th. 31, 5. Heó helode hire nebb ðæt hé hig ne mihte gecnáwan *she had covered her face that he might not know her*, Gen. 38, 15. Wé lǽraþ ðæt ǽnig geháded man his sceare ne helige *we enjoin that no man in orders conceal his tonsure*, L. Edg. C. 47; Th. ii. 254, 13. [*A.R*. helien: *Piers P*. helien, hylien; *pp*. helid, hiled: *Laym*. helede, *p*: *Wick*. hilide: *O. Sax*. bi-helian: *O. H. Ger*. hellen: *Ger*. hehlen.]

hell. v. hel.

hell-bend; *m. f. A hell-bond* :—Hellbendum fæst *fast in the chains of hell*, Beo. Th. 6137; B. 3072.

hell-cræft, es; *m. Hellish art*, Andr. Kmbl. 2205; An. 1104.

hell-cwalu, e; *f. Hell-torment*, Exon. 25 a; Th. 73, 15; Cri. 1190.

hell-deóful, es; *m. n. Orcus, Pluto*, Cot. 145, Lye.

hell-dor, es; *n. The gate of hell* :—Tó helldore in *infernum*, Ps. Th. 87, 3. Æt heldore, Exon. 40 b; Th. 135, 29; Gú. 531: Cd. 19; Th. 24, 20; Gen. 380: 23; Th. 29, 8; Gen. 447. [*O. Sax*. hell-dor.]

helle-. *In the case of at least some of the following words which are given as compounds, they might be taken as independent words, the first of which is the genitive of* hel. *For the meaning of such combinations the second word may be referred to.*

helle-bealu; *gen*. wes; *n. Hell-bale, woe of hell*, Exon. 28 b; Th. 87, 18; Cri. 1427.

helle-bróga, an; *m. The terror of hell* :—On hellebrógan gesette hí syndon *in inferno positi sunt*, Ps. Lamb. 48, 15. Of handa hellebrógan *de manu inferi*, 48, 16.

helle-bryne, es; *m. Hell-fire*, Judth. 10; Th. 23, 11; Jud. 116.

helle-ceafl, es; *m. The jaws of hell*, Andr. Kmbl. 3403; An. 1705.

helle-cinn, es; *n. The race of hell*, Exon. 31 b; Th. 99, 5; Cri. 1620.

helle-clam, -clom, Cd. 19; Th. 24, 6; Gen. 373. v. clam.

helle-deóful, -dióful, Exon. 75 a; Th. 280, 15; Jul. 629: Elen. Kmbl. 1799; El. 901: Andr. Kmbl. 2598; An. 1300. [Cf. hell-deóful.]

helle-dor, Exon. 121 a; Th. 464, 14; Hö. 87. [Cf. hell-dor.]

helle-duru, Elen. Kmbl. 2457; El. 1230.

helle-flór, Cd. 214; Th. 269, 9; Sat. 70.

helle-fýr, Bt. Met. Fox 8, 101; Met. 8, 51; Exon. 26 b; Th. 78, 6; Cri. 1270. On helle fýr *in gehennam ignis*, Mt. Kmbl. 18, 9. [*O. H. Ger*. hella-fiur *gehenna, tartarus*.]

helle-gást, -gæst, Exon. 72 a; Th. 269, 28; Jul. 457: 74 b; Th. 279, 17; Jul. 615: Beo. Th. 2552; B. 1274.

helle-geat, -gat, Homl. Th. i. 288, 1, 4.

helle-god, es; *n. A god of the infernal regions* :—Orfeus wolde gesécan hellegodu and biddan ðæt hí him ágeáfan eft his wíf *Orfeus would visit the gods of the infernal regions and pray them to give him his wife again*, Bt. 35, 6; Fox 168, 13. [*O. H. Ger*. hella-got *pluto, dis; pl. eumenides, manes*.]

helle-grund, Exon. 11 b; Th. 17, 4; Cri. 265: 16 a; Th. 35, 23; Cri. 562: Elen. Kmbl. 2608; El. 1305. [*O. Sax*. helli-grund: *O. H. Ger*. hella-grunt *tartarus*.]

helle-grut *the abyss of hell*, Hpt. Gl. 422. v. grut.

helle-, hylle-gryre, Cd. 223; Th. 291, 20; Sat. 433.

helle-hæft, Cd. 227; Th. 304, 16; Sat. 631.

helle-hæfta, Beo. Th. 1580; B. 788.

helle-hæftling, Andr. Kmbl. 2683; An. 1344: Exon. 69 a; Th. 257, 12; Jul. 246: Salm. Kmbl. 253; Sal. 126.

helle-heáf, Cd. 2; Th. 3, 19; Gen. 38.

helle-hinca, an; *m. The hell-limper, -hobbler, the devil lamed by his fall from heaven*, Andr. Kmbl. 2343; An. 1173. Grimm [Deutsche Mythologie, 944–5] speaking of the devil observes 'Am ersten fällt sein lahmer fuss auf, daher der *hinkende teufel* [diable boiteux], *hinkebein*, vom sturz aus dem himmel in den abgrund der hölle scheint er gelähmt, wie der von Zeus herabgeschleuderte Hephäst.' [Cf. *Icel*. hinka: *O. H. Ger*. hinkan *claudicare*.]

helle-hund, es; *m. A hell-hound* :—Sý hé Judas geféra Cristes belǽwendes and sý hé toren of hellehundes tóþum on ðám egeslícum hellewítum mid eallum deóflum bútan ǽlcum ende bútan hé hit ǽr his endedæge ribtlíce gebéte *may he be the companion of Judas the betrayer of Christ, and be torn by the teeth of a hell-hound in the awful torments of hell among all the devils without any end, unless he make due reparation before his last day*, Cod. Dipl. Kmbl. iii. 350, 18. [Cf. sceolde cuman ðære helle hund ongeán hine ðæs nama wæs Cerueruses *it was said that the hound of hell, whose name was Cerberus, came towards him*, Bt. 35, 6; Fox 168, 15.] v. Grimm. D. M. 948–9.

helle-hús, Exon. 42 b; Th. 142, 24; Gú. 649.

helle-líc; *adj. Infernal* :— Helelíc deópnes *barathrum, vorago, profunda*, Ælfc. Gl. 54; Som. 66, 97; Wrt. Voc. 36, 20. [*O. H. Ger*. helle-lích *tartareus*.] v. hel-líc.

helle-mere, es; *m. The lake of hell, Styx* :— Hellemere *hæc styx*, Ælfc. Gr. 9; Som. 14, 13. Helemere *Styx*, Ælfc. Gl. 54; Som. 66, 99; Wrt. Voc. 36, 22.

helle-níþ, Cd. 37; Th. 48, 13; Gen. 775.

helle-rúne, an; *f. One who is skilled in the mysteries of hell, the region of the dead, a sorceress, necromancer* :—Hellerúne *pythonissa*, Ælfc. Gl. 112; Som. 79, 102; Wrt. Voc. 60, 10. [*O. H. Ger*. hellirúna *necromantia*: v. Grm. D. M. 1175, 1178.] v. hell-rúna.

helle-scealc, Cd. 216; Th. 273, 8; Sat. 133.

helle-sceaþa, Elen. Kmbl. 1911; El. 957. v. hell-sceaþa.

helle-seáþ, es; *m. The pit of hell* :—Helleseáþ [Som. scead] *erebum*, Ælfc. Gl. 54; Som. 66, 98; Wrt. Voc. 36, 21: Exon. 71 b; Th. 267, 29; Jul. 422.

helle-þegn, Exon. 48 a; Th. 166, 14; Gú. 1042.

helle-wíte, es; *n. Hell-torment, punishment, hell* :—Hellewíte *tartara vel gehenna*, Ælfc. Gl. 54; Som. 66, 100; Wrt. Voc. 36, 23. Se for ðam méde onféhþ écum tintregum hellewítes *æternas inferni pœnas pro mercede recipiet*, Bd. 1, 7; S. 477, 40: Hy. 6, 36; Hy. Grn. ii. 286, 36. Mid heardum hellewítum *with hard pains of hell*, Soul Kmbl. 94; Seel. 47: 64; Seel. 32: Andr. Kmbl. 2106; An. 1054. [*O. Sax*. helliwíti *hell-torment*: *Icel*. hel-víti: *Dan*. helvede *hell*: *O. H. Ger*. hellawízi *gehenna, tartara*.]

hell-firen, e; *f. A hellish crime*, Exon. 98 a; Th. 366, 3; Reb. 6.

hell-fús; *adj. Bound for hell*, Andr. Kmbl. 99; An. 50: Exon. 24 a; Th. 69, 21; Cri. 1124.

hell-geþwing; es; *n. The restraint, constraint of hell* :—Se hellsceaða wiste ðæt hie sceoldon hellgeþwin[g] niéde onfón *the devil knew that they must needs receive the restraint of hell*, Cd. 33; Th. 43, 20; Gen. 696. [*O. Sax*. helli-geþwing.]

hell-heóþo; *indecl*; *f. Hell*, Cd. 228; Th. 308, 29; Sat. 700. v. heóþu.

hel-líc; *adj. Hellish, infernal* :— Ðeós hellíce súsl *hic tartarus*, Ælfc. Gr. 13; Som. 16, 29; Homl. Th. ii. 78, 20. Seó fæstnung ðære hellícan clýsinge ne geþafaþ ðæt hí ǽfre útábrecon *the fastening of the enclosure of hell does not permit them ever to break out*, i. 332, 20: ii. 80, 6. Wé wǽron mid eallum úrum fæderum on ðære hellícan deópnysse *we were with all our fathers in the deep of hell*, Nicod. 24; Thw. 12, 19. Ða hellícan fýnd *the fiends of hell*, Homl. Th. i. 380, 27.

hell-rúna, an; *m. One skilled in the mysteries of hell, a sorcerer, necromancer*, Beo. Th. 328; B. 163. v. helle-rúne.

hell-sceaða, an; *m. A hell-harmer, fiend, devil*, Cd. 33; Th. 43, 22; Gen. 694: Exon. 13 a; Th. 23, 5; Cri. 364: Byrht. Th. 137, 2; By. 180.

hell-træf, es; *m. A hellish, infernal building*, Andr. Kmbl. 3379; An. 1693.

hell-trega, an; *m. Hell-torment*, Cd. 4; Th. 5, 18; Gen. 73.

hell-waran; *pl. The inhabitants of hell* :— Ðás hellwaran hí *manes*, Ælfc. Gr. 13; Som. 16, 14. Him urnon ealle hellwaran ongeán *all the inhabitants of hell ran to meet him*, Bt. 35, 6; Fox 168, 29. Hlógan helwaran *the dwellers in hell laughed*, Exon. 120 a; Th. 460, 22; Hö. 21. Ðú míne sáwle álýsdest of helwarena hinderþeóstrum *eripuisti animam meam ex inferno inferiori*, Ps. Th. 85, 12: 140, 9. Helwarena stefn wæs gehýred *the voice of hell's people was heard*, Blickl. Homl.

87, 3. Cýðnise hellwarana *testamentum inferorum*, Rtl. 11, 9. Tô hell-warum *ad inferos*, 101, 16. Ne forlǽt ðû mîne sâwle mid hellwarum *leave not my soul in hell*, Blickl. Homl. 87, 33. v. hell-ware, -waru.

hell-ware, -wara; *pl. The inhabitants of hell* :—Ealle gesceafta heofon-wara eorþwara helwara onbûgaþ Criste *all creatures, those in heaven, those on earth, those in hell, bow to Christ*, Homl. Th. ii. 362, 1 : i. 36, 26. Ealle hellwara *all the inmates of hell*, Exon. 121 b; Th. 466, 18; Hö. 123. Wuldorweorudes and helwara *of the glorious host and of the dwellers in hell*, Exon. 12 a; Th. 18, 20; Cri. 286 : 114 a; Th. 437, 12; Rä. 56, 6. v. hell-waru.

hell-waru, e; *f. The body of inhabitants in hell* :— On ðam mycelan dôme ðǽr heofonwaru and eorþwaru and helwaru beóþ ealle gesomnode *in magno judicio ubi cælicolæ et terricolæ et inferi omnes congregabuntur*, L. Ecg. C. pref; Th. ii. 132, 22 : Hy. 7, 95; Grn. ii. 289, 95. Tô ðare helware [*or* ðara helwara (?)] stíðe pînnesse *to the severe torment of the people of hell*, Chart. Th. 369, 34. [Cf. burh-ware, -waru, -wara.]

HELM, es; *m.* **I.** *a* HELM, *helmet* :—Leðer helm *galea* : îren helm *cassis*, Ælfc. Gl. 51; Som. 66, 13, 14; Wrt. Voc. 35, 3, 4. Helmes camb *crista* : helmes bȳge *conus*, 53; Som. 66, 76, 77; Wrt. Voc. 36, 2, 3. Se hwîta, hearda helm, Beo. Th. 2900, 4502; B. 1448, 2255. **II.** *a crown, the top, overshadowing foliage of trees* :— Helm *corona*, Wrt. Voc. 64, 39. Mid þyrnenum helme his heáfod be-fêngon *encircled his head with a crown of thorns*, Homl. Th. ii. 252, 26; Mk. Skt. 15, 17. Ful oft unc holt wrugon wudubeáma helm *full oft the wood covered us the shady top of the forest trees*, Exon. 129 a; Th. 496, 2; Rä. 85, 8. Ðæt se stemn and se helm môte ðȳ fæstor and ðȳ leng standon *that the stem and top may stand the faster and longer*, Bt. 34, 10; Fox 148, 33 : Fox 150, 3. Hire hyrdeman sume âc âstâh and his orf læswode mid treowenum helme *her herdsman had ascended an oak and was feeding his cattle with its woody crown*, Homl. Th. i. 150, 31. Forðæm se þorn ðære gítsunga ne wyrþ forsearod on ðæm helme gif se wyrttruma ne biþ fǽrcorfen oððe forbærned æt ðæm stemne *si enim radix culpæ in ipsa effusione non exuritur, numquam per ramos exuberans ava-ritiæ spina siccatur*, Past. 45, 3; Swt. 341, 10 : Runic pm. 18; Kmbl. 342, 31; Hick. Thes. i. 135. **III.** *a covering* [in this sense the word is preserved in some dialects. Thus in Yorkshire and Lincolnshire Glossaries, English Dial. Soc. vols. ii. v. vi, *helm*, a hovel, an open shed for cattle, a shed built on posts] :—Wǽges helm [holm ?] *the covering made by the wave, the sea*, Elen. Kmbl. 459; El. 230. Under lyfte helm *under the air's covering*, Exon. 102 a; Th. 386, 19; Rä. 4, 64. Helme gedȳgled *concealed with a covering*, 1226; Th. 470, 10; Hy. 11, 33. **IV.** *in poetry the word is applied to persons, thus God and Christ are spoken of as* æþelinga, hæleþa, hâligra, duguþa, dryhtfolca, engla, gâsta, heofona, heofonrîces, wuldres helm and helm wera, ælwihta. *Similar phrases occur in speaking of earthly rulers*, æþelinga, heriga, lid-manna, wedra, weoruda helm and helm Scyldinga, Scylfinga. [*Goth.* hilms *a helmet* : *O. Sax.* helm : *Icel.* hjálmr : *O. H. Ger.* helm, *galea, cassis* : *Ger.* helm.] DER. bân-, grîm-, gûþ-, hæleþ-, heaþu-, heoloþ-, lyft-, mist-, niht-, sceadu-, sund-, wæter-helm.

Helma, an; *m. A* HELM, *rudder* :— Helma *clavus*, Ælfc. Gl. 104; Som. 77, 124; Wrt. Voc. 56, 42. Be ðæm is swîðe sweotol ðætte God ǽghwæs wealt mid ðæm helman his gôdnesse *Deus omnia bonitatis clavo gubernare jure credatur*, Bt. 35, 4; Fox 160, 14. [*Icel.* hjálm; *f.*]

helm-berend, es; *m. One who wears a helmet* :— Ne rôhte hê helm-berendra *he recked not of helmeted warriors*, Exon. 120 a; Th. 461, 18; Hö. 37. Gegrêtte hwate helmberend *he greeted the bold warriors*, Beo. Th. 5027; B. 2517 : 5277; B. 2642. [*O. Sax.* helm-berand : *and cf. the epithet* Hjálm-beri *helmbearer, given to Odin*.]

helmian; *p.* ode *To cover* :—Niht helmade beorgas steápe *night covered the high hills*, Andr. Kmbl. 2612; An. 1307.

helmiht; *adj. Full of leaves* or *boughs; frondosus*, Cot. 75, 198, Lye. v. helm II.

HELP, e; *f: also* es; *m.* HELP, *aid, succour* :—On ðæm burgum wæs getâcnad ðæt Crist is eáðmôdegra help *probans se esse conservatorem hu-milium*, Ors. 3, 2; Swt. 100, 25. Ðǽr is help gearu æt mǽrum manna gehwylcum *there is help ready at the hand of the mighty one for every man*, Andr. Kmbl. 1814; An. 909. Gionn helpe *præsta subsidium*, Rtl. 71, 37. Ða ðe hine helpe biddaþ *who ask him for help*, Ps. Th. 118, 2 : Andr. Kmbl. 2061; An. 1033. Gehȳr helpys bênan *exaudi me*, Ps. Th. 101, 2. Uton helpan aa ðam raðost ðe helpes betst behôfaþ *let us ever help him first who has most need of help*, L. C. S. 69; Th. i. 412, 3. Helpes bedǽled *deprived of help*, MS. Cott. Nero A. i. fol. 73. Helpes biddende *asking for help* : sumes helpes biddende *asking for some help*, Lchdm. iii. 365, col. 2. Hwá him tô hǽle and tô helpe on ðás world âstâg *who came down to this world as their salvation and help*, Blickl. Homl. 105, 32. Ðám burgwarum com mâra fultum tô ûtan tô helpe *more aid came from without to the citizens to help them*, Chr. 921; Erl. 107, 19. Rûmlícum helpe *benigno favore*, Rtl. 17, 35. Þurh ða gebedu gê mâgon on swîðe mycelan hylpe beón ge libbendum ge forþ-farenum *by prayers you may be of very great help both to the living and the departed*, L. E. I. 3; Th. ii. 404, 18 : 21; Th. ii. 414, 36. Nǽnige

helpe ðam byrnendan hûse gedôn mihton *nil ardenti domui prodesse va-lentes*, Bd. 3, 10; S. 534, 34. Ða nǽnig him ǽnige helpe findan mihte *cum nil salutis furenti superesse videretur*, 3, 11; S. 536, 25. Helpe úserne *adjutorium nostrum*, Rtl. 172, 23. Ðǽr mê wið lâþum lícsyrce mîn helpe gefremede *there against the foes my coat of mail afforded me help*, Beo. Th. 1107; B. 550. Gehýr mê and mê help freme *exaudi me*, Ps. Th. 68, 17 : Cd. 184; Th. 230, 20; Dan. 236. Dǽleþ help and hǽlo hæl-eþa bearnum *distributes help and salvation to the children of men*, 226; Th. 301, 15; Sat. 586 [*O. Sax.* helpa; *f : O. Frs.* helfe; *f : Icel.* hjâlp; *f : O. H. Ger.* helfa; *f. auxilium, adjutorium, subsidium, solatium : Ger.* hülfe.] v. helpe.

helpan; *p.* healp, *pl.* hulpon; *pp.* holpen; *v. trans. followed by gen. or dat. To help, aid, assist, succour* :—Ðû monegum helpst *thou helpest many*, Hy. 7, 44; Hy. Grn. ii. 288, 44. Wið fefre hylpþ marubis tô drincanne *for fever it helps to drink marrubium*, L. M. 1, 62; Lchdm. ii. 134, 27. Hê helpeþ þearfan *parcet pauperi*, Ps. Th. 71, 13. Ðonne helpe gê wel ðam ðe gê lǽraþ gif hî eówre lârum fyligean willaþ *then do ye well help those whom ye teach, if they will follow your teaching*, L. I. P. 21; Th. ii. 332, 21. Hê nyle helpan ðæs folces mid ðam ðe God his healp *ex muneribus quæ perceperit prodesse aliis non curat*, Past. 5, 2; Swt. 45, 5. Ðonne ðu hulpe mîn *when thou didst help me*, Ps. Th. 70, 20. Ða steortas hulpan ealle ðæs heáfdes *all the tails helped the head*, Shrn. 162, 16 : Exon. 27 b; Th. 83, 10; Cri. 1354. Help mê, Ps. Th. 60, 1. God ûre helpe. Amen *may God help us. Amen*, Swt. A. S. Rdr. 112, 225. Wê on ðisum lífe mâgon helpan ðam forþfarenum ðe on wîtnunge beóþ *we in this life may help the departed that are being punished*, Homl. Th. ii. 356, 11. Wê sceolon earmra manna helpan *we ought to help poor people*, 442, 14. Helpa *fovere*, Rtl. 122, 37. [*Chauc.* Piers. P. *p.* halp, help, *pl.* holpen; *pp.* holpen : *the pp.* holpen *occurs in the authorized version of the Bible* : *Goth.* hilpan : *O. Sax.* helpan : *O. Frs.* helpa : *Icel.* hjálpa : *O. H. Ger.* helfan : *Ger.* helfen.] DER. â-, ge-helpan.

helpe, an; *f. Help* :— Gif ðás fultumas ne sýn helpe *if these remedies are no help*, L. M. 2, 48; Lchdm. ii. 262, 15. [*Or should this be placed under* help ?] Hê him helpan ne mæg ænige gefremman *he can give him no help*, Beo. Th. 4888; B. 2448.

helpend, es; *m. A helper* :— Helpend *adjutor*, Rtl. 45, 18. Ealles middangeardes hǽlend and ealra sáula helpend *the saviour of all the earth and the helper of all souls*, Blickl. Homl. 105, 190. Helpend and hǽlend wið hellsceaðum *a helper and saviour against the harmers of hell*, Exon. 68 a; Th. 252, 2; Jul. 157. Helpend ne hafo ic *I have no helper*, Jn. Skt. Lind. 5, 7. Syððan hê ne hæbbe helpend ǽnne *quia non est qui eripiat eum*, Ps. Th. 70, 10. Helpendra leás *without helpers*, Exon. 28 b; Th. 86, 27; Cri. 1414.

helpend-bǽre; *adj. Helpful, assistant;* opifer, Cot. 148, Lye.

helpend-líc; *adj. Auxiliary.*

hêl-spure, an; *f. A heel* :— Unrehtwîsnis hêlspuran [hellspuran, Ps. Spl. 48, 5] mîne *iniquitas calcanei mei*, Ps. Stev. 48, 6. Hêlspuran [hellspuran, Ps. Spl. 55, 6] mîne *calcaneum meum*, 55, 7.

helto; *f. Haltness, lameness* :—Âfyrr ðû drihten from ðære stôwe blind-nesse and helto and dumbnesse *remove O Lord from the place blindness and lameness and dumbness*, Shrn. 101, 35.

helur-bledu, e; *f. The scale of a balance;* lanx, Cot. 26, Lye. v. bledu.

hem; *m. A hem, border* :— Hem *limbus*, Ælfc. Gl. 28; Som. 61, 7; Wrt. Voc. 26, 6. [*Laym.* þane hem : *Prompt. Parv.* hemme *fimbria, limbus.*] Cf. ham *an enclosure.*

hemlíc, hymlíc, es; *m* : hymlíce, an; *f. Hemlock* :— Hemlíc *cicuta*, Ælfc. Gl. 43; Som. 64, 47; Wrt. Voc. 31, 57. Hemlíc hâtte wyrt *a plant called hemlock*, L. M. 1, 77; Lchdm. ii. 150, 15. Wyrc híe of hemlíc *make the salve of hemlock*, 58; Lchdm. ii. 128, 7. Nim hemlíc *take hemlock*, 31; Lchdm. ii. 74, 6. Wyll nyoðerweardne hymlíc *boil the lower part of hemlock*, Lchdm. iii. 50, 17. Hymlíce *cicuta*, p. 331, col. 1. Dô tô hymlícan *put hemlock to it*, L. M. 1, 1; Lchdm. ii. 18, 27.

hemming, es; *m. A kind of shoe;* pero, Cot. 155, Lye.

hen. v. hæn.

hênan. v. hȳnan.

-hende. v. an-, ân-, ge-, of-, on-, spær-hende.

henge-clif, es; *n. A steep, precipitous cliff;* præruptum, Ælfc. Gl. 101; Som. 77, 38; Wrt. Voc. 55, 43.

hengen, e; *f.* **I.** *hanging* :— Eode and hî sylfe âheng . . . Se deófol hî tô hire âgenre hengene gelǽrde *she went and hung herself . . . The devil persuaded her to her own hanging* [*to hang herself*], Homl. Th. ii. 30, 24. Hêt hine hôn and mid hengene þráwan tô langere hwîle *bade hang him and for a long time torture him with hanging*, 308, 31. **II.** *that on which any one is hung, a gibbet, gallows, cross* :— Crist ðone ðe hî on hengene fæstnodon *Christ whom they fastened on a cross*, Homl. Th. ii. 256, 22 : 308, 30. Laurentius âstreht on ðære hengene þancode his Drihtne . . . Hê hêt âlýsan ðone diácon of ðære hengene *Lawrence stretched on the cross thanked his Lord . . . He ordered the deacon to be released from the cross*, i. 426, 32, 35. **III.** *prison, confinement, durance.* Schmid, p. 609, suggests a connection between this

meaning and that g´ven under I. in the following remark: 'Die grammatische Bedeutung des Wortes führt darauf, dass ursprünglich darunter das Anhängen an einen Block oder das Einspannen in den Stock, als die Art der Sicherung eines Gefangenen, der man sich bediente, wenn Gefängnisse fehlten, verstanden worden sei.' Accordingly he translates the following passage, L. Alf. pol. 35; Th. i. 84, 4: — Gif hé hine on hengenne [MS. B. hengene] álecgge 'wenn er ihn in den Stock legt,' which Thorpe renders *if he lay him in prison*. In the latter sense it is found L. C. S. 35; Th. i. 396, 27:—Gif freóndleás man swá geswenced weorþe ðæt hé borh næbbe ðonne gebúge hé hengenne [MS. B. hengene] and ðǽr gebíde óþ ðæt hé gá tó Godes ordále *if a friendless man be so distressed that he have no surety, then let him submit to prison, and there abide, until he go to God's ordeal*. Cf. L. H. 65, 5; Th. i. 568, 14, ponatur in *hengen*. [Cf. O. Sax. hie (Krist) welda thea werold alla mid is henginnia alósian, Hel. Heyne 5435: thuo sprak theró mannó óðer (the penitent thief) an thero henginna thâr hie geheftid stuod, 5591.]

hengen-wítnung, e; f. *The punishment of imprisonment* :—Gif forworht man friþstól geséce and þurh ðæt feorh geyrne ðonne sý þreóra án for his feore búte man bet geárian wille wergild éce þeówet hengenwítnung *if a man who has forfeited his life gain a sanctuary, and thereby secure his life, let there be one of three things instead of his life, unless he obtain remission more favourably, wergild, perpetual thraldom, imprisonment*, L. Eth. vii. 16; Th. i. 332, 18.]

hengest, es; m. *A gelding, horse, steed :* — Hengst *canterius*, Ælfc. Gl. 20; Som. 59, 46; Wrt. Voc. 23, 8. Án hundred wildra horsa and xvi tame hencgestas *a hundred wild horses and sixteen tame steeds*, Chart. Th. 548, 11. [*Laym.* hængest: *O. Frs.* hengst: *Icel.* hestr *a stallion, horse: O. H. Ger.* hengist *eunuchus, spado, cantarius, equus castratus: Ger.* hengst *a stallion*.] DER. brim-, faroþ-, fæt-, fríd-, mere-, sǽ-, sund-, wǽg-hengest.

Hengest, es; m. *Hengest*, Bd. 1, 15; S. 483, 28; Chr. 449; Erl. 13, 1–21: 455; Erl. 13, 22–25: 457; Erl. 12, 17–20: 465; Erl. 12, 21: 473; Erl. 12, 25: 488; Erl. 14, 3–4.

heng-wíte, es; n. *A fine to be paid for not keeping a criminal in custody so that he may be brought before the proper tribunal :* — Si quis latronem vel furem, sine clamore et insecutione ejus, cui dampnum factum est, ceperit, et captum ultra duxerit dabit x solid. de henwite [hengwite, French text], Andr. L. 4; Th. i. 469, 27.

henna, an; m. *A fowl :*—Gif swýn oðde henna ete of mannes líchaman *si porcus vel gallina de corpore hominis ederit*, L. Ecg. P. iv. 57; Th. ii. 220, 13. v. hæn.

henne-belle. v. hænne-belle.

hentan; p. te *To pursue, follow after, seize* [?] :—Gif hé man tó deáþe gefylle beó hé ðonne útlah and his hente mid hearme ǽlc ðara ðe riht wille *if he fell a man to death, let him then be an outlaw, and let every one that desires right pursue him with hue and cry* [?], L. E. G. 6; Th. i. 170. 10: L. C. S. 49; Th. i. 404, 11. Nime ðonne leáfe ðæt hé móte hentan æfter his ágenan *let him then take leave to follow after his own*, 19; Th. i. 386, 17. [*Chauc.* Piers P. hente *to seize, take, get :* Prompt. Parv. hentin *rapere*.] v. ge-hentan.

hénþ, hénþu. v. hýnþ, hýnþu.

heó. v. hé.

heó-dæg; adv. *To-day;* hodie, Cd. 30; Th. 41, 23; Gen. 661. [*O. Sax.* hiudu: *O. Frs.* hiudega, hiude: *O. H. Ger.* hiutu: *Ger.* heute: cf. *Goth.* himma daga.]

heóf, es; m. *Lamentation, grief, sorrow :* — Maximus mid micelum heófe gedréfed him tó com *Maximus troubled with great grief came to him*, Homl. Th. i. 414, 17. Sǽde ðæt hie hæfden bet gewyrht ðæt him mon mid heúfe [heófe MS. C.] ongeán cóme ðonne mid triumphan *Fabius oblatum sibi a senatu triumphum suscipere recusaret, quia luctus potius debebatur*, Ors. 2. 4; Swt. 70, 20. Heóf mínne *planctum meum*, Ps. Spl. 29, 13 [heáf, Ps. Th. 29, 11].

heófan; p. de *To lament, grieve, wail, mourn :* — Hungre heófeþ laments *for hunger*, Exon. 91 b; Th. 342, 30: Gn. Ex. 150. Heófaþ mid handum [Ps. Th. wépaþ and heówaþ] *plaudite manibus*, Ps. Spl. T. 46, 1 : 97, 8. Wé heófdon and gé ne weópon *lamentavimus et non plorastis*, Lk. Skt. 7, 32. Gif hé mid inweardre heortan heófe *if he heartily grieve*, L. Pen. 8; Th. ii. 280, 10. Heófende sprǽc *lamenting he spoke*, Andr. Kmbl. 3113; An. 1559. Álegdon ðá tó middes mǽrne þeóden hæleþ hiófende hláford leófne *warriors lamenting laid down in their midst the great prince, the lord beloved*, Beo. Th. 6275; B. 3142. [*Goth.* hiufan; *p.* hauf, v. Lk. 7, 32 : *O. Sax.* heobandi, hiobandi, *part. pres: O. H. Ger.* hiufit *luget;* hiufanti *luctuosus*.] v. heófian, heáfan.

heófe-líce; *adj. Lamentable, grievous ;* funebris, Som.

heofen. v. heofon.

heófian; p. ode *To lament, mourn, wail, bewail :* — Ic heófige *lugeo*, Ælfc. Gr. 26; Som. 28, 63. Gé hiófaþ and wépaþ *plorabitis et flebitis vos*, Jn. Skt. 16, 20. Hieremias heófode miclum ðæs folces synna swá swá his bóc ús segþ *Jeremiah lamented greatly the people's sins, as his book tells us*, Swt. A. S. Rdr. 70, 440. Ðá weópon hig ealle and heófodon hí *flebant autem omnes et plangebant illam*, Lk. Skt. 8, 52. Hí

heófodon folces synna *they bewailed people's sins*, Homl. Th. i. 540, 30. Wá eów ðe nú hlihgaþ gé sceolon heófian and wépan *woe to you that laugh now, ye shall mourn and weep*, 180, 15. Ðá ongann Ypolitus sárlíce heófian *then Hippolytus began sorely to lament*, 428, 12 : 408, 9 : L. E. I. prm ; Th. ii. 398, 36. Heófigende *lugens*, Ps. Spl. 34, 17. Heófiende *flebilis*, Bt. 2 ; Fox 4, 8. Of heófigendre menigu *from a mourning multitude*, Homl. Th. i. 86, 33. Mid heófigendum stemnum *with lamenting voices*, ii. 420, 16. v. heófan.

HEOFON, heofen, heofun, heofon, heben. hiofon, es; m. HEAVEN ; cǽlum :—Heofon and heofuna heofun and eorþe and ealle ða þing ðe sind on him sind Drihtnes *the heaven and the heaven of heavens is the Lord's, the earth with all that therein is*, Deut. 10, 14. Heofen and eorþe síde sǽflódas *cæli et terra, mare*, Ps. Th. 68, 35. Heofon and hel *heaven and hell*, Exon. 31 a ; Th. 97, 17 ; Cri. 1592. Heben til hrófe *heaven for a roof*, Swt. A. S. Rdr. 195, 13. Heofonas god *the god of heaven*, Hy. 3, 58 ; Hy. Grn. ii. 282, 58 : Andr. Kmbl. 3000 ; An. 1503. Hiofones leóhtes beorhto *the brightness of the light of heaven*, Bt. Met. Fox 21, 77 ; Met. 21, 39. Of hefene *from heaven*, Beo. Th. 3146 ; B. 1571. Mid his worde synt getrymede heofonas *verbo Domini cæli firmati sunt*, Ps. Th. 32, 5. Ðá wǽron fullfremode heofenas and eorþe *the heavens and the earth were finished*, Gen. 2, 1. Heofona ríce *regnum cælorum*, Mt. Kmbl. 13, 24. Of heofonum ðe of mannum *e cælo an ex hominibus*, 21, 25. Gif ic on heofenas up ástíge *si ascendero in cælum*, Ps. Th. 138, 6. [*O. Sax.* heban and himil : *Icel.* hifinn *and* himinn: *Goth.* himins : *O. Frs.* himul, himel: *O. H. Ger.* himil *cælum, lacunar : Ger.* himmel.] v. Grmm. D. M. 661.

heofon, heófon [?] :—Hergas on helle heofon ðider becom druron deófolgyld, Cd. 145 ; Th. 180, 17 ; Exod. 47. Grein translates *heofon* lamentation and *druron* mourned ; but may not *hergas* be from *hearg* q.v. and parallel to *deófolgyld*, and the passage be translated *the idols and false gods fell to hell and heaven came there?*

heofon-beácen, es; n. *A heavenly beacon or sign* [*the fiery pillar*], Cd. 148 ; Th. 184, 15 ; Exod. 107.

heofon-beorht; *adj. Heaven-bright, bright with the light of heaven*, Cd. 190 ; Th. 237, 21 ; Dan. 341 : Exon. 23 a ; Th. 63, 13 ; Cri. 1019.

heofon-býme, an; f. *A heavenly trumpet*, Exon. 21 b ; Th. 59, 8 ; Cri. 949.

heofon-candel, -condel, e; f. *A heavenly candle or light* [*the sun*], Andr. Kmbl. 488 ; An. 243 : [*the fiery pillar*] Cd. 148 ; Th. 184, 31 ; Exod. 115 : [*sun and moon*] Exon. 16 b ; Th. 38, 17 ; Cri. 608 : [*the stars*] 93 a ; Th. 349, 30 ; Sch. 54.

heofon-col, es; n. *The coal of the heavens :* — Brúne hátum heofoncolum *brown with the sun's heat* [*the Ethiopians*], Cd. 146 ; Th. 182, 5 ; Exod. 71.

heofon-cund; *adj. Heavenly, celestial :* — Heofuncund mett *manna*, Jn. Skt. Lind. 6, 31. Seó heofencunde weorþung *the heavenly honour*, Blickl. Homl. 165, 26. Heáh and hálig heofuncund þrýnes *O! high and holy heavenly Trinity*, Exon. 13 a ; Th. 24, 4 ; Cri. 379. Hý ðæs heofoncundan boldes bídaþ *they wait for the heavenly dwelling*, 33 b ; Th. 107, 6 ; Gú. 54: 35 a ; Th. 12, 11 ; Gú. 142. Ða beóþ ðǽre heofencundan Jerusalem burgware *who are citizens of the heavenly Jerusalem*, Bt. 5, 1 ; Fox 10, 7. [Cf. *Goth.* himina-kunds *cælestis*.]

heofon-cyning, es; m. *The king of heaven, heavenly king :*—God heáh heofoncyning *God high king of heaven*, Cd. 23 ; Th. 30, 7 ; Gen. 463. Ic eom heáhengel heofoncyninges *I am an archangel of the king of heaven*, Blickl. Homl. 201, 5 : Cd. 23 ; Th. 30, 28 ; Gen. 474: Andr. Kmbl. 184 ; An. 92. Heofoncining on heora heortum beran *to bear the king of heaven in their hearts*, Blickl. Homl. 79, 32. Heofoncyning hýhst *most exalted of heavenly kings*, Exon. 117 b ; Th. 451, 23 ; Dóm. 108. [*O. Sax.* heban-, himil-kuning: *O. H. Ger.* himel-chuning *superum regem (jovem)*.]

heofon-déma, an; m. *A heavenly judge*, Cd. 228 ; Th. 306, 4 ; Sat. 658.

heofon-dreám, es; m. *Heavenly joy, joy of heaven*, Ps. Th. 113, 11 ; Soul Kmbl. 206 ; Seel. 104 : Exon. 54 a ; Th. 190, 27 ; Az. 79.

heofon-duguþ, e; f. *A heavenly host*, Exon. 32 a ; Th. 101, 7 ; Cri. 1655.

heofone, an ; f. *Heaven :*—Heofone næs ná ǽr ǽrðan ðe se ælmihtiga wyrhta hí geworhte on anginne *heaven was not before the almighty workman wrought it in the beginning*, Hexam. 1 ; Norm. 4. Heofenan ríce *the kingdom of heaven*, Homl. Th. i. 68, 2 : 58, 4. God gesette hig on ðære heofenan ðæt hie scinon ofer eorþan *God set them in the firmament of heaven to give light upon the earth*, Gen. 1, 17, 14. On anginne gesceóp God heofenan and eorþan *in the beginning God created the heaven and the earth*, 1, 1.

heofon-engel, es; m. *An angel of heaven*, Exon. 15 a ; Th. 31, 8 ; Cri. 492 : 21 b ; Th. 57, 34 ; Cri. 928 : 75 a ; Th. 281, 7 ; Jul. 642 : Hy. 7, 13 ; Hy. Grn. ii. 287. 13.

heofon-feld, es; m. *A Northumbrian local name :*—Is seó stów on Englisc nemned Heofenfeld wæs heó geára swá nemned for tácnunge ðæra tóweardra wundra forðon ðe ðǽr ðæt heofonlíce sigebeácen árǽred beón sceolde and ðǽr heofonlíc sige ðam cyninge seald wæs *vocatur locus ille*

lingua Anglorum Hefenfelth, quod dici potest Latine cælestis campus, quod certo utique præsagio futurorum antiquitus nomen accepit significans nimirum quod ibidem cæleste erigendum trophæum, cælestis inchoanda victoria, Bd. 3, 2; S. 524, 33. Seó stów is geháten Heofonfeld oñ Englisc wið ðone langan weall ðe ða Rômániscan worhton *the place is called in English Heavenfield, by the long wall that the Romans made*, Swt. A. S. Rdr. 96, 33.

heofon-fugol, es; *m. A bird of the air, fowl of heaven* : — Heofon-fugelas healdaþ eardas *volucres cæli habitabunt*, Ps. Th. 103, 11 : Cd. 192; Th. 240, 16; Dan. 387 : 74; Th. 91, 21; Gen. 1515 : 10; Th. 13, 11; Gen. 201.

heofon-hæbbende *arcitenens, sagittarius*, Lye.

heofon-hálig; *adj. Heaven-holy, of celestial holiness*, Andr. Kmbl. 1455; An. 728.

heofon-hám, es; *m. A heavenly home, heaven* : — On heofonháme *in cælo*, Ps. Th. 102, 18 : 137, 6 : 148, 4 : Exon. 12 a; Th. 18, 33; Cri. 293. Ðũ ðe heofonhámas healdest and wealdest *qui habitas in cælo*, Ps. Th. 122, 1.

heofon-heáh; *adj. Heaven-high, reaching to heaven* :—Heofonheánne beám *a tree the height whereof reached unto heaven* [Dan. 4, 11], Cd. 202; Th. 250, 29; Dan. 554.

heofon-heall, e; *f. A heavenly hall* :—Ne hí swá fúle ne móton intó his fægeran heofonhealle *nor may they so foul enter into his fair heavenly hall*, L. Ælfc. P. 41; Th. ii. 382, 10.

heofon-hláf, es; *m. Heavenly bread, bread from heaven, manna* :—Hí heofonhláfe hálige gefylde *pani cæli saturavit eos*, Ps. Th. 104, 35. [Cf. *O. H. Ger.* himel-brot.]

heofon-hróf, es; *m.* I. *the roof of heaven, heaven* : — Under heofunhrófe *under the roof of heaven*, Exon. 58 a; Th. 209, 19; Ph. 173. II. *a roof, ceiling* :—Heofenhróf *lacunar*, Cot. 119, Lye. [Cf. *O. H. Ger.* himil *laqueare, lacunar, camera* : hiñilizi *lacunar, laquear*.]

heofon-hwealf, e; *f. The vault of heaven*, Andr. Kmbl. 1089; An. 545 : 2803; An. 1404.

heofonisc; *adj. Heavenly* : — Hũ ðæt heofenisce fýr forbærnde ðæt lond on ðæm wæron ða twá byrig on getimbred Sodome and Gomorre *how fire from heaven consumed the land in which were built the two cities Sodom and Gomorrah*, Ors. tit. 3; Swt. 1, 6. [Cf. *O. Sax. O. H. Ger.* himilisk : *O. Frs.* himelesk : *Icel.* hifneskr, himneskr.]

heofon-leóht, es; *n. Heavenly light*, Andr. Kmbl. 1948; An. 976. [Cf. *O. H. Ger.* himel-lieht.]

heofon-leóma, an; *m. A heavenly radiance, light*, Andr. Kmbl. 1675; An. 840. [Cf. *Icel.* himin-ljómi.]

heofon-líc; *adj. Heavenly* : — Mín se heofenlíca Fæder *Pater meus cælestis*, Mt. Kmbl. 18, 35 : Ps. Th. 67, 14. Ðín rihtwísnes is swá heáh swá ða heofonlícan muntas *justitia tua sicut montes Dei*, 35, 6. Heofon-lícæ þing *cælestia*, Jn. Skt. 3, 12. [Cf. *O. H. Ger.* himil-lîh *cælestis*.]

heofon-líce; *adv. From heaven, heavenly*; celitus, Ælfc. Gr. 38; Som. 42, 3.

heofon-ligende [lifigende?] *cælebs, virgo, quod vitam cælestem agat*, Som.

heofon-mægen, es; *n. Heavenly might* : — Bibodu hálgan heofon-mægnes *the commands of the holy heavenly power* [God], Exon. 118 a; Th. 454, 19; Hy. 4, 35. Heofonmægna God *God of the heavenly powers*, 256; Th. 75, 8 : Cri. 1218.

heofon-ríce, es; *n. The kingdom of heaven* : — Biþ him heofonríce ágiefen *to them shall be given the kingdom of heaven*, Exon. 26 a; Th. 77, 22; Cri. 1260. Heofenríces duru *the door of the kingdom of heaven*, Blickl. Homl. 9, 1. Heofonríces weard *auctorem regni cælestis*, Bd. 4, 24; S. 597, 20 : Cd. 69; Th. 82, 17; Gen. 1363. [*O. Sax* heban-ríki : cf. *O. Sax.* himil-ríki : *O. Frs.* himel-rík : *Icel.* himin-ríki : *Dan.* himme-rige : *O. H. Ger.* himil-ríchi : *Ger.* himmel-reich.]

heofon-steorra, an; *m. A star of heaven* :—Seó mænigeo mære wære swá heofonsteorran *the multitude should be great as the stars of heaven*, Cd. 190; Th. 236, 15; Dan. 321 : 192; Th. 239, 17; Dan. 371. Hreósaþ heofonsteorran *the stars of heaven shall fall*, Exon. 23 a; Th. 64, 27; Cri. 1044.

heofon-stól, es; *m. A heavenly throne*, Cd. 1; Th. 1, 15; Gen. 8.

heofon-þreát, es; *m. A heavenly band*, Cd. 218; Th. 278, 15; Sat. 222.

heofon-þrym, -mes; *m. Heavenly glory or majesty*, Andr. Kmbl. 962; An. 481 : 3436; An. 1722.

heofon-timber, es; *n. A heavenly structure*, Cd. 8; Th. 9, 23; Gen. 146.

heofon-torht; *adj. Heaven-bright*, Exon. 93 b; Th. 351, 1; Sch. 73 : Cd. 146; Th. 182, 19; Exod. 78 : Andr. Kmbl. 2035; An. 1020 : 2539; An. 1270 : Bt. Met. Fox 23, 6; Met. 23, 3.

heofon-tungol, es; *n. A heavenly body* :— Hádor heofontungol *the sun*, Bt. Met. Fox 22, 47; Met. 22, 24. Hædre heofontungol *bright heavenly bodies*, Exon. 18 a; Th. 43, 23; Cri. 693; 56 a; Th. 199, 28; Ph. 32 : Cd. 199; Th. 247, 23; Dan. 501. [Cf. *O. Sax.* himil-tungal : *Icel.* himin-tungl : *O. H. Ger.* himil-zungal *sidus*.]

heofon-ware; *pl. The inhabitants of heaven* : — Ealle gesceafta ge heofonware ge eorþware *all creatures, both those in heaven and those on earth*, Blickl. Homl. 11, 4. Ða hálgan heofenware *the holy dwellers in heaven*, 135, 17. v. next word.

heofon-waru, e; *f. The inhabitants of heaven* : — Hé dyde ðæt eal heofonwaru wundrode *he caused all the inhabitants of heaven to wonder*, Homl. Th. i. 442, 35; Hy. 7, 95; Hy. Grn. ii. 289, 95. Ealle heofon-wara and eorþwara on his andwerdnysse beóþ onstyred *all those in heaven and on earth shall be moved in his presence*, Chart. Th. 390, 10 : Homl. Th. ii. 360, 32. Bearn heofonwara *children of heaven-dwellers*, Salm. Kmbl. 930; Sal. 464. Ætforan heofonwarum and eorþwarum and hel-warum *before the inhabitants of heaven and of earth and of hell*, Homl. Th. ii. 604, 5. Cristes ácennednys gegladode heofenwara and eorþwara and helwara, i. 36, 25.

heofon-weard, es; *m. The guardian of heaven, God*, Cd. 6; Th. 8, 6; Gen. 120 : 86; Th. 107, 28; Gen. 1796. [*O. Sax.* heban-ward *an angel*.]

heofon-wolcen, es; *n. A cloud of heaven, of the sky* : — Of heofon-wolcnum *from the clouds of heaven*, Ps. Th. 147, 6. Ðær mec féddon hruse and heofonwolcn [? MS. wlonc] *where earth and rain from heaven fed me*, Exon. 126 b; Th. 485, 23; Rä. 72, 2. [Cf. *O. Sax.* himil-wolcan : *O. H. Ger.* himil-wolchen *nubes cæli*.]

heofon-wôma, an; *m. A heavenly sound, the sound heard at the day of judgment*, Exon. 20 a; Th. 52, 18; Cri. 835 : 22 b; Th. 62, 10; Cri. 999.

heofon-wuldor, es; *n. Heavenly glory*, Hy. 6, 12; Hy. Grn. ii. 286, 12.

heóf-sang, es; *m. An elegy*, Lye.

heófung, e; *f. Mourning, lamentation, grieving* :—Ðonne beóþ heora siblingas tó heófunge geneádode *then will their relations be forced to mourn*, Homl. Th. i. 88, 1. Mid micelre heófunge *with great lamentation*, ii. 516, 19. Biddende forgifennysse mid wópe and heófunge *asking forgiveness with weeping and lamentation*, H. R. 107, 27. On ðære wæron áwritene heófunga *scripta erant in eo lamentationes*, Ælfc. Gr. 48; Som. 49, 8, 9. Ær hé tó heófungum sôðre behreówsunge gecyrran mæge *before he can turn to the lamentations of true repentance*, Homl. Th. ii. 124, 13.

heófung-dæg, es; *m. A day of mourning* :—Ða heófungdagas wæron ðá gefyllede *completi sunt dies planctus*, Deut. 34, 8.

heófung-tíd, e; *f. A time of mourning* : — Fram ðisum dæge óþ eástron is úre heófungtíd *from this day until Easter is our time of mourning*, Homl. Th. ii. 86, 25.

heolca, an; *m.* [?] *Hoar-frost, rime* :—Swá swá bytte on heolcan *sicut uter in pruina*, Ps. Lamb. 118, 83.

heolfor, es; *n. Blood from a wound, gore*; cruor :—Blód út ne com heolfor of hreþre ðeáh mec bite stíðecg stýle *there came not out blood or gore from my breast though the steel with stiff edge bit me*, Exon. 130 a; Th. 499, 9; Rä. 88, 13. Heolfres þurstge *thirsty for gore*, 99 b; Th. 373, 24; Seel. 114. Flód blóde weól hátan heolfre *blood and hot gore bubbled up in the water*, Beo. Th. 2850; B. 1423 : 1702; B. 849 : 2609; B. 1302 : Andr. Kmbl. 2483; An. 1243 : 2555; An. 1279 : Cd. 166; Th. 206, 9; Exod. 449 : Th. 208, 1; Exod. 476.

heolfrig; *adj. Gory, bloody* :—Heolfrig herereáf *gory armour*, Judth. 12; Thw. 26, 8; Jud. 317 : 11; Thw. 23, 20; Jud. 130.

heoloran, holrian; *p. ede To weigh in a balance, to consider* : — Hé holrede *pensavit, cogitavit*, Mone B. 1604. Heolorende *librantes*, Cot. 123 : 180, Lye.

heoloþ-cynn, es; *n. A race living in a place of concealment* [?], *the devils in hell*, Exon. 30 b; Th. 94, 19; Cri. 1542. v. *next word*; *and* cf. heolstor.

heoloþ-helm, es; *m. A helm which conceals or makes invisible the wearer*, Exon. 97 a; Th. 362, 31; Wal. 45. [*Icel.* huliðs-hjálmr.] v. hæleþ-helm.

heolra, heolora, an; *m. The scale of a balance, a balance* [?] : — Twí-feald heolra *bilanx*, Lye. v. helur-blæd, heoloran.

heolstor, es; *n. That which covers or conceals, darkness, a veil, covering, place of concealment* :—Siðdan geára goldwine mínne hrusan heolstre biwráh *since long ago the veil of earth enwrapped my bounteous patron*, Exon. 76 b; Th. 287, 32; Wand. 23. Nágan wé ðæs heolstres ðæt wé ús gehýdan mágon *we have not the place of concealment to hide ourselves in*, Cd. 215; Th. 271, 5; Sat. 101. Gewitan him ðá gangan under beámsceade hýddon hie on heolstre ðá hie hálig word drihtnes ge-hýrdon *they retired then under the trees' shade, hid themselves in the darkness when they heard the holy word of the Lord*, 40; Th. 53, 12; Gen. 860. Ða com beácna beorhtost of heolstre *then came the sun out of darkness*, Andr. Kmbl. 485; An. 243 : Elen. Kmbl. 2223; El. 1113. Heolstre gehýded helme gedýgled þýstre ofersæðmed *with a veil hidden, with a covering concealed, with darkness enwrapped*, Exon. 122 b; Th. 470, 9; Hy. 11, 13 : 61 b; Th. 227, 4; Ph. 418 : 69 a; Th. 257, 2; Jul. 241 : Elen. Kmbl. 2161; El. 1082. Sume wuniaþ on wêstennum ge-sittaþ hámas on heolstrum *some dwell in deserts, occupy homes in hidden*

places, Exon. 33 b; Th. 107, 5; Gû. 54. [*Goth.* hulistr; *n. a veil*: cf. *Icel.* hulstr; *m. a sheath, case*: *Dut.* holster *holster*. *In Romaunt of Rose* hulstred *occurs* = *hidden* 'I wol herborow me There I hope best to hulstred be,' 6146.]

heolstor; *adj. Dark* :—Đǽr wunian sceal in đam heolstran hám hyhtwynna leás *there shall dwell in that dark abode reft of the joys of hope*, Judth. 10; Thw. 23, 14; Jud. 121.

heolstor-cófa, an; *m. A dark, concealed chamber, grave* :—Deáþræced hæleþa heolstercófan onhliden weorþaþ *the death houses, the graves of men shall be uncovered*, Exon. 56 b; Th. 200, 31; Ph. 49.

heolstor-hof, es; *n. A dark dwelling, hell*, Elen. Kmbl. 1524; El. 764.

heolstor-loca, an; *m. A dark enclosure, prison*, Andr. Kmbl. 288; An. 144: 2010: An. 1007.

heolstor-sceado; *f. A shadow that hides*, Cd. 5; Th. 7, 9; Gen. 103.

heolstor-scûwa, an; *m. Dark shadow, darkness*, Andr. Kmbl. 2508; An. 1255.

heolstrig; *adj. Látebrosus*, Cot. 169, Lye.

heona. v. heonan.

heonan, heonon, heonun, hionan; *adv. of place and time. Hence, from here* :—Heonon *abhinc*, Ælfc. Gr. 16; Som. 20, 4. Feor heonan *far from here*, Exon. 55 b; Th. 197, 19; Ph. 1. Ic mæg heonon geseón *I can see from here*, Cd. 32; Th. 41, 34; Gen. 666. Ær đû heonan môte *ere thou mayest go hence*, Exon. 72 a; Th. 269, 29; Jul. 457. Đis is mín ágen cýþ ic wæs ǽr hionan cumen *this is my own country, from here did I formerly come*, Bt. Met. Fox 24, 100; Met. 24, 50. Gáþ heonun *recedite*, Mt. Kmbl. 9, 24. Ásend đé heonun nyþer *mitte te hinc deorsum*, Lk. Skt. 4, 9. Ge heonon ge đanon *from here and there, from any quarter*, L. C. S. 19; Th. i. 386, 16. Ic forþ heonun đíne gewitnesse wel geheólde *I should henceforth keep thy testimony well*, Ps. Th. 118, 31, 24: Exon. 16 a; Th. 36, 27; Cri. 582. Heonon forþ and óþ on woruld *ex hoc nunc et usque in sæculum*, Blickl. Gloss: Gen. 8, 21. Gif hit sceal heonan forþ gódiende weorþan *if things from this time forward are to be improving*, Swt. A. S. Rdr. 105, 19. Mín feorh heonan on đisse eahteþan ende geséceþ *my life shall reach its end on the eighth day from this time*, Exon. 47 b; Th. 164, 10; Gû. 1009. [*Laym.* heonne, hinnes: *Piers P.* hennes: *O. Sax.* hinan: *O. H. Ger.* hinan, hinnan *hinc*: *Ger.* hennen.]

heonane, heonone; *adv. Hence* :—Far heonone *transi hinc*, Mt. Kmbl. 17, 20. Đú miht heonane gehýran *thou mayest hear from this place*, Cd. 37; Th. 49, 18; Gen. 794: 39; Th. 51, 24; Gen. 831. [*O. Sax.* hinana: *O. H. Ger.* hinana *hinc*.]

heonan-sîþ, es; *m. Departure, death*, Exon. 117 a; Th. 450, 12; Dôm. 86.

heonon-weard; *adj. Going hence, passing away* :— Đeós world is heononweard *this world is passing away*, Blickl. Homl. 115, 20: Cd. 71; Th. 86, 15; Gen. 1431.

heonu, heono, henu, hona; *interj. Lo, behold* :—Heonu [henu, Rush.] *ecce*, Mt. Kmbl. Lind. 11, 8. Heono, Jn. Skt. Lind. 1, 29. Hona lá mín hláford *ecce dominus meus*, Shrn. 60, 14.

heópa, an; *m. A briar. bramble* :— Ætt đæm heápe [heópe, Rush.] *secum rubum*, Lk. Skt. Lind. 20, 37. (*Or should this be placed under* heópe?) [*O. Sax.* hiopo: *O. H. Ger.* hiufo; *m. tribulus*.]

heóp-bremel, es; *m. A dog-rose, wild rose, bramble, briar* :—Heópbrymel *rubus*, Ælfc. Gl. 47; Som. 65, 22; Wrt. Voc. 33, 22. Heópbremles leáf *leaves of the dog-rose*, L. M. 2, 51; Lchdm. ii. 266, 8.

heópe, an; *f. A hip, seed-vessel of the dog-rose; also the plant on which the hip grows* [?] :— Heópe *butunus* [i.e. button, *Fr.* bouton, knob], Ælfc. Gl. 40; Som. 63, 90; Wrt. Voc. 30, 36. Genim brér đe hiópan on weaxaþ *take briar on which hips grow*, L. M. 1, 38; Lchdm. ii. 96, 15. [*Chauc.* hepe.] v. heópa.

heorcnian, hercnian; *p.* ode *To hearken, listen* :— Gúþlác eode sóna út and háwode and hercnode *Guthlac went out directly and looked and listened*, Guthl. 6; Gdwin. 42, 15. Ypolitus mid geþylde heora wordum heorcnode *Hippolytus listened to their words with patience*, Homl. Th. i. 442, 2. Maria gesæt æt Godes fótum his word heorcnigende *Mary sat at the feet of God hearkening to his words*, ii. 440, 16. Đæt hit tó hefigtýme ne þince đám heorcnigendum *that it may not seem too tedious to the listeners*, 72, 23. [*Orm.* herrcnenn: *A. R.* hercnen: *Laym.* hercnede; *p. Chauc.* herkneth.]

heorcnung, hearcnung, e; *f. Hearkening, listening, hearing, power of hearing* :—Wê sceolon úre eáran fram yfelre heorcnunge áwendan *we must turn away our ears from evil listening*, Homl. Th. i. 96, 23: ii. 564, 4: Ælfc. Gr. 1; Som. 2, 29. Hê forgeaf deáfum heorcnunge *he gave to the deaf hearing*, Homl. Th. i. 26, 13: ii. 16, 13. Hearcnunge, H. R. 7, 14. Drihten ic gehýrde heorcnunge đíne *Domine audivi auditionem tuam*, Cant. Abac. Lamb. fol. 189, 2.

HEORD, e; *f. A* HERD, *flock* :—Hiord *arimentum*, Wrt. Voc. 287, 53. Đǽr wæs án swýna heord *erat grex porcorum*, Mt. Kmbl. 8, 30. Ic hæbbe ôđre sceáp đa ne synt of đisse heorde *alias oves habeo quæ non sunt ex hoc ovili*, Jn. Skt. 10, 16. Hê dráf his heorde tó inneweardum đam wêstene *he led the flock to the backside of the desert*, Ex. 3, 1: L. R. S. 4;

Th. i. 434, 21. Rihtwîs hyrde ofer cristene heorde *a righteous shepherd over a christian flock*, L. I. P. 2; Th. ii. 304, 10. Of eówrum heordum *de gregibus tuis*, Ps. Th. 49, 10. Heora heorda wîslîce healdan *to keep their flocks wisely*, L. Eth. vi. 2; Th. i. 314, 14. Godcunde heorda *spiritual flocks*, L. C. E. 26; Th. i. 374, 34. [*Goth.* hairda: *Icel.* hjörð: *O. H. Ger.* herta *grex*: *Ger.* heerde.] v. hríđer-heord.

heordan '*hards of flax*; lini fila utiliora. Stuppa, Gl. C. 58 b. Naptarum heordena, Gl. Cleop. 65 c.' Lchdm. iii. 331, col. 1. [*Prompt. Parv.* hyrdys or herdys of flax, or hempe *stuppa, napta*. See note, p. 241. Hards, hurds tow, East Norfolk Gloss: Engl. Dial. Soc. vol. ii.]

heorde; *f. Care, guarding, custody* :— Hê út wæs gongende tô neáta scýpene đara heorde him wæs đære nihte beboden *egressus esset ad stabula jumentorum quorum ei custodia nocte illa erat delegata*, Bd. 4, 24; S. 597, 9. Forhwon beóþ ǽfre swǽ þríste đa ungelǽredan đæt hí underfón đa heorde đæs láreówdômes *ab imperitis ergo pastorale magisterium qua temeritate suscipitur*, Past. 1; Swt. 25, 17. Monige underfôþ heorde *nonnulli gregis curam suscipiunt*, 18, 5; Swt. 135, 25. [Cf. (?) *Icel.* hirð *a king's body-guard*: hirði- a prefix, tending, keeping.]

heorde. v. hirde.

heord-, hyrd-rǽden, e; *f. Guard, guardianship, care, keeping* :—Him is sinderlîce betǽht hyrdrǽden ofer eallum cristenum monnum *to him is especially committed the guardianship over all christian men*, Homl. Th. ii. 290, 26. Geþyld is wyrtruma and hyrdrǽden ealra háligra mægna *patience is the root and guard of all holy virtues*, 544, 5. Hí geswencaþ heora hláford þurh ymhîdignysse heordrǽdene *they distress their possessor through solicitude of guarding*, 92, 18. Gehwilc hæbbe him betǽhtne engel tô hyrdrǽdene *each has an angel assigned to him as guard*, i. 516, 32. Se stæf getácnaþ gýmene and hyrdrǽdene *the staff indicates care and guardianship*, ii. 280, 35. Tô heordrǽdene *ad custodiam*, Hymn. Surt. 11, 27. Đá gesette God æt đam infære engla hyrdrǽdene *then God set a guard of angels at the entrance*, Gen. 3, 24: Boutr. Scrd. 20, 32. Gê habbaþ heordrǽdenne *habetis custodiam*, Mt. Kmbl. 27, 65. Heordrǽdena se đe gesihþ swicunge hit getácnaþ *to see pickets betokens deception*, Lchdm. iii. 202, 13.

heóre, hýre; *adj. Gentle, mild, pleasant* :— Nis đæt heóru stów *it is a savage place*, Beo. Th. 2749; B. 1372. Culufre fótum stôp on beám hýre *the dove with her feet stepped on to the tree, gentle*, Cd. 72; Th. 88, 20; Gen. 1468. Đǽr se hýra gæst þíhþ an þeáwum *where the gentle spirit thrives in morals*, Exon. 38 a; Th. 126, 9; Gû. 368. [*Icel.* hýrr *sweet, smiling, mild*.] v. un-heóre.

heoro. v. heoru.

heorot, heort, es; *m. A hart, stag, male deer* :— Nán heort ne onscúnode nænne león *no hart shunned any lion*, Bt. 35, 6; Fox 168, 9. Heorot hornum trum *the hart firm of horns*, Beo. Th. 2742; B. 1369. Heorut *cervus*, Ps. Stev. 41, 1. Swá hwá swá slôge heort ođđe hinde hine man sceolde blendian *whoever killed hart or hind should be blinded*, Chr. 1086; Erl. 222, 27, 28. Mid heortes horne and mid ylpenbáne *with hart's horn and with ivory*, Herb. 131, 2; Lchdm. i. 244, 8: Med. ex Quadr. 2, 1, 2, 3; Lchdm. i. 334, 2, 5, 9. Heortas and hinda *harts and hinds*, Bt. Met. Fox 19, 33; Met. 19, 17. Heortas *cervos*, Coll. Monast. Th. 21, 31. [*Icel.* hjörtr: *O. H. Ger.* hiruz *cervus*: *Ger.* hirsch.]

heorot-berge, an; *f. Berry of the buckthorn*, Lchdm. iii. 331, col. 1. [hart-berries *vaccinium myrtillus*, Engl. Dial. Soc. No. 26.]

heorot-brembel, es; *m. Buckthorn*; rhamnus, Lchdm. ii. 391–2.

heorot-brér, e; *f. v.* [?] heorot-brembel :—Heortbrére *moro*, Lk. Skt. Rush. 17, 6.

heorot-, heort-clæfre, an; *f. Hart-clover*; medicago maculata, Lchdm. ii. 392.

heorot-crop *a bunch of the flowers of hartwort*, Lchdm. ii. 392.

Heorot-, Heort-ford, es; *m. Hertford* :—Æt Heorotforda [Heortforda MS. D.] *at Hertford*, Chr. 913; Erl. 102, 1: 673; Erl. 36, 2; 37, 2.

heorr, hior; *m. f. A hinge, cardinal point; cardo* :—Đeós heorr *hic cardo*, Ælfc. Gr. 9, 3; Som. 8, 61. Seó hior đe eall gôd on hwearfaþ *the hinge on which all good turns*, Bt. 34, 7; Fox 142, 35. Wæs đæt beorhte bold tôbrocen swíđe heorras tôhlidene *the splendid dwelling was sorely shattered, hinges were broken*, Beo. Th. 2002; B. 999. Heorras serras, Blickl. Gloss. Đis gesceád ys æfter đam feówor heorren *this distinction is according to the four cardinal points*, Lchdm. iii. 84, 11. [*Chauc.* 'no dore that he nolde heve of harre;' *Prompt. Parv.* herre of a lock *cardo*. v. note, p. 237: *Icel.* hjarri *a hinge*.]

heorra, an; *m.* [?] *A bar, hinge* [?] :—Hê gestrangode heorran geata đínra *confortavit seras portarum tuarum*, Ps. Lamb. 147, 2. v. heorr.

heort. v. heorot.

-heort. v. blíđ-, ceald-, earm-, gram-, grim-, hát-, heáh-, heard-, mild-, riht-, rûm-, sam-, stearc-, wulf-heort. [*Goth.* -hairts: *O. Sax.* -hert.]

heort-côđu, es; *f. A disease of the heart*, L. M. 2, 1; Lchdm. ii. 176, 13.

HEORTE, an; *f. The* HEART :—Gif đín heorte ace *if thy heart ache*, Lchdm. iii. 42, 1. Oþ đæt him heortan blôd foldan gesêceþ *until his heart's blood seek the earth*, Salm. Kmbl. 314; Sal. 156 Wyxþ wind

on ðære heortan *wind waxeth in the heart*, L. M. 1, 17 ; Lchdm. ii. 60, 92 a ; Th. 346, 10 ; Gn. Ex. 202. 7. Of ðære heortan cumaþ yfle geþancas *de corde exeunt cogitationes malæ*, Mt. Kmbl. 15, 19. Lustum heortena *desideriis cordum*, Ps. Th. 80, 12. [*Laym. A. R.* heorte: *Orm.* heorrte, herrte: *Chauc. Wick.* herte: *Goth.* hairto: *O. Sax.* herta: *O. Frs.* hirte: *Icel.* hjarta: *O. H. Ger.* herza: *Ger.* herz: *Lat.* cord-: *Grk.* καρδία.]

heort-ece, es ; *m. Pain at the heart* :—Heó wið heortece well fremaþ *it is very beneficial for heartache*, Herb. 18, 3 ; Lchdm. i. 110, 19 : *ad cardiacos*, 89, 3 ; Lchdm. i. 192, 16.

heorten ; *adj. Of a hart* :— Healfes pundes gewihte beran smeruwes and heortenes *of bear's grease and of hart's, by weight of half a pound*, Herb. 101, 3 ; Lchdm. i. 216, 15.

heort-gesida ; *pl. The entrails* ; enta, Lev. 3, 3.

HEORÞ, es ; *m. A* HEARTH, *fire-place ;* and taking the name of the whole from that of a part, *a house* :—Heorþ *foculare*, Ælfc. Gl. 30 ; Som. 61, 73 ; Wrt. Voc. 27, 2 : *arula*, Wrt. Voc. 63, 76. Hí ofslógon hine binnan his ágenan heorþæ *they slew him in his own house*, Chr. 1048 ; Erl. 177, 40. Hé sceolde bebeódan ðæt hí náman æt ælcum heorþe ánes geáres lamb *he was to command them to take a yearling lamb for every house*, Homl. Th. ii. 262, 27 : Chart. Th. 609, 7, 11, 30. Of ælcum heorþe, 27. Be ælcum frigan heorþe, L. Edg. I. 2 ; Th. i. 262, 17 : L. C. E. 11 ; Th. i. 366, 29 : L. In. 61 ; Th. i. 140, 14. Beþe hwílum ða sáran stówe æt heorþe *warm the sore place at times at the hearth*, L. M. 2, 59 ; Lchdm. ii. 280, 26. Genim ðæt sèleste hunig dô ofer heorþ *take the best honey, put it over the fire*, 2, 28 ; Lchdm. ii. 224, 17. Be heorþe, Lchdm. iii. 122, 21. Hweorfaþ æfter heorþe *they pass along the hearth* [*the floor of the fiery furnace*], Exon. 55 b ; Th. 196, 18 ; Az. 176. [*Prompt. Parv.* herthe, where fyre ys made *ignearium, focarium* : *O. Frs.* herth, hirth, herd : *O. H. Ger.* hert *arula* : *Ger.* herd.]

heorþa, herþa, an ; *m. A deer-skin* :—Heorþa *nebris*, Wrt. Voc. 86, 39.

heort-hama, an ; *m. A covering of the heart* :—Heorthama *bucleamen*, Ælfc. Gl. 75 ; Som. 71, 102 ; Wrt. Voc. 45, 9. Ðú nymst ðone hearthaman *thou shalt take the fat that covers the inwards*, Ex. 29, 22. [*O. Frs.* hert-hamo *præcordia*.]

heorþ-bacen ; *adj. Baked on the hearth* :—Heorþbacen hláf *subcinericius vel focarius*, Ælfc. Gl. 66 ; Som. 69, 64 ; Wrt. Voc. 41, 20. Mid heorþbacenum hláfe *with a loaf baked on the hearth*, Herb. 45, 2 : Lchdm. i. 148, 8. Abraham nam ðæt flæsc mid ðam heorþbacenum hláfum, Gen. 18, 8. Hí worhton þeorfe heorþbacene hláfas *they baked unleavened cakes*, Ex. 12, 39.

heorþ-cniht, es ; *m. A domestic, servant, attendant* :—Hió dyde sciella tô bisene his heorþcneohtum and ðus cwæþ *sub squamarum specie de ejus satellitibus perhibetur*, Past. 47, 3 ; Swt. 361, 18.

heorþ-fæst ; *adj. Having a house of one's own* :—Sý hé heorþfæst sý hé folgere *whether he have a house of his own or be the follower of another man*, L. C. S. 20 ; Th. i. 386, 23.

heorþ-geneát, es ; *m. A hearth-comrade, a follower who shares the hearth of his lord* :—Wé synt Hygeláces heorþgeneátas, Beo. Th. 528 ; B. 261: 3165 ; B. 1580: 4365 ; B. 2180: 6341 ; B. 3180; Byrht. Th. 137, 50 ; By. 204.

heort-hogu, e ; *f. Heart-care* :— Ðis mæg tô heorthoge æghwylcum bisceope *this may be care of heart for every bishop*, L. I. P. 5 ; Th. ii. 308, 27. v. hogu.

heorþ-pening, -peneg, es ; *m. A tax of a penny to be paid by every house* [e. g. *Peter's pence*] :— Be ðon heorþpeninge. Sý ælc heorþpenig ágifen be Petres mæsse dæge : and seðe hine tô ðam ándagan gelæst næbbe, læde hine tô Rôme, and ðær tô eácan xxx pænega and bringe ðonne swutelunge ðæt hé ðær swá micel betæht hæbbe. And ðonne hé hám cume gylde ðam cynge hundtwelftig scillinga *of the hearth-penny. Let every hearth-penny be paid up by St. Peter's mass day : and he who shall not have paid by that time, let him be led to Rome, and in addition thereto pay xxx pence, and then bring a certificate that he has there paid so much. And when he comes home let him pay the king a hundred and twenty shillings*, L. Edg. I. 4 ; Th. i. 264, 6-12. Sylle his heorþpænig on hálgan þunresdæg *let him pay his hearth-penny on holy Thursday*, L. R. S. 3 ; Th. i. 432, 26 : 4 ; Th. i. 434, 19. Heorþpenegas, Chart. Th. 432, 24.

heorþ-swæpe, an ; *f. A bridesmaid* ; pronuba, Som. [Cf. hád-swæpe.]

heorþ-werod ; *m. A band of household retainers, those who share the same hearth, a family* :—Ðá wearþ Jafeðe áféded heorþwerod suna and dôhtra *then for Japhet was reared a family of sons and daughters*, Cd. 78 ; Th. 96, 35 ; Gen. 1605. Se hálga hêht his heorþwerod wæpna onfôn *the holy man bade his retainers take their weapons*, 94 ; Th. 123, 4 ; Gen. 2039 : 95 ; Th. 125, 8 ; Gen. 2076 : Byrht. Th. 132, 30 ; By. 24.

heort-lufe, an ; *f. Love which comes from the heart*, Hy. 9, 29 ; Hy. Grn. ii. 292, 29.

heort-seóc ; *adj. Heart-sick* ; cardiacus, Cot. 209, Lye.

heort-seócnes *cardialgia*, Lye.

heort-wærc, es ; *m. Pain in the heart* :—Wið heortwærce *for pain in the heart*, L. M. 1, 17 ; Lchdm. ii. 60, 4.

heoru, heoro, hioro ; *m. A sword*, Beo. Th. 2574 ; B. 1285 : Exon. a ; Th. 346, 10 ; Gn. Ex. 202. The word is a poetical one both in English and Icelandic, and in these dialects, as in Old Saxon, is mostly used in compounds. [*Goth.* hairus: *O. Sax.* heru (in compounds only): *Icel.* hjörr.]

heoru-cumbul, es ; *n. A warlike ensign*, Elen. Kmbl. 213 ; El. 107.

heoru-dolg, es ; *n. A sword-wound, deadly wound*, Andr. Kmbl. 1883 ; An. 944.

heoru-dreór, es ; *m. Blood coming from wounds made by the sword, gore*, Beo. Th. 978 ; B. 487 : 1703 ; B. 849.

heoru-dreórig ; *adj.* I. *bloody with sword-wounds, gory*, Beo. Th. 1875 ; B. 935 : 3564 ; B. 1780 : 5434 ; B. 2720 : Andr. Kmbl. 1991 ; An. 998 : 2167 ; An. 1085 : Elen. Kmbl. 2427 ; El. 1215. [*O. Sax.* heru-drôrag.] II. *very sad, sad unto death*, Exon. 59 a ; Th. 212, 28 ; Ph. 217.

heoru-dryne ; *m. The sword's drink, blood flowing from a wound*, Beo. Th. 4706 ; B. 2358. [Cf. *Icel.* hjör-lögr (lögr *any liquid*) *blood*.]

heoru-fædm ; es ; *m. A deadly, hostile grasp* :—Wolde heoru [*huru* MS.] fædmum hilde gesceádan *meant with deadly grasps to decide the conflict*, Cd. 167 ; Th. 209, 24 ; Exod. 504. [Cf. wælfædmum, Th. 208, 9 ; Exod. 480.]

heoru-gífre ; *adj. Greedy, eager to destroy*, Exon. 22 a ; Th. 60, 29 ; Cri. 977 : 23 b ; Th. 65, 25 ; Cri. 1060 : 74 a ; Th. 276, 16 ; Jul. 567 : Th. 277, 25 ; Jul. 586 : Beo. Th. 3000 ; B. 1498.

heoru-grédig ; *adj. Greedy to destroy, bloodthirsty, savagely greedy*, Andr. Kmbl. 75 ; An. 38 : 158 ; An. 79.

heoru-grim ; *adj. Very fierce* or *cruel, savage*, Exon. 30 a ; Th. 93, 10 ; Cri. 1524 : 31 b ; Th. 98, 25 ; Cri. 1613 : 47 a ; Th. 161, 1 ; Gú. 952 : 53 a ; Th. 186, 29 ; Az. 27 : 111 a ; Th. 425, 12 : Rä. 41, 55 : Beo. Th. 3132 ; B. 1564 : 3698 ; B. 1847 : Elen. Kmbl. 237 ; El. 119 : Andr. Kmbl. 61 ; An. 31 : Cd. 189 ; Th. 235, 16 ; Dan. 307.

heoru-hóciht ; *adj. Furnished with sharp hooks, barbed*, Beo. Th. 2880 ; B. 1438.

heoru-scearp ; *adj. Terribly sharp*, Exon. 102 b ; Th. 388, 15 ; Rä. 6, 8.

heoru-sceorp, es ; *n. Warlike dress*, Exon. 120 b ; Th. 463, 20 ; Hö. 73.

heoru-serce, an ; *f. A war-shirt, coat of mail*, Beo. Th. 5072 ; B. 2539.

heoru-swealwe, an ; *f. A hawk*, Exon. 88 b ; Th. 332, 17 ; Vy. 86.

heoru-sweng, es ; *m. A blow with a sword*, Beo. Th. 3184 ; B. 1539 : Andr. Kmbl. 1903 ; An. 954.

heoru-wæpen, es ; *n. A weapon of war, a sword*, Judth. 12 ; Thw. 25, 16 ; Jud. 263.

heoru-weallende ; *part. pres. Boiling fiercely*, Beo. Th. 5556 ; B. 2781.

heoru-wearh ; *gen.* -wearges ; *m. A savage, bloody wolf*, Beo. Th. 2538 ; B. 1267.

heoru-word, es ; *n. A hostile, fierce word*, Exon. 81 a ; Th. 305, 7 ; Fä. 84.

heoru-wulf, es ; *m. A fierce wolf, a warrior*, Cd. 151 ; Th. 189, 7 ; Exod. 181. [Cf. here-wulf.]

heóþu, e ; *f. A room, hall* :—Hé on heóþe gestôd *he in the hall halted*, Beo. Th. 813 ; B. 404. [Dietrich in Haupt. x. 366 compares the word with κύτος : Heyne suggests a derivation from the root from which comes *heáh*, and translates as do Kemble and Thorpe *dais*, at the same time he gives the other etymology as a possible one.] v. hell-heóþo.

heow. v. hiw.

heowaþ, Ps. Th. 46, 1. v. heófan.

HÉR ; *adv.* HERE, *in this world, at this time* :—Hér hic, Ælfc. Gr. 38 ; Som. 40, 1. Ðá ic hér ærest com *when I first came here*, Cd. 129 ; Th. 164, 8 ; Gen. 2711. Hér gehýrþ Drihten ða ðe hine biddaþ and him sylleþ heora synna forgyfnesse. Hér is his mildheortnes ofer ús ac ðér is se éca dóm *in this world the Lord heareth those that ask him and giveth them forgiveness of their sins. In this world his mercy is upon us, but in the next is the eternal judgement*, L. E. I. prm ; Th. ii. 394, 4-16. Hér *in this year*, Chr. *passim*. [*Goth.* hér : *O. Sax.* hér, hier : *O. Frs.* hír ; *Icel.* hér : *O. H. Ger.* hiar, hier : *Ger.* hier.]

hér ; *adj. Noble, excellent, honourable, holy, sublime* :—Gehýr ðis hêre spel [herrespel, Thorpe], *hear this noble lay*, Exon. 93 a ; Th. 348, 32 ; Sch. 32. [*O. Sax.* hér : *O. H. Ger.* hér, hêre *almus, sanctus, magnificus* : *Ger.* hehr.]

héra, an ; *m. One who obeys another, a servant, follower* :—Héra t embehtmonn *minister*, Mk. Skt. Lind. 10, 43. Héra t þegn *minister*, Jn. Skt. Lind. 12, 26. Héro *ministros*, Rtl. 11, 35. Æþelinga hleó beorna beággifa hérna hildfruma *the shelter of princes, ring-giver of warriors, warlike chief of his followers*, Elen. Kmbl. 201 ; El. 101. v. ambeht-héra *and* hýran.

hér-æfter ; *adv. Hereafter* :— Swá swá wé eft hêræfter secgaþ *as we shall again hereafter say*, Bd. 3, 30 ; S. 562, 5.

héran. v. hýran.

hĕr-búende; *pl. People living in this world,* Cd. 52; Th. 66, 4; Gen. 1079: Judth. 10; Thw. 22, 38; Jud. 96: Bt. Met. Fox 29, 124; Met. 9, 62.

hĕr-bufan; *adv. Here above :*—Swã swã wē ǽr hĕrbiufan sǽdon on ðisse ilcan bēc *as we said before above in this same book;* sicut in priori hujus voluminis parte jam diximus, Past. 50, 4; Swt. 393, 2.

hĕr-cyme, es; *m. A coming here, coming to this world, advent :*—Þurh ðīnne hĕrcyme *through thy advent,* Exon. 11 b; Th. 16, 8; Cri. 250.

herd. v. heord.

herdan. v. hyrdan.

herde. v. hirde.

HERE; *gen.* heres, heriges, herges; *m. An army, a host, multitude, a large predatory band* [it is the word which in the Chronicle is always used of the Danish force in England, while the English troops are always the *fyrd*, hence the word is used for *devastation* and *robbery :*— Ne dohte hit nū lange inne nē úte ac wæs here and hunger bryne and blōdgyte *it is now long since matters were thriving at home or abroad, but there has been ravaging and famine, burning and bloodshed,* Swt. A. S. Rdr. 106, 68. Micel here *turba multa,* Mt. Kmbl. Lind. 14, 14. Here *legio,* Lk. Skt. Lind. 8, 30 : *exercitus,* 23, 11. Þeófas wē hátaþ ōð vii men from vii. hlōþ ōð xxxv siððan biþ here *up to seven men we call thieves, from seven to thirty-five a gang, after that it is an army,* L. In. 13; Th. i. 110, 14. [Cf. L. In. 15; Th. i. 112, 1, be herige; and L. Alf. 28; Th. i. 52, 2.] Hē gearo wǽre tō ðæs heres þearfe *he would be ready to supply the needs of the Danes,* Chr. 874; Erl. 76, 32: 878; Erl. 80, 3. Ðæs heriges hām eft ne com ǽnig tō láfe *of that host came no remnant back home,* Cd. 167; Th. 209, 30; Exod. 507: Elen. Kmbl. 410; El. 205. Herges, 285; El. 143. On Eást-Englum wurdon monige men ofslægene from ðam herige *in East Anglia many men were slain by the Danes,* Chr. 838; Erl. 66, 15: Andr. Kmbl. 2397; An. 1200. Herge, Cd. 4; Th. 4, 9; Gen. 51; Beo. Th. 2500; B. 1248. Se ðæm here waldeþ *who rules that host,* Bt. Met. Fox 25, 30; Met. 25, 15. Sió fierd ðone here geflíémde *the English force put the Danish to flight,* Chr. 894; Erl. 90, 26. Swã oft swã ða óðre hergas mid ealle herige ūt fōron ðonne fōron hie *as often as the other armies marched out in full force then they marched,* Erl. 90, 5. Tuelf hergas *duodecim legiones,* Mt. Kmbl. Lind. 26, 53. Hergia[s] *agmina,* Rtl. 115, 10. Ðý lǽs ǽfre cweðan óðre þeóda hǽðene herigeas *nequando dicant in gentibus,* Ps. Th. 78, 10: Andr. Kmbl. 1304; An. 652. Herigea mǽste *with the greatest of hosts,* 3001; An. 1503. Herega, Cd. 209; Th. 259, 29; Dan. 699. Heriga, Elen. Kmbl. 295; El. 148. Herga, 230; El. 115. Betwuh ðǽm twám hergum *between the two armies,* Chr. 894; Erl. 90, 9: Elen. Kmbl. 219; El. 110. Herigum, 811; El. 406. [Laym. Orm. here: Goth. harjis: O. Sax. heri: O. Frs. hiri, here: Icel. herr: O. H. Ger. hari, heri *exercitus, agmen :* Ger. heer.] DER. æsc-, egor-, flot-, forþ-, gúþ-, inn-, ísern-, sin-, scip-, þeód-, út-, wæl-here.

hĕre, e; *f. Dignity, majesty, greatness :*—Hwæt hiora hĕre búton se hlísa ān *what is their greatness but report alone,* Bt. Met. Fox 10, 107; Met. 10, 54. The prose, Fox 70, 10, has 'Hwæt is heora nū tō láfe bútan se lytla hlísa and se nama mid feáum stafum áwriten *signat superstes fama tenuis pauculis inane nomen litteris.'* [O. H. Ger. hĕre : f. *dignitas, majestas, magnitudo :* cf. O. H. Ger. hĕr-tôm *dignitas, auctoritas, principatus,* Grff. iv. 994 : O. Sax. hĕr-dôm.]

here-beácen, -beácn, es; *n. A military ensign, standard ;* also a *beacon, light-house :*—Herebeácn *farus :* upstandende herebeácn *pira,* Ælfc. Gl. 67; Som. 69, 93, 90; Wrt. Voc. 41, 45, 43. Herebeácen and segnas beforan mē lǽddon *cum signis et vexillis,* Nar. 7, 16. [O. H. Ger. heri-pouhan *vexillum, signum.*]

here-bleáþ; *adj. Fearful in fight, timorous :*—Flugon forhtigende woldon herebleáþe hámas findan *fearful they fled and shunning the battle would find their homes,* Cd. 166; Th. 206, 17; Exod. 453.

here-bróga, an; *m. The terror produced by an army* or *by war,* Beo. Th. 928; B. 462.

here-býme, an; *f. A war-trumpet,* Cd. 147; Th. 183, 29; Exod. 99. [Cf. Icel. her-horn, her-luðr *a trumpet :* O. H. Ger. heri-, her-horn *classicum, tuba.*]

here-byrne, an; *f. A war-corslet,* Beo. Th. 2890; B. 1443. [Laym. here-burne.]

here-cirm, es; *m. A war-shout, shout raised by a host,* Exon. 45 b; Th. 156, 9; Gú. 872.

here-cumbol, -combol, es; *m. A military signal :*—Wordum and bordum hōfon herecombol *with shouts and shields they raised the war-signal,* Elen. Kmbl. 49; El. 25. Cf. [?] Tacitus, Germania c. 3: 'As their line shouts, they inspire or feel alarm. It is not so much an articulate sound, as a general cry of valour. They aim chiefly at a harsh note and a confused roar, putting their shields to their mouths, so that, by reverberation, it may swell into a fuller and deeper sound.' [Icel. her-kuml *a war-token, arms on shields* or *helmets.*]

here-cyst, -cist, e; *f. A warlike troop,* Cd. 151; Th. 188, 32; Exod. 177: 156; Th. 194, 7; Exod. 257: 158; Th. 197, 3; Exod. 301.

here-draca, an; *A war-drake, an arrow :*—Herdracan, Hickes' Thes. p. 192. [Cf. hilde-nædre.]

here-feld, es; *m. A field, battle-field,* Elen. Kmbl. 537; El. 269: 251; El. 126: Andr. Kmbl. 19; An. 10: 35; An. 18.

here-feoh, gen. -feós; *n. Booty :*—Eal ðæt herefeoh forlēton *prædam amiserunt,* Ors. 3, 7; Swt. 118, 5.

here-fēða, an; *m. A martial band,* Exon. 22 b; Th. 63, 1; Cri. 1013.

here-fléma, an; *m. One who flees from battle,* Chr. 937; Erl. 112, 23; Æðelst. 23.

here-folc, es; *n. People forming an army,* Judth. 11; Thw. 24, 40; Jud. 234. [O. Frs. hiri-folk: Icel. her-fôlk *men of war.*]

here-fong, es; *m. An osprey;* ossifragus, Wrt. Voc. 280, 6.

Here-ford, es; *m. Hereford :*—Ða men of Hereforda *the men from Hereford,* Chr. 918; Erl. 102, 31.

here-fugol, es; *m. A bird which attends an army, eagle, vulture, raven,* Cd. 150; Th. 188, 2; Exod. 161. v. earn, hrefn.

here-gang, es; *m. An irruption, attack by an army :*—Tō wiðscúfanne swã rēþum heregange *ad repellendas tam feras inruptiones,* Bd. 1, 14; S. 482, 37, MS. B. [Laym. hire-ȝeong: Gen. and Ex. heregong *military expedition :* O. Frs. hiri-, heri-gong *an attack :* cf. Icel. her-ganga : f. *a march.*]

heregeat-land, es; *n. Heriot-land,* Chart. Th. 546, 37.

here-geatu; *gen.* -geatwe; *f.* I. *military equipment :*—Hí willaþ eów tō gafole gāras syllan ǽttrynne ord and ealde swurd ða heregeatu ðe eów æt hilde ne deáh *they will give you as tribute spears, the poisoned point and the swords they inherit, equipment for war that will not profit you in battle,* Byrht. Th. 133, 10; By. 48. Heregeatewa, MS. A : heregeatowe, B. wægeþ *it bears arms,* Salm. Kmbl. 106; Sal. 52. Ða beóþ mid gyldenum hyltsweordum and mid manigfealdum heregeatwum gehyrste *septos tristibus armis,* Bt. 37, 1; Fox 186, 6: Bt. Met. Fox 25, 17; Met. 25, 9. II. as a technical term, *heriot.* The amount of the heriot for various ranks is given L. C. S. 72; Th. i. 414, 4–20; further mention is also made in L. C. S. 71; Th. i. 412, 26–414, 2: 74; Th. i. 416, 3–18: 79; Th. i. 420, 13–17. The word also occurs in the following passages in wills, Chart. Th. 499, 29 : 512, 16 : 540, 5 ; 550, 28 : 573, 3. For the origin and nature of the heriot see Stubbs' Const. Hist. s. v. Kemble's Saxons in England, ii. 98. [Cf. Grmm. R. A. 372–3.]

heregend-líc. v. herigend-líc.

here-gild, es; *n. A war-tax, the Danegild, tax to support an army :*—Hēr wæs ðet heregeold gelǽst ðæt wǽron xxi þúsend punda and xcix punda *in this year the Danegild was paid, it was twenty-one thousand and ninety-nine pounds,* Chr. 1040; Erl. 167, 23. Swã fela sýðe swa menn gyldaþ heregyld oððe tō scipgylde *quotiens populus universus persolvit censum Danis, vel ad naves seu ad arma,* Chart. Th. 307, 23. Scotfrē fram heregeld *free from payment of the war-tax,* Cod. Dipl. Kmbl. iv. 224, 20.

here-grima, an; *m. A helmet,* Beo. Th. 797; B. 396: 4104; B. 2049: 5203; B. 2605.

heregung. v. hergung.

here-hand, a; *f. A hostile hand* or *power :*—Swã ðæt ne cyricum ne mynstrum seó herehand ne sparode ne árode *ita ut ne ecclesiis quidem, aut monasteriis manus parceret hostilis,* Bd. 4, 26; S. 602, 8.

here-hlóþ, e; *f. A hostile troop,* Exon. 48 a; Th. 166, 13; Gú. 1042.

here-hýþ, -húþe, e; *f. Spoil, booty, plunder :*— Hēr wæs mycel herehúþe [herehýþe, MS. C.] ðǽr genumen *in this year much spoil was taken at Bamborough,* Chr. 993; Erl. 133, 2. Hē his ðone feórþan dǽl and ðǽre herehýþe for Gode gesealde *quartam partem ejus et prædæ Domino daret,* Bd. 4, 16; S. 584, 10. Hiera heres ðone mǽstan dǽl hām sendon mid hiora herehýþe *præcipuam exercitus sui partem onustam præda domum revocant,* Ors. 1, 10; Swt. 46, 21. Mid ðǽre herehýþe [herehúþe, MS. E.], Chr. 885; Erl. 82, 30. Ða mycele herehúþe tō scipon brohton *they brought the great booty to the ships,* 1001; Erl. 137, 15. Ða herehýþþ ðe on helle genumen hæfde *the spoil that he had taken in hell,* Blickl. Homl. 89, 33. Genimon myccle herehýþ *to take great spoil,* 95, 2. Ymbe ða herehúþe hlemmeþ tōgædre grimme gōman *on the prey he snaps together his fierce jaws,* Exon. 97 b; Th. 363, 29; Wal. 61. Ðone here geflíémde and ða herehýþ áhreddon *put the Danes to flight and rescued the spoils,* Chr. 894; Erl. 90, 26. [O. H. Ger. heri-hunda, -hunta *preda.*]

here-láf, e; *f. The remnant of an army* or *people, what is left of an army after a battle, what is left after a battle, spoil :*—Se Chaldēa cyning com tō his earde mid ðǽre húþe and ðǽre herelāfe on ðǽre wæs Daniel se wītega and ða þrí cnihtas *the king of Chaldea came to his country with the spoil and the remnant of the people, among which was the prophet Daniel and the three children,* Swt. A. S. Rdr. 68, 380, 392. Gúþrum se hǽðene king twelf dages hēr on lande wunede and syððan gewende mid his herelāfe tō his ágenen earde *Guthrum the heathen king stopped twelve days in this land and afterwards returned with what remained of his army to his own country,* Shrn. 17, 8. Þurh gítsunge wearþ beswicen Sawl se cyning ðá ðá him leófran wǽron ða forbodenan herelāfa ðonne Godes willa *through avarice was king Saul betrayed when he preferred*

the forbidden spoils of the host [*of the Amalekites*, v. I Sam. xv. 9] *to the will of God*, Basil. admn. 9 ; Norm. 54, 8. Costontinus ne Ánláf mid heora hereláfum hlehhan ne þorftun *not Constantine nor Anlaf, with the remnants of their forces, had cause for laughing*, Chr. 937 ; Erl. 114, 13 ; Æðelst. 47.

here-líc ; *adj. Warlike, military* : — Ða herelícan *res militares*, Cot. 47, Lye.

here-lof, es ; *n. Praise gained in war, fame, glory* ; also *a trophy* ; rumor, fama, Hpt. Gl. 406, 511, 512 : 447·

here-mæcg, es ; *m. A man of war, warrior, man* [*used of the men of Sodom when attacking Lot*], Cd. 114 ; Th. 149, 31 ; Gen. 2483. [Cf. *Icel.* her-megir *warriors*.]

here-mægen, es ; *n. A warlike force, an army, a host, multitude*, Exon. 116 b ; Th. 447, 10 ; Dôm. 37 ; Andr. Kmbl. 1172 ; An. 586 : 1456 ; An. 728 : 2597 ; An. 1300 : 3299 ; An. 1652 : Elen. Kmbl. 339 ; El. 170.

here-man, -mann, es ; *m. A soldier* : — Heremenn *milites*, Lk. Skt. Lind. 7, 8. [*Icel.* her-maðr.] v. Grmm. R. A. 292.

hére-man. v. híre-man.

here-meðel, es ; *n. A warlike assembly* ; concio, Elen. Kmbl. 1096 ; El. 550.

here-nes, -nis, -ness, e ; *f. Praise* : — Herenes mín *laudatio mea*, Ps. Th. 103, 32 : 110, 8 : 117, 14. Herenis *laus*, Rtl. 30, 23 : 174, 31. In herenesse Godes *in laudem Dei*, Bd. 4, 24 ; S. 597, 17 : 599, 12 ; Ps. 55, 10. Hé geearnode ðæt hé ða hálgan hærenesse gehýrde *laudes beatas meruit audire*, Bd. 3, 19 ; S. 547, 35· v. here-word.

here-net, -nett, es ; *n. A war-net, coat of mail, corslet*, Beo. Th. 3110 ; B. 1553.

here-níþ, es ; *m. Hostility, enmity which is felt by those at war with one another*, Beo. Th. 4938 ; B. 2474.

here-nitig [?] *expeditio*, Cot. 73, Lye.

here-pád, e ; *f. A coat of mail*, Beo. Th. 4508 ; B. 2258.

here-, her-paþ, es ; *m. A road for an army, military road, road large enough to march soldiers upon* [occurs not unfrequently in charters] :— Ondlong herpoþes, Cod. Dipl. Kmbl. ii. 172, 18. Up tô herpaþe and fram ðam herpaþe súþrihte, 205, 20. On ðone brádan herpaþ, iii. 23, 35. Wísde herepoþ tô ðære heán byrig *shewed a road for his army to the lofty city*, Cd. 174 ; Th. 218, 12 ; Dan. 38. Hí swyrdum herpaþ worhton þurh láðra gemong *they with their swords wrought a road through the press of their foes*, Judth. 12 ; Thw. 26, 1 ; Jud. 303. DER. þeód-herpaþ.

here-ræswa, an ; *m. A chieftain*, Elen. Kmbl. 1987 ; El. 995.

here-reáf, es ; *n. Spoil, plunder, booty* :—Herereáf *spolia vel manubie vel prede*, Ælfc. Gl. 52 ; Som. 66, 52 ; Wrt. Voc. 35, 38 : *manubiæ, spolia*, Ælfc. Gr. 13 ; Som. 16, 16, 23. Achan behýdde of ðam herereáfe *Achan concealed some of the spoil*, Jos. 7, 1, 11. Ðú ús mycel herereáf gehéte *thou didst promise us much spoil*, Blickl. Homl. 85, 19. Hengest and Æsc gefuhton wið Walas and genámon unárímedlíco herereáf *Hengest and Æsc fought with the Britons and took countless spoils*, Chr. 473 ; Erl. 12, 26 : 584 ; Erl. 18, 25. Hé tôdælþ his herereáf *spolia ejus distribuit*, Lk. Skt. 11, 22. Ic geseah betwux ðam herereáfum sumne gildene dalc *I saw among the spoils a wedge of gold*, Jos. 7, 21.

here-rinc, es ; *m. A warrior*, Bt. Met. Fox 1, 141 ; Met. 1, 71 : [here-ric, MS.] Beo. Th. 2356 ; B. 1176. [*O. Sax.* heri-rink.]

here-sceaft, es ; *m. A war-shaft, spear*, Beo. Th. 675 ; B. 335·

here-sceorp, es ; *n. War-dress*, Fins. Th. 90 ; Fin. 45·

here-serce, -syrce, an ; *f. A coat of mail*, Beo. Th. 3027 ; B. 1511.

here-síþ, es ; *m. The journey of an army, a military expedition, march*, Elen. Kmbl. 265 ; El. 133 : Exon. 108 a ; Th. 411, 24 ; Rä. 30, 4 : 84 a ; Th. 317, 3 ; Môd. 60.

here-spéd, e ; *f. Success in war*, Beo. Th. 129 ; B. 64.

here-spel. v. hér.

here-stræl, es ; *m. An arrow*, Beo. Th. 2874 ; B. 1435.

here-stræt, e ; *f. A military road, one allowing the passage of an army, highway, high road* :—Léton ðone hálgan be herestræte swe ́fan on sibbe *they left the saint sleeping in peace by the highway*, Andr. Kmbl. 1662 ; An. 833. Ðanan on herestræt *thence to the high road*, Cod. Dipl. Kmbl. ii. 265, 30. [Cf. of ða wýdestræte, 32.] Wegas syndon drýge herestræta *the ways* [*through the Red Sea*] *are dry, the roads for the host*, Cd. 157 ; Th. 195, 29 ; Exod. 284. Ne mé herestræta ofer cald wæter cúþe sindon *nor are the highways over the cold water known to me*, Andr. Kmbl. 400 ; An. 200. Gegier ðæt ðíne willas iernan bí herestrætum *in plateis aquas divide*, Past. 48, 6 ; Swt. 373, 6. Æfter cyninga herestrætum *along king's highways*, 373, 18. Ic hí ádilgode swá swá wind dêþ dust on herestrætum *ut lutum platearum delebo eos*, Ps. Th. 17, 40. Omnes herestrete omnino regis sunt, L. H. 10, 2 ; Th. i. 519, 11. [*O. Frs.* hiri-strete : *O. H. Ger.* heri-stráza *via publica*.] Cf. here-paþ, -weg.

here-swêg, es ; *m. A martial sound*, Exon. 124 a ; Th. 477, 12 ; Ruin. 23.

here-teám, es ; *m.* I. *plundering, spoiling, devastation, taking part in a* '*here*,' i. e. *a predatory band of more than thirty-five members*

[v. here] :—Se ðe hereteáme betogen sý *he who is accused of taking part in a* '*here*,' L. In. 15 ; Th. i. 112, 2, MS. H. Heardlíc hereteám *fierce devastation*, Andr. Kmbl. 3100 ; An. 1553. II. *what is got by an army, plunder, booty, spoil* :—Ðæs hereteámes ealles teóþan sceat *a tithe of all the spoil*, Cd. 97 ; Th. 128, 4 ; Gen. 2121. Gewát hám síþian mid ðý hereteáme ðe him se hálga forgeaf *departed home with the spoil that the holy man gave him*, 98 ; Th. 130, 19 ; Gen. 2162.

here-têma, -týma, an ; *m. A leader of an army, of a people, a ruler, general* : — Se heretêma cyning selfa *the leader, the king himself* [*Theodoric*], Bt. Met. Fox 1, 62 ; Met. 1, 31. Se heretýma, caldéa cyning, Cd. 205 ; Th. 253, 30 ; Dan. 603. Ða cwæþ hé hwæs sunu is hit ða cwæþ se bisceop mínes heretêman *then said he 'whose son is it?' Then said the bishop 'my prince's'* [?], Shrn. 130, 9. Hé wearþ tô herete ́man *he became general*, Elen. Kmbl. 20 ; El. 10.

here-þreát, es ; *m. A troop, band of soldiers*, Cd. 170 ; Th. 214, 24 ; Exod. 574 : *cohortes*, Cot. 51, Lye.

here-þrym, es ; *m. A cohort*, Cot. 84, Lye.

here-toga, -toha, an ; *m. The leader of an army or of a people, a general* ; dux, consul : — Heretoga vel heorl *dux*, Ælfc. Gl. 68 ; Som. 70, 2 ; Wrt. Voc. 42, 11. Heretoga comes, Rtl. 193, 9. Of ðe forþgæþ se heretoga seðe recþ mín folc *ex te exiet dux, qui reget populum meum*, Mt. Kmbl. 2, 6. Consul ðæt wé heretoha hátaþ *consul which we call* '*heretoha*,' Bt. 1 ; Fox 2, 12 : 21 ; Fox 76, 4. Sum biþ heretoga fyrdwísa from *one is a leader, a good guide of the host*, Exon. 79 b ; Th. 297, 31 ; Crä. 76. Se heretoga Moyses *the leader Moses*, Homl. Th. i. 92, 25. Moises se mæra heretoga *Moses the great leader*, Num. 13, 1 : Jud. 1, 1 : Swt. A. S. Rdr. 60, 107. Uton ús gesettan heretogan *let us make a captain*, Num. 14, 4. Heora heretogan twe ́gen gebroðra Hengest and Horsa *duces eorum duo fratres Hengest and Horsa*, Bd. 1, 15 ; S. 483, 28. Heora heretogena sum ofslægen wearþ *one of their leaders was slain*, Chr. 794 ; Erl. 59, 21. Twelf heretogan hé gestríþ *twelve princes shall he beget*, Gen. 17, 20. De heretochiis, L. Ed. C ; Th. i. 456, note a. [*Laym.* here-toʒe : *O. Sax.* heri-togo : *Icel.* her-togi : *O. H. Ger.* heri-zoho, -zogo *dux, imperator* : *Ger.* herzog.] v. Stubbs' Const. Hist. s. v.

here-togen [?] ; *pp. Captive* : — Seó hereláf wunode ðæs heretogan [heretogenan ?] folces on Chaldéiscum earde *the remnant of the captive people dwelt in the land of Chaldea*, Swt. A. S. Rdr. 69, 393. [Cf. *Icel.* her-numinn, -tekinn *captive*.]

here-wæd, e ; *f. War-weed, armour*, Beo. Th. 3798 ; B. 1897. [*Icel.* her-váðir *armour*.] v. Grmm. R. A. 566-7.

here-wæpen, es ; *n. A weapon of war*, Ps. Ben. 34, 3 ; Ps. Grn. ii. 149, 3.

here-wæsmun :—Nó ic mé an herewæsmun hnágran talige gúþgeweorca ðonne Grendel hine, Beo. Th. 1358 ; B. 677. *Thorpe reads* wæstmum [*see the use of* wæstm *in the plural*] *and translates* '*in martial vigour.*' *Grein translates by* vis bellica *and refers the word to a nominative* wæsma, *comparing O. H. Ger.* wahsamo, wasmo, wasma *vigor, fructus, fertilitas*, Grff. i. 689. *Leo and Heyne connect with a root meaning rage, fury*, v. Leo. 494. *Taking either of the first the passage might be translated* '*I do not account myself worse in the warlike fruits of martial deeds than Grendel himself;*' *or an* herewæsmum *and* gúþgeweorca *might be taken as both dependent upon* hnágran.

here-wéða, an ; *m. A war-hunter, a hunter whose game is the enemy*, Judth. 11 ; Thw. 23, 17 ; Jud. 126 : Thw. 24, 5 ; Jud. 173. v. Grmm. Geschicht. D. S. 12 sqq.

here-weg, es ; *m. A highway, high road* :—Ealles hereweg *publica via*, Ælfc. Gl. 57 ; Som. 67, 52 ; Wrt. Voc. 37, 39. [*O. Frs.* heer-wei : cf. *Icel.* her-vegir *war-paths*.] v. here-paþ, -stræt.

here-weorc, es ; *n. A warlike deed or work*, Elen. Kmbl. 1308 ; El. 656.

herewian ; p. ode *To despise* : — Tô swíðe wé herewiaþ ús selfe *we despise ourselves too much*, Bt. 13 ; Fox 40, 12. Leófsunu herewade ðæs arcebiscopes gewitnesse *Leofsunu incepit vituperare archiepiscopum et testimonium ejus irritum facere*, Chart. Th. 273, 2. v. herwan.

here-wíc, es ; *n. An encampment, camp, dwelling* : — Míne welan ðe ic hæfde syndon ealle gewitene and míne herewíc syndon gebrosnode *my riches that I had are all departed and my dwellings are decayed*, Blickl. Homl. 113, 26. Him mon sægde ðæt ðær mon cymen wæs of Alexandres herewícum *he was told that a man was come from Alexander's camp*, Nar. 18, 9 : Cd. 95 ; Th. 123, 26 ; Gen. 2051.

here-wisa, an ; *m. The director, guide of an army, a leader, general*, Cd. 160 ; Th. 198, 15 ; Exod. 323.

here-wóp, es ; *m. The shout raised by an army*, Cd. 166 ; Th. 207, 2 ; Exod. 460. [*Icel.* her-óp *war-whoop, war-cry*.]

here-word, es ; *n. Praise, applause* : — Ða wolde Brihtríc geearnian him hereword *tunc cogitavit Brihtricus adquirere sibi laudem*, Chr. 1009 ; Erl 142, note 8. [*Laym.* hære-, here-word : A. R. 'a windes puf of worldes hereword, of mannes heriunge,' 148, 3.] v. here-nes, herian *to praise*.

here-wósa, an ; *m. One who is fierce in fight, a warrior* [?] : — Here-

wôsan hige *a warrior's soul*, Cd. 206; Th. 255, 24; Dan. 629. Siddan herewôsan heofon ofgǽfon *since those who fiercely fought gave up heaven*, 5; Th. 6, 7; Gen. 85. [Cf. ealo-wôsa, wudu-wâsa.]

here-wulf, es; *m. A war-wolf, warrior*, Cd. 94; Th. 121, 25; Gen. 2015.

herfest. v. hærfest.

hergan. v. herian.

hergaþ, hergoþ, es; *m. Harrying, plundering, making war :—*Hê wæs ðá útâfaren on hergaþ *he was then gone out a harrying*, Chr. 894; Erl. 91, 20: 911; Erl. 100, 25: 918; Erl. 102, 30. Faran on hergoþ *to wage war*, Thw. 162, 37.

heregend-líc. v. herigendlíc.

hergere, es; *m. One who praises*; laudator, Rtl. 124, 17.

hergian; *p.* ode; *pp.* od *To harry, pillage, plunder, ravage, waste, devastate, make an incursion* or *a raid, make war :—*Ða Cwénas hergiaþ hwílum on ða Norþmen ofer ðone môr hwílum ða Norþmen on hý *sometimes the Fins made incursions across the mountains on the Norwegians, sometimes the Norwegians on them*, Ors. 1, 1; Swt. 19, 3. Se here hergade on Peohtas *the Danes made raids upon the Picts*, Chr. 875; Erl. 78, 1. Fôr Willelm cyng into France mid fyrde and hergode uppan his ágene hláforde Philippe *king William marched with an army into France and made war upon his own lord Philip*, 1086; Erl. 220, 25: Homl. Th. ii. 58, 5. Wera hof hergode *laid waste the dwellings of men*, Cd. 69; Th. 83, 15; Gen. 1380. Ða hǽðenan on Norþhymbrum hergodon ðe *the heathens ravaged in Northumbria*, Chr. 794; Erl. 59, 20. Hie hergodon ofer Mercna land ôþ hie cômon tô Creccagelâde *they carried on their ravages across Mercia until they came to Cricklade*, 905; Erl. 98, 14. Mycel sciphere hider com and hergedon swíðe be Sefærn *a great fleet came to this country and committed great depredations along the Severn*, 910; Erl. 101, 7. Gif ǽnig sciphere on Engla lande hergie *if any fleet commit ravages in England*, L. Eth. ii. 1; Th. i. 284, 15, 18. Sǽdon ðæt hí woldan him sylfe niman and hergian ðǽr hí hit findan mihton *protestantur se cuncta insulæ loca vastaturos*, Bd. 1, 15; S. 483, 38. Hí sceoldan ealle ætgædere faran and hergian *they should go all together and harry*, Chr. 1014; Erl. 151, 2. Hê wæs heriende and feohtende fíftig wintra *arma foras extulit, cruentamque vitam quinquaginta annis bellis egit*, Ors. 1, 2; Swt. 28, 28. [*Laym.* hærþien : *Chauc.* haried, harwed : *Icel.* herja *to harry*; herjask á *to wage war on one another* : *O. H. Ger.* harion, herion *populare, vastare* : cf. *Ger.* verheeren.] DER. ge-, ofer-, on-hergian.

hergung, heregung, e; *f. Harrying, harrowing, plundering, devastation, waging war, an irruption, incursion, invasion, a raid, plunder :—*Seó hergung wæs þurh Alaricum Gotena cyning geworden *inruptio quæ per Alaricum regem Gothorum facta est*, Bd. 1, 11; S. 480, 11. Hêðenra manna hergung âdíligode Godes cyrican in Lindisfarena ee þurh reáflác and mansleht *the harrying of heathen men destroyed God's church at Lindisfarne by plundering and slaughter*, Chr. 793; Erl. 59, 11. Ðæt mǽste yfel ðe ǽfre ǽnig here dôn mihte on bærnette and hergunge and on manslihtum *the greatest evil that any army could do in the way of burning and plundering and manslayings*, 994; Erl. 133, 18. On ânre heregunge *in a single invasion*, Jos. 10, 40. Be his ǽriste and be his hergunga on helle *concerning his resurrection and his harrowing of hell*, Blickl. Homl. 83, 29. Hell oncneów Crist ðá ðá heó forlét hyre hæftlingas út þurh ðæs Hǽlendes hergunge *Hell acknowledged Christ when it let out its captives through the harrowing of Jesus*, Homl. Th. i. 228, 17. Hí hergodon and brohton tô ðam castele ða hergunge *they plundered and brought the plunder to the castle*, Chr. 1087; Erl. 224, 19. Ðá forlét hê his hergunga *then he left off his harryings*, 1016; Erl. 154, 10.

herian, hærian, hergan; *p.* ode, ede; *imper.* hera *and* here; *pp.* ed *To praise :—*Ðé ic herige swá swá wísne man *te laudo ut sapientem*, Ælfc. Gr. 15; Som. 17, 64: Ps. Th. 55, 4, 9. Ic herge, Exon. 41 b; Th. 138, 28; Gú. 583. Ðæt ðæt mon hereþ *hoc ipsum quod laudatur*, Past. 48, 5; Swt. 373, 2. Leofaþ sáwl mín and ðé hereþ *vivet anima mea et laudabit te*, Ps. Th. 118, 175. Heraþ, 101, 16. Weleras ðé míne heriaþ *labia mea laudabunt te*, 62, 3. Wê ðé hæriaþ *we praise thee*, Hy. 7, 116; Hy. Grn. ii. 289, 116. Hergaþ, Cd. 214; Th. 267, 33; Swt. 47. Ic nát for hwý gê ða tída swelcra brôca swá wel hergeaþ *I know not why ye praise so highly the times of such miseries*, Ors. 3, 7; Swt. 120, 4: Blickl. Homl. 89, 31. Hergaþ, Cd. 192; Th. 239, 24; Dan. 375. Heó Drihten herede *she praised the Lord*, Blickl. Homl. 13, 4: Lk. Skt. 16, 8. Ðæs cininges ealdormen heredon hig beforan him *the princes of Pharaoh commended her before Pharaoh*, Gen. 12, 15. Hit is áwriten ne hera ðú nænne man on his lífe *it is written 'Praise no man during his life*,' Homl. Th. ii. 560, 13. Ðé silfne ne hera *do not praise thyself*, Salm. Kmbl. 262, 21. Here ðú, Sion, swylce ðínne sôþne God *lauda Deum tuum, Sion*, Ps. Th. 147, 1. Mín hearpe herige Drihten *let my harp praise the Lord*, 56, 10. Herge, Beo. Th. 6333; B. 3177. Ðeáh hira híeremenn hie mid ryhte heregen *though their subjects with justice praise them*, Past. 19; Swt. 145, 22. Herian, Ps. Th. 65, 1. Hergen, Exon. 54 b; Th. 191, 27: Az. 94. Hie heofona helm herian ne cúðon *they did not know how to praise the heaven's protector*, Beo. Th. 367; B. 182. Hergan, Exon. 8 b; Th. 4,

8; Cri. 49. Herigean, Bd. 4, 24; S. 597, 20. Heó is ús tô herianne *she is to be praised by us*, Blickl. Homl. 11, 11. Tô herigenne, 63, 21. Tô hergenne, 223, 27. Se hálga wer hergende wæs metodes miltse *the holy man was praising the Lord's mercy*, Cd. 190; Th. 237, 8; Dan. 334. Herigende, Andr. Kmbl. 1314; An. 657. Ðú byst hered *perfecisti laudem*, Ps. Th. 8, 2; Blickl. Homl. 67, 4. [*Laym.* herien, hærien : *A. R.* herede; *þ : Chauc. Wick.* herie : *Spens.* herry, hery : *Goth.* hazjan *to praise*.]

herian [= herewian; cf. gearwian, gerian] *to despise :—*Agar ongan ágendfreán herian *Hagar despised her mistress* [cf. Gen. 16, 4 'her mistress was despised in her eyes'], Cd. 102; Th. 135, 5; Gen. 2238.

herigend-, hergend-líc; *adj. Praiseworthy, laudable :—*Ne biþ nán anginn herigendlíc bútan gôdre geendunge *no beginning is praiseworthy without a good ending*, Homl. Th. i. 56, 26: 212, 29. Hergendlíc in worlda world *laudabile in secula seculorum*, Blickl. Homl. 139, 11. Hergiendlíc *laudabilis*, Rtl. 181, 27. Ða giftu beóþ herigendlíce *that marriage is praiseworthy*, Homl. Th. ii. 54, 10.

herigend-, hergend-líce; *adv. Praiseworthily :—*Hê sylf herigendlíce leofode *he himself lived praiseworthily*, Homl. Th. ii. 118, 14. Hergiendlíce *laudabiliter*, Rtl. 105, 3. Hergeondlíce, Past. 7; Swt. 49, 193.

hér-inne; *adv. Herein*, Homl. Th. ii. 312, 4.

hér-, hǽr-líc; *adj. Noble, excellent :—*Næs ðæt hérlíc dǽd *that was no noble deed*, Bt. Met. Fox 9, 36; Met. 9, 18. Hǽrlíc, 1, 86; Met. 1, 43. [*O. H. Ger.* hér-líh *insignis*.] v. hér; *adj.*

hér-nis, herstan, hérsum. v. hýr-nis, hyrstan, hýrsum.

hér-ongemong; *adv. Here-among, amongst the rest, meanwhile :—*Gif wê Æfneres dǽda sume hérongemong secgaþ *si Abner factum ad medium deducamus*, Past. 40, 5; Swt. 295, 13. Gif wê Salamones cuida sumne hérongemong eówiaþ *si Salamonis ad medium verba proferantur*, 49, 5; Swt. 385, 33.

herra. v. hearra.

herþan; *pl. Testiculi*, Wrt. Voc. 65, 31. Wið hærþena sáre, L. Med. ex. Quadr. 8, 2; Lchdm. i. 358, 4 : Lchdm. iii. 116, 15; L. Alf. pol. 65; Th. i. 96, 25.

herþ-belig, -bylig, es; *m. Viscus, scrotum : —* Herþbelig, herþbylig *viscus*, Wrt. Voc. 283, 35 : 65, 13. Wið herþbylges sáre, L. Med. ex. Quadr. 5, 10; Lchdm. i. 350, 6.

herung, hering, e; *f. Praising, praise : —* Herung *laudatio*, Ps. Spl. 110, 10. For manna herunge *for the praise of men*, Homl. Th. i. 60, 33 : 38, 10 : 180, 20. On ðære heringe ðæs eádgan weres *in praise of the blessed man*, Past. 56, 7; Swt. 435, 18 : Bt. 27, 3; Fox 100, 4 : 30, 1; Fox 108, 22.

herwan. v. hyrwan.

HETE, es; *m.* HATE, *hatred, enmity, malignity, malice, spite :—*Hete *nequitia*, Mt. Kmbl. Rush. 22, 18. Ús hôl and hete derede swíðe þearle *slander and hate have injured us very sorely*, Swt. A. S. Rdr. 106, 70. Wæs his hete grim *fierce was its hate*, Exon. 109 a; Th. 416, 1; Rä. 34, 5 : Beo. Th. 5101 ; B. 2554 : 286 ; B. 142. Hê forseah and on hete hæfde odio *habebat et despiciebat*, Bd. 3, 21; S. 551; 25. Se wæs on hete heofoncyninges *he was hateful to the king of heaven*, Cd. 30; Th. 40, 32; Gen. 648. Ða Iudéiscan bôceras mid hete ðæt tældon *the Jewish scribes blamed that with malice*, Homl. Th. i. 338, 20. Ðú scealt hine álýsan of láþra hete *thou shalt release him from the hate of foes*, Andr. Kmbl. 1888; An. 946. Ðone mǽstan hete hê sent on eów *he shall pour upon you his fiercest hate*, Deut. 28, 59. Hete *malitiam*, Ps. Stev. 35, 5. Ic flýma wæs ðæt ic mé his hete berh and wearnode *qui vagabundus, hostium vitabam insidias*, Bd. 2, 12; S. 513, 28. Ða tô Sione hete hæfdon *qui oderunt Sion*, Ps. Th. 128, 3. Hete hæfde hê æt his hearran gewunnen *he had gained hate from his lord*, Cd. 16; Th. 19, 34; Gen. 301 : 37; Th. 47, 29; Gen. 768 : 103; Th. 137, 13; Gen. 2273. Mid fulryhte hete ic hie hatode *perfecto odio oderam illos*, Past. 46, 5; Swt. 353, 6. Mid inlíce hete *domestico odio*, Bd. 5, 23; S. 646, 38. Hetas *malitias*, Ps. Stev. 93, 23. [*Laym.* hete : *Orm.* hæte : *Prompt. Parv.* hate : *Goth.* hatis : *O.Sax.* heti : *Icel.* hatr : *O.H.Ger.* haz *odium : Ger.* hass.] DER. bil-, cumbol-, ecg-, leód-, môd-, morþor-, níþ-, scyld-, teón-, wǽpen-, wíg-hete.

hete-grim; *adj. Of malignant cruelty* or *fierceness*, Andr. Kmbl. 2789; An. 1397 : 3122; An. 1564. [*O. Sax.* heti-grim.]

hete-líc; *adj. Inspired by hate, hostile, malicious, evil :—*Heorowearh hetelíc *a wolf hostile and malignant*, Beo. Th. 2538; B. 1267. Mid hetelícum geþance *with evil intent*, H. R. 99. 4. Atregeas and Thigesþres hú hí heora fæderas ofslôgan and ymb hiora hetelícan forlignessa ic hit eall forlǽte *Atrei et Thyestis odia, stupra et parricidia dissimulo*, Ors. 1, 8; Swt. 42, 20. [*O. Sax.* heti-lík : *O. H. Ger.* haz-líh *invidus : Ger.* hässlich *ugly, wicked*.]

hete-líce; *adv. Fiercely, violently, vehemently : —* Hetelíce *mordicus*, Ælfc. Gr. 38; Som. 42, 5. Hine hetelíce swung [cf. Bd. 2, 6; S. 508, 13 mid grimmum swingum swong] *scourged him vehemently*, Chr. 616; Erl. 23, 3. Ús Godes yrre hetelíce on sitt *God's anger presses on us fiercely*, Swt. A. S. Rdr. 108, 109. Hit sáh hetelíce swíðe *it sank with great violence*, Homl. Th. ii. 508, 34. Hê hine hetelíce þídde *he stabbed him violently*, Jud. 3, 21 : Homl. Th. i. 452, 14 : H. R. 107, 7. **Hig**

hetelíce slóh and nán þing ne beléfde lybbende on him *smote them fiercely & and left no thing living among them*, Jos. 11, 8. Â hetelíce stýre ðam ðe þwyres willan *ever to punish those severely that desire perverseness*, L. I. P. 2; Th. ii. 304, 17.

hetend. v. hettend.

hete-níþ, es; *m. Enmity, hostility, malice, wickedness* :— Hí sprǽcon heteníþ *locuti sunt nequitiam*, Ps. Spl. T. 72, 8. Geheald ðú mé wið heteníþas and wið firenfulles folman *custodi me de manu peccatoris*, Ps. Th. 139, 4: Exon. 94 a; Th. 352, 22; Sch. 101. Grendel heteníþas wǽg *Grendel bore enmity*, Beo. Th. 307; B. 152.

hete-róf; *adj. Active in hate* or *hostility, hostile*, Andr. Kmbl. 2839; An. 1422.

hete-rún, e; *f. A charm causing hate* or *evil*, Exon. 109 a; Th. 416, 6; Rä. 34, 7.

hete-sprǽc, e; *f. Hostile* or *malicious speech*, Cd. 14; Th. 17, 22; Gen. 263.

hete-sweng, es; *m. A hostile blow*, Beo. Th. 4453; B. 2225.

hete-þanc, es; *m. A hostile thought*, Beo. Th. 955; B. 475: Exon. 70 a; Th. 261, 14; Jul. 315.

hete-þancol; *adj. Having hostile* or *evil designs*, Judth. 10; Thw. 23, 4; Jul. 105.

hetlen; *adj. Bearing hate, hostile, malignant*, Exon. 13 a; Th. 23, 5; Cri. 364.

hetol, hetel; *adj. Full of hate, hostile, malignant, evil* :—Se heáhengel ðe nú is hetol deófol *the archangel that now is a devil full of malice*, Boutr. Scrd. 17, 22. Maxentius ða burh geheóld mid hetelum geþance *Maxentius held the town with hostile intent*, Homl. Th. ii. 304, 21. Hí habbaþ nú ðone hetolan deófol him tó hláforde *they have now the malignant devil as their lord*, 254, 1: Swt. A. S. Rdr. 66, 327. Hér sind on earde cyrichatan hetole *here in the land are foes of the church full of malice*, 109, 154. [A. R. hetel: O. H. Ger. hazzal *malitiosus*.]

hettan; cf. hatian, *and see next word.*

hettend, hetend, es; *m. An enemy* :— Hettend lǽddon út mid ǽhtum abrahames mǽg *the enemy led forth Abraham's kinsman with his possessions*, Cd. 94; Th. 121, 17; Gen. 2011: 154; Th. 191, 4; Exod. 209: Chr. 937; Erl. 112, 10; Ǽdelst. 10: Andr. Kmbl. 61; An. 31. Hetend, Elen. Kmbl. 237; El. 119. Hettende, Exon. 62 a; Th. 228, 21; Ph. 441. Hetende, Beo. Th. 3660; B. 1828. Hettendra, Cd. 97; Th. 127, 13; Gen. 2110: Exon. 75 b; Th. 282, 14; Jul. 663. Hettendum, Beo. Th. 6000; B. 3004. Hetendum, Elen. Kmbl. 35; El. 18. [O. Sax. hettend, hetteand, hetand.] DER. eald-hettend.

hice-máse, an; *f. The blue titmouse* :—Hicemáse *vel* wrenna *parrax*, Ǽlfc. Gl. 38; Som. 63, 38; Wrt. Voc. 29, 56. [Cornish dialect, hickmal, hekky-mal *the blue titmouse*.] Cf. col-máse.

hicgan. v. hycgan.

híd, e; *f. A hide of land.* The form *híged*, which occurs Cod. Dipl. Kmbl. ii. 5, 25, seems to shew that the word is connected with *híwan, hígan*, and this etymology is supported by the use of *familia* and *híd* in the Latin and English versions respectively of Bede's Ecclesiastical History. The original meaning of the word would thus be 'as much land as will support one family.' v. Bd. 1, 25; S. 486, 19 : 2, 9; S. 87, 32 [Latin]: 3, 4; S. 106, 33 [Latin]: 4, 16; S. 584, 14. Further, in the charters, *híwisc* [q. v.] is used as equivalent to *híd*. The Latin words used as equivalent are *mansus, mansa, mansio, manens, cassatus, terra tributarii, familia*, Cod. Dipl. Kmbl. iii. xxx. See for further discussion of the word Kemble's Saxons in England, i. 4 : Stubbs' Const. Hist. s. v : Schmid. A. S. Gesetze, p. 610.

hídan. v. hýdan.

hider; *adv. Hither* :—Hider *huc*, Ǽlfc. Gr. 38; Som. 39, 65. Hideror *citerius*, Som. 41, 3. Sittaþ hér ðþ ðæt ic gá hider geond *sedete hic donec vadam illuc*, Mt. Kmbl. 26, 36. Hider and geond *huc illucque*, Bd. 5, 12; S. 629, 3. Hider and ðider *huc illucque*, Past. 9; Swt. 59, 5. Ne mæg hió hider ne ðider sígan ðé swíðor ðe hió symle dyde *it cannot decline to one side or the other more than it ever did*, Bt. Met. Fox 20, 328; Met. 20, 164. Sume hyder sume ðyder *some on one side, some on the other*, Elen. Kmbl. 1093; El. 548. [Chauc. Piers P. hider: Wick. hidir: Goth. hidre: Icel. heðra.] v. hidres.

hider-cyme, es; *m. A coming hither, to this world, advent* :— Ðín hidercyme *thy advent*, Exon. 13 a; Th. 23, 12; Cri. 367. Fram Cristes hidercyme *ab incarnatione Domini*, Bd. 1, 3; S. 475, 16 : 1, 4; S. 475, 26. On his hidercyme *in his coming hither* [*to Hell*], Blickl. Homl. 87, 2, 11. Hidercyme ðínne on wráþra geweald *thy coming hither into the power of enemies*, Andr. Kmbl. 2634; An. 1318: Exon. 10 a; Th. 9, 29; Cri. 142 : 62 a; Th. 227, 10; Ph. 421: 16 a; Th. 37, 2; Cri. 587.

hider-weard; *adj. Hitherward, in this direction* :— Híe ǽr fǽtte wǽron and beóþ hiderwearde *they were before fat and are still disposed this way*, L. M. 2, 36; Lchdm. ii. 242, 5.

hider-weard; *adv. Hitherward* :— On ðisum geáre menn sǽdon ðæt Cnut cyng fundade hiderward *in this year men said that king Cnut was making for this country*, Chr. 1085; Erl. 217, 40. [Laym. hider-ward, wardes: Piers P. hiderward.]

híd-gild, es; *n. A land tax, tax paid on every hide* :— Ðis mycel is gegolden of ðære cyricean W. cyninge syððan hé ðis land áhte wiðútan ðam hídgelde ðe nán man wiðútan Gode ánum átellan ne mæg *this much has been paid from the church* [*of Worcester*] *to king William since he owned this country, besides the hide-tax, which no one but God alone can reckon*, Chart. Th. 439, 22. [Cf. Chr. 1083; Erl. 217, 33-5, Se cyng lét beódan mycel gyld and hefelíc ofer eall Engla land ðæt wæs æt ælcere hýde twá and hundseofenti peanega.]

híd-mǽlum; *adv. By hides* :—Ðæt líþ hídmǽlum and æcermǽlum *it lies by hides and by acres*, Cod. Dipl. Kmbl. vi. 98, 4.

hidres; *adv. In the phrase* hidres ðidres *hither and thither* :— Ic ondrǽde ðæt ic ðé lǽde hidres ðidres on ða paþas of ðínum wege ðæt ðú ne mǽge eft ðínne weg áredian *verendum est, ne deviis fatigatus, ad emetiendum rectum iter sufficere non possis*, Bt. 40, 5; Fox 240, 21: Past. 22; Swt. 168, 13.

hie. v. hé.

hiénþo. v. hýnþ.

hiéran, etc. v. hýran, etc.

hierde. v. hirde.

hierstan. v. hyrstan.

hiertan. v. hyrtan.

hiéwe-stán, es; *m. A hewn stone* :—Ǽlcne hiéwestán tóbeátan *to beat to pieces every hewn stone*, Ors. 4, 13; Bos. 100, 10.

híf. v. hýf.

hig hay. v. heg.

hig they. v. hé.

hígan. v. híwan.

hige. v. hyge.

hí-gedryht, e; *A band of household retainers*, Exon. 94 b; Th. 353, 32; Reim. 21.

higera, higora, an; *m* : higere, an; *f. A magpie* or *a woodpecker;* see Exon. 106 b; Th. 406, 14; Rä. 25 where the name of a bird that can imitate various sounds is given by the runes G, A, R, O, H, I. Higera *picus*, Wrt. Voc. 62, 34. Higere *picus*, 281, 5 : *gaia vel catanus*, Ǽlfc. Gl. 37; Som. 63, 14; Wrt. Voc. 29, 37 : *cicuanus*, Cot. 34. Lye. [O. H. Ger. hehara, hehera *picus, attacus, orin*.] v. Grein, ii. 72.

higian; *p.* ode *To hie, hasten, strive* :—Ðonne hé higaþ tó ðæm godcundum þingum ánum *cum ad sola, quæ interiora sunt, nititur*, Past. 14, 3; Swt. 83, 14. Se ðonne se ðe suá higaþ tó andweardnesse his scippendes *qui igitur sic ad auctoris speciem anhelat*, 14, 6; Swt. 87, 10. Se ðe æfter ðæm higaþ ðæt hé eádig síe on ðisse worulde *qui festinat ditari*, Past. 44, 9; Swt. 331, 14. Higaþ ealle mægne ðæt hé wolde . . . *strives with all his might to . . .* , Bt. 30, 1; Fox 110, 4 : Bt. Met. Fox 13, 130; Met. 13, 65. Gehiéren ða reáferas ða ðe higiaþ wið ðæs ðæt hie willaþ óðre men bereáfian hwæt be him gecweden is *cum aliena rapere intendunt, audiunt, quod scriptum est*, Past. 44, 8; Swt. 329, 16. Ðætte suá hwelc suá inweard higige tó gangenne on ða duru ðæs écean lífes *ut, quisquis intrare æternitatis januam nititur*, 16, 5; Swt. 105, 14: Bt. 22, 2; Fox 78, 18 : 37, 2; Fox 118, 16. Ða ðe hé gesyhþ tó Gode higian *those that he sees striving towards God*, Blickl. Homl. 29, 22. Hé sceal simle higian ðæt hé weorþe geedniwad *he must ever strive to be renewed*, Past. 22, 1; Swt. 169, 10. [Orm. hiʒhenn: Laym. hiʒeden, *p. pl* : A. R. hien: Piers P. hyed, hiʒed, *p* : Wick. hiʒed, *pp*.]

higre [cf. higera] *or* hígre [cf. híwan] *verna*, Cot. 23, Lye : Gl. Epin. 663.

hig-scipe. v. híw-scipe.

hiht, hihtan. v. hyht, hyhtan.

hilc. v. hylc.

hild grace. v. hyld.

hild, e; *f.* [a poetical word] *War, battle;* pugna, prælium :— In the Scandinavian mythology Hildr is the name of one of the Valkyrias, and Grimm considers that the word occurs, denoting a person, in the Anglo-Saxon poetry, e. g. gif mec hild nime, Beo. Th. 909; B. 452 : 2967 ; B. 1481. v. Grmm. D. M. 392 sqq. Hild swedróde *war ceased*, Beo. Th. 1807; B. 901: 3180; B. 1585 : 3698; B. 1847 : Andr. Kmbl. 2840; An. 1422 : Elen. Kmbl. 36; El. 18 : 298; El. 149. Hyne Hetware hilde gehnǽgdon *him the Hetwaras conquered in battle*, Beo. Th. 5825; B. 2916 : 4159; B. 2076 : 4586; B. 2298 : Exon. 100 a; Th. 378, 10; Deór. 14 : Menol. Fox 493; Gn. C. 17 : Apstls. Kmbl. 41; Ap. 21: Cd. 150; Th. 188, 3; Exod. 162. Nǽfre hit æt hilde ne swác manna ǽngum *never had it failed in fight any man*, Beo. Th. 2925; B. 1460: 3322; B. 1659 : 5143; B. 2575 : 5361; B. 2684: Cd. 98; Th. 129, 25; Gen. 2149: Byrht. Th. 133, 24; By. 55 : 135, 24; By. 123: 138, 20; By. 223 : 140, 14; By. 324 : 131, 15; By. 8: Wald. 6; Vald. 1, 4: Andr. Kmbl. 823; An. 412: Salm. Kmbl. 320; Sal. 159: Fins. Th. 75; Fin. 37: Wald. 55; Vald. 1, 30 : Exon. 79 a; Th. 297, 5; Crä. 63 : 104 a; Th. 395, 7; Rä. 15, 4 : 120 a; Th. 461, 17; Hö. 37: Cd. 95; Th. 124, 11; Gen. 2061: 155; Th. 193, 5; Exod. 241: Elen. Kmbl. 63; El. 32 : 97; El. 49: 103; El. 52 : 129; El. 65. Ongenþeów hæfde Higeláces hilde gefrunen *Ongentheow had heard of Higelac's fighting*, Beo. Th. 5897; B. 2952 : 1299; B. 647 : 3984; B. 1990 :

Wald. 87; Vald. 2, 15: Exon. 16 a; Th. 35, 31; Cri. 566: Cd. 151; Th. 189, 3; Exod. 181: 167; Th. 209, 25; Exod. 504: Judth. 12; Thw. 25, 9; Jud. 251.   Heardre hilde *with hard fighting*, Elen. Kmbl. 165; El. 83: Judth. 12; Thw. 25, 36; Jud. 294. Fela ic gebâd heardra hilda *many hard battles have I experienced*, Fins. Th. 52; Fin. 26: Andr. Kmbl. 2980; An. 1493. [*O. Sax.* hild: *Icel.* hildr: *O. H. Ger.* hilt.] v. Grff. iv. 912.]

hilde-bedd, es; *n. Deathbed*, Andr. Kmbl. 2186; An. 1094.

hilde-bil, -bill, es; *n. Battle-blade, sword*, Beo. Th. 3337; B. 1666: 1118; B. 557: 3044; B. 1520: 5351; B. 2679.

hilde-bord, es; *n. A war-shield*, Beo. Th. 799; B. 397: 6270; B. 3139.

hilde-calla, an; *m. A war-herald*, Cd. 156; Th. 193, 26; Exod. 252.

hilde-corðor, es; *n. A warlike troop*, Apstls. Kmbl. 82; Ap. 41.

hilde-cyst, e; *f. Excellence in war, valour :*—Hildecystum *valorously*, Beo. Th. 5189; B. 2598.

hilde-deóful, es; *n. A devil, demon :* — Sindon ealle hæðene godu hildedeóful *omnes dei gentium dæmonia*, Ps. Th. 95, 5.

hilde-deór; *adj. Stout in war, brave*, Beo. Th. 629; B. 312: 1672; B. 834: 4220; B. 2107: 4372; B. 2183.   Hæle hildedeór *a warrior brave*, 3296; B. 1646: 3636; B. 1816: 6213; B. 3111: Andr. Kmbl. 2003; An. 1004: Elen. Kmbl. 1868; El. 936.   Hildedeóre *brave men*, Beo. Th. 6320; B. 3170. [Thorpe and Kemble take *deór* to be a noun.] v. hilde-freca.

hilde-freca. v. hild-freca.

hilde-frófor, e; *f. War-help, a weapon, sword* [?], *shield* [?] :—Hæfde him on handa hildefrófre [MS. frore] *had in his hand help for battle*, Vald. 2, 12.

hilde-gæst, -giest, es; *m. An enemy*, Exon. 113 b; Th. 436, 5; Rä. 54, 9.

hilde-geatwe; *pl. f. War equipments*, Beo. Th. 1353; B. 674: 4713; B. 2362.

hild-egesa, an; *m. Terror of battle*, Elen. Kmbl. 226; El. 113.

hilde-gicel, es; *m. A drop of blood*, Beo. Th. 3217; B. 1606.

hilde-grædig; *adj. Eager for battle*, Cd. 150; Th. 188, 3; Exod. 162.

hilde-gráp, e; *f. Hostile grasp*, Beo. Th. 2896; B. 1446: 5007; B. 2507. In the latter passage Thorpe and Kemble take *gráp* to be a verb.

hilde-hlem, -hlæm, mes; *m. Crash of battle*, Beo. Th. 4691; B. 2351: 5081; B. 2544: 4408; B. 2201.

hilde-leóma, an; *m. A hostile, warlike ray*, Beo. Th. 2291; B. 1143 [*a sword*]: 5159; B. 2583.

hilde-leóþ, es; *n. A battle-song, war-song*, Judth. 11; Thw. 24, 28; Jud. 211.

hilde-mæcg, es; *m. A warrior*, Beo. Th. 1603; B. 799.

hilde-méce, es; *m. A war-falchion*, Beo. Th. 4411; B. 2202.

hilde-nædre, an; *f. A war-adder, an arrow, dart, warlike missile*, Elen. Kmbl. 238; El. 119: 281; El. 141: Judth. 11; Thw. 24, 34; Jud. 222.

hilde-píl, es; *m. A dart, bolt, javelin*, Exon. 105 a; Th. 399, 5; Rä. 18, 6: 104 b; Th. 397, 33; Rä. 16, 28.

hilde-ræs, es; *m. A warlike onset*, Beo. Th. 605; B. 300.

hilde-rand, es; *m. A shield*, Beo. Th. 2489; B. 1242.

hilde-rinc, es; *m. A warrior*, Beo. Th. 2618; B. 1307: 2994; B. 1495: 3156; B. 1576: 6239; B. 3124: Byrht. Th. 136, 50; By. 169: Chr. 937; Erl. 114, 5; Æðelst. 39: Elen. Kmbl. 525; El. 263: Rood Kmbl. 122; Kr. 61: 143; Kr. 72.

hilde-sæd; *adj. Wearied with battle*, Beo. Th. 5439; B. 2723.

hilde-sceorp, es; *n. War-clothing*, Beo. Th. 4316; B. 2155.

hilde-scûr, es; *m. War-shower, flight of missiles*, Exon. 49 b; Th. 170, 24; Gú. 1116.

hilde-serce, an; *f. A war-shirt, corslet*, Elen. Kmbl. 468; El. 234.

hilde-setl, es; *m. A war-seat, saddle of a war-horse*, Beo. Th. 2082; B. 1039.

hilde-spell, es; *n. A warlike speech*, Cd. 170; Th. 214, 22; Exod. 573.

hilde-strengo; *f. Warlike strength*, Beo. Th. 4232; B. 2113.

hilde-swât, es; *m. Hostile vapour or steam*, Beo. Th. 5109; B. 2558.

hilde-swêg, es; *m. Sound of battle*, Cd. 93; Th. 120, 7; Gen. 1991.

hilde-þremma, an; *m. A warrior*, Exon. 66 b; Th. 246, 19; Jul. 64.

hilde-þrym, mes; *m. Warlike prowess*, Andr. Kmbl. 2064; An. 1034.

hilde-þryþ, e; *f. Strength in war*, Exon. 105 a; Th. 400, 6; Rä. 20, 4.

hilde-torht; *adj. Having warlike splendour*, Bt. Met. Fox 25, 18; Met. 25, 9.

hilde-tusc, -tux, es; *m. A battle-tusk, a tusk or tooth that serves as a weapon*, Beo. Th. 3026; B. 1511. [Cf. *Icel.* hildi-tannr.]

hilde-wǽpen, es; *m. A weapon of war*, Beo. Th. 77; B. 39.

hilde-wîsa, an; *m. A military leader, general*, Beo. Th. 2133; B. 1064.

hilde-wôma, an; *m. The crash and rush of battle*, Andr. Kmbl. 436; An. 218: Exon. 75 b; Th. 282, 15; Jul. 663: 67 b; Th. 250, 32; Jul. 136. v. Grmm. And. u. El. xxx.

hilde-wrǽsen, e; *f. A chain used to secure those taken in war* [?], Salm. Kmbl. 586; Sal. 292.

hilde-wulf, es; *m. A war-wolf, warrior*, Cd. 95; Th. 123, 25; Gen. 2051.

hild-freca, hilde-, an; *m. A warrior*, Beo. Th. 4721; B. 2366: 4416; B. 2205: Andr. Kmbl. 251; An. 126: 2141; An. 1072. v. freca.

hild-from; *adj. Stout or bold in war*, Andr. Kmbl. 2405; An. 1204.

hild-fruma, an; *m. A military chief or prince*, Elen. Kmbl. 19; El. 10: 201; El. 101: Exon. 65 b; Th. 243, 7; Jul. 7: Beo. Th. 3360; B. 1678: 5291; B. 2649: 5662; B. 2835.

hild-lata, an; *m. One sluggish in war, slow to fight, a coward*, Andr. Kmbl. 466; An. 233: Beo. Th. 5684; B. 2846.

hild-stapa, an; *m. One who steps to war, a warrior*, Andr. Kmbl. 2517; An. 1260.

hild-þracu; *gen.* -þræce; *f. Power, force in war*, Cd. 98; Th. 130, 9; Gen. 2157.

hil-hâma. v. hylle-hâma.

hill. v. hyll.

hilt, es; *m. n. Hilt, handle* [the plural, as in much later times, e. g. Shakspere's, is used of a single weapon]:—Ða wæs gylden hilt gamelum rince on hand gyfen *then was the golden hilt given into the old man's hand*, Beo. Th. 3358; B. 1677. Ðæs swurdes mid ðam sylfrenan hylte *the sword with the silver hilt*, Chart. Th. 558, 11. Ic ðæt hilt ðanon ætferede *I bore the hilt away from there*, 3341; B. 1668. Hylt, 3379; B. 1687. Blîcaþ ða hiltas *the hilt shines*, Salm. Kmbl. 446; Sal. 223. Ða hilt since fâge *the hilt many-coloured with treasure*, Beo. Th. 3233; B. 1614. Be hiltum *by the hilt*, 3152; B. 1574. [*Icel.* hjalt; *n. the boss or knob at the end of a sword's hilt; also the guard between the hilt and blade*. For some account of the hilts of old swords see Worsaae's Primeval Antiquities, pp. 29, 49.] DER. fealo-, fetel-, hroðen-, wreoðen-hilt. v. next word.

hilte, an; *f. A hilt, handle* :— Hilte *capulus, capulum* [?], Ælfc. Gl. 52; Som. 66, 47, 26; Wrt. Voc. 35, 34, 14.   Hiltan *capulum*, Wrt. Voc. 84, 21. Swâ ðæt ða hiltan eodon intó ðam innoþe *the haft went in after the blade*, Jud. 3, 22. Oþ ða hiltan *capulotenus*, Mone Gl. 432. [*O. H. Ger.* helza *capulus*.]

hilte-cumbor, es; *n. An ensign having a hilt*, Beo. Th. 2048; B. 1022.

hilted; *part. p. Provided with a hilt*, Beo. Th. 5966; B. 2987.

hilt-leás; *adj. Without a hilt* :—Hiltleás sweord *ensis*, Ælfc. Gl. 52; Som. 66, 46; Wrt. Voc. 35, 33.

hîna. v. hîne, hîwan.

hinan. v. heonan.

hinca. v. helle-hinca.

hind. v. hynd.

hind, e; *f. A hind, the female of the hart* :—Hind *cerva*, Ælfc. Gl. 19; Som. 59, 23; Wrt. Voc. 22, 64.   Hynd *cerva*, Wrt. Voc. 78, 27. Ðâ geseah se godes þeów wilde hinde melce *then the servant of God saw a wild hind in milk*, Shrn. 130, 3. Hê lægde laga ðæt swâ hwâ swâ slôge heort oððe hinde ðæt hine man sceolde blendian *he made laws that whoever should kill hart or hind should be blinded*, Chr. 1086; Erl. 222, 27. Sêcan heorotas and hinda *to hunt harts and hinds*, Bt. Met. Fox 19, 33; Met. 19, 17. [*Icel. Dan.* hind: *O. H. Ger.* hinta, hinda *cerva* : *Ger.* hinde, hindinn.]

hindan; *adv. From behind, at the back, in the rear, behind* :—Ðâ hêt hê gewrîðan ðone pâpan and ðone ôðerne preóst tô his hricge hindan *then he ordered the pope to be bound, and the other priest behind to his back*, Homl. Th. ii. 310, 31. Hindan þyrel *pierced from behind*, Exon. 129 b; Th. 497, 24; Rä. 87, 5.   Is him ðæt heáfod hindan grêne *its head is green at the back*, 60 a; Th. 218, 12; Ph. 293. Hie hindan ofrîdan ne meahte *could not overtake them*, Chr. 877; Erl. 78, 21: 894; Erl. 92, 22: Erl. 93, 7: 911; Erl. 100, 26: Ors. 6, 36; Bos. 131, 25. Heówan hereflêman hindan þearle *smote sorely the fugitives, pressing on their rear*, Chr. 937; Erl. 112, 23; Æðelst. 23.   Se cyng fêrde him æt hindan and offêrde hî *the king marched in their rear* [*pursued them*] *and overtook them*, 1016; Erl. 158, 1. Pharao fêrde him æt hindan *Pharao pursued after them*, Swt. A. S. Rdr. 63, 226. [*Goth.* hindana *beyond* : *O. Sax.* bi-hindan: *O. H. Ger.* hintana : *Ger.* hinten.]

hindan-weard; *adv. At the further end, hindwards* :—Sindon ða fiðru hwît hindanweard *the wings are white at the tips*, Exon. 60 a; Th. 218, 21; Ph. 298.

hind-berige, -berie, -berge, an; *f. A raspberry* :—Hyndberige *acimus, erimigio*, Wrt. Voc. 66, 59 : 67, 62. Genim hindbergean *take raspberries*, L. M. 2, 51; Lchdm. ii. 266, 8. Hindberge *ermigio*, Lchdm. iii. 302, col. 1. [hind-berry, hine-berry, v. English Plant Names, E. D. S. No. 26 : *O. H. Ger.* hind-beri: *Ger.* him-beere.]

hind-brér, es; *m. A raspberry plant; rubus idæus*, Lchdm. iii. 22, 31.

hind-cealf, es; *m. n. A fawn* :—Hindcealf *hinnulus*, Ælfc. Gl. 19; Som. 59, 26; Wrt. Voc. 22, 67: 78, 29. [*Halliw. Dict.* hind-calf, *a hind of the first year* : *O. H. Ger.* hint-kalb *hinnulus, damma, dammula*.]

hindema; *adj. Last :*—Hindeman sîðe *for the last time,* Beo. Th. &
4105; B. 2049: 5023; B. 2517. [Cf. *Goth.* hindumists.] Cf. next
word.

hinder; *adv. Back, on the further side, behind, down :*—Morðor món
sceal under eorþan befeolan hinder under hrusan *murder must be buried
under earth, down under ground,* Exon. 91 a; Th. 340, 24; Gn. Ex.
116. Hî mê âsetton on seáð hinder *posuerunt me in lacu inferiori,* Ps.
Th. 87, 6. Gengde on hinder *conversus est retrorsum,* 113, 3. On hinder
hê eode he [*the devil*] *went behind,* Homl. Th. i. 172, 35. Hê on hinder
gǽþ *he shall go back,* Salm. Kmbl. 254; Sal. 126. On hinder in helle hûs
down into hell, Exon. 42 b; Th. 142, 23; Gû. 648. [*Goth.* hindar *beyond: O. H. Ger.* hintar, hindar *retro, post: Ger.* hinter.]

hinder-geap, -gep; *adj. Crafty, cunning, guileful, deceitful :*—Hinder-
geap *versutus,* Ælfc. Gl. 84; Som. 73, 104; Wrt. Voc. 49, 11. Hindergepe
versuti, Coll. Monast. Th. 32, 29. [*Orm.* þatt mann iss fox and hinnderr-
ʒæp and forr will off ille wiless, 6646. Cf. *Goth.* hindar-weis *deceitful :*
hindar-weisei *guile.* Cf. *also* Carrais hine biðohte of ane hindere cræfte
[hiþer crafte, 2nd MS.]: *Laym.* 10489: Þe grune of hindre þat is of bi-
peching, *O. E. Homl.* ii. 213, 23: hinder-word, 59, 18: hinderfulle rede
consilium impiorum, 23.] v. geap.

hinder-hóc, es; *m. A stratagem, artifice, snare,* Exon. 83 b; Th.
315, 20; Môd. 34. [Cf. hinder-geap, hinder-scipe.]

hinderling, es; *m. A mean, base, contemptible person :*—Occidentales
Saxonici, scilicet execastre, habent in proverbio summi despectus, quod
summa ira commotus, unus vocat alterum hinderling, i. ab omni honestate
dejectum, L. Ed. C. 35; Th. i. 459, 36. [*Orm.* halde þe forr hinnderrling
and forr well swiþe unnwresste, 4860. Halliwell in his Dictionary says
under *hilding* 'the word is still in use in Devon, pronounced *hilderling* or
hinderling.']

hinder-scipe, es; *m. Wickedness;* nequitia, Hpt. Gl. 415.

hinder-þeóstru; *pl. Darkness in a remote* or *low place :*—Of helwarena hinderþeóstrum *ex inferno inferiori,* Ps. Th. 85, 12. v. hinder.

hinder-weard; *adj. Backward, slow :* — Nis hê hinderweard swâr ne
swongor swâ sume fuglas ða ðe late þurh lyft lâcaþ fiþrum *non tamen est
tarda, ut volucres quæ corpore magno incessus pigros per grave pondus
habent,* Exon. 60 a; Th. 220, 2; Ph. 314.

hinde-weard, -werd; *adj. Hindward :*—Mid hindewerdum ðam sceafte
aversa hasta, Past. 40, 5; Swt. 297, 10 13: 295, 17: L. Alf. pol. 36;
Th. i. 84, 17: Exon. 106 a; Th. 403, 29; Rä. 22, 15.

hind-fald, es [*or* -falda, an]; *m. A hind-fold,* Cod. Dipl. Kmbl. vi.
112, 33.

hind-hæleþe, -heolaþ, -heoloþe, -hioloþe, an; *f. Water agrimony :*—
Hyndhæleþe *ambrosia,* Wrt. Voc. 66, 60. Hindheolaþ, 79, 51. Genim
hindhæleþan, Lchdm. iii. 74, 4. Hindheoloþan, L. M. 2, 51; Lchdm.
ii. 266, 7: 1, 15; Lchdm. ii. 56, 21. Hindhioloþan, L. M. 1, 66; Lchdm.
ii. 142, 3: 1, 70; Lchdm. ii. 144, 22. v. Lchdm. iii. 331; col. 2.

hindrian; *p.* ede *To hinder, obstruct, keep back, repress :*—Â hê sceal
hǽðendôm hindrian *he must always repress heathenism,* L. I. P. 2; Th.
ii. 306, 7. [*Icel.* hindra: *O. H. Ger.* hintarian, Grff. iv. 704: *Ger.* hin-
dern.] v. ge-hindred.

hind-sîð. v. hin-sîð.

hîne [= (?) hînan *as* gehûse = gehûsan, hiwæ = hîwan *in the same verse*]
domesticos, Mt. Kmbl. Rush. 10, 25. Is this the word which gives later
English *hine,* Mod. E. *hind,* or are these taken from the gen. pl. of *hiwan,
hina,* which occurs most frequently in phrases *hina fæder,* etc., and which
may have come to be looked upon as an uninflected word used in such cases
as the first part of a compound? In v. 36 *domestici* is glossed *higu* i *hîne*
i *hîwen,* and 24, 34 *pater-familas = hîne-fæder* [but this may be for *hîna-
fæder*]. [*Laym.* children and hinen, 368: *O. E. Homl.* ðin owune hine,
i. 197, 112: *Chauc. Piers P.* hine.] v. hîwan.

hin-fús; *adj. Ready to go away* or *depart,* Beo. Th. 1514; B. 755:
Andr. Kmbl. 1223; An. 612.

hin-gang, -gong, es; *m. A going hence, departure, death,* Exon. 28 b;
Th. 86, 24; Cri. 1413: 30 b; Th. 95, 10; Cri. 1555: 44 b; Th. 150,
24; Gû. 783. [*O. H. Ger.* hina-gang *secessus.*]

hingrian. v. hyngrian.

hin-sîð, hinn-, hines-, es; *m. A journey hence, away, from this world,
departure, death,* Exon. 119 b; Th. 459, 29; Hö. 7: 87 a; Th. 328, 7;
Vy. 13: 97 b; Th. 364, 9; Wal. 68: 52 b; Th. 183, 22; Gû. 1331:
Cd. 33; Th. 44, 32; Gen. 718: Th. 45, 3; Gen. 74: Judth. 10; Thw.
23, 11; Jud. 117. Hindsîð, Blickl. Homl. 123, 6. [Cf. *O. Sax.* hin-fard:
O. H. Ger. hine-fart *exitus, obitus.*]

hinsîð-gryre, es; *m. Terror connected with death,* Cd. 223; Th. 293,
17; Sat. 456.

hió. v. hê.

hiofon. v. heofon.

hioful *the face :*—Ondwlita i hioful *facies,* Mt. Kmbl. p. 9, 11.

hion, e; *f. A bone of the head* [?] :—Gif sió ûterre hion gebrocen weorþeþ, L. Ethb. 36; Th. i. 12, 6, v. note, and cf. L. H. 93, 2; Th. i. 605,
12 si exterius os percussum sit.

hióp. v. heóp.

---

hior. v. heorr.

hiord. v. heord.

hioro. v. heoru.

hîr. v. hŷr.

hîran. v. hŷran.

hird *retinue, court :*—Hê fêrde tô Wudestoke and his biscopes and his
hird eal mid him *he [Henry] went to Woodstock, and his bishops and his
court all with him,* Chr. 1123; Erl. 249, 30. Dis geár heáld se kyng
Heanri his hird on Windlesoure *this year hing Henry held his court at
Windsor,* 1127; Erl. 255, 1. This form as it occurs in late specimens
may be merely a contraction of *hîrêd* [q. v.], or it may be a form influenced by the Danish *hirð.* In the former case it should be written *hîrd.*

hîrd-clerc. v. hîrd-preóst.

hirde, hierde, heorde, hiorde, hyrde, es; *m. A herd, shepherd, pastor,
guardian, guard, keeper :*—Hierde *arimentarius,* Wrt. Voc. 287, 52.
Crist ðú gôda hyrde *Christ, thou good shepherd,* Blickl. Homl. 191, 24.
Ic eom ðære stôwe hyrde *I am the guardian of the place,* 201, 9. Hire
âgenes hûses hirde *the keeper of her own house,* Bt. Met. Fox 13, 61;
Met. 13, 31. Rîces hirde *the guardian of a kingdom, a prince, king,*
26, 16; Met. 26, 8. Cilda hyrde *vel* lâreów *pædagogus,* Ælfc. Gl. 80;
Som. 72, 103; Wrt. Voc. 46, 60. Ic ðæs folces beó hyrde and healdend
I will be the people's keeper and preserver, Cd. 106; Th. 139, 25; Gen.
2315. Ne ic hyrde wæs brôðer mînes *I was not my brother's keeper,* 48;
Th. 62, 1; Gen. 1007. Heorde, Exon. 43 b; Th. 146, 33; Gû. 719.
Hiorde, Ps. Grn. ii. 279, 101. Rihtwîs hyrde ofer cristene heorde *a
righteous shepherd over a christian flock,* L. I. P. 2; Th. ii. 304, 9. Hie
settan him hyrdas tô *they set guards over him,* Blickl. Homl. 177, 26:
237, 18: Andr. Kmbl. 1986; An. 995. Úre ealdan fæderas wæron
ceápes hierdaʒ *antiqui patres nostri pastores,* Past. 17, 2; Swt. 109, 5.
Hyrdas *pastores ovium,* Gen. 46, 32. Hê hæfþ geset his englas ûs tô
hyrdum *he hath appointed his angels as our guardians,* Homl. Th. i.
170, 10. [*Goth.* hairdeis: *O. Sax.* hirdi: *Icel.* hirðir: *O. H. Ger.* hirti
pastor, custos: Ger. hirte.] DER. beór-, cú-, feorh-, gât-, grund-, hors-,
hrîðer-, neát-, sceáp-, swîn-hirde.

hirde-belg, -belig, es; *m. A shepherd's bag :*—Ðâ nam hê fîf stânas
on his herdebelig *then he took five stones in his shepherd's bag,* Blickl.
Homl. 31, 17.

hirde-bóc, hierde-, e; *f. Liber Pastoralis,* Past. Pref. Swt. 7, 19.

hirde-leás; *adj. Without a shepherd :*—Ne beóþ hî hyrdeleáse ðonne
hî ðê habbaþ *having thee they will not be without a shepherd,* Homl. Th.
i. 382, 23. Scêp heordeleáse *oves non habentes pastorem,* Mt. Kmbl.
Rush. 9, 36.

hirde-lîc; *adj. Pastoral :* — Ða byrðenne ðære hirdelecan giémenne
pastoralis curæ pondera, Past; Swt. 23, 11.

hirde-wyrt, e; *f.* I. the greater, *chlora perfoliata.* II. the
lesser, *erythæa centaureum,* Lchdm. iii. 332, col. 1.

hird-ness, hyrd-, e; *f. Guard, keeping, custody :*—Hê betǽhte hig ða
þrî dagas tô hirdnysse *tradidit ergo illos custodiæ tribus diebus,* Gen. 42,
17. Gif hwâ befæst his feoh tô hyrdnysse *si quis commendaverit pecuniam
in custodiam,* Ex. 22, 7. Swâ hî on niht hyrdnesse begangaþ *sicut custodia in nocte,* Ps. Th. 89, 5. On hyrdnyssa *in custodias,* Lk. Skt. 11, 12.

hîrd [=hîrêd] -preóst, es; *m. A domestic chaplain :* — Ælfrîc mîn
hîrdprêst, Cod. Dipl. Kmbl. iv. 269, 8: Chart. Th. 574, 10, 11.

hî-rêd, hîrd, es; *m. A household, house, family, the body of domestic
retainers of a great man* or *king, a court, the members of a religious
house, a company, band of associates :*—Hîrêd *vel* hîwrǽden *familia,* Wrt.
Voc. 72, 28. Se hâlga hŷrêd wæs wunigende ânmôdlîce on gebedum
the holy company continued with one accord in prayers, Homl. Th. i. 314,
4: Cd. 226; Th. 302, 1; Sat. 592: 221; Th. 288, 5; Sat. 376. Se
hîrd on Seynt Eádmundesbiri *the brotherhood at Bury St. Edmunds,* Chart.
Th. 574, 28, 33. Mîn ôwen hîrd *my own family,* 575, 21. Hîrêdes
fæder *paterfamilias,* Mt. Kmbl. 10, 25. Hîrêdes ealdor, 20, 1. Hîrêdes
hlâford, Wrt. Voc. 73, 20. Hîrêdes môder *materfamilias,* 73, 21. An
gewitnesse ðes hîrêdes æt Cristes cirican *with the witness of the brotherhood at Christchurch,* Cod. Dipl. Kmbl. ii. 3, 36. Gif hê stalie on gewitnesse ealles his hîrêdes gongen hie ealle on þeówot *if he steal with the
knowledge of all his household let them all go into slavery,* L. In. 7; Th.
i. 106, 17. Of Davides hûse and hîrêde *de domo et familia David.* Lk.
Skt. 2, 4. Tô ðæg is ðisum hîrêde hæl gefremmed *hodie salus domui
huic facta est,* Homl. Th. i. 582, 5. Cwæþ ðæt hê mid ðam Hælende on
hŷrêde wære *said that he was in company with Jesus,* ii. 248, 31. Hit
ne biþ nâ hûs bûton hit beó mid hîrêde âfylled *it is no house unless it be
filled with a household,* 582, 13. Ic wille ðæt mîne men bên frê on
hîrde and on tûne *I desire that all my men be free both in my household
and vill,* Cod. Dipl. Kmbl. iv. 269, 12. Ðam hîrêde intô ealdan mynstre
to the brotherhood at the old monastery, Chart. Th. 499, 14. Lucinius
bebeád ðæt nân cristen mon ne côme on his hîrêde *Licinius omnes
Christianos e palatio suo jussit expelli,* Ors. 6, 30; Swt. 282, 28. On
sumes cyninges hîrêde *in tanti patris familias dispositissima domo,* Bt. 36,
1; Fox 172, 18: 29, 2; Fox 104, 29: L. Edm. S. 4; Th. i. 248, 23:
L. C. S. 60; Th. i. 408, 14: L. R. 3; Th. i. 190, 20. Ðâ oferhogode

herodes hine mid hys hírēde *sprevit autem illum erodes cum exercitu suo,* ♱ 404, 4. Lk. Skt. 23, 11 : Cd. 222 ; Th. 290, 30 ; Sat. 423. God geswang Farao and ealne his hírēd *flagellavit Dominus Pharaonem et domum ejus,* Gen. 12, 17. Đone geset hys hláfurd ofer his hírēd *quem constituit dominus suus supra familiam suam,* Mt. Kmbl. 24, 45 : Cd. 106 ; Th. 139, 16 ; Gen. 2310. Gif hē beó tó đam gewelegod đæt hē hýrēd and ēht āge *if he be so enriched as to have a household and property,* L. Wg. 7 ; Th. i. 186, 23 [cf. 13 híwisc *landes*]. Fríone hierēd *a free monastery,* L. Alf. pol. 2 ; Th. i. 62, 1, v. note. Se cyng heóld đǣr his hírēd v dagas *the king held his court there five days,* Chr. 1085 ; Erl. 218, 18, 39. [O.E.Hom. hired : *Orm.* hird, hirrd : *Laym.* hiredes, *gen ;* hirde, *dat : A.R.* hird : *O.H.Ger.* hī-rāt *connubium : Ger.* heirath.] DER. iu-hírēd. v. hīwan.

hírēd-cniht, es ; *m. A man belonging to a 'hírēd,' a domestic :*—Þurh Paules bodunge gelýfdon đæs cāseres þegnas and hírēdcnihtas *through Paul's preaching the members of the emperor's household believed,* Homl. Th. i. 374, 34. [*Laym.* hird-cniht.]

hírēd-líc ; *adj. Familiaris,* Hpt. Gl. 463, 504.

hírēd-mann, hírd-man, es ; *m. A member of a 'hírēd' :*—Pharaones yldestan hírēdmen *senes domus Pharaonis,* Gen. 50, 7. His hírēdmen fērdon ūt mid feáwe maunan of đam castele and geslōgen and gelǣhton fíf hundred manna *the members of his household sallied out with few men from the castle, and slew and captured five hundred men,* Chr. 1087 ; Erl. 224, 29. Ongunnon đa hírēdmen heardlíce feohtan *the [earl's] household retainers began to fight stoutly,* Byrht. Th. 139, 28 ; By. 261. Hæbbe ǣlc hláford his hírēdmen [hírdmen (MS. A.)] on his āgenum borge *let every lord have the members of his household in his own 'borg,'* L. C. S. 31 ; Th. i. 394, 27 : L. Eth. i. 1 ; Th. i. 282, 9. [*Laym.* hired-, heredman (priveman, 2nd MS.)].

hírēd-wífmann, es ; *m. A female member of a household :*—Ic geann eallum mínum hírēdwífmannum *I give to all the women of my household,* Chart. Th. 531, 6.

hírēd-wist, e ; *f. Familiaritas,* Lye.

híre-man. v. hýre-man.

hirstan. v. hyrstan.

hírsum. v. hýrsum.

hiscan. v. hyscan.

hise. v. hyse.

hispan. v. hyspan.

hittan ; *p.* hitte *To hit upon, meet with :*—Đā com Harold ūre cyng on unwǣr on đa Normenn and hytte hī begeondan Eoforwīc æt Stemford brygge *then our king Harold came upon the Northmen unexpectedly and met with them beyond York at Stamford bridge,* Chr. 1066 ; Erl. 201, 26. [Borrowed from [?] *Icel.* hitta *to hit upon, meet with.*]

híw, hiow, e ; *f. Fortune :*—Swā hit oft gesǣleþ on đǣm sēlran þingum and on đǣm gesundrum đæt seó wyrd and sió hiow hie oft oncyrreþ *ut aliquid plerumque in secundis rebus fortuna obstrepit,* Nar, 7, 27.

híw, heow, hiow, heó, es ; *n. Shape, make, form, fashion, species, kind, appearance, symbol, hue, colour, beauty :*—Hiw *species,* Ælfc. Gl. 70 ; Som. 70, 45 ; Wrt. Voc. 42, 53. Hiw *figura, scema, specimen, forma, species,* Ælfc. Gr. 2 : 9 : 14 ; Som. 2, 45, 46 : 8, 22 : 9, 31 : 17, 19, 20. Hiw *figmentum,* Cd. 86. Đeós gerýnu is wedd and hiw *this mystery is a pledge and a symbol,* Homl. Th. ii. 272, 60. Sǣde hwylc đæs biscopes hiw wǣre *effigiem ejusdem Paulini referre esset solitus,* Bd. 2, 16 ; S. 519, 32 : Andr. Kmbl. 1449 ; An. 725. Heó is on onsýne ūtan yfeles heowes *outside it is in appearance of a very poor kind,* Blickl. Homl. 197, 11. Seó is brūnes heowes *it is of a brown colour,* 73, 22. Ānes hiwes *uniformis,* Ælfc. Gr. 49 ; Som. 59, 42. Hwǣlan hiwes *of a whale's shape,* Salm. Kmbl. 587 ; Sal. 263. Æt ānes heowes cý *from a cow all of one colour,* Lchdm. iii. 24, 13. Hiwes binotene *bereft of their [angelic] form,* Exon. 45 b ; Th. 156, 10 ; Gū. 872. On ōđrum hiwe *in alia effigie,* Mk. Skt. 16, 12. Hī ealle wurdon āwende of đam fǣgeran hiwe đe hí on gesceapene wǣron tó lāđlícum deóflum *they were all changed from the fair form in which they were created to loathly devils,* Homl. Th. i. 10, 30. On næddran hiwe *in the form of a serpent,* 16, 32 : 104, 23. On fýres hiwe *like as of fire,* 232, 15. On cuman hiwe *as a guest,* ii. 96, 35. Heowe, Blickl. Homl. 235, 29. Æfter his hiwe *secundum speciem suam,* Gen. 1, 12. Đū eart wlitig on hiwe *pulchra sis mulier,* 12, 11. Siđđan heó wunode mid fǣmnum on hira hiwe *afterwards she lived with women as a woman,* Shrn. 31, 16 : 52, 24. Se sunu onfēng mennisc hiw *the son took the form of a man,* Nar. 39, 32 ; Exon. 18 b ; Th. 45, 19 ; Cri. 721 : 46 a ; Th. 156, 28 ; Gū. 881. Heó, Elen. Kmbl. 12 ; El. 6. Tōcnāwan heofones hiw *faciem cæli dijudicare,* Mt. Kmbl. 16, 3. Scínende hiow and gewǣdu *shining face and garments,* Homl. Th. ii. 350, 18. Nū berþ Petrus đæt hiw odđe getācnunge đǣre hálgan gelaþunge *Peter is now the figure or symbol of the holy church,* 390, 14 : 406, 11. Weorþeþ sunne on blōdes hiw *the sun shall become the colour of blood,* Exon. 21 b ; Th. 58, 15 ; Cri. 936. Hiw *decorem,* Ps. Spl. C. 44, 13. Gimmas hwíte and reáde and hiwa gehwæs *gems, white and red and of every hue,* Bt. Met. Fox 19, 46 ; Met. 19, 23 : Exon. 95 b ; Th. 356, 31 ; Pa. 20. Behealdaþ eów wið leásum wītegum đe tó eów cumaþ on sceápa hiwum *take heed of false prophets that come to you as sheep,* Homl. Th. ii.

On mistlícum and mænigfealdum hiwum *of divers and manifold forms,* Lchdm. iii. 234, 13. [*Goth.* hiwi *form.*] v. feala-, scín-hiw.

híwan, hígan ; *pl. Members of a household, of a religious house, a family :*—Heora híwan *their household,* Cd. 133 ; Th. 168. 10 ; Gen. 2780. Hine ofslōgon his híwan [cf. hírēd] *the members of his household slew him,* Chr. 757 ; Erl. 53, 8. Híwan *members of a religious house,* L. Alf. pol. 5 ; Th. i. 64, 14. Đenewulf bisceop and đa híwan in Wintanceastre *bishop Denewulf and the brethren at Winchester,* Chart. Th. 151, 5. Hígen, Chart. Th. 47, 33 : 70, 33 : 461, 18, 33. Đa híwan đe on đam mynstre wǣron *qui erant in monasterio,* Bd. 3, 11 ; S. 535, 18. Hígo *familia,* Lk. Skt. Lind. 2, 4 : 12, 42. Hígo đa đe gihaldaþ *familia quæ abstinet,* Rtl. 16, 11 : 14, 30. Hígu *domestici,* Mt. Kmbl. Rush. 10, 36. Faderes hígna *patris familias,* Mt. Kmbl. Lind. 13, 27 ; Lk. Skt. Lind. 13, 25 : Chart. Th. 460, 9. Fæder hína, Mt. Kmbl. Rush. 20, 1 : 21, 33. Gehwilcne đe his hína wæs wǣpned cynnes *every one that of his family was of the male sex,* Cd. 107 ; Th. 142, 34 ; Gen. 2371. Đa hrýmde heó tó hire híwun *vocavit ad se homines domus suæ,* Gen. 39, 14. Mid hira híwun *cum domibus suis,* Ex. 1, 1. Gā tó đínum hūse tó đínum híwum *vade in domum tuam ad tuos,* Mk. Skt. 5, 19. Būton Noe and his seofan híwon *except Noah and the seven members of his family,* Homl. Th. ii. 58, 34 : i. 20, 34. On middum hire híwum *in medio eorum* [*the members of the monastery*], Bd. 4, 19 ; S. 588, 20 : Chart. Th. 468, 19 : L. Alf. pol. 2 ; Th. i. 62, 5. Giléf hígum đínum *concede famulis tuis,* Rtl. 30, 17. Ūt of earce híwan lǣd đū *lead thy family out of the ark,* Cd. 73 ; Th. 90, 3 ; Gen. 1489. Híwan [MS. A. munecas], Chr. 716 ; Erl. 45, 17. Híwæ *domesticos,* Mt. Kmbl. Rush. 10, 25. [*Ayenb. Chauc.* Piers P. hewe *a servant : Orm.* hiwenn *a family : cf. Goth.* heiwa-frauja οἰκοδεσπότης : *O. Sax.* hîwa *a wife : Icel.* hjú, hjún, hjón *man and wife, family, household : O. H. Ger.* hîwo *a married man ;* hîwa *a married woman.*] DER. gesam-, gesin-, in-, sam-, sin-híwan.

hiw-beorht, hiow- ; *adj. Bright of hue, beautiful in form* or *colour,* Elen. Kmbl. 145 ; El. 73 : Cd. 14 ; Th. 17, 27 ; Gen. 265.

híw-cūþ, heow- ; *adj. Familiar, well known :*—Híwcūþ *familiaris,* Ælfc. Gl. 115 ; Som. 80, 62 ; Wrt. Voc. 61, 36. Ic ne eom him suā híwcūþ *familiaritatis ejus notitiam non habemus,* Past. 10, 2 ; Swt. 63, 5 : Herb. 67, 1 ; Lchdm. i. 170, 13. Se đe hine selfne híwcūþne ne ongiet Gode *qui familiarem se ejus gratiæ esse nescit,* Past. 10, 2 ; Swt. 63, 8. Đa syndon heowcūþe đe wē geseón ne mǣgon *those things are familiar that we cannot see,* Blickl. Homl. 97, 23. Híwcūþe, Bt. Met. Fox 10, 122 ; Met. 10, 61.

híw-cūþlíce, hiew- ; *adv. Familiarly :*—Đa đe hine híwcūþlíce cūþan *qui eum familiariter noverunt,* Bd. 5, 2 ; S. 614, 27. Hē biþ hiewcūþlíce þeów đæm Godes feónde *hosti Dei familiarius servit,* Past. 47, 2 ; Swt. 361, 1. Hine God hiewcūþlícor on eallum þingum innan lǣrde đonne ōđre menn mid his gelōmlícran tōsprǣce *quem de cunctis interius per conversationem cum Deo sedulam locutio familiaris instruebat,* 41, 5 ; Swt. 304, 18.

híwcūþ-rǣdness, e ; *f. Familiarity, intimacy ;* familiaritas, Ælfc. Gl. 116 ; Som. 80, 66 ; Wrt. Voc. 61, 40.

hiwe ; *adj. Beautiful in form* or *colour,* Exon. 60 a ; Th. 218, 8 ; Ph. 291 : Th. 219, 4 ; Ph. 302. [Cf. twí-hiwe : *or is* hiwe *dative of* hiw ?]

híwen, es ; *n. A family, household :*—Đa þing đe eówre híwenu beþurfon *cibaria domibus vestris necessaria,* Gen. 42, 33. Tó mete eówrum híwenum *in cibum familiis,* 47, 24. Híwen *domestici,* Mt. Kmbl. Rush. 10, 36. [*Orm.* hiwenn *a family.*] v. híwan.

hiwene [?] *discoloration,* Lchdm. iii. 126, 8.

hiwere, es ; *m. One who pretends, a hypocrite :*—Hiwere *simulator,* Ælfc. Gr. 85 ; Som. 73, 105 ; Wrt. Voc. 49, 12. Wā eów hiwerum *woe to you hypocrites,* Homl. Th. ii. 404, 17.

híw-gedál, es ; *n. A separation of man and wife, divorce :*—Hē sylle hyre hyra híwgedáles bōc *det illi libellum repudii,* Mt. Bos. 5, 31 : 19, 7 ; Mk. Skt. 10, 4. [Cf. *Icel.* hjóna-skilnaðr *a divorce.*]

hiwian ; *p.* ode ; *pp.* od. *To form, fashion, shape, colour, feign, pretend :*—Hiwian *colorare,* Ælfc. Gl. 99 ; Som. 76, 112 ; Wrt. Voc. 54, 54. Ic hiwige *fingo,* Ælfc. Gr. 28 ; Som. 31, 61. Đū hiwast swilce đū đínum cildum hit sparige *you make as if you are saving it for your children,* Homl. Th. ii. 104, 8. Ealle đe hiwiaþ hí wiđūtan mid eáwfæstum þeáwum and wiđinnan sind geǣttrode mid árleásnysse *all that fashion themselves outwardly with pious manners, but inwardly are poisoned with impiety,* 404, 13. Sum fǣmne hí hiwode sárlíce seóce *some woman feigned herself very ill,* 506, 5. Herodes hiwode hine sylfne unrōtne *Herod pretended to be troubled,* i. 484, 26. Đū hiwodest *formasti,* Blickl. Gl. : Ps. Spl. C. 138, 4 : 93, 9. Hiwgende lang gebed *simulantes longam orationem,* Lk. Skt. 20, 47. v. ge-hiwian.

híwian ; *p.* ode ; *To marry :*—Hie forbiódaþ mannum đæt hie híwien *prohibentium nubere,* Past. 43, 9 ; Swt. 318, 1.

hiwing. v. hiwung.

híwisc, hígwisc, es ; *n. A family, household, house ; also a hide of land* [v. híd] :—Fæder hiogwuisc, hiowisc, hiuwisc *paterfamilias,* Lk. Skt. Lind. Rush. 12, 39 : 13, 25 : 14, 21. Gif hē hæbbe híwisc *landes if he have a hide of land,* L. Wg. 7 ; Th. i. 186, 13 [cf. l. 23]. On Cotenes-

felde ấn hÿwysce and þðđer đễl of Branok hyalf hĩwisce *in Cotensfield & one hide, and the other part of Branok half a hide,* Chart. Th. 107, 26–8. Hĩwisc, 428, 17. God bebeád Moyse đæt hễ and eall Israhễla folc sceoldon offrian æt ælcum hĩwisce Gode ấn lamb ấnes geáres *God commanded Moses that he and all the people of Israel should offer a lamb of the first year to God from every family* [*a lamb for an house,* Ex. 12, 3], L. In. 44; Th. i. 130, 5.

hiw-leás; *adj. Wanting in form* or *in colour :* — Hiwleás *deformis,* Wrt. Voc. 72, 16. Hū hiwleáse hie beóþ *how colourless the patients are,* L. M. 2, 36; Lchdm. ii. 242, 2.

hiwleás-ness *want of form;* deformitas, Som.

hiw-lĩc; *adj. Having good form* or *colour, shapely;* formosus :— Ansĩne hiwlĩce hine habban fultum getácnaþ *to see one's self with a handsome face betokens support,* Lchdm. iii. 204, 8. Reáf hiwlĩc habban blisse getácnaþ *to have a handsome robe betokens bliss,* 212, 6. Hiwlĩc *figuratus,* Hpt. Gl. 432.

hĩw-lĩc *matronalis,* Cot. 129, Lye.

hĩw-rǽden, e; *f. A family, household, house, a religious house :—* Hÿwrǽden *domus,* Ælfc. Gl. 106; Som. 78, 66; Wrt. Voc. 57, 45. Godes wĩngeard is Israhễla hĩwrǽden *God's vineyard is the house of Israel,* Homl. Th. ii. 72, 31 : Mt. Kmbl. 10, 6. Gang in tõ đam arce and eall đĩn hĩwrǽden *ingredere tu et omnis domus tua in arcam,* Gen. 7, 1 : 50, 8. For bễnum abbodes and đære heórǽdene æt Bercleá *for the prayers of the abbot and of the brethren at Berkeley,* Chart. Th. 129, 30: 168, 24. Sib sĩ đisse hĩwrǽddenne *pax huic domui,* Lk. Skt. 9, 5 : 19, 9 : Gen. 28, 2 : Ex. 2, 1. Hĩwrǽdene underféhþ *familiam susceperit,* L. Ecg. P. ii. 16; Th. ii. 188, 2.

hĩw-scipe, híg-, es; *m. A family, household, house :—* Hĩwscype *domus,* Ps. Lamb. 113, 17. Wæs sum hĩwscipes fæder and hĩna ealdor *erat paterfamilias,* Bd. 5, 12; S. 627, 9. Đã onfễng heó ænes hĩwscipes stõwe *accepit locum unius familiæ,* 4, 23; S. 593, 18. Ealle hĩwscipes þeóda *universæ familiæ gentium,* Ps. Lamb. 21, 28. [*O. E. Hom.* of elchan hiwscipe, i. 87, 8. v. Ex. 12, 3.] DER. sin-hĩwscipe.

hiwung, hiwing, e; *f. Forming, shaping, form, figure, pretence, feigning, hypocrisy, dissimulation :—* Hễ ne biþ đonne geleáfa ac biþ hiwung *it is not then belief but hypocrisy,* Homl. Th. i. 250, 21. Hywung, ii. 220, 32. Gễ sind wiđinnan ấfyllẹde mid hiwunge and unrihtwĩsnysse *within ye are filled with hypocrisy and unrighteousness,* 404, 21. Đa leásan licceteras đe mid hiwunge God sếcaþ *the false hypocrites that seek God with outward show,* i. 120, 2. Hễ com mid hiwunge *he came with dissimulation,* Chr. 1049; Erl. 172, 32. Mid đære hiwunga đe hió lícet đæt hió síe gõd *mendacium specie bonorum,* Bt. 29; Fox 72, 1. Hĩ on fruman tõ Godes hiwunga gesceapene wǽron *in the beginning they were created in the image of God,* Blickl. Homl. 61, 7. Þurh hiwwinge *per figuras,* Num. 12, 8. Đæt hluttre mõd đe Gode gelícaþ forsihþ đa hiwunga and healt sõđfæstnysse *the pure mind that pleases God despises pretences and holds the truth,* Basil admn. 5; Norm. 46, 8. Þurh deófles hiwunga *per diaboli figmenta,* L. Ecg. C. iii. 14; Th. ii. 202, 5. v. hiwian.

hĩwung, e; *f. Marriage :—* Mid his hĩwunge and his geférena *with the marriage of himself and of his companions,* Ors. 2, 2; Swt. 64, 24. v. hiwian.

hladan; *p.* hlõd; *pp.* hladen. I. *to heap, pile up, build, place, lade, load, freight :—* Ic mễ hrycg hlade đæt ic habban sceal *I load my back with what I am to have,* Exon. 102 a; Th. 386, 21; Rä. 4, 65. Wyrd wõp wecceþ weán hladeþ *fate awakens grief, heaps up misery,* Salm. Kmbl. 874; Sal. 436. Wễ gelíce sceolon leánum hleótan swã wễ weorcum hlõdun *we shall obtain rewards according as we built with our deeds* [cf. 1 Cor. 3, 12–14], Exon. 19 a; Th. 49, 12; Cri. 784. Hlõdan *they loaded,* 106 a; Th. 404, 19; Rä. 23, 10. Ongan đã ãd hladan *began then to build the pile,* Cd. 140; Th. 175, 25; Gen. 2901. Hÿ ne mõston on bæl hladan leófne mannan *they might not place the beloved man on the pile,* Beo. Th. 4259; B. 2126. Him on bearm hladan bunan and discas *to heap up in his bosom cups and dishes,* 5543; B. 2775. Naca hladen herewǽdum *the bark laden with war weeds,* 3798; B. 1897. Wæs wunden gold on wǽn hladen *twisted gold was laden on the wain,* 6260; B. 3134. Hærfest wæstmum hladen *autumn laden with fruits,* Menol. Fox 281; Men. 142. II. *to lade, draw* [*water*]; haurire :— Ic hlade *haurio,* Ælfc. Gr. 30; Som. 34, 40. Swã hwæt swã đũ hlætst of đam flõde *quidquid hauseris de fluvio,* Ex 4, 9. Hễ hlõd wæter mid ũs *hausit aquam nobiscum,* 2, 19. Đã mid ãne helme hlõd hit, Nar. 8, 3 : Homl. Th. ii. 118, 21. Đa þẽnas đe đæt wæter hlõdon *ministri qui haurierant aquam,* Jn. Skt. 2, 9. Hladaþ *haurite,* 8. Hlade đonne mid đære ylcan hand đæs wæteres mũþ fulne *let him then take up with the same hand a mouthful of the water,* Lchdm. iii. 68, 15 : 74, 16. Wæter tõ hladanne *ad hauriendam aquam,* Ex. 2, 16. Ne đũ næfst nãn þing mid tõ hladenne *neque in quo haurias habes,* Jn. Skt. MS. A. 4, 11. Gemễtte ænne ealdne munuc wæter hladende *found an old monk drawing water,* Homl. Th. ii. 180, 7. [*Orm.* lodenn. *p. pl*; lãdenn, *pp. to draw* (water) : *Ayenb.* lhade : *Prompt. Parv.* ladyñ i. *onero, sarcino*; ii. *vatilo* : *Goth.* hlaþan *to load* : *O. Sax.* hladan (*like A. Sax.*) : *O. Frs.* hlada *to lade :* *Icel.* hlaða *to lade, pile up, build :* *O. H. Ger.* hladan *onerare, ponere :* *Ger.* laden.] DER. ã-, ge-, tõ-hladan.

hladung, e; *f. A drawing,* haustus, Som.

hlæd, es; *n. A heap, pile, mound :—* Beraþ hiere hlæd tõ *comportabis aggerem,* Past. 21, 5; Swt. 161, 5 : 163, 10, 11. [*Icel.* hlað; *n:* hlaði; *m. a pile, stack.*] v. hladan.

hlædder. v. hlæder.

hlæd-disc, es; *m. A dish on which many things are heaped up* [?]; satura [MS. satira], Ælfc. Gl. 30; Som. 61, 69; Wrt. Voc. 26, 66. v. hlæd.

hlædel, es; *m. An instrument for drawing water, a ladle;* antlia, Hpt. Gl. 418. [*Chauc. Piers P.* ladel.] v. hladan.

hlæden *a vessel for drawing water, a bucket;* hauritorium, Ælfc. Gl. 25; Som. 60, 54; Wrt. Voc. 24, 50.

hlæder, hlædder, e; *f:* hlæddre, an [?]; *f. A ladder, flight of steps;* scala :— Đã geseah hễ on swefne standan ắne hlædre and godes englas up stĩgende and nyđer stĩgende on đære hlædre *viditque in somnis scalam stantem, angelos quoque dei ascendentes et descendentes per eam,* Gen. 28, 12, 13 : Past. pref; Swt. 23, 17 : Exon. 114 a; Th. 437, 11; Rä. 56, 6. On hlædran sittan, Lchdm. iii. 210, 23. Tõ heofnum up hlædra rǽrdon *they raised ladders up to the heavens,* Cd. 80; Th. 101, 1; Gen. 1675. Hie æfter hlæddrum ãstigon *they mounted by steps,* Blickl. Homl. 209, 7. [*Ayenb.* lheddre : *Piers P.* laddre : *O. Frs.* hladder, hleder : *O. H. Ger.* hleitar, leitara : *Ger.* leiter.]

hlæder-wyrt, hlædder-, e; *f. Ladder-wort, ladder to heaven, Jacob's ladder;* polemonium cæruleum or polygonatum multiflorum [v. E. D. S. No. 26, 'ladder to heaven'], Lchdm. iii. 8, 25.

hlæd-hweól, -weogl, -wiogl, es; *n. A wheel used in drawing water;* antlia, Cot. 9, 101, Lye.

hlæd-trendel, es; *m. A wheel used in drawing water;* rota hauritoria, Hpt. Gl. 418.

hlǽfdige, hlǽfdie, an; *f. A lady, mistress of a house;* after Bertric's time it is the title given to the wife of the West-Saxon king. v. William of Malm. bk. ii. c. 2 :—Hlǽfdige, *domina,* Wrt. Voc. 72, 79. Hĩredes hlǽfdige *materfamilias,* 73, 21. Gif hwylc wĩf hire wĩfman swingþ and heó þurh đa swingle wyrþ deád and heó unscyldig biþ fæste seó hlǽfdige vii geár *si mulier aliqua ancillam suam flagellis verberaverit et ex illa verberatione moriatur, et innocens sit, domina vii annos jejunet,* L. Pen. ii. 4; Th. ii. 184, 2. Cristes þegnas cweþaþ đæt đũ síe hlǽfdige wuldorweorudes *Christ's servants say that thou* [*the Virgin Mary*] *art the queen of the glorious host,* Exon. 124 a; Th. 18, 15; Cri. 284. Hlǽfdige mĩn *O lady mine!* Elen. Kmbl. 1309; El. 656. Đã com seó hlǽfdige hider tõ lande *then came the lady* [*Ethelred's wife*] *to this country,* Chr. 1002 : Erl. 137, 30 : 1013; Erl. 149, 29. Æþelflæd Myrcena hlǽfdige, 918; Erl. 103, 1 [cf. Henry of Hunt. 'Hæc igitur domina tantæ potentiæ fertur fuisse, ut a quibusdam, non solum domina vel regina sed etiam rex vocaretur']. On þÿs ilcan geáre forþférde seó ealde hlǽfdige Eádwerdes cinges mõder *in this same year departed the old lady, the mother of king Edward,* 1051; Erl. 176, 19. Cnut cyncg and Ælfgifu seó hlǽfdige, Chart. Th. 328, 20. Swã eágan gãþ earmre þeówenan đonne heó on hire hlǽfdigean handa lõcaþ *sicut oculi ancillæ, in manibus dominæ suæ,* Ps. Th. 122, 3 : Cd. 103; Th. 137, 13; Gen. 2273. Agar forseah hire hlǽfdian *Agar despexit dominam suam,* Gen. 16, 4. Đã forlét se cyng đa hlǽfdian seó wæs gehãlgod him tõ cwễne [*of Eward putting away his wife, Godwin's daughter*], Chr. 1048; Erl. 180, 20. Him tõ wĩfum dydon đa đe ǽr wǽron heora hlǽfdian *those who before had been their mistresses, they made their wives,* Ors. 4, 3; Bos. 80, 6. [*Laym.* lafdi, leafdi : *Orm.* lafidiȝ : *Ayenb.* lhevedi : *Chauc. Piers P.* lady, ladi.] v. hlãford.

hlǽnan; *p. de To cause to lean, to incline :—* Siđđan hÿ tõgædere gãras hlǽndon *after they had inclined their spears together,* Exon. 66 b; Th. 246, 18; Jul. 63. DER. ã-, bi-hlǽnan.

hlǽne; *adj. Lean, meagre;* macer :— Hlǽne *macer,* Ælfc. Gl. 89; Som. 74, 102; Wrt. Voc. 51, 15. Oxan fũle and swíđe hlǽne *boves fædæ confectæque macie,* Gen. 41, 3. Nũ wễ sind hlǽne *anima nostra arida est,* Num. 11, 6 : Ors. 4, 13; Bos. 100, 25. [*Laym. Piers P. Chauc.* lene.]

hlǽnian; *p.* ode *To make lean* or *to become lean :—*Đæt hễ his lĩchoman hlǽnige *ut caro maceretur,* Past. 14, 6; Swt. 87, 17. Đonne đæt flæsc hlǽnaþ *dum carnem macerant,* 43, 6; Swt. 313, 20. [*Prompt. Parv.* lenyñ *or* make lene *macero.*]

hlǽnnes, -ness, e; *f. Leanness :—*Hlǽnnes *macies* vel *tabitudo,* Ælfc. Gl. 89; Som. 74, 104; Wrt. Voc. 51, 17. Mõdes hlǽnnys *leanness of the mind,* Homl. Th. i. 522, 31.

hlǽnsian; *p.* ode *To make lean;* macerare, castigare, Hpt. Gl. 433. [Cf. *O. E. Hom.* 'Carnis maceratio fleises lensing. Mon lenseþ his fleis hwenne he him ȝefeđ lutel to etene,' i. 147.]

hlǽst, es; *n. Burden, freight, lading :—* Eów is holmes hlǽst and heofonfuglas and wildu deór on geweald geseald *into your power is given the ocean's freight* [*fishes*] *and the fowls of the air and wild beasts,* Cd. 74; Th. 91, 20; Gen. 1515. Hwã đæm hlǽste onfễng *who received that freight,* Beo. Th. 104; B. 52 : Cd. 71; Th. 85, 29; Gen. 1422.

Hlæst beran *to bear a burden*, Exon. 101 a ; Th. 381, 23 ; Rä. 2, 15. Ic ástíge mín scyp mid hlæstum mínum *ego ascendo navem cum mercibus meis*, Coll. Monast. Th. 26, 31. [*Chauc.* last : *Prompt. Parv.* leste, nowmbyr, as heryngys, and other lyke *legio* : *O. Frs.* hlest : *Icel.* hlass *a cart-load* : *Ger.* last *onus.*] v. hladan, brim-hlæst.

hlæstan. v. ge-hlæstan.

hléw, hláw, hláu, hléw, es ; *m.*   **I.** *a low or law* [*occurring in names of places*], *a rising ground, an artificial as well as a natural mound, a funeral mound* ; tumulus :—Wæs ðær on ðam eálande sum hláw mycel· ofer eorþan geworht, ðone ylcan men for feós wilnunga gedulfon and brǽcon *there was on the island a great mound made upon the earth, which same from the desire of treasure men had dug into and broken up*, Guthl. 4 ; Gdwin. 26, 5, 7 : Beo. Th. 2244 ; B. 1120. Dá hý ofer ðone hléw ridan *when they rode over the hill*, Lchdm. iii. 52, 14. Hátaþ hléw gewyrcean se sceal tó gemyndum mínum leódum heáh hlifian on Hrones næsse, ðæt hit sǽliðend syððan hátan Biówulfes biorh *bid them make a mound* ; *it shall as a memorial to my people tower high on Hronesness, so that hereafter may seafarers call it Beowulf's mount*, Beo. Th. 5597 ; B. 2802 : 6295 ; B. 3158 : 6319 ; B. 3170. Geworpene on wídne hléw *projecti in monumentis*, Ps. Th. 87, 5. On hwelcum hléwa hrusan þeccen bán Wélandes *in what tomb do Weland's bones cover the ground?* Bt. Met. Fox 10, 85 ; Met. 10, 43. Beorgas ðær ne muntas steápe ne stondeþ, ne stánclifu heáh hlifiaþ ne dene ne dalu ne dunscrafu hléwas ne hlincas *nec tumulus crescit, nec cava vallis hiat*, Exon. 56 a ; Th. 199, 13 ; Ph. 25. The word is found in local names, e.g. Cwicchelmes hléw, Chr. 1006 ; Erl. 140, 21 [for other examples see Cod. Dipl. Kmbl. iii. xxxi], and exists still in the forms -*low*, as Ludlow, Hounslow ; and -*law*, frequently applied to hills in Scotland. [Cf. *Icel.* haugr *a mound, funeral mound* ; *how* in local names.]   **II.** *the interior of a mound, a cave* :—Draca sceal on hléwe *a serpent shall dwell in a cave*, Menol. Fox 512 ; Gn. C. 26 : Beo. Th. 5539 ; B. 2773. Eorþsele hléw under hrusan *an earth-hall, a cave under ground*, 4813 ; B. 2411. [*Orm.* illc an lawe & illc an hill : *Havel.* lowe : *Goth.* hlaiw *a grave, tomb* ; hlaiwasna *grave, sepulchre* : *O. Sax.* hléwe (*dat.*) *grave* : *O. H. Ger.* hlaeo *mausoleum* ; laeo *acervus* ; hléo *agger* ; léuua *aggeres.*]

HLÁF, es ; *m. Bread, food, a loaf* :—Gehafen hláf *fermentacius panis* : ceorlisc hláf *cibarius* : geseórid hláf *acrizimus panis* : hwǽten hláf *siligeneus* vel *triticeus* : heorþbacen hláf *subcinericius* vel *focarius* : ofenbacen hláf *clibanius* : gehyrst hláf *frixius panis*, Ælfc. Gl. 66 ; Som. 69, 59–69 ; Wrt. Voc. 41, 15–23. Litel hláf *pastillus* : ofenbacen hláf *fermentum*, 31 ; Som. 61, 84, 94 ; Wrt. Voc. 27, 14, 24. Him hylpþ eác ofenbacen h'láf, L. M. 2, 27 ; Lchdm. ii. 222, 17. Smæl hláf *artolaganus*, Cot. 21, Lye. Tú hund greátes hláfes and þridde smales *two hundred* [*loaves?*] *of coarse bread, and a third of fine*, Chart. Th. 158, 25. Hwítes hláfes cruman *crumbs of white bread*, L. M. 1, 2 ; Lchdm. ii. 34, 21. Ne sý neáta cwyld ne ádl ne hláfes hungor *let there not be murrain among cattle, or disease, or lack of food*, Shrn. 104, 27. Sing ðis on ánum berenan hláfe and syle ðan horse etan *sing this over a barley loaf and give it the horse to eat*, Lchdm. iii. 68, 31 : Blickl. Homl. 179, 31 : Jn. Skt. 6, 9. Man sceolde dón dǽdbóte on hláfe and on wætere *poenitentia sit agenda in pane et aqua*, L. Ecg. C. 2 ; Th. ii. 134, 4. Úrne dæghwamlícan hláf syle ús tódæg *give us to-day our daily bread*, Mt. Kmbl. 6, 11. Mid Grécum diáconas ne móton brecan gehálgodne hláf *apud Græcos diaconis non licet frangere panem sanctum*, L. Ecg. C. 35 ; Th. ii. 160, 9 : L. M. 3, 41 ; Lchdm. ii. 334, 22 : L. Edg. C. 43 ; Th. ii. 254, 1. For hwon ne rǽcst ðú ús ðone hwítan hláf ðone ðú sealdest Saban *quare non nobis porrigis panem nitidum quem Saba dabas*, Bd. 2, 5 ; S. 507, 14. Cyse and drygne hláf *cheese and dry bread*, L. M. 2, 26 ; Lchdm. ii. 278, 21. Hláf wexenne *a wax plaster*, Lchdm. iii. 210, 1', 2. Gesufelne hláf, L. Ath. V. 8, 6 ; Th. i. 236, 36. Ðeorfe hláfas *unleavened loaves*, Homl. Th. ii. 264, 3. cxx. huǽtenra hláfa and xxx. clénra *one hundred and twenty wheaten loaves and thirty made without br·an*, 460, 16. cxx gesuffra hláfa, 32 : 469, 3. On xii mónþum ðú scealt sillan dínum þeówan men vii hund hláfa and xx hláfa búton morgenmetum and nónmetum *in twelve months thou shalt give thy slave-man seven hundred and twenty loaves, besides meals at morn and noon*, Salm. Kmbl. 192, 18. Cweþ ðæt ða stánas tó hláfum geweorþan *tell the stones to become loaves*, Blickl. Homl. 27, 7. [*Orm.* laf : *Laym.* laves, *pl* : *Ayenb.* lhove : *Goth.* hlaibs : *Icel.* hleifr : *O. H. Ger.* hlaiba, leib *panis, tortella* : *Ger.* laib.] DER. heofon-, offrung-hláf.]

hláf-ǽta, an ; *m. A loaf-eater, domestic, servant* :—Ceorles hláfǽta *a 'ceorl's' servant*, L. Ethb. 25 ; Th. i. 8, 10. [Cf. hláford, and *v.* (?) *under* hláf *the passage from* Salm. Kmbl. 192, 18.]

hláf-gang, es ; *m. The procession with the host*, L. Eth. vii. 27 ; Th. i. 334, 34.

hláf-gebrece, es ; *n. A fragment of bread* :—Swá hláfgebrece *sicut frustum panis*, Ps. Th. 147, 6.

hláf-gebroc, es ; *n. A fragment of bread* :—Ðara hláfgebroca wæs tó láfe twelf binna fulle *of the fragments there remained twelve baskets full*, Shrn. 48, 31.

hláf-hwǽte, es ; *m. Wheat for making bread*, Chart. Th. 144, 34.

hláf-leást, e ; *f. Lack of bread* :—For ðære hláfleáste ða eorþan ǽton *for lack of bread they ate the earth*, St. And. 34, 20.

hláf-mæsse, -messe, an ; *f. Lammas, a name for the first of August* :—Ðæt wæs on ðære tíde calendas Agustus on ðæm dæge ðe wé hátaþ hláfmæsse *it was on the first of August, on the day that we call Lammas*, Ors. 5, 13 ; Swt. 246, 17. On ðære nihte ðe gé hátaþ Hláfmesse *on the day that you call Lammas*, Homl. Th. ii. 384, 11. Bringeþ Agustus Hláfmæssan dæg *August brings Lammas-day*, Menol. Fox 277 ; Men. 140. Betwix hláfmæssan and middum sumera *be'ween Lammas and midsummer*, Chr. 921 ; Erl. 106, 5. Tóforan Hláfmæssan, 1101 ; Erl. 237, 24. Æfter hlámmessan, 1009 ; Erl. 142, 16. Tó Lámmæssan, 1085 ; Erl. 219. 3. [*Piers P.* lammasse : *Prompt. Parv.* lammasse *festum agnorum* vel *Festum ad vincula Sancti Petri.*] v. next word, and hláf-sénung.

hláfmæsse-dæg, es ; *m. Lammas-day, the first of August* : —Of ðam gehálgedan hláfe ðe man hálige on hláfmæssedæg *from the hallowed bread which is hallowed on Lammas-day*, Lchdm. iii. 290, 27. Ær hláfmæsse [dæge ?], L. M. 1, 72 ; Lchdm. ii. 146, 9. Æfter hlámmæssedæge. Chr. 1100 ; Erl. 235, 33.

hláford, es ; *m. A* LORD ; dominus, herus :—Hláford *heros*, Ælfc. Gl. 87 ; Som. 74, 46 ; Wrt. Voc. 50, 28. Scipes hláford *nauclerus*, 83 ; Som. 73, 66 ; Wrt. Voc. 48, 4. Hie cuædon ðæt him nǽnig mæg leófra nǽre ðonne hiera hláford *they said that no kinsman was dearer to them than their lord*, Chr. 755 ; Erl. 50, 20. Cwǽdon ðæt him nán leófra hláford nǽre ðonne heora gecynde hláford, 1014 ; Erl. 150, 6. Hé wæs ægðer mín mæg and mín hláford *he was both my kinsman and my lord*, Byrht. Th. 138, 23 ; By. 224. Ðæs þegenes lof is ðæs hláfordes wurþmynt. Sý lof ðam Hláforde ðe leofaþ on écnysse *the servant's praise is the Lord's honour. Praise be to the Lord that liveth for ever*, Homl. Th. ii. 562, 6. Sum sceal mid hearpan æt his hláfordes fótum sittan feoh þicgan *one shall sit with the harp at the feet of his lord, receive money*, Exon. 88 a ; Th. 332, 5 ; Vy. 80. Hine gecés tó hláforde Scotta cyning, *the king of Scots chose him as his lord*, Chr. 924 ; Erl. 110, 14. Tó hláforde geceósan *to elect king*, Ors. 3, 11 ; Bos. 74, 39. Óhthere sǽde his hláforde Ælfréde cyninge *Othhere said to his lord, king Alfred*, 1, 1 ; Bos. 19, 25. Ic geann mínum hláforde syxti mancusa goldes *I give to my lord sixty mancuses of gold*, Chart. Th. 516, 32. Úrum hláforde holde *loyal to our lord*, L. C. E. 20 ; Th. i. 372, 8. Ic mé be healfe mínum hláforde be swá leófan men licgan þence *beside my lord, by one so loved, I mean to lie*, Byrht. Th. 141, 7 ; By. 318 : Judth. 12 ; Thw. 25, 9 ; Jud. 251 : Andr. Kmbl. 823 ; An. 412. Heora hláford gewrecan *to avenge their lord*, Ors. 3, 9 ; Swt. 134, 30. Hé bebeád ðone hláford lufian swá hine selfne *he commanded to love the lord as himself*, L. Alf. 49 ; Th. i. 58, 13. Áhte ic fela wintra folgaþ tilne holdne hláford *I had for many years a good service, a gracious lord*, Exon. 100 b ; Th. 379, 26 ; Deór. 39. Álegdon ðá tómiddes mǽrne þeóden hláford leófne *they laid down in their midst the great prince, their beloved lord*, Beo. Th. 6276 ; B. 3142. Ða menn ða ðǽr hláfordas wǽron *the men that were lords there*, Chart. Th. 459, 16. Hláforda wín *honorarium vinum*, Ælfc. Gl. 32 ; Som. 62, 1 ; Wrt. Voc. 27, 67. Heó [*Hagar*] gewát hire hláfordum [*Abram and Sara*], Cd. 104 ; Th. 138, 21 ; Gen. 2295. [*Laym.* laverd : *Orm.* laferrd : *A. R.* loverd : *Proclam. H. III.* lhoaverd : *Ayenb.* lhord : *Piers P. Chauc.* lord.] DER. cyne-, eald-, hús-, worold-hláford.

hláford-dóm, es ; *m. Dominion, lordship* :—For Godes ege under ðæm geoke his hláforddómes þurhwunigen and hine for Godes ege weorþigen, suá mon hláford sceal *divino timore constricti ferre sub eis jugum reverentiæ non recusent*, Past. 28, 5 ; Swt. 197, 8. Se ðe on láreówes onlícnesse ða þenenga ðæs ealdordómes gecierþ tó hláforddóme *qui ex simulatione disciplinæ ministerium regiminis vertit in usum dominationis*, 17, 9 ; Swt. 121, 24. [*Orm.* laferrd-dom.]

hláford-gift *principatus*, Hpt. Gl. 412. [Cf. [?] Hlafordes gifu, L. Eth. iii. 3 ; Th. i. 292, 16, and see the Glossary.]

hláford-hyldo ; *f.* -hyld, -held [?] *m* ; or -hyldu, e ; *f. Fidelity to a lord, loyalty* :—Ac hí gecýðdon raðe ðæs hwylce hláford-hyldo hí þohton tó gecýðanne on heora ealdhláfordes bearnum *but soon after they shewed what kind of loyalty they intended to shew to the children of their late lord*, Ors. 6, 37 ; Bos. 132, 23. Eall ðæt wé ǽfre for riht-hláfordhelde dóþ *all that we ever do from true loyalty*, L. C. E. 20 ; Th. i 372, 10.

hláford-leás ; *adj. Lordless, not having a lord* :—Ætwítan mé ðæt ic hláfordleás hám síðie *to taunt me that I return home without my lord*, Byrht. Th. 139, 8 ; By. 251 : Exon. 105 b ; Th. 401, 35 ; Rä. 21, 22 : Beo. Th. 5863 ; B. 2935 : Andr. Kmbl. 810 ; An. 405. Be hláfordleásum mannum *concerning men who have no lord*, L. Ath. 1, 2 ; Th. i. 200, 4.

hláford-scipe, es ; *m. Lordship, rule* ; dominatio :—Hláfordscipe ðín *dominatio tua*, Ps. Spl. 144, 13. Hwí wæs Adame án treów forboden ðá ðá hé wæs ealles óðres hláford ? *why was one tree forbidden to Adam, when he was lord of every other? To the end that he might not exalt himself with so great lordship*, Boutr. Scrd. 17, 28. Ðú winsþ wið ðam hláfordscipe ðe ðú self gecure *you strive against the rule you have yourself chosen*, Bt. 7,

2 ; Fox 18, 30. Ðonne wē ágyltaþ wið ða hláfordas, ðonne ágylte wē wið ðone God ðe hláfordscipe gescóp *cum præpositis delinquimus, ejus ordinationi, qui eos nobis prætulit, obviamus*, Past. 28, 6 ; Swt. 201, 3 : 29 ; Swt. 201, 22. Dominationes sind hláfordscypas gecwedene, Homl. Th. i. 342, 32.

hláford-searu ; *f. n. Plotting against the life of a king or lord* :— Búton æt hláfordsearwe ðam hie náne mildheortnesse ne dorston gecwædan *except in cases of treason against a lord; to that they dared not assign any mercy*, L. Alf. 49 ; Th. i. 58, 9. Be hláfordsearwe. Gif hwá ymb cyninges feorh sierwie, síe hē his feores scyldig and ealles ðæs ðe hē áge *of plotting against a lord. If any one plot against the king's life, let him forfeit his life and all that he owns*, L. Alf. pol. 4 ; Th. i. 62, 14 : 1 ; Th. 60, 4 : L. Ath. i. 4 ; Th. i. 202, 1 : L. Edg. ii. 7 ; Th. i. 268, 23 : L. C. S. 26 ; Th. i. 392, 1. [Cf. L. Eth. v. 5 ; Th. i. 312, 5 : vi. 37 ; Th. i. 324, 16 : L. C. S. 58 ; Th. i. 408, 1.]

hláford-sócn, e ; *f. The 'seeking' a lord for the purpose of being in his service, and under his protection* [cf. hláford sēcan, L. Alf. pol. 37 ; Th. i. 86, 3 : L. Ath. iv ; Th. i. 220, 24] :—Ne dominus libero homini hlafordsoknam interdicat si eum recte custodierit, L. Ath. ii. 4 ; Th. i. 216, 25 : iii. 5 ; Th. i. 218, 25.

hláford-swica, an ; *m. A betrayer of his lord, a traitor to his lord* :— Se man ðe ðis gefæst ne þearf hē him ná ondrǽdan hellewítan bútan hē beó hláfordswica *the man that keeps this fast need not fear the pains of hell, unless he be a traitor to his lord*, Lchdm. iii. 228, 24. Hēr sind on earde on mistlíce wísan hláfordswican manige *here in the land are in divers manners many traitors*, Swt. A. S. Rdr. 107, 88 : 110, 176. [Laym. lauend-, louerd-swike *traitor*.]

hláford-swice, es ; *m. Treachery to a lord, treason* :—Ealra mǽst hláfordswice se biþ on worulde ðæt man his hláfordes sáwle beswíce and full mycel hláfordswice eác biþ ðæt man his hláford of lífe forrǽde oððe of lande lifigendne drife *the greatest treachery in the world against one's lord is to betray his soul, and very great treachery also is it to deprive him of life, or to drive him from the country alive*, Swt. A. S. Rdr. 107, 88. v. hláford-searu.

hláf-sēnung, e ; *f. Blessing of bread, which took place on August first or Lammas-day* : — On ðam ylcan dæge [Aug. 1] æt hláfsēnunga, Shrn. 112, 8. v. hláf-mæsse.

hlagol ; *adj. Apt to laugh*, Lye.

hlám-mæsse. v. hláf-mæsse.

hlanc ; *adj. Lank, lean, gaunt* :—Ðæs se hlanca gefeah wulf in walde *at that rejoiced the gaunt wolf in the wood*, Judth. 11 ; Thw. 24, 25 ; Jud. 205. Swá ðū on hríme setest hlance cylle *sicut uter in pruina*, Ps. Th. 118, 83.

hland, hlond, es ; *n. Urine*, Lchdm. i. 362, 18 : ii. 40, 20 : 156, 14. [Icel. hland.]

hláw. v. hlǽw.

hleahtor, hlehter, es ; *m. Laughter* : — Hleahter *risus*, Wrt. Voc. 83, 35. Ða gesíðas wóp and hleahtor *the comrades weeping and laughter*, Salm. Kmbl. 695 ; Sal. 347 : Beo. Th. 1226 ; B. 611. Hie habbaþ suá micle mēde óðerra monna gódra weorca suá wē habbaþ ðæs hleahtres ðonne wē hliehaþ gligmonna unnyttes cræftes *sic eis virtutum sanctitas, sicut stultis spectatoribus ludicrarum artium vanitas placet*, Past. 34, 2 ; Swt. 231, 6. Dū ús gesettest tó hleahtre and tó forsewennesse eallum ðǽm ðe ús ymbsittaþ *posuisti nos derisu et contemptu his qui in circuitu nostro sunt*, Ps. Th. 43, 15. Hē wæs heáfde becorfen for scondfulles gebeórscypes hleahtre *he [John the Baptist] had his head cut off for the amusement of a shameful feast*, Shrn. 123, 8. Be hleahtre ðe of milte cymþ *of laughter that cometh from the spleen*, L. M. 2, 36 ; Lchdm. ii. 142, 21. Ne sceal sprecan ýdelu word ða ðe unnytte hleahtor up áhebben ne hē eác sceal lufigean micelne and ungemetlícne cancettende hleahtor, L. E. I. 21 ; Th. ii. 416, 35. Se herewísa hleahtor álegde *the host's leader hath put away laughter* [*is dead*], Beo. Th. 6033 ; B. 3020. Hleahtor álegdon ðá hí swíðra oferstág weard *they put away laughter when a stronger guard had overcome them*, Exon. 35 b ; Th. 116, 1 ; Gū. 200. God mē worhte hlehter *risum fecit mihi deus*, Gen. 21, 6, Hwǽr beóþ ða ungemetlícan hleahtras *where are the immoderate laughings*, Blickl. Homl. 59, 18 : 195, 15. [Laym. lehtre : A. R. leihtres, pl : Icel. hlátr : O. H. Ger. hlahter *risus*.]

hleahtor-bǽre ; *adj. Given to laughter*, Lye.

hleahtor-full ; *adj. Scornful, derisive* :—Geþence ǽlc ðara tǽlendra and hleahterfulra *let every one that blames and derides reflect*, Guthl. prol. ; Gdwin. 2, 14.

hleahtor-líc ; *adj. Ridiculous* :—Gif hē hēr hwylc hleahterlíc word onfinde *if he here find any ridiculous word*, Guthl. prol. ; Gdwin. 2, 12.

hleahtor-smiþ, es ; *m. One who causes laughter, mirth, joy* :—Wóp wæs wíde worulddreáma lyt wǽron hleahtorsmiþum handa belocne *widespread was the wailing and little of this world's joys, the hands of those who wrought laughter were closed*, Cd. 144 ; Th. 180, 10 ; Exod. 43.

hleápan ; *p.* hleóp, *pl.* hleópon *and* hlupon [cf. Icel. hlupu]; *pp.* hleápen *To* LEAP, *jump, dance, run* :—Ic hleápe *salio*, Ælfc. Gr. 30 ; Som. 34, 45. Ðonne hleápþ se healta swá swá heort *the lame shall leap*

as a hart, Homl. Th. ii. 16, 18. Se ðe hleápeþ *he who dances*, Exon. 88 b ; Th. 332, 11 ; Vy. 83. Hē hleóp on ðæs cyninges stēdan *ascendens emissarium regis*, Bd. 2, 13 ; S. 517, 9 : 3, 9 ; S. 534, 3. Roger hēt án of heom se hleóp intó ðam castele æt Norþwíc *Roger was the name of one of them, he threw himself into the castle at Norwich*, Chr. 1087 ; Erl. 224, 34. „Hēr Eádwine eorl and Morkere eorl hlupon út and mislíce fērdon on wuda and feldon *in this year earl Edwin and earl Morcere fled away and went different ways through wood and open country*, 1072 ; Erl. 210, 26. Ðæt hie ne hliépen unwillende on ðæt scorene clif unþeáwa *per multa, quæ non appetunt, iniquitatum abrupta rapiuntur*, Past. 33, 1 ; Swt. 214, 7. Lege on ða wunde gyf heó tósomne hleápan wolde *lay on the wound if it be ready to close up* [cf. Icel. sáríð var hlaupit í sundr], Herb. 90, 13 ; Lchdm. i. 198, 2. Hwílum hleápan léton on geflit faran fealwe mearas *at times they made their fallow steeds run, contend on the course*, Beo. Th. 1733 ; B. 864. Hē á wæs gangende and hleápende *ambulans et exsiliens*, Bd. 5, 2 ; S. 615, 23. Heó him beforan hleápende wæs *the hind kept running before them*, Lchdm. iii. 426, 32. Herodes swór ðæt hē wolde ðǽre hleápendan dēhter forgyfan swá hwæt swá heó bǽde *Herod swore that he would give the dancing daughter whatever she asked*, Homl. Th. i. 452, 34. [Laym. lepen ; *p. pl.* leopen, lupen : Orm. læpen ; *subj. p.* lupe : Ayenb. lheape ; *p.* lhip : Piers P. lepen ; *p. pl.* lope : Chauc. lepe ; *p.* lep, leep : Goth. us-hlaupan *to leap up* : O. Sax. a-hlôpan : O. Frs. hlâpa : Icel. hlaupa *to leap*; also *to run* : O. H. Ger. hlaufan currere : Ger. laufen.] DER. á-, æt-, be-, ge-, ofer-hleápan.

hleápere, es ; *m. A leaper, dancer, runner, courier* :—Hleápere *saltator*, Wrt. Voc. 73, 70. Tuegen hleáperas Ælfrēd cyning sende mid gewritum king Alfred *sent two couriers with letters*, Chr. 889 ; Erl. 86, 23. [Prompt. Parv. lepare *or* rennare *cursor* : Scot. land-louper : Icel. hlaupari *a courser, charger* : O. H. Ger. loufari *circumcellio, cursor* : Ger. laufer.]

hleápestre, an ; *f. A dancer* ; *saltatrix*, Wrt. Voc. 73, 71.

hleápe-wince, an ; *f. The lap-wing* : — Hleápewince *cucurata*, Wrt. Voc. 62, 22 : *cucu*, 280, 27. [Ayenb. lhap-wynche : Gower. lappewinke : Prompt. Parv. lappe-wynge, lap-wynke *upipa* : Wick. lap-, leep-winke.]

hleápung, e ; *f. Leaping, dancing* : — Herodias swá mǽres mannes deáþ tó gife hire dēhter hleápunge underfēng *Herodias received as a gift for her daughter's dancing the death of so illustrious a man*, Homl. Th. i. 488, 3 : 480, 35.

hlec ; *adj. Having cracks or rents* :—Hlec, *rimosus, scissurosus*, Hpt. Gl. 529. Swíðe lytlum síceraþ ðæt wæter and swíðe dēgellíce on ðæt hlece scip, and deáh hit wilnaþ ðæs ilcan ðe sió hlúde ýþ dēþ on ðære hreón sǽ búton hit mon ǽr útáweorpe *hoc agit sentina latenter excrescens, quod patenter procella sæviens*, Past. 57, 1 ; Swt. 437, 15.

hlecan ; *p.* hlæc [?] *To join, unite, cohere* :—Swá eác his folgeras swá hie unwiðerweardran and gemódran beóþ swá hie swíður hlecaþ tósomne and eác fæstor tósomne beóþ gefégde tó gódra manna hiénþe *sequaces quippe illius, quo nulla inter se discordiæ adversitate divisi sunt, eo in bonorum gravius nece glomerantur*, Past. 47, 3 ; Swt. 361, 20.

hlēda, hlēde ; *m. A seat* : — Ðes hlēda, hlēde *sedile*, Ælfc. Gr. 9, 2 ; Som. 8, 26.

hleglende [= hlegiende, cf. (?) hlehhan *or* hlēgiende, cf. (?) hlówan] *sonans*, Cot. 24, Lye.

hlehhan, hlæhan, hlihhan, hlichan, hlihan, hlihgan ; *p.* hlóh ; *pl.* hlógan *To* LAUGH [*with gladness or contempt*], *to deride* :—Ic hliche *rideo*, Ælfc. Gr. 26, 3 ; Som. 28, 53 : 47 ; Som. 47, 15. Hē gedēþ ðæt wē hlihhaþ on morgen *ad matutinum lætitia*, Ps. Th. 29, 5. Eádgo ða ðe nū gie woepeþ forðon gie hlæheþ *beati qui nunc fletis quia ridebitis*, Lk. Skt. Lind. 6, 21. Wǽ iúh ða ðe hlǽhas forðon gie woepaþ *væ vobis qui ridetis nunc quia lugebitis*, 25. Hlihaþ, Homl. Th. i. 180, 14. Hlihaþ, Blickl. Homl. 25, 23. Hliehaþ, Past. 27 ; Swt. 187, 19. Ðonne wē hliehaþ gligmonna unnyttes cræftes *when we laugh at the useless art of gleemen*, 34, 1 ; Swt. 231, 7. Ne hlóh ic ná . . . . ac ðū hlóge *non risi . . . . sed risisti*, Gen. 18, 15. Ðū hlóge and ic weóp *thou didst laugh and I wept*, L. E. I. pref ; Th. ii. 398, 15. Se eorl wæs ðē blíðra hlóh ðá, Byrht. Th. 136, 6 ; By. 147 : Judth. 10 ; Thw. 21, 17 ; Jud. 23 : Cd. 33 ; Th. 45, 10 ; Gen. 724. Hlógun ł tēldon hine *deridebant eum*, Lk. Skt. Lind. 8, 53. Ða apostoli hlógon ðæra deófla leásunga and se ealdorman cwæþ mē stent ege ðysse andsware and gē hlihaþ *the apostles laughed at the devils' lying words, and the general said 'Fear comes upon me at this answer, and you laugh'*, Homl. Th. ii. 482, 25. Ealle geseónde mē hlógon on bysmor *omnes videntes me deriserunt me*, Ps. Lamb. 21, 8 : Exon. 120 a ; Th. 160, 22 ; Hö. 21. Deáh ðē mon hwylces hlihge and ðū ðē unscyldigne wite ne rēhst ðū hwæt hý rǽdon hý teóþ ðæt ðæs ðe hý sylfe habbaþ *though you are derided* [*or blamed?*] *for anything, and know yourself to be innocent, you shall not care what they say ; they accuse you of what they have themselves*, Prov. Kmbl. 12. Hē sǽde ðæt hē gesáwe crist selfne and ðæt hē him hlóge tó *he said that he saw Christ himself, and that he smiled upon him*, Shrn. 70, 9. Hlehhan ne þorftun *they had no*

need to laugh, Chr. 937; Erl. 114, 13; Æðelst. 47. Ne þorfton hlúde hlihhan, Cd. 4; Th. 5, 17; Gen. 73. Hwæt sceal ic ðonne búton hliehchan [Cot. MS. hliehhan] ðæs ðonne gē tô lose weorþaþ *what shall I do but laugh at it, when you come to ruin*; ego quoque in interitu vestro ridebo, Past. 36, 1; Swt. 249, 1. Forðon hí hlyhhan mǽgen *for this reason they can laugh*, L. M. 2, 36; Lchdm. ii. 242, 24. Ða deóflu sægdon hlúde hlihhende *the devils said, laughing loudly*, Homl. Th. ii. 350, 30 : i. 376, 5 : Herb. 9; Lchdm. i. 98, 27. Hlichende, Ælfc. Gr. 48; Som. 49, 18. Mid hlihendum múþe *with a smile on his lips*, Homl. Th. i. 428, 34: Elen. Kmbl. 1986; El. 995. Ðǽm hlæhendum *ridentibus*, Lk. Skt. p. 5, 7. [*Orm.* lahh3henn : *Laym.* leh3en, lih3en; *p.* loh, *pl.* lo3en : *A. R.* lauhwen : *Ayenb.* lhe33e : *Piers P. Chauc.* laughen : *Wick.* la3hen, lei3e; *p.* lei3ede : *Goth.* hlahjan; *p.* hlôh : *O. Sax.* hlahan; *p.* hlôg; *pp.* hlagan : *O. Frs.* hlaka; *p.* hlackade : *Icel.* hlæja; *p.* hló, *pl.* hlógu; *pp.* hleginn : *O. H. Ger.* hlahan; *p.* hlôc : *Ger.* lachen.] DER. ã-, be-, bi-hlehhan.

hlehter. v. hleahtor.

hlem, mes; *m. A sound, noise, crash :*—Nán monn ne gehiérde ne æxe hlem ne biétles suēg *absque mallei sonitu*, Past. 36, 5; Swt. 253, 17. [*Cf. Icel.* hlam; *n. a dull, heavy sound*; hlamman *crash, din.*] DER. hilde-, inwit-, uht-, wæl-hlem.

hlemman, *p.* de *To cause to sound, to clash :*—Hē ymbe ða herehúþe hlemmeþ tôgædre grimne gōman *about the prey he clashes his fierce jaws together*, Exon. 97 b; Th. 363, 30; Wal. 61. [*O. Sax.* hlamon : *Icel.* hlamma : *O. H. Ger.* hlamon *crepitare.*] v. hlimman.

hlenca or **hlence,** an; *m.* or *f. A link, a chain of links, a coat of mail formed with links or rings* [cf. hringlocen serce *and other compounds of* hring] :—Moyses bebeád frecan árísan habban heora hlencan beran beorht searo *Moses bade the warriors arise, take their coats of mail, bear their bright arms*, Cd. 153; Th. 191, 21; Exod. 218. Cf. L. M. 3, 55; Lchdm. ii. 342, 4; gif men sió heáfodpanne beó gehlenced *if a man's skull seem to be iron-bound.* [*Icel.* hlekkr; *m. a link, a chain of links* : *Dan.* lænke.] v. wæl-hlenca.

hlenor-teár, es; *m. Hyssop :*—Hlenorteáre *hyssopo*, Ps. Lamb. 508.

hleó. v. hleów.

hleomoc, hleomoce, an; *f. Brook-lime*, Lchdm. Gloss. ii. iii.

hleón. v. hleówan.

hleonaþ, hleonian, hleonung. v. hlinaþ, hlinian, hlinung.

hleór, es; *n. A cheek, face :*—Hleór *malæ*, Ælfc. Gl. 71; Som. 70, 79; Wrt. Voc. 43, 12. Hleór *maxilla*, Wrt. Voc. 70, 38: *facies*, 282, 37: Exon. 90 a; Th. 337, 18; Gn. Ex. 66 : 29 a; Th. 88, 5; Cri. 1435. On ðám nôsum oððe on ðam hleóre *on the nose or on the cheek*, Herb. 2, 18; Lchdm. i. 86, 2: L. Ethb. 46; Th. i. 14, 11. Dô his hleór xxx síðum tô eorþan *vultum suum xxx vicibus ad terram inclinet*, L. Ecg. C. 5; Th. ii. 138, 8 : Exon. 37 b; Th. 122, 13; Gú. 305 : Elen. Kmbl. 2195; El. 1099: Cd. 107; Th. 140, 33; Gen. 2337. [*Laym.* leores, *pl :* A. R. leor : *Piers P.* lere : *O. Sax.* hlior, hlier, hlear, hleor : *Icel.* hlýr *cheek.*]

-hleór; *suffix in adjectives* blác-, dreórig-, fæted-, swátig-, teárig-hleór.

hleór-bán, es; *n. Cheek-bone, temple :*—Þun-wængum ł hleórbánum *temporibus*, Ps. Lamb. 131, 5.

hleór-beran :—Eofor lic scionon [o]fer hleor beran gehroden golde fat [and] fyr heard ferh wearde heold, Beo. Th. 612-6; B. 303-5. Grein and Heyne take *hleor beran* as a compound, the former explaining ' *was auf dem Gesicht getragen wird, Helmvisier?* [*oder faciei munimentum?*]' the latter rendering it *cheek.* Thorpe reads *bæron*, Kemble *beran*, an infinitive after *scionon =* they seemed [?]. But may not the verb on which *beran* depends be *gewiton*, v. 607, vv. 608-11 be parenthetical, and *scionon* an adverb, the passage then being translated thus, *they went bearing above their faces the boar's shape, fairly* [scionon] *adorned with gold?*

hleór-bolster, es; *m. A cushion for the cheek, pillow*, Beo. Th. 1381; B. 688. [*Cf.* heáfod-bolster.]

hleór-dropa, an; *m. A tear*, Exon. 52 a; Th. 182, 24; Gú. 1315. [*Cf. Icel.* hlýra skúrir *tears.*]

hleór-sceamu, e; *f. Shame or confusion of face*, Ps. Th. 68, 8.

hleór-slæge, -slege, es; *m. A blow on the cheek or face :*—Hleórslægeas hē underfēng *alapas accepit*, Past. 36, 9; Swt. 261, 6.

hleótan; *p.* hleát, *pl.* hluton. I. *to cast lots :*—Ic hleóte *sortior*, Ælfc. Gr. 31; Som. 35, 55. Ðá hluton ðá consulas hwylc hiera ǽrest ðæt gewinn underfēnge. Ðá gehleát hit Quintus Flaminius *then the consuls cast lots which of them should first undertake that war. Then the lot fell to Quintus Flaminius*, Ors. 4, 11; Swt. 202, 33. Ðonne seó tíd gewinnes and gefeohtes com ðonne hluton hí mid tánum tô ðám ealdormannum and swá hwylc heora swá him se tán ætýwde ðonne gecuron hí ðone him tô heretogan *Satrapæ, ingruente belli articulo mittunt æqualiter sortes, et quemcumque sors ostenderit hunc tempore belli ducem omnes sequuntur*, Bd. 5, 10; S. 624, 24. Lēton tán wísian bluton hell-cræftum, Andr. Kmbl. 2205; An. 1104. Uton hleótan *sortiamur*, Jn. Skt. 19, 24. II. *to obtain by lot, get a share, share in, participate, obtain :*—Ðæs ðú gife hleótest háligne hyht gif . . . *for that shalt thou obtain grace and holy hope, if* . . . , Andr. Kmbl. 960; An. 480. Hē

feorhwunde hleát *he got a mortal wound*, Beo. Th. 4760; B. 2385. Hí ðím hlutan eádigne upwæg *they obtained glory, a blessed ascension*, Menol. Fox 382; Men. 192. Ða Godes þeówas on Israhēla þeóde náne landáre hleótan ne môston *to the servants of God among the people of Israel might not be allotted any landed possessions*, Homl. Th. ii. 224, 5. Hē sceolde þurh deáþes cyme dōmes hleótan *he was to gain glory through the coming of death*, Exon. 47 a; Th. 160, 18; Gú. 945 : 48 a; Th. 164, 20; Gú. 1014 : 74 b; Th. 280, 1; Jul. 622 : Runic pm. 1; Kmbl. 339, 6. Leánum hleótan *to obtain rewards*, Exon. 19 a; Th. 49, 10; Cri. 783. [*O. Sax.* hliotan : *Icel.* hljóta *to get* : *O. H. Ger.* hliozan *sortiri.*] DER. ge-hleótan.

hleóðo, hleóðu. v. hlið.

hleóðor, es; *n.* I. *hearing :*—Ontýn eárna hleóðor ðæt gehērnes hehtful weorðe on gefeán blíðse forþweard tô ðē *auditui meo dabis gaudium et lætitiam*, Ps. C. 50, 77; Ps. Grn. ii. 278, 77. [*Cf. Icel.* hljóð, *e.g.* gefa hljóð, biðja hljóðs *to give, ask for, a hearing* : *Goth.* hliuþ.] II. *what is heard, sound, noise, voice, speech, song :*—Ðá hleóðor cwom býman stefne ofer burhware *when the sound came of the voice of the trumpet over the city-dwellers*, Cd. 181; Th. 226, 29; Dan. 178: Exon. 86 b; Th. 325, 2; Vid. 105: 94 b; Th. 353, 46; Reim. 28: Andr. Kmbl. 3101; An. 1553. Heofonlíce hleóðor gehýred wæs a *heavenly voice was heard*, Exon. 52 a; Th. 181, 22; Gú. 1297: Cd. 162; Th. 204, 6; Exod. 417: Andr. Kmbl. 1478; An. 740. Hleóðor háligra *the voice of saints*, Exon. 65 b; Th. 241, 14; Ph. 656: 108 b; Th. 414, 9; Rä. 32, 17. Biþ ðæs hleóðres swēg eallum songcræftum swētra *the sound of its voice is sweeter than all singing*, 57 b; Th. 206, 24; Ph. 131: 52 a; Th. 181, 15; Gú. 1293. Heriaþ hine on hleóðre bēman *laudate eum in sono tubæ*, Ps. Th. 150, 3: 107, 2: Exon. 104 a; Th. 395, 8; Rä. 15, 4. Him brego sægde æt hleóðre hwæt hē freman wolde *in speech with him the Lord told him what he meant to do*, Cd. 64; Th. 78, 8; Gen. 1290. Ic onhyrge gúþfugles hleóðor *I imitate the war bird's* [*eagle's*] *voice*, 106 b; Th. 406, 22; Rä. 25, 5 : 81 b; Th. 307, 8; Seef. 20 : 49 b; Th. 171, 19; Gú. 1129 : 42 b; Th. 143, 7; Gú. 657. Hleóðra wyn *the delightful sound of the voices heard in heaven*, 56 a; Th. 198, 18; Ph. 12. Stefnum herigaþ hálgum hleóðrum heofoncyninges þrym *with voices and holy songs they praise the glory of heaven's king*, Andr. Kmbl. 1445; An. 723: Bt. Met. Fox 13, 94; Met. 13, 47 : Exon. 46 a; Th. 156, 22; Gú. 878: Cd. 81; Th. 102, 1; Gen. 1693. [*O. H. Ger.* hlioda *sonitus* : cf. *also Icel.* hljóð *sound* ; hljóðan *a sound, tune* : *Dan.* lyd : *Swed.* ljud.] DER. efen-, ofer-, swēg-, word-hleóðor.]

hleóðor-cwide, -cwyde, es; *m. A saying, vocal utterance, words, speech, discourse :*—Ic ðæt gehýre þurh ðínne hleóðorcwide ðæt . . . *I learn from thy words that* . . . , Exon 72 b; Th. 270, 7; Jul. 461: Beo. Th. 3962; B. 1979. Éces word hálges hleóðorcwide, Exon. 61 b; Th. 226, 1; Ph. 389: Andr. Kmbl. 1786; An. 895. Bodan þurh hleóðorcwide hyrdum cýðdon *messengers made known to the shepherds by speech*, Exon. 14 a; Th. 28, 21; Cri. 450. Hleóðorcwyde, Cd. 179; Th. 225, 16; Dan. 155: 190; Th. 236, 5; Dan. 316: 190; Th. 143, 20; Gen. 2382. Þurh hleóðorcwidas, Exon. 53 b; Th. 187, 18; Az. 32. Hleóðorcwydas, Cd. 107; Th. 141, 1; Gen. 2338. Wuton wuldrian weoraða Dryhten hálgan hleóðorcwidum *let us glorify the Lord of hosts with holy songs*, Hy. 8, 2; Hy. Grn. ii. 290, 2. Andreas herede hleóðorcwidum háliges láre *Andrew praised with his words the doctrine of the holy one*, Andr. Kmbl. 1637; An. 820. Æfter hleóðorcwidum *according to the words*, 3240; An. 1623. [*Cf.* meðel-cwide.]

hleóðor-cyme, es; *m. A coming that is attended with sound* [*of trumpets*; cf. hleóðor cwom býman stefne, v. hleóðor], *the coming of an army :*—Hie iudéa blǽd forbrǽcon billa ecgum and þurh hleóðorcyme herige genámon beorhte frætwe ða hie tempel strudon *they destroyed the glory of the Jews with the edge of the sword, and by their coming took with their host the bright ornaments, when they spoiled the temple*, Cd. 210; Th. 260, 15; Dan. 710. [*Cf.* þrym-cyme *a glorious coming.*] Thorpe and Bouterwek translate *oraculum, prophetia*; Grein takes *cyme* as a separate word, and as an adjective.

hleóðor-stede, es; *m. A place where words have been spoken*, Cd. 109; Th. 145, 1; Gen. 2299. [*Cf.* meðel-stede.]

hleóðrian; *p.* ode *To sound, make a sound* [*with the voice*], *to speak, sing, cry, exclaim, resound :*—Drihten hleóðraþ of heofonum and se hýhsta syleþ his stefne *intonuit de cælo Dominus et altissimus dedit vocem suam*, Bd. 4, 3; S. 569, 19: Ps. Spl. 17, 15. Ðonne hleóðriaþ hálge gæstas sáwla sóþfæste song áhebbaþ *when holy spirits shall lift up their voices, just souls raise a song*, Exon. 63 b; Th. 234, 12; Ph. 539. Ðá hleóðrade hlúdan stefne *then cried with a loud voice*, Andr. Kmbl. 2719; An. 1362 : 1073; An. 537. Hleóðrode, 921; An. 461. Geornlíce on gebede hleóðrede *obnixius orationi incumberet*, Bd. 4, 3; S. 569, 11. Azarias hleóðrade drihten herede and ðá word ácwæþ *Azarias cried out, praised the Lord, and these words then spake*, Cd. 188; Th. 233, 25; Dan. 281: Fins. Th. 2; Fin. 2. Ðæt lond hleóðrade for ðara wyrma hwistlunge *sibilabat tota regio*, Nar. 13, 21. Him þuhte ðæt hit eall betweox heofone and eorþan hleóðrode ðám egeslícum stefnum *it seemed to him that all between heaven and earth it resounded with those awful voices.* Hē

Guthl. 5; Gdwin. 36, 4. Ic gehýrde ðæt hit hleóðrode *I heard that it* ⳨ [*the cross*] *uttered a sound*, Rood Kmbl. 52 : Kr. 26. Hit hleóðrode ðä swíðe tóward Haraldes *the general voice was very much in favour of Harold*, Chr. 1036; Erl. 164, 28. Hyre stefn oncwæþ word hleóðrade *her a voice addressed, a word was heard*, Exon. 69 b ; Th. 259, 17; Jul. 283: Andr. Kmbl. 2860; An. 1432. Hé wæs ðæra worda wel gemyndig ðe hé hleóðrade tó Abrahame *memor fuit verbi quod locutus est ad Abraham*, Ps. Th. 104, 37. Fýnd ðíne hleóðrodon *inimici tui sonuerunt*, Ps. Spl. C. 82, 2. Hí ealle samod mid gedrémum sange Godes wuldor hleóðrodon *they all together with melodious song sounded the glory of God*, Homl. Th. i. 38, 7. Swá hleóðrodon *so spake*, Andr. Kmbl. 1383; An. 691. Eáran habbaþ ne hí áwiht mágon holdes gehýran ðeáh ðe him hleóðrige *aures habent et non audient*, Ps. Th. 134, 17. Hé sæde ðæt hé openlíce hí gehýrde betwyh óðer leóþ monig hleóðrian and singan *referre erat solitus, quod aperte eos inter alia resonare audiret*, Bd. 3, 19; S. 547, 37. Ongan ðá hleóðrian helle deófol hwæt is ðis lá manna *then exclaimed the devil of hell : Lo ! what man is this*, Elen. Kmbl. 1798; El. 901. Múþ habbaþ and ne mágon wiht hleóðrian *os habent, et non loquentur*, Ps. Th. 113, 13. Hleóðrian *increpare, redarguere*, Cot. 51: 105, Lye. Mid hleóðrigende dreáme *consona vocis harmonia*, Hpt. Gl. 467. Hleóðriyndum *sonantibus*, Ps. Spl. C. 150, 5. [Cf. Icel. hljóða *to sound, cry out*.]

hleóðrung, e; f. *Speaking, reproving, reproof :* — Ná hæbbende on múþe his hleóðrunga *non habens in ore suo redargutiones*, Ps. Spl. 37, 15.

hleów, hleó, es; n. *A shelter, protection, covering, refuge; often applied to persons :* —Dægscealdes hleó *the sun's* [*cf. Icel. himin-targa = the sun*] *covering, i. e. the pillar of cloud*, Cd. 146; Th. 182, 22 ; Exod. 79. God hleó þarfendra *deus, refugium pauperum*, Rtl. 40, 25. Con- stantínus æðelinga hleó, Elen. Kmbl. 198; El. 99. Beorna hleó éce ælmihtig, Exon. 69 b ; Th. 258, 28 ; Jul. 272. Duguþa hleó [*Guthlac*], 48 a ; Th. 165, 26; Gú. 1034. Wæs earmra hleó *be a refuge for the poor*, Cd. 203; Th. 252, 32; Dan. 587. Eorla hleó [*Beowulf*], Beo. Th. 1586; B. 791: *Hrothgar*, 2074; B. 1035: 3736; B. 1866 : Exon. 100 b; Th. 379, 30: Deór. 41. Tó ðam bisceope reordode : Ðú eorla hleó, Elen. Kmbl. 2145; El. 1074. Freónda hleó [*Guthlac*], Exon. 47 b ; Th. 162, 33; Gú. 985. Sóþne god gæsta hleó, 66 b ; Th. 245, 23; Jul. 49. Hæleþa hleó [*Byrhtnoth*], Byrht. Th. 133, 62; By. 74. Heriga helm wígena hleó [*Constantine*], Elen. Kmbl. 300; El. 150. Wígendra hleó [*Hrothgar*], Beo. Th. 863; B. 429: [*Sigemund*], 1803; B. 899 : [*Beowulf*], 3949; B. 1972: Andr. Kmbl. 1011; An. 506: [*Andrew*], 1792; An. 898. Ðú eart weoroda god wígendra hleó, Exon. 13 b; Th. 25, 31 ; Cri. 409. Wíggendra hleó Eádmund cyning, Chr. 942; Erl. 116, 18; Edm. 12. Ðonne hí tó his húse hleówes wilniaþ *when they desire shelter at his house*, Ps. Th. 108, 10. Under hleó *under shelter*, Cd. 209; Th. 259, 13; Dan. 691: Exon. 16 b; Th. 38, 13; Cri. 606 : 61 a; Th. 224, 11; Ph. 374: Andr. Kmbl. 1664; An. 834: Elen. Kmbl. 1011; El. 507. Ðe hé of hleó sende *whom he sent from the shelter* [*of heaven*], Cd. 5; Th. 7, 7; Gen. 102. Eallum tó hleó *as a refuge for all*, Exon. 25 a; Th. 73, 29; Cri. 1197: Andr. Kmbl. 221; An. 111 : 1133; An. 567. Uton gán on ðisne weald innan on ðisses holtes hleó *let us go into this wood, into the shelter of this grove*, Cd. 39; Th. 52, 7; Gen. 840: Exon. 62 a; Th. 227, 26; Ph. 429. Hé him beád his recedes hleów *he offered them the shelter of his house*, Cd. 112; Th. 147, 18; Gen. 2441. Ðæt hé ðonne stán nime wið hungres hleó hláfes ne gýme *that he should take a stone then as a protection against hunger, and care not for the bread*, Elen. Kmbl. 1228; El. 616. [O. Sax. hleo *in waldes hleo*: O. Frs. hlí : Icel. hlé; n. *lee* (*a sea-term*). Cf. also Icel. hlý *warmth*; hlýr *warm*; hlýja *to shelter*: Goth. hlija *a tent.*] DER. hús-, turf-hleów.

hleówan, hleón, hlýwan; p. de. I. *to make warm, cherish, protect, shelter :* —Ðære sunnan hæto ðe ðás eorðan hlýweþ *the heat of the sun which warms this earth*, Blickl. Homl. 51, 21. Wudubearwas eorþ- welan hleóþ [*cf. holtes hleó ; or is rén the subject of the verb?*] *the groves protect the earth's wealth*, Exon 54 a ; Th. 191, 8; Az. 85. Se king ðér sæt hleówwinde hine beo ðan fýre *the king sat there warming himself by the fire*, Shrn. 16, 16. [*Icel.* hlýja *to cover, shelter, make warm.*] II. *to become warm :* —Gif hit wæter sý hæte man hit óþ hit hleów tó wylme *if it be water let it be heated until it become so warm as to boil*, L. Ath. iv. 7 ; Th. i. 226, 14. v. hleów, hleówe, gehlýwan.

hleów-bord, es; n. *A board which serves for covering* or *protection* [*the binding of a book*], Exon 107 a ; Th. 408, 14; Rä. 27, 12.

hleów-burh; gen. -burge; f. *A city which affords shelter, protection*, Beo. Th. 1828; B. 912: 3467; B. 1731.

hleów-dryhten, es; m. *A lord who protects, a patron*, Exon. 86 a ; Th. 324, 19; Wíd. 94.

hleówe; adj. *Warm, sheltered :* —Gefere ðæne mannan on swíðe fæstne cleofan and wearmne gereste him swíðe wel hleówe ðær and wearme gléda bere man gelóme inn *carry the man into a room very fast shut and warm, let him rest himself there quite warm and snug, and let warm coals be often carried in*, L. M. 2, 59; Lchdm. ii. 280, 12. [Cf. Icel. hlúa að einum *to make one warm and snug* : hlýr, hlær, *warm, mild* : Wick. lew *lukewarm*.] DER. ge-, un-hleówe.

hleów-fæst; adj. *Sheltering, protecting :* —Heáh gæst hleófæst *exalted and sheltering spirit*, Exon. 13 a ; Th. 22, 27; Cri. 358.

hleów-feðer, e; f. *A sheltering wing :* — Gefór hleówfeðrum þeaht *journeyed covered by* [*his creator's*] *sheltering wings*, Cd. 131; Th. 165, 2; Gen. 2740.

hleów-hræscnes?:—Miclode ofor mé hleóhræscnesse ł forcæncednysse *magnificavit super me supplantationem*, Ps. Lamb. 40, 10.

hleów-leás; adj. *Not having* or *not affording shelter, protection, com- fort, cheerless :* —Ða ðe hleóleásan wíc wunedon *those who had occupied a cheerless dwelling*, Andr. Kmbl. 261; An. 131. Ne mótun hí on eorþan eardes brúcan ac hý hleóleáse háma þoliaþ *they may not enjoy a home on earth but shelterless lose their dwellings*, Exon. 35 b; Th. 115, 21; Gú. 193.

hleów-lora weorþan *to become unprotected*, Cd. 92; Th. 117, 14; Gen. 1953.

hleów-mǽg, es; m. *A near relation, one who is bound to offer shelter* or *help* [?], Cd. 48; Th. 61, 34; Gen. 1007: 75; Th. 94, 3; Gen. 1556: 78; Th. 96, 16; Gen. 1596: 76; Th. 95, 21; Gen. 1582: Exon. 81 b; Th. 307, 18; Seef. 25.

hleów-sceorp, es; n. *A protecting garment*, Exon. 103 a; Th. 391, 15; Rä. 10, 5.

hleów-stede, es; m. *A sheltered, warm place :* — Hleówstede *apricus locus*, Wrt. Voc. 86, 24.

hleów-stól, es; m. *A place of protection, one's native city :*—Síðedon fǽmnan and wuduwan freóndum beslægene from hleówstóle *damsels and widows bereft of friends journeyed from their sheltering home* [*of the people of Sodom driven from their city*], Cd. 94; Th. 121, 16; Gen. 2011.

hleówþ, hleóþ, hlíwþ, hlýwþ, e; f. *Shelter, protection, warmth :*— Hleówþ *apricitas*, Wrt. Voc. 86, 25. Ðonne him cælþ hé cépþ him hlýwðe *when he gets cold he looks out for warmth*, Hexam. 20 ; Norm. 28, 23. Tó neste bǽron heora briddum tó hleówþe *bore it to their nest to shelter their young*, Homl. Th. ii. 144, 23. Foresceáwian bigleofan and hleówþe *to provide food and shelter*, 462, 18. Hlýwþe, Basil admn. 9; Norm. 52, 23. Cold bæþ ongeán ða hlíwþe *a cold bath to atone for the warmth*, L. Pen. 16; Th. ii. 284, 5. Ða hlýwþe gódra weorca *the shelter of good works*, L. E. I. 32; Th. ii. 430, 24. Gé hyra hulpon and him hleóþ géfon *ye helped them* [*the poor*] *and gave them shelter*, Exon. 27 b; Th. 83, 11; Cri. 1354. [*Laym.* leoð *protection*.]

hleówung, hlýwing, e; f. *Shelter, refuge :*—Hlýwing *refugium*, R. Conc. 11, Lye.

hlét, hlíet, es; m. *A lot;* sors :— Missenlíce hlǽte *varia sorte*, Bd. 2, 20; S. 521, 10. Be hlǽte *sorte*, Hpt. Gl. 426. Hé hí hæþ oferstigene mid ðam hlíete his anwaldes *quos sorte potestatis excesserit*, Past. 17, 3; Swt. 111, 16. [*Goth.* hlauts; m. *a lot:* O. Sax. hlót; m: O. H. Ger. hlóz; m. n.] v. hlot, hlyt.

hlichan. v. hlehhan.

hlid, es; n. *A lid, cover, the opening which is closed by the cover :*— Hlidd *opertorium*, Ps. Spl. 101, 28. Ðá lédon ða þegenas ðone Hælend ðǽron and mid hlide belucon úre ealra Álýsend *then the thanes laid our Redeemer therein, and closed up with a cover the Redeemer of us all*, Homl. Th. ii. 262, 4. Se engel áwylte ðæt hlid of ðære þryh *the angel rolled away the cover from the tomb*, i. 222, 8. Hé tóáwylte mycelne stán tó hlide ðære byrgene *advolvit saxum magnum ad ostium monumenti*, Mt. Kmbl. 27, 60. [*Icel.* hlið, *gate, gateway:* O. H. Ger. hlit *operculum.*] DER. ge-hlid.

hlídan. v. be-, of-, on-, to-hlídan.

hlid-fæst; adj. *Having a lid :* — Hió becwyþ Eádmǽre áne hlidfæsþe cuppan *she bequeaths to Eadmer a cup with a lid*, Chart. Th. 536, 4.

hlid-geat, es; n. *A swing-gate, folding-door :*—On ðonæ stocc ðæ ðæt hlidgeat on hangodæ *to the post that the swing-gate hung on*, Cod. Dipl. Kmbl. ii. 176, 13. Of ðam hlidgeate, 236, 25. Hlidgata *valva*, Ælfc. Gl. 29; Som. 61, 36; Wrt. Voc. 26, 35.

hliépa. v. blýpa.

hliét. v. hlét.

hlifendre *minium*, Lye.

hlifian; p. ode *To stand out prominently, tower up, to be raised high :*— Ic hlifige under heofenum *I am high raised under the heavens*, Rood Kmbl. 167; Kr. 85. Se beorhta beág eádigra gehwam hlifaþ ofer heáfde *the bright crown rises o'er the head of each blessed one*, Exon. 64 b; Th. 238, 14; Ph. 604. Beorgas ne muntas steápe ne stondaþ ne stánclifu heáh hlifiaþ *nec tumulus crescit*, 56 a; Th. 199, 9, 27; Ph. 23, 32. Hlifiaþ eáran ofer eágum *ears stand up above my ears*, 104 b; Th. 396, 14; Rä. 16, 4. Wudubeám hlifode tó heofontunglum *the tree towered up to the stars of heaven*, Cd. 199; Th. 247, 22; Dan. 501. Hlifade, Beo. Th. 163, 3801; B. 81, 1898. Hlifodon, Andr. Kmbl. 1681; An. 843. Hlifedon, Cd. 146; Th. 183, 9; Exod. 89. Gesáwon salo hlifian *saw the halls towering up*, 109; Th. 145, 10; Gen. 2403: Exon. 113 b; Th. 435, 16; Rä. 54, 1. Heáh hlifian, Beo. Th. 5602; B. 2805. Hlifigan, Cd. 139; Th. 174, 12; Gen. 2877: 205; Th. 253, 29; Dan. 103. Hlifigean, 66; Th. 79, 25; Gen. 1321. Mid ðý ðe hé wæs

hlífigende ofer sǽs brim *whilst he was standing high up above the sea*, Blickl. Homl. 143, 5. DER. ofer-hlífian.

hlígan *or* hligan? *To allow one a reputation for anything, to give one glory* :—Ne forlét ðú ûsic éce drihten for ðám miltsum ðe ðec men hlígaþ *forsake us not, eternal Lord, because of those mercies for which men account thee glorious*, Cd. 190; Th. 235, 25; Dan. 311. Willaþ mid ðý gedón ðæt hie mon hlíge wísdómes *they desire thereby to make men allow them a reputation for wisdom;* doctrinæ sibi opinionem faciunt, Past. 48, 2; Swt. 367, 19. v. hlísa.

hligiung, e; *f. Laughing*, L. M. 2, 46; Lchdm. ii. 258, 20.

hligsa. v. hlísa.

hlihan, hlihhan. v. hlehhan.

hlimman ; *p.* hlamm ; *pl.* hlummon *To sound, roar* [*as the sea*], *clang, clash* :—Gársecg hlymmeþ *the ocean roars*, Andr. Kmbl. 784; An. 392. Hlimmeþ, Exon. 101 a; Th. 382, 2; Rä. 3, 5. Ðrǽd ne hlimmeþ *the thread makes no sound to me*, 109 a; Th. 417, 18; Rä. 36, 6. Scildas hlúde hlummon *loud clanged the shields*, Judth. 11; Thw. 24, 24; Jud. 205. Ic ne gehýrde bútan sǽ hlimman *I heard nought but the sea roaring*, Exon. 81 b; Th. 307, 4; Seef. 18. v. hlemman.

hlimme, an; *f. A torrent* :— Dó him swá ðú dydest Madiane and Sisare swylce Jabin ealle ða námon Ændorwylle and Cisone clǽne hlimme *fac illis sicut Madian et Sisaræ; sicut Jabin in torrente Cisson; disperierunt in Endor*, Ps. Th. 82, 8. Ða ðe on wege weorðaþ wǽtres and hlimman deópes ondrincaþ *de torrente in via bibet*, 109, 8. Oft úre sáwl swýðe frécne hlimman gedǽgde hlúdes wǽteres *torrentem pertransivit anima nostra*, 123, 4; 125, 4. v. hlimman; *and cf.* hlyn.

hlin. v. hlyn.

hlinaþ, hleonaþ, es; *m. A place to lie down in* :—Ic getimbre hús and hleonaþ, Exon. 36 a; Th. 117, 10; Gú. 222.

hlin-bedd, es; *n. A couch* :—Fundon on sande sáwulleásne hlin- [MS. hlim-] bed healdan *they found him without life occupying his couch*, Beo. Th. 6060; B. 3034. [*Cf. O. H. Ger.* hlína *recubitus, accubitus, reclinatorium.*]

hlinc, es; *m.*   I. *a link, linch, rising ground;* 'agger limitaneus, parœchias, etc. dividens,' Junius. The word occurs in the charters, e. g :—Of ðere díc on þornhlinch; ðanone on dynes hlinch; of ðam hlince, Cod. Dipl. Kmbl. iii. 223, 29. Ðanon on ðone miclan hlinc, Chart. Th. 160, 24. Fearnhlinc, landsore hlinc, sweord hlincas, wotan hlinc are other instances of its occurrence. In later times the word is given with a similar sense in provincial glossaries, e. g. in Suffolk some woods are called *links : linchets* grass partitions in arable fields, Lisle : *linch* a bawke or little strip of land, to bound the fields in open countries, Pegge's Kenticisms. v. E. D. S. Publications, and Halliwell's Dict.   II. *a hill, rising ground* :—Beorgas ne muntas steápe ne stondaþ ne stánclifu heáh hlífiaþ ne dene ne dalu ne dúnscrafu hlǽwas ne hlincas *nec tumulus crescit nec cava vallis hiat*, Exon. 56 a; Th. 199, 13; Ph. 25. Heá hlincas, 101 b; Th. 384, 7; Rä. 4, 24.

hlín-duru, a; *f. A door formed of lattice-work, a grated door* :—Helle hlínduru [cf. *Icel.* Hel-grindr], Exon. 97 b; Th. 364, 29; Wal. 78. Geseh hé fore hlíndura hyrdas standan *he saw guards standing before the grated door* [*of his prison*], Andr. Kmbl. 1985; An. 995. [*Cf. O. H. Ger.* hlínun, *pl.* cancelli, Grff. iv. 1096.]

hlinian, hleonian; *p.* ode *To lean, bend, lie down, recline, rest* :— Ic hlinige *cubo*, Ælfc. Gr. 24; Som. 25, 55. Ne ðǽr hleonaþ unsmeðes wiht *nor does aught unsmooth rest there*, Exod. 56 a; Th. 199, 14; Ph. 25. Ða ðe him godes egsa hleonaþ ofer heáfdum *those on whose heads rests the fear of God*, 33 b; Th. 106, 20; Gú. 44. Monige hleonigaþ mid Abraham *multi recumbent cum Abraham*, Mt. Kmbl. Rush. 8, 11. Hlionigaþ [hlinigaþ, Lind.] † restaþ *accumbent*, Lk. Skt. Rush. 13, 29. Hlionede hé in húse *discumbente eo in domo*, Mt. Kmbl. Rush. 9, 10. Hleonede [hlionade, Lind.], 26, 20. Án ðæra leorning cnihta hlinode on ðæs hǽlendes bearme *erat recumbens unus ex discipulis ejus in sinu iesu*, Jn. Skt. 13, 23. Æt ðæm uferran ende Drihten hlinode *Domino desuper innitente*, Past. 16, 3; Swt. 101, 20. Ánra gehwylc hleonade wið handa each one leaned on his hand, Cod. 222; Th. 291, 19; Sat. 433. Ne hlina [hliona, Rush.] ðú *non discumbas*, Lk. Skt. Lind. 14, 8. Hí sécaþ ðæt hie fyrmest hlynigen æt ǽfengieflum *primos in cœnis recumbiti quærunt*, Past. 1, 2; Swt. 27, 7. Hé fyrgenbeámas ofer hárne stán hleonian *funde he found the mountain trees resting on the grey rock*, Beo. Th. 2835; B. 1415. Ofer ða se hálga bisceop hlyniende forþférde *cui incumbens obiit*, Bd. 3, 17; S. 544, 18. Heó wæs hleonigende ofer hire ræste *she was lying on her bed*, Blickl. Homl. 145, 26. Fond hlingende freán *found his master lying in his bed*, Exon. 49 b; Th. 171, 2; Gú. 1120. [*Laym.* leonede, *p: A. R.* leonie, *subj: O. Sax.* hlinon : *O. H. Ger.* hlinen *obcumbere, incumbere, recumbere, inniti.*] DER. ge-, on-hlinian : v. hlǽnan.

hlín-rǽced, es; *n. A place with grated doors, a prison*, Andr. Kmbl. 2924; An. 1465: Exon. 69 a; Th. 257, 6; Jul. 243. [*Cf.* hlín-duru.]

hlín-scúa, -scúwa, an; *m. The darkness of a prison*, Andr. Kmbl. 2143; An. 1073: Exon. 73 b; Th. 275, 2; Jul. 544. v. preceding word.

---

hlinung, e; *f. Leaning, resting, a couch* :—Hlinunge wiðersæc *unfavourable to leaning*, L. M. 2, 46; Lchdm. ii. 258, 20. Ða forman hlininga *primos discubitos*, Lk. Skt. 20, 46.

hlísa, hligsa, hliosa, an; *m. Sound, rumour, report, reputation, renown, fame, glory* :—Hlísa *fama*, Wrt. Voc. 76, 1. Ða férde hys hlísa intó ealle Syriam *abiit opinio ejus in totam Syriam*, Mt. Kmbl. 4, 24. Ðes hlísa wearþ cúþ ðæra leóda cynegum ðe begeondan Iordane eardiende wǽron *this report became known to the kings of the nations that were dwelling beyond Jordan*, Jos. 9, 1. Hwæt is heora nú tó láfe bútan se lytla hlísa and se nama mid feáum stafum áwriten *signat superstes fama tenuis paucnlis inane nomen litteris*, Bt. 19; Fox 70, 10: 68, 21, 4. Hí wilnodon ðæs hlísan æfter heora deáþe, 18. 4; Fox 68, 9. Sume hí gebycgaþ weorþlícne hlísan ðisses andweardan lifes mid heora ágnum deáþe forþam hí wénaþ ðæt hí næbben nán oðer fioh ðæs hlísan [hliosan, Bod.] wyrðe bútan hiora ágnum fiore *nonnulli venerandum sæculi nomen, gloriosæ pretio mortis, emerunt*, Bt. 39, 11; Fox 228, 27. Ðeáh ðe monig mon herige ne gelýf ðú him tó wel: ac ðæs hlísan þenc ðé silf hwæt ðæs sóþes sý *though many men praise thee, do not believe them too much ; but thyself consider how much of this reputation is true*, Prov. Kmbl. 69. Gif wé mid hlýsan gódra weorca úrne Drihten sécaþ *if we come to our Lord with the fame of good works*, Homl. Th. i. 222, 4 : Exon. 34 b; Th. 111, 17; Gú. 128 : 33 a; Th. 105, 31; Gú. 31. Ðæt is ðonne ðæt mon his mearce brǽde ðæt mon his hligsan [hlísan, Hatt MS.] and his noman mǽrsige *terminum vero suum dilatare est opinionis suæ nomen extendere*, Past. 48, 2; Swt. 366, 13. Ða gehýrde heó Salomones hlísan *she heard of Solomon's fame*, Homl. Th. ii. 584, 8 : Exon. 54 a; Th. 191, 9; Az. 85. Ge gehýraþ gefeoht and gefeohta hlísan *audituri estis prælia et opiniones præliorum*, Mt. Kmbl. 24, 6. v. hlígan.

hlís-bǽre ; *adj. Famous, glorious*, Som.

hlís-eádig ; *adj. Successful in acquiring fame, famous, renowned* :—Biþ hlíseádigra se ðe hit selþ ðonne se ðe hit gaderaþ: eác ða welan beóþ hlíseádigran ðonne ðonne hie mon selþ ðonne hie beón ðonne hí mon gaderaþ. Ðe gítsung gedéþ heore gítseras láðe and ða cysta gedóþ ða hlíseádige *hæc effundendo magis quam coacervando melius nitent : avaritia odiosos, claros largitas facit*, Bt. 13; Fox 38, 11–17. Gif hé nǽre hlíseádig *egere claritudine*, 33, 1; Fox 120, 35.

hlíseádig-ness, e; *f. Renown, celebrity;* claritudo, Bt. 33, 1; Fox 122, 3.

hlís-ful ; *adj. Famous, of good repute, renown* :—Hlísful *famosus vel opinosus*, Ælfc. Gl. 82; Som. 73, 35; Wrt. Voc. 47, 39 : *famosus*, Wrt. Voc. 75, 71. Ðý læs ðe hé wurde tó hlísful on worulde and ðæs heofenlícan lofes fremde wǽre *lest he should become too famous in this world and be a stranger to the praise of heaven*, Homl. Th. ii. 142, 26. Cumlíðnys is swíðe hlísful þing *hospitality is a thing of very good repute*, 286, 16. Hlísfulle weras *men of renown*, Gen. 6, 4.

hlísful-líce ; *adv. Gloriously* :—Óswold cyning his cynedóm geheóld hlísfullíce *king Oswald maintained his kingdom gloriously*, Swt. A. S. Rdr. 99, 119.

hliþ, es; *n. A slope, declivity, hill-side, hill* :—Of hliþes nósan *from the promontory*, Beo. Th. 3789; B. 1892: Exon. 123 b; Th. 473, 28; Bo. 22. Beneoþan ðam hlíþe *under the hill*, Cod. Dipl. Kmbl. iii. 52, 15. Swá tó ðam westhlíþe, 123, 5. Hie be hlíþe heáre dúne eorþscræf fundon *they found a cavern on the slope of a lofty hill*, Cd. 122; Th. 156, 25; Gen. 2594. Weallsteápan hleoþu *hills steep as walls*, 86; Th. 108, 8 ; Gen. 1803. Hleoþo, 72; Th. 88, 3; Gen. 1459. Hleoþa, Exon. 101 a; Th. 382, 6; Rä. 3, 7. Hliþo, 130 a; Th. 498, 17; Rä. 88, 3. Beorgas steápe hleoþum hlifedon *steep hills rose high with their slopes*, Andr. Kmbl. 1681; An. 843. [*Icel.* hlíð *a slope, mountain side: O. H. Ger.* hlíta *clivus: Ger.* leite in cpds. Grff. iv. 1096.] DER. beorg-, burh-, fen-, heáh-, mist-, næs-, sand-, stán-, wulf-hliþ.

hlíwþ. v. hleówþ.

hlodd. v. hlot.

hlond. v. hland.

hlosnere, es; *m. A listener;* auscultator, Hpt. Gl. 461.

hlosnian ; *p.* ode *To listen, be silent in expectation of hearing, listen for the coming of a person, watch, await, be on the look out* :—Ða on sumere nihte hlosnode sum óðer munuc his færeldes and mid sleaccre stalcunge his fótswaðum filigde *then one night another monk was on the watch for his going, and with stealthy tread followed his footsteps*, Homl. Th. ii. 138, 5. Eoda ða tó mæssan and hlosnode georne be ðære líflícan onsægednesse *he went then to mass and waited eagerly for the living sacrifice*, Homl. Swt. 3, 157. Æfter ðissum wordum weorud hlosnode swígodon ealle *after these words the multitude listened* [*astonished or expectant*], *all were silent*, Andr. Kmbl. 1522; An. 762. Ðæt folc hlosnende wæs gehérde hine *populus suspensus erat audiens illum*, Lk. Skt. Lind. 19, 48. Hlosniend *attonitus*, Cot. 3, Lye. [*Cf. O. H. Ger.* hlosen *audire, attendere, obedire, auscultari:* hlosenti *adtonitus*.]

hlot, es; *n. A lot, portion, share* :—Ðis hlot *hæc sors*, Ælfc. Gr. 9, 44; Som. 13, 3. Hig wurpon hlot ðǽr ofer *sortem mittentes*, Mt. Kmbl. 27, 35. Hlott, Mk. Skt. Lind. 15, 24. Hlott † tán, Jn. Skt. Lind. 19, 24. Æfter gewunan ðæs sacerdhádes hlotes *secundum consuetudinem sacerdoti*

sorte, Lk. Skt. 1, 9. Sel mē dǣl ł hlodd [hlott, Rush.] striónes *da mihi portionem substantiæ*, Lind. 15, 12. Hie sendon hlot him betweónum *they cast lots among them*, Blickl. Homl. 229, 5. Hlotu wurpon *mittentes sortem*, Mk. Skt. 15, 24: Lk. Skt. 23, 34. v. hlēt, hlyt.

hlōþ, e; *f.* I. *spoil, booty :*—Hē yteþ hlōþe *comedet ł rædam*, Bd. 1, 34: S. 499, 27. Mycle hlōþe þurh his lāre and fulluhte ðam ealdan feónde āfyrde *magnas antiquo hosti preædas docendo et baptizando eripuit*, 2, 20; S. 522, 22. II. *a band, troop, company, gang, crew, body of robbers :*—Þeófas wē hātaþ ōþ vii men from vii hlōþ ōþ xxxv siððan biþ here, L. In. 13; Th. i. 110, 13. Ðȳ geáre gegadrode ān hlōþ wícenga *in that year a gang of vikings collected*, Chr. 879; Erl. 80, 28. Com ðā hǣðenra hlōþ hāliges neósan *then came a band of heathens visiting the saint*, Andr. Kmbl. 2777; An. 1391: 3085; An. 1545. Feónda hlōþ *a fiendish crew*, Exon. 46 a; Th. 157, 5; Gū. 887. Gif mon twȳhyndne mon unsynnigne mid hlōþe ofsleá gielde se ðæs sleges andetta sīe wer and ǣghwelc mon ðe on sīþe wǣre geselle xxx scill. tō hlōþbóte *if any one in company with others slay an unoffending 'twyhynde' man let him who acknowledges the blow pay 'wer' and 'wite;' and let every one who was engaged in the matter pay thirty shillings as fine*, L. Alf. pol. 29; Th. i. 80, 6–9. Ne cōman hig nā tō fiohtanne ac ðæt hig woldan mid hlōþe geniman *they did not come to fight, but with the intention of robbing*, Shrn. 38, 10. Geseh hē hǣðenra hlōþ, Andr. Kmbl. 1984; An. 994: 84; An. 42. Heó ðæt weorud āgeaf hlōþe of ðam hātan hreþre *she gave up that multitude, troops from her hot bosom*, Exon. 24 b; Th. 71, 29; Cr. 1163: 75 b; Th. 283, 6; Jnl. 676. Hē ðā his here on tū tōdǣlde sum ymb ða burg sætt and hē mid sumum hlōþum fōr and monega byrg bereáfode on Cheranisse *inde propter agendam prædam et curandam obsidionem divisit exercitum. Ipse autem cum fortissimis profectus, multas Cheronesi urbes cepit : profligatisque populis opes abstulit*, Ors. 3, 7; Swt. 116, 17: 3, 1; Swt. 100, 2. Fōran hie hlōþum *they went in bands*, Chr. 894; Erl. 90, 12: Exon. 45 b; Th. 156, 1; Gū. 868: 99 b; Th. 373, 23; Seel. 114. III. *the crime of taking part in the action of a* hlōþ :—Be hlōþe. Seðe hlōþe betygen sīe geswicne se hine be cxx hīda oððe swā bēte, L. In. 14; Th. i. 110, 15. DER. here-hlōþ.

hlōþ-bót, e; *f.* Compensation or fine to be paid by a member of a 'hlōþ' for the wrong committed by any one of them, L. Alf. pol. 29; Th. i. 80, 9. v. hlōþ.

hlōþere, es; *m.* A robber, spoiler; prædator, Cot. 170, Lye.

hlōþ-gecrod, es; *n.* A press of troops or bands :—Biersteþ hlúde heáh hlōþgecrod *with loud noise breaks the press of [cloud-] troops on high*, Exon. 102 a; Th. 386, 17; Rä. 4, 63.

hlōþian; *p.* ede To take booty, rob, spoil :—Ða ðe ǣlce geáre ofer ðone sǣ hlōþedon and hergedon *qui anniversarias prædas trans maria cogere solebant*, Bd. 1, 12; S. 481, 2. Ða ðe monige geár ǣr in onhergedon and hlōþedon *qui per multos annos prædas in terra agebant*, 1, 14; S. 482, 19.

hlōþ-sliht, es; *m.* Slaying by a member of a 'hlōþ,' L. Alf. pol. 29; Th. i. 80, 5. v. hlōþ.

hlówan; *p.* hleów To low, bellow, make a loud noise : — Oxa hlēwþ *bos mugit*, Ælfc. Gr. 22; Som. 24, 9. Hleówon hornboran *the trumpeters sounded*, Elen. Kmbl. 107; El. 54. Hlówendra fearras flǣsc *the flesh of lowing oxen*, Homl. Th. i. 590, 15. [*Icel.* hlōa *to roar (of streams): O. H. Ger.* hlōon *mugire, rudere.*]

hlówung, e; *f.* Lowing, noise : — Hlóweng *bombus*, Cot. 27, Lye. [*O. H. Ger.* hlōhunga *mugitus.*]

HLŪD; *adj.* LOUD, *sonorous :*—Heora stefn wæs swīðe hlúd *their voice was very loud*, Blickl. Homl. 149, 27 : Cd. 148; Th. 184, 14; Exod. 107. Hlimman hlúdes wæteres *torrentem*, Ps. Th. 123, 4. Hlúdre stefne *with a loud voice*, Blickl. Homl. 181, 18. Hlúddre stefne, 15, 19 : Cd. 227; Th. 302, 18. Hlúdan stefne, Andr. Kmbl. 2720; An. 1362. Hlúde wǣran hȳ ðā hȳ ofer ðone hlǣw ridan *loud were they when they rode over the hill*, Lchdm. iii. 52, 13. Francan wǣron hlúde *loud was the sound of the javelins*, Cd. 93; Th. 119, 20; Gen. 1982. Hlúddra sang *chorea*, Ælfc. Gl. 34; Som. 62, 47; Wrt. Voc. 28, 28. Ðæt ār ðonne hit mon slihþ hit biþ hlúdre ðonne ǣnig ōðer ondweorc *aes dum percutitur amplius metallis ceteris sonitum reddit*, Past. 37, 3; Swt. 267, 24. Hlúdast, Menol. Fox 467; Gn. C. 4. [*O. Sax. O. Frs.* hlúd: *O. H. Ger.* hlút: *Ger.* laut.]

hlúd-clipol; *adj.* Calling aloud, R. Ben. interl. 7.

hlúde; *adv.* Loudly :—Folc ðe hlúde singeþ *a people that sings loudly*, Blickl. Homl. 149, 30 : 217, 33. Ðæs cocces þeáw is ðæt hē micle hlúdor singþ on uhtan ðonne on dægrēd *gallus profundioribus horis noctis altos edere cantus solet*, Past. 63; Swt. 461, 2.

hlúd-stefn, -stemn; *adj.* Loud-voiced, Cot. 105, Lye.

hlúd-swēge; *adv.* With a loud voice :—Se hana sóna hlúdswēge sang *the cock straightway crowed with a loud voice*, Homl. Th. ii. 248, 33. Marcus swā swā leó hlúdswēge clipode, Ælfc. T. p. 25; Grn. 13, 8.

hlutor, hluttor; *adj.* Clear, pure, bright, sincere : — Hluttor wæter *limpha*, Ælfc. Gl. 97; Som. 76, 69; Wrt. Voc. 54, 13. Swīðe wynsum and hluttor wǣta *a very pleasant and pure stream*, Blickl. Homl. 209, 2. Hlutor, Bt. Met. Fox 5, 26; Met. 5, 13. Wæs hē hluttor and clǣne on

his life *he was pure and clean in his life*, Blickl. Homl. 217, 9 : Ps. Th. 72, 17. Ōþ ðæt byþ āhafen hluttor móna *donec extollatur luna*, 71, 7: Exon. 58 b; Th. 210, 9; Ph. 183. Gif ðín eáge biþ hluttor *si oculus tuus fuerit simplex*, Lk. Skt. 11, 34. xxx ambra hluttres ealoþ, L. In. 70; Th. i. 146, 17. Hlutres aloþ, Chr. 852; Erl. 67, 38. Ðæt hig drincon hluttor wín '*thou didst drink the pure blood of the grape*,' Deut. 32, 14. Genim ða ylcan sealfe hluttre *take the same salve clear*, L. Med. ex Quadr. 3, 3; Lchdm. i. 340, 2. Ōþ hlutturne dæg *usque ad ortum diei*, Bd. 4, 19; S. 588, 13. Þurh hlutterne dæg *during the daylight*, Exon. 105 b; Th. 401, 5; Rä. 21, 7. Hluttor píc *resin*, L. M. 1, 4; Lchdm. ii. 44, 24: 1, 31; Lchdm. ii. 72, 25. Dō on hluttor æg *add the white of an egg*, 2, 64; Lchdm. ii. 288, 9. Lǣt standan ōþ hit sȳ hluttor nim ðonne ðæt hluttre *let it stand till it be clear, then take the clear part*, Lchdm. iii. 4, 3. Weder hluttor gesihþ ceápes ferþrunge hit getācnaþ *if he sees clear weather, it betokens furthering of traffic*, 198, 17. Hluttre móde and bylehwíte *simplici et pura mente*, Bd. 4, 24; S. 599, 8 : Exon. 12 a; Th. 18, 34; Cri. 293. Mid hluttrum sáwlum *with pure souls*, Cd. 21; Th. 25, 21; Gen. 397. Mid hlutrum eágum *with clear eyes*, Bt. Met. Fox 21, 74; Met. 21, 37. Ðone hlutrestan stream *the stream most pure*, 23, 5; Met. 23, 3. [*Orm.* lutter. *Goth.* hlutrs *pure : O. Sax.* hluttar: *O. Frs.* hlutter: *O. H. Ger.* hlutar *clarus, lotus, purus, mundus : Ger.* lauter.] DER. glæs-hlutor.

hlutor-, hluttor-líce; *adv.* Clearly, plainly :—Hlutorlíce tōcnāwaþ *clearly distinguish*, Lchdm. iii. 440, 29. Gif hē him ðæt hluttorlíce gecȳðan wolde *hwæt hē wǣre si simpliciter sibi quis fuisset proderet*, Bd. 4, 22; S. 591, 37: 5, 13; S. 634, 2.

hlutor-, hluttor-ness, e; *f.* Clearness, purity :—Hū heora gecynd bútan ǣlcre besmitennysse on ēcere hluttornysse þurhwunaþ *how their nature continues without any pollution in eternal purity*, Homl. Th. i. 538, 29. Tō hluttornisse geleáfan *ad simplicitatem fidei*, Bd. 2, 5; S. 507, 42. On hluttornesse and on clǣnnesse *in sinceritate*, 4, 9; S. 576, 21: 2, 15; S. 518, 30.

hluttre, hlutre; *adv.* Clearly, brightly :—Heofon hluttre ongeat *heaven clearly perceived*, Exon. 24 b; Th. 71, 3 : Cri. 1150. Ðonne heofontungol hlutrost scíneþ *when the sun shines brightest*, Bt. Met. Fox 22, 48; Met. 22, 24. DER. dæg-hluttre.

hluttran [?] *to grow or make pure, clean, bright*, Exon. 54 a; Th. 191, 8; Az. 85. v. next word.

hluttrian; *p.* ode. I. *to become clear :*—Hit wile hluttrian *it will become clear*, Lchdm. iii. 76, 7. II. *to make clear, purify* [v. āhluttrian] — Morgenrēn hluttraþ [*or is the verb in the plural?] the morning rain purifies*, Exon. 54 a; Th. 191, 8; Az. 85.

hlȳd, es; *n. A sound :*—Losaþ gemynd heora mid hlȳde [MS. hlydne] *periit memoria eorum cum sonitu*, Ps. Spl. T. 9, 7. [*Laym.* mid lude.] v. ge-hlȳd.

hlȳda, an; *m. The month noisy with wind and storm, March :*—Hagolscúrum færþ geond middangeard Martius rēðe Hlȳda *with hail-showers passes through the earth rude March [which we call] Hlyda*, Menol. Fox 74; Men. 37. Mónaþ Martius ðe menn hātaþ hlȳda, Lchdm. iii. 152, 30. Ðæs mónþes ðe wē hātaþ Martius ðone gē hātaþ Hlȳda, Homl. Th. i. 100, 5. On Martius ðæt is on hlȳdan mónþe, Lchdm. iii. 152, 9; 250, 5. Se ǣresta frigedæg ðe man sceal fæsten is on hlȳdan *the first Friday to fast on is in March*, 228, 21. [Lide as a name for March is given in the E. D. S. East Cornwall Glossary.]

hlȳdan; *p.* de To sound, make a loud noise, to clamour, vociferate :—Ic hlȳde *strepo*, Ælfc. Gr. 28; Som. 30, 63. Ic hlȳde *garrulo*, 36; Som. 38, 29. Se tympano biþ geworht of drygum felle and ðæt fell hlȳt ðonne hit mon sliehþ *in tympano sicca et percussa pellis resonat*, Past. 46, 2; Swt. 347, 5. Ðíne fȳnd hlȳdaþ *inimici tui sonaverunt*, Jud. 5; Thw. 156, 1: Exon. 20 b; Th. 55, 14; Cri. 883. Se uncer hláford hlȳdde ðǣr úte *that master of ours was vociferating without*, Shrn. 43, 14. Hlóh and hlȳdde *he laughed and clamoured*, Judth, 10; Thw. 21, 18; Jud. 23. Ðā hlȳddon hig and cwǣdon *at illi invalescebant dicentes*, Lk. Skt. 23, 5. Ðā hē geseah hwistleras and hlȳdende menigeo *cum vidisset tibicines et turbam tumultuantem*, Mt. Kmbl. 9, 23. Hlȳdende *clamando*, Past. 15, 2; Swt. 91, 22, 23. Hlȳdende swíðust innan *sounding chiefly from within*, L. M. 2, 46; Lchdm. ii. 258, 19. Se ðe wylle drincan and dwæslíce hlȳdan drince him æt hām nā on Drihtnes húse *he who wants to drink and make a foolish noise let him drink at home, not in the Lord's house*, L. Ælfc. C. 35; Th. ii. 357, 40. Hēt hí mid handum sleán on ðæt hleór ðæt heó hlȳdan ne sceolde *he bade strike her with their hands on the face that she should not declaim*, Homl. Swt. 8, 70. [*O. Sax.* a-hlúdian: *O. H. Ger.* hlútian *sonare, clamare, concrepare : Ger.* lauten.]

hlȳden. v. hlȳd.

hlȳdend *garrulus*, Cot. 170, Lye. v. hlȳdan.

hlȳdig *garrulus*, Hpt. Gl. 439. [Cf. *O. H. Ger.* -hlútig -*sonus*, Grff. iv. 1098.]

hlȳd-mónaþ. v. hlȳda.

hlyn, hlin, es; *m.* [?] *The name of a tree, maple* [?], Exon. 114 a; Th. 437, 17; Rä. 56, 9. [*Icel.* hlynr *maple.*]

hlyn, hlynn, hlin, es; *m. A sound, noise, clamour, din :* — Tō ðon

N n

đonne hit hât wǽre and mon đa earman men oninnan dôn wolde hû se hlynn mǽst wǽre đonne hie đæt sûsl đǽron þrowiende wǽron *ut cum inclusus ibidem subjectis ignibus torreretur, sonum vocis extortæ capacitas concavi aeris augeret*, Ors. 1, 12; Swt. 54, 25. Hlynn wearþ on ceastrum *a great cry arose in the cities*, Cd. 119; Th. 153, 30; Gen. 2546. Hlyn scylda and sceafta *the din of shields and shafts*, 95; Th. 124, 12; Gen. 2061. Hlin, Exon. 101 a; Th. 381. 7; Rä. 2, 7. Hearpan hlyn *the sound of the harp*, 57 b; Th. 207, 1; Ph. 135; Cd. 52; Th. 66, 7; Gen. 1081: Beo. Th. 1227; B. 6, 11. DER. ge-hlynn.

hlynian; *p*. ode *To make a noise, roar* :—Wælfýra mǽst hlynode *the greatest of funeral fires roared*, Beo. Th. 2244; B. 1120.

hlynn, e; *f. A torrent* :—Ofer þah hlvnne *trans torrentem*, Jn. Skt. Rush. 18. 1. [*Scott*. lin. lyn, lynn *a cataract*.] v. *previous and following words, and cf.* hlimme *and* hlimman.

hlynnan; *p*. ede *To sound, make a noise, shout* :—Gûþwudu hlynneþ scyld scefte oncwyþ *the war-wood resounds, shield replies to shaft*, Fns. Th. 11; Fin. 6. Gârsecg hlynede *the ocean roared*, Andr. Kmbl. 476; An. 238. Hlynede and dynede *raised shout and din*, Judth. 10; Thw. 21, 18; Jud. 23. Stefn in becom hlynnan under hârne siân *the voice got in and sounded under the grey stone*, Beo. Th. 5099; B. 2553. Hlynnende hlûde streámas, *torrentes*, Ps. Th. 73, 15.

hlynsian, hlinsian; *p*. ode *To sound, resound* :— Reced hlynsode *the mansion resounded*, Beo. Th. 1545; B. 770. Hlinsade, Exon. 108 b; Th. 415, 26; Rä. 34, 3. Hôfan and hlynsadan hlûdan reorde *elevaverunt flumina voces suas*, Ps. Th. 92, 4. Hlynsodon, Andr. Kmbl. 3089; An. 1547.

hlýp, es; *m. A leap, jump* :—Hlýp *saltus*, Ælfc. Gl. 61; Som. 68, 49; Wrt. Voc. 39, 33: Ælfc. Gr. 11; Som. 15, 14. Se dæg is gehâten saltus lunæ đæt is đæs mônan hlýp *the day is called saltus lunæ, that is, the moon's leap*, Lchdm. iii, 264, 24: Exon. 18 b; Th. 45, 16, 29; 46, 1, 13; Cri. 720, 726, 730, 736. Hlýpum *by leaps*, Th. 46, 31; Cri. 747. Heorta hlýpum *leaping like the hart*, Cd. 203; Th. 252, 5; Dan. 574. [*Laym*. lupe, leope: *A. R*. lupes, *pl* : *Icel*. hlaup; *n* : *O. H. Ger*. louf *cursus*: *Ger*. lauf.]

hlýp, e; *f*. [?] :—Dis sind đa landgemǽra . . . of đære ealdan hæcce into Presta hlýpe . . . of đam æssce tô đære ældan hlýpe of đare hlýpe, Chart. Th. 394, 16: 395, 9, 34, 35.

hlýpa, hliépa, an; *m. That which helps in leaping, in leaping on or mounting a horse, a horse-block* :—Siđđan hê wæs đæm cyninge tô đon geset ôþ his lîfes ende, đæt hê sceolde swâ oft stûpian swâ hê tô his horse wolde, and hê đonne se cyning hæfde his hrycg him tô hliépan *hoc infamis officii continua donec vixit damnatione sortitus, ut ipse acclinis humi, regem super adscensurum in equo dorso adtolleret*, Ors. 6, 24; Swt. 274, 25. Æt hinde hlýpan, Cod. Dipl. Kmbl. ii. 249, 35. [v. Halliwell's Dict. ' leapingblock *a horse-block* : leaping *the operation of lowering tall hedges for the deer to leap over*.']

hlýp-geat, es; *n*. [?] :—Ondlang geardes on đæt hlýpgeat, Cod. Dipl. Kmbl. iii. 180, 28.

hlýrian *to puff out the cheeks as in blowing a trumpet, to blow* [*a trumpet*] :—Býmaþ ł hlýriaþ mid býman *buccinate tuba*, Ps. Lamb. 80, 4. v. hleór.

hlýsa. v. hlîsa.

hlyst, es; *m*: e; *f. The sense of hearing, hearing, listening* :—Hlyst *auditus*, Ælfc. Gr. 11; Som. 15. 15. Đa fîf andgitu . . . hlyst . . . *the five senses . . . hearing* . . . , Homl. Th. ii. 550, 11: i. 138, 27. Gif se hlyst ôþstande đæt hê ne mǽge gehiéran *if the hearing be stopped so that he cannot hear*, L. Alf. pol. 46; Th. i. 92, 23. Đâ wearþ hæleþa hlyst *then was there listening of men*, Cd. 181; Th. 226, 28; Dan. 178: Exon. 55 b; Th. 196. 5; Az. 169. On đæs folces hlyste *in aures plebis*, Lk. Skt. 7, 1. On hlyste *auditione*, Ps. Th. 111, 6; Ælfc. Gr. 1; Som. 2, 29. Lǽcedômas wiđ yfelre hlyste *leechdoms against bad hearing*, L. M. 1; Lchdm. ii. 2, 14. Gif [mon] yfelne hlyst hæbbe *if a man have bad hearing*, i. 3; Lchdm. ii. 40, 26. [*Laym*. lust: *O. Sax*. hlust *hearing* : *Icel*. hlust *the ear*.] DER. ge-hlyst.

hlystan; *p*. te *To list, listen to, hear, hearken* :—Hî gefeallaþ on đa heortan đe hiera hlyst *they fall on the heart that listens to them*, Past. 15, 6; Swt. 97, 1. Mid đam đe hê hlyste đæs heofonlîcan sanges *whilst he was listening to the heavenly song*, Homl. Th. ii. 98, 5. Ne hlyst đû nâ ungesceádwîses monnes worda *do not listen to the words of an indiscreet man*, Prov. Kmbl. 47: Nicod. 3; Thw. 2, 5. Hlyst hider *hearken*, Past. 49, 2; Swt. 381, 14. Sunu mîn hlyste mînre lâre *fili mi acquiesce consiliis meis*, Gen. 27, 8. Ne hliste đû his worda *non audies verba illius*, Deut. 13, 3. Hlystaþ hwæt ic secge *hear what I say*, L. I. P. 5; Th. ii. 310, 8. Hlyste hê gôdes rǽdes *let him hearken to good counsel*, Homl. Th. i. 54, 16. Wê biddaþ đê leóf đæt đû hlyste ûre sprǽce *oramus, domine, ut audias nos*, Gen. 43, 20. Man lâreówum hlyste *let teachers be listened to*, L. Eth. vii. 19; Th. i. 332, 26. Hig hlyston him *audiant illos*, Lk. Skt. 16, 29. Hê sceal bôclârum hlystan swýđe georne *he must pay diligent attention to the teaching of books*, L. I. P. 2; Th. ii. 306, 8. Hig fundon hine hlystende *they found him listening*, Lk. Skt. 2, 46: Past. 49, 5; Swt. 385, 23. [*Laym*. lusten: *Orm*. lisstenn : *Ayenb*. lheste: *Icel*. hlusta.]

---

hlystend, es; *m. A hearer, listener* : — On môde đære hlystendra *in the mind of the hearers*, Homl. Th. i. 362, 18.

hlystere, es; *m. A hearer, listener* :—Đæt âþweahþ his hlysteras from synna horewum *that washes its hearers from the foulnesses of sins*, Homl. Th. ii. 56, 7.

hlyt [*or* hlýt?], es; *m. A lot, portion* : — Hlyt *sors*, Ælfc. Gr. 9, 44; Som. 13, 3. Đû gedydest đæt wê mǽtan ûre land mid râpum and mîn hlyt gefeóll ofer đæt betste *funes ceciderunt mihi in præclaris*, Ps. Th. 15, 6. On handum dînum hlyt mîn *in manibus tuis sortes meæ*, Ps. Spl. 30, 18. Hlyt wîsode đǽr hie dryhtnes ǽ dêman sceoldon *the lot appointed where they should judge the Lord's law*, Apstls. Kmbl. 18; Ap. 9. On hlyte sorti, Ælfc. Gr. 38; Som. 41, 18. Đû hit tôdǽlst mid hlyte *tu eam sorte divides*, Deut. 31, 7. Mid hâligra hlyte wunigan *to dwell with the saints*, Elen. Kmbl. 1639; El. 821. Hî sendon hlyt *miserunt sortem*, Ps. Spl. 21, 17. Swâ him dryhten sylf hlyt getǽhte *as God himself assigned a lot to them*, Andr. Kmbl. 12; An. 6: 28; An. 14. Ne sêc đû þurh hlytas hû đê geweorþan scyle *do not seek by casting of lots what thy fate is to be*, Prov. Kmbl. 32. Gif hwâ hlytas begâ *si quis sortilegia exerceat*, L. Ecg. P. iv. 19; Th. ii. 210, 11. [*The Pastoral has the form* hliet (v. hlêt), *which seems to correspond with the Gothic* hlauts *and would suggest* ý *not* y *in* hlyt. *But compare Icel*. hlutr, hlaut, Cl. and Vig. Dict.] v. hlot.

hlyta, hlytta, an; *m. A diviner, one who divines by casting lots* :—Flaminius forseah đa sægene đe đa hlyttan him sǽdon đæt hê æt đæm gefeohte ne côme wiđ Gallie *Flaminius contemtis auspiciis quibus pugnare prohibebatur adversum Gallos*, Ors. 4, 7; Swt. 184, 26. Tânhlyta *sortilegus*, Ælfc. Gl. 112; Som. 79, 106; Wrt. Voc. 60, 13. v. efen-hlytta; hlyt.

hlytere. v. tân-hlytere.

hlyþran, Gen. 41, 27. v. lyþer.

hlytm *a parting or deciding by lot, an arranging of shares* :—Næs đâ on hlytme hwâ đæt hord strude *the part of each in despoiling the hoard was not carefully allotted [each took what he could]*, Beo. Th. 6243; B. 3126.

hlyttrian *to purify* :—Ic hlyttrige *liquo*, Ælfc. Gr. 37; Som. 39, 41.

hlyttrung, e; *f. A purifying, refining* ; *defecatio vel* purgatio, Ælfc. Gl. 100; Som. 77, 23; Wrt. Voc. 55, 27.

hlýwing. v. hleówung.

hlýwþ. v. hleówþ.

hnǽcan. v. nǽcan.

hnǽgan; *p*. de *To neigh* :—Ic hnǽge *hinnio*, Ælfc. Gr. 30, 5; Som. 34, 58. Hors hnǽgþ *equus hinnit*, 22; Som. 24, 9. [*Wick*. neʒen: *Prompt. Parv*. neyyñ *hinnio* : *Icel*. gneggja, hneggja.]

hnǽgan; *p*. de *To cause to bow, bring low, humble, humiliate* :—Ic bebeóde bearnum mînum đæt hie đê hnǽgon æt gûþe *I command my sons to humble thee in battle*, Andr. Kmbl. 2660; An. 1331. [*Goth*. hnaiwjan *to abase*: *Icel*. hneigja *to bow*: *O. H. Ger*. hneigjan *subjicere, inclinare* : *Ger*. neigen.] v. ge-hnǽgan, hnâh, hnîgan.

hnǽgan, Beo. Th. 2641; B. 1320. v. nǽgan.

hnǽgung, e; *f. Neighing* :—Horsa hnǽgung *neighing of horses*, Ælfc. Gr. 1; Som. 2, 35.

hnæpf, hnæpp, hnæp, es; *m. A cup, bowl* : — Hnæp *ciatus, anthlia*, Ælfc. Gl. 25; Som. 60, 50, 51; Wrt. Voc. 24, 46, 47. Hnæp *anaphus*, Wrt. Voc. 82, 43. Hnæpp *patera*, 290, 74. Of đam hnæpfe *from the bowl*, Chart. Th. 439, 31. ii gebonede hnæppas *two polished bowls*, 429, 30. [*Laym*. nap *a cup*: *A. R*. nep: *Du*. nap, *a cup, basin* : *O. H. Ger*. hnapf *cratera, patera*, Grff. iv. 1130: *O. French* hanap: *Low Lat*. hanapus, v. Skt. Dict. *hamper*.]

hnæppan *to strike* [?] :—Swâ swâ sió nafu simle biþ swâ gesund hnæppen đa felga on đæt đe hî hnæppen *if the nave is always quite safe the fellies may strike against what they will*, Bt. 39, 7; Fox 222, 26. [Cf. (?) nap *to strike the head sharply with a stick*, E. D. S. Mid-Yorkshire Glossary; knap *to strike*; nap *a stroke*, Halliwell Dict.]

hnæppan, hnæppung. v. hnappian, hnappung.

hnæsce. v. hnesce.

hnâh; *adj. Bent down, low, lowly, humble, abject, mean, poor* : — And hê huâh tô eorþan âleât wiđ đæs engles *adoravitque eum pronus in terram*, Num. 22, 31. Næs hió hnâh ne tô gnêþ gifa *she was not mean nor too sparing of gifts*, Beo. Th. 3863; B. 1929. Iudas cwæþ đæt hê wênde him trage [Kmbl. þrage] hnâgre *Judas said that he expected for himself humiliating pain*, Elen. Kmbl. 1333; El. 668. Wêndon hie wera cwealmes þrǽge hnâgran *they expected the death of men, a still worse time*, Andr. Kmbl. 3195; An. 1600. Nô ic mê hnâgran talige đonne Grendel hine *I think myself no worse man than does Grendel himself*, Beo. Th. 1359; B. 677. Ful oft ic leán teohhode hnâharan rince sǽmran æt sæcce *full oft have I appointed reward to a warrior inferior and of less worth in battle*, 1909; B. 952. [*Goth*. hnaiws *lowly, humble*.]

hnappian; *p*. ode *To slumber, sleep, doze* :—Ne slǽpþ ne ne hnappaþ se đe hylt Israhêl *non dormitabit neque dormiet qui custodit Israel*, Homl. Th. ii. 230, 6. Hnæppaþ, Ps. Spl. 120, 4. Se đe hnæppaþ *qui dormit*, 40, 9. Đa mǽdenu hnappiaþ *the maidens slumber*, Homl. Th. ii. 566, 26. Ne slǽpþ se nô fæsđe ac hnappaþ *non autem dormire ed dormitare est*, Past. 28, 4; Swt. 195, 8. Ac đonne hnæppiaþ ûre

bræwas *palpebræ vero dormitant*, 195, 2. Gif hē hwôn hnappode ðær-rihte hine drehton nihtlīce gedwimor *if he dozed a little, straightway nightly phantoms tormented him*, Homl. Th. i. 86, 18. Ic hnæppode *ego dormivi*, Ps. Spl. 3, 5. Ðā hnappedon hig ealle and slēpon *dormitaverunt omnes et dormierunt*, Mt. Kmbl. [MS. A.] 25, 5. Ne ne hnæppie se ðe healde ðē *neque dormiet qui custodit te*, Ps. Spl. 120, 3. Ne ne hnappigen dîne bræwas *ne dormitent palpebræ tuæ*, Past. 28, 4; Swt. 193, 24, 19. Hnappiende *dormiens*, Ps. Spl. 77, 71. [*A. R.* nappen: *Chauc. Wick. Piers P.* nappe: *Prompt. Parv.* nappyñ or slomeryñ *dormito*: cf. *O. H. Ger.* nafizan, Grff. ii. 1053.]

hnappung, hnæppung, e; *f. Slumbering, dozing, drowsiness* :—Æresð mon hnappaþ gif hē ðonne ðære hnappunge ne swîcþ ðonne hnappaþ hē ôþ ðæt he o wierþ on fæstum slæpe *dormitando vero oculus ad plenissimum somnum ducitur*, Past. 28, 4; Swt. 195, 11. Wið hnappunge *against drowsiness*, L. Med. ex Quadr. 8, 10; Lchdm. i. 358, 24. Hnæppunge *dormitationem*, Ps. Spl. 131, 4: hnappunga, Ps. Th. *and* Lamb. [*Wick.* napping: *Prompt. Parv.* nappynge or slomerynge *dormitacio*: *O. H. Ger.* naffezung *dormitatio*.]

hnātan; *p.* hneót *To strike together, clash*, Andr. Kmbl. 8; An. 4. v. hnītan.

hneáw; *adj. Stingy, near, niggardly* :—Ðȳ læs se hneáwa and se gîtsigenda fægnige ðæs ðætte menn wēnen ðæt hē sîe gehealdsum on ðæm ðe hē healdan scyle oððe dǽlan *ne aut cor tenacia occupet, et parcum se videri in dispensatione exultet*, Past. 20; Swt. 149, 17. Ic ðē hneáw ne wæs landes and lissa *I was no niggard to thee of land and favours*, Cd. 136; Th. 171, 5; Gen. 2823. [*Icel.* hnöggr *niggardly, stingy*: *Ger.* ge-nau.] DER. un-hneáw.

hneáw-līce; *adv. Sparingly, stingily* :— Him ðæs leán āgeaf nalles hneáwlīce *to him for that the Lord gave reward with no sparing hand*, Cd. 86; Th. 108, 20; Gen. 1809.

hneáw-ness, e; *f. Stinginess, parsimony, niggardliness* :— Monig mon dêþ micel fæsten, and hæfþ ðone hlîsan ðæt hē hit dô for forhæfdnesse and dêþ hit ðeáh for hneáwnesse and for feohgîtsunge *many a man fasts much, and has the reputation of doing it for abstinence, and yet does it for stinginess and avarice*; sæpe sub parsimoniæ nomine se tenacia palliat, Past. 20; Swt. 149, 6. Swā ða rūmmôdan fæsthafolnesse lǽren, swā hī ða uncystegan on yfelre hneáwnesse ne gebrengen *sic prodigis prædicetur parcitas, ut tamen tenacibus periturarum rerum custodia non augeatur*, 60; Swt. 453, 29.

HNECCA, an; *m. A* NECK, *nape of the neck, back of the head* :—Hnecca *cervix* vel *jugulum*, Ælfc. Gl. 72; Som. 70, 116; Wrt. Voc. 43, 44: Wrt. Voc. 70, 26. Wā ðæm ðe willaþ lecggean bolster under ælcne hneccan menn mid tô gefônne ... Ðonne biþ se hnecca underlēd mid bolstre *væ his qui faciunt cervicalia sub capite universæ ætatis ad capiendas animas ... Quasi cervicalibus caput jacentis excipitur*, Past. 19, 1; Swt. 143, 14. Gnīd ðone hneccan mid ðȳ *rub the back of the neck with it*, L. M. 1, 1; Lchdm. ii. 20, 25. Ðæt ðū næbbe nān þing hāles fram ðam fôtwolmum oþ ðone hneccan *sanari non possis a planta pedis usque ad verticem tuum*, Deut. 28, 35. [*Laym.* necke: *Chauc. Piers P. Prompt. Parv.* nekke *collum*: *O. Frs.* hnecka: *Icel.* hnakki *the nape of the neck, back of the head*: *O. H. Ger.* hnach *testa capitis, occiput, cacumen*: *Ger.* nacken.]

hnesce, hnæsce, hnysce; *adj. Nesh, soft, delicate, tender, effeminate* :—Hnysce hwîtel *linna*, Ælfc. Gl. 63; Som. 68, 113; Wrt. Voc. 40, 23. Hnesce on môde tô flǽsclîcum lustum *yielding easily to the lusts of the flesh*, Homl. Th. ii. 220, 4. Gefrēdan hwæt biþ heard hwæt hnesce *to feel what is hard, what soft*, 372, 32: Elen. Kmbl. 1226; El. 615. Heó is hnesce on æthrine *it is soft to the touch*, Herb. 15, 1; Lchdm. i. 108, 1. Sîe ðær eác lufu næs ðeáh tô hnesce *sit itaque amor, sed non emolliens*, Past. 17, 11; Swt. 127, 2. Hwæt getácnaþ ðonne ðæt flæsc būton unfæsd weorc and hnesce *quid enim per carnes nisi infirma quædam ac tenera, fuerit*, 34, 6; Swt. 235, 15. Ðonne hys twig byþ hnesce *cum ramus ejus tener fuerit*, Mt. Kmbl. 24, 32. Æghwæt hnesces oððe heardes, L. de Cf. 9; Th. ii. 264, 6: Salm. Kmbl. 574; Sal. 286. Ðonne geþafaþ him mon on ðære hnescean ôlecunge *eique mollities favoris adhibetur*, Past. 19, 1; Swt. 143, 21. Swā hē ðone hnescan þafettere on récelēste ne gebrenge *ut remissis ac lenibus non crescat negligentia*, 60; Swt. 453, 25. Ne gedafenaþ ūs ðæt wē symle hnesce beón on ūrum geleáfan *it befits us not to be ever delicate in our belief*, Homl. Th. i. 602, 12. Mann hnescum gyrlum gescrȳdne *hominem mollibus vestitum*, Mt. Kmbl. 11, 8; Lk. Skt. 7, 25. Heó biþ hnescuum leáfum *it is a plant with soft leaves*, Herb. 6, 1; Lchdm. i. 96, 14. Ic hæbbe hnesce litlingas *parvulos habeam teneros*, Gen. 33, 13. Syle him etan hnesce ægere *give him lightly boiled (?) eggs to eat*, Lchdm. iii. 134, 22. Ælc wuht biþ innanweard hnescost *every creature is softest inside*, Bt. 34, 10; Fox 150, 6. Drihten næfre ne forsyhþ ða eáþmôdan heortan ne ða hnescestan *the Lord never despises the humble heart nor the weakest*, Blickl. Homl. 99, 5. [*A. R.* nesche: *Orm.* nesshe: *Chauc.* nesh: *Goth.* hnaskwus *soft*.]

hnescian, hnexian; *p.* ode *To make*, or *to become, soft, to soften* :—Ic hnexiğe *mollio*, Ælfc. Gr. 30; Som. 34, 53. Lege ðonne on ðær hit heardige hnescaþ hyt sôna *apply where it is hard, it will at once soften*,

Herb. 2, 11; Lchdm. i. 84, 4. Ðonne hnescaþ se swile sôna *then the swelling will soften at once*, L. M. 2, 19; Lchdm. ii. 202, 10. Se hearda stān aðamans hnescaþ ongeán ðæt lîðe buccan blôd *durus adamas leni hircorum sanguine mollescit*, Past. 37, 4; Swt. 271, 4. Hī hnescodon sprǽca his *mollita sunt sermones ejus*, Ps. Spl. 54, 24. Ongunnon ða godes cempan hnexian *God's warriors began to yield*, Homl. Skt. 5, 48, 51: 8, 29. [*Orm.* nesshenn: *Ayenb.* nhesseþ, *pres*: *Prompt. Parv.* neschyñ or make nesche *mollifico*.] DER. ā-hnescian.

hnesc-līc; *adj. Effeminate* :—Hē wæs swîðe hnesclīc man *he [Sardanapalus] was a very effeminate man*, Ors. 1, 12; Bos. 35, 15. Hī beóþ hneslīce swā forlegene *hi sunt delicati ita fornicantes*, L. Ecg. P. iv. 68, 6; Th. ii. 228, 18.

hnesc-līce; *adv. Gently, softly, tenderly* :—Hē his hiéremonna yfelu tô hnesclīce forberan ne sceal *subditorum mala tolerari leniter non debent*, Past. 21, 5; Swt. 159, 25. Ðonne hē his wambe suā hnesclīce ôlecþ *dum ventri molliter serviunt*, 43, 5; Swt. 313, 12.

hnesc-ness, e; *f. Softness, delicacy, gentleness, weakness* :—Hnescnyss *mollities*, Ælfc. Gr. 12; Som. 15, 56. Ðære hnescnesse ūres flǽsces wē beóþ underþiédde *corruptionis nostræ infirmitatibus subjacemus*, Past. 21, 4; Swt. 159, 5. Genim ðyses wæstmes hnescnysse innewearde *take the inward soft part of this fruit*, Herb. 185, 2; Lchdm. i. 324, 9. Gif hwā for his hnescnysse ðæt fæsten āberan ne mæg *si quis præ mollitie sua jejunium perferre nequeat*, L. Ecg. P. iv. 60; Th. ii. 220, 24. Gif þurh his hnescnysse seó heord forwurþ *if through his want of vigour the flock perish*, L. I. P. 19; Th. ii. 326, 22.

hnifol, es; *m. The forehead* :—Hnifol *frons*, Wrt. Voc. 282, 46. Smire mid ða þunwangan and ðone hnifol and ufan ðæt heáfod *smear therewith the temples and the forehead and the top of the head*, L. M. 3, 1; Lchdm. ii. 306, 6.

hnifol-crumb; *adj. Cernuus*, Cot. 45, 56, Lye.

hnigan; *p.* hnāh; *pp.* hnigen *To bend, bow down, incline, descend, decline, sink* :—Ðonne hnīge eft under lyfte helm londe neár *then I bend again under the airy cover nearer the land*, Exon. 102 a; Th. 386, 18; Rä. 4, 63. Loth ðām giestum hnāh *Lot bowed to the guests*, Cd. 112; Th. 147, 15; Gen. 2440. Hnāg ic ðām secgum tô handa *I bowed down within the reach of the men*, Rood Kmbl. 118; Kr. 59. Hnigon ða mid heáfdum heofoncyninge tôgeánes *bent then their heads before heaven's king*, Cd. 13; Th. 16, 1; Gen. 237: 218; Th. 279, 18; Sat. 240: 225; Th. 298, 15; Sat. 533. Wit noldon hnîgan mid heáfdum hālgum Drihtne *we would not bend our heads to the holy Lord*, 35; Th. 46, 10; Gen. 742: 217; Th. 277, 22; Sat. 208. Ðā hē tô helle hnîgan sceolde *when he must sink to hell*, 221; Th. 288, 4; Sat. 375. [*Goth.* hneiwan *to bend downwards, decline*: *O. Sax.* hnîgan: *Icel.* hnîga *to bow down, sink, fall gently*: *O. H. Ger.* hnîgan *obstipare, adorare*.] DER. ge-, on-, under-hnîgan; *and see* hnǽgan.

hnigian; *p.* ode *To bend down* [*the head*] :—Ðonne uplang āsitte hnigie *let him sit up and bend his head downwards*, L. M. 1, 1; Lchdm. ii. 18, 16.

hnipend *humilis*, Hpt. Gl. 436. v. next word.

hnipian; *p.* ode *To bow the head* :—Biþ wuhta gehwilc onhnigen tô hrusan hnipaþ of dūne on weoruld wlîtaþ wilnaþ tô eorþan [cf. *in the prose version*, Fox 254, 28, ealle biôþ of dūne healde wið ðære eorðan] *prona tamen facies hebetes valet ingravare sensus*, Bt. Met. Fox 31, 26; Met. 31, 13. Ðā wearþ Cain suîðe hrædlîce irre and hnipode of dūne *iratusque est Cain vehementer, et concidit vultus ejus*, Past. 34, 5; Swt. 235, 6. [Þa nipeden hyo ealle *dormitaverunt omnes*, Mt. Kmbl. 25, 5, col. 2: *Laym.* þa sunne gon to nipen: cf. *Icel.* hnîpa *to be downcast, droop*: hnipna *to droop, despond*: *M. H. Ger.* nipfen: *Ger.* nippen *to nod*.]

hnītan; *p.* hnāt, *pl.* hniton; *pp.* hniten *To strike, thrust, push, come against with a shock* :—Ðonne hniton fēðan *in the shock of meeting hosts*, Beo. Th. 2659; B. 1327: 5082; B. 2544. Gif oxa hnite wer oððe wîf *si bos percusserit virum aut mulierem*, Ex. 21, 28. Ðonne ic hnîtan sceal hearde wið heardum *when I shall batter hard on the hard*, Exon. 129 b; Th. 497, 21; Rä. 87, 4. [*Icel.* hnîta *to strike, clash*.] DER. of-hnîtan.

hnitol; *adj. Given to striking, thrusting, pushing, having the head bent* [*as an animal when it butts* (?)] :—Hnitol vel eádmôd *cernuus, pronus* vel *inclinatus*, Ælfc. Gl. 9; Som. 56, 116; Wrt. Voc. 19, 1. Gif se oxa hnitol wære *si bos cornupeta fuerit*, Ex. 21, 29, 36: L. Alf. 21; Th. i. 48, 29.

hnitu, e; *f. A nit* :—Hnitu *lens* vel *lendix*, Ælfc. Gl. 23; Som. 60, 8; Wrt. Voc. 24, 12. Hnite and wyrmas on weg tô dônne ðe on cildum beóþ *to remove nits and worms that are on children*, L. Med. ex Quadr. 9, 15; Lchdm. i. 364, 6. [*Prompt. Parv.* nyte, wyrme *lens*: *Icel.* gnit; *f*: *O. H. Ger.* niz: *Ger.* niss.]

hnoc *mutinus*, Ælfc. Gl. 22; Som. 59, 83; Wrt. Voc. 23, 49. v. [?]hnot.

hnol, hnoll, es; *m. The top, crown of the head* :—Hnol *vertex*, Ælfc. Gl. 69; Som. 70, 32; Wrt. Voc. 42, 40: 64, 22. Eástdǽl his hnol heóld *the crown of his head held the east*, Homl. Th. ii. 256, 2. Fram ðam hnolle ufan oþ his fôtwylmas neoðan *from the crown of his head down to the soles of his feet*, 480, 12: 452, 26: 524, 2. On hnol his

in verticem ejus, Ps. Spl. 7, 17 : 67, 23.   [*Wick.* nol *cervix :* O. H. Ger. hnol *culmen, cacumen, vertex, sinciput.*]

hnoppa, an ; *m. Nap of cloth ;* villus, Som.   [*Prompt. Parv.* noppe of a clothe *villus, tomentum,* see note.]

hnossian ; *p.* ode *To beat, strike :*—Mec hnossiaþ homera lāfe *swords shall strike me,* Exon. 102 b ; Th. 388, 13 ; Rä. 6, 7.   [*Cf. Icel.* hnoss *an ornament.*]

hnot ; *adj. Bald, shaven, close-cut :* — Calu oððe hnot *glabrio,* Ælfc. Gr. 9, 3 ; Som. 8, 37. Hnot *mutilum, mutilatum,* Cot. 131, Lye. Tô ðon hnottan seale *to the pollard-willow,* Cod. Dipl. Kmbl. v. 193, 35. On ða hnottan dîc of ðære hnottan dîc *the dike without turf* (?), iii. 211, 24.   [*Chauc.* not-heed, Prol. 109 : *Dep. Rich.* not of his nolle, 3, 46: *Halliwell Dict.* not smooth, without horns; to shear, poll : see *Nares' Gloss.* nott, nott-pated, -headed.]

hnut-beám, es ; *m. A nut tree ;* corylus avellana : — Hnutbeám *nux* vel *nucarius,* Ælfc. Gl. 47 ; Som. 65, 38 ; Wrt. Voc. 33, 35. Hnutbeámes rind, L. M. i. 3, 6 ; Lchdm. ii. 42, 3 ; 52, 1. Hnutbeámes leáf, Lchdm. iii. 6, 15.   [O. H. Ger. hnuz-boum *amygdalus, nux, nucus, corylus :* Ger. nuss-baum.]

hnut-cyrnel, es ; *m. n. A nut kernel :*—Genim hnutcyrnla, L. M. 1, 2 ; Lchdm. ii. 34, 19.

hnutu, e ; *f. A nut :* — Hnutu *juglantis* vel *nux,* Ælfc. Gl. 45 ; Som. 64, 97 ; Wrt. Voc. 32, 32. For æppla and hnuta æte *from eating of apples and nuts,* L. M. 2, 39 ; Lchdm. ii. 246, 21. Hnute hula *culliole,* Ælfc. Gl. 31 ; Som. 61, 105 ; Wrt. Voc. 27, 34. Oðera hnutena cyrnlu *kernels of other nuts,* iii. 134, 23. Of frenciscen hnutu[m] *made of French nuts,* 122, 28. Cyrnlu of pîntrŷwenum hnutum *kernels out of pine tree nuts,* Herb. 134, 2 ; Lchdm. i. 250, 9. Gif heó gelôme eteþ hnyte *if she is often eating nuts,* iii. 144, 20. Hnyte somnian, gaderian *to gather nuts,* 174, 5 : 208, 18. On ðam ôðrum dæge wæs Aarones gyrd gemętt grôwende and berende hnyte *on the next day Aaron's rod was found growing and bearing nuts,* Homl. Th. ii. 8, 16, 18. Bringaþ ðam men lâc sumne dæl tyrwan and hunig and hnite *deferte viro munera, modicum resinæ et mellis et amygdalarum,* Gen. 43, 11.   [*Ayenb.* nhote : *Prompt. Parv.* note *nux, nucleus :* Icel. hnot ; *f. pl.* hnetr : O. H. Ger. hnuz, nuz *nux, migdola :* Ger. nuss.] DER. hæsel-, pîn-hnutu.

hnygela [or hnigela ?], hnygele, an ; *m. f. A shred, clipping :*—Hnygela *tomentum ;* seolce hnygele *platum* [= placium, Som.] Ælfc. Gl. 64 ; Som. 69, 3, 4 ; Wrt. Voc. 40, 37, 38. Hnyglan *putamina,* Cot. 152, Lye. [*Cf.* (?) nig *the clippings of money :* niggling *clipping :* niggler *a clipper,* Grose's Slang Dict : see also *Halliw. Dict.* niggle, niggling.]

hnŷlung, e ; *f. A kneeling, reclining ;* accubitus, Ælfc. Gl. 65 ; Som. 69, 52 ; Wrt. Voc. 41, 9.

hnyte. v. hnutu.

hô. v. hôh.

hô-banca, an ; *m. A couch, sofa ;* sponda, Wrt. Voc. 290, 13. v. hôh.

hoc ; *gen.* hocces *Hock, mallow :*—Hocces leáf, L. M. 3, 37 ; Lchdm. ii. 330, 3. Hocces moran, 41 ; Lchdm. ii. 334, 27. Hoc, Lchdm. iii. 22, 2. [In E. D. S. Plant Names 'hock *althæa rosea, malva sylvestris, malva rotundifloria.*' Skeat, Etymol. Dict. supposes the word was borrowed from Celtic : Welsh *hocys* mallows.]

HÔC ; *m. A* HOOK :— Hooc *arpago* vel *palum,* Ælfc. Gl. 3 ; Som. 55, 71 ; Wrt. Voc. 16, 43. Ic eom swâ swâ fisc on hôce *I am as the fish on the hook,* Nar. 40, 33. Ðonne biþ hê geteald tô ðære fŷrenan eá and tô ðam îsenan hôce *then shall he be assigned to the fiery river and the iron hook,* Blickl. Homl. 43, 25, 27. Wîngearda hôcas ðe hî mid bindaþ ðæt him nêhst biþ *capreoli* vel *cincinni* vel *uncinuli,* Ælfc. Gl. 59 ; Som. 68, 9 ; Wrt. Voc. 38, 59. Ðâ sôhtan heora gewinnan him sarwe and worhtan him hôcas *at contra non cessant uncinata hostium tela,* Bd. 1, 12 ; S. 481, 21 : Homl. Th. i. 362, 27. v. hinder-hôc.

hôced ; *adj. Shaped like a hook, curved :* — Oþ ðat hit cymþ tô ðan hôkedan gâran *until it comes to the curved strip of land,* Cod. Dipl. Kmbl. iii. 434, 10.

hôcer. v. hôcor.

hociht ; *adj. Full of mallows :*—Ærest onlong Foss on ða hocihtan dîc of ðere hocihtan dîc on ðone brâdan þorn *to the mallowy ditch,* Cod. Dipl. Kmbl. ii. 365, 25. [So Cockayne, Lchdm. iii. 332, col. 1, translates the word ; or should the word be written *hóciht* = with many bends ? Cf. hôced.]

hôciht. v. heoru-hôciht.

hoc-leáf, es ; *n. Mallow :*—Hocleáf *malva,* Wrt. Voc 79, 11. Hocleáf. Ðeós wyrt ðe man maluæ erraticæ oðrum naman hocleáf nemneþ byþ cenned æghwær on begānum stôwum *this plant, which is called malva erratica, and by another name hockleaf, is produced everywhere in cultivated places,* Herb. 41, 1 ; Lchdm. i. 142, 4 : L. M. 3, 8 ; Lchdm. ii. 312, 17. Hocleáf, Lchdm. iii. 48, 18.

hôcor, es ; *m.* [?] *Mockery, scorn, insult, derision :* — Tô oft man mid hôcere gôddæd hyrweþ *too often good deeds are depreciated with derision,* Swt. A. S. Rdr. 110, 162.   [O. E. Hom. to lusten hoker : *Laym.* hoker and scarn : *Chauc.* hoker and bissemare.]

hôcor-wyrde ; *adj. Using scornful, mocking language :*—Hêr sind on

earde hôcorwyrde æghwær *there are in the land here everywhere men of scornful speech,* Swt. A. S. Rdr. 109, 156.   [*Cf. Laym.* Sexisce men mine unhæle me atwiten mid heore hokerworden.]

hôd, es ; *m. A hood ;* cucullus, caputium, Cot. 31, Lye.   [*Laym.* A. R. hod : O. H. Ger. huot, hôt ; *m. mitra, tiara, cidaris :* Ger. hut.]

hoeg. v. heg.

hoelan = hēlan *to speak evil of, calumniate :*—Hoelende *calumniantes,* Mt. Kmbl. Rush. 5, 44.   [*Icel.* hæla *to praise, flatter, boast.*] v. hôl, hôlian.

hof, es ; *n. A house, hall, dwelling, building,* ædes, domus : — Lytel hof *ædicula,* Ælfc. Gl. 107 ; Som. 78, 84 ; Wrt. Voc. 57, 60. Cinges hof *basilica,* Som. 78, 86 ; Wrt. Voc. 58, 1. Hof sêleste *dwelling most excellent* [*the ark*], Cd. 69 ; Th. 84, 6 ; Gen. 1393 : 66 ; Th. 79, 25 ; Gen. 1316: 67 ; Th. 81, 15 ; Gen. 1345 : 73 ; Th. 90, 2 ; Gen. 1489. Gif hwâ hwylce hefige yfelnysse on his hofe geseó genime mandragoran on middan ðam hûse swâ mycel swâ hê ðonne hæbbe ealle yfelu hê ût ânŷdeþ *if any one see some grievous evil in his home, let him take mandragora into the middle of the house, as much as he has at the time, he will drive out all evils,* Herb. 132, 7 ; Lchdm. i. 248, 11: Cd. 76 ; Th. 94, 29 ; Gen. 1569 : 128 ; Th. 148, 13 ; Gen. 2456. Hê gewât from his âgenum hofe isaac lædan *he departed from his own house leading Isaac,* 139 ; Th. 173, 32 ; Gen. 2870. Him Hrôðgâr gewât tô hofe sînum rîce tô reste *Hrothgar had gone to his sleeping-chamber,* Beo. Th. 2477 ; B. 1236. Tô hofe sînum *to her dwelling,* 3019 ; B. 1507 : 3953 ; B. 1974. Se hâlga wæs tô hofe læded in ðæt dimme ræced *the saint was led to the building* [*prison*] *into that dark house,* Andr. Kmbl. 2616 ; An. 1309. Of ðam engan hofe, Exon. 73 b ; Th. 274, 12 ; Jul. 532 : Elen. Kmbl. 1420 ; El. 712. Tô hofe *to the* [*queen's*] *house,* 1111 ; El. 557. In ðam reónian hofe *underground,* 1664 ; El. 835. Him hof tæhte *pointed out to them the dwelling* [*of Hrothgar*], Beo. Th. 630 ; B. 312. Ðæt rædleáse hof *hell,* Cd. 2 ; Th. 3, 32 ; Gen. 44 : 217 ; Th. 276, 23 ; Sat. 193. Hofa *ædes,* Ælfc. Gl. 107 ; Som. 78, 83 ; Wrt. Voc. 57, 59. Hê ða hofa gehealdeþ and begŷmeþ *qui illa oppida maritima observat,* Nar. 37, 26. Hofu, Andr. Kmbl. 1676 ; An. 840 : Exon. 124 a ; Th. 477, 26 ; Ruin. 30. For Faraones hofun *in domos Pharaonis,* Ex. 8, 24. Hofum, Beo. Th. 3677 ; B. 1836.   [O. Sax. O. Frs. hof: Icel. hof *a temple :* O. H. Ger. hof *curtis, curta, atrium, aula, domus :* Ger. hof.] DER. ceaster-, gæst-, gnorn-, grorn-, heolstor-, mearc-, morþor-, sand-, stân-, sûsl-, ŷþ-hof.

HÔF, es ; *m. A* HOOF :—Hôf *ungula,* Ælfc. Gl. 72 ; Som. 71, 6 ; Wrt. Voc. 43, 59 : Wrt. Voc. 71, 76. Hors hôfum wlanc *the horse proud of hoofs,* Runic pm. Kmbl. 343, 5 ; Rûn. 19.   [*Icel.* hôfr : O. H. Ger. huof *ungula :* Ger. huf.]

hofding, es ; *m. A chief, captain, principal, ringleader :*—Rawulf eorl and Rogcer eorl wæron hofdingas [cf. yldast tô ðam unreode, l. 13] æt ðisan unræde *earl Ralph and earl Roger were ringleaders in this evil counsel,* Chr. 1076 ; Erl. 213, 31.   [Borrowed from *Icel.* höfðingi *a chief, leader, ringleader.*]

hofe. v. dim-hofe.

hôfe, an ; *f. Hove, alehoof* [v. English Plant Names, E. D. S.] ; glechoma hederacea :—Hôfe *viola,* Ælfc. Gl. 41 ; Som. 63, 132 ; Wrt. Voc. 31, 13. Genim hôfan *take hove,* L. M. 1, 1 ; Lchdm. ii. 20, 5. Brûne hôfe, Lchdm. iii. 292, 9. Genim ða reádan hôfan, L. M. 1, 2 ; Lchdm. ii. 34, 14. Mersc-hôfe, 1, 38 ; Lchdm. ii. 94, 10. Tûnhôfe, 3, 60 ; Lchdm. iii. 344, 2.

hofer, es ; *m.* [?] *A hump, swelling :* — Hofer *gibbus* vel *struma,* Wrt. Voc. 86, 71.   [O. H. Ger. houar, houer *gibbus.*]

hoferede ; *adj. Humpbacked :* — Hoferede *gybberosus* vel *strumosus,* Wrt. Voc. 86, 70 : 49, 7. Ðæt cild biþ hoforode *the child is humpbacked,* Lchdm. iii. 144, 26. Hoferede *gibbus,* Past. 11, 1, 3 ; Swt. 65, 4 ; 66, 12.   [O. H. Ger. houaradi *gibbus ;* hofaroht *gibberosus.*]

hoffing, es ; *m. A circle ;* orbis :— Hoffingas *orbes,* Lye.   [*Leo,* 40, 20 ; 197, 12, gives a gloss hôf-ring, hôf-hring *orbis,* explaining the word as a *horse-shoe.*]

hôf-rec, -ræc, es ; *n. Hoof-track :*—Sing on ðæt hôfrec *sing over the hoof-track,* Lchdm. i. 392, 9. Dryp on ðæt hôfræc ðæt wex *drop the wax into the hoof-track,* iii. 286, 4.

hof-rede ; *adj. Confined to the house ;* clinicus, Ælfc. Gl. 77 ; Som. 72, 30 ; Wrt. Voc. 45, 62.

hof-þela *tesqua,* Lye.

hof-weard, es ; *m. An ædile ;* ædilis, Ælfc. Gl. 8 ; Som. 56, 105 ; Wrt. Voc. 18, 54.

hog-. v. hoh-.

hoga ; *adj. Careful, thoughtful, prudent :*—Hoga *prudens,* Rtl. 105, 1. Geleáffull þegn and hoga *fidelis servus et prudens,* Mt. Kmbl. Lind. 24, 25. Wosas gê hogo *estote prudentes,* 10, 16. Hogum *prudentibus,* 11, 25. Gearnfulle ł hogo wosa *solliciti esse,* Lk. Skt. Lind. 12, 11.

hoga, an ; *m. Care,* R. Ben, 53, Lye.   v. ymb-hoga.

hoga-fæst ; *adj. Careful, prudent :*—Hogofæste, *prudentes,* Mt. Kmbl. Lind. 25, 2, 4. v. hoh-fæst.

hoga-scipe, es ; *m. Prudence, carefulness, thoughtfulness, wisdom :*—

Hogascip *prudentia*, Rtl. 81, 14. Hogascip *prudentia*, Lk. Skt. Lind. 2, 47. Tô hogascipe *ad prudentiam*, 1, 17.

hogde. v. hycgan.

hogian; *p.* ode *To employ the mind, to think, mind, consider, know, understand, care, be solicitous* or *anxious, to purpose, strive, intend, be intent on, resolve* :—Ymbe mîne mâgas ic hogige *erga propinquos curo*, Ælfc. Gr. 47; Som. 47, 29. Ðû hogast embe ðîne neóde *thou art busied about thy needs*, Homl. Th. i. 488, 23. Ne hogaþ hê be ðam heofenlîcan læcedôme *he is not anxious about the heavenly medicine*, ii. 470, 16. Hê hogaþ tô ðære betran wynne *he directs his mind to the better joy*, Exon. 95 a; Th. 355, 23; Reim. 81. Hogaþ *satagit*, Mone Gl. 356. Hogiaþ *satagunt*, 435. Hia hogaþ *sapiant*, Mt. Kmbl. p. 2, 5. For ðâm mannum ðe mid mâran gewilnunge ðæs æteorigendlîcan lîfes hogiaþ ðonne ðæs êcan *for those men whose minds are busied with a greater desire of the life that perishes than of the life eternal*, Homl. Th. ii. 368, 4: 342, 28. Ymbe ðîne handgeweorc ic hogode georne *in factis manuum tuarum meditabar*, Ps. Th. 142, 5. Mid ðý ic wæs lytel ic hogade swæ lytel *cum essem parvulus sapiebam ut parvulus*, Rtl. 6, 17. Ic ðæt hogode ðæt ic eówra leóda willan geworhte *I purposed to work your people's will*, Beo. Th. 1268; B. 632. Hwæt hogodest ðû hidercyme ðînne on wrâðra geweald *why didst thou resolve to come hither into the power of hostile men*, Andr. Kmbl. 2633; An. 1318. Ic on ðînre hælu hogode *I thought on thy salvation*, Ps. Th. 118, 81. Ðû ne hogodest *thou didst not consider*, Soul Kmbl. 83; Seel. 42. Hê on heortan hogode georne hû hê mid searuwe swylce âcwealde *he diligently considered in his heart how with cunning he might kill such*, Ps. Th. 108, 16: Swt. A. S. Rdr. 98, 92. Hê lythwon hogode ymbe his sâwle þearfe *he thought little about the needs of his soul*, 101, 201; Homl. Th. ii. 118, 15. Se feónd hogode on ðæt micle morþ men forweorpan *the foe intended to cast men into that great perdition*, Cd. 32; Th. 43, 14; Gen. 690. Hê tô friþe hogode *his purpose was to protect*, Andr. Kmbl. 1244; Án. 622. Ealle ðe mê yfel hogedon *qui cogitant mihi mala*, Ps. Th. 69, 3: 57, 2. Hî hine lufedan leáse mûþe ne ðæs on heortan hogedan âwiht *dilexerunt eum in ore suo, et lingua sua mentiti sunt ei*, 77, 35. Ðæt hî ðý læs ymb fleám hogodan *minus posse fugam meditari*, Bd. 3, 18; S. 546, 26. Hogedon âninga *their only purpose was*, Judth. 12; Thw. 25, 9, 22; Jud. 250, 273. Hogodon georne hwâ ðær mid orde ærost mihte on fægean men feorh gewinnan *they eagerly strove who there first with the sword's point might of the fey man win the life*, Byrht. Th. 135, 25; By. 123. Ne hoga ðû embe ðæt *be not anxious about that*, Homl. Swt. 3, 416. Hogiaþ *consider*, Homl. Th. ii. 124, 14. Hogiaþ *sapite*, Ps. Spl. C. 93, 8. Hogaþ gie *sapite*, Rtl. 13, 21: 25, 5. Hogige se yfela ðæt hê âstande *let the evil man be intent upon standing*, Homl. Th. i. 56, 23. Wê sceolon hogian embe ða bôte *we must busy ourselves about the reparation*, 274, 11. Wê sceolon carfullîce hogian ðæt wê ðone mâran gylt forfleón *we ought anxiously to endeavour to flee from the greater guilt*, 484, 5. Wê sceolon hogian hû wê hî begyton *we must consider how we may obtain it*, ii. 316, 25. Ne þurfon gê nô hogian on ðam anwealde ne him æfter þringan *ye need not aim at power nor press after it*, Bt. 16, 1; Fox 50, 29. Ne beó gê nâ hogiende ymb ða morgenlîcan neóde *nolite esse solliciti in crastinum*, Mt. Kmbl. 6, 34. Hogiende *cogitantes*, Mone Gl. 390. Hogiendum *nitentibus*, 420. [Laym. hoȝede, p: Icel. huga; pp. hugat: O.H.Ger. hugeta, hogeta, p.] DER. be-, for-, ge-, ofer-, wið-, ymb-hogian. v. hycgan.

hogo-. v. hoga-.

hogu, e; *f. Care, anxiety, solicitude* :—Habbon hî hoge ðæt hî sýn swilce ðæt hî wurþfullîce herigan mâgon *let them have a care that they be such that they may worthily praise*, Homl. Th. i. 446, 32. Hê næfþ nân andgit ne hoga embe Godes beboda *he hath no understanding nor cares about God's commandments*, 132, 13. [O. and N. hoȝe: R. Glouc. howe.] v. heort-hogu, hoga.

hogung, e; *f. Caring, care; cura*, Lye.

hôh, hô; *gen.* hôs; *m. A heel, hough* :—Hôh niþeweard *calx*, Wrt. Voc. 283, 75. Hô *calx*, Ælfc. Gr. 9, 72; Som. 14, 17, Hwæt is ðæs wîfes hô? ... Ðæs wîfes hô getâcnode ... *what is the woman's heel? ... The woman's heel signified* ..., Boutr. Scrd. 20, 13, 19. Hôs mînes *calcanei mei*, Ps. Spl. 48, 5. Dô on ðînne winstran scô under ðînum hô *put it into thy left shoe under thy heel*, Lchdm. i. 396, 2. Âhefþ hys hô ongeán mê *levabit contra me calcaneum suum*, Jn. Skt. 13, 18: Gen. 3, 15. Him on hôh beleác heofonrîces weard merehûses mûþ *God closed the door of the ark behind him*, Cd. 89; Th. 82, 16; Gen. 1363. Mînra hôa *calcanei mei*, Ps. Th. 48, 5. Pharao him filigde æt ðâm hôu *Pharaoh followed at their heels*, Homl. Th. ii. 194, 22. Hôs mîne *calcaneum meum*, Ps. Spl. 55, 6. [Cf. Icel. hâ-sin.]

hôh, hôgh, hô, hoo a form occurring in local names whose meaning is thus given by Kemble : 'Originally a point of land, formed like a heel, or boot, and stretching into the plain, perhaps even into the sea,' Cod. Dipl. iii. xxvi, where see the references to the various forms. Kemble's supposition is borne out by the following passage, in which the word occurs independently :—Wê ðâ fôron forþ be ðæm sæ and ðær ða heán hôs and dene and gârsecg ðone æthiopia wê gesâwon *promuntoria ad oceanum in*

ethiopia vidimus, Nar. 24, 9. [Cf. (?) over hil and hogh, Cursor Mundi 15826.]

hôh-, hog-fæst; *adj. Firm of mind, prudent, wise* : — Hogfæstum *prudentibus*, Mt. Kmbl. Lind. 11, 25. v. hoga-fæst.

hôh-fôt, es; *m. The heel* : — Hô ł hôhfôt *calcaneum*, Ps. Lamb. 55, 7.

hôh-, hog-ful; *adj. Mindful, careful, anxious, wise, prudent* : — Ic nû on sibbe gesitte on mînne cynestôl hohful embe ðæt hû ic his lof ârære *quiete pace perfruens, studiosus sollicite de laudibus Creatoris omnium occupor addendis*, Chart. Th. 240, 8. Ðâ wearþ ðæt mæden mycclum hohful hû heó æfre wæras wissian sceolde *then became the maiden very anxious how she was ever to direct men*, Homl. Skt. 2, 121. Ðâm ðe lufiaþ swîðor ða heálîcan clænnysse ðonne ða hohfullan gâlnysse *to those that love exalted chastity more than the wantonness which is full of care*, Homl. Th. ii. 324, 5. Hogfullum *prudentibus*, Mt. Kmbl. Lind. 11, 25. [Laym. hoh-fulle, pl : Orm. hoȝhe-full.]

hohful-ness, e; *f. Anxiety, care, trouble* : — Sæde ic mînum witun mînes môdes hohfulnysse *I told the anxiety of my mind to my 'witan,'* Cod. Dipl. Kmbl. iii. 349, 11.

hôh-hwyrfing, e; *f. A turning on the heel so as to describe a circle* [?] ; *orbis*, Som.

hohinge-rôd, e; *f. A cross, gibbet*, W. Cat. p. 294.

hoh-, hog-lîce; *adv. Prudently, thoughtfully* :—Hoglîce, *prudenter*, Lk. Skt. Lind. 16, 8.

hoh-môd; *adj. Having an anxious mind, anxious*, Lye.

hohmôd-ness, e; *f. Anxiety, trouble, care*, Som.

hôh-scanca, an; *m. The shank; crus* :—Sceápes hôhscancan, L. M. 1, 2; Lchdm. ii. 38, 8.

hôh-sinu, we; *f. Hough-sinew, ham-string, heel-sinew* :—Gif hôhsino forad sîe *if a heel-sinew be broken*, L. M. 1, 71; Lchdm. ii. 146, 3. Heora horsa hôhsina ðû ofcirfst *equos eorum subnervabis*, Jos. 11, 6. [Wick. houȝ-senu : Icel. hâ-sin : Dan. hase : cf. O.H.Ger. hahsanon subnervare, Grff. iv. 800.]

hôh-spor, es; *n. The heel; calx*, Ælfc. Gl. 75; Som. 71, 97: Wrt. Voc. 45, 5.

HOL, es; *n. A* HOLE, *hollow, cavern, den* :—Tô ðam ealdan hole; of ðam hole, Cod. Dipl. Kmbl. iii. 423, 22. Swâ swâ leó dêþ of his hole *quasi leo in cubile suo*, Ps. Th. 9, 29. Mec hæleþ ût týhþ of hole hâtne *a man draws me out hot from a hole*, Exon. 125 a; Th. 480, 6; Râ. 63, 7. On ðis dimme hol *into this dark den* [prison], Bt. Met. Fox 2, 21; Met. 2, 11. Ðæt cûðe hol, Exon. 112 b; Th. 431, 10; Râ. 45, 5. Wild deóra holl and denn *lustra*, Ælfc. Gl. 110; Som. 79, 38; Wrt. Voc. 59, 10. Hwelpas leóna on heora holum beóþ gelogode *catuli leonum in cubilibus suis collocabuntur*, Ps. Lamb. 103, 22. Foxas habbaþ holu *vulpes foveas habent*, Mt. Kmbl. 8, 20: Lk. Skt. 9, 58. Hola, Homl. Th. i. 160, 33. [Laym. hol: Chauc. hole: Prompt. Parv. hoole or pyt in an hylle *caverna*: O.Frs. O.Dut. Icel. O.H.Ger. hol *concavum, caverna spelunca, antrum*: cf. Goth. hulundi *spelunca*.] v. hola.

hol, es; *n. A covering* [?] :—Ân hol stæfes *apex*, Mt. Kmbl. Rush. 5, 18.

hol; *adj. Hollow* :—On middan hol *hollow in the middle*, Herb. 174, 1; Lchdm. i. 306, 9. Gif se weobud ufan hol nære *si in altari fossa non esset*, Past. 33, 2; Swt. 217, 21. Hol stân *fornix*, Cot. 93, Lye. Sca maria hine âcende on ðære nihte on ânum holum stânscræfe *St. Mary gave birth to him in a hollow cave*, Shrn. 29, 28. Ðæt wæter dranc of his holre hand *drank the water out of the hollow of his hand*, 50, 11. On ânne ealdne holne weg *to an old hollow way*, Chart. Th. 495, 8. Hole dene *convallem*, Ps. Spl. 59, 6. Hý beóþ innan hole *they are hollow within*, Herb. 180, 1; Lchdm. i. 316, 2. Gif heó hæfþ hole eágan *if she be hollow-eyed*, Lchdm. iii. 144, 7. [Prompt. Parv. hol *cavus, concavus*: York-dialect holl: O.Frs. hol: Icel. holr: O.H.Ger. hol *cavus, concavus*: Ger. hohl.] v. holh.

hol; *adj. Having a covering* or *crust* [?] :—Holne hlâf *tortam panis unius crustulam*, Ex. 29, 33. [Cf. hal-, heal-, healh-stân *crusta, crustulum*, Cot. 191, Lye.] v. also heal, healh; hol.

hôl, es; *n. Vain speech, evil speaking without cause, calumny, slander* :—Hôl and hete and rýþera reáflâc ûs derede *slander and hatred and the rapine of robbers hath harmed us*, Swt. A. S. Rdr. 106, 70. Hôl *calumnia*, Off. Episc. 8, Lye. Ne teó ic N. ne for hete ne for hôle [MS. H. hêle] ne for unrihte feohgyrnesse *I do not accuse N. from hate or with the intention of slandering him or from an unjust desire for money*, L. O. 4; Th. i. 180, 11. Ðæs deópne âþ Drihten âswôr and ðone mid sôðe swylce gefrymede ðæt hê hine for hôle ær ne âswôre gehêt Dauide swâ hê him dyde syððan *juravit Dominus David veritatem, et non frustrabitur eam*, Ps. Th. 131, 11; cf. Grff. iv. 849, huolian. [Icel. hôl *flattery, boasting*.] v. hôlunga, hoelan, hôlian.

hola, an; *m. A hole* : — Of ðam oterholan *from the otter hole*, Cod. Dipl. Kmbl. iii. 23, 30. [Prompt. Parv. hole *foramen*: Icel. hola; *f. a hole*: O.H.Ger. holi: Ger. höhle.]

holc, es; *n.* [?] *A hollow, cavity* :—Weaxcþ ðæt yfele blôd on ðâm holcum ðæs lîchoman *the evil blood increases in the hollow parts of the*

body, L. M. 1, 72; Lchdm. ii. 148, 7. On ðám holcum ðære lifre *in the hollows of the liver*, Lchdm. ii. 160, 26. [Cf. snikeð in ed te breoste holke, O. E. Homl. i. 251, 19: *Halliwell Dict.* holke, holket *hollow:* or is the meaning similar to that of *hylca*, q. v ?]

hold, es; *m. A title which seems to have been introduced by the Danes. It occurs several times in the Chronicle*, e. g. Ysopa hold and Óscytel hold, 905; Erl. 98, 34. Þurcytel eorl and ða holdas ealle, 918; Erl. 104, 22. Þurferþ eorl and ða holdas, 921; Erl. 107, 28. *It is the Norse* höldr *which is thus defined* 'sá er höldr er hann hefir óðol at erfðum tekit bæði eptir föður ok móður, þau er hans forellrar hafa átt áðr fyrir þeim,' see Cl. and Vig. Dict. höldr. *The importance of the hold in England is marked in the following passage:* — Holdes and cyninges heáhgeréfan wergild iiii þúsend þrymsa, L. Wg. 4; Th. i. 186, 8.

hold, es; *n. A carcase, body*:—Swá hwǽr swá hold byþ *ubicunque fuerit corpus*, Mt. Kmbl. 24, 28. Ðá woldon óðre fugelas fleón tó ðam holde *descenderunt volucres super cadavera*, Gen. 15, 11. Swá swá grǽdige ræmmas ðar ðar hí hold geseóþ *like greedy ravens, where they see a carcase*, L. Ælfc. P. 49; Th. ii. 386, 3: L. I. P. 19; Th. ii. 328, 5. Tódǽlon ðæs deádan hold him betwýnan *cadaver mortui inter se dispertient*, Ex. 21, 35. [Þu fule hold *olidum cadaver*, O. E. Homl. ii. 183, 15: *Icel.* hold *flesh*.]

hold; *adj. Kind, friendly, pleasant, favourable, gracious [of a prince to his subject], faithful, loyal, devoted, liege [of a subject to his prince]*:— Drihten gedyde ðæt ðæs cwearternes ealdor him wærþ swíðe hold *dominus dedit ei gratiam in conspectu principis carceris*, Gen. 39, 21. Hé wearþ cristnum monnum swíðe hold *benignus erga Christianos*, Ors. 6, 12; Swt. 266, 22. Swá hold is God mancynne ðæt hé hæfþ geset his englas us tó hyrdum *God is so gracious to mankind that he hath appointed angels as our guardians*, Homl. Th. i. 170, 9: Cd. 60; Th. 73, 10; Gen. 1202: 107; Th. 142, 26; Gen. 2367. Ðam byþ God hold ðe biþ his hláforde rihtlíce hold *God will be gracious to him who is rightly faithful to his lord*, L. C. E. 20; Th. i. 372, 12. Hé cwæþ ðæt hé heom hold hláford beón wolde, Chr. 1014: Erl. 150, 10. Ðonne biþ se holda þeówa geset ofer manegum gódum *then will the faithful servant be set over many goods*, Homl. Th. ii. 552, 23. Ic wille beón N. hold and getríwe *I will be faithful and true to N.*, L. O. 1; Th. i. 178, 4: Cd. 196; Th. 244, 4; Dan. 443: Beo. Th. 2463; B. 1229. Ic eom ðín hold scealc *tuus sum ego*, Ps. Th. 118, 94. Fram sóðum martirdóme ðæs hálgan weres his holdan pápan *from the true martyrdom of the holy man, his gracious pope*, Homl. Th. ii. 310, 29. Hé horn hefeþ holdes folces *exaltavit cornu populi sui*, Ps. Th. 148 14. Heriaþ hine on hleóðre holdre béman *laudate eum in sono tubæ*, 150, 3. Eáran habbaþ ne hí áwiht mágon holdes gehýran *ears have they but nought pleasing can they hear*, 134, 17. Holdum Gode ic sealmas singe *psallum Deo meo*, 145, 1. Ic gebócie sumne dǽl landes mínum holdan and getríwan þegne, Cod. Dipl. Kmbl. iii. 256, 8. Hé hí on hihte holdre lǽdde *deduxit eos in spe*, Ps. Th. 77, 53. Áhte ic holdne hláford *I had a gracious lord*, Exon. 100 b; Th. 379, 26; Deór. 39: Ps. Th. 150, 1: Cd. 106; Th. 139, 22; Gen. 2313. Ic geornlíce gode þegnode þurh holdne hyge *I diligently served God with loyal mind*, 28; Th. 37, 7; Gen. 586. Heó dyde hit ðeáh þurh holdne hyge *yet did she it with purpose kind*, 33; Th. 44, 12; Gen. 708: Beo. Th. 539; B. 267. Áhyld ðín eáre tó holde móde *graciously incline thine ear to me*, Ps. Th. 70, 2: 85, 6. Nele mé Israhél behealdan holde móde *Israel will not regard me with loyalty*, 80, 11; 118, 112. Ealle Rómáne wurdon cristnum monnum swá holde ðæt hie on monegum templum áwriten ðæt ǽlc cristen mon hæfde friþ *all the Romans shewed so much favour to the Christians that they wrote up in many temples that every Christian man should have protection*, Ors. 6, 13; Swt. 268, 19: Exon. 36 b; Th. 119, 7; Gú. 251. Holde frýnd mé sǽdon *faithful friends told me*, Homl. Th. 414, 7. Uton beón á úrum hláforde holde and getrýwe *let us ever be to our lord loyal and true*, L. C. E. 20; Th. i. 372, 8: Homl. Th. ii. 68, 9. Hí woldon him beón holde and gehýrsume *they [the monks] would be loyal and obedient to him [the abbot]*, Chr. 1083; Erl. 217, 6. Alle míne þegnas and míne holde freónd on Hertfordesíre *all my thanes and faithful friends in Hertfordshire*, Cod. Dipl. Kmbl. iv. 217, 5. Frýnd synd hie míne georne holde on hyra hygesceaftum ic mæg hyra hearra wesan, Cd. 15; Th. 19, 8; Gen. 288. We witon ðæt ǽghwylcum men biþ leófre swá hé hæbbe holdra freónda má *we know that the more faithful friends a man has the better he likes it*, Blickl. Homl. 123, 1: Beo. Th. 979; B. 487. Is sáwl mín symble on ðínum holdum handum *anima mea in manibus tuis semper*, Ps. Th. 118, 109. Holdost *most faithful*, Byrht. Th. 132, 31; By. 24. [*Laym.* þin holde mon: *Orm.* þin laferrd birrþ þe beon hold and trigg: *O. E. Homl.* mid holde mode: *O. Sax. O. Frs.* hold: *Icel.* hollr *gracious, faithful, wholesome*: *O. H. Ger.* hold *propitius, fidelis, devotus*: *Ger.* hold.] v. un-hold.

hold-áþ, es; *m. An oath of fealty*:—Hí wéron his menn and him holdáþas swóron ðæt hí woldon ongeán ealle óðre menn him holde beón *they did homage to him and swore oaths of fealty to him that they would be loyal to him against all other men*, Chr. 1085; Erl. 219, 7. Hé dyde ðæt ealle ða heáfodmæn on Normandig dydon manrǽden and holdáþas

his sunu Willelme, 1115; Erl. 245, 12. [*R. Glouc. Havel.* holde-, holdoþ.]

holde; *adv. Graciously, with devotion*, Ps. Th. 71, 2: 142, 6. v. hold.

holdigean *eviscerare*, Gl. Prud. 337.

hold-líce; *adv. Graciously, with kindness* or *friendliness, with devotion* or *attachment, faithfully, loyally*:—Holdlíce *affectuose* vel *devote*, Ælfc. Gl. 115; Som. 80, 50; Wrt. Voc. 61, 28. Hé cwæþ swíðe holdlíce be ús 'Fæder mín ic wille ðæt ða ðe ðú mé forgeáfe beón mid mé ðǽr ic beó' *he said very graciously concerning us 'My Father, I will that those whom thou hast given me be with me where I am,'* Homl. Th. ii. 368, 10: Cd. 220; Th. 283, 27; Sat. 311: Ps. Th. 54, 1: 58, 3. Holdlíce *kindly*, Exon. 27 b; Th. 83, 18; Cri. 1358. Hé mé holdlíce þegnade *he served me faithfully*, Ps. Th. 100, 6. Hwá ðás ǽlmesse holdlíce healde healde hine God, Chart. Th. 369, 29. Cwǽdon holdlíce hýran woldon *said they would listen devoutly*, Andr. Kmbl. 3276; An. 1641. Eádwearde hýrdon holdlíce *loyally obeyed Edward*, Chr. 1065; Erl. 196, 33; Edw. 14: Exon. 41 b; Th. 138, 14; Gú. 576. Ðæt Drihtne ful holdlíce hýran *ut serviant Domino*, Ps. Th. 101, 20.

hold-rǽden, e; *f. Faithfulness, loyalty, faithful discharge of duty to a superior*:— Hire hyrdeman þurh holdrǽdene sume ác ástáh *her herdsman in the discharge of his duty had ascended an oak*, Homl. Th. ii. 150, 30.

hold-scipe, es; *m. Loyalty, fealty, allegiance*:—Eallra ðæra manna land hí fordydon ðe wǽron innan ðæs cynges holdscipe *they destroyed the lands of all those men that were in allegiance to the king*, Chr. 1087; Erl. 224, 15. Sǽgdon ðæt hí hit dyden for ðes mynstres holdscipe *said that they did it on account of the loyalty of the monastery*, 1070; Erl. 209, 15.

holen, holegn, es; *m. Holly*:— Holen *acrifolius*, Ælfc. Gl. 47; Som. 65, 23; Wrt. Voc. 33, 23: *ulcia*, Wrt. Voc. 80, 12: *acrivolus*, 285, 37. Holegn *acrifolius*, Gl. Amplon. 131: Gl. Mett. 34 [Leo]. Holenrinde *holly-bark*, L. M. 1, 32; Lchdm. ii. 78, 12. Holenleáfa *holly leaves*, 3, 69; Lchdm. ii. 356, 11. Holen sceal in ǽled *holly shall to the fire*, Exon. 90 a; Th. 338, 17; Gn. Ex. 80. Se fealwa holen *the sere holly*, Exon. 114 a; Th. 437, 19; Rä. 56, 10. [*A. R.* holin, holie. For the form *hollen* (hollin, holyn) see E. D. S. Plant Names, p. 263.] v. cneówholen.

holenga. v. holunga.

holh, holg, es; *n. A hollow, cavity, hole*:—Hwæt tácnaþ ðæt holh on ðǽm weobude búton gódra monna geþyld? Forðam ðonne mon his mód geeáðmódgeþ ðæt hé wiðerweardnesse and scande forbere ðonne geeácnaþ hé sum holh on his móde swá swá ðæt weobud hæfþ on him uppan. Holh wæs beboden ðæt sceolde beón on ðæm weobude uppan . . . wel hit wæs gecueden ðæt ðæt holh sceolde beón on ðæm weobude ánre elne brád and ánre elne long *quod est altaris fossa, nisi bonorum patientia quæ, dum mentem ad adversa toleranda humiliat, quasi more foveæ hanc in imo positam demonstrat? Fossa ergo in altari fiat . . . Bene autem hæc eadem fossa unius cubiti esse monstratur*, Past. 33, 3; Swt. 219, 1–10. Ðǽr ðǽr se iil hæfde his holh *ibi habuit foveam ericius*, 35, 3; Swt. 241, 7. In ðæm wæs ðæt holg ðæs nearwan scræfes, Lchdm. iii. 365, col. 1. [*Laym.* holȝes, *pl.* and holh; *adj:* *R. Glouc.* holu, *sing. adj;* holwe, *pl:* *Chauc.* holwe *pl. adj.*]

holian; *p.* ode *To hollow out, make hollow, dig, make a hole; cavare*:— Hí ðá hwæthwega holodon and ðærrihte ðæt wæter swá genihtsumlíce út fleów ðæt hit arn streámrynes of ðam munte *they then hollowed out [the rock] a little, and straightway the water flowed out so abundantly that it ran streaming from the mountain*, Homl. Th. ii. 162, 7. [*A. R.* ne holieþ nout aduneward ase doþ þe uoxes: *Prompt. Parv.* holyn *cavo, perforo, terebro:* *Goth.* us-hulon *to excavate:* *Icel.* hola *to make hollow:* *O. H. Ger.* holian, holon *fodere, perforarare, excavare:* *Ger.* höhlen.] DER. á-holian.

hólian *to speak evil of, slander, calumniate*:—Ne sele ðú mé hóliendum mé *non tradas me calumniantibus me*, Ps. Lamb. 118, 121. [*Orm.* holen o þe laȝhe leod, 9319, *with which compare Goth.* holon *in* Lk. 3, 14: *cf. O. H. Ger.* huolian, Grff. iv. 849.] v. hól, hólunga.

hólinga. v. hólunga.

holl. v. hol.

holm, es; *m. A mound, hill, rising ground;* but in this sense, which belongs to the word in the Old Saxon, it is not found in English. **I.** Its most common use in the latter, in the poetry, is in reference to water with the meaning *wave, ocean, water, sea*:— Freá engla hét wesan wæter gemǽne ðá stód hraðe holm under heofonum síd ætsomne *the lord of angels bade the waters be together, then quickly stood ocean under heaven far-stretching continuously*, Cd. 8; Th. 10, 23; Gen. 161. Holm *the [Red] sea*, 157; Th. 195, 30; Exod. 284: 166; Th. 206, 9; Exod. 449. Holm *the water of the deluge*, 71; Th. 86, 15; Gen. 1431. Holm storme weól, Beo. Th. 2267; B. 1131. Holm heolfre weóll *[of the lake where Grendel dwelt]*, B. 2137: 3189; B. 1592. Wíde rád ofer holmes hringc hof séleste *[of the ark]*, Cd. 69; Th. 84, 5; Gen. 1393. Eów is holmes hlæst and heofonfuglas and wildu deór on geweald geseald *the fishes of the sea, the fowls of the air, and the beasts of the earth are*

delivered into your hand, 74; Th. 91, 20; Gen. 1515. Wið holme foldan sceldun guarded land against sea, Exon. 22 a; Th. 61, 4; Cri. 979. On holme, 97 a; Th. 363, 9; Wal. 51: Beo. Th. 1090; B. 543: 2875; B. 1435. Æt holme by the sea, 3832; B. 1914. Sealt wæter hreóh me holme besencte tempestas demersit me, Ps. Th. 68, 2. Ðá wæs heofonweardes gást ofer holm boren the spirit of God moved upon the face of the waters, Cd. 6; Th. 8, 7; Gen. 121. Léton holm beran they let the sea bear him, Beo. Th. 96; B. 48. Ofer wídne holm, Exon. 79 a; Th. 296, 23; Crä. 55. Ofer heánne holm, Elen. Kmbl. 1962; El. 983: Cd. 213; Th. 266, 4; Sat. 17: Exon. 77 b; Th. 291, 14; Wand. 82. Ðá ic on holm gestáh when I embarked, Beo. Th. 1269; B. 632: Andr. Kmbl. 858; An. 429. Heá holmas deep waters, Exon. 54 b; Th. 193, 17; Az. 123. Holmas dælde waldend ûre God divided the waters, Cd. 8; Th. 9, 24; Gen. 146: Exon. 93 a; Th. 349, 31; Sch. 54. Hider ofer holmas hither over the waves, Beo. Th. 485; B. 240. Windge holmas stormy seas, Exon. 20 a; Th. 53, 26; Cri. 856. Holma begang the way across the waters, Ps. Th. 138, 18: Andr. Kmbl. 390; An. 195: Bt. Met. Fox 11, 69; Met. 11, 30. Holma geþring, Beo. Th. 4271; B. 2132. Holma gelagu, Exon. 82 a; Th. 309, 28; Seef. 64. II. From the Scandinavian hólmr an islet especially in a bay, creek, lake, or river, it is used in English with the meaning land rising from the water, an island in a river, etc., holm [in local names]:—Ðý ilcan geáre wæs ðæt gefeoht æt ðam Holme Cantwara and ðara Deniscra, Chr. 902; Th. 180, col. 2. Hér fôr Cnut Cyng tô Denmearcon mid scipon tô ðam holme æt eá ðære hálgan, 1025; Erl. 163, 7. [Laym. holm : Prompt. Parv. holm, place besydone a water hulmus : of a sonde yn the see bitalassum vel hulmus. v. p. 243, note 2, and 244, note 2.] DER. sǽ-, wǽg-holm.

holm-ærn, es; n. A sea-house, vessel, ship:—Holmærna mǽst earc Noes, Cd. 71; Th. 85, 30; Gen. 1422.

holm-clif, es; n. A sea-cliff, cliff by the water-side:—On, fram ðam holmclife [the holm is the lake where Grendel dwelt], Beo. Th. 2846, 3274; B. 1421, 1635. Se ðe holmclifu healdan scolde he who had to guard the sea-cliffs, 465; B. 230. [O. Sax. holm-klif a hill.]

holmeg; adj. Oceanic :—Holmegum wederum with storms such as blow at sea, Cd. 148; Th. 185, 6; Exod. 118.

holm-mægen, es; n. The might of the ocean, the ocean, Exon. 101 a; Th. 382, 10; Rä. 3, 9.

holm-þracu; g. -þræce; f. The violence of the sea, the tossing of the waves, the ocean, Andr. Kmbl. 933; An. 467. Ðû geworhtest heofon and eorþan and holmþræce thou didst make heaven and earth and the sea with its tossing waves, Elen. Kmbl. 1453; El. 728: Exon. 17 b; Th. 42, 25; Cri. 678: 57 b; Th. 205, 19; Ph. 115.

holm-weall, es; m. A wall formed by the sea, Cd. 166; Th. 207, 16; Exod. 467.

holm-weard, es; m. One who keeps guard at sea, a sea-warder, Andr. Kmbl. 718; An. 359.

holm-weg, es; m. A way over the sea, Andr. Kmbl. 764; An. 382.

holm-wylm, es; m. The surge of the sea, Beo. Th. 4814; B. 2411.

holor, hólrian. v. heolora, heoloran.

HOLT, es; m. n. I. a HOLT, wood, grove, copse :—Holt lucus, Ælfc. Gr. 8; Som. 7. 30: nemus, 9, 32; Som. 12, 17: saltus, Ælfc. Gl. 45; Som. 64, 104; Wrt. Voc. 32, 39: nemus vel saltus, Wrt. Voc. 80, 34. Wildeóra holt, Salm. Kmbl. 116; Sal. 82. Holtes frætwe fruit, Exon. 57 a; Th. 202, 21; Ph. 73. Hê lét him ðá of handon fleógan hafoc wið ðæs holtes he let the hawk fly from his hands towards the wood, Byrht. Th. 131, 14; By. 8: Rood Kmbl. 58; Kr. 29. Uton gân innan on ðisses holtes hleó let us go within the shelter of this grove, Cd. 39; Th. 52, 7; Gen. 840: Exon. 62 a; Th. 227, 26; Ph. 429. Wulf holtes gehléða, Elen. Kmbl. 225; El. 113. Sum sceal on holte of heáhbeáme feallan, Exon. 87 b; Th. 328, 21; Vy. 21: Bt. Met. Fox 13, 103, 73; Met. 13, 52, 37. Gewiton áweg tô holte they went away to the wood, Homl. Th. ii. 516, 12. Holt ofgeáfon they left the wood, Beo. Th. 5685; B. 2846: 5190; K. 2598. Abraham ðá plantode ænne holt Abraham vero plantavit nemus, Gen. 21, 33. Ful oft unc holt wrugon wudubeáma helm, Exon. 129 a; Th. 496, 1; Rä. 85, 7. Ðû geond holt wunast thou shalt dwell among the woods, Cd. 203; Th. 252, 6; Dan. 574. II. wood; lignum :—Ic geseah holt hweorfende I saw wood moving, Exon. 114 a; Th. 438, 5; Rä. 57, 3. Holte bi[h]lænan to pile wood round, 74 a; Th. 277, 7; Jul. 577. [Laym. Chauc. holt : Prompt. Parv. holt, lytylle wode lucus, virgultum, p. 244, v. note : O. Frs. holt wood, stick : Icel. holt wood, coppice (nearly obsolete) : a rough stony hill : O. H. Ger. holz nemus, silva, saltus, arbor, lignum : Ger. holz.] DER. æsc-, firgen-, ofer-, wudu-holt.

holt-hana, an; m. A wood-cock; acegia, Gl. Mett. 41 : Gl. Amplon. 138.

hól-tihte, an; f. Calumny, slander :—Hóltihte vel teóne calumnia, Ælfc. Gl. 15; Som. 58, 36; Wrt. Voc. 21, 29.

holt-wudu, a; m. I. a wood; silva, nemus, Beo. Th. 2743; B. 1369 : Exon. 58 a; Th. 209, 16; Ph. 171. II. wood from a

holt, forest-wood; lignum, Beo. Th. 4669; B. 2340 : Rood Kmbl. 179; Kr. 91.

hólunga; adv. In vain, to no purpose, without cause, without intent :—Hólunga sine causa, Mt. Kmbl. Lind. 15, 9. Nales hólunge not without cause, Cd. 48; Th. 61, 14; Gen. 997. Nalles hólinga, Beo. Th. 2156; B. 1076. Wæs his fæder geléred in ða gerýno Cristes geleáfan ac hólinga pater ejus sacramentis Christianæ fidei imbutus est, sed frustra, Bd. 2, 15 ; S. 518, 29. Gif hê hit hólinga dô fæste i geár si casu fecerit, i annum jejunet, L. Ecg. P. iv. 68, 22; Th. ii. 230, 27. Ðære tíde wæs ðæt mæste wæll geworden on Norþanhymbra þeóde and cyrican. Ne wæs ðæt hólenga fordon óðer ðæra heretogena wæs hæðen óðer wæs ðam hæðenan grimra quo tempore maxima est facta strages in ecclesia vel gente Nordanhymbrorum, maxime quod unus ex ducibus paganus, alter erat pagano sævior, Bd. 2, 20; S. 521, 19. Mid ðý wê wið ðam winde and wið ðam sæ holonga campodan cumque cum vento pelagoque frustra certantes, 5, 1 ; S. 613, 27.

hom, hôme, homer. v. ham, ôme, hamer.

homela, homola, an; m. A word of uncertain meaning occurring in the following passage :—Gif hê hine on bismor tô homolan bescire mid x scitt. gebéte. Gif hê hine tô preóste bescire mid xxx scitt. gebéte, L. Alf. pol. 35; Th. i. 84, 5. See the note there; see also on cutting the hair as a mark of disgrace, Grimm's Deutsche Rechtsalterthümer, pp. 702-3. v. hamelian, and cf. [?] Scot. hummel, homyll having no horns.

hôn; p. hêng; pp. hangen To hang, suspend, crucify :—Gê hig hôþ crucifigetis, Mt. Kmbl. 23, 34. Hine man hêng ille suspensus est in cruce, Gen. 41, 13. Hig hine hêngon crucifixerunt eum, Lk. Skt. 23, 33. Ðone hêngon on heáne beám fæderas ûsse, Elen. Kmbl. 847; El. 424. Hôh hine crucifige eum, Mk. Skt. 15, 13. Hôh hyne hôh hyne; Ðá cwæþ pilatus tô him Nime gê hine and hôþ, Jn. Skt. 19, 6. Hôh on earm hang it on to the arm, Med. ex Quadr. 9, 12; Lchdm. i. 362, 27. Ðone ôderne hê hêt hôn on gealgan alterum suspendit in crucem, Gen. 40, 22. Hêt se wælhreówa hine hôn on heardre hengene, Homl. Th. ii. 308, 29. Ðær wæron geládde twegen sceaþan for heora synnum tô hônne there were brought two thieves to be crucified for their sins, 254, 22. Tô hôanne ad crucifigendum, Mt. Kmbl. Lind. 20, 19. Ic hæbbe mihte ðê tô hônne, Jn. Skt. 19, 10. Ðæm hôendum crucifigentibus, Lk. Skt. 11, 7. Frignan ongan on hwylcum ðara beám wealdendes hangen wǽre, Elen. Kmbl. 1701; El. 851. [Laym. hon; p. heng : Orm. Chauc. Piers P. heng, p: Goth. hahan; p. haihah: Icel. hanga; p. hêkk pendere : O. Frs. hua; p. heng; pp. huen: O. H. Ger. hahan; p. hieng figere, crucifigere, suspendere.] DER. a-, be-, bi-, ge-hôn.

hôn tendrils of a vine [?] :—Ðá geseah ic gyldenne wíngeard trumlícne and fæstlícne and ða twigo his hongodon geond ða columnan. ða wundrode ic ðæs swíðe. wæron in ðæm wíngearde gyldenu leáf and his hôn and his wæstmas wæron cristallum and smaragdus eác ðæt gimcyn mid ðæm cristallum ingemong hongode vineamque solidam auro argentoque inter columnas pendentem miratus sum. in qua folia aurea racemique cristallini ligis erant interpositi, distinguentibus smaragdis, Nar. 4, 31.

hona, hon-, hond, hongian. v. hana, heonu, han-, hand, hangian.

hôp. v. fen-, môr-hôp.

HOPA, an; m. HOPE :—Geleáffullum mannum mæg beón micel hopa tô ðam menniscum Gode Criste believing men may have great hope on the human God, Christ, Homl. Th. i. 350, 24. Ne bepǽce Ezechias eów mid leásum hopan let not Hezekiah deceive you with false hope, 568, 8. [Laym. Orm. A. R. hope : Du. hoop: Dan. haab : M. H. Ger. hoffe.] DER. tô-hopa.

hôp-gehnást, es; n. The dashing together of waves in a bay [?] :—Bídaþ stille stealc stánleópu streámgewinnes hôpgehnástes ðonne heáh geþring on cleofu crýdeþ the steep rocks await quietly the strife of the sea, the dash of the waves, when the press of waters towering up crowds on to the cliffs, Exon. 101 b; Th. 384, 13; Rä. 4, 27. [Cf. Icel. hóp a small landlocked bay or inlet : Scot. hope a haven.]

hopian; p. ode, ede To hope, have hope or confidence [in a person], expect, watch for [with gen.] :—Ic hopige tô him swá gôdan and swá mildheortan ðæt hê hit nylle sylf dôn I have confidence in him, so good and merciful, that he himself will not do it, Chart. Th. 548, 20. Ðû dysegost manna ðû hopast ðæt ðû hæbbe þoftrǽdene tô ðam áwyrigedan deófle thou most foolish of men, thou trustest that thou hast fellowship with the accursed devil, Homl. Th. ii. 416, 14. Swá eác ûre hiht ne becom ná tô ðam ðe hê hopaþ so also our hope has not arrived at that for which it hopes, i. 250, 25. Ðonne hê eall forsihþ eorþlícu gôd and hopaþ tô ðam écum, Bt. Met. Fox 7, 87; Met. 7, 44. Se synfulla hopaþ symle ðæs rihtwísan consideret peccator justum, Ps. Th. 36, 32. Ðæt ðæt Maria dyde tô ðam wê hopiaþ that which Mary did, for that we hope, Homl. Th. ii. 442, 33. Landfranc gewát of ðissum lífe ac wê hopiaþ ðæt hê fêrde tô ðæt heofonlíce ríce, Chr. 1089; Erl. 226, 15. Ic tô ðê hopode in e speravi, Ps. Th. 30, 17. Hê hopode ðæt hê gesáwe sum tácen sperabat signum aliquod videre, Lk. Skt. 23, 8. Hæbbende ðæs ðe wê ǽr hopedon, Homl. Th. i. 250, 35. Wê tô ðínum hidercyme hopodan and hyhtan, Blickl. Homl. 87, 11. Hopedon sperabamus, Lk. Skt. 24, 21. Ðá fíf cyningas hopodon tô lífe the five kings hoped to save their lives, Jos.

10, 16.   Ne hopige nán man tó ðyssere leásunge, Homl. Th. ii. 572, 21. Hit nys ró unnyt ðæt wé hopien tó Gode fordæm hé ne went swá swá wé dóþ *it is not vain for us to have hope in God; for he does not change as we do*, Bt. 42 ; Fox 258, 20.   Ðæt hí swá hopigen tó ðære forgiefnesse *ut sic de spe fiduciam habeant*, Past. 53, 5 ; Swt. 415, 19. Bebeódaþ ðæt hí ne hopian on heora ungewissum welan *bid them not to put their trust in their uncertain riches*, Homl. Th. i. 256, 25. Ne þearf hé hopian nó ðæt hé ðonan móte *he has no ground for hoping that he may go thence*, Judth. 10; Thw. 23, 12; Jud. 117. Ða hopiendan on ðé *sperantes in te*, Ps. Spl. 16, 8. [*M. H. Ger.* hoffen.] DER. tó-hopian.

hópig ; *adj. In hills and hollows* [*applied to the sea in reference to the deep depressions between high waves*; cf. Scot. hope *a sloping hollow between two hills, or the hollow that is formed between two ridges on one hill*] :—Com ic on sǽs hricg ðǽr mé sealt wæter hreóh and hópig holme besencte *veni in altitudinem maris; et tempestas demersit me*, Ps. Th. 68, 2.

hoppa. v. gærs-hoppa.

hóp-páda, an; *m. An upper tunic, cope* :—Hóppáda *ependeton* [= ἐπενδύτης], Ælfc. Gl. 112; Som. 79, 83; Wrt. Voc. 59, 52.

hoppe, an; *f. An ornament suspended from the neck, a bell* [?] *hung from a dog's neck* :—Hryðeres belle and hundes hoppe ǽlc biþ ánes sciłł. weorþ and ǽlc is melda geteald *an ox's bell and that on a dog's collar, each is worth a shilling and each is reckoned an informer*, L. Edg. H. 8; Th. i. 260, 16. Hie eall him gesealdon ðæt hie ðá hæfdon búton ðæt ǽlc wífmon hæfde áne yndsan goldes and án pund seolfres and ǽlc wǽpnedmon ǽnne hring and áne hoppan *ita ut nihil præter annulos singulos, bullasque sibi ac filiis, et deinde per filias uxoresque suas singulas tantum auri uncias, et argenti non amplius quam singulas libras relinquerent*, Ors. 4, 10; Swt. 196, 21.

hoppere, es; *m. A dancer*; saltator, Som.

hoppestre, an; *f. A female dancer* :—Ðæs mǽran wítegan deáþ ðære lyðran hoppystran tó méde forgeaf *rewarded that vile dancer with the death of the illustrious prophet*, Homl. Th. i. 484, 3. [*Chauc.* hoppestre.]

hoppetan ; *p.* te *To jump about* [*for joy*], *leap, rejoice, to throb* [*of a wound*] :—Swá benne ne burnon ne burston ne hoppetan *so that the wounds should neither burn nor burst nor throb*, L. M. 3, 63; Lchdm. ii. 352, 1. Ðæñe ðe méder on rífe hoppetende beclýsed Iohannes undergeat *quem matris alvo gestiens clausus Iohannes senserat*, Hymn. Surt. 51, 1. v. next word.

hoppian, *p.* ode *To hop, leap, dance* :—Ðá blissode mín cild on mínum innoþe and hoppode ongeán his Drihten *then rejoiced my child in my womb, and leaped towards his Lord*, Homl. Th. i. 202, 18. [*Chauc. Piers P.* hoppe *to dance, jump: Icel.* hoppa *to skip, bound: M. H. Ger.* hoppen: *Ger.* hüpfen.]

hopp-scýte, an; *f. A coverlet* [?] :—Ic geann ánes beddreáfes mid wahhryfte and mid hoppscýtan, Chart. Th. 529, 12.

hopu *lygustra*, Lchdm. iii. 332, col. 2.

horas. v. horh.

hora-seáþ, Bt. 37, 2; Fox 188, 1. v. horu-seáþ.

hór-cwene, an; *f. An adulteress, whore* : — Hórcwenan, L. E. G. 11; Th. i. 172, 21: L. Eth. vi. 7; Th. i. 316, 21: L. C. S. 4; Th. i. 378, 7. [*Icel.* hór-kona *an adulteress.*]

HORD, es; *n. m.* HOARD, *treasure* ; — Hord *thesaurus*, Wrt. Voc. 86, 47. Ðá wæs óþboren beága hord *then was borne off the hoard of rings*, Beo. Th. 4557; B. 2284: 6015; B. 3011. Hyrde ðæs hordes *keeper of the hoard*, Exon. 130 a; Th. 498, 7; Rä. 87, 9: Beo. Th. 1778; B. 887. Ðæs ðe heáh hlioþo horde onféngon *after the lofty hills had received the treasure* [*the ark*], Cd. 71; Th. 86, 32; Gen. 1439. Hæðnum horde, Beo. Th. 4438; B. 2216. Hord eald enta geweorc, 5540; B. 2773. Ðæt hord, 6244; B. 3126. Hord under hrusan [*the nails of the cross*], Elen. Kmbl. 2181; El. 1092. Hí ealgodon hord and hámas *they defended treasures and homes*, Chr. 937; Erl. 112, 10; Æðelst. 10. Hé ðæt fácen hafaþ in his heortan, hord unclǽne *he hath that deceit in his heart, a hoard unclean*, Frag. Recd. 11; Leás 6. Hord, heortan geþohtas, Exon. 23 a; Th. 65, 1; Cri. 1048: 23 b; Th. 65, 17; Cri. 1056. Breósta hord, Th. 66, 17; Cri. 1074. Breósta hord, gást *the breast's treasure, the spirit*, Cd. 79; Th. 97, 6; Gen. 1608. His synna hord onténde *he confessed his sins*, Ps. C. 50, 28; Grn. ii. 277, 28: 151, 155; Grn. ii. 280, 151, 155. Sáwle hord, Beo. Th. 4835; B. 2422. Hordas, gerýne *arcana*, Mone B. 4216 (v. gold-hord). [*Laym. Orm. A. R. Chauc.* hord: *Goth.* huzd; *n: O. Sax.* hord; *n: Icel.* hodd; *n.* (*but a late form* hoddar; *pl. occurs*) *in poetry only hoard, treasure: O. H. Ger.* hort; *n. thesaurus.*] DER. beáh-, bóc-, brand-, breóst-, feorh-, flǽsc-, gold-, greót-, líc-, máðm-, mód-, sáwl-, wamb-, word-, wyrm-hord.

hord-burh, -burg, e; *f. A city containing treasure*, Cd. 93; Th. 121, 9; Gen. 2007: Beo. Th. 938; B. 467.

hord-cleófa, -clýfa, an; *m. A treasure-chamber, treasury, store-room, closet* :—Hí gáþ in tó ðínum húse and tó ðínum bedde and tó ðínum hordclýfan *ingredientur cubiculum lectuli tui et super stratum tuum*, Exod. 8, 3. Ic hæbbe on mínum hordcleófan án wundorlíc weorc *I have in my*

---

treasury a wondrous work, Homl. Skt. 5, 260. Hí sóhton ðone behíddan mete on heora hordcleófan *they sought the hidden food in their closets*, Ælfc. T. 42, 14; Grn. 21, 13. v. next word.

hord-cófa, an; *m. A place for treasure, a retired chamber, closet, a place where the thoughts are stored* [v. hord], *the breast, heart* :—Ðá æfter ðon ðá cégde seó hálige Mariæ tó eallum apostolum on hire hordcófan *post hec vocavit Sancta Maria omnes apostolos in cubiculo suo*, Blickl. Homl. 143, 34. Ðæt hé his ferþlocan fæste binde healde [MS. healdne] his hordcófan *that he close fast his mind's coffer and preserve the treasury of his thoughts*, Exon. 76 b; Th. 287, 14 [cf. 22]; Wand. 14. Hine mid ealle innancundum heortum hordcófan helpe biddaþ *in toto corde exquirunt eum*, Ps. Th. 118, 2.

hordere, es; *m. A treasurer, steward, chamberlain* [v. Kemble's Saxons in England ii. 106] :—Hordere *cellerarius*, Wrt. Voc. 83, 6. Ðá hét hé his hordere ðæt glæsene fæt syllan ðam biddendan subdiácone. Se hordere cwæþ him tó andsware gif hé ðam biddendum sealde ðæt hé nán þing næfde his gebróðrum tó syllenne *then he bade his steward give the glass vessel to the requesting subdeacon. The steward said in answer, that if he gave it he should have nothing to give to his brethren*, Homl. Th. ii. 178, 22: Chr. 1131; Erl. 260, 12. Ðis forward wæs makid wid ordríc hordere, Chart. Th. 438, 3, 7. Cynges hordera oððe úra geréfena swilc, L. Ath. I, 3; Th. i. 200, 23, see note. Nán man ne hwyrfe nánes yrfes bútan ðæs geréfan gewitnesse ... oððe ðæs horderes, 9; Th. i. 204, 19. [*Ayenb.* hordier *treasurer.*]

hord-ern, -ærn, es; *n. A store-house, store-room, treasury* :—Hordern *cellarium*, Ælfc. Gl. 108; Som. 78, 100; Wrt. Voc. 58, 15: Lk. Skt. Lind. 12, 24. Cellaria uini id est hordern *promptuaria*, Blickl. Gl. 259, 5; Ps. Surt. 143, 13. Búton hit under ðæs wífes cǽglocan gebroht wǽre ðæt is hire hordern and hire cyste *unless it has been put into the places which the wife locks up, that is, her storeroom and her chest*, L. C. S. 77; Th. i. 418, 21. Hordærne neáh *near to the treasure-house*, Beo. Th. 5655; B. 2831. Hé is gód hordern on tó scǽwiene *it is a good day for examining a storeroom*, Lchdm. iii. 180, 6. Heora hordernu wǽron mid monigfealdum wlencum gefylde *their storehouses were filled with manifold riches*, Blickl. Homl. 99, 16. Hordærna sum, Beo. Th. 4548; B. 2279.

horder-wíce, an; *f. The office of a treasurer* or *steward*, Chr. 1137; Erl. 263, 14.

hord-fæt, es; *n. A vessel for holding treasure* : — Se Hálga Gást wunode on ðam æþelan innoþe and on ðam gecorenan hordfæte [*of the Virgin Mary*], Blickl. Homl. 105, 15: Hy. 11, 18; Hy. Grn. ii. 294, 18. Hí geopenodon heora hordfatu [cf. Mt. 2, 11 *apertis thesauris suis*] and him lác geoffrodon, Homl. Th. i. 78, 27: 116, 3. On heora hordfatum behíddon *absconderunt inter vasa sua*, Jos. 7, 11.

hord-geat, -gat, es; *n. A door through which a treasure is reached* :—Hwylc ðæs hordgates cǽgen cræfte ða clamme onleác *which, by the key's art, unlocked the fastenings of the door to the treasure*, Exon. 112 a; Th. 429, 28; Rä. 43, 11.

hord-gestreón, es; *n. Hoarded, accumulated wealth, that which has been acquired and now forms a 'hord'* :—Sum wæs ǽhtwelig in commedia heóld hordgestreón *there was one of large possessions, he kept in Nicomedia his stored-up wealth*, Exon. 66 a; Th. 244, 3; Jul. 12. Ne mót hé hionane lǽdan of disse worulde wuhte ðon mǽre hordgestreóna ðonne hé hider brohte, Bt. Met. Fox 14, 21; Met. 14, 11: Beo. Th. 6175; B. 3092. Mǽst hlífade ofer Hroþgáres hordgestreónum *the mast towered above the riches that had come from Hrothgar's hoard*, 3803; B. 1899. Næs him hyht tó hordgestreónum *no hope had they in hoarded wealth*, Andr. Kmbl. 2229; An. 1116.

hordian, *p.* ode *To* HOARD, *lay up* [*treasure*], *store* :— Ðæt hé for gýtsunge uncyste nánum óðrum syllan ne mæg ðæt hé hordaþ and nát hwam swá swá se wítega cwæþ 'on ídel biþ ǽlc man gedréfed se ðe hordaþ and nát hwam hé hit gegaderaþ' *what he from the vice of avarice can give to no other he hoards, and knows not for whom, as the prophet says 'In vain is every man troubled who hoards, and knows not for whom he gathers it,'* Homl. Th. i. 66, 3. Hordiaþ eówerne gol hord on heofenum *lay up your treasure in heaven*, ii. 104, 31. DER. ge d-hordian.

hord-loca, an; *m. A treasure-chest, coffer*, metaph. *the mind* [v. hord] :—Ðeáh ðe hé feohgestreón under hordlocan ǽhte *though he had wealth in his coffer*, Exon. 66 b; Th. 245, 11; Jul. 43. Heald hordlocan hyge fæste bind *keep thy thought's treasury, fast bind thy mind*, 122 a; Th. 469, 16; Hy. 11, 3: Andr. Kmbl. 1342; An. 671.

hord-máðmum, es; *m. A valuable present, jewel* :—Healsbeága mǽst, hordmáðmum, Beo. Th. 2400; B. 1198.

hord-mægen, es; *n. Abundance of wealth, riches*, Cd. 209; Th. 258, 13; Dan. 675.

hord-weard, es; *m. A guard of a hoard* or *treasure* : — Hordweard *the dragon which watched over the treasure*, Beo. Th. 4576; B. 2293: 4594; B. 2302: 5102; B. 2554: 5179; B. 2593. Hordweard hæleþa *the Danish king*, 2098; B. 1047: 3708; B. 1852. Hordwearda hryre [*of the death of the first-born in Egypt*], Cd. 144; Th. 179, 27; Exod. 35: [*of the destruction of the Egyptians in the Red Sea*], 169; Th. 210,

6; Exod. 511. Hordwearda gestreón *the wealth of the princes of Israel*, 174; Th. 220, 3; Dan. 65.

hord-wela, an; *m. Hoarded, stored-up wealth* :—Ðeáh ðe hordwelan heólde lange, Beo. Th. 4677; B. 2344.

hord-weorþung, e; *f. The honouring a person by bestowal of treasure*, Beo. Th. 1908; B. 952.

hord-wynn, e; *f. The delightful object that consists in hoarded treasure [applied to the treasure guarded by the dragon]*, Beo. Th. 4533; B. 2270.

hóre, an; *f. A whore, harlot*; meretrix, Hpt. Gl. 475, 484. [*Laym. A. R.* hore; *Icel.* hóra : *O. H. Ger.* huora : *Ger.* hure.]

horeht. v. horheht.

horh, horg, es; *m. n. A clammy humour, phlegm, rheum* :—Hrog [=horg] *phlegma*, Wrt. Voc. 64, 51. Horg *flegma*, 282, 67. Sió gífernes árist of ðæs hores wǽtan *the voracity arises from the humour of the phlegm*, L. M. 2, 16; Lchdm. ii. 196, 3. Wið langum sáre ðara tóþa þurh horh, 1, 1; Lchdm. ii. 24, 4. Gif him ofstondeþ on innan ǽnigu ceald wǽte ðonne spíwaþ hie ðæt horh . . . ðæt ofstandene þicce horh, 2, 16; Lchdm. ii. 194, 15–21. Ðonne spíwaþ hie sóna ðone þiccan horh, 2, 28; Lchdm. ii. 224, 15. Horas *pituita*, i. e. *minuta saliva*, Ælfc. Gl. 78; Som. 72, 55; Wrt. Voc. 46, 15. v. horu.

horheht; *adj. Full of phlegm, phlegmatic* :—Mid yfelre wǽtan horhehtre, L. M. 2, 28; Lchdm. ii. 224, 9: 2, 27; Lchdm. ii. 222, 26. v. horweht.

horian, Ps. Th. 27, 1, note. v. harian.

horig, horhig; *adj. Foul, dirty, defiled* :—Swá hit gedafenlíc is ðæt his reáf ne beó horig *so is it proper that his vestment be not foul*, L. Ælfc. C. 22 : Th. i. 350, 21. Næs his reáf horig, Homl. Th. i. 456, 20. Mid horium reáfe, 528, 24. Mid horhgum sicelse, Th. Ap. 13, 26. [*O. E. Homl.* þat brinþ hori to clene : *Wick.* hoori *unclean* : *Chauc.* horowe; *pl* : *O. H. Ger.* horig *lutulentus, cenosus.*]

hóring, es; *m. An adulterer, fornicator* :—Hér sindon miltestran and bearnmyrdran and fúle forlegene hóringas, Swt. A. S. Rdr. 110, 181. [Cf. *Goth.* hórs : *Icel.* hórr.]

HORN, es; *m. A* HORN, *a drinking-horn, a cupping-horn, a trumpet, the horn-shaped projection on the gable-end of a house* [v. Dasent's translation of Njála, plate 3, p. cvii], *a pinnacle* :—Oxan horn biþ x pæninga weorþ *an ox's horn shall be worth ten pence*, L. In. 58; Th. i. 138, 21. Se horn mínre hǽlo *cornu salutis meæ*, Ps. Th. 17, 3. Horn stundum song *sometimes the horn sounded*, Beo. Th. 2851; B. 1423. Hwílum teóh mid glǽse oððe mid horne *draw at times with a cupping-glass or horn*, L. M. 2, 18; Lchdm. ii. 200, 13. Sete horn on ða openan scearpan *put a cupping-horn on the open scarifications*, 1, 56; Lchdm. ii. 126, 21. Gif feorrancumen man oððe frǽmde búton wege gange and hé ðonne náwþer ne hrýme ne hé horn ne bláwe for þeóf hé biþ tó prófianne *if a man come from a distance, or a stranger, go out of the highway, and he then neither shout nor blow a horn, he is to be tried as a thief*, L. Wih. 28; Th. i. 42, 24. Syððan hie Hygeláces horn and býman galan ongeáton, Beo. Th. 5879; B. 2943. ii hnæppas and iiii hornas *two bowls and four drinking-horns*, Chart. Th. 429, 31. Ne býman ne hornas, Exon. 57 b; Th. 206, 30; Ph. 134. Ne hér ðisse healle hornas [horn næs, Th.] ne byrnaþ *nor here do this hall's gables burn*, Fins. Th. 7; Fin. 4. Ic wiht geseah wundorlíce horna ábitweónum húþe lǽdan *I saw a creature [the moon] wondrously bringing spoil between its horns*, Exon. 107 b; Th. 411, 19; Rä. 30, 2. Heorot hornum trum *the hart firm-antlered*, Beo. Th. 2742; B. 1369. Óþ wigbedes hornas *usque ad cornu altaris*, Ps. Th. 117, 25. [*Goth.* haurn; *n. a horn, drinking-horn, trumpet, husk* : *O. Sax.* horn-[seli] : *O. Frs.* horn; *n. cornu, tuba* : *Icel.* horn; *n. a horn, drinking-horn, trumpet* ; *a corner* : *O. H. Ger.* horn; *n. cornu, tuba, promontorium* : *Ger.* horn; *n.*] DER. blǽd-, drenc-, fyhte-, gúþ-horn. v. án-horn.

horn [horh?]-ádl, e; *f. A disease of foul humours in the stomach*, L. M. 2, 27; Lchdm. ii. 222, 31.

horn-bǽre; *adj. Horned, having horns*; corniger, Ælfc. Gr. 8; Som. 7, 20.

horn-bláwere, es; *m. A horn-blower, trumpeter* : — Hornbláwere *cornicen*, Wrt. Voc. 73, 63 : Ælfc. Gr. 9; Som. 9, 24. Ðær mihte wel bén ábúton twenti oðer þritte hornblaweres, Chr. 1127; Erl. 256, 36. [Cf. *Goth.* haurnja : *O. H. Ger.* horn-bláso *tubicen, cornicen.*]

horn-boga, an; *m. A bow with the ends curved like a horn* or *a bow made of horn [?]*, [cf. *Icel.* horn-bogi *a horn-bow*, Cl. and Vig. Dict.] :—Léton forþ fleógan hildenædran of hornbogan, Judth. 11; Thw. 24, 34; Jud. 222: Beo. Th. 4866; B. 2437. Ðær hé hornbogan [horn bogan ?] hearde gebendeþ *ibi confregit cornua arcuum*, Ps. Th. 75, 3.

horn-bora, an; *m. A horn-bearer, trumpeter*, Elen. Kmbl. 107; El. 54.

horn-fisc, es; *m. A garfish, a kind of pike* : — Hornfisc plegode glád geond gársecg, Andr. Kmbl. 740; An. 370. [*Icel.* horn-fiskr : *Dan.* hornfisk *garfish*, esox belone.]

horn-geáp; *adj. Having a wide extent between the 'horns'* [v. horn], *an epithet of a building* : — Tempel dryhtnes heáh and horngeáp, Andr.

Kmbl. 1335; An. 668 : Beo. Th. 164; B. 82. [Cf. under geápne hróf, 1677; B. 836.]

horn-gestreón, es; *n. An abundance of pinnacles*, Exon. 124 a; Th. 477, 11; Ruin. 23.

horn-píc, es; *n.* [?] *A pinnacle* : — Sette hine ofer hornpíc temples *statuit eum supra pinnam templi*, Lk. Skt. Lind. 4, 9.

horn-reced, es; *n. A house having 'horns'* [v. horn] or *pinnacles*, Beo. Th. 1412; B. 704.

horn-sǽl, es; *n. A hall having 'horns' in its roof* :—Hornsalu, Andr. Kmbl. 2318; An. 1160: Exon. 101 b; Th. 383, 10; Rä. 4, 8. v. horn-reced, -sele.

horn-sceaða, an; *m. A pinnacle* : — Ofer hornsceaðe temples *supra pinnaculum templi*, Mt. Kmbl. Lind. 4, 5. v. sceaða.

horn-scip, es; *n. A ship having a beak* [rostrum], *a ship with a horn-like projection in the bow*, Andr. Kmbl. 547; An. 274.

horn-sele, es; *m. A building having pinnacles*, Cd. 86; Th. 109, 11; Gen. 1821. [*O. Sax.* horn-seli.] v. horn-sǽl.

hornung-sunu, a; *m. A bastard*, Cot. 142. [*O. Frs.* horning *spurius, nothus* : *Icel.* hornungr *a bastard son*.] v. Grmm. R. A. 476, note.

horo-. v. horu.

hor-pyt, -pytt, es; *m. A dirt-pit, slough* [?] :—Tó ðæm horpytte, Cod. Dipl. Kmbl. iii. 37, 21: 162, 9. v. horu.

HORS, es; *n. A* HORSE :—Geþracan hors *mannus vel brunnicus* : hors of stéden *vel* of asrenne *burdo*, Ælfc. Gl. 5; Som. 56, 18, 19; Wrt. Voc. 17, 23, 24. Hors hófum wlanc, Runic pm. Kmbl. 343, 5; Rún. 19. Ne beó gé ná swylce hors *nolite fieri sicut equus*, Ps. Th. 31, 10. Ðá wæs Hróðgáre hors gebǽted wicg wundenfeax *then for Hrothgar was a horse bitted, a steed with plaited mane*, Beo. Th. 2803; B. 1399. Ne hé on horses hrycge cuman wolde ac hé his fótum geeode *non equorum dorso sed pedum incessu vectus*, Bd. 3, 5; S. 526, 28. Nis horses flǽsc forboden *caro equina non est prohibita*, L. Ecg. C. 38; Th. ii. 162, 16. Wið horses hreófle . . . dó on ðæt hors swá hit hátost mǽge *for a horse's leprosy . . . apply it to the horse as hot as possible*, L. M. 1, 88; Lchdm. ii. 152, 10. Gelícnes horses and monnes, Exon. 109 b; Th. 418, 26; Rä. 37, 11. Ðí byþ swíðe dysig se ðe getrúwaþ on his horses swiftnesse *falsus equus ad salutem*, Ps. Th. 32, 15. Cwæþ mid hospe horse mete is bere *said contemptuously 'Barley is food for a horse*,' Homl. Skt. 3, 216. Man his hors under him ofsceát *his horse was shot under him*, Ors. 5, 2; Bos. 101, 42. Ic seah sroh [*the word is written in runes*] hygewloncne, Exon. 105 a; Th. 400, 1; Rä. 20, 1. Horsa steal *carceres*, Ælfc. Gl. 61; Som. 68, 54; Wrt. Voc. 39, 37. Horsa hnǽgung *neighing of horses*, Ælfc. Gl. 1; Som. 2, 38. Hé wæs mid ðæm fyrstum mannum on ðæm lande nǽfde hé ðeáh má ðonne twentig hrýðera and twentig sceápa and twentig swýna ; ond ðæt lytle ðæt hé erede hé erede mid horsan *he [Ohthere] was among the first men of the country ; and yet he had not more than twenty oxen and twenty sheep and twenty swine ; and the little that he ploughed, he ploughed with horses*, Ors. 1, 1; Swt. 18, 12–15. Ða hors óþbær *it bore away the horses*, Exon. 106 a; Th. 404, 20; Rä. 23, 10. [*O. Sax.* hros; *n* : *O. Frs.* hars, hers, hors, ros; *n* : *Icel.* hross; *m* : *O. H. Ger.* hros; *n* : *Ger.* ross.] v. cræte-hors.

Horsa, an; *m. Horsa* :—On hiera dagum Hengest and Horsa gesóhte Bretene, Chr. 449; Erl. 12, 1. Hér Hengest and Horsa fuhton wið Wyrtgeorne ðam cyninge in ðære stówe ðe is gecueden Agǽlesþrep and his bróður Horsan man ofslóg, 455; Erl. 12, 13.

hors-bǽr, e; *f. A horse-bier*; feretrum caballarium, Bd. 4, 6; S. 574, 5. [*Laym. R. Glouc.* horse-bere : *Prompt. Parv.* hors-bere *lectica*, p. 247, see the note.]

horsc; *adj. Quick, ready, active, valiant*, applied generally to mental activity [cf. snel *active* : *Icel.* snjallr *eloquent*], *wise, sagacious, sharp, quick-witted* :—Horsc *prudens*, Cot. 191, Lye. Hwylc is hǽleþa ðæs horsc and ðæs hygecræftig ðæt ðæt mǽge ásecgan *who amongst men is so quick and cunning of mind as to be able to declare that*, Exon. 101 a; Th. 380, 36; Rä. 2, 1. Nis ǽnig ðæs horsc ne ðæs hygecræftig ðe ðin frumcyn mǽge fira bearnum sweotule gesēðan, 11 a; Th. 15, 24; Cri. 241. Horsc and hreðergleáw herges wísa *a guide of the host, prompt and prudent*, Cd. 143; Th. 178, 17; Exod. 13. On horscum wyllan *by the quick-flowing* [?] *spring*, Cod. Dipl. Kmbl. iii. 456, 15. Þurh horscne hád *through wisdom*, Exon. 8 b; Th. 4, 7; Cri. 49. Módum horsce *sagacious of mind*, 54 a; Th. 190, 12; Az. 72. Horsce mé heredon hilde generedon feóndon biweredon *the valiant praised me, from battle saved me, from foes defended me*, 94 a; Th. 353, 27; Reim. 19. [*O. Sax.* horsk (hugiskaft) : *Icel.* horskr *wise* : *O. H. Ger.* horsc *alacer, celer, præproperus, volucer, promtus, sagax*, v. Grff. iv. 1039–42.]

hors-camb, es; *m. A horse-comb, curry-comb*; strigilis, Wrt. Voc. 83, 34.

horsc-líce; *adv. Readily, promptly, with activity [bodily or mental], wisely, prudently* :—Biþ seó tunge tótogen forðon heó ne mæg horsclíce [MS. horslíce] wordum wrixlan wið ðone wergan gǽst *the tongue shall be rent asunder, therefore it will not be able to converse readily with the accursed spirit*, Exon. 99 b; Th. 373, 28; Seel. 116. [Hors[c]líce *prudenter*, Cot. 138, Lye. [*O. H. Ger.* horsc-lícho *naviter, strenue, agiliter*.]

hors-cræt, es; *n. A chariot*; biga, Lye.

hors-elene, -helene, an; *Elecampāne*; inula helenium, Lchdm. iii. 333, col. 1. Horshelene *helena*, Ælfc. Gl. 44; Som. 64, 68; Wrt. Voc, 32, 4. Horselene, Wrt. Voc. 79, 42. See horshele, E. D. S. Plant Names.

hors-ern, es; *n. A horse-house, stable :* — Horsern *æquiale*, Ælfc. Gl. 2; Som. 55, 33; Wrt. Voc. 16, 7.

hors-gærstūn, es; *m. A meadow for the pasturing of horses :*—Onbūtan ðone horsgærstūn, Cod. Dipl. Kmbl. iii. 414, 25.

hors-here, es; *m. A mounted force*; exercitus equestris, Lye. v. here.

hors-hirde, -hyrde, es; *m. A horse-keeper, groom :*—Horshyrde *pabulator*, Ælfc. Gl. 9; Som. 56, 123; Wrt. Voc. 19, 6. Horshyrde *agaso*, Ælfc. Gr. 9; Som. 8, 37.

hors-hwæl, es; *m. A walrus :* — Swīðost hē fōr ðider tōeácan ðæs landes sceáwunge for ðæm horschwælum for ðæm hie habbaþ swīðe ædele bān on heora tōþum *his principal object in going there, in addition to the observation of the country, was to get the walruses, for they have very excellent ivory in their tusks*, Ors. 1, 1; Swt. 17, 36. [*Icel.* hross-hwalr: *Ger.* wall-ross.]

horsian; *p.* ode *To horse, provide with horses :*—West Seaxe horsodon ðone here *the people of Wessex provided the Danes with horses*, Chr. 1015; Erl. 153, 1. Hē beád ðæt man sceolde his here metian and horsian, 1013; Erl. 148, 3: 1014; Erl. 151, 2. DER. be-, ge-horsian.

hors-minte, e; *f. Wild mint*; menthastrum, Lye. v. E. D. S. Plant Names, horse mint.

hors-syðða, an; *m: v.* hors-bær.

hors-þegn, es; *m.* **I.** *a groom :* — Horsþen *agaso*, Ælfc. Gl. 20; Som. 59, 42; Wrt. Voc. 23, 5: *mulio*, Hpt. Gl. 438: Gl. Mett. 516. **II.** *the title of an officer of the royal household* [cf. marescalcus *among the Franks*] :—Ecgulf cynges horsþegn, Wulfrīc cynges horsþegn, Chr. 897; Erl. 95, 5: 96, 16. v. Kemble's Saxons in England ii. 107–8.

hors-wægn, -wæn, es; *m. A chariot :*—Horswæn *carpentum, currus*, Ælfc. Gl. 48; Som. 65, 68; Wrt. Voc. 34, 3.

hors-wealh, es; *m. A servant that attends to horses* [*Thorpe takes* wealh *to mean one of British origin*, v. Glossary] :—Be cyninges horseweale. Cyninges horswealh se ðe him mæge geǽrendian ðæs wergield biþ cc scill., L. In. 33; Th. i. 122, 12.

hors-weard, e; *f. A taking care of horses :*—Horswearde healdan, L. R. S. 2; Th. i. 432, 17.

hors-weg, es; *m. A horse-road :* — Tō horsweges heale, Cod. Dipl. Kmbl. iii. 219, 2.

horu; *gen.* -wes; *m. Dirt, filth, foulness :*—Færmaþ gyf ðǽr hwæt horwes on biþ *cleanse if there be any foulness in it*, Herb. 9, 2; Lchdm. i. 100, 4. Horewes, Mone B. 3561. Gē mid horu speówdon on ðæs andwlitan *ye foully spat on his face*, Elen. Kmbl. 594; El. 297. Mīn flǽsc is ymscrýd mid dustes horwum *my flesh is clothed with the filth of dust*, Homl. Th. ii. 456, 10. On his blōde aþwogen fram synna horwum *washed in his blood from the impurities of sins*, Homl. Swt. 11, 297. Horewum, Homl. Th. ii. 56, 8. [*O. E. Homl.* horie, hore (of þe hore þat is cleped hordom): *O. Sax.* horu *dirt :* *O. Frs.* hore: *O. H. Ger.* horo; *gen.* horawes; *dat.* horowe, horewe, horwe, hore *limus, cenum, lutum, palustre.*] v. horh.

horu-seáþ, es; *m. A foul pit, sink :*—Gesihst ðū nū on hū miclum and on hū diópum and on hū þióstrum horaseáþe [MS. Cott. horoseáþa] ðara unþeáwa ða yfelwillendan sticiaþ *videsne igitur quanto in cœno probra volvantur*, Bt. 37, 2; Fox 188, 1.

horu-weg, es; *m. A dirty road, a lane* [?] :—Ðar horoweg ūtt sceát, Cod. Dipl. Kmbl. v. 173, 17. Horwegstige *devia semita*, Cot. 61, Lye.

horweht; *adj. Foul, filthy, dirty :*—Hine ðā lǽddon on ðone sweartan fenn and hine ðā on ða horwehtan wæter bewurpon *they led him then to the black fen and flung him into the foul water*, Guthl. 5; Gdwin. 36, 9. v. horheht.

hōs, e; *f. A bramble, thorn :* — Hōs *butrus*, Wrt. Voc. 285, 27: *rhamnus, vimen; butrus*, Cot. 25, 165, Lye. Twīgu ł hōsa *rhamnum*, Ps. Spl. C. 57, 9.

hōs, e; *f. A company, band :*—Mid mægþa hōse *with a band of maidens*, Beo. Th.1853; B. 924. [*Goth.* hansa *multitudo : O. H. Ger.* hansa *cohors :* cf. Hanse *applied to an association of towns.*]

hosa, an; *m.* [or hose; *f.*(?) v. next word, and cf. other dialects]. **I.** *a covering for the leg*, HOSE :—Hosa *caliga vel ocrea*, Wrt. Voc. 81, 48. [*Prompt. Parv.* hose *caliga*, p. 248, see note : *Laym.* hose, v. 15216: *R. Glouc.* (*in the corresponding passage*) hose: *A. R.* hosen; *pl : Chauc.* hosen : *Icel.* hosa; *f. a covering for the leg between the knee and the ankle, serving as a kind of legging or gaiter : O. H. Ger.* hose; *f. caliga : Ger.* hose; *f. breeches, hose.*] **II.** *a husk, a covering for a grain or seed* [or is this a different word?] :—Wilnade gefylle womb his of beánbælgum ł pīsum hōsum *cupiebat implere ventrem suum de siliquis*, Lk. Skt. Lind. 15, 16. v. Jamieson's Dict. hose *the seed-leaves of grain : vagina, the* hose *of corn*. See also E. D. S. Reprinted Glossaries, No. 5.

---

hose-bend, es; *m. A hose-band, garter :* — Hosebendas *periscelides*, Lye: Hpt. Gl. 517. [Cf. *Icel.* hosna-reim.]

hosp, es; *m. Reproach, opprobrium, contempt, contumely, insult, blasphemy :*—Hosp *opprobrium*, Ps. Spl. 14, 4 : 21, 5. Ða ðe forþgewēteþ of welerum mīnum nā ic dō hosp *quæ procedunt de labiis meis, non faciam irrita*, 88, 34. Hē geseah mīnne hosp áfyrran *respexit auferre opprobrium meum*, Lk. Skt. 1, 25. Nū tō dæg ic ádyde ðæra Egiptiscra hosp fram eówrum cynne *this day have I rolled away the reproach of Egypt from off you*, Jos. 5, 9. Hǽcenra hosp, Judth. 11; Thw. 24, 30; Jud. 215: Exon. 10 b; Th. 11, 16; Cri. 171: 29 a; Th. 88, 22; Cri. 1444. Hī mid hospe his lāre forsáwon *they with contumely despised his teaching*, Homl. Th. ii. 110, 5. Cwæþ mid hospe *said contemptuously*, Homl. Swt. 3, 216. Ða hrýmde Julianus mid hospe and earmlīce gewāt *then cried out Julian blaspheming and miserably died*, 275. Swā hwilcne swā hī tō hospe habban woldon hī cwǽdon be ðam ðæt hē wǽre Samaritanisc *whomsoever they wished to hold up to contempt, they said of him that he was a Samaritan*, Homl. Th. ii. 228, 32. Ðonne wurdon hī tō hospe gedōne *then were they made a reproach*, Ælfc. T. 12; Grn. 6, 22. Unrihtwīse habbaþ on hospe ða ðe him sindon rihtes wīsran *the unrighteous hold in contempt those that are better skilled in right than themselves*, Bt. Met. Fox 4, 87; Met. 4, 44. Hospe gereccan *to reproach opprobriously*, Exon. 70 a; Th. 260, 21; Jul. 300: 90 a; Th. 337, 17; Gn. Ex. 66. Menigfealde earfoþnyssa and hospas wolde gehwā eáðelīce forberan wið ðan ðæt hē mōste sumum rīcan men tō bearne geteald beón *anybody would put up with all kinds of hardships and affronts on condition that he might be accounted the son of some great man*, Homl. Th. i. 56, 11.

hosp-cwide, es; *m. Contemptuous, opprobrious, insulting language*, Elen. Kmbl. 1044; El. 523.

hosp-sprǽc, e; *f. Contemptuous, insulting language :* — Se eádmōda biscop ðe wē ymbe sprecaþ wæs swīðe geþyldig wið þwyrum mannum and him ne eglede heora hospsprǽc ac forbær blīðelīce ðeáh ðe him man bysmor cwǽde *the lowly-minded bishop that we are talking about was very patient with perverse people, and their contemptuous language did not vex him, but he cheerfully bore with it, though he was reviled*, Homl. Th. ii. 514, 11.

hosp-word, es; *n. A word expressing contempt, contumely, reproach, abuse :*—Án ðæra hospworda hē forbær suwigende *one of their reproaches he bore with in silence*, Homl. Th. ii. 230, 8. Ðā hēt martianus mid his hospwordum ðæt hē sǽde his sīþ him eallum *then Martianus bade him with expressions of contempt tell his journey to them all*, Homl. Swt. 4, 283: Exon. 68 b; Th. 253, 33; Jul. 189. Ongan tō ðam hālgan hospword sprecan *began to speak words of contempt to the saint*, Andr. Kmbl. 2632; An. 1317.

hoðma, an; *m. A covering* [?], *cloud* [?]. *darkness :*—Ðǽr wīsna fela wearþ inlīhted ðe ǽr under hoðman biholen lǽgon *there many things were illumined that before lay concealed in darkness*, Exon. 8 b; Th. 3, 32; Cri. 45. Rīdend swefaþ hæleþ in hoðman *knights and warriors sleep in the darkness* [*of death*], Beo. Th. 4907; B. 2458. [Cf. heóðu.]

hrā. v. hrǽw.

hrāca, an; *m. Expectoration, spittle, matter brought up when clearing the throat :*—Ðæs seócan mannes hrāca biþ maniges hiwes *the sick man's expectoration is many-coloured*, L. M. 2, 46; Lchdm. ii. 260, 13. Hyt gelīðigaþ ðone hrācan, Herb. 55, 2; Lchdm. i. 158, 10. Wið swiðlīcne hrācan, 146, 2; Lchdm. i. 270, 2. Mycelne hrācan, 158, 1; Lchdm. i. 284, 23. [*Icel.* hráki *spittle.*] v. hrǽcan.

hracca [hnacca?] *the back part of the head*; occiput, Som. [Cf. a rack of mutton, *dorsum ovile*, E. D. S. vol. 3, B. 18.]

hrace, an; *f.* hraca, an; *m. The throat :*—Hrace *gula*, Wrt. Voc. 283, 4: hracu, 64, 64. Ðǽr gýnude on ðare hracan swylce ðǽr hwylc seáþ wǽre *there yawned in the throat as if there had been a pit*, Lchdm. ii. 364, col. 1. Ne hī on hracan áwiht hlūde ne cleopiaþ *non clamabunt in gutture suo*, Ps. Th. 134, 19. Ne him gāst on hracan eardaþ *neque est spiritus in ore eorum*, 113, 16. Swille ða hracan *let him swill the throat*, L. M. 1, 1; Lchdm. ii. 24, 27. Stinge him on ða hracan ðæt hē mæge spīwan, 1, 18; Lchdm. ii. 62, 12. Hire man bestang sweord on ða hracan, Shrn. 56, 14. Fýrene tungan and gyldenne hracan *a fiery tongue and a golden throat*, Salm. Kmbl. 148, 32. Hracan [bracan, Som.] *fauces*, Ælfc. Gl. 72; Som. 70, 109; Wrt. Voc. 43, 37. [*O. H. Ger.* racho *sublinguium : Ger.* rachen *throat, jaws.*]

hracing, e; *f. A holding back, stopping, stay*; detentio, Rtl. 65, 27. [Cf. (?) *Icel.* hrakning *bad treatment, insult.*]

hracod *laceratus*, Som. [Cf. *Icel.* hrekja *to worry, vex.*]

hradian; *p.* ode *To quicken, hasten, accelerate, forward :*—Hreaða *accelera*, Ps. Stev. 30, 3. Hreaðedon *acceleraverunt*, 15, 4. DER. for-, ge-hradian.

hradung, e; *f. A hastening*; festinatio, acceleratio, Lye.

hréc. v. hreác.

hrǽcan; *p.* hrǽhte *To clear the throat, hawk, spit :*—Ic hrǽce oððe ic spǽte *screo*, Ælfc. Gr. 26, 6; Som. 29, 17. Hrǽce hió him on ðæt nebb foran *huic in faciem mulier expuat*, Past. 5, 2; Swt. 43, 15. Gif hwā blōd swīðe hrǽce *if any one spit much blood*, Herb. 40, 2; Lchdm. i.

142, 1. Wið ðæt man hefelíce hræce *for difficulty in clearing the throat in cases of cold*, 46, 1; Lchdm. i. 148, 12, 15. [*Icel.* hrækja *to hawk, spit :* cf. *O. H. Ger.* rachison *screare.*] v. hráca.

hrǽcea, an; *m. Clearing the throat, hawking :* — Þurh spátl and hrǽcean *by spittle and clearing the throat*, L. M. 1, 1; Lchdm. ii. 24, 8.

hrǽcetung, e; *f. Retching, eructation :*—Wið bitere hrǽcetunge, L. M. 2, 8; Lchdm. ii. 186, 26.

hrǽc-gebræc, es; *n. A cold in the chest, hoarseness :* — Hrǽc-gebræc *branchos* [= βράγχος], Ælfc. Gl. 10; Som. 57, 23; Wrt. Voc. 19, 29. v. bræc, gebræceo.

hrǽctan; *p.* te *To eructate, retch :*—Biþ sió wamb áþened and hrǽctaþ gelóme *the stomach is extended and they eructate frequently*, L. M. 2, 28; Lchdm. ii. 224, 12.

hrǽc-tunge, an; *f. The uvula :* — Biþ reád ymb ða hrǽctunga[n?], L. M. 1, 4; Lchdm. ii. 46, 10.

hrǽcung, e; *f. A clearing of the throat, hawking :*—Gelóme spǽtunga oððe hrǽcunga *frequent spittings or hawkings*, L. M. 2, 1; Lchdm. ii. 174, 21. DER. blód-, wyrs-hrǽcung.

hrǽd, hrǽd, hreð; *adj. Quick, swift, speedy, sudden, alert, rapid, prompt, active :*—Hrǽd oððe glæd *agilis :* hrǽddre *agilior :* ealra hrǽdost *agillimus*, Ælfc. Gr. 5; Som. 5, 6. Hrǽd oððe glæd *alacer*, 9, 18; Som. 9, 66. Tó hrǽd ierre *præceps ira*, Past. 13, 2; Swt. 79, 14, 11. Worda tó hrǽd, Exon. 88 a; Th. 330, 13; Vy. 50. Sum biþ hrǽd tæfle *one is quick at games of chance*, 79 a; Th. 297, 25; Crä. 73. Ðæt wæs hrǽd ǽrendraca se tylode tó secganne hys ǽrndunge ǽr ðon ðe hé lyfde *that was a quick messenger, who strove to tell his message before he lived*, Shrn. 95, 20. Se gást is hrǽd *spiritus promptus est*, Mt. Kmbl. 26, 41. Níþ godes hreð [hrẽd?] of heofonum *God's anger swift from heaven*, Cd. 206; Th. 255, 6; Dan. 620. Hrǽd and unlæt, Exon. 113 b; Th. 436, 9; Rä. 54, 11. Ðú ðe on hrǽdum fǽrelde ðone heofon ymbhweorfest *qui rapido cælum turbine versas*, Bt. 4; Fox 6, 31. On hrǽde sprǽce *in prosam*, Bd. 5, 23; S. 648, 22. Hrǽde weámetta *sudden sadnesses*, L. I. P. 10; Th. ii. 318, 32. Hrade [MS. T. hræþe; Ps. Th. hraðe] fót heora tó ágeótenne blód *veloces pedes eorum ad effundendum sanguinem*, Ps. Spl. 13, 6. Ða hradan ðonne sint tó manianne *præcipites admonendi sunt*, Past. 39, 1; Swt. 281, 20. Mé is fenýce fóre hreðre is dæs gores sunu gonge hrǽdra *more swift than I is the fen-frog in its course, the son of dirt [beetle] is more rapid in its walk*, Exon. 111 a; Th. 426, 9–12; Rä. 41, 71–2. [*Icel.* hraðr *swift, fleet :* O. H. Ger.* hrat, hrad *velox.*]

hrǽd-, hrǽd-bíta, an; *m. An insect which eats away clothes*, etc; *blata*, Wrt. Voc. 281, 44.

hrǽdding. v. hredding.

hrǽd-férness, e; *f. Quickness, rapidity :*—Behealdaþ ða hrǽdférnesse ðisses heofenes *respicite cæli celeritatem*, Bt. 32, 2; Fox 116, 6.

hrǽd-hýdignes, e; *f. Precipitancy, hastiness :*—Ðý læs hie unnytlíce forweorpen ðæt ðæt hie sellen for hira hrǽdhýdignesse *ne præcipitatione hoc quod tribuunt inutiliter spargant*, Past. 44, 2; Swt. 321, 18. Ðonne oncann hé hiene selfne for ðære hrǽdhýdignesse ðe hé ǽr tó fela sealde *occasionem contra se impatientiæ exquirit*, 4; Swt. 325, 16. For hrǽdhýdignesse *præcipiti festinatione*, 49, 1; Swt. 375, 16.

hrǽding, e; *f. Hurry, haste :*—Be ðisum þeófum ðe man on hrǽdinge fúle geáxian ne mæg and man eft geáxaþ ðe hé fúl biþ *concerning the thieves that are not at once found out to be guilty, and afterwards it is found on enquiry that he is guilty*, L. Æðelst. v. 9; Th. i. 238, 29. Hí burigdon swá swá heó líhtlucost mihten on swylce [h]rǽding *they buried him as best they could in such a hurry*, Th. An. 123, 22.

hrǽd-líc; *adj. Quick, hasty, sudden, speedy, precipitate :*—Hit wǽre tó hrǽdlíc gif hé ðá on cildcradole ácweald wurde *it had been precipitate, had he been slain then in the cradle*, Homl. Th. i. 82, 28. Æfter hrǽdlíce tíde *after a short time*, Ors. 1, 10; Swt. 44, 28. Hé wæs mid hrǽdlíce deáþe forgripen *morte immatura præreptus est*, Bd. 4, 23; S. 594, 36. Ðǽr forþférde Sideman bisceop on hrǽdlícan deáþe *died suddenly*, Chr. 977; Erl. 127, 36.

hrǽd-líce; *adv. Quickly, hastily, speedily, immediately, at once, forthwith :*—Hrǽdlíce *actutum*, Ælfc. Gr. 38; Som. 41, 64. Hrǽdlíce hé ástáh of ðam wætere *confestim ascendit de aqua*, Mt. Kmbl. 3, 16 : *continuo*, 13, 5, 20. Gif ðú wille mildheortnesse ús dón sǽge ús ðæt hrǽdlíce *if thou wilt do us kindness, tell us so at once*, Blickl. Homl. 233, 19. Him ðá áþas swóron ðæt hie hrǽdlíce of his ríce fóren *they swore oaths to him that they would speedily march out of his kingdom*, Chr. 876; Erl. 78, 11. Hé wæs æfter ðam swíðe hrǽdlíce gehálgod tó cyninge *very soon after that he was consecrated king*, 979; Erl. 129, 30. Hrǽdlícor *ocius;* hrǽdlícost *ocissime*, Ælfc. Gr. 38; Som. 42, 9. Se hit mæg hrǽdlícor geféran *he can perform the journey more quickly*, Blickl. Homl. 231, 24: Bd. 3, 14; S. 540, 19.

hrǽd-lícness, e; *f. Quickness, suddenness, rapidity, haste :*—Ða micclan welan ðe hig ǽrhwílon áhton hé geseh on hrǽdlícnysse ealle gewítan *the great riches that they formerly owned he saw all quickly pass away*, Guthl. 2; Gdwin. 14, 23.

hrǽd-ness, e; *f. Quickness, rapidity :*—Wundorlícre hrǽdnysse *with wonderful quickness*, Herb. 18, 4; Lchdm. i. 112, 1. Ond wē ðá mid wunderlícre hrǽdnysse porrum ðone cyning ofercwomon *mira celeritate poro rege devicto*, Nar. 4, 4. Se on hrǽdnesse swá mycele menigo heora fornom *quæ in brevi tantam ejus multitudinem stravit*, Bd. 1, 14; S. 482, 30.

hrǽd-sprǽce. v. un-hrǽdsprǽce.

hrǽd-wægn, -wǽn, es; *m. A swift chariot :*—Se stiórþ ðam hrǽdwǽne eallra gesceafta *volucrem currum regit*, Bt. 36, 2; Fox 174, 20 : Bt. Met. Fox 24, 81; Met. 24, 41.

hrǽd-wilness, e; *f. Precipitancy, haste :*—Sió hátheortness and sió hrǽdwilnes ðæt mód gebringþ on ðæm weorce ðe hine ǽr nán willa tó ne spón *mentem impellit furor, quo non trahit desiderium*, Past. 33, 1; Swt. 215, 9. Ðeáh for hrǽdwilnesse tó fóþ *tamen præcipitatio impellit*, 23, 2; Swt. 177, 15: 49, 1; Swt. 375, 20. [Cf. hrǽd-hýdigness.]

hrǽd-wyrde; *adj. Quick, hasty of speech :*—Ne sceal nó tó hátheort ne tó hrǽdwyrde *he must not be too passionate nor too hasty of speech*, Exon. 77 b; Th. 290, 17; Wand. 66.

hræfn, es; *m. A raven :*—Hrefn *corvus*, Wrt. Voc. 280, 33. Hræmn, Ælfc. Gr. 8; Som. 7, 35. Blac hræm *niger corvus*, 6; Som. 4, 21; Wrt. Voc. 77, 13. Noe ásende út ǽnne hremn se hremn fleáh ðá út and nolde eft ongeán cirran *Noe dimisit corvum, qui egrediebatur et non revertebatur*, Gen. 8, 7. Ðá wæs sum wild hrem . . . hé ðá wearp ðam hremme ðone gesǽttrodan hláf *there was a wild raven . . . he threw the poisoned bread to the raven*, Homl. Th. ii. 162, 21, 23. Se wanna hrefn wælgífre fugel, Judth. 11; Thw. 24, 25; Jud. 206: Beo. Th. 6041; B. 3024. Hrefn blaca, 3606; B. 1801. Se swearta hrefn, Soul Kmbl. 108; Seel. 54. Ðǽr him hrefn nimeþ heáfodsýne slíteþ salwigpád sáwelleásne *there shall the raven, dark-coated, pluck from him his eyes, shall tear him lifeless*, Exon. 87 b; Th. 329, 18; Vy. 36. Hræfen wan, Elen. Kmbl. 104; El. 52 : Fins. Th. 69; Fins. 34. Ðǽr wæs se gúðfana genumen ðe hí ræfen héton *there was the banner taken which they [the Danes] called the Raven* [see Asser's life of Alfred under the year 878 for an account of this banner; and see further references in Cl. and Vig. Icel. Dict. under *hrafn*], Chr. 878; Erl. 81, 3. Hrefnes briddum *pullis corvorum*, Ps. Th. 146, 10. His sunu hangaþ hrefne tó hróðre *his son hangs a solace for the raven*, Beo. Th. 4887; B. 2448. Saluwigpádan ðone swearten hræfn hyrnednebban *the black raven, dusky-coated, hard-beaked*, Chr. 937; Erl. 115, 10; Æðelst. 61. Hí læccaþ eallswá gýfre hremnas of holde dóþ *they seize just as greedy ravens do from a corpse*, L. I. P. 19; Th. ii. 328, 5. Swá swá grǽdige ræmmas, L. Ælfc. P. 49; Th. ii. 386, 3. Besceáwiaþ ða hrefnas *considerate corvos*, Lk. Skt. 12, 24. [*Laym.* rem : *Icel.* hrafn, hramn : *O. H. Ger.* hraban, hram *corvus, corax :* Ger.* rabe.] DER. niht-hræfn.

hræfn; *m. A crab :*—Se hrefn ðe sume menn hátaþ crabba *the 'hrefn' that some people call a crab*, Shrn. 162, 21. Hrefnes geallan and leaxes *a crab's gall and a salmon's*, L. M. 3, 2; Lchdm. ii. 308, 6; see note. Hræfnes geallan, Lchdm. iii. 2, 11. Genim cucune hrefn ádó ða eágan of and eft cucune gebring on wætre *take a live crab, put its eyes out, and put it back in the water alive*, L. M. 3, 2; Lchdm. ii. 306, 20. v. hæfern.

hræfn-cynn, es; *n. The raven-kind :*—Nán þing hrefncynnes, Lev. 11, 17.

hræfnes fót *ravensfoot; ranunculus gramineus*, see Lchdm. iii. 333, col. 1.

hræfnes leác *orchis*, see Lchdm. iii. 333, col. 1. v. Grmm. D. M. 1144.

hrægel, hrægl, es; *n. A garment, dress, robe, rail* [in *night-rail*] *clothing :*—Gerǽwen hrægel *segmentata vestis :* þicce gewefen hrægel *pavidensis :* þenne gewefen hrægel *levidensis :* purpuren hrægel *clavus* vel *purpura :* feala hiwes hrægel *polymita :* wógum bewerod hrægel *ralla* vel *rasilis :* geedniwod eald hrægel *interpola vestis :* geclútad hrægel *panucla :* gediht hrægel *acupicta :* þrýlen hrægel *trilicis*, Ælfc. Gl. 63; Som. 68, 99–109; Wrt. Voc. 40, 10–19. Hrægl and hringas *robe and rings*, Beo. Th. 2394; B. 1195. Sæt ðǽr sum þearfa nacod bæd hrægles and ælmessan *a beggar sat there naked asked for a garment and an alms*, Blickl. Homl. 213, 33. Hrægles þearfa ic mé leáfum þecce *lacking raiment I cover me with leaves*, Cd. 40; Th. 53, 25; Gen. 866. Ðisses hrægles neót *use this robe*, Beo. Th. 2439; B. 1217. Wíf móton under brúnun hrægle tó húsle gán *mulieribus licet sub nigro velamine eucharistiam accipere*, L. Ecg. C. 37; Th. ii. 162, 7. Wese hé hrægle gelíc *fiat ei sicut vestimentum*, Ps. Th. 108, 19. Mid mete and mid hrægle *with food and clothing*, Blickl. Homl. 41, 29. Se ðe mid ðon ánum hrægle wæs gegyrwed *who was dressed in that one garment*, 169, 1. On medmyclan hrægle gehealdene *moderate in dress*, 185, 17. Man hine forbærneþ mid his wǽpnum and hrægle *he is burnt with his arms and clothing*, Ors. 1, 1; Swt. 21, 8. Ðæt hrægl ðe hé ǽr ðæm þearfan sealde *the cloak that he had given to the beggar*, Blickl. Homl. 215, 18: 223, 8. Ongan his hrægl teran *began to rend his robe*, Judth. 12; Thw. 25, 28; Jud. 283. Ða hwítan hrægl ðara engla *the white robes of the angels*, Blickl. Homl. 121, 24. Sylle earmum mannum his ealde hrægl *let him give his old clothes to the poor*, 53, 13. Hie hæfdon manige glengas deórwyrþra hrægla *they had many ornaments of costly garments*, 99, 19. Beaduscrúda best hrægla sélest, Beo. Th. 912; B. 454. Án cild hreglum [hræglum, MS. C.] bewunden *infantem pannis involutum*, Lk. Skt. 2, 12. Mid godwebbenum hræglum *with purple raiment*, Blickl. Homl. 95, 20.

Hrægl *spolia*, Ps. Spl. 67, 13. [*O. Frs.* hreil, reil: *O. H. Ger.* hregil *indumentum, coturnus; pl. trophæa, spolia.*] DER. beadu-, beód-, brèc-, frum-, fyrd-, hrycg-, mere-, set-, setl-, wîte-hrægel.

hrægel-cist, e; *f. A clothes-chest, trunk :—*Ân hræglcysð *one clothes-chest*, Chart. Th. 538, 20.

hrægel-gefrætwodness, e; *f. Elegance* or *adornment of dress :—* Hwǽr is nû heora gold and heora hrægelgefrætwodnes? L. E. I. prm; Th. ii. 396, 27.

hrægel-gewǽde, es; *n. Dress, clothes*, Cot. 118, Lye.

hrægel-hûs, es; *n. A vestry;* vestiarium, C. R. Ben. 67, Lye. [Railhus *vestiarium*, Wrt. Voc. 93, 56.]

hrægel-talu, e; *f. A fund for providing vestments :—*Ic ðás land ǽcelîce sælle into sanctæ trinitatan ðám hîwum tô hira beódlandæ and tô hregltalæ *ego has terras dono æternaliter familiæ æcclesiæ sanctæ trinitatis ad refectorium fratribus et ad vestimenta*, Cod. Dipl. Kmbl. v. 218, 20.

hrægel-þegn, -þén, es; *m. An officer of the royal household* or *of a monastery :—*Ic Leófríc hrægelþén, Cod. Dipl. Kmbl. iii. 351, 16. Ælfríc wæs ðá hrǽlþen, Chart. Th. 170, 10. Hé scolde setten ðǽr prior of Clunni and circeweard and hordere and reilþein, Chr. 1131; Erl. 260, 12. Hræglþegn *vestiarius*, C. R. Ben. 55, Lye. [See Kemble's Saxons in England, ii. 106.]

hrægel-weard, es; *m. One who has charge of vestments :* — Hræglweard *vestiarius*, Wrt. Voc. 289, 69.

hrægl. v. hrægel.

hræglung, e; *f. Clothing;* vestitus, Ælfc. Gl. 62; Som. 68, 85; Wrt. Voc. 39, 68.

hrægn-loca = [?] brægn-loca *that which encloses the brain, the skull*, Exon. 126 b; Th. 487, 11; Rä. 72, 21.

hrǽm, hræmn. v. hræfn.

hrǽn *capreolus*, Som. v. hrán.

hrǽron. v. hreran.

hrǽtele, hrætel-wyrt *rattlewort*, Lchdm. iii. 333, col. 2.

hrǽð. v. hræd.

hrǽða. v. hréða.

hrǽðe. v. hraðe.

hrǽw, hráw, hreáw, hrá, es; *n. m. The body of a man living or dead, a corpse, carcase, trunk, carrion :* — Lîc *vel* hreáw *funus*, Ælfc. Gl. 85; Som. 74, 1; Wrt. Voc. 45, 25. Ðû earma nû ðû byst geworden ðæt fúleste hreáw and wyrma mete *thou miserable thing, now art thou become a very foul corpse and food for worms*, L. E. I. prm; Th. ii. 398, 16. Hrá wundum wérig *the body weary with wounds*, Andr. Kmbl. 2556; An. 1279: 2062; An. 1033: Exon. 36 b; Th. 119, 14; Gú. 254. Hé ðæt hrá gescóp *he created the body*, 8 a; Th. 2, 5; Cri. 14. Hrá biþ ácólad *the corpse is cooled*, 59 a; Th. 213, 22; Ph. 228: Elen. Kmbl. 1767: El. 885. Hrá wide sprong *far away sprang the trunk* [*as the head was severed from it*], Beo. Th. 3181; B. 1588. Ðonne flǽsc onginneþ hráw côlian *when the flesh, the body begins to grow cold*, Runic pm. 29; Kmbl. 345, 14. Wealdendes hrǽw *the ruler's* [*Christ*] *body*, Rood Kmbl. 106; Kr. 53: 144; Kr. 72. Ðá lócade hé on his ágenne lîchaman swá swá on uncûþne hreáw *he gazed on his own body as on an unknown corpse*, Shrn. 52, 4. Ða sticca Simones hreáwes *the pieces of Simon's carcase*, Homl. Th. i. 380, 34. Sang se wanna fugel hrǽs on wénan *the dusky fowl sang hoping for carrion*, Cd. 93; Th. 119, 25; Gen. 1985. Furseus ðá beseah tô his lîchaman swilce tô uncûþum hreáwe, Homl. Th. ii. 346, 7. Ða lîchoman heáhfædera hrá *the bodies, the patriarchs' corpses*, Andr. Kmbl. 1581; An. 792. Heora fædera hreáw *cadavera patrum*, Num. 14, 33. Hrǽ, hrǽw [*other MSS.* hráw, hrá] *corpses*, Chron. 937; Erl. 115, 9; Æðelst. 60. Reócende hrǽw *reeking carcases*, Judth. 12; Thw. 26, 7; Jud. 314. Hrǽwas ł ða deáþlícan ðínra þeówana *morticina servorum tuorum*, Ps. Lamb. 78, 2. Deádra hrǽwum *over the corpses of the dead*, Cd. 144; Th. 180, 6; Exod. 41. [*O. Sax.* hréo: *O. Frs.* hrê: *Icel.* hræ *a corpse, carrion: O. H. Ger.* hréo *cadaver, funus:* cf. *Goth.* hraiwadubo.]

hrǽw'raw. v. hreáw.

hrá-fyl, -fyll, es; *m. Slaughter*, Beo. Th. 559; B. 277.

hragan. v. ofer-hragan.

hrá-gîfre; *adj. Greedy for corpses, deadly :* — Hrágýfra *funestus*, Cot. 90, Lye. [Cf. wæl-gîfre.]

hrágra, an; *m. A heron :—*Hrágra *ardea*, Ælfc. Gl. 36; Som. 62, 111; Wrt. Voc. 29, 9: 63, 13. Hrágra *larum*, Shrn. 29, 18. [*O. H. Ger.* raiger, *regera ardea :* Ger. reiher *a heron*.]

hrá-lîc; *adj. Deadly* [?], *funereal* [?]; funebris, Cot. 88, Lye. [*O. H. Ger.* rê-lîh *funestus, funebris*.]

hramma, an; *m. Cramp, spasm :* — Hramma *spasmos*, Ælfc. Gl. 10; Som. 57, 12; Wrt. Voc. 19, 21. Gif hwylcum men hramma derige *if cramp annoy any man*, Herb. 94, 11; Lchdm. i. 206, 21. Wið hramman, 153, 5; Lchdm. i. 280, 5. [Cf. *Icel.* hrammr *that with which one clutches, a bear's paw*.] v. hremman.

hramsan; *pl. Ramsons, broad-leaved garlic;* allium ursinum, Lchdm. iii. 333, col. 2. [See Skeat, Etymol. Dict.]

hran, hron, es; *m. A whale, a mussel* [?] :—Hran *ballena*, Wrt. Voc. 65, 62. Hron *ballena* vel *pilina*, 281,55. Hran *musculus*, Ælfc. Gl. 102; Som. 77, 78; Wrt. Voc. 56, 1. On huntunge hranes *in venationem balenæ*, Coll. Monast. Th. 24, 25. Hér beóþ oft fangene seolas and hronas and mereswýn *capiuntur sæpissime et vituli marini, et delphines necnon et ballenæ*, Bd. 1, 1; L. 473, 16. Hronesnæs, Beo. Th. 5603, 6264; B. 2805, 3136.

hrán, es; *m. A reindeer :—*Se byrdesta sceall gyldan fíf hránes fell *a man of the highest rank has to pay five reindeer skins*, Ors. 1, 1; Swt. 18, 20. Ða deór hí hátaþ hránas; ðara wǽron syx stælhránas: ða beóþ swýðe dýre mid Finnum, forðæm hý fóþ ða wildan hránas mid *those deer they call 'rein;' six of them* [*Ohthere's*] *were decoys : those are very precious among the Fins, for they catch the wild reindeer with them*, 10–12. [*Icel.* hreinn, see Cl. and Vig. Dict.]

hrand-spearwa, an; *m. A sparrow :*—Hrondsparwas ł staras *passeres*, Mt. Kmbl. Lind. 10, 29.

hran-fisc, es; *m. A whale :*—Hronfixas, Beo. Th. 1085; B. 540.

hran-mere, es; *m. The whale-mere, the sea :*—Hronmere, Bt. Met. Fox 5, 19; Met. 5, 10.

hran-rád, e; *f. The whale-road, the sea :*—Ús bær on hranráde heáhstefn naca *us the high-stemmed bark bore on the sea*, Andr. Kmbl. 531; An. 266: 1267; An. 634. Geond hronráde *throughout the ocean*, Cd. 10; Th. 13, 19; Gen. 205: Beo. Th. 19; B. 10: Andr. Kmbl. 1641; An. 822.

hraðe, hræðe, hreðe; *adv. Quickly, immediately, at once, soon, forthwith, straightway :*—Gá hraðe on ða strǽta *exi cito in plateas*, Lk. Skt. 14, 21: 16, 6. Cûþ is ðætte hraðe Drihten ðæs ðe hé of ðam fulwihtes bæþe eode ðá fæstte hé sóna *it is known that the Lord directly after he came from baptism at once fasted*, Blickl. Homl. 27, 23. Ðá wæs hraðe geworden ðæt hé gelýfde *then immediately it came to pass that he believed*, 153, 13. Gif heó hraðe gǽþ *if she walks quickly*, Lchdm. iii. 144, 8. Hraðe æfter *directly after*, Ps. Th. 59, 3. Mé hraðe syððan gefultuma *ad adjuvandum me festina*, 69, 1. Tô hraðe *too soon*, Bt. 3, 1; Fox 4, 23. Hé wæs Godes bearn swá hraðe swá hé mannes bearn wearþ *he was the Son of God so soon as he became the Son of man*, Homl. Th. ii. 526, 1. Swíðe hræðe *repente*, Past. 21, 7; Swt. 166, 14. Héton ût hraðe æþeling lǽdan *they bade quickly lead out the noble one*, Andr. Kmbl. 2545; An. 1274: 3039; An. 1522. Ðû ealne hræðe hefon ymbhwearfest *rapido cælum turbine versas*, Bt. Met. Fox 4, 6; Met. 4, 3. Ðá wæs háten hreðe *then was bidden straightway*, Beo. Th. 1986; B. 991. Hreðe siððan *directly after*, Bt. Met. Fox 25, 94; Met. 25, 47. Ne scule gé hit nó ðý hraðor þurhteón *none the sooner shall ye accomplish it*, Ps. Th. 4, 5: Cd. 212; Th. 263, 2; Dan. 756. Nó hé fleótan meahte hraðor on holme *not more swiftly than I could he float on the ocean*, Beo. Th. 1090; B. 543. Hí hogedon hû hí unriht hraðost ácwǽdon *they considered how soonest they might utter iniquity*, Ps. Th. 72, 6. Swá hwilc swá gearo wearþ hraðost *whosoever was soonest ready*, Chr. 755; Erl. 51, 3. Hé árás swá hé hraðost meahte *he arose as quickly as ever he could*, Exon. 49 a; Th. 168, 24; Gú. 1082. And hraðost is tô cweðenne *in short*, Swt. A. S. Rdr. 106, 60. Ðæt is nú hraðost tô secganne, Bt. 7; Fox 60, 14. [Cf. *Icel.* ok er þat skjótast af honum at segja.] [*Laym. Orm. A. R. Piers P. Chauc.* raþe; *compar.* raþer: *Icel.* hratt *quickly; superl.* (sem) hraðast: *O. H. Ger.* hrado *celeriter, protinus, continuo; compar.* hrador; *superl.* hradost *contissime.*]

hraðer. v. hreðer.

hraðian. v. hradian.

hráw. v. hrǽw.

hrá-wérig; *adj. Wearied in body*, or *grievously wearied, wearied to death* [cf. hrá-lîc] :—Ic hæle hráwérig gewîte on longne síþ *I, a man sore wearied, shall depart on a long journey*, Exon. 63 b; Th. 235, 8; Ph. 554.

hreác, es; *m. A heap, stack, rick, reek* [in dialects, v. E. D. S. Old Country and Farming Words, ii, iii, and Halliwell's Dict.] :—Hreác *acervus*, Wrt. Voc. 89, 44. Healfne æcer gauolmǽde on hiora ágiere hwîle and ðæt on hreáce gebringan [*to mow*] *half an acre of 'gafol-meadow' in their own time and to bring the hay together in a reek*, Chart. Th. 145, 4. Hreácas *acervi*, Cot. 18, Lye. [*Prompt. Parv.* hreek *acervus: Chauc. Wick.* rekes; *pl : Icel.* hraukr *in* torf-hraukr *a peat-stack*.] v. hrycce.

hreác-copp, hreác-mete *food given to the labourers on completing a rick*, L. R. S. 21; Th. i. 440, 28, 27. The Latin version has *macoli summitas, caput macholi* for the former, and *firma ad macholum faciendum* for the latter. Thorpe in explanation of the passage quotes the following from Spelman ' Habetur macholum pro ipsa frugum seu garborum strue, quam hodie dicimus, *a reack or stack of corn*. Hujus olim ad constructionem epulari solebant agricolæ et messores.'

hreám, es; *m. A cry, outcry, hue and cry, crying, tumult, uproar :*—Ðæra Sodomitiscra hreám ys gemenigfyld *clamor Sodomorum multiplicatus est*, Gen. 18, 20: Past. 55; Swt. 427, 33: Cd. 229; Th. 309, 28; Sat. 717. Ðám hálgan were wæs geþuht ðæt ðæs gefeohtes hreám mihte beón gehýred geond ealle eorþan *it seemed to the holy man that the uproar of the conflict could be heard over all the earth*, Homl. Th. ii. 336, 17: Cd.

166; Th. 206, 10; Exod. 449: Beo. Th. 2608; B. 1302. Hreám and wôp *crying and weeping*, Blickl. Homl. 61, 36: 115, 15. Of ðam leahtre cymþ hreám dyslíc dyrstignys and mansliht *from that sin comes uproar, foolhardiness and manslaughter*, Homl. Th. ii. 220, 14. Hás ys for hreáme *raucus est præ clamatione*, Th. An. 19, 31. Julianus mid anþræcum hreáme forswealt *Julian with a horrible cry died*, Homl. Th. i. 452, 16. Ða heorde mid hreáme bewerian *to defend the flock with outcry*, L. I. P. 19; Th. ii. 326, 10. Gif hwá þeóf gemête and hine his þances áweg læte búton hreáme ... and gif hwá hreám gehýre and hine forsitte *if any one find a thief and voluntarily let him escape without hue and cry ... and if any one hear hue and cry and disregard it*, L. C. S. 29; Th. i. 392, 14-17: 170, 10 [MS. hearme]. [*Laym.* ræm, ream: *Orm.* ræm: *A. R.* ream: cf. *Icel.* hreimr (= hreymr?) *a scream, cry*: hraumi *a noisy fellow*.] v. hréman.

hreámig. v. hrémig.

hreán :—Wið hreán *for indigestion* [?], L. M. 2, 41; Lchdm. ii. 252, 16. Somner gives *phthisis*, but see hreáw, and cf. *Icel.* hrái *crudeness*.

hreáðe-mús, e; *f. A mouse ornamented, furnished with wings* [cf. hreóðan?], *a bat :* — Tôsnidenre hreáðemúse blôd *the blood of a bat cut up*, L. M. 2, 33; Lchdm. ii. 236, 17. Swilce eác cwôman hreáðemýs ... hæfdon hie eác ða hreáðemýs têþ in monna gelícnesse *sed et vespertilionum vis ingens ... habentes dentes in morem hominum*, Nar. 15, 5-8. [Cf. hrére-mús.]

hreáw *a body*. v. hráw.

HREÁW, hræw [*also written* hreów]; *adj.* RAW, *uncooked :*—Ne ne eton gê of ðam nán þing hreówes *non comedetis ex eo crudum quid*, Ex. 12, 9. Ne ete gê of ðam lambe nán þing hreáw, Homl. Th. ii. 264, 5. Syle etan oððe gesodene oððe hræwe *give* [*the plant*] *to eat either sodden or raw*, Herb. 136, 2; Lchdm. i. 254, 5. Ete ðara hundteóntig hreáwra *eat a hundred of them* [*lentils*] *raw*, L. M. 2, 13; Lchdm. ii. 190, 17. Meng wið hreáw ægru *mix with raw eggs*, I, 39; Lchdm. ii. 102, 7. Gif hí mon hreáwe swylgeþ *if they are swallowed raw*, L. Med. ex Quadr. 4, 10; Lchdm. i. 344, 16. Flæscmettas hreáwe *carnes crudas*, Coll. Monast. Th. 29, 13. [*Icel.* hrár *raw : Dan.* raa : *Swed.* rå : *Du.* raauw : *O. H. Ger.* rou *crudus : Ger.* roh.]

hreá-wíc, es; *n. A place of the dead, a place where people lie slain*, Beo. Th. 2432; B. 1214. [Cf. wæl-stôw.]

HREDDAN ; *p. de To* RID, *take away, save, liberate :*—God hí hredde wið heora fýnd *God rid them of, or saved them from, their enemies*, Homl. Th. i. 312, 9. Hrede t nere *eripe*, Blickl. Gl. Ps. 58, 2. Bûtan ðú úsic æt ðam leódsceaþan hreddan wille *unless thou wilt save us from the destroyer*, Exon. 11 b; Th. 17, 23; Cri. 274. Hwílum ic wráððum sceal stefne mínre forstolen hreddan *sometimes with my voice I shall save the stolen from enemies*, 104 a; Th. 396, 4; Rä. 15, 18. Óþ ðæt him god wolde þurh hryre hreddan heá frce *until god would take from him by death his exalted power*, Cd. 208; Th. 258, 5; Dan. 671. [*Orm.* redden: *O. Frs.* hredda, reda : *O. H. Ger.* rettan, Grff. 2, 471; *Ger.* retten.] DER. á-hreddan.

hredding, e; *f. Saving, salvation, liberation :*—Ús becom deáþ and forwyrd þurh wíf and ús becom líf and hredding þurh wimman *death and destruction came upon us by a woman, and by a woman came life and salvation*, Homl. Th. i. 194, 33. His ágen líf syllan for ðæs folces hreddinge *to give his own life for the redemption of the people*, 240, 14. Ongunnon for his hreddinge biddan *began to pray for his liberation*, 534, 27. Heó mid hreáme hyre hræddinge ofclypode *the result of her outcry was to save her*, Homl. Swt. 2, 219.

hréd-mónaþ. v. hréð-mónaþ.

hréfan ; *p. de To roof :*—Hé læt it réfen *he had it roofed*, Chr. 1137; Erl. 263, 8. v. ge-hréfan.

hrefl, Wrt. Voc. 66, 12. v. hrisil.

hrefn. v. hræfn.

hréh. v. hreóh.

hrem. v. hræfn.

hréman. v. hrýman. [*From the meaning the word would seem to correspond to O. Sax.* hrômian: *O. H. Ger.* hrômian, hruomian *gloriari, jactare ; but the adjective* hreámig, hrêmig, *though especially in the compound* sige-hrêmig *it agrees in meaning with the O. Sax.* hrômag: *O. H. Ger.* hrômag, hruomag *gloriosus :* siguhrômlîh *triumphalis, points to a connection with the noun* hreám: *the verb is therefore given under* hrýman, *the most usual form under which the verb connected with hreám in form and meaning occurs.*]

hrémig, hreámig; *adj. Clamorous* [*from joy or grief*], *exultant, lamenting, boasting, vaunting :*—Blissum hrêmig *exultant*, Andr. Kmbl. 3394; An. 1701: Elen. Kmbl. 2273; El. 1138; Exon. 48 b; Th. 168, 18; Gú. 1079: 57 b; Th. 206, 14; Ph. 126: 64 b; Th. 237, 19; Ph. 592. Gehþum hrémig *lamenting*, 98 a; Th. 367, 18; Seel. 9. Húþe hrémig *exulting in spoil*, Beo. Th. 248; B. 124: 3768; B. 1882: 4114; B. 2054: Elen. Kmbl. 297; El. 149: Andr. Kmbl. 1728; An. 866. Wuldrum hrémge *gloriously exulting*, Exon. 8 b; Th. 4, 17; Cri. 54. Wíges hreámige [*the e is written above the line*] *boasting of battle*, Chr.

937; Erl. 115, 8; Æðelst. 59. Hrêmge [*so the* MS.], Beo. Th. 4715; B. 2363. DER. sige-hrêmig. v. hréman.

hremman ; *p. de To hinder, obstruct, cumber :*—Forceorf hit tô hwí hremþ hit ðisne stede *cut it down; why cumbereth it this place?* Homl. Th. ii. 408, 4. Úre unlustas and leahtras ðe ús hremaþ *our evil desires and vices that hinder us*, i. 156, 12. Ði læs ðe seó smeáung ðæra æhta hí æt ðære láre hremde *lest the contemplation of the possessions should be a hindrance to them in learning*, 60, 30: 394, 14. Ne hremmaþ mínne martyrdôm *hinder not my martyrdom*, 592, 7. [Cf. *Icel.* hremma *to clutch.*]

hremming, e; *f. A hindering, hindrance, obstruction, obstacle, impediment :*—Nú is ðære eorþan sinewealtnys and ðære sunnan ymgang hremming ðæt se dæg ne byþ on ælcum earde gelíce lang *now the roundness of the earth and the course of the sun is an obstacle to the day being equally long in every country*, Lchdm. iii. 258, 11. Mycele swýðor sceal se sôþa Godes cempa bûton ælcere hremminge hræðe gehýrsumian Cristes sylfes bebodum *much more shall the true soldier of God, without any hindrance, at once obey the commands of Christ himself*, Basil admn. 2; Norm. 34, 23.

hremn. v. hræfn.

hrenian *redolere*, Scint. 28, Lye.

hreoce *rubellio, rutilus*, Lye. v. reohhe.

HREÓD, es; *n. A* REED :—Hwí férde gê on wêstene geseón ðæt hreód ðe byþ mid winde ástyred *quid existis in desertum videre harundinem vento moveri*, Lk. Skt. 7, 24: Mt. Kmbl. 11, 7. For cynegyrde him hreód forgeáfon *gave him a reed for a sceptre*, Homl. Th. ii. 252, 27. Hreódes spír *a spike of a reed*, L. M. 2, 51; Lchdm. ii. 266, 10. Grôwnys hreódes and ricsa *viror calami et junci*, Bd. 3, 23; S. 554, 23. Synd ðær manige eáland and hreód *there are there many islands and reeds*, Guthl. 3; Gdwin. 20, 6. [*O. Dutch* ried: *O. H. Ger.* reod, ried, riet *carectum, carex.*]

hreód-bedd, es; *n. A reed-bed :*—Ðá wæs ðær on middan ðam mere sum hreódbed *there was in the middle of the mere a reed-bed*, Guthl. 9; Gdwin. 50, 15. Heó ásette hyne on ánum hreódbedde be ðæs flôdes ôfre *exposuit eum in carecto ripæ fluminis*, Ex. 2, 3. Ðeós wyrt biþ cenned on dícon and on hreódbeddon *this plant* [*lion-foot*] *is produced in dikes and reed-beds*, Herb. 8, 1; Lchdm. i. 98, 13.

hreódeum [=hreódegum? cf. hreódiht] *reedy, covered with rough grass* [?] :—In heágum môrum and in hreódeum [*other MS.* hréþum] *in arduis asperisque montibus*, Bd. 4, 27; S. 604, 27.

Hreód-ford *Redbridge, Hants*, Bd. 4, 16; S. 584, 29.

hreódiht ; *adj. Reedy :*—On ðone hreódihtan môr, Cod. Dipl. Kmbl. iii. 121, 20.

hreód-wæter, es; *n. Fenny land where reeds are growing :*—Ðá wæs ðæt land eall swá wê geférdon ádrigad and fien and hreádwæteru *palus erat sicca et ceno habundans*, Nar. 20, 23.

hreód-writ, es; *n. A reed for writing, pen ;* calamus scribæ, Ps. Spl. C. 44, 2.

hreóf ; *adj. Rough, rugged, scabby, leprous :* — Hreóf *leprosus*, Mt. Kmbl. Rush. 8, 2. Ðonne biþ se líchoma hreóf ðonne se bryne ðe on ðæm innoþe biþ útáslihþ tô ðære hýde *fervor intimus usque ad cutis scabiem prorumpit*, Past. 11, 5; Swt. 71, 5. In hûse simonis ðæs hreófan *in domo Simonis leprosi*, Mt. Kmbl. Rush. 26, 6. Symones hreáfes, Mk. Skt. Lind. 14, 3. Lǽcedôm wið hreófum líce *a recipe for a scabby body*, L. M. 1, 32; Lchdm. ii. 78, 1. Is ðæs hiw gelíc hreófum stáne *it looks like a rough stone*, Exon. 96 b; Th. 360, 20; Wal. 8. Monige hreófe [hreáfo, Lind.] *multi leprosi*, Lk. Skt. Rush. 4, 27: 17, 12: Elen. Kmbl. 2428; El. 1215: Blickl. Homl. 177, 15. Hreófum, Andr. Kmbl. 1155; An. 578. [*Icel.* hrjúfr *rough, scabby : O. H. Ger.* riob *leprosus.*]

hreófl, hreófol, e; *f. Roughness of the skin, scabbiness, leprosy :*—Ðonne bí ðam sceabbe suíðe ryhte sió hreófl getácnaþ ðæt wôhhǽmed *in scabie fervor viscerum ad cutem trahitur, per quam recte luxuria designatur*, Past. 11, 5; Swt. 71, 4. Hreóful [Lind. hriófol] *lepra*, Mt. Kmbl. Rush. 8, 3. Hríofal [Lind. ríófol], Mk. Skt. Rush. 1, 42: Lk. Skt. Lind. 5, 13. Wer full hriófle *vir plenus lepra*, 12. Wið horses hreófle ... gif sió hreófol síe micel, L. M. 1, 88; Lchdm. ii. 156, 10, 13. Wið hreóf[l, L. Med. ex Quadr. 16, 10; Lchdm. i. 352, 18. Seðe ete his líchaman hreófel *qui corporis sui scabiem edit*, L. Ecg. P. iv. 52; Th. ii. 218, 30.

hreófl ; *adj. Leprous :*—Ðá brohte hé hig [his hand] forþ hreófle swá hwít swá snáw *quam protulit leprosam instar nivis*, Ex. 4, 6. v. next word.

hreófla, an; *m. A leper :*—Ðá geneálǽhte án hreófla tô him *ecce leprosus veniens*, Mt. Kmbl. 8, 2. On simones húse ánes hreóflan, Mk. Skt. 14, 3. Ðæs hreóflan, Mt. Kmbl. 26, 6. Moyses ǽ forbeád tô hrepenne ǽnigne hreóflan *the law of Moses forbade to touch any leper*, Homl. Th. i. 122, 5. Hreóflan synt gehǽlede *leprosi mundantur*, Lk. Skt. 7, 22.

hreófla, an; *m. Leprosy, scabbiness :*—Se hreófla him fram férde *lepra discessit ab illo*, Lk. Skt. 5, 13: Mt. Kmbl. 8, 3: Homl. Th. i. 120, 15. Swá mycel hreófla *tanta scabies*, Bd. 5, 2; S. 614, 44. Geseah ðæt hire

líchama wæs āfylled mid hreóflan *eam vidisset perfusam lepra*, Num. 12, 10. Wið sceápa hreóflan *against scab in sheep*, Lchdm. iii. 56, 19.

hreóflia. v. hreóf-lig.

hreóf-líc; *adj. Having elephantiasis*; elephantinus, Hpt. Gl. 519. v. next word.

hreóf-lig; *adj. Leprous:*—Đā com sum hreóflig *there came a certain leprous man*, Homl. Th. i. 120, 11. Se hreóflia *the leper*, 122, 10. Getācnode đes hreóflia man eal mancyn đe wæs ātelíce hreóflig . . . Lāđlíc biþ đæs hreóflian líc *this leper betokened all mankind that was foully leprous . . . Loathsome is the body of the leper*, 16–21 : 33. Wacode ealle đa niht mid đam wædlian hreóflian, Homl. Swt. 3, 486. Reóflium menn gelíc *like a leper*, Homl. Th. ii. 178, 13. Martinus getācnode ænne hreóflinne mannan, 512, 5.

hreóf-ness, e; *f. Leprosy:*—Hreófnis swā snāw *lepra quasi nix*, Num. 12, 10.

hreóh, hrēh; *n. Roughness of weather, storm, tempest:*—Flód ł hrēh midđý āwarþ *inundatione facta*, Lk. Skt. Lind. 6, 48. Sumne sceal hungor āhíđan sumne sceal hreóh fordrífan *famine shall waste one man, a storm drive another to destruction*, Exon. 87 a; Th. 328, 10; Vy. 15. Ic bíde đæs beornes đe mē bōte eft mindōm and mægenes hreóh *expectabam eum, qui me salvum faceret a pusillo animo et tempestate*, Ps. Th. 54, 7. v. hreóh-full, and next word.

HREÓH; *adj.* ROUGH, *fierce, savage, rough [of the weather, the sea, etc.], stormy, tempestuous, disturbed [of the mind]:* — Hreóh weder *tempestas*, Mt. Kmbl. 16, 3. Heom on becom swíđe hreóh weder, Chr. 1075; Erl. 212, 23. Hit wæs hreóh sǽ *mare exsurgebat*, Jn. Skt. 6, 18. Flód hreóh under heofonum, Cd. 69; Th. 83, 29; Gen. 1387; Andr. Kmbl. 933; An. 466: 3083; An. 1544. Hreóh wæter, Ps. Th. 68, 1. Ne wedra gebregd hreóh under heofonum *non ibi tempestas nec vis furit horrida venii*, Exon. 56 b; Th. 201, 18: Ph. 58. Brond hreóh onetteþ þa flame hurries fierce, 59 a; Th. 212, 19; Ph. 217. Hríoh biþ đonne seó đe ǽr gladu onsíéne wæs *rough then is the sea that before was smooth*, Bt. Met. Fox 5, 20; Met. 5, 10. Ān wiht is hreóh and rēđe *there is a creature fierce and fell*, Exon. 127 b; Th. 491, 20; Rä. 81, 2. Yrre gebolgen hreóh and hygeblind *angry, cruel and blind of mind*, 66 b; Th. 246, 13; Jul. 61: 74 b; Th. 278, 9; Jul. 595. Hreóh and heorogrim, Beo. Th. 3132; B. 1564. Wæs him hreóh sefa ege from đam eorle *troubled was his mind, he was in fear of the man*, Bt. Met. Fox 1, 142; Met. 1, 71. Ne mæg wērig mōd wyrde wiđstondan ne se hreó hyge helpe gefremman *a weary heart cannot withstand fate nor the troubled mind afford help*, Exon. 76 b; Th. 287, 18; Wand. 16: 94 b; Th. 354, 9; Reim. 43. Đā wæs beorges weard on hreóum mōde *then became the hill-ward of fierce mood*, Beo. Th. 5156; B. 2581. On đære hreón sǽ *turbato mari*, Past. 9; Swt. 59, 2. On hreón mōde *troubled*, Beo. Th. 2619; B. 1307. Wē geliden hæfdon ofer hreóne hrycg *we had sailed over a troubled sea*, Exon. 20 b; Th. 53, 31; Cri. 859. Hreó hæglfare *a hailstorm*, 78 a; Th. 292, 26; Wand. 105. Hreó wǽron ýđa *rough were the billows*, Beo. Th. 1101; B. 548: Andr. Kmbl. 1496; An. 749: Exon. 55 a; Th. 194, 19; Az. 141. Hreóra wǽga, 56 b; Th. 200, 24; Ph. 45. Đonne seó sǽ hreóhost byþ đonne wōt hē gewiss smelte wedere tōwæard *when the sea is roughest then he knows certainly that fair weather is to come*, Shrn. 179, 18. [*Laym.* reh, ræh : *O. Sax.* hrē.] v. hreów.

hreohehe = reohhe, q. v.

hreóh-full; *adj. Stormy:* — Hreóhfull geár *a stormy year*, Lye. v. hreóh.

hreóh-mód; *adj. Savage, fierce of mind, ferocious, troubled in mind:*— Hāt and hreóhmód *angry and savage*, Beo. Th. 4581; B. 2296. Hreóhmód wæs se hǽđena þeóden *fierce of heart was the heathen prince*, Cd. 186; Th. 231, 4; Dan. 242. Se þeóden hreóhmód *the prince with troubled heart*, Beo. Th. 4270; B. 2132. v. hreóh.

hreóhmód-ness, e; *f. Ferocity*, Som.

hreóh-ness, hreó-ness, e; *f. Roughness of the weather, of the sea, storm, tempest:*—Ofer eów cymeþ mycel storm and hreóhnes *tempestas vobis superveniet*, Bd. 3, 15; S. 541, 33. Hreánis *tempestas*, Mt. Kmbl. Rush. 16, 3. On ymbhwyrfte his hreóhnys strang *in circuitu ejus tempestas valida*, Ps. Spl. 49, 4: Homl. Th. ii. 18, 5. Micel hreóhnys on đære sǽ, 378, 14. Seó hreóhnys wearþ gestilled *the tempest was stilled*, i. 246, 10, 1. Ic geseó đæt đās brōđor synd geswencede of đisse sǽwe hreónesse *I see that these brethren are wearied from the roughness of the sea*, Blickl. Homl. 233, 26. On đissere cealdan hreóhnysse *in this cold storm*, Homl. Swt. 11, 187. Gif hwā hreóhnysse on rēwytte þolige . . . seó hreóhnys byþ forboden *if any one suffer stormy weather in rowing . . . the rough weather will be stopped*, Herb. 171, 3; Lchdm. i. 302, 5. Wið hagol and hreóhnysse . . . heó āwendeþ hagoles hreóhnysse, 176; Lchdm. i. 308, 10, 14, 16, 23. Hē dyde swíđe hreónesse đære sǽwe *he made the sea very rough*, Blickl. Homl. 235, 5. On đissere worulde hreóhnyssum *in the storms of this world*, Homl. Th. ii. 384, 26.

hreól *a reel*; alibrum, Ælfc. Gr. 111; Som. 79, 55; Wrt. Voc. 59, 26. [*Prompt. Parv.* reel, *womannys instrument alabrum*.]

Hreopa-, Hreope-, Hrypa-dūn, e; *f. Repton*, Chr. 755; Erl. 52, 1: 874; Erl. 76, 21: 875; Erl. 76, 33. Gūþlāc fērde tō mynstre đe ys

gecweden Hrypadūn and đǽr đa gerýnelícan sceare onfēng Sce Petres Guthlac *went to a monastery that is called Repton and there received the mystical tonsure of St. Peter*, Guthl. 23; Gdwin. 16, 20.

hreórig; *adj. Ruinous:*— Hrōfas sind gehrorene hreórge torras *the roofs are fallen, the towers ruinous*, Exon. 124 a; Th. 476, 6; Ruin. 3.

hreósan; *p.* hreás; *pl.* hruron; *pp.* hroren *To fall [rapidly, headlong], fall down, go to ruin*; ruere, corruere:— Ic hreóse *ruo*; tō hreósenne *ruiturus*, Ælfc. Gr. 28; Som. 30, 54. His weorc hrýst tō micclum lyre *his work falls to great perdition*, Homl. Th. i. 368, 25. Đā hrýsþ se stōl nyđer *then the throne falls down*, L. I. P. 4; Th. ii. 308, 2. On hærfest hrēst and fealuwaþ *in autumn it falls and fades*, Bt. Met. Fox 11, 116; Met. 11, 58. Twegen unþeáwas hreósaþ on ænne man *duorum vitiorum languor irruit*, Past. 62, 1; Swt. 457, 9. Wongas hreósaþ *the plains shall sink away*, Exon. 19 b; Th. 51, 5; Cri. 811. Hreósaþ tōbrocene burgweallas, 22 a; Th. 60, 30; Cri. 977. Hreósaþ heofonsteorran *the stars of heaven shall fall*, 23 a; Th. 64, 27; Cri. 1044. Đǽr ne hægl ne hrím hreósaþ tō foldan, 56 b; Th. 201, 23; Ph. 60. Heofon and eorþe hreósaþ tōgadore *heaven and earth shall rush together*, Andr. Kmbl. 2875; An. 1440. Ne hreósaþ hī tō hrusan hearde gebíged *non est ruina maceriæ*, Ps. Th. 143, 8. Swā đæt hē hreás and feóll on eorþan *ita ut corruens in terram*, Bd. 4, 31; S. 610, 13. Gomela Scylfing hreás blāc *the aged Scylfing fell down pale*, Beo. Th. 4969; B. 2488: 5654; B. 2831. Hie hrúron gāre wunde *they fell wounded by the spear*, 2153; B. 1074. Hruron him teáras *tears fell from him*, 3749; B. 1872. Hie onweg hruron *they plunged away [of the creatures on the top of the water which sank to the bottom on the appearance of Beowulf and his companions]*, 2865; B. 1430: Andr. Kmbl. 3199; An. 1602. Đæt se swā stronglíce hrure on đa circan *that it [the wind] beat so strongly on the church*, Shrn. 81, 22. Hreósan under heolstorhofu, Elen. Kmbl. 1525; El. 764: Exon. 28 b; Th. 86, 25; Cri. 1413. Gesihþ hreósan hrím and snāw, 77 a; Th. 289, 14; Wand. 48. Hit hreósan wile sígan sond æfter rēne, Bt. Met. Fox 7, 44; Met. 7, 22. Hió is mā hreósende for ealddōme đonne of ǽniges cyninges niéde *magis imbecillitate propriæ senectutis quam alienis concussæ viribus contremiscunt*, Ors. 2, 4; Swt. 76, 2. Đý læs cild sý hreósende đæt is fylleseóc *lest a child be falling, that is, be ill of the falling sickness [epilepsy]*, L. Med. ex Quadr. 5, 12; Lchdm. i. 350, 12. Hríđ hreósende *the storm rushing*, Exon. 78 a; Th. 292, 20; Wand. 102. Ongeán đam hreósendum treówe *towards the falling tree*, Homl. Th. ii. 508, 35. Synt swíđe hreósende đās gesǽlþa *these goods are very perishable*, Bt. 11, 2; Fox 34, 22. [*Laym.* reosen; *p.* rees; *pl.* ruren: *Icel.* hrjósa *to shudder.*] DER. ā-, be-, ge-, of-, ofer-, on-, tō-hreósan.

hreóse. v. wind-hreóse.

hreósende. v. hreósan.

hreósend-líc; *adj. Frail, perishable, ready to fall:*—Gē sēcaþ đære heán gecynde gesǽlþa and heore weorþscipe tō đam niđerlícum and tō đam hreósendlícum þingum *ab rebus infimis excellentis naturæ ornamenta captatis*, Bt. 14, 2; Fox 44, 30. Hreósendlíc *cassabundus, corruendus*, Hpt. Gl. 422, 459.

hreóđa. v. bord-, scild-hreóđa.

hreóđan. v. hroden.

hreóung, hríung, e; *f. Shortness of breath, hardness of breathing:*— Hríung *suspirium*, Ælfc. Gl. 10; Som. 57, 28; Wrt. Voc. 19, 34. Hreóung hlýdende swiđust innan *hard breathing sounding chiefly from within*, L. M. 2, 46; Lchdm. ii. 258, 19.

hreów *raw.* v. hreáw.

hreów, e; *f. Sorrow, regret, penitence, penance, repentance:* — Būton him seó sóþe hreów gefultmige *unless true penitence help them*, Blickl. Homl. 101, 7: Bt. Met. Fox 18, 21; Met. 18, 11. Ān hreów ys wydewan and fæmnan *viduæ et puellæ una est pœnitentia*, L. Ecg. P. iv. 68, 9; Th. ii. 228, 30. Ic đec lǽdan sceal tō đam hālgan hām đǽr nǽfre hreów cymeþ *I shall lead thee to that holy home where sorrow never comes*, Exon. 32 b; Th. 102, 20; Cri. 1675: Beo. Th. 4645; B. 2328. Hū langæ đū on hreówe ǽwunian sceole *quamdiu pœnitentiæ insistere*, Bd. 4, 25; S. 600, 11. On gōdre hreówe *in vera pœnitentia*, L. Ecg. C. 2; Th. ii. 136, 24. Mid synna hreówe *with repentance for sins*, L. Wih. 3; Th. i. 36, 18: 5; Th. i. 38, 8. From đære incundan hreówe *ab intentione pœnitentiæ*, Past. 53, 5; Swt. 415, 36. Būtan hreówe *without regret*, 44, 5; Swt. 324, 18. Dōn wē ūrum Drihtne sóþe hreówe and bōte, Blickl. Homl. 35, 36. Hreówe and dædbōte, 79, 5. Ne hē wihte hafaþ hreówe on mōde đæt him hālig gǽst losige *he hath not regret for the loss of his holy spirit*, Exon. 30 b; Th. 95, 16; Cri. 1558. Hreówa tornost *most grievous of sorrows*, Beo. Th. 4265; B. 2129. Hreówum gedreahte *afflicted with regrets*, Exon. 22 b; Th. 61, 34; Cri. 994. [*O. and N.* reowe: *O. H. Ger.* hriuwa, hriuwi *pœnitentia, pœnitudo, dolor: Ger.* reue.]

hreów; *adj.* In Andr. Kmbl. 2233; An. 1118 the alliteration seems to require reów. In the compounds blód-, wæl-hreów the second syllable seems to be hreóh [or is it reów, or may hreów be a confusion of the two forms?], as the form hreóh does not occur independently in the sense of

fierce. Grein separates *hreóh* [*hreów*] under two heads with the meanings *sævus, mæstus,* but this seems unnecessary, as the idea of mental disturbance may be derived from that of physical disturbance in *hreóh,* q. v. see also *hreówe.* However, as Ettmüller, p. 504, observes, perhaps the three forms *hreóh, hreów, hreáw* are sometimes confounded.

hreówan; *p.* hreáw *To rue, make sorry, grieve;* often *impers :—* Him nān yfel ue hrīwþ *quam mala nulla contristant,* Past. 53, 5; Swt. 417, 1. Hī hēr syngiaþ and hit him nō ne hreówþ *they sin in this world and are not sorry for it,* 55, 2; Swt. 429, 17. Hreóweþ, Exon. 44 b; Th. 150, 23; Gū. 783: Cd. 22; Th. 27, 31; Gen. 426. Ðonne hreóweþ hire ðæt heó hire gehāt ne gefylde *pœnitentia mota quod votum suum non impleverit,* L. Ecg. C. 33; Th. ii. 158, 7. Hreáw him *pœnituit eum,* Ps. Spl. 105, 42. Hreáw hine, Ps. Th. 105, 34: Cd. 64; Th. 77, 17; Gen. 1276. Gif ðū ongite ðæt him his synna hreówen *if you see that his sins cause him sorrow,* L. de Cf. 2; Th. ii. 260, 19. Swā swā hī læsse ongietad on him selfum ðæs ðe him hreówan þyrfe *cum minus se respiciunt habere quod defleant,* Past. 52, 9; Swt. 411, 5. For ðæm ðe hie ne māgon ealneg ealla on āne tīd emnsāre hreówan *neque enim uno eodemque tempore æque mens de omnibus dolet,* 53, 3; Swt. 413, 29. Ne hit him ne lǣt hreówan *does not let it trouble him,* Bt. 39, 12; Fox 232, 2: Cd. 38; Th. 50, 29, 36; Gen. 816, 819: Exon. 28 b; Th. 86, 28; Cri. 1415: 100 a; Th. 376, 5; Seel. 150. [*Laym.* reouwen: *Orm.* reoweþþ, *prs;* ræw, *p: Chauc.* reweþ: *Prompt. Parv.* ruwyñ *peniteo, penitet; compatior: O. Sax.* hrewan: *Icel.* hryggja, hryggwa *to distress, grieve: O. H. Ger.* [h]riuwan; *Ger.* reuen.] DER. ge-, of-hreówan.

hreów-cearig; *adj. Troubled, anxious, sorrowful :—* Hreðer innan swearc hyge hreówcearig *his soul grew dark within, his mind distressed,* Exon. 48 a; Th. 165, 9; Gū. 1026: 73 b; Th. 274, 21; Jul. 536: Rood Kmbl. 49; Kr. 25. Hreówcearigum *help to the troubled,* Exon. 13 a; Th. 23, 11; Cri. 367.

hreówe; *adj. Sad, grieved, sorrowful, penitent :—* Hreówum teárum lacrymis *pœnitentiæ,* Bd. 4, 25; S. 600, 15. [*O. Sax.* hriwi: *Icel.* hryggr *afflicted, grieved.*] v. hreów, *and for the form of the word cf.* treówe.

hreówian *to repent :—* Hreówigas *pœnitemini,* Mk. Skt. Lind. 1, 15. [*O. Sax.* hriwôn: *O. H. Ger.* hriuwôn.]

hreówig; *adj. Sad, mournful :—* Nū wit hreówige māgon sorgian for his sīþe *now may we mournful sorrow for his journey,* Cd. 38; Th. 49, 29; Gen. 799. [*O. Sax.* hriwig: *O. H. Ger.* [h]riuwag *pœnitens, compunctus corde.*]

hreówig-mód; *adj. Sad at heart :—* Wīf hreówigmód [*Eve*] Cd. 37; Th. 48, 5; Gen. 771. Hī hreówigmóde wurpon hyra wǣpen of dūne *they disconsolate flung down their weapons,* Judth. 12; Thw. 25, 33; Jud. 290. [*O. Sax.* hriwig-mód.]

hreów-, hrīw-līc; *adj. Grievous, miserable, pitiful, sad :—* Hreówlīc *calamitosus,* Hpt. Gl. 518. His wīf wyrþe wydewe hreówlīc *fiat uxor ejus vidua,* Ps. Th. 108, 9. Wāla ðæt wæs hreówlīc sīþ *alas! that was a miserable thing,* 1057; Erl. 192, 20. Wē geseóþ ðæt wē elles hrýwlīcum deáþe forwurþan sceolon *we see that otherwise we shall perish by a miserable death,* St. And. 36, 7. [*Laym.* reowlich: *R. Glouc.* rewlich.]

hreów-līce; *adv. Miserably, cruelly, grievously :—* Ða ðe swā hreówlīce ācwealde wǣron *crudeliter interemptos,* Bd. 1, 15; S. 484, 3: Chr. 1036; Erl. 164, 35. Blǣdran swīðe hreówlīce berstende *blisters bursting very painfully,* Ors. 1, 7; Swt. 38, 7. Māgon hie swā hreówlīce wēpan swā gē māgon ðara ōðra blīþelīce hlihhan, 3, 7; Swt. 120, 6. Earme menn sindon hreówlīce besyrwde *poor men are grievously ensnared,* Swt. A. S. Rdr. 106, 47. Hreówlīce gefærþ seðe hine sylfne ðus forþ forscyldigaþ and gesǣlig biþ hē ðeáh . . . *miserably does he fare who thus continues to incur guilt; and yet he will be happy* . . . , L. Pen. 12; Th. ii. 280, 28: Chr. 1096; Erl. 233, 22.

hreów-ness, e; *f. Penitence, repentance, sorrow, contrition :—* Æfter his dǣdbōte hreównysse *post pœnitentiæ contritionem,* L. Ecg. P. Th. ii. 170, 13. Hreównisse [hrēunisse, Rush.] *pœnitentiam,* Mt. Kmbl. Lind. 11, 21. Hreóuisse, 3, 8. Hreáwnise, 21, 29: Mk. Skt. Lind. 6, 12. Hreóunisse *pœnitentia,* Rtl. 8, 33.

hreów-ness. v. wæl-hreówness, *and* hreów.

hreówsian, hrýwsian; *p.* ode *To be sorry, grieve, repent, do penance :—* Ðæt hē æfre ne beþence ymbe ða hreówsunge ðe hē ǣr hreówsade and ðe him sint forgifena swīðe manega synna forðæmðe hīō swīðe hreówsade, Past. 52, 9; Swt. 411, 12. Hrýwsode *pœnituit,* Ps. Spl. C. 105, 42. Hreówsiaþ *pœnitemini,* Mk. Skt. Rush. 1, 15. Sume wyllaþ ðæt hē hreówsige *nonnulli volunt ut pœniteat,* L. Ecg. C. 24; Th. ii. 150, 9. Ðæt se rihtwīsa man hreówsige hine sylfne swylce hē wið God forwyrht sig *ut justus homo pœnitentiam agat eorum, quæ erga Deum deliquerit,* L. Ecg. P. i. 5; Th. ii. 174, 6. Heora synna hreówsian and dǣdbōte dōn, Ors. 6, 2; Swt. 256, 13. Ðā ongann hē hreówsian *pœnitentia ductus,* Mt. Kmbl. 27, 3. Mīnum hreówsiendan geþohte *to my sorrowing thought,* Bt. 3; Fox 4, 26. For hreówsigendum man *pro pœnitenti,* L. Ecg. C. 36; Th. ii. 160, 20. Fore hreósendum *pro pœnitentibus,* Rtl. 177, 7. [*Laym.* reousien: *O. H. Ger.* [h]riuwisôn.] DER. be-hreówsian.

---

hreówsung, e; *f. Sorrowing, sorrow, penitence, repentance :—* Hreówsung *pœnitudo,* Hpt. Gl. 510. Se apostol bebeád ðæt hī þrītig daga be hreówsunge dǣdbētende Gode geoffrodon *the apostle ordered that they for thirty days with penitence should offer to God doing penance,* Homl. Th. i. 68, 17. Gif hī hwæt gesyngodon hī hit eft mid hreówsunge gebēton *if they sinned in aught they should make amends therein with repentance,* Bt. 41, 3; Fox 248, 14. Hig hreówsunge dydon *pœniterent,* Lk. Skt. 10, 13. Ðæt hē þurh ða hreówsunga gemēte forgiefnesse beforan ðære sōþfæsdnesse *ut per lamenta veniam in conspectu veritatis obtineat,* Past. 21, 7; Swt. 165, 22. Forlǣtaþ eówre hreówsunga *cease your lamentations; capita vestra nolite nudare et vestimenta nolite scindere,* Lev. 10, 6. Be his sylfes heortan hreówsungum *according to the penitence of his own heart,* L. Pen. 3; Th. ii. 278, 11. [*Orm.* reowwsunnge.] v. be-hreówsung.

hrepian, hreopian; *p.* ode *To touch, treat :—* Se ðe eów hrepaþ hit mē biþ swā egle swylce hē hreppe ða seó mīnes eágan *he that touches you, it will be as painful to me as if he touches the apple of my eye,* Homl. Th. i. 392, 15: 516, 22. Seó hrepaþ swýðost ymbe Cristes godcundnysse it [*the gospel of St. John*] *treats chiefly of Christ's divinity,* 70, 1. Swā hraðe swā his sceadu hī· hreopode *as soon as his shadow touched them,* 316, 16: 492, 25. Hrepede, 176, 6. Gif hī his reáfes gefnædu hreppe . . . heó hreopode his reáfes fnædu . . . Hwā hreopode mē . . . ðū āxast hwā ðē hreopode . . . ðæt wīf hine hrepode, ii. 394, 10–18. Wē ne hrepodon ðone traht *we did not treat the exposition,* i. 104, 6. Ne hrepa ðū ðæs treówes wæstm *touch not the fruit of the tree,* 14, 1: Homl. Swt. 5, 302. Gōd bebeád ūs ðæt wē ðæt treów ne hrepodon *præcepit nobis deus ne tangeremus illud* [*lignum*], Gen. 3, 3. v. gehrepod, *and next word.*

hreppan *to touch, treat :—* Ic hreppe *tango,* Ælfc. Gr. 28; Som. 32, 56. Ic hreppe Pharao mid ānum wīte *una plaga tangam. Pharaonem,* Ex. 11, 1. Se ðe wudu hrepeþ *he who touches the wood,* Exon. 127 b; Th. 490, 7; Rä. 79, 7. Ða wē ne hreppaþ *those* [*nouns*] *we shall not treat of,* Ælfc. Gr. 9; Som. 12, 30. Ðeáh hī hwā hreppe heó hit ne gefrēt *though any one touch it* [*the soul*] *it does not feel it,* Homl. Swt. 1, 220. Ða rēðe deór ne dorston hī reppan *the fierce beasts durst not touch them,* 4, 405. Hire on beseón oðða hī hreppan *to look upon her or touch her,* 7, 151. Hwā dearr hī hreppan, Homl. Th. i. 458, 17. His eágan hreppan mid ðam seáwe *to touch his eyes with the juice,* Herb. 31; Lchdm. i. 128, 12. Moyses sǣ forbeád tō hrepenne ǣnigne hreófian *the law of Moses forbade to touch any leper,* Homl. Th. i. 122, 5. v. preceding word, and for such pairs of verbs see March's Anglo-Saxon Grammar, § 222. [*Icel.* hreppa *to reach, catch, obtain.*]

hrepsung, e; *f. The evening :—* Ǣfen oððe hrepsung *vesper,* Som.

hrepung, e; *f. Touch, touching :—* Hrepung *tactus,* Ælfc. Gr. 11; Som. 15, 15. Ða andgitu sint gehātene ðus . . . tactus hrepung on eallum limum *the senses are named thus . . . tactus touch, in all the limbs,* Homl. Swt. 1, 199: Homl. Th. ii. 372, 26. Hē mihte mid his worde hine gehǣlan būton hrepunge ac hē geswutelode ðæt his hrepung is swīðe hālwende geleáfullum *he could have healed him with his word without touching; but he shewed that his touch is very salutary to believers,* Homl. Th. i. 122, 9. Drihten gehǣlde ða untruman þurh his reáfes hrepunge *the Lord healed the sick by the touch of his garment,* ii. 394, 5.

hrēr; *adj. Rear* [*provincial*], *not thoroughly cooked, lightly boiled* [*of eggs*] :—Nim hrēr henne æg *take a hen's egg lightly boiled,* L. M. 2, 52; Lchdm. ii. 272, 16. [*Prompt. Parv.* rere, or nesche, as eggys *mollis;* see the note p. 430.] v. hrēren-brǣden.

hreran [?] *to fall :—* Ðæt ic hryre † gefealle [= ? hrure † gefeólle] *ut caderem,* MS. T: hī hrǣron, Ps. Spl. 117, 13.

hrēran; *p.* de *To move, shake, stir :—* Ic wudu hrēre *I move the wood,* Exon. 101 a; Th. 381, 9; Rä. 2, 8. Hrēra, 101 b; Th. 383, 9; Rä. 4, 8. Forhwī drēfe gē eówru mōd mid unrihte fióunge swā swā ȳða for winde ða sǣ hrēraþ *quid tantos juvat excitare motus,* Bt. 39, 1; Fox 210, 25: Bt. Met. Fox 27, 5; Met. 27, 3. Hig wegdan hrērdan heora heáfod *moverunt capita sua,* Ps. Th. 108, 25. Hrēr swīðe *stir thoroughly,* L. M. 1, 38; Lchdm. ii. 94, 4. Hrēr mid sticcan, 3, 26; Lchdm. ii. 322, 28. Hrēre ðonne swīðe *let it be thoroughly shaken,* 1, 36; Lchdm. ii. 88, 1: 38; Lchdm. 92, 4: 94, 13. Hē ne lǣtaþ mīne fēt lāðe hrēran *non dedit commoveri pedes meos,* Ps. Th. 65, 8. Hrēran mid hondum hrīmcalde sǣ *to row on the ice-cold sea,* Exon. 76 b; Th. 286, 21; Wand. 4. Sum mæg fromlīce ofer sealtne sǣ sundwudu drīfan hrēran holmþræce, 17 b; Th. 42, 25; Cri. 678. [*O. Sax.* hrórian: *Icel.* hræra *to move, stir: O. H. Ger.* hruorian *movere, agitare, tangere: Ger.* rühren.] v. on-hrēran. hrór.

hrēred-ness, e; *f. Agitation, haste, precipitation :—* Ealle word hrýrednesse *omnia verba præcipitationis,* Ps. Lamb. 51, 6.

hrēre-mūs, e; *f. A rear-, rere-mouse, bat :—* Hrēre-mūs *vespertilio,* Wrt. Voc. 77, 40. [See Nare's Gloss. rear-; rere-mouse, and cf. *Ger.* fleder-maus.] v. hreáðe-mūs: hrór.

hrēren-brǣden; *adj. Not thoroughly cooked :—* On ān hrērenbrǣden æg *over an egg lightly cooked,* Lchdm. iii. 294, 8. v. hrēr.

hrēr-ness, e; *f. Motion, disturbance, agitation, commotion, storm :—* Hroernis michelo geworden wæs in sǣ *motus magnus factus est in mari,* Mt. Kmbl. Lind. 8, 24. Gāst hrýrenesse † stormes *spiritus procellæ,* Ps.

Lamb. 106, 25. Eorþ hroernisse *terræ motu*, 27, 54. Swā ðū hí on yrre ehtest and drēfest ðæt hí on hrērnesse hraðe forweorþaþ *ita perse-queris illos in tempestate tua; et in ira tua conturbabis eos*, Ps. Th. 82, 11. v. eorþ-hrērness.

hresigende. v. hrisian.

hrēst = **hrȳst**, Bt. Met. Fox 11, 116; Met. 11. 58. v. hreósan.

hrētan. v. hrȳtan.

hreþ. v. hræd.

hrēð, es; *m.* [?] *Glory, fame, triumph, honour :*—Siððan him gesēlde sigorworca hrēð ðæt hē ealdordōm āgan sceolde ofer cynericu *afterwards fell to him the glory of victorious deeds, that he should have dominion over kingdoms*, Cd. 158; Th. 198, 2; Exod. 316. Him wyrd ne gescrāf hrēð æt hilde *fate ordained not for him triumph in battle*, Beo. Th. 5143; B. 2575. [*O. H. Ger.* hruodi (*in proper names*), Grff. iv. 1153: cf. *Icel.* hróðr *praise, fame.*] v. gūþ-, sige-hrēð; hrēðig, hrōðor.

hrēða, an; *m. A garment made of goat's skin;* melotes, Cot. 133, Lye. v. bord-, scild-hreóða [-hrēða].

hrēðan; *p. de To glory, triumph :* — Hrēðdon hildespelle *they triumphed with the song of* [*victorious*] *battle*, Cd. 170; Th. 214, 22; Exod. 573.

hrēðe; *adj. Fierce, cruel, savage, rough :*—Wearþ hire wrāþ on mōde heard and hrēðe *was wroth with her, harsh and cruel*, Cd. 103; Th. 136, 20; Gen. 2261. Deáþ neálǣcte strong and hrēðe, Exon. 49 b; Th. 170, 18; Gū. 1113. Hroeðo suīðe *sævi nimis*, Mt. Kmbl. Lind. 8, 28. In heágum mōrum and hrēðum *in arduis asperisque montibus*, Bd. 4, 27; S. 604, 27. Ðām hrēðestum feóndum *sævissimis hostibus*, Mone Gl. 346. v. rēðe.

hrēð-eádig; *adj. Glorious, noble, triumphant :*—Biþ ðǣr his þegna eác hrēðeádig heáp *there too shall be a triumphant band of his servants*, Exon. 21 b; Th. 58, 33; Cri. 945. Sum biþ on huntoþe hrēðeádigra deóra drǣfend *one is more famous in hunting, a chaser of wild beasts*, 78 b; Th. 295, 23; Crä. 37. [*Thorpe and Grein take* hrēðeádigra *as gen., but see* Th. 298, 1; Crä. *78 for another comparative.*] Hærfest biþ hrēðeádegost hæleþum bringeþ gēres wæstmas ða ðe him god sendeþ *autumn is most glorious, it brings to man the fruits of the year which God sends them*, Menol. Fox 475; Gn. C. 8. [*Cf. Icel.* hróðr-auðigr *famous.*]

hrēðe-mōnaþ. v. hrēð-mōnaþ.

hreðer, hræðer, hraðer, es; *m.* [?] *Breast, bosom :*—Hreðer innan wæs wynnum āwelled *the breast within was joyously agitated*, Andr. Kmbl. 2036; An. 1020. Hreðer [hreder, MS.] innan weóll beorn breóstsefa, Exon. 16 b; Th. 34, 9; Cri. 539: 46 b; Th. 158, 15; Gū. 910: Beo. Th. 4233; B. 2113. Hreðer innan swearc hyge hreówcearig *dark within grew his breast, troubled with care his mind*, Exon. 48 a; Th. 165, 8; Gū. 1025. Hreðer ǣðme weóll *his breast heaved with breathing*, Beo. Th. 5780; B. 2593. Is mē ǣnige gāst innan hreðres *anxiatus est in me spiritus meus*, Ps. Th. 142, 4. On breóston inne on hraðre, Bt. Met. Fox 25, 91; Met. 25, 46. Him of hræðre [hwæðre, MS.] gewāt sāwol *from his bosom departed the soul*, Beo. Th. 5631; B. 2819. Him on hreðre heáfodswīma heortan clypte *in his bosom stupor clasped his heart*, Cd. 76; Th. 94, 27; Gen. 1568. Ðe dryhtnes bebod heóldon on hreðre *who kept the lord's command in their breast*, Exon. 24 b; Th. 71, 23; Cri. 1160. Him wæs hreów on hreðre hygesorga mǣst, Beo. Th. 4645; B. 2328. Hē mē in hreðre bileác wísdōmes giefe, Exon. 51 a; Th. 176, 33; Gū. 1219; Andr. Kmbl. 138; An. 69: Cd. 161; Th. 201, 2; Exod. 366: Beo. Th. 2306; B. 1151. Ys mē on hreðre heorte gedrēfed *cor meum conturbatum est in me*, Ps. Th. 54, 4: 70, 8. Biþ on hreðre drepen biteran strǣle *is smitten in the breast with the bitter shaft*, Beo. Th. 3494; B. 1745. Æt helle duru dracan eardigaþ hāte on hreðre *at hell's door dwell dragons that send fire from within* [*firedrakes*], Cd. 215; Th. 271, 1; Sat. 99. Baðu hāt on hreðre *hot baths*, Exon. 124 b; Th. 478, 16; Ruin. 42: Beo. Th. 6287; B. 3148. Blōd ūt ne com of hreðre *blood came not from my breast*, Exon. 130 a; Th. 499, 9; Rä. 88, 13. Mē on hreðre heáfod sticade *in hẹr bosom she stuck my head*, 124 b; Th. 479, 9; Rä. 62, 5. Hālig heofonlíce gāst hreðer weardode æðelne innoþ *the holy heavenly spirit guarded her breast, her noble womb*, Elen. Kmbl. 2288; El. 1145: Exon. 49 a; Th. 169, 20; Gū. 1102. Him hildegrāp hreðre ne mihte aldre gesceððan *the hostile grasp could not harm his breast, his life*, Beo. Th. 2897; B. 1446. Hreðra gehygd *counsel*, 4096; B. 2045: Exon. 77 b; Th. 290, 28; Wand. 72. v. mid-hriðre.

hreðer-bealo; *n. Breast-bale, hurt to the mind, care, grief*, Beo. Th. 2690; B. 1343.

hreðer-cofa, an; *m. The breast*, Exon. 27 a; Th. 81, 25; Cri. 1329.

hreðer-gleáw; *adj. Prudent of mind*, Cd. 143; Th. 178, 17; Exod. 13.

hreðer-loca, an; *m. The breast*, Exon. 51 a; Th. 178, 1; Gū. 1237: 82 a; Th. 309, 17; Seef. 58: 23 b; Th. 65, 17; Cri. 1056: Elen. Kmbl. 172; El. 86.

hrēðig; *adj. Triumphant, exultant.* [*Goth.* hróþeigs *victorious, triumphant: Icel.* hróðugr *triumphant, glorious;* mod. *boasting.*] DER. eád-, eáð-, sige-, will-hrēðig.

hrēð-leás; *adj. Inglorious, joyless, without the joy of victory*, Exon. 46 a; Th. 156, 21; Gū. 878.

hrēð-, hrēð-mōnaþ, es; *m. March :*—On ðæm þriddan mōnþe on geáre biþ ān and þrittig daga and se mōnþ is nemned on lǣden martius and on ūre geþeóde hrēðmōnaþ *in the third month in the year are one and thirty days, and the month is called in latin* martius, *and in our language* hrēð-mōnaþ, Shrn. 59, 9. Ðonne se hrēðmōnaþ biþ āgān ðonne biþ seó niht twelf tīda lang and se dæg ðæt ilce *when March is past then the night is twelve hours long and the day the same*, 69, 7. Bede in his work ' De temporum ratione' c. 13 says ' Rhedmonath a dea illorum Rheda, cui in illo sacrificabant, nominatur.' Grimm quotes similar forms from other German sources, *Retmonat, Redimonet*, as names of March or February; and supposes an *O. H. Ger. Hruod, Hruoda* to correspond to the English *Hrêð, Hrêðe*, which would be connected with *hruod* [v. hrēð] *fame, glory*. See D. M. 267.

hrēð-ness, e; *f. Fierceness, roughness* [*of weather*], *cruelty :*—Hroeðnise *sævitiam*, Rtl. 122, 14. Hroeðnise *tempestatem*, Lk. Skt. Lind. 8, 24. [*Cf.* hreóh-ness.]

hreðor. v. hreðer.

hrēð-sigor, es; *m. Glorious victory*, Beo. Th. 5160; B. 2583.

hric, hricg. v. hrycg.

hricsc [= ? hrisc *or* hrics] *a rick, crick, a wrench accompanied with a small sound :*—Of fylle oððe of slege oððe of hricsca hwilcum *from a fall or from a blow or from any crick*, L. M. 1, 31; Lchdm. ii. 72, 23. [*Cf.* hriscan.]

hriddel, es; *n.* [?] *A riddle, sieve*, Som. [*Prompt. Parv.* rydyl *cribrum.*] v. hridian, hridder.

hridder, es; *n. A sieve, instrument for winnowing corn :*—Hridder *capisterium, taratantara*, Ælfc. Gl. 50; Som. 65, 116, 117; Wrt. Voc. 34, 45, 46. Ðā ābæd his fōstormōder ān hridder . . . Benedictus genam ða sticcu ðæs tōclofenan hriddores . . . hí ðæt hridder up āhēngon æt heora cyrcan geate, Homl. Th. ii. 154, 16–24. [*O. H. Ger.* ritra *cribrum, cribellum.*] v. hriddel.

hridrian; *p.* ode *To sift, winnow :*—Satanas gyrnde ðæt hē eów hridrude swā swā hwǣte *Satanas expetivit vos ut cribraret sicut triticum*, Lk. Skt. 22, 31. [*O. H. Ger.* ritaron *cribrare; Ger.* reitern *to sift.*]

hrif, rif, es; *n. The womb, belly;* uterus, venter :—Ðín ðæt fǣdmlíce hrif *thine enfolding womb*, Blickl. Homl. 7, 29. Hrif *uterus*, Mt. Kmbl. Lind. 1, 8: 19, 12: Rtl. 51, 27. Ðæt uferre hrif, L. M. 2, 28; Lchdm. ii. 224, 8. Rif *vel* seó inre wamb *alvus*, Ælfc. Gl. 74; Som. 71, 55; Wrt. Voc. 44, 38. Wið hrifes āþundennesse *for puffing of the visceral cavity*, Lchdm. iii. 70, 24. Of mōdur hrife mīnre *de utero matris meæ*, Ps. Th. 138, 11: 70, 5. Of hryfe *ex utero*, Ps. Spl. 21, 8. On hrife ðǣre ā clǣnan fǣmnan, Blickl. Homl. 33, 15. Bān biþ funden on heortes heortan hwílum on hrife *a bone is found in a hart's heart, sometimes in its belly*, L. Med. ex Quadr. 2, 17; Lchdm. i. 338, 6. Ácsedon hwider hie fleón woldon ðæt hie óðer gener næfden būton hie on heóra wífa hrif gewiton *quærentes, num in uteros uxorum vellent refugere*, Ors. 1, 12; Swt. 54, 4. Lācnung on ðæt hrif tō sendanne *to send medicine into the belly*, L. M. 2, 32; Lchdm. ii. 234, 19. Þurh mīnre mōdor hrif, Exon. 111 a; Th. 424, 27; Rä. 41, 44: 14 a; Th. 27, 4; Cri. 425. [*O. Frs.* rif, ref: *O. H. Ger.* href, ref *uterus.*] v. mid-hrif.

hrifþo; *f. Roughness of the skin, scurf :*—Heáfdes hrifþo, L. M. 2, 35; Lchdm. ii. 240, 20. v. hreóf.

hrif-wirc, -wærc, es; *m. A pain in the belly;* yleos, Ælfc. Gl. 10; Som. 57, 16; Wrt. Voc. 19, 24.

hrif-wund; *adj. Wounded in the belly :* — Gif [hē] hrifwund [hrif wund, Thorpe] weorþeþ xii scill. gebēte. Gif hē þurhþirel weorþeþ xx scill. gebēte *if he be wounded in the belly let twelve shillings be paid. If he be run through let twenty shillings be paid* [cf. the passage given in the note from Alamannic Laws, 'si in interiora membra transpunctus fuerit, quod hrefwunt dicunt, cum xii sol. componat. Si transpunctus fuerit cum xxiv sol. componat.' See, too, Graff. i. 897-8], L. Ethb. 61; Th. i. 18, 6.

hrig. v. hrycg.

hrilæcung [?] *ratiocinatio*, Som.

HRÍM, es; *m.* RIME, *hoar-frost :*—Hrím *pruina*, Ælfc. Gl. 94; Som. 75, 102; Wrt. Voc. 52, 52. Hrím and forst hāre hildstapan, Andr. Kmbl. 2516; An. 1259. Se hearda forst hrím heorugrimma, Exon. 111 a; Th. 425, 12; Rä. 41, 55. Hægel se hearda and hrím, 127 b; Th. 490, 11; Rä. 79, 9. Ðǣr ne hægl ne hrím hreósaþ tō foldan *nec gelido terram rore pruina tegit*, 56 b; Th. 201, 22; Ph. 60. Hrím hrusan bond hægl feóll on eorþan *frost bound the land, hail fell on earth*, 81 b; Th. 307, 31; Seef. 32. Ne hægles hryre ne hrímes dryre, 56 a; Th. 198, 27; Ph. 16. Mid herige hrímes and snāwes *with the legions of frost and snow*, Menol. Fox 406; Men. 204. On hríme *in pruina*, Ps. Th. 118, 83. Hríme gehyrsted *adorned with hoar-frost*, Menol. Fox 70; Men. 35: Exon. 77 b; Th. 291, 4; Wand. 77. Wineleás guma gesiþþ him beforan baðian brimfuglas brǣdan feðra hreósan hrím and snāw hagle gemenged *the friendless man sees before him the sea-birds bathe, and spread their wings, sees rime and snow fall mingled with hail*, 77 a; Th. 289, 14; Wand. 48. Nǣnig mōste heora hrórra hrím æpla gedígean *occidit moros*

eorum in pruina, Ps. Th. 77, 47. [Icel. hrím; n. hrími; m: O. H. Ger. rime gelu, Grff. ii. 506.]

hríman. v. hrýman.

hrím-ceald; adj. Icy cold: — Hrímcalde sǽ, Exon. 76 b; Th. 286, 22; Wand. 4. [Icel. hrím-kaldr.]

hrím-gicel, es; m. An icicle: — Bihongen hrímgicelum, Exon. 81 b; Th. 307, 1; Seef. 17.

hrímig; adj. Rimy, covered with hoar-frost: — Swíðe hrímige bearwas woods thickly covered with hoar-frost, Blickl. Homl. 209, 32: 207, 27 [?]. Winter biþ cealdost lencten hrímigost black frosts in winter, white frosts in spring, Menol. Fox 411; Gn. C. 6.

hrímig-heard; adj. Hard with frost, hard frozen, Exon. 130 a; Th. 498, 25; Rä. 88, 7.

hrínan; p. hrán; pp. hrinen To touch, reach, strike. **I.** with gen: — Ðú his hrínan meaht thou mayest touch it, Cd. 29; Th. 38, 34; Gen. 616. **II.** with dat: — Grundum ic hríne the depths I touch, Exon. 125 b; Th. 482, 22; Rä. 67, 5: 102 b; Th. 389, 8; Rä. 7, 4: 104 b; Th. 397, 31; Rä. 16, 28. Gif ic hríno wéde his if I touch his garment, Mt. Kmbl. Lind. 9, 21. Se hǽlend and hrán [or andhrán? cf. O. Sax. ant-hrínan] égum heora Iesus tetigit oculos eorum, Rush. 20, 34. Se hǽlend hrán him tangens eum, Mk. Skt. Rush. 1, 41: Exon. 110 a; Th. 421, 18; Rä. 40, 30. Hrinon hearmtánas drihta bearnum, Cd. 47; Th. 61, 4; Gen. 992. Ðeáh ðe hwæt wund hrine though the wound had touched him, Beo. Th. 5945; B. 2976. Ele synfulra ǽfre ne móte heáfde mínum hrínan oleum peccatorum non impinguet caput meum, Ps. 140, 7. Nǽnig wæter him hrínan ne mihte no water might reach him, Beo. Th. 3035; B. 1515: 1981; B. 988: Cd. 69; Th. 84, 11; Gen. 1396. Ðæt hý him mid hondum hrínan mósten, Exon. 38 b; Th. 127, 5; Gú. 381: 73 a; Th. 273, 7; Jul. 512. Ðé hondum hrínan, 36 b; Th. 119, 13; Gú. 254. Hrínande him tangens eum, Mk. Skt. Lind. 1, 41. **III.** with acc: — Ic hríno ðone hiorde percutiam pastorem, 14, 27. Gif hé mid his mihte muntas hríneþ qui tangit montes, Ps. Th. 103, 30: Exon. 106 b; Th. 406, 4; Rä. 24, 12. Hrín ða góman mid touch the fauces with it, L. Med. ex Quadr. 5, 3; Lchdm. i. 348, 10. Ne sceolon míne ða hálgan hrínan nolite tangere christos meos, Ps. Th. 104, 13. Wát ic Matheus þurh mǽnra hand hrínan heorudolgum, Andr. Kmbl. 1883; An. 944. **IV.** with object omitted: — Ðæt hé má wolde afrum onfengum earme gǽstas hrínan léton that he would further let the wretched spirits with their dire attacks touch him [Guthlac], Exon. 40 a; Th. 133, 17; Gú. 491. Swá hit him on innan com hrán æt heortan so it came within him, touched him at his heart, Cd. 33; Th. 45, 9; Gen. 724. Oþ ðæt deáþes folm hrán æt heortan until the hand of death touched him at his heart, Beo. Th. 4532; B. 2270. [A. R. rineð, prs: Orm. ran, p: O. Sax. hrínan: Icel. hrína to cleave, to hurt: O. H. Ger. hrínan tangere, obtrectare.] DER. æt-, and-, ge-, on-hrínan.

hrind. A word of doubtful meaning occurring in the following passage, 'Nis ðæt feor heonon ðæt se mere standeþ ofer ðæm hongiaþ hrinde bearwas wudu wyrtum fæst wæter oferhelmaþ,' Beo. Th. 2731; B. 1363. Thorpe translates barky, Kemble rinded, but in this case there should be no initial h. In Ælfc. Gl. 59; Som. 68, 5, 6; Wrt. Voc. 38, 56, 57 hrind translates caudex vel codex, and liber is translated seó inre hrind, but perhaps the better reading for the former would be rind = cortex. Otherwise hrinde bearwas might be [?] 'groves with [large-] stemmed trees.' Grein compares the word with forms given by Halliwell rind frozen to death, rinde to destroy, and suggests dead; Heyne takes hrinde = hrínende and compares with Icel. hrína sonare. Might hrinde = hringde in the sense 'placed in a ring or circle,' so that hrinde bearwas would be the trees placed round or encircling the mere?

hrindan; p. hrand, pl. hrundon To push, thrust: — Hé hrand [MS. rand], Exon. 113 b; Th. 436, 21; Rä. 55, 4. [Icel. hrinda to thrust.]

hrine, es; m. Touch: — Hrine tactus, Wrt. Voc. 282, 32. Drihten ðú ðe wé ne mágon ongytan mid hrine Lord thou whom we cannot perceive with the touch, Shrn. 166, 21. v. æt-hrine.

hrine-ness, e; f. Touching, contact: — Fram werelíce hrinenesse a viri contactu, Bd. 4, 19; S. 587, 37. Mid ða ylcan hrinenesse eodem tactu, 31; S. 610, 34. v. ge-hrineness.

HRING, hrincg, es; m. A RING, circle, circuit, cycle, orb, globe, festoon: — Ágymmed hrincg ungulus: geheáfdod hringce samothracius: lytel hring anelus, Ælfc. Gl. 65; Som. 69, 30, 31, 49; Wrt. Voc. 40, 59, 60; 41, 6. Hringc ansa, Wrt. Voc. 66, 34: 284, 7. Hring fibula, legula, sertum, Cot. 85, 186, 190, Lye. Án fýren hring globus ignis, Ors. 5, 10; Swt. 234, 3. Mon geseah ymbe ða sunnan swelce án gylden hring circulus ad speciem cælestis arcus orbem solis ambiit, 14; Swt. 248, 9. Ðæs seó hrincg circulus [pupillæ], Ælfc. Gl. 70; Som. 70, 64; Wrt. Voc. 42, 72. Se hring ealles geáres totius anni circulus, Bd. 4, 18; S. 586, 40. Hring útan ymbbearh the ring [armour formed of rings] protected him without, Beo. Th. 3011; B. 1503: 4513; B. 2260. Sunnan hring beága beorhtast the rainbow [?], Exon. 60 a; Th. 219, 11; Ph. 305. Ðone hálgan hringe betelðaþ flyhte on lyfte contrahit in cætum sese genus omne volantum, 60 b; Th. 221, 24; Ph. 339. Ðonne ðæt gecnáwaþ feónd ðætte fira gehwylc on his hringe biþ fæste geféged when the devil knows that any man is fast fixed

in his ring [fetters, chain or circle over which his power extends?], 97 a; Th. 362, 22; Wal. 40. Gim sceal on hringe standan the gem must stand in the ring, Menol. Fox 594; Gn. C. 22. Syllaþ him hring on his hand date anulum in manum ejus, Lk. Skt. 15, 22. Seðe his geleáfan hring mé lét tó wedde, Homl. Swt. 7, 30. Dyde him of healse hring gyldenne doff'd from his neck a golden ring, Beo. Th. 5611; B. 2809. Gewyrc ánne hring ymb ðone slite make a ring round the incision, L. M. 1, 45; Lchdm. i. 112, 1. Ðú geáres hring mid gyfe bletsast benedices coronæ anni benignitatis tuæ, Ps. Th. 64, 12. Ǽr sunne twelf mónþa hringc útan ymbgán hæbbe, Guthl. 21; Gdwin. 96, 5. Ofer holmes hrincg over the ocean's circuit, Cd. 69; Th. 84, 5; Gen. 1393. Hrincg ðæs heán landes, 137; Th. 172, 34; Gen. 2854. Wíngearda hringa [s] corimbi, Ælfc. Gl. 59; Som. 68, 11; Wrt. Voc. 38, 60. Hrægl and hringas raiment and rings, Beo. Th. 2394; B. 1195. Hringa hyrde, 4482; B. 2245: 3018; B. 1507: 4680; B. 2345. Heortan unhneáweste hringa gedáles the heart least niggardly in the giving of rings, Exon. 85 b; Th. 323, 4; Víd. 73. Hæft mid hringa gesponne bound with the linked chain, Cd. 35; Th. 47, 17; Gen. 762: 19; Th. 24, 14; Gen. 377. Hringum gehrodene adorned with rings, Judth. 10; Thw. 21, 27; Jud. 37: Beo. Th. 2187; B. 1091. Hringum gyrded, Exon. 129 b; Th. 497, 22; Rä. 87, 4. Hringan, 102 b; Th. 387, 8; Rä. 5, 2. Hé wolde ðæs beornes beágas gefecgan reáf and hringas, Byrht. Th. 136, 34; By. 161. Hringas dǽlan, Beo. Th. 3944; B. 1970: 6661; B. 3034. Ða nigontýnlícan hringas rihtra Eástrana and hét fordilgian ða gedwolan hringas feówer and hundeahtatig geára circuli Paschæ decennovenales oblitteratis erroneis octoginta et quatuor annorum circulis, Bd. 5, 21; S. 643, 26. [Icel. hringr a ring, ring of a coat of mail, circle: O. H. Ger. hring circulus, orbis, spira, sphæra, bulla, corona, sertum, torques, vinculum, laqueus: Ger. ring.] DER. bán-, bridels-, eág-, eáh-, eár-hring. v. beág.

hring, in the phrase wópes hring occurs four times, in poems by the same author: — Ðá cwom wópes hring þurh ðæs beornes breóst blát út faran weóll waðuman streám, Andr. Kmbl. 2558; An. 1281. Ðá wæs wópes hring hát heáfodwylm ofer hleór goten nalles for torne teáras feóllon, Elen. Kmbl. 2262; El. 1132. Ðǽr wæs wópes hring torne bitolden wæs seó treówlufu hát æt heortan hreðer innan weóll, Exon. 15 b; Th. 34, 5; Cri. 537. Him ðæs wópes hring torne gemonade teagor ýðum weól háte hleórdropan, 52 a; Th. 182, 21; Gú. 1313. The meaning given by Grein, sonus [cf. hringan], does not seem to suit the context very well, which, as in the second passage, where the phrase appears equivalent to hát heáfodwylm, points to shedding tears as the idea to be conveyed. Grimm explains fletus intensissimus, quasi circulatim erumpens, And. u. El. p. 130, and this seems to give the meaning though the connection with hring is not very evident.

hringan; p. de; v. trans. and intrans. To ring: — His searo hringeþ his armour rings, Salm. Kmbl. 534; Sal. 266. Byrnan hringdon their byrnies rang, Beo. Th. 660; B. 327. Hí ringden ða belle they rang the bells, Chr. 1131; Erl. 259, 37. Hringe tácn sonet signum, Lye. Yc gef leáua ðám munche tó hringinde hyre tíde I give leave to the monks to ring their hours, Chart. Th. 437, 13. [Laym. ringe; p. ringeden: 2nd MS. rongen: R. Glouc. Chauc. Piers P. ringe; p. rong: Icel. hringja.]

hring-bán, es; n. A circular bone, bone in the shape of a ring: — Hringbán ðæs eágan teuco, Ælfc. Gl. 70; Som. 70, 73; Wrt. Voc. 43, 6.

hring-boga, an; m. A serpent [from its being bent into coils (hring)], Beo. Th. 5115; B. 2561. [Cf. Icel. hring-laginn coiled up; hringa sik to coil (of a serpent).]

hringed; adj. Furnished with rings, formed of rings: — Hringedu byrne lorica, Cot. 121, Lye: Beo. Th. 2495; B. 1245: 5224; B. 2615. [Icel. hringa to furnish with a ring; and cf. hringa-brynja a coat of ring-mail: O. H. Ger. gi-ringotero hamata (lorica).]

hringed-stefna, an; m. A ship having its stern adorned with spiral or ring-shaped ornaments [?], or furnished with a ring or hook; or having a curved stern, Beo. Th. 64; B. 32: 3799; B. 1898: 2266; B. 1132. [Cf. wunden-stefna; hring-naca; and Icel. hring-horni the mythol. ship of the Edda.]

hring-fáh; adj. Of many colours, diversified with circular spots of colour [?]: — Hringfégh polimita vel oculata, Ælfc. Gl. 29; Som. 61, 29; Wrt. Voc. 26, 28. Hét wircean him hringfáge tunecan fecit ei tunicam polymitam, Gen. 37, 3. v. hring-wíse.

hring-finger, es; m. The ring-finger, the third finger: — Hringfinger anularis, Wrt. Voc. 283, 23. Mid þuman and mid hringfingre, L. Med. ex Quadr. 1, 5; Lchdm. i. 330, 21. v. Halliwell Dict. ring-finger.

hringian to surround, encircle. [Icel. hringja: cf. O. H. Ger. ga-hringian congyrare.] v. ymb-hringian.

hring-íren, es; n. The iron rings of a coat of mail: — Gúþbyrne scán heard hand-locen hringíren scír song in searwum the corslet shone, hard, hand-wrought, the bright iron rings rang in their armour, Beo. Th. 650; B. 222.

hring-loca, an; m. A coat of mail formed with rings, Byrht. Th. 136, 2; By. 145.

hring-mǽl; adj. Ornamented with inlaid rings [of a sword], Beo. Th.

3133; B. 1564. [Cf. *Icel.* mál *used of inlaid ornaments, e. g.* mála-sax *an inlaid sword;* and for ring ornaments see Worsaae's Primeval Antiquities, p. 40.]

hring-mǽled; *adj. Ornamented with inlaid rings:* — Hringmǽled sweord, Cd. 93; Th. 120, 10; Gen. 1992. v. preceding word.

hring-mere, es; *n. A round pool, a bath,* Exon. 124 b; Th. 478, 21; Ruin. 45.

hring-naca, an; *m. See* hringed-stefna, Beo. Th. 2728; B. 1862.

hring-nett, es; *n. A net-work of rings, a coat of mail formed of rings:*—Hringnet bǽron locene leoðosyrcan, Beo. Th. 3783; B. 1889. [Cf. *Icel.* hring-kofl, -serkr, -skyrta *a coat of mail;* hring-ofinn *woven of rings,* an epithet applied to such a coat.]

hring-sele; *m. A hall in which rings are distributed* or *stored up,* Beo. Th. 4024; B. 2010 [*Hrothgar's palace*]: 6008; B. 3053 [*the cavern where the dragon guarded the treasure*]: 5672; B. 2840. v. beág-sel, -sele.

hring-seta *circenses ludi,* Cot. 43, Lye.

hring-sete *circus,* Cot. 183, Lye.

hring-sittend *circumsedens, spectans,* Hpt. Gl. 407.

hring-stede *circulare stadium,* Lye.

hring-þegu, e; *f. Acceptance of rings, of gifts given by a lord:*—Ne biþ him tó hearpan hyge ne tó hringþege, Exon. 82 a; Th. 308, 24; Seef. 44. v. beág-þegu.

hring-weorþung, e; *f. Honouring by the gift of a ring:*—Ne mægþ habban on healse hringweorþunge *no maiden's neck shall be graced with a ring,* Beo. Th. 6027; B. 3017. v. hord-weorþung.

hring-windel *sphæra,* Lye.

hring-wíse, an; *f. In the phrase* on hringwísan *ring-wise, in rings:* — Hwítes hiowes and eác missenlíces wæs hió on hringwísan fág *candido versicolore in modum ranarum,* Nar. 16, 1. v. hringfáh.

hrínung, e; *f. Touch;* tactus: — In hrínung hláfes *intincti panis,* Jn. Skt. p. 7, 3. Mid ríning ł middÿ gehrán *tactu,* 8, 7.

hrís, es; *n. A twig, branch,* RISE :—Hrís *frondes,* Cot. 93, Lye. [Laym. O. and N. Chauc. ris: v. Halliwell Dict. rise: Icel. hrís; *n. shrubs, brushwood:* O. H. Ger. hrís *ramus, frondes, ramusculus:* Ger. reis *a twig, rod.*]

hríscan. v. hryscan.

hríseht; *adj. Bushy, bristly;* setosus, Cot. 186, Lye.

hrísel, hresl, es; *m.* [?] *A shuttle;* radius:—Hrisl *radiolum,* Ælfc. Gl. 110; Som. 79, 54; Wrt. Voc. 59, 25: *radium,* Wrt. Voc. 281, 75. Hresl [hresl, Wrt.] *radius,* 66, 12. Hrisil, Exon. 109 a; Th. 417, 20; Rä. 36, 7. v. hrisian, *and* cf. scytel.

hrisian; *p.* ede *To shake:*—Syrcan hrysedon *shook their coats of mail,* Beo. Th. 458; B. 226. Hrisedon heáfud *moverunt capita,* Ps. Surt. 21, 8: 108, 24. [Cf. Hresigende *febricitans,* Mk. Skt. 1, 30 (later MS.).] Stefn drihtnes hrysiendis wésten *vox Domini concutientis desertum,* Ps. Spl. T. 28, 7. [Laym. rusien: Ayenb. resie: Chauc. rese: Goth. hrisian: O. Sax. hrisian *to shake, tremble:* cf. Icel. hrista *to shake.*] v. á-hrisian.

hristenda [hriscenda?] *astridulus, stridulus,* Lye. v. hryscan, *or next word* [?].

hristlan *to rustle:*—Hristlend[e] *crepens,* Lye.

hristlung, e; *f. A rustling;* crepitus, strepitus, Lye.

hristung, e; *f. A quivering, spasmodic action:*—Ceolan hristung and hreóung hlÿdende swíðust innan [*or should* hristlung (v. *preceding word*) *be read?*], L. M. 2, 46; Lchdm. ii. 258, 18. Cockayne, *who explains as above, compares with Icel.* hrista *to shake. See also* hristenda.

hríð, e; *f. A storm, tempest:*—Hríð hreósende *the driving storm,* Exon. 78 a; Th. 292, 20; Wand. 102. [*Icel.* hríð; *f. a storm, snow-storm.*]

hrið, es; *m. Fever:*—Fefer ðæt is micel hǽto and hrið [MS. hruð], L. M. 2, 24; Lchdm. ii. 214, 7. [O. H. Ger. rito; *m. febris.*]

hrið-ádl, e; *f. A fever:*—Gif him hriðádl getenge biþ *if fever be upon him,* L. M. 2, 24; Lchdm. ii. 214, 16.

hriðer, hrÿðer, es; *n. Horned cattle, ox, cow, heifer:* — Jung hrÿðer *juniculus* [*anniculus?*], Ælfc. Gl. 22; Som. 59, 86; Wrt. Voc. 23, 45. Geong hrÿðer L. M. 2, 16; Lchdm. ii. 196, 24. Se hláford geáhsode ðæt ðæt hrÿðer [cf. fear, 7] geond ðæt wésten férde *the master learned that the bull was going through the desert,* Blickl. Homl. 199, 9, 11, 14, 19, 26. Ðǽr wǽron gecÿpe hrÿðeru and scép *there were for sale oxen and sheep,* Homl. Th. i. 406, 18. Hwílum hÿ him rǽredon on swá hrÿðro *sometimes they bellowed at him like oxen,* Shrn. 141, 10. Gif hrÿðera steorfan *if cattle are dying,* Lchdm. iii. 54, 31. Ðǽron næs orscynnes nán mǽre búton vii hruðeru, Cod. Dipl. Kmbl. iv. 275, 7: Ex. 34, 19. Bige mid ðam ylcan feó swá hwæt swá ðé lícige hrÿðera and sceáp *emes ex eadem pecunia quidquid tibi placuerit sive ex armentis sive ex ovibus,* Deut. 14, 26. Hrÿðera and scép, Jos. 6, 21. Næfde hé má ðonne twentig hrÿðera and twentig sceápa and twentig swÿna, Ors. 1, 1; Swt. 18, 14. Hrÿðera gehlów *the lowing of oxen,* Ælfc. Gr. 1; Som. 2, 35. Hine oftorfodon mid bánum and mid hrÿðera [hrÿðeres, MS. F: neáta, MS. D.] heáfdum *they stoned him to death with bones and heads of cattle,* Chr. 1012; Erl. 146, 18. Hrÿðra fald *bucetum,* Ælfc. Gl. 1; Som. 55,

23; Wrt. Voc. 15, 22. Of hriðerum *de armento,* Lev. 1, 3. Of nÿtenum ðæt ys of hriðerum and of sceápum *de pecoribus id est de bobus et ovibus,* 2. [*A. R.* reoðer: *Laym.* ruðeren, roðere; *pl: R. Glouc.* roþeren: *O. Frs.* hrither, rither, reder: cf. *O. H. Ger.* hrind *armentum, bos: Ger.* rind.] v. eald-hriðer.

hriðeren; *adj. Of cattle;* bovinus: — Genim hrÿðeren flǽsc *take ox-flesh,* L. M. 2, 7; Lchdm. ii. 186, 18. [Cf. *O. H. Ger.* rinderin *bovinus, bubula (caro).*]

hriðer-freóls *taurilia,* Hpt. Gl. 515.

hriðer-heáwere, es; *m. A butcher:* — Hrÿðerheáwere *bucida, qui boves mactat,* Ælfc. Gl. 33; Som. 62, 33; Wrt. Voc. 28, 16.

hriðer-heord, e; *f. A herd of cattle:*—Eówre sceáp and eówer hrÿðerheorda *oves tuæ et armenta tua,* Gen. 45, 10.

hriðer-hirde, es; *m. A neat-herd, herdsman:* — Amos hátte sum hrÿðerhyrde *Amos was the name of a certain herdsman,* Homl. Th. i. 322, 35. [Cf. *O. E. Hom.* Amos het a reoðer heorde.]

hriðian; *p.* ode *To shake, quake, have a fever:*—Sió wamb hryt *the stomach is fevered,* L. M. 2, 25; Lchdm. ii. 216, 20. Hie hriðiaþ *they are feverish,* 26; Lchdm. ii. 220, 5. Hé hriðode *he was sick with a fever,* Homl. Th. i. 86, 7. Hriðgende [cf. Lind. cuacende ł bifigende] *febricitantem,* Mt. Kmbl. 8, 14. Hriðigende, Mk. Skt. 1, 30. Hé biþ hriðende *he is feverish,* L. M. 2, 17; Lchdm. ii. 198, 21. [Cf. *O. H. Ger.* ridan *febricitare.*] v. hrisian.

hriðing, e; *f. Fever, feverishness:*—Mid hriðingum swíðe strangum *with very violent fevers,* L. M. 2, 46; Lchdm. ii. 258, 2.

hrið-suht [?], e; *f. Fever:*—Hál of ridesohte *the fever left her,* Mk. Skt. Rush. 1, 31. Perhaps the word is borrowed; cf. *Icel.* riðu-sótt *fever, ague.*

HRÓC, es; *m. A* ROOK, *a raven, a jackdaw:* — Hróc *graculus vel garrulus,* Ælfc. Gl. 38; Som. 63, 27; Wrt. Voc. 29, 47: 77, 44. Hróc *gralus, grallus,* 62, 31: 281, 1: *garrula,* Shrn. 29, 1. Se selþ nÿtenum mete and briddum hróca cígendum hine *qui dat jumentis escam ipsorum, et pullis corvorum invocantibus eum,* Ps. Spl. 146, 10. [*O. and N.* rok: *Prompt. Parv.* rook *frugella, graculus: O. Du.* rouca *garula : Icel.* hrókr: *O. H. Ger.* hruoh *graculus.*]

hroden; *pp.* of hreóðan *Laden, laden with ornaments, ornamented, adorned:* — Brÿd beága hroden *a bride adorned with rings,* Exon. 12 a; Th. 18, 31; Cri. 292. Ðá wæs heal hroden feónda feorum *then was the hall burdened with the lives of his foes* [*filled with the slain*], Beo. Th. 2307; B. 1151. Hroden ealowǽge *the ornamented ale-cup,* 995; B. 495: 2048; 1022. [Cf. *Icel.* hroðian *in* hroðit sigli.] v. beág-, ge-, gold-, sinc-hroden; on-hreóðan.

HRÓF, es; *m. A* ROOF, *the top, summit, highest part* [cf. Tennyson's 'Why should we only toil the *roof* and crown of things?']:—Góma *vel* hróf ðæs múþes *palatum vel uranon,* Ælfc. Gl. 71; Som. 70, 106; Wrt. Voc. 43, 35. Hróf *camara,* 290, 2. Se hróf hæfde mislíce heáhnysse *the roof was not all of one height,* Homl. Th. i. 508, 18. Ðæt héhste gód is hróf eallra óðra góda *the chief good is the roof and crown of all other goods,* Bt. 34, 7; Fox 142, 35. Wið ðæs heán hrófes ðæs héhstan andgites *in summæ intelligentiæ cacumen,* 41, 5; Fox 254, 16. Under fæstenne folca hrófes *under the firmament,* Cd. 8; Th. 10, 8; Gen. 153. Mec feredon under hrófes hleó *bore me under the shelter of the sky,* Exon. 107 b; Th. 409, 22; Rä. 28, 5. Martinus ástáh on ðam sticelan hrófe, Homl. Th. ii. 510, 7. Ðe ne beóþ tó ðam hrófe ðonne git cumen fulfremedra mægena *nondum ad extremam manum virtutum perfectione perductas,* Bt. 18, 1; Fox 60, 22. From hróf eardes *a summo terræ,* Mk. Skt. Lind. 13, 27. On hrófe gestód heán landes *he stopped on the summit of the mount,* Cd. 140; Th. 175, 20; Gen. 2898. Of he[um] heofnes hrófe *ex summa cæli arce,* Rtl. 101, 24. Hé gescóp eorþan bearnum heofon tó [h]rófe *qui filiis hominum cælum pro culmine tecti creavit,* Bd. 4, 24; S. 597, 22. Ðenden hé on ðysse worulde wunode under wolcna hrófe, Judth. 10; Thw. 22, 19; Jud. 67: Elen. Kmbl. 178; El. 89: Cd. 158; Th. 196, 28; Exod. 298. Ðæt wé tó ðam hÿhstan hrófe gestígan *that we may mount to heaven,* Exon. 18 b; Th. 47, 3; Cri. 749. Ðe ðæs húses hróf stadeliaþ *qui ædificant domum,* Ps. Th. 127, 1. Gif hwylc wíf seteþ hire bearn ofer hróf *si mulier aliqua infans em suam super tectum posuerit,* L. Ecg. C. 33; Th. ii. 156, 45. Ofer l eánne hróf, Beo. Th. 1970; B. 983: 1857; B. 926: 1677; B. 836. Under beorges hróf *in the cave,* 5504; B. 2755. Ðá gewát se engel up on heánne hróf heofona ríces, Cd. 196; Th. 244, 2; Dan. 442. Fiðru mid ðæm ic fleógan mæg ofer heáne hróf heofones ðisses *pennæ quæ celsa conscendant poli,* Bt. Met. Fox 24, 5; Met. 24, 3: Cd. 46; Th. 58, 34; Gen. 956. Ofer wealles hróf *super muros,* Ps. Th. 54, 9: Exon. 108 a; Th. 412, 1; Rä. 30, 7. Hylles hróf, 104 b; Th. 397, 30; Rä. 16, 27. Helmes hróf, Beo. Th. 2064; B. 1030. Under wǽtera hrófas [*of passing through the Red Sea*], Cd. 170; Th. 214, 18; Exod. 571. Bodiaþ uppan hrófum *prædicate super tecta,* Mt. Kmbl. 10, 27. [*Laym.* róf: *Orm.* rhof: *O. Frs.* hróf: *Icel.* hróf *a shed under which ships are built* or *kept.*] DER. heofon-, inwit-hróf.

Hrofes-, Hrofe-ceaster; *f. Rochester,* Chr. 741; Erl. 46, 31: 885; Erl. 82, 20. Tó Hrofeceastre *in civitate quam gens Anglorum a pri-*

mario quondam illius qui dicebatur Hrof, Hrofæs cæstræ cognominat, Bd. 2, 3; S. 504, 25.

hróf-fæst; *adj. Having the roof firmly fixed:*—Healle hróffæste, Bt. Met. Fox 7, 11; Met. 7, 6.

hróf-sele, es; *m. A hall having a roof:*—Nǽnig wæter him for hrófsele hrínan ne mihte *no water could touch him for the roofed hall*, Beo. Th. 3034; B. 1515.

hróf-stán, es; *m. A roof-stone, stone forming part of a roof:*—Of ðam hrófstáne, Homl. Th. i. 508, 33. [Cf. hróf-tigel: *Mod. E.* roof-tree.]

hróf-tigel, e; *f. A tile for roofing:*—Hróftigla *tegulæ, imbrices, lateres vel laterculi*, Ælfc. Gl. 58; Som. 67, 92; Wrt. Voc. 38, 18.

hróf-timber, es; *n. Material for roofing*, imbrex, Hpt. Gl. 459.

hróf-wyrhta, an; *m. A workman who works at roofs, a builder:*—Hrófwyrhta *sarcitector vel tignarius*, Ælfc. Gl. 9; Som. 56, 125; Wrt. Voc. 19, 8.

hromese *acitula*, Cot. 206. v. hramsan.

hron, hrond-. v. hran, hrand-.

hrop. v. rop.

hróp, es; *m. Crying, clamour, outcry:*—Ðǽr biþ â wóp and hróp *there shall be ever weeping and wailing*, Blickl. Homl. 185, 7. [*Laym.* rop: *Scot.* roup *an outcry, a sale by auction*; cf. *Goth.* hrôpei *clamor*; *Icel.* hróp; *n. scurrility, crying: O. H. Ger.* hruof; *m. clamor: Ger.* ruf.]

hrópan; *p.* hreóp *To cry out, clamour, make a noise, shout, scream:*—Hreópon friccan *the heralds shouted*, Andr. Kmbl. 2314; An. 1158: Elen. Kmbl. 108: 1097; El. 550. Hreópon mearcweardas *the warders of the border* [*the wolves*] *clamoured*, Cd. 151; Th. 188, 14; Exod. 168. On hwæl hreopon [MS. hwreopon] herefugolas *the birds of war wheeled about screaming*, 150; Th. 188, 1; Exod. 161. Wóp âhófun hreópun hwílum wédende swâ wilde deór, Exon. 46 a; Th. 156, 21; Gú. 878. Hrefnes briddum ðonne heó hrópende on cígeaþ *to pullis corvorum invocantibus eum*, Ps. Th. 146, 10. [*A. R.* ropeð, *prs: Scot.* roup *to cry, shout; to sell by auction: Goth.* hrôpjan; *p.* hrôpida *to cry out: O. Sax.* hrôpan; *p.* hreóp: *O. Frs.* hrôpa; *p.* rôp *and* rôpte: *Icel.* hrópa; *p.* hrópaði *to slander; to call aloud: O. H. Ger.* hruofan; *p.* hrief: hruofian; *p.* hruofta (Grff. iv. 1135) *clamare: Ger.* rufen; *p.* rief.]

hrór; *adj. Stirring, active, agile, nimble, vigorous, stout, strong:*—Hrór hægstealdmon *a stout fellow*, Exon. 113 b; Th. 436, 18; Rä. 55, 3. Sǽde ðæt his byrne ábrocen wǽre heresceorpum hrór [heresceorp unhrór, Th.] *said that his byrnie was broken, strong* [*though it was*] *as armour*, Fins. Th. 90; Fin. 45. Ðâ Israélas ǽhte gesǽtan hróres folces *et habitavit in tabernaculis eorum tribus Israel*, Ps. Th. 77, 56. Swâ seó strǽle byþ strangum and mihtigum hrórum on handa *sicut sagittæ in manu potentis*, 126, 5. Ðâ wæs of ðæm hróran [*Beowulf*] helm and byrne lungre álýsde, Beo. Th. 3262; B. 1629. Drihten his heáhsetl hrór timbrade *Dominus paravit sedem suam*, Ps. Th. 102, 18: 88, 26. Geseoh hróre meaht hysse ðínum *da potestatem tuam puero tuo*, 85, 15. Hróre stence *with strong perfume*, 132, 2. Ðæt hé folc gesceóp fægere Drihten heraþ holdlíce hróre geþance *populus qui creabitur laudabit Dominum*, 101, 16. Nǽnig móste heora hrórra hrím æpla gedígean *occidit moros eorum in pruina*, 77, 47. Hrórum neátum odðe unhrórum *mobilibus belluis aut immobilibus animantibus*, Bt. 41, 5; Fox 254, 14. [*O. Sax.* hrôr: cf. *O. H. Ger.* ga-hrôrig *viridis, floridus, florens: Ger.* rührig: cf. also *Prompt. Parv.* rooryñ or ruffelyñ *amonge dyuerse thyngys manumitto*; *and the epithet* roaring *as applied in the Elizabethan times to bullies*, v. Nares' Gloss.] v. fela-, un-hrór; *and* hréran.

hroren-líc; *adj. Ready to fall*; ruiturus, Som.

hróst, es; *m. A wooden framework* [*of a roof*], a ROOST:—Hróst *petaurum*; henna hróst *gallinarium*, Lye. [*Scot.* roost *the inner roof of a cottage, composed of spars reaching from the one wall to the other*: cf. *O. Sax.* he (*Christ*) ina kuman gisah thurh thes huses hrost (*of the man who was let down through the roof*): *O. Du.* roest *craticula, gallinarium: Ger.* rost '*craticula focaria, clathrum, fundamentum ædificii in cratis modum positum, clathrum galeæ*,' Grein: v. Grff. ii. 552, rôst; *m. craticula, arula, ratago, catasta*.]

hróst-beág [?] *the woodwork of a circular roof:*—Tigelum sceáðeþ hróstbeáges hróf [MS. hrost beages rof] *the woodwork of the roof parts from the tiles, the tiles fall off leaving the woodwork of the roof bare*, Exon. 124 a; Th. 477, 29; Ruin. 32.

hrot, es; *n. Thick fluid, scum, mucus:*—Gewyrc ðé lǽcedóm ðus of ecede and of hunige, genim ðæt sēleste hunig dó ofer heorþ áseóþ ðæt weax and ðæt hrot of *make yourself a medicine thus of vinegar and honey; take the best honey, put it over the fire, seethe* [*strain*?] *off the wax and the scum*, L. M. 2, 28; Lchdm. ii. 224, 17. [*O. H. Ger.* hroz, roz *mucca, mucus, vomen, phlegma, reuma; Ger.* rotz.]

Hróð- *in proper names, e. g.* Hróð-gâr, -mund, -wulf. [Cf. hréð, hréðig.]

hroð [or roð?]-**hund**, es; *m. Inutilis canis*, Ælfc. Gl. 21; Som. 59, 77; Wrt. Voc. 23, 36. v. roð-hund.

hróðor, es; *m. Solace, comfort, benefit, pleasure:* — Ic ðé Andreas

onsende tô hleó and tô hróðre *I will send Andrew to you to protect and comfort you*, Andr. Kmbl. 221; An. 111: 1133; An. 567. His sunu hangaþ hrefne tô hróðre *his son hangs a solace for the raven*, Beo. Th. 4887; B. 2448: Apstls. Kmbl. 190; Ap. 95. Ðú ðe cwôme heánum tô hróðre *thou* (*Christ*) *who hast come for a comfort to the humble*, Exon. 13 b; Th. 26, 7; Cri. 414. Feóndum tô hróðor *to the delight of thy foes*, 17 a; Th. 39, 16; Cri. 623. Hungrum tô hróðor [cf. Soul Kmbl. 224, hungregum tô frôfre], 99 b; Th. 373, 27; Seel. 116: 71 b; Th. 267, 17; Jul. 416. Tô hleó and tô hróðer, 25 a; Th. 73, 29; Cri. 1197: Elen. Kmbl. 32; El. 16: 2317; El. 1160. Forðon ðé hróðra oftiþþ grêne folde *therefore shall the green earth withdraw from thee her delights* [*fruits*], Cd. 48; Th. 62, 21; Gen. 1017. Gehwæðer óðrum hróðra gemyndig *each to other was mindful of benefits*, Beo. Th. 4349; B. 2171. Wêrigmód heán hróðra leás *wearied, humbled, comfortless*, Andr. Kmbl. 2733; An. 1369. Heánmôd hróðra bidǽled, Exon. 71 a; Th. 265, 33; Jul. 390. v. hréð.

hrúm, es; *m. Soot:*—Hrúm *cacobatus*, Wrt. Voc. 291, 24. Micelne sigelhearwan ðæm wæs seó onsýn sweartre ðonne hrúm *a great Ethiopian with a face blacker than soot*, Shrn. 120, 24. v. cetel-hrúm; hrýme.

hrúmig; *adj. Sooty*; fuliginosus, Cot. 31, Lye. v. be-hrúmig.

hrung, e; *f. A rung, staff, rod, beam, pole:*—Ongunnon stígan on wægn weras and hyra wicg somod hlôdan under hrunge ðâ ða hors oðbær wægn tô lande *the men mounted the wain and their steeds with them, they stowed them under the rung* [*the pole that supported the covering*?]; *then the wain bore the horses to land*, Exon. 106 a; Th. 404, 19; Rä. 23, 10. [*Chauc.* Piers P. rong (*of a ladder*): *Goth.* hrugga *a staff*: cf. *Icel.* Hrungnir *name of a giant*, v. Grmm. D. M. 494: *Ger.* runge *a pin, bolt*.] v. scil-hrung.

hruse, an; *f. The earth, ground:*—Beofaþ middangeard hruse under hæleþum *the world shall tremble, the earth under men*, Exon. 20 b; Th. 55, 13; Cri. 883: Beo. Th. 5110; B. 2558. Ðǽr mê siteþ hruse on hrycge *there the earth presses on my back*, Exon. 101 b; Th. 383, 5; Rä. 4, 6. Ic goldwine mínne hrusan heolstre biwráh *I buried my lord*, 76 b; Th. 287, 32; Wand. 23. Ligeþ him behindan hefig hrusan dǽl *there remains behind the heavy earthy part*, Bt. Met. Fox 29, 107; Met. 29, 53. Ne gelýfdon ðætte líffruma in monnes hiw from hrusan âhafen wurde *did not believe that the author of life had been raised from the ground in the form of a man*, Exon. 17 b; Th. 41, 19; Cri. 658. Ne hreósaþ hî tô hrusan *non est ruina maceriæ*, Ps. Th. 143, 18. Under hrusan *under ground*, Beo. Th. 4813; B. 2411: Elen. Kmbl. 435; El. 218. Wæs hungor ofer hrusan *there was a famine upon the earth*, Chr. 975; Erl. 126, 29; Edg. 55. Hreás on hrusan nalles æfter lyfte lácende hwearf, Beo. Th. 5654; B. 2831. Heofonas ðú wealdest hrusan swylce *tui sunt cæli et tua est terra*, Ps. Th. 88, 10: 120, 2: 133, 4. Under eorþan befeolan hinder under hrusan, Exon. 91 a; Th. 340, 24; Gn. Ex. 116. For ansýne êcean Drihtnes heofonas droppetaþ hrusan forhtiaþ *terra mota est; etenim cæli distillaverunt a facie Dei*, Ps. Th. 67, 9. Heofenas blissiaþ hrusan swylce gefeóþ *lætentur cæli et exultet terra*, 95, 11. Hyllas and hrusan and heá beorgas ðec wurðiaþ, Cd. 192; Th. 240, 7; Dan. 383. [Grimm D. M. p. 230 says 'mit *crusta* wird das ags. hruse genau verwandt sein.']

hrut *or* hrút *balidus*, Cot. 28, Lye. *Ettmüller suggests* balidus = balans *animal, and compares Icel.* hrútr *a ram: Ducange has the following* '*balidus fortasse pro validus, ad coitum aptus.*' *See* hryte.

hrútan; *p.* hreát, *pl.* hruton *To make a noise, to snore; stridere, stertere:*—Ic hrúte *sterto*, Ælfc. Gr. 28, 3; Som. 30, 64. Ne æt mê hrútende hrisil scríðeþ *nor does the shuttle come whizzing at me*, Exon. 109 a; Th. 417, 19; Rä. 36, 7. [*Prompt. Parv.* rowtyn, yn slepe *sterto: Chauc.* route *to snore, roar*, 'the wynde so loude kan to route:' *Wick.* route þ stertit: Piers P. rutte *snored:* E. D. S. Reprint. Gloss. B. 15, rute *to cry fiercely;* rowt, rawt *to low like an ox or cow: Icel.* hrjóta (*older* rjóta) *to snore: O. H. Ger.* riuzan; *p.* rôz, *pl.* ruzun *flere, plangere, stridere:* cf. also ruzian, ruzon *stertere:* ruzonti *stridulus, stridens*, Grff. ii. 562.] v. reótan.

hruð, hruðer. v. hrið, hríðer.

hruxl *a noise*; strepitus, Som. v. hryscan, ge-hruxl.

hryc. v. hrycg.

hrycce. v. corn-hrycce.

hrycg, es; *m.* I. *a back of a man* or *animal*; dorsum, spina:—Hricg *dorsum*, Ælfc. Gl. 74; Som. 71, 47; Wrt. Voc. 44, 30. Hricc, Blickl. Gl. Bæc ł hricc, Ps. Spl. 17, 42. Swylce mê wǽre se hrycg forbrocen *dum configitur* [*confringitur*, Ps. Surt.] *spina*, Ps. Th. 31, 4. Hiora hrycg simle gebíeged . . . se hrycg ðæt sint ða hieremenn . . . se hrycg fæþ æfter ǽlcre wuhte *dorsum illorum semper incurva . . . qui subsequenter inhærent dorsa nominantur*, Past. 1, 4; Swt. 29, 9–14. Hêt gewríðan ðone pâpan and ðone óðerne preóst tô his hricge hindan, Homl. Th. ii. 310, 31: 416, 10. Pricaþ innan ðân scüldru[m] and on ðan hrigge swilce ðǽr þornas on sý *there are prickings in the shoulders and back as if there were thorns in*, Lchdm. iii. 120, 10. Ne hê on horses hrycge cuman wolde *non equorum dorso vectus*, Bd. 3, 5; S. 526, 28. Se cyning hæfde his hrycg him tô hlíepan *ut ipse acclinis humi regem*

super adscensurum in equum dorso ad'olleret, Ors. 6, 24; Swt. 274, 24. Đonne went hē his hrycg tō him *jam terga in ejus faciem mittit*, Past. 52, 4; Swt. 407, 8: Lchdm. iii. 242, 13. Of hry[g]um *de spinis*, Mt. Kmbl. Lind. 7, 16. Hrygas *spinæ*, 13, 7. **II.** *a ridge, rigg* [of *barley*, etc; *see* Halliw. Dict. *rig*], *high line of continuous hills, an elevated surface :*—Anlang hrycges tō đære eorþburh *along the ridge to the earthen fort*, Cod. Dipl. Kmbl. iii. 411, 21. Eal būtan ánan hrycge, 19, 4. West đonan on đone hrycg, 416, 17. Ofer đæs temples hricg *supra pinnam templi*, Lk. Skt. 4, 9. Com ic on sǽs hricg *veni in altitudinem maris*, Ps. Th. 68, 2. Ofer sǽs hrygc, Lchdm. iii. 34, 16. Sende ic ofer wæteres hrycg ealde mádmas *I sent across the water old treasures*, Beo. Th. 947; B. 471. On wæteres hricg, Salm. Kmbl. 38; Sal. 19. Ǽr đon wē tō londe geliden hæfdon ofer breóne hrycg *ere to land we came across the rough sea*, Exon. 20 b; Th. 53, 31; Cri. 859. Rídan ýđa hrycgum *to ride on the crests of the waves*, 101 b; Th. 384, 25; Rä. 4, 33. [Laym. rugge: A. R. rug: Ayenb. reg: Havel. rig: Piers P. rugge: Prompt. Parv. rygge, of a lond *porca*: Icel. hryggr *back, spine*; *a ridge*: Dan. rug: O. H. Ger. hrucki *dorsum, tergum*: Ger. rücken.] v. stán-, sund-hrycg. The word under the forms *rig, ridge* may be found in many compounds among various dialects. See E. D. S. Reprinted Glossaries, Halliwell's Dictionary, and Jamieson's Scottish Dictionary.

hrycg-bán, es; *n. Back-bone, spine :*—Hrygcbán *spina*, Ps. Lamb. 31, 4. [Rygboon, v. Halliw. Dict. under *rig : Prompt. Parv.* ryggebone of bakke (rigbone or bakbone) *spina, spondile :* Dan. ryg-ben *backbone, spine :* O. H. Ger. hrucki-beini *spina*.]

hrycg-brædan [-brǽdan ?]; *pl. The parts of the back which stand out on the right and left side :*—Smyre ābūtan đane swyran and ābūtan đa hrigbræde *smear the neck and on either side of the spine*, Lchdm. iii. 118, 24. [Cf. lenden-brædena (*gen. pl.*) and *O. H. Ger.* ruggi-bratun *palæ, sunt dorsi leva dextraque eminentia membra.* v. Grff. iii. 284–5, *where see the remark under* brat *as to the vowel.*]

hrycg-hǽr, es; *n. Hair on the back of an animal :*—Gif đū hafast mid đē wulfes hrycghǽr and tæglhǽr đa ýtemestan on sīđfate būtan fyrhtu đū đone sīđ gefremest ac se wulf sorgaþ ymbe his sīđ *if you have with you on a journey hairs from a wolf's back and from the tip of its tail, without fear you will perform the journey ; but the wolf will have trouble about his journey*, L. Med. ex Quadr. 9, 3; Lchdm. i. 360, 20.

hrycg-hrægel, es; *n. A dorsal, mantle :*—Ic geann ānes hricghrægles đæs sēlestan đe ic hæbbe *I give one dorsal the best that I have*, Chart. Th. 529, 10, where Thorpe appends this note in explanation of the word, '"manteau très riche d'ornemens, qui n'étoit porté que par les gens de haute condition." Roquefort, *voce* Dossal. A dorsal is also a wall-hanging of tapestry, used chiefly in the church at the back of the stalls.' vii setl-hrægel and iii ricghrægel and ii wahræft, 429, 28.

hrycg-mearh *the spinal marrow.* [Dan. ryg-marv *spinal marrow*.] v. next word.

hrycgmearh-liþ, es; *n. The spine :*—Hrygmergliþ *spina*, Wrt. Voc. 283, 46.

hrycg-ribb, es; *n. A rib :*—Hricgrib *spondilia*, Wrt. Voc. 65, 22. Hrycrib, 283, 49.

hrycg-rible, -riple *the parts of the back which stand out on the right and left side :*—Ricgrible *pale*, Wrt. Voc. 65, 20. Hrycriple *palæ*, 283, 45. v. hrycg-brædan.

hrycg-teúng, e; *f. A spasm in the lower part of the back :*—Hrigteúng *vel* hrifwirc *yleos*, Ælfc. Gr. 10; Som. 57, 16; Wrt. Voc. 19, 24.

hrycg-weg, es; *m. A road running along a ridge* or *elevated piece of ground :*—On đone beorh tō đem ricgwege đonne eást andlang hricgweges *on to the hill to the road that runs along it, and then east along the road*, Cod. Dipl. Kmbl. iii. 427, 33.

hrycgian *to plough into ridges ;* resulcare, Gl. Prud. 716.

hryding, e; *f. A clearing, a patch of cleared land :*—Hryding *subcisiva*, Ælfc. Gl. 57; Som. 67, 71; Wrt. Voc. 37, 57. [Cf. *O. E. Homl.* þe schal ruden þine wei *qui præparabit viam tuam : E. D. S. Cumberland Gloss.* rid, rud *to uproot trees or hedges.* 'The frequent names of Ridding and Rudding applied to houses and fields have doubtless originated from this :' *Icel.* [h]ryðja *to clear land, a road, etc.*]

hrýfing, e; *f. Roughness, scab, crust of a healing wound :*—Smire mid hunige đæt đý đē raþor sió hrýfing of fealle, L. M. 1, 35; Lchdm. ii. 86, 4.

hrygile-búc, es; *m.* [?] Of đam æscene đe is ōđre namon hrygilebúc gecleopad, Chart. Th. 439, 26. [Cf. ridgil-back *a back having a rise or ridge in the middle*, Halliwell's Dict. According to this the word might mean 'having a prominent belly' and refer to the shape of the vessel.]

hrýman, hrēman; *p.* de *To call, cry out, to cry out* [*with exultation or in lamentation, complaint*], *boast, exult, lament, murmur :*—Ne hē ne hrýmþ *neque clamabit*, Mt. Kmbl. 12, 19. Wē biddaþ ł wē hrēmaþ *imploramus*, Rtl. 121, 1. Forhuon gie hrēmas *quid ploratis*, Mk. Skt. Lind. 5, 39. Đa hrýmaþ tō hyra efengelícon *clamantes coæqualibus*, Mt. Kmbl. 11, 16. Hig hrýmaþ, tō mē and ic gehíre hira hreám *vociferabuntur. ad me et ego audiam clamorem eorum*, Ex. 22, 23. Đa hrýmde heó tō hire híwun . . . đā hē gehírde đæt ic hrímde *vocavit mulier ad se homines domus suæ* . . .

cum ego succlamassem et audisset vocem meam, Gen. 39, 14, 15. Đā hrýmde sum wōd man and cwæþ, Homl. Th. i. 458, 2. Se cæsere wēdde and hrýmde dæges and nihtes *the emperor raved day and night*, Shrn. 139, 6. Ne đý hrađor hrēmde *nor the more vaunted*, Cd. 212; Th. 263, 2; Dan. 756. Israhēla bearn hrímdon and ongeán Moisen micclum ceorodon *the children of Israel murmured against Moses*, Num. 13, 31. Gaas đætte hrēme *vadit ut ploret*, Jn. Skt. Lind. 11, 31. Gif feorrancumen man odđe frǽmde būton wege gange and hē đonne nāwđer ne hrýme ne hē horn ne blǽwe *if a man from a distance or a stranger go off the high road and then neither call out nor blow a horn*, L. Wih. 28; Th. i. 42, 24. Đā ongunnon đa hrýman đe þurh đæs dracan blǽde álēfode wǽron, Homl. Th. ii. 294, 30. Wē sceolon hrýman swīđor and swīđor tō đam Hǽlende, i. 156, 22. Đā begann hē tō hrýmenne and cwæþ, 152, 15. Mid fleáme com on his cyþþe Constontinus hrēman ne þorfte *by flight Constantine got home, had little cause to boast*, Chr. 937; Erl. 114, 5; Æđelst. 39. Hrēmende *ululatus*, Mt. Kmbl. Lind. 2, 18 : *plorantem*, Jn. Skt. Lind. 11, 33. Mid micelre stemne hrýmende *crying with a loud voice*, Homl. Th. i. 46, 33. [*Laym. A. R.* remen : *Halliw. Dict.* reem, reme.] v. hreám, hrēmig.

hrýme *soot ;* fuligo, Cot. 83, Lye. v. hrúm.

hrýmpelle. v. rimpel.

hryre, es; *m. Fall, downfall, ruin, destruction, perdition, decay, decline, death :*—Hyrre *casus*, Ælfc. Gr. 11; Som. 15, 10 : *ruina*, Ps. Spl. 105, 28. His hryre wæs micel *fuit ruina ejus magna*, Mt. Kmbl. 7, 27. Hægles hryre *fall of hail*, Exon. 56 a; Th. 198, 26; Ph. 16. Đæt đæs folces sceolde micel hryre beón *that there should be a great destruction among the Romans*, Ors. 4, 1; Bos. 77, 45. Líces hryre *the fall of the body* [*death*], Exon. 48 b; Th. 167, 26; Gū. 1066 : 65 a; Th. 240, 27; Ph. 645 : Andr. Kmbl. 457; An. 229. Đǽr him næs ne lífes lyre ne líces hryre *there was for him* [*Adam*] *no loss of life, no bodily decay*, Exon. 44 b; Th. 151, 27; Gū. 801. Yfle preóstas bióþ folces hryre *laqueus ruinæ populi mei sacerdotes mali*, Past. 2, 1; Swt. 31, 9. Đætte hie đone spild đæs hryres him ondrǽden *ut præcipitem ruinam metuant*, 52, 5; Swt. 407, 21. Gif wē æfter đæm hryre ūrre scylda tō him gecierdon *nobis post lapsum redeuntibus*, 52, 3; Swt. 405, 16. Betwux đæra stána hryre betǽhte hē his fýnd Gode *whilst the stones were falling he commended his foes to God*, Homl. Th. i. 50, 23. Đis cild is gesett manegum mannum tō hryre *positus est in ruinam multorum*, 144, 18 : Bt. Met. Fox 9, 8; Met. 9, 4. Đa twā forman gesceapennyssa feóllon on hryre and seó þridde wæs on hryre ácenned, Homl. Th. ii. 8, 31. Ne fægnode ic on mínes feóndes hryre, 448, 22. On myclum hryre seó heord wearþ on sǽ besceofen *magno impetu grex præcipitatus est in mare*, Mk. Skt. 5, 13. Đone hryre đe se feallenda deófol on engla werode gewanode *the loss which the falling devil had caused in the host of angels*, Homl. Th. i. 32, 23, 28. Hordwearda hryre, Cd. 169; Th. 210, 6; Exod. 511: Exon. 76 b; Th. 287, 1; Wand. 7. Ne timbreþ hē nō healle ac hryre *non habitaculum sed ruina fabricatur*, Past. 49, 3; Swt. 383, 33. Mid gelōmlǽcendum hryrum *by frequent destructions*, Homl. Th. i. 578, 34. Hē gefylde hryras *implebit ruinas*, Ps. Spl. 109, 7. Hwilce hryras *quantas ruinas*, Bt. 16, 4; Fox 58, 1. v. leód-, líc-, wíg-hryre; *and cf.* dryre.

hryre ; *adj. Falling, decaying, perishing :*—Sōđlíce mid đisum wordum is geswuteloð đæt đises middangeardes wæstm is hryre. Tō đam hē wext đæt hē fealle *verily by these words is manifested that the fruit of this world is decaying* [or *a ruin* (?) *v. preceding word*]. *It grows that it may fall*, Homl. Th. i. 614, 8. [Cf. *for a similar relation in form between adj. and verb O. Sax.* luggi ; *adj. and* liogan.]

hrýred-ness, hrýre-mūs, hrýre-ness. v. hrēred-ness, hrēre-mus, hrēr-ness.

hrysc, hrysca *irruptio*, Som.

hryscan *to make a noise :*—Hriscan *stridere*, Hpt. Gl. 494. Hristenda [hriscende ?] *astridulus, stridulus*, Lye. v. hruxl.

hrysian. v. hrisian.

hrystan. v. hyrstan.

hrýtan ; *p.* te *To scatter :* — Se đe hrēt *qui sternit*, Prov. 10, Lye. [*Icel.* hreyta *to spread, scatter*.]

hryte or **hrýte ;** *adj. Balidinus*, Ælfc. Gl. 79 ; Som. 72, 94; Wrt. Voc. 46, 41. *The word occurs in a list of names of colours, but the meaning is uncertain. Ducange has* 'balidinus forte legendum badius vel balius nostris bay, bayard.' v. hrut.

hrýđer. v. hríđer.

hrýđig ; *adj. Dismantled ?* [cf. *Icel.* hrjóđa *to strip, clear*] or *tottering ?* [cf. hriđian], Exon. 77 b; Th. 291, 5; Wand. 77.

hryđđa. v. ryđđa.

hrýw-líc, hrýwsian. v. hreów-líc, hreówsian.

HÚ ; *adv. How.* **I.** *in direct questions :*—Hū mæg man ingán on stranges hūs *quomodo potest quisquam intrare in domum fortis ?* Mt. Kmbl. 12, 29 : 34. Hū ne synt gē sēlran đonne hig *nonne vos magis plures estis illis ?* 6, 26 : 25. Hū sculon wit nū libban *how are we to live ?* Cd. 38 ; Th. 50, 7; Gen. 805. **II.** *in exclamations* [*see also* **I**] :—Hū la ! ne gewearþ unc tō ánum peninge *how now ! was not our agreement for a penny ?* Th. An. 74, 20. Hū gōd is ēce God *quam bonus*

Deus, Ps. Th. 72, 1. Eálá gǽsta god hú ðú mid noman ryhte nemned wǽre emmanuhel *oh ! God of spirits, how rightly wast thou named by the name of Emmanuel !* Exon. 9 b; Th. 9, 6; Cri. 130: 11 a; Th. 14, 8; Cri. 216. Eálá on hú grimmum and on hú grundleásum seáðe swinceþ ðæt sweorcende mód, Bt. Met. Fox 3, 1, 2; Met. 3, 1. **III.** *in dependent clauses with indic. or subjunct :* — Nú wundraþ gehwá hú se deófol dorste geneálǽcan tó ðam Hǽlende *now every one will wonder how the devil durst come near Jesus*, Homl. Th. i. 166, 32. Wé gehírdon hú gé ofslógon twegen cynegas Seon and Og *audivimus quod interfecistis Sehon et Og*, Jos. 2, 10. Hí gehírdon hú seó hálige sprǽc, Judth. 11; Thw. 23, 37; Jud. 160. Wé gesáwon hú hé wæs on heofenas ástígende, Nicod. 18; Thw. 8, 39. Ús secgaþ béc hú ástág in middangeard bearn godes, Exon. 19 a; Th. 49, 15; Cri. 786. Ðá angan Thomas his spǽce hú hé com tó Cantuuarebyri and hú se arcebiscop áxode hýrsumnesse at him *then Thomas began his speech, how he had come to Canterbury, and how the archbishop had demanded obedience from him*, Chr. 1070; Erl. 208, 14. Ðá áxode se cásere ðone ǽnne preóst hú his nama wǽre oððe hú gefyrn hé gelýfde, Homl. Th. ii. 310, 15. Ðá wearþ ðæt mǽden hohful hú heó ǽfre wæras wissian sceolde, Blickl. Swt. 2, 122. Gefada embe hú ðú wylle *dispose of it how thou wilt*, 3, 285. Hycgaþ his ealle hú gé hí beswícen *consider of it all, how ye may entrap them*, Cd. 22; Th. 28, 9; Gen. 433. Ábídan sceal miclan dómes hú him metod scrífan wille *must abide the great doom, how the Lord will adjudge to him*, Beo. Th. 1962; B. 979. **IV.** *with a comparative* [cf. þý, swá] :— Lufade hine lenge hú geornor, Exon. 34 b; Th. 110, 18; Gú. 109. **V.** *qualifying, or in combination with, other words :* — Hú mycel scealt ðú *quantum debes?* Lk. Skt. 16, 5. Hú mycel gód is on gehýrsumnesse and hú mycel yfel on ungehýrsumnysse, Boutr. Scrd. 19, 26. On ðyssere dǽde is geswutelod hú micclum fremige ðǽre sóðan lufe gebed, Homl. Th. i. 50, 35. Hú micele swíðor *how much more?* 68, 24. On hú manegum wísum is Godes weorc? Boutr. Scrd. 18, 14. Hú fela se hǽlend him dyde *quanta sibi fecisset ihesus*, Mk. Skt. 5, 20. Hú fela sagena hig ongén ðé secgeaþ *quanta adversum te dicant testimonia*, Mt. Kmbl. 27, 13. Hú lange forbere ic eów *usque quo patiar vos?* 17, 17. Hú long tíd *quantum temporis*, Mk. Skt. 9, 21. Be gebróðrum hú gesibbe wíf hig habban móton *de fratribus quam prope cognatas uxores habere possint*, L. Ecg. C; Th. ii. 130, 8: 13. Hú héh and deóp hell seó, Cd. 228; Th. 309, 9; Sat. 707. Witan hú ðú ǽðele eart, Hy. 3, 14; Hy. Grn. ii. 281, 14. Mé com swíðe oft on gemynd hú gesǽliglíce tída wǽron giond Angelcynn *it has often come into my mind what happy times there were in England*, Past. Pref. Swt. 3, 4. Ðæt se lǽreów ðe him tela tǽce him sylf elles hú dó *that the teacher who teaches him well, himself act otherwise*, L. E. I. 21; Th. ii. 418, 4. Ne meg nú hú ælles beón it *cannot be otherwise*, Shrn. 195, 7. Hú geáres *according to the time of year*, L. M. 2, 34; Lchdm. ii. 238, 22. Swá hú swá hit gewurde *however it may have happened*, Homl. Th. i. 588, 29. Hí habbaþ æt Gode swá hú swá hí geearniaþ *they will have from God, in accordance with whatever they merit*, ii. 326, 30. [*Laym. Orm.* hu: *A. R.* hwu, hu: *Ayenb.* hou: *Goth.* hwê: *O. Frs.* hu, ho: *O. Sax.* hwô: *O. H. Ger.* hweó, v. Grff. iv. 1193: *Ger.* wie.] v. ge-hú; hú-meta, -hwega.

hú-. v. hw-.

hucs. v. husc.

húdenian *in the following passage :* —Húdenige ǽrest hine selfne, óþ hé wacige and áhrisige siððan óðre tó geornfulnesse gódra weorca *prius se per sublimia facta excutiant, et tunc ad bene vivendum alios sollicitos reddant*, Past. 64; Swt. 461, 16. [Sweet, in the note on this passage, suggests that the word may be from the same root as *quatio*, adding that Prof. Skeat compares the Scotch *houd* to shake. May not the word however be used from a misconception of the Latin word, by which *excutere* is considered as connected with *cutis = hýd?*]

húf, es; *m. Part of the mouth* or *upper part of the throat, a tumour affecting that part :*—Húf *sublinguium*, Ælfc. Gl. 71; Som. 70, 98; Wrt. Voc. 43, 28. Ad úfam. Ðes lǽcecræft deáh wyð ðone húf *ad uvam. This medicine is good for tumour on the epiglottis*, Lchdm. iii. 106, 6. Of ðan úve droppaþ uppan ða tunga, 138, 28.

húf, es; *m. A horned owl ;* bubo, Wrt. Voc. 63, 19. [The word occurs both in English and *O. H. Ger.* with and without initial *h*, húf, úf; húvo, úvo *bubo*.] v. úf.

húfe, an; *f. A covering for the head :*—Húfe *cidaris vel mitra*, Ælfc. Gl. 64; Som. 69, 11; Wrt. Vcc. 40, 45. Biscopes húf *flammeolum*, vel *flammeum*, 112; Som. 79, 88; Wrt. Voc. 59, 55. Húfan hættes *mitræ*, Lye. [*Chauc. Piers P.* houve : *Prompt. Parv.* howe, heed hyllynge *tena, capedulum, sidaris;* and see the note, p. 249 : *Scot.* how a *coif, hood :* *Icel.* húfa *a hood, cap, bonnet :* *O. H. Ger.* húba *mitra, thyara :* *Ger.* haube.]

húfian ; *p.* ode *To put on a hufe :*—Hé his suna húfode swá drihten bebeád *he put bonnets upon them, as the Lord commanded ;* imposuit mitras ut jusserat dominus, Lev. 8, 13.

Hugas ; *n. pl. The name of a people in the neighbourhood of West Friesland*, Beo. Th. 4998; B. 2502: 5820; B. 2914.

hugu. v. hwega.

hú-hwega, -hugu; *adv. About, somewhere about :*—Húhugu ymb ða teóþan tíd dæges *hora circiter decima diei*, Bd. 3, 27; S. 558, 12. Húhugu syx hund hída *familiarum circiter sexcentarum*, 4, 19; S. 590, 3. Húhwega ymb iii niht *somewhere about three days*, L. M. 2, 59; Lchdm. ii. 280, 16. Húhwego fíf hund manna, Blickl. Homl. 201, 14.

húilpa, an; *m. The name of a bird so called from its note* [cf. *Ger.* uhu *owl*] ?— Dyde ic mé tó gomene ganetes hleóþor and huilpan swég, Exon. 81 b; Th. 307, 9; Seef. 21.

hulc, es; *m.* [?] *A light ship, a hulk* [but in later times the word is applied to a heavy ship of clumsy make] ; liburna, Ælfc. Gl. 103; Som. 77, 104; Wrt. Voc. 56, 23. Si adveniat ceol vel hulcus, L. Eth. iv. 2; Th. i. 300, 9. [*Prompt. Parv.* hulke, shyppe *hulcus*, and see the note, p. 252 : *O. Du.* hulke *navis oneraria :* *O. H. Ger.* holcho *actuaria navis*.]

hulc, es; *m. A hut, hovel, cabin :*—Hulc *tugurium*, Ælfc. Gr. 8; Som. 7, 62 : Ælfc. Gl. 108; Som. 78, 116; Wrt. Voc. 58, 30 : 85, 74. Gyf hé his scip uppe getogen hæbbe oððon hulc geworhtne oððon geteld geslagen ðæt hé ðǽr friþ hæbbe and ealle his ǽhta *if he have drawn his ship ashore or have built a hut or pitched a tent, let him and all his property be unmolested*, L. Eth. i. 3; Th. i. 286, 9. Hé wolde geneálǽcan his hulce *he [the leper] wanted to reach his hut*, Homl. Th. i. 336, 10. On wáclícum screafum oððe hulcum lútigende *lurking in miserable dens or hovels*, 544, 30. [*Wick.* hulke, Is. 1, 8.]

hulfestre, an; *f. A plover ;* pluvialis [the word occurs in a list of names of birds], Ælfc. Gl. 38; Som. 63, 24; Wrt. Voc. 29, 44.

hulfstan *ciupella*, Wrt. Voc. 63, 24.

hú-líc, *pron. Of what sort ;* qualis :—Hé áhsode hwæt alexander se cyning dyde and húlíc mon hé wǽre and in hwylcere yldo *he asked what king Alexander was doing, and what sort of man he was, and of what age*, Nar. 1, 12. Nú ic wille secgan húlucu heó wæs *I will tell you what it [Carthage] was like*, Ors. 4, 13; Bos. 99, 57. Húlíc is ðes *qualis est hic?* Mt. Kmbl. Rush. 8, 27. Húlíc is se organ tó begonganne, Salm. Kmbl. 107; Sal. 53. Húlig, Lk. Skt. Lind. 1, 29. Gisih húlíce [húlco, Lind.] stánas and húlíc [huulig, Lind.] timber *aspice quales lapides et quales structuræ*, Mk. Skt. Rush. 13, 1. v. hwilc.

hulu, e; *f. A hull, husk :*—Hnute hula *culliole*, Ælfc. Gl. 31; Som. 61, 105; Wrt. Voc. 27, 34: Gl. Prud. 156: Hpt. Gl. 439. [*Prompt. Parv.* hoole or huske *siliqua ;* hoole of pesyn or benys or oðer coddyd frute *techa ;* see note, p. 242 : *Scot.* hule *a husk :* cf. *O. H. Ger.* hulsa *siliqua :* *Ger.* hülse.]

Humbre, an; or *indecl. f. The Humber :*—Óþ gemǽro Humbre [streámes] *ad confinium usque Humbræ fluminis*, Bd. 1, 25; S. 486, 17. Óþ Humbre stréam *Humbræ fluvio*, 2, 5; S. 506, 11. Behionan Humbre . . . begiondan Humbre, Past. Pref; Swt. 3, 14, 16. Be súþan Humbre, Chr. 827; Erl. 62, 33. Ofer Humbre múþan, 867; Erl. 72, 6. Humbra [MS. B. Humbran] eá, 942; Erl. 116, 10. Tó Humbran múþan, 993; Erl. 132, 12. Com Tostig eorl intó Humbran mid lx scipum, 1066; Erl. 201, 6.

hú-meta; *adv. How, in what manner ;* quomodo :—Húmeta eodest ðú in *quomodo intrasti?* Mt. Kmbl. 22, 12. Húmeta bitst ðú æt mé drincan *quomodo bibere a me poscis?* Jn. Skt. 4, 9. Húmeta bodaþ hé [Paul] Cristes geleáfan? Homl. Th. i. 388, 22. Nú is tó besceáwigenne húmeta se ælmihtiga God geþafaþ ðæt . . . *now it is to be considered how it is that the almighty God permits that . . .*, 486, 17. Ðú sǽdest ðæt ðú ne mihte witan húmeta hé his weólde oððe hú hé his weólde *you said that you could not see in what manner or by what means he governed it [the world] ;* quibus gubernaculis mundus regatur, Bt. 35, 2; Fox 156, 25.

hun [hún?], e; *f. Impurity* [?] ; tabes, Cot. 192. v. hunel.

Húnas and **Húne ;** *pl. The Huns :*—Húne *Hunni*, Bd. 5, 9; S. 622, 15. Húnas, Elen. Kmbl. 42; El. 21. Húna cyning, 64; El. 32: Chr. 443; Erl. 10, 22. Ætla weóld Húnum, Exon. 85 a; Th. 319, 26; Vid. 18: 85 b; Th. 322, 2; Víd. 57. [*Icel.* Húnar : *M. H. Ger.* Hiune.] v. Grmm. D. M. 489-91.

HUND, es; *m. A* HOUND, *dog; applied to persons as a term of abuse in English and in other dialects :*—Ðá hé ðider com ðá sceolde cuman ðǽre helle hund ongeán hine ðæs nama wæs Ceruerus *when he came thither, it is said, that then the dog of hell, whose name was Cerberus, came towards him*, Bt. 35, 6; Fox 168, 15. Wið hundes slite *for the bite of a dog*, Herb. 177, 2; Lchdm. i. 310, 8. Of ðæs hundes handa *de manu canis*, Ps. Th. 21, 18. Ðone hǽðenan hund *the heathen dog* [Holofernes], Judth. 10; Thw. 23, 7; Jud. 110. Swá hundas *ut canes*, Ps. Th. 58, 6. Dumbe hundas *canes muti*, Past. 5, 1; Swt. 89, 17. Hunda gebeorc *barking of dogs*, Ælfc. Gr. 1; Som. 2, 35. Nys hit ná gód ðæt man nime bearna hláf and hundum worpe *non est bonum sumere panem filiorum et mittere canibus*, Mt. Kmbl. 15, 26. [*Goth.* hunds : *O. Sax. O. Frs.* hund : *Icel.* hundr : *O. H. Ger.* hunt : *Ger.* hund.] DER. heáh-deór-, helle-, hroð-, wéde-hund.

hundes beó *a dog-fly,* Cot. 54, Lye.

hundes cwelcan *berries of the wayfaring tree ;* baccæ de viburno opulo, colocinthidæ, Lchdm. iii. 333, col. 2.

hundes fleóge *a dog-fly :* — Hundes fleóge *cinomia*, Ælfc. Gl. 21;

Som. 59, 119; Wrt. Voc, 23, 37.   Hundes fleógan *muscam caninam*, Ps. Th. 77, 45: Ors. 1, 7; Swt. 38, 1.   [*Wick.* hound-fleȝe: *O. H. Ger.* hunt-, huntes-fliuge *cynomia, musca canina*.]

hundes heáfod *snapdragon*, Lchdm. ii. 395, col. 2.

hundes lús *a dog-fly*; *cinomia*, Wrt. Voc. 77, 54.   [Cf. *Ger.* hunds-laus.]

hundes micge *cynoglossum officinale*, Lchdm. ii. 333, col. 2.

hundes tunge *hound's tongue*; *cynoglossum officinale*, Lchdm. ii. 333, col. 2.   [*O. H. Ger.* huntes-zunga *cynoglossa*.]   v. E. D. S. Plant Names, hounds-tongue.

hundes wyrm *a dog-worm*; *ricinus*, Ælfc. Gl. 24; Som. 60, 33; Wrt. Voc. 24, 33.

HUND; *n. A* HUNDRED; *centum* :—Gyf hwylc mann hæfþ hund sceápa *si fuerint alicui centum oves*, Mt. Kmbl. 18, 12.   Hund sestra . . . hund mittena hwǽtes, Lk. Skt. 16, 6, 7.   Senatum ðæt wæs án hund manna ðéh heora æfter fyrste wǽre þreó hund, Ors. 2, 4; Swt. 70, 36.   Mid án hund scipa, Bt. Met. Fox 26, 30; Met. 26, 15.   Sum hund scipa *some hundred ships*, Chr. 894; Erl. 91, 5.   Ðæt flód stód ða swá án hund daga and fíftig daga *obtinuerunt aquæ terram centum quinquaginta diebus*, Gen. 7, 24.   Æfter óðer healf hund daga *post centum quinquaginta dies*, 8, 3.   Mid penningum twǽm hundum *denariis ducentis*, Mk. Skt. Lind. 6, 37.   Ðǽr wǽron twá hund and eahta and feówertig wera, Blickl. Homl. 239, 14.   Mid ccl hunde [þridde healf hund, MS. E.] scipa, Chr. 893; Erl. 88, 25.   Ðá geceás Gedeon þreó hund manna, Jud. 7, 6.   Þreó hund manna and eahtatýne men, Gen. 14, 14.   Geseald tó þrim hunde penega *sold for three hundred pence*, Blickl. Homl. 69, 8: 75, 22.   Þriim hundum peninga, Jn. Skt. Lind. 12, 5.   Feówer hund geára, Gen. 15, 13.   Ðá ða hé wæs fíf hund geára, 5, 32.   Nigon hund wintra and lxxi, Blickl. Homl. 119, 2.   Hira monig hund ofslógon *slew many hundreds of them*, Chr. 895; Erl. 93, 28.   Hund síðon on dæge *a hundred times a day*, Homl. Th. i. 456, 21.   [*Goth.* hund: *Ó. Sax.* hund: *O. H. Ger.* hunt. This word is the representative of a fuller form which is seen in Gothic as *taihun-téhund*, *-taihund* [Lk. 15, 4: 16, 6], and which points to a primitive *dakan-dakanta* = ten-tenth = hundred.   The Latin *centum* shews a similar modification.]   v. next word.

hund- as a prefix to numerals from 70 to 120 is a shortened form of the word which appears in Gothic as *téhund*, *taihund* [v. preceding word], and may be explained *decade*.   *O. Sax.* prefixes *ant* [= *hund*?], in *O. Frs.* the prefix is *t*, and a trace of such forms is yet left in the Modern Dutch *t-achtig* = 80.   On these numerals March remarks 'Gothic has *sibun-téhund*.   The Anglo-Saxon form was once *hund-seofonta* [decade seventh], like *O. Sax.* ant-sibunta.   The -*ta* changed to -*tig* through conformation with the smaller numbers, and *hund-*, whose meaning had faded, was retained as a sign of the second half of the great hundred.' Grammar, p. 75.   See also Helfenstein's Comparative Grammar, p. 229.   For the great hundred [120] cf. Icel. *tólfrætt hundrað* as distinguished from *tírætt hundrað*.   See Cl. and Vig. Dict. hundrað.

hund-eahtatig; *num. Eighty*:—Hundeahtatig *octoginta*, Ælfc. Gr. 49; Som. 49, 44.   Heó wæs wudewe óþ feówer and hundeahtatig geára *hæc vidua usque annos octoginta quatuor*, Lk. Skt. 2, 37.   Mid hund-ehtatigum scipum, Chr. Erl. 5, 2.   Ær ðæm ðe Rómeburg getimbred wǽre iiii hunde wintrum and hundeahtatigum *anno ante urbem conditam* ccclxxx, Ors. 1, 10; Swt. 44, 4.

hundeahtatig-wintre; *adj. Eighty years old* :—Hundeahtatigwintre and sixwintre wæs Abram ðá ða Ager ácende Ysmael, Gen. 16, 16.

hunden; *adj. Of a dog, canine* :— Hundene *caninam*, Blickl. Gloss.   [*O. H. Ger.* huntin *caninus*.]

hund-endlefontig; *num. One hundred and ten* :—Feówer and hund-ǽndlæftig ealdra swína *one hundred and fourteen old swine*, Chart. Th. 163, 3.

hund-endleftigoða; *num. One hundred and tenth* :—On ðæm eahta and hundælleftiigoðan psalme *in the hundred and eighteenth psalm*, Past. 65, 5; Swt. 465, 23.

hundes beó, etc.   *See above after* hund.

hund-feald; *adj. Hundredfold* :—Hundfeald getel is fulfremed *the number a hundred is perfect*, Homl. Th. i. 338, 27.   Swá hwæt swá wé be ánfealdan Godes þearfum syllaþ hé hit ús forgylt be hundfealdum, ii. 106, 2.   Mid hundfealdum, i. 180, 26.   Sealdon wǽstm sum hundfealdne *dabant fructum aliud centesimum*, Mt. Kmbl. Ms. A. 13, 8.

hund-líc; *adj. Doglike, canine* :—Hundlíce [tēþ] *canini*, Wrt. Voc. 282, 74.   Nú sende hé hundas tó mé forðan ðe hé næfþ godcundlíce englas, ac hæfþ hundlíce *now has he sent dogs to me, for he has not divine angels, but he has doglike ones*, Homl. Th. i. 378, 3.

hund-nigontig; *num. Ninety* :—Hundnigontig *nonaginta*, Ælfc. Gr. 49; Som. 49, 44.   Se sumor hafaþ hundnygontig daga . . . Se winter hæfaþ tú and hundnigontig daga, Shrn. 83, 33; 146, 7.   Hundteóntig geára wæs Abraham and his gebedda hundnigontig *Abraham was a hundred years old and his consort ninety*, Homl. Th. i. 92, 21.   Nigon and hundnigontig *nonaginta novem*, Lk. Skt. 15, 4.   Mid þrim and hund-nigentigon scipum, Chr. 993; Erl. 132, 2.   Feówer hund geára and hundnigontig geára, Swt. A. S. Rdr. 71, 459.

hundnigontig-wintre; *adj. Ninety years old*, Gen. 17, 17.

hundred; *pl.* u; *n. A hundred* :—Getalu *vel* heápas *vel* hundredu *centurias*, Ælfc. Gl. 96; Som. 76, 25; Wrt. Voc. 53, 34.   Ðeáh ðe heora hundred seó *though there be a hundred of them*, Ps. Th. 89, 10.   On lxv and þreó hundræd hí beóþ tódǽlede *they are divided into three hundred and sixty-five*, Nar. 49, 25.   Seox hundred wintra and iii and hundseofenti wintra, Chr. 656; Erl. 33, 34.   Hundrað scillinga *centum denarios*, Mt. Kmbl. Lind. 18, 28.   On twegera hundred penega wurþe, Jn. Skt. 6, 7.   Wið þrim hundred penegon, 12, 5.   Mid twám hundred penegon, Mk. Skt. 6, 40.   Hí ðá sǽton hundredon and fíftigon *discubuerunt per centenos et per quinquagenos*, 37.   [*O. Frs.* hundred, hunderd: *Icel.* hundrað: *O.H. Ger.* hundert: *Ger.* hundert.   Two etymologies are suggested for the word; according to one *hunder-* corresponds to *Lat. centur-ia*; according to the other *-red* (*Icel.* rað) is a suffix akin to the *-rædr* which is found in *Icel.* átt-rædr, etc.   v. Grmm. Gesch. D. S. 175–6.]

hundred, es; *n. A hundred, a territorial division, the assembly of the men in such a division* :—Hú mon ðæt hundred haldan sceal.   Ǽrest ðæt hí heó gegaderian á ymb feówer wucan and wyrce ǽlc man óðrum riht *how the [assembly of the] hundred is to be held.   First, they [the men of the hundred] are to assemble themselves every four weeks; and each man is to do justice to other*, L. Edg. H; Th. i. 258, 2–4, and see the whole section.   Fó se hláford tó healfan and tó healfan ðæt hundred *let the lord take half, and the hundred half*, L. Edg. 2, 7; Th. i. 268, 20.   Gewitnys sý geset tó ǽlcere byrig and tó ǽlcum hundrode, L. Edg. S. 3; Th. i. 274, 8, 10.   Twegen þegenas innan ðam hundrede, L. Eth. i. 1; Th. i. 280, 11: L. C. S. 17; Th. i. 384, 30: 19; Th. i. 386, 12.   [Various explanations of the word have been given.   'It has been regarded as denoting simply a division of a hundred hides of land; as the district which furnished a hundred warriors to the host; as representing the original settlement of the hundred warriors; or as composed of a hundred hides, each of which furnished a single warrior,' Stubbs' Const. Hist. I, 97; see also following pages and pp. 71–3: Grmm. R. A. 532 sqq: Kemble's Saxons in England, c. ix: Schmid A. S. Gesetz. p. 613–4.]

hundredes ealdor, es; *m.*   I. *a centurion* :—Ðá geneáhlǽhte hym án hundredes ealdor *accessit ad eum Centurio*, Mt. Kmbl. 8, 5.   II. *the presiding officer of the court of the hundred* :—Gif se hundredes ealdor ðæt geáscoþ, L. Edg. S. 10; Th. i. 276, 8.   Cýðan hit ðæs túnes men ðam hundredes ealdre, 8; Th. i. 274, 28.

hundredes man *apparently the same as preceding word*, II : —Cýðe hit man ðam hundredes men, L. Edg. H. 2; Th. i. 258, 7.   v. hundred-mann.

hundred-gemót, hundredes gemót, es; *n. The assembly of the hundred* [v. hundred] :—Séce man hundredgemót swá hit ǽr geset wæs and ðǽr beó on scirebisceop and se ealdorman *let the hundredmoot be attended as was before appointed; and let the bishop of the shire and the alderman be there present*, L. Edg. ii. 5; Th. i. 268, 2–5.   Séce man hundredes gemót be wíte *let the hundredmoot be attended under penalty of a fine*, L. C. S. 17; Th. i. 386, 1.

hundred-mann, es; *m. The chief of a hundred men, a centurion* :—Ðá clypode hé ðæne hundredman *accersito centurione*, Mk. Skt. 15, 44.   Sette hig tó ealdrum and tó hundredmannum and tó fíftigesmannum and tó teóðingmannum *constitui eos principes, tribunos et centuriones et quinquagenarios et decanos*, Deut. 1, 15.   Þúsendmen and hundrydmen and fíftiesmen and teóðingmen *tribunos et quinquagenarios et decanos*, Ex. 18, 21.   [Cf. *O. H. Ger.* hunteri *centurio*.]

hundred-penig, es; *m.* 'A collection made for the support of his office by the sheriff or lord of the hundred :'—Hundredpenegas, Chart. Th. 432, 25: 433, 29.   v. Glossary.

hund-seofontig; *num. Seventy* :—Hundseofontig *septuaginta*, Ælfc. Gr. 49; Som. 49, 43.   Ealles hundseofontig manna *seventy men in all*, Homl. Th. ii. 190, 30.   His suna gestríndon twá and hundseofontig suna *his sons begot seventy-two sons*, Swt. A. S. Rdr. 61, 154.   Ne secge ic ðé óþ seofon síðas, ac óþ seofon hundseofontigon síðon *non dico tibi usque septies, sed usque septuagies septies*, Mt. Kmbl. 18, 22.

hundseofontig-feald; *adj. Seventy-fold* :—Septuagesima is hundseofontigfeald getel, Homl. Th. ii. 84, 28: 86, 2.

hundseofontig-wintre; *adj. Seventy years old* :—Ðá hé wæs seofonhundwintre and seofon hundseofontigwintre, Gen. 5, 31.

hund-teóntig; *num. A hundred* :—Hundteóntig *centum*, Ælfc. Gr. 49; Som. 49, 44.   Hundteóntig geára wæs Abraham *Abraham was a hundred years old*, Homl. Th. i. 92, 20.   Joseph leofode hundteóntig geára and tín tó eácan *Joseph lived a hundred and ten years*, Swt. A. S. Rdr. 63, 208.   Hundteóntig and twentig *a hundred and twenty*, Shr. 85, 12.   Hundteóntig and þreó and fíftig, Jn. Skt. 21, 11.   Fæder Abrahames wintra hæfde twá hundteóntig and fífe eác *and the days of Terah were two hundred and five years*, Cd. 83; Th. 104, 26; Gen. 1741.

hundteóntig-feald; *adj. Hundredfold* :—Tó hundteóntigfealdre méde, Blickl. Homl. 41, 19.

hundteóntigfeald-líc; *adj. Hundredfold* :—Ðæt hé on ðyssum lífe hundteóntigfealdlíce méde onfénge *ut in hac vita centuplum acciperet*, Bd. 5, 19; S. 636, 36.

hundteóntig-geáre ; adj. Aged a hundred :—Adam leofode hund-teóntigeáre and þrittegeáre, Gen. 5, 3.

hund-twelftig ; num. A hundred and twenty :—Hundtwelftig geára wæs Moses ðá ðá hé gewât Moyses centum et viginti annorum erat, quando mortuus est, Deut. 34, 7 : Cd. 64 ; Th. 76, 26 ; Gen. 1263. Se wudu is eástlang and westlang hundtwelftiges míla lang oððe lengra from east to west the wood is a hundred and twenty miles long, or longer, Chr. 893 : Erl. 88, 28.

hund-twentig ; num. A hundred and twenty :—Mid ðam ðe hé wæs on ylde hundtwentig wintra when he was a hundred and twenty years of age, Ælfc. T. Grn. 6, 1. Hé gean ðæra hundtwæntiga hída æt Wyrðæ he gives the hundred and twenty hides at Worth, Chart. Th. 526, 32.

hundtwentig-wintre ; adj. A hundred and twenty years old :—Ic eom tô-dæg hundtwentigwintre centum viginti annorum sum hodie, Deut. 31, 2.

hund-wealh, es ; m. A servant to attend to dogs :—Hundwæalh canum servitor, Ælfc. Gl. 8 ; Som. 56, 110 ; Wrt. Voc. 18, 58.

hund-wintre ; adj. A hundred years old :—Hé ;ylf wæs ðá hundwintre cum centum esset annorum, Gen. 21, 5. Wênst ðú ðæt sunu beó ácenned of hundwintrum men putasne centenario nascetur filius ? 17, 17.

hune, an ; f. Horehound ; marrubium vulgare :—Hunan seáw juice of horehound, L. M. 1, 3 ; Lchdm. ii. 42, 19. Nim hunan take horehound, 31 ; Lchdm. ii. 74, 8. Wyll ða háran hunan boil the horehound, Lchdm. iii. 48, 14. v. hár-hune.

Húne. v. Húnas.

hunel ; adj. Foul, wanton, impudent ; procax, protervus, immodestus, impudicus, Lye. v. hun.

HUNGOR, es ; m. HUNGER, famine :—Nis ðær hungor ne þurst slǽp ne swár leger ne sunnan bryne there is there neither hunger nor thirst, sleep nor grievous sickness, nor burning heat of the sun, Exon. 32 a ; Th. 101, 20 ; Cri. 1661. Beóþ ðé hungor and þurst hearde gewinnan, 36 b ; Th. 118, 27 ; Gú. 246. Hæfde hí hungor and þurst esurientes et sitientes, Ps. Th. 106, 4. Hér wæs se micla hungor on Angelcynne in this year was the great famine in England, Chr. 976 ; Erl. 127, 34. Hér on ðyssum geáre wæs se mycla hungor geond Angelcynn swilce nán man ǽr ne gemunde swá grimme, 1005 ; Erl. 139, 36. Hungor se hâta ne se hearda þurst, Exon. 64 b ; Th. 238, 32 ; Ph. 613. Se grimma hungor ne se hâta þurst, 112 a ; Th. 430, 5 ; Rä. 44, 3. Hunger se hearda hámsittendum wælgrim werum, Cd. 86 ; Th. 108, 32 ; Gen. 1815. Hungres on wênum blátes beódgæstes in expectation of hunger, pallid guest at the board, Andr. Kmbl. 2176 ; An. 1089. Hungre wæron þearle geþreátod swá se þeódsceaða hreów rícsode, 2230 ; An. 1116. Lǽtaþ cuelan hungre Cristes þearfan cum fame crucientur Christi pauperes, Past. 44, 6 ; Swt. 327, 6. Ic on hungre forwurðe fame pereo, Lk. 15, 17. Hungre ácwelan to die of hunger, Chr. 894 ; Erl. 92, 28 : 918 ; Erl. 104, 13. Hungre heófeþ wulf se grǽga the grey wolf howls for hunger, Exon. 91 b ; Th. 342, 30 ; Gn. Ex. 150. Hungur heaðugrimne heardne, Ps. Th. 145, 6. Mann-cwealman and hungras pestilentiæ et fames, Mt. Kmbl. 24, 7. [Goth. huhrus : O. Sax. hungor : O. Frs. hunger, honger : Icel. hungr : O. H. Ger. hungar fames : Ger. hunger.]

hungor-biten ; adj. Hunger-bitten, suffering from hunger :—Ac ðes folces ðe be Hungire fôr fela þúsenda ðær and be wæge earmlíce forfôran and fela hreówlíce and hungerbitene ongeán winter hám tugon but of the people that went by Hungary many thousands perished miserably there and by the way, and many came home towards winter in pitiful plight and suffering from hunger, Chr. 1096 ; Erl. 233, 22.

hungor-geár, es ; n. A year of famine : — Ðá hæfde se hálga wer ge-dǽled ðæs mynstres þing hafenleásum mannum for ðam hungergeáre the saint had distributed the provisions of the monastery to indigent men on account of the year of famine, Homl. Th. ii. 178, 20.

hungor-lǽwe ; adj. Hungry, famished :— Ða hungerlǽwan gefylde synt famelici saturati sunt, Ps. Lamb. Cantic. Annæ, 5.

hungrig ; adj. Hungry, famished :—Gewât se wilda fugol hungri, Cd. 72 ; Th. 88, 10 ; Gen. 1463. Ðæm hungrige esurienti, Rtl. 5, 22. Gif ðú ðissere hungrige ceasterwaran gehelpest if thou helpest this starving town, Th. Ap. 9, 18. Hungrig esuriens, Mt. Kmbl. Lind. 25, 37. Hý him hung-rige ymb hond flugon, Exon. 43 a ; Th. 146, 13 ; Gú. 709. Ða hungrian, Ps. Th. 106, 8. Hungrium, 35 : 131, 16. Hungregum tô frôfre, Soul Kmbl. 224 ; Seel. 116. [Orm. hunngri;: O. H. Ger. hungarag impastus, esuriens, famelicus : Ger. hungerig, hungrig.]

hunig, es ; n. Honey :—Ðær [Estland] biþ swýðe mycel hunig and fisc[n]aþ and se cyning and ða rícostan men drincaþ myran meolc and ða unspédigan and ða þeówan medo in that country there is very much honey and fishing ; and the king and the principal men drink mare's milk, and the poor and the slaves mead, Ors. 1, 1 ; Swt. 20, 15. Doran hunig dumbledore's honey, L. M. 1, 2 ; Lchdm. ii. 28, 20. [Cf. O. H. Ger. humbel-honag.] Englisces huniges of English honey, 2, 65 ; Lchdm. ii. 292, 23 : 3, 71 ; Lchdm. ii. 358, 10. Þynceþ þegna gehwelcum huniges bíbreád healfe ðý swétre gif hé hwéne ǽr huniges teáre bitres onbyrgeþ dulcior est apium mage labor, si malus ora prius sapor edat, Bt. Met. Fox 12, 17 ; Met. 12, 9. Swá þicce swá huniges teár as thick as honey

that drops from the comb, L. M. 1, 31 ; Lchdm. ii. 74, 4 : 2 ; Lchdm. ii. 28, 4. Tô ðam lande ðe eall flêwþ on riðum meolce and hunies . . . of ðam lande ðe weóll meolce and hunie in terram, quæ fluit rivis lactis et mellis . . . de terra, quæ lacte et melle manabat, Num. 16, 14, 13. Beón gif hí man ácwellaþ cwelle hig man raðe ǽr hí tô ðam hunige cumon, L. Ecg. C. 39 ; Th. ii. 164, 2. [Orm. huni;: A. R. huni : Ayenb. honi : O. Frs. hunig : Icel. hunang : O. H. Ger. honag, honig : Ger. honig.] v. wudu-hunig.

hunig-æppel, es ; m. Pastillus, Cot. 155, Lye.

hunig-bére ; adj. Mellifluus, Hpt. Gl. 408, 457.

hunig-camb, e ; f. Honey-comb :—Hunigcamb t^áres favum nectaris, Lchdm. ii. 396, col. 1.

hunig-flôwende ; adj. Flowing with honey, dropping honey, melli-fluous :—Wyrta geblôwene hunigflôwende, Exon. 51 a ; Th. 178, 26 ; Gú. 1250. [Cf. Icel. hunangs-fljótandi flowing with honey.]

hunig-gafol, es ; n. Rent paid in honey :—Syllan huniggafol to pay rent in honey, L. R. S. 4 ; Th. i. 434, 31. [Cf. mid ús is geræd ðæt hé (beó-ceorl) sylle v. sustras huniges tô gafole, 5 ; Th. i. 436, 1.]

hunig-smæc ; gen. -smæcces ; m. Taste or flavour of honey :—Hafaþ on gehátum hunigsmæccas use honeyed words in their promises, Frag. Kmbl. 53 ; Leás. 28.

hunig-súce, -súge, an ; f. Privet, a plant from which honey may be sucked :—Húnisúge ligustrum, Ælfc. Gl. 47 ; Som. 65, 31 ; Wrt. Voc. 33, 30. Hunisúce, Wrt. Voc. 68, 3.

hunig-swés ; adj. Like honey ; melleus, Hpt. Gl. 481.

hunig-swéte ; adj. Sweet as honey, mellifluous :—Hé hlôd ðá mid þurstigum breóste ða flôwendan láre ðe hé eft æfter fyrste mid hunig-swéttre þrotan bealcette, Th. An. 45, 4.

hunig-teár, es ; m.' Distillation from the comb, without squeezing, virgin honey ; mel purissimum, e favo sponte quod effluxit, mell stillativum,' Lchdm. ii. 396, col. 1:—Hunigteár nectar, Hpt. Gl. 468. Hunigteáres nectaris, Mone Gl. p. 384. Sý gemenged tôgædre hunigteár and wín let virgin honey and wine be mixed together, Lchdm. iii. 292, 16. Besmyra mid hunigteáre, 11. [Cf. O. E. Hom. swete al swá hunitíar felle upe ;iure hierte, i. 217, 27.]

hunig-teáren ; adj. Sweet as honey or nectar :—Hunigteárenne nec-tareum, Gl. Prud. p. 140.

hunigteár-líc ; adj. Like nectar ; nectareus, Cot. 138, Lye.

hún-spuran ' dolones ; great spars or staves with small heads of iron, and swords within,' Som. Lye gives hun-spera, -spura dolo, Cot. 62. v. hún-þyrel.

hunt, e ; Hunting :—Of hunte de venatione, Rtl. 117, 4. [Or is hunte for huntunge ?].

hunta, an ; m. A hunter :— Hunta venator, Ælfc. Gr. 36 ; Som. 38, 43 ; Wrt. Voc. 73, 43. Ænne cræft ic cann. Hunta ic eom unam artem scio. Venator sum, Coll. Monast. Th. 21, 1-6 : 22, 27. Wê lǽraþ ðæt preóst ne beó hunta ne hafecere we enjoin that a priest be not a hunter nor a hawker [cf. Chaucer's Monk : ' He ;af nat of that text a pulled hen, That seith, that hunters been noon holy men '], L. Edg. C. 64 ; Th. ii. 258, 7. Eal wêste bûton ðær huntan gewícodon oððe fisceras, Ors. 1, 1 ; Swt. 17, 29. Wêste land bûtan fiscerum and fugelerum and huntum, Swt. 17, 26. Bethsaida is gereht domus venatorum ðæt is huntena hûs, Shrn. 78, 9. [Ðá sôn ðæræfter ða sægon and hêrdon fela men feolde huntes hunten. Ða huntes wæron swarte and micele and ládlíce, Chr. 1127; Erl. 256, 28. Laym. hunte; pl. hunten: Orm. hunnte: Chauc. hunte.] v. hwæl-hunta.

hunta, an ; m. A hunting spider ; salticus scenicus or aranea taran-tula [?] :—Wið ðon gif hunta gebîte mannan ðæt is swiðra in case a hunting spider bite a man, that is the stronger, L. M. 1, 68 ; Lchdm. ii. 142, 18 [see the note]: 14, 19. Wið huntan bite, 144, 2, 5.

Huntan-dún, e ; f. Huntingdon :—Fôr se here of Huntandûne and of Eástenglum and worhton ðæt geweorc æt Tæmese forda and forlêton ðæt ôðer æt Huntandûne ... And ðá se firdstemn fôr hám ðá fôr ôðer út and gefôr ða burg æt Huntandûne and hie gebêtte and geedneowade ðær heó ǽr tôbrocen wæs be Eádweardes cyninges hæse, Chr. 921; Erl. 106, 16 : 107, 31. Tôward Huntendûne porte, 656; Erl. 31, 19.

Huntandûn-scir, e ; f. Huntingdonshire :—Tô Huntandûnscire, Chr. 1016; Erl. 154, 7.

huntaþ, es ; m. Hunting, game ; venatio :—On feáwum stôwum wíciaþ Finnas, on huntoþe on wintra and on sumera on fiscaþe be ðære sǽ, Ors. 1, 1; Swt. 17, 5. On huntoþe, Exon. 78 b; Th. 295, 22; Cri. 37. Tô huntaðe [a prayer] for hunting, Rtl. 117, 1. On ðæt geráð ðet ðenne ðæs neód biþ his men beón gearuwe tô huntoþe on the condition that, when there shall be need for it, his men may be ready for hunting, Chart. Th. 148, 3. Isaac lufode Esau for his huntoþe Isaac amabat Esau, eo quod de venationibus illius vesceretur, Gen. 25, 28. Bring mê of ðínum huntoþe affer mihi de venatione tua, 27, 7 : Homl. Th. ii. 576, 34. Huntaþ dôn gestreón getácnaþ to hunt betokens gain, Lchdm. iii. 212, 2. Môna se fíf and twentigoþa huntoþas begán nytlíc the five and twentieth moon is good for all sorts of hunting, 196, 1. [R. Glouc. Edgar an honteþ ywend was.] v. hwæl-huntaþ.

huntaþ-faru, e; f. *A hunting expedition, hunting :* — Cýpinga and folcgemóta and huntaþfara and woroldlícra weorca on ðam hálgan dæge geswíce man georne *let people diligently abstain from marketings and folk-moots and hunting expeditions and secular employments on the holy day* [*Sunday*], L. Eth. vi. 22; Th. i. 322, 12 : L. C. E. 15; Th. i. 368, 18. [Cf. the Icelandic law ' Maþr a at fiskja drottins dag eþa messu dag eþa veiþa annat ef hann vill. Hann scal hafa messu um morgininn aþr oc lata eigi veiþina standa fyrir tiþa socninni.']

hún-pyrel, es; n. *The hole in the mast-head through which the halyard went :* — Húnþyrlu *carchesia*, Wrt. Voc. 63, 49. [Icel. húnn *a knob at the end of a staff, at the top of a mast;* hún-bora *the hole in the mast-head through which the halyard went.*]

huntian; p. ode *To hunt :* — Ic ásende míne fisceras and hí gefixiaþ hí míne huntan and hí huntiaþ hí of ælcere dúne and of ælcere hylle *I will send for many fishers and they shall fish them; and after will I send for many hunters and they shall hunt them from every mountain and from every hill* [A. V. Jer. 16, 16], Homl. Th. i. 576, 28. Gif him þince ðæt hé huntige beorge him georne wið his fýnd *if he fancies that he is hunting, let him guard himself well against his foes*, Lchdm. iii. 172, 19. Ne canst ðú huntian búton nettum *nescis venari nisi cum retibus*, Coll. Monast. Th. 21, 21. Ic fare huntian *venatum pergo*, Ælfc. Gr. 24; Som. 25, 10. Huntigendra *venantium*, Ps. Spl. 90, 3 : 123, 6.

huntigestre, an; f. *A huntress :* —Huntigystran *venatrices*, Nar. 38, 3.

huntig-spere; n. *A hunting-spear, boar-spear :* — Bárspere *vel* huntigspere *venabulum*, Ælfc. Gl. 51; Som. 66, 23; Wrt. Voc. 35, 12.

huntnaþ, huntnoþ, es; m. *Hunting :*—Be huntnaþe. Ic wylle ðæt ælc man sý his huntnoþes wyrðe on wuda and on felda on his ágenan. And forgá ælc man mínne huntnoþ hwær ic hit gefriþod wille habban *Of hunting. I will that every man have the right to hunt in wood and in open country on his own property. And let every man leave my hunting alone where I wish to have it preserved*, L. C. S. 81; Th. i. 420, 23–6. Wǽre ðú tó-dæg on huntnoþe *fuisti hodie in venatione?* Coll. Monast. Th. 21, 35. Hé of huntnoþe com *venerat de venatu*, Bd. 3, 14; S. 540, 33. On fiscnoþum and on huntnoþum and on fugelnoþum *piscationibus, venationibus, aucupationibus*, Cod. Dipl. Kmbl. iii. 350, 9.

huntung, e; f. *Hunting :*—Mǽre on huntunge heorta and rána *cervorum caprearumque insignis*, Bd. 1, 1; S. 474, 41. Gyrstandæg ic wæs on huntunge *heri fui in venatione*, Coll. Monast. Th. 22, 3. Hwæt ðest ðú be ðínre huntunge? Ic sylle cync swá hwæt swá ic gefó *quid facis de tua venatione? Ego do regi quicquid capio*, 25–7. Of huntungum *de venationibus*, Rtl. 118, 39.

hup-bán, -seax. v. hype-bán, -seax.

húru; adv. *At least, at all events, at any rate, in any case, however, even, yet, only, indeed, certainly, especially :* —Húru gif ic hæfde ǽnne penig *saltim si haberem unum denarium*, Ælfc. Gr. 44; Som. 46, 35. Húru nú hæþ mín heáfod uppáhafen ofer míne fýnd *nunc autem exaltavit caput meum super inimicos meos*, Ps. 26, 7. Ðæt ic húru underfó sum fóstercild of hyre *si forte saltem ex illa suscipiam filios*, Gen. 16, 2. Beó ðú húru gehyrt *tu tantum confortare*, Jos. 1, 18, 17. Húru ðæt hig ofer niht ðæron ne wunigon *ita saltem ut non per noctem ibi restent*, L. Ecg. C. 39; Th. ii. 164, 2. Óðre lytle fugelas sind læssan ðonne heó sý and hwæðere hí ofsleáþ sum þing húru ðás fleógan *other little birds are less than it* [*the dove*] *is, and yet they kill something, at any rate these flies*, Homl. Th. ii. 46, 17. Woldon hine habban húru swá deádne *they would have him when he was dead at any rate*, 518. 23. Húru fíftene míla brád *at least fifteen miles broad*, Ors. 1, 1; Swt. 20, 8. Ðæt hé húru þreó þing ðananforþ healdan wille, L. Eth. v. 6; Th. i. 306, 8 : L. C. E. 19; Th. i. 370, 33. Be emnihte oððe húru be ealra hálgena mæssan *by the equinox or in any case by Allhallows' mass*, L. Eth. ix. 9; Th. i. 342, 22. Eallum cristenum gebyreþ ðæt hí riht lufian and húru [*certainly*] gehádode men scylon á riht rǽran, L. I. P. 7; Th. ii. 312, 34. Húru hit wyrþ ðonne egeslíc, Swt. A. S. Rdr. 104, 5. Gif hit on ǽnegum men ǽnige hwíle fæstlíce wunaþ se deáþ hit húru áfirreþ, Bt. 8; Fox 26, 4. Ðæt deáh tó ælcum and húru tó deópun dolgum *it is good for all, and especially for deep wounds*, L. M. 1, 45; Lchdm. ii. 114, 1. Ðæt man cristene men and unforworhte of earde ne sylle ne húru on hǽðene teóde *certainly not to a heathen nation*, L. Eth. v. 2; Th. i. 304, 15. Heora eáþmetto ne mihton náuht forstandan ne húru heora ofermetta *their humility could not avail aught, and certainly not their pride*, Bt. 29, 2; Fox 104, 34. [A. R. hure.]

húru-þinga; adv. *Especially, at least, at any rate :*—Húruþinga *presertim*, Ælfc. Gr. 38; Som. 41, 65. Hú ne scolde hine húruþinga sceamian seofon dagas *nonne debuerat saltem septem diebus rubore suffundi?* Num. 12, 14. Hyne bǽdon ðæt hig húruþinga his reáfes fnæd æthrinon *rogabant eum ut vel fimbriam vestimenti ejus tangerent*, Mt. Kmbl. 14, 36. Lǽtaþ mé fyrst óþ tómerigen húruþinga fyrst óþ tómerigen *allow me respite until to-morrow, allow me respite until to-morrow*, Homl. Th. i. 414, 23. Swilce hé swutelíce cwæde ' Gif gé noldon Gode lybban on cildháde, ne on geógoþe, gecyrraþ nú húruþinga on ylde tó lifes wege,' ii. 78, 13.

HÚS es; n. *A* HOUSE, *a family :*—Hic lar þis fýr on ánfealdum getele, and hit getácnaþ hús on mænigfealdum getele, *hi lares* ðás hús; ðanon is

gecweden *lardum* spic, forðan hit on húsum hangaþ lange, Ælfc. Gr. 9; Som. 9, 48. Baðiendra manna hús ðǽr hí unscrédaþ inne *apodyterium*, i. e. *domus qua vestimenta balneantium ponuntur*, Ælfc. Gl. 55 : Som. 67, 9; Wrt. Voc. 37, 6. Lytle hús of bredan *tabernæ vel gurgustia*, Wrt. Voc. 37, 8. Byþ gelíc ðam wísan were se hys hús ofer stán getimbrode *assimilabitur viro sapienti qui ædificavit domum suam supra petram*, Mt. Kmbl. 7, 24. Gewát neósian heán húses *went and visited the lofty house*, Beo. Th. 233; B. 116. Maria húse gesætt *Maria domi sedebat*, Jn. Skt. Lind, 11, 20. Lét fleógan hrefn of húse út [*out of the ark*], Cd. 71; Th. 87, 2; Gen. 1442. Se wilda fugel ofer heánne beám hús getimbreþ, Exon. 58 b; Th. 211, 24; Ph. 202. Ðæt fǽge hús *the corpse*, Elen. Kmbl. 1759; El. 881. Israhéla hús *domus Israel*, Ps. Th. 113, 18, 1, 19 : 134, 21. Nis nán wítega búton wurþscipe búton on his éðele and on his mægþe and on his húse *non est propheta sine honore nisi in patria sua et in cognatione sua et in domo sua*, Mk. Skt. 6, 4. [Goth. O. Sax. O. Frs. Icel. O. H. Ger. hús : Ger. haus.] DER. ambiht-, bán-, bed-, dóm-, eorþ-, feld-, feoh-, feorh-, friþ-, gæst-, geofon-, gift-, græf-, helle-, mán-, mere-, morðor-, nicor-, sáwel-, wíg-, wíte-hús.

húsa, an; m. *A member of a household :* —Fióndes menn húsa his *inimici hominis domestici ejus*, Mt. Kmbl. Lind. 10, 36. v. ge-húsa.

hús-bonda, -bunda, an; m. *The master of a house :*—Án his manna wolde wícian æt ánes bundan húse his unþances and gewundode ðone húsbundon and se húsbunda ofslóh ðone óðerne. Ðá wearþ Eustatius uppon his horse and his geféoran uppon heora and férdon tó ðam húsbundon and ofslógon hine binnan his ágenan heorþa *one of his men wanted to stop at a man's house against his will, and wounded the man of the house, and the man of the house slew the other. Then Eustace got on his horse and his companions on theirs, and went to the man of the house and slew him in his own home*, Chr. 1048; Erl. 177, 35–40. [O. E. Homl. þe husbonde þat is wit warneþ his hus þus, i. 247, 19 : *Laym.* of æverelche huse þat husbonde wunede, 31958 : *Prompt. Parv.* hose-, hus-bonde *paterfamilias;* also *maritus : Icel.* [*from which the word seems borrowed*] hus-bóndi [= -búandi] *a house-master; a husband.* Cf. *Chauc. Wick.* husbond-, housbonde-man *a householder.*]

hús-bonde, an; f. *The mistress of a house :*—Ða Israéliscan wíf biddaþ æt ðám Egitiscean wífon æt hira néhgebúron and æt hira húsbondum sylfrene fatu *postulabit mulier a vicina sua et ab hospita sua vasa argentea*, Ex. 3, 22.

hús-brice, es; m. *Housebreaking, burglary :*—Húsbrice [-brec, MS. A.] and bærnet æfter woruldlage is bótleás *housebreaking and arson are according to the secular law inexpiable*, L. C. S. 65; Th. i. 410, 5. Cf. quedam non possunt emendari, que sunt husbreche, et bernet, L. H. 12, 1; Th. i. 522, 27 : 47; Th. i. 546, 10. [O. Frs. hús-breke : cf. Icel. hús-brot *housebreaking, burglary :* and O. H. Ger. hús-prehho *prædator.*] v. brecan, á-brecan.

hús-bryne, es; m. *The burning of a house, a fire :*—Æt húsbryne ælc mon ánne pening *at the burning of a house let every man contribute one penny*, Chart. 614, 13. [Icel. hús-bruni : cf. O. Frs. hús-brand.]

husc, hucs, hux, es; m. [cf. hosp.] *Insult, scorn, scoffing, mockery :*—Abraham mid hucse bewand ða hleóðorcwidas on hige sínum [cf. *Sarah laughed within herself*, Gen. 18, 12], Cd. 107; Th. 140, 34; Gen. 2337 : 109; Th. 143, 21; Gen. 2382. Þurh hucx *per ironiam*, Cot. 186, Lye. [*Laym.* hux and hoker : O. L. Ger. hosc *subsannatio :* O. H. Ger. hosc *sugillatio.*] v. hux-líc.

hús-carl, es; m. [*A word apparently taken from the Scandinavians, as the English form would be hús-ceorl.*] *A member of the king's body-guard :* — Ðurstán mín húskarll *præfectus meus palatinus Þurstanus*, Cod. Dipl. Kmbl. iv. 202, 4. Urk mín húskarl, 221, 6. On gewitnesse eallra ðæs kynges húscarlan [-carla?], 291, 15. Ða Densca húscarles, Chr. 1070; Erl. 207, 25. Man gerǽdde ðæt Ælfgifu Hardacnutes módor sǽte on Winceastre mid ðæs cynges húscarlum hyra suna, 1036; Erl. 165, 5. [O. Frs. hús-kerl : Icel. hús-karl I. *a man-servant*, opposed to hús-bondi *a master;* II. *a member of the king's body-guard.* See Cl. and Vig. Dict.] v. Kemble's Saxons in England, ii. 118 sqq : Stubbs' Const. Hist. i. 150.

husc-word, es; n. *An insulting, scornful word or speech :*—Huscworde ongan ealdorsacerd hyspan, Andr. Kmbl. 1338; An. 669. [*Laym.* hux-word.]

HÚSEL, húsul, húsl, es; n. *The* HOUSEL, *consecrated bread and wine, the Eucharist :*—Ðæs hláfes wé onbyriaþ ðonne wé mid geleáfan tó húsle gáþ forðan ðe ðæt hálige húsel is gástlíce Cristes líchama *that bread we taste when we believingly go to the Lord's supper, for the consecrated bread is spiritually Christ's body*, Homl. Th. i. 34, 18. Hwí is ðæt hálige húsel gecweden Cristes líchama oððe his blód, gif hit nis sóþlíce ðæt ðæt hit geháten is? Sóþlíce se hláf and ðæt wín ðe beóþ þurh sacerda mæssan gehálgode óðer þing hí æteówiaþ menniscum andgitum wiðútan and óðer þing hí clypiaþ wiðinnan geleáffullum módum. Wiðútan hí beóþ gesewene hláf and wín ǽgðer ge on hiwe and on swæcce, ac hí beóþ sóþlíce æfter ðære hálgunge Cristes líchama and his blód þurh gástlícere gerýnu, ii. 268, 21–9. Ðæt húsel is Cristes líchama ná líchamlíce ac gástlíce ná se líchama ðe hé on þrowode ac se líchama ðe hé embe spræc ðá ðá hé bletsode hláf

and wín tó húsle ... and cwæþ be ðam gebletsodan hláfe Ðis is mín líchama and be ðam gehálgodan wíne Ðis is mín blód ... Understandaþ ðæt se Drihten dæghwamlíce bletsaþ þurh sacerda handa hláf and wín tó his gástlícan líchama and blóde *the housel is Christ's body, not bodily but spiritually; not the body that he suffered in, but the body that he spoke about when he blessed bread and wine for housel ... and said of the bread he had blessed: 'This is my body,' and of the hallowed wine: 'This is my blood' ... Understand that the Lord daily blesses, by the priest's hands, bread and wine so that they become his spiritual body and blood,* L. Ælfc. C. 36; Th. ii. 360, 15-24. Ðæm folce húsl syllan *Eucharistiam populo dare,* Bd. 2, 5; S. 507, 13. Hé frægn hwæðer hí ænig húsel ðærinne hæfdon. Ða andswaredon hí hwylc þearf is ðe húsles ... Cwæþ hé Beraþ mé hwæðere húsel tó *interrogavit, si Eucharistiam intus habērent. Respondebant, 'Quid opus est Eucharistia?' 'Et tamen' ait 'afferte mihi Eucharistiam,'* 4, 24; S. 598, 35-9: L. Ælfc. C. 36; Th. ii. 358, 16-38, 360, 5-15, 24-29. Tó húsle gán *to go to the sacrament,* Blickl. Homl. 207, 5: 209, 6. Húsle gereorded ðý æþelan gyfle *having been fed with the Eucharist, that noble meal,* Exon. 51 b; Th. 180, 4; Gú. 1274. [*The older meaning of the word is seen from the Gothic* hunsl *sacrifice;* hunslian *to offer;* hunsla-staþs *an altar,* see Grmm. D. M. 35. *The word is found in* Icel. húsl: *Swed.* husl: *Orm.* A. R. O. E. Hom. *husel:* R. Glouc. hosel: *Piers P. Chauc.* housel: *and for later use see Nares' Gloss.*]

húsel-bearn, es; n. *A person who may partake of the Eucharist:*—Hálig húsulbearn [*Guthlac*], Exon. 40 b; Th. 135, 28; Gú. 531.

húsel-disc, es; m. *Housel-dish, the plate for the consecrated bread, the paten:*—Húseldisc *patena,* Ælfc. Gl. 26; Som. 60, 91; Wrt. Voc. 25, 31: *patina,* Wrt. Voc. 81, 2. Ðis mon sceal wrítan on húsldisce and on ðone drenc mid háligwætere þweán and singan on *this is to be written on a paten and washed into the drink and sung over,* L. M. 1, 62; Lchdm. ii. 136, 3.

húsel-fæt, es; n. *A sacrificial vessel, [in Christian times] a sacramental vessel:*—Húselfatu *vasa sacra,* Bd. 1, 29; S. 498, 9. Subdiaconus is underdiácon se ðe ða fatu byrþ forþ tó ðam diácone and þénaþ under ðam diácone æt ðam hálgan weófode mid ðam huselfatum, L. Ælfc. C. 15; Th. ii. 348, 11. Húslfatu hálegu *the vessels of the temple,* Cd. 209; Th. 260, 5; Dan. 705: 212; Th. 262, 24; Dan. 749.

húsel-gang, es; m. *Attendance upon* or *partaking of the sacrament:*—Fulluht and synna forgyfenys húselgang sind eallum gemæne earmum and eádigum *baptism and forgiveness of sins, attendance at the sacrament, are common to all, to poor and rich,* Homl. Th. i. 64, 32: ii. 48, 29. Se ðe hit singþ æt his endedæge ðonne forstent hit him húselgang *he who sings it at his last day, for him it shall stand instead of receiving the Eucharist,* Lchdm. iii. 288, 16. Gearwige tó húslgange oft and gelóme gehwá hine sylfne, L. Eth. v. 22; Th. i. 310, 7. Gearwige hine tó húselgange húru þríwa on geáre, vi. 27; Th. i. 322, 7: L. C. E. 19; Th. i. 370, 32. v. next word.

húsel-genga, gengea, an; m. *One who goes to the Lord's supper, a communicant:*—Gif hé húslgengea síe, L. Wih. 23; Th. i. 42, 7: L. In. 19; Th. i. 114, 11. Be húslgengum, 15; Th. i. 112, 4.

húsel-hálgung, e; f. *The sanctifying that comes from receiving the Eucharist, attendance at the Eucharist:*—Ðreó heálíce þing gesette God mannum tó clænsunge án is fulluht óðer is húselhálgung þridde is dædbót ... Se húselgang ús gehálgaþ, Homl. Th. ii. 48, 27. Úre gástlícan lác sind úre gebedu and lofsang and húselhálgung *our spiritual gifts are our prayers and praise and attendance at the Eucharist,* i. 54, 27.

húsel-láf, e; f. *What is left of the housel:*—Man ne mót hálgian húsel on Langa Frigedæg ... Gange se preóst tó ðam weofode mid ðære húsellúfe ðe hé hálgode on Ðunresdæg *housel must not be hallowed on Good Friday ... Let the priest go to the altar with what remains of the housel that he hallowed on Thursday,* L. Ælfc. C. 36; Th. ii. 358, 22.

húsel-portic, es; m. *Sacristy:*—His líchoma wæs bebyriged beforan ðam húselportice *sepultus est corpore ante secretarium,* Bd. 2, 1; S. 500, 15.

húsel-þegn, es; m. *An acolyte:*—Acolitus ðæt is húslþén, L. Ecg. C. 41; Th. i. 166, 20.

húsel-wer, es; m. *One who may take the sacrament, a communicant:*—Húsulweras, Exon. 44 a; Th. 149, 28; Gú. 768.

hús-fæst; adj. *Having a house, being a householder:*—Ælc man húsfæst on his ówe land *every man having a house on his own land,* Chart. Th. 438, 5.

hús-heofon, es; m. *A ceiling:*—Húshefen *lacunar,* Cot. 119, Lye.

hús-hláford, es; m. *The master of a house:*—Secgeaþ ðam húshláforde *dicetis patrifamilias,* Lk. Skt. 22, 11.

hús-hleów, es; n. *Shelter afforded by a house:*—Gif[e] his húshleów and mete and munde ðam ðe ðæs beþurfe *let him give the shelter of his house and food and protection to him that needs it,* L. Pen. 15; Th. ii. 282, 25.

húsian; p. ode *To house, give shelter in a house:*—Féde þearfan and scrýde and húsige *let him feed the needy and clothe and house them,* L. Pen. 14; Th. ii. 282, 15. [*Icel.* húsa *to shelter;* hýsa *to house.*]

hús-incel, es; n. *A small house, a habitation;* domicilium, tabernaculum:—Husincil *tabernaculum,* Rtl. 181, 5, 15. In húsincle *in domicilio,* Ps. Surt: húsincyle, Ps. Spl. C. 101, 7. [*Cf. O. H. Ger.* húsili *domiciium, domuncula.*]

húsl. v. húsel.

húslian; p. ode *To housel, to administer the sacrament:*—Hý mihton wel habban wíf on ðám dagum forðan ðe hý næfre ne mæssodon ne menn ne húslodon *they might well have wives in those days for they never celebrated mass nor administered the Eucharist to men,* L. Ælfc. C. 7; Th. ii. 346, 8. Wé læraþ ðæt ælc preósta seóce men húslige ðonne heom þearf sí, L. Edg. C. 65; Th. ii. 258, 10. Diaconus mót ðæt folc húsligan, L. Ælfc. C. 16; Th. ii. 348, 14. [*Cf. Diaconus* mót hláf sillan, L. Ælfc. P. 34; Th. ii. 378, 12.] Gif man biþ tó húsligenne, 29; Th. ii. 352, 31. [*Orm.* huslenn: *Prompt. Parv.* howselyn wythe the sacrament *communico,* see note, p. 250: *Piers P. Chauc.* houseled; *pp:* cf. *Shaks.* un-houseled. *Goth.* hunslian *to offer:* Icel. húsla *to give the Corpus Domini to a sick person.*]

húslung, e; f. *The administration of the sacrament:*—Æfter ðære húslunge gewát tó ðam lífigendan gode, Homl. Swt. 3, 622: Homl. Th. ii. 548, 9.

hús-ræden, e; f. *A house, family:*—Húsræden israhéles *domus israel,* Ps. Lamb. 113 [2nd], 9. Húsrædenne hire *domus ejus,* 47, 14.

hús-stede, es; m. *The site of a building:*—Deós wyrt byþ cenned on ealdum hússtedum, Herb. 52, 1; Lchdm. i. 154, 25: 85, 1; Lchdm. i. 188, 12. [*O. Sax.* hús-stedi: *O. Frs.* hús-stede: *Icel.* húsa-staðr: *O. H. Ger.* hús-stat.]

hús-ting, es; n. *A word taken from the Scandinavians [Icel.* hús-þing *a council* or *meeting* to which a king, earl or captain summoned his people or guardsmen], *a meeting, court, tribunal,* apparently so called from its being held within a building when other courts were held in the open air. The word occurs in the following passages [Latin]:—Debet eciam in Londoñ, que caput est regni et legum, semper curia domini regis singulis septimanis die Lune hustingis sedere et teneri, L. Th. i. 457, 36. Ad folkemoth vel ad hustenge, 463, 11. Non on hustenge neque in folkesmote, 503, 3. Ad pondus Hustingie Londonensis, Chart. Th. 533, 10. It is found also in English:—Mid hundeahtigum marcan hwítes seolfres be hústinges gewihte, 329, 22. Hí [the Danes] leaddon ðone biscop tó heora hústinga, Chr. 1012; Erl. 146, 17.

hús-wist; e; f. *A house, household:*—Ic ingange on ðínum húswiste Í intó ðínum húse *introibo in domum tuam,* Ps. Lamb. 5, 8.

húðe [v. herehúðe], e; f. *Prey, spoil, booty:*—Húðe hrémig *exulting in spoil,* Elen. Kmbl. 297; El. 149: Beo. Th. 248; B. 124. Cómon tó Moyse mid micelre húðe *adduxerunt prædam ad Moysen,* Num. 31, 12. Se Chaldéa cining com ðá tó his earde mid ðære húðe, Ælfc. T. Grn. 8, 23: Cd. 174; Th. 220, 2; Dan. 65. Habbaþ nú ða húðe and ðæt orf eów gemæne *prædam vero et omnia animantia diripiens vobis,* Jos. 8, 2: Cd. 97; Th. 127, 19; Gen. 2113: 98; Th. 129, 24; Gen. 2149. Húða mæste *greatest of spoils,* Exon. 16 a; Th. 35, 35; Cri. 568. [*Cf. Goth.* hunths *captivity:* O. H. Ger. heri-hunda *præda.*] v. here-hýð.

húðe, tó *in portum,* Ps. Lamb. 106, 30. v. hýð.

hux-, husc-líc; adj. *Ignominious, involving shame, scorn, insult:*—Huxlíc *dedecor,* Ælfc. Gr. 9, 21; Som. 10, 34. Ða þuhte him tó huxlíc ðæt hé híran sceolde ænigum hláforde *it seemed to him too ignominious to obey any lord,* Ælfc. T. Grn. 2, 36. Ða þuhte ðam heáhgeréfan huxlíc ðæt heó óðerne tealde tóforan his gebyrdum, Homl. Swt. 7, 24. v. husc.

hux-líce; adv. *Ignominiously, disgracefully, unbecomingly:*—Ðone seó eorþlíce árleásnyss huxlíce tealde *whom earthly impiety had disgracefully calumniated,* Homl. Th. i. 48, 23. Ða ðe hí huxlíce hér on lífe gedrehton *those who shamefully afflicted them in this life,* Jud. 5; Thw. 156, 10. Gelædde ðone kining mid him swíðe huxlíce *carried the king with him very ignominiously,* Ælfc. T. Grn. 8, 20.

hwá; m. f.; hwæt; n. *Who; what.* I. *in direct questions* [*with indic.* or *subj.*]:—Quis hwá is werlíc hád *que hwilc is wíflíc, cujus* hwæs, *cui* hwam *a quo* fram hwam ... Gif ic cweðe *quis* interrogativum ðæt is áxigendlíc, Ælfc. Gr. 18; Som. 21, 12-27. Hwá hwylc mann swá Drihten ondræt *quis est homo qui timeat Dominum?* Ps. Th. 24, 10. Hwá is moncynnes ðæt ne wundrie *what man is there that does not admire?* Bt. Met. Fox 28, 10; Met. 28, 5. Hwá þegna, 86; Met. 28, 43. Hwæt is se gewuldroda cyning *quis est iste rex gloriæ?* Ps. Th. 32, 10. Hwæt hátte Noes wíf *what was Noe's wife called?* Salm. Kmbl. 184, 28. Hwæt wénst ðú hwæt is ðes *quis putas est iste?* Mk. Skt. 4, 41: Lk. Skt. 5, 21. Hwæt ys ðes mannes sunu? Jn. Skt. 12, 34. Hwæt sind ðás búton þrymsetl heora Scyppendes *what are these but thrones of their Creator?* Homl. Th. i. 346, 11. Hwæt sind ða strangan? Ða beóþ strange and trume ðe þurh geleáfan wel þeónde beóþ, ii. 390, 22. Ða cwæþ Isaac: Hwæt eart ðu? Hé andwirde: Ic eom Esau. Ða cwæþ Isaac: Hwæt wæs se ðe mé ær brohte of huntoþe? Gen. 27, 32-3. Hwæt is ðe slóh *quis est qui te percussit?* Mt. Kmbl. 26, 68. Hwæt eom ic manna ðæt ic mihte god forbeódan *what manner of man am I, that I could forbid God,* Homl. Swt.

10, 191: Elen. Kmbl. 1802; El. 903: Beo. Th. 479; B. 237. Hwæt is þinga ðe bitere sie *what thing is there that is bitterer?* Past. 21; Swt. 164, 1. Hwæt næddercynna sí on eorþan *how many kinds of snakes are there on the earth?* Salm. Kmbl. 204, 7. Hwæt suna hæfde Adam *what sons had Adam?* 184, 31. Hwæt synt ðínum esne ealra dagena *quot sunt dies servi tui?* Ps. Th. 118, 84. Hwæt gódes dó ic *quid boni faciam?* Mt. Kmbl. 19. 16. Hwæt þincþ eów be Criste hwæs sunu ys hé *quid vobis videtur de Christo? cujus est filius?* 22, 42. Hunta ic eom. Hwæs? *venator sum. Cujus?* Coll. Monast. Th. 21, 7. Hwæs wénaþ se ðe nyle gemunan *what does he expect that will not remember?* Exon. 25 b; Th. 74, 1; Cri. 1200. Tó hwam gá wé *ad quem ibimus?* Jn. Skt. 6, 68. Bí hwon scealt ðú lifgan *by what art thou to live?* Exon. 36 b; Th. 118, 23; Gú. 244. For hwan næron eorþwelan gedæled gelíce *why have not earth's treasures been equally divided?* Salm. Kmbl. 685, 693, 703; Sal. 342, 346, 351. For hwan gæst ðú swá búton wæstme ðínes gewinnes? St. And. 24, 15: Ps. Th. 73, 11: 113, 5. For hwon sécest ðú sceade? Cd. 42; Th. 54, 7, 12; Gen. 873, 876. On hwam mæg man geseón mannes deáþ *by what can one foresee a man's death?* Salm. Kmbl. 206, 10. On hwan *in quo?* Ps. Th. 118, 9. Tó hwæm willaþ gé þider faran *why will ye go thither?* St. And. 6, 18. Tó hwam, Salm. Kmbl. 894; Sal. 446. Tó hwan, Soul Kmbl. 39; Seel. 17. Hwæne séce gé *quem quæritis?* Jn. Skt. 18, 7. For hwí *quare?* Ps. Th. 113, 5: Coll. Monast. Th. 24, 19. For hwí swá *cur sic?* 27. Tó hwí stande gé ídele *why stand ye idle,* Homl. Th. ii. 74, 35. Hwý biþ his anwald áuhte ðý mára gif hé náh his selfes geweald *in what way will his power be at all the greater if he has not command over himself?* Bt. Met. Fox 16, 39; Met. 16, 20. II. *in dependent clauses:*—Gif ic cweþe *nescio quis hoc fecit* nát ic hwá ðis dyde ðon biþ se *quis infinitivum dæt* is ungeendigendlíc. Gif ic cweþe *tu scis quis hoc fecit* ðú wást hwá ðys dyde ðon biþ *quis relativum dæt* is edlesendlíc, Ælfc. Gr. 18; Som. 21, 27–30. Hogodon georne hwá ærost mihte on fægean men feorh gewinnan *strove eagerly who might first obtain the life of a 'fey' man,* Byrht. Th. 135, 26; By. 124. Men ne cunnon secgan hwá ðæm hlæste onféng, Beo. Th. 104; B. 52: Andr. Kmbl. 761; An. 381. Ic nú scortlíce secgan scyle hwá ðæs ordfruman wæron *I will now shortly tell who its authors were,* Ors. 5, 9; Swt. 232, 18. Næfdon hwæt hí æton *nec haberent quod manducarent,* Mk. Skt. 8, 1. Ne rædde gé ðæt hwæt dauid dyde ðá hine hingrede *nec hoc legistis quod fecit dauid cum esurisset,* Lk. Skt. 6, 3. Ðonne sceal gehwá him æteówian hwæt hé mid ðam punde geteolod hæfþ, Homl. Th. ii. 558, 10. Gehíéren hwæt áwriten is, Past. 44; Swt. 323, 7: 45; Swt. 341, 12: 52; Swt. 405, 29. Geþince gé hwæt gé síen and hwelce gé síen *pensa quod es,* Past. 21, 4; Swt. 159, 14: 1, 3; Swt. 27, 23. Hé sæde hyre hwæt heó man ne wæs *he told her how she was not a man,* Homl. Swt. 2, 78. Seó eorþe is tó wundrienne hwæt heó ærest oð, ðe gódra þinga cenne *mirandum est terra quantum aut bonarum rerum pariat,* Nar. 2, 12. Mé wæs uncúþ hwæt ðæs ðám lícian wolde ðe æfter ús wæren *I did not know how much of it would please those that should be after us,* L. Alf. 49; Th. i. 58, 22. Hit næs ná gesæd hwæt Pirruses folces gefeallen wære, Ors. 4, 1; Bos. 77, 30. [Ðæt is ungeliéfedlíc tó gesecganne] hwæt ðæs ealles wæs *what there was of it all,* 5, 12; Swt. 240, 16: Chr. 1046; Erl. 171, 3. Hé nyste hwæt ðæs sóðes wæs *he did not know how much truth there was in it,* Ors. 1, 1; Swt. 17, 33. Hý ne áhsedon hwæt ðæra gefarenra wære, ac hwæt heora ðonne tó láfe wære *they did not ask how many were dead, but how many of them were then left,* 4, 4; Bos. 80, 12. Ðá befran se sceaða hwæt hé manna wære, Homl. Th. ii. 502, 27: Cd. 64; Th. 77, 6; Gen. 1271. Saga hwæt ic hátte *say what I am called,* Exon. 102 b; Th. 387, 1; Rä. 4, 72. Ðæt hie geþencen hwæs folgeras hie sindon *ut cujus sint sequaces agnoscant,* Past. 47, 1; Swt. 357, 16. Wé cwædon hwæs se wyrðe wære ðe óðrum ryhtes wyrnde, L. Ed. 2; Th. i. 160, 10. Ic cýðe hwæs ic gean intó ealdan mynstre, Chart. Th. 333, 10: Andr. Kmbl. 290; An. 145. Swá wæs gemearcod hwam ðæt sweord geworht ærest wære *so was marked for whom that sword was first wrought,* Beo. Th. 3397; B. 1696. Ic ne can for hwam se streám ne mót stillan nihtes *I know not why the stream cannot rest at night,* Salm. Kmbl. 795; Sal. 397. Lyt ðú gemundest tó hwan ðínre sáwle þing siððan wurde *little didst thou mind to what thy soul's condition would come,* Soul Kmbl. 39; Seel. 20: Beo. Th. 4149; B. 2071. Sió hálige gesomnung þurh gesceádwísnesse gesiehþ of huan ælc costung cymeþ *sancta ecclesia, quæ ex causis singulis tentamenta prodeant, per discretionem conspicit,* Past. 11, 2; Swt. 65, 24. Ac ðú findst wið hwone ðú meaht flítan *sed contra quos valeatis vos extendere, semper invenitis,* 44, 8; Swt. 331, 5. Be hwý *according to what principle,* Chart. Th. 171, 7. Ic wundrige for hwý se góda God læte ænig yfel beón *I wonder for what reason the good God allows any evil to exist,* Bt. 36, 1; Fox 172, 4. For hwig, St. And. 32, 13. Frægn hí mid hwí hí gescildan heora hús *he asked them what they protected their house with,* Shrn. 90, 7. III. [an indefinite pronoun] *any one, some one; anything, something:*—Gif hwá on cirican hwæt þeófige *if any one steal anything in a church,* L. Alf. pol. 6; Th. i. 66, 2. Gyf hwá eów ænig þingc tócwyþ *si quis vobis aliquid dixerit,* Mt. Kmbl. 21, 3. Nellaþ hí ge-

lýfan ðeáh hwá of deáþe árise *they will not believe, though one rose from death,* Homl. Th. i. 334, 21: Bt. Met. Fox 10, 53; Met. 10, 27. Ðeáh ánra hwá ealles wealde ðæs íglandes *though any one rule all that island,* 16, 31; Met. 16, 16. Hwæt hwá óðrum tó wó gedó *what any one does wrongfully to another,* L. E. I. 35; Th. ii. 432, 26. Búton hwá þurh flánes flyht fyl genáme, Byrht. Th. 133, 56; By. 71. Gif hé næbbe hwæt hé selle *if he have not anything to give,* L. Alf. 24; Th. i. 50, 16. Ne furþum ne giémaþ hwæt hie dón oð ðe hwonne hie hwæt dón *qui nequaquam, quæ quando agant, inspiciunt,* Past. 39, 3; Swt. 287, 7. Ánes hwæt tó singanne *to sing something,* 46, 2; Swt. 347, 6: Beo. Th. 6013; B. 3010. Tó ðæm gleáw ðæt hé swelces hwæt tócnáwan cunne so *skilled that he can distinguish in a matter of such a kind,* Past. 52, 10; Swt. 411, 26. Blæc oð ðe won oð ðe swilces hwæt *pale or livid or something of that kind,* L. M. 1, 35; Lchdm. ii. 82, 13: Beo. Th. 1764: B. 880. Gif hwæt yfles on biþ, L. M. 2, 24; Lchdm. ii. 214, 13. Lytles hwæt, Ors. 3, 7; Swt. 120, 4: 3, 9; Swt. 136, 18. Gif friþgeard sí on hwæs lande *if a 'friþgeard' be on any one's land,* L. N. P. L. 54; Th. ii. 298, 16. Gif hwæs bróðor deád biþ *si cujus frater mortuus fuerit,* Mk. Skt. 12, 19. Ðonne ðæt mód hwæs wilnode tó witanne ðæs ðe hit ær for sweotole ongytan ne meahte, Shr. 164, 19. Ðeáh hwæm swá ne þince *though to any one it seem not so,* Bt. 20; Fox 70, 32. Rinca hwæm, Bt. Met. Fox 22, 56; Met. 22, 28. Oft hwæm gebyreþ ðæt hé hwæt mærlíces and wundorlíces gedéþ, Past. 4, 1; Swt. 39, 6: 40, 5; Swt. 297, 4. Hit biþ on ánes hwæm ðe unfæstre *impar quisque invenitur ad singula,* 4, 1; Swt. 37, 15. Sóna swá sacerda hwylc hwone on wóh gesyhþ *directly any priest sees any one in error,* L. E. I. 28; Th. ii. 424, 26. Ðeáh mon hwone gódra mid rihte herige, Bt. 30, 1; Fox 108, 8: Bt. Met. Fox 10, 1; Met. 10, 1: Beo. Th. 312; B. 155. IV. *in combination with swá, whosoever, whatsoever, whatever:*—Swá hwá *quicunque,* Ælfc. Gr. 18; Som. 21, 37: swá hwá *quisquis,* 34. Swá hwá swá ðe genýt þúsend stapa *quicunque te angariaberit mille passus,* Mt. Kmbl. 5, 41: Cd. 22; Th. 28, 20; Gen. 438: 24; Th. 31, 10; Gen. 483. Swá hwæt swá hig woldon *quæcumque voluerunt,* Mt. Kmbl. 17, 12: Cd. 35; Th. 47, 4; Gen. 755. [Hí mósten cěsen of clerchádes man swá hwam (acc.) swá hí wolden, Chr. 1123; Erl. 250, 11. V. *taking the place of the earlier* se:—Hé wið ðone cyng geworhte for hwan hine se cyng ealles benæmde *he acted against the king; on which account the king deprived him of everything,* 1104; Erl. 239, 31: 1110; Erl. 243, 15: 1117; Erl. 246, 21.] [Laym. wha; whæt, what, wat: *Orm.* wha; whatt: *A. R.* hwo; hwat: *O. and N.* hwo, wo; hwat, what, wat: *R. Glouc.* wo; wat: Ayenb. huo; huet: *Chauc.* Piers P. who; what: *Goth.* hwas, m: hwó; f: hwa; n: *O. Sax.* hwe; hwat: *O. Frs.* hwa; hwet: *Icel.* hvar; hvat: *O. H. Ger.* hwer; hwaz: *Ger.* wer; was: *Lat.* quis; quid.] v. hwæt, hwý; ge-whá.

hwæcca *a chest, hutch:* — Corn-hwæcca *arca frumentaria,* Lye. [*Piers P.* (A.) Til perneles porfyl be put in heore *whucche,* iv. 102: *Allit.* Poems Alle woned in the *whichche* (ark) þe wylde & þe tame, 49, 362: *Jos. of Arith.* Make a luytel *whucche,* 2, 39: *Prompt. Parv.* whyche or hoche, hutche *cista, archa,* pp. 242, 255, see note on latter page.]

hwæder, hweder; *adv.* *Whither:* — Hwæder gá ic *ego quo ibo,* Gen. 37, 30. Ic gesette him hwæder hé búgan sceal *constituam tibi locum, in quem fugere debeat,* Ex. 21, 13. Gif hé eów áxie hweder gé willon *si interrogaverit 'quo vadis?'* Gen. 32, 17. [*Goth.* hwadre *whither.*] v. hwider.

hwæg, hwæig, hweg, es; *n.* [?] *Whey:*—Hwæg *serum,* Wrt. Voc. 290, 36. Ðeówan wifmen hwæig on sumera tó a *servant maid shall be given whey in summer,* L. R. S. 9; Th. i. 436, 32. Sceáphyrdes riht is ðæt hé hæbbe ... blede fulle hweges oð ðe syringe ealne sumor, 14; Th. i. 438, 25. DER. cýse-, wring-hwæg.

hwæl, es; *m.* *A whale:* — Hwæl *balena* vel *cete* vel *cetus* vel *pistrix,* Ælfc. Gl. 101; Som. 77, 54; Wrt. Voc. 55, 57. Hwæl *cætus,* Ælfc. Gr. 8; Som. 7, 31. Se hwæl biþ micle læssa ðonne óðre hwalas *the walrus is much less than other whales,* Ors. 1, 1; Swt. 18, 3. On ðæs hwæles innoþe *in ventre ceti,* Mt. Kmbl. 12, 40. Hwæles ðel *the sea,* Andr. Kmbl. 548; An. 274: Exon. 82 a; Th. 309, 20; Seef. 60: Chr. 975; Erl. 126, 22; Edg. 48. Bí ðam miclan hwale *concerning the great whale,* Exon. 96 b; Th. 360, 10; Wal. 3. God ðá gegearcode ænne hwæl and hé forswealh ðone wítegan, Homl. Th. i. 246, 12. Wilt ðú fón sumne hwæl? Nic. For hwí? Forðam plyhtlíc þingc hit ys gefón hwæl *vis capere aliquem cetum? Nolo. Quare? Quia periculosa res est capere cetum,* Coll. Monast. Th. 24, 15–22. Hé gesceóp ðá micclan hwalas, Lchdm. iii. 234, 12. [*Icel.* hvalr: *O. H. Ger.* wal *balæna, cetus,* Grff. i. 839.] v. hors-hwæl.

hwæl:—On hwæl hreópon [hwreopon, MS.] herefugolas *the birds of war screamed as they wheeled round,* Cd. 150; Th. 188, 1; Exod. 161. [Cf. *Icel.* hvel.] v. hwél *in* hweogul.

hwæla, an; *m.* *A whale:*—Hé is on middan hwælan hiwes *he is of a whale's shape in the middle,* Salm. Kmbl. 527; Sal. 263.

hwæl-hunta, an; *m.* *A whale-hunter, whale-fisher, whaler:*—Hwæl-hunta *cetarius,* Ælfc. Gl. 101; Som. 77, 55; Wrt. Voc. 55, 59. Ðá wæs

hē swā feor norþ swā ða hwælhuntan firrest faraþ *was as far north as the whalers ever go*, Ors. 1, 1; Swt. 17, 12.

hwæl-huntaþ, es; *m. Whale-fishing, whaling :*—On his āgnum lande is se betsta hwælhuntaþ, Ors. 1, 1; Swt. 18, 5.

hwæl-mere, es; *m. The sea*, Exon. 101 a; Th. 382, 2; Rä. 3, 5: Andr. Kmbl. 739; An. 370.

hwæm *a corner*. v. hwem.

hwēne. v. hwēne.

hwænne. v. hwanne.

hwǽr [or hwær?], hwar; *adv. Where*. **I.** *in direct questions :*—Gyf ic cweþe *ubi posuisti meum librum*, hwǽr lēdest ðú mīne bóc ðonne is se *ubi interrogativum* ðæt is äxigendlíc, Ælfc. Gr. 38; Som. 40, 60. Hwǽr ys se Judēa cyning ðe ācenned ys *ubi est qui natus est rex Judæorum?* Mt. Kmbl. 2, 2. Hwǽr cwom mearg hwǽr cwom mago hwǽr cwom māððumgyfa *where is the steed gone, where the rider, where the giver of treasure?* Exon. 77 b; Th. 291, 34; Wand. 92: Cd. 213; Th. 267, 11; Sat. 36. **II.** *in dependent clauses :*—Gif ic cweþe *tu scis ubi liber tuus est* ðonne biþ *ubi relativum*. Gif ic cweðe *nescio ubi inveniam meum librum*, nát ic hwǽr ic finde mīne bóc, ðonne biþ se *ubi infinitivum*, Ælfc. Gr. 38; Som. 40, 61. Ic næbbe hwǽr ic mæge ealle mīne wæstmas gegaderian *I have not where I may gather together all my fruits*, Homl. Th. ii. 104, 16: Mt. Kmbl. 8, 20. Hí gesáwon hwǽr hē ða deádan tó lífe ārǽrde, Homl. Th. ii. 414, 8: Cd. 32; Th. 41, 35; Gen. 667. Hwǽr mon unsófte getilaþ on forewearde ða ádle *where the treatment is severe in the early stage of the disease*, L. M. 2, 46; Lchdm. ii. 260, 15. Lóca hwǽr ðæt blód útwealle *see where the blood wells out*, Lchdm. iii. 142, 15: 226, 13. Ðá frægn wuldres aldor cain hwǽr abel eorþan wǽre *the Prince of glory asked Cain where on earth Abel was*, Cd. 48; Th. 61, 26; Gen. 1003. Ic sēce mīne gebróðru hwar hig healdon hyra heorda, Gen. 37, 16. **III.** *indefinite, anywhere, somewhere :*—Gyf hý hwǽr hit tóbrǽcaþ *if they violate it anywhere*, L. Ælfc. C. 34; Th. ii. 356, 16: Homl. Th. i. 170, 18: 482, 26. Gif se líchoma hwǽr mid hefiglícre hǽto sý gebysgod, Herb. 2, 6; Lchdm. i. 82, 8. Swǽ gelǽrede biscepas swǽ swǽ nú wel hwǽr [or welhwǽr] siendon *bishops so learned as now are nearly everywhere*, Past. pref. Swt. 9, 5: Chr. 897; Erl. 95, 19. Elles hwǽr *elsewhere*, Beo. Th. 277; B. 138. Hý writon hwǽr ánne dóm hwǽr óðerne *they wrote at one place one doom, at another another*, L. Alf. 49; Th. i. 58, 16. **IV.** *combined with* swā, *wheresoever, wherever :*—Swā hwǽr swā hold biþ *ubicunque fuerit corpus*, Mt. Kmbl. 24, 28. Swā hwǽr swā hē on wíc oððe on tūnas eode *quocunque introibat in vicos vel in villas*, Mk. Skt. 6, 56. Swā hwǽr swā *ubicunque*, 14, 9. [*A. R.* hwar; *O.* and *N.* hwar, war: *Orm.* whær: *Laym.* whær, wher: *Chauc. Wick. Piers P.* wher: *Ayenb.* huer: *Goth.* hwar: *O. Sax.* hwár: *O. Frs.* hwēr: *Icel.* hvar: *O. H. Ger.* hwâr.] DER. â-, ǽ-, ǽg-, ge-, gewel-, nā-, nât-, ó-, wel-hwǽr.

hwǽr *a vessel*. v. hwer.

hwæs; *adj. Sharp, keen :*—Hí hwæsne beág ymb mín heáfod heardne gebýgdon *they encircled my head with a crown sharp and hard [the crown of thorns]*, Exon. 29 a; Th. 88, 23; Cri. 1444. [*Goth.* hwass-aba *sharply : Icel.* hvass *sharp*.] Cf. hwæt.

hwǽstrian, hwǽstrung. v. hwástrian, hwástrung.

hwæt; *neut.* of hwā, *used as an adv.* or *interj. Why, what! ah!*—Be ðæs folces heringe ic nát hwæt wē ðæs fægniaþ *as regards popular applause, I know not why we rejoice at it*, Bt. 30, 1; Fox 108, 22. Hwæt befealdest ðú folmum ðínum bróðor ðínne *why hast thou felled thy brother with thy hands?* Cd. 48; Th. 62, 6; Gen. 1010: Andr. Kmbl. 1257; An. 629. Hwæt ðú leóda feala forleólce and forlǽrdest *how many people hast thou deceived and seduced?* 2726; An. 1365: Beo. Th. 1064; B. 530. Hwæt iudas hēt ðá settan ðæt líc *ah! then Judas bade them put down the body*, H. R. 13, 26. Hwæt mē ðín hand ðyder lǽdeþ *etenim illuc manus tua deducet me*, Ps. Th. 138, 8. Hwæt ðá Sem and Jafeth dydon ánne hwítel on hira sculdra *at vero Sem et Japheth pallium imposuerunt humeris suis*, Gen. 9, 23. Hwæt ðú ēce God *O! thou eternal God*, Bt. Met. Fox 20, 7; Met. 20, 4: 20, 92; Met. 20, 46. Hwæt ðú eart se sylfa God ðe ús ádrife fram dóme *nonne tu Deus qui repulisti nos?* Ps. Th. 107, 10. Hwæt wē nú gehýraþ *ah! now we learn*, Cd. 45; Th. 57, 36; Gen. 939. Hwæt wē gefrunon twelfe tíreádige hæleþ *lo! we have heard of twelve glorious heroes*, Andr. Kmbl. 1; An. 1: Beo. Th. 1; B. 1: Cd. 143; Th. 177, 27; Exod. 1: Rood Kmbl. 1; Kr. 1. Eá lā hwæt! Bt. Met. 4, 49; Met. 4, 25. [*So O. Sax.* hwat: *Icel.* hvat: *O. H. Ger.* waz *cur, quid, quare*.] v. hwā, hū.

hwæt; *adj. Quick, active, vigorous, stout, bold, brave :*—Sum biþ tó horse hwæt *one is a bold rider*, Exon. 79 b; Th. 298, 7; Crä. 81. Nis mon ofer eorþan tó ðæs hwæt ðæt hē ā his sǽfóre sorge næbbe *there is no man on earth so bold as never to have anxiety for his journey on the sea*, 82 a; Th. 308, 16; Seef. 40. Ne scyle se hwata esne ymb ðæt gnornian *virum fortem non decet indignari, quoties increpuit bellicus tumultus*, Bt. 40, 3; Fox 238, 10: Beo. Th. 6048; B. 3028. Hwatum Heorowearde, 4328; B. 2161. Hwate Scyldingas, 3206; B. 1601: 4111; B. 2052. Hý beóþ heortum þý hwætran *they will be the stouter of heart*, Exon. 107 a; Th. 408, 30; Rä. 27, 20.

ðēh ðe Sciþþie hæfdon māran monmenie and self hwætran wǽron *cum Scythæ et numero et virtute præstarent*, Ors. 3, 7; Swt. 116, 25. Ðone cræftgestan dǽl and ða hwatestan men ealles ðises middangeardes *fortissimas mundi partes*, 1, 10; Swt. 48, 6. Of ðæm hwatestan monnum Germanie *from the bravest men of Germany*, Swt. 48, 14. [*O. E. Homl.* hwat, wat: *Laym.* whæt, wat: *Ayenb.* huet: *O. Sax.* hwat: *Icel.* hvatr.] DER. ār-, bearhtm-, blēd-, dǽd-, dóm-, flyht-, fyrd-, gold-, gūþ-, leód-, mód-, sund-, swíð-hwæt; *and see* hwæs, hwettan.

HWǼTE, es; *m.* WHEAT: — Hwǽte *triticum*, Wrt. Voc. 287, 17. Grǽg hwǽte *far*, Ælfc. Gr. 9, 17; Som. 9, 52. Þurh ða gemetgunge hwǽtes *per mensuram tritici*, Past. 63; Swt. 459, 13. Fyrsas ða ðe willaþ derian clǽnum hwǽte, Bt. Met. Fox 12, 9; Met. 12, 5. Hē hí fēdde mid hwǽte, Ps. 80, 15. Tó ðǽm ðæt hē him tó tíde gemetlíce gedǽle ðone hwǽte *ut det illis in tempore tritici mensuram*, Past. 63; Swt. 459, 13. Fullne hwǽte on ðam eare *plenum frumentum in spica*, Mk. Skt. 4, 28. Hwǽtas *frumenta*, Ælfc. Gr. 13; Som. 16, 10. On hwǽtum *frumento*, Ps. Th. 64, 14. [*Orm.* whæte: *Ayenb.* huete: *Piers P.* whete: *Goth.* hwaiteis: *O. L. Ger.* huēte: *Icel.* hveiti; *n : Dan.* hvede: *O. H. Ger.* hwaizi *triticum, frumentum : Ger.* weizen.] DER. hláf-hwǽte.

hwǽt-eádig; *adj. Successful in war* [cf. *other compounds of* eádig] :—Biþ se hwæteádig wíggeweorþod se ðe ðæt wicg byrþ *he shall be successful and honoured in war whom that steed bears*, Elen. Kmbl. 2388; El. 1195.

hwǽte-corn, es; *n. A grain of wheat* :— Genim hnutcyrnla and hwǽtecorn *take nut-kernels and grains of wheat*, L. M. 1, 2; Lchdm. ii. 34, 19. [*O. E. Homl.* hwete-corn: *Icel.* hveiti-korn.]

hwǽte-cynn, es; *n. Wheat-kind :*—Hē ðē gesadade mid ðý sēlestan hwǽtecynnes holde lynde *adipe frumenti satiat te*, Ps. Th. 147, 3.

hwǽte-god *Ceres*, Lye.

hwǽte-gryttan, *pl. Coarse wheaten meal :*—Hwǽtegryttan *apludes vel cantalna* [= *cantabra*], Ælfc. Gl. 50; Som. 65, 124; Wrt. Voc. 34, 53.

hwǽte-healm, es; *m. The straw* or *stalk of wheat :*—Genim hwǽtehealm and gebærn tó duste, L. M. 1, 60; Lchdm. ii. 130, 14.

hwǽte-land, es; *m. Wheat-land, land for growing wheat upon :*—Ðæt hæft se arcebisceop genumen tó hwǽtelande, Cod. Dipl. Kmbl. iii. 159, 23.

hwǽte-melu, wes; *n. Wheaten meal* or *flour :*—Mid hwǽtemelwe, L. M. 3, 65; Lchdm. ii. 354, 12. [*Icel.* hveiti-mjöl.]

hwǽten ; *adj. Wheaten :* — Hwǽten hláf *siligeneus vel triticeus panis*, Ælfc. Gl. 66; Som. 69, 63; Wrt. Voc. 41, 19. Ic secge eów ðæt hwǽtene corn wunaþ āna būton hyt fealle on eorþan and sý deád *dico vobis nisi granum frumenti cadens in terram mortuum fuerit ipsum solum manet*, Jn. Skt. 12, 24. Mid hwǽtenan meluwe, Herb. 184, 4; Lchdm. i. 322, 13. Of hwǽtenum mealte geworht, iii. 74, 3. Hwǽtenne hláf, L. M. 1, 53; Lchdm. ii. 126, 1: Ps. Th. 77, 25. On hwǽtene wyrte *in wheaten wort*, L. M. 2, 57; Lchdm. ii. 268, 12. Nim hwǽten corn, L. M. 1, 75; Lchdm. ii. 150, 8. cxx hwǽtenra hláfa, Chart. Th. 460, 15.

hwǽte-smedeme, an; *f. Fine wheaten flour :*—Hunig and hwǽtesmedman, Lchdm. iii. 18, 5.

hwǽte-wǽstm, es; *m.* [?] *Corn* ; frumentatio, Ps. Vos. 77, 29, Lye.

hwǽder; *pron.* **I.** *which of two :*—Hwǽder ðara twegra dyde ðæs fæder willan *whether of them twain did the will of his father?* Mt. Kmbl. 21, 31. Hwǽder ys māre ðe ðæt gold ðe ðæt templ ðe ðæt gold gehālgaþ *whether is greater, the gold or the temple that sanctifieth the gold?* 23, 17, 19. Hwǽder wǽre twegra strengra wyrd ðe warnung? Salm. Kmbl. 853; Sal. 426. Gebíde gē hwǽder sēl mǽge wunde gedýgan ucer twega, Beo. Th. 5054; B. 2530. Hwǽdres ðonne ðara yfelra is betre ǽr tó tilianne būton swǽdres swǽder frēcenlícre is *quæ igitur pestis ardentius insequenda est, nisi quæ periculosius premit?* Past. 62, 1; Swt. 457, 21. Hwǽdres biþ hira folgoþ betra? Salm. Kmbl. 740; Sal. 369. Hwǽderne wylle gē ðæt ic forgyfe eów of ðísum twām *whether of the twain will ye that I release unto you?* Mt. Kmbl. 27, 21. Ðá befran Pilatus hwǽderne hí gecuron Hǽlend oððe Barraban? Homl. Th. ii. 252, 12. Nāst ðú hwǽder beóþ ðæs rícan mannes bān hwǽder ðæs þearfan *thou knowest not which are the rich man's bones, which the poor one's*, Homl. Th. i. 256, 16. **II.** *one or other of two, either :*—Hie hit gesund begen āgifan swā hit hwǽder hiora ǽr onfēnge būton hiora hwǽder þingode ðæt ... *let them both return it sound as either of them may have before received it, unless either of them made a condition that* ..., L. Alf. pol. 19; Th. i. 74, 11: Bt. Met. Fox 5, 81; Met. 5, 41. Gif hwā tó hwǽdrum ðissa geníed síe *if any one be forced to either of these*, L. Alf. pol. 1; Th. i. 60, 3. Tó manigenne sint ða gesomhīwan ðeáh hira hwædrum hwæt-hwugu hwílum mislícige on óðrum ðæt hie ðæt geþyldelíce forberen *admonendi sunt conjuges, ut ea, in quibus sibi aliquando displicent, patientes invicem tolerent*, Past. 51, 3; Swt. 395, 32. **III.** *each of two, both :*—Hwǽder hāt and ceald hwílum mencgaþ *both heat and cold at times mingle*, Cd. 216; Th. 273, 5; Sat. 132. **IV.** *in combination with* swā, *whichever of two :*—Heora eáþmetto ne mihton nāuht forstanden ne hūru heora ofermetta dydon swā hwæþer swā hý dydon *their humility availed naught nor indeed did their pride, whichever course they followed*, Bt. 29, 2; Fox 106, 1. Bí swā hwæðerre efes swā hit ðonne fierdleás

wæs *on whichever border there was then no force*, Chr. 894; Erl. 90, 13. On swā hwæðere hond *on whichever hand*, Beo. Th. 1376; B. 686. Drihtenes āre oððe deófles þeówet swā hwæðer wē geearniaþ hēr on lífe, Hy. Grn. ii. 289, 99; Hy. 7, 99. [*Laym.* whaðer: *O. and N.* hweþer: *Chauc.* whether: *Goth.* hvaþar: *O. Sax.* hwedar: *O. Frs.* hweder: *Icel.* hvárr: *O. H. Ger.* hwedar.] DER. ā-, æg-, nā-, nō-hwæðer; *and see* swæðer.

hwæðer, hweðer; *conj. Whether.* **I.** *in direct questions* :— Hwæðer ic mōte lybban ōþ ðæt ic hine geseó *may I live till I see him?* Homl. Th. i. 136, 30. Hwæðer gē willen on wuda sēcan gold ðæt reáde? Bt. Met. Fox 19, 9, 29; Met. 19, 5, 15. Hwæðer ðe ðín eáge mánful ys forðam ðe ic gōd eom *an oculus tuus nequam est, quia ego bonus sum?* Mt. Kmbl. 20, 15. Hwæðer cweþe wē ðe ūre ðe ðæra engla *shall we say ours or the angels?* Homl. Th. i. 220, 20. Cwyst ðū hwæðer ic hyt sī *numquid ego sum?* Mt. Kmbl. 26, 25. **II.** *in dependent clauses* :—Lǣtaþ ðæt wē geseón hwæðer elias cume *sinite videamus si veniat helias*, Mk. Skt. 15, 36. Gregorius befran hwæðer ðæs landes folc cristen wǣre ðe hǣðen, Homl. Th. ii. 120, 23. Hī nysten hwæðer hē on Godes mihte ða þing worhte ðe þurh deófles cræft, Guthl. 17; Gdwin. 70, 17. Swīðe hræðe æfter ðon hē gecýðde hwæðer hē mǣnde ðe ðæs mōdes fōster ðe ðæs líchoman *qui hoc in loco pastionem cordis an corporis suaderet, aperuit*, Past. 18, 6; Swt. 137, 18. Hwæðer hit sig ðe sōþ ðe leás *utrum vera an falsa sint*, Gen. 42, 16. Josep ðode hig hwæðer hira fæder wǣre hāl oððe hwæðer hē lyfode *ille interrogavit eos dicens: Salvusne est pater vester? adhuc vivit?* 43, 27. Sceáwiaþ ðæt land hwæðer hit wæstmbǣre sī . . . and hwæðer ðæt landfolc sī tō gefeohte stranglíc oððe untrumlíc, feáwa on getele hwæðer ðe fela, Num. 13, 19–20. [*O. Sax.* hwedar: *O. Frs.* hweder: *Icel.* hvárt: *O. H. Ger.* hwedar.] v. *preceding word.*

hwæðere, hwæðre, hwæððre, hweðre; *adv. Yet, however, nevertheless* :—Ac nǣnig hwæðere him gelíce dōn ne mihte *but none however could do like him*, Bd. 4, 24; S. 596, 39. Hwæðere ðū meaht mē singan *attamen mihi cantare habes*, 597, 15. Hwæðere for fremsumnysse *tamen pro benignitate*, 1, 27; S. 493, 7. Hwæðere *verumtamen*, Ps. Th. 61, 5, 9: 67, 21. Ðeáh ðe . . . hwæðere *although . . . yet*, Beo. Th. 3441; B. 1718. Ne ðū hweðere on mōde milde weorþest eallum *non miseraris omnibus*, Ps. Th. 58, 5. Nō hweðere reste fand *did not find rest however*, Cd. 72; Th. 87, 30; Gen. 1456. Hwæþre hē getrymede heora geleáfan mid ðon heofonlícon weorce ðeáh hie ðæt word ðæs heofonlícan gerýnes ne ongeáton, Blickl. Homl. 17, 7. Hwæðre ðeáh *however*, Bt. Met. Fox 20, 108; Met. 20, 54. Hwæðre swā ðeáh, Beo. Th. 4876; B. 2442. Hwæððre, Past. 56, 2; Swt. 431, 26. Hweðre, Blickl. Homl. 125, 31: 207, 34. v. ðeáh-hwæðere.

hwæðere, hwæðre [=hwæðer]; *conj. Whether*, Exon. 37 b; Th. 123, 15; Gū. 323: Beo. Th. 2632; B. 1314.

hwæt-hwega, -hwigu, -hugu; *pron. and adv.* [cf. *use of* something *in* Shakspere.] *Something, somewhat, a little* :—Sing mē hwæthwegu *canta mihi aliquid*, Bd. 4, 24; S. 597, 12. Hwæthugu wundurlícre hálignesse *aliquid miræ sanctitatis*, 3, 9; S. 534, 1. Hwæthwego seldcúþes *something strange*, Bt. 34, 4; Fox 138, 28. Hwæthwygo *aliquid*, Nar. 1, 18. Ic hwæthwugo on bōcum geleornode, 39, 19. Huodhuoegu *aliquid*, Jn. Skt. Lind. 7, 4. Hwæthwega *paulisper, parumper*, Ælfc. Gr. 38; Som. 41, 65. Hē hwæthwego fram ðam wage ða limu āhōf, Guthl. 20; Gdwin. 82, 27: Homl. Th. ii. 90, 29. Hwæthwega ufor gān, 32, 22. Hwæthwegu tōdǣled *somewhat separated*, Bt. 34, 6; Fox 142, 14. Hwæthwiga *aliquantulum*, Ps. Th. 89, 15. Hwæthwygu, 93, 8. Hwæthwugu, Bt. Met. Fox 221; Met. 20, 111.

hwæt-hweganunges, -hwigununges, -huguninges, *adv. Somewhat* :—Hwæthweganunges [MS. Cot. -hwugununges] *aliquantum*, Bt. 11, 1; Fox 30, 27. Hwæthwegnunges, 11, tit; Fox xii. 10. Ða niétenu ðonne beóþ hwæthuguningas [MS. Cott. -hwugununges] *from eorþan āhæfen in animalibus vero jam quidem cogitationes aliquantulum a terra suspensæ*, Past. 21, 3; Swt. 155, 15.

hwæt-hwoegno; *pron. Anything, something*; aliquid, Jn. Skt. Rush. 7, 4. v. hwæt-hwega.

hwæt-líce; *adv. Quickly, speedily* :—Gehýr mē hwætlíce *exaudi me*, Ps. Th. 137, 4. Hwætlícor *citius*, Coll. Monast. Th. 31, 23. [*Icel.* hvat-liga *quickly*.]

hwæt-mōd; *adj. Stout-hearted, bold* :—Hæleþ hwætmōde *men stout of heart*, Elen. Kmbl. 2009; El. 1006: Exon. 55 b; Th. 197, 3; Az. 184.

hwæt-ness, e; *f. Quickness, agility* :—Seó fægernes and seó hwætnes ðæs líchoman geblissaþ ðone mon *pulcritudo atque velocitas videntur præstare celebritatem*, Bt. 24, 3; Fox 84, 6.

hwæt-rǣd; *adj.* [?] *Strong of purpose* or *counsel* :—Hwætrēd, hygerōf, Exon. 124 a; Th. 477, 5; Ruin. 20.

hwæt-scipe, es; *m. Quickness, boldness, bravery, valour* :—Oft mon biþ swīðe rempende and rǣsþ suīðe dollíce on ælc weorc and hrædlíce and ðeáh wēnaþ men ðæt hit sīe for arodscipe and hwætscipe *sæpe præcipitata actio velocitatis efficacia putatur*, Past. 20, 1; Swt. 149, 13. For hiora cræftum and for hiora hwætscipe iówra selfra anwald[es] eóweres

unþonces habban mehton *by their strength and valour might have had dominion over you against your will;* armis vindicare potuissent, Ors. 1, 10; Swt. 48, 21. Sinope tōeácan hiere hwætscipe and hiere monigfealdum duguþum hiere líf geendade on mægþhāde *Sinope singularem virtutis gloriam perpetua virginitate cumulavit*, Swt. 46, 24.

hwalf. v. hwealf.

hwall; *adj. Procax*, Cot. 171, Lye. [*O. H. Ger.* hwell *procax*; hwelli *pertinacia*.]

hwalwa [=hwalfa?] *devexus*, Cot. 67, Lye. v. hwealf.

hwamm, hwomm, es; *m. A corner* :—Heáfod hwommys *caput anguli*, Ps. Spl. C. 117, 21. Huommes, Mk. Skt. Lind. 12, 10:, Lk. Skt. Lind. 20, 17. Ðā eode ūt of ðæs karcernes hwomme swīðe egeslíc draca ðen came *a very horrible dragon out of a corner of the prison*, Nar. 43, 13. Hwommona heágost *caput anguli*, Ps. Th. 117, 21. On ðínes hūses hwommum *in lateribus domus tuæ*, 127, 2. In hwommum worþana *in angulis platearum*, Mt. Kmbl. Rush. 6, 5. Ofer ealle heá hwommas *super omnes angulos excelsos*, Past. 35, 5; Swt. 245, 7. v. hwemm.

hwam-stán, es; *m. A corner-stone* :—In heáfut huomstānes *in caput anguli*, Mt. Kmbl. Lind. 21, 42.

hwanan, hwanon, hwonan, hwanone; *adv. Whence.* **I.** *in direct questions* :—Interrogativa synd āxigendlíce, *unde* hwanan, Ælfc. Gr. 38; Som. 41, 58. Hwanon hæfde hē coccel *unde habet zizania?* Mt. Kmbl. 13, 27. Hwanun wāt ic þis *unde hoc sciam?* Lk. Skt. 1, 18. Hwanone sceoldest ðū specan on Hebréisc *how should you speak in Hebrew?* Nicod. 4; Thw. 2. 27. **II.** *in dependent clauses* :—Hī spyredan hwæt and hwonan hē wæs *investigantes unde vel quis esset*, Bd. 1, 33; S. 499, 12. Ic ne wāt hwonon his cyme sindon *I know not whence is his coming*, Exon. 50 b; Th. 175, 18; Gū. 1196. Hwanan, Beo. Th. 4798; B. 2403. Ðā næfde hē hwanon hē his wer āgulde *he had not means to pay his 'wer,'* Chart. Th. 207, 36. [*Laym.* whanene: *O. and N.* wanene, hwenene, hwenne: *Ayenb.* huannes: *Chauc.* whennes: *O. Sax.* hwanan: *O. H. Ger.* hwanan, hwanana: *Ger.* wannen.] DER. æg-, ge-, nā-, ō-hwonan.

hwanne, hwænne, hwonne; *adv. When.* **I.** *in direct questions* :—*Quando venisti* hwænne cōm ðū? is *interrogativum*, Ælfc. Gr. 38; Som. 40, 64. Hwonne ǣr beó deád oððe hwænne his nama āspringe *quando morietur, et peribit nomen ejus?* Ps. Th. 40, 5. **II.** *in dependent clauses* :—*Quando ero doctus* hwænne beó ic gelǣred, is *infinitivum*, Som. 40, 65. Sege ūs hwænne ðás þing gewurdon *dic nobis quando ista fient*, Mk. Skt. 13, 4: Mt. Kmbl. 2, 7. Þincþ him tō lang hwænne hē beó genumen of ðyses lífes earfoþnyssum *it seems to him too long* [*to the time*] *when he shall be taken from the troubles of this life*, Homl. Th. i. 140, 9. Lǣt gebídan beornas ðíne hwænne ðū eft cyme *let thy men await the time of thy return*, Andr. Kmbl. 800; An. 400. Ðā wæs ðæt hē sorgende bād hwonne seó ādl tō him cōme *qui cum sollicitus horam accessionis exspectaret*, Bd. 3, 12; S. 537, 6. Hit biþ long hwonne se hláford cume *moram facit Dominus meus venire*, Past. 17, 8; Swt. 121, 12. Hit earfoþe is ænegum menn tō witanne hwonne hē geclǣnsod síe *it is difficult for any man to know when he is cleansed*, 7, 2; Swt. 51, 5. Sǣles bídeþ hwonne ǣr heó cræft hyre cýðan mōte *it waits for the time for displaying its art* [cf. *O. Sax.* that werod bēd hwan ēr the frōdo man gifrumid habdi waldandes willeon], Exon. 108 b; Th. 413, 29; Rä. 32, 13. **III.** *indefinite, at some time* :—Se ilca ūs wile nū hwonne eft mid eallum egesan gesēcan *the same will visit us again at some time with all terror*, Blickl. Homl. 123, 32. [*Laym.* whenne, wonne: *Orm.* whanne: *O. and N.* hwanne, wonne: *A. R.* hwonne, hwon: *Wick.* whanne: *Chauc.* whan: *Goth. O. Sax.* hwan: *O. H. Ger.* hwanne, hwenne *quando, aliquando: Ger.* wann.]

hwar. v. hwǣr.

hwarne, Mt. Kmbl. Lind. 8, 30. v. hwergen.

hwast, es; *or* [?] hwasta, an; *m. An effeminate person* [?], *a eunuch* :—Hwastas *molles*, Som. Huastana *eunuchorum*, Mt. Kmbl. p. 18, 9.

hwāstrian, hwæstrian; *p.* ede *To whisper, murmur, mutter* :—Āgēn mē hwæstredun ealle fýnd míne *adversum me susurrabant omnes inimici mei*, Ps. Lamb. 40, 8. Huæstredon *murmurabant*, Mt. Kmbl. Lind. 20, 11: Lk. Skt. Lind. 19, 7: Jn. Skt. Lind. 6, 41, 61. Huæstria *murmurari*, 6, 43. Huæstrende *murmurantem*, 7, 32. [Cf. hwisprian, hwistlian; and *Icel.* hvískra, hvísla *to whisper*.]

hwāstrung, hwæstrung, e; *f. A whispering, murmuring, muttering* :—þurh hwāstrunge *per susurrationem*, Confess. Peccat. Huæstrung *micel murmur multus*, Jn. Skt. Lind. 7, 12. [Cf. *Icel.* hvískran *a whispering*.]

hwat, es; *n. Augury, divination* :—Ne gímon hwata ne swefna *non augurabimini nec observabitis somnia*, Lev. 19, 26. Wē lǣraþ ðæt preósta gehwilc forbeóde hwata and galdra *we enjoin that every priest forbid auguries and incantations*, L. Edg. C. 16; Th. ii. 248, 3. v. hwatung.

hwata; *adj.* v. hwæt.

hwata, an; *m. An augur, diviner* :—Warna ðē ðæt ðū ne gíme drýcræfta ne swefena ne hwatena *nec inveniatur in te, qui ariolos sciscitetur et observet somnia et auguria*, Deut. 18, 10. v. fugel-hwata.

hwætend *iris illyrica*, Lchdm. iii. 334, col. 1.

hwatung, e; *f. Divination* :—Ālýfed nys ídele hwatunga tō begánne

permissum non est vanas divinationes exercere, L. Ecg. P. ii. 23, title; Th. ii. 180, 36. Nis nā sōðlíce ālýfed nānum cristenum men ðæt hē ídele hwatunga begā swā hǽðene men dōþ ðæt is ðæt hig gelýfon on sunnan and on mōnan and on steorrena ryne and sēcon tída hwatunga hyra þing tō begynnanne *homini christiano certe non est permissum vana auguria facere, uti gentiles faciunt, id est, quod credant in solem et lunam, et in cursum stellarum; et auguria temporum exquirant, ad negotia sua incipienda*, 23; Th. ii. 190, 30–3. Gif hwā hwatunga begā *si quis divinātiones exerceat*, iv. 19; Th. ii. 210, 11. v. hwat.

hwealf, e; f. *An arched* or *vaulted covering* :—Under heofenes hwealf *under the vault of heaven*, Beo. Th. 1156; B. 576: 4034; B. 2015. Behealde hē hū wídgille ðæs heofenes hwealfa biþ *late patentes ætheris cernat plagas*, Bt. 19; Fox 68, 22. Hū wídgil sint heofenes hwealfe, Bt. Met. Fox 10, 13; Met. 10, 7. Hwalf *clima*, Cot. 56, Lye. [*Icel.* hválf; n. *a vault; the concavity of a shield*.] v. heofon-hwealf.

hwealf; adj. *Arched, vaulted, concave* [*of a shield*] : — Hwealfum lindum, Judth. 11; Thw. 24, 29; Jud. 214. v. preceding word.

hwealfian *to arch, vault*, Som. [*Icel.* hwelfa *to arch, vault*.]

hwearf, es; m. *A crowd, troop, band of people* :—Hwearfum þringan *to press in crowds*, Judth. 12; Thw. 25, 8; Jud. 249: Exon. 36 a; Th. 118, 3; Gū. 234. [*O. Sax.* hwarf *a crowd*. Cf. hwearfian; *and* gang *a number of people* (*in its connection with the verb* gangan).]

hwearf, hwerf, es; m. *A turn, space, change, exchange, that which is exchanged* : — Be hwearfe. Nān man ne hwyrfe nānes yrfes būtan ðæs gerēfan gewitnesse . . . Gif hit hwā dō fō se landhláford tō ðam hwearfe *Of exchange. Let no man exchange any property without the witness of the reeve . . . If any one do so let the lord take possession of the property exchanged*, L. Ath. i. 10; Th. i. 204, 16–21. In huarf *in spatio*, Lk. Skt. Lind. 24, 13. Huelc seles monn hwerf fore sáuel his *quam dabit homo commutationem pro anima sua*, Mt. Kmbl. 16, 26. Huoerf, Mk. Skt. Lind. 8, 37. Gif huerf gie sellas *si mutuum dederitis*, Lk. Skt. Lind. 6, 34. Ðæt wharfe and ðæt foreward *pactionem et commutationem*, Cod. Dipl. Kmbl. iv. 241, 37. [Cf. *O. Frs.* hwarf, werf (*with numerals*) achte werf *octies*: *O. H. Ger.* sibun warb *septies*; hwarba *motus, vicis*, Grff. iv. 1235. Cf. *the use of* síþ *in A. S. and the corresponding forms in other dialects, and the use of* gang *in Danish and Swedish, with numerals*.] v. ge-hwearf, hwearf-líce.

hwearf, es; m. *A wharf, bank, shore* : — Ðā gyrnde ðæt hē mōste macian foran gēn Mildryþe æker ǽnne hwerf wið ðon wōdan tō werianne *then he desired that he might make a bank opposite Mildred's field for protection against floods* [?], Chart. Th. 341, 7. v. mere-hwearf.

hwearf; adj. *Turning about, shifting, veering, changeable* :—Norþan wind heaþogrim and hwearf *a wind from the north deadly fierce and whirling in eddies*, Beo. Th. 1100; B. 548. Thorpe, Kemble, Heyne read andhwearf = came against [us]; Grein takes and hwearf, and compares *Icel.* hverfr *shifty*. The word may describe a strong wind often shifting its direction and whirling round with violent gusts. Cf. ge-hweorf; hwerf-líc.

hwearfan. v. hwerfan.

hwearfian; p. ode *To turn, change, roll about, revolve, wander, move, toss about* :—Ic nū giet hwearfige mē self on ðǽm ýdum mínra scylda *adhuc in delictorum fluctibus versor*, Past. 65, 7; Swt. 467, 22. Ǽlc gesceaft hwearfaþ on hire selfre swā swā hweól and tō ðam heó swā hwearfaþ ðæt heó eft cume ðǽr heó ǽr wæs *every creature turns on itself as a wheel, and it so turns to the end that it may come again where it was before* : repetunt proprios quæque recursus, rediituque suo singula gaudent, Bt. 25; Fox 88, 32: Bt. Met. Fox 13, 150; Met. 13, 75. Hē biþ fremede freán ælmihtigum englum ungelíc āna hwearfaþ *he shall be a stranger to the almighty Lord, unlike angels, alone shall he wander*, Salm. Kmbl. 70; Sal. 35. Drihtnes stíge hwearfaþ aa wísra gewyrdum *Ascension-day ever changes according to the rules of the learned*, Menol. Fox 131; Men. 65. Wē hwearfiaþ heánlíce *we wander abjectly*, Exon. 13 a; Th. 23, 21; Cri. 372. Hálige englas ðǽrābútan hwearfiaþ *holy angels hover round about the place*, L. C. E. 4; Th. i. 360, 34. Ðū wāst hū ða woruldsǽlþa hwearfiaþ . . . hwí ne hwearfost ðū mid him *thou knowest how worldly blessings change . . . why dost thou not change with them?* Bt. 7, 2; Fox 18, 6. Swā swā on wænes eaxe hwearfiaþ ða hweól as the *wheels turn on the axle of a waggon*, 39, 7; Fox 220, 32. Gūþ hwearfode *the battle rolled on* [or could guþ here be taken as a person, one of the Valkyrias, and *hwearfode* = hover about, as in the passage above, L. C. E. 4?], Cd. 149; Th. 187, 29; Exod. 159. Fana hwearfode on sceafte *the banner waved on its staff*, Bt. Met. Fox 1, 20; Met. 1, 10. Hwæt is ðē ðæt ðū ðǽrmid ne ne hwearfige *why shouldest thou not change with them?* Bt. 7, 3; Fox 22, 22. Nis ǽnegu gesceaft ðe ne hwearfige swā swā hweól dēþ, Met. Fox 13, 147; Met. 13, 74. Hwearfode, 20, 411; Met. 20, 206. Hwearfian, Bt. 33, 4; Fox 132, 11. Heán hwearfian *to wander abject*, Andr. Kmbl. 1781; An. 893. Fōran hwearfigende [hwearfiende, MS. Cott.] geond ðæt wēsten *they went wandering through the desert*; per vasta deserti evagatur, Ors. 6, 31; Swt. 286, 19. [*Goth.* hwarbōn *to go about*: *O. Sax.* hwarbōn: *Icel.* hvarfa *to wander about*: *O. H. Ger.* hwarbōn *versari*.]

---

hwearf-líce; adv. *In turn* :—Huoerflíce *vicissim*, Lk. Skt. p. 10, 6.

hwearflung. v. hwerflung.

hwearft, es; m. *A circuit, circle, revolution* :—Hwæt bíðaþ gē on hwearfte *why do ye stand round waiting?* Exon. 15 a; Th. 32, 12; Cri. 511. Under heofones hwearfte *under heaven's circuit*, 110 b; Th. 424, 3; Rä. 41, 33. Brādne hwearft *the broad expanse* [*of the sky*], 53 b; Th. 187, 29; Az. 38. Ymb wintra hwearft *after years have rolled on*, Th. 188, 5; Az. 41. v. ymb-hwearft, hwyrft.

hwearftlian; p. ode *To turn round, roll round, revolve, move about, rove* :—Ic hwearftlige *verso*, Ælfc. Gr. 37; Som. 39, 15. Ða eágan ðe nū þurh unālýfedlíce gewilnunga hwearftliaþ *the eyes that now rove through unallowed desires*, Homl. Th. i. 530, 31. Se cwyrnstān ðe tyrnþ singallíce and nǽnne færeld ne þurhtíhþ getácnaþ woruldlufe ðe on gedwyldum hwyrftlaþ and nǽnne stæpe on Godes wege gefæstnaþ *the millstone that is continually turning and makes no progress, betokens worldly love, that goes round and round in errors and takes no firm step in the way of God*, 514, 21. Micel trúwa hwearftlode on Petres heortan *great trust was revolving in Peter's heart*, 392, 34.

hwearfung, e; f. *A turning, revolution, change, exchange, barter* :—Ðē wæs ðeós hwearfung betere forðam ðe ðissa woruldsǽlþa tō wel ne lyste *this change was more tolerable to thee, because thou didst not take too much pleasure in temporal blessings*, Bt. 7, 3; Fox 22, 23. On midre ðisse hwearfunga, Fox 22, 19. Ðæt tácnaþ ceápunge and hwearfunge *that betokens chaffer and barter*, Lchdm. iii. 156, 6. Ne miht ðū ðara woruldsǽlþa hwearfunga onwendan *nor canst thou avert the revolutions of worldly happiness*, Bt. 7, 2; Fox 18, 37. v. hwerfung.

hweg. v. hwǽg.

hwega. v. hū-, hwæt-, hwilc-hwega.

hwelan, hwylan; p. hwæl *To roar, bellow* :—Streámwelm hwileþ *the surf roars*, Andr. Kmbl. 990; An. 495. [Cf. *Icel.* hvellr *a shrill sound*; hwellr *shrill*.] v. on-hwelan; hwelung.

hwelc. v. hwilc.

hwele *putrefaction*, Som. [*Prompt. Parv.* whele or whelke [whelle] *pustula*.] v. next word.

hwelian; p. ode, ede *To turn to matter; in pus converti* :—Ðanon se andiga hwelaþ *inde invidus contabescit*, Lchdm. iii. 365, col. 1. Gif ðæt líc heard sí ūtan lege on ðane lǽcedom ðe ðæt heard forðí hwelige and ðæt yfel ūt teó *if the body be hard on the outside apply such leechdom as the hard part may turn to matter thereby, and may draw out the mischief*, L. M. 2, 59; Lchdm. ii. 282, 23. [*Prompt. Parv.* whelyñ, as soorys *pustulo*.] v. ge-hweled; hwele.

HWELP, es; m. *A* WHELP, *a young dog, the young of other animals; catulus* :—Hund *canis*, hwylp *catulus*, Wrt. Voc. 78, 53. Hwelp *catulus* [*leonis*], Ps. Th. 16, 11. Ða hwelpas etaþ of ðām crumum ðe of hyra hláforda beódum feallaþ *catelli edunt de micis quæ cadunt de mensa dominorum suorum*, Mt. Kmbl. 15, 27: Mk. Skt. 7, 28. [*Laym.* whelp: *Orm.* (leness) whellp: *A. R.* hweolp: *Prompt. Parv.* whelp, lytyl hownde *catellus, catulus*: *O. Sax.* hwelp: *Icel.* hvelpr: *Dan.* hvalp: *O. H. Ger.* hwelf *the young of animals* (*lion, tiger, ape*).] DER. león-, wæl-hwelp.

hwelung, e; f. *Sound, noise* :—Hwelung *clangor tubæ*, Cot. 109, Lye. v. hwelan.

hwem, hwemm, es; m. *A corner, angle* :—Hwæt fremaþ ðære burhware ðeáh ðe ðæt port beó trumlíce on ælce healfe getimbrod gif ðǽr biþ ān hwem open forlǽten ðæt se onwinnenda here þurh ðam infær hæbbe *what does it avail the citizens, though the town be firmly built on every side, if a corner be left open, so that the assailing host may have entrance through it?* Homl. Th. ii. 432, 4. Hwæm *angulus*, Ps. Spl. T. 117, 21. Ða feówer hwemmas ealles middangeardes *the four corners of the whole world*, Homl. Th. i. 130, 21: ii. 252, 3. v. hwamm.

hwem-dragen; adj. *Sloping, not perpendicular* :—Wæs ðæt ilce hūs hwemdragen nalas æfter gewunan mennisces weorces ðæt ða wagas wǽron rihte ac git swíðor on scræfes onlícnesse ðæt wæs æteówed *that same house had sloping walls, not at all after the custom of men's work so that the walls should be perpendicular, but it appeared much more like a cave*, Blickl. Homl. 207, 17. v. next word.

hwemman; p. de *To slope, incline* :—Hí hwemdon ðā mid ðām scypon wið ðæs norþlandes *they inclined then with the ships towards the north shore*, Chr. 1052; Erl. 184, 25.

hwēne, hwǽne [= hwoene]; adv. *A little, somewhat* :—Hwēne ǽr *a little before*, Bt. 23; Fox 78, 27. Hwēne ǽror, Homl. Th. i. 358, 24. Hwēne wíddre ðonne bydenfæt *somewhat wider than a bushel measure*, Blickl. Homl. 127, 6. Hwēne rūmedlícor *paulo latius*, Past. 12; Swt. 75, 17. Nioþor hwēne *somewhat lower*, Beo. Th. 5392; B. 2699. Hwǽne heardor and strangor *paulo districtius*, Bd. 1, 27; S. 490, 12. Hwǽne ǽr, Shrn. 50, 13. Hwǽne gangende *progressus pusillum*, Mt. Kmbl. Rush. 26, 39. Hwoene lǽssan *paulo minus*, Ps. Stev. 8, 6. [*In Cumberland Dialect* wheen, whun *a few* : *Scot.* quheyne *few*; quhene *a small number*; wheen *a number*.] v. hwōn.

hweogul, hweowol, hweohl, hweól, es; n. *A wheel* :—Se firmamentum went on ðām twām steorrum swā swā hweogel [hweogul, MS. L; hweowul,

MSS. R. P.] tyrnþ on eaxe *the firmament turns on those two stars just as a wheel turns on an axle*, Lchdm. iii. 270, 22. Swá swá hweowol *ut rotam*, Ps. Spl. 82, 12. Wǽnes hweowol *a waggon-wheel*, Shrn. 32, 12. Swá swá yrnende hweowol, Hexam. 5; Norm. 8, 29. Ðǽre sunnan hweogul *solis rota*, Hymn. Surt. 22, 25. Hweól *rota*, Ælfc. Gl. 2; Som. 55, 48; Wrt. Voc. 16, 20. Ðæt hweól hwerfþ ymbútan, Bt. 39, 7; Fox 220, 29. Ðæt unstille hweól ðe Ixion wæs tó gebunden ðæt óþstód, 35, 6; Fox 168, 31. Ðæs hweohles [hweoles, MS. Cott.] felga, 39, 7; Fox 222, 19. On hweohle *in rota*, Ps. Spl. 76, 17. Hwél in hwélum *rota in rota*, Mt. Kmbl. p. 9, 20. [*A. R.* hweol: *Ayenb.* hueȝel: *Orm.* wheol, whel: *Icel.* hvel, hjól: *Dan. Swed.* hjul. Zacher in his ' Das Gothische Alphabet,' pp. 114–5, compares the two forms *hweol, hweogel* with the Greek κίρκος, κύκλος (=κύκλος) respectively, and so does not write hweól. See also Grmm. D. M. p. 664, where *hweol* is taken as corresponding to a Gothic *hwil*.]

hweóled; *adj. Provided with wheels*:—Héhhwiólod wǽn *a waggon having high wheels*, Lye.

hweól-fág; *adj. Circular and ornamented* [*applied to a dress*]:—Hwiólfág *cyclas*, Cot. 49, Lye.

hweop *a whip*; *flagellum*, Som.

hweorf. v. hwearf.

hweorfa, an; *m. Something which turns, a joint, a whorl* [*of a spindle*]:—Hweorfa *vertuba*, Wrt. Voc. 65, 16. Hwerfa *vertigo*, Ælfc. Gl. 74; Som. 71, 49; Wrt. Voc. 44, 32 [*in both cases the words occur among names of parts of the body—the two following are found among words connected with spinning*]. Hweorfa *verticillum*, Wrt. Voc. 66, 16: *vertelum*, 281, 72. Nim ðone hweorfan ðe wíf mid spinnaþ bind on his sweoran *take the whorl that women spin with, bind it on his neck*, L. M. 3, 6; Lchdm. ii. 310, 21. [Halliwell quotes Kennett's description of a *whorle*, 'the piece of wood put upon the iron spindle to receive the thread.' Cf. *O. H. Ger.* hwerbo *vortex, vorago*.] v. þeóh-hweorfa.

hweorfan, hweorfan, hwurfan *To turn, change, go, return, depart, go about, wander, roam, hover about*:—Nǽfre ic from hweorfe ac ic mid wunige áwa tó ealdre *I will never go from you, but I will dwell with you for ever*, Exon. 14 b; Th. 30, 8; Cri. 476. Ðú hweorfest of hénþum in gehyld godes *thou shalt pass from humiliations into the favour of God*, Andr. Kmbl. 233; An. 117. Mín folc hider hweorfeþ *revertetur huc populus meus*, Ps. Th. 72, 8: Exon. 76 a; Th. 284, 27; Jul. 703. Siððan heó ofer brim hweorfeþ *after it* [*the sun*] *had gone beyond the ocean*, 93 b; Th. 351, 17; Sch. 81: 110 a; Th. 422, 13; Rä. 41, 5. Gé tó mé on hyge hweorfaþ *ye turn to me in thought*, 98 a; Th. 366, 2; Reb. 6. On hinderling hweorfaþ *revertentur inimici mei retrorsum*, Ps. Th. 55, 8: 69, 3. On heora ágen dust æfter hweorfaþ *in pulverem suum revertentur*, 103, 27. Hí tówrecene wíde hweorfaþ *ipsi dispergentur*, 58, 15. Hweorfaþ æfter heorþe *they walk along the floor of the furnace*, Exon. 55 b; Th. 196, 18; Az. 176. Swá hweorfaþ gleómen so *gleemen roam about*, 87 a; Th. 326, 28; Víd. 135. Ðá seó scyld ðá tó his heortan hwearf *ad cor suum rediit*, Bd. 4, 25; S. 599, 35. Ierre hé hwearf ðonan tó his ágnum, Chr. 584; Erl. 18, 25. Hé ána hwearf mondreámum from *he went alone from human joys* [i. e. *died*], Beo. Th. 3433; B. 1714. Hwearf geond ðæt healreced Hæreðes dóhtor, 3965; B. 1981. Hé hwearf æfter wegum *he went along the roads*, Blickl. Homl. 199, 13: Beo. Th. 5657; B. 2832. Hwearf ðér Hróðgár sæt, 717; B. 356. Fæder ellor hwearf, 110; B. 55: Judth. 10; Thw. 23, 9; Jud. 112. Hwǽrf him ðá tó heofenum hálig drihten *the holy Lord returned to heaven*, Cd. 13; Th. 16, 7; Gen. 240. Hwearf eft tó his ágnum biscopdóme, Chr. 813; Erl. 60, 22. Hé hwearf be wealle *he went along the wall*, Beo. Th. 3150; B. 1573: 2380; B. 1188. Hengest hwearf him on láste *Hengest went after them*, Fins. Th. 35; Fin. 17. Gástas hwurfon sóhton engla éþel *spirits went and sought the angels' country*, Andr. Kmbl. 1280; An. 640. Hyssas hále hwurfon in ðam hátan ofne *the men walked unharmed in that hot furnace*, Cd. 188; Th. 233, 5; Dan. 271. Bláce hworfon sceaþan hwearfdon, 214; Th. 269, 11; Sat. 71. Ǽr hí on tú hweorfon *before they separated*, Andr. Kmbl. 2102; An. 1052. Hweorfon ða hǽðenan hæftas fram ðám hálgan cnihton *the heathen slaves went from the holy youths*, Cd. 187; Th. 232, 28; Dan. 267. Hweorfaþ eft tó mé *return to me*, Blickl. Homl. 235, 16. Him his gebed hweorfe tó fyrenun *oratio ejus fiat in peccatum*, Ps. 108, 6. Ðý læs hé for wlence of gemete hweorfe and forhycge heánspédigran *lest from pride he depart from moderation and despise the more scantily endowed*, Exon. 78 b; Th. 294, 35; Crä. 25. Ǽr hé on weg hwurfe gamol of geardum, Beo. Th. 534; B. 264. Hogedon georne ðæt ǽ godes ealle gelǽste and ne áwácodon wereda drihtne ne ðan má gén [(?) þan mægen, Th: heánmægen, Grein: mægenhwyrfe, Btwk.] hwyrfe in hǽðendom *they strove earnestly to perform all God's law, and not to be apostate from the Lord of hosts any more than to turn to heathendom*, Cd. 183; Th. 229, 22; Dan. 221. Hwonne se dæg cume ðæt hé sceolde ðæs ealles ídel hweorfan *when the day comes that he must depart having nothing of it at all*, Blickl. Homl. 97, 26. Ðæt ic meahte hweorfan ymbe ðínne ðone hálgan alter *circumdabo altare tuum*, Ps. Th. 25, 6: Cd. 32; Th. 42, 5; Gen. 669. Ðam þegne ongan his hige hweorfan *the man's mind*

began to change, 33; Th. 44, 8; Gen. 706. Hweorfan fram helltrafum tó fægeran gefeán, Andr. Kmbl. 3378; An. 1693. Hé lǽteþ hworfan monnes módgeþonc *he lets the mind of man roam*, Beo. Th. 3461; B. 1728. Hweorfan, Exon. 77 b; Th. 290, 29; Wand. 72. Hámleás hweorfan *to wander homeless*, 110 a; Th. 420, 25; Rä. 40, 9. Ic seah searo hweorfan giellende faran, 108 b; Th. 414, 29; Rä. 33, 3: Cd. 219; Th. 281, 11; Sat. 270: 215; Th. 272, 16; Sat. 120. On wræc hweorfan, 43; Th. 57, 15; Gen. 928: 48; Th. 62, 15; Gen. 1014. Of gesyhþe ðínre hweorfan *to go from thy presence*, 50; Th. 63, 21; Gen. 1035. Ðæt hé in ðone grimman gryre gongan sceolde hweorfan gehýned, Exon. 41 a; Th. 136, 20; Gú. 544. Com on sefan hwurfan swefnes wóma, Cd. 177; Th. 222, 25; Dan. 110. Hie wǽron eft hám hweorfende *they were returning home*, Blickl. Homl. 67, 10. Ðá wæs Maria eft hweorfende tó hire húse, 139, 3. Hie ymb ðæt fuhton on hweorfendum sigum *Samniticum bellum ancipiti statu gestum*, Ors. 3, 5; Swt. 106, 3. In the following passage the verb is transitive:—Fulwiaþ folc hweorfaþ tó heofonum *baptize people and turn them to heaven*, Exon. 14 b; Th. 30, 25; Cri. 485. [*Goth.* hwairban *to walk*: *O. Sax.* hwerban *to go, wander*: *O. Frs.* hwerva: *Icel.* hverfa: *O. H. Ger.* hwerban *redire, reverti, remeare, ambulare*.] DER. á-, æt-, be-, ge-, geond-, on-, tó-, ymbe-hweorfan; v. hwearfan. [Cf. *Mod. E.* walk, went.]

hweorf-, hwyrf-, hwer-bán, es; *n. A joint* [*of the back*], *vertebra*, [*of the knee*], *the knee-cap*:—Hwyrfbán *vertibulum*, Ælfc. Gl. 11; Som. 57, 43; Wrt. Voc. 19, 46. Hwerbán *vertibulum vel vertebra*, 74; Som. 71, 50; Wrt. Voc. 44, 33. Hweorbán *vertibula*, Wrt. Voc. 283, 38. Hwiorfbán, Lchdm. ii. 396, col. 1. [Cf. *Prompt. Parv.* whyrle-bone, or hole of a joynt *anca, vertebrum, vertibulum*, and see note, p. 524: *Scot.* whorle-bane *hip-joint*: *Ger.* wirbel-bein *vertebra*.]

hweoða. v. hwíða.

hweoðerian, hwoðerian; *p.* ode *To roar, be tempestuous*:—Se brym hwoðerode under his fótswaðum *the sea roared under his footsteps*, Homl. Th. ii. 388, 19. v. hwíða.

hweoðerung, e; *f. Murmuring*; murmuratio, Lye.

hweowol. v. hweogul.

hwer, es; *m. A kettle, pot, basin, caldron, cooking-vessel*:—Hwer *lebes*; cyperen hwer *cucuma*, Ælfc. Gl. 26; Som. 60, 84, 83; Wrt. Voc. 25, 24, 23. Moab mínes hyhtes hwer *Moab olla spei meæ*, Ps. Th. 59, 7. Ðá hét se cásere meltan on hwere leád and pic and hé hét ðone cniht on ðæs hweres welm ásetton *the emperor ordered lead and pitch to be melted in a caldron, and ordered the young man to be put into the boiling of the caldron*, Shrn. 91, 7. Áwyl ða wyrte on hwere *boil the plants in a pot*, L. M. 1, 32; Lchdm. ii. 76, 18. Ænne sylfrene hwer· on v pundon *a silver basin of five pounds*, Chart. Th. 558, 35. Ðǽr wǽron inne geseted hweras and pannan and hé clypte ða hweras and cyste ða pannan ðæt hé wæs eall sweart and behrúmig *pots and pans had been put in there, and he embraced the pots and kissed the pans, so that he was all black and sooty*, Shrn. 69, 27, 30. [*Icel.* hverr *a caldron, boiler*; hverna *a pan, basin*.]

hwer-bán. v. hweorf-bán.

hwerf, hwerfa. v. hwearf, hweorfa.

hwerfan, hwierfan, hwirfan, hwyrfan; *p.* de; *pp.* ed. I. *to turn, revolve, move about, go, return, depart*:—Óþ ðæt ðú eft hwyrfest tó him *until thou shalt return to him*, Blickl. Homl. 233, 29. Mannes sáwl hweóle gelícost hwærfeþ ymbe hý selfe *man's soul, just like a wheel, revolves about itself*, Bt. Met. Fox 20, 422; Met. 20, 211. Hwærfþ, 434; Met. 20, 217. Hwerfeþ, 28, 30; Met. 28, 15. Hwyrfeþ, Exon. 103 b; Th. 394, 3; Rä. 13, 12. Hægl hwyrft of heofones lyfte *hail whirls down from the sky*, Runic pm. 9; Kmbl. 341, 5. Hí hám hwyrfaþ *domum redeunt*, L. Ecg. P. i. 14; Th. ii. 178, 6. Cynna gehwylcum ðara ðe cwice hwyrfaþ *for every race that living moves*, Beo. Th. 197; B. 98. Hig eft syððan tógædere hwyrfdon *postea iterum se conjunxerint*, L. Ecg. P. iv. 8; Th. ii. 206, 8. Hie eft hwirfdon tó hiora ealdormannum *they returned to their rulers*, Blickl. Homl. 239, 29. Hwearfdon geond ðæt atole scref *roamed through that horrid den*, Cd. 214; Th. 269, 13; Sat. 72. Gehwá hám hwyrfe *let every one return home*, L. E. I. 24; Th. ii. 422, 1. On gemynd hwyrfe unrihtwísnys fædera his *in memoriam redeat iniquitas patrum ejus*, Ps. Spl. 108, 13. Hwyrf eft on ða ceastre *go again to the city*, Blickl. Homl. 249, 8. Wæs eft hwyrfende *was returning*, 199, 6: 207, 30: 249, 12. Ðæt hwerfende hweól *the revolving wheel*, Bt. 7, 2; Fox 18, 35. II. *to turn, change* [*trans. and intrans.*]:—Hé hwierfde his stemne nales his mód *vocem, non mentem mutavit*, Past. 36, 7; Swt. 257, 18. Adame his hyge hwyrfde and his heorte ongann wendan tó hire willan *Adam's mind changed, and his heart began to turn to her desire*, Cd. 33; Th. 44, 28; Gen. 716. Ðeáh ðe his leóht gelómlíce hwyrfe *though its light change frequently*, Lchdm. iii. 242, 16. Hwærfe hia *convertantur*, Mt. Kmbl. Lind. 13, 15. Hiora heortan hé ongan hwyrfan *convertit cor eorum*, Ps. Th. 104, 21. Hwý ðú woldest ðæt seó wyrd swá hwyrfan sceolde *cur tantas lubrica versat fortuna vices?* Bt. 4; Fox 8, 12. III. *to exchange, barter* [*with gen.*]:—Aðelwold bisceop and Wulfstán Uccea hwyrfdon landa on Eádgáres cyninges gewytnesse *bishop Athelwold and Wulfstan Uccea exchanged lands with the witness of king Edgar*, Chart. Th. 230, 1. Nán man ne

hwyrfe nānes yrfes būtan đæs gerēfan gewitnesse *let no man exchange any property without the witness of the reeve*, L. Ath. i. 10; Th. i. 204, 17. Nán man ne bycge ne hwyrfe [hwirfe, MS. H.] būton hē gewitnesse hæbbe *let no man either buy or barter unless he have a witness*, L. Eth. 1, 3; Th. i. 282, 26. Huerfa *mutuari*, Mt. Kmbl. Lind. 5, 42. [*Laym.* whærven; *p.* whærfde: *Orm.* wherrfedd *perverse*: *O. Sax.* gi-hwerbian *to turn, change*: *Icel.* hverfa; *p.* hverfði *to turn*: *O. H. Ger.* hwarbian; *p.* hwarpta *versare, rotare, redire, convertere, revertere*, Grff. iv. 1233.] DER. ā-, be-, for-, ge-, on-, ymb-hwerfan; *and see* hweorfan, hwearfian.

hwerfel. v. sin-hwerfel. [*O. H. Ger.* sin-hwerbal *rotundus, teres.*]

hwerfere, es; *m: A changer, trader.* [*O. H. Ger.* werbare *negotiator.*] v. pening-hwerfere.

hwerf-līc; *adj. Changeable, shifting, not enduring:*—Hū hwerflīce đås woruldsǽlþa sint *quam sit mortalium rerum misera beatitudo*, Bt. 11, 1; Fox 32, 37. [*O. H. Ger.* hwarb-, hwerb-līh *versatilis, volubilis:* cf. *Icel.* hwerfull *shifty, changeable.*] v. hwearf; *adj.*

hwerflung, e; *f. Wandering, error:*—Hwærflung *error*, Mt. Kmbl. Lind. 24, 24. [Cf *Icel.* hvarfla *to wander.*] v. hwurf.

hwerfung, e; *f. Change, mutation, vicissitude:* — Hwæt singaþ đa leóþwyrhtan ōðres be đisse woruld būton mislīca hwerfunga đisse worulde *quid tragœdiarum clamor aliud deflet, nisi indiscreto ictu fortunam felicia regna vertentem?* Bt. 7, 3; Fox 22, 21. v. hwearfung.

hwergen; *adj. Somewhere:*—Elles hwergen *elsewhere, somewhere else*, Beo. Th. 5173; B. 2590. [*O. Sax. O. H. Ger.* hwergin *usquam, alicubi:* cf. *Icel.* hwargi *wheresoever.*]

hwer-hwette, an; *f. A cucumber:*—Hwerhwette *cucumer*, Ælfc. Gl. 40; Som. 63, 99; Wrt. Voc. 30, 47. Hwerwette, L. M. 1, 23; Lchdm. ii. 66, 9. Hwerhwettan gesihþ on swefnum untrumnysse getācnaþ *if a man sees in dreams a cucumber it betokens illness*, Lchdm. iii. 200, 16.

hwerwe *a plant name,* perhaps *colchicum autumnale:*—Đa greátan wyrt hwerwe hātte, L. M. 2, 52; Lchdm. ii. 268, 22. [Cf. Ðeós wyrt đe man hieribulbum and ōðrum naman greáte wyrt nemneþ, Herb. 22, 1; Lchdm. i. 118, 13. v. Lchdm. ii. 396, col. 1.]

hwēsan; *p.* hweós *To wheeze, make a noise in breathing, to breathe hard:* — Gif hē mid earfoþnysse hwēst *if he breathes with difficulty*, Lchdm. iii. 122, 3. Hē hwēst swýðe hefelīce, 126, 9. Hē egeslīce hweós *he wheezed terribly*, Homl. Th. i. 86, 1. [*Icel.* hvæsa *to hiss.*]

hwet-stān, es; *m. A whetstone:*—Hwetstān cos, Ælfc. Gl. 58; Som. 67, 100; Wrt. Voc. 38, 25: Ors. 4, 13; Bos. 100, 30. Nim đonne hwetstān brādne *then take a broad whetstone*, Lchdm. iii. 16, 21. [*O. H. Ger.* wezi-stein, cos: *Ger.* weþ-stein.]

hwettan; *p.* te *To* WHET, *sharpen, instigate, urge, incite, excite:*—Ic hwette *acuo*, Ælfc. Gr. 28; Som. 30, 48: Exon. 103 b; Th. 393, 1; Rä. 12, 3. Se lǽce his seax *hwæt the physician sharpens his knife*, Past. 26, 3; Swt. 187, 5. Ūsic lust hwæteþ *desire urges us*, Andr. Kmbl. 571; An. 286. Ðurh đæt his mōd hweteþ *by that means excites his mind*, Salm. Kmbl. 988; Sal. 495: Exon. 82 a; Th. 309, 26; Seef. 63: 83 b; Th. 314, 23; Mōd. 18. Hwettaþ hyra blōdigan tēþ *they whet their bloody teeth*, L. E. I. prm; Th. ii. 396, 6. Ic hig hwette tō fleánne *I instigated her to fly*, Shrn. 41, 25. Swā đīn sefa hwette, Beo. Th. 985; B. 490. Hwetton higerōfne, 413; B. 204. Hȳ hwetton *exacuerunt*, Blickl. Gloss. [*Laym.* whætte; *p*: *Icel.* hvetja *to whet, incite:* *O. H. Ger.* wezzen *acuere, exacuere, provocare:* *Ger.* weþen.] DER. ā-, ge-hwettan.

hwī. v. hwý.

Hwiccas, Hwicceas, *and* **Hwiccan** [?] *or* [?] Hwicce [cf. Seaxe]; *pl. The people of a small state which extended over Gloucestershire, Worcestershire, and part of Warwickshire:*—Đæt is geseted in Huicca mægþe in đære stōwe đe mon hāteþ Weogernaceaster *it is situated in the province of the Hwiccas, in the place that is called Worcester*, Chart. Th. 28, 31. Fērde đā in Hwicca mægþe đær wæs đā Ōsrīc cyning *divertit ad provinciam Huicciorum cui tunc rex Osric præfuit*, Bd. 4, 23; S. 594, 22. Wilfrid is Hwicna biscop *provinciæ Huicciorum Vilfrid episcopus*, 5, 23; S. 646, 22. Ðȳ ilcan dæge rād Æþelmund aldorman of Hwiccium [Hwiccum MS. E.] ofer æt Cynemǽres forda. Đā mētte hine Weoxtan aldorman mid Wilsætum, Chr. 800; Erl. 60, 5. Seó cwēn đære nama wæs Æbbe on hire mægþe đæt is on Hwyccum wæs gefullad *regina nomine Eabæ in sua, id est, Huicciorum provincia fuerat baptizata*, Bd. 4, 13; S. 582, 16.

hwider; *adv. Whither* [*in direct interrogation, or in dependent clauses*]:—Hwider wylt đū *quo vadis?* Gen. 16, 8: Deut. 1, 28. Hwyder gǽst đū *quo vadis?* Jn. Skt. 13, 36. Đū nāst hwanon hē cymþ ne hwyder hē gǽþ *non scis unde veniat et quo vadat*, 3, 8: 12, 35. [*O. and N.* hwider, wider: *Ayenb.* huider: *Laym.* whuder, woder: *Gen. and Ex.* quider: *Wick.* whidir.] v. hwæder.

hwig. v. hwý.

HWĪL, e; *f. A* WHILE, *space of time:*—Wæs seó hwīl micel *it was a great while*, Beo. Th. 295; B. 146. Đā wæs hwīl dæges ǽr hē đone grundwong ongytan mihte *it was a day's space ere he might feel the bottom*, 2995; B. 1495. Ǽr dæges hwīle *before day-time*, 4630; B. 2320. On dæges hwīle *in the day-time*, Cd. 191; Th. 238, 4; Dan. 349. Crist on đære hwīle tō helle gewende *Christ during that time* [*while in the tomb*] *went to hell*, Homl. Th. i. 26, 35. Iu hwīle tīde *in momento temporis*, Lk. Skt. Lind. 4, 5. Tō hwīle lǽn *momentum*, Ælfc. Gl. 15; Som. 58, 47; Wrt. Voc. 21, 36. Bētan tō hwīle *to make better for a time*, L. M. 3, 62; Lchdm. ii. 348, 21. Tō langre hwīle *for a long while*, Cd. 24; Th. 31, 22; Gen. 489. Tō litelre hwīle, Homl. Th. i. 64, 14. Tō suīðe scortre hwīle, Past. 36, 6; Swt. 255, 11. Đa hwīle his līfes *vivendi spatia*, 2; Swt. 249, 25. Đa hwīle đisses andweardan līfes *the time of this present life*, Bt. 18, 3; Fox 66, 4. Đa hwīle đe his līf [tīma, l. 20] wæs, Chr. 1016; Erl. 155, 18. Sume hwīle *some time*, 1055; Erl. 190, 12. Gōde hwīle đone here geflīemde *put the Danes to flight for a good while*, 837; Erl. 66, 8. Nū is đīnes mægnes blǽd āne hwīle *for a while*, Beo. Th. 3528; B. 1762. Ǽnige hwīle, 5090; B. 2548. Ealle hwīle *all the while*, Byrht. Th. 140, 47; By. 304. Nū hwīle *just now*, Blickl. Homl. 109, 6. Grendel wan hwīle wið Hrōđgār *Grendel strove for a time with Hrothgar*, Beo. Th. 306; B. 152: 211; B. 105. Đa đe on carcerne hwīle wunedon, Andr. Kmbl. 262; An. 131. Man gīslade đa hwīle *hostages were given the while*, Chron. 994; Erl. 133, 29. Đa hwīle đe đū eart on wege mid him *dum es in via cum eo*, Mt. Kmbl. 5, 25. Đā besǽt sió fierd hie đǽr ūtan đa hwīle đe hie đǽr lengest mete hæfdon *the English force besieged the Danes there as long as ever they had provisions there*, Chr. 894; Erl. 90, 29. Hwīle mid weorce hwīle mid worde hwīle mid geþohte *at one time with deed, at another with word, at another with thought*, Hy. 3, 44–5; Hy. Grn. ii. 282. 44–5. Đæs ungeendodan līfes hwīla *æternitatis infinita spatia*, Bt. 18, 3; Fox 66, 15. [*O. and N.* hwīle: *A. R.* hwule: *Orm.* while: *Laym.* while, wile: *Goth.* hweila: *O. Sax.* hwīla: *O. Frs.* hwīle: *O. H. Ger.* hwīla *hora, momentum:* *Ger.* weile: cf. *Icel.* hvíla *a bed;* hvila *rest.*] DER. bearhtm-, dæg-, earfoþ-, gesceap-, gryre-, hand-, langung-, orleg-, rōt-, sige-, þræc-, wræc-hwīl. v. hwīlum.

hwilc, hwylc, hwelc; *pron.* **I.** *which, who, of what kind,* [*in direct questions*]:—Quis hwā is werlīc hād, *que*, hwilc is wīflīc, *quod*, hwilc nis nāðres cynnes; *cujus* hwilces; *cui* hwilcum; *quem virum laudas* hwilcne wer herast đū; *a quo* fram hwilcum. *Pluraliter qui* hwilce; *quorum* hwilcera; *quibus* hwilcum; *quos laudas* hwilce herast đū; *a quibus* fram hwilcum ... *Qualis* hwilc getācnaþ þreó þingc *interrogationem* and *infinitionem* and *relationem*. Gif ic cweđe *qualis est rex* hwilc is se cingc, đon biþ hē *interrogativum* ... Đū cwyþst *qualis est ille* hwilc is hē, ic cweđe *talis est* swilc is hē, Ælfc. Gr. 18; Som. 21, 12–18, 57–63. Hwylc man is of eów *quis est ex vobis homo?* Mt. Kmbl. 7, 12. Hwylc þearf is đē hūsles *quid opus est Eucharistia?* Bd. 4, 24; S. 598, 37. Hwā is ūre Fæder? Se Ælmihtiga God. And hwilcera manna Fæder is he? Swutelīce hit is gesǽd, yfelra manna. And hwilc is se Fæder? *who is our Father? The Almighty God. And of what sort of men is he Father? It is plainly said, of evil men. And of what kind is the Father?* Homl. Th. i. 254, 5–8. Hwylc is mihtig God būtan ūre se mǽra God *quis Deus magnus sicut Deus noster*, Ps. Th. 76, 11. Hwylc is wīsra đe đās mid gehygde healdan cunne *quis sapiens et custodiet hæc?* 106, 42. Hwylces đæra sufona byþ đæt wīf *cujus erit de septem uxor?* Mt. Kmbl. 22, 28. Hwylcum bigspelle widmete wē hit *cui parabolæ cumparabimus illud?* Mk. 4, 30. **II.** [*in dependent clauses*]:—Gif ic cweđe *nescio qualis est rex* nāt ic hwilc se cyngc is, đon is se *qualis infinitivum.* Gif ic cweđe *tu scis bene qualis est* đū wāst wel hwilc hē is, đon biþ hit *relativum*, Ælfc. Gr. 18; Som. 21, 59–61. Geseó hē hwylc se man sig oððe đæt neát *videat qualis homo sit vel pecus*, L. Ecg. C. 14; Th. ii. 142, 19. Gē habbaþ gehȳred hwilc đes god is đe gē wēndon đæt eów gehēlde, Homl. Th. i. 464, 10. Hwelc se bión sceal đe tō reccendōme cuman sceal *qualis quisque ad regimen venire debeat*, Past. 10; Swt. 61, 5. Bæd đæt hē him geswutelode hwylc basilius wǽre on wurðscype mid him *prayed that he would reveal to him what manner of man Basil was in honour as compared with himself*, Homl. Swt. 3, 498. Sege ūs hwilc tācn sī đīnes tōcymys *dic nobis quod signum adventus tui*, Mt. Kmbl. 24, 3. Hēt sēcan hwilc đære geógoþe gleáwost wǽre *bade seek which of the youth was most skilled*, Cd. 176; Th. 220, 34; Dan. 81: Andr. Kmbl. 821; An. 411. Cwēn frignan ongan on hwylcum đara beáma bearn wealdendes hangen wǽre, Elen. Kmbl. 1698; El. 851. Dō mē wegas wīse đæt ic wite on hwylcne ic gange *notam mihi fac viam, in qua ambulem*, Ps. Th. 142, 9. Geþence gē hwæt gē sīen and hwelce gē sīen *pensa quod es*, Past. 21, 4; Swt. 159, 14. Đā onfunde se mōdiga, hwilce his mihta wēron *then the proud spirit found out what his powers were*, Ælfc. T. Grn. 2, 47. **III.** *indef. pron. any one, any, of any kind, some:*—Oððe gif hwylc cyningc wyle faran *aut quis rex iturus*, Lk. Skt. 14, 31. Hwæt wēnstū nū, gif hwelc forworht monn cymþ and bitt ūrne hwelcne đæt wē hine lǽden tō sumum rīcum menn and him geþingien *si enim fortasse quis veniat, ut pro se ad intercedendum nos apud potentem quempiam virum ducat*, Past. 10, 2; Swt. 63, 1. Ne hig ne gelȳfaþ đeáh hwylc of deáþe ārise *neque si quis ex mortuis surrexerit credent*, Lk. Skt. 16, 31. Swelc ic wǽre hwelc folclīc mon and mē wǽre mete and wīnes þearf *ut vini et carnis quidam emptor*, Nar. 18, 4. Wēn is đæt hwilc wundor ineode on đæt carcern, St. And. 14, 28. Manslyht oððe elles hwilc đara heáfodlīcra leahtra *manslaughter or any other of the capital crimes*, L. E. I. 26; Th. ii. 422, 5. Sōna swā sacerda hwylc hwone on wōh gesyhþ *as soon as any priest sees any one in error*, 28; Th. ii.

424, 25. Gif mînra þegna hwilc, Cd. 22 ; Th. 27, 7 ; Gen. 414. Ânra hwilc *each one*, Bt. Met. Fox 20, 129 ; Met. 20, 65. Gif him þince ðæt hé on hwylcere fægerre stôwe sî *if it seems to him that he is in some fair place*, Lchdm. iii. 174, 26. Æt mæstra hwelcre misdæde *for almost every misdeed*, L. Alf. 49 ; Th. i. 58, 6. Gyf hwylce ðǽr beóþ ðara ðe hwæt æbylhþa wið ôðre habbaþ *if there are any there who have any grudges against others*, L. E. I. 36 ; Th. ii. 434, 7. Wê gesâwon oft in cyrcean ǽgðer ge corn ge hig ge hwylce woroldlîcu þing beón gehealdene *we have often seen in churches corn and hay, and any kind of secular things kept*, 8 ; Th. ii. 406, 31. Gif hwâ biþ mid hwelcum welum geweorþod and mid hwelcum deórwyrþum ǽhtum gegyrewod, Bt. 14, 3 ; Fox 46, 11. **IV.** *combined with* swâ :—Quisquis swâ hwâ, quæque swâ hwilc, quodquod swâ hwilc ; quicunque swâ hwâ, quæcunque swâ hwilc, Ælfc. Gr. 18 ; Som. 21, 35, 37. Ðæs cyninges þegnas ðider urnon swâ hwelc swâ ðonne gearo wearþ *the king's thanes ran thither, whichever of them was ready*, Chr. 755 ; Erl. 50, 3. Swâ hwylc swâ sylþ ánne drinc *quicumque potum dederit*, Mt. Kmbl. 10, 42. Swâ hwylcum manna swâ him gemet þuhte, Beo. Th. 6106 ; B. 3057 : 1890 ; B. 943. Swâ hwylce daga *in quacumque die*, Ps. Th. 137, 4. Ðæt git ne lǽstan wel hwilc ǽrende swâ hê sendeþ *that ye will not perform what business soever he sends*, Cd. 26 ; Th. 35, 15 ; Gen. 555. **V.** *correlative of* swilc [v. I] :—Hit is scondlîc ymb swelc tô sprecanne hwelc hit ðâ wæs *it is shameful to talk about such a state of things as it then was*, Ors. 1, 10 ; Swt. 48, 4. [O. E. Homl. hwilche : Laym. whilc, whulc : Orm. whillc : R. Glouc. wuch : Piers P. Chauc. which : Goth. hwêleiks, hwileiks : O. Sax. hwilîk : O. Frs. hwelîk, hwelk, hulk, hwek : Icel. hvílíkr : O. H. Ger. hwelîh : Ger. welcher.] DER. ǽg-, ge-, wel-hwilc.

hwilc-hwega, -hwugu, -hugu [*in the Northern Gospels the whole form is declined, elsewhere only* hwilc] ; *pron. Some, any, some one* :—Gehrân mec huoelchuoege *tetigit me aliquis*, Lk. Skt. Lind. 8, 46. Hwilc-æthwega yfel wǽte *some evil humour*, L. M. 2, 59 ; Lchdm. ii. 284, 27. Brôðer huoelchuoeges *frater alicujus*, Lk. Skt. Lind. 20, 28. Swâ hê sîe mid hwilcre-hwega byrþenne gehefegod *as if he is weighted with some burden*, L. M. 2, 23 ; Lchdm. ii. 212, 11. Gif man forleóse gehâlgodne mete hwylcne-hwugu dǽl *si quis perdiderit cibi consecrati aliquantulum*, L. Ecg. P. iv. 52, note ; Th. ii. 218, 23. Hwelcne-hugu dǽl, Ors. 3, 7 ; Swt. 110, 13. Hwelce-hwugu gerisenlîce leáfe dyde *he gave some suitable leave*, Past. 51, 4 ; Swt. 397, 25. Heó geþingode tô gode sumre hǽðenre fǽmnan gǽste hwylce-hwegu rǽste in ðǽre ēcan worulde, Shrn. 133, 16. Ðe hwilce-hwega gefēlnesse hæbbe, L. M. 1, 35 ; Lchdm. ii. 82, 30. Hafaþ ðæt môd hwylce-hugu scyldo *habet animus aliquem reatum*, Bd. 1, 27 ; S. 496, 42. Hwylce-hugu tîd *aliquanto tempore*, 4, 22 ; S. 591, 31. Hwælchuoego *quid*, Mk. Skt. Lind. 13, 15. Huoelchuoegu *aliquid*, Rtl. 146, 23. [Cf. hwæt-, hû-hwega ; and next word.]

hwilc-hwéne, -hwône ; *pron. indef. Some, some one* :—Bēcon hwelchuoene *signum aliquid*, Lk. Skt. Lind. 23, 8. Wið huelchûône *adversus aliquem*, Mk. Skt. Lind. 11, 25.

hwilc-ness, e ; *f. Quality* :—Sume synd *qualitātis* ðe getácniaþ hwilcnysse, Ælfc. Gr. 38 ; Som. 40, 31. [Cf. O. H. Ger. hweolîhi, hweolîhnissi *qualitas*.] v. ge-hwilcness.

hwilen ; *adj. Lasting only for a time, transitory, brief* :—Uton sibbe tô him on ðás hwílnan tîd hǽlu sēcan *let us seek in this brief season* [*the present life*] *peace and salvation from him*, Exon. 97 b ; Th. 365. 10 ; Wal. 87. [O. H. Ger. hwîlin *temporalis*.] v. un-hwílen.

hwîlend-lîc ; *adj. Lasting only for a time, of time, temporal, temporary, transitory* :—Þrió þing sindon on ðis middanearde. Ân is hwîlendlîc... Oðer þing is ēce... Ðridde þing is ēce *three things there are in this world. One is of time... the second... and the third are of eternity*, Bt. 42 ; Fox 256, 15. Ðâ se cyning wæs ceasterwara gefremed ðæs ēcan rîces and wolde eft ðæt ēþel sēcan his hwîlendlîcan rîces *rex æterni regni jam civis effectus, temporalis sui regni sedem repetiit*, Bd. 3, 22 ; S. 552, 33. Mid ðýs hwîlendlîcan onwalde *temporali potentia*, Past. 17, 4 ; Swt. 113, 11. Mid ðissum hwîlendlîcum þingum *temporali sollicitudine*, 18, 7 ; Swt. 139, 7. Ðû næfst ða hwîlendlîcan ārwyrþnessa ðe ðû ǽr hæfdest *thou hast not those temporary dignities that thou hadst before*, Bt. 8 ; Fox 24, 31. v. hwîlwendlîc.

hwil-fæc *a space of time*, Lye.

hwîlon. v. hwîlum.

hwíl-stycce, es ; *n. A fragment* or *short portion of time* :—Æghwæ ðæs ðe hie on ænegum hiora hwílsticcum geearnian mægen *all that they can earn in any of their fragments of time*, L. Alf. pol. 43 ; Th. i. 92, 12.

hwíl-tídum ; *dat. pl. as adv. At times, sometimes* :—Hwíltídum oððe nû ðâ *modo*, Ælfc. Gr. 38 ; Som. 41, 37 ; *aliquando*, Past. 57, 1 ; Swt. 437, 3 ; Lchm. iii. 240, 23 : 242, 18. Eác hê sceal hwíltídum geara beón on manegum weorcum tô hláfordes willan *also at certain times he must be prepared for many kinds of work at the lord's pleasure*, L. R. S. 5 ; Th. i. 436, 3. Ðeós woruld ðeáh ðe heó myrige hwíltídum geþuht sý *this world though sometimes it appear joyous*, Homl. Th. i. 154, 17. Seó sǽ is hwíltídum smylte and myrige on tô rôwene, hwílon eác swíðe hreóh and egeful on tô beônne, 182, 32. [Cf. O. H. Ger. stunt-hwîla *momentum*, Grff. iv. 1226]

hwîlum, hwîlon ; *dat. pl. as adv. At times, for a time, sometimes, whilome* :—Hwîlon ic dyde swâ *aliquando feci sic* ... Dudum gefyrn, quandam hwîlon, and *olim* getácniaþ þreó tída, forþgewitene and andwerde and tôwerde, Ælfc. Gr. 38 ; Som. 39, 62–4. Ic wiste ðæt ðû hwîlon lufodest God *scivi te aliquando amasse Deum*, 24 ; Som. 25, 9. Ða ðe on horsum hwîlon wǽron *qui ascenderunt equos*, Ps. Th. 75, 5. Hwîlum tô gebede feóllon *sometimes they fell to praying*, Cd. 37 ; Th. 48, 18 ; Gen. 777 : 38 ; Th. 50, 17 ; Gen. 810. Hî hwîlum gelýfaþ *qui ad tempus credunt*, Lk. Skt. 8, 13. Hwîlon ǽr wē wǽron hēr and bohton ûs hwǽte *jam ante descendimus, ut emeremus escas*, Gen. 43, 20. Ic secge ðæt ic hwîlon ǽr forsuwode *I say what I sometime before passed over in silence*, Boutr. Scrd. 18, 27. Hwîlan ǽr, Bt. Met. Fox 29, 106 ; Met. 29, 53. Hwîlum on áne healfe hwîlum on ǽlce healfe *now on one side, now on every side*, Chr. 891 ; Erl. 88, 20. Ða hâlgan lâreówas hwîlon sprecaþ be ðam Ælmihtigan Fæder and his Sunu, hwîlon swutollîce embe ðǽre Hâlgan Ðrynnesse, Homl. Th. ii. 56, 26 : Cd. 216 ; Th. 273, 7–12 ; Sat. 132–5. [O. Sax. hwîlun : O. H. Ger. hwîlon *paulatim, nunc* ; hwîlom ... hwîlom *modo* ... *modo* : M. H. Ger. hwîlont, Grff. iv. 1225 : Ger. weiland *formerly*.]

hwîl-wende ; *adj. Temporary, lasting for a time, not eternal* :—Him fremede tô ēcere hǽlþe seó hwîlwende ehtnys *the persecution that lasted but for a time, helped him to the salvation which lasts for ever*, Homl. Th. ii. 528, 7. Hê hî mǽrsaþ on ðǽre ēcan worulde for heora hwîlwendum geswince ðises sceortan lîfes, 562, 3. Ðæt hî gelýfon tô geágenne ða ēcan welan, ða ðe for his naman ða hwîlwendan spēda forhogiaþ, i. 64, 20. [Cf. Goth. hweila-hwairbs *lasting only for a time*.]

hwîlwend-lîc ; *adj. Temporary, lasting only for a time, not eternal* :—Hit is hwîlwendlîc *est temporalis*, Mt. Kmbl. 13, 21. Þreó þing synd on middanearde ân is hwîlwendlîc ... ôðer þing is ēce ... þridde þing is ēce, Homl. Swt. 1, 25. Manna freóndscipe biþ swîðe hwîlwendlîc *the friendship of men lasts but a very short time*, Blickl. Homl. 195, 26. Se ælmihtiga se ðe is ēce leóht ǽrest ðæt hwîlwendlîc leóht geworhte, Boutr. Scrd. 19, 5. Hwîlwendlîc lîf ... ēce lîf *the life of time ... the life of eternity*, Homl. Th. ii. 240, 15–20. Nalæs ðæt ān ðæt hê hî fram yrmþum ēcre niðerunge ac swylce eác fram ðam mânfullan wæle hwîlwendlîcre forwyrde generede *non solum eam ab ærumna perpetuæ damnationis, verum et a clade infanda temporalis interitus eripuit*, Bd. 4, 13 ; S. 582, 27. Hê swanc for heofonan rîce swîðor ðonne hê hogode hû hê geheólde on worulde ða hwîlwendlîcan geþincþu, Swt. A. S. Rdr. 98, 93. Ðæt wē ða heofonlîcan þinga mid ðam eorþlîcum and ða ēcelîc mid ðam hwîlwendlîcum geearniaþ, L. Ath. i. prm ; Th. i. 196, 27. [Orm. hwilwendlic : cf. O. H. Ger. wîlwendige *fortuna*, Grff. i. 763.] v. hwîlend-lîc.

hwîlwend-lîce ; *adv. Temporarily, for a time only* :—Beóþ blôwende and welige hwîlwendlîce ðæt gē ēcelîce wǽdlion *be flourishing and wealthy for time that ye may be beggars for eternity*, Homl. Th. i. 64, 15 : 162, 15 : ii. 384, 26.

hwînan ; *p. hwân* ; *pl. hwinon To make a whistling, whizzing sound* [*as an arrow, etc. in its flight*] :—Ful oft of ðam heápe hwînende fleág giellende gâr *full oft from that band flew whistling the shrieking javelin*, Exon. 86 b ; Th. 326, 12 ; Vîd. 127. [Prompt. Parv. whynyñ, as howndys or oþer beestys *ululo, gannio* : Chauc. for as an hors I coude bite and whine : Icel. hvína ; *p.* hvein *to give a whizzing sound* [*as an arrow*], e. g. örvarnar flugu hvínandi yfir höfuð þeim : Dan. hvine *to whistle* (*of the wind*) ; hvin *a piercing shriek*.]

hwîoð. v. hweóð.

hwioð. v. hwið.

hwirfan. v. hwerfan.

hwirfel, es ; *m. A whirl-pool* [?] :—On ðone hwyrfel, Cod. Dipl. Kmbl. iii. 412, 8. [Cf. Icel. hvirfill *a ring* ; *the crown of the head* ; *a top, summit* : Dan. hvirvel *a whirl-pool* ; *the top of the head* : O. H. Ger. hwirvil *turbo* : Ger. wirbel.]

hwirf-pôl, es ; *m. A whirl-pool* :—Hwyrfepôle *vorago, syrtis*, Cot. 59, Lye.

hwisprian ; *p. ode, ede To mutter, murmur*, WHISPER : — Alle hwispredon *omnes murmurabant*, Lk. Skt. Rush. 19, 7. Hwispradun, Jn. Skt. Rush. 6, 41 : *murmurarent*, 61. Nallaþ gê hwispriga *nolite murmurari*, 43. [Prompt. Parv. whysperyñ *mussito* : O. Du. wisperen : O. H. Ger. hwispalôn *sibilare* : Ger. wispern.] v. following words.

hwisprung ; *f. A muttering, murmuring*, WHISPERING :—Hwisprung murmur, Jn. Skt. Rush. 7, 12.

hwistle, an ; *f. A pipe, flute*, WHISTLE : — Hwistle oððe pîpe *musa* ; hwistle *fistula*, Wrt. Voc. 73, 60, 65. Mið hwistlum *tibiis*, Lk. Skt. Lind. 7, 32. [Chauc. so was hire joly whistle wel ywette.]

hwistlere, es ; *m. A piper, player on a flute* :—Pîpere oððe hwistlere *tibicen*, Ælfc. Gr. 9 ; Som. 9, 25. Ðâ hê geseah hwistleras *cum vidisset tibicines*, Mt. Kmbl. 9, 23.

hwistlian ; *p. ode ede To make a hissing sound, to hiss, whistle* : — Hê hwystlode stranglîc[e] stemne *he* [*the devil*] *made a great hissing*, Nar. 43, 17. [Wick. whistlen *hiss* (A. V.) : Piers P. whistlen (*to birds*).]

hwistlung, e; f. A hissing, WHISTLING, piping, music:—Ðeós hwistlung hic sibilus, ðás hwystlunga hæc sibila, Ælfc. Gr. 13; Som. 16, 28. Hwistlung sibilatio, Ælfc. Gl. 79; Som. 72, 67; Wrt. Voc. 46, 24. Huislung simphonia, Lk. Skt. Lind. 15, 25. Ic beswíce fugelas mid hwistlunge decipio aves sibilo, Coll. Monast. Th. 25, 15. Suā suā mid liðre wisdlunga mon hors gestilleþ suā eác mid ðære illcan wistlunga mon mæg hund ástyrigean lenis sibilus equos mitigat, catulos instigat, Past. 23; Swt. 173, 21.

HWÍT; adj. WHITE, bright, clear, fair, splendid:—Hwít albus; amineus vel albus, Ælfc. Gl. 79; Som. 72, 71-2; Wrt. Voc. 46, 28-9. His reáf hwít scínende vestitus ejus albus refulgens, Lk. Skt. 9, 29. Wlitescýne hwít and hiwbeorht hæleþa náthwylc some man beauteous, shining and bright of hue, Elen. Kmbl. 145; El. 73. Hwít heard stán creta vel cimolia, Ælfc. Gl. 56; Som. 67, 40; Wrt. Voc. 37, 29. Se hwíta stán mæg wið stice the white stone is effective against stitch, L. M. 2, 64; Lchdm. ii. 290, 9. Se hwíta helm the shining helm, Beo. Th. 2900; B. 1448. Ðú ne miht ænne locc gedón hwítne oððe blacne non potes unum capillum album facere aut nigrum, Mt. Kmbl. 5, 36. Hæfde hé hine swá hwítne geworhtne gelíc wæs hé ðám leóhtum steorrum so splendid had he formed him he was like the bright stars, Cd. 14; Th. 17, 4; Gen. 254. Leóht hwít clear light, 29; Th. 38, 2; Gen. 616. Ðone hwítan hláf panem nitidum, Bd. 2, 5; S. 507, 14. Fæst ælce dæge and forgang hwít jejuna quotidie et abstine te ab albo, L. Ecg. C. prm; Th. ii. 132, 5. Gedó æges hwít tó add white of egg, L. M. 3, 59; Lchdm. ii. 342, 18. Dó æges ðæt hwíte tó, 1, 13; Lchdm. ii. 56, 6: 25; Lchdm. ii. 66, 21: Homl. Th. i. 40, 27. His reáf wǽron swá hwíte swá snáw vestimenta ejus facta sunt alba sicut nix, Mt. Kmbl. 17, 2. Ða scíran dagas hwítan the clear bright days, L. M. 2, 41; Lchdm. ii. 252, 10. Hwíte metas lacticinia [cf. Icel. hvítr matr milk, curds, etc. opposed to flesh], Lye. Wǽron on ðyssum felda unríme gesomnunge hwíttra manna and fægera erant in hoc campo innumera hominum albatorum conventicula, Bd. 5, 12; S. 629, 25. Engla and deófla, beorhtra and blacra, hwítra and sweartra, Exon. 21 a; Th. 56, 9; Cri. 898. Hire þuhte hwítre heofon and eorþe heaven and earth seemed brighter to her, Cd. 29; Th. 38, 7; Gen. 603. Engla scýnost and hwíttost most beautiful and most splendid of angels, 18; Th. 22, 11; Gen. 339. [Goth. hweits: O. Sax. O. Frs. hwít: Icel. hvítr: O. H. Ger. hwíz albus, candidus, lacteus: Ger. weiss.] v. eall-, geolu-, healf-, snáw-hwít.

hwíta. v. sweord-hwíta.

hwítan to make white, to polish, Exon. 95 a; Th. 354, 48; Reim. 62. v. hwítian.

Hwít-círice, an; f. A local name, WHITCHURCH:—Æt Hwítcíricean, Chr. 1001; Erl. 136, 7.

hwít cwidu, cudu, es; n. v. cwudu.

hwítel, es; m. A WHITTLE, a cloak, mantle, blanket:—Hwítel sagum, Ælfc. Gl. 27; Som. 60, 111; Wrt. Voc. 25, 51. Hnysce hwítel linna, 63; Som. 68, 112; Wrt. Voc. 40, 23. Seó wimman mid hire hwítle bewreáh hine she covered him with a mantle; opertus ab ea pallio, Jud. 4, 18: L. M. 1, 32; Lchdm. ii. 76, 23. Ðá hét Beñedictus beran ða tócwysedan lima on ánum hwítle intó his gebedhúse, Homl. Th. ii. 166, 21. Sem and Jafeth dydon ánne hwítel on hira sculdra Sem et Japheth pallium imposuerunt humeris suis, Gen. 9, 23. Ðá eode ðes bróðor sume dæge ðæt hé wolde his reówan and hwítlas ða ðe hé on cumena búre brúcende wæs on sǽ wacsan and feormian hic cum quadam die lenas sive saga quibus in hospitale utebatur in mari lavasset, Bd. 4, 31; S. 610, 10. [A. R. (MSS. C. T.) hwitel (other MS. kurtel): Piers P. for when he streyneþ hym to strecche þe straw is hus whitel, C-text 17, 76: Halliwell Dict. whittle 'a blanket. Kennett says "a coarse shagged mantle." The whittle, which was worn about 1700, was a fringed mantle, almost invariably worn by country women out of doors': Icel. hvítill a white bed cover.] v. gafol-hwítel.

Hwít-ern, es; n. Whitherne in Galloway:—His mynster is æt Hwíterne, Chr. 565; Erl. 19, 7. [Cf. Bd. 3, 4:—Qui locus ad provinciam Berniciorum pertinens, vulgo vocatur Ad candidam casam, eo quod ibi ecclesiam de lapide, insolito Brittonibus more fecerit. See also 5, 23:—On ðære stówe ðe is gecíged æt Hwítan earne quæ candida casa vocatur, S. 646, 31.]

hwít-fót; adj. Having white feet:—Hwítfót albipedius, Wrt. Voc. ii. 6, 48. Huítfoot, 99, 71.

hwíða, hweoða, an; m: hweoðu, e; f. A breeze:—Hwíða oððe weder aura, Wrt. Voc. 76, 43: Ælfc. Gl. 94; Som. 75, 109; Wrt. Voc. 52, 59. Hwioðan oððe oreþe aura, ii. 6, 56. Ælc hwiða windes every breath of wind, Past. 42. 1; Swt. 306, 6. Hé ýste mæg eáðe oncyrran ðæt hí windes hweoðu weorþeþ smylte statuit procellam in auram, Ps. Th. 106, 29. On lyftu í tó hwiðan í tó wedere in auram, Ps. Lamb. 106, 28. On lyste [MS. C. wedyre í hweoðan], Ps. Spl. 106, 29. [Icel. hvíða a squall of wind.]

hwítian; p. ode To be or become white, to whiten:—Ic hwítige albeo, albesco, Ælfc. Gr. 35; Som. 38, 6: albo, albico, 36; Som. 38, 29-30. Ðæt ðæt fel hwítige that the skin may become white, L. M. 1, 38; Lchdm. ii. 96, 6. [A. R. hwiteþ prs. becomes white: Piers P. whitten to make white: Prompt. Parv. whytoñ or make whyte dealbo, candido: Goth. ga-hweitjan to make white: O. H. Ger. hwízên to become white; ga-hwízit albatus: Ger. weissen to whiten.]

hwíting, e; f. Whiting, chalk and size:—Of hwítingmelwe, L. M. 3, 39; Lchdm. ii. 332, 20.

hwíting-treów, es; n. Whitten tree; pirus aria:—Hwítingtreów variculus, Ælfc. Gl. 47; Som. 65, 25; Wrt. Voc. 33, 25. v. Lchdm. iii. 334. col. 1.

hwít-leác, es; n. Onion; allium cæpe:—Hwítleác poletis, Ælfc. Gl. 41; Som. 63, 118; Wrt. Voc. 30, 61.

hwít-loc; adj. Having white or bright, shining hair:—Exon. 112 a; Th. 429, 12; Rä. 48, 3. v. next word.

hwít-locced; adj. Fair-haired, having bright hair, Exon. 127 a; Th. 489, 7; Rä. 78, 4.

hwít-ness, e; f. Whiteness:—Seó reádnes ðære rósan and seó hwítnes ðære lilian, Blickl. Homl. 7, 30: Homl. Th. i. 444, 14. His gewǽda scinon on snáwes hwítnysse his raiment shone with the whiteness of snow, ii. 242, 7.

Hwít-sand Wissant near Calais, Chr. 1095; Erl. 231, 5.

hwít-stów is the translation of Libanus, Ps. Spl. 71, 16.

hwom. v. hwamm.

hwón; adj. Little, few [but the word occurs for the most part only in the neuter acc. with a substantive or adverbial force = a little]:—Dó huniges hwón tó put a little honey to it, L. M. 1, 2; Lchdm. ii. 32, 15. Hwón buteran, 8; Lchdm. ii. 54, 3. Hwón buteran and pipores hwón and hwón sealtes, 2, 52; Lchdm. ii. 268, 25-6. Swá hwæt swá hé læs and hwón hæfde geearnunge si quid minus haberet meriti, Bd. 4, 29; S. 608, 1. Bealosíþa hwón, Exon. 81 b; Th. 307, 24; Seef. 28. Dó hwón on ðíne tungan put a little on to your tongue, L. M. 2, 52; Lchdm. ii. 272, 18: 1, 59; Lchdm. ii. 130, 7. Genim hwón sealt take a little salt, 2; Lchdm. ii. 32, 3. Ácrind and hwón wermód gecnua pound oak rind and a little wormwood, 52; Lchdm. ii. 124, 22. Huón aron ða ðe onfindes ða ilco pauci sunt qui inveniunt eam, Mt. Kmbl. Lind. 7, 14. Ofer lytla í huón super pauca, 25, 21. Huón í unmonige paucos, p. 15, 7. Búta hwón untrymigo gehælde nisi paucos infirmos curavit, Mk. Skt. Lind. 6, 5. Huónum paucis, Lk. Skt. p. 7, 19. Ðanon hwón ágán progressus inde pusillum, Mk. Skt. 1, 19. Uton ús hwón restan requiescite pusillum, 6, 31. Huón paululum, Lind. 14, 35. Hine hwón fram ðám cnihtum geweænde, Ap. Th. 21, 27. Gif huidir huón ic sægde quominus dixissem, Jn. Skt. Lind. 14, 2. Gif hé hwón hnappode if he dozed a little, Homl. Th. i. 86, 18. Ðá hwón onslép, Shrn. 60, 17. Hwónn, Bd. 3, 9; S. 534, 11. Ðám mannum ðe mágon hwón gehýran for those people who can hear but little, L. Med. ex Quadr; Lchdm. i. 362, 20. Mót ic nú cunnian hwón ðínne fæstrædnesse pauculis rogationibus, Bt. 5, 3; Fox 10, 34. Hé wæs hwón giernende ðissa worolðþinga and micelra onwalda vir tranquillissimus, Ors. 6, 30; Swt. 280, 28. Hé ðær bád westanwindes and hwón norþan he there waited for a wind rather from the north of west, 1, 1; Swt. 17, 15. Hwón lange rather long, Herb. 152, 1; Lchdm. i. 276, 24. Hwón weredre swæce of a rather sweet taste, 151, 1; Lchdm. i. 276, 9. Tó hwón God andrǽdeþ fear God too little; minime, Past. 17, 2; Swt. 109, 15: 63, 7; Swt. 417, 35. [O. E. Hom. wan: Laym. whon.] v. lyt-hwón; hwón-líc, -líce; hwéne.

hwonan. v. hwanan.

hwón-líc; adj. Little, slight, small:—Gif wé eów ða gástlícan sǽd sáwaþ hwónlíc biþ ðæt wé eówere flǽslícan þing rípon if we sow the spiritual seeds for you, it is a slight matter that we reap your fleshly goods, Homl. Th. ii. 534, 26. Ic wearþ belocen on ánre lytlan byrig mid hwónlícum fultume I was shut up in a little town with an inconsiderable force, Homl. Swt. 7, 347.

hwón-líce; adv. Little, slightly:—Ða hwílwendlícan geþincþu ðe hé hwónlíce lufode the temporal dignities that he loved but little, Swt. A. S. Rdr. 98, 94. Nú gé habbaþ hwónlíce tó geswincenne, Homl. Th. ii. 78, 14. Hé byþ hwónlíce biter on byrgincge it is a little bitter of taste, Herb. 140, 1; Lchdm. i. 260, 9. Hwónlíce þyrnihte, 161, 1; Lchdm. i. 288, 16. Heó hwónlíce undergǽþ ðære geendunge it goes a little below the horizon, Lchdm. iii. 260, 6: 134, 3. Him hwónlíce speów he had but little success, Homl. Skt. 7, 94. Mid ðære sceall seó sáwul ealle þing gemætegian ðæt hit tó swíðe ne sý ne tó hwónlíce therewith shall the soul moderate all things, that there be not error by excess or by defect, 1, 162. Hwónlícor minus, Ælfc. Gr. 38; Som. 40, 47. On ðám máran ðe swýðor syngaþ, on ðám læssan ðe hwónlícor syngaþ, Homl. Th. i. 460, 27. Hwónlícost minime, Ælfc. Gr. 38; Som. 40, 49.

hwón-lotum; adv. A little while:—Huónlotum parumper, Wrt. Voc. ii. 116, 46.

hwonne, hwonon. v. hwanne, hwanon.

hwópan; p. hweóp To threaten:—Ne ondrǽd ðú ðé ðeáh ðe elþeódige egesan hwópan heardre hilde fear not though strangers threaten terror and cruel war, Elen. Kmbl. 164; El. 82. Bælegsan [bell egsan, MS.] hweóp hátan líge ðæt hé on wéstenne werod forbærnde nymðe hie moyses hýrde with terror of fire, with hot flame it [the pillar of fire] threatened that it would consume the host in the wilderness, unless they hearkened to Moses, Cd. 148; Th. 185, 12; Exod. 121. Geofon deáþe hweóp the ocean threatened death, 166; Th. 206, 6; Exod. 447: Th. 208, 3; Exod. 477. Ongan ðá þurh swefn sprecan tó ðam æþelinge and him yrre hweóp then did God speak in a dream to the prince and in anger threatened him, 125; Th. 159, 18; Gen. 2636. Ðonne hý him yrre hweópan frécne fýres wylme, Exon. 35 a; Th. 113, 22; Gú. 161. Ðǽr ǽnig ne mæg lǽþþum hwópan there cannot any threaten injuries, 64 a; Th. 236, 31; Ph. 582. [Goth. hwópan to boast.]

P p

hworfan. v. hweorfan.

hwōsan. v. hwēsan.

hwōsta, an; *m. A cough* :—Hwōsta *tussis*, Wrt. Voc. 289, 5 : Ælfc. Gr. 9 ; Som. 14, 33. Hwōsta and nearones breósta, L. M. 2, 21 ; Lchdm. ii. 204, 26. Hine dreceþ þyrre hwōstan and him on ðam, hwōstan hwīlum losaþ sió stemn *he is troubled with a dry cough and at times during the cough he loses his voice*, 51 ; Lchdm. ii. 264, 13. Wið hwōstan hū hē missenlíce on mon becume and hū his mon tilian scyle *for cough, in what different ways it comes on a man and how it must be treated*, 1, 15 ; Lchdm. ii. 56, 13. [*Prompt. Parv.* hosse, host, hoost *tussis : Scott.* host, hoast, hoist *a cough : Icel.* hósti : *O. H. Ger.* huosto *tussis : Ger.* husten.]

hwōstan, *p.* te *To cough :* — Hwōstaþ [hwosaþ, MS.] gelóme *they cough frequently*, L. M. 2, 46 ; Lchdm. ii. 258, 7. [*Prompt. Parv.* hostyñ, or rowhyñ, or cowghyñ *tussio, tussito : Scott.* host, hoist *to cough : Icel.* hósta : *Dan.* hoste : *O. H. Ger.* huostōn : *Ger.* husten.]

hwoðerian. v. hweoðerian.

hwu. v. hū.

hwugu. v. hwega.

hwurf *a going about, wandering, error :* — Huurf *error*, Mt. Kmbl. Lind. 27, 64.

hwurfan. v. hweorfan.

hwurf-bán, Lchdm. iii. 98, 16. v. hweorf-bán.

hwurf-líc ; *adj. Changeable ;* mutabilis, Hpt. Gl. 470, 62. v. hwerf-líc.

hwurfling, es ; *m. That which turns :*—Hwurflinces *orbis*, Hpt. Gl. 453.

hwurful ; *adj. Changeable, fickle :*—Hwæt getácniaþ ða truman ceastra būtan hwurfulu mód *what do the strong cities betoken but fickle minds ;* quid per civitates munitas nisi suspectæ mentes, Past. 35, 5 ; Swt. 245, 7.

hwurful-ness, e ; *f. Changeableness, mutability :* — Ða twigu ðære hwurfulnesse *genimina mutabilitatis*, Past. 42, 3 ; Swt. 308, 1. Hió hit gecýþ self mid hire hwurfulnesse ðæt hió biþ swíðe wancol *se instabilem mutatione demonstrat*, Bt. 20 ; Fox 70, 34.

HWÝ, hwí ; *inst. of* hwæt. WHY.   I. *in direct questions :*— *Interrogativa* synd áxigendlíce *cur* hwí, Ælfc. Gr. 38 ; Som. 40, 58. Hwí didest ðū ðæt *quare hoc fecisti ?* Gen. 3, 13 : Mt. Kmbl. 9, 11. Hwý sceal ic æfter his hyldo þeówian ? Cd. 15 ; Th. 18, 33 ; Gen. 282.   II. *in dependent clauses :*—Se wísa Augustinus smeáde hwí se hálga cýðere cwæðe ... *the wise Augustine inquired why the holy martyr said* ..., Homl. Th. i. 48, 10. Eall ðæra Iudéiscra teóna árás þurh ðæt hwí Drihten Crist seðe æfter flæsce sóðlíce is mannes sunu eác swilce wære gecweden Godes sunu *all the quarrel of the Jews had its origin from this, why Christ, who according to the flesh is truly the son of man, should also be called the son of God*, 16. Ða áscade hē Æðelm hwý hit him ryht ne þuhte ðæt wē him gereaht hæfden *then he asked Æthelm why that did not seem right to him which we had arranged for him*, Chart. Th. 171, 12. Ða óðre ða ðe ðǽr nǽron þurh gewrite atíwdon hwí hí ðǽr beón ne mihton *the others who were not there shewed by letter why they could not be there*, Chr. 1070 ; Erl. 206, 6. v. hwá.

Hwyccas, hwyder, hwylc. v. Hwiccas, hwider, hwilc.

hwylca, an ; *m. A swollen vein ;* varix, Ælfc. Gl. 76 ; Som. 71, 129 ; Wrt. Voc. 45, 32.

hwyrfan, hwyrf-bán, hwyrfel, hwyrfere, hwyrfolung, hwyrf-pól. v. hwerfan, hweorf-bán, hwerfel, hwerfere, hwerflung, hwirf-pól.

hwyrf-ness, e ; *f. Giddiness :*—Wið brægenes hwyrfnesse, Lchdm. iii. 70, 20.

hwyrft, es ; *m. A turn, revolution, going, course, orbit, circuit, orb, circle :*—Ða ðe ofercumaþ allum hwyrfte *quæ superveniunt universo orbi*, Lk. Skt. Rush. 21, 26. Hwá ne wundraþ ðætte sume tunglu habbaþ scyrtran hwyrft ðonne sume habban *who does not wonder that some stars have a less orbit than others ?* Bt. 39, 3 ; Fox 214, 18. Heofonsteorran bebúgaþ brádne hwyrft *the stars of heaven encompass a spacious circle* [*the earth*], Cd. 190 ; Th. 236, 16 ; Dan. 322. Geþancmeta on hwilce healfe ðú wille hwyrft dón cyrran mid ceápe *consider on which side thou wilt bend thy course, turn with thy cattle*, 91 ; Th. 115, 12 ; Gen. 1918. Gif ic on helle gedó hwyrft ænigne *si descendero in infernum*, Ps. Th. 138, 6. Helle hlínduru nágon hwyrft ne útsíþ æfre *never is there return or passage out through the grated doors of hell*, Exon. 97 b ; Th. 364, 30 ; Wal. 78. Náhton máran hwyrft *they could go no further*, Cd. 154 ; Th. 191, 6 ; Exod. 210. Náh ic hwyrft weges [*Grein reads* hwyrftweges] *I cannot return*, Exon. 101 b ; Th. 383, 6 ; Rä. 4, 6. Sóna æfter ðæm wordum helle hæftas hwyrftum scríðaþ þúsendmǽlum *straightway after those words shall the captives of hell by thousands bend thither their steps*, Cd. 227 ; Th. 304, 17 ; Sat. 631 : Beo. Th. 329 ; B. 163. Ða wæs ágangen geára hwyrftum *then had passed in course of years*, Elen. Kmbl. 2 ; El. 1. DER. ed-, ymb-hwyrft. v. hwearft.

hwyrftlian. v. hwearftlian.

hý. v. hē.

hycgan, hycgean ; *p.* hogde.   I. *to employ the mind, take thought, be mindful, think, consider, meditate :*—Bēc bodiaþ ðam ðe wiht hycgeþ *books tell to him that thinks at all*, Salm. Kmbl. 476 ; Sal. 238. Hycgeþ ymbe se ðe wile *he shall think about it who will*, Bt. Met. Fox 19, 2 ;

Met. 19, 1. Ðám ðe mid heortan hycgeaþ rihte *his qui recto sunt corde*, Ps. Th. 72, 1. Ða inwit and fácen hycgeaþ on heortan *qui cogitaverunt malitias in corde*, 139, 2, 8. Gif gē teala hycgaþ, Andr. Kmbl. 3223 ; An. 1614. Hwæt hē on hyge hogde heortan geþoncum *what he meditated in his mind with the thoughts of his heart*, Exon. 51 a ; Th. 177, 14 ; Gū. 1227. Ðú wið Criste wunne hogdes wið hálgum *thou didst strive with Christ, didst plot against the saints*, 71 b ; Th. 267, 28 ; Jul. 422. Hycgaþ his ealle hū gē hí beswícen *all think of this, how ye may deceive them*, Cd. 22 ; Th. 28, 8 ; Gen. 432. Hicgeaþ on ellen *let your thoughts be of valour*, Fins. Th. 21 ; Fin. 11. Ðæt seó forlǽtene cyrice ne hycgge ymb ða ðe on hire neáwiste lifgeaþ *that the forsaken church will take no thought for those that live in her neighbourhood*, Blickl. Homl. 43, 1. Hū ðú ymb módlufan mínes freán on hyge hycge *how thou mayest think in thy mind of the love of my lord*, Exon. 123 a ; Th. 473, 5 ; Bo. 10. Hū gód biþ ðætte bróður on án hicgen *how good it is that brothers should be unanimous*, Ps. 132, 1. Wærwyrde sceal wísfæst hæle breóstum hycgan *a man cautious of words and wise must keep his thoughts to himself*, Exon. 80 b ; Th. 303, 24 ; Fä. 58. Uton wē hycgan hwǽr wē hām ágen and ðonne geþencan hū wē ðider cumen *let us consider where we may have a home, and then devise how we may come thither*, 83 a ; Th. 312, 30 ; Seef. 117. Á sceal snotor hycgean ymbe ðisse worulde gewinn *ever must the prudent man meditate about the struggle of this world*, Menol. Fox 570 ; Gn. C. 54. Iç mid heortan ongann hycggean *meditatus sum cum corde meo*, Ps. Th. 76, 6. Hycgan on ellen, Cd. 154 ; Th. 191, 22 ; Exod. 218. Micel is tó hycganne wísfæstum menn hwæt seó wiht sý *to a sagacious man it is a great subject for thought what the creature may be*, Exon. 107 b ; Th. 411, 14 ; Rä. 29, 13. Hycgenne, 108 b ; Th. 414, 21 ; Rä. 32, 23. Hycgende mon *a man who thinks*, 92 b ; Th. 347, 10 ; Sch. 10. Wē sculon á hycgende hǽlo rǽdes gemunan sigora waldend *mindful of saving counsel must we ever remember the disposer of victories*, 84 b ; Th. 318, 13 ; Mód. 82. Gemune ūs on módsefan forþ hycgende folces ðínes *remember us, being continually mindful of thy people ; memento nostri in beneplacito populi tui*, Ps. Th. 105, 4.   II. *to direct the mind* [*to an object*], *to be intent upon, to intend, purpose, determine, endeavour, strive :*—Ic hicge molior, Ælfc. Gr. 31 ; Som. 35, 51 : nitor, 36 ; Som. 38, 53. Ic mid ealre mínre heortan hige hycge swíðe ðæt ic ðín bebod átredde *ego in toto corde meo scrutabor mandata tua*, Ps. Th. 118, 69. Ic hycge ðæt ic sóðne dóm symble healde *statui custodire judicia justitiæ tuæ*, 106 : 146. Hió hogde georne ðæt hire mægþhád clǽne geheólde *she earnestly determined to keep her maidenhood pure*, Exon. 66 a ; Th. 244, 18 ; Jul. 29. Freóndrǽdenne heó from hogde *her mind revolted from relationship with him* [i. e. *she determined not to marry*], Th. 244, 28 ; Jul. 34. Hicg þegenlíce *viriliter age*, Jos. 1, 18. Hycge swá hē wille ne mæg wērigmód wyrde wiðstondan *strive as he will the weary-hearted cannot withstand fate*, Exon. 76 b ; Th. 287, 15 ; Wand. 14. Ne hycge tó slǽpe se ðe heoldeþ ðē *neque obdormiet qui custodit te*, Ps. 120, 3. Hēt ðá hyssa hwæne hicgan tó handum *he bade then each of his men look to the arms in their hands*, Byrht. Th. 131, 6 ; By. 4. Ongunnon ðæt ðæs monnes mágas hycgan þurh dyrne geþoht ðæt hý tódǽlden unc *this did the man's kinsmen through dark design endeavour, to part us two*, Exon. 115 a ; Th. 442, 12 ; Kl. 11. Wē ðæs sculon hycgan georne ðæt ... *we must therefore earnestly endeavour to* ..., Cd. 19 ; Th. 25, 22 ; Gen. 397 : 226 ; Th. 302, 6 ; Sat. 594.   III. *to direct the mind with a feeling of confidence, to hope :*—Ic on ðē geare hycge *sperabo in eum*, Ps. Th. 90, 2. Ic hycge tó ðē in te speravi, 142, 8. Hycge him hálig folc hǽlu tó Drihtne *sperate in eum, omnis conventus plebis*, 61, 8. Wē cunnon hycgan and hyhtan ðæt ... *we can hope that* ..., Frag. Kmbl. 83 ; Leas. 44. [*Goth.* hugjan : *O. Sax.* huggian : *Icel.* hyggja *to think, intend, purpose : O. H. Ger.* huggen *meditari, sperare*, Grff. iv. 786.] DER. á-, be-, for-, ge-, ofer-, on-, wið-hycgan ; *and see* hogian.

-hycgende. v. bealu-, deóp-, gleáw-, gram-, heard-, morðor-, níþ-, rǽd-, stíð-, swíð-, þanc-, þríst-, wís-, wiðer-hycgende.

HÝD, e ; *f.* HIDE, skin :—Hýd *cutis* vel *pellis ; corium* vel *tergus*, Ælfc. Gl. 73 ; Som. 71, 31, 32 ; Wrt. Voc. 44, 17, 18. Getannede hýd *subacta coria* vel *medicata* vel *confecta*, 17 ; Som. 58, 103 ; Wrt. Voc. 22. 19. Hiora hýd biþ swíðe gód tó sciprápum *their* [*walruses*] *hide is very good for ship-ropes*, Ors. 1, 1 ; Swt. 18, 2. Him seó hýd áheardod wæs on ðǽm cneówum swá olfendan cneó beóþ *the skin on his knees had got as hard as a camel's knees are*, Shrn. 93, 10. Þurh ðære hýde wunda ádwæscte his módes wunda *through the wounds of his skin extinguished the wounds of his mind*, Homl. Th. ii. 156, 31. Twegen sciprápas óðer of hwæles hýde geworht óðer of sioles, Ors. 1, 1 ; Swt. 18, 22. Se bát wæs geworht of þriddan healfre hýde *the boat was made of two hides and a half*, Chr. 891 ; Erl. 88, 9. Þinch him [*cattle*] genóg on ðam ðe hí binnan heora ægenre hýde habbaþ tóeácan ðam fódre ðe him gecyndelíc biþ, Bt. 14, 2 ; Fox, 44, 23. Gif mon óðrum rib forsleá binnan gehálre hýde geselle x scill. tó bóte gif sió hýd síe tóbrocen ... *if a man fracture another's rib without breaking the skin let him pay ten shillings in compensation ; if the skin be broken* ..., L. Alf. pol. 70 ; Th. i. 98, 11–13. Ða heó [*the snake*] gefylled wæs hē hēt hý behyldan and ða hýde tó

Rôme bringan . . . heó wæs hundtwelftiges fôta lang, Ors. 4, 6; Bos. 85, 1. Hě healde iii niht hýde [*of an ox*] and heáfod and sceápes eallswā. And gif hě ða hýde āweg sylle gilde xx ôran, L. Eth. iii. 9; Th. i. 296, 18. Hwæt sind gescý būton deádra nýtena hýda? Homl. Th. ii. 280, 30. Ic bicge hýda and fell *ego emo cutes et pelles*, Coll. Monast. Th. 27, 29. Horses hýda hí habbaþ him tô hrægle *pelliculas equorum ad vestimentum habentes*, Nar. 38, 2. In the Laws the word is used in technical phrases relating to flogging [cf. colloquial ' to give one a *hiding* '] :—Wealh gafolgelda cxx scill. . . . weales hýd twelfum *the ' wer ' of a tenant of British race is one hundred and twenty shillings . . . the ' hide-gild ' of a man of British race is twelve shillings* [the ' hide-gild ' of a *þeów* (v. infra), whose *wer* was half that of a *wealh*, was six shillings ; if the same proportion was kept, the *weales hýd* would be, as here, twelve shillings], L. In. 23; Th. i. 118, 4. Þeówman þolie his hýde oððe hýdgyldes *let a slave be flogged or pay the ' hide-gild*,' L. E. G. 7, 8; Th. i. 172, 1, 7: L. C. S. 45: 47; Th. i. 402, 16, 26. Ðara hyrda ælc þolige ðære hýde, L. Edg. S. 9; Th. i. 276, 3. Gif þeów deóflum geldaþ vi scill. gebête oððe his hýd *if a slave offer to devils let him pay six shillings or be flogged*, L. Wih. 13: 15: 10; Th. i. 40, 8, 11: 38, 22. Gif hwā his hýde forwyrce and cirican geierne síe him sió swingelle forgifen *if any one be liable to flogging* [lit. *forfeit his hide*] *and escape into a church, let the scourging be forgiven him*, L. In. 5; Th. i. 104, 15. Se ðe ænig ðissa dô, gilde wîte, frîman xii ôr, þeówman ða hýde, L. N. P. L. 56; Th. ii. 298, 25. v. Grm. R. A. 703. [*Laym. A. R. O. and N.* hude: O. Frs. hûd, hêd : Icel. húð *a hide*: also a law term as above, e. g. fyrirgöra húð síuni *to forfeit one's hide*; leysa húð sína *to redeem one's hide*; cf. hýða *to flog*: O. H. Ger.* hût *cutis, corium, pellis, tergus, birsa : Ger.* haut.]

-hýd = -hygd, q. v.

HÝDAN; *p.* de *To* HIDE, *conceal* :—Ic mě wið heora hete hýde *absconderem me ab eo*, Ps. Th. 54, 12. Se lǽce hýd his îsern wið ðone monn ðe hé snîðan wile *the surgeon hides his knife from the man that he means to cut*, Past. 26, 3; Swt. 185. 25. Hýt *abscondit*, Swt. 187, 9. Se ðe his hwæte hýtt *qui abscondit frumenta*, 49, 1; Swt. 377, 13. Hýdeþ, Exon. 82 b; Th. 311, 34; Seef. 102. Hí on holum hýdaþ hí *in cubilibus suis se collocabunt*, Ps. Th. 103, 21. Fleóþ ðonne tô muntum and hié hýdaþ for ðara engla onsýne, Blickl. Homl. 93, 26 : Past. 15, 1; Swt. 89, 15. Ic on mínre heortan hýdde *in corde meo abscondi*, Ps. Th. 118, 11 : Bt. Met. Fox 29, 109; Met. 29, 55. Ðe hælend hine hýdde *Iesus abscondit se*, Jn. Skt. Rush. 8, 59. Hýddon hié *they hid themselves*, Cd. 40; Th. 53, 12; Gen. 860. Hýde se ðe wylle *hide who will*, Beo. Th. 5526; B. 2766. Ne sylþ hé hit ûs tô ðon ðæt wé hit hýdon, Blickl. Homl. 53, 17. Crist hêt hine hýdan ðæt hearde îsen [*put up his sword*], Homl. Th. ii. 246, 24. Nô ðý mínne þearft hafelan hýdan [*bury*], Beo. Th. 896; B. 446. Hwær se wuldres beám under hrusan hýded wære *where the tree of glory* [*the cross*] *under ground was hidden*, Elen. Kmbl. 436; El. 218. Ðær ða æðelestan hýdde wæron, 2214; El. 1108. [*Orm.* hidenn; *Laym. A. R. O.* and *N.* huden: *Ayenb.* hede: *Chauc.* hide.] DER. ā-, be-, bi-, for-, ge-hýdan.

hýdd-ern. v. hêdd-ern.

hýdels, es; *m. A place of concealment, hiding-place, cavern* :—Hýdels þeáfana *spelunca latronum*, Mk. Skt. Rush. 11, 17. Gif hit on hýdelse funden sý *if it be found in a place of concealment*, Lv. Ath. iv. 6; Th. i. 226, 4. [*Laym.* an hudlese wuneden *lived in caverns : A. R.* inę hudles *in secret : Trev.* break out of his hydels (hudels, huydels) *de latibulo suo erumpens : Wick.* in hidils (hudlis) *in abscondito.*]

hýd-gild, es ; *n. A payment made to escape the punishment of flogging.* v. hýd.

hýdig ; *adj. Made of hide, leathern* :—Hýdig fæt *bulga*, Ælfc. Gl. 29; Som. 61, 28; Wrt. Voc. 26, 28. [Cf. leðer-coddas *bulgæ*, 16; Som. 58, 58; Wrt. Voc. 21, 45.]

hýdig = hygdig, q. v.

HÝF, e ; *f. A* HIVE :—Hýf *canistrum vel alvearium*, Ælfc. Gl. 25; Som. 60, 60; Wrt. Voc. 25, 2. Hýf *alvearia*, Wrt. Voc. 284, 40. Hýfe *alvearii*, ii. 4, 64. Hýfi *alvearia*, 100, 1. Wið ðæt beón æt ne fleón genim ðäs ylcan wyrte ðe wé veneriam nemdon and gehôh tô ðære hýfe ðonne beóþ hý wungynde *that bees may not fly away, take this same plant that we called veneria and hang it to the hive, then will they be stationary*, Herb. 7, 2; Lchdm. i. 98, 1. Mæderecíþ on ðínre hýfe ðonne ne āsponþ nān man ðíne beón ne hí man ne mæg forstelan ða hwíle ðe se cíþ on ðære hýfe biþ [*put*] *a plant of madder in your hive ; then nobody will lure away your bees, nor can they be stolen while the plant is in the hive*, Lchdm. i. 397, 2-4. [Hé wunede eall riht swā dräne dôþ on hîue *he lived exactly as drones do in the hive*, Chr. 1127; Erl. 256, 20. *Rel. Ant.* huive: *M. L. Ger.* huve.]

-hygd. v. ge-, for-, wan-hygd,-hýgd.

hygdig, hýdig ; *adj. Disposed, minded, careful, considerate, chaste, modest* : — Þancolmôd wer þeáwum hýdig *a man of thoughtful mind, virtuously disposed*, Cd. 82; Th. 102, 17; Gen. 1705. Hygdig *casta*, Rtl. 68, 12. Hygdign friódôm *casta libertas*, 105, 1. Hygdego, 109, 35. [*O. Sax.* hugdig, hûdig (*in compounds*).] DER. ān-, bealu-, deóp-, fæst-, gleáw-, gram-, læt-, lytel-, nîþ-, ofer-, reðe-, stíð-, þríst-, un-, wan-, wîs-, wiðer-hygdig, -hýdig.

---

hygdig-lîce ; *adv. Chastely* : — Hia seolfa hia hygdiglige beheóldon *seipsos castraverunt*, Mt. Kmbl. Lind. 19, 12.

hygdig-ness, e ; *f. Chastity, modesty* : — Hygdignisse *castitatis*, Rtl. 77, 33 : 103, 40. Hygdignisse *pudore*, 110, 5.

hyge *the upper part of the throat, fauces* :—Hyge *faus* [*faux* or *fauces* ?] Wrt. Voc. 282, 78 : ii. 36, 46.

hyge, es ; *m. Mind, heart, soul* :—Cwæþ ðæt hine his hige speóne ðæt hě wyrcean ongunne getimbro *he said that his heart lured him to attempt making buildings*, Cd. 15; Th. 18, 17; Gen. 274. Ôþ hine his hyge forspeón and his ofermetta ealra swîðost *until his heart seduced him, and his pride most of all*, 18 ; Th. 22, 34; Gen. 350. Hyge Euan wîfes wâc gebóht *the mind of Eve, weak thought of woman*, 30 ; Th. 40, 34; Gen. 648. Ðam þegne ongan his hige hweorfan *the man's mind began to change*, 33; Th. 44, 8; Gen. 706. Næs him blîðe hige *no cheerful mind was his*, 178; Th. 223, 10; Dan. 117. Wearþ him hýrra hyge ðonne gemet wære *haughtier grew his soul than was meet*, 198 ; Th. 247, 2; Dan. 491. Him wæs geómor sefa hyge murnende *mournful was their mind, sorrowing their soul*, Exon. 15 a ; Th. 31, 24; Cri. 500. Forðon is mín hyge geómor, 115 a ; Th. 442, 24; Kl. 17. Se hreó hyge, 76 b ; Th. 287, 18; Wand. 16. Ys mínre heortan hige hluttor and clǽne *quia delectatum est cor meum*, Ps. Th. 72, 17. Hyge wearþ mongum blissad *the heart of many was made glad*, Exon. 24 b; Th. 71, 30; Cri. 1163. Hlihende hyge *a gladsome mind*, Elen. Kmbl. 1986; El. 995. Hyge wæs him hinfús *he was minded to flee away*, Beo. Th. 1514; B 755. Ne biþ him tô hearpan hyge *no mind hath he for the harp*, Exon. 82 a ; Th. 308, 23; Seef. 44. Ne wæs him bleáþ hyge *no coward heart had he* [cf. *Icel.* hug-blauðr *timid*; hug-bleyði *cowardice*], Andr. Kmbl. 462; An. 231. Ðā wæs hyge onhyrded *then was his heart confirmed*, Elen. Kmbl. 1678; El. 841. Se hearda hyge wunade *the stout heart continued*, Exon. 40 b; Th. 134. 31; Gû. 517. Hyge sceal heardum men *a bold man must have courage*, 92 a ; Th. 346, 15; Gn. Ex. 205. Hige sceal ðê heardra heorte ðê cênre ðê ûre mægen lytlaþ *the firmer must courage be, braver the heart, the more our force dwindles*, Byrht. Th. 140, 62; By. 312. Hyge weallende *a mind agitated by violent emotions*, Andr. Kmbl. 3415; An. 1711. Weóll him on innan hyge ymb his heortan, Cd. 18; Th. 23, 5; Gen. 354. Mín hyge dreóseþ bysig æfter bôcum : hwílum hyge heortan neáh hreáw wealleþ, Salm. Kmbl. 122-6; Sal. 60-2. Ðú wâst ðæt ic eom unwís hyges *tu scis insipientiam meam*, Ps. Th. 68, 6. Ic mín gehât mid hyge gylde ðæt míne weleras ǽr wíse gedǽldan *reddam vota mea, quæ distinxerunt labia mea*, 65, 12: 102, 19. Ic andette ðê mid hyge ealle heortan mínre *confitebor tibi in toto corde meo*, 110, 1: 118, 69 : 94, 10. Wesan ðíne eáran gehýrende mid hige on eall gebedd esnes ðínes *fiant aures tuæ intendentes in orationem servi tui*, 129, 2. Mid hyge þencan *to think with the mind*, Exon. 82 b ; Th. 311, 23; Seef. 96. Wese heorte mín on hige clǽne *fiat cor meum immaculatum*, Ps. Th. 118, 80. On mínum hyge hreóweþ *I am grieved to think*, Cd. 22; Th. 27, 31; Gen. 426. Ne nieahte hê æt his hige findan ðæt hê wolde þeódne þeówian *he could not find it in his heart to serve his prince*, 14 ; Th. 18, 1; Gen. 266. Hálig on hige *holy of thought*, 133 ; Th. 168, 9; Gen. 2780: Exon. 73 b ; Th. 274, 14; Jul. 533. On heardum hyge *in my hard heart*, Elen. Kmbl. 1614; El. 809. Hêt hicgan tô hige gôdum *bade them see to it that they were of good courage*, Byrht. Th. 131, 7; By. 4. Hí on heofon setton hyge hyra mûþes *posuerunt in cælum os suum*, Ps. Th. 72, 7. Hæfde hyge strangne *he had a strong heart*, Cd. 23 ; Th. 29, 9; Gen. 447. Heardrǽdne hyge, 107 ; Th. 141, 21; Gen. 2348. Ic georulíce gode þegnode þurh holdne hyge *diligently I served God with loyal heart*, 28 ; Th. 37, 7; Gen. 586: Beo. Th. 539; B. 267. Þurh yrne hyge *in anger*, Exon. 16 b ; Th. 39, 10; Cri. 620: Andr. Kmbl. 1941; An. 973. Ðinne hyge gefæstna *strengthen thine heart*, Exon. 93 a ; Th. 348, 33; Sch. 37: Andr. Kmbl. 2427; An. 1215. Dôþ eówre heortan hige hále and clǽne *effundite coram illo corda vestra*, Ps. Th. 61, 8. Nyllan gé eów on heortan ða hige staðelian *nolite cor apponere*, 11. [*Laym.* huʒe: *Orm.* hiʒ: *Goth.* hugs: *O. Sax.* hugi: *O. Frs.* hei: *Icel.* hugi, hugr: *O. H. Ger.* hugu, hugi *animus, sensus, affectus*.] DER. hāt-hyge.

hyge-bend, es ; *m* : e ; *f. A tie or bond which is furnished by the mind* :—Hygebendum fæst *fixed firm by the mind's chains*, Beo. Th. 3761; B. 1878.

hyge-blind ; *adj. Having the mind blinded*, Exon. 66 b ; Th. 246, 13 ; Jul. 61.

hyge-blîðe ; *adj. Glad at heart*, Andr. Kmbl. 3378; An. 1693 : Exon. 107 a ; Th. 408, 31; Rä. 27, 20.

hyge-clǽne ; *adj. Pure in mind*, Ps. Th. 104, 3.

hyge-cræft, es ; *m. Mental power, intellect, wisdom* : — Ealle þeóde ēcne Drihten mid hygecræfte herigan *let all nations praise the Lord with the powers of their minds*, Ps. Th. 116, 1 : 118, 61, 73. Gif ðū mé ðínne hygecræft hylest and ðíne heortan geþohtas *if thou dost conceal from me thy wisdom and thy heart's thoughts*, Exon. 88 b ; Th. 333, 12 ; Gn. Ex. 3. Wísdôm higecræft heáne, Cd. 176; Th. 222, 1; Dan. 98. Hygecræftum, Hy. 6. 3 ; Hy. Grn. ii. 286, 3.

hyge-cræftig ; *adj. Having mental power, wise, sagacious*, Exon.

11 a ; Th. 15, 25 ; Cri. 241 : 92 b ; Th. 348, 8 ; Sch. 25 : 101 a ; Th. 380, 37 ; Rä. 2, 1.

hý-gedriht. v. hí-gedryht.

hyge-fæst ; *adj. Firm of mind, prudent, wise*, Exon. 112 a ; Th. 429, 33 ; Rä. 43, 14. [*Icel.* hug-fastr *steadfast.*] Cf. hoga-fæst.

hyge-fród ; *adj. Wise of mind, prudent*, Cd. 92 ; Th. 117, 13 ; Gen. 1953.

hyge-frófor, e ; *f. Comfort for the mind* or *heart*, Elen. Kmbl. 709 ; El. 355 : Hy. 9, 13 ; Hy. Grn. ii. 291, 13.

hyge-gælsa ; *adj. Slow, sluggish :*—Nis hé swongor swá sume fuglas ða ðe late þurh lyft lácaþ fiþrum *non tamen est tarda, ut volucres quæ corpore magno incessus pigros per grave pondus habent*, Exon. 60 b ; Th. 220, 3 ; Ph. 314. v. gælan.

hyge-gál ; *adj. Light-minded, wanton*, Exon. 103 b ; Th. 394, 2 ; Rä. 13, 12. v. gál.

hyge-gár, es ; *m. A dart of the mind, a wile, device*, Exon. 83 b ; Th. 315, 21 ; Mód. 34.

hyge-geómor, -giómor ; *adj. Sad in mind, mournful, sorrowful*, Cd. 42 ; Th. 54, 18 ; Gen. 879 : Andr. Kmbl. 2175 ; An. 1089 : 3112 ; An. 1559 : Exon. 49 b ; Th. 171, 20 ; Gú. 1129 : Beo. Th. 4807 ; B. 2408 : Exon. 21 a ; Th. 55, 29 ; Cri. 891. Hygegeómorne, 115 a ; Th. 442, 28 ; Kl. 19. Hygegeómre, 10 a ; Th. 10, 17 ; Cri. 154 : 22 b ; Th. 61, 33 ; Cri. 994 : 70 b ; Th. 262, 4 ; Jul. 327 : 45 b ; Th. 155, 8 ; Gú. 857 : 46 a ; Th. 157, 31 ; Gú. 900 : Elen. Kmbl. 2429 ; El. 1216.

hyge-gleáw ; *adj. Wise, prudent, having clear mental vision*, Exon. 25 a ; Th. 73, 23 ; Cri. 1194 ; Chr. 975 ; Erl. 126, 25 ; Edg. 51 : Elen. Kmbl. 665 ; El. 333.

hyge-grim ; *adj. Cruel of mind, fierce, savage*, Exon. 74 b ; Th. 278, 9 ; Jul. 595.

hyge-leás ; *adj. Thoughtless, careless, foolish :*—Ne geríseþ biscopum ne æt hám ne on síðe tó higeleás [iuncg'íc, MS. G] wíse ac wísdóm and weorþscipe gedafenaþ heora háde *a too thoughtless manner is not seemly for bishops, neither at home nor when travelling, but wisdom and dignity are becoming to their rank*, L. I. P. 10, note ; Th. ii. 318, 41. Higeleás plega *senseless play*, Homl. Th. ii. 220, 6. Hygeleáse *lacking wisdom* [*the rebellious angels*], Cd. 3 ; Th. 4, 10 ; Gen. 51. Leahtra hegeleásra *of sins committed thoughtlessly*, Ps. C. 50, 144 ; Ps. Grn. ii. 280, 144. [*Icel.* hug-lauss *fainthearted.*]

hyge-leást, e ; *f. Thoughtlessness, foolishness, folly, want of wisdom, heedlessness :*—Eálá gē cildra gáþ út bútan hygeleáste tó claustre oððe tó leorninge *O vos pueri egredimini sine scurrilitate in claustrum vel in gymnasium*, Coll. Monast. Th. 36, 9. Ne ús ne gedafenaþ ðæt wē úrne líchaman ðe Gode is gehálgod mid unþæslícum plegan and higleáste ge-scyndan *it doth not beseem us to put our body, that is sanctified to God, to shame with indecent play and folly*, Homl. Th. i. 482, 12. Wē sceolon blissian on úrum Drihtne ná on higleáste *we ought to rejoice in our Lord, not in folly*, ii. 292, 32. Englas wæron befeallene on ða hátan hell þurh hygeleáste and þurh ofermétto *angels had fallen into the hot hell through folly and through pride*, Cd. 18 ; Th. 21, 29 ; Gen. 331. Biscopum gebiraþ wísdóm . . . ne gerísaþ heom micele ofermétta ne ǽnige higelíste, L. I. P. 10, note ; Th. ii. 318, 32. [Cf. *Icel.* hug-leysa *timidity.*]

hyge-mǽd, e ; *f. Honour that is shewn with the heart* or *mind, reverence ;* or *fitness that is determined by the mind* [?] :—Wíglaf healdeþ higemǽdum [hige mēdum, Th.] heáfodwearde *Wiglaf keeps guard reverently* [or *duly*], Beo. Th. 5810 ; B. 2909. v. mǽd.

hyge-méðe ; *adj. Wearying the heart* or *mind*, Beo. Th. 4875 ; B. 2442.

hyge-róf ; *adj. Stout, strong of mind* or *heart, magnanimous*, Exon. 124 a ; Th. 477, 6 ; Ruin. 20 : 46 b ; Th. 159, 13 ; Gú. 926 : Andr. Kmbl. 465 ; An. 233 : 2009 ; An. 1007 : Beo. Th. 413 ; B. 204 : Cd. 82 ; Th. 102, 32 ; Gen 1709 : 75 ; Th. 93, 22 ; Gen. 1550 : Exon. 15 b ; Th. 33, 31 ; Cri. 534 : Judth. 12 ; Thw. 26, 1 ; Jud. 303.

hyge-rún, e ; *f. A secret of the mind* or *heart :*—Cyriacus hygerúne ne máþ gástes mihtum tó Gode cleopode *Cyriacus did not conceal the secret of his heart, but with the powers of the spirit cried to God*, Elen. Kmbl. 2196 ; El. 1099. v. Grmm. A. u. E. 139. [*Icel.* hug-rúnar *magical runes with a power of wisdom.*]

hyge-sceaft, e ; *f. Mental constitution, mind, disposition, heart :*—Frýnd synd hié míne georne holde on hyra hygesceaftum *they are my zealous friends, loyal in their hearts*, Cd. 15 ; Th. 19, 8 ; Gen. 288. [*O. Sax.* hugi-skafti ; *pl.*]

hyge-snottor ; *adj. Wise of mind, prudent, sagacious*, Exon. 49 a ; Th. 168, 23 ; Gú. 1082 : 71 a ; Th. 265, 24 ; Jul. 386 : Bt. Met. Fox 10, 14 ; Met. 10, 7.

hyge-sorh, -sorg, e ; *f. Mental care, anxiety*, Cd. 94 ; Th. 122, 31 ; Gen. 2035 : Exon. 10 b ; Th. 11, 21 ; Cri. 174 : 47 b ; Th. 162, 28 ; Gú. 982 : 50 a ; Th. 174, 15 ; Gú. 1178 : 51 a ; Th. 176, 32 ; Gú. 1219 : Cd. 37 ; Th. 48, 16 ; Gen. 776 : Beo. Th. 4646 ; B. 2328.

hyge-teóna, an ; *m. Deliberate injury* or *offence :*—Ic him hygeteónan hwítan seolfre béte *with white silver will I make reparation to him for injury*, Cd. 130 ; Th. 165, 13 ; Gen. 2731 : 69 ; Th. 83, 16 ; Gen. 1380. Higeteónan spræc on fæmnan *from her heart spoke injuriously against the woman*, Cd. 103 ; Th. 136, 21 ; Gen. 2261.

hyge-þanc, es ; *m. Thought*, Andr. Kmbl. 1634 ; An. 818 : Exon. 27 a ; Th. 81, 30 ; Cri. 1331 : 109 a ; Th. 417, 14 ; Rä. 36, 4 : Elen. Kmbl. 311 ; El. 156 : Ps. Th. 74, 5.

hyge-þancol ; *adj. Thoughtful*, Andr. Kmbl. 681 ; An. 341 : Cd. 176 ; Th. 221, 26 ; Dan. 94 : Judth. 11 ; Thw. 23, 20 ; Jud. 131.

hyge-þrymm, es ; *m. Strength of heart* or *mind*, Beo. Th. 683 ; B. 339.

hyge-þrýþ, e ; *f. Pride of heart* or *mind, insolence :*—Higeþrýþe wæg was insolent, Cd. 102 : Th. 135, 6 ; Gen. 2238.

hyge-þyhtig ; *adj. Doughty of heart*, Beo. Th. 1497 ; B. 746.

hyge-treów, e ; *f. Faith deliberately pledged*, Cd. 107 ; Th. 142, 25 ; Gen. 2367.

hyge-wælm, es ; *m. Agitation of the mind, violent emotion*, e. g. anger, Cd. 47 ; Th. 60, 12 ; Gen. 980.

hyge-wlanc ; *adj. Proud, elated in mind*, Exon. 105 a ; Th. 400, 1 : Rä. 20, 2 : 112 b ; Th. 431, 21 ; Rä. 46, 4.

hyht, es ; *m.* [*f.* Ps. Th. 77, 53.] *Hope, joyous expectation, joy :*—Hiht on Gode *hope in God*, Homl. Th. ii. 602, 11. Ðære gástlícan strenge mycel hyht *the great hope of spiritual strength*, Blickl. Homl. 135, 26. Mé is hálig hyht on hine *spes mea in Deo est*, Ps. Th. 61, 7 : 70, 4. Ðú eart hyht ealra ðe on ðysse eorþan útan syndon *spes omnium finium terræ*, 64, 6. Hwílum hié gehéton æt heargtrafum wigweorþunga bædon ðæt him gástbona geóce gefremede. Swylc wæs þeáw hyra hǽðenra hyht *sometimes they vowed in their temples idolatrous honours, prayed that the destroyer of souls would afford them help. Such was their custom, such the hope of the heathens*, Beo. Th. 360 ; B. 179. Ðú eart mín se sóða hiht *tu es spes mea*, Ps. Th. 141, 5. Ælc hyht lífes *omnis spes vitæ*, Rtl. 3, 28. Ðú cægst his noman Iohannes and ðé biþ ðonne hyht and gefeá *vocabis nomen suum Johannem et erit gaudium tibi et exultatio*, Blickl. Homl. 165, 10. Lífes hyht and ealles leóhtes gefeá, Exon. 16 a ; Th. 36, 32 ; Cri. 585 : 42 a ; Th. 141, 23 ; Gú. 631. Ðær is hyht and blis *there is joy and bliss*, Exon. 18 b ; Th. 47, 5 ; Cri. 750 : 15 b ; Th. 33, 22 ; Cri. 529 : 46 b ; Th. 159, 14 ; Gú. 926. Ne biþ him tó hearpan hyge ne tó wífe wyn ne tó worulde hyht *he hath no mind for the harp, nor delight in woman, nor joy in life*, 82 a ; Th. 308, 26 ; Seef. 45. Næs him tó hǽlende wyn hyht tó hordgestreónum, Andr. Kmbl. 2229 ; An. 1116. Sigbég hyhtes *corona spei*, Rtl. 1, 15. Hygtes, 3, 26. Is mé Moab mínes hyhtes hwer *Moab olla spei meæ*, Ps. Th. 59, 7. Ðære hǽlo ðe hé ús tó hyhte forgeaf *for the salvation which he hath given us to hope for*, Exon. 16 b ; Th. 38, 28 ; Cri. 613. Hæbbe ic mé tó hyhte heofonríces weard *I have the guardian of the kingdom of heaven as my hope*, 68 b ; Th. 255, 10 ; Jul. 212. Hæfdon hym tó hyhte helle flóras beornende bealo *they had the bottom of hell and burning torments to look forward to*, Cd. 214 ; Th. 269, 8 ; Sat. 70. Nabbaþ wé tó hyhte nymþe weán and wítu *we have nothing to expect but woe and punishments*, 220 ; Th. 285, 9 ; Sat 335. Se beorn wæs on hyhte *the man was in good hopes* [*of performing his journey*], Andr. Kmbl. 478 ; An. 239 : 1274 ; An. 637. Ic eom wunderlícu wiht wífum on hyhte *I am a wondrous creature giving joy to women*, Exon. 106 b ; Th. 407, 7 ; Rä. 26, 1 : Runic pm. Kmbl. 342, 16 ; Rún. 16. Hé hí on hihte holdre lædde *deduxit eos in spe*, Ps. Th. 77, 53. Ic háligne gást hyhte belúce emne swá écne *I believe the Holy Ghost to be just as eternal*, Hy. 10, 41 ; Hy. Grn. ii. 293, 41. Hé him forgeaf éces lífes hyht, Blickl. Homl. 137, 7. Hí on God ǽnne heora hyht gesetton *they placed their hope on God only*, 185, 15 : Ps. Th. 113, 20. Beón ða ófdrædde ða ðe sint ofsette mid flǽsclícum lustum, and nabbaþ nǽnne hiht tó engla werode *let those be afraid that are oppressed with fleshly lusts and have nothing to hope for from the angelic host*, Homl. Th. i. 222, 29. Ðæt hí gleáwne hiht tó Gode hæfdan *ut ponant in Deo spem suam*, Ps. Th. 77, 9. Ic hiht on ðon hæbbe georne *exultabo*, 62, 7. Hyhta leáse helle sóhton *hopeless they sought hell*, Exon. 75 b ; Th. 283, 18 ; Jul. 682. Hyhtum tó wuldre *with hopes of glory*, 116 b ; Th. 448, 3 ; Dóm. 48. [*O. E. Homl.* huht, hiht : *Orm.* hihht : *O. and N.* hihte, histe.] DER. tó-, woruld-hyht.

hyhtan ; *p.* te *To hope, trust, look forward to with hope* or *joy, rejoice :*—Ic under ðínum fiðrum sceal on pennis ejus sperabis, Ps. Th. 90, 4. Ic ðé hihte tó *sperantem in te*, 85, 2. Hihte ic tó ðínra handa hálgum dǽdum *in operibus manuum tuarum exultabo*, 91, 3. Heorte mín and flǽsc hyhtaþ georne on ðone lifgendan Drihten *cor meum et caro mea exultaverunt in Dominum vivum*, 83, 2. Hé hyhte tó him *in me speravit*, 91, 14. Ðám [ðe] longe his hyhtan hidercyme *to those who had long hoped for his advent*, Exon. 10 a ; Th. 9, 29 ; Cri. 142. Sione bearn symble hihtan *filii Sion exultant*, Ps. Th. 149, 2. Se þeóda láreów lǽrde ða rícan ðæt hí heora hiht ne besetton on ðám swicelum welum, ac hihton on God ðæra góda syllend *the teacher of the gentiles taught the rich that they should not set their hope on deceitful riches, but should hope in God, the giver of good things*, Homl. Th. ii. 328, 1. Wē cunnon hyhtan ðæt wē heofenes leóht ágan móton *we can hope that we may possess the light of heaven*, Fragm. Kmbl. 84 ; Leás. 44. Ic ellen wylle habban and hlyhhan and mé hyhtan tó *I will have courage, and laugh and look forward with hope*, Exon. 119 a ; Th. 456, 22 ; Hy. 4, 70 : 12 b ; Th. 21, 26 ; Cri. 340. Gód ys on Dryhten tó hyhtanne *bonum est confidere in Domino*, Ps. Th. 117, 9. [*O. and N.* hihte.] v. ge-hyhtan.

hyht-ful; adj. Full of hope or joy, joyous, exultant, glad, pleasant :—
Ic þurh Judas ǽr hyhtful gewearþ and nū gehýned eom þurh Judas eft through Judas formerly I became exultant, and now again through Judas am I humiliated, Elen. Kmbl. 1842; El. 923. Ontýn eárna hleóðor ðæt mín gehérnes hehtful weorþe auditui meo dabis gaudium, Ps. C. 50, 78; Ps. Grn. ii. 278, 78. Him on lǽste beleác hihtfulne hám hálig engel a holy angel closed behind them the pleasant abode [paradise], Cd. 45; Th. 58, 14; Gen. 946. Wē hyhtfulle hǽlo gelýfaþ we, filled with hope, trust the salvation [or hyhtfulle may agree with hǽlo], Exon. 9 b; Th. 8, 17; Cri. 119.

hyht-gifa, an; m. One who gives hope or joy [an epithet of Christ], Elen. Kmbl. 1700; El. 852.

hyht-gifu, e; f. A gift which causes hope or joy, Exon. 94 b; Th. 353, 31; Reim. 21.

hyhting, e; f. Exultation, joy :—Hihting exultatio, lætitia, Wrt. Voc. ii. 146, 30.

hyht-leás; adj. Without hope [of that which is promised], joyless :—Áhóf brýd Abrahames hihtleásne hleahtor Abraham's wife laughed incredulously [without hope that the promise of a son would be fulfilled], Cd. 109; Th. 144, 9; Gen. 2387. v. hyht-ful.

hyht-líc; adj. Giving, or having, cause for hope or joy, hopeful, pleasant, joyous, exultant :—Hyhtlíc heorþwerod a hopeful family, Cd. 78; Th. 96, 35; Gen. 1605. Beóþ ðonne eádge ðe ðǽr in wuniaþ hyhtlíc is ðæt heorþwerud happy are they that dwell therein, joyous is that band, Exon. 93 b; Th. 352, 1; Sch. 91: Cd. 95; Th. 125, 8; Gen. 2076. Hyhtlíc heofontimber the pleasant frame of heaven, Th. 9, 23; Gen. 146: Exon. 116 a; Th. 446, 18; Dóm. 24. Hyhtlícra hám, Cd. 218; Th. 278, 3; Sat. 216: 216; Th. 273, 17; Sat. 138. Ðonne biþ hyhtlícre . . . biþ ðæt ǽrende eádiglícra, Soul Kmbl. 250; Seel. 129. Háma hyhtlícost, Andr. Kmbl. 207; An. 104. [O. E. Hom. hihtliche bure a pleasant chamber: cf. Laym. un-huhtlic.]

hyht-plega, an; m. Joyous play, sport, Exon. 18 b; Th. 46, 14; Cri. 737: 105 b; Th. 402, 12; Rä. 21, 28.

hyht-willa, an; m. Desire accompanied by hope or joy :—Hyhtwillan leás without hope of attaining any good, Cd. 216; Th. 274, 25; Sat. 159.

hyht-wynn, e; f. Joy of hope :—Ne þearf hē hopian nó ðæt hē ðonan móte ac ðǽr wunian sceal hyhtwynna leás no need has he to hope that he may go thence, but there shall he dwell hopeless and joyless, Judth. 10; Thw. 23, 14; Jud. 121.

hylc, es; m. A bend, turn, winding :—Ábrocen land vel hilces anfractus, Ælfc. Gl. 100; Som. 77, 9; Wrt. Voc. 55, 12. Wóge hylcas anfractus, reflectus, Hpt. Gl. 448. Hylcas anfractus, 486. Hylcum anfractibus, 493.

hyld, held, es; m. Favour, protection, grace [of a superior to an inferior], loyalty, allegiance [of the inferior to the superior] :—Ic hálsige eów for ðæs cáseres helda ðæt gē mē secgon I adjure you by your allegiance to the emperor that you tell me, Nicod. 8; Thw. 4, 7. Gecýþe ðæt on Godes helde and on 'hláfordes let him declare that on his faith towards God and the lord, L. C. S. 23; Th. i. 388, 23. On gesyhþe ðara háligra ðe ðínne held curan in the sight of the saints that chose thine allegiance [chose thee as their lord]; ante conspectum sanctorum tuorum, Ps. Th. 51, 8. Ðe his hyld curon, Cd. 198; Th. 246, 19; Dan. 481. Gē ðe úres ðæs hálgan Godes held begangeþ ye who practise loyalty towards our holy God, Ps. Th. 133, 2. Ðeáh ðe ic on mínes húses hyld gegange si introiero in tabernaculum domus meæ, 131, 3. Hyld hæfde his ferlorene he had lost the favour of his chief, Cd. 16; Th. 20, 1; Gen. 301. Hæfde wuldres beám werud gelǽded on hild godes the pillar of glory had conducted the host into the favour of God, 170; Th. 214, 13; Exod. 568. On gástes hyld, 195; Th. 243, 29; Dan. 440. Hylda leáse without favours, Exon. 53 a; Th. 186, 20; Az. 21. Ðē ǽfre on fullum hyldum hold and on fulre lufe faithful to thee with full faith and with full love, Chart. Th. 598, 31. For eówrum hyldum ðe gē mē symble cýddon for your fidelity that you have ever shewn me, L. Edg. 5, 12; Th. i. 276, 19. v. helde, hyldu, gehyld, hold; Grmm. R. A. 252.

hyldan, heldan; p. de; trans. and intrans. To bend, incline, heel, tilt :—Ðú gestaþoladest eorþan swíðe fæstlíce ðæt heó ne helt on náne healfe thou hast fixed earth very firmly, so that it does not incline to any side, Bt. 33, 4; Fox 130, 36. Heldeþ, Bt. Met. Fox 20, 327; Met. 20, 164. Hylde hine hleór bolster onfēng he bent himself [to the couch] and the pillow received his cheek, Beo. Th. 1380; B. 688. Ða hig hyra andwlitan on eorþan hyldan cum declinarent vultum in terram, Lk. Skt. 24, 5. Hié tō gebede hyldon they bent down to pray, Andr. Kmbl. 2054; An. 1029. Ne hyld ðú míne heortan ut non declines cor meum, Ps. Th. 140, 5. Ic hyldan mē ne dorste I dare not bow myself [the Ruthwell cross has hælda ik ni darstæ], Rood Kmbl. 90; Kr. 45. Is mín feorh tō helldore hylded geneahhe vita mea in infernum appropinquavit, Ps. Th. 87, 3. [Laym. scipen gunnen helden : A. R. helden win ine wunden : Prompt. Parv. heldyñ or bowyñ inclino, flecto, deflecto, p. 234, see note : Wick. Piers P. helde fundere : O. Sax. af-heldian : Icel. halla to lean or turn sideways ; hella to pour out : Dan. hælde to incline : Swed. hälla : O. H. Ger. halden vergere, recubare ; haldian, heldian inclinare, declinare.] DER. á-, on-hyldan.

hyldan ; p. de To flay, take off the skin :—Hyldeþ discoriat, Wrt. Voc. ii. 140, 78. And hyldon ða offrunge detractaque pelle hostiæ. Lev. 1, 6. [Laym. Wick. hilde : Icel. hylda to slash.] v. be-, on-hyldan ; hold and hyldere.

hyld-áþ, es ; m. An oath of fealty or fidelity :—Ðus man sceal swerigean hyldaþas in this manner are oaths of fealty to be sworn, L. O. 1; Th. i. 178, 2 : see 252, 5. v hold-áþ.

hylde, an ; f. The slope of a hill :—Óþ ðæs clifas norþ hyldan to the north side of the cliff, Cod. Dipl. Kmbl. iii. 418, 24. [Icel. hallr ; m. a slope, hill : O. H. Ger. halda ; f. clivus.] v. hyldan, held.

-hylde. v. earfoþ-, on-hylde and heald.

hylde-mǽg, es ; m. A near and dear kinsman, Cd. 52 ; Th. 67, 1; Gen. 1094 : 94 ; Th. 122, 25 ; Gen. 2032.

hyldere, es ; m. A flayer, butcher :— Hyldere oððe cwellere oððe flǽsctawere lanio vel lanista vel carnifex vel macellarius, Ælfc. Gl. 113; Som. 79, 121 ; Wrt. Voc. 60, 27. From hylderum a lanionibus, Wrt. Voc. ii. 10, 2. v. hyldan, hold.

hylding, e ; f. A bending, inclination ; curvatura, Wrt. Voc. ii. 23, 66.

hyld-rǽden, e ; f. Fidelity :—Ǽlc ōðrum áþ on háligdóme sealde sōðre heldrǽdenne each should give to other on the relics an oath of true fidelity, Chart. Th. 610, 32. v. hold-rǽden.

hyldu, e ; hyldo ; indecl. f. Kindness, favour, affection, friendship, grace, fidelity, loyalty [v. hold.] :—Ys mē heortan gehygd hyldu Drihtnes Deus cordis mei, Ps. Th. 72, 21. His hyldo is unc betere tō gewinnanne ðonne his wiðermēdo his favour is better for us to gain than his hostility, Cd. 30 ; Th. 41, 20 ; Gen. 659. Ðē wæs leófra his sibb and hyldo ðonne ðín sylfes bearn his [God's] peace and grace were dearer to thee than thine own child, 141 ; Th. 176, 33 ; Gen. 2921. Unc is his hyldo þearf we need his favour, 32 ; Th. 41, 30 ; Gen. 664 : Judth. 9 ; Thw. 21, 3 ; Jud. 4. Hyldo tō wedde as a pledge of favour, Beo. Th. 5989 ; B. 2998. For ealdre hyldo from old friendship ; amicitia vetus, Ors. 3, 9 ; Swt. 130, 28. Eallum monnum nánuht swá gōd ne þuhte swá hié tō his hyldo becóme to all men nothing seemed so good as to obtain his favour, 5, 15 ; Swt. 250, 18. Ácwæþ hine fram his hyldo, Cd. 16 ; Th. 20, 6 ; Gen. 304. Hyldo affectum, Wrt. Voc. ii. 1, 12. Swá ic áge Pharaones hyldo so may I possess the favour of Pharaoh ; per saltem Pharaonis, Gen. 42, 15. Wē hraðe begytan hyldo ðíne cito anticipet nos misericordia tua, Ps.Th. 78, 8. Wutun úrum Hǽlende hyldo gebeódan jubilemus Deo salutari nostro, 94, 1. Ic hyldo sōhte I sought grace, 118, 123. Englas ðe on godes hyldo gelǽston angels who were loyal to God, Cd. 17 ; Th. 21, 9 ; Gen. 321 : Ps. Th. 55, 10 : 84, 8. [O. Sax. huldi grace, favour, devotion : O. Frs. helde, hulde : Icel. hylli favour, grace : O. H. Ger. huldi gratia, favor, devotio, fides : Ger. huld.] DER. hláford-, un-hyldu ; and see helde, hyld.

hyll. v. hel.

hyll, es ; m. e ; f. A hill :—Hyll collis, Ælfc. Gl. 97 ; Som. 76, 62 ; Wrt. Voc. 54, 6. Þurþ hylles hrōf through the top of the hill, Exon. 104 b ; Th. 397, 29 ; Rä. 16, 27. Stondende on lytlum hylle, Shrn. 70, 14. Hí huntiaþ hí of ǽlcere hylle they shall hunt them from every hill, Homl. Th. i. 576, 28. Hyllas montes, Ps. Spl. C. 71, 3. Hyllas and heá beorgas, Cd. 192; Th. 240, 7; Dan. 383. Hyllas and cnollas, Exon. 18 a; Th. 45, 11; Cri. 717. Gebiðe synt hylla middaneardes incurvati sunt colles mundi, Cant. Abac. 6. Dúna and hylla montes et colles, Hymn. T. P. 75. Ðonne hie cweþaþ tō ðǽm dúnum and tō ðǽm hyllum tunc incipient dicere montibus et collibus, Blickl. Homl. 93, 33 : Lk. Skt. Lind. 23, 30. [Laym. A. R. hul : Orm. hill : Ayenb. hell : Prompt. Parv. hylle.] v. sand-hyll.

hyll-háma, an ; m. A cricket :—Hilháma cicada, Ælfc. Gl. 37 ; Som. 63, 7 ; Wrt.Voc. 29, 29. Hylleháma oððe gærstapa cicada, ii. 21, 54. v. háma.

hyll-wyrt, e ; f. Hill-wort :—Hylwurt samum, Ælfc. Gl. 40 ; Som. 63, 82 ; Wrt. Voc. 30, 34. Hylwyrt pollegia, 44 ; Som. 64, 83 ; Wrt. Voc. 32, 19. v. Lchdm. ii. 392, col. 2. In E. D. S. Plant Names hillwort is given as [1] mentha pulegium ; [2] thymus serpyllum.

hyl-song a timbrel :—On hylsongæ in tympano, Ps. Spl. T. 150, 4.

hylsten. v. hilsten (Appendix).

hylte, es ; m. A wood, shrubbery :—Scoom hylti frutices, Wrt. Voc. ii. 39, 60.

hymblícae cicuta, Ep. Gl. 7 d, 8. v. hemlíc.

Hymbre. v. Norþan-, Norþ-, Súþ-hymbre.

hymele, an ; f. The hop plant ; humulus lupulus, Lchdm. ii. 392, col. 2. [Icel. humall : Dan. humle hop-plant.]

hymen, es ; m. A hymn :—Be ðam hymene ðe wē be hire geworhton of the hymn that we composed about her, Bd. 4, 19 ; S. 587, 16.

hýnan, hénan ; p. de To abuse, humiliate, rebuke, correct, treat with insult or contumely, despise, oppress, afflict, ill-treat, bring or lay low, subject :—Ðám ilcan monnum ðe hē ðǽr þreátaþ and hénþ ipsis fratribus qui corriguntur, Past. 17, 7 ; Swt. 117, 16. Ða ðe hē ðǽr hýnþ those whom he subjects there, 23 ; Swt. 218, 19. Hē hermcweðend hýneþ humiliabit calumniatorem, Ps. Th. 71, 5. Seðe iuih gehéneþ mec hénes seðe wutedlíce mec hénes gehéneþ ðone seðe mec sende qui me sende me spernit, qui autem me spernit spernit eum qui me misit, Lk. Skt. Lind. 10, 16. Ðæt se bealofulla hýneþ heardlíce the baleful one cruelly afflicts it, Exon. 11 b ; Th. 16, 27 ; Cri. 260. Hí hýnaþ da heorde ðe hí sccoldan healdan they ill-treat the flock that they ought to keep, L. I. P. 12 ;

Th. ii. 320, 17 : Swt. A. S. Rdr. 109, 135. Ic hiora fýnd fylde and hýnde *ad nihilum inimicos eorum humiliassem*, Ps. Th. 80, 13. Hē Godes hálgan hýnde mid wítum *he oppressed God's saints with torments*, Homl. Th. ii. 310, 25. Hē bebeád ðæt hié mon on ælce healfe hiénde *he ordered that they should be treated with insult on every side*, Ors. 6, 3; Swt. 258, 6. Se gúþsceaþa Geáta leóde hatoðe and hýnde, Beo. Th. 4627; B. 2319. Hē heów and hýnde *he smote and felled*, Byrht. Th. 141, 18; By. 324. Hī Godes cyrican hýndan and bærndon *they evilly entreated and burned the churches of God*, Chr. 684; Erl. 41, 22. Hý ða slógon and hýndon ðe ealle Rómáne friþian woldon, Ors. 4, 1; Bos. 79, 4. Hefe ðū ðīne handa and hýn hiora oferhygd *raise thine hand and humble their pride*; leva manum tuam in superbiam eorum in finem, Ps. Th. 73, 4. Ne hēn ðū *ne despicias*, Rtl. 43, 13. Hergian and hýnan *to ravage and ill-use*, Ors. 4, 1; Bos. 79, 1. Of ðæs handum ðe hine hýnan wolde *from the hands of him that would have laid him low*, Homl. Th. ii. 510, 23. Hē sceal rýperas and reáferas hatian and hýnan *robbers and plunderers he must hate and humble*, L. I. P. 2; Th. ii. 304, 20. Ic wolde helpan ðæs ðe unscyldig wære and hēnan ðone ðe hine yfelode, Bt. 38, 6; Fox 208, 17. Hēnan ða yfian and fyrþrian ða gódan *to bring the evil low and to promote the good*, 39, 2; Fox 212, 22. Ic eom frymdi tō ðē ðæt hī helsceaþan hýnan ne móton *I am suppliant to thee that fiends of hell may not evil entreat it* [*the soul*], Byrht. Th. 137, 3; By. 180. [*O. E. Hom.* stala and steorfa swiðe eow scal hene: *Laym.* hænen and hatien: *Goth.* haunjan to humiliate: *O. Frs.* hēna : *O. H. Ger.* hônjan *debilitare, illudere* : *Ger.* höhnen.] DER. ā-, for-, ge-hýnan; *and see* heán.

-hynde. v. six-, twelf-, twý-hynde.

hynden, e ; *f. A legal association of one hundred men. It will appear from the following passage that the hynden was an association of ten tithings :—* Ðæt wē tellan ā x. menn tōgædere and se yldesta bewiste ða nigene tō ælcum ðara gelāste ðara ðe wē ealle gecwædon and syððan ða hyndena heora tōgædere and ænne hyndenman ðe ā x. mynige tō ūre ealre gemæne þearfe and hig xi. healdan ðære hyndene feoh [*resolved :*] *that we always count ten men together, and that the chief one should direct the nine in each of those duties that we have all agreed upon; and then groups of ten tithings and* [*in each such group*] *one chief man* [hyndenman] *who may admonish the ten* [*chiefs of tithings*] *to the common benefit of us all; and let these eleven keep the money of the hynden to which they belong*, L. Æðelst. v. 3; Th. i. 230, 22–23, 3. On ðære hyndenne, L. In. 54; Th. i. 136, 11. v. next word; and see for a discussion of the term Kemble's Saxons in England, i. 242, sqq.

hynden-mann, es ; *m. The head man of a hynden :—* Ðæt wē ūs gegaderian a emban ænne mōnaþ gif wē mágon and æmtan habban ða hyndenmenn and ða ðe ða teóþunge bewitan ... and habban ða xii [xi ?] menn heora metscype tōgædere [*resolved :*] *that we gather to us once every month, if we can and have leisure, the* hyndenman *and those who direct the tithings ... and let these eleven* [*the* hyndenman *and one from each tithing in the* hynden *of which he was the head*] *have their refection together*, L. Æðelst. v. 8; Th. i. 236, 1–6. v. preceding word; and cf. hundred-mann.

hyngrian, hyngran ; *p.* ode, ede *To hunger.* **I.** *with nom. of person :—*Eádige synd gē ðe hingriaþ nū *beati qui nunc esuritis*, Lk. Skt. 6, 21. Eádige ða ðe rihtwísnesse hingriaþ *beati qui esuriunt justitiam*, Mt. Kmbl. 5, 6. Hingrian is of untrumnysse ðæs gecynnes *esurire ex infirmitate naturæ est*, Bd. 1, 27; S. 494, 14. Hwænne gesáwe wē ðē hingrigendne *quando te vidimus esurientem*, Mt. Kmbl. 25, 37. Ðane hingriendan *famelicum*, Wrt. Voc. ii. 34, 27. Gē gēson hingrendum hláf, Exon. 27 b; Th. 83, 12; Cri. 1355. God gefylþ ða hingrigendan mid his gódum, Homl. Th. i. 202, 35. **II.** *with dat. or acc. of person :—*Siððan him hingrode *afterwards he hungered*, 166, 12. Him nán þing ne hingrode, 168, 19. Hine hingrede *esuriit*, Lk. Skt. 4, 2. Mē hingrode *esurivi*, Mt. Kmbl. 25, 35. Ða ongan hyne syððan hingrian *postea esuriit*, 4, 2. [*Piers P.* þe hungreþ: *Goth.* huggrjan *impers. with acc.* : *O. Sax.* gihungrian : *O. Frs.* hungera : *Icel.* hungra : *O. H. Ger.* hungarian *pers. and impers. with acc. esurire* : *Ger.* hungern.] v. ge-hyngran.

hyngrig, *adj. Hungry :—*Ic wæs hingcgrig *esurivi*, Mt. Kmbl. Lind. 25, 35. v. hungrig.

hýn-ness, e ; *f. Humiliation, abasement, proscription :—*Unsceaþþiendra hýnnysse *proscriptionibus innocentum*, Bd. 1, 6; S. 476, 25, note. v. heán, hýnan.

hynni-laec *ascolonium*, Ep. Gl. 2 d, 6. v. enne-leác.

hýnþ, e ; hýnþu [-o] ; *indecl. f. Humiliation, abasement, disgrace, contempt, injury, harm, loss :—*Hýnþ *vel* lyre *vel* hearm *dispendium vel damnum vel detrimentum*, Ælfc. Gl. 81; Som. 73, 24; Wrt.Voc. 47, 29. Mycel hýnþ and sceamu hyt ys men nelle wesan ðæt ðæt hē ys and ðæt hē ys ne wesan sceal *magnum damnum et verecundia est homini nolle esse quod est, et quod esse debet*, Coll. Monast. Th. 32, 3. Hēnþa *detrimentum, damnum*, Wrt. Voc. ii. 140, 69. Sorh is mē tō secganne hwæt mē Grendel hafaþ hýnþo gefremed *a grief it is to me to say what harm Grendel hath done me*, Beo. Th. 954; B. 475 : 1190; B. 593. Undóm dēman earmum tō hýnþe *to judge unjust judgment to the injury of the poor*, L. I. P. 11; Th. ii. 318, **24.** Hī willaþ geinnian ða æftran hīnþe mid ðám uferan gestreónum *they*

desire to supply the consequent loss with the heavenly gains, Homl. Th. i. 340, 33. Hýnþu and hráfyl *injury and slaughter*, Beo. Th. 559; B. 277. Wē hēnþo geþoliaþ *we shall suffer humiliation*, Cd. 222; Th. 289, 18; Sat. 399. Helle hiénþu heofones mærþu *the disgrace of hell, the glory of heaven*, Exon. 16 b; Th. 37, 10; Cri. 591. Hýnþu unrim *ills unnumbered*, Cd. 37 ; Th. 48, 15; Gen. 776. Fela heardra hýnþa *many cruel injuries*, Beo. Th. 334; B. 166. Hēnþa, Bt. Met. Fox 12, 41; Met. 12, 21. Ná beóþ ða eádige ðe for hýnþum oððe lirum hwílwendlícra hýðða heófiaþ *they are not blessed, who mourn for losses of temporal comforts*, Homl. Th. i. 550, 28. Eall gē ðæt mē dydon tō hýnþum *ye did all that against me*, Exon. 30 a ; Th. 92, 24 ; Cri. 1514. Hié in hýnþum sculon wergþu dreógan *in abject state shall they undergo damnation*, Elen. Kmbl. 420; El. 210. Ðū hweorfest of hēnþum in gehyld godes *thou shalt go from humiliations into the grace of God*, Andr. Kmbl. 233; An. 117. Ðæt wē on ðam tōweardan life hýnþa forbúgan mágon *that in the life to come we may escape disgrace*, H. R. 17, 29. Hēnþa, Dóm. L. 6, 88. Ic heóld nū nigon geár wið ealle hýnþa ðínes fæder gestreón *I have kept now nine years thy father's wealth from all losses*, Homl. Skt. 9, 42. [*O. E. Hom.* hend : *O.H.Ger.* hônida *contumelia, ignominia, calumnia, dedecor, crimen, humilitas.*] v. heán, hýnan.

HYPE, es ; *m. The* HIP, *haunch, upper part of the thigh :—*Hype *clunis*, Wrt. Voc. 71, 49: *ilia*, ii. 110, 54. Ánra gehwylc hæfde sweord ofer his hype for nihtlícum ege *every man had his sword upon his thigh because of fear in the night* [Song of Sol. 3, 8], Blickl. Homl. 11, 18. Dó his sweord tō his hype *ponat vir gladium super femur suum*, Past. 49, 2 ; Swt. 383, 2. Hypas *clunes*, Ælfc. Gl. 74; Som. 71, 70; Wrt. Voc. 44, 52. [*A. R.* R. Glouc.* hupe : *Wick. Chauc.* hipe, hippe : *Goth.* hups; *m* : *Icel.* huppr : *m* : *O. H. Ger.* huf; *f. femur, coxa, clunis* : *Ger.* hüfte.]

hýpe, an; *f. A heap :—*Hýpe *acervus*, Wrt. Voc. 74, 70. Hí beóþ gegaderode tō micelre hýpan gif wē hí weaxan lætaþ *they will be gathered together into a great heap, if we let them grow*, Homl. Th. ii. 466, 7. Goldes and seolfres ungerime hýpan, i. 450, 21. [*Cf. O. H. Ger.* húfo; *m. strues, acervus, tumulus, congeries.*] v. mold-hýpe, heáp.

hype-bán, es ; *n. The hip-bone*, Ælfc. Gl. 74; Som. 71, 54 ; Wrt. Voc. 44, 37. Hupbán *catacrinis*, ii. 22, 63. Hupbánan *lumbi*, 54, 11.

hýpel, es ; *m. A heap :—*On hýpel *in cumulum, in augmentationem*, Hpt. Gl. 465. Hypplas *congeries*, 499. On reáde hýplas *in rubicundas congeries*, 449. Cf. scald-hýflas *vel* sond-hyllas *alga*, Wrt. Voc. ii. 99, 73. [Hupel *acervus*, Wrt. Voc. 89, 44 : *Wick.* hipil : *Trev.* huples ; *pl.* Cf. also *Wick.* hipilmelum *acervatim*.]

hype-, hup-seax, es ; *n. A knife hanging at the hip, a dagger, short sword :—*Lytel sweord *vel* hypesex *pugio vel clunabulum*, Ælfc. Gl. 52; Som. 66, 50; Wrt. Voc. 35, 37. Helm oððe hupseax, Exon. 79 a; Th. 297, 6; Crä. 64. Helmas and hupseax, Judth. 12 ; Th. 26, 15 ; Jud. 328.

hype-werc, es ; *m. Pain in the hip, sciatica :—*Hipwerc *sciascis*, Ælfc. Gl. 11; Som. 57, 42 ; Wrt. Voc. 19, 45.

HÝR, e ; *f.* HIRE, *payment for service done* or *money lent, interest :—*Ne nim ðū ná máre æt him tō hýre ðonne ðū sealdest. Ne syle ðū ðín feoh tō hýre *quare non dedisti pecuniam meam ad mensam*, Lk. Skt. 19, 23. Ðe hyra feoh lǽnaþ tō hýre *qui pecuniam suam mutuam dant fænore*, L. Ecg. P. iii. proem; Th. ii. 194, 31. [*Laym.* hure: *A. R.* hure, huire: *Piers P.* hyre: *Wick.* hire: *Du.* huur *wages* : *Dan.* hyre *hire* : *O. Frs.* hére *a lease.*]

hýra, an; *m. A hired servant, hireling :* — Se hýra se ðe nis hyrde *mercenarius qui non est pastor*, Jn. Skt. 10, 12, 13 : Homl. Th. i. 238, 14 : 240, 15. Hýrena þeúwe gē fleóþ . . . swá se hýra ðonne hē ðone wulf gesyhþ *ye flee after the manner of hirelings . . . as the hireling does when he sees the wolf*, Past. 15, 1; Swt. 38, 14.

hýra, an; *m. One who is subject to another :—*Æþelbryhtes hýra *sub potestate positus Ædilbercti*, Bd. 2, 3 ; S. 504, 21.

HÝRAN, hēran, hiéran ; *p.* de [*with acc., with infin., and with acc. and infin.*] **I.** *to* HEAR, *hear of :—*Morgensteorran ðe wē ōðre naman æfensteorra nemnan hēraþ *the morning star which we hear called evening star by another name*, Bt. Met. Fox 4, 29; Met. 4, 15. Nænigne ic sélran hýrde hordmáðmum *no better treasure did I ever hear of*, Beo. Th. 2399; B. 1197. Æfre ic ne hýrde ðon cymlícor ceól gehladenne heáhgestreónum *never have I heard of a bark any fairer laden with treasures*, Andr. Kmbl. 720; An. 360. Wundorlícor ðonne æfre byre monnes hýrde *more wonderfully than ever child of man heard*, Exon. 57 b; Th. 206, 19; Ph. 129. Ic londbūend secgan hýrde *I have heard the people of the country say*, Beo. Th. 2697; B. 1346. Ne hýrde ic idese lædan mægen fægerre *I have not heard of a queen leading a fairer force*, Elen. Kmbl. 480; El. 240. Hýrde ic ðæt hē ðone healsbeáh Hygde gesealde *I have heard that he gave the collar to Hygd*, Beo. Th. 4350; B. 2172. **II.** *to listen to, follow, serve, obey, be subject to, belong to :—*Ic hēro *servio*, Lk. Skt. Lind. 15, 29. Se port hýrþ in on Dene *the port belongs to the Danes*, Ors. 1, 1; Swt. 19, 24. Ic gean ðæs landes æt Holungaburnan and ðæs ðe ðærtó hýrþ *I grant the land at*

Hollingbourn and what belongs thereto, Chart. Th. 558, 27. Se haga æt Wiltûne ðe hýrþ intô Wilig, Cod. Dipl. Kmbl. iii. 415, 4. Hê mínum lǽrum hýreþ [MS. hyraþ] *he listens to my teachings*, Exon. 71 a ; Th. 264, 29 ; Jul. 371. Ða men ðe hîraþ intô heora mynstre *the men that belong to their minster*, L. Ælfc. P. 49 ; Th. ii. 384, 4. Ða ígland ðe in Denemearce hýraþ, Ors. 1, 1 ; Swt. 19, 31. Ðás land eall hýraþ tô Denemearcan, 36 ; 20, 4. Inc hýraþ eall *all shall be subject to you two*, Cd. 10 ; Th. 13, 20 ; Gen. 205. Gif gê hýraþ mê *if ye obey me*, 106 ; Th. 139, 26 ; Gen. 2315. Ðû tunglu genêdest ðæt hie ðê tô hêraþ *legem pati sidera cogis*, Bt. Met. Fox 4, 10 ; Met. 4, 5. Hýrde on ðam ða bysene ðæs ǽrestan hyrdes Godes cyricean *in quo exemplum sequebatur primi pastoris ecclesiæ*, Bd. 2, 4 ; S. 505, 11. Hié cwǽdan ðæt se ân wǽre sôþ God se ðe Martinus hýrde *they said that he alone was true God whom Martin followed*, Blickl. Homl. 231. 1. Englas hêrdon him *angeli ministrabant illi*, Mk. Skt. Rush. 1, 13. Ðæt Israélisce folc hýrdon gode and Moise his þeôwe *populus crediderunt domino et Moysi servo ejus*, Ex. 14, 31. Tô ðǽm landum eallum ðe ðǽrtô hiérdon *to all the lands that thereto belonged*, Chr. 912 ; Erl. 100, 32. Filgan hî ðam láfordscipe ðe ðæt land tô hýre *let them follow the lordship that the land belongs to*, Chart. Th. 549, 33. Eal ðæt folc ðe ðê hîran sceal *omnis populus qui subjectus est tibi*, Ex. 11, 8. Ne mæg nán mon twám hláfordum hiéran *nemo potest duobus dominis servire*, Past. 18, 2 ; Swt. 129, 24. Hêra, Mt. Kmbl. Lind. 6, 24. Gif hê Gode wile rihtlíce hýran, L. Edg. C. 60, note ; Th. ii. 256, 36. Him ǽghwilc hýran scolde gomban gyldan *him each one had to obey, to him pay tribute*, Beo. Th. 20 ; B. 10. Hêran, Bt. Met. Fox 1, 61 ; Met. 1, 31. Holdlíce hýran woldon *were ready loyally to obey*, Andr. Kmbl. 3277 ; An. 1641. Hǽðengild hýran wig weorþian *to follow false Gods, to worship idols*, Apstls. Kmbl. 94 ; Ap. 47. [*Goth.* hausjan : *O. Sax.* hôrian *to hear, obey* : *O. Frs.* hêra : *Icel.* heyra *to hear, hearken* : *belong to* : *Dan.* höre : *O. H. Ger.* hôrian *audire, auscultare, obedire, pertinere* : *Ger.* hören.] DER. ge-, mis-, ofer-hýran.

hýran *to hire.* v. hýrian.

hýrcnian *to hearken*, Andr. Kmbl. 1307 ; An. 654 : Exon. 47 b ; Th. 162, 21 ; Gû. 979. v. heorcnian.

hyrdan, herdan, hierdan ; *p.* de *To make hard, strong or bold, to embolden, encourage, brace* :—Sôna æfter ðon suíðe lîðelîce hierde [hirde, Cott. MS.] ða ðe hê unfæstráde wisse *caute monendo postmodum, quæ infirma sunt, roborat*, Past. 32, 2 ; Swt. 213, 8. Tô ðam wǽge gesǽg heafelan onhylde hyrde ðá gêna ellen on innan tô ðe wall he sank, *bowed his head, yet within did he brace up his strength*, Exon. 51 a ; Th. 178, 15 ; Gû. 1244. Hyrde hine georne *diligently encouraged him*, Wald. 1 ; Vald. 1, 1. Herd hyge ðînne heortan staðola *make thy soul strong, firm fix thine heart*, Andr. Kmbl. 2427 ; An. 1215. [*Goth.* ga-hardjan *to harden* : *O. Sax.* gi-herdian *to make strong, firm* : *O. Frs.* herda : *Icel.* herða *to harden, temper* [*iron*] ; *to exhort, cheer* : *O. H. Ger.* hartian *confortare*.] DER. â-, for-, ge-, on-hyrdan. v. heardian.

hyrde. v. hirde.

hyrdel, es ; *m. A hurdle, a frame of intertwined twigs* or *bars* :— Hyrdel *cleta, cratis*, Ælfc. Gl. 29 ; Som. 61, 44 ; Wrt. Voc. 26, 43 : *cratis* i. *flecta*, 49 ; Som. 65, 88 ; Wrt. Voc. 34, 20. Ðá forlêt se cásere ðone hálgan lîchaman uppon ðam ísenan hyrdle *then the emperor left the holy body [of St. Lawrence] on the iron hurdle*, Homl. Th. i. 430, 23. Hyrþil *cratem, flecta*, Wrt. Voc. ii. 105, 45. Hyrdlas *crates*, 80, 22. [*Prompt. Parv.* hyrdel *plecta, flecta, cratis* : *R. Glouc.* an chyrche of herdles and of 3erden : cf. *Goth.* haurds *a door* : *Icel.* hurð *a door, a hurdle* : *O. H. Ger.* hurt *crates, craticula* : *Ger.* hürde *a hurdle*.]

hyrd-ness. v. hird-ness.

hyrd-rǽden. v. heord-rǽden.

hyrdung, e ; *f. Strengthening, restoring* :— Hyrdung *constructio vel instructio* : ealdere timbrunga bôte *instructio* : niwe timbrung *constructio*, Ælfc. Gl. 62 ; Som. 68, 74–6 ; Wrt. Voc. 39, 57–9. v. hyrdan, â-hyrding.

hýre *hire.* v. hýr.

hýre-borg, es ; *m. Interest, usury* :—Hiéreborg (or? hiére, borg) *fenus*, Wrt. Voc. ii. 88, 18.

hý-rêd. v. hî-rêd.

hýred-ness *fame, report* ; *fama*, Lye.

hýre-gilda, an ; *m. One who receives (?) pay for service, a mercenary* :— Hýregildan *mercedarii*, Ælfc. Gl. 8 ; Som. 56, 96 ; Wrt. Voc. 18, 46.

hyrel [?] :—Andlang ðære fyrh ðæt hit cymþ tô hyrel ; ðonne þwyres ofer hyrel on ða furh ofer clǽnan dûne, Cod. Dipl. Kmbl. iii. 435, 9.

hýre-, hiére-, hýr-mann, es ; *m. One who obeys*, or *is subject to, another, a subject, follower, servant, subordinate, [as an ecclesiastical term] a parishioner, a hearer* :—Forðon oft for ðæs láreówes unwîsdôme misfaraþ ða hiéremenn and oft for ðæs láreówes wîsdôme unwîsum hiéremonnum biþ geborgen *for often from the ignorance of the teacher the followers go astray, and often from the wisdom of the teacher the followers are preserved* ; *per pastorum ignorantiam hi, qui sequuntur, offendant*, Past. 1, 4 ; Swt. 29, 5. Ðe hrygc ðæt sint ða hiéremenn *hi, qui subsequenter inhærent, dorsa nominantur*, Swt. 29, 12. On ôðre wîsan sint tô monianne ða ealdormen on ôðre wîsan ða hiéremenn *quomodo admonendi subditi et prælati*, 28 ; Swt. 189, 13. Bist ðû ûre cyning oððe beóþ wê ðîne

hýrmen *rex noster eris aut subjiciemur ditioni tuæ?* Gen. 37, 8. Wê beódaþ eác úrum hîremannum *we also command our subjects*, L. Ædelst. v. 8, 7 ; Th. i. 238, 1. Æt his hýremannum *from his subordinates*, 11 ; Th. i. 240, 16. Eówrum hýremonnum cýðon *to make known to your parishioners*, L. E. I. 26 ; Th. ii. 422, 20. His hiéremonnum *auditores suos*, Past. 8 ; Swt. 53, 17. His hýrmen, L. Ælfc. P. 46 ; Th. ii. 384, 22. Is gehwylcum mæssepreóst micel þearf ðæt hê his hýremen georne lǽre, L. E. I. 25 ; Th. ii. 422, 6 ; 28 ; Th. ii. 424, 33. v. hýran.

hýr-geoht, es ; *n. A hired yoke of oxen* :—Be hýrgeohte, L. In. 60 ; Th. i. 140, 7.

hýrian ; *p.* ode : hýran ; *p.* de ; *pp.* ed *To hire* :—Ús nán man ne hýrode *nemo nos conduxit*, Mt. Kmbl. 20, 7. Nán man ús ne hýrde, Homl. Th. ii. 76, 5. Seðe wolde hýrian wyrhtan, 72, 19. [*A. R.* huren : *Prompt. Parv.* hyryū *conduco* : *Laym.* hureden, *p. pl* : *Piers P.* huyred, *pp* : *O.Frs.* hêra : *O.Dut.* hueren : *M.L.Ger.* huren.] v. â-hýrian, ge-hýran.

hýrian ; *p.* ede *To imitate* :—Hió hyrigaþ monnum *they imitate men*, Bt. 41, 5 ; Fox 252, 26. v. æfter-, on-hyrian.

hýrig-mann, es ; *m. A subject, follower, parishioner* :—Hýrigmonnum, L. E. I. 26 ; Th. ii. 422. 27. Hýrigmen, 28 ; Th. ii. 424, 16. v. hýre-mann.

hýrig-mann, es ; *m. A person hired to work* :—Ðá gewearþ ðam hláforde and ðám hýrigmannum wið ánum peninge *an agreement to work for a penny was made between the lord and the workmen he had hired*, Th. An. 73, 30.

hýrling, es ; *m. A hireling, one who works for hire* :—Hî heora fæder on scipe forlêton mid hýrlingum *relicto patre suo in navi cum mercenariis*, Mk. Skt. 1, 20.

hýr-mann, es ; *m. One who works for hire* :—Hýrman *mercenarius*, Wrt. Voc. 86, 40. Mid ðǽm hýremonnum *cum mercenariis*, Mk. Skt. Rush. 1, 20. [Hurmon, Wrt. Voc. 95, 51.]

hýr-mann. v. hýre-mann.

hyrnan ; *p.* de *To project in the shape of a horn* or *wedge* :—Andlang ðæs streámes on ðone mǽdham ðe hyrnþ into Scylftûne and fram Scylftûne andlang streámes ðæt it cymþ tô ðam mylewere ðe hyrnþ intô duceling dûne *along the stream to the meadow-enclosure that projects wedge-shaped into Scylfton ; and from Scylfton along stream until it comes to the mill-weir that juts out into Ducklingdown*, Cod. Dipl. Kmbl. iv. 92, 29. [Cf. 'Herne, a nook of land, projecting into another district, parish, or field,' Forby.] v. hyrne.

hyrne, an ; *f. A horn, corner, angle* :—Hyrne *angulus*, Wrt. Voc. 80, 73. Ðæt wæter ðe man ða bán mid áþwôh binnan ðære cyrcan wearþ ágoten on ânre hyrnan *the water that the bones were washed with in the church was poured away in a corner*, Swt. A. S. Rdr. 100, 162. Tô ðæs hegges hyrnan *to the corner of the hedge*, Cod. Dipl. Kmbl. iii. 423, 18. Tô môrmǽde norþ hyrnan, 449, 19. On strǽta hyrnum *in angulis platearum*, Mt. Kmbl. 5, 6. On ðæs weofodes hyrnan *super cornua altaris*, 29, 12 : Lev. 4, 18 : 8, 15. On ða feówer hyrnan ðære earce *per quatuor arcæ angulos*, Past. 22, 1 ; Swt. 169, 21. [*A. R.* hurne : *R. Glouc.* hurne : *Prompt. Parv.* hyrne *angulus*, see note, p. 241 : *Chauc.* herne : *Piers P.* huirne, hirne, hyrne : *O. Frs.* herne : *Icel.* hyrna *one of the horns* or *points of an axe-head ; a mountain peak.*] v. horn.

-hyrne. v. ân-, þreó-hyrne.

hyrned ; *adj. Provided with a horn* or *beak [of a ship], having angles* or *corners* :—Ða hyrnedan nǽddran *the horned snakes*, Homl. Th. i. 102, 7. Hyrnde ciólas *ships having horn-shaped prows*, Bt. Met. Fox 26, 46 ; Met. 26, 23. v. ân-, eahta-, ofer-, six-hyrned.

hyrned-nebba ; *adj. Horny- or hard-beaked [epithet of raven and eagle]*, Judth. 11 ; Thw. 24, 28 ; Jud. 212 : Chr. 937 ; Erl. 115, 11 ; Ædelst. 62.

hyrnen ; *adj. Made of horn* : — On stefne bêmen hyrnenre [Ps. Spl. býman hyrnendre ; Ps. Stev. hornes hyrnes ; *Wick.* þe hornene trumpe] *voce tubæ corneæ*, Ps. Lamb. 97, 6. [*O. H. Ger.* hurnin *corneus*.]

hýr-ness, e ; *f. Obedience, subjection, a district in subjection to secular or ecclesiastical authority* :—His môd biþ âfêdd mid ðære smeáunga ðære wilnunga ôðerra monna hiérnesse *in occulta meditatione cogitationis ceterorum subjectione pascitur*, Past. 8, 2 ; Swt. 55, 6. Hê underþeódde and him tô hêrnysse geteáh *subjecit*, Bd. 3, 24 ; S. 557, 33. Wæs hê ǽrest arcebiscopa ðæt him eall Angelcynn hýrnysse geþafode *is primus erat in archiepiscopis, cui omnis Anglorum ecclesia manus dare consentiret*, 4, 2 ; S. 565, 22. Underþeódde on hêrnysse *subjecti*, 30 ; S. 561, 36. Ne spane nán mæssepreóst nánne mon of ôðre cyrcean hýrnysse tô his cyrcan ne of ôðre preóstscyre lǽre ðæt mon his cyrcan gesêce *let no priest entice any man from the parish of another church to his church, nor persuade any one to come from another district to attend his church*, L. E. I. 14 ; Th. ii. 410, 31. Eall Beorcleá hýrnesse hî áwæston, Chr. 1087 ; Erl. 224, 21. From hwǽm ondfóaþ gǽfle oððe hêrnisse *a quibus accipiunt tributum vel censum?* Mt. Kmbl. Rush. 17, 25. v. mis-hýrness ; hýran, hýre-mann.

hyrnetu, hyrnet, e ; *f. A hornet* :—Hyrnet *crabro*, Ælfc. Gl. 22 ; Som. 59, 107 ; Wrt. Voc. 23, 63. Hyrnetu *crabro*, ii. 16, 25. Hurnitu, 105, 46. Ic ásende hyrnytta *mittam crabrones*, Ex. 23, 28. [*O. H. Ger.* hornuz *crabro, scabro* : *Ger.* horniss.]

hyrn-ful ; *adj. Full of corners* ; *angulosus*, Hpt. Gl. 409.

hyrn-stán, es ; *m. A corner-stone* :—Hê is se hyrnstán ðe gefêgþ ða twegen weallas tôgædere *he is the corner-stone that joins together the two*

walls, Homl. Th. i. 106, 12, 23. [*Orm.* he wass himm sellf þatt hirnestan þatt band ta twezzinge wazhess.]

hýr-oxa, an; *m. A hired ox*, L. In. 60; Th. i. 140, 7, note.

hyrst, e, *f. An ornament, a decoration, jewel, anything of value, trapping, equipment, armour, implement*:—Hyrsta *falerarum*, Wrt. Voc. ii. 36, 74. Hryste *farelas*, 108, 34. Hyrsta scýne bord and bräd swyrd brúne helmas *beautiful equipments, shield and broad sword, brown helms*, Judth. 12; Thw. 26, 9; Jud. 317: Fins. Th. 41; Fin. 20. Bég and siglu eall swylce hyrsta swylce on horde ær men genumen hæfdon *ring*[s] *and jewels, just such ornaments as before men had taken in the hoard*, Beo. Th. 6309; B. 3165. Íren byrnan heard swyrd hilted and his helm häres hyrste *the iron byrnie, the hard and hilted sword, and his helm, the hoary one's equipments*, 5968; B. 2988. Hyrste [hyrsta, Soul Kmbl. 114] ða reádan ne gold ne seolfor [*not*] *the red ornaments, nor gold nor silver*, Exon. 99 a; Th. 370, 15; Seel. 57. Hwílum mec ähebbaþ hyrste míne *sometimes my trappings* [*wings*] *raise me up*, 103 a; Th. 390, 1; Rä. 8, 4: 103 b; Th. 392, 16; Rä. 11, 8: Th. 392, 24; Rä. 12, 1. Hyrste gerîm rodores tungel *number* [*heaven's*] *ornaments, the stars of the firmament*, Cd. 100; Th. 132, 7; Gen. 2189. Ðeáh ðe hyrsta unrîm ǽhte *though he owned jewels unnumbered*, Exon. 66 b; Th. 245, 12; Jul. 43. Ne mót hé ðara hyrsta hionane lǽdan wuhte ðon máre hordgestreóna ðonne hé hiðer brohte *defunctum leves non comitantur opes*, Bt. Met. Fox 14, 17–22; Met. 14, 9–11. Fyrnmanna fatu hyrstum behrorene *vessels of men of old, deprived of their ornaments*, Beo. Th. 5517; B. 2762. Hilderincas hyrstum gewerede, Elen. Kmbl. 526; El. 263. Hyrstum frætwed wliitig on wäge, Exon. 104 a; Th. 395, 22; Rä. 15, 11: 108 b; Th. 413, 15; Rä. 32, 20: 113 b; Th. 454, 7: 129 a; Th. 495, 22; Rä. 85, 7. [*O. H. Ger.* hrusti, Grff. ii. 546.] DER. ge-, wîg-hyrst.

hyrst, es; *m. A hurst, copse, wood.* The word occurs most frequently in compounds, e. g. *hnut-hyrst, æsc-hyrst*, etc., and is still found as *hurst* in names of places. See Cod. Dipl. Kmbl. iii. xxxii, and Leo's Anglo-Saxon Names, p. 107:— In hyrst sciofingden, Cod. Dipl. Kmbl. i. 273, 6. Wermód hér on hyrstum heasewe standeþ *wormwood stands dusky here in the woods* [Grein takes *hyrstum* under the previous word], Exon. 111 a; Th. 425, 24; Rä. 41, 61. v. horst, hurst, Grff. iv. 1042.

hyrstan, hrystan; *p.* te; *pp.* ed *To ornament, decorate, deck*:—Beón hyrst *comi*, Wrt. Voc. ii. 23, 43. Hyrsted sweord, Beo. Th. 1349; B. 672. Helm hyrsted golde, 4503; B. 2255. Hyrsted gold *gold fairly wrought*, Cd. 98; Th. 130, 5; Gen. 2155. Hyrstedne hróf hálgum tunglum *the* [*heavenly*] *canopy adorned with holy stars*, 46; Th. 58, 34; Gen. 956. Beorc byþ on helme hyrsted [hrysted, MS.] fægere *the birch at its top is fairly adorned*, Runic pm. Kmbl. 342, 32; Rún. 18. [*O. H. Ger.* hrusten *ornare*, Grf. ii. 546.] v. ge-hyrstan; ísen-hyrst.

hyrstan, hierstan; *p.* te; *pp.* ed *To fry, roast*:—Ic herste *frigo*, Ælfc. Gr. 28; Som. 31, 64. Hwæt is þinga ðe bietere síe on ðæs láreówes móde oððe hit suíður hierste *quid vero acrius doctoris mentem frigit?* Past. 21, 6; Swt. 165, 2. Nim áne clǽne panne and hyrste hý mid ele *take a clean pan and fry them with oil*, Lchdm. iii. 136, 4. Hé hine hét áþenian on írenum bedde and hine cwicne hirstan and brǽdan and swá hine mon má hirste wæs hé fægera on ondwlitan *he ordered him to be stretched on an iron bed and roasted alive; and the more he was roasted the fairer was his face*, Shrn. 116, 3–5. v. ge-hyrstan.

hyrste *a little gridiron; craticula*, Wrt. Voc. ii. 136, 53.

hyrste-panne. v. hearste-panne.

hyrst-geard, es; *m. An enclosed wood* [?]:—In ðone hyrstgeard, Cod. Dipl. Kmbl. iii. 19, 1.

hyrsting, hiersting, e; *f. Frying, burning, a frying-pan* [?]:—Hyrstincg *cremium*, Ps. Lamb. 101, 4. Hyrstyngc[-panne?] *frixorium*, Wrt. Voc. 82, 69. Hyrstung *frixorium*, Ælfc. Gr. 28; Som. 31, 65. Hyrsting *frixura*, Wrt. Voc. ii. 150, 84. Mid ðisse pannan hierstinge wæs Paulus onbærned *Paulus hujus sartaginis urebatur frixura*, Past. 21, 6; Swt. 165, 3. [Cf. *O. H. Ger.* harsta *frixura*.]

hyrsting-hláf, es; *m.Crust*:—Herstinghláfum *crustis*, Wrt.Voc.ii.18,51.

hyrsting-panne, an; *f. A frying-pan*:— Hyrsting [dyrsting, MS.] panne *sartago vel frixorium*, Ælfc. Gl. 25; Som. 60, 59; Wrt. Voc. 25, 1.

hyrsudon [?], Bd. 3, 14; S. 540, 11, note.

hýr-, heár-sum; *adj. Obedient, compliant*:—Se ðe him hýrsum beón wolde hé gehét *qui sibi obtemperantibus promitteret*, Bd. 1, 25; S. 486, 26. Him hýrsum beón *ei obtemperare*, 2, 12; S. 574, 16. Hit biþ his láreówum hýrsum *it is obedient to its teachers*, Salm. Kmbl. 798; Sal. 398. Wé beóþ hírsume *erimus obedientes*, Ex. 24, 7. Nemne ic gode sylle hýrsumne hige *unless I give to God an obedient mind*, Exon. 37 b; Th. 124, 13; Gú. 340. Heársume, 42 b; Th. 144, 13; Gú. 677: 43 a; Th. 145, 19; Gú. 697. [*O. E. Homl.* her-sum: *Orm.* herr-summ: *Laym.* hær-sum: *O. H. Ger.* hôr-sam.] v. ge-hýrsum.

hýrsumian; *p.* ode, ede *To be obedient, obey, serve*:—Windas and sǽ him hýrsumiaþ *venti et mare obediunt ei*, Mt. Kmbl. 8, 27: Homl. Th. ii. 368, 28. Hýrsumiaþ *ancillantur*, Ælfc. Gl. 100; Som. 77, 6; Wrt. Voc. 55, 9. Wé ðe on ðissum ne hérsumiaþ *we shall not obey thee in this*, Blickl. Homl. 243, 19. Ða hálgan heofonware him hýrsumedon 135,

17. Hé ðæm bebodum heársumede, Bd. 2, 6; S. 508, 41. [*O. E. Homl.* hersunian: *O. H. Ger.* hôrsamôn *obedire*.] v. ge-hýrsumian.

hýrsum-ness, e; *f. Obedience, subjection*:—Myrcna cyninge on hýrsumnesse underþeódded syndon *Merciorum regi subjectae sunt*, Bd. 5, 23; S. 646, 27. Þurh ða hýrsumnysse ðe wé heom hýrsomiaþ *through the obedience with which we obey them*, L. Edg. S. 1; Th. i. 272, 21. [*O. E. Homl.* hersamnise: *Laym.* hersumnesse.] v. ge-hýrsumnys.

hyrtan, hiertan; *p.* te *To* HEARTEN, *encourage, animate*:—To heora ágenne þearfe hyrteþ *ad propriam eorum necessitatem animat*, L. M. I. P. 13; Th. ii. 266, 8. Hyrt *cohortat*, Wrt. Voc. ii. 136, 5. Mid óðrum worde hé hierte mid óðrum hé brégde *favet ergo ex desiderio, et terret ex praecepto*, Past. 8, 1; Swt. 53, 11. Hyrte hyne hordweard *the hoardward* [*dragon*] *took courage*, Beo. Th. 5179; B. 2593. [*Laym.* hirten: *Prompt. Parv.* hertyñ *animo*.] v. ge-hyrtan.

hyrwan, hyrwian; *p.* de, ede *To speak ill or contemptuously of any one, blaspheme, despise, condemn, treat ill, oppress, vex, harass*:—Óðerne herweþ *alterum contemnet*, Mt. Kmbl. Rush. 6, 24. Ða earman ðe nú Godes bebodu hyrwiaþ beóþ cwylmede *the miserable men that now despise God's commandments shall be tormented*, L. E. I; Th. ii. 396, 36. Ðú heruwdest Godes bebodu, Blickl. Homl. 49, 36. Hé hyrwde godes naman and wirigde hine *cum blasphemasset nomen et maledixisset ei*, Lev. 24, 11. Ðá hyrwdon hí ealle hine *omnes condemnaverunt eum*, Mk. Skt. 14, 64. Hié hyrwdon ðé *they despised thee*, Elen. Kmbl. 710; El. 355. Gé gewritu herwdon *ye despised the scriptures*, 774; El. 387. Ne hyrw ðú úre godas *blaspheme not our gods*, Homl. Th. i. 424, 13. Ne hyrwe gé útancymenne man *non exprobretis advenae*, Lev. 19, 33. Sceal wís cyning cristendóm miclian and mærsian and á hé sceal hǽðendóm hindrian and hyrwan *a wise king must extend and magnify christianity, and ever must he hinder and harass heathendom*, L. I. P. 2; Th. ii. 306, 7. [*O. H. Ger.* harwian *exasperare*, Grff. iv. 1043.] v. ge-hyrwan.

hyrwe *name of a tree; torriculum*, Wrt. Voc. 285, 50.

hyrwend, es; *m. A blasphemer*:—Léd út ðone hirwend *educ blasphemum*, Lev. 24, 14.

hyrwend-líc; *adj. Contemptible, despicable*:—Heruuendlícae *contemptum*, Ep. Gl. 7 d, 9. Heuuendlíce, Wrt. Voc. ii. 104, 31. Ða hirwendlícan *contemtibiliora*, 15, 62.

hyrw-ness, e; *f. Contempt, reproach*:—Hirwnessæ *contemptus*, Ps. Spl. T. 118, 141. Gefylled wé synd hirwnesseum *repleti sumus despectione*, 122, 4. Ða earman ðe on ða godcundan láreówas, Wulfst. 235, 25.

hyscan; *p.* te *To mock, deride, taunt, reproach*:—Hé hiscþ geþeahtas ealdra *reprobat consilia principum*, Ps. Lamb. 32, 10. Seðe earðaþ on heofonum hyseþ [hyscþ ?] hý *qui habitat in caelis irridebit eos*, Ps. Spl. T. 2, 4. Ðonne hyscte hé on ða godcundan láreówas, Wulfst. 235, 25. Hyhsan *conviciari*, Gl. Prud. 696. Hihsendes *subsannantis*, Hpt. Gl. 524. v. husc, ge-, in-hyscan.

hyse, es; *m. A young man, warrior*:— Hyse cwom gangan *there came a young man*, Exon. 113 b; Th. 436, 14; Rä. 55, 1. Him be healfe stód hyse unweaxen cniht on gecampe *by his side stood a youth not yet grown up, a boy in battle*, Byrht. Th. 136, 17; By. 152. Hyse [*Beowulf*], Beo. Th. 2438; B. 1217: Andr. Kmbl. 1190; An. 595: 1622; An. 812: Elen. Kmbl. 1043; El. 523. Hé lét his francan wadan þurh ðæs hysses hals *he pierced the man's neck with his javelin*, Byrht. Th. 135, 60; By. 141. Hysse ðínum *puero tuo*, Ps. Th. 85, 15. Tó Abrahame his ágenum hysse *ad Abraham puerum suum*, 104, 37. Ðissum hysse hold *gracious to this man*, Andr. Kmbl. 1099; An. 550. Hysas, Byrht. Th. 135, 24; By. 123. Beornas feóllon, hyssas lágon, 135, 2; By. 112. Noldon ða hyssas hýran lárum hǽðnum *the youths would not listen to heathen lore*, Cd. 183; Th. 229, 14; Dan. 217: 184; Th. 230, 11; Dan. 231. Hét hyssa hwǽne *bade each man*, Byrht. Th. 131, 2; By. 2: 135, 34; By. 128: Fins. Th. 96; Fin. 48. v. þegn-hyse.

hyse-beorþor, -berþor, -borþor, es; *n. The bearing of male offspring, the offspring itself, a young man*:—Hyseberþor *puerperium*, Mone B. 3894. Hyseborþor, 4975. Hysebeorþ[or], Wrt. Voc. ii. 94, 42. Woldon on ðam hysebeorþre [cf. 2253, se geonga] heafolan gescénan *they would hurt the head of the man*, Andr. Kmbl. 2285; An. 1144. v. beorþor.

hyse-berþling, es; *m. The bearing of a male child, a male child* [?]; *puerperium*, Ælfc. Gl. 5; Som. 56, 8; Wrt. Voc. 17, 16.

hyse-cild, es; *n. A male child*:— Ǽlc hysecild betwux eów beó ymbsniden *circumcidetur ex vobis omne masculinum*, Gen. 17, 10. Gif hit hysecild byþ *si masculus fuerit*, Ex. 1, 16. Hyscild *mas*, Ælfc. Gl. 86; Som. 74, 22; Wrt. Voc. 50, 6. Beó hit hysecild beó hit mǽdencild *sit masculus infans, sit femina*, L. Ecg. P. ii. 21; Th. ii. 190, 21: L. M. cont. 2, 60; Lchdm. ii. 172, 17. Ðá féddon hié ða mǽdencild and slógon ða hysecild *mares enecant, feminas nutriunt*, Ors. 1, 10; Swt. 46, 11: Homl. Th. i. 30, 15.

hyseþ, Ps. Spl. T. 2, 4. v. hyscan.

hyse-wíse; *f. The manner of young men*:—Hysewíse *hircitallo*, Wrt. Voc. ii. 43, 26. This gloss is sufficiently explained by the following quotation from Paulus' epitome of Festus, ed. Müller, p. 101:—*Hirquitalli pueri primum ad virilitatem accedentes, a libidine scilicet hircorum dicti.* Further, in the notes to this word is added, *hirquitalli* βούπαιδες; *irquitalus* νηπιώτατος.

hyspan; *p.* te *To mock, scorn, taunt, revile, insult, reproach* :—Drihten hispeþ hŷ *Dominus subsannabit eos*, Ps. Spl. 2, 4. Hû lange hyspeþ feónd *usque quo improperabit inimicus?* 73, 11. Se ðe hespþ *qui calumniatur*, Kent. Gl. 497. Hŷ mē hyspaþ *exprobraverunt me*, Ps. Th. 41, 12. Tógeánes mē hyspton ealle fŷnd mîne *adversum me susurrabant omnes inimici mei*, Ps. Spl. C. 40, 8. Hyne hyspdun *improperabant ei*, Mt. Kmbl. 27, 44. Ðone hyspton *quod exprobraverunt*, Blickl. Gl. Hié Cristes bebod hyspton and hit forsáwon *they scorned Christ's commandment and despised it*, Ors. 6, 3; Swt. 256, 25. Hysptun hearmcwidum *mocked opprobriously*, Exon. 24 a; Th. 69, 15; Cri. 1121. Ðæt nā hyspen *ut non insultent*, Blickl. Gl. Hyspan *exprobrare*, Mt. Kmbl. 11, 20. Ongan hine hyspan and hearmcwiddigan, Bt. 18, 4; Fox 66, 33: Andr. Kmbl. 1341; An. 671. Fram stemne hyspendes *a voce exprobrantis*, Ps. Spl. 43, 18. Hyspendra *exprobrantium*, Blickl. Gl. v. hosp, ge-hyspan.

hyspend. v. hyspan.

hysping, e; *f. Reproach, reviling, contumely* :—Sidðan hē his hyspinge gehēred hæfde *acceptaque contumelia*, Bt. 18, 4; Fox 66, 35.

hysp-ness, e; *f. Reproach, opprobrium* :—Ðû settest ûs hyspnesse neáhgebûrum ûrum *posuisti nos opprobrium vicinis nostris*, Ps. Spl. T. 43, 15.

HŶÐ, e; *f.* '*A* HITHE, *or place that receives the ship, etc., on its landing ; a low shore, fit to be a landing place for boats, etc.*,' *a port, haven* :—Hŷð *angiportus, i. refrigerium navium*, Ælfc. Gl. 5; Som. 56, 32; Wrt. Voc. 17, 36: *confugium, i. statium, portus*, ii. 131, 51. Hŷð *portus*, Ælfc. Gr. 11; Som. 15, 8. Seó ân hŷð byþ simle smyltu æfter eallum ðām ŷstum ûrra geswinca *hic portus placida manens quiete*, Bt. 34, 8; Fox 144, 27: Bt. Met. Fox 21, 21, 25; Met. 21, 11, 13. Ðæt hie wilnigen ðære hŷðe ðæs gesinscipes *ut conjugii portum petant*, Past. 51, 8; Swt. 401, 33. Martha swanc ðā swilce on rēwette and Maria sæt stille swilce æt ðære hŷðe, Homl. Th. ii. 440, 32. Hera ðone steórman ac nā ærðan ðe hē becume gesundful tô ðære hŷðe, 560, 22. Cômon ðær þrý men tô ðære hŷðe *three men came to the landing-place*, Guthl. 11; Gdwin. 54, 24. Ðær æt hŷðe stód ædelinges fær, Beo. Th. 63; B. 32: Elen. Kmbl. 495; El. 248: Exon. 52 a; Th. 182, 8; Gú. 1307. Hē hî on hælo hŷðe gelædde *eduxit eos in portum*, Ps. Th. 106, 29: Exon. 20 b; Th. 53, 34; Cri. 860: Salm. Kmbl. 489; Sal. 245. [*Prompt. Parv.* hyþe, where *bootys ryve to londe, or stonde* stacio. 'Hithe occurs in the names of seaports, and also landing-places on rivers, far from the coast', p. 242, note 1. Kemble, Cod. Dipl. iii. xxxii, notes 'Rotherhithe (hrŷðra hŷð) the place where oxen were landed; Clayhithe, near Cambridge; Erith, in Kent and Cambridge, Eárhŷð; Cwēnhŷð, Queenhithe.']

hŷð; *gen.* hŷðe; *f. Advantage, gain, profit, benefit* :—Hŷð *vel* freme *commodum, questus*, Ælfc. Gl. 81; Som. 73, 25; Wrt. Voc. 47, 30. Gif feohbót ārîseþ ðæt gebyreþ rihtlîce tô þearfena hŷðe *if a money-fine arises, it is properly applied for the benefit of the needy*, L. Eth. vi. 51; Th. i. 328, 6. Uton dôn þearfum sume hŷðe ûre gôda *let us do some good to the needy with our wealth*, Homl. Th. ii. 100, 35. Ða ðe for lirum hwîlwendlîcra hyðða heófiaþ *those who mourn for losses of temporary advantages*, i. 550, 29. On earmra manna hyððum *for the advantage of poor men*, L. I. P. 19; Th. ii. 328, 11. Se hŷra smeáþ embe ða woruldlícan hŷðða and lǽt tô gýmeleáste ðæra sceápa lyre *the hireling inquires after worldly advantages, and leaves to neglect the loss of the sheep*, Homl. Th. i. 240, 29.

hŷdan; *p.* de *To despoil, plunder, lay waste, pillage, ravage* :—Hîðeþ and tô hām tŷhþ *it plunders and brings home*, Exon. 109 a; Th. 416, 25; Rä. 35, 4. Hŷðaþ wîde gifre glēde *widely shall the greedy flames lay waste*, 23 a; Th. 64, 28; Cri. 1044. Hit feor and wîde hŷðde and hergode *longe lateque devastans*, Bd. 3, 16; S. 542, 17. Cwæþ ðæt hē mid his gesîðum wolde hŷðan eal heofona rîce *said that with his comrades he would ravage all the kingdom of heaven*, Salm. Kmbl. 909; Sal. 454. Hîðende lēg *the wasting flame*, Exon. 22 a; Th. 60, 23; Cri. 974: 130 b; Th. 499, 28; Rä. 88, 22: 109 a; Th. 416, 5; Rä. 34, 7. Hîðendum *grassantibus*, Wrt. Voc. ii. 41, 49. [Cf. *O. H. Ger.* far-hundit *captivus*, Grff. iv. 965.] v. hûð, ā-hŷðan.

hŷðegung, e; *f. Profit, advantage*; *commodum*, Lye. v. ge-hŷðegod.

hŷðe-líc; *adj. Convenient, advantageous* :—Ðæt wæs hŷðelíc *that was convenient*, Exon. 124 b; Th. 478, 17; Ruin. 42. v. hŷð, ge-hŷðelíc, be-hŷðelíce.

hŷð-gild, es; *n. A port-due* [?] :—Hŷðgilda *portunalia*, Hpt. Gl. 515.

hŷð-líc; *adj. Relating to a port* :—Ða hŷðlícan *portunalia*, Wrt. Voc. ii. 67, 19.

hŷð-scip, es; *n. A pirate-ship* :—Hîðscip *myoparo*, Ælfc. Gl. 103; Som. 77, 100; Wrt. Voc. 56, 21. Hŷdscip *nioparo*, ii. 59, 26. v. hŷðan.

hŷð-weard, es; *m. One who guards a hithe*, Beo. Th. 3833; B. 1914.

hyw. v. hiw.

hŷwyt *hewn, cut*; ðolatum, i. incisum, planum, Wrt. Voc. ii. 141, 63.

I

THE Runic character | for this vowel was named *is* :—Îs byþ oferceald ungemetum slidor; glisnaþ glæshluttur gimmum gelîcust, Runic pm. Kmbl. p. 341.

The short *i* generally corresponds to Gothic *i*. e. g. *in*, Goth. *in*,

---

biddan, Goth. *bidjan*; the long *i*, which is sometimes written *ii*, é. g. *riiknæ* on the Ruthwell Cross, to Gothic *ei*, e. g. *isern*, Goth. *eisarn*, *bidan*, Gothic *beidan*. In early West Saxon MSS., however, *i*, *í* are found arising from other sources. Thus the mutation of the breaking *ea* is written *i*, e. g. *ildu, irmþu* from *eald, earm*; and the mutations of *eó, eá* are written *í*, e. g. *onlîhtan, híran*. In such cases, however, instead of *i* the diphthong *ie* is very often found; and not only in such, but also in those where the root-vowel is *i* or *í*, e. g. *ongietan, wietan* [=*witan*]; even in the place of *ŷ*, e. g. *ieðegende*. In the later MSS. instead of *i* or *ie*, *y* is found very commonly; indeed even in the earlier MSS. *y* has in some instances already made its way into the place of *i*, thus *ryht* is the form regularly used in Alfred's translation of Gregory's Pastoral Care. In the case of *niht* in the earliest times, in that of *niht* and its compounds in later, *i* takes the place of original *a*.

Initial *i* before *a, o, u* is found where most generally *ge* is used; for examples see below.

iá; *adv. Yea* :—Æt ðû tôdæg? Iá ic dyde *manducasti hodie? Etiam feci*, Ælfc. Gr. 31; Som. 40, 17. Eart ðû Esau mîn sunu? And hē cwæþ : Iá leóf ic hit eom *tu es filius meus Esau? Respondit : Ego sum*, Gen. 27, 23. Se kyng befealh georne hire brēðer oþ ðæt hē cwæþ já wið *the king pressed her brother eagerly until he said yes in reply*, Chr. 1067; Erl. 204, 23. v. geá.

iáces sûre, Wrt. Voc. 286, 21. v. geác.

iacincð, es; *m. Jacinth* :—Iacincðe [iacinte, Cot. MS.] *ex hyacintho*, Past. 14, 4; Swt. 87, 3.

Ianuarius; *m. January* :—Forma mônaþ folc mycel Ianuarius hēton *the Romans called the first month January*, Menol. Fox 19; Men. 10.

IC; *pron. of 1st pers. s. I* :—Ic Æðelstân cyninge cŷðe *I, king Athelstan, proclaim*, L. Ath. 1, prm; Th. i. 194, 2. Ic hyt eom *it is I*; ego sum, Mt. Kmbl. 14, 27. Ic sylf hit eom *ipse ego sum*, Lk. Skt. 24, 39. Ic eom Gabriel ic ðe stande beforan gode *ego sum gabrihel qui adsto ante deum*, 1, 19. For Wulfgáres sáwle ðe ic hit selle *for Wulfgar's soul* [*I*] *who give it*, Chart. Th. 496, 24. [*Laym. O. and N.* ic, ich, ihc: *Orm.* icc, I: *Chauc.* ich, I: *Goth. O. Frs. O. Sax.* ik: *Icel.* ek: *Dan.* jeg: *Swed.* jag: *O. H. Ger.* ih: *Ger.* ich: *Lat.* ego; *Gk.* ἐγώ.] For other forms in the declension of the pronoun of the first person, see the several words.

ícan, iécan, îcean, ŷcan; *p.* îhte, ícte *To* EKE, *increase, add to, augment* :—Ðû ŷcest ðîne yrmþo *thou dost increase thy misery*, Andr. Kmbl. 2381; An. 1192. Hwæt is ðis manna ðe íceþ ealdne nîð *what man is this that adds to ancient hate?* Elen. Kmbl. 1806; El. 905. Ŷceþ, Exon. 89 a; Th. 335, 9; Gn. Ex. 31. Sunne and môna iécaþ eorþwelan *sun and moon increase the wealth of earth*, 16 b; Th. 38, 23; Cri. 611. Ŷcaþ, 119 a; Th. 457, 32; Hy. 4, 93. Ðá îhte hē eft his synna *auxit peccatum*, Ex. 9, 34. Ðær eác ŷcte tô *also he added thereto*, Bd. 4, 16; S. 584, 15. Iécte, Cd. 55; Th. 68, 25; Gen. 1122: 108; Th. 143, 9; Gen. 2376. Icte, 59; Th. 72, 22; Gen. 1190. Sidðan wôcan ða îcton mægburh Caines *afterwards were born those who increased the kindred of Cain*, 52; Th. 65, 13; Gen. 1065. In eallum hî ðissum îhtan synne *in omnibus his peccaverunt adhuc*, Ps. Th. 77, 31. Ac ða hwîle ðe hē giernþ ðæt hē his welan iéce hē ágiémeleásaþ ðæt hē forbúge his synna *profecto enim, qui augere opes ambit, vitare peccatum negligit*, Past. 44, 9; Swt. 331, 16. Hwylc eówer mæg þencende ícan âne elne tô his anlícnesse *quis vestrum cogitando potest adjicere ad staturam suam cubitum unum?* Lk. Skt. 12, 25. Ðû gehēte ðæt ðû hŷra frumcyn ícan wolde *thou didst promise that thou wouldest increase their race*, Cd. 190; Th. 236, 8; Dan. 318. Hí sculon ælce dæg eácan [Cott. MS. ŷcan] ðæt mon ælce dæg wanaþ, Bt. 26, 2; Fox 94, 1. Ŷcan, Judth. 11; Thw. 24, 11; Jud. 183: Exon. 108 a; Th. 413, 3; Rä. 31, 9. Ŷcean *augmentare*, Bd. 2, 4; S. 505, 16. Ŷced *increased*, Exon. 53 b; Th. 187, 25; Az. 36. [*Laym.* æchen, eche: *Orm.* ekenn: *R. Glouc. Chauc.* eche: *O. Sax.* ôkian: *O. H. Ger.* auhhôn *augere, adjicere.*] v. eác, ēcan, eácan.

ice. v. yce.

ícend, es; *m. One who increases or augments* :—Ðon hē cymþ of ðam worde *augeo* ic geíce and hē getácnaþ geeácnunge ðon macaþ hē *hic auctor* ðes ícend and *hæc auctrix* ðeós ícestre *when it comes from the word* augeo *I increase, and indicates augmentation, then it makes* hic auctor *this augmenter, and* hæc auctrix *this augmentress*, Ælfc. Gr. 9, 21; Som. 10, 42-4.

ícestre, an; *f.* v. preceding word.

icge gold, Beo. Th. 2219; B. 1107. The translation of this phrase is difficult. Thorpe has 'moreover,' Kemble 'heaped up;' Heyne suggests comparison with *Sskr.* iç *dominare, imperare*, and gives 'Schatzgold, reiches gold;' Grein's note is as follows: 'Sollte vielleicht zu *icg* das Altn. *yggr* [terror] zu halten sein, da das Gold Altn. auch *ôgnar ljómi* [splendor terroris] heisst? oder sollte sich etwa der Begriff Sühngold herausbringen lassen?' Grundtvig suggests the reading *éce-gold*, i.e. gold given in addition on the occasion of a solemn reconciliation.

Iclingas; *pl. The name of a Mercian family to which St. Guthlac belonged* :—Hē was ðæs yldestan and ðæs ædelstan cynnes ðe Iclingas wǽron genemnede he [*Guthlac's father*] *was of that chiefest and noblest race that were called Iclings*, Guthl. 1; Gdwin. 8, 4. [Icelingtûn (*Ickleton in Cambridgeshire?*) occurs Cod. Dipl. Kmbl. iv. 300, 24; *and there is Icklingham in Suffolk.*]

í-dæges; *adv. On the same day :*—Se ðe sleá his ágenne þeówne esne and hê ne sý ídæges deád *he who smites his own slave, and he die not on the same day*, L. Alf. 17 ; Th. i. 48, 13. Hí ne môston metes þicgan gif hí igdæges tô mynstre gecyrran mihton *they were not allowed to partake of food if they could return to the monastery on the same day*, Homl. Th. ii. 166, 32. Swá hraþe swá hê him tô com ýdæges swá gewât hê of ðisum andwerdum lífe *as soon as he came to him, on the same day, he departed from this present life*, 176, 3. [Cf. í-sídes.]

ÍDEL ; *adj.*   I. *empty :*—Tô hwan mæg ðis eorþlíce hús gif hit ýdel stent ? Hit ne biþ nâ hús búton hit beó mid híréde áfylled *what purpose can this earthly house serve, if it stand empty ? It is not a house unless it be filled with a household*, Homl. Th. ii. 502, 12. Is nû forðí gehwilcum men tô hogienne ðæt hê ýdel ne cume his Drihtne tôgeánes on ðam gemǽnelícum ǽriste *now is it therefore for every man to take care that he come not empty-handed to meet his Lord at the general resurrection*, 558, 18. Ðonne se geohsa of ðære ídlan wambe cymþ *when the hiccup comes from the empty stomach*, L. M. 1, 18 ; Lchdm. ii. 60, 28. Ídelne híne forlêton *dimiserunt eum inanem*, Lk. Skt. 20, 10, ñ1. Sáwle ídle *animam inanem*, Ps. Th. 106, 8. Hê forlêt ða rícan ídele, Homl. Th. i. 204, 6.   II. *not possessing, destitute, void, devoid* [*with gen.*] :—Londrihtes môt monna ǽghwilc ídel hweorfan *every man must wander destitute of land-right*, Beo. Th. 5768 ; B. 2888. Se deófol on sumum uncystum gebringþ ðone ðe hê gemêt ídelne ǽlces gôdes weorces *the devil brings into some vices him whom he finds devoid of every good work*, L. E. I. 3 ; Th. ii. 404, 13. Ða ðe ídle beóþ swelcra giefa *those who are devoid of such gifts*, Past. 9 ; Swt. 59. 17.   III. *vain, useless, idle, to no purpose :*—Seó eorþe wæs ýdel and ǽmtig *terra erat inanis et vacua*, Gen. 1, 2. Ídel sangere *temelici*, Ælfc. Gl. 61 ; Som. 68, 57 ; Wrt. Voc. 39, 40. Eall eówer geswinc biþ ídel *consumetur incassum labor vester*, Lev. 26, 20. Ýdel biþ se lǽcedom ðe ne mæg ðone untruman gehǽlan ; swá biþ eác ýdel seó lár ðe ne gehǽlþ ðære sáwle leahtras *vain is the medicine that cannot heal the sick; so also is the doctrine vain that does not heal the sins of the soul*, Homl. Th. i. 60, 11. Wese wíc heora wêste and ídel *fiat habitatio eorum deserta*, Ps. Th. 68, 26. Unnyt oððe ýdel *supervacuus*, Ælfc. Gr. 47 ; Som. 48, 46. Oft biþ swíðe ídel and unnyt ðara yfelena manna hreówsung *plerumque mali inutiliter compunguntur*, Past. 54, 4 ; Swt. 431, 11. Ðes wída grund stód ídel and unnyt, Cd. 5 ; Th. 7, 14 ; Gen. 106 : Beo. Th. 830 ; B. 413 : 293 ; B. 145. Man byþ merwe gesceaft mihtum ídel *homo vanitati similis factum est*, Ps. Th. 143, 5. Ídel gelp him on ne rícsode *vanæ gloriæ contemptorem*, Bd. 3, 17 ; S. 545, 9. Ídel gylp *vanitas*, Ps. Th. 51, 6. Ídel searu, 138, 17. Ídel gielp *inanis gloria*, Past. 62, 1 ; Swt. 457, 20. Ídel wuldor *vainglory*, Exon. 33 a ; Th. 107, 12 ; Gû. 57. Hê nǽfre nôht leásunga ne ídeles leóþes wyrcean ne mihte *nihil unquam frivoli et supervacui poematis facere potuit*, Bd. 4, 14 ; S. 596, 42. Ða bodan ðæs ídlan fætes *the messengers of the useless vessel*, Past. 47, 3 ; Swt. 361, 16. Híg ðâ æfter ridon ídelan færelde *they rode after, but their journey was to no purpose*, Jos. 2, 7. Guman geþancas ídle synt *cogitationes hominum vanæ sunt*, Ps. Th. 93, 11. Ðǽr ðæt heáfod biþ unhâl eall ða limu bióþ ídelu *languente capite membra incassum vigent*, Past. 18, 2 ; Swt. 129, 8. Ýdele spellunga *fabulæ*, Ælfc. Gr. 50, 29 ; Som. 52, 2. Ne hí ðǽr ǽnig unnit ne geþafian ne ídele spǽce ne ídele dǽde, L. Edg. C. 26 ; Th. ii. 250, 6 : Hy. 7, 108 ; Hy. Grn. ii. 289, 108. Ídel word *idle words*, Exon. 37 a ; Th. 120, 30 ; Gû. 279. On ídel *in vain* ; nequiquam, Ælfc. Gr. 38 ; Som. 41, 55. On ídel gê swincaþ and eówre fýnd his brúcaþ *frustra seretis sementem, quæ ab hostibus devorabitur*, Lev. 26, 16. Ne nemne gê drihtnes naman on ídel, Deut. 5, 11. Ne sint híg eów on ídel beboden *non incassum præcepta sunt vobis*, 32, 47. On ídel hí mê wurðiaþ *in vanum me colunt*. Mk. Skt. 7, 7 : Ps. Th. 62, 8.   IV. *idle, unemployed :*—Hê geseah ôðre on strǽte ídele standan *vidit alios stantes in foro otiosos*, Mt. Kmbl 20, 3. Hwí stande gê hér eallne dæg ídele, 6 : Exon 92 a ; Th. 345, 6 ; Gn. Ex. 184. [*Orm. Piers P. Chauc.* on idel *in vain: O. Sax.* ídal : *O. Frs.* ídel : *O. H. Ger.* ítal *vanus, inanis : Ger.* eitel.] v. mân-ídel.

ídel, es ; *n. Idleness, vanity, futility, frivolity :*—Ðæt ýdel fêt unþeáwas *idleness nourishes bad habits*, Prov. Kmbl. 1. Ælc ýdel fêt unhǽlo, 61. Wê lǽraþ ðæt preóstas ðǽr ne geþafian ne ídele spǽce ne ídele dǽde ne ǽnig ídel *we enjoin that priests do not permit there* [*in the church*] *idle talk or action or any frivolity*, L. Edg. C. 26 ; Th. ii. 250, 27. Gif ðú gesihst manega gêt ýdel getácnaþ *if thou seest many goats it betokens frivolity*, Lchdm. iii. 214, 1. Nys eác mid ídele tô forlǽtenne ðæt wundor ðæt þurh wítedômes cræft hê wiste *nor is the miracle, that he knew things by prophetic power, to be lightly dismissed*, Guthl. 17 ; Gdwin. 70, 2 [cf. 76, 10]. Ða ídlo *vanitates*, Rtl. 162. 32. v. preceding word.

ídel-georn ; *adj. Fond of idleness, lazy, inert :*—Ne beó ðú tô slǽpor ne tô ídelgeorn forðan ðe slǽp and ðæt ýdel fêt unþeáwas and unhǽlo ðæs líchoman *be not too fond of sleep or idleness, for sleep and idleness nourish bad habits and bad health in the body*, Prov. Kmbl. 1. Eálá gê eargan and ídelgeornan *ah! ye sluggish and lazy ones ;* inertes, Bt. 40, 4 ; Fox 238, 30.

ídel-gild, es ; *n. False worship, idolatry :*—Híg mê tirigdon mid hira ídelgildum *ipsi me provocaverunt in eo qui non erat deus et irritaverunt in vanitatibus suis*, Deut. 32, 21. v. ídelness.

ídelgild-offrung, e ; *f. An offering to an idol :*—Ídelgildoffrung *idolothytum*, Ælfc. Gl. 18 ; Som. 58, 109 ; Wrt. Voc. 22, 25.

ídel-hende ; *adj. Empty-handed, empty :*—Ne cum ðú tô mínum húse ídelhende *nec apparebis in conspectu meo vacuus* ; *none shall appear before me empty*, Ex. 34, 20. Gif hê cume ídelhende tô *si vacuus appropinquat*, Past. 49, 2 ; Swt. 379, 21. Hê biþ ealra his ǽhta ídelhende *he shall be destitute of all his possessions*, Blickl. Homl. 49, 26. Nô ídelhende bona of ðam goldsele gongan wolde, Beo. Th. 4169 ; B. 2081. Ne lǽt ðú hine gân ídelhende fram ðe *nequaquam vacuum abire patieris*, Deut. 15, 13. Forleórton hine ídelhende *dimiserunt eum inanem*, Lk. Skt. Lind. 20, 10, 11. Ðonne gê út faraþ ne fare gê ídelhende *cum egrediemini, non exibitis vacui*, Ex. 3, 21. [Cf. *Ayenb.* idel-honded.]

ídel-ness, e ; *f. Idleness, vanity, frivolity, uselessness, futility, emptiness, falseness :*—Seó ýdelnes is ðære sáwle feónd *idleness is an enemy of the soul*, L. E. I. 3 ; Th. ii. 404, 11. Ælces libbendes mannes mægen and anwald is ídelnes *universa vanitas omnis homo vivens*, Ps. Th. 38, 6. Ðonne hí mid fulle gesceáde ongietaþ ðæt ðæt wæs leás and ídelnes ðæt hí ǽr heóldon *cum certo judicio deprehenderint falsa se vacue tenuisse*, Past. 58, 1 ; Swt. 441, 19. Sebastianus cwæþ ðis is swutol gedwyld and leás ýdelnyss, Homl. Skt. 5, 274. Sanctus Paulus cwæþ ðæt sió gítsung wǽre hearga and ídelnesse geféra *avaritia quæ est idolorum servitio*, Past. 21, 3 ; Swt. 157, 6. On ídelnisse gê fæstniaþ eówer môd on him *incassum cor figitis*, 51, 2 ; Swt. 395, 29. Ne minne noman ne cíg ðú on ídelnesse, L. Alf. 2 ; Th. i. 44, 7. Hierusalem winþ for rihtwísnysse and Babilonia winþ ongeán for unrihtwísnysse seó óðer for sóðfæstnysse óðer for ýdelnysse *Jerusalem fights for righteousness, and Babylon fights in opposition for unrighteousness : the one for truth, the other for falsehood*, Homl. Th. ii. 66, 31. Ða gímeleásan men ðe heora líf ádrugon on ealre ídelnisse *the careless men who passed their lives quite idly*, Ælfc. T. Grn. 1, 13. Nys eác mid ídelnysse tô forlǽtenne ðæt wundor ðe ðes hálga wer foreséde *nor is the wonder which this holy man foretold to be lightly dismissed*, Guthl. 19 ; Gdwin. 77, 10 [cf. 70, 2]. Ne ðú manna bearn tô ídelnesse geworhtest *non vane constituisti filios hominum*, Ps. Th. 88, 40 : Bd. 4, 3 ; S. 567, 27. Forhwan gê mid ídelnesse ealle áríseþ ǽrdon leóht cume *in vanum est vobis ante lucem surgere*, Ps. Th. 126, 3. Hwí lufige gê ídelnessa and sécaþ leásuncga *quid diligitis vanitatem, et quæritis mendacium ?* 4, 3. Hê forlêt ða ídelnesse deófolgylda *relictis idolorum superstitionibus*, Bd. 2, 15 ; S. 518, 26. Ðis synt ða ídelnyssa ðisse worlde *hæ sunt vanitates hujus mundi*, L. Ecg. P. 1, 8 ; Th. ii. 174, 32. On ídelnyssum heora *with their vanities*, Cant. Moys. ad fil. 21. [*O. Frs.* tô ýdelnisse *in vain: O. H. Ger.* ídalnissa *desolatio*.]

ídel-sprǽce ; *adj. Talking idly, vainly :*—Ða felaídelsprǽcan *multiloquio vacantes*, Past. 23 ; Swt. 175, 25.

ides, e ; *f. A woman* [it is a word little used except in poetry, and it is supposed by Grimm to have been applied, in the earliest times, like the Greek νύμφη, to superhuman beings, occupying a position between goddesses and mere women, v. D. M. 372] :—Ides *virgo*, Kent. Gl. 1196. Freólecu mæg ides ǽwiscmód [*Eve*], Cd. 42 ; Th. 55, 18 ; Gen. 896. Freólecu mæg ides eaforan fêdde [*Cain's wife*], 50 ; Th. 64, 22 ; Gen. 1054. Wlitebeorht ides [*Sarah*], 82 ; Th. 103, 34 ; Gen. 1728. Monig blächleór ides [*the women of Sodom and Gomorrah*], 92 ; Th. 118, 24 ; Gen. 1970. Freólecu mæg ides egyptisc [*Hagar*], 101 ; Th. 134, 19 ; Gen. 2227. Ides ælfscínu [*Judith*], Judth. 9 ; Thw. 21, 11 ; Jud. 14. Ides Helminga beághroden cwên [*Wealtheow, Hrothgar's queen*], Beo. Th. 1245 ; B. 620. Ides Scyldinga, 2341 ; B. 1168. Idese onlícnes *a woman's form*, 2706 ; B. 1351. Him brýda twá idesa eaforan fêddon [*Lamech's wives*], Cd. 52 ; Th. 65, 34 ; Gen. 1076. Weras and idesa, Exon. 50 b ; Th. 176, 7 ; Gû. 1205. Eorlas and hira idesa mid, Andr. Kmbl. 3275 ; An. 1640. A weak form occurs in Hpt. Gl. 456, 76 :—Tô, on ydesan *in juvenculam*, Ælfc. Gr. Sax. idis : *O. H. Ger.* itis *matrona* ; itis-líh *matronalis*, Grff. i. 159. Grimm D. M. 373 takes the Icel. *dís* to be the same word, and compares the phrase from the Edda *dís skjöldunga* with the similar phrase given above from Beowulf.]

idig [?] ; *adj. Busy, active :*—Tôþas idge *busy teeth* [*referring to the eating of the forbidden fruit by Adam and Eve*], Exon. 61 b ; Th. 226, 18 ; Ph. 407. [Cf. *Icel.* iðja *activity :* iðinn *assiduous, diligent ;* iðja *to be active, busy.* The passage is somewhat uncertain, as the MS. has *to þas*, and Thorpe prints as if there were a gap between *þas* and *idge*.]

ídiso, ýddisc, es ; *pl.* e ; *m. n* [?]. *Property, household stuff :*—Ýddisc *supellex*, Ælfc. Gl. 27 ; Som. 80, 98 ; Wrt. Voc. 25, 38. Ýddisce *supplex*, Wrt. Voc. 83, 28. Ne forlǽte gê nân þing of eówrum ýddisce *nec dimittatis quidquam de supellectili vestra*, Gen. 45, 20. Ágíf ðises ceorles ýddysce [cf. ǽhta l. 1, þing, l. 23] *give up this fellow's property*, Homl. Th. ii. 180, 27. DER. in-ídisc. v. eád, ǽdisc.

ídlian ; *p.* ode *To become vain or idle, come to nought, to make vain or empty :*—Him hyge brosnaþ ídlaþ þeódscype *their mind corrupts, discipline comes to nought*, Exon. 81 a ; Th. 304, 13 ; Fä 69. Ídlodon on ídelnyssum heora *irritaverunt in vanitatibus suis*, Cant. Moys. ad. fil. 21. Wæs ídlod *cassaretur*, Hpt. Gl. 515. Ídelude *exinanita*, Ps. Spl. T. 74, 8. [Cf. *O. H. Ger.* ki-ítallent *adnullabunt : Ger.* ver-eiteln.] DER. á-, ge-ídlian.

ídol, es; *n. An idol*:—Hǽðenscype biþ ðæt man ídola [idol, MS. 13; deófolgyld, MS. G.] weorðige *it is heathendom, to worship idols*, L. C. S. 5; Th. i. 378, 18. Ídola wurðing *worship of idols*, L. N. P. L. 48; Th. ii. 298, 1.

ié *gen. dat.* of eá, Ors. 1, 1; Swt. 8, 10, 11, 14.

ie, ié. *For words beginning with these combinations look under* i, î, *and see the preliminary remarks under the letter* I.

IFIG, ifegn, es; *n. Ivy*:—Ifig *eder*, Wrt. Voc. 286, 2. Ifegn *eder*, ii. 106, 78. Yfig. Ðeós wyrt ðe man hederam crysocantes and óðrum naman ifig nemneþ is gecwedeu crysocantes forðý ðe heó byrþ corn golde gelíce *Ivy. This plant, which is named hedera crysocantes, and by another name ivy, is called crysocantes, because it bears berries like gold*, Herb. 121; Lchdm. i. 234, 1–4. Nim ðæt ifig ðe on stáne weaxe *take the ivy, which grows on stone*, L. M. 3, 30; Lchdm. ii. 326, 3. Ifies seáw *juice of ivy*, 1, 3; Lchdm. ii. 40, 26. Weal se is mid ifige bewrigen *a wall that is covered with ivy*, Shrn. 139, 27. [*O. and N.* ivi: *Prompt. Parv.* ivy *edera*: *O.H.Ger.* ebah *hedera*, Grff. i. 91.] DER. eorþ-ifig.

ifig-crop, -cropp, es; *m. A cluster of ivy berries*:—Ifigcrop *corymbus*, Wrt. Voc. 68, 2.

ifig-croppa, an; *m. A cluster of ivy berries*:—Ifigcroppena fíf and xx *five and twenty bunches of ivy berries*, L. M. 2, 24; Lchdm. ii. 214, 18.

ifig-leáf, es; *n. An ivy leaf*:—Nim ifigleáf ðe on eorþan wixþ *take leaves of ivy that grows on the ground*, L. M. 3, 31; Lchdm. ii. 326, 11.

ifig-tearo; *n*: -tara, an; *m. Ivy tar, gum that comes from ivy when it is cut*:—Nim sciptearo and ifigtearo, L. M. 1, 76; Lchdm. ii. 150, 12. Dó clǽne ifigtaran ðǽr on gif ðú hæbbe [cf. dó gódne sciptaran tó, 326, 14], 3, 26; Lchdm. ii. 322, 27.

ifiht; *adj. Covered with ivy*:—On ðonæ ifihtan stoc *to the ivy-covered post*, Cod. Dipl. Kmbl. iii. 176, 8. In ða ifihtan ác, 379, 29. On ðone ibihtan alr; of ðam ibihtan alre, v, 124, 27.

-ig *a suffix connoting possession of an object denoted by the stem, used in the formation of adjectives, and represented in modern English by* y. *Early English and cognate forms may be seen in the following examples*: *Orm.* modiȝ: *Laym.* modi: *A.S.* módig: *Goth.* módags: *O.Sax.* módag, módig: *Icel.* móðugr, móðigr: *O.H.Ger.* muotig, muotich, muodic: *Ger.* müthig: *Orm.* mahhtiȝ: *Laym.* mæhti: *A.S.* meahtig: *Goth.* mahteigs: *O.Sax.* mahtig: *O.Frs.* machtich: *Icel.* máttugr, máttigr: *O.H.Ger.* mahtig: *Ger.* mächtig: *A.S.* hálig: *Icel.* heilagr: *Goth.* haudugs.

íg, e; *f. An island*:—Wulf is on íege ic on óðerre fæst is ðæt églond fenne biworpen sindon wælreówe weras ðǽr on íge *the wolf is on one island, I on another; closely is that island surrounded with fen, fierce men are there on the island*, Exon. 100b; Th. 380, 6–11; Rä. 1, 4–6. *The word occurs in names of places*:—Án ígland ðæt is Meresíg háten, Chr. 895; Erl. 93, 24. Hér hǽðne men on Sceápíge sǽtun, 855; Erl. 68, 23. Æt Æðelinga íge [eigge, MS. A.], 878; Erl. 81, 5. Of Ceortesíge, 964; Erl. 124, 3. On Beardanigge, 716; Erl. 44, 14. [*Icel.* ey *frequent in local names, e.g.* Fær-eyjar *the Faroe islands*, Orkneyjar *the Orkneys*: *Dan.* öe: *Swed.* ö.]

íg-búend, es; *m. A dweller in an island, an islander*:—Hí ígbúend óðre worde Baðan nemnaþ *island-dwellers by another name call it Bath*, Chr. 973; Erl. 124, 12. Ðis ǽrendgewrit Agustinus ofer sealtne sǽ súðan brohte iégbúendum *this letter Augustine brought across the salt sea from the south to the islanders*, Past. Pref; Swt. 9, 8. [Cf. *Icel.* ey-búi *an islander*.] v. ég-búend *and next word*.

íg-búende; *part. Dwelling in an island*:—Swá hine cígaþ ígbúende Engle and Seaxe weras mid wífum *so call it the island-dwellers, Angles and Saxons, men and women*, Menol. Fox 367; Men. 185. v. *preceding word*.

ig-dæges. v. í-dæges.

-íge -eyed. v. -eáge.

ígeoþ, ígoþ, iggaþ, iggoþ, es; *m. An eyot, ait, islet, small island*:—Ðá ásende hé hine on wræcsíþ tó ánum ígeoþe ðe is Paðmas gecíged *then he sent him away into exile to an island that is called Patmos*, Homl. Th. i. 58, 31. Binnan ánum ígoþe Pathmos gehäten, Ælfc. T. Grn. 16, 23. Binnan iggoþe, Cod. Dipl. Kmbl. iii. 61, 7. Hié flugon up be Colne on ánne iggaþ *they fled up along the Colne on to an island*, Chr. 894; Erl. 90, 28. Ðus feale synden ðere ýgetta ðe liggeþ intó Chertesēge *so many are the islets that belong to Chertsey*, Cod. Dipl. Kmbl. v. 17, 30.

igil, íl, es; *m. A hedgehog, porcupine, an urchin*:—Se mára igil *istrix* [=ὕστριξ], Ælfc. Gl. 24; Som. 60, 29; Wrt. Voc. 24, 30. Íl *yricius vel equinacius*, Wrt. Voc. 78, 21. Se læssa íl *iricius*; se mára íl *istrix*, ii. 49, 52, 53. Hé wæs ðara [strǽla] swá full swá igl biþ byrsta *he* [*St. Sebastian*] *was as full of arrows as a hedgehog is of bristles*, Shrn. 55, 9. Se iil ǽrdǽm hé gefangen weorðe mon mæg gesión ægðer ge his fét ge his heáfod ac sóna swá hiene mon geféhþ swá gewint hé tó ánum cliewene and tíhþ his fét swá hé inmest mæg and gehýt his heáfod *ericius cum apprehenditur, ejus et caput cernitur, et pedes videntur; sed mox ut apprehensus fuerit, semetipsum in sphæram colligit, pedes introrsus subtrahit, caput abscondit*, Past. 35, 3; Swt. 241, 9–12. Íl, Swt. 243, 6. Ðonne biþ ðæs íles heáfud gesewen *caput enim ericii cernitur*, 241, 16. Hé [Eádmund] all wæs biset mid heoræ scotungum swylce ýles burstæ swá swá Sebastianus wæs, Th. An. 122, 17. Íles byrsta, Homl. Skt. 5, 428. Stán is gener

iglum [Blickl. Gl. ílum] *petra est refugium erinaceis*, Ps. Lamb. 103, 18. [*A.R.* ylespilles felles *hedgehogs' skins*: *Trev.* iles piles *ericii*: *Icel.* ígull *a sea-urchin*; ígul-köttr *a hedgehog*: *O.H.Ger.* igil *erinacius*: *Ger.* igel *hedgehog, urchin*.]

íg-land, es; *n. An island*:—Brittene ígland is ehta hund míla lang and twá hund brád. And hér sind on ðis íglande fíf geþeóde *the island of Britain is eight hundred miles long and two hundred broad. And at present there are five languages in this island*, Chr. pref; Erl. 3, 1. Heora cyng him gesealde ðæt ígland ðe man Ii nemnaþ, 565; Erl. 18, 1. Sió wunode on ðam íglande, Bt. 38, 1; Fox 194, 21. Hié cómon on án ígland ðæt is úte on ðære sǽ ðæt is Meresíg háten, Chr. 895; Erl. 93, 24: Bt. 38, 1; Fox 184, 11. Ðæt íland ðe wé hátaþ Thyle, 29, 3; Fox 106, 23. [*Laym.* i-lond: *Icel.* ey-land.] v. eá-, ég-, eig-land.

ígoþ. v. ígeoþ.

-iht *an adjective suffix having much the same meaning as* -ig, *or as the Latin* -osus, *e.g.* stǽniht: *O.H.Ger.* steinaht: *Ger.* steinicht *petrosus*. *Icel. has a suffix* -óttr.

íht, e; *f. Increase*:—Ic sóhte hwylc wǽre elnes oððe iéhte eorlscipes se Pater Noster *I sought what in respect of power or increase of valour the Pater Noster might be*, Salm. Kmbl. 22; Sal. 11. v. ícan.

Ii, Hii, *Iona*:—Heora cyng him gesealde ðæt ígland ðe man Ii nemnaþ . . . Nu sceal beón ǽfre on Ii abbod and ná biscop and ðan sculon beón underþeódde ealle Scotta biscopas forðan ðe Columban was abbod ná biscop *their king gave him* [*Columba*] *the island that is called Iona . . . Now there must always be in Iona an abbot and not a bishop, and to him all the bishops of the Scots must be subject, for Columba was abbot, not bishop*, Chr. 565; Erl. 18, 1–8. Wæs hé sended of ðam eálande and of ðam mynstre ðe Hii is nemned *de insula quæ vocatur Hii*, Bd. 3, 3; S. 526, 11.

iil, íl. v. igil.

ilca; *pron.* [*occurs in the weak declension only*]. *The same*:—Hé sylf oððe se ylca *ipse*; heó sylf oððe seó ylce *ipsa*; hí sylfe oððe ða ylcan *ipsi*, Ælfc. Gr. 15; Som. 18, 53–4. Ðú byst se ilca se ðú ǽr wǽre *tu idem ipse es*, Ps. Th. 101, 24. Se ilca hét ácwellan ða rícostan witan *the same man* [*Nero*] *ordered the greatest senators to be killed*, Bt. Met. Fox 9, 47; Met. 9, 24. Hætþ se ilca god eorþan and wætere mearce gesette *the same God hath appointed a limit to earth and water*, 11, 127; Met. 11, 64. Ðis is se ilca ealwalda god ðone on fyrndagum fæderas cúðon, Andr. Kmbl. 1501; An. 752. Seó ylce bóc *idem libellus*, Bd. 4, 10; S. 578, 16. Hé weorþan sceolde eft ðæt ilca ðæt hé ǽrdon wæs *it should become again the same, that it was before*, Exon. 61 a; Th. 224, 21; Ph. 379. Hié cwǽdon ðæt tæt ilce hiera gefērum geboden wǽre *they said that the same offer had been made to their comrades*, Chr. 755; Erl. 50, 22. On ðisse ylcan tíde *hac ipsa hora*, Ex. 9, 18. On ðære ylcan tíde *eadem hora*; Wick. *in the same hour*, Lk. Skt. 24, 33. Hí smeágaþ unriht and on ðam ilcan forweorþaþ *scrutati sunt iniquitatem; defecerunt scrutantes scrutinio*, Ps. Th. 63, 5. Gelíce ðisse ilcan ðe wé ymb sprecaþ *like the very one we are talking about*, Bt. Met. Fox 26. 5; Met. 26, 3. Ðisne ilcan þreát *this same band*, Exon. 16 a; Th. 36, 2; Cri. 570. Ðyssum ylcum tídum *his temporibus*, Bd. 5, 7; S. 621, 14. Swá ðam ilcum byþ ðe nellaþ ðínre æt bebod healdan *so shall it be with those, who will not keep thy law*, Ps. Th. 118, 36. [*Ilk is used as late as the time of Chaucer, and remains yet in the phrase ' of that ilk,' but its place was gradually occupied by* same (*the Icelandic* sami) *which occurs once in the Ormulum*.]

ilce; *adv. In the same way*:—Hú ne eom ic monn suá ilce suá ðú *am I not a man the same as you are?* Past. 17, 6; Swt. 115, 12. Eft swá ilce *again in the same way*, Bt. 16, 1; Fox 50, 10. [Cf. swilce.]

ild, e; *f. I. an age, period of time; ævum, sæculum*:—Yld *ævum*, Ælfc. Gl. 94; Som. 75, 118; Wrt. Voc. 52, 68. Hér wæs seó forme yld ðissere worulde and seó óðer yld wæs óþ Abrahames tíman . . . Seó þridde yld wæs ðá wuniende óþ David *at this time was the first age of this world, and the second age was till Abraham's time . . . The third age was lasting then till David*, Ælfc. T. Grn. 4, 5, 34. Hé com on ðære syxtan ylde, Blickl. Homl. 71, 26. Se eahtoþa dæg getácnode ða eahtoþan ylde ðyssere worulde, Homl. Th. i. 98, 8. Be ðám syx yldum, Bd. 4, 20; S. 648, 15. II. *age, time of life, years; ætas*:—Eádig is heora yld seó ðe ðá gyt ne mihte Crist andettan and móste for Criste þrowian *blessed is their* [*the children of Bethlehem*] *age, which as yet could not confess Christ, and might suffer for Christ*, Homl. Th. i. 84, 3. Ealle wé cumaþ tó ánre ylde on ðam gemǽnelícum ǽriste ðeáh ðe wé nú on myslícere ylde of ðyssere worulde gewíton *we shall all come at one age at the general resurrection, though now we depart from this world at different ages*, 23–5. Deóplícor mid ús ðú smeágast ðonne yld úre anfón mæge *profundius nobiscum disputas quam ætas nostra capere possit*, Th. An. 33, 11. Hé wæs ðá sixhund geára on ylde *he was six hundred years of age*, Gen. 7, 6. Ðá was ágán his ielde xxiii wintra *he was then twenty-three years of age*, Chr. prm; Erl. 4, 19. Hé leng ne leofaþ ðonn on midre ilde *he will not live beyond middle age*, Lchdm. iii. 162, 21. Ǽrdǽmðe hé self wǽre fulfremedre ielde *nisi perfecta ætate*, Past. 49, 5; Swt. 335, 19. Hundehtatig ylda *octoginta anni*, Ps. Th. 89, 11. III. *mature or old age, eld; senectus, vetustas*:—Yld *senectus*, Ælfc. Gr. 9; Som. 12, 28. Seó nóntíd

biþ úre yld forðan ðe on nóntíde ásíhþ seó sunne and ðæs ealdigendan mannes mægen biþ wanigende *the ninth hour is our old age, for at the ninth hour the sun sinks, and the force of the man that grows old is diminishing*, Homl. Th. ii. 76, 20. Geswenced yld *wearied age*, Dôm. L. 16, 255. Ðonne mê ylde tíd on gesíge *in tempore senectutis*, Ps. Th. 70, 8. On hyre ylde ácende sunu *peperit filium in senectute sua*, Gen. 21, 2. Cild ðæt ðe heó Abrahame on his ylde ácende *filium quem peperit ei [Abraham] jam seni*, 7: Beo. Th. 43; B. 22. Sume beóþ gelædde on cildhâde tô rihtum lîfe, sume on cnihthâde, sume on geþungenum wæstme, sume on ylde, sume on forwerodre ealdnysse, Homl. Th. ii. 76, 26. Ðǽr is geógoþ búton ylde *there is youth without age*, Blickl. Homl. 65, 17: Exon. 32 a; Th. 101,6; Cri. 1654. Gód sceal wyð yfele geógoþ sceal wið ylde sacan, Menol. Fox 562; Gn. C. 50. Nǽron eówre gescí mid ylde fornumene *nec calceamenta pedum vestrorum vetustate consumpta sunt*, Deut. 29, 5. Gesceádlíce tôsceádan ylde and geóguþe *to distinguish discreetly between age and youth*, L. Cf. 4; Th. ii. 262, 5. **IV.** *age, old people, chief people* [v. eald] :— Seó yld hí gebæd and seó iúguþ wrát *age prayed and youth wrote*, Homl. Th. ii. 506, 21. Ðǽr wærþ Eást-Engla folces seó yld ofslagen *there the principal men of the East Angles were slain*, Chr. 1004; Erl. 139, 33. [Goth. alds, alþs *an age, generation* : O. Sax. O. L. Ger. eldi [*old*] *age*; *antiquitas, senectus* : O.Frs. elde : Icel. elli *old age* : O. H. Ger. alti, elti *ætas, ævum, senium, senectus, vetustas*.] v. eld, æfter-yld; ildu.

ilda. v. ildu.

ildan; *p.* de *To delay, tarry, defer, put off, postpone, procrastinate. delay the notice of anything, connive at, dissimulate* :— Tô hwon yldestû middangeard tô onlýhtenne *why dost thou delay to enlighten the world?* Blickl. Homl. 7, 33. Tô hwon yldest ðú ðæt ðú raðost dô ðæt man ðâs menn wítnige and cwelle *why dost thou delay at once to cause these men to be punished and killed*, 183, 1. Seó hâlige cyrice sum þing þurh sceáwunge yldeþ and swâ âbireþ and ældeþ ðæt oft ðæt wiðerwearde yfel âberende and yldende beweraþ *sancta ecclesia quædam per considerationem dissimulat, atque ita portat et dissimulat, ut sæpe malum quod adversatur portando et dissimulando compescat*, Bd. 1, 27; S. 491, 29–32. Ðâ se brýdguma ylde *moram faciente sponso*, Mt. Kmbl. 25, 5. Hê ilde [Cott. MS. ielde] and þasode ða scylda *dissimulavit culpas*, Past. 21, 1; Swt. 151, 22. Hê ða gewilnunge nâht ne ylde *he did not long that desire*, Th. Ap. 1, 17. Ne ylde hê hit ðâ leng *nec exinde distulit*, Bd. 2, 12; S. 512, 34. Hê ylde ðâ gyt *distulit*, Ps. Th. 77, 23. Ne yld ðæt ðú mê árie *ye tardaveris*, 39, 21. Ðeáh ðe ic hit læng ylde *though I should longer delay to notice the matter*, Chr. 1100; Erl. 236, 11. Ne ðæt se agláca yldan þohte *nor did the wretch mean to delay that*, Beo. Th. 1483; B. 739: 4471; 2239. Yldan *dissimulare*, Wrt. Voc. ii. 27, 37. Yldende tô andettenne *differentes confiteri*, Bd. 5, 12; S. 630, 5. Ðonne se lâreów ieldende sêcþ ðone tíman ðe hê his hiéremonn sidelíce on þreátigean mæge *cum tempus subditis ad correptionem quæritur*, Past. 21, 2; Swt. 153, 5. [O.H.Ger. altian *differre*; altôn *dissimulare*; altinôn *differre, dissimulare, elongare*.] v. ældan, ildcian, ildian; for-ildan.

ildcian; *p.* ode *To delay* :— Se dysega ungeþyldega all his ingeþonc hê geypt ac se wísa hit ieldcaþ and bítt tíman *totum spiritum suum profert stultus, sapiens autem differt et reservat in posterum*, Past. 33, 4; Swt. 220, 10. v. elcian, eldcung.

ilde; *pl. m. Men* [a poetical term] :— Hâtaþ ylde eorþbúende fison *men, earth-dwellers, call it Pison*, Cd. 12; Th. 14, 19; Gen. 221. Yldo ofer eorþan, 163; Th. 205, 15; Exod. 436. Nædran ða aspide ylde nemnaþ, Ps. Th. 57, 4. Ylda ǽghwilc *every man*, Cd. 24; Th. 31, 4; Gen. 480. Ylda gehwilc, Ps. Th. 77, 4. Earmlíc ylda cwealm *miserable slaughter of men*, Andr. Kmbl. 363; An. 182: 3108; An. 1557. Ylda Waldend God, Beo. Th. 3327; B. 1661. Ilda cyn *the race of men*, Elen. Kmbl. 1040; El. 521, Ylda bearn *the children of men*, Cd. 113; Th. 149, 6; Gen. 2470: 177; Th. 222, 17; Dan. 106. Sceal mid yldum wesan ismahel hâten *shall be called among men Ishmael*, 104; Th. 138, 3; Gen. 2286: Beo. Th. 154; B. 77. Ðæt wæs yldum cúþ, 1415; B. 705: Ps. Th. 144, 9. Niht becwom ôðer tô yldum, Beo. Th. 4240; B. 2117: Menol. Fox 174; Men. 88: Elen. Kmbl. 1581; El. 792. [O. Sax. eldi; *pl. men*; eldeo barn *children of men* : Icel. öld; aldir; *pl.* [*in poetry*] *men*; alda börn *children of men*.]

ildend, es; *m. One who delays* :— Næs ðâ nænig yldend [ylding?] tô ðam ðæt syððan hí on ðæt hús cômon hí ðâ sóna ðone hâlgan wer gebundon *there was no one, after they had got into the house, who delayed at once to bind the holy man*, Guthl. 5; Gdwin. 36, 5. *See note, where the other reading ylding is given.*

ildend-líc; *adj. Tardy, dilatory* :— Eldendlíce *morosa*,Wrt.Voc.ii.54,58.

ildest; *superl.* of eald. **I.** *eldest, oldest* :— Úre ieldesta mǽg *parens primus* [Adam], Past. 43, 5; Swt. 313, 15. Hê sôhte fram ðam yldestan ôþ ðone gingestan *quos scrutatus, incipiens a majore usque ad minimum*, Gen. 44, 12. Ða yldestan *senes*, Ps. Th. 104, 18. Ða yldestan chus and cham hâtene wǽron *the eldest were named Cush and Ham*, Cd. 79; Th. 97, 22: Gen. 1616. **II.** As the *oldest* might be supposed best fitted to fill the highest positions the word gets the meaning *principal, chief, greatest* :— Se yldesta *cardinarius, i. primarius*, Ælfc. Gl. 48; Som. 65, 66; Wrt. Voc. 34, 1. Yldest byrla *magister calicum*, 113:

Som. 79, 130; Wrt. Voc. 60, 34. Hê wæs ieldest [*summus*] ofer ða hâlgan cirican, Past. 17, 6; Swt. 115, 16. Hwylc hyra yldest wǽre *quis eorum major esset*, Lk. Skt. 9, 46, 22, 24. Ieldesta bisceop *pontifex maximus*, Ors. 5, 4; Swt. 224, 2. Tyrus hêt him tô clypian ðone ðe on ðam scype yldost wǽre *Tyrus bade call to him the principal man on the ship*, St. And. 28, 6. Hê clipode him tô his yldestan gerêfan *dixit ad servum seniorem*, Gen. 24, 2. Aaron and ða yldestan men *tam Aaron quam principes synagogæ*, Ex. 34, 31. Ða ieldestan men ðe tô Bedanforda hiérdon, Chr. 918; Erl. 104, 23. Ða yldestan witan gehâdode and leáwede Angelcynnes, 1012; Erl. 146, 7: 978; Erl. 127, 9. Ða yldestan þægenas, 1015; Erl. 151, 19. Ealle ða yldestan menn on West-Seaxon *all the principal men of Wessex*, 1036; Erl. 165, 1. Ða ðe ieldeste wǽron *equites*, Ors. 6, 4; Swt. 260, 24. Ða yldstan setl on gesamnungum *the highest seats in the synagogues*; primas cathedras in synagogis, Lk. Skt. 20, 46. Ic hit rehte ðâm yldostan Egiptan witun *I told it to the chief wise men of Egypt*, Gen. 41, 24. [Laym. ældeste : Ayenb. eldeste : Icel. ellztr : O. H. Ger. altist, altost *primus, primogenitus*; thie altoston thes folkes *seniores*. For the use similar to that given under **II.** of a word denoting in the first instance age, cf. Goth. þai sinistans (lit. *eldest*) manageins *by which Ulfilas translates* οἱ πρεσβύτεροι τοῦ λαοῦ; *and the passage in Ammianus Marcellinus* 'sacerdos omnium maximus apud Burgundios vocatur *sinistus*.'] v. ildra.

ildian; *p.* ode *To delay, defer, put off* :— Nis forðí nânum synfullum tô yldigenne âgenre gecyrrednysse ðýlæs ðe hê mid sleacnysse forleóse ða tíd Godes fyrstes *it is not, therefore, for any sinner to delay his own conversion, lest by remissness he lose the time of God's respite*, Homl. Th. i. 350, 14. v. ildan.

ilding, e; *f. Delay, putting off, deferring, prolonging, delaying to notice anything, connivance* :— Ylding *tricatio*, Wrt. Voc. ii. 88, 19. Ne wæs ðâ ylding tô ðon ðæt hí heápmælum côman *non mora ergo confluentibus catervis*, Bd. 1, 15; S. 483, 31. Ne wæs ðâ ylding ðæt monige gelýfdon *quid mora? crediderunt nonnulli*, 1, 26; S. 487, 39 : 3, 9; S. 533, 38. Ðâ hit mycel ylding wæs *cum mora multa fieret*, Mk. Skt. 6, 35. Hwæt is ðæt líf elles ðysses middangeardes búton lytelu ylding ðæs deáþes *what else is the life of this world but a little deferring of death?* Blickl. Homl. 59, 27. Hit biþ deáþes ylding swíðor ðonne lífes *it is rather the deferring of death, than the prolonging of life*, 32. Beó ðú on tíd gearu ne mæg ðæs ærendes ylding wyrðan *be thou at the time ready, the errand may not brook delay*, Andr. Kmbl. 430; An. 215. Ðâ bæd hê hine yldinge and fyrstes *petens inducias*, Bd. 4, 1; S. 564, 7. Bútan ǽnigre yldinge *sine ulla dilatione*, 1, 27; S. 493, 30. Búton yldinge, Homl. Th. i. 84, 34. Búton ælcere yldinge, Blickl. Homl. 87, 4. Be ðære ildinge [MS. Cott. ieldinge] suíðe wel Drihten þreáde Iudéas *qua dissimulatione bene Iudæam Dominus corripit*, Past. 21, 1; Swt. 151, 19. Ðæt ic yldinge onfô tô lifianne *ut inducias vivendi accipiam*, Bd. 3, 13; S. 538, 34. Ieldinga *morarum*, Wrt. Voc. ii. 54, 57. v. ildan, eldung.

ildo. v. ildu.

ildra; *m.* ildre; *f. n. comp.* of eald. **I.** *elder, older, grand* [in grand-father, cf. eald-fæder, -môder] :— Ældra *senior*, Wrt.Voc. ii. 120, 48. Seó yldre hâtte Lia and seó gingre Rachel *nomen majoris Lia, minor vero appellabatur Rachel*, Gen. 29, 16. Hys yldra sunu wæs on æcere *erat filius ejus senior in agro*, Lk. Skt. 15, 25. Mín yldra mǽg *my elder brother*, Beo. Th. 940; B. 468. Yldra brôðor, 2653; B. 1324. Oþ ðæt hê yldra wearþ *until he got older*, 4746; B. 2378. Ic eom micle yldra *I am much older*, Exon. 111 a; Th. 424, 20; Rä. 41, 42. Ældra fæder *avus*, Wrt.Voc. ii. 101, 22. Yldra fæder *avita*, 78, 3. Geornful tô witanne ðætte ǽr wæs ǽr ðú ácenned wǽre oððe furðum ðín yldra fæder geboren wǽre *desirous to know what was before you were begotten, or even before your grandfather was born*, Shrn. 198, 29 : Elen. Kmbl. 872; El. 436. For míne sâwle and for mínes fæder and for mínes ieldran fæder *for my soul, and for my father's, and for my grandfather's*, Chart. Th. 496, 21: 497, 15. Þurh heora yldran môdor lâre hí gelýfdon gode *through their grandmother's teaching they believed on God*, Shrn. 53, 10, 16, 21. Ða gingran árísaþ wið ðam yldrum *the younger shall arise against the elder*, Blickl. Homl. 171, 23. Swelce snytro swylce manegum ôðrum ieldran gewittum oftogen is *such wisdom as is withheld from many older minds*, Bt. 8; Fox 24, 28. **II.** *greater, superior* [v. yldest II.] :— Hwæðer ys yldra ðe se ðe þenaþ ðe se ðe sitt *quis major est qui recumbit an qui ministrat?* Lk. Skt. 22, 27. Gewurþe hê swâ swâ gingra seðe yldra ys betwux eów *qui major est in vobis fiat sicut junior*, 26. Ða ðe synt yldran habbaþ anweald on him *qui majores sunt, potestatem exercent in eos*, Mt. Kmbl. 20, 25. [Orm. eldre : Laym. ældre, eldre : O. Sax. aldiro (*as a noun*) : Icel. ellri : O. H. Ger. altero.] v. next word.

ildra, an [*but the singular rarely occurs*]; *m. A parent, ancestor, father, forefather, predecessor, elder* :— Ðâ mê yldra mín âgeaf andsware ðæt mê reordode *then my father answered me and spake* [cf. 872: El. 436: 891; El. 447: 906; El. 454], Elen. Kmbl. 921; El. 462. Hí forgeten hæfdon ðara wundra heora yldran on lôcadan *obliti sunt mirabilium quæ ostendit coram patribus eorum*, Ps. Th. 77, 13. Úre ieldran ða ðe ðâs stôwa ǽr hióldon hie lufodon wísdom *our forefathers, who formerly held these places, loved wisdom*, Past. Pref.; Swt. 5, 14 : Exon.

47 a; Th. 160, 20; Gú. 946. Úre yldran swultan and swíðe oft ús from wendan *our parents have died and very often gone from us*, Blickl. Homl. 195, 26. Wǽron his yldran fæder and módor hǽðne *his parents, father and mother, were heathens*, 211, 19: 213, 2. Úre yldrena lage *traditionem seniorum*, Mt. Kmbl. 15, 2. Twegen gebróðru ðe hæfdon behwyrfed eall heora yldrena gestreón on deórwyrþum gymstánum *two brothers who had converted all their parents' wealth into precious stones*, Homl. Th. i. 60, 23. Bebirge mé mid mínum yldrum *condas me in sepulchro majorum meorum*, Gen. 47, 30. Eafora æfter yldrum *the son after the parents*, Cd. 56; Th. 69, 1; Gen. 1129. Suna ic lǽrde ðæt hié hýrdon heora yldrum *I taught sons to obey their parents*, Blickl. Homl. 185, 20. Nolde hé him geceósan welige yldran *he [Christ] would not choose wealthy parents for himself*, 23, 25. [Laym. aldren, ældere, eldre *forefathers*: R. Glouc. eldren: Piers P. Chauc. eldres: O. Sax. aldiro *a forefather*; pl. *parents*; eldiron, pl. *parents*: O. Frs. alder, elder, aldera, ieldera *father*, *parent*: O. H. Ger. altiron, eldiron *parentes*: Ger. ältern, eltern *parents*.] v. eldran.

ildu; *indecl. f.* **I.** *an age*; *ævum*:—Nis ðæt tó geortrýwianne ðæt on úre yldo ðæt beón mihte ðæt forþgongendre yldo oft geworden getreówe spell secgaþ *nec diffidendum est nostra etiam ætate fieri potuisse, quod ævo præcedente aliquoties factum fideles historiæ narrant*, Bd. 4, 19; S. 587, 32: 3, 27; S. 558, 31. **II.** *age, time of life*; *ætas*:—Óþ nigon and fíftig wintra mínre yldo *usque ad annum ætatis meæ quinquagesimum nonum*, 5, 24; S. 647, 32. On ðære ǽrestan yldo his lífes *in prima ætate*, 5, 13; S. 633, 32. Mid ðí ðe heó bicom tó giftelícre yldo *when she arrived at a marriageable age*, Th. Ap. 1, 10. Ða ðe nabbaþ náwþer ne ildo ne wísdóm *quos vel imperfectio vel ætas prohibet*, Past. 49, 3; Swt. 383, 21. **III.** *age, old age*; *senectus*:—Seó yldo and se ende ðæs heora lífes *their old age and the end of their life*, Blickl. Homl. 163, 5. Heora ylda gelíffæsted wæs, 18. Him æfter ðý yldo ne derede *after that age should not harm him*, Cd. 23; Th. 30, 24; Gen. 471. Nis ðǽr on ðam londe yldu ne yrmþu *there is not in that land old age nor misery*, Exon. 56 b; Th. 201, 6; Ph. 52. On geóguþe . . . on yldo, 88 a; Th. 330, 32, Vy. 60. Geógoþ búton yldo, Blickl. Homl. 103, 35. Heó hire on ylda ðá wǽre *she was in her old age*, 163, 10. Nú gyt syndan manige manna swylce ðe hiom yldo gebídan ǽr tó genihte *adhuc multiplicabuntur in senecta uberi*, Ps. Th. 91, 13. Ða yldu wendan tó lífe to turn old age to life, Exon. 58 b; Th. 210, 23; Ph. 190. Míne yldo beóþ ǽghwǽr genihtsum *senectus mea in misericordia uberi*, Ps. Th. 91, 9. [Orm. A. R. R. Glouc. Ayenb. Piers P. Chauc. Wick. elde *age, old age*, eld: Icel. öld *an age*. v. ild *for other related words*.] v. æfter-, ǽr-, frum-yldo; ældo, eldo, ild.

ile, es; *m.* **I.** *the sole of the foot*:—Ile [? cf. 283, 75 hela *calx*, occurring in a very similar list] *calx*, Wrt. Voc. 65, 47. Ilas, wearras *calces*, ii. 127, 45. From his hnolle ufewerdan óþ his ilas neoþewerde *from the crown of his head to the soles of his feet*, Homl. Th. ii. 452, 27. Mid íseum pílum heora ilas gefæstnode *fastened the soles of their feet with iron nails*, Homl. Skt. 5, 388. **II.** *hard skin* [*such as comes on the sole of the foot?*], *callosity*:—Ile *callus*, Ælfc. Gl. 78; Som. 72, 51; Wrt. Voc. 46, 11. Weorras vel ill *callos*, ii. 103, 16. Him weóxon ylas on olfendes gelícnysse on his cneówum *callosities grew on his knees, just as on a camel's*, Homl. Th. ii. 298, 26. [To þe yle of hire helen, Marh. 10, 19: O. Frs. ili, ile, il *hard skin*: Icel. il; *gen.* iljar; *f. the sole of the foot.*]

ilf, e; *f. An elf*:—Ðanon untydras ealle onwócon eotenas and ylfe *thence sprang all monstrous things, giants and elves*, Beo. Th. 224; B. 112. Gif hit wǽre ésa gescot oððe hit wǽre ylfa gescot oððe hit wǽre hægtessan gescot *if it were Æsir's shot, or elves' shot, or witches' shot*, Lchdm. iii. 54, 10. [Cf. *Scot.* elf-shot; elf-arrow, *Halliw. Dict*; Grmm. D. M. 429: *Prompt. Parv.* elfe *lamia*, 138, see note: M. H. Ger. elbe; *f.* see Grmm. D. M. 411.] v. ælf, -elfen.

ilfette, an; ilfetu, e; *f. A swan*:—Aelbitu *olor*, *cicnus*, Wrt. Voc. ii. 115, 47: *tantalus*, 98, 30. Ilfatu *alvor*, 6, 55. Ilfetu *olor*, 63, 40. Ylfete *cignus*, Ælfc. Gl. 36; Som. 62, 105; Wrt. Voc. 29, 3. Elfetu, Wrt. Voc. 62, 5. Ylfette *olor vel cignus*, 77, 25. Ylfete song *the song of the swan*, Exon. 81 b; Th. 307, 6; Seef. 19. Sume fugelas beóþ langsweorede swá swá ylfettan *some birds are long-necked, such as swans*, Hexam. 8; Norm. 14, 17. [*Icel.* álpt, álft 'the common Icel. word for *swan*; svan is only poët:' O. H. Ger. albiz, alpiz, elpiz *olor*, Grff. i. 243.]

ilfig; *adj. Affected by elves* [?], *mad, frantic*:—Fanaticus, i. minister templi, futura præcinens, vel ylfig, Wrt. Voc. ii. 147, 40. Ylfie vel mónaþseóce *comitiales*, i. e. garritores, 132, 26. Comitiales, lunaticos wanseóce i. garritores, ylfie, Hpt. Gl. 519, 44.

illeracu, e; *f. A surfeit*: crapula, Wrt. Voc. ii. 21, 62. v. ge-illerocaþ.

ilnetu *ciciris* [? v. DuCange 'cicurris *domesticus sus'*], Wrt. Voc. ii. 16, 15.

im-byrdling. v. in-byrdling.

impe [?], an; *f. An imp, scion, graft, shoot*:—Ðæt is sió hálige gesomnung Godes folces ðæt eardaþ on æppeltúnum ðonne hie wel begáþ hira plantan and hiera impan óþ hié fulweaxne beóþ *ecclesia quippe in hortis habitat, quæ ad viriditatem intimam exculta plantaria virtutum servat*, Past. 49, 2; Swt. 381, 17. [Gunge impen me bigurt mid þornes, A. R. 378, 24: Yzet mid guode ympen. Þe ilke ympen byeþ þe virtues, Ayenb. 94, 34: I was the coventes gardyner, for to graffe ympes, Piers P. 5,

137: *Prompt. Parv.* impe or graffe *surculus*: cf. O. H. Ger. impitunga *insertio*: ga-impitón *inserere*, Grff. i. 262: and see Skeat's Etym. Dict. imp.]

IN; *prep. cum dat. inst. acc.* 'In is not found in Alfred's Metres, in the Runic poem, or in Byrhtnoþ; it occurs twice in the metrical Psalms, three times in Cædmon's Genesis; elsewhere in the poetry in and on freely interchange; but in prevails in the North, on in the South. The distinctive on has a vertical element [up or down], which easily runs to against or near,' March, p. 163. **I.** *with dat. inst. In, on*:—Wé sceolan on ðisse sceortan tíde geearnian éce ræste ðonne móton wé in ðære engellícan blisse gefeón mid úrum Drihtne *we must in this short time earn eternal rest, then may we in angelic bliss rejoice with our Lord*, Blickl. Homl. 83, 2. On sumre stówe hé wæs ðæt man mid his handa neálíce geræcean mihte in sumre eáðelíce mid heáfde gehrínan *in one place the roof was so that it could hardly be reached with the hand, in another it could easily be touched with the head*, 207, 22. Hé wæs on Pannania ðære mǽgðe ǽrest on woruld cumen, in Arrea ðæm túne. Wæs hé hweðre in Italia áféded, in Ticinan ðære byrig, 211, 16-18. Ðara monna ðe in ðam here weorþuste wǽron *of the men that were most distinguished in the army*, Chr. 878; Erl. 80, 21. In woruldháde *in sæculari habitu*, Bd. 4, 23; S. 592, 42: 4, 7; S. 574, 34. In regollíces lífes láre swýðe geornful *regularis vitæ institutioni multum intenta*, 4, 23; S. 593, 33. Eall ða hé in gehérnesse geleornian mihte *cuncta quæ audiendo discere poterat*, 4, 24; S. 598, 5. Hafaþ in hondum heofon and eorþan, Exon. 42 a; Th. 140, 32; Gú. 619. Wé sculon á gemunan in móde ðone sigora waldend *we must ever keep in mind the disposer of victories*, 84 b; Th. 318, 15; Mód. 83. Lifgan fracoþ in folcum *to live vile among nations*, 10 b; Th. 12, 33; Cri. 195. Ðú ðe in dryhtnes noman cwóme *thou who didst come in the name of the Lord*, 13 b; Th. 26, 5; Cri. 413. In hwítum hræglum gewerede *clad in white raiment*, 14 a; Th. 28, 15; Cri. 447: Cd. 154; Th. 191, 10; Exod. 212. Wuniaþ in wynnum *they dwell in delights*, 224; Th. 296, 26; Sat. 508. Þafaþ in geþylde *allows in patience*, Exon. 79 a; Th. 297, 20; Crä. 71. Ic on unrihtum eác ðan in synnum geeácnod wæs *I was conceived in iniquity and in sin*, Ps. C. 50, 60; Ps. Grn. ii. 278, 60: Bd. 2, 12; S. 574, 9. In campe *in battle*, Beo. Th. 5003; B. 2505. In Caines cynne ðone cwealm gewræc éce Drihten *the eternal Lord avenged that death among the race of Cain*, 214; B. 107. Ne móste Efe ðá gyt wlítan in wuldre *Eve might not as yet look on glory*, Cd. 222; Th. 290, 2; Sat. 409. Ne hafu ic in heáfde hwíte loccas *I have not white hairs on my head*, Exon. 111 b; Th. 427, 28; Rä. 41, 98. Ábídan sceolan in sinnihte *they shall abide in eternal night*, 31 b; Th. 99, 29; Cri. 1632. In grimmum sǽlum *in rough seasons*, 89 b; Th. 336, 20; Gn. Ex. 52. In lífdagum *in lifetime*, Cd. 163; Th. 204, 22; Exod. 423. In geárdagum *in days of yore*, Beo. Th. 2; B. 1. [Cf. On fyrndagum, Andr. Kmbl. 2; An. 1.] On stówe seó is gecíged in Hripum, Bd. 5, 19; S. 638, 38. In ðýs ginnan grunde *in this wide world*, Judth. 9; Thw. 21, 1; Jud. 2. **II.** *with acc. into, in, to*:—Ǽr ðon ðe hé in heofenas ástige *before he ascended into heaven*, Blickl. Homl. 125, 16. Genáman his líc and in ða stówe ásetton ðe Vaticanus hátte *they took his body and put it into the place called the Vatican*, 191, 33. Ðá eode hé in ða cetan *then he went into the cell*, 219, 14. Gúþlác sette hyht in heofenas, Exon. 39 a; Th. 128, 18; Gú. 406. Heó hine in ðæt mynster onféng . . . Hé eall in ðæt swéteste leóþ gehwyrfde *susceptum in monasterium . . . Ipse cuncta in carmen dulcissimum convertebat*, Bd. 4, 24; S. 598, 3-7. Ðá gewát heó in Eást-Engla mǽgþe *secessit ad provinciam Orientalium Anglorum*, 4, 23; S. 593, 8: Exon. 96 b; Th. 361, 7; Wal. 16. Ne inlǽd úsih in [West Sax. on] costunge *ne inducas nos in temtationem*, Mt. Kmbl. Lind. 6, 13: Hy. 6, 28; Hy. Grn. ii. 286, 28. Beraþ forþ scíre helmas in sceaþena gemong *bear forth your bright helms into the press of the foes*, Judth. 11; Thw. 24, 17; Jud. 193. Héton ǽðeling lǽdan in wráðra geweald, Andr. Kmbl. 2547; An. 1275. Ðá wæs eft geseted in aldordóm babilone weard *the king of Babylon was restored to sovereignty*, Cd. 208; Th. 256, 16; Dan. 641. Ðá hié ðá in ðone heofon lócodan æfter him *as they looked after him unto heaven*, Blickl. Homl. 121, 21. Se ágend upárǽrde reáde streámas in randgebeorh *the Lord hath raised up the waters of the Red Sea as a protection*, Cd. 156; Th. 196, 24; Exod. 296. Gelǽred in ða gerýno Cristes geleáfan, Bd. 2, 15; S. 518, 28. In ða tíd bád ðone écan sige *ipso tempore coronam exspectabat æternam*, Bd. 4, 23; S. 593, 14: 2, 3; S. 504, 20. In áne tíd on one hour, Andr. Kmbl. 2183; An. 1093. In ðín dóm wunaþ in ǽlce tíd *thy glory lasteth to all time*, Exon. 13 b; Th. 25, 26; Cri. 406. In ealle tíd, Exon. 83 a; Th. 313, 15; Seef. 124: 95 b; Th. 356, 25; Pa. 17. In woruld weorulda *in sæcula sæculorum*, Elen. Kmbl. 901; El. 452. **III.** *In sometimes follows its case*:—Ðǽr se eádga mót eardes neótan, wyllestreáma wuduholtum in, wunian in wonge, Exon. 61 a; Th. 223, 20; Ph. 362. Blǽd wíde sprang Scyldes eaferan Scedelandum in, Beo. Th. 38; B. 19. [*Goth. O. Frs. O. H. Ger. Ger.* in: *Icel.* í: *Lat.* in: *Grk.* ἐν.]

in [*adv. and noun*]. v. inn.

in-, inn-. In the case of some of the verbs where *in* is given as a prefix perhaps it should be separated; the passages may then be taken as illustrating the adverb *inn*.

in-áberan; *p.* -bær *To bring in:*—Be ðam hunde ðe his hand eft innábær *of the dog that brought his hand in again,* Homl. Th. ii. 520, 14.

in-ádl, e; *f. An internal disease:*—Sára inádle, L. M. 2, 1; Lchdm. ii. 174, 28. Wið eallum inádlum, 2, 41; Lchdm. ii. 252, 6.

in-ǽlan; *p.* de *To kindle:*—Eów wæs ád inǽled *for you a pile was kindled,* Exon. 42 a; Th. 142, 6; Gú. 640. v. on-ǽlan.

in-ásendan; *p.* de *To send in:*—Hí inásendan ðæt bed *summiserunt grabatum,* Mk. Skt. 2, 4.

in-áwritting, e; *f. An inscription:*—Innáwritting *inscribtio,* Lk. Skt. Lind. 20, 24.

in-bærniss, e; *f. Incense, frankincense:*—Inbærnis *tus,* Wrt. Voc. 289, 54. Inbernisse *incensum,* Ps. Surt. 140, 2. v. an-, on-bærniss; and cf. in-récels.

in-belǽdan; *p.* de *To lead in, introduce:*—Ðú inbelǽdst hig *introduces eos,* Cantic. Moys, 17.

in-belgan; *p.* -bealg; *pp.* -bolgen *To exasperate:* — Ða inbolgeno *aspirando,* Rtl. 15, 40. v. á-belgan.

in-belúcan; *p.* -leác *To shut:*—Ðá ða duru inbeleác æfter him *then he shut the door after them,* Blickl. Homl. 217, 26.

in-bend; *m. f. An internal bond:*—Wæs se báncofa ádle onǽled inbendum fæst *his body was inflamed with disease, fast with the fetters within,* Exon. 46 b; Th. 159, 18; Gú. 928.

in-beódan; *p.* -bead; *pp.* -boden *To announce, declare, proclaim:*—Inboden fæsten *indicto jejunio,* Mt. Kmbl. p. 9. 5. v. on-beódan.

in-beornan; *p.* -bearn *To burn, be on fire:*—Inbiorne wē *inardescamus,* Rtl. 95, 27.

in-beran; *p.* -bær *To bring in,* Beo. Th. 4310; B. 2152.

in-berdling. v. in-byrdling.

in-berþ. v. in-byrd.

in-bestingan; *p.* -stang *To pierce, penetrate, make a thrust which enters but does not go quite through:*—Gif hē þurhstinþ .vi. scill. gebéte. Gif man inbestinþ .vi. scill. gebête, L. Ethb. 64; Th. i. 18, 12.

in-bewindan; *p.* -wand *To wrap up, enwrap:*—Innbewand *involvit,* Lk. Skt. Lind. 23, 53. Inbewunden *involutum,* 2, 12.

in-bewreón; *pp.* -wrigen *To cover up:*—Heora andwlitan inbewrigenum *with their faces covered up,* Cd. 77; Th. 95, 28; Gen. 1585.

in-bindan; *p.* -band *To unbind:*—Án sceal inbindan forstes fetre *one shall unbind the fetters of frost,* Exon. 90 a; Th. 338, 8; Gn. Ex. 75. v. an-, on-bindan.

in-birding. v. in-byrding.

in-birigan; *p.* de *To taste:*—Inberigde *gustavit,* Jn. Skt. Rush. 2, 9. v. on-birian.

in-bláwan; *p.* -bleów *To inspire, breathe upon:*—Inbleów on hine *insuflavit,* Jn. Skt. Rush. 20, 22. Ðec inbláwende *te inspirante,* Rtl. 103, 32.

in-borh; *gen.* -borges; *m. A security required in cases where property had been stolen, bail:*—Gif hwá þíſþe betogen sý ... ðonne niman ða ðe hit tógebyreþ on his æhtan inborh *if any one be accused of theft ... then let those to whom it appertains take security from his property,* L. Ed. 6; Th. i. 162, 20. Ðonne sette mon inborh *let security be given* [the property in dispute is þeófstolen, v. l. 12], L. O. D. 8; Th. i. 356, 10. [Cf. L. H. 1; Th. i. 589, 19, de suo aliquid pro *inborgo* retineatur. Heore godfaderes scullen beo *inborȝes* for hem, O. E. Homl. i. 73, 32. Inboreges, ii. 17, 20.]

in-brengan; *p.* -brohte *To bring in* or *to, present:*—Húmicele hefigra biþ se wēnenda deáþ ðonne se inbrohta *how much more grievous is death when it is expected than when it is presented to us,* Shrn. 42, 31. v. next word.

in-bringan *to bring in, present:*—Ðá hí ne mihton hine inbringan *cum non possent offerre eum illi,* Mk. Skt. 2, 4.

in-bryne, es; *m. A fire, burning:*—Inbyrno *incendia,* Rtl. 64, 12.

in-bryrdan; *p.* de *To stimulate, instigate, incite, animate, inspire:*—Inbryrdende Godes gefe *God's grace instigating me,* Chart. Th. 129, 25. Breóstum inbryrded *animated in spirit,* Exon. 73 b; Th. 274, 18; Jul. 535. Breóstum inbyrrded tó ðam betran hám, 42 a; Th. 141, 12; Gú. 626. Ðá wæs, þurh ðæt hálige treó, imbryrded breóstsefa, Elen. Kmbl. 1680; El. 842. Inbyrrded breóstsefa, 2089; El. 1046. v. on-bryrdan.

in-bryrdniss, e; *f. Inspiration, animation, compunction, feeling:*—Mid ða mǽstan swētnesse and inbryrdnisse [inbyrdnisse, MS.] *maxima suavitate et compunctione,* Bd. 4, 24; S. 596, 34: 3, 19; S. 549, 21. Tó inbryrdnesse [inbyrdnesse, MS.] and tó gemynde ðære æfterfyligendra *ad instructionem memoriamque sequentium,* 17; S. 585, 16, note. v. on-bryrdniss.

in-búan *to inhabit:*—Seðe inbyeþ in ðæm *qui inhabitat in ipso,* Mt. Kmbl. Lind. 23, 21.

in-búend, es; *m. An inhabitant, native:*—Inbúend *colonus, incola, inquilinus,* Wrt. Voc. ii. 134, 25.

in-burh; *gen.* -burge; *f. A hall, vestibule:*—Inburh *atrium,* Wrt. Voc. 84, 35.

inburh-fæst; *adj. Stationed in a hall;* atriensis; scil. atrii janitor, seneschallus, lictor, Lye.

in-byrde; adj. *Born in a master's house:*—Dunne wæs inbyrde tó Hǽðfelda *Dunne belonged by birth to Hatfield,* Chart. Th. 650, 28. Wifús and Dunne and Seoloce syndan inbyrde tó Hǽðfelda, 649, 33. Ða inberðan menn tó Eblesburnan, 152, 8. See next two words, and Kemble's Saxons in England, i. 203 sqq. [Cf. *Icel.* inn-borinn *native:*

O. H. *Ger.* in-burto *oriundus;* in-burtig *indigena:* O. L. *Ger.* in-burdig *indigena.*]

in-byrding, es; *m. A slave born in a master's house:*—Inbirding *vernaculus,* Ælfc. Gl. 8; Som. 56, 103; Wrt. Voc. 18, 52. v. next word.

in-byrdling, es; *m. A slave born in a master's house:*—Inberdling *vel* fóstorling *verna vel vernaculus,* Ælfc. Gl. 86; Som. 74, 34; Wrt. Voc. 50, 17. Inbyrdlinge *vernaculus,* Wrt. Voc. 72, 82. Sicul inberdli[n]c, sicilisc inhyrdlincg (= -byrdling) *siculus indigena,* Hpt. Gl. 499. Mín inbyrdling biþ mín yrfenuma *vernaculus meus heres meus erit,* Gen. 15, 3. Ælc werhádes man on eówrum mǽgðum and inbyrdlingum and geboht þeówa *omne masculinum in generationibus vestris, tam vernaculus quam emptitius,* 17, 12. Ealle werhádes men his inhírēdes ǽgðer ge imbyrdlingas ge gebohte þeówan *omnes viri domus illius, tam vernaculi quam emptitii,* 27.

inc; *dat:* inc, incit; *ac:* incer; *gen. of dual of pronoun of 2nd person:*—Inc ágēnyrnþ sum man *occurrit vobis homo,* Mk. Skt. 14. 13. Nys mē inc tó syllanne *non est mean dare vobis,* Mt. Kmbl. 20, 23. Hwî gewearþ inc swá ðæt gyt dorston fandian Godes *why have ye* [Ananias and Sapphira] *agreed to tempt God?* Homl. Th. i. 316, 33. Bǽm inc *to you both,* Exon. 13 a; Th. 22, 26; Cri. 357. Inc bám twám, Cd. 27; Th. 35, 30; Gen. 562. Neótaþ inc ðæs ōðres ealles wariaþ inc wið ðone wæstm ne wyrþ inc wilna gǽd, 13; Th. 15, 18-21; Gen. 235-6. Incit, 130; Th. 165, 16; Gen. 2732: 139; Th. 174, 19; Gen. 2880. Incer twega *of you two,* Exon. 123 b; Th. 475, 14; Bo. 47. Yncer ǽgðer ofslyhþ ōðerne and hundas licciaþ eówre blód and fugelas fretaþ incer flǽsc and yncer wíf beóþ on ānum dæge wudewan, Shrn. 148, 1-4. Gehwæðer incer *either of you two,* Beo. Th. 1173; B. 584. [*Laym.* 1st MS. inc selven; 2nd MS. ȝou seolve: *Marh.* inc baðen: *Orm.* ȝunnc baþe; gunkerr baþre.] v. git, incer.

inca, an; *m. Doubt, question, cause of complaint, offence, ill-will* or *fear:*—Inca *apporia,* Wrt. Voc. ii. 10, 8: *occasio,* R. Ben. 38, Lye. Ðá ongan hē mē ácsian hwæðere ic wiste hwæðer ic on riht bútan incan gefullad wǽre *cœpit me interrogare, an me esse baptizatum absque scrupulo nossem,* Bd. 5, 6; S. 619, 45. Ðá frægn hē hwæðer hí ea]le smylte mód and bútan eallum incan blíðe tó him hæfdon. Ðá andswaredon hí ealle ðæt hí nǽnigne incan tó him wiston *then he asked them, whether they all were peaceably and kindly disposed to him without any cause of complaint. Then they all answered that they knew no cause of complaint again't them;* interrogavit, si omnes placidum erga se animum, et sine querela controversiæ ac rancoris haberent. Respondebant omnes, se mentem ad illum ab omni ira remotam habere, 4, 24; S. 598, 39-41. Ðú mē scealt edwít mín of áwyrpan ðæt mē tó incan áhwǽr gangeþ *thou shalt cast from me my reproach, which everywhere goes as a cause of fear to me;* amputa opprobrium meum, quod suspicatus sum, Ps. Th. 118, 39. Ðeáh ðe ic nō [MS. on] ingcan wiste hú ic míne heortan heólde mid sōðe *though I did not know any cause of complaint, as to the manner in which I had kept my heart truly;* ergo sine causa justificavi cor meum, 72, 11. Ne ic culpan in ðē incan ǽnigue ǽfre onfunde womma geworhtra *I found not fault in thee, nor cause of complaint for sins committed,* Exon. 10 b; Th. 11, 29; Cri. 178. Incan *scrupulum,* Wrt. Voc. ii. 85, 7. Incan *causas,* 130, 13.

-incel *a diminutive suffix,* e. g. ráp-incel, scip-incel, hús-incel.

in-cempa, an; *m. A member of a household capable of bearing arms:*—Incempa, gescota *commanipularius, collega, miles,* Wrt. Voc. ii. 132, 48. v. in-hirdman, in-híred, in-cniht.

incer; *adj. pron. 2nd person dual. Of* or *belonging to you two:* — Ic nū ðás þing wríte tó ðē gemǽnelíce and tó mínre mēder and mínum geswustrum forðon incer lufu sceal beón somod gemǽne *nunc tibi et matri mee sororibusque meis de singulis regni mei commodis scribebam, que tibi et illis communia esse arbitror,* Nar. 3, 6-9. Sý inc æftyr incrun [MS. A eowrum] geleáfan *secundum fidem vestram fiat vobis,* Mt. Kmbl. 9, 29. Ðý læs gyt láð gode incrum [Adam and Eve] waldende weorðan þyrfen, Cd. 27; Th. 36, 25; Gen. 577. Tó incre andsware, Th. 35, 19; Gen. 557. Biddaþ incerne [Moses and Aaron] god, Ex. 10, 17. Dǽlan somwist incre [Hagar and Sarah], Cd. 104; Th. 137, 27; Gen. 2280. Fyllaþ eorþan incre [Adam and Eve] cynne, 10; Th. 13, 4; Gen. 197. [*Laym.* 1st MS. incker moder inc hateþ; 2nd MS. ȝoure moder ȝou hoteþ: Gen. a. Ex. gunker: Goth. iggkwar.] v. git, inc.

incge, *in the phrase* incge láfe [a sword], Beo. Th. 5747; 13, 2577, *appears to be a proper name.* Ing *occurs in stanza 22 of the Runic poem, and Ing-winas is a name of the Danes in Beowulf.*

in-cígan; *p.* de *To invoke:*—Ic incēgo *invoco,* Rtl. 119, 5.

in-cígung, e; *f. Invocation:*—Innceigungum *invocationibus,* Rtl. 121, 26. Innceiginge *invocationem,* 122, 22. Inceigence, 172, 8.

incit. v. inc.

in-cleofa, an; *m. An inner chamber, closet, bed-chamber, den, cave:*—Incleofa *cellarium,* Wrt. Voc. ii. 130, 56: *camera,* 127, 79. Incleofe *spelunca,* Ps. Spl. T. 9, 10. Forþ of hire inclifan *out of her closet,* Chart. Th. 230, 17. On inclifum [bedcliofum, MS. T.] eówrum *in cubilibus vestris,* Ps. Spl. 4, 5. On incleofum [bedcliofum, MS. T.] his *in cubili suo,* 35, 4. On incleofum [bedclyfum, MS. T.] heora *in cubilibus suis,* 149, 5: Blickl. Gl. Ácende eorþe heora froggan on inclyfum heora

cyninga *edidit terra eorum ranas in penetralibus regum ipsorum*, Ps. Lamb. 104, 30. Hwelpas leóna on incleofum heora hí gesomniaþ *catuli leonum in cubilibus suis collocabuntur*, Ps. Spl. 103, 23.

in-cnapa, an; *m. A domestic servant*, Lye. v. next word.

in-cniht, es; *m. A servant in a house, household* or *domestic servant:*— Incniht *cliens* vel *clientulus*, Wrt. Voc. 72, 80. Incniht *parasitus, cliens, domesticus*, Hpt. Gl. 427, 483, 514. Se hláford gegaderode micele menigu his incnihta *the master gathered together a great many of his household servants*, Homl. Th. i. 502, 13. [*O. L. Ger.* in-kneht *apparitor*: *O. H. Ger.* in-kneht *vernaculus, servus* vel *domigena, verna, inquilinus, apparitor*.]

in-cofa, an; *m. An inner chamber*, [*metaph.*] *the breast, heart:*—On his incofan ł on his clyfan *in cubili suo*, Ps. Lamb. 35, 5. On díglum ł on incofan ł on eówrum clyfum *in cubilibus vestris*, 4, 5. Eal ðæt hé hæfde on his incofan *all that he had in his breast*, Bt. Met. Fox 22, 35; Met. 22, 18. v. breóst-cofa.

in-coðu, e; *and* an; *f. An internal disease:*—Wið incoðe, L. M. 2, 55; Lchdm. i. 276, 6. Fela incoða hé gehælde untrumra sáwla mislícra manna *many diseases of sick souls of diverse men he healed*, Homl. Th. ii. 560, 33. Incoða *infirmitates*; incoðe *fibras* [= *febris?*], Hpt. Gl. 453. Incoðan *melancholias*, 478. [Cf. in-ádl.]

in-cuman; *p. -com To come in, enter:*—Ðonne gé incumaþ on ðæt lond ðe ic eów sille *cum ingressi fueritis terram, quam ego dabo vobis*, Lev. 23, 10. On swá hwilcum húse swá gé incumaþ *whatever house you enter*, Homl. Th. ii. 534, 8. Gá hé út mid swilcum reáfe swilce hé incom *cum quali veste intraverit, cum tali exeat*, Ex. 21, 3. Ðá hié tósamne incóman *when they entered together*, Blickl. Homl. 173. 5. Ðær næfre nænig dæl regnes incuman ne mæg *never can any rain enter there*, 125, 33. Incuma *introire*, Mk. Skt. Lind. 1, 45.

in-cund; *adj. Internal, inward, intimate:* — Ða óðre werod brúcaþ ðære incundan embwlátunge his godcundnysse swá ðæt hí náteshwón fram his andweardnysse ásende ne gewítaþ *the other hosts enjoy the closest contemplation of his divinity, so that on no account do they depart on any mission from his presence*, Homl. Th. i. 348, 7. Ðære þeóde sáwla þurh ða ýttran wundra beóþ getogene tó ðære incundan gife *the souls of that people are drawn by those outward miracles to the inward grace*, ii. 132, 3. Ðonne hé ða úterran þing dôn sculon, ðæt hié ne síen ðam incundum ingeþance áfirrede . . . hié lætaþ ácólian ða incundan lufan *ne, dum cura ab eis exterior agitur, ab interna intentione mergantur . . . ab intimo amore frigescunt*, Past. 18, 7; Swt. 138, 5–9. Wið æghwylcum incundum earfoþnyssum *for all internal difficulties*, Herb. 90, 11; Lchdm. i. 196, 21. Tó incundum *ad intima*, Kent. Gl. 999. v. innan-, inne-cund.

in-cúð; *adj. Strange, not friendly, grievous:* — Hé wolde eác swylce þurh ðone regul oncnáwaþ ða wíslícan gefadunge ðe snotorlíce geset is be incúðra þinga endebyrdnesse *he wished also to know by means of the Rule [of Benedict] the wise arrangement, that is prudently appointed concerning the disposition of strange matters*, Lchdm. iii. 440, 26. Hé hálegra cyricena land incúðum reáferum tódælde *he [Edwy] distributed the lands of holy churches to strangers and robbers*, 436, 1. v. next word.

in-cúðlíce; *adv. Grievously, sorely:*—Ðá begann se ealda incúðlíce siccetan and mid wôpe wearþ ofergoten *then the old man began to sigh grievously and became suffused with tears*, Ælfc. T. Grn. 18, 1.

in-dælan; *p. de To impart, infuse:*—Ðæt léht scínende indæl heartum úsum *illud lumen splendidum infunde cordibus nostris*, Rtl. 2, 13. Indælde *infudit*, 47, 1.

Indea, India *India:*—Ðæt sint India gemæro *in his finibus India est*, Ors. 1, 1; Swt. 10, 15. Hé fór on Indie *Indiam petit*, 3, 9; Swt. 132, 4. Ðá wilnode ic Indeum innwearde tô geseónne *interiorem indiam perspicere cupiens*, Nar. 5, 17. On Indea *to India*, Chr. 883; Erl. 83, 17.

Indeas; *pl. Indians:*—Ðæm strengstan Indea cyninge *fortissimo Indorum rege*, Ors. 3, 9; Swt. 132, 17. Tô Indéum, Apstls. Kmbl. 85; Ap. 43; Bt. 29, 3; Fox 106, 22. Oþ Indeas, Bt. Met. Fox 16, 35; Met. 16, 18.

in-dípan; *p. te To dip in, immerse:*—Ðætte indépe útaweard fingres in wætre *ut intinguat extremum digiti in aquam*, Lk. Skt. Lind. 16, 24. [Cf. *Goth.* daupjan.]

Indisc; *adj. Indian:*—Ðone gársecg mon Indisc *e qua oceanus Indicus vocari incipit*, Ors. 1, 1; Swt. 10, 8. On indisc sprecende *indice loquentes*, Nar. 25, 16. Indisce mýs *mures indici*, 16, 5. Indiscum wordum *indico sermone*, 29, 8.

in-drencan; *p. te To soak, saturate, inebriate:*—Hí ðá sylfe betweónum indrencton mid ðám cerenum ðære gôdspellícan swêtnysse *they mutually saturated each other with the wines of evangelic sweetness*, Guthl. 17; Gdwin. 72, 7. [Cf. *Ger.* ein-tränken *to soak, impregnate*.] v. indrincan.

in-drífan; *p. -dráf To impel, send forth, utter:*—Hé in wîtum word indráf *in torments he spoke impetuously*, Cd. 214; Th. 269, 29; Sat. 80.

in-drincan; *p. -dranc To imbibe, drink:*—Indranc *inbibit*, Mt. Kmbl. p. 1, 7. Indrungno [*Rush.* indruncne] *inebriati*, Jn. Skt. Lind. 2, 10.

in-dryhten; *adj. Noble, courtly, befitting one who belongs to a king's body-guard* [cf. *Icel.* inn-drótt *a king's body-guard*]:—Ðæt biþ in eorle indryhten þeáw ðæt hé his ferþlocan fæste binde *it is a noble habit in a man, to bind fast his mind's casket*, Exon. 76 b; Th. 287, 11; Wand. 12. Ic eom indryhten and eorlum cúð *I am noble and known to men*, 130 b;

Th. 500, 3; Rä. 89, 1. Ic wát indryhtne giest, 112 a; Th. 430, 1; Rä. 44, 1. *Does* indryhten wicg *ippus* (= ? ἵππος), Wrt. Voc. ii. 48, 37 *belong here?*

in-dryhto; *f. Nobleness, honour, glory:*—Blæd is gehnæged eorþan indryhto ealdaþ and searaþ *glory is laid low, earth's honour grows old and withers*, Exon. 82 b; Th. 311, 8; Seef. 89. Gehwone wyrta wynsumra ðe wuldercyning ofer eorþan gescóp tô indryhtum ælda cynne *every pleasant plant that the king of glory created on earth as honours for the race of men*, 58 b; Th. 211, 15; Ph. 198.

Ine, es; *m. Ine, king of the West Saxons from* A. D. 688 *to* 726:—Hér Ine féng tó Wesseaxna ríce and heóld xxxvii wint., Chr. 688; Erl. 42, 4. Hér Ine férde tô Róme and ðær his feorh gesealde, 728 [726, MS. E]; Erl. 44, 33. Ine wæs Cénréding. pref; Erl. 4, 10. The laws of Ine are given in Thorpe's Ancient Laws and Institutes of England, vol. i. pp. 102–150.

in-éddisc. v. in-ídisc.

in-elfe. v. in-ylfe.

in-erfe. v. in-irfe.

in-fær, es; *n. An entrance, ingress:*—Ðá gesette God æt ðam insære engla hyrdrædene *then God set a guard of angels at the entrance*, Gen. 3, 24. Mid ðam innfære mid ðam ðe hé inn áfaren wæs *by the entrance at which he had entered*, Homl. Th. i. 178, 2. Hé hæfþ gerýmed rihtwísum mannum insær tô his ríce *he hath opened to righteous men an entrance to his kingdom*, 28, 13. Geopenige úre sárnys ús insær sóðre gecyrrednysse *let our affliction open to us an entrance to true conversion*, ii. 124, 7. Of inferum *ex aditis*, i. *ex ingressibus*, Wrt. Voc. ii. 144, 49. v. in-faru.

in-færeld, es; *n. An entrance:*—Úre gást forhtode tô eówrum insærelde *elanguit cor nostrum ad introitum vestrum*, Jos. 2, 11. Infæreld *introitus*: insærelda *vestibula, introitus*, Hpt. Gl. 498.

infangeneþeóf '*the right to judge one's own thief when taken within the jurisdiction, and the privilege consequent upon that jurisdiction, viz. the receiving of the mulct, or money-payment for the crime*,' Cod. Dipl. Kmbl. i. xlv. The word, which does not occur in the earlier laws, is thus defined in those of Edward the Confessor:—De infangeneþef. Justicia cognoscentis latronis sua est de homine suo, si captus fuerit super terram suam, L. Ed. C. 22; Th. i. 452, 4. In the preceding chapter, '*descripcio libertatum diversarum*,' it is said the lords 'haberent eos [*their men who had committed crime*] ad rectum in curia sua, si haberent sacham et socham, tol et theam, et infangene thef.' Other passages in which the word is found are L. Wil. I. 2; Th. i. 467, 27, Si quis eorum, qui habent soche et sache et tol et them et infangene theof, implacitetur in comitatu; and L. H. xx. c; Th. i. 528, 9, Archiepiscopi, episcopi, comites, et alie potestates in terris proprie potestatis sue sacam et socnæm habent tol et theam et infangentheaf. The word also occurs in the following charters of Edward the Confessor:—Concedo eis in omnibus terris suis prænominatis, consuetudines hic Anglice scriptas, scilicet, infangene þeóf, etc. Chart. Th. 359, 3. A similar enumeration occurs in 384, 25 and in 411, 32. In 369, 13 the word occurs in an Anglo-Saxon charter. See also Cod. Dipl. Kmbl. iv. 227, 9, where is the form 'mid infangenum þeófe.'

in-faran; *p. -fôr To go into, enter:*—Ic infare on húse dînum *introibo in domum tuam*, Ps. Spl. 5, 8. Innfæreþ *ingredietur*, Jn. Skt. Lind. 10, 9. Infôr se cingc on ða sǽ *ingressus est pharao in mare*, Cantic. Moys. 19. Ðis synd Israhéla naman ðe infôron on Egipta land *hæc sunt nomina filiorum Israel, qui ingressi sunt in Ægyptum*, Gen. 46, 8. Infaraþ tô his cafertúnum *introite in atria ejus*, Ps. Lamb. 95, 8. Ne mæg hé infaran on godes ríce *non potest introire in regnum dei*, Jn. Skt. 3, 5. Ðæt hé ælmessan underfêncge æt ðám infarendum *that he might receive alms from those entering*, Homl. Skt. 10, 27.

in-faru; *f. Invasion, march into a country, inroad:* — Se cyng bæd hine faran intó Cent . . . ac se eorl nolde ná geþwærian ðære infare *the king bade him [Godwin] march into Kent . . . but the earl would not assent to the invasion*, Chr. 1048; Erl. 187, 11.

in-feccan *to fetch in:*—Ðá hêht hé ðone drý infeccan beforan hine *he ordered the sorcerer to be fetched into his presence*, Blickl. Homl. 175, 1.

in-féran; *p. de To enter:*—Infoerden *ingrediun'ur*, Mk. Skt. Lind. 1, 21. Ge in giwinne hiora infoerdun *vos in laborem eorum introistis*, Jn. Skt. Rush. 4, 38.

in-fiht, -feoht, es; *n. An attack made upon a person by one inhabiting the same dwelling*; *it was a breach of the peace for which a fine had to be paid to the head of the house if he were competent to exercise jurisdiction:*—Infiht [infiftht, MS.] vel *insocna est quod ab ipsis qui in domo sunt contubernales agitur*; *hoc eciam wita emendabitur patrifamilias, si questionem habent querentem vel quesitam*, L. H. 80, 12; Th. i. 587, 25.

in-findan; *p. -fand To find, discover:*—Soecaþ gé and gé .infindes *quærite et invenietis*, Mt. Kmbl. Lind. 7, 7. Infund restende *invenit vacantem*, 12, 44. Ic ne infand in him intinga *ego non invenio in eo causam*, Jn. Skt. Rush. 19, 6. Infunden wæs *inventa est*, Mt. Kmbl. Lind. 1, 18. v. on-findan.

in-flæscness, e; *f. Incarnation*, Lye.

in-fléde; *adj. Full of water [of a stream]:*—Tigris eá infléde *Tigris*, Cd. 12; Th. 15, 12; Gen. 232. Lǽt nú stréamas weallan, eá infléde, Andr. Kmbl. 3006; An. 1506. v. fléde.

in-fóster, es; *n. Rearing, breeding* :—Hit mín ágen æht is and mín infóster *it is my own property and my rearing*, L. O. 3 ; Th. i. 180, 7.

in-fród ; *adj. Very old* or *very wise* :—Hé him helpe ne mæg eald and infród ænige gefremman *old and stricken in years he can afford him no help*, Beo. Th. 4889 ; B. 2449. Him wæs wén ealdum infródum, 3752 ; B. 1874.

-ing *a suffix of feminine nouns denoting action.*

-ing.   I. *a patronymic suffix* :—Sume naman syndon patronymica, dæt synd fæderlíce naman, æfter Grēciscum þeáwe, ac seó Lēdenspræc næsþ ða naman ; hí sind swá ðeáh on Engliscre spræce, Penda, and of ðam Pending, Ælfc. Gr. 5 ; Som. 4, 52–4. Ælfrēd Æþelwulfing *Alfred the son of Ethelwulf*, Chr. 871 ; Erl. 76, 3. The use of this suffix is well shown by the genealogies in the Chronicle, e. g. pref ; Erl. pp. 2, 4 : 855 ; Erl. 68, 69, with which may be compared similar lists in Icelandic where -*son* is used. See also Lk. Skt. Lind. 3, 23–38 where the suffix is used with the foreign names, e. g. *Seth Adaming* Seth son of Adam. In a rather extended sense the suffix is found in the names of families or peoples, who are regarded as descendants of a common ancestor, and traces of this use remain in many place-names in England. 'The Wælsings, in Old Norse Völsungar, reappear at Walsingham in Norfolk, Wolsingham in Northumberland, and Woolsingham in Durham. The Billings at Billing, Billingham, Billinghoe, etc. Such local names are for the most part irregular compositions, of which the former part is the patronymic -*ing*, declined in the genitive plural. The second portion is a mere definition of the locality, as -geat, -hyrst, -hám, -wíc, -tún, -stede, and the like. In a few cases the patronymic stands alone in the nominative plural, as Tôtingas, Tooting, Surrey ; Wôcingas, Woking, Surrey ; Meallingas, Malling, Kent. . . . In dealing, however, with these names, some amount of caution is necessary : it is by no means enough that a name should end in -*ing*, to convert it into a genuine patronymic. On the contrary it is a power of that termination to denote the genitive or possessive, which is also the generative case : and in some local names we do find it so used : thus Æðelwulfing lond [Cod. Dipl. No. 179, a. 801] is exactly equivalent to Æðelwulfes lond, the estate of a duke Æðelwulf, not of a family called Æðelwulfings. So again, ðæt Folcwining lond [Cod. Dipl. No. 195, a. 811], ðæt Wynhearding lond [Cod. Dipl. No. 195, a. 811], imply the land of Folcwine, of Wynheard, not of marks or families called Folcwinings, and Wynheardings. [Cf. Cāsering † caseres gafel *didrachma*, Mt. Kmbl. Lind. 17, 24.] Woolbedington, Wool Lavington, Barlavington, are respectively Wulfbæding tún, Wulflāfing tún, Beórlāfing tún, the tún or dwelling of Wulflāf, Wulfbæd, and Beórláf. Between such words and genuine patronymics the line must be carefully drawn, a task which requires both skill and experience ; the best security is, where we find the patronymic in the genitive plural. . . . Changes for the sake of euphony must also be guarded against, as sources of error : thus Abingdon in Berks would impel us strongly to assume a family of Abingas ; the Saxon name Æbban dún convinces us that it was named from an Æbba [*m.*] or Æbbe [*f.*]. Dunnington is not Duning tún, but Dunnan tún.' Kemble's Saxons in England, i. 59, nn ; see also the text in the following pages, and Taylor's Names and Places, pp. 82–3, 89. As was seen above in *Adaming*, the native suffix could be applied in the case of individuals to foreign names : it was so also in the case of peoples. Thus in the Rushworth Gloss, Mt. 8, 28, 'in lond geransinga' translates *in regionem Gerasenorum* ; in 10, 15 'eorðe sodominga and gomorringa' is the rendering of *terra Sodomorum et Gomorræorum*, and above in v. 5 of the same chapter we have 'cæstra samaringa' for *civitates samaritanorum*. These may be compared with the forms in the Chronicle, West Kentingas, 999 ; Erl. 134, 28 ; Eást Centingas, 1009 ; Erl. 142, 19 ; Centingas, 1011 ; Erl. 144, 27.   II. The suffix is also found in nouns formed from adjectives with a force which may be seen in the following examples :—æðeling *a prince* : earming *a wretch.*

ing *the name of the nasal guttural* ᚾ *ng, in the Runic alphabet. In the Gothic the name seems to have been iggws*, see Zacher, Das Gothische Alphabet, p. 3. *In the Runic poem* 22 ; Kmbl. 343, 27 *it is taken as the name of a prince of the East Danes* :—Ing wæs ærest mid Eást Denum gesewen secgum ; oþ hé siððan eft ofer wæg gewât. Ðus heardingas ðone hæle nemdon. This name [cf. Gothic form] may be the same as that found in a genealogy in the Chronicle a. 547 :—Esa wæs Inguing Ingui Angenwitting, Erl. 16, 11. As a proper name or as part of a proper name Ingi occurs in Icelandic, e. g. Ingi-björg, Ing-veldr, Ingi-mundr, Ingólfr : 'many more compounds are found in the Swedish-Runic stones as this name was national among the ancient Swedes ; cf. also Yngvi and Ynglingar.' Cl. and Vig. Ingi. For the Rune see Zacher, pp. 30, 56–7 : Taylor's Greeks and Goths, pp. 31, 82 : and for the name Grmm. D. M. pp. 320–1.

ing, e ; *f. A meadow, an ing* [in dialects of north and east, see E.D.S. Reprinted Glossaries, Nos. 2, 15, 16, 17]. The word occurs in local names, e. g. Ing-ham, Ing-thorpe, Ink-set, Ink-pen ; see Cod. Dipl. Kmbl. vi. 306. [*Icel.* eng ; *f. a meadow* ; engi ; *n. meadowland, a meadow : Dan.* eng : *Swed.* äng.]

in-gán ; *p.* -eode *To go in, enter* :—On swá hwylce burh swá gē ingáþ . . . Ðonne gē ingán on ðæt hús *in quamcumque civitatem intraveritis . . . Intrantes in domum*, Mt. Kmbl. 10, 11, 12. Ðá hé ineode *ingresso*, Gen. 48, 3. Hé on ðæs gesiðes hús ineode, Bd. 5, 4 ; S. 617, 16. Hú mæg man ingán on stranges hús *quomodo potest quisquam intrare in*

domum fortis, Mt. Kmbl. 12, 29 : Lk. Skt. 8, 51. Hé nolde ingán *nolebat introire*, 15, 28.

in-gang, es ; *m. Entrance, entry, ingress, entrance-fee* :—þurh ðē sceal beón se ingang eft geopenod *through thee* [*the Virgin Mary*] *shall the entrance* [*to heaven*] *be again opened*, Homl. 9, 8. Hundteóntiga swína ingang *right of entry into a pasture for a hundred swine*, Cod. Dipl. Knbl. iii. 283, 12. Ingong and útgong *ingress and egress*, Chart. Th. 578, 26. Ðæt beó gelæst binnan twám dagum be ðæs inganges wíte *let that be done within two days, under penalty of forfeiting the entrance-fee*, 606, 10, 20. Gebēte hē be his ingange, 25. Gylde his ingang, 35. Be útgonge Israhēla folces of Ægypta lande and be ingonge ðæs gehātlondes *de egressu Israel ex Ægypto et ingressu in terram repromissionis*, Bd. 4, 24 ; S. 598, 11. Him ôðres lífes ingang gegearwode *vitæ alterius ingressui paravit*, S. 599, 2. Ingang ðín and útgang ðín *thy going out and thy coming in*, Ps. Spl. 120, 8. Inngang, Ps. Th. 117, 19. [O. E. Homl. A. R. in-3oug : Laym. in-3eong : Piers P. in-gong, -gang *entrance : O. Frs.* in-gong, -gung : *Icel.* inn-ganga, -gangr *entrance, entering : O. H. Ger.* in-gang *introitus, aditus, vestibulum, janua : Ger.* ein-gang.]

in-gangan ; *p.* -gēng *To enter, go in* :—Ic ingange *ingredior*, Ælfc. Gr. 29 ; Som. 33, 47. Ic on unscyldignyssa mínre ic ingange *ego in innocentia mea ingressus sum*, Ps. Spl. 25, 1. Ingangeþ cyninge wuldres *introibit Rex gloriæ*, 23, 7. Oþ ðæt ic ingange on hāligra godes *donec intrem sanctuarium Dei*, 72, 17. Gē nū þyder ingongaþ *do ye now enter in*, Blickl. Homl. 207, 2. Cyricean duru ingangan *ecclesiæ januam ingredi*, Bd. 5, 14 ; S. 634, 19. Wæs ingangende on ðære hālgan Marian hús *entered the house of the Holy Mary*, Blickl. Homl. 147, 1. Ingongende, 4. Ðæt deófol genam mid him ôðre seofon deóflo and ingangende on ðæt carcern, 243, 5. Ðonne is ôðer ingangendum ðam mônþe ðe wē agustus hātaþ se æresta mônan dæg *the second day is at the beginning of the month that we call August, the first Monday*, Lchdm. iii. 76, 16. Ðæt ða ingangendan leóht geseón *ut intrantes videant lumen*, Lk. Skt. 8, 16.

in-geat [?] *cubiculum*, Lye.

in-gebed, es ; *n. Hearty, earnest prayer* :—Gange mín ingebed [or gebed in ?] on ðín gleáwe gesihþ *intret oratio mea in conspectu tuo*, Ps. Th. 87, 2. [From the Latin *intret* the *in* might be expected to belong to the verb ; if so it should occupy some other place.]

in-gebyrigan ; *p.* de *To taste* :—Ingeberigde *gustavit*, Jn. Skt. Lind. 2, 9.

in-gedón *to put in* :—Hé on ðæt gemynegade mynster ingedón wæs *monasterio supra memorato inditus*, Bd. 5, 12 ; S. 631, 9.

in-gefeoht, es ; *n. Intestine* or *civil war* :—Ðætte Bryttas sume tíd gestildon fram útgefeohte and hie sylfe þræston on ingefeohtum *ut Brittones quiescentibus ad tempus exteris, civilibus sese bellis contriverint*, Bd. 1, 22 ; S. 485, 12. [Cf. in-gewinn.]

in-gefolc, es ; *n. A native race*, Cd. 149 ; Th. 186, 22 : Exod. 142. [Cf. in-geþeóde.]

in-gehrif, es ; *n. The womb* :—Of ingerife *ex utero*, Ps. Spl. T. 21, 8. v. hrif.

in-gehýgd, -hýd, e ; *f*: es ; *n. Thought, mind, intent, sense, knowledge, understanding, conscience, intention, purpose* :—Hwæt fremaþ ðē ðæt ðín cyst stande ful mid gódum and ðín ingehýd beó æmtig ælces gódes *what doth it profit thee that thy chest stand full of good things, and thy mind be empty of every good thing?* Homl. Th. ii. 410, 11. Ðæs mannes wísdom is árfæstnys and sôð ingehýd ðæt heó yfel forbúge *the fear of the Lord, that is wisdom ; and to depart from evil is understanding*, Homl. Skt. 1, 237. Æfæstre ingehýde *religiosæ intentionis*, Bd. 4, 28 ; S. 605, 10. Treów ingehýdes gódes and yfeles *lignum scientiæ boni et mali*, Gen. 2, 9. Gē ætbrudun ðæs ingehýdes cæge *tulistis clavem scientiæ*, Lk. Skt. 11, 52. Cherubin is gecweden gefyllednys ingehýdes oððe gewittes, Homl. Th. i. 544, 3. Ure wuldor is seó gecýðnys úres ingehýdes *our glory is the testimony of our conscience*, ii. 564, 32. Mid ealle inngehýgde heortan mínre *in toto corde meo*, Ps. Th. 118, 145. Ðá onwende heó hine fram ðære yfelan ingehygde his môdes *revocavit eum illa ab intentione*, Bd. 2, 12 ; S. 574, 37. Wæs se ylca munuc mid hluttre ingehýde ðæs upplícan edleánes *erat idem monachus pura intentione supernæ retributionis*, 4, 3 ; S. 567, 18. Ðá andwyrde eugenia and cwæþ mid ðisum ingehýde ðæt ða gewylnunga ðissere andweardan worulde synt swíðe swicole *then answered Eugenia and spoke to this effect, that the desires of this present world are very deceitful*, Homl. Skt. 2, 163. Mid ingehýgde *conscientia*, Ps. Stev. ii. 203, 11. Se Hālga Gæst him forgeaf ingehýd ealra gereorda *the Holy Ghost gave them knowledge of all languages*, Homl. Th. i. 318, 13. Sumum men hé forgifþ wísdom sumum gód ingehýd *to one man he gives wisdom, to another good knowledge* [cf. 1 Cor. xii. 8], 322, 26. Wæter getácnaþ on ðyssere stôwe mennisc ingehýd, ii. 280, 2. Ðá [*the seven gifts of the Holy Spirit*] sind wísdôm and andgit, ræd and strengþ, ingehýd and ærfæstnys ; Godes ege is se seofoða, 292, 23. Hé heold his þeáwas swá swá heálíc biscop and his munelíce ingehýd swá þeáh betwux mannum *he behaved as an exalted bishop, and yet to all intents and purposes was a monk among men*, 506, 13. Hí hæfdon ðæt góde ingehýd on heora heortan ðæt hí woldon Gode ánum gecwēman and ná cēpan dysegra manna herunge *they had the good sense in their hearts, to wish to please God only, and not to care for the*

praise of foolish men, 564, 29. Seó geladung geopenaþ Criste hire ingehýd and ða dígelan geþohtas on sóðre andetnysse *the church opens her mind and secret thoughts to Christ in true confession*, 586, 20. Ðus áfandaþ God his gecorenan, ná swilce hé nyte heora ingehýd, Boutr. Scrd. 23, 7. Ðæt gold getácnode úrne geleáfan and úre góde ingehíd ðe wé Gode offrian sceolon, Gen. pref. Thw. 3, 33. On ðam is godcundnesse wén ðe manna ingehygd wát and can, Blickl. Homl. 179, 26. Ða eorþlícan sorga hie forléton and ða ingehýd heora heortan ful fæstlíce on ðone heofonlícan hyht gestaþelodon *they dismissed earthly cares, and fixed full firmly the intents of their heart on the heavenly hope*, 135, 29. Ingehýd *conscientias*, Hymn. Surt. 127, 8. Ic ingehygd eal geondwlíte *I survey all his mind*, Exon. 71 b; Th. 266, 16; Jul. 399. Sió swíðe gedræíþ sefan ingehygd monna gehwelces *sorely does it trouble the thought of every man's mind*, Bt. Met. Fox 25, 84; Met. 25, 42. God ingehýda drihten is *Deus scientiarum dominus est*, Cantic. An. 3. [Cf. in-geþanc.]

in-gehygdness, e; *f. Intention, purpose :* — Ic ontýne on sealmlofe ingehygdnessa ł foresetnysse *aperiam in psalterio propositionem*, Ps. Lamb. 48, 5.

in-geledan; *p. de To lead* or *bring in, introduce :* —Ingelédde ofer hie Drihten weter séwe *the Lord brought upon them the water of the sea*, Cantic. Moys. 23; Thw. notæ, p. 30. Oþ ðæt ic ðé ingelæde on mínes Fæder hús *until I bring thee unto my father's house*, Blickl. Homl. 191, 19. Ingelæded *introducta*, Bd. 4, 9; S. 576, 37.

in-geladian; *p. ode To invite :* —Se ðe ðé ingeladode *is qui te vocavit*, Lk. Skt. 14, 9, 10. Ðá sæde hé sum bigspel ðám ingeladudan *dicebat ad invitatos parabolam*, 7.

in-gemynd, es; *n : e; f. Memory, mind, remembrance :* — Ic ðæs wuldres treówes oft hæfde ingemynd *oft had I remembrance of the tree of glory*, Elen. Kmbl. 2504; El. 1253. Húlíc is organ ingemyndum tó begonganne ðam ðe his gást wile ásceádan of scyldum *of what nature is the Pater Noster for use by the mind, in the case of him who will separate his spirit from guilt*, Salm. Kmbl. 108; Sal. 53. v. in-gehygd.

in-gemynde; *adj. Recollected, remembered, in mind, in memory :* — Ðá wæs ðam folce on ferhþsefan ingemynde swá him á scyle wundor ða ðe worhte weoroda dryhten *then did the people remember in mind, as is ever their duty, the miracles which the Lord of Hosts wrought*, Elen. Kmbl. 1788; El. 896.

in-genga, an; *m. An aggressor, invader :* — Seoððan Grendel wearþ ingenga *since Grendel became my aggressor*, Beo. Th. 3557; B. 1776.

in-geótan; *p. -geát To pour in :* — Hí on ælce healfe inguton *they poured in on every side*, Guthl. 5; Gdwin. 34, 18.

in-geóting, e; *f. A pouring in, purification :* —Yngeóting *lustramentum*, Hpt. Gl. 483.

in-gerec, es; *n. A tumult :* — Hé ðá eác on ðam ingerece óðerne cyninges þeng mid ðý mánfullan wæpne ácwealde *in ipso tumultu etiam alium de militibus sica nefanda peremit*, Bd. 2, 9; S. 511, 26. v. ungerec, gerec.

in-gerif. v. in-gehrif.

in-gesteald, es; *n. Household goods :* — Tó scypum feredon eal ingesteald swylce hie æt Finnes hám findan meahton sigla searogimma, Beo. Th. 2314; B. 1155.

in-geswell, es; *n. An internal swelling; empus [= ἔμπυος], Ælfc. Gl. 10; Som. 57, 30; Wrt. Voc. 19, 36.

in-geþanc, es; *m. n. Thought, thinking, cogitation, intent, mind, heart, conscience :* —Seaxes ord and seó swíðre hond eorles ingeþonc and ord somod *the knife's point and the right hand, the mind of man and the point combined*, Exon. 123 a; Th. 472. 8; Rä. 61, 13. Ðæt ingeþonc ælces monnes ðone líchoman lít [læt?] ðider hit wile *the mind of every man bends [leads?] the body whither it will*, Bt. Met. Fox 26, 235; Met. 26, 118. Gif hé his ingeþances anweald næíþ *if he has not power over his mind*, Bt. 29, 3; Fox 106, 26. Eft sint tó manigenne ða geþyldegan ðætte ðæt hie mid hiera wordum and dædum forgiefaþ ðæt hie ðæt eác on hiera ingeþonce forgifen ðý læs hé mid ðý nîðe yfles ingeþonces tóweorpe ða mægenu ðæs gódan weorces ðe hé Gode útan anwealglíce forgeaf *contra admonendi sunt patientes, ne in eo, quod exterius portant, interius doleant : ne tantæ virtutis sacrificium, quod integrum foras immolant, intus malitiæ peste corrumpant*, Past. 33, 5; Swt. 220, 19. Mid eáðmóde ingeþonce ðú mé ciddesð *me humili intentione reprehendis*, prm; Swt. 22, 10. Suelcum ingeþonce geríst *cujus intentioni bene congruens*, 10, 1; Swt. 61, 9. Se Déma se ðe ðæt inngeþonc eall wát hé eác ðæm inngeþonce démþ *intus quippe est qui judicat, intus, quod judicatur*, 4, 2; Swt. 39, 11. Geleornigen eác ða bearn ðæt hí suá hiéren hira iúðran suá suá hie selfe wieten on hira inngeþonce beforan ðæs diéglan Déman eágum ðæt hí hit for Gode dón *illi discant, quomodo ante occulti arbitri oculos sua interiora componant*, Past. 191, 2. Of úrum ágnum ingeþonce *a nobismet ipsis*, 49, 4; Swt. 385, 9. Mid ealles módes geornfullan ingeþance hige *with diligent thought of the whole mind strive*, Bt. 22, 2; Fox 78, 18. Agustinus worhte twá béc be his eágnum ingeþance *Augustine composed two books about his own mind*, Shrn. 164, 16. Ðú ongitst ðín ágen ingeþanc ðæt hit biþ micele beorhtre ðonne seó sunne, Bt. 35, 1; Fox 154, 28. God besceáwaþ ælces mannes

inngeþanc *Deus intuetur cujuslibet hominis cogitationem*, L. Ecg. P. i. 2; Th. ii. 172, 13. Hyra ingeþanc hig forleósaþ on hyra wege *they lose their conscience on their way*, L. E. I. 35; Th. ii. 432 ,22. Nú ic wilnige ðæt ðeós spræc stigge on ðæt ingeþonc ðæs leorneres *ut ad lectoris sui animum gradiatur*, Past. prm; Swt. 23, 16. Se dysega ungeþyldega all his ingeþonc hé geypt *totum spiritum suum profert stultus*, 33, 4; Swt. 220, 10. Drync se onwende gewit wera ingeþanc *a drink that perverted the wit, the mind of men*, Andr. Kmbl. 70; An. 35. Næfdon hí máre monnum gelíces ðonne ingeþonc; hæfde ánra gehwylc his ágen mód, Bt. Met. Fox 26, 188; Met. 26, 94. Hie forgytaþ ðæt hie hwéne ær ymbhygdigum eárum and ingeþancum gehýrdon reccean *they forget what they a little before with anxious ears and minds have heard related*, Blickl. Homl. 55, 27. Ðá azarias ingeþancum hleóðrade *then did Azariah sing full thoughtfully*, Cd. 188; Th. 233, 24; Dan. 280. Ingeþoncum beofiaþ *they tremble at heart*, Exon. 22 b; Th. 63, 4; Cri. 1014. Hiorte geclánsod and geeádméded ingeþancum, Ps. C. 50, 128; Ps. Grn. ii. 279, 128. Oþ ðæt hé ongeat ðæs módes ingeþancas *until he understood the mind's thoughts*, Bt. 7, 1; Fox 16, 5. Hie behealdaþ ealle ða ingeþoncas hiora módes *tota illud mentis intentione custodiunt*, Past. 21, 5; Swt. 161, 14. Uncláene ingeþoncas *impure thoughts*, Exon. 27 a; Th. 80, 34; Cri. 1316. Uton word and weorc rihtlíce fadian and úre ingeþanc clænsian georne *let us order our words and works aright, and purify our thoughts diligently*, Swt. A. S. Rdr. 111, 218. Gesamnige swá hé swíðost mæge ealle tó ðæm ánum his ingeþonc *let him collect, as far as possible, all his thoughts to that one object*, Bt. Met. Fox 22, 24; Met. 22, 12.

in-geþeóde; *pl. Peoples, nations :* —Dryhten is ofer ealle ingeþeóde [? MS. inca þeode.] se heáhsta *excelsus super omnes gentes Dominus*, Ps. Th. 112, 4; Cd. 163; Th. 205, 30; Exod. 443.

in-gewinn, es; *n. An intestine struggle :* —Scortlíce ic hæbbe nú gesæd hiora ingewinn *I have now shortly related their intestine struggles*, Ors. 2, 6; Swt. 88, 29. [Cf. in-gefeoht.]

in-gewitness, e; *f. Knowledge, knowing, consciousness, conscience :* —Besmitene syndon ge heora mód ge heora ingewitnys *coinquinata sunt et mens eorum et conscientia*, Bd. 1, 17; S. 494, 42. Ða wyrstan ingewitnesse mé ic geseó *pessimam mihi scientiam præ oculis habeo*, 5, 13; S. 632, 32.

Ingwine; *pl. A name of the Danes*, Beo. Th. 2092; B. 1044: 2642; B. 1319. v. Grmm. D. M. 320-1; *and see* Ing.

in-heald *interrasilis*, Wrt. Voc. ii. 46, 24.

in-hebban *to raise, remove*, Exon. 12 a; Th. 20, 6; Cri. 313.

in-heord, e; *f. A herd belonging to the lord and kept on his estate :* —Æhteswáne ðe inheorde healt gebyreþ . . . *servo porcario, qui dominicum gregem curie custodit, pertinet* . . . , L. R. S. 7; Th. i. 436, 22.

in-here, es; *m. A native army, the army of a country, home-force :* —Se here férde swá hé sylf wolde and se fyrdinge dyde ðære landleóde ælcne hearm ðet him naðor ne inhere ne úthere ne ðæm Danes *went as they liked, and the English levy did every kind of harm to the people of the country, so that neither the native nor the foreign army did them any good*, Chr. 1006; Erl. 140, 13.

in-hirdmann, es; *m. A member of a retinue or body-guard :* —þegnas ł innheardmenn *milites*, Mt. Kmbl. Lind. 8, 9. v. hird.

in-híréd, es; *m. Household, family, house :* —Tirus wæs on Cryst gelýfende hé sylf and eall hvs ynhýréd *Tyrus believed on Christ, he himself and all his household*, St. And. 30, 15. Inhýredes *clientelæ*, Hpt. Gl. 523. Ealle werhádes men his inhírédes ægðer ge inbyrdlingas ge gebohte þeówan *omnes viri domus illius, tam vernaculi quam emptitii*, Gen. 17, 27. Ðá wearþ gefullod fæder and sunu mid heora inhýréde *then was baptized the father and son with their household*, Homl. Skt. 5, 308. v. in-híwan.

in-hirness, e; *f. A belonging to any one :* —Ðe Æðelréd cyning geúðe God elmihtigum and his hálgan apostolan Petre and Paule on éce inhýrnesse *which king Ethelred granted to Almighty God and to his holy apostles Peter and Paul to belong to them for ever*, Cod. Dipl. Kmbl. vi. 136, 14.

in-híwan, -hígan; *pl. Members of a household, of a convent, domestics :* —Gif gesíþcund mon þingaþ wið cyning for his inhíwum *if a 'gesithcund' man make terms with the king for his household*, L. In. 50; Th. i. 134, 3. Ælce gære áne dægfeorme inhiowum *every year one day's provision for the members of the convent*, Chart. 509, 14. Gie aron inhígo godes *estis domestici Dei*, Rtl. 82, 33. [Cf. Al mi nestfalde cun beoð me meast feondes and mine *inhinen* alre meast hearmen, Jul. 33, 5.] v. híwan.

in-hoh; *adj. Evidens, manifestus*, Hpt. Gl. 523.

in-hold; *adj. Thoroughly loyal, loyal from the heart :* —Abbodissum wé tæcaþ ðæt hí inholde sín and ðæs hálgan regoles gebodum eallum móde þeówigen *we teach abbesses to be heartily loyal, and to be subservient to the commands of the holy rule with all their mind*, Lchdm. iii. 442, 28.

in-ídisc, es; *m. n.* [?] *Household furniture :* —Inéddisc *vel* inorf *entheca g. suppellex*, Ælfc. Gl. 58; Som. 67, 90; Wrt. Voc. 38, 16.

in-ilve. v. in-ylfe.

in-irfe, es; *n. Household stuff* or *goods :* —Se ðe micel inerfa [MS.

Cott. innierfe] and mislíc ágan wile hé beþearf eác micles fultumes *pluribus adminiculis opus est ad tuendam pretiosæ supellectilis varietatem*, Bt. 14, 2; Fox 44, 10. v. in-orf; *and cf.* O.Frs. in-bold, in-gód *household furniture.*

in-lád, e; *f. A way in, bringing in, introduction, entrance-fee* [? v. in-gang]:—Æhtu óra seulfres tó inláde *eight oras of silver as entrance-fee*, Jn. Skt. p. 188, 9. Mid inláde and útláde *cum inductione et eductione*, Cod. Dipl. Kmbl. iv. 209, 5. v. lád.

in-lǽdan; *p.* de *To lead* or *bring in, introduce* :—Ne inlǽd úsih in costunge *ne inducas nos in temtationem*, Mt. Kmbl. Lind. 6, 13. Se ðe má manna inlǽde ðonne hé sceole *he who introduces more men than he ought*, Chart. Th. 606, 32. Mid ðý inlǽddon ðone cnæht aldro his *cum inducerent puerum parentes ejus*, Lk. Skt. Lind. 2, 27.

in-lǽnde, -lǽndisc. v. in-lende, -lendisc.

in-lagian; *p.* ode *To restore an outlaw to the protection of the law* :—Ærest ðæt hé his ágenne wer gesylle ðam cyninge and Criste and mid ðam hine sylfne inlagige *first, that he* [*a man who has committed manslaughter in a church*] *pay his own ‘wer’ to the king and to Christ, and therewith inlaw himself*, L. Eth. ix. 2; Th. i. 340, 13. Inlagie, L. C. E.; Th. i. 520, 11. Inlagie, L. C. E.; Th. i. 360, 3. Cf. Si rex paciatur ut qui in ecclesia fecerit homicidium ad emendacionem veniat, primo episcopo et regi precium nativitatis sue reddat, et ita se *inlegiat*, L. H. 11, 1; Th. i. 520, 11. v. ge-inlagian.

in-land, es; *n.* ‘ *Demesne land, that part of a domain which the lord retained in his own hands, in contradistinction to* út-land *terra tenementalis, signifying land granted out for services* ; terra dominicalis, pars manerii dominica ’:—Wulfége ðæt inland and ælfége ðæt útland, Chart. Th. 502, 13. Sex æceras innlondes ǽgðer ge mǽdlondes ge eyrþlondes, Cod. Dipl. Kmbl. ii. 95, 16. xxx hída .ix inlandes and xxi. hída gesettes landes . . . is sum inland sum hit is tó gafole gesett *thirty hides, nine of ‘inland’ and twenty-one hides of let land . . . some is ‘inland,’ some of it is let*, iii. 450, 11–18. Ægðer ge of þegnes inlande ge of geneátlande, L. Edg. i. 1; Th. i. 262, 8. Ðat inlond ðe Leófríc hædde for his eádmódre hérsumnesse, Cod. Dipl. Kmbl iii. 256, 11. His hláfordes inland, L. R. S. 3; Th. i. 432, 27.

in-laðian; *p.* ode *To invite* :— Ðá cwæþ hé tó ðam ðe hine inlaðode *dicebat ei qui se invitaverat*, Lk. Skt. 14, 12. Ic wæs cuma and gé mé inlaðodon *hospes eram, et collegistis me*, Mt. Kmbl. 25, 35.

in-lenda, an; *m. A native*—Inlenda *indigena*, Ælfc. Gl. 8; Som. 56, 102; Wrt. Voc. 18, 51: ii. 49, 47. Inle[n]da *accola, habitator*, Hpt. Gl. 490, 52. Inlendan *accolas*, Hymn. Surt. 57, 10. v. next word.

in-lende; *adj. Native, indigenous* :—Inlænde ic eam on eorþan *incola ego sum in terra*, Ps. Lamb. 118, 19. Ðǽr on fyrd hyra færspell becwom óht inlende *there to their host came tidings sudden and terrible, fear of the men of the land* [*the Israelites hearing of the pursuit by the Egyptians*], Cd. 148; Th. 186, 9; Exod. 136. David mǽnde tó Drihtne be his feóndum ǽgðer ge inlendum ge útlendum *David complained to the Lord about his enemies, both of his own land and of other lands*, Ps. Th. 2, Arg. [*Icel.* inn-lendr *native: cf.* O. Frs. in-lendes: O. H. Ger. in-lenti *patria*, Grff. ii. 238.]

in-lendisc; *adj. Native, indigenous* :— Inlendisc *indigena vel incola*, Wrt. Voc. 74, 63. Sí hé gemang eów swá inlendisc *sit inter vos quasi indigena*, Lev. 19, 34. Ðǽr útlendisc man inlendiscan derie *where a foreigner injures a native*, L. O. D. 6; Th. i. 354, 29. Se forsǽda bisceop angan tó befrínenne sume inlendisce ymbe ðæs íglondes gewunan *the aforesaid bishop began to ask some of the natives about the customs of the island*, Lchdm. iii. 432, 28. Hæbben for ðí ða ungelǽredan inlendisce ðæs hálgan regules cýððe þurh ágenes gereordes anwrigennesse *the unlearned natives therefore may have knowledge of the holy Rule, through an explanation in their own language*, 442, 8. [*Icel.* inn-lenzkr *indigenous:* Ger. in-ländisch.]

in-lendiscness, e; *f. Incolatus, peregrinatio*, Lye.

in-líc; *adj. Inner, internal, inward* :—Inlíca *intimus*, Hymn. Surt. 66, 13. Se inlíca déma *internus arbiter*, Bd. 3, 15; S. 541, 19. Mid ðone inlícan gewitan *apud internum testem*, 5, 6; S. 618, 32. Mid inlíce hete *domestico odio*, 5, 24; S. 646, 38. Fram ðam inlícum bendum ðara synna *internis peccatorum vinculis*, 4, 25; S. 600, 2. [O. Frs. in-lēk, -lík: O. H. Ger. in-líh *internus*.]

in-líce; *adv. Inwardly, internally, thoroughly, heartily* :— Hé hine bæd and hét ðæt hé inlíce ðam biscope freónd wǽre *amicum episcopo fieri petiit et impetravit*, Bd. 5, 19; S. 641, 8. Ðú miht openlíce ongiton ðæt ðæt is for inlíce gód þing ðæt . . . *you can plainly perceive that that is a very thoroughly good thing that* . . ., Bt. 34, 12; Fox 152, 32. [*Piers P.* in-liche: O. H. Ger. in-lího *medullitus*.]

in-líchamung, e; *f. Incarnation* :—Inlíchomung *incarnatio*, Rtl. 44, 40: 66, 27.

in-líhtan; *p.* te *To illumine, enlighten* :— Ðú tída gehwane inlíhtes *thou dost enlighten every season*, Exon. 9 b; Th. 7, 29; Cri. 108. Inléhteþ ðec *inluminabit te*, Lk. Skt. Rush. 11, 36. Inlíhteþ *inluminat*, Jn. Skt. Lind. 1, 9. Hine inlýhte *he enlightened him*, Exon. 34 a; Th. 108, 9; Gū. 70. Ðæt ðú inleóhte *that thou illumine*, 9 b; Th. 8, 9; Cri, 115. Inlíhte *inluminare*, Lk. Skt. Lind. 1, 79. Inlíhted, Exon. 8 b; Th. 3, 29; Cri. 43. Inlýhted, 42 a; Th. 141, 14; Gū. 817. v. on-líhtan.

in-líhtend, es; *m. One who enlightens* :—Inlíhtend *inluminator*, Rtl. 2, 11.

in-líhtian; *p.* ode *To illumine, enlighten* :—Inléhtaþ ðec *inluminabit te*, Lk. Skt. Lind. 11, 36. Inlíchtade *inluminasset* : inlíchtet *inluminatus*, Jn. Skt. p. 6, 1, 2.

in-liþewác; *adj. Inflexible, intractable* ; *intractabilis*, Wrt. Voc. ii. 48, 72. v. un-liþewác.

in-líxan, -líxian *to shine, grow light* :—Sunnadæg inlíxade [wæs inlíxende, Rush.] *sabbatum inlucescebat*, Lk. Skt. Lind. 23, 54.

in-merca *inscriptio*, Mk. Skt. Lind. 12, 16.

INN, es; *n. A dwelling, house, chamber, lodging* :—Næs Beówulf ðǽr ac wæs óðer in ǽr geteohhod *Beowulf was not there, but other lodging had before been assigned to him*, Beo. Th. 2604; B. 1300. Ðá eode hé tó his inne ðǽr hé hine restan wolde *intravit cubiculum, quo dormire disponebat*, Bd. 2, 12; S. 513, 18: Cd. 76; Th. 94, 25; Gen. 1567: Judth. 10; Thw. 22, 21; Jud. 70. Hé com tó his inne *venit in domum*, Mt. Kmbl. 13, 36. Sóna swá hí út of ðam inne eodon *directly they went out of the house*, Guthl. 11; Gdwin. 54, 16. Ðá lǽdde heó hine on ða cyrcan . . . and on ðam ylcan inne hé oncneów hwæt ðǽr inne wæs *then she led him into the church . . . and in the same house he recognized what was therein*, 22; Gdwin. 96, 23–98, 5. Ðá hé tó his inne com hé hine ǽnne ðǽr inne beleác and hine sylfne ofslóh *when he came to his house, he shut himself in alone, and slew himself*, Ors. 4, 5; Bos. 81, 39: Homl. Th. ii. 490, 10. Se steorra him ðæs cildes inn gebícnode *the star pointed out to them* [*the Magi*] *the child's lodging*, Homl. Th. i. 110, 16. Ðǽr Petrus inn hæfde *where Peter lodged*, 372, 34. [*Laym.* he hafde an in iȝarked toȝeines him: *Orm.* þær he wass at *inne : A. R.* in: *Piers. P.* where dowel was at *inne : Icel.* inni; *n. abode, home.*]

inn-. v. in-.

inn, in; *adv. In, within* :—Ic wæs cuma and gé mé ne in ne gelaðodun *I was a stranger, and ye did not invite me in*, Mt. Kmbl. 25, 43. Waciaþ and gebiddaþ eów ðæt gé in ne gán on costunge *vigilate et orate ut non intretis in temtationem*, 26, 41. Gangaþ inn þurh ðæt nearwe geat *intrate per angustam portam*, 7, 13: Ps. Th. 117, 19. Ðæne se geatweard lǽt in *whom the porter lets in*, Jn. Skt. 10, 3. Hé áwearp ða scyllingas in on ðæt templ *he cast the money into the temple*, Mt. Kmbl. 27, 5. Æt hám gebring and nǽfre in on ðone mon *bring it home and never into the man's presence*, L. M. 2, 65; Lchdm. ii. 292, 26. Ðæt land beág ðǽr súþryhte oððe seó sǽ in on ðæt land, Ors. 1, 1; Swt. 17, 18. Héht óðre dæge hie ealle þrý in beforan hine *next day he ordered them all three in before him*, Blickl. Homl. 175, 18. Ðǽr gedydon twá weofedu in *they put two altars in there*, 205, 15. Duru ðæt mannes heáfod ge ða sculdro mágan in *a door so that a man's head and shoulders may get in*, 127, 9. Ðá heó ðá in tó ðære hálgan Elizabethe eode *when she went in to the holy Elizabeth*, 165, 28. Ðá eode Simon in tó Nerone, 175, 10. Ðá eodan hí in tó swǽsendum, Bd. 3. 14; S. 540, 31. Hreóh wæter tó mínum feore inn flóweþ and gangeþ *introierunt aquæ usque ad animam meam*, Ps. Th. 68, 1. Ðá mé gerýmed wæs síð inn under eorþweall *when a road was cleared for me in under the earthwall*, Beo. Th. 6171; B. 3090. [*Goth.* inn: O. Sax. O. Frs. in: *Icel.* inn: O. H. Ger. in, Grff. i. 287: Ger. ein.]

inna [?], an; *m. The womb* :—In inna *in utero*, Lk. Skt. Lind. 1, 15, 31, 41: 2, 21. Inna *vulvam*, 2, 23.

innan; *adv. and prep. gen. dat. acc. In, into, within, from within.* I.—Gé synt innan fulle reáfláces *intus estis pleni rapina*, Mt. Kmbl. 23, 25. Hig synt innan fulle deádra bána *intus plena sunt ossibus mortuorum*, 23, 27. Heorot innan wæs freóndum áfylled, Beo. Th. 2039; B. 1017. Breóst innan weóll þeóstrum geþoncum *his breast was agitated within by dark thoughts*, 4652; B. 2331. Smire mid ða eágan innan *smear the eyes therewith inside*, L. M. 3, 2; Lchdm. ii. 308, 5. Innan of manna heortan yfele geþancas cumaþ *abintus de corde hominum malæ cogitationes procedunt*, Mk. Skt. 7, 21. Innan and útan, Cd. 66; Th. 80, 1; Gen. 1322: Exon. 22 b; Th. 62, 21; Cri. 1005: 60 a; Th. 219, 2; Ph. 301. II. *with gen* :—Is mé ænige gást innan hreðres *anxiatus est in me spiritus meus*, Ps. Th. 142, 4. Hie hiora onweald innanbordes [cf. *Icel.* innan-borðs] gehióldon *they maintained their power at home*, Past. pref; Swt. 3, 7. Innabordes *intus*, Rtl. 2, 21. III. *with dat* :—Ðá hé sæt innan húse *discumbente eo in domo*, Mt. Kmbl. 9, 10. Hé ádráf út ealle ða ðe ceápodon innan ðam temple *ejiciebat omnes vendentes et ementes in templo*, 21, 12. Gif hé ǽr on ðæs ofermódan engles wísan innan his geþance of Godes gesiehþe ne áfeólle *nisi more superbientis angeli a conspectu conditoris prius intus aversione mentis caderet*, Past. 47, 1; Swt. 359, 1. Hé wæs bebyrged innan ðære cyrican *he was buried inside the church*, Chr. 789; Erl. 57, 32. Ródetácn wearþ æteówed innan ðære dæȝenge *a cross appeared at dawn*, 806; Erl. 60, 24. IV. *with acc* :—Feall innan ða sǽ *jacta te in mare*, Mt. Kmbl. 21, 21. Ne gá gé innan samaritana ceastre *in civitates Samaritanorum ne intraveritis*, 10, 5: Andr. Kmbl. 2350; An. 1176. Innan ðæs týd Gifemund forþférde and Brihtwald gehálgode Tobian on his steall *at this time* [or *meanwhile*] *Gifemund died and Brihtwald consecrated Tobias in his place*, Chr. 693;

Erl. 43, 17. Hēr fōr se here innan Mierce *in this year the Danes marched into Mercia*, 868; Erl. 72, 21. **V.** *in combination with* in, on [cf. *O. Sax.* an innan], geond, be :—Đā hēt ic feá strǣla sendan in đa burh innan *paucas in civitatem dejici sagittas imperavi*, Nar. 10, 22. In đone ofn innan, Cd. 184; Th. 230, 24; Dan. 238; Exon. 58 b; Th. 211, 19; Ph. 200. On đæt morþer innan, Cd. 18; Th. 22, 18; Gen. 342. Burgum in innan, Beo. Th. 3941; B. 1969. In innan *intrinsecus*, Mt. Kmbl. Rush. 7, 15. Eardode ic in innan, Exon. 98 a; Th. 368, 31; Seel. 33. Ne wæs mē feorh đā gēn, ealdor in innan, 103 a; Th. 391, 10; Rä. 10, 3. Innan on đisses holtes hleó, Cd. 39; Th. 52, 7; Gen. 840. On innan đē *in te*, Ps. Th. 147, 2. Geond woruld innan, Exon. 14 b; Th. 29, 28; Cri. 469 : 95 b; Th. 355, 43; Pa. 4. Geond Bryten innan, 45 b; Th. 155, 5; Gū. 855. Be innan đam carcerne, Bt. 1; Fox 4, 2. [*Laym.* inne: *A. R.* inne, ine : *Ayenb.* ine : *Goth.* innana ; *adv. and prep. with gen :* *O. Sax.* innan *adv. and prep. with dat. acc :* *O. Frs.* inna, ina; *id : Icel.* innan ; *adv. and prep. with gen :* *O. H. Ger.* innan, innana ; *adv. and prep. gen. dat. acc.* Grff. i. 296 : *Ger.* innen.] v. innane.

innan-bordes. v. innan, **II.**

innan-burhware ; *pl. Those living within a town* —Đa gefērscipas innanburhwara and ūtanburhwara *the fellowships of the in-townsmen and of the out-townsmen,* Chart. Th. 510, 31.

innan-cund ; *adj. Inward, internal, not superficial, thorough, earnest, genuine, sincere :*—Đonne deáh hit wiđ ǣghwylcre innancundre unhǣlo *then it does for every internal complaint,* Herb. 2, 22; Lchdm. i. 86, 18 : Lchdm. iii. 44, 27. Ic đē mid ealre innancundre heortan sēce *in toto corde meo exquisivi te,* Ps. Th. 118, 10, 2. v. in-, inne-cund.

innane ; *adv. Within :*—Hig beóþ innane reáfigende wulfas *intrinsecus sunt lupi rapaces,* Mt. Kmbl. 7, 15. v. innan.

innan-onfeall. v. oufeall.

innan-weard ; *adj. Inward, internal, interior :*—Ǣlc wuht cwices biþ innanweard hnescost *mollissimum quodque, sicuti medulla est, interiore semper sede reconditur,* Bt. 34, 10; Fox 150, 6. Flet innanweard *the interior of the hall,* Beo. Th. 3957; B. 1976 : 1987; B. 991. Breóst innanweard *the breast within,* Andr. Kmbl. 1294; An. 647 : Exon. 71 b; Th. 266, 19; Jul. 400. Eal innanweard wæs wynsumra đonne hit in worulde mǣge stefn āreccan *all the interior of the dwelling was more delightful than any voice in the world can declare,* 52 a; Th. 181, 16; Gū. 1294. Mec īsern innanweardne bennade *iron wounded me within,* 130 a; Th. 499, 6; Rä. 88, 11. [*Icel.* innan-verðr.] v. inne-weard.

inne ; *adv. In, within, inside, in-doors :*—Đonne đǣr biþ man deád hē liþ inne unforbǣrned mid his freóndum … and ealle đa hwīle đe đæt līc biþ inne đǣr sceal beón gedrync and plega *when there is a man dead, he lies unburnt in the house among his friends … and all the while that the body lies inside, there has to be drinking and playing,* Ors. 1, 1; Swt. 20, 20-6 : Bd. 5, 4; S. 617, 7. Gif man inne feoh genimeþ se man iii gelde gebēte *if a man take property within* [i. e. *in a house*] *let that man pay a threefold compensation,* L. Ethb. 28; Th. i. 10, 1 [cf. *Icel.* brenna inni *to be burnt to death in a house*]. Hwæđer đe ūte đe inne *utrum intus an foris.* Bd. 2, 12; S. 513, 39. Ne mæg đe deófol sceþþan inne ne ūte *the devil cannot harm thee in-doors nor out,* L. M. 3, 58; Lchdm. 342, 15. Sīe se drenc đǣr inne đǣr se seóca man inne sīe *let the drink be in the same place that the sick man is in,* 3, 64; Lchdm. ii. 352, 15. On đam scyran đe ordríc abbud hæfþ land inne *in those shires that abbot Ordric has land in,* Cod. Dipl. Kmbl. iv. 228, 5. Alle đa đe đǣr inne eardedon *all who dwelt therein,* Chr. 491; Erl. 14, 6. Hie sume inne wurdon *some of them got inside* [*York*], 867; Erl. 72, 14. Đone here mētton đǣr on đam geweorce and hine inne besetton *they found the Danes there in the fort, and besieged them inside,* 868; Erl. 73, 25. Đǣr wǣron fīf wucan inne *they were in there five weeks,* 910; Erl. 100, 15. Seó ān inne āwunode, Bd. 5, 12; S. 627, 16. Bēte swā seó dōmbóc sæcge gif hit sȳ hēr inne. Gif hit sȳ eást inne gif hit sȳ norþ inne bēte be đam đe đa friþgewritu sæcgan *let him make ' bót' as the law says, if it be in this part of the country. If it be in the east or north let him make ' bót' according to what the treaties say,* L. Ed. 8; Th. i. 164, 7. Inne on đære þeóde, Bt. 18, 3; Fox 64. 31. On breóstum inne *within their breasts,* Bt. Met. Fox 25, 90; Met. 25, 45. Hēr inne *herein,* Cd. 22; Th. 28, 16; Gen. 436. Hié đǣr inne fulgon *they got in,* Chr. 755; Erl. 50, 27; Fox 2567; B. 1281. [*Goth,* inna : *O. Sax. O. Frs.* inne : *Icel.* inni *in-doors :* *O. H. Ger.* inna, inni, inne *adv. and prep. intus, intra.*] v. innor, innemest.

inne-cund ; *adj. Internal, inward :*—Is geornlīce tō behealdenne đonne hie đa ūterran þing dōn sculon đæt hie ne sīen đæm innecundan ingeþonce āfierrede … hí đonne lǣtaþ ācólian đa innecundan lufan *est vigilanter intuendum, ne, dum cura ab eis exterior agitur, ab interna intentione mergantur … ab intimo amore frigescunt,* Past. 18, 7; Swt. 139, 5-8. v. in-, innan-cund.

inne-fare, an ; *f. The intestines :*—Wiđ wambe cōđe and wiđ inneforan sāre *for dysentery,* L. M. 2, 30; Lchdm. ii. 228, 22. Sió filmen biþ þeccende đa wambe and đa innefaran *the film covers the stomach and the inwards,* 2, 36; Lchdm. ii. 242, 17.

innemest ; *adv. A superlative form from* inne :— Innemest *intime,* Ǣlfc. Gr. 38; Som. 42, 13.

innemest ; *adj. Inmost :*—Ealle đa innemestan geþohtas *all the inmost thoughts ; omnia cogitationum interiora,* Past. 21, 3; Swt. 155, 7.

innera, innra ; *adj. Inner, interior :*— Seó inre hrind *liber,* Ǣlfc. Gl. 59; Som. 68, 6; Wrt. Voc. 38, 57. Se innra man đæt is seó sāwl *interior homo, id est anima,* L. Ecg. P. iv. 63; Th. ii. 224, 6. Se inra wind, Homl. Th. ii. 392, 32. Þurh đa twā pund wæs getācnod ǣgđer ge đæt ȳttre andgit ge đæt inre *by the two pounds was signified both the external and the internal sense,* 554, 34. Se leó gewāt on đæt inre wēsten *the lion departed into the interior of the desert,* Glostr. Frag. 110, 22. Eall mīn inneran *omnia interiora mea,* Ps. Th. 102, 1. Ealle mīne đa inneran, Blickl. Homl. 89, 2. Đeáh hē mē đara ūterrena gewinna gefreóde đeáh winnaþ wiđ mē đa inran unrihtlustas *though he has freed me from outward struggles, yet the inner lusts strive with me,* Ps. Th. 15, 7. On đām inneran gōdum ge on đām ūttran *interioribus bonis et exterioribus,* Bd. 4, 13; S. 582, 39. [*O. Frs.* inra : *Icel.* inri, iðri : *O. H. Ger.* innero, Grff. i. 297.]

inne-weard ; *adj. Inward, internal, interior ;* the word may generally be rendered by the phrase *the inner part of* [the noun with which it agrees]. In the neut. sing. and pl. it is used as a noun, *intestines, viscera, the inward part :*—Inneweard þeoh *femen,* Ǣlfc. Gl. 75; Som. 71, 78; Wrt. Voc. 44. 60. Đes windiga sele eall inneweard *all the interior of this windy hall,* Cd. 216; Th. 273, 15; Sat. 137. Hū hēh and deóp hell inneweard seó, 228; Th. 309, 10; Sat. 707 : Beo. Th. 2000; B. 998. Tō inneweardum đam wēstene *ad interiora deserti,* Ex. 3, 1. Đā com of inneweardre đære byrigenne swā mycel swētnysse stenc *tantæ fragrantia suavitatis ab imis ebullivit,* Bd. 3, 8; S. 532, 17. Of inneweardre heortan *intimo ex corde,* 2, 1; S. 501, 14 : 3, 27; S. 559, 4. Mid inneweardum mōde *with all my mind,* Bt. 2, 17; Fox 76, 7, 24. Inneweard *intestina,* Ǣlfc. Gl. 74; Som. 71, 62; Wrt. Voc. 44. 44. Innoþes innewearde *viscera,* 75; Som. 71, 99; Wrt. Voc. 45, 7. Đā gewand him ūt eall his innewearde *all his intestines came out,* Homl. Th. i. 290, 19. Etaþ đæt heáfod and đa fēt and đæt innewearde, ii. 264, 6 : 280, 7. Etaþ his heáfod and his fēt and innewærde *caput cum pedibus ejus et intestinis vorabitis,* Ex. 12, 9. Innewerde, 29, 17. v. innan-, in-weard.

innian ; *p.* ode *To get within, put in, bring in, put up, lodge :*—Hē werodaþ syđđan hē innaþ *interius recepta dulcescant,* Bt. 22, 1; Fox 76, 31. Đā hī đider cōmon đā woldon hī innian hí đǣr heom sylfan gelícode *when they came thither then they wanted to put themselves up, where it pleased themselves,* Chr. 1048; Erl. 177, 35. [*Me* nuste wære hem inny *people did not not know where to lodge them,* R. Glouc. 336, 14. Þe kyng lette lede hem to a feir old court and *innes* hem þere, Jos. 174 Theseus *ynned* hem, everich at his degre, Chauc. Kn. T. 1334. *O. Frs.* innia *to harbour, lodge :* *O. H. Ger.* innôn *recipere, suscipere, adjungere, afferre,* Grff. i. 298.] v. inne, ge-innian.

innihte ; *adv. Within certain limits :*—Innihte beborene *municipales,* Wrt. Voc. ii. 59, 16.

in-niwian ; *p.* ode *To renew :*—Inniwa *innova,* Rtl. 168, 23.

innon. v. innan.

innor ; *adv. cpve of* inne :—Innor *interius,* Ǣlfc. Gr. 38; Som. 42, 13. [*O. H. Ger.* innor *interius.*]

INNOÞ, innaþ, es ; *m. f.* [?] *The inner part of the body, the inside, stomach, womb, bowels, the breast, heart :*—Innoþ *alvus ;* wīfes innoþ *uterus,* Ǣlfc. Gr. 8; Som. 7, 52, 30 : *viscus,* 9; Som. 12, 12. Wīfmannes innoþ *matrix, uterus,* Ǣlfc. Gl. 74; Som. 71, 56; Wrt. Voc. 44, 39. Eádig is se innoþ đe đē bær *beatus venter qui te portavit,* Lk. Skt. 11, 27. His innoþ tōfleów *his bowels gushed out,* Homl. Th. ii. 250, 26. Đætte hira mōdes innaþ yfele and hefiglíce mid gefylled wæs *quæ mentis intima deprimebat,* Past. 54, 1; Swt. 419, 32. Sió his innaþ wan wætere gelíc *intravit sicut aqua in interiora ejus,* Ps. Th. 108, 18. Wiđ innoþes sār *for sore of inwards,* Herb. 11, 2; Lchdm. i. 102, 11. Wiđ innoþes fæstnysse *for costiveness,* 62; Lchdm. i. 164, 16. Wæstm đe of his innaþe āgenum cwōme *de fructu ventris tui,* Ps. Th. 131, 12. Đa litlingas fuhton on hire innoþe *collidebantur in utero ejus parvuli,* Gen. 25, 22. Hē biþ swīđe líþe on đam innoþe *it is very mild in the stomach,* Bt. 22, 1; Fox 76, 31. Ealle đās yfelu of đam innoþe cumaþ *omnia hæc mala ab intus procedunt,* Mk. Skt. 7, 23. Đē ic andette mid mūþe and mid mīnre heortan and mid eallum innoþe ic đē gewilnige *with my mouth and with my heart I confess thee, and with all that is within me I desire thee,* Homl. Skt. 7, 237. Hālig gāst hređer weardode ædelne innoþ, Elen. Kmbl. 2289; El. 1146. Mæg hē eft cuman on his mōdor innoþ *numquid potest in ventrem matris suæ iterato introire?* Jn. Skt. 3, 4. Inneþas *viscera,* Wrt. Voc. 283, 76. Eádige synt đa innoþas đe ne cendun *beati ventres qui non genuerunt,* Lk. Skt. 23, 29. Wiđ innoþa wræc *for pain of intestines,* L. Med. ex Quadr. 2, 18; Lchdm. i. 338, 9. Wiþ tōbrocenum innoþum *for ruptured bowels,* L. M. 2, 33; Lchdm. ii. 236, 23. On innoþas his *in interiora ejus,* Ps. Spl. M. 108, 17. Đæt sār hwyrfde on hire innoþas *converso ad interanea dolore,* Bd. 4, 23; S. 595, 26. Innaþa *viscera,* Rtl. 13, 33. [*O. E. Homl.* inneþ : *O. L. Ger.* innethron *viscera :* *O. H. Ger.* innod *uterus, viscera ;* innodili *viscera.*]

innoþ-tyderness, e ; *f. A weakness of the intestines :* — Wiđ eallum innoþtydernessum, L. M. 2, 64; Lchdm. ii. 288, 24.

innoþ-wund, e; *f. A wound of the intestines :*—Wið innoþwundum, L. M. 2, 33; Lchdm. ii. 236, 18, 21.

innung, e; *f. A putting or getting in, what is put or got in :*—Se heofon is betera and fægera ðonne eall his innung būton monnum ānum *the heaven is better and fairer than all it includes, except men only*, Bt. 32, 2; Fox 116, 10. Ðes tūnes cýping and seó innung [*the getting in, or revenue?*] ðara portgerihta gange intó ðere hālgan stówe *villæ merci-monium censusque omnis civilis sanctæ æcclesiæ deserviat*, Cod. Dipl. Kmbl. iii. 138, 10.

in-orf, es ; *n. Household goods :*—Inéddisc *vel* inorf *entheca*, g. *suppellex* Ælfc. Gl. 58; Som. 67, 90; Wrt. Voc. 38, 16. Gif hit sý innorf *if it be goods from a house* [*that are taken*], Lchdm. iii. 286, 5. For hwilcum gylta férdest ðú ðus æfter mē and tówurpe eall mīn inorf *quam ob culpam meam sic exarsisti post me et scrutatus es omnem supellectilem meam?* Gen. 31, 36. v. in-irfe.

inra. v. innera.

in-rǽsan; *p. de To rush upon :*—Inrǽsdon *inruerunt*, Mt. Kmbl. Lind. 7, 25. Inrǽsan *inrumpere*, Wrt. Voc. ii. 44, 84.

in-récels, es ; *n. Incense :*—Inrécels, *incensum*, Lk. Skt. Rush. 1, 9.

in-sǽte ; *adj. Belonging to one who is ' settled in ' the household of the lord, one who lives close to the lord's mansion* [?] :—Insǽte hús *vel* lytel hús *casa vel casula*, Ælfc. Gl. 108 ; Som. 78, 113; Wrt. Voc. 58, 28. v. -sǽta, -sǽte.

in-sceáwere, es ; *m. An inspector :*—Ofer-insceáweras *super-inspectores*, Rtl. 194. 25, 29.

in-sceáwung, e; *f. Inspection*, Mt. Kmbl. p. 4, 6.

in-segel, es ; *n. A seal, signet :*—Insegel *sigillum vel bulla*, Ælfc. Gl. 29; Som. 61, 31; Wrt. Voc. 26, 30. Insegl *sigillum*, Wrt. Voc. 83, 4. Geþenc nū gyf ðīnes hláfordes ǽrendgewrit and his insegel tó ðē cymþ hwæðer ðú mǽge cweþan ðæt ðú hys willan ðǽr on gecnáwan ne mǽge *consider now, if your lord's letter and his seal come to you, whether you can say that you cannot recognise his pleasure in them*, Shrn. 176, 10. Insegle *signaculo*, Hpt. Gl. 504, 37. Ðá com Sparhafoc tó him mid ðæs cynges gewrite and insegle, Chr. 1048; Erl. 177, 20. Swā hwæðer swā heó beó fúl swā clǽne binnan ðam insegle *whether it* [*the hand*] *be foul or clean within the seal*, L. Æðelst iv. 7 ; Th. i. 226, 32. Ðá sende se cyning his insegel tó ðam gemóte, Chart. Th. 288, 22. [Ðet insēil þe þe deofel ne mei nefre tobreocan, O. E. Homl. i. 127, 33. He haueð his merke on me iseilet wið his inseil, Marh 5, 16. Bisett wiþþ seffne inseʒʒless, Orm. O. Frs. in-sigel, -sigil *a seal* : Icel. inn-sigli *a seal, a seal-ring*; also the *wax* affixed to a deed: O. H. Ger. in-sigili *sigillum, signaculum, lunula, annulus, moneta : Ger.* in-siegel.] v. insigle.

in-seglian ; *p. ode To seal, place a seal upon :*—Hig innseglodon ðone stán *signantes lapidem*, Mt. Kmbl. 27, 66. Inseglige man ða hand *let a seal be put upon the hand*, L. Æðelst. iv. 7 ; Th. i. 226, 30. [*Icel.* inn-sigla *to seal* : O. H. Ger. in-siglian *signare*.] v. ge-inseglian.

in-seglung, e; *f. A sealing, seal :*—Ic bidde ðē for godes lufan ðæt ðú mē unlýse ða insæglunge *I pray thee for the love of God that thou unloose for me the seal*, Homl. Skt. 3, 537. [*Icel.* inn-siglan *sealing.*]

in-sendan; *p. de To send in :*—Insendes *inmittit*, Mt. Kmbl. Lind. 9, 16. Insende engel dryhten *inmittit angelum Dominus*, Ps. Surt. 33, 8 : 39, 4.

in-setness, e ; *f. A rule, regulation, institute :*—Insetnissum *institutis*, Rtl. 34, 14. Insætnissum, 18, 21.

in-settan ; *p. te To appoint, institute :*—Insette *instituit*, Bd. 4, 23; S. 593, 38.

in-sigle, es ; *n. A seal, signet :*—Hē brohte insigle tó mē ... Ðá ágeaf ic ðæt insigle ðē *he brought a signet to me ... Then I gave the signet to thee*, Chart. Th. 173, 8, 11. Wyrðe arð onfóа bóc and untýne insigloe his *dignus es accipere librum et aperire signaculum ejus*, Rtl. 29, 19. v. in-segel.

in-siht; e; *f. An account, narrative, argument :*—Onginneþ insiht æfter iohannem *incipit argumentum secundum Johannem*, Jn. Skt. p. 1, 1. [*Goth.* in-sahts *narrative.*]

in-sittende; *part. Sitting within :*—Ealra wǽron fīfe eorla and idesa insittendra, Exon. 112 b; Th. 432, 3; Rä. 47, 7.

in-smoh; *gen.* -smós [?] ; *m. A slough :*—Hē ágeaf ðone clǽnan gást and ðæs līchaman insmoh [*exuvias*] forlēt monnum tó mundbyrde *he gave up the clean spirit, and left the slough of the body as a protection for men*, Shrn. 126, 2. v. smúgan; and cf. O. Frs. in-smuge *a creeping in.*

in-spinn, es; *n. An instrument for spinning, a spindle :*—Inspinn *netorium*, Ælfc. Gl. 110 ; Som. 79, 46; Wrt. Voc. 59, 17. Inspin, Wrt. Voc. 66, 15. [*Netorium fusus quo netur : fusum, fusile*, Du Cange.]

in-stæppan ; *p. te To step in, enter :*—Ic ne instæppe ðide ingá oðde ic ne fare *non introibo*, Ps. Lamb. 25, 4. Insteppaþ oðde ingáþ on gesihþe his *introite in conspectu ejus*, 99, 2. On unscyldignysse mīure instæppende ic eom *in innocentia mea ingressus sum*, 25, 11. Hī sume gesáwon englas instæppende *some of them saw angels entering*, Homl. Th. ii. 546, 23.

in-stæpe, es ; *m. Entrance :*—Hí gemēttou ðæt éce líf on instæpe ðæs andweardan lífes *they found the life eternal at the entrance of the present life*, Homl. Th. i. 84, 7. [O. Frs. in-stap, in-steppi *entrance.*] v. next two words.

in-stæpe, -stepe ; *adv. At the outset, at once, directly, immediately :*—

Instæpe *confestim*, Bd. 2, 12 ; S. 514, 21 : *extemplo*, 4, 25 ; S. 601, 30. Árás hē instæpe *surrexit continuo*, 5, 5 ; S. 618, 14. Hī instæpe fram mīure gesihþe gewiton *statim disparuerunt*, 5, 13 ; S. 633, 15. Ðonne wǽre mīn blód instæpe ágoten *then had my blood been at once shed*, Shrn. 39, 17. Seó strǽl instepe wearþ eft gecyrred, Blickl. Homl. 199, 21. v. next word.

in-stæpes, -stepes; *adv. At once, immediately :*—Se mon se ðe óðerne ácwelþ and instæpes hine sylfne ongyteþ ðæt hē mycel mán gedón hæbbe *the man who kills another, and at once perceives himself to have done a great wrong*, Blickl. Homl. 65, 5. Hē ðá sóna instæpes geseh *he then immediately saw*, 15, 27. Ðēh gē sóna instæpes ðǽre mēde ne ne onfón, 41, 13. Instepes, 33, 19. Ðæt fæsten wæs ongunnen instepes ðæs ðe ... *the fast was begun directly after* ..., 35, 5. Hī flugon instæpes *they fled forthwith*, Elen. Kmbl. 254; El. 127.

in-standan; *p.* -stód *To be near or present* ; *instare :*—Éce instondaþ wuldur *perennis instat gloria*, Rtl. 165, 7. Instond[end]um *instantibus*, 69, 11.

in-standendlíc; *adj. Present, of to-day :*—Hláf úre instondenlíce sel ús tó dæge *give us to-day our daily bread*, Mt. Kmbl. Rush. 6, 11.

in-stede, -styde [or in stede; cf. *Icel.* í-stað *on the spot, at once*] ; *adv. On the spot, at once, immediately :*—Instyde *continuo*, Mt. Kmbl. Rush. 27, 48. Instyde *statim*, Mk. Skt. Rush. 1, 28 : 2, 12.

in-stice, es ; *m. An inward stitch, a pricking sensation within :*—Wið instice, L. M. 2, 54; Lchdm. ii. 274, 27.

in-stihtian; *p. ode To arrange, regulate, dispose :*—Instihtade ł dihtade *instigante*, Lk. Skt. p. 2, 6. v. stihtian.

in-sting, es ; *m. Authority :*—Nán ðere biscope ne habbe nán insting on ðæt mynster *let no bishop have any authority in that monastery*, Chart. Th. 348, 12. v. on-sting.

in-swán, es; *m. The herd who had charge of the lord's swine :*—Ǽlc gebúr sylle .vi. hláfas ðam inswáne ðonne hē his heorde tó mæstene drífe *omnis geburus det vi. panes porcario curie quando gregem suum minabit in pastinagium*, L. R. S. 4; Th. i. 434, 21.

in-swápen. v. swápan.

in-swógenness, e ; *f. A rushing in with a loud sound, violent entrance :*—Hē mid ðæs unclǽnan gástes inswógennisse þrycced wæs *spiritus inmundi invasione premebatur*, Bd. 2, 5 ; S. 507, 4. v. swógan.

inðer ; *adv. Apart* ; *seorsum*, Mt. Kmbl. Rush. 17, 1.

in-þicce ; *adj. Gross, thick :*—Inþicce is hearta folces ðisses *incrassatum est cor populi hujus*, Mt. Kmbl. Lind. 13, 15.

in-þinen, e; *f. A female domestic servant* ; *incola*, Germ. 401, 125.

in-timbrian; *p. ede, ode To instruct :*—Hē hī intimbrade and gelǽrde *he instructed and taught them*, Bd. 4, 16; S. 584, 34. Intimbrede, 4, 27 ; S. 603, 45. In cyriclícum þeódscipum and in mynsterlícum heálíce intimbred *ecclesiasticis ac monasterialibus disciplinis summe instructus*, Bd. 5, 8 ; S. 621, 35 : S. 622, 2. v. on-timbrian.

in-tinga, an; *m. A cause, sake, plea, case, occasion, matter, affair, business :*—Intinga *pragma*, Ælfc. Gl. 12 ; Som. 57, 93; Wrt. Voc. 20, 34 : *negotium*, 81; Som. 73, 17 ; Wrt. Voc. 47, 24 : *causa vel negotium*, 90; Som. 74, 115; Wrt. Voc. 51, 28 : *causa*, Wrt. Voc. 83, 62. Ðysse þeóde wæs se ǽresta intinga tó onfónne Cristes geleáfan ðæt . . . *huic genti occasio fuit percipiendæ fidei, quod* . . ., Bd. 2, 9 ; S. 510, 18. His intinga wæs geondsóhte beforan Agaþone *causa ejus ventilata est præsente Agathone*, 5, 19 ; S. 639, 28. Se forma intinga mennisces forwyrdes wæs ðá ðá se deófol ásende óðerne deófol tó Evan *the first cause of man's perdition was when the devil sent another devil to Eve*, Homl. Th. i. 194, 30. Ðæt mín sáwul lybbe for ðínum intingan *ut vivat anima mea ob gratiam tui*, Gen. 12, 13. For hwilcum intingan *quam ob causam*, 19. Tó ðisum is genumen se grēcisca y for intingan grēciscra namena *to these* [*the vowels*] *is added the Greek y for the sake of Greek names*, Ælfc. Gr. 2; Som. 2, 51. For his intingan hē hit dēþ *sui causa facit*, 17 ; Som. 20, 50 : Homl. Th. i. 84, 2. Ic ongann be ðam intingan hwæthwega geornlícor smeágan *I began to inquire somewhat more diligently about the matter*, ii. 32, 23. Gif hiú of cealdum intingan cymþ ðonne sceal mon mid hátum lǽcedómum lácnian *if it* [*the disease*] *comes from a cold cause then it is to be cured with hot medicines*, L. M. 1, 1; Lchdm. ii. 22, 5. Búton intingan *sine causa*, Ps. Spl. 3, 7. Bútan intingan hig mē wurðiaþ *sine causa colunt me*, Mt. Kmbl. 15, 9. Ðá hí ðá heora intingan him wēpende sǽdon ðá wæs hē sóna mid mildheortnysse gefylled *when with tears they had told him their business, he was at once filled with pity*, Guthl. 12 ; Gdwin. 58, 25. Tósceáð intingan mīnne *discerne causam meam*, Ps. Spl. 42, 1. Dēm intingan ðīnne *judica causam tuam*, 73, 23. Ne finde ic nänne intingan on ðysum men *nihil invenio causæ in hoc homine*, Lk. Skt. 23, 4, 14. Hē nolde syllan intingan ðām Iudēiscum ðæt hē hī forsáwe ðe Godes ǽ heóldon, and ðæt hǽdene folc him tó getuge *he would not give the Jews cause to complain, that he despised those who kept God's law, and drew to him the heathen people*, Homl. Th. ii. 112, 5. Fordon misenlíce intingan gelimpeþ *quia diversæ causæ impediunt*, Bd. 4, 5 ; S. 573, 7.

in-tó; *prep. Into*. **I.** *with dat :*—Ðú gæst intó ðam arce . . . and twegen gemacan ðú lǽtst intó ðam arce *ingredieris arcam* . . . *et bina induces in arcam*, Gen. 6, 18, 19. Noe eode intó ðam arce *ingressus est Noe in arcam*, 7, 7. Ic gange intó ðǽre byrig *in urbem vado*. Ic

gange intó ðínum huse *introibo in domum tuam*. Intó ðære ceastre rád se kyning *in civitatem equitavit rex*, Ælfc. Gr. 47; Som. 48, 15-7. Ðá se hǽlend com intó ðæs ealdres healle *cum venisset Iesus in domum principis*, Mt. Kmbl. 9, 23. Sume urnon intó cyrcean and belucan ða duran intó heom *some ran into the church and shut the doors upon them*, Chr. 1082; Erl. 217, 13. **II.** *with acc :*—Férde his hlísa intó ealle Syriam *abiit opinio ejus in totam Syriam*, Mt. Kmbl. 4, 24. Wið feó sealdon wíde intó leódscipas *they sold them far and wide into various nations*, Blickl. Homl. 79, 23. **III.** *with inst :*—Ðá ongeáton hie ðæt se eádiga Michael him sylfa ðæt tácn ðæs siges gecýðde intó ðý swiðan slǽpe *then they perceived that the blessed Michael had himself made known that token of victory in the deep sleep*, 205, 4.

in-trahtnung, e; *f. Explanation, interpretation :* — Sóþ intrahtnung *vera interpretatio*, Mt. Kmbl. p. 2, 6.

in-trifelung, e; *f. Intritura*, Cot. 109, Lye.

in-wǽte, an; *f. An inward humour :*—Gif hit biþ cumen of yfelre inwǽtan *if it is come of an evil inward humour*, L. M. 2, 46; Lchdm. ii. 258, 27.

in-weard; *adj. Inward, inner, internal :* — Gif gé hine mid inweardre heortan séceaþ *si toto corde quæsieris*, Deut. 4, 29. Biddaþ mid inweardre heortan ðysne Godes apostol, Homl. Th. i. 68, 8. Ðá wilnode ic indeum innewearde tó geseónne *interiorem indiam perspicere cupiens*, Nar. 5, 17. v. innan-, inne-weard.

in-weard; *adv. Within :* — Ðætte inweard is *quod intus est*, Lk. Skt. Lind. 11, 39. Ðá hig inweard fóron ðá gemytton hig twegen ealde weras *when they went in, they met two old men*, Nicod. 31; Thw. 18, 3. [Let þene lust gon inward, A. R. 272, 8. Inwardes, 92, 6.]

inweard-líc; *adj. Inward, internal :*—Innweardlíc *interius*, Rtl. 4, 20. On heora inweardlícum stówum *in their inward parts*, L. Med. ex Quadr. 3, 1; Lchdm. i. 338, 19, MS. H.

inweard-líce; *adv. Inwardly, thoroughly, heartily, earnestly :* — Heroðes innweardlíce gelearnade from him *Herodes diligenter didicit ab eis*, Mt. Kmbl. Lind. 2, 7, 8. Innweardlíce cliopaþ hine *invocate eum*, Rtl. 10, 26. Is ðæt for inweardlíce riht racu *that is a very thoroughly right explanation*, Bt. 40, 1; Fox 236, 9. Se ðe æfter rihte mid gerece wille inweardlíce æfterspyrian swá deóplíce ðæt hit tódrífan ne mæg monna ǽnig *quisquis profunda mente vestigat verum, cupitque nullis ille deviis falli*, Bt. Met. Fox 22, 3; Met. 22, 2. Wearþ ðá him inweardlíce gelufod *he was heartily loved by him*, Homl. Th. i. 58, 18. Ða ðe tó ge-leáfan cyrden hé ða inweardlýcor lufade *credentes arctiori dilectione amplecteretur*, Bd. 1, 26; S. 488, 16.

in-weorud, es; *n. A band of domestics* or *courtiers, a household :*—Ðæt wæs innweorud Earmanríces, Exon. 86 b; Th. 325, 13; Víd. 111. [Cf. in-híréd.]

inwid, inwit, es; *n. Fraud, guile, deceit, evil, wickedness :*—Inwid *dolus*, Ps. Spl. T. 14, 3. Ne beó nænig man hér on worldríce bregda tó full ne inwit tó leóf *let no man in this world be too full of wiles, nor let guile be too dear to him*, Blickl. Homl. 109, 29. Ne wæs ǽfre fácen ne inwid on his heortan *nor was never deceit nor guile in his heart*, 223, 31. Gramlíc inwit *nequitia*, Ps. Th. 54, 15. Mán and inwit, 9. Forðan mé inwit næs on tungan *quia non est dolus in lingua mea*, 138, 2. Mán inwides *dolus*, 54, 10. For inwite *propter dolos*, 72, 14. Mið inwite [mit fácne, A. S.] *dolo*, Mt. Kmbl. Lind. 26, 4. Gé on heortan hogedon inwit *in corde iniquitates operamini*, Ps. Th. 57, 2. Hió ðá inwit feala ýwdan on tungan *locuti sunt adversum me lingua dolosa*, 108, 2. His esnum inwit fremedan *dolum facerent in servos ejus*, 104, 21. Ða inwit and fácen hycgeaþ on heortan *qui cogitaverunt malitias in corde*, 139, 2. Hie sprecaþ fácen and inwit, Cd. 109; Th. 145, 31; Gen. 2414. Inwit syredon *they plotted evil*, Andr. Kmbl. 1220; An. 610. Hwǽr áhangen wæs waldend þurh inwit, Elen. Kmbl. 413; El. 207. [O. Sax. inwid: cf. Goth. inwindiþa *injustice*.]

inwid-. v. inwit-.

inwidda, inwit; *adj. Guileful, deceitful, evil, wicked, malicious :*—Gelpan ne þorfte eald inwidda [inwitta, MSS. B. C. inwuda MS. D.], *no cause to boast had he, old and crafty*, Chr. 937; Erl. 114, 12; Æðelst. 46. Swá se inwidda ofer ealne dæg dryhtguman síne drencte mid wíne *so the evil one [Holofernes] all through the day his men drenched with wine*, Judth. 10; Thw. 21, 20; Jud. 28. Ealle weleras inwiddæn *universa labia dolosa*, Ps. Spl. T. 11, 3. Wordum inwitum *with guileful words*, Cd. 229; Th. 310, 22; Sat. 731. [Cf. Goth. inwinds *unjust, perverse*.]

in-wise, an; *f. A condiment :*—Ðæt hit síe on ða onlícnesse geworht ðe senop biþ getemprod tó inwisan *that it may be made like mustard when it is mixed for a condiment*, L. M. 2, 6; Lchdm. ii. 184, 22.

inwit. v. inwid, inwidda.

inwit-feng, es; *m. A wily* or *malicious grasp*, Beo. Th. 2898; B. 1447.

inwit-flán, es; *m. A treacherous shaft*, Exon. 83 b; Th. 315, 27; Mód. 37.

inwit-full; *adj. Deceitful, guileful, malicious, evil :*—Inwitfull *dolosus, insidiosus, fraudulentus, callidus*, Wrt. Voc. ii. 141, 66. Ne mæg ðær inwitfull ǽnig geféran womscyldig mon *there may none guileful come,*

none guilty of sin, Cd. 45; Th. 58, 18; Gen. 498. From ðære inwit-fullan yflan tungan *a lingua dolosa*, Ps. Th. 119, 3. Hé áfylleþ ða inwitfullan word of his tungan *he causes deceitful words to fall away from his tongue* [cf. Ps. Th. 14, 3, non egit dolum in lingua sua], Blickl. Homl. 55, 16. Ðá geseah sigora waldend hwæt wæs monna mánes and ðæt hí wǽron inwitfulle *then saw, the Lord of victories what the wickedness of men was, and that they were full of deceit*, Cd. 64; Th. 77, 10; Gen. 1273. Synfulra and inwitfulra múþas *os peccatoris èt dolosi*, Ps. Th. 108, 1.

inwit-gǽst, es; *m. A guileful, evil guest*, Beo. Th. 5333; B. 2670.

inwit-gecynd, es; *n. A malicious, evil nature*, Salm. Kmbl. 660; Sal. 329.

inwit-gyren, e; *f. A treacherous snare :*—Forhýddon mé oferhýdge inwitgyrene *absconderunt superbi laqueos mihi*, Ps. Th. 139, 5.

inwit-hlemm, es; *m. A stroke treacherously* or *maliciously given*, Rood Kmbl. 93; Kr. 47.

inwit-hróf, es; *m. A deceitful, evil roof [the fire-drake's den]*, Beo. Th. 6238; B. 3123.

inwit-net, es; *n. A net of treachery* or *malice*, Beo. Th. 4340; B. 2167.

inwit-níþ, es; *m. Malicious, treacherous enmity*, Beo. Th. 3720; B. 1858: 3898; B. 1947: Hy. 3, 46; Hy. Grn. ii. 282, 46. [O. Sax. inwid-níð.]

inwit-rún, e; *f. Malicious, guileful counsel*, Exon. 74 b; Th. 279, 7; Jul. 610.

inwit-scear, es; *m. Slaughter effected by craft*, Beo. Th. 4949; B. 2478. [Cf. gúþ-scear.]

inwit-searo; *n. Malicious* or *treacherous artifice*, Beo. Th. 2206; B. 1101.

inwit-sorh; *gen.* -sorge; *f. Sorrow brought about by malice* or *guile*, Beo. Th. 1666; B. 831: 3477; B. 1736.

inwit-spell, es; *n. A tale of evil*, Cd. 94; Th. 122, 9; Gen. 2024.

inwit-stæf, es; *m. Evil, wickedness, malice;* nequitia, Ps. Th. 54, 15: 140, 5.

inwit-þanc, es; *m. Evil, malicious, deceitful thought* or *purpose*, Andr. Kmbl. 1339; An. 670: 1118; An. 559: Elen. Kmbl. 616; El. 308: Bt. Met. Fox 9, 16; Met. 9, 8: 27, 46; Met. 27, 23: Beo. Th. 1502; B. 749.

inwit-wrásen, e; *f. A chain of guile* or *malice*, Andr. Kmbl. 126; An. 63: 1892; An. 948.

in-wreón; *p.* -wráh; *pl.* -wrigon *To uncover, reveal :*—Ðú mé inwrige wyrda gerýno *thou hast revealed to me the mysteries of fate*, Elen. Kmbl. 1621; El. 813. v. on-wreón.

in-writting, e; *f. An inscription;* inscriptio, Mt. Kmbl. p. 4, 5.

in-wund, e; *f. An inward wound :*—Wið inwunde magan *for an inward wound of the stomach*, L. M. 2, 9; Lchdm. ii. 188, 11. [Cf. O. Frs. in-werdene *internal injury*.]

in-wuneness, e; *f. Persistence, perseverance;* instantia, Wrt. Voc. ii. 47, 41.

in-wunung, e; *f. Habitation, dwelling*, Lye.

in-ylfe, es; *n. A gut, bowel :* — Inelfe *intestinum*, Wrt. Voc. 65, 55. Inilve, 284, 2. Inelve *interamen*, 286, 60. Ðý læs ðæt innelfe útsíge *lest the matrix prolapse*, L. M. 3, 37; Lchdm. ii. 328, 25. Gif men síe innelfe úte . . . gedó ðæt innelfe on ðone man *if a man's bowel protrude . . . put the bowel into the man*, 3, 73; Lchdm. ii. 358, 23-5. Inelfe *viscera*, Wrt. Voc. 65, 32. Inilve, 285, 58. Sume ninnaþ hwelpes innylfe *some take a whelp's intestines*, L. Med. ex Quadr. 9, 5; Lchdm. i. 362, 7. [Icel. inn-yfli, -ylfi; *n. pl. entrails, bowels :* O. H. Ger. inn-uveli, -oveli *viscera*.]

Iob, es; *m. Job :*—Sum wer wæs geseten on ðam lande ðe is geháten Hus, his nama wæs Iob, Homl. Th. ii. 446, 10. Iobes dóhtra, 458, 32. Tó mínum þeówan Iobe, 456, 30. Be ðan eádigan were Iob, 446, 4.

Iob, es; *m. Jove, Jupiter :*—Iob Saturnes sunu, Bt. 35, 4; Fox, 162, 5. Ercules Iobes sunu, 16, 2; Fox 52, 34. Iobes templ, Nar. 37, 23. v. Iofes.

ioc. v. iuc, geoc.

Iofes, es; *m. Jove :*—Ðanc hafa ðú, Iofes, Ors. 4, 1; Bos. 77, 37. Hyra héhstan godes hús Iofeses, 4, 2; Bos. 79, 11. v. Iob.

Iól *Yule*, Chart. Th. 423, 5. v. Geól.

iór, es; *m. The name of the rune ✳; also of a fish, perhaps the eel :*— ✳ byþ eáfixa [sum] and ðeáh á brúceþ fódres on faldan *eel is a river-fish, and yet ever eats food on the ground*, Runic pm. 28; Kmbl. 345, 4. See Zacher's Das Gothische Alphabet, p. 26; Taylor's Greeks and Goths, pp. 97-8.

Iotas, Iutan; *pl. The Jutes :*—Ðá cómon ða men of þrim mégðum Germanie of Ald-Seaxum of Anglum of Iotum. Of Iotum cómon Cantwara and Wihtwara ðæt is seó mégð ðe nú eardaþ on Wiht and ðæt cyn on West Sexum ðe man nú git hǽt Iutna cyn *then came the men from three tribes of Germany, from old Saxons, from Angles, from Jutes. From the Jutes came the people of Kent and Wight, that is, the tribe that now lives in Wight and the race among the West Saxons that is to the present time called the Jutes' race*, Chr. 449; Erl. 13, 10-14. The Anglo-Saxon version of Bede, i. 15, has *Geat* for *Iot*, but in 4, 16 *Iutorum provincia* is rendered *Eota land*. See Grmm. Gesch. D. S. 511 sqq. [Icel. Iótas *Jutes*.]

iów, iówian, ιówih. v. eów, eówian, eówic.

ir; adj. Angry:—Yr on móde, Cd. 4; Th. 4, 33; Gen. 63. v. irre, ir-scipe.

Íra-land, es; n. Land of the Irish, Ireland:—Gewitan him ða Norþmenn Dyflen sécean eft Íraland [Yraland, hira land], Chr. 937; Th. 206, col. 2, l. 15; Æðelst. 56. In Ors. 1, 1; Swt. 19, 15, 16 Íraland is doubtful. In the Anglo-Saxon version of Bede's History Hibernia generally is rendered by Hibernia Scotta eáland. v. Ír-land.

Íras; pl. The Irish [v. Íra-land]:—Férde twelf geár bodiende betwux Yrum and Scottum and siððan ofer eal Angelcyn he went twelve years preaching among the Irish and Scotch, and afterwards over all England, Homl. Th. ii. 346, 35. But the people of Ireland are often spoken of as Scottas, e. g:—Pyhtas cóman ǽrost on norþ Ybernian up and ðǽr bǽdon Scottas ðæt hí ðér móston wunian, Chr. Erl. 3, 9. Scotta sum ðǽl gewát of Ybernian on Brittene, 18. Þrie Scottas cuómon tó Ælfréde of Hibernia, 891; Erl. 88, 5. So in Alfred's Orosius it is said Igbernia ðæt wé Scotland hátaþ, 1, 1; Swt. 24, 16. [Icel. Írar.]

íren, es; n. Iron, an iron weapon [cf. use of steel in modern English], a sword, blade:—Ðæt swurd, drihtlic íren, Beo. Th. 1788; B. 892. Gif ðæt gegangeþ ðæt ádl oððe íren nimeþ ealdor ðínne if it come to pass, that disease or sword take off thy prince, 3700; B. 1848. Mé sceal wǽpen niman, ord and íren, Byrht. Th. 139, 12. Áres and írenes ǽris et ferri, Bd. 1, 1; S. 473, 23, note. Heardes írenes grindlas gratings of hard iron, Cd. 19; Th. 24, 25; Gen. 383. Ðeáh hé wǽre mid írne ymbfangen, 224; Th. 297, 15; Sat. 513. Héht his sweord niman, leóflic íren, Beo. Th. 3622; B. 1809. His sweord, írena cyst, 1350; B. 673: 1609; B. 802. Bite írena, 4511; B. 2259. Íren ecgheard, Andr. Kmbl. 2363; An. 1183. [Icel. járn: Dan. jern: Swed. järn.] v. hring-íren; ísen, ísern.

íren; adj. Of iron, iron:—Ecg wæs íren the edge was of iron, Beo. Th. 2922; B. 1459: 5549; B. 2778. Hé hine hét áþenian on írenum bedde and hine cwicne hirstan he bade stretch him on an iron bed, and roast him alive, Shrn. 116, 2. Mid írenum gyrdum with iron rods, 115, 24: Salm. Kmbl. 55; Sal. 28: 942; Sal. 470. Scyttelas ýrenne hé forbræc vectes ferreos confregit, Ps. Spl. 106, 16. v. eal-íren, the following compounds, and ísen, ísern.

íren-bend, es; m. An iron bond or band:—Licgaþ mé ymbe írenbendas, Cd. 19; Th. 24, 2; Gen. 371. Írenbendum fæst, Beo. Th. 2001; B. 998. [Cf. Goth. eisarna-bandi.]

íren-byrne, an; f. An iron byrnie:—Námon írenbyrnan, heard swyrd hilted, and his helm, Beo. Th. 5965; B. 2986.

íren-gelóma, an; m. An iron implement:—Ða írengelóman ferramenta, Nar. 9, 19. v. gelóman.

íren-heard; adj. Iron-hard, Beo. Th. 2227; B. 1112.

íren-helm, es; m. An iron helmet:—Írenhelm [or íren helm; but cf. preceding compounds] cassis, Ælfc. Gl. 51; Som. 66, 14; Wrt. Voc. 35, 4. [Cf. Icel. járn-hattr a kind of helmet.]

íren-þreát, es; m. A band having iron armour, Beo. Th. 666; B. 330.

ire-þweorh; adj. Having the mind perverted by rage, Exon. 67 a; Th. 248, 3; Jul. 90.

irfan; p. de To inherit:—Yrfan hí swá hí wyrðe witan let the land devolve upon such as they know to be worthy or entitled [v. wyrðe], Chart. Th. 578, 9. v. [?] Cod. Dipl. Kmbl. i. xxxiii–v on the leases of church lands for lives, in which such phrases as the following occur:—His dæg forgeaf, and æfter his dæg twám yrfeweardum. Such lives were sometimes named in the instrument setting forth the grant. [O. Frs. ervia to inherit: O. L. Ger. gi-ervan hereditare: Icel. erfa to honour with a funeral feast; mod. to inherit: O. H. Ger. erbet hæreditabit: Ger. erben.]

irfe, ierfe, yrfe, es; n. Inheritance, property:—Gewriten yrfe legatum, Ælfc. Gl. 13; Som. 57, 96; Wrt. Voc. 20, 37. Ungewriten yrfe intestata hereditas, Som. 57, 101; Wrt. Voc. 20, 41. Yrfe drihtnes hereditas Domini, Ps. Spl. 126, 4. Yrfe sceal gedǽled deádes monnes a dead man's property must be divided, Exon. 90 a; Th. 338, 18; Gen. Ex. 80. Ne wilna ðú ðínes néhstan ierfes mid unrihte covet thou not thy neighbour's goods, L. Alf. 9; Th. i. 44, 21. Þolige his wǽpna and his ierfes let him forfeit his weapons and his property, L. Alf. pol. 1; Th. i. 60, 14. Gif hwá gefeohte on cyninges húse sie hé scyldig ealles his ierfes, L. In. 6; Th. i. 106, 3. Ðonne is riht ðæt heó sý healfes yrfes wyrðe and ealles gif hý cild gemǽne hæbban then is it right that she be entitled to half the property, and to all if they have children together, L. Edm. B. 4; Th. i. 254, 15. Ðú ðínes yrfes æðele gyrde álýsdest liberasti virgam hæreditatis tuæ, Ps. Th. 73, 3. Malalehel wæs æfter iarede yrfes hyrde fæder on láste Mahalaleel was after Irad the guardian of the heritage, in succession to his father, Cd. 52; Th. 65, 17; Gen. 1067. Nelle ic from mínum hláforde ne from mínum wífe ne from mínum bearne ne from mínum ierfe I will not go from my lord, nor from my wife, nor from my child, nor from my goods, L. Alf. 11; Th. i. 46, 9. Ne sylle gé ðæt land on éce yrfe terra non vendetur in perpetuum, Lev. 25, 23. His yrfe forhogode hæreditatem suam sprevit, Ps. Th. 77, 62. Gif hé wite hwá ðæs deádan ierfe hæbbe tiéme ðonne tó ðam ierfe and bidde ða hond ðe ðæt ierfe hafaþ ðæt hé him gedó ðone ceáp unbeceásne oððe gecýðe ðæt se

deáda nǽfre ðæt ierfe áhte if he know who has the property of the dead, let him then vouch the property to warranty, and demand of the hand which has that property, that he make the chattel uncontestable to him; or prove that the dead man never owned that property, L. In. 53; Th. i. 136, 4–8. Him on láste heóld land and yrfe malalehel, Cd. 58; Th. 71, 8; Gen. 1167. [Under the single form yrfe two words seem to be comprised; the one just given, also written ærfe, erfe, and another, which would correspond with a Gothic aurbi, connected with orf, with the meaning cattle. With the former may be compared Goth. arbi; n. heritage, inheritance: O. Sax. erbi; n: O. L. Ger. ervi; n. hæreditas: O. Frs. erve; n: O. H. Ger. arbi, erbi, arpi; n. possessio: Ger. erbe; n: Icel. arfr; m. inheritance; erfð; f. inheritance. See yrfe, orf]. v. sundor-irfe; and Grmm. R. A. pp. 466–7; 565.

irfe-béc; pl. f. A will, testament:—Uncwedene yrfebéc ruptum testamentum: forswiged yrfebéc suppressum testamentum: underne yrfebéc nuncupatio: samhíwna yrfebéc jus liberorum, Ælfc. Gl. 13; Som. 57, 102–8; Wrt. Voc. 20, 42–6: ii. 49, 14. Áwégune yrfebéc irritum testamentum: unárlíce yrfebéc inofficiosum testamentum, 49, 15–18.

irfe-first, es; m. A delay before entering upon an inheritance; cretio, Ælfc. Gl. 13; Som. 57, 106; Wrt. Voc. 20, 44.

irfe-gedál, es; n. A division of an inheritance or property:—Yrfegedál familiæ erciscundæ, Ælfc. Gl. 13; Som. 57, 109; Wrt. Voc. 20, 47. Yrfegedál familia erciscundæ, quia ærciscunda enim apud veteres divisio nuncupabatur, ii. 39, 26.

irfe-geflit, es; n. A dispute about inheritance: — Ðá gehýrde wé manegu yrfegeflitu then did we hear of many disputes about the inheritance, Chart. Th. 486, 12.

irfe-gewrit, es; n. Writing concerning an inheritance, a will, testament:—Ac hit gelamp ðæt Æðelréd cingc gefór ðá ne cýðde mé nán mann nán yrfegewrit ne náne gewitnesse ðæt hit ǽnig óðer wǽre bútan swá wit on gewitnesse ǽr gecwǽdon but it happened that king Ethelred died; then no man made known to me any testament or any witness that it was any other than as we two before with witness agreed, Chart Th. 486, 7. On ðam yrfegewrite in the testament, 32.

irfe-hand, a; f. One who manages the estate of a deceased person, an administrator [?]:—Se mann se tó londe fóe ágefe hire erfehonda xiii pund pendingæ and heó forgifeþ xv pund for ðý ðe mon ðǽs feorme ðý soel gelǽste let the man who succeeds to the land give to her administrator thirteen pounds of pennies; and he will give fifteen pounds, in order that this refection may be the better provided, Chart. Th. 474, 9. v. hand.

irfe-láf, e; f. An hereditary relic, heirloom, what is left of an inheritance, inheritance, heir:—Hé fédeþ folc Iacobes and Israhéla yrfeláfe pascere Jacob servum suum, et Israel hæreditatem suam, Ps. Th. 77, 70. Æghwylcum máððum gesealde yrfeláfe to each he gave a gift, an heirloom, Beo. Th. 2110; B. 1053. Hé bátwearde swurd gesealde, ðæt hé syððan wæs máððum ðý weordra, yrfeláfe, 3810; B. 1903. Wolde líge gesyllan his swǽsne sunu ángan ofer eorþan yrfeláfe he [Abraham] was ready to give to the flame his dear son, the only heir that was left him on earth, Cd. 162; Th. 203, 14; Exod. 403.

irfe-land, es; n. Land that passes as an inheritance, heritable land:—Ic cýðo hú mín willa is ðæt mín ærfelond fére ðe ic gebohte on éce ærfe I declare how my will is that my heritable land shall go, that I bought in perpetual inheritance, Chart. Th. 476, 12. Hie dydon mín land him selfum tó ierfelonde dederunt terram meam sibi in hereditatem, Past. 50, 2; Swt. 387, 30. Gebletsa ðín yrfeland benedic hæreditati tuæ, Ps. Th. 27, 10. Sealde heora eorþan on yrfeland dedit terram eorum hæreditatem, 135, 22.

irfe-numa, an; m. One who takes an inheritance, an heir:—Ðes and ðeós yrfenuma hic et hæc heres, Ælfc. Gr. 6; Som. 5, 33. Mín inbyrdling biþ mín yrfenuma . . . Ne byþ ðes ðín yrfenuma ac ðone ðú háfast tó yrfenuman ðe of ðé sylfum cymþ vernaculus meus heres meus erit . . . Non erit hic heres tuus, sed qui egredietur de utero tuo, ipsum habebis heredem, Gen. 15, 3–4: 21, 10. Ðes ys yrfenuma hic est heres, Mt. Kmbl. 21, 38: Mk. Skt. 12, 7. Hit wǽre geþuht ðæs ðe máre gemynd ðæs fæder, ðá ðá se sunu, his yrfenuma, wæs gecíged ðæs fæder naman, Homl. Th. i. 478, 11. Fæderas and móddru bestandaþ heora bearna líc and heora yrfenuman him sylfum tó forwyrde forestæppaþ fathers and mothers stand about the corpses of their children, and their heirs precede them to destruction, ii. 124, 18. Se ðe sitte uncwydd and uncrafod on his áre on life ðæt nán man on his yrfenuman ne spece æfter his dæge he who sits without contest or claim on his property during life, that no one bring an action against his heir after his day, L. Eth. iii. 14; Th. i. 298, 10. Gif hwá tó deádan týme, búton hé yruenoman hæbbe ðe hit clǽnsie, ii. 9; Th. i. 290, 9. Gif se bónda ǽr hé deád wǽre beclypod wǽre ðonne andwyrdan ða yrfenuman swá hé sylf sceolde ðeáh hé líf hæfde if the man of the house before his death were cited; then let the heirs answer as he himself would have had to do if he had lived, L. C. S 73; Th. i. 416, 1. Se mann ðe on fyrdunge ætforan his hláforde fealle, beón ða heregeata forgyfene, and fón ða yrfenuman tó lande and tó ǽhtan, 78; Th. i. 420, 16. [Goth. arbi-numja an heir: O. H. Ger. arpi-, erpi-nomo hæres: Ger. erb-nehmer: cf. Icel. arf-takari, arf-taki, arf-tökumaðr an heir.]

irfe-stól, es; *m. An hereditary seat :*—Se burgstede, eádges yrfestól, Exon. 52 a; Th. 181, 14; Gú. 1293. Eafora chuses yrfestóle weóld, Cd. 79; Th. 98, 13; Gen. 1629. Ne þearf ic yrfestól eaforan bytlian ǽnegum minra . . . ne sealdest ðú mé sunu *I need not build an hereditary seat for any descendant of mine . . . thou hast not given me a son,* 99; Th. 131, 14; Gen. 2176.

irfe-weard, es; *m. The guardian of an inheritance, an heir, possessor of a property :*—Hér ys se yrfeweard [erfuard, Lind: erfeword, Rush.] *hic est heres,* Lk. Skt. 20, 14. Ðæt mé gifeðe ǽnig yrfeweard æfter wurde *if any heir to follow me had been granted me,* Beo. Th. 5455; B. 2731: Cd. 83; Th. 103, 33; Gen. 1727. Óðres ne gýmeþ tó gebídanne yrfeweardas *cares not to await another heir,* Beo. Th. 4897; B. 2453. Wæs swá mycel mancwealm ðæt manige land binnan ðære byrig wǽran bútan ǽlcum yrfewearde *there was so great a pestilence that many lands within the city were without any to inherit them,* Ors. 5, 2; Bos. 102, 13. Ús is swíðe uncúþ hwæt úre yrfeweardas and lástweardas dón willon æfter úrum life *we are very ignorant of what our heirs and successors will do after our life,* Blickl. Homl. 51, 35. Ða ðe God bletsiaþ beóþ eorþan yrfeweardas *benedicentes eum possidebunt terram,* Ps. Th. 36, 21. Geréfa mín mynteþ ðæt mé æfter síe eaforan síne yrfeweardas *my steward supposes that after me his children shall be heirs,* Cd. 100; Th. 131, 29; Gen. 2183. Hwæðer freá wille ǽnigne ðé yrfeweardum on woruld lǽtan, 101; Th. 134, 26; Gen. 2230. Ða sylfan wilniaþ him tó yrfeweardum tó habbanne *ipsos habere heredes quærunt,* Bd. 1, 27; S. 490, 18. Ic landes sumne dǽl sumum wífe hiere dæg forgæaf and æfter hiere dæge twám yrfeweardum *I granted a certain portion of land to a certain woman for her life, and after her death to be held for two other lives,* Cod. Dipl. Kmbl. iii. 5, 10. See i. xxxiv. Forlét hé ðæs hwílenlecan ríces yrfeweardas his suna þrý *tres suos filios regni temporalis heredes reliquit,* Bd. 2, 5; S. 507, 8. [*Gen. and Ex.* er(f)ward: *O. Sax.* erƀi-ward: *Icel.* (poët.) arf-vörðr *an heir.*]

irfe-, irf-weardness, e; *f. An inheritance :*—Yrfeweardnes *hereditas,* Ælfc. Gl. 13; Som. 57, 95; Wrt. Voc. 20, 36. Drihtnes dǽl wæs his folc and Iacob his yrfeweardnis *pars domini populus ejus, Jacob funiculus hereditatis ejus,* Deut. 32, 9. God cwæþ ðæt hé sylf wǽre heora yrfweardnys, Homl. Th. ii. 224, 7. Ðonne biþ úre seó yrfeweardnes *nostra erit hereditas,* Mk. Skt. 12, 7.

irfe-, irf-weardian; *p.* ode *To inherit, possess an inheritance :*—Ðú yrfweardast on eallum þeódum *tu hæreditabis in omnibus gentibus,* Ps. Spl. 81, 7. Hí yrfweardiaþ eorþan *hæreditabunt terram,* 36, 11. Ðæt ðú yrfweardige eorþan, 36. DER. be-irfeweardian.

irfeweard-wrítere, es; *m. One who specifies his heir in writing, a testator :*—Yrfeweardwrítere *legatarius,* Ælfc. Gl. 13; Som. 57, 99; Wrt. Voc. 20, 39.

irfe-wrítend, es; *m. One who writes concerning the disposition of his property, one who makes a will :*—Yrfewrítend *testator,* Ælfc. Gl. 13; Som. 57, 100; Wrt. Voc. 20, 40.

irf-. v. irfe-.

irgþ, e; irgþu, irgþo; *indecl* ; *f. Sluggishness, cowardice, timorousness, pusillanimity :*—Wé witon georne ðæt wé for iergþe náðer ne durran ne swá feor friþ gesécan ne furþon hie selfe æt hám hie werian *we know well that they from cowardice dare neither seek peace at such a distance, nor even defend themselves at home,* Ors. 3, 9; Swt. 136, 28. For eówre forhtnysse and yrhþe ðe eów eglaþ *propter cordis tui formidinem, qua terreberis,* Deut. 28, 67. Se man ðe ætfleó fram his hláforde oððe fram his geféran for his yrhþe sý hit on scipfyrde sý hit on landfyrde þolige ealles ðæs ðe hé áge and his ágenes feores *the man that flies from his lord, or from his comrade, from cowardice, be it on an expedition by sea or by land, let him lose all that he owns and his own life,* L. C. S. 78; Th. i. 420, 8. Ða héton hí secgan ðysses landes wæstmbǽrnysse and Brytta yrgþo *nunciatum est simul et insulæ fertilitas, ac segnitia Brittonum,* Bd. 1, 15; S. 483, 15. Þurh lyðre yrhþe Godes bydela ðe clumedon mid ceaflum ðær hí scoldon clipian *through the vile sluggishness of God's messengers, who mumbled with their mouths when they should have cried aloud,* Swt. A. S. Rdr. 111, 202. [*Laym.* Arður, ærhðe bideled, 23546: *O. and N.* he for arehþe hit ne forlete, 404: *O. H. Ger.* argida *hebitudo, ignavia.*] v. earg.

irhþ. v. irgþ.

Iringes weg *via secta,* Wrt. Voc. ii. 123, 50. v. Grmm. D. M. 332.

Ír-land, es; *n. Ireland :*—Ðrie Scottas cuómon tó Ælfréde cyninge of Ýrlande, Chr. 891; Erl. 88, 6, note. Tó Írlande, 918; Erl. 104, 15: 1051; Erl. 176, 18. Se preóst cwæþ ðæt án we wǽre on Írlande geléred, Swt. A. S. Rdr. 101, 200. Hé férde geond eal Yrrland, Homl. Th. ii. 346, 28. v. Íra-land.

ir-líc; *adj. Angry :*—Hé swíðe irlícum andwlitan beseah tó ðam iungan cnyhte *he looked at the young man with a very angry countenance,* Th. Ap. 4, 6: 5, 3.

irman; *p.* de *To make miserable* or *wretched, to afflict, vex :*—Ðá ongunnan twá þeóda Pyhtas norþan and Scottas westan hí onwinnan and heora æhta niman and hergian and hí fela geára yrmdon and hýndon *then began two peoples, the Picts from the north, the Scots from the*

west, to attack them, and to take their possessions, and to harry, and afflicted and vexed them many years ; Brittania denique subito duabus gentibus transmarinis vehementer sævis, Scottorum a circio, Pictorum ab aquilone, multos stupet gemitque per annos, Bd. 1, 12 ; S. 480, 24. Hé hæfde him tó gamene hú hé eorþcyninges yrmde and cwelmde *he* [*Nero*] *made it his sport, how he could vex and torment the kings of this earth,* Bt. Met. Fox 9, 94; Met. 9, 47. Ic mæg sleán and ierman míne heáfodgemæccan *I can beat and vex my companions,* Past. 17, 8; Swt. 121, 12. [Cf. *O. H. Ger.* ki-ermit uuerdemes *aporiamur,* Grff. i. 423.] v. for-, ge-yrman.

irmen, yrmen; *adj.* A word occurring mostly as a prefix with the idea of *greatness, universality.* In the following passages it occurs independently :—Faraþ geond ealne yrmenne grund *go through the whole earth,* Exon. 14 b; Th. 30, 18; Cri. 481. Ofer ealne yrmenne grund, 66 a; Th. 213, 14; Jul. 10. [*O. Sax.* irmin-: *Icel.* jörmun-, e. g. jörmungrund *the earth.* See Grmm. D. M. 104-7: 325, sqq.] v. eormen-.

irmen-þeóde; *pl. The peoples of the earth :*—Bringeþ Agustus yrmenþeódum hláfmæssan dæg *August brings Lammas day to all the nations of earth,* Menol. Fox 276; Men. 139. [Cf. *O. Sax.* ik allun skal irminthiodun dómôs adélian *I shall judge all the nations of the world,* Hel.3316.]

irming, es; *m. A poor, mean, wretched, miserable person, a wretch :*—Ic eom ána forlǽten yrming *unicus et pauper sum ego,* Ps. Th. 24, 14. Ic eom yrming and þearfa *ego egenus et pauper sum,* 39, 20. Ðú eart ðé godes yrming *as to thee, thou art God's pauper,* Exon. 36 b; Th. 118, 22; Gú. 243. Betere is ðé ðæt ðé sceamige nú hér beforan mé ánum yrmingce ðonne eft beforan Gode on ðam mycelan dóme *melius est tibi nunc hic coram me solo misero pudefieri, quam posthac coram Deo in magno judicio,* L. Ecg. C. prm; Th. ii. 132, 20. Ðá ðá iermingas ðe ðǽr tó láfe wurdon út of ðǽm holan crupon ðe heó on lutedan *when the wretched people that remained crept out of the holes that they had lurked in,* Ors. 2, 8; Swt. 92, 29. Se ðe ǽnigne ðissa ierminga besuícþ *qui scandalizaverit unum de pusillis istis,* Past. 2, 2; Swt. 30, 17. Ðæt is sió friþstów and sió frófor án eallra yrminga æfter ðissum weoruldgeswincum *that alone is the asylum and the comfort of all the wretched after these labours in the world,* Bt. Met. Fox 21, 33; Met. 21, 17. [Makede him *erming* þer he was er king, O. E. Homl. 2, 62: þu *erming* þu wrecche gost, O. and N. 1111: Agag þe king, þu ært an *ærming,* Laym. 16690: *Icel.* armingi *a poor fellow, a wretch : O. H. Ger.* arming *pauper.*] v. earming, erming.

irmþ, e; irmþu, irmþo; *indecl. f. Poverty, penury, misery, wretchedness, calamity, distress, disorder :*—Yrmþ *miseria,* Ælfc. Gr. 33; Som. 37, 24. Nis ðǽr on ðam londe yldu ne yrmþu *in that land there is not age or misery,* Exon. 56 b; Th. 201, 6; Ph. 52: 64 b; Th. 238, 34; Ph. 614. Him gewearþ yrmþu tó ealdre *upon them* [*Adam and Eve*] *came misery for ever,* 73 a; Th. 272, 24; Jul. 504: 119 a; Th. 457, 15; Hy. 4, 84. Ne biþ him hyra yrmþu án tó wíte ac ðara óðerra eád tó sorgum *nor alone shall their own misery be torment, but the bliss of the others shall be a grief,* 26 b; Th. 79, 19; Cri. 1293. For yrmþe unspédig[ra] *propter miseriam inopum,* Ps. Spl. 11, 5. Ðeós of hyre yrmþe eall ðæt heó hæfde sealde *hæc de pænuria sua omnia quæ habuit misit,* Mk. Skt. 12, 44. Ðonne sende hé him fultum þurh sumne déman ðe hí álísde of heora yrmþe *then he sent them help by some judge, who released them from their misery,* Ælfc. T. Grn. 6, 26. Wið ðæs migðan yrmþe *for disorder of the urine,* Herb. 163, 3; Lchdm. i. 292, 7. Ic ádreáh feala yrmþa ofer eorþan *I suffered many miseries on earth,* Andr. Kmbl. 1939; An. 972: Exon. 26 b; Th. 78, 5; Cri. 1269. Ic eom gefylled mid iermþum *saturatus sum miseria,* Past. 36, 5; Swt. 253, 8. Seðe hine fram swá monigum yrmþum and teónum generede *qui se tot ac tantis calamitatibus ereptum,* Bd. 2, 12; S. 514, 19. Ðæt hí ðám yrmþum â ne wiðstanden *in miseriis non subsistent,* Ps. Th. 139, 10. Gif hé ðære tíde yrmþo beswicode *si temporis illius ærumnis exemptus,* Bd. 2, 12; S. 512, 36. Ðus hí heora yrmþo árehton *ita suas calamitates explicant,* 1, 13; S. 481, 43. Disse worlde yrmþa *the miseries of this world,* Blickl. Homl. 61, 3. Yrmþo, 203, 20. Dreógan yrmþu bútan ende *to suffer endless misery,* Elen. Kmbl. 1902; El. 953. Ðú scealt ēcan ðíne yrmþu, Andr. Kmbl. 2767; An. 1386. Yrmþo, 2381; An. 1192. Ides yrmþe gemunde *the woman remembered her misery,* Beo. Th. 2523; B. 1259. Hé ða yrmþu oncyrde ðe wé ǽr drugon *he averted the miseries that before we suffered,* Exon. 16 b; Th. 38, 29; Cri. 614. [*O. E. Homl.* ermðe *poverty : Laym.* ærmðe *misery : O. H. Ger.* armida *paupertas, inopia, penuria.*] v. ermþu, earmþu, eormþu, weoruld-irmþu.

irnan; *p.* arn, *pl.* urnon; *pp.* urnen *To run :*—Ic yrne *cucurri,* Ps. Spl. T. 118, 32. Seó eá Danai irnþ ðonan súþryhte *the river Don runs thence due south,* Ors. 1, 1; Swt. 8, 17. Æspringe irneþ wið his eardes. *ipse ad armentum cucurrit,* Gen. 18, 7. Ðonne orn hé eft inn tó ðæm temple *ad templum recurrit,* Past. 16, 3; Swt. 103, 4. Ðú urne mid him *simul currebas cum eo,* Ps. Th. 49, 19. Ðá urnon him tógenes twegen ðe hæfdon deófolseócnesse *occurrerunt ei duo habentes dæmonia,* Mt. Kmbl. 8, 28. Gangende ðǽr urnon, Mk. Skt. 6, 33: Jn. Skt. 20, 4. Tó ðam ylcan ryne ðe hié ǽr urnon, Bt. 21; Fox 74, 12. Ðæt hí mægen iernan and fleón tó ðæs láreówes móde *ut ad pastoris mentem recurrant,* Past.

16. 4; Swt. 103, 22. Hé sceal yrnan forþ *he must run forth*, Exon. 128 b; Th. 494, 9; Rä. 82, 5. Seó [eá] is irnende of norþdǽle, Ors. 1, 1; Swt. 8, 15. Ac hí forweorþan wǽtere gelícost ðonne hit yrnende eorþe forswelgeþ *ad nihilum devenient, velut aqua decurrens*, Ps. Th. 57, 6. Óþ ðæt wintra biþ þúsend urnen *until a thousand years are passed*, Exon. 61 a; Th. 223, 23; Ph. 364. DER. á-, be-, ge-, geond-, ofer-, on-, óþ-, tó-, þurh-, under-, up-, ymb-irnan. v. rinnan.

irnere. v. fore-irnere.

irre, es; *n*. *Anger, wrath, ire, rage*:—Ðonne tyht hie ðæt ierre [Cott. MS. irre] ðæt hie wealwiaþ on ða wédenheortnesse . . . Ðonne ðæt ierre æfþ anwald ðæs monnes hé self nát hwæt hé on ðæt irre déþ *impellente ira in mentis vesaniam devolvuntur . . . Quos cum furor agit in præceps, ignorant quidquid irati faciunt*, Past. 40, 1; Swt. 289, 5–10. Godes yrre ys ofer hig *egressa est ira a domino*, Num. 18, 46. Nú is gefylled ðæt mycelle hátheort and ðæt mycelle yrre ðyses ealdermannes *now is completed the great rage and anger of this ruler*, Blickl. Homl. 151, 11. Síe ǽlc monn lætt tó iorre iorra forðon weres sóþfæst godes ne giwyrcaþ *sit omnis homo tardus ad iram; ira enim viri justitiam Dei non operatur*, Rtl. 28, 21: 40, 35: 41, 3. Seó gesceádwísnes sceal wealdan ǽgðer ge ðære wilnunga ge ðæs yrres *reason must rule both desire and anger*, Bt. 33, 4; Fox 132, 9. Hé him weg worhte wráðan yrres *viam fecit semitæ iræ suæ*, Ps. Th. 77, 50. Ic bidde ðé, hláford, ðæt ic móte bútan yrre wið ðé sprecan *oro, domine mi, loquatur servus tuus verbum in auribus tuis et ne irascaris*, Gen. 44, 18. Wurdon mé on yrre yfele and hefige *in ira molesti erant mihi*, Ps. Th. 54, 3. Ðæt gé fleón fram ðam tówerdan yrre *fugite a ventura ira*, Lk. Skt. 3, 7. Ðá cwæþ se hláford mid yrre *tunc iratus paterfamilias dixit*, 14, 21. Mid miclum wylme and yrre onstyred *nimio furore commotus*, Bd. 1, 7; S. 477, 41. Ðá wæs hé mid yrre swíðlíce onstyred, Blickl. Homl. 199, 16. Ic ondréd his graman and his yrre *timui indignationem et iram illius*, Deut. 9, 19. Ágeót ofer hí ðín ðæt grame yrre *effunde super eos iram tuam*, Ps. Th. 68, 25. Ða hine on yrre gebringaþ *qui in ira provocant*, 65, 6. Ðé læs gé habban godes yrre *ne super omnem coetum oriatur indignatio*, Lev. 10, 6. Hé gearwe wiste ðæt hie godes yrre habban sceoldon, Cd. 33; Th. 43, 24; Gen. 695: Exon. 61 b; Th. 226, 20; Ph. 408. Godes yrre bær *the wrath of God was upon him*, Beo. Th. 1427; B. 711. [Godess irre iss upponn himm, Orm. 18000: O. E. Homl. A. R. eorre: *Reliq. Antiq.* urre.] v. eorre, *and next word*.

irre, yrre; *adj*. **I.** *Gone astray, wandering, confused, perverse, depraved*:—Ðæt wæs earfoðcynn yrre and réðe *genus pravum et peramarum*, Ps. Th. 77, 10. Óþ ðæt his eáge biþ æfþancum ful yrre geworden *until his eye is filled with evil thoughts and gone astray*, Salm. Kmbl. 994; Sal. 498. Sumum méces ecg yrrum ealowósan ealdor óþþringeþ *the edge of the sword crushes the life out of one, confused* [or *angry?*] *and mad with drink*, Exon. 87 b; Th. 330, 10; Vy. 49. Ealle synt yrre ða ðe unwíse heora heortan hige healdaþ mid dysige *turbati sunt omnes insipientes corde*, Ps. Th. 75, 4. **II.** *angry, enraged, wrathful, indignant*:—And ierre hé hwearf ðonan *and he went away in a rage*, Chr. 584; Erl. 18, 25. Iorra *iratus*, Rtl. 179, 36. Hwí eart ðú yrre *quare iratus es?* Gen. 4, 6. Se cyning wæs yrre wið mé, 41, 10. Hé wæs mé yrre, Deut. 1, 37. Ðá wearp yrre god and ðam werode wráþ, Cd. 2; Th. 3, 12; Gen. 34. Ne hine nǽnig man yrne ne grammódne ne funde *nor did any man find him angry or cruel*, Blickl. Homl. 223, 33. Þurh yrne hyge *with cr... purpose*, Exon. 16 b; Th. 36, 10; Cri. 620. Hé hine on yrre mód gebrohtan *exacerbaverunt eum*, Ps. Th. 77, 40. Ða irran [Cott. MS. ierran] nyton hwæt hie on him selfum habbaþ and eác ðætte wierse is ðætte hie ful oft wénaþ ðæt hie hierre [Cott. MS. ierre] síe ryhtwíslíc anda *ignorant quidquid a semetipsis patiuntur irati; nonnunquam vero, quod est gravius, iræ suæ stimulum justitiæ zelum putant*, Past. 40, 1; Swt. 289, 10. Hie wǽron tó ðon hátheortlíce yrre ðæt hie woldan ðone cásere cwicenne forbærnan *they were so furiously enraged, that they wanted to burn the emperor alive*, Blickl. Homl. 191, 11. Yrre wǽron begen réðe *angry were both and fierce*, Beo. Th. 1543; B. 769. [Ford wende þe eorl ire [2nd MS. yr] on his mode, Laym. 18597: þe eorre Demare *iratus Judex*, A. R. 304, 24: Goth. airzis wisan *or* wairþan *to go astray, err*; airzei, airziþa *error*; airzjan *to lead astray*: O. Sax. irri *angry*; irrian *to disturb, confuse*: O. L. Ger. irrôn *errare, commovere*: O. H. Ger. irri *vagus, lascivus*; irre isn *errare*; irra-heit *error*; irrado *impedimentum*; irran *impedire, confundere*; irrôn *errare, apostatare*: Ger. irre *confused, wandering*; irren *to err, go astray*. Cf. irsian, *and see* Diefenbach i. 21: Grff. i. 449 sqq.] v. eorre.

irre-mód; *adj*. *Of angry mood, angry-minded*:—Eode yrremód, him of eágum stód líge gelícost leóht unfæger, Beo. Th. 1456; B. 726.

irre-weorc, es; *n*. *A work undertaken in anger*:—Engla drihten wile uppe heonan sáwla lǽdan and wé seoddan á ðæs yrreweorces hénþo geþoliaþ *the Lord of angels will up from hence lead souls, and we ever after shall suffer the humiliation of that angry feat* [*the harrowing of Hell*], Cd. 222; Th. 289, 17; Sat. 399.

irringa, irrenga; *adv*. *Angrily, in anger*:—Be ðæm ilcan hé cwæþ eft ierrenga *hinc iterum iratus dicit*, Past. 56, 7; Swt. 435, 11. Ðá tó evan god yrringa spræc, Cd. 43; Th. 56, 27; Gen. 918. Seó beó sceal losian ðonne heó hwæt yrringa stingþ *the bee shall perish when she stings any-*

thing in anger, Bt. 31, 2; Fox 112, 26: Bt. Met. Fox 18, 13; Met. 18, 7. Yrrenga, 26, 167; Met. 26, 84. Se brǽda sǽ of clomme bræc up yrringa on eorþan fæðm *the broad sea from durance broke up angrily on to earth's bosom*, Exon. 24 b; Th. 70, 31; Cri. 1147. Gé mec yrringa up gelǽddon ðæt ic of lyfte londa getimbru geseón meahte, 39 b; Th. 131, 13; Gú. 455. Hé yrringa slóh *he angrily smote*, Beo. Th. 3135; B. 1565: 5921; B. 2964. v. eorringa.

ir-scipe, es; *m*. *Anger*:—Æfter mycelnes[se] his irscipes *secundum multitudinem iræ suæ*, Ps. Lamb. second 9, 4.

irsian; *p*. ode. **I.** *to be angry, to rage*:—Hú lange yrsast ðú on ðínes esnes gebed *quousque irasceris in orationem servi tui*, Ps. Th. 79, 5. Synfull yrsaþ *peccator irascetur*, 111, 9. Ðonne ús ðara manna mód yrsade and ús wiðerwearde wǽron *cum irasceretur animus eorum adversum nos*, 123, 3. Swá him yrsade se for ealle spræc feónda mengu *so did he, who spake for all the multitude of fiends, rage against him* [*Guthlac*], Exon. 35 a; Th. 114, 11; Gú. 171. Moises ðá yrsode and áxode *iratusque Moyses ait*, Num. 31, 14. His gebróðru yrsodon swíðe wið hine *invidebant ei fratres sui*, Gen. 37, 11. Ne yrsa ðú wið mé, Nar. 43, 7. Yrsiaþ *irascimini*, Ps. Lamb. 4, 5. Ic bidde ðæt ðú ne yrsie *obsecro ne irascaris*, Gen. 18, 32. Yrre is ðære sáwle forgifen tó ðý ðæt heó yrsige ongeán leahtres *anger is given to the soul that it may be angry against vice*, Homl. Skt. 1, 104. Ðæt ðe hió mid ryhte irsian sceall *that with which rightly it must be angry*, Past. 40, 4; Swt. 293, 13. Ðær ðær ðú neóde irsian scyle gemetiga ðæt ðeáh *in case you needs must be angry, still be moderate*, Prov. Kmbl. 24. Ûþwitan secgaþ ðæt sió sáwul hæbbe þrió gecynd án is ðæt heó biþ wilnigende óðer ðæt hió biþ irsiende þridde ðæt hió biþ gesceádwís *philosophers say that the soul hath three natures, one is that it desires, the second that it is angry, the third that it is rational*, Bt. 33, 4; Fox 132, 4. Ðæt irsigende mód hé gegremeþ and wierse ierre [Cott. MS. irre] hé ástyreþ *irati animus ad deteriora provocatur*, Past. 10, 3; Swt. 63, 13. Hwæthwugu biþ betweoh ðǽm irsiendan and ðǽm ungeþyldgan . . . ða iersigendan hit tó getióþ ðæt ðætte hie eáþe bútan bión meahton *in hoc ab impatientibus iracundi differunt . . . isti, quæ tolerentur, important*, 40, 4; Swt. 293, 15. Ða Iudéiscan yrsigende cwǽdon tó Criste *the Jews being angry said to Christ*, Homl. Th. ii. 236, 4. **II.** *to make angry, to anger, provoke*:—Hí yrsodon moyses *irritaverunt Moysen*, Ps. Spl. 105, 16.

irsigend-líc; *adj*. *Capable of anger*:—Ûþwytan secgaþ ðæt ðære sáwle gecynd is þrýfeald. Án dǽl is on hire gewylnigendlíc óðer yrsigendlíc þrydde gesceádwíslíc *philosophers say that the nature of the soul is threefold. There is one part in her capable of desire, a second capable of anger, a third is rational* [cf. Bt. 33, 4; Fox 132, 4], Homl. Skt. 1, 97.

irsung, e; *f*. *Anger, readiness to anger, irascibility*:—Twá ðara gecyndu habbaþ nétenu swá same swá men óðer ðara is wilnung óðer is irsung *two of those natures beasts have the same as men, one of them is desire, the other is anger*, Bt. 33, 4; Fox 132, 6. Yrsung, Bt. Met. Fox 20, 371; Met. 20, 185. Oft ungemetlícu irsung biþ gelícet ðæt menn wénaþ ðæt hit síe ryhtwíslíc anda *sæpe effrenata ira spiritalis zeli virtus æstimatur*, Past. 20, 1; Swt. 149, 11. Sió gesceádwísnes sceal on gehwelcum waldan semle irsunge [cf. wealdan ðæs yrres, Fox 132, 10], Bt. Met. Fox 20, 397; Met. 20, 199. Of irsunge wyxt seófung and of ðære geþwǽrnesse lufu *from anger grows sighing, and from gentleness love*, Prov. Kmbl. 23. Gé yldran ne sceolan gé eówru bearn tó yrsunge geciegean *ye parents, ye shall not provoke your children to anger*, L. E. I. 33; Th. ii. 430, 39. Hé hyne sceal forhabban wyð yrsunga *he shall restrain himself from anger*, Lchdm. iii. 140, 27. Ac ða irsunga [Cott. MS. iersunga] sindun swíðe ungelíca óðer biþ swelce hit síe irres anlícnes . . . óðer biþ ðæt ierre ðæt mon síe gedréfed on his móde bútan ǽlcre ryhtwísnesse óðer ðara irsunga biþ tó ungemetlíce átyht on ðæt ðe hió mid ryhte irsian sceall óðer on ðæt hió ne sceal biþ ealneg tó swíðe onbærned *sed longe alia est ira, quæ sub æmulationis specie subripit, alia, quæ turbatum cor et sine justitia prætexta confundit. Illa enim in hoc, quod debet, inordinate extenditur; hæc autem semper in his, quæ non debet, inflammatur*, Past. 40, 4; Swt. 293, 9–14.

irþ, e; *f*. **I.** *ploughing, tilling*:—For yrþe *for ploughing*; ad arandum, L. R. S. 21; Th. i. 440, 27. **II.** *the produce of arable land, a crop*:—Ðæt ðæs wæstmes yrþ ðǽr má upyrnende wǽre. Ðá him ðá ðæt sǽd broht wæs ofer ealle tíd tó sáwenne and ofer eallne hiht wæstm tó beranne ðe hé on ðam ylcan land seów ðá georn ðǽr sóna upp genihtsumlíc yrþ and wæstm *ut illius frugis ibi potius seges oriretur. Quod dum sibi adlatum, ultra omne tempus serendi, ultra omnem spem fructificandi, eodem in agro sereret; mox copiosa seges exorta est*, Bd. 4, 28; S. 605, 38–602, 1. Ic sello ðás land mid cwice erfe and mid earþe and mid eallum þingum ðe tó londum belimpaþ *I give these lands with the live stock, and crops and all things that belong to the lands*, Chart. Th. 481, 3. Rípe yrþe *maturam segetem*, Bd. 1, 12; S. 480, 35, note. **III.** *ploughed land*:—Cf. on ða foryrþe eástewerde, Cod. Dipl. Kmbl. iii. 449, 32 *where* Kemble *translates* foryrþ '*the land which is first ploughed*,' xlii. [On erthes *aracionibus*, Pall. 4, 68: Scott. earth *the act of earing* or *ploughing*.] v. gærs-, gafol-, lencten-yrþ; *and* erian.

irþ-land, es; *n*. *Arable land*:—Ierþland *arva*, Wrt. Voc. 285, 6. Yrþland

arva, 289, 77. Đanon up andlang yrþlandes, Cod. Dip. Kmbl. iii. 23, 31. Ic áwéste ðinne buruh and gewyrce tó yrþlande *I will lay waste thy city and make it into ploughed land*, Homl. Skt. 3, 224. Đonne is ðes londes ðe ic hígum selle xvi gioc ærþelandes and médwe *now of the land that I give to the convent there are sixteen acres of arable land and meadow*, Chart. Th. 477, 26.

irþling, es; *m.* I. *a husbandman, farmer, ploughman*:—Yrþlingc *arator*, Wrt. Voc. 73, 34: Ælfc. Gr. 41; Som. 44, 8. Noe ðá yrþling began tó wircenne ðæt land *coepitque Noe vir agricola exercere terram*, Gen. 9, 20. Móna se twentigoþa cild ácenned yrþlincg *a child born on the twentieth day of the moon will be a husbandman*, Lchdm. iii. 194, 6. Hwæt sægest ðú Yrþlingc *quid dicis tu, Arator?* Coll. Monast. Th. 19, 11. Hwilce ðé geþuht betwux woroldcræftas heoldan ealdordóm? Eorþtilþ forðam se yrþling ús ealle fétt *qualis tibi videtur inter seculares artes retinere primatum? Agricultura, quia arator nos omnes pascit*, 30, 23-8. Sume synt yrþlincgas sume scéphyrdas sume oxanhyrdas *alii sunt aratores, alii opiliones, quidam bubulci*, 19, 3. *Laboratores* sind yrþlingas and æhtemen, tó ðam ánum betæhte, ðe hig ús bigleofan tiliaþ, Ælfc. T. Grn. 20, 19. II. *the name of a bird, a cuckoo* [?]:—Irþling *cucuzata*, Wrt. Voc. 281, 14: *birbicariolus*, 281, 22. Ærþling *tanticus*, 29, 63. Geác *cuculus*, eorþling *birbicaliolus*, 63, 3-4. Yrþling *berbigarulus* vel *tanticus*, Wrt. Voc. ii. 12, 60. Erdling *bitorius*, 102, 1. Erþling *enistrius*, 143, 57. In connection with the cuckoo it may be noticed that *cucusare* is given in DuCange as the verb properly used of the note of the cuckoo; and see Grmm. D. M. 640, sqq. on the cuckoo as associated with a particular season of the year. However, in Wrt. Voc. 62, 22 the *lapwing* is glossed by *cucurata*.

is is. v. eom.

ÍS, es; *n.* I. ICE:—Ís *glacies*, Ælfc. Gl. 94; Som. 75, 103; Wrt. Voc. 52, 53. Hwí ne wundriaþ hí hwí ðæt ís weorþe *why do not they wonder why ice comes?* Bt. 39, 3; Fox 214, 35. Ofer eástreámas ís brycgade *the ice formed a bridge over the streams*, Andr. Kmbl. 2524; An. 1268: Exon. 90 a; Th. 338, 4; Gn. Ex. 73. Íses gicel *stiria, stillicidia*, Ælfc. Gl. 16; Som. 58, 68; Wrt. Voc. 21, 55. Hit eal gemealt íse gelícost *it all melted just like ice*, Beo. Th. 3221; B. 1608. Đá code hé sumre nihte on íse unwærlíce *dum incautius forte noctu in glacie incederet*, Bd. 3, 2; S. 525, 1. Styccum healfbrocenra ísa *semifractarum crustis glacierum*, 5, 12; S. 631, 26. II. *the name of the Rune* |=i:—| byþ oferceald ungemetum slidor *ice is exceedingly cold and excessively slippery*, Runic pm. 11; Kmbl. 341, 14. [O. Frs. O. H. Ger. ís; *n:* Icel. íss; *m:* Ger. eis; *n.*]

-isc, modern *-ish*, *a suffix of adjectives, connoting the quality of the object denoted by the stem*, e. g. ceorl-isc *churl-ish*, cild-isc *child-ish*; *also connotes origin from a place or stock*, e. g. Engl-isc, Grēc-isc, Iudē-isc. The suffix may be seen in the cognate dialects in the following words, Goth. þiud-isk-o *after the manner of the Gentiles*; Iudaiw-isk-s: *O. Sax.* menn-isk *human*: *O. Frs.* mann-isk: *Icel.* bern-sk-r *childish*: En-sk-r *English*: *Dan.* Engel-sk *English*: *O. H. Ger.* diut-isc: *Ger.* deut-sch.

ís-ceald; *adj. Ice-cold*:—Ísceald sǽ, Exon. 81 b; Th. 306, 28; Seef. 14: 307, 5; Seef. 19: Bt. Met. Fox 27, 6; Met. 27, 3.

isen, iesen, iesend. v. Lchdm. iii. 361, col. 2; *and* gesen *in the appendix*.

ÍSEN, es; *n. Iron, steel, an implement made of iron*:—Ísen *ferrum*, Wrt. Voc. 85, 13: Ælfc. Gr. 5; Som. 4, 58. Đis ýsen *hic calibs*, 9; Som. 13, 18. Eorþe swilce ísen *terra ferrea*, Deut. 28, 23. Đá wæs se ofen onhǽted ísen eall þurhgléded *then was the furnace heated, the iron made red hot*, Cd. 186; Th. 231, 8; Dan. 244. Ísenes scearpnyss *acumen*, Ælfc. Gr. 9; Som. 9, 31. Gemeng tógædere mid glówende ísene *mix together with a glowing iron*, L. M. 2, 24; Lchdm. ii. 216, 1. Ne delfe nán man ða moran mid ísene *let no man dig up the roots with iron*, Lchdm. iii. 30, 24. Bútan ǽlcan ísene genumen *gathered without using any iron implement*, Lchdm. iii. 4, 29 [cf. Grmm. D. M. 1148, sqq. as to the use of iron in getting plants]. *The two following passages refer to the ordeal* [v. ísen-ordál] *by hot iron*:—Gif hé hine ládian wille ðonne gá hé tó ðam hátum ísene and ládige ða hand mid ðe man týhþ *if he be willing to clear himself, then let him go to the hot iron, and clear the hand therewith that is accused*, L. Ath. i. 14; Th. 206, 23. Ǽlc tiónd áge geweald swá hwæðer hé wille swá wæter swá ísen, L. Eth. iii. 6; Th. i. 296, 4. Ácéle ðú wealhát ísen ðonne hit furþum síe of fýre átogen *cool very hot iron when it is just drawn from the fire*, L. M. 2, 45; Lchdm. ii. 256, 15. [*Ayenb.* izen (*but the general form in middle English is that with* r): *O. H. Ger.* ísen: *Ger.* eisen.] v. íren, ísern; brand-, delf-, gád-, ordál-ísen.

ísen; *adj. Iron, made of iron*:—Ísen *ferreus*, Ælfc. Gr. 5; Som. 4, 58. Seó gyrd wæs eal ísen *the rod was all iron*, Homl. Th. ii. 312, 17. Hig hyne on ánum ýsenum scrýne gebrohton on ðære byrig Damascus *they brought him in an iron chest to the city of Damascus*, St. And. 38, 8. Drihten sett ísen geoc on eówerne swuran *dominus ponet jugum ferreum super cervicem tuam*, Deut. 28, 48. Ísene bendas *vincula ferrea*, Ps. Th. 149, 8. Ádrífan ísene næglas þurh ða handa, Homl. Th. i. 146, 11. Đá wurdon hrædlíce forþ áborene ísene clútas and ísene clawa and ísen bedd ... Decius cwæþ \Lecgaþ ða ísenan clútas háte glówende tó his sídan,' 424, 18-35. v. íren, ísern.

ísen-grǽg; *adj. Iron-grey*:—Ísengrǽg *ferrugo*, i. *color purpuræ subni-*

grǽ: ísengrǽgum blóstme *ferrugineo flore* vel *purpureo*, Wrt. Voc. ii. 147, 63-67. Đa ísengrǽgan *ferrugineas*, 38, 44. [*Icel.* járn-grár: *Ger.* eisen-grau.]

ísen-hearde, an; *f. Ironhard*; *centaurea nigra*, Lchdm. iii. 4, 28: 22, 31: 334, col. 2. See Plant Names in E. D. S. Pub. iren-harde, iron-heads, iron-weed.

ísen-hyrst; *adj. Fitted with iron*:—Ǽrest of ísenhyrste gate ... eft in on ísenhyrsten geat *first, from the gate fitted up with iron ... back to the same gate*, Cod. Dipl. Kmbl. iii. 130, 27 ... 131, 19. [Cf. *Icel.* járnsleginn *mounted with iron*.]

ísenian; *p.* ode. *To furnish* or *cover with iron (armour)*: Đa ísnodan truman *ferratas acies*, Wrt. Voc. ii. 147, 52.

ísen-ordál, es; *n. The ordeal by hot iron*, in which the accused who wished to clear himself had to bear, on the naked hand, a piece of red hot iron. The passages from which the following extracts are taken will illustrate this mode of trial:—Gif hit sý ýsenordál beón þreó niht ǽr man ða hand undó *if it be the ordeal by hot iron, let it be three days before the hand be undone*, L. Ath. i. 23; Th. i. 212, 3. Wé cwǽdon ... ðæt man ... myclade ðæt ordálýsen ðæt hit gewege þrý pund ... and hæbbe se teónd cyre swá wæterordál swá ýsenordál swá hwæðer him leófre sý *we have ordained that the ordeal-iron be increased so that it weigh three pounds ... and let the accuser have the choice of ordeal by water or by iron, whichever he prefer*, iv. 6; Th. i. 224, 12-16. See too, Dóm be hátan ísene, 7; Th. i. 226, 7, sqq; *and* Schmid A. S. Gesetz. p. 419. [Cf. *Icel.* bera járn, járn-burðr in Cl. and Vig. Dict. *and see* Grmm. R. A. 915, sqq.] v. ordál.

ísen-panna, -panne, an; *m. f. A frying-pan*:—Ísenpanna *sartago*, Wrt. Voc. 82, 68. Ísenpanne, Ælfc. Gl. 26; Som. 60, 94; Wrt. Voc. 25, 34. See other compounds of íren, ísen, ísern.

ísen-smiþ, es; *m. An iron-smith, worker in iron, blacksmith*:—Tubalcain wæs ǽgðer ge goldsmiþ ge ísensmiþ *Tubalcain fuit malleator et faber in cuncta opera æris et ferri*, Gen. 4, 22. Ic hæbbe smiþas ísen[e]smiþas goldsmiþ seolforsmiþ ársmiþ *habeo fabros, ferrarios, aurificem, argentarium, ærarium*, Coll. Monast. Th. 29, 35. [Cf. *Wick.* iren-smiþ: *Icel.* járn-smiðr *a blacksmith*: *O. H. Ger.* ísarn-smid *faber ferrarius*: *Ger.* eisen-schmied.] v. ísen-, ísern-wyrhta.

ísen-swát, es; *m.* [?]:—Smít on ísenswát, L. M. 2, 65; Lchdm. ii. 296, 18. *See* iii. 366, col. 1.

ísen-tanga, an; *m. A pair of snuffers*:—Candel *candela*; ísentanga *munctorium*, Wrt. Voc. 81, 34-5. v. tang, tange.

ísen-wyrhta, an; *m. A worker in iron, blacksmith*:—Ísenwyrhta *ferrarius*, Wrt. Voc. 73, 28. v. ísen-smiþ, ísern-wyrhta.

ísern, es; *n. Iron, an instrument* or *weapon made of iron*:—Sweord sceal on bearme drihtlíc ísern *the sword shall lie in the lap, the noble steel*, Menol. Fox 511; Gn. C. 26. Oft mec ísern scód sáre on sídan *oft has iron harmed me sorely in the side*, Exon. 126 a; Th. 485, 14; Rä. 71, 13: 130 a; Th. 499, 5; Rä. 88, 11. Áres and ísernes *æris, ferri*, Bd. 1, 1; S. 473, 23: Cd. 52; Th. 66, 23; Gen. 1088. Ísernes dǽl, Exon. 114 b; Th. 439, 25; Rä. 59, 9. Wið slege ísernes oððe stenges *for a blow from iron* [*sword*] *or stick*, Herb. 32, 8; Lchdm. i. 132, 4. Wið wunda som hý sýn of íserne som hý sýn of stenge, 63, 3; Lchdm. i. 166, 9. Achilles mid ðysse sylfan wyrte (*yarrow*) gehǽlde ða ðe mid íserne geslegene and gewundude wǽran, 90, 1; Lchdm. i. 194, 8. Đú swyltst nalles mid íserne ácweald swá ðú wénst ac mid átre *morieris, non ferro quod suspicaris, sed veneno*, Nar. 31, 27. Gebundene on íserne *ligatos in ferro*, Ps. Th. 106, 9. Íserne wund, Exon. 102 b; Th. 388, 2; Rä. 6, 1. Purh ðæt ísern ðæt mægen ðara þreátunga is getácnod *per ferrum increpationis fortitudo signatur*, Past. 21, 6; Swt. 163, 24. Se lǽce hýd his ísern wið ðone moun ðe hé sníðan wile *the surgeon hides his knife from the man he wants to cut*, 26, 3; Swt. 185, 25. [*Goth.* eisarn *iron, an iron fetter*: *O. Sax.* ísarn: *O. L. Ger.* ísarn *chalybs*: *O. Frs.* ísern: *Icel.* ísarn *(occurs five times in old poetry; the usual form is* járn): *O. H. Ger.* ísarn.] v. íren, ísen; hóc-, leóht-, mearc-, stemping-ísern.

ísern; *adj. Iron, made of iron*:—Hé him tǽhte ðæt hé him genáme áne íserne hearstepannan and sette betweoh hine and ða burg for íserne weall *et tu sume tibi sartaginem ferream, et pones eum murum ferreum inter te et inter civitatem*, Past. 21, 5; Swt. 161, 7: Cd. 186; Th. 231, 16; Dan. 248. Íserne steng *vectes ferreos*, Ps. Th. 106, 15. Hét gebindan beám ðone miclan ǽrenum clammum and ísernum *he bade bind that great tree with brazen bands and with iron*, Cd. 200; Th. 248, 29; Dan. 520. [*Goth.* eisarneins: *O. L. Ger. O. H. Ger.* ísarnin: *Ger.* eisern.] v. íren, ísen.

ísern-byrne, an; *f. An iron byrnie* or *corslet*:—Hé him of dyde ísernbyrnan, Beo. Th. 1347; B. 671. v. íren-byrne.

ísern-gelóman. v. gelóman.

ísern-here, es; *m. An iron-clad host*:—Ísernhergum án wísode, Cd. 160; Th. 199, 33; Exod. 348.

ísern-wyrhta, an; *m. A worker in iron, a blacksmith*; *ferrarius*, Ælfc. Gl. 2; Som. 55, 46; Wrt. Voc. 16, 18. v. ísen-wyrhta.

ís-gebind, es; *n. A bond of ice*:—Winter ýðe beleác ísgebinde *winter locked up the wave with icy bond*, Beo. Th. 2270; B. 1133.

ís-geblǽd, es; *m.* [?] *A blister that is produced by ice*:—Wið ýsgeblǽd Lchdm. iii. 36, 22.

îs-gicel. v. gicel.

îsig; adj. Icy, covered with ice :—Ðær stôd hringedstefna îsig and ûtfûs, Beo. Th. 65; B. 33. v. eall-îsig.

îsig-feðera; adj. Having ice on the wings :—Stearn îsigfeðera, Exon. 81 b; Th. 307, 15; Seef. 24.

î-sîðes; adv. At that time, at once, directly :—Man îsîðes sôna ðæræfter swytelaþ it is immediately thereafter manifested, L. I. P. 24; Th. ii. 338, 11. [Cf. î-dæges.]

Ismahéli; pl. m. Ishmaelites, Bedouins :—Ðâ þiccodan ðider semninga ða ismahéli on horsum and on olfendum then crowded thither on a sudden the Bedouins, on horses and camels, Shrn. 38, 4.

Ismahélitas; pl. m. Ishmaelites : — Ismæhélita, Ismahelitum, Ps. Th. 82, 6. Ysmahélitum Ismaelitis, Gen. 37, 28.

Ismahélitisc; adj. Ishmaelite :— Æt ðâm Ismahélitiscum mannum de manu Ismaelitarum, Gen. 39, 1.

îs-mere, es; m. A mere covered with ice :— Scîneþ sunne sôna îsmere weorþeþ tô wætre the sun shines, at once the icy lake turns to water, Bt. Met. Fox 28, 123; Met. 28, 62.

Ispania Spain; Hispania, Ors. 1, 1; Swt. 24, 1, 7, 9.

-isse. This suffix, Lat. -issa, which in later English became the common suffix to mark the feminine gender, is found before the Norman Conquest in the word abbud-isse abbess. [Cf. -estre.]

Îstas; pl. m. The Esthonians :—Ic wæs mid Îstum, Exon. 86 a; Th. 323, 31; Vid. 87. [Icel. Eistir.] v. Êste; and see Grmm. Gesch. D. S. 499, sqq.

istoria history :—Istoriam Indêa rîces, Salm. Kmbl. 7; Sal. 4.

-istre. v. -estre.

Italie, a; pl. The Italians or Italy :—Pencentes Italia folc, Ors. 4, 2; Swt. 160, 27. Pirrus fôr of Italium (ab Italia), 4, 1; Swt. 158, 30 : 154, 32. îð, iéð, ŷð; adv. compve. More easily :—Ðæt hie hiera godum ðe iéð blôtan mehten that they might the more easily sacrifice to their gods, Ors. 2, 2; Swt. 64, 29. Hwâ meahte iéð monnum rêdan bûtan scylde ðonne se ðe hí gescôp quis principari hominibus tam sine culpa, quam is, qui hos nimirum regeret, quos ipse creaverat? Past. 3. 1; Swt. 33, 16. Ðŷ ŷþ, Exon. 120 b; Th. 463, 6; Hö. 66. v. eáðe, êð.

îðan; p. de To lay waste, desolate, destroy :—Ic ŷðde eotena cyn and on ŷdum slôg niceras nihtes, Beo. Th. 846; B. 421. Ŷðde disne eard-geard ælda scyppend the creator of men laid waste this world, Exon. 77 b; Th. 291, 20; Wand. 85. Ŷðan, 126 a; Th. 484, 13; Rä. 70, 7. Îðende depopulis, Wrt. Voc. ii. 27, 27. [Icel. eyða to lay waste, destroy, waste desolate : O. H. Ger. ôdian desolare, Grff. i. 150 : Ger. ver-öden.] v. â-îðan.

îðast, îðost; adv. superl. Most easily :—Ŷðast meahtan frôfre findan might find comfort most easily, Exon. 19 b; Th. 50, 15; Cri. 800. Ŷðæst, 26 b; Th. 79, 1; Cri. 1284. Ŷðost, Hy. 7, 3; Hy Grn. 287, 3. v.îð.

îð-belig; adj. Easily made angry :—Ne wê tô ŷðbelige [eáðbylige, MS. D.] ne sŷn, ne tô langsum yrre hæbben, Wulfst. 253, 11.

îð-dæde; adj. Easy to do :—Hit wæs Gode ŷðdæde, ðâ hê hit swâ gedôn habban wolde, Wulfst. 15, 18. v. eáð-dæde.

îðe; adj. Easy, pleasant :—Nô ðæt ŷðe byþ tô befleónne that is not easy to flee from, Beo. Th. 2009; B. 1002: 4822; B. 2415. On his heortan hê Gode þancie ealles ðæs ðe hê him forgeaf ægðer ge ŷdran ge unŷdran in his heart let him thank God for all that he has given him, both pleasant and unpleasant, L. E. I. 29; Th. ii. 426, 11. Ûs ðis se æðeling ŷðre gefremede this the prince has made easier for us, Exon. 17 a; Th. 39, 25; Cri. 627. v. eáðe, êðe, un-îðe.

îðe-lîce; adv. Easily :—Iéðelîce and scortlîce ic hæbbe nû gesæd hiora ingewinn without making the account difficult or long I have now related their intestine struggle, Ors. 2, 6; Swt. 88, 28. Iéðelîce forneáh bûton ælcon gewinne easily, almost without any struggle, 3, 7; Swt. 112, 28. Ŷðelîce, Beo. Th. 3116; B. 1556. Forðæm se lytega feónd swâ micle iéðelîcor ðæt môd gewundaþ swâ hê hit ongiet nacodre ðare byrnan wær-scipes quia hostis callidus tanto liberius pectus percutit, quanto nudum a providentiæ lorica deprehendit, Past. 56, 1; Swt. 431, 10. v. eáðe, êðe-lîce, un-îðelîce.

îð-fynde; adj. Easy to find :—Ŷðfynde, Andr. Kmbl. 3092; An. 1549. v. eáð-, êð-fynde.

îð-gesŷne; adj. Easy to see :—Ŷðgesŷne, Beo. Th. 2493; B. 1244. v. êð-gesŷne.

îð-ness, e; f. Easiness, freedom, ease, satisfaction, delight : — Hwelce îðnesse hæfþ God æt ûrum wîtum neque Deus nostris cruciatibus pascitur, Past. 54, 5; Swt. 425, 11. v. un-îðness, êðness.

iú. v. geó.

iuc, ioc a yoke; jugum, Wrt. Voc. 284, 54: ii. 46, 37 : juger, 38. Ioc jugum, Ælfc. Gl. 3; Som. 55, 58; Wrt. Voc. 16, 30: Ps. Spl. C. 2, 3: Mt. Kmbl. Rush. 11, 29, 30: Rtl. 108, 21. [These examples should be given under geoc.] v. geoc.

iuc-boga, an; m. The bow or curved part of a yoke :—Iucboga jujula [among things connected with vehicles], Wrt. Voc. 284, 50: jugula, ii. 46, 36.

iucian; p. ode To join, yoke :—Ic iucige jungo, Ælfc. Gr. 28; Som. 31, 53. v. ge-iukod.

iuc-sticca, an; m. The bar of a yoke :—Ioc-sticca obicula, Ælfc. Gl. 3; Som. 55, 61; Wrt. Voc. 16, 33. (Cf. O. H. Ger. iuh-rota pertica, Grff. ii. 491.]

iuc-têma, an; m. An animal yoked with another :—Ioctêma jugalis, Ælfc. Gl. 3; Som. 55, 59; Wrt. Voc. 16, 31. [cf. ge-týme.]

Iudan burh Jedburgh, Chr. 952; Erl. 118, 26.

iú-dæd, e; f. A deed done of old or formerly :—Gû-dæda, Exon. 64 a; Th. 235, 12; Ph. 556. Iúdædum, 76 a; Th. 284, 26; Jul. 703: Cd. 217; Th. 276, 10; Sat. 186.

Iudéa Judea : — Fram Iudéa de Judæa, Mt. Kmbl. 4, 25. On ðam wêstene Judéæ in deserto Judææ, 3, 1.

Iudéas; gen. a; pl. m. The Jews; Judæi, Jn. Skt. 2, 20. Eal Iudéa þeód omnis Iudæa Mt. Kmbl. 3. 5. Betwux ðâm Iudêum, Jn. Skt. 10, 19.

Iudéisc; adj. Jewish :—Ðâ stôd ân Iudéisc wer, ðæs nama wæs Nichodemus, Nicod. 11; Thw. 5, 38 : Jn. Skt. 18, 35. Crist cwæþ be ðâm ungeleáffullum Iudéiscum wâ eów Christ said of the unbelieving Jews ' Woe to you,' Ælfc. Gr. 48; Som. 49, 5.

iú-geára; adv. Formerly :—Breoton wæs iúgeára Albion hâten Brittania cui quondam Albion nomen fuit, Bd. 1, 1; S. 473, 8. v. geó-geára.

iugian to join, yoke :—Ic iugie hí tô syl jungo eos ad aratrum, Coll. Monast. Th. 19, 15. v. iucian.

iúgoþ youth, young people, Jos. 5, 5 : Homl. ii. 506, 21 : Homl. Skt. 6, 2. v. geógoþ.

Iúla, an; m. December or January : — Mônaþ Decembris, ærra Iúla, Menol. Fox 439; Men. 221. v. Geóla.

iú-leán, es; n. A reward for something done long ago : — Iúleán ðæs ðe hine of nearwum Widia ût forlêt a reward, because in time past Widia released him from straits, Wald. 2, 7.

iú-mann, es; m. A man of old, of a former time : — Iúmonna gold, Beo. Th. 6096; B. 3052. v. gió-mann.

iú-meówle, an; f. One who was a maiden long ago, an old woman :— Ió-meówlan, Beo. Th. 5854; B. 2931.

iung; adj. Young :—Sum iung man, Th. Ap. 3, 23 : 4, 7 : Bd. 2, 12; S. 514, 27 : Ælfc. Gl. 45; Som. 64, 106; Wrt. Voc. 32, 41: 64, 03; Wrt. Voc. 32, 28. v. geong.

iung-lîc; adj. Youthful : — Iunglîcre ylde, Ælfc. T. Grn. 16, 40. v. geong-lîc.

iung-ling, es; m. A youth :—Iunglingc juvenis, Wrt. Voc. 73, 19 : Gen. 4, 23. Sum iungling him fyligde adulescens quidam sequebatur eum, Mk. Skt. 14, 51 : Homl. Th. ii. 312, 16. v. geong-ling.

Iútan, Iútas. v. Iótas.

iú-wine, es; m. A friend of old or former times : — Wât his iúwine eorþan forgiefene knows that his friends of old are committed to earth, Exon. 82 b; Th. 311, 15; Seef. 92.

ÎW, es; m. Yew :—Îw taxus, Ælfc. Gl. 46; Som. 64, 131; Wrt. Voc. 32, 65: Wrt. Voc. 79. 74: 285, 49. Se hearda îw, Exon. 114 a; Th. 437, 18; Rä. 56, 9. On ðone ealde îw ðonan of ðon îwe to the old yew; thence from the yew, Cod. Dipl. Kmbl. iii. 218, 35. In proper names, vi. 306, col. 2; 307, col. 1. [Chauc. ew : Icel. ýr a yew, a bow : O. H. Ger. îwa taxus : Ger. eibe.] v. eow.

îwan; p. de To show, bring before the eyes, display, reveal :—Ŷweþ and yppeþ shews and reveals, Salm. Kmbl. 985; Sal. 494. Ðâ ŷwde hê ðær synne wisan culpam esse demonstravit, Bd. 1, 27; S. 496, 2. Ðâ ŷwde ic him sôna ða ylcan bôc ðara reogola quibus statim protuli eundem librum canonum, 4, 5; S. 572, 25. Mid his sylfes dæde ŷwde and cŷdde propria actione præmonstraret, 4, 27; S. 604, 40. Ŷwaþ mê ânne peninc ostendite mihi denarium, Lk. Skt. 20, 24. Wênþ gif hê hit him iéwe ðæt hê him nylle geþafigean ðæt hê hine snîðe he expects, if he show it [the knife] to him, that he will not allow him to cut him, Past. 26, 3; Swt. 185, 25. Ðîne miltse ŷwe show thy mercy, Exon. 11 b; Th. 15, 32; Cri. 245. Ðæt land ðe ic ðê ŷwan wille the land that I will show thee, Cd. 83; Th. 105, 11; Gen. 1751. Ord and ende ðæs ðe him ŷwed wæs the beginning and end of what was revealed to him, 180; Th. 225, 31; Dan. 162. DER. æt-, ge-, oþ-îwan [-ŷwan]; and see eáwan, eówan.

K.

THE letter k appears to have had no distinct duty to perform in the oldest English, but to have been a mere variant of c. In the MSS. (more particularly the Cotton) of Alfred's translation of Gregory's Pastoral Care, where in the words kyning, kynn &c. it occurs not unfrequently, this writing is not uniform. Thus in Sweet's edition Angelkynn is found p. 2, ll. 3, 13, but Angelcynn l. 4; whilst in each case the Hatton MS. has c. So in the following page in l. 10, kynn, in l. 20, cynn. On pp. 2, 3, l. 1 kyning is the writing of both MSS. while pp. 34, 35, l. 14 it is cyning : p. 32, 20–1 we find kyning, kynehad, the Hatton MS. in the same passage has c : p. 38, ll. 13, 18 kyning, kynestol, where the Hatton MS. has cyning, kynestol : pp. 6, 7, l. 18 both have kynerice : p. 84, ll. 10, 12, 13 kynelic occurs four times, in the Hatton MS. it is twice written with c, twice with k. On p. 212, l. 15 is found Crist, while the Hatton writes Krist; on p. 152, line 5 the Cotton MS. has kræft, the Hatton MS. cræft. On p. 459, ll. 29, 31, 32 (Hatton MS.) occur the forms kokka, kokkum, kok. So in the Chronicle. Erl. p. 8, l. 15 kyning; but

p. 6, l. 23 *cyning*: p. 24, l. 1, *kyning*; 26, 1, *cyning*. The later use with regard to the letter may be, to some extent, illustrated from the concluding years. For many years previous to 1111 the form is *cyng*, in that year we have *Kyng* Henri; again until 1122 the opening line of each annual contains the phrase *Cyng* Henri, then until the end the spelling is *k*.

Words beginning with *k* are to be looked for under *c*.

L

In the later specimens of the West Saxon dialect those words in which the vowel *a* immediately preceded a combination of consonants beginning with *l* are generally found to have undergone a change which was represented by writing *ea* instead of *a*. This change does not occur to the same extent in the earlier specimens, and seems not to occur at all in the Northumbrian dialect, or in the kindred languages. Thus in the translation of Gregory's Pastoral Care and in the Parker MS. of the Chronicle *alle*, *onwald* are found as well as *ealle*, *onweald*, while in Ælfric's Homilies they are regularly written in the latter form. So the West Saxon forms, *healdan*, *sealt*, *healf*, are found in the Northumbrian Gospels as *halda*, *salt*, *half*, and in Gothic, O. Sax., Icel., O. H. Ger. the vowel also is *a*.

In the Runic alphabet the character, which in name and form agrees with the Scandinavian rune ↑, *lögr*, was ↑ *lagu*. The same name seems to have been given to the corresponding letter in the Gothic alphabet, though it occurs only in a corrupt form *laar = lagus*. The meaning of this word may be seen from the verses in the Runic poem that are devoted to the letter:

Lagu byþ leódum	*water to wanderers*
langsum geþuht	*wearisome seemeth*
gif hí sculun néþan	*if they must venture*
on nacan tealtum	*on vessel unsteady*
and hí sǽyþa	*and them the sea-waves*
swýðe brégaþ	*sorely affright*
and se brimhengest	*and the sea-horse*
bridles ne gýmþ	*steering despiseth.*

Runic pm. 21; Kmbl. 343, 19-26.

lá. I. *interj.* *Lo! Oh! Ah!*:—Lá næddrena cyn *Oh! generation of vipers*, Mt. Kmbl. 3, 7: 12, 34. Lá ðú líccetere, 7, 5. Lá freónd *amice*, 22, 12. Lá Drihten *Domine*, Ps. Th. 21, 17: 118, 176. Lá hú oft hí gremedon hine *quotiens exacerbaverunt eum*! Ps. Spl. 77, 45. Áfæst lá and hí lá hí and wel lá well and ðyllíce óðre syndon englisc interjectiones, Ælfc. Gr. 48; Som. 49, 28. Weg lá weg lá *euge, euge*, Ps. Th. 69, 4. Wá lá se tówyrpþ ðæt tempel *ua qui destruit templum*, Mk. Skt. 15, 29. Wá lá áhte ic mínra handa geweald *alas! had I power over my hands*, Cd. 19; Th. 23, 32; Gen. 368. Wá lá wá *heu, proh dolor!* Bd. 2, 1; S. 501, 14. Wei lá wei, [cf. Chauc. *weilawey*: Shakspere *welladay*] Bt. 35, 6; Fox 170, 12, Cott. MS. Wel lá men wel *oh! men*, 34, 8; Fox 144, 23. Wel lá, Bt. Met. Fox 21, 1; Met. 21, 1. **II.** *Enclitic particle used to emphasise interrogation, exclamation, entreaty, affirmation, negation*:—Understenst ðú lá *sentisne*, wylt ðú lá *visne*, Ælfc. Gr. 44; Som. 45, 47. Is ðǽr genoh lá *satisne est*, Som. 46, 40. Hú lá ne wurpe wé þrý cnihtas intó ðam fýre *why, did not we cast three youths into the fire?* Homl. Th. ii. 20, 12. Wénst ðú lá ðæt ðú beó álýsed fram ðisum tintregum *do you suppose then that you will be released from these torments?* Homl. Th. i. 424, 29. Ðá cwæþ ic hwæt is ðæt lá then said I '*what then is that?*' Bt. 34, 5; Fox 140, 14. Hwæt is ðæt lá þinga? 38, 3; Fox 200, 2. Hwæt is ðis lá manna? Elen. Kmbl. 1802; El. 903. Hwæt biþ hit lá elles búton flǽsc seoððan se écea dǽl of biþ hwæt biþ lá elles seó láf búton wyrma mete *why, what else is it but flesh when the eternal part is away? what else then is the remnant but worms' food?* Blickl. Homl. 111, 31. Hwǽr biþ lá ðonne se ídla lust? hwǽr beóþ ðonne ða symbelnessa? 58, 16. Is ðis lá wundorlíc and winsum spell *this is indeed a wonderful and delightful speech*, Bt. 34, 5; Fox 140, 10. Ðæt lá mæg secgan se ðe sóð and riht fremeþ *that indeed may he say who does truth and right*, Beo. Th. 3404; B. 1700: 5720; B. 2864. Ðæt lá wæs fæger, Cd. 223; Th. 293, 18; Sat. 457. Uton lá geþencan *let us then determine*, 227; Th. 305, 9; Sat. 644. Ac feor ðæt lá sí ðæt . . . *sed absit ut* . . . Bd. 1, 27; S. 490, 24. Ic ðæs lá wísce ðæt wegas míne on dínum willan weorðan gereahte *I do indeed wish that my ways may be directed according to thy will*; utinam dirigantur viæ meæ, Ps. Th. 118, 5. Bidde ic ðé lá gif . . . *precorque si* . . . Bd. 3, 13; S. 538, 40: 4, 3; S. 568, 27: Dóm. L. 6, 65. Nese lá nese, Bt. 27, 2; Fox 96, 27. v. eálá.

lác; *generally neuter, but occasionally feminine* [v. Shrn. pp. 3-4], *or masculine, as in the compound* lyb-lác q. v. The idea which lies at the root of the various meanings of this and of the next word seems to be that of motion. Thus *lácan* and Icel. *leika* are used to describe the motion of a vessel riding on the waves, the flight of a bird as it rises and falls in the air, the flickering, wavering motion of flame, and the like; while Gothic *laikan* renders σκιρτᾶν in Luke i. 41, 44: vi. 23. From this idea of activity we pass to that of *games, playing, dancing* &c.; and so Gothic *laiks* = χορός in Luke xv. 25; in Icel., where the meaning *play, sport* is the prevailing one (see also compounds in which *leik-* occurs), *leikr* is used of *dancing, athletics, various games, music*, as in *strengleikr*, *leika* = to play, to *lake* in the dialect of the North of England. In O. H. Ger. the application is generally to music, *leih, leich* = modus, modulus, carmen versus, but in *rang-leih* = wrestling the meaning is similar to the Icelandic (see Grff. ii. 152-3.) And just as *plega* is used, by itself or in its compounds, of war and battle, so in the Icelandic poetry we have *Hildar leikr*, *sverða leikr* = battle (see Cl. and Vig. Dict. p. 382, col. 2), and in English *lác* could be applied in the same way. But in the latter language the more frequent meanings are those of *offering, gift*, and to connect these with the preceding ones Grimm notes the association of dancing and playing with offerings and sacrifices. From this special meaning of *offering* the more general one of *gift, present* might easily come. To quote his words '*Das wort* (*lác*) *scheint einer wurzel mit dem goth. laiks* (saltatio) *ahd. leih* (ludus, modus) *altn. leikr*, *ursprünglich also tanz und spiel, die das opfer begleiteten, allmählich die gabe selbst zu bezeichnen,*' D. M. 35. The passages which follow will shew the English use of the word. **I.** *battle, struggle*:—Wíga unlæt láces *a warrior not slow to fight* (*referring to death which was approaching Guthlac*), Exon. 47 b; Th. 164, 5; Gú. 1007. **II.** *an offering, sacrifice, oblation*:—Gode onsægdnesse tó beranne ðæs hálgan láces *ad offerendas Domino victimas sacræ oblationis*, Bd. 4, 22; S. 592, 26. Hí him sculon láces lof lustum bringan *sacrificent sacrificium laudis*, Ps. Th. 106, 21. Ic ðé láces lof lustum secge *tibi sacrificabo hostiam laudis*, 115, 7. Ic ðé lustum láce cwéme *voluntarie sacrificabo tibi*, 53, 6. And bærnon uppan ðam weofode drihtne tó láce *adolebuntque super altare in oblationem domino*, Lev. 3, 5. Offrian tó láce *to offer as a sacrifice*, Ælfc. T. Grn. 4, 27. Hie drihtne lác begen brohton *they both brought an offering to the Lord*, Cd. 47; Th. 60, 2; Gen. 975. Se rinc Gode lác onsægde, 85; Th. 107, 21; Gen. 1792. Onbleót ðæt lác Gode, 142; Th. 177, 21; Gen. 2933. Ðú scealt blótan sunu, and leófes líc forbærnan, and mé lác bebeódan, 138; Th. 173, 9; Gen. 2858. Ðú dínne lác offrige, Homl. Skt. 7, 119. Þurh lác ðære hálwendan onsægdnesse *per oblationem hostiæ salutaris*, Bd. 4, 22; S. 592, 22. Mára is allum cwicum lácum and sægdnissum *majus est holocaustomatibus et sacrificiis*, Mk. Skt. Rush 12, 33. Ǽnig ðæra þinga ðe gedwolgodum tó lácum betǽht biþ *any thing that is appointed to false gods for sacrifices*, Swt. A. S. Rdr. 105, 30. Nemne hé lufige mid lácum ðone ðe gescóp heofon and eorþan *unless by offerings he shew his love to him that created heaven and earth*, Exon. 67 a; Th. 249, 13; Jul. 111. Mid háligra lofsanga lácum cóman *with offerings of holy hymns they came*, Blickl. Homl. 207, 9. Gode lác onsægdon, 201, 13: Guthl. 20; Gdwin 32, 13. On ðám lácum geleáfsumra *fidelium oblationibus*, Bd. 1, 27; S. 488, 38. Geoffrode lác *obtulit holocausta*, Gen. 8, 20. Genimaþ eów lác and ingangaþ on his wíctúnas *tollite hostias et introite in atria ejus*, Ps. Th. 95, 8. Seó cwén Sabæ geseah ða lác ðe man Gode offrode *the queen of Sheba saw the offerings that were made to God*, Homl. Th. ii. 584, 16. Hé fræt fíftýne men and óðer swylc út offerede láðlícu lác *he* (*Grendel*) *devoured fifteen men and as many bore away, horrid sacrifices*, Beo. Th. 3172; B. 1584. **III.** *a gift, present, grace, favour, service; a present or offering of words, a message*:—Lác *munus*, Ælfc. Gr. 9, 22; Som. 12, 14. Lác *munus vel zenia*, Ælfc. Gl. 35; Som. 62, 77; Wrt. Voc. 28, 55. Lác *elogia*, i.e. *munus*, Wrt. Voc. ii. 143, 19: 29, 24: *xenium, donum*, Hpt. Gl. 496: *munificentia*, 414. Gúþlác se nama ys on rómánisc *belli munus*, Guthl. 2; Gdwin 10, 23. Leóht wé geseóþ láce *lumen videmus muneris*, Hymn. Surt. 43, 17. Behátenre fæderes láce *promisso Patris munere*, 95, 27. Láce *eulogiæ, benedictionis*, Hpt. Gl. 496. Tóforan ðære cynelícan láce ðe hé hire geaf, Homl. Th. ii. 584, 31. Sende tó láce *sent it as a present*, Elen. Kmbl. 2398; El. 1200. Hé ðære mægeþ sceolde láce (*acc. fem.?*) gelǽdan láþspel tó sóþ *he to the maiden must bring the message, the grievous tale too true*, Exon. 52 a; Th. 182, 28; Gú. 1317. Tíd is ðæt ðú fére and ða ærendu eal biþence ófestum lǽde swá ic ðé ǽr bibeád lác tó leófre *time is that thou go and think about those errands* [cf. Th. 173, 24 sqq. where Guthlac speaks of his burial], *with speed bring, as I before bid thee* [cf. Th. 172, 31 sqq], *the message to my dear sister*, 51 b; Th. 179, 35; Gú. 1272. Heó lác weorðade ðe hire brungen wæs *she honoured the gift* [*the nails of the cross*] *that was brought her*, Elen. Kmbl. 2272; El. 1137. Cwæþ hé his sylfes suna syllan wolde . . . Hie ða lác hraðe þégon tó þance *he said he would give his own son . . . They that gift soon accepted thankfully*, Andr. Kmbl. 2224; An. 1113. Ða hálgan þrynnesse georne biddan ðæt heó ðæt lác ðæt hie þurh ðone hálgan heáhengel ǽrest æteówde mannum wundorlíc tácn ðæt hie ðæt mannum tó fylgenne oncýðde *earnestly to entreat the holy Trinity that the grace of shewing by the holy archangel a wondrous token to men, that that it would make known to men for their guidance*, Blickl. Homl. 205, 30. Ðonne onfóþ hí from Gode máran méde ðonne hí from ǽnigum óðrum lácum dón *then shall they receive from God greater reward than they do from any other gifts*, 45, 34. Him lácum cwémaþ *dona adducent*, Ps. Th. 71, 10. Lácum, þeódgestreónum, Beo. Th. 86; B. 43. Him eorla hleó gesealde máþmas xii.

hét hine mid ðǽm lácum leóde sécean, 3740; B. 1868. Culufre gewāt fleógan eft mid lácum hire (the olive branch), Cd. 72; Th. 88, 28; Gen. 1472. Hí geopenodon heora hordfatu and him lác geoffrodon gold and récels and myrram they opened their treasures, and presented unto him gifts; gold, and frankincense, and myrrh [Mt. 2, 11], Homl. Th i. 78, 27. Lác gifan, Exon. 100 b; Th. 380, 2; Rä. 1, 1. Bringan lác and luftácen to bring gifts and love-tokens, Beo. Th. 3730; B. 1863. Lǽc munera, Ps. Spl. T. 14, 6. IV. medicine :—Heofendlícere lác [= heofenlícere lác] cælestis medicinæ, Hpt. Gl. 415, 36. Lác medicamine, 507, 77. Lác medicamenti, 527, 18. [Laym. 1st MS. lac, 2nd MS. lock gift: Orm. lac a sacrifice, offering: Gen. a. Ex. loac; Piers P. laik a game.] v. ag-, æfen-, beadu-, berne-, brýd-, cwic-, feoht-, freó-, ge-, hæmed-, headu-, lyb-, mæsse-, reáf-, sǽ-, scín-, wed-, wíf-, wíte-lác. It also occurs in proper names, e. g. Gûþ-lác, Hyge-lác.

lácan; p. leólc, léc; pp. lácen. I. to swing, wave about, move as a ship does on the waves, as a bird does in its flight, as flames do :— Ic láce mid winde I wave about with the wind, Exon. 108 a; Th. 412, 17; Rä. 31, 1. Sum láceþ on lyfte one swings in the air [of the man who is hung on a tree], 87 b; Th. 328, 25; Vy. 23. Is ðæt frécne stream ýða ofermǽta ðe wē hēr on lácaþ perilous is the stream, huge the waves, on which here we toss, 20 a; Th. 53, 24; Cri. 855. Hie ofer feorne weg ceólum lácaþ, Andr. Kmbl. 506; An. 253. Fuglas ða ðe late þurh lyft lácaþ fiðrum birds which slowly through the air move with their pinions, Exon. 60 b; Th. 220, 7; Ph. 316. Brondas lácaþ on ðam deópan dæge fires shall flame up on that solemn day [cf. to play applied to flame, and Icel. logi lēk um þá v. Cl. and Vig. Dict. leika II. 2], 116 b; Th. 448, 23; Dóm. 58. Ða ðe lácaþ ymb eaxe ende those stars that revolve about the pole, Bt. Met. Fox 28, 44; Met. 28, 22. Leólc on lyfte he took his flight through the air [of the lost angel who was to tempt Adam], Cd. 23; Th. 29, 10; Gen. 448: Exon. 114 a; Th. 438, 15; Rä. 57, 8. Hē leólc ofer laguflód he bounded o'er the water, 75 b; Th. 283, 2; Jul. 674. Fugel uppe sceal lácan on lyfte up in the air must the bird wing its flight, Menol. Fox 537; Gn. C. 39. Hwylc hyra [the seraphim] nēhst mǽge nergende flihte lácan, Exon. 13 b; Th. 25, 11; Cri. 399. Dû meahtes ofer rodorum feðerum lácan, feor up ofer wolcnu windan, Bt. Met. Fox 24, 17; Met. 28, 9. Heofonfuglas ða ðe lácende geond lyft faraþ, Exon. 55 a; Th. 194, 24; Az. 144: Beo. Th. 5657; B. 2832: Elen. Kmbl. 1797; El. 900. Lagu lácende the tossing waves, Andr. Kmbl. 873; An. 437. Lácende líg the leaping flame, Cd. 197; Th. 246, 8; Dan. 476: Exon. 31 a; Th. 97, 23; Cri. 1595: Elen. Kmbl. 1156; El. 580: 2219; El. 1111. II. to play [as in 2. Sam. 2, 14 'Let the young men play before us ... And every one thrust his sword in his fellow's side,' cf. æsc-plega], make use of a weapon, fight : Ða ne dorston ǽr dareðum lácan on hyra mandrýhtnes miclan þearfe who before had not dared at their lord's dire need to play in the javelin-play, Beo. 5689; B. 2848. III. to play [a musical instrument] :—Hió dumb wunaþ hwædre hyre is on fôte fæger hleóþor; wrætlíc mē þinceþ hū seó wiht mǽge wordum lácan þurh fót neoþan dumb does it dwell, yet in its foot hath a fair voice; wondrous it seems to me how the wight can play with words by its foot from below, Exon. 108 b; Th. 414, 13; Rä. 32, 19. [Orm. to þeowwtenn Godd and lakenn [sacrifice], 973; þa þre kingess lakedenn [presented] Crist wiþþ þrinne kinne lakess, 7430: Havel. leike; p. leikede to play: Piers P. laike to play: Goth. laikan; p. lailak: Icel. leika; p. lēk: M. H. Ger. leichen.] DER. be-, for-, geond-lácan: daroþ-, faroþ-, lyft-lácende. v. lǽcan, ellen-lǽca, and preceding word.

lâc-dǽd, e; f. Munificence; munificentia, Hpt. Gl. 496.

lâc-gifa, an; m. One who gives gifts :—Drihten is lácgeofa manna bearnum dominus dedit dona hominibus, Ps. Th. 67, 18.

lacing (?) :—Ðis sint ða landgemǽra. ǽrest of cealcforda on ealdan lacing ... ðoñ tô smalan wege and on lacing, Cod. Dip. Kmbl. ii. 317, 22–26. [Cf. (?) lacu.]

lâc-lîc; adj. Sacrificial, having the nature of a sacrifice or offering :— Swâ oft swâ hí offrodon ða láclícan lác ðe ðá gewunelíce wǽron as often as they offered the sacrificial offerings that were then customary, L. Ælfc. P. 39; Th. ii. 380, 18.

lâcnian; p. ode To heal, cure, tend, take care of, treat, dress (a wound) :—Ic lácnige medeor, Ælfc. Gr. 33; Som. 36, 47. Se lǽce ðonne hē on untíman lácnaþ wunde hió wyrmseþ secta immature vulnera deterius infervescunt, Past. 21, 2; Swt. 153, 3. Ðæt lácnaþ ðone milte that heals the milt, L. M. 2, 38; Lchdm. ii. 246, 11. Hē mid ælmessan sáwla lácnaþ, Exon. 122 a; Th. 467, 30; Alm. 9. Betwyh ðon ðe hine mon lácnode inter medendum, Bd. 4, 26; S. 603, 15. Lácnode fomentat, Wrt. Voc. ii. 37, 17. Lǽcnode, 91, 39. Hē hine lácnude curam ejus egit, Lk. Skt. 10, 34. Lécnade monigo curavit multos, Mk. Skt. Lind, 1, 34. Ne ða wanhálan gē ne lácnedon neque ægras sanavistis, L. Ecg. P. iii. 16; Th. ii. 202, 26. Ðonne ðæt dolh open sý genim ða ylcan wyrte unsodene ... lácna ða wunde ðǽrmid ðonne byþ heó sóna hál when the incision (made by a snake) is open, take the same plant unsodden ... dress the wounds therewith; it will soon be well, Herb. 90, 16; Lchdm. i. 198, 16. Lácna mid ðý, L. M. 1, 30; Lchdm. ii. 70, 19. Lá lêce lécna ðec solfne medice cura te ipsum, Lk. Skt. Rush. 4, 23. Cymeþ and

lécnigaþ venite et curamini, 13, 14. Ðonne sceal man mid cealdum lǽcedómum lácnian it must be cured with cold medicines, L. M. 1, 1; Lchdm. ii. 22, 4. Ðan scealt ðû hine ðus lácnigean, Lchdm. iii. 126, 12. Freónd ðe his gýmenne dyde and his wunda lácnian wolde amicos qui sui curam agerent, Bd. 4, 22; S. 591, 2. Ðis is þearf ðæt se se ðe wunde lácnian (Hatt. MS. lácnigean) wille géote wîn on necesse est, ut, quisquis sanandis vulneribus praeest, in vino morsum doloris adhibeat, Past. 17, 10; Swt. 124, 11. Se lácnigenda the physician, 21, 2; Swt. 153, 4. Lácnod wæs fram his wundum curabatur a vulneribus, Bd. 4, 16; S. 584, 30. [O. E. Homl. lechinen : Laym. lechinien (2nd MS. lechnie), lacnien (2nd MS. lechni): A. R. lecnen : Piers P. lechnede (other MS. lechede), p.: Goth. lēkinon, leikinon to cure, heal : O. L. Ger. lácnôn mederi : Icel. lækna : O. H. Ger. láhinon mederi, fomentare, temperare.] v. ge-lácnian, lǽcnan; lǽce.

lâcnigend-lîc; adj. Medical, surgical : —Lácnigendlíc tôl a surgical instrument, Hpt. Gl. 478.

lâcnung, lǽcnung (v. sealf-lǽcnung), e; f. Healing, cure, remedy, medicine :—Lácnung medicamen, R. Ben: medicamentum, Hpt. Gl. 478. On gôdan lǽce biþ gelang seóces mannes lácnung the sick man's cure depends on a good doctor, L. Pen, 1; Th. ii. 278, 4. Ða hē gehǽlde ðe lácnunga beþorftun eos qui cura indigebant sanabat, Lk. Skt. 9, 11. Gebête wið hine ða wunde and begyte him ða lácnunge compenset ei vulnus, et sanationem ei comparet, L. Ecg. P. iv. 22; Th. ii. 210, 25 [O. E. Homl. hit (Christ's blood) beo mi lechnunge, i. 202, 16: Jul. ne mahte he wið ute þe lechnunge of hire luue libben, 7, 4: Icel. lækning a cure, medicine; the art of healing: Dan. lægning healing: O. H. Ger. láchenunga medicine.]

lacra, Fins. Th. 68; Fin. 34. v. læc.

lâc-sang, es; m. A song made when offering (?) :—Lácsang (MS. lane sang) offertorium, Ælfc. Gl. 34; Som. 62, 62; Wrt. Voc. 28, 42.

lactuca, an; f. This word seems to retain its Latin form in the nominative, but otherwise conforms to English usage, and is generally treated as a weak noun. The form lactucas, however, occurs in the Leechdoms, which, though it looks like a strong plural masc., seems to be singular :—Lactuca hætte seó wyrt ðe hí etan sceoldon mid ðám þeorfum hláfum heó is biter on þigene lettuce was the name of the herb that they were to eat with the unleavened loaves; it is bitter in the eating, Homl. Th. ii. 278, 26. Nim lactucan áne hand fulle take a hand full of lettuce, Lchdm. iii. 114, 13. Eton þeorfe hláfas mid ðære lactucan ðe on felda wixþ edent azymos panes cum lactucis agrestibus, Ex. 12, 8. Etan þeorfe hláfas mid feldlícere lactucan, Homl. Th. ii. 264, 3. Lácnian innan mid lactucan to cure by the internal application of lettuce, L. M. 2, 37; Lchdm. ii. 244, 16. Mid feldlícum lactucum, Homl. Th. ii. 278, 19. Him is tô sellanne lactucas lettuce is to be given him, L. M. 2, 33; Lchdm. ii. 212, 7. Him is nyt ðæt hē hláf þicge and lactucas ðæt is leahtric it is beneficial for him to eat bread, and lactucas, that is, lettuce, 16; Lchdm. ii. 194, 6. [O. H. Ger. ladducha, latoch, lattouch lactuca, Grff. ii. 202.]

lacu, e; f. A pool, pond, piece of water, lake :—Oþ ðæt seó lacu ût scýt—ðæt norþ andlang lace to the point where the water runs out of the lake ... then along the lake, Cod. Dip. Kmbl. ii. 250, 26. Ðonne of exa[n] on ða smala[n] lace of ðære lace eft on exan then from the Exe to the small pool, from the pool again to the Exe, ii. 205, 10. Tô æscwylles lace heáfdon, 24. Tô æscwylles lace, 20. On Suttûninga lace, iii. 211, 23. Andlang foslace, 25, 19. On ða ealdan lace ; andlang lace on ða norþeá, vi. i. 20. Laca lacos, Wrt. Voc. ii. 51, 52. [Meres and laces, Chr. 656; Erl. 31, 19 : Laym. ouer þen lac (2nd MS. þe lake) of Siluius and ouer þen lac (2nd MS. þan lake) of Philisteus : Prompt. Parv. lake lacus. It might be supposed that lacu was taken from Latin lacus, and the fact that the gender of the Latin is not that of the English word does not disprove the supposition; for feminine porticus gives masculine portic, and masculine versus gives neuter fers. And in the specimens of later English just quoted (in Laym. it will be observed the gender is no longer feminine) it may have been to Latin that the English word is due; but there may have been at an earlier time a native word: cf. leccan to water, and O. H. Ger. lacha; cf. palus, botinus, Grff. ii. 100.]

lâd, e; f. I. a course, way :—Micel is lád ofer lagustreám great is the way across the water, Andr. Kmbl. 845; An. 423: Exon. 94 a; Th. 353, 17; Reim. 14. Brimwudu láde fûs the ship swift in its course, 52 a; Th. 182, 6; Gû. 1306. Ne lǽt ðû ðec síðes getwǽfan láde geletan lifgende monn do not thou let living man divert thee from thy journey, hinder thee from thy way, 123 b; Th. 474, 3; Bo. 24: Beo. Th. 1142; B. 569. Hû lomp eów on láde ðá ðû gehogodest sæcce sécean ofer sealt water, 3978; B. 1987. Ic freóndu beþearf on láde ðonne ic sceal langne hám áua gesécan I need friends on my way, when alone I must seek my long home, Apstls. Kmbl. 183; Ap. 92: Andr. Kmbl. 551; An. 276. Noe tealde ðæt hē (the raven) hine, gif hē on ðære láde land ne funde, sécan wolde, Cd. 72; Th. 87, 5: Gen. 1444. Se ûs ðás láde sceóp who shaped this course for us, 89; Th. 110, 21; Gen. 1841. II. a lode, watercourse (as a component in local names) :—Mariscem quam circumfluit Iægnlaad, Cod. Dip. Kmbl. i. 190, 6. Ad aquæ ripam Iaenláde, 163, 16. Cappelád, Wodelád are other instances occurring in the Charters. III. carrying, carriage, bringing (see lǽdan) :—Sunnandæges cýpinge wē forbeódaþ and ælc weorc and ælce láde ægðer ge on wǽne

ge on horse ge on byrdene *we forbid Sunday traffic and all work and all carrying (of goods,* &c.) *both by waggon and by horse and by the man himself,* L. N. P. L. 55; Th. ii. 298, 22. [The word *lád* in this passage can hardly be translated 'journeying;' for, in the first place, such a meaning does not well suit the phrase *on byrdene,* and, next, some journeying was allowed. Thus, L. E. I. 24; Th. ii. 420, 21–, it is said no secular work was to be done 'bútan hwam gebyrige ðæt hē nýde faran scyle'; ðonne mót hē swá rídan swá rówan swá swilce færelde faran swylce tó his wege gebyrige.' The threefold division of the means of carriage seems to be that found in the Icelandic law where, dealing with the observance of Sunday, it is said of the amount that might be carried in journeying on that day 'er rētt at bera â sjálfum ser (= on byrdene) eþa fara â skipi eþa bera â hrossi.'] On sumon hē sceal lǽde lǽdan *on some lands the 'geneát' has to furnish means of carriage,* L. R. S. 2; Th. i. 432, 14. Cf. 436, 5–6:—Hē sceal beón gehorsad ðæt hē mǽge tó hláfordes seáme ðæt syllan oððe sylf lǽdan. The word used in both cases in the Latin translation is *summagium,* in reference to which, and to the English words which it translates, may be quoted Thorpe's explanation in his glossary : ' Lád, seám, summagium. A service, which consisted in supplying the lord with beasts of burthen, or, as defined by Roquefort (*voce* somey): " Service qu'un vassal devoit à son seigneur, et qui consistoit à faire faire quelques voyages par ses bêtes de somme." *See* Spelman *sub voce, and* Du Cange *voce* Sagma.' The phrase *láde lǽdan* occurs in a similar passage, dealing with the duties of the 'geneát,' in Cod. Dip. Kmbl. iii. 450, 31–:—Se geneát [at *Dyddanham*] sceal wyrcan swá on lande, swá of lande, hweðer swá man být and rídan, and auerian, and láde lǽdan, dráfe drífan, and fela oðra þinga dón. The later English *lode* seems to keep this meaning. Thus *Prompt. Parv.* 310, loode or caryage *vectura ;* lodysmanne *vector, lator, vehicularius :* the verb *lead* is found with the sense of *carry,* e. g. p. 62 cartyn or *lede* wythe a carte ; and in the note, and again in a note on p. 293, we have the phrases ' to *lede* dong,' ' to *lede* wheet,' &c. See also scip-lád. IV. *Sustenance, provision, means of subsistence :*—Ne sceal se dryhtnes þeów in his módsefan máre gelufian eorþan ǽhtwelan ðonne his ánes gemet ðæt hē his líchoman láde hæbbe *nor shall the servant of the Lord love more of earth's possessions, than a sufficiency for himself, that he may have sustenance for his body,* Exon. 38 a; Th. 125, 27; Gū. 360. With this use of *lád* may be compared the later English *lif-lode* which, besides the meaning *conduct,* has that of *sustenance :*—Heo tilede here lyflode . . . heo fonden hem sustynance ynow, R. Glouc. 41, 22 : *Prompt. Parv.* lyvelode *victus ;* lyflode or warysone *donativum.* So *O. H. Ger.* lîb-leita *victus, annona, alimonium.* [In further illustration of *lád* the following native and foreign words are given. *Orm.* þe steoressmann aʒʒ lokeþþ till an steorrne þatt stannt aʒʒ still . . . forr þatt he wile follʒhenn aʒʒ þatt illke steorrness *lade (guidance) ;* o lade *on the way : A. R.* lode *burthen* (v. III) : *Mod. E.* lode-star : *Icel.* leið. I. *a way, course, road.* II. *a levy : O. H. Ger.* leita, *funus, ducatus ; pl. exequiæ ; see also compounds of* leiti, Grff. ii. 187]. DER. brim-, eá-, ge-, in-, lagu-, lif-, mere-, sǽ-, scip-, út-, ýð-lád.

lád, e ; f. I. *excuse, defence against a charge :*—Nú hí nabbaþ nâne láde be hyra synne *nunc excusationem non habent de peccato suo,* Jn. Skt. 15, 22. Ðætte hē nâne láde ne mǽge findan ac síe súa mid his âgnum wordum gebunden *et in nulla sui defensione se exerceat, quam sententia proprii oris ligat,* Past. 26, 3; Swt. 185, 16. Ða nǽnige láde gedón ne mágon on dómes dæge ah sceolon mid deóflum in ēce wîte gefeallan *those will not be able to make any defence at the day of judgment, but will have to fall with devils into everlasting punishment,* Blickl. Homl. 57, 20. II. as a technical term in the laws, *purgation, exculpation, the clearing one's self from a charge or accusation.* The accused might clear himself by his own oath, supported by the oaths of a certain number of compurgators, or he might undergo some form of ordeal. The *lád* varied with the character of the deed with the commission of which the accused was charged. In the *ánfeald lád,* if the purgation were by oath, the oaths of the accused, and two others were necessary, in the *þrýfeald lád,* the accused was to bring five compurgators ; if the ordeal was used, in the former case the iron weighed one pound, in the latter, three. Other passages than those cited below, which may illustrate the terms *ánfeald, þrýfeald,* are the following :—Wē cwǽdon ðe ðam morþslyhtum ðæt man dýpte ðone áþ be þrýfealdum and myclade ðæt ordálísen ðæt hit gewege þrý pund, L. Ath. iv. 6; Th. i. 224, 12–14. Gange hē tó ðam þrýfealdan ordále ; and ofgá man ðæt þrýfealde ordál ðus : nime fífe and beó hine sylfa syxta, L. C. S. 30; Th. i. 394, 3–5 : 44; Th. i. 402, 7. The term 'lád,' it will be seen from the following passages, does not, as Schmid observes, occur in the laws before Ethelred's time, *canne* and *andsæc* being used previously :—Gyf mon ðone hláford teó . . . nime him fíf þegnas tó and beó him sylf syxta and láde hine ðæs. And gif seó lád forþcume beó hē ðæs weres wyrðe *if the lord be accused . . . let him take to himself five thanes, and be himself the sixth, and clear himself of the charge. And if he be successful in clearing himself, let him be entitled to the 'wer,'* L. Eth. i. 1; Th. i. 282, 7 : L. C. S. 30; Th. i. 394, 22. Gif him seó lád byrste *if the attempt to clear himself fail,* L. Eth. i. 1; Th. i. 282, 14 : L. C. S. 8; Th. i. 380, 21 : 31; Th. i. 396, 5. Gif lád for-

berste, 54; Th. i. 406, 10. Ðeáh lád teorie, L. O. D. 4; Th. i. 354, 14: 6; Th. i. 354, 31. Ne stent nân óðer lád æt tihtlan búte ordál betweox Wealan and Englan búte man þafian wille *no other method of clearing a man upon accusation is valid between Welsh and English but the ordeal, unless it be permitted,* L. Eth. i. 354, 1. Láde wyrðe beón *to be entitled to clear one's self (by oath or by ordeal),* L. C. S. 20; Th. i. 386, 21. Sý ælc getrýwa man ðe tihtbysig nǽre and nâðor ne burste ne áþ ne ordál ânfealdre láde wyrðe *let every true man that has not previously been accused, and in whose case neither oath nor ordeal has failed, be entitled to single purgation,* 22; Th. i. 388, 11. Dúnstan gedēmde ðæt se mæssepreóst nǽre, gif hē wíf hæfde, ǽnigre óðre láde wyrðe, bútan eallswá lǽwede sceolde ðe efenboren wǽre, gif man mid tihtlan ðæne belēde, L. Edg. C. 60, note ; Th. ii. 256, 38. Gebyreþ ðæt mon óðrum riht wyrce ge æt láde ge æt ælcre sprǽce ðe him betweox biþ *it is proper for men to do right to one another both as regards clearing themselves of charges and as regards any suits that there are between them,* L. O. D. 2; Th. i. 352, 17. Gif æt láde mistíde dēme se bisceop *if the attempt to clear himself miscarry, let the bishop pass sentence,* L. C. S. 57; Th. i. 406, 27. Geládige hine mid fulre láde, 42; Th. i. 400, 25. Geládige swá mid þrýfealdre swá mid ânfealdre láde be ðam ðe seó dǽd sí, L. C. E. 5; Th. i. 364, 2 : L. Eth. ix. 27; Th. i. 346, 15. Ládige hine mid þrýfealdre láde, L. C. S. 8; Th. i. 380, 20 : 48; Th. i. 404, 3. Ofgá man ânfealde láde mid ânfealdan foraþe and þrýfealde láde mid þrýfealdan foraþe [*the Latin version has the following in explanation :*—Qui autem conquirere debet simplicem púrgationem, simplici sacramento hoc faciat, hoc est, accipiat duos et sit ipse tertius, et sic jurando conquirat. Triplex vero juramentum sic conquiratur ; accipiat quinque et ipse sit sextus, et sic jurando acquirat triplex judicium aut triplex juramentum'], 22; Th. i. 388, 14. Se gerēfa namige ða láde *let the reeve name the compurgators,* L. Eth. iii. 13; Th. i. 388, 15. Se ðe ofer ðæt láde geþafie oððe se ðe hý sylle gilde vi healfmarc *he that admits, or he that offers, purgation after that, shall pay six half-marks,* Th. i. 298, 7. Hēr swutelaþ an (ðissum gewrite) ðæt Godwine hæfþ gelǽd fulle láde æt ðan unrihtwífe ðe Leófgâr bisceop hine tihte and ðæt wæs lǽd æt Licitfelda *in this writing is declared that Godwine has fully cleared himself of the charge in the matter of the woman about whom bishop Leofgar accused him : and he cleared himself at Lichfield,* Chart. Th. 373, 31. *See* wer-lád, cor-snǽd, ordál, ládian ; Stubb's Const. Hist. i. 609–; Grmm. R. A. 856, 859–; Du Cange *sub voce* lada ; Richthofen's Altfries. Wört. lēde, láde.

ládian, *p.* ode. I. *to excuse, clear* [*one's self of a charge*], *exculpate, defend :*—Ðe hit symle lytiglíce ládaþ *sese callide defendentis,* Past. 35, 3; Swt. 244, 9. For ðan dú tôwyrpest ðíne fýnd and ealle ða ðe unrihtwísnesse ládiaþ and scyldaþ *ut destruas inimicum et defensorem,* Ps. Th. 8, 3. Ðære leóhtmódnesse sanctus Paulus hine ládode ðá hē cwæþ . . . *a mentis levitate se alienum Paulus fuisse perhibuit, cum dicit . . .* Past. 42, 3; Swt. 308, 7. Ðá ládode hē hine *ille se excusans,* Bd. 3, 7; S. 530, 26. Ðá cwæþ Petrus wǽre ðú mid ðínum fæder ðá hē mē swá ládode ðæt hie mē ne gegripon *then said Peter ' Wast thou with thy father when he made such excuse for me that they did not seize me?'* Blickl. Homl. 151, 26. Him Rômâne his forwierndon and hit under ðæt ládedon for ðon ðe hē ǽr æt ðæm óðrum cirre sige næfde *the Romans refused it [the triumph] to him, and excused [the refusal] under the pretext that before on the other occasion he had not gained the victory,* Ors. 5, 2; Swt. 216, 31. Ic bidde ðē ðæt ðú mē ládige *I pray thee to excuse me,* Homl. Th. ii. 374, 10. Ðæt synfulle môd ðe hit simle wile ládian *peccantem animam excusantemque se,* Past. 35, 3; Swt. 241, 7. Hú mæg ic ládigan láðan sprǽce oððe andsware ǽnige findan wrâðum tôwiðere *how can I clear myself of the hateful speech, or find any answer in reply to my foes?* Exon. 10 b; Th. 12, 9; Cri. 183. II. as a technical legal term [lád, II.] *to clear from an accusation.* [Amongst instances in which suspicion of crime is removed by the oath of the suspected party and the oaths of compurgators, may be taken that of King Alfonso who, when suspicion rested on him of complicity in the murder of his brother Sancho, cleared himself by the oaths of himself and twelve of his vassals. See the account in the Cronica del Cid. cc. 76–79.] :—Gif se húshláford hit nât ládie hine [*shall clear himself by oath*] si latet fur, dominus domus . . . jurabit, quod non extenderet manum in rem proximi sui, Ex. 22, 8. Gif hē hine ládian wille gâ hē tó ðam hâtum îsene and ládige ða hand mid ðe man týhþ ðæt hē ðæt fácen mid worhte *if he be willing to clear himself, then let him undergo the ordeal by hot iron, and therewith clear the hand with which he is accused of committing the fraud,* L. Ath. i. 14; Th. i. 206, 22–4. Gyf mon ðone hláford teó, nime him fíf þegnas tó, and beó him sylf syxta, and ládie hine ðæs [*by his own oath and the oaths of five compurgators clear himself of that charge*], L. Eth. i. 1; Th. i. 282, 4–6, 13. Hē hine twelfa sum ládige ðæt hē ða sôcne nyste *let him clear himself by his own oath, supported by the oaths of eleven others, from the charge of having known that the slain man had sought sanctuary,* L. Ath. iv. 4; Th. i. 224, 2. Gif man hwilcne man teó ðæt hē ðone man fēde ðe úres hláfordes griþ tôbrocen habbe ládige hine mid þrinna xii (cf. Icel. þrennar tylftir), L. Eth. iii. 13; Th. i. 296, 29. Mæssepreóst ládige hine on ðam húsle . . . Diacon nime six his gehâdan and ládige mid ðâm . . . &c. I.

Eth. ix. 19-27; Th. i. 344, 346: L. C. E. 5; Th. i. 362, 364. Bûtan hê hine ládian mæge ðæt hê him nân fâcn on nyste *unless he can clear himself from the charge of having known of any fraud in the man*, L. Ath. iv. 4; Th. i. 224, 6. Bûtan hê hine ládian durre be ðæs flýman were [*the degree* of *lád to be determined by the status of the fugitive*] ðæt hê hine flýman nyste, i. 20; Th. i. 210, 13. Ládian be ðæs cynges wergilde oððe mid þrýfealdan ordâle, L. Eth. v. 30; Th. i. 312, 6. Ládian be ðam deópestan âþe oððe mid þrîfealdan ordâle, vi. 37; Th. i. 324, 18. Gif mon cyninges þegn beteó manslihtes, gif hê hine ládian dyrre, dô hê ðæt mid xii cyninges þegnum, L. A. G. 3; Th. i. 154, 6. Gif se hláford hine ládian wylle mid twâm gódum þegenum, L. Eth. iii. 4; Th. i. 294, 12. DER. â-, be-, ge-ládian; *see previous word.*

ládigend-lîc; *adj. Excusable:*—Ládiendlíce excussabile, Wrt. Voc. ii. 146, 19.

lád-mann, es; *m. A leader, guide:*—Ðû canst wegas geond ðæt wêsten beó ûre ládmann *thou knowest the ways through the desert; be our guide;* eris ductor noster, Num. 10, 31. Abram fêrde of Egipta lande and Farao him funde ládmen *præcepit Pharao super Abram viris et deduxerunt eum*, Gen. 12, 20. [Cf. *Laym.* ȝe scullen habben lædesmen and ford ȝe scullen liðen (2nd MS. lodesmen forþ ȝou to lede): *Ayenb.* þe ssipmen yhyerþ þane smite of þe lodesmanne: *Prompt. Parv.* p. 311, n. lodesman *pilot.*]

lád-rinc, es; *m. A word of uncertain meaning occurring in the following passage:*—Gif cyninges ambihtsmiþ oððe laadrinc mannan ofslehþ meduman leódgelde forgelde *if the king's smith or 'ládrinc' kill a man, let him pay for it with a half fine* [cf. § 21; Th. i. 8, 3), L. Ethb. 7; Th. i. 4, 8. The word, as Schmid observes, might have the same meaning as *lád-mann* q. v. just as Layamon uses the compound *lod-cniht*, 'biforen rad heore *lod-cniht*' 25730; or taking *lád* in the sense of journey the reference may be to a messenger of the king, cf. L. In. 33; Th. i. 122, 13 where it speaks of 'Cyninges horswealh se ðe him mæge geǽrendian.' But there is another use of lád [v. lád, III) which perhaps is that in the passage; then the *lád-rinc* would be the king's carrier, one who did for the king similar service to that which the *geneát* does for his lord. In the Prompt. Parv. *lodysmanne* is rendered by *vector, lator, vehicularius.*

ladsar *laserwort;* laserpitium:—Nim ladsar, Lchdm. iii. 88, 20.

lád-scipe, es; *m. Leadership, command;* ducatus, Wrt. Voc. ii. 72, 70.

lád-teáh, lât-têh; *gen.* -teáge, -tége; *f. A leading-rein:*—Lǽtteh *ducale*, Ælfc. Gl. 21; Som. 59, 64; Wrt. Voc. 23, 24.

lád-teów, es; *m. A leader, guide, conductor, a leader in war, general:*—Ænne of þám þrím englum ða ðe him on æghwæðere gesihþe ládteów wæs *unum de tribus angelis, qui sibi in tota utraque visione ductores adfuerunt*, Bd. 3, 19; S. 548, 31. Ðæt hê his ládteów beón sceolde on Breotone *ut ipse eum perduceret Brittaniam*, 4, 1; S. 564, 15. Hengest se ðe wæs ǽrest ládteów and heretoga Angelcynnes on Breotene *Hengist qui Brittaniam primus intravit*, 2, 5; S. 506, 34. Hê sende fyrd ðære wæs Beorht ládteów and heretoga *misso cum exercitu duce Bercto*, 4, 26; S. 602, 5. Ládteáw, Bt. tit. 36; Fox xviii. 4. Lâteáu, Kent. Gl. 131. Ládtow *dux*, Ps. Surt. 30, 4: 54, 14. Mîn ládþeów *dux mihi*, Ps. Th. 30, 4: Ps. Spl. C. 54, 14. Ðû eart ǽgðer ge weg ge ládþeów *tu semita, dux*, Bt. 33, 4; Fox 132, 37. Lâtteów *dux*, Ælfc. Gr. 33; Som. 37, 49. Heretoga and lâtteów *dux*, Bd. 1, 16; S. 484, 18. Lâtteów wæs ðara leóda *duces eorum*, Ps. Th. 67, 25. Ic eom ealdor and lâtteów drihtnes heres *sum princeps exercitus domini*, Jos. 5, 14. Wilferþ bæd ðæt hê him ðæs siiþfǽtes lâtteów wǽre *Vilfridum ducem sibi itineris fieri rogaret*, Bd. 4, 5; S. 571, 35: 2, 20; S. 521, 41. Lâtteów ðæs weges, Ælfc. T. Grn. 18, 11. God, lífes lâtteów, Elen. Kmbl. 1037; El. 520: 1794; El. 899. Lífes lâtþeów, Cd. 147; Th. 184, 8; Exod. 104. Wæs ðæt se mîn lâtþeów se ðe mê ǽr lædde *ille erat ipse qui me ante ducebat*, Bd. 5, 12: S. 629, 8. Lâtþeów *ductor*, S. 629, 40. Lâtþeów *dux*, Ps. Spl. 54, 14. Lífes lâdteów *the guide of life*, Dôm. L. 52, 9. Ðes and ðeós lâteów oððe heretoga *hic et hæc dux*, Ælfc. Gr. 9; Som. 14, 9: Wrt. Voc. 72, 60. Drihten ðe eówer lâteów ys *dominus qui ductor est vester*, Deut. 31, 8. Ðæt hê ðæs lâtteówes lârum hýre *that he listen to the guide's instructions*, Exon. 37 b; Th. 124, 5; Gú. 335: Elen. Kmbl. 2417; El. 1210. Hê sôhte hine him tô lâtðeówe on ðæm wege *ducem requirebat in via*, Past. 41, 5; Swt. 305, 5. Seó leó gif heó blôdes onbirigþ âbît ǽrest hire lâdteów *the lioness, if she tastes blood, will first rend her keeper;* primusque lacer dente cruento domitor rabidas imbuit iras, Bt. 25; Fox 38, 14. Þurh sume ða Wyliscean ðe him tô wǽron cumen and his lǽdteówas wǽron *by means of some of the Welsh who had come to him and were his guides*, Chr. 1097; Erl. 233, 39. Hig synt blinde and blindra lâtteówas (Lind. lâtuas) *cæci sunt, duces cæcorum*, Mt. Kmbl. 15, 14. Wǽron heora lâtteówas and heretogan twegen gebrôðra Hengest and Horsa, Bd. 1, 15; S. 483, 27. Ic mê ðâ mid genom .cc. lâdþeówa and eác .l. ðe ða gênran wegas cûðan ðara sîðfato *acceptis .cl. ducibus qui brevitates itinerum noverant*, Nar. 6, 7. Gê preóstas synd gesette tô ládþeówum and tô lâreówum ofer Godes folc. L. Ælfc. P. 5; Th. ii. 366, 4. Him ðâ Rômâne æfter ðæm lâdteówas gesetton, ðe hie consulas hêton, Ors. 2, 2; Swt. 68, 2. Ealle mîne lâdþeówas ðe mec on swelc earfeðo gelǽddon *locorum demonstratores qui nos in insidias deducebant*, Nar. 16, 25. In Mt. Kmbl. Lind. 2, 6: Rtl. 38, 15: 193, 15, the form *látwa*

with pl. *látuas*, Mt. 15, 14, occurs; also *látwu*, Rtl. 193, 17, 19; and in 2, 5 *látuan* glosses *ducere*. [O. E. Homl. latteu *a guide: Jul.* lauerd, liues lattow: *cf. Icel.* leið-togi *a guide.*] v. under-lâdteów.

lâdteów-dôm, es; *m. Leadership, guidance, conduct:*—Mid engla lâdþeówdôme *ducentibus angelis*, Bd. 4, 3; S. 568, 41. Ðÿlæs hî underfô ðone lâdteówdôm (Hat. MS. lâtteówdôm) ðæs forlores *ne ducatum suscipiat perditionis*, Past. 3, 1; Swt. 32, 9 Ðone lâdteówdôm (Hat. MS. lâttiówdôm) ðæs folces *plebium ducatum*, 7, 2; Swt. 50, 18. Lâdteówdôm (Hat. MS. lâtteówdôm) geearwian *ducatum præbere*, 18, 7; Swt. 138, 16. Lâdteówdôm *magisterium, pædagogium*, Hpt. Gl. 477.

lâdung, e; *f.*   I. *An excusing, a clearing of* or *defending against a charge, an apology, excuse, a defence, exculpation:*—Lâdung *apologia*, Ælfc. Gl. 106; Som. 78, 64; Wrt. Voc. 57, 43: *excussatio*, Wrt. Voc. ii. 146, 15. God lǽt him fyrst ðæt hê his mândǽda geswîce gif hê wile: gif hê nele ðæt hê beó bûtan ǽlcere lâdunge swîðe rihtlíce tô deófles handa âsceofen *God allows the wicked man time, that he may, if he will, cease from his wicked deeds: that, if he will not, he may, having nothing to plead in his defence, very justly be thrust into the hands of the devil*, Homl. Th. i. 270, 1. Môd ymbtrymedu mid lytelícre lâdunge *mentes fallaci defensione circumdatæ*, Past. 35, 5; Swt. 245, 8. Hî simle sêceaþ endeléase lâdunga *semper improbas defensiones quærunt*, 35, 2; Swt. 239, 8.   II. *as a legal term, purgation, the clearing himself on the part of an accused person, by oath or by some form of ordeal, of the charge made against him:*—And stande betwux burgum án lagu æt lâdunge, L. C. S. 34; Th. i. 396, 22. Bisceop sceall æt tihtlan lâdunge gedihtan ðæt ænig man ôðrum ænig wôh beódan ne mæge aðor oððe on âþe oððe on ordâle *when accusation is made, the bishop shall so order the proceedings by which the accused is to clear himself, that no man may be able to offer wrong to another in the matter of taking oath or of undergoing the ordeal*, L. I. P. 7; Th. ii. 312, 15. v. lád, ládian, be-lâdung.

lǽ *hair:*—Lǽ wiffex *cæsaries*, Wrt. Voc. ii. 16, 46. [*Icel.* lá *hair: cf.* ló, lóð *shagginess;* also *a flock of wool.*] Perhaps we may compare here *lee* of threde, Prompt. Parv. 291, where the following note is given. 'Forty threads of hemp-yarn are termed in Norfolk a lea. The "lea" by which linen yarn was estimated at Kidderminster, contained 200 threads.' Halliwell gives as a northern word '*lea* the seventh part of a hank or skein of worsted.'

lǽc *a gift.* v. lác.

lǽc; *adj.* The word, if this be the true form of it, occurs only once, in the following passage:—Gârulf gecrang ealra ǽrest . . . ymb hyne gôðra fela hwearf lacra hrǽr hrǽfn wandrode sweart and sealobrún, Fins. Th. 64-70; Fin. 33-5. All the editors for *hrær*, which Hickes gives, read *hræw*, but in the MSS. *r* (ꞃ) and *s* (ſ) are so nearly alike that perhaps *hræs*, the genitive of *hrá*, was the original word. With regard to *lacra* various explanations have been given. Kemble and Conybeare print *hwearflacra*, Ettmüller reads *hwearfllicra*, Thorpe *hwearf láðra*, Grein *hwearf lacra*. Taking the word to be independent, and retaining the reading of Hickes, we may compare it with *Icel. lakr* lacking, defective, and render it by *weak, failing (from wounds), wounded*. Another form that attracts comparison is given by Graff ii. 100, *lah*, which has reference to cutting, and this suggests the rendering *wounded*. With the reading *hræs* for *hrær* the passage might be translated '*first of all sank down Garulf . . . around him moved many a stout man weak or wounded in body: the raven wheeled round swart and dusky.*' Ettmüller p. xxiv, giving a meaning to *wandrian* which it will hardly bear, translates the doubtful part of the passage '*volubilium (=mortuorum) cadavera corvus conculcavit.*' Similarly, as regards the first part, Conybeare has '*circa illum fortes multi caduci moriebantur.*'

lǽca, an; *m. A leech, doctor, physician:*—Se lǽca ðe sceal sâre wunda wel gehǽlan hê môt habban gôde sealfe ðǽrtô *the doctor who has to make a good cure of painful wounds, must have good salve for the purpose*, L. Pen. 4; Th. ii. 278, 15: 5; Th. ii. 278, 20. v. lǽce.

-lǽca. v. ag-, ellen-, lyb-, scîn- lǽca.

lǽcan; p. lǽhte, lǽcte *To move quickly, spring, leap* [as *flame*]:—Hwîlum se wonna lêg lǽhte wið ðes lâþan *at times the lurid flame leaped towards the fiend*, Cd. 229; Th. 309, 25; Sal. 716. DER. Æfen-, dyrst-, ed-, efen-, geán-, gedyrst-, geneá-, geriht-, geþríst-, lof-, neá-, riht-, sumor-, þríst-, winter-lǽcan; *and see* lácan.

lǽccan, lǽccean; p. lǽhte; pp. lǽht *To take, grasp, seize, catch, apprehend, capture:*—Lǽdeþ hine and lǽceþ and hine geond land spaneþ *leadeth and taketh him, and through the land lures him*, Salm. Kmbl. 989; Sal. 496. Hî lǽccaþ of manna begeatum hwæt hî gefón mǽȝn eallswá gýfre hremnas of holde dôþ *they seize of men's gettings what they can grasp, just as greedy ravens do from a corpse*, L. I. P. 19; Th. ii. 328, 4. Hî gærs ǽton georne and ǽlc lǽhte of ôðrum gif hê hwæt litles hæfde *they eagerly ate grass, and each seized from the other, if he had any little bit*, Ælfc. T. Grn. 21, 10. Heora ǽgðer uppon ôðerne tûnas bærnde and eác menne lǽhte *in their struggle they burned one another's towns and captured one another's men*, Chr. 1094; Erl. 230, 13. Ðætte ðíostro iuih ne lǽcga *ut non tenebræ vos compræhendant*, Jn. Skt. Lind. 12, 35. Allswǽ tô þeáfe gié foerdon mið suordum and stengum tô lǽccanne mec *tam-*

✠

quam ad latronem existis cum gladiis et lignis comprehendere me, Mk. Skt. Lind. 14, 48. Ðæt wíf wearþ ðá læht and gelǽd tó ðam cininge *sublata est mulier in domum Pharaonis,* Gen. 12, 15. [*Orm.* to lacchenn þurrh trapp; bikahht and lahht (*pp.*): *A. R.* lecche; *p.* lahte: *O. and N.* grine þe for to lacche: *Piers P.* to lacche foules; *p.* lauȝte: *Gen. and Ex.* lagt *pp.*] v. ge-læccan.

LÆCE, es; *m.* I. A LEECH, [Shakspere uses the word once, and even now it has not quite died out, but perhaps, in prose at least, its meaning is usually that given by Bailey in his Dictionary 'a Farrier or Horse-Doctor,' a doctor rather for animals than men], *doctor, physician :—Lǽce medicus,* Wrt. Voc. 74, 4. Eálá lǽce gehǽl ðé sylfne [lá lǽce lécne ðec seolfne, Lind.] *medice cura te ipsum,* Lk. Skt. 4, 23. Cyneferþ lǽce se æt hire wæs ðá heó forþférde *medicus Cynifrid, qui morienti illi adfuit,* Bd. 4, 19; S. 588, 41. Hálig lǽce [*the Deity*] Hy, 7, 62; Hy. Grn. ii. 288, 62. Hé [*the Pater Noster*] is lamena lǽce, Salm. Kmbl. 155; Sal. 77. Lǽteþ flint brecan his sconcan ne biþ him lǽce gód *he shall cause the stones to break his legs, no doctor shall avail him,* 206; Sal. 102. Nys hálum lǽces nán þearf *non est opus valentibus medico,* Mt. Kmbl. 9, 12: Lk. 5, 31: Exon. 89 b; Th. 336, 8; Gn. Ex. 45. Hé hine gelǽdde on his lǽcehús and hine lácnode and brohte óðrum dæge twegen penegas and sealde ðam lǽce *duxit illum in stabulum et curam ejus egit, et altera die protulit duos denarios et dedit stabulario,* Lk. Skt. 10, 34-5. Oððe hí lǽceas (Ps. Spl. lǽcas) weccean *aut medici suscitabunt,* Ps. Th. 87, 10. Ðeáh ða woroldlecon lǽceas [Hat. MS. lǽcas] scomaþ ðæt hí onginnen ða wunda lácnian ðe hí gesión ne mágon ... hwílon ne scomaþ ða ðe ðæs módes lǽceas bión sceoldon ðeáh ðe hí náne wuht ongitan ne cunnon ðara gǽstlecena beboda ðæt hí him onteóþ ðæt hí sín heortan lǽceas *tamen sæpe qui nequaquam spiritalia præcepta cognoverunt, cordis se medicos profiteri non metuunt: dum qui pigmentorum vim nesciunt, videri medici carnis erubescunt,* Past. 1, 1; Swt. 24, 19-26, 2. Witodlíce ne mágon lǽceas [MS. B. lǽcas] náht mycel hǽlan bútan ðisse wyrte *certainly, doctors cannot heal much without this plant,* Herb. 20, 4; Lchdm. i. 114, 22. Lǽcas lǽraþ ðisne lǽcedóm, L. M. 2, cont. 18, 20; Lchdm. ii. 160, 17, 22. Lǽceas secgaþ, 19; Lchdm. ii. 160, 19. Seó cóðu ðe lǽcas hátaþ paralisin, Homl. Th. ii. 546, 29. Gelácna ðú hý forðan ðú éðest miht ealra lǽca, Hy. 1, 6; Hy. Grn. ii. 280, 6. Fram manegum lǽcum *a compluribus medicis,* Mk. Skt. 5, 26. Is seó geoluwe swá ðeáh swíðost lǽceon [MS. B. lǽcan] gecwéme *the yellow is however most suitable for doctors,* Herb. 165, 1; Lchdm. i. 294, 11. Josep beád his þeówan lǽcon *Joseph præcepit servis suis medicis,* Gen. 50, 1. Seó fordǽlde on lǽcas eáll ðæt heó áhte *in medicos erogaverat omnem substantiam suam,* Lk. Skt. 8, 43. Lǽceas, Ep. Gl. 18 b, 21. [*O. E. Homl.* lache, leche: *Orm.* læche: *A. R.* leche: *Chauc. Piers P.* leche: *Prompt. Parv.* leche *aliptes, empiricus, medicus, cirurgicus,* a surgion; p. 291 note, q. v.: *Goth.* lēkeis, leikeis: *O. Frs.* leza, letza, leischa: *O. H. Ger.* láhhi, láche *medicus: Dan.* læge: cf. *Icel.* laknari, læknir.] v. heáh-lǽce. II. *a leech* (species of worm):—Lǽce *sanguisuga* vel *hirudo,* Ælfc. 23; Som. 60, 5; Wrt. Voc. 24, 9: *sanguisuga,* Wrt. Voc. ii. 11, 17. Lýces *sanguissuge,* Kent. Gl. 1085. [*Prompt. Parv.* leche.] -lǽcea. v. ag-lǽcea.

lǽce-bóc, e; *f. A book on medicine, book of recipes* :—Ðonne sceal him mon blód lǽtan on ðás wísan ðe ðeós lǽcebóc segþ *then shall he be let blood in these ways that this book on medicine sayeth,* L. M. cont. 2, 42; Lchdm. ii. 168, 12. [*Dan.* læge-bog *a medical book.*]

lǽce-cræft, es; *m. The art of medicine, a particular instance of the application of this art, a remedy, recipe, medicine* :—Swá gedéþ se lǽcecræft ðæt se mon biþ lǽce *medicina medicos facit,* Bt. 16, 3; Fox 54, 31. Ic ðé wille nú secgan hwelc se lǽcecræft is mínre láre hé is swíðe biter on múþe *I will now tell thee of what kind the medicine of my teaching is. It is very bitter in the mouth,* Bt. 22, 1; Fox 76, 28. Ðes lǽcecræft ys áfandud *this remedy is a proved one,* Herb. 183, 1; Lchdm. i. 320, 9. Brúce ðysses lǽcecræft[es] *use this remedy,* Lchdm. iii. 126, 20. Ðis sceal ðan manna tó lǽcecræfte *this shall be a remedy for the men,* 22. Wé habbaþ hwæðere ða bysne on hálgum bócum ðæt mót se ðe wile mid sóðum lǽcecræfte his líchaman getemprian *we have however the examples in holy books that he who will may cure his body with true leechcraft* [cf. wiccecræft l. 22], Homl. Th. i. 474, 34. Lǽcecræftas and dolgsealfa and drencas wið eallum wundum *medicines and unguents and potions for all wounds,* L. M. cont. 1, 38; Lchdm. ii. 8, 26. Lǽcecræftas be lifre ádlum *recipes for diseases of the liver,* L. M. cont. 2, 17; Lchdm. ii. 160, 10. Be wylddeóra lǽcecræftum *of medicines obtained from wild animals,* Lchdm. i. 326, 9. On ðissum ǽrestan lǽcecræftum gewritene sint lǽcedómas wið eallum heáfdes untrymnessum *in these first recipes are written remedies for all infirmities of the head,* L. M. 1, 1; Lchdm. ii. 18, 21. [Ne þurh nenne lǽcecræft ne mihte he líf habben, Laym. 7616: Þurrh Crisstenndomess lǽchecrafft, Orm. 1869: he ne secheð nout leche ne lechecraft, A. R. 178, 13: þe kyng lette do under lechecraft hem þat ywonded were, R. Glouc. 141, 6: lered lechecraft his lyf for to save, Piers P. 16, 104: *Dan.* læge-kraft *healing power.*] cf. lǽce-dóm.

lǽce-cræftig; *adj. Skilled in medicine* :—Arestolobius wæs háten án cing hé wæs wís and lǽcecræftig hé ðá gesette forðon gódne morgendrænc

wið eallum untrumnessum ðe mannes líchoman iond styriaþ *there was a king named Arestolobius, he was wise and skilled in medicine, for which reason he composed a good-morning drink for all infirmities that stir throughout man's body,* Lchdm. iii. 70, 16.

lǽce-cynn; *n. The race of physicians* or *surgeons* :—Nǽfre [ic] lǽcecynn on folcstede findan meahte ðara ðe mid wyrtum, wunde gehǽlde *never could I find on the battlefield the leeches, those who with herbs my wounds would heal,* Exon. 102 b; Th. 388, 20; Rä. 6, 10.

lǽce-dóm, es; *m. Medicine, a medicine, remedy, cure* :—Lǽcedóm *medicina,* Wrt. Voc. 74, 5: Lchdm. ii. 16, 9-27. Lécedom, Kent. Gl. 148. Lǽcedóm *malagma,* Wrt. Voc. ii. 75, 59: *cura,* 92, 61. In untrymnisse wæs ðú lécedóme *in infirmitate sis medicina,* Rtl. 105, 13. On ðare smyrunge biþ lǽcedóm and sinna forgifnes and ne biþ ná hádung *unction is medicinal, and in it there is forgiveness of sins, but there is no ordination,* L. Ælfc. P. 48; Th. ii. 384, 32. Ýdel biþ se lǽcedóm ðe ne mæg ðone untruman gehǽlan *vain is the medicine that cannot heal the sick,* Homl. Th. i. 60, 34. Búton hé ðone tíman ǽrédige ðæs lǽcedómes ðonne biþ hit swutol ðæt se lácnigenda forliésþ ðone cræft his lǽcedómes *nisi cum tempore medicamenta conveniant, constat procul dubio, quod medendi officium amittant,* Past. 21, 2; Swt. 153, 3-5. Hwí ne bidst ðú ðé lífes lǽcedómes æt lífes freán, Dóm. L. 6, 81. Mycel wund behófaþ mycles lǽcedómes *grande vulnus grandioris curam medelæ desiderat,* Bd. 4, 25; S. 599, 40. Tó lǽcedóme and tó hǽle untrumra manna *ad medelam infirmantium,* 3, 10; S. 534, 24. For hwylcum lǽcedóme *pro aliquo remedio,* L. Ecg. C. 21; Th. ii. 156, 14. Becuman tó ðam sóþan lǽcedóme *pervenire ad veram medelam,* L. Ecg. P. i. 4; Th. ii. 174, 4: Blickl. Homl. 107, 15. Ne hogaþ hé be ðam heofenlícan lǽcedóme, Homl. Th. ii. 470, 16. Wið untrumnysse lǽcedóm sécan *medicamentum contra ægritudines explorare,* Bd. 1, 27; S. 494, 18. Him lǽcedom bǽron *illis solent adferre medelam,* 4, 6; S. 574, 10. Ðá sóhte Colemannus ðysse unsibbe lǽcedóm *quæsivit Colmanus huic dissensioni remedium,* 4, 4; S. 571, 6. Ic wolde ymbe ðone lǽcedóm ðara dínra lára hwéne máre gehýran *remedia audiendi avidus vehementer efflagito,* Bt. 22, 1; Fox 76, 17. Ús is néðþearf ðæt wé sécan ðone lǽcedóm úre sáuwle, Blickl. Homl. 97, 31. Þurh his lǽcedóm *by means of the remedy he has provided,* Cd. 226; Th. 301, 30; Sat. 589. Lǽcedóm findan, Exon. 31 a; Th. 96, 13; Cri. 1573. Lǽcedómas, see Lchdm. ii. pp. 2-16: pp. 158-174. Hí tó ðam dweoligendum lǽcedómum deófolgylde éfeston *ad erratica idolatriæ medicamina concurrebant,* Bd. 4, 27; S. 604, 7. Tó lécedómum écum *ad remedia æterna,* Rtl. 23, 20. Untrymnessa lǽcedómas onfergon *languorum remedia conquisiere,* Bd. 3, 17; S. 544, 47. Lege on lǽcedómas ða ðe út teón ða yfelan wǽtan *apply remedies that may draw out the evil humour,* L. M. 1, 4; Lchdm. ii. 46, 26. [*O. E. Homl.* ȝif he lechedom con, i. 111, 2: *Orm.* Drihhtiness læchedom and sawless ȝife sallfe, 1851: *O. H. Ger.* láh-tuom *medicina, medicamentum, fomentum:* cf. *Icel.* læknis-dómr *medicine: Dan.* læge-dom *medicine, healing power, cure.*]

lǽcedóm-ness, e; *f. A plaster* :—Lǽcedómnessa oððe sealfe *cataplasma,* Wrt. Voc. ii. 18, 30.

lǽce-feoh; *g.* -feós; *n. A physician's fee, money paid to a doctor* :—Swá hwylc man swá óðrum womwlite ongewyrce forgylde him ðone womwlite and his weorc wyrce óþ ðæt seó wund hál sig and ðæt lǽcefeoh ðam lǽce gylde, *quicunque homo alio vulnus in faciem inflixerit, emendet ei vulnus, et opus ejus operetur, donec vulnus sanetur, et mercedem medico solvat,* L. Ecg. C. 22; Th. ii. 148, 19. [Cf. Si vulneravit quis alium, et satisfacere debeat, in primis reddat ei *lich-fe* quantum scilicet in curam vulneris impendit, L. W. I. 1, 10; Th. i. 471, 25. Cf. *Icel.* læknis-fé.]

lǽce-finger, es; *m. The leech-finger, the fourth finger* [though in one gloss it seems to be the *little-finger*] :—þuma *pollex,* scytelfinger *index,* middelfinger *medius,* lǽcefinger *medicus,* éarefinger *auricularius,* Wrt. Voc. 71, 30-34. At p. 44. 7-8 the names are different :—Goldfinger *medicus* vel *annularis,* lǽcefinger *auricularis,* Ælfc. Gl. 73; Som. 71, 22. Sing on díne lǽce-finger paternoster, Lchdm. i. 394, 2. [In later times it was the fourth finger e. g. Halliwell in his Dictionary quotes from a MS. of the 15th cent.

Ilke a fyngir has a name, als men thaire fyngers calle,
The lest fyngir hat *lityl man,* for hit is lest of alle;
The next fynger hat *leche man,* for qwen a leche dos oȝt,
With that fynger he tastes all thyng. howe that hit is wroȝt.

In Prompt. Parv. p. 291 note the reason for the name is given differently. 'The fourth finger was called the leech finger, from the pulsation therein found, and supposed to be in more direct communication with the heart, as in the tract attributed to Joh. de Garlandiâ . . . it is said '*Stat medius* [medylle fyngure] *medio, medicus* [leche fyngure] *jam convenit* [accordyt] *egro.*'' See too in the same writer's *Dictionarius,* Wrt. Voc. p. 121, 35 '*medicus* dicitur digitus eo quod illo medici imponunt medicinam.' Cf. *Icel.* læknis-fingr.]

lǽce-hús, es; *n. A hospital, a house where the sick are tended by a leech* :—Hé hine gelǽdde on his lǽcehús [Lind. léchús] and hine lácnude *And brohte óðrum dæge twegen penegas and sealde ðam lǽce and ðus cwæþ* Begým hys *illum duxit in stabulum et curam ejus egit. Et altera*

die protulit duos denarios et dedit stabulario et ait curam illius habe, Lk. Skt. 10, 34-5. [The translator seems not to have kept close to the text, but to have rendered the passage in accordance with the part played by the Good Samaritan. A more literal translation is given Past. 17, 10; Swt. 125 where *in stabulum* is rendered *tó ðæm giesðhúse*.] [*Prompt. Parv.* a leche house *laniena, quia infirmi ibi laniantur*, p. 291, note 4.]

læce-sealf, e; *f. A medicinal salve* or *ointment, a plaster*; malagma, Wrt. Voc. ii. 87, 77.

læce-seax, es; *n. A surgeon's knife*:—Se læce hýt ðonne his læceseax under his cláðum *medicus abscondit igitur ferrum medicinale sub veste*, Past. 26, 3; Swt. 187, 9.

læce-wyrt, e; f.　　I. *a herb having medicinal virtue*:—Se wísa Augustinus cwæþ ðæt unpleólíc sý ðeáh hwa læcewyrte þicge ac ðæt hé tælþ tó unálýfedlícere wiglunge gif hwá ða wyrta on him becnitte búton hé hí tó ðam dolge gelecge *the learned Augustine said, that it is not dangerous, though any one eat a medicinal herb; but he considers it as unlawful sorcery, if any one bind the herbs on himself, unless he lay them to the wound*, Homl. Th. 1. 476, 4.　　II. *the name of a particular plant*:—Læcewyrt *quinquenerina* [*quinquenervia*], Wrt. Voc. 286, 39. Lêciwyrt *quinquenervia*, Wrt. Voc. ii. 118, 57. Læcewyrt. Deós wyrt ðe man lichanis stefanice and óðrum naman læcewyrt nemneþ *this plant which is named λύχνις στεφανική and by another name leechwort* [Cockayne Lchdm. ii. 396, col. 2 suggests *campions* or *ragged robin* or one of that kindred as the plant here meant], Herb. 133, 1; Lchdm. i. 248, 15-7. Læcewyrt *plantago lanceolata*, L. M. 1, 32; Lchdm. ii. 78, 7: 1, 38; Lchdm. ii. 96, 14. See Cockayne as above where he gives lákeblad *plantago major*, in West Gothland. [*Dan.* læge-urt *medicinal plant* : cf. *Icel.* læknis-gras *a healing herb*.]

læcing, e; *f. Blame, reproof*; redargutio, Somner. [Cf. *Chauc. Piers P.* to lakke *to blame, dispraise, speak ill of*; *Prompt. Parv.* lakkyn *vitupero, culpo*; lacke or blame *vituperium*, p. 285, note 3, where this line from Lydgate, besides other instances, is given 'with lawde or *lack* liche as they have deserved': *O. Frs.* laking *impugnatio*; lakia *impugnare*.]

læcnan *to tend*:—Læcnende *procurans*, Wrt. Voc. ii. 90, 72. v. lácnian.

læcnung. v. lácnung.

læcung, e; *f. Healing, remedy.* [*O. E. Homl.* hit beo mi *lechunge* hit beo mi *bote*, i. 187, 35: *O. H. Ger.* láhunka *remedium*.] v. sealf- -læcung, *and* cf. lácnung.

læd, Chart. Th. 166, 21. v. læwed.

lǽd. v. un-lǽd.

LǼDAN; *p.* de; *pp.* læded, læd *TO LEAD, conduct, take, carry, bring, bring forth, produce* [the word translates the Latin verbs *ducere, ferre* with many of their compounds]:—Ic naman Drihtnes herige and hine mid lofsange lǽde swylce *laudabo nomen Dei mei cum cantico, et magnificabo eum in laude*, Ps. Th. 68, 31. Twegen gemacan ðú lǽtst in tó ðam arce *bina induces in arcam*, Gen. 6, 19. Se wísa mon eall his líf lǽt on gefeán [cf. orsorg líf lǽdaþ woruldmen wíse, Bt. Met. Fox 7, 80; Met. 7, 40] *duces serenus ævum*, Bt. 12; Fox. 36, 24. Se blinda gyf hé blindne lǽt *cæcus si cæco ducatum præstet*, Mt. Kmbl. 15, 14. Lǽt, Dóm. L. 18, 294. Se ðe nimeþ ł lǽdeþ synne middangeardes *qui tollit peccatum mundi*, Jn. Skt. Lind. 1, 29. Gé cunnon hwæt se hláford is se ðisne here lǽdeþ, Exon. 16 a; Th. 36, 11; Cri. 574. Man ða moldan nimeþ and men wíde geond eorþan lǽdaþ tó reliquium *the earth is taken, and men carry it far and wide over the world as relics*, Blickl. Homl. 127, 16. Hí hergiaþ and tó scipe lǽdaþ *they harry and carry off the plunder to their ships*, Swt. A. S. Rdr. 109, 137. Hí Crist heriaþ and him lof lǽdaþ *Crist they laud and to him bring praise*, Hy. 7, 25; Hy. Grn. ii. 287, 25. Ic wille ácwellan cynna gehwylc ðara ðe lyft and flód lǽdaþ and fédaþ *I will destroy every kind that air and water produce and nourish*, Cd. 65; Th. 78, 25; Gen. 1298. Wæstme tydraþ ealle ða on Libanes lǽdaþ [MS. lǽdeþ] on beorge cwice cederbeámas ða ðú sylfa gesettest *cedri Libani quas plantasti*, Ps. Th. 103, 16. Ða men mon lǽdde tó Winteceastre tó ðæm cynge *the men were brought to Winchester to the king*, Chr. 897; Erl. 96, 10. Se deófol hine genam and lǽdde hine on swíðe heáhne munt *assumpsit eum diabolus in montem excelsum valde*, Mt. Kmbl. 4, 8: Blickl. Homl. 27, 16. Ða cwæþ hé tó ðam engle ðe hine lǽdde *then said he to the angel that conducted him*, 43, 32. Eal ðæt folc hine lǽdde mid gefeán, 249, 21. Ecgbryht lǽdde fierd wið Norþanhymbre *Egbert led a force against the Northumbrians*, Chr. 827; Erl. 64, 7. Hé wæs ofslegen mid ealle ðý weorude ðe hé lǽdde, Bd. 1, 34; S. 499, 34. Hé unbeád ðæt hé of Róme cóme and ðæt betste ǽrende lǽdde *mandavit se venisse de Roma ac nuncium ferre optimum*, 1, 25; S. 486, 26. Hé ancorlíf lǽdde *vitam solitariam duxerat*, 4, 27; S. 603, 28. Hé lǽde *eduxit*, Blickl. Gl. Hé hine lǽde forþ tó ðon cafortúne ðæs húses. Blickl. Homl. 219, 20. Lǽde mon hider tó ús sumne untrumne mon. Ða lǽdde mon forþ sumne blindne mon of Angelcynne. Wæs hé ǽrest lǽded tó Brytta biscopum *adducatur aliquis æger . . . Allatus est quidam de genere Anglorum, oculorum luce privatus; qui oblatus Brittonum sacerdotibus*, Bd. 2, 2; S. 502, 21-5. Ða Abraham ǽhte lǽdde of Egypta éðelmearce, Cd. 90; Th. 112, 20;

Gen. 1873. Hé hêt smiðian áne lytle róde ða hé lǽdde on his swiðran *he ordered a little cross to be forged, that he laid upon his right hand*, Homl. Th. ii. 304, 16. His ðegnas lǽddon him tó ðone eosol *his disciples brought the ass to him*, Blickl. Homl. 71, 6. On hæftnéd lǽddon *led into captivity*, 79, 22. Ða fíf cyningas mit húðe lǽddan (*predati sunt*) Loth gebundenne, Prud. 2 a. Mé lǽddon *me deduxerunt*, Ps. Spl. 42, 3. Ða ilcan ðe ǽr landgemǽre lǽddon *the same that before had marked the boundaries of the land*, Chart. Th. 376, 19. Hettend lǽddon út mid ǽhtum abrahames mǽg of Sodoma byrig, Cd. 94; Th. 121, 17; Gen. 2011. Ne lǽd ðú ús in costunge *lead us not into temptation*, Hy. 6. 27; Hy. Grn. ii. 286, 27. Lǽd út mid ðé *educ tecum*, Gen. 8, 17. Ða cwæþ hé tó his geréfan lǽde in ðás menn and gearwa úre þenunga *præcepit dispensatori domus suæ dicens: Introduc viros domum, et instrue convivium*, 43, 16. Ða cwæþ hé lǽde hig tó mé *adduc, inquit, eos ad me*, 48, 9. Lǽdaþ hig forþ and forbearnaþ hig *producite eam ut comburatur*, 38, 24. Fare gé tó eówrum húse and lǽde eówerne gingstan bróðor tó mé *vos abite in domos vestras et fratrem vestrum minimum ad me adducite*, 42, 20. Gáþ and lǽdaþ út ðæt wíf *producite eam*, Jos. 6, 22. Lǽde seó eorþe forþ cuce nîtenu *producat terra animam viventem*, Gen. 1, 24. Lǽdæ þrounc *tollat crucem*, Mk. Skt. Lind. 8, 34. Hé his ða menniscan gecynd on heofenas lǽdon wolde *he would take his human nature into heaven*, Blickl. Homl. 127, 24. Hé hêt his lîchoman up ádón and lǽdon tó Wintonceastre *translatus in Ventam civitatem*, Bd. 3, 7; S. 529, 24: Blickl. Homl. 193, 10. Hé forðon cóme ðæt hé sceolde mete lǽdan *propter victum adferendum*, Bd. 4, 22; S. 591, 8. Hí hæfdon ǽrend ðe hí him lǽdan sceolden *haberent aliquid legationis quod deberent ad illum perferre*, 5, 10; S. 624, 22. Ne dorste siððan nán Scotta cininga lǽdan here on ðás þeóda, Chr. 603; Erl. 21, 16. Sceal ic lǽdan ðînne sunu eft tó ðam lande ðe ðú of férdest? Beó wǽr æt ðam ðæt ðú nǽfre mînne sunu ðyder ne lǽde *numquid reducere debeo filium tuum ad locum, de quo egressus es? Cave, ne quando reducas filium meum illuc*, Gen. 24, 4-5. Wíf lǽdan *to take a wife*, Lchdm. iii. 190, 5: 212, 8. Þuhte mé ðæt ic gesáwe treów on lyft lǽdan *methought that I saw a tree borne aloft*, Rood Kmbl. 9; Kr. 5. Wudu mót him weaxan tánum lǽdan *wood may grow, be productive of twigs*, Exon. 119 b; Th. 458, 23; Hy. Grn. ii. 285, 105. Ecbyrht munuclíf wæs lǽdende on Hibernia, Bd. 3, 27 tit.; S. 558, 8. Hé wæs eft swá ǽr lof lǽdende *he was again as before bringing forth praise*, Andr. Kmbl. 2952; An. 1479. Se ána ealra beáma up lǽdendra *it alone of all trees that bear on high their branches*, Exon. 58 b; Th. 209, 30; Ph. 178. Sagaþ Matheus ðætte se Hǽlend wǽre lǽded on wésten, Blickl. Homl. 27, 4. Ða wæs geond ða werþeóde wíde lǽded mǽre morgenspel *then was a mighty report carried far and wide among the people*, Elen. Kmbl. 1935; El. 969. Feorran lǽded *brought from far*, Exon. 107 b; Th. 411, 2; Rä 29, 6. Ðæt wæs lǽd æt Licitfelda *that [the exculpation from the charge] was produced at Lichfield*, Chart. Th. 373, 34. Tó ðam écan setle ðæs heofonlícan ríces lǽded wæs *ad æternam regni cælestis sedem translatus est*, Bd. 2, 1; S. 500, 11. Forðon of Breotone nædran on scipum lǽdde wǽron *nam de Brittania adlati serpentes*, Bd. 1, 1; S. 474, 34. [*Laym.* læden *to lead, take : Orm.* ledenn ȝuw *to conduct yourselves : A. R.* lede lif : *Gen. and Ex.* leden song *to sing : O. Sax.* lêdian *to lead, bring, bear : O. L. Ger.* lêdian, leidan *ducere, deducere : O. Frs.* lêda *to lead, conduct : Icel.* leiða : *O. H. Ger.* leitan : *Ger.* leiten.] v. lád, III. á-, an-, for-, ge-, in-, on-, óþ-, út-, wið-lǽdan.

lǽdan, *to excuse.* v. lǽdend.

Lǽden, es; *n.*　　I. *Latin, the Latin tongue*:—Is ðæt Lêden on smeáunge gewrita eallum ðam óðrum gemǽne *quæ* [i. e. *lingua Latinorum*] *meditatione scripturarum cæteris omnibus est facta communis*, Bd. 1, 1; S. 474, 4. Swá gelǽred ðæt hé Grécisc gereord of miclum dǽle cúþe and Lêden him wæs swá cúþ swá swá Englisc *in tantum institutus, ut Græcam linguam non parva ex parte, Latinam non minus quam Anglorum noverit*, 5, 20; S. 641, 34. Wé ne durron ná mǽre áwrítan on Englisc ðonne ðæt Lîden hæfþ, ne ða endebirdnisse áwendan búton ðam ánum ðæt ðæt Lêden and ðæt Englisc nabbaþ ná áne wísan on ðære sprǽce fadunge [fandunge, Thw.]. Ǽfre se ðe áwent of Lêdene on Englisc, ǽfre hé sceal gefadian hit swá ðæt ðæt Englisc hæbbe his ágene wísan, elles hit biþ swíðe gedwolsum tó rǽdenne ðam ðe ðæs Lêdenes wísan ne can, Ælfc. Gen. Thw. 4, 5-11. Hé Grécisc geleornode mid Lêdene *Græcam cum Latina didicit linguam*, Bd. 5, 23; S. 645, 16. Of Lǽdene on Englisc áreccean *to translate from Latin into English*, Past. pref; Swt. 3, 15. Of Lǽdene tó Engliscum spelle gewendan, Bt. pref; Fox viii, 9. *Glossa* is ðonne man glêsþ ða earfoþan word mid eáðran lédene *faustus* is on óðrum lédene *beatus* ðæt is eádig *fatuus* is on óðrum lédene *stultus* ðæt is stunt *a gloss is when the difficult words are explained with easier Latin; another Latin word for* faustus *is* beatus i. e. *happy; another Latin word for* fatuus *is* stultus i. e. *foolish*, Ælfc. Gr. 50; Som. 51, 43-4. Ða bóc ðe is genemned on Lǽden Pastoralis, and on Englisc Hierdebóc, Past. pref.; Swt. 7, 19. Hér is geleáfa lǽwedum mannum ðe ðæt lêden ne cunnon, Homl. Th. ii. 596, 2. Gitrahtad on læden [Lind. in Latin] *interprætatum*, Mk. Skt. Rush. 5, 41. On læden [Lind. læddin] *latine*, Jn. Skt. Rush. 19, 20. Didymus, gemi-

nus in lætin, Lind. 20, 24, margin. Hí beóþ oft óðres cynnes on léden, and óðres cynnes on englisc; wé cweþaþ on léden *hic liber*, and on englisc ðeós bóc, Ælfc. Gr. 6; Som. 5, 37–40. On léden *latine* and *latialiter*, 38; Som. 41, 32. Gelǽrede on léden and on grécisc, Homl. Skt. 2, 44: Bd. 4, 1; S. 564, 11. Sum mæssepreóst cúðe be dǽle Lýden understandan *a certain mass-priest could understand Latin partially*, Ælfc. Gen. Thw. p. 1, 20. [Cf. Icel. Látína; *f.*] II. *any tongue, speech, language* :—Spasmus ðæt ys on úre leódene hneccan sár σπασμός, *that is in our language, a pain at the back of the neck*, Lchdm. iii. 110, 1. Mara ðæt ys on úre lýden biternys, Ex. 15, 23. Ealle hig sprecaþ án lýden *est unum labium omnibus*, Gen. 11, 6. [Laym. *cerno* an Englisc leoden, ich iseo, 29677; *Marh.* þe moneþ ðæt on ure ledene is ald englisch esterlið inempnet, 23, 6: *A. R.* on ebrewische ledene, 136, 24; on englische leodene, 170, 9: *Piers P.* I leve his ledne be in owre lordes ere lyke a pyes chiteryng, 12, 253: *Chauc.* every thing that any foul may in his ledene seyn, F. 435 [see Skeat's note in the Clarendon Press edition]. For the extended use of forms in Romance from *latinus* cf. the passage, given in that note, of Dante's Canzone beginning ' Fresca rosa novella,' ' Cantino gli augelli ciascuno in suo *latino*;' Parad. iii. 63 si che 'l raffigurar m'è piu latino [*clear*]; Convito bk. 2, c. 3 a piu *latinamente* veder la sentenza. In Old Spanish *ladino* is explained ' el que sabe otra lengua o lenguas ademas de la suya.' Is it possible that in the case of English the forms geþeóde, þeód may have had some influence in giving currency to *lýden* in the general sense of *language*, by suggesting a connection of this latter form with *leód* ?]

Léden; *adj. Latin :*—Léden *latinus*, Ælfc. Gr. 38; Som. 41, 32. Ealle naman lédenre sprǽce [lédensprǽce, MS. O.] ðe on a geendiaþ *all latin nouns that end in a*, 7; Som. 6, 55. Ða gemetu gebyriaþ tó lédenum leóðcræfte *metres belong to latin poetry*, 50; Som. 51, 66. On lédenum gereorde, Homl. Skt. 6, 367. Lédene láreówas maciaþ on sumum namum accusativum on *im*, Ælfc. Gr. 9; Som. 14, 32. Stafum créciscum and lǽdenum [latinum, Lind.] *litteris græcis et latinis*, Lk. Skt. Rush. 23, 38. See the compounds of which *Léden* forms the first part.

Léden-bóc; *f. A Latin book :*—Nán man næfþ lédenbóca angit be fullon búton hé ðone cræft cunne *no man perfectly understands Latin books, unless he know that art* [grammar], Ælfc. Gr. 50; Som. 50, 65. Áwriten on lédenbócum *written down in Latin books*, Homl. Skt. p. 4, 48. Ða ealdan lǽces gesetton on lédonbócum, Lchdm. iii. 152, 1.

lédend, es; *m. One who leads* or *brings :*—Se wæs ǽ bringend, lára lǽdend, Exon. 10 a; Th. 9, 27; Cri. 141.

lédend, es; *m. One who excuses :*—Ne hyld ðú míne heortan ðæt ic lǽdend wese láðra firena *ut non declines cor meum ad excusandas excusationes in peccatis*, Ps. Th. 140, 5. Cf. ládian.

Léden-gereord, -gereorde, es; *n. Latin, the Latin language :*—Of lǽdengereorde on englisc, Lchdm. iii. 440, 27.

Léden-geþeóde, es; *n. The Latin language :*—Lǽre mon furður on Lǽdengeþióde ða ðe mon furður láran wille . . . Ða ic ðá gemunde hú sió lár Lǽdengeþiódes áfeallen wæs giond Angelcynn *let those to whom it is desired to give further instruction, be instructed in Latin . . . When I remembered how the teaching of Latin was decayed throughout England*, Past. Pref; Swt. 7, 13–17.

Lédenisc; *adj. Latin :*—On Lédenisc gereorde ge on Grécisc, Bd. 4, 1; S. 563, 33. On Lédennisc, 4, 2; S. 565, 28: 5, 8; S. 622, 1. On lǽddin þ lédinisc *latine*, Jn. Skt. Kmbl. 19, 20. [Cf. *O. H. Ger.* in latinisgon *latine*.]

Léden-nama, an; *m. A Latin noun :*—Gif ðú nâst sumne lédennaman [lǽden- MS. H] hwylces cynnes hé sý *if you do not know some Latin noun, of what gender it is*, Ælfc. Gr. 50; Som. 51, 35.

Léden-sprǽc, e; *f. The Latin speech* or *language :*—Ealle naman lédensprǽce [also lédenre sprǽce] *all Latin nouns*, Ælfc. Gr. 7; Som. 6, 55. On lédensprǽce, 2; Som. 2, 47. Hálige láreówas hit áwriton on lédensprǽce, Homl. Skt. p. 6, 51. Se cræft geopenaþ lédensprǽce [MS. H. lǽden-] Ælfc. Gr. 50; Som. 50, 65.

Léden-stæf, es; *m. A Latin letter :*—Hit wæs áwriten grécisceon and lédenstafon *erat scribtum græce et latine*, Jn. Skt. 19, 20. [Cf. *Icel.* Látínu-stafr.]

Léden-ware; *pl. The Latins, the Romans :*—Lédenware wendon hié ealla on hiora ágen geþeóde *the Romans turned them all into their own language*, Past. Pref; Swt. 6, 3. On Lédenwara gereorde *lingua Latinorum*, Bd. 1, 1; S. 474, 4. Sui næfþ nænne nominativum náðer ne mid Grécum ne mid Lédenwarum, Ælfc. Gr. 15; Som. 18, 5. Firgilius wæs mid Lédenwarum sélest *amongst the Romans Virgil was best*, Bt. 41, 1; Fox 244, 5.

Léden-word, es; *n. A Latin word :*—Ðás word ne beóþ ná lédenword gif se r byþ áweg gedón, Ælfc. Gr. 19; Som. 22, 54.

lédere, es; *m. A leader, guide*, Cant. Moys. [?], Lye. [Piers P. leder: *O. Frs.* folk-lédera: *O. H. Ger.* leitari *dux*.]

léd-ness, e; *f. A bringing forth, production :*—On ðæs tuddres lédnysse *in prolis prolatione*, Bd. 1, 27; S. 493, 21 note. v. forþ-lædness.

léd-teów. v. lád-teów.

léf. v. láf, leáf.

---

lǽfan; *p. de.* I. *to leave :*—Ic lǽfe eów sibbe *pacem relinquo vobis*, Jn. Skt. 14, 27. Ic lǽfe *lego*, Wrt. Voc. ii. 49, 66. Gif hwæs bróðor deád biþ and lǽfþ his wíf *si cujus frater mortuus fuerit et dimiserit uxorem*, Mk. Skt. 12, 19. Hig ne lǽfaþ on ðé stán ofer stáne *non relinquent in te lapidem super lapidem*, Lk. Skt. 19, 44. Se forma lǽfde his bróðer his wíf *primus reliquit uxorem suam fratri suo*, Mt Kmbl. 22, 25. Ða men ðe hé beæftan him lǽfde ǽr *those men that before he had left behind him*, Chr. 755; Erl. 50, 13. Hié begeáton welan and ús lǽfdon *they got wealth and left it us*, Past. pref; Swt. 5, 15. Swá hit his yldran begeáton and létan and lǽfdon ðam tó gewealde ðe hý wel úðan, L. O. 14; Th. i. 184, 3. Ðínum mágum lǽf folc and ríce *leave to thy kinsmen people and power*, Beo. Th. 2361; B. 1178. Ðonne him forþsíð gebyrige gýme his hláford ðæs hé lǽfe *when his death happens, let his lord take charge of what he leaves*, L. R. S. 4; Th. i. 434, 28 : 5; Th. i. 436, 9. Ne biþ lǽfed stán uppan stáne, Mt. Kmbl. 24, 2. Ân byþ genumen and óðer byþ lǽfed, 24, 41. Heora landáre ðe him lǽfed wæs *their landed property that was left them*, Homl. Skt. 4, 82. Ná lǽfedum sǽde *non relicto semine*, Mk. Skt. 12, 20. Ðæt ða bán áne beón lǽfed *so that the bones only are left*, L. Med. ex Quad. 3, 11; Lchdm. i. 340, 26. II. *to remain, be left remaining :*—Gif hwæt lǽfde *if anything remained*, Homl. Th. ii. 40, 14. Hia lǽfdun *superaverunt*, Jn. Skt. Rush. 6, 12. [Goth. bi-laibjan *to remain* : O Sax. farlébian *to remain*; lébôn *to be left* : O. Frs. léva *to leave* : Icel. leifa *to leave* : O. H. Ger. leibjan *relinquere* : leibén *to remain*.] DER. be-, ge-, ofer-lǽfan.

lǽfan *to allow.* v. lífan.

lǽfel, es; *m. A cup, vessel, bowl :*—Lǽfel *sciffus*, Wrt. Voc. 85, 66. Lǽvel, 25, 18. Lǽvil *manile*, 290, 69. Lǽuel *aquemanile*, Wrt. Voc. ii. 7, 14. Label *aquemale*, 100, 60. Lebil *manile*, 113, 43. Lebl *triplia*, 122, 62. Se lǽfyl ðe gé forstǽlon wæs mínum hláforde swíðe dýre *scyphus, quem furati estis, ipse est, in quo bibit dominus meus*, Gen. 44, 5. Ǽren fæt, lǽfel oððe céc, Lchdm. iii. 292, 9. Of ðǽm hlæfle, Chart. Th. 439, 30. Ðonne gesealde Aðelwold biscop his cynehláforde ánne sylfrenne lefel on fíf pundum *dedit autem Athelwoldus episcopus regi quoddam vas argenteum quinque libras appendens*, 236, 11. Nym mínne sylfrenan lǽfyl *scyphum meum argenteum*, Gen. 44, 2. viii lǽflas *eight cups*, Chart. Th. 429, 36. [Laym. water me brohte- mid guldene lǽflen : O. L. Ger. lavil *pelvis* : O. H. Ger. label, lapei *labium, concha, pelvis*, Grff. ii. 78–9.]

lǽfend, es; *m. One who misleads* [*a traitor*, = lǽwend ?]; *seductor*, Ælfc. Gl. 85; Som. 73, 110; Wrt. Voc. 49, 17.

lǽfer, e; *f.* I. *a rush :*—Lǽfer *pirus* [l. *papyrus*], *gladiolus*, Ælfc. Gl. 47; Som. 65, 15; Wrt. Voc. 33, 15 : *scirpio*, Wrt. Voc. 69, 9 : *scirpia*, 289, 44. Lebr *scirpea*, Wrt. Voc. ii. 119, 81. Eórisc, leber *scirpea*, 120, 17. Genim lǽfre neoðowearde *take the lower part of a bulrush*, Lchdm. i. 382, 21. II. *a thin plate of metal :*—Gylden lǽfr *bractea*, Ælfc. Gl. 52; Som. 67, 111; Wrt. Voc. 38, 34. Xerxes beworhte ða bígelsas mid gyldenum lǽfrum *Xerxes wrought over the arches of the roof with golden plates*, Homl. Th. ii. 498, 3. Mid lǽfrum *liscis*, Wrt. Voc. ii. 51, 9. [See E. Ð. S. Plant Names s. v. levers: Grff. ii. 80 leber *scirpus, herba rotunda*.]

lǽfer-bedd, es; *n. A bed of rushes :*—Lǽferbed *pirorium* [v. lǽfer], Ælfc. Gl. 47; Som. 65, 14; Wrt. Voc. 33, 14. [Cf. *liver-ground* the place where the plant grows, E. D. S. Plant Names s. v. levers.]

-læg. v. or-læg.

lǽl, lél, e; *f.* I. *a pliant twig, withe, whip, switch :*—Lǽl *vimen*, Ælfc. Gl. 46; Som. 65, 13; Wrt. Voc. 33, 12 : *vibex*, Wrt. Voc. ii. 88, 4 : 96, 35. Lǽla *mastigias* [mastigia *flagrum, flagellum, virga*, Ducange], 55, 25. Lélan *vibice*, 123, 68. II. *a weal, mark left on the flesh by a stroke from a rod, stripe, mark, bruise, swelling :*—Sylle wunde wið wunde lǽl wið lǽle *reddat vulnus pro vulnere, livorem pro livore*, Ex. 21, 25 : L. Ælfc. 19; Th. i. 48, 22. Ne sý him blódig wund líces lǽla ac gé hine gesundne ásettaþ ðær gé hine genóman *let there be no bloody wound on him, no stripes on his body, but do you put him down sound, where you took him*, Exon. 42 b; Th. 143, 34; Gú. 671. Léla *livor*, Kent. Gl. 763. Lǽla *nevorum*, Wrt. Voc. ii. 59, 50. Wið láðum lǽlum and wommum *ad perniones*, L. Med. ex Quad. 2, 20; Lchdm. i. 338, 15. Wið ðæt man lǽla and óðre sár of líchaman gedó *in order that weals and other sores may be removed from the body*, Herb. 102, 2; Lchdm. i. 216, 21. Wið yfele lǽla *oldήματa*, 153, 4; Lchdm. i. 280, 1. Ðá eode se mæssepreóst tó ðam byscope and hym eówde ða lǽla ðæra swyngellan ðe hé from dryhtne onféng *then the priest went to the bishop and shewed him the marks of the scourging that he had received from the Lord*, Shrn. 98, 18. [(?) Scot. leill *a single stitch in marking on a sampler*.] v. lǽlan, lǽlian; and cf. [*for the double use*] walu.

lǽlan *to become black and blue with blows, to be bruised :*—Geseoh nú seolfes swæðe swá ðín swat ágeat blódige stíge líc lǽlan *see now thy track, where thy blood hath poured forth, a bloody path, see thy body bruised*, Andr. Kmbl. 2884; An. 1445. *Kemble and Grimm read* líelǽlan spots [*of blood*] *on the body, but cf. lǽlian.*

lǽlian *to become black and blue*; *livescere*, Wrt. Voc. ii. 50, 41.

laembis lieg. v. lendis lieg.

lǽmen; *adj. Made of clay, earthen:*—Lǽmen fæt *lagena*, Ælfc. Gl. 26; Som. 60, 93; Wrt. Voc. 25, 33. Lēmen fet *vas fictile*, Kent. Gl. 1001. Lǽmen crocca *testa*, Ps. Th. 21, 13. Lǽmen fæt *a vessel of earth*, Exon. 74 a; Th. 277, 2; Jul. 574: L. Ath. iv. 7; Th. i. 226, 15. Lǽmene fatu *fictilia vel samia*; readde lǽmene fatu *aretina* [MS. *alsierina*], Ælfc. Gl. 66; Som. 69, 94, 95; Wrt. Voc. 41, 46, 47. Lǽmene fatu beóþ on ofne āfandode, Homl. Th. i. 554, 33. Leomo lǽmena *limbs of clay*, Exon. 8 a; Th. 2, 6; Cri. 15. Lǽmina *fictilia*, Wrt. Voc. ii. 36, 35. [*O. H. Ger.* leimin *fictilis, luteus*.]

lǽn, lān [*v. under* lǽn-land], e; *f.*   I. *a loan, grant, gift:*—Lǽn *commodum*, Ælfc. Gl. 14; Som. 58, 5; Wrt. Voc. 21, 1. Lǽn *commodum, lucrum*, Wrt. Voc. ii. 132, 1 : *depositum* i. e. *commendatum*, 139, 1. Borg *vel* lǽn *fenus* i. e. *lucrum, usura*, 148, 24. Tô hwîle lǽn *momentum*, Ælfc. Gl. 15; Som. 58, 47; Wrt. Voc. 21, 36. Ðæt hridder tôbærst on ðære lǽne. Seó fôstermôdor weóp for ðære āwyrdan lǽne *the sieve broke in two during the loan. The foster-mother wept for the injured loan*, Homl. Th. ii. 154, 16. Sum man sceolde āgyldan healf pund and wæs ðearle geswenct for ðære lǽne *a certain man had to pay back half a pound, and was exceedingly harassed on account of the loan*, 176, 35. Hê tô ðære lǽne fācn ne wiste *he knew of no ill-design in the loan* [*of arms*], L. Alf. pol. 19; Th. i. 74, 7. Se ðe æt his nēhstan hwæt tô lǽne ābit *qui a proximo suo quidquid mutuo postulaverit*, Ex. 22, 14. Ðē biddaþ manega þeóda dînes þinges tô lǽne and ðū ne bitst nānne *foenerabis multis gentibus, et ipse a nullo foenus accipies*, Deut. 28, 12. Tô lǽne syllan *mutuum dare*, 15, 8. Tô lǽne beón to be lent, Past. pref; Swt. 9, 7. Lǽne syllaþ *mutuum dare*, Lk. Skt. 6, 35. Ðā meahte heó wîde geseón þurh ðæs lāðan lǽn *then could she widely see through the fiend's gift*, Cd. 29; Th. 38, 3; Gen. 601. Lǽn Godes, ælmihtiges gife, 32; Th. 43, 18; Gen. 692. Ðeáh hê him nānra ôðerra lǽna [*but* Cott. MS. leana] ne wēne *though he expect no other benefits*, Bt. 24, 3; Fox 84, 1.   II. [*in connection with land*] *a grant that may be recalled, lease, fee, fief:*—Landes lǽn *precarium*, Ælfc. Gl. 14; Som. 58, 6; Wrt. Voc. 21, 2. Mon gerehte ðæt yrfe cinge forðon hê wæs cinges mon and Ordlāf fēng tô his londe forðon hit wæs his lǽn ðæt hê onsǣte hê ne meahte uā his forwyrcan *the property went to the king because he* [*Helmstan*] *was the king's man; and Ordlaf took the land, for the land that he* [*Helmstan*] *occupied was held in fee from Ordlaf, so he* [*Helmstan*] *could not forfeit it*, Chart. Th. 173, 4. v. Cod. Dip. Kmbl. i. lix. Ðā oferbād Ælfeh his brôðor and Ælfeh tô his lǽne [cf. geúðe hê him &c. 9–12] *then Ælfeh survived his brother, and resumed the lands he had granted to him*; mortuo Ælfrico Ælfegus statim omnia præstita sua, quæ fratri suo viventi præstiterat [resumpsit], 272, 13, 21. Ælcne man lyst siððan hê ǽnig cotlýf on his hlāfordes lǽne getimbred hæfþ ðæt hê hine môte hwîlum ðar on gerestan and his on gehwilce wîsan tô ðære lǽnan [*is this a form* lǽne, *an*; *f.* = lǽn, *or can it be the adj.* lǽne *transitory, as the opposite of which* ǣce *occurs afterwards, with the noun that it qualifies omitted?*] tilian ôþ ðone fyrst ðe hê bôcland and ǣce yrfe gearnige *every man, after he has built any cottage on land granted him by his lord, desires that he may rest himself therein at times, and in some fashion provide for himself from the grant* [?], *until the time that he has gained a freehold and a perpetual possession*, Shrn. 164, 2–8. Æþelwald and Alhmund his sunu hit woldon habban on his lǽne and hîna *Ethelwald and his son Alhmund would hold it* [*certain land*] *of an him* [*the bishop*] *and of the convent in fee*, Chart. Th. 140, 32. Denewulf biscop and ða hŷwan on Wintanceastre ænlǽnan Ælfrēde his deg xl hîda landes æt Alresforda æfter ðære lǽna ðe Tūnbryht bisceop ǽr ālēnde his yldran *bishop Denewulf and the convent at Winchester lease to Alfred for his life xl hides of land at Alresford, according to the lease that Tunbryht granted before to his parents*, 147, 29. [*O. E. Hom.* se riche *lane* as beoð þeos sustren, i. 257, 22 : *A. R.* *Ayenb.* lone *what is lent*: *Piers P.* lone, loone, lene. The double form of the word in later English may be partly owing to Scandinavian influence. Icelandic has both lēn; *n. a fief, fee, grant*, and lān; *n.* (though an older feminine is indicated) *a loan, fief*. *O. Frs.* lēn; *n. a grant, fee, fief* : *O. H. Ger.* lēhan; *n. fœnus, beneficium, usura, præstatio* : *Ger.* lehen; *n. fief, fee*.] v. Kemble's Saxons in England, i. 310.

lǽnan; *p.* de *To lend, grant, lease:*—Lǽnþ *commodat*, Ps. Spl. 36, 27 : 111, 5 : Blickl. Gl. Lēnþ *fenerator*, Kent. Gl. 699. Gif gē lǽnaþ ðām ðe gē eft æt onsôþ hwilc þanc is eów sôþlice synfulle synfullum lǽnaþ *si mutuum dederitis his a quibus speratis accipere quæ gratia est vobis? nam et peccatores peccatoribus fœnerantur*, Lk. Skt. 6, 34. Hig lǽnaþ eów and gē ne lǽnaþ him *ipse fœnerabit tibi et tu non fœnerabis ei*, Deut. 28, 44. Ðæt hê hæbbe ðæt land æt Ludintûne iii. geár for ðām þreóm pundum ðe hê lǽnde, Chart. Th. 434, 33. Lǽn mē þrý hláfas *commoda mihi tres panes*, Lk. Skt. 11, 5. Lǽne mē ða bôc tô rǣdenne *commoda mihi librum ad legendum*, Ælfc. Gr. 24; Som. 25, 20. Ne lǽne dînum brêðer nān þing tô hîre *non fœnerabis fratri tuo ad usuram pecuniam*, Deut. 23, 19. Ys forboden ðæt hê his feoh tô nānum unrihtum gafole ne lǽne *prohibitum est, pecuniam suam ullo injusto fœnore mutuam dare*, L. Ecg. P. ii. 30; Th. ii. 194, 16. Ymb ðæt land ðæ ðū mē firmdlg tô wǣræ ðæt ic dǣ ēndæ *de terra illa, de qua egisti apud me, ut ego eam tibi commodarem*,

---

Chart. Th. 162, 15. Him drihten mihte spēde lǽnan *the Lord could grant him success*, Cd. 95; Th. 124. 8; Gen. 2059. Hlǽnan *mutuare*, Wrt. Voc. ii. 56, 10. [*Laym.* lenen *to grant* :—þis lond he hire lende, 228: *Orm.* lenen: *Chauc.* lene : *Prompt. Parv.* leendyñ *presto, fenero* : *O. Fris.* lēna *to lend*: *Icel.* lēna *to grant*; lāna *to lend* : *O. H. Ger.* lēhanon *mutuari* : *Ger.* lehnen.] DER. ā-, be-, ge-, on-lǽnan.

lǽn-dagas; *pl. m. The days granted to a man in which to live, the time during which a man lives:*—Sceolde lǽndaga [MS. þend daga] æþeling ende gebîdan worulde lîfes *the end of the days that had been granted, of life in this world, was to come upon the prince*, Beo. Th. 4672; B. 2341. Swā sceal ǽghwylc mon ālǽtan lǽndagas, 5175; B. 2591. Cf. lǽne.

lǽnding. v. lending.

lǽne, an; *f.* = [?] lǽn, Shrn. 164, 6. v. lǽn II.

lǽne; *adj. Granted as a* lǽn [q. v.]. *granted for a time only, not permanent, transitory, temporary, frail* [generally used as an epithet of things of this world when they are contrasted with those of the next]:—Ac ic wolde witan hweðer ðē þuhte be ðam ðe ðū hæfst hweðer hyt wǽre ðe lǽne ðe ǣce þu? *I would know whether you thought of what you have, that it was temporary or eternal*, Shrn. 176, 29. Hēr biþ feoh lǽne hēr biþ freónd lǽne hēr biþ mon lǽne *in this world shall not wealth endure, or friend, or man*, Exon. 78 a; Th. 292, 32; Wand. 108: Elen. Kmbl. 2539; El. 1271. Ðis lǽne lîf ðe wē lifiaþ on *this transitory life in which we live*, Ps. Th. 62, 3. Ðis deáde lîf, lǽne on londe, Exon. 82 a; Th. 309, 32; Seef. 66. Ðeós lǽne gesceaft [*the world*], 20 a; Th. 52, 34; Cri. 843. Ðis is lǽne dreám [*the present life*], Cd. 169; Th. 211, 25; Exod. 531. Lǽnes landes bryce *fructus*, Wrt. Voc. ii. 39, 31. Lǽnan lîfes leahtras, Exon. 62 b; Th. 229, 16; Ph. 456: Cd. 156; Th. 194, 29; Exod. 268. Ende him on becom ðisses lǽnan lîfes *there came upon him an end of this life which is but for a moment*, Blickl. Homl. 113, 8. Ne biddan wē ūrne Drihten ðyses lǽnan welan, ne ðyssa eorþlîcra geofa ðe hrædlîce from monnum gewîtaþ, 20, 11. Se dæg wæs fruma ðyses lǽnan leóhtes, and hē biþ fruma ðæs ēcan æfterfylgendan, 133, 10. Mā dereþ monna gehwylcum môdes unþeáw ðonne mettrymnes lǽnes lîchoman, Bt. Met. Fox 26, 225; Met. 26, 119. Eádgār ðis wāce forlēt lîf ðis lǽne, Chr. 975; Erl. 124, 32. Suelce hê cwǣde ic eów onlǣne ðās gewîtendan and ic eów geselle ða þurwuniendan. Gif ðonne ðæs monnes môd and his lufu biþ behleápen eallunga on ða lǽnan sibbe ðonne ne mæg hê nǣfre becuman tô ðære ðe him geseald is *relinquo scilicet transitoriam, do mansuram. Si ergo in ea cor, quæ relicta est, figitur, nunquam ad illam quæ danda est, pervenitur*, Past. 46, 5; Swt. 350, 12–16. Monnes lîfdagas lǽne syndan, Ps. Th. 102, 14. Ǽghwilc þing ðe on ðîs andweardan lîfe lîcaþ lǽnu sindon eorþlîcu þing ā fleóndu *everything that pleases in this present life, transient are they, earthly things ever fleeting*, Bt. Met. Fox 21, 58; Met. 21, 29. Sceoldon sēcan dreám æfter deáþe, and ðás lǽnan gestreón, îdle ǣhtwelan forhogodon, Apstls. Kmbl. 166; Ap. 83. Gylt gefremmaþ þurh lîchaman lǽne geþohtas, Ps. C. 50, 15 ; Ps. Grn. ii. 277, 15. [*O. E. Homl.* 3if we forleosað þas lenan worldþing, i. 105, 30: *O. Sax.* lēhni (fehu, werold).] DER. un-lǽne.

lǽne-, lǽn-lic; *adj. Transitory, transient, not enduring:*—Cwæþ se godspellere Martha and Maria getācniaþ ðis lǽnelîce lîf and ðis gewîtendlîce, Blickl. Homl. 73. 9. Hēr is seó lǽnlîc winsumnes ac ðǽr is seó syngale nearones *in this world is the delight that endures not, but in the next is the anxiety that continues for ever*, L. E. I. pref; Th. ii. 394, 7.

lǽnend, es; *m. A creditor, lender:*—Lǽnend *fenerator vel commodator vel creditor, redditor*, Wrt. Voc. ii. 148, 26. Twegen gafolgyldon wǽron sumum lǽnende *duo debitores erant cuidam feneratori*, Lk. Skt. 7, 41.

lǽnend-lîc; *adj. Transitory, transient:*—Uton geþencan hū lǽnendlîc ðeós woruld ys, Wulfst. 136, 27. v. lǽne-lîc.

lǽnere, es; *m. A creditor, lender:*—Lǽnere *creditor*, Ælfc. Gl. 113; Som. 79, 124; Wrt. Voc. 60, 29. [*Ayenb.* lenere : *Wick.* leenere : *Prompt. Parv.* lendare *fenerator, creditor*: *O. Frs.* lēner : *O. H. Ger.* int-lēhenari *fœnerator*.]

lǽnian, Gen. 50, 15. v. leánian.

lǽn-land; *n. Land let on lease, which was never out of the possession of the lessor:*—Ðonne is ðæs landes iii hîda ðe Ōswald arcebisceop bôcaþ Eádrîce his þegne swā swā hê hit ǽr hæfde tô lānlande *there are three hides of land that archbishop Oswald conveys by charter to the possession of Eadric his thane, such as before he held by lease*, Cod. Dip. Kmbl. iii. 165, 5. Fîf hîda ðe Ōswald bôcaþ Eádrîce swā swā hê hit ǽr hæfde tô lǽnlande, 217, 20. Wē wrîtaþ ðæt hê hæbbe hit swā rūm tô bôclande swā hê ǽr hæfde tô lǽnlonde, 258, 29. Eall ðæt yrfe ðæ ic hæbbe on lǽnelondum, v. 333, 21. v. lǽn; and see Cod. Dip. Kmbl. i. lxii: Kemble's Saxons in England, i. c. xi.

lǽnung. v. feoh-lǽnung.

lǽpeldre *a dish, platter:*—Lǽpeldre fæt *paropsis vel catinus*, Ælfc. Gl. 26; Som. 60, 89; Wrt. Voc. 25, 29. Se ðe bedypþ on disce mid mē his hláf on lǽpeldre *qui intingit mecum manum in parapside* [Mt. 26, 23], Homl. Th. ii. 244, 4.

-lǽpped *having laps or lobes:*—Fîf-lǽppedu, Lchdm. ii. 160, 12.

lǽppa, an; *m. A skirt* [*of a garment*], *lappet, lobe* [*of the ear &c.*],

lap [in dew-*lap*; cf. also *lop*-eared], *a detached portion, a district:*— Læppa oððe ende *ora*, Ælfc. Gr. 50; Som. 51, 30. Hý môstan ðam læppan friþ gebicgean ðe hý under cyngces hand oferhæfdon *they might purchase peace for that district which, subject to the king, they ruled over,* L. Eth. ii. 1; Th. i. 284, 13. On læppan his hrægles *in oram vestimenti ejus,* Ps. Spl. C. 132, 3. Hê genam his loðan ænne læppan tô tâcne ðæt hê his geweald âhte, Past. 3, 2; Swt. 36, 6. Hê forcearf his mentles ænne læppan *oram chlamydis ejus abscidit,* 28, 6; Swt. 197, 21: 199, 11, 17. Læppan *vel* fnado *fimbria,* Ælfc. Gl. 64; Som. 68, 128; Wrt. Voc. 40, 33. Lappan *lacinia,* Wrt. Voc. ii. 51, 51. Lifre læppan *vel* þearmas *fibræ,* 76; Som. 71, 110; Wrt. Voc. 45, 16. Hér sint tâcn âheardodre lifre ge on ðâm læppum and filmenum *here are symptoms of a hardened liver both on the lobes and the membranes,* L. M. 2, 21; Lchdm. ii. 204, 4. Sió lifer hæfþ fíf læppan, 2, 17; Lchdm. ii. 198, 1. [*P. L. S.* þe lappe of oure loverdes cloþ, 21, 29: *Laym.* leyde uppe his lappe [1st MS. bærm], 30261: *Chauc.* lappe: *Piers P.* he shal lese for hir loue a lappe of caritatis, 2, 35: *Prompt. Parv.* lappe, skyrte *gremium,* p. 287 where see note: *O. Frs.* lappa: *Dut.* lap *a remnant, patch: Dan.* lap *a patch: O. H. Ger.* lappa *lacinia: Ger.* lapp.] DER. eár-, fræt-, lifer- læppa.

lǽr. v. lâr.

lǽran; *p.* de *To teach, instruct, educate, to give religious teaching, to preach, to teach a particular tenet* or *dogma, to enjoin a rule, to exhort, admonish, advise, persuade, suggest:*—Ic lǽre *instruo,* Ælfc. Gr. 29; Som. 32, 4: *erudio,* 30; Som. 34, 60. Ic tý oðde lǽre *imbuo,* 28, 3; Som. 32, 46. Ic eów lǽre Godes ege *timorem Domini docebo vos,* Ps. Th. 33, 11. Ânra manna gehwylcne ic myngie and lǽre ðæt ânra gehwylc hine sylfne ongyte *I admonish and exhort every man to understand himself,* Blickl. Homl. 107, 11. Ic lǽre *persuadeo,* Ælfc. Gl. 99; Som. 76, 107; Wrt. Voc. 54, 50. For ðon ic lǽre ðæt ðæt tempel wê on fýre forbærnon *unde suggero ut templa igni contradamus,* Bd. 2, 13; S. 516, 33. Ne mæg ic ðæt dôn ðæt þ mê lǽrest *non hoc facere possum quod suggeris,* 2, 12; S. 513, 24. Ðû lǽrst ûs *tu doces nos,* Jn. Skt. 9, 34. Paulus ðæt ilce lǽreþ, Blickl. Homl. 175, 13. Lêrþ *erudit,* Kent. Gl. 470. Wê lǽraþ ðæt . . . *we enjoin that,* L. Edg. C; Th. ii. 244–258. Gyf se dêma ðiss geáxaþ wê lǽraþ hyne and gedôþ eów sorhleáse *si hoc auditum fuerit a præside, nos suadebimus ei et securos vos faciemus,* Mt. Kmbl. 28, 14. Ic lǽrde sibbe ymb manige þeóda; ǽrest ic lǽrde ðæt men lufodan hié him betweónan . . . Fæderas ic lǽrde ðæt hié heora bearnum ðone þeódscipe lǽrdon Drihtnes egsan, Blickl. Homl. 185, 10–20. Se Hâlga Gâst hié ǽghwylc gôd lǽrde, 131, 30. Lǽrde Paulinus Godes word *prædicabat Paulinus verbum,* Bd. 2, 16; S. 519, 18. Ðæt Agustinus Brytta biscopas lǽrde and monade *ut Augustinus Brittonum episcopos monuerit,* 2, 2; S. 502, 2. Lǽrde hine and manede *ammonens,* 2, 12; S. 514, 37; Blickl. Homl. 19, 36; Chr. 1042; Erl. 169, 16. Ðâ lǽrde se câsere hine ðæt hê forlête Cristes geleáfan *the emperor advised him to leave the faith of Christ,* Shrn. 83, 14. Ðâ gewunode se cyning ðæt hê hine trymede and lǽrde *solebat eum hortari,* Bd. 3, 22; S. 552, 10. Hî lǽrde tô healdenne reogollîces lifes þeódscipe *disciplinam vitæ regularis custodire docuit,* S. 553, 10. Hê ful baldlîce beornas lǽrde *full boldly he exhorted the warriors,* Byrht. Th. 140, 61; By. 311. Se wiðermêda wordum lǽrde folc tô gefeohte, Andr. Kmbl. 2392; An. 1198. Leóde lǽrde on lifes weg *he brought people by his teaching into the way of life,* 339; An. 170. Hê lǽrde men geornlîce tô Godes geleáfan *he urged men in his teaching to a belief in God,* Shrn. 125, 8. Se bisceop hié lǽrede ðæt hié sendon tô ðæm pâpan *the bishop advised them to send to the pope,* Blickl. Homl. 205, 18. Scottas lǽrdon geonge and ealde on reogollîcne þeódscipe *imbuebantur præceptoribus Scottis parvuli Anglorum, una cum majoribus, studiis et observatione disciplinæ regularis,* Bd. 3, 3; S. 526, 9. Hî hî on metercræfte and on tungolcræfte and on gramatiscræfte týdan and lǽrdon, 4, 2; S. 565, 26. Hî him lifes weg bodedon and lǽrdon *verbum ei vitæ prædicarent,* 1, 25; S. 487, 8. Ða ðe bododan and lǽrdon *qui dogmatizabant,* 5, 19; S. 639, 34. Hî hî trymedon and lǽrdon ðæt hî fæsten worhtan, 1, 12; S. 480, 31. Ðâ cleopedon his þegnas him tô and hine bǽdon and geornlîce lǽrdon ðæt hê hine ofslôge *cum eum viri sui ad feriendum Saul accenderent,* Past. 28, 6; Swt. 197, 18. Mê bǽdon and lǽrdon Rômâne ðæt ic gewât heonon onweg, Blickl. Homl. 191, 13. Hine hys yldran tô woruldfolgaþe tyhton and lǽrdon, 211, 28. Sume lǽrdon ðæt hine mon onweg âcurfe *quidam abscidendum esse dicebant,* Bd. 4, 32; S. 611, 20. Lǽr ûs *doce nos,* Lk. Skt. 11, 1: Ps. Th. 118, 12. Ðû ðê lǽr be ðon, Beo. Th. 3449; B. 1722. Lǽraþ ðæt hig healdon ealle ða þing ðe ic eów bebeád, Mt. Kmbl. 28, 20. Lǽre Pharao *ut suggeras Pharaoni,* Gen. 40, 14. Ða men ðe bearn habban lǽran hié ðâm rihtne þeódscipe, Blickl. Homl. 109, 17. Heora scriftbêc tǽcan and lǽran, 43, 8. Se mæg hine sylfne be ðare bysene lǽran *he may teach himself by this example,* 101, 6. Hê ongan lǽran tô healdenne ða þing, Bd. 4, 5; S. 571, 41. Lange sceal leornian se ðe lǽran sceal *long must he learn who is to teach,* L. Ælfc. P. 46; Th. ii. 384, 15: L. I. P. 14; Th. ii. 322, 8. Mid brôðorlîce lufan hî manigean and lǽran *eis fraterna admonitione suadere,* Bd. 2, 2; S. 502, 8. Wæs ic seald tô fêdanne and tô lǽranne *datus sum educandus,* Bd. 5, 24; S. 647, 22. Cyningas and rîce men sendon heora dôhtor ðider [*France*] tô lǽranne *filias suas erudiendas mittebant,* 3, 8; S. 531,

18. Ðæt willsume weorc ðâm þeódum godspell tô lǽranne *desideratum evangelizandi gentibus opus,* 5, 11; S. 625, 33. Godcunde lâre tô lǽranne on Angelþeóde, 2, 2; S. 502, 10. Tô lǽrenne, Blickl. Homl. 233, 17. Lǽrende and strangende hira heortan, 249, 17. Lǽrendum Athamnano *instante Adamnano,* Bd. 5, 15; S. 635, 10. Cneohtas and geonge men týdde and lǽrde wǽron *pueri erudirentur,* 3, 18; S. 546, 1: Elen. Kmbl. 345: El. 173. [*Orm. Laym.* lǽren: *A. R.* learen, leren: *Gen. and Ex.* leren *to learn: R. Glouc.* lere *to learn: Piers P.* lere: *Chauc.* lere *to learn: Prompt. Parv.* lerýn or techýn another *doceo;* lerýn or receyue lore of anothere *addisco: Goth.* laisjan *to teach: O. Sax., O. L. Ger.* lêrian: *O. Frs.* lêra: *Icel.* læra *to teach;* but in modern usage *to learn: Dan.* lære *to teach, learn: O. H. Ger.* lêran *docere, instruere, monere, redarguere: Ger.* lehren.] DER. â-, for-, ge-lǽran.

-lǽred. v. ge-, sam-, þurh-, un-lǽred.

lǽrest, least:—Æt ðam lǽrestan wlitewamme iii scillingas and at ðam mâran vi scill. *for the smallest disfigurement of the face iii shillings, for the greater vi shillings,* L. Ethb. 56; Th. i. 16, 15. As this seems to be the only instance in which this form occurs, Schmid suggests that *r* is wrongly written for *s,* but the O. Frs. *lerest* may justify the presence of the *r.* v. lǽssa.

lǽrestre, an; *f. A female teacher, an instructress, preceptress:*—Lǽrestre *doctrix,* Ælfc. Gr. 9, 64; Zup. 71, 8. Siððan clypode heó hire tô ða ylcan lǽrestran, Homl. Th. ii. 543, 8. [*Cf. Wick.* lerere: *Prompt. Parv.* lerare *doctor, instructor: Goth.* laisareis: *O. H. Ger.* lêrari: *Ger.* lehrer; but the form to which these point, and which would be a masculine corresponding to *lǽrestre,* seems not to occur, the usual word being lâreów.]

lǽr-gedéfe:—Leorna lâre lǽr gedéfe wene ðec in wisdóm. Exon. 806; Th. 303, 31; Fä. 61. *In this passage Ettmüller and Grein take* lǽrgedéfe *as an adj. but* lǽr *may well be, as Thorpe takes it, the imperat. of* lǽran, *which verb naturally accompanies* leornian.

lǽrig. A word of doubtful meaning occurring only twice:—Bærst bordes lǽrig, Byrht. Th. 140, 6; By. 284. Ne him bealubenne gebiden hæfdon ofer linde lǽrig, Cd. 154; Th. 192, 29; Exod. 239. Grein suggests comparison with λαισήϊον *a buckler, target,* and that the word like *rand* may mean *the rim of the shield* and also *the body of the shield.* Either rendering is admissible so far as the sense is concerned.

lǽring, e; *f. Instruction, teaching* [see next two words and *Icel.* læring *teaching, learning: O. H. Ger.* lêrunga *institutio, doctrina.*]

lǽring-mǽden, es; *n. A girl who is receiving instruction, a female pupil:*—Nim nû lâreów and bryng ðínum lǽringcmǽdene *take now, master, and bring them to thy pupil,* Th. Ap. 20, 13. [*Cf. Icel.* læri-mær *a female disciple.*]

lǽring-mann, es; *m. A disciple;* discipulus, R. Ben. 5, Lye. [*Cf. Icel.* læri-sveinn *a disciple.*]

lǽre; *adj. Empty* [see next word.] [*O. and N.* lere house: *R. Glouc.* was þis lond of Romaynes almest lere, 81, 1: *O. Sax.* lâri: *O. H. Ger.* lâri *inanis: Ger.* leer.] v. ge-lǽr.

lǽr-ness, e; *f. Emptiness:*—Se cymþ of tô micelre lǽrnesse *it* [*hiccup*] *comes of too much emptiness,* L. M. 1, 18; Lchdm. ii. 60, 20.

lǽs, we, e; *f. A pasture, leasow* [still found in local names]:—Lǽs *pascua;* gemǽne lǽs *compascuus ager,* Ælfc. Gl. 96; Som. 76, 44, 47; Wrt. Voc. 53, 51, 54. Se wudu and seó lǽs is gemǽne tô ðâm ân and twentigum hidum, Cod. Dip. Kmbl. v. 319, 28. Sceáp lǽswe ðínre *oves pascuæ tuæ,* Ps. Spl. 73, 1: 78, 14: 94, 7: 99, 4: 22, 1. Hit is gescræpe on lǽswe sceápa and neáta *alendis apta pecoribus ac jumentis,* Bd. 1, 1; S. 473, 14. Ic wylle hî healdan on genihtsumere lǽse *I will keep them in an abundant pasture,* Homl. Th. i. 242, 15. On gemǽnre lǽse, L. Edg. 5, 8; Th. i. 274, 26: 9; Th. i. 276, 1: L. R. S. 12; Th. i. 438, 14. Ic drîfe sceáp mîne tô heora leáse *mino oves meas ad pascua . . .* Ic lǽde hig tô lǽse *ego duco eos* [*boves*] *ad pascua,* Coll. Monast. Th. 20, 13, 27. Ne land ne lǽsse [lǽswe? MS. H. lǽse], L. O. 14; Th. i. 184, 7. Fint lǽse [lêsua, Lind: lêswe, Rush.] *pascua inveniet,* Jn. Skt. 10, 9. Waldon ða swängerêfan ða lǽswe forður gedrîfan, Chart. Th. 70, 20. Lǽswe *pascua,* Wrt. Voc. 80, 49. Lǽsa *pascua,* Ælfc. Gr. 13; Som. 16, 24. Mid heora fæder heordum on lǽsum *in pascendis gregibus patris,* Gen. 37, 12. [*O. E. Homl.* leswe; *acc:* Laym. leswa [2nd MS. lesewes]; *pl:* A. R. leswe: *Wick.* leswe, lesewe: *R. Glouc.* lese. Tusser uses *lease* = *pasture.*] v. eten-lǽs.

lǽs, e; *f. A letting* [of blood]:—Blôdes lǽs *bloodletting,* L. M. 3, 47, cont; Lchdm. ii. 302, 23. Cf. þurh ða blôdlǽse geclǽnsad, 2, 23; Lchdm. ii. 210, 18; and v. blôd-lǽswu, lǽtan: cf. ǽs *and* etan.

LÆS; *adv. also used in conjunctional phrases* **Less, lest:**—Hió mê lytle lǽðe woldan ðisses eorþweges ende gescrîfan *paulominus consummaverunt me in terra,* Ps. Th. 118, 87. Nôht ðon lǽs *nihilominus,* Bd. 2, 14; S. 516, 6. Nôhte ðon lǽs, 3, 6; S. 528, 10. Nôhte ðý lǽs unârǽenedlîc *non minus intolerabile,* 5, 12; S. 627, 38. Cýð ðis folc ðæt hig ne gân ofer ða gemǽro ðê lǽs hig swelton *contestare populum ne forte velit transcendere terminos et pereat,* Ex. 19, 21, 24: Ps. Th. 68, 14. Ðê les ne, Kent. Gl. 161. Wê hine mid swâ micle mâran unryhte oferhycgeaþ swâ hê læs forhogaþ ðæt hê ûs tô him spane, Past. 52, 4; Swt. 407, 18. Ðý læs ðe, Homl. Th. i. 88, 32: Gen. 32, 11. Ân læs

R r 2

twentig *undeviginti*, twām læs twentig *duodeviginti*, Ælfc. Gr. 49; Som. 50, 41. Hē rīxode twā læs xxx geára, Chr. 641; Erl. 27, 16. Oðrum healfum læs ðe xxx wintra, 901; Erl. 96, 24. Ðȳ ilcan sumera forwearþ nō læs ðonne xx scipa *that same summer no less than twenty ships were lost*, 897; Erl. 96, 14. Ða wæs āgangen fíf þúsend geára and ðan geáre læs ðonne twā hund, Shrn. 29, 34. Gif læs manna beó *sin minor est numerus*, Ex. 12, 4. Swā man māre sprycþ swā him læs manna gelýfeþ *the more a man speaks, the fewer men believe him*, Prov. Kmbl. 38. Swā mid læs worda swā mid mā *whether with fewer words or with more*, Bt. 35, 5; Fox 166, 12. Forðon hit næs þeáw on ðǽm tídum ðæt mon ǽnig wæl on ða healfe rímde ðe ðonne wieldre wæs būton ðǽr ðȳ læs ofslagen wǽre *quia scriptorum veterum mos est, ex ea parte quæ vicerit occisorum non commemorare numerum: nisi forte cum adeo pauci cadunt*, Ors. 4, 1; Swt. 156, 22. Ðes dæg is geweorþod mid manegum godcundum geofum næs ðara gifena læs ðonne Drihtnes ǽrist and eác ðonne seó gifu ðæs Hálgan Gāstes *this day is distinguished by many divine gifts, no less gifts than the Lord's resurrection, and also than the gift of the Holy Ghost*, Blickl. Homl. 133, 3. Áhte ic holdra ðȳ læs, Beo. Th. 929; B. 487: 3897; B. 1946: Exon. 103 a; Th. 391, 27; Rä. 10, 11. Ða ðe læs águn 33 b; Th. 106, 33; Gū. 50. Læsast brúcan, 37 b; Th. 122, 22; Gū. 309. Licgende beám læsest gróweþ, 91 b; Th. 343, 19; Gn. Ex. 159. Ðonne hí læst wēnaþ *when they least expect*, Ps. Th. 13, 9: 10, 2: Homl. Th. ii. 104, 12: Bd. 4, 25; S. 601, 30. Hū gé fullecost mágon Gode þiówian ðæt eów læst þing mierþ *quod facultatem præbeat sine impedimento Domino observiendi*, Past. 51, 7; Swt. 401, 17. [O. Sax. les.]

læsast, læsest. v. læs, læssa.

læs-boren; adj. *Of inferior birth* :—Wē lǽraþ ðæt ǽnig forþboren preóst ne forseó ðone læsborenan *we enjoin that any highborn priest do not despise the one of inferior birth*, L. Edg. C. 13; Th. ii. 246, 21.

læs-hosum = [?] læst-hosan; *pl. Some species of covering for the foot, socks without soles* :—Fōt-leáste [= -læste], læshosum [= læsthosan] *cernui* ['*cernui* socci sunt sine solea,' Ducange], Ælfc. Gl. 28; Som. 61, 17; Wrt. Voc. 26, 16. v. læst, læst-wyrhta.

læsian. v. læswian.

læssa; adj. cpve. *Less* :—Se ðe læssa ys ys on heofena ríce him mǽre *qui autem minor est in regno cælorum, major est illo*, Mt. Kmbl. 11, 11. Hwæðere hē ðǽm ðe on sceare māran wǽron on ðām mægnum eáþmódnesse and hȳrsumnesse nōhte ðon læssa wæs *verum eis quæ tonsura majores sunt virtutibus humilitatis et obedientiæ non mediocriter insignitus*, Bd. 5, 19; S. 637, 18. Ðæt mǽre leóht and ðæt læsse leóht *luminare majus et luminare minus*, Gen. 1, 16. Gaderodon sum māre sum læsse *collegerunt, alius plus, alius minus*, Ex. 16, 17. Ne eart ðū læst [læsæst, Rush.] *nequaquam minima es*, Mt. Kmbl. 2, 6. Nis ðæt læsast, Exon. 43 b; Th. 148, 7; Gū. 741. Ðara ánum ðeáh hit se læsta and se heánosta to one of them, *though it were the least and the humblest*, Blickl. Homl. 169, 22. Ðæt læste fæc *parvissimum spatium*, Bd. 2, 13; S. 516, 20. Ðone læstan dǽl þunges *the least bit of aconite*, L. M. 2, 52; Lchdm. ii. 268, 31. Ðara læstena worda hreówsian *se de tenuissima verbi laceratione reprehendunt*, Past. 28, 6; Swt. 199, 15. Se ðe tōwyrpþ ān of ðysum læstum [leasestum, Lind : læsest, Rush.] bebodum ... se biþ læst [leasest, Lind: se læsesta, Rush.] genemned on heofonan ríce *qui solverit unum de mandatis istis minimis, ... minimus vocabitur in regno cælorum*, Mt. Kmbl. 5, 19. Æt læstan l scypa *at least 50 ships*, Chr. 1049; Erl. 173, 15. [O. Frs. lessa.] v. lærest.

læst. v. læssa, læs.

læst *a track*. v. lást.

læst, e; *f. A covering for the foot, a boot* :—Læste *ocreæ*, Ælfc. Gl. 29; Som. 61, 25; Wrt. Voc. 26, 24. v. læs-hosum, læst-wyrhta, lást, *and* Icel. leistr; *m. a short sock* : O. H. Ger. leist *calopodium, forma* : Ger. leisten.

læst *act, performance* (?) :—Nalles hige gehyrdon háliges láre siððan leófes leóþ læste neár swēg swiðrode *they did not neglect the holy one's [Moses] teaching, after the loved one's lay, when the time drew nearer for action [crossing the Red Sea], and his voice died away*, Cd. 158; Th. 197, 17; Exod. 308. v. Bouterwek's Cædmon i. 321; *and cf.* ful-læst, ge-lást.

læstan; *p.* te.   I. *to follow, attend, accompany, do suit and service* :—Sōna ða beótunge dǽdum læstan *neque seguius minas effectibus prosequuntur*, Bd. 1, 15; S. 483, 39, MS. C. Allum ðám ðe him læstan woldon *with all those who would follow him*, Chr. 874; Erl. 76, 31. Gif hí leódfruman læstan dorsten, Bt. Met. Fox 1, 54; Met. 1, 27. Him se líchoma læstan nolde *the body would not do him service* [*of Grendel powerless in the grasp of Beowulf*], Beo. Th. 1629; B. 812.   II. *to do, perform, observe, carry out, execute, discharge* [*a debt* or *duty*] :—Ic lufan symle læste wið eówic *I will ever love you*, Exon. 14 b; Th. 30, 10; Cri. 471. Gif hē læst mína lára *if he does my teachings*, Cd. 29; Th. 39, 3; Gen. 619. Ðenden ðū míne láre læstest, 99; Th. 130, 32; Gen. 619. Læstes, 27; Th. 36, 15; Gen. 572. Forðon hē ða godspellícan bebode heóld and læste *quod evangelica præcepta servaret*, Bd. 3, 22; S. 553, 23: 4, 25; S. 600, 20. Ðæt hí ða ungewergadre geornfullnysse fylidon and læston *ut instituta indefessa instantia sequerentur*, 4, 3; S. 568, 15. Ðæt mid dǽdum læston ða ðe hí ongitan mihton *ut ea quæ intelligere poterant, operando sequerentur*, 4, 27; S. 604,

18. Wē sōðfæstes swaðe folgodon, læston lárcwide, Andr. Kmbl. 1347; An. 674. Læstun, Exon. 25 b; Th. 75, 21; Cri. 1225. Bibeád ic eów ... earge gē ðæt læstun *my command was ... ill have ye performed it*, 30 a; Th. 92, 3; Cri. 1503. Leófa Beówulf læst eall tela, Beo. Th. 5320; B. 2662: Cd. 106; Th. 139, 4; Gen. 2304. Læste ðū georne his ambyhto *do diligently his messages*, 25; Th. 33, 9; Gen. 517. Ðū læstan scealt ðæt his bodan bringaþ, Th. 32, 26; Gen. 509. Geongordōm læstan, 30; Th. 41, 26; Gen. 663. Gif ðū wilt his wordum hȳran and his bebodu læstan *if thou wilt hear his words, and do his commands*, Blickl. Homl. 185, 1: Exon. 45 a; Th. 152, 28; Gū. 815. [Ic an six marc silures and ðat schal Godríc míne bróðer lēsten (*pay*), Chart. Th. 566, 23, 31.]   III. *to continue, last* :—Ðonne him dagas læstun *in their life-time*, Exon. 26 b; Th. 79, 12; Cri. 1289. [Ðæt fír læste swa lange þ hit wæs liht ofer eall, Chr. 1122; Erl. 249, 25. ð lastede þa xix wintre wile Stephne was king, 1137; Erl. 262, 19.] [*Gen. and Ex.* lesten *to perform*: *Orm.* lasstenn *to last* : *Laym.* læsteþ *lasts* : *O.E. Homl.* lasteþ : *A.R.* lesteþ : *Marh.* leasteþ : *Mand.* laste *to perform* : *Goth.* laistjan *to follow* : *O. Sax.* lēstian *to do, perform* : *O. Frs.* lásta, lēsta, *to do, perform, pay* : *O.H. Ger.* leistan *reddere* : *Ger.* leisten.]

lēste, es; *m* [?] *A shoemaker's last* :—Læste *musticula* ['*mustricola* machina ad stringendos pedes, hoc est, ad calceum suendum qui pedes tegit et stringit : forma in qua calceus suitur,' Ducange]. Ælfc. Gl. 29; Som. 61, 25; Wrt. Voc. 26, 24. Læste *vordalium* [?], Wrt. Voc. 287, 37. [lest *formipedia*, Wrt. Voc. 181, 13 : *Prompt. Parv.* leste, sowtarys forme *formula, calopodia* : *Dan.* læst *a last* : *O.H. Ger.* leist *calopodium, forma* : *Ger.* leisten.] v. læst.

lēstend, es; *m. One who performs* or *executes* :—Ðara þinga ðe hē óðre lǽrde tō dōnne hē sylfa wæs se wylsumesta fyllend and læstend *eorum quæ agenda docebat erat executor devotissimus*, Bd. 5, 22; S. 644, 4. Fylgend and læstend, 4, 3; S. 568, 15, note.

Læsting, Læsting eá *Lastingham in Yorkshire* :—Fram ðám bróðrum ðæs mynstres ðe Læstinga eá is nemned, Bd. pref : S. 472, 17. Getimbrede ðǽr mynster ðæt is nū gecýged Læstinga eá, 3, 23; S. 555, 3. Hē gewát tō his mynsterscire ðæt is on Læstinga eá, 5, 19; S. 639, 14. On Læstinge, 4, 3; S. 566, 28.

lēst-wyrhta, an; *m. A shoemaker* :—Læstweorhta *caligarius*, Ælfc. Gl. 28; Som. 61, 23; Wrt. Voc. 26, 22. Læstwyrhta, Wrt. Voc. ii. 127, 66. v. læst.

lēswian, lēsian: *p.* ode, ede, *trans. and intrans. To pasture, feed, graze* :—Ic læswige *pasco*, Ælfc. Gr. 35; Som. 38, 13. Ic læsewige, læswige, 28 : Som. 30, 33. Ic hí læswige on dōme and on rihtwísnysse *I will feed them in judgement and righteousness*, Homl. Th. i. 242, 18. Hig man læswode on mōrium lande *pascebantur in locis palustribus*, Gen. 41, 2. Ða læswede heó hire fēstermódor sceápum *then she fed her foster-mother's sheep*, Shrn. 101, 14. Ða læswede hē mid his fæder sceápum, 108, 31. Ða assan wið hí læswodon *the asses were grazing by them*, Homl. Th. ii. 450, 6. Læswa míne scēp *feed my sheep*, 290, 30. Oxanhyrde mōt læswian ii. oxan oððe mā on gemǽnre læse *bubulco licet adherbare duos boves, et alicubi plus, in communibus pascuis*, L. R. S. 12; Th. i. 438, 13. Heord læswiende *grex pascens*, Mt. Kmbl. 8, 30. Læsgende, Mk. Skt. 5, 11. Hwylc eówer hæfþ þeów scēp læsgende [lēsuande, Lind.] *quis vestrum habens servum pascentem*, Lk. Skt. 17, 7. Heord swýna læsiendra [lēsuandra, Lind.] *grex porcorum pascentium*, 8, 32. On læswigendum eówdum *in pascendis gregibus*, Ælfc. Gr. 26; Som. 28, 20. [O. E. Homl. lesewep, *prs.* 3 : *A. R.* leswe, *imper* : *Wick.* leseweden, *p.* 3.]

læt, es; *m. One of a class that was inferior to that of the ceorl but above that of the slave. The word occurs only in the following passage* :—Gif[man] læt ofslæhþ ðone sélestan lxxx scill. forgelde gif ðane óðerne ofslæhþ lx scillingum forgelde ðane þriddan xl scillingum forgelden *if any one slay a 'læt' of the highest class, let him pay eighty shillings; if he slay one of the second, let him pay sixty shillings; let them pay for one of the third with forty shillings*, L. Ethb. 26; Th. i. 8, 12–14. See Stubbs' Const. Hist. s. v : Grmm. R. A. 305–309 : Grff. ii. 190 : Thorpe's Glossary : Kemble's Saxons in England, i. c. 8 : Lappenberg's Hist. ii. 321.

læt; *adj. Late, slow, sluggish, tardy* :—Wundrodon ðæt hē on ðam temple læt wæs *mirabantur quod tardaret ipse in templo*, Lk. Skt. 1, 21. Hræd tō gehiéranne and læt tō sprecenne *velox ad audiendum, tardus ad loquendum*, Past. 38, 8; Swt. 281, 6. Hlæt, Rtl. 28, 19. Nalas elnes læt *not slow of courage*, Beo. Th. 3063; B. 1529. Ne sceal se tō sǽne beón, ðissa lárna tō læt, seðe him wile lifgan mid Gode, Exon. 117 a; Th. 450, 17; Dōm. 89: Apstls. Kmbl. 66; Ap. 33. Se mæssepreóst se ðe biþ tō læt ðæt deófol of men ádrífe *that the priest who is too slow in driving the devil from a man*, Blickl. Homl. 43, 22: Exon. 74 a; Th. 276, 29; Jul. 573: 76 a; Th. 285, 11; Jul. 712. Heora behreówsung wæs tō læt *their repentance was too late*, Homl. Th. ii. 572, 15. Nis seó stund latu ðæt ... *the time does not tarry, when* ... Andr. Kmbl. 2422; An. 1212: Exon. 46 a; Th. 156, 16; Gū. 875. Nis seó tíd latu, 51 a; Th. 178, 4; Gū. 1239. Be latre meltunge *of sluggish digestion*, L. M. 2, 33; Lchdm. ii. 238, 6. Wið latre meltunge, 2, 34;

Lchdm. i. 238, 27. Late gange *gradu lento*, Wrt. Voc. ii. 41, 76. Læte dissides, i. *tardi*, 141, 6. Ne beóþ æfre tô late *numquam sunt sera*, 62, 18. Swæfna gewisse synt oft late *dreams are certain, but often late of fulfilment*, Lchdm. iii. 186, 27. Ðæt hî ne beón ne wordes ne weorces, ne ealles tô hræde ne tô swíðe læte, L. I. P. 10; Th. ii. 318, 36. Nalæs late wæron eorre æscberend tô ðam orlege, Andr. An. 46. On heortan læte [hlatto, Lind.] tô gelýfenne *tardi corde ad credendum*, Lk. Skt. 24, 25. Ðæt hê ðý lætra biþ tô uncystum *that it [the body] be the less ready to vices*, L. E. I. 3; Th. ii. 404, 20. And â swâ hit forþwerdre beón sceolde swâ wæs hit lætre *and ever as things ought to have been more forward, did they go on more slowly*, Chr. 999; Erl. 134, 33. Siððan ðû spræce tô ðínum þeówe ic hæfde ðê lætran tungan *ex quo locutus es ad servum tuum, impeditioris et tardioris linguæ sum*, Ex. 4, 10. [*Goth.* lats *slothful*: *O. Sax.* lat: *O. Frs.* let: *Icel.* latr: *O. H. Ger.* laz *piger, segnis, stupidus, tardus: Ger.* lass.] v. unlæt, lata; lætemest, lætest.

-læta. v. freó-, frig-, scyld-læta.

LÆTAN, *p.* lêt, leórt; *pp.* lǽten. The ellipsis of a verb in the infinitive, the meaning of which may be inferred from the context, not unfrequently takes place after *lǽtan;* and the connection of many of the meanings which follow with the simple one seems explainable in this way. **I.** *to* LET, *allow, permit, suffer* :—God læt him fyrst ðæt hê his mándǽda geswíce *God allows him time that he may cease from his crimes*, Homl. Th. i. 268, 32. Ðonne ne lǽteþ hê ûs nô costian ofer gemet *then he will not let us be tempted beyond measure*, Blickl. Homl. 13, 8. Gif Drihten ðê lǽteþ ðone teóþan dǽl ânne habban *if the Lord lets thee have only the tenth part*, 51, 3. God lǽt hí habban âgenne cyre, Homl. Th. i. 10, 19. Ne leórt ǽnigne monno tô fylgenne hine *non admisit quemquam sequi se*, Mk. Skt. Lind. 5, 37. Ðâ onlýsde hê hine and lêt hine féran æfter ðam biscope *absolvit eum, et post Theodorum ire permisit*, Bd. 4, 1; S. 565, 3. Se dêma lêt ða môdor tô ðam suna on synderlíce clýsingce *the judge allowed the mother to come to the son in a chamber apart*, Homl. Skt. 4, 342. Se êca Drihten hine sylfne lêt lǽdon on ða heán dûne *the Lord eternal allowed himself to be led on to the high mountain*, Blickl. Homl. 33, 10. Drihten ealle ða gefylde ða ðe hié on eorþan lêton hingrian and þyrstan for his naman *the Lord had filled all those who let themselves, or were content to, hunger and thirst for his name's sake* [cf. in Icel. *láta* with a reflex. infin.], 159, 17. Lǽt beón ealne dæg *let it be all day*, L. M. 2, 22; Lchdm. ii. 106, 25. Lǽtaþ ægðer weaxan *sinite utraque crescere*, Mt. Kmbl. 13, 30. Lête pateretur, Wrt. Voc. ii. 67, 4. Hê lifde bûton synnum ðeáh ðe hê hine lête costian, Blickl. Homl. 33, 17. Lǽtan nǽnne lybban *to let none live*, Ex. 14, 5. Se ðe mýn blôd nolde lǽtan âgeótan *he that would not suffer my blood to be shed*, Nicod. 20; Thw. 10, 17. Lêton, Exon. 46 b; Th. 152, 3; Gû. 921. Gif ðû ðê wilt dôn manegra beteran ðonne scealt ðû ðê lǽtan ânes wyrsan *if thou wilt make thyself the superior of many, thou must allow thyself to be the inferior of one*, Bt. 32, 1; Fox 114, 14. **II.** *to let* [*alone*], *let go, give up, dismiss, leave, forsake, let* [*blood*] :—Ne recce ic hwæt hí dêman. Ic lǽte tô ðínum dôme mâ ðonne tô hiora *I care not what judgements they make. I give myself up, or trust, to your judgement more than to theirs*, Bt. 38, 5; Fox 206, 14. Hwý nelt ðû géman ðæt mín sweostor mê lǽt âne þegnian *why dost thou not heed that my sister leaves me to serve alone?* Blickl. Homl. 67, 31. Hê lǽt his hláfordes gebod tô gíemeléste *he leaves to neglect* [*neglects*] *his lord's command*, Past. 17, 8; Swt. 121, 14. Lǽtt ðonne ân ðæt gefeoht sume hwíle *he lets the battle alone then for some time*, 33, 7; Swt. 227, 10. Hê cwæþ tô him lǽtaþ ðæt nett on ða swiðran healfe ðæs réwettes ... hig lêton *dixit eis mittite in dexteram navigii rete ... miserunt*, Jn. Skt. 21, 6. Ðæt ic sylf ongeat ne lêt ic ðæt unwriten *what I myself knew, I did not leave unwritten*, Bd. pref; S. 472, 26. God hine lêt frigne *God left him free*, Homl. Th. i. 18, 29. Ic lêt míne wylne tô ðê *ego dedi ancillam meam in sinum tuum*, Gen. 16, 5. Se arcebiscop lêt hit eall tô heora âgene rǽde *the archbishop left it all to their own discretion*, Chart. Th. 341, 11. God hî hǽðenum leódum lêt tô anwealde *God left them to the power of heathen nations*, Jud. 1, 8. Hê lǽt hí tô handa Madian *tradidit illos in manu Madian*, 6, 1. Ðâ gyrnde se cyng ealra ðæra þegna ðe ða eorlas ǽr hæfdon, and hí lǽtan hî ealle him tô handa, Chr. 1048; Erl. 180, 9. Hê forsâwon eall min geþeaht and lêton eów tô gíemeléste ðonne ic eów cídde *despexistis omne consilium meum et increpationes meas neglexistis*, Past. 36, 1; Swt. 247, 22. Swâ swâ hit his yldran lǽtan and lǽfdan ðam tô gewealde *as his parents left and bequeathed it to him at his disposal*, L. O. 14; Th. i. 184, 3; Lchdm. iii. 286, 15. Hine eft ðǽm mannum hâlne and gesundne âgeaf ðám ðe hine ǽr deádne lêton *gave him back safe and sound to the men who before had left him dead*, Blickl. Homl. 219, 24. Lǽt ðíne lâc beforan ðam altare *relinque munus tuum ad altare*, Mt. Kmbl. 5, 24. Lǽt ðû him blôd on ǽdre *let blood for him from a vein*, L. M. I, 4; Lchdm. ii. 46, 22. Beó ðû be ðínum and lǽt mê be mínum *be thou with thine, and leave me with mine*, Lchdm. iii. 288, 8. Gif hê tôþ of âsleá hwâ frige dentem si excusserit, dimittet eos liberos, Ex. 21, 27. Lǽte hig frige, 26. Wê lǽraþ ðæt man ǽnig ne lǽte unbiscpod tô lange, Wulfst. 120, 15. Lǽte [ðæt feoh] ân and fô se âgend tô *let him give up* [*the property*], *and*

let the owner take it, L. H. E. 7; Th. i. 30, 9: 12; Th. i. 34, 12. Hwilce hwíle hwile wille Drihten hêr on worlde lǽtan *how long the Lord will leave him in this world*, Blickl. Homl. 125, 9. Hê sceal lǽtan his unnyttan geþancas of his môde *he must dismiss his idle thoughts from his mind*, Wulfst. 234, 26. Ðonne hê hî nyle lǽtan tô hiera âgnum wilnungum *quos in sua desideria non relaxat*, Past. 50, 4; Swt. 391, 22. Hê nô ðe ðǽm ânum lǽtan wolde ac ofer ðone gârsecg ðone ylecan leóman ðæs fullan geleáfan âspringan lêt *he would not leave off when that* [*the spreading of the gospel over part of the world*] *only was done, but caused the same beam of the perfect faith to spring forth across the ocean* [*to England*], Lchdm. iii. 432, 16. Wið poccum swíðe sceal mon blôd lǽtan, L. M. I, 40; Lchdm. ii. 106, 3. Nis him blôd tô lǽtanne, 35; Lchdm. ii. 82, 16. **III.** *to let, cause, make, get, have, cause to be, place* :—Ic hine symble gehýre and mine mildse ofer ðone lǽte *I will ever hear him, and my mercy shall be upon that man*, Wulfst. 264, 11. Swâ hí hiora lufe neár Gode lǽtaþ swâ hí bióþ orsorgu *the nearer to God they place their love, the more free are they from care*, Bt. 39, 7; Fox 222, 24: 40, 7; Fox 242, 26–28. Hê lêt betwux him and mínum feóndum ðæt he nǽfre gesewen [wæs] fram him *posuit tenebras latibulum suum*, Ps. Th. 17, 11. Ðâ fôr hê norþryhte be ðæm lande lêt him ealne weg ðæt wêste land on ðæt steórbord *then he sailed due north along the coast : he had the waste land all the way on his starboard*, Ors. 1, 1; Swt. 17, 10. Ðâ hê lêt standan beforan ymbeûtan ða eardungstôwe *quos stare fecit circa tabernaculum*, Num. 11, 24. Hê sette scole and on ðǽre hê lêt cnihtas lǽran *he set up a school, and had boys taught in it*; instituit scholam in qua pueri literis erudirentur, Bd. 3, 18; S. 545, 45. Se cing lêt gerídan ealle ða land ðe his môdor âhte him tô handa, Chr. 1042; Erl. 169, 19: 1023; Erl. 162, 35: 1035; Erl. 164, 22. Wit ðæt ðâ lêtan and unêþelíce þurhtugan ðæt hê ðæs geþafa wolde beón *with difficulty we got him to assent to it*, Bd. 5, 4; S. 617, 17. Hí lêton hig hâdian tô bisceopum *they got themselves ordained bishops*, 1053; Erl. 188, 14. **IV.** *to make a thing appear* [*so and so*], *make as if, make out, profess, pretend, estimate, consider, suppose, think* :—Ic lêto *existimabo*, Lk. Skt. Lind. 13, 18. Hê lêttes *arbitretur*, Jn. Skt. Lind. 16, 2. Ða lǽtaþ wê *non dissimulamus*, Wrt. Voc. ii. 62, 22. Ealle wê lǽtaþ efendýrne Engliscne and Deniscne *we estimate all at the same amount, Englishman and Dane*, L. A. G. 2; Th. i. 152, 12. [Cf. *Icel.* manngjöld skyldi jöfn látin ok spora-höggit.] Ðonne wê ðisses middangeardes welan foresettaþ and ûs leófran lǽtaþ ðonne ða lufan ðara heofonlícra eádignessa *cum mundi divitias amori cælestium præponimus*, Bd. 3, 19; S. 548, 16. Fela is ðæra ðe embe bletsunga oððe unbletsunga leóhtlíce lǽtaþ *many are there that esteem lightly of blessings or cursings* [cf. *Piers P.* iv. 160–161 moste peple ... *leten* mekenesse a maistre and Mede a mansed schrewe. Loue *lete* of hir liȝte and lewte ȝit lasse: *Orm.* 7523– uss birrþ *lætenn* unnorneliȝ and litell off uss sellfenn and *lætenn* wel off oþre menn], L. I. P. 6; Th. ii. 310, 36. Gê beótlíce lǽtaþ *ye boast*, Wulfst. 46, 15. Hý þencaþ and lǽtaþ ðæt tô warscype, ðæt hý ôðre mâgan pæcan, 55, 2. Hê lêt ðæt hyt Dryhtnes sylfes andwlyta wǽre *he supposed that it was the face of the Lord himself*, St. Andr. 42, 9. Ðâ sendun hig mid searwum ða ðe rihtwíse lêton *observantes miserunt insidiatores qui se justos simularent*, Lk. Skt. 20, 20. Ðâ lêton hý sume ðæt ðæt mycel unrǽd wǽre *some of them considered it a very bad plan*, Chr. 1052; Erl. 179, 32. Manige lêton ðæt hit cometa wǽre *many supposed that it was a comet*, 1097; Erl. 234, 13. Ðæs ðe men lêton *as men supposed*, Erl. 234, 17. Hê hit selfe lêton ægðer geo for heáne ge for unwræste *they considered themselves as abject and undone*; ultima propemodum desperatione tabuerunt [cf. *Piers P.* xv. 5 somme *leten* me for a lorel], Ors. 3, 1; Swt. 98, 22. Lêton ðâ gedwealde men, swylce Simon Godes sylfes sunu wǽre, Wulfst. 99, 7. Ðæt man þurh ðæt lǽte ðæt hê sí ðæs legeres wyrðe *so that for that reason it be considered that he is worthy of such burial*, L. Edg. C. 29; Th. ii. 250, 17. Ðæt hê ða ðe him underþiedde síen lǽte him gelíce *æqualem se subditis deputet*, Past. 17, 1; Swt. 107, 15. Ðæt cild ðe læg on cradele ða gýtseras lêton efenscyldig and hit gewittig wǽre, L. C. S. 77; Th. i. 420, 2. Ic wælle lêta *æstimabo*, Lk. Skt. Lind. 13, 20. Se ealdormonn sceal lǽtan hine selfne gelícne his hiéremonnum, Past. 17, 1; Swt. 107, 8. **V.** *to behave towards, treat* :—Ðam elþeódigan and ûtancumenan ne lǽt ðû nô uncûþlíce wið hine *as regards the alien and foreigner do not behave unkindly towards him* [cf. *Icel.* björn lætr allblítt við hana], L. Alf. 47; Th. i. 54, 20. **VI.** *to let* [*land, &c.*] :—Eádward cyning and ða hiwan (in Wintanceastre lǽtaþ tô Dænewulfe bisceope twentig hída landes, Chart. Th. 158, 7. Ðâ com sum ôðer and beád mâre ðonne ðe ôðer ǽr sealde and se cyng hit lêt ðam menn ðe him mâre beád *then some other man came and offered more than the other had before given, and the king let it to the man that offered more*, Chr. 1086; Erl. 220, 10. Ðâ hý lǽtan him tô ðæt land æt Eádburge byrig, Cod. Dip. Kmbl. iv. 76, 5: Chart. Th. 151, 6. **VII.** *with adverbs* :—Ðâs ôðre lǽtaþ ðone n âweg on sopinum *these others let the n fall away in the supine*, Ælfc. Gr. 28; Som. 31, 60. Ælmǽr abbod hí lǽtan âweg *they let abbot Aylmer go away*, Chr. 1011; Erl. 145, 13. Ðâ cwǽdon ða witan ðæt betere wǽre ðæt man ðene aþ âweg lête ðonne hine man

sealde . . . Ða lēt hē ðone áþ áweg *then the witan said that it would be better that the oath should be dispensed with than that it should be taken . . . Then he omitted the oath*, Chart. Th. 289, 24–30. Æt ealre ðære hergunge and æt eallum ðám hearmum ðe ǽr ðam gedón wǽre ǽr ðæt friþ geset wǽre man eall onweig lǽte and nán man ðæt ne wrǽce ne bóte ne bidde *as regards all the harrying and all the injuries that were done before the peace was made, let it all be dismissed, and let no man avenge it or ask for compensation*, L. Eth. ii. 6; Th. i. 288, 3. Petrus cnucode óþ ðæt hī hine inn lēton *Peter knocked until they let him in*, Homl. Th. i. 382, 23. Hē lǽt him eáþelíce ymbe ðæt *he takes it easily*, Wulfst. 298, 30: Homl. Skt. 4, 342. [Goth. lētan: O. Sax. látan: O. Fries. lēta: Icel. láta: O. H. Ger. lâzan.] DER. á-, for-, ge-, of-, on-, tó- lǽtan.

lǽt-byrd, e; f. *A late or slow birth:—*Se wífman se hire cild áfédan ne mæg gange tó gewitenes mannes birgenne . . . and cweþe ðás word ðis mē tó bóte ðære lápan lǽtbyrde *let the woman who cannot nourish her [unborn] child go to the grave of a dead man . . . and say these words: 'May this help me with the troublesome late birth,'* Lchdm. iii. 66, 21.

-lǽte. v. á-lǽte.

lǽtemest; *a double superlative of lǽt. Last:—*In ðǽm lǽtemestan dæge *in novissimo die*, Jn. Skt. Rush. 6, 44: 39, 40. Stówe ða lǽtemestu *novissimum locum*, Lk. Skt. Rush. 14, 9, 10. Monige wutudlíce bióþun ǽrist ða foerþmestu and ða lǽtemestu foerþmest *multi autem erunt primi novissimi et novissimi primi*, Mk. Skt. Rush. 14, 31. Ða endo ꞇ lǽtmesta *novissima*, Mt. Kmbl. Lind. 12, 45.

lǽtemest; *adv. Lastly, at last, finally:—*Lǽtemest (lǽtmest, Lind.) *novissime*, Mk. Skt. Rush. 16, 14.

lǽtere. v. blōd-lǽtere.

lǽtest; *superl. of lǽt. Last:—*Ðe lǽtest [ða lǽtmesta, Lind.] *the last*, Mt. Kmbl. Rush. 22, 27. [O. Eng. Homl. latest: Orm. latst: A. R. Laym. last.]

lǽð, es; *n. Land:—*Dō swá ic lǽre beó ðé [Lchdm. ðú] be ðínum and lǽt mē be mínum ne gyrne ic ðínes ne landes ne sace ne sócne ne ðú mínes ne þearft *do as I advise; be thou with thine and leave me to mine; I desire nothing of thine, neither lea nor land, neither 'sac' nor 'socn'; nor needest thou mine*, L. O. 14; Th. i. 184, 15; Lchdm. iii. 288, 8. The Icelandic has the same alliterative phrase, e. g. 'deyr fé; deyja frændr; eyðisk land ok láð.'

lǽð *a lathe* [e. g. Kent is divided into six *lathes*], *a district containing several hundreds*, v. Stubbs' Const. Hist. i. 100. The word occurs in the Latin laws of Edward the Confessor:—In quibusdam vero provinciis Anglice vocabatur *leð*, quod isti dicunt tithinge [or trihinge], Th. i. 455, n. 3. In L. Hen. I. viii. 2 occurs amongst the names of other officials *leidegrevei* = *láðgeréfan*, Th. i. 514, note 1. Cf. Icel. leið, leiðangr *a levy*: Dan. leding. Skeat, Etymol. Dict. under *lathe*, suggests that lǽð = lægð, in which case perhaps it may be compared with *Dan.* lægd a *levying district*.

lǽðan; *p. de To speak ill of, accuse, abuse, execrate, detest, hate:—*Man eall hyrweþ ðæt man scolde herian and láðeþ [lǽðeþ?] ðæt man scolde lufian *people scorn what they ought to praise, and hate what they ought to love*, Swt. A. S. Rdr. 110, 167. Gif hwelc cymiþ tó mē and ne lǽdes [lǽðes, Lind] fæder his *si quis venit ad me et non odit patrem suum*, Lk. Skt. Rush. 14, 26. Middý iuih lǽðeþ menn *cum vos oderint homines*, Lind. 6, 22. Ða ðe lǽdes ꞇ lǽðedon *qui oderunt*, Mt. Kmbl. Lind. 5, 44. Ða ðe lǽðdon, Lk. Skt. Lind. i, 71. Hý wǽron ealle ánspræce ðonne hý mē leahtrodon and lǽðdon *loquebantur simul*, Ps. Th. 40, 7. [Cf. Icel. leiða *to make a person loathe a thing*: O. Sax. a-lēðian *to disgust*: O. H. Ger. leidan *accusare, detestari*; leiden *execrari, odiosum facere*.] v. be-lǽðan, láðian.

lǽðð[u], e; lǽððo; indecl.; f. *An injury, offence, hatred, enmity, malice:—*Lǽððe *livoris*, Wt. Voc. ii. 50, 16. Mið lǽðo hæfe ðū fiónd ðinne *odio habebis inimicum tuum*, Mt. Kmbl. Lind. 5, 43. Lǽððo *odio*, 24, 10. Seðe unlage rære oððe undóm gedéme heononforþ for lǽððe oððe for feohfange *he that from this time forth shall set up unjust law, or judge unjust judgement on account of malice or of bribery*, L. C. S. 15; Th. i. 384, 9. Þurh Pendan lǽððe hyra cyninges, Bd. 3, 18; S. 546, 14. Ðæt is ðonne ðæt ǽrest ðæt man tó óðrum lǽððe hæbbe *now first it is murder, that a man hate another*, Blickl. Homl. 63, 36. Ne dóm ic ðe laxðo *non facio tibi injuriam*, Mt. Kimbl. Lind. 20, 13. Ðæt hié ongietan ðæt ðæt sindon ða forman lǽððo ðe hié Gode gedoon mǽgen *ut noverint, quod hanc primam injuriam faciunt Deo*, Past. 45, 2; Swt. 339, 7. Ðara lǽðða ðe gē lange drugon *for the injuries that ye have suffered long*, Judth. 11; Thw. 23, 36; Jud. 158. Hē mid lǽððum ús eglan móste, Thw. 24, 12; Jud. 185. Ðone Jacobum Judǽa leorneras otslógan for Cristes lǽððum *that James the disciples of the Jews slew from hatred to Christ*, Shrn. 93, 12. Lǽððum hwópan *to threaten injuries*, Exon. 64 a; Th. 236, 31; Ph. 582. [Cf. Icel. leiða; f. *irksomeness*: O. H. Ger. leida; f. *accusatio*.]

lǽt-hýdig; *adj. Slow-minded, slow of thought, dull:—*Nis mon on moldan . . . ðæs læðhýdig ðæt hine se árgifa ealles biscyrge módes crǽfta *no man is there on earth so dull, that the bounteous giver hath quite cut him off from powers of mind*, Exon. 78 b; Th. 294. 5; Crä. 10.

lǽt-líce; *adv. Slowly:—*Ða andswarode hē him lǽtlíce *then he an-*

swered him slowly, Guthl. 20; Gdwin, 80, 12. Lǽtlícor *more slowly* Exon. 118 a; Th. 454, 16; Hy. 4, 33.

lǽtmest. v. lǽtemest.

lǽt-rǽde; *adj. Slow of counsel, deliberate:—*Oft mon biþ suíðe wandigendre æt ǽlcum weorce and suíðe lǽtrǽde and wēnaþ menn ðæt hit síe for suármódnesse and for unárodscipe and biþ ðeáh for wisdóme and for wǽrscipe *often a man will be very hesitating in every action, and very deliberate, and men suppose that it is from stupidity and from cowardice, and yet it is from wisdom and caution*; the Latin however has 'sæpe agendi tarditas gravitatis consilium putatur,' Past. 20, 1; Swt. 149, 14.

lǽtsum; *adj. Slow, late:—*Wæs suíðe lǽtsum geár on corne and on ælces cynnes wæstmum *it was a very late year for corn and crops of every kind*, Chr. 1089; Erl. 226, 18.

lǽtt, e; f. *A lath:—*Lǽtta *asseres*, Ælfc. Gl. 29; Som. 61, 42; Wrt. Voc. 26, 41. Latta *vel* reaftetes *asseres*, 108; Som. 78, 123; Wrt. Voc. 58, 35. [Hic *asser* a lath, Wrt. Voc. 235, 37: Prompt. Parv. lathe latthe, laththe *tignus, tignum, tigillum*: O. H. Ger. latta, lata *tignum, asser, tegula: Ger.* latte *a lath*.]

lǽuw. v. leów.

lǽwa, an; *m. A betrayer, traitor:—*Lǽwa *proditor* vel *traditor*, Wrt. Voc. 85, 43. Judam scarioð se wæs lǽwa [hlēga, Lind] *iudam scarioth qui fuit proditor*, Lk. Skt. 6, 16. His lǽwa him tácen sealde *dederat traditor ejus signum*, Mk. Skt. 14, 44: Homl. Th. ii. 246, 10. Mid Judan ðe Cristes lǽwa wæs, Cod. Dip. Kmbl. iii. 138, 21. Hér is ðæs lǽwan hand *ecce manus tradentis me*, Lk. Skt. 22, 21.

lǽwan; *p. de To betray:—*Ðonne lǽweþ bróðer óðerne hǽðnum on deáþ and sunu se lǽweþ his fæder ðone *then one brother shall betray another to the heathen to death, and a son he shall betray his father*, Blickl. Homl. 171, 21. [Goth. lēwjan, *to betray*: O. H. G. gi-lâti; *p. (he) betrayed*.] v. be-lǽwan.

lǽwed, lēud, es; *m. A layman:—*Gif man lēud ofsleá on þeófþe licge bútan wyrgelde *if a layman be slain while thieving, let no wergild be paid for the slaying*. L. Wih. 25; Th. i. 42, 13. v. next word.

lǽwede; *adj. Lay, laic, not learned, not of the church;* by gradual change of meaning it has become the later *lewd:—*Lǽwede man *laicus*, Wt. Voc. 72, 8. Ðara manna sum wæs bescoren preóst sum wæs lǽwede sum wæs wífmon *e quibus hominibus quidam erat adtonsus ut clericus, quidam laicus, quædam femina*, Bd. 5, 12; S. 628, 35. Hī underfēngon ða ðigelnyssa ðǽre láre ðe ðæt lǽwede folc undergitan ne mihte *they [the apostles] received the mysteries of the doctrine that the unlearned people could not understand*, Homl. Th. i. 190, 13. Búton ða láreówas screádian symle ða leahtras þurh heora láre áweg ne biþ ðæt lǽwede folc wæstmbǽre on gódum weorcum, ii. 74, 17. Hē munuclíce leofode betwux ðam lǽwedan folce *he lived as a monk among laymen*, 97, 67. Sum wer wæs on lǽwedum háde *fuit vir in laico habitu*, Bd. 5, 13; S. 632, 7. Ðeáh ðe hē ðá gyt on lǽwedum háde beón sceolde . . . hē munuclífe gyta swíðor lifde ðonne ðonne lǽwedes mannes, Blickl. Homl. 213, 9–11. Ðæt hit nǽfre on lǽdu hand ne wende *that it should never pass to a lay hand*, Chart. Th. 166, 21. Ealle ge bescorene ge lǽwede, Bd. 3, 5; S. 526, 36: 5, 7; S. 621, 14. Ða ðe mid him wǽron swíðust lǽwde *qui cum ipso erant, maxime laici*, 5, 6; S. 618, 42. Ða witan ealle ge hádode ge lǽwede *all the witan both churchmen and laymen*, Chr. 1014; Erl. 150, 4. Ne úre nǽnig his líf ne fadode swá swá hē scolde, ne gehádode regollíce ne lǽwede lahlíce, Swt. A. S. Rdr. 107, 78. þurh gelǽredra regolbryce and þurh lǽwedra lahbryce *through breach of [monastic] rule by the learned and breach of law by the unlearned*, Swt. A. S. Rdr. 111, 199. [In the later English the *lewed* are contrasted with the *lered*, e. g. Orm. ȝa lǽwedd follc, ȝa lǽredd; and Robert Manning writes 'not for þe *lerid* but for the *lewed*:' Prompt. Parv. lewde *illitteratus, inscius, ignarus, laicus*.]

lǽwend, es; *m. One who betrays, a traitor:—*Lǽwend *proditor*, Ælfc. Gl. 85; Som. 73, 125; Wrt. Voc. 49, 18: Wrt. Voc. ii. 68, 75. Lēwend, Kent. Gl. 1156.

lǽwerce. v. láwerce.

Lǽwes, Lǽwe *Lewes in Sussex:—*Tó Lǽwe [other MS. Lǽwes] *at Lewes*, L. Ath. i. 14; Th. i. 208, 1. Æt Hamme wið Lǽwe, Cod. Dip. Kmbl. ii. 388, 18. Hamme juxta Lǽwes, vi. 46, 11.

láf, e; f. I. *what is left, remnant, remains, relic, remainder, rest, lave* [in northern dialects]:—Láf *superstes*, Ælfc. Gr. 9; Som. 11, 7. Healmes láf *stipulae*, Ælfc. Gl. 59; Som. 67, 131; Wt. Voc. 38, 51. Ðǽr wæs ungemetlíc wæl geslægen and sió láf wið ðone here friþ nam *there was immense slaughter, and those who were left made peace with the Danes*, Chr. 867; Erl. 72, 17: 894; Erl. 93, 1. Seó wǽpna láf *the weapons' leavings, the survivors of a battle*, Cd. 93; Th. 121, 5; Gen. 2005. Secg gára láf se ða gúþe genæs, 94; Th. 121, 32; Gen. 2019. Ða Norþmen dreórig daraþa láf, Chr. 937; Erl. 115, 3; Æðelst. 54. Seoððan se ēcea dǽl of biþ ðæt is seó sáwl hwæt biþ elles seó láf búton wyrma mete *when the eternal part, that is the soul, is gone, what else is the rest but food for worms?* Blickl. Homl. iii. 32. Ic beó tó láfe *resto*, Ælfc. Gr. 24; Som. 25, 62. Ne wearþ ðǽr forþon án Bret tó láfe *there [at Anderida] was not even one Briton left*, Chr. 491; Erl. 14, 7. Ðæs folces ðe ðǽr tó

lâfe wæs, Blickl. Homl. 79, 20. Betǽcan eów on hǽðenra hand heries lâfe *to deliver you into the hands of the heathen, all that is left of* or *by a host*, Wulfst. 295, 20. Sumes þinges lâfe *reliquiæ*, Ælfc. Gr. 13; Som. 16, 19. Lâfa ârleásra forwurþaþ *reliquiæ impiorum interibunt*, Ps. Spl. 36, 40. Wætra lâfe *the survivors of the flood*, Cd. 75; Th. 93, 21; Gen. 1549. Hî nâmon ða lâfa *tulerunt reliquias*, Mt. Kmbl. 14, 20. II. *used in poetry of weapons with the gen. of the implement employed in making them* :—Ic eom wrâðra lâf fŷres and feóle *I am the leaving of foes, of fire and of file* [a sword, forged in the fire and sharpened by the file], Exon. 126 a; Th. 484, 6; Rä. 70, 3. Homera lâfa *swords*, Beo. Th. 5651; B. 2829: Exon. 102 b; Th. 388, 14; Rä. 6, 7: Chr. 937; Erl. 112, 6; Ædelst. 6. III. *what is left as an inheritance, legacy, heirloom* [of armour or weapons: 'das schwert ist des mannes gröztes kleinod, das nur auf seinen nächsten männlichen erben übergeht' Grmm. Gesch. D. S. p. 12] :—Beaduscrúda betst ðæt mîne breóst wereþ; ðæt is Hrædlan lâf, Welandes geweorc, Beo. Th. 913; B. 454. Gomel swyrd Eánmundes lâf *an ancient sword, an heirloom from Eanmund*, 5216; B 2611: 5250; B. 2628. Ðǽr brægd eorl Beówulfes ealde lâfe, 1595; B. 795: 2981; B. 1488. Hêt in gefetian Hrêðles lâfe; næs sincmâððum sêlra on sweordes hâd, 4389; B. 2191. IV. *a relict, widow* :—Lâf *vel* forlǽten wíf *derelicta*, Ælfc. Gl. 88; Som. 74, 65; Wrt. Voc. 50, 46. Ne nime ðæs forþfarenan lâf nânne óðerne man bûton his bróður *uxor defuncti non nubet alteri, sed accipiet eam frater ejus*, Deut. 25, 5. And ǽfre ne geweorþe ðæt Cristen man gewîfige on ðæs lâfe ðe swâ neáh wǽre on woroldcundre sibbe *and never let it happen that a Christian man marry the relict of him who was so near* [within the prohibited degrees] *in worldly relationship*, L. Eth. vi. 12; Th. i. 318, 15: L. C. E. 7; Th. i. 364, 23. Se forlêt his fulluht and lifode on hǽðenum þeáwe swâ ðæt hê heafde his feder lâfe tô wífe, Chr. 616; Erl. 21, 40. Paulinus genam Ædelburge Eádwines lâfe and gewât on scipe tô Cent, 633; Erl. 25, 21. Ðâ gewât Eádríc ... Ðâ hæfde Eádríc lâfe and nân bearn *then Eadric died ... Eadric left a widow but no child*, Chart. Th. 272, 22. [*Goth.* laiba *a remnant:* O. Frs. lâva : O. Sax. lêba : *Icel.* leif: O. H. Ger. leiba.] DER. ege-, ende-, eormen-, here-, hûsel-, met-, sǽ-, un-, weá-, ŷð-, yrfe-lâf.

lafian; *p.* ode *To lave, bathe, pour water on* :—Nim ðone wǽtan and wyrm and lafa ðin heáfod mid *take the liquor and warm it and lave thy head with it*, Lchdm. iii. 48, 7. Wyrc ðæt bæþ of ðâm ilcum wyrtum on cealdum wyllewætre gecnuwa ða wyrta swîðe wel lege on ðæt wæter lafa on ðone swile *make the bath of the same herbs in cold spring-water, pound the herbs very thoroughly, lay on, pour the water on to the swelling*, L. M., i, 31 ; Lchdm. ii. 74, 29. Genim beren eár beseng lege on swâ hât and hât wæter lafa on *take a barley ear, singe it, apply it as hot as possible, and pour hot water on*, 1, 51 ; Lchdm. ii. 124, 18. [O. H. Ger. labian, labên, labôn *reficere, refocillare:* Ger. laben.] v. ge-lafian.

lafor, es; *m.* A *leopard* [so Cockayne, but ought not the word in the following passage to be *eoforas*?] :—Swelce eác laforas ðǽr cwôman unmǽtlîcre micelnisse and monig óðer wildeór and eác tigris *nec minus apri ingentis forme mixti maculosis lincibus tygribusque*, Nar. 15, 1.

-lafte. v. twî-lafte.

lag-. v. lah-.

laga, an; *m. Law* :—Stande ân laga, L. C. S. 34; Th. i. 396, 22, MS. B. Rǽde gē forþ lagan fyrþor ic wolde gif mē tô anhagode *proceed further in determining laws; I would, if it were convenient for me*, Wulfst. 275, 11. v. riht-, woruld-laga.

-laga. v. ân-, ût-laga.

lagian; *p.* ode *To make a law, ordain* :—Lagiaþ gôde woruldlagan and lecgaþ ðǽrtôeácan ðæt ûre cristendóm stande *ordain good secular laws, and add thereto the establishment of our christianity*, Wulfst. 274, 7. [*Kath.* lahede *ordained*.] v. ge-in-, in-, ût-lagian.

lago-. v. lagu-.

lagu, e; *f. Law, statute, decree, regulation, rule, fixed custom* :—Lagu *jus*, Ælfc. Gr. 9; Som. 12, 22. God him sette ǽ ðæt ys open lagu ðam folce tô steóre *God appointed them law, that is a plain rule, for the guidance of the people*, Ælfc. T. Grn. 5, 36. *Deuteronomium* ðæt ys óðer lagu, 39. Gif hē hine lâdian wille dô ðæt be ðam deópestan âðe on Engla lage and on Dena lage be ðam ðe heora lagu sî *if he will clear himself, let him do it by the most solemn oath in the district under English law; in that under Danish, by what their law may be*, L. Eth. vi. 37; Th. i. 324, 20. Manna gehwilc óðrum beóde ðæt riht ðæt hê wille ðæt man him beóde and ðæt is swŷðe riht lagu *let every man offer that justice to another that he wishes to be offered to himself, and that is a very just rule*, 49; Th. i. 326, 32. Nû is seó ealde lagu geendod æfter Cristes tôcyme and men ne ceósaþ nû on ðissere cristenan lage of nânum biscopcynne óðerne biscop ac of ǽlcum cynne *now the old law is ended after Christ's advent, and men do not now under the Christian law choose a bishop from an episcopal race, but from any race*, L. Ælfc. P. 40; Th. ii. 380, 24. Hig gesceótaþ tô Aarones dǽle and his suna ēcre lage *cedent in partem Aaron et filiorum ejus jure perpetuo*, Ex. 29, 28. Hwî forgŷmaþ ðíne leorningcnihtas ûre yldrena lage? ... Gē for nâht dydon Godes bebod for eówre lage *quare discipuli tui transgrediuntur traditionem seniorum? ... Irritum fecistis mandatum Dei propter traditionem vestram*,

Mt. Kmbl. 15, 2, 6. Ðǽr hæfþ âne lage earm and se welega *there poor and rich shall have one law*, Dóm. L. 12, 163. Godes lage healdan, Swt. A. S. Rdr. 105, 36, 23. Hē niwade ðǽr Cnutes lage, Chr. 1064; Erl. 196, 2. Ðǽr þegen âge twegen costas lufe oððe lage *where a thane has two alternatives love or law* [i. e. where a case may be arranged amicably or by appeal to law], L. Eth. iii. 13; Th. i. 298, 6. Ðis synd ða bebodu and dômas and laga ðe drihten gesette *hæc sunt judicia atque præcepta et leges quas dedit dominus*, Lev. 26, 46. Ðis ys seó ǽ ðe Moises foresette and laga and dômas *ista est lex quam proposuit Moyses, et hæc testimonia et ceremoniæ atque judicia*, Deut. 4, 44-45. Ic wille ðæt hig beón swâ gôdera lagana wurðe swâ hig best wǽran on ǽniges cynges dæge *I will that they be entitled to as good laws as there ever have been in any king's day*, Chart. Th. 416, 24. And ic wille ðæt woruldgerihta mid Denum standan be swâ gôdum lagum swâ hŷ best geceósen mǽgen, L. Edg. S. 2; Th. i. 272, 30. Hwilc óðer þeód is swâ mǽre ðæt hæbbe laga and rihte dômas and ealle ǽ *quæ est alia gens sic inclyta, ut habeat ceremonias justaque judicia et universam legem*, Deut. 4, 8. Hē lægde laga ðæt swâ hwâ swâ slôge heort oððe hinde ðæt hine man sceolde blendian, Chr. 1086; Erl. 222, 26. Ic wylle ðæt man rihte laga upp ârǽre and ǽghwilce unlage georne âfylle, L. C. S. 1; Th. i. 376, 7. In the phrases on Engla, Dena, &c. lage, which may be compared with the *Icel. i þrænda lögum, lagu* is nearly equivalent to 'district in which certain [English, Danish, &c.] laws prevail,' and in Cl. & V. Dict. [v. lög ii.] *lög* is rendered 'law community, communion, also a law district.' So in L. E. G. 7; Th. i. 172, 3 it is said :—Gif hwylcað his þeówan freólsdæge nŷde tô weorce gylde lahslitte inne on Deone lage and wîte mid Englum. These laws are the first in which *lagu* or *lah-* occurs, afterwards these forms are not unfrequent, and are continued in the Laws of William the Conqueror 'en Dene lahe, en Merchene lahe, en West Sexene lahe,' Th. i. 466, and in L. H. I. ' in Denelaga,' 566. From the time of the appearance of the word it would seem that its use was due to Scandinavian influence. v. Steenstrup's *Normannerne*, iv. 15 sqq. In Icelandic the word is used in the sense of law only in *pl.* lög: *Dan.* lov. v. land-, inǽg-, riht-, þegen-, un-, woruld- lagu; laga *and* lah.

lagu, lago; *m.* I. *sea, water* :—Ðæt gelimpan sceal ðætte lagu flôweþ ofer foldan *it shall come to pass that the sea shall flow over the earth* [at the last day], Exon. 115 b; Th. 445, 1; Dóm. 1. Lagu, wæter under wolcnum, Beo. Th. 3265; B. 1630. Lagu lâcende *the tossing water*, Andr. Kmbl. 873; An. 437. Lyft and lagu [cf. *Icel.* lopt ok lögr] land ymbclyppaþ gârsecg embegyrt gumena rîce *air and sea embrace earth, ocean girds round the kingdom of men*, Bt. Mt. Fox 9, 72; Met. 9, 40. Stille þynceþ lyft ofer londe and lagu swîge, Exon. 101 b; Th. 383, 16; Rä. 4. 11. Lagu land gefeól lyft wæs onhrêred *sea fell to earth, air was stirred* [of the destruction of the Egyptians in the Red Sea], Cd. 167; Th. 208, 12; Exod. 482. Ðâ gesundrod wæs lago wið lande, 8; Th. 10, 27; Gen. 163. Lago yrnende, 12; Th. 13, 32; Gen. 211. Willflôd ongan lytligan eft, lago ebbade [of the subsiding deluge], 71; Th. 85, 12; Gen. 1413. Mid lande and mid loge mid wude and mid felde *cum terra et cum aqua, cum sylva et cum agro*, Cd. Dip. Kmbl. iv. 202, 1. Under lyft ofer lagu, Exon. 57 a; Th. 204, 21; Ph. 101. Â hafaþ longunge seðe on lagu fundaþ, 82 a; Th. 308, 30; Seef. 47. Ne lagu drêfde ne of [on?] lyfte fleág *it troubled not water, nor flew it in air*, 106 a; Th. 404, 31; Rä. 23, 16. Ic ymb sîþ spræce and on lagu þence, 119 a; Th. 458, 9; Hy. 4, 97. II. *the name of the Rune* ⌐ :—Lagu byþ leódum langsum geþuht gif hî sculon nêðan on nacan tealtum *water to men wearisome seemeth, if they must venture on vessel unsteady*, Runic Poem. Kmbl. 343, 19; Rún. 21. Swâ ⌐ tôglîdaþ, Elen. Kmbl. 2536; El. 1269. [*Goth.* (see the name of Gothic *l*) lagus: O. Sax. lagu (in cpds.): *Icel.* lögr; *m.* sea, water, liquid; also name of Rune ⌐: O. H. Ger. lagu name of Runic letter.]

lagu-cræftig; *adj. Skilled in matters connected with the sea* :—Lagucræftig mon, Beo. Th. 423; B. 209.

lagu-fædm, es; *m.* A *watery embrace* :—Ŷð sió brûne lagufæðme beleólc *the dark wave played round me with its watery embrace*, Exon. 122 b; Th. 471, 26; Rä. 61, 7.

lagu-fæsten, es; *n.* A *water-fastness, sea, ocean* :—Ofer lagufæsten, Andr. Kmbl. 796; An. 398: 1650; An. 826: Elen. Kmbl. 2031; El. 1017. Lagofæsten, 497; El. 249.

lagu-flód, es; *m. Sea, ocean, stream, wave, water* :—Laguflód *unda*, Wrt. Voc. ii. 130, 33. Lyfthelm and laguflód *air and sea*, Menol. Fox 553; Gn. C. 46. Swâ wē on laguflóde ofer ceald wæter ceólum lîðan geond sîdne sǽ, Exon. 20 a; Th. 53, 16; Cri. 851. Heliseus leólc ofer laguflód on swonrâde, 75 b; Th. 283, 2; Jul. 674. Fereþ oft lagoflód on lyfte *oft bears water aloft*, 114 b; Th. 440, 3; Rä. 59, 12. Ǽr gescôp ēce dryhten laguflóda bigong *before had the Lord eternal created the course of the waters*, 54 b; Th. 193, 29; Az. 129: Bt. Met. Fox 20, 345; Met. 20, 173. Twelf sîþum ðæt tîrfæste lond geondláce laguflóda wynn *fons duodecies undis irrigat omne nemus*, Exon. 56 b; Th. 202, 16; Ph. 70. Lageflôdum þodenum *ceruleis turbinibus*, Wrt. Voc. ii. 133, 38. ⌐ flôdum bilocen, Exon. 19 b; Th. 50, 26; Cri. 807. Ofer lagoflôdas, Andr. Kmbl. 487; An. 244.

lagu-lád, e; *f. A way across water,* Exon. 76 b; Th. 286, 19; Wand.
3: Andr. Kmbl. 627; An. 314. [Cf. *O. Sax.* lagu-líðandi *a seafarer.*]

lagu-mearh, -mearg; *m. A sea-steed, ship,* Exon. 52 a; Th. 182, 7;
Gú. 1306. [Cf. *Icel.* lög-dýr, -fákr *a ship.*]

lagu-síþ, es; *m. A sea-journey* :—Ðære láfe lagosíþa *for those who
are left after sea-journeyings* [*those who were saved in the ark*], Cd. 67;
Th. 81, 11; Gen. 1343. Lagosíþa rest *rest from sea-journeyings* [*on
coming out of the ark*], 73; Th. 89, 26; Gen. 1486.

lagu-stræt, e; *f. A sea-road, the sea* :—Ofer lagustræte, Beo. Th.
483; B. 239.

lagu-streám, es; *m. Sea, stream, river, water* :—Folde and lagustreám *earth and sea,* Bt. Met. Fox 11, 86; Met. 11, 43. On lagustreáme [*the Danube*], Elen. Kmbl. 273; El. 137. Lyft wið lagustreám
air with water, Exon. 93 b; Th. 351, 22; Sch. 84. Lád ofer lagustreám,
Andr. Kmbl. 845; An. 423: Bt. Met. Fox 26, 31; Met. 26, 16. Ðær
lagustreámas wyllan onspringaþ *fons in medio est,* Exon. 56 b; Th. 201,
27; Ph. 62. Lagustreáma full *full of water,* 102 a; Th. 385, 1; Rä.
4, 38. Álýs mé and genere wið lagustreámum manegum wæterum *eripe
me, et libera me de aquis multis,* Ps. Th. 143, 8: Cd. 91; Th. 115, 21;
Gen. 1923. Ofer lagustreámas [*the waters of the deluge*], 161; Th. 201,
5; Exod. 367. Ofer lagustreámas *across the sea,* Beo. Th. 599; B. 297.
[*O. Sax.* lagu-ström.]

lagu-swimmend, es; *m. A creature that swims, a fish* :—Lagu-
swimmendra, Salm. Kmbl. 580; Sal. 289.

lah; *n.* (?) *Law* :—Ælc mynetere ðe betihtlad sí bicge him lah mid xii
óran [cf. bicge him lage, Th. i. 294, 8] *let every minter that is accused
buy himself law with xii ores* [v. lah-ceáp], L. Eth. iii. 8; Th. i. 296,
16. [*Icel.* lög, is neuter.] v. lagu; lah-ceáp.

lah-breca, an; *m. A law-breaker,* Scint. 2, Lye.

lah-bryce, es; *m. A breach of the law* :—Ðæt wæs geworden ðæs
ðe hé sæde þurh geléredra regolbryce and þurh læwedra lahbryce *that
happened, according to him* [*Gildas*], *through the violation of their rule
by ecclesiastics, and through the breaking of the law by laymen,* Swt.
A. S. Rdr. 111, 199. Deóflíce dæda on mistlícan lahbrycan [MS. D. lag-
brycan] on hádbrycan and on æwbrycan *devilish deeds in the shape of
diverse violations of law, of holy orders and of marriage,* L. Eth. v.
25; Th. i. 310, 18: vi. 28; Th. i. 322, 18. Wearþ ðes þeódscipe
swíðe forsyngod þurh lahbrycas and þurh æswicas þurh hádbrycas and
þurh æwbrycan, Swt. A. S. Rdr. 109, 147.

lah-ceáp, -cóp, es; *m. Payment made for re-entry into legal rights
which have been lost;* redemptio privilegiorum quæ per utlagationem
fuerint amissa :—Lahceáp, L. N. P. L. 67; Th. ii. 302, 5. Lahcóp, L.
Eth. iii. 3; Th. i. 294, 1. In the note on the latter passage an illustration is quoted from old Danish Law, where ' bylagh ' [*town law*] being
lost under certain conditions after an absence of a year and a day, a man
' bör at köbe sigh thet igen a ny.' The term is found in Old Sleswick
Law :—' Rex habet quoddam speciale debitum in Slæswick, quod dicitur
Læghköp, quo redimitur ibi hereditas [quorundam] morientium.' In the
same passage occurs the phrase ' emere lagh.' v. lah.

lah-líc; *adj. Lawful,* Scint. 9, Lye. v. next word.

lah-líce; *adv. Lawfully, according to law* :—Ne úre nænig his líf ne
fadode swá swá hé scolde ne gehádode regollíce ne læwede lahlíce *nor hath
any one of us ordered his life as he should, neither those ordained
according to their rules nor the laymen according to the law,* Swt. A.
S. Rdr. 107, 78. Ðæt hí læran ðæt gehádode menn regollíce libban and
læwede lahlíce heora líf fadian, L. I. P. 18; Th. ii. 324, 27.

lah-mann, es; *m. A man acquainted with, and whose duty it was
to declare, the law* :—xii lahmenn scylon riht tæcean Wealan and Ænglan vi Engliscne and vi Wylisce. Þolien ealles ðæs hý ágon gif hí wóh
tæcen oþþe geládian hí ðæt hí bet ne cúþon *xii lawmen shall declare the
law to Welsh and English, vi English and vi Welsh. Let them forfeit
all they own if they declare wrong; or clear themselves* [*on the
ground*] *that they knew no better,* L. O. D. 3; Th. i. 354, 9. In L. Ed.
C. 38; Th. i. 461, 21 the latinized form of the word occurs :—Postea
inquirat justicia per *lagemannos,* et per meliores homines de burgo vel
hundredo vel villa. See Cl. & Vig. Dict. *sub voce* lögmaðr.

lah-riht, es; *n. Legal right* :—Æghwylc lahriht ge burhriht ge landriht *every legal right, both of town and country,* L. I. P. 7; Th. ii.
312, 19. Gif hwá openne wiðercwyde ongean lahriht Cristes oþþe
cyninges gewyrce *if any one act in open contradiction to the legal right
of Crist or of the king,* L. Eth. v. 31; Th. i. 312, 9.

lah-slit; n [?]; -sliht, -slite, es; *m;* -slitt, e: *f. According to its component parts the word means a breach or violation of the law;* in the
Laws however it is applied to *the fine payable for the breach,* and is used
only with reference to the Danes, the corresponding term among the
English being *wíte* :—Beó se wið ðone cyninge hundtwelftig scill. scyldig
on Engla lage . . . and on Dena lage lahslites scyldig, L. C. S. 15; Th. i.
384, 15. Gebéte ðæt be ðæm ðe seó dæd sý swá be wíte swá be lahslitte
[lahslite, MS. B.] *let him make 'bot' for that according to what the deed is,
either by 'wíte'* [*if English*] *or by 'lahslit'* [*if Danish*], L. E. G. 3; Th.
i. 168, 6. Gif preóst fulluhtes forwyrne ðam ðe ðæs þearf sý, gylde wíte

mid Englum and mid Denum lahslit, ðæt is twelf óran, 10–13. Gylde
swá wíte swá lahslitte [lahslite, MS. B], 2; Th. i. 168, 9. Gylde lahslitte inne on Deone lage and wíte mid Englum, 7; Th. i. 172, 3. Lahslite, 8; Th. i. 172, 7. Lahslit, 9; Th. i. 172, 11. Ðonne gilde hé
lahsliht, L. N. P. L. 51 : 52 : 53; Th. ii. 298, 9 : 12 : 15. The word is
continued in the Laws of William the Conqueror :—In Danelahe erit in
forisfactura de suo *laslite* [*laxlite* in French], Th. i. 483, 24. In Th. i.
168, note a, a passage is quoted from old Swedish law in which ' lagsliht '
occurs. See also Grmm. R. A. 623 : Steenstrup's Normannerne, iv. 264 sqq.

lah-wita, an; *m. One who has a knowledge of law, a lawyer* :—
Cyningan and bisceopan eorlan and heretogan geréfan and déman lárwitan and lahwitan gedafenaþ mid rihte ðæt hí Godes riht lufian *it rightly
befits kings and bishops, nobles and generals, sheriffs and judges, those
who have learning and those who know law, to love God's justice,* L.
I. P. 5; Th. ii. 308, 14.

lám, es; *n. Clay, mud, mire, earth* :—Laam *argilla,* Ælfc. Gl. 56;
Som. 67, 35; Wrt. Voc. 37, 25: Wrt. Voc. ii. 100, 66. Lám a[r]gella,
Wrt. Voc. 285, 7: *limus,* Ælfc. Gr. 13; Som. 16, 4: Wrt. Voc. ii. 112,
81. Lámes gelícnes *the body* [*after death*], Exon. 98 a; Th. 368, 9;
Seel. 19. God gesceóp man of ðære eorþan láme *formavit dominus deus
hominem de limo terræ,* Gen. 2, 7: Homl. Th. i. 12, 29: 236, 15.
Áfæstnod ic eom on láme grundes *I sink in deep mire;* infixus sum in
limo profundi, Ps. Spl. 68, 2. Genera mé of láme *deliver me out of the
mire;* eripe me de luto, C. 68, 18. Láme bitolden *covered with earth*
[*buried*], Exon. 64 a; Th. 235, 11; Ph. 555 : 50 a; Th. 173, 27; Gú.
1167 : 117 b; Th. 451, 5; Dóm. 99. Ic áworpe ða myht fram mé ðe
mé fram ðé geháten ys swá ðæt lám ðe ic myd mýnum fótum ontrede
*I cast away from me the power that is promised me by thee, as the
dirt that I tread upon with my feet,* Shrn. 151, 22. [*O. L. Ger.* lêmo,
leimo *limus* : *O. H. Ger.* leim *argilla, limus, lutum* : *Ger.* lehm.]

lama, loma, lame; *adj. Lame, disabled in the limbs, maimed, crippled,
weak, paralysed, palsied, paralytic* :—On sídan lama *pleuriticus,* Ælfc.
Gl. 10; Som. 57, 25; Wrt. Voc. 19, 31. Lame *debilis* vel *enervatus,*
77; Som. 72, 22; Wrt. Voc. 45, 55. *Conclamatus* i. *commotus, convoc-
atus, desperatus, vel* loma, Wrt. Voc. ii. 136, 28 : *conclamatus,* 105, 20.
Áune man se wæs lama *hominem qui erat paraliticus,* Lk. Skt. 5, 18. Ic
eom lama þearfa *egenus et pauper sum,* Ps. Th. 108, 22. Ðá læg ðær
sum creópere lama fram cildháde *then lay there a cripple lame from his
childhood,* Homl. Skt. 10, 25. Ánne bædrydan for eahte geárum lama *a
bedridden man paralysed for eight years,* 42. Man ne mót nán þing
gehǽlan on restedagum þéh hyt lama beó nú hǽlþ hé ǽgðer ge healte ge
blynde ge deáfe ge dumbe ge gebýgede laman and deófolseóce, Nicod. 2;
Thw. 1, 29. Án mǽden seó wæs lama *puella paralytica,* Bd. 3, 9; S. 533,
5. Hé wæs lama and eallra his lima þénunge benumen *deficiente penitus
omni membrorum officio,* 5, 5; S. 617, 37. Mid langre ádle laman legeres
swíðe gehefigod *longo paralysis morbo gravatam,* 3, 9; S. 534, 5. Oft
him feorran tó laman liomseóce cwómon healte hreófe and blinde *oft to
him from far came the lame, the crippled, the halt, the leprous, and the
blind,* Elen. Kmbl. 2425; El. 1214. Lamena [lamana, MS. B.] hé is
lǽce *of the lame it is the leech,* Salm. Kmbl. 155; Sal. 77. Iii hit oftræd and hié tó loman gerénode ðæt hié mec ænigre note nytte beón ne
meahton *duos et l. calcatos inutiles fecit,* Nar. 15, 26. Laman *paraly-
ticos,* Mt. Kmbl. 4, 24. [*O. Sax.* lamo : *O. Frs.* lam, lom : *Icel.* lami,
lama : *Dan.* lam *lame, palsied, paralytic* : *O. H. Ger.* lam *claudus,
mancus, debilis, paralyticus.*] v. ád-, lim- lama.

lamb, es; *and* lamber; *n. A lamb* :—Ðæt lamb sceal beón ánwintre *erit
agnus anniculus,* Ex. 12, 5. Hér is Godes lamb *ecce agnus dei,* Jn. Skt. 1, 29.
Swá plegende lamp *quasi agnus lasciviens,* Kent. Gl. 214. Hé gefullode
ðone wulf and geworhte tó lambe *he baptized the wolf and made it a lamb,*
Homl. Th. i. 390, 26. Godes lomber folgian, Exon. 48 a; Th. 164, 22;
Gú. 1015. Nyme ælc mann án lamb *tollat unusquisque agnum,* Ex. 12, 3.
Swá swá lamb *sicut agni,* Ps. Spl. 113, 4. Swá sáceóne lambru, Ps. Th. 113,
4, 6. Lambra, Ps. Spl. 113, 6. Mid lamba rysle *cum adipe agnorum,*
Deut. 32, 14. Abram gesette seofon lamb on sundron *statuit Abram sep-
tem agnas seorsum,* Gen. 21, 28. Ic eów sende swá swá lamb [lombro,
Lind : lombor, Rush.] betwux wulfas *ego mitto vos sicut agnos inter lupos,*
Lk. Skt. 10, 3. Heald míne lamb [lombor, lomboro, Lind : lombor,
Rush.] *pasce agnos meos,* Jn. Skt. 21, 15, 16. [*Orm.* lammbre; *pl* : Ayenb.
Piers P. lambren : *Goth.* O. Sax. Icel. O. H. Ger. lamb; in *O. H. Ger.*
lember, lembir *as well as* lamb *are found in pl.* v. Grff. ii. 214.]

lambes cerse, an; *f. Lamb's cress; cardamine hirsuta* :—Cersan
sǽdes sume men hátaþ lambes cersan, L. M. 1, 1; Lchdm. i. 24, 16. v.
E. D. S. Plant names.

Lamb-, Lambe-hýþ, e; *f. Lambeth in Surrey* :—Hér forþferde Hardacnut æt Lambhýþe, Chr. 1041; Erl. 167, 30. Ðis synd ða landgemǽre
intó Lambehýþe, Cod. Dip. Kmbl. iv. 158, 4. v. hýþ.

lam-byrd, e; *f. A lame, weak, imperfect birth,* Lchdm. iii. 66. 22. v.
læt-byrd.

lám-fæt, es; *n. A vessel of clay, the body,* Exon. 74 a; Th. 277, 9,
Jul. 578 : 100 a; Th. 375, 4; Seel. 133.

lamprede, an; *f. A lamprey* [Low Latin *lampreda.*] :—Hwilce fixas

gefĕhst ŏú? Lampredan *quales pisces capis? murænas*, Coll. Monast. Th. 23, 35. [*O.H.Ger.* lampreda, lantprida *murenula*, Grff. ii. 241.]

lăm-pytt, es; *m.* *A clay-pit*:—Swā andlang mearce on lāmpyttas, Cod. Dip. Kmbl. iii. 252, 24.

lăm-seáðe [?], an; *f.* *A clay-* [or *mud-*] *pit*:—Of sceadwellan in lāmseáðan; of lāmseáðan in ledene, Cod. Dip. Kmbl. iii. 80, 14.

lăm-wyrhta, an; *m.* *A worker in clay, a potter*:—Lāmwyrhte [-wrihta, Lind.] *figuli*, Mt. Kmbl. Rush. 27, 7. Lāmwyrhtæ [-wrihtæs, Lind.], 10.

LAND, es; *n.* I. LAND *as opposed to water or air, earth*:— Wē ŏec in lyft gelæddun oftugon ŏē landes wynna *we led thee aloft, earth's pleasures withdrew from thee*, Exon. 39 b; Th. 130, 15; Gú. 438. Đā siŏŏan tófērdon ŏa apostolas wíde landes geond ealle ŏás world *then afterwards the apostles separated and went far and wide on earth, throughout all this world*, L. Ælfc. P. 21; Th. ii. 372, 6: Wulfst. 105, 6. Monigra folca ceápstŏw of lande and of sǽ cumendra *multorum emporium, populorum terra marique venientium*, Bd. 2, 3; S. 504, 19. Úsic æt lande gebrohte, 5, 1; S. 614, 10. Hig tugon hyra scypo tó lande *subductis ad terram navibus*, Lk. Skt. 5, 11. Đā cōmon hié tó londe on Cornwalum, Chr. 891; Erl. 88, 11. Wǽron ŏa menn uppe on londe of ágáne, 897; Erl. 95, 24. Án scip flotigende swā nēh ŏan lande swā hit nýxt mǽge, 1031; Erl. 162, 7. Đá gesundrod wæs lago wiŏ lande, Cd. 8; Th. 10, 27; Gen. 163. Com ŏá tó lande swiŏmód swymman, Beo. Th. 3250; B. 1623. Stille þynceþ lyft ofer londe, Exon. 10 b; Th. 383, 15; Rä. 4, 11. Lifigende ŏa ŏe land tredaþ *living creatures that walk the earth*, Cd. 10; Th. 13, 16; Gen. 203. II. *a land, country, region, district, province*:—Đæs landes golda ys golda sēlost *aurum terræ illius optimum est*, Gen. 2, 12. Is seó cirice on Campania ŏæs landes gemǽro *the church is on the borders of the land of Campania*, Blickl. Homl. 197, 19. Úres landes mann *nostras:* eówres landes mann *vestras*, Ælfc. Gr. 15; Zup. 94, 8: 102, 21. Ne nim ŏú nāne sibbe wiŏ ŏæs landes menn *ne ineas pactum cum hominibus illarum regionum*, Ex. 34, 15. Twegen landes menn and án ælþeódig, Homl. Th. ii. 26, 20. Twegen sacerdas ŏe ǽr on lífe wǽron his landes menn *two priests who before, when living, had been his countrymen*, 342, 3. Đá cōmon ŏa landes menn [*the Northumbrians*] tōgeánes him and hine ofslógon, Chr. 1068; Erl. 205, 2. Hí wǽron of Galiléam ŏǽm lande, Blickl. Homl. 123, 21. Hē leng on ŏam lande gewunian ne mihte *he could not live longer in that country*, 113, 11. On Lindesse lande *in provincia Lindissi*, Bd. 3, 27; S. 558, 34. Andreas sette his hand ofer ŏara wera eágan ŏe ŏǽr on lande wǽron *Andrew placed his hand upon the eyes of the men who were there in that country*, Blickl. Homl. 239, 3. Ceólwulf and Eádbald of ŏǽm londe áfōron *Ceolwulf and Eadbald left the country*, Chr. 794; Erl. 58, 6. Ælþeódige mǽn of lande mid héora ǽhtum and mid synnum gewiten *let foreigners depart from the country with their goods and with their sins*, L. Wih. 4; Th. i. 38, 2. Perh ōŏer woeg eft gecerrdon in lond hiera *per aliam viam reversi sunt in regionem suam*, Mt. Kmbl. Lind. 2, 12. Mid ŏý hí ŏider cōman on land *cum illo advenissent*, Bd. 5, 10; S. 624, 1. Đæt wǽron ŏa ǽrestan scipu Deniscra monna ŏe Angelcynnes lond gesōhton, Chr. 787; Erl. 56, 16. Ǽlc ŏæra landa ŏe ǽnigne friþige ŏæra ŏe Ængla land hergie *every land that affords protection to any of those that harry England*, L. Eth. ii. 1; Th. i. 284, 17. Đá lædde hē mē on fyrran lænd *cum me in ulteriora produceret*, Bd. 5, 12; S. 628, 9. In ŏa nēsta gemǽro and londo [lond, Rush.] *in proximas villas et vicos*, Mk. Skt. Lind. 6, 36. III. *land, landed property, estate, cultivated land, country* [*as opposed to town*]:— Gesāwen æcer *vel* land *seges*, Ælfc. Gl. 97; Som. 76, 48; Wrt. Voc. 53, 55. Land *solum vel tellus vel terra, vel arvum*, 98; Som. 76, 98; Wrt. Voc. 54, 42. Đis land *hoc rus*, Ælf. Gr. 9; Som. 12, 21. Land *agellum*, Bd. 4, 12; S. 581, 5. Se árfæsta bigenga ŏæs gástlícan landes *pius agri spiritalis cultor*, 5; S. 519, 8. xii hída gesettes landes *xii hides of cultivated land*, L. In. 64: 65; Th. i. 144, 6: 9. Be gyrde londes *of a yard of land*, 67; Th. i. 146, 1. Þolige landes and lífes *let him lose land and life*, L. C. E. 2; Th. i. 358, 21. On lande *ruri*, Ælfc. Gr. 38; Som. 41, 18. Sý hit binnan byrig sý hit upp on lande, L. C. S. 24; Th. i. 390, 5. Ge on lande ge on ōŏrum þingum ge on ōŏrum gestreónum *consisting of land and of other things and of other acquisitions*, Blickl. Homl. 51, 7. Noe began tó wircenne ŏæt land *cæpit Noe exercere terram*, Gen. 9, 20. Búton earmre wudewan ŏe næfde nán land *except a poor widow that had no land*, L. Ath. v. 2; Th. i. 230, 20. Færende on lond *euntes in villam*, Mk. Skt. Lind. 16, 12. Heora wlenca wǽron swíŏe monigfealde on landum and on wíngeardum, Blickl. Homl. 99, 15. Hér geswutelaþ on ŏissum gewrite ŏæt Leófríc eorl and his gebedda habbaþ geunnen twá land *hac inscriptione manifestatur Leofricum comitem et Godgivam comitissam duas villas concessisse*, Cod. Dip. Kmbl. iv. 72, 20. Ǽlc ŏe forlæt land [londo, Lind.] *omnis qui reliquit agros*, Mt. Kmbl. 19, 29: Bd. 5, 19; S. 636, 35. Feówer land hē forgeaf ælþeódigum tó andfencge and tó ælmesdædum *he gave four estates for the reception of strangers and for deeds of charity*, Homl. Skt. 7, 386. Byrig and land þurhfēran *oppida et rura peragrare*, Bd. 3, 28; S. 560, 32: 3, 30; S. 562, 13. [The word occurs in all the Teutonic languages.] DER. burg-,

eá-, eard-, eást-, ég-, el-, ele-, éŏel-, feld-, feor-, folc-, gehlot-, heáfod-, heáh-, íg-, in-, irfe-, irþ-, lín-, mearc-, mór-, omer-, sand-, síd-, sundor-, sundor-geref-, tún-, þeód-, un-, út-, wea[lh]-, wíd-, wyn-, wyrŏe-land.

-landa. v. ge-landa.

land-ádl, e; *f.* *Nostalgia* [so Cockayne, but cf. lond-iuil *epilepsy*, Prompt. Parv.]:—Wiŏ londádle, L. M. 2, 65; Lchdm. ii. 296, 13.

land-ælf, e; *f.* *A land-elf*:—Landælfe *ruricolas musas*, Wrt. Voc. ii. 88, 83.

land-ágend, es; *m.* *A land-owner, one of those to whom a country belongs, a native*:—Hí wǽron on myclum ege ŏám sylfan landágendum ŏe hí ǽr hider laþedon *ipsis qui eos advocaverant indigenis essent terrori*, Bd. 1, 15; S. 483, 34 note. [Cf. *Icel.* land-eigandi *a land-owner.*]

land-ágende; *adj.* *Owning land*:—Gif gesíþcund mon landágende forsitte fyrde, geselle cxx scill. and þolie his landes, L. In. 51; Th. i. 134, 8. Landágende man, L. N. P. L. 49: 52; Th. ii. 298, 4: 10. Landágende men ic lǽrde ŏæt hié heora gafol mid gehygdum águldon, Blickl. Homl. 185, 21. DER. un-landágende.

land-ár, e; *f.* *Property in land, landed estate*:—Of Seint Petres landáre *in territorio Sancti Petri*, Cod. Dip. Kmbl. iv. 242, 16. Hē him ŏa landáre forgeaf ŏe hē ŏæt mynster on getimbrade *quo concedente et possessionem terræ largiente, ipsum monasterium fecerat*, Bd. 4, 18; S. 586, 35. Wilniende ŏætte heó him funden swylce londáre swylce hē mid árum on beón mehte *desiring that they should provide him such an estate as he might reside on with dignity*, Chart. Th. 47, 21. Đæt land æt Boccinge intó Cristes cyrcean, and his ōŏre landáre intó ōŏran hálgan stōwan, 540, 26. Nimaþ ŏis gold and bicgaþ eów landáre, Homl. Th. i. 64, 12. Đa ŏe landáre hæfdon hí hit beceápodon, 316, 10: ii. 224, 5. Ic wille ŏæt se cyng beó hláford ŏæs mynstres ŏe ic getimbrede, and ŏǽre landára ŏe ic ŏyderinn becweden hæbbe, Chart. Th. 547, 31.

land-begenga, an; *m.* I. *a cultivator of land, husbandman, farmer*:—Se mǽra landbegenga [londbegengea, MS. Cott.] *magnus colonus*, Past. 40, 3; Swt. 293, 2. Gif hit on Wōdnes dæg þunrige ŏæt tácnaþ landbigencgena cwealm and cræftigra *if it thunder on Wednesday, that betokens death of husbandmen and craftsmen*, Lchdm. iii. 180, 14. Ágæf ŏa ŏǽm londbigencgum [-bigengum, Rush.] *locavit eam agricolis*, Mk. Skt. Lind. 12, 1. II. *an inhabitant of a country, a native*:—Đá sǽgdon mē ŏa londbigengan *mihi locorum incole affirmabant*, Nar. 20, 16. Đám sylfan landbigengum *ipsis indigenis*, Bd. 1, 15; S. 483, 34. Ealle ŏa landbigengan útamǽran *omnes indigenas exterminare*, 4, 16; S. 584, 6. [*O.H.Ger.* lant-pikengeo *accola, indigena.*]

land-begang, es; *m.* *Cultivation of land, or habitation in a land*:—Londbigonges mínes *incolatus mei*, Ps. Surt. 118, 54. v. preceding word.

land-bōc; *f.* *A charter in which land is granted*:—Đis is ŏara xxv hída landbōc ŏe Eádgár cyng gebōcede Gode and Sca. Marian intó Abbandúne, Cod. Dip. Kmbl. iii. 29, 10. Ic wylle ŏæt man ágyfe ŏám híwum æt Domrahamme hyra landbēc, ii. 116, 35. Landbēc *donatio*, Wrt. Voc. ii. 141, 77.

land-brēce, es; *m.* *Breaking up or ploughing of* [*fallow*] *land*:—Landbrǽce *proscissio*, Ælfc. Gl. 1; Som. 55, 20; Wrt. Voc. 15, 20. [Cf. *O.H.Ger.* brāhha *aratio prima;* brāhhōn *proscindere*, Grff. iii. 268: *Ger.* brachen *to plough a field after it has been lying fallow.*]

land-búend, es; *m.* I. *a cultivator of the land, husbandman*:—Fæder mín londbúend [-býend, Rush.] is *pater meus agricola est*, Jn. Skt. Lind. 15, 1. Đa landbúendo *agricolæ*, Mt. Kmbl. Lind. 21, 38. Đǽm londbúendum *agricolis*, 33: 40: Mt. Skt. Lind. Rush. 12, 2. Đæt scipmannum is beboden gelíce and ŏǽm landbúendum ŏæt ealles ŏæs ŏe him on heora ceápe geweaxe hig Gode ŏone teóþan dǽl ágyfen *it is commanded to those who trade with ships, just as to those who cultivate land, that they give to God the tenth part of all their increase*, L. E. I. 35; Th. ii. 432, 28. Sende ŏa londbúend *misit agricolas*, Mt. Kmbl. Lind. 21, 34. II. *an inhabitant of a country, a native, a dweller on earth*:—Hæleþ wǽron irre landbúende *the men were angry, the inhabitants of the land*, Judth. 11; Thw. 24, 36; Jud. 226. Ælda bearn, londbúendra, Exon. 130 b; Th. 500, 23; Rä. 89, 11. Gesette sunnan and mōnan leóman tó leóhte landbúendum, Beo. Th. 191; B. 95. Londbúendum, Exon. 78 b; Th. 295, 7; Crä. 29: 87 a; Th. 326, 22; Víd. 132. Londbúendum [*the Jews*], Judth. 12; Thw. 26, 7; Jud. 315. Ic ŏæt londbúend leóde míne secgan hýrde *I heard the land's inhabitants, my people, say*, Beo. Th. 2694; B. 1345. v. next two words.

land-búend, e; *f.* *A settlement, colony*:—Seó landbúend *colonia*, Nar. 33, 8. v. note p. 78.

land-búende; *adj.* *Inhabiting a country, living on the earth*:—Hwá ŏæs leóhtes londbúende brúcan mōte *who that lives on land may enjoy that light*, Exon. 93 b; Th. 351, 15; Sch. 80.

land-búness; e; *f.* *A settlement, colony*:—Seó landbúness is swíŏost cýpemonnum geseted *hæc colonia est maxime negotiatorum*, Nar. 33, 15. Londbúnes *colonia*, 35, 18.

land-ceáp; es; *m.* *A fine or tax paid when land was purchased*:—Landcōp, L. Eth. iii. 3; Th. i. 292, 16. Landcēáp, L. N. P. L. 67; Th. ii. 302, 5. Ego Berchtwulf cyning sile Forŏrēde mínum þegne nigen higida lond ... hē salde tó londceápe xxx mancessan and nigen

hund sciłł. wið dæm londe *I, King Berchtwulf, sell my thane Forthred nine hides of land . . . he gave xxx mancusses as fine at the purchase, and nine hundred shillings for the land,* Cod. Dip. Kmbl. ii. 5, 24–31. [Cf. *Icel.* land-kaup ; *n. the purchase of land* ; in Norse, *a fine to be paid to the king* by one exiled or banished : *O. Frs.* land-kâp.] v. lah-ceáp.

land-cofa, an ; *m. A translation of Sicima* [Shechem], Ps. Lamb. 59, 8.

landes mann. v. land.

land-fæsten, es ; *n. A land-fastness, a strong military position on land, a pass :*—Leoniða on ánum nearwan londfæstenne him widstôd *Leonida in angustiis Thermopylarum obstitit,* Ors. 2, 5 ; Swt. 80, 14.

land-feoh ; *gen.* -feós ; *n. 'A recognitory rent for land,'* Cod. Dip. Kmbl. v. 143, 22. v. Kemble's Saxons in England ii. 328–9.

land-fird, e ; *f. An expedition, journey by land, a land-force :*—Ne him tô ne dorste sciphere on sǽ ne landfyrd *the fleet durst not approach them at sea nor the land force* [on land], Chr. 1001 ; Erl. 137, 18. Man sceolde mid scypfyrde and eác mid landfyrde hym ongeán faran, 999 ; Erl. 134, 30. Se man ðe ætfleó fram his hláforde sý hit on scypfyrde sý hit on landfyrde þolige ealles ðæs ðe hê áge and his ágenes feores, L. C. S. 78 ; Th. I. 420, 9. Ðæt is fyrdfara sig hit on scipfyrde sig hit on landfyrde *scilicet expeditio, sive sit in navali collectione, sive in pedestri,* Chart. Th. 333, 20. Ðâ gestihtade hê ðæt hê wolde landfyrde ðider gelǽdan *terrestri itinere illo venire disponebat,* Bd. 3, 15 ; S. 541, 26.

land-folc, es ; *n. The people of a land* or *country :*—Hwæðer ðæt landfolc sî tô gefeohte stranglîc oððe untrumlîc *populum, utrum fortis sit an infirmus,* Num. 13, 20. Ðet landfolc hardlîce widstôd *the people resisted stoutly,* Chr. 1046 ; Erl. 171, 4 : 1070 ; Erl. 207, 22. On sumere tíde com micel hungor on ðam lande and gehwǽr ðæt landfolc micclum geangsumode *at one time a great famine came on the land and very much afflicted the people everywhere,* Homl. Th. ii. 170, 32 : 164, 19.

land-fruma, an ; *m. A prince of a country :*—Leóf landfruma, Beo. Th. 61 ; B. 31.

land-gafol, es ; *n. Rent for land :*—Hê sceal landgafol syllan *he must pay rent,* L. R. S. 2 ; Th. i. 432, 13. Hê sceal . . . his láforde wyrcan . . . ne þearf hê landgafol syllan *he must work for his lord, then he need not pay rent,* 3 ; Th. i. 432, 23. v. gafol-land.

land-gehwearf, es ; *n. An exchange of land* ; commutatio terræ :—Ðis is seó gerǽdnes ðe Byrhtelm biscop and Aþelwold abbod hæfdon ymbe hira landgehwerf . . . Se biscop gesealde ða hîda æt Cenintûne and se abbud gesealde ðæt seofontŷne hŷda æt Crydanbricge, Chart. Th. 191, 6.

land-gemaca, an ; *m. A neighbour :*—*Vicinum* landgemacena, (*in margin*) *affinium* landgemaca, Hpt. Gl. 480, 18–20.

land-gemǽre, es ; *n. A boundary, confine :*—Ligeþ ðæt londgemǽre [*of Asia and Africa*] sûþ ðonan ofer Nilus ða eá, Ors. 1, 1 ; Swt. 8, 29. Cirus fôr ofer ðæt londgemǽre ofer ða eá ðe hâtte Araxis *Cyrus passed the boundary, the river that was called Araxis,* 2, 4 ; Swt. 76, 6. Ðis syndon ðara twegra hîda landgemǽru *these are the boundaries of the two hides,* Cod. Dip. Kmbl. iii. 206, 25. Landgemǽro, 207, 34. The word is of frequent occurrence in the Charters. Sî se man âwirged, ðe forhwyrfe his freóndes landgemǽro *maledictus, qui transfert terminos proximi sui,* Deut. 27, 17. Ofer landgemǽru *extra terminum,* Ælfc. Gr. 47 ; Som. 47, 29. Ðâ côman hî mid sciphere on heora landgemǽro *advecti navibus inrumpunt terminos,* Bd. 1, 12 ; S. 480, 34 : Ps. Th. 45, 8.

land-gemirce, es ; *n. A boundary :*—Se westsûþende Europe landgemirce is in Ispania westeweardum et ðæm gârsecge *Europæ in Hispania occidentalis oceanus terminus est,* Ors. 1, 1 ; Swt. 8, 23. Ðǽr Affrica and Europe hiera landgemircu tôgædre licgaþ, 10. Affrica and Asia hiera landgemircu onginnaþ of Alexandria, 28. Landgemyrcu, Beo. Th. 424 ; B. 209.

land-gesceaft, es ; *n. The earthly creation, created things on earth :*—Bǽdon bletsian eall landgesceaft êcne drihten *they called upon all created things on earth to bless the Lord eternal,* Cd. 191 ; Th. 238, 25 ; Dan. 360.

land-geweorc, es ; *n. The principal stronghold of a country, one which it has been the work of the country to build* [cf. Beo. Th. 135–152 ; B. 67–76], Beo. Th. 1880 ; B. 938.

land-gewyrpe, es ; *n. A heap of earth thrown up* [?] :—Andlang ðare landgewirpa, Cod. Dip. Kmbl. iii. 453, 30. On ða landgewyrpu . . . andlang ðara landgewyrpa, 434, 2–4.

land-hæbbende ; *adj.* **I.** *owning land :*—Monnes landhæbbendes, L. In. 45 ; Th. i. 130, 10. Cf. landâgende. **II.** *holding a country as a ruler :*—Landhæbbende ł his cynnes lâtwa *tribunus,* Rtl. 193, 15.

land-hæfen, e ; *f. Property in land :*—Be Wilisces monnes londhæfene. Gif Wylisc mon hæbbe hîde londes, his wer biþ cxx scill., L. In. 32 ; Th. i. 122, 8.

land-here, es ; *m. A military force which acts on land* [opposed to *sciphere*], or *which belongs to the land* [opposed to a foreign force] :—Æfter ðam gegadorode micel here hine of EástEnglum ǽgðer ge ðæs landheres ge ðara wîcinga ðe hié him tô fultume âspanen hæfdon *after that a great force collected from East Anglia, both of the native force and of the vikings that they had allured to their assistance,* Chr. 921 ; Erl.

107, 15. Hér fôr Æþelstân in on Scotland ǽgðer ge mid landhere ge mid scyphere, 933 ; Erl. 110, 27. [*Icel.* land-herr *people of the land.*]

land-hláford, es ; *m.* **I.** *a land-lord, an owner of land, lord of the manor :*—Tôdǽle man ða eahta dǽlas on twâ and fô se landhláford tô healfum tô healfum se bisceop sý hit cynges man sý hit þegnes [cf. H. I. 11 ; Th. i. 520, 18–20 *reliquam in duas partes dividant, dimidium habeat dominus, dimidium habeat episcopus, sit homo regis vel alterius,* L. Edg. i. 3 ; Th. i. 264, 3 : L. Eth. ix. 8 ; Th. i. 342, 19 : L. C. E. 8 ; Th. i. 366, 9. Healde se landhláford ðæt forstolene orf ôþ ðæt se ágenfrigea ðæt geácsige *let the lord keep the stolen cattle until the owner get to hear of it,* L. Edg. S. 11 ; Th. i. 276, 14 : L. Eth. i. 3 ; Th. I. 282, 27. And nân man ne hwyrfe nânes yrfes bûtan ðæs gerêfan gewitnesse oððe ðæs mæssepreóstes oððe ðæs landhláfordes, L. Ath. i. 10 ; Th. i. 204, 18. **II.** *the lord of a country :*—Hû stîde se landhláford sprǽc wið hig, and hig cwǽdon se landhláford wênde ðæt wê wǽron sceáweras *locutus est nobis dominus terræ dure et putavit nos exploratores esse,* Gen. 42, 30.

land-lagu, e ; *f. Law* or *regulation prevailing in a district :*—Ðeós landlagu stænt on suman lande *hæc consuetudo stat in quibusdam locis,* L. R. S. 4 ; Th. i. 434, 29. Landlaga sŷn mistlîce swâ ic ǽr sǽde *leges et consuetudines terrarum sunt multiplices et varie, sicut prelibavimus,* 21 ; Th. i. 440, 19.

land-leás ; *adj. Landless, not having land :*—Be landleásum mannum. Gif hwylc landleás man folgode on ôðre scire, L. Ath. i. 8 ; Th. i. 204, 4.

land-leód, es ; *pl.* e, an [cf. Seaxe, Seaxan] ; *m. An inhabitant of a country :*—Landleód *accola,* Wrt. Voc. ii. 3, 76 : [*in*]*digena,* 28, 59. Eft hê frægn hwæðer ða ylcan landleóde Cristene wǽron *rursus interrogavit, utrum iidem insulani Christiani essent,* Bd. 2, 1 ; S. 501, 12 : 4, 26 ; S. 602, 8. Ac hii ða londleóde tiolode mâ ûssa feónda willan tô gefremmanne ðonne ûrne *sed illi* [*periti regionum*] *majorem hosti quam mihi favorem accommodarent,* Nar. 6, 19. Ðâ wurdon ða landleóde his ware and him wið gefuhton, Chr. 917 ; Erl. 102, 16. Ðæt folc eal ðæt ðǽr tô lâfe wæs ðara landleóda beág tô Eádwearde cyninge *the people, all that remained of the inhabitants of the district, submitted to king Edward,* 921 ; Erl. 108, 1. Hê wæs ðæs cynges swica and ealra landleóda, 1055 ; Erl. 189, 4. Hié from ðam londleódum þurh seara ofslægene wurdon *conspiratione finitimorum per insidias trucidantur,* Ors. 1, 10 ; Swt. 44, 28. Hê betealde hine wið Eádward cyng his hláford and wið ealle landleódan *he cleared himself to his lord king Edward and to all the people,* Chr. 1052 ; Erl. 187, 20.

land-leód, es ; *m.* [?] : e ; *f. The people of a country :*—Se wer gebiraþ mâgum and seó cynebôt ðam leódum ; *other reading :*—Ðam were habbaþ ða mǽgas and ðam cynebôt se [seó?] landleód, L. Wg ; Th. i. 190, 9, and note 14. Schmid p. 396 gives the further reading :—Ðæt cynebôt tô ðam landleód. [These passages seem corrupt, so that much reliance perhaps cannot be placed upon them for determining the gender, but it may be noticed that *O. H. Ger. lant-liut* is masc. v. Grff. ii. 195.] Se fyrdinge dyde ðære landleóde ǽlcne hearm *the levy did the people of the country every kind of harm,* Chr. 1006 ; Erl. 140, 12. Ealle ðâs landleóda belicgaþ ûs *all these people will surround us,* Jos. 7, 9.

land-lyre, es ; *m. Loss of land :*—For his landlyre hér on lande *on account of his loss of land in this country,* Chr. 1105 ; Erl. 240, 11.

land-mann, es ; *m. A native of a country :*—Nâh nâðer tô farenne ne Wylisc man on Ænglisc land ne Ænglisc on Wylisc ðê mâ bûtan gesettan landmen se hine sceal æt stæðe underfôn and eft ðǽr bûtan fâcne gebringan. Gyf se landman ǽniges fâcnes gewita sý ðonne sý hê wîtes scyldig, L. O. D. 6 ; Th. i. 354, 23–7. Landmanna cyme *the coming of the men of the country,* Cd. 151 ; Th. 189, 4 ; Exod. 179. v. landes mann *under* land. [*O. H. Ger.* lant-man *patriota.*]

land-mearc, e ; *f. Boundary of an estate* or *of a country :*—Seó landmearce liþ þ Terstân upp be Hohtûninga mearce, Cod. Dip. Kmbl. iii. 189, 5. Londmearce neáh *near to the land's boundary,* Exon. 75 a ; Th. 280, 27 ; Jul. 635. [*O. H. Ger.* lant-marcha *funiculum.*] Cf. landgemǽre, -gemirce.

land-mearc ; *adj. Belonging to the boundaries of a country :*—Mîn is se landmearca and mîn is mannaseisca landsplot *meus est galaad* (= heap of witness) *et meus est mannases,* Ps. Lamb. 59, 9.

land-openung, e ; *f. Breaking up of land* ; proscissio, Ælfc. Gl. 57 ; Som. 67, 68 ; Wrt. Voc. 37, 54.

land-rǽden, ne ; *f. Institution, disposition, ordinance of a district* or *country :*—Hêde se ðe scîre healde ðæt hê wite â hwæt eald landrǽden sý and hwæt þeóde þeáw *videat qui scyram tenet, ut semper sciat que sit antiqua terrarum institutio, vel populi consuetudo,* L. R. S. 4 ; Th. i. 434, 33.

land-rest, e ; *f. A land-couch, grave :*—Lǽtan landreste *to leave the grave,* Andr. Kmbl. 1561 ; An. 782.

land-rica, an ; *m. A powerful man in a district, a landed proprietor, a land-lord* ; the term seems equivalent to *land-hláford,* q. v. :—Heáh landrîca *ierarchon,* Wrt. Voc. ii. 44, 23. Fô se landrîca tô healfan, and tô healfan ðæt hundred, L. Edg. S. 8 ; Th. i. 274, 30. Gif cyninges þegn oððe ǽnig landrîca hit forhæbbe, gilde x. healf-mearc, healf Criste healf cynge, L. N. P. L. 58, 59 ; Th. ii. 300, 3 : 6, 7. Fare ðæs cinges

geréfa tô, and ðæs bisceopes, and ðæs landrícan [cf. landhláford, 11], L. C. E. 8; Th. i. 366, 8 : L. Eth. ix. 8; Th. i. 342, 16. Healf landrícan, healf wǽpentake, L. Eth. iii. 3; Th. i. 294, 8, 9. Hê ðeáh gange ðam landrícan tô ordále, 4; Th. i. 294, 20. Healf landrícan, healf cinges geréfan binnan port, 7; Th. i. 296, 8. Gylde ðam cyninge oððe landrícan, L. C. S. 37; Th. i. 348, 13. Healf Criste and healf landrícan, L. N. P. L. 49; Th. i. 298, 5. Gif hwá borhleás orf hæbbe, and landrícan hit befón, ágife ðæt orf and gilde xx óran, L. Eth. iii. 5; Th. i. 296, 1.

land-ríce, es; *n.* A territory, region, estate :—Bócland *vel* landríce *fundos,* Wrt. Voc. ii. 152, 18. Hê ðágiet lytel landríce hæfde búton ðære byrig ánre *he had as yet little territory except the town only,* Ors. 2, 2; Swt. 66, 14. Ic hæbbe gesǽd ymb ða þrié dǽlas ealles ðises middangeardes ac ic wille nú ðara þreóra landríca gemǽre gereccan *tripartiti orbis divisiones dedi, ipsarum quoque partium regiones significare curabo,* 1, 1; Swt. 10, 5.

land-riht, es; *n.* I. *the law of the land, the rights and privileges belonging to the inhabitant of a country* or *to the owner of land* [?]:—Londrihtes môt ðære mǽgburge monna ǽghwilc ídel hweorfan *shall each man of the family wander lacking the rights of those who live in the land,* Beo. Th. 5765; B. 2886. Grimm, R. A. 731 q. v. quotes in illustration from Saxo the order of Frotho : 'Si quis in acie primus fugam capesseret, *a communi jure alienus* existeret.' See also pp. 39–42. Mid rihtum landrihte swá hit on lande stonde *in accordance with the regular law of the land, as it stands in the land,* Cod. Dip. Kmbl. iii. 435, 35. Unc môdige ymb mearce sittaþ . . ne willaþ rúmor unc landriht heora *round our border sit bold ones, who will not more largely allow us their landright,* i. e. *will not allow us to possess more land in their country,* Cd. 91; Th. 114, 28; Gen. 1911. Hê landriht geþah *he received landright, he was settled in the country with the right of a native,* 161; Th. 200, 10; Exod. 354. Áhte ic fela wintra folgaþ tilne holdne hláford ôððæt Heorrenda nú leóðcræftig mon londryht geþah *þæt mé eorla hleó ǽr gesealde good service had I for many a winter, a kind lord; until now Heorrenda, a man skilled in song, has received land right; the prince had before given me that,* i. e. *H. was now admitted, as Deór had been before, to the rights of a native, and had succeeded in attracting to himself the favour before shown to Deór,* Exon. 100 b; Th. 379, 29; Deór. 40. II. *that which is due from land or estates :*—Ðegenes lagu is ðæt hé þreó þinc of his lande dó . . . Eác of manegum landum mǽre landriht áríst tô cynges gebanne *the law as regards the thane is that he do three things for his land. Also for many lands or estates, more extensive dues arise upon decree of the king,* L. R. S. 1; Th. i. 432, 6. [O. Sax. land-reht *law of the land* e. g. irô aldironô ēo, therô liudiô landreht : O. Frs. land-riucht : O. H. Ger. lant-reht *jus, lex* : Ger. land-recht *common law.*]

land-séta, an; *m. One settled in a country, a colonist :*—Ôðres eardes landséta *colonus,* Ælfc. Gl. 8; Som. 56, 100; Wrt. Voc. 18, 49. [O. L. Ger. land-sêtio : Ger. land-sass.]

land-sceap, es; *n. A district, tract of country, land :*—Swá hê on landsceape stille stande ðǽr hine storm ne mæg wind áwecgan *as if it [the vessel] stand still on land, where storm or wind cannot move it,* Andr. Kmbl. 1002; An. 501. v. land-scipe.

land-scearu, e; *f.* I. *a share, division,* or *portion of land, land, country :*—Sume hine lǽtaþ ofer landscare ríðum tôrinnan. Nis ðæt rǽdlíc þing gif swá hlutor wæter tôflôweþ æfter feldum ôð hit tô fenne werþ *some let it [spring of water] run away over their land in rills. It is not a wise thing if water so pure disperses itself along the fields, until it becomes a marsh,* Past. 65; Swt. 469, 5. Héton lǽdan ofer landsceare . . . drôgon æfter dúnscræfum ymb stánhleoðo efne swá wíde swá wegas tôlǽgon innan burgum strǽte stánfáge *they bade lead him over the country . . . they dragged him by mountain caves, across rocky slopes, far as the roads stretched, within the towns, the streets with many-coloured stones,* Andr. Kmbl. 2460; An. 1231. II. *a boundary of land* [cf. Icel. skör *a rim, edge*]. With this meaning the word occurs in charters which Kemble [Cod. Dip. iii. xii.] notices as being of comparatively late date and belonging to the extreme south of England :—Ðis his ðara fíf hída landscaru tô westtúne [*then follow the boundaries* : cf. landgemǽra *in such phrases*], Cod. Dip. Kmbl. iii. 338, 4. Of ðam hlince tô ðam beorge tô Ælfrêdes landscare ; ðonne is hit ðær feówer furlanga brád bútan feówer gyrdan ; ðonne gǽþ hit ðær niðer be ðara wyrhtena landscare, 420, 25-7. Ðonne eást andlang hricgweges tô Brytfordinga landsceare, 302, 16. The word also occurs in compounds landscar-hlinc [*also* landscare hlinc], landscar-ác. [Halliwell in his Dictionary gives *land-share* as a Devonshire word, meaning 'headland of a field': he also gives the word *land-score*]

land-scipe, es; *m. A tract of land, region :*—Ic á ne geseah lâðran landscipe *never saw I a more hateful region,* Cd. 19; Th. 24, 11; Gen. 376. [O. Sax. land-skepi : Icel. land-skapr *a region* : O. H. Ger. lantscaf *regio, provincia, patria.*]

land-seten, e; *f.* I. *Land in possession* or *occupation, an estate :*—Ðis his sió landseten æt Stántúne ðe Cénwold hæfde [*then follow the boundaries*], Cod. Dip. Kmbl. iii. 403, 24. [cf. ii. 143 where it is said ' Æþelwulf suo fideli ministro nomine Cenwold jure hereditario possiden-

dam condonavit terram in loco ubi a ruricolis Stantun nominatur.' And 144, ' Territoria istius agelli his terminibus circumdata esse videntur.' II. *occupation of land :*—Gebyreþ ðæt him man tô landsetene sylle ii oxan and i cú and vi sceáp *moris est ut ad terram assidendam dentur ei ii boves, et i vacca, et vi oves,* L. R. S. 4; Th. i. 434, 23.

land-setla, an; *m. An occupier of land, a tenant :*—Ic an míne landseðlen here toftes tô ôwen áchte *I give to my tenants their tofts into their own possession,* Cod. Dip. Kmbl. iv. 282, 29. [O. H. Ger. land-sidilo *accola, colonus, indigena,* Grff. vi. 310 : also *a tenant.* v. Grmm. R. A. 317 : cf. Icel. land-seti *a tenant.*]

land-sidu, a; *m. Custom of a country :*—Gemacaþ ðæt his ege wierþ tô gewunan and tô landsida *he causes the fear of him to become a habit and custom of the country,* Past. 17, 9; Swt. 121, 25. Be landside *according to the usage of the district,* L. R. S. 8; Th. i. 436, 27. Ealle landsida ne sýn gelíce *omnium terrarum instituta non sunt equalia,* 4; Th. i. 434, 30. [O. Sax. land-sidu.]

land-sittende; *adj. Occupying land :*—Hê lētt gewrítan hú mycel ǽlc man hæfde ðe landsittende wæs innan Englalande on lande oððe on orfe and hú mycel feós hit wǽre wurþ *he [William I.] caused to be written how much every man that was in the occupation of land in England, had in land or in cattle, and how much money it was worth,* Chr. 1085; Erl. 218, 32.

land-sôcn, e; *f. Search for land* or *country :*—Tôfaran on landsôcne *to separate in search of land* [*of the dispersion at the tower of Babel*], Cd. 80; Th. 100, 17; Gen. 1665 : 81; Th. 102, 12; Gen. 1699.

land-spéd, e; *f. Property in land :*—Ða munecas tô biscopan gewurdan ðære cyrcean landspéde [*substantiam aecclesiae*], Cod. Dip. Kmbl. iii. 349, 24.

land-spédig; *adj. Rich in landed property, having large estates :*—Landspédig *locuples,* Ælf. Gl. 88; Som. 74, 72; Wrt. Voc. 50, 52. Ðes and ðeós landspédiga *hic et hæc locuples,* Ælfc. Gr. 9, 27; Som. 11, 22.

land-splott, es; *m. A small portion,* or *plot, of ground :*—Mín is mannaseisca landsplot *meus est mannases,* Ps. Lamb. 59, 9. Ðisne landsplot becwæþ Æþelwine intô Abbendúne [it is spoken of before as *parva ruris particula, ruris particula*], Cod. Dip. Kmbl. iv. 39, 12.

land-stede, es; *m. Land, country,* Exon. 115 a; Th. 442, 22; Kl. 16.

land-stycce, es; *n. A small portion of land :*—Him gebyreþ sum landstycce for his geswince *convenit, ut aliquam terre portiunculam habeat pro labore suo,* L. R. S. 18; Th. i. 440, 8. Him man hwilces landsticces geann, 19; Th. i. 440, 14.

land-waru, e; *f. The people of a country, country,* Beo. Th. 4631; B. 2321. [Cf. burh-, ceaster-waru.]

land-weard, es; *m. The guard of a country, prince, ruler,* Beo. Th. 3785; B. 1890.

land-wela, an; *m. The wealth of this earth,* Exon. 63 a; Th. 232, 11; Ph. 505.

lane, an; *f. A lane, a narrow and bounded path, a street in a town :*—Hit cymeþ on ægles lonan ; ondlang ðære lonan ðæt hit cymeþ eft in ða burnan, Cod. Dip. Kmbl. iii. 33, 7. On ða ealdan lanan, 456, 3. Ðínne líchoman geond ðisse ceastre lanan hié tôstenceaþ *thy body shall they scatter through the streets of this city,* Blickl. Homl. 237, 5 : 241, 21, 25. [O. Frs. lona, lana.] v. norþ-lane.

lane-sang. v. lác-sang.

lang *length of time.* v. leng.

LANG; *adj.* LONG, *tall :*—Hê sǽde ðæt ðæt land síe swíðe lang norþ ðonan *he said that the land stretches thence far to the north,* Ors. 1, 1; Swt. 17, 4. Se wudu is eástlang and westlang hundtwelftiges míla lang oððe lengra *the wood, measuring from east to west, is a hundred and twenty miles long, or longer,* Chr. 893; Erl. 88, 28. Ðæt is þrittiges míla lang eást and west *habet ab oriente in occasum triginta circiter milia passuum,* Bd. 1, 3; S. 475, 19. Ðæt hê wǽre lang on bodige *quod esset vir longæ staturæ,* 2, 16; S. 519, 33. Ðæt is nú ðæs líchoman gôd ðæt mon síe fæger and lang and brád, Bt. 34, 6; Fox 140, 32. Eádweard se langa, Byrht, Th. 139, 53; By. 273. Se biþ lang lífes and welig *he shall be long-lived and wealthy,* Lchdm. iii. 156, 18. Næs lang tô ðý ðæt his brôðor ðyses lǽnan lífes tíman geendode *it was not long before his brother died,* 434, 24. Nis hit lang tô ðon, Bd. 4, 24; S. 599, 5. Hié tealdon ðætte Israhéla ríce sceolde beón hér on eorþan mycel and lang *they reckoned that the kingdom of Israel should be great and lasting here on earth,* Blickl. Homl. 117, 18. Tô langum gemynde *as a lasting memorial,* Homl. Skt. pref. 51. Langere tíde *tanto tempore,* Bd. 1, 25; S. 487, 11. Mid langre ádle *longo morbo,* 3, 9; S. 534, 5. Ofer swá langne weg sǽs and landes *per tam prolixa terrarum et maris spatia,* 2, 18; S. 520, 36. Ealle ðás naman habbaþ langne .o. on eallum. casum *all these nouns have long o in all cases,* Ælfc. Gr. 9; Som. 8, 52. Ða andswarode hê ymbe long *then answered he after long,* Bt. 39, 2; Fox 214, 8. Lange tíde *multis temporibus,* Lk. Skt. 8, 27, 29. Hiwgende lang gebed *simulantes longam orationem,* 20, 47. Tô langre fiftiges elna lange and ða mǽstan fiftiges elna lange. Ors. 1, 1; Swt. 18, 6. Ða ðe tô lang tô secgenne syndon *which are too long to narrate,* Bd. 3, 8; S. 532, 12. Wæs se líchoma sponne lengra ðære þrýh *corpus mensura palmi longius erat*

sarcofago, 4, 11; S. 580, 5. Ne biþ hē lengra ðonne syfan elna lang, Ors. 1, 1; Swt. 18, 4. Ðis eálond hafaþ mycele lengran dagas on sumera ðonne ða sūþdǽlas middangeardes, Bd. 1, 1; S. 473, 32. Ðā bebeád hē ðæt him mon lengran cwidas beforan cwǽde *præcepit eum sententias longiores dicere,* 5, 2; S. 615, 14. Ða onfóþ lengestne dóm *hi accipient prolixius judicium,* Mk. 12, 40. [The word occurs in all the Teutonic dialects.] DER. and-, dæg-, ealdor-, ge-, morgen-, niht-, sumor- lang; *it also is found in combination with the words denoting the points of the compass,* eást-lang, &c.

Langa-Frige-dæg *Good-Friday:*—Ðes passio gebyreþ on Langa-Frigadæg, Jn. Skt. 18, 1, rubric. Man ne mót hālgian hūsel on Langa-Frigedæg forðan ðe Crist þrowode on ðone dæg for ūs *the eucharist must not be consecrated on Good Friday, for Christ suffered for us on that day,* L. Ælfc. C. 36; Th. ii. 358, 16. [On langfridæi him on rode hengen, Chr. 1137; Erl. 263, 25.] [*Icel.* Langi-frjádagr : *Da.* Lang-fredag. In the E. D. S. Holderness Glossary Lang-Friday is given as the first Friday in Lent.]

Langa-land, es; *n. Langeland* an island in the Baltic belonging to Denmark:—On bæcbord him.wæs Langaland . . . and ðás land eall hȳraþ tó Denemearcan; Ors. 1, 1 : Swt. 19, 35. [*Icel.* Langa-land.]

langaþ. v. langoþ.

Lang-beardas, -beardan; *m. pl. The Lombards:*—Ða Gallie ðe mon nū hǽt Longbeardas, Ors. 4, 7; Swt. 180, 25. Tó Longbeardna londe, Chr. 887; Erl. 86, 9. Longbeardum, Exon. 85 a; Th. 320, 21; Víd. 32 : 86 a; Th. 323, 18; Víd. 80. [*Icel.* Lang-barðar.] v. Grmm. Gesch. D. S. c. xxv; cf. Heaðo-beardan.

lange; *adv. Long, a long time, far :*—Lange *diu*; leng *diutius*; ealra lengst *diutissime,* Ælfc. Gr. 38; Som. 42, 10. Longe *procul,* Wrt. Voc. ii. 66, 71 : *penitus,* 72. Ðā hē ðā lange and lange hearpode *when then he had harped a long, long time,* Bt. 35, 6; Fox 170, 5. Hū longe *how long,* Past. pref; Swt. 9, 4. Hū langæ, Bd. 4, 25; S. 600, 10. Nóht longe æfter ðon *not long after that,* Shrn, 105, 9. Swa lange swā gē dydon ānum of ðysum mínum læstum gebróðorum swa lange gē hyt dydon *me quamdiu fecistis uni de his fratribus meis minimis, mihi fecistis,* Mt. Kmbl. 25, 40 : Blickl. Homl. 169, 21. Genóh lange *long enough,* Deut. 1, 6. Hwæt mæg ic leng dón *ultra quid faciam?* Gen. 27, 37. Hwider mæg ic nū leng fleón *quo enim nunc fugiam?* Bd. 2, 12; S. 513, 27. Swaðer uncer leng wǽre [lifede, 38] *which of us two lived the longer,* Chart. Th. 485, 29. Ðænne ðū lengc ne móst lífes brúcan, Dóm. L. 32, 61. Lencg, Lk. Skt. 16, 2. Leng swā swíðor, Cd. 47; Th. 60, 30; Gen. 989. Swā leng swā swíðor, Exod. 19, 19. Nā leng heó ne gebád ðonne hit dæg wǽs *she waited only till it was day,* Apol. Th. 19, 2. Ðone aldormon ðe him lengest wunode *the alderman that stopped with him longest,* Chr. 755; Erl. 48, 21.

lang-fǽre; *adj. Lasting, enduring, old :*—Nānwuht nis langfǽres on ðis andweardan lífe *there is nothing lasting in this present life,* Bt. 38, 2; Fox 198, 6. On langfǽre ylde bet hē dēþ *at an advanced age he will do better,* Lchdm. iii. 188, 26. Eác ða treówa ðe beóþ āheáwene on fullum mōnan beóþ heardran wið wyrmǽtan and lengfǽrran [langferran, MS. L.], 268, 10. Swā eác treówa gif hí beóþ on fullum mōnan geheáwene hí beóþ heardran and langfǽrran tó getimbrunge *so too trees, if they are cut down at the full moon, are harder and more lasting for building,* Homl. Th. i. 102, 23. [*O. H. Ger.* lanc-fāri *longævus,* Grff. 3, 574.]

lang-first, es; *m. A long space of time :*—Nolde fæder engla in ðisse lífe longfyrst ofer ðæt wunian léton *the father of angels would not let him remain in this life a long space after that,* Exon. 46 b; Th. 159, 2; Gū. 920.

langian; *p.* ode *To grow long :*—Ðonne se dæg langaþ ðonne gǽþ seó sunne norþweard óþ ðæt heó becymþ tó ðam tācne ðe is gehāten Cancer, Lchdm. iii. 250, 9. Se langienda dæg, 252, 6, 9. Eft on langiendum dagum hē ofergǽþ ðone sūðran sunnstede, 14.

langian; *p.* ode : *v. impers. with acc. of pers. To cause longing, desire, discontent,* or *pain in a person :*—Langaþ ðē āwuht *dost thou desire aught?* Cd. 25 : Th. 32, 1; Gen. 496. Hæleþ langode hwonne hié of nearwe stæppan mósten *the men longed for the time when they might step from durance,* 71; Th. 86, 16; Gen. 1431. Hine ðæs heardost langode hwanne hē of ðisse worlde móste, Blickl. Homl. 227, 1. Mec longade *I was ill at ease,* Exon. 115 a; Th. 442, 18; Kl. 14. Longiga *tædere,* Mk. Skt. Lind. 14, 13. Ðæt ūs nū æfter swelcum longian mǽge swelce ðā wǽron *that we should now long for such times as then were,* Ors. 2, 5; Swt. 84, 27. Ðā ongan hine eft langian on his cýððe *then he began to long again for his native land,* Blickl. Homl. 113, 15. [*O. Sax.* langón (*with acc. of pers.*) : *O. H. Ger.* langēn, langón (mih langet *desidero*).)]

langian; *p.* ode *To summon, call :*—Godes æncgel cwæþ ðæt hē sceolde ðe him tó langian [MS. U. langean] *God's angel said that he was to summon thee to him,* Homl. Skt. 10, 121. v. ge-langian.

langian; *p.* ode *To belong, pertain :*—Alle ða land ðe longen intó ðære hālagen stówe *all the lands that belong to the holy place,* Cod. Dip. Kmbl. iv. 215, 4. [*O. H. Ger.* ge-langón *pertingere.*] v. lengan *to belong.*

lang-líce; *adv. Long, at length, for a long time :*—Langlíce *tractim,* Ælfc. Gr. 38; Som. 41, 12. Hét ðone diácon langlíce swingan, Homl. Th. i. 426, 13 : ii. 490, 5. Langlíce on gebedum læg, 160, 35 : 510, 25. Langlíce bæd, i. 66, 23. [Cf. *O. H. Ger.* lang-líh *long* (of time) : *Icel.* lang-liga *for a long time past.*]

lang-líf -líf; *adj. Long-lived :*—Langlífe *longævus,* Ælfc. Gl. 35 : Som. 62, 95; Wrt. Voc. 28, 72. Langlíf [MS. C. langlífe, Zup. 320, 1] *longævus,* Wrt. Voc. 85, 59. Ðæt ðū sí langlífe *ut longo vivas tempore,* Deut. 5, 16 : 4, 1. Longlífe and gileáffull suǽ Sarra *longeva et fidelis ut Sarra,* Rtl. 109, 39. Langlífe hē biþ *he shall live long,* Lchdm. iii. 184, 4. [*Icel.* lang-lífr : *O. H. Ger.* lanc-líp *longævus,* Grff. 2, 46.]

lang-mód; *adj. Patient, long-suffering :*—Longmód *longanimis,* Ps. Stev. 7, 12. [*Ps.* 102, 8 lang-mode : *O. H. Ger.* lanc-mót *longanimis :* cf. *Ger.* lang-müthig *patient, long-suffering.*]

lang-ness, e; *f. Length :*—Brādnyss langnyss heáhnyss and deópnyss *breadth, length, height and depth,* Homl. Th. ii. 408, 21. Langnysse dagena ic gefylle hine *longitudine dierum replebo eum,* Ps. Spl. 90, 16. Ðonne sceal man ðysne wyrttruman gedrígean and ða langnysse tóceorfan on pysena gelícnysse *this plant is to be dried, and its length cut up into pieces about the size of peas,* Herb. 140, 1; Lchdm. i. 260, 15. Ealle óðre dagas on twelf mónþum habbaþ mislíce langnisse *all other days in the twelve months have various lengths,* Lchdm. iii. 258, 2.

langoþ, es; *m. Longing, desire, discontent,* or *weariness that arises from unsatisfied desire :*—Æfter men dyrne langaþ born *a secret longing for the man burned within him,* Beo. Th. 3763; B. 1879. Hine ne meahte longaþ gelettan, Exon. 37 b; Th. 123, 29; Gū. 330. Ic ǽfre ne mæg ðære módceare mínre gerestan ne ealles ðæs longaþes ðe mec on ðissum lífe begeat *never can I be at rest from my grief of mind, nor from all the weariness that in this life hath laid hold on me,* 115 b; Th. 444, 2; Kl. 41. Wā biþ ðam ðe sceal of [on?] langoþe leófes ābídan *woe to him that must wait, with unsatisfied longing, for one that he loves,* Th. 444, 26 : Kl. 53. Hæfde him tó gesíþþe sorge and langaþ *he had for company sorrow and discontent,* 100 a; Th. 377, 14; Deór. 3. Forðon mec longeþas lyt gegrétaþ *therefore longings visit me little,* 37 a; Th. 121, 11; Gū. 287. Forlēt longeþas lǽnra dreáma *he gave up desires for transitory delights,* Th. 122, 5; Gū. 301.

lang-sceaft; *adj. Having a long shaft :*—Mid longsceaftum sperum *longas hastas habebamus hastas,* Nar. 13, 24. Mid longsceaftum sperum *venabulis,* 15, 28. [Cf. *Icel.* lang-skeptr.]

lang-scip, es; *n. A long-ship, a large war-ship :*—Ðā hēt Alfréd cyng timbran langscipu [*other MSS.* lange scipu] ongēn ða æscas, Chr. 897; Erl. 95, 11. [*Icel.* lang-skip.]

lang-strang *glosses* longanimis *in* Ps. Lamb. 102, 8.

lang-sum; *adj. Long, taking a long time, prolix, lasting a long time, long-enduring, long-suffering :*—Nis mē ðæs þearf tó secgenne forðon hit longsum is and eác monegum cūþ *nec per ordinem nunc retexere nostrum est, quia et operi longum et omnibus notum videtur,* Ors. 1, 11; Swt. 50, 16. Ða tó talanna longsum is *quos enumerare longissimum est,* Mt. Kmbl. p. 7, 7 : Andr. Kmbl. 2962; An. 1484. Hū langsum wæs him se hlísa *how lasting was that fame for him?* Bt. 18, 4 ; Fox 68, 5 : Beo. Th. 3076; B. 1536. Hwæt gif ic bíde merigenes se ebréisca cwæþ ne biþ hit swā langsum '*What if I last till morning?*' *The Jew said* '*It will not be so long,*' Homl. Skt. 3, 585. Ðonne seó āheardung ðære lifre tó langsum wyrþ *when the hardening of the liver lasts too long,* L. M. 2, 22; Lchdm. ii. 210, 4 : Beo. Th. 268; B. 134 : Homl. Skt. 4, 128. On ðam tíman wæs swíðe langsum líf on mancynne *at that time life lasted long among men,* Homl. Th. ii. 460, 3. Lufu langsumu *lasting love,* Cd. 91; Th. 114, 18; Gen. 1906. Langsum *longanimis,* Ps. Spl. 102, 8. Mid heora langsuman gebede *sub obtentu prolixæ orationis,* Mk. Skt. 12, 40; Hpt. Gl. 500, 25. Ðam þeódscype tó langsuman rǽde *to the lasting advantage of the nation,* L. I. P. 4; Th. ii. 308, 5 : Cd. 219; Th. 280, 4; Sat. 250. Gehǽlede fram heora langsumum bróce *healed from their long sickness,* H. R. 105, 2. Him and his gebeddan tó langsumum gemynde *as a lasting memorial for him and his consort,* Chart. Th. 605, 12. His sāwle tó gescyldnesse on langsuman sȳðe *as a protection to his soul on its long journey,* Chr. 959; Erl. 121, 7. Ða þrý cyningas hæfdon langsume sprǽce wið ðone gedrehtan joh, Homl. Th. ii. 456, 24. Langsume *longanimem,* Wrt. Voc. ii. 53, 52. Tó langsumum wýtum, Homl. Skt. 4, 120. Him ēce geceás langsumre líf *he chose for himself a more enduring, an eternal life,* Apstls. Kmbl. 39; Ap. 20. Ús selfum betst word and longsumast æt ūrum ende gewyrcan *to gain for ourselves the best and most enduring fame at our death,* Ors. 2, 5; Swt. 82, 2. [*O. Sax., O. H. Ger.* langsam *longus, diuturnus, prolixus : Ger.* langsam *slow.*]

langsum-ness, e; *f. Length :*—Langsumnysse daga *longitudinem dierum,* Ps. Spl. 20, 4. Swā ðæt hí ne beón þurh ða deópnysse ǽmóde ne þurh ða langsumnysse ǽþrytte *so that they be not discouraged by the deepness, nor wearied by the length,* Homl. Th. ii. 446, 8. Ealle óðre dagas on twelf mónþum habbaþ mislíce langsumnysse, Lchdm. iii. 258, 2 note. Ða brādsumnessa and ða langsumnessa, Wulfst. 244, 27.

lang-sweored, -swyred; *adj. Having a long neck, long-necked :*—Sume fugelas beóþ langsweorede swā swā swanas *some birds are long-*

necked, such as swans, Hexam. 8 ; Norm. 14, 16. Ða beóþ langswyrede ðe lybbaþ be gærse swá swá olfend and assa, 9 ; Norm. 16, 2.

lang-twidig; *adj. Granted for a long time :*—Ðú scealt tó frófre weorþan eal langtwidig leódum ðínum *thou, granted for long to them, shalt prove a comfort to thy people*, Beo. Th. 3420 ; B. 1708.

langung, e ; *f. Longing, desire, weariness or grief that comes from unsatisfied desire :*—Hié langung beswác eorþan dreámas éces rǽdes *the longing for the joys of earth cheated them of eternal good*, Cd. 173 ; Th. 217, 28 ; Dan. 29. Hé for ðære langunga and for ðære geómrunga ðæs óðres úeáþes leng on ðam lande gewunian ne mihte . . . him nǽfre seó langung ne geteorode *for grief and sorrow at the other's death he could not live in that land any longer . . . his grief never wore itself out*, Blickl. Homl. 113, 10–14. Ðá wæs him micel langung and sorh on heora heortan ðá hié ðæt ongeáton ðæt hé leng mid him líchomlíce wunian nolde, 135, 21. Ða myclan byrþenne áberan ðære mycclan langunga heora ðæs leófes Hláfordes *to bear the great burden of the great longing after their dear [departed] Lord*, 135, 8. Tó frófre for ðære miclan langunga Drihtnes framfundunga *as a comfort for the great grief at the Lord's departure*, 131, 14. For longunge *præ tædio*, Ps. Spl. C. 118, 28. Longunge fús *longingly eager*, Exon. 119 a ; Th. 458, 8 ; Hy. 4, 97. Á hafaþ longunge se ðe on lagu fundaþ *ever hath he weariness whose way is on the water*, Exon. 82 a ; Th. 308, 29 ; Seef. 47. Langunga habban æfter ðam freóndum *to think with grief of dead friends*, Blickl. Homl. 131, 26.

langung, e ; *f. Lengthening, prolonging, delay :*—Longunga *prolixae*, Mk. Skt. Lind. 12, 40 ; *prolixa* [in both cases = prolixe], Jn. Skt. p. 7, 18. On ælcre longunge geþyldige *patient in every delay*, Past. 5, 1 ; Swt. 41, 16.

langung-hwíl, e ; *f. A time of longing or weariness :*—Feala[ic] ealra gebád langunghwíla, Andr. Kmbl. 249 ; An. 125.

lann, lonn, e ; *f. A bond, fetter :*—Licgeþ lonnum fæst *lies fast in fetters*, Salm. Kmbl. 531 ; Sal. 265. Fæste gebindan, lonnum belúcan, 557 ; Sal. 278. [Grein refers to Grff. 2, 217 'Lanna *lamina* (among words referring to weaving).']

lapian; *p.* ode *To lap, lick :*—Ic lapige *lambo*, Ælfc. Gr. 28 ; Som. 32, 25. Gedó ðonne on glæsfæt and ðonne mid hláfe oððe mid swá hwilcum mete swá ðú wille lapa on *then put it into a glass vessel, and then, with bread or with whatever food you will, lap it up*, L. M. 2, 6 ; Lchdm. ii. 184, 24. Lapien on hunig *let them lap up honey*, 16. [Cf. *Icel.* lepja *to lap* as a dog : *O. H. Ger.* laffan ; *p.* luof *lambere*.]

lappa. v. læppa.

LÁR, e ; *f.* I. LORE, *teaching, instruction, learning, knowledge, cunning, science, preaching, doctrine, dogma, precept :*—Lár *disciplina : doctrina*, Ælfc. Gl. 80 ; Som. 72, 100, 101 ; Wrt. Voc. 46, 57, 58. Folclíc lár *omilia*, 35 ; Som. 62, 75 ; Wrt. Voc. 28, 53. Lár *dogma*, Ælfc. Gr. 9 ; Som. 8, 24. On ðam wæs áwriten Lár and Sóðfæstnys *in quo erat Doctrina et Veritas*, Lev. 8, 8. Seó hálige lár *sancta prædicatio*, Bd. 1, 27 ; S. 495, 40. Seó rihtgelýfde lár wæs dæghwamlíce weaxende *crescente per dies institutione catholica*, 3, 28 ; S. 560, 39. Bisceopes dægweorc biþ . . . lár oððon leornung *a bishop's daily work . . is . . . teaching or learning*, L. I. P. 8 ; Th. ii. 314, 19. Him tó fultume godcundre láre *sibi adjutorem evangelizandi*, Bd. 2, 4 ; S. 505, 14. Mynster tó timbrianne ðam monnum ða ðe Scotta láre fyligdon *ad construendum monasterium his qui Scottos sequebantur*, 5, 19 ; S. 638, 39. Láre *gravitate*, Wrt. Voc. ii. 40, 34. Ic mé gúþbordes sweng láre gebearh *I warded off the blow from me by cunning*, Cd. 128 ; Th. 163, 7 ; Gen. 2693. Hé sceal habban láre ðæt he máge Godes folc mid wísdóme léran *he must have learning, that he may be able to instruct God's people with wisdom*, Homl. Th. i. 206, 26. Hú giorne ða godcundan hádas wǽron ægðer ge ymbe láre ge ymbe liornunga . . . and hú man útanbordes wísdóm and láre hieder on lond sóhte *how diligent the clergy were about teaching and learning . . . and how wisdom and instruction were sought here by foreigners*, Past. pref ; Swt. 3, 9–12. Ne sceolan ða láreówas ágímeleásian ða láre, Blickl. Homl. 47, 29 ; 7, 11. Tó bodigenne godcunde láre *ad prædicandum*, Bd. 2, 3 ; S. 504, 16. Þurh his láre *docendo*, 2, 20 ; S. 522, 22. Hé godspellíce láre lǽrde *opus evangelizandi exsequens*, 3. 19 ; S. 547, 9. Háliges láre [cf. langsum leornung, 2962] *the story of the saint*, Andr. Kmbl. 2955 ; An. 1480. Lǽre *disciplinam*, Ps. Spl. 118, 66. Bodigende his láre *prædicans præceptum ejus*, Ps. Lamb. 2, 6. Hálige lára *dogmatum*, Wrt. Voc. ii. 27, 58. Ic wolde ymbe ðone lǽcedóm ðara ðínra lára hwéne máre gehýran *I would hear a little more of the medicine of those instructions of thine*, Bt. 22, 1 ; Fox 76, 17. Lárna, Exon. 117 a ; Th. 450, 17 ; Dóm. 89 ; Andr. Kmbl. 964 ; An. 482. Gif wé óðre men teala lǽraþ, and hié be úrum lárum rihtlíce for Gode libbaþ, ðonne bringe wé Drihtne swétne stenc on úrum dǽdum and lárum, Blickl. Homl. 75, 14. Hig lǽraþ manna lára *docentes doctrinas hominum*, Mt. Kmbl. 15, 9. Betwih óðre láre tó lifigeanne *inter alia vivendi documenta*, Bd. 3, 5 ; S. 526, 20. Wið ALreum and his láre *contra Arium et ejusdem dogmata*, 4, 17 ; S. 585, 44 : 586, 1. Wé sceolan healdan ða lára ðara feówer godspellera *we must keep the precepts of the four evangelists*, Blickl. Homl. 35, 11. II. *exhortation, admonition, counsel, suggestion, instigation, persuasion :*—Mid his getrymnesse and láre *ejus hortatu*, Bd. 1,

33 ; S. 498, 35. Mid his dæghwamlícre láre *quotidiana exhortatione*, 2 9 ; S. 510, 37. Láre *hortamentis*, Wrt. Voc. ii. 42, 55. Ealle ða men Julius hét ofsleán ðe æt ðære láre wǽron ðæt mon Pompeius ofslóg *Julius ordered all the men to be killed who advised that Pompey should be slain*, Ors. 5, 12 ; Swt. 242, 23. Hé wið his hláford wan for óðra manna láre *he fought against his lord at the instigation of other men*, 6, 35 ; Bos. 131, 11. Hlyste mínre láre *acquiesce consiliis meis*, Gen. 27, 8. Þurh Wulfheres láre *suggerente rege Wulfhere*, Bd. 4, 13 ; S. 582, 7. Wes ðú ús lárena gód *be liberal to us of thy counsels*, Beo. Th. 544 ; B. 269. Lárum *hortamentis*, Bd. 2, 2 ; S. 502, 14. Hié swýðor fylgaþ deófles lárum *they rather follow the suggestions of the devil*, Blickl. Homl. 25, 10 : 61, 13. Ðín ríce for his lárum gefealleþ *thy kingdom will fall because of his counsels*, 181, 34. [*O. Sax.* léra : *O. Frs.* láre : *O. H. Ger.* léra *doctrina, dogma, sermo, præceptum, exhortatio, consultum :* *Ger.* lehre.] DER. bóc-, folc-, freónd-, mis-, un- lár.

lár-bóc ; *f. A book which conveys instruction :* Swá swá Beda áwrát, Engla þeóde láreów, on his lárbócum, Chart. Th. 241, 20.

lár-bysn, e ; *f. An example, proof, specimen :*—Lárbysn *documentum vel specimen*, Ælfc. Gl. 80 ; Som. 72, 104 ; Wrt. Voc. 46, 61.

lár-cræft, es ; *m. Knowledge, science :*—Ic íglanda eallra hæbbe lárcræftas onlocen. Salm. Kmbl. 5 ; Sal. 3.

lár-cwide, es ; *m. Precept, doctrine :*—Wé sóðfæstes lǽston lárcwide, Andr. Kmbl. 1347 ; An. 674.

láreów, es ; *m. A teacher, master, preacher :*—Láreów *doctor vel imbutor, vel eruditor : dogmatista*, Ælfc. Gl. 80 ; Som. 73, 98, 102 ; Wrt. Voc. 46, 55, 59. Cilda láreów *pædogogus*, Som. 73, 103 ; Wrt. Voc. 46, 60. Láreów *dogmatista*, Wrt. Voc. ii. 28, 50. Wé cildra biddaþ ðé eálá Láreów ðæt ðú tǽce ús sprecan *nos pueri rogamus te, Magister, ut doceas nos loqui*, Coll. Monast. Th. 18, 1. Ne gyrne gé ðæt eów man Láreówas nemne áu ys eówer Láreów *nolite vocari Rabbi : unus enim est Magister vester*, Mt. Kmbl. 23, 8. Hé is ordfruma and láreów ealre clǽnnesse *he is the origin and teacher of all purity*, Blickl. Homl. 13, 21. Heó æfter ðon wæs magister and láreów ðæs mynstres *deinde magistra exstitit*, Bd. 3, 24 ; S. 557, 5. On ðære heó mihte Gode willsumra wífmonna láreów and féstermódur gestandan *in quo ipsa Deo devotarum mater ac nutrix possit existere feminarum*, 4, 6 ; S. 574, 17. Wæs se Columba-se ǽresta láreów ðæs cristenan geleáfan *erat Columba primus doctor fidei christianæ*, 5, 9 ; S. 622, 40. Be ðære láre mínes láreówes, Blickl. Homl. 185, 8 : Exon. 14 b ; Th. 29, 6 ; Cri. 458. Hí sendon Aidan ðone biscop Angelþeóde tó láreówe *ad prædicationem gentis Anglorum Aidanum miserant antistitem*, Bd. 5, 22 ; S. 644, 25 : 3, 5 ; S. 527, 29. Hé ða hálgan láreówas hider onsende *hlios prædicatores mittens*, 1, 7 ; S. 501, 36. [*Orm.* lárew : cf. *O. Sax.* léreo.] DER. heáh-láreów.

láreów-dóm, es ; *m. The office of a teacher, mastership, governance, teaching :*—Forðonðe nán cræft nis tó lǽranne ðæm ðe hine ǽr geornlíce ne leornode forhwon beóþ ǽfre suǽ þriste ða ungelǽredan ðæt hí underfón ða heorde ðæs láriówdómes ðonne se cræft ðæs láreówdómes biþ cræft ealra cræfta *nulla ars doceri præsumitur, nisi intenta prius meditatione discatur. Ab imperitis ergo pastorale magisterium qua temeritate suscipitur, quando ars est artium regimen animarum*, Past. 1, 1 ; Swt. 25, 15–19. Ne hí scoldon ne underfón ða áre ðæs láreówdómes *ne locum regiminis subeant*, 2, arg ; Swt. 29, 19. Ðæt biþ ðæs recceres ryht ðæt hé þurh ða stemne his láriówdómes ætiéwe ðæt wuldor ðæs uplícan éðles *debitum rectoris est supernæ patriæ gloriam per vocem prædicationis ostendere*, 21, 5 ; Swt. 159, 22. Tó Criste hé Angle gehwyrfde mid árfæstnysse láreówdómes *ad Christum Anglos convertit pietate magistra*, Bd. 2, 1 ; S. 500, 28. Mid ealdorlícnesse láreówdómes *auctoritati magistri*, 4, 27 ; S. 603, 44. Wæs on his láreówdóme áféded *erat in magisterio illius educatus*, 4, 3 ; S. 569, 6 : 5, 19 ; S. 638, 15 : L. Ælfc. P. 10 ; Th. ii. 368, 3. Ðætte unlǽrde ne dyrren underfón láreówdóm *ne venire imperiti ad magisterium audeant*, Past. 1, arg ; Swt. 25, 14 : Homl. Th. ii. 320, 12.

láreów-líc ; *adj. After the manner of a teacher :*—Léreówlíc *exhortatorium*, Hpt. Gl. 512, 45. Láreówlícum cræftum *gymnicis* (gl. *magisterialis*) *artibus*, 405, 8.

láreów-setl, es ; *n. The seat of a teacher* or *doctor :*—Ofer Moyses láreówsetl *super cathedram Mosi*, Mt. Kmbl. 23, 2.

lár-hlystend, es ; *m. One who listens to instruction, a catechumen*, Mone B. 2802.

lár-hús, es ; *n. A house for instruction, a school ; gymnasium*, Hpt. Gl. 405, 11.

lár-leást, -lýst, e ; *f. Lack of learning or instruction :*—Þurh lárleáste hí ne cunnon ne lǽdan ne láran hí *through want of knowledge they cannot guide or teach them*, L. I. P. 19 ; Th. ii. 326, 28. Wé sceolon bodigan ðam lǽwedum ðý læs ðe hý for lárlýste losian sceoldan *we must preach to the laymen, lest for lack of instruction they should perish*, L. Ælfc. C. 23 ; Th. ii. 352, 1 : Wulfst. 79, 19.

lár-líc ; *adj. Instructive :*—Sume Godes þeówan mid lárlícre sprǽce óðre getrymmaþ *some servants of God confirm others with instructive discourse*, Homl. Th. i. 346, 22. Hit is swíðe gedafenlíc ðæt gé sume lárlíce word æt eówerum láreówum gehýron, ii. 282, 31.

lár-smiþ, es; *m.* *A wise man, a counsellor* :—Lársmiþas, Elen. Kmbl. 406; El. 203. Lársmeoþas, Andr. Kmbl. 2441; An. 1221.

lár-spell, es; *n.* *A discourse, sermon, homily, treatise* :—God cwæþ be láreówum on his lárspelle *God said of teachers in his sermon*, Homl. Th. ii. 320, 25. Se bisceop ðam folce sǽde lárspell, Homl. Skt. 3, 141. Ic gesett hæbbe wel feówertig lárspella *I have composed quite forty homilies*, Ælfc. T. Grn. 13, 45. Swá swá wē áwriton ǽror on óðrum lárspellum, 4, 15. Ða apostoli gesetton eác swilce lárspell [*the epistles*] tō ðám leódscipum ðe tō geleáfan bugon, 14, 3. [*Laym. Orm.* lar-spell *a sermon*.]

lár-swic, es; *m. n.* [?] *Deception, seduction, delusion, treachery* :—Mycel is nýdþearf manna gehwylcum, ðæt hē wið deófles lárswice warnige symle, Wulfst. 309, 14.

lár-wita, an; *m.* *A learned man* :—Lárwitan and lahwitan, L. I. P. 5; Th. ii. 308, 14.

laser, es; *m. n.* [?]. *A tare, cockle* :—Laser, zizania, Ælfc. Gl. 101; Som. 77, 29; Wrt. Voc. 55, 34. Lasur *lolium*, Wrt. Voc. ii. 54, 15. Ætan ł lasor zizania, 72, 61.

lást, lǽst, leást, es; *m.* *A step, footstep, sole of the foot, track, trace* :—Lǽst *solum*, Ælfc. Gl. 75; Som. 71, 98; Wrt. Voc. 45, 6. Ðú ðás werþeóde wræccan láste feorran gesóhtest *from far with the foot of an exile this people hast thou sought*, Cd. 114; Th. 149, 22; Gen. 2478. Sarran brýde láste beddreste gestáh, 129; Th. 164, 15; Gen. 2715. Of láste *e vestigio, statim*, Wrt. Voc. ii. 144, 33. On láste *e vestigio*, 107, 41. Him on láste setl wíde stódan *behind them heaven stood spacious*, Cd. 5; Th. 6, 10; Gen. 86. Malalehel wæs æfter Jarede yrfes hyrde fæder on láste *Mahalaleel was after Jared the guardian of the heritage in succession to his father*, 52; Th. 65, 18; Gen. 1068. Him on láste fór sweót Ebrēa *on their track marched the band of Hebrews*, Judth. 12; Thw. 25, 38; Jud. 298. Yldran ússe án forlēton ðone wlitigan wong on láste *our parents left that beauteous plain behind*, Exon. 62 a; Th. 228, 18; Ph. 440. Frǽtwe lēton licgan on láste, 104 a; Th. 394, 30; Rä. 14, 11. Ðá wearþ forht ferþ manig folces on láste *then was the mind of many a man of that folk left in fear*, Andr. Kmbl. 3191; An. 1598. Hié ðæs láðan lást sceáweodon *they marked the track of the foe*, Beo. Th. 265; B. 132. Lást weardian [cf. lást-weard] *to guard the track of one gone before, to remain behind; also to follow in the steps of another*. Cyning úre gewát þurh ðæs temples hróf ðǽr hý tō sēgun ða ðe leófes lást weardedun [*of the disciples watching the ascension of Christ*], Exon. 15 a; Th. 31, 16; Cri. 496. Se ðe his mondryhten life bilidene lást weardian wiste *who knew his lord, of life bereft, remained behind*, 52 a; Th. 182, 19; Gú. 1312. Sceal se líchoma leást weardigan eft on eorþan *the body shall again be left in the ground*, Bt. Met. Fox 20, 482; Met. 241. Hē his folme forlēt lást weardian, Beo. Th. 1947; B. 971. Hýrde ic ðæt ðám frætwum feówer mearas lást weardode *I heard that four steeds followed those trappings*, 4335; B. 2164. Him arn on lást þýstre genip *dark cloud succeeded it*, Cd. 8; Th. 9, 8; Gen. 138. Him fleáh on lást earn ǽtes georn, Judth. 11; Thw. 24, 27; Jud. 209. Geseoh nú seolfes swæðe . . . Ðá on lást beseah leóflíc cempa '*see now thine own track*.' . . . Then the good warrior looked behind, Andr. Kmbl. 2880-90; An. 1443-48. On lást faran *to return*. Beo. Th. 5883; B. 2945. Wesseaxe on lást legdun láþum þeódum *the West Saxons hung on the track of the foe*, Chr. 937; Erl. 112, 22; Æðelst. 22. On lást [cf. Icel. á lesti] *at last*. Ðú sárgige on lásð *gemas in novissimis*, Past. 36, 2; Swt. 249, 13. Hit on lást of his tungan útábirst tō openum bismere *ad extremum usque ad apertas lingua contumelias erumpat*, 38, 7; Swt. 279, 8. Ðæt mód him ǽrest ná ne ondrǽt ða lytlan scylda, ne ðonne on lást ða miclan, 57, 2; Swt. 437, 28 : Bt. 7, 20; Fox, 16, 11; Fox 72, 7. Lástas wǽron wíde gesýne, gang ofer grundas, Beo. Th. 2809; B. 1402. Ic sume in bryne sende ðæt him lásta wearþ síðast gesýne *some have I sent into the fire, so that no trace of them was left*, Exon. 72 b; Th. 270, 33; Jul. 474. Blódgum lástum, 36 b; Th. 119, 25; Gú. 260. Ðonne is ðǽr geworht emb ða lástas . . . ðæt man mæg tō ðǽm lástum onhnígan and mænige men ða moldan neomaþ on ðǽm lástum *the footsteps are built about, yet so that people can stoop down to the footsteps, and many men take the earth from the footsteps*, Blickl. Homl. 127, 5-11, 15, 19. Ðæt nǽnig man ða lǽstas sylfe ufan oferwrycean ne mid golde ne mid seolfre *so that no man might overlay the footsteps themselves, neither with gold nor with silver*, 125, 35. Sceáwian láðes lástas, Beo. Th. 1686; B. 841. Lástas lecgan [cf. colloquial *to make tracks*] *to journey, travel*. Ic lástas sceal wíde lecgan *wide must I wander*, Cd. 49; Th. 63, 3; Gen. 1026. Gewít ðú féran, lástas lecgan, 137; Th. 172, 26; Gen. 2850 : 118; Th. 153, 9; Gen. 2536 : 109; Th. 145, 3; Gen. 2400. [*Goth.* laists *a footstep*.] DER. æf-, feorh-, fēt-, fēðe-, fót-, sweart-, úríg-, wíd-, wræclást. v. lǽst.

lást. v. ge-lást.

lástian. v. wræc-lástian.

lást-weard, es; *m.* *One who keeps in the steps of another, a successor, pursuer* :—Ðone lástweard, his swǽsne sunu [*Isaac*], Cd. 162; Th. 203, 7; Exod. 400. Wræcmon gebád láðne lástweard *the fugitive awaited the foe that followed*, 148; Th. 186, 13; Exod. 138. Ús is swíðe uncúþ hwæt úre yrfeweardas and lástweardas getreówlíces dón willon efter úrum

life *it is quite unknown to us how faithfully our heirs and successors will act after our death*, Blickl. Homl 51, 36. Ic ne míne lástweardas *neither I nor my successors*, Chart. Th. 29, 12.

lást-word, es, *n.* *Report, reputation* :—Eorla gehwam lástworda betst *the best reputation for every man*, Exon. 82 b; Th. 310, 12; Seef. 73.

lata, an; *m.* *One who is late or slow* :—Ðeáh heó ðæs bearnes lata wǽre *though she were late in bearing the child*, Blickl. Homl. 163, 8. [*Icel.* lati *the lazy one*.] v. hild-lata.

late; *adv. Slowly, late, at length, at last* :—Alexander late unweorðlícne sige gerǽhte [*anceps*] *pugna tandem tristem pene victoriam Macedonibus dedit*, Ors. 3, 9; Swt. 134, 8. Hú ne cymþ se deáþ ðeáh ðe hē late cume and áðēþ eów of ðisse worulde *sera vobis rapiet hoc etiam dies*, Bt. 19; Fox, 70, 16. Gif wit ðæt ealle sculon ásmeágan ðonne cume wit late tō ende ðisse bēc oððe nǽfre, 42; Fox, 256, 22. Hú late hī on ðysne middangeard ácennede wurdon and hú raþe hī him eft of gewítan sceolan, Blickl. Homl. 59, 23. Late on geáre *late in the year*, Chr. 867; Erl. 72, 11. Late mylt gǽten flǽsc *goat's flesh digests slowly*, L. M. 2, 16; Lchdm. ii. 196, 16. Gif heó gǽþ late . . gif heó hraþe gǽþ, Lchdm. iii. 144, 7 : Exon. 49 b; Th. 172, 2; Gú. 1137. Ic ðæt gecneów tō late *too late I perceived it*, 72 a; Th. 269, 2; Jul. 444 : Elen. Kmbl. 1412; El. 708. Síð and late *at last*, Judth. 12; Thw. 25, 24; Jud. 275. Ǽr oðð lator *prius aut posterius*, Athan. 25. Lator *tardius*, Bd. 4, 9; S. 577, 10. Ðæt ðæt lator biþ, ðæt hæfþ anginn, Homl. Th. i. 284, 7. Onbútan Martines mæssan and gyt lator, Chr. 1089; Erl. 226, 20. Ðæt hit hraþost weaxan mæg, and latost wealowigan, Bt. 34, 10; Fox 148, 22. Sý ágifen be emnihte oððe latest be ealra hálgena mæssan *let it be paid by the equinox, or at latest by All-Hallows' Mass*, Wulfst. 208, 5.

láteów. v. lád-teów.

láð, es; *n.* *What is hateful or harmful, harm, evil, injury, hurt, trouble, grief, pain, annoyance, enmity* :—Ðætte monnum hēh is laaþ [*adj.?*] is mið Gode *quod hominibus altum est, abominatio est apud deum*, Lk. Skt. Lind. 16, 15. Hit sóna nǽnig láð ne biþ *it [the pain] will soon be no annoyance*, Herb. 1, 11; Lchdm. i. 74, 10. Hē mē nówiht láðes ætýwde *ille mihi nil inimicitiarum intulerit*, Bd. 2, 12; S. 513, 25. Ðæt hē ðe nánwiht láðes ne dó *ut nec ipse tibi aliquid mali faciat*, 514, 3. Ðæt him mon nóht láðes gedón dorste *ne qui prædicantibus quicquam molestiæ inferret*, 5, 10; S. 624, 6. Ic eom mid ðæs láðes sáre swíðe ofþrycced *I am sorely oppressed with the pain of this trouble*; insitus animum mœror praegravat, Bt. 8; Fox 24, 14. Ða ungeþyldegan ne mágon áberan nánwuht ðæs láðes ðe him mon on legþ oððe mid wordum oððe mid dǽdum *the impatient cannot bear any annoyance that is put upon them either by word or deed*; impatientes ab aliis illata non tolerant, Past. 40, 4; Swt. 293, 16. Ðeáh hié nán mann mid láðe ne grēte hié sēceaþ ða ðe hié fleóþ *though no man attacks them, they seek those that flee from them*; iracundi se declinantes insequuntur, 293, 19. Hié hit tō nánum fácne ne tō nánum láðe nǽfdon ðætte ða earman wífmen hié swá tintredon *nec tamen miseriæ hominum pressura temporum deputata est*, Ors. 1, 10; Swt. 48, 13. Wið ðæm ðe hié of ðæm londe mósten búton láðe *ut tutum et incolumem exercitum a locorum periculo liberaret*, 6, 32; Swt. 286, 28. Mid lufe ge mid láðe *with what is pleasant and what is unpleasant*, Blickl. Homl. 45, 8. Nis hit gód ðæt hié sīen on ðam láðe *it is not good that they be in that durance [the fiery furnace]*, Cd. 193; Th. 243, 2; Dan. 430. Ne dó ic him ná láð *I will not harm them*, Gen. 18, 30 : Nar. 16, 22. Eálá hwæt ðú mē mycel yfel and láð dēst mid ðínre ærninge *O quam magnum væ facis mihi sic equitando*, Bd. 5, 6; S. 619, 14 : Cd. 21; Th. 25, 11; Gen. 392. Wið eal ðæt láð ðe intó land fare *against all the harm that comes into the land*, Lchdm. i. 388, 14. Ðonne hié láð gedóþ hié sculon lufe wyrcean *when they do evil, they must act so as to regain love*, Cd. 29; Th. 39, 11; Gen. 624. Ðú míne sáwle of deáþes láðum wiðlǽddest *eripuisti animam meam de morte*, Ps. Th. 55, 11. [*O. Sax. O. Frs.* lēð : *O. H. Ger.* leid *dolor, moeror, injuria, malum, execratio* : *Ger.* leid.]

láð; *adj.* I. *Causing hate, evil, injury, annoyance; hateful, hated, loathed, loth, displeasing, injurious, grievous* :—Láth *ingratus*, Ep. Gl. 12 b, 16. Laath *invisus*, 12 f, 5. Ðá wæs ic swíðe onscúniende and mē láð wæs *multum detestatus sum*, Bd. 5, 12; S. 630, 32. Ðeáh hit láð wǽre, Chr. 1006; Erl. 141, 7. Him wæs láð tō ámyrrene his ágenne folgaþ, 1048; Erl. 178, 11. Fram allum mannum hē biþ láð *he shall be hated of all men*, Lchdm. iii. 162, 19. Se wæs láð Gode, on hete heofoncyninges, Cd. 30; Th. 40, 31; Gen. 647. Swá láð wæs Pēna folc Scipian *so hateful were the Carthaginians to Scipio*, Ors. 4, 10; Swt. 198, 15. Mánswara láð leóda gehwam, Exon. 10 b; Th. 12, 31; Cri. 194. Leófost on lífe láð biþ ðænne *what is dearest in this life, shall then be hateful*, Dóm. L. 16, 243. Láð biþ ǽghwǽr wineleás hæle *he is everywhere unloved, a friendless man*, Exon. 87 b; Th. 329, 9; Vy. 31. Wæs ðæt gewinn tō láð and longsum *that strife was too grievous and long*, Beo. Th. 268; B. 134. Hē mē álýsde of láðan grine huntum unholdum *ipse liberavit me de laqueo venantium*, Ps. 90, 3. Lǽdan on láðne síþ *to lead to hell*, Exon. 118 b; Th. 455, 20; Hy. 4, 52. Ðec gelegdon on láðne bend *they put thee into grievous captivity*, Cd. 225; Th. 298, 27; Sat. 539. Ða fuglas ús nǽnige láðe ne yfle ne

wǽron *aves non nobis perniciem ferentes*, Nar. 16, 18. Ða rihtwísan sint láðe and forþrycte *the righteous are hated and oppressed*, Bt. 3, 4 ; Fox 6, 23. Hé hæfde fela ǽhta ðe him wǽron láðe tó forlǽtenne *he had many possessions that he was loth to leave*, Basil admn. 9 ; Norm. 56, 7. Gě habbaþ ús gedón láðe Pharaone, Ex. 5, 21. Láð gewidru *grievous storms*, Beo. Th. 2754 ; B. 1375. Næs ic him láðra ówihte ðonne his bearna hwylc *I was not a whit less dear to him than any of his children*, 4856 ; B. 2432. Ic ǎ ne geseah láðran landscipe *never saw I scene more hateful*, Cd. 19 ; Th. 24. 11 ; Gen. 376. Sege ðínum leódum miccle láðre spell *tell to thy people a tale that will please much less*, Byrht. Th. 133, 15 ; By. 50. Gnornsorga mǽst wyrda láðost *greatest of griefs, most grievous of fates*, Elen. Kmbl. 1953 ; El. 978. Ðǽr ðě láðast biþ, Exon. 41 a ; Th. 137, 17 ; Gú. 560. Áne ða mǽstan synne and Gode þa láðustan *one of the greatest sins and most displeasing to God*, Ex. 32, 21. II. *bearing hate to another, hostile, malign, inimical :*—Ne leóf ne láð *nor friend nor foe*, Beo. Th. 1026 ; B. 511. Láð wið láðum *foe with foe*, 884 ; B. 440. Láðe cyrmdon *the foes shouted*, Cd. 166 ; Th. 207, 3 ; Exod. 461. Wið láðra lygesearwum *against false wiles of foes*, Exon. 19 a ; Th. 48, 23 ; Cri. 776 : Judth. 12 ; Thw. 25, 38 ; Jud. 304. Ðæt on land Dena láðra nǽnig sceððan meahte, Beo. Th. 490 ; B. 242. Láðan fingrum *with hostile fingers*, 3015 ; B. 1505. Láðum eágan, Cd. 151 ; Th. 189, 3 ; Exod. 179. Láðum wordum, Exon. 28 a ; Th. 84, 17 ; Cri. 1376. Álýs mě fram láðum *libera me a persequentibus*, Ps. Th. 141, 7. Ðæt hé ðě ne forlǽte láðum tó handa, Dóm. L. 30, 29. Hé ne lǽteþ míne fét láðe hrèran, Ps. Th. 65, 8. [O. Sax. lèð : *Icel.* leiðr : O. H. Ger. leid *exosus, odiosus, invisus, tristis, malignus, ingratus* : Ger. leid.] v. þurh-láð.

láð-bite, es ; *m. A wound :*—Blód ætsprang láðbite líces, Beo. Th. 2248 ; B. 1122.

láðe ; *adv. With hatred* or *enmity, in detestation :*—Hió mě lytle læs láðe woldon ðisses eorþweges ende gescrífan *paulominus consummaverunt me in terra*, Ps. Th. 118, 87. Ðis ungesǽlige geár gyt tó-dæg láðe wunaþ *this miserable year still continues in detestation to-day*, Bd. 3, 1 ; S. 523, 33. [O. H. Ger. leido *invise, odiose*.]

láðettan ; *p.* te *To be odious or hateful, be hated, be hostile, to abominate, hate :*—Láðetteþ *detestantur*, Wrt. Voc. ii. 26, 8. Man láðette tó swýðe ðæt man scolde lufian *people hated too much what they ought to love*, Wulfst. 168, 13. Uncer láðette ǽgðer óðer óðrum ðe hé hít óðrum ne sǽde *each of us hated the other, though he did not say so to the other*, Shrn. 39, 22. Ðás gyltas ne mǽgon úre sáwla ofsleán ac hí mágon hí áwlǽtan and Gode láðettan *these sins cannot destroy our souls, but they can pollute them and be hateful to God*, Homl. Th. ii. 590, 29. Hundas beorcynde gesihþ oððe him láðhetan *if a man sees dogs barking, or be hostile to him*, Lchdm. iii. 200, 26. Olfendas geseón and fram him gesihþ láðhetan *to see camels and if he sees himself to be hated by them*, 31. [O. H. Ger. leidezan, leidezzan *detestari, abominari, aversari, inhorrescere*, Grff. 2, 177.] v. láðian.

láð-geníðla, an ; *m. A foe, enemy*, Exon. 56 b ; Th. 201, 3 ; Ph. 50 : 69 a ; Th. 256, 15 ; Jul. 232.

láð-geteóna, an ; *m. One who does evil, an enemy*, Beo. Th. 1953 ; B. 974 : 1123 ; B. 559.

láð-gewinna, an ; *m. A hated opponent, an enemy*, Exon. 104 b ; Th. 397, 33 ; Rä. 16, 29.

láðian ; *p.* ode *To invite, call, call upon :*—Hwílum ic rincas láðige tó wíne *at times I invite men to wine*, Exon. 104 a ; Th. 395, 32 ; Rä. 15, 16. Ðyder ðe unc láðaþ anð cěgþ uncer Drihten *whither our Lord invites and calls us*, Blickl. Homl. 187, 26 : Cd. 226 ; Th. 301, 29 ; Sat. 589. Loth hig·láðode geornlíce *Lot compulit illos oppido*, Gen. 19, 3. Hé hí láðede ðæt hí onféngon ðam gerýno Cristes geleáfan *ad fidei suscipiendæ sacramentum invitaret*, Bd. 3, 5 ; S. 526, 31. Mě of weorulde cígde and láðode *me de sæculo evocare dignatus est*, 4, 3 ; S. 568, 18. Heora ða leásan godas hié him láðodon on fultum *they called upon their false gods to help them*, Blickl. Homl. 201, 31. Hé hêht hám láðian Mellitum and Iustum *revocavit Mellitum et Justum*, Bd. 2, 6 ; S. 508, 33. Ðá hêt hé Willfriþ tó ðam sinoþe láðian *vocari jussit Vilfridum*, 5, 19 ; S. 639, 35. Hé sende his þeówan tó láðigenne mancynn tó ðære ècan feorme, Homl. Th. ii. 372, 5. [Goth. laþôn : O. Sax. lathian : O. Frs. lathia : Icel. laða : O. H. Ger. ladôn : Ger. laden.]

láðian ; *p.* ode *To be hateful* or *loathed :*—Heora fela wǽron mid olfendes hǽrum tó líce gescrýdde and ðǽr láðode sóftnys *many of them were clad with camel's hair next to the body, and there softness was hateful*, Homl. Th. ii. 506, 24. Hió ðǽm folce láðode *she was hateful to the people*, Ors. 3, 11 ; Swt. 148, 15. [þe schal laðin his luue, Jul, 16, 6 : þat te schal laði þi lif, H. M. 9, 2 : him loðie, A. R. 324, 27 : us lotheth þe lyf, Piers P. prol. 155 : O. Sax. lèðôn : O. H. Ger. leidôn.] v. lǽðan, láðettan.

láð-leás ; *adj. Innocent, harmless, free from harm* or *annoyance :*—Gif hé láðleás [MS. H. ladleas] beó sêce swylcne hláford swylcne hé wille forðý ðe·ic an ðæt ǽlc ðara ðe láðleás [MS. H. ladleas] beó folgie swylcum hláforde swylcum hé wille, L. Ath. iv. 1 ; Th. 1, 220, 24–222, 1. Láðleáse *immunes*, Wrt. Voc. ii. 43, 68.

láð-líc ; *adj. Hateful, loathsome, disgusting, unpleasant, detestable,*

abominable, horrible :—Láðlíc *detestabile*, Wrt. Voc. ii. 26, 5. Láðlíc biþ ðæs hreófsian líc mid menigfealdum springum *the leper's body is loathsome with manifold ulcers*, Homl. Th. i. 122, 21. Ðæt is láðlíc líf ðæt hí swá maciaþ *it is an abominable life that they do so*, L. I. P. 14 ; Th. ii. 322, 26 : Exon. 266 ; Th. 78, 19 ; Gri. 1276. Þincþ his neáwist láðlíco and unfæger *his [the dead man's] nearness seems disgusting and displeasing*, Blickl. Homl. 111, 30. Nis ðér ne se láðlíca cyle ne láðlíc storm, Dóm. L. 16, 259, 262 ; Soul Kmbl. 306 ; Seel. 157. Hine mon ðér láðlíce deáþe ácwealde *eum detestanda omnibus morte interfecit*, Bd. 3, 14 ; S. 539, 46 : 541, 10. Láðlíc wíte, Elen. Kmbl. 1038 ; El. 520. Hér æfter sint lungenádla láðlícu tácn *here follow the unpleasant symptoms of lung disease*, L. M. 2, 51 ; Lchdm. ii. 264, 9. Ða láðlecan *obscena*, Wrt. Voc. ii. 63, 12. [*Prompt. Parv.* lothli *abominabilis ;* O. Sax. lèð-líc : *Icel.* leiði-ligr : O. H. Ger. leid-líh *detestabilis, execrabilis, exosus, horrendus.*]

láð-líce ; *adv. Hatefully, detestably, horribly, unpleasantly :*—Ongunnon láðlíce rýnan *they began to roar horribly*, Bt. Met. Fox 26, 166 ; Met. 26, 83. Wit gewídost lifdon láðlícost *we should live as far apart as possible, and in most grievous sort*, Exon. 115 a ; Th. 442, 17 ; Kl. 14.

láð-scipe, es ; *m. A painful condition, calamity :*—Abram wolde Loth álynnan of láðscipe [*when Lot was carried off captive*], Cd. 95 ; Th. 123, 20 ; Gen. 2048.

láð-searu *a fell device*, Cd. 195 ; Th. 243, 14 ; Dan. 436.

láð-síþ *a painful journey*, Cd. 144 ; Th. 180, 12 ; Exod. 44.

láð-spell, es ; *n. A painful, grievous story :*—Hié ealle ðér ofslógon búton ánum se ðæt láðspel æt hám gebodade *omnes ibidem trucidati sunt ; uno tantum ad enunciandam cladem reservato*, Ors. 2, 4 ; Swt. 72, 19 : Andr. Kmbl. 2160 ; An. 1080 : Exon. 52 b ; Th. 182, 29 ; Gú. 1317.

láð-treów *a fell, harmful tree* [*the tree of knowledge*], Cd. 30 ; Th. 40, 25 ; Gen. 644.

laðu, v. freónd-, neód-, word-laðu.

laðung, e ; *f. A calling, invitation ;* vocatio, Past. 52, 4 ; Swt. 405, 23. [O. H. Ger. ladunga *vocatio, evocatio, ecclesia.*] v. ge-laðung.

láð-wende ; *adj. Evilly disposed, evil, hostile, malignant :*—Wæs láðwendo ongan wíð Sarran winnan *Hagar was evilly disposed and began to strive with Sarah*, Cd. 102 ; Th. 135, 7 ; Gen. 2239. Gyf mon mête ðæt hé gæt geseó ðonne mæg hé wênan ðæs láðwendan feóndes him on neáwyste *if a man dream that he sees goats then may he expect the devil in his neighbourhood*, Lchdm. iii. 176, 3. Láðwende here [*the fallen angels*], Cd. 4 ; Th. 5, 7 ; Gen. 68. Ludon láðwende rěðe wæstme *fruits evil and dire sprang forth*, 47 ; Th. 60, 29 ; Gen. 989. Láðwende men *evil men*, Exon. 31 a ; Th. 97, 24 ; Cri. 1595. [Cf. O. H. Ger. leid-wentige *calamitas*, Grff. 1, 763.]

láðwende-mód ; *adj. Evilly* or *hostilely disposed*, Cd. 23 ; Th. 29, 11 ; Gen. 448.

láð-weorc, es ; *n. An evil work, work that is hateful to another :*—Leornedan láðweorc Gode, Ps. Th. 105, 26. [O. Sax. lèð-werk : and cf. O. H. Ger. leid-tát *supplicium.*]

latian ; *p.* ode *To be slow, to linger, loiter, delay :*—Ic latige on sumere stówe *moror*, Ælfc. Gr. 25 ; Som. 27, 14. Hwí latast ðú swá lange ðæt ðú ðě lǽce ne cýðst *why dost thou delay so long to show thyself to the leech?* Dóm. L. 6, 66. Lataþ *tardat*, Wrt. Voc. ii. 138, 48. Deáþ ne lattaþ *mors non tardat*, Rtl. 11, 7. Eall líchoma hefegaþ and latiaþ ða fét *all the body grows heavy, and the feet are sluggish*, L. M. 2, 25 ; Lchdm. ii. 216, 23. Ic latode *distuli*, Cant. M. ad f. 27. Ðeáh ðe hé ðá get latode on ðissum líchomlícum gebyrde *though his birth was still deferred*, Blickl. Homl. 167, 7. Hit is swytol ðæt man ðæs latode ealles tó lange, Wulfst. 168, 2. Ne lata ðú *ne cuncteris*, Wrt. Voc. ii. 60, 34. Ne yld ðú ł ne lata ðú *non tardaveris*, Ps. Spl. 39, 24 : Ps. Th. 69, 7 : Exon. 13 a ; Th. 23, 23 ; Cri. 373. Smeáge húru georne gehwá hine sylfne and ðæs ná ne latige tó lange *at any rate let every one examine himself, and not delay in that too long*, Swt. A. S. Rdr. 111, 192. Nó latiendum *non cunctante*, Wrt. Voc. ii. 61, 22. [*Icel.* lata *to be slow :* O. H. Ger. lazôn *tardare.*]

latta. v. lætt.

lát-téh, -teów. v. lád-teáh, -teów.

latu. v. word-latu.

látwa. v. lád-teów.

laur, lawer, es ; *m. Laurel, bay :*—Laures croppan, seáw, blêda, leáf, Lchdm. ii. 20, 17 : 226, 2 : 228, 25 : 230, 3. Mid lawere gebeágod *crowned with laurel*, Blickl. Homl. 187, 27.

laur-beám, es ; *m. Laurel :*—Laurbeám *daphnis vel laurus*, Ælfc. Gl. 45 ; Som. 64, 110 ; Wrt. Voc. 32, 45. Lauwer [lawer] beám *laurus*, Wrt. Voc. 79, 78. Laurbeáme gelíce *similes lauro*, Nar. 36, 30.

laur-berige, an ; *f. A berry of the laurel :*—Lauberigan, Lchdm. iii. 122, 22 : 6, 16. Laurberigie, 106, 1. Lauwinberigean, 136, 28. Lauwerberian, i. 376, 6.

laur-treów, es, *n. Laurel :*—Laurtreówes leáf, Lchdm. iii. 88, 10. Of lawertreówe, i. 174, 11.

lawer, laber *laver* [a plant. v. E. D. S. Plant Names], Lchdm. i. 254, 1, 2.

láwerce, an ; *f. A lark, laverock :*—Láuerce *alauda*, Ælfc. Gl. 37 ;

Som. 62, 127; Wrt. Voc. 29, 22. Lâwerce *tilaris*, Wrt. Voc. 62, 42: *laude*, Wrt. Voc. ii. 50, 49. Lǽwerce *caradrion*, 13, 46. Lâuricae *allauda*, 100, 9. Lâurice *laudæ*, 112, 26. Lâfercan beorh *occurs several times in charters.* v. Cod. Dip. Kmbl. vi. 307. Cf. *O. H. Ger.* Lêrichanvelt. [*Icel.* lævirki: *O. H. Ger.* lêrahha *caradrius, caradrion, aloda, laudula*: *M. H. Ger.* lêrche: *Ger.* lerche.]

leác, lǽc, lêc, es; *n.* Generally, *a garden herb* [as in leác-tún, &c.], *an alliaceous plant* [v. compounds], *a leek* :—Ðis lêc *hoc cepe*: ðis leác *hoc porrum*, Ælfc. Gr. 13; Som. 16, 32, 35. Leác *ambila*, Wrt. Voc. 284, 24: Wrt. Voc. ii. 8, 49. Lâec, Ep. Gl. 2 d, 8. Leáces heáfod *cartilago*, 17, 40. Ðæt greáta crȃuleác; nim ðes leáces heáfda, Lchdm. i. 376, 3. On ðære mycele ðe leáces, Herb. 49; Lchdm. i. 152, 16. Leáces sǽd, Lchdm. i. 104, 26. Gebeát ðæt leác [*garlic*]. L. M. 2, 32; Lchdm. ii. 234, 24: Leác, i, 32; Lchdm. ii. 78, 7: iii. 16, 10. Nim forcorfen leác and cnuca hyt, 102, 13. v. brâde-, crâw-, crop-, enne-[ynne-], gâr-, hol-, hwîte-, por-, secg-leác. [*Icel.* laukr; *m.*: *O. H. Ger.* louch *cepa, porrum.*]

leác-cærse, an; *f.* 'A cress with an onion-like smell,' alliaria officinalis' E. D. S. Plant Names. Cockayne says 'erysimum alliaria,' Lchdm. ii. 318, 7: 320, 3. In Wrt. Voc. ii. 60, 40, leáccærse *id est* tûncærse glosses *nasturcium*.

leác-, leáh-tric, es; *m. A lettuce* :—Leáhtric *lactuca*,Wrt.Voc. 67,47: ii. 50,57.*Lactucas* ðæt is leáhtric, L.M. 2,16; Lchdm. ii. 194,6: 3, 8; Lchdm. ii. 312,20. Ðâ geseah heó ǽnne leáhtric ðâ lyste hî ðæs and hine genam and forgeat ðæt heó hine mid Cristes rôdetâcne gebletsode *then she saw a lettuce and had a longing for it, and took it and forgot to bless it with the sign of the cross*, iii. 336, col. 1. Wudu-lêctric *lactuca silvatica*, Herb. 31; Lchdm. i. 128, 6, 8.

leác-trog, -troc, es; *m. A bunch of berries* :— Leáctrogas *corimbos*, Wrt. Voc. ii. 14, 78: 104, 70. Leáctrocas *corimbus*, Ep. Gl. 8 f, 34. Cockayne, Lchdm. iii. 336,col. 1, puts this with the preceding word.

leác-, leáh-, lêh- tûn, es; *m. A garden of herbs, a kitchen-garden* :—Leáhtûn *ortus olerum*, Wrt. Voc. 285, 76: ii. 64, 9 Ðêr wæs lêhtûn *ubi erat hortus*, Jn. Skt. Lind. 18, 1: 19, 41. Nân man on ðysne dæg wyrte in lêhtûne ne fatige, Wulfst. 227, 8: 231, 18. Monn sende in lêhtûne his *homo misit in hortum suum*, Lk. Skt. Lind. 13, 19. [*Misc.* leyhtun *a garden.*] Cf. wyrt-tûn.

leáctûn-weard, es; *m. A gardener* :—Lêctûnweard *olitor*, Ælfc. Gl. 31; Som. 61, 82; Wrt. Voc. 27, 12. [*Misc.* leyhtunward *a gardener.*]

leác-weard, es; *m. A gardener* :— Leácweard *holitor*, Wrt. Voc. ii. 42, 57. Lêcueard *hortulanus*, Jn. Skt. Lind. 20, 15. Lêcword, p. 8, 4. Cf. wyrt-weard.

LEÁD, es; *n. Lead* :—Leád *plumbum*, Wrt.Voc. 85, 11. Ðæt leád is hefigre ðonne ǽnig ôðer andweorc *plumbum ceteris metallis est gravius*, Past. 37, 3; Swt. 269, 7. Írenes and leádes ða men on ðâm londum wædliaþ and goldes genihtsumiaþ *ferro et plumbo egent, auro habundant*, Nar. 31, 4: Bd. 1, 1; S. 473, 23. Beworhte mid leáde, Homl. Skt. 3, 532. Ðû herast ðone mancgere ðe begytt gold mid leáde, Homl. Th. i. 254, 26.

leáden; *adj. Leaden* :— Leáden *plumbeus*, Ælfc. Gr. 5; Som. 4, 60. Sî ðæt âlfæt îsen oððe ǽren leáden oððe lǽmen, L. Ath. iv. 7; Th. i. 226, 15: Nar. 46, 3. Mid leádenum swipum swingan, Homl. Th. i. 426, 13.

leád-gedelf, es; *n. A lead-mine* :—Eft in leádgedelf; of leádgedelfe, Cod. Dip. Kmbl. iii. 401, 7.

leád-stæf, es; *m. A scourge* [cf. *last entry under* leáden] :—Leádstafum *mastigiis*, Wrt. Voc. ii. 54, 75.

LEÁF, es; *n. A* leaf *of a tree, of a book, a shoot* :—Leáf hys ne fylþ *folium ejus non defluet*, Ps. Spl.; his leáf and his blǽda ne fealwiaþ ne ne seariaþ *folium ejus non decidet*, Ps. Th. 1, 4. Leáf *antes*, Wrt. Voc. ii. 9, 16. Leáf *folia*, Mt. Kmbl. 21, 19: 24, 32: Mk. Skt. 13, 28: Bt. Met. Fox 11, 114; Met. 11, 57. Man scóf ðara bóca leáf ðe of Hibernia côman and ða sceafþan dyde on wæter *rasa folia codicum qui de Hibernia fuerant, et ipsam rasuram aquæ immissam*, Bd. 1, 1; S. 474, 37. Mid grênum leáfum *virentibus foliis*, Gen. 8, 11. [*Goth.* laufs; *m.*: *O. Sax.* lôf: *O.Frs.* lâf: *Icel.* lauf: *O. H. Ger.* laub *folium, frons*: *Ger.* laub.] DER gold-leáf.

LEÁF, e; *f.* LEAVE, *permission, license* :—Leáf *licentia*, Ælfc. Gr. 33; Som. 37, 17. Lôciaþ ðæt ðiós eówru leáf ne weorðe ôðrum monnum tô biswice *videte, ne forte hæc licentia vestra offendiculum fiat infirmis*, Past. 59, 6; Swt. 451, 32. Gif him lîf seald wǽre, Bd. 1, 23; S. 486, 8, note. Ða seofan cnihtas ðe ðe ðînre leáfa lyfedan bûton ehtnisse *the seven youths that by your leave lived without persecution*, Homl. Skt. 4, 255. Se Englisca be fulre leáfe hine werige *Anglicus plena licentia defendat se*, L. Wil. ii. 2; Th. i. 489,13. Hê sæt on ðam biscoprîce ðe se cyng him ǽr geunnan hæfde be his fulre leáfe, Chr. 1048; Erl. 177, 27. Be ðæs cynges lǽfe and rǽda, 1043; Erl. 169, 25. Bûtan ðæs cyninges leáfe and his witena, 901; Erl. 96, 28. Bûton ðæs bisceopes leáfe *absque permissu episcopi*, Bd. 4, 5; S. 573, 4. Ða ðe willaþ grîpan on leáfe *for hiera gîtsunge* hié ðôþ him tô leáfe ðone cwide ðe sanctus Paulus cwæþ *qui præesse concupiscunt, ad usum suæ libidinis instrumentum apostolici sermonis arripiunt, quo ait*, Past. 8, 1; Swt. 53, 7. Hê

begeat ðâ leáfe ðæt hê of ðam lande môste *he got leave to go out of the country*, Homl. Skt. 5, 328. Hî habbaþ leáf [Cott. MS. leáfe) yfel tô dônne *they have leave to do evil*, Bt. 38, 4; Fox 204. 13. Hæbbe hê fulle leáfe swâ tô dônne, L. Wil. ii. 1; Th. i. 489, 8. Ðâ ðâ Aulixes leáfe hæfde ðæt hê ðonan môste, Bt. Met. Fox 26, 42; Met. 26, 21. Mê ða leáfe forgyf tô geopenienne ðone ingang ðînre hâlgan cyrcan, Glostr. Frag. 106, 13. Leáfe syllan *to give leave*, Gen. 50, 5: Lchdm. iii. 424, 27. Hî bǽdon lǽfa æt mê *they asked leave of me*, Guthl. 14; Gdwin 62, 13. [Cf. *Icel.* leyfi, *leave*: *O. Sax.* or-lôf: *Icel.* or-lof: *O. H. Ger.* urlaup *licentia, permissus.*]

leáfa, an; *m. Belief, faith* :—Hû mæg se leáfa [other MS. geleáfa] beón forþgenge gif seó lâr and ða lâreówas âteoriaþ *how can belief be prosperous if teaching and teachers fail*, Ælfc. Gr. pref; Som. 1, 37. Leáfa *fides*, Mt. Kmbl. Lind. 8, 10: 15, 28. Leáfo, 21, 21. [*O. H. Ger.* laubo.] v. ge-leáfa.

leáfa [?], an; *m. Leave* :—Be his leáfan ârǽrde mynster *with his leave raised a monastery*, Homl. Skt. 6, 145.

leáf-full; *adj. Believing, faithful* :—Leáffull *fidelis*, Mt. Kmbl. Lind. 25, 21: Jn. Skt. Lind. 20, 27. Ic cȳðe on ðissan gewrite eallum leáffullum mannum hwet ic gerǽd habbe wið mîne arcebiscópes, Chart. Th. 347, 26. God cwæþ tô Moysen ðæt hê wolde cuman and hine ætforan ðam folce gesprecan ðæt hî ðȳ leáffulran wǽron *God said to Moses that he would come and talk with him before the people, that they might be the more believing* [v. Exod. 19, 9], Homl. Th. ii. 196, 18.

leáf-helmig; *adj. Having a leafy top*; frondicoma Germ. 390.

leáf-hlystend, es; *m. A catechumen* :—[Ge ?] leáfhlestend *catechumenus*, Hpt. Gl. 457, 12. v. geleáfhlystend.

leáf-, lêf-, lȳf-ness; e; *f. Leave, permission, licence* :—Gif him lêfnys seald wǽre *if leave had been given him*, Bd. 1, 23; S. 486, 8. Lȳfnes *licentia*, 4, 18; S. 586, 34: 2, 1; S. 501, 32: 5, 19; S. 640, 10. Bûtan heora leáda geþafunge and leáfnysse *absque suorum consensu ac licentia*, 2, 2; S. 502, 35. Bûtan kyninges lêfnesse [MSS. B. H. leáfe], L. Alf. pol. 8; Th. i. 66, 16. Mid his lêfnysse *accepta ab eo licentia*, Bd. 1, 25; S. 486, 11. Mid Ebrinum lȳfnysse, 4, 1; S. 564, 44. Heó his leáfnysse hæfde ðæt . . . *she would have his permission to* . . . 1, 25; S. 486, 34. Nymðe þurh leáfnysse his âgenes abbudes *nisi per demissionem proprii abbatis*, 4, 5; S. 572, 38. Hî mâran lêfnysse onfêngon tô lǽranne *majorem prædicandi licentiam acciperent*, 1, 26; S. 488, 5. Lȳfnesse, 5, 11; S. 625, 30. Lȳfnesse sealde ðæt . . . *gave leave to* . . . 1, 25; S. 487, 20. Him lȳfnesse sealde tô farene, 4, 1; S. 564, 34. Heó freó lêfnesse sealdon, 2, 5; S. 507, 10. Forgeaf him lȳfnesse, 4, 22; S. 592, 9. v. leáf.

leáf-scead, es; *n. A place made shady by leaves* or *foliage*, Exon. 58 b; Th. 212, 4; Ph. 205.

leáf-wyrm, es; *m. A canker, caterpillar* :— Hê sealde leáfwyrme (MS. C. treowyrme) wæstm heora *he gave their increase unto the caterpillar* (A.V.), Ps. Spl. 77, 51.

leágung, e; *f. Lying* :—Ðȳ lǽs on mê mǽge îdel spellung oððe scondlîc leágung [leásung?] beón gestǽled *ne aut fabulæ aut turpi mendacio dignus efficiar*, Nar. 2, 21.

leáh; *g.* leás; *m. A lea, meadow, open space, untilled land* :—Ðanne is ðêr se leáh ðe man ðæt lond mid friþe haldan scæl an eásthealfe sió ealdæ strǽt &c. *now there is the open space* (?) *by which the land is protected; on the east side the old road &c.*, Cod. Dip. Kmbl. ii. 71, 20. Ðonne geúðe ic Ælfwine and Beorhtulfe ðæs leás and ðæs hammes be norþan ðære lytlan dîc *I granted Alfwine and Beorhtulf the meadow and the enclosure to the north of the little dike*, 249, 33. Æt ðam leá ufeweardan, 36. Tô ðam leá . . . on eásteweardan ðam leá . . . tô fealuwes leá ðæt on fealuwes leá . . . fram fealuwes leá, 250, 2, 16, 29, 32. Æt Eardulfes leá . . . tô Aþelwoldes leá, Chart. Th. 291, 19, 22. Ðæt intô Eardulfes leá; of ðan leá, ðæt tô ðære greátan dîc, 292, 4. Þurh ðone leá tô ðam miclan hæslwride, Cod. Dip. Kmbl. 250, 34. Betweox ða twegen leás, 21. Lytle leás *amarcas*, Wrt. Voc. ii. 10, 14. v. next word.

leáh; *g.* leáge; *f. A lea*, as a termination of local names -*leigh*, -*ley*, -*ly*; it occurs frequently in the charters :—Hrîdra leáh *campus armentorum*, Cod. Dip. Kmbl. i. 232, 21. Ðis syndon ða landgemǽro tô madanleáge (cf. 120, 28 madan lieg) ǽrest on witena leáge, iii. 121, 13-4. On mapodorleáge; be eástan ðære leáge . . . eft on Heortleáge westeweardre, 407, 7, 8, 13. On hemlêclêge, 437, 4. Ðonne on ðæt (ða ?) lêge . . . ðonon on gerihte on riscleáge, 10, 24-5. Of ðam clyfe on heán leáge; ðæt on lungan leáge . . ðonne on Swonleáge, 48, 6, 7. On Wytleáhe; of Wytleáge, 14, 6. Of ðæt lêge, 406, 27. [*Piers P.* bad hym eryen his leyes, 7, 5: *Promp. Parv.* lay, londe not telyd, see note 2, p. 285; cf. *Pol. Songs Wrt.* mi lond leye liþ and leorneþ to slepe, 152, 10: ley *tere freche*, Wrt. Voc. 153, 4. *O.H.Ger.* v. Grmm D. M. 1202, has lôh; *m. lucus*, which occurs also in local names, Hohenlohe, Grff. 2, 127-8: the same suffix is found in Water-loo.] v. preceding word.

leáh; *g.* leáge; *f. Lye, a mixture of ashes and water* :—Lǽg *læxiva*, Wrt. Voc. ii. 112, 28. Leáh *lexiva* 50, 50: 'lixa, 52, 13. On ðare bisteran lêge, L. Med. Ex. Quad. 9, 14; Lchdm. i. 364, 5. Ofergeót ða ascen mide, mac swâ tô lêga, 378, 11. Wyrc him leáge of ellenahsan, L. M. 3, 47; Lchdm. ii. 338, 25. [*Ayenb.* we byeþ alle ywesse of onelepi

leȝe, 145, 22: *Prompt. Parv.* ley for waschynge *lixivium*, 294, see note: *O.H. Ger.* louga *lixivia.*]

leahan. v. leán.

leáh-hrycg, es; *m.* *The ridge of a lea:*—Tó ðæm ealdan lǽghrycge, Cod. Dip. Kmbl. iii. 437, 17.

leáh-mealt-wurt *some kind of wort:*—Lēhmealtwurt *lexinum* (? *lixivum*, cf. *lixivum mustum* the wine that runs out of the grapes before they are pressed), Ælfc. Gl. 33; Som. 62, 23; Wrt. Voc. 34, 6.

leahter, es; *m.* **I.** *a moral defect, a crime, fault, offence, sin, vice, disgraceful* or *shameful act, reproach, opprobrium, blame, disgrace:*—Leahter *crimen,* Ælfc. Gr. 9; Som. 9, 29. Hosp, lehter *probrum,* Wrt. Voc. ii. 67, 35. Ǽghwilc mennisc leahter on ðæm eádigan Sancte Johanne cennendum gestilled *wæs every human vice was stilled in the blessed St. John's parents,* Blickl. Homl. 163, 15, 1. Būtan leahtre *sine crimine,* Ælfc. Gr. 47; Som. 48, 3: Mt. Kmbl. 12, 5. Hié eodan on eallum Drihtnes bebodum būtan leahtre *they walked in all the commandments of the Lord blameless,* Blickl. Homl. 161, 31. Būtan ǽlcon womme and swá clǽne fram ǽlcon leahtre *stainless and pure from every vice,* Nicod. 28; Thw. 16, 31. *Vitia* ðæt synd lehtras on lēdensprǽce, Ælfc. Gr. 50; Som. 51, 53. Swā sceal wīsdómes bodung healdan manna heortan wið brosnunge fūlra leahtra, Homl. Th. ii. 536, 21. Ic mē synnum and leahtrum þeódde *vitiorum implicamentis solebam servire,* Bd. 3, 13; S. 538, 30. Hē unscyldig and būtan leahtrum wæs clǽne gemēted *absque crimine inventus est,* 5, 19; S. 639, 30. Bysmrian leahtrum belecgan *to revile and load with opprobrium,* Andr. Kmbl. 2591; An. 1297. Hē begann tó lufienne leahtras tó swīðe *he began to love vices too much,* Ælfc. T. Grn. 17, 13. Leahtras *noxas* (cf. gylt *noxam,* 50), Wrt. Voc. ii. 61, 41. Ýdel byþ seó lār ðe ne gehǽlþ ðǽre sáwle leahtras (v. II.) and unþeáwas, Homl. Th. i. 60, 35. Wið ða heáfodlícan leahtras *against the deadly sins,* Blickl. Homl. 37, 3. **II.** *a bodily defect, disease, disorder, hurt, malady:*—Hyt áfeormaþ ðone leahtor ðegrēcas hostopyturas hâtaþ, ðæt ys, scurf ðæs heáfdes, Herb. 184, 4; Lchdm. i. 322, 15. Hyt ealne ðone leahtor genimeþ *it takes away all the malady,* 13, 3; Lchdm. i. 106, 2. Heó ðone leahtor [*cancer*] gehǽlan mæg, 32, 3; Lchdm. i. 130, 14. Leahtras *noxas* (cf. dare *noxam,* 64], Wrt. Voc. ii. 61, 41. Wið leahtras ðæs mūþes *for blotches of the mouth,* Herb. 145, 3; Lchdm. i. 268, 13. Wið misenlíce leahtras ðæs bæcþearmas, 165, 3; Lchdm. i. 294, 15. DER. syn-leahter.

leahter-cwide, es; *m. Opprobrious, insulting, injurious speech, blasphemy:*—Æfter leahtorcwidum, Exon. 68 b; Th. 254, 18; Jul. 199.

leahter-full; *adj. Vicious, seductive:*—Leahterfulle þeáwas *vitiosos mores,* Bd. 3, 13: S. 538, 32. Leahte[r]fulle *decipulosa* i. *inlecibrosa,* Wrt. Voc. ii. 138, 1.

leahter-leás; *adj. Faultless, free from defect, free from sin, innocent:*—Forðon nis nán man leahtorleás *quoniam nemo vitiorum expers est,* L. Ecg. P. i. 9; Th. ii. 176, 16. Ðonne ðū ōderne man tǽle, ðone geþenc ðū ðæt nán man ne byþ leahterleás, Prov. Kmbl. 3. Ic ða meorde wāt leahtorleáse *I know the reward to be faultless,* Exon. 48 b; Th. 167, 14; Gū. 1060. Hié freóndrǽdenne fæste gelǽston leahtorleáse *firmly should they friendship maintain, free from offence,* Elen. Kmbl. 2415; El. 1209.

leahter-líce; *adv. Viciously, noisomely :*— Ðæt deáde flǽsc rotaþ leahtorlíce ðonne se deádlíca líchama þeówaþ gálnysse *the dead flesh rots noisomely when the mortal body is a slave to lust,* Homl. Th. i. 118, 13.

leahter-wyrþe. v. un-leahterwyrþe.

leahtrian; *p.* ode. **I.** *to charge with crime, impeach, accuse, blame, revile, reproach:*—Ic leahtrige *criminor;* ic leahtrode *criminatus sum,* Ælfc. Gr. 25; Som. 26, 61. Man godfyrhte lehtreþ ealles tó swīðe *godfearing men are reviled far too much,* Swt. A. S. Rdr. 110, 163. Ða ðe ða tída ūres cristendómes leahtriaþ *hi qui de temporibus Christianis murmurant,* Ors. 2, 1; Swt. 62, 33. Ðā herede hé and nánuht ne leahtrade *laudavit,* 6, 1; Swt. 254, 14. Hý wǽran ealle ánsprǣce ðonne hý mē leahtrodon and lǣþdon *loquebantur simul,* Ps. Th. 40, 7. Ðæt hié ðás tída leahtrien, Ors. 3, 9; Swt. 136, 31. Gif se midwinter byþ on Seternes deag ða clēnan beóþ leahtrode *if midwinter be on a Saturday the guiltless will be accused,* Lchdm. iii. 164, 12. Leahtrian *insimulare,* Hpt. Gl. 506, 3. **II.** *to corrupt, vitiate:*—Lehtriende *inficians,* Wrt. Voc. ii. 48, 7. v. ge-leahtrian.

leáh-tric. v. leác-tric.

leahtrung, e; *f. Accusation, blame, detraction:*—Lehtrung *derogatio,* Ælfc. Gl. 61; Som. 68, 44; Wrt. Voc. 39, 28.

leáh-tún. v. leác-tún.

leán, es; *n. Reward, recompense, remuneration, requital, retribution :*— Leán *meritum laboris,* Wrt. Voc. ii. 143, 40. Se ðe ðæt gelǣsteþ him biþ leán gearo, Cd. 22; Th. 28, 14; Gen. 435. Him ðæs grim leán becom *terrible retribution befel them for that,* 2; Th. 3, 36; Gen. 46. Gif hé eal wel gefriðaþ [ðe] hé wealdan sceal ðonne biþ hé gódes leánes ful wel weorðe *if he protects well all that he has to keep, then is he quite entitled to good pay,* L. R. S. 20; Th. i. 440, 18. Ic ðē tó leánes ðinne noman mǣrsige *in recompense I will magnify thy name,* Lchdm. iii. 436, 28. Hwæt ðēst ðū ūs ðæs tó leáne *what recompense will you give us for that?* Homl. i. 392, 33: Cd. 135; Th. 170, 27; Gen. 2819. Sigores tó leáne *as a reward of victory,* Beo. Th. 2047; B. 1021. Be

hundfealdon hē onfēhþ leán *centuplum accipiet,* Mt. Kmbl. 19, 29. Wē sceolan habban ánfald leán ðæs ðe wē on lífe ǽr geworhtan, L. C. E. 18; Th. i. 370, 21. Gebyreþ ðæt man his geswinces leán gecnáweþ *it is proper that the reward of his labour be acknowledged* [i. e. *he be rewarded for his labour*], L. R. S. 20; Th. i. 440, 12. Ðǽr leán cumaþ werum bī gewyrhtum *there rewards come to men according to their deserts,* Exon. 27 b; Th. 84, 2; Cri. 1367. Sǣgde leána þanc and ealra ðara ðe him sīð and ǽr gifena drihten forgifen hæfde, Cd. 142; Th. 177, 22; Gen. 2933. Gē eów ondrǽdaþ ðæt gē onfón tó lytlum leánum *you are afraid of receiving too little reward,* Blickl. Homl. 41, 21. Leánum mīne gife gyldan *to requite my gift,* Cd. 22; Th. 27, 4; Gen. 412. Nealles ic ðām leánum forloren hæfde, mægnes mēde, Beo. Th. 4296; B. 2145. Ðonne forliést gód man his leánum ðonne hē his gód forlǽt *tum suo praemio carebit, cum probus esse desierit,* Bt. 37, 2; Fox 189, 26. Ðæt edleán is ofer ealle ōðre leán tó lufienne, Fox 190, 1. [*Goth.* laun: *O. Sax.* lón: *O. Frs.* lán: *Icel.* laun; pl.: *O.H.Ger.* lôn *praemium, merces, stipendium, remuneratio: Ger.* lohn.] DER. æfter-, and-, dǽd-, drinc-, ed-, eft-, ende-, feorh-, fóstor-, hand-, iú-, morþor-, sige-, sigor-, wiðer-, word-, wuldor-leán.

leán; *p.* lóg [*a weak form also occurs* (cf. *Icel.*):—Se ðe wolde leógan oftost on his wordon, ealle hine *leádan,* ða ðe God lufedan, Wulfst. 168, 17.] *To blame, reproach, find fault with, disapprove, scorn :*—Ne leá ic ðē ná ðæt ðū ǽgðer lufige *I blame thee not for loving either,* Shrn. 197, 2. Hý nǽfre man lyhþ se ðe secgan wile sóð æfter rihte *a man that will rightly tell the truth will never blame them,* Beo. Th. 2101; B. 1048. Ða ðe ðæt unliéfde leáþ and swā ðeáh dóþ *qui accusant prava, nec tamen devitant,* Past. 55, 1; Swt. 427, 12. Paulus ðæt yfel ðære forlegnesse swā manegum áwiergdum leahtrum lôh *Paulus fornicationis vitium tot criminibus execrandis inseruit,* 51, 8; Swt. 401, 26. Hē him lôh ðæt hē hæfde his bróðor wíf him tó ciefse *he reproached him with having his brother's wife as his concubine,* Shrn. 123, 1. Nales wordum lôg mēces ecge *he brought no word of blame against the blade's edge,* Beo. Th. 3627; B. 1811. Ðara monna ðe mē ðæt lôgon ðæt ic ðǣm wegum fērde *hominum qui dixerant mihi ne festinarem,* Nar. 6, 27. Ðone sīðfæt him snotere ceorlas lythwôn lôgon *prudent men a little blamed him for that journey,* Beo. Th. 408; B. 203. Ne hié winedrihten wiht ne lôgon, 1729; B. 862. Ne ðē silfne ne hera ne ðē silfne ne leah *neither praise thyself, nor blame thyself,* Prov. Kmbl. 36. Herigaþ oft suā suīðe suā hié hit leán scoldon *plerumque laudant etiam, quod reprobare debuerant,* Past. 17, 3; Swt. 111, 6. Ða dēman beóþ swīðor tó herigenne ðonne tó leánne, Blickl. 63, 21. Eal swilc is tó leánne nǽfre tó lufianne, L. Eth. vi. 29; Th. i. 322, 22. Bóclāre leánde and unriht lufiende *scorning booklearning and loving wrong,* Wulfst. 82, 2. [*Goth.* laian; *p.* lailô *to revile; O. Sax.* lahan; *p.* lôg: *Icel.* lâ; *p.* lâði *to blame: O.H.Ger.* lahan; *p.* luog *vituperare.*] v. be-leán.

leán-gifa, an; *m. One who gives recompense* or *reward :*—Swylce se rihtwísa leángyfa nó mid wordum ac mid dǣdum ðus cwǣde *as if the righteous Recompenser had said not with words but with deeds,* Lchdm. iii. 436, 23.

leánian; *p.* ode *To reward, recompense, requite, pay :*—Ic ðē ða fǣhþe leánige ealdgestreónum *I will recompense thee for the strife with ancient treasures,* Beo. Th. 2765; B. 1380. Ðū ūs leánest unfreóndlíce *thou dost requite us unkindly,* Cd. 127; Th. 162, 29; Gen. 2688. God mǣrlíce leánaþ ǣghwylcum ðære ðe him gód behét and ðæt eft fullíce gelǣst, Lchdm. iii. 436, 16: Exon. 20 a; Th. 52, 4; Cri. 828: 113 a; Th. 434, 12; Rä. 51, 9. Gūþlāce God leánode ellen mid árum, 39 a; Th. 129, 13; Gū. 420. Mē ðone wǣlrǣs wine Scyldinga leánode manegum mádmum, Beo. Th. 4211; B. 2102. Lofe leánige, Exon. 54 b; Th. 193, 13; Az. 121. Ðæt hió him leánige ðæt hē ǽr tela dyde *that it may reward him for having done well,* Bt. 40, 1; Fox. 236, 4. Ðām gódum leánian hiora gód *to reward the good for their goodness,* 39, 12; Fox 230, 25. Nū ic wolde ðē ðone unþanc mid yfele leánian *valet manus mea reddere tibi malum,* Gen. 31, 29. Ðā cwæþ heó ðæt heó ne dorste him swā leánian swā hē hire tó geearnud hæfde *then said she, that she dared not requite him as he had deserved of her,* Chart. Th. 202, 21. Ǽghwylcum ánum men gyldan and leánigean æfter his sylfes weorcum, Blickl. Homl. 123, 34. [*O.Sax.* lônôn: *O.Frs.* lânia: *Icel.* launa: *O.H.Ger.* lônôn *retribuere, munerare, reddere : Ger.* lohnen.] v. ge-leánian.

leánung, e; *f. Reward, recompense :*—Leánung [? leasung, Wrt.] *hostimen,* Wrt. Voc. ii. 43, 20. v. ed-leánung.

leáp, es; *m.* **I.** *a basket, a basket containing a certain amount,* [*two-thirds of a bushel ? 'Lepe* quod est tertia pars duorum bussellorum ;' in Sussex, time of Ed. I.] *a weel for catching fish :*—Leáp *corbis,* Wrt. Voc. ii. 23, 6: *calatus,* 127, 73. Leóht leáp *imbilium,* Wrt. Voc. 287, 27: ii. 46, 40. Leáp *vel* wilige *cophinus,* Ælfc. Gl. 101; Som. 77, 32; Wrt. Voc. 55. 37. Leáp *vel* bogenet *nassa,* Som. 73, 90; Wrt. Voc. 48, 28. Sǣdere gebyreþ ðæt hē hæbbe ǣlces sǣdcynnes ǣnne leáp fulne, L. R. S. 11; Th. i. 438, 9. Leápas *corbes,* Wrt. Voc. ii. 20, 52. Ðā bær man up of ðan ðe hí lǣfdon twelf leápas fulle, Wulfst. 293, 32. **II.** *trunk* [*of the body*], Judth. 10; Thw. 23, 8; Jud. 111. [The word is to be found among English dialects, see the note in Prompt. Parv. p. 296; also the following reference in E. D. S. Publications '*Leáp* a

S s

large deep basket; a' chaff basket, B. 2. *Leap* or *lib* half a bushel [in Sussex], B. 16, 18. *Lep* a large wicker basket, Gloss. of old farming words, vi. *Leap* a wicker basket for catching eels, Lincoln. *Icel.* laupr *a basket of lattice work.*] v. sǽd-leáp.

leás; *adj.*   I. *loose, free from, destitute* or *void of, without :—* Hé wæs ealra fyrena leás *he was free from all sins,* Blickl. Homl. 135, 2 : Exon. 9 b ; Th. 8, 25 ; Cri. 123. Wer womma leás *a man spotless,* Cd. 188 ; Th. 233, 29 ; Dan. 283. Land leóhtes leás and líges full *a land without light and full of flame,* 18 ; Th. 21, 32 ; Gen. 333. Ríces leás *powerless,* 19 ; Th. 24, 4 ; Gen. 372. Búendra leás *without inhabitants,* 5 ; Th. 6, 16 ; Gen. 89. Alles leás ēcan dreámes *void of all eternal joy,* 217 ; Th. 276, 1 ; Sat. 182 : Beo. Th. 1705 ; B. 850. Nǽge wē nāne þearfe ðæt wē ðyses weorþan leáse ac utan dón swā ūs þearf is gelǽstan hit georne *we have no need to fail in this ; but let us do, as there is need for us, diligently perform it,* Wulfst. 38, 13.   II. *vain, false, lying, deceitful, deceptive, faulty :—* Leás *pellax,* Wrt. Voc. ii. 95, 60. *Solocismus* biþ sum leás word on ðam verse, Ælfc. Gr. 50; Som. 51, 51. Ðonne sægde Petrus ðæt hē wǽre leás drý *then said Peter that he was a false sorcerer,* Blickl. Homl. 175, 7. Hit is swīðe leás tōhopa *falsus equus ad salutem,* Ps. Th. 32, 15. Hwæðer hit sig ðe sōð ðe leás ðe gē secgaþ *utrum vera an falsa sint, quæ dixistis,* Gen. 42, 16. Se leása gewita *the false witness,* Deut. 19, 19. Se leása. gylp *vainglory,* Blickl. Homl. 59, 18. Mid leásre gecýðnesse *with false witness,* 173, 35. Ne beó ðū on liésre gewitnysse ongēn ðīnne nēhstan *non loqueris contra proximum tuum falsum testimonium,* Exod. 20, 16 : Wulfst. 40, 11. Leáse mūðe *with lying mouth,* Ps. Th. 77, 35. Sume sǽdon leáse cýðnesse āgēn hine *quidam falsum testimonium ferebant adversus eum,* Mk. Skt. 14, 57. Leáse sybbe ne sceal mon syllan *feigned friendship must not be formed,* Glostr. Frag. 112, 14. Ðonne cumaþ leáse Cristas and leáse wītegan *surgent enim pseudo-cristi et pseudo-prophetæ,* Mt. Kmbl. 24, 24. Ðās leásan spell *hæc fabula,* Bt. 35, 6 ; Fox 170, 15. Fram leásum wītegum *a falsis prophetis,* Mt. Kmbl. 7, 15. Wiðsacaþ ðām leásum welum *renounce the deceitful riches,* Blickl. Homl. 53, 23. Ða leásan godas *false gods,* 201, 30. Fiscere ðone leásostan *a fisherman most false,* 179, 14. [*R. Glouc.* les : *Prompt. Parv. Chauc.* lees : *Goth.* laus *empty, vain :* O. Sax. lôs *free from ; false :* O. Frs. lās : *Icel.* lauss *loose, free, void :* O. H. Ger. lôs *levis, turpis :* Ger. los.]

leás es ; *n. Falsehood, falseness :—* Hī ongietaþ ðæt ðæt wæs leás and īdelnes ðæt hī ǽr heóldon *they perceive that that was falsehood and vanity that they formerly held ;* deprehenderint falsa se vacue tenuisse, Past. 58, 1 ; Swt. 441, 18. Ðæt leás, Elen. Kmbl. 1157 ; El. 580. Gif gē ðisum leáse leng gefylgaþ *if longer ye follow this falsehood,* 1148 ; El. 576. Būtan leáse *truly,* Bt. 41, 1 ; Fox 244, 12 : Bt. Met. Fox 30, 36 ; Met. 30, 18. [*A. R.* leas *falsehood,* 82, 16 : *Laym.* buten lese.]

-leás *a frequently occurring suffix used to form adjectives, having the force of* without [v. leas I.], *modern* -less. *It is found in the cognate dialects.* v. leás.

leás-bregd, -brēd ; *adj. False, deceitful, cheating :—* Ðū leásbrēda feónd and fācnes ordfruma, Homl. Skt. 6, 314.

leás-bregd, -brēd, es ; *m. Deceit, fraud, a trick, cheat, wile :—* Hē hiwode þurh drýcræft fela leásbregda *he performed many tricks by magic,* Wulfst. 99, 16. Swicol on dǽdum and on leásbregdum, 107, 2. Þurh his leásbregdas, 252, 19. Mid leásbregdum earmum mannum derian *to harm poor men with tricks,* L. I. P. 12 ; Th. ii. 320, 25.

leás-bregdende, -brēdende ; *adj. Wily, deceitful :—* Hund sīðon līhþ se leásbrēdenda *centies mentitur versipellis,* Ælfc. Gr. 49 ; Som. 50, 31.

leás-bregdnes, -brēdness es ; *f. Deception, falsehood,* Leo. 220, 22.

leás-cræft, es ; *m. A false art, deception :—* Hē hié getýhþ tō eallum uncystum and tō ðære lufan ðisse worlde mid his leáscræftum *he draws them to all vices and to the love of this world with his false arts,* Blickl. Homl. 25, 12.

leásere, es ; *m.*   I. *a false person, hypocrite :—* Leáseras i̔ lēgeras *falsos,* Mt. Kmbl. p. 15, 8.   II. *one who feigns* or *acts, a buffoon, jester :—* Se wæs ǽrest sumes lǽses mima, ðæt is leásere and sang beforan him scandlīcu leóþ *first he was some emperor's mima, that is, jester, and sang obscene songs before him,* Shrn. 121, 9. Ða gesealde hē ða fǽmnan his leáserum, 154, 23. Ða hēt hē his leáseres hine lǽdan tō ðæm wuda, 83, 18. [*O. H. Ger.* lôsare *dolosus.*]

leásettan ; *p.* te *To feign, pretend :—* Leásetende ðæt hī woldon hine eft tō lífe ārǽran *pretending that they would raise him to life again,* Homl. Th. ii. 474, 10.

leás-ferhþnes, e ; *f. Inconstancy, falseness, folly :—* Hū micel leóhtmódnes and leásferþnes *quanta mentis levitas,* Past. 43, 5 ; Swt. 313, 10. v. leás-líc.

leás-fyrhte (= (?) leás-ferhþ), *false :—* Leásfyrhte is unrihtwīsnys here *mentita est iniquitas sibi,* Ps. Spl. 26, 18.

leás-gewitnes, e ; *f. False witness :—* Leásgewitnyssa, Homl. Th. ii. 592, 5.

leás-gilp, es ; *m. Vain-glory :—* Ðæt hié ne wilnigen leásgielpes ne *inanem gloriam quaerant,* Past. 48, 2 ; Swt. 367, 24.

leásian ; *p.* ode *To lie :—* Leásiaþ ðē fýnd ðīne *mentientur tibi inimici tui,* Ps. Spl. C. 65, 2.

leásing, es ; *m. A false person* [cf. earming] :— Nǽfre ðū gelǽrest ðæt ic leásingum dumbum and deáfum deófolgieldum gaful onhāte *never shalt thou persuade me to promise tribute to false creatures, to dumb and deaf idols* [or *is* leásingum = *with lies, falsely.* v. leásung], Exon. 68 a ; Th. 251, 23 ; Jul. 149.

leás-líc ; *adj. False, vain, frivolous :—* Wēnþ ðæt hit hæbbe sum heálic gōd gestrýned . . . ond mē þincþ ðæt hit hæbbe geboht sume swīðe leáslíce mǽrþe *it supposes that it has gained some exalted good . . . and methinks it has purchased a very false greatness,* Bt. 24, 3 ; Fox 82, 24. Leáslíce cristene *false christians,* Wulfst. 93, 8. Leóhtlícu weorc and leáslícu *levitas operis,* Past. 43, 1 ; Swt. 309, 1. Mid leáslícum wordum hī hine beswīcaþ *with false words they deceive him ;* blandientes sermone ut decipiant eos, Nar. 37, 5. Ða leáslícan ceápas binnan ðam Godes hūse geþafedon *they allowed false bargains within God's house,* Homl. Th. i. 406, 15.

leás-líce ; *adv. Falsely, deceptively :—* Leáslíce *falso,* Ælfc. Gr. 38 ; Som. 41, 35. Leáslíce geclypode oððe ǽwritene *pronounced or written wrongly,* 50 ; Som. 51, 52. Hit biþ swīðe leáslíce on siolufres hiewe [*stannum*] *argenti speciem mentitur,* Past. 37, 3 ; Swt. 269, 3 : Bd. 2, 9 ; S. 511, 20 note.

leás-lícettan ; *p.* te *To dissemble, feign :—* Leáslíccettan *dissimulari,* Wrt. Voc. ii. 27, 38.

leás-lícettung, e ; *f. Dissimulation, pretence :—* Næs hē begangende leáslícetunge *he did not practise dissimulation,* Guthl. 2 ; Gdwin 12, 18.

leásmód-ness, e ; *f. Inconstancy, want of stability :—* Ðære leóhtmódnesse and ðære leásmódnesse sanctus Paulus hine lādode *a quibus* [*mentis levitas, cogitationum inconstantia*] *se alienum Paulus fuisse perhibuit,* Past. 42, 3 ; Swt. 308, 6.

leás-ness, e ; *f. Levity, fickleness ; falseness, lying :—* Þurh leásnesse *per mendacium,* Confess. Peccat. Ðæt ic swā wǽre ālýsed frɔm ðære scylde ðære swýðe īdlan leásnesse *ut sic absolvar reatu supervacuæ levitatis,* Bd. 4, 19 ; S. 589, 30.

leás-ólecung, e ; *f. Flattery, cajolery :* — Leásólecung *lenocinia,* Wrt. Voc. ii. 49, 68.

leás-sagol ; *adj. Saying what is false, mendacious :—* Se ðe wǽre leássagol weorðe se sōðsagol *he that told lies, let him tell the truth,* Wulfst. 72, 16.

leás-spanung, e ; *f. Seduction, allurement, enticement :—* Leássponunge *nec lenonum* [*lenocinium ?*], Wrt. Voc. ii. 59, 71.

leás-spell, es ; *n. A false story, fiction, fable :* — Leásspel *figmenta,* Wrt. Voc. ii. 34, 43. Be swylcum menn leásspell secgaþ *de qualibus fabulæ ferunt,* Bd. 4, 22 ; S. 591, 26.

leás-spellung, e ; *f. Idle, vain,* or *false talking :—* Leásspellunga *fabulationum,* Bd. 4, 25 ; S. 601, 14. Leásspellunga *nenias,* Wrt. Voc. ii. 59, 74. Sōna swā hit forlǽt sōðcwidas swā folgaþ hit leásspellunga *ut quoties abjecerint veras, falsis opinionibus induantur,* Bt. 5, 3 ; Fox 14, 16.

leást. v. lǽst.

-leást, -liést, -lēst, -lýst *a termination of nouns formed from adjectives in* -leás.

leás-tyhtan ; *p.* te *To wheedle, flatter :—* Leástyhtendum *lenocinantibus,* Wrt. Voc. ii. 50, 15.

leás-tyhtung, e ; *f. Wheedling, flattery, cajolery :—* Leástihtinge *lenocinia,* Wrt. Voc. ii. 49, 68.

leásung, e ; *f. Leasing, lying, vain* or *frivolous speech, fiction, false witness, falsehood, falseness, hypocrisy, deception, deceitfulness, artifice :—* Leásung *vel* faam *famfaluca* (Ital. fanfaluca, *a whim, trifle,* and see Ducange, s. v.), Ep. Gl. 9 d, 17. Leásung ðissa woruldwelena *fallacia divitiarum.* Mt. Kmbl. 13, 22. Leásung *falsitas,* Rtl. 37, 31. Heóra leásung wæs gecyrred tō heom sylfum *mentita est iniquitas sibi,* Ps. Th. 26, 14. Ðæs forwyrd and leásung and forleornung swīðe raþe cymþ tō him ðe hē hine sylfne dēþ tō ðon ðe hē nis *for this reason destruction and lying and error come quickly to him, that he makes himself out what he is not,* Blickl. Homl. 183, 34. Sōðfæstnysse feóung and seó lufu líges and leásunge *odium veritatis amorque mendacii,* Bd. 1, 14 ; S. 482, 24. *Nebulonis* heowunga ; *fallacis* scūan i̔ leásunge, Hpt. Gl. 459, 14. Hē nǽfre nōht leásunga ne īdeles leóþes wyrcean ne mihte *nihil unquam frivoli et supervacui poematis facere potuit,* 4, 24 ; S. 596, 52. Fulle mid leásunge *pleni hypocrisi,* Mt. Kmbl. Lind. 23, 28. Ne beó ðū leás gewita. Ðis bebod wiðcweþ leásunge '*Thou shalt not be a false witness.*' *This commandment forbids leasing-making* [cf. Scott. ' leasing-making ' *the crime of uttering falsehood against the king to the people or vice versa*], Homl. Th. ii. 208, 27. Ða Judēiscan noldon gehýran Cristes sōðfæstnysse, forðan ðe hī wǽron āfyllede mid heora fæder leásunga, 226, 24. Ðonne glād ðæt deófol ūt mid his leásunge swā swā smýc æt his eágdura *then the devil by his artifice stepped out at his eye in the form of smoke,* Shrn. 52, 33. Wrec ðē gemetlíce ðý lǽs ðe men [man ?] leásunga teó ðæt ðū ðíne cysta cýðe *revenge thyself in moderation, lest the charge be falsely made, that thou display thy virtues,* Prov. Kmbl. 46. Leásunga *frivola,* Wrt. Voc. ii. 34, 55. Leásunga *factiones,* Hpt. Gl. 472, 3. Leásunga *leonum,* 500, 55. On leásungum *in mendaciis,* Coll. Monast. Th. 32, 29. Se hlísa ðe hē ǽr mid leásungum wilnode *the reputation* (*of philosopher*)

that he had before desired under false pretences, Bt. 18, 4; Fox 68, 5. Nelle wē eác mid leásungum þyllíc líccetan, Homl. Skt. pref. 49. Leásingum beswícen ðæt hē wēneþ furþon ðæt hē man ne sý [so] *deceived by false notions that he thinks even that he is not man*, Blickl. Homl. 179, 5: Elen. Kmbl. 2243; El. 1123. For dínum leásungum *on account of thy falsehoods*, Cd. 214; Th. 268, 28; Sat. 62. Ðú fordēst ða ðe symle leásinga specaþ *thou shalt destroy them that speak leasing* [A. V.], Ps. Th. 5, 5. Onscúna ðú á leásunga, L. Ælf. 44; Th. i. 54, 14: Homl. Th. ii. 482, 25. Fācen and leásunga from úrum heortum ádoon *to remove deceit and falseness from our hearts*, Blickl. Homl. 95, 27. Þurh áðbrycas and þurh weddbrycas and þurh mistlíce leásunga, Swt. A. S. Rdr. 109, 151. Búton ðú forlǽte ða leásinga, weohweorðinga, Exon. 68 a; Th. 253, 13; Jul. 179: Elen. Kmbl. 1375; El. 689. Ðyllíce leásunga hí worhton and mihton eáþe secgan sóþsped gif him ða leásunga nǽron swētran, Bt. 35, 4; Fox 162, 14: 38, 1; Fox 196, 8. [*Prompt. Parv.* leesynge *mendacium;* lesynge *nuga: Icel.* lausung *lying, falsehood.*]

leásung-spell, es; *n. A false or foolish story, a fable:*—Ðá hæfdon mouige unwíse menn him tó worde and tó leásungspelle ðæt sió hǽte nǽre for hiora synnum ac sǽdon ðæt hió wǽre for Fetontis forscapunge *ex quo quidam, dum non concedunt Deo potentiam, suas inanes ratiunculas conquirentes, ridiculum Phaetontis fabulam texuerunt*, Ors. 1, 7; Swt. 40, 8.

leáþor, es; *n* [?]. *A kind of nitre used for soap, lather:*—Leáþor *nitrum*, Wrt. Voc. ii. 62, 3. Of leáþre *nitria*, 61, 27. Gníd swíðe ðæt heó sý eall gelēþred þweah mid ðý leáþre ðæt heáfod gelōme rub strongly so that it may be all lathered, wash the head frequently with the lather*, Lchdm. iii. 2, 4. [*Icel.* lauðr; *n. froth or foam of the sea water; a kind of nitre or soap.*]

leáþor-wyrt, e; *f. Lather-wort, soap-wort;* saponaria officinalis:—Leáþorwyrt, *borith, erba fullonum*, Wrt. Voc. ii. 12, 47: 38, 43: L. M. 1, 3; Lchdm. ii. 42, 22.

leáw-finger, e; *m. The forefinger:*—Leáwfinger *index*, Ps. Th. 72, 11. [Cf. [?] O. H. Ger. gi-lou *versutus, sollers, gnarus*, Grff. 2, 35.]

leax, læx, lex, es; *m. A salmon, lax* [Scott.]:—Lex *salmo* vel *esocius*, Ælfc. Gl. 102; Som. 77, 65; Wrt. Voc. 55, 70. Leax *ysox*, 65, 66: *esox*, Wrt. Voc. ii. 30, 48. Laex *isic*, 112, 8. Leax sceal on wǽle mid sceóte scríðan *swiftly shall the salmon in the stream's eddy move*, Menol. Fox 538; Gn. C. 39. Leaxes geallan, L. M. 3, 2; Lchdm. ii. 308, 6. Hwý gē nú ne settan on sume dúne fiscnet eówru, ðonne eów fón lysteþ leax? Bt. Met. Fox 19, 23; Met. 19, 12. Hwæt fēhst ðú on sǽ? Hæringcas and leaxas *quid capis in mari? Aleces et isicios*, Coll. Monast. Th. 24, 9. Ðis is seó gerǽdnes ... gesyllan ælce geare xv. leaxas *this is the agreement ... that they give xv salmon every year*, Cod. Dip. Kmbl. iii. 295, 34: L. In. 70; Th. i. 146, 19. [*Icel.* lax *a salmon: O. H. Ger.* lahs *salmo, esox: Ger.* lachs.]

leax-heáfod, es; *n.* ?:—Lex heáfod *capital*, Wrt. Voc. ii. 128. 43.

leber, lebr. v. lǽfer.

lec *rimosus*, Germ. 400. v. hlec.

lec (?), *sweet:*—Lec *dulcia*, Hpt. Gl. 411, 47.

lêc. v. leác.

lêc, es; *m. Look, sight:*—Wē sceolon áwendan úrne lêc fram yfelre gesihþe, urne hlyst fram yfelre sprǽce, Homl. Th. ii. 374, 3. v. on-lêc.

leccan; *p.* lehte, leohte *To moisten, wet:*—Ic lecce *rigabo*, Ps. Spl. 6, 6. Hæglas and snáwas and se oftrǽda rén leccaþ ða eorþan on wintra *hiemem defluus irrigat imber*, Bt. 39, 13; Fox 234, 16: Met. Fox 29, 128; Met. 29, 64: Exon. 56 b; Th. 202, 4; Ph. 64. Sumu twigu hē lehte mid wætere *some twigs he watered*, Past. 40, 3; Swt. 293, 7. His eágospind mid teárum leohte *wetted his cheeks with tears*, Guthl. 20; Gdwin 82, 4. Leohte ðæt ̒liðe land lago yrnende, Cd. 12; Th. 13, 30; Gen. 210. Seó wæs wætrum weaht and wæstmum þeaht lagostreámum leoht *it was refreshed by the waters, covered with various growths, irrigated by running streams*, 91; Th. 115, 21; Gen. 1923. Leccende *rigans*, Ps. Surt. 103, 13. [*O. H. Ger.* lekjan; *p.* lacta *rigare, irrigare: Ger.* lecken: cf. *Icel.* leka; *p.* lak *to drip.*] DER. ge-, geond-leccan.

leccing, e; *f. Watering, moistening:*—Leccinc *inrigatio*, Kent. Gl. 33.

lêce. v. lǽce.

lecg, e; *f. Some part of a weapon, the cross bar in the hilt* [?]:—Án handsex and [an?] ðæræ lecge is hundeahtati mancussa goldæs, Chart. Th. 527, 9. Leo takes *lecg* = gift, legacy, and then a dish of three pounds and a cup of equal amount would go to make up the amount of eighty mancusses. As regards the value of a *handseax*, Chart. Th. 501, 5 may be quoted, where one worth eighty mancusses is mentioned. [Cf. *ledge*, a bar E. D. S. Publ. B. 20: *ledge* the horizontal bar of a gate, Lincolnshire. In *Prompt. Parv.* legge, ouer twarte byndynge *ligatorium*, occurs: other words that suggest themselves by their form for comparison are M. H. Ger. lecke *leiste, saum: O. H. Ger.* legge *tornaturus, intransversum ligna tornata: Icel.* lögg *the ledge or rim at the bottom of a cask.*]

LECGAN; *p.* legde, lægde, lēde *To cause to lie.*   I. *to lay, place, put, lay* [a dead body in the grave.]:—Syððan hē ðanne grundweall legþ *postea quam posuerit fundamentum*, Lk. Skt. 14, 29. Ða ungeþyldegan ne mágon áberan nánwuht ðæs láðes ðe him mon on legþ *impatientes ab aliis illata non tolerant*, Past. 40, 4; Swt. 293, 17. Wá ðǽm ðe

willaþ under ælcne elnbogan lecggean pyle ... Se legeþ pyle under ælces monnes elnbogan seðe ... *væ his qui consuunt pulvillos sub omni cubito manus ... Pulvillos sub omni cubito manus ponere, est ...* 19, 1; Swt. 143, 14. Cwēn mec hwílum hond on legeþ, Exon. 127 a; Th. 489, 8; Rä. 78, 4. Ða land ðe hig ðiderin lecgeaþ beón ða ðám gebrōðran ðe ðær binnan beóþ tó fódnoþe and tó scrúde *let the lands, that they assign thereto, be for the feeding and clothing of the brethren there*, Chart. Th. 370, 25. Sege mē hwar ðú hine lēdest *dicito mihi ubi posuisti eum*, Jn. Skt. 20, 15. Se cyng lægde hí wið Eádward kyng hire hláforde *the king laid* [*buried*] *her by King Edward her lord*, Chr. 1075; Erl. 214, 12. Lēde him ætforan *posuit coram eis*, Gen. 18, 8. Hē nam stánas and lēde under his heáfod, 28, 11. Hine betellan æt ælc ðæra þinga ðe him man on lēde *to clear himself from every thing that was laid to his charge*, Chr. 1048; Erl. 180, 12. Abraham legde hleór on eorþan, Cd. 107; Th. 140, 32; Gen. 2336. Se mec wrǽde on æt frumsceafte legde *who at the beginning binding laid on me*, Exon. 101 b; Th. 383, 22; Rä. 4, 14. Wē on bearm lægdon *we put them into our laps*, Salm. Kmbl. 864; Sal. 431. Gē on his wergengan wíte legdon *ye imposed pain upon his pilgrim*, 43 a; Th. 144, 29; Gú. 685. Ðæt folc geald heom swá mycel swá hí heom on legden *the people paid as much as they imposed*, Chr. 1052; Erl. 183, 15. Hig lægdon ærende on hine tó ðam cynge *they commissioned him to the king*, 1064; Erl. 194, 24. Ðá lægdon hí fýr on *they set fire to it*, 1083; Erl. 209, 1. Lege hit hēr beforan dínum freóndum *pone hic coram fratribus tuis*, Gen. 31, 37. Lecgaþ ðærtōeácan *add thereto*, Wulfst. 274, 7. Sleá mon hine and on fúl lecge *let him be slain and buried in unconsecrated ground*, L. Eth. i.̒ 4; Th. 284, 2: vi. 21; Th. i. 320, 6: L. C. S. 33; Th. i. 396, 17. Hwá wolde gelýfan ðæt Sarra sceolde lecgan cild tó hyre breóste tó gesoce *quis crederet, quod Sara lactaret filium*, Gen. 21, 7. Josue hēt lecgan him on uppan ormǽte weorcstánas *præcepit, ut ponerent super os ejus saxa ingentia*, Jos. 10, 27. Lecgan ðone mæst *to lower the mast*, Bt. 41, 3; Fox 250, 15. Ægru lecgan *to lay eggs*, Lchdm. iii. 204, 30. Lástas lecgan *to go, journey*, Cd. 109; Th. 145, 3; Gen. 2400: 118; Th. 153, 9; Gen. 2536: Exon. 82 a; Th. 309, 14; Seef. 57.   II. *to cause to lie* [dead. v. licgan], *to slay:*—Hine lecge for þeóf seðe him tó cume *let him that comes at him slay him for a thief*, L. Ath. i. 2; Th. i. 200, 10. Gif hine hwá lecge, L. Eth. iv. 4; Th. i. 222, 9. Se ðe mid þeófe stande and mid feohte, lecge hine man mid ðam þeófe. v. 1, 3; Th. i. 228, 23. Ðæt hine man lecgan ne mōste, Th. i. 230, 6. [*Goth.* lagjan: *O. Sax.* leggian: *O. Frs.* leia: *Icel.* leggja: *O. H. Ger.* legjan: *Ger.* legen.] DER. a-, be-, ge-, of-, tō-, under-, wið-lecgan.

lecþ, e; *f.* ?:—Lecþ [=? legþ] *peana*, Wrt. Voc. 287, 29. Ducange gives '*peanius lignum tectis conficiendis aptum;*' Spanish has *peana* a pedestal, a frame put at the foot of an altar to tread upon.

lecþa, an; *m. The lowest part of a ship, in which bilge water collects:*—Sentina lechta *ubi multae aquæ colliguntur in navem*, Ep. Gl. 23 d, 15. Lechta *sentina*, Wrt. Voc. ii. 120, 27. Cf. (?) lec, hlec.

Lēden. v. Lǽden.

lêf [or lef?]; *adj. Weak, injured, infirm:*—Lêf *debilis*, Germ. 389. On fýre hí ne lyst lócian gif se æppel lêf biþ *men do not like to look at fire if the apple of the eye be injured*, Bt. 38, 5; Fox 204, 29. Lêf mon lǽces behōfaþ *a sick man needs a doctor*, Exon. 89 b; Th. 336, 8; Gn. Ex. 45. On fēðe líf seonobennum seóc *weak for walking, sick with sinew-wounds*, 87 b; Th. 328, 16; Vy. 18. Oft him feorran tó laman liomseóce lêfe cwōmon *oft from far to him the paralytic, the cripple, the infirm came*, Elen. Kmbl. 2426; El. 1214. See note to Grmm. A. u. E. p. 166. [*O. Sax.* O. Frs. lêf: *Dut.* loof.] v. â-, ge-lêfan; *adj; n;* lêfung.

lêf, es; *n. Hurt, damage, injury:*—Ðeore feórþan niht gif wind byþ lêf byþ litel *if there is wind on the fourth night, the damage will be little*, Lchdm. iii. 164, 17.

lêfan *to permit.* v. lífan.

lefel. v. læfel.

lêf-ness. v. leáf-ness.

lêft, e; *f. A vow;* votum, Ps. Spl. T. 64, 1: 65, 12. [Cf. (?) *Icel.* leyfd *praise.*]

lêfung, e; *f. Weakening, laming, lameness, paralysis:*—Ðí læs ðe hí ðás lêfunge on heora limum gebrohton *lest they should bring this paralysis* [*want of power to speak, walk and see*] *upon their limbs*, Homl. Th. ii. 486, 18.

lêg. v. líg.

leger, es; *n.*   I. *a lying:*—Hys spêda hý forspendaþ mid ðan langan legere ðæs deádan mannes inne *they squander his wealth with the long lying of the dead man in the house*, Ors. 1, 1; Swt. 21, 9.   II. *a lying sick or dead, sickness, death:*—Nis ðær hungor ne þurst ne slǽp ne swár leger *there is neither hunger nor thirst nor sleep nor grievous sickness*, Exon. 32 a; Th. 101, 21; Cri. 1662: 56 b; Th. 201, 15; Ph. 56. On ðam sixtan dæge his legeres *on the sixth day of his illness*, Homl. Th. ii. 186, 28. Mid langre ádle laman legeres swíðe gehefigod *longo paralysis morbo gravatam*, Bd. 3, 9; S. 534, 6. Moyses and Aaron geendodon heora líf swaðeáh búton legere *Moses and Aaron ended their lives, yet without sickness*, Homl. Th. ii. 212, 13. Se preóst sceal smyrigan ða seócan symble on legere *the priest must always anoint the sick in ill-*

ness, L. Ælfc. C. 32 ; Th. ii. 354, 14. Tó hæbbenne and tó syllanne for life and for legere *to have and to give during life and at death*, Chart. Th. 208, 3. Ðá cwæþ se cyng ðæt mihte beón geboden him wið clǽnum legere *then the king said, the offer might have been made to him, if the death had been by fair means* [it was by drowning], 31. **III.** *a place to lie in, a couch, a lair, a place where the dead lie, a grave* :— Hálig leger [legerstów (?)] *cimiterium*, Ælfc. Gl. 49 ; Som. 65, 74 ; Wrt. Voc. 34, 9. Þolige hé clǽnes legeres and Godes mildse *let him forfeit a hallowed grave and God's mercy*, L. N. P. L. 62, 63 ; Th. ii. 300, 19, 22 : Wulfst. 39, 19. Wé lǽraþ ðæt man innan circan ǽnigne man ne birige búton . . . hé sí ðæs legeres wyrðe *we enjoin that no man be buried within a church, unless he be worthy of such a place of burial*, L. Edg. C. 29 ; Th. ii. 250, 17. On gehálgodan legere licgan to be buried in consecrated ground, 22 ; Th. ii. 248, 20. Ge on lífe ge on legere *both alive and in the grave*, L. Eth. v. 9 ; Th. i. 306, 22 : vi. 5 ; Th. i. 316, 14 : ix. 28 ; Th. i. 346, 19. Unsac hé wæs on lífe beó on legere swá swá hé móte, i. 184, 13 ; Lchdm. iii. 288, 6. Líchoman, se ðe on legre sceal weorþan wyrme tó hróðor, Exon. 71 b ; Th. 267, 15 ; Jul. 415. Be ðǽre róde ðe ǽr in legere wæs lange bedyrned [*of the cross that had been buried*], Elen. Kmbl. 1200 ; El. 602 : 1442 ; El. 723. Líc legere fæst, 1762 ; El. 883. Se wæs fíftiges fótgemearces lang on legere *he was fifty feet long in the place where he lay*, Beo. Th. 6078 ; B. 3043. ¦Leger ðis *lectum istum*, Rtl. 111, 24. On legir *in lectum*, 181, 7. Frýnd leger weardiaþ ðonne ic on úhtan ána gonge *my friends rest in their couches, when ere the dawn I go solitary*, Exon 115 b ; Th. 443, 23 ; Kl. 34. [O. E. Homl. (to) leire *couch* : O. Sax. legar : O. Frs. legor : O. H. Ger. legar *cubile, lustrum, accubitus, concubitus* : Ger. lager : Goth. ligrs ; *m. a couch.*]

leger-bǽre ; *adj. Suffering from sickness* :— Bútun hé on hláfordes neóde beó oððe legerbǽre *unless he be on his lord's necessary business, or suffering from sickness*, Chart. Th. 611, 20.

leger-bedd, es ; *n. A sick-bed, bed of death, grave* :— Sum mǽden hé gehǽlde ðæt ðe langlíce læg on legerbedde seóc *a maiden he healed that had long been confined to her bed by sickness*, Homl. Th. ii. 510, 25. Árís nú and ber hám ðín legerbed, i. 472, 25. Ðæt ðú ðus láðlíc legerbed cure *that thou shouldst choose so loathily a couch* [*the grave*], Soul Kmbl. 307 ; Seel. 157 : Wulfst. 187, 12. Sceal ðis sǽwelhús legerbedde fæst wunian wælræste, Exon. 47 b ; Th. 164, 2 ; Gú. 1005 : Beo. Th. 2019 ; B. 1007. [O. Sax. legar-bed.]

-legere. v. for-legere.

leger-fæst ; *adj. Sick, ill*, R. Ben. 39, Lye. [O. Sax. legar-fast.]

legerian ; *p.* ode *To be ill, afflicted with sickness.* v. ge-legerian.

leger-stów, e ; *f. A burial-place, cemetery* :—Hálig leger [legerstów ?] *cimiterium, poliandrium*, Ælfc. Gl. 49 ; Som. 65, 74 ; Wrt. Voc. 34, 9. Cyricean ðe legerstów on sý *a church at which there is a burial-place*, L. Edg. i. 2 ; Th. i. 262, 12 : L. C. E. 11 ; Th. i. 366, 24 : 3 ; Th. i. 360, 23. Ðæt hí þolian woroldǽhta and gehálgodre legerstówe *that they forfeit worldly possessions and a consecrated burial-place*, L. Edm. E. 1 ; Th. i. 244, 14 : 4 ; Th. i. 246, 6. Ypolitus bebyrigde ðone hálgan líchaman on ðære wudewan legerstówe *Hippolytus buried the holy body in the burial-place of the widow*, Homl. Th. i. 430, 26. [Laym. leir-stow.]

leger-teám, es ; *m. Matrimony, sexual intercourse* [*lawful or unlawful*] :—Matheus him sægde ðæt hé wǽre swá synnig wið God gif hé ða gehálgodan fǽmnan tó legerteáme onfénge swá se þeów wǽre se ðe fénge on kyninges quéne tó unryhtum hǽmde *Matthew said to him, that he would be as guilty against God, if he received the consecrated virgin as his wife, as the slave would be who took a king's queen to commit adultery with her*, Shrn. 132, 4. Legerteám *flagitium*, Wrt. Voc. ii. 39, 34.

leger-wíte, es ; *n. A fine for lying with a woman*, L. H. 23 ; Th. i. 529, 23 : 81 ; Th. i. 589, 3. [Trev. leir-wite *fine for lying with a bond-woman*.]

légetu *lightning.* v. lígetu.

Legra ceaster. v. Ligora ceaster.

léh *lye.* v. leáh.

léhtan *to alleviate.* v. líhtan.

lehter *disgrace.* v. leahter.

léh-tric, -tún. v. leác-tric, -tún.

lél. v. lǽl.

leloþre [*error for* geloþre *according to Cockayne.* v. gelod-wyrt], *A kind of dock* :—Lelodrae *lapatium* (= λάπαθον ; cf. uude docce *lapatium*, Lchdm. iii. 303, col. 2), Ep. Gl. 13 f, 31. Lelothras *radinape*, 22 b, 32. Leloþre *lapadium*, Wrt. Voc. 69, 14 : ii. 54, 24. Lelodrae *lapatium*, 112, 35. Lelothrae *rodinope*, 119, 24.

lemian ; *p.* ede *To lame, cripple, enfeeble, strike* [?] :—Swá wildu hors ðonne wé hié ǽresð gefangnu habbaþ wé hié stráciaþ mid brádre handa and lemiaþ *equos indomitos blanda prius manu tangimus*, Past. 41, 4 ; Swt. 303, 11. Hine sorhwylmas lemedon [MS. lemede] tó lange *the waves of care had crippled him too long*, Beo. Th. 1814 ; B. 905. [Icel. lemja *to beat so as to lame* or *disable, to suppress* : O.H.Ger. lemian *debilitare* : Ger. lähmen.]

lempedu, e ; *f. A lamprey* :—Lempedu *lemprida*, Wrt. Voc. ii. 53, 42.

lemp-healt, laempi-halt ; *adj.* The word occurs in Wrt. Voc. ii. 51, 20, and in Ep. Gl. 13 f, 4 as the gloss of *lurdus* which Ducange explains as *foul*, cf. Ital. *lordo*, or *stupid*, cf. Fr. *lourde, lourdand*. Lye quotes without reference *lempe lenitas* ; *Icel.* has *lempiligr* pliant, could the word mean 'unable to bend, stiff, awkward?'

lencg ; *adv. Longer.* v. lange.

lencten, lengten, lenten, es ; *m. Spring, Lent* :—Lencten *ver* : foreweard lencten *vel* middewǽrd lencten *ver novum*: æfterwǽrd lencten *ver adultum*, Ælfc. Gl. 95 ; Som. 76, 7, 12-14 ; Wrt. Voc. 53, 21, 26, 27. Swá nú lencten and hærfest ; on lencten hit gréwþ, and on hærfest hit fealwiaþ, Bt. 21 ; Fox 74, 22. Gif middes wintres messedeg biþ on sunnandeg, ðonne biþ gód winter and lengten windi, Lchdm. iii. 162, 26. Winter biþ cealdost, lencten hrímigost, Menol. Fox 471 ; Gn. C. 6. Wæs ðá lencten ágán bútan vi. nihtum ǽr sumeres cyme on Maias Kł., Elen. Kmbl. 2452 ; El. 1227. Ðæs sylfan lentenes hé fór tó Róme *in the course of the same spring he went to Rome*, Chr. 1048 ; Erl. 177, 13. Ðá com Æðelréd cyning innan ðam lenctene hám tó his ágenre þeóde, 1014 ; Erl. 150, 17. Sunnan glǽm on lenctenne lífes tácen weceþ *the sun's gleam in spring wakes signs of life*, Exon. 59 b ; Th. 215, 16 ; Ph. 254. Ðé má ðe man mót on lenctene flǽsces brúcan *any more than flesh may be eaten in Lent*, Wulfst. 305, 25. Sumor ðú and lencten swylce geworhtest *æstatem et ver tu plasmasti ea*, Ps. Th. 73, 16. Ðone lencten wǽron him on Cent *during the spring they were in Kent*, Chr. 1009 ; Erl. 143, 14. Nis nán blódlǽstíd swá gód swá on foreweardne lencten *there is no time for letting blood so good as in the early spring*, L. M. 1, 72 ; Lchdm. ii. 148, 3 : 2, 30 ; Lchdm. ii. 228, 8. Gif mon in lencten hálig ryht in folce bútan leáfe álecgge gebéte mid cxx. scił *if any one in Lent suppress holy law among the people without leave, let him make amends with cxx shillings*, L. Alf. pol. 40 ; Th. i. 88, 13. Ðú dydes sumer and lenten, Ps. Surt. 73, 17. [Piers P. lenten : Prompt. Parv. lente : cf. O. H. Ger. lengiz *and* lenzo *ver* : Ger. lenz. v. Grmm. D. M. 715.]

lencten-ádl, e ; *f. A fever, typhus fever, tertian fever* :— Lengtenádl *tipus*, Ælfc. Gl. 10 ; Som. 57, 24 ; Wrt. Voc. 19, 30. Lenctenádl *tertiana*, 289, 58. Lenctinádl *tertiana*, ii. 122, 20. Án lytel cniht fram lengtenádle wæs gelácnod . . . sum cniht on langre lengtenádle wæs hefiglíce geswenced *puerulus e febre curatus sit . . . puerulus quidam longo febrium incommodo graviter vexatus fuit*, Bd. 3, 12 ; S. 537, 2-5. Ða ðe on lengtenádle wǽron *febricitantes*, 4, 6 ; S. 574, 6. Wið lenctenádle, L. M. 1, 62 ; Lchdm. ii. 134, 28 : 3, 11 ; Lchdm. ii. 306, 12.

lencten-bryce, es ; *m. A breach of the Lenten fast* :—Gif hwá openlíce lengctenbryce gewyrce, L. C. S. 48 ; Th. i. 402, 29.

lencten-dæg, es ; *m. A day in Lent* :—Lengctendagum, L. C. E. 17 ; Th. i. 370, 3 : Wulfst. 117, 15.

lencten-eorþe, an ; *f. Land ploughed in the spring* ; *veractum*. Ducange gives '*veractum* champ reonné' and refers to *warectum* 'terra novalis, seu requieta, quia alternis requiescit, sic dicta, inquit Edw. Cokus quasi novo victum, vel subactum.'], Ælfc. Gl. 1 ; Som. 55, 16 ; Wrt. Voc. 15, 16.

lencten-fæsten, es ; *n. The fast of Lent*, L. Alf. pol. 5 ; Th. i. 64, 25 : 40 ; Th. i. 88 ; 12 : L. C. E. 16 ; Th. i. 368, 22 : Wulfst. 117, 9.

lencten-líc ; *adj. Vernal, lenten* :—Lengtenlíc dæg *dies vernalis*, Ælfc. Gl. 95 ; Som. 76, 11 ; Wrt. Voc. 53, 25. Manegra manna cwyddung is ðæt seó lenctenlíce emniht gebyrige rehtlíce on Marian mæssedæge, Lchdm. iii. 256, 4. Ða clǽnan tíd lenctenlíces fæstenes *the pure time of the Lenten fast*, Homl. Th. ii. 98, 24. Ðæs lænctenlíces emnihtes dæg *the day of the vernal equinox*, Lchdm. iii. 238, 17. Ebréi healdaþ heora geáres annginn on lenctenlícre emnihte, 246, 17. On lenctenlícre tíde *in spring time*, Hexam. 4 ; Norm. 8, 3. Nú is ús álýfed ðæt wé dæghwomlíce on ðyssere lenctenlícan tíde úre líchaman gereordigan mid forhæfednysse and clǽnnysse. Stuntlíce fæst se lenctenlíc fæsten, se ðe on ðisum clǽnum tíman hine sylfne mid gálnysse befýlþ, Homl. Th. ii. 100, 13-17.

lencten-sufel, es ; *n. Food for the spring* or *for Lent* :—Syster beána tó lǽngtensufle *i. sester fabe ad quadrigesimalem convictum*, L. R. S. 9 ; Th. i. 436, 31.

lencten-tíd, e ; *f. Spring-time, spring, Lent* :—Ver is lenctentíd, Lchdm. iii. 250, 9. Hit wæs lenctentíd *erat vernum tempus*, Gen. 48, 7. On lengtentíde mónþes tíde *mense verni temporis*, Ex. 34, 18. Nǽfre on lenctentíde *never in Lent*, Wulfst. 305, 24. Hé on lenctentíd gesceóp ðone forman dæg ðyssere worulde ðæt is xv cl. Aprilis *he in spring created the first day of this world, that is the 18th of March*, Hexam. 4 ; Norm. 8, 4 : Bt. Met. Fox 29, 135 ; Met. 29, 68.

lencten-tíme ; *adj. Vernal* :—Lenctentíme *vernali* (s. *tempore*) Hpt. Gl. 496, 44.

lencten-wicu, an ; *f. A week in Lent* :—Ðys sceal on þursdæg on ðære óðre lenctenwucan *this shall be read on Thursday in the second week in Lent*, Rubc. Jn. Skt. 5, 30.

-lenda, -lende. v. in-, ut-lenda, -lende.

lendan ; *p.* de *To arrive, come to land* :—Man hine lǽdde tó Eligbyrig . . . sóna swá hé lende on scype man hine blende *he was brought to Ely . . . as soon as he arrived he was blinded on board ship*, Chr. 1036 ; Erl. 165, 27 ; Ælf. Tod. 14. [Icel. lenda *to come to land, get to* : O.H. Ger. lantian *applicare*.] v. ge-lendan.

lenden-bán, es ; *n. The loin-bone* :—Lendenbán neoþeweard *sacra spina*, Ælfc. Gl. 74 ; Som. 71, 52 ; Wrt. Voc. 44, 35. [Cf. *Misc.* 12, 360, leigeð his skinbon on oðres *lendbon*.]

lenden-, lende-bræd, e, f: -bræda, an; m. A loin:—Lendebræde lumbulos, Wrt. Voc. ii. 51, 31. Lendebrēde, 113, 35. Wið lendenbrǣdena sáre against lumbago [?], Herb. 1, 10; Lchdm. i, 74, 3. Sió helt ða lendenbrǣdan it [the liver] has a hold on the false ribs, L. M. 2, 17; Lchdm. ii. 198, 1. [Cf. O. H. Ger. lenti-práto; m. ren, renunculus, lumbulus, lumbus, Grff. 3, 285: Ger. lenden-braten loin, sirloin: and see hrycg-brǣdan.]

lenden-reáf, es; n. A covering for the loins, an apron:—Lenden-, síd-reáf lumbare vel renale, Ælfc. Gl. 63; Som. 68, 112; Wrt. Voc. 40, 22.

lendenu; pl. The loins, reins:—Lendenu renes vel lumbi, Ælfc. Gl. 74; Som. 71, 53; Wrt. Voc. 44, 36. Lændenu lumbi, 65, 26. Lendena renes, 71, 41. Laendino rien, Wrt. Voc. ii. 119, 17. Lendene renes, Ps. Spl. T. 15. 7. Beón eówer lendena ymbgyrde ... On ðám ymbgyrdum lendenum is se mægþhád tó understandenne let your loins be girded ... By the girded loins virginity is to be understood, Homl. Th. ii. 564, 25. Beóþ eówre lændenæ ymbgirde ... on ðám lændenum is getácnad swá swá wē leorniaþ on bócum seó fúle gálnes, L. Ælfc. P. 13–14; Th. ii. 368, 32–35. Begyrdaþ eówer lendenu renes vestros accingetis, Ex. 12, 11: Homl. Th. ii. 264, 8. Se Johannes hæfde fellennę gyrdel embe hys lendenu ipse Joannes habebat zonam pelliciam circa lumbos ejus, Mt. Kmbl. 3, 4. [Cf. O. L. Ger. lenda; f. ren: Icel. lend; f: O. H. Ger. lenti: f: Ger. lende; f.]

lenden-wǣro, es; m. A disease of the kidneys; nefresis [nefritis?], Ælfc. Gl. 10; Som. 57, 39; Wrt. Voc. 19, 42.

lending, e; f. Landing, landing-place:—Ic ann ealle ða lændinge and ða gerihte of ðam ilkan wætere concedo omnes exitus ejusdem acquæ, Chart. Th. 317, 22. [Icel. lending landing, landing-place.]

lendis lieg bofor, Wrt. Voc. ii. 102, 12. Laembis lieg, 11, 28.

-lendisc. v. dūn-, eówer-, in-, up-, ūre-, ūt-lendisc.

leng; adv. Longer. v. lange.

leng, e; f. Length [of time or space], height, stature:—Mannes leng statura, Ælfc. Gr. 43; Som. 45, 4. Nǣfre ne sý se hálga eásterdæg gemǣrsod ǣr ðan ðe ðæs dæges lenge [lencge MS. P; længe, MS. L.] ofersíge ða niht never let the holy Easter-day be celebrated, before the length of the day exceed the night, Lchdm. iii. 256, 13. Swá micel swá seó sǣ heó mǣst widteóhþ and git ánes mannes lenge ðe healt ánne spreót on his hand and strecþ hine swá feor swá hē mæg árǣcan intó ðere sǣ quantum mare plus se retraxerit, et adhuc statura unius hominis tenentis lignum quod Angle nominant spreót, et tendentis ante se quantum potest, Chart. Th. 318, 10. Lenge proceratitis, Wrt. Voc. ii. 66, 8. Hú lang wæs Adam on lenge gesceapen how tall was Adam created? Salm. Kmbl. 180, 19. Hwilc eówer mæg geícan áne elne to his lenge? Homl. Th. ii. 464, 2. Forneán on lenge ungeendod almost infinite in length, 350, 7. Þreóhund fæðma biþ se arc on lenge, Gen. 6, 15. Far geond ðis land on lenge and on brǣde perambula terram in longitudine et in latitudine sua, 13, 17: Nar. 33, 22. Leáf on fingeres længe leaves of the length of a finger, Herb. 147, 1; Lchdm. i. 270, 22. On fingres lencge, 150, 1; Lchdm. i. 274, 14. Seó sunne stód stille ánes dæges lencge [længce, MS. M.] the sun stood still for the length of one day, Lchdm. iii. 262, 9. Dó ðus ða lange ðe hit beþurfe do thus for the length of time that is necessary, 114, 18. Tele ða lenge ðǣre hwíle ... compare the length of time ... Bt. 18, 3; Fox 66, 6. On ðínum handum synd ða lenge mínra tída in manibus tuis tempora mea, Ps. Th. 30, 17. [O. H. Ger. lengi: Ger. länge.] v. lengu.

lengan; p. de To make or to become long, protract, delay, extend, lengthen:—Lengeþ, Exon. 107 b: Th. 411, 6; Rä. 29, 8. Ðá lengde hit man swá lange it was so long delayed, Chr. 1052; Erl. 183, 10. Ne lengde ðá leóda aldor wítegena wordcwyde on hí wíde beád metodes mihte the prince was not slow to heed the prophet's words, but widely proclaimed the might of the Lord, Cd. 208; Th. 256, 25; Dan. 646. Hyre lof lengde geond londa fela her praise extended through many lands, Exon. 86 a; Th. 324, 23; Vid. 99. Giestas lisse lengdon the guests prolonged their pleasure, 94 a; Th. 353, 13; Reim. 12. Hí lengdon (prolongaverunt) unrihtwísnyssa heora, Ps. Spl. 128, 3. [Havel. lenge to prolong: Ayenb. lenge to delay: Piers P. lenge to delay, tarry: Icel. lengja to lengthen, prolong: O. H. Ger. lengjan protrahere, differre.] DER. gelengan.

lengan; p. de To pertain, belong:—Ðonne heó byþ ii and xx niht eald ðæt ðú gesihst hit lengeþ tó góde and gefeán when the moon is twenty-two nights old, what thou seest belongs to good and to joy, Lchdm. iii. 160, 9. v. lenge, langian, ge-lang.

lenge; adj. Belonging, related:—Him biþ lenge húsel to them belongs the housel, Exon. 326; Th. 103, 9; Cri. 1685. Gód biþ wið God lenge good hath affinity with God, 91 a; Th. 341, 5; Gn. Ex. 121. v. preceding word, and ge-lenge.

lengian; p. de v. impers. To long:—Lengaþ hine hearde sorely doth he long, Salm. Kmbl. 542; Sal. 270. [Cf. Icel. lengjask mjök to long exceedingly.]

lengeo, lengo. v. lengu.

leng-færra. v. lang-fære.

lengten. v. lencten.

lengþ, e; f. Length:—On lengþe mid him hē begeat ealle ða eástlond at length with them he gained all the east country, Ors. 3, 11; Swt. 144,

1. [Hit weáx on lengþe it grew in length, Chr. 1122; Erl. 249, 22.] [Icel. lengd length.]

lengu; indecl. f. Length:—Gerisenlícre lengo tó gemete ðæs líchoman congruæ longitudinis ad mensuram corporis, Bd. 4, 11; S. 580, 14. Seó wæs ungeendodre lengo infinitæ longitudinis, 5, 12; S. 627, 36. Hí tóætýcton lengeo ðære þrýh twegra fingra gemet addiderunt longitudini sarcofagi quasi duorum mensuram digitorum, 4, 11; S. 580, 6. Ðæs lengo ne his heánesse ǣnig ende gesewen wæs cujus neque longitudini neque altitudini ullus esse terminus videretur, 5, 12; S. 629, 13. Tó lengo his ad staturam suam, Mt. Kmbl. Lind. Rush. 6, 27. Lengu dæga longitudine dierum, Ps. Surt. 90, 16. Lengu, Lk. Skt. Rush. 12, 25. Se ðe lífa gehwæs lengu wealdeþ he who determines the length of every life, Exon. 40 a; Th. 133, 2; Gū. 483. Tele nū ða lengu ðære hwíle, Bt. 18, 3; Fox 66, 6 note. v. leng.

lent, e; f. A lentil:—Lent legumen (cf. lentis, legumen, Ep. Gl. 13 e, f, 8), Germ. 390. [Cf. O. H. Ger. linsi; f. lens: M. H. Ger. linse.]

leó, g. león; [a dat. leóne and acc. f. leó are found as well as regular forms león: the dat. pl. leónum is put under leóna q. v.] m. f. A lion, lioness:—Leó leo, Wrt. Voc. 77, 78. Leó leo, leena, Wrt. Voc. ii. 53, 47, 49. Ðæt nǣfre míne fýnd ne grípen míne sáwle swá swá leó nequando rapiat ut leo animam meam, Ps. Th. 7, 2: 21, 11. Ðá ongan seó leó fægnian ... Seó leó mid hire earmum scrǣf geworhte, Glostr. Frag. 110, 7, 15. Ðonne seó leó bringþ his hungregum hwelpum hwæt tó etanne, Ors. 3, 11; Swt. 142, 24. Seó leó ðeáh hió wel tam sē and hire magister swíðe lufige, Bt. 25; Fox 88, 9. Etan león flǣsc ... Nim león gelynde to eat lion's flesh ... take lion's suet, L. Med. ex Quad. 10, 12; Lchdm. i. 364, 22, 24. Gefríða mē of ðæs león mūðe libera me de ore leonis, Ps. Th. 21, 19. Of león hwelpum, 56, 4: 103, 20. León hwelpas leunculi, Wrt. Voc. ii. 51, 42. Griffus fiðerfóte fugel, leóne gelíc on wæstme, Wrt. Voc. 78, 2. Hió sceolde forsceoppan tó león and ðonne seó sceolde sprecan ðonne rýnde hió she turned into a lioness, and when it ought to have spoken, then she roared, Bt. 38, 1; Fox 194, 33. Nán heort ne onscúnode nǣnne león, 35, 6; Fox 168, 9. Hē gelǣhte áne león be wege, Jud. 14, 5. Ða wildan leó hē gewylde the wild lion he subdued, Ælfc. T. Grn. 7, 16. Ðú miht tredan león and dracan conculcabis leonem et draconem, Ps. Th. 90, 13: Glostr. Frag. 110, 3. Ús symle león and beran ūre ehtan incursantibus leonibus ursisque, Nar. 12, 3. Ða ðe león wǣron ongunnon rýnan, Bt. Met. Fox 26, 165; Met. 26, 68. Tólýseþ leóna mægen molas leonum confringet, Ps. Th. 57, 5. Hwelpas leóna catuli leonum, Ps. Spl. 103, 22. Hý mon sende in wildra deóra menigo, in leóna and in berena, Shrn. 133, 10: Wulfst. 200, 23. Hwænne áhredst [dū] míne ángan sáwle æt ðǣm leóum (leóm, Ps. Surt.) restitue a leonibus unicam meam, Ps. Th. 34, 17. Hē hēt gelǣdan león and beran, manega and mycele, Homl. Skt. 4, 403. [In Orm. and Laym. leo occurs as well as leon. Icel. léo; m: O. H. Ger. lio, leuuo; g. leuuen, Grff. 2, 31.]

leód, es; pl. [which is more frequent] leóde; m. A man, poet. a prince [cf. Icel. álfa ljóði]; in pl. men, people, people of a country, country [cf. the use of proper names, e. g. hē gewát intó Galwalum he departed into Gaul, Chr. Erl. 5, 14]:—Leód Ebréa [Abraham], Cd. 136; Th. 171, 28; Gen. 2835. Ebréa leód, 98; Th. 130, 21; Gen. 2163. Wedera leód [Beowulf], Beo. Th. 687; B. 341: 702; B. 348: 1254; B. 625. Gif hwá his geneáte leód [MS. H. leód] bebycgge if any one sell his own countryman, L. In. 11; Th. i. 110, 3. Ðá hatedon hine his leóde cives autem ejus oderant eum, Lk. Skt. 19, 14. Ða leóde ðá flugon ðá hié ðone here tóweardne wiston the people fled when they knew the army was coming, Blickl. Homl. 79, 12. Ðá flugon ða hǣðnan leóde, 203, 16. Lifigende leóde, Cd. 205; Th. 255, 3; Dan. 618. Leóde ne cúðan módblinde men meotud oncnáwan people, men mind-darkened, could not their maker recognize, Exon. 25 a; Th. 73, 10; Cri. 1187. Wedera leóde, Beo. Th. 455; B. 225. Wē synt gumcynnes Geáta leóde by race are we men of the Gauts, 526; B. 260. Hē ealle ða landbigengan wolde ūtāmǣran and his ágenra leóda mannum gesettan omnes indigenas exterminare, ac suæ provinciæ homines pro his substituere contendit, Bd. 4, 16; S. 584, 7. Hit ná geweorþan sceolde ðæt se wǣre leóda cyning se ðe ǣr wæs folce þeów it ought not to be, that he that had been a servant to a people, should be a king of men, Ors. 4, 6; Swt. 178, 11. Leóda lífgedál Lothes gehýrde brýd Lot's wife heard the death of men, Cd. 119; Th. 154, 25; Gen. 2561. Leóda ǣnigum nytte of use to any man, Beo. Th. 1591; B. 793. Láþ leóda gehwam, Exon. 10 b; Th. 12, 31; Cri. 194. Hæleþa eðel, leóda gesetu, Andr. Kmbl. 2519; An. 1261. Wē ðissa leóda land gesóhton, 535; An. 268. Ðǣr wæs þreó þúsend ðæra leóda there was three thousand of the people, Elen. Kmbl. 570; El. 285. Leóda bearn [cf. O. Sax. liudi-barn] the children of men, Exon. 24 a; Th. 69, 11; Cri. 1119: Exon. 975; Erl. 124, 32; Edg. 24. Leóda [MS. leode] þeódum, Ps. Th. 80, 12. Geáta leóda cempan warriors of the men of the Gauts, Beo. Th. 416; B. 205. Ic eówra leóda willan geworhte, 1273; B. 634. Næs ðǣr má sínra leóda nemne elleffne orettmæcgas, Andr. Kmbl. 1326; An. 663. Wæs hē eallum his leódum leóf ipse [Oswin] amabilis omnibus præfuit, Bd. 3, 14; S. 539, 33. Bæd hē Theodor ðæt hē him and his leódum bisceop funde [sibi suisque], 4, 3; S. 566, 25. Tó nytnysse his leódum utilitati suæ gentis, 2, 16; S. 520, 3. Ðæt Súþseaxna

mǽgþ sceolde habban ágenne bisceop on heora leódum *ut provincia Au- stralium Saxonum ipsa proprium haberet episcopum*, 5, 18; S. 636, 14. Æþelwulf tó his leódum cuom, Chr. 855; Erl. 68, 31. Ælþeódige men ... swǽse men in leódum *aliens ... natives of the country*, L. Wih. 4; Th. i. 38, 3. [Cf. below, Beo. Th. 3741.] Ðǽm Cristenum leódum com Godes engel on fultum *God's angel had come to the Christians as a help*, Blickl. Homl. 203, 25, 20: Cd. 24; Th. 31, 22; Gen. 489: 157; Th. 195, 16; Exod. 277. Hié wíf tó Denum feredon lǽddon tó leódum *they bore her to Denmark*, Beo. Th. 2322; B. 1159. Wǽron æþelingas eft tó leódum fúse tó farenne *the nobles were eager to go back to their people*, 3613; B. 1804. Gif cyning his leóde tó him geháteþ and heom mon ðǽr yfel gedó *if a king summon his people to him and evil is done to them there*, L. Ethb. 2; Th. i. 2, 8. Ceadwealla slóh ða Norþhymbran leóde æfter heora hláfordes fylle, Swt. A. S. Rdr. 95, 9. Leóda, 96, 40. Leóde hogode on ðæt micle morþ, men forweorpan, Cd. 32; Th. 43, 14; Gen. 690: Andr. Kmbl. 339; An. 170. Leóde, Judéa cyn, Elen. Kmbl. 416; El. 208. Hét hine leóde swǽse sécean *bade him seek his own people*, Beo. Th. 3741; B. 1868: 2677; B. 1336. Land and leóde, Andr. Kmbl. 2643; An. 1323: Chr. 1065; Erl. 198, 6; Edw. 25. Gif ðú ðæt gerǽdest ðæt ðú ðíne leóda lýsan wille *if you decide to save thy men*, Byrht. Th. 132, 56; By. 37. [O. L. Ger. liud; *m*; *pl*. liudî: O. Sax. liudî: O. Frs. liode, liude: *Icel.* lýðir; *pl*. [e. g. af lýðum sínum *by his people*]: O. H. Ger. liuti *homines*: Ger. leute.] v. burh-, eást-, ge-, land- leód; *and next word*.

leód, e; *f*. A people, nation, race, district occupied by a people [v. pre- ceding word, and cf. mǽgþ], *country* :—Hit wæs hwílum on Engla lagum ðæt leód and lagu fór be geþincþum *at one time it was in the laws of the English, that the people and the law went according to ranks*, L. R. 1; Th. i. 190, 11. Ðæt leód and lagu trumlíce stande, Wulfst. 74, 8. Feówer folccyningas, leóde rǽswan, Cd. 95; Th. 125, 6; Gen. 2075. Ða fǽhþe eówer leóde *the hostility of your people*, Beo. Th. 1197; B. 596. Tó fela Deniga leóde, 1396; B. 696: 1202; B. 599. Se wæs Cantwara leóde *oriundus de gente Cantuariorum*, Bd. 3, 14; S. 539, 27. Moyses leóde *from the Israelites*, Cd. 149; Th. 187, 16; Exod. 152. Wæs his gewuna ðæt hé his ágene leóde Norþanhymbra mǽgþe sóhte *solebat suam, id est, Nordanhymbrorum provinciam revisere*, Bd. 3, 23; S. 554, 6. Hé wæs ealle ða land and leóde þurhfærende *omnia pervagatus*, 3, 30; S. 562, 13. Úres hláfordes gerǽdnes is ðæt man cristene menn of earde ne sylle ne húru on hǽðene leóde *our lord's ordinance is, that Christian men be not sold out of the land, certainly not into a heathen country* [or leóde = men, preceding word], L. Eth. v. 2; Th. i. 304, 16: Beo. Th. 387; B. 192. Ðone Denisca leóda lufiaþ swýðost him [Thor] *the Scan- dinavian peoples love most*, Wulfst. 106, 23. Beneuentium and Sepontium hátton ða twá leóde *Benevento and Sepontus were the two places called*, Blickl. Homl. 201, 22. Ealle him leóda lácum cwemaþ *all nations shall make offerings to please him*, Ps. Th. 71, 10. [O. Sax. liud-: O. Frs. liod: *Icel.* ljóð-; lýðr; *m*. *people, common people*: O. H. Ger. liut; *m*. *n*. populus, plebs.] v. land-leód, *and preceding word*.

leód, es; *m*. Fine for slaying a man [cf. leudus, id est weregildus; *and see other passages in* Grmm. R. A. 652] :—In xl nihta ealne leód forgelde *let him pay the whole fine within forty days*, L. Ethb. 22; Th. i. 8, 6. Healfne leód, 23; Th. i. 8, 7. v. leód-geld, wer-geld.

leóda, an; *m*. A man, one of a people or country :—Gif hwá his ágenne geleód [MS. B. leódan] bebycgge *if any one sell a man of his own people*, L. In. 11; Th. i. 110, 3. Be leódan bygene *concerning the sale of a man of one's own country*, Th. i. 110, 1 note.

leódan; *p*. leád; *pl*. ludon To spring, grow :—Swá Libanes beorh líðeþ and gróweþ *sicut cedrus Libani multiplicabitur*, Ps. Th. 91, 11. Of ðam twige ludon rǽde wæstme *from that branch sprang dire fruits*, Cd. 47; Th. 60, 29; Gen. 989. [Goth. liudan: O. Sax. liodan: O. H. Ger. ar-, fram-liutan.] DER. á-, ge-leódan.

leód-bealu, wes; *n*. Harm or bale which affects a people, Beo. Th. 3448; B. 1722: 3896; B. 1946.

leód-biscop, es; *m*. A bishop of a district, province, or diocese, a bishop subordinate to an archbishop, a suffragan. *The* leódbiscop *ranks with the* ealdorman, *the* arcebiscop *with the* æþeling. In Rtl. 194, 34–40 occurs the following 'Chore episcopi; Grece core, Latine vicari, episcopi: hii in vicis et villis constituti habentes licentiam constituere gradum mi- norem, non presbiterum neque diaconum, propter scientiam episcopi ·in cujus regione est.' The Greek form is here glossed by *liódbiscop*, the Latin by *scirebiscop*. Ercebiscop *archiepiscopus*; leódbiscop, *episcopus*, Wrt. Voc. 71, 70, 71. Se hálga Cúðbertus Lindisfarnensiscere gelap- unge leódbiscop [cf. hé wæs tó biscope gecoren ðære cyricean æt Lindis- farena eá, Bd. 4, 28; S. 606, 7], Homl. Th. i. 148, 22. Gif hwá arcebiscopes oðða æþelinges borh abrece ... Gif hwá leódbiscopes oðða ealdormannes, L. C. S. 69; Th. i. 408, 8–10. Ðæt Turonisce folc hine geceás him tó leódbiscope *the people of Tours chose him as their bishop*, Homl. Th. ii. 506, 3: Chr. 971; Erl. 125, 34. Bútan hit be- foran cyninge oðða leódbiscope oðða ealdorman beó, Chart. Th. 612, 13. Séce man tó ðam leódbiscope; and gif man furþor scule tó ðam arce- biscope; and syððan tó ðam pápan, Wulfst. 275, 6. Gif hé sóhte leódbiscop oðða ealdorman ðonne áhte hé vii nihta griþ, L. Eth. vii. 5;

Th. i. 330, 14. Ða bǽdon ealle ða leódbisceopas ðone hálgan apostol ðæt hé ða feórþan bóc gesette *then all the provincial bishops asked the apostle to compose the fourth gospel*, Homl. Th. i. 70, 6. Hé lætt gewrítan hú mycel landes his arcebiscopas hæfdon and his leódbiscopas and his abbodas and his eorlas, Chr. 1085; Erl. 218, 30. [Mid arce- biscopes and leódbiscopes and abbotes, 1125; Erl. 254, 8. Ealle ða leódbiscopes ða ða wǽron on Englalande, 1129; Erl. 258, 10.] [*Icel.* adopts from English ljóð-, lýð-biskup *a suffragan bishop*.] Cf. scír-biscop.

leód-burh; *f*. A people's town, a town of a country, town occupied by a people :—Of ðysse leódbyrig [Sodom], Cd. 116; Th. 150, 33; Gen. 2501. Hé eaferum lǽfde lond and leódbyrig *he to his children left his land and its towns*, Beo. Th. 4933; Th. 2471.

leód-cyning, es; *m*. The king of a people :—Beówulf Scyldinga leóf leódcyning, Beo. Th. 107; B. 54. [*Laym.* leod-king.]

leóde; *pl*. people. v. leód.

leód-fruma, an; *m*. The first in time of a people, the founder of a people, a patriarch; the first in rank among a people, a prince, chieftain, king :—Him wæs án fæder leóf leódfruma *one father had they, founder beloved*, Cd. 161; Th. 200, 9; Exod. 354. Leódfruma [St. Andrew], Andr. Kmbl. 3318; An. 1662: [Constantine], Elen. Kmbl. 382; El. 191. Mín leódfruma *my lord*, Exon. 115 a; Th. 442, 5; Kl. 8. Sethes cynn, leófes leódfruman, Cd. 63; Th. 75, 26; Gen. 1246. Of ðam leódfruman brád folc cumaþ *from that patriarch* [Isaac] *shall come nations wide-spreading*, 106; Th. 140, 24; Gen. 2332. Gif hí leód- fruman lǽstan dorsten *if they durst follow their chief*, Bt. Met. Fox 1, 53; Met. 1, 27. Cyning, leófne leódfruman, Exon. 60 b; Th. 222, 7; Ph. 345: [Hrothgar], Beo. Th. 4266; B. 2130: [St. Andrew], Andr. Kmbl. 1977; An. 991.

leód-geard, es; *m*. The dwelling of a people, country :—Sunu æfter heóld leódgeard, Cd. 62; Th. 74, 20; Gen. 1225. Ethiopia land and leódgeard, Th. 15, 6; Gen. 22: 85; Th. 106, 18; Gen. 1773. [Cf. *Icel.* ljóð-heimar *the people's abode, the world*.]

leód-gebyrga, an; *m*. The protector of a people, a prince, chief man :— Se æþeling, leódgebyrga [Constantine], Elen. Kmbl. 405; El. 203. Hláf- ord ðinne, leódgebyrgean [Hrothgar], Beo. Th. 543; B. 269. Leód- gebyrgean *the chief men of the city* [cf. ceastre weardas *applied to the same persons in* v. 767], Elen. Kmbl. 1108; El. 556.

leód-geld, es; *n*. The fine paid for slaying a man, L. Ethb. 21; Th. i. 8, 4: 7; Th. i. 4, 9. v. Grmm. R. A. 653, *and* leód.

leód-geþyncþ, es; *f*. Rank existing amongst a people :—Be leódge- þincþum, L. R.; Th. i. 190, 10.

leód-gewinn, es; *n*. Strife :—Lǽt sace restan, láð leódgewin, Exon. 68 b; Th. 254, 22; Jul. 20.

leód-gryre, es; *m*. Terror affecting a people, Salm. Kmbl. 558; Sal. 278.

leód-hata, an; *m*. A tyrant :—Nalæs swá swá sigefæst cyning ac swá swá leódhata *non ut rex victor sed quasi tyrannus*, Bd. 3, 1; S. 523, 29. Bana, láð leódhata [the angel that destroyed the first-born in Egypt], Cd. 144; Th. 180, 4; Exod. 40. For wédenheortnesse ðæs leódhatan Brytta cyninges *propter vesanam Brittonici regis tyrannidem*, Bd. 3, 1; S. 524, 2: Bt. 16, 2; Fox 52, 30. Láðne leódhatan [Holofernes], Judth. 10; Thw. 22, 22; Jud. 72. Hér sind on earde leódhatan grimme ealles tó manege *here in the land are fierce tyrants all too many*, Swt. A. S. Rdr. 109, 155. Áwyrgede womsceaðan, leáse leódhatan, Elen. Kmbl. 2597; El. 1300. Cyningas ða habbaþ under him mænigfealde leódhatan *reges sub se multos habentes tyrannos*, Nar. 38, 19.

leód-hete, es; *m*. Hate or enmity felt by a people, Andr. Kmbl. 2278; An. 1140: 224; An. 112: 2300; An. 1151.

leód-hryre, es; *m*. Fall or destruction of a people, Beo. Th. 4771; B. 2391: 4064; B. 2030.

leód-hwæt; *adj*. Very brave [cf. leud a *prince* ?] :—Se leódhwate lind- geborga, Elen. Kmbl. 21; El. 11. [Grein suggests lindhwata leódge- borga; cf. leód-gebyrga.]

leód-mǽg, es; *m*. A kinsman as being one of the same race, tribe or people, a man of the same nation with one's self :—Hí fundon fíf hund leódmǽga *they found five hundred of their race*, Elen. Kmbl. 759; El. 380. Leódmágum feor *far from my kinsmen* [Abraham in Egypt], Cd. 128; Th. 163, 6; Gen. 2694.

leód-mægen, es; *n*. The might of a people, its fighting men :—Ðæt leódmægen, gúþrófe hæleþ, eorlas æscrófe, Elen. Kmbl. 544; El. 272. Lofige hine eall his leódmægen *laudate eum omnes virtutes ejus*, Ps. Th. 148, 2. Leódmægnes worn *a host of warriors*, Cd. 151; Th. 190, 7; Exod. 195: Th. 188, 13; Exod. 167.

leód-mearc, e; *f*. A people's territory, a country, Andr. Kmbl. 572; An. 286: 1554; An. 778.

leód-riht, es; *n*. Public law, common law, the law which affects a whole people, law of the land; jus publicum :—Mid rihtum landrihte and leódrihte swá hit on lande stonde *in accordance with the common law of the land*, Cod. Dip. Kmbl. iii. 435, 35. Bútan leódrihte, Andr. Kmbl. 1357; An. 679. v. folc-, land-riht.

leód-rúne, an; *f*. A witch, wise woman [cf. burh-rúne *furia*; helle-

rûne *pythonissa*: Grmm. D. M. 375 *on the forms of feminine names in* -rûn, -rûna]:—Wið ǽlcre yfelre leódrûnan . . . eft óðer dust and drenc wið leódrûnan, L. M. 1, 64; Lchdm. ii. 138, 23, 26. Cockayne translates the word 'heathen charm.' Cf. *Laym.* 9121 seolcuðe leodronen [tocke, 2nd. MS.]: leoten weorpen & fondien leodrunen [*incantations*], 15499, 15511: leodrunen [deorne rouning, 2nd MS.], 14553.

leód-scearu, e.; *f. A people, nation,* Cd. 160; Th. 199, 12; Exod. 337. Cf. folc-scearu.

leód-sceaða, an; *m. A harmer of men, a public enemy:*—Láð leódsceaða [*the serpent*], Cd. 43; Th. 56, 24; Gen. 917. Æt ðam leódsceaðan hreddan *to save from the devil,* Exon. 11 b; Th. 17, 20; Cri. 273. Ic ðam leódscaðan [*Grendel*] hondleán forgeald, Beo. Th. 4193; B. 2093. Hearmcwide láðra leódsceaðena [*the Mermedonians who abused St. Matthew*], Andr. Kmbl. 159; An. 80. [*O. Sax.* liud-skaðo (*the devil*).] cf. folc-sceaða.

leód-scipe, es; *m. A people, nation, country occupied by a people:*—Ðe ðes leódscype longe bieode *whom this people have long worshipped,* Exon. 68 b; Th. 255, 2; Jul. 208. Of ðam leódscipe ðe is Siria geháten *from the country that is called Syria,* Homl. Th. i. 400, 7: Exon. 64 a; Th. 236, 30; Ph. 582. Eallum his leódscipe tó ðearfe *for the behoof of all his people,* L. Edg. pref; Th. i. 262, 4: L. Eth. ii. 1; Th. i. 284, 10. Woruldrihta ic wille ðæt standan on ælcum leódscipe [*English and Danish and British, see the rest of the section*], L. Edg. S. 2; Th. i. 272, 23: Beo. Th. 4400; B. 2197. On ðam leódscipe [*the Greeks*], Bt. Met. Fox 30, 3; Met. 30, 2. Hwæt tó bóte mihte æt ðæm færcwealme ðe his leódscipe swýðe drehte, L. Edg. S. 1; Th. i. 270, 10: Chr. 1014; Erl. 150, 9: Reo. Th. 5495; B. 2751: Bt. Met. Fox 1, 135; Met. 1, 68. Ðrý leódscipas sind gehátene India, Homl. Th. i. 454, 11. Hí cyning habban woldon swá swá óðre leódscipas hæfdon *they wanted to have a king, as other nations had,* Ælfc. T. Grn. 6, 45. Tó ðam leódscipum ðe tó geleásan bugon, 14, 3. Ða cynegas ðe eardodon on ðam leódscipum *reges Amorrhæorum et Chanaan,* Jos. 5, 1. Bodigende geleáfan ðam leódscipum ðe sind gecwedene Galatia, Cappadocia, Bithinia, Asia, Italia, Homl. Th. i. 370, 26: L. I. P. 23; Th. ii. 334, 28. Hé wið feó sealdon wíde intó leódscipas *sold them into distant countries,* Blickl. Homl. 79, 23. [*O. Sax.* liud-skepi *a people: O. H. Ger.* liut-scaf.] Cf. þeód-scipe.

leód-stefn, es; *m. A race, family, people,* Ps. Th. 82, 7. [*O. H. Ger.* liut-stam : cf. *O. Sax.* liud-stemni; *adj. belonging to a people.*]

leód-þeáw, es; *m. Custom of a people* or *country:*—Ðá hé tó mé cwom ðá grétte hé mé sóna 'and [h]álette his leódþeáwe *cum me more rituque salutaret,* Nar. 27, 3. Ne wolde ðám leódþeáwum Loth onfón *Lot would not adopt those customs of the country,* Cd. 92; Th. 116, 18; Gen. 1938.

leód-weard, e; *f. The guard* or *government of a people* or *country,* Cd. 59; Th. 72, 1; Gen. 1180: 60; Th. 72, 3; Gen. 1196: 145; Th. 181, 6; Exod. 57.

leód-wer, es; *m. A man of a nation:*—Leódweras [*the Egyptians*], Cd. 89; Th. 110, 5; Gen. 1833. Ofer leódwerum [*the Israelites*], 148; Th. 184, 20; Exod. 110.

leód-werod, es; *n. The host formed by a people:*—Wolcen lædde leódwerod [*the Israelites*], Cd. 146; Th. 182, 17; Exod. 77.

leód-wita, an; *m. A man of intelligence in a people:*—Ða wæron þeódwitan [leódwitan, MS. H.] weorþscipes wyrþe, eorl and ceorl, þegen and þeóden, L. R. 1; Th. i. 190, 12. v. Grmm. R. A. 267.

leód-wynn, e; *f. Joy that comes from being among one's own people:*—Leódwynna leás, wineleás wrǽcca, Exon. 119 a; Th. 457, 25; Hy. 4, 89.

leóf, *used as a form of address to one* or *to many, cf. modern 'dear sir':*—Wé biddaþ ðé leóf ðæt ðú hlyste úre sprǽce *oramus, domine, ut audias nos,* Gen. 43, 20: 3, 10: Ælfc. Gen. Thw. 1, 5, 14. Ðá cwæþ ðæt wíf tó him leóf ðæs mé þingþ ðú eart wítega *dicit ei mulier domine video quia propheta es tu,* Jn. Skt. 4, 19. Hí cwǽdon, leóf, wé wyllaþ geseón ðone hælend, 12, 21. Seó gegaderung his leorningcnihta cwæþ Drihten leóf wilt ðú nú gesettan ende ðysre worulde *the assembly of his disciples said, Lord, wilt thou now put an end to this world,* Homl. Th. i. 294, 24. Ic bidde eów leóf ðæt gé gecirron tó mínum húse *obsecro, domini, declinate in domum pueri vestri,* Gen. 19, 2. Gefyrn ic hine cúðe leóf . . . La leóf nele hé gelýfan mínum wordum *long ago I knew him, Sir . . . Ah! Sir, he will not believe my words,* Glostr. Frag. 2, 10, 19. Lá leóf O Lord, Gen. 18, 23, 25, 28, 30, 31. Hí cwǽdon tó ðám apostolon lá leóf hwæt is ús tó dónne *they said to the apostles, Sirs, what shall we do?* Homl. Th. i. 314, 33. v. next word.

LEÓF; *adj.* LIEF, *desirable, pleasant, acceptable, loved, beloved, dear; used substantively, one who is dear, a friend, loved one:*—Se ðe gód onginneþ and ðonne áblinneþ ne biþ hé Godes leóf on ðæm néhstan dæge *he who begins good and then ceases, will not be God's friend at the last day,* Blickl. Homl. 21, 35. Wæs hé eallum his geférum leóf *he was dear to all his companions,* 213, 12: Cd. 4; Th. 5, 30; Gen. 79. Hé wæs leóf Gode, 130; Th. 165, 26; Gen. 2737. Ealre his þeóde leóf heora ríce tó habbanne and tó healdenne *totæ suæ genti ad tenenda servandaque regni sceptra exoptatissimus,* Bd. 5, 19; S. 636, 33. Ne ǽnig mon ne leóf ne láð *no man, neither friend nor foe,* Beo. Th. 1026; B. 511. Gode is

swíðe leóf ðæt gé earmum mannum syllon *it is very acceptable to God, that you give to poor men,* Blickl. Homl. 53, 28. On ða tíd wæs mannum leóf ofor eorþan and hálwende *at that time it was pleasant for men upon earth, and healthful,* 115, 8. Ðá cwæþ Petrus and Andreas tó Johanne ðú leófa drihten gecýðe ús hwylce gemete ðú cóme tódæg tó ús *then said Peter and Andrew to John, 'Dear Sir, tell us how thou camest to us to-day,'* 141, 20. Brúc disses beáges, Beówulf leófa, mid hǽle, Beo. Th. 2437; B. 1216. Eálá leóf hláford, O, mi domine, Coll. Monast. Th. 19, 13. Hér is mín leófa sunu *hic est filius meus dilectus,* Mt. Kmbl. 17, 5. Matheus mín se leófa, beheald on mé, Blickl. Homl. 229, 30. Forþférde Gode se leófa fæder Agustinus *defunctus est Deo dilectus pater Augustinus,* Bd. 2, 3; S. 504, 30. Se leófa cuma and se lufigendlíca *hospes ille amabilis,* 4, 3; S. 568, 16. Mé sealde sunu on leófes stæl ðæs ðe Cain oflslóh *he gave me a son in place of the loved one, him whom Cain slew,* Cd. 55; Th. 68, 7; Gen. 1113. Leófes and láðes *of friend and foe,* Beo. Th. 5813; B. 2910. Fela sceal gebídan leófes and láðes *he shall experience much pleasure and pain,* 2126; B. 1061. Ic ðé wolde leófum lofsang cweþan, Ps. Th. 118, 164. Áledon leófne þeóden on bearm scipes, Beo. Th. 68; B. 34. Hláford leófne, 6276; B. 3142. Leófe ðíne *dilecti tui,* Ps. Th. 59, 4. Míne bróðru leófon *my dear brethren,* Bd. 4, 24; S. 598, 43. Ðǽr ne biþ leófra gedál ne láðra gesamnung *there shall not be parting of friends there, or meeting of foes,* Blickl. Homl. 65, 20. Ðá cwǽdon hié ðæt him nǽnig mǽg leófra nǽre ðonne hiera hláford *then they said that no kinsman was dearer to them than their lord,* Chr. 755; Erl. 50, 19. Hí cwǽdon ðæt him nán hláford leófra nǽre ðonne hiora gecynda hláford, 1014; Erl. 150, 25. Leófre mé ys ðæt ic hig sylle ðé ðonne óðrum men *melius est, ut tibi eam dem, quam alteri viro,* Gen. 29, 19. Ic wylle and mé leófre sig gif ðú mǽge *volo et multum delector, si potes,* Bd. 5, 3; S. 616, 31. Ǽghwilcum men biþ leófre swá hé hæbbe freónda ðe má *the more true friends he has, the better every man likes it,* Blickl. Homl. 121, 36. Ús biþ ðonne leófre ðonne eal eorþan wela gif hé ús miltsian wile *if he will shew us mercy, shall we not prefer that to all the wealth of earth?* 51, 29. Ǽnne tíman ðonne ús wǽre leófre ðonne eall ðæt on middanearde is, ðæt wé áworhtan georne Godes willan, L. C. E. 18; Th. i. 370, 18. Ne dém ðú óðerne dóm ðam liófran and óðerne ðam láðran, L. Alf. 43; Th. 1, 54, 12. Him wǽron ær his ǽhta leófran tó hæbbenne ðonne Godes lufu *he would rather have his possessions than God's love,* Blickl. Homl. 195, 9. Eall forlǽteþ ðæt him wæs leófost tó ágenne and tó hæbbenne, 111, 26. For oft hit wyrþ raðost forloren ðonne hit wǽre leófost gehealden *too often it is most quickly lost, when keeping it would be most pleasant* [or leófost *adv.*?], Wulfst. 109, 4. Ðes is mín leófesta sunu *hic est filius meus carissimus,* Mk. Skt. 9, 7. Ðú leófesta [Hat. MS. leófusta] bróður *frater carissime,* Past; Swt. 22, 9. Ic sende grétan ðone leófastan cyninge Ceólwulf, Bd. ded; S. 471, 8. Míne gebróðra ða leófostan *my dearest brethren,* Homl. Th. ii. 4, 19. Men ða leófostan, 188, 25: Blickl. Homl. 165, 33. Leófestan, 9, 13. Ða word ðe hé wénþ ðæt him leófoste sýn tó gehýrenne *the words that he thinks will be most pleasant for him to hear,* 55, 20. [*Goth.* liubs: *O. Sax.* liof: *O. Frs.* liaf, lief: *Icel.* ljúfr: *O. H. Ger.* liub, liob, lieb *gratus, desiderabilis, carus, optatus, amicus: Ger.* lieb.] DER. fela-, mód-, ofer-, un-leóf. The word occurs forming part of proper names, e.g. Leóf-ríc, Leóf-sunu, Leóf-wine; so in other dialects.

leófan; *p.* leáf; *pl.* lufon. Grein suggests that this verb is found in the following passage:—Éðelweardas lufan lífwelan ðenden hié lét metod, Cd. 174; Th. 219, 17; Dan. 56. Is it possible however that a verb such as *hæfdon* should be supplied, and that *lufan* is the accusative after it?

leófen. v. lifen.

leófian. v. lifian.

leófian, *p.* ode *To be dear* or *pleasant, to delight:*—Him leófedan londes wynne bold on beorhge *the pleasures of the country were dear to him, the house on the hill,* Exon. 34 b; Th. 110, 19; Gú. 110. [Cf. *O. L. Ger.* ge-lievan *delectari, delectare: O. H. Ger.* liubjan *diligere, affectare, commendare.*]

leóf-líc; *adj. Lovely, beautiful, delightful, pleasant, lovable, dear:*—Wíglaf leóflíc lindwíga *Wiglaf, warrior dear,* Beo. Th. 5199; B. 2603. Leóflíc cempa, Andr. Kmbl. 2891; An. 1448. Leóflíc wíf, Elen. Kmbl. 572; El. 286. Eafora leóflíc on life, Cd. 82; Th. 103, 4; Gen. 1713. Leóflíc geþwǽrnes *fair concord,* Dóm. L. 18, 270. Ðone wlitigan wong and wuldres setl leóflíc *the beauteous plain and the pleasant seat of glory,* Exon. 62 a; Th. 228, 18; Ph. 440. Hié Sarran wlite heredon óð ðæt hé lǽdan héht leóflíc wíf tó his selfes sele, Cd. 89; Th. iii. 16; Gen. 1856. His sweord leóflíc íren *his sword, weapon of price,* Beo. Th. 3622; B. 1809. Lofiaþ leóflícne *they laud the beloved (God),* Exon. 13 b; Th. 25, 13; Cri. 400. [*Goth.* liuba-leikr *lovely* (Phil. 4, 8): *O. Sax.* liof-lík: *O. Frs.* liaf-lík: *O. H. Ger.* liub-líh *amoenus, venustus, pulcher, gratus, elegans, splendidus: Ger.* lieb-lich.]

leóf-líce; *adv. Kindly, graciously, gladly, lovingly:*—Deáh ðe ic scyle ealle wucan fæstan ic ðæt leóflíce dó *though I have to fast all the week, I will do it gladly,* Bd. 4, 25; S. 600, 7. Hé leóflíce lífes ceápode moncynne *graciously he purchased life for mankind,* Exon. 24 a; Th. 67,

29; Cri. 1096. Fore onsýne éces déman lǽddon leófsíce *before the face of the eternal judge they led him lovingly*, 44 a; Th. 149. 3; Gú. 756. [*O.H. Ger.* liub-lího *gratifice, perfloride, evitaliter.*]

leóf-spell, es; *n. A pleasant message :*—Leófspell manig, Elen. Kmbl. 2032; El. 1017.

leóf-tǽle, -tǽl; *adj. Loving, dear, desirable, estimable, grateful, pleasant, gracious :*—Hé biþ freónd and leóftǽl lufsum and líðe *he (Christ) shall be friendly and gracious, kind and gentle*, Exon. 21 a; Th. 57, 4; Cri. 913. Hé is monþwǽre, lufsum and leóftǽl, 96 a; Th. 357, 21; Pa. 32. Óðer biþ unlǽde on eorþan óðer biþ eádig swíðe leóftǽle mid leóda duguþum *one will be miserable on earth, the other fortunate, high in favour with the best of men*, Salm. Kmbl. 733; Sal. 366. Nán cræft nis Gode deórwyrðra ðonne sió lufu ne eft ðam deófle nán cræft leóftǽlra ðonne hié mon slíte *nil pretiosius est Deo virtute dilectionis, nil est desiderabilius diabolo extinctione caritatis*, Past. 47, 2; Swt. 359, 24. Ða welan beóþ hlíseádigran and leóftǽlran ðonne ðonne hié mon seþ ðonne hié beón ðonne hí mon gadraþ. Seó gítsung gedéþ heore gítseras láðe ǽgðer ge Gode ge monnum and ða cysta gedóþ ða simle leóftǽle and hlíseádige *divitiæ effundendo magis quam coacervando melius nitent: siquidem avaritia semper odiosos, claros largitas facit*, Bt. 13; Fox. 38, 13–17.

leóf-wende; *adj. Pleasing, gracious, acceptable, amiable, estimable :*—Nó liófwende *non gratus*, Wrt. Voc. ii. 61, 62. Sum biþ leófwende hafaþ mód and word monnum geþwǽre *one man is amiable, he hath mind and speech in accord with men*, Exon. 79 b; Th. 298. 13; Crä. 84. Ne beó ðú nó tó tǽlende ac beó leófwende *be not too ready to blame, but be amiable*, 81 a; Th. 305, 22; Fä. 92. Ðæt ic meotud ðínum lǽrum leófwendum lyt geswíce *that I, O Lord, little desert thy pleasant precepts*, Andr. Kmbl. 2581; An. 1292. Wuton wuldrian weorada dryhten lufian liófwendum lífes ágend *let us glorify the Lord of hosts, gratefully love the disposer of life*, Hy. 8, 3; Hy. Grn. ii. 290, 3: Exon. 14 b; Th. 29, 31; Cri. 471. Ðeáh hit gód seó and deóre ðeáh biþ hlíseádigra and leófwendra se ðe hit selþ ðonne se ðe hit gaderaþ and on óðrum reáfaþ *though it (gold) be good and precious, yet will he be of better repute and esteem who gives it, than he who collects it and robs it from another*, Bt. 13; Fox 38, 12.

LEÓGAN; *p.* leáh; *pl.* lugon *To lie, tell a lie, say falsely, break one's word, play false, deceive, feign :*—Ic leóge *mentior*, Ælfc. Gr. 31; Som. 35, 53. Eal hit is swá, ne leóge ic, Blickl. Homl. 179. 3. Ðú líhst ðæt ðú God sý *thou sayest falsely that thou art God*, Homl. Th. i. 378, 7. Seó orsorge wyrd simle líhþ and lícet *prospera fortuna semper mentitur*, Bt. 20; Fox 70, 30: Ælfc. Gr. 49; Som. 50, 30. Hé líhþ him sylfum,Wulfst 66, 3. Se ðe lýhþ oððe ðæs sóðes ansaceþ, Salm. Kmbl. 362. Sal. 181. Má sceamigan ðonne fagnian ðonne hí geheóraþ ðæt him man on líhþ *qui falso prædicantur, suis ipsi necesse est laudibus erubescant*, Bt. 30, 1; Fox 108, 8. Ic geseó tó sóðe nales mé sefa (MS. selfa) leógeþ *I do indeed see, my mind deceives me not*, Cd. 193; Th. 242, 9; Dan. 416. Ðíne feóndas ðé fǽcne leógaþ (lǽgaþ, Ps. Surt.) *mentientur tibi inimici tui*, Ps. Th. 65, 2: 80, 14. Nú cwǽdon gedwolmen ðæt deófol gesceópe sume gesceafta, ac hí leógaþ, Homl. Th. i. 16, 20. Oft ða unþeáwas leógaþ and lícettaþ ðæt hí sién góde þeáwas *plerumque vitia virtutes se esse mentiuntur*, Past. 20; Swt. 149, 2. Ðá ðá hé leág *fefellisset*, Wrt. Voc. ii. 34, 26: Exon. 84 b; Th. 318, 12; Mód. 81. Hér begann se deófol tó reccanne hálige gewrita and hé leáh mid ðære race *here the devil began to expound holy writ, and he spake falsely in his exposition*, Homl. Th. i. 170, 4. Ðá swóran hié swíðe ðæt hié sóð sægdon and nóht lugon ðara þinga *quibus jurantibus se nichil falsi commiscere*, Nar. 25, 28. Sǽdon ðæt hí wǽran on Criste gelýfede, ac hí lugon swá ðeáh, Homl. Skt. 2, 303. Hig hym fæla ongeán lugon *they brought many false charges against him*, Nicod. 34; Thw. 19, 39. Gé tó dæge wǽron Somnitum þeówe gif gé him ne álugen (other MS. lugon) *iówra wedd hodie Romani Samnio servirent, si fidem fæderis ipsi Samnitibus servavissent*, Ors. 3, 8; Swt. 122, 13. Ne leóh ðú leng *noli ultra fallere*, Ex. 8, 29. Ne leóh ðú *non mentiemini*, Lev. 19, 11. Swá wénaþ manige men, ðæt ðes diáccon leóge be ðam fýre, Wulfst. 206, 13. Ðone ilcan geþang ic ðé ǽr sǽde, ǽr hé leóge, ðæt hé ðé leógan ne durre, Blickl. Homl. 179, 29. Búton Priscianus luge *unless Priscian have made a mistake*, Ælfc. Gr. 17; Som. 20, 49. Se ðe wolde leógan on his wordon, Wulfst. 168, 17. Ðonne onginþ him leógan se tóhopa ðære wræce *then the hope of revenge begins to deceive them*, Bt. 37; Fox 186, 23: Bt. Met. Fox 25, 100: Met. 25, 50: Exon. 90 a; Th. 337, 27; Gn. Ex. 71. Ðæne nǽnig mæg leógan *quem nemo potest fallere*, Hymn. Surt. 33, 15. Ðú leógende sagast, Blickl. Homl. 179, 22. Ðonne hí secgeaþ ǽlc yfel ongén eów leógende *cum dixerint omne malum adversum vos mentientes*, Mt. Kembl. 5, 11. Gé sind leógende *mentita es*, Past. 21, 1; Swt. 151, 21. [*Goth.* liugan: *O. Sax.* liogan: *O. Frs.* liaga: *Icel.* ljúga: *O. H. Ger.* liugan *mentiri, fallere, fingere: Ger.* lügen.] DER. á-, for-, ge-, of- leógan.

leógere, es; *m. A liar, one who speaks or acts falsely, a false witness :*—Up árísaþ leáse leógeras, Wulfst. 79, 4. Leógeras, L. C. S. 5; Th. i. 380, 5. Ðá cómon twegen ðæra leógera *venerunt duo falsi testes*, Mt. Kmbl. 26, 60. [*Icel.* ljúgari *a liar: O. H. Ger.* liugari *fictor.*]

LEÓHT, líht, es; *n.* LIGHT, *a light :*—Geweorþe leóht and leóht wearþ geworht *fiat lux, et facta est lux*, Gen. 1, 3. Tweóne leóht *crepusculum :*

tweónul leóht *maligna lux* vel *dubia*, Ælfc. Gl. 94; Som. 75, 122, 125; Wrt. Voc. 53, 3, 6. Ðæt leóht ðe wé dægréd hátaþ *the light that we call dawn*, Lchdm. iii. 234, 28. Ic geseó ðis hús mid swá mycele leóhte gefylled ðætte ðæt eówer blácern and leóht mé is eallinga þýstre gesewen *domum hanc tanta luce impletam esse perspicio, ut vestra illa lucerna mihi omnimodis esse videatur obscura*, Bd. 4, 8; S. 576, 3. Ðære sunnan beorhtnys and ðæs mónan leóht and ealra tungla, Homl. i. 64, 29: Blickl. Homl. 91, 23. Ðenden him leóht and gǽst somod fæst seón *whilst he lives*, Exon. 31 a; Th. 96, 27; Cri. 1580. Ðú eart dóhtor mín míura eágna leóht *thou art my daughter, the light of mine eyes*, 67 a; Th. 248, 14; Jul. 95. Leóhtes leóhting *lucubrum*, Ælfc. Gl. 67; Som. 69, 89; Wrt. 41, 42. Se blinda bæd his eágena leóhtes *the blind man asked for his eye-sight*, Blickl. Homl. 21, 6: Elen. Kmbl. 596; El. 298. Se dæg wæs fruma ðyses lǽnan leóhtes *the day was the beginning of this transitory light*, Blickl. Homl. 133, 10. Godes cyrcan mid leóhte and lácum gelóme gegrétan *to visit God's church frequently with candles and offerings*, Wulfst. 308, 28. Of ðissum leóhte álǽded *de hac vita subtractus*, Bd. 3, 20; S. 550, 23. Ða ðe hí of ðissum leóhte foreode *qui eas ex hac luce præcesserant*, 4, 7; S. 575, 4. Se sacerd forbærnþ ða drihtne tó leóhte and tó wynsumum stence *adolebit ea sacerdos in holocaustum et suavem odorem domino*, Lev. 1, 9. On lifgendra leóhte *in lumine viventium*, Ps. Th. 55, 11. Ic tó ðé æt leóhte gehwam wacie *ad te de luce vigilo*, 62, 1. Ðe dæg lǽdeþ tó líhte ðǽr hí líf ágon á tó aldre, Cd. 221; Th. 287, 2; Sat. 361. Geearnian leóht ðæs écan lífes, Blickl. Homl. 17, 21. Ðæt þridde ne geseah ðære sunnan leóht nǽfre, Glostr. Frag. 8, 27. Hé Godes leóht geceás *he died*, Beo. Th. 4930; B. 2469: Exon. 52 b; Th. 184, 13; Gú. 1343. Eádgár ceás him óðer leóht, Chr. 975; Erl. 124, 30; Edg. 22. Drihten nam in óðer leóht Agustinus, Menol. Fox 191; Men. 97. Beó nú leóht on ðære heofenan fæstnysse *fiant luminaria in firmamento cœli*, Gen. 1, 14. Gesceóp God twá miccle leóht *God created two great lights*, Lchdm. iii. 234, 7. Wé sceolon on ðisum dæge beran úre leóht tó cyrcan and lǽtan hí ðǽr bletsian *we must on this day carry our lights to church and have them blessed there*, Homl. Th. 1, 150, 27. [*Goth.* liuhaþ: *O. Sax.* lioht: *O. Frs.* liacht: *Icel.* ljós: *O. H. Ger.* lioht, lieht *lux, candela, lucerna, lumen: Ger.* licht.] DER. ǽfen-, fýr-, heofon-, morgen-leóht.

leóht, léht, líht; *adj. Light, bright, cheerful* (perhaps the passages in which the word has the meaning of *cheerful* should be put under the next word v. leóht-mód), *shining, clear :*—Cwæþ ðæt his líc wǽre leóht and scéne *he said that his body was bright and beautiful*, Cd. 14; Th. i. 26; Gen. 265. Léht (líht, Rush.) biþ all líchoma ðín *lucidum erit totum corpus tuum*, Mt. Kmbl. Lind. 6, 23. Bebod drihtnes leóht *præceptum dominum purum*, Ps. Spl. C. 18, 9. Him wæs leóht sefa ... blíðheort wunode *his soul was unclouded by sorrow ... blithe of heart he continued*, Andr. Kmbl. 2504; An. 1253. Him wæs leóht sefa, ferhþ gefeónde, Elen. Kmbl. 346; El. 173. Swá leóhtes andwlitan men *tam lucidi vultus homines*, Bd. 2, 1; S. 501, 15. Ðam ðe ic ofonn leóhtes geleáfan *to whom I grudge clear belief*, Exon. 71 a; Th. 265, 8; Jul. 378: Apstls. Kmbl. 131; Ap. 66. Leóhte gesihþe *lucidus aspectu*, Bd. 5, 12; S. 627, 32. Æt leóhtum fýre *at a bright fire*, L. M. 1, 2; Lchdm. ii. 30, 7. Se ðe reáfaþ man leóhtan dæge *he who robs a man in daylight*, L. Eth. iii. 15; Th. i. 298, 11. Ða þiostro ðínre heortan willaþ míure leóhtan láre wiðstondan, Bt. Met. Fox 5, 43; Met. 5, 22. Mid leóhtum andgite *with clear understanding*, Blickl. Homl. 105, 31: Wulfst. 252. 5. Gé syttaþ ealle niht and drincaþ óð leóhtne dæg, and swá áwendaþ dæg tó niht and niht tó dæge, 297, 28. Be leóhtne dæg *in matutino*, Ps. Th. 72, 11. Dó ðíne ansýne esne ðínum leóhte *faciem tuam illumina super servum tuum*, 118, 135. Ðonne wurþaþ ðín eágan swá leóht *then shall thine eyes become so clear*, Cd. 27; Th. 35, 34; Gen. 564. Gelíc wæs hé (*Lucifer*) ðám leóhtum steorrum, 14; Th. 17, 7; Gen. 256. Ðæt wé mágon oft leóhtum dagum geseón *quam sæpe lucidioribus diebus aspicere solemus*, Bd. 1, 1; S. 474, 15. Leóhte nihte on sumera hafaþ *lucidas æstate noctes habet*, S. 473, 29. Se heofon mót brengon leóhte dagas, Bt. 7, 3; Fox 20, 21. Wurde ðín líchoman leóhtra micle, Cd. 25; Th. 32, 13; Gen. 502. Eác wǽre ðam earman leóhtre on móde gif hé ðæs rícan mannes welan ne gesáwe *also the poor man would have been more cheerful, if he had not seen the rich man's wealth*, Homl. Th. i. 330, 11. Benedictus ðe ús bóc áwrát leóhtre be dǽle ðonne Basilius *Benedict who wrote us a book clearer in some respects than Basil did*, Basil prm; Norm. 32, 9. Ingeþonc leóhtre and beorhtre ðonne se leóma sunnan on sumera, Bt. Met. Fox 22, 43; Met. 22, 22: Ors. 5, 14; Swt. 248, 11. Léga leóhtost *brightest of flames*, 9, 33; Met. 9, 17. [*O. Sax.* lioht: *O. Frs.* liacht: *Icel.* ljóss: *O. H. Ger.* lioht, lieht *lucidus: Ger.* licht.]

leóht, léht, líht [*from comparison with other dialects the proper spelling would seem to be* líht, *but* leóht (*or* leoht?), *in West-Saxon at least, is the regular form*]; *adj. Light, not heavy, inconsiderable; not slow, quick, ready, nimble, fickle, easy :*—Mín byrðyn ys leóht (Lind. léht; Rush. líht) *onus meum leve est*, Mt. Kmbl. 11, 30. Leóht and leoþuwác *nimble and supple*, Exon. 79 b; Th. 298, 12; Crä. 84. Hé is snel and swift and leóht *levis et velox est*, 60 b; Th. 220, 9; Ph. 317: 52 a; Th. 182, 6;

Gú. 1306. Mē leóht slǣp oferarn *levis mihi somnus obrepsisset*, Bd. 5, 9; S. 622, 33. Leóht drenc *a light drink*, L. M. 2, 51; Lchdm. ii. 264, 26. Leóht wýn, Lchdm. iii. 122, 1. Hwílum ða leóhtan scylda beóþ beteran tó forlǣtenne *aliquando leviora vitia relinquenda sunt*, Past. 62; Swt. 457, 7. Hý habbaþ swýðe lytle scypa and swíðe leóhte *they have very little ships and very light ones*, Ors. 1, 1; Swt. 19, 8. Ðæt sió wamb ðý ðē leóhtre síe *by it the stomach may be relieved*, L. M. 2, 25; Lchdm. ii. 218, 1. Wið módes (?innoþes, MS.) hefignesse . . . sóna biþ ðæt mód leóhtre, Lchdm. iii. 50, 23. Leóhtre ic eom micle ðonne ðes lytla wyrm *I am much lighter than this little worm*, Exon. 111 b; Th. 426, 19; Rä. 41, 76. Líhtre *tolerabilius*, Mt. Kmbl. Lind. 10, 15. Wē underfóþ scortne ryne ðæs leóhtran gewinnes *we have a short course of the easier conflict*, Homl. Th. i. 418, 10. Mid nánum leóhtran þinge gebēte ðonne him mon áceorfe ða tungan of, L. Alf. pol. 32; Th. i. 80, 21. Hý habbaþ ðæs ðē leóhtran gang *they shall walk the easier for it*, L. Med. ex Quad. 3, 15; Lchdm. i. 342, 12. Se hæfde moncynnes leóhteste hond *he had of all men the readiest hand*, Exon. 85 b; Th. 323, 1; Wíd. 72. [*Goth.* leihts: *O. Sax.* líht(-líc): *O. Frs.* licht: *Icel.* léttr: *O. H. Ger.* líhti *levis, facilis*: *Ger.* leicht.]

leóhtan; *p.* te *To give light, to illumine, make light, cause to shine*:—Ðær leóhtes ne leóht lytel sperca earmum ǣnig *there doth not any little spark give light to the miserable ones*, Dóm. L. 14, 218. Hē lofe leóhteþ leófe ða hálgan *hymnus omnibus sanctis ejus*, Ps. Th. 148, 14. Ðíne lígetta leóhteþ and beorhteþ, 143, 7. Beorhte leóhte ðínne andwlitan *illuminet vultum suum*, 66, 1. v. líhtan.

leóht-bǣre; *adj. Luminous, brilliant, splendid*:—Hyra leóhtbǣran ryne *their (the stars) luminous course*, Lchdm. iii. 272, 10. Á ðæs dóm áge leóhtbǣre lof se ús ðis líf giefeþ *ever therefore may he have glory, splendid praise, who giveth us this life*, Exon. 80 a; Th. 299, 34; Crä. 112.

leóht-beámed; *adj. Having bright beams* or *rays*:—Sind sume steorran leóhtbeámede, fǣrlíce árísende and hrǣdlíce gewítende, Homl. Th. i. 610, 2.

leóht-berend, es; *m. Lucifer*:—Leóhtberend *Lucifer*, Ælfc. Gr. 8; Som. 7, 19. Ðá wæs ðæs teóþan werodes ealdor swíðe fæger and wlitig gesceapen swá ðæt hē wæs geháten Leóhtberend, Homl. Th. i. 10, 22. Se hätte Lucifer, ðæt ys Leóhtberend, Ælfc. T. Grn. 2, 35.

leóht-berende; *adj. Light-bearing, Lucifer, luminous*:—Lucifer hâten, leóht-berende, Cd. 221; Th. 287, 15; Sat. 367. Swylce án ofen eall smóciende and leóhtberende fýr férde ofer ða lác *apparuit clibanus fumans et lampas ignis inter divisiones illas*, Gen. 15, 17.

leóht-brǣdness, e; *f. Illumination*:—Leóhtbrǣdnesse *facibus*, Hpt. Gl. 515, 11.

leóhte; *adv. Brightly, clearly*:—Leóhte and beorhte scínaþ *clearly and brightly they shine*, Blickl. Homl. 127, 35: Exon. 116 a; Th. 446, 10: Dóm. 20: 26 a; Th. 76, 14; Cri. 1239: Elen. Kmbl. 2229; El. 1116: Bt. Met. Fox 9, 25; Met. 9, 13. Leóhte oncnáwan *clearly recognise*, Exon. 24 a; Th. 69, 12; Cri. 1119: Elen. Kmbl. 1929; El. 966. Wæs se blâca beám bôcstafum âwriten beorhte and leóhte, 183; El. 92. Wearþ mē on hige leóhte *my mind was enlightened*, Cd. 32; Th. 42, 20; Gen. 676. Scýnan leóhtor *to shine more brightly*, Exon. 21 a; Th. 56, 18; Cri. 902.

leóhte; *adv. Lightly, easily, gently*:—Líhte *lento*, Wrt. Voc. ii. 49, 62. Ða wæs heó gesewen þurh twegen dagas ðæt hire leóhtor wǣre *videbatur illa per biduum aliquanto levius habere*, Bd. 4, 19; S. 589, 3. [*O. Sax.* liohto: *O. H. Ger.* líhto *leviter, leniter*.]

leohte; *p.* leoht; *pp.* v. leccan.

leóht-fæt, es; *n. A lamp, light, lantern*:—Leóhtfæt *lucernarium*, Ælfc. Gl. 30; Som. 61, 55; Wrt. Voc. 26, 54. Ðínes líchaman leóhtfæt is ðín eáge *lucerna corporis est oculus*, Mt. Kmbl. 6, 22. Leóhtfatu *lampades*, 25, 1. Judas com mid leóhtfatum *Judas venit cum lanternis*, Jn. Skt. 18, 3: Homl. Th. ii. 246, 9. Hē leóhtfatu (lehtfeatu, Ps. Surt.) micel geworhte *fecit luminaria magna*, Ps. Th. 135, 7: Hymn. Surt. 126, 12. [*O. Sax.* lioht-fat: *O. H. Ger.* lioht-faz *lucerna, lampas, luminarium (cœli), lanterna*.]

leóht-fruma, an; *m. The author* or *origin of light* (cf. lucis auctor, Exon. 65 b; Th. 242, 3; Ph. 667):—Lífes leóhtfruma *God*, Cd. 9; Th. 11, 14; Gen. 175: 43; Th. 57, 10; Gen. 926: Exon. 41 a; Th. 137, 26; Gú. 565: 41 b; Th. 138, 24; Gú. 581: Ps. C. 50; Ps. Grn. ii. 277, 46: Bt. Met. Fox 11, 143; Met. 11, 72.

leóht-gesceot, -gescot, es; *n. Contribution made to furnish the church with lights*. The various regulations respecting it may be seen in the following passages:—Gif hwá leóhtgescot ne gelǣste, gylde lahslit mid Denum, wíte mid Englum, L. E. G. 6; Th. i. 170, 4. Gelǣste man leóhtgescot þriwa on geáre, L. Eth. V. 11; Th. i. 308, 2: vi. 19; Th. i. 320, 3. Leóhtgescot gelǣste man tó Candelmæssan; dó oftor se ðe wylle, ix. 12; Th. i. 342, 31. Leóhtgescot þriwa on geáre: ǣrest on Eásteræfen, healfpenigwurþ wexes ǣt ǣlcre híde; and eft on Ealra Hálgena mæssan eall swá mycel; and eft tó ðǣm Sanctam Mariam clǣnsunge eal swá, L. C. E. 12; Th. i. 366, 31. Leóhtgescot þreówa on geáre: ǣrest healfpeningwurþ wexes tó Candelmæssan, and eft on Eásteræfen and þriddan síþe tó Ealra Hálgena mæssan. Wulfst. 116, 6. Leóhtgescota, 113.

11. Leóhtgescot gelǣste man be wíte tó Cristes mæssan and tó Candelmæssan and tó Eástron; dó oftor se ðe wylle, 311, 9. [Cf. *Icel.* ljóstollr *fee to a church for lighting*.]

leóhtian; *p.* ode *To give light*:—Leóma leóhtade leóda mǣgþum *a ray gave light to the tribes of men*, Exon. 11 a; Th. 15, 10; Cri. 234. v. líhtan.

leóhtian, *p.* ode *To grow light, become less heavy*, or *easy, be relieved*:—Ðonne leóhtaþ him se líchoma *his body will be relieved of the pain*, Herb. 1, 16; Lchdm. i. 76, 2. v. líhtan *to ease*.

leóhting, e; *f. Lighting*:—Leóhtes leóhting *lucubrum*, Ælfc. Gl. 67; Som. 69, 89; Wrt. Voc. 41, 42. v. líhting.

leóht-ísern, es; *n. A candlestick*:—Lēhtísern *candelabrum*, Mt. Kmbl. Lind. 5, 15: Mk. Skt. Lind. 4, 21: Lk. Skt. Lind. 8, 16.

leóht-leás; *adj. Without light*:—Hē sǣde ðæt hē wǣre gelǣd tó leóhtleásre stówe *he said that he was conducted to a place without light*, Homl. Th. ii. 504, 29.

leóht-líc; *adj. Light, bright, shining*:—Lyftfæt leóhtlíc [*the moon*], Exon. 108 a; Th. 411, 21; Rä. 30, 3.

leóht-líc; *adj. Light, of little weight* or *value*:—Leóhtlícu weorc *levitas operis*, Past. 43, 1; Swt. 309, 1. [*O. Sax.* líht-lík: *Icel.* létt-ligr: *O. H. Ger.* líht-líh *levis, infimus, humilissimus*.] v. next word.

leóht-líce; *adv. Lightly, slightly, gently, without trouble* or *effort, easily, quickly*:—Swá swá leóhtlíce gebylged *quasi leviter indignata*, Bd. 4, 9; S. 577, 24. Swá swá hē leóhtlíce onslǣpte *quasi leviter obdormiens*, 4, 11; S. 580, 2. Geswēt swíðe leóhtlíce mid hunige *sweeten very slightly with honey*, L. M. 1, 2; Lchdm. ii. 36, 3: 1, 19; Lchdm. ii. 62, 20. Hí forlēton ða scipo ðus leóhtlíce . . . and lēton ealles þeódscipes geswincg ðus leóhtlíce forwurþan *they abandoned the ships thus lightly . . . and let all the nation's labour thus lightly come to nought*, Chr. 1009; Erl. 142, 10–13. Ða weras mon sceal hefiglecor lǣran and ða wíf leóhtlecor *illis* [*men*] *graviora, istis* [*women*] *injungenda sunt leviora*, Past. 24; Swt. 179, 16. [Swá swá heó líhtlucost mihten *as quickly as they could*, Th. An. 123, 21.] [*Icel.* létt-liga *lightly, easily, readily*: *O. H. Ger.* líht-líhho *leniter, levius*: *Ger.* leicht-lich.]

leóht-mód; *adj. Of light* or *cheerful mind, light-hearted, easy-tempered; light-minded* (v. next word), *inconstant, fickle*, Exon. 90 a; Th. 338, 30; Gn. Ex. 86. [Cf. *Icel.* létt-látr *cheerful*: létt-lyndr *easy-tempered*; létt-úð *light-heartedness*; mod. *levity, frivolity*: *O. H. Ger.* líht-mótig *levis*.]

leóhtmód-ness, e; *f. Lightness of mind, want of gravity* or *steadiness, levity, frivolity, inconstancy*:—Gif ǣresd se wyrtruma biþ forcorfen ðæt is sió leóhtmódnes . . . Mon hine bewarige wið ða leóhtmódnesse . . . Paulus cwæþ 'Wēne gē nú ðæt ic ǣnigre leóhtmódnesse brúce' . . . þ ðǣre leóhtmódnesse unþeáwes nánwuht næfde *cum prius radicem levitatis abscidunt . . . Mentis levitas caveatur . . . Paulus dicit 'Numquid levitate usus sum?' . . . levitatis vitio non succumbo*, Past. 42, 3; Swt. 308, 2–11: 32, 2; Swt. 215, 2: 33, 6; Swt. 225, 12. For hira leóhtmódnesse *levitate cogitationum*, 42, 1; Swt. 305, 17. [Cf. *O. H. Ger.* líht-móti *levitas*.]

leóht-sceáwigend *light-seeing*; *lucivída*, Wrt. Voc. ii. 51, 56.

leóht-sceot. v. leóht-gescot.

leólc. v. lácan.

leóma, an; *m. Light, radiance, sheen, splendour, lightning, ray* or *beam of light*:—Ðes leóma *hoc jubar*, Ælfc. Gr. 9; Som. 9, 43. Candeles leóma *lampas*, Ælfc. Gl. 67; Som. 69, 88; Wrt. Voc. 41, 41. Leóma *globus*; leómum *globis*, Wrt. Voc. ii. 40, 74, 75: 109, 73: *globis, luminibus*, Hpt. Gl. 472, 27. Fýres leóma *illuminatio ignis*, Ps. Th. 77, 16. Sunnan leóma *the light of the sun*, Exon. 21 a; Th. 56, 16; Cri. 901. Swegles leóma *the radiance of the sky*, 57 a; Th. 204, 26; Ph. 103. Berhtre ðonne se leóma sunnan on sumera *brighter than sun-light in summer*, Bt. Met. Fox 22, 46; Met. 22, 23. Stôd se leóma him of swylce fýren þecele ongeán norþdǣle middangeardes *the brightness [tail of a comet] proceeded from them [two comets] as a fiery torch towards the north*, Bd. 5, 23; S. 645, 29; Beo. Th. 5532; B. 2769. God eástan sende leóhtne leóman *God from the east sent bright radiance*, Judth. 11; Thw. 24, 16; Jud. 191: Cd. 223; Th. 294, 11; Sat. 469. Ðæt nánes mannes gesihþ ðæs leóhtes leóman sceáwian ne mihte, Homl. Th. i. 76, 11. Fýrleóht geseah blácne leóman beorhte scínan *he saw the firelight, a pale gleam, shine brightly*, Beo. Th. 3038; B. 1517. Seó sunne byþ swá feorr súþ ágán ðæt hyre leóman ne mágon tó ðam lande gerǣcan *the sun is gone so far south, that its rays cannot reach that land*, Lchdm. iii. 260, 10: Cd. 148; Th. 184, 25; Exod. 112. Leóman *fulgura*, Hymn. T. P. 73: Ps. Lamb. 134, 7. Leómena leás *blind*, Exon. 87 a; Th. 328, 13; Vy. 17. Leómum inlýhted *illumined with his rays*, 42 a; Th. 141, 14; Gú. 627. Seó sunne behýdde hire hátan leóman *the sun hid its hot beams*, Homl. Th. ii. 256, 34. [*O. Sax.* liomo: *Icel.* ljómi *radiance, a ray*.] DER. æled-, beadu-, bryne-, fýr-, ge-, heofon-, hilde-, sweord- leóma.

leomu *limbs*. v. lim.

león. v. leó.

león; *p.* láh. *To lend, grant for a time*:—Mín lond ðe ic hæbbe, and mē God láh, Chart. Th. 469, 25: Beo. Th. 2916; B. 1456. Líh mē þreó hláfas *commoda mihi tres panes*, Lk. Skt. Lind. 11, 5. [*Goth.*

leihwan : *O.Sax.* far-líhan : *O.L.Ger.* lían : *O.Frs.* lía : *Icel.* ljá : *O.H.Ger.* líhan *commodare, fenerare, mutuare : Ger.* leihen.] v. on-léon.

león-fót, es ; *m. Lion's foot* [plant name] ; alchemilla vulgaris :—Leónfót *leontopodium,* Wrt. Voc. 67, 50 : Herb. 8, 1 ; Lchdm. i. 98, 12. Liónfót *leontopedium,* Wrt. Voc. ii. 53, 48. [*Icel.* ljóns-fótr *alchemilla.*]

leóna, an ; *m. A lion* or *lioness* :—Zosimus tó ðam leónan cwæþ : Eálá ðu mæsta(e) wildeór [cf. l. 15 seó leó mid hire earmum], Glostr. Frag. 110, 9. Oft hálige men wunedon on wëstene betwux wulfum and leónum, Homl. Th. i. 102, 5 : 488, 4 : 572, 13 : ii. 192, 24. Fram leónum *a leonibus,* Ps. Spl. 34, 20. [*Icel.* leóna *a lioness :* león, ljón ; *gen.* ljóns : *m. n. a lion.*] v. leó.

leonian. v. linian.

leópard, es ; *m. A leopard :*—Fore hundum tigros and leópardos hí fêdaþ *pro canibus tigres et leopardos nutriunt,* Nar. 38, 4. [*Icel.* leópardr, hlébardr : *O. H. Ger.* lëbarto, lëbard, leóparto.]

leóran, *p. de To go, depart, pass, pass away :*—Ic ne leóru *non emigrabo,* Ps. Surt. 61, 7. Leoreþ *transeat,* 56, 2. Wið ða hwíle lióres [geleóreþ, Rush.] heofon and eorþo *donec transeat cælum et terra,* Mt. Kmbl. Lind. 5, 18. Hê leórde ðonan *transiit inde,* 11, 1 : Andr. Kmbl. 247 ; An. 124. Hê tó drihtne mid sibbe leórde *he departed in peace to the Lord,* Glostr. Frag. 110, 30. Hê leórde tó heófonum *migravit ad cælos,* Bd. 2, 7 ; S. 509, 36. Of ðissum leóhte leórde, 3, 20 ; S. 550, 26. Leórde *transivit,* 4, 23 ; S. 592, 39. Ðe of weorulde leórdan *qui de sæculo migraverant,* 4, 22 ; S. 592, 27. Ðá leórdon ða gástas tó êcum gefeán, Shrn. 134, 7. Lungre leórdon, nalas leng bidon, Andr. Kmbl. 2085 ; An. 1044. Leór † gewít heonan *transi hinc,* Mt. Kmbl. Rush. 17, 20. Ðê gedafenaþ ðæt ðú leóre on ðíne bære, Blickl. Homl. 149, 11. Leóre from mê ðes calic *transeat calix iste,* Mt. Kmbl. Rush. 26, 39. Ðætte munecas ne leóran of stówe tó óðre *ut monachi non migrent de loco ad locum,* Bd. 4, 5 ; S. 572, 37. Leóran *transire,* Mt. Kmbl. Rush. 26, 42. Leórendum dagum *in the transitory days* [*of this life*], Exon. 118 a ; Th. 454, 9 ; Hy. 4, 30. DER. â-, forþ-, be-, fore-, ge-, ofer-, þurh-leóran.

leóred-ness, e ; *f. Migration, departure, exit, vision :*—Liórednesse *visione spiritali,* Hpt. Gl. 486, 30. v. ge-leóredness.

leornere, es ; *m. A learner, disciple scholar, learned person, reader :*—Be ðam wrât Beda se leornere *of him the scholar Bede wrote,* Shrn. 155, 25. Gif leornere geþeh þurh láre ðæt he hád hæfde and þênode Criste *if a scholar succeeded by learning so that he had holy orders and served Christ,* L. R. 7 ; Th. i. 192, 12. Brýde beág bêc leornere *a ring for a bride, books for a scholar,* Exon. 91 a ; Th. 341, 25 ; Gn. Ex. 131. Ðone leornere ic nú bidde *lectorem obsecro,* Bd. pref ; S. 472, 31. Swá leorneras secgaþ *as scholars say,* Shrn. 63, 10 : Exon. 62 a ; Th. 227, 17 ; Ph. 424. Se Hælend tóbræc ða hláfas and sealde his leornerum, Homl. Th. ii. 400, 21 : Blickl. Homl. 131, 20. Johannes gesende twægen leorneras his *Joannes mittens duos de discipulis suis.* Mt. Kembl. Rush. 11. 2, 1 : 10, 1. Ealle ða geláredestan men and ða leorneras *multis doctioribus viris,* Bd. 4, 24 ; S. 597, 30. v. stæf-leornere.

leór-ness, e ; *f. Going, departure, withdrawal :*—Dægas leórnisse his *dies assumptionis ejus,* Lk. Skt. Rush. 9, 51. Leornisse *transmigrationis,* Ps. Surt. ii. 191, 3. In leórnisse *in secessu* [cf. gang Il.], Mt. Kmbl. Rush. 15, 17. v. ge-, ofer-leórness.

leornesse [?] Bd. 6, 5 ; S. 527, 16, *other MSS.* have geornesse.

leornian ; *p.* ode *To learn, study, read :*—Swá swá in ðære bêc his lifes gemêteþ swá hwylc swá hí rædeþ and leornaþ *sicut in volumine vitæ ejus quisque legerit inveniet,* Bd. 4, 31 ; S. 611, 7. Ælc ðe gehýrde æt fæder and leornode *omnes qui audivit a patre et didicit,* Jn. Skt. 6, 45. Fram ðám hê ðæt gemet leornode regollíces þeódscipes *a quibus normam disciplinæ regularis didicerat,* Bd. 3, 23 ; S. 554, 35. Hê hálige gewritu leornade and smeáde *scripturis legendis operam daret,* S. 555, 29. Ða ðe hê on gewritum leornode tó dônne *ea quæ in scripturis agenda didicerat,* 3, 28 ; S. 560, 16. Gê ne leornodan *non legistis,* 4, 3 ; S. 560, 16. Leorna ðæt ðú ondræde drihten *ut discas timere dominum,* Deut. 14, 23. Leorneaþ æt mê *discite a me,* Mt. Kmbl. 11, 29. Leornigeaþ bigspell ðe ðam fictreówe *ab arbore fici discite parabolam,* 24, 32. Syle andgit ðæt ic ðíne gewitnesse wel leornige *da mihi intellectum ut sciam testimonia tua,* Ps. Th. 118, 125. Ic hit tór ðære hælo ðe hit leornige oððe gehýre áwrât *ob salutem legentium, sive audientium narrandam esse putavi,* Bd. 5, 13 ; S. 634, 2. Lange sceal leornian se ðe læran sceal *long must he learn who has to teach,* L. Ælfc. P. 46 ; Th. ii. 384, 15 : L. I. P. 14 ; Th. ii. 322, 8. Bêc on tó leornianne *libros ad legendum,* Bd. 3, 27 ; S. 558, 27. Ða hús ða ðe on tó gebiddenne and tó leornigenne geworhte wæron *domunculæ quæ ad orandum vel legendum factæ erant,* 4, 25 ; S. 601, 12. Ealswá David dyde leornigendum móde [*with docile mind*], Wulfst. 172, 22. [*O. Frs.* ge-lerna, -lirna : *O. H. Ger.* lernên, lirnên *discere, meditari : Ger.* lernen. *Goth.* has leisan, *and* ga-laisjan sik : *O. Sax.* linôn : *mod.* Scandinavian dialects use forms corresponding to læran.] v. ge-leornian.

leornung, e ; *f. Learning, study, meditation, reading :*—Lár oððe leornung *teaching* or *learning,* L. I. P. 8 ; Th. ii. 314, 20 : Past. pref ; Swt. 3, 10. Micel is tó secganne langsum leornung ðæt hê in life ádreág *much is it to tell, lengthy the reading, what he in life underwent,* Andr. Kmbl. 2962 ; An. 1484. Geleoso ðære godcundan leornunge *studia di-*

vinæ lectionis, Bd. 3, 13 ; S. 538, 29. On smeáwunge and on leornunge háligra gewrita . . . ðonne hí on heora leornunge wæron and heora bêc ræddon and beeodan *meditationi scripturarum . . . cum illi intus lectioni vacabant,* 4, 3 ; S. 567, 29-34. On leornunge úra stafa *nostrarum lectione litterarum,* 5, 14 ; S. 635, 8. On leornunge *in discendo,* Coll. Monast. Th. 18, 18. Gáþ út tó claustre oððe tó leorninge *egredimini in claustrum vel in gymnasium,* 36, 9. Hí hiene nieddon tó leornunga ðeh hê gewintred wære *they compelled him to go to school, though he was an old man,* Ors. 6, 31 ; Swt. 284, 21. Tó liornunga óðfæste, Past. pref ; Swt. 7, 12. Ðú hatodest leornunga *tu odisti disciplinam,* Ps. Th. 49, 18. Mid ða leornunga ðissa bôca *hujus* [*libri*] *lectione,* Bd. 5, 18 ; S. 636, 4. Tó begangenne his leornunge *lectioni operam dare,* 5, 2 ; S. 614, 35. Hê micle gýminge hæfde háligra leorninga *curam non modicam lectionibus sacris exhibebat,* 3, 19 ; S. 547, 27. On hálgum leornungum, 4, 2 ; S. 565, 33. [*O. H. Ger.* lirnunga, lernunga *disciplina, industria, doctrina, lectio.*]

leornung-cild, es ; *n. A scholar, pupil, disciple :*—Benedictus bemænde ðæt his leorningcild Maurus ðæs óðres deáþes fægnian sceolde, Homl. Th. ii. 164, 10.

leornung-cniht, es ; *m. A youth engaged in study, scholar, disciple :*—Leorningcniht *discipulus vel mathites,* Ælfc. Gl. 80 ; Som. 72, 99 ; Wrt. Voc. 46, 56. Nys se leorningcniht ofer his láreów *non est discipulus super magistrum,* Mt. Kmbl. 10, 24. Sí ðú his leorningcniht, wê synt Moyses leorningcnihtas, Jn. Skt. 9, 28. [The word occurs frequently in the Gospels, as it regularly translates *discipulus.*] Monige ðeáh ðe hí næfre leorningcnihtas næren wilniaþ ðeáh láreówas tó beónne *plerique qui, quæ non didicerint, docere concupiscunt,* Past. proem ; Swt. 25, 8. Ða undergeat se preóst ðæt hê ne mihte ðone hálgan wer líchamlíce ácwellan, and wolde ðá his leorningcnihta sáwla fordón, Homl. Th. ii. 162, 30.

leornung-cræft, es ; *m. Learning, erudition :*—Ða ðe leornungcræft hæfdon *scholars,* Elen. Kmbl. 760 ; El. 380.

leornung-hús, es ; *n. A house for study, a school :*—Leorninghús *gymnasium,* Ælfc. Gl. 107 ; Som. 78, 76 ; Wrt. Voc. 57, 54 ; Wrt. Voc. ii. 46, 56.

leornung-mann, es ; *m. A learner, pupil, scholar, student, disciple :*—Sum leorningman well geláred on gewritum *scholasticus quidam doctus studio literarum,* Bd. 3, 13 ; S. 538, 18. Ærest discipula and leorningmon reogollíces lífes *primo discipula regularis vitæ,* 3, 24 ; S. 557, 4. Ic wylle tó him gecyrran and biddan ðæt ic môte heononforþ his leorningman beón, Homl. Th. ii. 414, 15. Hí [*Martha and Mary*] wæron ðæs Hælendes leorningmen, 438, 18. Mæssepreóstas sceolon symble æt heora húsum leorningmonna sceole habban, and gif hwylc gódra wile his lytlingas hiom tó láre befæstan, hig sceolon swíðe lustlíce hig onfón and him êstlíce tæcan, L. E. I. 20 ; Th. ii. 414, 7-10.

leórt. v. lætan.

leósan. v. be-, for-leósan.

LEÓÞ, es ; *n. A song, poem, ode, lay, verses :*—Ðis leóþ *hoc carmen,* Ælfc. Gr. 9 ; Som. 9, 28. Leóþ *poema,* Ælfc. Gl. 112 ; Som. 79, 98 ; Wrt. Voc. 60, 6. Sárlíc leóþ *tragædia,* Wrt. Voc. ii. 82, 37. Leóþ wæs âsungen *the song was recited,* Beo. Th. 2323 ; B. 1159. Leóþ Gode úrum *carmen Deo nostro,* Ps. Spl. 39, 4. Hê for ðon næfre nóht leásunga ne ídeles leóþes wyrceanne mihte *unde nihil unquam frivoli et supervacui poematis facere potuit,* Bd. 4, 24 ; S. 596, 42. Ðý betstan leóþe geglenged *optimo carmine compositum,* S. 597, 37. Ðone leóþ singan *dicere carmen,* 597, 31. Ðis leóþ him answaraþ for gewitnysse and ðæt leóþ ne ádiligaþ nán man of ðínes ofspringes múþe *respondebit ei canticum istud pro testimonio, quod nulla delebit oblivio ex ore seminis tui,* Deut. 31, 21. Ic geworhte hí eft tó leóþe *I made a poetical version of it,* Bt. proem ; Fox viii, 10. Ðá ic ðis leóþ âsungen hæfde *when I had recited these verses,* 3, 1 ; Fox 4, 16. Leóþ *odai,* Wrt. Voc. ii. 64, 63. Leóþa gleáw *skilled in songs,* Exon. 79 a ; Th. 296, 16 ; Crä. 52. Omerus wæs mid Crêcum leóþa cræftgast, Bt. Met. Fox 30, 4 ; Met. 30, 2. Ic lióþa fela sang, 2, 1 ; Met. 2, 1 : Exon. 91 b ; Th. 344, 8 ; Gn. Ex. 170. Leóþum and spellum leódum reahte *in songs and stories he related to men,* Bt. Met. Fox 30, 15 ; Met. 30, 8. Ða lióþ ðe ic geó lustbærlíce song *carmina qui quondam studio florente peregi,* Bt. 2 ; Fox 4, 6. Hê gewunode gerisenlíce leóþ wyrcean ða ðe tó æfæstnesse and tó árfæstnesse belumpon *carmina religioni et pietati apta facere solebat,* Bd. 4, 24 ; S. 596, 31. Ne wäne ælda cynnes ðæt ic lygewordum leóþ somnige wríte wóðcræfte *let none imagine of the race of men that with lying words my lays I compose, writing in verse,* Exon. 63 b ; Th. 234, 29 ; Ph. 547. [*Goth.* awi-liuþ : *Icel.* ljóð : *O. H. Ger.* leod, lied *carmen : Ger.* lied.] DER. æfen-, bismer-, brýd-, byrgen-, dæg-, dryht-, fús-, fyrd-, galdor-, gift-, gryre-, gúþ-, hearm-, hilde-, líc-, sæ-, sige-, sorg-, wíg-, wóp-leóþ.

leóþ-cræft, es ; *m. The art of poetry, poetry, verse, a poem :*—Ðes leóþcræft *hoc poema :* ðás leóþcræftas *hæc poemata* [all the other cases are also given], Ælfc. Gr. 9 ; Som. 8, 16-21. Hê biþ swá ðeáh on leóþcræfte ægðer ge lang ge sceort it [*i of the genitive in certain words*] is however in poetry both long and short, 18 ; Som. 21, 51. Ða gemetu gebyriaþ tó feáldum leóþcræfte *metres pertain to Latin poetry,* 50 ; Som. 51, 66. Sixfealdum leóþcræfte *exametro heroico,* Wrt. Voc. ii. 144, 47. Hê ðone leóþcræft geleornode *canendi artem didicit,* Bd. 4, 24 ; S. 596, 40.

leóþ-cræftig; *adj. Skilled in poetry:*—Leóþcræftig mon, Exon. 100 b; Th. 379, 28; Deór. 40.

leóþ-cwide, es; *m. A poem:*—Ic nát for hwí eów sindon ða ærran gewin swá lustsumlíce on leóþcwidum tó gehiéranne *I do not know why the earlier contests are so pleasant for you to hear in poems*, Ors. 3, 7; Swt. 120, 2.

leóþ-gidding, e; *f. A poem, song*, Andr. Kmbl. 2956; An. 1481.

leóþian. v. á-leóþian, liþian.

leóþian; *p.* ode *To sing, sound:*—Wód óðer ne lythwón leóþode ðonne in lyft ástág ceargesta cirm *a second cry sounded, nor weakly, when to the heavens rose the wail of the troubled spirits*, Exon. 38 a; Th. 125, 32; Gú. 363. Folcum ic leóþode *to peoples I sang*, 94 b; Th. 354, 4; Reim. 40. [*Goth.* liuþôn *to sing : O. H. Ger.* liudôn *canere, jubilare.*]

leóþ-líc; *adj. Poetical:*—Beda ðises hálgan líf ægðer ge æfter ánfealdre gereccednysse ge æfter leóþlícere gyddunge áwrát *Bede wrote this saint's life both in prose and in verse*, Homl. Th. ii. 134, 1.

leoþa. v. leoþu.

leóþ-sang, es; *m. A song, poem:*—In swinsunge leóþsanges *in modulationem carminis*, Bd. 4, 24; S. 597, 35. For his leóþsongum *cujus carminibus*, S. 596, 36.

leoþu. v. liþ.

leoþu [?]:—Wæs on lagustreáme lád ðær mé leoþu ne biglád [cf. (?) *Icel.* lið *a host, people*, or lið *a ship*], Exon. 94 a; Th. 353, 18; Reim. 14.

leoþu-bend; *m. f. A fetter, bond :*—Ic ðé álýse of ðyssum leoþubendum *I will release thee from these bonds*, Andr. Kmbl. 200; An. 100 : 2746; An. 1375 : 327; An. 164. Of leoþobendum, 2066; An. 1035 : 3127; An. 1566. Lioþobendum, Cd. 19; Th. 24, 23; Gen. 382. [*O. Sax.* liðo-bend.]

leoþu-bíge, -bíg; *adj. Flexible at the joints, humble, meek :*—Ðá wearþ ðæt hálige líc hál on eorþan gemét liþebíge on limum *the holy body was found in the earth sound, and with the limbs not yet stiff*, Homl. Th. ii. 152, 33. Ic gesette eów sóðe gebysnunge, ðæt eówer ælc sceole óðres fét áþweán, swá swá ic láreów eów liþebíg [*humble*] áþwóh, 242, 28.

leoþu-cæge, an; *f. A limb-key, key which consists of limbs :*—Ðé [*the Virgin Mary*] æfter him engla þeóden eft unmæle lioþucægan bileác, Exon. 12 b; Th. 21, 13; Cri. 334.

leoþu-cræft, es; *m. Bodily skill, skill in the use of the limbs :*—Se gedæleþ missenlíce leoþucræftas londbúendum, Exon. 78 b; Th. 295, 6; Crä. 29. Segn eallgylden hondwundra mæst gelocen leoþocræftum [*skilfully*; or leoþocræftum, (cf. leóþcræft *and next word*) *with charms, magically*; cf. the Danish banner, the Raven, supposed to be woven by the daughters of Ragnar, and to which extraordinary qualities were attributed. See also Burnt Njal, c. 156.]

leóþu-cræft, es; *m. Poetic art or skill*, Elen. Kmbl. 2499; El. 1251. v. preceding word.

leoþu-cræftig; *adj. Skilful with the limbs*, Exon. 59 b; Th. 216, 14; Ph. 268.

leoþu-fæst; *adj. Firm of limb, strong, able :*—Sum biþ bóca gleáw, lárum leoþufæst, Exon. 79 b; Th. 298, 34; Crä. 95.

leoþu-geþynd. v. leoþu-sár.

leoþu-líc; *adj. Belonging to the limbs, bodily :*—Leoþolíc and gástlíc, Andr. Kmbl. 3254; An. 1630. [*O. H. Ger.* lido-líh.]

leóþu-rún, e; *f. Counsel conveyed in verse*, Elen. Kmbl. 1042; El. 522.

leoþu-sár, es; *n. A pain of the limbs or joints :*—Leoþusár *vel* geþind *condolomata articula*, Wrt. Voc. ii. 135, 67.

leoþu-sirce, an; *f. A coat of mail :*—Locene leoþosyrcan, Beo. Th. 3014; B. 1505 : 3784; B. 1890.

leoþu-wác; *adj. With pliant joints, flexible, pliant, supple :*—Liþowác *habile*, Wrt. Voc. ii. 42, 67. Leoþuwác, 110, 25. Leúht and leoþuwác *nimble and supple*, Exon. 79 b; Th. 298, 12; Crä. 84. Swilce liðewácum *velut lentescente*, Hpt. Gl. 520, 36. Liðewácum tagum [?tánum] *lentis viminibus*, 514, 69. [*O. H. Ger.* lido-weih *flexible*; *lentus.*] v. un-leoþuwác.

leoþuwác-ness. v. un-leoþuwácness.

leoþu-wácung, e; *f.* In Ps. Spl. T. 78, 11 *compeditorum* is glossed by liþewácunga.

leoþuwæcan, liþewæcan; *p.* -wæhte *To become or to make soft, or pliant, to grow calm, to assuage, soften :*—Liþewæcaþ brymmas sæs *the surges of the sea become still*, Hymn. Lye. Liþewæhte *lentesceret*, Hpt. Gl. 479, 30. Liþewæhtan *mollescerent, delenirent*, 481, 13. Leoþewæce *mitigare, pacificare*, 495, 22. v. ge-liþewæcan.

leóþ-weorc, es; *n. Song-making, poetry*; *poesis*, Ælfc. Gl. 112; Som. 79, 99; Wrt. Voc. 60, 7.

leóþ-wíse, an; *f. A poetical manner, verse :*—Mycel Englisc bóc on leóþwísan geworht *a large English book composed in poetry*, Chart. Th. 430, 24. Áwend of Lédene on Englisce on leóþwísan, Homl. Th. ii. 520, 10.

leóþ-word, es; *n. A word in a poem*, Andr. Kmbl. 2975; An. 1490.

leóþ-wyrhta, an; *m. A poet :*—Leóþwyrhta *poeta vel vates*, Ælfc. Gl. 112; Som. 79, 100; Wrt. Voc. 60, 8 : 73, 68. Hleot [=leóþ] wyrhta *melopius*, 291, 26. Leódwyrhta *melopius*, Wrt. Voc. ii. 56, 50. Ælfréd cyning Westsexna leóþwyrhta, Bt. Met. Fox introduc. 5; Met. Einl. 3.

leów, es; *pl.* (?) leówer, leówera; *n. A thigh, ham :*—Án hríðres læuw

---

a ham of beef, Cod. Dipl. Kmbl. ii. 355, 7. Leówer *pernas*, Lchdm. i. lxix, 13. Léwera, lxxiii, 31. [*Or is* leower *a different word.* Cf. *Icel.* lær *thigh*, and see Lchdm. iii. 366, col. 1.]

leówe, an; *f. A league, a mile :*—Leóuue *miliarium*, Ælfc. Gl. 57; Som. 67, 81; Wrt. Voc. 38, 7. ['*Lat.* leuca, leuga *a Gallic mile of* 1500 *Roman paces*; a word of Celtic origin.' Skt. Etym. Dict. under *league*.]

lepeþ:—Sum sceal wildne fugel átemian . . . fédeþ on feterum . . . lepeþ lyftswiftne lytlum giefum óþ ðæt se wælisca his ætgiefan eáþmód weorþeþ, Exon. 88 b; Th. 332, 14-27; Vy. 85-91. Grein compares with *M. H. Ger.* erlaffen *languefacere*, the passage would then mean that the hawk's fierceness and wildness were subdued by giving it little to eat. Might we however for **lepeþ** read *léfeþ* [cf. *léf*] or *léweþ* [cf. *ge-léwan*] = weakens, which would give very much the same meaning?

les. v. ge-les.

lesan; *p.* læs; *pl.* læson; *pp.* lesen *To lease* [= *glean* dialect.], *gather, collect :*—Se eorþlíca anweald næfre ne sæwþ cræftas ac lisþ and gadraþ unþeáwas *earthly power never sows virtues, but collects and gathers vices*, Bt. 27, 1; Fox 94, 25. Gif gé lesaþ wyrte on Sunnandæg *if ye gather herbs on Sunday*, Wulfst. 231, 18. Ic læs *I collected*, Elen. Kmbl. 2474; El. 1238. Hí læson æfre forþ mid heom ealle ða butsecarlas ðe heó gemétton *they kept on all the while collecting and joining to themselves all the sailors they found*, Chr. 1052; Erl. 184, 15. Ne gé ne gaderion ða eár ðe bæftan eów beóþ ac lætaþ þearfan and tíðacymene hig lesan *nec remanentes spicas colligetis, sed pauperibus et peregrinis dimittetis eas*, Lev. 23, 22. [*Piers P. Wick.* lese *to glean : Goth.* lisan : *O. Sax.* lesan : *O. L. Ger.* lesan *to read : O. Frs.* lesa : *Icel.* lesa : *O. H. Ger.* lesan *legere, colligere : Ger.* lesen.] DER. á-lesan.

lésan *to loose*. v. lísan.

lesu; *indecl. f. Numen :*—Leso *numine*, Wrt. Voc. ii. 62, 19.

lesu; *adj.* v. lysu.

letanía, an; *also with pl.* -as; *m. A litany :*—Ðæt hé ðysne letanían sungan *quia hanc litaniam modularentur*, Bd. 1, 25; S. 487, 24. Mid reliquium and mid letanían, Wulfst. 170, 18. Cristes folc mærsiaþ letanías, Shrn. 79, 28.

leter. v. eald-leter.

LEÐER, es; *n. Hide, skin, leather.* [The word is found chiefly, if not exclusively, in compounds. So in Icelandic, though frequent in modern usage, it is not found in old writers except in compounds. *O. H. Ger.* leder *corium.*] v. geweald-, heals-, weald-leðer.

leðer-codd, es; *m. A leather bag :*—Leðercoddas *bulgæ*, Ælfc. Gl. 16; Som. 58, 58; Wrt. Voc. 21, 45.

leðeren, liðeren, leðern, leðren; *adj. Leathern, of leather :*—Leðern *scorteus*, Ælfc. Gl. 99; Som. 76, 126; Wrt. Voc. 54, 66. Leðren fæt *scortia*, 16; Som. 60, 75; Wrt. Voc. 25, 15. Liðerene trymsas *asses corteas*, Wrt. Voc. ii. 7, 18. Lidrinae *scorteas*, Ep. Gl. 2 b, 10. [*O. H. Ger.* lidirin *pellicea : Ger.* ledern.]

leðer-helm, es; *m. A leather helmet*; *galea*, Ælfc. Gl. 51; Som. 66, 13; Wrt. Voc. 35, 3.

leðer-hose [-hosu?]; *f. A leather covering for the leg, gaiter :*—Leðerhose [-hosan?] *caligas*, Coll. Monast. Th. 27, 33. [*Icel.* ledrhosa; *f. a gaiter : O. H. Ger.* leder-hosa; *f. ocrea, cenarga.*] v. hosa.

leðer-wyrhta, an; *m. A tanner, currier*; byrseus, byrsarius, Wrt. Voc. ii. 11, 49 : 102, 38 : 127, 31. Lediruuyrcta, Ep. Gl. 6 d, 13.

lēðran. v. liðran.

letig. v. lytig.

lettan; *p.* te *To cause to be slow* [læt], *to let, hinder, impede, delay :*—Ne leteþ *non tricaverit*, Wrt. Voc. ii. 60, 75. Ðæs andwearda wela ámerþ and læt [MS. Cot. let] ða men ðe beóþ átihte tó ðám sóþum gesælþum, Bt. 32, 1; Fox 114, 3. Ðæt flæsc oft lett [MS. Hat. lætt] ða geornfulnesse and ðone willan ðæs þeóndan módes hér on worulde. Swæ swæ mon oft lett fundiende monnan and his færelt gælþ, swá gælþ se líchoma ðæt mód, Past. 36, 7; Swt. 256, 4-6. Óþ oreldo hí hine hwílum lettaþ *they sometimes defer it (death) until extreme old age*, Bt. 41, 2; Fox 246, 10. Ðæt syððan ná brimliþende láde ne letton *so that afterwards they did not hinder seafarers from their course*, Beo. Th. 1142; B. 569. Ac ic ðé hálsige ðæt ðú mé nó leng ne lette *tu modo quem excitaveris ne moreris*, Bt. 36, 3; Fox 174, 32. Gyf ðonne ðissa þreóra þinga ænig hwylcne man lette, ðæt hine tó ðam fæstene ne onhagie *if any of these three things hinder any man, so that the fast be inconvenient to him*, Wulfst. 285, 4. Hwí wille gé lettan úre síþfæt *why will ye hinder our journey?* Homl. Th. ii. 336, 11. Wé ðé ðæs nú nellaþ lettan ðæs ðú ær geþoht hæfdest *we will not hinder thee from that which thou didst before purpose*, Guthl. 5; Gdwin. 30, 24. [*O. Sax.* lettian : *O. Frs.* letta : *Icel.* letja : *O. H. Ger.* lezjan *retardare.*] v. ge-lettan, latian.

letting, e; *f. Letting, hindering, obstruction, delay, retarding :*—Ðeós yl[d]fulle letting *hæc morosa tricatio*, Hpt. Gl. 529, 6. Lettinge *obstaculo*, 523, 16. Lettincge *offendiculo*, 429, 35. On ðære lettinge his færeltes *in ejus itineris retardatione*, Past. 36, 7; Swt. 254, 20. [Se cyng scipa út on sæ sende his bróðer tó dære and tó lættinge, Chr. 1101; Erl. 237, 19.] Blindne se ðe hine gesihþ lettincge getácnaþ *if a man [in a dream] sees himself blind, it betokens hindrance*, Lchdm. iii. 200, 14 : 202, 3 : 204, 2.

leu, leuw. v. leów.

léwsa, an; m. *Weakness, infirmity, misery:*—Eágan míne sárgodon for léwsan *oculi mei languerunt præ inopia*, Ps. Spl. T. 87, 9. v. léf, ge-léwan.

lib-. v. lyb-.

LIBBAN; *p.* lifde *To* LIVE:—For ðam ic lybbe and gé lybbaþ *quia ego vivo et vos vivetis*, Jn. Skt. 14, 19. Ne lybbe ic, ac Crist leofaþ, Blickl. Homl. 165, 23. Wé lybbaþ mislíce on twelf mónþum; nú sceole wé lybban Gode, wé ðe óðrum tíman ús sylfum leofodon, Homl. Th. i. 180, 17. Godes þeówas ðe be gódra manna ælmessan libbaþ *God's servants who live by the alms of good men*, Wulfst. 120, 4. Hié ðe úrum lárum libbaþ *they live according to our instructions*, Blickl. Homl. 75, 15. Eal his líf hé lifde búton synnum, 33, 16. Hé on wynsumnesse lifde, 113, 7: Bd. 3, 27; S. 559, 27. Hé on ællþeódignesse lifde *exulabat*, S. 559, 30. Hé hér on eorþan engelíce lífe lifde, Blickl. Homl. 167, 33: 213, 11. Se þeódcyning þeáwum lyfde *the king lived virtuously*, Beo. Th. 4295; B. 2144. Wynnum lifde *lived joyously*, Exon. 111 b; Th. 428, 13; Rä. 41, 107. Wé ealne ðysne geár lifdon mid úres líchoman willan *we have lived all this year as it was pleasing to our body*, Blickl. Homl. 35, 27. Æfter ðon ðe hí lǽrdon hí sylfe þurh eall lifdon *secundum ea quæ docebant ipsi per omnia vivendo*, Bd. 1, 26; S. 487, 37. Hí ðágyt on hǽðennysse gedwolum lifdan *paganis adhuc erroribus essent implicati*, 2, 1; S. 501, 13. Ðá námon hí him wintersetl on Temesan and lifdon [lifedon, MS. E.] him of Eást Seaxum *they took up their winter quarters on the Thames and got their provisions out of Essex*, Chr. 1009; Erl. 143, 4 note. Swá ða drihtguman dreámum lifdon, Beo. Th. 199; B. 99. Swá swá diácon ðe regollíf libbe, L. Eth. ix. 21; Th. i. 344, 21. Swínes scearn ðæs ðe on dúnlande and wyrtum libbe, L. M. 1, 20; Lchdm. ii. 62, 28. Ne hié selfe ðý beteran ne taligen ðe ða óðre ðeáh ða óðre be him libben *ne se meliores æstiment, quia contineri per se ceteros vident*, Past. 44, 1; Swt. 319, 19. Ða niétenu onlútaþ tó ðære eorþan fordon hié sculon be ðære libban, 21, 3; Swt. 154, 17. *Laboratores* syndon weorcmen, ðe tilian sceolon ðæs ðe eal þeódscipe big sceal lybban, Wulfst. 267, 15. Uton libban ðam lífe ðe scrift ús wísige, 112, 18: 150, 13. Lífe swilcum libban *vitam talem vivere*, Hymn. Surt. 90, 13. Hé sǽde ðæt hé wolde óðer oððe ðǽr libban oððe ðǽr lecgan *he said that he would either live [conquer] there or die there*, Chr. 901; Erl. 96, 33: Ors. 3, 10; Swt. 158, 32. Hé hié ealle geniédde ðæt hié áþas swóran, ðæt hié ealle ætgædere wolden oððe on heora earde licggean, oððe on heora earde libban, 4, 9; Swt. 190, 27. Hé cwæþ 'Ðú eart ðæs lifigendan Godes sunu.' Se is lybbende God ðe hæfþ líf þurh hine sylfne, Homl. Th. i. 366, 33. Eall ðæt ic hæbbe on libbandan and on licgendan *all the live and dead stock that I have*, Chart. Th. 548, 12. Nán man nán þing ne bycge ofer feówer peninga weorþ ne libbende ne licgende, L. C. S. 24; Th. i. 390, 3. Hý hit be ðám libbendan habban *let them have it during their lifetime*, Chart. Th. 491, 25. [*Goth.* liban: *O. Sax.* libbian: *O. Frs.* libba: *Icel.* lifa: *O. H. Ger.* lebên.] v. lifian.

libn. v. lifen.

líc, es; *n. A body [living or dead]* generally the latter; the word remains in *lich*-gate, *lyke*-wake:—Líc oððe líchama *corpus*, Ælfc. Gr. 9, 32; Som. 12, 16. Líc ǽgðer ge cuces ge deádes *corpus*; líc oððe hreáw *funus*; líc oððe hold *cadaver*, Wrt. Voc. 85, 51-54: 49, 25. Næs nán hús on eallum Egipta lande ðe líc inne ne lǽge *neque erat domus, in qua non jaceret mortuus*, Ex. 12, 30. Ealle ða hwíle ðe ðæt líc biþ inne, ðér sceal beón gedrync and plega, Ors. 1, 1; Swt. 20, 25. Ðǽr ðæs hǽlendes líc áléd wæs *ubi positum fuerat corpus iesu*, Jn. Skt. 20, 12. Cwæþ ðæt his líc wǽre leóht and scéne, Cd. 14; Th. 17, 25; Gen. 265. Ðendan bu somod líc and sáwle lifgan móte *whilst both soul and body may live together*, Exon. 27 a; Th. 81, 21; Cri. 1327. Líc and gǽst, 46 b; Th. 160, 8; Gú. 940: 50 a; Th. 172, 25; Gú. 1149. Næs fýre gemǽled ne líc ne leoþu *neither body nor limbs were marked by the fire*, 74 a; Th. 278, 3; Jul. 592. Líc sáre gebrocen, bánhús blódfág, Andr. Kmbl. 2808; An. 1406. Ðé is gedál witod líces and sáwle, Cd. 43; Th. 57, 20; Gen. 931. Sweostor mín líces mǽge *my sister, kinswoman according to the flesh*, Th. 110, 4; Gen. 1833. Líces lustas *lusts of the flesh*, Exon. 71 b; Th. 267, 2; Jul. 409: 26 b; Th. 79, 28; Cri. 1297. Gang tó ciricean tó ðæs hálgan Oswaldes líce and site ðær *ingredere ecclesiam, et accedens ad sepulcrum Osualdi, ibi reside*, Bd. 3, 12; S. 537, 9. Stód se biscop æt ðam líce, 4, 11; S. 580, 13: L. Edg. c. 65; Th. ii. 258, 13. Bæþ wið ðam miclan líce *a bath for elephantiasis*, L. M. 1, 32; Lchdm. ii. 78, 18. Mynte ðæt hé gedælde ánra gehwylces líf wið líce *meant to part the life of each one from the body*, Beo. Th. 1470; B. 733. Hé ðæt andweorc of Adames líce áleoþode, Cd. 9; Th. 11, 18; Gen. 177. Hé sceáf reáf of líce, 76; Th. 94, 21; Gen. 1565. Forþ gewát Cham of líce *Ham died*, 79; Th. 97, 35; Gen. 1623. Hí his líc námon and hine on byrgene lédon, Mk. Skt. 6, 29: Beo. Th. 4261; B. 2127: L. Eth. v. 12; Th. i. 308, 5: vi. 21; Th. i. 320, 6. Ðæs mynstres bróðra ðydon scé. Cúþberhtes líc of eorþan, and hí ðæt gemétton swá gesund swá hé ðágyt lifde, Shrn. 82, 14. Se ús líf forgeaf, leomu, líc and gǽst, Exon. 19 a; Th. 48, 25; Cri. 777. His [*the Phœnix*] líc, 59 b; Th. 216, 14; Ph. 268. Hé wearp hine ðá on wyrmes líc, Cd. 25; Th. 31, 26; Gen. 491. Eowre líc sceolon sweltan on ðisum wéstene *vestra cadavera jacebunt in*

solitudine, Num. 14, 32. Ðǽr ðara arcebisceopa líc bebyrigde syndon *ubi archiepiscopi Cantiæ sepeliri solent*, Bd. 4, 1; S. 565, 5. Forleósan líca gehwilc ðara ðe lífes gást fǽdmum þeahte, Cd. 64; Th. 77, 26; Gen. 1281. Lícu *cadavera*, Hymn. Surt. 52, 27. [*Goth.* leik: *O. Sax.* *O. Frs.* lík: *Icel.* lík: *Dan.* lig: *Swed.* lik: *O.H.Ger.* líh: *Ger.* leiche.] DER. eofor-, wyrm-líc.

-líc. v. ge-líc, *and the numerous adjectives of which* -líc [*modern* -ly] *forms the last part*.

-líca. v. efen-, ge-, man-, swín-líca.

lícan *to please:*—Ne lícaþ him ðeáh his earfoþu *his troubles do not please him*, Ps. Th. 40, 1. Wel lícaþ Drihtne ða ðe hine him ondrǽdaþ *beneplacitum est Domino super timentes eum*, 146, 12. [*Goth.* leikan: *O. H. Ger.* líchên.] v. lícend-líc; lícian.

líc-beorg [beorg; *m. a hill, funeral mound*; or beorg; *f. protection*; or beorg *connected with* beorgan *to taste, eat, a literal reproduction of* sarcophagus?] *a sarcophagus:*—Lícbeorg *sarcofago*, Wrt. Voc. ii. 119, 50

líc-bysig; *adj. Of active body, active with the body:*—Ic eom lícbysig láce mid winde *active am I of body, move hither and thither with the wind*, Exon. 122 b; Th. 470, 22; Rä. 31, 1.

líccettan. v. lícettan.

liccian; *p.* ode *To lick:*—Ic liccige *linguo*, Ælfc. Gr. 28; Som. 31, 57: *lambo*, 32, 25. Seó lyft liccaþ and átýhþ ðone wǽtan of ealre eorþan and of ðære sǽ, and gegaderaþ tó scúrum, Lchdm. iii. 276, 12. Fýnd his eorþan liccaþ [liccigeaþ, Th.] *inimici ejus terram lingent*, Ps. Spl. 71, 9. Ða rǽdan deór heora liþa liccodon mid liðran tungan, Homl. Skt. 4, 407: Lk. Skt. 16, 21. Liccedon *linxerunt*; liccigan *lincxere*, Wrt. Voc. ii. 51, 54, 55. His fétlástas licciende. Glostr. Frag. 110, 3. [*O. Sax.* likkôn: *O.H.Ger.* lechôn *lambere, lingere: Ger.* lecken.] v. ge-liccian.

liccung, e; *f. Licking:*—Hundes liccung gehǽlþ wunda *a dog's licking heals wounds*, Homl. Th. i. 330, 23.

-líce *a frequent adverbial termination, modern* -ly.

lícend-líc; *adj. Pleasing, pleasant:*—Fordon on his folce is fægere Drihtne wel lícendlíc *quia beneplacitum est Domino in populo suo*, Ps. Th. 149, 4. v. líciend-líc.

lícend-líce; *adv. Pleasingly:* — Fordon mín gebed nú gyt bécnum standeþ ðæt him on wísum is wel lýcendlíce *quoniam adhuc est oratio mea in beneplacitis eorum*, Ps. Th. 140, 8.

Licetfeld, a; *m. Lichfield in Staffordshire:*—Hæfde hé bisceopsetl on ðære stówe ðe gecýd is Licitfeld, Bd. 4, 3; S. 646, 14. Licetfelda bisceop, 5, 24; S. 646, 14. Æt Licetfelda, Shrn. 59, 20. On Licetfelda, Chr. 716; Erl. 45, 14. Æt Licetfelda, Chart. Th. 373, 34.

lícettan; *p.* te *To feign, pretend, profess falsely, simulate:*—Fordam seó orsorge wyrd simle liþþ and lícet ðæt mon scyle wénan ðæt heó is sió sóþe gesǽlþ *illa [prospera fortuna] enim specie felicitatis, cum videtur blanda, mentitur*, Bt. 20; Fox 70, 30. Oft ða unþeáwas leógaþ and lícettaþ ðæt hí síen góde þeáwas *plerumque vitia virtutes se esse mentiuntur*, Past. 20; Swt. 149, 2. Hé lícette hine selfne ðæt hé wǽre ungeleáffull *in se personam infidelium transfigurans*, 16, 2; Swt. 101, 8. Job lícette ðæt hé sceolde bión se hêhsta god *Jove feigned to be the supreme god*, Bt. 28, 1; Fox 104, 13. Lícetton *scemmatizarunt*, Wrt. Voc. ii. 84, 49. Ðá lícettan hí fleám beforan him *simulantibus fugam hostibus*, Bd. 4, 26; S. 602, 19. Ðæt is wísdóm ðæt wís man lícette dysig *it is wisdom for a wise man to feign folly*, Prov. Kmbl. 37. Ðeáh hé lícete untrymnesse, Ps. Th. 40, 9. Nelle wé mid leásungum ðyllíc lícetan *we will not feign such things with falsehoods*, Homl. Skt. pref. 49. Monige tíde ðe mon sceal wærlíce lícettan *nonnulla prudenter dissimulanda sunt*, Past. 21, 1; Swt. 151, 13. Biþ gód tó lícettanne suelce hé hit nyte *it is good to make as if he did not know it*, 151, 9. Lícettende *scemmatizans*, Wrt. Voc. ii. 94, 70. Ðus mid wordum lícettende *offering the following pretext*, Homl. Th. i. 400, 18. [*O. H. Ger.* líhizan *simulare, fingere*.] v. -ge-, leás-lícettan.

lícettere, es; *m. One who feigns, a hypocrite:* — Lícetere *ypochrita*, Wrt. Voc. 85, 39: *fictor vel hipocrita*, 49, 13. Swylce leáse líceteras [lícetteras, Rush.] *sicut hypocritæ*, Mt. Kmbl. 6, 16. Wá eów lícceteras *væ vobis, hypocritæ*, 23, 13. Lícetteras, 23, 15. Líceteras and leógeras Godes graman habban búton hig geswícan *may those who are false in deed and in word have the wrath of God, unless they desist*, L. C. S. 7; Th. i. 380, 5. Wel wítegod Isaias be eów lícceterum *bene prophetavit Esaias de vobis hypocritis*, Mk. Skt. 7, 6. [*O. H. Ger.* líhizari *hypocrita*.] DER. riht-, þeód-lícettere.

lícettung, e; *f. Feigning, pretence, false representation, simulation, hypocrisy*:—Lícetung *hypocrisis*, Lk. Skt. 12, 1. Hwǽr com seó manigfealde lícetung heora freónda *what is become of the manifold flattery of their friends?* Blickl. Homl. 99, 33. Innan gé synt fulle lícettunge [MSS. A. B. lícetunge] *intus pleni estis hypocrisi*, Mt. Kmbl. 23, 28. Búton lícetunge *sine hypocrisi*, Coll. Monast. Th. 33, 7. Ðæt his gesacan on miclum dǽle lícettunge and leáse wið hine syredon and onsǽgdon *accusatores ejus nonnulla in parte falsas contra eum machinasse calumnias*, Bd. 5, 19; S. 640, 14. Þurh lícetunge *per simulationem*, Confess. Peccat. [*O.H.Ger.* líhizunga *dissimulatio*.] v. leás-lícettung.

líc-fæt, es; *n. The body*, Exon. 48 b; Th. 167, 20; Gú. 1063.

LICGAN; *p.* læg: *pl.* lǽgon; *pp.* legen. I. *To* LIE, *be at rest,*

be in bed, lie dead, lie low, fail :—Árís nú hwí líst ðú neowel on eorþan *surge! cur jaces pronus in terra?* Jos. 7, 10. Hwæt ligst ðú on horwe? Dóm. L. 6, 77. Mín cnapa líþ on mínum húse lama *puer meus jacet in domo paralyticus*, Mt. Kmbl. 8, 6. Gif hine on iii nihte ealdne mónan gestandeþ se líþ fæste and swylt *if sickness attack him when the moon is three days old he will be confined to his bed and will die*, Lchdm. iii. 182, 8. Ðonne ðín flǽsc ligeþ *when thou art dead*, Cd. 100; Th. 132, 5; Gen. 2188. Nú se wyrm ligeþ *the serpent is dead*, Beo. Th. 5484; B. 2745. Ðonne wind ligeþ weder biþ fæger *when the wind is at rest the weather is fair*, Exon. 58 b; Th. 210, 7; Ph. 182. Swá ðín blǽd líþ *so shall thy glory lie low*, Cd. 202; Th. 251 13; Dan. 563. Ða creópendan licgeaþ mid ealle líchoman on eorþan *creeping things lie on the earth with all the body*, Past. 21, 3; Swt. 155, 17. Heora líchoman licggaþ on eorþan and beóþ tó duste gewordne, Blickl. Homl. 101, 2. Á ðǽr hé læg [*in his bed*] hé hæfde his handa upweardes, 227, 16. Hé læig æt forþsíðe *he lay at the point of death*, Homl. Th. i. 128, 7; Homl. Skt. 3, 301. Nǽfre on óre læg [*failed*] wídcúþes wíg, Beo. Th. 2088; B. 1041. Ðǽr se cyning ofslægen læg, Chr. 755; Erl. 50, 14. Hié simle feohtende wǽron óð hié alle lǽgon *they kept on fighting until they all lay dead*, 50, 7. On carcern-um lǽgon *they lay in prison*, Ors. 5, 1; Swt. 214, 18. Hí eallne ðone geár an monncwealme lǽgan *all that year they suffered from a pestilence*, 3, 5; Swt. 106, 10. Lige on ða sídan *lie on the side*, L. M. 1, 47; Lchdm. ii. 118, 10. Licge bútan wyrgelde *let him lie [dead] without wergeld*, L. Wlh. 25; Th. i. 42, 13. Gif hine mon ofsleá licgge hé orgilde, L. Alf. pol. 1; Th. i. 60, 15. Hine wulfas ábiton ðǽr hé ástifod lǽge, Blickl. Homl. 193, 8. Hié gemétton ða seofon hyrdas deáde licgan, 239, 25. Tó tácne ðæt hié óðer woldon oððe ealle libban oððe ealle licgan *parato animo, ni vincant, mori*, Ors. 3, 10; Swt. 138, 32 : Chr. 901 : Erl. 96, 28. Gif hé nylle hit geþafian léton hine licgan *if he will not allow it, they shall kill him*, L. Ath. i. 20; Th. i. 210, 9. Ðǽr ða scipu sceoldan licgan *the ships were to lie there*, Chr. 1009; Erl. 141, 24. Hwæt hé gefélde cealdes æt his sídan licgean, Bd. 3, 2; S. 525, 15. Hé má ge-wunode on his smiþþan dæges and nihtes sittan and licgean *magis in officina sua die noctuque residere consuerat*, 5, 14; S. 634, 16. Ðæt mægn ðæs licgendan *the virtue of the dead man*, Glostr. Frag. 110, 7. Hana ða licgenda[n] áwecþ *gallus jacentes excitat*, Hymn. Surt. 6, 36. Licgende feoh *dead [as opposed to live] stock, other property than cattle, ready money* :—Heó beceápode ða scínendan gymmas and eác hire landáre wið licgendum feó *she sold the shining gems and her landed property too, for ready money*, Homl. Skt. 9, 54. Eall ðæt ic hæbbe on libbandan and on licgendan *all the live and dead stock that I have*, Chart. Th. 548, 13. Ðǽr wæs xx M horsa gefangen ðeh hié ðǽr nán licgende feoh ne métten *pecorum magna copia abducta, auri atque argenti nihil repertum*, Ors. 3, 7; Swt. 116, 32. And nán man nán þing ne bygcge ofer feówer peninga weorþ ne libbende ne licgende, L. C. S. 24; Th. i. 390, 3. See Grmm. R. A. pp. 491 sqq. **II. to lie, be situated** [*of a place*]*, go or run* [*of a road or stream*] :—On ðam wege ðe líþ tó Euphrate *in via, quæ ducit Euphratam*, Gen. 35, 19. Swá swá se weg líþ, wé faraþ, Num. 21, 22. Ðam wege ðǽr eást ligþ . . . on ðone wege ðe líþ tó Stánleáge . . . ðam wege ðe tó Stanleáge ligþ, Cod. Dip. Kmbl. iii. 409, 2–17. Sió stów ðe se weg tó ligþ, Bt. 33, 4; Fox 132, 37. Seó Wisle líþ út of Weonodlande and líþ in Estmere . . . and ligeþ of ðæm mere west and norþ on sǽ, Ors. 1, 1; Swt. 20, 7–12. On Swalewan streáme se ligþ be Ceterehtúne *in fluvio Sualua, qui vicum Cataractam præterfluit*, Bd. 2, 14; S. 518, 15. Lindesse ligeþ út on sǽ *Lindissi, pertingens usque ad mare*. 2, 16; S. 519, 19 : 1, 25; S. 486, 21. On his gehlotland ðe líþ on Ephraim dúne *in finibus possessionis suæ, quæ est sita in monte Ephraim*, Jos. 24, 30. Saulus ríce swá hit súþ licgeþ ymbe Gealboe, Salm. Kmbl. 382; Sal. 190. Seó forme India líþ tó ðara Silhearwena ríce seó óðer líþ tó Medas, seó þridde tó ðam micclum gársecge *the first India extends to the kingdom of Ethiopia, the second to Media, the third to the great ocean*, Homl. Th. i. 454, 12. On ðam wege ðe læg tó Thamnaþa *in bivio itineris, quod ducit in Thamnam*, Gen. 38, 14. Ðá lǽg ðǽr án micel eá up in on ðæt land, Ors. 1, 1; Swt. 17, 20. Hé wolde fundian hú longe ðæt land norþryhte lǽge *he wanted to try how far the land extended due north*, Swt. 17, 8. **III. with prep. or adv.** :— Se hláford ðe ryhtes wyrne and for his yfelan man licge *the lord who refuses justice, and makes his wrong doing man's cause his own*, L. Ath. i. 3; Th. i. 200, 15. Godwine eorl and ealle ða yldestan menn on West Seaxon lágon ongeán swá hí lengost mihton *earl Godwin and all the chief men in Wessex opposed as long as ever they could*, Chr. 1036; Erl. 165, 2. xiiii æceras and ða mǽde ðe ðǽr tó líþ *xiiii fields and the meadow belonging thereto*, Cod. Dip. Kmbl. ii. 3, 34. Mid eallon ðam þingon ðe ðǽr tó læg forðam ðe his witan him sǽdon ðæt hit hwílon ǽr læg ðiderin *with everything that belonged thereto; for his witan told him that in former times it had belonged to that place*, vi. 190, 20. Ǽlc ðæra landa ðe on mínes fæder dæge læg intó Cristes cyrcean, iv. 232, 10. [Ic wille ðæt ðæt ligge intó sainte Petre 219, 26 : 220, 19.] [*Goth.* ligan : *O. Sax.* liggian : *O. Frs.* liga : *Icel.* liggja : *O. H. Ger.* ligan, liggan : *Ger.* liegen.] DER. á-, æt-, be-, dyrn-, for-, ge-, tó-licgan.

líc-hama, an; *m. The body* [*generally of a living person*]*, the corpo-*real, *in contrast to the spiritual, part of man* :— Se líchoma biþ líchoma ða hwile ðe hé his lima ealle hæfþ, Bt. 34, 9; Fox 148, 6. Is ðæs monnes líchoma betera ðonne ealle his ǽhta . . . seó sáwl betere ðonne se líchoma, 32, 2; Fox 116, 11–13 : Mt. Kmbl. 6, 25. Hire líchama wæs áfylled mid hreóflan, Num. 12, 10. Ðæt ðín líchama sí eallum fugelum tó mete *sit cadaver tuum in escam cunctis volatilibus cæli*, Deut. 28, 26. Ðis is mín líchaman [líchama, MS. A.] *hoc est corpus meum*, Mt. Kmbl. 26, 26. Án líchama mid his fæder wæs *una caro cum patre fuit*, Bd. 1, 27; S. 491, 15. Cépecnihtas hwítes líchoman and fægeres andwlitan *pueros venales candidi corporis, ac venusti vultus*, 2, 1; S. 501, 7. Ðínes líchaman leóhtfæt is ðín eáge, Mt. 6, 22. Hé wearþ ðá mann gesceapen on sáwle and on líchaman *he became then man formed of soul and body*, Homl. Th. i. 12, 30. Ðonne betǽcþ Crist ða mánfullan mid líchaman and mid sáwle intó helle-wíte *then will Christ deliver the wicked, body and soul, into hell*, ii. 608, 7. Hí tú beóþ on ánum líchoman *erant duo in carne una*, Bd. i, 27; S. 491, 14. Hé wæs álǽded of líchaman *raptus est e corpore*, 3, 19; S. 547, 33 : 4, 3; S. 569, 46. Ne beó gé brégyde fram ðám ðe ðone líchaman ofsleáþ and nabbaþ syððan hwæt hig má dón, Lk. Skt. 12, 4. In ðam ealra ærcebiscopa líchoman syndon bebyrged bútan twegra, heora líchaman sindon on ðære cyricean sylfre gesette, Bd. 2, 3; S. 504, 36. Wé nán ðing nabbaþ búton land and líchaman, Gen. 47, 18. Hé healdeþ ða deádan líchoman ungemolsnode *he keeps the dead bodies undecayed*, Shrn. 82, 21. [*O. Sax.* lík-hamo : *O. Frs.* líkkoma, lícma : *Icel.* líkami, líkamr : *O. H. Ger.* líchamo.]

lícham-leás; *adj. Without a body, incorporeal* :—Englas líchamleáse, Ælfc. T. Grn. 2, 25.

lícham-, lícum-líc [*cf. cognates under* líc-hama]; *adj. Bodily, corporeal, material, carnal, not spiritual* :— Seó [heofene] is geháten firmamentum seó is gesewenlíc and líchamlíc *it* [*heaven*] *is called the firmament; it is visible and material*, Lchdm. iii. 232, 14. Hyre líchomlíce dóhtor *filia ipsius carnalis*, Bd. 5, 3; S. 616, 3. His lícumlíce untrumness *corporea infirmitas*, 4, 1; S. 564, 5. Oswald hæfde lícumlícre yldo xxxvii wintra *anno ætatis suæ trigesimo octavo*, 3, 9; S. 533, 13. Hí wilnodon ðæs líchomlícan deáþes . . . wið ðan écan lífe, Bt. 11, 2; Fox 36, 3; Blickl. Homl. 103, 10. Se hálega gást ástáh líchamlícre ansýne *corporali specie*, Lk. Skt. 3, 22. Wæs hé líchomlícre gebyrdo æþeles cynnes *erat carnis origine nobilis*, Bd. 2, 7; S. 509, 15. Lícumlícre gegaderung *copulæ carnalis*, 2, 9; S. 511, 1. Ealle ða líchamlícan gód biþ forcúþran ðonne ðære sáwle cræftas, Bt. 24, 3; Fox 84, 5. Ne geseó wit unc ofer ðæt líchomlícum eágum *we shall never see one another after that with our bodily eyes*, Bd. 4, 29; S. 607, 21 : Blickl. Homl. 21, 20. [*Icel.* líkamligr : *O. H. Ger.* líhham-líh *corporalis, carnalis*.] DER. un-líchamlíc.

lícham-, lícum-líce; *adv. Bodily, in the body* :—Ðeáh ðe hé líchamlíce on heora slege andwerd nǽre *though he was not present in the body at their slaughter*, Homl. Th. i. 82, 33. Líchamlíce *corporaliter*, Ælfc. Gr. 38; Som. 41, 6. Seó stów ðe Drihten líchomlíce nëhst on stód on middangearde, Blickl. Homl. 125, 15. Hié hine líchomlíce gesáwon *they saw him with their bodily eyes*, 135, 19. Ðeáh ðe hé lícumlíce æfward wære *quamvis corporaliter absens*, Bd. 3, 15; S. 542, 6.

líc-hord, es; *n; The inner parts of the body*, Exon. 46 b; Th. 159, 19; Gú. 929 : 47 b; Th. 163, 31; Gú. 1002.

líc-hrægel, es; *n. Winding-sheet* :— Hí dydon scë Cúþberhtes liic of eorþan . . . ðá bǽron hí ðæs líchrægles dǽl tó Eádberhte ðǽm biscope, Shrn. 82, 16.

líc-hryre, es; *m. Fall of the body, death*, Cd. 52; Th. 67, 11; Gen. 1099.

lícian; *p.* ode *To please* :—Ic lícige *placebo*, Ps. Spl. T. 114, 9. Ne mæg nán man hine sylfne tó cynge gedón ac ðæt folc hæfþ cyre tó ceósenne ðone tó cyninge ðe him sylfum lícaþ *no man can make himself king, but the people have the option of choosing him as king who pleases them*, Homl. Th. i. 212, 8. Hé mé wel lícaþ, ii. 40, 5. Ðë lícaþ se almihtiga God bet ðonne þeodisius, Shrn. 196, 35. Ealle ða þing ðe hér líciaþ sint eorþlice, Bt. 34, 8; Fox 144, 35. Hit lícode Herode, Mt. Kmbl. 14, 6 : Mk. Skt. 6, 22. Swá heó wiste ðæt his fæder lícode, Gen. 27, 14. Ac mé swá ðeáh nó ne lícade on him ðæt hé ða weorþunge Eástrena on riht ne heóld *however I did not like in him his not keeping Easter rightly*, Bd. 3, 17; S. 545, 2. For ðí sceolde ǽlc mon beón on ðam wel gehealden ðæt hé on his ágenum earde lícode *erit igitur pervagata inter suos gloria quisque contentus*, Bt. 18, 3; Fox 64, 28. Hí cwǽdon ðæt ðú Gode lície, Blickl. Homl. 64, 34. Ac lícige swá hit lícige *but please as it may*, Wulfst. 191, 21. Ǽghwylc man þurh góde dǽda Gode lícian sceal, Blickl. Homl. 129, 34. Hé ðam cyninge wæs líciende, Bd. 5, 13; S. 632, 9. Him silfan lícigende, Lchdm. iii. 190, 24. [*O. Sax.* líkôn : *O. Frs.* líkia : *Icel.* líka.] DER. ge-, mis-lícian ; v. lícan.

líciend-líc; *adj. Pleasing, pleasant* :—Sie Gode wel líciendlíc *beneplacitum est Deo*, Ps. Th. 67. 16. Teala líciendlíc, 68. 13. v. lícend-líc.

líc-lǽlan. v. lǽlan.

líc-leóþ, es; *n. A funeral song, dirge; epicedion*, Wrt. Voc. ii. 31, 2.

líc-mann, es; *m. A person having to do with a corpse* :—Ealle ða lícmenn wurdon áfyllede mid ðam wynsumum stence, Homl. Th. ii. 98, 8 : 334, 31. His líc læg ealle ða niht inne beset, ac hé árás of deáþe. Ða

lícmenn ðá ealle flugon âweg, 348, 20 : 548, 15. Ðá bær sum wuduwe hire suna líc tô bebyrgenne . . . Seó dreórige môdor mid ðám lícmannum hí âstrehte æt ðæs hâlgan apostoles fôtum . . . Johannes ofhreów ðære mêder and ðæra lícmanna dreórignysse, i. 66, 15–21. [*Icel.* lík-maðr.]

líc-ness, e; *f. Likeness, form, image, stature* :—Lícnessa *imaginis*, Mt. Kmbl. p. 19, 5. Tô lícnesse *ad staturam*, 6, 27. an-, ge-, un-ge-lícness.

líc-pytt, es; *m. A grave* :—Lícpytt [MS. ic pytt] *scrobs*, Ælfc. Gr. 9, 51; Som. 13, 17.

líc-rest, e; *f. A place of rest for a dead body, tomb, sepulchre* :—Hé hæfde ðæt land syððan him sylfon tô lícreste *he had the land afterwards for his own burial place*, Gen. 23, 20. On líchryste *in cœmeterio*, Hpt. Gl. 507, 67. Man slôh ân geteld ofer ða hâlgan bân binnan ðære lícreste, Swt. A. S. Rdr. 100, 150. Heó hyre lícreste geceás on êlíg byrig *she chose her burial place in Ely*, Lchdm. iii. 430, 17. [*Laym.* þu hit scalt leden to ðere lichraste . . . þer þine wines liggeþ.]

lícsan. v. lixan.

líc-sang, es; *m. A funeral song, dirge* :—Wôpleóþ † birisang † lícsang *tragœdiam, miseriam, luctum*, Hpt. Gl. 488, 56. [*Icel.* líksöngr.]

líc-sár, es; *n. A body-wound, a mortal wound* [?], Beo. Th. 1635; B. 815 : Exon. 28 b; Th. 87, 25; Cri. 1430.

líc-sirce, an; *f. A coat of mail*, Beo. Th. 1105; B. 550.

líc-þegnung, -þénung, e; *f. Last offices done to the dead, funeral, exequies* :—Ic mæg habban ârwurþfulle lícþénunge of heófigendre menigu *I may have honourable service done to my corpse by a mourning multitude*, Homl. Th. i. 86, 33. Ðá ðá his frýnd ða lícþénunge gearcodon *when his friends were performing the last offices for the dead*, ii. 28, 3. Ða fæmnan dedan hire lícþénunge and læddon hí tô byrgenne, Shrn. 87, 27. Lícþénunga *exsequiæ*, Ælfc. Gr. 13; Som. 16, 17.

líc-þeóte, an; *f. A pore* :—Lícþeótan *pori i. spiramenta unde sudor emanat*, Ælfc. Gl. 73; Som. 71, 41; Wrt. Voc. 44, 25.

líc-þrowere, es; *m. A leper, one suffering from ulcers on the body* :—Lícþrowere *leprosus*, Ælfc. Gl. 78; Som. 72, 32; Wrt. 45, 64. Lazarus wæs lícþrowere [*ulceribus plenus*], Homl. Th. i. 328, 15 : Homl. Skt. 3, 480. On Simones húse ðæs lícþroweres *in the house of Simon the leper*, Blickl. Homl. 73, 2. Manega lícþroweras *multi leprosi*, Lk. Skt. 4, 27 : H. R. 105, 2. [*Cf. Icel.* lík-þrá *leprosy*.]

líc-tún, es; *m. An enclosure in which to bury people, a grave-yard, cemetery* :—Hí woldon ðæt heora líctún wære geseted *cimeterium fieri vellent*, Bd. 4, 7; S. 574, 37 : Glostr. Frag. 8, 20. On ðæra brôðra líctune wæs bebyriged *in cœmeterio fratrum sepultum est*, Bd. 3, 17; S. 543, 46 : 4, 10; S. 578, 2, 17, 28 : Chart. Th. 157, 23. Hé næfre binnan nânum gehâlgodum líctúne ne licge *let him never lie in a consecrated graveyard*, L. Ath. i. 25; Th. i. 212, 20 : L. C. E. 22; Th. i. 372, 35.

lícum-líc. v. lícham-líc.

lícung, e; *f. Pleasing, pleasure, gratification* :—Ðætte hié for ðære lícunga ðære heringe ðe hié lufigeaþ eác geþafigen ða tælinge *ut dura admittunt favores, quos diligunt, etiam correptiones recipiant*, Past. 41, 4; Swt. 303, 19. Wel gedafonaþ ðætte ða gôdan recceras wilnigen ðæt hié monnum lícigen, forðæm ðætte þurh ða lícunga hí mægen gedôn ðætte hiera Dryhten lícige ðæm folce, 19, 3; Swt. 147, 7. Ne sylþ Gode lícungæ his *non dabit Deo placationem suam*, Ps. Spl. T. 48, 7. Lícongum *libitos*, Wrt. Voc. ii. 52, 34.

líc-wiglung, e; *f. Necromancy*, L. Edg. C. 16; Th. ii. 248, 3.

líc-wund, e; *f. A wound*, Cd. 154; Th. 193, 1; Exod. 239. [*O. Sax.* lík-wunda.]

líc-wyrþe; *adj. Fit to please, pleasant, well-pleasing, acceptable, agreeable, estimable, sterling* (*of money*) :—Ne mæg heó nân ðæra þinga gedôn ðe Gode lícwyrþe beó *nequit quidquid eorum facere quæ Deo grata sunt*, L. Ecg. P. ii. 16; Th. ii. 188, 5 : Wulfst. 279, 17. Lícworþe, Shrn. 170, 31. On ðære lícwyrþe is Gode eardian *in quo beneplacitum est Deo habitare*, Ps. Lamb. 67, 17. Suæ wæs lícwyrþe before ðec *sic fuit placitum ante te*, Mt. Kmbl. Lind. 11, 26. Ðê micle mâ lícwyrþe se gehnysta gâst *much more pleasing to thee is the contrite spirit*, Ps. C. 50, 126; Ps. Grn. ii. 279, 126. Ne læt ðú unlofod ðæt ðú swutele ongite ðæt lícwyrþe sý *leave not unpraised what you clearly see is estimable*, Prov. Kmbl. 62. Hwæt biþ ðær ðonne lícwyrþes búton his gôd and his weorþscipe ðæs gôdan cyninges *quid in eis aliud, quam probitas utentium, placet?* Bt. 16, 1; Fox 50, 16. iiii pund lícwyrþes feós *four pounds of sterling money*, Cod. Dip. Kmbl. iii. 254, 15. For his lícweorþan feó, 255, 11. Ðínre ðære lícwurþan mundbyrdnesse *to thine acceptable protection*, Glostr. Frag. 108, 16. Him swâ gecwême and lícwyrþe folc, Lchdm. iii. 434, 5. Hié Gode swíðe lícwyrþe forhæfdnesse brengaþ *placentem Deo abstinentiam offerunt*, Past. 43, 8; Swt. 314, 21.

lícwyrþ-ness, e; *f. Good pleasure* :—On ðínre lícwyrþnysse *in beneplacito tuo*, Ps. Lamb. 88, 18.

lid, es; *n. A vessel, ship* :—On lides [*the ark*] bôsme, Cd. 67; Th. 80, 21; Gen. 1332 : 71; Th. 85, 6; Gen. 1410 : Chr. 937; Erl. 112, 27; Aðelst. 27. Tô lides stefne, Erl. 112, 34; Aðelst. 34 : Andr. Kmbl. 806; An. 403 : 3411; An. 1709. Seó [*the dove*] eft ne com tô lide [*the ark*] fleógan, Cd. 72; Th. 89, 11; Gen. 1479. Læt nú geferian flotan úserne,

lid tô lande, Andr. Kmbl. 795; An. 398. [*Icel.* lið; *n. a ship* (almost exclusively in poetry.)] v. liþ.

lida, an; *m. A sailor, traveller* :—Lida biþ longe on síþe, Exon. 90 b; Th. 339, 34; Gn. Ex. 104. [*Icel.* liði *a sailor, traveller.*] v. sæ-, sumor-, ýð-lida; *and* líðan.

líðeþ, Ps. Th. 91, 11. v. leóðan.

lid-mann, es; *m. A sailor, seaman* :—Wícinga werod . . . lidmen, Byrht. Th. 134, 44; By. 99. Lidmanna sum, 136, 41; By. 164. Lidmanna helm (*Beowulf*), Beo. Th. 3251; B. 1623. Lidmonna freá [*Ulysses*], Bt. Met. Fox 26, 126; Met. 26, 63. [*Cf. Icel.* liðs-maðr.]

lid-weard, es; *m. One who guards a ship* :—Lidweardas on merebâte, Andr. Kmbl. 487; An. 244.

lid-wérig; *adj. Weary of being on shipboard*, Andr. Kmbl. 963; An. 482.

Lid-wiccas, Lid-wícingas; *pl.* The people of Brittany [or using the name of the people for the country] *Brittany* :—Carl fêng tô eallum ðam westríce . . . bútan Lidwiccium *Charles took all the western kingdom . . . except Brittany*, Chr. 885; Erl. 84, 13. Two other MSS. have Lidwícingum, Th. 154, 155, and this form occurs in the Scop's Tale :—Ic wæs mid Lidwícingum, Exon. 86 a; Th. 323, 17; Víd. 80. Micel sciphere com súþan of Lidwicum, Chr. 910; Erl. 101, 32. Lidwiccum, 918; Erl. 102, 22. The word seems to contain the British name for Armorica, *Llydaw*. v. notes to the passages from the Cod. Exon. and from the Chron. 918.

LÍF, es; *n.* LIFE [the opposite of death], *mode of life, period during which a man lives* :—Hwæt is ðæt líf elles ðysses middangeardes búton lytelu ylding deáþes, Blickl. Homl. 59, 27. Twâ líf sind sôðlíce . . . ðæt ân líf is deádlíc, ðæt ôðer undeádlíc, Homl. Th. i. 224, 14–16. Ðis andwarde líf manna on eorþan, Bd. 2, 13; S. 516, 14. Lífes treów *lignum vitæ*, Gen. 2, 9. Lífes wæter *aqua viva*, Jn. Skt. 4, 10. Lífes weg, Blickl. Homl. 17, 19. Lífes bæþ, Bd. 2, 5; S. 507, 19. For heora lífes geearnunge geþungon ðæt hí wæron abbudissan *on account of the merit of their lives succeeded in becoming abbesses*; præ merito virtutum, 3, 8; S. 531, 23. Seó þearlwísnes ðæs heardan lífes *districtio vitæ arctioris*, 4, 25; S. 599, 32. Reogollíces lífes þeódscipe, 3, 22; S. 553, 10. On ðære béc Cúþberhtes lífes, 4, 30; S. 609, 32. Ealle hig wæron hâliges lífes menn, Wulfst. 270, 15. Hé geendode his dagas æfter mycclum geswince his lífes, Chr. 1016; Erl. 155, 3. On ðam ýtemestan dæge his lífes, Bd. 3, 17; S. 543, 19, col. 1. Lífes *alive* :—Ætýwde ðæt hé lífes wæs *quia viveret demonstrans*, 5, 19; S. 640, 24. Geáxlan hwæðer hé lífes wære, Homl. Th. ii. 186, 1 : L. Eth. ii. 9; Th. i. 290, 14 : Chart. Th. 471, 34 : Cod. Dipl. Kmbl. i. 234, 28. 32. Ðær belifon swâðeáh lífes on ðam mynstre feówer and twentig muneca, Homl. Skt. 6, 351. Gif hé biþ vi nihta eald and hine âdl gestandeþ se biþ lífes [*he will survive*], Lchdm. iii. 182, 12. [*Icel.* lífes *alive.*] Sume hit ne gedýgdan mid ðam líse *some did not get off with their lives*, Chr. 978; Erl. 127, 13. Heó of deáþe fêrde tô lífe *she went from death unto life*, Bd. 4, 23; S. 595, 32. Hé forþfêrde of ðyssum lífe and fêrde tô ðam uplícan lífe, 2, 1; S. 500, 13. On ðis lífe, Dóm. L. 32, 80. Hé nære nâ man geþuht, gif hé mannes líse ne lyfode, Homl. Th. i. 150, 8 : Blickl. 167, 33. Se hâlga Augustinus be his hâlan líue hine hâdode tô biscope [*while alive and in health*], Chr. 616; Erl. 22, 27. Ðearfendum lífe wunedon *pauperem vitam agebant*, Bd. 1, 15; S. 484, 8. Be nuneca líse *de vita monachorum*, 2, 4; S. 505. 33. On munuclícum lífe geseted, 4, 27; S. 603, 24 : 5, 1; S. 613, 6. Seó bôc þe is âwriten be his lífe, 3, 19; S. 547, 32. Seó freólsbôc ealra ðære landa ðe in tô ðæm mynechina lífe [*nunnery*, v. munuc-líf] æt Wiltúne forgifene sint, Cod. Dip. Kmbl. iii. 117, 25. On, tô lífe [*Icel.* á lifi *alive*] *alive, living* :—Ðá hé on líse wæs *adhuc vivens*, Mt. Kmbl. 27, 63. Hé wæs on líse eorþlíc cing, hé is nú æfter deáþe heofonlíc sanct, Chr. 979; Erl. 129, 9. Ða hwíle ðe hig on líse beón *quamdiu in vivis erunt*, L. Ecg. P. ii. 19; Th. ii. 188, 28. Hwí hig heóldon ða wífmenn tô líse *why they kept the women alive*, Num. 31, 15. Hé læfde æenne tô líse, Wulfst. 106, 8. Se deáþ cymeþ ðæt hé ðæt líf âsyrre, Bt. 8; Fox 26, 7. Sylle líf wið líse *reddat animam pro anima*, Ex. 21, 23. Deáh hé líf hæfde *if he had been alive*, L. C. S. 73; Th. i. 416, 1. Wê ús nyton witod líf æt æfen, Wulfst. 151, 17. Líf and land werian, 274, 17. Preóstas and nunnan heora líf rehtan *let priests and nuns order their lives*, 269, 15. Liif, Bd. 3, 18; S. 545, 42, col. 2. Nis mê tíd mín líf tô onwendenne *there is no time for me to change my life*, 5, 14; S. 634, 32 : Past. 17, 4; Swt. 111, 23. Seó Cúþburh ðæt lýf [*monastery*] æt Winburnan ârærde, Chr. 718; Erl. 45, 19. [*O. Sax. O. Frs. Icel.* líf : *O. H. Ger.* líp *vita, conversatio, habitus.* In Icelandic the word has also the meanings *body* [e. g. líf ok sála] *person*, and the latter use is found in Piers. P. e. g. no *lyf* elles. In *O. H. Ger.*, v. Grff. ii. 44, it is seldom, if ever, used with the meaning of the modern *leib.* DER. ancor-, edwít-, ende-, feorh-, munuc-, mynster-, regol-, sundor-, woruld-líf.

líf, *permission.* v. leáf.

líf; *adj.* v. léf.

lifat [?], Lchdm. iii. 82, 13.

lífan, léfan, lýfan; *p. de To give leave, allow, † permit* :—Ða feówer ic eów lýfe tô sæde and tô mete *quatuor reliquas permitto vobis in sementem*

et in cibum, Gen. 47, 24. Ic ðe selfes dôm life *I allow you to decide*, Cd. 91; Th. 115, 7; Gen. 1916. Moyses lýfde eów eówer wíf tó forlǽtenne *Moyses permisit vobis dimittere uxores vestras*, Mt. Kmbl. 19, 8. God lýfde Adame, ðæt hē môste brúcan ealra wæstma, Wulfst. 9, 6: Blickl. Homl. 189, 22. Ðá bǽdon hý ðæt hē lýfde him on ða gán. Þá lýfde hē him, Lk. Skt. 8, 32. Ðá se cing lýfde eallon Myrceon hám and hig swá dydon *then the king gave leave to all the Mercians to go home, and they did so*, Chr. 1049; Erl. 172, 37. Wē hit ne selfe ne lufedon ne eác ôðrum monnum ne lîfdon [lêfdon, Hat. MS] *we did not love it ourselves nor allow it to other men*, Past. pref; Swt. 4, 6. Ic bidde ðæt ðú mē lýfe ofer ðín land tó férenne *obsecro, ut transire mihi liceat per terram tuam*, Num. 21, 22. Tó ðam dyrstig, ðæt hē ǽfre líse ǽnigan men ðis fæsten tó ábrecenne, Wulfst. 174, 60. Gif priúst lǽfe unrihthǽmed, L. Wih. 6; Th. i. 38, 9. Gif eów Crist lýfan wylle, ðæt ... Exon. 41 a; Th. 137, 27; Gú. 565. [Icel. leyfa *to permit*] v. â-, ge-lîfan.

lîfan, lêfan, lýfan; *p.* de *To believe :* — Ðá lýfde Simplicus and fulwihte onféng, Shrn. 146, 18. Ða dysegan men ðe ðysum drýcræftum lýfdon, Bt. Met. Fox 26, 197; Met. 26, 99. Swá is tó lýfenne ðæt ... Blickl. Homl. 11, 12. [*Goth.* laubjan.] v. ge-lîfan.

lîfan *to remain.* v. be-lîfan.

líf-bryegung, e; *f. Life, intercourse;* conversatio, Rtl. 7, 29.

líf-bysig; *adj. Busy about saving life, struggling for life, anxious about life :* — Ðæt hē for mundgripe mínum scolde licgean lífbysig bútan his líce swice *that for my handgrip he should lie struggling for life, unless his body should escape*, Beo. Th. 1936; B. 966.

líf-cearu, e; *f. Care or anxiety about life*, Andr. Kmbl. 2856; An. 1430: Cd. 42; Th. 54, 17; Gen. 878.

líf-dæg, es; *m. A day of life, any portion of the time that a person lives :* — Ðín geleáfa in lífdæge úrum môde þurhwunige *may belief in thee while we live continue in our hearts*, Hy. 6, 8; Hy. Grn. ii. 286, 8. Swá his lífdagas lǽne syndon, Ps. Th. 102, 14. Ic on lífdagum healde ðínra worda waru *vivam et custodiam sermones tuos*, 118, 17: 139, 8: Cd. 162; Th. 203, 25; Exod. 409: Elen. Kmbl. 880; El. 441. On hyra lífdagum *in the days of their life*, Exon. 25 b; Th. 75, 22; Cri. 1225: 97 b; Th. 364, 23; Wal. 75: Bt. Met. Fox 15, 11; Met. 15, 6. Ic him lífdagas lange sylle *longitudine dierum replebo eum*, Ps. Th. 90, 16: Chart. Th. 372, 18. Gyf God ne gescyrte ðæs þeódscaþan lífdagas, Wulfst. 86, 17. Sumon dægbôte and sumon mâ daga and sumon ealle his lífdagas, L. Pen. 3; Th. ii. 278, 14: Cd. 43; Th. 56, 10; Gen. 910. Hē him lífdagas leófran ne wisse ðonne hē hýrde heofoncyninge *no pleasanter time in his life did he know, than when he obeyed heaven's king*, 162; Th. 203, 25; Exod. 409. Oflét lífdagas *died*, Beo. Th. 3248; B. 1622. [*Icel.* líf-dagar.]

lifen, leofen, e; *f. That by which one lives, support, sustenance :* — Libn *vicatum* [= *victum*]ₙ, Wrt. Voc. ii. 123, 51; Ep. Gl. 28 b, 17. Lífes tô leofne *for. the support of life*, Andr. Kmbl. 2247; An. 1125. [Cf. *Goth.* libains *life.*] v. and-lifen.

LIFER, e; *f. The* LIVER :— Lifer *jecur*, Wrt. Voc. 65, 50: 71, 6. Lifre læppan *fibræ*, Ælfc. Gl. 76; Som. 71, 110; Wrt. Voc. 45, 16. Ðære lifre nett *reticulum jecoris*, Ex. 29, 13. Ealle ða þing ðe tô ðære lifre clifiaþ *cuncta, quæ adhærent jecori*, Lev. 1, 8. Se vultor sceolde forlǽtan ðæt hē ne slât ða lifre Tyties ðæs cyninges, Bt. 35, 6; Fox, 170, 3. [*Icel.* lifr : O. H. Ger. libara : *Ger.* leber.]

lifer *a level surface* [?]; libramentum, Wrt. Voc. ii. 50, 78. v. [?] læfer.

lifer-ádl, e; *f. Disease of the liver*, L. M. Cont. 2, 23; Lchdm. ii. 162, 2.

lifer-býl, e; *f. A prominence on the liver*, L. M. 2, 21; Lchdm. ii. 204, 20. v. next word.

lifer-hol, es; *n. A hollow in the liver :* — Hwæðer on ðám liferbýlum ðe on ðám liferholum, L. M. 2, 21; Lchdm. ii. 204, 20.

lifer-læppa, an; *m. A lobe of the liver :* — Liferlæppa *fibra* i. *vena*, Wrt. Voc. ii. 148, 55. Librlæppan *fibræ*, 108, 54.

lifer-wærc, es; *m. Pain in the liver :* — Wið eallum liferwærcum, M. Cont. 2, 24; Lchdm. ii. 162, 5.

lifesne, Bd. 4, 27; S. 604, 9. v. lybesn.

líf-fadung, e; *f. The ordering or regulating of one's life :* — Be gehádodra manna lîffadunge *of the ordering of the life of men in orders*, L. Wilk. 82, 22.

líf-fæc, es; *n. The time during which life lasts, life :* — On lǽnan líffæce, L. Eth. vii. 21; Th. i. 334, 4. Æfter heora líffæce, Wulfst. 4, 6 : 5, 5.

líf-fæst; *adj. Living, having life, quickened :* — Ðæt hē onfôn wolde ðam gerýne ðære liffæstan rôde Cristes *ad suscipiendum mysterium vivificæ crucis*, Bd. 2, 12; S. 512, 29: Glostr. Frag. 108, 4. Ic mid ða liffæstan ðe þurgoten wæs *vitali unda perfusus sum*, Bd. 5, 6; S. 620, 17. Ða liffæstan leoþu, Exon. 37 a; Th. 327, 19; Vy. 6. v. next word.

líf-fæstan; *p.* te *To give life, quicken, vivify :* — Ðonne hine God liffæsteþ *when God shall quicken him; Deo vivificante*, Bd. 2, 1; S. 500, 20. Gǽst is se liffæsteþ *spiritus est qui vivificat*, Jn. Skt. Rush. 6, 63. Hē is se liffæstenda God, Homl. Th. i. 280, 23: ii. 598, 7. v. ge-liffæstan.

liffettan. v. lyffettan.

líf-freú, an; *m. The Lord of life* [epithet of God], Exon. 8 a; Th. 2, 7, 30; Cri. 15, 27: Beo. Th. 32; B. 16: Cd. 40; Th. 53, 28; Gen. 868: 1; Th. 2, 9; Gen. 16: 86; Th. 108, 18; Gen. 1808: 156; Th. 195. 3; Exod. 271: 192; Th. 240, 33; Dan. 396.

líf-fruma, an; *m. The author of life*, [*Christ*], Exon. 17 b; Th. 41, 16; Cri. 656: [*God*], 23 a; Th. 64, 25; Cri. 1043: Andr. Kmbl. 2570; An. 1286: [*Christ*] 1124; An. 562: Elen. Kmbl. 670; El. 335: Exon. 15 a; Th. 31, 31; Cri. 504: 42 a; Th. 140, 13; Gú. 609: [*God*], Cd. 208; Th. 256, 20; Dan. 643.

líf-gedál, es; *n. Parting with life, separation from life, death*, Beo. Th. 1687; B. 841: Exon. 87 b; Th. 330, 2; Vy. 45: 48 a; Th. 164, 29; Gú. 1019: Cd. 119; Th. 154, 25; Gen. 2561.

líf-gesceaft, e; *f. A condition of life as ordered by fate*, Beo. Th. 3910; B. 1953: 6120; B. 3064.

líf-getwinnan; *pl. m. Twins*, Salm. Kmbl. 284; Sal. 141.

LIFIAN, leofian ; *p.* ode *To* LIVE :— Ne swelte ic ac ic lifige *non moriar, sed vivam*, Ps. Th. 117, 17: 118, 93. Ðú eádig leófast, 127, 2. Ðenden ðú hér leofast, Cd. 43; Th. 57, 29; Gen. 935. Ðǽr hit lifaþ swá unnyt swá hit wæs *where it continues as useless as it was before*, Beo. Th. 6316; B. 3168. Lyfaþ *vivet*, Ps. Th. 71, 15. Þurh Godes fultum, ðe lyfaþ and ríxaþ â bútan ende, Blickl. Homl. 131, 6. Leofaþ, 13, 29. Ða gástlícan láre, ðe úre sául big leofaþ, 57, 9. On gewinne and on swáte hé leofaþ, 59, 36. Se ðe him sylfum leofaþ *he who lives to himself*, Homl. Th. ii. 78, 4. Bé heora ágénum handgewinne lifigeaþ *proprio labore manuum vivant*, Bd. 4, 4; S. 571, 22: 4, 28; S. 605, 16. Be ðæm balzamum ða men in ðæm londe lifgeaþ *opobalsamo vescuntur*, Nar. 31, 6. Godes is ðæt yrfe ðe wē big leofiaþ, Blickl. Homl. 51, 18. Ðǽm mannum ðe be his lárum lifiaþ, 61, 13. On hwylcum geswince hié lifiaþ, 59, 25. Gif wē ða dagas fulfremedlíce for Gode lifgeaþ, 35, 25. Ða hwíle ðe wē lifgaþ hér on worlde, 35, 35. Se cyning Eglippus leofode his líf on eáwfæstre drohtnunge, 476, 16. Se hálga swá leofode swá hē tǽhte, Homl. Th. ii. 186, 19. Se æþeling lyfode [other MS. leofode] ðá gyt, Chr. 1036; Erl. 165, 21. Hē ðǽr sum fæc on forhæbbendum lífe lifede *aliquandiu continentissimam gestit vitam*, Bd. 5, 11; S. 626, 16. Hí for heofonan ríces lufan on ellþeódignesse lifedon *pro æterna patria exulaverant*, 5, 10; S. 624, 12. Ðú leofa bútan mé gif ðú mǽge *live without me, if you can*, Wulfst. 259, 5. Ic beó láreów georn ðæt hē monþeáwum mínum lifge *I am diligent in teaching him to live according to my customs*, Exon. 71 b; Th. 267, 5; Jul. 410. Ðone geleáfan ðý Cristenan þeáwe lifigean and ðone wel healdan *fidem more christiano servare*, Bd. 2, 9; S. 510, 31. Hú hí mid heora geférum drohtian and lifigean scylon *qualiter cum suis clericis conversentur*, 1, 27; S. 488, 37. Leofigean, S. 489, 21. Hē wolde his líf on ælþeódignysse lyfian *peregrinus vivere vellet*, 3, 27; S. 559, 9. Se líchoma bútan mete and drence leofian ne mæg, Blickl. Homl. 57, 10. On forhæfdnesse lifigean, 35, 21. Hē ongan lifgean ongeán Gode ǽrdon ðe hē him sylfum lifgean mihte, 165, 22. Gif seó upplíce árfæstnys mé ǽnig fæc tó lifianne forgifan wylle, Bd. 3, 13; S. 538. 31. Ic symle tilode tó lifigenne tó ðínes múþes bebode, 4, 29; S. 607, 28. On dwolan lifigende, 2, 15; S. 518, 42. Be Diocletiane lyfgendum *vivente Diocletiano*, 1, 8; S. 479, 28: Chart. Th. 485, 33. Sume forlǽtaþ ða hig ǽr hæfdon and be lifiendre cwenan eft ôðre nimaþ *some leave the wives they had before, and while the wife is still living, take another*, Wulfst. 269, 23. Sum ðéh hē forlǽte ða hē ǽr hæfde, hē be lifiendre ðære eft ôðere nimþ, L. Eth. vi. 5; Th. i. 316, 10. Hine þurh ðone lifigendan Drihten hálsedon, Bd. 4, 28; S. 606, 14. [O. *Frs.* livia.] v. â-, ge-lifian ; cwic-, un-lifigende ; libban.

líf-lád, e; *f. Conduct of life, way of life, life*, R. Ben. 1, Lye [cf. hwa so eauer boc writ of mi *liflade*, Marh. 20, 16: heo goð mid gode *liflode* toward þe riche of heouene, A.R. 350, 4. It is also used to mean that by which life is supported, *livelihood :* — Heo tilede here *lyflode*, R. Glouc. 41, 22: lyvelode or lyfhode *victus*, Prompt. Parv. 308. So O. H. Ger. líb-leita *victus, annona, alimonium, alimentum.*]

líf-leás; *adj. Lifeless, without life :* — Ðú bist deád and ða ðe ðé tó lóciaþ beóþ lífleáse eác *morte morieris tu et omnia quæ tua sunt*, Gen. 20, 7. Fela templa árǽrdon and mid andgitleásum and lífleásum anlícnyssum áfyldon *erected many temples, and filled them with images that were without sense and without life*, Homl.Th. ii. 574, 28.

líf-leást, -lǽst, e; *f. Loss of life, death :* — On ǽlcum ðara daga gif man ǽnige ǽddran geopenaþ on ðara tíde ðæt hit biþ lífleást oðde langsum sár *on each of those days, if a vein be opened at that hour, it is death or long disease*, Lchdm. iii. 152, 5. Bendas oððe dyntas hwílum líflǽsta *bonds or blows, at times death*, L. Pen. 3; Th. ii. 278, 27.

líf-líc; *adj. Pertaining to life, living, causing life, vital :* — Líflíc *vitalis*, Ælfc. Gr. 9, 28; Som. 11, 36. Líflíc ys blôd lǽtan *to let blood* [at this time] *is as much as a man's life is worth*, Lchdm. iii. 190, 28. Is hwæðere swá tó lǽtanne swá ðæt líflíce mægen ne áspringe *blood however is to be let so that vital power be not dissipated*, L. M. 2, 42; Lchdm. ii. 254, 12. Wyll líflíc *fons vivus*, Hymn. Surt. 92, 15. Ic eom se líflíca hláf ðe of heofenum ástáh *I am the living bread, that came down from heaven*, Homl. Th. ii. 202, 5. Ðæra nǽddrena geslit wæs deádlíc Cristes deáþ wæs líflíc *the bite of the serpents brought death; Christ's death brought*

life, 238, 31. Lîflîc onsægednys *a living sacrifice*, i. 358, 18 : 482, 12. Adylegode of ðære liflican bêc *blotted from the book of life*, 68, 11. Liflîcum blôde *vivido sanguine*, Hymn. Surt. 80, 21. God âbleów on his ansŷne liflîcne blǽd, Hexam. 11 ; Norm. 18, 26.

lîf-lîce ; *adv. Vitally, so as to infuse life :*— Hê genam ðâ hlâf and hine liflîce hâlgode, Homl. Th. ii. 244, 10.

lîf-lyre, es ; *m. Loss of life :*—Gif liflyre wurþe *if loss of life occur*, L. E. B. 2 ; Th. ii. 240, 11.

lîf-neru, e ; *f. Support of life, food :*—Tô lifnere Andr. Kmbl. 2180 ; An. 1091. [*O. Sax. O. L. Ger.* lîf-nara *sustenance : O. H. Ger.* lîb-nara *victus, alimonia.*]

lifnes, Bd. 4, 27 ; S. 604, 9 note. v. lyfesn.

lifrig ; *adj. Connected with the liver :*—Dæt þiccæ and lifrige blôd, L. M. 2, 40 ; Lchdm. ii. 250, 10.

lift *the air.* v. lyft.

lift, lyft *a grant, allowance :*—Ús bôceras beteran secgaþ lengran lŷft wynna *learned men tell us of a better and longer grant of joys*, Cd. 169 ; Th. 211, 24 ; Exod. 531. Cf. lîfan. [*Bouterwek suggests* lyst (= lust) wynna, Grmm. Gr. ii. 466 *would read* lyftwynn *recreatio in aere ; Thorpe suggests* lîf *for* lyft.]

lîf-weard, es ; *m. A guardian of life* [*Christ*], Elen. Kmbl. 2069 ; El. 1036.

lîf-weg, es ; *m. A way which leads to life, way of life, one's path in life :*—Lîfweg [*the road followed by the Israelites under the guidance of the pillar of cloud*], Cd. 147 ; Th. 184, 9 ; Exod. 104. Uton nû ealle ûre lîfwegas geornlîce rihtan *let us diligently amend our ways*, Wulfst. 75, 22. Lîðe lîfwegas, Exon. 43 b ; Th. 148, 5 ; Gû. 740.

lîf-wela, an ; *m. Riches that confer* or *possess life, heavenly riches, wealth belonging to this,* or *to the next, life :*— Him wæs wuldres dreám, lîfwela leófra ðonne ðâs leásan godu, Apstls. Kmbl. 97 ; Ap. 49. Ða lîfwelan, swǽse swegldreámas, Exon. 27 b ; Th. 82, 33 ; Cri. 1348. Lîfwelan *the wealth of this world*, Cd. 174 ; Th. 219, 17 ; Dan. 56.

lîf-welle ; *adj. From a living spring :*—Lîfwelle wæter [wæter cwicwelle, Rush.] *aquam vivam*, Jn. Skt. Lind. 4, 10.

lîf-wraðu, e ; *f. A support of life*, Beo. Th. 1946 ; B. 971 : 5746 ; B. 2877.

lîf-wynn, e ; *f. A pleasure* or *joy of life :*—Hê lytle hwîle lîfwynna breác *a little while he enjoyed the pleasures of life*, Beo. Th. 4201 ; B. 2097 ; Exon. 19 b ; Th. 50, 27 ; Cri. 807 : Elen. Kmbl. 2535 ; El. 1269.

lîg, lêg, es ; *generally masc. but ðæt* lêg *occurs. Flame, lightning :*—Lîg *flamma*, Wrt. Voc. 76, 49 : 82, 52. Lêg, 284, 12. Dæt fŷr and ðæt lêg [se lîg MS. C.] swîðe weóx ... Ða fôr se wallenda lêg ... ðær se lêg mǽst wæs, Bd. 2, 7 ; S. 509, 19–24. Se lêg ongan sleán ongeán ðone wind, Blickl. Homl. 221, 12. Wonna lêg *the pale flame*, Beo. Th. 6221 ; B. 3115. Hlemmeþ hâta lêg, Exon. 21 b ; Th. 58, 9 ; Cri. 933. Reáða lêg, 19 b ; Th. 51, 2 ; Cri. 810. Sweart lîg, Cd. 110 ; Th. 145, 33 ; Gen. 2415. Ne biþ ðær [heaven] nânes lîges gebrasl, Dôm. L. 16, 259 : Beo. Th. 166 ; B. 83. Wylm ðæs wæfran lîges *the heat of the flickering flame*, Cd. 185 ; Th. 231, 2 ; Dan. 241. Ligges leóma, 190 ; Th. 237, 25 ; Dan. 343. Ðone deópan grund ðæs hâtan lêges and ðæs heardan lêges [*hell*], Blickl. Homl. 103, 15. For ðæs lêges [*lightning*] bryne, 203, 11. Lêges blæstas, Andr. Kmbl. 3103 ; An. 1554. Biscopas mid folcum mid îserne and lîge fornumene wæron *presules cum populis ferro et flammis absumebantur*, Bd. 1, 15 ; S. 484, 2. Fŷres lîge wæs fornumen, 4, 25 ; S. 599, 20. On fîres lîge *in flamma ignis*, Ex. 3, 2. On brâdum ligge, Wulfst. 188, 3. Ligge gelîcost, Beo. Th. 1458 ; B. 727. For dracan lêge *for the flame that was sent forth by the dragon*, 5092 ; B. 2549. Blâcan lŷge, Andr. Kmbl. 3081 ; An. 1543. Ûre synna lîg, Wulfst. 287, 9. Hî wǽron on ǽnne unmǽtne lêg gesomnade *in immensom adunati sunt flammam*, Bd. 3, 19 ; S. 548, 21. Ðǽr [*hell*] hê hæfþ weallendne lêg, and hwîlum cŷle ðone grimmestan, Blickl. Homl. 61, 35. On ðæt lîg tô ðê hweorfan, Cd. 35 ; Th. 46, 33 ; Gen. 753. Lîga *fulminum*, Hpt. Gl. 509, 30. On fŷrenra lêga onlîcnesse *in the form of flames of fire*, Blickl. Homl. 133, 20. Lêgea, 135, 3. Monige heápas sweartra lîgea *crebri flammarum tetrarum globi*, Bd. 5, 12 ; S. 628, 16, 22. Lêga leóhtost, Bt. Met. Fox 9, 33 ; Met. 9, 17. Wununga âfyllede mid brastligendum lîgum, Homl. Th. i. 68, 5 ; Bd. 2, 7 ; S. 509, 34. Brand and brâde lîgas, Cd 18 ; Th. 21, 16 ; Gen. 325 : 36 ; Th. 47, 20 ; Gen. 763. Mellitus ða lîgeas his byrnendre ceastre gebiddende âdwæscte *Mellitus flammas ardentes suæ civitatis orando restinxerit*, Bd. 2, 7 ; S. 509, 2. Lîgeas gemonigfealdaþ and hî gedrêfeþ *fulgura multiplicavit et conturbavit eos*, 4, 3 ; S. 569, 20. Lîas *flammas*, Hymn. Surt. 10, 29. [*O. E. Homl.* leies ; *pl.: Piers P.* leye : *A. R.* leie : *Jul.* ley : *Icel.* leygr *a flame* [poet.] : *O. H. Ger.* louch, loug *flamma :* cf. *O. Sax.* lôgna : *Icel.* log, logi *a flame, lowe.*]

lîg-bǽre ; *adj. Flame-bearing, flaming, fiery :*—Lîgbǽrum *flammifera*, Hpt. Gl. 433, 71. Ligbǽrum scridum *flammigeris quadrigis*, Wrt. Voc. ii. 149, 13.

lîg-berend ; *adj. Flame-bearing, fiery :*—Lîgberend *flammiger*, Wrt. Voc. ii. 149, 9. Lêgberend, 36, 52.

lîg-bryne, es ; *m. Burning of flame, fire :* — Æfter lîgbryne, Exon. 64 a ; Th. 236, 20 ; Ph. 577. Lêgbryne, 22 b ; Th. 62, 15 ; Cri. 1002.

---

lîg-cwalu, e ; *f. Torment,* or *death by fire*, Elen. Kmbl. 591 ; El. 296.

lîg-draca, an ; *m. A fire-drake, dragon vomiting flames*, Beo. Th. 4655 ; B. 2333. Lêgdraca, 6073 ; B. 3040.

Lige *the river Lea.* v. Lyge.

lige, ligen, *a lie.* v. lyge, lygen.

lîg-egesa, an ; *m. Fear caused by fire*, Beo. Th. 5554 ; B. 2780.

lîgen ; *adj. Flaming, fiery :*—Lîgen *flammaticus*, Wrt. Voc. ii. 149, 5. Ðǽr wæs lîgen swurd gelogod æt ðam ingange *there was placed a flaming sword at the entrance*, Hexam. 19 ; Norm. 28, 1. Lêgene sweorde, Elen. Kmbl. 1511 ; El. 757. Heofon lîgenne gesihþ *if he sees the heavens fiery*, Lchdm. iii. 200, 14. Lîgen ðære sunnan hweogul *flammeam solis rotam*, Hymn. Surt. 22, 23. Se ealda deófol hine æteówode mid byrnendum mûþe and lîgenum eágum, Homl. Th. ii. 164, 23. [*O. H. Ger.* laugin *flammeus.*]

lîget, es ; *m. n. :* lîgetu, e ; *f. Lightning, a flash of lightning :*— Lîgit *fulgor* vel *fulmen*, Wrt. Voc. 52, 46. Hys ansŷn wæs swylce lîgyt, Mt. Kmbl. 28, 3 : 24, 27. Ðǽr begann tô brastligenne micel þunor and lîget sceótan, Homl. Th. ii. 196, 23. Swâ hâttra sumor swâ mâra þunor and lîget, Lchdm. iii. 280, 10. [Swâ stor þunring and lǽgt wes, swâ ðæt hit âcwealde manige men, Chr. 1085 : Erl. 219, 22.] Æfter ðæm wolcne cymeþ lêgetu and þunor, Blickl. Homl. 91, 33. Lêgitu, Ps. Surt. ii. 196, 19. Lêgite *fulgoris*, 190, 15. Men sweltaþ for ðæs þunres ege ânum and ðære lîgette, Wulfst. 207, 26. Ðæt fŷr âbyrst ût þurh lîgett [lîgette, MS. R. P.], Lchdm. iii. 280, 7. Hê lǽdeþ wind and lîget, Ps. Th. 134, 7. Gif lîgette and þunorrâde eorþan and lyfte brêgdon *si corusci ac tonitrua terras et aera terrerent*, Bd. 4, 3 ; S. 569, 12. Lîgette *coruscationes*, Ps. Th. 76, 15. Lîgetta, 143, 7 : *fulgura*, Ex. 19, 16 : Exon. 54 b ; Th. 192, 15 ; Az. 102. Ða flugon ða lîgetu swylce fŷrene strǽlas, Blickl. Homl. 203, 9. Lîgetu, Ps. Th. 17, 12 : Cd. 192 ; Th. 240, 2 ; Dan. 380. Lîgetas, Bd. 4, 3 ; S. 569, 22. Lîgettas *fulgura*, Ps. Lamb. 17, 15. Lîgetta [lŷgytu, MS. C.] *fulgura*, Ps. Spl. 134, 7. Hê gemanigfealdode his lîgeta *fulgura multiplicavit*, Ps. Th. 17, 14. Lêgite, Ps. Surt. ii. 197, 34. [Cf. *Goth.* lauhatjan *to lighten : O. H. Ger.* laugazan : lôhazan *rutilare, micare, coruscare.*]

lîget-ræsc, es ; *m. Lightning :*— Ic geseah Satanan swâ swâ lîgetræsct of heofone feallende, Lk. Skt. 10, 18, MS. A. v. lîg-ræsc.

lîget-slîht, ð ; *f. A flash of lightning :*—Ða com þunerrâd and lêgetsleht and ofslôh ðone mǽstan dǽl ðæs hǽðnan folces, Shrn. 57, 35. Lêgedslæht *fulgor*, Lk. Skt. Lind. 10, 18.

lîg-fémende, -fâmblâwende, -fŷrberende [-ferbærnde, MS.] *vomiting flame ;* flammivomus [the words are those used by the several MSS. in] Bd. 5, 12 ; S. 630, 12.

lîg-fŷr, es ; *n. Flaming fire*, Cd. 146 ; Th. 182, 18 ; Exod. 77.

lîg-hrægel [?] :—Lîgrægel *orbiculata* [*vestis*], Ælfc. Gl. 29 ; Som. 61, 30 ; Wrt. Voc. 26, 29. Cf. Ducange ' duo *orbicularia* de opere ad acum ' *under* orbiculare.

lîg-locc ; *adj. Having flaming locks :* — Lîgloccum *flammicomis*, Wrt. Voc. ii. 149, 12.

lîg-loccod ; *adj. Furnished with fiery locks :* — Lîgloccode [-liccode, Wrt.] *flammicomos*, Wrt. Voc. ii. 149, 10.

lignian ; *p.* ede *To deny :*—Ðû lignest nû ðæt sîe lifgende se ofer deófium dugeþum wealdeþ, Cd. 212 ; Th. 263, 18 ; Dan. 764. Hû hine [*Christ*] lŷgnedon leáse on geþoncum, Exon. 24 a : Th. 69, 13 ; Cri. 1120. [*Goth.* laugnjan : *O. Sax.* lôgnian *to deny : O. H. Ger.* laugauian *negare, diffiteri, inficiari : Ger.* laugnen.]

lîg-ræsc, es ; *m. Lightning, a flash of lightning, bright light :*—Lîgræsc *coruscatio* i. *fulgor*, Wrt. Voc. ii. 136, 2. His ansŷn wæs swylce lîgræsc, Nicod. 15 ; Thw. 7, 20. Ic geseah Satanan swâ swâ lîgræsc of heofone feallende *videbam Satanaň sicut fulgor de cælo cadentem*, Lk. Skt. 10, 18 : 17, 24. Ðæt leóhtfæt ðæs lîgræsces *lucerna fulgoris*, 11, 36. For lîgræsce *præ fulgore*, Ps. Spl. 17, 14. Lîgrascas *coruscationes*, 76, 18. Lîgræscas *fulgura*, 96, 4. Lêgræscas, 17, 16 : *coruscationes*, Blickl. Gl. Ðæt ðû âwende hagolas and lîgræscas [-ræceas, MS. O.] *that thou avert hail and lightning*, Herb. 176 ; Lchdm. i. 308, 23. Lîgræsceas gesihþ *orsorhnesse hit getâcnaþ*, Lchdm. iii. 202, 17. Lîghræscas, Ps. Lamb. 134, 7.

lîg-ræscetung, e ; *f. Lightning :*—Lîgrescetunga *fulgura*, Ps. Lamb. 17, 15.

lîg-spiwol ; *adj. Vomiting flame :*—Ðær beóþ ða welras gefylde lîgspiwelum bryne, Dôm. L. 14, 209 ; Wulfst. 139, 9.

lîg-pracu ; *gen.* -þræce ; *f. Violence* or *tumultuous movement of flames :*—Æfter lîgþræce *after the fire has spent its force*, Exon. 59 a ; Th. 213, 15 ; Ph. 225.

lîg-ŷp, e ; *f. A wave of flame*, Beo. Th. 5338 ; B. 2672.

lîht. v. leóht.

lihtan ; *p.* te *To shine, lighten, give light :*—Hit lîht *fulminat*, Ælfc. Gr. 22 ; Som. 24, 7. Ðæt leóht lŷht on þŷstrum *lux in tenebris lucet*, Jn. Skt. 1, 5. Se môna lîht on niht, Bt. 21 ; Fox. 74, 25. Swâ swâ ðæt leóhtfæt lîeht on nieht ûrum eágum, ðætte ðá gewritu on ðæg liehten ûrum môde, Past. 48, 1 ; Swt. 365, 15. Lihteþ *luceat*, Ps. Surt. ii. 202, 11 ; Exon. 64 a ; Th. 237, 9 ; Ph. 587. Wedercondel (*the sun*) wearm weorodum lŷhteþ, 58 b ; Th. 210, 18 ; Ph. 187. Ne hêr dæg lŷhteþ *day shines not here*,

Cd. 215; Th. 271, 14; Sat. 105. Líhte *auroresceret*, Wrt. Voc. ii. 88, 54. Ðá dæg lýhte *at dawn*, 180; Th. 225, 23; Dan. 158: Andr. Kmbl. 2794; An. 1399: Exon. 21b; Th. 58, 21; Cri. 939. Swá se lígræsc lýhtende scínþ *sicut fulgor coruscans fulget*, Lk. Skt. 17, 24. Hé wæs byrnende leóhtfæt and lýhtende *ille erat lucerna ardens et lucens*, Jn. Skt. 5, 35. Sumre niwre gyfe líhtendre *nova quadam relucente gratia*, Bd. 5, 22; S. 644, 26. [*Goth*. liuhtjan: *O. Sax*. lióhtian: *Icel*. lýsa: *O. H. Ger*. liuhtjan: *Ger*. leuchten.] DER. á-, geond-, in-, on- líhtan; *and see* leóhtan.

líhtan; *p. te*. **I.** *to make light or easy, to alleviate, relieve, assuage* :— Líht ðæt ðone swencendan magan *that relieves the labouring stomach*, L. M. 2, 7; Lchdm. ii. 186, 20: 2, 44; Lchdm. ii. 256, 13. Gif ðǽr hwylc wíteþeówman sý bútan ðyson hió gelýfþ tó hyre bearnon ðæt hí hine willon lýhtan for hyre sáulle *if there be any penal slave besides these, she trusts to her children that they will relieve (release*, v. líhting*) him for her soul's sake*, Chart. Th. 535, 38. Ðá wolde ic mínne þurst léhtan *sitim levare cupiens*, Nar. 8, 28. **II.** *to relieve of a burden, to light, alight* :— Hé lýhte of his horse *he alighted from his horse*, Bd. 3. 22; S. 553, 32. Ðá líhte se eorodman, 3, 9; S. 533, 33: H. R. 103, 17: Byrht. Th. 132, 28; By. 23. [*Icel*. léhta *to lighten, ease, leave off what is laborious: O. H. Ger*. ga-líhtjan *lenire, levare, relevare*.] DER. á-, ge- líhtan; *and see* leóhtian *to grow light*.

líhte. v. leóhte.

líhting, e; *f. Lighting, shining, illumination, giving light* :— On líhtinge fýres *in illuminatione ignis*, Ps. Spl. 77, 17. God geworhte ðæt máre leóht tó ðæs dæges líhtinge, Gen. 1, 16. Ða steorran sint tó nihtlícere líhtinge gesceapene, Homl. Th. i. 110, 15. Mid sóðre sunnan líhtincge úre heortan álíhte, Btwk. 196, 17. Nú is ǽlc dæg of ðære sunnan lýhtinge, Lchdm. iii. 234, 18. Hí (*the stars*) nabbaþ náne lýhtinge for ðære sunnan andwerdnysse, 236, 1. Se móna næfþ náne líhtincge *the moon shall not give her light*, Wulfst. 137, 12. Ðæt swearte fýr him náne líhtinge ne déþ '*from those flames no light*,' Homl. Th. i. 132, 17. Healde man ælces sunnandæges freólsunga fram nóntíde ðæs Sæternes dæges óþ ðæs mónandæges líhtinge, L. Edg. i. 5; Th. i. 264, 20: Wulfst. 117, 4: 207, 12. Ðæt ða gesceaftu gesewenlíce wurdon þurh ðæs dæges líhtinge, Hexam. 4; Norm. 8, 3. Líhtunge *coruscationes*, Ps. Spl. T. 76, 18. DER. á-, on-líhting. v. leóhting.

líhting, e; *f. Lightening, alleviation, relief, mitigation, release* :— Ðis is seó líhtincge ðe ic wylle eallon folce gebeorgan ðe hig ǽr ðyson mid gedrehte wǽron ealles tó swýðe *this is the relief that I will secure to all folk in regard to matters with which they were ere this all too much harassed*, L. C. S. 70; Th. i. 412, 18. Ðonne biþ him gesæld his synna líhtincge *then shall a release from his sins be given him*, L. Pen. 18; Th. ii. 286, 3. Gif ðæt riht tó hefig sý séce siððan ða líhtinge tó ðam cynge, L. Edg. ii. 2; Th. i. 266, 12.

líhting-ness, e; *f. Lightness of taxation; levitas tributi*, L. I. P. 3; Th. ii. 306, 22.

líht-líce. v. leóht-líce.

líht-ness, e; *f. Lightness, brightness* :— Se sunnandæg is wuldorlíc dæg and líhtnesse dæg, Wulfst. 230, 12. [*O. H. Ger*. liuht-nissa *illuminatio*.] DER. á-líhtness.

lilie, lilige, an; *f. A lily* :— Lilie *lilium*, Ælfc. Gl. 39; Som. 63, 60; Wrt. Voc. 30, 10. Liliæ. Ðás wyrt man lilie and óðrum naman *lilium* nemneþ, Herb. 109; Lchdm. i. 222, 5. Lilige, Lchdm. iii. 24, 9. Genim ða twá wyrta, ðæt is, lilie and róse; bet tó bearneácenum wífe . . . gif heó nimþ lilian, heó cenþ cnyht; gif heó nimþ rósan, heó cænþ mǽden, 144, 10–13. Ðeáh ðe lilie sý beorht on blóstman ic eom betre ðonne heó, Exon. 110b; Th. 423, 24; Rä. 41, 27. Drince hé lilian wyrttruman áwylledne on wíne, L. M. 1, 37; Lchdm. ii. 90, 13. Genim neoþeweard e lilian, Lchdm. i. 374, 6. Godes gelaþung hæfþ on sibbe lilian, ðæt is clǽne drohtnung; on ðam gewinne, rósan, ðæt is martyrdóm, Homl. Th. ii. 546, 2. Besceáwiaþ æcyres lilian, Mt. Kmbl. 6, 28: Lk. Skt. 12, 27. [*O. Sax*. lilli: *Icel*. lilja: *O. H. Ger*. lilia; *f*; lilio; *m*: *Ger*. lilie.]

LIM, es; *n*. (but it also occurs with adj. fem. :). *A limb, joint, member of a body, branch of a tree* :— Án lim *membrum*; má lima *membra*, Wrt. Voc. 70, 20, 21. Gif men cíne hwylc lim, genim regen mela, dó on ðæt lim, L. M. 1, 73; Lchdm. ii. 148, 22. Be ðæs limes (*the finger*) micelnysse, Homl. Th. ii. 204, 6. Limes dǽl *commata* (*commota*, Wrt.) Wrt. Voc. ii. 20, 25. On ælcre lime, L. M. 2, 64; Lchdm. ii. 288, 22. Wið foredum lime *for a broken limb*, 1, 25; Lchdm. ii. 66, 22, 26. Ne biþ nán tó ðæs lytel liþ on lime áweaxen, Soul Kmbl. 192; Seel. 96. Ic nán lim onstyrian ne mihte *I could not stir a limb*, Bd. 5, 6; S. 619, 26. Hafa ðúne niéxtan swá swá ðín ágen lim, Basil Admn. 5; Norm. 44, 24. Monegu limu beóþ on ánum men, and weorþaþ ðeáh ealle tó ánum líchoman, Bt. 34, 6; Fox 140, 25. Gif wé tó lange sittaþ, slapaþ ða lima, Homl. Th. i. 490, 1. Gé sindon Cristes líchama and leomu (cf. *Icel*. Guðs, fjándans limir), ii. 276, 19. Unrihtwíse syndon deófles leomo, Blickl. Homl. 33, 8. Ðæs biscopes leoma on ðysse byrigenne syndon betýned, Bd. 2, 1; S. 500, 22. Leomu gnornian *the (leafless) branches (shall) mourn*, Exon. 89a; Th. 334, 35; Gn. Ex. 26. Án dínra lima *unum membrorum tuorum*, Mt. Kmbl. 5, 29: Bd. 4. 9; S. 577, 17. Hé

biþ Cristes lima án, Wulfst. 37, 5. Þurh deófles oððe his lima láre, Cod. Dip. Kmbl. iii. 138, 16. Leoma, Blickl. Homl. 147, 15. Leomena, Salm. Kmbl. 205; Sal. 102. Fram árleásum deófles limum, Homl. Th. i. 556, 8: Wulfst. 37, 7: Ps. Th. 21, 15. Ic geseó óðre ǽ on mínum leomum . . . synne ǽ seó is on mínum limum, Bd. 1, 27; S. 497, 35–37. Leomum, Blickl. Homl. 33, 11: 167, 2. Leomum and leáfum *with branches and leaves*, Beo. Th. 194; B. 97. Hé ongan his limu þræstan, Bd. 3, 11; S. 536, 15. Hé his lima gesette and hine gerestan wolde, 4, 11; S. 579, 32. Limo 4, 24; S. 597, 10. Leomu, 2, 6; S. 508, 11. Ádyde ða leomu and ðæt heáfod on weg ðæs sceápes, Blickl. Homl. 183, 24. Leomo lǽmena, Exon. 8a; Th. 2, 6; Cri. 15. [*Icel*. limr; *m. a limb, a joint (of an animal)*: lim; *n. a branch*; limar; *pl. f. branches of a tree*.] DER. gecynd-, sceam- lim.

LÍM, es; *m*. LÍME, *material which causes adhesion, cement, mortar, glue, gluten, bird lime, thick substance made of curds, paste* :— Ánes cynnes lím *bitumen*, Ælfc. Gl. 56; Som. 67, 43; Wrt. Voc. 37, 31. Lím tó fugele *gluten*; eglím *glara*, Ælfc. Gl. 80, 81; Som. 72, 118, 119; Wrt. Voc. 47, 1, 2. Gebærnd lím *calcis viva*, Wrt. Voc. ii. 127, 49. Lím *cementum, i. cesura lapidis*, 130, 62: *bitumen*, 11, 8: *cola*, 20, 24: *gluten*, 40, 25: *glus*, 40, 72. Lím, caluuer *galmilla*, 109, 55. Lím, molecgn *galmilla*, 40, 62. Liim, molegn, Ep. Gl. 10f, 32. Lím *calmilla*, Wrt. Voc. 290, 35: *gluten*, Ælfc. Gr. 9, 12; Som. 9, 30. Swá lím gefæstnaþ fell tó sumum brede *as glue fastens a skin to a board*, 44; Som. 45, 25. Límes *calcis*, Wrt. Voc. ii. 19, 52. Áfæstnod ic eom on líme grundes *infixus sum in limo profundi*, Ps. Spl. 68, 2. Ic beswíce fugelas mid líme *decipio aves glutino*, Coll. Monast. Th. 25, 13. Eorþan líme . . . ðæt is syndrig cynn, symle biþ ðý heardra ðé hit swearte sǽstreámas swíðor beátaþ, Cd. 66; Th. 80, 2–10; Gen. 1322–1326. Þurh lím *per cola*, Hpt. Gl. 411, 7. [*Icel*. lím; *n. lime, glue, paste: O. H. Ger*. lím *bitumen, gluten, viscus: Ger*. leim; *m*.] DER. æg-, fugel-, stán-lím; *and see* ge-líman, -límian.

limb-stefning, e; *f. An awning, curtain; peripetasma*, Ælfc. Gl. 116; Som. 80, 69; Wrt. Voc. 61, 46.

lim-gelecg, es; *n. The disposition or arrangement of the limbs, form, shape* :— Limgelecg *liniamento*, Wrt. Voc. ii. 52, 31.

lim-hál; *adj. Sound of limb*, Exon. 42b; Th. 143, 14; Gú. 661.

límian. v. ge-límian (*Appendix*).

líming, e; *f. Daubing, plastering, cementing* :— Líming *liture*, Wrt. Voc. ii. 52, 43. Líminge *lituræ*, Hpt. Gl. 509, 54. [*A. R.* limung *joining: Icel*. líming *glutinatio*.]

lim-lǽw, e; *f. Injury to the limbs, mutilation* :— Bendas oððe dyntas . . . hwílum lim-lǽwa and hwílum liflǽsta *bonds and blows . . . at times mutilations of the limbs, and at times deprivation of life*, L. Pen. 3, note; Th. ii. 278, 27. v. next word.

lim-lǽweo; *adj. Maimed or injured in the limbs* :— Gif limlǽweo (other MS. -læpeo) lama ðe forworht wǽre weorþe forlǽten and hé æfter ðam þreó niht álibbe siððan man mót hylpan *if a criminal that has been mutilated be left, and he live after that three days, then he may be helped*, L. E. G. 10; Th. i. 172, 16. v. léf, ge-léfan, léwsa *and preceding word*.

lim-lama; *adj. Lame in the limbs, crippled* :— Manege ðǽr wurdan hále, ðe ǽr wǽran limmlaman, Wulfst. 4, 12.

lim-leás; *adj. Without limbs* :— His (Christ's) gástlíca líchama, ðe wé húsel hátaþ, is of manegum cornum gegaderod, búton blóde and bâne, limleás and sáwulleás, Homl. Th. ii. 270, 22.

lim-mǽlum; *adv. Limb-meal* (used by Shakspere in Cymbeline), *limb by limb, a limb at a time* :— Limmǽlum *membratim*, Wrt. Voc. ii. 54, 55: *membratim, particulatim*, Hpt. Gl. 443, 3: *membratim, per singula membra*, 486, 44. [*Laym*. he hine limmele todroh.]

lim-nacod; *adj. With uncovered limbs, naked* :— Se eádega wer [Noah] him selfa sceáf reáf of líce; læg ðá limnacod, Cd. 76; Th. 94, 23; Gen. 1566.

-limp. v. ge-, mis-limp.

limpan; *p*. lamp, *pl*. lumpon *To befall, happen, fall (to one's share), pertain, belong, affect, concern* :— Ða yfelan habbaþ gesælþa, and him gelimpþ (Cott. MS. limpþ) oft æfter heora ágnum willan, Bt. 39, 2; Fox 214, 5. Ða unrihtwísan ne beóþ ná swylce ne him eác swá ne limpþ *non sic impii, non sic*, Ps. Th. 1, 5. Eádig biþ ðæt folc ðe him swá on foldan fægre limpeþ *beatum populum, cui hæc sunt*, 143, 19: Exon. 81b; Th. 306, 26; Seef. 13. Hwæt limpeþ ðæs tó ðé of hwylcum wyrtruman ic ácenned sí *quid ad te pertinet qua sim stirpe genitus?* Bd. 1, 7; S. 477, 27. Sorgaþ ymb óðerra monna wísan ðe him náuht tó ne limpþ *is busied about other men's affairs, that do not all concern it*, Past. 53, 5; Swt. 415, 21. Ðis sind ða landgemǽra ðæs londes ðe lynnþ tó Stúre *these are the boundaries of the land that belongs to Stour*, Cod. Dip. Kmbl. iii. 81, 34. Hú lomp eów on láde *what hap was yours by the way?* Beo. Th. 3978; B. 1987. Twegra sceopa ðǽrtó ðe limpende beóþ *of two ships that are thereto pertaining*, Chart. Th. 28, 26. [*O. H. Ger*. limphan, limfan *convenire*.] DER. á-, be-, ge-limpan.

limp-líce; *adv. Fitly, opportunely, conveniently* :— God swíðe limplíce geset ðæt gewrixle eallum his gesceaftum, Bt. 21; Fox 74, 21.

lim-ræden, e; f. A cloak (?). In Hpt. Gl. 465, 72 limrǽdenne is given as a marginal reading against chlamide.

lim-seóc; adj. Having diseased limbs, Andr. Kmbl. 1157; An. 579: Elen. Kmbl. 2425; El. 1214.

lim-wǽd, e; f. A garment :—Swá limwǽdum sicut vestimento, Ps. Th. 103, 2.

lim-wæstm, es; m. Limb-growth, stature, size of body :—Ic eom limwæstmum ꝺæt ic gelutian ne mæg so large am I of limb, that lie hid I cannot, Cd. 216; Th. 273, 2; Sat. 130.

lim-wérig; adj. Having the limbs wearied :—Álédon hié ꝺǽr limwérigne, Rood Kmbl. 125; Kr. 63.

lín, es; n. Flax, linen, something made of linen :—Flæx ꝼ lín linum, Mt. Kmbl. Rush. 12, 20. Lín manitergium, Wrt. Voc. ii. 113, 44. Besweópun hine miꝺ líne ligaverunt eum linteis, Jn. Skt. Rush. 19, 40. Miꝺ ꝺý onféng ꝺæt lín cum accepisset linteum, 13, 4. Bohte lín and hine biwand in líne mercatus sindonem eum involvit sindone, Mk. Skt. Rush. 15, 46: Lk. Skt. Rush. 23, 53. Gisæh ꝺa lín gisetedo vidit linteamina posita, Jn. Skt. Rush. 20, 6. [Goth. lein linen : O. Sax. lín : Icel. lín flax, linen : O. H. Ger. lín linum : Ger. lein.] DER. biscop-, heáfod-, breóst-, hand-, swát- lín.

lind, e; and linde, an; f. I. the linden or lime-tree :—Lind seno vel tilia, Ælfc. Gl. 45; Som. 64, 111; Wrt. Voc. 32, 46. Linde tilie, Wrt. Voc. ii. 75, 29. In ꝺa greátan lindan; of ꝺære lindan, Cod. Dip. Kmbl. iii. 79, 24. On ꝺa gemearcodan lindan; of ꝺære gemearcodan lindan, vi. 182, 2. Ꝺonon in áne linde, iii. 392, 1. II. what is made of the wood of the tree, a shield (in poetry) :—Wisse hé gearwe, ꝺæt him holtwudu helpan ne meahte, lind wiꝺ líge, Beo. Th. 4671; B. 2341. Ofer linde lǽrig, Cd. 154; Th. 192, 29; Exod. 239. Under linde protected by the shield, Andr. Kmbl. 91; An. 46. Leófsunu his linde áhóf, Byrht. Th. 138, 63; By. 244. Rond, geolwe linde, Beo. Th. 5213; B. 2610. On fyrd wegan fealwe linde, Cd. 94; Th. 123, 14; Gen. 2044. Under lindum, 154; Th. 192, 7; Exod. 228 : 155; Th. 193, 23; Exod. 251. Bordum beꝺeahte, hwealfum lindum, Judth. 11; Thw. 24, 30; Jud. 214. Beraꝼ linde forꝼ, Thw. 24, 16; Jud. 191. Scyldas wégon, linde bǽron, Byrht. Th. 134, 45; By. 99: Beo. Th. 4719; B. 2365. Hwíte linde, Cd. 158; Th. 107, 4; Exod. 301. [Icel. lind a lime-tree; poet. a shield, a spear: O. H. Ger. linta tilia : Ger. linde.] DER. heaꝺu-lind; and see linden.

lind-croda, an; m. Shield-press, battle, Cd. 93; Th. 120, 21; Gen. 1998.

linden; adj. Made of the lime-tree :—Scyld, leóht linden bord, Exon. 90 b; Th. 339, 16; Gn. Ex. 95.

Lindisfaran, pl. Name of people settled in part of Northumbria (the word occurs generally with eá or eá-land) :—Oswald Aidanum on Lindesfarona eálonde biscopsetl forgeaf (in insula Lindisfarnensi) : on Lindesfearona eá, Bd. 3, 3; S. 525, 20, 35. Lǽdde mon his líchoman tó Lindisfarena eá, 3, 17; S. 543, 37, col. 2. Mid ꝺám bróꝺrum ꝺære cyricean æt Lindisfarena a fratribus ecclesiæ Lindesfarnensis, pref; S. 472, 29. Is Cynebyrht Lindisfarena biscop provinciæ Lindisfarorum Cyneberct episcopus præest, 5, 24; S. 646, 22. Hér forꝼferde Higbald Lindisfarna biscop, Chr. 803; Erl. 61, 22. Hé wæs on ꝺam munuclífe ꝺe is Lindisfarneá geháten, Homl. Th. ii. 142, 6.

Lindisfarnensisc; adj. Of Lindisfarne :—Se hálga Cúꝼberhtus, Lindisfarnensiscere gelaꝼunge leódbiscop, Homl. Th. ii. 148, 21.

Lindesse, Lindisse, Lindesíge Lindsey, the northern part of Lincolnshire [Lat. Lindi colonia] :—Lǽrde Scs. Paulinus Godes word on Lindesse : seó mǽgꝼ is seó nýhste on súꝼhalfe Humbre streámes, ligeꝼ út on sǽ, Bd. 2, 16; S. 519, 18. On Lindesége mǽgꝼe, 519, 16. On Lindese, 3, 11; S. 535, 14. On Lindesse and on Eást-Englum, Chr. 838; Erl. 66, 13 : 873; Erl. 76, 19 : 874; Erl. 76, 21. Lindisse, 627; Erl. 25, 5. On Lindesíge ge on Norꝼhymbran, 993; Erl. 133, 4: 1013; Erl. 147, 20 : 1014; Erl. 151, 2. His lýchama resteꝼ on Lyndesse mǽgꝼe Shrn. 155, 24.

Lindis-ware; pl. The people of Lindsey :—Man gehálgode Lindiswarum tó biscope Eádhéd; se wæs on Lindissi ǽrost biscopa, Chr. 678; Erl. 41, 8.

lind-geborga, an; m. A protector bearing a shield, a warlike protector [?]. v. leóд-hwæt.

lind-gecrod, es; n. A shield-bearing crowd, Andr. Kmbl. 2442; An. 1222.

lind-gelác, es; n. A shield-conflict, battle, Apstls. Kmbl. 151; Ap. 76.

lind-gestealla, an; m. A companion in arms, Beo. Th. 3950; B. 1973.

lind-hæbbende; part. as noun. Shield-bearer, warrior, Beo. Th. 495; B. 245 : 2808; B. 1402.

lind-hóh; gen. -hós; m. A hóh [q. v.] where lime-trees are growing [?] :—On lindhóh; of lindhó, Cod. Dip. Kmbl. iii. 76, 33.

lind-hrycg, es; m. A ridge on which lime-trees are growing [?] :—On lindrycg; of lindrycge, Cod. Dip. Kmbl. iii. 79, 20.

lind-plega, an; m. Shield-play, battle, Beo. Th. 4085; B. 2039: [MS. hild-] 2151; B. 1073.

lind-weorud, es; n. A band armed with shields, Elen. Kmbl. 283; El. 142.

lind-wíga, an; m. A warrior armed with a shield, Beo. Th. 5199; B. 2603.

lind-wígend, -wiggend, es; m. A warrior armed with a shield, Bt. Met. Fox 1, 25; Met. 1, 13 : Judth. 10; Thw. 22, 1; Jud. 42: Elen. Kmbl. 539; El. 270.

líne, an; f. I. a line, rope, a coil of rope :—Lange línan with a long line, Salm. Kmbl. 589; Sal. 294. Línan spiræ, Ælfc. Gl. 104; Som. 78, 14; Wrt. Voc. 56, 60. II. a line, row, line for guidance, rule, canon :—Þurh ꝺæs cantices cwide, Cristes línan [the rule laid down by Christ in the Lord's Prayer], Salm. Kmbl. 34; Sal. 17. Ꝺǽr sceal wesan se torhta æsc án an línan ácas twegen hægelas swá some 'æ' must occur once, 'a' and 'h' twice [in forming the words hæn, hana], Exon. 112 a; Th. 429, 25; Rä. 43, 10. [Icel. lína a line (cord), line (mathem.): O. H. Ger. línna linea.] DER. sceát-, steding-, sund-, toh-líne.

línen; adj. Made of flax, linen :—Línen lineum : línen wearp linostema, Ælfc. Gl. 62, 63; Som. 68, 97, 98; Wrt. Voc. 40, 6, 8. Línnin rýhae villa, Ep. Gl. 28 d, 19. Línen byssina, Hpt. Gl. 526, 31. Hig bewundon hine mid línenum claꝼe [línnínum hræglum, Lind.] ligaverunt eum linteis, Jn. Skt. 19, 40. Mid línenum reáfe subucula linea, Lev. 8, 7 : Past. 14, 4; Swt. 83, 23. Línen hrægel linteum, Jn. Skt. 13, 4. Hió becwiꝼ línnenne cyrtel oꝺꝺe línnen web she bequeathes a linen kirtle or a piece of linen, Chart. Th. 537, 24. Ne hé wyllenra hrægla breác ac línenra ealra, Shrn. 93, 8. [O. H. Ger. línin lineus : Ger. leinen.]

línen-werd; adj. Dressed in linen :—Hé wæs línenwerd and his lændena wéron ymbgirde he was clothed in linen, and his loins were girded, L. Ælfc. P. 17; Th. ii. 370, 11. [Cf. wolleward dressed in woollen garments, Piers P. B. 18, 1.]

línete, an; f. A linnet [for connection with lín cf. Ger. hanf hemp, hänfling linnet] :—Línete cardella, Wrt. Voc. 62, 46.

líne-twige, -twigle, an; f. A linnet :—Línetwige carduelis, Wrt. Voc. ii. 13, 43 : 103, 13 : fronulus, 36, 3. Línetuigle fronulus, 109, 14. Cf. ꝼisteltuige cardella, 102, 76. [Cf. Scot. lyntquhit a linnet.]

-ling. v. deór-, eorꝼ-, geong-, hæft-, hýr-, níd-ling.

-ling, -linga, -lunga. v. bæc-, ears-, hinder-ling; bæc-, grund-, hand-linga.

lín-héwen; adj. Flax-coloured [?] :—Þurh línhǽwenne claꝼ, Lchdm. iii. 2, 23 : 4, 22.

linian, leonian to leave [?] :—Ic leonige óꝺrum eorꝼcyningum tó bysne ꝺæt hié witen ꝺý gearwor ꝺæt mín ꝼrym and mín weorꝼmynd máran wéron ꝺonne ealra óꝺra kyninga ꝺe in middangearde ǽfre wéron I leave it [an account of my exploits] as an example to other kings, that they may the better know that my glory and honour were greater than all other kings that ever were in the world, Nar. 33, 2. v. á-linian.

lín-land, es; n. Land where flax grows :—Ꝺæt lytle línland, Cod. Dip. Kmbl. iii. 19, 4. [Cf. Icel. lín-akr.]

linnan; p. lann, pl. lunnon To cease, leave off, desist, part from, lose :—Blǽd his blinniꝼ blisse linniꝼ [-aꝼ MS.] listum [lissum?] linneꝼ his glory comes to an end, he ceases from joy, desists from delights, Exon. 95 a; Th. 354, 30; Reim. 53. Lunnon sáwlum they parted from their souls i. e. they died, Cd. 167; Th. 209, 9; Exod. 496. Ealdre linnan to die, Exon. 88 a; Th. 330, 21; Vy. 54: An. 2277; An. 1139: Beo. Th. 2960; B. 1478. Ealdres linnan, 4878; B. 2443. [Goth. af-linnan : Icel. linna : O. H. Ger. bi-linnan.] DER. á-linnan, blinnan [= be-linnan], á-, ge-blinnan.

lín-sǽd, es; n. Linseed :—Línsǽd elimos vel lini semen, Wrt. Voc. 69, 32. Mid línsǽde, Herb. 39, 3; Lchdm. i. 140, 13.

lín-wǽd, e; f. A linen garment, linen cloth :—Hé drígde hig mid ꝺære línwǽde ꝺe hé wæs mid begyrd coepit extergere linteo quo erat praecinctus, Jn. Skt. 13, 5. Hé geseah ꝺa línwǽda licgan videt posita linteamina, 20, 5 : Lk. Skt. 24, 12. [O. H. Ger. lín-wât linteamen.]

lín-wyrt, e; f. Flax, L. M. 1, 25; Lchdm. ii. 66, 17 : 3, 65; Lchdm. ii. 354, 10.

lippa, an; m. A lip :—Ufeweard lippa labium : foreweard feng ꝺære lippena tógædere rostrum, Ælfc. Gl. 71; Som. 70, 93-95; Wrt. Voc. 43, 24-26. Lippan labia, Hpt. Gl. 481, 24. Áwergode beón heora tungan and lippan, Wanl. Catal. 137, 51. Wiꝺ lippe sár. Eft sóna ꝺes lǽcedóm sceal ꝺan manne ꝺa hyra lippa beóꝼ sáre oꝺꝺe hyra tunga .. smire mid ꝺa lippa, Lchdm. iii. 100, 15-21. [O. Frs. lippa; m : cf. O. L. Ger. lepor : O. H. Ger. leffur labium : lefs labium.]

lira, an; m. Fleshy part of the body without fat or bone, brawn :—Lira pulpa vel viscum, Ælfc. Gl. 73; Som. 71, 37; Wrt. Voc. 44, 21. Lira pulpa, Wrt. Voc. 65, 12 : 290, 48: ii. 76, 10. Sár ꝼeóh and lira the thigh and the fleshy parts are sore, L. M. 2, 51; Lchdm. ii. 264, 11. Ꝺa liran ꝺara lendena sáriaꝼ the fleshy parts of the loins get sore, 2, 25; Lchdm. ii. 216, 24. [Toleac liꝺ ba and lire broke both joints and flesh, Jul. 59, 10. Lire the flesh of an animal or rather the increasing substance as it grows bulky, E. D. S. Whitby Glossary. See also Halliw. Dict. Scot. lire flesh or muscles, as distinguished from the bones.] DER. ears-, spear-lira.

lireht; adj. Brawny, fleshy :—Hí habbaꝼ lirehte fét, L. M. 2, 36; Lchdm. ii. 242, 14.

lísan, lýsan; *p.* de *To loosen, release, redeem, deliver*:—Mín sáwl ða ðú sylf lýsdest *anima mea, quam redemisti*, Ps. Th. 70, 21. Se sylfa cyning mid síne líchoman lýsde of firenum, Exon. 25 b; Th. 74, 22; Cri. 1210. Gif hé ða hand lésan [álýsan, MS. H; lýsan, MS. B.] wille . . . gelde swá tó his were belimpe, L. Alf. pol. 6; Th. i. 66, 5: Byrht. Th. 132, 57; By. 37: Elen. Kmbl. 592; El. 296: Rood Kmbl. 82; Kr. 41. [*Goth.* lausjan: *O. Sax.* lósian: *O. Frs.* lésa: *Icel.* leysa: *O. H. Ger.* lósen: *Ger.* lösen.] v. á-, ge-, on-, tó-lísan; untólísende.

lísian *to release, redeem*:—Gif hé on hand gán wille dó hine man on carcern swá hit æt Greátanleá gecweden wæs and hine be ðam ylcan lýsige *if he is ready to submit, let him be put in prison, as it was determined at Greatanlea* [v. Th. i. 198], *and according to the same let him be redeemed*, L. Æthelst. v. 12; Th. i. 240, 33.

lísing, es; *m. A freedman*:—Lísingas and þeówe, Chart. Th. 592, 1. Búton ðam ceorle ðe on gafollande sit, and heora [*the Danes*] liésingum [lýsingum]; ða syndan efendýre, ægðer tó cc. sciłł., L. A. G. 2; Th. i. 154, 3. [*Icel.* leysingi, leysingr *a freedman*.]

lísing, e; *f. A loosing, releasing, redemption*:—Lésing *redemtio*, Lk. Skt. Lind. 2, 38. v. á-, crism-lísing.

lisne, Ps. Th. 52, 6. v. [?] lyswen.

lís-ness, e; *f. Redemption, release, deliverance*; *redemtio*, Mk. Skt. Lind. Rush. 10, 45: Lk. Skt. Lind. Rush. 1, 68: 2, 38. DER. á-, tó-lísness.

lisnian. v. be-lisnian.

liss, e; *f. Mildness, lenity, mercy, kindness, favour, grace, delight, joy*:—Hé bæd ðæt Lazarus móste his tungan drýpan ac him næs getíðod ðære lytlan lisse *he prayed that Lazarus might put a drop of water on his tongue; but that little favour was not granted to him*, Homl. Th. i. 330, 30. Ic ðé biddan wile lífes and lisse *I will ask thee for life and favour*, Ps. C. 50, 69; Ps. Grn. ii. 278, 69. Hé þancode lífes leóhtfruman lisse and ára, Cd. 90; Th. 113, 19; Gen. 1889. Hé him ðære lisse leán forgildeþ *he will requite him for that grace* [*honouring God*], Exon. 14 a; Th. 27, 21; Cri. 434. Ða eádigan ceasterwaran gefeóþ and wynsumiaþ on lisse and on blisse and on écum gefeán, Wulfst. 265, 12. Lifgan in lisse *lucis et pacis to live in the delight of light and peace*, Exon. 656; Th. 242, 12; Ph. 672. Hé onfón sceal blisse mínre lufan and lisse *he shall receive my joy, my love and my favour*, Cd. 106; Th. 140, 23; Gen. 2332: 190; Th. 237, 19; Dan. 340. Forgif mé tó lisse bitre bealodæde *in mercy to me forgive my evil deeds*, Exon. 118 a; Th. 453, 21; Hy. 4, 18. Lífes tó lisse *to save life*, Andr. Kmbl. 2223; An. 1113. Lisse ic gelýfe leahtra gehwylces *I believe in the forgiveness of sins*, Hy. Grn. ii. 294, 54. Se þinc on líchoman lisse sóhte *Enoch while yet in the body sought* [*heaven's*] *joy*, Cd. 60; Th. 73, 14; Gen. 1204. Ðé is édelstól gerýmed, lisse on lande, 73; Th. 89, 25; Gen. 1486. Ic ðé lissa lífigendum giet læte brúcan, 126; Th. 161, 10; Gen. 2663: 136; Th. 171, 6; Gen. 2824: Exon. 13 a; Th. 23, 24; Cri. 373: Beo. Th. 4306; B. 2150. Wilna biscirede, lufena and lissa, Exon. 48 b; Th. 166, 27; Gú. 1049. Lufum and lissum, Cd. 130; Th. 165, 25; Gen. 2737. Wé ðé getǽhton land tó lissum ðú ús leánest nú unfreóndlíce *we assigned thee land for thy delight, now dost thou repay us in fashion unfriendly*, 127; Th. 162, 28; Gen. 2688. Lissum *kindly, graciously*, Andr. Kmbl. 1735; An. 870. v. liðs.

lissan *to soften, weaken, tame, subdue*:—Yldo beoþ on eorþan ǽghwæs cræftig . . . lisseþ eal ðæt heó wile beám heó ábreóteþ . . . friteþ wildne fugol . . . heó oferwígeþ wulf *on earth age has power over everything . . . she subdues all that she will; the tree she destroys . . . the wild bird she devours . . . the wolf she conquers*, Salm. Kmbl. 590; Sal. 294. [Cf. I trowe my peyne shalle never *lisse*, Chauc. R. R. 4128: it shulde *lisse* me, Gow. iii. 82, 19: hire care to *lisse*, Will. 631. Jamieson gives the verb in his Scottish Dict. lis *to ease, assuage*; liss *to cease, stop*.] v. liss.

list, es; *m.: list*, e; *f. Art, skill, craft, cunning, artifice*:—Lot sceal mid lyswe list mid gedéfum *cunning goes with evil, skill with things proper, i. e. lot and list are the names for a corresponding vice and virtue*, Exon. 92 a; Th. 345, 17; Gn. Ex. 189. Ðú miht león and dracan liste gebýgean *conculcabis leonem et draconem*, Ps. Th. 90. 13. List *art* (*of poetry*), Bt. Met. Fox Introd. 5; Met. Einl. 3: Exon. 79 a; Th. 296, 13; Crä. 50. Þurh calle list, 27 a; Th. 81, 5; Cri. 1319. Ðæs líchoman listas and cræftas of ðæm móde cumaþ *the arts and powers of the body come from the mind; intus est hominum vigor arce conditus abdita*, Bt. Met. Fox 26, 216; Met. 26, 108. Hé fela onginþ leornian lista *many arts doth he learn*, 28, 153; Met. 28, 77: Cd. 13; Th. 16, 5; Gen. 239. Mid listum speón idese on ðæt unriht *with wiles he lured the woman to that wrong*, 28; Th. 37, 12; Gen. 588: 32; Th. 43, 8; Gen. 687. Listum *skilfully, craftily, cunningly*:—Him listum áteáh rib of sídan *skilfully drew a rib from his side*, 9; Th. 11, 19; Gen. 177: 77; Th. 95; 29; Gen. 1586: Judth. 10; Thw. 23, 2; Jud. 101: Bt. Met. Fox 13, 84; Met. 13, 42: 1, 118; Met. 1, 59: Beo. Th. 1566; B. 781: Ps. Th. 87, 10. Wyl tógædere listum *boil them skilfully together*, L. M. I, 1; Lchdm. ii. 24, 11: 1, 2; Lchdm. ii. 26, 8. Hé ðé hét lista lǽran *he bade teach thee arts*, Cd. 25; Th. 33, 8; Gen. 517. [*Goth.* lists: *O. Sax.* list: *O. Frs.* lest: *Icel.* list; *f.: O. H. Ger.* list; *m. f. ars, ingenium, astutia, peritia: Ger.* list; *f.*]

líste, an; *f. A list, hem, border, selvage*:—Líste *lembus*, Wrt. Voc. ii. 113, 1: 50, 68. Lístan *lembum*, 112, 54. Lístum, *lembus*, 50, 69. [*Icel.* lista; *f.*: listi; *m.* list, border: *O. H. Ger.* lísta; *f. limbus, fimbria: Ger.* leiste.]

list-hendig; *adj. Having skilful hands*:—Sum biþ listhendig tó áwrítanne wordgerýnu, Exon. 79 b; Th. 299, 1; Crä. 95.

listig-, liste-líce; *adv. Skilfully: Seóð æt leóhtum fýre listelíce, L. M. 1, 2; Lchdm. ii. 30, 7.

list-wrenc, es; *m. Wile, artifice*, Lye. v. lot-wrenc.

lítan [*from* lútan, *as* bígan *from* búgan] *to cause to bow, to bend, incline*:—Ðæt ingeþonc ǽlces monnes ðone líchoman lít ðider hit wile *the mind of every man inclines the body whither it will*, Bt. Met. Fox 26, 237; Met. 26, 119.

líte-líce. v. lytig-líce.

liþ, es; *m. n. A joint, lith* [Scott. e. g. the Laird of Auchinleck to Johnson, Cromwell 'gart kings ken they had a *lith* in their necks'], *member of the body, limb*:—Liþ *artus: lytel liþ articulus*, Wrt. Voc. 283, 16, 17: Soul Kmbl. 191; Seel. 96. Ðætte sum man fram deáþes liþe wæs gehǽled *ut sit quidam a mortis articulo revocatus*, Bd. 3, 13; S. 538, 3. Ðæt hé dyppe his fingres liþ on wætere *that he may dip the tip of his finger in water*, Lk. Skt. 16, 24. On ðone liþ ðæra eaxla, L. M. 2, 36; Lchdm. ii. 242, 12. On ðæt liþ, 1, 61; Lchdm. ii. 132, 6. Liþu *artus*, Wrt. Voc. 64, 77. Ða máran liþa *artus*, Ælfc. Gl. 72; Som. 71, 4; Wrt. Voc. 43, 58. Gif men his leoþu acen, Herb. 3, 1; Lchdm i. 86, 21. Foxes leoþu, L. Med. ex Quad. 3, 1; Lchdm. i. 338, 20: Exon. 87 a; Th. 327, 18; Vy. 6: 74 a; Th. 278, 3; Jul. 592. Sint mé leoþ tólocen líc sáre gebrocen, Andr. Kmbl. 2807; An. 1406. Býgendlíc on gebeódnessum his liþa *flexilibus artuum compagibus*, Bd. 4, 30; S. 608, 38. Betwyh liþum *inter femora*, L. Ecg. P. iv. 68, 6; Th. ii. 228, 24. Hé ðé worhte of liþum mínum, Cd. 38; Th. 50, 33; Gen. 818. Leoþum onfón, Exon. 23 a; Th. 64, 3; Cri. 1032. Liþa *articulus*, Hpt. Gl. 443, 61. Bígdon heora heáfda tó ðære hálgena fótum and heora liþa liccodon, Homl. Skt. 4, 407. Of láme ic ðé leoþe gesette, Exon. 28 a; Th. 84, 31; Cri. 1382. Leoþo, Andr. Kmbl. 1562; An. 782. Leomena liþ, Salm. Kmbl. 205; Sal. 102. [*Goth.* liþus; *m. a limo, member: O. Sax.* lið; *m: O. Frs.* lith; *n: Icel.* liðr; *m. a joint, limb: O. H. Ger.* lid; *m. n. artus, articulus, membrum: Ger.* g-lied.] v. hrycgmearh-liþ, leoþu-.

liþ, es; *n. Strong drink*:—Ðæt ðæt liþ gescired wæs *digesto vino*, Past. 40, 4; Swt. 295, 6. Ðam men ðe hine ne lyst his metes ne liþes *for the man that does not care for his meat or drink*, L. M. 1, 19; Lchdm. ii. 62, 16. Of mistlícum dryncum ðæs liþes *from various strong drinks*, Bt. 37, 1; Fox 186, 17. Se ðe ús oferdrencþ mid ðæs écan líþes liþe *aeterna nos dulcedine inebrians*, Past. 36, 9; Swt. 261, 15. Ðá bær unc mon liþ forþ *oblato poculo*, Bd. 5, 3; S. 616, 31. [*Goth.* leiþus οἴκερα: *O. Sax.* líð: *O. Frs.* lith: *Icel.* lið cider: *O. H. Ger.* lid, liþ *potus, liquor, poculum, fiala, sicera: it remains in some provincial German words*, e. g. leit-haus *an ale-house*; leit-geber *keeper of an ale-house*.]

liþ, es; *n.* [*The Scandinavian form of* lid q. v.] *a fleet*:—Ðæs sumeres com ðæt liþ of Humbran *in the course of the summer the fleet came from the Humber*, Chr. 1070; Erl. 210, 4: 1052; Erl. 183, 12: 1069; Erl. 207, 12. [*Icel.* lið *a host* by land or sea.]

líð; *adj.* v. líðe.

Líða, an; *m. Name of the months June and July*:—Se mónaþ is nemned on lǽden Iunius, and on úre geþeóde se ǽrra Líða, for ðon seó lyft biþ ðonne smylte and ða windas. Ond monnum biþ ðonne gewunelíc ðæt hí líðaþ ðonne on sǽs bryme, Shrn. 87, 34. Se ǽrra Lýða, 99, 11. Ǽrra Líða, Junius, Menol. Fox 213; Men. 108. Mónaþ ðone wé nemnaþ on lýden Iulius . . . ðone mónaþ wé nemnaþ on úre geþeóde se æftera Lýða, Shrn. 99, 26: 110, 24. [iþe moneþ þ on ure ledene is ald englisch efterlið inempnet iulius o latin, Marh. 23, 6.] v. Grmm. Gesch. D. S. 56 sqq.

liþ-ádl, e; *f. Gout*:—Liþádl *articterius vel artriticus*, Ælfc. Gl. 11; Som. 57, 44; Wrt. Voc. 19, 47. Wið liþádle, L. Med. ex Quad. 3, 11; Lchdm. 1, 340, 25.

líðan; *p.* láð *To go* [generally by sea], *sail*:—Ic tólíðe, ic líðe *applicabo*, Wrt. Voc. ii. 4, 54. Monnum biþ gewunelíc ðæt hí líðaþ ðonne [ǽrra Líða, *June*] on sǽs bryme, Shrn. 88, 1. Ða ðe sǽ séceaþ mid scipe líðaþ *qui descendunt mare in navibus*, Ps. Th. 106, 22. Hé ofer sǽ láð in Gallia ríce *navigavit Galliam*, Bd. 3, 19; S. 550, 1: Shrn. 60, 5. Se cyning sylfa and se hálga bisceop líðan on ðæt eálond *rex ipse cum sanctissimo antistite insulam navigavit*, Bd. 4, 28; S. 606, 12. Nú ðon gelícost swá wé on laguflóde ofer cald wæter ceólum líðan, Exon. 20 a; Th. 53, 18; Cri. 852. Líðan cymeþ *comes sailing*, Exon. 90 b; Th. 340, 11; Gn. Ex. 109: 108 b; Th. 415, 23; Rä. 34, 1: Andr. Kmbl. 512; An. 256: Bt. Met. Fox 26, 119; Met. 26, 60. Líðendum wuda *a ship*, Exon. 103 b; Th. 392, 9; Rä. 11, 5. Ða líðende land gesáwon *those sailing saw land*, Beo. Th. 447; B. 221. Ða wæs sund liden *then was the sea passed* [cf. *Icel.* líða as a transitive verb], Beo. Th. 452; B. 223. Dóhtor mín eácen up liden *my daughter, great and grown up* [?], Exon. 109 a; Th. 416, 13; Rä. 34, 11. [*Goth.* ga-leiþan: *O. Sax.* líðan: *Icel.* líða: *O. H. Ger.* ga-lídan *peregrinari, cedere, evanescere*.] DER. be-, for-, ge-, ofer-, tó-, ymb-líðan; brim-, eá-, headu-, mere-, sǽ-, scip-, wǽg-líðende.

lîðan *to suffer loss* [?] :—Beám sceal leáfum lîðan *a tree must lose its leaves*, Exon. 89 a ; Th. 334, 34 ; Gn. Ex. 26. [Cf. (?) *O. H. Ger.* lîdan *to suffer.*]

lîðan *to assuage, mitigate, soften* :—Ðæt se hié lîðe and hǽle *foveantur sananda*, Past. 17, 10 ; Swt. 124, 12. v. lîðian.

lîðe, lîð ; *adj.* *Lithe, soft, gentle, meek, mild, serene, benign, gracious, pleasant, sweet* :—Swâ fæder þenceþ his bearnum milde weorþan swâ ûs God ðâm ðe hine lufiaþ lîðe weorþeþ *sicut miseretur pater filiis, ita misertus est Dominus timentibus se*, Ps. Th. 102, 13. Leorniaþ æt mê ðæt ic eom lîðe and swîðe eáðmód *discite a me, quia mitis sum et humilis corde*, Homl. Th. i. 210, 18. Hê biþ ðâm gódum lufsum and lîðe, Exon. 21 a ; Th. 57, 5 ; Cri. 914. Óðer [wæstm] wæs swâ wynlîc wlitig and scêne lîð *the other [fruit] was so delightful, beauteous and fair, delicate*, Cd. 23 ; Th. 30, 17 ; Gen. 468. Hwæðer him cume ðe rêþu wyrd ðe lîðu *whether fortune foul or fair come to him*, Bt. 40, 3 ; Fox 238, 9. Lîðe t̄ smilte *serenum*, Hymn. Surt. 24, 15. Hê forlêt eall ðæt ðǽr lîðes wæs and swêtes *vino epulisque deseruit*, Ors. 2, 4 ; Swt. 76, 14. Cumb fulne lîðes aloþ *a coomb full of mid ale*, Cod. Dip. Kmbl. i. 203, 8 : Chart. Th. 105, 12. Dreám lîðes lîfes *the joy of the serene life* [*of heaven*], Exon. 32 a ; Th. 100, 7 ; Cri. 1638. Mid lîðran mulsa, Hpt. Gl. 481, 14. Mid lîðra tungan *with lithe tongue*, Homl. Skt. 4, 407. Mid lîðran gesceafte [*water*], Boutr. Scrd. 22, 30. Andwlitan mid lîðan *vultu sereno*, Hymn. Surt. 22, 11 : 143, 2. Mid lîðere sprǽce *with gentle speech*, Ap. Th. 2, 25. Mid lîðre wisdlunga mon hors gestilleþ *lenis sibilus equos mitigat*, Past. 23 ; Swt. 173, 21. On lîðum wîne, Herb. 57, 1 ; Lchdm. i. 160, 1 : 80, 2 ; Lchdm. i. 182, 19. Lîðne (*lenis*) drenc, Bt. 39, 9 ; Fox 226, 12. Ðæt lîðe land *the pleasant land*, Cd. 12 ; Th. 13, 31 ; Gen. 211. Eádige beóþ ða lîðan . . . Ða synd lîðe and gedéfe, ða ðe ne wiðstandaþ yfelum, ac oferswýðaþ mid heora goodnesse ðone yfelan, Homl. Th. i. 550, 19 : Mt. Kmbl. 5, 5. Lîðe æppla *mitia poma*, Ælfc. Gr. 47 ; Som. 48, 26. Nû ic freónda beþearf lîðra on lâde *now need I gracious friends on my course*, Apstls. Kmbl. 183 ; Ap. 92. Hearda wunda beóþ mid lîðum beðengum gehnescode *dura vulnera per lenia fomenta mollescunt*, Past. 26, 2 ; Swt. 183, 20. Swîðe lîðum wordum *with very gentle words*; *humanitatis lege eos mulcens*, Nar. 25, 10 : Exon. 37 b ; Th. 124, 3 ; Gû. 334. Mid lîðum styrungum *with gentle gestures*, Glostr. Frag. 110, 8. Lagu lâcende sceal lîðra wyrðan *the tossing wave shall become calmer*, Andr. Kmbl. 874 ; An. 437. Oft byþ ðæt brocc lîðre *the disease is often less severe*, Wulfst. 12, 5. Ðǽr syndon lýðran wedera ðonne on Brettania *coeli solique temperie magis utilis*, Ors. 1, 1 ; Swt. 24, 19. Se sceortigenda dæg hæfþ lîðran gewederu ðonne se langienda dæg, Lchdm. iii. 252, 9. Lîðesta *mittissime*, Hymn. Surt. 65, 11 : 126, 2. Manna mildust, leódum lîðost, Beo. Th. 6346; B. 3183. [*O. Sax.* lîði : *Icel.* linr : *O. H. Ger.* linde lind *lenis, mollis* : *Ger.* ge-lind, -linde.] DER. cum-, gæst-, uncum-, un- lîðe ; v. lîðig.

lîðe-bíge. v. leoþu-bíge.

lîðeg. v. lîðig.

lîðe-lîc ; *adj.* *Gentle, mild, soft*: — Lîðelîce stefne *lena voce*, Nar. 36, 21. Mid lîðelîcum wordum *with gentle words*, Past. 30, 2 ; Swt. 205, 8.

lîðe-lîce ; *adv.* *Gently, mildly, softly, kindly, graciously* :—Lîðelîce, fægere *pedetemtim*, Wrt. Voc. ii. 64, 49. Hê hié lîðelîce hǽlan wolde *graciously he would heal them*, Blickl. Homl. 105, 26. Lîðelîce hé âdlaþ *he will have a mild attack of illness*, Lchdm. iii. 186, 15. Hwîlum lîðelîce tô þreátianne hwîlum suîðlîce and strǽclîce tô þrafianne *aliquando leniter arguenda, aliquando vehementer increpanda*, Past. 21, 1 ; Swt. 151, 11. Ðû scealt lîðelîce monian *suadendo, blandiendo*, Bd. 1, 27 ; S. 492, 22. Sume þearflícor synd gerihte *quidam districtius, quidam levius corrigantur*, S. 490, 11 : L. C. S. 69 ; Th. i. 412, 5.

lîðend, es ; *m.* *A traveller, sailor* :—Lîðend brohte elebeámes twig in tô handa *the traveller (the dove) brought home an olive-branch* [Bouterwek takes lîðend to be a dative ; if it is, the word refers to Noah], Cd. 72 ; Th. 88, 29 ; Gen. 1472. v. sǽ-lîðend, lîðan.

lîðercian ; *p.* ode *To soften, charm, flatter* :—Lîðercaþ, ôleccaþ *adulatur*, Wrt. Voc. ii. 127, 7. Lîðercade *promulserit*, 117, 72. Lîðircadae, Ep. Gl. 17 f, 30.

lîðere, an ; *f.* lîðera, an ; *m.* *A sling* :—Lîðere *funda*, Wrt. Voc. 84, 34. Lyðre, 35, 30. Lîðre, Wrt. Voc. ii. 109, 41. Leðera *funda* : lîðeran *fundibulæ*, 36, 23, 24. Swâ mycelre brǽdo swâ mon mæg mid lîðeran geworpan *amplitudinis quasi jactus fundæ*, Bd. 4, 13 ; S. 583, 11. Mid his lîðeran ofwearp ðone geleáfleásan ent, Ælfc. T. Grn. 7, 18. Of blacere lîðran, Salm. Kmbl. 54 ; Sal. 27. v. stæf-lîðere.

lîðeren. v. leðeren.

lîðer-lîc ; *adj.* *Of a sling* :—Lîðerlîcum swêge *fundali stridore*, Wrt. Voc. ii. 152, 16.

lîðe-wâc, lîþewǽcan. v. leoþu-wâc, leoþuwǽcan.

lîþ-geat. v. hlid-geat.

lîðian ; *p.* ode *To be, become,* or *make lîðe* [q.v.] :— Miltsige man for Godes ege and lîðige man georne *let mercy be shewn for fear of God, and let kindness be diligently shewn*, L. Eth. vi. 53 ; Th. i. 328, 28. Swâ hwæt swâ gê gebindaþ hêr ofer eorþan eall hit wyrþ on heofenan mid Godes yrre gebunden bûtan gê lîðian *whatsoever ye bind on earth shall all*

be bound in heaven with God's anger, unless ye be gracious, Wulfst. 178, 4. Biþ ðæs innoþes sâr lîðigende ðæt hit sóna nǽnig lâð ne biþ *the disease of the stomach will grow easier, so that soon it will be no annoyance*, Herb. 1, 11 ; Lchdm. i. 74, 10. [*Icel.* lina *to soften, alleviate, abate* : *O.H.Ger.* lindian *mollire, blandiri.*] DER. ge-, on-lîðian ; v. lîðigian, lîðan.

lîðig ; *adj.* *Lithe, pliant, supple, flexible, soft, yielding* :—Heó biþ lîðig swâ clâþ ongeán deófles lâre *it [a man's heart] is pliant as cloth to the devil's teaching*, Wulfst. 234, 22. Ðâ gelǽhte Petrus hire lîðian [lîðigan, MSS. U. B.] hand *then Peter took her supple hand*, Homl. Skt. 10, 73. On his lîðegum cneówum, Homl. Th. ii. 298, 27. His lîðegan fingeras, 512, 1. Ðæt ðû lîðegie *ut mitiges*, Ps.Spl. 93, 13. v. ge-lîðian.

lîðigian, lîðegian ; *p.* ode *To make,* or *be soft* or *yielding, to assuage, calm* :—Gewylc ýða his ðû lîðegast *motum fluctuum ejus tu mitigas*, Ps. Spl. 88, 10. Se ðe on ðam ǽrran tôcyme lîðegode se dêmþ stîðne dôm æt ðam æfteran tôcyme *he that was mild at the first advent shall judge stern judgement at the second*, Homl. Th. i. 320, 17. Ðæt ðû lîðegie *ut mitiges*, Ps.Spl. 93, 13. Uton lîðegian ûre môde *leniamus animum nostrum*, L. Ecg. P. iv. 66 ; Th. ii. 226, 26. v. ge-lîðian.

lîþ-incel, es ; *n.* *A little joint* ; *articulum*, Wrt. Voc. 283, 17 : ii. 8, 3.

lîþ-lîc. v. riht-lîþlíc.

lîd-mann. v. lid-mann.

lîð-ness, e ; *f.* *Softness, gentleness, mildness, lenity, kindness* : — Hî sind gesewene mid lîðnysse ac heora lîðnys is sóðlîce âsolcennys *they appear with gentleness, but their gentleness is really sluggishness*, Homl. Th. ii. 46, 11. On lîgette is ôga and on snâwe lîðnyss ðǽre beorhtnysse *in lightning is the terror of brightness, in snow its mildness*, i. 222, 32. Hê forbær manna yfelnysse þurh his lîðnysse *he endured the evil of men by reason of his gentleness*, 320, 16. Swâ is tô mengenne ða lîðnesse wið ða rêðnesse *miscenda ergo est lenitas cum severitate*, Past. 17, 11 ; Swt. 124, 13. DER. cum-, gæst-lîðnesse.

lîþrian ; *p.* ede *To lather, smear* :—Lêðrede *unxit*, Jn. Skt. Lind. 11, 2. Lýþre mid sâpan, L. M. 1, 50 ; Lchdm. ii. 124, 5. [*Icel.* leyðra *to wash.*]

lîðs, e ; *f.* *Gentleness, calm, ease, pleasure* :—Lîðsa and wynna hâm *a home of pleasures and of joys* [*Eden*], Cd. 45 ; Th. 58, 13 ; Gen. 945. Lîðsum gewunedon *they lived at ease*, 80 ; Th. 100, 28 ; Gen. 1671. v. liss.

lîþ-seáw, es ; *n.* *The oily matter between the joints, synovia* :—Gif mon biþ on eaxle wund ðæt ðæt lîþseáw ût flôwe gebête mid xxx scill., L. Alf. pol. 53 ; Th. i. 94, 22. Manegum men lîþseáu sýhþ . . . wið lîþseáwe, L. M. i. 61 ; Lchdm. ii. 132, 10–13.

lîðs-, lits-mann *a sailor* :—Ða lîðsmenn [*the Danes*], Chr. 1036 ; Erl. 164, 14. Litsmanna, 1047 ; Erl. 175, 11. [*Icel.* lîðs-maðr.] v. lid-mann.

lîþule [= lîþ-ele, Cockayne, Lchdm. ii. 398, col. 1] *synovia* :—Gif lîþule ût yrne, L. M. 1, 61 ; Lchdm. ii. 134, 3, 8. v. lîþ-seáw.

lîðung, e ; *f.* *Relieving, alleviation, relief* : — Hê ongit ðæs innoþes lîðunge *he will find relief for the stomach*, Herb. 18, 4 ; Lchdm. i. 112, 2.

lîþ-wǽrc, es ; *m.* *Pain in the joints* :— Wið lîþwǽrce, L. M. 1, 61 ; Lchdm. ii. 132, 2, 4.

lîþ-wǽge, es ; *A drinking-cup, wine-cup*, Beo. Th. 3969; B. 1982.

lîð-wyrt ; *f.* *Dwarf elder* :—Lýðwyrt. Ðeós wyrt ðe man ostriage and ôðrum naman lýðwyrt nemneþ, Herb. 29, 1 ; Lchdm. i. 124, 13. Lîð-wyrt, L. M. 1, 61 ; Lchdm. ii. 132, 13. Lîðwyrt *ostriago*, Wrt. Voc. 69, 26 : *eripheon*, 68. 12 : *ostriago*, ii. 65, 48. v. Gloss. to Lchdms. ii. iii.

litel, litig. v. lytel, lytig.

lîxan, lîcsan ; *p.* te *To shine, glitter, gleam* :—Seó reádnes ðǽre rôsan lîxeþ on ðê, and seó hwîtnes ðǽre lilian scîneþ on ðê, Blickl. Homl. 7, 30. Môna lîxeþ, Exon. 18 a ; Th. 44, 6 ; Cri. 698. Ðæt nebb lîxeþ swâ glæs oððe gim *the beak glitters like glass or gem*, 60 a ; Th. 218, 24 ; Ph. 299. Sôðfæste scînes t̄ lîxeþ swâ sunna *justi fulgebunt sicut sol*, Mt. Kmbl. Lind. 13, 43. Lîxaþ, 64 b ; Th. 238, 15 ; Ph. 604. Lîxte *fulminavit*, Wrt. Voc. ii. 37, 18 : Exon. 15 a ; Th. 31, 34 ; Cri. 505 : Beo. Th. 627 ; B. 311. Ðonne dæg lixte, 975 ; B. 485. Sumum scinan ða scilla and lîxtan swylce hié wǽron gyldene *auri fulgori similes*, Nar. 13, 19 : Elen. Kmbl. 46 ; El. 23 : 180 ; El. 90 : 2229 ; El. 1116 : Cd. 148 ; Th. 185, 20 ; Exod. 125. Hié gesâwon eóred lîxan *they saw the host glitter*, 149 ; Th. 187, 28 ; Exod. 157 : Exon. 57 a ; Th. 204, 8 ; Ph. 94. Ðonne lîgette lîxan cwôman *illuxerunt coruscationes tuæ*, Ps. Th. 76, 15 : Bt. Met. Fox 9, 25 ; Met. 9, 13. Lîxende *fulgens*, Lk. Skt. Lind. 24, 4 : *lucens*, Jn. Skt. Lind. 5, 35. Lîcxændum *coruscantibus*, Rtl. 3, 1. Lîxende lîgetta, Exon. 54 b ; Th. 192, 14 ; Az. 106. Lîxende lof *brilliant praise*, 93 a ; Th. 349, 20 ; Sch. 49. v. in-lîxan.

lîxende ; *adv.* *Splendidly* :— Fegerlîce t̄ lîcsendo *splendide*, Lk. Skt. Lind. 16, 19.

lîxung, lîcsung, e ; *f.* *Splendour, brightness* : —Lîxung *splendor*, Mt. Kmbl. p. 14, 11 : Rtl. 3, 13. Lîcsung, 38, 29.

lobbe ; *f.* *A spider* :—Ûre gêr swâ swâ lobbe oððe rynge beóþ âsmeáde *anni nostri sicut aranea meditabuntur* [cf. Ps. Th. 89, 10, anlîcast geongewefran ðonne hió geornast biþ ðæt heó âsêre fleógan on nette], Ps. Lamb. 89, 9. Mistlîce þreála gebyriaþ for synnum bendas oððe dyntas carcernþýstra lobban *various punishments are proper for sins, bonds or blows, prison darkness, spiders*, L. Pen. 3 ; Th. ii. 278, 26. Cf. (?) *Icel.* lubbi *a shaggy longhaired dog.*]

loc, es; *n.* **I.** *A lock, bolt, bar, that by which anything is closed, an enclosed place, enclosure, fold* :—Loc *clausura*, Wrt. Voc. 81, 17. Locc *mandra vel ovile*, 23, 55. Loc *caula*, 85, 73. Gâta loc *titula*, 288, 20. Loce † fæstene *clustello*, Hpt. Gl. 527, 72. In scípa locc *in ovile ovium*, Jn. Skt. Lind. 10, 1 : p. 6, 2. Ic scitte sum loc oððe hæpsige *sero*, Ælfc. Gr. 37; Som. 39, 21. Uton belúcan ðás circan and ðæt loc inseglian, Homl. Skt. 3, 329. Sceápa locu *caule*, Ælfc. Gl. 2; Som. 55, 21; Wrt. Voc. 16, 6. Ða locu feóllan, clûstor of ðâm ceastrum, Exon. 120 a; Th. 461, 22; Hö. 39. Ealle ða îsenan scyttelas helle loca wurdan tôbrocene, Blickl. Homl. 87, 5. Hwylc manna is ðæt his âgene sâwle fram helle locum generige *quis eruet animam suam de manu inferi*, Ps. Th. 88, 41. Tô helle locum gelæded beón sceolde *ad inferni claustra raperetur*, Bd. 3, 13; S. 538, 22 : 5, 13; S. 633, 20. Mid ðâm trumestum locum getimbrade *seris instructa firmissimis*, 1, 1; S. 473, 27. Ðonne wê sittaþ innan ceastre ðonne wê ûs betýnaþ binnan ðæm locum ûres môdes *in civitate quippe considemus, si intra mentium nostrarum nos claustra constringimus*, Past. 49, 4; Swt. 385, 6 : L. E. I. 45; Th. ii. 442, 13. Heó hêht ða rôde in seolfren fæt locum belúcan, Elen. Kmbl. 2051; El. 1027. Locu *mandras, caulas*, Hpt. Gl. 476, 30. Loca *caulas*, Coll. Monast. Th. 20, 17. Godes engel undyde ða locu ðæs cwearternes, Homl. Th. i. 572, 27 : Exon. 12 b; Th. 20, 21; Cri. 321. **II.** *A close, conclusion, settlement* :—Loces *syllogismi, conclusionis*, Hpt. Gl. 481, 65. And ðises loces ærendracan wæran . . . Ðonne is hér seó gewitnes ðe æt ðisum loce wæs *and of this settlement the commissioners were . . . Here are the witnesses that were at this settlement*, Chart. Th. 303, 12–19. Mid ðâm ilcan mannan ðe ær ðæt loc makedon *with the same men that had before made the settlement*, Chr. 1094, Erl. 230, 3. [*Icel.* lok *a conclusion;* loka *a lock, latch.*] DER. âr-, clûster-, word-loc; v. loca.

lôc, lôca *look, see, look you;* the word often occurs in connection with a pronominal form, and seems equivalent to a suffixed *-ever*, loca hû *however*, &c. :—Efne oððe lôca nû hêr hit is *en*, Ælfc. Gr. 38; Som. 40, 56 : Homl. Th. i. 358, 9. Hig cwædon Lôca nû hû hrædlîce ðæt fîctreów forscranc *dicentes: quomodo continuo aruit ficulnea*, Mt. Kmbl. 21, 20. Lôca nû hû hê hyne lufode *ecce quomodo amabat eum*, Jn. Skt. 11, 36. Þreá hig lôca hû ðú wylle *punish her, look you, as you will*, Gen. 16, 6. Ðú hæfst ðæt feoh mid ðê, gefada embe, lôca, hû ðú wylle, Homl. Skt. 3, 285 : 4, 262. Hî ferdon lôc [MSS. C.D. lôca] hû hî wolden *they went however they liked*, Chr. 1009; Erl. 142, 26. Lôca, hwâ ðâ gange, l*ige* hê ofslagen, Jos. 2, 19. Lôca, hwâ ðære mihte âge, hê môt gehæftne man âlýsan [*whoever has the power*], Wulfst. 294, 32. Lôca hwylc cristen man sý ungesibsum, 295, 4. Hlystan lôca hwæt ða lâreówas tæcan, 294, 26. Dôn lôc hwæt wê mâgon, 141, 28 : 150, 11. Lôc hwæt eald sî *hîc et hæc et hoc vetus*, Ælfc. Gr. 9, 32; Som. 12, 9. Lôc hwæt hæbbe týn fêt *decempes*, 49; Som. 50. 49. Bide mê lôce hwæs ðú wille *ask me for whatever you will*, Homl. Th. ii. 576, 10. Lôc hwær ic hit gefriþod wille habban *wherever I will have it protected*, L. C. S. 81; Th. i. 420, 26. Lôca hwonne *whenever*, Wulfst. 199, 16. Swâ ðæt lôc hwenne ðæt flôd byþ ealra hêhst, Chr. 1031; Erl. 162, 5. Lôc hweðer ðæra gebrôðra ôðerne oferbide wære yrfeweard ealles Englalandes *whichever of the two brothers should survive the other, should inherit all England*, 1101; Erl. 237, 31 : Chart. Th. 605, 27. v. lôcian.

loca, an; *m. That which closes or shuts, a bar, bolt, lock, an enclosed place, locker* :—Hepse † loca *clustella, serra*, Hpt. Gl. 500. Âlýsde leóda bearn of locan deófla [*hell*], Elen. Kmbl. 362; El. 181. Under helle cinn under líges locan, Exon. 31 b; Th. 99, 7; Cri. 1621 : 72 b; Th. 270, 32; Jul. 19. Se ðe healdeþ locan who *guards the lock*, 8 a; Th. 2, 14; Cri. 19: Salm. Kmbl. 371; Sal. 185. DER. bân-, brægn-, breóst-, burg-, feorh-, ferhþ-, ferþ-, fýr-, fyrhþ-, gewit-, hearm-, heolstor-, hord-, hreðer-, hring-, nîþ-, þeóster-, word-loca; v. loc.

loca, an; *m. A lock of wool* :—Loca *floccus*, Wrt. Voc. ii. 35, 71, cf. locc.

loc-bore, an; *f. One wearing long hair, a free woman* :—Frî wîf locbore, L. Ethb. 73; Th. i. 20, 7. See the note there, and Grmm. R. A. 286, 239.

locc, es; *m. The hair of the head, a hair, a lock of hair, a curl, ringlet* :—*Comatus* se ðe hæfþ loccas, *coma* is locc, Ælfc. Gr. 43; Som. 45, 9. Locc unscoren *coma vel cirrus*, Wrt. Voc. 42, 45. Locc *uncinus*, 42, 48. Loc *coma*, 70, 33 : *cicinnus* i. *vinnus*, ii. 131, 12 : *cirrus, crinus*, 24. Ne ân loc of eówrum heáfde forwyrþ *not a hair of your head shall perish*, Blickl. Homl. 243, 33 : Andr. Kmbl. 2845; An. 1425. Locces *cincinni*, Hpt. Gl. 526, 44. Se deófol lædde hine ût of ðære cyrican be ðam locce, Wulfst. 236, 10. Ðû ne miht ænne loc gedôn hwîtne oððe blacne *non potes unum capillum album facere aut nigrum*, Mt. Kmbl. 5, 36. Se scînenda lîg his locc up âteáh *the shining flame drew up his hair*, Homl. Th. ii. 514, 3. Wîfmannes loccas *crines :* loccas vel unscoren hær *comæ*, Wrt. Voc. 42, 49, 64. Loccas *capilli*, 64, 27. Loccas oððe feaxeácan *antiæ frontis*, Wrt. Voc. ii. 3, 66. Winde loccas *cincinni*, 20, 43 : 14, 23. Locca *criniculorum*, Hpt. Gl. 435, 27. Cyrpsum loccum *crispantibus*, 435, 11. Loccum *cirris*, Wrt. Voc. ii. 18, 70. Hî ne scoldon hira loccas lætan weaxan *non comam nutrient*, Past. 18, 7; Swt. 139, 13. Teóh him ða loccas and wringe ða eáran and ðone wangbeard twiccige, L. M. 2, 16; Lchdm. ii. 196, 13. Hê hæfde crispe loccas *capillis crispis*, Bd. 5, 2; S.

&. 615, 30. Fýrene loccas, Cd. 148; Th. 185, 10; Exod. 120. Wundne loccas *curled locks*, Exon. 111 b; Th. 428, 7; Rä. 41, 98. [*Icel.* lokkr : *O. H. Ger.* loc, locc *cincinnus, capillus, crinis* : *Ger.* locke.] DER. eár-, hêr-locc.

-locc, **-locced**, **-locked**. v. hwît-, líg-, wunden-locc, hwît-, líg-locced.

loccian. v. ge-loccian.

locen, *an enclosed place* (?), Cd. 220; Th. 283, 6; Sat. 300.

locer *a carpenter's tool, a plane* :—Locor *runcina*, Wrt. Voc. 287, 12. Locer, sceaba, ii. 119, 32. Locaer *vel* scraba, Ep. Gl. 22 b, 23.

loc-feax, es; *n. Hair* :—Ðæs wonges locfeax *cæsaries*, Wrt. Voc. ii. 22, 57.

loc-gewind, es; *n. Hair* :—Locgewind *vel* fexnes *capillatura*, Wrt. Voc. ii. 128, 38.

LÔCIAN; *p.* ode *To LOOK, see, gaze, observe, regard, take heed, look (to), belong, pertain* :—Gif ic on ealle ðîne bebodu lôcie *dum respicio in omnia mandata tua*, Ps. Th. 118, 6. Ðú eádmôdra lôcast *humilia respicit Dominus*, 137, 6. Ðâs sælâc ðe ðú tô lôcast *these offerings from the sea that thou dost look at*, Beo. Th. 3313; B. 1654. Hê on ðâs eorþan ealle lôcaþ *qui respicit terram*, Ps. Th. 103, 30. Lôcaþ unhióre *looks fiercely*, Salm. Kmbl. 532; Sal. 265. Ealles ðæs ðe mê ðâ tô lôcaþ *all that there belongs to me*, Chart. Th. 542, 11. Ðás ii bêc lôciaþ intô Ryppel, Cod. Dip. Kmbl. iii. 19, 22 : 256, 31. Búton Raab âna libbe and ða ðe lôciaþ tô hire *sola Rahab vivat cum universis, qui cum ea in domo sunt*, Jos. 6, 17 : 8, 1. Hié simle lôcigeaþ tô ðære eorþan *they (animals) always look to the earth;* ad terram semper inclinentur, Past. 21, 3; Swt. 155, 20. Hwæt stondaþ gê hér and up on ðysne heofon lôcaþ? Blickl. Homl. 123, 22. Ðâ lôcode Petrus tô Paule, 187, 34. Lôcode ðâ up wið Simones, 189, 6. Hê forþ lôcade of his ðam heán hâlgan setle *prospexit de excelso sancto suo*, Ps. Th. 101, 17. Hê on heofon lôcode *intuens in cælum*, Mk. Skt. 6, 41. Oþ hê on ðone æþeling lôcude *until his eyes fell on the atheling*, Chr. 755; Erl. 48, 34. Hié lôcodan æfter him, Blickl. Homl. 121, 22. Blinde men gehælde ðæt hié lôcodan *healed blind men so that they saw*, 173, 28 : Wulfst. 5, 1. Ðær men tô lôcedon *where men were looking on*, 98, 21. Lôcæ feónd mînne *respice inimicos meos*, Ps. Spl. T. 24, 20. Lôca nû *receive thy sight*, Blickl. Homl. 15, 26. 'Lôca hider;' ðâ lôcade hê ðider, Wulfst. 236, 20. Lôciaþ brâde and nân þing gecnâwaþ *look far and wide, and understand nothing*, 47, 13. Gâþ and lôciaþ *ite et videte*, Mk. Skt. 6, 38. Lôciaþ nû ðæt ðiós eówru leáf ne weorðe ôðrum monnum tô biswice *videte, ne forte hæc licentia vestra offendiculum fiat infirmis*, Past. 59, 6; Swt. 451, 32. Fore cyningum ðær hig eágum on lôcian *in conspectu regum*, Ps. Th. 118, 46. Ic ræhte mîne hond tô eów nolde iówer nân tô lôcian *extendi manum meam, et non fuit qui aspiceret*, Past. 36, 1; Swt. 247. 22. God hêt hyne lôcian tô heofonum *suspice cælum*, Gen. 15, 5. Hié ongeán lôcian ne mihton for ðæs léges bryne, Blickl. Homl. 203, 11. Swîðe fæger an tô lôcianne, Ors. 2, 4; Swt. 74, 13. Matheus ðâ lôciende geseah Drihten Crist, Blickl. Homl. 229, 30. Forðam ðe lôciende hig ne geseóþ *quia videntes non vident*, Mt. Kmbl. 13, 13. v. lôc.

locor. v. locer.

loddere, es; *m. A beggar, poor person* :—Se rîca besihþ on his pællenum gyrlum and cwyþ 'Nis se loddere mid his tættecon mîn gelîca *but the rich man looks at his purple robes and says 'the beggar with his rags is not my fellow,'* Homl. Th. i. 256, 8. [*Icel.* loddari *a tramp, juggler :* cf. *O. H. Ger.* lotar *cassus, vanus, inanis*.] v. lodrung.

lodrung, e; *f. Nonsense, triviality* :—Lodrung *nenias*, Wrt. Voc. ii. 71, 51. [cf. *O. H. Ger.* loter unde unreht *iniquitas* ; lotarum sprácha *nenias*, Grff. ii. 204.] v. loddere.

lof, es; *n. m. Praise, glory, a song of praise, hymn* :—Ðam Dryhtne sý lof and wuldor *to the Lord be praise and glory*, Blickl. Homl. 53, 3. Sý ðê þanc and lof ðînre mildse, Hy. 7, 58; Hy. Grn. ii. 288, 58. Ðær biþ gehýred ðîn hálige lof, 7, 32; Hy. Grn. ii. 287, 32. Ðis lof *hic pean*, Ælfc. Gr. 9, 11; Som. 9, 21. Be ðam Fortunatus on fæmnena lofe cwæþ *de quo Fortunatus in Laude Virginum ait*, Bd. 1, 7; S. 476, 32. Be ðam is gecweden on ðære brýde lofe, Past. 11, 2; Swt. 65, 22. Gecwedenum lofe *hymno dicto*, Mk. Skt. 14, 26. Wê cweþaþ lof ymb hié, Blickl. Homl. 149, 32. Drihtnes lof singende, 231, 9. Lof secgean, Ps. Th. 106, 31. Eall folc Gode lof sealde, Lk. Skt. 18, 43. Wê herigaþ hira cræftas and ðeáh nyllaþ hî habban forðæm wê hiera nabbaþ nân lof *we praise their arts, and yet do not wish to have them, for we get no credit from them*, Past. 34, 2; Swt. 231, 8. Ic eów sylle mîne sibbe þurh mîn ðæt hêhste lof (*the Holy Ghost*), Blickl. Homl. 157, 30. Gegân longsumne lof *to earn lasting praise*, Beo. Th. 3076; B. 1536. Lofa ic cweþe ðê *laudem dixi tibi*, Ps. Spl. 118, 164. Lofu † herunga *præconia, laudes, favores*, Hpt. Gl. 500, 2 : *melos*, Hymn. Surt. 5, 31. [*O. Sax., O. L. Ger., O. Frs., Icel.* lof : *O. H. Ger.* lob *laus, favor, hymnus* : *Ger.* lob.] v. here-lof.

lôf, es; *m.* ?—Hæfde sigora weard on ðam wangstede wære betolden leófne leódfruman mid lôfe sînum, Andr. Kmbl. 1978; An. 991. Grimm A. u. E. 989 would translate 'lôf' *hand*, comparing *Goth.* lôfa : *Icel.* lôfi (*Scott.* loof) *the palm of the hand*. In Hpt. Gl. 525, 8 *redimicula* is

glossed 'wrǽdas oððe cynewiððan, lofas;' would this be the same word as that in the above passage?

lof-bǽre; *adj. Laudatory, giving praise:* — Lofbǽrum werodum *hymniferis choris,* Hymn. Surt. 57, 12.

lof-dǽd, e; *f. A deed deserving praise,* Beo. Th. 48; B. 24.

lof-georn; *adj. Desirous of praise:* —Se ðe wǽre lofgeorn for ídelan weorþscype weorþe se carfull hū hē swýðast mǽge gecwēman his drihtne *he that was eager for praise on account of empty honour, let that man be careful how best he may please his Lord,* Wulfst. 72, 10. Manna lofgeornost of all men most desirous to deserve praise (*Beowulf*), Beo. Th. 6347; B. 3183. [*Jactantia* þet is idelʒelp on englisc, ðeine mon biþ lofʒeorn and deþ for ʒelpe mare þenne for godes luue, O. E. Homl. i. 103, 29. *Icel.* lof-gjarn: cf. *O. H. Ger.* lob-gerni *jactantia.*]

lof-herung; e; *f. Praising, commendation:* —Ic áʒylde lofherunga ðē *reddam laudationes tibi,* Ps. Lamb. 55, 12.

lofiah; *p.* ode *To praise, value, put a price upon:* —Míne weleras gefeóþ wynnum lofiaþ ðonne ic ðē singe *gaudebunt labia mea dum cantavero tibi,* Ps. Th. 70, 21. Wē ðē hǽriaþ and lofiaþ *we laud and praise thee,* Dōm. L. 48, 116; Cd. 192; Th. 240, 33; Dan. 396: Elen. Kmbl. 904; El. 453: Exon. 13 b; Th. 25, 13; Cri. 400. Job herede helm wera, hǽlend lofede, 17 a; Th. 40, 6; Cri. 634. Song áhōfun, lofedun liffruman, 15 a; Th. 31, 31; Cri. 504. Hē gehýrde hū hī God lofodon and heredon, Bd. 3, 19; S. 547, 36. Lofa *lauda,* Ps. Lamb. 147, 1. Ðec mihtig God gāstas lofige, Cd. 192; Th. 239, 21; Dan. 373. Lofigen, Exon. 54 b; Th. 192, 2; Az. 100. Ic gehýrde hine ðīne dǽd and word lofian, Cd. 25; Th. 32, 24; Gen. 508. [*O. Sax.* lofōn: *Icel.* lofa: *O. H. Ger.* lobōn *hymnizare, glorificare, commendare, magnificare: Ger.* loben.] v. ge-lofian; lofung.

lof-lác, es; *n. An offering made to do honour:* —Ða hǽðenan him brohton oft mistlíce loflác *the heathens often brought him offerings of divers kinds to do him honour,* Wulfst. 107, 6.

lof-lǽcan; *p.* -lǽhte *To praise:* — Sāwle mín lōflǽceþ [MS. -aþ] ł heraþ ðē *anima mea laudabit te,* Ps. Lamb. 118, 175.

lof-líc; *adj. Praiseworthy, laudable, honorable:* —Of lofflícere *laudabili, honorabili,* Hpt. Gl. 498, 45. [*Icel.* lof-ligr: *O. H. Ger.* lobe-líh *laudabilis: Ger.* löb-lich.]

lof-líce; *adv. Honorably, gloriously:* —Uton wē gehýran hū swíðe loflíce Sanctus Johannes wæs mid ðæs Hálgan Gāstes mǽgenum gefylled, Blickl. Homl. 165, 16. [*Icel.* lof-liga *gloriously.*]

lof-mægen, es; *n. Abundance* or *greatness of praise:* —Hwylc mæg spēdlíce eall Drihtnes lofmægen leóde gehýran *quis auditas faciet omnes laudes Domini,* Ps. 105, 2.

lof-sang, es; *m. A song of praise, hymn, psalm,* as an ecclesiastical term *lauds:* —Lofsang *ymnus,* Ælfc. Gl. 34; Som. 62, 45; Wrt. Voc. 28, 26. Fram ðǽre tíde ðæs ūhtlícan lofsanges *a tempore matutinæ laudis,* Bd. 3, 12; S. 537, 23. Ðā se sealmsang gefylled wæs ðæs ūhtlícan lofsanges *expletis matutinæ laudis psalmodiis,* 4, 7; S. 575, 3. Mid lofsange *cum cantico,* Ps. Th. 68, 31: Ex. 15, 21. Mid þysum lofsange *with this psalm* (v. Ps. Th. 53, 1), Homl. Skt. 11, 89. Moises sang Gode lofsang *cecinit Moyses carmen hoc Domino,* Ex. 15, 1. Ðā hig hǽfdon heora lofsang gesungenne *hymno dicto,* Mt. Kmbl. 26, 30. Lofsang cweþan *laudem dicere,* Ps. Th. 118, 164. Æfter ða hālgan lofsangas and mæssan gefyllede wǽron *after the holy psalms and masses were completed,* Blickl. Homl. 207, 29. God heriaþ mid gāstlícum lofsangum, Ælfc. Gr. 48; Som. 49, 11. Him lofsangum cwēmdan *cantaverunt laudes ejus,* Ps. Th. 105, 11. 'Gloria in excelsis Deo' sungon englas . . . Nū forlǽte wē ðæs lofsangas, Homl. Th. ii. 88, 3. Ic sang ūhtsang æfter ðā wē sungon dægrēdlíce lofsangas *cantavi nocturnam, deinde cantavimus matutinales laudes,* Coll. Monast. Th. 33, 27; Bd. 4, 7; S. 575, 5. [*O. Sax.* lof-sang: *Icel.* lof-sōngr: *O. H. Ger.* lob-, lobe-sang *hymnus.*]

lof-singende *hymning, hymn-singing:* —Lofsingende *hymnizantes,* Hpt. Gl. 519, 9.

lof-sum; *adj. Deserving praise, excellent, noble:* —Wæstm wæs lofsum, Cd. 23; Th. 30. 17; Gen. 468. [*O. Sax.* lof-sam; *O. H. Ger.* lob-sam *probabilis, meritus: Ger.* lobe-sam.]

loft *air:* —Heó ne liþ on nānum þinge ac on lofte heó stynt *it* (*the earth*) *does not rest on anything, but stands in the air,* Hexam. 6; Norm. 10, 20. v. lyft.

lofung, e; *f. Praising, appraising:* —Næfþ Godes ríce nānes wurþes lofunge ac biþ gelofod be ðæs mannes hǽfene. Heofenan ríce wæs álǽten ðisum gebrōðrum for heora nette and scipe and ðam rícan Zacheo tō healfum dǽle his ǽhta and sumere wudewan tō ānum feorþlinge and sumum menn tō ānum wæteres drenc *God's kingdom hath no fixed price, but a price is put upon it according to a man's property. The kingdom of heaven was allowed to these brothers for their net and ship, and to the rich Zacheus for half his possessions, and to a certain widow for a farthing, and to a certain man for a drink of water,* Homl. Th. i. 580, 21–26. Lofunga ł herunga *laudationes,* Ps. Lamb. 9, 15.

-loga. v. áð-, treów-, wǽr-, wed-, word-loga.

lōgian (v. lōh); *p.* ode *To lodge, place, put in order, arrange, frame:* —Tō þreágenne gē lōgiaþ eówere sprǽce *ye frame your speech to reprove,*

Homl. Th. ii. 454, 25. Hí on heora scype heora nett lōgodon *in navi componentes retia,* Mk. Skt. 1, 19. Wē lǽraþ ðæt man intō circan ǽnig þinga ne lōgige ðæs ðe ðartō ungedafenlíc sí *we enjoin that nothing be lodged in the church that is unsuitable for the place,* L. Edg. c. 27; Th. ii. 250, 11. [Ne neuer se stede ne uurþe lōged mid ōðere hōdes manne ðanne mid moneke, Cod. Dip. Kmbl. iv. 231, 9.] v. ge-logian.

logðor, logeðer *plotting mischief, wily, crafty:* —Logðor *cacomicanus,* Wrt. Voc. ii. 102, 77: 127, 35. Logðer, 13, 31. Logeðer *marsius* (cf. (?) 'Marsi homines, quibus naturalem vim contra serpentes inesse olim creditum, incantatores,' Ducange, v. wyrm-galere), 55, 58.

lōh; *gen.* lóges; *n. A place, stead:* —Gehādode Tobias on his lōh (on his steall, two other MSS.), Chr. 693; Thorpe 67, 9 col. 3. [*O. Frs.* lōch; *dat.* lōge *a place: O. H. Ger.* luog *specus, cubile: M. H. Ger.* luoc *locus.* v. Grmm. R. A. 955.]

lōh-sceaft, es; *m. A bolt, bar* (?): —Gaderode mē kigelas and stuþan sceaftas and lōhsceaftas, Shrn. 163, 6.

Loidis *Leeds:* —Ða æftran cyninigas him botl worhton on ðam lande ðe Loidis[is] hāten *reges posteriores fecere sibi villam in regione quæ vocatur Loidis,* Bd. 2, 14; S. 518, 21: 3, 24; S. 557, 12.

lōma *a tool.* v. and-, andge-, ge-lōman.

lomb, lond, long. v. lamb, land, lang.

lōm-lǽcan; *p.* -lǽhte *To use often, repeat, frequent:* —Lōmlǽhtan *frequentabant,* Hpt. Gl. 457, 44. v. ge-lōmlǽcan.

loppe, an; *f. A flea* (?), *a spider* (?); also *a silk-worm:* —Furþum ðeós lytle loppe hine hwílum deádne gedēþ *even this little flea sometimes kills him,* Bt. 16, 2; Fox 52, 13. Seolucwyrm oððe sīdwyrm oððe loppe *bombix,* Wrt. Voc. ii. 12, 23 (or is this a different word, corresponding to another meaning of *bombix,* 'silk or fine wool;' cf. *Icel.* lyppa *wool drawn into a long hank before being spun?* In Ps. Lamb. 38, 12 *sicut araneam* is glossed 'swā swā ætterloppan'; if this is not a mistake for 'ættercoppan,' by which the word ʾis rendered in Ps. Spl. 38, 15, 'loppe' would be rather a *spider* than a *flea,* and the same word might be used for the silk worm, as both insects are spinners. And in Wrt. Voc. 24, 1 *loppe* (apparently however intended to be a Latin word) is given as the equivalent of 'fleónde næddre *vel* āttorcoppe.') [Lop *a flea,* in some dialects, v. E. D. S. Reprinted Gloss. B. 15, 22; C. 1; and gloss. of Mid-Yorkshire and Holderness: *Dan.* loppe.]

lopystre, an; *f. A lobster, a locust:* —Loppestre *polypus,* Ælfc. Gl. 102; Som. 77, 77; Wrt. Voc. 56, 2: 77, 69. Lopust *locusta,* ii. 113, 11. Hwæt fēhst ðū on sǽ? Crabban and lopystran *quid capis in mari? Cancros et polypodes,* Coll. Monast. Th. 24, 13. Lopestro (loppestra, Rush.) *lucustas,* Mk. Skt. Lind. 1, 6.

lor, es; *n.* (v. ðæt forlor, Past. Swt. 403, 13). *Loss, destruction:* —Ðæt tō lore weorþe ān ðíne lioma *ut pereat unum membrorum tuorum,* Mt. Kmbl. Rush. 5, 29: 9, 17: 10, 6. Ðæt nǽniges mannes feorh tō lore wearð for ðam ofslægenan cyninges brēðer *ut nullius anima hominis pro interfecto regis fratre daretur,* Bd. 4, 21; S. 590, 23. Ðās heán mihta hēr on worulde áfeallaþ and tō lore wurþaþ *these lofty powers here on earth decay and perish,* Wulfst. 149, 4: 262, 17. Éðel ðe nǽfre tō lore ne weorþeþ *nunquam amittenda hæreditas,* Past. 36, 6; Swt. 255, 4. Ðonne hié him ǽr tíde tō tiōþ ðæt hí ne mágon, ðonne is him tō ondrǽdenne ðæt him weorþe tō lore ðæt hié tō ryhtre tíde gefolgian meahton, ðæt is se wisdōm, ðe hié ǽr tíde wilniaþ and eówiaþ, ac hē him wyrþ ðonne swíðe ryhtlíce tō lore *admonendi ne, cum arripiunt intempestive, quod non valent, perdant etiam quod implere quandoque tempestive potuissent: atque scientiam, quia incongrue conantur ostendere, juste ostendantur amisisse,* 49, 3; Swt. 383, 25–28. Ðonne ðín líchoma beó tō lore gedōn and ðín flǽsc gebrosnod *quando consumseris carnes et corpus tuum,* 36, 2; Swt. 249, 13. Tō hwon sceolde ðeós smyrenes ðus beón tō lore gedōn *why should this ointment be thus wasted?* Blickl. Homl. 69, 7. DER. for-lor; *and see* los.

lorh, lorg, e; *f. A pole, a weaver's beam:* —Lorh *vel* webbeám *liciatorium* [lignum in quo licium involvitur, et laqueus qui de filo solet fieri, Ducange], Ælfc. Gl. 110; Som. 79, 48; Wrt. Voc. 59, 19. Lorg *amitis* [amis lignum bifurcatum, per quod venatores expandunt retia, ad capiendas feras, Ducange], 285, 17: ii. 8, 38. Loerge *amites,* Ep. Gl. 1 b, 3.

lorian. v. losian.

los, es; *n. Loss, destruction:* —Ða þing tō lose wurdon ðe on ðam scipe wǽron *perditis his quæ in navi erant rebus,* Bd. 5, 9; S. 623, 20. Ðonne gē tō lose [Cott. MSS. lore] weorþaþ *in interitu vestro,* Past. 36, 1; Swt. 249, 1. Weg ðiú lǽdas tō lose *via quæ ducit ad perditionem,* Mt. Kmbl. Lind. 7, 13. Ðæt tō lose weorþe *ut pereat,* Rush. 18, 14: 5, 30. Hū hine mæhtes tō lose gedōa *quomodo eum perderent,* Lind. 12, 14. [*Icel.* los *looseness, breaking up.*] v. lor.

los-, lose-wist; *e; f.: es, m.* [?] *Hurt, loss, destruction, waste:* —Tō huon losuist ðiós smirinisse áworden wæs *quid perditio ista ungenti facta est,* Mk. Skt. Lind. 14, 4. Loswist [losewest, Rush.] walana *deceptio divitiarum,* 4, 19. Suna losuistes [loswest, Rush.] *filius perditionis,* Jn. Skt. Lind. 17, 12. Of losuist ðe *interitu,* Rtl. 169, 33. Sáules loswist geþolas *animæ detrimentum patiatur,* Mt. Kmbl. Lind. 16, 26. Losuist [losewest] gedōe, Mk. Skt. Lind. 8, 36: Lk. Skt. Lind. Rush. 9, 25.

lose [?] *frutectum, locus ubi ponunt,* Wrt. Voc. ii. 109, 23.

losian; *p.* ode *To perish, be lost, stray, escape:*—'Drihten ic losige.' Cweþ 'ic losige' ðý læs ðe ðú losige '*Lord, I perish.*' *Say* '*I perish,*' *lest thou perish,* Homl. Th. ii. 394, 1–2. Hwílum losaþ sió stemn *sometimes the voice is lost,* L. M. 2, 51; Lchdm. ii. 264, 14. Hwæt losaþ æfre ðam ælmihtigan Gode *what is ever lost to the Almighty God?* Homl. Skt. 11, 278. Gif hwylc mann hæfþ hund sceápa and him losaþ án of ðam . . . ða nigon and hundnigontig ðe ná ne losedon *si fuerint alicui centum oves et erraverit una ex eis . . . nonaginta novem, quæ non erraverunt,* Mt. Kmbl. 18, 12, 13: Homl. Th. i. 338, 27. Nó hé on helm losaþ *she shall not escape into shelter,* Beo. Th. 2789; B. 1392. Ealra ðæra sáwla ðe þurh ðæt losiaþ *all the souls that perish through that,* L. I. P. 19; Th. ii. 328, 37. Ðæt sæd ðe feóll be ðam wege mid twýfealdre dare losode [*perished*], Homl. Th. ii. 90, 14. Ðá losade hió him sóna *she was at once lost to him,* Bt. 35, 6; Fox 170, 15. Hé onweg losade *he escaped,* Beo. Th. 4199; B. 2096. Fíftig þurh fleám onweg losedon *quinquaginta fuga lapsos esse,* Bd. 2, 2; S. 504, 6. Ðý læs ðe ðú losige *ne tu pereas,* Gen. 19, 15. Gif hé losige and hine mon eft gefó *if he escape and he be caught a second time,* L. Alf. pol. 7; Th. i. 66, 11: 1; Th. i. 60, 17. Gif hit [feoh] him losige, 20; Th. i. 74, 17. Gaderiaþ ða láfe and hí ne losion *gather the remnants, and let them not be lost,* Homl. Th. i. 182, 21. Ne sceal hé for ðam læssan losian *he shall not be lost for the lesser sins,* ii. 336, 22. Swá swá seó beó sceal losian, ðonne heó hwæt yrringa stingþ, Bt. 31, 2; Fox 112, 26. Ðætte nú foraldod is ðæt is forneáh losad *quod enim antiquatur, prope interitum est,* Past. 30; Swt. 205, 9. Ðonne ðé mon ærest secge ðæt ðín ceáp sý losod, Lchdm. iii. 60, 9: L. Eth. ii. 8; Th. i. 288, 15. Mé syndon losode fóta gangas *effusi sunt gressus mei,* Ps. Th. 72, 1. *v.* ge-losian.

losigend-líc; *adj. Ready to perish, in danger of destruction:*—Ða tóweardan frecednyssa ðises losigendlícan middangeardes, Homl. Th. ii. 538, 7. Se ðe ða losigendlícan buruhware [*people of Jerusalem*] bemǽnde, i. 408, 6.

losing, e; *f. Loss, perdition:* Tó lose ł losing *ad perditionem,* Mt. Kmbl. Lind. 7, 13.

lot, es; *n. Deceit, guile, fraud, craft, cunning:*—Náuht ne deregaþ monnum máne áþas ne ðæt leáse lot ðe beoþ mid ðám wrencum bewrigen *nil perjuria, nil nocet ipsis fraus, mendaci compta colore,* Bt. 4; Fox 8, 17: Exon. 92 a; Th. 345, 16; Gn. Ex. 189 [v. list]. Mid his lote bewunden *encompassed with his deceit,* Past. 35, 3; Swt. 243, 1: 46, 3; Swt. 347, 19. þurh ðara scuccena lot *dæmonum solertia,* Fox 220, 14 note. *v.* lytig.

lotendra? *madendum,* Wrt. Voc. ii. 57, 46.

loða, an; *m. A cloak, upper garment:*—Loða *lodix,* Wrt. Voc. ii. 52, 58: *lacerna,* 53, 65: *sandalium,* 119, 55: *sagulum,* 119, 58: *colobium, dictum quia longum est, et sine manicis,* 134, 37. Hé genom his loðan ǽnne læppan *he took a skirt of his robe,* Past. 3, 2; Swt. 37, 5. Loðan *clamidem,* Wrt. Voc. ii. 21, 31. Hloðan, gegirelan *liniamento,* 50, 4. Heora andwlitan bewrigenum under loðum *their faces wrapped under their cloaks,* Cd. 77; Th. 95, 29; Gen. 1586. [*Icel.* loði *a fur cloak;* cf. loðinn *shaggy:* O. H. Ger. ludo, lodo *birrus, penula, lodix, genus vestimenti.*]

lot-wrenc, es; *m. Deceit. deception, cunning, fraud, device, wile, craft:*—Lotwrænc *deceptio, fraus,* Wrt. Voc. ii. 138, 13. Mid hwelcum lotwrence hit deófla dydon *with what deception devils did it,* Ors. 3, 3; Swt. 102, 18. Philippus mid his lotwrence áliéfde ðæt heora anwaldas móston standan swá hié ǽr dydon *Philip, with his craftiness, allowed their powers to stand as they did before,* 3, 7; Swt. 118, 9. Gif hwá mid his lotwrencum óðres mannes folgere fram him ápǽce *si quis versutiis suis alius hominis pedisequam ab eo allexerit,* L. Ecg. P. ii. 14; Th. ii. 186, 22. Ðá wearþ se mann mid deófles lotwrencum bepǽht, Homl. Th. i. 192, 11: 376, 9: Wulfst. 84, 19. For his lotwrencium, Past. 30, 1; Swt. 203, 19. þurh ðara scuccena mislíce lotwrencas *dæmonum varia solertia,* Bt. 36, 6; Fox 220, 14. Hé heora lotwrencas [-wrencceas, MS. B.] wiste *sciens versutiam eorum,* Mk. Skt. 12, 15. Ða ðe ðisse worulde lotwrenceas cunnon *sapientes hujus seculi,* Past. 30, 1; Swt. 203, 5: Swt. 205, 17.

lotwrenc-ceást, e; *f. Wiliness, cunning:*—Hé heora lotwrencceáste *sciens versutiam eorum,* Mk. Skt. 12, 15.

lox, es; *m. A lynx:*—Lox *linx,* Ælfc. [Gl. 19; Som. 59, 14; Wrt. Voc. 22, 55. Aristoteles sǽde ðæt deór wǽre ðæt mihte ǽlc wuht þurhseón ge treówa ge furþum stánas; ðæt deór wé hátaþ lox, Bt. 32, 2; Fox 116, 22. [*O. H. Ger.* luchs, lohs *lynx, pardus, panthera: Ger.* luchs.]

loxe *in* loxanwudu, Cod. Dip. Kmbl. v. 345, 5. Cf. [?] Grff. ii. 163 luhsa, Linsa [*sylvestris, Gottheit*]'; or loxan wudu = *lynx-wood.*'

lúcan; *p.* leác, *pl.* lucon; *pp.* locen *To close, conclude, fasten, lock:*—Ðæt hé leác on háre tungon *qui statim conclusit et omnino confirmavit totum quod pater suus in sua fecerat,* Chart. Th. 272, 5. On ðæt gerád ðe ðæt stande ðe wit beforan ðam ealdormen lucan *on the condition that that arrangement stand which we concluded before the alderman,* 597, 32. Hrím and forst lucon leóda gesetu *rime and frost shut up men's dwellings,* Andr. Kmbl. 2519; An. 1261. Ðǽr com flówende flód æfter ebban lucon lagustreámas *there came flowing flood after ebb, the streams inter-*twined *or closed up* [*the surface of the water shewing a network of lines from the varying currents, as the tide flowed up the river*], Byrht. Th. 133, 46; By. 66. Siððan ða ýslan eft onginnaþ lúcan tógædere geclungne tó cleowenne *afterwards the ashes begin to close up again, pressed to a ball; in massam cineres coactos,* Exon. 59 a; Th. 213, 16; Ph. 225. Lúcan eorþan cíðas (*frost shall*) *lock up the germs of earth,* 90a; Th. 338, 6: Gn. Ex. 74. Lúcan [onlúcan?], Cd. 220; Th. 283, 5; Sat. 300. Sincgim locen *the jewel fastened in its setting,* Elen. Kmbl. 528; El. 264. Locen *is applied to coats of mail, which were formed of* [*interlacing*] *rings fastened on to some material to which they might be sewn, see* hring *with its compounds, and cf.* brogden byrne; *also Icel.* hring-ofin:—Locene leoþosyrcan, Beo. Th. 3014; B. 1505: 3784: 1890. Locen beág *a closed ring* [*not a spiral* wunden beág], 5982; B. 2995: Andr. Kmbl. 605; An. 303. [*O. Sax.* ant-, bi-lúkan: *O. Frs. Icel.* lúka: *O. H. Ger.* lúhhan.] DER. á-, be-, ge-, on-, tó-, un-lúcan.

lúcan; *p.* leác *To pull up:*—Swá swá londes ceorl of his æcere lýcþ yfel weód monig, Bt. Met. Fox 12, 55; Met. 12, 28 [*E.D.S. Mid-York. Gloss.* louk, louk *to weed: Holderness Gloss.* lookers *weeders in a cornfield;* look *to hoe weeds in a field of young corn:* lowker *runcinator,* Wrt. Voc. 218, col. 2: *O. H. Ger.* ar-, úz-liuhhan *evellere,* Grff. ii. 138.] *v.* á-lúcan.

lud-geat, es; *n. A back door, postern:*—þurh ludget *per seudoterum* [ψευδοθυρον], Wrt. Voc. ii. 67, 72. þorh ludgæt, 116, 70: Ep. Gl. 18 b, 16.

lufe. *v.* lufu.

lufe-líc. *v.* luf-líc.

lufen, e; *f. Hope* [?] :—Sceal eall eðelwyn eówrum cynne lufen álicgean [lufena licgean, MS.] *all delight in their country and hope shall fail your kin,* Beo. Th. 5764; B. 2886. [Grein who emends thus compares *lufen* with Gothic *lubains;* Grimm takes *lufen = leofen* victus, R. A. 731.]

lufestice, es, *also,* an; *m. Lovage:*—Lufestice *lubestica,* Ælfc. Gl. 39; Som. 63, 79; Wrt. Voc. 30, 27: 69, 23. Lubestica *conixe,* 67, 40. Lufestice *libestica,* 79, 2. Genim lubastican wyrttruman, Herb. 146, 3; Lchdm. i. 270, 7. Lufestices sǽd, L. M. 3, 12; Lchdm. ii. 314, 20: iii. 128, 22. Genim lufestice, 4, 10.

lufestre, an; *f. A sweetheart:*—Lufestran *amatricis,* Hpt. Gl. 509, 70.

lufian; *p.* ode *To love, feel affection for, shew love to:*—Simon lufast ðú mé . . . hé cwæþ tó him ðú wást ðæt ic ðé lufige *Simon diligis me . . . dicit ei tu scis quia amo te,* Jn. Skt. 21, 15. Se ðe lufaþ his sáwle forspilþ hig *qui amat animam suam perdet eam,* 12, 25. Lufiaþ mid lácum ða ðe læs águn *shew their love with gifts to those that have less,* Exon. 33 b; Th. 106, 32; Gú. 50. Hé ágsode hý, hwá wolde on ðære gefǽrrædenne beón ðe hé wǽre, and ðæt lufian ðæt hé lufode, L. Eág. 3; Th. i. 162, 6. Hé mé mid syndrige lufan lufode, Bd. 5, 6; S. 619, 33. Hú ús wuldres weard wordum and dǽdum lufode in life, Andr. Kmbl. 1193; An. 597. Ðú mé on ðínum weorcum lufadest *delectasti me in factura tua,* Ps. Th. 91, 3. Hí hine lufedan leáse múðe *dilexerunt eum in ore suo,* 77, 35. Lufigean his néhstan swá hine sylfne, Mk. Skt. 12, 33. Ðæt is tó lufigenne on ðysse wyrte ðæt heó hafaþ gehwǽdne wyrttruman *it is an excellent property of this plant, that it has a small root,* Herb. 140, 1; Lchdm. i. 260, 5. Ðeo lufigenda wer *hic amans vir;* ðis lufigende wíf *hæc amans fœmina,* Ælfc. Gr. 5; Som. 3, 49. Hé wæs fram eallum mannum lufad, Bd. 3, 14; S. 540, 11: 5, 19; S. 637, 19. *v.* ge-lufian.

lufiend, lufigend, es; *m. A lover:*—Amans Deum, lufigende God, is *participium,* and *amans Dei* is nama, ðæt is, *amator Dei,* Godes lufigend, Ælfc. Gr. 43; Som. 44, 61. Swá swíðe se cyning wæs geworden lufiend ðæs heofonlícan ríces, Bd. 3, 18; S. 546, 5 col. 2. Ic hæbbe óðerne lufiend *I have another lover,* Homl. Skt. 7, 27. Se wísdom gedeþ his lufiendas wíse, Bt. 27, 2; Fox 98, 1. Lufigendas, Homl. Th. ii. 392, 27.

lufiend-, lufigend-líc; *adj. Lovely, lovable, amiable:*—Lufigendlíc *amabilis,* Ælfc. Gr. 9, 28; Som. 11, 40. Lufigendlíc miht *amanda virtus,* 26; Som. 28, 19. Luffendlíc stede *amenus locus,* Ælfc. Gl. 48; Som. 65, 63; Wrt. Voc. 33, 59. Swíðe lufigendlíc and leóf æghwæðere þeóde *utrique provinciæ multum amabilis,* Bd. 4, 21; S. 590, 16: 4, 3; S. 568, 16. Swíðe lufiendlíce sind geteld ðín *quam amabilia sunt tabernacula tua,* Ps. Surt. 83, 2.

luf-líc; *adj. Lovely, lovable, worthy of love, amiable, dear:*—Luflíc *amabilis,* Hymn. Surt. 38, 5. Cild ácenned god luflíc *a child born at this time will be good and amiable,* Lchdm. iii. 190, 5. Hú luflíce geteld ðín *how amiable are thy tabernacles;* quam dilecta tabernacula tua, Ps. Spl. 83, 1.

luf-líce; *adv. Amiably, kindly, dearly, with good will* or *love, willingly:*—Luflíce *affabiliter,* Wrt. Voc. ii. 5, 11. Hé luflíce him hýrde *libenter eum audiebat,* Mk. Skt. 6, 20. Ðæt hé luflíce swá gedyde *libentissime se facturum,* Bd. 4, 11; S. 579, 31: Blickl. Homl. 203, 33. Ælfréd cyning háteþ grétan Wærferþ biscep his wordum luflíce and freóndlíce *with love and friendship,* Past. Swt. 3, 1; Blickl. Homl. 199, 36. Hú luflíce hé ús gesóht hider on middangeard *with how great love he visited us here on earth,* 129, 11: Wulfst. 204, 16. Ic wylle cýpan luflícor ðonne ic gebicge *volo vendere carius quam emi,* Coll. Monast. Th. 27, 19.

luf-rǽdenn, e; *f. Love:*—Hig gesetton hatunge for lufrǽddenne mínre *posuerunt odium pro dilectione mea,* Ps. Lamb. 108, 5.

luf-sum; *adj. Amiable, pleasant, lovable:*—Lufsum swǽ Rahel *amabilis ut Rachel*, Rtl. 109, 37. Lufsum and líðe leófum monnum *amiable and kind to the men that are dear to him*, Exon. 21 a; Th. 57, 5; Cri. 914: 96 a; Th. 357, 21; Pa. 32.

lufsum-líce; *adv. Kindly, graciously:*—Ðá sende Vitalianus se pápa cyninge lufsumlíce ǽrendgewrit, Bd. 3, 29; S. 561, 18.

lufsum-ness, e; *f. Amiability, pleasantness, love, kindness:*—Lufsumness *delectatio*, Wrt. Voc. ii. 138, 56. Lufsumnisse *dilectionis*, Rtl. 3, 24: 13, 21. Lufsumnisse *jocunditatem*, 45, 33.

luf-tácen, es; *n. A token of love*, Beo. Th. 3730; B. 1863.

luf-tíme; *adj. Giving rise to love, pleasant, grateful:*—Gregorius ðæt luftýme weorc gefremode *Gregory performed that grateful work [the conversion of the English]*, Homl. Th. ii. 126, 26.

LUFU, e *and* an [v. Anglia vi. 176]; *f.* LOVE:—*Te amo* ðě ic lufige, ðon befylþ mín lufu on ðé and ðú miht cweþan *amor a te* ic eom gelufod fram ðé, Ælfc. Gr. 19; Som. 22, 36. Gif ðonne ðæs monnes mód and his lufu biþ behleápen on ða lǽnan sibbe *si ergo in ea [pace] cor quæ relicta est figitur*, Past. 46, 5; Swt. 351, 14. Swá mycel lufu tó godcundre láre *tantus amor persuadendi*, Bd. 4, 27; S. 604, 20. Sǽde hire ðá his lust and his willan ðæt his lufu wǽre ðæt hé ða stówe neósode ðara eádigra apostola *indicavit ei desiderium sibi inesse beatorum apostolorum limina visitandi*, 5, 19; S. 637, 30. On ðæm welme ðære sóþan lufan, Blickl. Homl. 29, 10: Exon. 107 a; Th. 409, 7; Rä. 27, 25. Mid bróðorlíce lufan hí lǽran ðæt hí rihte sibbe and lufan betwih him hæfdon, Bd. 2, 2; S. 502, 8. For Godes lufon *pro Domino*, 3, 19; S. 547, 16. For úre lufan *for love of us*, Blickl. Homl. 23, 35. Mid luse ge mid láðe, 45, 8. For hylde and lufe *affectu*, Wrt. Voc. ii. 3, 65. Gif hé secge ðæt hé hæbbe hire freóndscipe ðæt ys he lufe *si dicat se amicitiam ejus habere, id est, amatorie*, L. Ecg. P. iv. 68, 17; Th. ii. 230, 17. Hié sceolan lǽran Godes lufan and manna, Blickl. Homl. 77, 20. Godes ege and his lufe fæstlíce on úrum heortum healdan, 131, 3. And ðar þegen áge twegen costas lufe odðe lage and hé ðonne lufe geceóse *and where a thane has a choice of two courses, love or law [an amicable settlement or appeal to law] and he choose the former*, L. Eth. iii. 13; Th. i. 298, 5. Ðeós woruld nǽre wyrðe ðæt man tó hire lufe hæfde ealles tó swíðe *this world does not deserve to be loved too much*, Wulfst. 273, 14. For Godes ege and for his lufu, 302, 27. Ðér wé sib and lufu samod gemétaþ, Hy. 7, 30; Hy. Grn. ii. 287, 30. Lufena tó leáne, Exon. 119 b; Th. 459, 11; Hy. 4, 115. Deáh monn good onginne for sumes wíteð ege, hit mon sceal ðeáh geendigean for sumes gódes lufum, Past. 37, 1; Swt. 265, 7. For ðǽm lufum ðe hí tó him habbaþ *per caritatem*, 52, 7; Swt. 409, 13. Ðone mon lufaþ for lufum, Bt. 24, 3; Fox 82, 34. For ðínum lufum, 22, 2; Fox 78, 12. Hé onféng ða ilcan gecynde for úrum lufon *he received the same nature for our sakes*, Blickl. Homl. 23, 24. For mínum lufan, Wulfst. 231, 17. Lufum *voluntariis*, Hpt. Gl. 435, 64. DER. bearn-, brýd-, eald-, eard-, feoh-, freónd-, fyrhþ-, gást-, heáh-, heort-, mǽg-, man-, mód-, ofer-, sib-, sorg-, treów-, wíf-lufu.

luf-wende; *adj. Beloved, amiable, pleasant:*—Cild ácenned lufwende *a child born [at that time will be] amiable*, Lchdm. iii. 186, 24. Mid lufwendum módes willan *cum benevolo animi affectu*, Lye. Ða lufwende eardas *dilecta rura*, Wrt. Voc. ii. 140, 42.

lufwend-líc; *adj. Amiable; amabilis*, Lye.

luh (a borrowed word apparently, Welsh *llwch*; cf. *pól* and Welsh *pwll*]; *n. A loch, lough:*—Ofer ðæt luh *trans fretum*, Mt. Kmbl. Lind. 14, 34: Mk. Skt. Lind. 8, 13: Lk. Skt. Lind. 8, 22. Ofer luh ł lytel sǽ, Mt. Kmbl. Lind. 14, 22: Mk. Skt. Lind. Rush. 5, 1.

Lunden *London:*—Hé bebohte hine on Lundenne *he sold him in London*, Bd. 4, 22; S. 592, 3: Chr. 839; Erl. 66. 16: 898; Erl. 96, 20. Of Eástenglum and of Lunden, 992; Erl. 131, 33.

Lunden-burh; *f. London:*—Ðes geáres forbarn Lundenburh, Chr. 1077; Erl. 215, 12. Ða Bryttas forlēton Kentland and mycclum ege flugon tó Lundenbyrig, Chr. 456; Erl. 13, 29: 872; Erl. 76, 15. Ðý ilcan geáre gesette Ælfréd cyning Lundenburg, 886; Erl. 84, 26.

Lunden-ceaster, e; *f. London:*—Is heora [*East Saxons*] ealdorburh nemmed Lunden-ceaster on ofre geseted ðæs foresprecenan streámes [*the Thames*] . . . Ðá hét Æþelbyrht on Lundenceastre cyricean getimbrian and ða gehálgian Sce. Paule, Bd. 2, 3; S. 504, 17–23. Se wæs Lundenceastre biscop, 2, 7; S. 509, 8. Eác swylce Eást-Seaxum hé gesette Ercenwold biscop in Lundenceastre, 4, 6; S. 573, 43.

Lundenisc; *adj. Belonging to London:*—Lundenisc *Lundoniensis*, Ælfc. Gr. 5; Som. 4, 28.

Lunden-waran, -ware; *pl. The people of London:*—Mellitum ðone biscop Lundenwaran onfón ne woldon *Mellitum Lundonienses episcopum recipere noluerunt*, Bd. 2, 6; S. 508, 37. Ðá wurdon Lundenware hǽdene, Chr. 616; Erl. 23, 10.

Lunden-wíc, es; *n. London:*—Æþelbyrht gesealde Mellite biscopsetle on Lundenwíc, Chr. 604; Erl. 21, 22.

lund-laga, an; *m. Rein, kidney:*—Ðes lundlaga *hic rigen* odðe *ren*, Ælfc. Gr. 9; Som. 9, 34. Lundlaga *lien*, Wrt. Voc. 45, 14: *renunculus*, ii. 118, 72. Lundlaga *renunculi*, Wrt. Voc. 44, 67: *renunculæ*, 65, 58. Ðú nymst twegen lundlagan *sumes duos renes*, Ex. 29, 13: Lev. 8, 25. [Cf. Icel. lundir; *pl. f. the flesh along the back: O.H. Ger.* lunda *arvina*; *and see* gelynd, gelyndu.]

lungen, e; *f. A lung:*—Lungen *pulmo*, Wrt. Voc. 45, 11: *pulmon*, 65, 51. Lungena *pulmones*, 71, 5. Ðone man ðe biþ lungenne wund, L. M. 1, 38; Lchdm. ii. 92, 21. [*Icel.* lungu; *pl. n. the lungs*; *O.H. Ger.* lunga, lungina; *f. pulmo: Ger.* lunge.]

lungen-ádl, e; *f. Disease of the lungs*, Lchdm. iii. 20, 24: 22, 8.

lungen-sealf, e, *f. A salve for the lungs:*—Ðás wyrte sculon tó lungensealfe, Lchdm. iii. 16, 6.

lungen-wyrt, e; *f. Lung-wort*, Lchdm. ii. 398, col. 1: iii. 337, col. 1.

lungre; *adv. Quickly, soon, at once, straightway, speedily:*—Loth eode lungre út *Lot went out straightway*, Cd. 113; Th. 148, 24; Gen. 2461: Beo. Th. 5480; B. 2743. Cyning álýsde hine lungre, Ps. Th. 104, 16. Wén is ðæt hí ús lifigende lungre wyllen, snióme forsweolgan, 123, 2. Ðǽr him lífgedál lungre weorþeþ *there the parting with life shall happen to him suddenly*, Exon. 87 b; Th. 330, 3; Vy. 45: 10 a; Th. 11, 8; Cri. 167. Hié lungre ǽr feorh álēton *just before they had lost their lives*, Andr. Kmbl. 3255; An. 1630. Næs him gewemmed wlite ne wlóh of hrægle lungre álýsed *his beauty was not spoiled nor a fringe of his garment even loosened* [Grimm would translate *lungre* here *acriter, fortiter*], 2942; An. 1474. [Cf. *O. Sax.* lungar *strong* : *O. H. Ger.* lungar *strenuus*.]

lús; *f. A louse:*—Lús *pediculus vel sexpes*, Wrt. Voc. 24, 11. Swínes lús *usia*, 24, 34. Luus *peducla*, ii. 117, 8. Hine byton lýs, Hexam. 17; Norm. 24, 30. Hé áfylde eal heora land mid froggon, and siðdan mid gnættum, eft mid hundes lúsum, Homl. Th. ii. 192, 21. [*Icel.* lús: *f*; *pl.* lýss: *O. H. Ger.* lús *pediculus: Ger.* laus.]

LUST, es; *m.* LUST, *desire, pleasure, voluptuousness:*—Epicurus sǽde ðæt se lust wǽre ðæt hēhste gód *Epicurus summum bonum voluptatem esse constituit*, Bt. 24, 3; Fox 84, 23. Swá mycel hǽto and lust Cristes geleáfan *tantus fervor fidei et desiderium*, Bd. 2, 14; S. 518, 4. Him wæs metes micel lust *he had a craving for food*, Homl. Th. i. 86, 6. Lust odðe gǽlsa *luxus*, Ælfc. Gr. 11; Som. 15, 11. Hwǽr beóþ ðonne se ídla lust and seó swétnes ðæs hǽmeðþinges ðe hé ǽr hátheortlíce lufode, Blickl. Homl. 59, 16. Lustes *veneris*, Wrt. Voc. ii. 92, 79. Luste *oblectamento*, Hpt. Gl. 525, 68. Mid ungeswencedlíce luste heofonlícra góda *infatigabili cælestium bonorum desiderio*, Bd. 5, 12; S. 631, 35. Nú is ðín folc on luste *now is thy people desirous*, Andr. Kmbl. 2046; An. 1025: Elen. Kmbl. 276; El. 138. Wedres on luste *glad on account of fair weather* or [?] *desirous of fair weather*, Exon. 97 a; Th. 361, 28; Wal. 26. Of luste flǽsces *ex voluntate carnis*, Jn. Skt. Rush. 1, 13. In lust *in luxum*, Wrt. Voc. ii. 47, 3: Hpt. Gl. 514, 5. Him sǽde his willan and his lust *ei indicasset desiderium suum*, Bd. 2, 15; S. 519, 7. Ofer lust mínne *a desiderio meo*, Ps. Th. 139, 8. Ðonne hafaþ hé micelne lust *ita ingentem libidinem haberet*, L. Med. ex Quad. 8, 8; Lchdm. i. 358, 20. Plegan, lustas *ludrica*, Wrt. Voc. ii. 52, 64. Him sweðraden synna lustas *sinful lusts were stilled in him*, Exon. 34 a; Th. 109, 3; Gú. 84. Se man hine forhabban sceal on manegum þingum his lífes lusta *homini a multis vitæ suæ libidinibus abstinendum sit*, L. Ecg. P. 1, 5, arg; Th. ii. 170, 10. Hé hine ætbrǽd ðám flǽsclícum lustum, Homl. Th. i. 58, 19. Hé fulgǽþ his lustum and his plegan *he follows his desires and his pleasure*, 66, 12. Tó ðám upplícan lustum *ad superna desideria*, Bd. 4, 29; S. 607, 15. Of lustum ðiss lífes *voluptatibus vitæ*, Lk. Skt. 8, 14. Lustum *joyfully, gladly, voluntarily*, Cd. 1; Th. 2, 8; Gen. 16. Ic ðé lustum láce cwéme *voluntarie sacrificabo tibi*, Ps. Th. 53, 6. Néde odðe lustum, Bt. Met. Fox 9, 88; Met. 9, 44. Mid lustum, Dóm. L. 6, 70. Wesan on lustum *to live joyously*, Cd. 23; Th. 30, 26; Gen. 473. Here wæs on lustum *joyous were the people*, Judth. 11; Thw. 23, 38; Jud. 162. Ne heora lustas ne heora willan gefyllan *nec desideria vel vota complere*, Bd. 1, 7; S. 477, 38. Líces lustas, Exon. 71 b; Th. 267, 2; Jul. 409. [*Goth.* lustus: *O. Sax. O. Frs. O. H. Ger.* lust *luxus, appetitus, venus, delectatio, concupiscentia: Ger.* lust.] DER. firen-, syn-, un-lust.

lust; *adj.* (?) *Pleased, glad, desirous:*—Ðæt ðú ne gehýre lustum móde ðæra twýsprǽcena word *that thou be not glad or desirous to hear the words of the double-tongued* [or does *lustum* belong to the preceding word?], Wulfst. 246, 10.

lust-bǽre; *adj. Producing* or *having desire* or *pleasure, desirous, desirable, pleasant, agreeable:*—Lustbǽre *libens*, Ælfc. Gr. 33; Som. 37, 18: 44; Som. 46, 32. Lustbǽre on gesihþe *aspectu delectabile*, Gen. 3, 6. Ic wæs swíðe lustbǽre hine tó gehýranne *me audiendi avidum*, Bt. 22, 1; Fox 76, 7. Sió hǽlu hine geðéþ lustbǽrne *salubritas videtur præstare voluptatem*, 24, 3; Fox 84, 9. Wǽron lustbǽre for ðone leófsan drihten wíta tó þrowienne, Homl. Skt. 4, 116. Ðás word sind lustbǽre tó gehýrenne *these words are pleasant to hear*, Homl. Th. i. 130, 16.

lustbǽr-líc; *adj. Desirable, pleasant:*—Eálá hú lustbǽrlíce tída on ðam dagum wǽron *O tempora desiderio dignissima!* Ors. 2, 5; Swt. 84, 25.

lustbǽr-líce; *adv. With delight, pleasure, eagerness, pleasantly:*—Ða leóþ ðe ic geó lustbǽrlíce asungen hæfde *hæc cum philosophia leniter suaviterque cecinisset, qui quondam studio florente peregi*, Bt. 2; Fox 4, 7. Ðá se wísdóm ðis leóþ lustbǽrlíce ásungen hæfde, 36, 1; Fox 170, 25.

lustbǽr-ness, e; f. Desire, pleasure, pleasantness:—Lustbǽrnes delectatio, Wrt. Voc. ii. 138, 56. Ða bereáfodon ǽlcere lustbǽrnesse they robbed me of every pleasure, Bt. 2 ; Fox 4, 11. Wě ðonne ne beóþ onǽlde mid ðære lustbǽrnesse úres módes ðonne bistilþ sió slǽwþ on ús óþ ðæt heó ús áwyrtwalaþ from ǽlcere lustbǽrnesse gódra weorca ipsa quippe mentis desidia, dum congruo fervore non accenditur, a bonorum desiderio funditus convalescente furtim torpore mactatur, Past. 39, 1 ; Swt. 283, 3. Hit biþ onstyred mid ðære lustbǽrnesse ex delectatione pulsatur, 53, 6 ; Swt. 417, 13. Þurh Evan lustbǽrnesse oferswíðed delectatione superatus, 53, 7 ; Swt. 417, 28. Hí náne lustbǽrnisse nabbaþ hí tó sécanne they have no desire to seek them, Bt. 32, 3 ; Fox 118, 23. Lustbǽrnesse nimþ cupidinem contrahat, L. Ecg. P. iii. 14 ; Th. ii. 202, 4 : Wrt. Voc. ii. 33, 72.

lust-full; adj. Desirous :—Gif his hwá síe lustfull máre tó witanne sěce him ðonne self ðæt if any one be desirous to know more of it, let him seek it himself, Ors, 3, 2 ; Swt. 100, 27.

lustfullian ; p. ode To rejoice, be glad, take pleasure [in] : — Swá ic lustfullige on ðisum láðum wítum, swá swá ðe gesihþ ðone ðe hě gewilnode, Homl. Skt. 8, 116. Heó lustfullode-on hire fóstormóder húse, Nar. 40, 12. Se cyning ongan lustfullian ðæt clǽneste líf háligra and heora ðám swētestan gehátum [rex] ipse delectatus vita mundissima sanctorum, et promissis eorum suavissimis, Bd. 1, 26 ; S. 488, 8. Mid ðý se líchoma ongynneþ lustfullian cum caro delectari cœperit, 1, 27 ; S. 497, 22. Wě witan ðæt se líchoma ne mæg lustfullian bútan ðam móde cum caro delectare sine animo nequeat, 497, 28. Ða ongan hě lustfullian ðæs biscopes wordum, 2, 9 ; S. 511, 34. Ða ongan se biscop lustfullian his wíslícra worda, 5, 19 ; S. 637, 46. Evan swá swá líchoma wæs lustfulliende Eva velut caro delectata est, 1, 27 ; S. 497, 15 : 5, 12 ; S. 630, 32. Lustfulligende, 4, 25 ; S. 600, 22. DER. ge-lustfullian.

lustful-líce ; adv. With joy or pleasure, joyfully, gladly :—Lustfullíce libenter, Bd. 4, 27 ; S. 604, 30. Se mildheorta Drihten onfehþ swíðe lustfullíce eallum ðæm gódum ðe ǽnig man gedéþ his ðæm néhstan, Blickl. Homl. 37, 25.

lustful-ness, e; f. Pleasure, delight, desire :—Lustfulnes oblectamenta, Wrt. Voc. ii. 62, 49. Seó lustfulnys biþ þurh líchoman delectatio fit per carnem, Bd. 1, 27 ; S. 497, 13, 10, 18, 12, 30 : Past. 53, 6 ; Swt. 417, 7, 8, 21, 24, 25. Drihten eallum geleáffulum monnum heora gong gestaþelade tó lífes wege ðæt hié mágon þurh ða lustfulnesse heora módes mid gódum dǽdum geearnian leóht ðæs ěcan lífes the Lord established for all believers their passage to the way of life, that they may through the ardent desire of their mind earn with good deeds the light of everlasting life, Blickl. Homl. 17, 20.

lustfullung, e ; f. Pleasure, delight :—Of ýdelum gylpe biþ ácenned lustfullung leásre herunge from vainglory is born a delight in false praise, Homl. Th. ii. 220, 33. Lustfullunge oblectamento, Hpt. Gl. 525, 68.

lustgeorn-ness, e ; f. Desire, concupiscence :—Lustgeornnisse fornicationis, Mt. Kmbl. p. 14, 16. Lustgiornisses concupiscentiæ, Mk. Skt. Lind. 4, 19.

lust-grin, e ; f. Snare set by pleasure, Soul Kmbl. 46 ; Seel. 23. [The MS. has lustgryrum for which Grein proposes to read lustgrynum.]

lús-þorn, es ; m. The spindle tree ; euonymus Europæus :—On lúsþorn ; of lúsþorne, Cod. Dip. Kmbl. iii. 77, 19. [v. E. D. S. Plant Names louse-berry tree : Dutch luizen-boom.]

lust-líce ; adv. With pleasure, gladly, willingly : — Lustlíce libenter, Ælfc. Gr. 44 ; Som. 46, 32. Lustlíce onfón libenter excipere, Bd. 3, 11 ; S. 535, 18 : 3, 3 ; S. 525, 30. For ðě wé wolden lustlíce sweltan for thee we would gladly die, Ap. Th. 26, 6. Ðe nú lustlíce sibbsumes friþes æt eów biddende sindon who now are willing to ask a friendly peace from you, Ors. 1, 11 ; Swt. 48, 22. Ða godcundan láre lustlíce gehýran, Blickl. Homl. 47, 28 : 49, 32. v. for-lustlíce.

lustmoce, an ; f. Lady's smock ; Cardamine pratensis :—Lustmoce croppan, L. M. 1, 38 ; Lchdm. ii. 92, 23. Lustmocan crop, Lchdm. ii. 92, 8. Genime lustmocan, 1, 30 ; Lchdm. ii. 70, 17.

lustsum-líc ; adj. Pleasant, delectable :—Ic nát for hwí eów sindon ða ǽrran gewin swá lustsumlíce on leóþcwidum tó gehíeranne, Ors. 3, 7 ; Swt. 120, 2. [Cf. O. H. Ger. lustsam amoenus, dulcis, delectabilis.]

LÚTAN ; p. leát ; pl. luton ; pp. loten To lout, bow, bend forward, stoop, fall down before one :—Hé lúteþ æfter he boweth after it, Salm. Kmbl. 806 ; Sal. 402. Leótt [hleát, Lind.] tó fótum his procidit ad pedes ejus, Mk. Skt. Rush, 5, 22. Hé árás and ðá tó eorþan leát he rose up, and then bowed to the ground, Guthl. 17 ; Gdwin. 74, 7. Hé leát tó ðæs cáseres eáre he bent down to the emperor's ear, Homl. Th. i. 376, 28. Ðæt heofonlíce wolcn leát wið his and hine genam the cloud from heaven stooped towards him, and received him, 296, 2. Hé forþ leát on his andwlitan procideret in faciem, Bd. 4, 3 ; S. 569, 11. Hé leát forþ ðæt him man áslóh ðæt heáfod of he bent forward so that his head was struck off, Ors. 6, 34 ; Bos. 130, 16. Hé leát forþ tó ðæm men ðe hine sleán mynte, Blickl. Homl. 223, 7. Gásta unclǽnra lutun tó him spiritus inmundi procidebant ei, Mk. Skt. Rush. 3, 11. Loð and Josue luton wið heora (the angels they saw), Homl. Th. i. 38. 21. Deáh heó onsíge and lúte tó ðære eorþan though she [the sun] sink and stoop to the earth, Bt. 25 ; Fox 88, 25. Forþ lúten wě procidamus, Ps. Surt.

94, 6. [R. Glouc. Chauc. Piers P. loute : Icel. lúta to bow down.] DER. á-, ge-, on-, under-lútan.

luðer-. v. lyðer-.

lutian ; p. ode To lie hid, be concealed, lurk, skulk, be latent : — Sum gedwyld lutaþ ðǽr aliquis latet error, Ælfc. Gr. 44 ; Som. 45, 46. Of ðam fýre ðe him on lutaþ from the fire that is latent in it, Lchdm. iii. 274, 4. Hú moniga dígla costunga ðæs ealdan feóndes lutigeaþ on ðýs andweardan life quanta in hujus vitae itinere tentamenta antiqui hostis lateant, Past. 21, 5 ; Swt. 159, 24. Ðú lutodest óþ ðis on ðam láðum cristendóme thou hast skulked until now in that detestable Christianity, Homl. Skt. 5, 413. Ða iermingas út of ðæm holan crupon ðe heó on lutedan the wretched creatures crept out of the holes that they had lurked in, Ors. 2, 8 ; Swt. 92, 30. Ða óðre ðe lutedon on ðære dígelnisse insidiæ, quæ latebant, Jos. 8, 19. Lutiaþ ðǽr þrý dagas ibi latitate tribus diebus, 2, 16. Eal ðæt gehýddes lutige omne, quod clausum latet, Past. 21, 3 ; Swt. 153, 15. Nys hyt swá stearc winter ðæt ic durre lutian æt hám for ege hláfordes mínes non est tam aspera hyems ut audeam latere domi prae timore domini mei, Coll. Monast. Th. 19, 17. Fěrde ðá lutigende geond heges and weges geond wudes and feldes swá ðæt hé [king Alfred] gesund becom tó Æþelingěge, Shrn. 16, 11. Dígelne leahter on menniscre heortan lutigende secret sin lurking in the human heart, Homl. Th. i. 496, 18. Cwæþ ðæt hě god wǽre on mannes hiwe lutiende said that he was a god concealed in the form of a man, ii. 474, 22. [Laym. Trev. Piers P. Chauc. lotie to lie hid : O. H. Ger. luzên latere, Grff. ii. 322.] Cf. lot, lytig.

lybb, es ; n. Medicine, drug, simple, in a bad sense poison ; the word often implies the use of witchcraft, see the compounds ; as Grimm says ' aus der bedeutung des erlaubten φάρμακον gieng hernach die des schädlichen, zauberhaften hervor,' D. M. 1103 :—Lyb obligamentum, Wrt. Voc. ii. 65, 31. Lybb, Ep. Gl. 17 b, 13. Ðæt biþ lyb wið eágena dimnesse that is a medicine for dimness of eyes, L. M. 1, 2 ; Lchdm. ii. 30, 14. Oxna lyb green or black hellebore, Lchdm. ii. 34, 28. Ðis ðě lib be cyrneles this may be a medicine for thee for churnel, iii. 62, 21. [O. L. Ger. lubbe ; dat. suco : cf. lubbian medicare : Icel. lyf ; f. also n. a herb, simple, esp. with the notion of healing, witchcraft, or supernatural power ; cf. ú-lyfjan poison ; lyfja to heal : O. H. Ger. luppi ; n. maleficium, succus lethiferus ; luppôn medicare : cf. Goth. lubja-leisei φαρμακεία.] v. cýs-lybb, un-lybbe, lybesn.

lybbestre, an ; f. A witch, sorcerer :—Lybbestran carios, Wrt. Voc. ii. 129, 12. v. lybb, lyb-lǽca ; and cf. O. H. Ger. luppari veneficus, maleficus.

lyb-corn, es ; n. ' A grain of purgative effect, especially the seeds of various euforbias, probably also of some of the gourds, as momordica elaterium, cucumis colocynthis,' Cockayne Lchdm. ii. 397, col 2 :—Libbcorn catharticum, Wrt. Voc. 67, 8. Libcorn lacyride, 67, 73 : tytymalosca, 68, 55. Lybcorn cartomo, ii. 14, 14 ; lattyride, 54, 23 : cartam, 103, 53 ; chartamo, 76 : catarticum, potus, 129, 43. Wyrc útyrnendne drænc genim fíf and hundeahtatig lybcorna make a purgative drink thus ; take eighty-five purgative seeds, Lchdm. iii. 18, 12 : 20, 1. Wyrc óðerne [spíwdrænc] of beóre and of feówertig lybcorna, 20, 10.

lyb-cræft, es ; m. Magic, witchcraft, skill in the use of lybb :—Hió him sealdon áttor drincan ðæt mid myclen lybcræfte wæs geblanden, Blickl. Homl. 229, 12.

lybesn, lyfesn, lybsen, e ; f. A charm, an amulet :—Lyb, lybsn obligamentum, Wrt. Voc. ii. 115, 23. Lyb, lyfesn, 63, 23. Lybsin lustramenta, 82, 10. Lyfesna filacteria, 36, 72 : 73, 16. Lybesne strenas, 121, 36. Swá swá hí ðæt sende wíte fram Gode scyppende þurh heora galdor oððe lifesne oððe óðre dígolnesse deófolcræftes bewerian mihte quasi missam a Deo conditore plagam per incantationes, vel fylacteria, vel alia dæmonicæ artus arcana cohibere valerent, Bd. 4, 27 ; S. 604, 9.

lyb-lác, es ; n. m. Sorcery, witchcraft, the art of using drugs or potions for the purpose of poisoning, or for magical purposes :—Ðis synt ða ídelnyssa ðisse worulde . . . lyblác . . . scíncræft hæ sunt vanitates hujus mundi . . . maleficium . . . ars magica [cf. Gal. 5, 20 where Gothic has lubjaleisei = φαρμακεία, A. V. witchcraft], L. Ecg. P. i. 8 ; Th. ii. 174, 34. Hér ys seó bót hú ðú meaht ðíne æceras bētan gif ðǽr hwilc ungedéfe þing on gedón biþ on drý oððe on lyblǽce, Lchdm. 1, 398, 3. Gif hí hwilc man niman wile oððe hyra æthríneþ ðonne forbærnaþ hí sona eall his líc ðæt syndon ungefrægelícu lyblác if any man wants to catch them [certain fowls] or touches them, then at once they consume all his body : those are most extraordinary cases of witchcraft, Nar. 34, 3. Wið ealra bealwa gehwylc ðara lybláca against every harm from sorceries, Lchdm. i. 402, 11. Wě cwǽdon be ðǽm wiccecræftum and be liblácum gif ðǽr man ácweald wǽre . . . we have ordained concerning witchcrafts and sorceries, if in such cases any one were killed . . ., L. Ath. i. 6 ; Th. i. 202, 10. Be liblácum. Ða ðe lyblác wyrcaþ sýn hí á fram ǽlcum Godes dǽle áworpene, búton hí tó rihtre dǽdbóte gecyrran, L. Edm. E. 6 ; Th. ii. 246, 13-16. Bebeorh ðě wið lyblácas and áttorcræftas cave tibi a maleficiis et veneficiis [cf. ne unrihtlyblácas ne ongynne wě, Wulfst. 253, 11, MS. D.], L. Ecg. C. prm ; Th. ii. 132 9. DER. unriht-lyblác. v. next word.

lyb-lǽca, an; *m. A sorcerer :*—Lyblǽcan *caragios* [*caragius* sortilegus, præstigiator qui characteribus magicis utitur, Ducange], Wrt. Voc. ii. 13, 53. v. fugel-hwata.

lyb-wyrhta. v. unlyb-wyrhta.

lýcþ, Bt. Met. 12, 55; Met. 12, 28. v. lúcan.

lýden. v. lǽden.

lýfan. v. lífan.

lyfesn. v. lybesn.

lyffetere, es; *m. A flatterer :*—Lyffetere *adulator*, Wrt. Voc. 85, 40. Liffetere, 49, 14. Đonne ádumbiaþ đa ýdelan lyffeteras *then shall the vain flatterers be dumb*, Homl. Th. ii. 570, 35. Faraþ tô đǽm lyffeterum đe eów ǽr leáslíce ôlǽhton *go to the flatterers that before fawned on you falsely*, 570, 23 : i. 494, 10.

lyffettan; *p.* te *To flatter, pay court to :*—Ic lyffytte *adulor*, Ælfc. Gr. 25; Som. 26, 63. Đa byrþeras đe hine tô byrgenne feredon synd ôlǽcunga lyffetyndra geférena *the bearers who carried him to the grave are the blandishments of flattering companions*, Homl. Th. i. 492, 28. Lyffetyndra tungan gewríđaþ manna sáwla on synnum *the tongues of flatterers bind the souls of men in sins*, 494, 6.

lyffetung, e; *f. Flattery, adulation, paying court to :*—Lyffetung *adulatio*, Wrt. Voc. 85, 41. Liffetung, 49, 15. Herige hine ná on đisum lífe, ac æfter his geendunge, đonne ne deraþ nán lyffetung đǽm herigendum, and nán upǽhefednys ne costnaþ đone geheredan, Homl. Th. ii. 560, 19. Ne hlyste gê heora geswǽsan lyffetunge, 404, 29. Heora nán ne gedyrstlǽce đæt heó Godes landâre woroldrícum sellen for lyffetunge *let none of them dare to give God's lands to the powerful of the earth as a means of paying court to them*, Lchdm. iii. 442, 32. Đæt mǽden ne mihte beón bepǽht þurh ǽnige lyffetunge fram hire leófan drihtne, Homl. Skt. 7, 86. Hwǽr beóþ đa líđan lyffetunga đe hine forlǽddon ǽror *where are the fair flatteries that formerly seduced him ?* Basil admn. 8; Norm. 50, 27. Đa smêđan lyffetunga, Homl. Th. ii. 572, 1. Lyffetungum befangen, i. 492, 32.

lýf-ness. v. leáf-ness.

lyft, es, e; *m. f. n. Air, atmosphere, breeze, sky, heavens, cloud :*—Lyft *aer*, Wrt. Voc. 52, 55. Lybt *sudum*, ii. 121, 66. Stemn is geslagen lyft . . . ǽlc stemn biþ geworden of đæs múþes clypunge and of đǽre lyfte cnyssunge; se múþ drýfþ út đa clypunge and seó lyft biþ geslagen mid đǽre clypunge, Ælfc. Gr. 1; Som. 2, 31–35. Ân đǽra [*the elements*] is eorþe, ôđer wæter, þridde lyft, feówrþe fýr, Bt. 33, 4; Fox 128, 30. Đeós lyft đe wê on libbaþ is án đǽra feówer gesceafta . . . Lyft is swýđe þynne, seó ofergǽþ ealne middangeard, and up ástíhþ forneán ôþ đone mônan, on đam fleóþ fugelas . . . Ne mihte heora nán fleón nǽre seó [đæt MS. R.] lyft đe hí byrþ. Ne nán man nǽþ nâne orþunge bûton þurh đa lyfte [đæt lyft MS. M.], Lchdm. iii. 272, 12–22. Seó lyft đonne heó ástyred is byþ wind, 274, 10. Se storm and seó stronge lyft *the storm and the strong blast*, Exon. 22 b; Th. 61, 28; Cri. 991. Seó hǽwene lyft *the azure air*, Cd. 166; Th. 207, 33; Exod. 476. Đeós lyft scínþ unwederlíce *rutilat triste cælum*, Mt. Kmbl. 16, 3. Lyft *nubes, aer*, Hpt. Gl. 493, 52. Seó lyft hí ofersceadewude and stefn com of đǽre lyfte *facta est nubis obumbrans eos et venit vox de nube*, Mk. Skt. 9, 7. Đǽre lyfte fugelas, Gen. 1, 28. Laguflôda gelâc lyfte and tungla *the movement of waters, of air and of stars*, Bt. Met. Fox 20, 346; Met. 20, 173. Under lyfte helm, Exon. 102 a; Th. 386, 19; Rä. 4, 64. Líxeþ lyftes mægen, 116 b; Th. 448, 16; Dôm. 55. On genipum lyftes *in nubibus aeris*, Ps. Spl. 17, 13. Se giem jacintus, se is lyfte onlícusđ on hiwe, Past. 14; Swt. 85, 5. Beorc byþ lyfte getenge *the birch towers to the sky*, Runic pm. Kmbl. 343, 2; Rún. 18. Hægl hwyrft of heofones lyfte, 341, 5; Rún. 9: Exon. 116 a; Th. 446, 10; Dôm. 20. Nân wolcn næs on đǽre lyfte gesewen *no cloud was seen in the sky*, Homl. Th. ii. 182, 35. Leólc on lyfte *sported in air*, Cd. 23; Th. 29, 10; Gen. 448. On lyfte cumende *venientem in nube*, Lk. Skt. 21, 27. On lofte heó stynt *it* [*the earth*] *rests in the air*, Hexam. 6; Norm. 10, 20. Under lyfte *sub divo*, Wrt. Voc. ii. 83, 34: Andr. Kmbl. 839; An. 420. Nalles æfter lyfte lácende hwearf *he went not sporting through the air*, Beo. Th. 5656; B. 2832. Hê gesette storm his on lyfte *statuit procellam ejus in auram*, Ps. Spl. 106, 29. Hê gesceóp đæt upplíce lyft, Hexam. 4; Norm. 6, 24. Đæt lyft hê gesceóp, Norm. 8, 17. Sôna swá hí [*snakes*] đæs landes [*Ireland*] lyft gestuncan, swá swulton hí, Bd. 1, 1; S. 474. 35. Đonne lígette and þunorráde eorþan and lyfte brêgdon, 4, 3; S. 569, 13. Swá oft swá hê lyft onstyrige, 569, 29. Hí fleóþ geond đás lyft, Homl. Th. ii. 90, 21: Elen. Kmbl. 1464; El. 734. On lyft ástáh *rose into the air*, 1796; El. 900. Đû þurh lyft lǽtest leódum tô freme mildne morgenrén *for the benefit of men thou dost let the gentle morning rain fall through the air*, Exon. 54 a; Th. 190, 30; Az. 81. Fugel under lyft ofer lagu lôcaþ georne, 57 a; Th. 204, 21; Ph. 101. Áhafen on đa heán lyft *raised aloft*, Cd. 69; Th. 84, 22; Gen. 1401. Hátwendne lyft *the torrid air*, 146; Th. 182, 12; Exod. 74. Đonne gê geseóþ đa lyfte cumende on westdǽle *cum videritis nubem orientem ab occasu*, Lk. Skt. 12, 54. Hibernia on smyltnysse lyfta is betere mycle đonne Breotone land *Hibernia serenitate aerum multum Brittaniæ præstat*, Bd. 1, 1; S. 474, 30. Geleht lyftum *moistened by the clouds*, Bt. Met. Fox 20, 195; Met. 20, 98. Lyftu

æthera, aera, Hpt. Gl. 457, 48. Geond lyftu *per aera*, Hymn. Surt. 66, 5. Đás lyfta and windas hê ástyraþ, Wulfst. 196, 6. [*Goth.* luftus; *m* : *O. Sax.* luft; *m. f* : *Icel.* lopt; *n* ; *O. H. Ger.* luft; *f. n.* : *Ger.* luft; *f.*]

lýft. v. lift.

lyft-ádl, es; *f. Palsy, paralysis :*—Mid đa ádle đe Grécas nemnaþ *paralysis*, wê cweþaþ lyftádl, Bd. 4, 31; S. 610, 17. Fram lyftádle gehǽled *a paralysi sanatus*, 610, 2. Wiđ lyftádle, L. M. 1, 59; Lchdm. ii. 130, 1.

lyft-edor, es; *m. An enclosure formed by clouds* [? v. lyft] :—Síđboda lyftedoras brǽc *the pillar of fire broke through the clouds*, Cd. 155; Th. 193, 24; Exod. 251.

lyften; *adj. Aërial, airy :*—Hwí is đæt tâcn on đǽre lyftenan heofonan gesewen *why is that sign* [*the rainbow*] *seen in the aërial heaven?* Boutr. Scrd. 21, 23. Hí sind genumene tô lyftenne heofenan ná tô rodorlícre *they* [*Enoch and Elijah*] *are taken to the aërial heaven, not to the etherial heaven*, Homl. Th. i. 308, 3. Lyftene gnættas *the gnats of the air*, Hexam. 17; Norm. 24, 30. [*O. H. Ger.* luftin *aëreus.*]

lyft-fæt, es; *n. An aërial vessel* [*the moon*], Exon. 108 a; Th. 411, 21; Rä. 30, 3.

lyft-fleógend, es; *m. That which flies in the air, a bird :*—Lyftfleógendra, Salm. Kmbl. 579; Sal. 289.

lyft-floga, an; *m. A flier in the air* [*a dragon*], Beo. Th. 4619; B. 2315.

lyft-gelác, es; *n. Motion in or of the air :*—þurh lyftgelác on land becwom [*he was borne through the air*], Andr. Kmbl. 1653; An. 828. þurh lyftgelác léges blǽstas weallas ymbwurpon [*the winds blew the flames*], 3102; An. 1554.

lyft-geswenced; *adj. Weather-beaten :*—Ceól lyftgeswenced on lande stôd, Beo. Th. 3830; B. 1913.

lyft-helm, es; *m. The air, atmosphere, cloud :*—Lyfthelm and laguflôd air [or *cloud?*] *and water*, Menol. Fox 553; Gn. C. 46. Wǽron land heora lyfthelme beþeaht *their lands were covered with cloud*, Cd. 145; Th. 181, 13; Exod. 60.

lyft-lácende *sporting* or *playing in the air, moving hither and thither in the air :*—Ic bidde đæt đú mê gecýđe hwæt đes þegn sý lyftlácende, Exon. 69 b; Th. 259, 12; Jul. 281. Forlǽt rêc ástígan lyftlácende, Elen. Kmbl. 1588; El. 796. Síđ tugon lyftlácende *took their way in flight through the air* [*of evil spirits*], Exon. 34 b; Th. 110, 31; Gú. 117. Hefonfugelas lyftlácende, Cd. 192; Th. 240, 17; Dan. 388.

lyft-sceaþa, an; *m. The robber of the air* [*the raven*], Exon. 87 b; Th. 329, 24; Vy. 39.

lyft-wundor, es; *n. A wonder of the air* [*the pillar that conducted the Israelites*], Cd. 146; Th. 183, 11; Exod. 90.

lyft-wynn, e; *f. The pleasantness of the air :*—Lyftwynne heóld *enjoyed himself* [*the dragon*] *by flying through the air*, Beo. Th. 6079; B. 3043.

Lyge, an; *f. The river Lea :*—Úre landgemǽra up on Temese and đonne up on Ligan [Ligean, 2nd text] and andlang Ligan [Ligean] ôþ hire ǽwylm *our* [*English and Danes*] *boundaries : up on the Thames, then up on the Lea, up to its source*, L. A. G. 1; Th. i. 152, 9. Đa Deniscan tugon hira scipu up on Temese, and đá up on Lygan, Chr. 895; Erl. 93, 35. Se foresprecena here worhte geweorc be Lygan, 896; Erl. 93, 35. Lygean, 913; Erl. 102, 2.

lyge, es; *m. A lie, lig* [*provincial*], *falsehood :*—Ic eów tô sôþe secgan wille and đæs in lífe lyge ne wyrþeþ *in truth I will tell you, and never shall it prove false*, Elen. Kmbl. 1147; El. 575. Sôþfæstnysse lyge and seó lufu liges and leásunge *odium veritatis amorque mendacii*, Bd. 1, 14; S. 482, 24. Liges fýr *mendacii ignis*, 3, 19; S. 548, 13. Bûta lyg *verumtamen*, Mt. Kmbl. Lind. 11, 24. Hí on lige lange feredon *de mendacio compellantur*, Ps. Th. 58, 12. Mengan lyge wiđ sôđe, Elen. Kmbl. 613; El. 307. Đú ǽr sægdest sôþlíce and nú on lyge cyrrest, 1329; El. 666. Đú ús gelǽrdæst þurh đínne lyge *thou didst persuade us through thy falsehood*, Cd. 214; Th. 268, 11; Sat. 53. Hwæđer him mon sôþ đe lyge sagaþ, Exon. 27 a; Th. 80, 16; Cri. 1307. Ic đê tô sôþe secgan wille, nelle ic lyge fremman, 67 b; Th. 250, 27; Jul. 133. Mân on môde, in múþe lyge, 80 b; Th. 302, 13; Fä. 35. [*Icel.* lygi; *f. a lie* : *O. H. Ger.* lugi; *f. mendum, falsum, figmentum, fabula : Ger.* lüge.]

lyge, lycce; *adj. Lying, mendacious, false :*—Sôhtun lyge gewitnisse wiđ đone hǽlend . . . đonne monige lyge [leáse ł lycce, Lind.] gewitu cwômun ætnǽhste đá cwôman twægen lyge [leáso ł liycce, Lind.] gewitu *quærebant falsum testimonium contra Jesum . . . cum multi falsi testes accessissent novissime autem venerunt duo falsi testes*, Mt. Kmbl. Rush. 26, 59–60. Monige lyge ł leáse wîtga *multi pseudoprophetæ*, 24, 11 : 24, 44. Behaldeþ eów wiđ lyge ł leáse wîtgu *attendite a falsis prophetis*, 7, 15. [*O. Sax. O. L. Ger.* luggi: *O. H. Ger.* luggi, lucki *mendax, falsus.*] *See also the compounds of which lyge is the first part.*

lyge *a plant name*, sicalia, Wrt. Voc. 68, 72.

Lygean-burh, *Lenborough, near Buckingham :*—Hér Cúþwulf genom Lygeanburg, Chr. 571; Erl. 18, 13. See Green's Making of England, pp. 118 sqq.

Lyge-tún, Lyg-tún *Leighton, in Bedfordshire :*—Đæt rád út wiđ Lygtúnes, Chr. 917; Erl. 102, 16. Æt Lygetúne, Cod. Dip. Kmbl. i. 196, 3.

lygen, e; *f. A lie, falsehood :*—Ðær lyt geháta biþ ðær biþ lyt lygena *where there are few promises, there are few lies,* Prov. Kmbl. 7. Mid lignenum *with lies,* Cd. 25 ; Th. 31, 36 ; Gen. 496 : 26 ; Th. 34, 2 ; Gen. 531 : 28 ; Th. 37, 11 ; Gen. 588. Lygenum, Th. 37, 31 ; Gen. 598. [O. Sax. lugina : O. H. Ger. lugina *mendacium.*]

lygen-word, es ; *n. A lying word, lie, falsehood :*—Mid ligenwordum, Cd. 33 ; Th. 43, 32 ; Gen. 699. Cf. lyge-word.

lyge-searu, wes ; *n. A false trick, artifice, wile, snare, lying art :*—Hý ligesearwum áhófun hearmstafas *with lying arts they stirred up mischiefs,* Exon. 35 b ; Th. 115, 34 ; Gú. 199 : Elen. Kmbl. 415 ; El. 208. Lygesearwum, Exon. 19 a ; Th. 48, 23 ; Cri. 776.

lyge-spell, es ; *n. A false speech :*—Mid ligespelle me[n]dosa mandata, Wrt. Voc. ii. 58, 32. [Cf. Icel. lygi-saga *a lying story, false report.*]

lyge-synnig ; *adj. Guilty of lying, false :*—Lygesynnig feónd, Elen. Kmbl. 1795 ; El. 899.

lyge-torn, es ; *n. Feigned anger* or *grief* [?] :—Ne biþ cwénlíc þeáw ðæt freoþuwebbe feores onsæce æfter ligetorne leófne mannan *it is no womanly fashion that a peaceweaver* [*woman*] *attack a loved man's life, having only a pretended cause for anger against him* [? *Thorpe reads* lígtorn *burning anger*], Beo. Th. 3890 ; B. 1943.

lyge-word, es ; *n. A lying word, lie, falsehood :*—Lygeword spǽcon *locuti sunt falsa,* Ps. Th. 57, 3 : Cd. 210 ; Th. 261, 3 ; Dan. 720. Ne wéne ǽnig ðæt ic lygewordum leóþ somnige, Exon. 63 b ; Th. 234, 28 ; Ph. 547. [*Icel.* lygi-orð.]

lyge-wyrhta, an ; *m. A liar, a forger of lies :*—Mid ðám ligewyrhtum *with the forgers of lies,* Fragm. Kmbl. 19 ; Leás. 11.

lyg-ness, e ; *f. Deceitfulness, falseness :*—Lygnisse weolan *fallacia divitiarum,* Mt. Kmbl. Rush. 13, 22.

lýgnian. v. lígnian.

lýhtan. v. líhtan.

lynd, e ; *f. Grease, fat, fatness :*—Lind *arvina,* Wrt. Voc. 65, 14. Lynde [a]rvina, 284, 6. Hé hí fédde mid fætre lynde hwǽte *cibavit eos ex adipe frumenti,* Ps. Th. 80, 15 : 147, 3. [O. H. Ger. lunda *arvina.*]

lyni-bór [v. Wrt. Voc. ii. 98, 7 boor *dasile*] *a gimlet, auger :*—Lynibór *terebellus,* Wrt. Voc. ii. 287, 14. v. next word.

lynis, es ; *m. An axletree :*—Spácan *radii :* felg *canti :* lynis *axedo :* eax *axis,* Wrt. Voc. 284, 47-51. Lynis axsedo : lynisas axsedones, ii. 7, 52, 51. [*Wm. of Shoreham* linses *axles :* cf. *O. H. Ger.* lun *obex : Du.* luns : *Ger.* lünse *a linch-pin : Dan.* lun-stikke *a linch-pin.* Linch-, lin-pin *is earlier spelt* lins-pin.]

lypen-wyrhta, an ; *m. A tanner, currier :* — Lypenwyrhta *byrseus,* Wrt. Voc. 288, 14. Leðerwyrhta oððe lypenwyrhta *byrseus,* ii. 11, 49.

lyre, es ; *m. Loss, damage, destruction, detriment :*—Lyre *jactura,* Wrt. Voc. 74, 51. Hýnþ *vel* lyre *vel* hearm *dispendium* vel *damnum* vel *detrimentum,* 47, 29. Hire lima lyre [*of a person paralysed*], Homl. Th. ii. 546, 31. 'Ic wille ofgán æt ðé his blód' ðæt is his lyre '*I will require at thy hands his blood,*' *that is, his destruction,* i. 6, 27. Lífes lyre *death,* Exon. 44 b ; Th. 151, 26 ; Gú. 801. Ne se enga deáþ, ne lífes lyre, 56 b ; Th. 201, 8 ; Ph. 53. Ne biþ ðær wædl ne lyre ne deáþes gryre, Dóm. L. 16, 265 ; Wulfst. 139, 32. Hé macode heora líf tó lyre *he destroyed them,* 106, 6. Hwílum forlidenesse ic þolie mid lyre ealra þinga mínra *aliquando naufragium patior, cum jactura omnium rerum mearum,* Coll. Monast. Th. 27, 1. On lyre *in perditione,* Ps. Lamb. 87, 12. Lyre *jacturam, damnum,* Hpt. Gl. 480, 43. Ná beóþ ða eádige ðe for hýnþum oððe lirum hwílwendlícra hyðða heófiaþ, Homl. Th. i. 550, 28. DER. feorh-, land-, líf-lyre ; *and see* lor.

lýsan, lýsing. v. lísan, lísing.

lyssen. v. lyswen.

LYSTAN ; *p. te To* LIST, *cause pleasure* or *desire* [*with dat.* or *acc. of person in whom the feeling is caused, and gen. of the thing,* or *infin.*]:—Mé ne lyst *piget,* Ælfc. Gr. 33 ; Som. 37, 23. Mé lyst rǽdan *lecturio,* 34 ; Som. 37, 56. Hine ne lyst his willan wyrcean, Blickl. Homl. 51, 16. Hú ne biþ ǽlc mon genóg earm ðæs ðe hé næfþ ðonne hit hine lyst habban *is not every man poor enough as regards that which he has not, when he desires to have it?* Bt. 26, 1 ; Fox 92, 2. Ne him nǽfre genóg ne þincþ ǽr hé hæbbe eall ðæt hine lyst, 33, 2 ; Fox 124, 7. Wel mé lícode ðæt ðú ǽr sǽdest and ðises mé lyst nú get bet *I liked well what you said before, and am still better pleased with this,* 35, 4 ; Fox 162, 3 ; 34, 6 ; Fox 142, 12. Ðam men ðe hine ne lyst his metes *for the man who has no appetite for his food,* L. M. 1, 19 ; Lchdm. ii. 62, 15. Ðonne hine ǽtes lysteþ, Exon. 97 a ; Th. 363, 12 ; Wal. 52 : Bt. Met. Fox 10, 27 ; Met. 10, 14. Se leahtor déþ ðæt ðam men ne lyst nán þing tó góde gedón *that sin causes a man to have no desire to do anything to good purpose,* Homl. Th. ii. 220, 22. Him lyste ðær on dígolnysse his gebedu begangan, Bd. 3, 16 ; S. 542, 33. Hine lyste mid him etan and drincan *ipse delectaretur manducare et bibere cum eis,* 5, 5 ; S. 618, 16 : Beo. Th. 3591 ; B. 1793. Hine nánes þinges ne lyste on disse worulde *he cared for nothing in this world,* Bt. 35, 6 ; Fox 168, 12 : Bt. Met. Fox 26, 142 ; Met. 26, 71. Se gesceádwíslíca willa ðæt hine ðara twega lyste *the rational will which delights in them both,* Bt. 14, 2 ; Fox 44, 26 : Bt. Met. Fox 10, 2 ; Met. 10, 1. Hé sceal syllan

his gód on ða tíd ðe hine sylfne sélest lyste his brúcan, Blickl. Homl. 101, 20. [Cf. *Goth.* lustón (*with gen.*) *to desire : O. Sax.* lustean (*acc. of pers., gen. of thing*) : *Icel.* lysta (*acc. of pers.*) : *O. H. Ger.* lustjan (*acc. of pers., gen. of thing,* or *infin.*); cf. *also* lustón *to desire : Ger.* lüsten (*impers.*)] DER. ge-, of-lystan.

lystere (=? hlystere) :—Lysteres *fautoris,* Hpt. Gl. 5714, 40.

lysu ; *adj. Depraved, corrupt, evil, dishonourable, shameful, profligate :* — Lyswe lársmeoþas *corrupt counsellors,* Andr. Kmbl. 2441 ; An. 1222. Cf. lyswen.

lysu, wes ; *n. What is depraved* [v. preceding word] :—Gif cyning æt mannes hám drincæþ and ðær man lyswæs hwæt gedó ii bóte gebéte *if the king be entertained at a man's house, and any evil be done there, let a double fine be paid,* L. Ethb. 3 ; Th. i. 4, 2. Gif frí wíf leswæs hwæt gedéþ xxx scill. gebéte, 73 ; Th. i. 20, 7. Lot sceal mid lyswe, list mid gedéfum [v. list], Exon. 92 a ; Th. 345, 16 ; Gn. Ex. 189.

lyswen, lyssen ; *adj. Full of matter, corrupt, purulent; depraved* [?] :—Ðonne se swile tóbyrst ðonne biþ seó micge lyswen swilce worms, L. M. 2, 17 ; Lchdm. ii. 198, 26. [In Ps. Th. 52, 6 the word *lisne* occurs ; can this be the adverb from this adjective, taken in the sense given to *lysu :*—Manna bán mihtig Drihten lisne tósceádeþ *scatters with shame* or *dishonour* ?] v. lysu *and next word.*

lyswen, lyssen *matter, purulence :*—On ðære þrotan biþ swyle and lyssen, L. M. 1, 4 ; Lchdm. ii. 46, 14.

lyt ; *indecl. used as subst. adj. and adv. Few, little :*—Ðæra is nú tó lyt ðe wile wel tǽcan *there are now too few of those that will teach well,* Homl. Th. i. 6, 22. Ðæra biþ ealles tó lyt, ðe hé ne beswíce, Wulfst. 97, 7. Is swíðe lyt monna ðæt ne sý mid ðæm sumum besmiten *there are very few men that are not defiled with some of them,* L. E. I. 31 ; Th. ii. 428, 4. Wóp wæs wíde, worulddreáma lyt, Cd. 144 ; Th. 180, 9 ; Exod. 42. Ðé eádes tó lyt þuhte, Exon. 28 a ; Th. 86, 1 ; Cri. 1401. Wergendra tó lyt þrong ymbe þeóden, Beo. Th. 5758 ; B. 2882. Ðæt lyt manna þáh *it succeeded with few,* 5665 ; B. 2836. Hé on folce lyt freónda hæfde, Cd. 124 ; Th. 158, 32 ; Gen. 2626. Cyning hæfde wígena tó lyt, Elen. Kmbl. 126 ; El. 63. Hé mid lyt wordum ac geleáffullum his hǽle begeat *he obtained his salvation with words few but full of faith,* Dóm. L. 6, 61. Ne sceal hé tó lyt þancian heora ælmessan *he shall not be too sparing of thanks for their alms,* Blickl. Homl. 43, 13. Forðon hé lyt genihtsumede on smeáwunge and on leornunge háligra gewrita hé ðý má mid his handum wonn and worhte *nam quo minus sufficiebat meditationi scripturarum, eo amplius operi manuum studium impendebat,* Bd. 4, 3 ; S. 567, 29. Hé lyt ongeat ðæt him swá earme gelamp. Cd. 76 ; Th. 94, 32 ; Gen. 1566. Ðæt eów swá lyt gespeów, Andr. Kmbl. 2688 ; An. 1346. [*O. Sax.* lut (werodes).]

lyteg. v. lytig.

LYTEL ; *adj.* LITTLE :—Nú gyt is án lytel fyrst *adhuc modicum,* Jn. Skt. 14, 19. Hwæt is ðæt líf elles búton lytelu ylding ðæs deáþes, Blickl. Homl. 59, 27. Lytulu sprǽc, Exon. 116 a ; Th. 445, 16 ; Dóm. 8. Se lytla finger, L. Alf. pol. 60 ; Th. i. 96, 7. Lá lytle heord *pusillus grex,* Lk. Skt. 12, 32. On swá lytlum fæce *in such a little space,* Elen. Kmbl. 1917 ; El. 960. Ælfréd cyning gefeaht wið alne ðone here lytle werede, Chr. 871 ; Erl. 76, 5. Lytle læs *paulo minus,* Ps. Th. 118, 87. Lytle ǽr, Elen. 1325 ; El. 664. Lytle lengre ðonne seofon fóta, Lchdm. iii. 220, 4. Lytle mǽre ðonne feówer, 220, 12. Ðæt lytle ðæt hé erede, hé erede mid horsan, Ors. 1, 1 ; Swt. 18, 15. Lytle hwíle sceolde hé his lífes niótan, Cd. 24 ; Th. 31, 16 ; Gen. 486. Se lícette litlum and miclum, gumena gehwylcum, Bt. Met. Fox 26, 72 ; Met. 26, 36. On ǽlcum þingum ðe ðær unbecweden biþ, on bócum and on swylcum lytlum, Chart. Th. 536. 26. On swíðe lytlon hiera hæfþ seó gecynd genóg *paucis minimisque natura contenta est,* Bt. 14, 1 ; Fox 42, 10. Ða lytlan *parvulos,* Ps. Th. 114, 6. Lytlum *by little, by degrees, in little pieces, a little at a time :*—Lytlum *paulatim,* Ælfc. Gr. 38 ; Som. 40, 30. Tóbrec hig lytlum *divides eos minutatim,* Lev. 2, 6. Sele ðæt lytlum súpan, L. M. 2, 52 ; Lchdm. ii. 270, 1. Hé gewýt swá lytlum and lytlum fram Gode *so little by little he departs from God,* Ælfc. Gr. pref ; Som. 1, 35 : Past. 39, 1 ; Swt. 283, 9. Ic geseah weaxende blósman litlum and litlum *videbam crescere paulatim in gemmas,* Gen. 40, 10. [*Goth.* leitils ; *O. Sax.* luttil : *Icel.* lítill : *O. H. Ger.* luzil, luzzil.]

lytel ; *neut. of adj. used as subst. or adv. A little :*—Dó lytel sealtes tó *put a small quantity of salt to it,* Herb. 2, 19 ; Lchdm. i. 86, 7. Hwerhwette niþewearde án lytel *the lower part of cucumber, a little,* L. M. 3, 41 ; Lchdm. ii. 336, 4. Mycel *multum,* lytel *parum,* Ælfc. Gr. 38 ; Som. 40, 34. Ymbe lytel *post pusillum,* Mk. Skt. 14, 70. Ymbe án lytel gé mé ne geseóþ and eft ymbe lytel gé mé geseóþ *modicum non videbitis me et iterum modicum et videbitis me,* Jn. Skt. 16, 16.

lytel-fóta ; *adj. Having small feet :*—Litelfóta *petilus,* Ælfc. Gl. 76 ; Som. 71, 132 ; Wrt. Voc. 45, 35.

lytel-hygdig-, hýdig ; *adj. Small-minded, pusillanimous :* — Mon ðæs lytelhýdig ne ðæs læthýdig *no man of mind so small and so sluggish,* Exon. 78 b ; Th. 294, 4 ; Crä. 10.

lyte-líc. v. lytig-líc.

lytel-mód ; *adj. Of little courage, faint-hearted, pusillanimous :*—Se

mec hâlne dyde from lytelmôduṁ *qui me salvum faceret a pusillanimo*, Ps. Surt. 54, 9. Đa lytelmôdan and đa unþrîstan đonne hié ongietaþ hiera unbældo and hiera unmiehte hié weorþaþ oft ormôde *pusillanimes dum nimis infirmitatis suæ sunt conscii, plerumque in desperationem cadunt*, Past. 32, 1; Swt. 209, 7.

lytel-ne; *adv.* All but, almost, nearly :— Hê lytelne [lytesne?] Breotona rîce forlêt *Brittaniam pene amisit*, Bd. 1, 3; S. 475, 22.

lytel-ness, e; *f.* Littleness —Sume [*adverbs*] syndon *quantitatis ;* đa getâcniaþ mycelnysse ođđe lytelnysse, Ælfc. Gr. 38; Som. 40, 34.

lytes-nâ, lytes-ne, lytest-ne; *adv. Almost, nearly, within a little :*— Lytesnâ *concedam*, Wrt. Voc. ii. 104, 49. Lytisnâ, 14, 65: Ep. Gl. 7 d, 31. Wæs his rîce brâd wîd ofer werþeóde lytesnâ ofer ealne yrmenne grund *his realm was broad, wide over mankind, almost over all the world*, Exon. 66 a; Th. 243, 13; Jul. 10. Lytestne eall his weorod ofslegen wæs *omnis pene ejus est cæsus exercitus*, Bd. 1, 34; S. 499, 32 : 3, 24; S. 556, 30. Lytesne [*pene*] of ealre Lindesse stôwum, 1, 11; S. 535, 25. Lytesne of eallum *de cunctis prope*, 3, 14; S. 540, 11. Bôc lytestne unâberendlîcre byrþenne *codicem ponderis pene importabilis*, 5, 13; S. 633, 6.

lyđer-, luđer- full; *adj.* Base, vile, dissolute, depraved :— Leófan men ne beón gê nâđor ne leáse ne luđer- [lyđer- MS. B] fulle, ne fûle ne fracode, ne on ænige wîsan tô lehterfulle, Wulfst. 40, 5.

lyđer-lîc; *adj. Sordid, mean, vile :* — Se cyning self mid swîđe lyđerlîcum gegierelan *ipse imperator sordida servilique tunica discinctus*, Ors. 4, 5; Swt. 166, 16. [The word comes to mean *lazy* in later times. Cf. Tusser 'some *litherly* lubber leaveth undone that another will do.']

lyđer-lîce; *adv. Wickedly, vilely :*—Luđerlîce *pessime*, Ælfc. Gl. 99; Som. 76, 101; Wrt. Voc. 54, 45. [Leiden swa *luđerliche* on hire lichđæt hit brec oueral, Marh. 5, 21: A. R. 290, 8. A clerk hath *litherly* byset his while Bot if he cowde a carpenter byggle, Chauc. Miller's Tale, 113.]

lyđre; *adj. Evil, wicked, base, mean, poor, sordid, vile, lewd, depraved :* — Đæt Godes feoh ne ætlicge and hê beó lyđre þeówa gehâten *that God's money be not idle, and he be called a wicked servant*, Ælfc. Gr. pref; Som. 1, 30. Lytel is se fyrst đyses lîfes and lyđre is *few and evil are the days of this life*, Wulfst. 109, 2. Hû læne and hû lyđre đis lîf is on tô getrûwianne, 189, 3. Eálâ đû lyđra þeówa *serve nequam*, Mt. Kmbl. 18, 32 : Lk. Skt. 19, 22 : Homl. Th. ii. 552, 6. Ic eom se lytla for đê and se lyđra man, se syngige swîđe genehhe, Hy. 3, 41; Hy. Grn. ii. 282, 41. Eówre lyđre môd *incircumcisa mens*, Lev. 26, 41. Gif hwylc wîf for hwylcum lyđrum andan hire wîfman swingþ *si mulier aliqua, ex prava aliqua invidia, ancillam suam flagellis verberaverit*, L. Ecg. P. ii. 4; Th. ii. 182, 32 : L. M. I. P. 12; Th. ii. 268, 11. Se đe Crist belæwde for lyđrum sceatte *who betrayed Christ for filthy lucre*, Homl. Th. ii. 244, 26 : Wulfst. 297, 26. Đæs mæran wîtegan deáþ đære lyđran hoppestran [*the daughter of Herodias*] tô mêde forgeaf, Homl. Th. i. 484, 3. Lyđerne earhscype *base cowardice*, Wulfst. 53, 12. Þurh lyđre wrhþe, 166, 26. Đa seofon hlyđran ear *septem spicæ tenues*, Gen. 41, 27. Ođre lyđre cynn *cetera adulterina genera*, Ælfc. Gl. 101; Som. 77, 31 ; Wrt. Voc. 55, 36. Lyđra bearn *filii excussorum*, Ps. Th. 126, 5. Se Hælend geþafode lyđrum mannum đæt hî hine ofslôgon, Homl. Th. i. 168, 6. Se ealdorman hî betæhte liđrum mannum tô behealdenne *the aldorman entrusted it to base* [cf. đa wæron yfele and earge l. 27] *men to hold*, Ors. 6, 36; Bos. 131, 23. Eár lyđre and forscruncene *spicæ tenues et percussæ uredine*, Gen. 41, 6. Þurh lîchaman leđre geþohtas *through the wicked thoughts of the body*, Ps. C. 50, 41; Ps. Grn. ii. 277, 41. [A. R. Laym. luđer : Piers P. luþer, liþer : Prompt. Parv. lyder or wyly *cautus* [see note for *lither* = lazy in later English] : cf. Ger. lüder-, lieder-lich.]

lyđre; *adv. Badly, vilely :*—Habbaþ wê alle for đînum leásungum lyđre gefêred *we have all fared miserably for thy falsehoods*, Cd. 214; Th. 268, 29 ; Sat. 62.

lyt-hwôn; *subst.* and *adv.* A little [*space, time, quantity*] : — Meng lythwôn wið hunig *mix a little with honey*, L. M. 1, 1; Lchdm. ii. 22, 20. Lythwôn becom cwicera tô cýđđe *few living reached their country*, Judth. 12 ; Thw. 26, 5; Jud. 311 : Elen. Kmbl. 284 ; El. 142. Đâ hê wæs lythwôn đanon âgân *progressus pusillum*, Mt. Kmbl. 26, 39 : Mk. Skt. 14, 35. Hê his eágan lythwôn fram đære eorþan up âhôf, Glostr. Frag. 104, 13. Đara đe lythwôn rêccaþ embe bôca beboda, L. I. P. 6; Th. ii. 310, 34 : Swt. A. S. Rdr. 101, 200 : Beo. Th. 408; B. 203. Ne lythwôn *not a little*, Exon. 38 a ; Th. 125, 32 ; Gû. 363. Đâ geswîgode heó lythwôn *parumper reticuit*, Bd. 4, 9; S. 577, 22. v. lyt.

lytig, lyteg ; *adj. Cunning, astute, sly, artful, crafty, wily :*—Litig *procax*, Wrt. Voc. ii. 67, 48. Se lytega sætere *seductor callidus*, Past. 65, 2; Swt. 463, 11. Hû manega costunga đæs lytegan feóndes *quanta hostis callidi tentamenta*, 21, 5 ; Swt. 161, 18. Forđæm him [*a simple person*] is micle iedre tô gestiganne on đone ryhtan wîsdom, đonne đæm lytegan sîe tô anbûganne, for đæm đe hê biþ ær upâhæfen for his lotwrencium, 30, 1 ; Swt. 203, 18. Marius đone consul â swâ lytigne swâ hê wæs *Marii consulis, qui non minore pene quam ipse præditus erat astutia*, Ors. 5, 7; Swt. 228, 32. Đone leásan lytegan đû scealt hâtan fox *insidiator occultis surripuisse fraudibus gaudet? vulpeculis exaequetur*, Bt. 37, 4 ; Fox 192, 17. On leásungum lytige *in mendaciis vafri*, Coll. Monast. Th. 32, 29. Đa lytegan *sapientes hujus seculi*, Past. 30, 1 ; Swt. 203, 6, 24 : 205, 3.

lytigian; *p.* ode To act cunningly :—Ongunnon lytegian đâ lâđe gystas *began then to act guilefully the hateful guests*, Byrht. Th. 134, 18 ; By. 86. v. be-lytigian.

lytig-, lyte-lîc ; *adj. Deceitful, false :*—Ymbtrymedu mid lytelîcre lâdunge *fallaci defensione circumdatæ*, Past. 35, 5 ; Swt. 245, 8.

lytig-, lyte-lîce ; *adv. Cunningly, artfully, craftily :*—Đe hit symle lytiglîce lâdaþ *sese callide defendentis*, Past. 35, 3 ; Swt. 241, 8. Litelîce *callide*, Ex. 32, 12. Đa woruldsælþa mid swîđe manigre swêtnesse swîđe lytelîce ôleccaþ đæm môdum, Bt. 7, 1; Fox 16, 10. Hû lytelîce hý đonne deófol bepæhte, Wulfst. 11, 9, 16. Ne weorþeþ on worulde lytelîce swicolra đonne hê wyrþeþ *none in the world is more craftily deceitful than he*, 54, 22. Se đe litelîcost cûđe leáslîce hiwian unsôþ tô sôþe *he that most cunningly could make untruth appear truth*, 128, 9.

lytig-ness, e ; *f. Cunningness, craftiness, astuteness :* — Đære nædran lytignes *astutia serpentis*, Past. 35, 1 ; Swt. 237, 22.

lytlian; *p.* ode To make or to become little, to lessen, diminish :—Gidæfnaþ đæt ih lytlige *oportet me minui*, Jn. Skt. Rush. 3, 30. Đonne lytlaþ him se tôhopa đe hê hæfde đâ hê synful wæs *spem, quæ esse potuit de peccatore, subtraxit*, Past. 58, 10; Swt. 447, 14. Heorte sceal đe cênre môd đê mâre đe ûre mægen lytlaþ *heart shall the hardier be, courage the more, the fewer our forces*, Byrht. Th. 140, 65. Lytlaþ đæt his anweald and êcþ his ermþa *it lessens his power, and increases his miseries*, Bt. 29, 1 ; Fox 102, 19. Drenc đe lytlaþ đa yfelan wætan, L. M. 2, 59; Lchdm. ii. 282, 10. Đonne lyttlaþ hê đæt fæsten *tunc breviabit jejunium*, L. Ecg. P. Add. 19; Th. ii. 234, 18. Cristes lage wanedon and cyninges lage lytledon *Christ's laws waned, and the king's laws were weakened*, L. Eth. ix. 37; Th. i. 348, 19. Lytligen đa grambæran hiera gedrêfednesse *damnent iracundi perturbationem*, Past. 40, 2 ; Swt. 291, 2. Willfiôd ongan lytligan, Cd. 71 ; Th. 85, 11 ; Gen. 1413. Hý mon sceal lytlian *they shall be lessened*, L. M. 2, 1 ; Lchdm. ii. 178, 12. Se đe hit þence tô litlianne, gelitlige hine God elnihtig hêr on worulde, Cod. Dipl. Kmbl. iv. 171, 21. Biþ se ece litliende [litligende, MS. B], Herb. 3, 3, 4 ; Lchdm. i. 88, 2, 7.

lytling, es ; *m.* A little one, a young person, child :—Se đe underfêhþ ænne lytling on mînum naman *he that receives one little one in my name*, Homl. Th. ii. 286, 30. Lyttlingas, i. 512, 21. Furþon litlincgas nellaþ forbîgean mê *nec parvuli nolunt præterire me* [*the baker*], Coll. Monast. Th. 29, 1. Đa litlingas fuhton on hire innoþe, Gen. 25, 22. Ænue of đyssum lytlingum *unum de pusillis istis*, Mt. Kmbl. 18, 6 : Homl. Th. i. 84, 11. His efenealdan lytlingas [*the children killed in Bethlehem*], 88, 12. Ic hæbbe hnesce litlingas *parvulos habeam teneros*, Gen. 34, 13 : 50, 21. Gif hwylc gôdra wile his lytlingas hiom [*priests*] tô lâre befæstan, hig sceolon swîđe lustlîce hig onfón, and him tæcan, L. E. I. 20; Th. ii. 414, 8.

lytluc[c], es ; *m.* A bittock, small piece :—Lytluccas (MS. lyttuccas) *segmenta, particulas*, Germ. 400, 531.

lytlum. v lytel.

M

Original *m*, generally speaking, is preserved in Anglo-Saxon, and is found corresponding to *m* in the Gothic and other cognate dialects, e. g. *mê, manna, dôm* ; Goth. *mik, manna, dôms*. When, however, *m* is not initial, the correspondence is not always maintained ; thus, A.S. *fîf*, but Goth. *fimf*; A.S. *sôfte*, O.H.Ger. *samfto*. Also for earlier *fn* is found *mn*, as in *emn* along with *efn*, Goth. *ibn* ; *stemn* and *stefn*, Goth. *stibna*. In some inflexions *m* is no longer found ; so in the 1st pers. sing. pres. indic. *eom* is the only instance in which the old person-ending has maintained itself; though *beón*, *dôn*, and *gân* offer occasional instances of its retention in the Northern Gospels; while the *m* which is found in the plural of the Gothic and O. H. Ger. conjugations has left no trace. In declensions *n* in the later times began to take the place of *m* in the dative, so *đan* for *đam*.

The form of the Runic letter, whose name was *man*, was ᛗ, but from the similarity to the *d*-rune (*đæg*) ᛞ, the two seem to be sometimes confounded. In each case the symbol was sometimes employed, after the runes had been generally supplanted by the Latin letters, to express the word which was its name ; thus in the Durham Ritual *quis* is glossed *ænig* ᛗ, *nemo, ne ænig* ᛗ : the same symbol being also used to gloss *dies*. The form of the rune accompanying the Runic poem is ᛗ, Kmbl. plate 16, fig. 11, and the verse attached to it the following :—

Man byþ on myrgþe	Men will be cheerful!,
his mâgan leóf	dear to their friends,
sceal đeáh ânra gehwylc	shall yet each one
ôđrum swîcan	depart from other,
forđam dryhten wile	for the Lord will
dôme sînum	by his doom
đæt earme flæsc	the 'vile body'
eorþan betæcan.	commit to earth.
	Kmbl. 343, 11-18.

má; *indecl. cpve. used as subst. and adj. More.* I. *as subst.*:— Sume naman sind *omonima; ða getácniaþ má þinga mid ánre clypunge,* Ælfc. Gr. 5; Som. 4, 13. Seó þridde declinatio hefþ eahta and hundseofontig geendunga oððe má, 9; Som. 8, 15: Elen. Kmbl. 1264; El. 634. Hé hæfþ weána má ðonne æniges mannes gemet sý ðæt hié áríman mæge, Blickl. Homl. 61, 36: 213, 28. Æghwylcum men biþ leófre swá hé hæbbe holdra freónda má, 123, 1. Mid ðý eówer má is *cum sitis numero plures,* Bd. 2, 2; S. 503, 13. Ne gehérde ða ondsware má manna ðonne ða míne getreówestan freónd, Nar. 32, 15. Má ðæra Iudéiscra ealdra embe Cristes cwale smeádon, Homl. Th. i. 88, 28. Næfde hé má ðonne twentig swýna, Ors. 1, 1; Swt. 18, 14. Nó ðé láðes má gedón môton *no worse may they do thee,* Andr. Kmbl. 2885; An. 1446. Ða habbaþ twegen mislíce casus and ná má on gewunan . . . nis ðér ná má mislíce casa *they have two different cases, and no more generally . . . there are no more different cases,* Ælfc. Gr. 14; Som. 17, 3–7: 15; Som. 17, 38: Blickl. Homl. 35, 24. Donatus tép gyt má tô ðysum . Gyt synd má ðyssera æfter Priscianus, Ælfc. Gr. 44; Som. 46, 6–10. Gyt má wæs ðe ðæt dôn ne wolde *there were yet more who would not do that,* Bd. 1, 14; S. 482, 17. Swá ðær má beáh tô ðam sôðan geleáfan, Homl. Th. ii. 540, 27. Ða geneáléhton má hine meldigende, 248, 32. Nabbaþ syððan hwæt hig má dôn *non habent amplius quod faciant,* Lk. Skt. 12, 4. Hwæt sceal ic ðonne má secgean fram Sancte Johanne, Blickl. Homl. 169, 24: Bd. 3, 27; S. 559, 22: Ps. Th. 125, 2. Gif hé má wille, drince hé hát wæter, L. M. 2, 59; Lchdm. ii. 284, 5. Be ðam man mæg gecnáwan and be má þinga, Wulfst. 5, 4. Swá má læs worda, swá mid má, Bt. 35, 5; Fox 166, 12. Hé ne úðe ðæt ænig ôðer man æfre mærða ðon má gehédde ðonne hé sylfa *he would not allow that any other man should have any more distinctions than he himself had,* Beo. Th. 1012; B. 504. Wát ic sorga ðý má, Cd. 42; Th. 54, 33; Gen. 886. Mæ wundra *plura signa,* Jn. Skt. Lind. 7. 31. II. *as adj.*:— Seó sáwul ys má ðonne se líchama and se líchama má ðonne ðæt reáf *anima plus est quam esca, et corpus quam vestimentum,* Lk. Skt. 12, 23. Má wén is ðæt ðú onsende ðínne engel *there is more hope if you send your angel,* Blickl. Homl. 231, 23. Má wæter of ðínum mûþ ðú ne send, 247, 7. Ic nelle nán word má of ðínum mûþe gehýran, Nar. 45, 23. Ic wæs sixtýne síðum on sæbáte . . . is þys áne má *I have been sixteen times in a sea boat...this is once more,* Andr. Kmbl. 984; An. 492. Ðæt wæs má cræft ðonne hit eorþbúend ealle cúþan [cf. use of *mikil* in O. Sax. kûðean kraft mikil], Exon. 13b; Th. 26, 24; Cri. 421. Ne synd ná má namanspeligende bútan ðás fífténe *there are no more pronouns than these fifteen,* Ælfc. Gr. 15; Som. 17, 46. v. *next word, and* mæst.

má, mæ; *adv. More, rather, further*:— Mæ *amplius,* Ps. Surt. 50, 4. Gáþ má tô ðam sceápum *potius ite ad oves,* Mt. Kmbl. 10, 6: 28. Ælces monnes æþelo bióþ má on ðam môde ðonne on ðam flæsce, Bt. 30, 1; Fox, 110, 2: Past. 17, 9; Swt. 121, 22. Nis him blôd tô lætanne ac má hira man sceal tilian mid wyrtdrencum *he is not to be let blood, but rather the symptoms are to be treated with drinks made from herbs,* L. M. 1, 35; Lchdm. ii. 82, 16. Hé ðone ná eft ne wyrge, ac hine má bletsige, L. E. I. 21; Th. ii. 416, 12. Forðon ðe Godes willa is ðæt tô Columban mynstre hé má fære and lære *Dei enim voluntatis est ut ad Columbæ monasteria magis pergat docenda,* Bd. 5, 9; S. 622, 39. Hé má geceás ðæt hé wæs eft hám hweorfende *he preferred to return home,* 5, 2; S. 615, 33. Him wíslícre and gehyldre wære ðæt hí má hám cyrdan ðonne hí ða eallreordan þeóde gesécan sceoldan, 1, 23; S. 485, 32. Ðæt hié má mehten heora weras wrecan *that they might better avenge their husbands,* Ors. 1, 10; Swt. 46, 4. Gyt má oððe gyt swíðor *immo,* Ælfc. Gr. 38; Som. 42, 18: Bt. 32, 1; Fox, 114, 17. Ne ðonne má *nor further,* 16, 3; Fox, 54, 29. Ongunnon hí Moyses má bysmrian, Ps. Th. 105, 14. Se má eallum Angelcyningum Brytta þeóde fornom *qui plus omnibus Anglorum primatibus gentem vastavit Brittonum,* Bd. 1, 34; S. 499, 19. Wénestû recce hé hire æfre má *numquid revertetur ad eam ultra,* Past. 52, 3; Swt. 405, 12: Cd. 216; Th. 273, 21; Sat. 140. Ne synga ðú næfre má, Jn. Skt. 8, 11. Ðæt ðú má ne síe mínra gylta gemyndig, Elen. Kmbl. 1630; El. 817. Má of heora mûþe hit ne eode *it (water) no longer came out of its mouth,* Blickl. Homl. 247, 9. Sægdon ðæt hí nó má ne mihton swencte beón *they said that they could not be troubled any more,* Bd. 1, 27; S. 481, 3. Ðam mycle má hé scrýt eów *quanto magis vos vestit,* Mt. Kmbl. 6, 30. Mycle má, 7, 11. Swá mycele má, Lk. Skt. 12, 28. Hwæt is ðæt ðé má ðæt ænig man mæge ôðrum dôn ðæt hé ne mæge him dôn ðæt híe *quid autem est, quod in alium facere quisquam potest, quod sustinere ab alio ipse non potest,* Bt. 16, 2; Fox 52, 27. Ða clypodon hig ðæs ðé má [*so much the more,* cf. O. H. Ger. des diu mêr: Ger. desto mehr], Mt. Kmbl. 20, 31: Mk. Skt. 6, 51: 10, 48. Hit ðær ne weaxt ðé má ðe gimmas weaxaþ on wíngeardum *it does not grow there any more than jewels grow in vineyards,* Bt. 32, 3; Fox 118, 10: 34, 1: Fox 134, 15. Ðæra máðma ne róhte ðé má ðe reócendes meoxes, Homl. Skt. 7, 20: L. Edg. C. 7; Th. ii. 280, 6. Gelpan ne þorfte Costontinus an Anláf ðý má *no need had Constantine to boast, no more had Anlaf,* Chr. 937; Erl. 114, 12; Æðelst. 46. Næs him se swég tô sorge ðon má ðe sunnan scíma *the noise (of the flames) was not troublesome to them any more*

than sunshine, Cd. 187; Th. 232, 23; Dan. 264. Hié ðæs ne onmunden ðon má ðe eówre geféran, Chr. 755; Erl. 50, 25. Ða ne wolde se pápa ðæt geþafigean ne ða burhware ðon má *then the pope would not permit it, no more would the citizens;* et si pontifex concedere illi quod petierat voluit, non tamen cives potuere permittere, Bd. 2, 1; S. 501, 33: Ps. Th. 93, 13: Salm. Kmbl. 436; Sal. 218. Má and má *magis magisque,* Bd. 4, 29; S. 607, 15. Weaxan á má and má, Past. 37, 1; Swt. 263, 18. Se wela ðe [hí] him dæghwamlíce gesamnodan má and má, Blickl. Homl. 99, 29. [Mo, moe *remains down to Shakspere's time.* O. Frs. má; *adv. and subst.; other dialects have forms which contain the comparative suffix:* Goth. mais: *adv.;* ni þana mais *no more;* O. Sax. mêr: *subst. and adv.;* þan mêr *any more:* Icel. meir: *adv.;* O. H. Ger. mêr; *adv.*]

maca, an; *m. A make, mate, match:*—Fadores æc gimaca ðæm maca *patrisque compar unice* (the glosser seems to have misunderstood *unice*), Rtl. 165, 11. [*Make is used by Ben Jonson. Icel.* maki *a match, mate: Dan.* mage.] v. ge-maca, ge-mæc.

maca-, macca-líc; *adj. Fit, suitable, convenient:*—Mið ðý dæg maccalíc [macalíc, Rush.] gecuom *cum dies opportunus accidisset,* Mk. Skt. Lind. 6, 21 [*Scot.* makly *seemly: Icel.* mak-ligr *meet, becoming, fitting.*] v. ge-mæc, *and preceding word.*

MACIAN; *p. ode To* MAKE, *do, act:*—Ic macige ðé mycelre mægþe *faciam te in gentem magnam,* Gen. 12, 2. Seó forme declinatio macaþ hire genitivum on *ae,* Ælfc. Gr. 7; Som. 6, 4: 24; Som. 24, 24. Ðæt is ðæt hêhste gôd ðæt hit eall swá mehtiglíce macaþ *that is the highest good, which does everything so mightily,* Bt. 35, 4; Fox 162, 1. Ne swincaþ á ymbe ænige þearfe ac maciaþ eall be luste and be éþnesse . . . Ðæt is láþlíc líf ðæt hí swá maciaþ *they never labour at any necessary matter, but do all for pleasure and ease . . It is a detestable life, that they act so,* L. I. P. 14; Th. ii. 322, 23–26. Sweriaþ mê ðæt gê dôn wið mê swilce mildheortnisse, swá ic macode wið eów, Jos. 2, 12. Ða befrán heó ðæt cild hú hit macode on eallum ðam fyrste *then she asked the child what it had been doing in all the time,* Homl. Th. i. 566, 20. Swá hé hit macode on his líf *such was his practice in his life,* ii. 354, 24. Jubal wæs fæder ðæra ðe organan macodun, Gen. 4, 21. Forðan hí macodon mæst ðet unseht betweónan Godwine eorle and ðam cynge, Chr. 1052; Erl. 187, 27. Ðæt ic macige mete ðínum fæder ðær of, Gen. 27, 9. Ðæt ða cristenan hine tô martyre ne macion *that the Christians may not make a martyr of him,* Homl. Skt. 5, 460. Hé (*Lucifer*) wolde hine macian tô gode, Ælfc. T. Grn. 2, 43. Bædon sume ðæt Samson môste him macian sum gamen *some asked, that Samson should make sport for them,* Jud. 16, 25. Riht is ðæt mynecena mynsterlíce macian *it is right that nuns that should practise the rules of their monasteries,* L. I. P. 15; Th. ii. 322, 32. Gestihtode hú men sceoldon ðærinne hit macian *qualiter debeant conversari dispensat,* Past. 16, 1; Swt. 98, 11. Se wísdóm sæde him hú hit macian sceolde gif hé heora þegen beón sceolde, Bt. tit. 7; Fox x, 16. [O. Sax. makôn: O. Frs. makia: O. H. Ger. machôn: Ger. machen.] v. ge-macian.

má-cræftig; *adj. Very (?) skilled or powerful:*—Hwanon cômen gê ceólum líðan mácræftige menn, Andr. Kmbl. 513; An. 257. Næfre ic sælidan sélran mêtte, mácræftigran, 943; An. 472. [Grimm in a note on the former passage suggests that *má* in this compound may be a substantive from the same root and with the same meaning as *mere.*]

macung, e; *f. Making, doing, action:*—þurh ðes macunge mæst se eorl Rotbert ðises geáres ðis land mid unfriþe gesôhte *it was mostly his doing that Earl Robert attacked this country in the course of this year,* Chr. 1101; Erl. 238, 1.

mád (v. ge-maad *vecors,* Wrt. Voc. ii. 123, 36); *adj. Unreasoning, foolish, mad:*—Þrinteþ him on innan ungeméde mád mód *within him (one guilty of oferhygd) swells a mind displeasing by its folly,* Exon. 83b; Th. 315, 2; Môdd. 25. v. ge-mæd.

mádm. v. máðm.

mæ, *more.* v. má.

mæc; *adj. Well-matched, equal, agreeable (?):*—Hár hildering hréman ne þorfte mæcan (*other MSS.* mecca, meca, mecga) geménan *the grey-haired warrior had no need to boast of well-matched intercourse,* i. e. *would not boast of being a match for those against whom he fought, and by whom he had been defeated,* Chr. 937; Erl. 114, 6; Æðelst. 40. [*Prompt. Parv.* make *or* fyte *and* mete; mak, fyt, esy *aptus, conveniens: Icel.* makr *suitable, easy to deal with.*] v. ge-mæc.

mæced [= má-éced? cf. má-geéct] *glosses* mactus, Wrt. Voc. ii. 79, 53.

Mæcedonie; *pl. The Macedonians:*—Philippus Mæcedonia cyning, Ors. 4, 11; Swt. 204, 5. Gewin wið Mæcedonie, Swt. 202, 33.

Mæcedonisc; *adj. Macedonian:*—Ðæt Mæcedonisce gewin, Ors. 4, 11; Swt. 208, 5.

mêce-fisc. v. mêce-fisc.

mæcg, mecg, es; *m. A man:*—Ic meþelcwide mæcges (*the angel that visited Guthlac*) ongeat, Exon. 37b; Th. 175, 9; Gú. 1192. Mægþ and mæcgas, 45 a; Th. 153, 29; Gú. 833: 113 a; Th. 434, 7; Rä. 51, 7. Frêfra ðíne mæcgas (*the disciples of St. Andrew*), Andr. Kmbl. 843; An. 422. Mæcga misgehýd *men's evil intent,* 1543; An. 773. Mæcgea (mecga, MS. C.) mundbora (*Edmund*), Chr. 942; Erl. 116, 8,

Mecga (*those in hell*) gnornunge, Cd. 220; Th. 285, 8; Sat. 334. Mæc-gum (*the children in the fiery furnace*), 187; Th. 232, 24; Dan. 265. Adam iécte siddan mægþum and mæcgum mægburg síne *Adam afterwards increased his family with daughters and sons*, 55; Th. 68, 26; Gen. 1123. DER. ambeht-, earfoþ-, eóred-, Geát-, gigant-, here-, hilde-, oret-, wræc-mæcg.

mæcga, an; *m. A man*, Exon. 88 a; Th. 330, 16; Vy. 52. v. gúþ-, ofer-, wræc-mæcga.

mæcige, Lchdm. iii. 126, 19. v. mecgan.

mæctor. v. mæte.

MÆD, e *and* we; mædwe, an; *f. also* (?) mædwa, an; *m. A MEAD, meadow :*—Mæd *pratum*, Ælfc. Gl. 57; Som. 67, 75; Wrt. Voc. 38, 1: 96; Som. 76, 45; Wrt. Voc. 53, 52. XII æcras an westhealfe dære stræte and án mêdwa beneoþan dære hlípe *xii acres on the west side of the road, and one meadow beneath the hill*, Cod. Dipl. Kmbl. iii. 52, 15. vi æcras mæde on da geréfmæde, 53, 2. xvi gioc ærþelandes and mêdwe, i. 316, 26. On Wíferþes mæduan hege *to the hedge of Wiferth's meadow*, iii. 78, 21. Andlang heges on Eomeres mæduan (cf. on Eomeres mêdwa, 405, 24); of dam mæduan . . . andlang burnan on Hereferþes mæduan, 78, 6–9. Tó wudumædwan; of dæm mædwan, 246, 22. (In the last two passages perhaps the forms are plural as in) Tó dæm mædwum wid súdan da mædwa, 169, 2–3. [Mid læswe and mid mêdwe, Chr. 777; Erl. 55, 12.] Gelíce and mon mæd mâwe *just as one mows a meadow*, Ors. 2, 8; Swt. 92, 15. XIIII æceras and da mæde de dær tó líþ Dúnstán gebohte æt Uhtlufe *xiiii acres and the meadow pertaining thereto Dunstan bought of Uhtlufu*, Cod. Dipl. Kmbl. ii. 3, 34. Norþrihte on mære mæde westewearde, iii. 416, 18. Of dere ealdan díc dæt on wylihte mædwan; of wylihte mædwan, 235, 16. On rýdmædwan ufewarde, 378, 14. Eahta æceras mêdwa . . . xii æceras mædwa, 4, 12–13. Mæda *prata*, Hpt. Gl. 409, 38. Deós wyrt biþ cenned on mædum *this plant is produced in meadows*, Herb. 1, 1; Lchdm. i. 70, 2. [Cf. *Ger.* mähde *a meadow*.] v. gafol-, geréf-, mór-mæd; mæþ.

mæd, mædan. v. ge-mæd, ge-mædan.

mæden. v. mægden.

mæder (?), *a measure :*—Ofgeót mid. iii. mædrum ealoþ, Lchdm. iii. 28, 16.

mædere, an; *f. Madder :*—Mæddre *vermiculi, rubia*, Ælfc. Gl. 42; Som. 64, 13, 19; Wrt. Voc. 31, 24, 29. Mædere *anchorum*, 67, 38: *veneria*, 68, 38: *sandix* (*herba*), Hpt. Gl. 524, 41. Deós wyrt de man gryas and ôdrum naman mædere nemneþ, Herb. 51, 1; Lchdm. i. 154, 12: L. M. 2, 51; Lchdm. ii. 268, 15. [*Icel.* madra.] v. feld-mædere.

mædere-cíp, es; *m. A sprig of madder*, Lchdm. i. 397, 2.

mæd-land, es; *m. Meadow-land, grass-land which is mown :*—Ægder ge mêdlondes ge eyrþlondes *both of land for mowing and of arable land*, Cod. Dipl. Kmbl. ii. 95, 16. Mêdlandes, vi. 219, 4. v. mæd-land.

mæd-mæwect, *the mowing of a meadow :*—Eác hê sceal hwíltídum geara beón on manegum weorcum tó hláfordes willan tóeácan . . . mædmæw-ecte *also he shall at times be ready for labour of many kinds at his lord's pleasure, besides* . . . *mowing his meadows*, L. R. S. 5; Th. i. 436, 3–5.

mæd-rædenn, e; *f. A mowing, grass mown on a piece of land :*—Seó mædræden beniþan díc betweónan cealdan lace and cullig, Cod. Dipl. Kmbl. vi. 153, 10. Cf. wudu-rædenn.

mæd-splott, es; *m. A plot of meadow-land :*—Ænne mædsplot, Cod. Dip. Kmbl. iv. 72, 7.

mædwa. v. mæd.

mædwe-land, es; *n. Meadow-land, land where grass that is to be mown grows :*—Hió sellaþ him dæt mêdweland bí westan Sæferne . . Éc twelf æceras gôdes mædwelandes, Cod. Dipl. Kmbl. ii. 150, 10–18: vi. 219, 3. v. mæd-land.

MÆG, es; *m. A relative, kinsman :*—Mæg *propinquus*, Wrt. Voc. 72, 45: Ælfc. Gr. 5; Som. 4, 51. Hwylc þyncþ dê dæt sý dæs mæg de on da sceadan befeóll *quis videtur tibi proximus fuisse illi qui incidit in latrones?* Lk. Skt. 10, 36. Meig *contribulius*, Wrt. Voc. ii. 104, 26. Meeg, Ep. Gl. 6 f, 17. Se wæs his mæg and his freónd and hæfde his sweoster tó wífe *qui erat cognatus et amicus ejus, habens sororem ipsius conjugem*, Bd. 3, 21; S. 551, 6: Blickl. Homl. 113, 22. Him cýþdon dæt hiera mægas him mid wæron . . And dâ cuædon hié dæt him nænig mæg leófra nære donne hiera hláford . . and dâ budon hié hiera mægum dæt hié gesunde from eodon, Chr. 755; Erl. 50, 17–21. Hér Æþelherd cining forþférde and fæng Cûdred his mæg tó West-Seaxna ríce, 740; Erl. 47, 33: 754; Erl. 49, 18: 962; Erl. 120, 2. Abrahames mæg (*Lot*), Cd. 94; Th. 121, 19; Gen. 2012. Higeláces mæg (*Beowulf*), Beo. Th. 820; B. 408. Úre ieldesta mæg *our first parent*, Past. 43, 5; Swt. 313, 15. Ne hæme nân man wid his mæges (*fratris*) wíf, Lev. 18, 16. Mæges *filii*, Cd. 140; Th. 176, 5; Gen. 2907. Moises heóld his mæges (*soceri*) sceáp, Ex. 3, 1. Moises gecirde tó his mæge, 4, 18. Abrahame, mæge Lothes, Cd. 141; Th. 177, 2; Gen. 2923. Cênwalh gesalde Cûþréde his mæge (*fratrueli*), Chr. 648; Erl. 26, 15. Ne bysmra dû dinne mæg *non facies calumniam proximo tuo*, Lev. 19, 13. Gif man gehâdodne man odde ælþeódigne forræde donne sceal him cyningc beón for mæg and for mundboran, L. C. S. 40; Th. i. 400, 6. Ne his mâgas (*fratres*) ne gelýfdon on hyne, Jn. Skt. 7, 5. His eorþlícan mâgas *his kinsmen according to the flesh*, Chr. 979; Erl. 129, 12. His mâgas and his frýnd *cognati atque amici*, L. Ecg. C. 36; Th. ii. 160, 22. Gif

bana of lande gewîteþ his mâgas healfne leód forgelden, L. Ethb. 23; Th. i. 8, 7. Bócland him his mâgas (MS. B. his yldran) leáfden, L. Alf. pol. 41; Th. i. 88, 16. Hine môton his mægas (MS. B. mâgas) unsyngian *his kindred may exculpate him*, L. In. 21; Th. i. 116, 8. Sunu odde mægas (MS. B. mâgas), 23; Th. i. 116, 15. Mâga *affinium*, Hpt. Gl. 480, 18. Ænig dînra mâga odde yldrena *aliquis de tuis parentibus aut cognatis*, Bd. 2, 12; S. 514, 15. Mid gýmenne mínra mâga *cura propinquorum*, 5, 24; S. 647, 22. Se wæs ædelboren of æwfæstum mâgum *he was nobly born of pious parents*, Homl. Skt. 4, 3. Suna ic lærde dut hié hýrdon heora yldrum and heora mâgum, Blickl. Homl. 185, 21. Súþ-Seaxe and Eást-Seaxe from his mâgum (*ancestors*) ær mid unryhte ânídde wærun, Chr. 823; Erl. 62, 23. Gê beóþ gesealde fram mâgum and gebrôdrum and cûdum and freóndum *trademini a parentibus et fratribus et cognatis et amicis*, Lk. Skt. 21, 16. Lêraþ eówre suna and eówre mâgas *docebis filios ac nepotes tuos*, Deut. 4, 9. Mâgos *propinquos*, Kent. Gl. 368. Bearn árísaþ ongén mâgas *insurgent filii in parentes*, Mt. Kmbl. 10, 21. Ymbe míne mâgas ic hogige *erga propinquos curo*, Ælfc. Gr. 47; Som. 47, 29. Dîne leófostan frýnd fæder and môdor and dîne mâgas *patrem tuum et matrem et omnem cognationem tuam*, Jos. 2, 18: Ps. Th. 73, 8. Ealle wyrd forsweóp míne mâgas, Beo. Th. 5622; B. 2815: Blickl. Homl. 139, 16. [*Laym. mæi a cousin: Goth.* mégs *a son-in-law: O. Sax.* mâg *a relation: O. Frs* mêch: *Icel.* mâgr *a father-in-law: O. H. Ger.* mâg *cognatus, affinis*.] v. cneó-, fæderen-, freó-, fride-, heáfod-, hleó-, hylde-, leód-, mêdren-, neáh-, wine-, woruld-mæg; un-mæg; ge-mâgas.

mæg, e; *f. A woman, kinswoman :*—Freólecu mæg (*Eve*), Cd. 42; Th. 55, 17; Gen. 895: (*Cain's wife*), 50; Th. 64, 21; Gen. 1053: (*Hagar*), 101; Th. 134, 18; Gen. 2226. Drihtlícu mæg (*Sara*), 89; Th. 111, 2; Gen. 1850: 133; Th. 168, 12; Gen. 2781. Mæg ælfsciéno (*Sara*), 86; Th. 109, 23; Gen. 1827: 130; Th. 165, 11; Gen. 2730. Seó eádge mæg, sancta Maria, Exon. 9 a; Th. 6, 21; Cri. 87. Seó æþele mæg (*Juliana*), 68 a; Th. 253, 4; Jul. 175. Seó wuldres mæg, 74 b; Th. 278, 20; Jul. 600. Cáseres mæg (*Elene*), Elen. Kmbl. 660; El. 330: 1335; El. 669. [*Laym.* may: *Orm.* maȝȝ: *Chauc.* mai.] v. eád-, wyn-mæg.

mæg *may*. v. magan.

mæg-bana, an; *m. A destroyer of one's kinsmen :*—Hit (*surfeiting*) biþ mægbana, and hit ne murneþ for nânum men, ne for fæder ne for mêder ne for brôder ne for swuster ne for nânum gesibban men, Wulfst. 242, 5.

mæg-bót, e; *f. The 'bót' paid to the kinsmen of a slain man for the slaying of the latter*. It seems to be used only in the case of the spiritual relationship of godfather and godchild :—Gif hwâ ôdres godsunu sleâ odde his godfæder síe sió mægbót and sió manbót gelíc. Weaxe sió bót be dam were swâ ilce swâ sió manbót dêþ de dam hláforde sceal . . . Gif hê on done geonbyrde de hine slôg donne ætfealle sió bót dæm godfæder swâ ilce swâ dæt wíte dam hláforde dêþ *if any one slay another's godson or his godfather, let the compensation to the godfather or godson and that to the lord of the dead man be alike. Let them both increase in proportion to the 'wer'* . . . *If he* (*the slain man*) *strove against him that slew him, then let there be no 'bót' to the godfather just as there is no 'wíte' to the lord*, L. In. 76; Th. i. 150, 13–20. Ægder ge mægbóte ge manbóte fullíce gebête, L. C. E. 2; Th. i. 360, 7.

mæg-burh; *gen.* -burge; *f. Kindred, family, relatives, tribe :*—Mæg-burg *cognatio*, Wrt. Voc. ii. 15, 70. Weóx under wolcnum mægburh Semes, Cd. 82; Th. 102, 20; Gen. 1703: 100; Th. 132, 14; Gen. 2193: 81; Th. 102, 4; Gen. 1695. Ne weorþeþ sió mægburg gemicledu eaforan mínum, Exon. 105 b; Th. 401, 31; Râ. 21, 20. Heó ongan his mægburge men geîcean sunum and dôhtrum, Cd. 56; Th. 69, 7; Gen. 1132: 101; Th. 134, 5; Gen. 2220: Beo. Th. 5766; B. 2887. Hê hit ne môste sellan of his mægburge *he might not sell it* (bócland) *out of the family*, L. Alf. pol. 41; Th. i. 88, 18. Wes mægburge mínre ârfæst *be kind to my kindred*, Cd. 136; Th. 171, 8; Gen. 2825: Exon. 88 a; Th. 331, 3; Vy. 62. Gielden siddan his mægas done wer gif hê mægburg (-borh, MS. B.: -burh, MS. H.) hæbbe freó *let his kinsmen afterwards pay the wergild, if he have free kindred*, L. In. 74; Th. i. 148, 19. Mægburge míne *my children*, Exon. 104 b; Th. 397, 15; Râ. 16, 20. Iécte mægburg síne, Cd. 55; Th. 68, 27; Gen. 1123. Mægburh, 52; Th. 65, 14; Gen. 1066. Cûde æghwilc mægburga riht *each one knew the rights of the tribes*, 161; Th. 200, 5; Exod. 352. Da de mægburge mæst gefrunon frumcyn feora fæderæþelo gehwæs *those who were best informed as to families, as to the origin of men, and the ancestry of each*, Th. 200, 21; Exod. 360.

mæg-cild, es; *n. A young kinsman :*—Hine âhsode hwær hê his mægcildum cumen hæfde de hê him forstolen hæfde *asked him what he had done with his young kinsmen* (*cousins*) *whom he had stolen away from him*, Lchdm. iii. 424, 37. Dý læs ænig man cwede dæt ic míne mægcild mid wô fordêmde *lest any man say that I wrongfully decided against my kinsmen* (*nephews*), Chart. Th. 486, 27.

mæg-cûd; *adj. Related :*—Mægcûdre sibbe *cognate propinquitatis*, Wrt. Voc. ii. 133, 34.

mæg-cwealm, es; *m. Murder of a father or kinsman :*—Mægcualm *parricidio*, Wrt. Voc. ii. 116, 53.

mǽg-cynren, es; *n. Race, family* :—Macynnere [=(?) mægcynrene] *prosapia*, Hpt. Gl. 437, 11.

mægden, mǽden, es; *n. A maiden, girl, virgin* :—Mǽden oððe geong wífman *puella*, Wrt. Voc. 73, 5. Nis ðis mǽden nā deád ac heó slǽpþ ... Hé nam ðæs mǽdenes mōdor, Mk. Skt. 5, 39–40. Ðū nū sceáwa ðínes mæg(d)enes (*the Virgin Mary*) eáþmōdnesse, Blickl. Homl. 159, 4. Ðā wearþ ðæs mægdnes mōd miclum geblissad, Exon. 74 b; Th. 279, 3; Jul. 608. Hit ᵹealde ðam mǽdene (*the daughter of Herodias*), and ðæt mǽden hit sealde hire mēder, Mk. Skt. 6, 28. Gif hwá mǽden nýdnǽme *si quis violenter virginem opprimat*, L. C. S. 53; Th. i. 406, 3. Ne nýde man nāðer ne wíf ne mǽden tō ðam ðe hyre sylfre mislícige *let no woman, whether she have been married before or not, be forced to a marriage which she dislikes*, 75; Th. i. 416, 20: L. Edm. B. 1; Th. i. 254, 2. Mǽdenu *virgines*, Ps. Th. 44. 15. Tō abbudissan gehādod ofer mā ðonne twām hund mǽdenum, Homl. Th. ii. 476, 20. Mǽdenu niman on þeáwe gōdne tíman getácnaþ, Lchdm. iii. 208, 28. [*O. H. Ger.* magatín : *M. H. Ger.* magetîn.]

mægden-ǽw, e; *f. Marriage with a virgin* :—Ðæt biþ rihtlíc líf ðæt cniht þurhwunige on his cnihthāde ōþ ðæt hé on rihtre mǽdenǽwe gewífige and hæbbe, ða syððan and nǽnige ōðre ða hwíle ðe seó libbe *that is right life, that a young man remain a bachelor until in lawful matrimony he take a maiden to wife, and let him have her afterwards and no other while she lives*, L. I. P. 22; Th. i. 332, 29.

mægden-cild, es; *n. A female child, girl* :—Gif hit hysecild byþ ofsleáþ ðæt gif hit sí mǽdencild healdaþ ðæt *si masculus fuerit, interficite eum, si femina reservate*, Ex. 1, 16. Ðonne ða wíf heora bearn cendon, ðonne fēddon hié ða mǽdencild and slōgon ða hysecild, and ðǽm mǽdencildum hié fortendun ðæt swíðre breóst foran, Ors. 1, 10; Swt. 46, 10–12. Tǽcende ðǽm mǽdencildum *docendo puellas*, Ælfc. Gr. 26; Som. 28, 16.

mægden-hád, es; *m. Maidenhood, virginity* :—Ðeáh wæs hyre (*the Virgin Mary*) mægdenhád ǽghwæs onwalg, Exon. 28 b; Th. 87, 5; Cri. 1420. Gif ǽnig wer oððe wíf gehāte ðæt hé wylle mǽdenhád gehealdan *si quis vir aut mulier voverit virginitatem servare*, L. Ecg. C. 19; Th. ii. 146, 1. v. mægþ-hád.

mægden-, mǽden-heáp, es; *m. A virgin band, troop of maidens*, Dōm. L. 18, 288.

mægden-líc; *adj. Maidenly, girlish, virginal* :—Mǽdenlíc *puellaris, virginalis*, Ælfc. Gr. 5; Som. 5, 23. Seó mǽdenlíce clǽnnys *virginalis castitas*, Hymn. Surt. 118, 21. Mǽdenlícere *virginalis*, Hpt. Gl. 506, 38. Godes sunu þurh mǽdenlícne innoþ ácenned wearþ, Homl. Th. i. 458, 33.

mægden-mann, es; *m. A maid, virgin* :—Mǽdenman *virgo*, Wrt. Voc. 73, 6. Gā ān mǽdenman tō, and hó hit on his sweoran, Lchdm. iii. 42, 9. Gif hwylc mǽdenman on gefērrǽdene mid gehādodum wunaþ *si puella aliqua in societate cum ordinatis habitet*, L. Ecg. P. ii. 17; Th. ii. 188, 9. Gif man wið cyninges mægdenman geligeþ, L. Ethb. 10; Th. i. 6, 4. For ðon Mesiane noldon ðæt Lǽcedemonia mægdenmenn mid heora ofreden and heora godum onsægden *propter spretas virgines suas in solemni Messeniorum sacrificio*, Ors. 1, 14; Swt. 56, 16. [*Orm.* Sannte Marᵹe wass æfre maᵹᵹdennmann.] v. mægþ-mann.

mǽge, an; *f. A kinswoman* :—Elizabeth ðín mǽge (MSS. A. B. mage.) *Elisabeth cognata tua*, Lk. Skt. 1, 36. Hér sit Leóflǽd mín mǽge, Ðurcilles wíf, Chart. Th. 337, 30. Cwæð ðæt heó wǽre gramena mǽge, Deáðes dōhtor, Homl. Skt. 2. 173. Saga ðæt ðū síe sweostor mín, líces mǽge, Cd. 89; Th. 110, 4; Gen. 1833 : 127; Th. 162, 18; Gen. 2683. In Dauides dýrre mǽgan (*the Virgin Mary*), Exon. 9 a; Th. 7, 5; Cri. 96. v. māge, mǽg.

Mǽgelan, Mǽgelang, *Milan* :—Tō Mǽgelan [Mǽgolange, MS. C.] *apud Mediolanum*, Ors. 6, 36; Swt. 294, 30.

MÆGEN, es; *n.* I. MAIN, *might, strength, force, power, vigour, efficacy, virtue, faculty, ability* :—Úrum líchoman cymþ eall his mægen of ðam mete ðe wē þicgaþ *all its strength comes to our body from the food that we take*, Bt. 34, 11; Fox 150, 34. Ðæt mycle mægen mínra handa *the mighty power of my hands*, Ps. Th. 80, 13. Micel drihten úre and micel mægen his *Magnus Dominus noster, et magna virtus ejus*, Ps. Spl. 146, 5. Ðǽm monnum ðe him mægen and cræft wiexþ eác hwílum eákiaþ æfter ðam mægenum ða costunga *crescente virtute plerumque bella tentationis augentur*, Past. 21, 5; Swt. 163, 8. Se wæs moncynnes mægenes strengest *he was mightiest among men*, Beo. Th. 395; B. 196. Nǽnne man ðæs ne tweóþ ðæt se seó strong on his mægene ðe mon gesihþ ðæt stronglíc weorc wyrcþ *nemo dubitat esse fortem, cui fortitudinem inesse conspexerit*, Bt. 16, 3; Fox 54, 28. Ǽr hí geseón Godes ríce on mægne cuman *donec videant regnum dei veniens in virtute*, Mk. Skt. 9, 1. Hé sealde ǽghwylcum be hys āgenum mægene *dedit unicuique secundum propriam virtutem*, Mt. Kmbl. 25, 15. Lufa ðinne drihten mid eallum mægne *diliges dominum tuum ex tota fortitudine tua*, Deut. 6, 5. Of eallum ðínum mihtum and of eallum ðínum mægene *ex omnibus viribus tuis et ex omni mente tua*, Lk. Skt. 10, 27. Eallon mǽgene tilian, Bt. 24, 2; Fox 82, 6. Wiðstandan ealle mægene, Past. 15, 1; Swt. 91, 1 : Beo. Th. 5328; B. 2667. Ðū ne wēnst ðæt heó mǽge swā mycel mægen habban *you will not expect that the plant*

can have so great efficacy, Herb. 12, 4; Lchdm. i. 104, 12. Hé moncynnes mǽste hǽfde mægen and strengo, Cd. 79; Th. 98, 19; Gen. 1632. Ða ðe snyttro mægn and mōdcræft mǽste hæbben *those who in the greatest degree have wisdom, ability and mental power*, Elen. Kmbl. 815; El. 408. Ðonne hí ðæt mægen ðære unmǽtan hǽto áræfnan ne mihton *cum vim fervoris immensi tolerare non possent*, Bd. 5, 12; S. 627, 41. Mægyn and mihta (*angeli*) *potentes virtute*, Ps. Th. 102, 19. Eall his bearna mægen *omnes virtutes ejus*, 20. Seó sýfernes and ōðre mægnu *sobrietas et alie virtutes*, Prud. 54 a : 64 a. Ða sōðan welan ðæt sind hálige mægnu *the true riches, they are holy virtues*, Homl. Th. ii. 88, 310. Mægenu, Basil admn. 2; Norm. 38, 9. Mægno and cræftas, Bt. 32, 1; Fox 116, 1. Wísdóm mōdur eallra mægena *virtutum omnium nutrix*, 10; Fox 26, 24. Mycelre mægna fǽmne *magnarum virgo virtutum*, Bd. 3, 8; S. 531, 12. Geleáfa is e-alra mægena fyrmest, Homl. Th. i. 134, 2. Geþyld is wyrtruma ealra háligra mægna, and ungeþyld is ealra mægna tōstencednys, ii. 544, 6. Þurh ðínra mægna spéd *through the abundance of thy powers*, Bt. Met. Fox 20, 516; Met. 20, 258 : Cd. 1; Th. 1, 6; Gen. 3. Eallum hire mihtum and mægenum, L. M. 3, 63; Lchdm. ii. 352, 5. Ða ðe faraþ fram leahtrum tō mæignum *those who pass from vices to virtues*, Homl. Th. ii. 54, 26. Mægnum, Prud. 28 a. Ða ongunnon hí mōd and mægen niman ... Mōd and mægen Bryttas onféngon *ceperunt illi vires animosque resumere ... vires capessunt Brittones*, Bd. 1, 16; S. 484, 15–19. Ðeáh ðe ic nū gyt ða ǽrran mægen ne hæbbe *etsi necdum vires pristinas recepi*, 5, 3; S. 616, 34 : 5, 4; S. 617, 25. Heó hæfþ ðás mægnu it (*henbane*) *has these virtues*, Herb. 5, 1; Lchdm. i. 94, 10. Megene vires, Kent. Gl. 930. II. *an exercise of power, effort, a mighty work, miracle* :—Mægene *conamine*, Wrt. Voc. ii. 24, 57. Hé ne mihte ǽnig mægen wyrcan *non poterat virtutem ullam facere*, Mk. Skt. 6, 5. Ān mægen and ān wundor of monegum āsecgan *unum e pluribus virtutis miraculum enarrare*, Bd. 3, 2; S. 524, 28. Monige mægen and hǽlotácen gefremede wǽron *innumeræ virtutes sanitatum noscuntur esse patratæ*, S. 524, 28. On him synd mægenu geworht, Mk. Skt. 6, 14. Ða burga on ðām wǽrun gedōne manega hys mægena, Mt. Kembl. 11, 20. III. *a force, military force* :—Gif ðet full mægen ðǽre wǽre ne eodan hí nǽfre eft tō scipon *if the full force had been there, they would never have got back to the ships*, Chr. 1004; Erl. 139, 34. Úre mægen lytlaþ *our force lessens*, Byrht. Th. 140, 65; By. 313. Mægen, folc Ebréa, Judth. 12; Thw. 25, 15, 10; Jud. 261, 253. Werod, mōdigra mægen, Cd. 147; Th. 184, 2; Exod. 101 : 158; Th. 197, 1; Exod. 300. Mægen forþgewāt, 160; Th. 199, 30; Exod. 346. Mægen (*the Egyptian army*) wæs ádrenced, 166; Th. 206, 28; Exod. 458. Seó sibgedriht bād mǽran mægenes *the Israelites awaited the greater force of the Egyptians*, 154; Th. 191, 15; Exod. 215. Mægenes wísa (*Belshazzar*), 979; Th. 260, 2; Dan. 703. Se wæs mid his dǽdum snelra ðonne hé mæ[ge]nes hæfde *he was quicker in his actions than in proportion to the force he had*; celeritate magis quam virtute fretus, Ors. 2, 5; Swt. 78, 27. Hé self fór ðǽrtō mid eallum ðæm mægene ðe he ðǽrtō gelǽdan mehte *he himself marched thither with all the troops that he could lead there*, Swt. 80, 24. Martyra mægen unlytel *no small host of martyrs*, Andr. Kmbl. 1752; An. 878: Beo. Th. 894; B. 445. Mægen unríme *hosts innumerable*, Elen. Kmbl. 121; El. 61. [*O. Sax.* megin : *Icel.* magn *and* megin: *O. H. Ger.* magan, megin, *robur, vigor, vis, virtus, fortitudo.*] DER. beadu-, deáþ-, eal-(æl-), eorþ-, eorl-, folc-, gæst-, gesíþ-, hand-, here-, heáh-, heofon-, holm-, hord-, leód-, lof-, ofer-, rǽd-, tōþ-, þeód-mægen.

mægen-ágende; *adj. Possessing strength, mighty*, Beo. Th. 5666; B. 2837.

mægen-byrðenn, e; *f. A mighty burden*, Beo. Th. 3254; B. 1625 : 6174; B. 3091.

mægen-corþer, es; *n. A powerful band*, Cd. 93; Th. 119, 27; Gen. 1986.

mægen-cræft, es; *m. Main force, great power or might, mighty power* :—Mægencræft ðe him meotud engla forgiefen hæfde *the power which the Lord of angels had given him*, Exon. 49 a; Th. 170, 1; Gū. 1105. Is ðæt mægencræft micel mōda gehwylces ofer líchoman (cf. hit is micel cræft ðæs mōdes for ðone líchoman, Bt. 38, 1; Fox 196, 10), Bt. Met. Fox 26, 209; Met. 26, 105. Ðæt hé þrittiges manna mægencræft on his mundgripe hæbbe, Beo. Th. 765; B. 380. Mircne mægencræft, Exon. 26 b; Th. 78, 26; Cri. 1280. [*O. Sax.* megin-kraft: *O. L. Ger.* megin-craft *majestas*: *O. H. Ger.* magan-kraft *majestas*.]

mægen-cyning, es; *m. A chief, mighty or powerful king* :—Mægencyning (*God*), Elen. Kmbl. 2493; El. 1248: Exon. 116 b; Th. 448, 21; Dóm. 57: (*Christ*), 21 a; Th. 57, 11; Cri. 917. Mægencyninga meotod *the Lord of mighty kings*, 21 b; Th. 58, 29; Cri. 943: 116 a; Th. 445, 12; Dóm. 6. [Cf. *Icel.* megin-dróttning (*the Virgin Mary*): megin-skjöldungr (*Christ*).]

mægen-dǽd, e; *f. A mighty deed, an action requiring strength*, Exon. 78 b; Th. 294, 9; Crä. 12.

mægen-eáca, an; *m. An increase of strength, succour* :—Monnum tō mægeneácan *a succour for men*, Exon. 55 a; Th. 194, 14; Az. 138.

mægen-eácen; *adj. Endowed with strength, powerful* :—Mōde mægen-

eácen, Exon. 79 b; Th. 299, 7; Crä. 98. Mægeneácen folc (*the victorious Hebrews*), Judth. 12; Thw. 25, 35; Jud. 293.

mægen-earfeþe, es; *n. A great labour* or *hardship:*—Nales fore lytlum geómre, ac fore ðam mǽstum mægenearfeþum, Exon. 22 a; Th. 60, 4; Cri. 964. Mægenearfeþu, sár and swár gewin and sweartne deáþ, 28 b; Th. 86, 20; Cri. 1411.

mægen-ellen, es; *n. Mighty valour*, Beo. Th. 1323; B. 659.

mægen-fæst; *adj. Strong, vigorous, firm:*—Sealde him snyttru mægen-fæste gemynd *he gave him wisdom, vigorous thought*, Exon. 39 b; Th. 130, 28; Gú. 445. Ǽlc líchamlíce gesceaft ðe eorþe ácenþ is fulre and mægenfæstre on fullum mónan ðonne on gewanedum *every bodily creature that earth produces is more complete and more vigorous at the full moon than when the moon has waned*, Homl. Th. i. 102, 21. [Cf. mægen-leás.]

mægen-folc, es; *n. A mighty people:*—Mægenfolc micel (cf. *O. Sax.* meginfolk mikil *the multitude that flocked about Christ*) *a people mighty and vast* (*the good at the day of judgment*), Exon. 20 b; Th. 55, 1; Cri. 877.

mægen-fultum, es; *m. A powerful help:*—Næs ðæt mǽtost mægen-fultuma (*the sword lent to Beowulf by Hunferth*), Beo. Th. 2915; B. 1455.

mægen-heáp, es; *m. A powerful band:*—Mægenheápum, Cd. 151; Th. 190, 11; Exod. 197.

mægen-heard; *adj. Very strong, powerful:*—Ðam ðe sitteþ on ufan meare mægenheardum, Runic pm. 5; Kmbl. 340, 5.

mægenian, mægnian; *p.* ode *To gain strength:*—Mód mægnode *mind gained might*, Exon. 94 b; Th. 353, 55; Reim. 33. v. ge-mægened.

mægen-leás; *adj. Without strength, powerless, weak, feeble:*—Mægen-leás *enervis*, Wrt. Voc. i. 46, 6: *elumbis*, Germ. 396, 216. Seó sáwul, gif heó næfþ ða hálgan láre, heó biþ ðonne weornigende and mægenleás, Homl. Th. i. 168, 33. [*Icel.* megin-lauss.]

mægenleás-líce; *adv. Feebly, impotently:*—Mægenleaslíce *eviscerando*, Germ. 398, 122.

mægen-leást, e; *f. Weakness, feebleness, impotence:*—Ða ofhreów ðam munece ðæs hreóflian mægenleást (*inability to walk*), Homl. Th. i. 336, 11. Módes mægenleást *weakness of mind*, ii. 220, 5. Hí ne mihton for heora mægenleáste ða meniu bewerian (*of the Jews reduced by famine during the siege of Jerusalem*), Ælfc. T. Grn. 21, 8.

mægen-rǽs, es; *m. A mighty* or *violent attack:*—Mægenrǽs forgeaf hilde bille (*Beowulf attacking Grendel's mother*), Beo. Th. 3043; B. 1519.

mægen-róf; *adj. Of great power:*—Módig and mægenróf mid ðære miclan hand (*applied to God*), Cd. 156; Th. 195, 11; Exod. 275. Þegn, mægenrófa ínan, Exon. 109 b; Th. 419, 9; Rä. 38, 3.

mægen-scipe, es; *m. Power, might:*—Metodes mægenscipe, Cd. 173; Th. 217, 9; Dan. 20.

mægen-spéd, e; *f. Abundance of strength, strength* (cf. on ðínes mægenes miclum spédum *in virtute tua*, Ps. Th. 73, 13), *power, virtue:*—Ic ðé sceal meotudes mægenspéd gesecgan *to thee am I to tell the Maker's abundant might*, Exon. 92 b; Th. 348, 6; Sch. 24. Hé mec for miltsum and mægenspédum næfre wille án forlǽtan *on account of his mercy and his might he will never forsake me*, 42 a; Th. 140, 17; Gú. 611: Andr. Kmbl. 2572; An. 1287. Mín múþ sægeþ ðíne mægenspéde *os meum pronuntiabit justitiam tuam*, Ps. Th. 70, 14.

mægen-stán, es; *m. A mighty stone* or *rock:*—Him on innan felþ muntes mægenstán (cf. ðǽr micel stán wealwiende of ðam heáhan munte on innan fealþ, Bt. 6; Fox 14, 28), Bt. Met. Fox 5, 31; Met. 5, 16. Ðis synd ðæra xx hída gemǽro .. andlang wægæs óþ ðonæ mægenstán, Cod. Dip. Kmbl. v. 112, 18.

mægen-strang; *adj. Strong in power:*—Hú ðú mǽre eart mihtig and mægenstrang *how great thou (Christ) art, how mighty and strong in power*, Hy. 3, 21; Hy. Grn. ii. 282, 21. Ðú eart se miccla and se mægenstranga, 3, 38; Hy. Grn. ii. 282, 38. Mægenstrong, Exon. 129 a; Th. 495, 5; Rä. 84, 3.

mægen-strengo; *indecl. f. Main strength, great force:*—Gúþcyning (*Beowulf*) mægenstrengo slóh hilde bille *with mighty force the warrior-king smote with his battle-blade*, Beo. Th. 5350; B. 2678. Sum biþ gleáw módes cræfta sum mægenstrengo onfêhþ *one is skilled in the arts of the mind, another receives great bodily strength*, Exon. 78 b; Th. 295, 15; Crä. 33.

mægen-strengþu; *indecl.*: -strengþ, e; *f. Great strength, power:*—Hí ðíne mægenstrengþu mǽrsien wíde *magnitudinem tuam narrabunt*, Ps. Th. 144, 6. Ic siges mihte and mægenstrengþe swá micele eów sille ðæt gê eów tó gamene feónda áfillaþ swá fela swá gê reccaþ *I will give you so great victorious might and power, that it shall be sport to you to slay as many foes as you can count*, Wulfst. 132, 19.

mægen-þegen, es; *m. A mighty minister* (an angel), Exon. 49 a; Th. 169, 23; Gú. 1099.

mægen-þreát, es; *m. A mighty band*, Cd. 174; Th. 218, 26; Dan. 45: 169; Th. 210, 8; Exod. 512.

mægen-þrymm, es; *m.* (The word is used almost exclusively in reference to the Deity). **I.** *Majesty, greatness, glory:*—Se myccla mægenþrym *the great majesty* (of *Christ*), Blickl. Homl. 179, 8. Mægen-þrymmes God *Deus majestatis*, Ps. Th. 28, 3. Mægenþrymmes ðínes

majestatis tuæ, 144, 5. His mægenþrymmes *magnitudinis ejus*, 150, 2. His ríces ongin, ne his mihte, ne his mægenþrymmes nǽfre gewonad ne weorþeþ, Blickl. Homl. 9, 17. Hé (*Christ*) hine ungyrede ðæs godcundan mægenþrymmes, and gegyrede hine þeówlíce, 105, 3. Ðonne se heofenlíca déma cymþ on egeslícum mægenþrymme, Homl. Th. ii. 558, 9: Lk. Skt. 9, 26, 31, 32. Mid ðý mǽstan mægenþrymme, Exon. 22 b; Th. 62, 30: Cri. 1009. Johannes on Godes mægenþrymme hí gebletsode, Homl. Th. i. 64, 4. Wé gesáwon Godes mægenþrim and his micelnisse (*majestatem et magnitudinem suam*), Deut. 5, 24. **II.** (*using the attribute for the person*), *Christ:*—Mægenþrym árás, sigefæst and snottor, Exon. 120 a; Th. 420, 25; Hö. 22. **III.** *great power, might:*—Gê geseóþ mannes Bearn sittende on ða swýðran healfe Godes mægenþrymmes *videbitis filium hominis sedentem a dextris virtutis Dei*, Mt. Kmbl. 26, 64. Ic sóhte hwylc wǽre mægenþrymmes oððe elnes se Pater Noster, Salm. Kmbl. 20; Sal. 10. In ðam mægenþrymme mid ðam sý áhefed heofon and eorþe *in that mighty power with which is uplifted heaven and earth*, Exon. 93 b; Th. 351, 31; Sch. 88. Hé hine of sáwle deáþe áwehte þurh ðone mægenþrym *he raised him from the death of the soul through divine power*, Blickl. Homl. 77, 10. Mægen-þrymmum mǽst *mightiest*, Cd. 160; Th. 199, 35; Exod. 349. **IV.** *an instance in which the divine glory or power is displayed:*—Eftwyrd cymþ, mægenþrymma mǽst, dæg dǽdum fáh (*doom's day*), Cd. 169; Th. 212, 16; Exod. 540. **V.** *the glory of heaven, heaven, the angels who inhabit heaven:*—Wuldres ealdor middangeardes and mægen-þrymmes *the prince of glory, of earth and of heaven*, Exon. 68 a; Th. 251, 33; Jul. 154. Hé is cyning middangeardes and mægenþrymmes, wuldre biwunden, 65 b; Th. 241, 33; Ph. 665: 16 a; Th. 35, 13; Cri. 557. Ufan of roderum, of his mægenþrymme, 98 a; Th. 368, 24; Seel. 29. Hêht sigores fruma his heáhbodan hider (*to earth*) gefleógan of his mægenþrymme, 12 a; Th. 19, 5; Cri. 296. Næs ǽnig ðá giet engel geworden ne ðæs miclan mægenþrymmes nán *was not any angel then created, nor any of that great and glorious band*, 12 b; Th. 22, 16; Cri. 352.

mægenþrym-ness, e; *f. Majesty, magnificence, glory:*—His mægen-þrymnes (-þrymmes, MS.) micellíc standeþ *magnificentia opus ejus*, Ps. Th. 110, 2. Mæg[en]þrymnysse *majestatis*, Hpt. Gl. 486, 18. Ælmihtig God, ánes gecyndes, and ánre mægenþrymnisse on ánre godcundnysse, Hexam. 2; Norm. 4, 23. Ðonne sit hé on dómsetle his mægenþrym-nysse, Wulfst. 287, 31. Mæg[n]þrumnysse *majestati*, Hpt. Gl. 416, 52. God sylf se ðe ǽfre þurhwunode on his miclan wuldre and on his mægen-þrimnisse, Ælfc. T. Grn. 2, 4. Ða ðe gesáwon míne mægenþrimnisse *qui viderunt majestatem meam*, Num. 14, 22.

mægen-þyse, an; *f. Violence, force:*—Sóna ðæt onfindeþ se ðe mec fêhþ ongeán and wið mægenþisan mínre genǽsteþ ðæt hé hrycge sceal hrusan sécan *soon doth he find that fights against me, and with my force comes into conflict, that with his back he must visit the earth*, Exon. 107 b; Th. 410, 24; Rä. 28, 10. [Cf. *Icel.* þysja *to rush.*]

mægen-weorc, es; *n. A mighty work:*—Hú micle synt ðíne mægen-weorc *quam magnificata sunt opera tua*, Ps. Th. 91, 4. [*Icel.* megin-verk; *pl. mighty works.*]

mægen-wísa, an; *m. The leader of a force* or *army*, Cd. 170; Th. 213, 17; Exod. 553.

mægen-wudu, es; *m. A mighty spear-shaft:*—Þegn Hróðgáres cwehte mægenwudu mundum *Hrothgar's thane shook his mighty shaft with his hands*, Beo. Th. 477; B. 236.

mægen-wundor, es; *n. A very great wonder* (of the circumstances attending the day of judgement), Exon. 21 b; Th. 57, 31; Cri. 927.

mæger; *adj. Meagre, lean:*—Ða men beóþ mægre and bláce on onsýne ðeáh ðe hié ǽr fætte wǽron *the men will be lean and pale of aspect, though before they were fat*, L. M. 2, 36; Lchdm. ii. 242, 3. [*Icel.* magr: *Dan., Swed., Du.* mager: *O. H. Ger.* magar *macilentus: Ger.* mager.]

mægerian; *p.* ode *To macerate, emaciate, make lean:*—Mægeregan *macerare*, Wrt. Voc. ii. 57, 16: 96, 34. [*Icel.* megra *to emaciate: O. H. Ger.* magarian *macerare, macrescere;* cf. magar fleiski *pulpa: Ger.* magern.]

mægeþ, mǽgeþ. v. mægþ, mǽgþ.

mægeþe *name of a plant.* v. nageþe.

mǽg-gemót, es; *n. A meeting of kinsmen:*—Hé bebeád ofer ealne middangeard ðæt ǽlc mǽgþ tógædere cóme, ðæt ǽlc man ðý gearor wiste hwǽr hé gesibbe hæfde. Ðæt tácnode ðæt on his dagum sceolde beón geboren se se ðe ús ealle tó ánum mæggemóte gelaþaþ, Ors. 5, 14; Swt. 248, 18.

mǽg-gewrit, es; *n. A writing containing a list of kinsmen, a genea-logical table, pedigree*, Cot. 213, Lye.

mǽg-gildan (?) *to pay part of the wergild for a homicide committed by a kinsman:*—Ne þearf se frigea mid ðam þeówan mǽggieldan (*or should this be mǽge gieldan? cf. MS. B. which has* mid ðam þeówan men gyldan. But the word is supported by L. H. i. 70, 5:—Non cogitur liber cum servo *meggildare*), L. In. 74; Th. i. 150, 1.

mǽg-hǽmed, es; *n. Incest:*—Nǽnig mǽghǽmed ne uncláene fremme *nullus incestum faciat*, Bd. 4, 5; S. 573, 15.

mǽg-hand, a; *f. A relation, kinsman:*—Nis Eðelmóde ǽnig mǽghond

neór ðes cynnes ðanne Eádwald *there is no nearer relative to Ethelmod in the family than Eadwald*, Chart. Th. 466, 1. Wes hit becueden his bróðar suna and siððan nëniggra mëihanda má ðes cynnes, 465, 20. Cf. ða nýhstan hand më, 491, 13.

mægister. v. magister.

mǽg-lagu, e; f. *Law regulating the duties and responsibilities of kinsmen* (mǽgas), e. g. in the matter of paying or receiving certain parts of the wergild if one of their number slew or was slain :—Hë (mynstermunuc) gǽþ of his mǽglage ðonne hë gebýhþ tó regoⅼlage, L. Eth. ix. 25 ; Th. i. 346, 2. v. mǽgþ-lagu *and* lagu.

mǽg-leás ; adj. *Without kinsmen* :—Gif hë sí mǽgleás *if he have no kinsmen*, L. Eth. ix. 24 ; Th. i. 344, 28 : L. In. 23 ; Th. i. 116, 16 : L. C. E. 5 ; Th. i. 362, 24. Fædrenmǽga mǽgleás mon *a man having no kinsmen on the father's side*, L. Alf. pol. 27 ; Th. i. 78, 20.

mǽg-líc ; adj. *Belonging to kinsmen* :—Hë hine lufode ná swá micclum for ðære mǽglícan sibbe *he loved him, not so much because they were relations*, Homl. Th. i. 58, 4. Næfde hë ðæt andgit þurh mǽglíce láre *he did not have that intelligence through the teaching of his parents*, 368, 10.

mǽg-lufu, an ; f. *Love* :—Heó sagaþ ðæt heó mǽglufan mínre ne gýme *she* (*Juliana*) *says that she cares not for my* (*Heliseus', who wished to marry Juliana*) *love*, Exon. 66 b ; Th. 246, 31 ; Jul. 70.

mǽg-morðor, es ; n. *Murder of a kinsman* :—Mǽgmorðor *parricidium*, Hpt. Gl. 519, 74. Mǽgmorðres wítnung *parricidii actio*, Ælfc. Gl. 14 ; Som. 58, 15 ; Wrt. Voc. 21, 10. [Cf. *O.H. Ger.* mág-mord *parricidium*.]

mǽg-myrðra, an ; m. *One who murders a kinsman, a parricide* :—Mǽgmyrðra *parricida*, Wrt. Voc. ii. 67, 15 : Hpt. Gl. 509, 72.

mægn. v. mægen.

mǽg-racu, e ; f. *The account of a family, a genealogy* :—Ðis is seó bóc Adames mǽgrace *hic est liber generationis Adam*, Gen. 5, 1. Gif ðú telst ða mǽgrace fram Judan ðonne findst ðú fíf mǽgþa *if you reckon the genealogy from Judah, then you will find five generations*, Boutr. Scrd. 22, 19.

mǽg-rǽdenn, e ; f. *Kinship, relationship* :—Gesibbere mǽgrǽdene *consanguinitatis*, Hpt. Gl. 472, 20. Hë (*Julius Cæsar*) hiene (*Octavianus*) for mǽgrǽdenne gelǽrde, Ors. 5, 13 ; Swt. 244, 24. Næfre ic ðæs þeódnes þafian wille mǽgrǽdenne *I will never consent to marry the prince*, Exon. 67 a ; Th. 249, 9 ; Jul. 109.

mǽg-rǽs, es ; m. *An attack by men upon their kinsmen* :—Wearþ ðes þeódscype swýðe forsyngod þurh morððǽda and þurh mándǽda . . þurh mǽgrǽsas and þurh manslihtas *this nation is sunk in sin through deeds fell and foul . . through attacks of kinsmen upon kinsmen and through manslaughters*, Wulfst. 164, 4.

mǽg-scír, e ; f. *A division of a people, containing the kinsmen of a particular family* :—Teá monna látwu ofer tëno oððe of mǽgscíre is *decanus super x. vel decurio* (the glosser seems to have taken *de* as a separate word) *est*, Rtl. 193, 19.

mǽg-sibb, e ; f. **I.** *kinship, relationship* :—Eva hine hálsode for scá Marian mǽgsibbe ðæt hë hire miltsade. Heó cwæþ tó him gemyne mín drihten ðæt heó wæs bán of mínum bánum and flǽsc of mínum flǽsce *Eve conjured him* (*Christ*) *on account of her kinship to St. Mary to pity her. She said to him 'Remember, my Lord, that she was bone of my bone and flesh of my flesh,'* Shrn. 68, 15. Hë (*Christ*) hym (*men*) his mildse onwreáh and his mǽgsibbe gecýdde. Ǽr ðam wë wǽron steópcild gewordene, Wulfst. 252, 9 : Blickl. Homl. 107, 2. Wel is tó warnianne ðæt man wite ðæt hý (*the man and woman about to be married*) þurh mǽgsibbe tó gelænge ne beón (i. e. *are not within the prohibited* (*seven*) *degrees*), L. Edm. B. 9 ; Th. i. 256, 9. Seó hálige lé forbeódeþ ða sceondlícnysse onwreón mǽgsibba (ðære mǽgsibbea, MS. B.) *sacra lex prohibet cognationis turpitudinem revelare*, Bd. 1, 27 ; S. 491, 7. **II.** *Love between kinsmen, affection* :—Mǽgsibbe *affectui* vel *dilectione*, Wrt. Voc. ii. 99, 52. Mǽgsibbe, Ep. Gl. 3 b, 9.

mǽgsib-líc ; adj. *Of kin, related* :—Mǽgsiblícum *contribulibus*, Wrt. Voc. ii. 20, 18.

mǽg-slaga, an ; m. *The slayer of a kinsman* :—Mǽgslaga *parricida*, Ælfc. Gl. 85 ; Som. 73, 114 ; Wrt. Voc. 49, 21 : Ælfc. Gr. 7 ; Som. 6, 46. Se mǽgslaga Cain *the fratricide Cain*, Homl. Th. ii. 58, 28. Hër syndan mannslagan and mǽgslagan, Wulfst. 165, 27 : 266, 26.

mǽg-sliht, es ; m. *The slaughter of a kinsman* :—Wearþ ðes þeódscipe swíðe forsingod þurh manslihtas and þurh mǽgslihtas, Wulfst. 130, 2. [*O. H. Ger.* mág-slaht ; f. *parricidium*.]

mǽgþ, mǽgeþ ; *without inflection in the sing. and in the n. ac. pl.*, f. *A maid, virgin, girl, maiden, woman* (almost confined to poetry) :—Gif man mǽgþ gebigeþ ceápe geceápod sý gif hit unfácne is *if a man make terms for his marriage with* (lit. *buys with a price,* cf. Icel. kona mundi keypt) *a woman, let the bargain stand, if it be without fraud*, L. Ethb. 77 ; Th. i. 22, 1. Wæs seó fémne geong, mǽgþ mánes leás (*the Virgin Mary*), Exon. 8 a ; Th. 3, 14 ; Cri. 36. On fǽmnan, mǽgeþ unmǽle, 18 b ; Th. 45, 18 ; Cri. 721 : 122 b ; Th. 470, 14 ; Hy. 11, 16. Þa torhtan mǽgþ (*Judith*), Judth. 10 ; Thw. 22, 1 ; Jud. 35. Mǽgþ scýne *maiden fair*, Beo. Th. 6025 ; B. 3016. Ofer mǽgþ giunge, Bt.

Met. Fox 26, 134 ; Met. 26, 67. Þurh Judithe láre, mǽgþ módigre, Judth. 12 ; Thw. 26, 18 ; Jud. 335. Mǽgeþ, brýde dínre (*Sarah*), Cd. 134 ; Th. 169, 10 ; Gen. 2797. Hë ðære mǽgeþ (*Guthlac's sister*) sceolde láce gelǽdan lád spel, Exon. 52 a ; Th. 182, 27 ; Gú. 1316. Mǽgþ and mæcgas, 45 a ; Th. 153, 29 ; Gú. 833. Mǽgeþ and mæcgas, 113 a ; Th. 434, 7 ; Rä. 51, 7. Him tó nimaþ mǽgeþ tó gemæccum *take to themselves maidens as mates*, Cd. 64 ; Th. 76, 18 ; Gen. 1259. Mǽgþa síd *the maidens' coming*, 123 ; Th. 157, 11 ; Gen. 2604 : Beo. 1853 ; B. 924. Swá hwylc mǽgþa swá ðone magan cende, 1890 ; B. 943. Mǽgþa cynnes *of womankind*, Exon. 73 b ; Th. 275, 16 ; Jul. 551. Mǽgþum and mæcgum, Cd. 55 ; Th. 68, 26 ; Gen. 1123. [*Goth.* magaþs *a maid, virgin : O. Sax.* magað : *O. Frs.* megith : *O. H. Ger.* magad *virgo : M. H. Ger.* maget : *Ger.* magd.] v. heals-mǽgeþ.

mǽgþ, e ; f. *Importunate desire, ambition* :—Ðæt mód sǽde ðæt him nǽfre seó mǽgþ and seó gítsung forwel ne lícode, Bt. tit. 17 ; Fox xii, 24. Cf. Ðú wást ðæt më nǽfre seó gítsung and seó gemǽgþ ðisses eorþlícan anwealdes forwel ne lícode *scis ipsa minimum nobis ambitionem mortalium rerum fuisse dominatum*, 17 ; Fox 58, 23. v. máh, ge-mǽhþ (*with which* ge-mǽgþ *in the above passage should be put*).

mǽgþ, mǽgeþ, e ; f. *A collection of* mǽgas. **I.** *with a more limited extent, a family, stock, race* :—Mǽgþ oððe styb *styrps,* Ælfc. Gr. 3 ; Som. 3, 17. Mǽgþ *progenies,* Wrt. Voc. 72, 48 : *cognatio,* Ps. Spl. 73, 9. Mýgþ *propinquus,* Kent. Gl. 876. Ða wæs án mǽgþ ðe nǽfre ne ábeáh tó nánum deófolgylde . . Seó mǽgþ ásprang of Noes eltstan suna . . And ðyssere mǽgþe God sealde ǽ . . forðan ðe hë wolde of ðyssere mǽgþe him módor geceósan, Homl. Th. i. 24, 5–20. Woldon ofsleán Claudius for Gaiuses þingum ðæs ǽrran cēsares and ealle ða de ðære mǽgþe wǽron *evertenda penitus Caesarum universa familia decrevissent,* Ors. 6, 4 ; Swt. 258, 25. Rím miclade monna mǽgþe, Cd. 63 ; Th. 75, 22 ; Gen. 1244. Mǽgþe dínre (*Abraham's*), 84 ; Th. 105, 34 ; Gen. 1763. Nis nán wítega búton wurþscipe búton on his ǽðele and on his mǽgþe (*cognatione*) and on his húse, Mk. Skt. 6, 4. Ða hwíle ðe ǽnig man wǽre on hira mǽgþe ðe godcundes hádes beón walde *as long as there was any man of their stock that was willing to take orders,* Chart. Th. 166, 16. **II.** *as a technical term in the laws, relatives, kindred, the* mǽgas *who were living at the same time, and to whom the* mǽg-lagu *applied* :—Gá seó mǽgþ him on borh *let the family go bail for him* (*the thief*), L. Ath. i. 1 ; Th. i. 198, 24. Gif ðonne ðæt gebyrige ðæt ǽnig mǽgþ tó ðan strang sý . . ðæt ðonne þeóf foran forstande, V. 8, 2 ; Th. i. 236, 9 : 12, 2 ; Th. i. 242, 3 : L. Edm. S. 1 ; Th. i. 248, 5. Bëte ðam cyninge swá ilce swá ðære mǽgþe *let amends be made to the king in the same way as to the kindred,* L. In. 76 ; Th. i. 150, 17 : L. Ath. i. 2 ; Th. i. 200, 7. Ealle of ǽgðere mǽgþe, L. E. G. 13 ; Th. i. 174, 21. Se slaga wille bëtan wið mǽgþe, L. Edm. S. 7 ; Th. i. 250, 15. Gebëte wið ða mǽgþe, L. C. S. 39 ; Th. i. 398, 27 : L. Edm. S. 4 ; Th. i. 248, 25. **III.** *in a wider sense, descendants of a common ancestor living at the same time, a generation* :—Ðë ic geseah sóðlíce rihtwísne ætforan më on ðissere mǽgþe *te enim vidi justum coram me in generatione hac,* Gen. 7, 1. On ealræ mǽgþe *in omni generatione,* Ps. Spl. 44, 19. Hwí is áwriten on ðære bëc Genesis ðæt Abrahames cynn sceolde gecyrran ongeán fram Aegypta lande on ðære feórþan mǽgþe and seó óðer bóc Exodus sǽgþ ðæt hí fërdon of Aegyptan lande on ðære fíftan mǽgþe ? . . Gif ðú telst ða mǽgrace fram Iudan ðonne findst ðú ðǽr fíf mǽgþa, and gif ðú telst fram Leui ðonne findst ðú ðǽr feówer mǽgþa, Boutr. Scrd. 22, 16– 20 : Homl. Th. ii. 458, 34. Noe wæs rihtwís wer on his mǽgþum *Noe vir justus fuit in generationibus suis,* Gen. 6, 9 ; 9, 12. **IV.** *with wider limits than those implied by family,* (*a*) *a tribe, subdivision of a people* :—Mǽgþ *tribus,* Wrt. Voc. 72, 48 : Ælfc. Gr. 11 ; Som. 15, 23. Gegaderiaþ eów tó mǽgþum [and gange] ðæt gehlot fram mǽgþe tó mǽgþe and be manna híwrǽdenum *accedetis singuli per tribus vestras, et quamcumque tribum sors invenerit, accedit per cognationes suas,* Jos. 7, 14. Of Asseres mǽgþe *de tribu Asser,* Lk. Skt. 2, 36. Leóda mǽgþe *the tribes of men,* Cd. 80 ; Th. 100, 16 ; Gen. 1665. Ðæra mǽgþa ealdras *principes tribuum,* Num. 1, 4. Of ðám twelf mǽgþum, 13, 3 : Blickl. Homl. 155, 30. (*b*) *a people, nation* :—Ðære mǽgþe monwísan *the manners of the people* (*of Sodom*), Cd. 92 ; Th. 116, 20 ; Gen. 1939. Ná dyde hë swylc ǽlcre mǽgþe *non fecit taliter omni nationi,* Ps. Spl. 147, 9 : 49, 7. Gebannan manigre mǽgþe geond ðisne middangeard, Beo. Th. 150 ; B. 75. Ðonne hë ys tóweard on micelre mǽgþe and ða strengstan mǽgþe nú ealra eorþan mǽgþ beóþ on him gebletsode *cum futurus sit in gentem magnam ac robustissimam et benedicendae sint in illo omnes nationes terræ,* Gen. 18, 18. Fremde þeóde, óðre mǽgþe, Ps. Th. 88, 43. Hæfdon ða mǽgþa ǽlcne for ëcne god *the nations held each to be god eternal,* Bt. Met. Fox 26, 98 ; Met. 26, 49. Mǽgþa tída *tempora nationum,* Lk. Skt. 21, 24 : Cd. 124 ; Th. 158, 12 ; Gen. 2616 : Beo. Th. 49 ; B. 25 : 9 ; B. 5. (*c*) *as in the case of proper names the word for the people is used for their country, so province, country* :—Seó mǽgþ West-Seaxna *provincia occidentalium Saxonum,* Bd. 3, 7 ; S. 529, 2. Seó ylce mǽgþe ǽrest ðysne biscop ágenne onfëng *hunc primum eadem provincia proprium accepit praesulem,* 4. 12 ; S. 581,

24. Willferþ bisceop Sûþ-Seaxna mǽgþe (*provinciæ*), 4, 13; S. 581, 37. From Armoricano ðære mǽgeþe, 1, 1; S. 474, 7. Mid his mǽgþe Eást-Englum, 2, 15; S. 518, 27. On Beornicia mǽgþe, 2, 14; S. 518, 14. Hé fêrde geond ealle Angelcynnes mǽgþe *perlustrans universa*, 4, 2; S. 566, 1. Him twâ mǽgþe (*duas provincias*) forgeaf, 4, 13; S. 582, 10. Ða mǽgþe ðe mon hâteþ Gallia Belgica, 1, 1; S. 473, 12. On Palestina ðære mǽgþe, Shrn. 100, 26. On Tiro ðære mǽgþe, Th. Ap. 3, 24; Blickl. Homl. 211, 16; Andr. Kmbl. 528; An. 264. [*Orm.* off Asæress ma33þe.] v. fæderen-, folc-, ge-, mêdren-, sûþ-, wer-mǽgþ.

mǽgþa, an; *m. Maithen, may-weed;* anthemis cotula:—Mǽgþa *herba putida*, Ælfc. Gl. 42; Som. 64, 11; Wrt. Voc. 31, 22: *caluna* (= *calmia*, v. Lchdm. ii. 398, col. 2); 39; Som. 63, 71; Wrt. Voc. 30, 19. Him mon mǽgþan tô mete gegyrede, Lchdm. iii. 34, 11. v. mageþe.

Mǽgþa land *the Polish province of Mazovia* (?) :—Be norþan Horiti is Mǽgþa land; and be norþan Mǽgþa londe Sermende ôþ ða beorgas Riffen, Ors. 1, 1; Swt. 16, 21.

mǽgþ-, mǽgeþ-blǽd, es; *n. Pudendum muliebre:*—Mǽgeþblǽdd *virginal*, Germ. 400, 8. Leo 508, 9 says on this word ' Dieselbe Bedeutung hat Blatt noch in der deutschen Jägersprache : das Blatt einer Ricke, einer Hinde.'

mǽgþ-bôt, e; *f. The fine to be paid by an unmarried woman:*—Mǽgþbôt sî swâ friges mannes *let the fine to be paid by an unmarried woman be the same as that by a free man (for the same offence)*, L. Ethb. 74; Th. i. 20, 9. This regulation follows one that settles the fine to be paid by ' frî wîf locbore.'

mǽgþ-, mǽgeþ-hâd, es; *m.*   **I.** *maidenhood, virginity, celibacy, chastity:*—Ðû cennest cyning ealra clǽnnessa and ðînne mǽgþhâd nô ne gewemmest, Blickl. Homl. 7, 36: Exon. 12 a; Th. 18, 25; Cri. 289: 9 a; Th. 6, 16; Cri. 85: Homl. Th. i. 460, 4. Mǽgþhâd is ǽgðer ge on wǽpmannum ge on wîfmannum. Ða habbaþ rihtne mǽgþhâd ða ðe fram cildhâde wuniaþ on clǽnnesse, 148, 13. Mǽigþhâd, 7. Ðæt sindan ða ða ðe mid wîfum ne beóþ besmitene, and hira mǽgeþhâd habbaþ gehealdenne, Past. 52, 7; Swt. 409, 7. Mǽ[g]þhâdes *virginitatis, puritatis*, Hpt. Gl. 411, 32: *castitatis*, 441, 69: *celibatus, pubertatis*, 453, 56. Hê sceal foresceáwian ðam mǽdene hire mǽgþhâdes wurþ (*pretium pudicitiæ*), Ex. 21, 10: L. Alf. 12; Th. i. 46, 18. Án man ðe sý mǽgþhâdes man, cnapa oððe mǽgeþ, Herb. 104, 2; Lchdm. i. 218, 21. Hire meiþhâdes *pupertatis sue*, Kent. Gl. 26. Ic bidde ðê for Scam. Marian mǽgþhâde, Bt. Fox 260, 3.   **II.** *a body of young persons:*—Mǽgeþhâde *pedagogio*, Wrt. Voc. ii. 77, 30. [*Mark.* meidhad : *Orm.* ma33phadd: *O. H. Ger.* magad-heit *virginitas, pubertas, coelibatus*.] v. mǽgden-hâd.

mǽgþ-hâd, es; *m. Kinship, relationship:*—'Se ðe his brôðor ne lufaþ hê wunaþ on deáþe.' Ealle wê sind gebrôðra ðe on God gelýfaþ and wê ealle cweþaþ ' Ûre Fæder ðe eart on heofonum.' Ne gedyrstlǽce nân man be mǽgþhâde bûtan sôðre lufe ' *he who loveth not his brother continueth in death*' . . . *All we are brethren that believe on God, and we all say* ' *our Father that art in heaven.*' *Let no man presume on kinship without true love*, Homl. Th. i. 54, 6–11.

mǽgþhâd-lîc; *adj. Virgin, virginal:*—Mǽg[þ]hâdlîcre sidefulnysse *pudicitiæ virginalis*, Hpt. Gl. 440, 65.

mǽgþ-lagu = mǽg-lagu q.v., L. C. E. V; Th. i. 362, 28.

mǽgþ-leás; *adj. Belonging to no family, not of distinguished family; ignobilis*, Wrt. Voc. ii. 138, 73.

mǽgþ-mann, es; *m. A maiden, virgin:*—Gif man mǽgþman nêde genimeþ *if a maiden be carried off by force (to be married)*, L. Ethb. 82; Th. i. 24, 3. v. mǽgden-mann.

mǽgþ-sibb, e; *f. Kindred:*—Mǽgþsybbe *parentelæ*, Hpt. Gl. 523, 10. v. mǽg-sibb.

mǽg-tudor, es; *n. That which is produced from the same stock:*—Mǽgtuðre *cognatæ*, Hpt. Gl. 469, 52. Cf. magu-tudor.

mǽg-wine, es; *m. A kinsman and friend:*—Mon mænig be his mǽgwine *many a man standing by his kinsman* (of the people at the tower of Babel), Cd. 80; Th. 100, 9; Gen. 1661. Mǽgwinas mîne, Beo. Th. 4951; B. 2479. Mǽgwinum, Cd. 149; Th. 187, 4; Exod. 146: 158; Th. 197, 28; Exod. 314: Salm. Kmbl. 719; Sal. 359. [*O. Sax.* mâg-wini.]

mǽg-, mêg-wlite, es; *m. Appearance, form, species:* species, forma, aspectus:—Mêgwlit *aspectus*, Mt. Kmbl. Lind. 28, 3. Mǽgwlit (mêg-wlitt, Rush) onsióne his *species vultus ejus*, Lk. Skt. Lind. 9, 29. Tô mǽgwlite andgytes *ad formam sensus*, Bd. 5, 24; S. 647, 34. Ðæt ðû meahte mînum weorþan mǽgwlite gelîc, Exon. 28 b; Th. 87, 30; Cri. 1433. Gedyde ic ðæt ðû onsýn hæfdest, mǽgwlite mê gelîcne, 28 a; Th. 84, 35; Cri. 1384: Andr. Kmbl. 1711; An. 858. Ne mêgulit (mǽgwlit, Rush.) his gesêgon *neque speciem ejus vidistis*, Jn. Skt. Lind. 5, 37. Mêgwlite, Rtl. 2, 7. Mêgewlit Godes *majestatem Dei*, 1, 19. Mon ne mǽge ða lâstas on ôðerne mǽgwlite oncyrran; ah hié â beóþ on ðære ilcan onsýne *the footsteps cannot be changed into another form; but they always appear the same*, Blickl. Homl. 127, 19. Ælc hafaþ mǽgwlite metodes and engla, Cd. 75; Th. 92, 17; Gen. 1530. Monge mǽgwlitas *many species*, Exon. 43 a; Th. 146, 7; Gû. 706: Bt. Met. Fox 31, 9;

Met. 31, 5. Woroldgife monige on misenlîcum mǽgwlitan *dona in diversis speciebus perplura*, Bd. 1, 32; S. 498, 21.

mǽgwlitian *to form, shape:*—Oferhiuad t [ofer] mǽgwlitgad *transfiguratus*, Mt. Kmbl. Lind. 17, 2.

mǽgwlit-lîce; *adv. Figuratively:*—Mêgwlitlîce *figuraliter*, Mk. Skt. p. 4, 10.

mæhe (*for mæhte?*) *dicione*, Wrt. Voc. ii. 27, 75.

mæht, mæhtig. v. meaht, meahtig.

mǽl, mâl, mêl, es; *n. m.* (?)   **I.** *a measure:*—Dô wînes þrié mêl on *pour three measures of wine on*, L. M. 1, 45; Lchdm. ii. 110, 26. v. cucler-mǽl, dæg-mǽl, fot-mǽl, mǽl-tange; *and* cf. *Icel.* mâl *a measure: Dan.* maal.   **II.** *a mark, sign, cross, crucifix:*—Hêr ôþiewde reâd Cristes mǽl on hefenum *in this year a red cross appeared in the sky*, Chr. 773; Erl. 52, 23. Mid ðâm wæs sum mycel gylden Cristes mǽl *in quibus crucem magnam auream*, Bd. 2, 20; S. 522, 9. Hê ðæt Cristes mǽl hrǽde weorce geworhte . . and ðæt Cristes mǽl genam and on ðone seáþ sette, 3, 2; S. 524, 16–18. Bǽron Cristes rôde tâcen sylfrene Cristes mǽl *crucem pro vexillo ferentes argenteam*, 1, 25; S. 487, 3. Ænne sylfrene mǽle on V. pundon *a silver crucifix of five pounds*, Chart. Th. 558, 33. Ðon on ealdan Cristes mǽle ; of ðam Cristes mǽle, Cod. Dip. Kmbl. vi. 66, 34. Ealle hit writen mid Cristes mǽl *all signed it with a cross*, Chr. 963; Erl. 123, 25. v. fýr-mǽl, ge-mǽl, grǽg-mǽl: *O. Sax.* hobid-mâl *head on a coin* and cf. *Icel.* mâl *applied to the inlaid ornamenting of weapons: and English* hring-, wunden-mǽl. *The word is also used for the sword itself* brogden mǽl, Beo. Th. 3236; B. 1616: 3338; B. 1667: Elen. Kmbl. 1574; El. 759. v. mâl-sweord.   **III.** *fixed, suitable, appointed time, season, occasion:*—Mǽl is mê tô fêran *it is time for me to go*, Beo. Th. 637; 316. Ðâ wæs sǽl and mǽl ðæt tô healle gang Healfdenes sunu, 2021; B. 1008. Ðâ ðæs mêles wæs mearc agongen *then was the appointed time past*, Cd. 83; Th. 103, 16; Gen. 1719: 224; Th. 296, 12; Sat. 501. Ic ðæt mǽl geman ðonne wê gehêton ûssum hlâforde *I remember the time when we promised our lord*, Beo. Th. 5259; B. 2633. Ǽlce mǽle *on each occasion*, Exon. 119 a; Th. 457, 30; Hy. 4, 92. Se geweald hafaþ sǽla and mǽla *he hath power over times and seasons*, Beo. Th. 3226; 1611. Efne swylce mǽla swylce . . . *just at such times as . . .*, 2502; B. 1249. Mǽla gehwylce *on every occasion*, 4121; Ps. Th. 118, 62. Ðû him mete sylest mǽla gehwylce *and ðæs tîdlîce tîd gemearcast *tu das escam illis in tempore opportuno*, 144, 16: 21. Ðæt ǽr feala mǽla behýded wæs *which long before was hidden*, Elen. Kmbl. 1971; El. 987. Ærran mǽlum *on former occasions*, Beo. Th. 1819; B. 907: 4466; B. 2237: 6062; B. 3035.   **IV.** *the time for eating, a meal:*—Ðâs hâlgan lenctenlîce tîde gehealdan mid clǽnum fæstene ǽlce dæge tô ânes mǽles (*having only one meal a-day*, cf. *Icel.* fasta einmælt), Wulfst. 285, 2. Hê gereordade æt ânum mǽle fîf þûsend manna *he fed at one (meal) time five thousand men*, 293, 27. Yfel biþ ðæt man rihtfæstentîde ǽr mǽle ete, L. C. S. 47; Th. i. 402, 24: Homl. Th. ii. 590, 25. Gifernys biþ ðæt se man ǽr tîman hine gereordige oððe æt his mǽle tô micel þicge *it is greediness when a man eats before the time or takes too much at his meal*, 218, 30. Ne fæsþ se nô Gode ac him selfum se ðe ðæt nyle þearfum sellan ðæt hê ðonne on mǽle lǽsþ ac wile hit healdan est tô ôðrum mǽle *non Deo, sed sibi quisque jejunat, si ea quæ ventri ad tempus subtrahit, non egenis tribuit sed . . custodit*, Past. 43, 8; Swt. 317, 4. Mûþa gehwylc mete þearf mǽl sceolon tîdum gongan *every mouth needs meat; meals must there be at times*, Exon. 91 a; Th. 341, 13; Gn. Ex. 125. [*Laym. Orm.* mæl: *O. E. Homl. A. R.* mel: *Chauc.* mel, meel *a meal : Prompt. Parv.* meel *pastus: Goth.* mêl *a time: Icel.* mâl *time, meal-time, season : O. H. Ger.* mâl *time, occasion : M. H. Ger.* mâl: *Ger.* ein-mal, etc.: *M. H. Ger.* mâl *time for eating, meal: Ger.* mahl.] v. -mǽlum.

mǽl, e; *f. A speech, talk, conversation:*—Gemuna ða mǽla ðe wê oft æt meodo sprǽcon *think of the talks that we oft had at table*, Byrht. Th. 137, 66; By. 212. [*Icel.* mâl; *n. speech, colloquy, talk.*] v. mǽlan.

mǽl, es; *n. A cause, suit, action* (?):—Ðû symle furðor feohtan sôhtest mǽl ofer mearce *thou didst ever press on to fight, didst pursue thy cause* (i. e. *carry on war*) *over the border*, Wald. 1, 33; Vald. 1, 19. Cf. *Icel.* mâl *a suit, cause;* sǽkja mâl *to prosecute* (as a law term). Stephens takes mǽl here = *mark, goal* : Rieger (quoted by Grein) takes it = *gemôt, concio*, so figuratively battle. v. mâl.

mǽlan; *p.* de *To speak:*—Se stân mǽlde for mannum *the stone spake before men*, Andr. Kmbl. 1533; An. 768. Wîcinga âr wordum mǽlde, Byrht. Th. 132, 35; By. 26: 133, 1; By. 43: 137, 63; By. 210. Hyre se feónd oncwæþ, wordum mǽlde, Exon. 70 b; Th. 263, 18; Jul. 351. Be eów Essaias for weorodum wordum mǽlde, Elen. Kmbl. 702; El. 351. Him ðâ tô wuldorgâst wordum mǽlde, Cd. 141; Th. 176, 16; Gen. 2913. Him Andreas wið, wine þearfende, wordum mǽlde, Andr. Kmbl. 600; An. 300. Him ðâ tôgênes ða gleáwestan wordum mǽldon, Elen. Kmbl. 1072; El. 537. Hwæt mê God on mînum môdsefan mǽlan wille *quid loquatur in me dominus*, Ps. Th. 84. 7. [*Orm.* mælenn: *Havel.* mele: *Icel.* mǽla *to speak.*] v. ge-, on-mǽlan.

mǽlan *to mark.* [*Goth.* mêljan *to write: O. Sax.* mâlon *to mark* (of

a wound made by a sword): *O. H. Ger.* mālôn, mālēn *pingere: Ger.* malen *to paint.*] v. hring-, scîr-mǽled; mǽl **II.**

mǽlan *to spot, blemish.* v. ge-mǽlau, mál, un-mǽle.

mǽl-cearu, e; *f. Care or trouble belonging to a particular time:*— Swā ða mǽlceare maga Healfdenes singala seáþ *so did Healfdene's son ever brood over the trouble of that time,* Beo. Th. 380; B. 189.

mǽl-dæg, es; *m. A day, season, an appointed time:*—Hē dæs mǽldæges self ne wēnde dæt him Sarra bringan meahte on woruld sunu *he himself never hoped for the day when Sarah could bring him a son into the world,* Cd. 107; Th. 141, 4; Gen. 2339. Hē moncynnes mǽste hæfde on dǽm mǽldagum mægen and strengo, 79; Th. 98, 18; Gen. 1632.

mǽl-dropa, an; *m. Phlegm:*—Mǽldropa *flegma.* i. *saliva,* Wrt. Voc. ii. 149, 39.

mǽl-dropiende *phlegmatic;* flegmaticus, Ælfc. Gl. 77; Som. 72, 13; Wrt. 45. 47.

Mǽldūn MALDON *in Essex,* Chr. 913; Erl. 102, 5: 920; Erl. 104, 32: 993; Erl. 132, 5.

mǽle *spotted.* v. un-mǽle.

mǽl-gesceaft, e; *f. That which happens at its appointed time in accordance with the decrees of fate:*—Ic bâd mǽlgesceafta *I waited for that which in due time fate would assign me,* Beo. Th. 5467; B. 2737.

mǽl-mete, es; *m. Food to eat:*—Ne biþ ðec mǽlmete nymþe môres græs *no food shall there be for thee but the grass of the moor,* Cd. 203; Th. 252, 7; Dan. 575. [Grein, *quoting Dietrich, would read* mǽl mēte (= *obvius*), v. Hpt. Zeitsch. x. 358.]

mǽl-sceafa, an; *m. A canker:*—Mǽlscæafa *eruca,* Ælfc. Gl. 23; Som. 60, 3; Wrt. 24, 7. Mǽlsceafa *caniglata,* Wrt. Voc. ii. 128, 19. Mǽlsceafa *eruca,* Wrt. Voc. 78, 66; Zup. 310, 5. In the last reference one MS. (v. Wrt. Voc. 91, 23) has *mæslesceafe;* in Wrt. Voc. 161, 23 *maseles* translates *rugeroles* (see also Skeat's Dict. s.v. *measles*), so *mǽl,* in this word, would mean a *spot.*

mǽl-tange, an; *f.* -tang, es; *m.* (?) *A pair of compasses:*—Mǽltange *circinum,* Ælfc. Gl. 49; Som. 65, 70; Wrt. Voc. 34, 5: 62; Som. 68, 78; Wrt. Voc. 39, 61. Mǽltanges *pinca centrum,* 39, 62.

-mǽlum *-meal* (in *piece-meal*). v. æcer-, bit-, dǽl-, drop-, flocc-, folc-, fôt-, heáp-, hîd-, lim-, nam-, sceáf-, stæp-, stund-, stycce-, þrag-, þreát-, þūsend-, worn (wearn)-, wrǽd-mǽlum.

mǽnan; *p.* de *To mean.* **I.** of persons (*a*) *to intend to convey a certain sense:*—Gif hē of wege ǽnigne gebrohte .. dæt is dæt ic mǽne gif hē ǽnigne man on synne bespeóne *if he have brought any man out of the way ..., what I mean, is, if he have lured any man to sin,* L. Pen. 16; Th. ii. 284, 12. Hwet mǽnde Crist ðā cwæþ: 'Ða unrihtwîsan faraþ on ǽce wītu,' Shrn. 197, 18. God ðā geopenude Abrahame hwæt hē mid dǽre sprǽce mǽnde, Gen. 18, 20. (*b*) *to intend to indicate a certain person or thing without direct statement:*—Cweþan swā hē tô ânum sprece and hwædre ealle mǽneþ *to say, as if he speaks to one and yet means all,* Exon. 28 a; Th. 84, 24; Cri. 1378. Hē gecýdde dæt hē ne mǽnde (*indicaret*) ðis andwearde līf, Past. 50, 2; Swt. 389, 22. Hwylc beren mǽnde hē ðonne elles bûton heofona rîce, Blickl. Homl. 39, 27. Crist mǽnde ðone ēcan deáþ . . . ða Iudēiscan mǽndon ðisne andweardan deáþ, Homl. Th. ii. 232, 20. Ne mǽnde ûre Drihten mid ðisum wordum ða treówa ðe on appeltûne wexaþ, 406, 9. (*c*) *to mean, purpose, have as an object to which the mind is directed, intend:*—Gif hē ðara nân ne dēþ ðonne nât hē hwæt hē mēnþ (Cott. MS. mænþ) *if he does none of these, then he does not know what he means,* Bt. 38, 2; Fox 198, 28. Ðā ongon hē sprecan swîðe feorran ymbûtan swilce hē nā ða sprǽce ne mǽnde, 39, 5; Fox 218, 12. Hwæt ðū ðonne mǽne mid dǽre gîtsunge dæs feós *what do you mean by the greed of money?* 32, 1; Fox 114, 7. **II.** (of things) *to signify, have a certain signification or purpose:*—Saga hwæt ic mǽne, Salm. Kmbl. 472; Sal. 236: Exon. 124 b; Th. 479, 18; Rä. 62, 9. Oft gehwā gesihþ fægre stafas and nât hwæt hî mǽnaþ, Homl. Th. i. 186, 3. Hwæt mǽnde dæt syxtig wera strongera? Blickl. Homl. 11, 22: Homl. Th. ii. 234, 31. Faraþ and leorniaþ hwæt dæt mǽne: 'Ic wylle mildheortnysse, and nā offrunge,' 470, 18. Geleornian hwæt fulluht mǽne, Wulfst. 123, 4. Understandan hwæt ða twā word mǽnan, *abrenuntio* and *credo,* 38, 8. [*O. Sax.* mēnian: *O. Frs.* mēna: *O. H. Ger.* meinian *dicere: Ger.* meinen.] v. ge-mǽnan.

mǽnan; *p.* de *To tell of, relate, declare:*—Ne wyrneþ word lofes, wîsan mǽneþ mîne for mengo (cf. *O. Sax.* thû fora thesaro thiod telis, mahtig mēnis), Exon. 105 b; Th. 401, 14; Rä. 21, 11. Hæleþ hý hospe mǽnaþ *men speak of her contemptuously,* 90 a; Th. 337, 17; Gn. Ex. 66. Secgas nemnaþ, mǽnaþ mid mûþe modugâles gedrinc, 88 a; Th. 330, 26; Vy. 57. Ðý læs dæt weras gieddum mǽndan be mē lîfgendum *lest men should tell of it in songs during my lifetime,* 50 b; Th. 176, 9; Gū. 1206. Ic mæg singan and secgan, spell mǽnan, hū mē cynegôde cystum dohten, 85 b; Th. 321, 32; Wîd. 55: Beo. Th. 2139; B. 1067. Ðǽr wæs Beówulfes mǽrþo mǽned *there was told Beowulf's greatness,* 1718; B. 857. [*O. Sax.* mēnian, gi-mēnian *to make known: O. H. Ger.* meinian *dicere:* ga-meinian *dicere, dicare,* Grff. ii. 785, 788.]

mǽnan; *p.* de *To lament, mourn, complain.* **I.** *intrans.:*—Ðū simle mid wôpe and mid unrôtnesse mǽnst gif ðē ǽnies willan wana biþ

(*tu*) *qui abesse aliquid tuae beatitudini tam luctuosus atque anxius conquereris,* Bt. 11, 1; Fox 30, 22. Ðā hē gehiérde dæt dæt folc mǽnde tô him Arone ymb hiera earfeðo *Moyses cum contra se et Aaron conqueri populum cognovisset,* Past. 28, 6; Swt. 201, 4. Ealfē wordum mǽndon, Cd. 222: Th. 288, 24; Sat. 386. **II.** *followed by a clause:*— Ða welan ðe dū mǽndest dæt dū forlure *the wealth which you complain of having lost,* Bt. 7, 3; Fox 20, 18. Bonan mǽndon dæt hý monnes bearn oferþunge, Exon. 38 b; Th. 128, 8; Gū. 401. **III.** *with acc.:*—Hū Boetius his earfoðu tô Gode mǽnde, Bt. tit. cap. 4. His tungan he mǽnde swiðost *he complained most of his tongue,* Homl. Th. i. 330, 31. Basilius mǽnde dæt unriht, Homl. Skt. 3, 322. Hē misbeád his munecan and ða munecas hit mǽndon lufelîce, Chr. 1083; Erl. 217, 4. Hî mǽndon mondryhtnes cwealm *they mourned their lord's death,* Beo. Th. 6289; B. 3149. Ic wundrige hwæt ðē seó oððe hwæt ðū mǽne *admiror cur aegrotes,* Bt. 5, 3; Fox 12, 11. Hū miht ðū mǽnan dæt wyrse nū ðū dæt leófre hæfst gehealden *poterisne, meliora quæque retinens, de infortunio jure caussari?* 10; Fox 28, 10. Cyning mǽnan *to mourn their king,* Beo. Th. 6324; B. 3172. Ic gehēre gnorniende cynn grundas mǽnan (*the devils in hell*), Cd. 216; Th. 273, 10; Sat. 134. Ðæt ic sceal teárum mǽnan *that I must mourn with tears,* Exon. 76 a; Th. 285, 10; Jul. 712. v. bemǽnan.

mǽne; *adj.* **I.** *mean, wicked, false, evil:*—Synna lustas mǽne môdlufan *the pleasures of sin, vicious love,* Exon. 71 a; Th. 364, 26; Jul. 370. Hygeleáse mǽne *mad and false* (*the rebel angels*), Cd. 4; Th. 4, 11; Gen. 52. Þurh mǽnra hand searonettum beseted, Andr. Kmbl. 1882; An. 943. **II.** the word however occurs most often in reference to oaths:—Se ðe his þances mǽnne āþ swerige and hē wite dæt hē mǽne biþ æfter ðam *qui sua sponte perjuraverit et postea scit quod perjurus est,* L. Ecg. C. 34; Th. ii. 158, 20, 14, 16. Gif hwā swereþ and se āþ beó mǽne . . se ðe mǽne āþas begā *si quis juraverit et perjurium sit . . . Qui perjuria commiserit,* L. Ecg. P. iv. 68; Th. ii. 228, 7–9: L. Edg. C. 8; Th. ii. 262, 31. Gif mæssepreóst stande on leásre gewitnesse oððe on mǽnan āþe *if a masspriest be concerned in false witness or perjury,* L. Eth. ix. 27; Th. i. 346, 9: L. C. E. 5; Th. i. 362, 30. Se ðe mānāþ (*other MS.* mǽnne āþ) swerige, L. Ath. i. 25; Th. i. 212, 18. Be mǽnan āþe. Gif hwā mǽne āþ swerige, L. C. S. 36; Th. i. 398, 3–4. Gebēte ðone mǽnan āþ, L. In 35; Th. i. 124, 13. Swerian mǽnne āþ þurh swā miclan mægenþrymme, Wulfst. 214, 15. Eall yfel forlǽtan ge on manslihte ge on mǽnum āþum, 228, 21. v. un-mǽne; mān.

mǽne; *adj. Common:*—Mǽna lǽse *common pasturage,* Cod. Dip. Kmbl. iv. 284, 8. v. ge-mǽne.

mǽngan, Mǽn-īg, mǽnig, mǽnigeo, mǽnnisc. v. mengan, mon-īg, manig, menigu, mennisc.

mǽnoe. v. mene.

mǽnsumian; *p.* ode. **I.** *to have the companionship of a person, to marry:*—Ne hiá mǽnsumiaþ (mǽnsumigaþ, Rush.) ne hiá biþon gemǽnsumad (*i. ne ceorl hæfis wifes gemǽna ne wîf hæfis ceorles*) *neque nubent neque nubentur,* Mk. Skt. Lind. 12, 25. **II.** *to share with another, to communicate:*—Mǽnsumede *participavit, communicavit,* Hpt. Gl. 467, 2.

mǽnsumung, e; *f.* **I.** *communion, admission to fellowship with others* (opp. of *excommunication*):—Benedictus cwæþ dæt hî unāmánsumode wǽron . . Hî underfēngon ða hálgan mǽnsumunge æt Gode þurh his þeówan Benedicte, Homl. Th. ii. 174, 31. **II.** *participation:*—Hē ûs forgeáfe dǽl on his rîce, and mǽnsumunge on his godcundnysse, 174, 10, 11.

mǽntel. v. mentel.

mǽr. v. wudu-mǽr.

mǽra, mera, an; *m. An incubus:*—Mera † satyrus *incuba,* Ep. Gl. 12 f, 14. v. mǽre.

mǽr-āc, e; *f. An oak which serves as part of a boundary* (?):—Of ðǽre āc in ða mǽrāc, Cod. Dip. Kmbl. iii. 379, 31. v. mǽr-brôc, mearc-bēce.

mǽran, mǽran; *p.* de *To make known, celebrate, declare, proclaim:*— Mîn mūþ sægeþ ðîne mægenspēde and ðîn sôþfæst weorc mǽreþ *os meum pronuntiabit justitiam tuam,* Ps. Th. 70, 14. Songe lofiaþ mǽraþ môdigne meaglum reordum *they praise with song and with powerful voices celebrate the noble bird,* Exon. 60 b; Th. 221, 21; Ph. 338. For cyning mǽraþ leófne leódfruman *they proclaim the loved chief as king,* Th. 222, 6; Ph. 344. Swylce mîn tunge tîdum mǽrde ðîn sôþfæst weorc *sed et lingua mea tota die meditabitur justitiam tuam,* Ps. Th. 70, 22. Ðæt hî heora bearnum budun and sægdun and cinn ôðrum cýðden and mǽrden *ut notam faceret eam filiis suis; ut cognoscat generatio altera,* 77, 7. Gē scyldigra synne secgaþ, sôþfæstra nô môd and monþeáw mǽran willaþ, Exon. 40 a; Th. 132, 26; Gū. 478. Hit nǽnig mon ðû cýðan ne môste, ðý læs ða elreordigan kyningas on ðæt fǽgon, dæt ic swā lytle hwîle lifgean môste. Ne hit ǽnig mon ðǽre ferde ðûn mā ût mǽran môste, ðý læs hié for ðon ormôde wǽren on, Nar. 32, 22. [*Goth.* mērjan *to proclaim, announce: O. Sax.* mārian: *Icel.* mǽra *to praise: O. H. Ger.* mârian *diffamare, declarare, clarificare, praedicare.*] v. ge-mǽran.

mǽr-apeldre, an; *f. An apple-tree which serves as a boundary:*— Hit cymeþ tô mǽrapeldran, Cod. Dip. Kmbl. iii. 390, 5.

mǽr-bróc, es; m. *A brook which forms a boundary*, cf. mearc-bróc :—Tó mǽrbróce; of mǽrbróce, Cod. Dip. Kmbl. iii. 79, 5 : 438, 27 : v. 284, 29 (*where* mǽr-bróc *is the same as* merc-bróc *of* l. 13). v. mere *and* mǽre *a boundary*.

mǽrc. v. mearc, mearh.

mǽr-díc, e ; f. *A boundary dike* :—On ða mǽrdíc, Cod. Dip. Kmbl. iii. 378, 24. On ða ealdan mǽrdíc, 449, 10.

mǽre *a mere.* v. mere.

mǽre, mare, mere, an ; f. *A night-mare, a monster oppressing men during sleep* (cf. passage quoted in Cl. and Vig. under *mara* : 'En er hann hafði lítt sofnat, kallaði hann ok sagði at mara trað hann. Menn hans fóru til, ok vildu hjálpa honum ; en er þeir tóku uppi til höfuðsins, þá trað hón fótleggina svá at nær brotnuðu. þá tóku þeir til fótanna, þá kafði hón höfuðit, svá at þar dó hann') :—Mǽre *faecce*, Wrt. Voc. ii. 108, 44 : *incuba*, 111, 46. Mere *fecce*, 35, 26. Gif mon mare ríde, L. M. 1, 64 ; Lchdm. ii. 140, 9. Hí beóþ góde wið nihtgengan and maran, 3, 1 ; Lchdm. ii. 306, 12. [*Prompt. Parv.* mare or nyȝhte mare *epialtes* ; mare or wyche *magus, maga, sagana*, and see note, p. 326 : *Icel.* mara : *M.H.Ger.* mare : *Ger.* mahr : cf. *French* cauchemar.] v. mǽr, mǽra.

mǽre, es ; n. *A boundary, limit, confine, border* :—Ondlong ðæs mǽres (meres?) heges, Cod. Dip. Kmbl. iii. 32, 30 : ii. 250, 7 (?). In mǽre Judéana *in fines Judæa*, Mt. Kmbl. Rush. 19, 1. In mǽrum *in villas*, Mk. Skt. Lind. 6, 56. In mǽrum (mǽro, Rush.) *in vicos*, Lk. Skt. Lind. 14, 21. [Cf. *Icel.* mǽrr *a border-land*.] v. ge-mǽre ; mǽr-ác, -apeldre, -bróc, -díc, -heg, -stán, -þorn, -weg.

mǽre ; adj. *Great, excellent, distinguished, illustrious, sublime, splendid, celebrated, famous, widely known* (of persons or things) :—Mǽre *clarus, insignis, nobilis, perspicuus*, Wrt. Voc. ii. 131, 66 : *inclytus*, 46, 10, 11. Mǽre weard *percrebuit*, Ep. Gl. 18 b, 10. Mǽre *celeber*, Ælfc. Gr. 9, 18 ; Zup. 44, 10. Mǽrne *celebre*, Hpt. Gl. 525, 45. Beorht ł mǽre *præclara, splendida*, 436, 43. Mǽr[re] *illustrius*, 460, 25. I. (*of persons and* (a) *in a good sense*) :—Dryhten ys mǽre God and mihtig *Dominus est deus magnus et potens*, Deut. 10, 17. Ðú eart mǽre God, and Jacobes God se mǽra, Ps. Th. 83, 8 : 103, 23. God mǽre (*excelsus*) álysend heora is, Ps. Spl. 77, 39. Freá ælmihtig, mǽre þeóden, Cd. 40 ; Th. 52, 34 ; Gen. 853. Se mǽra Fæder (*God*), L. Ælfc. C. 3 ; Th. ii. 344, 4. Hé byþ mǽre beforan Drihtne *erit magnus coram domino*, Lk. Skt. 1, 15 : 32. Ðeáh hé on ðam lande seó mǽre ðonne biþ hé on óðrum unmǽre *though he be famous in one country, he is not in another*, Bt. 30, 1 ; Fox 108, 15. Wæs hé (*St. Martin*) swíðe mǽre geond middangeard, Blickl. Homl. 221, 1. Mǽru cwén *the illustrious queen* (*Wealhtheow*), Beo. Th. 4037 ; B. 2016. Sunu se ðe biþ góde mǽre *a son* (*Isaac*) *who shall be great in goodness*, Cd. 100 ; Th. 132, 24 ; Gen. 2198 : Beo. Th. 3909 ; B. 1952. Mihtum mǽre *great in power*, Elen. Kmbl. 679 ; El. 340. Marian mǽrre meówlan *of Mary, maiden illustrious*, Exon. 14 a ; Th. 28, 13 ; Cri. 446. Smeágende cwidas and dǽda ðara mǽrena (*illustrium*) wera úre þeóde, Bd. pref. ; S. 471, 13. Ðes ys mǽrra (*major*) ðonne ðæt templ, Mt. Kmbl. 12, 6. Nis betwux wífa bearnum nán mǽrra wítega ðonne Johannes, Lk. Skt. 7, 28. Nán man ne biþ for óðres góde nó ðý mǽrra ne nó ðý geheredra *splendidum te aliena claritudo non efficit*, Bt. 30, 1 ; Fox 108, 27. David wæs hearpera mǽrost, Ps. C. 50 ; Ps. Grn. ii. 276, 4. Ðás mánfullan men wǽron getealde for ða mǽrostan godas, Wulfst. 106, 17. (*b*) *in a bad sense, notorious, distinguished by evil deeds* ; insignis :—Hæfdum ǽnne gebundenne mǽrne (mérne, Lind.) monn se wæs háten Barrabas (cf. *O. Sax.* mári meginthiof) *habebat vinctum insignem qui dicebatur Barabbas*, Mt. Kmbl. Rush. 27, 16. Grendel, þære mearc-stapa, Beo. Th. 206 ; B. 103 : 1528 ; B. 762 (?). II. (*of things*) :—Sum deófolgild ðe mid ðǽm hǽðenum mannum swíðe weorþ and mǽre wæs *a certain idol that was held in high honour and esteem among the heathens*, Blickl. Homl. 221, 7. Swíðe nǽre burh se is háten Sepontus *a very famous town which is called Sepontus*, 197, 20. On ðam mǽran (*industri*) túne, se is nemned æt Walle, Bd. 3, 21 ; S. 551, 11 : Cd. 205 ; Th. 254, 10 ; Dan. 609. Tó ðære mǽran byrig (*the heavenly Jerusalem*), 227 ; Th. 304, 4 ; Sat. 624. Tempel heáhst and háligost, hæleþum gefrǽgost, mǽst and mǽrost (*Solomon's temple*), 162 ; Th. 202, 28 ; Exod. 395. Ðæt wæs ðæt mǽreste hús ðe on eorþan geworht wurde *that* (*the temple*) *was the most splendid house that was built in the world*, Wulfst. 278, 1. Mǽre wurdon his wundra geweorc wíde and síde *far and wide spread the fame of the wonders he wrought*, Exon. 45 b ; Th. 155, 1 ; Gú. 853. Eall ðeós mǽre gesceaft *the universe*, Rood Kmbl. 24 ; Kr. 12. Mǽre wundur *mirabilia*, Ps. Th. 106, 30 : 110, 3. Sunne mǽre tungol *the sun, resplendent star*, Chr. 937 ; Erl. 112, 14 ; Æðelst. 14. Mǽrost tungla, Exon. 57 b ; Th. 205, 28 ; Ph. 119. In dege mǽrum *in die insigni*, Ps. Surt. 80, 4. Ðone mǽron symbeldæg Drihtnes upstige, Blickl. Homl. 131, 10 : Cd. 8 ; Th. 10, 11 ; Gen. 1550. Seó mǽre tiid (*Easter*), Menol. Fox 114 ; Men. 57. Se mǽra dæg *the great and terrible day of the Lord*, Exon. 23 b ; Th. 65, 16 ; Cri. 1055. Ðæt is mǽre spell *no common tale is that*, Cd. 119 ; Th. 155, 2 ; Gen. 2566 : Elen. Kmbl. 1936 ; El. 970. Æfter ðisse dǽde his noma wes weorþ and mǽre geworden *after this deed his name became honoured and*

famous, Blickl. Homl. 219, 4 : Exon. 107 a ; Th. 409, 11 ; Rü. 27, 27. Is wuldur ðín wíde and síde ofer ðás eorþan ealle mǽre *in omnem terram gloria tua*, Ps. Th. 56, 6. Se mǽresta hlísa *fama celeberrima*, Bd. 3, 13 ; S. 538, 37. Ðæt is mǽro wyrd *that is a tremendous event* (*the deluge*), Cd. 69 ; Th. 84, 18 ; Gen. 1399. Ðín mægen is swá mǽre, swá ðæt ǽnig ne wát eorþbúende ða deópnesse Drihtnes mihta, Hy. 3, 31 ; Hy. Grn. ii. 282, 31. (*In a bad sense*) Caudenes Furcules seó stów gewearþ swíðe mǽre for Rómana bismere *Caudinas furculas satis celebres et famosas Romanorum fecit infamia*, Ors. 3, 8 ; Swt. 120, 21. [Cf. *Goth.* waila-mérs *of good report* ; wailaméreins *good report* : *O. Sax.* mári : *Icel.* mærr : *O. H. Ger.* mári *memorabilis, famosus, illustris, insignis, clarus.*] v. efen-, folc-, fore-, forþ-, freá-, frǽ-, heaðo-, un-, wíd-mǽre.

mǽre *pure, in the phrase* mǽre peningas = *Lat.* meri denarii i.e. coins made of pure silver, v. Ducange s.v. *merus*, quoted by Schmid. The passage in which the word is found occurs in L. Alf. pol. 3 ; Th. i. 62, 10 :—Mid V. pundum mǽrra pæninga. With this may be compared the following passage :—For his lícweorðan feó, ðæt is ii pund mérehwítes seolfres, Cod. Dip. Kmbl. iii. 255, 12.

mǽrels, mǽrels, es ; m. and **mǽrels-ráp**, es ; m. *A rope for mooring a ship* ; pronesium [v. Ducange : '*pronexium* funis quo navis religatur ad palum'] :—Mǽrelsráp *pronesium*, Ælfc. Gl. 105 ; Som. 78, 21 ; Wrt. Voc. 57, 3. Mǽrels *prosnesium*, 63, 62. [Both words occur in lists giving the names of ships, and their various parts. Cf. *Du.* marlijn, *also* marl-reep = mar-reep *a marline, a small cord used for binding large ropes, to protect them* : *O. Du.* maren *to tie knots*, which occurs in English in the phrase *to moor a ship. Also* cf. marlyñ *illaqueo*, marlyd *illaqueatus*, Prompt. Parv. 327, and note.] v. scip-mǽrels.

mǽre-torht. v. mere-torht.

mǽrh. v. mearh.

mǽr-heg, es ; m. *A boundary* (?) *hedge* :—Ondlong ðære búrnan óþ hit cymeþ tó ðæm mǽrhege ; ondlong ðæs mǽres heges ðæt hit cymeþ up on ða dúne, Cod. Dip. Kmbl. iii. 32, 29. Cf. gemǽr-haga.

mǽr-hlísa, an ; m. *Great fame, celebrity* :—Mid mǽrhlísan *cœlebri*, Wrt. Voc. ii. 23, 74.

mǽrian ; p. ode *To become great, be distinguished* :—Swá mǽregend[iend]um cýðere *tanto prestanti martiri*, Hymn. Surt. 46, 3.

mǽring *a plant name* :—Hwít mǽringc (Cockayne suggests *sweet basil*), Lchdm. iii. 2, 21.

mǽr-líc ; adj. *Great, magnificent, glorious, splendid, illustrious* (of persons or things) :—Mǽrlíce *magnificas*, Gl. Wülck. 254, 11. I. (*of persons*) :—Mǽrlíc (*God*) on hálignysse *magnificus in sanctitate*, Cant. Moys. 11. Ðæt wæter feóll ofer Pharaones mǽrlícum riddum *the water fell upon Pharaoh's splendid knights*, Ælfc. T. Grn. 5, 31. II. (*of things*) :—Mýrlíc cynehelm *corona inclita*, Kent. Gl. 67. Gabrihel bodade Zacharian his mǽrlícan drohtnunge *Gabriel announced to Zacharias his* (*John's*) *glorious life*, Homl. Th. i. 352, 26. Ðá hæfde ðæt cild swíðe mǽrlíce stemne *the boy had a magnificent voice*, Wulfst. 152, 11. Hwæðer má mǽrlecra dǽda gefremed hæfde ðe Philipus ðe Alexander *which had performed more splendid deeds, Philip or Alexander*, Ors. 3, 9 ; Swt. 130, 27. Hwæðer ðe ðonne þynce unweorþ and unmǽrlíc seó gegaderung ðara þreóra þinga . . oððe hwæðer hit ðé þince eallra þinga weorþlícost and mǽrlícost *obscurumne hoc, atque ignobile censes esse, an omni celebritate clarissimum*? Bt. 33, 1 ; Fox 120, 31. [*O. Sax.* már-lík : *O. H. Ger.* mári-líh.] v. fore-, un-mǽrlíc.

mǽrlíce ; adv. *Magnificently, excellently, nobly, splendidly, with distinction* :—Mǽrlíce *insigniter*, Wrt. Voc. ii. 85, 81 : Hpt. Gl. 512, 47. Ðam sý mǽrlíce mægen and wurðment bútan ænde *cui sit magnifice virtus et honor sine fine*, Hymn. Surt. 47, 32 ; Hy. 7, 19 ; Hy. Grn. ii. 287, 19. Hé mǽrlíce weorhte *magnifice fecit* (he hath done excellent things, A. V.), Cant. Es. 5. Sum welig man . . dæghwamlíce mǽrlíce (*splendide*) leofode, Homl. Th. i. 328, 13. Joseph leofode on ðam lande (*Egypt*) mǽrlíce, Ælfc. T. Grn. 5, 8. Hwæt is ðes mihtiga ðe ðus mǽrlíce færeþ (*Christ entering Jerusalem*), Blickl. Homl. 71, 14. Mǽrlíce ðæt líc behwurfon mid micelum wópe *celebrantes exequias planctu magno*, Gen. 50, 10. Healdaþ ðisne dæg on eówerum gemynde and freólsiaþ hine mǽrlíce, Homl. Th. ii. 264, 15. Swá hé ús mǽrlícor gifeþ swá wé him mǽrlícor þancian scylon *the more excellent his gifts are, the more excellent ought our thanks to be*, Wulfst. 261, 20. [*O. Sax.* mǽr-líko.]

mǽr-ness, e ; f. *Greatness, distinction, celebrity* :—Mycelnesse ł mǽrnesse *magnitudinis*, Ps. Lamb. 144, 3. Mǽrnesse *insignia*, Wrt. Voc. ii. 45, 12. Mǽrnessa *preconia*, 66, 39. v. fore-mǽrness.

mǽr-pytt, es ; m. *A pit that forms part of a boundary* (?) :—On ðone mǽrpyt ; of ðam pytte, Cod. Dip. Kmbl. iii. 439, 1. Eást tó mǽrpytte, ii. 250, 5.

mǽrsere, es ; m. *One who proclaims or makes widely known, a herald* :—Mérseris *preconis*, Rtl. 56, 35.

mǽrsian ; p. ode. I. *to make great, extend* :—Hig tóbrǽdaþ hyra heálsbǽc and mǽrsiaþ heora reáfa fnadu *dilatant philacteria sua, et magnificant fimbrias*, Mt. Kmbl. 23, 5. II. *to make known, spread the knowledge of anything, declare, proclaim, announce, celebrate* :—Ic mǽrsige *insignio*, Ælfc. Gr. 30 ; Som. 34, 60. Mǽrsaþ

tunge mín spǽce ðíne *pronuntiabit lingua mea eloquium tuum*, Ps. Lamb. 118, 122. Wē mērsiaþ *prædicamus*, Rtl. 71, 25 : 6, 11. Ðíne mægenstrengþu mǽrsien wíde *magnitudinem tuam narrabunt*, Ps. Th. 144, 6. Ðǽr gǽsta gedryht Hǽlend hergaþ, and heofoncyninges meahte mǽrsiaþ, singaþ Metude lof, Exon. 64 b; Th. 239, 6; Ph. 617. Sceal manna gehwylc weorc Godes wíde mǽrsian (*annuntiaverunt*), Ps. Th. 63, 8. Wuldur ðín wíde mǽrsian (*cantare*), 70, 7. Mǽrsiga ðæt word *diffamare sermonem*, Mk. Skt. Lind. 1, 45. Ðætte hiá ne mǽrsades hine *ne manifestarent eum*, 3, 12. Ðæt is ðæt mon his mearce brǽde ðæt mon his hlísan and his naman mǽrsige *terminum suum dilatare, est opinionis suæ nomen extendere*, Past. 48, 2; Swt. 367, 14. Mǽrsedon *celebrabant*, Hpt. Gl. 514, 21. Mǽrsud [wearþ] *crebruit*, Wrt. Voc. ii. 23, 71. Ðǽr hǽlo untrumra manna and neáta mǽrsode syndon *sanitates infirmorum et hominum et pecorum celebrari non desinunt*, Bd. 3, 9; S. 533, 19. **III.** *to celebrate* (*a particular event, season, &c.*):— His symbeldæg wē mērsiaþ *ejus natalitia celebramus*, Rtl. 44, 30. Be ðisse hálgan tíde (*birthday of John the Baptist*) weorþunga ðe wē nú tódæg mǽrsian sceolan . . . swíðe ús is ðes dæg tó mǽrsienne . . nǽniges Godes háligra gebyrd ciricean ne mǽrsiaþ, nemþe Cristes sylfes and ðyses Johannes, Blickl. Homl. 161, 4–11: Bd. 5, 10; S. 625, 19: Homl. Th. i. 324, 8. Wē ðe his ǽriste mǽrsiaþ, Blickl. Homl. 91, 8. Swēg mǽrsiendes *the voice of one celebrating a festival*; *sonus epulantis*, Ps. Lamb. 41, 5. **IV.** *to celebrate, perform a rite, ceremony, &c. with due solemnity*:— Ða hálgan gerȳne mǽrsian *sacra mysteria celebrare*, Bd. 1, 27; S. 496, 23. Ða symbelnysse tó mǽrsianne massæsanges *missarum sollemnia celebrandi*, S. 497, 1: 2, 5; S. 507, 12. **V.** *to magnify, exalt, praise, glorify*:— Clypa mē on dæge ðínre gedréfednysse and ic ðē áhredde and ðū mǽrsast mē *invoca me in die tribulationis; eripiam te, et magnificabis me*, Homl. Th. ii. 126, 8. Mǽrsa ðínne Sunu ðæt ðín Sunu ðē mǽrsige *clarifica filium tuum ut filius tuus clarificet te*, 360, 8. Mín sáwl mǽrsaþ Drihten *magnificat anima mea dominum*, Lk. Skt. 1, 46. Ic onginne ðē tó mǽrsigenne *incipiam exaltare te*, Jos. 3, 7. Ðǽr Sicilia sǽstreámum in êþel mǽrsaþ *where Sicily, the sea streams among, her land makes illustrious*, Bt. Met. Fox 1, 32; Met. 1, 16. [*O. L. Ger.* ge-mârson *mirificare*.] v. ge-, wíd-mǽrsian; mǽran.

mǽr-stán, es; *m. A boundary-stone*:— Ðis syndon ða landgemǽro . . . On mǽrstán; of mǽrstáne on ðone ealdan gáran, Cod. Dip. Kmbl. iii. 438, 28.

mǽrsung, e; *f.* **I.** *a making known, report, rumour*:— Spranc mǽrsung ðiús (*fama hæc*) in alle eorþo, Mt. Kmbl. Lind. 9, 26. Gefehto and mǽrsungo (*opiniones*) ðara gefehto, Mk. Skt. Lind. 13, 7. **II.** *fame, renown, celebrity*:— Gesprang mǽrsung his in alle Syria *abiit opinio ejus in totam Syriam*, Mt. Kmbl. Lind. 4, 24. Herodes gehērde mǽrsung (*famam*) Hǽlendes, 14, 1. Gesprang mǽrsung (*rumor*) his in all lond, Mk. Skt. Lind. 1, 28. **III.** *celebration* (*of a rite, festival, &c.*):— Gibedes ðisses gêrlícre mǽrsunge *observationis hujus annua celebritate*, Rtl. 9, 21. Mǽrsung his gebyrdtíde *the celebration of his birthday*, Homl. Th. i. 480, 34. Ðás fíftig daga sind ealle gehálgode tó ánre mǽrsunge, 312, 23. On ðære Eástrena mǽrsunge *in celebratione Paschæ*, Bd. 3, 17; S. 545, 21. Mid ða mǽrsunga ðara heofonlícra gerȳna, 2, 9; S. 510, 37: 4, 22; S. 591, 21. **IV.** *a making great, magnifying, glorification*:— Se Fæder hine sette tó his swíðran on heofenan ríce . . Deós is Cristes mǽrsung æfter ðære menniscnysse, Homl. Th. ii. 360, 28. Mid eálre þoncunga and mǽrsunga hine herian *to praise him with giving thanks and glory to him*, Blickl. Homl. 31, 21. **V.** *Greatness, magnificence, excellency, honour, favour*:— Syllaþ mǽrsunge Gode úrum *date magnificentiam deo nostro*; ascribe ye greatness to our God (*A. V.*), Cant. M. ad f. 4. Mǽrsunge *favore*, Rtl. 8, 40. Ofer gesamnunge is his mǽrsung *his excellency* (*magnificentia*) *is over Israel*, Ps. Lamb. 67, 35: Ps. Spl. 110, 3: 70, 23. Ðæt ic synge ealne dæg mǽrsunga (*magnitudinem*) ðíne, 70, 9. Stefn Drihtnes on mǽrsungum *the voice of the Lord is full of majesty*, 28, 4. v. cyric-, ge-mǽrsung.

mǽrsung-tíma, an; *m. A time of celebration* or *glorification*:— Ðá wæs his mǽrsungtíma, ðæt se Fæder hine mǽrsode swá ðæt hē hine sette tó his swíðran on heofenan ríce, and him forgeaf andweald on heofenan and on eorþan, and eác ofer hellwarum, Homl. Th. ii. 360, 25.

mǽrþ *a weasel*. v. mearþ.

mǽr-þorn, es; *m. A hawthorn tree which serves as a boundary*:— Of ðæm pytte on ðone díc, ðæt on mǽrþorne; of ðæm þorne norþ on ðone hwítan stán, Cod. Dip. Kmbl. iii. 168, 33.

mǽrþu, mǽrþo; *indecl.*: mǽrþ, e; *f.* **I.** *greatness, honour, glory, fame*:— Gesprang nierþu his in all lond Galilæe *processit rumor ejus in omnem regionem Galilaeae*, Mk. Skt. Lind. 1, 28. Lof wíde sprang, miht and mǽrþo, ofer middangeard, þeodnes þegna, Apstls. Kmbl. 13; Ap. 7. Ðær wæs Beówulfes mǽrþo mǽned *there was celebrated Beowulf's glory*, Beo. Th. 1718; B. 857 : 1322; B. 659. Mǽrþo fremman *to achieve glory*, 4274; B. 2134. Ðæt hié him tó mǽrþe burh geworhte *that they should build a city in their own honour*, Cd. 80; Th. 100, 12; Gen. 1663. Ðú ongunne ætȳwan ðíne mǽrþe (*magnitudinem*), Deut. 3, 24 : Ps. Lamb. 150, 2. Sillaþ mǽrþe (*magnificentiam*) úrum Gode, Deut. 32, 3. Dryhtne ðe hyre weorþmynde geaf mǽrþe *to the Lord that gave her honour and glory*, Judth. 12; Thw.

26, 25; Jud. 344. Geceósan swá helle hiénþu swá heofones mǽrþu, Exon. 16 b; Th. 37, 11; Cri. 591. Mē þincþ ðæt hit hæbbe geboht sume swíðe leáslíce mǽrþe, Bt. 24, 3; Fox 82, 24. Ic ongite ðæt . . ða mǽstan mǽrþa ne sint on ðysse woruldgylpe *video .. nec celebritatem gloria posse contingere*, 33, 1; Fox 120, 4. Mǽrþa gesǽligost *most blessed of glories*, Salm. Kmbl. 136; Sal. 67. Mǽrþa ðíne hig tellaþ *magnitudinem tuam narrabunt*, Ps. Lamb. 144, 6. Eálá mín drihten . . mǽrþum gefrǽge, Bt. Met. Fox 20, 4; Met. 20, 2. Hine God trymede mǽrþum and mihtum *him God confirmed with glory and with might*, Elen. Kmbl. 29; El. 15. **II.** *a great, honourable, glorious action, a wonderful thing, mighty work*:— Hē hēt ða hȳde tó Rôme bringan and hié ðǽr tó mǽrþe áþenian for ðon heó wæs hundtwelftiges fóta lang *corium* (*serpentis*) *Romam devectum* (*quod fuisse centum viginti pedum spatio ferunt*) *cunctis miraculo fuit*, Ors. 4, 6; Swt. 174, 16. Sceoldon hiera senatus ða menn beforan him drífan gebundene ðe ðǽr gefongene wǽron, ðæt heora mǽrþa sceoldon ðȳ þrymlícran beón, 2, 4; Swt. 70, 30. Ðǽr syndon ða micclan mǽrþa ðæt syndon ða geweorc ðe Alexander hēt gewyrcean *ibi sunt illa magna insignia que Alexander operari jusserat*, Nar. 33, 20. Mǽrþa georne *eager to do great things*, Cd. 80; Th. 101, 5; Gen. 1677. Hæbbe ic mǽrþa fela ongunnen, Beo. Th. 827; B. 408 : 5284; B. 2645 : Exon. 82 b; Th. 310, 34; Seef. 84. Dú hit worhtes eall . . ðeáh ðē nǽnegu nédþearf wǽre ealra ðara mǽrþa *thou didst make it all .. though thou didst not need all those mighty works*, Bt. Met. Fox 20, 51; Met. 20, 26. Mǽrþa fruma *God*, Chr. 975; Erl. 126, 15; Edg. 41. Standaþ and geseóþ Drihtnes mǽrþa (*magnalia*), Ex. 14, 13 : Hy. Surt. 96, 36. Mǽrþa, Ps. Spl. 105, 21. Ic wylle fǽhþe sêcan, mǽrþum (*gloriously, nobly*) fremman, Beo. Th. 5021; B. 2514. Hæfdon neowne gefeán mǽrþum (*wondrously, miraculously*) gemêted, Elen. Kmbl. 1738; El. 871. [*Goth.* mêritha *fame, report*: *O. Sax.* mârida : *O. H. Ger.* mârida *fama, opinio, rumor, praeconium, claritudo*.] v. ellen-mǽrþu.

mǽr-weg, es; *m. A boundary* (?) *road*:— On ðone mǽrweg; ondlong ðæs mǽrweges, Cod. Dip. Kmbl. iii. 32, 33. Ondlong ðæs lytlan weges ðæt hit cymeþ on ðone norþran mǽrweg; ondlong ðæs mǽrweges, 33, 5 : 77, 26. [Cf. mearc-weg, 202, 5; *but also* on piddes meres weg, 77, 14.]

mǽr-weorc, es; *n. A great, splendid work*, Ps. Th. 110, 4.

Mǽs, e; *f. The Maes* or *Meuse*; *Mosa*:— Hēr fôr se here up onlong Mǽse feor on Froncland, Chr. 882; Erl. 82, 7. [*O. H. Ger.* Masa : *Ger.* Maas.]

mǽscre, an; *f. A mesh of a net*:— Mǽscre *macula*, Wrt. Voc. ii. 59, 5. v. masc.

mǽsen [*for* (?) mǽseren]; *adj. Of maple*:— Vi mǽse[r]ne sceala *vi vessels of maple*, Chart. Th. 429, 29. [Cf. *Icel.* mösur-skál *a vessel of maple*; 'such bowls are frequently mentioned in inventories of churches; cp. mid. H. G., where *maser* is even used of *a chalice, a maple-wood cup.*' Cl. and Vig. Dict. See also *Prompt. Parv.* masere *murrus*, p. 328 and note there. The noun perhaps occurs in Maser-feld, Chron. 641; Erl. 27, 8.]

mǽslen, mǽsling v. mæstling.

mǽsle-sceafe. v. mǽl-sceafa.

mǽsse, messe, an; *f.* **I.** *a service of the church, mass*:— Mǽsse *missa*, Wrt. Voc. ii. 59, 8. Ǽne þrowade Crist, ac swáðeáh dæghwomlíce biþ his þrowung geedníwod þurh gerȳnu ðæs hálgan húsles æt ðære hálgan mǽssan; forðí fremaþ seó hálige mǽsse miclum ge ðám lybbendum ge ðám forþfarenum, Homl. Th. ii. 376, 10–13. Nú is seó mǽsse gemynd Drihtnes þrowunge, L. Ælfc. P. 31; Th. 6, 13. Mǽssan singan *to celebrate mass*, Bd. 1, 27; S. 496, 23 : 4, 22; S. 592. 8. Mǽssan dôn, 4, 22; S. 591, 29 note. Se biscop and se mǽssepreóst sceolan húru embe seofon niht mǽssan gesingan for eal cristen folc ðe ǽfre ácenned wæs, Blickl. Homl. 45, 31. Æfter ðon ðe ðǽr wǽron ða hálgan lofsangas and mǽssan gefyllede, 207, 59. **II.** *a festival day when a solemn mass was celebrated, -mas in Christmas, Michaelmas, &c.*:— Temples mǽssa *scenopegia*, Jn. Skt. Lind. 7, 2. Æfter Andréas mǽssan, Ælfc. Gr. 9, 18; Som. 9, 56. Tó sanctae Michaheles mǽssan, Blickl. Homl. 197, 2. Tó sancte Martines mǽssan, 211, 11. Ǽr ealra háligra mǽssan, Chr. 901; Er!. 96, 22. Tó Cristes mǽssan, 1104 : Erl. 239, 13. Wē Marian mǽssan healdaþ, Menol. Fox 40; Men. 20 : L. Alf. pol. 43; Th. i. 92, 7. [From Low Latin *missa* v. Skeat's Dict. s.v. mass, for the meaning. *Icel.* messa : *O. H. Ger.* messa, missa : *M. H. Ger.* messe : *Ger.* messe.] v. candel-, capitol-, hláf-mǽsse; mǽsse-dæg.

mǽsse-ǽfen, es; *m. The eve of a festival, e.g. Christmas Eve*:— On sēe Michaeles mǽsseǽfan, Chr. 1014; Erl. 151, 13. Fǽstaþ ðæra háligra martyra mǽsseǽfenas, Wulfst. 136, 19.

mǽsse-bóc; *gen.* -béc; *f. A mass-book, missal*:— Saltere and pistol-bóc, godspellbóc and mǽssebóc, sangbóc and handbóc, gerím and pastoralem, penitentialem and rǽdingbóc, ðás béc sceal mǽssepreóst néde habban, L. Ælfc. C. 21; Th. ii. 350, 13; Chart. Th. 430, 7. On ðǽm ealdan sacramentorium, ðæt is on ðǽm ealdan mǽssebócum, Shrn. 88, 5. [*Orm.* Havel. messe-bok : *O. H. Ger.* missi-puoh *missalis* : *Icel.* messu-bók.]

mǽsse-créda, an; *m. The creed used in the service of the mass, the Nicene creed*:— On ðam sinoþe (on ðære ceastre Nicēa) wǽron gesette

ða hálgan cyricþénunga, and se mæssecréda, L. Ælfc. C. 4; Th. ii. 344, 9. *The mæssecréda is given in* Homl. Th. ii. 596, 24-598, 14.

mæsse-dæg, es; *m. A festival* (v. mæsse, **II.**):—Uton sécan úre cyrcean Sunnandagum and mæssedagum *frequentemus ecclesias nostras diebus Dominicis, et diebus festis*, L. Ecg. P. iv. 66; Th. ii. 226, 29: Blickl. Homl. 47, 27. Be mæssedaga freólse, L. Alf. pol. 43; Th. i. 92, 1. November onginþ on ealra hálgena mæssedæg, Ælfc. Gr. 9, 18; Som. 9, 56. Uppon sce Laurent mæssedæg, Chr. 1103; Erl. 239, 5. [*Orm.* messeda33 to freollsenn: *Ayenb.* messeda3es *holidays.*]

mæsse-gierela, an; *m. Vestment used at the celebration of the mass*, Past. 14, 6; Swt. 87, 19.

mæsse-hacele, an; *f. A cope:*—Mæssehacele *casula*, Wrt. Voc. 81, 42. [Ic án þeódréd mín wíte massehakele ðe ic on Pauie bouhte, Chart. Th. 515, 16: 512, 30. Messehacel, Chr. 963; Erl. 123, 16. Mæssehakeles, 1070; Erl. 207, 35: 1122; Erl. 249, 8.] [*Icel.* messu-hökul *a cope: O. H. Ger.* missa-hachul *casula.*]

mæsse-hrægel, es; *n. A surplice:*—Se sacerd scolde beón fæste bewǽfed on bǽm sculdrum mid ðæm mæssehrægle *in utroque humero sacerdos velamine superhumerali adstringitur*, Past. 14, 3; Swt. 83, 9. Ðes pápa gesette ðæt mæssepreóstas ne sceoldon brúcan gehálgodra mæssehrægla búton on cyrcean ánre, Shrn. 112, 19.

mæssian; *p.* ode *To say mass:*—Be ðam sacerde ðonne hé mæssaþ hwæt hé on him hæbbe *de iis quibus indutus esse debet sacerdos, cum missam celebrat*, L. Edg. C. tit. ix.; Th. ii. 128, 19. Mæssode se apostol ðam folce, Homl. Th. ii. 478, 14. For mé gelómlíce mæssaþ *pro me missas crebras facit*, Bd. 4, 22; S. 591, 29. For hreówsigendne man man mót mæssian ymb. xxx nihta, L. Ecg. C. 36; Th. ii. 160, 21. Hý mihton wel habban wíf on ðam dagum forðan ðe hý næfre ne mæssodon, L. Ælfc. C. 7; Th. ii. 346, 8. Wé lǽraþ ðæt preóst on ǽnigum húse ne mæssige, búton on gehálgodre cirican, L. E. B. 30; Th. ii. 250, 18. (*For other regulations see* §§ 31-33, 35, 37; and L. N. P. L. 13, 14, 16, 18; Th. ii. 292, 16-24.) Benedictus ásende áne ofeletan, and hét mid ðære mæssian, Homl. Th. ii. 174, 27. Ymbe underntíd ðá ðá se bróðor wæs gewunod tó mæssigenne, 358, 21. [*Icel.* messa.]

mæsse-lác, es; *n. The mass-offering, the host:*—Messelác *fertum*, Wrt. Voc. ii. 39, 41: 147, 76. Messelác, Ælfc. Gl. 34; Som. 62, 61; Wrt. Voc. 28, 41. [v. Ducange: '*fertum genus panis, in Glossis MSS.* Isidoro et Papiæ dicitur oblatio, quæ ad altare fertur et sacrificatur a Pontificibus, a quo offertorium nominatur. In Festus; *fertum genus libi dictum, quod crebrius ad sacra ferebatur altero genere libi.*']

mæsse-niht, e; *f. The night which precedes a festival* (mæsse-dæg):—Ðis sceal on mydde-wyntres mæssenyht (i.e. *on* Christmas morning) tó ðære forman mæssan, Lk. 2, 1 (rubric). Nágan lǽwede men wífes gemánan mæssenihtum, Wulfst. 305, 23.

mæsse-preóst, es; *m.* **I.** *A priest not of the Christian church:*—Melchisedec wæs cyningc and mæssepreóst, Prud. 5 a. Ðá cwǽdon ða ealdras and ða mæssepreóstas tó Pilate, Nicod. 10; Thw. 5, 22: 11; Thw. 6, 2. **II.** *a priest of the Christian church, who had attained the last of the seven appointed orders, and might celebrate the mass. His orders were the same as those of the bishop, but the latter alone could ordain priests, confirm children, and consecrate churches. He might be a regular or not. There is the* mæssepreóst *ðe regollíce libbe or the folcisc* mæssepreóst *ðe regollíf næbbe*, L. Ælfc. ix. 19, 21; Th. i. 344, 11, 21; *but he was forbidden to marry. As compared with the laity his oath was equal to that of a thane, and he was worthy of thane-right.* [v. mæsse-þegen.] *His presence was necessary at a wedding, and he was one of those who were proper witnesses when property was exchanged. For manslaughter and other crimes he might be deprived of his orders. See the passages below taken from the Laws.* Mæssepreóst *presbiter*, Wrt. Voc. 42, 21: 71, 75. Swá hwæðer ðú sý swá mæssepreóst swá munuc, Coll. Monast. Th. 31, 35. Ælc mæssepreóst sceal beón swá hé geháten is *sacerdos*, ðæt is on Léden *sacrum dans* .. Hé sceal syllan hálignysse ðam folce ðe hé tó láreówe biþ geset, L. Ecg. P. iii. 16; Th. ii. 202, 16. *Presbiter* is mæssepreóst oððe ealdwita; on ðæt ǽlc eald sý, ac ðæt hé eald sý on wísdóm. Se hálgaþ Godes húsel, L. Ælfc. C. 17; Th. ii. 248, 20. Beggen sind on ánum háde, se biscop and se mæssepreóst, ðeáh on ðam seofoþan ciricháde, L. Ælfc. P. 35; Th. ii. 378, 14. Nis ná máre betwyx mæssepreóste and biscop búton se bisceop biþ gesett tó hádigenne preóstas, and tó bisceopgenne cild, and tó hálgyenne cyrcan, and tó gýmenne Godes gerihta, L. Ælfc. C. 17; Th. ii. 348, 25. Mæssepreóstes áþ and woruldþegenes is on Engla lage geteald efendýre; and for ðám seofon cirichádan ðe se mæssepreóst geþeáh ðæt hé hæfde, hé biþ þegenrihtes wyrðe, L. O. 12; Th. i. 182, 14. *For the books necessary for the* mæssepreóst *and for rules to be observed by him in celebrating mass see passages given under* mæsse-bóc, mæssian *respectively.* Æt ðám giftan sceal mæssepreóst beón mid rihte, L. Edm. B. 8; Th. i. 256, 6. Nán man ne hwyrfe nánes yrfes bútan ðæs geréfan gewitnesse, oððe ðæs mæssepreóstes, oððe ðæs landhláfordes oððe ðæs horderes, oððe óðres ungelygenes mannes, L. Ath. i. 10; Th. i. 204, 18. Gif hwá ðonne ða teóþunge gelǽstan nelle, fare ðæs cynges geréfa and ðæs bisceopes, and ðæs mynstres mæssepreóst, L. Edg. i. 3

Th. i. 262, 25. Mæssepreóstum and diáconum is eallunge forboden ǽlc hǽmed. Þreó hund biscopa and eahtatýne gesetton canon, ðæt nán mæssepreóst oððe diácon on his wununge wífhádes mann næbbe, búton hit sý his móder, oððe sweoster, oððe faðu, oððe módrie; and gif hé ðearnunge oððe eáwunge wífes brúce, ðæt hé his hádes þolige, Homl. Th. ii. 94, 27-33: L. Ecg. P. iii. 1; Th. ii. 196, 12: iii. 6; Th. ii. 198, 7. *But the rule is still stricter in* L. E. I. 12; Th. ii. 410, 7. Nis hyt ryht ðæt ǽnig wífmon mid mæssepreóste on húsum wunige. *Other regulations which concern the* mæssepreóst *follow q.v.* Gif mæssepreóst manslaga wurðe oððe elles mánweorc tó swíðe gewurce, ðonne þolige hé ægðres ge hádes ge eardes, L. Eth. ix. 26; Th. ii. 346, 4: L. Ecg. P. iii. 3; Th. ii. 196, 23: iv. 2; Th. ii. 204, 10. For other crimes and their punishment see L. Eth. ix. 27; Th. i. 346, 8-16: L. Ecg. P. iv. 7; Th. ii. 206, 1. Ic Ælfríc munuc and mæssepreóst, Homl. Th. i. 2, 12. Arrius se mæssepreóst *Arius presbyter*, Ors. 6, 30; Swt. 282, 33. Mamméa sende æfter Origenise ðæm gelǽredestan mæssepreóste, 6, 18; Swt. 270, 27. [*Icel.* messu-prestr.] v. efen-mæssepreóst.

mæssepreóst-hád, es; *m. The orders of a mass-priest:*—Of ðære tíde ðæs ðe ic mæssepreóstháde onfeng *ex quo tempore accepti presbyteratus*, Bd. 5, 24; S. 647, 32: 5, 11; S. 613, 12.

mæssepreóst-scír, e; *f. The district attached to the church at which a masspriest officiated:*—Gif man hwylc metrum cild tó mæssepreóste bringe, sý of swylcre mæssepreóstscýre swylce hyt sý, L. E. I. 17; Th. ii. 412, 21. Cf. Ne spane nán mæssepreóst nánne mon of óðre cyrcan hýrnysse tó his cyrcan, ne of óðre preóstscýre lǽre ðæt mon his cyrcan geséce, and him heora teóþinge syllan, and ða geryhtu ðe hig ðam óðrum syllan sceoldan, 14; Th. ii. 410, 30-33.

mæsser-bana, an; *m. One who slays a priest:*—Mæsserbanan (MS. C. sacerdbanan), Wulfst. 165, 28.

mæssere, es; *m. One who says mass, a mass-priest:*—Mæssere *presbiter*, L. Ecg. C. 7; Th. ii. 140, 1: Exon. 55 a; Th. 194, 34; Az. 149.

mæsse-reáf, es; *n. Vestment used when celebrating mass:*—Wé lǽraþ ðæt ǽlc preóst hæbbe corporalem ðonne hé mæssige, and subuculam under his alban and eal mæssereáf wurdlíce behworfen, L. Edg. C. 33; Th. ii. 250, 28: L. Ælfc. C. 22; Th. ii. 350, 19. Ic geann ánes mæssereáfes mid eallum ðam ðe ðærtó gebyreþ, Chart. Th. 529, 8.

mæsse-sang, es; *m. The service of the mass:*—Ða symbelnysse tó mærsianne mæssæsanges *missarum sollemnia celebrandi*, Bd. 1, 27; S. 497, 1. Mæssesong dón *missas facere*, 1, 26; S. 488, 4. Gewuna mæssesonga *consuetudo missarum*, 1, 27; S. 489, 33. On mæssesangum and on sealmsangum, L. Edg. C. 14; Th. ii. 282, 17.

mæsse-þegen, es; *m. A mass-priest:*—Mæsseþegnes and woruldþegnes wergild ii þúsend þrymsa, L. Wg. 5; Th. i. 186, 10. v. mæsse-preóst.

mæsse-tíd, e; *f. A time at which mass was said:*—Æt mæssetídum *tempore missæ*, L. Ecg. C. 9; Th. ii. 140, 20.

mæsse-wín, es; *n. Wine used in the service of the mass:*—Messewín *infertum vinum*, Ælfc. Gl. 32; Som. 61, 126; Wrt. Voc. 27, 52. [*Icel.* messu-vín.]

mæst, es; *m. A pole to support a sail, a mast:*—Mæst *malus* vel *artemo: artemon* vel *malus*, Ælfc. Gl. 83, 104; Som. 73, 81: 77, 126; Wrt. Voc. 48, 19: 56, 43. Mest *malus*, 63, 47. Mæstum *malis*, Wrt. Voc. ii. 57, 15. Mæst (?) *columbarium*, 134, 61 (cf. ár-locu *columbaria*, Wrt. Voc. 63, 41). Segelgyrdena, mæsta *antennarum*, Hpt. Gl. 529, 20: Menol. Fox 508: Gn. C. 24: Beo. Th. 71: B. 36: 3801: B. 1898: 3814; B. 1905: Andr. Kmbl. 929; An. 465. Hé hǽt fealdan ðæt segl, and eác hwílum ðone mæst, Bt. 41, 3; Fox 250, 15; Ors. 4, 6; Swt. 172, 5. [*O. H. Ger.* mast *malus.*]

mæst, es; *m. Mast, fruit of forest trees e.g. oak, beech, used for feeding swine:*—Ðrím hunde swína mæst, ond se biscop and ða hígen áhten twǽde ðæs wuda ond ðæs mæstes, Cod. Dip. Kmbl. i. 279, 3. Mid wude and mid felde mid mæste *cum sylva et cum agro, cum porcorum esca*, iv. 202, 2. Micle beámas ða ðe mæst and wæstm mannum bringaþ *ligna fructifera*, Ps. Th. 148, 9. [*O. H. Ger.* mast *sagina.*] v. mæsten, mæstan. **mæst.** v. micel.

mǽst; *adv.* **I.** *most, chiefly, especially:*—Se westsúþende Europe landgemirce is in Ispania westeweardum and mǽst (*maixme*) æt ðæm íglande ðætte Gades hätte, Ors. 1, 1; Swt. 8, 24. Ðara nýtena meolc ðe hý mǽst bí libbaþ, 1, 2; Swt. 30, 10. Geond ealle world, and ðeáh mǽst in Thasalia, 1, 6; Swt. 36, 8. Swá hié mǽst mehten *as much as ever they could*, 6, 5; Swt. 260, 32: Past. 28; Swt. 190, 9. Ealles mǽst *maxime*, Bd. 2, 4; S. 505, 7. Preóst oftor ne mæssige ðonne þríwa mǽst ðara þinga (*at the utmost*), L. Edg. C. 37; Th. ii. 252, 4. **II.** *with the adj.* eall, *almost, nearly:*—Hit is eal mǽst mid háligra manna naman geset *it is almost all occupied with holy men's names*, Homl. Th. ii. 466, 22. Ðæt him sealde mǽst eal his sunu *almost all of which his son gave him*, Chart. Th. 271, 33. Wígheard and mǽst ealle (*omnes pene*) his geféran, Bd. 4, 1; S. 563, 25. Hié mǽst ealle ofslægene wurdon, Ors. 2, 5; Swt. 80, 22. Swá swá ealle mǽst ðyssere declinunge, Ælfc. Gr. 9, 7; Som. 9, 9. Ða óðre ealle mǽst *almost all the others*, 9, 4; Som. 10, 24. Ealle mǽst ðás word, 30; Som. 38, 35. v. má, *and* micel.

mæstan; *p.* mæste; *pp.* mæsted, mæst *To fatten* :—Maestun *saginabant*, Wrt. Voc. ii. 119, 61 : Ep. Gl. 24 b, 27. Ic wylle ðæt man mæste mínum wífe twá hund swína, Chart. Th. 596, 21. Is mæst *saginatur*, *nutritur*, Hpt. Gl. 489, 43. Weorþaþ mæsted *pinguescent*, Ps. Th. 64, 13. [*Prompt. Parv.* Mastyñ beestys *sagino, impinguo* ; mast-hog, mastid swyne *maialis* : *O. H. Ger.* mastian *to feed* ; ge-mestet, ge-mast *fattened*, v. Grff. ii. 882 : *Ger.* mästen.] v. â-, ge-mæstan.

mæst-cist, e ; *f. The hole in which the mast is fixed* :—Mest *malus* : mastcyst *modius*, Wrt. Voc. 63, 48, 49. Mæstcyst *modius*, ii. 59, 27. [' dicitur *modius* cavum illud in navi cui arbor institit', Forcellini.]

mæstel-bearh; *gen.* -bearges ; *m. A fattened barrow pig* :—Ante *porcos*, before bergum ; ðæt sindon ða mæstelbergas ; ðæt aron ða gehádade menn, and ða góde menn, and ða wlonce menn forhogas Godes bebod and godspelles, Mt. Kmbl. Lind. 6, 6 note.

mæsten [n], es ; *m. Mast-pasture, pasture for swine, consisting of the fruit of forest trees* :—Man mæste mínum wífe twá hund swína, ðænne ðær mæsten sý, Chart. Th. 596, 23 : Cod. Dip. Kmbl. iv. 20, 5. Be unáliéfedes mæstennes onfenge. Gif mon on his mæstene unáliéfed swín geméte, L. M. 49; Th. i. 132, 11. Ðonne hé [se inswán] his heorde tô mæstene drífe, L. R. S. 4 ; Th. i. 434, 21. [Ðis geár wæs gæsne on mæstene, Chr. 1116 ; Erl. 245, 36.] v. mæsten-ræden.

mæsten-treów, es ; *n. A tree producing mast* :—Mæstentriów *suberies* (*suberes* ?) Ælfc. Gl. 45 ; Som. 64, 102 ; Wrt. Voc. 32, 37.

mæsten-ræden [n], e ; *f. The right to feed swine in places where there was mast* :—[Hæbbe] mæstenrædene ðonne mæsten beó, Cod. Dip. Kmbl. iii. 451, 10. v. mæst-ræden.

mæst-land, es ; *n. Land on which mast is produced* :—Eall ðæt wudulond ðæt Æþelbald gesealde tô mæstlonde, Chart. Th. 140, 2.

mæstling, mæsling, mæslen [n], es ; *n.* **I.** *A kind of brass.* The word is used to gloss *aes*, *aurichalcum*, and *electrum* :—Mæstlingc ær and tin *aurichalcum, aes et stannum*, Coll. Monast. Th. 27, 11. Mæstlinc, grêne âr *auricalcos*, Wrt. Voc. 286, 66. Cwicseolfer *vel* mæstling *electrum* i. *sucus arboris*, ii. 142, 78. Mæslen *aes*, Mk. Skt. Lind. Rush. 6, 8. Ðæt mæslenn (mæslen, Rush.), 12, 41. Mæslen, Jn. Skt. Lind. Rush. 2, 15. **II.** *a vessel made of the metal* (? v. *Halliw. Dict.* ' Plater, disse, cop and *maseline*) :—Calicea frymþa and ârfata and mæstlinga *baptismata calicum et urceorum et eramentorum et lectorum*, Mk. Skt. 7, 4. Gedôn on cyperen fæt oððe mæstling [-fæt ?] oððe bræsen, Lchdm. iii. 292, 17. [*A. R.* copper, *mestling*, breas : al is icleopet or : *Halliw. Dict.* bras, *maslyn*, yren and stel ; where also *mastelyn* panne : *R. Glouc.* mastling : cf. also *Icel.* mersing, messing *brass* : *M. H. Ger.* messinc : *Ger.* messing.] v. gold-mæstling.

mæstling-, mæslmg-smiþ, es ; *m. A worker in brass* :—Mæstlincsmiþ *aerarius*, Ælfc. Gl. 81 ; Som. 73, 7 ; Wrt. Voc. 47, 14. Mæslingcsmiþ, 73, 32.

mæst-lôn (?) *pulleys at the top of the mast over which the ropes are drawn* :—Carceria, mæstlôn, *sunt in cacumine arboris trocliae, quasi flicteria, per quas funes trahuntur*, Wrt. Voc. ii. 128, 59.

mæst-ræden [n], e ; *f. The right of feeding swine in places where mast is produced* :—Hé næfre hine bereáfian wolde ðære mæstréddene ðe he him âléfed hæfde on Longan hrycge, Chart. Th. 140, 35. v. mæsten-ræden.

mæst-ráp, es ; *m. A rope fastening a sail to a mast*, Cd. 146; Th. 182, 27 ; Exod. 82.

mæst-twist, es ; *m. A rope to support a mast, a stay* :—Mæsttwist *parastates*, Ælfc. Gl. 104; Som. 77, 127 ; Wrt. Voc. 56, 44. Mæstwist, 63, 48.

mæt = mete, q.v.

mǽtan; *p.* te *To dream* (with dat. or acc. of person ; cf. *Icel.* dreyma which takes acc. of dreamer and of dream) :—On ânre nihte ealdne mônan, swá hwæt swá ðé mæteþ ðæt cymþ tô gefeán, Lchdm. iii. 154, 15. Gyf mon (*acc.* cf. l. 27) mêteþ ðæt hé geseó . . . , 168, 8. Gyf man mǽte ðæt hé hæbbe . . , 176, 2. Ongitan swelce eów mǽte, Bt. 26, 1 ; Fox 90, 4 : tit. 26; Fox xiv, 16. Hit gelamp ðæt hine mǽtte, Gen. 37, 5. Mín swefen ðe mé mǽtte, 37, 6. Oðer swefen hine mǽtte, 37, 9 : 41, 5, 11 : 42, 9. Gif hé secge ðæt him mǽtte swefen, Deut. 13, 1. Ðære Perpetuan mǽtte ðæt heó wǽre on weres hiwe, Shrn. 60, 28. [*Chauc.* meten.] v. ge-mǽtan.

mǽte; *adj. Moderate, mean* (*between two extremes*), *small, poor, bad* ; in the *cpve. inferior*, applied to persons, *of a middle* or *lower class* :—Reste hé ðær mǽte weorode (*alone*), Rood. Kmbl. 138; Kr. 69. *So again* Ic ána wæs mǽte werede, 245; Kr. 124. Unrîm ealra cwycra, mycelra and mǽtra (*pusilla et magna*), Ps. Th. 103, 24: 113, 21: Exon. 33 a ; Th. 105, 16; Gû. 24. Ic ðé feáwe dagas mînra mǽttra môde secge *I will tell thee the fewness of my days poor and evil* ; paucitatem dierum meorum enuntia mihi, Ps. Th. 101, 21. Ðe mǽtu sprecaþ ofer mé *qui maligna loquuntur super me*, Ps. Spl. T. 34, 30. Biþ seó sîþre tîd sǽda gehwylces mǽtræ in mægne (*inferior in virtue*), Exon. 33 a ; Th. 105, 2 ; Gû. 17. Gif hió biþ gôd drenc, biþ on peninge; gif mǽtra, biþ on ôðrum healfum oððe on twám ; and gif isel þrîm, ac ne mǽ, L. M. 2, 52 ; Lchdm. ii. 272, 24. Hors tô healfan punde gif hit swá gôd sý ; and gif hit mǽtre sý, gilde be his wlites wyrþe, L. Ath. V. 6 ; Th. i. 232, 25. Nalæs ðæt án ðætte ða mǽttran (mǽteran, MS. B.) . . ac eúc swylce cyningas and ealdormen *non solum mediocres . . sed*

---

etiam reges et principes, Bd. 4, 23 ; S. 593, 43 note. Eall ðás getimbro ge ða máran ge ða mǽttran *cuncta hæc ædificia publica vel privata*, 4, 25 ; S. 600, 33. Micle and mǽttran (MS. and micle mǽttan), Chart. Th. 510, 32. Mǽtran, Bt. 39, 7 ; Fox 222, 11 note. Næs ðæt mǽtost mægenfultuma *not poorest of aids was that*, Beo. Th. 2914; B. 1455. Mǽtestum *pessimi*[*s* ?], Kent. Gl. 711. v. ge-, ofer-, or-, un-, unge-mǽte.

mǽþ, e ; *f.* (*but ofer ðínne mǽð*, Prov. Kmbl. 27.) **I.** *measure, degree, proportion* :—Gilde be ðære giftan mǽþe *reddet pecuniam juxta modum dotis*, Ex. 22, 17 ; L. Ecg. P. i. 11 ; Th. ii. 176, 28. Be ðære synne mǽþe *secundum peccati gradum*, tit. i ; Th. ii. 170, 5 : Ors. 1, 12 ; Swt. 56, 4. Be dǽde mǽþe, L. C. E. 5 ; Th. i. 364, 1. Beó seó æht gescyft swíðe rihte wíse and cildan and nêhmágon ælcum be ðære mǽþe ðe him tô gebyrige *let the property be shared among the wife and children and near relatives with strict justice, to each according to the proportion that is proper for him*, L. C. S. 71 ; Th. i. 414, 2. **II.** *the measure* or *extent of power, ability, capacity, efficacy* :—Nis ná eówer mǽþ tô witenne ðone tíman *it is not for you to know the time* (Acts 1, 7), Homl. Th. i. 298, 22. Úre mǽþ nis ðæt wé ealle Godes gecorenan eów gereccan, ii. 72, 1 : 188, 28. Nis ǽfre æniges mannes mǽþ ðæt hé cunne God swá forþ geherian swá hé wyrþe is *it is never within any man's power to praise God to the extent he deserves*, Btwk. 194, 15. Ðeáh hit úre mǽþ ne síe ðæt wé witan hwæt hé síe, wé sculon ðeáh be ðæs andgites mǽþe ðe hé ûs gifþ fundigan, Bt. 42 ; Fox 256, 2. Ælc winþ be his andgites mǽþe *each strives according to the measure of his understanding*, 41, 4 ; Fox 250, 26 : Homl. Th. i. 344, 22. Crist dǽlþ his gyfe his limum be gehwylces mannes mǽþe *according to each man's ability*, ii. 526, 8. Gif ðú oncnǽwst ðínne Drihten mid ðínum æhtum be ðínre mǽþe, i. 140, 30. Gôdne dǽl ælces be ðære mǽþe (*efficacy* of the ingredient), Lchdm. iii. 12, 20. Dô ðǽrtô ða huniges mǽþe, 76, 9. Gôde sind ðás þing (*bread, fish, &c.*) be heora mǽþe *these things are good as far as they go*, Homl. Th. i. 252, 26. Ofer mǽþe úre ðú forþtýhst sprǽce *ultra ætatem nostram protrahis sermonem*, Coll. Monast. Th. 32, 11. Ðeáh wé nú ofer úre mǽþ þencen *sive mente excedimus*, Past. 16, 2 ; Swt. 101, 11. Ðæt môd ðe ofer his mǽþ biþ upáhæfen *animus qui extra se in elationem ducitur*, 36, 7 ; Swt. 255, 18. Ðú scealt gelýfan on ðone lifigendan God and ná ofer ðíne mǽþe môtian be him, Hexam. 3 ; Norm. 6, 17. Ðú bǽde ofer míne mǽþe *thou hast asked beyond my power*, Homl. Skt. 3, 515. Ne wilna ðú ofer ðínre mǽd tô witanne ymbe ða heofonlícan þing, Prov. Kmbl. 27. Manna gehwylc mæg be his mǽþe, mid ðám lácum ðe hé hæfþ, Gode eáðe gecwêman, forðam ne gewilnaþ hé ná máran ðonne ðæs mannes mǽþa beóþ, Wulfst. 280, 27. **III.** *degree, rank, status, condition* :—' Ne onwreáh ðé flæsc ne blôd ðisne geleáfan.' Flæsc and blôd is gecweden his flǽsclíce mǽiþ '*flesh and blood did not reveal this belief to thee.*' *His fleshly condition is called flesh and blood*, Homl. Th. i. 368, 9. Ða wǽron þeódwitan weorþscipes wyrþe, ælc be his mǽþe, eorl and ceorl, þegen and þeóden, L. R. 1 ; Th. i. 190, 13. Eallum cristenum mannum gebyraþ ðæt hí háda gehwylcne weorþian be mǽþe, L. C. E. 4 ; Th. i. 360, 28 : L. Eth. vii. 3 ; Th. i. 330, 8. **IV.** *due measure, right* :—Hé þeáh swá hit mǽþ wæs fægere forþwerd *he made good progress, as was right and fit*, Wulfst. 17, 8. Manna gehwilc ôðrum beóde ðæt riht ðæt hé wille ðæt man him beóde, be ðam ðe hit mǽþ sí, L. Eth. vi. 49 ; Th. i. 326, 31. Mahna má ðonne hit ænig mǽþ wǽre *more men than was at all right*, Byrht. Th. 137, 33. Ofer mǽþe *justo amplius*, Ger. 395, 58. **V.** *due measure in regard to others, honour, respect* (v. mǽþ-full) :—Hwílum wǽron heáfodstedas and heálíce hádas micelre mǽþe and munde wyrþe, and gríþian mihton ða ðe ðæs beþorftan and ðǽrtô sôhtan aa be ðære mǽþe ðe ðǽrtô gebyrede *formerly chief places and high orders were entitled to much respect, and to the right of giving protection, and they could afford sanctuary to those that needed it, and repaired thereto, ever according to the dignity that thereto belonged*, L. Eth. vii. 3 ; Th. i. 330, 7. Se wæs ðonne mǽþe and munde swá micelre wurþe, swá ðonne ðam háde gebirede, L. R. 7 ; Th. i. 192, 13. Ðæt Godes circan beón beteran mǽþe and munde wyrþe, Wulfst. 266, 9. Godes þeówas syndan mǽþe and munde gewelhwar bedǽlde, 157, 19. Man sceal mǽþe on háde gecnáwan *people must feel respect for the clergy*, L. C. E. 4 ; Th. i. 362, 4 : L. I. P. 19 ; Th. ii. 328, 26. Ælc cristen man áh mycele þearfe ðæt hé on ðam griþe mycle mǽþe wite (*shew great respect to*), 25 ; Th. ii. 338, 38 : Wulfst. 161, 2. Se hæfþ árfæstnysse ðe mǽþe carn on ôðrum mannum . . and nele forseón ôðerne, 51, 30. Deófol sendeþ árleásnesse ðæt ungesǽlig man mǽðe ne geseó on his underþeóddum ne on his efenlícan *shews no respect for his subordinates or equals*, 53, 24. [*Orm.* mett and mæþ i claþess : *Allit. Pms.* in mesure and meþe.]

mǽþ, es ; *n.* (?) *Math* in after-*math, mowing, hay-harvest* :—Freóh ælces weoruldcundes þeówetes búton þreom þingum án is circsceat and ðæt hé mid eallum cræfte twuga on geáre [wyrce ?] ǽne tô mǽþe and ôðre síþe tô rípe *free from every secular service except three things ; one is church scot, and (the other two) that he [work] with all his might twice a-year, once at hay-harvest, the other time at corn-harvest*, Cod. Dip. Kmbl. ii. 400, 30. [*O. H. Ger.* mâd : *M. H. Ger.* mât ; *gen.* mâdes ; *n.* also *f.* : *Ger.* mahd ; *f.*] v. mǽðere.

Mḗðas, Mḗðe, Mēðas, Mēdas *the Medes* :—Siððan hæfdon Mḗðe onwald : ofer Mēðas ðæt lond : Asiria anwald gehwearf on Mēðas : Mḗða rīce, onwald : on ðara Mḗða anwalde : Mēða ealdorman : betuh Mḗðum : Mḗðum gafol guldon : cyning in Mēðen, Ors. 1, 12 ; 2, 1 ; Swt. pp. 52, 54, 60. Mēða māððumselas, Salm. Kmbl. 379 ; Sal. 189 : Cd. 209 ; Th. 259, 7 ; Dan. 688. Mēdum, Th. 258, 26 ; Dan. 681. v. Mḗðisc.

mǽðel, medel, meðel, es ; *n.* **I.** *an assembly, a deliberative* or *judicial meeting, council* :—In mædle *in curia*, Wrt. Voc. ii. 111, 45 : Ep. Gl. 12 d, 35. An medle oððe an þinge, L. H. E. 8 ; Th. i. 30, 12. Sum in mǽðle mǽg mōdsnottera folcrǽdenne forþ gehycgan, ðǽr witena biþ worn ætsomne, Exon. 79 a ; Th. 295, 30 ; Crä. 41 : 128 b ; Th. 494, 16 ; Rä. 83, 2. On medle, Elen. Kmbl. 1088 ; El. 546 : 1182 ; El. 593. Se þeóden ongan geþinges wyrcan .. and ðā on ðam medle bebeád, Cd. 197 ; Th. 245, 28 ; Dan. 470. Upp āstōdon manige on medle *many stood up in the assembly*, Andr. Kmbl. 3250 ; An. 1628. Æt medle on ðam miclan dæge *at the assembly on that great day (of judgment)*, 2870 ; An. 1438 : Exon. 63 b ; Th. 234, 10 ; Ph. 538. Mǽðel hēgan *to hold a meeting, take counsel, consult, address* (cf. *Icel.* heyja þing) :—Ðā mōdigan mid him mǽðel gehēdon (*took counsel together*), Andr. Kmbl. 2100 ; An. 1051. Hē wið ǽnne ðæra (*pillars*) mǽðel gehēde (*addressed*), 2991 ; An. 1498. **II.** *speech, address, harangue, conversation* :—Ðū gehȳrdest ðone hālgan wer Moyses on medle (cf. *Icel.* vera â mâli *to converse*) *thou didst hear the holy man Moses when conversing with him*, Elen. Kmbl. 1568 ; El. 78 b. Mōdiges medel monige gehȳrdon *many heard the proud one's harangue (of Moses addressing the Israelites when pursued by Pharaoh)*, Cd. 156 ; Th. 194, 3 ; Exod. 255. [*Goth.* maþl ἀγορά : *O. H. Ger.* madal *in cpds.* v. Grff. ii. 706 : cf. *O. Sax. O. H. Ger.* mahal *concio.*] v. mǽdlan, maðelian, here-meðel.

mǽðel-ǽrn, -ern, es ; *n. A house of meeting for speaking or for consulting* :—In mǽðlern *in preterium* (l. *pretorium*), Wrt. Voc. ii. 46, 52 : 74, 23.

mǽðel-cwide, es ; *m. Discourse, converse* :—Ic ðæs þeódnes word medelcwide ongeat gǽstes sprǽce *I the words of the prince, his discourse, have heard, the guest's speech*, Exon. 50 b ; Th. 175, 9 ; Gū. 1192. Hyrcnigan hālges lāra mildes medelcwida *to listen to the instructions of the holy man, the discourses of the kind one*, 47 b ; Th. 162, 23 ; Gū. 980. Meaht ðū medelcwidum worda gewealdan *are words at thy command for discourse*, Th. 163, 4 ; Gū. 988. Ðonne wē on geflitum sǽton medelcwidas mengdon *when we sat in discussion, and now one, now another spoke*, Salm. Kmbl. 865 ; Sal. 432.

mǽðel-hēgende ; *part. pres. Attending, holding* or *addressing an assembly* or *council, consulting, conversing* (cf. *Icel.* þing-heyjandi 'the law term for any person who visits a þing, or a summons to perform any public duty,' Cl. and Vig.) :—Biscopas and bóceras and ealdormen mǽðel-hēgende (*in council*), Andr. Kmbl. 1217 ; An. 609. Beornas cōmon mǽðelhēgende .. Ðā wæs tō ðam þingstede þeód gesamnod *men came who had to attend the meeting .. Then was the people collected at the meeting-place*, 2194 ; An. 1098. Hwæt se manna wæs medelhēgendra *who of men that speak was he*, 524 ; An. 262. Hēht gebeódan medelhēgende on gemōt cuman, ða ðe deóplicost Dryhtnes gerȳno reccan cūðon, Elen. Kmbl. 557 ; El. 279. v. mǽðel.

mǽðel-hergende ; *past. pres. Speech-praising, esteeming conversation highly* :—Monige beóþ mǽðelhergendra, sittaþ æt symble, wordum wrixlaþ, Exon. 83 b ; Th. 314, 13 ; Mōd. 13.

Mǽð-hild, e ; *f. A woman's name, Matilda* :—Wē ðæt Mǽðhilde gefrugnon, Exon. 100 a ; Th. 378, 10 ; Deór. 14. *Grein would read* mǽð hilde, *comparing* mǽð *with Icel.* meiða *to injure, spoil.*

mǽðel-stede, es ; *m.* **I.** *A place of assembly, place where a meeting is held* (cf. þing-stede) :—Tō ðam medelstede manige cōmon snottere selerǽdend, Andr. Kmbl. 1315 ; An. 658 : 1393 ; An. 697. Swā him Offa ǽr āsæde on ðam medelstede ðā hē gemōt hæfde, Byrht. Th. 137, 40 ; By. 199. Is eów rǽdes þearf on medelstede (*in the queen's palace*), mōdes snyttro, Elen. Kmbl. 1104 ; El. 554 : Cd. 179 ; Th. 224, 33 ; Dan. 145. Tō ðam medelstede (*Mount Moriah*), 162 ; Th. 203, 1 ; Exod. 397. On ðam medelstede *the place of the last judgment*), 169 ; Th. 212, 20 ; Exod. 542. **II.** *a place of hostile meeting, a battle-place* :—Hē ne meahte on ðǽm medelstede wið Hengeste wiht gefeohtan, Beo. Th. 2169 ; B. 1082. [Cf. *O. H. Ger.* mahal-stat *curia*.]

mǽðel-word, es ; *n. A word used in a formal address* :—Þegn Hróðgāres medelwordum frægn (*of the question put by the coast-guard to Beowulf on his landing*), Beo. Th. 478 ; B. 236.

mḗðere, es ; *m. A mower* :—Sīþberend *vel* mḗðre *falcarius*, i. *falciferens vel falcifera*, Wrt. Voc. ii. 146, 80. Mḗðeras *fenisece*, 148, 21. [*O. H. Ger.* mādari *feniseca, messor.*]

mǽþ-full ; *adj, Shewing respect to others, courteous, humane* (v. mǽþ, **V.**) :—Mǽðfull *humanus*, Ælfc. Gr. 45 ; Som. 41, 42. v. mǽþ-líc, mǽþian.

mǽþian ; *p.* ode *To regard, respect* :—Hē sylf ārleásnysse ðæt hē ne ārige ne eác ne mǽþige his underþeóddum ne his gelícum *the devil gives*

---

pitilessness, so that the man neither spares nor regards his subordinates or his equals, Wulfst. 59, 17. v. mǽþ, **V** ; ge-mǽðian.

Mḗðisc, Mēdisc ; *adj. Of the Medes* :—Mycel fyrd Mēdiscra monna, Nar. 17, 8. v. Mēðas.

mǽdlan, medlan, *a word occurring only in poetry, to speak* :—Ðǽr (*at the day of judgment*) hē (*Christ*) tō ðam eádgestum ǽrest mǽðleþ, Exon. 27 b ; Th. 82, 14 ; Cri. 1338. Gehȳreþ cyning mǽdlan, sprecan rēðe word, 19 b ; Th. 50, 9 ; Cri. 797. Ic God mǽdlan gehȳrde, Cd. 26 ; Th. 33, 23 ; Gen. 524. Ongan wordum mǽdlan, 101 ; Th. 134, 2 ; Gen. 2218 : Exon. 27 b ; Th. 83, 30 ; Cri. 1364 : 50 a ; Th. 174, 10 ; Gū. 1175. Mǽdlan, Andr. Kmbl. 2879 ; An. 1442. v. maðelian.

mǽþ-leás ; *adj. Without moderation, greedy* :—' Ðás fugelas habbaþ feónda gelícnysse, ðe menn grǽdelíce grípaþ tō grimre helle.' Ðā hēt Martinus ða mǽþleásan fugelas ðæs fixnoþes geswīcan, Homl. Th. ii. 516, 11.

mǽþ-líc ; *adj. Moderate, in accordance with due measure, proper to a person's degree, having regard to others* (v. mǽþ-líce) :—Beón ða heregeata swā hit mǽþlíc sȳ *let the heriots be as is proper to the several degrees* (*earl's, king's thane, &c.*), L. C. S. 72 ; Th. i. 414, 4. Gif hwilc forwyrht man hiówan gesǽce, bió se þingad swā hit medlíc sió be ðæs geltes mǽðe *if any criminal betake himself to the convent, let terms be made for him, as may be fit and proper according to the measure of the crime*, Chart. Th. 509, 23. v. mǽþ, un-mǽþlíc.

mǽþ-líce ; *adv. With due regard to others, courteously* :—Mǽþlíce *humaniter*, Ælfc. Gr. 45 ; Som. 41, 43 : 42, 6.

mǽþrian ; *p.* ode *To shew respect to, honour* :—Būton hē hwæne furþor gemǽþrian (mǽðrian, MS. A. gemǽðian, MS. B.), and hē him ðæs weorþscipes geunne, L. C. S. 12 ; Th. i. 382, 15.

mḗting, e ; *f. A dream* :—On xxii nihta seó mǽtinga biþ eall costunge full ; ne biþ ðæt nā gōd swefen, Lchdm. iii. 156, 7. Gē mǽtinge míne ne cunnon, Cd. 179 ; Th. 224, 24 ; Dan. 141.

mḗt-líc. v. ofer-, un-ge-mǽtlíc.

mḗt-ness. v. or-, un-mǽtness.

mḗw, meáu, mǽu, es ; *m. A sea-mew, gull* :—Mǽw *alcedo* vel *alcion*, Ælfc. Gl. 37 ; Som. 63, 1 ; Wrt. Voc. 29, 24 : 62, 13 : *alacid*, Wrt. Voc. ii. 7, 62 : *alcido*, 10, 31. Meáu *alcido*, 100, 2 : *gabea*, 109, 56 : *larus*, 112, 35. Mēu *larus*, 50, 59. Mēu *vel* mēg *larum*, Shrn. 29, 2. Se grǽga mǽw, Andr. Kmbl. 742 ; An. 371. Mǽw singende, Exon. 81 b ; Th. 307, 11 ; Seef. 22. Mǽwes song, 106 b ; Th. 404, 25 ; Rä. 25, 6. Mere, mǽwes ēðel, 123 b ; Th. 474, 6 ; Bo. 25. [*Icel.* már : *Dan.* maage : *Du.* meeuw : *O. H. Ger.* mēh : *Ger.* mōwe.]

maffa, an ; *m. A caul* ; omentum, Wrt. Voc. ii. 63, 43 : Ep. Gl. 17 d, 23.

maga, an ; *m. The* MAW, *stomach* :—Maga *stomachus*, Ælfc. Gl. 76 ; Som. 71, 114 ; Wrt. Voc. 45, 19 : 65, 54 : Wrt. Voc. ii. 121, 40. *Fleumon*, magan *untrymness*, 39, 12. Magan *masdi*, 56, 9. Gif se maga āþened sīe, L. M. 2, 2 ; Lchdm. ii. 158, 4. Be geswelle ðæs magan, 158, 6. Hū ðone cealdan magan ungelíclíce mettas lyste, 2, 16 ; Lchdm. ii. 160, 7. Hū ðone magan ealne āfeormaþ, Herb. 70 ; Lchdm. i. 162, 19. Lege ofer ðone magan, L. M. 2, 15 ; Lchdm. ii. 192, 20. [*H. M.* mahe : *A. R. Chauc. Piers. P.* mawe : *Icel.* magi : *Dan.* mave : *O. H. Ger.* mago : *Ger.* magén.] v. mage.

maga ; *adj. used as subst. Powerful, strong, a powerful person* :—Ic lǽre ǽlcne ðara ðe maga sí *I advise every one that is powerful*, Shrn. 163, 12. Ne derige se maga ðam unmagan *let not the strong injure the weak*, L. I. P. 7 ; Th. ii. 314, 1. Se maga and se unmaga nā gelíce byrdene āhebban, L. Edg. C. 4 ; Th. ii. 262, 2 : L. Eth. vi. 52 ; Th. i. 328, 160. Ne mǽg se unmaga ðam magan gelíce byrdene āhebban, L. C. S. 69 ; Th. i. 412, 7. v. dirn-, un-maga.

mǽga, an ; *m.* (cf. nið *for similar division of meanings*) **I.** *a relative*, v. heáfod-, níd-mǽga ; mǽge. **II.** *a son* :—Mǽga Healfdenes (*Hrothgar*), Beo. Th. 381 ; B. 189 : 2953 ; B. 1474 : 4293 ; B. 2143. Mǽga Ecgþeówes (*Beowulf*), 5168 ; B. 2587. Ic (*Christ*) sylf gestāg mǽga in mōdor, Exon. 28 b ; Th. 87, 4 ; Cri. 1420. Fǽder eft lǽrde mǽgan, 80 a ; Th. 301, 32 ; Fä. 28. Ðonne mōdor mǽgan cenneþ, Salm. Kmbl. 742 ; Sal. 370. On mǽgan, ðīn āgen bearn, Cd. 109 ; Th. 144, 26 ; Gen. 2395. Mǽgan (*Isaac*) gelǽdde Abraham, 162 ; Th. 203, 2 ; Exod. 397. Se eorl wolde sleán eaferan sínne, mǽgan, Th. 204, 2 ; Exod. 413. **III.** *a man* :—Se mǽga geonga (*Wiglaf*), Beo. Th. 5343 ; B. 2675. On ðǽre mǽgþe mǽga wæs hāten Tubal Cain, Cd. 52 ; Th. 66, 11 ; Gen. 1082. Mǽga cystum eald *a man old in virtues*, Exon. 80 a ; Th. 300, 7 ; Fä. 2. Se mǽga (*Christ*), Andr. Kmbl. 1278 ; An. 639 : 1630 ; An. 816 : (*St. Andrew*), 1967 ; An. 986 : 1249 ; An. 625. Mǽga māne fāh (*Grendel*), Beo. Th. 1960 ; B. 978. v. gūþ-, wuldor-mǽga.

MAGAN (*the infin. does not occur in W. S. but* mǽge *glosses* posse, Mk. Skt. p. 3, 1 ; *and* **magende** (cf. *Icel.* megandi) = *quiens*, Ælfc. Gr. 41 ; Som. 44, 21. Megende *valens*, Kent. Gl. 189 : *the later English forms seem to point to* mugan, *Gen. and Ex.* mugen : *Orm.* muȝenn : *Chauc.* mowen : *Wick.* mowe : *Prompt. Parv.* mown. *Icel.* has mega : *O. H. Ger.* magan *and* mugan : *M. H. Ger.* mugen, mügen : *Ger.* mögen) ; *prs.* ic, hē mæg ðū meaht, mæht, meht, miht ; *pl.* mǽgon, māhau, mǽgon (*or* magon ?) : *Goth.* keeps α

throughout: *Icel.* megum: *O. Sax. O. Frs.* mugan: *O. H. Ger.* (sie) magun, mugun (*later* mugen); *þ.* meahte, mæhte, mehte, mihte (*Goth.* mahta: *O. Sax.* mahta, mohta; *O. Frs.* machte: *Icel.* mátti: *O. H. Ger.* mahta, mohta: *M. H. Ger.* mohte: *Ger.* mochte); *subj. prs.* mæge, mâge, mêge, meige (*or* mæge? *Icel.* megi: *O. Sax.* mugi: *O. H. Ger.* megi, mugi) **I.** *to be strong, efficacious, to avail, prevail, be sufficient*:—Gif ðú meht si *vales*, Kent. Gl. 52. Wel mæg ðæm dæg wêrignise his *sufficit diei malitia sua*, Mt. Kmbl. Lind. 6, 34: Mk. Skt. Lind. Rush. 14, 41. Ne meg mon *non prævaleat homo*, Ps. Surt. 9, 20. Ne mâgon úre woruldfrýnd ús ðonne ǽnigum gôde *our friends will avail us nothing then*, Wulfst. 151, 12. Helle gatu ne mâgon ongên ða *portæ inferi non prævalebunt adversum eam*, Mt. Kmbl. 16, 18. Magan tô *to serve a purpose, be good for, have an effect, be the cause of*:—Ne mæg tô nâhte *ad nihilum valet*, 5, 13. Biþ men ful lytle ðý bet ðeáh ðe hé gôdne fæder hæbbe, gif hé self tô nâuhte ne mæg, Bt. 30, 1; Fox 108, 30. Tô hwan mæg ðis eorþlíce hús, gif hit ýdel stent, Homl. Th. ii. 582, 12: 432, 15: Past. Swt. 7, 12. Him mæg tô sorge ðæt hé nât hwæt him tôweard biþ *it causes him anxiety that he knows not what will happen to him*, Bt. 11, 1; Fox 32, 12. Wæs geworden ðætte seó ylce eorþe mihte tô hǽle *factum est ut ipsa terra gratiæ salutaris haberet effectum*, Bd. 3, 11; S. 535, 34: Exon. 21 b; Th. 57, 21; Cri. 922: 1001, 22; Th. 374, 17; Seel. 127: 82 b; Th. 311, 30; Seef. 100. Magan wið (cf. *Icel.* mega við) *to prevail with* or *against, to be efficacious against* (of a medicine) *to be good for* (a disease):—Gif ic swâ wel wið ðé mæg *if I am so influential with thee*, Homl. Skt. 3, 176. Wið ǽlcum ǽttre mâgon *contra venenum valent*, Bd. 1, 1; S. 474, 36. Ðeós wyrt mæg wið manega untrumnyssa, Herb. 171, 1; Lchdm. i. 300, 24: L. Med. ex Quad. 5, 3; Lchdm. i. 348, 9: L. M. 2, 64; Lchdm. ii. 290, 10. Ðis mæg horse wið ðon ðe him biþ corn on ða fêt, Lchdm. iii. 62, 24. Migtigra wîte wealdeþ ðonne hé him wið mæge *one too mighty for him to withstand is the disposer of punishment*, Cd. 200; Th. 249, 1; Dan. 523. **II.** *to be strong, be in good health* (so *Icel.* mega vel, &c.):—'Hú mæg hê?' Hig cwǽdon ðæt hé wel mihte '*sanusne est?*' '*Valet,*' *inquiunt*, Gen. 29, 6. Ðâ sǽde se cnapa ðæt hé swíðe wel mihte, Homl. Skt. 3, 435. Ðonne ðú mé getrymedest, ðæt ic teala mihte, Ps. Th. 70, 20. **III.** *to be able*, *may* (because a thing is possible):—Ic mæg *queo*; magende *quiens*, Ælfc. Gr. 41; Som. 44, 21. Ic mæg *queo*, ðú miht *quis*, hê mæg *quit*; ic mihte *quivi*, 30; Som. 35, 5. (1) *With infin.*:—Ic mid handum ne mæg heofon geræcan, Cd. 275, 9; Sat. 169. Hér ys seó bôt hú ðú meaht ðíne æceras bêtan, Lchdm. i. 398, 1: Cd. 27; Th. 36, 1; Gen. 565. Ðú .. ðe æghwylc miht wundor gewyrcean, Ps. Th. 76, 11. Hú mæg ðæt yfel beón ðætte ælces monnes ingeþanc wênþ ðætte gôd síe, Bt. 24, 4; Fox 86, 12. Ðæt mæg engel ðín eáþ geféran, 387; An. 194. Eall ðis mâgon him sylfe geseón ... mâgun leóda bearn oncnáwan, Exon. 24 a; Th. 69, 5-12; Cri. 1115. Hí ne mâgon ðone earman gefyllan, Bt. 11, 1; Fox 34, 1. Him ða stormas derian ne mâhan (mǽgon, Cott. MS.), 7, 3; Fox 22, 6. Wê ðæt sôþ mǽgon secgan, Cd. 94; Th. 121, 21; Gen. 2013. [Beo ðan wê mugen understanden, Shrn. 17, 26.] Ðæt hê âna mǽge geríman, Cd. 163; Th. 205, 21; Exod. 439. Ic mæge, Mt. Kmbl. Lind. 26, 61. Ic mêge *possim*, Ps. Surt. 70, 8. Ðú meige *possis*, Kent. Gl. 958. Ðæt ic mâge geseón, Homl. Th. i. 152, 22. Cunnige mâge man of eágum teáras geræcan *try whether tears can be drawn from their eyes*, L. P. M. 3; Th. ii. 288, 4. Gif wê hit mǽgen aþencan, Cd. 21; Th. 26, 2; Gen. 400: 226; Th. 302, 11; Sat. 597. Uê mǽgi, Rtl. 45, 3. Mǽgi hiá, 95, 16. Wîddra ðonne befǽdman mǽge foldan sceattas, 163; Th. 204, 32; Exod. 428. [Ðæt heó þurh ða mugen tô lífes wege becumen .. ðæt ða ðe ðǽr ingáþ mugen ðone leóme geseón, Shrn. 12, 10-13.] Ne meahte hê æt his hige findan, Cd. 14; Th. 18, 1; Gen. 266: Beo. Th. 3322; B. 1659. Mehte, 2168; B. 1082. Eáþe heó mehte beón geseald, Blickl. Homl. 69, 7. Swâ swâ mihte beón fíf þúsend werum, Homl. Th. i. 182, 16. Ðú meahtes geseón ægðer ge fêt ge heáfod, Past. 35; Swt. 241, 14: St. And. 10, 22: Exon. 39 b; Th. 130, 19; Gú. 440. Mihtest, Blickl. Homl. 175, 28. Ða ne meahton âsecgan, 145, 13: Cd. 115; Th. 150, 14; Gen. 2491. Wê ðæt deór gewundigan ne meahte, Nar. 21, 4. Maehtun, Ps. Surt. 20, 12. Mehton, Blickl. Homl. 15, 13. Mihton, 79, 16. Ðæt lâðra nǽnig sceððan ne meahte, Beo. Th. 243. B. 243. Oþ ðæt ðú meahte .. forsíon, meahtes .. lácan, Bt. Met. Fox 24, 11-17; Met. 24, 6-9. Mihte, Blickl. Homl. 45, 27. Swâ hit men fægrost geþencean meahton, 125, 23: Elen. Kmbl. 648; El. 324. Meahten, Exon. 64 a; Th. 236, 13; Ph. 253. Meahte, 39 a; Th. 128, 14; Gú. 404. Mehten, Ors. 3, 1; Swt. 98, 3. Mihtan, Blickl. Homl. 45, 14: 137, 1. Mihten, Cd. 224; Th. 298, 11; Sat. 500 Mihton, Blickl. Homl. 49, 10. Mihte, Ps. Th. 77, 1. (2) *followed, by a clause*:—Hwâ mæg ðæt hê ne wundrige, Bt. 34, 10; Fox 150, 9. (3) *with ellipsis of the infin.* (a) *of a verb which occurs elsewhere in the sentence*:—Gelácna ðú hý forðan ðú édest miht (gelácnian), Hy. 1, 6; Hy. Grn. ii. 280, 6. Nelle ic aldre beneótan, ðeáh ic eáðe mǽge, Beo. Th. 1365; B. 680. Ðæs ofereode, ðisses swâ mæg, Exon. 100 a; Th. 377, 22; Deór. 7. Telle ðás steorran, gif ðú mǽge, Gen. 15, 5: Bd. 5, 3; S. 616, 31. Forlǽte swâ hê oftost mǽge, Bt. Met. Fox 22, 18;

Met. 22, 9: 27, 58, 66; Met. 27, 29, 33. Árás swâ hé hraðost meahte, Exon. 49 a; Th. 168, 24; Gú. 1080. Wolde ic freóndscipe ðínne, gif ic mihte, begitan, Andr. Kmbl. 958; An. 479. (b) *of a verb whose place is taken by* swâ:—Wolde freádrihtnes feorh ealgian, ðǽr hié meahton swâ, Beo. Th. 1599; B. 797. Cwǽdon ðæt heó ríce âgan woldon, and swâ eáðe meahtan, Cd. 3; Th. 4, 4; Gen. 48. Wyllen forsweolgan, gif hí swâ mâgon, Ps. Th. 123, 2. (c) *of a verb to be inferred from the context* (i) *verbs of motion*:—Nó ðý ǽr fram meahte (*might escape*), Beo. Th. 1513; B. 754. Ic ne mæg of ðissum lioþobendum, Cd. 19; Th. 24, 22; Gen. 381. Ne mæg hé on ðæt *non intrabit in illud*, Mk. Skt. 10, 15. On ðone forecwedenan portic mâ ne mihte *prædicta porticus plura capere nequivit*, Bd. 2, 3; S. 504, 38. Ðæt ic up heonon mǽge, Cd. 222; Th. 291, 3; Sat. 425. (ii) *other verbs* (see also **I**):—Wel ðæt swâ mæg *that may well be so*, Bd. 2, 1; S. 501, 18. Þuhte heom ðæt hit mihte swâ, ðæt hié wǽron seolfe swegles brytan, Cd. 213; Th. 266, 15; Sat. 22: Andr. Kmbl. 2786; An. 1395. Wolde hyre búr âtimbrian, gif hit swâ meahte, Exon. 108 a; Th. 411, 28; Rä. 30, 6. Wísdóm sǽde ðæt men mihton (*could understand*) be Gode swelce hí mǽte, Bt. tit. 26; Fox xiv. 16. Ne mâgon ðam breahtme býman ne hornas (*cannot equal*), Exon. 57 b; Th. 206, 29; Ph. 134. **IV.** *may* (*because a thing is permissible* or *lawful, because there is sufficient cause*):—Ðú miht ðæs habban þanc, ðæt ðú mínra gifa wel bruce. Ne miht ðú nó gereccan ðæt ðú ðínes âuht forlure, Bt. 7, 3; Fox 20, 12. Hú miht (mæht, Lind.) ðú secgan ðínum brêðer, Lk. Skt. 6, 42. Ðú meaht ðé forþ faran, Cd. 26; Th. 34, 25; Gen. 543. Hié leng ne mâgon healdan heofonríce, 35; Th. 45, 24; Gen. 731. Nú wit mâgon sorgian for his síðe *we have good cause to rue his journey*, 38; Th. 49, 29; Gen. 799; Exon. 9 b; Th. 8, 34; Cri. 127. Hwæðer sél mæge wunde gedýgan, Beo. Th. 5054; B. 2530. Hit ne meahte swâ *that was not allowed*, Exon. 41 a; Th. 136, 29; Gú. 548. **V.** *in the Northumbrian Gospels the verb is used as an auxiliary in the translation of the Latin subjunctive, or fut. indic.*:—Synngiga mæge *peccabit*, Mt. Kmbl. Lind. 18, 21. Wê habbas t mâgon habba *habebimus*, 21, 38. (*Also the W. S. version in Mt. Kmbl. 26, 54, has* hú nâgon beón gefyllede *quomodo implebuntur*.) Hú hine mæhtes tô lose gedôa *quomodo eum perderent*, 12, 14. Ðatte hiá éton t mæhton eata *quod manducarent*, Mk. Skt. Lind. 8, 1. Huu hine hiá âcuoella mæhton (mæhtun, Rush.) *perderent*, 11, 18. Mæghton (mæhtun, Rush.), Lk. Skt. Lind. 22, 4.

magdala-treów, es; *n.* *An almond-tree*; amigdala *vel* nutida, Ælfc. Gl. 47; Som. 65, 36; Wrt. Voc. 33, 34.

mage, an; *f.* *The belly*; ventriculus, Ælfc. Gl. 74; Som. 71, 43; Wrt. Voc. 44, 26. v. maga.

mâge, an; *f.* *A kinswoman*:—Elizabeþ ðín mǽge (mâge, MSS. A. B.) *cognata tua*, Lk. Skt. 1, 36. Seó cwên his mâge *regina propinqua illius*, Bd. 3, 24; S. 557, 24. Ða landes ðe hire mǽge hire geúþe, Chart. Th. 338, 14: 337, 27. From bearme ânre mâgan, Exon. 112 b; Th. 430, 25; Rä. 44, 14. Grendles mâgan (*mother*) gang, Beo. Th. 2786; B. 1391. Be hire mâgan (*propinqua*), Bd. 3, 8; S. 531, 3. Ne hǽme nân man wið his mâgan ne wið his mǽges wíf, Lev. 18, 16. Se wolde niman his mâgan (*cousin*) tô wífe, Homl. Th. ii. 476, 19. Menn hæfdon on frymþe heora mâgan tô wífe, Homl. Skt. 10, 215. v. mǽge, mâga.

mâ-geéct (mâ = *magis*, ge-écan = *augere*), *mactus* (= *magis auctus*):— Ða mâgeéctan *macta*, Wrt. Voc. ii. 55, 3. Cf. 54, 71. Mâgeécte *morota* (*macta?*), 57, 24.

Mage-sǽte, -sǽtan; *pl.* *The people of Herefordshire*, Chr. 1016; Erl. 158, 4.

mageþe, an; *f.* *A plant-name, maythe, chamomile, ox-eye*:—Mageþe beneolentem [*camemelon*], Wrt. Voc. i. 67, 27: obtalmon, 68, 50. Magoþe *optalmon*, 16, 52. Ðás wyrte ðe man *camemelon*, and ôðrum naman mageþe nemneþ, Herb. 24; Lchdm. i. 120, 14. Wildre magþan wyrttruman (*matricaria chamomilla*) L. M. 2, 22; Lchdm. ii. 206, 15. Magoðe, L. M. 1, 64; Lchdm. ii. 140, 7. Ða reádan mago-þan (*anthemis tinctoria*) 140, 4. [Maiþe *camomilla*, Wrt. Voc. i. 140, 27. Mathen (maythe) *ameroke*, 162, 20. Maythe *embroca*, 190, 51. See Lchdm. ii. 398, col. 2, iii. 337, col. 1, and E. D. S. Plant-names under *mathes* and *May-weed*.] v. mægþa.

magister, mægister, es; *m.* *A master*:—Se magister, Past. 61; Swt. 455, 20. Byrla magister (cf. byrla ealdor, v. 20), Gen. 40, 21. Mægister, Wrt. Voc. i. 75, 6. Mín mægister Euripides, Bt. 31, 1; Fox 112, 20. For his magistre, Bd. 1, 7; S. 477, 10. Ðeáh hió hire magister lufige, Bt. 25; Fox 88, 10. His âgenne mægistre, 29, 2; Fox 104, 19. Magistra betst, Bt. Met. Fox 30, 8; Met. 30, 4. Hí hæfdan magistras, Bd. 4, 2; S. 565, 34. Mægestras, Ex. 1, 11.

magu, a; *m.* **I.** *A child, son*:—Ðâ wearþ eafora fêded, mago Caines, Malalahel, Cd. 58; Th. 70, 28; Gen. 1160. Mago Ecglâfes (cf. Ecglâfes bearn, 1003), Beo. Th. 2935; B. 1465. Mago Healfdenes (cf. sunu Healfdenes, 541), 3738; B. 1867: 4027; B. 2011. Eald fæder ongon his mago monian, Exon. 80 b; Th. 303, 28; Fä. 60. Ðínum magum (mágum?) lǽf folc and ríce, Beo. Th. 2361; B. 1178.

II. *a young person, a servant* (cf. cniht, cnapa, geongra):—Ongan his

magu frignan (cf. ombehtþegn, l. 9), Exon. 47 b; Th. 162, 30; Gū. 983. **III.** *a young, strong man, a man* (cf. cniht):—Hwǽr cwom mearg hwǽr cwom mago *where is the steed gone? where his rider?* 77 b; Th. 291, 34; Wand. 92. Mago Ebrēa (*Abraham*), Cd. 100; Th. 132, 34; Gen. 2203: 109; Th. 145, 25; Gen. 2411: 127; Th. 161, 32; Gen. 2674. Maga gemēdu, Beo. Th. 499; B. 247. [*Goth.* magus παῖς (*puer, servus*): *O. Sax.* magu *child*: *Icel.* mögr *a son, a man.*]

magu-dryht, e; *f. A band of young men*:—Oþ ðæt seó geóguþ geweóx, magodriht micel, Beo. Th. 134; B. 67.

magu-geóguþ, e; *f. Youth,* Exon. 28 b; Th. 87, 23; Cri. 1429. [Cf. *O. Sax.* magu-jung *young*].

magu-rǽdend, es; *m. One who advises men*:—Woldon cræfta gehygd magorǽdendes (*St. Andrew*) mōd oncyrran, Andr. Kmbl. 2920; An. 1463.

magu-rǽswa, an; *m. A leader of men, a chief*:—Se magorǽswa mǽgþe sīnre dōmas sægde, Cd. 79; Th. 98, 2; Gen. 1624. Se ðe lǽdde, mōdig magorǽswa (MS. -rǽwa), 145; Th. 181, 2; Exod. 55: 143; Th. 178, 25; Exod. 17.

magu-rinc, es; *m. A child, young man, a man, warrior*:—Se magorinc sceal wesan Ismahel hāten, Cd. 104; Th. 138, 2; Gen. 2285: (*Isaac*), 106; Th. 140, 15; Gen. 2328. Ða magorincas (*youths*), Abraham and Loth, 82; Th. 103. 6; Gen. 1714: (*Cato and Brutus*), Bt. Met. Fox 10, 111; Met. 10, 56. Cwom LX monna . . ne meahton magorincas ofer mere feolan, Exon. 106 a; Th. 404, 9; Rä. 23, 5. Magorinca heáp (*the men in Hrothgar's hall*), Beo. Th. 1464; B. 730. Magorinca mōd, Bt. Met. Fox 1, 51; Met. 1, 26.

magu-þegn, *m. A thane, vassal, follower, retainer, warrior, servant*:—Ic eom Higelāces mǽg and magoþegn, Beo. Th. 820; B. 408: (*Beowulf's follower, Wiglaf*), 5507; B. 2757. Mǽrum maguþegne (*a retainer of Hrothgar*), 4164; B. 2079: (*God's servant, Matthew*), Andr. Kmbl. 188; An. 94: (*St. Andrew*), 2416; An. 1209. His engel, mǽrne maguþegn, 731; An. 366. Ic maguþegnas (*servants*) mīne hāte flotan eówerne healdan, Beo. Th. 591; B. 293. Mōdige maguþegnas (*the Mermedonians*), Andr. Kmbl. 2281; An. 1142: 3028; An. 1517: Exon. 77 a; Th. 290, 8; Wand. 62: Judth. 12; Thw. 25, 1; Jud. 236. Magoþegna ðone sēlestan (*Æschere s. vv. 2654 sqq.*), Beo. Th. 2815; B. 1405.

magu-timber, es; *n.* **I.** *A child*:—Ðā heó wæs magotimbre eácen worden *when she was with child*, Cd. 101; Th. 134, 36; Gen. 2235. Mē sealde sunu sigora waldend, and mē cearsorge mid ðȳs magotimbre of mōde āsceáf, 55; Th. 68, 10; Gen. 1115. [Cf. *Icel.* manns-efni (efni *material, stuff*) *a promising young man.*] **II.** *progeny, all those who are born*:—Ne sȳ ðæs magutimbres gemet ofer eorþan gif hī ne wanige se ðās worulde teóde *there would be no bounds upon earth to those who are born, if they waned not through him that created the world*, Exon. 89 a; Th. 335, 13; Gn. Ex. 33.

magu-tudor, es; *n. Offspring*:—Ær ðȳ magotudre mōdor wǽre eácen be eorle, Cd. 132; Th. 167, 13; Gen. 2765. Ús ðis se æþeling gefremede . . monnes magutudre *for us, the human race, the prince* (*Christ*) *did this*, Exon. 17 a; Th. 39, 28; Cri. 629. Cf. magu-timber.

māh; *adj. Wicked, wanton,* Exon. 95 a; Th. 354, 47; Reim. 62. v. ge-māh.

māl, es; *n. A mole, spot, mark*:—Fūll maal on [h]rægel *stigmentum*, Ælfc. Gl. 28; Som. 61, 13; Wrt. Voc. 26, 12. Māl *maculam*, Wrt. Voc. ii. 57, 9: 92, 19. [*Goth.* mail *spot, blemish*: *O. H. Ger.* meil.]

māl, es; *n.* **I.** *An action, suit, cause*:—Māl *clasma* (cf. clasma clam oðde wed oðde wæra. 'This barbarous word meant in medieval Latin, an action at law, for a bond or other obligation,' 21, 2), Wrt. Voc. ii. 83, 42: Hpt. Gl. 496, 4. [*Icel.* māl *an action*: *O. H. Ger.* mahal *concio, pactio, fœdus.*] **II.** occurring late in the chronicle and borrowed from Icelandic (?):—Ðǽr bær Godwine up his māl (*case*) (cf. *Icel.* bera upp māl), Chr. 1052; Erl. 187, 19. Eádwerd scylode ix scypa of māle (= *Icel.* skilja af māli) *put an end to the agreement with, paid off, nine ships,* 1049; Erl. 174, 38. Hē sette ealle ða litsmen of māle, 1050; Erl. 176, 13. Se cyng sealde his lande swā deóre tō māle swā heó deórost mihte *made as hard terms as ever he could,* 1086; Erl. 220, 8. [*Icel.* māl *a case; terms, agreement.*] v. mǽl and next word.

māl-dæg, es; *m. An agreement, covenant, settlement* (?) (*Icel.* māldagi) or *a day on which terms are fixed* (?) (*O. H. Ger.* mahal-tag *dies sponsionis*) *a day when the dowry was settled*:—Ic an mīne wīfe al þe þing þe ic haue on Norfolke so ic hire gaf tō mund and tō māldage, Chart. Th. 574, 1. v. mǽl-dæg.

māletung, e; *f. Verbosity*:—Hlȳdig gewyrd, malelung (maletung?) *garrula verbositas,* Hpt. Gl. 439, 60.

malscra. v. next word.

malscrung, e; *f. Bewitching, fascination*:—Malscrung *fascinatus,* i. *laudatis stultæ,* Wrt. Voc. ii. 35, 7: *fescinatio,* 108, 23. Wið malscrunge, Lchdm. iii. 36, 13. Wið feóndes costunga and nihtgengan and maran and malscra (malscrunga?), L. M. 3, 1; Lchdm. ii. 306, 13. [*O. H. Ger.* mascrunc *fascinatio, laus stulta*: cf. *Goth.* untila-malsks προπετής: *O. Sax.* malsk *proud*: *Allit. pms.* þe mon malskred (*fascinated, spell-bound*) in drede; þat malscrande mere: *Will.* hou he

hade . . malskrid (*wandered as under the influence of a charm, mazed*) aboute.]

māl-sweord, es; *n. A sword with inlaid ornament*:—Ic geann ðæs mālswurdes, Chart. Th. 560, 33. [Cf. *Icel.* māla-sax *an inlaid sword.*]

malt, malu. v. mealt, mealu.

Mame-ceaster, e; *f. Manchester*:—Mameceaster on Norþhymbrum, Chr. 923; Erl. 110, 4.

mamme, an; *f. A teat, breast* (*Lat.* mamma):—An mamman *in papillas,* Germ. 401, 77.

mamor, es; *m. Deep sleep, unconsciousness*:—Mamor *soporem,* Kent. Gl. 695. Momna (= mamor?) *sopor,* Wrt. Voc. ii. 120, 82. v. next word.

mamorian, mamrian *to be deep in thought about anything* (?):—Hī mamriaþ mān and unriht *they are plunged in thought of crime and wrong;* scrutantes scrutinio, Ps. Th. 63, 5. [*Somner gives* mamerung *dormitio, dormitatio:* cf. *later English* mammering:—He sits now in a *mammering,* As one that minds it not. Halliw. Dict. q. v. See also Nare's Glossary.]

man, mon; *indef. pron.* (*originally nom. of noun* mann q. v.; cf. *French* on *from* homo). *One, anyone, they, people;* it is often used with the active voice where modern English would take the passive:—Man brohte his heáfod on ānum disce and sealde ðam mǽdene *allatum est caput ejus in disco, et datum est puellæ,* Mt. Kmbl. 14, 11. Tō middyre nihte man hrȳmde *media nocte clamor factus est,* 25, 6. His brōþur Horsa man ofslōg, Chr. 455; Erl. 12, 15. Man gehālgode ii. biscopas on his stal, 678; Erl. 41, 7. Hine man hēng . . Hyne man dyde up and hine man efosode and scrȳdde hine and brohte hine tō ðam cynge *ille suspensus est in cruce. Eductum de carcere Joseph totonderunt, ac veste mutata obtulerunt regi,* Gen. 41, 13, 14. Ne ete man his flǽsc *non comedentur carnes ejus,* Ex. 21, 28. Gif hē nǽbbe hwæt hē wið ðære stale sylle sylle man hine wið feó. Gif man cucu finde ðæt hē stæl *si non habuerit, quod pro furto reddat, ipse venundabitur. Si inventum fuerit apud eum, quod furatus est, vivens,* 22, 3, 4. Hū mæg man (*quisquam*) ingān on stranges hūs, būton hē gebinde ǽrest ðone strangan, Mt. Kmbl. 12, 29. Worhte man hit him tō wīte, Cd. 17; Th. 21, 2; Gen. 318. Hit gedēfe biþ ðæt mon his winedryhten herge, Beo. Th. 6332; B. 3176. [*Later English* me: *Du.* men: *Ger.* man.]

mān, es; *n. A bad, shameful action, a crime, crime, guilt, wickedness*:—Maan *facinus,* Ælfc. Gl. 84; Som. 73, 98; Wrt. Voc. 49, 5. Mān, Wrt. Voc. ii. 34, 54: *piaculum,* 68, 68. Mān and inwit *guilt and guile,* Ps. Th. 54, 9. Mān and unriht *iniquitas,* 118, 69. Mān, yfel endeleás, Andr. Kmbl. 1388; An. 694. Mān and morðor (cf. *O. Sax.* mēn endi morðwerk), misdǽda worn (v. Fox 58, 2, hwilc mān hē weorhte), Bt. Met. Fox 9, 13; Met. 9, 7. Mānes *fraudis,* Wrt. Voc. ii. 33, 44. Mānes wyrhtan *peccatores,* Ps. Th. 100, 8. Māne *piaculo,* Hpt. Gl. 432, 50: Lev. 19, 29. Mid manegum māne *with many a crime* (cf. eác ðam wæs unrīm ōðres mānes, Met. 1, 44), Bt. Fox 1, 10. Gē mid māne men ongunnon *irruitis in homines,* Ps. Th. 61, 3: Cd. 16; Th. 19, 30; Gen. 299. For þȳ māne (*the murder of Abel*), Beo. Th. 220; B. 110. Māne fāh *stained with crime,* 1960; B. 978. Mān *nequitiam,* Ps. Spl. 72, 8: Ps. Th. 140, 4. Tō ðam ilcan men (*Achan*) ðe ðæt mān (*taking of the forbidden spoil*) gefremode, Jos. 7, 17: Cd. 10; Th. 12, 22; Gen. 189. Ne swera ðū mān (cf. *O. Sax.* ni thū mēnes ni sweri) *non perjurabis,* Lev. 19, 12. Se man ðe swereþ mān, 5, 4. For ǽghwæðerum ðyssa māna *utroque scelere,* Bd. 2, 5; S. 506, 40. Hī geclǽnsian ðæra ǽrrena māna *a pristina flagitiorum sorde purgare,* 3, 23; S. 554, 28. On manegum mānum (*flagitias*) hī sylfe besencton, 1, 22; S. 485, 12. Ealle ða mān (*scelera*) ðe ic ǽfre gefremede, 5, 13; S. 633, 8. [*Orm.* man inn aþess and i wittness: *O. Sax.* mēn: *O. H. Ger.* mein *nefas, inlicitum:* *Icel.* mein *hurt, harm.*] v. next word.

mān; *adj. Wicked, false, base*:—Mān inwitstæf *nequitia,* Ps. Th. 54, 15. Heora mǽnige māne swultan *many a wicked one of them died,* 77, 30. Nāuht ne deregaþ monnum māne āþas *nil perjuria nocet ipsis,* Bt. 4; Fox 8, 16. Mānum treówum woldon hié ðæt feorhleán, fācne gyldan, Cd. 149; Th. 187, 11; Exod. 149. [*Icel.* meinn *mean, base:* *O. Frs.* mēn *false* (*oath*): *O. H. Ger.* mein.] v. mǽne *and preceding word.*

man-. v. mann-.

mān-āþ, es; *m. A false oath, perjury*:—Se ðe mānāþ [*other reading* mǽnne āþ] swerige *he who commits perjury,* L. Ath. i. 25; Th. 212, 18. [*Orm.* þatt tu ne swere nan manaþ: *O.E. Homl.* man-að: *O.Sax. O.L. Ger. O. Frs.* mēn-ēð: *Icel.* mein-eiðr: *Da.* meen-ed: *O. M. Mod. H. Ger.* mein-eid.] v. mān; *adj.,* mǽne.

mān-bealu, wes; *n. Wicked injury,* Cd. 174; Th. 218, 27; Dan. 45.

mān-bryne. v. mann-bryne.

mancus, es; *m. A mancus, the eighth of a pound, the sum of thirty pence*:—Fīf penegas gemacigaþ ǽnne scillingc and xxx penega ǽnne mancus (*other* MSS. manccus, mancs), Ælfc. Gr. 50; Som. 52, 8. *In Cnut's laws the heriot of an earl included* twā hund mancus goldes (*which is rendered in a Latin version by* quinquaginta marcas auri, v. Schmid. p. 309, *so that the* mancus *is the fourth of a* marc), L. C. S. 72; Th. i. 414, 8. Cf. *for an instance of the manner in which this might be*

paid the will of an ealdorman *where the heriot included* feówer beágas twegen on hundtwelftigum mancosum and twegen on hundeahtatigum, Chart. Th. 500, 3. *The value of the mancus is also seen from* L. Ath. v. 6, 2 ; Th. i. 234, 1 :—Oxan tô mancuse *compared with* Th. i. 232, 7 *where an ox is rated at thirty pence*, be xxx pænega oððe be ánum hrýðere. The word occurs not unfrequently in the charters. Gedæle hê ælcum mæssepreóste binnan Cent mancus goldes, Chart. Th. 471, 19. Ágyfe man mînra (*king Alfred*) ealdormanna ælcum ân hund mancgusa . . . and Æðeréde ealdormenn ân sweord on hundteóntigum mancusum, 489, 29–33. Ic geann ælcum bisceope v. mancessa goldes, 544, 8. Ân hund mancosa, 596, 9. Mancussa, 530, 13. Ænne beáh on þrittigan mancysan, 501, 9. Ânes beáges on sextigum mancussum goldes, 529, 4 : 531, 4. Mid xvi. mancussum reádes goldes, 536, 21. Týn mancusas goldes, v. mancusas goldes, 544, 11–14. [*O. H. Ger.* mancusa, manchusa, manchussa (*nummos*) *aureos, philippos, solidos*, Grff. ii. 808 : *O. L. Ger.* mancusi *aureos*.]

mand, mond, e ; *f. A basket, mand, maund* (archaic or dialectic v. E. D. S. Pub. Gloss. B. 1 : 15 : 16 : Mid-Yorkshire and Lincolnshire Gloss. *Prompt. Parv.* mawnd, skype *sportula*, p. 300, see the note for other examples) :—Mand *corvis*, Wrt. Voc. i. 291, 20 : *cophinus*, ii. 74, 47 : 104, 62 : *qualus*, 118, 47 : *corben*, 104, 42. Manda *coffinos*, 17, 47 : 72, 68. Twælf monde fulle *duodecim cophinos plenos*, Mt. Kmbl. Rush. 14, 20 : 16, 9. Hû monig monda *quot sportas*, Lind. 16, 10. Mondo, Mk. Skt. Lind. 8, 8. Huu monig mondo (monde, Rush.) *quot cophinos*, 19.

mán-dǽd, e ; *f. An evil deed, crime, sin* :—Mándǽd *crimen, peccatum*, Wrt. Voc. ii. 137, 3. Mándǽda *scelera*, 149, 29. Hê sume mándǽde (*aliquid sceleris*) gefremede, Bd. 4, 25 ; S. 599, 34. Mándǽda forlǽtan *intermissis facinoribus*, S. 601, 27. His synne and mándǽde *scelera sua*, 5, 13 ; S. 632, 12 : Exon. 62 b ; Th. 229, 18 ; Ph. 457. Mándǽda *facinorum, peccatorum*, Hpt. Gl. 415, 14 : 469, 9 : *flagitiorum*, 529, 73 : Ors. 1, 8 ; Swt. 42, 17. Ðá ðá hê ða môdigan preóstas for heora mándǽdon ðanan ût ádrǽfde and ðérinne munecas gelógode, Chart. Th. 227, 21. Wolde mid mándǽdum menn beswîcan, Cd. 23 ; Th. 29, 16 ; Gen. 451. [*O. E. Homl.* man-dede : *O. Sax.* mên-dâd : *O. H. Ger.* mein-tât *scelus, flagitium, facinus, piaculum* : cf. *Icel.* mein-görð *offence*.]

mán-dǽde ; *adj. Doing evil, wicked, flagitious* :—Hê sceal mándǽde men þreágean þearle *he must sharply rebuke evil-doers*, Wulfst. 266, 24 : L. I. P. 2 ; Th. ii. 304, 18. Ealles tô ídele ælcere gôddǽde and tô mándǽde *far too deficient in every good deed and too ready to do evil*, 14 ; Th. ii. 322, 14. [Cf. *O. Sax.* mên-dâdig : *O. H. Ger.* mein-tâtig *flagitiosus, sacrilegus*.]

mán-deorf ; *adj. Labouring to do evil, wicked* :—Ne mæg se yfela preóst mid his yfelnysse, ðeáh hê mándeorf sý and mánful on dǽdum, ne mæg hê nǽfre Godes þênunge gefilan, naðer ne ðæt fulluht, ne ða mæssan, L. Ælfc. P. 41 ; Th. ii. 382, 12. v. deorfan.

mán-drinc, es ; *m. An evil, poisonous drink* :—Ðone mándrinc (*the poison from an arrow*, cf.ǽttren l. 7), Exon. 106 b ; Th. 406, 6 ; Rä. 24, 13.

manetian (?), *to admonish, reprove* :—Gê monetigaþ Godes êce bearn (cf. vv. 1331 sqq. *for the speech of the* ealdorsacerd), Andr. Kmbl. 1492 ; An. 747. Cf. manian.

mán-fǽhþu ; *f. Guilt, wickedness* (cf. máne fá, morþorscyldige, Andr. Kmbl. 3196 ; An. 1601 : also Beo. Th. 1960 ; B. 978) :—Mánfǽhþu bearn (*those who were drowned by the deluge*), Cd. 69 ; Th. 83, 11 ; Gen. 1378.

mán-feld, es ; *m. The field of crime* :—Mon hǽtt ðæt lond Mánfeld ðér hié mon byrgde *obruta est in campo, qui nunc Sceleratus vocatur*, Ors. 3, 6 ; Swt. 108, 20.

mán-folm, e ; *f. A hand that does evil* :—Alýs mê and genere wið mánfolmum fremdra beorna, Ps. Th. 143, 8.

mán-fordǽdla, an ; *m. One who wickedly destroys* :—Mánfordǽdlan (*the sea monsters that attacked Beowulf*), Beo. Th. 1130 ; B. 563.

mán-forwyrht, es ; *n. Sin, crime* :—Fore moncynnes mánforwyrhtum, Exon. 24 a ; Th. 67, 28 ; Cri. 1095.

mán-freá, an ; *m. The prince of evil, the devil* :—Morðres mánfreá, Andr. Kmbl. 2627 ; An. 1315 : Elen. Kmbl. 1880 ; El. 942 : Exon. 73 b ; Th. 275, 6 ; Jul. 546.

mán-fremmende ; *part. Doing evil, working wickedness* :—Mid mannum mánfremmendum *cum hominibus operantibus iniquitatem*, Ps. Th. 140, 6 : Exon. 67 b ; Th. 250, 34 ; Jul. 137 : 29 a ; Th. 88, 9 ; Cri. 1437 : Elen. Kmbl. 1810 ; El. 907.

mán-full ; *adj. Evil, wicked, flagitious, producing an evil effect, dire* :—Mánful *profanus*, Ælfc. Gl. 84 ; Som. 73, 101 ; Wrt. Voc. 49, 8 : *infandum*, Wrt. Voc. ii. 111, 2 : *flagitiosus, criminosus*, 149, 27. Mánfull *nequam*, Ælfc. Gr. 9, 78 ; Som. 14, 30 : Mt. Kmbl. 6, 23. Mánful, 20, 15. Ðæt mánfulle wuht *the devil*, Blickl. Homl. 31, 7. Mánfulles *fanaticae*, Hpt. Gl. 467, 61. Mánfulles scínláces *fanaticæ superstitionis, infandæ vanitatis*, 488, 40 : 509, 38. Becom ðæt tô eáran ðæs mánfullan (*nefandi*) ealdormannes, Bd. 1, 7 ; S. 477, 6. Ðone mánfullan *flagitiosum*, Wrt. Voc. ii. 33, 52. Mánfulle and synfulle *publicani et peccatores*, Mt. Kmbl. 9, 10, 11 : Mk. Skt. 2, 15, 16.

Ðonne ûs mánfulle menn onginnaþ *cum insurgerent homines in nos*, Ps. Th. 123, 2 ; Andr. Kmbl. 359 ; An. 180. Mánfulre wurte *dirorum* (*nefandorum*) *graminum*, Hpt. Gl. 450, 9. Sodoman and Gomorran ðæra mánfulra þeóda, Gen. 14, 10 : Andr. Kmbl. 84 ; An. 82 : Salm. Kmbl. 298 ; Sal. 148. Ða mánfullan *infandas*, Wrt. Voc. ii. 47, 69. Eác mycle mánfullran (*sceleratiora*) fremedon, Bd. 4, 25 ; S. 601, 29. [*O. Sax.* mên-ful : *O. H. Ger.* mein-fol *profanus, flagitiosus, nefarius, funestus.*]

mánful-lîc ; *adj. Evil, wicked* :—Hê sǽwþ mánfullîce geþohtas intô ðæs mannes heortan, Boutr. Scrd. 20, 17.

mánful-lîce ; *adv. Wickedly*, Scint. 4.

mánful-ness, e ; *f. Wickedness* :—Git Martianus for his mánfulnysse nolde on God gelýfan, Homl. Skt. 4, 389. Hê leornode ǽfre máran and máran on his mánfulnysse and ne lêt nánne his gelícan on yfele, Ælfc. T. Grn. 17, 28.

mán-genga, an ; *m. One conversant with* or *practising evil, a sacrilegious person* :—Ðone mángengan and ðone wiðfeohtend *rebellem ac sacrilegum*, Bd. 1, 7 ; S. 477, 18.

mán-genîþla, an ; *m. A wicked, evil persecutor* :—Ðæt ne môton mángenîþlan, grame grynsmiþas, gáste gescedðan, Andr. Kmbl. 1832 ; An. 918.

mangere, es ; *m. A monger* (in iron-*monger*, cheese-*monger*, &c.), *merchant, trader, dealer* :—Mangere *mercator* vel *negotiator*, Wrt. Voc. i. 73, 72. Hwæt sægst ðu, mancgere (*mercator*)? Coll. Monast. Th. 26, 23. Ne preóst ne beó mangere *a priest shall not be mangere* (cf. *Icel.* prestar skulu eigi fara með mangi nê okri), L. Ælfc. C. 30 ; Th. ii. 354, 1. Wê lǽraþ ðæt preósta gehwilc tilige him rihtlîce and ne beó ǽnig mangere mid unrihte, L. Edg. C. 14 ; Th. ii. 246, 24. Heofena ríce is gelíc ðam mangere (*negotiatori*), Mt. Kmbl. 13, 45. Ðû herast ðone mancgere ðe begytt gold mid leáde, Homl. Th. i. 254, 25. [*Icel.* mangari : *O. H. Ger.* mangari, mengari ; *Graff quotes an O. L. Ger.* fleiscmengere.] v. flǽsc-mangere.

mán-gewyrhta, an ; *m. A worker of wickedness*, Ps. Th. 77, 38.

mangian ; *p. ode To trade, traffic, act as a monger* :—Ic mangige *mercor*, Ælfc. Gr. 25 ; Som. 27, 12. Mid sceápum hê mangaþ *he traffics with sheep*, Homl. Th. i. 412, 6. Gif man mid cirican mangie, bête be lahslite, L. N. P. L. 20 ; Th. ii. 292, 28. Hwæt forstent ǽnigum menn ðæt ðeáh hê mangige ðæt hê ealne ðisne middangeard âge gif hê his sáule forspildt *what does it benefit any man, though he come to own all this world by his trading, if he destroys his soul*, Past. 44, 10 ; Swt. 333, 9. [*A. R.* mangen : *O. Sax.* mangôn : *Icel.* manga *to trade* : cf. *Du.* mangelen *to barter*.] v. ge-mangian.

mangung, e ; *f. Trade, traffic, business, commerce, dealing ; also merchandise* :—Mangung *mercimonium*, gestreón i. *commercium*, Hpt. Gl. 500, 44. Mid mangunge ł gestreóne *commercio*, 478, 31. Fram mangunge *a negotio*, Ps. Lamb. 90, 6. Hig férdun, sum tô his tûne, sum tô his manggunge (*negotiatione*), Mt. Kmbl. 22, 5. Se fǽrþ embe his mangunge (cf. sume tô heora ceápe, l. 9), Homl. Th. i. 524, 12. [Cf. *Icel.* mang *traffic*.]

mangung-hûs, es ; *n. A house for traffic* :—Ne wyrce gê mínes feder hûs tô mangunghûse (*domum negotiationis*), Jn. Skt. 2, 16.

mán-hûs, es ; *n. A house of wickedness, hell* :—Mánhûs fæst under foldan, ðér biþ fýr and wyrm, open scræf yfela gehwylces, Cd. 169 ; Th. 212, 7 ; Exod. 535.

manian, manigean, monian ; *p. ode.* **I.** *to bring to mind what ought to be done, to urge upon one what ought to be done, to admonish, exhort, instigate* :—Ðonne manige ic ðæt gê eów âlêsan of eówrum synnum, Blickl. Homl. 51, 32. Ic myngige and manige manna gehwylcne ðæt hê his âgene dǽda georne smeáge, 109, 11. Manaþ *cohortatur, ammonet*, Hpt. Gl. 451, 52. Uton forhradian Godes ansýne on andettnysse, swá swá se wîtega ûs manaþ, Homl. Th. ii. 124, 24. Monaþ môdes lust tô fêran, Exon. 82 a ; Th. 308, 7 ; Seef. 36. Ealle ða gemoniaþ môdes fûsne fêran tô sîþe . . . swylce geác monaþ, Th. 309, 6 ; Seef. 53. Menede *instigavit, monuit*, Hpt. Gl. 511, 30. Hê manode hig georne ðæt hig Moyses ǽ heóldon, Jos. 23, 1. Manade, Bd. 5, 13 ; S. 632, 11. Agustinus Brytta biscopas for rihtgeleáffulra sibbe lǽrde and monade (*monuerit*), 2, 2 ; S. 502, 3. Hine mid ðisum wordum manode, Homl. Th. ii. 130, 33. Hî hî manedon and lǽrdon ðæt hî him wǽpno worhton, Bd. 1, 12 ; S. 481, 5. Ongan hî manigean and lǽran ðæt hî sibbe hǽfdon, 2, 2 ; S. 502, 8. Manian, Byrht. Th. 138, 31 ; By. 228. Maniende *instigantes, incitantes, cohortantes*, Hpt. Gl. 416, 23. **II.** *to bring to mind what should not be forgotten, to admonish, remind, suggest, prompt* :—Forþon ic eów manige ealle ðæt *therefore I remind you all of it*, Blickl. Homl. 143, 7. Hêr ûs manaþ and mynegaþ be (*we are here reminded of*) ðisse hálgan tíde weorþunga, 161, 3. Manaþ swá and myndgaþ sárum wordum Beo. Th. 4120 ; B. 2057. Mec ðæs þearf monaþ, micel môdes sorg, Exon. 76 a ; Th. 285, 21 ; Jul. 717. **III.** *to tell what ought to be done, to teach, instruct, advise* :—Hê hié mid ðissum wordum lǽrde and manode *he taught them what they should do in these words*, Blickl. Homl. 169, 12. Hê ûs lǽrde and monade, hû wê ûs gebiddan sceoldan, 19, 36. Hê dyde swá swá hê manede, Homl. Th. i, 238, 23. God bebeád Moyse ðæt hê

manode đæt folc, đæt swâ hwâ swâ âbiten wǽre, besáwe up tô đǽre
ǽrenan næddran, ii. 238, 17. Heó lǽrde hine and manede, đæt đæt ne
gedafenade, đæt hê sceolde his freónd on gold bebycgean, Bd. 2, 12;
S. 514, 37. Fæder ongon his mago monian (cf. l. 13 lǽrde), Exon. 80 b;
Th. 303, 28; Fä. 60. **IV. to claim of a person** (acc.) *what is due*
(gen); in jus vocare (cf. *the Frankish* ad mallum mannire, *and the use
of* monere *in the laws*. v. Grmm. R. A. 842; *Mod. Ger.* mahnen *to ask
payment of a debt: Icel.* mana *to provoke, challenge*):—Hwane manaþ
God mâran gafoles đonne đone biscop *of whom will God demand more
tribute, than of the bishop?* Blickl. Homl. 45, 16. Drihten manaþ
ǽghwylcne man đæs đe hê him hêr syleþ, 49, 31. Đam đe Drihten
micel syleþ, mycles hê hine eft manaþ, Wulfst. 261, 22: 148, 18. Forgield mê đîn iht...đæs lífes ic manige, Exon. 29 b; Th. 90, 24; Cri. 1479.
Láþ se đe londes monaþ, leóf se đe mǽre beóđeþ, 89 b; Th. 337, 5; Gn.
Ex. 60. Đâ cwæþ se đe đæs feós manode, Shrn. 127, 30. Mana đone
đæs ângyldes, L. In. 22; Th. i. 116, 11. [*O. Sax.* manôn: *O. Frs.*
monia *to admonish, to claim* (with gen.): *O. H. Ger.* manôn, manên
monere, suggerere with acc. of person (and gen. of thing)]. v. â-, fore-,
ge-manian; maniend, manung.

mân-ídel; *adj. Wicked and vain:*—Đara mûþas sprecaþ mânídel
word *quorum os locutum est vanitatem,* Ps. Th. 143, 9, 13.

maniend, es; *m. One who claims* (*debts* &c.):—Se wæs ǽrest theloniarius đæt is gafoles moniend *he* (*St. Matthew*) *was first theloniarius, that
is a tax-gatherer,* Shrn. 131, 24.

MANIG, maneg, monig, mænig; *adj.* **I. with a noun or adjective,** MANY, (with sing. noun) *many a:*—Đǽr biþ swýđe manig burh,
Ors. 1, 1; Swt. 20, 14. Đâ wæs ymb đa gifhealle gûþrinc monig, Beo.
Th. 1681; B. 838. Manig man cwyþ *multi dicunt,* Ps. Th. 4, 7. Geong
manig, Beo. Th. 1712; B. 857. Monig, 345; B. 171. With a plural
verb:—Wlanc manig on stæþe stôdon, Elen. Kmbl. 461; El. 231.
Maniges þinges hê wilniaþ, Bt. 34, 7; Fox 142, 32. Đises hí wundriaþ
and manies þyllíces, 39, 3; Fox 214, 31. Mid manegum mâne, 1; Fox
2, 10. Manegum men þuhte, 11, 3; Fox 32, 24. Swîđe manigne
hláford and swîđe manigne mundboran, Shrn. 35, 32. Mid monige wîte,
101, 23. Dê biddaþ manega þeóda, Deut. 28, 12. Hû đa monegan
yflan wundor wurdon on Rôme, Ors. 4, 2, tit; Swt. 3, 25. Ic sceal đara
monegena gewinna geswîgian, 5, 2; Swt. 218, 20. Đû bist manegra
þeóda fæder, Gen. 17, 4. Hê sende Agustinum and ôđre monige munecas,
Bd. 1, 23; S. 485, 27. **II. used absolutely:**—On manig dǽlan, Bt.
33, 1; Fox 120, 11. Đû tôsyndrodest hig on manega, Hy. 7, 65;
Hy. Grn. ii. 288, 65. Mænego, 9, 21; Hy. Grn. ii. 291, 21. Đyllícu
þing and ôđre manega, Shrn. 35, 28. Mænige gesêþ manega, Coll.
Monast. Th. 25, 1. Hwí árîsaþ swâ mænige wiđ mê, Ps. Th. 3, 1. Đǽr
môdlíce manega sprǽcon, Byrht. Th. 137, 43; By. 200. Hié witon đæt
đæt ilce yfel ofereode, swâ đa monegan ǽr dydan, Ors. 5, 2; Swt. 218, 3.
Manigra sumne *one of many,* Beo. Th. 4188; B. 2091. **III. with a genitive:**—Moniges breác wintra, Cd. 62; Th. 74, 31; Gen. 1230. Heáfod hê
gebreceþ hæleþa mæniges, Ps. Th. 109, 7. Heora manigne ofslôg, Bt.
35, 4; Fox 162, 25. Monige sint cwucera gesceafta unstyriende, 41, 5;
Fox 252, 20. Monige đara brôđra sǽdon, Bd. 3, 8; S. 532, 4. Geseah
hê rinca manige, Beo. Th. 1461; B. 728. [*Goth.* manags: *O. Sax.*,
O. H. Ger. manag: *O. Frs.* monich: *Ger.* manch.] v. un-manig.

manig-brǽde (?); *adj. Consisting of many things:*—Mænibrǽde dôm
satura lex (*lanx?*), Ælfc. Gl. 13; Som. 57, 111; Wrt. Voc. 20, 49.
Cf. (?) brǽdan *to roast.*

manig-feald; *adj.* **I. Manifold, multifarious, of many kinds,
various, consisting of many parts, complex:*—Mænigfeald *multiplex,* Ps.
Th. 67, 17. Ys mænigfeald *multiplicata est,* 118, 69. Đes pistol is
swîđe menigfeald ûs tô gereccenne *this epistle is very complex for us to
expound,* Homl. Th. i. 448, 7. Ûs þincþ tô manigfeald đæt wê swîđor
ymbe đis sprecon, Lchdm. iii, 276, 8. Manigfealde *multifariam,* Wrt.
Voc. ii. 57, 51. Manigfealdne *multimodam,* 58, 20: Exon. 17 b; Th.
41, 27; Cri. 662. On swâ manigfeald gedǽled, Bt. 34, 9; Fox 146, 17.
Wê swâ monigfeald witon, alra tácna gehwylc, Elen. Kmbl. 1284; El.
644. Đa manigfealdan mîne geþohtas, Exon. 18 a; Th. 453, 1; Hy. 4, 8.
Þurh monigfealdra mægna gerýno, 16 b; Th. 38, 7; Cri. 603: 42 a;
Th. 140, 26; Gû. 616. For đâm mistlícum and manigfealdum weor-
uldbisgum, Bt. prooem; Fox viii, 5. Hit sceal heonanforþ mænigfealdre
weorþan, Wulfst. 83, 19. Monigfealdran, Exon. 51 a; Th. 177, 2; Gû.
1221. Wæs đǽr seó monigfealdeste wôl, mid moncwealme, ge eác đætte
ne wîf ne niéten ne mehton nánuht libbendes geberan, Ors. 4, 1; Swt.
158, 17. **II. Manifold, numerous, abundant;** as a grammatical term,
plural:—Menifeld *augmentatus,* Hpt. Gl. 440, 51. *Numerus is getel,
singularis* anfeald, and *pluralis* menigfeald, Ælfc. Gr. 15, 59.
Sume naman maciaþ heora mænigfealdan dativum on *-bus,* 7; Som. 8, 54.
On hyra menigfealdan spǽce *in multiloquio suo,* Mt. 6, 7. Manifealde
copiosa, Hpt. Gl. 468, 5. Mid mænifealdre *crebra,* 512, 34. Heora
ǽhta wǽron menifælde, Gen. 13, 6. Hî cômon swâ manigfealde swâ
swâ sandceosol, Jos. 11, 4. Mænigfealdum þenungum *exequiis pluribus,*
Wrt. Voc. ii. 144, 78. [*Goth.* manag-falþs: *O. Sax.*, *O. H. Ger.* manag-
fald *multiplex, frequens, varius.*]

---

manigfeald-líc; *adj. Manifold, having many parts, of many kinds,
various:*—Đeáh hit ûs manigfealdlíc þince, sum gôd, sum yfel, hit is
đeáh him ânfeald gôd, Bt. 39, 6; Fox 220, 8. Forđon wǽron swâ
manigfealdlíce sorga Cristes þegnum *therefore Christ's servants had such
manifold sorrows,* Blickl. Homl. 135, 18. Sangeras and mæssepreóstas and
manigfealdlíce ciricean þegnas *Church ministers of many kinds,* 207, 32.

manigfeald-líce; *adv. Manifoldly, in many ways;* as a grammatical
term, *in the plural:*—Monigfaldlíce *multipliciter,* Ps. Surt. 62, 2. Wê
mihton be eallum đam ôđrum stafum mænigfealdlíce sprecan *we might
speak of all the other letters under various heads,* Ælfc. Gr. 2; Som. 3,
10. Mænigfealdlíce *pluraliter,* 5; Som. 3, 42: 13; Som. 16, 9, 12.
Se ealda mænegfealdlíce bæd *the old man made many prayers,* Glostr.
Frag. 110, 18. Mænifealdlíce, Menol. Fox 185; Men. 94. [*O. H. Ger.*
managfalt-lího *multifariam.*]

manigfeald-ness, e; *f. Multiplicity, complexity; abundance, great
number:*—Manifealdnes *perplexitas,* Wrt. Voc. ii. 68, 20. Of monig-
faldnise *ex habundantia,* Lk. Skt. Lind. 6, 45. On mænigfealdnysse
in multitudine, Ps. Spl. 65, 2: 68, 20: Cant. Moys. 7. [Cf. *O. H. Ger.*
managfaltí *multitudo, affluentia.*]

manigfildan; *p. de To multiply:*—Ic mænigfylde *multiplico,* Ælfc.
Gr. 24; Som. 25, 55. [Cf. *O. H. Ger.* managfalton *multiplicare.*] v. ge-
mænigfyldan.

manig-síđes; *adv. Many times, often:*—Manisíđes swutelaþ đæt man
wile on ǽnne God gelýfan, Wulfst. 144, 11.

manig-teáw, -tíwe; *adj. Skilful, dexterous:*—Mænigtíwe *sollers,*
Wrt. Voc. 73, 49. [Menituwe, 88, 48.] Mænigtýwe, Ælfc. Gr. 9, 43;
Som. 12, 67. Mænigteáwum *sollerti,* Hpt. Gl. 512, 29. Đære mæni-
teáwestan *sollertissimæ,* 407, 65. v. æl-teáw.

manigteáw-ness, e; *f. Skill, dexterity:*—Mæniteáwnys *sollertia,*
Hpt. Gl. 428, 3. Meniteáwnysse *sollertiam,* 407, 7.

MANN, man, mon, es; *m.* **I.** MAN, *a human being of either
sex:*—Hic et hæc homo ǽgþer is mann ge wer ge wíf, Ælfc. Gr. 9;
Som. 8, 54. Đes mann *iste homo,* đises mannes *istius hominis,* dat. đisum
menn, acc. đysne mann, abl. fram đisum menn; *pl. n. acc.* đâs menn,
gen. đyssera manna, *dat.* đisum mannum, 15; Som. 18, 25-28. Uton
wircean man (*hominem*) tô úre andlícnisse..God gesceóp man tô his
andlícnisse, Gen. 1, 26, 27. Se man (*homo*) wæs geworht on libbendre
sáwle, 2, 7. Wást đû hwæt mon síe. Đâ cwæþ ic: Ic wát đæt hit is
sáwl and líchoma. Đâ cwæþ hê: Hwæt đû wást đæt hit biþ mon đa
hwíle đe seó sáwl and se líchoma undǽlde beóþ; ne biþ hit nán mon
siđđan hí tôdǽlde biôþ, Bt. 34, 9; Fox 148, 3-6. Hû Hanna ân mon
wæs onwaldes giernende, Ors. 4, 5 tit; Swt. 3, 32. Hiene ofslóg Othon
ân mon, 6, 6; Swt. 262, 9. Hê geceás him tô fultume Traianus đone
mon, 6, 10; Swt. 264, 18. Hê ofslôg Albínus đone mon, 6, 15; Swt.
270, 10: 6, 26; Swt. 276, 23: 6, 31; Swt. 284, 20. Gif hund mon
tôslíte, L. Alf. pol. 23; Th. i. 78, 2. Gif mon swâ gerádne mon ofsleá, 28;
Th. i. 80, 2. Syxhynde mon, 30; Th. i. 80, 11. Gif mon cierliscne
mon gebinde, 35; Th. i. 84, 2. Hwæne secgeaþ menn đæt sý mannes
sunu *quem dicunt homines esse filium hominis?* Mt. Kmbl. 16, 13.
Hwæt eom ic manna đæt ic mihte God forbeódan '*what was I, that I
could withstand God?*' Homl. Skt. 10, 191. Đâ đû ǽrest tô monnum
becôme *cum te matris ex utero natura produxit,* Bt. 7, 3; Fox 20, 10.
Englas hê worhte, đa sind gâstas, and nabbaþ nǽnne líchaman. Menn
hê gesceóp mid gâste and mid líchaman. Nýtenu hê gesceóp on flǽsce
bûtan sáwle. Mannum hê gesealde uprihtne gang, đa nýtenu hê lêt gân
álotene, Homl. Th. i. 276, 1-5. *Used of a male:*—Đeós biþ gecíged
fǽmne, for đam đe heó ys of were genumen. For đam forlǽt se man fæder
and môdor and geþeót hine tô his wífe, Gen. 2, 23-24. Gelíc đam
dysigan men (*viro,* cf. wîsan were, 24), Mt. Bos. 7, 26. Hê sǽde hyre
hwæt heó man ne wæs *he told her that she* (*Eugenia*) *was no man* (cf.
vv. 48-53 from which it is seen that Eugenia was dressed as a man),
Homl. Skt. 2, 78. *Used of a female,* cf. wíf-man:—Đæt se mon (*woman*)
swǽte swíđe, L. M. 3, 38; Lchdm. ii. 332, 1. Ercongota háli fémne
and wundorlíc man, Chr. 639; Erl. 27, 5. Agathes clypode: '*Mín drihten
đe mê tô menn gesceópe,' Homl. Skt. 8, 185. His môdor wæs cristen,
swíđe gelýfed mann, Homl. Th. ii. 306, 4. *Used of both:*—Twegen
men, wer and wíf (*Adam and Eve*), 206, 21: Hexam. 17; Norm. 24,
24: Cd. 33; Th. 45, 18; Gen. 728. **II. a man who is under the
authority of another** (cf. mann-rǽden), *a servant, vassal, liege-man;* as
an ecclesiastical term, *a parishioner:*—Se cyng Melcolm griđede wiđ
đone cyng Willelm and his man wæs, Chr. 1072; Erl. 211, 6. Sý hit
cynges man, sý hit þegnes, L. Edg. i. 3; Th. i. 264, 4. Sý đæs mannes
man đe hê sý, L. C. S. 13; Th. i. 382, 20. Nân man his men fram
him ne tǽce, đet hê clǽne sý ǽlcere sprǽce, 28; Th. i. 392, 11. Ne
underfô nán man ôđres mannes man bûtan đæs leáfe đe hê ǽr fyligde, L.
Ed. 10; Th. i. 164, 16: L. Ath. i. 22; Th. i. 210, 20. Ealle đa land-
sittende men ofer eall Englaland, wǽron đæs mannes men đe hí wǽron.
And ealle hí bugon tô him and wǽron his menn, Chr. 1086; Erl. 219,
4-6. Se đe hý feormige ođđe hyra manna ǽnigne, L. Ath. iv. pref.;
Th. i. 220, 12. Eác is mæssepreóstum micel þearf đæt hig hyra mannum
cýđen, L. E. I. 27; Th. ii. 422, 34. **III. the name of the Rune**

for M, which is sometimes used instead of writing the word *man*, e. g. ǽnig ᚻ *quis*, Rtl. 11, 41. Ne ǽnig ᚻ *nemo*, 13, 25, 29. ᛗ byþ on myrgþe, Runic pm. Kmbl. 343, 11; Rún. 20. *So the compound* mann-dréam *is written with the rune* in, Exon. 124 a; Th. 477, 14; Ruin. 24. The word forms the second part of very many compounds. [Cognate forms are found in all the Teutonic dialects, but in Gothic a nominative occurs only in the weak form, and in *Icel.* the nom. takes the form *maðr*.] v. man, manna.

manna, monna, an; *m. Man, a man:*—Hwæt is se manna *quid est homo?* Ps. Th. 143, 4. On mannan mód, 117, 8. For ðissum earfoþnessum ðe wé ðissum mannan dydon, Blickl. Homl. 247, 18. Ic ádílige ðone mannan *delebo hominem*, Gen. 6, 7. God geworhte ǽnne mannan of láme, Homl. Th. i. 12, 29. Ðá wolde God wyrcan mannan, Hexam. 11; Norm. 18, 9. Gif man frigne mannan ofsleahþ, L. Ethb. 6; Th. i. 4, 6. Eorlcundne mannan, L. H. E. 1; Th. i. 26, 8. Gif frigman mannan forstele, 5; Th. i. 28, 10. Abraham, leófne mannan, Cd. 121; Th. i. 156, 11; Gen. 2587. Geongne monnan, Exon. 89 b; Th. 336, 9; Gn. Ex. 45. Fremde monnan, 90 b; Th. 339, 32; Gn. Ex. 103. [*Goth.* manna: *Icel.* manni.] v. mann.

manna, monna; *indecl. Manna:*—Nemdon ðone mete manna, Ex. 16, 31: Ps. Spl. T. 77, 28: Num. 11, 9. Monna, Past. 17, 11; Swt. 125, 19.

mann-bǽre; *adj. Productive of men:*—Ic tówurpe ðás burh and tó yrþlande áwende, swá ðæt heó biþ cornbǽre swíðor ðonne mannbǽre, Homl. Th. i. 450, 12.

mann-bót, e; *f. A fine to be paid to the lord of a man slain.* Its amount was regulated by that of the 'wer':—Sió sió mægbót and sió manbót gelíc. Weaxe sió [mæg]bót be ðam were swá ilce swá sió manbót dêþ ðe ðam hláforde sceal, L. In. 76; Th.i. 150, 14–16. Æt twýhyndum were mon sceal sellan tó monbóte xxx. sciłł, æt vi. hyndum Lxxx. sciłł, æt twelfhyndum cxx., 70; Th. i. 146, 13–15; L. Edm. S. 7; Th. i. 250, 21: L. E. G. 13; Th. i. 174, 27: L. C. E. 2; Th. i. 360, 7; L. W. I. 7; Th. i. 471, 11: L. H. I. 43; Th. i. 543, 27. [*Icel.* mannbætr; *pl.*]

mann-bryne, es; *m. A fire in which men lose their lives* (?):—Ðá wæs swíðe micel mancwealm, and se micela manbryne wæs on Lundene, and Paules mynster forbarn, Chr. 962; Erl. 120, 6. [Thorpe with previous translators renders the word by *fever;* Earle would read *mánbryne* =destructive fire. If *mánbryne* be taken perhaps an *incendiary fire* is meant.]

mann-cwealm, es; *m. Death of men, pestilence, mortality, slaughter:*—Mancwealm *pestilentia*, Bd. 1, 14, tit; S. 482, 14. On ðǽm dagum wæs se mǽsta mancwealm (*pestes plurimas dirosque morbos*), Ors. 1, 6; Swt. 36, 15. Se micla moncwealm *ingens pestilentia*, 3, 3; Swt. 102, 4. Ðý ilcan geáre wæs micel mancwealm, Chr. 664; Erl. 34, 21. Wæs swíðe micel mancwealm (cf. se færcwealm ðe his (*Edgar*) leódscipe swýðe drehte and wanode, L. Edg. 5; Th. i. 270. 9), 962; Erl. 120, 5. On ða tíd ðæs mancwealmes *tempore mortalitatis*, Bd. 3, 30, tit; S. 561, 31. Mec ongan hreówan ðæt moncynnes tuddor sceolde mancwealm seón, Exon. 28 b; Th. 86, 33; Cri. 1417. Hú monege missenlíce moncwealmas gewurdon *quantae clades gentium fuere*, Ors. 1, 12; Swt. 52, 11. Manncwealmas (*pestilentiæ*) beóþ, Mt. Kmbl. 24, 7.

mann-cwealmness, e; *f. Man-slaying, homicide:*—Monncualmniss *homicidium*, Mk. Skt. Lind. (moncwælmnisse, Rush.) 15, 7.

mann-cwild, e; *f. Mortality, pestilence:*—On ða tíd ðæs miclan wóles and moncwylde *tempore mortalitatis*, Bd. 3, 13; S. 538, 15.

mann-cynn, es; *n.* I. *mankind, men, the human race:*—Engla hláf ǽton mancynn *panem angelorum manducavit homo*, Ps. Th. 77, 25. Sende se Fæder his áncennedan sunu tó cwale for mancynnes álýsednysse, Homl. Th. ii. 6, 17. For ealles mancynnes hǽle, Blickl. Homl. 129, 14. Ord moncynnes (*Adam*), Cd. 55; Th. 68, 2; Gen. 1111. Drihten of deáþe árás mancynne tó bysene, Blickl. Homl. 83, 21. Hié sceoldan geond ðysne middangeard mancynne bodian, 121, 4. Hine on woruld tó moncynne módor brohte, Cd. 132; Th. 167, 23; Gen. 2770. Hine feor forwræc Metod mancynne fram *the Lord drove him away far from men*, Beo. Th. 221; B. 110. Hé wolde mancyn lýsan, Rood Kmbl. 82; Kr. 41: Blickl. Homl. 71, 26. Hé ealle eáðmódnysse wið mancynne gecýðde, 123, 31. II. *a race of men, a people, men* (a limited number):—Ðonne is sum eáland on ðære Reádan Sǽ ðǽr is moncynn (*hominum genus*) ðæt is mid ús Donestre genemned, Nar. 37, 1. Æfter ðam ðe Iosue ðæt mankyn (*the Israelites*) gebrohte tó ðam behátenan earde, Jud. pref. 3. Hé ða burg gewann and eall ðæt moncynn ácwealde *he took the town and slew all the inhabitants*, Ors. 3, 7; Swt. 112, 16. Micel ðæs moncynnes sum ácwealde sum on Mæcedonie lǽdde *magnam Romanorum praesidiorum multitudinem partim occidit, partim in Macedoniam duxit*, 4, 11; Swt. 208, 15. [*Laym.* mon-kun: *Orm.* mann-kinn: *Ayenb.* man-kende: *O. Sax.* man-kunni: *Icel.* mann-kyn: *O. H. Ger.* man-chunni *humanum genus, generatio.*]

mann-dréam, es; *m. Human joy, joyous life among men, joyous noise:*—Ðú ne gemyndgast æfter mandreáme, ne wást bútan wildeóra

þeáw *thy mind shall not be according to human life, nor shalt thou* (*Nebuchadnezzar*) *know aught but the habit of wild beasts*, Cd. 203; Th. 251, 30; Dan. 37: Andr. Kmbl. 74; An. 37. Cain fág gewát mandreám fleón, Beo. Th. 2533; B. 1264. Lifde and lissa breác Malalehel mondreáma hér, Cd. 59; Th. 71, 26; Gen. 1176. Meodo heall moni g ᛗ dreáma full, Exon. 124 a; Th. 477, 14; Ruin. 24. Hé ána hwearf mondreámum from, Beo. Th. 3435; B. 1715. [*Laym.* þa aras þe mondrem þat þe uolde dunede aȝen.]

mann-dryhten, es; *m. A lord of men, liege lord* (cf. mann, II.):—Mandryhten, Beo. Th. 3961; B. 1978. Úre mandryhten (*Beowulf*), 5287; B. 2647. Mondryhten, 5722; B. 2865. Mondrihten, 876; B. 436. Æfter mandrihtne, æfter ðam ædelinge (*Nebuchadnezzar*), Cd. 207; Th. 256, 8; Dan. 637. Ðá ic ðæt wíf (*Sarah*) gefrægn wordum cýðan hire mandrihtne (*Abraham*), 102; Th. 135, 15; Gen. 2243. Hé fore his mondryhtne módsorge wæg (*of Guthlac and his disciple*), Exon. 48 a; Th. 165, 5; Gú. 1024: (cf. onbehtþegn, Th. 170, 29) 49 b; Th. 171, 10; Gú. 1124. [*O. Sax.* Mattheus warð im úses drohtines man, kós . . milderan medgebon than ér is mandrohtin wâri an thesero weroldi, 1200.]

mann-eáca, an; *m. An increase of human beings:*—Ðæt hié wǽron ortríewe hwæðer him ǽnig moneáca cuman sceolde *ut defectura successio crederetur* (*on account of pestilence no children were born alive*), Ors. 4, 1; Swt. 158, 20.

mann-faru, e; *f. A going of men* or *a moving band of men*, v. faru:—Wé ðás wíc mágun fótum áfyllan, meara þreátum and monfarum, Exon. 36 b; Th. 119, 20; Gú. 257. [Cf. *Laym.* al mi mon-uerde (2nd MS. alle mine cnihtes), 16453: he sende after man-ferde (1st MS. monweored), 10747.]

mann-fultum, es; *m. Military force, troops:*—Hié ǽr tweóde hwæðer hiene mon mid ǽnige monfultume gefíéman mehte *they before doubted whether he* (*Hannibal*) *could be routed by any troops*, Ors. 4, 9; Swt. 192, 16: 5, 7; Swt. 230, 9. Hié gegaderodon máran monfultum ðonne Philippus hæfde *they got together a greater force than Philip had*, 3, 7; Swt. 118, 16.

mannian; *p.* ode *To supply with men, to garrison:*—Heora ǽlc férde tó his castele and ðone mannoden and metsoden swá hig betst mihton *every one of them went to his castle and garrisoned and provisioned it as well as ever they could*, Chr. 1087; Erl. 224, 16. v. ge-mannian, fullmannod.

mann-leás; *adj. Without men, uninhabited, deserted:*—Rófleáse and monleáse ealde weallas *parietinae*, Ælfc. Gl. 110; Som. 79, 35; Wrt. Voc. 59, 8. [*Icel.* mann-lauss.]

mann-líca, an; *m. A human form, image of a man, statue:*—Æfre siððan se monlíca (*the pillar of salt into which Lot's wife was turned*) stille wunode, Cd. 119; Th. 155, 1; Gen. 2566. Eall Adames cynn ðe módor gebær tó manlícan *all the race of Adam that mother gave the form of man to at birth*, Wulfst. 137, 26: Dóm. L. 131. Ǽnne manlícan (*the golden image which Nebuchadnezzar set up*), gyld of golde árǽrde, Cd. 180; Th. 226, 20; Dan. 174. Hé þurh dreócræft worhte stǽnene manlícan and ærene, and hié hié styredan, Blickl. Homl. 173, 23. Twegen manlícan (*images in the sick man's eyes of the observer*) beóþ on mannes eágum; gif ðú ða ne gesihst, ðonne swilt se man, and biþ gewiten ǽr þrim dagum, Salm. Kmbl. p. 206, 11. v. Grmm. D. M. 1133. [*Goth.* man-leika *imago: O. H. Ger.* man-líha *statua, imago, figura, effigies: Icel.* mann-líkan *a human image, idol, being in human shape.*]

mann-líce; *adv. Manfully, in a manner becoming to a man, nobly:*—Swá manlíce mǽre þeóden heaðoræsas geald mearum and máðmum, Beo. Th. 2096; B. 1046. [*Icel.* mann-liga: cf. *O. H. Ger.* man-líh *virilis.*]

mann-lufu, an; *f. Love of men:*—Woldun ðæt him tó móde fore monlufan sorg gesóhte, ðæt hé síþ tuge eft tó eþle *they desired that for love of men care would visit his mind, that he might take his journey back to his country* (*and not remain as a hermit*), Exon. 37 b; Th. 123, 18; Gú. 324.

mann-mægen, es; *n. A force of men, a troop of men, cohort:*—Ðæt monnmægen ł þegna uorud *cohortem*, Jn. Skt. Lind. 18, 3. [Cf. *O. Sax.* man-kraft *a host of people.*]

mann-menigu; *f. A multitude of people;*—Manmenio (*the tribe of Reuben*), Cd. 160; Th. 199, 5; Exod. 334. [*Grein reads* mán menio *but there seems no reason to apply such an epithet to the* menio *in question.*] Ðéh ðe Sciþþie hæfdon máran monmenie *cum Scythae numero praestarent*, Ors. 3, 7; Swt. 116, 24.

mann-mirring, es; *f. Destruction of men:*—Ac man þǽr ne gespædde bútan manmyrringe *they did not succeed without loss of men*, Chr. 1096; Erl. 233, 29.

mann-rǽdenn, -rǽden, e; *f.* I. *homage, the condition of being another's man* (v. mann, II.):—Ðá cwǽdon úre frínd ðæt wé cómon tó eówre manrǽdene *then our friends said that we should come and make submission to you*, Jos. 9, 11. Ealle hig bugon tó Israéla manrǽdene, 13, l. 5: Th. An. 120, 27. Sum man deófle mannrǽdene befæste *a certain man sold himself to the devil*, Homl. Th. i. 448, 15. [Hé dyde ðæt

ealle ða heáfodmæn on Normandig dydon manræden his sunu Willelme, Chr. 1115; Erl. 245, 12. Cf. Hí hadden him manrēd maked, 1137; Erl. 261, 32. *Laym.* he heora monredne onfeng.]   II. *service or dues paid by the tenant to the owner :*—Ðæt is ǽrest of ðam lande æt Nigon hídon seó mannrēdden intó Tantún, cirhsceattas..., Chart. Th. 432, 22.

mann-rím, es; *n. A number of men :*—Ðínre mægþe monrím, Cd. 84; Th. 105, 35; Gen. 1763. Monrím mægeþ (mægþa?) *a number of women (the Egyptian women spoken of before as* freó *and* þeówe), 131; Th. 166, 15; Gen. 2748. Hwæt ðǽr eallra wæs on manríme... deádra gefeallen, Elen. Kmbl. 1296; El, 650.

mann-scipe, es; *m. Humanity, kindness, civility :*—Manscipes weldǽdum underþeódde *humanitatis officiis deditos*, Cod. Dip. Birch 154, 38. Manscipe gyfan beþearfendum and ælþeódigum *humanitatem peregrinis et egentibus impendere*, 155, 5.

mann-silen, e; *f. The wrongful selling of men into slavery :*—Þurh mannsylena, Wulfst. 164, 1. Mansilena, 130, 1. Leódhatan ðe þurh mansylene bariaþ ðas þeóde, 310, 5. Cf. earme men wǽron út of ðisan earde gesealde swýðe unforworhte fremdum tó gewealde, 158, 13. *And see* L. Eth. v. 2; Th. i. 304, 14.

mann-slaga, an; *m. A homicide, man-slayer :*—Manslaga *homicida*, Wrt. Voc. i. 85, 44: L. Edm. E. 4; Th. i. 246, 7. Ne beó ðú manslaga *non occides*, Deut. 5, 17: L. Eth. ix. 1; Th. i. 340, 8: L. C. S. 41; Th. i. 400, 13. Gē sind manslagan *ye are murderers*, Homl. Th. i. 46, 24. Ðyder sculan mannslagan, Wulfst. 26, 14. [*O.H.Ger.* man-slago.]

mann-slege, es; *m. Man-slaying, homicide :*—Gif þeóf brece mannes hús nihtes and hē weorðe ðǽr ofslegen, ne síe hē (*the slayer*) nā mansleges scyldig. Gif hē æfter sunnan upgonge ðis dēþ, hē biþ mansleges scyldig, and hē ðonne self swelte, L. Alf. 25; Th. i. 50, 18–21: Blickl. Homl. 189, 34. Be manslege. Gif Ænglisc man Deniscne ofsleá gylde hine mid xxx pundum, oððon mon ðone handdǽdan āgyfe, L. Eth. i. 5; Th. i. 286, 20.

mann-sliht, -slieht, -slæht, sleht, es; *m. Manslaughter, homicide, murder :*—Ða heáfodleahtras sind, mansliht..., Homl. Th. ii. 592, 4. Ðonne mæg hē beón orsorg ðæs monnslihtes (monnsliehtes, Hatt. MS.) *reus perpetrati homicidii non tenetur*, Past. 21, 7; Swt. 166, 20. Manslehtes beteón, L. A. G. 3; Th. i. 154, 5. Be monslihte (monnslyhte, MS. H.), L. In. 34; Th. i. 122, 15: L. Edm. E. 3; Th. i. 246, 1: L. Edm. S. 1; Th. i. 248, 1. Be ðǽm monnum ðe heora wǽpna tó monslyhte lǽnaþ. Gif hwā his wǽpnes óðrum onlǽne ðæt hē mon mid ofsleá, L. Alf. pol. 19; Th. i. 74, 1–4. Manslyht gewyrcan *to commit murder*, Mk. Skt. 15, 7. Heðenra manna hergung ādiligode Godes cyrican þurh reáflác and mansleht, Chr. 793; Erl. 59, 12. Manslæht, Confess. Peccat. Ðis synt ða ídelnyssa ðisse worlde.. manslehtas (*homicidia*), L. Ecg. P. i. 8; Th. ii. 174, 34: Wulfst. 164, 4. Ðǽr wǽron swā micle monslihtas on ǽgðere healfe ðæt hie mon bebyrgan ne mehte *inhumatas strages reliquit*, Ors. 4, 6; Swt. 176, 30. Ungetíma ǽgder ge on monslehtum ge on hungre, 1, 11; Swt. 50, 19: Chr. 994; Erl. 133, 18. [*Laym.* monslæht: *A.R.* mon-sleiht: *Gen. and Ex.* manslagt: *O.Sax.* man-slahta: *O.Frs.* mon-slachta: *O.H.Ger.* man-slaht.]

mann- (mān-?) **swica**, an; *m. A traitor :*—Ðyder (*to hell*) sculan mannslagan and ðider sculan manswican, Wulfst. 26, 15.

mann-þeáw, es; *m. A manner, custom, practice :*—Gē scyldigra synne secgaþ sóþfæstra nō monþeáw mǽran willaþ *ye rehearse the sin of the guilty, the practice of the just ye will not celebrate*, Exon. 40 a; Th. 132, 25; Gū. 478. Ðæt hē monþeáwum mínum lifge *that he live according to my customs*, 71 b; Th. 267, 4; Jul. 410. Hē forlǽteþ lāre ðíne, and manþeáwum mínum folgaþ, Elen. Kmbl. 1856; El. 930. In monþeáwas, Exon. 55 b; Th. 197, 15; Az. 190. [Cf. þe hwile hit (*a child*) is lutel ler him *monþewes*, Morris Spec. i. 152, 432.] Cf. mann-wíse.

mann-þeóf, es; *m. A man-stealer :*—Manigu wítu [wǽron] mǽran ðonne óðru; nū sint ealle gelíce bútan manþeófe, cxx sciłł, L. Alf. pol. 9; Th. i. 68, 7. Cf. Gif mon *forstolenne* man befó æt óðrum, L. In. 53; Th. i. 134, 16. Gif þeówne man man *forstǽle*, L. Æðelst. v. 6; Th. i. 234, 4. Man-stealing is dealt with in Theodore's Liber Penitentialis: 'si quis servum alterius, vel quemcunque hominem, furtu quolibet in captivitatem duxerit aut transmiserit, vii annos pæniteat, ii in pane et aqua,' xxiii. 13. See also xlii. 5.

mann-þwǽre; *adj. Gentle, mild, meek, not harsh, courteous :*—Manþwǽre *cicur*, i. *mansuetus, placidus*, Wrt. Voc. ii, 131, 35; *cicur*, 17, 12: i. 288, 46. Cyningc ðín cymeþ ðē monnþwǽre (*mansuetus*), Mt. Kmbl. Rush. 21, 5. Milde and monþwǽre, Blickl. Homl. 71, 4. Earmum mannum milde and manþwǽre *pauperibus benignus et mitis*, L. Ecg. C. pref.; Th. ii. 132, 14. Manþwǽre (*propitius*) heora fyrendǽdum, Ps. Th. 77, 37. Mildheort and manþwǽre *misericors et miserator*, 144, 8: Bt. 42; Fox 258, 9. On þeáwum monþwǽre *moribus civilis*, Bd. 3, 14; S. 540, 8. On óðre wísan sint tó manienne ða monþwǽran on óðre ða grambǽran *quomodo admonendi mansueti et iracundi*, Past. 40; Swt. 287, 20: Ps. Th. 33, 2: 149, 4. God geriht ða manþwǽran (*mites*) on dómum, 24, 7. Manna mildust and monþwǽrost *most gentle and courteous of men*, Beo. Th. 6345; B. 3182.

mann-þwǽrness, e; *f. Gentleness, meekness, courtesy :*—Forðam oft gebyreþ ðæm monþwǽran ðonne hē wierþ ríece ofer óðre menn ðæt hē for his monnþwǽrnesse āslāwaþ and wierþ tó unbeald forðǽm sió unbieldo and sió monnþwǽrnes bióþ swíðe anlíce *nonnunquam enim mansueti, cum praesunt, vicinum et quasi juxta positum torporem desidiae patiuntur*, Past. 40, 1; Swt. 287, 24. Manþwǽrnes *mansuetudo*, Ps. Th. 89, 12: 131, 1. Mycelre monþwǽrnysse (*mansuetudinis*) mon, Bd. 3, 3; S. 525, 31. On his hāðeortnesse (*fervor*) and on his monþwǽrnesse (*mansuetudo*), Past. 21, tit; Swt. 151, 6. Scearpnyssa beóþ āwende tó smēðum wegum, ðonne ða yrsigendan mōd, and unlíþe gecyrraþ tó manþwǽrnysse, Homl. Th, i. 362, 30: ii. 226, 9: Blickl. Homl. 33, 29.

mann-werod, es; *m. A band of people, an assembly :*—Ða Philippuse gebyrede ðæt hē for ðæm plegan út of ðæm monweorode ārād, Ors. 3, 7; Swt. 118, 33. Gemun ðín mannweorod *memento congregationis tuæ*, Ps. Th. 73, 2. [*Laym.* mon-weored: *O.Sax.* man-werod.]

mann-weorþ, es; *n. The value or price of a man :*—Gif mannes esne eorlcundne mannan ofslæhþ.. se āgend āgefe ðone banan, and dō ðǽr þrió manwyrþ tó. Gif se bana óþbyrste feorþe manwyrþ hē tó gedó, L. H. E. 1–2; Th. i. 26, 8–28, 1: 3–4; Th. i. 28, 4–8.

mann-weorþung, e; *f. The worshipping human beings :*—Wē lǽraþ ðæt preósta gehwilc forbeóde wilweorþunga.. and manweorþunga, L. Edg. C. 16; Th. ii. 248, 3.

mann-wíse, an; *f. Custom, fashion, usage, manner of men :*—Æfter monwísan *after the manner of men*, Exon. 9 a; Th. 5, 30; Cri. 77. Hē ðære mægþe monwísan fleáh *he shunned the customs of that country*, Cd. 92; Th. 116, 21; Gen. 1939.

mān-sceaða, -scaða, an; *m.*   I. *A wicked and harmful person :*—Se mánsceaða (*the firedrake*), Beo. Th. 5022; B. 2514. Se mánsceaða (*Grendel*), 1428; B. 712: 1479; B. 737: (*Grendel's mother*), 2682; B. 1339. Míne myrðran and mánsceaðan (*evil spirits*), Exon. 42 a; Th. 141, 5; Gū. 622: 46 a; Th. 156, 27; Gū. 881: (*the giants before the flood*), Cd. 64; Th. 77, 2; Gen. 1269: (*the Egyptians who oppressed the Israelites*), 154; Th. 179, 31; Exod. 37.   II. *a sinner, one who wickedly does wrong :*—Ðonne mánsceaða fore Meotude forht on ðam dōme standeþ, Exon. 30 b; Th. 95, 20; Cri. 1560. Ðǽr fýr maansceaðan ða synfullan forbærnde *flamma combussit peccatores*, Ps. Th. 105, 16. [*O.Sax.* mēn-skaðo *applied to the devil and to the Jews.*]

mān-sceatt, es; *m. Usury, unjust gain :*—Of mánsceatte and of māne *ex usuris et iniquitate*, Ps. Th. 71, 14.

mān-scyld, e; *f. Guilt, sin :*—Ðū eart ðæt hālige lamb ðe mánscilde middangeardes tōwurpe, Hy. 8, 23; Hy. Grn. ii. 290, 23. [*O Sax.* alāt ūs managorō mēnskuldio *forgive us our trespasses.*]

mān-scyldig; *adj. Guilty of crime :*—Mē mánscyldigne (*Cain*) Cd. 49; Th. 63, 7; Gen. 1028: 50; Th. 64, 11; Gen. 1048.

mān-slagu, e; *f. A wicked blow :*—Ne mōton hié ðinne líchoman lehtrum scyldige deáþe gedǽlan, ðeáh ðū drype þolige, myrce mánslaga (*or* manslagan *in apposition to* scyldige?), Andr. Kmbl. 2437; An. 1220.

mānsumian. v. ā-mānsumian.

mānsumung, e; *f. Anathema :*—Nellaþ ða apostoli nǽnne rihtwísne mid heora mánsumunge [āmánsumunge?] gebindan, Homl. Th. i. 370, 10. v. ā-mánsumung.

mān-swara, -swora, an; *m. A perjurer, one who swears falsely :*—Gif man mannan mánswara hāteþ, L. H. E. 11; Th. i. 32, 4: Exon. 10 b; Th. 12, 30; Cri. 193. Mánswaran, Blickl. Homl. 61, 13: 63, 13. Mánsworan, Wulfst. 26, 16: Exon. 31 b; Th. 98, 23; Cri. 1612: L. Ed. 3; Th. i. 160, 18, 19: L. E. G. 11; Th. i. 172, 19, 20. [*Icel.* mein-særi: *O.H.Ger.* mein-swero *perjurus.*]

mān-swaru, e; *f. Perjury :*—Mánswara *perjuria*, Wrt. Voc. ii. 96, 70: L. Eth. v. 25; Th. i. 310, 15: vi. 28; Th. i. 322, 15. [*Laym.* monsware: cf. *Icel.* mein-særi; *n.*]

mān-swerian; *p.* swōr; *pp.* -sworen *To swear falsely, commit perjury, forswear :*—Gif man wāt ðæt óðer mánsweraþ (*or* māu sweraþ, cf. se man ðe swereþ mān, v. 2), Lev. 5, 1. Be mánsworum. Ða ðe mánsweriaþ, L. Edm. S. 6; Th. i. 246, 14. Ne swerige hē dýlæs hē mánswerige, L. E. I. 8; Th. i. 416, 8. Ða mánsweriendan *perjurantes*, Hpt. Gl. 472, 8. [*Laym.* þ he weore touward his lauerd manswore: *Scott.* to mansweir *to perjure*; manswearing *perjury*: Mid. York. Gl. main-swear *to forswear.*]

manung, e; *f.*   I. *monition, admonition, advice :*—Seó monung ðære godcundan ārfæstnesse *admonitio divinæ pietatis*, Bd. 4, 25; S. 599, 24. Ða sealdon hí strange manunge *dant fortia monita*, 1, 12; S. 481, 13. Tō onfōnne and tō ongitanne ða monunge ðære hālwendan lāre *ad suscipienda et intelligenda doctrinæ monita salutaris*, 2, 12; S. 512, 26.   II. *a claiming or exaction of debt, tribute, &c. :*—Gafules manung *exactio*, Wrt. Voc. ii. 30, 10. Ic beóde ðæt hý nān man ne brocie mid feós manunge, Chart. Th. 472, 10.   III. *the place where toll is demanded, the district in which a power of summoning or exacting is exercised :*—Monno sittende æt gæflæs monunge *hominem sedentem in teloneo*, Mt. Kmbl. Rush. 9, 9. Nemne man on ælces gerēfan manunge swā fela manna swā man wite ðæt ungelygne sýn, L. Ath. iv. 1; Th. i. 222, 9. Ðæt wē rídan be eallum tó mid ðam gerēfan ðe hit on his

monunge sȳ, v. 8, 2; Th. i. 236, 13. **IV.** *the people residing in such a district, and bound to answer his summons :*—Fô se gerêfa tô mid his monunge, and ádrífe ðæt spor ût of his scîre, v. 8, 4; Th. i. 236, 22. v. manian.

mân-wamm, es; *m. A blot caused by sin :*—Mânwomma gehwone geseón on ðâm sáwlum *to see every guilty stain in the souls*, Exon. 26 b; Th. 78, 27; Cri. 1280.

mân-weorc, es; *n. A wicked work, crime :*—Gif mæsse preóst mânweorc tô swîðe gewurce, L. Eth. ix. 26; Th. i. 346, 4: L. C. S. 41; Th. i. 400, 14. Ðæt hȳ môstun mânweorca tôme lifgan, Exon. 25 b; Th. 74, 25; Cri. 1211: 72 b; Th. 270, 2; Jul. 459. Ðæt ic in mânweorcum môd oncyrre, 72 a; Th. 268, 28; Jul. 439. Ær man áweódige ða unriht and ða mânweorc ðe man wîde sæwþ, Wulfst. 243, 19. [*O. Sax.* mên-werk.] Cf. mân-dǽd.

mân-weorc; *adj. Doing evil, wicked :*—Ðæt ðû mê swâ mânweorcum inwrige wyrda gerȳno, Elen. Kmbl. 1621; El. 812. v. mân-wyrhta.

mân-word, es; *n. A wicked word :*—Ys hyra mûðes scyld mânworda feala ða hî mid welerum âsprǽcan *delicta ores eorum sermo labiorum ipsorum*, Ps. Th. 58, 12.

mân-wyrhta, an; *m. A worker of wickedness, a sinner :*—Mânwyrhtan *peccatores*, Ps. Th. 93, 3: *qui operantur iniquitatem*, 118, 3.

mapulder (-dur, -dor); *m.(?) f.(?) A maple tree :*—Mapuldur *acerabulus*, Ep. Gl. 26, 14: Wrt. Voc. ii. 99, 14. Mapuldor, 4, 26: L. M. 1, 36; Lchdm. ii. 86, 6. Mapulder *acer*, Ælfc. Gl. 46; Som. 65, 1; Wrt. Voc. 33, 1. Mabuldor *acerabulos*, 285, 35. On ðære (ðæne ?) ealdan mapolder, Chart. Th. 146, 26. Tô ðon reádleáfan mapuldre; of ðam mapuldre, Cod. Dip. Kmbl. v. 298, 16. The word is found in several place-names in the Charters v. Cod. Dip. vi. 313, and still occurs, e. g. Mappledurwell in Hampshire, Mapplederham in Oxfordshire. v. mapultreów, *and* cf. apulder.

mapulderen; *adj. Made of maple :*—Mapuldern *acernum*, Ælfc. Gl. 46; Som. 65, 1; Wrt. Voc. 33, 1. On mapoldren geat, Cod. Dip. Kmbl. iii. 81, 18.

mapul-treów (*it is made masc. in the following*) :—In ðonne mapultrê . . from ðam mapoltrê, Cod. Dip. Kmbl. iii. 381, 1–2. v. mapulder.

mára, *more*. v. micel.

mâran. v. mǽran.

marc, es; *n. A mark, half a pound* (in the laws only the half-mark occurs) :—Swîðe strang gyld, ðæt wæs viii. marc, Chr. 1040; Erl. 166, 21. Six marc silures . . âne marc goldes, Chart. Th. 566, 21–29. ii marc gold, 567, 33. Tô marc goldes tô ðe kynges heregete and half-marc goldes ðe erl Harold and half-marc goldes Stîgand bisscop, 573, 10–14. Wið x marcun goldes, Wanl. Cat. 150, 11. Gilde x healfmarc, L. N. P. L. 48; Th. ii. 298, 2. (See also several of the following paragraphs.) Tô viii. healfmarcum âsodenes goldes, L. A. G. 2; Th. i. 154, 1. [*O. Frs.* merk, mark; *f:* Icel. mörk; *f:* M. Lat. marca.]

mare, mârels. v. mǽre, mǽrels.

mare, an; *f. Silverweed*, L. M. 1, 37; Lchdm. ii. 74, 9. [*Icel.* mara. v. Lchdm. ii. 399, col. i.]

margen. v. morgen.

marian. v. â-marian *and* mirran.

market, es; *n. Market :*—Ðat market æt Dûnhâm *mercatum de Dunham*, Chart. Th. 422, 20 (a charter of Edward the Confessor). [Market and toll. Ic wille ðat markete beó in þe selue tûn, Chr. 963; Erl. 124, 5–18.] [*O. Frs.* merked, market: Icel. markaðr: *O. H. Ger.* markat *mercatus, forum*; all from Latin *mercatus*.] v. geár-market.

marma, an; *m. Marble :*—Heó hæfþ hwîtes marman (marbran, MS. H.) bleoh *it has the colour of white marble*, Herb. 51, 1; Lchdm. i. 154, 14. [Cf. *Icel.* marmari: *O. H. Ger.* marmul.] v. marman-stân.

marman-stân, es; *m. Marble, a piece of marble :*—Gehêr ðû marmanstân, Andr. Kmbl. 2994; An. 1500. Þrûh of marmanstâne, Homl. Th. i. 564, 20. On ðam marmanstâne, 506, 11: Blickl. Homl. 203, 35: 207, 13. [Cf. *Icel.* marmara-steinar *slabs of marble*.]

marm-stân, es; *m. Marble, a piece of marble :*—Ðes marmstân *hoc marmor*, Ælfc. Gr. 9, 21; Som. 10, 31: Wrt. Voc. i. 85, 19. Of marmstâne geworht, Chart. Th. 241, 12. On mearmstâne, Exon. 60 b; Th. 221, 12; Ph. 333. Of ðiterscîtum marmstânum geworht *made of squared blocks of marble*, Homl. Th. ii. 496, 35. [*Laym.* mearm-stân, marbre-ston: *R. Glouc.* marbre-ston: *O. E. Homl.* marbel-ston: cf. *O. H. Ger.* marmulstein *marmor*.]

marmstân-gedelf, es; *n. Marble-quarrying :*—Mâ ðonne twâ þûsend cristenra manna ðe tô marmstângedelfe gesette wǽron, Homl. Th. i. 560, 32.

Maroara; *The people of Moravia :*—Hié Maroara habbaþ bewestan him Þyringas ... Be eástan Maroara londe is Wisle lond, Ors. 1, 1; Swt. 16, 10–17.

martyr, martyre, es; *m. A martyr :*—Se strengesta martyr *martyr fortissimus*, Bd. 1, 7; S. 478, 33. Wæs se martyre from moncynnes synnum âsundrad, Exon. 40 a; Th. 133, 5; Gû. 485. Hê wilnade ðæt hê mid ðone martyr þrowian môste, Bd. 1, 7; S. 478, 18. Hî cóman tô ðæs martyres hûse, S. 477, 9. Ðǽr martiras meotode cwêmaþ, Cd. 228;

Th. 305, 30; Sat. 655. Hê gemynegode ðara eádigra martyra, Bd. 1. 7; S. 476, 33: Andr. Kmbl. 1751; An. 878. Martira gemynd, Menol. Fox 137; Men. 69. Æfter gerisenre âre martyrum, Bd. 5, 10; S. 625, 17. [*O. L. Ger.* martir: *O. Frs.* martir, martil: *O. H. Ger.* martyr.]

martyr-dôm, es; *m. Martyrdom :*—Mid sige martyrdômes, Homl. Th. i. 374, 24. Hê (*Stephen*) is fyrmest on martyrdôme, ii. 34, 22. His martyrdôme wyrþe *ejus martyrio condigna*, Bd. 1, 7; S. 479, 7. Hê gearcodon heora môd tô ðam martyrdôme, Homl. Skt. 5, 150. Martyrdôm (*martirium*) þrowiende, Bd. 5, 10; S. 623, 36: Menol. Fox 249; Men. 126: 287; Men. 145. [*O. H. Ger.* martar-toam *martyrium*.]

martyr-hâd, es; *m. Martyrdom :*—Se ðe rǽdeþ bôc mînes martirhâdes, Nar. 47, 11. Hê martyrhâd gelufade, Exon. 39 b; Th. 130, 24; Gû. 443. Ne heora martyrhâda wona wǽron heofonlîcu wundru *nec martyrio eorum cælestia defuere miracula*, Bd. 5, 10; S. 625, 4.

martyrian. v. ge-martyrian.

martyrung, e; *f. Suffering as a martyr :*—Ymbe his martyrunga *de passione Christi*, Ors. 6, 2; Swt. 254, 24. [*O. H. Ger.* martirunga *passio*.]

masc, max, es; *n. A mesh, a net, toil :*—Ic wyrpe max mîne on eá *pono retia mea in amne*, Coll. Monast. Th. 23, 9: 21, 13. On ðâm maxum *in retibus*, 21, 19. [*Prompt. Parv.* maske of a nette *macula*: *Scott.* mask a crib *for catching fish*; to mask *to catch in a net*: cf. *Icel.* möskvi a mesh: *O. L. Ger.* *O. H. Ger.* masca a mesh; mascun; *pl.* retia, plagæ, maculæ.] v. mǽscre.

mǽsc-, max-wyrt, e; *f. 'Mash-wort, the wort in the mash-tub.* On the malt boiling water is poured and allowed to stand three quarters of an hour; the liquid is wort, or mash-wort,' Lchdm. ii. 399, col. i:—Mâxwyrte amber fulne, L. M. 1, 41; Lchdm. ii. 106, 16. Wylle swîðe on mâxwyrte, 1, 36; Lchdm. ii. 86, 14. Dô picce mâxwyrt on gemang, 1, 38; Lchdm. ii. 96, 18. [Cf. *Prompt. Parv.* maschyn yn brewynge *misceo*, maschynge *mixtura*: *Scott.* to mask *to infuse*; mask-fat *a vat for brewing*: *Dan.* mask grains: *Swed.* mäsk: *Ger.* meisch *mash*; meisch-fass *mash-tub*.]

mâse, an; *f.* (*Mouse in*) *tit-mouse :*—Mâsae *parrula*, Ep. Gl. 20 b, 13. Mâse *parula*, Wrt. Voc. ii. 67, 62: 116, 36. [*O. and N.* mose: *O. H. Ger.* meisa *parus, parix:* *Ger.* meise: *Du.* mees: *Icel.* meisingr.] v. col-, cum-, frǽc-, hice-, spic-mâse.

masian. v. â-masian.

massere, es; *m. A merchant :*—Gif massere geþeáh ðæt hê fêrde þrige ofer wîdsǽ be his âgenum cræfte, se wæs ðonne syððan þegenrihtes weorþe, L. R. 6; Th. i. 192, 9. Ne beó ǽnig mangere mid unrihte, ne gîtsigende massere, L. Edg. C. 14; Th. ii. 246, 24: L. Ælfc. C. 30; Th. ii. 354, 1.

maða, an; *m. A grub, worm, maggot :*—Maþa *tomus* (=*tarmus*), Ælfc. Gl. 23; Som. 60, 12; Wrt. Voc. 24, 16. Maða (maðu ?) *cimex*, Wrt. Voc. ii. 131, 44. His gesceapu maðan weóllon, Homl. Th. i. 86, 10. Cf. Eorþ-mata (-maða ?) *vermis*, Wrt. Voc. ii. 123, 44. [York. Gl. mad *an earthworm: Prompt. Parv.* make, maþe, wyrm yn þe fleshe *tarmus: O. E. Homl.* meaðen i forrotet flesch, i. 251, 19: *Goth.* maþa a worm: *O. L. Ger.* matho lignorum et lardi vermis: *O. H. Ger.* mado tarmus, tarmes: *Ger.* made: cf. *Icel.* maðkr grub, worm.] v. maðu.

maðelian; *p. ode To speak, harangue, make a speech, declaim :*—Maðelaþ *concionatur*, i. conclamat, loquitur, contestatur in populo, Wrt. Voc. ii. 135, 34. Maðalode *contionatur, declamat*, Wülck. Gl. 15, 36. Satan maðelode, sorgiende spræc, Cd. 18; Th. 22, 27; Gen. 347. Abraham maðelode .. ongan his brȳd wordum lǽran, 86; Th. 109, 9; Gen. 1820: Beo. Th. 701; B. 348: 747; B. 371. Byrhtnoþ maðelode, wordum mǽlde, Byrht. Th. 132, 66; By. 42. Byrhtwold maðelode, hê ful baldlîce beornas lǽrde, 140, 60. Elene maðelade, and fore eorlum sprǽc, Elen. Kmbl. 807; El. 404. Wîdsîþ maðolade, wordhord onleác, Exon. 84 b; Th. 318, 19: Vid. 1. Maðeliendra *concionatorum, rhetorum*, Hpt. Gl. 460, 76. v. mǽðlan.

maðelere, es; *m. One who speaks* or *harangues :*—Maðelere *contionator*, Wrt. Voc. ii. 24, 72. Môtere *vel* maðelere *concionator*, i. locutor, 135, 32.

maðelig; *adj. Tumultuous, inciting to tumult* as in the case of one who harangues people (?) :—Maðeli *tumultuosa*, Kent. Gl. 725.

maðelung, e; *f. Loquacity, garrulity :*—Maðelunge *garrulitatis, verbositatis, loquacitatis*, Hpt. Gl. 475, 42.

mâðm. v. mâðum.

maðu, e; *f. A bug, maggot* (?) :—Maðu *cimex*, Ælfc. Gl. 23; Som. 60, 9; Wrt. Voc. 24, 13: 78, 69. [*Prompt. Parv.* mathe *cimex, tarmus*.] v. flǽsc-maðu.

mâðum, mâðm, mâðm, mâððum, es; *m. A precious* or *valuable thing* (often refers to gifts), *a treasure, jewel, ornament :*—Gylden mâðm, sylofren sincstán, searogimma nân, middangeardes wela môdes eágan ne onlȳhtaþ, Bt. Met. Fox 21, 40; Met. 21, 20. Mâððum ôðres weorþ gold mon sceal gifan *treasure shall change hands, gold must be given*, Exon. 91 b; Th. 343, 11; Gn. Ex. 155. Næs him tô mâðme wynn, hyht tô hordgestreónum, Andr. Kmbl. 2228; An. 1115. Deórum mâðme (*a sword*), Beo. Th. 3060; B. 1528. Æghwylcum eorla drihten mâð-

dum gesealde *to each the lord of earls (Hrothgar) gave a rich present,* 2109; B. 1052. Hé ðone máððum byreþ ðone ðe ðú mid rihte rǽdan sceoldest *he the jewel bears, that of right should be thine,* 4117; B. 2055. Máðm, *goldhilted sword,* Exon. 114 a; Th. 437, 26; Rä. 56, 13. Ðis synd ða mádmas ðe Æðelwold sealde intó ðam mynstre .. ón Cristes bóc mid sylure berénod, and iii. róde eác mid sylure berénode, ii. sylure candelsticcan and ii. ouergylde, Cod. Dip. Kmbl. vi. 101, 21–26. Fato ł mádmas *vasa,* Mt. Kmbl. Lind. 12, 29. Hió hyre ða betstan mádmas tó Cantwaran cyricean brohte, Lchdm. iii. 422, 14. Heora dýre gold ne biþ náhte wurþ wið ða foresǽdan mádmas (*St. Swiðhun's bones*), Glostr. Frag. 2, 30. Hí be hyra gate tó sǽ eodon, and mádmas ofer L míla fram sǽ fǽttan, Chr. 1006; Erl. 140, 27. Ic (*Hrothgar*) ðǽm gódan (*Beowulf*) sceal mádmas beódan, Beo. Th. 776; B. 385. Mádmas, 3739; B. 1867. Ealde mádmas (*the spoil of the Egyptians drowned in the Red Sea*), Cd. 171; Th. 215, 19; Exod. 585. Welan þicgan, mádmas and meoduful, Exon. 88 a; Th. 331, 2; Vy. 62. Gehét unrím mádma and cynelícra gyfena *promisit se ei innumera ornamenta regia vel donaria largiturum,* Bd. 3, 24; S. 556, 8. Ða ciricean giond eall Angelcynn stódon mádma and bóca gefýldæ, Past. pref.; Swt. 5, 10. Unc sceal worn fela mádma gemǽnra *many a precious thing will we share,* Beo. Th. 3572; B. 1784: 5590; B. 2799. Ðǽr wæs mádma fela, frætwa gelǽded, 72; B. 36. Mádma, 81; B. 41. Dýrwurþre eallum mádmum *omnibus ornamentis pretiosior,* Bd. 2, 12; S. 514, 41. Ðæt se fénge ǽgðer ge tó lande ge tó mádmum and tó eallum his ǽhtum *that he should succeed to the land and to the valuables and to all his possessions,* Chart. Th. 486, 1. On circlícum máðmum (*then follows a list of crucifixes, chalices and other valuables connected with a church*), 429, 11. [Se cyng sende his dóhter mid mænigfealdan mádman ofer sǽ, Chr. 1110; Erl. 242, 33.] Rúmheort beón mearum and mádmum, Exon. 90 a; Th. 339, 2; Gn. Ex. 88; Beo. Th. 3900; B. 1898: 2100; B. 1048. Wine Scyldinga fættan golde fela leánode, manegum mádmum, 4212; B. 2103. [*Laym.* maðmes; *pl.* (2nd MS. godes): *Orm.* maddmess; *pl.* (*the gifts brought by the Magi*): *Goth.* maiþms ðῶρον: *O. Sax.* mêdmôs; *pl. gifts, precious things:* *Icel.* meiðmar; *pl. gifts, presents.*] v. dryht-, gold-, hord-, ofer-, sinc-, þeóden, wundor-máðum.

máðum-ǽht, e; *f. A costly possession, valuable, treasure:*—Ne nom hé máðmǽhta má, ðeáh he monige geseah, búton ðone hafelan and ða hilt somod since fáge *more things of price he took not, though many he saw, than the head and the hilt gay with gold,* Beo. Th. 3230; B. 1613. Draca máðmǽhta wlonc *the dragon proud of his treasures,* 5659; B. 2833.

máðum-cist, e; *f. A treasure-chest, treasury:*—Nys hyt ná ályfed ðæt wé ásendon hyt on úre máðmcyste (*in corbanan,* cf. *Goth.* kaurban, þatei ist maiþms, Mk. 7, 11), Mt. Kmbl. 27, 6.

máðum-fæt, es; *n. A costly vessel:*—Máððumfæt mǽre, Beo. Th. 4801; B. 2405. Ðá genam hé ða máðmfatu, gyldene and sylfrene, binnon Godes temple, Hml. Th. ii. 432, 25. Ða máðmfatu ðæs temples ungeríme, gyldene and sylfrene, mid óðrum goldhordum, 66, 7. [Ða Ælfréd king forlét his mádmes and máðmfaten, Shrn. 16, 10.]

máðum-gesteald, es; *n. Treasure, riches:*—Eall ðæt máððumgesteald ðe in ðæs æðelinges ǽhtum wunade, Exon. 66 a; Th. 244, 32; Jul. 36.

máðum-gestreón, es; *n. Treasure:*—Næs heó tó gnéað gifa Geáta leódum, mádmgestreóna, Beo. Th. 3866; B. 1931.

máðum-gifa, an; *m. A giver of costly gifts, a liberal prince:*—Hwǽr cwom máððumgyfa? Exon. 77 b; Th. 292, 1; Wand. 92. [*O. Sax.* mêðom-gibo (*Christ*).]

máðum-gifu, e; *f. A costly gift:*—Æfter máððumgife, Beo. Th. 2606; B. 1301.

máðum-hirde, es; *m. A treasurer:*—Ða mádmhyrdas ðe ðæt feoh heóldon ðe mon ðam ferdmonnum on geáre sellan sceolde, Bt. 27, 4; Fox 100, 13.

máðum-hord, es; *n. Treasure:*—Máðmhorda mǽst (*the Ark with its contents*), Cd. 161; Th. 201, 6; Exod. 368. [*O. Sax.* mêdom-hord.]

máðum-hús, es; *n. A treasure-house, treasury:*—Mádmhús *gazophilacium,* Ælfc. Gl. 81; Som. 73, 11; Wrt. Voc. 47, 18. Máðmhús, 86, 48: *erarium,* Wrt. Voc. ii. 30, 42. On ðæs cynges máðmhúse *in ærarium regis,* Gen. 47, 14: Ors. 6, 3; Swt. 258, 13. Gesæt se Hǽlend binnan ðam temple ætforan ðam máðmhúse, Homl. Th. i. 582, 12. Hé lǽdde ða ællþeódgan ǽrendracan on his máðmhús and hiм geiéwde his goldhord, Past. 4, 1; Swt. 39, 3. Ðá fór Julius and ábræc hiera máðmhús (*ærarium*), Ors. 5, 12; Swt. 240, 15.

máðum-sele, es; *m. A hall in which a prince gives costly gifts, or a hall containing costly things* (cf. gold-sele):—Méda máððumselas, Salm. Kmbl. 379; Sal. 189.

máðum-sigle, es; *n. A costly jewel:*—Geseah máððumsigla fela, Beo. Th. 5508; B. 2757.

máðum-sweord, es; *n. A costly sword:*—Mǽre máððumsweord, Beo. Th. 2050; B. 1023.

máðum-wela, an; *m. Wealth consisting of costly things:*—Æfter máððumwelan (*the contents of the fire-drake's cave*), Beo. Th. 5493; B. 2750.

matt, meatt, e; meatte, an; *f. A mat:*—Matte *spiato* (= *psiato*), Wrt. Voc. ii. 121, 7. Meatte *matta,* i. 82, 20. Meatta *storia vel psiata,* i. 41, 30. [*Prompt. Parv.* matte *matta, storium:* *O. H. Ger.* matta, madda *psiatum, matta.*]

mattuc, mattoc, mettoc, meottic, es; *m. A mattock, kind of pickaxe:*—Mattuc *ligonem;* mattucas *lagones,* Wrt. Voc. ii. 51, 35, 36. Mettac *tridens,* i. 289, 59. Mettocas *ligones, rastros,* Ep. Gl. 22 d, 29: *lagones,* 13 b, 20: *ligones,* 13 f, 1: Wrt. Voc. ii, 50, 77: *rastros,* 118, 68. Meottoc *tridens,* 122, 64. Meotticas *ligones,* 112, 66. Ðonne hét hé hiene (*the rock*) mid fýre onhǽtan and siððan mid mattucun heáwan *rupes igni ferroque rescindit,* Ors. 4, 8; Swt. 186, 19. [*Mattok bidens,* Wrt. Voc. 234, 10: *Prompt. Parv.* mattok, pykeye or twybyl *ligo, marra.* Welsh matog, *a hoe.*]

máwan; *p.* meów [cf. *Laym.* medewen heo *meowen* (2nd MS. *mewen*)]; *pp.* máwen *to mow:*—Ðǽr nænig mann heg ne máweþ, Bd. 1, 1; S. 474, 32. Gelíce ond mon mǽd máwe, Ors. 2, 8; Swt. 92, 15: Ps. Th. 128, 5. Rípan and máwan, L. R. S. 2; Th. i. 432, 15. Máwenum hege, Ps. Th. 102, 14. [*O. H. Ger.* májan: *Ger.* mähen.]

max, máx-wyrt. v. masc, másc-wyrt.

mé; *dat.:* mé, mec, meh, mech; *acc. of pronoun of first person. Me:*—Ealle þing mé synt gesealde *omnia mihi tradita sunt,* Mt. Kmbl. 11, 27. Ælcne ðe mé (Lind. meh; Rush. mec) cýð *omnis qui confitetur me,* Mt. Kmbl. 10, 32. Ða ðe swencaþ mec *qui tribulant me,* Ps. Surt. 3, 2, 5, 6. Hálne mé dóa *salvum me fac,* 3, 7; 4, 2. Se ðe geléfes on mech (mec, Rush.) *qui credit in me,* Jn. Skt. Lind. 6, 35. Ne hæfes ðú dǽl mech (mec, Rush.) mið *non habes partem mecum,* 13, 8. Hé mé habban wile dreóres fáhne, gif mec deáþ nimeþ, Beo. Th. 897, 899; B. 446, 447. [*Goth.* mis; *dat.;* mik; *acc.;* *O. Sax.* mi; mi, mik: *O. Frs.* me; mi: *Icel.* mér; mik: *O. H. Ger.* mir; mih.]

meagol, megol; *adj. Earnest, strenuous, firm:*—Ðæt ic Gode and Sancta Marian meaglum móde on éce yrfe geseald hæbbe *what I, with mind immovable, have given as a perpetual inheritance to God and St. Mary* (cf. the form 'Ego donationem *indeclinabiliter* consensi,' 322, 6), Cod. Dip. Kmbl. v. 331, 5. Mandryhten holdne gegrétte meaglum wordum *the lord (Hygelac) greeted his liege (Beowulf, on his return) with earnest words, gave him a hearty greeting,* Beo. Th. 3964; B. 1980: Exon. 43 a; Th. 146, 8; Gú. 706. Fugla cyn hine weorþedon meaglum stefnum, 46 a; Th. 157, 13; Gú. 891: 60 b; Th. 221, 22; Ph. 338. v. un-meagol *and following words.*

meagol-líce; *adv. Earnestly, strenuously:*—Hié ðone lifgendan God and ðone hálgan heáhengel Michael meagollíce (cf. Homl. Th. i. 504, 7 *where in the same narrative* geornlíce bǽdon *occurs*) gebǽdon *they earnestly prayed to the living God and the holy archangel Michael,* Blickl. Homl. 201, 13. Hé hafaþ wíslícu word, wile meagollíce módum tǽcan, Cd. 169; Th. 211, 16; Exod. 527.

meagol-mód; *adj. Of earnest mind, earnest, strenuous:*—Ic synful bydde ðæt ðú onsende in mé (mé in?) heortan meagolmód gemynd and gedéfe hreówe and sóðe ondetnesse ealra mínna synna *I, sinful, pray that thou send into my heart an earnest mind, and suitable penitence, and the true confession of all my sins,* Wanley Cat. 246, 9.

meagolmód-ness, e; *f. Earnestness, diligence:*—Hé sang ǽghwylce dæge mæssan Gode tó lofe myd swýðe mycelre meagolmódnysse and myd wépendum teárum *every day he sang mass to the praise of God with very great earnestness, and with tears,* Shrn. 98, 3. Ðæs wé sceolan mid ealre heortan meagolmódnesse úrum Drihtne þanc secgan, Blickl. Homl. 123, 16. v. next word.

meagol-ness, e; *f. Earnestness:*—Lufian wé hine mid eallre úre heortan megolnesse *let us love him in all earnestness of heart,* Blickl. Homl. 65, 23. v. preceding word.

meaht, maht, mæht, meht, mieht, miht, e; *f.* (but mihtes, Ps. Th. 70, 18). I. *Might, power, virtue, ability:*—Meaht eorþlíces ríces *potestas terreni imperii,* Bd. 2, 9; S. 510, 13. Seó godcunde meht, Blickl. Homl. 19, 20. Gif hǽto oððe meht ne wyrne lǽt him blód *if heat, or his ability to bear it do not forbid, let him blood,* L. M. 2, 42; Lchdm. ii. 254, 4. Miht is Drihtnes *potestas Dei est,* Ps. Th. 61, 12. Meahte *opis,* Wrt. Voc. ii. 65, 26: *potentatus,* 77, 78. Mihte lufigend amans *virtutis,* Ælfc. Gr. 43; Zup. 255, 10. His ríces ongin, ne his mehte, ne his mægenþrymmes næfre gewonad ne weorþeþ, Blickl. Homl. 9, 17. Ðínes mihtes þrym *potentiam tuam,* Ps. Th. 70, 18. Meahte nutu, Wrt. Voc. ii. 60, 78: 91, 31. Ungeléredne fiscere, náwðer ne on worde ne on gebyrdum mid nænigre mihte (*ability*) gewelgode, Blickl. Homl. 179, 15. Hé on mihte (mæhte, Lind.) and on mægene unclǽnum gástum bebýt (*in potestate et virtute*), Lk. Skt. 4, 36. Bútan ðínre miht *abs te,* Ps. Th. 138, 10. Maht *potentiam,* Ps. Surt. 144, 4. Meahte *numen,* Wrt. Voc. ii. 61, 25. Ðín wuldor ús gecýð, cræft and meaht, Exon. 53 b; Th. 188, 11; Az. 44. Swá swá mæht hæbbende *sicut potestatem habens,* Mt. Kmbl. Rush. 7, 29. Ða mæhte (ðæt mæht, Lind.) seðe eode from him *virtutem quæ exierat de eo,* Mk. Skt. Rush. 5, 29. Hé nænige mehte wið ús nafaþ, Blickl. Homl. 31, 33. Þurh his godcunde meht, 121, 15. Ðín mægen is áterod and ða mihte ðú næfst, Homl. Skt. 3, 611. Se weard hafaþ miht and strengþo, Cd. 45; Th. 58, 22; Gen. 950. *Virtutes* sind

gecwedene mihta, þurh ða wyrcþ God fela wundra, Homl. Th. i. 342, 27. His meahte synt *powers are his*, Ps. Th. 98, 10. Þurh ðínra mehta spéd *through the abundance of thy powers*, Bt. Met. Fox 4, 64; Met. 4, 32. His mihta name *nomen majestatis ejus*, Ps. Th. 71, 19. Ðú sǽs wealdest mihtum *tu dominaris potestati maris*, 88, 8. Gástes miehtum, Hy. 8, 12; Hy. Grn. ii. 290, 12. Eallum hire mihtum and mægenum *with all her might and main*, L. M. 3, 63; Lchdm. ii. 352, 5. Eallum mihtum, L. C. E. 20; Th. i. 372, 9. Mid eallum mægene and eallum mihtum *ex omni virtute, et omnibus viribus*, L. Ecg. C. pref.; Th. ii. 132, 13. On hyre yldrena mihtum *in potestate parentum suorum*, 27; Th. ii. 152, 15. Ðæt geþyld oferswíððum leahtrum sprecþ tó ðám mihton (mægnum, 28 a) *patientia devictis vitiis ad virtutes loquitur*, Prud. 28 b. On díne ða myclan mihte *in potentias Domini*, Ps. Th. 70, 15. Mihta strange, 102, 6. II. *an exercise of power, mighty work :*—Swilce mihta (mæhto, Lind.: mæhte, Rush.) ðe þurh his handa gewordene synd *virtutes tales quæ per manus ejus efficiuntur*, Mk. Skt. 6, 2. Ne dyde mæhto ð mæguo monigo *non fecit virtutes multas*, Mt. Kmbl. Lind. 13, 58: 14, 2. [O. E. Homl. maht: Laym. mæht, miht: Orm. mahht, mihht: Ayenb. miȝt: Goth. mahts: O. Sax. maht: O. Frs. macht, meht: Icel. máttr: O. M. H. Ger. maht: Ger. macht: Du. magt.] v. eall-, heáh-, un-meaht.

meaht; *adj.* I. *mighty, powerful :*—Se meahta moncynnes fruma, Exon. 61 a; Th. 224, 17; Ph. 377. Se micla dæg meahtan Dryhtnes, 20 b; Th. 54, 16; Cri. 869. Ealle ðínes múðes meahte dómas, Ps. Th. 118, 13. II. *possible :*—Alle mæhte sindun mið God *omnia possibilia sunt apud Deum*, Mk. Skt. Rush. 10, 27. [*Goth.* mahts *possible.*] v. æl-miht.

meahte-, meaht-líc; *adj. Possible :*—Gode synt mihtelíce ða ðing ðe mannum synt unmihtelíce *quæ impossibilia sunt apud homines possibilia sunt apud Deum*, Lk. Skt. 18, 27. Ealle þing synd gelýfedum mihtlíce (MS. A. myhtelíce), Mk. Skt. 9, 23. [Cf. *Icel.* máttu-ligr *mighty*; *possible:* O. H. Ger.* maht-líh *possibilis.*] v. un-mihtelíc.

meahte-, meaht-líce; *adv. Mightily, powerfully, with power, in power :*—Mihtelíce *potenter*, Hy. Surt. 26, 4. Myhtylíce *potentialiter*, 29, 11. Mihtlýce *potenter*, 49, 19. Sǽ oncneów ða Cristofer hyre ýða mihtelíce eode *the sea acknowledged him, when Christ in his might walked over the waves*, Homl. Th. i. 108, 17. Mid ðám he ðý mihtlícor wiðscúfan mihte *quibus potentius confutare posset*, Bd. 5, 21; S. 642, 39. Meahtelícor, Exon. 111 a; Th. 425, 27; Rä. 41, 62. [Cf. *Icel.* máttu-liga *mightily.*] v. meahtig-líce.

meahtig, mæhtig, mehtig, mihtig; *adj.* I. *mighty, powerful, able :*—Meahtig God, Ps. Th. 98, 9: Exon. 44 a; Th. 149, 12; Gú. 760: Hy. 4, 108; Hy. Grn. ii. 285, 108. Dryhten strong and maehtig (*potens*), Ps. Surt. 23, 8: 71, 12: Mk. Skt. Lind. 9, 29. Mæhtih, Lk. Skt. Lind. 24, 19. Meahtig God, Ps. C. 50; Ps. Grn. ii. 278, 89. Cyning ríce and mihtig *rex potentissimus*, Bd. 1, 25; S. 486, 16. Wyrta módor, innan mihtigu, Lchdm. iii. 32, 8. Heó was swá mihtegu wið God ðæt heó sealde blindum gesihþe, Shrn. 31, 12. Meotud biþ meahtigra ðonne ænges monnes gehygd, Exon. 83 a; Th. 312, 28; Seef. 116. Migtigra, Cd. 200; Th. 248, 33; Dan. 522. Allra mæhtigust is snytro *omnium potentior est sapientia*, Rtl. 81, 9. On ðysum eahta dælum (*parts of speech*) synd ða mǽstan and ða mihtigostan *nomen* and *verbum*, Ælfc. Gr. 5; Som. 4, 5. II. *possible :*—Mæhtiga *possibilia*, Mk. Skt. Lind. 9, 23: Lk. Skt. Lind. 18, 27. Cf. meaht; *adj.* and meahte-líc, meahte-líce. [*Goth.* mahteigs: *O. Sax. O. L. Ger. O. H. Ger.* mahtig: *Ger.* mächtig: *O. Frs.* machtich: *Icel.* máttigr.] v. eal-, efen-, fela-, fore-, ofer-, swíð-, tír-, un-meahtig.

meahtig-líce; *adv. Mightily, powerfully, with might :*—Ðæt is ðæt héhste gód ðæt hit eall swá mihtiglíce macaþ, Bt. 35, 4; Fox 160, 32. Mihtiglíce hé mihte mid his worde hine gehǽlan búton hrepunge *by an exercise of power he could have healed him with his word, without touching*, Homl. Th. i. 122, 8. Ðás seofonfealdan gifa wunodon on Criste æfter ðære menniscnysse swíðe mihtiglíce, Wulfst. 57, 9. [*O. Sax.* mahtiglík *mighty.*] v. meahte-líce.

meaht-leás; *adj. Powerless :*—Ðonne (*at the day of judgement*) stent ealra hergea mǽst heortleás and earh, mihtleás and áfǽred, Wulfst. 137, 23. [*Icel.* mátt-lauss *weak.*]

meaht-mód, es; *n. Strong feeling, passion :*—Wǽron heaðowylmas heortan getenge mihtmód wera *fierce rage pressed on the heart, and the mighty passions of men*, Cd. 149; Th. 187, 10; Exod. 149.

meala. v. melu.

Mealdumes burh *Malmsbury :*—Aldhelme abbode æt Mealdumesbyrig, Cod. Dip. Birch 154. 6. Æt Meldum, ðæt is óðrum naman Maldunes buruh geclypud, 24, Binnon Mealdelmes byrig, Chr. 1015; Erl. 152, 3: Cod. Dip. Kmbl. vi. 312, col. 2.

meale-hús. v. melu-hús.

mealm, es; *m. Sand, chalk* (?) (see next two words). [*Goth.* malma: *m.* ἄμμος: *O. Sax. O. H. Ger.* melm; *m. pulvis: Icel.* málmr; *m. sand* (in names of places).]

mealmmiht; *adj. Sandy, chalky* (?) :—Tó mealmehtan leáhe (*the land lay in Surrey*), Cod. Dip. Kmbl. iii. 394, 13. [E. D. S. Ellis' Farming

Words, 'The chalk and mould were so mixed together, that in Hertfordshire we call it a *maumy (malmey)* earth.' 'A chalk or a *maume*.' 'Chalk, *maume*, or loam.']

mealm-stán, es; *m. Maum-stone.* 'In agro Oxoniensi lapidem invenies friabilem, quem *maum* vocant indiginæ.' E. D. S. Gloss. B. 15. A correspondent of Dr. Bosworth's writes: 'The *Maumstone* is to be found, more or less, all over Wiltshire, especially towards Stonehenge. It is used for the foundation of walls, and the poor people use it for whitening, in keeping their hearth-stones clean. It is not so white as chalk, and is much more brittle.'—Mon heardlíce gníde ðone hnescestan mealmstán, Ors. 4, 13; Swt. 212, 28.

mealt; *adj. Cooked, boiled* (?) :—On gewylledre mealtre meolce (mealtre = gewylledre ? Cockayne says the word should be struck out), Lchdm. iii. 6, 17. v. miltan.

mealt, malt, es; *n. Malt :*—Malt *bratium*, Ep. Gl. 6 b, 2: Wrt. Voc. 102, 18. Mealt, malt, 11, 44: 127, 15: *macetum*, 58, 13. [*Icel.* malt; *n. O. H. Ger.* malz *brasium.*] v. alo-malt.

mealt-gescot, es; *n. A contribution of malt :*—Sceóte man swá hwæt swá witan gerǽdan, hwílum weaxgescot, hwílum mealtgescot, Wulfst. 171, 2.

mealt-hús, es; *n. A malt-house*; brationarium, Ælfc. Gl. 108; Som. 78, 127; Wrt. Voc. 58, 38. [Cf. *Icel.* malt-hlaða.]

mealt-wyrt, -wurt, e; *f. Malt-wort :*—Maltwyrt *acinum*, Wrt. Voc. ii. 10, 37, 54. Mealtwurt, i. 28, 7. v. leáh-mealtwurt.

mealwe, an; *f. Mallow :*—Malwe *malva*, Ælfc. Gl. 42; Som. 64, 31; Wrt. Voc. 31, 41. Mealewe, 67, 56. Wildre mealwan seáw, L. M. 2, 24; Lchdm. ii. 214, 14. Hé hláf þicge and mealwan, 16; Lchdm. i. 194, 6: 33; Lchdm. ii. 238, 14. [From *Lat.* malva.] v. mersc-mealwe.

mearc, e; *f.* I. *a mark, sign* made upon a thing :—Tácon ł merca *titulus*, Mk. Skt. Lind. 15, 26. Cf. onmerca *inscribtio*, 12, 16. Merce ł stæfes heafud *apicem*, Lk. Skt. Lind. 16, 17. Mearce *caracteres*, Wrt. Voc. ii. 23, 81. II. *a mark, ensign :*—Hé nam ðone stán and árærde hine tó mearce (*in titulum*; for a pillar, A. V.), Gen. 28, 18. Moyses getimbrode twelf mearca (*titulos*; pillars, A. V.), Ex. 24, 4. Nimaþ ða sigefæstan mearca *victricia tollite signa*, Ælfc. Gr. 9, 64; Som. 13, 66. [*O. Frs.* merke; *f. a mark*; macula: *Icel.* mark; *n. a mark, sign*; merki; *n. a mark, landmark*; *standard: O. H. Ger.* marcha, marca *titulus: M. H. Ger.* marc; *n. a sign.*]

mearc, e; *f.* I. *a limit, bound, term* (of time) :—Ðá ðæs mǽles wæs mearc agongen *then was the limit of the time passed*, Cd. 83; Th. 103, 17; Gen. 1719: 224; Th. 296, 13; Sat. 501. Him ðær tó mearce wearþ hé ðær feorhwunde hleát *that proved his life's limit; there his death-wound he got*, Beo. Th. 4758; B. 2384. II. *a limit, boundary* (of place), (a). :—Beó ðær gemeten nygon fét of ðam stacan tó ðære mearce (*the limit up to which the hot iron had to be carried*; cf. Grmm. R. A. 918), L. Ath. iv. 7; Th. i. 226, 13. Hé hæfþ heora mearce swá gesette ðæt hié ne mót heore mearce gebrǽdan ofer ða stillan corþan *ut fluctus avidum mare certo fine coerceat, ne terris liceat vagis latos tendere terminos*, Bt. 21; Fox 74, 27: Bt. Met. Fox 11, 129, 139, 146; Met. 11, 65, 70, 73: 20, 177; Met. 20, 89. Swá ðæt heora nán óðres mearce ne ofereode, Bt. 33, 4; Fox 128, 32. (b) *a boundary* (= gemǽre) *of a particular estate :*—Ðis is eástmærc tó stánmere . . . swá tó Rithmærce, Cod. Dip. B. 280, 18, 12. Swá be mearce . . . ðonon súð andlang mearce, 148, 31–37. His metis rus hoc gyratur . . . forþ on ða mearce . . . andlang mearce . . . ðonon tó Æðelbirhtes mearce . . . ðonan forþ on ða mearce tó Beonetlǽga gæmære . . . ðonan west on ða mearce ðǽr Ælfstán líþ on hǽðenan byrgels . . . ðonan Wulfstanes mearce, Cod. Dip. Kmbl. iii. 130, 26–131, 13. Be rihtre mearce (cf. be gerihtum gemǽre, l. 22) tó ðæm gemǽrþornan ; ðæt tó ðære reádan róde ; swá forþ be ealdormonnes mearce ; á be mearce ðæt hit cymþ on Icenan, 404, 31–405, 2. Heallingwara mearc, 400, 24. (c) *a boundary, confine of a district, border :*—Sí swá hwǽr swá hý, swá be norþan mearce, swá be súþan, á of scíre on óðre, L. Ath. v. 8, 4; Th. i. 236, 26: 4; Th. i. 232, 19. Cépeman oððe óðerne ðe sió ofer mearce cuman, L. H. E. 15; Th. i. 32, 17: L. Wih. 8; Th. i. 38, 17. (Thorpe in the last two examples would take mearc to be the limit of an estate.) Ðú symle furðor feohtan sóhtest mǽl ofer mearce, Wald. 1, 33; Vald. 1, 19. Ðæt is ðonne ðæt mon his mearce brǽde . . . hira mearce mid tó rýmanne *terminum suum dilatare est . . . ad dilatandum terminum suum* (cf. getryman hira landgemǽru, 4), Past. 48, 2; Swt. 367, 13–15: Cd. 136; Th. 171, 19; Gen. 2830. Unc módige ymb mearce sittaþ (*sit on our borders*), 91; Th. 114, 21; Gen. 1907. Merce gemǽrde wið Myrgingum, Exon. 85 a; Th. 321, 6; Víd. 42. Hé sume on wræcsíð forsende sume on óðra mearca gesette *alios avulsos a sedibus suis, alios in extremis regni terminis statuit*, Ors. 3, 7; Swt. 114, 34. III. *the territory within the boundaries*; *fines :*—Hit wæs geond ealle Rómána mearce ðæt *it was the custom throughout all the Roman territories* (cf. O. Sax. thero marka giwald égan *to succeed to the throne*), Bt. 37, 4; Fox 100, 13. Hwílum wycg byreþ mec ofer mearce, hwílum merehengest fereþ ofer flódas, Exon. 104 a; Th. 395, 11; Rä. 15, 6. Mearce healdan (*or* II. c), Cd. 98; Th. 128, 32; Gen. 2135. Nǽfre on his weorþige weá áspringe mearce má scýte mán

X x

inwides *non defecit de plateis ejus usura, et dolus*, Ps. Th. 54, 10. [*Goth.* markōs; *pl. borders* (of a country): *O. Sax.* marka *border, district* : *O. L. Ger.* marka *district* : *O. Frs.* merke *limit, district* : *Icel.* mörk *a forest* ; in compounds, *a border-land, district* : *O. H. Ger.* marcha, marka *limes, confinium, terminus, fines* : *Lat.* margo.] v. édel-, first-, land-, leód-, tæl-, þeód-, Weder-mearc ; ge-mearc, ge-mirce, *and 'the following compounds with* mearc- ; *and cf. these with compounds of* mær-. On the *mark* see Stubbs' Const. Hist. i. 49-52, and Kemble's Saxons in England, vol. i.

mearc-béce, an ; *f. A beech-tree which forms part of a boundary* :—Ðis synd ðæra viii. hída landgeméra ... tó ðære mearcbécean ; of ðære bécean, Cod. Dip. B. i. 295, 9. On ða ealdan mearce bécan, 296, 26.

mearc-beorh ; *gen.* -beorges ; *m. A hill which forms part of a boundary* :—Prædicta tellus his terminis circumcincta. Ǽrest on æsc-woldes hláw : ðonne on gemótbiorh ... ðonne on mearcbiorh, Cod. Dip. Kmbl. ii. 195, 14. Ǽt ðæne mearcbeorh, iii. 175, 35. Cf. gemǽr-beorh, iii. 403, 27. [Kemble says 'the *mearcbeorh* appears to denote the hill or mound which was the sité of the *mearc-mót*.' Saxons in England, i. 56.]

mearc-bróc, es ; *m. A brook which serves as a boundary* :—Andlang Ecclesburnon tó ðam mearcbróce, Cod. Dip. Kmbl. v. 193, 31. Ðis synd ða landgeméra. Ǽrest ðér mercbróc scýt on Seolesburnan ; of mearc-bróce ... swá andlang burnan eft on mérbróce, 284, 12-30.

mearc-denu, e ; *f. A valley which serves as a boundary* :—Tó mearc-dene, Cod. Dip. Kmbl. iii. 404, 23.

mearc-díc, e ; *f. A ditch which serves as a boundary* :—On ða ealdan mercdíc, Cod. Dip. B. i. 295, 7.

mearcere, es ; *m. A notary, writer* :—Mǽrcerum, wrí[terum] *notariis*, Hpt. Gl. 528, 67.

mearc-hof, es ; *n. A dwelling in a mark or country*, Cd. 145 ; Th. 181, 14 ; Exod. 61.

mearcian ; *p.* ode (mearc *a mark*). **I.** *to make a mark on any-thing* :—Hé byreþ blódig wæl ... mearcaþ (*marks with blood*) mórhopu, Beo. Th. 904 ; B. 450. Mearciaþ on marmstáne hwonne se dæg and seó tíd geeáwe *in marmore signant titulo remque diemque*, Exon. 60 b ; Th. 221, 11 ; Ph. 333. Mearcode *sulcaret t scriberet t labararet*, Hpt. Gl. 465, 6. Hé mearcode ða stówe, Homl. ii. 160, 35. Mearca ðé sylfne mid tácne ðære hálgan róde, i. 534, 22. Mearcie (*brand*) man hine (þeówman) æt ðam forman cyrre, L. C. S. 32 ; Th. i. 396, 9. Mercande *signantes*, Mt. Kmbl. Lind. 27, 66. **II.** *to mark out, design* :—Ǽlc cræftega þencþ and mearcaþ his weorc on his móde ǽr he it wyrce *every artificer considers and marks out his work in his mind before he does it*, Bt. 39, 6 ; Fox 220, 4. Him tó gingran metot mearcode *the Lord marked them out for his servants*, Cd. 23 ; Th. 29, 33 ; Gen. 459. [*O. Sax.* markón *to mark out* : *O. Frs.* merkia : *Icel.* marka *to mark* ; *mark out, design* ; merkja *to mark* : *O. H. Ger.* marchón *significare, notare* ; markjan, markén *notare, designare*.] v. ge-, tó-mearcian ; fore-mearcod.

mearcian ; *p.* ode (mearc *a limit*) *To fix the bounds or limits of a place* :—Se mearcode ða stówa ðe gé eówre geteld on sleán sceoldon *metatus est locum, in quo tentoria figere deberetis*, Deut. 1, 33. [*O. H. Ger.* marchón *definire, collimitare*.]

mearc-ísen, es ; *n. A branding-iron* :—Mearcísen *cauterium*, Wrt. Voc. ii. 13, 18. Mearcísene *cauterio*, Hpt. Gl. 453, 22. Hé sǽde ðæt hé gesége ðæt ic wǽre gemearcod mid deófles mearcísene, Shrn. 37, 13. v. next word.

mearc-ísern, es ; *n. A branding-iron* :—Mearcísern *cauterium*, Ep. Gl. 8 d, 35 : Wrt. Voc. ii. 129, 76 : *ferrum quo note pecudibus inuruntur*, 3. Mercíseren, 102, 58.

mearc-land, es ; *n.* **I.** *a border-land, waste land lying outside the cultivated* :—Se mylenhám and se myln and ðæs mearclandes swá mycel swá tó þrím hídon gebyraþ, Cod. Dip. Kmbl. iii. 189, 11. v. Kemble's Saxons in England, i. 50. Mearclonde (*the sea coast*) neáh, Exon. 101 b ; Th. 384, 6 ; Rä. 4, 23. Him ðe feára sum mearclond gesæt (*of Guthlac when he retired to his hermitage. Cf. what is said before of his dwelling place* :—Wæs seó londes stów bimiðen fore monnum, óððæt meotud onwráh beorg on bearwe, 34 b ; Th. 110, 32-35), Exon. 35 a ; Th. 112, 17 ; Gú. 145. Héht ymbwícigean Æthanes byrig mearc-landum on *bade them encamp about Etham's town, in its borders*, Cd. 146 ; Th. 181, 27 ; Exod. 67. **II.** *a district, country, territory* :—Ðæt mearcland, folcstede gumena, hæleþa éðel, Andr. Kmbl. 37 ; An. 19. Geweoton ða wítigan mearcland tredan, 1603 ; An. 803. v. Kemble's Saxons in England, i. 46 sqq. [*Icel.* mark-land *forest-, border-land*.]

mearc-mót, es ; *n. The place where the assembly* (mót) *of a district* (mearc) *was held* :—Ðis syndon ða landgeméra ... tó mercemót ; fram mercemóte, Cod. Dip. Kmbl. iii. 71, 31. v. Saxons in England, i. 55.

mearc-pæð, es ; *m. n.*(?) *A path leading through a country* :—Be mearcpaðe, strǽte neáh, Andr. Kmbl. 2124 ; An. 1063. Ic síðade wíð-dor mearcpaðas (*paths across the marches*?) træd, móras pæðde, Exon. 126 a ; Th. 485, 7 ; Rä. 71, 10. Gewát hé ða féran ofer mearcpaðu (-paðum ? -waðu, Grimm, Kemble), ðæt hé on Membre becom, Andr. Kmbl. 1575 ; An. 789. v. mearc-wæd.

mearc-stapa, an ; *m. One who wanders about the desolate mark or border-land* :—Grendel, mǽre mearcstapa, Beo. Th. 206 ; B. 103. Hié gesáwon swylce twegen micle mearcstapan móras healdan ; óðer wæs idese onlícnes óðer on weres wæstmum wrǽclástas træd ... Hié dýgel lond wari-geaþ, wulfhleoþu, windige næssas, frécne fengelád, 2698-2722 ; B. 1347-1359. v. Kemble's Saxons in England, i. 48.

mearc-stede, es ; *m. Desolate, border-land* :—Saga mé from ðam lande ðér nǽnig fira ne mæg fótum gestæppan ... Hé on ðam felde geslóg xxv dracena, .. forðan ðás foldan ne mæg fira ǽnig, ðone mearcstede, mon gesécan, fugol gefleógan, ne ðon má foldan neát, Salm. Kmbl. 418-436; Sal. 209-218. v. preceding word.

mearc-þreát, es ; *m. A band of men occupying the frontier of a country* :—Manna þengel mearcþreáte rád (cf. Th. 187, 33 : 188, 14), Cd. 151 ; Th. 188, 25 ; Exod. 173.

mearc-treów, es ; *n. A tree serving as a boundary* :—Ðonne tó mearc-treówe, Cod. Dip. Kmbl. iii. 434, 18. Cf. gemǽr-treów.

mearcung, e ; *f.* **I.** *a marking, mark* :—Nota ðæt is mearcung Ðæra mearcunga sind manega, Ælfc. Gr. 50 ; Som. 51, 19. Mǽrcunge *characteres*, Hpt. Gl. 473, 13. **II.** *a marking out, description, arrangement, disposition* :—Mercung *descriptio*, Lk. Skt. Rush. 2, 2. Mearc-ung *capitulatio*, Wrt. Voc. ii. 128, 40. Mearcunge *constellationem, constellationes*, Hpt. Gl. 468, 1, 3. [*O. H. Ger.* marchunga *propositum, institutio*.] v. fore-, ge-, on-mearcung.

mearc-wǽd, es ; *n. Boundary-water, the water by the shore* :—Wlanc monig on stæþe stódon stundum wrǽcon ofer mearcwaðu and ða gehlódon hildesercum wǽghengestas *many a proud one stood on the shore ; now and again they pressed over the border-floods, and then laded the wave-steeds with their war-shirts* (but cf. mearc-pæð), Elen. Kmbl. 465 ; El. 233.

mearc-weard, es ; *m. A mark-warden, a wolf*, Cd. 151 ; Th. 188, 14 ; Exod. 168.

mearc-weg, es ; *m. A road that forms part of a boundary* :—Andlang mearcweges, Cod. Dip. Kmbl. v. 40, 3. On mearcwei, iii. 202, 5. Cf. mǽr-, gemǽr-weg.

meard. v. meord.

mear-gealla, an ; *m. A kind of gentian* :—Mergelle, Lchdm. iii. 24, 1. Wyl mergeallan on meolcum, L. M. 2, 65 ; Lchdm. ii. 296, 18. v. mersc-meargealla.

mearh, mærh, es ; *n. m. Marrow, pith* ; *also a sausage*. Cf. mearh-gehæcc :—Mearh *medulla*, Wrt. Voc. i. 65, 23. Mearg, 283, 48. Mǽrh, 70, 47. Merg, ii. 114, 3. Mearh *lucanica* (*lucanica* genus farciminis ex porcinis carnibus concisis a Lucanis populis, a quibus Romani milites primum didicerunt, Forcellini), 51, 55 : *amilarius*(?), 6, 59 : 100, 19. Mǽrh, 113, 22. Meargh, Wrt. Voc. i. 286, 53 (given amongst words de *suibus*). Mearh *medulla* vel *lucanica*, 44, 42. Mid mearche *cum medulla*, Cant. M. ad fil. 14. Wuduþistles ðone grénan mearh ðe biþ on ðam heáfde, L. M. 3, 70 ; Lchdm. ii. 358, 1. Gedó ðæt mearh on ða eágan, 1, 2 ; Lchdm. ii. 38, 9. Heortes smeoruw oððe ðæt mearh, Herb. 96, 3 ; Lchdm. i. 208, 22. Nim foxes smero and ráhdeóres mearh, Lchdm. iii. 2, 25. Wulfes mearh, L. Med. ex Quad. 9, 6 ; Lchdm. i. 362, 9. Heortes mearg, 10, 4 ; Lchdm. i. 366, 4. Nim mærc, sápan (MS. mærcsápan) and hinde meolc, Lchdm. iii. 4, 1. Mearga *medullas*, Germ. 397, 493. [*O. L. Ger.* marg : *O. Frs.* merg : *Icel.* mergr ; *m.* : *O. H. Ger.* marag, marg, mark : *Ger.* mark ; *n.*]

mearh ; *g.* meares ; *m. A horse, steed* :—Mearh moldan træd, Elen. Kmbl. 109 ; El. 55. Cyninges mearh, 2383 ; El. 1193. Se swifta mearh burhstede beáteþ, Beo. Th. 4521 ; B. 2264. Hwǽr cwom mearg, hwǽr cwom mago, Exon. 77 b ; Th. 291, 34 ; Wand. 92. Sum biþ meares gleáw *one is skilful in the management of a steed*, 79 a ; Th. 297, 17 ; Crä. 69. Tomes meares, 91 a ; Th. 342, 13 ; Gn. Ex. 142. Ðá hé on meare rád, on wlancan ðam wicge, Byrht. Th. 138, 54 ; Elen. Kmbl. 2349 ; El. 1176. Ðe him mænigne mear gesealde, Byrht. Th. 137, 19 ; By. 188. Eahta me…ras, Beo. Th. 2075 ; B. 1035. Fealwe mearas, 1735 ; B. 865. Mearas æppelfealuwe, 4333. Meara and máðma, 4338 ; B. 2166. Mearum and máðmum, 3800 ; B. 1898. Beornas cómon wiggum gengan on mearum módige, Andr. Kmbl. 2193 ; An. 1098. [*Icel.* marr *a steed* (in poetry ; used in compounds, e. g. vág-marr *wave-steed*, of ships : *O. H. Ger.* marah, march *equus*.] v. lagu-, sǽ-, ýð-mearh.

mearh-cofa, an ; *m. A marrow-chamber, a bone* :—Mearhcofan *ossa*, Ps. Th. 101, 3.

mearh-gehæcc, es ; *n. A kind of pudding, a sausage* :—Mearhgehæcc *isica* (*insicia* genus farciminis, seu obsonii ex carne concisa, Forcellini), Wrt. Voc. ii. 48, 35. Mǽrhgehæc (-hæt, Wrt.) *isicia*, i. 27, 22. [Halliwell gives ' *hack* the lights, liver, and heart of a boar or swine : *hackin* a pudding made in the maw of a sheep or hog : *hack-pudding* a mess made of sheep's heart, chopped with suet and sweet fruits : *hatcher* a dish of minced meat.] v. next word and haccian.

mearh-hæccel es ; *n. A sausage, hog's-pudding* :—Gehæcca oððe mearhæccel *farcimen* (*farcimen* intestinum varie ac minutim concisa carne refertum, Forcellini), Wrt. Voc. ii. 39, 77. v. preceding word.

mearh-, mearg-líc ; *adj. Marrowy, fat* :—Onsegdnisse merglíce ic

offriu *holocausta medullata offeram*, Ps. Surt. 65, 15. [Cf. *O. H. Ger.* marag-haft (*in same passage*).]

mearrian; *p.* ode *To err, go astray* :—Ne þyncþ ðeáh ðám monnum ðæt hí áuht mearrigen ðe ðæs wilniaþ tô begitanne ðæt hí máran ne þu. fon tilian *num enim videntur errare hi, qui nihilo indigere nituntur?*, Bt. 24, 4; Fox 86, 1. v. ge-mearr, mirran.

mearþ, es; *m. A marten, a kind of weasel* :—Mearth *furuncus*, Ep. Gl. 9 d, 11. Mearþ, Wrt. Voc. ii. 36, 21 : *furo, idem deminutive furunculus*, 39, 58 : *ferunca* vel *ferunculus*, i. 22, 51. Mærþ *feruncus*, 78, 17 : *rumusculus*, ii. 76, 36. Merþ *ferunca*, 40, 12. Se byrdesta sceall gyldan xv mearþes fell (cf. *Icel.* marð-skinn.), Ors. 1, 1 ; Swt. 18, 20. Ofer mearþes hrycg (*in an enumeration of boundaries*), Cod. Dip. Kmbl. iii. 391, 20. [*Icel.* mörðr.]

mearu, mæru, meru, myru; *adj. Tender, soft, delicate* :—Ðonne his twig biþ mearu (*tener*), Mk. Skt. 13, 28. Merwe, Mt. Kmbl. Rush. 24, 32. Mearuwe *delicatus i. tenerus*, Wrt. Voc. ii. 138, 40. Gyf se lîchoma mearu (MS. B. mearuw) sý *if the body be tender (with sores)*, Herb. 102, 2 ; Lchdm. i. 216, 24. Hwæðer sió gecynd ðæs lîchoman sîe heard ðe hnesce and mearwe, L. M. 1, 35 ; Lchdm. ii. 84, 14. Man byþ merwe gesceaft, Ps. Th. 143, 5. Myra *tenellus*, Kent. Gl. 62. Se myrwa *mactus*, Wrt. Voc. ii. 54, 71. Ðære mærwan cyrican weaxnesse *tenellis ecclesiæ crementis*, Bd. 2, 5 ; S. 506, 37. Blôd fleów of hire ðæm merwan lîchoman, Shrn. 101, 22. Genim ðás wyrte swâ mearwe *take this plant as young and tender as possible*, Herb. 89, 1 ; Lchdm. i. 192, 8, 12. Mearawa *tenera, gracilia*, Hpt. Gl. 457, 42. Ne gedafenaþ ûs ðæt wê symle hnesce beón on ûrum geleáfan swâ swâ ðás merwan cild, Homl. Th. i. 602, 13. Þurh ða myrwan *per tenera*, Wrt. Voc. ii. 66, 23. Hí (*the leaves*) beóþ mearwran (MS. H. mearuwran), Herb. 153, 1 ; Lchdm. i. 278, 15. Ða hwîtan lîchoman beóþ mearuwran and tedran ðonne ða blacan, L. M. 1, 35 ; Lchdm. ii. 84, 21. Mærwost, 2, 14 ; Lchdm. ii. 190, 21. On mearwis[tum?] *in tenerrima, gracillima*, Hpt. Gl. 444, 69. Merewistan *gracillima*, 521, 29. [*A. R.* meruwe (*of young trees*) : *O. H. Ger.* marawi, maro *tener, delicatus ; there is besides* muruwi, murwi *with same meaning* : *M. H. Ger.* mürwe : *Ger.* mürbe.]

mearuw-ness, e; *f. Tenderness, delicacy* :—Hira môdes mearuwnesse (Cott. MSS. meruwenesse) *eorum teneritudinem*, Past. 32, 2 ; Swt. 211, 18. Marenysse *teneritudine*, Hpt. Gl. 441, 35.

meatt. v. matt.
meáu. v. mǽw.
mec. v. mê.

mêce, es; *m. A sword, falchion, blade* :—Mêce *machera*, Hpt. Gl. 470, 44 : 424, 30 : Wrt. Voc. ii. 54, 47 : *mucro*, 114, 35. Mêcha *aciem gladii, vim gladii*, 98, 36. Mêche *frameam*, Ps. Spl. T. 16, 14. Mêces ecge, Beo. Th. 3628 ; B. 1812. Mid áwendenlîcum mêce *romphæa versatili vel volubili ancipíti, utraque parte acutus*, Hpt. Gl. 433, 70. Slôh fâgum mêce, bloody blade : 10 ; Thw. 23, 4 ; Jud. 104. Scîrne mêce *a bright blade*, Exon. 79 a ; Th. 297, 8 ; Crä. 65. Heardne mêce, Byrht. Th. 136, 47 ; By. 167. Mêcea gemânan, Chr. 937 ; Erl. 114, 6 ; Æðelst. 40. Mêcum mylenscearpan, Erl. 112, 24 ; Æðelst. 24. [*Laym.* mæche : *Goth.* mêki (*acc.*) : *O. Sax.* mâki : *Icel.* mækir.]

mêce-fisc, es; *m. A mullet* :—Mêce-(mêce-)fisc *mugil*, Ælfc. Gl. Zup, 308, 5. Cf. gâr-fisc.

mecg. v. mæcg.

mecgan; *p.* mægde (?) *To stir, mix* :—Cnuca eall ðás tôgadere and magce tôgadere *pound all these together, and stir together*, Lchdm. iii. 134, 8. Nîme ðæt dust and mæcige mid ðan æge *take the dust and stir it up with the egg*, 126, 19. Stréam sceal mecgan mereflôde *the river shall stir up (as it pours in) or mix with, the ocean*, Menol. Fox 507 ; Gn. C. 24.

mechanisc; *adj. Mechanical* :—Án wurþlíc weorc on mechanisc geweorc, Homl. Skt. 5, 251.

mêd, e; *f. Meed, reward* :—Mêd *merces*, Ælfc. Gr. 9, 27 ; Som. 11, 25 : Wrt. Voc. i. 61, 45 : *merx*, ii. 58, 41. Ðín mêd byþ swîðe micel, Gen. 15, 1 : Lk. Skt. 6, 35. Hwæt byþ ûs tô mêde, Mt. Kmbl. 19, 27 : Judth. 12 ; Thw. 26, 19 ; Jud. 335. Elles næbbe ge mêde mid eówrum fæder ðe on heofenum ys, Mt. Kmbl. 6, 1. Mêde onfón, 6, 5. Hê mê mêde gehêt, Beo. Th. 4275 ; B. 2134. Ðê sind gehealdene ðíne mêda gewisse, Homl. Th. ii. 516, 24 : Cd. 19 ; Th. 130, 29 ; Gen. 2167. Ðú mêdum scealt onfón, 141 ; Th. 176, 24 ; Gen. 2916 : Bd. 4, 3 ; S. 568, 34. [*O. Sax. O. L. Ger.* mêda, miéda : *O. Frs.* mêde, meide, mîde : *O. H. Ger.* mieta, miata : *Ger.* miete.] v. meord.

mêdan. v. on-mêdan.
Mêdas, Mêdisc. v. Mǽdas, Mǽdisc.
mêdder, mêddern. v. môdor, mêdren.
med-drosna ; *pl. f. Dregs of mead*, L. M. 1, 56 ; Lchdm. ii. 126, 15.
-mêde ; *subst. and adj.* v. eáþ-, ge-, ofer-, unblîðe-, unge-, wiðer-mêde (-mêdu).
medel. v. mæðel.
medeme. v. medume.
mêderce. v. mýdrece.
mêderen. v. mêdren.

---

meder-wyrhta. v. meter-wyrhta.

Medeshámstede, es ; *m. Peterborough* :—Abbud ðæs mynstres ðe. gecweden is Medeshámstyde on Gyrwan lande, Bd. 4, 6 ; S. 573, 41. Nama hit gáuen Medeshámstede, forðan ðet ðǽr is án wæl ðe is gehâten Medeswæl, Chr. 654 ; Erl. 29, 9. Hê geaf hit ðâ tô nama Burch ðe ǽr hêt Medeshámstede, 963 ; Erl. 123, 34. See also Cod. Dip. Kembl. vi. 312.

mêd-gilda, an ; *m. One who receives pay, a needy person* :—Wædla l̄ mêdgylda *mendicus*, Ps. Lamb. 39, 18. Se hýra oððe se mêdgylda *the hireling or the mercenary*, Homl. Th. i. 242, 5. Swâ swâ mêdgildan (*hireling's*) dagas, ii. 454, 27. Nafa ðû ðînne nêhstan for weal and for mêdgildan *non fratrem tuum opprimes servitute famulorum*, Lev. 25, 39.

-medla. v. an-, on-, ofer-medla.
medlen. v. midlen.
mêd-lîc. v. mǽþ-lic.

med-micel; *adj.* **I.** *not great, moderate, small* (of time, space, quantity) :—Se medmicla fyrst *modica illa intercapedo*, Bd. 5, 1 ; S. 614, 14 : Blickl. Homl. 111, 24. Is on westan medmycel duru, 127, 8. Se yfela dêma onsêhþ medmycclum feó, 61, 30. Ðâ fêng hê tô medmycclan bigleofan, ðæt wæs tô ðam berenan hlâfe, Guthl. 5 ; Gdwin. 34, 5. Hæfde hê medmycel (*permodicum*) mynster, Bd. 4, 13 ; S. 582, 21. Cærenes gôdne bollan fulne, and ecedes medmicelne, L. M. 1, 1 ; Lchdm. ii. 24, 20. Midmycle (*other MS.* medmycle), Bd. 2, 16 ; S. 519, 34. Medmiclu and miclu *pusilla et magna*, Blickl. Gl. *Used as a noun* :— Dô medmicel on ða eágan *put a little into the eyes*, 1, 2 ; Lchdm. ii. 36, 8. Medmicel pipores, 2, 44 ; Lchdm. ii. 256, 5. Medmicel hlâfes, Bd. 3, 27 ; S. 559, 35. Ðæs medmâsta (*or* medmasta ?) *from* medume. v. *also under* **II, III**) geleáfe *minime fidei*, Mt. Kmbl. Rush. 10, 30. **II.** *not great, trifling, venial, not important* :—Gif man medmycles (*exigui*) hwæthwega deóflum onsægþ, fæste i. geár ; gif hê mycles hwæt onsecge, fæste x winter, L. Ecg. C. 32 ; Th. ii. 156, 15. Medmycel ǽrende wê ðyder habbaþ, Blickl. Homl. 233, 11. Ða gôd ðe ic ǽfre dyde wǽron swîðe feáwe and medmicle (*nimium pauca et modica*), Bd. 5, 13 ; S. 632, 38. Ne mâgon wê bûton ðæm medmyclum synnum beón, Blickl. Homl. 37, 10. On mycclum gyltum oððe on medmycclum, 107, 14. Micclum þingum and medmiclum, Cod. Dip. Kmbl. ii. 304, 12. Ðæt ic on ðam medemǽstan (medemæstan ?) geþohte gesyngode *quæ tenuissima cogitatione peccavi*, Bd. 5, 13 ; S. 633, 10. **III.** *not great, lowly, mean, poor* :—On medmyclum hrægle gehealdene *content with mean apparel*, Blickl. Homl. 185, 17. On ðone medmycclan innoþ ðære á clǽnan fǽmnan *into the lowly womb of the ever clean virgin*, 5, 18, 33 : 23, 23. Æt ânum of ðissum medmǽstan *unum de pusillis istis*, L. Ecg. P. Add. 23 ; Th. ii. 236, 10.

medmicel-ness, e ; *f. Smallness* :—Medmicelnysse gástæs *pusillanimitate spiritus*, Ps. Spl. 54, 8.

medmicle ; *adv. Humbly, meanly* :—Oft wîc beóþ on manegum stôwum medmyccle gesette ; seó ceaster ðonne wæs hêh and aldorlîc, Blickl. Homl. 77, 24.

medo. v. medu.

mêdren, mêdern, mêddern; *adj. Maternal, (of lineage) on the mother's side* :—Eádweard his brôðor on mêdren (cf. *Icel.* môðerni *the mother's side*), Chr. 1041 ; Erl. 166, 28. Þurh mêdderne *per maternam*, Hpt. Gl. 440, 70. Of mêdernum hrîfe *de vulva* ; mêdernum *maternis*, 441, 41, 25. Of mêddernum geeácnungum *partubus*, 480, 9. v. *following words* and ge-mêdred.

mêdren-cynn, es ; *n. Maternal kin, kin by the mother's side* :—Ælfrêdes reht meódrencynn *Alfred's direct maternal kin*, Chart. Th. 483, 5. Ðæt wê ðín mêdrencynn môtan cunnan, nû wê áreccan ne mâgon ðæt fædrencynn, Exon. 11 b ; Th. 15, 34 ; Cri. 246.

mêdren-gecynd, es ; *n. Nature derived from the mother* :—Hê wæs sôð man þurh his mêdrengecynd (mêddrengecynd) *he was very man in the nature derived from his mother*, Wulfst. 17, 7.

mêdren-mǽg ; *m. A kinsman by the mother's side, maternal kinsman* :—Mêddernmágas *cognati*, Wrt. Voc. i. 51, 80. Ðara mêdrenmǽga (mêddrenmága, MS. H.) dǽl, L. Alf. pol. 8 ; Th. i. 66, 21. Gif hê mêdrenmǽgas nâge, 27 ; Th. i. 78, 21.

mêdren-mǽþ, e ; *f. Kindred by the mother's side* :—Gebyriaþ twelf men tô werborge, viii fæderenmǽgþe, and iii. mêdrenmǽgþe, L. E. G. 12 ; Th. i. 174, 19.

med-rîce ; *adj. Of little power, not powerful, of the lower as opposed to the higher classes* :—Medrîca gesetnyssa *plebisscita* ; rîccra gesetnes *senatus consultum*, Wrt. Voc. i. 20, 65–66.

med-sǽlþ, e ; *f. Bad fortune, ill success* :—Ðæt hié môsten gefandian hweðer hié heora medsêlþa oferswîðan mehte, Ors. 4, 4 ; Swt. 164, 28.

mêd-sceatt, es ; *m.* **I.** *payment in reward of service done, a reward, wages, fee* :—Ne onfêng hê ðæt tô mêdsceatte *he did not accept it as a fee*, Shrn. 135, 24. Hê ne sealde Gode nânne mêtsceat for his sáule . . . Ðæt is ðonne se mêdsceat wið his sáule ðæt hê him gielde gôd weorc *non dabit Deo pretium redemtionis animæ suæ. . . Pretium namque redemtionis dare, est opus bonum reddere*, Past. 45, 2 ; Swt. 339, 9–11. Swelce hié ða mêtsceattas rîmen ðe hié Gode sellen . . . Ac hié sceoldon

gehiéran ðone cwide ðe áwriten is : ' Se ðe médsceattas gaderaþ hé legeþ hié on þyrelne pohchan.' An þyrelne pohchan se legþ ðæt hé tó métsceatte sellan þencþ *quasi mercedem numerant . . . Audiant, quod scriptum es :* ' *Qui mercedes congregavit, misit eas in sacculum pertusum.' In sacculo pertuso videtur, quando pecunia mittitur,* 45, 4; Swt. 343, 16-21. II. *payment for service or favour expected* (generally in a bad sense), *a gift, present, a bribe :*—Sí se áwirged ðe unscildigne man beléowe wið médscette *maledictus, qui accepit munera, ut percutiat animam sanguinis innocentis,* Deut. 27, 25. Ælc wóh for lyðran médsceatte geléataþ tó rihte, Wulfst. 297, 26. Se man ðe bringþ médsceat ðam geréfan, se geǽrendaþ bet ðonne se ðe nænne ne bringþ, 238, 8. Gif hwá æt þeófe médsceatt nime, L. Ath. i. 17; Th. i. 208, 14. Swylc geréfa swylc médsceat nime, and óðres ryht þurh ðæt álecge, iv. 1; Th. i. 222, 5: L. E. I. 16; Th. ii. 412, 12. Médsceattas *munera propriæ,* Wrt. Voc. ii. 59, 9. Médsceattas áblendaþ wísra manna geþancas, Deut. 16, 19. Swýðre heora gefylled is of médsceattum (*muneribus*), Ps. Spl. 25, 10 : L. Alf. 46; Th. i. 54, 17 : L. Ed. 7; Th. i. 162, 25.

med-spédig ; *adj. Unprosperous, poorly provided :*—Ne biþ ǽnig ðæs earfoþsǽlig mon on moldan, ne ðæs medspédig ðæt hine se árgifa ealles biscyrge módes cræfta *no man upon earth is there of such hard fortune or so meanly endowed, that the gracious giver quite cuts him off from powers of mind,* Exon. 78 b; Th. 294, 3; Crä. 9.

med-strang ; *adj. Of moderate means, of middle rank :*—Ic lærde wlance men and heáhgeþungene . . . Ic lærde eác ða medstrangan men (cf. Homl. Th. i. 370, 20, *see under* medume) . . . and þearfum ic lærde, Blickl. Homl. 187, 13-17.

med-, met-trum ; *adj.* I. *not strong in health, infirm, weak, ill :*—Hwá biþ medtrum ðæt ic ne síe for his þingum seóc *quis infirmatur, et ego non infirmor ?* Past. 21, 6; Swt. 165, 4. Se mettruma líchoma *debile corpus,* 61, 2; Swt. 455, 27. Sint tó manianne ða mettruman (*ægri*), 36, 4; Swt. 251, 20. Manega wurdon mettrume gehǽlede, Homl. Th. ii. 512, 7. Mettrumra *ægrotorum,* Hpt. Gl. 415, 20. II. *of inferior position*(?) :—Nalæs ðæt án ðætte ða metruman (MSS. O. T. mǽttran: MS. B. mǽteran) men ymb heora nédþearfnesse wǽron ac eác cyningas and ealdormen from hire geþeaht sóhton *non solum mediocres in necessitatibus suis, sed etiam reges ac principes ab ea quærerent consilium,* Bd. 4, 23; S. 593, 43. Cf. med-strang.

med-, met-trum-, -trym-ness, e ; *f. Infirmity, ill-health, sickness, illness :*—Seó lange mettrumnes ðæs seócan mannes, ðonne hine God forlǽtan nele éþelíce lifian, ne hé swyltan ne móte, Blickl. Homl. 59, 28. Hwílum ofþrycþ ðone líchoman ungemetlícu mettrymnes (*languor*). Ongeán swelce metrymnesse món beþorfte strónges lǽcedómes . . . swá hé mǽge ða mettrymnesse (*morbum*) mid geflíéman, Past. 61, 2; Swt. 455, 26-30. Se ðe biscephád underféhþ hé underféhþ ðæs folces mettrymnesse *quasi ad ægrum medicus accedit,* 9 ; Swt. 59, 23. Hé gefór on ðære mettrymnesse, Ors. 6, 30; Swt. 282, 21. Ða gehældon hié sum wíf of micelre medtrumnesse, Shrn. 135, 16. Mettrumnesse, Ps. Th. 5, arg : 6, arg : 15, arg : Guthl. 20; Gdwin. 82, 13. Ða lǽcas cunnon heora medtrumnesse ongitan, Bt. 39, 9; Fox 226, 16. Mettrymnysse *infirmitates,* Ps. Spl. C. 15, 3. Metrymnisse *ægrotationes,* Mt. Kmbl. Rush. 8, 17. Wíf sceolon gemunan hyra mettrumnessa and hyra hádes tyddernessa *women must remember their infirmities and the weaknesses of their sex,* L. E. I. 6; Th. ii. 406, 12.

medu, meodn, a ; *m. : wes ; n. Mead, a drink made from honey :*—Medu *medo* vel *medus,* Wrt. Voc. i. 27, 41. Meodu *medo,* 82, 30. Medo *mulsum,* 290, 60. Medo, geswét *vel* weall *defrutum,* i. *vinum,* ii. 138, 24. Meodu, Andr. Kmbl. 3051, An. 1528. Medewes *defruti,* Hpt. Gl. 480, 74. Ða mǽla ðe wé oft æt meodo sprǽcon, Byrht. Th. 137, 66; By. 212. Tó medo, Beo. Th. 1212; B. 604. Ða wé medu þégon, 5260; B. 2633. Ða þeówan drincaþ medo, Ors. 1, 1; Swt. 20, 17. Wylle swá swýðre medo, L. M. 2, 52; Lchdm. ii. 270, 7. Gedó on ðone drenc swíðe gód medo, 2, 53; Lchdm. ii. 274, 15. Hwítne medu, Fins. Th. 78; Fin. 39. Ðær hý meodu drincaþ, Exon. 105 b; Th. 401, 16; Rä. 21, 12. Medewa, wín *defruta, decocta vina,* Hpt. Gl. 468, 38. [*Icel.* mjödr ; *m : O. H. Ger.* meto, mito *mulsum, medum : Ger.* meth : *Lithuan.* middus : *Gk.* μέθυ.]

medu-ærn, es ; *n. A house in which mead is drunk, a banqueting-house :*—Meodoærn micel, Beo. Th. 138; B. 69.

medu-benc, e ; *f. A bench in a banqueting-hall :*—Medubenc monig, Beo. Th. 1556; B. 776. On ðære medubence, 2108; B. 1052. Medobence, 4376; B. 2185. Meodobence, 3808; B. 1902. Meodubence, Exon. 87 b; Th. 330, 9; Vy. 48.

medu-burh ; *f. A city in which mead is drunk, one in which mead-drinking warriors live :*—On ðære medobyrig, Judth. 11 ; Thw. 24, 2; Jud. 167. On meoduburgum, Exon. 123 a; Th. 473, 18; Bo. 16.

medu-dreám, es ; *m. Joy attending mead-drinking, festivity :*—Ne seah ic medudreám máran, Beo. Th. 4036; B. 2016. Meododreáma, Exon. 123 b; Th. 475, 8; Bo. 44.

medu-drenc, es ; *m. Mead :*—Ðonne biþ heom heora meoodudrenc wín and beór eall tó écum þurste áwend *then shall their mead and wine and beer all be turned for them to eternal thirst,* Wulfst. 245, 4.

medu-drinc, es ; *m. Mead-drinking :*—Fore medodrince *instead of mead-drinking,* Exon. 81 b; Th. 307, 12; Seef. 22.

medu-full, es ; *n. A mead-cup :*—Meoduful, Exon. 88 a ; Th. 331, 2 ; Vy. 66. Medoful, Beo. 1253; B. 624 : 2034; B. 1015.

medu-gál ; *adj. 'Flown with wine,' excited with mead :*—Holofernus módig and medugál, Judth. 10; Thw. 21, 19; Jud. 26 : Cd. 209; Th. 260, 1 ; Dan. 703. Meodugál, Exon. 88 a; Th. 330, 16; Vy. 52. Meodugáles gedrinc, 330, 27 ; Vy. 57.

medu-heall, e ; *f. A mead-hall, banqueting-hall :*—Ðeós (*Hrothgar's*) medoheal, Beo. Th. 972; B. 484. Meodoheall, Exon. 124 a; Th. 477, 13 ; Ruin. 24. In meoduhealle, 76 b; Th. 288, 6 ; Wand. 27 : 79 a ; Th. 297, 16; Crä. 69 : 85 b; Th. 321, 33 ; Víd. 55. In medohealle, Elen. Kmbl. 2515 ; El. 1259.

meduma, meoduma, an ; *m. A weaver's beam :*—Meoduma *insubula,* Wrt. Voc. i. 66, 33 : 282, 18. (Cf. Webbeámas *insubulæ,* 59, 43.) Meodoma, ii. 46, 33.

medume, medeme, meodume ; *adj.* I. *middling, moderate, common :*—Medeme *mediocer,* Ælfc. Gr. 9, 18; Som. 9, 67. Gif hwylc man forstele deórwurþe þing . . . Gif hwylc man medeme þing (*rem mediocrem*) stele, L. Ecg. P. ii. 25 ; Th. ii. 192, 17-20. II. *occupying the middle* or *mean position as regards (a) size, amount,* etc. :—Medume leódgeld *a half fine* (cf. *medietas leudis,* and other examples, Grmm. R. A. 653), L. Ethb. 7; Th. i. 4, 9: 21 ; Th. i. 8, 3. Hé hæfþ medemne wæstm *he is of middle height,* Homl. Th. i. 456, 18. Heáfdu medumra manna *heads of average, ordinary men,* Salm. Kmbl. 525; Sal. 262. Gehwar gebürrihta sýn hefige, gehwar medeme (*moderate*), L. R. S. 4; Th. i. 434, 5. Se mǽsta segl *acateon* ; se medemesta segl *epidromas* ; se lesta *dalum,* Wrt. Voc. i. 56, 51-53. (*b) place, rank, means :*—Medemra þegna heregeata *the medial thanes' heriots,* L. C. S. 72 ; Th. i. 414, 12. Ic tǽhte ðam rícan . . . ic tǽhte ðam medeman mannum . . . Ic bebeád þearfum, Homl. Th. i. 378, 20. Heáfodmynstres griþbryce . . . medemran mynstres . . and ðonne git læssan, L. Eth. ix. 5 ; Th. i. 342, 1 : L. C. E. 3 ; Th. i. 360, 21. Ðæs medemestan lífes (*the life mid-way between the best and worst,* cf. mon forlǽt ðæt wyrreste líf and ne niæg git cuman tó ðæm betstan, 10), Past. 51, 6; Swt. 399, 15. (*c) age :*—Mínre yldstan dehter . . . ðære medeman . . . ðære gingstan, Chart. Th. 488, 28-32 : 489, 23-25. III. *observing the just mean, perfect, meet, fit, worthy :*—Hé wæs þurh eall meodum (MS. B. medeme: MS. O. medum) *erat dignus per omnia,* 8, 3 ; S. 567, 19. Meoduma, Mt. Kmbl. Rush. 10, 37. Hwelc se beón scolde ðe medome (*dignus*) hierde bión sceolde, Past. 11, 7 ; Swt. 73, 20. Medeme, Blickl. Homl. 129, 35. Hé wyrþ ǽlces cræftes medeme (*fit for, capable of*) . . . ǽlces þinges swá medeme swá hé ǽfre medemast (medomist, MS. Cott.), Bt. 38, 5 ; Fox 206, 25-29. Hwylc ðæt medeme gód wæs hwylc ðæt unmedeme *quæ sit imperfecti, quæ perfecti boni forma,* 35, 1 ; Fox 134, 4. Medeme fæsten *a proper fast,* L. E. I. 39 ; Th. ii. 436, 35. Medeme lác, Blickl. Homl. 37, 32. Ful medome wæstm, 55, 5. Drihtne tó geearnienne medome folc (' *a prepared people,'* Lk. 1, 17), 165, 15. Ne gedéþ se anweald gódne ne meodumne (MS. Cott. medomne) *power makes him neither good nor worthy,* Bt. 16, 3 ; Fox 56, 20. Góde and medeme, Blickl. Homl. 129, 23, 32. Mid medemum wæstmum hreówe *dignis pænitentia fructibus,* Bd. 4, 27 ; S. 604, 24 : Mt. Kmbl. 3, 8. Medeme þinc *res dignas,* Kent. Gl. 396. Drihten ðú ðe eall medemne geworhtest and náht unmedemes, Shrn. 165, 31. Ne mágon wé nánwuht findan betere (MS. Cott. medemre) ðonne God, Bt. 34, 4 ; Fox 138, 26. Nis meodumre ne mára ðonne *it is not too good nor too great for,* Exon. 38 a ; Th. 125, 16 ; Gú. 355. Ðæt medemæste *the best,* Bt. 24, 4 ; Fox 86, 10. Ða medumestan ealdras *exspectabiles senatores,* Wrt. Voc. ii. 145, 51. [*O. H. Ger.* metam, metem.] v. un-medume.

medumian, medemian, medmian ; *p.* ode. I. *to fix the measure of anything :*—Dóm æfter dǽde medemige man be mǽðe *according to the deed let the measure of doom be fixed in proportion,* L. Eth. vi. 10 ; Th. i. 318, 6 : vi. 53 ; Th. i. 328, 17. Man sceal medmian and gescádlíce tóscádan ylde and geógoþe *youth and age must have their proper place assigned them, and be discreetly distinguished,* vii. 52 ; Th. i. 328, 18. Medmian (medemian), L. C. S. 69 ; Th. i. 412, 8. II. *to deem worthy* (v. medume, III.), *respect, esteem :*—Ic gemedemige (*other MSS.* medemige) ðé tó ðam þinge *dignor te illa re,* and medemigende ðé tó ðam þinge *dignans te illa re,* Ælfc. Gr. 41; Zup. 250, 9-10. Weofoðþéna mǽðe medemige man, L. Eth. ix. 18; Th. i. 344, 9. [*O. H. Ger.* metamón *temperare, moderare, dimidiare.*] v. ge-medemian.

medum-líc ; *adj.* I. *middling, moderate, small :*—Gehwǽdum † medemlícum *mediocri,* Hpt. Gl. 505, 55. Hé hæfþ medemlíce nosu (cf. medmicle neosu þynne *naso pertenui,* Bd. 2, 16; S. 519, 34) *he has a slender nose,* Homl. Th. i. 456, 18. II. *worthy, honourable :*—Medomlícan *dignitosam,* Wrt. Voc. ii. 28, 64. Medomlíce *dignitosa,* 106, 55 : 140, 27.

medum-líce ; *adv.* I. *moderately, in a small degree, imperfectly :*—Medomlíce *mediocriter,* Wrt. Voc. ii. 140, 27. Wé cunnon ðære leóde gereord, ná medemlíce ac fulfremedlíce, Homl. Th. ii. 474, 3. II. *worthily, fitly, kindly* (cf. mǽþ-líce, medum-ness):—Hí ne

mágon medomlíce (Cott. MSS. medumlíce) þénian *ministrare digne nequeunt*, Past. 1, 2 ; Swt. 27, 10. Suíðe medomlíce Iacobus his stírde *hinc pie Iacobus prohibet*, 3, 1 ; Swt. 33, 9. Medomlíce *benigniter*, Wrt. Voc. ii. 11, 3. Meodomlíce *digne*, Rtl. 2, 41.

medumlíc-ness, e ; *f. Smallness :*—Gehwǽdnys ł medemidlícnys (medemlícnys?) *mediocritas, parvitas*, Hpt. Gl. 467, 14.

medum-ness, e : *f.* **I.** *worth, dignity :*—Medumnes (Cott. MSS. medomnes) *dignitas*, Bt. 16, 3 ; Fox 56, 25. Nán man for his ríce ne cymþ tó cræftum and tó medemnesse ac for his cræftum and for his medumnesse hé cymþ tó ríce *non virtutibus ex dignitate, sed ex virtute dignitatibus honor accedat*, 16, 1 ; Fox 50, 20–22. Gé underþíoðaþ eówre héhstan medemnesse under ða eallra nyðemestan gesceafta *vos dignitatem vestram infra infima quæque detruditis*, 14, 2 ; Fox 44, 34. Ðæt gé nǽfre swá heálíce medumnesse (*the priestly office*) ne forwyrcen, L. E. I. 1 ; Th. i. 402, 27. Ealdordómes medomnysse, Shrn. 151, 19. **II.** *kindness, condescension, appreciation of worth in others* (cf. mǽþ, V) :—Medemnysse ðínre *benignitatis tuæ*, Blickl. Gl. ; Ps. Spl. 64, 12. Medumnysse *benignitatem*, 51, 3 : Blickl. Homl. 145, 33. Cf. medumlíce, II.

medumung, e ; *f.* **I.** *the fixing of the measure of anything :*—Á sceal dóm æfter dǽde and medemung be mǽðe *ever shall doom be according to deed, and fine be fixed with fair measure*, L. Eth. ix. 5 ; Th. i. 342, 5 : L. E. B. 10 ; Th. ii. 242, 11. **II.?** :—Ðonon á be ecge on ða medemuncga (medemunga) ; of ðære medemuncge (mǽdemunge) on ðone ealdan widig, Cod. Dip. Kmbl. iii. 25, 21–23 : v. 286, 31–33. [*O. H. Ger.* metemunga *temperies, temperamentum.*] v. medumian.

medu-rǽden[n], e ; *f. Strong drinks, cellar* (in the sense of the liquors contained in it) :—Rúmheort beón meodorǽdenne *liberal with liquors*, Exon. 90 a ; Th. 339, 3 ; Gn. Ex. 88.

medu-scenc, es ; *m. A draught* or *cup of mead :*—Meoduscencum hwearf geond ðæt healreced (cf. Ymbeode ides Helminga óððæt heó Beówulfe medoful æt bær, 1244–), Beo. Th. 3965 ; B. 1980.

medu-seld, es ; *n. Mead-house, house in which feasting takes place*, Beo. Th. 6123 ; B. 3065.

medu-setl, es ; *n. A mead-seat, a seat in a banqueting-hall*, Beo. Th. 10 ; B. 5.

medu-stíg, e ; *f. Path to the mead-hall :*—Cyning of brýdbúre treddode ... and his cwén mid him medostíg gemæt ... Hróðgár tó healle geóng, Beo. Th. 1845–1855 ; B. 920–925.

medu-wǽge, an: -wǽg, e ; *f. The Medway :*—Sint dæs londes gemǽra : an westhealfæ Scipflíót, an norþhalfe Meodowǽge, Cod. Dip. Kmbl. ii. 71, 25. Miodowǽge, iii. 400, 26. Partem fluminis Meduwaeian, i. 135, 34. Andlang Medwǽge, 283, 4. Andlang Medwǽgan, Chr. 999 ; Erl. 134, 24. In tó Medewǽge, 1016 ; Erl. 157, 4. Óþ mediwǽgan sindan ða gemǽra. Fram Miadawegan, Cod. Dip. Kmbl. ii. 86, 24. Óþ Miodowegan, 17. In flumen Medewiæge, iii. 386, 26. Óþ ða eá Medewegan, 400, 31.

medu-wang, es ; *m. A mead-plain, the ground surrounding the house where mead is drunk :*—Tó sele comon feówertýne Geáta gongan, módig (*Beowulf*) on gemonge meodowongas træd. Ða com ingán ealdor þegna, Beo. Th. 3291 ; B. 1643.

medu-wérig ; *adj. Sated with feasting*, Judth. 11 ; Thw. 24, 38 ; Jud. 229 : 12 ; Thw. 25, 6 ; Jud. 245.

medu-wyrt, e ; *f. Meadow-sweet, also mead-sweet :*—Meodowyrt *melleuna*, Wrt. Voc. ii. 59, 43 : L. M. 1, 38 ; Lchdm. ii. 94, 14. Medowyrt, Lchdm. ii. 96, 17 : i. 44 ; Lchdm. ii. 108, 11. Medewyrt *malletina* (?), Wrt. Voc. i. 31, 1 : Lchdm. iii. 6, 12 : 16, 9. Meodeuyrt *mellauna*, papamo, 304, 1, 35. [*Scott.* med-uart : *Dan.* mjöd-urt.]

med-wís ; *adj. Not wise, dull, foolish :*—Ða medwísan *hebetes*, Past. 30, 1 ; Swt. 203, 6, 15, 21 ; 205, 2, 4, 17. Sume wísran sume medwísran *quosdam sapientes, quosdam tardiores*, 30, 2 ; Swt. 205, 7. Medwísum men, Exon. 102 b ; Th. 387, 24 ; Rä. 5, 10.

még, megen, megende. v. mǽg, mǽw, mǽgen, magan.

meh, meht. v. mé, meaht.

méi, meig. v. mǽg.

mela. v. melu.

melc, meolc ; *adj. Giving milk, milch :*—Melc *foetus*, Wrt. Voc. i. 287, 57 : *fetus*, ii. 36, 33. Melce and tydrende *foetus*, 36, 32. Hé geseah wilde hinde melce and se geþyrsta mon meolcode ða hinde, Shrn. 130, 3. Wið tittia sár wífa ðe beóþ melce, Herb. 19, 4 ; Lchdm. i. 112, 16. Meolce breóst *ubera*, Wrt. Voc. i. 44, 14. [*Icel.* mjólkr *giving milk : O. H. Ger.* melch *foetus : Ger.* melk.]

melcan ; *p.* mealc, *pl.* mulcon ; *pp.* molcen *To milk :*—Ic melce *mulgeo*, Ælfc. Gr. 26, 3 ; Som. 28, 55. Melke, Coll. Monast. Th. 20, 17. Se ðe melcþ *qui emulget*, Kent. Gl. 1121. Milciþ *morgit* (?), Ep. Gl. 14 f, 16. Milcet, Wrt. Voc. ii. 55, 73. Milcit, 114, 17 : *mulgit*, Wülck. 33, 26. Hé éwa mealc, Shrn. 61, 19. Ðæt fæt ðe ðú wille on melcan, L. M. 1, 67 ; Lchdm. ii. 142, 9. Nige molcen, 2, 27 ; Lchdm. ii. 222, 13 : 2, 25 ; Lchdm. ii. 218, 22. [*O. H. Ger.* melchan.] v. meolcian.

melcing-fæt, es ; *n. A milk-pail :*—Melcingfata *mulctra*, Germ. 390, 66. v. meolc-fæt.

---

meld, e ; *f.* [*O. H. Ger.* melda ; *f. delatura, delatio, proditio*] *Declaration, proclamation :*—Hé wíde beád Metodes mihte ðǽr hé meld áhte *he declared the Lord's power widely, where he could proclaim it*, Cd. 208 ; Th. 256, 30 ; Dan. 648.

melda, an ; *m.* **I.** *a narrator, an informer, announcer :*—Ðæs ðe ic ǽfre on ealdre ǽngum ne wolde monna ofer moldan melda weorþan *what I would never relate to any man upon earth*, Exon. 50 b ; Th. 176, 3 ; Gú. 1203 : 73 b ; Th. 275, 28 ; Jul. 557. Sió æsc biþ melda, nalles þeóf *the axe is an informer, not a thief* (i. e. the noise made by hewing with an axe would attract the attention, which a thief would certainly shun, v. Grmm. R. A. 47), L. In. 43 ; Th. i. 128, 23 : L. Edg. H. 8 ; Th. i. 260, 17. Þurh ðæs meldan hond ; se sceolde wong wísian, Beo. Th. 4802 ; B. 2405. Ic tó melda wearþ *I turned informer* (cf. Th. 259, 28 sqq., 270, 10 for the narrative forced from the devil by Juliana : cf. also Jul. pp. 39 sqq.), Exon. 74 b ; Th. 279, 30 ; Jul. 621. Ðæt wé ðæs morþres meldan ne weorþen *that we be not informers of the crime*, Elen. Kmbl. 856 ; El. 428. **II.** *a betrayer :*—Gé sind meldan and manslagan (*betrayers and murderers*, Acts vii. 52), Homl. Th. i. 46, 24. [Cf. *O. L. Ger.* meldari *sponsor : O. H. Ger.* meldari *delator, proditor.*]

meldan ; *p.* ede *To announce, declare :*—Ús frunon fǽcnum wordum meldedan *they questioned us, with crafty words declared*, Ps. Th. 136, 3. Ic ne mæg word sprecan, mældan for monnum, Exon. 105 a ; Th. 399, 18 ; Rä. 19, 2. Meldan, 109 b ; Th. 411, 13 ; Rä. 29, 12. v. tó-meldan, meldian.

melde, an ; *f. Orach*, a plant-name :—Melde, Lchdm. iii. 6, 11. Nim meldon ða wyrt, 54, 23. [*Dan.* meld : *O. H. Ger.* malta *beta* ; melda *atriplex : Ger.* melde.] v. tún-melde.

meld-feoh, gen. -feós ; *n. Fee paid for giving information :*—Se ðe hit (forstolen flǽsc) ofspyraþ, hé áh ðæt meldfeoh, L. In. 17 ; Th. i. 114, 4. v. Grmm. R. A. 656.

meldian ; *p.* ode, ede. **I.** *to declare, announce, tell :*—Múþ habbaþ and ne meldiaþ wiht *os habent, et non loquentur*, Ps. Th. 134, 16. Hí sprecaþ unnyt sæcgeaþ and wóh meldiaþ *pronuntiabunt et loquentur iniquitatem*, 93, 4. Ælfréd cræft meldode *Alfred displayed his art*, Bt. Met. Fox Introd. 4 ; Met. Einl. 2. Ic sceal mód meldian swá ðú mé beódest *I must tell all my mind, as thou dost bid me*, Exon. 72 b ; Th. 270, 10 ; Jul. 463. Ongan meldigan ðone hálgan wer *the devil began to tell who the holy man was,*. Andr. Kmbl. 2341 ; An. 1172. Ðá geneálǽhton má hine meldigende (*declaring that Peter was with Jesus*), Homl. Th. ii. 248, 32. **II.** *to inform against, accuse :*—Oft mec ísern scód sáre on sídan, ic swígade, nǽfre meldade monna ǽngum (*never accused any man* (?) *or told no man*), Exon. 126 a ; Th. 485, 17 ; Rä. 71, 15. Meldadun *vel* wroegdun *defferuntur*, Wrt. Voc. ii. 106, 17. Meldedun, 25, 26. *Desequunt vel* meldadan i. *accusabant*, 139, 15. Hé nolde meldian on his geféran ðe mid him sieredon *he would not inform against his companions who had plotted with him*, Bt. 16, 2 ; Fox 52, 20. [*O. Sax.* meldón *to declare, betray, proclaim : O. H. Ger.* meldén, meldón *prodere, deferre, producere : Ger.* melden.] v. ge-meldian, meldan.

meldung, e ; *f. Information* (against a person), *betrayal :*—Hé swýðe mánfullíce ácweald wæs þurh meldunga his ágenes wífes *multum nefarie peremptus est proditione conjugis suæ*, Bd. 3, 24 ; S. 557, 39. [*O. H. Ger.* meldunga *proditio, delatura : Ger.* meldung.]

méle, mæle, es ; *m. A cup, bowl, basin :*—Meeli *aluium*, Ep. Gl. 26, 38 : Wrt. Voc. ii. 99, 72. Méli *avum* (= *alvium?*), 101, 31. Méle *albium*, 8, 27 : i. 285, 9 : *patera*, 24, 39. Mélas *karchesia*, 24, 42 : *ciatos*, ii. 22, 44. Dó méle fulne buteran on, L. M. 1, 36 ; Lchdm. ii. 86, 17. [Halliw. Dict. meles and payles.] v. wæter-méle (-mǽle).

mele-, mil-deáw, es ; *n. m. Honey-dew, nectar :*—Hunig [deáw] oððe mildeáw *nectar*, Wrt. Voc. ii. 61, 38. Nó hé fóddor þigeþ nete on moldan nemne meledeáwes dǽl gebyrge se dreóreþ oft æt middre nihte *non illi cibus est nostro concessus in orbe, ambrosios libat cælesti nectare rores, stellifero teneri qui cecidere polo*, Exon. 59 b ; Th 215, 29 ; Ph. 260. [Swetter is munegunge of þe þen mildeu o muðe, *O. E. Homl.* i. 269, 5. In *Prompt. Parv.* and *Wick.* the word has the modern sense *blight, uredo, aurugo*; so *O. H. Ger.* mili-tou : *M. H. Ger.* mili-tou : *Ger.* mehl-thau. The first part of the word seems to mean *honey*, cf. milisc and *Goth.* miliþ *honey*. Grmm. D. M. p. 607, gives another etymology, connecting it with *Icel.* mél *bit* (*of a bridle*), the dew being the foam which fell from the bit of the horse Hrímfaxi.]

melsc. v. milisc.

meltan ; *p.* mealt, *pl.* multon ; *pp.* molten. **I.** *to melt, become liquid, be consumed, dissolved :*—Ic mylte *liqueo*, Ælfc. Gr. 35 ; Som. 38, 7. Mylt *dissolvitur*, Wrt. Voc. ii. 147, 25. Swá weax melteþ, Ps. Th. 57, 7. Mylteþ, 67, 2. His sylfes hám brynewylmum mealt (*was consumed*), Beo. Th. 4642 ; B. 2326. Multon meretorras (*when the waters of the Red Sea fell upon the Egyptians*), Cd. 167 ; Th. 208, 16 ; Exod. 484. Ðonne mé mægen mylte *dum defecerit virtus mea*, Ps. Th. 70, 8. Ne sceal ánes hwæt meltan (*be consumed on the pile*), Beo. Th. 6014 ; B. 3011. Weax miltende *cera liquescens*, Ps. Spl. 21, 13. Myltende *liquidas*, Hpt. Gl. 470, 73. **II.** of food, *to digest :*—Late mylt gǽten flǽsc *goat's flesh digests slowly*, L. M. 2, 16 ; Lchdm. ii. 196, 16, 25. Ða scearpan þing unyþelíce meltaþ, 2, 23 ; Lchdm. ii.

212, 2. Wið ðon ðe men mete untela melte, 2, 29; Lchm. ii. 226, 5. Ða ðe on ðære uferan wambe gewuniaþ and ne mágon meltan, 1, 2; Lchdm. ii. 26, 17. Myltan, 2, 27; Lchdm. ii. 222, 18. Wel meltende mettas, 2, 16; Lchdm. ii. 196, 21. v. for-, ge-meltan; miltan.

meltung, e; *f. Melting* (of food), *digestion:*—Ðara metta meltung, L. M. 2, 17; Lchdm. ii. 198, 3. Hió næfþ góde meltunge *it* (*the stomach of a watery nature*) *hath not good digestion,* 2, 27; Lchdm. ii. 220, 27. v. un-meltung.

melu, melo, mela, meolu, mealu, wes; *n. Meal, flour:*—Melu oððe offrung *ador,* Ælfc. Gr. 9, 21; Som. 10, 32: *farina,* Wrt. Voc. i. 83, 17: ii. 38, 70. Swá swá mon melo (Cott. MS. meolo) sift, ðæt melo (meolo) þurhcrýpþ ælc þyrel, Bt. 34, 11; Fox 152, 2. Ðæt mela biþ gód, L. M. 1, 38; Lchdm. ii. 94, 2. Genim hwætenes meluwes smedman, L. M. 1, 61; Lchdm. ii. 134, 4. Melwes (Lind. mælo) *farinæ,* Mt. Kmbl. 13, 33. Melues *similæ,* Lev. 6, 20. Melewes smedma *simila,* 83, 65. Melewes *polline,* mealewes *farinæ,* Hpt. Gl. 497, 36, 37. Ðrittig mittan clǽnes melowes (*fine flour*) and sixtig mittan óðres melowes, Homl. Th. ii. 576, 32. Meolwes, Chart. Th. 40, 10. *Pollis* smedma, *pollinis* of melowe, Ælfc. Gr. 9, 28; Som. 11, 48. Windlas mid meluwe *canistra farinæ,* Gen. 40, 16. Of rigenum melwe, L. M. 2, 32; Lchdm. ii. 236, 9. Genim beren mela gód, L. M. 1, 5; Lchdm. ii. 50, 3. Beren meala, Lchdm. iii. 8, 15. [*Icel.* mjöl: *O. H. Ger.* melo *farina, polenta, pulvis: Ger.* mehl.] v. ed-melu.

melu-gescot, es; *n. A contribution* or *payment made in meal:*—Hwílum weaxgescot, hwílum mealtgescot, hwílum melagescot, Wulfst. 171, 2 note. [Cf. *Icel.* mjöl-skuld *rent to be paid in meal.*]

melu-hús, es; *n. A house in which to keep meal:*—Mealehús *farinale,* Wrt. Voc. i. 58, 41.

men *in* nim sealtes, þrý men *take of salt three parts,* L. M. 1, 50; Lchdm. ii. 124, 4. [*Cockayne compares the word with Swedish* mån *apart.*]

mend-líc (?); *adj. Moderate, small:*—Tó medmyclum (MS. C. mendlícum) fæce *ad modicum,* Bd. 2, 13; S. 516, 21.

mene, myne, es; *m. A necklace, an ornament:*—Maenoe *crepundia,* Wrt. Voc. ii. 105, 44. Mene *lunules,* 71, 1. Myne *crepundium* i. *monile gutturis,* 136, 68. Myne *vel* sweorbéh *monile vel serpentinum,* i. 40, 50: 74, 58. Ðes myne *hoc monile,* Ælfc. Gr. 9, 2; Som. 8, 28. Brósinga mene, Beo. Th. 2403; B. 1199. (v. Grmm. D. M. 283.) Menas *monilia,* Wrt. Voc. i. 16, 60: *crepundia, ornamenta, monilia,* Hpt. Gl. 419, 30: 517, 29. Mynas, 481, 43: *lunulas,* 458, 30. Menum *monilibus,* 434, 71. Mynum *lunulis,* Wrt. Voc. ii. 49, 71. [*O. Sax.* hals-meni: *Icel.* men; *n. a necklace: O. H. Ger.* menni; *pl. monilia.*] v. heals-mene.

menen, mennen, minnen, es; *n. A female servant, bondwoman, hand-maid:*—Án menen ł þeówæ *ancilla,* Mt. Kmbl. Rush. 26, 69: *vernacula,* Wrt. Voc. ii. 123, 37. Mennen *ancilla,* 2, 39. Sunu menenes ðínes *filius ancillæ tuæ,* Ps. Surt. 115, 16: 122, 2. Minenes, p. 200, 6. Be ceorles mennenes niédhǽmede. Gif mon ceorles mennen tó nédhǽmde geþreáteþ, L. Alf. pol. 25; Th. i. 78. 11–12: Cd. 103; Th. 136, 14; Gen. 2258: 97; Th. 128, 13; Gen. 2126. Deáh hwá bebycgge his dóhtor on þeówenne ne síe hió ealles swá þeówu swá óðru mennen is, L. Alf. 12; Th. i. 46, 13. CCL ðara monna, esna and mennena (*servos et ancillas*), Bd. 4, 13; S. 583, 20. [Cf. *Icel.* man; *n. a bondman* or *bondwoman: O. H. Ger.* mana-houpit = *a servant;* v. Grmm. R. A. 301.] v. drunc-, mere-, þeów-menen (-mennen).

menen-líc (= ?), myniend-líc *hortandus, ammonendus,* Hpt. Gl. 485, 64.

mene-scilling, es; *m. A coin worn as an ornament:*—Menescillingas *lunules,* Ep. Gl. 13 b, 37: Wrt. Voc. ii. 113, 15. Mynescillingas, 49, 72.

mengan, mængan, mencgan; *p. de.* **I.** *to mix, mingle, combine:*—Ic menge *mango* (?), Wrt. Voc. ii. 58, 42. Mengio, 113, 59: Epl. Gl. 156, 36. Mænge *margo* (*mango?*), Wrt. Voc. ii. 58, 48. Menget *confundit,* 105, 11. Ic mǿnne drinc mengde wið teárum *potum meum cum fletu temperabam,* Ps. Th. 101, 7. Ðú wið fýre foldan mengdest, Bt. Met. Fox 20, 223; Met. 20, 112. Ðara blód Pilatus mengde (*miscuit*) mid hyra offrungum, Lk. Skt. 13, 1. Ðonne wé medelcwidas mengdon *when we conversed,* Salm. Kmbl. 865; Sal. 432. Hí hí wið mánfullum megndan þeóde *commisti sunt inter gentes,* Ps. Th. 105, 26. Hí mínne mete mengde wið geallan, 68, 22. Meng ða blisse wið ða unrótnesse, Prov. Kmbl. 71. Fífleáfon seáw mencg (mængc, MS. B) tó wíne, Herb. 3, 6; Lchdm. i. 88, 12. Menge mon wið áseowen hunig, L. M. 2, 26; Lchdm. ii. 220, 10. Nánne wǽtan hí ne cúþon wið hunige mengan, Bt. 15; Fox 48, 10. Mængan, Bt. Met. Fox 8, 48; Met. 8, 22. Mengan lyge wið sóðe, Elen. Kmbl. 612; El. 306. *Of sexual intercourse:*—Is eác bewered ðæt mon hine menge wið his bróðor wífe *cum cognata misceat prohibitum est,* Bd. 1, 27; S. 491, 16, 10. **II.** *intrans:*—Hú ceald hwílum mencgaþ, Cd. 216; Th. 273, 6; Sat. 132. **III.** *to mingle together, stir up, disturb:*—Mengan merestreámas, Exon. 123 b; Th. 475, 3; Bo. 42. Meregrundas mengan, Beo. Th. 2903; B. 1449. [Cf. his mod him gon mengen, Laym. 3407: wraþþe meinþ þe heorte blod, O. and N. 945. *Prompt. Parv.* mengyn *misceo: O. Sax. O. L. Ger.* mengian: *O. Frs.* mengia: *O. H. Ger.* chi-menghid; *pp.: Ger.* mengen.] v. ge-, geond-mengan.

mengung, mencgung, e; *f. Mixture, preparation, composition:*—Mencingc *confectio,* Hpt. Gl. 250, 30. [*Prompt. Parv.* mengynge *mixtura, commixtio.*] v. ge-mengung.

menian, menig. v. mynian, manig.

menigdu; *f. A multitude, a body of people:*—Menigdu *manum,* Wrt. Voc. ii. 58, 26. [*O. H. Ger.* managoti; *f. manus.*]

menigu, mengu, menigeo; *indecl.: also gen.* e; *f. A many, multitude, crowd, great number:*—Seó menigu ðara freónda, Bt. 29, 2; Fox 106, 6. Menigo, Andr. Kmbl. 898; An. 449. Menego, Cd. 214; Th. 270, 1; Sat. 83. Menigeo (MS. A. mænigeo) *turba,* Mk. Skt. 2, 13. Mænigeo (MS. A. mænio), Mt. Kmbl. 9, 8. Mænegeo, Cd. 121; Th. 156, 14; Gen. 2588. Mengu, Elen. Kmbl. 450; El. 225. Mengeo, Cd. 80; Th. 100, 13; Gen. 1663. Mengio, Bt. 14, 1; Fox 42, 20. Menio, Cd. 223; Th. 294, 25; Sat. 476. Mænieo, 173; Th. 216, 12; Dan. 5. Ðære menigo þeáw, Andr. Kmbl. 354; An. 177. Menego, Cd. 220; Th. 284, 14; Sat. 321. On menigeo *in multitudine,* Ps. Th. 65, 2. Mænigeo, 68, 13. Mid manigeo, Rood Kmbl. 300; Kr. 151. From mengu *a multitudine,* Ps. Surt. 63, 3: Exon. 66 b; Th. 245, 16; Jul. 45. Mid mengo, Elen. Kmbl. 754; El. 377. For ðære meniu, Gen. 16, 10. For ðære miclan menige, Ors. 3, 9; Swt. 124, 36. Of menge wetra *de multitudine aquarum,* Ps. Surt. 17, 17. For ðære mænige, Rood Kmbl. 221; Kr. 112: Kr. 112. Mt. Fox 26, 121; Met. 26, 61. Ic álýse ealle ða menigo, Andr. Kmbl. 201; An. 101. Menigeo (MS. A. mænio) *turbam,* Mt. Kmbl. 9, 25. Mænegu (Rush. mengu), 15, 33. Mænego, Cd. 91; Th. 116, 7; Gen. 1932. Manegu, Hy. 10, 8; Hy. Grn. ii. 293, 8. Mengu *multitudinem,* Ps. Surt. 9, 25. Mengo, Exon. 128 b; Th. 493, 12; Rä. 81, 29. Mengeo, Cd. 83; Th. 103, 30; Gen. 1726. Meniu *exercitum,* Wrt. Voc. ii. 106, 46. God ða miclan Pharones menge gelytlode, Ors. 1, 7; Swt. 38, 27: Cd. 56; Th. 69, 8; Gen. 1132. Cômon menigu (MS. A. mænigu) *conveniunt turbæ,* Mk. Skt. 10, 1. Ða menigeo (MS. A. mænio: B. mænigeo: Rush. menigu: Lind. menigo) *turbæ,* Mt. Kmbl. 12, 23. Forlǽt ðás mænegeo (MS. A. mænygeo: B. mænegu: Rush. mengu) *demitte turbas,* 14, 15. Ða eargan mengo *fugaces turmas,* Wrt. Voc. ii. 151, 48. [*Goth.* managei: *O. Sax. O. L. Ger.* menegî, menigî: *O. Frs.* menî: *O. H. Ger.* managî, menigî *multitudo, turba, legio, caterva: Ger.* menge.] v. mann-menigu.

menisc, men-lufigende, mennen. v. mennisc, menn-lufigende, menen.

mennisc; *adj. Human:*—Nán mennisc man *no human being,* Bt. 33, 2; Fox 122, 15. Ne gegrípe eów nǽfre nán costung búton menniscu *tentatio vos non apprehendat, nisi humana,* Past. 11, 5; Swt. 71, 12. Þus mǽrsode se mennisca Crist his heofenlícan Fæder, Homl. Th. ii. 362, 11. Ðá getreówde hé in godcundre fultom ðǽr se mennesca wan wæs, Bd. 2, 7; S. 509, 23. Anginn mennisce álýsednysse . . . intinga mennisces forwyrdes, Homl. Th. i. 194, 27–30. Mennisce handa hit ne mihton tówurpan, Homl. Th. ii. 510, 13. Hæleþa forlor, menniscra morþ, Cd. 33; Th. 45, 5; Gen. 722. [*Goth.* mannisks: *O. Sax.* mennisk, mannisk: *O. Frs.* mannisk: *Icel.* mennskr: *O. H. Ger.* mennisc.]

mennisc, es; *n. Men, people:*—Ðis is ðæt mennisc ðe ealle míne dǽda mid heora wordum onwendan, Blickl. Homl. 175, 24. Ðonne eówre wǽrgaþ mennisc *when men curse you,* Mt. Kmbl. Rush. 5, 11. Gif ðær óðer mennisc borh síe *if other people be surety,* L. Alf. pol. 1; Th. i. 60, 19. Ðá wearþ micel mennisc geweaxen *then men began to multiply,* Homl. Th. i. 20, 21. Ðǽr wæs mycel mennisc tóweard *there was a great multitude of people coming,* 182, 5. Deáh eal mennisc wǽre gegaderod *though all men were gathered together,* 26, 26. Ðære þeóde mennisc swá wlitig wǽre *the men of that nation were so beautiful,* in 120, 22. Ðæt ðú ne nyme wíf mínum suna of ðisum menisce (*de filiabus Chananæorum,* Gen. 24, 3. Josue ofslóh eall ðæt mennisc ðe on muntum wunode (*omnem terram montanam,* Jos. 10, 40: Thw. 161, 37. Á-cwealde ðæt earme mennisc, Homl. Th. ii. 474, 7. [Cf. *O. H. Ger.* mannisco, mennisco *homo: Ger.* mensch.]

mennisc-líc; *adj. Human:*—Mennisclíc *humanus,* Ælfc. Gr. 38; Som. 41, 42. Mennisclíc (*humanum*) is ðæt mon on his móde costunga þrowige, Past. 11, 5; Swt. 71, 13. [*O. H. Ger.* manisc-, menisc-, mennisc-líh *humanus: Ger.* mensch-lich.]

mennisc-líce; *adv. Humanly, after the manner of men;* humaniter, humanitus, Ælfc. Gr. 38; Som. 41, 43: 42, 6.

menniscness, e; *f.* **I.** *humanity, human nature* (generally in reference to Christ), *incarnation:*—Crist becom on hire innoþ and þurh hí on menniscnysse wearþ ácenned (*was born a man*), Homl. Th. i. 194, 8. Ne wearþ se Fæder mid menniscnysse befangen, 284, 23. Wé wurþiaþ úres Hǽlendes ácennednysse æfter ðære menniscnysse. Hé wæs ácenned mid líchaman and mid sáwle, se ðe wæs ǽfre mid ðam Fæder wunigende on ðære godcundnysse, i. 4, 20. Úre Hǽlend Crist underfêng menniscnysse, 600, 6. Fram Drihtnes menniscnysse *ab incarnatione Domini,* Bd. 1, 5; S. 476, 5. Æfter ðære drihtenlícan menniscnysse, 1, 6; S. 476, 16. **II.** *humaneness, humane behaviour:*—Hí syndon fremfulle (*benigni*) menn, and gyf hwylc mann tó him cymeþ ðonne gyfaþ hí him wíf ǽr hí hine on weg lǽtan. Se Macedonisca Alexander ðá ðá hé him tó com ðá wæs hé wundriende hyra menniscnysse (*miratus*

est eorum humanitatem), Nar. 38, 25. [*O. H. Ger.* mannisc-nissa; *and*
cf. mennisg-heit *humanitas, incarnatio.*]

menniscu, e; *f. Humanity, state of man :*—Hē forleás his mennisce
ut homo esse perderet, Past. 4, 2; Swt. 39, 24. [*Mid. E.* menske *honour :
O. Sax.* menniskî *humanitas : Icel.* menska : *O. H. Ger.* mennisgî.]

mentel, es; *m. A mantle, cloak :*—Mentel colobium, Wrt. Voc. ii. 134,
38. Hē forcearf his mentles ǽnne lǽppan *oram chlamydis ejus abscidit*,
Past. 28, 6; Swt. 197, 21. Mid twyfealdum mentle *diploide*, Ps. Spl.
108, 28. Hyre beteran mentel, Chart. Th. 537, 32. [*Lat.* mantellum :
Icel. möttull : *O. H. Ger.* mantel, mandal *chlamys, pallium.*]

mentel-preón, es; *m. A mantle-pin, brooch :*—Hió becwiþ hyre
mentelpreón, Charl. Th. 533, 33.

meó; *gen.* meón *A shoe or sock covering the foot :*—Meó *pedula*, Wrt.
Voc. i. 26, 2. Meón *pedulos* (cf. Wülck. 601, 19-21 '*pedules*, pars cali-
garum que pedem capit, *a vampey : pedulus* a pynson, or a sok '), 82, 1 :
calsus (cf. *Fr.* chausser : *Span.* calzar *to put on shoes*), ii. 127, 71.

meodu-, meocs, meohs, meolc ; *adj.* v. medu-, meox, melc.

meolc, meoluc, milc, e; *f. Milk :*—Ðeós meolc *hoc lac*, Ælfc. Gr. 9,
76; Som. 14, 21 : Wrt. Voc. i. 283, 31. Sûr meolc *oxygala, acidum
lac :* þicce meolc *colustrum*, 28, 2-3. Âwilled meolc *juta*, 290, 45. Hē
(*the Pater Noster*) biþ sâwle hunig and môdes meolc, Salm. Kmbl. 135;
Sal. 67. Meoluc, Wrt. Voc. i. 65, 9 : Ps. Th. 118, 70. Of ðam lande
ðe weóll meolce and hunie . . . ðe flêwþ on riðum meolce and hunies,
Num. 16, 13-14. Mid þynre meolce *with skim milk*, Bd. 3, 27; S.
559, 35. Mid lytle meolc (MS. B. meoloce) wætere gemengedre *cum
parvo lacte aqua mixto*, 3, 23; S. 554, 33. Ðe flêwþ meolece and
hunie, Ex. 3, 8. Abraham nam meoloc, Gen. 18, 8. Meoluc, Deut.
32, 14. Dô on þeorfe meoluc *put into skim milk*, L. M. 2, 52; Lchdm.
ii. 272, 1. Ða rícostan men drincaþ myran meolc, Ors. 1, 1; Swt. 20,
17. Is ðæt eálond welig on meolcum *dives lactis insula* (*Hibernia*),
Bd. 1, 1; S. 474, 40. Wyl on meolcum *boil in milk*, L. M. 2, 65;
Lchdm. ii. 296, 19. Mid cû. meolcum, 2, 25; Lchdm. ii. 218, 22.
From milcum âððen *ablactatus*, Blickl. Gl. [*Goth.* miluks : *O. Frs.*
melok : *Icel.* mjólkr : *O. H. Ger.* miluh.] v. frum-meolc.

meolc-fæt, es; *n. A vessel for holding milk, a milk-pail :*—Meolcfæt
mulctrale vel *sinum* vel *mulctrum*, Wrt. Voc. i. 25, 13. [*O. H. Ger.*
melich-faz *multra*.] v. melcing-fæt.

meolc-hwít; *adj. Milk-white :*—Of meolchwýttre *lacteo*, Germ. 389,
70. Meolchwítum *lacteis*, 397, 32.

meolcian; *p.* ode. I. *to milk, take milk from an animal :*—Se
geþyrsta mon meolcode ða hinde and dranc ða meolc, Shrn. 130, 4.
Nân wîf hire yrfe ne meolcige, bûtan heó ða meolc for Godes lufan
syllan, Wulfst. 227, 10. Hyt biþ gôd ceáp tô milcian, Lchdm. iii. 178,
30. II. *to give milk, to suckle* (v. ge-milcian) :—Ða breóst ða ðe
nǽfre meolcgende nǽron, Blickl. Homl. 93, 32. [*Icel.* mjólka *to milk ;
also to give milk.*] v. melcan.

meolc-sûcend, es; *m. A suckling :*—Meolocsûcendra *lactantium*,
Wrt. Voc. ii. 51, 71. Meolcsucgendra, 73, 9.

meolc-teónd, es; *m. A suckling :*—Of mûðe cilda and milcdeóndra
ex ore infantium et lactentium, Ps. Surt. 8, 3.

meolu, meoluc. v. melu, meolc.

meord, meorð, meard, e; *f. Reward, pay :*—Byþ ðē meorð wið God,
Andr. Kmbl. 550; An. 275. Meard *premium*, Rtl. 165, 5. Leán ł
meard (mearða, *pl.* Lind.) *merces*, Mt. Kmbl. Rush. 5, 12. Leán ł
mearde *mercedem*, 6, 2; (meard, Lind.), 10, 41. Geld him meard *redde
illis mercedem*, Lind. 20, 8. Meorde (mearda, Lind.) *mercedem
accipit*, Jn. Skt. Rush. 4, 36 : Exon. 48 b ; Th. 167, 13; Gû. 1059 : 62 b ;
Th. 230, 15; Ph. 472 : 76 a ; Th. 286, 9; Jul. 729. Meorda hleótan,
gingra geafena, 48 a ; Th. 164, 20; Gû. 1014. Ðē síe þonc meorda and
miltsa *to thee be thanks for rewards and mercies*, 118 b ; Th. 456, 15; Hy.
4, 67. Morða, 95 a ; Th. 355, 24; Reim. 82. [*Goth.* mizdô ; *Gk.* μισθός.]

meoring, e; *f. Obstacle, impediment, hindrance :*—Moyses ofer ða
fela meoringa fyrde gelǽdde *Moses with many hindrances led the army
across them*, Cd. 145; Th. 181, 16; Exod. 62. [Cf. *O. H. Ger.* marunga
impedimentum.] v. mirran.

meornan; *p.* mearn, *pl.* murnon; *pp.* mornen *To care, feel anxiety,
trouble one's self about anything, reck :*—Nalles for ealdre mearn *he
recked not of life*, Beo. Th. 2889; B. 1442. Nalas for fǽhþe mearn
for fear of the feud was not troubled, 3079; B. 1537. Nô mearn fore
fyrene *he cared not for the crime he committed*, 273; B. 136. Lyt
ǽnig mearn ðæt hié ût geferedon dýre mâðmas *little anxiety did any feel
about bringing out the precious treasures*, 6250; B. 3129. Wôdon
wælwulfas for wætere ne murnon (*cared nought for water*), Byrht. Th.
134, 39; By. 96. v. be-meornan and murnan.

meós, es; *m. n.* (?) *Moss :*—Treówes meós *muscus*, Wrt. Voc. ii. 57,
72. Ragu and meós fornymþ eówres landes wæstmas *omnes fruges
terræ tuæ rubigo consumet*, Deut. 28, 42. Sumne dǽl ealdes meóses ðe
on ðam hâlgan treówe geweaxen wæs (*aliquid de veteri musco*), Bd. 3, 2;
S. 525, 10 : Swt. A. S. Rdr. 96, 30. Meóse *muscum*, Wrt. Voc. ii. 59,
38. Cf. meós môr, Cod. Dip. Kmbl. iii. 81, 29. [*O. H. Ger.* mios :
M. H. Ger. mies; *m. n.*] v. mos *and next word*.

meós; *adj. Mossy :*—Innon meóson môre; of meóson môre, Cod. Dip.
Kmbl. iii. 384, 23.

meóse, meotud, meottuc. v. mêse, metod, mattuc.

meoto *thought* (?) *in :*—Site nû tô symle and onsǽl meoto secgum swâ
ðîn sefa hwette *sit now at the feast, let loose thy thoughts to men, as thy
mind prompts thee*, Beo. Th. 983 ; B. 489.

meówle, an ; *f. A maid, damsel, virgin, woman :*—Ǽnlîcoste meówle
juvencula pulcherrima, Hpt. Gl. 456, 39. Seó hâlige meówle (*Judith*),
Judth. 10; Thw. 22, 10; Jud. 56. Him brýd sunu, meówle (*Mahalaleel's
wife*) tô monnum brohte, Cd. 58; Th. 71, 17; Gen. 1172. Afrisc
meówle, 171; Th. 215, 7; Exod. 579. Meówle, seó hyre bearn gesihþ
brondas þeccan, Exon. 87 b ; Th. 330, 5; Vy. 46. Secg oððe meówle
man or maid, 102 b ; Th. 387, 15; Rä. 5, 5. Ceorles dôhtor, môdwlonc
meówle, 107 a ; Th. 407, 18; Rä. 26, 7. Freólîcu meówle *a damsel
fair*, 124 b ; Th. 479, 2; Rä. 62, 1. Marian, mǽrre meówlan, 14 a ;
Th. 28, 13; Cri. 446. In wîfes lufan, fremdre meówlan, 80 b ; Th.
302, 20; Fä. 39. Wið ða hâlgan mægþ, Metodes meówlan (*Judith*),
Judth. 12; Thw. 25, 15; Jud. 261. [*Goth.* mawilo *a damsel, girl.*]
v. iû-meówle.

meox, mix, myx, es ; *n. Muck, dung, ordure, dirt :*—Meox *stercus*,
Ælfc. Gr. 9, 32; Som. 12, 17 : *coenum*, 13; Som. 16, 6 : *rudera vel
ruina*, Wrt. Voc. i. 22, 12. Fugeles meox *avium stercus*, L. Ecg. P.
add. 10; Th. ii. 232, 32. Ðæt treów biþ bedolfen and mid meoxe
beworpen . . . ðæt meox is ðæt gemynd his fûlan dǽda . . . Hwæt is
fûlre ðonne meox? Homl. Th. ii. 408, 29-33 : Lk. Skt. 13, 8. Licgaþ
forsewene swâ swâ meox (Cott. MS. miox) under feltûne, Bt. 36, 1;
Fox 172, 11 : Homl. Skt. 2, 241. Heó eall forseah on meoxes gelícnysse,
8, 38. Ða nýtenu forrotedon on heora meoxe, Homl. Th. i. 118, 15.
Bûton hē ǽrest ârîse of ðam reócendum meoxe, ii. 320, 23. Ðone hlâf
ðe biþ tô meoxe âwend, i. 258, 2. Tô meohxe, Ps. Th. 82, 8. Meoxe
(meoxene?) *sterquilinio*, Hpt. Gl. 488, 21. Mixe, horwe *ceno*, i. *luto*,
Wrt. Voc. ii. 130, 70. Of myxe dustes *de fece pulveris*, Hy. Surt. 136, 1.
Meoxa *stercorum*, 484, 22. [*Mid. E.* mix, mex : *Frs.* miux : cf. *Goth.*
maihstus : *O. H. Ger.* mist.]

meox-bearwe, an ; *f. A dung-barrow, basket for carrying dung :*—
Wylige oððe meoxbearwe *corbis* vel *cofinus*, Wrt. Voc. i. 86, 2. v. meox-
wilige.

meoxen. v. mixen.

meox-force, an ; *f. A fork used for removing dirt :*—Myxforce *rotabulum* (*rotabulum* furca cum quo ignis movetur in fornace
causa coquendi : et dicitur sic, quia rotat et proruit ignem furni gratia
coquendi vel *stercora* purgandi), Wrt. Voc. i. 16, 34.

meox-wilige, an ; *f. A basket for carrying dung :*—On meocswilian
in cophino, Ps. Lamb. 80, 7. v. meox-bearwe.

merc, meocs, Mercisc, merce, mercels. v. mearc, Mirce, Mircisc,
merece, mircels.

mere, mǽre, es ; *m. f* (?). I. *the sea* (*mer* in *mer*-maid) :—Mere
swîðe grâp on fǽge folc (*of the waters of the deluge*), Cd. 69; Th. 83,
18; Gen. 138. Mere (*the Red Sea*) stille bâd, 158 ; Th. 197, 2 ;
Exod. 300 : 166 ; Th. 206, 27; Exod. 458. Mere sweoðerade, ýða
ongin eft oncyrde, Andr. Kmbl. 930 ; An. 465. Æt meres ende *on the
shore*, 442 ; An. 221. Ofer wîdne mere, 566 ; An. 283. Ofer sealtne
mere, Menol. Fox 203 ; Men. 103. Mere sêcan, mǽwes êþel, Exon.
123 b ; Th. 474, 5; Bo. 25. II. *a mere, lake :*—Meri *stagnum*,
Ep. Gl. 25 b, 16. Mere *stagnum*, Wrt. Voc. i. 54, 15 : ii. 121, 28.
Nis ðæt feor heonon ðæt se mere standeþ, Beo. Th. 2729; B. 1362.
In eálonde ðæs myclan meres (*stagni*), Bd. 4, 29; S. 607, 10. Seó
menigeo ðe stôd begeondan ðam mere, Jn. Skt. 6, 22. On culfran mere,
of ðæm mere . . . On weorces mere ; of ðære mere, Cod. Dip. Kmbl. iii.
76, 37-77, 3. Wið ðone mere *secus stagnum*, Lk. Skt. 5, 1, 2 : 8, 22.
Ðæt wē fundon sumne swîðe micelne mere in ðǽm wǽre fersc wæter,
Nar. 11, 26. On mære *in stagnum*, Blickl. Gl. Be norþan hodes mære
. . . ðonon up on ðone mære, Cod. Dip. Kmbl. iii. 10, 19-26. Ofer
burnan ge ofer meras and ofer ealle wæterpyttas *super rivos ac paludes et
omnes lacus aquarum*, Ex. 7, 19. III, *an artificial pool, cistern :*—
On Syloes mere *in natatoria Siloae*, Jn. 9, 7, 11. Drinc ðæt wæter of
ðînum âgenum mere *bibe aquam de cisterna tua*, Past. 48, 5 ; Swt. 373,
4, 8. [*Goth.* marei; *f. : O. Sax. O. L. Ger.* meri ; *f. : Icel.* marr ; *m.:
O. H. Ger.* mari, meri ; *m. n. : Ger.* meer ; *n. : Lat.* mare.] v. fisc-, hran-,
hring-, hwæl-, îs-, sund-, wîn-, ýð-mere.

mere, myre, an ; *f. A mare :*—Mere *equa*, Wrt. Voc. i. 23, 7. Mire,
287, 78. Myre, ii. 30, 42 : Ælfc. Gr. 7 ; Som. 7, 2. Myran meolc, Ors.
1, 1 ; Swt. 20, 16. Ðǽre myran sunu, Bd. 3, 14 ; S. 540, 30. On myran
rîdan, 2, 13 ; S. 517, 7. [*Icel.* merr : *O. H. Ger.* meriha, marha : *Ger.*
mähre.] v. ass-, stôd-mere.

mêre. v. mǽre.

mere-bât, es ; *m. A sea-boat*, Andr. Kmbl. 492 ; An. 246.

mere-candel, e ; *f. The sea-candle, the sun which rises from*, or *sets
in the sea*, Bt. Met. Fox 13, 114 ; Met. 13, 57. Cf. heáðu-sigel.

merece, merce, es ; *m. Marche* (a plant), *smallage ; apium graveolens :*—
Merici *apio*, Ep. Gl. 1 f, 4. Merice, Wrt. Voc. ii. 100, 46. Merce, 8, 44 :

i. 286, 5: *apium*, 30, 37: 66, 69. Swînes mearce *apiaster*, ii. 7, 7. Merce *merculiaris*, 59, 45: *apiaster*, Ælfc. Gr. 8; Som. 7, 16. Merces sæd, Herb. 97, 1; Lchdm. i. 210, 8. Grênes merces leáf, L. M. 1, 39; Lchdm. ii. 98, 23. Genim merce nioðoweardne, 1, 61; Lchdm. ii. 134, 3. Merece (meric, Lind.) *mentam*, Lk. Skt. Rush. 11, 42. [*Dan.* mærke *smallage, water-parsley.*] v. stân-, wudu-merece (-merce).

mere-cist, e; *f. A sea-chest* :—Noe ongan wyrcan micle merecieste (*the ark*), Cd. 66; Th. 79, 26; Gen. 1317.

mere-deáþ, es; *m. Death in the sea, death by drowning*, Cd. 169; Th. 210, 9; Exod. 512. Meredeáþa mæst (*the destruction of the Egyptians in the Red Sea*), 166; Th. 207, 9; Exod. 464.

mere-deór, es; *n. A sea-beast*, Beo. Th. 1120; B. 558. [*O. L. Ger.* meri-dier *a water-fowl*: *O. H. Ger.* meri-tier.]

mere-fara, an; *m. A sea-farer*, Beo. Th. 1008; B. 502.

mere-faroþ, es; *m. Sea-waves* :—On merefaroþe *on the waves*, Andr. Kmbl. 577; An. 289: 701; An. 351: Exon. 122 b; Th. 471, 16; Rä. 61, 2.

mere-fisc, es; *m. A sea-fish* :—Wæs merefixa môd onhrêred, Beo. Th. 1102; B. 549. [*O. H. Ger.* mere-uisc *piscis maris.*]

mere-flôd, es; *m.* I. *a flood of water, deluge* :—Mereflôd *diluvium*, Exon. 56 b; Th. 200, 18; Ph. 42: Cd. 67; Th. 81, 7; Gen. 1341. Streám fleów ofer foldan . . . miclade mereflôd, Andr. Kmbl. 3050; An. 1528. II. *a body of water, flood, ocean* :—Mereflôdes ýþa, Bt. Met. Fox 27, 4; Met. 27, 2: Cd. 167; Th. 209, 23; Exod. 503. On mereflôde middum *in the midst of the waters*, 8; Th. 9, 21; Gen. 145. Bisencte on mereflôde *drowned in ocean*, Exon. 72 b; Th. 271, 10; Jul. 480: 82 a; Th. 309, 19; Seef. 59.

mere-grot, es; *n. A pebble* or *stone of the sea, a pearl* :—Ne forlǽte ic ðê nǽfre, mín meregrot! Blickl. Homl. 149, 2. Is heofena ríce gelíc ðam mangere ðe sôhte ðæt gôde meregrot. Ðá hê funde ðæt ân deórwyrðe meregrot ðá bohte hê ðæt meregrot, Mt. Kmbl. 13, 45–46. Bergean swylce meregrota (*margaritæ*), Nar. 37, 29. Gefrætwod swá swá mid meregrotum, Homl. Th. i. 596, 8. [Cf. *O. H. Ger.* meri-grioz *margarita, unio.*] v. next word.

mere-grota, an; *m. A pearl* :—Meregrota *margarita*, Wrt. Voc. i. 85, 24. On ðám beóþ oft gemêtte ða betstan meregrotan *quibus inclusam sæpe margaritam optimam inveniunt*, Bd. 1, 1; S. 473, 18. [Cf. *O. Sax.* meri-grita, -griota.]

mere-grund, es; *m. The bottom of a sea* or *lake*, Beo. Th. 2902; B. 1449: 4207: B. 2100.

mere-hengest, es; *m. A sea-steed, a ship*, Exon. 104 a; Th. 395, 12; Rä. 15, 6: Bt. Met. Fox 26, 49; Met. 26, 25.

mere-hrægel, es; *n. A sea-garment, a sail* :—Merehrægla sum, segl sále fæst, Beo. Th. 3815; B. 1905.

mere-hûs, es; *n. A sea-house (Noah's ark)*, Cd. 65; Th. 78, 34; Gen. 1303: 69; Th. 82, 18; Gen. 1364.

mere-hwearf, es; *m. A sea-wharf, sea-shore*, Cd. 169; Th. 210, 16; Exod. 516.

mére-hwít. v. mǽre *pure.*

mere-lâd, e; *f. A sea-way, the road which the sea furnishes*, Exon. 123 b; Th. 474, 9; Bo. 27.

mere-líðende, *a sea-faring, a sea-faring person*, Cd. 71; Th. 84, 34; Gen. 1407: Beo. Th. 515; B. 255: Andr. Kmbl. 705; An. 353. [Cf. *Icel.* mar-líðendr; *pl.* sea-farers.]

mere-men[n], e; *f. A siren* :—Meremen *sirena*, Wrt. Voc. i. 289, 6. Meremenna *sirenarum*, Hpt. Gl. 498, 65. [Brutus iherde siggen þurh his sæmonnen of þan ufele ginnen þe cuðen þa mereminnen, Laym. 1337: *O. H. Ger.* mer-min *siren*; meri-meni, -menni *scylla.*] v. next word and Grmm. D. M. 404-407.

mere-menen, -mennen, e; *f. A siren* :—Meremenin *sirina*, Wülck. 47, 7. Meremennena *sirenarum*, Wrt. Voc. i. 84, 12. [Cf. *Icel.* mar-mennill; *m. a sea-goblin.*] Cf. mere-wíf.

mere-næddra, an; *m.* -nædre, an; *f. A sea-adder, a lamprey* :—Mere-næddra *murena vel murina vel lampreda*, Wrt. Voc. i. 55, 65. Myre-næddra, 77, 72. Merenædre, ii. 59, 23.

mere-smylte; *adj. Having the sea calm* :—Meresmylta wíc, Bt. Met. Fox 21, 24; Met. 21, 12.

mere-strǽt, e; *f. The road which the sea furnishes*, Elen. Kmbl. 483; El. 242: Beo. Th. 1032; B. 514.

mere-streám, es; *m. A sea-stream, the sea, water of the sea*, Cd. 39; Th. 51, 27; Gen. 833: 154; Th. 191, 5; Exod. 210: 166; Th. 207, 17; Exod. 468. Merestreám ne dear ofer eorþan sceát eard gebrǽdan (cf. sǽ, Bt. Fox 74, 26), Bt. Met. Fox 11, 130; Met. 11, 65: 20, 228; Met. 20, 114. Oþ merestreámas *unto the waters of the sea*, Cd. 199; Th. 247, 27; Dan. 503: Bt. Met. Fox 28, 65; Met. 28, 33. Manegum merestreámum *de aquis multis*, Ps. Th. 143, 12. [*O. Sax.* meri-strôm.]

mere-strengu; *f. Strength in the sea, strength for swimming* :—Ic merestrengo mâran âhte, earfeþo on ýðum, ðonne ǽnig ôðer man, Beo. Th. 1070; B. 533.

mere-swín, es; *n. A sea-pig, porpoise, dolphin* :—Ðes mereswín *hic delfin*, Ælfc. Gr. 9, 14; Som. 9, 37: Wrt. Voc. ii. 26, 15: i. 281, 56.

Mereswín *bacharus*, 281, 57: 65, 61: *delphin* vel *bocharius* vel *simones*, 55, 60. Mereswýn *bacharus*, 21, 46. Meresuín *bacarius*, ii. 102, 11. Ǽlc seldfynde fisc ðe weordlíc byþ, styria and mereswýn, Cod. Dip. Kmbl. iii. 450, 28. Nim mereswînes fel, L. M. 3, 40; Lchdm. ii. 334, 1. Mereswýn and stirian *delphinos et sturias*, Coll. Monast. Th. 24, 9: Bd. 1, 1; S. 473, 17. [*Icel.* mar-swín: *O. H. Ger.* meri-suín: *Ger.* meerschwein *dolphin, porpoise.*]

mere-þyssa, an; *m. A sea-rusher, a ship* :—On mereþyssan, Andr. Kmbl. 892; An. 446. On mereþissan, 514; An. 257. [Cf. *Icel.* þysja *to rush;* þyss *uproar.*]

mere-torht; *adj. Bright from bathing in the sea* (epithet of morning) :—Sió sunne brencþ eorþwarum morgen meretorhtne *the sun rising from the sea brings bright morn to men*, Bt. Met. Fox 13, 121; Met. 13, 61. Becwom ofer gârsecges [begong] morgen mæretorht [*or* mæretorht *splendidly bright*, cf. *O. H. Ger.* mâri-mihil], Cd. 160; Th. 199, 29; Exod. 346. Cf. mere-candel.

mere-torr, es; *m. A tower formed by the sea (the walls formed by the waters of the Red Sea)*, Cd. 167; Th. 208, 16; Exod. 484.

mere-weard, es; *m. A sea-ward, one who keeps guard in the sea* :—Se mereweard (*the whale*), Exon. 97 a; Th. 363, 13; Wal. 53.

mere-wêrig; *adj. Weary of journeying on the sea* :—Merewêrges môd *the mind of the sea-weary man*, Exon. 81 b; Th. 306, 23; Seef. 12.

mere-wíf, es; *n. A water-witch, woman living in a lake (Grendel's mother)*, Beo. Th. 3042; B. 1519. [*O. H. Ger.* meri-wíb *sirena.*]

mergen. v. merigen.

merian; *p.* ede; *pp.* ed *To purify, refine* :—Ðam ðe his gâst wile mergan (MS. B. merian) of sorge âsceádan of scyldum *for him who will purify his spirit from the dross of care, separate it from guilt*, Salm. Kmbl. 112; Sal. 55. v. â-merian.

merig. v. mirig.

**merigen, merien, mergen, es; *m.* I. *morning* :—Úres andgites merigen is úre cildhâd, Homl. Th. ii. 76, 14. Ðá se mergen geworden wæs *when it was morning*, St. And. 10, 3. Mergen þridda, Cd. 8; Th. 10, 11; Gen. 155: Beo. Th. 4213; B. 2103: 4255; B. 2124. Merien *mane*, Wrt. Voc. i. 76, 53. On mergenne *mane*, Ps. Spl. 91, 2: Ps. Th. 54, 17: 89, 16: Beo. Th. 1134; B. 565. In merne *mane*, Mt. Kmbl. Lind. 20, 1: 21, 18. Tô merne, 16, 3. On ðam dæge worhte God merigen and æfen, Homl. Th. i. 100, 5. On mergen *mane*, Ps. Spl. 89, 6. II. *the morning of the next day, morrow* :—Ðú ðe nâst hwæðer ðú merigenes gebíde *thou that knowest not whether thou wilt live to see the morrow*, Homl. Th. ii. 104, 26. Hwæt gif ic bíde merigenes, Homl. Skt. 3, 585. In merne *in crastinum*, Mt. Kmbl. Lind. 6, 34. On merne, Jn. Skt. Lind. 1, 43: 12, 12. Tô merne *cras*, Lk. Skt. Lind. 13, 32. On mergen *in crastinum*, Jn. Skt. 1, 43: 12, 12. On merien, Homl. Th. ii. 502, 16. Wê nyton hwæt tô merigen biþ tôweard, 82, 17: i. 374, 21: 462, 3. Tô merigen *cras*, Ælfc. Gr. 38; Som. 39, 59. v. ǽr-, ǽrne-mergen, *and* morgen.

merigen-, mergen-dæg, es; *m. Morrow* :—Hê ðæs mergendæges gebídan môste, Blickl. Homl. 213, 25. v. morgen-dæg.

merigen-, mergen-líc; *adj.* I. *belonging to the morning* :—Se merigenlíca tilia *the labourer who came to work in the morning*, Homl. Th. ii. 74, 29. Se mergenlíca steorra *the morning star*, Blickl. Homl. 137, 32. II. *belonging to the morrow* :—Ðam ne fyligþ merigenlíc dæg, forðan ðe him ne forstôp se gysternlíca, Homl. Th. i. 490, 19. Ðýs mergenlícan dæge, Blickl. Homl. 143, 21: 147, 29. v. morgen-líc.

merigen-, mergen-tíd, e; *f. Morning-time, morning* :—Fram ðære mêran mergentíde oþ ðæt æfen cume *a custodia matutina usque ad noctem*, Ps. Th. 129, 6. v. morgen-tíd.

merisc. v. mersc.

merne. v. merigen.

merra, merran, merring. v. mirra, mirran, mirring.

mersc, es; *m. A marsh* :—Mersc *calmetum*, Wrt. Voc. ii. 13, 42: 103, 10: 127, 55. Tô mærsce, Cod. Dip. Kmbl. iii. 175, 32. Ðæt lond at Ðorpe mid mêdwe and mid merisce, iv. 295, 7. On sealtum mersce, Ps. Spl. 106, 34. Hê ða weaxendan wende eorþan on sealtne mersc (*in salsuginem*), Ps. Th. 106, 33: Blickl. Gl.: Cd. 160; Th. 199, 4; Exod. 333. Ne fersc ne mersc, Lchdm. iii. 286, 21. Sumra wyrta eard biþ on merscum *alias herbas ferunt paludes*, Bt. 34, 10; Fox 148, 23. On feldum and on mǽdum and on sealtum merscum, Cod. Dip. Kmbl. iii. 350, 8. Mersc *Romney Marsh*, Chr. 796; Erl. 58, 11.

mersc-land, es; *n. Marsh-land* :—Forneáh ǽlc tilþ on mersclande forfêrde, Chr. 1098; Erl. 235, 12.

mersc-mealwe, an; *f. Marsh-mallow* :—Merscmealewe *althea*, Wrt. Voc. i. 67, 20. Merscmealwe *hibiscum*, ii. 43, 3. Merscmealuwe. Ðeós wyrt ðe man *hibiscum* and ôðrum naman merscmealwe (mealuwe, MS.) nemnaþ, Herb. 39; Lchdm. i. 140, 3–5. Merscmealwan crop, L. M. 3, 63; Lchdm. ii. 350, 24. Nim merscmealwan, 3, 8; Lchdm. ii. 312, 12.

mersc-mear-gealla, an; *m. A kind of gentian*; gentiana pneumonanthe :—Nim merscmeargeallan, L. M. 1, 39; Lchdm. ii. 100, 5: 1, 50; Lchdm. ii. 124, 1.

Mersc-ware; *pl. The inhabitants of marshy land* :—Myrcena cining oferhergode Cantware and Merscware (*men of Romney Marsh*), Chr. 796;

Erl. 59, 40. Monige on Merscwarum *many of the men of the fens*, 838;
Erl. 66, 12.

mertze (?):—Mertze *merx*, Wrt. Voc. ii. 113, 82. [Cf. *O. H. Ger.*
merzi *merx*, Grff. ii. 861.]

mes (?) *dung*:—Gesomna cûe mesa *collect cow-dung*, L. M. 1, 38;
Lchdm. ii. 98, 5. [' Mes *stercus, fimus* (Kilian),' Cockayne.]

mêsan *to feed, eat*:—Ic mêsan mæg meahtelícor ealdum þyrse *I can
eat mightier meals than an old giant*, Exon. 111 a; Th. 425, 26; Rä.
41, 62. v. môs.

mêse, meóse, mîse, mýse, an; *f. A table; also what is on a table*:—
Mîse (MS. T. mêse) *mensa*, Ps. Spl. 68, 27. Meóse *rrensorium* (*mensorium quod est in mensa, ut mantile, et vas escarium*), Wrt. Voc. i. 26, 61.
Mýse t beód *mensa*, 82, 21 ... Ða hwelpas etaþ of ðam crumon ðe feallaþ
of heora hláfordes mýsan ... Seó mýse is bóclíce lár ... Ðe ðære mýsan
cwæþ se wîtega: Drihten ðû gegearcodest mýsan on mînre gesihþe,
Homl. Th. ii. 114, 24–28: i. 330, 31, 34: Ps. Spl. 127, 4: Mk. Skt. 7,
28: Lk. Skt. 12, 21, 30. [*Goth.* mês: *O. H. Ger.* mias, meas *mensa.*]
met. v. ge-, tæl-met.

metan; *p.* mæt, *pl.* mæton; *pp.* meten. **I.** *to mete, measure*:—
Ic mete *metior*, Ælfc. Gr. 31; Som. 35, 32. Ic meotu *metibor*, Ps.
Surt. 59, 8: 107, 8. Ælfc ðæra þinga ðe man met on fate *everything
that is measured in a vessel*, Ælfc. Gr. 13; Som. 16, 8. On ðam ylcan
gemete ðe gê metaþ eów byþ gemeten *qua mensura mensi fueritis,
remetietur vobis*, Mt. Kmbl. 7, 2. Hwílum mid folmum [hê] mæt weán
and wîtu, Cd. 229; Th. 309, 22; Sat. 714. **II.** *to measure out,
mark off, assign the bounds of a place*:—Se geleáfa and seó lufu mæton
ðone stede hwær hió drihtnes tempel ræran woldan, Prud. 80. Ðû
gedydest ðæt wê mæton ûre land mid rápum, Ps. Th. 15, 6. Wîcsteal
metan *castra metari*, Cd. 146; Th. 183, 16; Exod. 92. **III.** *to
measure by paces, to traverse, pass over*:—Him eoh fore mílpaðas mæt,
Elen. Kmbl. 2523; El. 1263. Férdon forþ ðanon, fêðelástum foldweg
mæton, Beo. Th. 3271; B. 1633: 1032; B. 514: 1838; B. 917. Forþ
gesáwon lifes látþeów lífweg (liftweg?) metan, Cd. 147; Th. 184, 9;
Exod. 104. **IV.** *to measure one thing by or with another, to compare*:—Se swêg wæs be winde meten *the sound was compared to the
wind*, Blickl. Homl. 133, 31. Hê mæt ðone welan tô ðære winestran
handa *he compared wealth to the left hand*, Past. 50, 2; Swt. 389, 18.
Ne sint hî nô wiþ eów tô metanne *they are not to be compared with you*,
Bt. 13; Fox 40, 10: 39, 8; Fox 224, 5: Bt. Met. Fox 21, 83; Met. 21,
42. Tô metenne wið ðæt mód, Bt. 16, 2; Fox 52, 6: 32, 2; Fox
116, 7. Tô mettanne, 18, 1; Fox 62, 4. [*Goth.* mitan: *O. L. Ger.*
metan: *O. Frs. Icel.* meta: *O. H. Ger.* mezan: *Ger.* messen.] v. á-, be-,
ge-, wið-, wiðer-metan.

mêtan; *p.* te *To paint*:—Ic mête *pingo*, Ælfc. Gr. 28, 5; Som. 31,
60. Swá mêteras mêtaþ on anlícnyssan *as painters paint in likenesses*,
Wrt. Voc. i. 41, 5. Seó ðe mêtan sceall *pictura*, Ælfc. Gr. 43; Som.
45, 3. Mêtton ofergeweorke *depicto mausoleo*, Coll. Monast. Th. 32,
35. [Fe33 haffdenn liccness metedd, Orm. 1047. Cf. *Goth.* maitan *to
cut: Icel.* meita *to cut*; meitill *a chisel: O. H. Ger.* meizan *to cut*; meizil
a chisel.] v. á-, ge-mêtan, *and* mêting.

mêtan; *p.* te *To meet with, come upon, come across, find*:—
Ealle ðe hê mildheorte mêteþ and findeþ, Ps. Th. 75, 6. For ðý hî hit
ne gemêtaþ (MS. Cott. mêtaþ) ðe hî hit on riht ne sêcaþ, Bt. 36, 3;
Fox 178, 4. Gê unæþelne ænigne [ne] mêtaþ (gê nânne ne mágon
mêtan unæþelne, Bt. 30, 2; Fox 110, 16), Bt. Met. Fox 17, 34; Met.
17, 17. Moette *offendit*, Wrt. Voc. ii. 115, 41. Mêtte, 63, 35. Ðá
eode hê furþor ôþ hê gemêtte (MS. Cott. mêtte) ða Parcas *then he went
on until he came upon the Fates*, Bt. 35, 6; Fox 168, 24. Ðá mêtte hê
ðane man forþfêredne *he found the man departed*, Blickl. Homl. 217, 17.
Hê ne mêtte mundgripe máran, Beo. Th. 1506; B. 751: Andr. Kmbl.
942; An. 471: 1106; An. 553. Hê þreó mêtte rôda ætsomne *he came
upon three crosses together*, Elen. Kmbl. 1663; El. 833. Hí mêtton
invenerunt, Ps. Spl. 106, 4. Nime se ðe hit on his æcere mête, L. In.
42; Th. i. 128, 14. Swá ær swá hê hádes wyrþne mon mêtan mihte *as
soon as he could meet with a man worthy of the (episcopal) rank*, Bd. 3,
29; S. 561, 26. Ðær byþ sôþ symble mêted *truth is ever found there*,
Ps. Th. 118, 160. Ðæt sigorbeácen mêted wære, funden in foldan, Elen.
Kmbl. 1969; El. 986. [*Goth.* ga-môtjan: *O. Sax.* môtian: *O. Frs.*
mêta: *Icel.* mœta.] v. ge-mêtan.

met-cund (? meter-cund, q. v.); *adj. Metrical*:—Ðý metcundan (dymetcunda, Wrt.), Wrt. Voc. ii. 75, 30. v. next word.

metcund-lic; *adj. Metrical*:—Metcundlícere getincnesse *metrica facundia*, Hpt. Gl. 409, 17. v. preceding word.

METE, mæte, es; *m.* MEAT, *food*:—Mete *cibus*, Wrt. Voc. ii. 22,
80. Mín mete (mett, Lind. Rush.) is ðæt ic wyrce ðæs willan ðe mê
sende, Jn. Skt. 4, 34. Gesoden mæt on wætere *elixus cibus*, Wrt. Voc. i.
27, 17. Swête mete *dapis*, ii. 28, 29. Ðû scealt mid earfoþnyssum ðé
metes tilian *thou shalt with hardships get thyself food*, Homl. i. 18, 15.
Ðæt hig beón eów tô mete *ut sint vobis in escam*, Gen. 1, 29: Cd. 38;
Th. 50, 25; Gen. 814. Gá hyt eft in tô ðam hálegan mynstre mid
mete and mid mannum *let it revert to the holy monastery with meat and*

with men, Chart. Th. 379, 21. Wyt æton swêtne mete (*dulces cibos*),
Ps. Th. 54, 13. Ðæt ic macige mete ðînum fæder *ut faciam escas patri
tuo*, Gen. 27, 9. Gif hý him syððan ne dôþ mete ne munde *if they
afterwards give him neither food nor favour*, L. Edm. S. 1; Th. i. 248, 7.
Ðær mæte þygde, Bd. 5, 4; S. 617, 11. Mettas *cibaria*, Wrt. Voc. ii.
15, 71: *dapes*, 28, 1: *fercula*, Hpt. Gl. 492, 75. Ða mettas (*cibos*) ðe
God self gesceóp, Past. 43, 9; Swt. 319, 1. Mínum þeówum ic sylle
mettas, Ælfc. Gr. 15; Som. 18, 65. Se ðe mettas (*escas*) hæfþ, Lk.
Skt. 3, 11. Earmra hungur hê oferswýþde mid mettum, Bd. 2, 1; S.
500, 24. Mid cynelícum mettum (*regalibus epulis*) gefylled, 2, 6; S.
528, 14. Fram swêttrum mettum *a cibis luculentioribus*, Wrt. Voc. ii.
6, 25. [*Goth.* mats: *O. Sax.* meti: *O. Frs.* mete: *Icel.* matr: *O. H. Ger.*
maz; *n. esca.*] v. æfen-, côcor-, dæg-, êst-, flæsc-, hreác-, mæl-, morgen-,
nôn-, pan-, undern-, wyrt-mete.

mete-ærn, es; *n. A room for taking meals in*:—Gemæne metern
cœnaculum, Wrt. Voc. i. 58, 50.

mete-áfliúng, e; *f. Atrophy*; atrophia, Wrt. Voc. i. 19, 44.

mete-bælg, es; *m. A bag for food, wallet*:—Bûta metbælge (metbælig, Lind.) *sine pera*, Lk. Skt. Rush. 22, 35.

mete-corn, es; *n. Corn for food*:—Ílk habbe his metecû and his
metecorn, Chart. Th. 580, 7. v. next word.

mete-cû, e; *f. A cow that is to furnish food*:—Ânan esne gebyreþ tô
metsunge xii pund gôdes cornes and i gôd metecû, L. R. S. 8; Th. i.
436, 27. v. preceding word.

mete-fæt, es; *n. A dish*:—Micel and rûm metfæt *graves et ampla
parabsis*, Germ. 403, 18.

mete-fætels, es; *m. A wallet*:—Metefætels *sitarchia*, Wrt. Voc. i. 16, 39.

mete-fisc, es; *m. An edible fish*:—Ðes metefisc *hic mugil*, Ælfc. Gr.
9, 8; Som. 9, 10.

mete-gafol, es; *n. Tax or rent paid in food*:—On sumen lande gebûr
sceal syllan huniggafol, on suman metegafol, on suman ealugafol, L. R. S.
4; Th. i. 434, 32.

mete-gearwa; *pl. f. Preparations of food*:—Ôðre hwætene (MS.
wætan) metegearwa sint tô forbeódanne *other preparations of wheaten
food are to be forbidden*, L. M. 2, 23; Lchdm. ii. 210, 26.

mete-gird. v. met-gird.

metegian, metegung. v. metgian, metgung.

mete-láf, e; *f. A remnant of food*:—Dælon ealle ða metelâfe *let them
distribute all the remnants of food*, L. Æðelst. v. 8, 1; Th. i. 236, 7.
On ðîne metelâfa *in reliquias ciborum tuorum*, Ex. 8, 3. Ða metlâfo
reliquias, Mt. Kmbl. Lind. 14, 20.

mete-leás; *adj. Without food, lacking food*:—On sumere tîde wæs
micel menigu mid ðam Hælende on ânum wêstene meteleás (*nec haberent,
quod manducarent*), Homl. Th. ii. 396, 1: Homl. Th. i. Elen. Kmbl. 1220; El. 612:
1392; El. 698. Heó wunode seofon niht meteleás *she remained seven
days without food*, Homl. Skt. 10, 283. [*Icel.* mat-lauss.]

mete-leást, -líest, -læst, -lêst, -líst, e; *f. Want of food*:—Him ofhreów ðæs folces meteleást, Homl. Th. ii. 396, 19. Ðá wæron hié mid
metelíeste gewægde *they were reduced by want of food*, Chr. 894; Erl.
92, 27. For metelíeste heora líf álætan, Ors. 3, 8; Swt. 120, 30.
Metelæste *inedia*, Hpt. Gl. 480, 34. Metelêste, 497, 31. Metelæste
cibi inopia, 416, 75. Murnende môd nales metelíste, Exon. 101 a;
Th. 380, 29; Rä. 1, 15. For meteleáste mêde, Andr. Kmbl. 77; An.
39: 2315; An. 1159. [Cf. *O. Sax.* meti-lôsi: *Icel.* mat-leysa *lack of food*.]

metend, es; *m. One who measures or metes*:—Him leán âgeaf metend
(God), Cd. 86; Th. 108, 21; Gen. 1809. Middangeardes metend *ex
Ormista* (the A. S. gloss seems to be intended as a translation of the title
commonly given to Orosius' History, [H]Ormesta Mundi, and is *the
measurer or describer of the world*, i. e. a general history of the world),
Wrt. Voc. ii. 30, 18. Cf. metod, metten.

metend-lîce, meten-ness. v. á-metendlíce, wið-metenness.

meter, es; *n. Metre*:—Missenlíce metre *diverso metro*: eroico metre
heroico metro, Bd. 5, 24; S. 648, 36, 37. [*O. H. Ger.* meter; *n.*]

meter-cræft, es; *m. The art of versification*; ars metrica, Bd. 4, 2;
S. 565, 25.

meter-cund; *adj. Relating to metre*:—Metercund *catalecticus, ubi in
pede versus una sillaba deest*, Wrt. Voc. ii. 129, 41. Ðý metercundum
catalectico, 17, 67.

mêtere, es; *m. A painter*:—Mêtere *pictor*, Wrt. Voc. i. 46, 72: 75,
18. Sîd reáf swylce mêtere[s] wyrceþ on anlícnysse *toga*; scrûd swá
mêteras mêtaþ on anlícnyssan *cinctus gabinus*, 41, 3, 5. Ælfnôþ ðe
mêtere, Cod. Dip. Kmbl. iv. 261, 20. v. mêtan, mêting.

meter-fers, es; *n. Hexameter verse*:—Be his life wê âwriton ge
meterfers ge gerædre spræce *de vita illius et versibus heroicis et simplici
oratione conscripsimus*, Bd. 4, 28; S. 605, 13. Meterfersum *versibus
hexametris*, 5, 18; S. 636, 6.

meter-geweorc, es; *n. Verse*:—Paulinus bêc of metergeweorce on
geráde spræce ic gehwyrfde *I turned Paulinus' books from verse into prose*,
Bd. 5, 23; S. 648, 21.

meter-lîc; *adj. Metrical, poetical*:—Mid meterlícum fôtum *pedibus
poeticis*, Hpt. Gl. 411, 3. [*O. H. Ger.* meter-líh.]

met-ern. v. mete-ærn.

meter-wyrhta, an; *m. A verse-maker, poet :*—Mederwyrhta *metricus*, Wrt. Voc. ii. 114, 7. Meterwyrhta, 55, 64. [Cf. *O. H. Ger.* meter-wurcha *poetica musa.*]

mete-, met-[?]sacca, an; *m. A kind of measure :*—Metesacca *legula* (*ligula* mensuræ genus quod alio nomine cochlea dicitur et est octava pars cyathi) vel *coclea*, Wrt. Voc. i. 26, 62.

mete-seax, es; *n. A meat-knife, knife used in cutting food, dagger :*—Hiene mid heora metseacsum ofsticedon, Ors. 5, 12; Swt. 244, 18. [*O. H. Ger.* maz-sahs *cultellum.*]

mete-sócn, e; *f. Desire for food, appetite :*—Of đæs magan ádle cumaþ ungemetlíca metesócna, L. M. 2, 1; Lchdm. ii. 174, 27.

mete-swamm, es; *m. An edible mushroom :*—Metteswam *fungus* vel *tuber*, Wrt. Voc. i. 31, 52.

mete-þearfende, *part. Wanting food :*—Hié æghwylcne ellþeódigra dydon him tô môse meteþearfendum *they made every foreigner food for themselves in want of meat*, Andr. Kmbl. 54; An. 27: 272; An. 136.

mete-þegn, es; *m. An officer whose duty it is to see after food, a sewer*, Cd. 148; Th. 185, 31; Exod. 131. [Cf. disc-þegn.]

mete-útsiht, e; *f. A disease which causes food to pass the bowels without digestion :*—Meteútsiht *lienteria* (λειεντερία), Wrt. Voc. i. 19, 54. Meteútsihþ, ii. 53, 75.

met-fæt. v. mete-fæt *and* gemet-fæt.

metgian, metegian, metian; *p.* ode. **I.** *to assign due measure* (with dat.) :—Đonan metgaþ ǽlcum be his gewyrhtum *thence assigns to each due measure according to his deserts ;* quid unicuique conveniat, agnoscit, et, quod convenire novit, accomodat, Bt. 39, 9; Fox 226, 23. **II.** *to moderate, regulate* (with acc.) :—Se ilca God se đæt eall metgaþ *the same God who regulates all that*, Bt. Met. Fox 11, 188; Met. 11, 88. **III.** *to measure in the mind, consider, meditate upon* (cf. *Goth.* mitôn *to consider*) :—Ic đíne gewitnysse on môde metegie *georne testimonia tua meditatio mea est*, Ps. Th. 118, 24. Đæt ic ǽ đíne metige *lex tua meditatio mea est*, 118, 174. Ic ǽ đíne on môde metegade, 118, 97, 143: 142, 5. Ic on đínre sôđfæstnesse symble meteode (*meditabor*), 118, 16. Ic metegian ongan mænigra weorca *meditatus sum in omnibus operibus tuis*, 76, 10. v. ge-metgian.

met-gird, -geard, -gyrd, e; *f. A rod for measuring, a rod, perch :*—Metgeard *pertica*, Wrt. Voc. i. 38, 5. Riht is đæt ne beó ǽnig metegyrd lengre đonne ôđer, L. I. P. 7; Th. ii. 314, 6. Đonne is đæs imbganges ealles þrió furlanges and þreó metgeurda, Chart. Th. 157, 27. Twegræ metgyrda brád, 232, 17.

metgung, metegung, e; *f.* **I.** *moderation, temperance :*—Wísdóm is se hêhsta cræft, and se hæfþ on him feówer ôđre cræftas, đara is ân wærscipe, ôđer metgung, þridde is ellen, feórþe rihtwísnes, Bt. 27, 2; Fox 96, 34. **II.** *meditation :*—Mê is metegung hû ic ǽ đíne efnast healde *lex tua meditatio mea est*, Ps. Th. 118, 77. v. ge-metgung.

Mêđas, međel. v. Mǽđas, mæđel.

mêđe; *adj.* **I.** *weary, exhausted* (with labour, hunger, disease, etc.) :—Hé hine đær hwíle reste, mêđe æfter đam miclan gewinne, Rood Kmbl. 129; Kr. 65. Mêđe and meteleás, Elen. Kmbl. 1220; El. 612: 1392; El. 698: Exon. 90 b; Th. 340, 15; Gn. Ex. 111. Mêđe for đam miclan bysgum *exhausted by disease*, 49 a; Th. 168, 25; Gú. 1083. Mê swâ mêđum (*exhausted from want of food*), Elen. Kmbl. 1620; El. 812. Mêđne *fessum*, Wrt. Voc. ii. 38, 26: Exon. 47 b; Th. 163, 3; Gú. 988: 49 b; Th. 171, 23; Gú. 1131. Mêđe stôdon, hungre gehæfte, Andr. Kmbl. 2316; An. 1159: 78; An. 39. Hié slǽp ofereode mêđe be mæste, 929; An. 465. **II.** *weary in mind, troubled, sad :*—Đé unrôtne, mêđne, môdseócne, Exon. 51 a; Th. 177, 30; Gú. 1235. Hyge geómurne, mêđne môdsefan, 52 a; Th. 182, 16; Gú. 1311. Ongunnon sorhleóþ galan, đá hié woldon síđian mêđe fram đam mǽran þeódne, Rood Kmbl. 137; Kr. 69. Mêđra frêfrend *comforter of the weary-hearted*, Exon. 62 a; Th. 227, 13; Ph. 422. **III.** *troublesome, causing weariness :*—Nelle đú mê moeđe í hefig wosa *noli mihi molestus esse*, Lk. Skt. Rush. 11, 7. [*O. Sax.* môđi: *Icel.* môđr *weary, exhausted* : *O. H. Ger.* muodi *fessus, fatigatus, lassus* : *Ger.* müde.]

međema = (?) meduma :—Međema persa (wersa,Wrt.) *tramarium*,Wrt. Voc. i. 59, 27.

mêđian *to grow weary :*—Wiđ miclum gonge ofer land . . . mucgwyrt nime him on hand ôđđe dô on his scô đý læs hé mêđige *for much walking over the country . . . let him take mugwort into his hand, or put it into his shoe, lest he grow weary*, L. M. 1, 86; Lchdm. ii. 154, 10. [*O. H. Ger.* muođên *fatiscere, lassari:* cf. *Icel.* mœđa *to weary, trouble.*] Cf. ge-mêđgian.

mêđig; *adj. Weary, exhausted :*—Hié hiene mêđigne on cneówum sittende mêtten, Ors. 3, 9; Swt. 134, 31. Đa đe tô láfe beón môston wæron tô đæm mêđie đæt hié ne mehton đa gefarenan tô eorþan bringan *the survivors (of the pestilence) were exhausted to such a degree, that they could not inter the dead*, 2, 6; Swt. 86, 28. v. mêđe.

metian *to supply with food :*—Đa beád hé đæt man sceolde his here metian (MS. C. mettian) and horsian *he ordered that his army should be supplied with food and with horses*, Chr. 1013; Erl. 148, 3. v. metsian.

mêting, e; *f. A painting, picture :*—Mêtincg *pictura*, Ælfc. Gr. 28, 5; Som. 31, 61. Mêtingc, Wrt. Voc. i. 46, 73: 75, 19. Swâ swâ on mêtinge biþ forsewen seó blace anlícnys, đæt seó hwíte sý beorhtre gesewen, Homl. Th. i. 334, 12. On ôđre wîsan wê sceáwiaþ mêtinge, and on ôđre wîsan stafas. Ne gǽþ nâ mâre tô mêtinge bûton đæt đú hit geseó and herige, 186, 5–7. v. mêtan.

met-líc. v. un-metlíc.

metod, metud, meotud, meotod, es; *m. A word found only in poetry* (the phrase *se metoda drihten* occurs twice in Ælfric's Homilies, but in alliterative passages). The earlier meaning of the word in heathen times may have been *fate, destiny, death* (cf. metan), by which Grein would translate *metod* in Wald. 1, 34; Val. 1, 19 :—Đý ic đê metod ondrêd đæt đú tô fyrenlíce feohtan sôhtest (Stephens here takes *metod* as vocative with the meaning of prince) ; in this sense it seems to be used in its compounds, and in the Icelandic *mjötuđr* weird, bane, death (Cl. and Vig. mjötuđr, II). Could this be the meaning in the phrase *se metoda drihten* used of Christ in the following passages ?—Ne dorston đa deóflu, đá đá hí ádrǽfde wǽron, intô đam swýnum, gif hê him ne sealde leáfe, ne intô nánum men forđan se metoda drihten úre gecynd hæfde on him sylfum genumen, Homl. Th. ii. 380, 4–7. Gemyndig on môde hû se metoda drihten cwæþ on his godspelle be his godcundan tôcyme, 512, 27. But the word, which occurs frequently, is generally an epithet of the Deity as the O. Sax. *metod;* so too Icel. *mjötuđr* (Cl. and Vig. mjötuđr, I) is applied to heathen gods :—Metod engla, lífes brytta, Cd. 6; Th. 8, 9; Gen. 136. Blíđheort cyning, metod alwihta monna cynnes, 10; Th. 12, 29; Gen. 193. Hine forwræc metod mancynne fram, Beo. Th. 220; B. 110. Metud *O Lord!* Elen. Kmbl. 1634; El. 819. Middangeardes meotud, Exon. 116 b; Th. 449, 2; Dôm. 65. Cyninga wuldor, meotud mancynnes, Andr. Kmbl. 343; An. 172. Sôđfæst meotud, 772; An. 386. Meotod hæfde miht đá hê gefestnade foldan sceátas, Cd. 213; Th. 265, 3; Sat. 2. Meotod mancynnes, 223; Th. 293, 22; Sat. 459. Meotod alwihta, 228; Th. 308, 24; Sat. 697. Mægencyninga meotod, Exon. 21 b; Th. 58, 29; Cri. 943. Cf. metend, metten.

metod-gesceaft, e; *f. Decree of fate, death :*—Sum sceal seonobennum seóc sâr cwânian, murnan meotudsceafte (*approaching death*), Exon. 87 b; Th. 328, 19; Vy. 20. [*O. Sax.* hie iro mundoda wiđer metodigiskeftie (*the death of her son*).] v. next word.

metod-sceaft, e; *f. Decree of fate, doom, fate after death :*—Ealle Wyrd forsweóp míne mâgas tô metodsceafte (*to their doom*), Beo. Th. 5623; B. 2815. Gâst onsende Matheus his tô metodsceafte (*to the fate appointed to it*), in êcne gefeán, Menol. Fox 342; Men. 172. Weccaþ of deápe dryhtgumena bearn tô meotudsceafte *the children of men shall awake from death to doom*, Exon. 21 a; Th. 55, 24; Cri. 888. Hê forþ gewât metodsceaft seón *he died*, Cd. 83; Th. 104, 31; Gen. 1743: Beo. Th. 2364; B. 1180. Heó metodsceaft (*the death of her kinsmen*) bemearn, 2158; B. 1077.

metod-wang, es; *m. The plain where the decrees of fate are executed, a battlefield :*—Đonne rond and hand on herefelda helm ealgodon, on meotudwange, Andr. Kmbl. 21; An. 11.

met-ráp, es; *m. A line for sounding the depth of water :*—Sundgyrd on scipe *vel* metráp *bolidis* (βολὶs), Wrt. Voc. ii. 126, 46: 11, 17.

met-seax. v. mete-seax.

met-scipe, es; *m. Food, refection :*—Habban đa xii heora metscype tôgædere, and fêdan hig swâ swâ hig sylfe wyrđe munon, and dǽlon ealle đa metelâfe, L. Æđelst. v. 8, 1; Th. i. 236, 6. [*Icel.* mat-skapr *victuals, food.*]

metsian; *p.* ode. **I.** *to feed :*—Đú metsast ûs *cibabis nos*, Ps. Spl. 79, 6. Hê metsode hí *cibavit illos*, 80, 15: *nutriebat*, Hpt. Gl. 466, 28 : *saginaverit*, 493, 9. Đú ûs geþafodest him tô metsianne swâ swâ sceáp, Ps. Th. 43, 13. **II.** *to furnish with provisions :*—Heora ǽlc fêrde tô his castele and đone mannoden and metsoden swâ hig betst mihton *each of them went to his castle and manned and provisioned it as well as ever they could*, Chr. 1087; Erl. 224, 16. Him man metsod *they were furnished with provisions*, 1006; Erl. 141, 11. v. ge-metsian.

metsung, e; *f. Provision, food :*—Be manna metsunge. Ânan esne gebyreþ tô metsunge xii pund gôdes cornes, L. R. S. 8; Th. i. 436, 25. Hí tô metsunge fêngon and tô gafle *they accepted provisions and tribute*, Chr. 1002; Erl. 137, 26. Đa gerǽdde se cyng đæt man him gafol behête and metsunge, 994; Erl. 133, 23 : 1006; Erl. 141, 10. Beád đá Swegen full gild and metsunga tô his here, 1013; Erl. 149, 3. Heom man geaf gíslas and metsunga, 1052; Erl. 184, 6.

mettoc. v. mattoc.

metten, e; *f. One of the Fates :*—Đa graman gydena (MS. Cott. mettena) đe folcisce men hâtaþ Parcas, Bt. 35, 6; Fox 168, 24. Cf. metend, metod.

mêtto, met-trum, metud, mêu. v. eáþ-, ofer-mêtto, med-trum, metod, mǽw.

micel; *adj. Mickle, great.* **I.** *of size ; magnus :*—Mycel *magnus*, Wrt. Voc. i. 83, 54, 67. Mycel belle *campana*, 81, 39. Geswollne mid đære ádle đæs myclan líces (*elephantiasis*), Lchdm. ii. 399, col. 2. Micel *grandem*, Wrt. Voc. ii. 41, 70. Đa miclan tân *alloces*, 5, 18.

God geworhte twá micele leóht, ðæt mǽre leóht tô ðæs dæges líhtinge, and ðæt læsse leóht tô ðære nihte líhtinge, Gen. 1, 16. Se læssa íl *iricius*; se mǽra íl *istrix*, Wrt. Voc. ii. 49, 52, 53. Ic tôwurpe míne bernu and ic wyrce mǽran (*majora*), Lk. Skt. 12, 18. Hit is ealra wyrta mǽst *majus est omnibus holeribus*, Mt. Kmbl. 13, 32. Feldhûsa mǽst, Cd. 146; Th. 183, 3; Exod. 85. Of mǽstan dǽle *maxima ex parte*, Bd. 5, 13; S. 633, 2: Ors. 1, 1; Swt. 21, 2. Ða geseah ic beforan unc ðone mǽstan weal, 5, 12; S. 629, 13. Ða têþ of áðó ða ðe hé mǽste hæbbe *remove the biggest teeth it has*, L. Med. ex. Quad. 1; Lchdm. i. 326, 13. II. *of quantity*, *much*, *many*; multus:—Mycel *multum*, Wrt. Voc. i. 83, 67. Ðá com micel wynsum stenc, Shrn. 91, 28. Gê sáwaþ micel sǽd and rípaþ litel *sementem multam jacies in terram et modicum congregabis*, Deut. 28, 38. Him fyligdon mycele menigu (*turbæ multæ*), Mt. Kmbl. 4, 25. Eálá sáwel ðú hæfst mycele gôd (*multa bona*), Lk. Skt. 12, 19. Ðes man wyrcþ mycele tácna (*multa signa*), Jn. Skt. 11, 47. Him mon sôhte mǽstra daga ǽlce *they were attacked most days*, Chr. 894; Erl. 90, 15. His fultum mihte mǽstra (MS. C. mǽstne) ǽlcne heora flâna on heora feóndum áfæstnian, Ors. 6, 36; Bos. 132, 10. III. *great in a metaphorical sense*:—God, ðú eart se miccla kyning, Hy. 3, 38; Hy. Grn. ii. 282, 38. Ic ne eom swá micel swelgere *I am not so great a glutton*; non sum tam vorax, Coll. Monast. Th. 34, 35. Ðá wæs geworden mycel (*loud*) stefn of heofonum, Blickl. Homl. 145, 14: Mt. Kmbl. 27, 46. Micel sído and Rômwarum wæs ðæt ðær nâne ôðre on ne sǽton bûton ða weorþestan (*a custom carefully observed*), Bt. 27, 1; Fox 96, 1. Micel is ðæt and wundorlíc ðæt ðú gehǽtst *magna promittis*, 36, 3; Fox 174, 30. Micel óga him becom, Gen. 15, 12. Biþ ðær seó miccle milts áfyrred, Exon. 28 a; Th. 84, 9; Cri. 1371. On ðam miclan dæge (*the day of judgment*), 23 a; Th. 65, 7; Cri. 1051. On hyra mandryhtnes miclan þearfe, Beo. Th. 5691; B. 2849. Mǽre ł miclu weorc drihtnes *magna opera domini*, Ps. Lamb. 110, 2. Se lícette litlum and miclum, gumena gehwylcum, Bt. Met. Fox 26, 72; Met. 26, 36. Ne árás betwyx wífa bearnum mâra Johanne Fulwihtere, Mt. Kmbl. 11, 11. Ðes is mâra ðonne Saolmon, 12, 42. Nys ôðer mâre bebod, Mk. Skt. 12, 31. Ne þorfte hê nâ mâran fultumes ðonne his selfes, Bt. 26, 2; Fox 92, 23 : 33, 1; Fox 120, 13. Se hæfþ mâran synne se ðe mê sealde, Jn. Skt. 19, 11. Ægðer ge on ðæm mâran (*main*) landum ge on ðæm íglandum, Ors. 1, 1; Swt. 16, 25. Ðonne ðæt gefeoht mǽst wǽre *when the fight was hottest*, 4, 11; Swt. 206, 18. Se mǽsta *precipuus*, Wrt. Voc. ii. 81, 66. Drihten is on Sion dêma se mǽsta, Ps. Th. 98, 2. Manege tellaþ ðæt tô mǽstum gôde and tô mǽstere gesǽlþe ðæt mon síe simle blíðe, Bt. 24, 2; Fox 82, 12. On ðam mǽstan dæge (*the day of judgment*), Exon. 115 b; Th. 445, 11; Dôm. 6. Pirrusan ðone mǽstan feónd Rômánum, Ors. 3, 5; Swt. 106, 4. On ðâm wǽron ða ǽrestan and ða mǽstan (*primi et præcipui*), Bd. 1, 29; S. 498, 7. IV. *neuter used substantively* (a) *with gen.*:—Ic nât nâht gewislíce hwæðer ðæs feós swâ micel is, ne ic nât ðeáh his mâre sý, Chart. Th. 490, 15. Heora heriges wæs mycel ofslægen, Bd. 3, 18; S. 546, 35. Hé wæs wilniende ðæt hê ðæs gewinnes mehte mâre gefremman *he was desirous to carry on the struggle*, Ors. 2, 5; Swt. 82, 8. Hit mâre ðæs landes forbærnde ðonne hit ǽrfþ ǽr dyde, 5, 2; Swt. 220, 16. Ðæt hí þurh ðæt mǽge mǽst bearna begitan, Bt. 24, 3; Fox 82, 25. Ðǽr manna wese mǽst ætgædere, Ps. Th. 78, 10. Se ðissum herige mǽst hearma gefremede, Andr. Kmbl. 2397; An. 1200. (b) *without gen.*:—On swâ miclum heó hæfþ genóg swâ wê ǽr sprǽcon. Gif ðú heore mâre selest . . ., Bt. 14, 1; Fox 42, 11. Ðæt hê mid swâ lytle weorode swâ micel anginnan dorste, Ors. 3, 9; Swt. 124, 16. Hû mycel scealt ðú *quantum debes?* Lk. Skt. 16, 5. Hû mycel hê dyde mínre sáwle, Ps. Th. 65, 14. Ðæt hê genóg hæbbe and nó mâran ne þurfe, Bt. 26, 1; Fox 92, 10. Ðǽm ðe ǽnigre wuhte mâre habbaþ . . . swâ hê mâre hæfþ swâ hê mâ monna ôleccan sceal, 26, 2; Fox 92, 29–33: 26, 3; Fox 94, 16. Ic sceal erian fulne æcer oððe mâre . . . Hwæt mâre dêst ðú? Gewyslíce mâre ic dó, Coll. Monast. Th. 19, 23–35. Ðonne hí mǽst tô yfele gedôn hæfdon, ðonne nam man grið and frið wið hí, Chr. 1011; Erl. 145, 2. V. *oblique cases used adverbially*:—Se lǽce biþ micles tô beald (*much too bold*), Past. 9; Swt. 61, 2. Ðara micles tô feala winþ wiþ gecynde, Bt. Met. Fox 13, 32; Met. 13, 16. Micles on æþelum wíde is geweorðod hâligra tíd, Menol. Fox 236; Men. 119. Hié God wolde onmunan swâ micles, Andr. Kmbl. 1789; An. 897. Micclum *nimium*, Ælfc. Gr. 38; Som. 40, 46. Ne cweþe ic nâ ðæt ðeós bôc mâge micclum tô lâre fremian, pref.; Som. 1, 43: Herb. 17, 2; Lchdm. i. 110, 10. Ealle micclum ðæs wundrodon, Homl. Th. i. 42, 16, 21: Ps. Th. 103, 14. Ne him mycelum ondrǽdeþ, 111, 6. Swâ man æt mêder biþ miclum fêded, 130, 4: Andr. Kmbl. 244; An. 122: Bt. Met. Fox 13, 40; Met. 13, 19. Micel ic gedeorfe *multum laboro*, Coll. Monast. Th. 20, 25. Oftor micle *much oftener*, Bt. Met. Fox 19, 37; Met. 19, 19. Hê wæs micle ðê blíðra, 9, 63; Met. 9, 32. Swíðe micle scyrtran ymbhwearft, 28, 14; Met. 28, 7. Nôht micle ǽr, Bd. 4, 23; S. 593, 21. Ðam mycle mâ (*quanto magis*) hê scrýt eów, Mt. Kmbl. 6, 30. Ic þegnum ðínum dyrnde and sylfum ðê swíðost micle *I concealed it from thy servants, and from thee much the most*, Cd. 129; Th. 164, 12; Gen. 2713. [Laym. O. E. Homl. A. R.

Chauc. Ayenb. muchel, mochel: Orm. Havel. mikel; Gen. a. Ex. mikel, michel: Goth. mikils: O. Sax. O. L. Ger. mikil: Icel. mikill: O. H. Ger. michil.] v. efen-, frǽ-, mis-, ofer-, wuldor-micel, *and* mâ.

micel-ǽte; adj. *Eating much, gluttonous:*—Ic geseó dæighwamlíce ðæt ðú mycelǽte eart, Shrn. 16, 20. Cf. ofer-ǽte.

micel-dóend; adj. *Doing great things*; magnificus, Rtl. 45, 14.

micel-heáfded; adj. *Having a great head:*—Mycelheáfdode *capitosus*, Wrt. Voc. i. 45, 34. Micelheáfdede, ii. 22, 69.

micelian, miclian, miclian; p. ode. I. *to become great, to increase in size or in quantity:*—Micelaþ *grandescit, crescit*, Wrt. Voc. ii. 42, 42. Rím miclade, Cd. 63; Th. 75, 21; Gen. 1243; Andr. Kmbl. 3050; An. 1528. Wæter micladon *the waters waxed*, 3105; An. 1555. Ðæt folc ongan weaxan and myclian (*grandescere*), Bd. 1, 15; S. 483, 33. On ðǽm dagum wæs ðæt norþmeste (*ríce*) micliende, Ors. 6, 1; Swt. 252, 12. II. *to make great, to increase the size or quantity of a thing:*—Man myclade ðæt ordálýsen *the ordeal-iron should be increased in weight*, L. Æðelst. iv. 6; Th. i. 224, 13. Ðæt ic mǽgburge môste ðínre rím miclian, Cd. 101; Th. 134, 7; Gen. 2221. III. *metaphorically, to extol, magnify:*—Miclaþ sáwel mín drihten *magnificat anima mea dominum*, Lk. Skt. Rush. 1, 46. Mycclaþ, Blickl. Homl. 7, 2. Ic micliu *magnificabo*, Ps. Surt. 68, 31. Wê micliaþ *magnificabimus*, 11, 5. Eal ðæt folc his noman myccledon, Blickl. Homl. 16, 29. Myclian wê his noman, 13, 7. [*Jul.* muchelin, mucli: A. R. muchelen: Ps. mikel: Goth. mikiljan: Icel. mikla: O. H. Ger. michilên.] v. ge-miclian.

micel-líc; adj. *Great, grand, magnificent, splendid, illustrious:*—Micellíc *magnificum*, Wrt. Voc. ii. 54, 64. Wæs se wer for Gode and for mannum micellíc (*magnificus*), Bd. 5, 20; S. 641, 38. Hú his mægenþrynnes mycellíc standeþ, Ps. Th. 110, 2. Hwæt ðæt síe mǽrlíces and micellíces ðæt git mec gehâtaþ *quid sit illud quod mihi tam illustre et tam magnificum pollicemini*, Nar. 25, 12: Bt. 18, 1; Fox 62, 21. Hú micellíce (*magnificata*) sind werc ðín, Ps. Surt. 91, 6. [Icel. mikilligr: O. H. Ger. michil-líh illustris, magnificus.]

micel-líce; adv. I. *greatly, grandly, splendidly:*—Singaþ dryhtne forðon micellíce (*magnifice*) dyde, Ps. Surt. p. 184, 15. II. *greatly, exceedingly:*—Micellíce intimbred *multipliciter instructus*, Bd. 5, 8; S. 622, 2. Micellíce gelǽred *doctissimus*, 5, 23; S. 645, 13. [Icel. mikilliga: O. H. Ger. michil-lího magnifice, magnopere, exaggerative.]

micel-môd; adj. *Having a great mind, magnanimous:*—Nis his micelmôdes mægenes ende *magnitudinis ejus non est finis*, Ps. Th. 144, 3. [O. H. Ger. michil-muot magnanimus, animosus.]

micel-ness; e; f. I. *greatness, bigness, size:*—Stânas on pysna mycelnesse *stones the size of peas*, Herb. 180, 1; Lchdm. i. 314, 22: Blickl. Homl. 181, 21. Se clâð wæs swíðe gemǽte hire micelnysse *the garment was exactly adapted to her size*, Homl. Skt. 7, 157. His micelnesse ne mæg nâu man âmetan, Bt. 42; Fox 258, 12. II. *greatness (of quantity), multitude, abundance:*—Ðá wæs geworden mid ðam engle mycelnes (*multitudo*) heofonlíces werydes, Lk. Skt. 2, 13. Ne meahton âsecgan for ðæs leóhtes mycelnesse, Blickl. Homl. 145, 14. Æfter micelnisse ðínre mildheortnisse, Num. 14, 19. III. *greatness, magnificence:*—Micylnys *magnificentia*, Ps. Spl. C. 8, 2. In micelnisse *in magnificentia*; in mikelnes, Ps. Surt. 28, 4. On mycelnysse earmes ðínes *in magnitudine brachii tui*, Cant. Moys. 16: Ps. Spl. 78, 12. Ús weorþ þuruh ðíne mycelnesse milde and blíðe, Ps. Th. 66, 1. Sancte Johannes mycelnesse se Hǽlend sylfa tâcn sægde, Blickl. Homl. 167, 17. [Wick. michelnes: O. H. Ger. michil-nessi majestas.]

micel-sprecende; adj. *Talking big, boasting:*—Tungan micelsprecende *linguam magniloquam*, Ps. Lamb. 11, 4.

micelu, e; f. *Size:*—On ðære mycele ðe leáces *of the size of a leek*, Herb. 49, 1; Lchdm. i. 152, 16. [Goth. mikilei greatness: O. H. Ger. michili magnitudo, quantitas.]

micelung, miclung, e; f. *A doing of great things*; magnificentia:—Miclung ł mǽrsung weorc his *magnificentia opus ejus*, Ps. Lamb. 110, 3. v. ge-miclung.

micga, an; m. *Urine:*—Hlond *vel* micga *lotium*, Wrt. Voc. i. 21, 63: urina, 46, 8. Drince buccan micgan . . . sélost ys se micga ðæt hê sý oftost mid fêded, L. Med. ex Quad. 6, 16; Lchdm. i. 354, 12, 15. Fûles hlondes, miggan *foetentis lotii*, Hpt. Gl. 483, 19. Stingendum miggan *putenti lotio* (*urina*), 487, 65. [A. R. migge.] v. micge.

micge, an; f. *Urine:*—Gesceáwa ǽlce dæge ðæt ðín útgong and micge síe gesundlíc. Gif sió micge síe lytelu . . ., L. M. 2, 30; Lchdm. ii. 226, 20. Ðonne onginþ ðære hǽto welm wanian þurh ða micgean, 2, 23; Lchdm. ii. 212, 7 : 1, 37; Lchdm. ii. 88, 20.

micgern. v. mycgern.

micgþa. v. migþa.

micgung, e; f. *Making water:*—Miggung *minctio*, Wrt. Voc. i. 46, 9.

micle, micles, miclum; miclung. v. micel; micelung.

MID, (in Gloss. Ep. and Lindisfarne Gospels) mið; *prep. with dat. acc. inst. With*; at the root of the various meanings lies the idea of association, of being together. I. having very nearly the same force as *and*, (a) *with dat. or inst.*:—Hig lǽddon hí of ðære byrig mid eallum

hire māgum (*Rahab et cunctum cognationem illius*), Jos. 6, 23. Wē sungon seofon seolmas mid letanian, Coll. Monast. Th. 33, 29. Se feónd mid his gefērum eallum feóllon of heofnum, Cd. 16; Th. 20, 10; Gen. 306. Đū scealt friþ habban mid sunum đīnum *thou and thy sons shall be protected*, 65; Th. 78, 28; Gen. 1300. Æđelinga bearn, weras mid wīfum, 83; Th. 104, 20; Gen. 1738. (b) *with acc.*:—Wes đū hāl mid đās willgedryht, Andr. Kmbl. 1828; An. 916. **II.** with the idea of joint action or companionship, *in conjunction with, in company with, along with*, (a) *with dat.* or *inst.*:—Ic sang ūhtsang mid gebrōđrum *cantavi nocturnam cum fratribus*, Coll. Monast. Th. 33, 25. Mittan wītegan clypige, R. Ben. 29, 6. Mit đam wītegan cweđan, 31, 16. Đā fērde se Hælend mid him, Lk. Skt. 7, 6. Mycel menegu wæs mid hyre, 7, 12. Đā bebeád se fæder đæm consule đæt hē mid his fierde angeán fōre, and hē beæftan gebād mid sumum đæm fultume, Ors. 3, 10; Swt. 140, 19. Gefeaht Æþelhelm wiđ Deniscne here mid Dornsǽtum, Chr. 837; Erl. 66, 8. Se winterlīca wind wan mid (*in league with*) đam forste, Homl. Skt. 11, 144. Ic fleáh mid fuglum, Exon. 126 b; Th. 487, 16; Rä. 73, 3. Hē fulluhtes gerýno onfēng mid his þegnum đe mid hine wǽron, Bd. 3, 3; S. 525, 27. Đa eágan ... ætgædere mid đæs martyres heáfde on eorþan feóllan, 1, 7; S. 478, 38. (b) *with acc.*:—Đē dǽlnimende gedēþ mid þone, 2, 12; S. 515, 29. Hē bæd đæt hē mid þone martyr þrowian mōste, 1, 7; S. 478, 18: 1, 23; S. 485, 27. Nemþe hē Cristes geleáfan onfēnge mid đa þeóde đe hē ofer cyning wæs, 3, 21; S. 551, 1. Hē gewāt mid cyning engla, Cd. 60; Th. 73, 26; Gen. 1210: Beo. Th. 1329; B. 662. Đæt mīnne līchaman mid mīnne goldgyfan glēd fæđmie, 5297; B. 2652. Ic mid mec gelædde mīne frýnd, Nar. 29, 26. Mid dryhten rūne besǽton, Andr. Kmbl. 1252; An. 626. (c) *with inst.*:—Eode hē in mid āne his preósta, Bd. 3, 5; S. 527, 4. His hand mid đý earme đe of his līchoman āslegen wæs hē hēt tō āhōn, 3, 12; S. 537, 34. Mid medmycele werede hē fērde, 3, 24; S. 556, 20. **III.** with the idea of reciprocal action:—Hē wolde mid his freóndum sprǽce and geþæht habban, Bd. 2, 13; S. 515, 36. **IV.** expressing the relation between animate and inanimate things, (a) *with dat.* or *inst.*:—Đā đa wīfmen urnon mid stānum wiđ đara wealla *cum matronae currerent, et convehere in muros saxa gestirent*, Ors. 4, 10; Swt. 194, 11. Twelf stānas hī hæfdon forþ mid him, Jos. 4, 8. Faran tō eá mid scype mīnum, Coll. Monast. Th. 24, 23. Ic āstīge mīn scyp mid hlæstum mīnum, 26, 31: Beo. Th. 250; B. 125. Hælend cymeþ mid wolcnum, Cd. 227; Th. 303, 5; Sat. 608. Hī fērdon mid đý hālgan Cristes mǽle, Bd. 1, 25; S. 487, 22. (b) *with acc.*:—Đa (*these things*) mid hine brohte, 2, 4; S. 505, 38. Mid đa nōþe niđer gewīteþ, Exon. 97 a; Th. 361, 31; Wal. 28. **V.** with the idea of an association which affords protection or help:—For đan đe ic beó mid đē on eallum đām đe đū tō færst, Jos. 1, 9; Mt. Kmbl. 28, 20. Theodosius hæfde đone wind mid him, đæt his fultum mehte mǽstra ǽlcne heora flāna on hiora feóndum āfæstnian, Ors. 6, 36; Swt. 294, 26. **VI.** with the idea of permanent association, (residing) *with, at,* (when the relation expressed is that of one to many) *among*; apud, penes, (a) *with dat.*:—Elles næbbe gē mēde mid eówrum Fæder (*apud patrem vestrum*), Mt. Kmbl. 6, 1. Bæd æt Gode đæt hē him geswutelode hwylc Basilius wǽre on wurđscype mid him (*in what estimation he was with God*), Homl. Skt. 3, 498. Eallum ūs leófre ys wīkian mid (*apud*) đam yrþlinge đonne mid (*apud*) đē, Coll. Monast. Th. 31, 1. Ys seó mildheortnes mid (*apud*) đē, Ps. Th. 129, 4. Albanus hæfde đone andettere mid (*penes*) him, Bd. 1, 7; S. 477, 7. Mid mannum ic eom *apud homines sum*, mid đam biscope hē wunaþ *apud episcopum manet* .. mid eów hē is *penes vos est*, mid đēmum *penes judices*, Ælfc. Gr. 47; Som. 47, 23-47. Ic wæs mid Englum, Exon. 85 b; Th. 322, 10; Víd. 61 (and often). Ic hæfde đē mid đam fyrmestan đe mīnum hȳrēde folgodon *I held thee among the first who followed my court*, Homl. Skt. 5, 412: Ors. 1, 1; Swt. 18, 13. Gefrugnen mid folcum *known among nations*, Exon. 11 a; Th. 14, 26; Cri. 225. (b) *with acc.*:—Is mīn hyht mid God, 37 a; Th. 121, 16; Gū. 289: 39 a; Th. 128, 27; Gū. 410. Sibb sý mid eówic, 75 b; Th. 282, 25; Jul. 668. Wuna mid ūsic, Cd. 130; Th. 164, 29; Gen. 2722. **VI a.** *between*:—Dēma mid unc twih *a judge between us two*, 102; Th. 136, 5; Gen. 2253. **VII.** expressing an accompanying circumstance, the phrase being often equivalent to an adverb of manner, (a) *with dat.*:—Mid gódum willan fæstan, Blickl. Homl. 37, 27 : 35, 27. Mid his sylfes willan, willum *ultro*, Bd. 1, 7; S. 477, 22, 15. Mid mycelre willsumnysse bodian *magna devotione predicare*, 3, 3; S. 526, 4. Hē hæfde hī mid mycelre āre mid him, 4, 1; S. 564, 33. Wæs sió fǽmne mid hyre fæder willan beweddad, Exon. 66 a; Th. 244, 24; Jul. 32. Brūc đisses beáges mid hǽle, Beo. Th. 2438; B. 1217. Ic eów mid gefeán ferian wille, Andr. Kmbl. 693; An. 347. Winnan mid māne (*criminally*), Cd. 16; Th. 19, 30; Gen. 299. Mid swāte and mid sorgum libban, 24; Th. 31, 8; Gen. 482. Wīf đonne heó mid cylde biþ *mulier gravida*, L. Ecg. C. 28, tít; Th. ii. 130, 14. Heó wæs mid bearne (cf. *Icel.* ganga međ barni), Shrn. 60, 33. Đā heó mid đam bearne wæs, 149, 1. Swā heó đam cilde wearþ, Homl. Th. i. 460, 7. (b) *with acc.*:—Đæt hē mid đa mǽstan swētnesse (*maxima suavitate*) geglencde, Bd. 4, 24; S. 596, 34.

(c) *with inst.*:—Đā ongan hē mid gleáwe mōde þencan, 3, 10; S. 534, 20 : Past. 9, 1; Swt. 55, 20. **VIII.** expressing the idea of instrumentality, *by, through*, (a) *with inst.* or *dat.*:—Hié wǽron gebrocede ... mid đæm đæt manige đara sēlestena cynges þēna forþfērdon *they suffered from the death of many of the best king's thanes*, Chr. 897; Erl. 94, 32. Ne canst đū huntian būton mid nettum ? ... Mid swiftum hundum ic betǽce wildeór, Coll. Monast. Th. 21, 21-27. Đū đæt land tōdǽlst mid hlyte (*sorte*), Deut. 31, 7. Mid đissum woruldgesǽlþum and mid đīs andweardan welan mon wyrcþ oftor feónd đonne freónd, Bt. 24, 3; Fox 84, 2-4. Mid his handum gesceóp, Cd. 14; Th. 16, 30; Gen. 251. Hié heora līchoman leáfum beþeahton, weredon mid đȳ wealde, 40; Th. 52, 19; Gen. 846. Stōd bewrigen folde mid flōde, 8; Th. 10, 15; Gen. 157. Ofgeót mid scīre wīne ealde, L. M. 2, 11; Lchdm. ii. 188, 20. Mid monige wīte þreágan, Shrn. 101, 23. Mid đȳ blōde gewurþad, Bd. 1, 7; S. 478, 24. Mid deáþe fornumen, forgripen, 1, 27; S. 492, 30: 3, 8; S. 532, 27. Mid his lāre *by means of his teaching*, 3, 28; S. 560, 38. Mid gȳmenne mīnra māga *by the care of my kinsmen*, 5, 24; S. 647, 22. Dǽle hē swā mycel feoh for hyne swā hē ǽr mid him nam (*as much as he got with him*, i. e. *by selling him*), L. Ecg. P. iv. 26; Th. ii. 212, 12. Eom ic leóhte geleáfan and mid lufan gefylled, Exon. 42 a; Th. 141, 9; Gū. 624. Hē frægn hī mid hwī hī gesceldan heora hūs wiđ đæs fȳres frēcennysse, Shrn. 90, 7. Gewiton mid đȳ wǽge in forwyrd sceacan *carried by the wave they hurried to destruction*, Andr. Kmbl. 3186; An. 1596: Cd. 12; Th. 14, 5; Gen. 214. (b) *with acc. (and inst.*):—Hē mid hī fēran sceolde tō đon đæt hē đa fǽmnan æghwæđer ge mid đa (đære, MS. B.) mǽrsunge heofonlīcra gerýna ge mid his dæghwamlīcre lāre trymede, Bd. 2, 9; S. 510, 37. Hē monige ... mid đa leornunga đissa bōca gelǽdde, 5, 18; S. 636, 4: Cd. 100; Th. 133, 9; Gen. 2208. Se mihtiga slōh mid hālige hand, 167; Th. 208, 18; Exod. 485. **IX.** having reference to time, *with, at*:—On ūhtan mid ǽrdæge, Beo. Th. 253; B. 126: Andr. Kmbl. 2776; An. 1390: 3048; An. 1527: Cd. 121; Th. 155, 19; Gen. 2575. **X.** *giving direction*:—Onlong brōces mid streáme *along the brook in the direction in which it runs*, Cod. Dip. Kmbl. vi. 226, 20. **XI.** in adverbial or conjunctional phrases, (a) *with* eallum, ealle:—Hyne myd scrȳne myd eallum on feastum cwearterne beclȳsdon *they shut him up cage and all in prison*, St. And. 38, 9. Mid ealle *penitus*, Ælfc. Gr. 38; Som. 40, 46. Mid stybbe mid ealle *stirpitus*; mid wyrttruman mid ealle *radicitus*, Som. 42, 3-4. Hié āsettan hī on ǽnne sīþ ofer mid horsum mid ealle, Chr. 893; Erl. 88, 24 (cf. *Icel.* međ öllu). (b) *with dat. or inst.* case of the demonstrative, denoting that the two actions expressed by the verbs in the connected clauses are in close association, being either simultaneous, or the one following upon, and being regarded, more or less, as the result of the other, *when, since, seeing that*; cum:—Mid đam đe se apostol stōp intō đære byrig, đā bær man him tōgeánes ānre wydewan līc, Homl. Th. i. 60, 11. Mid đam đe hē hig geseah đá ēfste hē *quos cum vidisset, cucurrit*, Gen. 18, 2. Hū yfele mē dōþ manege woruldmenn, mid đam đæt ic ne mōt wealdan mīnra āgenra þeówa *how ill do many men act towards me, when I may not rule my own servants*, Bt. 3; Fox 20, 19. Mid đȳ đe heó gehȳrde ... đa cwæþ heó, Blickl. Homl. 7, 19 : 15, 6. Mid đī đe hié cōmon ... hié gemētton seofon hyrdas standan, 237, 17. Mid đȳ đe (*dum*) hē hine geseah on singalum gebedum ... đā wæs hē semninga mid đam godcundan gyfe gemildsad, Bd. 1, 7; S. 476, 37. Mittē *dum*, Ps. Surt. 67, 8. Mid đȳ (*cum*) Peohtas wīf næfdon, bǽdon him fram Scottum, Bd. 1, 1; S. 474, 19. Gif hē eów ne wyllan ārīsan tōgeánes, mid đȳ eówer mā is (*cum sitis numero plures*), 2, 2; S. 503, 13: 1, 27; S. 493, 42. Mid đī hē đis cwæþ, hē āstāh on heofenas, Blickl. Homl. 237, 15. **XII.** *used after its case* or *as an adverb*:—On đam clifian đe him gód mid worhte *cleave to him who did good with them*, Bt. 16, 3; Fox 56, 10, 12. Đa him mid scoldon *which were to go with him*, Beo. Th. 82; B. 41. Đara đe hē him mid hæfde, 3255; B. 1625: 1783; B. 889: Homl. Th. ii. 490, 24. Manega ōđre đe him mid (*simul cum eo*) fērdon, Mk. Skt. 15, 41. Mid fērdan *comeant, simul pergebant*, Wrt. Voc. ii. 132, 45. Hē his heres þriddan dǽl gehȳdde and him self mid wæs, Ors. 3, 7; Swt. 116, 27. Hine mid wunode ān ombehtþegn, Exon. 47 a; Th. 162, 8; Gū. 972. Biddan đone ele đæt đū Adam myd smyrian mōte *to ask for the oil, to anoint Adam with*, Nicod. Thw. 13, 23. Smyre đone man mid, Herb. 54, 3; Lchdm. i. 158, 2. Đā sceolde hē sendan lȳgetu and windas, and tōwyrpan eall hira geweorc mid, Bt. 35, 4; Fox 162, 14. Ic wilnode andweorces đone anweald mid tō gereccenne, 17; Fox 60, 8: 20; Fox 72, 24. Se forma hád and se ōđer hád beóþ ǽfre ætgedere ... se þridda hád is hwīlon mid, hwīlon on ōđre stōwe, Ælfc. Gr. 15; Som. 17, 39. Đonne se mon nō his āgenne gielp mid ne sēcþ, Past. 59; Swt. 451, 15. Gif he nōht geseón ne mǽge mid, L. Alf. pol. 47; Th. i. 94, 6. Hē hæfde mildheortnysse đa þearfan mid tō frēfrigenne, Bd. 3, 17; S. 545, 13. Geond đone ofen eodon and se engel mid, Cd. 191; Th. 238, 14; Dan. 354. Đæt wæs Satane and his gesīđum mid, Exon. 30 a; Th. 93, 7; Cri. 1522. Ælc đara đe mid stande *every one that stands by* (*assists*) *him*, L. Ath. i. 1; Th. i. 200, 3. [*Mid* occurs in Piers P., and still remains in *mid*-wife: *Goth.* miþ, mid;

O. Sax. midi, mid: O. Frs. mith, mit, mei: O. L. Ger. mid, mit, met: Icel. með: Swed. Dan. med: O. H. Ger. miti, mit: Ger. mit: Du. met.]

midd; adj. with superl. midemest, midmest Mid, middle. I. of place:—Seó burh wæs on midre ðære eá (in medio amne), Nar. 10, 11. Ðá wé wǽron on midde ðære sǽ (in medio mari), Bd. 5, 1; S. 613, 23. Is on middre ðære cyricean, 2, 3; S. 504, 39. Hire (the axis) midore ymbe (cf. ymb ða eaxe middewearde, Bt. 39, 3; Fox 214, 23), Bt. Met. Fox. 28, 46; Met. 28, 23. On middum ðínum temple in medio templi tui, Ps. Th. 47, 8. On mereflóde middum, Cd. 8; Th. 9, 22; Gen. 145. Gáþ from geate tó geate þurh midde ða ceastre (per medium castrorum), Past. 49, 2; Swt. 383, 3: St. And. 14, 17. On middum ðǽm úrum wícum in media castrorum parte, Nar. 12, 24. Ða gesettan scép in middum wulfum (in medio luporum), Bd. 2, 6; S. 508, 16. Hé mé lǽdde betweoh midde ða þreátas inter choros medios, 5, 12; S. 629, 26. Hǽlettend midemest finger salutarius; ʒewiscberend midmesta finger impudicus, Wrt. Voc. i. 283, 21–22. Gif hí ðone midmestan weg áredian willaþ, Bt. 40, 3; Fox 238, 23. Ða sélestan men ... ða midmestan ... swá bióþ ða midmestan men, 39, 7; Fox 222, 1–10, 15. II. of time:—Tó middes dæges Crist wæs on róde áþened, Btwk. 216, 14. On middes wintres mæssenint, Chr. 827; Erl. 62, 30. Swá hé in swoloþan middes sumeres wǽre quasi in mediae aestatis caumate, Bd. 3, 19; S. 549, 30. Sunnon upgong æt middan sumere ortum solis solstitialem, 5, 12; S. 627, 35. Hwæt án lengten foran tó middan wintra (ante Natale Christi) ... fæste ii lengtenu, án tóforan middan sumera (ante mediam æstatem), óðer foran tó middan wintra, L. Ecg. P. iv. 22, 23; Th. ii. 210, 25–28. Tó middan (middum, MS. B.) wintre, L. Ath. iv. pref.; Th. i. 226, 5. Tó middyre (MS. A. myddre) nihte media nocte, Mt. Kmbl. 25, 6. Æt midre niht, Ps. Th. 118, 62. Æt middre nihte, Exon. 59 b; Th. 216, 2; Ph. 262. Æt middere niht, Cd. 144; Th. 179, 32; Exod. 37. Hé leng ne leofaþ ðonn on midre ilde he will not live beyond middle age, Lchdm. iii. 162, 21; Ps. Th. 54, 24. On midne dæg meridie, Ælfc. Gr. 38; Som. 41, 47. Seó seofoþe tíd dæges, ðæt is án tíd ofer midne dæg, Bd. 5, 6; S. 619, 27. On midne winter, Chr. 878; Erl. 78, 28. Ofer ðone midne sumor after midsummer, 1006; Erl. 140, 5. Ofer midne sumor, Lchdm. iii. 74, 11. On midde niht, Bd. 4, 8; S. 575, 40. [Goth. midjis: O. Sax. middi: O. Frs. midde: Icel. miðr: O. H. Ger. mitti.] v. on-middan, tó-middes.

mid-dæg, es; m. Mid-day:—Middæg sexta, Wrt. Voc. i. 53, 12: Coll. Monast. Th. 33, 33: Jn. Skt. 4, 6. Middæg meridies, Ælfc. Gr. 12; Som. 15, 46: Hymn. Surt. 16, 29. Ðæs middæges gereord, R. Ben. 65, 20 Tó middæges, 65, 18. Tó middæge at midday, Lchdm. iii. 218, 4, 6, 9, etc. On ðæm sumerlícan sunnstede on middæge (MS. R. middan dæge), 258, 15. [O. Frs. mid-dei: Icel. mið-dagr: O. H. Ger. mitti-tag: Ger. mit-tag.] v. middel-, midne-dæg.

middæg-líc; adj. Midday, meridian:—Ðære middæglícan sunnan scíman beorhte solis meridiani radiis præclarior, Bd. 5, 12; S. 629, 23. Fram deófle middæglícum ab daemonio meridiano, Ps. Spl. C. 90, 6. [O. H. Ger. mittitaga-líh.]

middæg-sang, es; m. The midday service:—Úhtsang and prímsang, undernsang and middægsang, nónsang and ǽfensang, and nihtsang, L. Ælfc. C. 19; Th. ii. 350, 7. De officio sextae horae. Middægsang. On midne dæg wé sculon God herian, Btwk. 216, 13: R. Ben. 39, 19: 40, 7.

middæg-tíd, e; f. The midday hour, noon; meridies, Wrt. Voc. ii. 58, 66.

middandæg-líc, adj. Midday, meridian:—Fram middendægiícum deófle ab daemonio meridiano, Ps. Lamb. 90, 6.

middan-eard, es; m. The middle dwelling, the abode of men, the earth, the world (in a physical sense)—De mundo. Middaneard is ʒeháten eall ðæt binnan ðam firmamentum is ... Seó heofen and sǽ and eorþe synd geháte middaneard, Lchdm. iii. 254, 6–9. Hé sǽde, ðæt al ðes middaneard nǽre ðe mǽre dríges landes ofer ðone mycelan ǽrsecg, ðonne wan þæen prican áprície on ánum brádum brede. And ys ðes middaneard búton swylce se seofoþa dǽl ofer ðone mycelan ǽrsecg, se ðe mid his ormǽtnysse ealle ðas eorþan útan emblíþ, Wulfst. 46, 19–24. Middaneardes gewissast ðú ðe getimbrunge mundi regis ui fabricam, Hymn. Surt. 91, 21. Ðone eard Asiam, se ðe is geteald ʒ healfan dǽle middaneardes, Homl. Th. i. 68, 35. Eálá middaneard! ʒlá dæg leóhta! eálá upheofon! Cd. 216; Th. 275, 2; Sat. 165. ᵹme sceolon hweorfan geond hælepa land ... geond middaneard, 219; ʒh. 281, 16; Sat. 272. Geond eorþan ... ofer middaneard, Ps. Th. 137, : 144, 12. Ðú miht on ánre hand befealdan ealne middaneard, Hy. 7, ᵹo; Hy. Grn. ii. 290, 120. II. the world, mankind:—Ealle ðe ᵹriaþ ... eall middaneard, 9, 38; Hy. Grn. ii. 292, 38. Middaneardes ᵹélynd salvator mundi, Jn. Skt. 4, 42. Ic eom middaneardes leóht ða ᵹíle ðe ic on middanearde eom, 9, 5: 8, 12. [Laym. midden-erd; cf. ᵹym. Orm. Gen. and Ex. Havel. middel-erd, -ærd.] v. middan-geard d next word.

ᵹiddaneard-líc; adj. Earthly. I. in a physical sense:—Ðæt ᵹ mid hyre hǽtan middaneardes (other MSS. middaneardlíce) wæstmas

ne forbærne, Lchdm. iii. 250, 17. II. as distinguished from spiritual or heavenly, worldly, mundane, earthly:—Godes sunu becom tó ðissum middanearde tó ðí ðæt hé mid his hálgan láre middaneardlíc gedwyld (human error) ádwæscte, Homl. Th. ii. 90, 13: 366, 9. On middaneardlícum lustum in worldly pleasures, 368, 3. Ealle middaneardlíce þing forhogiende despising all the things of this world, 130, 1. Middaneardlíce genipu mundana nubila, Hymn. Surt. 74, 3: 91, 23: Homl. Skt. 2, 241.

middan-geard, es; m. I. the middle dwelling (between heaven and hell), the earth, world:—Middangeard chosmos, Wrt. Voc. ii. 16, 36. Se læssa middangeard microchosmos, 56, 22. On Godes onwealde is eal ðes middangeard, and ðás windas and ðás regnas syndon ealle his, and ealle gesceafta syndon his, Blickl. Homl. 51, 19. Ðes middangeard wæs tó ðon fæger, ðæt hé teáh men tó him þurh his wlite, 115, 10. Ðes middangeard daga gehwylce fealleþ and tó ende éfsteþ, 59, 26: Exon. 77 a; Th. 290; Wand. 62. Cwealmdreóre swealh middangeard earth drank gore, Cd. 47; Th. 60, 23; Gen. 986. Gefylled wearþ eall ðes middangeard monna bearnum, 75; Th. 93, 30; Gen. 1554. Beofaþ middangeard, hrúse under hæleþum, Exon. 20 b; Th. 55, 12; Cri. 882. Ealne ðisne ymbhwyrft ðises middangeardes swá swá Oceanus útan ymblígeþ orbem totius terrae, Oceani limbo circumseptum, Ors. 1, 1; Swt. 8, 1. Middangeardes, eorþan sceátta, Beo. Th. 1507; B. 751. Ríce middangeardes ðǽr nó men búgaþ hunc orbem, mors ubi regna tenet, Exon. 58 a; Th. 208, 17; Ph. 157. Rícsian on ðiosan middangearde, Ors. 1, 2, tit.; Swt. 1, 4. Seó ród biþ árǽred on ðæt gewrixle ðara tungla, seó nú on middangearde áwergede gástas flémeþ, Blickl. Homl. 91, 24. Ðæt nǽre nǽfre nǽnig tó ðæs hálig mon on ðissum middangearde, ne furþum nǽnig on heofenum, 117, 26. Swá hwǽr swá ðys godspel byþ gebodud on eallum myddangearde (in toto mundo), Mt. Kmbl. A. 26, 13. Geond ealne middangeard, Blickl. Homl. 69, 19. Ðá ic wíde gefrægn weorc gebannan manigre mǽgþe geond ðisne middangeard, Beo. Th. 151; B. 75: Exon. 33 a; Th. 104, 1; Gú. 1: 95 b; Th. 355, 37; Pa. 1. God ðysne middangeard tócleófeþ, Blickl. Homl. 109, 35: Andr. Kmbl. 322; An. 161. II. the world and they that dwell therein, mankind:—Se middangeard ús wæs lange underþeóded, and ús deáþ mycel gafol geald, Blickl. Homl. 85, 11. Him æteówde eal eorþan ríce and ídel wuldor ðisses middangeardes, 27, 17: 65, 15. Líf ðysses middangeardes this present life, 59, 27. Gé synt middangeardes (-geardes, MS. A.) leóht vos estis lux mundi, Mt. Kmbl. 5, 14. Ða hwatestan men ealles ðises middangeardes, Ors. 1, 10; Swt. 48, 6. Hú gesǽlig seó forme eld wæs ðises middangeardes, Bt. 15; Fox 48, 3. Heofones waldend, ealles waldend middangeardes, Exon. 16 a; Th. 35, 12; Cri. 557: 65 b; Th. 241, 32; Ph. 665: Andr. Kmbl. 453; An. 227. Middangeardes weard (Nebuchadnezzar), Cd. 205; Th. 253, 17; Dan. 597. Gecýþ nú middangearde blisse, Blickl. Homl. 87, 24. Hé getácnaþ ðysne middangeard, se wæs synna and mána full, 75, 5. Hé com on ðære syxtan ylde on ðysne middangeard mancyn tó álýsenne, 71, 26: Homl. Th. i. 62, 11. [Goth. midjun-gards οἰκουμένη: O. H. Ger. mittan-, mittin-gart: cf. myddellyard the world, Chest. Plays 1, 67: O. Sax. middel-gard: O. H. Ger. mittil-gart orbis: Icel. mið-garðr. 'The Icel. Edda has preserved the true mythical bearing of the word.— The earth (miðgarð), the abode of men, is seated in the middle of the universe, bordered by mountains and surrounded by the great sea (úthaf); on the other side of this sea is the Út-garð, the abode of giants; the Miðgarð is defended by the Ás-garð (the burgh of the gods), lying in the middle (the heaven being conceived as rising above the earth). Thus the earth and mankind are represented as a stronghold besieged by the powers of evil from without, defended by the gods from above and from within.'—Cl. and Vig. Dict. s.v. See also Grmm. D. M. 754.] v. middan-eard.

middangeard-líc; adj. Terrestrial, physical as opposed to spiritual:—Forðon hé oft stormas ðara werigra gásta fram his sylfes scépenisse and his geférena mid bedum wiðsceáf, wæs ðæt ðæs wyrþe ðæt hé wið ðam middangeardlícum windum and lígum swíðian mihte (ventus flammisque mundialibus), Bd. 2, 7; S. 509, 34. v. middaneard-líc.

middan-sumor, -winter. v. under midd, II, where perhaps in the instances in which middan occurs that word is to be taken as the first part of a compound. Cf. midde-sumor, -winter, and middandæg-líc.

midde, an; f. The middle (only in the phrase on middan):—Se fugel hafaþ iiii heáfdu ... and hé is on middan hwælan hiwes the bird hath four heads ... and in the middle it is of a whale's shape, Salm. Kmbl. 526; Sal. 262. Forwrát hé wyrm on middan, Beo. Th. 5404; B. 2705. Múð wæs on middan, Exon. 108 b; Th. 415, 10; Rä. 33, 9. On æge biþ gioleca on middan, Bt. Met. Fox 20, 339; Met. 20, 170. [O. Sax. middea (an middean): Icel. miðja (í miðju): cf. O. H. Ger. mittí (in mittí): Ger. mitte.] v. on-middan.

middel, es; middela (?), an; m. The middle, centre:—In middle in centro, Wrt. Voc. ii. 92, 13. On middele (Ps. Lamb. midle) innoþes mín in medio ventris mei, Ps. Spl. 21, 13, 21. Hé ánne cnapan gesette on hyra middele (in medio eorum), Mk. 9, 36. Se ðe álǽdde Israhel of middele heora ... þurh middele his, Ps. Spl. 135, 11, 14. Of midle ex centro, Wrt. Voc. ii. 31, 47. Of ðæs wuda midle, Exon. 56 b; Th. 202,

6; Ph. 65. Hió is gesceapen on ðam midle, betwux ðære drýgan and ðære cealdan eorþan and ðam hátan fýre, Bt. 33, 4; Fox 128, 37. Hió is on midle fýres and eorþan, Bt. Met. Fox 20, 163; Met. 20, 82. On midle mínra dagena, Ps. Th. 101, 21. Is ðis eálond geseted ongeán midle Súþ-Seaxna (contra medium Australium Saxonum), Bd. 4, 16; S. 585, 1. Intó ðam middelan (intó middan, other MS.) ðere strǽte, Cod. Dip. Kmbl. iii. 385, 9. On middel ðæs unmǽtan cyles, Bd. 5, 12; S. 627, 42: 628, 1. On ðone middel ðære mǽran byrig, Elen. Kmbl. 1724; El. 864. Hié gegripan on hire middel laid hold of her waist (cf. Laym. 28069, þa leo iueng me bi þan midle: Piers. P. 5, 358, B. text), Blickl. Homl. 141, 29. [Cf. Icel. á, í meðal among; á, í milli (from midli) between: M. H. Ger. mittel.] v. next word.

middel; superl. midlest; adj. Middle:—Be midelen streáme in mid stream, Cod. Dip. Kmbl. iii. 385, 15. Se midlesta finger the middle finger, L. Alf. pol. 58; Th. i. 96, 3. Be ðam midlæstan (the third in a list of five names) is nú tó secgenne, Bd. 4, 23; S. 594, 15. Swá biþ ðám midlestan monnum so it is with men of an intermediate class (between the best and the great majority of mankind), Bt. 39, 7; Fox 222, 4 (v. midd). [O. Sax. middil-gard: O. Frs. middel; superl. midlest, -ost, -ast: Icel. meðal-, in cpds.: O. H. Ger. mittil: Laym. Gen. and Ex. A. R. Ayenb. have superl. midlest.] Middel is found as the first part of many names of places, e.g. Middel-tún Middleton, Middel-hám Middleham, etc., Cod. Dip. Kmbl. vi. 315; see also following words.

middel-dæg, es; m. Mid-day:—Syle drincan middeldagum, Lchdm. iii. 74, 6: L. M. 1, 15; Lchdm. ii. 56, 22. Hé ðonne on middeldagum inne gewunode, 1, 72; Lchdm. ii. 146, 13. [Cf. O. H. Ger. mittila-tagun meridianus (ventus).] Cf. middel-niht.

middel-dǽl, es; m. The middle:—Ongén ðæm middeldǽle (other MS. middele) on ðæm eástende ad mediam frontem orientis, Ors. 1, 1; Swt. 10, 6.

Middel-Engle, a; pl. The Middle Angles, the Angles of Leicestershire (v. Green's Making of England, pp. 74–80):—Of Engle cóman Eást-Engle and Middel-Engle and Myrce and eall Norþhembra cynn de Anglis Orientales Angli, Mediterranei Angli, Merci, tota Nordanhymbrorum progenies . . sunt orti, Bd. 1, 15; S. 483, 25. Middel-Angle, Chr. 449; Erl. 12, 12. Middel-Engla mǽgþ . . . wæs cristen geworden. Ðissum tídum Middel-Engle Cristes geleáfan onféngon, Bd. 3, 21; S. 550, 36–39. Ðá wæs Déma biscop geworden Middel-Engla and eác Myrcna samod . . . hé forþférde on Middel-Englum on ðam þeódlande ðe is nemned on Feppingum, S. 551, 32–36: 3, 24; S. 557, 17. [When the Middle Angles had a bishop of their own the see was at Leicester.] Færpinga þreó hund hýda is in Middel-Englum, Cod. Dip. B. i. 414, 27. Ðone Ceaddan se ercebiscop æsænde Myrceon tó biscope and Middel-Englum and Lindesfarum, Shrn. 59, 14.

middel-finger, es; m. The middle finger:—Middelfinger medius vel impudicus, Wrt. Voc. i. 44, 6: 71, 32: ii. 58, 5. Gif man middelfinger of áslæhþ iv. scill. gebéte, L. Ethb. 54; Th. i. 16, 11.

middel-fléra, an; m. -flére, an; f. A partition (?; it occurs as an alternative with words meaning) the gristle of the nose, bridge of the nose:—Middelfléra interpinnium, Wrt. Voc. ii. 49, 48. Nose grystle vel middelflére internasus vel interfinium vel interpinium, i. 43, 20. [v. interfinium the grystell of the nose, Wülck. 590, 15: bryg of the nese, 634, 9: 675, 25.]

middel-fót, es; m. The middle of the foot, the instep:—Middelfót subtel, Wrt. Voc. i. 45, 3.

middel-geméru; pl. n. A middle or central district:—On Filistina middelgemǽrum in the centre of the land of the Philistines, Salm. Kmbl. 509; Sal. 255.

middel-gesculdru, -gescyldru; pl. n. The part between the shoulders:—Middelgesculdru interscapilium, Wrt. Voc. i. 44, 29. Middel-gescyldru interscapulum, ii. 49, 49. [Cf. Icel. mið-herðar mid-shoulders.]

middel-niht, e; f. Mid-night:—Nalles æfter lyfte lácende hwearf middelnihtum, Beo. Th. 5658; B. 2833: 5557; B. 2782: Bt. Met. Fox 28, 93; Met. 28, 47: Exon. 129 b; Th. 498, 4; Rä. 87, 7. Cf. middel-dæg and mid-niht.

Middel-Seaxe, -Seaxan; pl. The Middle-Saxons, Saxons who settled in the district west of London, and whose name is preserved in the present Middlesex: they appear to have been an offshoot of the East Saxons. v. Green's Making of England, p. 111, note:—Hér Middel-Seaxe (but MS. E. Middal-Engla, v. under Middel-Engle) onféngon ryhtne geleáfan, Chr. 653; Erl. 26, 24. Hí hæfdon ðá ofergán i. Eást-Engle, and ii. Eást-Sexe, and iii. Middel-Sexe, 1011; Erl. 144, 33. In provincia quæ nuncupatur Middel-Seaxan, Cod. Dip. Kmbl. i. 59, 20 (the charter is of a king of Essex). In Middil-Saexum, 142, 7.

middes. v. tó-middes.

midde-sumor, es; m. Mid-summer:—Ðis godspel gebyraþ on middesumeres mæsseðǽn, Lk. Skt. 1, 1, rubric. On middesumeres dæg, Herb. 4, 5; Lchdm. i. 90, 17. [Icel. mið-sumar.] v. midde-winter, mid-sumor.

midde-weard; adj. Mid-ward, middle of (the noun with which the word agrees):—Middeweard hand vola vel tenar vel ir, Wrt. Voc. i. 43, 54. Middewǽrd lencten vel foreweard lencten ver novum, 53, 26.

Middeweard hit mæg bión þrítig míla brád oððe brádre Norway may be thirty miles or more across the middle, Ors. 1, 1; Swt. 18, 31. Andlangæs bróces middesweardes along the middle of the brook, Cod. Dip. B. i. 295, 31. On middeweardum (-an, MSS. R. L.) hyre ryne, Lchdm. iii. 250, 26. On middeweardre sǽ in medio mari, Cant. Moys. 8. Ymb ða eaxe middewearde hwearfaþ they revolve about the middle of the axis, Bt. 39, 3; Fox 214, 23. Seó eá is irnende þurh middewearde Babylonia burg mediam Babyloniam interfluentem, Ors. 2, 4; Swt. 74, 3: 1, 3; Swt. 32, 6. As a noun:—On middeweardan innoþes mínes in medio ventris mei, Ps. Lamb. 21, 15.

midde-winter, es; m. Mid-winter, Christmas:—Ðis sceal on Sunnan-dæg betweox myddewintres mæssedæge and twelftan dæge, Lk. Skt. 2, 33, rubric. Ne miht ðú wín wringan on midne winter (meddewinter, MS. Bod.), Bt. 5, 2; Fox 10, 32. v. midde-sumor, mid-winter.

mid-eard, es; m. The world:—Mideardes ordfruman mundi originem, Hymn. Surt. 13, 30. Seó sunne ðe onlíht ealne mideard, Homl. Skt. 1, 72. v. middan-eard.

mid-fæsten, es; n. Mid-Lent:—Wæs mycel gemót tó midfestene, Chr. 1047; Erl. 175, 11. [Cf. Icel. mið-fasta mid-Lent.]

mid-feorh, gen. -feores; m. n. The period of middle age:—Midferh juventus, Wrt. Voc. ii. 112, 17. Oft biþ on hálgum gewrietum genemned midfeorh (MS. mid feorwe) tó giúguþháde aliquando adolescentia juventus vocatur, Past. 49, 5; Swt. 385, 31. [Cf. Ps. Th. 54, 24 on middum feore: O. Sax. (man) mid-firi: O. H. Ger. mitti-uerha dimidio (dierum meorum).] v. next two words.

mid-ferhþ, es; m. n. Middle life or age:—On cnihtháde . . . swá forþ ealne giógoþhád . . . and ðonne lytle ǽr his midferhþe, Bt. 38, 5; Fox 206, 25.

mid-ferhtness, e; f. Middle age:—Seó heora iúgoþ and seó mid-fyrhtnes bútan ǽgwylcum leahtre gestanden, hwylc talge wé ðæt seó yldo and se ende ðæs heora lífes wǽre? Blickl. Homl. 163, 3–6.

mid-help, es; m: e; f. Help, assistance:—Tó miðhelpe adjuvando, Rtl. 29, 36.

mid-hrif, es; n. m. [mid middle, hrif ventus] The mid-riff, the diaphragm, separating the heart from the stomach, etc.; also the entrails:—Midrif disseptum, Wrt. Voc. i. 44, 51: exta, 44, 49. Wið ðæt mannes midrif ace, Herb. cont. 3, 6; Lchdm. i. 6, 21. Midrife, Lchdm. i. 88, 11. On ðam uferan hrife oððe on ðam midhrife, L. M. 2, 46; Lchdm. ii. 260, 20. Of ðam midhrife, se is betweox ðære wambe and ðære lifre, 2, 56; Lchdm. ii. 278, 10. [O. Frs. mid-ref.] v. next word, and see hrif.

mid-hriðere, -hridir, es; n. The membrane enclosing the entrails:—Midhridir, nioþanweard hype ilia, Wrt. Voc. ii. 110, 54. Midhriðre omentum, i. 65, 56. Midhryðre, 284, 3. Midhryðere, ii. 64, 4. [O. Frs. midrithere membrana qua jecor et splen pendent; cf. also mid-rede, -rith the mid-riff: mydrede diafragma, Wrt. Voc. i. 208, 31.]

midl, es; n. I. a bit, curb (of a bridle):—Midl frenum vel lupatum: brídles midl chamus, Wrt. Voc. i. 23, 21, 22. Midlum lupatis (repagulis), Hpt. Gl. 406, 27. Of ísenum midlum I brídlum ferratis salivaribus (repagulis), 458, 3: Homl. Th. i. 360, 19: Elen. Kmbl. 2349; El. 1176: 2384; El. 1193. Midlum, Wrt. Voc. ii. 119, 49. II. the thong which bound the oar to the pin:—Midla strupiar, Wrt. Voc. i. 57, 6. Midlu, 63, 65. [Cf. in the same list of words connected with ships ár-widðe struppus, 56, 37.]

midlen, es; n. The middle, midst, centre:—Of midlene ex centro i. ex medio, Wrt. Voc. ii. 145, 66. On medlene in meditullio i. in medio, Hpt. Gl. 405, 37. Ic eom on eówrum midlene, Lk. Skt. 22, 27. On fýres midlene de medio ignis, Deut. 4, 15: 5, 24. Hé eardode in hǽðenra midlene . . . on þorna midlynæ, Shrn. 125, 7–8. Ða englas ásyndriaþ ða yfelan of ðæra gódra midlene, Mt. Kmbl. 13, 49. Ic eom on hyra midlene, 18, 20. On midline in dimidio, Blick. Gl. Se Hǽlend gesette ǽnne lytling on hyra midlen, 18, 2. Ðá férde hé þurh hyra midlen, Lk. Skt. 4, 30.

mid-lencten, es; n. m. Mid-Lent:—On mydlenctenes Sunnandæg, Jn. Skt. 6, 1, rubric. Tó midlængtene, Chart. Th. 349, 28.

midlest. v. middel.

midl-hring, es; m. The ring of a bit:—Midlhringas armillae, Wrt. Voc. ii. 10, 18.

midlian; p. ode To bridle, curb, restrain:—Forðæm is sió tunge gemetlíce tó midliganne (midlianne, Cot. MSS.) lingua itaque discrete frenanda est, Past. 38, 5; Swt. 275, 11. v. ge-midlian; á-, un-, unge-midled.

midlian; p. ode To mediate. [Icel. miðla to mediate.] v. midligend and ge-midlian.

mid-lifiend, es; m. One co-existent with another:—Uppstige ðæs midlifiendes [ðæs lifigendan, MS. Ca.], Bd. 3, 17; S. 545, 24, note. v. next word.

midligend, es; m. A mediator:—Uppstige ðæs midligendes Godes ascensionem mediatoris, Dei, Bd. 3, 17; S. 545, 24, note.

midlung, e; f. The middle, midst:—Of midlunge hwelpa de medio catulorum, Ps. Lamb. 56, 5. Of midlunge ðínum bósme de medio sinu

tuo, 73, 11. On midlunge sceaduwe dǽþes *in medio umbrae mortis*, Ps. Spl. 22, 4: Cant. Moys. 19 : Cant. Abac. 2 : Ps. Lamb. 73, 12. On midlunga, 81, 1. v. next word.

midlunga; *adv. To a moderate* or *middling degree, intermediate between muċh and little:*—Sam hē hine miclum lufige, sam hē hine lytlum lufige, sam hē hine mydlinga lufige, Shrn. 194, 14. v. preceding word.

midmest. v. midd.

midne-dæg, es; *m. Mid-day:*—Se rehta geleáfa swǽ swǽ midnedæg *fides velut meridies*, Ps. Surt. ii. 201, 25. Cf. ǽrne-mergen *in another version of the same hymn:*—Clǽnnyss sȳ swā swā ǽrnemergen, geleáfa swā swā middæg, Hymn. Surt. 16, 27.

mid-ness, e; *f. Middle, midst:*—In midnesse ðæs mynstres ... wit wǽron on midnesse miccles eges; ðā genāmon wit on midnysse ðæs eówdes twegen buccan, Shrn. 41, 20–27.

mid-niht, e; *f. Mid-night:*—Seó niht hæfþ seofan dǽlas ... feórþa is *intempestum*, ðæt is midniht, Lchdm. iii. 244, 3 : Wrt. Voc. ii. 49, 32. Midniht *intempestum vel intempesta nox*, i. 53, 5. On middre nihte wearþ clypung gehȳred ... Hwæt getácnaþ seó midniht būton seó deópe nytennys, Homl. Th. ii. 568, 4. [Cf. *Icel.* mið-nætti : *O. H. Ger.* mittinaht : *Ger.* mitter-nacht.] v. middel-niht; midd, II.

mid-rād, e; *f. A riding with another:*—Ðæt ǽlc man wǽre óðrum gelāstfull ge æt spore ge æt midráde (*in accompanying the other in following the trace of the lost property*), L. Æðelst. v. 4 ; Th. i. 232, 12. [*Icel.* með-reið.]

mídrece. v. mȳdrece.

mid-rif. v. mid-hrif.

mid-singend, es; *m. One who sings with another;* concentor, Wrt. Voc. i. 28, 23.

mid-sīðian; *p.* ode *To accompany:*—Hū ne midsīðgadest ðū *comitarisne tū?* Midsīðige *comitatur, sequitur*, Wrt. Voc. ii. 132, 34–38. Midsīðudu *comitata*, 23, 39. v. ge-midsīðian.

mid-spreca, an; *m. One who speaks on behalf of another:*—Paulus wæs midspreca and bewerigend ðǽre ealdan ǽ *Paul was an advocate and defender of the old law*, Homl. Th. i. 388, 32. [Cf. *Icel.* með-mæli *the speaking a good word for one.*]

mid-sumor, es; *m. Mid-summer:*—Ǽr midsumeres mæsseǽfen, Chr. 1052 ; Erl. 182, 5. v. midde-sumor.

[**midsumor-dæg,** es; *m. Midsummer-day:*—Tō midsumer dæi, Chr. 1131 ; Erl. 259, 34.]

mid-weg, es; *m. Mid-way:*—Segor stód on midwege betweox ðǽm muntum and ðǽm merscum, Past. 51, 5 ; Swt. 399, 13.

mid-winter, es; *m. Mid-winter, Christmas:*—Gif se (seo, MS.) midwinter biþ on Wódnesdæg, ðonne biþ heard winter and grim ... Gif heó byoþ on Ðunresdæg, ðonne byoþ gód winter ... Gif se midwinter byþ on Frigedæge, ðonne byþ onwendǽdlíc winter ... Gif se midwinter byþ on Seternesdæg, ðonne byþ winter gedréfedlíc, Lchdm. iii. 164, 1–10. On ðǽre hālgan midwintres tíde, L. C. E. pref.; Th. i. 358, 7. [*O. Frs.* mid-winter.] v. mid-sumor, midde-winter.

mid-wist, e; *f. The being with others, presence, society:*—Þurh fonthālgunge gewyrþ sóna Godes midwist *by the hallowing of the font God becomes at once present*, Wulfst. 36, 2. Ǽlc ðe gewita oððe gewyrhta sí ðǽr útlendiscan man inlendiscan derie gelādie ðǽre midwiste *let every one that is cognisant or co-operating, where a stranger injures a native, clear himself of the participation*, L. O. D. 6 ; Th. i. 354, 29. Snottre men lufiaþ midwist míne, Exon. 130 b ; Th. 500, 17; Rä. 89, 8. [*O. H. Ger.* mite-wist *consortium, participatio.*]

mid-wunung, e; *f. Dwelling with others:*—Þusend þusenda þénodon wealdende, and tēn þusend síðan hundfealde þusenda him mid wunodon. Óðer is þénung, óðer is midwunung, Homl. Th. i. 348, 5. Éce líf and midwununcg mid Gode, R. Ben. 133, 18. Ðæt wé on ðam tóweardan lífe diófla midwununga forbúgan mǽgon, H. R. 17, 29.

mid-wyrhta, an; *m. One who works with others, a co-operator:*—On ðæt gerād ðæt hē wǽre his midwyrhta ǽgðer ge on sǽ ge on lande *on the condition that he would co-operate with him by sea and by land*, Chr. 945; Erl. 116, 31 : Past. 38, 8; Swt. 279, 25. Hyt áwriten hys, ðæt ǽlcum welwyrcendum God myd beó mydwyrhta, Shrn. 179, 29.

mígan; *p.* māh, *pl.* migon *To make water:*—Ic míge mingo, Ælfc. Gr. 28, 5; Som. 31, 63. Ic míge meio; míge gē meite; mígan meire, 33; Som. 37, 44–45. Ðæt hē mȳhþ (*mingit*), byþ sweart, Lchdm. iii. 140, 22. Ðām ðe under hȳ mígaþ, L. Med. ex Quad. 8, 12; Lchdm. i. 360, 8. [*Laym.* mæh, meh; *p.: Icel.* míga: *M. L. Ger.* mígen.] v. ge-mígan.

miggung. v. micgung.

míging; e; *f. A making water;* minctio, Wrt. Voc. ii. 58, 10.

migol; *adj. Diuretic:*—Ðām monnum synd tō sellanne migole drincan, L. M. 2, 22; Lchdm. ii. 206, 27 : 208, 7. Mid wyrtdrencum ūtyrnendum oððe migolum, 1, 35; Lchdm. ii. 206, 17.

migoþa, migþa, micgþa, an; *m. Urine:*—Gif se micgþa ætstanden sȳ, Herb. 7, 3; Lchdm. i. 98, 5. Heó earfoþlícnysse ðæs migþan ástyreþ, 143, 1; Lchdm. i. 266, 3. Mid his selfes migoþan, ii. 42, 1. Swā hwæt

---

swā ðæne migþan gelet, 4, 6; Lchdm. i. 90, 26 : 7, 3; Lchdm. i. 98, 8 : 152, 1; Lchdm. i. 278, 4. v. micga, micgung, cū-migoþa.

miht. v. meaht.

míl, es; *n. Millet:*—Miil *milium*, Wrt. Voc. ii. 114, 9. Míl, 55, 68.

míl, e; míle (?), an; *f. A mile:*—Ālecgaþ hit on ánre míle ðone mǽstan dǽl fram ðæm tūne, ðonne óðerne ... ðū ðe hyt eall āléd biþ on ðære ánre míle, Ors. 1, 1 ; Swt. 20, 30–32 : Blickl. Homl. 129, 4. Leóuue, míle *milliarium*, Wrt. Voc. i. 38, 7. Twelf míla, Blickl. Homl. 197, 23. Of ðære burnan tō míla stāne, Cod. Dip. Kmbl. iii. 382, 22. Hund þusenda míla, Cd. 229; Th, 310, 9; Sat. 724. Ehta hund míla lang, Bd. 1, 1; S. 473, 11. On nygan mílum, 4, 27; S. 603, 30. [*Icel.* míla : *O. H. Ger.* míla, mílla.]

milc. v. meolc.

milcen; *adj. Of milk:*—Mylcen mete *food made of milk*, L. M. 1, 67; Lchdm. ii. 142, 14.

milcian. v. meolcian, melcan.

mild-beorht; *adj. Mildly bright, serene:*—Miltbeorhtum leóhte *luce serena*, Hpt. Gl. 484, 29.

MILDE; *adj.* I. MILD, *gentle, meek, benign, liberal* (?):—Se wæs milde wer and monþwǽre *vir omnium mansuetissimus ac simplicissimus*, Bd. 4, 27; S. 603, 35. Heora cining cymeþ milde and monþwǽre (*mansuetus*, cf. Mt. 21, 5), Blickl. Homl. 71, 4. Ic eom milde and eáþmódre heortan *mitis sum et humilis corde*, Bd. 2, 2; S. 503, 4. Ðæt milde mód (*Guthlac*), Exon. 43 b; Th. 146, 17; Gū. 711. Of árfæstre heortan and mildre, Blickl. Homl. 37, 27. Milde *mitia*, Wrt. Voc. ii. 57, 43. Spræc mildum wordum, Beo. Th. 2348; B. 1172. Mildre *indulgentior*, Ælfc. Gr. 43; Som. 44, 49. Manna mildost (*Moses*), Cd. 170; Th. 213, 8; Exod. 549. Cwǽdon ðæt hē wǽre manna mildust and monþwǽrost *they said that he was kindest and most courteous of men*, Beo. Th. 6344; B. 3182. Se leó gewāt swā ðæt mildoste lamb, Glostr. Frag. 110, 22. II. *of the more towards the less powerful, merciful, clement, propitious:*—Biddende ðæt Drihten him árfæst and milde wǽre *Dominum sibi propitium fieri precabatur*, Bd. 4, 31; S. 610, 31. God beó ðū milde (*propitius*) mé synfullum, Lk. Skt. 18, 13: Ps. Lamb. 98, 8; Blickl. Homl. 47, 32. Mé milde weorþ *miserere mei*, Ps. Th. 56, 1. His milde gehigd *misericordia sua*, 56, 4. Cyning cystum gód, clǽne and milde (*clement*), Chr. 1065; Erl. 199, 6. Ðam mildestan cyninge Wihtrǽde ríxigendum *in the reign of the most clement king Wihtræd*, L. Wih. pref.; Th. i. 36, 4. [*Goth.* milds: *O. Sax.* mildi : *O. Frs.* milde : *Icel.* mildr *mild*; also *munificent*: *O. H. Ger.* milti *mansuetus, largus, munificus.*] v. un-milde.

milde; *adv. Mercifully, graciously:*—Ús milde æteów ðinne andwlitan, Ps. Grn. 79, 18 : Ps. C. 50, 72; Ps. Grn. ii. 278, 72 : Hy. 6, 35; Hy. Grn. ii. 286, 35 : Exon. 11 b; Th. 16, 7; Cri. 249. [*O. Sax.* mildo.]

mil-deáw. v. mele-deáw.

milde-líc; *adj. Merciful, clement, propitious:*—Mildelíc *propitius*, Rtl. 37, 19. [*Icel.* mild-ligr *gentle.*]

milde-líce; *adv. Graciously, kindly, mercifully:*—His se cyning mildelíce onféng *the king received him kindly*, Ors. 1, 8; Swt. 40, 18. Swā mildelíce wæs Rómeburg on fruman gehālgod mid bróðor blóde, 2, 2; Swt. 66, 4. Hāwa mildelíce on ðās earman eorþan, Bt. 4; Fox 8, 20. Mildelíce *propitiatus*, Rtl. 120, 9. [*O. H. Ger.* milt-líhho *largiter:* *Icel.* mild-liga *gently.*]

mild-heort; *adj.* I. *kind-hearted, of gentle disposition, meek:*—Leorniaþ æt mé forðon ðe ic eom mildheort and eáþmód (*mitis et humilis corde*, Mt. 11, 29), Blickl. Homl. 13, 19. Uton beón eáþmóde and mildheorte and ælmesgeorne, 95, 26. Ðā weóp hē eác sylf ... swā hē wæs manna mildheortost, 225, 23. II. *merciful, compassionate, gracious, clement:*—Ðū God mildheort (*misericors*), Ps. Spl. 85, 14: *miserator*, 102, 8. Beóþ mildheorte swā eówer fæder is mildheort, Lk. Skt. 6, 36: Blickl. Homl. 97, 32. Ðín mildheort mód *misericordia tua*, Ps. Th. 107, 4. Mid mildheortum weorcum *with works of mercy*, Blickl. Homl. 37, 19. Cyng ðū mildheortesta *rex clementissime* (*Christ*), Hymn. Surt. 86, 29 : Ors. 6, 30; Bos. 126, 39 note. Hē wæs eallra monna mildheortast *he was most compassionate of all men*, 5, 12; Swt. 242, 20. [*O. H. Ger.* milt-herzi *misericors.*]

mildheort-líce; *adv. Kindly, compassionately, mercifully:*—Mildheortlíce *misericorditer*, L. Ecg. P. i. 9; Th. ii. 176, 15 : ii. 2; Th. ii. 182, 27 : Past. 44, 1; Swt. 319, 12, 14 : Blickl. Homl. 101, 36. Mildheortlícor *clementius*, Hymn. Surt. 138, 1.

mildheort-ness, e; *f. Mercy, compassion, pity, clemency:*—Hys mildheortnes *misericordia ejus*, L. Skt. 1, 50. Drihtnes mildheortnesse, Blickl. Homl. 49, 24. Úre sáula smerian mid mildheortnesse ele, 73, 24. Þurh mildheortnesse weorc, 97, 2. Mid ánre mildheortnyssa *sola clementia*, Hymn. Surt. 115, 27 : Bd. 3, 17; S. 545, 13. Mildheort God ... ðū ðe gehilst mildheortnysse *Deus misericors ... qui custodis misericordiam*, Ex. 34, 6. Hí náne mildheortnesse ne gearnodon, Bt. 38, 4; Fox 202, 28. Godes módor hire mildheortnisse ðære burhware gecȳðde, Chr. 994; Erl. 133, 15. Ðǽr beóþ gegearwode Godes mildheortnessa,

Blickl. Homl. 193, 20 : 103, 18. Hē him lytle mildheortnesse gedyde, Ors. 3, 9 ; Swt. 128, 15.

mild-hleahtor, es ; *m. Gentle laughter :*—Bysmrodon mē mildleahtre (*or* mid hleahtre ?) *subsannaverunt me subsannatione,* Ps. Spl. 34, 19.

mildian ; *p.* ode *To become mild :*—Mildode *mansuescit,* Germ. 399, 435. v. ge-mildian.

milds, mildsian, mildsiend, mildsung. v. milts, miltsian, miltsiend, miltsung.

milescian. v. miliscian.

mîl-gemearc, es ; *n. Space of a mile* or *distance measured by miles :*— Nis ðæt feor heonon mîlgemearces ðæt se mere standeþ *it is not far hence, measuring by miles, that the mere lies,* Beo. Th. 2728 ; B. 1362. Cf. fôt-, geár-gemearc, *and* mîl-getæl.

mîl-gemet, es ; *n. A mile-measure, a mile-stone :*—On ðæt mîlgemæt, Cod. Dip. Kmbl. iii. 252, 21.

mîl-getæl, es ; *n. The number of paces in a mile, a mile :*—On rîme ðæs læssan mîlgetæles ðe *stadia* hâtte fîf hund and ðæs miclan mîlgetæles ðe *leuua* hâtte þreó hund and eahta and syxtig *reckoning according to the smaller mile, which is called* stadia, *it is five hundred miles, and according to the great mile, which is called* leuua (*league*), *it is* 368, Nar. 33, 9–11.

milisc ; *adj. Honeyed, sweet, mellow,* (of drink) *mulled :*—Milisc apuldor *melarium :* milisc æppel *metianum,* Wrt. Voc. i. 285, 54, 55. (Melarium, *pomarium melis* (μῆλοις), *hoc est malis, consitum,* Du Cange : the Anglo-Saxon glosser seems to connect the word with *mel ?*.) Milisc æppel *nicalalbum,* 289, 74 : ii. 60, 42. Ðære miliscan *mulsæ,* 32, 66 : 54, 35. Myliscre, Hpt. Gl. 520, 39. Drince mylsce drincan, sió gebēt ða biternesse ðæs geallan, L. M. 1, 42 ; Lchdm. ii. 108, 2. Milscra (milscre, Wrt.) treówa blôsman qui[n]tinas, g. *caducas* (Du Cange quotes Isidore : ' Flores malorum (*punicorum*) a Græcis appellati sunt quintinæ. Latini caducum vocant '), Wrt. Voc. i. 22, 16. Melsc appla *nicolaos* (cf. *nicolaus = dactulus,* Wrt. Voc. ii. 75, 79 ; *nicolatis* palmæpla, 60, 67;) Hpt. Gl. 496, 65. Genim milsce æppla (*dates ?*), L. M. 2, 4 ; Lchdm. ii. 182, 19. Mylsce æppla, 2, 16 ; Lchdm. ii. 194, 9. [Cf. *Icel.* milska *a honeyed beverage ;* milska *to mix* (a beverage) : *Goth.* miliþ *honey.*] v. next word.

miliscian *to become sweet* or *mellow :*—Milescian *mitescere,* Wrt. Voc. ii. 55, 8.

mîl-pæþ, es ; *m. A road along which miles are reckoned :*—Wlance þegnas mæton mîlpaþas meara bôgum *proud thanes traversed the roads on their steeds,* Cd. 151 ; Th. 188, 20 ; Exod. 171 : Elen. Kmbl. 2523 ; El. 1263 : Runic pm. Kmbl. 340, 16 ; Rûn. 5.

miltan, mieltan, meltan ; *p.* te.   I. *trans.* (a) *To melt :*—Nim heortes mearg mylt *take heart's marrow, melt it,* L. Med. ex Quad. 10, 4 ; Lchdm. i. 366, 4. Mylt buteran, Lchdm. iii. 6, 22. Beó ælc calic geworht of myldendum antimbre (*of fusible material*), gilden oððe seolfren, glæsen oððe tinen ; ne beó nâ hyrnen, ne hûru treówen, L. Ælfc. P. 45 ; Th. ii. 384, 6. (b) *to digest :*—Sió wamb seó ðe biþ hâtre gecyndo melt mete wel . . . Seó ðe biþ wæterigre gecyndo næfþ gôde meltunge, swiðost on ðâm mettum ðe uneáþe melte beóþ, L. M. 2, 27 ; Lchdm. ii. 220, 22–28. (c) *to refine by melting :*—Ðæm ðe his gâst wile meltan (MS. B. miltan) wið morðre âsceádan of scyldum *by him who will refine his spirit from the dross of crime, separate it from sins,* Salm. Kmbl. 111 ; Sal. 55.   II. *intrans.* (=meltan) *To melt, become liquid :*—Ic mylte *liqueo,* Ælfc. Gr. 35 ; Som. 38, 8. Ðonne mē mægen and môd mylte *dum defeceret virtus mea,* Ps. Th. 70, 8. Weax miltende *cera liquescens,* Ps. Spl. 21, 13. Myltende *madens,* Wrt. Voc. ii. 57, 56. Myltende[s] *liquidas,* Hpt. Gl. 470, 73. [*Icel.* melta *to digest.*] v. ge-miltan, meltan.

milt-coðu, e *and* an ; *f. Disease of the spleen ;* lienosis, Wrt. Voc. ii. 53, 74.

MILTE, es ; *m.*: an ; *f. The* MILT, *spleen :*—Milti, Ep. Gl. 256, 24. Milte *lien,* Wrt. Voc. ii. 53, 67 : 112, 71 : *splen,* i. 45, 12 : *splena,* 65, 52. Se milte biþ emlang ðære wambe, L. M. 2, 36 ; Lchdm. ii. 242, 15, 22, 28. Þeós milte *hic splen,* Ælfc. Gr. 9, 13 ; Som. 9, 34. Hyt gelamp hwîlon ðæt man þearmas mid ðære miltan uppan ðâs wyrte gescearp, ðâ geclufde seó milte tô ðysse wyrte and heó hrædlîce ða miltan fornam . . hý beón bûtan miltan gemētte, Herb. 57, 1 ; Lchdm. i. 160, 3–10. Wið miltan sâre . . heó ðæt sær fornimþ ðære miltan, 32, 6 ; Lchdm. i. 130, 22 : L. Med. ex Quad. 2, 8 ; Lchdm. i. 334, 23. Wið ðam wætan yfle ðæs miltes . . . ðæt lâcnaþ ðone milte, L. M. 2, 38 ; Lchdm. ii. 246, 9–11, 18. Of milte, Lchdm. ii. 248, 1. Wið âswollenum milte, 2, 45 ; Lchdm. ii. 256, 16. [*O. Frs.* milte ; *f.: Icel.* milti ; *n.: O. H. Ger.* milzi ; *n.: Ger.* milz ; *f.*]

milte-seóc ; *adj. Splenetic :*—Milteseóc *lienosus,* Wrt. Voc. i. 19, 41. Wið milteseócum men, him mon sceal sellan eced, L. M. 2, 39 ; Lchdm. ii. 248, 9 : 2, 41 ; Lchdm. ii. 252, 5.

milte-wærc, milt-wræc, es ; *m. Pain in the spleen :*—Be miltewærce, L. M. 2, 36 ; Lchdm. ii. 242, 1 : 3, 16 ; Lchdm. ii. 318, 9. Wið milt-wræce, L. Med. ex Quad. 9, 5 ; Lchdm. i. 262, 5.

miltestre, an ; *f. A harlot :*—Myltestre *meretrix* vel *scorta,* Wrt. Voc.

i. 86, 72 : Gen. 38, 15. Ne lǣt ðû ðîne dohtor beón myltestre *ne prostituas filiam tuam,* Lev. 19, 29. Beclypte seó myltestre ðæt clǣne mǣden, Homl. Skt. 2, 169 : 7, 178. Cômon tô ânre miltistran hûse *ingressi sunt domum mulieris meretricis,* Jos. 2, 1. Melt[r]estran hûs *lupanar,* Hpt. Gl. 500, 61. Myltistryna hûs, Ælfc. Gr. 9, 16 ; Som. 9, 45 : Homl. Skt. 7, 148. Oððe ðû mid mǣdenum ðîne lâc geoffrige, oððe ðû lâðum myltestrum scealt beón gefêrlǣht, 7, 119. Mânfulle and myltystran *publicani et meretrices,* Mt. Kmbl. 21, 31, 32.

miltestre-hûs, es ; *n. A brothel :*—Myltestrehûs *lupanar,* Wrt. Voc. i. 58, 53.

milts, milds, e ; *f.*   I. *mildness, kindness, favour, mercy* (most commonly with reference to the Deity) :—Mid ðec nilds is *apud te propitiatio est,* Ps. Surt. 129, 4. Ðonne wurþe ûs eallum Godes milts ðē gearuwre, L. C. E. 19 ; Th. i. 372, 5 : L. C. S. 85 ; Th. i. 424, 23 : Past. 44 ; Swt. 325, 13. Biþ ðær seó miccle milts âfyrred . . . ðæs Ælmihtigan, Exon. 28 a ; Th. 84, 10 ; Cri. 1371. Ûs wæs â syððan Merewioingas milts ungyfeþe, Beo. Th. 5835 ; B. 2919. Þolige hē clǣnes legeres and Godes mildse, L. N. P. L. 62 ; Th. ii. 300, 19. Ðû mid mildse mînre fêrest *thou shalt depart with my favour,* Andr. Kmbl. 3344 ; An. 1676. Hē Drihtnes mildheortnysse gecýgde and ða mildse bæd monna cynne *misericordiam Domini invocaret, et eam generi humano propitiari rogaret,* Bd. 4, 3 ; S. 569, 9. Miltse gecýðan, onwreón, Blickl. Homl. 39, 23 : 107, 2. Hæbbe hē Godes miltse (mildse), L. Eth. v. 9 ; Th. i. 306, 20 : L. N. P. L. 64 ; Th. ii. 300, 24. Gemyne mildsa ðînra *reminiscere miserationum tuarum,* Ps. Surt. 24, 6 : 68, 17. Secggan wē him þanc ealra his miltsa, Blickl. Homl. 103, 26 : 109, 10. Âsecggan ða miltsa ðe hē wið ðis mennisce cynn gecýdde, 103, 19. For his miltsum *by his mercies,* Exon. 88 b ; Th. 333, 6 ; Vy. 98 : 42 a ; Th. 140, 16 ; Gû. 611.   II. *meekness, humility* (?), *joy* (?), (cf. *O. H. Ger.* milti *hilaritas*) :—Ðec Anananias and Azarias and Misahel miltsum [*humbly* (?), *joyously* (?)] hergaþ, Exon. 55 a ; Th. 195, 11 ; Az. 154 : Th. 194, 29 ; Az. 146 : 54 b ; Th. 193, 8 ; Az. 118.

miltsian, mildsian ; *p.* ode *To have* or *take pity upon a person, shew mercy, be merciful, pity.*   I. *not followed by an object :*—Ic miltsige *indulgeo,* Ælfc. Gr. 26, 3 ; Som. 28, 54 : *ignosco,* 28, 1 ; Som. 30, 31. Miltsige (mildsige, MS. B.) man for Godes ege *for fear of God let mercy be shewn,* L. C. S. 68 ; Th. i. 410, 22 : L. Eth. vi. 53 ; Th. i. 328, 28. Cum and mildsa, Hy. 7, 27 ; Hy. Grn. ii. 287, 27.   II. *with dative :*—Ic miltsige ðē *misereor tui* . . . miltsa ûs Drihten *miserere nostri Domine,* Ælfc. Gr. 41 ; Som. 43, 63–64. Ðû eallum miltsast ðæm ðe on ðē gelýfaþ, Blickl. Homl. 145, 19. Hē bæd ðæt Hǣlend him miltsade, 19, 13. Hē ðînum mândǣdum miltsade eallum *qui propitiatur omnibus iniquitatibus tuis,* Ps. Th. 102, 3. Mon mildsige ðâm yfelum, Bt. 39, 1 ; Fox 212, 7 : 38, 7 ; Fox 210, 18. Gebrôðru, miltsige eów God, Homl. Th. ii. 158, 24. Eálâ ! ðû man, miltsa ðē, L. E. I. pref.; Th. ii. 394, 30. Miltsa mē *miserere mei,* Mk. Skt. 10, 48. Miltsa eallum ðînum wiðerwinnum, and âgyld gôd for yfele, Homl. Th. ii. 344, 2. Mildsa monna cynne, Hy. 8, 32 ; Hy. Grn. ii. 290, 32. Him wile git God miltsian, Blickl. Homl. 47, 7. Gif hē ûs ârian and miltsian wile, 51, 30. Ðe hió sôna ûs efenþrowiende and hraðe miltsiende, 19, 30. Hǣlend wæs miltsiende Adame, 87, 35.   III. *with genitive :*— Hē þearfendra miltsude, Ps. Th. 106, 40. Miltsa mîn *miserere mei,* 56, 1. Tîd tô mildsiende his *tempus miserendi ejus,* Ps. Surt. 101, 14. Miltsigende ðîn *miserens tui,* miltsigende his *miserens illius,* Ælfc. Gr. 41 ; Som. 43, 63.   IV. *with a preposition,* v. miltsiend. v. ge-miltsian.

miltsiend, mildsiend, es ; *m. One who takes pity :*—Ðû wǣre miltsiend ofer heora cild, Blickl. Homl. 249, 6. Mildheort and mildsiend *miserator et misericors,* Ps. Spl. 102, 8. Mildsiend *miserator,* Ps. Lamb. 85, 15. Milsend, Rtl. 69, 7 : 170, 9. v. ge-mildsiend.

miltsigend-lîc ; *adj. To be pardoned, venial :*—Miltsigendlîc *propitiabilis,* Germ. 401, 130. Hwî wæs ðæs heáhengles syn unmiltsigendlîc and ðæs mannes miltsigendlîc ? Boutr. Scrd. 17, 21.

miltsung, mildsung, e ; *f. Mercy, pity, compassion, a shewing mercy, pardon, indulgence :*—Hit is rihtre ðæt him mon mildsige ðæt is ðonne hiora mildsung ðæt mon wrece hiora unþeáwas *it is more fitting that mercy be shewn them. Now this it is to shew them mercy, to punish their vices,* Bt. 38, 7 ; Fox 210, 18. Ealle for miltsunge stefne uton sellan *omnes pro indulgentia vocem demus,* Hymn. Surt. 37, 22. Swâ micclum swâ ðæs mannes gecynd unmihtigre wæs swâ hit wæs leóhtre tô miltsunge *the weaker was man's nature, the easier was it to pardon,* Boutr. Scrd. 17, 24. Bûtan forgifenysse I miltsunge (milsunge) *sine respectu,* Hpt. Gl. 487, 53. Hē ûs mid his miltsunge (*sua miseratione*) gescylde, Bd. 3, 2 ; S. 524, 24. Petrus tîhþ ða geleáffullan þurh þingrǣdene þurh miltsung him forgyfenre mihte *Peter draws the faithful by intercession, by the merciful exercise of the power given to him,* Homl. Th. ii. 292, 2. Crist mæg ðîne nytennysse þurh his miltsunge onlîhtan, Homl. Skt. 5, 200. Gemune miltsunga ðînra (*miserationum tuarum*), Ps. Spl. 24, 8 : 50, 2. v. un-miltsung.

milt-wræc. v. milte-wærc.

mimor. v. ge-mimor *and next word.*

mimorian; *p.* ode *To keep in the memory, remember :*—Pater noster and crēdan mymerian (mynegian, MS. C.) ða yldran and tǽcan heora gingran, Wulfst. 74, 15.

min; *adj.* I. *small :*—Ne ðē sunne on dæge ne gebærne ne ðē mōna on niht min ne geweorþe *may the sun not burn thee by day, nor the moon withhold her light from thee by night*, Ps. Th. 120, 6. II. *mean, vile :*—Hwīlum cyrdon eft minne mánsceaþan on mennisc hiw *at times the vile criminals turned into human form*, Exon. 46 a; Th. 156, 27; Gū. 881. [The positive does not occur in the other Teutonic dialects, but comparative and superlative forms are found in Gothic, O. Frs., O. Sax., Icel. and O. H. Ger. Cf. also Lat. *minor, minimus.*] *v.* minsian, min-dōm.

mīn; *pron. gen.* of ic *Of me :*—Beó ðū mīn gemyndig, Ps. Th. 24, 6. Miltsa mīn, 56, 1. Ne æthrīn ðū mīn, Jn. Skt. 20, 17. Ic sprece *ego loquor*, mīn sprǽc *mei locutio*, Ælfc. Gr. 15; Som. 17, 56. Ǽr ðū ða miclan meaht mīn oferswīðdest, Exon. 73 a; Th. 273, 25; Jul. 521. Ne wât ic hygeþoncum mīn, 109 a; Th. 417, 14; Rä. 36, 4. Hē wæs mīn on ða swīðran, Elen. Kmbl. 694; El. 347. Mīn sylfes gást wæs ōrmod worden, Ps. Th. 76, 4. Mīn sylfes weorc hī gesáwon, 94, 9. (Cf. next word, V.) [*Goth.* meina : *O. Sax. O. Frs. Icel. O.H. Ger.* mīn.]

mīn; *adj. pron.* Mine, my. I. *with a noun :*—Mīn cnapa līþ on mīnum hūse lama . . . Ne eom ic wyrðe ðæt ðū ingange under mīne þecene . . . Ic cweþe tō mīnum þeówe, Mt. Kmbl. 8, 6–9. Hwylc is mīn mōdor and hwylce synt mīne gebrōðra, 12, 48. Fæder mīn! 26, 39. Ðis is mīnes fæder willa, Jn. Skt. 6, 40. Mīnre faðan yldre mōder, Wrt. Voc. i. 52, 19. On mīnre gesiþþe, Ps. Th. 88, 31. Ne cunne gē mē ne mīnne fæder, Jn. Skt. 8, 19. Nimaþ mīn geoc ofer eów, Mt. Kmbl. 11, 29. Mīne fearras and mīne fuglas synt ofslegene, and ealle mīne þing synt gearwe, 22, 4. Mid lyre ealra þinga mīnra, Coll. Monast. Th. 27, 1. Hū gelýfe gē mīnum wordum, Jn. Skt. 5, 47. II. *as predicate :*—Eall eorþe ys mīn, Ex. 19, 5. Ealle ða þing synd mīne, Gen. 31, 43. Ðīne twegen suna beóþ mīne, 48, 5. III. *used substantively :*—Wlwine habbe ðat lond ðe hē mīnes hafde, Chart. Th. 580, 24. Ic heóld mīn tela, Beo. Th. 5468; B. 2737. Gif ic mōt mīne wealdan, Cd. 102; Th. 136, 1; Gen. 2251. Ealle mīne synt ðīne, and ðīne synt mīne, Jn. Skt. 17, 10. Ðū mundbora wǽre mīnum, Exon. 120 b; Th. 463, 25; Hö. 75. Ða mīnan, Cd. 224; Th. 296, 19; Sat. 504. IV. *with a pronoun :*— Hēr is mīn se gecorena sunu *hic est filius meus dilectus*, Mt. Kmbl. 3, 17. Ðes mīn sunu, Lk. Skt. 15, 24. Se mīn wine, Exon. 115 b; Th. 444, 21; Kl. 50. Mīn se ēca dǽl in gefeán fareþ, 38 a; Th. 125, 11; Gū. 352. Mīn se swētesta sunnan scīma; 68 a; Th. 252, 20; Jul. 510. Bi ðam bitran deáþe mīnum, 29 b; Th. 90, 18; Cri. 1476. Ic mid mec gelǽdde mīne þrié ða getreówestan frýnd, Nar. 29, 27. Mīne ða hálgan, Ps. Th. 104, 13 : 121, 8. Ða manigfealdan mīne geþohtas, Exon. 118 a; Th. 453, 1; Hy. 4, 8. V. *with self* (a) *agreeing with the noun* (see also preceding word) :—On mīnne sylfes dōm, Beo. Th. 4301; B. 2147. (b) *agreeing with* self :—Mīnes sylfes mūþ on mehum, Ps. Th. 77, 2. Mīnes sylfes gebed *oratio mea*, 140, 2. Mīnre sylfre sīþ, Exon. 115 a; Th. 441, 20; Kl. 2. VI. *with* ágen :—Ic ne mōt wealdan mīnra ágenra þeówa, Bt. 7, 3; Fox 20, 20. [*Goth.* meins : *O. Sax. O. Frs. O.H. Ger.* mīn : *Icel.* mīnn.]

min-dōm, es; *m. Smallness, abjectness, pusillanimity :*—Ic bīde ðæs beornes ðe mē bōte (? bēte) eft mindōm *expectabam eum qui me salvum faceret a pusillo animo*, Ps. Th. 54, 7. *v.* min, minsian.

mine, es; *m. A minnow :*—Mȳne *vel* ǽlepūte *capito*, Wrt. Voc. i. 55, 75. Mynas and ǽlepūtan *menas et capitones*, Coll. Monast. Th. 23, 33.

mīn-līce; *adv. In my way, in my manner :*—Mīnlīce *meatim* (= *meo more*, Wülck. 32, 20), Wrt. Voc. ii. 58, 46.

minna (?) *a sheaf :*—Ða minnan gaderaþ *qui manipulos colliget*, Ps. Spl. T. 128, 6.

minsian; *p.* ode *To lessen, diminish, become small :*—Wlite minsode, Cd. 187; Th. 232, 30; Dan. 268. Minsade, Exon. 94 a; Th. 353, 48; Reim. 29. Cf. Ne mæg ǽnig man Godes mihta ne his mǽrþa geminsian, Wulfst. 35, 3. [*O. Sax. O. L. Ger.* minsôn *to make less:* cf. *Icel.* minnka *to make less.*] *v.* next word.

minsung, e; *f. Parsimony :*—Forhæuednys *parsimonia*; minsong *abstinentia*, Hpt. Gl. 494, 41.

minte, an; *f. Mint :*—Minte *menta*, Wrt. Voc. i. 31, 11 : ii. 98, 18 : *mentha*, i. 67, 65. Eal mintan cyn *mentastrum*, ii. 56, 34. Gē ðe teóðiaþ mintan, Lk. Skt. 11, 42 : Mt. Kmbl. 23, 23. *v.* brōc-, feld-, fen-, hors-, sǽ-, tūn-minte.

mirc-apuldor *a dark apple-tree :*—Mircapuldur *melarium* (as if from μέλας?), Wrt. Voc. ii. 113, 78. *v.* milisc.

Mircan. *v.* next word.

Mirce, Mierce, Myrce; *pl. The Mercians,* (and as the name of the people is used where modern English uses the name of their country) *Mercia* [see Green's The Making of England, p. 85]:—Hēr Mierce wurdon Cristne, Chron. 655; Erl. 28, 1. Ða nāmon Mierce (Myrce, MS. E.) friþ wið ðone here, 872; Erl. 76, 16. Of Engle cōman Eást-Engle and Middel-Engle and Myrce (*Merci*) and eall Norþhembra cynn,

Bd. 1, 15; S. 483, 25. Miercna cyning, land, rīce, Chr. 853; Erl. 68, 7 : 877; Erl. 78, 26 : 794; Erl. 58, 7. Mircena cining, 704; Erl. 43, 30. Mercna land, rīce, cyningcynn, 905; Erl. 98, 14 : 655; Erl. 28, 4 : Bd. 2, 20; S. 521, 8. Myrcna cynn, mǽgþ, þeód, 3, 21; S. 551, 23 : 4, 3; S. 566, 24 : 2, 12; S, 515, 7. Myrcna landes is þrittig þūsend hýda ðǽr mon ǽrest Myrcna hǽt, Cod. Dip. B. i. 414, 15. Myrcena cining, land, Chr. 792; Erl. 59, 1 : 796; Erl. 59, 39 : L. Alf. 49; Th. i. 58, 25 : L. Eth. i. pref.; Th. i. 280, 4. Ða fēng Æðelbald tō rīce on Mercium (Myrcum, MS. E.), Chr. 716; Erl. 44, 14. In Mercum preóst, 731; Erl. 47, 10. On Myrcean, L. C. S. 14; Th. i. 384, 1. On West-Sexan and on Myrcan and on Eást-Englan, 72; Th. i. 414, 14 : Swt. A. S. Rdr. 100, 146. Hine on Mierce (Myrce, MS. E.) lǽddon, Chr. 796; Erl. 58, 12. Hē fōr ofer Mierce on Norþ-Walas, 853; Erl. 68, 10. Innan Mierce (Myrce, MS. E.) tō Snotengahām, 868; Erl. 72, 21. Of Wesseaxum on Merce, 853; Erl. 68, 22. *v.* Norþ-, Sūþ-Mirce; *and* mearc.

mirce; *adj.* I. *dark, murky :*—Ða mircan gesceaft (*Hell*), Exon. 116 a; Th. 446, 23; Dōm. 26. Gang ofer myrcan mōr *her course o'er the dark moor*, Beo. Th. 2814; B. 1405. II. in a metaphorical sense (of sin, crime, etc.) *dark, black, evil :*—Mircne mægencræft mánwomma gehwone *dark power, each sinful stain*, Exon. 26 b; Th. 78, 26; Cri. 1280. Ðeáh ðū drype þolige, myrce mánslaga, Andr. Kmbl. 2437; An. 1220. Leahtras mirce mándǽde *crimes, black deeds of wickedness*, Exon. 62 b; Th. 229, 18; Ph. 457. Mircast mánweorca *blackest of crimes*, 73 a; Th. 272, 26; Jul. 505. [*Havel.* mirke : *Chauc. Piers P.* merke : *Prompt. Parv.* myrke *obscurus, tenebrosus :* *O. Sax.* mirki : *Icel.* myrkr : *Dan. Swed.* mörk.] *v.* mirc-apuldor, æl-myrca, Gūþ-myrce.

mirce, es ; *n. Darkness :*—Se ðe hié of ðam mirce (*the fiery furnace*) generede, Cd. 196; Th. 244, 15; Dan. 448. Myrce (*or adv.?*) gescȳrded *shrouded with darkness*, Andr. Kmbl. 2628; An. 1315. [*Piers P.* men þat in merke sitten : *Scot.* mirk : *Icel.* myrkr ; *n. darkness* ; mjörkvi *darkness, thick fog :* *Dan.* mörke.]

mircels, es; *m.*: e ; *f.* I. *a sign, mark, token :*—Ðū ásettest ðīnes wuldres myrecels on worlde, sete nū ðīn wuldres tācn in helle, Blickl. Homl. 87, 16. II. *a mark to aim at :*—Hē miste mercelses, and his mǽg ofscēt, Beo. Th. 4869; B. 2439. Hī setton hine tō myrcelse, and heora flán him on áfæstnodon, Homl. Skt. 5, 426. III. *a signet, seal :*—Gehealdenre mercelse *salvo signaculo*, Hpt. Gl. 501, 27. Insegle, mercelse *signaculo*, 504, 37. IV. *an ensign, a trophy :*—Ðā hēt se hǽðena cyning his heáfod of ásleán and his swiðran earme, and settan hī tō myrcelse, Swt. A. S. Rdr. 99, 135. Ðā ðū gehēte ðæt ðec hálig gǽst wið earfeþum eáðe gescilde for ðam myrcelse ðe (ðec?) monnes hond from ðīnre onsýne áhwyrfde *when thou didst promise, that the Holy Spirit would easily shield you from troubles, on account of the ensign (the cross?) that would turn man's hand from thy face*, Exon. 39 a; Th. 129, 30; Gū. 429. V. *a marked spot :*—Hē hēt ða gebrōðru ádelfan ǽnne pytt, ðǽr ðǽr hē ǽr gemearcode . . . Ða gebrōðru ðā eodon tō ðam mercelse, Homl. Th. ii. 162, 1–6.

Mircisc; *adj. Mercian :*—Be Merciscan áðe, L. O. 13 ; Th. i. 182, 18.

mire *a mare.* *v.* mere.

mirgan; *p.* de *To be merry, to rejoice, be glad :*—Fægniaþ and myrgaþ Gode mid wynsumre stemne *jubilate Deo in voce exultationis*, Ps. Th. 46, 1.

mirgen *that which causes delight, poetry* (?) :—Him wæs lust micel ðæt hē ðiossum leódum leóþ spellode, monnum myrgen *great his (Alfred) delight was lays to relate, matter of mirth for men*, Bt. Met. Fox introd. 9; Met. Einl. 5. Cf. miriginess.

mirhþ. *v.* mirigþ.

mirige; *adj. Pleasant, delightful, sweet :*—Myrige leóþ *dulce carmen*, Hymn. Surt. 55, 17. Ðeós woruld ðeáh ðe heó myrige hwīltīdum geþuht sȳ *this world, though it seem at times pleasant*, Homl. Th. i. 154, 17. Ðeós woruld is hwīltīdum myrige on tō wunigenne, 182, 24. Gærs myrige on tō sittenne, 182, 15. Wǽre hit ðonne murge mid monnum, Bt. Met. Fox 11, 203; Met. 11, 102. Eall se eard wæs mirige (*or adv.?*) mid wætere gemenged, Gen. 13, 10. Dōmes dæg, ðæt is se myriga dæg, Wulfst. 244, 15. Hwæt ða woruldlustas myreges (myrges, MS. Cott.) brengaþ *quid habeat jucunditatis*, Bt. 31, 1; Fox 112, 4. Ne geleofaþ man náht miriges ða hwīle ðe mon deáþ ondrǽt *one gets no pleasure from life, while one fears death*, Prov. Kmbl. 16. Mid merigum lofsange *dulci ymno*, Hymn. Surt. 141, 38. Him ða twigu þincaþ swā merge *the boughs seem so pleasant to them*, Bt. Met. Fox 13, 89; Met. 13, 45. Ða mergen *amoena*, Hpt. Gl. 409, 36. [*Laym. A. R.* murie : *Gen. and Ex.* mirie : *Prompt. Parv.* myry yn chere *letus, jocundus* ; myry, mery weder *malacia :* *Chauc. Piers P.* murie, merie.] *v.* next word.

mirige; *adv. Pleasantly, sweetly, gladly :*—His mōdor gehýrde hū myrge hē sang mid ðām munecum and hyre wæs myrge on hyre mōde *his mother heard how sweetly he sang with the monks, and she was glad at heart*, Wulfst. 152, 11–13.

mirig-ness, e ; *f. Pleasantness, sweetness* (of sound), *music :*—Myrgnis *musica*, Wrt. Voc. ii. 114, 45. *v.* mirgen, mirige.

mirigþ, mirgþ, mirhþ, myrþ, e ; *f. Pleasure, joy, delight, sweetness* (of

sound):—Dæg byþ myrgþ eádgum and earmum *day is a delight to rich and to poor*, Runic pm. Kmbl. 344, 12; Rún. 24. Wā him ðære mirigþe búte hē ðæs yfeles ǽr geswîce *alas for his delight, unless first he leave evil*, Hy. 2, 6; Hy. Grn. ii. 281, 6. Hē ádrǽsed wæs of neorxena wanges myrþe (*paradisum voluptatis*), Gen. 3, 24. For ðære mirínte (mergþe, MS. Cott.) ðæs sónes, Bt. 35, 6; Fox 168, 11. On heofonan ríces mirhþe, Ælfc. T. Grn. i, 11. Myrhþe, Homl. Th. i. 58, 4. Đa heorde tô heofonlícre myrhþe (myrþe, MS. B.) lǽdan, L. C. S. 85; Th. i. 424, 11. Man byþ on myrgþe (*joyous*), Runic pm. Kmbl. 343, 11; Rún. 20. Đû ðǽr náne myrhþe on næfdest ðá ðá ðû hié hæfdest *thou hadst no pleasure in them, when thou hadst them*; nec habuisse te in ea pulcrum aliquid, Bt. 7, 1; Fox 16, 17. Đín ríce ðǽr wē gemétaþ ealle mirhþe, Hy. 7, 31; Hy. Grn. ii. 287, 31. Đǽr (*heaven*) syndan mihta, mǽrþa and myrhþa, Wulfst. 5, 5 : 167, 9 : 28, 7. Adam wearþ of myclum myrhþum bescofen tô hefigum geswincum, 104, 1. v. myrige, un-mirigþ.

mirra, merra, an; *m. One who leads astray, a deceiver*:—Merra *seductor*, Mt. Kmbl. Lind. 27, 63.

mirran, mierran, merran; *p. de.*   **I.** *to be a stumbling-block to, to hinder, obstruct*:—Đe ðone ungesceádwîsan mirþ (*scandali occasionem praebere*), Past. 59, 6; Swt. 453, 4. Sió ofersméaung mirþ (*is a hindrance to*) ða unwîsan, 15, 5; Swt. 97, 17. Đæt eów lǽst þinga mierþ *sine impedimento*, 51, 7; Swt. 401, 17. Đæs andwearda wela ámerþ and lǽt (MS. Cott. myrþ and let) ða men ðe beóþ átihte tô ðám sôþum gesǽlþum, Bt. 32, 1; Fox 114, 3. Merþ, tit. 32; Fox xvi, 12. Seó ungesceádwîsnes heora eágena hí myrþ (ámerraþ, Cott. MS.), 32, 2; Fox 116, 26. Gyf hí ðé myrraþ and lettaþ, Shrn. 185, 5. Hwî mirraþ git ðis folc fram heora weorcum *quare sollicitatis populum ab operibus suis?* Ex. 5, 4. God nolde ðæt hié ðone Cristendóm mierde leng *God would not that they should longer obstruct Christianity*, Ors. 6, 7; Swt. 262, 21. Gif hwā Godes lage wyrde oððe folclage myrre, L. I. P. 2; Th. ii. 306, 12.   **II.** *to waste, squander*:—Đý læs mon unnytlîce mierre ðæt ðæt hē hæbbe *ne, quae possident, inutiliter spargant*, Past. 44, 4; Swt. 325, 3. Ne myr ðú eal ðæt ðú hæbbe, ðý læs ðe geþearfe tô ôðres mannes ǽhtum, Prov. Kmbl. 73. Gif ðú ðín ágen myrre, ne wît ðú hit nā Gode, 51. Se hordere nā mynstres ǽhta ne ýte, ne nā myrre, R. Ben. 55, 4.   **III.** *intrans. To err*:—Gié merras ł geduellas *erratis*, Mt. Kmbl. Lind. 22, 29. [*Goth.* marzjan σκανδαλίζειν: *O. Sax.* merrian (*trans. and intrans.*): *O. Frs.* meria: *O. H. Ger.* marrian *impedire, scandalizare*.] v. ā-, ge-myrran.

mirrelse, an; *f. A hindrance, stumbling-block*:—Gif sôþfæstra þurh myrrelsan môd ne ôðcyrreþ *if the mind of the righteous, through rock of offence, turn not aside*, Exon. 70 b; Th. 262, 25; Jul. 338.

mirring, e; *f.*   **I.** *hindering, leading astray*:—Merrunga *seductiones*, Mk. Skt. p. 5, 8. Cf. mirra. [*O. Frs.* meringa *hindrance: O. H. Ger.* marunga *impedimentum*.]   **II.** *waste, squandering* (v. mírran, II):—Oððe se gielpna for his gôda mierringe (mirringe, Cott. MSS.) gielpe and wêne ðæt hē síe kystig and mildheort *aut cum effuse quid perditur, largum se glorietur*, Past. 20, 2; Swt. 149, 20. Đa uncystgan cysta læteð he, swā hé ða cystgan on merringe ne gebringe *sic tenacibus infundatur tribuendi largitas, ut prodigis effusionis frena minime laxentur*, 60; Swt. 453, 27. v. mann-mirring.

mirt, myrt *a mart, market*:—Cêping *mercatum*: scipmanne myrt þe (Wrt. se) cêping *teloneum*, Wrt. Voc. i. 37, 9–10.

mis-, miss-, mist-, misse- a prefix denoting *defect, imperfection, Goth.* missa- (*for* mihto- *a participial form connected with root meaning* to lose): *O. Sax. O. L. Ger. O. Frs.* mis-: *Icel. Da. Swed.* mis-: *O. H. Ger.* missa-, missi-: *M. H. Ger.* misse-: *Ger.* mis-, miss-.

mis-begán *to cultivate badly, waste, disfigure*:—Misbegáas onsióne hiora *exterminant facies suas* (cf. unrôtlîce dôþ *exterminant*, Wrt. Voc. ii. 30, 64), Mt. Kmbl. Lind. 6, 16.

mis-beódan; *p.* -beád, *pl.* -budon; *pp.* -boden *To do wrong to, to offend, abuse, ill-use*:—Hē misbeád his munecan on fela þingan *he ill-used his monks in many things*, Chr. 1083; Erl. 217, 3. Đé læs ǽnig man misbeóde *lest any do wrong to other*, L. I. P. 7; Th. ii. 312, 22 : Chart. Th. 320, 13 : 416, 13. Ne misbeóde ǽnig ôðrum, forðam eal ðæt ǽnig man ôðrum on unriht tô hearme gedêþ, eal hit sceal eft mænigfealdlîce deóran him sylfum, Wulfst. 112, 7–11. Misbeódan, 157, 20. Gif him ǽnig man heálîce misboden hæbbe (cf. Who hath yow misboden, or offended, Chaue. Kn. T. 51), L. Edg. C. 5; Th. ii. 244, 18. [*Piers P.* mysbede nou3te þi bondemen: *Icel.* mis-bjóða *to ill-use, offend*.]

mis-boren; *pp.*   **I.** *mis-born, mis-shapen at birth, abortive*:—Gif cild misboren sý, Herb. 115, 3; Lchdm. i. 228, 10. (Cf. H. M. 33, 34 : 3if hit (the child) is mis-born, as hit ilome limpeð.)   **II.** *degenerate*:—Misboren *degener*, Germ. 393, 130. v. mis-byrd, -byrdo.

mis-bregdan *to remove, draw aside* (?):—Misbroden [ic eom?] *disto* ł *differo*, Wrt. Voc. ii. 141, 50. [Cf. *Icel.* mis-brigði *deviation*.]

mis-byrd; *f. A mis-birth, abortion*:—Misbyrd *abortus*, Wrt. Voc. ii. 4, 13 : 98, 17 : Ep. Gl. 2 f, 4. [*Da.* mis-byrd *miscarriage, abortion*.]

mis-byrdo; *f. indecl. Imperfect nature* or *quality*:—Be wambe missenlîcre gecyndo oððe ðære misbyrdo, L. M. 2, 27; Lchdm. ii. 220,

14. Sió wamb sió ðe biþ cealdre gecyndo oððe misbyrdo, 222, 3. [Cf. *Da.* mis-byrd *mean birth.*]

mis-bysnian; *p.* ode *To set a bad example*:—Gif ða láreówas wel tǽcaþ and wel bysniaþ ðonne beóþ hí gehealdene; gif hí mistǽcaþ, oððe misbysniaþ, hí forpǽraþ hí sylfe, Homl. Th. ii. 50, 3–5.

miscan, miscean *to injure, afflict*:—Hwî lǽtst ðú mē gān unrótne ðonne mē mysceaþ mîne fýnd *quare tristatus incedo, dum affligit me inimicus?* Ps. Th. 44, 11. [Cf. *Icel.* miski *a misdeed, offence*.]

mis-cealfian; *p.* ode *To cast a calf*:—Miscalfaþ *abortabit*, Wrt. Voc. ii. 62, 1.

miscian; *p.* ode *To mix, to mix in due proportion*:—Hē of ðǽm heán hrófe hit eall gesihþ and ðonan miscaþ and metgaþ ǽlcum be his gewyrhtum *qui, cum ex alta providentiae specula respicit, quid unicuique conveniat, agnoscit, et, quod convenire novit, accommodat*, Bt. 39, 9; Fox 226, 22. Gehwæðeres sceal mon nyttian and miscian ðæt ðone líchoman hǽle *each method (treatment by hot or by cold remedies) shall be used and applied in due proportion, that the body may be cured*, L. M. I, 1; Lchdm. ii. 22, 7. [*Ö. H. Ger.* misken *mis ere: M. H. Ger. Ger.* mischen.]

mis-cirran *to pervert*:—Oft ic miscyrre cúðe sprǽce, Bt. Met. Fox 2, 15; Met. 2, 8. v. mis-fôn.

mis-crocettan *to make a horrible noise*:—Hí (*evil spirits*) miscrocetton on hásrúnigendum stefnum, Guthl. 5; Gdwin. 36, 1. v. cræcetung.

mis-cweðan.   **I.** *to speak amiss* or *incorrectly*:—Miscweden word *barbarismus*, Wrt. Voc. ii. 12, 46 : Ælfc. Gr. 50, 21 ; Som. 51, 49. *Solocismus*, ðæt is miscweden word on endebyrdnysse ðære rǽdinge of ðam rihtan cræfte, 50, 22; Som. 51, 49.   **II.** *to curse*; *maledicere* :—Se ðe miscweðes feder ł moeder *qui maledixerit patri aut matri*, Mk. Skt. Rush. 7, 10. Miscuêdon him *maledixerunt ei*, Jn. Skt. Lind. 9, 28. [Cf. *Goth.* missa-kwiss *dissension: Icel.* mis-kviðr *a slip in pleading*.]

mis-dǽd, e; *f. A mis-deed, evil action, transgression, offence, injury*:—Mîne misdǽda bióþ simle beforan mē *delictum meum coram me est semper*, Past. 53, 2; Swt. 413, 18. God him geunne ðæt his gôde dǽda swýðran wearþan ðonne misdǽda, Chr. 959; Erl. 121, 6. Gif hund mon tóslíte ǽt forman misdǽde geselle vi scitt ... Gif ǽt ðissa misdǽda hwelcere se hund losige ... Gif se hund mā misdǽda gewyrce, L. Alf. pol. 23; Th. i. 78, 3–6. Menn scamaþ for gôddǽdan swýðor ðonne for misdǽdan, Wulfst. 164, 16. Forsyngod þurh mænigfealde synna and þurh fela misdǽda, 163, 20 : L. Eth. vi. 52 ; Th. i. 328, 15 : L. Alf. pol. 14; Th. i. 70, 16. Gif hwā lengctenbryce gewyrce ... þurh ǽnige heálîce misdǽda, L. C. S. 48; Th. i. 404, 1. [*Goth.* missa-dêds : *O. L. Ger.* mis-dât *delictum: O. Frs.* mis-dêd : *Da.* mis-daad : *O. H. Ger.* missa-, mis-tât *offensio, delictum, culpa, injuria: Ger.* misse-that.]

mis-dôn *to act wrongly, offend, transgress*:—Gif hit geweorðeþ ðæt man unwilles ǽnig þing misdêþ, nā biþ ðæt nā gelíc ðam ðe sylfwilles misdêþ, and eác se ðe nýdwyrhta biþ ðæs ðe hé misdêþ, L. Eth. vi. 52 ; Th. i. 328, 21 : L. Edg. C. 4; Th. ii. 262, 6. Se ðe misdôeþ *qui male agit*, Jn. Skt. Lind. 3, 20. Se ðe misdyde, hē hit gebéte, L. I. P. 19; Th. ii. 328, 15. Tô fela is ðæra ðe misdydan, Wulfst. 270, 30. [Durste nān man misdôn wið ôðer on his tîme, Chr. 1135; Erl. 261, 7.] [*O. Frs.* mis-dûa: *O. H. Ger.* missa-, mis-tuon *delinquere, offendere, culpare*.]

mis-efesian *to cut the hair improperly* (of the tonsure):—Wē lǽraþ ðæt ǽning gehádod man his sceare ne helige, ne hine misefesian ne lǽte, L. Edg. C. 47; Th. ii. 254, 13.

mis-endebyrdan *to arrange improperly, put in wrong order*:—Gif preóst misendebirde ciriclîce geárþenunga, L. N. P. L. 38; Th. ii. 296, 7.

misen-líc. v. missen-líc.

mis-fadian *to misconduct, order wrongly*:—Gif hē his líf misfadige *if he do not order his life aright*, L. Eth. ix. 29; Th. i. 346, 20. Gif preóst ordál misfadige, L. N. P. L. 39; Th. ii. 296, 9.

mis-fadung, e; *f. Misconduct, irregularity*:—For oft hit getímaþ ðæt sacu and ungeþwǽrnessa on mynstre áspringaþ þurh ðæs profostes misfadunge, R. Ben. 124, 5. Þurh ðis beóþ áwecte saca and tala, ungeþwǽrnessa and misfadunga, 124, 18. Misfadunga *exordinationes*, Wrt. Voc. ii. 145, 78.

mis-faran.   **I.** *to go astray, to err, transgress*:—Oft for ðæs láreówes unwísdóm misfaraþ ða híeremenn *per pastorum ignorantiam hi, qui sequuntur, offendant*, Past. 1, 4; Swt. 29, 4. Đæt men for nytennysse misfaran ne sceolon, Homl. Th. ii. 314, 5. [Cf. If Joseph sag hise breðere misfaren His fader he it gan unhillen and baren, *Gen. and Ex.* 1911.]   **II.** *To fare badly, have ill success*:—Sume secgaþ ðæt hí (*certain animals*) þurh bletsunge misfaraþ, and þurh wyrigunge, geþeóþ, Homl. Th. i. 100, 31. Þurh deófol fela þinga misfôr *by the devil's agency many things have gone on badly*, Wulfst. 104, 22. Se ðe Gode nele hýran, witod hé sceal misfaran, 178, 21. [*O. Frs.* mis-fara *to act falsely: Icel.* mis-fara *to go astray, to transgress*; mis-farask *einum to go badly with one: Da.* mis-fare *to miscarry: O. H. Ger.* missa-faran *to transgress*.] v. mis-féran.

mis-fédan *to feed improperly*:—Misfédeþ glosses *de-pascet in* Ps. Spl. T. 48, 14.

mis-féran *to go astray, transgress*:—Hē (*Saul*) ðæt folc bewerode

wið ða hǽðena leóda, ðeáh hē misférde on manegum óðrum þingum, Ælfc. T. Grn. 7, 4. [*Laym.* mis-ferde; *p. wandered: Havel.* mis-ferde; *p. acted ill.*] v. mis-faran.

mis-fón *to fail to take, to mistake:*—Ic hwílum gecoplíce funde ac ic nū gerádra worda misfó *once I readily invented, but now I fail to get appropriate words,* Bt. 2; Fox 4, 9. Be ðǽm ðe on cyricean misfón. Gif hwylc bróðor wǽgþ and misfēhþ (*makes a mistake*) on boduncge sealma, R. Ben. 71, 4–5. Wīn gedēþ, ðæt furþon witan oft misfóþ and fram rihtum geleáfan būgan, 65, 5. Ðý lǽs ǽnig ðǽre tale brúce ðæt hē ðý dǽge misfēnge (*mistook the day*), Lchdm. iii. 442, 3. [Mine songe þah he beó god me hine mai misfonge (*mis-apply, take wrongly*), O. and N. 1374: cf. *Icel.* mis-fangi *a taking one thing for another.*]

mis-gedwild, es; *n. Error:*—Ðæt wē sóðfæstra, þurh misgedwield, mód oncyrren, Exon. 70b; Th. 262, 1; Jul. 326.

mis-gehygd, es; *n. Evil mind* or *thought,* Andr. Kmbl. 1543; An. 773. [Cf. *Icel.* mis-huga *to think evil.*]

mis-gelimp, es; *n. Mishap, misadventure:*—Hē sende misgelimpu on manna bearn, Wulfst. 211, 30.

mis-gemynd, e; *f. Evil memory* or *memorial:*—Ýweþ him earmra manna misgemynda *shews him the evil memories of wretched men,* Salm. Kmbl. 987; Sal. 495.

mis-gewider, es; *n. Bad weather:*—Hwanan sió ádl cume be misgewiderum, L. M. 2, 36; Lchdm. ii. 244, 11. v. mis-wider.

mis-gíman *to fail to take care, to neglect:*—Gif preóst sceare misgýme beardes oððe feaxes, L. N. P. L. 34; Th. ii. 294, 27.

mis-grétan *to affront, insult:*—Se gylda ðe óðerne misgrēt ... gebēte hē ðæt wið ðone man ðe hē mysgrētte, Chart. Th. 606, 22–27. Gif hwilc gegilda óðerne misgrēte, 612, 18. Cf. mis-beódan.

mis-hæbbende *being ill:*—Alle mishæbbende *omnes male habentes,* Mt. Kmbl. Lind. 8, 16. Cf. yfel-hæbbende.

mis-healdsumness, e; *f. Want of observance, negligence:*—Be muneces mishealdsumnysse *de monachi inobservantia,* L. Ecg. P. iii. tit. 11; Th. ii. 196, 3.

mis-híran *to pay no attention to a person speaking, to disobey:*—Se ðe eów gehýrþ, hē gehýrþ mē, and se ðe eów mishýrþ, hē mishýrþ mē, R. Ben. 19, 23. Mid ðám murcnerum ðe Gode mishýrdon, 21, 5. Mancynn Gode mishýrde, Wulfst. 104, 23. Mishýrdan, 13, 13. Ūre bisceopas ðe wē nǽfre mishýran ne scylon on nán ðara þinga ðe hī ūs tǽcaþ, L. Edg. S. 1; Th. i. 272, 19.

mis-hírness, e; *f. Disobedience, act of disobedience:*—Forlǽt mē hý on wíta lǽdan, and ða mishērnessa gewrecan, ðe hý wið ðē forworhtan, Wulfst. 256, 4.

mis-hwirfed; *pp. Perverted:*—Swá hit is mishweorfed *sic rerum versa conditio est,* Bt. 14, 2; Fox 44, 18. Mishwyrfedre *praepostero,* Hpt. Gl. 496, 41 : 518, 19. v. next word.

mis-hworfen; *pp. Perverted, inverted:*—Tó mishworfenum *depravandam,* Wrt. Voc. ii. 26, 73; 85, 61. Mishworvenre tíde *tempore praepostero,* Hpt. Gl. 496, 42. [Cf. *O. H. Ger.* missa-huarpida *eversio;* missa-huarpari *eversor,* Grff. iv. 1236, 1237.]

mis-lǽdan *to mislead, lead astray:*—Gif hē lāre ne can, ne hē leornian nele, ac mislǽt his hýrmen and hine silfne forþ mid, L. Ælfc. P. 46; Th. ii. 384, 22.

mis-lǽran *to teach wrongly, to persuade a person to do what is wrong:*—Ðá ongunnon heora mágas behreówsian ðæt hí ǽfre ða martyras mislǽran woldon, Homl. Skt. 5, 119. [Luþer men ðat hine mislerede, Laym. 4311.]

mis-lár, e; *f. Bad teaching* or *doctrine,* Scint. 21 : 78.

mis-libban *to lead a bad life:*—Biþ mannum sceamu ðæt hí mislybban sceolon, and ða nýtenu healdaþ heora gesetnysse, Homl. Th. ii. 324, 18.

mis-, mist-, misse-líc; *adj.* **I.** *wanting in likeness* or *unity, unlike, diverse, various:*—Sorh manig and mislíc, Frag. Kmbl. 2; Leás. 2. Hū ne sǽdon wē ðæt ðis andwearde líf nǽre nó ðæt hēhste gód, forðam hit wǽre mistlíc (MS. Cott. mislíc), Bt. 34, 9; Fox 146, 17. Mistlíc *promiscuum, mixtum,* Hpt. Gl. 497, 5. Mistlíc bleó *discolor,* Wrt. Voc. i. 46, 35. Mistlíces bleós *discolor,* 77, 5. Gescý mistlíces cynnes *calceamenta diversi generis,* Coll. Monast. Th. 27, 31. Se hróf wæs on mislícre heánesse *the roof was of varying height,* Blickl. Homl. 207, 21. Se ðe micel inerfe and mislíc ágan wile, Bt. 14, 2; Fox 44, 10. Synna beóþ mislíce, Blickl. Homl. 43, 17. Mistlíce wōge wegas *divortia, diverticula,* Wrt. Voc. i. 37, 44. Mistlícra (*variarum*) cræfta biggenceras, Coll. Monast. Th. 30, 1. Misselícum *sweccum variis odoribus,* Kent. Gl. 1016. Mistlícum *diversis,* Hpt. Gl. 522, 73. Ðæt geár wæs hefigtýme on manegum þingum and mislícum .. þurh mistlíce coða, Chr. 1041; Erl. 169, 5–9. Mistlíce *varios, multimodos,* Hpt. Gl. 524, 33. **II.** *diverging from the usual course* (?), *erratic* (v. mis-líce, **II**):—Mistlícum *errabundis, vagabundis,* Hpt. Gl. 493, 20. [*Goth.* missa-leiks *various: O. Sax.* mis-lîc: *O. H. Ger.* missa-, mis-lîh *varius, diversus, dispar, multiplex, multifarius.*]

mis-, mist-líce; *adv.* **I.** *diversely, variously, in different ways:*—Godwine his geféran mislíce ofslóh *Godwine killed his companions in different ways,* Chr. 1036; Erl. 164, 33; Ælf. Tod. 2: Exon. 107b; Th. 411, 13; Rä. 29, 12. Hí his mistlíce (Cott. MS. mislíce) willnigen,

Bt. 36, 3; Fox 176, 26. **II.** *in an irregular manner* (v. mis-líc, **II**):—Eádwine eorl and Morkere eorl hlupon ūt and mislíce sērdon (*went wandering about*) on wuda and on feldon óþ ðæt Eádwine wearþ ofslægen fram his ágenum mannum, Chr. 1072; Erl. 210, 20. [Cf. Laym. 6270: fulle seouen ʒere heo misliche foren (*wandered about*).]

mis-lícian *to displease:*—Gif heó mislícaþ (*displicuerit*) ðam hláforde, Ex. 21, 8. Se ðe him sylfum mislícaþ tó ðí ðæt hē Gode gelícige, Homl. Th. i. 512, 35. Ðonne eów mislícíaþ ða mettrumnessa ðe gē on óðrum monnum geseóþ, Past. 21, 4; Swt. 159, 13. Hē him sylfum mislícade, Bd. 5, 13; S. 632, 10. Ðeós ūre mynegung wile mislícian eów wel manegum, L. Ælfc. P. 2; Th. ii. 364, 14. [*Icel.* mis-líka: *O. H. Ger.* misse-líchen *displicere.*]

mis-, mist-lícness, e; *f. Diversity, variety:*—Be swefena mistlícnysse *de somniorum diversitate,* Lchdm. iii. 198, 4. Mislícnysse *varietate,* Ps. Spl. 44, 11. Mistlícnesse *varietates, diversitates,* Hpt. Gl. 431, 75. Ðás ylcan mislícnyssa ðæra foresǽdra tída, Homl. Th. ii. 76, 12.

mis-limp, es; *n. A mishap:*—Mislimp *excessus,* Wrt. Voc. ii. 145, 67. Mislimp tearte *casus asperos,* Hymn. Surt. 16, 5.

mis-limpan *to turn out unfortunately:*—Æfter ðæm ðe him swá oftrǽdlíce mislamp hié angunnan hit wítan heora lǽtteówum *iterum infelicius victi sunt; propter quod ducem suum exsulare jusserunt,* Ors. 4, 4; Swt. 164, 24. Nis nán wundor ðeáh ūs mislimpe *it is no wonder, though we have ill success,* Wulfst. 163, 16. Gif hit geweorðe ðæt folce mislimpe þurh here oððon hunger, L. I. P. 18; Th. ii. 324, 28. [*O. E. Homl.* him mai sone mislimpe.]

mis-micel; *adj. Wanting in greatness* or *quantity* (?), *few:*—On feorhgebeorh hæfde eallum eorþcynne ēce láfe frumcneów gehwæs fæder and móder tuddorteóndra geteled ríme mismicelra (*misselícra or misselícra?*) ðonne men cunnon *to preserve the life of all that lives on earth Noah had an everlasting remnant (one from which an endless line of descendants would come), an original pair, father and mother, of every one of the offspring-producers, few in number, (fewer indeed) than men know,* (or *? of many kinds when reckoned up, more so than men know*), Cd. 161; Th. 201, 16; Exod. 373.

mis-rǽcan *to reach* or *touch wrongly,* metaph. *to apply abusive language to a person:*—Ðæt man biddendne þearfan misrǽce *to abuse a needy person who begs* (*is one of the lighter offences*), Homl. Th. ii. 590, 25. v. ge-rǽcan (*the last example there given*).

mis-rǽd, es; *m.* **I.** *evil advice* or *direction, mis-guidance:*—Hí beóþ geyrmede þurh unwíse cyning on manegum ungelimpum for his misrǽde *they (a people) are made miserable through an unwise king, by many mischances, on account of his misguidance,* Homl. Th. ii. 320, 3. **II.** *evil conduct:*—God hí (*the Israelites*) betǽhte ðam hǽðenan folce feówertig geára for heora misrǽde, Jud. 13, 1. [Cf. *Icel.* mis-rǽði *an ill-advised deed.*]

mis-rǽdan *to counsel amiss, give bad advice:*—Gif geférræden ðæne rǽd on gemǽnum geþeahte misrǽdaþ (-rǽdaþ) and feáwa witena ðæs geféres ða þearfe wíslícor tōcnáwaþ stande ðara rǽd ðe mid Godes ege and wísdóme ða þearfe geceósaþ *if the society in a general council act ill-advisedly (in the choice of an abbot), and a few wise men of the society with greater wisdom recognize what is necessary, let their counsel prevail, who with the fear and wisdom of God choose what is necessary,* R. Ben. 117,19. [Cf. Laym. ʻwe adreded ðat heo him-ræden.ʼ Þa answerede þe abbed: ʻNæi ac heo him radeþ god,ʼ 13130: *Ayenb.* me him gyleþ and misret, 184, 31: *Icel.* mis-rǽðit *ill-advised.*]

missan; *p.* miste. **I.** *to miss, fail to hit* (with gen. of object):—Hē miste mercelses, Beo. Th. 4869; B. 2439. **II.** *to escape the notice of a person* (with dat.):—Beó se canon him ætforan eágum, beseó tó gif hē wille, ðý lǽs ðe him misse (*lest any part be omitted by him*), L. Edg. C. 32; Th. ii. 250, 25. [*Laym.* missen *to notice the absence of a person: Gen. and Ex.* missen *to lose, fail: Prompt. Parv.* missyñ *careo: O. Frs.* missa *to be without: Icel.* missa *to fail to hit, to lack, to omit, to lose: O. H. Ger.* missan *carere.* The verb governs the gen. in the cognate dialects.]

mis-scrence; *adj. Shrivelled up, distorted:*—Hí (*demons*) hæfdon wōge sceancan and misscrence tán, Guthl. 5; Gdwin. 36, 1. Cf. ge-screnge.

mis-scrýdan *to clothe improperly:*—Bindaþ ðone misscrýddan (*the man who had not on the wedding garment*), Homl. Th. i. 530, 13.

missen-, misen-, missend-líc; *adj. Dissimilar, different, diverse, various, divers:*—Hwítes hiowes and eác missenlíces *candido versicolore,* Nar. 16, 1. Draca missenlíces hiwes, 43, 13. For missenlíce heora feaxes hiwe óðer wæs cweden se bleaca Heáwold óðer se hwíta (*pro diversa capillorum specie*), Bd. 5, 10; S. 624, 16. Misenlíco wilddeór him cōmon tó, Shrn. 88, 16. Wið misenlíce (misendlíce, MS. B.) leahtras, Herb. tit. 165, 3; Lchdm. i. 62, 8. Missendlíce cynno *diversitatem gentium,* Rtl. 32, 1. Hū gedǽleþ missenlíce (*or adv.?*) leoþocræftas londbúendum, Exon. 78b; Th. 295, 4; Crä. 28. Hē ūs sylþ missenlíc mód, 89a; Th. 334, 8; Gn. Ex. 13. Ealle yfelhæbbende missenlícum ádlum (*variis languoribus*), Mt. Kmbl. 4, 24. Mid eallum missenlícum áfeddum blóstmum *with all the various flowers that are brought forth,* Blickl. Homl. 7, 31. For missenlícum intingan *diversis ex causis,* Bd.

4, 1; S. 564, 17. Mid missendlícum blôstmum *variis floribus*, 1, 7; S. 478, 22. v. mis-líc.

missen-líce; *adv. Variously, diversely, differently:*—Ðeáh hê hié mannum missenlíce dǽle, Blickl. Homl. 39, 18: Exon. 88 a; Th. 331, 6; Vy. 64: 79 b; Th. 299, 18; Crä. 104.

missenlíc-ness, e; *f. Variety, diversity:*—Ðanon him wæs eágena missenlícnes geseald *thence was given him variety of eyes*, Salm. Kmbl. 180, 14. Ðeós wyrt is gecweden *iris illyrica* of ðære missenlícnysse (*variegated character*) hyre blôstmena, for ðý ðe is geþuht ðæt heó ðone heofonlícan bogan mid hyre bleó geefenlǽce, Herb. 158, 1; Lchdm. i. 284, 14. Missenlícnesse *varietatibus*, Ps. Spl. T. 44, 16.

missere, missare, es; *n. A period of half a year* [cf. *Icel.* ár heitir tvau misseri, *but the word also means* a year: *as in the following examples the Icelandic word* (*also written* missari) *occurs generally in the plural*. v. Grmm. D. M. 716]:—Swā ic Hring-Dena hund missera (*fifty years*) weóld, Beo. Th. 3543; B. 1769: 3001; B. 1498. Fela missera *many a year*, 309; B. 153: 5234; B. 2620: Cd. 145; Th. 180, 23; Exod. 49. Hê forþ gewāt misserum frôd (*well stricken in years*), 83; Th. 104, 30; Gen. 1743. Missarum frôd, 107; Th. 141, 16; Gen. 2345.

mis-spówan *to succeed badly:*—Hê sǽde ðæt hit ðæm cyninge læsse edwit wǽre, gif ðæm folce bûton him misspeówe *if it went ill with the people when he was not with them*, Ors. 2, 5; Swt. 82, 34.

mis-sprecan *to murmur:*—Misspreca *murmurari*; missprêcon *murmurabant*, Jn. Skt. Lind. 6, 43, 41.

mist, es; *m. Mist, dimness:*—Mist *vel* genip *nebula*, Wrt. Voc. i. 52, 61. Dymnys oððe myst *caligo*, Ælfc. Gr. 9, 3; Som. 8, 58. Ðā slôh ðǽr micel mist *facta est caligo tenebrosa*, Gen. 15, 17. Ǽr se þicca mist þinra weorðe, Bt. Met. Fox 5, 11; Met. 5, 6. Woruld miste oferteáh *covered the world with mist*, Exon. 51 b; Th. 178, 35; Gú. 1254. Tôdríf ðone mist ðe nū hangaþ beforan ûres môdes eágum, Bt. 33, 4; Fox 132, 32. Ðone sweartan mist, môdes þióstro, Bt. Met. Fox 23, 9; Met. 23, 5. Ða mistas ðe ðæt môd gedrêfaþ, Bt. 5, 3; Fox 14, 17. On ðás sweartan mistas (*hell*), Cd. 21; Th. 25, 9; Gen. 391. *Dimness (of sight)*:—Lǽcedômas wið eágna miste, L. M. 1, 2; Lchdm. ii. 26, 6. Of wlǽtan cymþ eágna mist, Lchdm. ii. 28, 1. Ðeós eáhsealf mæg wið ælces cynnes broc on eágon ... wið mist, Lchdm. iii. 292, 2. [Cf. *Icel.* mistr *mist*.] v. eáh-, gedwol-, wæl-mist; mistian, mistrian.

mis-tǽcan *to teach wrongly:*—Gif ða láreówas wel tǽcaþ, ðonne beóþ hí gehealdene; gif hí mistǽcaþ, hí forpǽraþ hí sylfe, Homl. Th. ii. 50, 4. [*Gen. and Ex.* mis-tagte *mis-directed*.]

mistel, es; *m* (?). I. *basil*:—Mistel *ocimum*, Wrt. Voc. i. 68, 37: ii. 65, 51. Genim ðás wyrte ðe man *ocimum*, and ôðrum naman mistel nemneþ, Herb. 119, 1; Lchdm. i. 232, 11. Heó hafaþ leáf neáh swylce mistel, 137, 1; Lchdm. i. 254, 12. v. eorþ-mistel. II. *mistletoe*:—Mistil *viscus*, Ep. Gl. 28 d, 21. Mistel, Wrt. Voc. ii. 123, 59. v. ác-mistel. [*Da.* mistel: *O. H. Ger.* mistil; *m. viscus*.] v. next two words.

mistel-lám, es; *n. Bird-lime made from the berries of the mistletoe:*—Mistellám *viscus*, Wrt. Voc. i. 289, 65.

mistel-tán, es; *m. Mistletoe:*—Mistiltán *viscarago*, Wrt. Voc. i. 31, 66. [*Icel.* mistil-teinn: *Da.* mistel-ten.] v. tán *a twig*.

mist-glóm *darkness caused by mist:*—Helle sêceþ grundleásne wylm under mistglôme *seeks hell, bottomless burning, amid the misty gloom*, Exon. 97 a; Th. 363, 1; Wal. 47.

mist-helm, es; *m. A veil or covering of mist:*—Oft ic misthelme forbrægd eágna leóman *oft have I drawn a misty veil before the light of their eyes*, Exon. 72 b; Th. 270, 25; Jul. 470.

mis-þeón; *p.* -þáh *To succeed badly, to fail to improve, to degenerate:*—Ic misþeó *degenero*, Wrt. Voc. i. 39, 29. Misþíhþ *degenerat*, ii. 138, 36. Misthágch *degeneraverat*, 106, 30. Misþáh, 25, 36: Exon. 95 a; Th. 354, 39; Reim. 58. [*O. H. Ger.* missi-díhan *deprimi*.] v. geþeón.

mist-hlíþ, es; *n. A mist-covered hill-side:*—Ðá com of môre under misthleoþum Grendel gongan *then came from the moor, under the misty slopes, Grendel walking*, Beo. Th. 1425; B. 710. Ðis leóhte beorht (*the sun*) cymeþ morgna gehwam ofer misthleoþu wadan ofer wǽgas, Exon. 93 a; Th. 350, 8; Sch. 60.

mistian; *p.* ode *To grow dim:*—Mê mistiaþ mîne eágan *caligo*, Ælfc. Gr. 36; Som. 38, 48. v. mistrian.

mis-tídan; *p.* de (*used impersonally*) *To turn out badly:*—Gif æt láde mistíde *if the attempt at exculpation prove a failure*, L. C. S. 57; Th. i. 406, 27. [Cf. *O. and N.* þu miht wene þat þe mistide, 1501.] Cf. mis-tímian.

mistig; *adj. Misty, covered with mist:*—Ofer môr mistig *super montem caliginosum*, Rtl. 18, 38. Hê heóld mistige môras, Beo. Th. 326; B. 162.

mis-tímian; *p.* ode *To happen amiss, to do amiss* (with dat. of person):—Gif ðu hwene gesihst geþeón on gôde blissa on his dǽdum and gif him hwæt mistímaþ besárga his unrôtnysse *if you see any one*

---

flourish in goodness, rejoice at his deeds, and if any mischance befall him (or *if he do anything amiss?*) *sorrow for his disquietude*, Basil admn. 5; Norm. 44, 30. [Gyf ǽnie prusten mistímide on áþaran mynstre ne fôre hê náwider ac gesôhte hê his nágabûras and him þingadan *if there were misconduct on the part of any priest in either monastery, he would go no whither, but would seek his neighbours, and they would mediate for him*, Chart. Th. 324, 8. *A. R.* þe ueorðe is Gledschipe of his vuel, lauhwen oðer gabben, gif him misbiueolle (mistimes, MS. T.; mistimeð, MS. C.), 200, 21.]

mist-líc. v. mis-líc.

mistran; *p.* ede *To grow dim:*—His eágan ne mistredon *non caligavit oculus ejus*, Deut. 34, 7. v. mistian.

mis-tríwan *to mistrust, be diffident:*—Wê mistríwaþ *difidimus*, Rtl. 39, 32. [Cf. *Icel.* mis-trûa *to mistrust*: *O. H. Ger.* missa-trûên *diffidere*: *Ger.* miss-trauen.]

mis-tucian *to maltreat:*—Ðe abbot wolde hí (*the monks*) mistukian, and sende æfter lǽwede mannum, and hî cômon intô capitulan fullgewêpnede, Chr. 1083; Erl. 217, 9.

mis-tyhtan *to incite or persuade to what is wrong, dissuade:*—Hig ðæt folc mistihton *murmurare fecerant multitudinem*, Num. 14, 36. Hê cwæþ tô ðám mágum ðe ða martyras mistihton (*urged them to renounce Christianity*), Homl. Skt. 5, 69. v. next word.

mistyhtend-líc; *adj. Dissuasive:*—Sume (*adverbs*) synd *deortativa*, ðæt synd forbeódendlíce oððe mistihtendlíce, Ælfc. Gr. 38; Som. 40, 8.

mis-weaxan *to grow in an improper way:*—Ðæt hí symle ða misweaxendan bôgas of áscreádian, Homl. Th. ii. 74, 12.

mis-wendan; *p.* de. I. *trans. To pervert, apply to a wrong use, abuse:*—Ðá miswendon sume ða englas heora ágenne cyre, and hý sylfe tô deóflum geworhton *then some of the angels made an ill-use of the choice that was theirs, and made themselves devils*, Homl. Th. i. 112, 7. Hê begann tô þreágenne ða gebrôðru ðe miswende wǽron *he began to rebuke the two brothers who were perverted*, 66, 34. Mid þweorum ðú bist miswend *cum perverso perverteris*, Ps. Lamb. 17, 27. II. *intrans. To turn in a wrong direction, be perverted:*—Gif seó gewylnung miswent, ðonne ácenþ he[ó] gýfernesse and forlygr and gítsunge, Homl. Skt. 1, 102. [*Ayenb.* hwanne he miswent and went to þe worse half al þet he yherþ, 62, 15: *O. H. Ger.* missa-wenten *evertere*; missa-wentit *transversus, obliquus*.]

mis-weorc, es; *n. An evil deed:*—Miswerc *mala opera*, Jn. Skt. Rush. 3, 19. [*Icel.* mis-verk.]

mis-weorþan *to turn out badly* (for a person, dat.):—Gif ða penegas teóþ swiðor ðonne ðæt gold ðonne miswyrþ ðam men hraðe *if the pennies weigh more than the gold, then will it soon prove a bad thing for the man*, Wulfst. 240, 4.

mis-weorðian, -wurðian *to dishonour, treat disrespectfully:*—Gif preóst circan miswurðige, ðe eal his wurðscipe of sceal árísan, gebête ðæt, L. N. P. L. 25; Th. ii. 294, 10.

mis-wider, es; *n. Bad weather, storm:*—Gif hwæt fǽrlíces on þeóde becymþ, beón hit hererǽsas, beón hit miswyderu oððon unwæstmas, Wulfst. 271, 2. v. mis-gewider.

mis-wissian *to mis-direct:*—Gif mæssepreóst folc miswissige æt freólse and æt fæstene, gylde xxx scill. mid Englum, L. E. G. 3; Th. i. 168, 8.

mis-wrítan *to write incorrectly, make a mistake in writing:*—Barbarismus, ðæt is ánes wordes gewæmmednyss, gif hit biþ miswriten, Ælfc. Gr. 50, 21; Som. 51, 48. On manegum wîsum miswritene, 50, 23; Som. 51, 54.

míte, an; *f. A small insect, a mite:*—Míte *tamus*, Wrt. Voc. i. 24, 16. [*Chauc.* These wormes, ne these mothes, ne these mites Upon my paraille fret hem never a del: *O. Du.* mijte *acarus*: *O. H. Ger.* mîza *culex*.]

mið. v. mid.

míðan; *p.* máð, *pl.* miðon; *pp.* miðen. I. *to conceal, dissemble* (a) *with gen.*:—Ðú mê tǽldesð forðon ic mín máð and wolde fleón ða byrðenne ðære hirdelecan giémenne *pastoralis curae me pondera fugere delitescendo voluisse reprehendis*, Past. proem.; Swt. 23, 11. Mê nǽfre næs ealles swā ic wolde ðeáh ic his miðe *it was never with me just as I would, though I dissembled the fact*, Bt. 26, 1; Fox 90, 28. (b) *with acc.*:—Ic on môde máð, monna gehwylcne, þeódnes þrymcyme, Exon. 51 a; Th. 177, 18; Gú. 1229. Hê ða wyrd ne máð, fǽges (*Guthlac*) forðsíð, 52 b; Th. 182, 33; Gú. 1319. Ðá hié ûs gesáwon hié selfe sôna in heora hûsum deágollíce hié miðan *visis nobis continuo inter tectorum suorum culmina delituerunt*, Nar. 10, 18. Ne sceal ic mîne onsýn forð eówere mengu míðan, 43 a; Th. 144, 18; Gú. 680. Ic míðan sceal monna gehwylcum síðfæt mînne, 127 b; Th. 491, 12; Rä. 80, 13. Ic monnan funde heardsǽligne môd míðendne *I found a man of hard fortunes, his thoughts concealing*, 115 a; Th. 442, 29; Kl. 20. (c) *case undetermined*:—Míðiþ *dissimulat*, Wrt. Voc. ii. 106, 42. Míðeþ, 25, 51. Fela gê fore monnum míðaþ, ðæs ðe gê in môde gehycgaþ, Exon. 39 a; Th. 130, 10; Gú. 436. Cyriacus hygerûne ne

mǽð tô Gode cleopode *Cyriacus concealed not the secret of his mind, but cried to God,* Elen. Kmbl. 2196; El. 1099. Hwîlum biþ gôd tô mîðanne his hiéremonna scylda *aliquando subjectorum vitia prudenter dissimulanda sunt,* Past. 21, 1; Swt. 151, 8. Miðene *concealed,* Bd. 4, 27; S. 604, 24. II. *intrans. To be concealed, lie hid:*—Ðonne biþ sôna sweotol æteówod on him ðæt ǽr deágol mǽð *then at once will be made manifest in him what before lay hid,* L. M. 2, 66; Lchdm. ii. 298, 8. Monig þing ge egeslíce ge willsumlíce ðe ôðre men miðon *multa, quae alios laterent, vel horrenda, vel desideranda,* Bd. 5, 12; S. 627, 30. Miðende *dilitiscendo,* Wrt. Voc. ii. 140, 39. III. *to avoid, refrain from, forbear (with inst. (?) dat. (?) or intrans.) :*—Ic þurh mûþ sprece ... hleoðre ne miðe *I speak with my mouth ... refrain not from sound,* Exon. 103 a; Th. 390, 20; Rä. 9, 4. Wulf on walde wælrûne ne mǽð, Elen. Kmbl. 56; El. 28. Ne mîð ðú for menigo *forbear not on account of the multitude,* Andr. Kmbl. 2419; An. 1211. Ne mæg ic ðý mîðan, Exon. 125 a; Th. 481, 1; Rä. 64, 10. [*Havel.* his sorwe he couþe ful wel miþe (*conceal*), 948: *Gen. and Ex.* ðog ðis folc miðe (*forbore*) a stund, 3807: *O. Sax.* mîðan (*with gen. acc. and intrans.*) *to avoid, forbear:* O. H. Ger. mîdan *vitare, cavere, latere, latitare, occultare, erubescere:* Ger. meiden: O. Frs. for-mîtha.] v. be-mîðan.

mitinc. v. mitting.

mitta, an; m. *A measure, both dry and liquid, as for corn, meal, ale, honey;* according to one passage it seems equal to two 'ambers':—Under mittan *sub modio,* Wrt. Voc. ii. 85, 9: Hpt. Gl. 505, 4. Under mitte (mytte, Rush), Mt. Kmbl. Lind. 5, 15. Mitta, Mk. Skt. 4, 21: mitto, Lk. Skt. Lind. 11, 33. Selle mon xxx ombra gôdes Welesces aloþ, ðet limpaþ tô xv mittan, and mittan fulne huniges, oððe twegen wînes, Chart. Th. 460, 22–28. Mittan *bata,* Wrt. Voc. ii. 11, 52: *chori,* 55, 82. His bigleofa wæs ælce dæg þrittig mittan clǽnes melewes and sixtig mittan ôðres melewes 'Solomon's provision for one day was thirty measures of fine flour, and threescore measures of meal' (1 Kings 4, 22), Homl. Th. ii. 576, 31–32. Hund mittena *centum choros,* Lk. Skt. 16, 7. Wíf gehýdeþ in meolo mitto þrió *mulier abscondit in farinae sata tria,* Lind. 13, 21. [Cf. *Goth.* mitaþs, mitaþjo *a measure:* O. H. Ger. mezzo: Ger. metze.] v. an- (on-), cyric-, hand-mitta.

mittan; p. te *To meet with, find:*—Ne meahton ceastre weg cûðne mittan *viam civitatis non invenerunt,* Ps. Th. 106, 3. v. ge-mittan.

mitting, e; f. *A meeting:*—Ðonne habbaþ wê gecweden ðæt úre mytting síe þriwa on xii mônþum *we have agreed that our meeting be thrice a year,* Chart. Th. 613, 25. Se mæssepreóst â singe twâ mæssan æt ælcere mittinge, 614, 5. v. gâr-, ge-, word-mitting.

mix. v. meox.

mixen, [n]e; f. *A mixen, dung-heap;* also *dung:*—On ðære nyðemestan flêringe (*of the ark*) wæs heora gangpyt and heora myxen, Boutr. Scrd. 21, 7. Meoxine *sterculii,* Germ. 397, 449. Job sæt on his mixene, Homl. Th. ii. 452, 28. Nis hyt nyt ne on eorþan ne on myxene (mixen, Lind.: mixenne, Rush.) *neque in terram neque in sterculinium utile est,* Lk. Skt. 14, 35. Ðeós wyrt biþ cenned on ealdum myxenum (myxennum, MS. H.), Herb. 14, 1; Lchdm. i. 106, 12. Meoxena *sterquilinia,* Hpt. Gl. 504, 2. Ic sendo micxseno (mixenne, Rush.), *mittam stercora,* Lk. Skt. Lind. 13, 8.

mixen-plante, an; f. *The mixen-plant;* 'solanum nigrum, which is morella minor, and is often found on mixens.* Otherwise *night-shade,'* Lchdm. iii. 338, col. 2:—Of ðære wyrte ðe man hâteþ myxenplante, L. M. 1, 58; Lchdm. ii. 128, 23.

môd, es; n. I. *the inner man, the spiritual as opposed to the bodily part of man,* e. g. ða ryhtæþelo biþ on ðam môde, næs on ðam flǽsce, Bt. 30, 2; Fox 110, 19. Ðone blindan ðe on lîchoman wæs gehǽled ge eác on môde, Blickl. Homl. 21, 10. Like the English *spirit, soul* it can be used to denote a person, e. g. ðæt ǽðele môd (*St. Andrew*), Andr. Kmbl. 2486; An. 1244: (*St. Juliana*), Exon. 68 b; Th. 255, 4; Jul. 209. Ðæt milde môd (*St. Guthlac*), 43 b; Th. 146, 17; Gû. 711; and throughout Alfred's translation *ðæt môd* represents Boethius, e. g. ðâ ðæt môd ðillic sâr cweþende wæs se wîsdôm him blíþum eágum on lôcude and hê for ðæs môdes geómerunge næs nâuht gedréfed *haec ubi continuato dolore delatravi, illa vultu placido, nihilque meis questibus mota,* Bt. 5, 1; Fox 8, 23–26. (a) with more especial reference to intellectual or mental qualities, *mind:*—Gesceád *ratio,* môd *mens,* Ælfc. Gr. 5; Som. 4, 48. Môd *vel* geþanc *animus,* Wrt. Voc. i. 42, 33. Seó sâwul *is animus,* ðæt is môd, ðonne heó wât; heó is *mens,* ðæt is môd, ðonne heó understent, Homl. Skt. 1, 184: Blickl. Homl. 229, 14, 18. Nû ic wât tela and ic onféng gewit mînes môdes, Bd. 3, 11; S. 536, 34. Hit is ælces môdes wîse ðæt sóna swâ hit forlǽt sôþcwidas swâ folgaþ hit leásspellunga *eam mentium constat esse naturam, ut quoties abjecerint veras, falsis opinionibus induantur,* Bt. 5, 3; Fox 14, 15. Hê ongeat ðæs môdes ingeþancas, 7, 1; Fox 16, 5. Hâles môdes *sane mentis,* Mk. Skt. 5, 15. Hê ðâ cwices môdes (*animi vivacis*) geornlîce leórnade, Bd. 5, 19; S. 637, 37. Môdes snyttru, Exon. 17 b; Th. 41, 28; Cri. 662: 78 b; Th. 295, 14; Crä. 33: Cd. 52; Th. 66, 16; Gen. 1085. Heó cwæþ on hyre môde *dicebat intra*

se, Mt. Kmbl. 9, 21. Nis mê on geþance *vel* on môde *non mihi est cordi,* Wrt. Voc. i. 54, 47. Ic hæfde mê êce geár ealle on môde *annos aeternos in mente habui,* Ps. Th. 76, 5. Gleáw on môde, Cd. 107; Th. 143, 2; Gen. 2373: 213; Th. 266, 14; Sat. 22. Môde gegrípan *to comprehend,* Exon. 92 b; Th. 348, 10; Sch. 26. Môd *mentes,* Wülck. 253, 30. (b) with reference to the passions, emotions, etc., *soul, heart, spirit, mind, disposition, mood:*—God biþ ðonne þearlwîsra ðonne ǽfre ǽnig môd gewurde *God shall then be more severe than ever any soul might be,* Blickl. Homl. 95, 31. Ðâ weóp hê sylf, and his môd wæs onstyred, 225, 22: Cd. 35; Th. 47, 10; Gen. 758. Him wæs murnende môd *sad hearts had they,* Beo. Th. 99; B. 50. Hî lǽrdon ðæt hî him wǽpno worhton and môdes strengþo nâman *they (the Romans) urged them (the Britons) to make themselves weapons and to take courage,* Bd. 1, 12; S. 481, 5. In môdes heánnesse *in extasi,* Wrt. Voc. ii. 47, 20. On gnornunga môdes *in merore animi,* Kent. Gl. 517. Môdes heánes *loftiness of soul,* Blickl. Homl. 119, 20: 31, 34. Ðæt is ðînes môdes willa *the desire of thy heart,* 225, 19. Ða ðe betran môdes wǽron *those who were better disposed,* 215, 11. His þegnas wǽron flǽsclices môdes (*carnally minded*), 17, 5: Ors. 4, 13; Swt. 212, 25: 5, 3; Swt. 222, 2: Ps. Th. 118, 60: 144, 5. Lufa ðînne drihten mid ealre ðînre heortan and mid eallum môde (*ex tota anima tua*), Deut. 6, 5: 13, 3. Forseó ðisse worulde wlenco gif ðû wille beón welig on ðînum môde; forðam ða ðe ðæs welan gîtsiaþ, hî biþ wædlan on hyra môde, Prov. Kmbl. 50. Hê wæs â on ânum môde and heofonlíce blisse mon mihte â on his môde ongytan *he was always the same, and heavenly joy might ever be seen in him,* Blickl. Homl. 223, 34. Ðâ wǽron hié swîðe erre on heora môde *then were they very angry in their hearts,* 149, 28: Cd. 3; Th. 4, 33; Gen. 63: 16; Th. 20, 2; Gen. 302. God onsende on ðara brôðra môd ðæt hî woldan his bân geniman *God put it into the hearts (in animo) of the brethren to take his (Cuthbert's) bones,* Bd. 4, 30; S. 608, 28. Bêgan wê úre môd from ðære lufan ðisse worulde, Blickl. Homl. 57, 22. Is mê nû swîðe earfeþe hiera môd tô âhwettanne, nû hit nâwþer nyle beón, ne scearp ne heard, Ors. 4, 13; Swt. 212, 30. Hî tíon on yrre môd gebrohtan *in ira concitaverunt eum,* Ps. Th. 77, 40: Cd. 3; Th. 4, 28; Gen. 60: 21; Th. 26, 7; Gen. 403. Hý se sylfa cyning lýsde þurh milde môd, Exon. 25 b; Th. 74, 23; Cri. 1211. Ða tydran môd, 43 b; Th. 147, 19; Gû. 729. Drihtnes weg gearwian tô heora môdum, Blickl. Homl. 81, 8. Hê ûs syleþ missenlícu môd (*different dispositions*), Exon. 89 a; Th. 334, 8; Gn. Ex. 13. Môde, *inst.* with much the same force as the Romance suffix -mente, -ment:—Unforhte môde *fearlessly,* Blickl. Homl. 67, 1. Untweógende môde *undoubtingly,* 171, 13. Erre môde, 189, 25. Sorgiende môde, Bd. 1, 15; S. 484, 8. Mid freó môde, 2, 5; S. 507, 32. II. a special quality of the soul, (a) in a good sense, *Courage, high spirit:*—Æfter ðam ðe his môd wæs mid ðam bismre âhwæt hê fôr eft on Perse and hî geflýmde *after his courage had been sharpened by this disgrace, he again marched against the Persians, and put them to flight,* Ors. 6, 30; Bos. 126, 17. Heorte sceal ðê cénre môd sceal ðê mâre ðe úre mægen lytlaþ *heart shall the braver be, courage the higher, as our force dwindles,* Byrht. Th. 140, 64; By. 313. Ðâ ongunnon hî môd niman *then they began to take courage,* Bd. 1, 16; S. 484, 15. Hê hæfde môd micel, Beo. Th. 2338; B. 1167. Woldon ellenrôfes môd gemiltan, Andr. Kmbl. 2785; An. 1395. (b) in a bad sense, *Pride, arrogance :*—Ðæs engles môd, Cd. 1; Th. 3, 2; Gen. 29. Hyre môd âstâh *her (Hagar's) pride mounted up,* 101; Th. 134, 35; Gen. 2235: 205; Th. 253, 18; Dan. 597: Exon. 42 a; Th. 141, 27; Gû. 633. Cf. Hê wæs on swâ micle ofermêtto âstigen *efferatus superbia,* Ors. 6, 9; Swt. 264, 8. Næs mê for môde *it was not from pride in me,* 28 b; Th. 87, 22; Cri. 1429. Him se mǽra môd getwǽfde, þæt forbîgde, Cd. 4; Th. 4, 14; Gen. 53. Þurh ðîn (*Lucifer's*) micle môd, 35; Th. 46, 2; Gen. 738. III. applied to inanimate things, *Greatness, magnificence, pride :*—Heriaþ hine æfter môde his mægenþrymmes *laudate eum secundum multitudinem magnitudinis ejus,* Ps. Th. 150, 2. Mycel môd and strang ðínes mægenþrymmes *magnificentiam majestatis tuae,* 144, 5. Ne mihton forhabban werestreámes môd *they could not restrain the pride of the flood (of the Egyptians drowned in the Red Sea),* Cd. 167; Th. 208, 24; Exod. 448. [*Goth.* môds *anger:* Icel. môðr *wrath, grief:* O. Sax. O. Frs. môd *mind, heart, courage:* O. H. Ger. muot *mens, animus, anima, cor:* Ger. muth.] v. ofer-môd.

-môd in composition of adjectives. v. âcol-, an-, ân-, ǽttren-, ǽwisc-, blíðe-, deór-, dreórig-, eád-, eáð-, forht-, freórig-, gâl-, gealg-, geómor-, gewealden-, glæd-, gleáw-, gûþ-, heáh-, heán-, heard-, hreóh-, hreówig-, hwæt-, irre-, lâðwende-, leóht-, meagol-, meaht-, micel-, ofer-, or-, reomig-, reónig-, rêðe-, rêðig-, rûm-, sârig-, sceóh-, stíð-, styrn-, swíð-, þancol-, þearl-, til-, torht-, torn-, wêrig-, wrâð-môd.

môd-blind; adj. *Having the mind's eye darkened, undiscerning:*—Leóde ne cûðan, môdblinde men, Meotud oncnâwan, Exon. 25 a; Th. 73, 11; Cri. 1188: Andr. Kmbl. 1627; An. 815: Elen. Kmbl. 611; El. 306. [Cf. *O. H. Ger.* muot-plinti *coecitas animi.*]

môd-blissiende *rejoicing at heart:*—Môdblissiendra *laetantium,* Ps. Th. 67, 17.

mód-bysgung, e; *f. Anxiety of mind:*—Ðam ðe his synna sáre geþenceþ módbysgunge micle dreógeþ *to him who his sins with sorrow remembers, much anxiety suffers of mind*, Exon. 117 a; Th. 450, 7; Dóm. 84.

mód-cearig, *adj. Anxious at heart*, Exon. 76 b; Th. 286, 18; Wand. 2. [*O. Sax.* mód-karag.]

mód-cearu, e; *f. Sorrow of heart, grief:*—Ðæt gelumpe módcearu mægum, Exon. 35 a; Th. 114, 1; Gú. 166. Ic æfre ne mæg ðære módcéare mínre gerestan, 115 b; Th. 443, 34; Kl. 40. Dreógeþ mín wine micle módceare, Th. 444, 22; Kl. 51. Hygesorge wæg, micle módceare, 47 b; Th. 162, 29; Gú. 983 : 52 a; Th. 182, 26; Gú. 1316: Beo. Th. 3560; B. 1778 : 3989; B. 1992. Higum unróte nódceare mændon mondryhtnes cwealm *troubled in mind they mourned with sorrow of soul their lord's decease*, 6289; B. 3149. [*Laym.* heo þolede modkare, 3115: *O. Sax.* mód-kara.]

mód-cræft, es; *m. Mental power or skill:*—Ða ðe snyttro mid eów and módcræft habben, Elen. Kmbl. 815; El. 408. Módcræfte séc þurh sefan snyttro ðæt ðú wite, Exon. 14 a; Th. 28, 4; Cri. 441.

mód-cræftig, *adj. Possessing mental power, intelligent, skilled:*—Módcræftig smiþ, Exon. 79 a; Th. 297, 2; Crä. 62.

mód-cwánig, *adj. Sad at heart:*—Mengo módcwánige, Elen. Kmbl. 754; El. 377. v. cwánian.

móddor, móddrige. v. módor, módrige.

mód-earoþ, es; *n. Travail of soul, distress of mind:*—Ic wonn (MS. þonc) módearroþa má, Exon. 119 a; Th. 457, 19; Hy. 4, 86.

móde-líc, -wæg, móder. v. módig-líc, -wæg, módor.

mód-full; *adj. Proud, arrogant:*—Cild ácenned [biþ] weallende módful *a child born (on the eleventh day of the moon) will be turbulent and arrogant*, Lchdm. iii. 188, 26. [Oswi hæfde emes sunen þe weoren swiðe þrute gumen, and ma of his cunne þe weoren modfulle, Laym. 31464.]

mód-gehygd, es; *n. Thought:*—Ic tó ðé mid módgehygde clypade *I cried to thee in thought*, Ps. Th. 87, 13. Hine fyrwyt bræc módgehygdum *his thoughts were distracted by curiosity*, Beo. Th. 471; B. 233.

mód-gemynd, es; *n.*: e; *f. Mind, thought, intelligence:*—Ðá wæs módgemynd miclum geblissod hyge onhyrded *then was his mind much rejoiced, his heart confirmed*, Elen. Kmbl. 1676; El. 840. Ða ðe leornungcræft þurh módgemynd hæfdon *those who had knowledge through intelligence*, 761; El. 381: Andr. Kmbl. 1375; An. 688: Exon. 96 b; Th. 360, 9; Wal. 3.

mód-geómor, *adj. Sad at heart, of mournful mind:*—Ðæt eorlwerod módgiómor sæt, Beo. Th. 5779; B. 2894. Þeód wæs módgeómre, Andr. Kmbl. 2227; An. 1115: 3412; An. 1710.

mód-geþanc, es; *m. n. Mind, thoughts, thought:*—Hé mid his eágum up tó heofenum lócade ðyder his módgeþanc á geseted wæs *with his eyes he looked up to heaven, whither his thoughts were ever directed*, Blickl. Homl. 227, 17: Exon. 50 a; Th. 173, 33; Gú. 1170. Módgeþonc, R. Met. Fox 31, 37; Met. 31, 19. Næron gé swá eácne ofer ealle men módgeþances *ye were not so gifted above all men with understanding*, Cd. 179; Th. 224, 16; Dan. 137. Mætra on módgeþanc *more humble in mind*, 207; Th. 256, 3; Dan. 635. Nú gé fyrhþsefan and módgeþanc mínne cunnon, Elen. Kmbl. 1067; El. 535. Nú wé sceolan herigean metodes módgeþanc (-gidanc) *nunc laudare debemus creatoris consilium*, Bd. 4, 24; S. 597, 20. Monnes módgeþonc, Beo. Th. 3462; B. 1729: Bt. Met. Fox 5, 45; Met. 5, 23. Ne þearf hé gefeón módgeþance *he need not rejoice in his heart*, Cd. 75; Th. 92, 5; Gen. 1524. On hige sínum, módgeþance, 107; Th. 141, 3; Gen. 2339. Ðá þeahtode þeóden úre módgeþonce, 5; Th. 6, 23; Gen. 93. Swá monig beóþ men ofer eorþan swá beóþ módgeþancas *quot homines, tot sententiae*, Exon. 91 b; Th. 344, 4; Gn. Ex. 168: 91 a; Th. 341, 11; Gn. Ex. 124.

mód-geþoht, es; *m. Mind, thought:*—Mihtigne on his módgeþohte *mighty of mind*, Cd. 14; Th. 17, 1; Gen. 253. [*O. Sax.* módgiþaht.]

mód-geþyldig; *adj. Patient of soul*, Andr. Kmbl. 1962; An. 983.

mód-gewinna, an; *m. A foe of the mind, care, anxiety:*—Læt ðé áslúpan sorge of breóstum, módgewinnan, Cd. 134; Th. 169, 9; Gen. 2797.

mód-glæd; *adj. Of gladsome mind*, Exon. 49 b; Th. 171, 23; Gú. 1131.

mód-gleáw; *adj. Wise of mind*, Salm. Kmbl. 361; Sal. 180.

mód-hete; es; *m. Hate:*—Ic hine wergþo on míne sette, and módhete, Cd. 83; Th. 105, 21; Gen. 1756.

mód-hord, es; *n. m. The mind:*—Módhord onleác weoruda dryhten and ðus wordum cwæþ, Andr. Kmbl. 344; An. 172.

mód-hwæt; *adj. Strong of soul, courageous, brave:*—Mægeþ módhwatu *a maiden strong of soul*, Exon. 122 b; Th. 470, 14; Hy. 11, 16. Nymðe hié módhwate Moyses hýrde *unless they with courage good obeyed Moses*, Cd. 148; Th. 185, 17; Exod. 124. Ða módhwatan *the courageous ones*, 191; Th. 238, 20; Dan. 357.

módig; *adj.* **I.** *of high* or *noble spirit, high-spirited, noble-minded:*—Ðis is se écea God módig and mægenróf *this is the eternal God, noble and mighty*, Cd. 156; Th. 195, 11; Exod. 275: Exon. 18 b; Th. 46, 32; Cri. 746: Rood Kmbl. 81; Kr. 41. Ðæt wæs módig cyn *that was a high-spirited race*, Cd. 173; Th. 216, 16; Dan. 7. Se fugel engla eard gesóhte, módig, meahtum strang, Exon. 17 a; Th. 40, 31; Cri. 647. Is se wyrhta módig meahtum spédig *of noble mind is the maker, abundant in might*, 56 a; Th. 198, 14; Ph. 10: 42 b; Th. 143, 26; Gú. 667. Ðæt is módig wuht *it (the bull) is a high-spirited creature*, Runic pm. Kmbl. 339, 12; Rún. 2: Elen. Kmbl. 2524; El. 1263. Hlóh ðá módi man (*Byrhtnoth*), Byrht. Th. 136, 6; By. 147. Se módiga (*Holofernes*), Judth. 10; Thw. 22, 7; Jud. 52. Se módega mæg Higeláces (*Beowulf*), Beo. Th. 1630; B. 813. Se módga (*the Phenix*), Exon. 59 b; Th. 216, 3; Ph. 262. Geáta leód trúwode módgan (*Beowulf's*) mægnes, Metodes hylde, Beo. Th. 1344; B. 670. Unc módige ymb mearce sittaþ, þeóda þrymfæste, Cd. 91; Th. 114, 20; Gen. 1907. Módge maguþegnas, Exon. 77 a; Th. 290, 8; Wand. 62. **II.** *bold, brave, courageous* (physically or morally):—Wæs from se ðe lædde, módig magoræswa, Cd. 145; Th. 181, 2; Exod. 55. Gæþ se ðe mót tó medo módig *he that may shall go bold to the mead*, Beo. Th. 1212; B. 604: Andr. Kmbl. 481; An. 241. Ðæt wæs módig secg *a brave man was he*, Beo. Th. 3629; B. 1812: 3021; B. 1508. Næs ænig ðæs módig mon ofer eorþan . . . ðæt mec ðus bealdlíce bendum bilegde, Exon. 73 a; Th. 273, 8; Jul. 513. Sió hand gebarn módiges mannes, Beo. Th. 5329; B. 2698. Beówulfes síþ, módiges merefaran, 1008; B. 502. Hægsteald módige, wígend unforhte, Cd. 160; Th. 198, 24; Exod. 327. **III.** *proud, arrogant:*—Módig *superbus* . . . eádmód *humilis*, Wrt. Voc. i. 76, 25, 27. Ne beó rænig man hér on worldríce tó módig, Blickl. Homl. 109, 27. Módig and medugál '*flown with insolence and wine*,' Judth. 10; Thw. 21, 19; Jud. 26. Mære and módig (*Nebuchadnezzar*), Cd. 177; Th. 222, 15; Dan. 105. Æfter ðæra módigra gásta hryre, Homl. Th. ii. 82, 11. Hé tóstæncte ða módigan *dispersit superbos*, Cant. Mar. 51. **IV.** *hearty, earnest, impetuous;* in a bad sense, *bold, headstrong, stubborn, wilful:*—Bidde ic monna gehwone . . . ðæt hé mec neódful . . . gemyne módig *I pray every man that diligently and heartily he bear me in mind*, Exon. 76 a; Th. 285, 28; Jul. 721. Merestreám módig *the impetuous flood* (v. mód, III; *and* módigian), Cd. 166; Th. 207, 17; Exod. 468. Módig *contumax*, Ælfc. Gr. 9, 60; Som. 13, 42. Gif ænig man hæbbe módigne sunu and rancne *si genuerit homo filium contumacem et protervum*, Deut. 21, 18. On óðre wísan sint tó manianne ða módgan (*protervi*), on óðre ða unmódgan (*pusillanimes*), Past. 32, 1; Swt. 209, 4. [*Goth.* módags *angry: Icel.* móðugr: *O. Sax.* módag: *M. H. Ger.* muotec: *Ger.* muthig.] v. fela-, ofer-, til-, un-módig.

módigian, módigan; *p.* ode. **I.** *to be* or *become proud, to glory, exult:*—Se unwæra oft módegaþ on gódum weorcum *the heedless is often proud of good works*, Homl. Th. ii. 222, 4. Se ríca módegode on his welum *the rich man gloried in his wealth*, i. 328, 19. Se deófol ðe módegode *the devil who grew proud*, 138, 11. Swá módgade wuldres cempa *thus exulted the soldier of glory* (*Guthlac*), Exon. 37 a; Th. 121, 25; Gú. 294. Bebeódaþ ðám rícum ðæt hí ne módigan on heora ungewissum welan, Homl. Th. i. 256, 25. Ðá begann hé (*Lucifer*) tó módigenne for ðære fægernesse ðe hé hæfde, 10, 22. Wá lá wá ðæt ænig man sceolde módigan swá, hine sylf áhebban, and ofer ealle men tellan, Chr. 1086; Erl. 222, 36. **II.** *to take offence through pride:*—Sum æþelboren cild heóld leóht ætforan his mýsan, and ongann módigian ðæt hit on swá wáclícum þingum him wícnian sceolde. Se hálga underget his módignysse, Homl. Th. ii. 170, 25. **III.** *to bear one's self proudly, impetuously:*—Flota módgade (*moved proudly*), Cd. 160; Th. 198, 32; Exod. 331. Ðær ær wegas lágon mere módgode (v. módig, IV) *where before ran the roads, now raged the sea*, 166; Th. 206, 27; Exod. 458. v. ofer-módigian.

módig-líc; *adj.* **I.** of persons, *Noble-mind, high-souled, courageous, brave:*—Eálá mín drihten! ðæt ðú eart ælmihtig, micel, módílíc, Bt. Met. Fox 20, 3; Met. 20, 6. Módiglíce menn síðfrome *brave men, bold in travel*, Andr. Kmbl. 491; An. 246. Ne seah ic elþeódige men módiglícran *no braver men from foreign lands have I seen*, Beo. Th. 680; B. 337. **II.** of things (v. mód, III), *Superb, magnificent:*—Nænig man nafaþ tó ðon módelíco gestreón hér on worlde, Blickl. Homl. 111, 24: 113, 6.

módiglíce; *adv. Boldly, bravely:*—Modelíce manega sprǽcon ðe eft æt þearfe (MS. þære) þolian noldon *many used brave words, who would fail at need*, Byrht. Th. 137, 42; By. 200. [ȝho mihhte modiȝlike onnȝæn Anndswerenn þuss, Orm. 2035.]

módig-ness, e; *f.* **I.** in a bad sense, *Pride:*—Módignys *superbia*, Wrt. Voc. i. 76, 26. Se eahteoða heáfodleahter is módignyss (þe ehtuðe sunne is ihatan *superbia*, þet is on englisc modinesse, O. E. Homl. i. 103, 33), Homl. Th. ii. 218, 22. Flǽsces tóbryte módignesse *carnis terat superbiam*, Hymn. Surt. 9, 22. Ða heofenlícan myrhþe ðe ða englas þurh módignysse forluron, Homl. Th. i. 360, 28. **II.** in a good sense, *Highmindedness, magnanimity, greatness of mind which does not resent injury:*—Eahta sweras syndon ðe rihtlícne cynedóm up wegaþ:

sóþfæstnys, módignes (*patientia*), L. I. P. 3; Th. ii. 306, 28. [Þatt wære módiȝnesse ꝥ idell ȝellp, Orm. 12040: stiȝþ on heh þurh modinesse, O. and N. 1405.]

módig-wæg, es; *m. An impetuous wave:*—Módewæga mæst (*the water that overwhelmed the Egyptians*), Cd. 167; Th. 209, 14; Exod. 499. v. módig, IV.

mód-leás; *adj. Spiritless, dull;* excors, Kent. Gl. 400.

mód-leást, e; *f. Want of courage, pusillanimity:*—Ðá wearþ se wælhreówa wódlíce geancsumod, ꝥ his mágas ne mihton his módleáste ácuman, ac héton ácwellan ꝥæt mæden, Homl. Skt. 9, 125. [Þe sixte unþeau is þet þe ðe to lauerd bið iset þet he for modleste ne mei his monnan don stere, O. E. Homl. i. 111, 24.]

mód-leóf; *adj. Dear to the heart, beloved:*—Fæder lærde módleófne mágan, Exon. 80 a; Th. 301, 32: Fä. 28.

mód-lufu, an; *f. Heart's love, affection,* Beo. Th. 3650; B. 1823: Exon. 26 a; Th. 77, 25; Cri. 1262: 71 a; Th. 264, 26; Jul. 370: 76 a; Th. 284, 18; Jul. 699: 123 a; Th. 473, 3; Bo. 9. [*O. H. Ger.* mót-luba *affectu.*]

módor; *gen.* módor, méder; *dat.* méder; *f. A mother* (of human beings or of animals):—Heó is ealra libbendra módor, Gen. 3, 20. Hér is ðín módor, Mk. Skt. 3, 32. Ánes cildes módor *mater;* manigra cilda módur *materfamilias,* Wrt. Voc. ii. 59, 20, 21. Fæder and módor, Exon. 103 a; Th. 391, 8; Rä. 10, 2. Módur, Gen. 37. 10: Ps. Th. 108, 14. Ðæt is móddor monigra cynna, Exon. 112 a; Th. 428, 16; Rä. 41, 2: 128 a; Th. 492, 13; Rä. 81, 15. Þridde móder *proavia:* feówerþe móder *abavia:* fífte móder *tritavia,* Wrt. Voc. i. 51, 56, 58, 60. Wynburge þridde módor, Chart. Th. 650, 23. Of his módor (móderes, Lind.: moeder, Rush.) innoþe, Lk. Skt. 1, 15. Of módur hrife, Ps. Th. 70, 5. From bearme móddor, Exon. 112 b; Th. 430, 27; Rä. 44, 15. Þurh geleáfan ꝥæs fæder and ꝥære méder, Homl. Th. ii. 52, 2: 50, 35: 116, 13: i. 66, 21. Hé mín ne ræcþ ne ꝥære méder, Homl. Skt. 4, 313. Þurh þingunge his ꝥære eádigan méder, Bd. 5, 19; S. 640, 42. Segþ his fæder and méder, Mk. Skt. 7, 11: Ps. 130, 4: Wulfst. 119, 3: Cd. 50; Th. 64, 10; Gen. 1048: Exon. 8 b; Th. 3, 15; Cri. 36. Riht is ꝥæt ꝥæt bearn médder folgige, L. H. E. 8; Th. i. 30, 4: 99 a; Th. 370, 7; Seel. Ex. 53. Nim ꝥæt cild and his módor, Mt. Kmbl. 2, 13. Gif mon cú oððe stódmyran forstele, and folan oððe cealf of ádrífe forgelde . . . and ða móder be hiora weorðe, L. Alf. pol. 16; Th. i. 72, 1. Ealle fæderas and móddru, Homl. Th. ii. 34, 32. 124, 17. Heáp móddra *caterva matrum,* Hymn. Surt. 52, 5. Ðé læs hé ofsleá ðás módra, Gen. 32, 11. [*The Gothic uses* aiþei, *the other dialects use a form corresponding to the English. O. Sax.* módar: *O. Frs.* móder: *Icel.* móðir: *O. H. Ger.* muotar: *Lat.* mater: *Grk.* μήτηρ.] v. beó-, eald-, fóstor-, steóp-módor.

módor-cynd, e; *f. The nature derived from the mother:*—Hé wæs sóþ man þurh his médrengecynd (módercynde, MS. H.), Wulfst. 17, 7.

módor-leás; *adj. Motherless:*—Fylstan fæderleásum and móderleásum cildum, Wulfst. 228, 22.

módor-líc; *adj. Maternal:*—Móderlíc *maternus,* Ælfc. Gr. 5; Som. 4, 57. Móderlícere stæððinysse *materna gravitate,* Hpt. Gl. 469, 37.

módor-slaga, an; *m. A matricide;* matricida, Wrt. Voc. i. 85, 46.

módren, móddren; *adj. Maternal:*—Móddrenum flǽsce ic brúce *materna carne vescor,* Ap. Th. 4, 12. v. médren.

módrige, móderge, móddrige, an; *f.* **I.** *an aunt:*—Mín móddrige *matertera mea,* Wrt. Voc. i. 52, 25: 51, 53: Bd. 3, 8; S. 532, 21. Módriæ, Kent. Gl. 1190. Bisceop næbbe on his húse nænne wífman búton hit sý his módor . . . oððe módrige, L. Ælfc. C. 5; Th. ii. 344, 14. Módrige, Homl. Th. ii. 94, 32. Módrigan sunu *fratrueles,* Wrt. Voc. ii. 39, 53: 55, 31. Módrian sunu *consobrinus,* Ors. 3, 9; Swt. 130, 21: Ælfc. T. Grn. 16, 9. Módergan sunu, Shrn. 93, 3. Móddrian sunu, Homl. Th. i. 58, 5: Wrt. Voc. i. 52, 2, 27, 28. **II.** *a cousin:*—Móderge *consobrinus,* Wrt. Voc. ii. 105, 31. Mínre móddrigan móder *matertera mea materna,* 55, 33. Tó ðære hire (*the Virgin Mary*) móddrian ðære hálgan Elizabethe, Blickl. Homl. 165, 28. [His moddrie sune, Laym. 30644.]

mód-sefa, an; *m.* [*a poetical word with much the same meaning as* mód, e. g. Swá bióþ ánra gehwæs monna módsefan áwegede of hiora stede, Bt. Met. Fox 7, 47; Met. 7, 24 = swá ꝥæt mennisce mód biþ áweged of his stede, Bt. 12; Fox 36, 17: *and* Gif heora módsefa meahte weorþan staþolfæst gereaht, 11, 195; Met. 11, 98 = gif heora mód wære gestaþelod, Bt. 21; Fox 74, 40.] *The inner man, mind, spirit, soul, heart:*—Ðæt ðín módsefa mára wurde and ðín líchoma leóhtra micle *that thy mind would be mightier and far fairer thy body,* Cd. 25; Th. 32, 10; Gen. 501. Ðá wæs módsefa miclum geblissod *greatly then was his heart gladdened,* Andr. Kmbl. 1783; An. 894: Elen. Kmbl. 1748; El. 876. Wæs módsefa áfýsed on forþwege *my soul longed to be gone,* Rood Kmbl. 246; Kr. 124. Mé ðín módsefa lícaþ *you please me,* Beo. Th. 3711; B. 1853. Ne gemealt him se módsefa *his heart did not fail,* 5249; B. 2628. Helle gemundon in módsefan *hell had they in mind,* 362; B. 182. Ic ne métte on módsefan máran snyttro, Andr. Kmbl.

1107; An. 554. Ne sceal se Dryhtnes þeów in his módsefan (*in his heart*) mǽre gelufian eorþan æhtwelan, Exon. 38 a; Th. 125, 22; Gú. 358: 66 b; Th. 247, 1; Jul. 72. Man cweþeþ on his módsefan *dicet homo,* Ps. Th. 57, 10. On módseofan, 115, 2. Módsefan ásecgan *to open one's heart to another,* Exon. 76 b; Th. 287, 6; Wand. 10. Hé his módsefan fæste trymede *he his soul surely stablished,* 46 b; Th. 159, 26; Gú. 933: Andr. Kmbl. 2420; An. 1211. Syððan hé módsefan mínne cúðe *after he knew my heart,* Beo. Th. 4028; B. 2012: Exon. 54 a; Th. 188, 24; Az. 50. Beóþ módsefan dálum gedǽled, sindon dryhtguman ungelíce, 83 b; Th. 314, 29; Mód. 21. [*O. Sax.* módsebo: *Icel.* móð-sefi.]

mód-seóc; *adj. Sick at heart, with mind diseased, distressed:*—Unrótne, módseócne, Exon. 51 a; Th. 177, 30: Gú. 1235. [*O. H. Ger.* muot-siuh: cf. *Icel.* hug-sjúkr *distressed.*]

mód-seócness, e; *f. Disease of the stomach:*—Módseócnes vel [mód-] unmiht *morbus cordis (cardiacus),* Wrt. Voc. ii. 128, 66.

mód-snotor, -snottor; *adj. Prudent of mind, wise, sagacious:*—Fród fæder freóbearn lærde, módsnottor, Exon. 80 a; Th. 300, 6; Fä. 2. In mǽðle módsnottera, 79 a; Th. 295, 31; Crä. 41: 100 a; Th. 374, 19; Seel. Ex. 128. Módsnotra, Soul Kmbl. 249; Seel. Verc. 128.

mód-sorh; *gen.* -sorge; *f. Care or sorrow of mind, sorrow of soul:*—Eác is hearm Gode, módsorg gemacod, Cd. 35; Th. 47, 3; Gen. 755. Hé módsorge wæg hefige æt heortan *sorrow of soul bore he heavy at heart,* Exon. 48 a; Th. 165, 6; Gú. 1024: Elen. Kmbl. 122; El. 61. [Mid muchele modsorȝe (sorewe, 2nd MS.), Laym. 8692.]

mód-staþol, es; *m. The foundation on which the mind rests:*—Stedefæst módstaþol biþ witena gehwilcum weorþlícre micle ðonne hé his wísan fágige tó swíðe *a firm foundation for the mind is much more honourable for every man of counsel, than an excessive variation of manners,* L. I. P. 10; Th. ii. 318, 38.

mód-staþolfæstness, -staþolness, e; *f. Stability of mind:*—Ongeán módstaþolnysse (-staþolfæstnesse, MS. C.) and módes strencþe se mánfulla deófol sendeþ wácmodnysse and lyðerne earhscype, Wulfst. 53, 10.

mód-swíð; *adj. Strong of mind* or *soul*:—Wec ðú in mé módswíðne geþanc *crea in me spiritum rectum,* Ps. C. 50, 89; Ps. Grn. ii. 278, 89.

mód-þracu; *gen.* -þræce; *f. Impetuosity of mind, impetuous* or *daring courage:*—Ic ðæm gódan (*Beowulf*) sceal for his módþræce mádmas beódan, Beo. Th. 775; B. 385. [*O. Sax.* mód-thraka *conflict of mind, grief:*—Sind that módthraka manno gehwilikumu, that hé farlátan skal liobana herron, Hel. 4775.]

mód-þreá; *gen.* -þreán; *m. f. Pain* or *torment of mind:*—Egsa micel módþrea *terror, great torment of mind,* Exon. 102 a; Th. 385, 25; Rä. 4, 50.

mód-þrýðu (o); *indecl. f. Violence of mind:*—Módþrýðo wæg folces cwén *a violent heart bore the queen of the people,* Beo. Th. 3867; B. 1931.

mód-þwǽre; *adj. Gentle, meek, mild:*—Hé gerehþ módþwǽre on dóme *diriget mansuetos in judicio,* Ps. Lamb. 24, 9.

mód-þwǽrness, e; *f. Gentleness, meekness, patience:*—Módþwǽrnes (*patientia vel geþyld,* MS. E.), Wulfst. 69, 1.

mód-unmeaht, -miht. v. mód-seócness.

mód-welig; *adj. Rich in spiritual* or *mental gifts:*—Gregorius, Rómwara betest, monna módwelegost, Past. Swt. 9, 12.

mód-wén, e; *f. Hope entertained by the mind:*—Forþ áscúfan ꝥæt mínes freán módwén (ꝛ, MS.) freoþaþ middelnihtum *to push on what my lord's hopes favour at midnight* (*to carry out the plans which are thought on at night, and in which he hopes to succeed?*), Exon. 129 b; Th. 498, 3; Rä. 87, 7.

mód-wlanc; *adj. Proud, haughty, of high courage:*—Nis ꝥæs módwlonc mon ofer eorþan ꝥæt hé á his sǽfóre sorge næbbe *no man upon earth is of courage so high, as on his sea-journey ne'er to feel fear,* Exon. 82 a; Th. 308, 13; Seef. 39. Módwlonc meówle *haughty maiden,* 107 a; Th. 407, 18; Rä. 26, 7.

mohþe. v. moððe.

molcen, es; *n. Curdled milk:*—Molcen lac coagolatum, Wrt. Voc. i. 290, 29: ii. 52, 7. Swá þicce swá molcen, L. M. 3, 39; Lchdm. ii. 332, 18. Nim súr molcen, 1, 39; Lchdm. ii. 98, 25.

mold-ǽrn, es; *n. An earth-house, a grave:*—Þeáh mín líc scyle on moldærne molsnad weorþan, Exon. 64 a; Th. 235, 28; Ph. 564: Rood Kmbl. 130; Kr. 65: Andr. Kmbl. 1604; An. 803.

molda or **molde,** an; *m.* or *f. The top of the head:*—Ðæt galdor man sceal singan ǽrest on ðæt wynstre eáre ðænne on ðæt swíðre eáre ðænne ufan ðæs mannes moldan *the charm must first be sung into the left ear, then into the right ear, then on the top of the man's head,* Lchdm. iii. 42, 9. [Cf. Trev. v. 369, 7: þe Longobardes used to schere of þe heere of hir heed from þe molde to þe nolle (from the toppe un to the hynder parte, MS. Harl.) *comam capitis a cervice usque ad occipitium tondebant.* Halliwell gives mold the suture of the skull.]

molde, an; *f.* **I.** *mould, dust, sand, earth:*—Molde *sabulum,* Wrt. Voc. i. 37, 24: *sablo,* ii. 119, 39: 89, 36. Of ðære moldan (*pulvere*) ðæs flóres monige untrume men gehǽlede wǽron. Ond heó

bæd ðæt hyre man sumne dǽl ðǽre hálwendan moldan (*pulveris*) sealde, Bd. 3, 11; S. 536, 5-8: 3, 10; S. 534, 23, 29. Ða ðe for hund wintrum mid eorþan moldan (*pulvere terrae*) bewrogene wǽron, L. Ecg. P. iv. 66; Th. ii. 226, 23. Ðonne hit (*cadaver*) biþ on ða byrgenne set, ðonne wyrpeþ man moldan ofer hit, L. Ecg. C. 36; Th. ii. 162, 3. His þegnas mid moldan hit (*a cross*) gefæstnedon *adgesto a militibus pulvere, terrae figeretur*, Bd. 3, 2; S. 524, 19. Be moldan ða ðe on ðǽre stöwe genumene wǽron, 3, 9; S. 533, 27.   **II.** *ground, earth, land:*—Molde *vel* land *humus, rus, arvum*, Wrt. Voc. i. 41, 61: *humus*, 70, 12: Ælfc. Gr. 8; Som. 7, 53. Of ðǽre moldan tyrf *from the grass of the ground*, Exon. 56 b; Th. 202, 8; Ph. 66. God forþ áteáh of ðǽre moldan (*de humo*) ǽlces cynnes treów, Gen. 2, 9. Þeóda wealdend árǽs of moldan (*rose from the grave*), Hy. 10, 34; Hy. Grn. ii. 293, 34: Exon. 120 a; Th. 460, 24; Hö. 22. Ðonne of ðisse moldan men onwecniaþ, deáde of duste árísaþ, Cd. 227; Th. 302, 22; Sat. 604. Ða moldan ðe meolce and hunige flēwþ *humum lacte et melle fluentem*, Num. 14, 8. Mearh moldan træd *the steed trod the ground*, Elen. Kmbl. 109; El. 55.   **III.** *earth* (the dwelling place of men):— Ne mihte ða on moldan man gerïman *no man on earth might number them*, Ps. Th. 104, 30: 127, 5: Cd. 202; Th. 251, 21; Dan. 567: Exon. 99 a; Th. 371, 13; Seel. 75. Of moldan on ða mǽran gesceaft *from earth to heaven*, Bt. Met. Fox 20, 561; Met. 20, 281. Men ofer moldan *men upon earth*, Rood Kmbl. 23; Kr. 12: Hy. 3, 12; Hy. Grn. ii. 281, 12: Exon. 50 b; Th. 176, 2; Gú. 1203. Meotud ða moldan gesette, 56 a; Th. 198, 15; Ph. 10. [*Goth.* mulda *dust*: *Icel.* mold *mould, earth*: *O. H. Ger.* molta *pulvis, humus, solum, terra.*]   v. græs-molde.

mold-corn, es; *n.* ' *The granular tuber of saxifraga granulata, and the plant itself,*' Cockayne:—Moldcorn *vulnetrum*, Wrt. Voc. i. 69, 8: Lchdm. iii. 18, 8.

mold-grǽf, es; *n.* *A grave:*—Wæs lǽded líc tö moldgræfe, Exon. 75 b; Th. 284, 1; Jul. 690. Ǽnra gehwylc from moldgrafum sēceþ Meotudes dóm, 63 b; Th. 233, 13; Ph. 524.

mold-hrērende *moving upon earth:*—Nis ðæt monnes gemet moldhrērendra *it is not within the compass of man, of those who move upon earth*, Exon. 92 b; Th. 348, 13; Sch. 27.

mold-hýpe, an; *f.* *A heap of earth* or *dust:*—Ðonne biþ hit swylce hē sý mid sumere moldhýpan ofhroren *it is as though he be overwhelmed by a heap of dust*, Homl. Th. i. 492, 33.

mold-stöw, e; *f.* *A place on the earth, a site,* or *a place in the earth, a grave:*—Moldstöwe, stöwlícere moldan *situ i. sepulcro*, Germ. 391, 195.

mold-weg, es; *m.* *A way upon earth, earth:*—Gif wē on moldwege fundne weorþen *if we are found on earth*, Exon. 70 b; Th. 262, 18; Jul. 334: 48 a; Th. 164, 15; Gú. 1012: Elen. Kmbl. 931; El. 467.

mold-wyrm, es; *m. An earth-worm, a worm in the grave:*—Ðec (*the body*) sculon moldwyrmas monige ceówan, Exon. 99 a; Th. 371, 7; Seel. 72. [*O. H. Ger.* molt-wurm *stellio*.]

molegn, es; *n.* (?) *A thick substance made of curds:*—Molegn *calmum* (occurs under the heading *de mensa*), Wrt. Voc. i. 290, 34: ii. 17, 20: *galmum*, 40, 63: Ep. Gl. 10 f, 15: *galmilla*, 10 f, 32. Molegen *galmilla*, Wrt. Voc. ii. 40, 64. Moling *galmum*, Wülck. 24, 4.

molegn-stycce, es; *n.* *A portion of* molegn (?):—Molegnstycce *galmulum*, Wrt. Voc. ii. 109, 54.

molsnian; *p.* ode *To moulder, become corrupt, decay:*—Söna hē molsnaþ and wyrþ tö ðære ilcan eorþan ðe hē ǽr of gesceapen wæs *soon it (the body) suffers corruption, and turns to the same earth from which before it was made*, Blickl. Homl. 21, 28. Ðonne hit (húsl) molsnaþ tö þicgenne *cum prae mucore percipi non potest*, L. Ecg. P. iv. 48; Th. ii. 218, 8. Ðeáh mïn líc scyle on moldærne molsnad weorþan, Exon. 64 a; Th. 235, 29; Ph. 564.   v. ā-, for-, ge-molsnian.

momna, Wrt. Voc. ii. 120, 82.   v. mamor.

mon.   v. man.

mön in the phrase full mön *plenilunium:*—Fullum möne *plenilunio*, Wrt. Voc. ii. 67, 42. [Cf. *O. H. Ger.* -máni *in* niu-máni *neomenia*; uol-máni *plenilunium*: unter-máni *interlunium*, Grff. 2, 795.]

möna, an; *m.: but also* niöne, an; *f.*   **I.** *the moon:*—Se möna and ealle steorran underföþ leóht of ðære miclan sunnan, Lchdm. iii. 236, 19. Se möna wæs æt fruman on ǽfen gesceapen, 264, 26. Sunna and möne (*but* næs se möna ðägyt uppe, 29, 22), Nar. 28, 20: Bt. Met. Fox 29, 73; Met. 29, 37. Ðæs sunnan ásprungnis oððe ðære mönan, Nar. 28, 10. Ðæs mönan trendel *the moon's disc*, Lchdm. ii. 242, 4.   **II.** *moon* as in new, full moon, the reference being to the stage reached in a lunar month:—Níwe möna *neomenia*, Wrt. Voc. i. 16, 51. Se níwa möna, Lchdm. iii. 264, 26. Möna se forma, se öðer, se þridda, etc., pp. 184-196. [Ful]les mönan *plene lunae*, Kent. Gl. 210. Næfre būton on níwum mönan, Lchdm. iii. 242, 23. On ānre nihte ealdne mönan . . . on tweigra nihta mönan, etc., 154, 15-28, 156, 1-16. Hē gesette ðone mönan fulne, 238, 27. Ðæt geár hæfþ twelf níwe mönan, 248, 25-26. [*Goth.* mēna; *m.: Icel.* máni; *m.: O. Sax.* máno; *m.:*

O. Frs. möna; *m.: Du.* maan; *f.: O. H. Ger.* máno; *m.: M. H. Ger.* máne; *m. also f.;* mánt, mánde: *Ger.* mond; *m.*]

Mönan-ǽfen, es; *m. Monday-eve, the evening of Sunday:*—Gif esne ofer dryhtnes hǽse þeówweorc wyrce an Sunnanǽfen efter hire setlgange öþ Mönanǽfenes setlgang, L. Wih. 9; Th. i. 38, 19.   v. Mönanniht.

Mönan-dæg, es; *m. Monday:*—Ūtgangendum ðam mönþe ðe we Aprelis hátaþ, se nýhsta Mönandæg & ingangendum ðam mönþe ðe we Agustus hátaþ se ǽresta Mönandæg . . . se ǽresta Monandæg æfter ūtgange ðæs mönþes Decembris *the last Monday in April . . . the first Monday in August . . . the first Monday after the end of December*, Lchdm. iii. 76, 14-18. On Mönandæg, Rubc. Jn. Skt. 2, 12: 7, 32. [*O. Frs.* möna-, mönan-dei: *O. H. Ger.* mäno-tag: *Ger.* mon-tag: *Icel.* mäna-dagr: *Dan.* man-dag.]   v. Mön-dæg.

Mönan-niht, e; *f. Monday eve, the evening of Sunday:*—Hē ús ðonne myngaþ ðæs Sunnandæges weorces and ðæs Sæternesdæges ofer nön and ðǽre Mönannihte, Wulfst. 210, 10.   v. Mönan-ǽfen.

mönaþ, mönþ, es; *pl.* mönaþ, mönþas; *m. A month, lunar* or *calendar:*—Ǽlce mönþe seó sunne yrnþ under án ðæra tácna . . . Ǽlc ðæra twelf tácna hylt his mönaþ, and ðonne seó sunne hí hæfþ ealle underurnen, ðonne byþ án geár ágán. On ðam geáre synd getealde twelf mönþas . . . Ðæs mönan mönaþ is ðonne hē gecyrþ níwe fram ðære sunnan öð ðæt hē eft cume hyre forne ágeán, eald and áteorod, and eft þurh hī beó ontend. On ðam mönþe synd geteald nigon and twentig daga and twelf tída, ðis is se mönelíca mönaþ . . . Se mönelíca mönaþ hæfþ ǽfre on ānum mönþe xxx nihta, and on öðrum nigon and xx. On swá hwilcum sunlícum mönþe swá se möna geendaþ, se byþ his mönaþ. Ic cwede nū gewislícor; gyf se ealda möna geendaþ twám dagum binnan Hlýdan mönþe, ðonne byþ hē geteald tö ðam mönþe, Lchdm. iii. 244-250. Ðá án mönuþ ágan wæs, Gen. 29, 14. Fullne mönoþ, Num. 11, 20. Se teóþa mönþ, October, Menol. Fox 360; Men. 181. On ðone seofenteoðan dæg ðæs mönþes, Gen. 7, 11: Lev. 23, 5. Healfum mönþe se möna biþ weaxende, healfum hē biþ wanigende, Homl. Th. i. 154, 27. Ðý syxtan mönþe ðæs ðe Sanctus Johannes on his mödor bösm onfangen wæs, Blickl. Homl. 165, 24. Æfter nigan mönþa fæce, 9, 29. Feola mönþa, Bd. 5, 19; S. 638, 19. On xII mönþum, Chart. Th. 433, 10. Fíf, syx mönþas, Lk. Skt. 1, 24: 4, 25. Feówer, eahta, seofon, nigon, twelf, feówertýne mönaþ, Ors. 6, 28; Swt. 278, 8: 6, 31; Swt. 286, 2: Blickl. Homl. 193, 13: 89, 19: 39, 15: Homl. Th. ii. 490, 25. The names of the months are as follows: Se æftera Geóla *January*, Sol-mönaþ *February*, Hrēd- or Hlýd-mönaþ *March*, Éaster-mönaþ *April*, Þrïmilci *May*, se ærra Líða or Sear-mönaþ *June*, se æftera Líða or Mǽd-mönaþ *July*, Weód-mönaþ *August*, Hálig- or Hærfest-mönaþ *September*, Winterfylliþ *October*, Blöt-mönaþ *November*, se ærra Geóla *December*. See the several words for references, and Grmm. Gesch. D. S. c. VI for the month-names in Anglo-Saxon and related dialects. [*Goth.* mēnoþs: *Icel.* mänuðr: *Dan.* maaned: *Swed.* monad: *O. L. Ger.* mänuth: *O. Frs.* mönath: *O. H. Ger.* mänod: *Ger.* monat.]

mönaþ-ádl, e; *f. A disease that occurs at intervals of a month:*—Ða ðe ðonne on gewunon mönaþ ádle numene beóþ . . . Ðæt wíf mid ðý heó ðone gewunan þrowaþ mönaþ ádle *cum in suetis menstruis detinentur . . . Mulier dum consuetudinem menstruam patitur*, Bd. i. 27; S. 493, 40-43.

mönaþ ádlig; *adj. Suffering from* mönaþádl:—Gif hwylc man gangeþ tö mönaþ ádligum wífe *si quis vir ad menstruatam mulierem accedat*, Bd. 1, 27; S. 493, 42.

mönaþ-blöd, es; *n. Menstruum:*—Mönaþblöd *menstrum*, Wrt. Voc. ii. 59, 22: *menstrua*, i. 46, 13. [Cf. *O. H. Ger.* mänod-blöti *menstruus.*]

mönaþ-böt, e; *f. Penance extending over a month:*—Sumon geárbóte, sumon mā geára . . .; sumon mönþbóte, sumon mā mönþa; sumon wucnbóte, sumon mā wucena, L. Pen. 3; Th. ii. 278, 12.

mönaþ-fyllen, e; *f. The time of full moon:*—Mönaþfylene *plenilunio*, Hpt. Gl. 525, 63.

mönaþ-gecynd, e; *f. Menstruum:*—Gif wífe tö swíðe of flöwe sió mönaþgecynd, L. M. 3, 38; Lchdm. ii. 330, 26, 13.

mönaþ-líc; *adj.*   **I.** *monthly:*—Ða mönaþlecan *menstrua*, Wrt. Voc. ii. 57, 34.   **II.** *lunar:*—Mönoþlíces clywnes *lunaris luminis*, Hpt. Gl. 418, 15. [*O. L. Ger.* mönoþ-líc: *O. H. Ger.* mänod-líh *menstruus.*]   v. symbel-mönaþlíc.

mönaþ-seóc; *adj.*   **I.** *lunatic, epileptic :*—Mönaþseóc *lunaticus*, Wrt. Voc. i. 45, 65. *Comitiales i. garritores* ylfie *vel* mönaþseóce, ii. 132, 26 (v. ilfig). Mönaþseóce *lunaticos*, Mt. Kmbl. 4, 24: Herb. 10, 2; Lchdm. i. 100, 18.   **II.** *suffering from* mönaþádl:—Bearneacnigende wíf and mönaþseóc, Homl. Th. ii. 94, 4. [*O. H. Ger.* mänodsiuh *lunaticus: and* cf. mänod-suhtig *menstruata.*]   v. mön-seóc.

mönaþseóc-ness, e; *f. Lunacy:*—Wið mönoþseócnysse, gyf man ðás wyrte ðam mönoþseócan þonne gyt ádligendum ofer áleþ, söna hē hyne sylfne hálne up áhefþ, Herb. 66, 2; Lchdm. i. 170, 4.

mond=(?) möd, Exon. 40 b; Th. 134, 26; Gú. 514.

Món-dæg, es; m. Monday :—Ælce Móndæge, L. R. S. 3; Th. i. 432, 21. v. Mónan-dæg.

móne-, món-líc; adj. Lunar :—Ðis is se mónelíca [mónlíca, MS. P.] mónaþ, Lchdm. iii. 248, 20 : 250, 1. Seó sunne biþ hwíltídum þurh ðæs mónelícan trendles underscyte áþýstrod, Homl. Th. i. 608, 32.

Mon-íg, e; f. The Isle of Man or Anglesey ; Mona :—Ðá gehergodon hí Moníge [Mæníge] then they harried the Isle of Man, Chr. 1000 (ed. Thorpe). Moníge Brytta eáland Angelcynnes ríce hé underþeódde Mevanias insulas imperio subjugavit Anglorum, Bd. 2, 9; S. 510, 16. [Icel. Mön; gen. Manar Isle of Man.]

món-seóc; adj. Lunatic, epileptic :—Mónsék (fylleseóc, W. Sax.) hé is lunaticus est, Mt. Kmbl. Rush. 15. Mónsékæ lunaticos, 4, 24. v. mónaþ-seóc.

mór, es; m. I. a moor, waste and damp land :—Moor uligo, Wrt. Voc. i. 37, 23. Móres græs the grass of the field (which Nebuchadnezzar was to eat), Cd. 203; Th. 252, 8; Dan. 575. On ðone hreódihtan mór; of ðon móre, Cod. Dip. Kmbl. iii. 121, 21 : Beo. Th. 1424; B. 710. Ofer myrcan mór, 2814; B. 1405. Ys on Breotonlande sum fenn un-mætre mycelnysse . . . Ðær synd unmæte móras, Guthl. 3; Gdwin. 20, 1–4. Fennas and móras paludes, Bt. 18, 1; Fox 62, 14. Sumra wyrta eard biþ on dúnum sumra on merscum sumra on mórum aliae herbae montibus oriuntur, alias ferunt paludes, 34, 10; Fox 148, 24. Ofer burna and ofer móras super rivos et paludes, Ex. 8, 5. Mistige móras, Beo. Th. 326; B. 162 : 207; B. 103. II. high waste ground, a mountain :—Licgaþ wilde móras wið eástan . . . on ðæm mórum eardiaþ Finnas . . . Ðær hit (Norway) smalost wære, hit mihte beón þreora míla brád tó ðæm móre; and se mór syððan, on sumum stówum, swá brád swá man mæg on twám wucum oferféran . . . Ðonne is tómnes ðæm lande súðeweardum, on óðre healfe ðæs móres, Sweóland (Ohthere's description of Norway), Ors. 1, 1; Swt. 18, 27–34, 19, 1–2. Ne munt ne mór, Salm. Kmbl. 845; Sal. 422 : 681; Sal. 340. In mór héh in montem excelsum, Mt. Kmbl. Lind. 4, 8 : 5, 1. Swá unefne is eorþe þicce, syndon ðás móras myclum ásprotene, Ps. Th. 140, 9. Ungeféredra móra inaccessorum montium, Bd. 4, 26; S. 602, 20. In heágum mórum and in hrédum in arduis asperisque montibus, 4, 27; S. 604, 27 : 3, 23; S. 554, 20. Of ðissum wéstum wídum mórum a desertis montibus, Ps. Th. 74, 6. Waldend scóp wudige móras, Exon. 54 b; Th. 193, 12; Az. 120. [O. H. Ger. M.H.Ger. muor; n. a marsh, bog.]

móraþ, mórod, es; n. A drink formed by boiling down and sweetening wine (with mulberries), a decoction of wine and herbs :—Móraþ carenum (cf. carenum æþele alu, ii. 23, 1), Wrt. Voc. i. 27, 64. Ne ete fersce gós . . . ne fersc swín ne náht ðæs ðe of mórode cume. Gif hé hwilc ðissa ete síe ðæt sealt do not let him eat fresh goose or fresh pork or aught of that which comes out of a decoction of wine and herbs (has been cooked with wine and herbs?). If he eat any of these, let it be salted, Lchdm. ii. 88, 9. Áwylle on ealdum mórode, 88, 14 : 122, 16. Nim eald mórod, iii. 14, 8. [M.H.Ger. móraz mulberry wine. v. Du Cange, moratum.]

mór-beám, es; m. A mulberry tree or blackberry bush :—Mórbeám morus vel rubus, Wrt. Voc. i. 32, 60 : murus, 80, 26. Márbeámas moros, Ps. Surt. 77, 47. [Cf. Wick. mór-tree.]

mór-denu, e; f. A swampy or fenny valley :—Of ðam stocce inn on mórdene; of mórdene inn on ðere saltstrét, Cod. Dip. Kmbl. iii. 384, 30. Cf. mór-fæsten.

more, moru, an; f. (also mora in cpds. q. v.) An (edible) root, a carrot, parsnip :—Bétan more a root of beet, Lchdm. iii. 6, 19. Wylisc moru carrot . . . Englisc moru parsnip, L. M. 3, 8; Lchdm. ii. 312, 16, 21. Eolonan moran dust, doccan moran dust, 1, 54; Lchdm. ii. 126, 6. Mintan broþ oððe moran (carrot), 1, 18; Lchdm. ii. 62, 6 : 2, 28; Lchdm. ii. 224, 25. Nim celeþonian moran and glædenan moran and hocces moran, 3, 41; Lchdm. ii. 334, 27. Ete wælwyrte moran, Lchdm. i. 354, 13. Nim Englisce moran, L. M. 1, 2; Lchdm. ii. 38, 15. Moran pastinace, Wrt. Voc. i. 69, 13. Genim ðæs scearpan þistles moran, L. M. 3, 12; Lchdm. ii. 314, 11. [O.H.Ger. moraha, morach pastinaca, cariota : Ger. möhre.] v. feld-, weal-, weald-more; æg-moran.

mór-fæsten, es; n. A place secure from attack from the swampy character of the country :—Hé (Alfred) lytle werede uniéþelíce æfter wudum fór, and on mórfæstenum, Chr. 878; Erl. 78, 34.

morgen, es; m. I. morning, morn :—Ðá hyt morgen wæs mane facto, Mt. Kmbl. 27, 1 : Blickl. Homl. 235, 18. Syððan morgen com, Beo. Th. 2159; B. 1077 : Cd. 160; Th. 199, 29 : Exod. 346. On morgene mane, Ps. Th. 91, 2. On morgene in matutino, 100, 8. Æt ðære þriddan tíde on morgenne, Blickl. Homl. 201, 35 : 203, 2. On morgne at morn, Exon. 50 b; Th. 175, 10; Gú. 1192 : Th. 176, 29; Gú. 1217. On marne mane, Ps. Surt. 5, 4, 5 : 54, 18 : Bd. 2, 6; S. 508, 23. Bringþ morgen tó mannum Decembris, Menol. Fox 435; Men. 219. On morgen mane, Gen. 28, 18 : Blickl. Homl. 69, 28 : 231, 36. Swíðe ær on morgen, Ps. Th. 18, 5. Morgena gehwilce every morning, Cd. 40; Th. 52, 23; Gen. 848 : Ps. Th. 58, 16. Morgna gehwam, Exon. 93 a; Th. 350, 7; Sch. 60. Morna, Beo. Th. 4892; B. 2450. Drince þrý morgenas let him drink three mornings, Lchdm. i.

88, 13. Nigon morgenas, ii. 118, 5. viiii morgnas . . . viii morgnas, 294, 1. Morghenas, iii. 6, 17. II. the morning of the next day, morrow :—Gá and cum tó morgenne go, and come to-morrow, Past. Swt. 325, 1. On morgne on the morrow, Beo. Th. 4961; B. 2484. On morne, Bd. 2, 6; S. 508, 7. Tó morgen cras, Ex. 8, 23 : Mt. Kmbl. 6, 30; Kent. Gl. 54 : Cd. 111; Th. 147, 12; Gen. 2438. Tó morhgen (morgen, MS. A.), Lk. Skt. 13, 32, 33. [Gen. and Ex. morgen, morwen : A. R. morwen : Ayenb. morȝen : Chauc. Piers P. morwe : Laym. morȝen, marȝen, morwe : Goth. maurgins : Icel. morginn : O. Sax. O. L. Ger. O. H. Ger. morgan : O. Frs. morn : Dan. Du. Ger. morgen : Swed. morgon.] v. ær-morgen, ærne, and mergen.

morgen-ceald; adj. Chilled with the cold of early morning :—Sceal gár wesan monig morgenceald, Beo. Th. 6036; B. 3022.

morgen-colla, an; m. Dread (?) or rage (?), furious attack (?) which comes in the morning :—Him færspel bodedon, morgencollan, atolne ecgplegan, Judth. 12; Thw. 25, 6; Jud. 245. v. collen-ferhþ.

morgen-dæg, es; m. I. morning, day-light :—Ðá hit wæs tóforan dæges ðá cwóman fugelas . . . hí eft gewiton. Ðá hit on morgendæg wæs ðá . . ., Nar. 16, 24. II. the morrow :—Be ðan morgendæge þencean, Blickl. Homl. 213, 22. v. mergen-dæg.

morgen-drenc, es; m. A drink or potion to be taken in the morning :—Hé gesette gódne morgendrænc wið eallum untrumnessum, Lchdm. iii. 70, 17. [Cf. Icel. morgin-drykkja.]

morgen-gifu, e; f. The gift made by the husband to the wife on the morning after the consummation of the marriage :—Morgengifu dos, Wrt. Voc. i. 20, 53. Hit (five hides of land) wæs hire morgengifu ðá heó ærest tó Aðulfe com, Chart. Th. 170, 24. Gif heó (a widow) binnan geáres fæce wer geceóse, ðonne þolige heó ðære morgengyfe, L. C. S. 74; Th. i. 416, 8 (cf. 522, 3 : 576, 2). Ic cýðe hwæt ic mínum wífe tó morgengife sealde, ðæt is Beadewan and Burgestede and Strætford and ða þreó hýda æt Heánhealan, Chart. Th. 596, 31. Hig ðone cincg bædon ðæt heó móste gesyllan hire morgengife intó Cristes cyrcean, 540, 18. Gif hió bearn ne gebyreþ fæderingmágas ágan morgengyfe, L. Ethb. 81; Th. i. 24, 2. [Gen. and Ex. morgen-giwe : A. R. marhen-, marech-, morh-giue : Laym. mor-, mær-ȝeue douarie : Prompt. Parv. mor-yve dos : Icel. morgun-gjöf : Dan. morgen-gave : O.H.Ger. morgangeba : Ger. morgen-gabe.] v. Grmm. R. A. 441.

morgen-lang; adj. Having a long morning :—Eorlwerod morgenlongne dæg módgiómor sæt sad at heart sat the warriors through a day whose evening seemed as if would never come, Beo. Th. 5780; B. 2894.

morgen-leóht, es; n. The morning light, morning, Beo. Th. 1213; B. 604 : 1839; B. 917. [Laym. morȝen-, more-liht : O.H.Ger. morganlioht mane.]

morgen-líc; adj. I. morning :—Morgenlíc matutinus, Wrt. Voc. ii. 116, 67. From gehæld morgenlícum a custodia matutina, Rtl. 181, 1. Tó morgenlícum tídum ad matutinas horas, 36, 35. Ic beó ðýs morgenlícan dæge (on the morning of this day: St. Mary's death seems to have taken place on the day when she says this) gongende of líchoman, Blickl. Homl. 143, 2 : 139, 18. II. of to-morrow :—Se morgenlíca dæg crastinus dies, Mt. Kmbl. 6, 34. [Icel. morgun-ligr matutinus : O. H. Ger. morgan-líh matutinus.] v. mergen-, myrgen-líc.

morgen-mete, es; m. A morning meal, breakfast :—On xii mónþum ðú scealt sillan ðínum þeówan men vii hund hláfa and xx hláfa, búton morgenmetum and nónmetum, Salm. Kmbl. p. 129, 19. [ȝief he frend were me sceolde ȝief him his morȝemete (cf. 231, 19 where it is called forme mete) þat he þe bet mihte abide þane more mete, O. E. Homl. i. 237, 33.]

morgen-regn, es; m. Rain that falls in the morning :—Ðú þurh lyft lætest, leódum tó freme, mildne morgenrén, Exon. 54 a; Th. 191, 2; Az. 82.

morgen-seóc; adj. Sick in the morning :—Him biþ â sefa geómor, mód morgenseóc, Exon. 119 a; Th. 458, 4; Hy. 4, 95.

morgen-spell, es; n. A story or narrative told in the morning :—Ðá wæs wíde læded mære morgenspel . . . ðæt Cristes ród funden wære, Elen. Kmbl. 1936; El. 970.

morgen-spréc, e; f. The periodical assembly of a guild held in the morning, or on the morrow after the guild-feast :—Se gegilda ðe ne geséce his morgenspæce gilde his syster huniges the member of a guild, who does not attend the assembly of the guild, shall pay a sester of honey, Chart. Th. 613, 7. [Cf. And if any broþer be somound to any morwe-speche . . . and wil nouht come, he scal paye a pound of wax, English Guilds (E. E. T. S.), p. 54. See also the Glossary for other references to the word, and Introduction, pp. xxxii–xxxiii, for remarks upon it. In the Promptorium morow-, morwe-, mor-speche = crastinum colloquium; cf. English Guilds, p. 30, where a meeting is held 'on morwe aftyr þe gylde day.']

morgen-steorra, an; m. The morning star :—Ðone beorhtan steorran ðe wé hátaþ morgensteorra Lucifer, Bt. 4; Fox 8, 3 : 39, 13; Fox 234, 3 : Bt. Met. Fox 4, 26; Met. 4, 13. [Prompt. Parv. morow-;

morwyn-sterre *Lucifer:* cf. *Icel.* morgun-stjarna: *Ger.* morgen-stern.] v. æfen-steorra.

morgen-swég, es; *m. A sound made in the morning:*—Ðá wæs on úhtan Grendles gúþcræft gumum undyrne. Ðá wæs æfter wiste wóp up áhafen, micel morgensweg, Beo. Th. 258; B. 129.

morgen-tíd, e; *f. Morning-tide, morning:*—In morgentíd *in matutinis,* Ps. Surt. 100, 8. On morgentíd, Beo. Th. 973; B. 484: 1041; B. 518: Chr. 937; Erl. 112, 14. On ða morgentíd, Judth. 12; Thw. 25, 1; Jud. 236. Útgong margentíde *exitus matutini,* Ps. Surt. 64, 9. Tó margentíde *ad matutinum,* 29, 6. In margentíd *in matutino,* 72, 14. [*Gen. and Ex.* morgen-tid: *O.Sax.* morgan-tíd: *Icel.* morgun-tíðir *matins.*] v. mergen-tíd.

morgen-torht; *adj. Bright with the brightness of morning* (applied to the sun), Andr. Kmbl. 482; An. 241.

morgen-wacian; *p.* ode *To get up early in the morning:*—Morgen-wacode *manicabat* (v. Lk. 21, 38), Wrt. Voc. ii. 73, 72: 56, 58.

mór-hǽþ, e; *f. A mountain-heath:*—Swá líg freteþ mórhǽþ *velut flamma incendat montes,* Ps. Th. 82, 10.

mór-heald (?):—Wǽron land heora lyfthelme beþeaht mearchofu mórheald, Cd. 145: Th. 181, 14; Exod. 61. *Grein takes the word to be an adjective=placed on a mountain slope,* cf. heald; *adj. But the word might be a noun,* cf. *O.H.Ger.* halda; *f. clivus: Icel.* hallr; *m. a slope, 'their march-dwellings were the mountain-slope.' Or perhaps* heald, ge-heald *in the sense of* keeping *might be compared, as also* hald *fermum,* Wrt. Voc. ii. 147, 71, so mór-heald=*mountain-hold or fastness. Yet again,* heald *may be* [*a northern form* (?) *of*] *the verb*=heóld, '*the mountain guarded their march-dwellings.' Bouterwek and Thorpe read thus.*

mór-hop, es; *n. A pool in a marsh:*—Hé byreþ blódig wæl ... mearcaþ mórhopu *he* (Grendel) *will bear the bloody corse ... will mark the marshy pools* (with the blood), Beo. Th. 904; B. 450. Cf. fen-hop.

mórig; *adj. Marshy, fenny:*—On mórium lande *in locis palustribus,* Gen. 41, 2. v. mór-mǽd.

mór-land, es; *n. Moor-land, wild hilly country:*—Se ðe on wéstenne, méðe and meteleás, mórland trydeþ, Elen. Kmbl. 1221; El. 612. Hé wunede on ðám mórlandum (*in montanis*), Bd. 4, 27; S. 604, 33. Se ǽresta láreów on ðám mórlandum ða ðe syndon tó norþdǽle Pehta ríces *primus doctor transmontanis Pictis ad aquilonem,* 5, 9; B. 622, 40. Ofer alle mórlonda *super omnia montana,* Lk. Skt. Lind. Rush. i. 65.

mór-mǽd, e; *f. A marshy meadow:*—Tó mórmǽde norþhyrnan, Cod. Dip. Kmbl. iii. 449, 19. v. mórig.

morne, mórod. v. morgen, móraþ.

mór-pytt, es; *m. A marshy pool:*—On mórpyt, Cod. Dip. Kmbl. iii. 381, 9.

mór-sceaþa, an; *m. A bandit, a robber who takes refuge in the moors* (v. mór):—Ðone mórsceaþo (*Barabbas*), Mk. Skt. Lind. Rush. 15, 15. 11. Wæs Barabbas mórsceaþe (sceaþa, Rush.) *erat Barabbas latro,* Jn. Skt. Lind. 18, 40. Swá tó mórsceaþe (scaþe, Rush.) gié cwómun (*ad latronem*), Mt. Kmbl. Lind. 26, 55. Tuoge mórsceaþo *duo latrones,* Mk. Skt. Lind. 15, 27: Lk. Skt. Lind. 23, 33.

mór-seáþ, es; *m. A boggy, marshy pit,* Cod. Dip. Kmbl. iii. 378, 13.

mór-secg, es; *m. n. Sedge:*—Bedde hys bed myd mórsecge, Lchdm. iii. 140, 25.

mór-stapa, an; *m. A moor-stepper, traverser of the moors:*—Mǽre mórstapa (*the bull*), Runic pm. Kmbl. 339, 11; Rún. 2.

mortere, es; *m. A mortar mortariola,* Wrt. Voc. ii. 58, 28. Se ealra mǽsta mortere *girba,* 42, 22: i. 20, 25. Gepuna eall tósomne on ánum mortere, Lchdm. i. 216, 13: 142, 18.

morþ, es; *n. m.* I. *death, destruction, perdition:*—Hit wæs hæleþa forlor menniscra morþ ðæt hié tó mete dǽdon ofet unfǽle *it was men's ruin, our race's destruction, that for their food they took that evil fruit,* Cd. 33; Th. 45, 5; Gen. 722. Mid morþes cwealme *with death's pang,* 35; Th. 47, 9; Gen. 758. Ðæt micle morþ (*death which followed the eating of the forbidden fruit*), 30; Th. 40, 16; Gen. 640. Nys ús ná tó secgenne ðone sceamlícan morþ ðe ðǽr gedón wæs (*the mortality, attended with so many horrible circumstances, that happened at the siege of Jerusalem*), Ælfc. T. Grn. 21, 15. II. *that which causes death:*—Ðú (*the evil soul*) wǽre ðǽr (*in this world*) morþ and myrþra, ac ðú ne miht hér (*in the next world*) swá beón, Wulfst. 241, 9. Ic bidde ðæt man ðæs morþres (*deadly sin, marriage by men in orders*) heononforþ geswíce, L. I. P. 23; Th. ii. 334, 23. Hé (*the devil*) hógode on ðæt micle morþ (*the eating of the forbidden fruit*) men forweorpan, forlǽran and forlǽdan, Cd. 32; Th. 43, 15; Gen. 691. Man téh ðæt morþ (*apparently an image of the intended victim whose destruction was being attempted through witchcraft by a widow and her son,* v. III *and* morþ-dǽd) *forþ of hire inclífan.* Ðá nam man ðæt wíf and ádrencte hí æt Lundenebrigce, Chart. Th. 230, 17. III. *murder;* (a) *as a technical term, slaying with an attempt at concealment of the deed.* Cf. *the distinction in Icelandic law between* morþ *murder and* víg *man-slaughter, 'þat er morþ ef maðr leynir eða hylr hræ ok gengr eigi í gegn,' but if declaration* (*lýsing*) *were made it was* víg. v. Cl. & Vig. Dict. *and* Grmm. R. A. 625. Schmid. A. S. Gesetz. p. 633, *suggests that morþ*

has particular reference to death caused by witchcraft or by poison, and refers to the connection in which the compounds morþ-dǽd-, weorc-, -wyrhta *occur:* see the passages given under those words. See also the last passage under II:—Gif open morþ weorþe ðæt man sý ámyrdred ágife man mágum ðone banan and gif hit tihtle sý and æt láde mistíde déme se bisceop *if there be a death and it afterwards appear that the man was murdered, the* (*supposed*) *murderer being discovered, let the latter be given up to the kinsmen* (*of the slain man*), *and if the accusation be brought, and the attempt of the accused to clear himself fail, let the bishop pass sentence,* L. C. S. 57; Th. i. 406, 25. Ǽbere morþ æfter woruldlage is bótleás *slaying, which is proved to be murder, according to the secular law, cannot be compounded for,* 65; Th. i. 410, 5. (b) *as a general term, murder, homicide:*—Hí swylc geblót and swylc morþ dónde wǽron (*of Busiris sacrificing strangers to the gods,* Ors. 1, 8; Swt. 40, 26. Ðæs ðe hé blódgyte, wælfyll weres wǽpnum gespédeþ, morþ mid mundum, Cd. 75; Th. 92, 13; Gen. 1528. [*Laym.* morþ *destruction: O. Sax.* morð: *O. Frs.* morth: *Icel.* morð: *O. H. Ger.* mord: *Lat.* mort-.] v. morþor.

morþ-bealu, wes; *n. Deadly harm, murder,* Beo. Th. 272; B. 136. v. morþor-bealu.

morþ-crundel. v. crundel.

morþ-dǽd, e; *f. A deed which causes destruction,* (a) *of the body:*—Be ðǽm wiccecræftum and be liblácum and be morþdǽdum, gif man ðǽr ácweald wǽre (v. *last passage under* morþ, II, *and* morþ-weorc), L. Ath. i. 6; Th. i. 202, 11. (b) *of the soul, deadly sin, evil deed:*—Hé gewenede swá hine sylfne tó heora synlícum þeáwum and tó márum morþdǽdum mid ðam mánfullum flocce ... Swá férde se cniht on his fraceþum dǽdum and on morþdǽdum micclum gestrangod on orwénnysse his ágenre hǽle, Ælfc. T. Grn. 17, 18–24. Wearþ ðes þeódscype swýðe forsyngod ... þurh morþdǽda and þurh mándǽda, Wulfst. 163, 21. [Ðonne scalt þu (*the body*), erming, up arisen imete þine morþdeden, Fragm. Phlps. 7, 37.]

morþor, es; *n. m.* I. *murder:*—Manige men wénaþ ðæt morþor sý seó mǽste synne; ac ús is tó witenne ðæt þreora cynna syndon morþras. Ðæt is ðonne ðæt ǽreste, ðæt man tó óðrum lǽþþe hæbbe, and hine hatige ... Ða æfstigan men, ðéh hí sýn ðæs morþres scyldige, hí hit him tó nánre synne ne gelýfaþ, Blickl. Homl. 63, 34–65, 11. Ðara banena byre morþres gylpeþ, Beo. Th. 4116; B. 2055. Ðeáh hié (*cannibals*) morþres feala gefremed habben, Andr. Kmbl. 1950; An. 977. Morþres on luste, 2282; An. 1142. Draca morþre swealt (*the dragon perished by the sword,* Beo. Th. 1789; B. 892. Ic on morþor ofslóh minra sumne hyldemága, Cd. 52; Th. 66, 32; Gen. 1093. Morþor sceal mon under eorþan befeolan, ðe hit forhelan þenceþ, Exon. 90 b; Th. 340, 23; Gn. Ex. 115. Morþer *homicidium* ... fore morþre *propter homicidium,* Lk. Skt. Rush. 23, 19, 25. Ne ðú morþur ne fremme *non homicidium facies,* Mt. Kmbl. Rush. 19, 18: Lind. 27, 16. Morþur *homicidia,* 15, 19. II. *mortal sin, great wickedness:*—Wælhreówes árleásta fela, mán and morþor, misdǽda worn (cf. hwilc mán and hwilce ǽrleásnesse Neron weorhte, Fox 58, 2), Bt. Met. Fox 9, 13; Met. 9, 7. Morþres brytta (*Holofernes*), Judth. 10; Thw. 22, 33; Jud. 90 (*the devil*), Andr. Kmbl. 2342; An. 1142. Ðæt wé ðæs morþres meldan ne weorþen, hwǽr ðæt hálige treó beheled wurde, Elen. Kmbl. 855; El. 428: 1248; El. 626. Ðǽre synwrǽce sceoldon, morþres ongyldan, Exon. 45 a; Th. 153, 30; Gú. 833. Hú lange mánwyrhtan morþre gylpaþ *usque quo peccatores gloriabuntur,* Ps. Th. 93, 3. Seó sáwl sceal mid deóflum drohtnoþ habban in morþre and on máne, Wulfst. 187, 18. Morþor (*adultery*), Exon. 10 b; Th. 12, 29; Cri. 193. Ic andette mínes módes morþor, L. de Cf. 8; Th. ii. 262, 31: Salm. Kmbl. 82; Sal. 41. III. *torment, deadly injury, great misery:*—Swá hwæt swá wit morþres þoliaþ, hit is Adame forgolden, Cd. 35; Th. 47, 4; Gen. 755. Se hié of ðam morþre álýsde (*from the fiery furnace*), 196; Th. 244, 23; Dan. 452. God wearp hine on ðæt morþer innan (*into hell*), 18; Th. 22, 18; Gen. 342. Heó his mǽg-winum morþor fremedon (*greatly afflicted*), 149; Th. 187, 5; Exod. 146. Sceolde his wíte habban, ealra morþra mǽst, 16; Th. 19, 26; Gen. 297. Ðe ús monna mǽst morþra gefremede, sárra sorga, Judth. 11; Thw. 24, 10; Jud. 181. [*Goth.* maurþr φόνος.] v. morþ.

morþor-bealu, wes; *n. Deadly hurt, murder:*—Geseón morþorbealo mága, Beo. Th. 2162; B. 1079: 5477; B. 2742. v. morþ-bealu.

morþor-bedd, es; *n. The bed of death. the bed where a murdered man lies:*—Wæs ðam yldestan mǽges dǽdum morþorbed stréd (*of a man shot by his brother*), Beo. Th. 4864; B. 2436.

morþor-cofa, an; *m. A prison,* Andr. Kmbl. 2008; An. 1006.

morþor-cræft, es; *m. Deadly or murderous art or power:*—Ðǽr sylfǽtan (*the cannibal Mermedonians*) éðel healdaþ morþorcræftum, Andr. Kmbl. 353; An. 177.

morþor-cwealm, es; *m. Murder, slaughter,* Exon. 91 b; Th. 343, 4; Gn. Ex. 152.

morþor-hete, es; *m. Murderous, deadly hate,* Beo.Th. 2214; B. 1105.

morþor-hof, es; *n. A place of torment or extreme misery* (*hell*), Elen. Kmbl. 2603; El. 1303.

morþor-hús, es; *n. A house of torment* (*hell*), Exon. 31 b; Th. 99, 15; Cri. 1625.

morþor-leán, es; *n. Recompense of sin* or *a terrible recompense* :— Ðǽr (*in hell*) sceolan þeófas and þeódsceaþan, leáse and forlegene, lífes ne wénan, and mánsworan morþorleán seón, Exon. 31 b; Th. 98, 24; Cri. 1612.

morþor-scyldig; *adj. Guilty of murder* or *of grievous sin*, Andr. Kmbl. 3197; An. 1601.

morþor-slaga, an; *m. A murderer, homicide* :—Morþorslago homicidas, Mt. Kmbl. Lind. 22, 7. v. morþ-slaga.

morþor-slagu (?), e; *f. Murder, homicide* :—Morþurslaga homicidium, Mt. Kmbl. p. 14, 13. Morþorslago (morþurslagu, Rush.) homicidia, Mk. Skt. Lind. 7, 21.

morþor-slege, es; *m. Murder, homicide* :—Swá hwylc swá morþorslege þafaþ quicunque ad homicidium consenserit, L. Ecg. C. 22; Th. ii. 148, 14.

morþor-sliht, es; *m. Slaughter, the slain* :—Hwæt wæs on manríme morþorslehtes, deádra gefeallen, Elen. Kmbl. 1297; El. 650. v. morþ-sliht.

morþor-wyrhta, an; *m. A worker of iniquity* or *of murder* :—Hér syndan mánsworan and morþorwyrhtan, Wulfst. 165, 30. v. morþ-wyrhta.

morþ-slaga, an; *m. A murderer, an assassin* :—Sý ǽlc morþslaga áwirged maledictus, qui clam percusserit proximum suum, Deut. 27, 24. Oferfyll biþ mǽgbana and morþslaga, Wulfst. 242, 6. [*O. E. Homl.* morð-slaga : *pl.*] v. morþor-slaga.

morþ-sliht, es; *m. Murder, assassination* :—Be morþslihtum, L. Æðelst. iv. 6; Th. i. 224, 11. v. morþor-sliht.

morþ-weorc, es; *n. An act which causes death* (*by witchcraft* or *poison*) :—Hǽðenscipe biþ ðæt man ... wiccecræft lufige oððe morþweorc gefremme (*causes death by witchcraft or poison*, v. morþ, III), L. C. S. 5; Th. i. 378, 21. Deóflíce dǽda on morþweorcum and on manslihtan, L. Eth. v. 25; Th. i. 310, 15 : vi. 28; Th. i. 322, 16. [*O. Sax.* morð-werk.] Cf. morþ-dǽd and next word.

morþ-wyrhta, an; *m. One who causes death* (*by witchcraft* or *poison*) :—Wiccan oððe wigleras, mánsworan oððe morþwyrhtan, L. E. G. 11; Th. i. 172, 20 (see note): L. Eth. vi. 7, 36; Th. i. 316, 21, 324, 11 : L. C. S. 4; Th. i. 378, 7 : Wulfst. 266, 25. v. morþ, III.

moru. v. more.

mór-wyrt, e; *f. Moor-wort* :—Wyrc hié (*a salve*) of ðære smalan mórwyrte (*drosera rotundifolia*, Cockayne), Lchdm. ii. 128, 8.

mos, es; *n. A moss, a marshy place* :—In ðæt micle mos ; of ðæm mose, Cod. Dip. Kmbl. iii. 121, 19. Cf. Tó mossetena gemǽre, and swá big mossetena gemǽre ... Ðis syndon ðæs landes gemǽre æt mosleáge, Cod. Dip. B. ii. 56, 22, 28. [*N. of England and Scott.* moss (*as in* moss-trooper): *O. H. Ger.* mos *palus*: cf. *Icel.* mosi *a moss*: *Dan.* mose *a bog, moor.*]

mós, es; *n. Food, nourishment* :—Gé oftugon hrægles nacedum, móses meteleásum, Exon. 30 a; Th. 92, 11; Cri. 1507. Tó móse ł ǽte *ad edulium*, Hpt. Gl. 494, 66. Ðú his heáfod sealdest tó móse (*in escam*), Ps. Th. 73, 14. Tó móse *manducare*, 77, 25; Andr. Kmbl. 53; An. 27: 271; An. 136: Salm. Kmbl. 576; Sal. 287. Móse fédan, Exon. 36 b; Th. 118, 26; Gú. 245. Wista ł mósa *epularum*, Hpt. Gl. 481, 15. [*O. L. Ger.* muos, mós *esca, cibus*: *O. H. Ger.* muos, mós *cibus, esca, edulium, coena, alimonia*: *Ger.* mus : cf. ge-múse.]

mot, es; *n. A mote, an atom* :—Mot *attomos*, Wrt. Voc. i. 284, 37: ii. 8, 10. Mote *atomo*, 9, 62. Tó hwí gesihst ðú ðæt mot (*festucam*) on ðínes bróðor égan, Mt. Kmbl. 7, 3, 5. Ðú gesáwe gehwǽde mot on ðínes bróðor eáge, R. Ben. 12, 3. Ðæt lytle mot ... ðone mot, Lk. Skt. Lind. 6, 41, 42.

mót *a meeting, court.* v. folc-, ge-mót, *and compounds in which* mót *forms the first part.*

mót, e; *f.* (?) *Toll, tax* :—Mót ðæs cyninge[s] *nomisma census*, Mt. Kmbl. Lind. 22, 19. [*Goth.* móta *toll, custom*: cf. *Icel.* múta *a fee*: *O. H. Ger.* múta *toll* : *Ger.* mauth.]

mótan = (?) métan :—Gif man óðerne sace tihte and hé ðane mannan móte (*meet with*; Price translates *cite*, see his note) an medle oððe an þinge, L. H. E. 8; Th. i. 30, 11.

[**mótan**;] ic, hé mót, ðú móst; wé móton; *p.* móste (*from* mót-te). **I.** *to be allowed, may, mote,* (a) *with an infinitive* :—Mót ic drincan *licet mihi bibere*, ic móste *mihi licuit*, gif wé móstan *si nobis liceret*, beón álýfed *licere*, Ælfc. Gr. 33; Som. 37, 15. Wé móton *nobis licet*, ðú móstest *tibi licuit*, 44; Som. 46, 29. Ðú móst heonon húðe lǽdan, Cd. 98; Th. 129, 25; Gen. 2148: Beo. Th. 3347; B. 1671. Monna gehwylc geceósan mót swá helle hiénþu swá heofones mǽrþu, Exon. 16 b; Th. 37, 9; Cri. 590. Gif hé ús geunnan wile ðæt wé hine grétan móton, Beo. Th. 700; B. 347. Ne mágon hié and ne móton (*are not able and are not permitted*) ðinne líchoman deáþe gedǽlan, Andr. Kmbl. 2431; An. 1217. Ðæt hié on ðæt fǽgon, ðæt ic swá lytle hwíle lífgean móste, Nar. 32, 21. Ðæt ðú wilwega wealdan móstest, Ps. Th. 90, 11. Móstes, Exon. 28 a; Th. 85, 10; Cri. 1389. Hé him álýfde ðæt hí

ǽrnan móstan, Bd. 5, 6; S. 618, 42. Ðæt ic gást mínne ágifan móte, Andr. Kmbl. 2832; An. 1418. Ðæt ðú móte írætwa dǽlan, Cd. 136; Th. 171, 15; Gen. 2828. Ðæt hé ða yldu móte wendan tó lífe, Exon. 58 b; Th. 210, 24; Ph. 190. Ðǽr wé móton sécan, 65 b; Th. 242, 8; Ph. 670. Mótan, 11 b; Th. 16, 1; Cri. 246. Móten, 13 a; Th. 23, 30; Cri. 376. (b) *with ellipsis of infinitive*, (1) *to be supplied from preceding clause* :—Ða ic for God wille gemundbyrdan gif ic mót, Cd. 114; Th. 149, 12; Gen. 2473. Blǽd biþ ǽghwǽm ðǽm ðe Hǽlende héran þenceþ, and wel is þam ðe ðæt mót, 221; Th. 287, 11; Sat. 365. Uton fleón ða hwíle ðe wé móton, Homl. Th. ii. 124, 20. Nú cweþaþ oft preóstas ðæt Petrus hæfde wíf: fulsóþ hý secgaþ, forðam ðe hé swá móste ðá, L. Ælfc. C. 6; Th. ii. 344, 23. (2) *to be inferred otherwise* :—Ic him yfle ne mót *I may not be harmful to him*, Exon. 127 b; Th. 491, 5; Rä. 80, 9. Ðú of néde móst (*mayst go*), Andr. Kmbl. 230; An. 115. Nǽfre hió tó helle mót, Exon. 110 a; Th. 421, 19; Rä. 40, 20. Hé begeat leáfe ðæt hé of ðam lande móste, Homl. Skt. 3, 328. Ðæt Metellus tó Róme móste, Ors. 5, 9; Swt. 232, 25. Ðæt hé móste mid ðǽm sunu wið Somnitum, 3, 10; Swt. 140, 17. **II.** *to be obliged, must* :—Man mót on eornost mótian wið his drihten, Ælfc. T. Grn. 15, 3. Londríhtes mót monna ǽghwylc ídel hweorfan, Beo. Th. 5765; B. 2886. Ðæt hit sceaðen mǽl scýran móste, 3883; B. 1939. [This verb is one of the small class of verbs called preterite-present. The infin. does not occur in any of the dialects, but in the forms which are found the conjugation is the same as that of the A. S. verb. *Goth.* ga-mót; *p.* -mósta : *O. Sax.* mót; *p.* mósta : *O. Frs.* mót; *p.* móste : *O. H. Ger.* muoz, móz; *p.* muosi, muoste.]

mót-ærn, -ern, es; *n. A court-house* :—Mótern *praetorium*, Jn. Skt. Lind. 18, 28. v. gemót-ærn.

mót-bell, e; *f. A bell rung to call an assembly together* :—Debent statim pulsatis campanis, quod Anglice vocant *mótbel*, convocare omnes et universos, quod Anglice dicunt *folcmóte*, L. Edw. Conf. Schmid, p. 509, § 4.

mótere, es; *m. One who addresses a meeting* :—Mótere *vel* maþelere *concionator*, i. *locutor*, Wrt. Voc. ii. 135, 31. On mótera ford; of mótera forde andlang mótera lace, Cod. Dip. Kmbl. iii. 313, 24. [*Prompt. Parv.* motare or pletare *disceptor*, p. 345, *and see note.*] v. mótian, II, III.

mót-geréfa, an; *m. The geréfa who presides at a court* or *mót* :—Swá ðæt nán scýrgeréfe oððe mótgeréfe ðǽr habban ǽne sócne oððe gemót búton ðæs abbudes ágen hǽse (*nullus vicecomes vel praepositus*), Cod. Dip. Kmbl. iv. 200, 9. v. Kemble's Saxons in England, ii. 181, 155, note 2.

moððe, an; *f. A moth* :—Moððe *tinea*, Wrt. Voc. i. 24, 15 : 78, 70. Ðǽr moððe (mohða, Lind. Rush.) ł mouþe, mouȝte, Wick.) hit forninþ *ubi tinea demolitur*, Mt. Kmbl. 6, 19, 20: Lk. Skt. 12, 33. Moððe word frætt, Exon. 112 b; Th. 432, 4; Rä. 48, 1. Ðǽr moððan hit áwéstaþ, Wulfst. 286, 32. [*H. M.* mohðe: *Prompt. Parv.* mouȝte: *Chauc.* mouhtes; *pl.* : *Icel.* motti : *Ger.* motte.]

mót-hús, es; *n. A house where a court* or *assembly is held* :—Dómhús *vel* móthús *epicausterium*, Wrt. Voc. i. 57, 52. Móthúses *þrod[r]omi*, Hpt. Gl. 476, 61.

mótian; *p.* ode. **I.** *to address one's self, speak* (*to a person*), *converse* (v. mótung) :—Man mót on eornost mótian wið his Drihten se ðe wyle ðæt wé sprecon mid weorcum wið hine *the Lord, who will have us speak to him by our deeds, must be addressed in all seriousness*, Ælfc. T. Grn. 15, 3. Ne hiwa ðú swilce ðú mid bilewitnysse mǽge ðé gán orsorh tó mǽdena húsum and wið hí mótian ðæt ðín mód ne beó yfele besmiten þurh ða ýdelan spellunga *do not pretend, as if in innocency you can go secure to maidens' houses and converse with them, and your heart not be defiled through the idle conversations*, Basil admn. 7; Norm. 48, 11. Gif se munuc wyle gán tó wífmanna húsum and wið hý mótian, and gif ðǽm mǽdenum líkiaþ hyra luftýman sprǽce, 48, 15. [Cf. Stille beo þu, ne schaltu motin wið me na mare, Marh. 17, 26.] **II.** *to address an assembly* (cf. mótere) :—Heródes hæfde gemót ... Mid ðam ðe hé swíðost mótode, on his dómsetle sittende (cf. Acts 12, 21 : *Herod sat upon his throne, and made an oration*), Homl. Th. ii. 382, 30. **III.** *to discuss, dispute, moot a question* (cf. a *moot point*) :—Ðú scealt gelýfan on ðone lifigendan God, and ná ofer ðíne mǽðe mótian be him, Hexam. 3; Norm. 6, 17. [Cf. ge-mótod, *and Prompt. Parv.* mootyn *discepto, placito* : mótynge *disceptacio.*]

mót-lǽdu *in* Chart. Th. 433, 22. *The word occurs in a list of services due from the tenant of certain land, and seems to mean ‘ courts, assemblies’* :—Þreó mótlǽdu ungeboden on xii mónþum *the tenant must attend three courts a year without summons. In the same charter, in similar lists, occur two phrases which seem identical in meaning with that just given,* þriwa sécan gemót on xii mónþum, 433, 9, *and* iii gemót on geáre, 433, 32. *The charter is later than* 1066, *perhaps the Icel.* leið *an assembly,* may be compared. Cf. also kynnis-leið *a visit to relations.*

mót-stów, e; *f. A place of assembly, forum* :—Mótstów on burge *forus* (*forum* ?) *vel prorostra*, Wrt. Voc. i. 36, 43 : 47, 22. v. gemót-stów.

mótung, e; *f. Conversation, discourse :*—Of mótunge *colloquio, sermocinatione*, Hpt. Gl. 511, 26. v. mótian, I.

mót-weorþ; *adj. Entitled to attend a* mót :—Ealle ða men ða beón mótwurðe, Cod. Dip. Kmbl. iv. 208, 32.

mucg-, mug-wyrt, e; *f.* A plant name *mug-wort*, (Scott.) *muggart, muggon*, also called *mother-wort*. In the Herbarium, Lchdm. i, three kinds of *mug-wort* are mentioned :—Mugcwyrt. Ðeós wyrt ðe man *artemisiam* and óðrum naman mucgwyrt nemneþ (*Artemisia vulgaris*), 102, 1–3. *Herba artemisia tragonthes* ðæt is nugcwyrt (*Artemisia dracunculus tarragon*), 102, 18. Mucgwyrt. Ðeós wyrt þridde ðe wé *artemisiam leptefilos*, and óðrum naman mucgwyrt nemdon (*Artemisia Pontica*), 104, 15–18. Mugwort was supposed to prevent weariness on a journey, v. Lchdm. i. 102, 3–7: ii. 154, 8–12. Mugwyrt *artemisia vel matrum herba*, Wrt. Voc. i. 36, 51: 66, 61. Mucgwyrt, ii. 8, 36. Mugwyrt *gagantes* (see above, Lchdm. i. 102, 18), i. 68, 78. Mucgwyrt, ii. 42, 40. See Lchdm. iii. 339 for other references, and Grmm. D. M. 1152.

muexle, múdrica. v. muscle, mýdrece.

múga, múha, múwa, an; *m. A mow* (as in barley-*mow*), *a heap* (of hay, corn) :—Múha *aceruus*, Wülck. 3, 10. Múwan *acervum*, Wrt. Voc. ii. 6, 10. Múwan, hreácas *acervos*, 9, 55. Gif fýr bærne múgan oððe standende æceras *si ignis comprehenderit acervos frugum sive stantes segetes in agris*, Ex. 22, 6. [Cf. Wrt. Voc. i. 154, 23 a mowe (reke, MS. Camb.) *une moye:* Sparewen grupen in þen mu3en, Laym. 29280: *Icel.* múgi *a swathe.*]

múl, es; *m. A mule :*—Múl *mulus*, Wrt. Voc. i. 23, 25: 78, 10: 287, 49: ii. 56, 40. Ne beó gé ná swylce hors and múlas, Ps. Th. 31, 10. [*From Lat.* mulus. *Icel.* múll: *O. H. Ger.* múl: *Ger.* maul (-thier, -esel).]

múl-hirde, es; *m. A mule-keeper :*—Múlhyrde *mulio*, Ælfc. Gr. 9, 3; Som. 8, 37.

munan (a pret. pres. verb); ic, hé man, ðú manst, wé munon; *p.* munde.   I. *to remember, be mindful of, to be careful of :*—Til mon tiles and tomes meares *a good man thinks of, is careful of, a good and quiet horse*, Exon. 91 a; Th. 342, 12; Gn. Ex. 142. [Cf. *Icel.* muna *to remember with feelings of gratitude, hate,* etc.]   II. *to consider, think :*—Fédan hig swá swá hig sylfe wyrþe munon *let their meal be such as they consider suitable*, L. Ath. v. 8; Th. i. 236, 7. Ðæt hine God ðæs cynedómes weorþne munde, Ps. C. 50, 150; Ps. Grn. ii. 280, 150. [*Goth.* ga-munan; *prs.* -man, *pl.* -munum; *p.* -munda *to remember : O. Sax.* far-munan; *prs.* -man, *pl.* -munun; *p.* -munsta *to despise: Icel.* muna; *prs.* man, *pl.* munuþ; *p.* mundi *to remember.*] v. á-, ge-, of-, on-munan.

mund, e; *f.*   I. *a hand :*—Hé cwehte mægenwudu mundum, Beo. Th. 477; B. 236: 6037; B. 3022. Merestræta mundum brugdon (*swam*), 1033; B. 516. Mundum brugdon scealcas of sceáðum scír-mæled swyrd, Judth. 11; Thw. 24, 38; Jud. 229. Gif monna hwelc mundum sínum aldre beneóteþ, Cd. 50; Th. 63, 31; Gen. 1040. Ic geféng mid mundum mægenbyrdenne, Beo. Th. 6173; B. 3091.   II. *a hand* (as a measure) :—Stænen bedd þrým mundum hiérra ðonne ðæs húses flór, Shrn. 69, 4.   III. (a) *protection* (cf. *to be in a person's hands, and* v. hand) :—Wé woldon gesettan ðás bóc mannum tó getrymminge and tó munde ús sylfum *we wished to compose this book to encourage other men, and to secure ourselves*, Homl. Skt. pref. 71. Gé orsorge wuniaþ on lande under mýnre munde, Wulfst. 132, 16. Ða hæðenan mid lácum heora leásra goda munde and gescyldnysse bædon, Homl. Th. i. 504, 19. Munde *patrocinium*, Hpt. Gl. 425, 19. Gif hý him syððan ne dóþ mete ne munde *if afterwards they do not feed or shelter him*, L. Edm. S. 1; Th. i. 248, 7. Gif mete and munde ðam ðe ðæs beþurfe, L. Pen. 15; Th. ii. 282, 25: Hy. 7, 48; Hy. Grn. ii. 288, 48. Hwí wénst ðú ðæt hý habban nænege munde heora freónda on ðisse weorulde *why do you think that they (the good who are dead) afford no protection to their friends in this world*, Shrn. 202, 25. (b) in a technical sense, *Guardianship :*—Ða betæhte Ecgferþ land and bóc on cynges gewitnesse Dúnstáne arcebiscope tó mundgenne his láfe and his bearna. Ða hé geendod wæs ða rád se bisceop tó ðam cynge myngude ðære munde and his gewitnesse *then Ecgferth delivered land and charter, with the witness of the king, to archbishop Dunstan, that he might act as guardian in respect to them, on behalf of his widow and children. When he died, the bishop rode to the king, and reminded him of the guardianship and his witness*, Chart. Th. 208, 10–18. (c) in a personal sense, *A protector, guardian* (cf. mund-bora, mundbyrdness, II) :—Ðæt hé beó ðærtó geheald and mund under mé, Chart. Th. 391, 17. Ic wile ðæt Ælthelm sý hire mund and ðæs landes, 545, 23. Ic wille ðæt Ælfríc and Ælfhelm bén mund and freúnd intó ðære stówe, 547, 37. Ic eom ðæs mynstres mund and upheald, Cod. Dip. Kmbl. iv. 232, 7. [Bé Alfríc and Toñi and Ðrunni ðese quides mundes, Chart. Th. 567, 1.]   IV. as a technical term in the laws, (a) *protection, guardianship* extended by the king to the subject, *the king's peace*, by the head of a family to its members :—Gif man his mæn freóls gefe freólsgefa áge munde ðære hína *if a man give his slave freedom, let him who gives the*

freedom be the guardian of the freedman's family, L. Wih. 8; Th. i. 38, 16. Ðonne ðæt gedón sý ðonne rære man cyninges munde ðæt is ðæt hý ealle gemǽnum handum of ǽgðere mægþe on ánum wǽpne ðam sémende syllan ðæt cyninges mund stande *when that is done, then let the king's peace be declared, that is, that they all of either kindred, with their hands in common upon one weapon, engage to the mediator that the king's peace shall not be broken*, L. E. G. 12; Th. i. 174, 20–22: L. Edm. S. 7; Th. i. 250, 19. Be munde. Hwílum wæron heáfodstedas and heálíce hádas micelre mǽþe and munde wyrþe and gridian mihton ða ðe ðæs beþorf[ton] (*they were entitled to afford protection, and might give 'grið' to those that needed it*), L. Eth. vii. 3; Th. i. 330, 7: Wulfst. 157, 19. Se ærcebiscop spæc tó mé ymbe Xpes circean freóls, ðæt heó hæfþ nú læsse munde ðonne hió hwílan ǽr hæfde, Chart. Th. 308, 20. [Ich wille ðæt hié habben alsuá hiere rigte ðane tún mid alsuá muchele munde alsuá on méseluen stant, Cod. Dip. Kmbl. iv. 204, 7.] (b) *the fine paid for violation of* mund, cf. mund-bryce, mund-byrd :—Mund ðare betstan widuwan eorlcundre, L. scillinga gebéte, L. Ethb. 75; Th. i. 20, 10. Gif man widuwan unágne genimeþ, ii gelde seó mund sý, 76; Th. i. 20, 14. Heáfodmynstres gridbryce béte man be cyninges munde, ðæt is mid .v. pundum (*let the fine be as in the case of breach of the king's* mund, cf. gif hwá cynges mundbrice gewyrce, gebéte ðæt mid .v. pundum, L. Eth. vii. 11; Th. i. 330, 29), L. Eth. ix. 5; Th. i. 342, 1: L. C. E. 3; Th. i. 360, 19. Gif hwá folces fyrdscip áwyrde, gebéte ðæt georne, and cyninge ða munde, L. Eth. vi. 34; Th. i. 324, 5. [*O. Sax.* mund *hand: Icel.* mund; *f.* hand (mostly poetry); also *hand* (a measure): *O. Frs.* mund *guardianship;* also *a guardian: O. H. Ger.* munt *palmus, cubitus; protectio; protector*, Grff. ii. 815: 813. v. Grmm. R. A. 447.] v. féðe-mund. The word also is found in proper names, e.g. Eád-mund.

mund (?) :—Hú ic fǽmnanhád mund inne geheóld and eác módor gewearþ Meotodes suna, Exon. 9 a; Th. 6, 32: Cri. 93.

mund-beorh, -beorges; *m. A sheltering hill :*—Hí (*Jerusalem*) synd mundbeorgas micle ymbútan, Ps. Th. 124, 2.

mund-bora, an; *m.*   I. *one who can give protection* (mund), *a protector, patron, guardian, advocate :*—Forspeca *vel* mundbora *advocatus, patronus vel interpellator*, Wrt. Voc. i. 57, 42: Mundbora *patronus*, ii. 67, 24: *subfragator*, 121, 55; Ep. Gl. 24 b, 31: *advocatus*, Hpt. Gl. 466, 73. (a) applied to the Deity :—Se ðe (*Christ*) is úre mundbora, Homl. Th. i. 350, 25: Exon. 120 b; Th. 463, 24; Hö. 75: 68 a; Th. 251, 36; Jul. 156. Drihten ðín mundbora *Dominus protectio tua*, Ps. Th. 120, 5. Úres mundboran (*Christ*) láre folgian, Blickl. Homl. 169, 17; (*God*), Exon. 40 b; Th. 134, 25; Gú. 514: 8 a; Th. 2, 33; Cri. 28. (b) to angels or saints :—Tó ðæm heáhengle Michaele, swá tó ðæm getreówestan mundboran, Blickl. Homl. 201, 27. Hé (*Dives*) ðone wolde habban him tó mundboran, ðam ðe hé nolde ǽr his cruman syllan, Homl. Th. i. 330, 27. (c) to earthly kings :—Wes ðú (*Hrothgar*) mundbora mínum magoþegnum, Beo. Th. 2964; B. 1480. Eádmund cyning, mága mundbora, Chr. 942; Edm. 2. Eádgár, West-Seaxena wine, Myrcene mundbora, 975; Erl. 125, 17. Eást-Engla cyning and seó þeód gesóhte Ecgbryht him tó mundboran, 823; Erl. 62, 25: 921; Erl. 108, 14. Sceal him (*an ecclesiastic or a foreigner who was wronged*) cyng beón oððon eorl and bisceop for mæg and for mundboran, L. E. G. 12; Th. i. 174, 8: L. Eth. ix. 33: Th. i. 348, 6: L. C. S. 40; Th. i. 400, 6.   II. *a guardian* (of things) :—Ðara máðma mundbora wæs, Beo. Th. 5552; B. 2779. [*O. Sax.* mund-boro: *O. L. Ger.* mundboro *municeps: O. H. Ger.* munt-poro *patronus, protector*.]

mund-bryce, es; *m.*   I. *a breach of* mund (v. mund, IV) :—Wé cwædon be munábrice, se ðe hit dó, ðæt hé þolige ealles ðæs ðe hé áge, L. Edm. S. 6; Th. i. 250, 9. Gif hwá cynges mundbrice gewyrce, gebéte ðæt mid v. pundum, L. Eth. vii. 11; Th. i. 330, 29. On Centlande æt ðam mundbryce (*for the offence*), v. pund ðam cingce, and þreó ðam arcebiscope, L. C. E. 3; Th. i. 360, 20.   II. *the fine paid for the offence to the authority whose* mund *was violated :*—Ðis syndon ða gerihta ðe se cyning áh ofer ealle men on Wessexan, ðæt is, mundbryce ..., L. C. S. 12; Th. i. 382, 13. Gif hwá folces fyrdscip ámyrre ðæt hit ǽnote weorþe forgilde hit fullíce and cyninge ðone mundbrice (*pay the fine to the king for the offence*), L. Eth. vi. 34; Th. i. 324, 7. Béte cynincge be fullan mundbryce, 42; Th. i. 400, 24: L. C. E. 2; Th. i. 360, 5. On Cantwara lage cyning and arcebiscop ágan gelícne and efendýrne mundbryce, L. Eth. vii. 6; Th. i. 330, 18. Myndbræcas and ǽlces wýtes, Chart. Th. 333, 33.

mund-byrd, e; *f.* (v. mund, mund-bora).   I. *protection, patronage, aid :*—Mundbyrd *suffragium*, Ep. Gl. 24 b, 32: *patrocinium*, Wrt. Voc. i. 288, 59: ii. 66, 53: 116, 3: Hpt. Gl. 497, 59. Hé þancaþ Gode his mundbyrde, ðonne hé hwylcum eacforþum álysed hæfþ, Ps. Th. 17, arg. Se ðe him écean Godes tó mundbyrde miht gestreóneþ *qui sperat in Domino*, 83, 13: Cd. 83; Th. 105, 14; Gen. 1753. Mundbyrde and fultome *presidio*, Wrt. Voc. ii. 67, 41. Under mundbyr[d]e *sub pretextu*, 79, 84: 84, 15. Ic mundbyrd on ðé hæfde *tu es meus protector*, Ps. Th. 70, 5. Heó funde mundbyrd æt ðam mǽran þeódne, Judth. 9; Thw. 21, 2; Jud. 3: Andr. Kmbl. 1447; An. 724: Exon.

35 a; Th. 113, 11; Gû. 113. Gif ðû ðê tô swá mildum (*heathen gods*) mundbyrd sécest, 68 a; Th. 252, 29; Jul. 170. Ða mundbyrde (*patrocinium*) ðæs fêrendan fæder tô Drihtne, Bd. 5, 22; S. 644, 41. Geornlíce mundbyrde gelýfaþ tô ðære stówe (*a church*), Blickl. Homl. 207, 3. Ðæt folc beág tô Eádwearde cyninge and sôhton his friþ and his mundbyrde, Chr. 921; Erl. 108, 2. Ús gehǽl mid mundbyrdum *nos salva patrociniis*, Hymn. Surt. 111, 44. II. *the fine paid for a violation of mund* (v. mund, IV a, b; mund-bryce, II):—Cyninges mundbyrd .L. scillinga, L. Ethb. 8; Th. i. 6, 1: 15; Th. i. 6, 12. Ciricean mundbyrd .L. scill. swá cinges, L. Wih. 2; Th. i. 36, 17. Scyldig (*liable to pay*) cyninges mundbyrde, L. Alf. pol. 5; Th. i. 64, 11. Forgylde ðem mæn his mundbyrd (*the fine for violating the man's mund by fighting in his house*), L. H. E. 14; Th. i. 32, 15: L. Ath. iv. 4; Th. i. 224, 1. [O. Sax. O. L. Ger. mund-burd: O. H. Ger. mundi-burd.]

mundbyrdan. v. ge-mundbyrdan.

mundbyrdness, e; f. I. *protection*:—Ic fare swá hwyder swá ðû mê tô mundbyrdnysse gerecst *I will go whithersoever thou dost direct for my protection*, Glostr. Frag. 106, 24. II. *in a personal sense* (v. mund, III b), *A protector, patron, advocate*:—Ic ðê tô mundbyrdnysse geceóse wið ðín ágen bearn *I choose thee for my advocate with thy own child*, 106, 19. Swá swá ic ǽr cwæþ ðínre ðære lícwurþan mundbyrdnysse, 108, 16. III. *a protection of rights granted by charter*:—Ic wille ðæt ðeós mundbyrdnesse beó strang *volo ut haec confirmatio vim obtineat*, Cod. Dip. Kmbl. iv. 202, 20: 205, 7. Icc nelle ðat any man ðás mundbyrdnesse tôbreke, 213, 19.

mund-cræft, es; m. *Power of hand or power to protect*:—Cunne ic his mihta, his mægen, and his mihta, and his mundcræftas, Lchdm. i. 384, 13.

mund-gripe, es; m. *Hand-gripe, grasp*:—Ðæt hê þrittiges manna mægencræft on his mundgripe hæbbe, Beo. Th. 766; B. 380. Strenge getrúwode, mundgripe mægenes, 3072; B. 1534. Æfter mundgripe, 3880; B. 1938. Ðæt hê ne mêtte middangeardes on elran men mundgripe máran, 1510; B. 753.

mund-heáls, -háls, e; f. (?) *Safety which comes from the protection* (mund) *afforded by another* (?):—Ðá se ælmihtiga ácenned wearþ siððan hê Marian mundheáls geceás *when Christ was born, after he had chosen a safe retreat in Mary's protecting womb*, Exon. 14 a; Th. 28, 14; Cri. 446.

mundian; p. ode. I. *to protect, shelter, guard*:—Se ðe ðê mundaþ swá swá fæder, Homl. Th. i. 274, 6: Exon. 36 a; Th. 117, 28; Gû. 231. Baldwine geaf Ælfgife wununge on Bricge and hê hí mundode and heóld ða hwíle ðe heó ðǽr wæs, Chr. 1037; Erl. 167, 4. Cristenum cyninge gebyreþ ðæt hê Godes áre mundie, Wulfst. 266, 17. II. *in a technical sense, To act as guardian.* v. mund, III b. [O. Sax. mundón: O. H. Ger. muntón *defendere*.] v. â-, ge-mundian.

mundiend, es; m. *A protector, guardian*:—Ic hine bidde ðæt hê mín fulla freónd and mundiend beó on mínum dege, Chart. Th. 525, 8.

mund-leów, (-leáw?), -laú, -leú, e; f. *A basin for washing the hands*:—Mundlaú *vescada* (among things belonging to the table), Wrt. Voc. i. 290, 68. Mundleú, ii. 123, 22: *conca* (cf. *Ital.* conca *a laver*: *Span.* cuenca *a wooden bowl*), 105, 7. Mundleów *conca, coclea*, 136, 15. [*Icel.* mund-laug *a basin for washing the hands, especially before and after a meal*.]

mund-róf; adj. *Ready or active with the hands*:—Þegn mægenstrong and mundróf, Exon. 129 a; Th. 495, 5; Rä. 84, 3.

munec, munecian, múnecenu. v. munuc, munucian, mynecenu.

munt, es; m. [*from Lat.* mons] *A mount, hill, mountain*:—Munt *mons*, Wrt. Voc. i. 54, 4. Wæs se munt Garganus bifigende, Homl. Th. i. 504, 28. Tô Oliuetes muntes nyðerstige, Lk. Skt. 19, 37. Ofer ðæs muntes cnæpp, 4, 29: Ex. 19, 20. Ne mæg hús on munte lange gelæstan, Bt. Met. Fox 7, 36; Met. 7, 18. Munte *promontorio*, Hpt. Gl. 420, 6. Munt *Scyllam*, 529, 20. Âbútan ðone munt, Ex. 19, 12. Ðæra munta cnollas, Gen. 8, 5. Tô ðám muntum, 14, 10. On heán muntum heortas wuniaþ, Ps. Th. 103, 17. On heálīcum muntum, Homl. Th. ii. 160, 29. Ðá ðá hê com tô muntum, ðá gemêtten hine sceaþan, 502, 24. Tô Alpes ðǽm muntum, Ors. 4, 8; Swt. 186, 16. Ofer ða muntas ðe Caucaseas wê hátaþ, Bt. 18, 2; Fox 64, 10: Gen. 8, 4. v. fore-munt.

munt-ælfen, e; f. *A mountain-nymph*:—Muntælfen *oreades*, Wrt. Voc. i. 60, 14.

munt-geóf, -ióf, -gióp, es; m. *The Alps*:—Muntiófes clifu *Alpes*, Wrt. Voc. ii. 9, 41. From muntgióp óð ðone mæran wearoþ (cf. betwux ðám muntum and Sicilia, Bt. 1; Fox 2, 4), Bt. Met. Fox 1, 27; Met. 1, 14. Ðá wæs ofer muntgióp monig ǽtyhted, 1, 15; Met. 1, 8. Hê com tô Alpis ðæm muntum . . . and ðone weg geworhte ofer munt Ióf, Ors. 4, 8; Swt. 186, 18. Muntgeófa *Alpium*, Wrt. Voc. ii. 2, 27.

munt-land, es; n. *A hilly country*:—Fêrde on muntland *abiit in montana*, Lk. Skt. 1, 39.

munuc, munec, es; m. [*Lat.* monachus] *A monk*:—Munuc *monachus*, Wrt. Voc. i. 42, 19. Ic Ælfríc munuc and mæssepreóst, Homl. Th. i. 2, 12: Bd. 5, 12; S. 630, 41. Be ðám ðe munecum heora feoh bútan leáfe befæstaþ. Gif mon óðres monnes munuce feoh óðfæste, bútan ðæs

munuces hláfordes lêfnesse, L. Alf. pol. 20; Th. i. 74, 13-16. Swá swá dafnaþ munuce, Coll. Monast. Th. 35, 5. Ic eom geanwyrde monuc *professus sum monachum*, 18, 28. Godes þeówas, biscopas, abbudas, munecas, preóstas, L. Eth. v. 4; Th. i. 304, 26. Wê willaþ ðæt munecas regollícor libban ðonne hí nú ǽr ðisan on gewunan hæfdon, ix. 31; Th. i. 346, 27. Muneca gehwylc ðe úte sý of mynstre and regoles ne gŷme . . . gebúge georne intô mynstre, v. 5; Th. i. 306, 1. Be munuca cynne. Feówer synt muneca cyn, R. Ben. 9, 2-3. Syx synt niuneca cynerena, 134, 3. Hê beád, ðæt nán his bearna ðæt menster leng mid preóstan gesette, ac ðæt hit êfre mid munecan stóde, Chart. Th. 227, 17. Hê sende Godes þeów Agustinum and óðre monige munecas, Bd. 1, 23; S. 485, 27. [*Icel.* múnkr: O. H. Ger. munich.] v. mynster-munuc.

munuc-cild, es; n. *A boy that is being brought up to be a monk*:—Sum munuccild drohtnode on his mynstre, and hæfde micele lufe tô his fæder and tô his mêder. Swíðor for ðære sibbe ðonne for Godes dǽle wearþ ðá oflangod, and arn of mynstre tô his mágum, Homl. Th. ii. 174, 33. Ân munuccild wunode on Mauricius mynstre . . . hæfde ðæt munuccild swíðe mǽrlíce stemne, Wulfst. 152, 7-11: 22.

munuc-gegerela, an; m. *A monastic dress*:—Gegyrede hine mid his munucgegyrelan, Bd. 1, 7; S. 477, 10.

munuc-hád, es; m. *Monk-hood, the monastic state* (of women as well as of men):—Munuchád and abbudhád syndon on óðre wísan (*different from the seven orders previously mentioned*), L. Ælfc. C. 18; Th. ii. 348, 31. Ægðer ge preósthádes ge munuchádes menn *both the secular and regular clergy*, Homl. Th. ii. 126, 16. Wæs sum mæssepreóst munuchádes *quidam monachus*, Bd. 5, 12; S. 630, 41, MS. B. Hê weoruldhad forlête and munucháde (*habitum monachicum*) onfênge, 4, 24; S. 598, 2. Of munucháde on bisceopháde gecorene *de monachorum collegio in episcopatus gradum adsciti*, 4, 12; S. 581, 21: Blickl. Homl. 219, 32. Seó ǽrest wífa is sǽd in Norþanhymbra mǽgþe ðæt heó munucháde and hǽligrifte onfênge *quae prima feminarum fertur in provincia Nordanhymbrorum propositum vestemque sanctimonialis habitus suscepisse*, Bd. 4, 23; S. 593, 23.

munuc-heáp, es; m. *A band of monks, the monks of a monastery*:—Ân abbod . . . mid eallum his munucheápe, Anglia viii. 325, 43.

munucian; p. ode *To make a person a monk*:—Hê hine môt munecian *se monachum potest facere*, L. Ecg. C. 27; Th. ii. 152, 13.

munuc-líc; adj. *Monastic*:—On munuclícre drohtnunge *in monachica conversatione*, Swt. A. S. Rdr. 96, 46: Bd. 4, 11; S. 579, 2: 4, 27; S. 603, 24. Hê wolde árǽran on his biscopríce munuclícne regol, Homl. Skt. 6, 59. Healdan his munuclíce scrúdware, L. Eth. v. 6; Th. i. 306, 9. Hê heóld his munelíce ingehýd swá ðeáh betwux mannum *he preserved the habit of mind which he had when a monk though mixing with men*, Homl. Th. ii. 506, 13. On munuclícum hádum *in monachico habitu*, Bd. 5, 19; S. 636, 21.

munuc-líce; adv. *Monastically, after the manner of a monk*:—Hê munuclíce leofode betwux ðám lǽwedan folce, Swt. A. S. Rdr. 97, 67.

munuc-líf, es; n. I. *the monastic life*:—Monige of Breotone for intingan munuclífes (*monachicae conversationis gratia*) gewunedon sêcan Francna mynstro, Bd. 3, 8; S. 531, 17. Hê in heardnesse munuclífes lifde *in monachica districtione vitam duxit*, 4, 26; S. 602, 40. Man on munuclífe gelǽred *viro monachica vita instituto*, 3, 21; S. 551, 40: (of a woman), 4, 23; S. 593, 1. Hê munuclífe swíðor lifde ðonne lǽwedes mannes, Blickl. Homl. 213. 10. Hê árǽrde mynster and munuclíf *he established a monastery and monastic discipline*, Homl. Skt. 6, 146. Munuclíf lǽdan, dôn *monachicam vitam ducere, agere*, Bd. 3, 27; S. 558, 7: (of a woman), 4, 23; S. 593, 19. Hê sundorlíf and munuclíf wæs foreberende *vitam privatam et monachicam praeferens*, 4, 11; S. 579, 8. II. *the place in which the monastic life is lived, a monastery*:—Hê árǽrde him munuclíf . . . Ðæt mynster hê gelógode mid wellybbendum mannum, ðæt wǽron hundeahtatig muneca, Homl. Th. ii. 506, 14. Hê árǽrde six munuclíf on Sicilia lande, and ðæt seofoþe binnan Rômána burh getimbrode, on ðam hê sylf regollíce under abbodes hǽsum drohtnode, 118, 27: Ors. 6, 34; Swt. 290, 4. Munuclífa *coenobiorum, monasteriorum*, Hpt. Gl. 412, 22. Hê Aþelwold biscop æft ða láre (*Latin*) on munuclífum árǽrde, Ælfc. Gr. pref.; Som. 1, 42. [Cf. *Icel.* múnk-lífi *a monastery*.]

munuc-regol, es; m. I. *the rule of a monastic order*:—Basilius áwrát munucregol, Homl. Skt. 3, 145. II. *the monastic order which observes a certain rule*:—Ic geann intô ǽlcum munucregole .i. pund, Chart. Th. 544, 12.

munuc-stów, e; f. *A place for monks*; *locus monachorum*, Bd. 3, 24; S. 556, 42.

munuc-wíse, an; f. *The manner of monks*:—On munucwísan gescrýd, Homl. Skt. 6, 247.

múr, es; m. *A wall*:—Burstan múras and stánas, Exon. 24 b; Th. 70, 23; Cri. 1143. [O. Sax. O. L. Ger. múra; f.: O. Frs. múre; f.: O. H. Ger. múra, murî; f.: M. H. Ger. múre, múr; f.: Ger. mauer; f.: *Icel.* múrr; m. all from Latin murus.]

murcen (?); adj. *Sad, complaining*:—Ða ðe murcne ǽr hungur heardne geþoledan, Ps. Th. 145, 6. [v. murcian, murcnian, *and cf. for*

similar relation murnan *and* un-murne, Ps. Th. 75, 4; *also* weoren Bruttes bliðe an mode þæ ær weoten *murne*, Laym. 16159.]

murcian; *p.* ode *To grieve, complain, repine*:—Hwî murcnast (MS. Bod. murcas) ðu wið mîn *quid tu reum me quotidianis agis querelis?* Bt. 7, 3; Fox 20, 3. Murcaþ forðý ðæt hê Gode nolde þeówian *gemunt homines quod Deo servire noluerunt*, Past. 36, 3; Swt. 250, 16. Ðæt hî him ondræden and murkien for hira unfullfremednesse *ut imperfectionis suae taedio tabescant*, 65, 6; Swt. 467, 13. Sóna swâ ic ðê on ðisse unrôtnesse geseah ðus murciende (Cott. MS. murcniende) *cum te moestum lacrymantemque vidissem*, Bt. 5, 1; Fox 8, 27. v. murcung, murcnian.

murcnere, es; *m.* *One who murmurs*:—On ēcum wîte mid ðâm murcnerum, R. Ben. 21, 5.

murcnian; *p.* ode *To murmur, complain, repine, grieve*:—Hwæt murcnast ðu æfter ðæm ðe ðu forlure oððe tô hwon fagnast ðu ðæs ðe ðu ær hæfdest *quid est, quod vel amissis doleas, vel laeteris retentis?* Bt. 14, 2; Fox 42, 31: 7, 3; Fox 20, 3 (v. murcian). Hî suâ murcniaþ ł geómriaþ *murmurabunt*, Ps. Spl. 58, 17. Gê murcnodon *murmurastis*, Deut. 1, 27. Ne murcniaþ, Jn. Skt. 6, 43. Ðâ ongunnon hig murcnian ongên ðone hîrēdes ealdor, Mt. Kmbl. 20, 11. Ðonne onginþ hê tô murcnienne, and þincþ him tô lang hwænne hê beó genumen of ðyses lifes earfoþnyssum, Homl. Th. i. 140, 19. Ða Pharisēi gehýrdon ða menigeo ðus murcnigende be him, Jn. Skt. 7, 32: Bt. 5, 1; Fox 8, 27 (v. murcian). v. be-murcnian.

murcnung, e; *f.* *Complaint, murmuring*:—Ðâ gehýrde Drihten folces murcnunge (*murmurationes*), Ex. 16, 11. Ic syngede þurh tale and þurh murcnunge (*per detractionem et per murmurationem*), Confess. Pecc. Wôplîcum murcnungum *flebilibus questibus*, Hpt. Gl. 518, 26. Hiófum, murcnungum *questibus*, 472, 64.

murcung, e; *f.* *Complaint, grief, murmuring*:—Hwæt is eówer murcung (*murmur*) wið unc? Past. 28, 6; Swt. 201, 5. Mid suâ micelre murcunga his âgen môd gedrêfþ *tanto mentem moerore conturbat*, 33, 7; Swt. 227, 19. Ðæt hié weorþen on murcunga and on ungeþylde *ad impatientiae murmurationem proruunt*, 45, 3; Swt. 341, 3. Hý ðê willaþ on murcunga gebringan ðonne hié ðê fram hweorfaþ *fortuna cum discesserit allatura moerorem*, Bt. 7, 2; Fox 18, 19 note.

murge. v. mirige.

murnan; *p.* de.   **I.** *intrans. To mourn, be sad, be anxious*:—Gif ðu ðonne heora þegen beón wilt and ðê heora þeáwas lîciaþ tô hwon myrnst ðu swâ swîðe *si probas, utere moribus, ne queraris*, Bt. 7, 2; Fox 18, 7. Sêlre biþ æghwæm ðæt hê his freónd wrece, ðonne hê sela murne, Beo. Th. 2775; B. 1385. Ðæt mín murnende môd, Bt. 3, 1; Fox 4, 18: Beo. Th. 99; B. 50: Andr. Kmbl. 3332; An. 1669: Exon. 101 a; Th. 380, 28; Rä. 1, 15. Geómor sefa, hyge murnende, 15 a; Th. 31, 24; Cri. 500. Cwom seofð murnende Maria, 119 b; Th. 459, 33; Hö. 9: 121 a; Th. 464, 22; Hö. 91. Bonan gnornedon, mændon murnende, 38 b; Th. 128, 8; Gú. 401. Murnan on môde *to be sad at heart*, Cd. 35; Th. 45, 31; Gen. 735: Judth. 11; Thw. 23, 33; Jud. 154. Hî murnaþ on môde, Cd. 169; Th. 212, 6; Exod. 535. Ne beó ðu on sefan tô forht, ne on môde ne murn *be not fearful of mind, nor anxious of heart*, Andr. Kmbl. 197; An. 99.   **II.** *with prepositions* for, æfter:—Ne mæg nâ for feore murnan se ðe wrecan þenceþ freán *not for life must he care that his lord will avenge*, Byrht. Th. 139, 25; By. 259. Ne murn ðu for ði mêce ðe wearþ mâðma cyst, Wald. 1, 44; Vald. 1, 24. Hyge wæs oncyrred ðæt hié ne murndon æfter mandreáme *the mind was o'erthrown, so that after the glad life of men they longed not*, Andr. Kmbl. 73; An. 37.   **III.** *trans.* (a) *To mourn, lament*:—Sum sceal murnan meotudgesceaft môde gebysgad *the Maker's decree shall one mourn, troubled in mind*, Exon. 87 b; Th. 328, 19; Vy. 20: Salm. Kmbl. 971; Sal. 485. (b) *to care about, regard*:—Se ðe hiora welt ne murnþ nâuþer ne friénd ne fiénd ðê mâ ðe wêdende hund *he that rules them regards neither friend nor foe any more than he would a mad dog* (cf. se hlâford ne scrîfþ freónde ne feónde, Met. 25, 15), Bt. 37, 1; Fox 186, 7. [*A.R.* murnen; *p.* murnede: *Laym.* murnede; *p.*: *Piers P.* mornede; *Goth.* maurnan μεριμνᾶν: *O.Sax.* mornôn: *Icel.* morna: *O.H.Ger.* mornên *moereo*; *part.* mornênti *moestus.*} v. be-murnan, meornan.

murnung, e; *f.* *Grief, anxiety*, Bt. 7, 2; Fox 18, 19. v. murcung (last passage).

murra, myrra, an; *m.* *Cicely*:—Murra hâtte wyrt, Lchdm. ii. 18, 3. Nim murran ða wyrt, iii. 8, 1. Myrran, 14, 20.

murre *myrrh.* v. myrre.

mús, e; *f.*   **I.** *a mouse*:—Muus *mus*, Wrt. Voc. ii. 114, 41. Mús *sorex*, i. 23, 31: *mus vel sorex*, 78, 23. Deós mús *hic mus*, Ælfc. Gr. 9, 33; Som. 12, 20. Gif gê gesáwen hwelce mús ðæt wære hláford ofer ôðre mýs, Bt. 16, 2; Fox 52, 2. Mýs *sorices*, Wrt. Voc. ii. 87, 73. [Ðæt gewrit beó geworpen músen tô gnagene, Chart. Th. 318, 28.]   **II.** *a muscle*:—Mús ðæs earmes *torus vel musculus vel lacertus*, Wrt. Voc. i. 43, 48. [*Icel.* mús, *pl.* mýss *a mouse*; *also a muscle: O.H.Ger.* mús *mouse; muscle:* *Ger.* maus *mouse; muscle: Gk.* μῦς *mouse; muscle.*] v. hreáðe-, hrēre-, scirfe-, sise-mús; múse-pise.

muscelle, muscle, muxle, musle, an; *f.* [from Latin] *A muscle* or *mussel, a shell-fish*:—Muscle *muscula*, Wrt. Voc. ii. 57, 76. Muxle,

---

i. 77, 71: *geniscula*, 281, 62. Mucxle, 65, 68: ii. 41, 19. Musclan scil *conca*, 15, 35. Of muscellan *de conca*, 26, 39: 75, 71: 89, 35. Musclan, Hpt. Gl. 417, 9. Hêr beóþ oft numene missenlîcra cynna muscule (muslena, note), Bd. 1, 1; S. 473, 17. Muslan *musculos*, Coll. Monast. 24, 11. [*O.H.Ger.* muscula; *f.*]

musc-fleotan. v. must-fleóge.

múse-pise, an; *f.* *Mouse-pea, a vetch*:—Músepise *vicia*, Wrt. Voc. i. 38, 55.

mús-fealle, an; *f.* *A mouse-trap*:—Muusfalle *muscipula*, Wrt. Voc. ii. 114, 34. Músfealle *pelx*, 71, 28. [*Prompt. Parv.* mows-falle: *O.H.Ger.* mús-falla; *f. muscipula: Ger.* mäuse-falle.]

mús-fealu; *adj. Mouse-coloured*:—Músfealu, bleóreád *myrteus*, Wrt. Voc. ii. 58, 8. [*Ger.* mäuse-fahl.]

mús-hafoc, es; *n. A mouse-hawk*:—Múshafoc *siricarius*, Wrt. Voc. i. 62, 17: *suricarius*, 280, 21. Múshabuc *soricarius*, ii. 120, 81.

must, es; *m.* (?) *Must, new wine*:—Must *mustum* (cf. nîwe wîn *mustum*, 27, 47), Wrt. Voc. i. 82, 36. Must mid hunig gemenged *inomellum*, 27, 45. Heortan manna must and wîndrinc myclum blissaþ *vinum laetificet cor hominis*, Ps. Th. 103, 14. Ne miht ðu wîn wringan on midne winter, ðeáh ðê wel lyste wearmes mustes, Bt. 5, 2; Fox 10, 32. Ðâs men sindon mid muste fordrencte ('*these men are full of new wine*,' Acts 2, 13), Homl. Th. i. 314, 21. [*O.H.Ger.* most; *m.*: *Ger.* most; *m.* From Latin.]

must-fleóge, an; *f. A small fly found in wine*; bibio, parva musca quae in vino nascitur:—Mustfleógan (muscfleotan, Wrt.) *bibiones, mustiones*, Wrt. Voc. i. 23, 74. Cf. bibulus musti bibiones (Anglice *myntys*) arcet amurca, 176, 24.

múþ, es; *m.*   **I.** *of persons*, (a) *The mouth*:—Múþ *os*, Wrt. Voc. i. 64, 52. Múþes hróf *palatum*, 64, 58. Gân[i]gende múþe *hiulco rostro*, ii. 79, 34. Hê for ðý sâre ne mihte his hand tô múþe gedón *could not put his hand to his mouth*, Bd. 3, 2; S. 525, 4. Eall ðæt on ðone múþ gæþ, gæþ on ða wambe, Mt. Kmbl. 15, 17. Múþum *buccis*, Wrt. Voc. ii. 12, 16. (b) *the mouth* as an instrument of speech:—Be ælcon worde ðe of Godes múþe gæþ, Mt. Kmbl. 4, 4. Hê æt his sylfes múþe gehýrde, Bd. 3, 27; S. 558, 40. Múþas ealle ða unriht sprecaþ *os loquentium iniqua*, Ps. Th. 62, 9. (c) *the face*:—Ic sprece tô him múþe tô múþe, Num. 12, 8.   **II.** *of things*, *A mouth, opening, orifice*:—Ælces kynnes múþ *orificium*, Wrt. Voc. i. 19, 57: Exon. 108 b; Th. 415, 10; Rä. 33, 9. Duru sceal on healle, rûm recedes múþ, Menol. Fox 533; Gn. C. 37. Gif mon biþ on hrif wund . . . gif hê þurhwund biþ, æt gehwæðerum múþe twentig scitt., L. Alf. pol. 61; Th. i. 96. 12. Beleác heofonrîces weard merehúses múþ (*the door of the ark*), Cd. 69; Th. 82, 18; Gen. 1364. [*Goth.* munþs: *Icel.* munnr, múðr: *O.Sax.* múð: *O.Frs.* muth, mund: *O.H.Ger.* mund.]

múþa, an; *m.*   **I.** *the mouth of a river*:—Ðær ligeþ se múþa út on ðone gârsecg ðære ié ðe mon hâteþ Gandis (*ostia fluminis Gangis*) . . . Be súþan ðæm múþan is se port Caligardamana . . . be norþan ðæm Gandes múþan is se port Samera. Be norþan ðæm porte is se múþa ðære ié . . . Ottorogorre, Ors. 1, 1; Swt. 10, 6–13. On Limene múþan . . . Se múþa is on eástweardre Cent . . . On ða eá hî tugon up hiora scipu ôþ ðone weald, iiii míla fram ðæm múþan útanweardum, Chr. 893; Erl. 88, 25–32. Ælc ceápscip friþ hæbbe ðe binnan múþan cuman, L. Eth. ii. 2; Th. i. 284, 20: ii. 3; Th. i. 286, 6. Ofer Humbre múþan, Chr. 867; Erl. 72, 7. On súþhealfe Sæfern múþan . . . ôþ Afene múþan, Erl. 104, 4–5: Bd. 5, 24; S. 647, 20. Ofer ðone múþan *trans fretum*, Mt. Kmbl. 18, 28. On hwelcum wæterum and on æghwelcra eá múþum hî sculun sēcan fiscas, Bt. 32, 5; Fox 118, 19.   **II.** *an opening, door*:—Recedes múþan, Beo. Th. 1452; B. 724. [*Icel.* munni *mouth* (of a cave, etc.).] v. ge-mýþe.

múþ-ádl, e; *f. A mouth-disease*:—Múþádl on gôman *mentedra* vel *oscedo*, Wrt. Voc. i. 43, 64: ii. 58, 7. v. múþ-coþu.

múþ-bana, an; *m. One who destroys with the mouth*:—Him Grendel wearþ tô múþbonan, leófes mannes líc eall forswealg, Beo. Th. 4165; B. 2079.

múþ-bersting, e; *f. A breaking out about the mouth*:—Múþberstinge (*in a list of diseases*) *frenus* (cf. frenusculi, ulcera circa rictum oris, similia his quae fiunt jumentis asperitate frenorum, Isid. 4, orig. 8), Wrt. Voc. i. 20, 14. Múþbersting, ii. 39, 17. Múþberstung, 150, 56.

múþ-coþu, e; *f. A mouth-disease*; oscedo (= oris ulcus), Wrt. Voc. i. 20, 13: ii. 61, 32.

múþ-freó; *adj. At liberty to speak*:—Hwí ne synt wê múþfreó? hú ne môton wê sprecan ðæt wê willaþ, Ps. Th. 11, 4.

múþ-hæl, es; *n. Salutary words pronounced by the mouth*:—Môdiges (*Moses*) múþhæl (cf. êce rædas Moyses sægde, Th. 210, 15–17), Cd. 170; Th. 213, 14; Exod. 552.

múþ-hróf, es; *m. The roof of the mouth, palate*:—Múþhrófe *palato*, Hpt. 414, 22.

múþ-leás, *adj. Without a mouth*:—Ic sceolde múþleás sprecan, Exon. 123 a; Th. 472, 1; Rä. 61, 9.

mútian. v. bi-mútian.

mútung, e; *f. A loan* (?):—Mútung *vel* wrixlung *mutuum*, Wrt. Voc.

ii. 58, 60. Cf. læn *commodum;* wrixlung *mutuum,* i. 21, 1–3: *and* tô borge *mutuum,* Kent. Gl. 817.

múwa, muxle. v. múga, muscelle.

mycel. v. micel.

mycg, mygg, es; *m.* : mycge (?), an; *f. A midge :*—Mygg *culix,* Wrt. Voc. ii. 105, 60: *sciniphes,* 120, 9. Mycg *culix,* 15, 55. Mygc, i. 281, 36. Micge (micgc?) *culex,* 24, 17. Mycgæs *cynomya,* Ps. Spl. T. 104, 29. Wið gnættas and micgeas, Lchdm. i. 54, 14. Heó gnættas and micgeas (micgas, MS. B.) ácwelleþ, 266, 2. [*O. L. Ger.* muggia; *f. culex* : *O. H. Ger.* nucca, mugga; *f. culex, conopis, scinifes* : *Ger.* mücke : *Icel.* mý; *n.* : *Dan.* myg : *Swed.* mygg.]

mycgern *fat about the kidneys :*—Micgern *exugium,* Wrt. Voc. i. 46, 10: *exugia,* ii. 30, 13. Micgerne *exugia* i. *minctura,* 146, 31. Rysele, mycgern *axungia;* micgern *arvina,* i. *adeps* ł *pinguedo,* Hpt. Gl. 471, 4–7. [Leo suggests borrowing from Welsh *mychiryn lard.*]

mycg-nett, es; *n. A mosquito-net :*—Fleóhnet *vel* micgnet *conopeum,* Wrt. Voc. i. 57, 24.

mydd, es; *n. A bushel;* modius:—Hannibal sende tô Cartaina þrió mydd gyldenra hringa his sige tô tâcne *Annibal in testimonium victoriae suae tres modios annulorum aureorum Carthaginem misit,* Ors. 4, 9; Swt. 190, 12. [*O. L. Ger.* muddi: *O. H. Ger.* mutti *modius.*]

mýdrece, an; *f. A chest :*—Mýderce (mêderce, MS. J.) oððe cyst *loculus,* Ælfc. Gl. Zup. 313, 15. Ðæs synt twâ micle mýdercan, and ân hræglcysð, and ân lytulu towmýderce, and eác twâ ealde mýdercan, Chart. Th. 538, 19–22. Heó becwiþ him twâ mýdrecan, and ðær aninnan ân bedreáf, eal ðæt tô ânum bedde gebyreþ, 536, 24: 537, 26. vi. mîdreca, 430, 2. Mûdrica *loculos,* Jn. Skt. Lind. 12, 6.

mýgþ. v. mægþ.

myl *dust :*—Ðæt ðære ylcan stôwe myl wið fýre wæs freomigende *ut pulvis loci illius contra ignem voluerit,* Bd. 3, 10, tit.; S. 534, 16. [*Prompt. Parv.* mul *pulvis,* p. 348, and note. Cf *Icel.* mylja *to crush.*]

myldan, myldende. v. be-myldan, miltan I (a).

mylen, es; *m. A mill :*—Myln *molendenum,* Wrt. Voc. i. 83, 7. Mylen *mula,* ii. 58, 16: 'R. Ben. 127, 6. Se mylenham and se myln ðærtô, Cod. Dip. Kmbl. iii. 189, 10. Of Eádweardes mylne, 438, 26: 439, 2. Ne mylnum nis âlýfed tô eornenne (*on Sunday*), Wulfst. 227, 11. [Myln *molendinum,* Wrt. Voc. i. 235, 60: *A. R.* mulne: *Wick.* milne: *Icel.* mylna: *O. H. Ger.* mulîn ; *f.* : *Du.* molen.]

mylen-brôc, es; *m. A mill-brook :*—On mylenbrôc; ðonne andlang streámes, Cod. Dip. Kmbl. v. 198, 30.

mylen-ham[m], es; *m. An enclosure in which a mill stands :*—Hit (*the boundary*) cymþ nyðer tô ðam mylenhammæ and se mylenham and se myln ðærtô, Cod. Dip. Kmbl. iii. 189, 10.

mylen-hweogul, es; *n. A mill-wheel :*—Seó heofon æfre tyrnþ onbûtan ûs swiftre ðonne ænig mylenhweól (-hweowul, MS. P.), Lchdm. iii. 232, 19.

mylen-púl, -pôl, es; *m. A mill-pool :*—On mylepûl; of mylenpûlle in Afene streám, Cod. Dip. Kmbl. iii. 401, 8. In ðone mylenpôl; of ðam pôle tô ðære portstræte, Cod. Dip. B. i. 418, 1.

mylen-scearp; *adj. Ground sharp :*—Heówan mêcum mylenscearpum, Chr. 937; Erl. 112, 24; Æðelst. 24. v. next word.

mylen-stân, es; *m. A stone for grinding :*—Feól oððe mylenstân *lima,* Wrt. Voc. ii. 49, 75: i. 287, 2.

mylen-steall, es; *m. A mill :*—Tô myllnstealle, Cod. Dip. Kmbl. iii. 4, 14. Mylenstall, 169, 9. v. next word.

mylen-stede, es; *m. A mill-stead, mill :*—Ðysne mylenstede ðe ðærtô gebyreþ æt Leóferes hagan, Cod. Dip. Kmbl. vi. 243, 10.

mylen-stíg, e; *f. A path to a mill :*—Æfter ðam grênan wege in tô ðære mylnstîge; of ðære mylenstîge, Cod. Dip. Kmbl. iii. 389, 9.

mylen-troh, -trog, es; *n. A mill-trough, the channel in which water comes to a mill-wheel :*—Mylentroh *canalis,* Wrt. Voc. ii. 128, 16.

mylen-waru, e; *f. A mill-dam* (? cf. *Icel.* vörr; *f. a fenced-in landing place*) :—Andlang streámes on ða mylenware; of ðare mylenware tô ðare swête apuldre, Cod. Dip. Kmbl. iii. 454, 7. Cf. mylen-wer.

mylen-weard, es; *m. A miller :*—Mylenwyrd *molendinarius vel molinarius,* Wrt. Voc. i. 34, 35. Myleweard *molendarus,* ii. 58, 17.

mylen-wer, es; *m. A mill-weir, mill-dam :*—Andlang streámes ðæt it cymþ tô ðam mylewere, Cod. Dip. Kmbl. iv. 92, 30.

mylma, an; *m. A retreat* (?); recessus, Germ. 398, 150.

myltan, mylte, myltestre. v. miltan, milte, miltestre.

-mynd. v. freónd-, ge-, weorþ-mynd.

myndgian; *p.* ode. **I.** *to bear in mind, recollect :*—Gê sweltaþ deáþe nymþe ic dôm wite sôðan swefnes ðæs mín sefa myndgaþ *ye shall die unless I know the import of the true dream, of which my mind is still conscious,* Cd. 179; Th. 224, 31; Dan. 144. Wê ðæs hereworces myndgiaþ (*recollect*), and ða wiggþræce on gewritu setton, Elen. Kmbl. 1311; El. 657. **II.** *to bring to the mind of another, recall, remind :*—Manaþ swâ and myndgaþ mæla gehwylce sârum wordum, Beo. Th. 4120; B. 2057. Ic wolde ðê nû myngian (Cott. MS. myndgian) ðære manigfealdan lâre ðe ðû mê ær gehête, Bt. 40, 5; Fox 240, 11. v. gemyndigian, mynegian *and next two words.*

myndgiend, es; *m. One who reminds :*—Gyf Frysna hwylc ðæs morþorhetes myndgiend wære, Beo. Th. 2215; B. 1105.

myndgung, e; *f. A reminding one of anything, admonition :*—Sió myndgung ðara hâligra gewrita *divinae admonitiones verba,* Past. 22, 1; Swt. 169, 8.

myndig; *adj. Mindful :*—Myndig wæs Petrus wordes ðætte cweden wæs him, Mk. Skt. Rush. 14, 72. v. ge-myndig.

mynd-leás; *adj. Senseless, foolish :*—Se wîsdôm hine sylfne ætbret fram myndleásum geþohtum, Homl. Th. ii. 326, 4. v. ge-myndleás.

myne, es; *m.* **I.** *the mind :*—Môd mægnade, mine fægnade, Exon. 94 b; Th. 353, 56; Reim. 33. **II.** *mind* (as in to have a *mind* for anything), *purpose, desire :*—Læssan hwîle ðonne his myne sôhte *for a less time than he would have desired,* Beo. Th. 5138; B. 2572. Wæs him ût myne fleón fealone streám *they had a mind to escape, to flee the yellow stream,* Andr. Kmbl. 3073; An. 1539. Gê holdlîce hyge staþeladon mid môdes myne (*with full purpose of heart*), Exon. 27 b; Th. 83, 20; Cri. 1359. Hê lârum wile, þurh môdes myne, mínum hýran, 71 a; Th. 265, 10; Jul. 379: 74 a; Th. 282, 2; Jul. 657. Nô hê ðone gifstôl grêtan môste for Metode ne his myne wisse *he might not approach the throne because of the Lord, and knew not his purpose,* Beo. Th. 341; B. 169. **III.** *love :*—Hwær ic feor oððe neáh findan meahte ðone ðe in meoduhealle mine wisse (*would feel love, would love*), oððe mec frêfran wolde, Exon. 76 b; Th. 288, 7; Wand. 27. [Do þu þis mid gode mune (*intent*), þenne eart þu godes sune, O. E. Homl. i. 57, 53. *Goth.* muns *purpose, device, readiness : Icel.* munr *the mind; mind, longing; love.*] v. wîf-myne.

myne. v. mene, mine.

mynecenu, e; *f. The feminine form corresponding to masc.* munuc :—Mynecenu *monacha vel monialis,* Wrt. Voc. i. 42, 20: Homl. Th. ii. 26, 28. Munuc and mynecenu ðe Gode sylfum beóþ gehâlgode, and hyra gehât Gode gehâten habbaþ, L. Ecg. P. iii. 11; Th. ii. 198, 32. Seó mynecynu *monacha,* iv. 9; Th. ii. 206, 16: Homl. Th. ii. 184, 1. Bysn be sumere mynecyne, 546, 26. Gif hwâ mynecene, ðe Godes brýd biþ gehâten, him tô wîfe nimþ, beó heó âmânsumad, L. Ecg. P. ii. 19; Th. ii. 188, 21. Godes þeówas, munecas and mynecena, preóstas and nunnan, L. Eth. v. 4; Th. i. 304, 26. Munecas and mynecena, canonicas and nunnan, vi. 2; Th. i. 314, 17: L. C. E. 6; Th. i. 364, 7. Be mynecenan. Riht is ðæt mynecena mynsterlîce macian, efne swâ wê cwædon ǽror be munecan (v. *next paragraph where* preóstas and nunnan *are taken together*), L. I. P. 15; Th. ii. 322, 31–33. Eugenia hæfde âsteald mynecena mynster, Homl. Skt. 2, 311. Munecena mynstru, R. Ben. 136, 4. Ða forlǽtenan mynstru mid munecum gesettan and eác mid mynecenum, Chart. Th. 240, 17. Basilisca wearþ môdor ofer manega mynecena, Homl. Skt. 4, 85. Mynecǽna, Lchdm. iii. 440, 15. [Ealra ðare landa ðe intô ðæ mynechina lîfe æt Wiltûne forgifene synt, Cod. Dip. Kmbl. iii. 117, 25. *Laym.* munechene: *Piers P.* monchen: *Trev.* minchin.]

mynegian, myngian; *p.* ode (*with acc. of person and gen. of thing, or with a clause*). **I.** *to bring to one's own mind, recall :*—Dauid myngode ðæra gyfa ðe God his fædrum and his foregengum sealde, Ps. Th. 43, arg. **II.** *to bring to another's mind,* (a) *to remind :*—Drihten ûs ðonne myngaþ ðæs Sunnandæges weorces *the Lord will remind us then of the work done on Sunday,* Wulfst. 210, 9. Mec ðæra nægla fyrwet myngaþ, Elen. Kmbl. 2156; El. 1079. Ic ðê ǽr mynegode (Cott. MS. myndgode) ðære ilcan spræce, Bt. 35, 3; Fox 160, 7. Hû ne mynegodest (Cott. MS. myndgodest) ðú mê ðære ilcan spræce, 35, 2; Fox 156, 14. Ic wolde ðê mynegian (Gott. MS. myndgian) ðære manigfealdan lâre ðe ðû mê ǽr gehête, 40, 5; Fox 240, 11. Wê willaþ eów myngian, ðæt hit ne gange eów of gemynde, Homl. Th. i. 220, 3. (b) *to bring a duty to the mind, to admonish, exhort :*—Eów ic mynegie *vos moneo,* Ælfc. Gr. 15; Som. 18, 3. Mîne wylna ic mynegige *meas ancillas moneo,* 19, 6. Ic myngige and manige manna gehwylcne, Blickl. Homl. 109, 11. Ic myngie and lǽre, 107, 10. Manaþ ûs and myngaþ seó âr and seó eádignes, 197, 3. Mynegaþ, 161, 3. Menegaþ *instigat,* Hpt. Gl. 526, 63. Eádweard cyning myngode his wytan ðæt hý smeádon hû heora friþ betere beón mæhte, L. Ed. 4; Th. i. 160, 23. Minga hine *hunc exhortare,* Deut. 1, 38. Ǽlc biscop ðone cyning myngige (MS. B. myndgige) ðæt ealle Godes cyrcan sýn wel behworfene, L. Edm. E; Th. i. 246, 11. Ænne hyndenman, ðe ða .x. mynige tô ûre ealre gemæne þearfe, L. Æðelst. v. 3; Th. i. 232, 2. Wê willaþ mynegian freónda gehwilcne, ðæt gehwâ hine sylfne beþence, L. Eth. vi. 42; Th. i. 326, 6. (c) *to remind of a debt, to ask for payment.* v. manian :—Myngaþ *exigit,* Wrt. Voc. ii. 144, 81. Sǽde on heortan hys ne myngeþ (*requiret*), Ps. Spl. T. 9, 15. Gif hê gelômlîce þurh his bydelas hys gafoles myngaþ *if he by his messengers often asks for his tribute,* L. Edg. S; Th. i. 270, 20. Heáhberht oft ðæs myngode, oððe ðæs landes bæd, Chart. Th. 167, 6. Se ðe nimþ ða þing ðe ðîne synt ne mynega ðû hyra (*ne repetas*), Lk. Skt. 6, 30. **III.** *to have in the mind, to purpose, intend, determine :*—Menegiaþ, hogiaþ *conati sumus, decrevimus,* Hpt. Gl. 527, 66. [*A. R.* munegen: *Marh.* munegin:

Laym. munegie: *Piers P.* munge, menewe: *O. H. Ger.* bi-munigŏn.] v. ge-mynegian.

mynegung, e; *f.* I. *admonition, exhortation* (v. mynegian, II b):—Mynegung *monitus,* Ælfc. Gr. 11; Som. 15, 16. Mynigung, 43; Som. 44, 53. Mynegunge *monitionem,* 15; Som. 18, 4. Þurh Albinus myngunge (*hortatu*), Bed. pref.; S. 472, 8. 'Ne ondræde gĕ eów' hĕ cwæþ . . . þurh đás minegunge . . ., L. Ælfc. P. 13; Th. ii. 364, 26. Þurh đæs apostoles mungunge (myngunge, MSS. O. F.; minegunge, MS. T.), R. Ben. 53, 1. Heó wolde þurh his mynegungum hire mŏd getrymman, Homl. Th. ii. 146, 10. Æfter mynegungum Æđeluuoldes đe mĕ oft manode, Chart. Th. 240, 30. Menegungum *hortamentis,* Hpt. Gl. 485, 52. II. *a demand for payment of what is due, a claim* (v. mynegian, II c):—Þurh đa gedurstegnysse đe folces men wiđhæfton dære gelómlícan mynegunge (myngunge, MS. F.) . . . đe úre láreówas dydon ymbe đæt neádgafol úres Drihtnes, L. Edg. S; Th. i. 270, 25. Ne forléte hĕ đa mynegunge *let him not relinquish the claim,* L. Æđelst. v. 7; Th. i. 234, 26.

mynele, an; *f. Desire, longing:*—Đæt hĕ tó his earde ǽnige nyste mŏdes mynlan *so that he (Ulysses) felt no heart's desire for his native land,* Bt. Met. Fox 21, 133; Met. 26, 67. v. myne.

myne-líc; *adj. Pleasant, desirable:*—Oft hĕ geþah mynelícne máþþum, Exon. 84b; Th. 318, 25; Víd. 4. [*O. Sax.* muni-líh: *Icel.* mun-ligr *pleasant.*] v. myne.

mynet, es; *n.* I. *a coin:*—Mynet *nummisma,* Wrt. Voc. i. 73, 48. Mynit *nomisma,* ii. 114, 75. Mynete *nummismate,* 61, 14: 96, 80. Genim pipores swilce ân mynet gewege, diles sǽdes swilce iiii mynet gewegen, Lchdm. ii. 192, 14. Ætgýwaþ mĕ đæs gafoles mynyt, Mt. Kmbl. 22, 19. Đæt hĕ sceoldon đæt gyldene mynet (*aureum illud numisma*) mid him geniman, Bd. 3, 8; S. 532, 1. Hĕ hĕt đæm cwelre syllan .xxv. gyldenra myneta, Shrn. 129, 12. II. *coinage, money:*—Đæt ân mynet sý ofer eall đæs cynges onweald, L. Ath. i. 14; Th. i. 206, 18: L. Edg. i. 8; Th. i. 268, 27. Ân mynet gange ofer ealle đás þeóde bútan ǽlcon false, L. Eth. vi. 32; Th. i. 322, 28: L. C. S. 8; Th. i. 380, 15: Wulfst. 272, 2. [*O. L. Ger.* munita: *f. nomisma, moneta*: *O. H. Ger.* muniza, munizza; *f.*: *Ger.* münze. *From Latin* moneta.]

mynet-cípa, an; *m. A money-dealer:*—Se đe him sylfum teolaþ on Godes gelaþunge, and ne caraþ ymbe Cristes teolunge, se biþ mynetcýpa getalod, Homl. Th. i. 412, 16.

mynetere, es; *m.* I. *a moneyer, a money-changer, money-dealer:*—Mynetere *nummularius,* Wrt. Voc. i. 47, 15: *trapezita,* 57, 33: trapezeta vel *nummularius,* 73, 47. Miyniteri *numularius, nummorum praerogator,* ii. 115, 2. Mynetere *trapezita,* Ælfc. Gr. 7; Som. 6, 43. Mynetera *nummulariorum,* Wrt. Voc. ii. 60, 51. Munetera, 73, 8, 41. Đa setl đara mynetera *the seats of the money-changers,* Blickl. Homl. 71, 19. Hyt gebyrede đæt đú befæstest mín feoh mynyterum, Mt. Kmbl. 25, 27: Homl. Th. ii. 554, 8. Hĕ gemĕtte sittende myneteras, Jn. 2, 14. II. *a minter, one who coins:*—Mynetere *monetarius,* Wrt. Voc. i. 57, 33. Be myneterum . . . Nân man ne mynetege bútan on porte. And gif se mynetere fúl wurþe, sleá man of đa hand đe hĕ đæt fúl mid worhte, and sette upp on đa mynetsmiđđan . . . On Cantwara byrig .vii. myneteras, L. Ath. i. 14; Th. i. 206, 17–26. Ælc mynetere đe man tíhþ đæt fals feoh slóge . . . gif hĕ fúl beó, sleá hine man, L. Eth. iii. 8; Th. i. 296, 12–15. Đa myneteras đe inne wuda wyrcaþ ođđe elles hwǽr; đæt đa bión heora feores scyldige, iii. 16; Th. i. 298, 13. Godes feoh biþ befæst myneterum tó sleánne, Homl. Th. ii. 554, 14. Ic habbe geunnen Baldewyne abbode ŏnne meonetere wiđinne Sæint Eáđmundes byrg, Cod. Dip. Kmbl. iv. 223, 6. [*O. Sax.* muniteri *a money-changer*: *Icel.* myntari *a minter*: *O. H. Ger.* munizari, munizzari *numularius, monetarius, trapezita*: *Ger.* münzer.]

mynetian; *p.* ode *To mint, coin:*—Nân man ne mynetege bútan on porte, L. Ath. i. 14; Th. i. 206, 19. [*O. H. Ger.* munizŏn *cudere.*]

mynet-smiđđe, an; *f. A mint, place for coining.* v. mynetere, II.

myngian. v. mynegian.

mynian; *p.* ede (cf. myne, II) *To have as the object of desire or purpose, to intend, direct one's course to an object:*—Đǽr mín hyht myneþ tó gesécenne *my heart's desire is to visit there,* Exon. Th. 167, 17; Gú. 1601: Andr. Kmbl. 583; An. 294. Ic lǽre ǽlcne đara đe maga sí and manigne wǽn hæbbe đæt hĕ menige tó đam ilcan wuda *I advise every one that is able and has many a waggon, to direct his steps to that same wood,* Shrn. 163, 13.

mynster, es; *n.* I. *a monastery, a place where a body of monks or of nuns resided:*—Gif hit beón mæg, swâ sceal mynster beón gestaþelod, đæt ealle neádbehéfe þing đær binnan wunian, đæt is wæter-scype, mylen, wyrtún and gehwylce misenlíce cræftas đe synd góde tó begânne, R. Ben. 127, 4–7. Wæs se ǽrest abbod đæs ylcan mynstres Petrus háten, Bd. 1, 33; S. 499, 5: 2, 2; S. 502, 40. Mynstres aldor, L. Wih. 17; Th. i. 40, 13. Gif hwá gefeohte on cyninges húse sie hĕ scyldig ealles his ierfes . . . Gif hwá on mynstre gefeohte, hundtwelftig sciłł. gebéte, L. In. 6; Th. i. 106, 4. Gif hwá gefeohteþ on mynstre bútan circean gebéte . . . be mynstres mǽđe, L. Eth. vii. 10; Th. i. 330, 26.

Muneca gehwylc đe úte sý of mynstre . . . gebúge georne intó mynstre, v. 5; Th. i. 306, 1–3. Gif hwá nunnan of mynstre út álǽde, L. Alf. pol. 8; Th. i. 66, 15. Wæs heó . . . on đam mynstre đe on Franclande wæs getimbrad fram dære abbadissan đe Fara hátte . . . forđon on đa tíd ne wǽron monige mynstra getimbrade on Angelþeóde; forđon monige of Breotone gewunedon sécan Francna mynstro, Bd. 3, 8; S. 531, 12–17. Mid đý đe wǽn đá com, đe đa bán on lǽded wǽron, in đæt foresprecene mynster, đá ne woldan đa híwan đe on đam mynstre wǽron him lustlíce onfón, 3, 11; S. 535, 17. Se munuc đe mynster næbbe, L. Eth. v. 6; Th. i. 306, 6. On mynstrum fæste gewunian and regollíce libban (*said of abbots*), ix. 32; Th. i. 348, 1. In mynsterum, Exon. 38b; Th. 127, 16; Gú. 387. Coleman twá mynstro geworhte, Bd. 4, 4; S. 570, 30. Twá ǽđele mynstere, 4, 6; S. 574, 12. Mynstru, R. Ben. 139, 4. II. *a church, minster* (v. mynster-clǽnsung):—Ne sín ealle circan ná gelícre mǽđe worldlíce wyrđe . . . Heáfodmynstres griþbryce . . . bête man be cyninges munde . . . and medemran mynstres mid hundtwelftigan sciłł., L. Eth. ix. 5; Th. i. 342, 1: L. C. E. 3; Th. i. 360, 21. Man ágife ǽlce teóþunge tó đam ealdan mynstre (*ad matrem ecclesiam*) đe seó hýrnes tó hýrþ, L. Edg. i. 1; Th. i. 262, 7. Đæs mynstres mæssepreóst, i. 3; Th. i. 262, 25. (See also sections 2 and 5.) Ôswold fullworhte on Eferwíc đæt ǽnlíce mynster đe his mǽg Eádwine ǽr begunnen hæfde, Swt. A. S. Rdr. 98, 90. [*Laym.* munster *a monastery*: *Orm.* i þeȝȝre minnstre (*the temple,* cf. i þe kirrke, 1099), 1017: *O. H. Ger.* munustiri *monasterium.* From the Latin.] v. heáfod-, nunn-mynster.

mynster-clǽnsung, e; *f. Purification of a minster (within whose walls a man has been slain):*—Đonne bête man đæt ciricgriþ intó đære circan . . . and đa mynsterclǽnsunge begite (cf. gif ǽnig man Godes ciricgriþ swá ábrece đæt hĕ binnon ciricwágum manslaga weorþe, ll. 6–8), L. Eth. ix. 3; Th. i. 340, 18: L. C. E. 2; Th. i. 360, 6.

mynster-gang, es; *m. Going into a monastery, entering on a monastic life:*—Heó đonne mót gif heó wile đæt forlǽtan and hyre mynstergang geceósan *tunc, si velit, licebit ei id derelinquere, et vitam monasticam sibi eligere,* L. Ecg. C. 20; Th. ii. 146, 23.

mynster-hám, es; *m. A monastic house, monastery:*—Gif hwá đara mynsterháma hwelcne, for hwelcre scylde geséce, đe cyninges feorm tó belimpe, ođđe ŏđerne freóne hiéred, L. Alf. pol. 2; Th. i. 60, 23. Đone oferécan mon gedǽle gind mynsterhámas tó Godes ciricum in Súđregum and in Cent, Chart. Th. 482, 18.

mynster-hata, an; *m. A hater or enemy of monasteries:*—Hér syndan sacerdbanan and mynsterhatan, Wulfst. 165, 28.

mynster-líc; *adj. Monastic:*—Man árǽrde cyrcan on his ríce geond eall and mynsterlíce gesetnyssa (*monastic institutions*), Swt. A. S. Rdr. 97, 71. [*O. H. Ger.* munistri-líh *monasterialis.*]

mynster-líce; *adv. Monastically, in a manner suitable to a monastery:*—Riht is đæt mynecena mynsterlíce macian (*act in accordance with monastic rules*), L. I. P. 15; Th. i. 322, 32. Hĕ æþele mynster getimbrede. Đá hĕ đá đæt hæfde mynsterlíce ge þeáwlíce gesett, Bd. 3, 19; S. 549, 37.

mynster-lif, es; *n.* I. *monastic life:*—Gif hláford nylle hire mynsterlífes geunnan, ođđa hiá siolf nylle, Chart. Th. 471, 2. Hĕ mynsterlíf đam weoruldlífe forbær *monasticam saeculari vitam praetulit,* Bd. 5, 19; S. 637, 7. Hĕ him sendan sceolde sume eáwfæste munecas đe him mynsterlíf ástealdon, Homl. Skt. 6, 57. II. *a place in which the monastic life is lived:*—Mynsterlíf *coenobium* (cf. *hec cenobium* an abbay, i. 230, 8), Wrt. Voc. ii. 19, 47: 93, 32: *gurgustia,* 93, 33. Ic wille đæt đær ǽfre beó mynstrelíf and samnung (*a monastery and brotherhood*), Chart. Th. 391, 29. Cf. munuc-líf.

mynster-mann, es; *m. A man who lives in a monastery, a monk:*—Gif hit mynsterman sig *si monasticus sit,* L. Ecg. C. 40; Th. ii. 166, 10. Đás bóc be đæra hálgena lífe đe mynstermenn mid heora þenungum wurđiaþ, Homl. Skt. pref. 44: Swt. Rdr. 100, 148. Đæt forme muneca cyn is mynstermanna, đe gemǽnan lífe drohtniaþ on mynstre, R. Ben. 134, 5: 9, 3. Mynstermannum gedafenaþ đæt hí on stilnysse heora líf ádreógan, Homl. Th. ii. 342, 29: Ælfc. Gr. pref.; Som. 1, 38.

mynster-munuc, es; *m. A monk who lives in a monastery:*—Ne þearf ǽnig mynstermunuc mid rihte fǽhþbóte biddan, L. Eth. ix. 25; Th. i. 346, 1. Đa mynstermunecum urnon tó, Homl. Th. ii. 176, 23. Benedictus mid his mynstermunecum, 178, 33: i. 532, 33.

mynster-prafost, es; *m. The provost of a monastery:*—Ælfnŏd mynsterprauost, Chart. Th. 434, 4.

mynster-preóst, es; *m. A priest who conducts service in a minster:*—Wĕ lǽraþ đæt mæssepreósta ođđe mynsterpreósta ǽnig ne cume binnan circan dyre, ne binnan weohstealle bútan his oferslipe, L. Edg. C. 46; Th. ii. 254, 8.

mynster-scír, e; *f. The management of a monastery:*—Hĕ gewát tó his mynsterscíre *ad monasterii sui curam secessit,* Bd. 5, 19; S. 639, 13.

mynster-stów, e; *f. A place where there is a minster, a town:*—Hĕ férde geond ealle ge þurh mynsterstówe ge þurh folcstówe *per cuncta et urbana et rustica loca,* Bd. 3, 5; S. 526, 27.

mynster-þeáw, es; *m. A monastic custom:*—Cyriclíce þeáwas ođđe

mynsterþeáwas *ritus ecclesiastici sive monasteriales*, Bd. 5, 19; S. 637, 24.

mynster-þegnung, e; *f. Service done in a monastery:*—Ðeós foresceáwung sý gehealden ... on eallum mynstres þénungum (mynster-þénungum, Wells Frag.), R. Ben. 85, 17.

mynster-wíse, an; *f. A custom or manner followed in a monastery:*—Se abbod ongeat sume ða mynsterwísan tó gerihtanne *the abbot managed to correct some of the abuses practised in the monastery*, Glostr. Frag. 110, 27.

myntan; *p.* te. I. *to mean, intend, purpose, determine*, (a) *with infin.:*—Se ðe Gode mynteþ bringan beorhtne wlite, Exon. 23 b; Th. 65, 22; Cri. 1058. Mynte ic hié háton yfsian *I had a mind to order them to be punished*, Nar. 25, 27. Heó hí mynte for hý tó abbudissan gesettan *abbatissam eam pro se facere disposuerat*, Bd. 5, 3; 616, 19. Hé mynte hine sleán, Blickl. Homl. 223, 7, 9, 11, 16. Hé mynte mid his discipulum tó his mynstre féran, 225, 11: Beo. Th. 1428; B. 712. Ðá mynton wé ús gerestan, Nar. 14, 25: Bt. Met. Fox 26, 143; Met. 26, 72. (b) *with infin. to be supplied:*—Gif ðú seó riht cyning swá ðú ǽr myntest, Cd. 228; Th. 308, 8; Sat. 688. Mynte se mǽra hwǽr hé meahte ðanon fleón *the mighty one designed (to get) where he could flee thence*, Beo. Th. 1528; B. 762. [Cf. *Prompt. Parv.* myntyñ or amyñ towarde *attempto.*] (c) *with a clause introduced by* ðæt:—Geréfa mín mynteþ ðæt mé æfter síe eaforan síne yrfeweardas *my steward means his children to be heirs after me*, 100; Th. 131, 27; Gen. 2182. Hé mynte ðæt hé gedǽlde líf wið líce, Beo. Th. 1466; B. 731. (d) *with a case:*—Wit sculon sécan ðæt ðæt wit ǽr mynton *sed quae proposuimus intueamur*, Bt. 35, 3; Fox 158, 11. Hí him sylfum ríce mynton, Wulfst. 145, 26. II. *to think, suppose:*—Mynton ealle, ðæt se brego and seó mægþ wǽron ætsomne, Judth. 12; Thw. 25, 10; Jud. 253. v. ge-myntan.

mynung (?) *admonition:*—Úre hálige fæderes mid gelómrǽdre menunge ús gemenegiþ, Chart. Th. 316, 27. v. mynegung, manung.

Myrce, myrce, myrcels, myre, myrhþ, myrgan, myrige. v. Mirce, mirce, mircels, mere, miriȝþ, mirgan, mirige.

myrgen-líc; *adj. Morning:*—Ðýs myrgenlícan dæge heó biþ gongende of líchoman *she will depart before evening*, Blickl. Homl. 141, 33. v. morgen-líc.

myrran, myrrelse, myrring. v. mirran, mirrelse, mirring.

myrre, myrra, an; *f. Myrrh:*—Hí him lác brohton; ðæt wæs gold and récels and myrre, Mt. Kmbl. 2, 11. Seó myrre getácnode ðæt hé wæs deádlíc, Homl. Th. i. 116, 10. Myrra déþ ðæt ðæt deáde flǽsc ne rotaþ, 118; 11. Murre *myrra*, Ps. Spl. 44, 10. Wín gemenged mid myrran *myrratum vinum*, Wrt. Voc. i. 27, 59. Uton him bringan myrran, Homl. Th. i. 116, 25: 118, 17. [O. Sax. O. H. Ger. myrra.]

myrt. v. mirt.

myrten, es; *n. Flesh of animals that have died a natural death:*—Ne ǽnig man myrtenes ǽfre ne ábíte, Wulfst. 71, 1. Gif hé myrten ete *si morticinam ederit*, L. Ecg. C. 15; Th. ii. 142, 26. v. next word.

myrten; *adj. That has died by disease:*—Gif swýn etaþ myrten flǽsc *si porci carnem morticinam ederint*, L. Ecg. C. 40; Th. ii. 164, 18. Grécas myrten flǽsc nǽnigum men ne lýfaþ ac ða hýda ðæra myrtenra neáta hý heom dóþ tó scón *Graeci carnem morticinam nulli permittunt, de pellibus tamen morticinorum animalium calceamenta sibi faciunt*, Th. ii. 166, 29–31.

myrþ. v. miriȝþ.

myrþra, an; *m. A murderer, homicide:*—Se man biþ myrþra (*homicida*), se ðe his bróþor hataþ, L. Ecg. C. 24; Th. ii. 150, 10. Gif hwylc man for his mǽges wrǽce man ofsleá, dó (*do penance*) hé swá myrþra .vii. geár oððe .x., L. Ecg. P. iv. 68, 18; Th. ii. 230, 19, 21: Bd. 2, 9; S. 511, 37. Ðú (*the soul*) wǽre ðǽr (*in the world*) morþ and myrþra, Wulfst. 241, 9. Ðonne biþ hé ealra ðara manna deápes sceldig and myrþra beforan ðæs écan Déman heáhsetle, Blickl. Homl. 53, 7. Myrþran and mánswaran, 61, 13. Míne myrþran and mánsceaþan (*the devils*), Exon. 42 a; Th. 141, 4; Gú. 622. Myrþra *homicidas*, Mt. Kmbl. Rush. 25, 7. [Goth. maurþrja: O. H. Ger. murdreo *latro*.] v. bearn-, mǽg-, self-myrþra.

myrþrian *to murder.* [Goth. maurþrjan: O. H. Ger. murdrian *jugulare.*] v. for-, of-myrþrian.

myrþrung, e; *f. Murder, homicide:*—Myrþrunge *parricidium*, Wrt. Voc. ii. 67, 30.

myrwa. v. mearu.

mysci; *pl. Flies:*—Sóna cwómon mysci manige *venit cynomyia*, Ps. Th. 104, 27. [From *Lat.* musca.]

mýse *a table.* v. mése.

mýðe (?); *pl. The mouth of a stream:*—Andlang bróces on ða mýðy; of ðes gemýðon on Ceahhanmere, Cod. Dip. Kmbl. iii. 48, 25. v. gemýðe.

mýðe (?), an; *f. The mouth of a stream:*—Ǽrest fram mýðan in cyrstilmǽl ác ... eft in ða mýðan, Cod. Dip. Kmbl. iii. 379, 20–380, 7. v. múða.

myxen. v. mixen.

N.

N, like *m* (q. v.), in Anglo-Saxon generally corresponds to *n* in Gothic and in other cognate dialects, e. g. *net, hand, án*; Goth. *nati, handus, ains*; O. H. Ger. *nezzi, hant, ein*; O. Sax. *net, hand, ên*; but, like *m*, it falls away before ð and *s*, and the vowel which preceded the *n* is lengthened, e. g. *cúð, tóð, óðer, múð, húsel, est*; Goth. *kunþs, tunþus, anþar, munþs, hunsl, ansts*; O. H. Ger. *chund, zand, andar, mund, anst*; O. Sax. *kúð, tand, óðar, múð, anst*. If, however, *n* and *s* come together by the loss of an intervening vowel the *n* remains, e. g. *winstre*; O. H. Ger. O. Sax. *winistar*. The character which appears in the Runic poem is ᚾ, and the verse, in which the name (cf. Icel. *nauð*) is given, is the following:—

Nýd byþ nearu on breóste   oft tó helpe
niða bearnum,   and tó hǽle gehwæðre
weorðeþ heó ðeáh   gif hí his hlystaþ ǽror.
                 Runic pm. Kmbl. 341, 8–13.

ná, nó; *adv. No, not*; non. I. *qualifying a verb expressed or implied*, (a) *without any other negative particle:*—Ná cunne *nesciat*, Wülck. Gl. 257, 28. Fela gódra háma ðe wé genemnan ná cunnan, Chr. 1001; Erl. 136, 29. Fremde ná heom God setton on gesyhþe, Ps. Th. 53, 3. Ná ðú andwlitan ðinne áwend fram mé, 101, 2. Swá sceal man dón, ðonne hé gegán þenceþ longsumne lof, ná ymb his líf cearaþ, Beo. Th. 3077; B. 1536. Ealle hí scínaþ, ná hwæðre ðeáh ealle efenbeorhte, Bt. Met. Fox 20, 460; Met. 20, 230. Gewíte ðes calic fram mé, ðeáh-hwæðere ná swá swá ic wylle, Mt. Kmbl. 26, 39. Ða habbaþ twegen casus and ná má on gewunan, Ælfc. Gr. 14; Som. 17, 3. Ðæt is se eþel ðe nó geendad weorþeþ, Exon. Th. 100, 12; Cri. 1640. Ic gelýfe nó ðæt him eorþwelan éce stondeþ, 309, 33; Seef. 66. Nó ðæt ðín aldor wolde Godes goldfatu in gylp beran, ne ðý hraðor hrémde ... ac ðæt oftor gecwæþ ... ðæt hé wǽre ána Drihten, Cd. Th. 263, 34; Dan. 754. Nó seoððan ðæt hí mósten in ðone écan andwlítan, 288, 8; Sat. 377: 304, 23; Sat. 634. (b) *with other negatives:*—Hyt ná ne feóll *non cecidit*, Mt. Kmbl. 7, 25. Ne eom ic ná Crist, Jn. Skt. 1, 20. Ráde ðe mon ná ne rímde, Chr. 871; Erl. 76, 12. Nǽs ðæt ná ðæt hé nyste, Blickl. Homl. 19, 33. Ne hé hine ná ne onstyreþ, 21, 27. Ðæt ðás láreówas ne sceolan Godes dómas náwðer ne ná wanian, ne ne écan, 81, 4. Ne wandige ús ðe mǽssepreóst nó for ríces mannes ege, 43, 9. Nǽron gé nó mín gemunende, ne gé nó geþohton, Past. 21; Swt. 151, 21. Swá nán óðer ná déþ, Menol. Fox 392; Men. 197. Nabbaþ ðás naman ná óðre gebígednysse, Ælfc. Gr. 11; Som. 15, 24. Nis ná má casa on gewunan ... nis ná má mislícra casa, 14; Som. 17, 4–7. Ne behófaþ náðor ðyssera pronomina ná má stemna búton twegra, 15; Som. 17, 38. Ne synd ná má namanspeligende búton ðás fíftíne, Som. 17, 46: Blickl. Homl. 35, 24. Nó má, Exon. Th. 441, 25; Kl. 4. Telle ic ða weorþ-mynd ðæm wyrhtan, næs ná ðé, Bt. 14, 1; Fox 42, 19. Nǽs ná for ðam ðe ðæs landes swá fela wǽre, ac for ðam ðe se Wendelsǽ hit hæfþ swá tó-dǽled, Ors. 1, 1; Swt. 24, 25. Lufian wé hine ... næs nó on gesundum þingum ánum, ac eác swylce on wiðerweardum þingum, Blickl. Homl. 13, 7. II. *qualifying* (a) *an adjective:*—Mid langum scipum ná manegum, Chr. Erl. 3, 7. Ðý ilcan sumera forwearþ nó læs (= *not a smaller number*) ðonne xx scipa, 897; Erl. 96, 14. Wíse sweltende samod ná wís *sapientes morientes, simul insipiens*, Ps. Spl. 48, 9. (b) *an adverb or adverbial phrase:*—Nis nó ðæt án ... ac eác *not only ... but also*, Blickl. Homl. 85, 15. Nǽs hit ná ðæt án ðæt ðú on ungemetlícum ungesǽlþum wǽre, ac eác ðæt ðú fulneáh mid ealle forwurde, Bt. 5, 3; Fox 14, 6. Ðe ná ðæt án mé, ac eác swylce míne geféran, mæg besencan, Coll. Monast. Th. 24, 31: L. Ecg. P. iii. 1; Th. ii. 196, 13. Ná ðá git *non dum*, Wrt. Voc. ii. 59, 55. Ná swá *numquam ita*, Wülck. Gl. 248, 9. Ná lancge *non diu*, Coll. Monast. Th. 28, 31. Ná elles *haud secus*, Ælfc. Gr. 38; Som. 42, 3. Ðú hit ná hú elles begitan ne miht, Bt. 32; Fox 114, 8. Ne mæg hé nó ðé raþor, Bt. tit. 32; Fox xvi, 15. Nǽs him nó ðý læs underþeóded eall ðes middangeard, 16, 4; Fox 58, 10. Ðá ná ðé læs beseah Lothes wíf underbæc, Scrd. 22, 42. Nó ðý fægra wæs, Cd. Th. 203, 6; Exod. 399. Nó ðý sél dyde, 246, 35; Dan. 489. Ne sý ná tó ðæs hwón (*on no account*) geendod náðer ne dægrédsang ne ǽfensang búton ðam drihtlícan gebede, R. Ben. 38, 14: 87, 1, 10: 95, 7. Ná tó hwón (ná tó ðæs hwón, MS. T.), 111, 19. v. lytes-ná.

nabban (= ne habban, *the verb is conjugated throughout*) *not to have, to be without:*—Nǽbbe ic synne gefremed, Cd. Th. 160, 15; Gen. 2650. Nǽbbe ic welan, Andr. Kmbl. 601; An. 301. Nafast hláfes wiste, 621; An. 311. Ðú næfst nán þing, Jn. Skt. 4, 11. Nǽft ðú, Bt. Met. Fox 20, 71; Met. 20, 36. Hé nǽnige mehte nafaþ, Blickl. Homl. 31, 33. Hé wilnaþ ... ðæs ðe hé næft, Bt. 11, 1; Fox 34, 2. Nǽþ, Ps. Th. 71, 12. Wé nabbaþ, Mk. Skt. 9, 13. Earmra manna gehelpan ðe sylfe nabbaþ and ðæra myhta nabbaþ ðæt hié wyrcen mágon, L. E. I. 3, Th. ii. 404, 22. Hí heora nabbaþ má ðonne hí heora habban, Bt. 26, 1;

Fox 90, 19. Ðonne ðú hæfdest ðæt ðú noldest oððe næfdest ðæt ðú woldest, Fox 90, 31. Næfde heó nôht on hire, Blickl. Homl. 147, 15. Næfde gé, Jn. Skt. 9, 41. Nafa ðú fremde godas, Deut. 5, 7. Ðonne gé faran næbbe gé mid eów hláf, Blickl. Homl. 233, 17. Gif hé wíf næbbe, Ex. 21, 4. Ne mæg ðæt ná beón ðæt ða bearn langunga nabban, Blickl. Homl. 131, 26. Næbben, Beo. Th. 3705; B. 1850. Hêt mê fremdne god hergan, oððe hí nabban, Exon. Th. 247, 12; Jul. 77. Sint hí ðé pliólícran hæfd ðonne næfd, Bt. 14, 1; Fox 42, note 10. [O. Frs. combines the negative with the verb in the same way.] v. ge-næfd.

nabo-gár, nabula. v. nafo-gár, nafola.

naca, an; m. A boat, bark, ship, vessel:—Ne hié scip fereþ, naca, Exon. Th. 439, 17; Rä. 59, 5. Sægeáp naca, Beo. Th. 3797; B. 1896. Heáhstefn naca, Andr. Kmbl. 532; An. 266. Of nacan stefne, 582; An. 291: Exon. Th. 306, 14; Seef. 7. On bearm nacan, Beo.Th. 433; B. 214. Nêðan on nacan tealtum, Runic pm. 343, 22; Rún. 21. Flotan níwtyrwydne, nacan, Beo. Th. 596; B. 295: Exon. Th. 474, 31; Bo. 39. [O. Sax. nako: O. H. Ger. nacho: Icel. nökkvi.] v. hring-, sæ-, ýð-naca.

nacian; p. ode To strip (the clothes off a person):—Ðá hé ðæt nolde hé wæs nacod and on carcern onsænded when he would not do that (deny Christ), he was stripped and sent to prison, Shrn. 51, 12. [The shenship of his flesh he shal nakyn, Wick. Lev. xx, 19; he nakide (later version, made nakid) the hous of the pore man, Job xx, 19: O nice men, whi nake ye youre bakkes, Chauc. Boeth. l. 4288: Prompt. Parv. nakyñ nudo, denudo, v. p. 351, note 1. The verb to nake occurs as late as Tourneur who has 'nake your swords;' v. Skeat Dict. s. v. naked.] v. be-nacian, nacod.

nacod, næcad; adj. I. naked, bare; nudus:—Nacod exertum, Wrt. Voc. ii. 144, 70. Næcad exerta, 107, 78. (a) of persons, without clothing:—Nacod and ceald nuda, 61, 65. Nacod plegere gymnosophista, i. 17, 10. Ic eom nacod (nudus), Gen. 3, 10, 11. Ðá sæt ðér sum þearfa nacod, bæd him hrægles, Blickl. Homl. 213, 33: Cd. Th. 255, 32; Dan. 633. Ic wæs nacud and gé mê scrýddon, Mt. Kmbl. 25, 36: Cd. Th. 207, 29; Exod. 474. Gif ðú earm gewurðe, geþenc ðú ðæt ðín môder ðé nacodne gebær, Prov. Kmbl. 15. Nacode wé wæron ácennede, and nacode wé gewítaþ, Homl. Th. i. 64, 28. Gé géfon hrægl nacedum, Exon. Th. 83, 13; Cri. 1355. Nacode scrýdan, Blickl. Homl. 213, 18. Se feónd swá micle iédlícor ðæt môd gewundaþ swá hé hit ongiet nacodre ðære byrnan wærscipes, Past. 56; Swt. 431, 10. (b) of an animal, unsáddled, bare-backed:—Hé nolde on nacedum assan rídan, Homl. Th. i. 210, 27. (c) of a sword, naked, unsheathed:—Him ne hangaþ nacod sweord ofer ðam heáfde, Bt. 29, 1; Fox 102, 27: Beo. Th. 1082; B. 539. II. bare in a metaph. sense, (a) of persons, destitute, stripped of property:—Se nacoda wegférend vacuus viator, Bt. 14, 3; Fox 46, 29. Ðú (Adam) scealt on wræc hweorfan, nacod niédwædla, neorxna wanges dugeþum bedæled, Cd. Th. 57, 16; Gen. 929. Ðú (Laban) mê (Jacob) woldest forlætan nacodne, Gen. 31, 42. (b) of words, not accompanied by deeds:—God nele ðæt ðú hine lufie mid nacodum wordum ac mid rihtwísum dædum, Basil admn. 4; Norm. 40, 18. [Goth. nakwaþs: Icel. nökviðr: O. Frs. nakad: O. H. Ger. nachot, nahhut: Ger. nackt.] v. eall-, lim-nacod; nacian.

nacodian. v. ge-nacodian. [O. H. Ger. gi-, ant-nachatôn.]

nǽcan to kill:—Ic nǽce (other MSS. knǽce, nǽte) oððe ic ácwelle neco, Ælfc. Gr. 24; Som. 25, 56. [Cf. O. H. Ger. neihan immolare, Grff. 2, 1015.]

nǽced, e; f. Nakedness:—Gif hwylc man stele mete oððe cláðas and hine hungor oððe næced ðærtó drife (fames vel nuditas eum coegerit), L. Ecg. P. iv. 25; Th. ii. 212, 4. Drihten ásent hungor on eów and þurst and næcede, Deut. 28, 48. [Goth. nakwadei: Icel. nekt nakedness.]

nǽcedness, e; f. Nakedness:—Swá ðæt hig ne gesáwon heora fæder næcednesse, Gen. 9, 23. Ðé ne sceamaþ ðínre næcednysse, Homl. Th. i. 432, 5.

nǽct, næder-, næddre, nǽdel. v. niht, næder-, nædre, nǽdl.

nǽder-bíta, an; m. An ichneumon:—Nædderbíta hinc neomon (= ichneumon), Wrt. Voc. ii. 43, 49. Nædderbíta cicidemon, 131, 40.

nǽder-winde, an; f. The name of a plant, adder-wort:—Nædderwinde viperina, Wrt. Voc. i. 63, 26. v. next word.

nǽder-, nǽdre-wyrt, e; f. Adder-wort; polygonum bistorta:—Nædderwyrt uiperina, Wülck. Gl. 300, 23. Nædrewyrt. Ðeós wyrt ðe man uiperinam and ôðrum naman nædderwyrt nemneþ, Lchdm. i. 96, 11. Nædderwyrt. Ðeós wyrt ðe man basilisca and ôðrum naman nædder- (nædre-, MS. O) wyrt nemneþ, 242, 7: iii. 8, 24. Genim næderwyrte, ii. 110, 25. [v. E. D. S. Pub. Plant Names, adderwort, and Lchdm. ii. Glossary.]

nǽdl, e; f. A needle:—Nædl acus, Wrt. Voc. i. 85, 4: Ælfc. Gr. 11; Som. 15, 18. Hwanon seámere nædl unde sartori acus, Coll. Monast. Th. 30, 33. Þurh nædle (nédle, Rush.) eáge per foramen acus, Mt. Kmbl. 19, 24: Lk. Skt. 18, 25. Þurh nédle þyrel, Mk. Skt. 10, 25: Wrt. Voc. ii. 73, 1. Nédle sceorpran, Soul Kmbl. 230; Seel. 116. Mið nédle acu, Wrt. Voc. ii. 117, 37. Mið naedlae, Ep. Gl. 19 f, 30. [Goth. neþla:

O. Sax. nádla: O. Frs. nédle: O. H. Ger. nádala: Icel. nál.] v. feax-, hǽr-nǽdl.

nǽdre, nǽddre, an; f. Any kind of serpent, adder, viper:—Nædre gipsa, Wrt. Voc. ii. 41, 55: natrix, 97, 36: 60, 77. Snaca oððe nædre coluber, 16, 75. Gerumpenu, gehyrnedu nædre coluber cerastis, 15, 68: 16, 2. Mê nædre beswác, Cd. Th. 55, 20; Gen. 897. Næddre vipera vel serpens vel anguis, Wrt. Voc. i. 78, 55. Fleónde næddre, 24, 1. Ðære nædran basilisci, ii. 12, 2: 86, 58. Efter gelícnisse nedran (serpentes), Ps. Surt. 57, 5: Cd. Th. 271, 8; Sat. 102. Ðære scortan næddran spalangii, Hpt. Gl. 450, 25. Nedran colubri, Kent. Gl. 1095. God cwæþ tó ðære næddran (ad serpentem), Gen. 3, 14. Nædran celidrum, Wrt. Voc. ii. 21, 21. Nædran hilidros, i. celidros, 43, 38. Swá swá Moyses áhôf ða næddran . . . Ðá sende God fýrene næddran . . . God bebeád Moyse ðæt hé geworhte áne ærene næddran, and sette up tó tácne, and ðæt hé manode ðæt folc ðæt swá hwá swá fram ðám næddrum ábiten wære, besáwe up tó ðære ærenan næddran, Homl. Th. ii. 238, 4–19. Nædran serpentes, Ps. Th. 139, 3. Gif mon hine (gagates) on fýr déþ, ðonne fleóþ ðær neddran onweg, Bd. 1, 1; S. 473, 25. Nædrena draconum, Wrt. Voc. ii. 27, 71. Nædrena áttor venenum aspidum, Deut. 32, 33. Lá næddrena (ætterna, Lind.; nedrana, Rush.) cyn progenies viperarum, Mt. Kmbl. 3, 7: 12, 34. Hig wurpon ealle hira gyrda nyðer and hí wurdon tó næddrum (versae sunt in dracones), Ex. 7, 12. [Goth. nadrs: Icel. naðr (in poetry); m.; naðra; f.: O. Sax. nadra: O. H. Ger. natra, natara; f.: Ger. natter.] v. hilde-, mere-, wæter-nædre.

nǽfde = ne hæfde. v. nabban.

nǽfig; adj. Not having means, poor:—Þarfa t næfga (næfge, Lind.) mendicus, Jn. Skt. Rush. 9, 8. Næfgum (næfigum, Lind.) egenis, 13, 29. Næfigum, Lind. 12. 5.

nǽfre (= ne ǽfre); adv. Never. I. alone:—Næfre ætýwde swylc, Mt. Kmbl. 9, 33. Næfre ic máran geseah eorl ofer eorþan, Beo. Th. 500; B. 247. Næfre gé mid blôde beóðgereordu eówre þicgeaþ, Cd. Th. 91, 26; Gen. 1518. Eádig biþ se ðe in his éþle geþíhþ; earm se him his frýnd geswícaþ; néfre (?) sceal se him his nest áspringeþ (never shall he thrive whose provision fails him (?). Grein takes nefre = infirmus), Exon. Th. 335, 23; Gn. Ex. 38. II. with another negative:—Ne hit næfre ne gewurðe nec unquam fiat, Ælfc. Gr. 38; Som. 40, 14. Ðæt hí næfre ne gedóþ, Bt. 14, 2; Fox 44, 15. Næfre siðan Rômáne ne rícsodon on Bretone, Chr. 409; Erl. 10, 7. Hié næfre his banan folgian noldon, 755; Erl. 50, 20. Nán man ne dorste sleán óðerne man, næfde hé næfre swá mycel gedón wið ðone óðerne, 1086; Erl. 222, 6.

nǽgan, negan; p. de To address, accost, speak to:—Nigeþan síþe nægde se gomola, sægde eaforan worn, Exon. Th. 304, 5; Fä. 65. But generally the verb is accompanied by wordum:—Ðú mê wordum nægest, fúsne frignest, 175, 26; Gú. 1200. Hine weroda God wordum nægde, Cd. Th. 179, 4; Exod. 23. Hé ðone wísan wordum nægde (hnægde, MS.) freán Ingwina, Beo. Th. 2641; B. 1318. Ongan ðá wíf weras wordum nêgan, Elen. Kmbl. 574; El. 287: 1115; El. 559. v. ge-nægan.

nǽgel, nægl, es; m. I. the nail of a finger or toe:—Nægel unguis; næglas ungues, Wrt. Voc. i. 43, 60. Fingras digiti . . . nægel ungula, 65, 4. Nægl, 283, 25. Nægl unguana, ii. 124, 10. Gif nægl of honda weorðe if a nail come off a hand, Lchdm. iii. 58, 7: ii. 80, 20. Gif þuman nægl of weorðeþ, .iii. scill. gebéte . . . Æt ðám nægclum gehwylcum scilling if a thumb-nail come off (from a blow) the bót shall be iii shillings . . . For each finger-nail a shilling (cf. L. Alf. pol. 56–60; Th. i. 94, 96 where the bót for the thumb-nail is 5 shillings, for the nail of the fore-finger and for that of the ring-finger 4 shillings each, for that of the middle finger 2 shillings, and for that of the little finger one shilling), L. Ethb. 54, 55; Th. i. 16, 9–14. Wið scurfedum nægle; nim gecyrnadne sticcan, sete on ðone nægl, Lchdm. ii. 150, 4: i. 570, 9: iii. 114, 21. Deóplíc dædbót biþ . . ., ðæt íren ne cume on hǽre ne on nægle, L. Edg. C. 10; Th. ii. 280, 21. God of ðam láme flǽsc worhte and blôd, bán and fell, fex and næglas, Homl. Th. i. 236, 16. II. a nail, peg:—Nægl clavus, Wrt. Voc. ii. 12, 10. Nægl paxillum, palum, 116, 27. Nægles epigri vel clavi, i. 39, 63. Nægle cuspide, ii. 21, 24. Ðǽr hýdde wǽron næglas (the nails by which Christ was fastened to the cross) on eorþan, Elen. Kmbl. 2216; El. 1109: 2227; El. 1115: 2344; El. 1173. Ne gelýfe ic búton ic geseó ðara nægela (clavorum) fæstnunge on his handa, and ic dô mínne finger on ðære nægela stede, Jn. Skt. 20, 25. Ðæt fýr eode andlang ðara nægla ðe seó studu mid gefæstnad wæs tó ðam wáge, Bd. 3, 17; S. 544, 31, col. 1. Mid næglum þurhdrífan ða hwítan honda, Exon. Th. 68, 27; Cri. 1110: Rood Kmbl. 91; Kr. 46. Hié námon treówu, and slôgon on óðerne ende monige scearpe ísene næglas, Ors. 4, 1; Swt. 158, 5. Heó lǽdde tó hire suna ða ísenan næglas ðe wǽron ádrífene þurh Cristes folman, Homl. Th. ii. 306, 15. Nægelas geseón anxsumnysse getácnaþ, Lchdm. iii. 212, 24. III. an instrument for striking the strings of a harp, v. hearpe-nægel, Exon. Th. 332, 12; Vy. 84. [O. Sax. nagal (in both senses): O. Frs. neil (in both senses): O. H. Ger. nagal unguis, clavus. paxillus: Icel. nagl unguis; nagli clavus: cf. Goth. ga-nagljan.] v. hearpe-, scóh-, steór-nægel.

nægel-seax, es; *n.* *A knife for cutting the nails* :—Næglsex *novaculum,* Wrt. Voc. i. 35, 22 : *novacula,* 86, 22. [*Laym.* nail-sax (-sex).]

nægen = ne mægen :—Gedó ðæt hý nægen dón ðæt yfel ðæt hý þencaþ *make them unable to do the evil that they devise;* decidant a cogitationibus suis, Ps. Th. 5, 11.

nægled-bord. v. næglian.

nægled-cnearr, es; *m.* *A vessel the planks of whose sides are nailed together* :—Gewitan him ða Norþmen nægledcnearrum, Chr. 937; Erl. 115, 2; Æðelst. 53.

næglian; *p.* ode, ede *To nail, fasten with nails* :—Hí dulfon ł nægledun handa míne and fét míne *foderunt manus meas, et pedes meas,* Ps. Lamb. 21, 17. Síæ nægled on róde *crucifigatur,* Mt. Kmbl. Rush. 27, 23. Ne hié scip fereþ naca nægled bord (or nægled-bord; adj.?) *nor does ship carry her, vessel, nailed plank* (or *with nail-fastened sides*), Exon. Th. 439, 17; Rä. 59, 5. Siððan nægled bord, fær séleste, flód up áhóf, Cd. Th. 85, 22. Hwonne hié of nearwe ofer nægled bord stæppan mósten *when from durance over the vessel's (the ark) nail-fastened side they might step,* 86, 20; Gen. 1433. Hió [næ]gled sinc hæleþum sealde (*bracelets fastened with rivets* or *studs*), Beo. Th. 4051; B. 2023. Næglede (ætlede, Th.) beágas, Exon. Th. 474, 22; Bo. 34. Nægledne, 400, 7; Rä. 20, 5. [*Goth.* ga-nagljan : *Icel.* negla : *O. Sax.* neglian; negilid sper, neglit skip : *O. H. Ger.* nagalian.] v. ge-nægled.

Nægling *the name of Beowulf's sword* :—Nægling forbærst, sweord Beówulfes, Beo. Th. 5354; B. 2680.

néh, næht, néht, næm, næman, næmne. v. neáh, niht, náht, níd-næm, be-, míd-næman, nemne.

néming, e; *f. Acceptance, agreement, bargain* :—Ceáp *distractio;* sala *venditio;* næmingce *contractio* vel *contractus,* Wrt. Voc. i. 55, 54–56.

nænig (= ne ænig). **I.** used as an adjective, *not any, none, no,* (a) without another negative :—Nænig óðer hý æfre má eft onlúceþ, Exon. Th. 20, 27; Cri. 324. Deáh ðe nænegu néðþearf wære, Met. 20, 25. Ðær nængu biþ niht on sumera, 16, 13. Naenge earbeðe *nullo negotio,* Wrt. Voc. ii. 115, 5. Nænigne ic sélran hýrde hordmáðmum, Beo. Th. 2398; B. 1197. Hafaþ tóþ nænigne, Exon. Th. 439, 24; Rä. 59, 8 : Cd. Th. 272, 20; Sat. 122. Him ðæs nænige bót dydon, Blickl. Homl. 201, 23. (b) with other negatives :—Nænig mon ne sceal lufian ne ne géman his gesibbes, 23, 16. Ðæt wíte ðe næfre nænig ende ne becymeþ, 51, 31. Ne hé nænigne man unrihtlíce fordémde, ne hine nænig man yrne ne funde, 223, 32. Ðær him nænig wæter wihte ne scepede, Beo. Th. 3032; B. 1514. Óðer nænig sélra nære, 1723; B. 859. Nis nænigu gecynd, Salm. Kmbl. 839; Sal. 419. Ne sý eów nænigu cearo, Blickl. Homl. 145, 8. Ne hié nænigo firen ne gewundode, 161, 33. Næs nænig ylding, 87, 17. Nis nænig máre mægen, 31, 30. Eów nænig wuht ne deraþ, And. 14, 8. God ðonne ne gýmeþ nænges mannes hreówe, Blickl. Homl. 95, 29. Nænges þinges máre þearf nære, 175, 8. Hé næfre nænigne woruldrícum men onbúgan nolde, 223, 27. Warna ðé ðæt ðú hyt nænegum men ne secge, Mt. Kmbl. 8, 4. Ðæt hé nænigum óðrum men ne sæde, Bd. 5, 9; S. 623, 3 : Blickl. Homl. 211, 16. Hié eów tó nænigre áre ne belimpeþ, Blickl. Homl. 41, 23 : 179, 15. Hí ne mihtan ðære heorde nænige góde beón, 45, 16. Hé nænige mehte wið ús nafaþ, 31, 33 : 79, 7. Ne bideþ hé æt ús nænig óðer edlean, 103, 21. Ne ðær nænige þingunga ne beóþ, 95, 30 : 157, 13 : 185, 9. Hé næfre nænige godcunde englas næfde, 181, 28. **II.** as a substantive, *no one, not any one,* (a) without another negative :—Nænig bihelan mæg wom unbéted, Exon. Th. 80, 23; Cri. 1311 : 294, 20; Crä. 18 : Beo. Th. 3870; B. 1933. Nænig óðerne freóþ, Frag. Kmbl. 69; Leás. 36 : Exon. Th. 491, 29; Rä. 81, 6. Nænegum þuhte dæg on þonce, Met. 12, 15. Se ðe nængum scód, Exon. Th. 90, 1; Cri. 1467. Nænige *neminem,* Hpt. Gl. 457, 57. Ðær hé nænige forlét bendum fæstne, Andr. Kmbl. 2074; An. 1039. (b) with other negatives :—Nis nænig swá snotor nymþe God seolfa, Cd. Th. 286, 8; Sat. 349. Ðone nænig heonon ne sceáwaþ, Blickl. Homl. 31, 9. Nænigne tweógean ne þearf, 83, 9. **III.** with partitive gen. (a) without another negative :—Nænig fíra ðæs fród leofaþ, Exon. Th. 351, 6; Sch. 76. Nænig wera gewiste, 412, 13; Rä. 30, 13. Nænig manna is, Andr. Kmbl. 1088; An. 544 : Salm. Kmbl. 120; Sal. 59. Him nænig wæs ælærendra óðer betera, Elen. Kmbl. 1008; El. 505. Nænig heora þohte, Beo. Th. 1385; B. 691. Nænegum áraþ leóde Deniga, 1201; B. 598. Ic nængum scedde burgsittendra, Exon. Th. 407, 9; Rä. 26, 2. Mid ðý se cyning nænige þinga (*nulla tenus*) his bénum geþafian wolde, Bd. 3, 24; S. 556, 11. (b) with other negatives :—Ne ðær nænig wihte wénan þorfte, Beo. Th. 316; B. 157 : 490; B. 242. Nænig gumena ongitan ne mihte, Andr. Kmbl. 1971; An. 988 : Salm. Kmbl. 867; Sal. 433. Næniges Godes háligra gebyrd, ne his heáhfædera . . . ciriecan ne mærsiaþ nemþe . . ., Blickl. Homl. 161, 9. Nis ðæt mín miht ne næniges úres (= úre næniges), 151, 29. Be ðare nænigum gecweden beón ne mihte, 161, 22. Ne eart ðú ðon leófre nænium lifgendra, Exon. Th. 370, 5; Seel. Ex. 54. Deós dæd nænige þinga forholen ne wurþe, Lchdm. iii. 60, 24 : Met. 10, 16 : 19, 37. Ne sculon mæssepreóstas náteshwón nænig þinga bútan óðrum mannum mæssan syngan, L. E. I. 7; Th. ii. 406, 21.

nænig-wiht; *adv.* *Nothing, not, not at all* :—Andreas nænigwuht ðú gefirnodest *Andrew, thou hast nothing sinned,* And. 10, 20. v. nán-wiht.

næniht. v. nán-wiht.

næp, es; *m.* *Turnip, rape* :—Næp *napus, rapa,* Wrt. Voc. i. 31, 44, 51 : *napis,* 68, 18 : 286, 26 : ii. 114, 56. Wilde næp *nap silvatica,* i. 31, 27 : *diptamnus vel bibulcos,* 32, 5. Nim Ængliscne næp, Lchdm. iii. 12, 14. Nim smælne næp, 40, 5. Healde hine wið næpas, and wið ða þing ðe windigne æþm on men wyrcen, ii. 214, 3. [*Nepe bacar,* Wrt. Voc. i. 191, 39 : nepe *coloquintida, cucurbita,* Prompt. Parv. 353. See also E. D. S. Plant Names, nape, nep : *Icel.* næpa; *f. a* turnip.]

næp-sæd, es; *n.* *Seed of turnip* or *of rape* :—Genim senepes sædes dæl and næpsædes, Lchdm. ii. 24, 15. Nim senepsæd and næpsæd, iii. 88, 15.

næpte, nære, næron. v. nepte, næs.

næs = ne wæs *was not* :—Wære ðú tódæg on huntnoþe? Ic næs, Coll. Monast. Th. 22, 1 : 34, 9. Ðú nære mildsiend ofer heora cild, Blickl. Homl. 249, 6. Man næs, ðe ða eorþan worhte, Gen. 2, 5. Næron ða welige hámas, ne diórwyrþra hrægla hí ne girndan, forðam hí ða git næran, Bt. 15; Fox 48, 4–6. Ða cyningas Rómeburg begeáton ðær Mutius nære (*if it had not been for Mucius*), Ors. 2, 3; Swt. 68, 20. Gif hé nære yfeldæde, ne sealde wé hine ðé, Jn. Skt. 18, 30. Hié wýscaþ ðæt hié næfre næron ácennede Blickl. Homl. 93, 28. [*O. Frs.* nas = ne was; nére = ne wére.]

næs; *adv.* *Not.* **I.** alone :—Búton hit riht spræc sý and behéfe næs ídel *nisi recta locutio sit et utilis, non anilis,* Coll. Monast. Th. 18, 16. Ic wylle mildheortnesse næs onsægdnesse, Mt. Kmbl. 9, 13. Gif hit fæger is, ðæt is of heora ágnum gecynde, næs of ðínum; heora fæger hit is, næs ðín, Bt. 14, 2; Fox 42, 33. ·Heó wæs ful cweden, næs æmetugu, Blickl. Homl. 5, 5. Ic cýðe mid dædum, næs mid wordum ánum, 181, 25 : Ps. Th. 48, 12. Næs hié ðære fylle gefeán hæfdon, Beo. Th. 1128; B. 562 : 6140; B. 3074. **II.** with another negative :—Ábréd of ða fiðeru, næs ne cerfe, Lev. 1, 17. Ic ondræde ðæt hé wirige mé, and næs ná bletsige, Gen. 27, 12. Ðonne telle ic ða weorþmynd ðæm wyrhtan, næs ná ðé, Bt. 14, 1; Fox 42, 19. Gif ðú gesáwe þeóf, ðú urne mid him, næs ná ongeán hine, Ps. Th. 49, 19. Gesceapene tó ðon écan lífe, næs ná tó ðon écan deáþe, Blickl. Homl. 61, 8. Næs ná mid golde, ac mid gódum dædum, 95, 19. Lufian wé hine næs nó on gesundum þingum ánum, ac eác swylce on wiðerweardum þingum, 13, 7. [*O. Frs.* nas.]

næsc *fawn-skin* :—Fel *pellis,* hýd *cutis* vel *corium;* næsc *nebris,* Wrt. Voc. i. 86, 37–39. Gefóh fox, ásleah of cucum ðone tuxl, læt hleápan áweg, bind on næsce, hafa ðé on, Lchdm. ii. 104, 13 : 140, 10. Dó on næsc, 36, 8. Naescum *tractibus* (cf. tracta), *pl.* in *mulomedicina emplastrum ex variis medicamentis compositum, et in tela linea distentum,* Forcellini.), Wrt. Voc. ii. 122, 77.

næse. v. nese.

næs-gristle *the gristle* or *cartilage of the nose* :—Naesgristlae *cartilago,* Ep. Gl. 7 b, 5. Naesgristle, Wrt. Voc. ii. 102, 45. Næsgristle, 13, 10. [Þe laðe helle wurmes þe freoteð ham ut te ehnen ant te nease gristles, O. E. Homl. i. 251, 16.] v. nos-gristle.

næss, ness, es; *m.* **I.** *a ness, land running out into water, headland, promontory.* [The word *ness* found in English local names is mostly of Scandinavian origin, *Icel.* nes; but, in a charter of 778, Cod. Dip. Kmbl. iii. 382, 28, Tucingnæs occurs, and in another of 801 is the passage, 'adjecto uno piscatorio on Taemise fluuio ubi dicitur Fiscnaes,' i. 216, 25. Other instances in the charters are, Herces næs, 437, 1 : on scearpan næsse, 438, 22. Earna næs *Eagles-ness,* Beo. Th. 6055; B. 3031, Hrones næs *Whales-ness,* 5603; B. 2805, are examples of the word in foreign local names] :—Æt brimes næsse *at the sea-headland,* Andr. Kmbl. 3417; An. 1712. Beorh wæterýðum neáh, be næsse, Beo. Th. 4478; B. 2243. Gesæt on næsse cyning, 4825; B. 2417. Wearþ on næs (*of a lake*) togen wundorlíc wægbora, 2883; B. 1439 : 3205; B. 1600. Se ðe næs (*by the sea*) geráð, 5789; B. 2898. Windige næssas *wind-swept headlands,* 2721; B. 1358. Neowle næssas *headlands that plunge into the water,* 2826; B. 1411. Hié Geáta clifu ongitan meahton, cúþe næssas, 3828; B. 1912. **II.** in connection with *under, niðer,* and often in pl. *ground* (as in *under-ground*) :—Ongan ða eorþan delfan, ðæt hé on twentigum fótmælum feor funde behelede under neólum niðer næsse gehýdde in þeóstorcofan (*he found the cross hidden twenty feet underground*), Elen. Kmbl. 1661; El. 832. Gæst ellor hwearf under neowelne næs (*underground,* i. e. *to hell*), Judth. Thw. 239; Jud. 113. Sunne gewát tó sete glídan under niflan næs (*sink beneath the horizon*), Andr. Kmbl. 2611; An. 1307. Fyrgenstreám under næssa genipu niðer gewíteþ (*the stream disappears in a dark chasm*), flód under foldan, Beo. Th. 2724; B. 1360. Hí (*the fallen angels*) gedúfan sceolun niðær undær nessas (*to hell*) in ðone neowlan grund, Cd. Th. 266, 32 : Sat. 31 : 270, 15; Sat. 91. Ingong in ðæt atule hús (*hell*) niðer under næssas, neóle grundas, Exon. Th. 136, 2; Gú. 535. v. sæ-næss *and next word.*

næsse, an; *f.* *A headland, promontory, cape* :—Óþ ða norþmestan næssan on eorþan *to the most northerly cape on earth,* Met. 9, 43. Næssun (-an?) *litora,* Germ. 400, 488. v. næss.

næss-hliþ, es; _n._ _The slope of a headland :_—Gesáwon on næshleoþum nicras licgean, Beo. Th. 2858; B. 1427.

nǽstan. v. ge-nǽstan.

nǽster _cancale_ (=? καυκαλίς _wild carrot_), Wrt. Voc. ii. 129, 74.

næs-þyrel, -þyrl, es; _n._ _A nostril:_—Næsþyrel _pennula,_ Wrt. Voc. i. 282, 66; _nares,_ ii. 62, 5. Dô on ðæt næsþyrl, Lchdm. i. 352, 4. On næsþyrl bestungen, 348, 4. His (_the dead man_) næsþyrlo beóþ belocene, Blickl. Homl. 59, 14. Wið næsþyrla (næsþurla, 14, 11) sáre, Lchdm. i. 114, 19. Blôdryne of næsþyrlon, 282, 12. Mid hundes lúsum, ða flugon intô heora múðe and heora næsþyrlum, Homl. Th. ii. 192, 22. Hit gǽþ þurh eówre næsþyrlu _exeat per nares véstras,_ Num. 11, 20 : Ps. Spl. 113, 14. Dô on ða næsþyrlu, Lchdm. i. 72, 21. [_Wick.: Prompt. Parv._ nese-þirl.] v. nos-þyrel.

nǽtan; _p._ te _To trample upon, crush, subdue :_—Oft ic cwice bærne nǽte mid nîþe _oft the living I burn, painfully oppress them,_ Exon. Th. 389, 7; Rä. 7, 4. Hé sceal weorðan his lîfe tô nytte mid ðý ðæt hé nǽte his unþeáwas _mores pravorum premere, vitae prodesse,_ Past. 46, 5 ; Swt. 353, 10. Nǽtendne _proterentem,_ Wrt. Voc. ii. 118, 3 : Ep. Gl. 18 b, 27. v. ge-nǽtan _and next word._

nǽting, e; _f._ _Blaming, upbraiding :_—Ac hú wêne wê hú micel scyld ðæt síe ðæt monn áþreóte ðære nǽtinge yfelra monna and nime sume sibbe wið ða wierrestan _pensandum ergo est, quando ab increpatione quiescitur, quanta culpa cum pessimis pax tenetur,_ Past. 46, 6 ; Swt. 353, 11. [Cf. _Goth._ naiteins _blasphemy._]

nafa _a nave._ v. nafu : nafa = ne hafa. v. nabban.

nafela, an; _m._ _The navel :_—Nabula _umbilicus,_ Wülck. Gl. 54, 13. Navela, Wrt. Voc. i. 44, 50. Ðínum nafelan, Kent. Gl. 32. Hé (_Minutius_) hiene (_the elephant_) on ðone nafelan ofstang, Ors. 4, 1 ; Swt. 156, 11. [_O. Frs._ navla : _Icel._ nafli : _O. H. Ger._ nabalo.]

nafel-sceaft, e; _f._ _The navel :_—Ðisne lǽcedôm man sceal dô ðan manne se his nafulsceaft in týhþ, Lchdm. iii. 124, 22.

nafeþa, an; _m._ _A nave :_—Naveþa _modiolus,_ Wrt. Voc. i. 16, 22.

nafu, e; _f._; nafa (?), an; _m._ _A nave :_—Nafu _modialis,_ Wrt. Voc. i. 284, 55. Sió nafa (nafu, Cott.) nêhst ðære eaxe, Bt. 39, 7 ; Fox 220, 29. Sió nafu, Fox 222, 1. Se nafa, 222, 12. Fæst on ðære nafe, 222, 3, 8, 9, 11, 12. [_Icel._ nöf : _O. H. Ger._ naba _modiolus._]

nafu-gár, es; _m._ _An auger :_—Nabogaar _terebellus,_ Wrt. Voc. ii. 122, 21. Nabogár _rotrum,_ 119, 31. Nafogár _foratorium,_ 149, 74 : _foratorium vel terebellum,_ 38, 50. Navegár _terebrum,_ i. 16, 12 : 84, 63. [Wymble, nauger _terere,_ 170, 17 : _O. H. Ger._ naba-gêr _terebellus, terebellum, terebrum : Icel._ nafarr : _Du._ ave-gaar.]

-nág. v. ge-nág.

nágan = ne ágan. I. _not to have,_ (a) with acc. :—Náh se sacerd náne þearfe (_sacerdoti non opus est_), ðæt hé forwyrne ðam men rihtre andetnysse, L. Ecg. P. i. 2 ; Th. ii. 172, 11. Gif hé náh his selfes geweald, Met. 16, 21. Helle hlinduru nágon hwyrft, Exon. Th. 364, 29 ; Wal. 78. Ðeáh ðú hî nǽfre náhtest, Bt. 14, 2 ; Fox 44, 1. Hé náhte his lîchoman geweald, Blickl. Homl. 203, 11. Náhton hié náðer ne mete ne freónd, Ors. 2, 8 ; Swt. 92, 34. Sî on cynges dôme hwæðer hé lîf áge ðe náge, L. Eth. vii. 9 ; Th. i. 330, 25. (b) with gen. :—Nágan wê ðæs heolstres, ðæt wê ús gehýdan mágon, Cd. Th. 271, 5 ; Sat. 101. II. _not to be allowed, ought not :_—Náh náðer tô farenne ne Wylisc man on Ænglisc land, ne Ænglisc man on Wylisc, L. O. P. 6; Th. i. 354, 23. Náge hé hié út on elþeódig folc tô bebycgganne _it shall not be allowable for him to sell her abroad into a foreign people,_ L. Alf. 12 ; Th. i. 46, 13. On ða gerád ðæt hine náge nán man of tô áceápienne, Chart. Th. 151, 13. Ðæt hit náge nán man fram ðære stôwe tô dǽlanne, 157, 6.

náht. v. ná-wiht.

ná-hwǽr, -hwár, -wér; _adv._ I. _no-where, in no place :_—Náhwǽr _nusquam,_ Ælfc. Gr. 38 ; Som. 41, 55. Ðú ne ætstande náhwár on ðisum earde _nec stes in omni circa regione,_ Gen. 19, 17. Hé sôhte his wǽpnu, ac hé ne geseah hî náhwǽr, Homl. Skt. 3, 257 : Blickl. Homl. 59, 20 : 181, 23. Ðeáh hé hire náwér ne geneálǽce on ǽlcere stôwe hé is hire emnneáh _though the sky nowhere approach the earth, it is every-where equally near to it,_ Bt. 33, 4 ; Fox 130, 22. II. _in no case, never :_—Ðás prepositiones ne beóþ náhwár ána, ac beóþ ǽfre tô sumum ôðrum worde gefêgede, Ælfc. Gr. 47 ; Som. 48, 50. Ne heard sweopu húse ðínum náhwǽr sceþþan [mágon], Ps. Th. 90, 10. Ðú mê náhwǽr forlǽte _thou didst never forsake me,_ Homl. Th. i. 74, 32. III. _in no respect, not at all :_—Eall moncynn and ealle nêtenu ne notigaþ náwér neáh feórþan dǽles ðisse eorþan _men and animals do not use anywhere near a fourth part of this earth,_ Bt. 18, 1 ; Fox 62, 8 : 18, 2 ; Fox 64, 6. Nese lá nese ne náwér neáh, Shrn. 106, 28. Ne trúige ús swá wel, ne náwér neáh swá ðám, 197, 13. [Cf. _Icel._ hvergi nær.]

ná-hwǽrn (?), -wérn; _adv._ _No-where :_ — Náwern _non usquam,_ Wrt. Voc. ii. 61, 9 : 95, 8. Cf. ǽgwêrn, Ors. Swt. 154, 22.

ná-hwæðer, náwðer, náðer, nóðer; _pron._ _Neither :_—Náðor _neuter, neutra, neutrum ;_ náðres _neutri ;_ náðrum _neutro,_ Ælfc. Gr. 18 ; Som. 21, 49. Getácnigende oððe sum þyngc tô dônne, oððe sum þingc tô þrowigenne, oððe náðor, 19 ; Som. 22, 23, 25. Náuðær næ sîe tô ðon gedurstig ne cyning næ bisceop ne nánes hádes man _nullus rex aut episcopus, vel_

aliquis alius potens, sit tam audax, Cod. Dip. Kmbl. v. 218, 26. Ne fornime nôðer ôðer ofer will _let neither of you deprive the other against his or her will,_ Past. 51 ; Swt. 399, 34. Hî gecýðaþ ðonne hié endiaþ ðæt hié náwðer ne bióþ, 16, 3 ; Fox 56, 27. Ða þing ða ðe náuðer ne sint, ne getrêwe tô habbenne ne eác êðe tô forlǽtanne, 7, 2 ; Fox 18, 15. Dydon swá hwæðer swá hý dydon ne dohte him náwðer whichever of the two they did, neither did them any good, Bt. 29, 2 ; Fox 106, 2 : Exon. Th. 12, 22 ; Cri. 189. His rihtwîsnys nolde hî neádian tô náðrum Homl. Th. i. 112, 3. Godes gelaþung nis búton náðrum ðæra (_the strong and the weak_), ii. 390, 29. Swá mîn sáwl bád ðæt ðú swylce heó for náhwæðer nôwiht hǽle _sicut expectavit anima mea, pro nihilo salvos facies eos,_ Ps. Th. 55, 6. Ðæt se yfela mǽge dôn yfel ðeáh hé gôd ne mǽge, and se deáda ne mǽge náuðer dôn, Bt. 36, 7 ; Fox 182, 25. Hié náðer (náwðer, Cott. MS.) ne mágon, ne ðîn helpan ne heora selfra, 14, 1 ; Fox 42, 9. Hié náðer næfdon siððan, ne heora namon ne heora anweald, Ors. 3, 1 ; Swt. 98, 7. Se ðe náðor nele, ne leornian ne tǽcan, Ælfc. Gr. pref. ; Som. 1, 34. v. next word.

ná-hwæðer, náwðer, náðor; _conj._ _Neither :_—Ðá ðá wê hit nô-hwæðer ne selfe ne lufodon ne eác ôðrum monnum ne lêfdon _when we neither loved it ourselves nor allowed it to other men,_ Past. 5, 6. Wê nôhwæðer ne hit witan nyllaþ ne hit bêtan nyllaþ, ne furðum ne rêcaþ hwæðer wê hit ongieten, 28; Swt. 195, 5. Náwðer ne ða wôhhǽmendan, ne ða ðe diófulgieldum þiówiaþ, ne ða unfæstrǽdan, ne ða þiófas, ne ða gîtseras, ne ða reáferas Godes rîce ne gesittaþ, 51 ; Swt. 401, 26. Nǽron náwðer ne on Fresisc gescæpene, ne on Denisc, Chr. 897 ; Erl. 95, 15 : Blickl. Homl. 45, 14. Lāreówas ne sceolan Godes dômas náwðer ne ná wanian, ne ne êcan, 81, 4. Hié náwðer ne him sylfum helpan ne mihton, ne nánum ðara ðe tô him áre wilnodan, 223, 2 : Bt. 29, 2 ; Fox 106, 5. Ðá nolde hé ásendan náðor ne engel, ne heáhengel, ne wîtegan, ne apostolas, Homl. Th. ii. 6, 15. v. preceding word.

ná-hwanon; _adv._ _From nowhere :_— Sió his gesǽlþ him náhwonan útane ne com, ac wæs simle on him selfum, Bt. 34, 7 ; Fox 144, 20.

ná-hwider; _adv._ _No-whither, to no place :_—Hý náhwider faraþ bútan ðæs abbodes rǽde, R. Ben. 137, 10.

nalas (-læs, -les), nalles. v. nealles.

nám, e; _f._ _Seizure of property belonging to one which is in the hands of another :_—Be naame. Ne nime nán man náne náme, ne innan scíre ne ût of scíre, ǽr man hæbbe þríwa on hundrede his rihtes gebeden ; (_but on the failure of legal means_) nime ðonne leáfe ðæt hé môte hentan æfter his ágenan, L. C. S. 19 ; Th. i. 386, 9–17. Cf. Nullus _namium_ capiat . . . accipiat licenciam _namium_ capiendi, L. W. I. 45 ; Th. i. 485, 13–17 : L. H. I. 29, 2 ; Th. i. 533, 7. Nulli sine judicio vel licencia _namiare_ liceat alium in suo vel alterius, 51, 3 ; Th. i. 550, 5. [Cf. _Icel._ land-nám in Norse law _an unlawful holding of another man's land,_ and hence a _fine_ for trespassing on another man's land ; in Icel. _the taking possession of land_ as a _settler_ : nes-nám _in phrase_ nema nesnám _to land on a ness and seize cattle :_ nám _a seizing by the mind, learning :_ O. H. Ger. náma ; nôt-náma _rapina._]

nama, an; _m._ I. _a name :_—Sumum men, ðam is Æþelm nama, Cod. Dip. Kmbl. ii. 383, 24. Wæs ðǽm hæftmêce Hrunting nama, Beo. Th. 2919; B. 1457. Ðǽre (eá) is Geon noma, Cd. Th. 15, 9 ; Gen. 230. _Ego hoc feci,_ ic dyde ðis, ðon stent se ic on ðînes naman stede, Ælfc. Gr. 5 ; Som. 3, 33. Naman _titulo,_ Hpt. Gl. 509, 4 : _vocabulo,_ 517, 61. Hé nemþ his ágene sceáp be naman _propias oves vocat nominatim,_ Jn. Skt. 10, 3. Be naman cígean, Ps. Th. 146, 4. Ðone ilcan wê hátaþ ôðre naman ǽfensteorra, Bt. 4; Fox 8, 3 : 33, 4 ; Fox 128, 27. Ðú nemdest eall mid áne noman, Met. 20, 56. Him se pápa Petrus tô noman scóp, Bd. 5, 7 ; S. 620, 43. God him sette naman Adam, Homl. Th. i. 12, 31. Hî him naman gesceópon, 92, 27. Hit ofetes noman ágan sceolde, Cd. Th. 44, 34; Gen. 719. II. _a noun_ = _Nomen_ is nama, mid ðam wê nemnaþ ealle þing . . . _Pronomen_ is ðæs naman speliend . . . _Amans_ lufigend cymþ of ðam worde _amo,_ ic lufige; ðon nymþ hé of ðam naman him ealle ða six casus, Ælfc. Gr. 5 ; Som. 3, 26–46. Sume synd ágene naman, swá swá is Eádgár, Dúnstán. Sume gemênelíce, kynincg, biscop, Som. 4, 10–11. [_Goth._ namô : _Icel._ nafn : _O. Sax._ namo : _O. Frs._ noma : _O. H. Ger._ namo.] v. freó-, heáh-nama.

nam-bóc; _f._ _A book in which names are written, a register :_—Nombéc _albo,_ Wrt. Voc. ii. 3, 1.

nam-bred, es; _n._ _A tablet on which names are written, a register :_—Nombred _albo,_ Wrt. Voc. ii. 81, 35.

nam-cúþ; _adj._ _Having the name well-known, celebrated, famous, of note, of renown :_—Nabochodonossor se namcúþa cining, Ælfc. T. Grn. 8, 15. Ælcre namcúþre wyrte dǽl _a bit of every well-known plant,_ Lchdm. i. 398, 9. Twegen sacerdas ðe ǽr on lîfe wǽron swîðe namcúþe, Homl. Th. ii. 342, 3. Heáhfæderas namcúþe weras (_the twelve patriarchs_), Ælfc. T. Grn. 5, 2 : R. Ben. 33, 20. On ðam gemótan ðeáh rǽdlíce wurðan on namcúðan stówan _in those assemblies, though advisedly they were made in places of note,_ L. Eth. ix. 37 ; Th. i. 348, 18. Se rîca biþ namcúðre on his leóde ðonne se þearfa _the name of the rich man is better known in his country than that of the poor man,_ Homl. Th. i. 330, 5. [_Sodome and Gomorre, and alle þe_ nomecuðe buruhwes (_famous cities_) A. R. 334, 25. Cf. _Icel._ nafn-kunnigr _famous._]

namcúþlíce; *adv. By name :*— Úre mǽþ nis ðæt wé ealle Godes gecorenan eów namcúðlíce gereccan *it is not within our power to recount to you by name all God's elect,* Homl. Th. ii. 72, 2. Hé gehwilce eardas namcúðlíce on gemynde hæfde, i. 558, 25. [Þurh him and ðurh ealle his freónd namcúðlíce, Chr. 1127; Erl. 256, 12.]

namian; *p.* ode. **I.** *to name, mention the name of, mention :*— Git ðú namast Crist *dost thou still name the name of Christ?* Homl. Skt. 8, 165. Ða twá tabelan getácnodon ða twá bebodu ðe ic nú namode, Homl. Th. ii. 204, 21. On ðære ylcan byrig ðe wé ǽr namodon, 296, 32. Namedon, Ælfc. Gr. 8; Som. 7, 7. Ðæt ðú nánne brýdguman nǽfre mé ne namige *that you never mention the name of any bridegroom to me,* Homl. Skt. 9, 37. Ðeáh ðe wé ðás sinderlíce namian *though we mention the names of these in particular,* Homl. Th. ii. 432, 23. **II.** *to name, appoint by name to a particular duty, nominate :*— Gif hé ne mehte, ðonne namede him man six men, L. Ed. 1; Th. i. 158, 21. Beforan his witum ðe se cyng silf namode, L. Ædelst. v. 10; Th. i. 240, 6. **III.** *to name, give a name to :*— Hwí namode Crist Abel rihtwísne? Boutr. Scrd. 18, 6. [*O. Frs.* nomia : *O. H. Ger.* namón.] v. ge-namian.

nam-mǽlum; *adv. Name by name :*—Nammǽlum *nominatim, per singula nomina,* Hpt. Gl. 427, 28.

namnian; *p.* ode *To name, call by name :*— Se namnode ðone Hǽlend be his naman, Ælfc. T. Grn. 10, 16. [*O. Frs.* namna, nanna.] v. nemnan.

nám-rǽden[n], e; *f. Learning, erudition :*—Námrǽdenne *litterature,* Wrt. Voc. ii. 50, 19. [*Cf. Icel.* nám *learning, study.*]

nán [=ne án], *pron.* **I.** as adjective, *not one, none, no,* (a) without other negatives :—Nán mǽrra man wurde ácenned, Menol. Fox 319; Men. 161. Hit is nánum men getiohhod ac is eallum monnum *it is not intended for one man, but for all men,* Bt. 37, 2; Fox 188, 15. (b) with other negatives :—Ne nán heora án nis ná lǽsse ðonne eall seó þrynnys *and no one of them is less than all thé Trinity,* Homl. Th. i. 284, 1. Nán heort ne onscunode nǽnne león, ne nán hara nǽnne hund, ne nán neát nyste nǽnne andan ne nǽnne ege tó óðrum, 35, 6; Fox 168, 9–11. Nán swylc ne cwom ... brýd, Exon. Th. 18, 28; Cri. 290. Swá nán óðer ná déþ mónaþ, Menol. Fox 392; Men. 197. Nán þing ðæs folces belyfen nǽs *there was nothing left of the people,* St. And. 34, 13. Nǽs ðæt nán þing wundor ðæt ... *it was no wonder that,* Deut. i. 37. Ða cild ðe niton nánes þinges nán gesceád ne gódes ne yfeles, i, 39. Seó leáse wyrd ne mæg ðam men dón nǽnne dem, forðam heó nis nánes lofes wyrðe, Bt. 20; Fox 70, 22–24. Ne cyning næ bisceop ne nánes hádes man *nullus rex aut episcopus, vel aliquis alius potens,* Cod. Dip. Kmbl. v. 218, 28. Hié nǽfre tó nánum men ne becumaþ, Bt. 11, 1; Fox 30, 27. Hé on náne wísan ne mæg forbúgan *he can in no wise avoid,* 16, 2; Fox 54, 5. Ða ne mihton hig him nán word andswarian ne nán ne dorste hyne nán þing máre áxigean, Mt. Kmbl. 22, 46. Ne sǽdon hyt mé náne swá sóðfeste men, Shrn. 204, 22. **II.** as predicate :—Forhwí ðé hátan dysige men wuldor nú ðú náne eart (nán neart, MS. Cott.) *why do foolish men call thee glory, when thou art none,* Bt. 30, 1; Fox 108, 3. **III.** as substantive, (a) absolutely, *none, no man, nothing :*—Nán mihtigra ðé nis, ne nán ðín gelíca, 33, 4; Fox 128, 11. Nán in nearowe néþan móste, Exon. Th. 436, 12; Rä. 54, 13. Ðam wæs nán tó gedále, Cd. Th. 84, 20; Gen. 1400. Hé nolde nǽnne forlǽtan ðe him fylgian wolde, Hy. Grn. 10. 38. Deáþ náne forlét, Met. 10, 66. (b) with partitive genitive :—Næs ǽnig engel geworden, ne ðæs miclan mægen-þrymmes nán, Exon. Th. 22, 17; Cri. 352. Him ne mæg ealdfeónda nán átre sceþþan, 229, 2; Ph. 449. Gúþbilla nán, Beo. Th. 1610; B. 803: 1980; B. 988. Næs heora neáta nán geyfelad *jumenta eorum non sunt minorata,* Ps. Th. 106, 37. Ne þearf hæleþa nán wénan, Met. 7, 6. Hwæt wille gé cueðan hwæs oððe hwæs gé síen? gé habbaþ gecýðed ðæt gé úres nánes (= úre nánes) ne siendon *quid vos hujus vel illius dicitis, qui nullius vos esse monstratis?* Past. 32, 1; Swt. 211, 14. Nánne ne sparedon ðæs herefolces, Judth. 11; Thw. 24, 40; Jud. 233. Náne þinga beór ne drince *on no account let him drink beer,* Lchdm. ii. 88, 10. [*Icel.* neinn : cf. *O. Sax.* nén : *O. H. Ger.* nein (*particle of negation*).] v. nǽnig.

nán-wiht, nán-uht. **I.** as subst. *nothing :*—Nánwiht *nihil,* Wrt. Voc. i. 47, 32. Heó hire self gecýþ ðæt heó nánwuht ne biþ *she herself shews that she is nothing,* Bt. 20; Fox 70, 24. Ðú wéndest ðæt ðé nánwuht unrihtlíces on becuman ne mihte, 7, 3; Fox 22, 15 : 16, 3; Fox 56, 31 : 38, 2; Fox 198, 6. Ðæt gecynd nyle nǽfre nánwuht wiðerweardes lǽtan gemengan, 16, 3; Fox 54, 36. Hió nánwuht elles ne lufaþ bútan ðé, 10; Fox 28, 24. Hé nánwuht ealles (*nothing at all*) næbbe ymbe tó sorgienne, 11, 1; Fox 32, 12. Hié hiora nánwuht ongiotan ne meahton, Past. Swt. 5, 22. Ðone ðe ðú nánwiht yfles on nystest, Blickl. Homl. 85, 36. Nánuht berendes, ne wíf ne niéten, ne mehton nánuht libbendes geberan, Ors. 4, 1; Swt. 158, 18. Nánuht ágiefan nolde ðæs ðe hié béna wǽron, 3, 11; Swt. 146, 35. Gé nánuht nabbaþ fæstes ne stronges, 2, 4; Swt. 74, 28. *The Northern gospels have* nǽniht (*from* nǽnig?) :—Nǽniht unmæht biþ *nihil impossibile erit,* Mt. Kmbl. Lind. 17, 20. Tó nówihte ł nǽnihte *ad nihilum,* 5, 13. Bibeód him ðæt nǽniht (ne ǽniht, Lind.)

hiǽ gilǽdde on woeg *praecepit ne quid tollerent in via,* Mk. Skt. Rush. 6, 8. Nǽneht ł ne óht (nǽniht ł nóht, Rush.) *nullam,* Lk. Skt. Lind. 23, 22. **II.** as an adverb, *nothing, not at all, no whit :*—Hé his godcundnesse nánwiht ne gewanode *he no whit diminished his divinity,* Blickl. Homl. 91, 9. Ne gefyrenodest ðú nánwuht *thou hast done no sin,* 235, 34. v. ná-wiht.

náþan. v. ge-náþan *and* níþan.

nard, es; *m. Spikenard* ; *nardus :*—Sealfbox deórwyrþes nardes *alabastrum ungenti nardi spicati praetiosi,* Mk. Skt. 14, 3. Nardys, Lchdm. i. 184, 19. Ete nardes eár, 354, 12. Ele ðe sý of nardo, 246, 20. Nardes stenc, Exon. Th. 423, 28; Rä. 41, 29. [*Goth.* nardus : *O. H. Ger.* narda. *From Latin.*]

naru. v. ealdor-, feorh-, líf-naru.

nást. v. nytan.

nasu; *f. The nose :*—Nasu *naris* ; eall seó nasu *columpna* ; forewerd nasu *pirula,* Wrt. Voc. i. 282, 63–65. Gif man þyrel weorþ, L. Ethb. 45; Th. i. 14, 10 : 48; Th. i. 14, 13. Gif man óðerne mid fyste in naso slæhþ, 57; Th. i. 16, 17. [*Icel.* nös : *O. H. Ger.* nasa.] v. nosu.

nát. v. nytan.

nátes-hwón; *adv. Not at all, by no means :*—Haud, adverbium, ðæt is on Englisc nátes-hwón, Ælfc. Gr. 50, 16; Som. 51, 25. Náteshwón *haud, minime, nullatenus,* 38; Som. 40, 13–15 : *nequaquam,* Som. 41, 55 : *nequaquam, nullo modo,* Hpt. Gl. 433, 60 : *haud,* 466, 70 : *minime,* 470, 24. Ne eart ðú náteshwón wacost burga *thou art by no means least of towns,* Homl. Th. i. 78, 14. Ne mæg ic náteshwón búton mynstre nihtes wunian, ii. 182, 33 : 80, 16. Sume teolunga sind ðe man earfoþlíce mæg oððe náteshwón (*hardly or not at all*) búton synnum begán, 288, 22 : Homl. Skt. 7, 104. Hé ne áwyrpþ náteshwón his wǽpna him fram, ǽr ðam ðe ðæt gewinn wurðe geendod, Basil admn. 2 ; Norm. 36, 9. Ne sculon mæssepreóstas náteshwón bútan óðrum mannum mæssan syngan, L. E. I. 7; Th. ii. 406, 21. v. ná-wiht.

náðor. v. ná-hwæðer.

nát-hwǽr; *adv. In some place unknown,* Exon. Th. 480, 8; Rä. 63, 8 : 407, 14; Rä. 26, 5.

nát-hwæt; *pron. indef. Something unknown :*—Rúwes náthwæt *something rough, but what I know not,* Exon. Th. 479, 17; Rä. 62, 9 : 436, 23; Rä. 55, 5 : 499, 25; Rä. 88, 21.

nát-hwilc; *pron. indef. Some one, I know not who :*—Hæleþa náthwylc *some man, I know not who,* Elen. Kmbl. 146; El. 73. Hé, gumena náthwylc, Beo. Th. 4459; B. 2233. Ðara banena byre náthwylces *the child of one of those murderers, but I know not of which,* 4113; B. 2053: 4451; B. 2224. Þurh náthwylces ..., Exon. Th. 12, 21; Cri. 189. Hé in níþsele náthwylcum wæs *in some unknown hall was he,* Beo. Th. 3031; B. 1513. (Cf. sceaða ic nát hwylc, 554; B. 274.)

nául. v. ná-wiht.

náwa [=ne áwa]; *adv. Never :*—Ðæt is swíðe strang ðam ðe ðæt náwa ǽr þigde *it is very strong for him who never before tasted it,* Lchdm. ii. 252, 14. [*Cf. O. Sax. O. H. Ger.* néo : *Goth.* ni aiw.] v. áwa.

ná-wérn. v. ná-hwǽrn.

ná-wiht, nó-wiht, ná-uht, náwht, náht, nóht. **I.** as subst. *with gen.* es; *n.* (a) *nothing, naught, a thing of no value, an evil thing :*—Is tó cýðanne hwelc nánwuht (náuht, Cott. MSS.) ðes woruldgielp is *intimandum est, quam sit nulla temporalis gloria,* Past. 41, 1; Swt. 299, 6. Náwuhtes cearu ofer ða ryhtwísnesse *care for nothing besides righteousness,* Swt. 302, 9. Ðú hí miht tó náwihte (*ad nihilum*) forniman, Ps. Th. 72, 16, 17 : 107, 12. Spoede míne swé swé nówiht beforan ðé biþ *substantia mea tanquam nihil ante te est,* Ps. Surt. 38, 6. Tó nówihte, 14, 4 : 80, 15. Fore nówihte *pro nihilo,* 55, 8. Hé nówiht ne fremede *nec ipse aliquid profecisset,* Bd. 5, 9 tit. ; S. 622, 6. Hé nówiht elles ne dyde, 2, 14; S. 518, 8. Yfel is náuht. Ðǽr yfel áuht wǽre, ðonne mihte hit God wyrcan. For ðý hit is náuht, Bt. 35, 5 ; Fox 164, 10–11. Heore þinyt eall náuht (nóht, Cott. MS.) ðæt heó hæfþ, 10; Fox 28, 28. Hú ne is se anweald ðǽr náuht? 16, 2; Fox 54, 7. Hú ne wást ðú ðæt hit nis náuht gecynde ne náuht gewunelíc ðæt ǽnig wiðerweard þing bión gemenged wið óðrum wiðerweardum *do you not know that it is not a natural or usual thing, for contraries to be mingled with other contraries,* 16, 3; Fox 54, 11. Ne eart ðú nó eallunga tó náuhte gedón *thou art not altogether brought to naught,* 10; Fox 30, 4. Weorðan tó náuhte *to come to naught,* Met. 11, 87. For náuht tó habbene *to be considered worthless,* Bt. 30, 1; Fox 108, 17. Mon ongiet mid hwelcum stæpum ðæt náwht (náuht, Cott. MSS.) wæs þurhtogen *quibus vestigiis nequitia sit perpetrata,* Past. 35, 3; Swt. 241, 18. Náht *nichil,* Wrt. Voc. i. 83, 68 : *nihili,* 47, 33 ; *nihil,* Ælfc. Gr. 9, 8; Som. 9, 13. Náht mé wana biþ *nihil mihi deerit,* Ps. Spl. 22, 1. Nis ðæs mannes fæsten náht, ðe hine sylfne on forhæfednysse dagum fordrencþ, Homl. Th. ii. 608, 23. Heora dýre gold ne biþ náhte wurð wið ða foresǽdan mádmas *their precious gold will be worth nothing in comparison with the aforesaid treasures,* Glostr. Frg. 2, 29. Tó náhte *ad nihilum,* Ps. Spl. 14, 5 : Ps. Th. 59, 11. On náhte nyt ne biþ *it is to no purpose,* Blickl. Homl. 57, 5. Hig tellaþ mín wedd for náht *irritum facient pactum meum,* Deut. 31, 20. For náhtum *pro nihilo,* Ps. Lamb. 80, 15. Ungeleáfsumum nóht biþ ciǽne *infidelibus*

nihil est mundum, Bd. 1, 27; S. 494, 40. Mon nôhtes wyrþe his sâule ne dêþ ne his goldes ne his seolfres *a man does not make his soul worthy of any-thing, of his gold or of his silver*, Blickl. Homl. 195, 5. Næfdon heó nôht on hire, búton dæt ân dæt heó hæfde mennisce onlícnesse, 147, 15. Ne forstent hit him nôht, ne him nôhte don mâ ne beóþ forlætna his âgna synna, Past. 21; Swt. 163, 19. ¶ genitive used as predicate :—Ða sælþa de hê ǽr wênde dæt gesælþa beón sceoldan, nâuhtas næran (*were worthless*), Bt. 10. tit.; Fox xii. 6. Eówer godas ne synd nâhtes, Homl. Skt. 7, 205. (b) with a genitive :—Eallinga nâwiht mægenes hæfeþ seó ǽfæstnys *nihil omnino virtutis habet religio illa*, Bd. 2, 13; S. 516, 3. Ic dæs nôwiht wât, Exon. Th. 393, 5; Rä, 12, 5. Ealles nâuht *nothing at all*, Bt. 36, 6; Fox 182, 8. Nâuht elles *nothing else*, 3, 2; Fox 6, 11. Hê ne mæg ûtane nâuht âgnes habban, 27, 2; Fox 98, 8. Ðes nâht yfeles ne dyde *hic nihil mali gessit*, Lk. Skt. 23, 41. Nâht elles búton *nothing but*, Blickl. Homl. 215, 3. Nôht elles ne wunaþ, búton dæt ân, 101, 4. Gif wê yfles nôht gedôn habbaþ, Exon. Th. 262, 8; Jul. 329. II. as an adverb, *not* :—Hit gelamp neâht micelre tîde æfter his slæge (*non multo exacto tempore*), Bd. 3, 9; S. 533, 30. Nâht feor eást, 2, 13; S. 517, 15 : Blickl. Homl. 43, 26. Mannum de nâht swîde God ne lufiaþ, 53, 18: Wrt. Voc. ii. 55, 27. Ic wât dæt dû nâht (âuht, Cott. MSS.) ne forslâwodest, Bt. 10; Fox 28, 15. Heó nôht lata ne wæs, Blickl. Homl. 163, 8. Ne þurfan gê nôht besorgian hwæt gê sprecan, 171, 18. Ne wæs hê nôht feor on oferhygd âhafen, 215, 32. Nôht longe ofer dis, Exon. Th. 172, 15; Gú. 1144. Æfter nôht langre tîde, Bd. 5, 11; S. 626, 10. Ic nôht don de dære ærninge blon, 5, 6; S. 619, 15. [O. Frs. nâ-wet: O. Sax. O. H. Ger. neô-wiht.] v. nâtes-hwôn *and following words*.

nâwiht-, nâht-fremmend, es ; *m. One who does evil* :—Genere mê fram nîde nâht-fremmendra *eripe me de operantibus iniquitatem*, Ps. Th. 58, 2.

nâwiht-, nâwht-, nâuht-gîtsung, e ; *f. Wicked avarice* :—Ðonne hié wilniaþ þurh da nâwhtgîtsunga (nâuhtgîdsunga, Cott. MSS.) dæt hié hira woruldspêda îcean *dum per avaritiae nequitiam multiplicari appetunt*, Past. 44, 10; Swt. 333, 5.

nâwiht-, nâht-lîc ; *adj. Good for nothing, worthless, naughty* :—Seó hæfde nigon dohtra, nâhtlîce and fracode, Homl. Skt. 8, 11. Manna rǽdas syndon nâhtlîce ongeán Godes gepeaht *men's plans are of no avail against God's counsel*, Chr. 979; Erl. 129, 27. Ða hê geceás de dyse-lîce and nahtlîce gepuhte synt *he chose those that seemed foolish and of no account*, R. Ben. 138, 30. v. next word.

nâwiht-, nôht-lîce ; *adv. Worthlessly, evilly* :—Ðætte nôhtlîce dû dôe *ut nequiter facias*, Ps. Surt. 36, 8, 9.

nâwiht-, nâht-ness, e ; *f. Worthlessness, cowardice* :—Heom seggan Brytwalana nâhtnesse (MS. E. nâhtscipe. Cf. secgan Brytta yrgþo (*segnitia*), Bd. 1, 15; S. 483, 15), Chr. 449; Erl. 12, 6.

nâwiht-, nâuht-wela, an; *m. False wealth, wealth that is not really wealth* :—Gê wênaþ dæt eówre nâuhtwelan (nôht-, Cott. MS.) sîen eówra gesælþa, Bt. 14, 2; Fox 44, 37.

nâwder. v. nâ-hwæder.

Nazarenisc, Nazaresc; *adj. Of Nazareth* :—Se Nazareniscea (Nazaresca, Lind.) Hælend, Mk. Skt. 10, 47: 14, 67. Ðone Nazareniscean (Natzarenisca, Lind.) Hælend, Jn. Skt. 18, 5.

ne. I. *adv.* (a) *Not;* non, ne :—Ic ne dyde *non feci*, Ælfc. Gr. 38; Som. 40, 13. Nis hit swâ hit nys *non, non*, 40, 23. Warna dæt dû dæt ne dô *cave ne hoc facias*, 40, 9. Hwî forbeád God eów, dæt gê ne ǽton of ǽlcum treówe? Gen. 3, 1. Hî nyllaþ geswîcan dæt hî ôdre men ne reáfigen, Past. 45, tit; Swt. 335, 4. Ne gǽst dû danone *non exies inde*, Lk. Skt. 12, 59. Ne sleh ðu. Ne synga ðû. Ne stel ðû, Ex. 20, 13–15. (b) *no, nay* :—Ne secge ic eów *I tell you, Nay*, Lk. Skt. 12, 51: 13, 5. Ne secge ic nâ, 13, 3. II. *conj. Nor, neither;* ne, neque, nec :—Ne tunge ne handa odde eágan syngion *ne lingua nec manus oculive peccent*, Ælfc. Gr. 44; Som. 45, 47. Ne ic ne herige ne ic ne tǽle *nec laudo, nec vitupero*, 45, 49. Ne ic ne dyde ne ic ne dô *neque feci, neque faciam*, 38; Som. 44, 9. Ne fare gê ne ne fyliaþ, Lk. Skt. 17, 23. Ne hig ne cwedaþ *neque dicent*, 17, 21. Ne him eác næfre genôg ne þincþ ǽr hê hæbbe eall dæt hine lyst, Bt. 33, 2; Fox 124, 6. Suelcum ingeponce gerîst dæt hê for lícuman tiedernesse ne for woroldbismere ânum wid da scîre ne winne, ne hê for sie giétsiende odˈerra monna ǽhta, Past. 10; Swt. 61, 9–11. Ða de nôhwæder ne odˈerra monna ne wilniaþ, ne hiora âgen nyllaþ sellan, Past. 45, tit.; Swt. 335, 1. *The word often occurs with other negatives.* v. nâ, nâ-wiht, nâ-hwæder; *it also coalesces with many words beginning with a vowel, with* w *or with* h. v. nabban, nâgan, næs, neom, nic, nyllan, nytan ; nân, nâ-, nât-, næfre, nænig. [*Goth*. ni: O. Sax. ne, ni: O. Frs. ne : O. H. Ger. ni.]

nê-, neâ-, neáh-, neô-, neáh-, nîd.

neádian; *p.* ode (v. nîd, VI) *To force, compel, constrain* :—Neádaþ forlǽtan *cogit intermittere*, Hymn. Surt. 56, 13 : 84, 17. Útlagan ûs wêpan neádiaþ *exules nos flere cogunt*, 56, 3. Se de ôderne neádaþ ofer his mihte tô drincenne, Ælfc. T. Grn. 21, 31. God hine ne neádode on nâdre healfe, ac lêt hine habban his âgene cyre, Hexam. 15; Norm. 22, 30. Ne neádige hine man tô fæstene *ne cogatur ad jejunium*, L. Ecg. P. iv. 25; Th. ii. 212, 5: L. Ælfc. C. 29; Th. ii. 352, 29. Neádede

cogeret, Hpt. Gl. 519, 19. Neádiendum *cogente*, 503, 39. His deópe rihtwîsnys nolde hî neádian tô nâdrum, Homl. Th. i. 112, 3. v. ge-, of-neádian, nîdan.

neádian, neódian ; *p.* ode (v. nîd, IV) *To be necessary* :—On cealdum eardum neódaþ (is neód, W. F.), dæt dæs reáfes mâre sý, on hleówfæstum læs. Ðæs abbodes forsceáwung sceal beón be dysum, hû dæs neódige, R. Ben. 89, 6, 8.

neádigness, e ; *f. Obligation* :—Neádinysse t neóde *debitum*, Hpt. Gl. 456, 14.

neádlunga; *adv. Forcibly, against one's will* :—Manega gewilniaþ ôdres mannes wôlîce and hî beóþ benǽmede neádlunga hyra âgenes *many covet another man's goods, and they shall be forcibly deprived of their own*, Basil admn. 9; Norm. 52, 20. v. nîdlinga *and next two words*.

neádung, e ; *f. Force* or *violence used against any one, compulsion, necessity* :—Ðeós neádung *haec vis*, Ælfc. Gl. 9, 29 ; Som. 11, 62. Of disum leahtre (gîtsung) beóþ âcennede reáflâc, stala, unmǽþlic neádung, Homl. Th. ii. 220, 12. Hê nolde geniman ûs neádunge of deófles anwealde, i. 26, 30. Hine betellan swilce hê neádunge gefremode dæt fâcn *to excuse himself, as if he committed that crime of necessity*, H. R. 105, 26. Neádunge *vim*, Hpt. Gl. 435, 70. [*Icel*. nauðung *compulsion*.] v. next word.

neádunga(-inga) ; *adv. Forcibly, not willingly, under compulsion, of necessity* :—Hê nolde niman mancyn neádunga of dam deófle búton hê hit forwyrhte *he would not have taken mankind by force from the devil, unless he had forfeited it*, Homl. Th. i. 216, 5. Ðone cniht de hê neádinga genam (*rapuisset*), Ors. 1, 8; Swt. 42, 10. Hî hine neádunga mid him lǽddon *invitum duxerunt*, 3, 18; S. 546, 22. Gif lǽweda man neádinga (*invite*) man ofsleá, L. Ecg. P. ii. 1; Th. ii. 182, 16. Neádunga, L. M. I. P. 6; Th. ii. 266, 27. Gif hê (*man*) wǽre neádunga (*without power of choice, necessarily*) Gode underþeódd, donne næfde hê nân wuldor for gôdum weorcum, Boutr. Scrd. 17, 26. Sió leáse gesælþ tîhþ on lâst neádinga (*inevitably*) da de hiere tô gepeódaþ from dǽm sôdum gesælþum; seó widerweardnes full oft ealle da de hiere underþeódde bióþ, neádinga getîhþ tô dǽm sôdum gesælþum, swâ swâ mid angle fisc gefangen biþ, Bt. 20; Fox 72, 7–11. v. nîdinga *and preceding word*.

neáh ; *adj.* I. *nigh, near* :—On dam neáhgum mynstre [neáh-nunnan-mynstre] *de vicino virginum monasterio*, Bd. 4, 1 ; S. 564, 4 note. Neágum *proximis* (cf. O. L. Ger. nâan *proximum*), Germ. 399, 409. Seó ûs neárre Ægyptus, Ors. 1, 1 ; Swt. 14, 3. Seó ûs neárre Ispania, Swt. 22, 31 : 24, 9. Sîe se lâreów eallum monnum se niéhsta and eallum monnum enþrowiende on hira gesuincum *sit rector singulis compassione proximus*, Past. 16; Swt. 97, 22. Seó mǽgþ is seó nýhste on sûþhalfe Humbre streámes *provincia (Lindissi) quae est prima ad meridianam Humbrae fluminis ripam*, Bd. 2, 16; S. 519, 19. Nîhsta *proxima*, Ps. Spl. 21, 10. Sió nêste hond *the nearest relative*, Chart. Th. 481, 21. Gif hwylc man wîfige on his nêhstan mâgan (*proximam cognatam*), L. Ecg. P. ii. 18; Th. ii. 188, 16. Hiera niéhstan friénd, Past. 49 ; Swt. 377, 1. Heora nýhstan mâgas, L. Eth. ii. 6; Th. i. 286, 32. II. in cpve. *later, latter* : superl. *last, latest.* v. ende-nêhst :—Se æftera t nærra *novissimus*, Mt. Kmbl. Rush. 21, 31. His da nêrran tîde wǽron wyrsan dǽm ǽrran *habuit posteriora pejora prioribus*, Bd. 2, 15; S. 518, 31. Cedd and Adda and Bete and Dema, se nýhsta wæs Scyttysces cynnes, 3, 21 ; S. 551, 15. Dis is Byrhtrîces nîhsta cwide (*last will*), Chart. Th. 500, 24. Oder is se ǽresta apostol, ôder se nêhsta, Blickl. Homl. 171, 9. On dǽm nêhstan dæge *on the last day*, 21, 35. On da nêhstan tîd disse worlde, on dômes dæge, 123, 32. Oþ da nýhstan orþuncge *until his latest breath*, L. Ælfc. E.; Th. ii. 392, 9. From Ninuse hiora ǽrestan cyninge ôþ Sardanapolim heora nîhstan, Ors. 6, 1; Swt. 252, 8. Be dǽm neáhstan twâm is æfter tô cweþanne *de ultimis infra dicendum est*, Bd. 4, 23 ; S. 594, 12. Monige beóþan da ǽrestu nêhstu and da nêhstu ǽrestu, Mt. Kmbl. Rush. 19, 30. From dǽm nêhstum ôþ de ǽrestu m, 20, 8. Æt nêhstan *postremo*, Bd. 2, 6 ; S. 508, 19. Æt nêhstan (Rush. nîhsto) *novissime*, Mt. Kmbl. 25, 11: Blickl. Homl. 85, 1. Æt nêxtan, Homl. Th. i. 66, 23. Æt niéhstan, Cd. Th. 84, 19 ; Gen. 1400. Æt nîhstan, Ors. 4, 9; Swt. 192, 35. Æt nýhstan, Bd. 2, 12; S. 513, 4. Æt nýxtan, Chr. 994 ; Erl. 133, 20 : 1010; Erl. 144, 9. v. nîhsta *and next word*.

neáh, nêh *nigh, near.* I. *as adv.* (1) *of place* :—Ealle hîre mâgas da de dǽr neáh wǽron, Blickl. Homl. 139, 16. Ic wât heáhburh hêr âne neáh, Cd. Th. 152, 9; Gen. 2517. Feor odde neáh, 63, 8; Gen. 1029. Ge neáh ge feor, Bd. 4, 4 ; S. 571, 7. Ge nêh ge feor, Andr. Kmbl. 1083; An. 542. Gâ hider neár *accede huc*, Gen. 27, 21. Mid dý ic dâ wolde neár geseón *quos cum adire vellemus vicinius*, Nar. 22, 11. Swa fyr swâ nŷr, L. I. P. 21 ; Th. ii. 332, 16. Ðær dǽr hê niéhst rŷmet hæfde, Chr. 894; Erl. 90, 9. (2) *of time* :—Disses middangeardes ende (*or dat.?*) swîde neáh is, Blickl. Homl. 107, 23. Eall dâs getimbro neáh is dæt is eall fŷr fornimeþ and on axsan gehwyrfeþ *cuncta haec aedificia, in proximo est ut ignis absumens in cinerem convertat*, Bd. 4, 25; S. 600, 33. Nemnan dæt ûs neáh (*lately*) geweard gecŷþed, Exon. Th. 107, 26; Gú. 64. Ðâ ic hine nêhst geseáh *when I last saw him*, Cd.

Th. 34, 12; Gen. 536. Ðonan hý God nýhst eágum segun, Exon. Th. 34. 1; Cri. 535. (3) of degree, *near, nearly, about :*—Heó hafaþ leáf neáh swylce mistel, Lchdm. i. 254, 12. Ða Finnas and ða Beormas sprǽcon neáh án geþeóde, Ors. 1, 1; Swt. 17, 34. Hié æt níhstan hæfdon ealra ðara anwald ðe ǽr nēh heora hæfdon *in the end they had dominion over all those who before nearly had dominion over them,* 4, 9; Swt. 192, 35. Swā neáh wæs þúsend áurnen *so nearly had a thousand years passed (all but twenty-seven),* Chr. 973; Erl. 124, 23. Swīðe neáh ðū ongeáte ðæt riht, Bt. 34, 12; Fox 154, 10. Nihtscūwan neáh ne mihton (*could not nearly*) heolstor āhȳdan, Col. Th. 184, 29; Exod. 114. Hī ne notigaþ nāwēr neáh feórþan dǽles ðisse eorþan *they do not use anywhere near a fourth part of the earth,* 18, 1; Fox 62, 9. Ðæt gē dōn ne māgon, ne furþum nāwēr neáh, 18, 2; Fox 64, 6: Shrn. 196, 28: 197, 13. Ne mæg hió ðeáh gescīnan āhwǽrgen neáh ealla gesceafta *the sun cannot reach with its rays anywhere near all creatures,* Met. 30, 10. Ūs is þearf ðæt wē geþencen hwæt Dauid cwæþ and eác ðon dōn swā wē nýhst mǽgan *we must consider what David said, and besides that act as nearly as we can accordingly,* L. E. I. 30; Th. ii. 426, 38. **II.** *as prep. with dat.* (1) *of place :*—Neáh helle *secus infernum,* Ps. Spl. 140, 9. Neáh [Lind. Rush. nēh] ðam tūne *juxta praedium,* Jn. Skt. 4, 5. Seó flōweþ neáh ðǽre ceastre wealle, Bd. 1, 7; S. 478, 5. Ðām wæs engel neáh, Exon. Th. 112, 14; Gū. 143. Ðæt nān ne sǽte hiere x mīlum neáh *that no one should settle within ten miles of it (Carthage),* Ors. 4, 13; Swt. 210, 22. Tō ðæm tūne nēh Oliuetes dūne, Blickl. Homl. 69, 33. Gang ne neár hider *come hither nearer to me,* 179, 30. Seó sió Ægyptus ðe ūs neár is, Ors. 1, 1; Swt. 12, 16. Swā hē biþ ðære sunnan neár swā biþ hire fyrr *whether it (the moon) is nearer to the sun or farther from it,* Shrn. 64, 32. Swā neár ende ðyssere worulde swā māre ehtnys ðæs deófles *the nearer to the end of the world, the greater the devil's persecution,* Homl. Th. ii. 370, 15. Ne biþ hió merestreáme ðē neár ðe oṅ midne dæg, Met. 28, 37. Ðā se swēg mē nýr wæs, Bd. 5, 12; S. 628, 31. Swā swā sió nafu fērþ nēhst (Cott. MS. neáhst) ðære eaxe, Bt. 39, 7; Fox 222, 1. Sceall beón se lǽsta dǽl nýhst ðæm tūne, Ors. 1, 1; Swt. 20, 33. (2) *of time :*—Ne ðīnre forþfore swā neáh is *neque mori adhuc habes,* Bd. 4, 24; S. 598, 37, 32: 3, 8; S. 531, 36. Ðǽre tȳde is neáh, ðæt Godes cyrce hafaþ sybbe on eorþan, Shrn. 154, 33. Biþ nēh ðæm seofoþan dæge, Blickl. Homl. 95, 11. Hié wēndon ðæt hit neár worulde endunge wǽre ðonne hit wǽre, Past. 32; Swt. 213, 6. (3) *of manner :*—Ic dō neáh ðam ðe ðū cwǽde *juxta verbum tuum faciam,* Ex. 8, 10. Neáh (*juxta*) eallon ðām þingum, ðe Drihten bebeád, Num. 1, 54: 8, 20. Neáh andefene *prope modum,* Wrt. Voc. ii. 66, 73. Āgnung biþ nēr ðam ðe hæfþ ðonne ðam ðe æfter sprecþ *possession is nine points of the law,* L. Eth. ii. 9; Th. i. 290, 20. Ūs sylfe gerihtlǽcan swā neáh swā wē nýhst māgon ðam rihte *to direct ourselves as much according to right as we possibly can,* Chart. Th. 615, 24. [**Goth.** nēhw; *adv.*; *cpve.* nēhwis: *Icel.* nā- (*in cpds.*); nær; *adv.* (*pos. and cpve.*): *O. Sax.* nāh; *adv. prep. with dat.* (*cpve.* nāhor: *O. Frs.* nī, nei; *adv.: O. H. Ger.* nāh; *adv. prep. with dat.*; nāh; *adj. contiguus, vicinus.*] v. efen- (emn-), uṅ-neáh, *the preceding word, the cpds. with* neáh-, *and* neáwung.

-neah. v. be-, ge-neah.

neáh-būend, es; *m. A near-dweller, a neighbour :* — Ic eom neáhbúendum nyt, Exon. Th. 407, 8; Rä. 26, 2.

neáh-búr. v. neáh-gebūr.

neáh-dūn, e; *f. A neighbouring hill :* — Of ðæm neáhdūnum and scrafum *ex vicinis montium speluncis,* Nar. 14, 6.

neáh-eá; *f. A neighbouring river :* — Hié of ðæm neáhéum and merum ða hronfiscas up tugon and ða áeton, Nar. 22, 9.

neáh-eáland, es; *n. A neighbouring island :* — On ðysum neáheálande ðæt is nemned Ulcani, Shrn. 86, 1.

neáh-freónd, es; *m. A near friend or relation :*—Ðǽr wæs mycel menigo manna gegaderod his mága and eác óðra his nēhfreónda, Guthl. 12; Gdwin. 56, 22. [*Icel.* nā-frændi *a near kinsman.*]

neáh-, nēh-gebūr, nēhche-, nēhhe-, nēche-, nēhe-būr, es; *m. A neighbour :*—Nēhgebūr *adfinis,* Wrt. Voc. ii. 9, 68: *convical,* 135, 56. Mīne frýnd and mīne mágas and mīne neáhgebūras *amici mei, et proximi mei,* Ps. Th. 37, 11. His neáhgebūras (nēhebūras, Lind.: nēhgibūras, Rush.) *vicini,* Jn. Skt. 9, 8. Hyre nēhchebūras (nēhhebūras, MS. A.: nēhebūras, Lind.: nēhgibūras, Rush.), Lk. Skt. 1, 58, 65. Nēhhebūras, 14, 12: 15, 6. Neapolite ða heora nēhgebūras, Blickl. Homl. 201, 19. Nēchebūrena gefeoht *intestinum bellum,* Wrt. Voc. i. 35, 16. V. men his neáhgebūra (nēhbūra), L. Ath. i. 9; Th. i. 204, 11. On his nēhebūra gewitnesse, v. 8, 7; Th. i. 238, 3. Se ðe æfter ǽnegum ceápe rīde, cȳþe his neáhbūrum ymbe hwæt hē rīde, L. Edg. S. 7; Th. i. 274, 20: Ps. Th. 30, 13. Ne laða ðū ðīne welegan neáhgebūras, Past. 44; Swt. 323, 21. [*Cf. Icel.* nā-búi *a neighbour.*] v. next two words.

neáh-gebȳrild, es; *m. A neighbour :*—Nēhebȳrildas *vicinas,* Lk. Skt. Lind. 15, 9. v. next word.

neáh-gebȳren, e; *f. A neighbour :*—Heó clypaþ hyre frýnd and nēhhebȳryna (-byrna, MS. A.) *convocans amicas et vicinas,* Lk. Skt. 15, 9. v. neáh-gebūr.

---

neáh-gehúsa, an; *m. A neighbour :*—Nēhgehūsum mīnum *vicinis meis,* Ps. Surt. 30, 12: 78, 4, 12: 79, 7.

neahhige; *adv. Abundantly, frequently,* Ps. 138, 9. v. geneahhie.

neáh-, neá-lǽcan; *p.* -lǽhte. -lǽcte *To draw nigh, approach :*—Ðis fȳr mē swīðe neálǽceþ *ignis mihi adpropinquat,* Bd. 3, 19; S. 548, 24: Exon. Th. 164, 4; Gū. 1006. Deáþ neálǽcte, 170, 16; Gū. 1112. Hē neálǽhte *accessit,* Gen. 27, 27. On ðǽre tīde ðe neálǽhte nidða bearnum, Cd. Th. 77, 32; Gen. 1284: Judth. Thw. 21, 25; Jud. 34. Hī neáhlǽhton tō ðære ceastre *adpropinquantes civitati,* Bd. 1, 25; S. 487, 21. Hī ðam mynstre neálǽctan, 4, 25; S. 600, 28. Tō him neálǽcan, 4, 3; S. 567, 43. v. ge-, tō-neáhlǽcan.

neáh-, neá-lǽcung, e; *f. A drawing nigh, approach :*—Ðā ðā hē gefrédde his deáþes neálǽcunge *when he was sensible of the approach of his death,* Homl. Th. i. 88, 8. Hȳ sylfe fram manna gesyhþe āscyriende ðara manna neálǽcynge nā underfōþ *cutting themselves off from the sight of men they do not admit the approach of men,* R. Ben. 135, 1. v. ge-neálǽcing.

neáh-līc; *adj. Near :*—Unrōtnysse neáhlīce *tribulatio proxima,* Ps. Lamb. 21, 12. [*Icel.* ná-ligr *near, close at hand.*] v. next word.

neáh-, neá-līce; *adv. Nearly, about :*—Hié neálīce swā fela (*tot þene*) þearfena ofsleáþ swā hié īðelīce mid hiera ælmessan gehelpan meahton, Past. 45, 1; Swt. 335, 15. Hȳ blōwaþ ðonne neálīce (*just about when*) ōðre wyrta scrincaþ, Lchdm. i. 204, 13. [*Icel.* ná-liga *nearly, almost: O. H. Ger.* nāh-līcho *ferme.*] v. ge-nēhlīce.

neáh-, neá-, nēh-mǽg, es; *m. A near kinsman :*—His gebrōðru and his neámágas *fratres ejus omnisque cognatio illa,* Ex. 1, 6. Neáhmága *adfinium,* Wrt. Voc. ii. 3, 8. Se man leóf his nēhmágum and his worldfreónda, Blickl. Homl. 113, 9. Se man leóf his nēhmágum and his worldfreóndum, 111, 27. Wīfe and cildan and nēhmágon (MS. B. neáhmágum), L. C. S. 71; Th. i. 414, 1. [*Icel.* ná-mágr *a near kinsman by marriage.*]

neáh-mǽgþ, e; *f. A neighbouring province* (v. mǽgþ, **IV.** c) :—On ða neáhmǽgþe *in proximam provinciam,* Bd. 4, 16; S. 584, 23. Ðæra neáhmǽgþa *finitimarum provinciarum,* 3, 24; S. 557, 15.

neáh-, nēh-mann, es; *m. A neighbour :*—Him se gesīþ eác fultumade and ealle ða neáhmenn *juvante etiam comite ac vicinis omnibus,* Bd. 4, 4; S. 571, 14. Ða nēhmen *vicini,* 1, 33; S. 499, 10. Ūrum neáhmannum *vicinis nostris,* Ps. Th. 79, 6: Shrn. 73, 35.

neáh-munt, es; *m. A neighbouring mountain :*—Of ðæm neáhmunte (*ex vicino monte*) wealleþ wæter, Nar. 31, 7.

neáhness, e; *f. Nearness, neighbourhood :*—Hwylc tōweard yfel ðū ðē on neáhnysse forhtast *quae ventura tibi in proximo mala formidas,* Bd. 2, 12; S. 514, 1. On nēhnesse his cytan *in vicinia cellae illius,* 5, 12; S. 630, 42.

neáh-nunnan-mynster. v. neáh, **I.**

-neáhsen. v. ge-neáhsen.

neáh-sibb, e; *f. Relationship :* — Nēhsibbe *propinquitatis,* Wrt. Voc. ii. 66, 36.

neáh-sibb; *adj. Related :*—Wē lǽraþ ðæt ǽnig cristen man ǽfre ne gewīfie on his mǽges lāfe ðe swā neáhsib (neáh sib, Th.) wǽre, L. C. E. 7; Th. i. 364, 24. Nān man ne wīfige on neáhsibban (neáh sibban, Th.) nēr (m', Th.) ðonne wiðūtan ðam .iiii. cneówe *let no one take a wife among his relations nearer of kin than beyond the fourth degree,* L. N. P. L. 61; Th. ii. 300, 14.

neáhsta a *neighbour.* v. nīhsta.

neáh-stów, e; *f.* **I.** a *neighbouring place :*—Ealle ða neáhstōwa ðǽr ymbūtan, Bt. 15; Fox 48, 22. **II.** *neighbourhood :*—On ðære circean oððe on hire neáhstōwe, Shrn. 81, 24.

neaht. v. niht.

neáh-þeód, e; *f. A neighbouring people :*—Europe ne Asia ne ealle ða neáhþeóda, Ors. 1, 10; Swt. 46, 28. Ǽgðer ge hié self wēndon ge ealle ða neáhþeóda ðæt hié ofer hié ealle mehte anwald habban, 3, 1; Swt. 96, 6.

neáh-tīd, e; *f. A time close at hand :*—Ðæt heó tō ðon ðider com ðæt heó hire sǽde ða neáhtīde hire geleórnesse *quod ipsa ei tempus suae transmigrationis in proximum nunciare venisset,* Bd. 4, 9; S. 577, 33.

neáh-tūn, es; *m. A neighbouring town :*—Sum eald man wæs in ðam nēhtūne ðær ic wæs ðæs nama wæs Malchus *there was an old man in the town near where I was, whose name was Malchus,* Shrn. 36, 6.

neáh-wæter, es; *n. A piece of water that is near :*—Wē gewīcodon be ðæm neáhwætrum, Nar. 22, 24.

neáh-west, -wist, e; *f.:* es; *m.* **I.** *nearness, neighbourhood :*—Hē ne dorste his neáwiste geneálǽcan *he dare not come into his neighbourhood,* Homl. Th. i. 88, 21. Ungewuniendlic for ðære sunnan neáweste *uninhabitable on account of the nearness of the sun,* Lchdm. iii. 260, 21. On ðære neáwiste næs nān wæterscipe, Jud. 15, 18. In ðara neáwiste *in quorum vicinia,* Bd. 5, 14; S. 634, 28. Hig on neáwiste (*in vicino*) eardodon, Jos. 9, 16: Elen. Kmbl. 133; El. 67. Wæs ðær on neáwiste (*in proximo*) hūs, Bd. 4, 24; S. 598, 27: Blickl. Homl. 197, 20: Chr. 924; Erl. 110, 12. Ðā wīcode se cyng on neáweste ðære byrig, 896; Erl. 94, 5. Swā feala earmra manna swā on ðæs rīcan

neáweste swéltaþ, Blickl. Homl. 53, 5. Đa ðe on hire neáwiste lifgeaþ, 43, 2. Ne [mágon] húse ðínum on neáweste náhwǽr sceþþan *flagella non appropiabunt tabernaculo tuo*, Ps. Th. 90, 10. Ealle ða wǽpnedmen ðe him on neáweste wǽron, Ors. 1, 10; Swt. 46, 2. Ealle ða rícu ðe him under beóþ oððe on neáweste, Bt. 16, 1; Fox 50, 3. Tó ðæs ríces neáwiste belimpeþ seó stów *ad cujus vicina pertinet locus ille*, Bd. 5, 12; S. 630, 22. II. *the being with another, presence, society, fellowship :—* Hwæt is betere ðonne ðæs cyninges folgaþ and his neáwest (cf. ðæs cyninges geférræden, l. 2) *what is better than to serve and be with the king?* Bt. 29, 1; Fox 102, 7. Hwelc is ǽngum men mǽre daru ðonne hé hæbbe on his geférrǽdenne and on his néweste feónd on freóndes anlíc-nesse, 29, 2; Fox 106, 14. Þincþ his neáwist (*the presence of the dead body*) láþlíco and unfæger, Blickl. Homl. 111, 30. Ne cume hé ná on ðæs cyninges neáwiste (ansýne, MS. H.), L. Edm. E. 3; Th. i. 246, 3. Se sacerd démde ðæt hé sceolde beón áscyred fram manna neáwiste *the priest judged that he (the leper) should be separated from the society of men*, Homl. Th. i. 124, 25. Hé férde tó folces neáwiste and bodade, 352, 11. From alre néweste geleáfulra sýn heó ásceádene, Chart. Th. 29, 19. Mid ðý ic wæs him on neáwiste, hé ðus wæs sprecende, Bd. 3, 13; S. 538, 23. Forlǽt mec englas geniman on ðínne neáwest (*into thy presence*), Exon. Th. 455, 13; Hy. 4, 49. Ic forboden ǽlcon bisceope and mæsse-preóste ðæt hig nánes wífmannes neáweste mid him næbbon (*ne mulieris alicujus societatem secum habeant*), L. Ecg. P. ii. 6; Th. ii. 198, 8. Wé lǽraþ ðæt ǽnig preóst ne lufige wífmanna neáwiste, L. Edg. c. 60; Th. ii. 256, 21. Hí wífes neáwiste forléton, L. Ælfc. C. 1; Th. ii. 342, 14: Homl. Skt. 10, 204. Libia and Agrippina wurdon swá gelýfede ðæt hí forbugon heora wera neáwiste, Homl. Th. i. 374, 33. Ne can ðara idesa ówðer þurh gebedscipe beorna neáwist, Cd. Th. 148, 36; Gen. 2467. [*Laym.* ne-, neo-weste (-uste): *Icel.* ná-vist *presence*: *O. H. Ger.* náh-wist *praesentia.*]

nealles, nalles, nallæs, nallas, nales, nalæs, nalas; *adv. Not, not at all :—*(a) in the second clause of a sentence. Đonne telle ic ða weorþ-mynd ðǽm wyrhtan, nealles ðe *I ascribe the honour to the maker, not to thee*, Bt. 14, 1; Fox 42, 19 note. Swá sceal mǽg dón, nealles inwitnet óðrum bregdan, Beo. Th. 4340; B. 2167: 4365; B. 2179. Hé spræc þurh feóndscipe, nalles hé hié freme lǽrde, Cd. Th. 38, 22; Gen. 610: 14, 2; Gen. 212. Hét hine ðære sweartan helle grundes gýman, nalles wið God winnan, 22, 26; Gen. 346. Nallæs, Soul Kmbl. 206; Seel. Verc. 104. Hwæðere him on ferhþe geáf breósthord blóðreów, nallas beágas geaf, Beo. Th. 3443; B. 1719: 3503; B. 1749. Ic feówer men geseó tó sóðe, nales mé selfa [sefa?] leógeþ, Cd. Th. 242, 9; Dan. 416. Ðær heó brynewelme bídan sceoldon, nales swegles leóht habban, 266, 27; Sat. 28. Waraþ hine wrǽclást, nales wunden gold; ferþloca freórig, nalæs foldan blǽd, Exon. Th. 288, 16-19. Ðis ic cweþe æfter forgife-nysse nalæs æfter bebode *hoc autem dico secundum indulgentiam, non secundum imperium*, Bd. 1, 27; S. 495, 45. Hí áwendan áweg, nalæs wel dydan, Ps. Th. 77, 57: Andr. Kmbl. 92; An. 46. Ðú eart geong, nalas wintrum fród, 1012; An. 506: Beo. Th. 2991; B. 1493: Blickl. Homl. 207, 17. (b) in the first clause :—Nealles him handgestealan ymbe gestódon, ac hý on holt bugon, Beo. Th. 5185; B. 2596: 4296; B. 2145: 4446; B. 2222. Nealles ... hwæðre, 5738; B. 2873. Heó nalles on goldes wlite ne scíneþ, ac on sundorweorþunge heó gewuldrad stondeþ, Blickl. Homl. 197, 8: Cd. Th. 173, 19; Gen. 2863: 249, 14; Dan. 530. Nales, Exon. Th. 60, 1; Cri. 963: 111, 3; Gú. 121. Nalæs ... ðægyt, Beo. Th. 85; B. 43. Nalas ... ah, Blickl. Homl. 121, 11: Andr. Kmbl. 3180; An. 1593. Nales ðæt án ðæt ... ac eác swelce *not only ... but also*, Ors. 1, 2; Swt. 30, 27: 1, 7; Swt. 40, 4. Nalæs ðæt án ðæt ... ac swylce eác, Bd. 3, 13; S. 538, 4: 1, 14; S. 482, 24: 4, 29; S. 608, 17. Nalæs ðæt án ... ac eác, 2, 12; S. 514, 8. Écan gesélþa sóhtan nallas þurh ðæt án ðæt hí wilnodon ðæs líchomlícan deáþes, ac eác manegra sárlícra wíta hié gewilnodon, Bt. 11, 2; Fox 36, 3. (c) with an adjective or adverb :—Nealles swǽslíce, Beo. Th. 6169; B. 3089. Nalles hólinga, 2156; B. 1076. Nalles hneáwlíce, Cd. Th. 108, 20; Gen. 1809. Nales feám síðum, Elen. Kmbl. 1633; El. 818. Nales hólunge, Cd. Th. 61, 14; Gen. 997. Nales swá wíde, Wrt. Voc. ii. 60, 55. Nales [nalles, 60, 69] ungeráde *non dissona*, 86, 12. Monge, nales feá, Exon. Th. 72, 11; Cri. 1171. Nalæs æfter myclum fæce *non multo post*, Bd. 1, 14; S. 482, 33. Nalæs æfter mycelre tíde, 4, 23; S. 593, 24. Oft, nalæs seldan, Ps. Th. 74, 4.

neán; *adv.* I. *from near :—*Neán and feorran *from near and far*, Cd. Th. 14, 28; Gen. 225. Feorran oððe neán, 64, 8; Gen. 1047. Somnaþ súþan and norþan, eástan and westan, faraþ feorran and neán, Exon. Th. 220, 26; Ph. 326: Beo. Th. 1683; B. 839. Ic eów wísige ðæt gé genóge neón sceáwiaþ beágas *I will guide you so that from near ye may gaze on rings in abundance*, Beo. Th. 3104. II. *near, close at hand :—*Gif ðú Grendles dearst neán bídan *if thou durst here await Grendel*, 1061; B. 528. Wæs ðæs wyrmes wíg wíde gesýne, neán and feorran, 4624; B. 2317. Hí ðære eaxe útan ymbhwearf, ðone norþende neán ymbcerraþ (cf. hí sint swá neáh ðam norþende ðære eaxe, Fox 214, 20), Met. 28, 14. III. *nearly, about :—*Neán twelfwintre *fere annorum duodecim*, Lk. Skt. 8, 42. Ða wæs geworden

æfter ðam wordum neán (MS. A. neáh) eahta dagas, 9, 28. Ðá wæs neán seó syxte tíd *erat autem fere hora sexta*, 23, 44. Wé ðæs here-weorces neán myndgiaþ *we bear that warlike deed in mind nearly as it happened, have an accurate remembrance of it*, Elen. Kmbl. 1311; El. 657. v. for-neán.

neap *a cup*, Lchdm. i. 374, 23. v. hnæpf.

Neapolite; *pl. The Neapolitans*, Blickl. Homl. 201, 19.

neár; neara, nearo. v. neáh; nearu.

nearu; *adj.* I. *narrow, strait, confined, not spacious :—*Neara scræf *gurgustulum*, Wrt. Voc. i. 58, 29. Neare pyt *puteus angustus*, Kent. Gl. 901. Gangaþ inn þurh ðæt nearwe (MS. B. nearuwe: Lind. nearuo: Rush. naarwe) geat ... Eálá hú neara (MS. A. nearu: Lind. naru: Rush. naru) is ðæt geat *intrate per angustam portam ... Quam angusta porta*, Mt. Kmbl. 7, 13-14: Lk. Skt. 13, 24. Se sǽ ðe ǽgðer is ge nearo ge hreóh, Ors. 1, 1; Swt. 28, 12. Alexander him ðæt ondréd for ðære nearwan stówe ðæt hé ðá on wæs *timens angustias quibus inerat locorum*, 3, 9; Swt. 124, 25. In án nearo fæsten ungeféredra móra *in angustias inaccessorum montium*, Bd. 4, 26; S. 602, 20. Nearo wíc *mansionem angustam*, 4, 28; S. 605, 23. Tóbrǽdan ofer ða nearwan eorþan (cf. ofer ðás nearowan eorþan sceátas, Met. 10, 16), Bt. 19; Fox 68, 25. Binnon nearwum gemǽrum *intra fines angustiores*, Bd. 4, 26; S. 603, 9. Nearewum *artis*, Wrt. Voc. ii. 5, 67. Mid ða nearwan *arta*, 5, 57. Ofereode stíge nearwe, enge ánpaðas, Beo. Th. 2823; B. 1409. II. *narrow, limited, poor, restricted :—*Hú ne ongite gé hú neara (Cott. MS. nearo) se eówer hlísa beón wile, Bt. 18, 2; Fox 64, 14. Swíðe nearewe (Cott. MS. nearwa) sent and swíðe heánlíce ða menniscan gesélþa, 11, 1; Fox 30, 25. Hét hié from hweorfan neorxna wange on nearore líf, Cd. Th. 58, 11; Gen. 944. III. *strait, oppressive, causing anxiety* (of that which restricts free action of body or mind) :—Nýd byþ nearu on breóste níða bearnum *need straitens the breast of man*, Runic pm. Kmbl. 341, 8; Rún. 10. Nearo nihtwaco *the anxious night watch*, Exon. Th. 306, 13. In hæft under nearone clom (*under confining fetter*), Exon. Th. 138, 2; Gú. 570. Ðone nearwan níþ onfón, Cd. Th. 43, 27; Gen. 697: 304, 22; Sat. 634. Of ðæm nearwum bendum, Homl. Skt. 3, 197: Exon. Th. 435, 6; Rä. 53, 3. Under nearwum clommum, 134, 22; Gú. 511. Hié wilnodan ðæt hé hié of ðæm nearwan þeóstrum álésde, Blickl. Homl. 103, 13. IV. *oppressed, not having free action :—*Wið nearwre sworetunge *for difficult breathing*, Lchdm. i. 340, 11. Hym beóþ on hyra brósten nearuwe (*people with asthma*), iii. 116, 23. V. *strict, severe :—*Ðæt hié ne þyrfen bión gesewene æt ðæm nearwan dóme *ut a districta judice videri non debeant*, Past. 53, 2; Swt. 413, 16. [*O. Sax.* naru.]

nearu, we; nearu (o); *indecl. f.* I. *confinement, durance, prison :—*Hwonne hié of nearwe stæppan mósten, of enge út æhta lǽdan (*when they might come out of the ark*), Cd. Th. 86; Gen. 1433. Hió bebeád ðæt hine man of nearwe and of nýdcleofan, fram ðam engan hofe forléte, Elen. Kmbl. 1418; El. 711. Nǽglas of nearwe scínende *the nails shining from the hole where they had been hidden*, 2227; El. 1115. Neb wæs mín on nearwe *my face was in confinement*, Exon. Th. 392, 1; Rä. 11, 1. Siððan ðæt nioþan upweardne on nearo fégde *afterwards fixed me upside down in durance*, 479, 12; Rä. 62, 6: 480, 8; Rä. 63, 8. II. *a strait, difficulty :—*On nearwe *in a strait*, Elen. Kmbl. 2203; El. 1103. Nearwe genýddon on norþwegas wiston him be súþan Sigelwara land *the difficulties of the situation forced them to the north for they knew that to the south of them lay the land of the Ethiopians*, Cd. Th. 181, 29; Exon. 68. Nearu, nearo þrowian *to be in straits*, Andr. Kmbl. 828; An. 414: Beo. Th. 5182; B. 2594. Hé ǽr sela nearo néþende níða gedígde *from many straits and strifes had he come safely*, 4689; B. 2350. Hine of nearwum út forlét, Vald. 2, 8. In nearowe néþan *to venture into difficulties*, Exon. Th. 436, 12; Rä. 54, 13.

nearu-bregd, es; *n. A wile or trick that brings others into straits* (v. *preceding word and* nearu, III) :—Néþde ic nearobregdum, ðǽr ic Neron biswác, ðæt hé ácwellan hét Cristes þegnas, Exon. Th. 260, 24; Jul. 302.

nearu-cræft, es; *m. An art that confines or imprisons* (?) :—Beorh wunode on wonge nearocræftum fæst ðǽr on innan bær eorl gestreóna ... feá worda cwæþ: Hold ðú nú hrúse eorla æhte *the mound stood on the plain firm in its prisoning powers* (*able to keep in durance the treasure entrusted to it*); *therein bore the earl treasures ... few words he spake :* Hold thou now, earth, the possessions of earls, Beo. Th. 4475-4488; B. 2241-2248. v. nearu; *f.* I.

nearu-fáh; *adj. Disastrously hostile, bearing enmity the result of which is to reduce others to straits :—*Wæs ðæs wyrmes wíg wíde gesýne, nearofáges níð, hú se gúþsceaþa Geáta leóde hatode and hýnde, Beo. Th. 4623; B. 2317.

nearu-gráp, e; *f. A close grasp :—*Án wiht is ... hreóh and reþe hafaþ ryne strongne ... and be grunde faraþ ... neól is nearográp, Exon. Th. 491, 28; Rä. 81, 6. v. gráp.

nearu-líc; *adj. Oppressive, distressing, grievous :—*Feala mé se Hǽlend hearma gefremede, níða nearolícra, Elen. Kmbl. 1822; El. 913. v. next word.

nearulíce; *adv.* I. *narrowly, within narrow limits, briefly :—*

Nearolíce *strictim*, Ælfc. Gr. 38; Som. 41, 60. Ys seó foresǽde bóc (*Genesis*) on manegum stówum swíðe nǽrolíce gesett (*is a mere narrative of events*), and ðeáh swíðe deóplíce on ðam gástlícum andgite, Ælfc. Gen. Thor. 4, 3. II. *oppressively, grievously*:—Ða ðe nearwlícast cúðan swician *those who knew how to cheat in most oppressive manner*, L. I. P. 12; Th. ii. 320, 24. III. *narrowly, exactly, strictly*:—Manegu díglu þing siudon nearolíce (*subtiliter*) tó smeáganne *many secret things are to be narrowly examined*, Past. 21, 3; Swt. 153, 13. Swá swýðe nearwelíce hé hit lett út áspyrian (*of the enquiry which was made when Doomsday Book was compiled*), Chr. 1085; Erl. 218, 34. [He nule nout so neruhliche demen ase ʒe siggeð, A. R. 334, 14.]

nearuness, e; *f.* I. *a strait*:—Mid longre nearonesse be eástan Constantinopolim ligeþ *juxta Constantinopolim longae mittuntur angustiae*, Ors. 1, 1; Swt. 8, 21. II. *oppression, distress* (of body):—Nearones breósta *oppression of the chest*, Lchdm. ii. 204, 27. Ðæt (*asthma*) ys nearunyss ... and breóst byþ innc 1 mid micle nearnysse, iii. 116, 23–26. III. *distress* (of mind), *anxiety, tribulation, trouble, grief*:—Hér is seó lǽnlíce winsumnes ac ðǽr is seó syngale nearones *in this world is the transient delight, in the next is the perpetual distress*, L. E. I. introd.; Th. ii. 394, 8. Hú ne witon wé ðæt nán nearewnes ne nán unrótnes nis nán gesǽlþ *nam non esse anxiam, tristemque beatitudinem quid attinet dicere?* Bt. 24, 4; Fox 86, 20. Hé on swá micelre nearanesse becom *he fell into so great trouble (was imprisoned)*, 1; Fox 2, 27. Swá hwá swá ða flǽsclícan unþeáwas forlǽtan wile hé sceal geþolian micele nearanesse *corporis voluptatum appetentia plena est anxietatis*, 31, 1; Fox 110, 26. Seó hreówsung ne beoþ ná bútan sorge and bútan nearonesse, Fox 110, 29. On swá micelre módes unréto and nearonisse *anxietate*, Nar. 30, 24. Nearonessa *angustia*, Ps. Th. 118, 143. On mínum earfoþum and nearonessum *in tribulatione*, 4, 1. Of nearonessum heora *de necessitatibus eorum*, Ps. Lamb. 106, 6. Nearonessum módes mentis *angoribus*, Bd. 2, 12; S. 513, 33 note.

nearu-níd, -néd, e; *f. Sore need, grievous trouble*:—Ða menigo ðe ðé mid wuniaþ on nearonédum [*or* (?) on nearo nédum *in confinement by force*], Andr. Kmbl. 203; An. 102. From naronéðe *de angusta* (as if *angustia?*), Lk. Skt. p. 8, 6.

nearu-searu, we; *f. A wile that causes restraint* or *confinement* (?):—Hýdde wǽron þurh nearusearwe næglas on eorþan (*of the nails in the cross that had been buried*), Elen. Kembl. 2215; El. 1109.

nearu-sorh, -sorg, e; *f. Oppressive care, grievous trouble*:—Nearu-sorge dreáh, Elen. Kmbl. 2520; El. 1261.

nearu-þanc, es; *m. Illiberal thought, wickedness*:—Feóndlícra nearaþanca *spiritalium nequitiarum*, Hpt. Gl. 426, 61. Syle heom æfter nearuþancum (nearoþancnysse, Ps. Lamb. 27, 4), wiðmétednyssa heora *da illis secundum nequitiam adinventionum ipsorum*, Ps. Spl. 27, 5.

nearu-þancness. v. preceding word.

nearu-þearf, e; *f. Pressing need*:—Ic on ýðum slóg niceras nihtes, nearoþearfe dreáh, Beo. Th. 849; B. 422: Exon. Th. 5, 14; Cri. 69.

nearu-wrenc, es; *m. A trick* or *wile that causes anxiety* or *trouble*, Exon. Th. 316, 5; Mód. 44.

nearwe; *adv.* I. *straitly, strictly, closely*:—Nearwe gebunden *straitly bound*, Exon. Th. 463, 2; Hö. 64. Hyne sár hafaþ nearwe befongen, Beo. Th. 1957; B. 976: Elen. Kembl. 2550; El. 1276: Met. 21, 5. II. *narrowly, strictly, exactly* (of enquiry):—Þeódcwén ongan georne sécan nearwe, tó hwan hió ða næglas gedón meahte, Elen. Kmbl. 2313; El. 1158: 2476; El. 1240. III. *oppressively, forcibly*:—Ðonne hiue æt niéhstan nearwe stilleþ G, Salm. Kmbl. 268; Sal. 133. Nearwe gebéged, Cd. Th. 292, 26; Sat. 446. IV. *anxiously, in a manner causing trouble*:—Hyge gnornende nihtes nearwe *the mind mourning in anguish at night*, Exon. Th. 174, 25; Gú. 1183. Ferþ gebysgad, nearwe genǽged, 162, 35; Gú. 986. Ðá heó nearwe beswác yldran ússe *when the serpent deceived our first parents to their hurt*, 226, 30; Ph. 413: Frag. Kmbl. 51; Leás. 27.

nearwe-líce. v. nearu-líce.

nearwian; *p.* ode. I. *to make narrow, straiten, compress*:—Se ðe mec nearwaþ, Exon. Th. 407, 25; Rä. 26, 10. II. *to become narrow, contracted*:—Sefa nearwode (*of Noah when drunk*), Cd. Th. 94, 32; Gen. 1570. Sinc searwade, sib nearwade, Exon. Th. 353, 63; Reim. 37. v. ge-nearwian, nirwan.

neát, es; *n. A neat, an ox* or *a cow, cattle, beast, animal*:—Gif neát mon gewundige, weorpe ðæt neát tó honda oððe foreþingie, L. Alf. pol. 24; Th. i. 78, 9. Nán neát nyste nǽnne andan tó óðrum, Bt. 35, 6; Fox 168, 10. Ne ligeþ hé eallinga on ðǽre eorþan suá ða creópendan wuhta, ac biþ hwæthwugu upáhæfen suá ðæt neát from eorþan, Past. 21, 3; Swt. 157, 1. Fugel oððe fisc on sǽ oððe eorþan neát, feldgongende feoh bútan snyttro, Exon. Th. 371, 23; Seel. 80. Foldan neát, Salm. Kmbl. 436; Sal. 218. Ic eom anlíc ánum neáte *ut jumentum factus sum*, Ps. Th. 72, 18. Sealde heora neát (*jumenta*) hæglum, 77, 48. Deór and neát *bestiae et universa pecora*, 148, 10. Ða dumban neát, Andr. Kmbl. 134; An. 67. Tó neáta scypene *ad stabula jumentorum*, Bd. 4, 24; S. 597, 9. Nǽnig mann scypene his neátum ne timbreþ, 1, 1; S. 474, 32. Ðám monnum ðe beóþ neátum gelíce, Bt. 14, 1; Fox

42, 3 note: 41, 5; Fox 254, 5. [*O. Frs.* nát: *Icel.* naut *cattle, oxen*: *O. H. Ger.* nôz *jumentum*.] v. sleg-neát.

Neátan-leáh (?) *Netley*, Chr. 508; Erl. 14, 18.

neáten, neá-west. v. níten, neáh-west.

neáwung, e; *f. Nearness, coming near*:—In neáwung síe sumer *in proximo sit aestas*, Mk. Skt. Lind. 13, 28.

nebb, es; *n.* I. *a neb* (dialect.), *nib, a beak, a beak-shaped thing*:—Neb *rostrum*, Wrt. Voc. ii. 119, 25. Ðæt nebb (*of the Phænix*) líxeþ swá glæs oððe gim, Exon. Th. 218, 24; Ph. 299. Neb (*of a ship*), 392, 1; Rä. 11, 1. Neb (*of a plough*), 403, 1; Rä. 22, 1. Nebb (*of a rake*), 416, 23; Rä. 35, 3. Neb (*of a musical instrument*), 413, 16; Rä. 32, 6. Ic (*a key*) bregde nebbe, 498, 6; Rä. 87, 8. Ic (*a helm*) hæbbe heard nebb, 489, 29; Rä. 79, 1. II. *a nose, the gristle of the nose*:—Neb *internasum* (cf. nose gristle *internasus*, 43, 20), Wrt. Voc. i. 64, 50. Gif mon óðrum ðæt neb (nebb, MSS. B. H.: næb, MS. G.) of ásleá, gebéte him mid lx. sciłł., L. Alf. pol. 48; Th. i. 94, 8. III. *the face, countenance*:—Neb *facies*, Wrt. Voc. i. 42, 51: *vultus*, Hpt. Gl. 475, 6. Hys nebb (*facies*) wæs mid swátlíne gebunden, Jn. Skt. 11, 44: Neb, Met. 31, 23. Be blǽdrum ðe on mannes nebbe sittaþ .. smyre ðæt neb mid, Lchdm. i. 86, 5–8. Mid ðam wlitegostan nebbe, Homl. Th. i. 430, 14. Ðonne wé wendaþ úre neb tó eástdǽle, 262, 10: ii. 102, 16. Heó helode hire nebb (*vultum*), Gen. 38, 15: Ex. 3, 6. Spǽte ðæt wíf on his nebb (*faciem*), Deut. 25, 9: Num. 12, 14. Ðá forceáw hé his ágene tungan and wearp hine on ðæt neb foran (*in os tyranni abjecit*), Bt. 16, 2; Bt. 52, 25. Hý habbaþ twá neb on ánum heáfde *duas in uno habentes capite facies*, Nar. 35, 24. [*Icel.* nef (*gen. pl.* nefja) *the nose; the beak of a bird*.]

nebbian; *p.* ode *To turn the face towards anyone* (?), *to retort upon anyone* (?):—Se ríca besihþ on his pǽllenum gyrlum, and cwyþ: 'Nis se loddere mid his tættecon mín gelica.' Ac se apostol Paulus hine nebbaþ mid ðisum wordum (*retorts upon him, meets him, with these words*): 'Ne brohte wé nán þing tó ðisum middangearde, ne wé nán þing heonon mid ús lǽdan ne mágon, Homl. Th. i. 256, 7–12.

neb-corn, es; *n. A pimple on the face*:—Gif nebcorn on wífmannes nebbe weaxan ... hit áfeormaþ of ealle ða nebcorn, Lchdm. i. 118, 22–25.

neb-gebrǽc, es; *n. A defluxion from the head, mucus of the nose*:—Nebgebrǽc *coriza* (= κορυζα), Wrt. Voc. i. 19, 28: ii. 135, 77.

Nebrond, es; *m. Nimrod*:—Freónd Nebrondes, Salm. Kmbl. 426; Sal. 213. v. Nefrod.

nebwlát-ful; *adj. Bold, impudent, shameless*:—Nebwlátful, scamleás *frontosa*, Hpt. Gl. 506, 78. v. *next word* and wlátian.

neb-wlátung, e; *f. Boldness, impudence*; *frontositas*, Lye (from a vocabulary in the Cotton library). v. preceding word.

neb-wlatung, e; *f. Dejection*; *vultus demissio*, Lye.

neb-wlitu, e; *f. The form of the face, the face, countenance*:—Heora nebwlitu sceán swá swá sunne, Homl. Th. ii. 426, 10. Ic ne mæg on his nebwlite beseón, Homl. Skt. 7, 104. Hí gesáwon his nebwlite swylce sumes engles ansýne, Homl. Th. i. 46, 5. Sege ús his nebwlite *describe his face to us*, 456, 15. Ne behealde gé heora nebwlite, ii. 404, 28.

néchebúr, necte-gale, néd-. v. neáh-gebúr, nihte-gale, níd-.

nediende (?) *abominandum, execrandum*, Hpt. Gl. 515, 40.

nefa, an; *m.* I. *a nephew*; nepos:—Bróðer sune vel suster suhe ðæt is nefa, Wrt. Voc. i. 51, 71. Neva *nepos*, 72, 35. Hlóþhere Ægelbrhytes nefa (cf. hé him onsende Leutherium his nefan (*nepotem*), Bd. 3, 7; S. 530, 29), Chr. 670; Erl. 34, 29: 789; Erl. 57, 34: Ælfc. Gr. 9, 31; Som. 11, 69. Eám and nefa, Exon. Th. 431, 35; Rä. 47, 6. Heó wæs Édwines nefan (*nepotis*) dohtor, Bd. 4. 23; S. 593, 2. Hé swylces hwæt secgan wolde eám his nefan, Beo. Th. 1766; B. 881. II. *a grandson*:—Nefena bearnum *pronepotibus, filiis nepotum*, Hpt. Gl. 50. Ealdra nefena *pronepotum*, 445, 56. III. *a step-son*:—Nefa *prifignus*, Wülck. Gl. 41, 28. [*Icel.* nefi *a cognate kinsman, a nephew*: *O. Frs.* neva: *O. H. Ger.* nefo *nepos, sobrinus*: *Ger.* neffe.] v. for-, ge-nefa.

nefene, an; *f. A niece* or *grand-daughter*:—Bróðer dochter vel suster dohter, nefene *neptis*, Wrt. Voc. i. 51, 72. Nefenu[m] *nepotibus* (*neptibus?*), Hpt. Gl. 485, 42.

nefne, nemne. I. *conj. connecting clauses, Unless, except*:—Hé hyra má ácwellan wolde, nefne him witig God forstóde, Beo. Th. 2116; B. 1056: 6101; B. 3054: Exon. Th. 340, 5; Gn. Ex. 106: 345, 11; Gn. Ex. 186. Hí sǽdon, nemne (*nisi*) hí him máran andlyfne sealdon, ðæt hí woldan him sylfe niman, Bd. 1, 15; S. 483, 37. Hé læg swá swá deád mon nemne ðynre éðunge ánre ætýwde ðæt hé lífes wǽre *quasi mortuus jacebat, halitu tantum pertenui quia viveret demonstrans*, 5, 19; S. 640, 24. Nymne, 1, 27; S. 493, 38. Nó hé fóddor þigeþ, nemne meledeáwes dǽl gebyrge, Exon. Th. 215, 29; Ph. 260: 124, 12; Gú. 338: 249, 10; Jul. 109: Beo. Th. 3108; B. 1552: 5302; B. 2654. Næfne, 506; B. 250. On weres wæstmum, næfne (*except that*) hé wæs mára ðon ǽnig man óðer, 2710; B. 1353. Hwæt hæfde seó godcunde þurh ða menniscan nemne búton ðæt heó mihte beón ácenned, Blickl. Homl. 19, 22. II. *connecting words in the same case* (contracted clauses, the verb of the second clause

being the same as that in the first, and not expressed :—Ne gehŷrde nǽnig man on his mūþe ôht elles nefne Cristes lof and nytte sprǽce, 223, 36 : Exon. Th. 308, 28 ; Seef. 46. Nǽneg dorste nefne sinfreá, Beo. Th. 3873 ; B. 1934. Ic lyt hafo heáfodmāga nefne đec, 4309 ; B. 2151. Đæt unc ne gedǽlde nemne deáþ āna ôwiht elles, Exon. Th. 442, 34 ; Kl. 23 : Andr. Kmbl. 1327 ; An. 664. III. *prep. Except :*—Nemne feáum ānum, Beo. Th. 2167 ; B. 1081.

nefre. v. *nǽfre.*

Nefrod, es ; *m. Nimrod :*— Nefrod se gigant ; se Nefrod wæs Chuses sunu, Bt. 35, 4 ; Fox 162, 17. v. *Nebroud.*

nefte. v. *nepte.*

nē-fugol, nēgan, nēh, nēhsta, neht, nele, nellan. v. neó-fugol, nǽgan, neáh, nīhsta, neht, nyllan.

nemnan, nemnian ; *p.* nemde. **I.** *to name, give a name to a person* or *thing :*—Đū cennest sunu đone đū nemnest Hǽlend, Blickl. Homl. 7, 19, Đū nemdest mid āne noman ealle tōgædere woruld, Met. 20, 55. Đa hē nemde (*nominavit*) apostolas ; Simonem đǽne hē nemde (*cognominavit*) Petrus, Lk. Skt. 6, 13-14. Hē đone yldestan Noæ nemde, Cd. Th. 75, 4 ; Gen. 1235. Hig nemdon (*vocant*) hyne hys fæder naman Zachariam, Lk. Skt. 1, 59. Đysne dæg hié nemdon siges dæg, Blickl. Homl. 67, 13. **II.** *to use such and such a name* or *title in speaking of a person* or *thing :*—Đone wē wifel wordum nemnaþ *which we call beetle, when we speak of it,* Exon. Th. 426, 14 ; Rä. 41, 73. Hine tō sylfcwale secgas nemnaþ *men speak of him as a suicide,* 330, 25 ; Vy. 56. Eác hī ôđre worde beornas Bađan nemnaþ *men also use the name Bath in speaking of it,* Chr. 973 ; Erl. 124, 13 : 975 ; Erl. 124, 32. David sylf nemde hine drihten *ipse David dicit eum dominum,* Mk. Skt. 12, 37. Heó sylf hié þeówen nemde, Blickl. Homl. 13, 13. Drihten đa cynelīcan burh forhogodlīce naman nemde *the Lord used a contemptuous name (wīc) in speaking of the royal city,* 77, 23, 26. Đis andwerde lîf hé nemde for weg *this present life he spoke of as a way,* L. E. I. 35 ; Th. ii. 432, 23. Đone hwîtan hlâf (*the eucharistic bread*) đone đū sealdest Saban ûssum fæder nemdon heó hine swā (*sic eum appellare consuerant*), Bd. 2, 5 ; S. 507, 15. Ne gyrne gē đæt eów man Lāreówas nemne *nolite vocari Rabbi,* Mt. Kmbl. 23, 8 : Ps. Th. 82, 4. Đêh đe gewrito oft nemnen eal đa lond Mêdia, Ors. 1, 1 ; Swt. 10, 24. Đeáh mon anweald and genyht tō twǽm þingum nemne đeáh hit is ān *though power and abundance be spoken of as two things, yet are they one,* Bt. 33, 1 ; Fox 120, 21. Hī gewunedon hī mōder cȳgean and nemnian (tō hātenne and tō nemnenne, MS. B.) *quam matrem vocare consueverant,* Bd. 4, 23 ; S. 594, 39. Đæs fæder wæs Wôden nemned, 1, 15 ; S. 483, 30 : Blickl. Homl. 81, 1. On đǽm bōcum đe nemned is *Actus Apostolorum,* 133, 11 : 137, 31. Đæt wæs swîđe heálīc nama đæt Sanctus Johannes engel wæs nemned, 167, 32. Đam is tō naman nemned Drihten *Dominus nomen est ei,* Ps. Th. 67, 4. **III.** *to call upon the name of, address by name, to invoke :*—Ne nemn đū Drihtnes naman on ȳdel ne byþ unscyldig se đe his noman on ȳdel nemþ *non assumes nomen Domini Dei tui in vanum ! nec enim habebit insontem Dominus eum, qui assumpserit nomen Domini frustra,* Ex. 20, 7. Ic naman đinne nemde, Dryhten, Ps. Th. 118, 55. Hē nemde mē mînne noman *vocavit me nomine meo,* Bd. 5, 6 ; S. 619, 37. Hine se ār be naman nemde, Elen. Kmbl. 155 ; El. 78. Se nemde God niþþa bearna ǽrest ealra, Cd. Th. 69, 13 ; Gen. 1135. God nemdon and hine bǽdon, 48, 22 ; Gen. 779. Ongan swegles weard be naman nemnan, Judth. Thw. 22, 27 ; Jud. 81. **IV.** *to mention by name, to mention, relate :*—For mîne brōđru ic bidde, and mîne đa neáhstan nemne swylce, Ps. Th. 121, 8. Đǽm unþeáwum đe ic ǽr nemde, Met. 25, 62. Ealle đa ôđru gōd đe wē ǽr nemdon, Bt. 24, 3 ; Fox 84, 24 : Cd. Th. 288, 20 ; Sat. 383. Sege hwæt ic þence, nemn gif đū hit gereccean mæge, Blickl. Homl. 181, 14. Mágun wē nemnan *we can tell,* Exon. Th. 107, 25 ; Gū. 64. Đeáh đe ic hȳ nîhst nemnan sceolde *though I should mention their names last,* 326, 10 ; Vîd. 126. *Pronomen spelaþ đone naman đæt đū ne þurfe tuwa hine nemnan the pronoun represents the noun so that you need not mention it (the noun) twice,* Ælfc. Gr. 5 ; Som. 3, 30. Swā on đǽre ilcan lāre nemned (*mentioned*) is, Blickl. Homl. 133, 34. **V.** *to name, nominate :*—Gif landāgende man ætsace, đonne nemne man him his gelîcan ealswā micel Wente swā cyninges þegne, L. N. P. L. 52 ; Th. ii. 298, 10. [*Goth.* namnjan : *O. Sax. O. H. Ger.* nemnian : *Icel.* nefna.] v. ge-nemnan, namnian, namian.

nemne. v. *nefne.*

nemnigend-līc ; *adj. Nominative :*—*Nominativus* is nemnigendlīc, mid đam casu wē nemnaþ ealle þing, Ælfc. Gr. 7 ; Som. 6, 17.

nempe. v. *nimþe.*

Nen *the river Nen in Northamptonshire :*—Đæt water, đæt man cleopeþ Nen, Chr. 963 ; Erl. 122, 17.

neó *a corpse.* v. dryht-nē, neó-bedd, -fugol, -sîđ. [*Goth.* naus : *Icel.* nār.]

neó-bedd, es ; *n. A bed for a corpse :*—Ic in mînum neste neóbed ceóse '*I shall die in my nest*' (A. V.), Exon. Th. 235, 7 ; Ph. 553. God wearp hine niđer on đæt neóbedd (*that couch of corpses,* Hell ; cf. Milton '*that fiery couch*'; and Icel. nā-strönd *the place where the dead came, who had not fallen in battle*), Cd. Th. 22, 19 ; Gen. 343.

neód, nēd, niéd, nȳd, e ; *f. Desire, eagerness, diligence, earnest endeavour :*—Wæs him neód micel đæt hié tôbrugdon fira flǽschoman him to fôdderþege *great was their desire to rend the bodies of men for their repast,* Andr. Kmbl. 316 ; An. 158. Biþ him neód micel đæt hē đa yldu môte wendan tô lîfe feorg geong onfôn *it is most eager to turn old age to life, to receive youth,* Exon. Th. 210, 22 ; Ph. 189 : 228, 3 ; Ph. 432. (Cf. *O. Sax.* was im niud mikil that sie selbon Krist gisehan môstin *they desired eagerly to see Christ.*) Ús is eallum neód đæt wē đīn mēdrencynn môtan cunnan *we all desire to know thy descent on the mother's side,* 15, 33 ; Cri. 245. Wundorlîc is geworden đīn wîsdôm ne mæg ic him on neóde ā neáh cuman (*I cannot with all my endeavours come near it*), Ps. Th. 138, 4. Noe tealde đæt hē on neód hine gif hē land ne funde sēcan wolde *Noah reckoned that if the raven did not find land it would eagerly seek him,* Cd. Th. 87, 4 ; Gen. 1443. Hié God herigaþ, and him be namon gehwam on neód (*earnestly*) sprecaþ, 242, 25 ; Dan. 424. Ic đinne naman on neód secge *confitebor nomini tuo,* Ps. Th. 137, 2. Sôđfæste đinne naman willaþ þuruh neód herigean *justi confitebuntur nomini tuo,* 139, 13. Se đe naman đinne þurh neód forhtaþ *he that is earnest in reverencing thy name,* 60, 4. ¶ *The instrumental with adverbial force occurs very frequently in the Psalms.* Neóde, nēde, niéde, nȳde *earnestly, diligently, eagerly :*—Weoroda mǽst fore Waldende gǽþ neóde and nŷde (*the good will go eagerly, the wicked only on compulsion*), Ps. Th. 138, 16, 15 ; Cri. 1072. Oft hē hǽþengield gesôhte neóde geneahhe (*very diligently*), 244, 7 ; Jul. 24 : Ps. Th. 82, 12, 13. His naman neóde heriaþ, 67, 4. His naman neóde lufiaþ, 68, 37. Hī hyrdnesse neóde begangaþ *they diligently keep watch,* 89, 5 : 112, 2 : 121, 6. Nēde, 105, 36 : 118, 55. Nŷde, 118, 132 : 114, 4. Niéde, Ps. Ben. 43, 27. Þurh đinra neóda (niéda, MS. Verc.), *lust by the pleasure of thy passions,* Exon. Th. 369, 29 ; Seel. Ex. 48. Hē ús on hæft nimeþ ofer ûsse neóde lust (*contrary to our desires*), 16, 30 ; Cri. 261. Wē đǽrinne andlangne dæg nióde nāmon *in the hall the live long day we took our pleasure,* Beo. Th. 4238 ; B. 2116. [*O. Sax.* niud : *O. Frs.* niod : *O. H. Ger.* niot ; *m. desiderium, cupido.*] v. *next word.*

neód (= neád) *necessity. The distinction in form between the word* = *Goth.* nauþs, *and the preceding word seems not to have been observed in A. S. MSS. See the passages under* nīd.

neód-fracu, e ; *f. Desire, appetite, the object of desire* or *of appetite :*—Wuhta gehwilc hnipaþ of dūne, wilnaþ tô eorþan, sume nēdþearfe, sume neódfræce (cf. ealle beóþ of dūne healde wiđ đære eorþan and đider wilniaþ ođđe đæs đe hī lyst ođđe đæs đe hī beþurfon, Bt. 41, 6 ; Fox 254, 28), Met. 31, 15.

neód-freónd. v. *nīd-fréond.*

neód-ful ; *adj. Earnest, zealous :*—Bidde ic monna gehwone đe đis gied wrǽce, đæt he mec neódful gymne, Exon. Th. 285, 26 ; Jul. 720.

neódian *to be necessary.* v. *neádian.*

neód-laðu ; *f. Earnest, hospitable invitation :*—Hē frægn gif him wǽre æfter neódlaðu niht getǽse *Beowulf asked if to Hrothgar the night had been pleasant after the hospitality of the preceding evening (?)* (cf. him wæs ful boren and freóndlaðu wordum bewægned, 2389 ; B. 1192), Beo. Th. 2644 ; B. 1320.

neód-līce ; *adv. Diligently, sedulously, zealously, eagerly, earnestly :*—Smire đa sîdan mid đŷ neódlîce *smear the sides with it diligently,* Lchdm. ii. 262, 11. Đā êfste se abbud wiđ đæs muneces, and neódlîce (*eagerly, anxiously*) cwæþ : 'Hwǽr is se đe đū feredest?' Homl. Th. i. 336, 22 : ii. 26, 5. Lustlîce gehyrdon đa đe him gelǽrde wǽron and eác swylce neódlîce mid dǽdum lǽston đa đe hī ongitan mihton *libenter ea quae dicerentur, audirent; libentius ea quae intelligere poterant, operando sequerentur,* Bd. 4, 27 ; S. 604, 18. Gôd is đæt man neódlîce Drihtnes naman āsinge, Ps. Th. 91, 1 : 128, 6 : 133, 3 : 148, 12. Neódlîce on naman đînum ealle eorþbüend egsan habbaþ *greatly do all dwellers on earth stand in awe of thy name,* 101, 13. Mē neódlîce tô forsceape scŷhte, Cd. Th. 53, 21 ; Gen. 897. Nǽnig đînra þegna neódlucor ne gelustfullīcor hine sylfne underþeódde tô ûra goda bigange đonne ic *nullus tuorum studiosius quam ego culturae deorum nostrorum se subdidit,* Bd. 2, 13 ; S. 516, 5. Nŷdlîcor *libentius,* 4, 13 ; S. 583, 4. [*O. Sax.* niud-lîko.]

neód-lof, es ; *n. Diligent praise :* — Herian naman Drihtnes mid neódlofe (cf. hebbaþ neódlîce eówre handa on hālig lof, 133, 3), Ps. Th. 148, 12.

neód-spearuwa, an ; *m. An active, restless sparrow* (cf. (?) sparuwe is a cheaterinde brid ; cheatereð euer ant chirmeð, A. R. 152), Ps. Th. 123, 6.

neód-weorþung, e ; *f. Great honouring :*—For naman đînes neódweorþunge *propter nomen tuum,* Ps. Th. 142, 11.

neó-, nē-fugol, es ; *m. A bird that feeds on carrion, a vulture* or *crow :*—Nēfuglas sittaþ þeódherga wæl þicce gefylled *carrion-birds sit gorged with the slain,* Cd. Th. 130, 12 ; Gen. 2158.

neól. v. *neowol.*

neom, neam, nam = ne eom *am not,* nis = ne is *is not :*—Đæs gescŷ neom (nam, Lind. : nǽm, Rush.) ic wyrđe tô berenne, Mt. Kmbl. 3, 11. Neam ic *non sum,* Ps. Surt. 118, 30. Sî eówer sprǽc : Hyt ys, hyt ys :

nyt nys, hyt nys, Mt. Kmbl. 5, 37. Nis álýfed *it is not allowed,* Homl. Th. i. 94, 29.

neómian (?) *to produce harmonious sounds :*—Nægl (*plectrum*) neómigende (MS. neome cende), Exon. Th. 332, 12; Vy. 84. [Grein compares the word with *O. H. Ger.* niumôn *jubilare, psallere ;* there is also the noun niumo *modulatio, sonus, canticum.*]

neón. v. neán.

neorxna wang, es; *m. Paradise :*—Paradisum ðæt wē hátaþ on Englisc neorxna wang, Hexam. 16; Norm. 24, 5 : Cd. Th. 13, 26; Gen. 208 : Blickl. 17, 15 : Homl. Th. i. 12, 32. Gif hē beget and yt rinde sió ðe cymþ of neorxna wonge, ne dereþ him nán ätter. Ðonne cwæþ se ðe ðas bóc wrät hió wǽre torbegete, Lchdm. ii. 114, 4. Neorxena wang, Gen. 2, 9 : 3, 8. Neorxnewong, Hpt. Gl. 447, 2. Nearxnewang, Hy. Surt. 64, 25. Nerxnewang, 47, 12. Nercsna ([n]erexna, Rush.) wong, Lk. Skt. Lind. 23, 43. Neirxna wong, Mt. Kmbl. p. 8, 5 : Rtl. 124, 7. Nerxna wong, 124, 3.

neósan; *p.* de (?) *with gen. acc. or clause.*   **I.** *to search out, find out by enquiry :*—Wolde ic ānes tō ðē cræftes neósan ðæt ðū mē getǽhte hū ðū sǽhengeste sund wīsige *one art would I find out by enquiry of thee ; that thou wouldest teach me how for the sea-horse thou guidest its swimming,* Andr. Kmbl. 968; An. 484. Hý neósan cwōman, hwæðere him ðæs wonges wyn swedrade, Exon. Th. 123, 12; Gū. 321.   **II.** *to seek, visit* (a) a place :—Gewāt his beddes neósan *Holofernes sought his couch,* Judth. Thw. 22, 15; Jud. 63 : Beo. Th. 3587; B. 1791. Setles neósan, 3576; B. 1786. Ceóles neósan, 3617; B. 1806 : Andr. Kmbl. 620; An. 310. Ēðles neósan, 1660; An. 832 : 2050; An. 1027. Burga neósan, Elen. Kmbl. 304; El. 152. Wīca neósan, Beo. Th. 251; B. 125 : Exon. Th. 184, 5; Gū. 1339. Þýstra, wīta neósan *to seek hell,* 275, 23; Jul. 554 : 280, 18; Jul. 631. Hāmes niósan, Beo. Th. 4722; B. 2366 : 4765; B. 2381. Ēce staðulas neósan, Cd. Th. 207, 30; Exod. 474. (b) a person :—Ūser neósan, Beo. Th. 4155; B. 2074. Com ða hæðenra hlōþ háliges neósan, Andr. Kmbl. 2778; An. 1391 : Exon. Th. 170, 30; Gū. 1119. Ic his neósan wille, 145, 8; Gū. 691. Word āres oft neósendes (ðīn), 175, 6; Gū. 1190.   **III.** *to seek with hostile intent* (cf. sēcan) :—Wyrm yrre cwom fiónda niósan, laðra manna, Beo. Th. 5336; B. 2671. [*Goth.* bi-niuhsjan *to spy out : Icel.* nýsa *to pry, enquire : O. Sax.* niusian : *O. H. Ger.* niusian *niti, conari.*] v. next word.

neósian; *p.* ode *with gen. acc. or clause.*   **I.** *to search out, find out by enquiry* or *inspection, to inspect :*—Wolde neósian Nergend, hwæt his bearn dyde, Cd. Th. 53, 2; Gen. 855. Gewāt neósian heán hūses hū hit Hring-Dene gebūn hæfdon *he came and inspected the lofty house, how the Hring-Danes had ordered it,* Beo. Th. 230; B. 115.   **II.** *to seek, visit* (a) a place :—Wæs his gewuna ðæt Norþanhymbra mǽgþe sōhte and neósode *solebat Nordanhymbrorum provinciam revisere,* Bd. 3, 23; S. 554, 7. Gewiton him wīgend wīca neósian, Frysland geseón, Beo. Th. 2255; B. 1125. (b) a person :—Mannes sunu ðe ðū neósast (*visitas*), Ps. Th. 8, 5. Neósode hē mīn eft *me revisens,* Bd. 5, 6; S. 619, 43. Se hine æghwylce daga neósade, Exon. Th. 162, 11; Gū. 974. Hwīlum mennisce āras neósedon (hine or his), 157, 16; Gū. 892. Ic wæs on ðæm carcerne and gē mīn neósodon, L. E. l. 32; Th. ii. 428, 29. Se leófa cuma se ðe gewunade ūre brōðer neósian (*visitare*), Bd. 4, 3; S. 568, 17. Ðone ðe hī untrumne neósian cōman, 4, 11; S. 579, 40: R. Ben. 17, 2. Ðonne Drihten ūre hwylces neósian wille, Blickl. Homl. 125, 13.   **III.** *to seek with hostile intent* (cf. sēcan), *to visit with calamity, disease,* etc. :—Leomu hefegedon, hē gecneów ðæt hine ælmihtig ufan neósade (cf. the phrase *the visitation of God*), Exon. Th. 159, 24; Gū. 931. Ðǽr Ongenþeów Eofores niósade (MS. niosað), Beo. Th. 4966; B. 2486. [*O Sax.* niusôn.] v. ge-neósian *and preceding word.*

neó-síþ, es; *m. Death :*—Se sceal æfter neósíþum wunian wītum fæst Exon. Th. 316, 27; Mōd. 55.

neósung, e; *f. A visiting, visitation :*—Synna forgyfenys, hūselgang and Godes neósung sind eallum gemǽne, Homl. Th. i. 64, 32. Johannes wearþ on ðysum ðære tō heofenan rīces myrhþe þurh Godes neósunge genumen, 58, 4. Mid ðȳ ðā æfter langre tīde com tō him for neósunge intingan (*gratia visitationis*), Bd. 4, 3; S. 569, 41. Būton niósunga *absque visitatione,* Kent. Gl. 710. v. ge-neósung.

neótan, niótan; *p.* neát, *pl.* nuton *To enjoy, have the benefit of, make use of,* (a) with gen. :—Brūc ðisses beáges and ðisses hrægles neót,'Beo. Th. 2439; B. 1217. Níotaþ inc ðæs ōðres ealles *all other take for your use,* Cd. Th. 15, 18; Gen. 235. Līfes, feores neótan tō hyw. Hwylc is manna ðæt feores neóte *quis est homo, qui vivet,* Ps. Th. 88, 41 : Exon. Th. 328, 14; Vy. 17. Níótan, Cd. Th. 31, 17; Gen. 486 : 26, 4; Gen. 401. Mīnes ēðelríces eádig neótan, Exon. Th. 89, 25; Cri. 1462 : 223, 18; Ph. 361 : 356, 14; Pa. 11. Geofona neótan, 225, 5; Ph. 384 : 152, 6; Gū. 804. Willum neótan blǽdes and blissa, 184, 21; Gū. 1347 : 82, 26; Cri. 1344. Ðæt hē ðær brūcan mōt wonges mid willum, and welan neótan līfes and lissa, 208, 2; Ph. 149. Sēcan swegles dreámas and (ðara dreáma) willum neótan, Andr. Kmbl. 1620; An. 811. Wǽpna neótan *to make good use of his weapons,* Byrht. Th. 140, 55; By. 308. (b) with acc. :—Ic ðē on ða fægran foldan gesette tō neótenne neorxna

wonges beorhtne blǽdwelan, Exon. Th. 85, 14; Cri. 1391. [*Gòth.* niutan : *O. Sax.* niotan : *Icel.* njóta : *O. Frs.* niata : *O. H. Ger.* niuzan (*with gen. and acc.*) *uti, frui.*] v. be-, (bi-)neótan.

neóten. v. níten.

neoþan; *adv. Down, beneath, from beneath :*—Nyþan (niþan, neoþan) *dedeorsum,* Ælfc. Gr. 38; Zup. 238, 10. On heofenum and on eorþan neoþan *in coelo sursum et in terra deorsum,* Jos. 2, 11. Ealle stōwa hē neoþan underwrepeþ, Blickl. Homl. 23, 20. Ðæt wæter wæs sweart under ðæm clife neoþan, 211, 2 : Cd. Th. 20, 18; Gen. 311. Wrætlīc is seó womb neoþan, Exon. Th. 219, 14; Ph. 307 : 392, 2; Rä. 11, 1 : 407, 14; Rä. 26, 5 : 414, 14; Rä. 32, 20. Ðū mē of neowelnesse neoþan álýsdest, Ps. Th. 70, 19 : 103, 7 : Elen. Kmbl. 2228; El. 1115. Neoþan, Exon. Th. 479, 11; Rä. 62, 6. v. be-, wið-neoþan, *and next word.*

neoðane; *adv. Beneath, below :*—Hēr is fýr micel ufan and neoþone, Cd. Th. 24, 8; Gen. 375. Ufane and neoþane, Met. 20, 141. [*O. Sax.* niðana : *O. H. Ger.* nidana *subtus, subtu.*]

neoþan-weard; *adj. Low in position :*— Nioþanweard hype *ilia,* Wrt. Voc. ii. 110, 54.

neoþemest. v. *next word and* neoþor.

neoþera, niþera; *adj.* (*without a positive form*) *Lower :*—Neoþera welor *albrum* (= *labrum*), Wrt. Voc. ii. 7, 79 : i. 282, 71. Niþera lippe *labrum,* 43, 25. Ðū gencredest mīne sáule of ðære neoþeran helle, Blickl. Homl. 89, 28. Neoþran, Ps. Spl. 85, 12. On seáðe ðam neoþeran *in lacu inferiori,* 87, 6. Cyng āh ðone uferan and bisceop ðone nyþeran, L. E. G. 4; Th. i. 168, 16. On nyþerum eorþan *in inferioribus terrae,* Ps. Spl. 138, 14. On ða neoþran eorþan, 62, 9. On ðás niþeran ðǽlas ðisse ceastre, Blickl. Homl. 239, 6. Yfemest is eallra gesceafta fýr ofer eorþan, folde neoþemest, Met. 20. 85. On ðære nyþemystan (*lowest*) bytminge, Homl. Th. i. 536, 10. Ða niþemestan ic gebrenge æt ðam hēhstan, and ða hēhstan æt ðam niþemestan, Bt. 7, 3; Fox 20, 35. Gē underþióðaþ eówre hēhstan medemnesse under ða eallra nyþemestan gesceafta, 14, 2; Fox 44, 34. On ða neoþemestan helle wītu, Blickl. Homl. 185, 6.

neoþe-, nioþo-, niþe-weard; *adj. Low, situated beneath, bottom of* (the noun with which the adjective agrees) :—Niþeweard fōt *planta ;* hōh niþeweard *calx,* Wrt. Voc. i. 283, 73, 75. Is se hals grēne nioþoweard and ufeweard, Exon. Th. 218, 23; Ph. 299. On nyþewerdum ðam munte *ad radices montis,* Ex. 19, 17. Hē (*Noah's ark*) wæs on nyþeweardan wið, and on ufeweardan nearo, Homl. Th. i. 536, 9. Wyrc hié of nioþoweardre netlan, Lchdm. ii. 128, 6. Wyl neoþewearde netlan, 312, 5. Lege on ðone pyt neoþeweardne *lay it at the bottom of the pit,* i. 398, 22. Neoþouard *crepidinem,* Wrt. Voc. ii. 98, 5. Tósliten of ufewerdum ōþ neoþewerd (nioðuord, Lind. : nioþawordum, Rush.), Mk. Skt. 15, 38. Nyþeweard (nioþaweard, Lind. : neoþewearde, Rush.), Mt. Kmbl. 27, 51. Of neoþeweardum *imis,* Wrt. Voc. ii. 43, 57. Fram his hnolle ufewerdan ōþ his ilas neoþewearde *from the crown of his head to the soles of his feet,* Homl. Th. ii. 452, 27.

neoþor, nioþor, niþor; *adv.* (*without a positive form*) *Lower, in an inferior position :*—Niþor *inferius,* nyþemyst *inferius,* Ælfc. Gr. 38; Som. 42, 14. Se ðe wæs neoþor on endebyrdnysse wearþ fyrmest on þrowunge he (*Stephen*) *that was lower in order, was first in suffering,* Homl. Th. i. 50, 4. Ðá heó ðá hié in ðæm gefeohte neoþor gesēgon *qui dum se inferiores in bello conspicerent,* Bd. 3, 18; S. 546, 16. Ðē læs ðe ðæt mōd sý neoþer ðonne se līchoma, Homl. Skt. 1, 58. Nioþoror, Bt. 41, 6; Fox 254, 31. Sió eorþe is nioþor ðonne ænig ōðru gesceaft, Bt. 33, 4; Fox 130, 20. Nioþor hwæne, Beo. Th. 5392; B. 2699. Ðæt mōd glīt nioþor and nioþor (niþor and niþor, Hatt. MS.) stæpmǽlum, Past. 38, 7; Swt. 278, 2. Hine nyþor äsette Metod *the Lord humbled him* (*Nebuchadnezzar*), Cd. Th. 247, 7; Dan. 493.

neówan, neówe, neowel, neówian, neówinga, neówness. v. nīwan, nīwe, neowol, nīwian, nīwinga, nīwness.

neowol, nifol, nihol, nihold, neól, niwol; *adj.*   **I.** *prone, prostrate :*—Nihol *pronus,* Ep. Gl. 20 b, 2. Nihold, Wrt. Voc. ii. 118, 20. Hwī līst ðū neowel on eorþan *cur jaces pronus in terra?* Jos. 7, 10. Hē feóll niwel on ða eorþan, Gen. 33, 3. Niwol, Bt. 1; Fox 4, 3. Neowol, Met. 1, 80. Ðærrihte fērde eall seó heord myclum onrǽse niwel on sǽ *ecce impetu abiit totus grex per praeceps in mare,* Mt. Kmbl. 8, 32. Neól ic fēre, Exon. Th. 403, 2; Rä. 22, 1. Hīt swá niowul (*prostrate*) up ārǽrde, Bt. 3, 1; Fox 4, 26. Neowle nihtscūwan *the shades of night that had settled down upon earth,* Cd. Th. 184, 28; Exod. 114. Ða neowlan *cernua,* Wrt. Voc. ii. 78, 59. Neóle *cernuas,* 83, 3. Nióle, 18, 42. Nifle nædran cynn *serpentes,* Ps. Th. 148, 10.   **II.** *deep down, low, profound* (v. neowolness) :—Niól *infima,* 110, 73. Under neólum niþer næsse *deep underground,* Elen. Kmbl. 1660; El. 832. In ðam neólan scrǽfe *in that deepest den* (*hell*), Exon. Th. 283, 23; Jul. 684. In ðissum neowlan genipe (*hell*), Cd. Th. 271, 7; Sat. 102. In ðone neowlan grund *to that profound abyss,* 267, 1; Sat. 31 : 270, 16; Sat. 91. In ðis neowle genip, 275, 31; Sat. 180 : 292, 25; Sat. 446. Drihten for ðē of ðæm heán heofone on ðas neowlan gesceaft niðer ästäh *for thee the Lord descended from the high heaven to this lower world,* L. E. I. prm. ; Th. ii. 396, 2. Gē beóþ foræltene on ðone neowlan helle seáð *ye shall be dismissed to the bottomless*

pit, 396, 18. Gǽst ellor hwearf under neowelne næs, Judth. Thw. 23, 9; Jud. 113. Sunne gewât tó sete glídan under nifian næs, Andr. Kmbl. 2611; An. 1307. Nyþer gefeallaþ under neowulne grund *descendunt usque ad abyssos*, Ps. Th. 106, 25. Neowle næssas *low-lying headlands*, Beo. Th. 2826; B. 1411; Niþer under næssas, neóle grundas (*hell*), Exon. Th. 136, 3; Gú. 535.

neowol-líc; *adj. Profound :*— Hé siccetunga teáh of niwellícum breóste *he heaved sighs from the depths of his breast*, Homl. Skt. 7, 66.

neowolness, e; *f. A deep place, an abyss :*—Neowelnys *abyssus*, Ps. Spl. 35, 6. Seó neólnes cliopaþ tó ðære neólnesse *abyssus abyssum invocat*, Ps. Th. 41, 8. Ealle wyllspringas ðære micelan niwelnesse, Gen. 7, 11 : 1, 2. Of neowelnesse *de abyssis terrae*, Ps. Th. 70, 19. In neólnesse, in súsla grund, Elen. Kmbl. 1882; El. 943. Ealle neowelnessa *omnes abyssi*, Ps. Th. 148, 7. Neowelnyssa, Cant. Moys. 5. Neólnessa, Blickl. Homl. 93, 12. On þa neowolnesse ðæs seáþes *in profunda*, Bd. 5, 12; S. 628, 21. Neólnisse *abyssos*, Ps. Surt. 32, 7, Nywolnessa, Ps. Th. 103, 7.

nép; *adj. Lacking, scanty* (?) :—Mægen wæs on cwealme fæste gefeterod forþganges *nép the force of the Egyptians was fast fettered in death, they could make no advance (when they were overwhelmed in the Red Sea*), Cd. Th. 207, 20; Exod. 469. v. next word.

nép-flód, es; *m. A neap-tide, a very low tide :*—Népflód *vel* ebba *ledona*, Wrt. Voc. i. 57, 11: *ledo*, 63, 74: ii. 98, 22. On ǽlcùm ânum geáre weaxeþ ðæt flód ðæs sǽs feówer and twentigum síða, and swá oft swáþ; fylleþflód biþ némned on lǽden *malina*, and se népflód *ledo*, Shrn. 63, 31. [Cf. Eng. Gilds (E. E. T. S.), p. 425, 30, where ' *neep* sesons ' are mentioned, the times of neap-tides.]

nepte, nefte, an; *f. Nep or nip* (v. E. D. S. Plant-Names), *cat's mint :*—Nepte *nepita*, Wrt. Voc. ii. 62, 40. Næpte, i. 30, 21. Nepte. Ðás wyrte man *nepitamon*, and óðrum naman nepte nemneþ, and eác Grécas hý *mente orinon* hátaþ, Lchdm. i. 208, 7–9. Nefte, ii. 122, 13 : 316, 5 : 318, 12. Neptan sǽd, iii. 72, 11. Wyl neftan, ii. 62, 25 : 76, 19 : 142, 3 : 266, 11. [*Prompt. Parv.* nepte *nepta*.]

ner, es; *n. A refuge :*—Geworden is [Dryhten] ner oððe rótnes ðam þearfan *factus est Dominus refugium pauperi*, Ps. Lamb. 9, 10. v. ge-ner.

nergend, nerigend, neriend, es; *m. A saviour, preserver :* — Dec, mihtig God, nergend, Cd. Th. 239, 24; Dan. 375. Crist nergend, Hy. Grn. ii. 291, 39. Dryhten God, nerigend fira, Andr. Kmbl. 2573; An. 1288. Neregend, 581; An. 291. Se Godes cwide is folces nerigend (MS. B. neriend), Salm. Kmbl. 162; Sal. 80. Nergendes hǽs *God's command*, Cd. Th. 173, 19; Gen. 2863. Nergende leóf, 77, 35; Gen. 1285. Ealra fǽmnena cwén cende ðone sóþan Scyppend and ealles folces Fērfrend, and ealles middangeardes Hǽlend, and ealra gâsta Nergend, and ealrasáula Helpend, Blickl. Homl. 105, 18. [*O. Sax.* neriand (*Christ*).]

nerian; *p.* ede *To save :*—Wyrd oft nereþ unfǽgne eorl *if a man's death be not doomed, oft destiny saves him,* Beo. Th. 1149; B. 572. Of neádum heora hé nerode (*eripuit*) hig, Ps. Spl. 106, 6. Hié hálig God nerede, Cd. Th. 84, 13; Gen. 1397: 90, 6; Gen. 1491. Hí freá nerede fram hellcwale, Exon. Th. 73, 14; Cri. 1189. Ðín ealdor nere, Cd. Th. 151, 2; Gen. 2502. Ðæt ðú nerige (*eruas*) mê, Ps. Spl. 39, 18. Se ðe wyle oððe sceall nerian *eruiturus*, Ælfc. Gr. 41; Som. 44, 26. Hwæt God wolde nergan wið níþum, Exon. Th. 135, 10; Gú. 525. Gewiton feorh heora fleáme nergan, Cd. Th. 120, 126; Gen. 2000. Nergean, 151, 16; Gen. 2509. Tó nergenne, 234, 1; Dan. 285. Tó nerganne, Exon. Th. 185, 11; Az. 6. Neriende Crist (cf. *O. Sax.* neriendi Krist), Hy. Grn. ii. 286, 4, 28. Nerigende, Cd. Th. 238, 15; Dan. 355. Nergende Crist, 300, 25; Sat. 570. [*Goth.* nasjan : *O. Sax.* nerian : *O. Frs.* nera : *O. H. Ger.* nerian *alere, pascere, sustentare, salvare* : *Ger.* nähren : cf. *Icel.* næra *to nourish.*] v. ge-nerian.

nering, e; *f. Protection, defence :*—Nerin[ge] *presidio, protectionis,* Hpt. Gl. 527, 68.

Neron, es; *m. Nero :*—Neron cwæþ, Blickl. Homl. 175, 33. Nerones wíf Libia, 173, 13. Tó Nerone, 173, 10.

nerwet. v. nirwett.

nesan; *p.* næs; *pl.* nǽson; *pp.* nesen *To be saved from, to escape from :*—Ðam ðe mid sceolon mereflód nesan *those who are to be saved with you from the flood (the living creatures in the ark with Noah*), Cd. Th. 81, 7; Gen. 1341. v. ge-nesan.

nese (=ne sí); *adv. No* (*the opposite of* gese) :—Wylt ðú ðis? Nese *vis hoc*? *Non*, Ælfc. Gr. 38; Som. 40, 13. Wylt ðú wé gadriaþ hig? Ðá cwæþ hé, Nese (*non*), Mt. Kmbl. 13, 29. Syllaþ ûs of eówrum ele ... Ðá andswarudun ða gleáwan, Nese, 25, 9. Ðá cwæþ hé : Nese (Lind. næsæ) fæder Abraham, Lk. Skt. 16, 30. Sume cwædon, he is gód; óðre cwædon, nese (Lind. næse), ac hé beswícþ ðis folc, Jn. Skt. 7, 12. Næsi, Jn. Skt. Lind. 21, 5. Hwæðer ðú swelces áuht geworhtes habbe. Nese, nese, Bt. 14, 1; Fox 40, 26, 33.

-nes[s], -nes[s], -nys[s], a frequently occurring suffix of feminine abstract nouns, cf. *Goth.* -assus, e. g. ufar-assus : *O. H. Ger.* -nessî; *f.* nessi; *n.;* -nissa, -nissi; *f.* -nissi; *n.* v. Grimm. Gram. ii. 321 sqq.

ness. v. næss.

nest, es; *n. A nest;* also *the young birds in the nest;* nidus :—Nest *nidus*, Wrt. Voc. i. 77, 39 : Ælfc. Gr. 8; Som. 7. 30. Ic in mínum neste neóbed ceóse, ' *I shall die in my nest*,' Exon. Th. 235, 6; Ph. 553 : 212, 25; Ph. 215. Nest timbran, gearwian, getimbran, wyrcan *to build a nest*, 210, 20, 21; Ph. 189 : 228, 2; Ph. 432 : 229, 6; Ph. 229. Ðíne bearn gegaderian swá se fugel deþ his nest (*nidum*) under his fiðerum, Lk. Skt. 13, 34. Heofones fuglas habbaþ nestþ (MS. A. nest : Lind. nesto) *volucres coeli habent nidos*, 9, 58. Nest (Lind. nestas ł nesto), Mt. Kmbl. 8, 20 : Homl. Th. i. 160, 34. [*O. H. Ger.* nest *nidus.*] v. nistian, nestlian.

nest, es; *n.* I. *provisions, victuals :*—Se him his nest áspringeþ *he whose provisions fail him*, Exon. Th. 335, 23; Gn. Ex. 38. Sum sceal on feorwegas gongan, and his nest beran, 329, 3; Vy. 28. On ðæm sǽtelse ðe hyre foregenga hyra begea nest þyder lǽdde, Judth. Thw. 23, 19; Jud. 128. II. *provisions served out at fixed times, rations :*—Nest *epimenia* (ἐπιμήνια, cf. fóstraþas epimoenia, 32, 41. *Epimenia* expensae vel exennia vel tributa quae dantur per singulos menses, Ducange), Wrt. Voc. ii. 107, 32. Ða cempan cwǽdon : Hwæt dó wê? Ðá sǽde hé him : Beóþ eðhylde on eówrum andlyfenum (Lind. Rush. nestum = *stipendiis*), Lk. Skt. 3, 14. [*Icel.* nest; *n. provisions*: *O. H. Ger.* wega-, fart-nest *viaticum.*] v. weg-nest, nest-pohha.

nést, nésta. v. neáh, níhsta.

nestan; *p.* te *To spin :*—Ne wynnes and ne nestas *non laborant neque nent*, Mt. Kmbl. Lind. 6, 28. Nestaþ, Lk. Skt. Lind. Rush. 12, 27. [Cf. *Icel.* nista *a pin*; nista *to pin*: *O. H. Ger.* nestilo, nestila *vitta, funiculus, redimiculum, vitta, fibula, ansa* : *Ger.* nestel.]

nestlian; *p.* ede *To make a nest :*—Ðár spearwan nestliaþ *illic passeres nidificabunt*, Ps. Lamb. 103, 16. [Þar nestleþ (1st MS. næstieþ) hearnes, Laym. 21753. Nestlyñ *nidifico*; nestlyd *nidificatus*; nestelynge *nidificatio*, Prompt. Parv. 354.] v. nistian, nistlan.

nest-pohha, an; *m. A bag for food, wallet :*—Nestpoha *pera*, Mt. Kmbl. Lind. 10, 10. [Cf. *Icel.* nest-baggi *a wallet.*]

neta, an; *m. A caul :*—Inilve *intestinum* ; midhryþre *onentem* ; neta *disceptum* ; blind þearm *cecum*, Wrt. Voc. i. 284, 2–5. v. nette.

netele, netle, netel, an; *f. A nettle :*—Netele *urtica*, Wrt. Voc. i. 289, 43 : ii. 65, 49. Netle, i. 31, 60 : 68, 25. Netel (netele, netle) *urtica* ; blind netel (netele, netle) *archangelica*, 79, 30, 31. Netle, blinde netle, Lchdm. ii. 66, 4. Netele. Genim ðysse wyrte seáw ðe man *urticam*, and óðrum naman netele nemneþ, i. 310, 14–16. Seó reáde netele *lamium purpureum*, iii. 52, 11 : ii. 58, 10 : 92, 10. Netelan sǽd, i. 228, 24 : ii. 94, 12. Of nioþoweardre netlan, 128, 7. Nim netelan, 152, 10 : 312, 5. Ða greátan netlan (*urtica dioica*), 86, 12. Smale netelan (*urtica urens*), 68, 4. Netlan *verticeta*, Wrt. Voc. ii. 124, 20. [*O. H. Ger.* nezila] v. worþig-netele.

néten, neteness. v. níten, nytenness.

néðan; *p.* de *To have courage to do, to dare to do, to venture :*—Néþeþ hwílum meówle ðæt heó on mec grípeþ *the maiden has at times the courage to lay hold on me*, Exon. Th. 407, 15; Rä. 26, 5. Néðde ðær ic Neron beswâ : *I dared to go where I deceived Nero*, 260, 24; Jul. 302. Hé in ðæt búrgeteld néðde *he ventured into the pavilion*, Judth. Thw. 25, 25; Jud. 277. Git on deóp wæter aldrum néðdon *ye ventured into deep water at the risk of your lives*, Beo. Th. 1024; B. 510 : 1080; B. 538. Ic néðan gefrægn hæleþ tó hilde *I have heard that warriors dared to do battle*, Cd. Th. 124, 9; Gen. 2060. Néðan on nacan tealtum *to venture upon the unsteady vessel*, Runic pm. Kmbl. 343, 21; Rún. 21. In nearowe néðan, Exon. Th. 436, 13; Rä. 54, 13. [*Goth.* ana-nanþjan *to be bold* : *Icel.* nenna *to strive, have a mind to* : *O. H. Ger.* nendjan *insurgere* ; gi-nendjan *audere.*] v. ge-néðan.

néðing, e; *f. Daring, audacity :*—Ðæt hé þurh néðinge wunne, Exon. Th. 109, 33; Gú. 99. Ða swá swíðe hiene ondrédan ðe on westeweardum ðisses middangeardes wǽron ðæt hié on swá micle néðinge ... hiene æfter friþe sóhton on eástweardum ðeosan middangearde *those who were in the west of this earth feared him (Alexander) so much, that they had the courage to visit him in search of peace in the east of this earth*, Ors. 3, 9; Swt. 136, 24. [*Icel.* nenning *activity, energy*: cf. *O. H. Ger.* nendigî *audacia.*]

net-gearn, es; *n. Net-yarn, string for making nets :* — Án cliwen gódes nettgernes, Cod. Dip. Kmbl. iii. 451, 7.

net-ráp, es; *m. A toil :*—Netrápas *plagas*, Wrt. Voc. i. 48, 26: 57, 21.

nett, es; *n.* I. *a net* (for fowling, fishing, or hunting) :—Net *rete*, Wrt. Voc. i. 285, 16. Nyt, 73, 41. Ned *cassis*, ii. 14, 3. Hyra net wæs tóbrocen, Lk. Skt. 5, 6. Ures fisceres nett *nostri piscatoris rete*, Ælfc. Gr. 15; Som. 19, 57. Feallaþ on nette his *cadent in retiaculo ejus*, Ps. Spl. 140, 11. Ic mín net út lǽte *laxabo rete*, Lk. Skt. 5, 5 : Mt. Kmbl. 4, 18. Lǽtaþ ðæt nett on ða swíðran healfe, Jn. Skt. 21, 6. Ic brǽde nett *plecto*, Ælfc. Gr. 28; Som. 32, 8. Oþ ðæt ðe hig (wildeór) cuman tó ðám nettan ... Ne canst ðú huntian búton mid nettum? Coll. Monast. Th. 21, 15–21: 22, 11. On feala wísan ic beswíce fugelas, hwílon mid nettum, 25, 11. Hí forlēton hyra nett (netta, Lind.) *relictis retibus*, Mt. Kmbl. 4, 22 : Homl. Th. i. 578, 21. II. *a mosquito-net :*—Nette, fleógryfte *conopio*, Wrt. Voc. ii. 19, 18. III. net-

work, web :—Swâ tedre swâ swâ gangewifran nett, Ps. Th. 38, 12. Ðonne hió (*the spider*) geornast biþ ðæt heó âfære fleógan on nette, 89, 10. Folc gescylde hâlgan nette (*with a net-work of clouds*), Cd. Th. 182, 11 ; Exod. 74. [*Goth.* nati : *O. Sax.* netti, (fisk-)net : *O. Frs.* nette : *Icel.* net ; *gen. pl.* netja : *O. H. Ger.* nezzi *rete.*] v. æl-, boge-, breóst-, deór-, dræg-, feng-, fisc-, fleóh-, here-, hring-, inwit-, mycg-, searo-, wæl-nett, *and next word.*

nette, an; *f. The net-like caul :*—Nette (*under the heading* de membris hominum) *disceptum* i. *reticulum* (cf. hoc reticulum, pinguedo circa jecur, 704, 7), Wülck. Gl. 293, 6. Nettae *oligia*, 35, 34. Nytte *obligia*, Wrt. Voc. i. 45, 18. Nette, ii. 63, 39 : *disceptum*, 26, 19. [*Icel.* netja *the caul :* cf. *O. H. Ger.* nezzi *adeps intestini* ; *pl. intestina.*] v. neta.

neurisn, e; *f. A kind of paralysis :*—Wið paralisin and wið neurisne, Lchdm. i. 12, 21 : 130, 11.

newe-, niwe-, nu-seóða, an; *m. The pit of the belly :*—Be ðam nafolan and bæcþearme and neweseóðan, Lchdm. ii. 232, 1. Niweseóðan, 164, 8. Sió biþ on ða swiðran sîdan âþened óþ ðone neweseóðan, 198, 1 : 242, 19 : 258, 6. Nuseóðan, 160, 12. Cf. (?) seód.

né-west, nêxt, nêxta. v. neáh-west, neáh, nîhsta.

ní-. v. nîw-.

nic=ne ic *not I :*—Wilt ðû fôn sumne hwæl? Nic *vis capere aliquem cetum? Nolo,* Coll. Monast. Th. 24, 17. Eart ðû wîtega ? Hê cwæþ nic, Jn. Skt. 1, 21. Eart ðú ðysses leorningcnihtum ? Ðâ cwæþ hê : Nicc, ne eom ic, 18, 17.

ní-cend, -cumen. v. nîw-cenned, -cumen.

nicor, es; *m.* I. *a hippopotamus :*—Him wæron ða breóst gelíce niecres breóstum *hypopotami pectore,* Nar. 20, 29. Nicoras *hypopotami,* 11, 11. II. *a water-monster :*—Sanctus Paulus wæs geseónde on norþanweardne ðisne middangeard, ðær ealle wætero niþer gewítaþ, and hê ðær geseah ofer ðæm wætere sumne hârne stân ... and under ðæm stâne wæs niccra eardung and wearga. And hê geseah ... manige swearte sâula ... and ða fýnd on nicra onlícnesse heora grípende wæron ... gewitan ða sâula niþer and him onfêngon ða nicras, Blickl. Homl. 209, 29–211, 5. On nicera mere, Beo. Th. 1695 ; B. 845. Ic on ýðum slóg niceras nihtes, 848; B. 422 : 1154; B. 575. Nicras, 2859; B. 1427. [*Icel.* nykr *a sea-goblin; a hippopotamus :* *O. H. Ger.* nichus *a crocodile.* v. Grmm. D. M. 135, 146.]

nicor-hûs, es ; *n. The abode of a 'nicor,'* Beo. Th. 2827 ; B. 1411.

níd, neád, nêd, neód, nied, nýd, es ; *n. :* e ; *f.* I. *necessity, inevitable-ness :*—Neód (nêd, Lind. Rush.) ys ðæt swycdômas cumon *necesse est ut veniant scandala,* Mt. Kmbl. 18, 7 : Homl. Th. i. 514, 33. Gif ðæt nýd âbædeþ *cum ipsa necessitas compellit,* Bd. 1, 27 ; S. 497, 1. Nenne hwylc nýd máre âbædde, 3, 5 ; S. 526, 28. Swâ hyne nýd fordrâf, Judth. Thw. 25, 25 ; Jud. 277. Nýd biþ wyrda heardost, Salm. Kmbl. 622; Sal. 310. Eádfriþ for neóde (neáde, MS. T. : nýde, MS. B.) tó Pendan gebeáh *Eadfrid necessitate cogente ad Pendam transfugit,* Bd. 2, 20 ; S. 521, 15 ; Chr. 1016; Erl. 154, 11. Mid nýde gebæded *necessitate cogente,* Bd. 3, 24 ; S. 556, 7. Nýde genýdde *forced by necessity,* Beo. Th. 2014; B. 1005. II. *necessity, need, urgent requirement :*—Ne nêd is ðê ðætte hwelc ðec gifregne *non opus est tibi ut quis te interroget,* Jn. Skt. Rush. 16, 30. Mê is neód *necesse habeo,* ic habbe neóde *necesse habeo,* Ælfc. Gr. 38 ; Som. 41, 38. Nis Gode nân neód ûre æhta, Homl. Th. i. 140, 24. Nis Gode nân neód ðæt wê gód wyrcan, ne hê nân þing ne hæft for his âgenre neóde, Homl. Skt. 11, 299. Seó þearlwîsnes ðæs heardan lîfes him ærest of nýde becom for bôte his synna ac forþgangendre tîde ðæt hê ðæt nýd on gewunon gecyrde ... *ex necessitate obvenerat, sed ... necessitatem in consuetudinem verterat,* Bd. 4, 25 ; S. 599, 32. Nabbaþ hí neóde tó farenne, Mt. Kmbl. 14, 16 : Lk. Skt. 14, 18. III. *a necessary business, duty :*—Neád *debita,* Wrt. ii. 139, 68. Ûs is neód (*it is our bounden duty*) ðæt wê ða hâlgan eástertîde be ðam sôðan regole healdon, Lchdm. iii. 256, 17. Ûs is twyfeald neód on bóclícum gewritum. Ânfeald neód ûs is, ðæt wê ða bóclícan lâre mid carfullum môde smeágan ; ôðer ðæt wê hî tó weorcum âwendan, Homl. Th. ii. 284, 23. Hê fêrde embe sumere neóde *he was going about some necessary business,* 508, 15. Nolde Maurus of ðam mynstre faran for nânre neóde, bûtan hê nýde sceolde, Homl. Skt. 6, 290. Eádsige hine wel lærde and tó his âgenre neóde and ealles folces manude (*exhorted him with regard to his duty as king*), Chr. 1043; Erl. 168, 5. Hê wolde gân embe his neóde forþ, Homl. Th. i. 290, 18. Gafele ꝥ nêdde (neáde?) *debito, necessitate,* Hpt. Gl. 440, 29. Neóde *debitum,* 456, 14. On ðam tôweardan lîfe ne beóþ ðâs neóda (*the duties of feeding the hungry, etc.*), ne ðâs þênunga Homl. Th. ii. 442, 18. Neódum *causis,* Hpt. Gl. 412, 57. IV. *need, what one wants :*—Ðæt man underfô mâre ðonne his lîchaman sý, Homl. Th. ii. 590, 21. Mid ðý hí ðâ ðæt scyp gehlæsted hæfdon mid ðâm þingum ðe swâ mycles sîþfætes nýd âbædde *cum navi imposuissent quae tanti itineris necessitas poscebat,* Bd. 5, 9 ; S. 623, 18. Ne lufode hê woruldlíce æhta for his neóde âna (*to supply his own needs only*), ac tó dælenne eallum wædliendum, Homl. Th. ii. 340, 21. Gylde se tûnscipe tó ðære muneca neóde (*ad usus*), Chart. Th. 307, 26. Tó ðæs minstres neóde, 362, 7. Ðû hogast embe ðîne neóde, Homl. Th. i. 488, 24. Ðâm mannum ðe heora neóde habbaþ *who have what they*

want, ii. 106, 18. Hê sylþ him his neóde, *he gives him what he wants,* Lk. Skt. 11, 8. God dæghwamlíce ûs deþ ûre neóde *God daily supplies our needs,* Basil Admn. 4 ; Norm. 40, 29 : Homl. Th. i. 516, 9. Ealle ûre neóda ægðer ge gâstlíce ge lîchamlíce, 272, 16. V. *necessity, need, difficulty, hardship, distress :*—Lust hæfþ wîte and neád wuldor-beáh gegearwaþ *pleasure hath punishment and hardship is a preparation for a crown,* R. Ben. 26, 9. Ðâ cwædon hié ðæt him leófre wære ðæt hié an swelcan niéde deáþ fornôme ðonne hié mid swelcan niéde friþ begeáte *tutius rati sese armatos mori quam miseros vivere,* Ors. 4, 6 ; Swt. 174, 25–27. Sume men ða wætan for ðæm nýde þigdon *suam urinam vexatos ultimis necessitatibus haurientes,* Nar. 9, 22. Moises sæde Drihtne ðæs folces neóde, Ex. 15, 25. Of neádum mînum genera mê *de necessitatibus meis erue me,* Ps. Spl. 24, 18 : 106, 6. Nêdum, 30, 9. Niédum, Andr. Kmbl. 2754 ; An. 1379. VI. *force, compulsion :*—Rîccra manna need *vis potestatis,* Wrt. Voc. i. 21, 28. Ne eom ic nânre neáde gecnéwe, Chart. 296, 1. Him beóþ ealle mid nêde (*by force*) on genumene, Blickl. Homl. 49, 26. Mid nænigum nêde gebæded, 83, 32. Ða kyningas ðe ic mid nêde tó hýrsumnesse gedyde, Nar. 32, 19. Mâ hreósende for ealddôme ðonne of æniges cyninges niéde, Ors. 2, 4 ; Swt. 76, 3. Heofena ríce þolaþ neád (*vim*), Mt. Kmbl. 11, 12. Nêd, Ps. Surt. 37, 13. Nýd, Ps. Spl. 37, 12. VII. *the name of the rune,* ᚾ, N ; hence the symbol is sometimes put instead of writing the word, Runic pm. Kmbl. 341, 8 ; Rûn. 10: Exon. Th. 429, 22 ; Rä. 43, 8 : 50, 14 ; Cri. 800 : 284, 28 ; Jul. 704: Elen. Kmbl. 2519; El. 1261. [*Goth.* nauþs: *Icel.* nauð, neyð : *O. Sax.* nôd: *O. Frs.* nêd : *O. H. Ger.* nôt *vis, violentia, exactio, necessitas, tribulatio, angor.*] v. hæft-, nearo-, ôht-, þeów-, þreá-níd ; nêde, nêdes, *and* neód.

nídan ; *p.* de *To force, compel, urge :*—Ic nýde *cogo,* Ælfc. Gr. 28 ; Som. 32, 14. Hê ûs ne nêt (Cott. MS. nêd) tó ðam ðæt wê nêde scylen gód dôn, Bt. 41, 4 ; Fox 252, 3. Hê nýt (*compellet*) eów ðæt gé faron ût, Ex. 11, 1. Hié hié selfe nídaþ (Cott. MSS. niédaþ) tó healdonne swîgean, Past. 38, 1 ; Swt. 271, 16. Se pâpa nêdde Adrianus ðæt hê biscophâde onfênge, Bd. 4, 1 ; S. 564, 6. Gif gê gesâwen hwelce mûs ðæt wære hláford ofer ôðre mýs and nêdde (Cott. MS. nêdde) hié æfter gafole (*exacted tribute from them*), Bt. 16, 2 ; Fox 52, 3. Hê nýdde his leorningcnihtas on scyp stîgan, Mk. Skt. 6, 45. Ne nýdde hê nâ ðæt folc tó his cwale *he did not force the people to kill him,* Homl. Th. i. 216, 1. Ðâ nýdde hê ðone unclænan gâst ût, Lk. Skt. 9, 42. Hê hié nýdde in fæðm fýres, Cd. Th. 230, 14 ; Dan. 233. Ða Egiptiscan nýddon (*urgebant*) ðæt folc ût of hira lande, Ex. 12, 33. Ðâ nýddon hine his yldran tó ðæm ðæt hê sceolde woroldlícum wæpnum onfón, Blickl. Homl. 213, 1. Ðone hig nýddon ðæt hê bære hys rôde, Mt. Kmbl. 27, 32. Nýd *compelle,* Lk. Skt. 14, 23 : Homl. Th. ii. 376, 14. Ne niéde (MS. H. nýde) ðû hine *you shall not press him* (*the debtor*), L. Alf. 35 ; Th. i. 52, 22. Hwæðer seó godcunde foretiohhung oððe sió wyrd ûs nêde tó ðam ðe hí willeu, Bt. 40, 7 ; Fox 242, 15. Nêdendum dôme *urgente decreto,* Hpt. Gl. 488, 68. Ic eom nýded, Bd. 3, 13 ; S. 538, 26. [*Goth.* nauþjan: *Icel.* neyða : *O. Sax.* nôdian : *O. Frs.* nêda : *O. H. Ger.* nôtian.] v. ge-nýdan, neádian.

níd-bâd, e ; *f. An exaction, a due, toll :*—Ic Æðelbald Myrcna cincg wæs beden from bisceope Milrêde ðæt ic him âlêfde alle nêdbâde tuegra sceopa, Chart. Th. 28, 25 : 29, 8. Hê nymeþ nýdbâde *he* (*Grendel*) *takes toll,* Beo. Th. 1200; B. 598. v. bâdian, bædan, níd *debitum,* níd-gafol, *and next word.*

níd-bâdere, es ; *m. One who exacts toll :*—Ic him âlýfde alle nêdbâde tuegra sceopa, ða ðe ðær âbædde beóþ from ðæm nêdbâderum in Lundentunes hýðe, ond næfre ic ne mîne lâstweardas ne ða nêdbâderas geþrîstlæcen ðæt heó hit onwenden, Chart. Th. 29, 7–14. v. preceding word.

níd-bebod, es ; *n. An urgent command, mandate :*—Healdeþ nýdbibod hâlgan Dryhtnes, Exon. Th. 350, 32 ; Sch. 72.

níd-beháfe ; *adj. Necessary, needful :*—Neádbehêfe *necessarium,* Hpt. Gl. 524, 65. Ân þing is niédbehêfe, Lk. Skt. 10, 42. Is ðeáh niéd-behêfe ungeláredum woroldmonnum, Lchdm. iii. 440, 32. Se man wæs ðam dêman þearle nýdbehêfe, Homl. Skt. 4, 144. On eallum ðissum þingum is geþyld nýdbehêfe, Homl. Th. i. 470, 31. Seó hand getácnaþ ûrne nýdbehêfan freónd, ðe ûs ûre neóde deþ, 516, 8. Synd gesealde from ðam abbode ealle neádbehêfe þing, R. Ben. 92, 2 : 127, 5. In Godes lofe and in nýdbehêfum weorcum wê sceolon gewunigan, L. E. I. 42 ; Th. ii. 438, 31. Wê habbaþ ða nýdbehêfestan ânunga âwritene, Boutr. Scrd. 23, 12. v. níd-behôf *and next word.*

níd-behêfe (?) *necessity :*—Ðâ hnêdbihoefe (Lind. nêd) hæfde *quando necessitatem habuit,* Mk. Skt. Rush. 2, 25.

níd-behôf ; *adj. Necessary, needful :*—Ân þing is nýdbehôf, Homl. Th. ii. 440, 9. v. níd-behêfe.

níd-behôflíc ; *adj. Necessary :*—Hê sæde ðæt him wære his líf nýdbehôflíc *quia multum necessaria sibi esset vita ipsius,* Bd. 5, 5 ; S. 618, 3.

níd-beþearf ; *adj. Necessary :*—Sumæ bêc ða ðe niédbeþearfosta (níd-beþyrfesta, Cott. MSS.) sîen, Past. præf. ; Swt. 7, 7.

níd-boda, an; *m. One who announces violence or distress* (v. níd, V,

VI):—Sincalda sǽ, nýdboda (*the Red Sea which overwhelmed the Egyptians*), Cd. Th. 207, 29; Exod. 474.

níd-brýce, es; *m. Necessary use, requirement, need :*—Ðá wolde se hálga sum hús timbrian tô his nêdbrícum, Homl. Th. ii. 144, 31.

níd-bysig; *adj. Troubled by distresses :*—Ðǽr (*in hell*) ðú (*the devil*) nýdbysig fore oferhygdum eard gesôhtes, Exon. Th. 267, 31; Jul. 423.

níd-bysigu, -bysgu; *f. Distress, trouble :*—Nýdbysgum neáh, Exon. Th. 354, 11; Reim. 44.

níd-clamm, -clomm, es; *m. Necessity, need, distress :*—Of neádclammum heora hê álǽdde *de necessitatibus eorum eduxit*, Ps. Lamb. 106, 28.

níd-cleofa, -clafa (?), an; *m. A prison :*—Ðæt hine man of nearwe and of nýdcleofan fram ðam engan hofe up forlête, Elen. Kmbl. 1419; El. 711. In nêdcleofan nearwe geheaðrod, 2249; El. 1276. Ðá wæs carcernes duru behliden . . . symle heó wuldorcyning herede in ðam nýdclafan, Exon. Th. 256, 31; Jul. 240. v. next word.

níd-cofa, an; *m. A prison :*—Se hálga wæs lǽded in ðæt dimme ræced, sceal ðonne in neádcofan nihtlange fyrst wunian, Andr. Kmbl. 2619; An. 1311.

níd-costing, e; *f. A distressing trial, affliction :*—Nearwum genǽged nýdcostingum, Exon. Th. 171, 14; Gú. 1126.

níd-dǽda, an; *m. One who does something under compulsion :*—Gif hê æfter sunnan upgonge ðis dêþ (*kills the housebreaker*), hê biþ mansleges scyldig, and hê ðonne self swelte, búton hê niéddǽda (nýd-, MS. H.) wǽre (*unless he were forced to do it in self-defence*), L. Alf. 25; Th. i. 50, 21. Cf. Se ðe hine nêdes ofslôge oððe unwillum oððe ungewealdes, 13; Th. i. 46, 22. v. níd-wyrhta.

níde, neáde, neóde, níðe, niéde, nýde; *adv.* (*a case of* níd, q. v.). I. *of necessity, as a natural, inevitable consequence, from force of circumstances :*—Gif gê neáde swá dôn sceolon (*si sic necesse est*), dôþ swá gê wyllon, Gen. 43, 11. Wegfêrende môton for neóde mete neáde ferian and for unfriþe man môt freólsǽfenan nýde fulfaran betweonan Eferwîc, and six míla gemete *travellers may, when compelled by circumstances, carry food to supply their needs; and on account of war, a man may, on the eve of a festival, when compelled by circumstances, travel between York and a distance of six miles*, L. N. P. L. 56; Th. ii. 298, 25–27. Forðamðe wê witon ðæt ân wealdend is eallra þinga wê sceolon beón nêde geþafan (*we must inevitably assent to the conclusion*) ðæt hê sie se hêhsta hróf eallra gôda, Bt. 34, 12; Fox 154, 7. Ðes middangeard nêde (*as the result of natural, inevitable laws*) on ðás eldo endian sceal, Blickl. Homl. 117, 35. Wæs his fæder cininges þegna aldorman. Ðá sceolde Sanctus Martinus nêde (*as an inevitable result*) beón on his geógoþháde on ðære gefærǽdenne cininges þegna, 211, 22. Niéde sceal bión gebrocen ðæt môd ðara hiéremonna, gif se láreów ágiémleásaþ ðæt hê hiera ûtan ne helpe, Past. 18; Swt. 137, 13: Ors. 5, 2; Swt. 218, 20. Ðǽr ðǽr ðú neóde irsian scyle, gemetiga ðæt ðeáh, Prov. Kmbl. 24. Hit is on worulde swá leng swá wyrse, and swá hit sceal nýde ǽr Antecristes tôcyme yfelian stýde, Wulfst. 156, 4: 157, 8. II. *of necessity, because a law, natural, moral or human, is to be satisfied :*—Ðis sceal se mæssepreóst nêde bebeódan *the priest is bound by his duty to proclaim this*, Blickl. Homl. 49, 6. Ðone andleofan ðe hê nêde big lifgean sceolde (*the provision that nature required*), 213, 20. Ðâs bêc sceal mæssepreóst nêde habban (*these books are indispensable*), and hê ne mæg bútan beón, L. Ælfc. C. 21; Th. ii. 350, 15. Niéde hê sceolde him forgyfan ânne (*custom required it*), Lk. 23, 17. III. *from force, under compulsion, without free-will :*—Nân nyle onginnan ðæt ðæt hê nele, búton hê nêde scyle (*unless he is forced*), Bt. 36, 3; Fox 176, 9: 41, 4; Fox 252, 3. Sceal nêde riht wyrcean se ðe ǽr nolde, L. O. D. 3; Th. i. 354, 9. Nêde oððe lustum hêran, Met. 9, 44. Niéde sceoldon gombon gieldan, Cd. Th. 119, 10; Gen. 1977. v. next word.

nídes; *adv. Of necessity, not willingly :*—Se ðe hine nêdes (nýdes, MS. G.) ofslôge, L. Alf. 13; Th. i. 46, 22. v. níd-dǽda *and preceding word*.

níd-, nýd-fara, an; *m. One who journeys under compulsion, who is forced to march*, Cd. Th. 191, 1; Exod. 208. v. *next word and* níd-genga, -dǽda.

níd-faru, e; *f. A journey one is forced to take, death :*—Fore there neidfaerae naenig uuirthit thoncsnotturra than him tharf sîe, Archaeologia, vol. xxviii. p. 357. v. níd-gedâl.

níd-freónd, es; *m. One closely connected by relationship or friendship :*—Hê wæs pápan ǽhte bifealden Enagrius his neódfreóndes, Shrn. 36, 4. [*O. H. Ger.* nôt-friunt; *pl. necessarii*.] v. níd-gesteatla, -mǽg.

níd-gafol, es; *n. A tax that must be paid, tribute :*—Nêdgaefel ðæm cásere *tributum Caesari*, Mt. Kmbl. p. 18, 2. Ymbe ðæt neádgafol ûres Drihtnes, ðæt sýn ûre teóþunga and cyricsceattas, L. Edg. S.; Th. i. 270, 26, 13. v. níd-gild.

níd-gedâl, es; *n. An inevitable parting, the parting of body and soul :*—Nis nú swíðe feor ðam ýtemestan endedôgor nýdgedâles, Exon. Th. 172, 9; Gú. 1141. Se Dryhtnes dôm wîsade tô ðam nýhstan nýdgedâle, 129, 5; Gú. 416. Þurh nýdgedâl, 158, 9; Gú. 906.

níd-genga, an; *m. One who is forced to go or one who goes in misery :*—Nacod nîdgenga (*Nebuchadnezzar*), Cd. Th. 255, 32; Dan. 633. v. níd-fara.

níd-gestealla, an; *m. One who is closely bound to another by the ties of comradeship :*—Hié â wǽron æt níða gehwam nýdgesteallan, Beo. Th. 1769; B. 882. v. níd-freónd.

níd-geweald, es; *n. Power that is forcibly exercised or that causes distress, tyranny :*—Of deófles nýdgewealde genered, Exon. Th. 89, 2; Cri. 1451.

níd-gewuna, an; *m. A necessary, suitable custom* (v. nêd, IV):—Neádgewuna *debitus usus, i. congruus*, Wrt. Voc. ii. 139, 72.

níd-gild, es; *n. Enforced payment, tribute, exaction :*—Scandlíce nýdgild ûs sind gemǽne, Wulfst. 162, 11. [*Icel.* nauð-gjald.] v. nídgafol.

níd-gilda, an; *m. One who is forced to pay :*—Neádgilda *debitor i. obnoxius, reus*, Wrt. Voc. ii. 139, 74.

níd-gripe (?) *a violent grasp :*—Hyne (*Grendel*) sâr hafaþ in nídgripe (MS. mid gripe; Th. nîþgripe) nearwe befongen, Beo. Th. 1956; B. 976.

níd-hâd, es; *m. Force, compulsion :*—Neádhâde *vim*, Wrt. Voc. ii. 72, 54.

níd-, nêd-, niéd-, nýd-hǽmed, es; *n. Rape*, L. Alf. pol. 25, 26; Th. i. 78, 11–18.

níd-hǽmestre, an; *f. A woman who has been violated, a mistress :*—Nêdhǽmestran *amatricis*, Hpt. Gl. 509, 70.

níd-hǽs, e; *f. A command which is attended by compulsion :*—Man for cyning gebidde and hine bûton neádhǽse heora willum weorðigen *let people pray for the king, and honour him without injunction, of their own accord*, L. Wih. 1; Th. i. 36, 16.

níd-help; *m. f. Help in need, needful help :*—On wîsum scrifte biþ swíðe forþ gelang forsyngodes mannes nýdhelp, L. Pen. 1; Th. ii. 278, 3.

níd-hírness, e; *f. Enforced obedience, servitude :*—In nêdhêrnesse ic bêgo *in servitutem redigo*, Rtl. 6, 9.

níding. v. neádung *and next word*.

nídinga (-unga); *adv. By force, against a person's will :*—Nêdunga *violenti*, Mt. Kmbl. Lind. 11, 12. Ðý læs nêdunga genom Crist menn *ne raperet Christus homines*, Rtl. 197, 35. Woldon hine dôn niédenga (nîdenga, Cott. MSS.) tô cyninge, Past. 3, 1; Swt. 33, 14. Ðâ tugon heó hine nýdinga of ðam mynstre *illum invitum monasterio duxerunt*, Bd. 3, 18; S. 546, 22. Gif hwâ mǽden nýdinga nimþ *si quis puellam invitam ceperit*, L. Ecg. P. ii.130; Th. ii. 186, 20. v. neádunga, nídlinga. -nídla, -nýdlíc. v. þreá-nídla, -nídlíc.

nýd-líce. v. neód-líce.

nídling, es; *m.* I. *one who serves of necessity, a slave, bondman :*—Gif ðû fioh tô borge selle ðínum geféran ðe mid ðé eardian wille, ne niéde ðû hine swá niédling (MS. H. nýdling), L. Alf. 35; Th. i. 52, 22. Hié on cnihtháde wǽron oðerra manna niédlingas *in youth they had been the bondmen of others*, Ors. 2, 2; Swt. 66, 17. Se æðeling bebeád ðæt hié ðâ consulas and witan him beforan drifen swá swá niédlingas, ðæt heora bismer ðý mâre wǽre, 3, 8; Swt. 122, 7. Hý ealle tô nýdlingum him gedydon, 1, 5; Swt. 34, 34. Wæterberere oððe nêdlungum *lixarum*, Wrt. Voc. ii. 52, 73. II. *one who has to serve on board ship, a sailor :*—Nêdling *nauta :* nêdlingas *nauticos*, 60, 30, 29. Ðâ ongunnon ðâ nýdlingas and ðâ scypmen ðâ ancras on ðone sǽ sendan woldon ðæt scyp mid gefæstnian *tentabant navitae anchoris in mare missis navem retinere*, Bd. 3, 15; S. 541, 40.

nídlinga; *adv. By force, against a person's will :*—Gif hwâ mǽden nýdlinga nimþ *si quis puellam invitam ceperit*, L. Ecg. P. ii.13; Th. ii. 186, 20 note. v. neádunga.

níd-mǽg, es; *m. A near kinsman, a cousin :*—Iohannes ûres Drihtenes nýdmǽg, L. Ælfc. P. 9; Th. ii. 366, 37. v. níd-mâge, -freónd, -sibb; *and cf. Icel.* nauð-maðr *a near kinsman*.

níd-mǽgen, es; *n. Force, violence :*—Nêdmægn *vim*, Rtl. 117, 25.

níd-mâge, an; *f. A near kinswoman, a cousin :*—Ǽfre ne geweorðe ðæt cristen man gewífige on ðæs wífes nýdmâgan ðe hê ǽr hæfde, L. Eth. vi. 12; Th. i. 318, 16. Nêdmâgan, L. C. E. 7; Th. i. 364, 24. v. níd-mǽg.

níd-micel; *adj. Very important, urgent :*—Nêdmycel (medmycel, MS. B.) ǽrende wê ðider habbaþ, and ûs is þearf ðæt wê hit gefyllon, St. Andr. 6, 20.

níd-ném, e; *f. A taking by force, rapine :*—Nǽnigum biscope âlýfed sî ôwiht of heora ǽhtum þurh nýdnǽme him on geniman (*violenter abstrahere*), Bd. 4, 5; S. 572, 36. Gif hwâ binnan ðâm gemǽrum ûres ríces reáflâc and niédnǽme dô, L. In. 10; Th. i. 108, 9. [*Cf. O. H. Ger.* nôt-numft *violentia, rapina*.] v. next word.

níd-néman; *p. de To take by force, to force a woman, to ravish :*—Gif hwâ nunnan gewemme oððe wydewan nýdnǽme, L. Eth. vi. 39; Th. i. 324, 25: L. C. S. 53; Th. i. 406, 2, 3. v. níd-niman.

nídness, e; *f. Necessity :*—Ðeáhhwæðere mid nýdnysse hire man môt lýfan ðæt heó mid ðam sig *tametsi si necesse est, licet viro ejus ei permittere secum esse*, L. Ecg. C. 33; Th. ii. 158, 10.

níd-nima, an; *m. One who takes by force :*—Nēdniomu *violenti*, Mt. Kmbl. Rush. 11, 12. Nēdnioma (-niomo, Lind.) *raptores*, Lk. Skt. Rush. 18, 11. [Cf. *O. H. Ger.* nôt-nemo *rapidus* ; nôt-numeo *raptor.*]

níd-niman ; *pp.* -numen *To take by force, ravish :*—Đeáh heó nýdnumen (neád-, MS. B.) weorđe, þolige đæra æhta, būton heó fram đam ceorle wille eft hām ongeán and næfre eft his ne weorđe, L. C. S. 74; Th. i. 416, 13. v. níd-nimung, -næman.

níd-nimu (?), e; *f. A taking by force, rapine :*—Fulle sint nēdnima (-nimende, Rush.) *pleni sunt rapina*, Mt. Kmbl. 23. 25. Full is mid nēdnime, Lk. Skt..Rush. 11, 39. Nēdnioma *rapinam*, Rtl. 21, 18. v. níd-næm.

níd-nimung, e; *f. A taking by force, rapine :*—Wísa nýdnimung *stuprum, raptum*, Wrt. Voc. i. 21, 32. Full is mid nēdnimincg *plenum est rapina*, Lk. Skt. Lind. 11, 39.

níd-riht, es; *n.* (v. níd, III). I. *a duty that must be performed, service, office ;* officium, debitum :—Nēdreht *debitum*, Rtl. 89, 26. Godcund þeówdōm is gesett on cyriclícum þēnungum æfter canoneclícan gewunan tô niédrihte eallum gehādedum mannum. On ælcne tíman man sceal God herian . . . Ac đeáhhwæđere sindon gesette tíman synderlíce tô đam ānum, đæt gif hwā for bisgan oftor ne mæge, đæt hē hūru đæt niédriht dæghwamlíce gefylle, Btwk. 194, 3-8. II. *a due, what must be paid :*—Eallum æhtemannum gebyreþ midwintres feorm and Eástorfeorm . . . tōeácan heora nýdrihte, L.R.S. 9; Th. i. 438, 2. v. next word.

níd-scyld, e; *f. Bounden duty :*—Sôna swā hē tô đære áre cymþ, swā þyncþ him đæt se hié him niédscylde sceolde se se hié him sealde *as soon as he comes to the honour, it seems to him that he who gave it him was bound to grant it as a matter of right ;* repente pervenies jure sibi hoc debitum, ad quod pervenerit, putat, Past. 9; Swt. 57, 6. v. preceding word.

níd-sibb, e; *f. Relationship :*—Neádsibba *necessitudinum*, Wrt. Voc. ii. 61, 15. v. níd-mæg.

níd-syn[n] (?), e; *f. A sin of violence :*—Hū ic becwom in đis neowle genip nídsynnum (MS. mid synnum : Grein, níþsynnum) fāh, Cd. Th. 275, 32; Sat. 180.

níd-syndrig, adj. *Quite apart* (?) :—Hí sylfe đa munecas nædsyndrige (*monachos*) *seipsos*, Cod. Dip. B. i. 154, 12.

níd-þearf, e; *f.* I. *necessity, inevitableness :*—Sum hit sceal geweorþan unáwendendlíce, đæt biþ đætte úre nýdþearf (nēd-, Cott. MS.) biþ, and his willa biþ. Ac hit is sum swā gerád đæt his nis nān neódþearf (nēd-, Cott. MS.), and đeáh ne deraþ nô đeáh hit geweorþe, Bt. 41, 3; Fox 250, 1-4. II. *necessity, constraint :*—Đē nān neódþearf ne lǽrde tô wyrcanne đæt đæt đū worhtest, ac mid đínum ágenum willan đū ealle þing geworhtest, 33, 4; Fox 128, 11. III. *need* (*for something*) :—Nis him nānes þinges nédþearf, 42; Fox 258, 8. Him biþ nídþearf (niéd-, Cott. MSS.) đæt hē fleó, Past. 21; Swt. 167, 16. Is suíđe micel niédþearf đæt . . ., Swt. 159, 2. Đē heora nān nýdþerf nis eft on mē tô nimene *bonorum meorum non eges*, Ps. Th. 15, 1. Nēdþearf, Met. 20, 20. Mycel is nýdþearf manna gehwylcum đæt . . ., Wulfst. 157, 10. Hē wæs fram him eallum āræfned fore nýdþearfe his úttran weorca (*ob necessitatem operum ipsius exteriorum*), Bd. 5, 14; S. 634, 13. Him nānes ne biþ wana, ne hé nānes neódþearfe næsþ, Bt. 41, 7; Fox 80, 22. Wē habbaþ nēdþearfe đæt wē ongyton, Blickl. Homl. 23, 1 : 81, 36. IV. *a necessary thing, what a person needs :*—Wuhta gehwylc wilnaþ tô eorþan, sume neódþearfe (cf. ealle đider (*earthwards*) wilniaþ . . . đæs đe hí beþurfon, Bt. 41, 6; Fox 254, 29) sume neódfræce, Met. 31, 15. Hē wirþ swā earm đæt hé næfþ furþum đa neódþearfe āne (*fit ut necessariis egeat*), đæt is wist and wæda, wilnaþ đonne đære neádþearfe, næs đæs anwealdes, Bt. 33, 2; Fox 124, 15-18. Seó gítsung ne cann gemet, ne næfre ne biþ gehealden on đære nídþearfe, ac wilnaþ simle māran đonne hē þurfe, 26, 2; Fox 94, 6. Se cyning his lāreówum sealde heora nýdþearfe on missenlícum æhtum (*necessarias in diversis speciebus possessiones*), Bd. 1, 26; S. 488, 20. V. *need, distress, trouble :*—Hwí noldest đū cuman tô ús tô đære tíde đe ús nýdþearf wæs *quid recessisti longe in tribulatione*? Ps. Th. 9, 20. Đæt gē mē ne forseón on đisse mycclan nēdþearfe tíde, Blickl. Homl. 151, 23. Gedô đū mē gefriđie æt mínre nýdþearfe *de necessitatibus meis eripe me*, Ps. Th. 24, 15. Fylston eów æt nýdþearfe *in necessitate vos protegant*, Deut. 32, 38. Seldon būtan māran nýdþearfe (*praeter arctiorem necessitatem*) mā đonne āne síþe on dæge đæt heó wolde mete þicgan, Bd. 4, 19; S. 588, 11. VI. *a necessary business :*—Đā cwædon hí đæt hí wæron on heora nýdþearfum swýđe geswencte, Guthl. 14; Gdwin. 64. 3. [O. H. Ger. nôt-duruft *necessitas, necessarium : Ger.* noth-durft.] v. níd and next word.

níd-þearf ; *adj. Necessary, needful :*—Ys cræft mín behēfe þearle eów and neódþearf *est ars mea utilis valde vobis et necessaria*, Coll. Monast. Th. 27, 27 : 29, 17. Is eallum mannum nēdþearf and nytlíc đæt hié heora fulwihthādas wel gehealdan, Blickl. Homl. 109, 25. Behōfíc ł nēdþarf *necessarius*, Mk. Skt. Lind. Rush. 11, 3. Ān is nēdþarf ł behōfíc *unum est necessarium*, Lk. Skt. Lind. Rush. 10, 42. Nēdþærfo tído ymbhuoerfnise undercymende *necessaria temporum vicissitudine*

succedente, Rtl. 37, 35. Habban gôde geféran and þearle neódþearfe (*necessarios*), Coll. Monast. Th. 29, 31. v. *preceding word* and nídþearfness.

nídþearf-líc ; *adj. Necessary, needful, useful :*—Neádþearflíc *operae pretium*, Hpt. Gl. 433, 25 : 506, 29 : *operae pretium, necessarium, utile, justum*, 499, 78 : *debitum, necessarium*, 424, 51. Is swýđe nýdþearflíc (*necessaria*) gesceád, Bd. 1, 27; S. 496, 35. Gif ic síe đínum folce nédþearflíc tô hæbbene, Blickl. Homl. 225, 26. Būtan tô his neódþearflíce þēnunge *nisi ad usum necessarium*, Bd. 2, 16; S. 520, 8. Be monigum sócnum đa đe him nýdþearflíce (*necessariae*) gesewen wæron, 1, 27; S. 488, 33. Đa þing đe heora andlyfene nédþearflíco gesawen wæron, 1, 26; S. 487, 35. Nýdþearflícu, 5, 9; S. 622, 26 : 4, 3; S. 567, 31. Be đām nýdþearflícan þingum, intingum, Bd. 1, 27; S. 488, 24 : 2, 4; S. 505, 30. Neádþearflícum gestreónum *debito emolumento*, Hpt. Gl. 432, 69. Đa nēdþearflícan hús, Bd. 4, 28; S. 605, 25.

nídþearflíce ; *adv. Necessarily :*—Nēdþearflíce (níd-, nýd-, neád-) *necessario*, Ælfc. Gr. 38; Som. 41, 37.

nídþearfness, e; *f.* I. *necessity, compulsion :*—Mid ríhtre nýdþearfnysse gebæded *justa necessitate compulsus*, Bd. 2, 2; S. 502, 27. II. *necessity, need* (*for something*) :—Mycel nýdþearfnys is đæt đæt gesceád . . . *necessaria est magna discretio*, 1, 27; S. 497, 17. III. *need, trouble, distress :*—Wæs biddende đætte hē on swā mycelre neódþearfnysse his bigengum gehulpe *deprecatus est ut in tanta rerum necessitate suis cultoribus succurreret*, 3, 2; S. 524, 15 : Cant. M. ad fil. 38. Of nédþearfnessum *de necessitatibus*, Ps. Surt. 30, 8 : 24, 17. On neádþearfnessum *in opportunitatibus*, Ps. Lamb. second 9, 1. Nýdþearfnyssum, Ps. Spl. C. 106, 6. Ymb heora nēdþearfnesse *in necessitatibus suis*, Bd. 4, 23; S. 594, 1.

níd-þeów, es; *m. A slave, thrall :*—Wē đē, Hælend, biddaþ, đæt đū gehýre hæfta stefne đínra niédþiówa, Exon. Th. 22, 33; Cri. 361. Ne derige se hláford his mannum, ne forđan his nýdþeówan, L. I. P. 7; Th. ii. 314, 3. v. next word.

níd-þeówetling, es; *m. One who is forced into slavery* (*for an unsatisfied claim*) :—Hēr kýđ on đissere béc đæt Ælfríc wolde þeówian Putraele him tô nýdþeówetlinge (*the enslavement was abandoned at the intercession of Bora, Ælfríc's brother, on payment to Ælfríc of eight oxen ; Bora received sixty pence for his mediation*), Chart. Th. 628, 11-26.

níd-þeówian ; *p.* ode *To reduce to servitude, to compel service from :*—Gif man cirican nýdþeówige (cf. ænig man heonan forþ cirican ne þeówige, L. Eth. v. 10; Th. i. 306, 27 : vi. 15; Th. i. 318, 26), L. N. P. L. 21 ; Th. ii. 294, 1.

níd-þing, es; *n. A necessary thing :*—Ealra neádþinga hē (*the monk*) sceal hihtan and wilnigan fram his mynstres fæder (*the abbot*), R. Ben. 57, 3.

níd-wædla, an; *m. A needy person :*—Đū scealt on wræc hweorfan nacod niédwædla, neorxna wanges dugeþum bedæled, Cd. Th. 57, 16; Gen. 929.

níd-wís ; *adj. Necessary, due :*—Lof .neádwís *laus debita*, Hy. Surt. 49. 29. Neádþearflíc ł neádwís *debitum, necessariat*, Hpt. Gl. 424, 52. Swā swā se líchama biþ ontend þurh unālýfede lustas, swā eác þurh seó sāwul þurh neádwís wíte, Homl. Th. ii. 338, 19. Neádwísum ł neádþearflícum gestreónum *debito emolumento*, Hpt. Gl. 432, 68. Þances hit āgylde neádwíse *grates rependat debitas*, Hy. Surt. 27, 21. Lofu neádwíse *laudes debitas*, 86, 33. v. níd, III.

nídwíslíce ; *adv. Of necessity :*—Hē sylf wæs đære hālgan ǽ underþeód, đæt hē đa álýsde đe neádwíslíce đære ǽ underþeódde wæron, Homl. Th. i. 94, 16.

nídwísness, e; *f. Necessity :*—Neádwísnysse *debitum*, Hpt. Gl. 462, 69.

níd-wraca, an; *m. One who is forced to be an avenger, who avenges an affront :*—Gif ænig gilda hwilcne man of스tleá, and hē neádwraca sí, and his bismer béte, fylste ælc gegylda, Chart. Th. 611, 29. v. níd-dæda, -fara.

níd-wracu, e; *f. Violence, misery caused by violence :*—Wæs đæt gewin lāđ and longsum, đe on đa leóde becom, nýdwracu níþgrim, nihtbealwa mæst, Beo. Th. 388; B. 193. Hyne God wolde nergan wiđ níþum, and hyra nýdwræce deópe dēman, Exon. Th. 135, 17; Gū. 525.

níd-wyrhta, an; *m. One who acts from necessity, an involuntary agent :*—Se đe nýdwyrhta biþ đæs đe he misdéþ, se biþ đȳ beteran dómes symle wyrđe, đe hē nýdwyrhta wæs đæs đe hē worhte, L. Eth. vi. 52 ; Th. i. 328, 23-25. On mænigre dæde đonne man biþ nýdwyrhta, đonne biþ se gebeorges đe betwyrđe đe hē for neóde dyde đæt đæt hē dyde, L. C. S. 69; Th. i. 412, 12-14. v. níd-dæda.

níéd, níeten. v. níd, níten.

nifol. v. neowol.

nift, e; *f. A niece, grand-daughter, or a step-daughter :*—Nift *privigna, filia sororis*, Ep. Gl. 18 b, 6. Nift *privigna*, Wrt. Voc. ii. 117, 80. Seó wæs nift đæs hína ealdres (*neptem patris familias*), Bd. 3, 9; S. 534, 5. Ic an míne lāuedy half marc goldes and míne nifte ānn ōre wichte goldes, Chart. Th. 556, 27. [*Prompt. Parv.* nypte, nifte *neptis* ;

nypt, broderys douter *lectis*: Rebecca was forð nefte (*great niece*) of Abraham, Gen. a. Ex. 1386: O. Frs. nift *niece*: Icel. nipt *a female relative, sister, daughter, niece*: O. H. Ger. nift *neptis, privigna*.] Cf. nefa.

níg-. v. níw-.

nígan (?):—Ðonne ic búgendre stefne styrme, stille on wícum siteþ nígende (*one who listens* [?]), Exon. Th. 390, 27; Rä. 9, 8.

nigon *nine.* I. *as subst.* :—Hwær synt ða nigone (nygene, MS. A: nigona, Lind.: nióne, Rush.), Lk. Skt. 17, 17. Ðá hét se cyng faran mid nigonum ðara níwena scipa, Chr. 897; Erl. 95, 20. II. *as adj.* :—Harold wes gewend mid nigon scipon, 1052; Erl. 183, 18. Nigon nihtum ǽr middum sumere, 898; Erl. 96, 19. Ic ofslóh niceras nigene, Beo. Th. 1154; B. 575. [*Goth. O. H. Ger.* niun: O. Sax. O. Frs. nigun: *Icel.* níu.]

nigon-feald; *adj. Nine-fold* :—Nigonfeald *novenarius*, Ælfc. Gr. 49; Som. 50, 17.

nigonteóþa *nineteenth* :—Se niganteóþa getælcircul *circulus decennovenalis*, Wrt. Voc. ii. 131, 33. Nigonteóþe healf geár, Chr. Erl. 4, 7: 855; Erl. 68, 33. Ðý nigonteóþan geáre mínes lífes, Bd. 5, 24; S. 647, 28.

nigontig *ninety* :—Ofer nigon and nigontigum rihtwísra, Lk. Skt. 15, 7. v. hund-nigontig.

nigontíne *nineteen* :—Embe nigontýne niht, Menol. Fox 141; Men. 71.

nigontín-líc; *adj. Containing the number nineteen* :—Ða nigontýnlícan hringas rihtra Eástrana *circuli Paschas decennovenales*, Bd. 5, 21; S. 643, 26.

nigon-wintre; *adj. Nine years old* :—Ðá hé nigonwintre cniht wæs *cum esset novem annos natus*, Ors. 4, 8; Swt. 186, 10.

nigoþa *ninth* :—Embe ða nigoþan tíde, Mt. Kmbl. 20, 5. Fram ðære sixtan tíde óþ ða nigoþan tíde. Ymbe ða nigoþan tíd clypode se Hǽlend, 27, 45, 46. Ðý nigeþan dæge, Bd. 5, 23; S. 645, 9. Nigend half *eight and a half*, Cod. Dip. Kmbl. iv. 194, 11. Nigende, vi. 203, 15. Nióþa, Mt. Kmbl. p. 3, 16: 11, 8.

nihol, nihold. v. neowol.

níhsta, an; *m. A neighbour*; *proximus* :—Se ðe his neáhstan yfeles nán þing ne dyde, and se ðe hosp on his neáhstan ne sette, R. Ben. 3, 20–22. Ne girn ðú ðínes neáhstan wífes (*uxorem proximi tui*), Deut. 5, 21. Gif ðú wed nime æt ðínum nǽhstan, Ex. 22, 26. Gif hwá ofslihþ his néhstan, 21, 14. Lufa ðínne néhstan (Lind. nésta), Mt. Kmbl. 19, 19. Hwylc is mín néhsta (neestæ, Lind.)? Lk. Skt. 10, 29. Lufa ðínne néxtan (néste, Lind: néxstan, Rush.), Mt. Kmbl. 5, 43. Hwá is úre néxta? Homl. Th. ii. 318, 1. Hwelc ðara niéhstena (níhstena, Cott. MSS.) ðæs ofslægenan, Past. 21; Swt. 167, 3. Tó nýhstan his, Ps. Spl. 11, 2: Ps. Th. 11, 2. v. neáh.

nihstig, nistig, nestig; *adj. Fasting* :—Gedrinc his on niht nistig, Lchdm. i. 74, 1, 6: 76, 7, 13. Nyhstig, iii. 48, 2. Nibstig, 48, 15: 50, 21: i. 82, 14: 84, 16. v. niht-nihstig.

nihstnig *fasting* (?) :—Eft hý (*monks*) gaderiaþ hý on nixtnig, ðæt hý raca gehýren æt heora fæder ... Hý siððan heora líchoman gereordaþ, R. Ben. 138, 2–8.

niht, næht, næct, neaht, neht, nyht, e; *f.* : *but also with gen. es.* I. *night* (as opposed to day) :—Niht is gesett mannum tó reste on ðysum middanearde ... Úre eorþlíce niht (nyht, MS. M.) cymþ þurh ðære eorþan sceade ... Seó niht hæfþ seofon dælas fram ðære sunnan settlunge óþ hire upgang. Án ðæra déla is *crepusculum*, óðer is *vesperum*, þridde is *conticinium*, feórþa is *intempestum*, ðæt is midniht, fífta is *gallicinium*, syxta is *matutinum vel aurora*, seofoþa is *diluculum*, Lchdm. iii. 240, 10–244, 5. Hé hine micelre tíde ðære deáhlan neahte swong, Bd. 4; S. 508, 13. Scínaþ þurh ða scíran neaht, Met. 20, 229. Niht (næht, Lind. Rush.) cymþ ðonne nán man wyrcan ne mæg, Jn. Skt. 9, 4: 13, 30. Fira bearnum neálǽhte niht seó þýstre, Judth. Thw. 21, 25; Jud. 34. Hé com tó him ánes nihtes, Shrn. 16, 27. Næs nǽnig man ðe ǽfre nihtes tídum dorste on ðære ciricean cuman, Blickl. Homl. 207, 34. Wacana nǽhtes *vigilia noctis*, Lk. Skt. Lind. Rush. 2, 8: Mt. Kmbl. Lind. Rush. 14, 25. Swá swá se beorhta dæg tódrǽfþ ða dimlícan þeóstru ðære sweartan nihte, Homl. Th. i. 604, 2. On dæge and on nihte, 36, 28. Hé fealh ðære ilcan niht of ðǽm bendum, Ors. 5, 11; Swt. 236, 12: Bd. 1, 33; S. 499, 9: Blickl. Homl. 215, 15. Tó niht *hac nocte*, Num. 22, 19. On næht nocte, Ps. Surt. 16, 3. On niht ǽr hé ræste, Blickl. Homl. 47, 18. Feówurtig daga and feówurtig nihta (Lind. nǽhta), Mt. Kmbl. 4, 2. Þreó niht and dagas, Cd. Th. 20, 12; Gen. 307. Dagum and nihtum, Met. 20, 213. II. *night, darkness* (as opposed to light) :—Seó swearte niht ðære écan geniþerunge, Homl. Th. i. 530, 23. Dryhten ðe ús of duste geworhte, nergend of nihtes sunde, Salm. Kmbl. 675; Sal. 337. III. *night* (as in se'n-*night*, fort-*night*; cf. Tacitus' Germania, c. xi : 'Instead of reckoning by days as we do, they reckon by nights.') :—Be ánre nihtes (MS. B. nihte) þiéfþe, L. In. 73; Th. i. 148, 11. Hé fór ymb áne niht tó Igleá, and ðæs ymb áne tó Eþandúne ... and ðǽr sæt xiiii niht ... and hé was xii niht mid ðam cyninge, Chr. 878; Erl. 80, 12–24. Embe seofon niht, Blickl. Homl. 45, 31.

Emb tén niht, 117, 16. On twám nihtum biþ mannes sunu geseald on synfulra hand, 73, 1. For tén nihtum *ten days ago*, 131, 10. Mid ðon dæge wæs gefylled se dæg ðe is nemned Pentecosten ymb fíftig nihta æfter ðære gecýþdan ǽriste, 133, 14. [*Goth.* nahts: O. Sax. naht: O. Frs. nacht: *Icel.* nátt, nótt: O. H. Ger. naht.] v. Eáster-, efen- (emn-, em-), Frige-, mæsse-, mid-, middel-, mónan-, sæter-, sin-, sunnan-, þunres-, Tíwes-, Wódnes-niht; nihtes.

niht-bealu, wes; *n. Bale* or *hurt that comes at night*, Beo. 389; B. 193.

niht-buttorfleóge, an; *f. An insect that flies at night*; *blatta*, Wrt. Voc. i. 23, 65.

niht-eáge, -ége; *adj. Able to see at night* :—Nihteáge *nyctalmus*, Wrt. Voc. i. 62, 23. Nihtége *nictalmus*, i. 20, 8.

niht-eald; *adj. A day old* :—Gif hit biþ nihteald þiéfþ *if it is a theft a day old*, i. e. *if a day passes between the commission of the crime and the capture of the thief*, L. In. 73 ; Th. i. 148, 10.

niht-egesa, an; *m. Terror by night* :—Ne ðú ðé nihtegsan ondrǽdest *non timebis a timore nocturno*, Ps. Th. 90, 5. [Cf. O. H. Ger. naht-forhta *timor nocturnus*.]

nihte-gala, an; *m. A nightingale* :—Nihtegala *luscinia*, Wrt. Voc. i. 62, 25. v. next word.

nihte-gale, an; *f. A bird whose note* (v. galan) *is heard at night.* I. *the night-raven* :—Naechthraebn, *ali dicunt* nectigalae *noctua*, Ep. Gl. 16 b, 15; *but more generally* II. *the nightingale* :—Naectegale *luscinia*, Wrt. Voc. ii. 113, 30: *roscinia*, 119, 23. Nectægalae *roscinia*, Ep. Gl. 22 b, 27. Nictigalae *achalantis*, 1 f, 6. Nehtegale *achalantis* vel *luscinia* vel *roscinia*, Wrt. Voc. ii. 99, 3. Nihtegale, 4, 24: *luscinia*, 51, 27: *philomela*, 1. 63, 23. Nightegale *luscinia* vel *philomela*, 29, 12. [O. L. Ger. nahti-gala *luscinia*, *acredula*: O. H. Ger. nahti-gala *luscinia*, *filomela*; also *corax*, *nocticorax*, *noctua*.]

nihte-líc. v. niht-líc.

nihterne, neahterne; *adj. Nocturnal* :—Þurh nihterne besmitenesse *per nocturnam pollutionem*, Confess. Peccat. v. next word.

nihterne, neahterne, nihternum; *adv. For a night* :—Ðæs gást wæs neahterne of líchoman álǽded *his* (*Fursey*) *spirit was for a night taken from his body*, Shrn. 51, 30. Lǽt standan neahterne, Lchdm. ii. 24, 21: 32, 25. Lǽt licgean neahterne, 66, 12. Bind on ða eágan nihterne, 34, 23. Lǽt beón nihterne, 74, 14: 270, 8. Lǽt standan nyhternum, iii. 16, 17. v. *preceding and following words, and* dægþerne.

nihternness, e; *f. Night-time* :—Ðonne gescylt ðé God wið unswefnum ðe nihternnessum on menn becumaþ *then will God protect thee against evil dreams that come to men at nights*, Lchdm. iii. 288, 22. v. preceding word.

nihtes (*gen.* of niht, q. v.); *adv. At night, night* :—Ne mæg ic búton mynstre nihtes wunian *I cannot stop out of the monastery at night*, Homl. Th. ii. 182, 34. Ðá gestód hé æt ánum éhþyrle óþ forþ nihtes (*far on into the night*), 184, 27. Hys leorningcnihtas cómon nihtes (*nocte*), Mt. Kmbl. 28, 13. Ðæra eágan scínaþ nihtes, Nar. 34, 14. Se biþ dæges hát and nihtes ceald, 36, 27. Dæges and nihtes *die et nocte*, Ps. Th. 1, 2: Mk. Skt. 4, 27: Blickl. Homl. 47, 11: 127, 30: 137, 22. Deges and naehtes, Ps. Surt. 31, 4. [O. Sax. nahtes: O. H. Ger. nahtes: Ger. nachts.]

niht-feormung, e; *f. Entertainment for the night* :—Hé (*Lot*) ðám rincum (*the angels*) beád nihtfeormunge, Cd. Th. 147, 2; Gen. 2433.

niht-genga, an; *m. A creature that goes at night, a goblin, evil spirit* :—Wið feóndes costunga and nihtgengan and maran, Lchdm. ii. 306, 12. Wyrc sealf wið nihtgengan, 342, 1. Wið ælfcynne and nihtgengan and ðám mannum ðe deófol mid hǽmþ, 344, 7. Gif men hwylc yfel costung weorþe oððe ælf oððe nihtgengan, 344, 16. Hió (*betony*) hyne scyldeþ wið unhýrum nihtgengum and wið egeslícum gesihþum and swefnum, i. 70, 5.

niht-genge, an; *f. A night-goer, an animal that prowls at night, a hyena* :—Naectgenge *hyna*, Wrt. Voc. ii. 110, 41. Nihtgenge *hyna*, 43, 6.

niht-gerím, es; *n. Reckoning by days* (v. niht, III), *number of days* :—Æfter seofontýnum nihtgerímes *after seventeen days*, Menol. Fox 52; Men. 26: 110; Men. 55: Andr. Kmbl. 229; An. 115: 315; An. 158. Ealra hæfde v. and syxtig ðá hé forþ gewát and nigon hund eác nihtgerímes *in all the number of his days when he died was nine hundred and sixty-five years*, Cd. Th. 72, 28; Gen. 1193. [Cf. dógorgerím, *and Icel.* náttar-tal: *Chauc.* nighter-tale.]

niht-gild, es; *n. A service, sacrifice celebrated at night* :—Nihtgild *nyctilia*, Wrt. Voc. ii. 61, 18. Blóstmfreólsas and nihtgilda *floralia nictelia* (cf. blóstmgeld *floralia*, Wrt. Voc. ii. 37, 52), Hpt. Gl. 515, 18.

niht-glóm, es; *m.* (?) *The darkness of night* :—Wæs ðam báncofan æfter nihtglóme (*when the shades of night prevailed*) neáh geþrungen, Exon. Th. 158, 27; Gú. 916. v. æfen-glóma.

niht-helm; *m. The covering of night, night's curtain* :—Nihthelm geswearc deorc ofer dryhtgumum *night's curtain dark was drawn over men*, Beo. Th. 3583; B. 1789.

niht-hræfn, es; _m._ _The night-raven, night-jar, night-owl_ :— Naehthraebn _noctua, nocticorax,_ Ep. Gl. 16 b, 15, 18. Naehthraefn _noctua,_ Wrt. Voc. ii. 114, 76. Nihthræfn _nycticorax,_ i. 63, 12. Nihthrefn, 281, 42. Nihthremn, ii. 60, 36. Nihtremn, i. 29, 35. Nihthrefne (nihtrefen, Ps. Spl.; næhthrefn, Ps. Surt.; nihthræm, Ps. Lamb.) gelíc, Ps. Th. 101, 5. [_Icel._ nátt-hrafn : _O. H. Ger._ naht-hraban _nocticorax, noctua, bubo._]

niht-hróc, es; _m._ _The night-rook, raven_ :— Nihtróc _nycticorax,_ Ps. Lamb. 101, 7.

niht-lang; _adj._ _Night-long, a night in length_ :— Nafa ðú nánes þearfan wedd mid ðé nihtlange fyrst _si pauper est proximus tuus, non pernoctabit apud te pignus,_ Deut. 24, 12 : Cd. Th. 191, 2 ; Exod. 208: Andr. Kmbl. 1668 ; An. 836 : 2620 ; An. 1311 : Elen. Kmbl. 134; Sl. 67. Nihtlongne fyrst, Beo. Th. 1060; B. 528. [_Icel._ nátt-langt _for a night._] v. next word.

niht-langes; _adv._ _For the night_ :— Ic bidde eów ðæt gé gecirron tó mínum húse and ðær wunion nihtlanges, Gen. 19, 2. [Ne moste nihtlonges istonden, Laym. 15564.] v. preceding word.

niht-líc; _adj._ _Nightly, of the night, nocturnal_ :— Fram ege nihtlícum _a timore nocturno,_ Ps. Spl. 90, 5. For nihtlecum ege, Past. 56; Swt. 433, 11. Hí swuncon on nihtlícum rewette, Homl. Th. ii. 384, 24. Ða steorran sint mannum tó nihtlícere líhtinge gesceapene, i. 110, 15. On nihtlícre tíde _at night,_ Lchdm. iii. 234, 21.: 270, 26. On nihtlícre gesyhþe _in a vision of the night,_ Bd. 5, 10; S. 625, 12. Hine drehton nihtlíce gedwimor, Homl. Th. i. 86, 18. Ðæs synfullan líf is widmeten nihtlícum þeóstrum, ii. 200, 33. Nihtlícum tídum _in the night seasons,_ Bd. 3, 11; S. 536, 11. Þerh næhtlíco mysto _per nocturnas caligines,_ Rtl. 171, 39. [_Icel._ nátt-ligr : _O. H. Ger._ naht-líh _nocturnus._]

niht-nihtig, -nestig; _adj._ _Fasting for a night_ :— Sele nihtnestig drincan, Lchdm. ii. 64, 18. Gedrinc ælce dæge neahtnestig, 30, 26. Drince iii morgenas neahtnestig, 296, 12. Mid his selfes nihtnestiges migoþan, 42, 1. Sele nihtnestigum drincan, 64, 9, 19: 186, 5. Syle on morgenne ðam seócum men neahtnestigum, 286, 11.

niht-rest, e; _f._ _The couch on which one rests at night_ :— Abram síne nihtreste ofgeaf, Cd. Th. 173, 18; Gen. 2863.

niht-rím, es; _n._ _A number of days_ :— Nihtrím scridon, Exon. Th. 167, 35; Gú. 1070.

niht-sang, es; _m._ **I.** _the service at the seventh of the canonical hours, compline_ :— Nú gebyraþ mæssepreóstum ðæt hí ða seofon tídsangas gesyngon . . . nihtsang seofoþan, L. Ælfc. C. 19; Th. ii. 350, 3–7 : R. Ben. 40, 7. Hwænne wylle gé singan nihtsange (_completorium_), Coll. Monast. Th. 34, 3. **II.** _a copy of the service_ :— Sind .ii. fulle sangbéc and .i. nihtsang . . . Hé ne funde ná má búton áne capitulare and .i. forealddone nihtsang . . ., Chart. Th. 430, 8, 28. [_Icel._ nátt-söngr.]

niht-scada (-sceadu ?) _night-shade_ (plant name) :— Nihtscada _strumus_ vel _uva lupina,_ Wrt. Voc. i. 31, 18.

niht-scúa, -scúwa, an; _m._ _The darkness, shades of night_ :— Ðonne nípeþ nihtscúa, Exon. Th. 292, 24; Wand. 104 : 307, 29; Seef. 31. Æfter nihtscúan, 162, 5; Gú. 971. Under nihtscúwan, Cd. Th. 124, 10; Gen. 2060. Neowle nihtscúwan, 184, 28; Exod. 114.

niht-slǽp, es; _m._ _Sleep during the night_ :— Ðæt ilce geþanc ðe heom amang ðam nihtslǽpe wæs on heora heortan, eall ða hí áwac. Odon hí ðæt sylfe geþohton, Homl. Skt. 23, 442.

niht-wacu (o); _f._ _A night-watch_ :— Mec oft bigeat nearo nihtwaco æt nacan stefnan, Exon. Th. 306, 13; Seef. 7. v. next word.

niht-wæcce, an; _f._ _A night-watch, vigil_ :— Nihtwæccan _vigiliae,_ Wrt. Voc. i. 18, 22. Hyrdas wǽron waciende and nihtwæccan (-wæcan, MS. C.) healdende _pastores erant vigilantes et custodientes vigilias noctis,_ Lk. Skt. 2, 8. [_Icel._ nátt-vaka : _O. H. Ger._ naht-wacha _vigilia._]

niht-waru, e; _f._ _Night-wear_ :— Genóh byþ ðam munuce ðæt hé hæbbe twá cúlan and twegen syricas for ðære nihtware and for ðæs reáfes þweále, R. Ben. 90, 4.

niht-weard, es; _m._ _A guard who keeps watch at night_ :— Heofoncandel (_the fiery pillar_) barn, níwe nihtweard, Cd. Th. 185, 1; Exod. 116.

niht-weorc, es; _n._ _A work done at night_ :— Nihtweorce (_the defeat of Grendel_) gefeh, Beo. Th. 1659; B. 827.

nillan. v. nyllan.

nima. v. níd-nima.

niman; _p._ nam, _pl._ námon; _pp._ numen (_kept in the slang word_ nim = _steal._ Cf. Shakspere's Corporal Nym). **I.** _to take, receive, get;_ sumere, accipere :— Nimþ _sumpserint,_ Kent. Gl. 1056. Hwær nime wé (hwonon ús tó niomane, Rush.) swá fela hláf? Mt. Kmbl. 15, 33. Cristes onsægdnesse ðe wé æt ðæm weofode nimaþ, Blickl. Homl. 77, 5. Ðæt (_food_) hé ǽr tó blisse nam, 57, 7. Ða nam Petrus and ða óðre apostolas hié (_Mary_), and hié ásetton ofer hire bǽre, 149, 5. Hié náman blówende palmtwigu and bǽron him tógeánes, 69, 30. Nim and telle Israhéla folc, Num. 1, 2. Nim ǽnne oððe twegen tó ðé _adhibe tecum unum vel duos,_ Mt. Kmbl. 18, 16. Nim ðé ðis ofæt on hand, Cd. Th. 33, 11 ; Gen. 518. Ðæt ðú nǽfre ne nyme wíf mínum suna of ðisum mennisce, Gen. 24, 3. Nán man ne sceal sceattas niman for Godes cyrcan,

Homl. Th. ii. 592, 21. Hé lǽrde tó healdenne reogollíces lífes þeódscipe swá swíðe swá ða níwan Cristenan hit niman (_capere_) mihton, Bd. 3, 22; S. 553, 11. Héht his sweord niman, Beo. Th. 3621; B. 1808. **II.** _to take, keep, hold;_ tenere :— Nimþ mé seó swýðre ðín _tenebit me dextera tua,_ Ps. Th. 138, 9. Ðú nǽme (_tenuisti_) hand ða swýðran, 72, 23. Hé ðæt wolcn him beforan nam _he had the cloud before him,_ Blickl. Homl. 121, 14. Hí oferhygd nam (_tenuit_), Ps. Th. 72, 5. Hí hí be handum nóman _junctis manibus,_ Bd. 4, 13; S. 582, 31. **III.** _to take, catch_ :— Hér beóþ oft numene missenlícra cynna weolcscylle, I, 1; S. 473, 17. **IV.** _to contain_ :— Nó swá ðæt heó (_the coffin_) ðone líchoman neoman mihte, 4, 11; S. 580, 7. **V.** _to take (with one), carry, bring_ :— Ða nam hé fíf stánas on his herdebelig, Blickl. Homl. 31, 17. Námon wé hláfas mid ús, Mt. Kmbl. 16, 7. **VI.** _to take (to one), give_ :— Hát ðé niman Pilatus ǽrendgewrit, Blickl. Homl. 177, 2. **VII.** _to take forcibly, seize, take away, carry off;_ tollere, capessere, auferre, rapere :— Ic nyme _tollo,_ Ælfc. Gr. 28; Som. 32, 53. Ðam ðe ðíu reáf nymþ _qui auferet tibi vestimentum,_ Lk. Skt. 6, 29. Se ðe hine deáþ nimeþ, Beo. Th. 887; B. 441. Gúþ nimeþ freán eówerne, 5066; B. 2536. Þeófas ðe on mannum heora æhta on wóh nimaþ, Blickl. Homl. 61, 22. Manige men ða moldan neomaþ on ðæm lǽstum, 127, 11. Nimaþ _capessunt,_ Wrt. Voc. ii. 23, 33. Nam _capessit,_ 20, 8. Hé nam _tulit,_ Kent. Gl. 209. Nam mid handa rinc on ræste, Beo. Th. 1496; B. 746. Ðæt hé móste niman ðæs Hǽlendes líchaman . . . Ða com hé and nam ðæs Hǽlendes líchaman _ut tolleret corpus Jesu . . . venit ergo et tulit corpus Jesu,_ Jn. Skt. 19, 38. Ðæt flód com and nam (_tulit_) hig ealle, Mt. Kmbl. 24, 39. Nimaþ ðæt pund fram him _auferte ab illo mnam,_ Lk. Skt. 19, 24. Nis nánum men álýfed ðæt hé nime on his þeówe ǽnig feoh _nemini licet servo suo pecuniam aliquam auferre,_ L. Ecg. P. addit. 35 note; Th. ii. 238, 11. Gif mec hild nime, Beo. Th. 909; B. 452. Mé sceal wǽpen niman, Byrht. Th. 139, 11; By. 252. Ne biþ álýfed æt ðam þeówan his feoh tó nimanne _non licet pecuniam suam servo auferre,_ L. Ecg. P. addit. 35; Th. ii. 238, 6. **VIII.** in phrases in a metaphorical sense :— Andan niman _to take umbrage, offence_ :— Ða nam ðæt folc micelne andan ongeán his láre, Homl. Th. i. 26, 21. Tó ðon ealdfeóndas ondan nóman, Exon. Th. 115, 14; Gú. 189. Bysne niman be, æt _to take example by, from._ Nime heó bysne be ðisre wudewan, Homl. Th. i. 148, 5. Hí námon ða bysne ðæs fæstenys æt ðam Niniveiscan folce, 244, 23. Casum niman _to take a case (of the government of verbs),_ Ælfc. Gr. 41; Som. 43, 57. Eard niman _to take up one's abode._ Ðær ic eard nime _hic habitabo,_ Ps. Th. 131,15. Heofones cyning sylf cymeþ, nimeþ eard in ða, Exon. Th. 5, 1; Cri. 63. Freóndrædene niman _amicitias jungere,_ Ex. 34, 12. Friþ niman wið _to make peace with,_ Chr. 867; Erl. 72, 17: 868; Erl. 72, 29 (often in the Chronicle). Geleáfan niman _to believe,_ Cd. Th. 41, 2 ; Gen. 650. Geþeódrǽdene niman wið _to associate with._ Gif hwylc bróðor gedyrstlǽcþ ðæt hé on ǽnige wísan geþeódrǽdene nime wið ðone ámánsumedan, R. Ben. 50, 11. Graman niman _to take offence, feel angry._ Ða nam hé micelne graman and andan tó ðam mannum, Homl. Th. i. 16, 30. Láre niman _to accept teaching._ Hý leng mid him láre ne námon, Salm. Kmbl. 926; Sal. 462. Lufe niman tó _to take an affection for._ Máran lufe nimþ se heretoga tó ðam cempan, ðe æfter fleáme his wiðerwinnan þegenlíce oferwinþ, Homl. Th. i. 342, 2. Mód niman _to take courage._ Þegenlíce mód niman, Bd. 1, 16; S. 484, 15. On niman _to take effect on,_ Lchdm. ii. 84, 6 : 234, 5 : 282, 22. On gemynd niman _to bear in remembrance,_ Elen. Kmbl. 2464; El. 1233. On hæft niman _to take captive,_ Exon. Th. 16, 29; Cri. 260. Sibbe niman wið _to make terms with._ Ne nim ðú náne sibbe wið ðæs landes menn _ne ineas pactum cum hominum illarum regionum,_ Ex. 34, 15. Sige niman _to gain the victory,_ Chr. 871 ; Erl. 74, 8 (and often). Tó gemæccan niman _to take to wife,_ Th. 76, 17; Gen. 1258. Tó suna niman _to adopt as a son,_ Ors. 1, 12; Swt. 52, 16. Wǽpna niman _to take up arms,_ 1, 10; Swt. 44, 32. Ware niman _to take care._ Hé ne nom náne ware húlíce hié wǽron, 5, 4; Swt. 224, 21. Weg niman _to take, go one's way,_ Cd. Th. 80, 16; Gen. 1329. Wícstówa niman _to pitch a camp,_ Ors. 4, 10; Swt. 200, 8. [_Goth._ niman : _O. Sax._ niman : _O. Frs._ nima, nema : _Icel._ nema : _O. H. Ger._ neman _tollere, carpere, vellere, rapere, capere._] v. á-, æt-, be- (bi-), dǽl-, for-, ge-, of-, ofer-, under-niman.

nimþe, nemþe, nymþe; _conj._ _Unless, except_ :— Nimþe _nisi,_ Wülck. Gl. 249, 9. Nimþe wén wǽre _ni forsan,_ Wrt. Voc. ii. 61, 4 : 93, 3. **I.** connecting clauses :— Ne hine mon on óðre wísan his bêne týþigean wolde, nemþe hé Cristes geleáfan onfênge, Bd. 3, 21; S. 550, 43. Ne sceal nǽfre his torn tó rycene beorn ácýðan, nemþe hé ǽr ða bóte cunne, Exon. Th. 293, 9; Wand. 113. Nymþe mé Drihten gefultumede, wéninccga mín sáwl sóhte helle, Ps. Th. 93, 16: Beo. Th. 3321; B. 1657. **II.** connecting words in the same case (contracted clauses, the verb of the second clause being the same as that in the first, and not expressed) :— Nǽnig óðerne freóþ in fyrhþe, nimþe feára hwylc (freóþ), Fragm. Kmbl. 71; Leas. 37. Næs monna gemet, ne mægen engla ðæt eów mihte helpan, nimþe Hǽlend God, Cd. Th. 295, 27; Sat. 493. Nǽniges Godes háligra gebyrd ciricean ne mǽrsiaþ, nemþe Cristes sylfes and ðyses Johannes, Blickl. Homl. 161, 11. Ne wé ús náht elles ne

3 A

wēnden nemþe deáþes sylfes, Bd. 5, 1; S. 613, 26. Unc gemǣne ne sceal elles āwiht nymþe lufu langsumu, Cd. Th. 114, 17; Gen. 1905: 252, 8; Dan. 575. Nis đē wiđerbreca man on moldan, nymþe Metod āna, 251, 22; Dan. 567. Hwā is đæt đē cunne, nymþe ēce God, 266, 7; Sat. 18. Nis nǣnig swā snotor, nymþe God seolfa, 286, 11; Sat. 350. Ic nǣngum sceþþe nymþe bonan ānum, Exon. Th. 407, 11; Rä. 26, 3. Nabbaþ wē tō hyhte nymþe cyle and fŷr, Cd. Th. 285, 10; Sat. 335. Eaforan syndon deáde nymþe feá āne, 128, 30; Gen. 2134. Cf. nefne.

nimung, e; f. A taking, plucking:—Niming hēra vulsio spicarum, Lk. Skt. p. 5, 3. v. nīd-nimung.

nió-bedd, nіód, nіól, nіowol, nіótan, nіoþan, -nіp. v. neó-bedd, neód, nīd, neowol, neótan, neoþan, ge-nip.

nip (?):—Nipum rudente, Germ. 399, 451.

nīpan; p. nāp, pl. nipon; pp. nipen To grow dark:—Đonne won cymeþ, nīpeþ nihtscúa, Exon. Th. 292, 24; Wand. 104. Nāp nihtscúa, 307, 29; Seef. 31. Nīpende niht, Beo. Th. 1098; B. 547: 1302; B. 649. v. ge-nīpan, ge-nip.

nirwan, nirwian; p. de, ode To constrain, repress, blame, threaten:—Moyses onfēng scīnendum wulderhelme forđon hē symle đa nyrugde đe God oferhogodan Moses received a shining crown, because he ever repressed those that despised God, Blickl. Homl. 49, 12. Hī fŷrene tangan him on handa hæfdon and mē nyrwdon and mē tōbeótodan đæt hī mē mid đām gegrīpon woldon forcipibus igneis quos tenebant in manibus, minitabantur me comprehendere, Bd. 5, 12; S. 628, 43. Ne ne on đīnum yrre ne nyrwa đū mē neque in ira tua corripias me, Ps. Spl. 37, 1. v. ge-nyrwian, nearwian.

nirwett, es; m. I. narrowness:—Ic hit gesēran ne mehte for đara wega nerwette (propter angustas semitas), Nar. 25, 5. II. a narrow place, pass:—Se engel eode intó ānum nyrwette angelus ad locum angustum transiens, Num. 22, 26. III. oppression of the chest, difficulty of breathing:—Hit frēmaþ myclum gedruncen wiđ nyrwyt, Lchdm. i. 140, 1: 144, 17. v. nearuness and next word.

nirwþ, e; f. Confinement, a prison:—Nirwþa ergastula, Wrt. Voc. ii. 33, 24.

nis. v. neom.

nistan, nistian to build a nest:—Đǣr sperwan nistiaþ illic passeres nidificabunt, Ps. Spl. 103, 18. Nistaþ (MS. nistađađ), Ps. Surt. 103, 17. [O. H. Ger. nistian, nistón nidificare.] v. nistlan.

nistig. v. nihstig.

nistlan to make a nest:—On đam spearwan nystlaþ, Ps. Th. 103, 16. Nistlaþ, 83, 3. v. nestlian, nistan.

nitan, nitenness. v. nytan, nytenness.

nîten, nіéten, neáten, nŷten, es; n. An animal, beast, cattle:—Ǣlc cuce þing vel nŷten animal: ǣlces kynnes nŷten pecus, jumentum, Wrt. Voc. i. 22, 37, 38: 78, 49. Rēđe nŷten feralis bestia, ii. 147, 54. Đis nŷten haec pecus, Ælfc. Gr. 9, 32; Som. 12, 10. Đis nŷten istud animal: đyses nŷtenes istius animalis; đǎs nihtenu ista animalia, 15; Som. 18, 34–36. Swā nŷten geworden eom ut jumentum factus sum, Ps. Spl. 72, 22. Sum nŷten is đe wē nemnaþ broc, Lchdm. i. 326, 11. Nēten, Met. 20, 191. Nіéten, Salm. Kmbl. 44; Sal. 22. Hē hine on his nŷten (nētne, Lind.) sette, Lk. Skt. 10, 34. Nŷtenu and deór fixas and fugelas God gesceóp on flǣsce būtan sāwle. Đa nŷtenu hē lēt gān ālotene, and hē forgeaf đām nŷtenum gærs, Homl. Th. i. 276, 3–6. Men and nŷtenu sweltaþ homines et jumenta morientur, Ex. 9, 19. Swā stunte nŷtenu sicut bruta animalia, Coll. Monast. Th. 32, 19. Đa ungesceádwîsan neótena, Bt. 14, 2; Fox 44, 21. Nētenu, 34, 11; Fox 152, 6. Đām monnum đe beóþ neátenum gelîce, 14, 1; Fox 42, 3. Lǣde seó eorþe forþ cuce nîtenu, Gen. 1, 24. v. neát.

nîten-lîc; adj. Animal, after the manner of a brute:—Gē nētelîcan (nētenlîcan, Cott. MS.) men O! terrena animalia, Bt. 16, 2; Fox 50, 35. **nîþ[þ]** a man. v. niþþas.

nîþ, es; n. A place low down, an abyss:—Lǣdaþ in đæt sceađena scræf, scúfaþ tō grunde in đæt nearwe niþ, Cd. Th. 304, 22; Sat. 634. v. niþ-sele, -wundor.

nîþ, es; m. I. envy, hatred, enmity, rancor, spite, ill-will, jealousy:—Đis synt đa îdelnyssa đisse worlde . . . nîþ and ǣfēsta and hāðheortnys hae sunt vanitates hujus mundi . . . odium et invidiae et furor, L. Ecg. P. i. 8; Th. ii. 174, 32. Ne mehte se nîþ būton him twǣm gelicgean the enmity between the two could not die out, Ors. 3, 11; Swt. 152, 14. Gif him þince đæt hē næddran geseó đæt biþ yfeles wîfes nîþ if he fancies he sees a snake, that means a bad woman's spite, Lchdm. iii. 174, 17. Blātene nîþ livid envy, Cd. Th. 60, 14; Gen. 981: Andr. Kmbl. 1536; An. 769. Nîþ wiđ God enmity with God, Exon. Th. 302. 23; Fä. 40. Āhredde mē hefiges nîþes feónda mînra eripe me de inimicis meis, Ps. Th. 58, 1. Genere mē fram nîþe nāhtfremmendra, Lchdm. iii. Paulus ehte cristenra manna, nā mid nîþe (rancorously), swā swā đa ludēiscan dydon, ac hē wæs bewêrigend đære ealdan ǣ, Homl. Th. i. 388, 31. Hió mid wîflîce nîþe wæs feohtende on đæt underiende folc she (Semiramis) with a woman's rancor was carrying on war against that harmless people, Ors. 1, 2; Swt. 30, 19. Đæt gē eówer mód gemetgien on đæm nîþe ut in increpationis zelo se spiritus temperet, Past. 21, 4;

Swt. 159, 15. Đǣr is friþ būtan ǣfēstum, sib būtan nîþe, Exon. Th. 101, 18; Cri. 1660. Hē slóh hildebille đæt hit on heafolan stód nîþe genŷded he smote with his battle-blade, that, forced on by hate, it stuck in the (dragon's) head, Beo. Th. 5353; B. 2680. Đa hwîle đe hē nŷþ odđe andan hæbbe on his heortan wiđ đone nēhstan quamdiu invidiam vel malitiam in corde suo cum proximo suo habet, L. Ecg. P. ii. 27; Th. ii. 192, 27. Sidđan genam Saul micelne nîþ tō Davide, Homl. Th. ii. 64, 16. Hannibal gecŷþde đone nîþ and đone hete (odium) đe hē beforan his fæder swór, Ors. 4, 8; Swt. 186, 9. Hē him forgeaf đone nîþ đe hē tō him wiste he (Augustus) forgave them (the Germans who had slain Varus) the ill-will he felt towards them, 5, 15; Swt. 250, 15. Hē ne rōhte heora eallra nîþ, Chr. 1086; Erl. 222, 32. Ic hine on sette mōd-hete, longsumne nîþ, Cd. Th. 105, 22; Gen. 1757: 47, 30; Gen. 768. Nîþa gebêded forced by feelings of hatred, Exon. Th. 254, 27; Jul. 203: 270, 9; Jul. 462. II. action which arises from hatred, strife, war, hostility:—Hē cwæþ nîþes ofþyrsted đæt hē on norþdǣle heáhsetl heofena rîces āgan wolde all too eager for strife he said that in the north of heaven a throne he would own, Cd. Th. 3, 7; Gen. 32: 120, 15; Gen. 1995. Gūþbill nacod æt nîþe the blade bared in battle, Beo. Th. 5163; B. 2585. Nîþe rōf bold in battle, Judth. Thw. 22, 7; Jud. 53. Nîþ āhebban wiđ to strive against, Elen. Kmbl. 1672; El. 838. Nîþa ofercumen, fǣge and geflŷmed, Beo. Th. 1694; B. 845. Æt nîþa gehwam nŷdgesteallan, 1768; B. 882. Nîþa cræftig, 3929; B. 1962: 4346; B. 2170. Hē nîþa gedîgde, hildehlemma, 4690; B. 2350: 4785; B. 2397. III. the effect of hatred, persecution, trouble, vexation, annoyance, affliction, tribulation, grief:—Đā wæs wyrmes wîg wîde gesŷne nearofāges nîþ (the disastrous effects of the dragon's malice), 4623; B. 2317: Cd. Th. 83, 22; Gen. 1383. Oft đǣr brōga cwom ealdfeónda nîþ oft came terror there, trouble from the hate of ancient foes, Exon. Th. 110, 24; Gū. 112: 125, 29; Gū. 361: 345, 25, 30; Gn. Ex. 195, 197: 346, 4; Gn. Ex. 200. Him leófre wǣre đæt hié an swelcan nîđe (MS. C. nēđe) deáþ fornōme đonne hié mid swelcan nіéde friþ begeáte cum intolerabiles conditiones pacis audissent, tutius rati sese armatos mori quam miseros vivere, Ors. 4, 6; Swt. 174, 26. Hæfde hē sele Hróđgāres genered wiđ nîþe, Beo. Th. 1658; B. 827: Andr. Kmbl. 2073; An. 1039. Hié habban sceoldon hellgeþwin, đone nearwan nîþ, Cd. Th. 43, 27; Gen. 697: 48, 13; Gen. 775. Ic wræc Wedera nîþ, Beo. Th. 850; B. 423. Scyld đū đe nú đū đysne nîþ genesan mōte, Lchdm. iii. 52, 17. Đæt đū mē generige nîþa gehwylces eripe me, Ps. Th. 118, 170: Exon. 230, 8; Ph. 469. Fela mē se Hǣlend hearma gefremede, nîþa nearolîcra, Elen. Kmbl. 1822; El. 913. Nîþa georn, bealwes beald, Blickl. Homl. 109, 28. Hǣle wiđ deófla nîþum, 171, 30. Hē mec wile wiđ đām nîþum genergan, Exon. Th. 116, 24; Gū. 212: 140, 34; Gū. 620. Ofer đa nîþas đe wē nú dreógaþ, 105, 8; Gū. 20. Ic mē forhtige fyrenfulra fǣcne nîþas conturbatus sum a tribulatione peccatoris, Ps. Th. 54, 2. Helle heáfas, hearde nîþas, Cd. Th. 3, 20; Gen. 38. IV. evil, wickedness, malice:—Nîþ synfulra nequitia peccatorum, Ps. Surt. 7, 10. On đara ācorenra monna heortan sceal đære nædran lytignes and hire nîþ đære culfran biliwitnesse gescirpan in electorum cordibus debet simplicitatem columbae astutia serpentis acuere, Past. 35, 1; Swt. 237, 22. Mid đŷ nîþe yfles ingeþonces malitiae peste, 33, 5; Swt. 220, 19. Nîþe nequitiae, Ps. Spl. 54, 17. Đara nædran nîþ through the serpent's malice, Cd. Th. 290, 8; Sat. 412: Exon. Th. 226, 29; Ph. 413. Nîþa geblonden (Holofernes), Judth. Thw. 21, 25; Jud. 34. Nîþa efter nîþum teolunge heara secundum nequitias studiorum ipsorum, Ps. Surt. 27, 4. Æfter nîþas, Ps. Spl. C. 27, 5. [A. R. Orm. Laym. nîþ: Gen. and Ex. nið and strif: Goth. neiþ φθόνος: O. Sax. nîd: O. Frs. nîth: Icel. nîð a libel, lampoon: O. H. Ger. nîd invidia, rancor, discidium, invidentia, iniquitas.] v. bealu-, fǣr-, gār-, helle-, hete-, hete-, inwit-, orleg-, searo-, sin-, spere-, wæl-nîþ.

nîþ; adj. (?) Vexatious, rancorous:—Ǣfǣstum onǣled, nîþum nearowrencum (or nîþum from preceding word?), Exon. Th. 316, 5; Mód. 44. [Cf. he fell off heffne dun Inntill nîþ hellepine, Orm. 13677.]

nîþ-cwalu; e; f. Grievous destruction:—Hē hŷ generede from nîþcwale, and eác forgeaf ēce dreámas, Exon. Th. 77, 18; Cri. 1258.

nîþ-cwealm, es; m. Violent death, destruction:—Heora neát nîþcwealm forswealh jumenta eorum in morte conclusit, Ps. Th. 77, 50.

nîþ-draca, an; m. A hostile, malicious dragon, Beo. Th. 4538; B. 2273.

nîþemest. v. neoþera.

nîþer; adv. Down, beneath, below:—Nîþer deorsum, Ælfc. Gr. 38; Som. 40, 6. Đē wearþ helle seáþ nîþer gedolfen beneath was the pit of hell dug for thee, Exon. Th. 267, 30; Jul. 423. Đā hē nyþer ābeáh cum se inclinasset, Jn. Skt. 20, 5. Ic nyþer ālǣte submitto, Ælfc. Gr. 28; Som. 31, 41. Hē nyþer ālēde deposuit, Lk. Skt. 23, 53. Nyþer āsceótan to cast down, Homl. Th. i. 170, 23. Hē niþer āsette đā mihtigan deposuit potentes, Cant. Mar. 52. Đonne heó nyđer byþ āstigen, Anglia viii. 319, 19. Đæt hî hine nyþer bescufon ut praecipitarint eum, Lk. Skt. 4, 29. Se đe nyþer com of heofonum qui descendit de caelo, Jn. Skt. 3, 13. Đū niþer færst (descendes) ōþ helle, Mt. Kmbl. 11, 23. Niþer feallaþ procident, Ps. Lamb. 71, 9: 94, 6. Niþer fylþ decidat, 89,

6. Gá nyþer *descende*, Mt. Kmbl. 27, 40. Hé nyþer ne eode, Ex. 32, 1. Gang niþer, Deut. 9, 12. Niþer gewítan *descendere*, Bd. 5, 12; S. 628, 21. Hwearf him eft niþer (*to hell*) boda bitresta, Cd. Th. 47, 18; Gen. 762. Niþer stígan *descendere*, Ps. Lamb. 27, 1: Lchdm. iii. 210, 17. Hé sceal má þencan up ðonne nyþer *he must direct his thought upwards rather than downwards*, Bt. 41, 6; Fox 254, 31. [*Laym*. niðer: *O. E. Homl. Marh*. neoþer: *Gen. and Ex*. neðer: *O. Sax*. niðar: *O. Frs*. nither: *Icel*. niðr: *O. H. Ger*. nidar *deorsum*: *Gen*. nieder.] v. niþre.

niþera. v. neoþera.

niþer-bogen *down-bent*:—Tó ðære niþerbogenan ác, Cod. Dip. Kmbl. iv. 72, 1. [Cf. *Icel*. niðr-bjúgr.]

niþer-dǽl, es; m. *A lower part*:—On niþerdǽlum eorþan *in inferioribus terrae*, Ps. Th. 138, 13.

niþre, niþre; adv. *Down, beneath, below*:—Læg mín flǽschoma in foldan bigrafen, niþre gehýded in byrgenne, ðæt ðú meahte beorhte uppe on roderum wesan, Exon. Th. 89, 34; Cri. 1467. Uppe ge niþre, 360, 3; Pa. 74. [*O. H. Ger*. nidare, niðere: *Icel*. niðri.]

niþer-gang, es; m. *Descent*:—Mid hyra upgange oððe nyþergange, Lchdm. iii. 246, 8. [*Icel*. niðr-gangr, -ganga.]

niþer-heald; adj. *Bent downwards*:—Nis ðæt gedafenlíc, ðæt se módsefa ǽniges niþerheald wese, and ðæt neb upweard, Met. 31, 23. [Cf. Hie mugen lihtliche cumen mid þare niðerhelde (*the downward slope*), O. E. Homl. ii. 230, 347: *O. H. Ger*. nidar-haldig *reclinus*.]

niþerian, niþrian; p. ode. I. *to bring low, humiliate*:—Se ðe hine nyþeraþ *qui se humiliat*, Lk. 14, 11. Hí nyþerodon *humiliaverunt*, Ps. Spl. 93, 5. II. *to accuse, condemn*:—Ne ǽnig mon ðec niþraþ (*condemnavit*), Jn. Skt. Rush. 8, 10. Hí niþriaþ *condemnabunt*, Ps. Spl. 93, 21. Niþrigaþ *condemnabunt*, Mt. Kmbl. Rush. 12, 41. Ðæt hiǽ niþradun (*accusarent*) hine, Mk. Skt. Rush. 3, 2. Niþrad *damnatus*, Mt. Kmbl. Rush. 27, 3. Wæs neþored *damnatur*, Hpt. 495, 2. [*Orm*. niþþrenn: *Laym*. neoþered: *Icel*. niðra *to put down, lower*: *O. H. Ger*. niderren *humiliare, accusare, condemnare*.] v. ge-niþerian.

niþerigend-líc; adj. *Deserving condemnation*:—Þurh gódne willan herigendlíc oððe of yflum willan nyþergendlíc, Boutr. Scrd. 20, 1.

niþer-líc; adj. I. *low* (of position):—Án þeósterful dene swíðe niþerlíc, Homl. Th. ii. 338, 5. Heortan niþerlícan *cordis ima*, Ps. Surt. ii. p. 202, 5. On nyþerlícum eorþan *in inferioribus terrae*, Ps. Lamb. 62, 10. Tó nyþerlícum *ad inferos*, Cant. An. 6. Hé his eágan bígde on ðás nyþerlícan þing *oculos in inferiora deflectens*, Bd. 3, 19; S. 548, 8. II. *low, humble, inferior*:—In ðisse nyþerlícan worulde *in this lower world*, Shrn. 123, 10. Ða nyþerlícan *humilia*, Blickl. Gl. Gé sécaþ ðære heán gecynde gesǽlþa tó ðám niþerlícum and tó ðám hreósendlícum þingum, Bt. 14, 2; Fox 44, 30: Homl. Th. ii. 522, 30.

niþerness, e; f. *Lowness, a low position, the bottom*:—Ða geseah hé swá þýstre dene under him in nyþernesse gesette *vidit quasi vallem tenebrosam subtus se in imo positam*, Bd. 3, 19; S. 548, 9.

niþer-scyfe, es; m. *A pushing down, falling down, hasty downward movement*:—Niþerscyfe *per praeceps*, Hpt. Gl. 468, 74. v. scyfe.

niþer-sige, es; m. *A going down, setting*:—Sunne oncneów niþersige (*occasum*) hire, Ps. Lamb. 103, 19: 113, 3. Ofer niþersi[g]e *super occasum*, 67, 5.

niþer-stige, es; m. *A descent*:—Se upstige and se niþerstige *the ascent and descent* (*of the angels seen by Jacob in his dream*), R. Ben. 23, 7. Tó Olivetes muntes nyþerstige *ad descensum montis Oliveti*, Lk. Skt. 19, 37. [Cf. *Icel*. niðr-stiga *a descent*: *O. H. Ger*. nidar-stiga.]

niþerung, e; f. I. *a bringing low, humiliation, overthrow*:—Ic salde..iów mæhte hénnisse-t niþrunge ofer nedre *dedi vobis potestatem calcandi supra serpentes*, Lk. Skt. Rush. 10, 19. Ǽttrige niþerunge *venenata detrimenta*, Hpt. Gl. 450, 39. II. *damnation, condemnation*:—Niþrung *damnatio*, Lk. Skt. Lind. 23, 40: 24, 20. Hé hí fram yrmþum écre niþerunge generede, Bd. 4, 13; S. 582, 26: 5, 13; S. 633, 14. In niþrunge *in condemnatione*, Rtl. 24, 19. For ðæs dæges nyþerunge *ad damnationem diei*, L. Ecg. P. add. 22; Th. ii. 236, 4: Bd. 5, 14; S. 635, 2. [*O. H. Ger*. nidarunga *damnatio*: cf. *Icel*. niðran *degradation*.]

niþer-weard; adj. *Downward, turned downwards*:—Neb is mín niþerweard, Exon. Th. 403, 1; Rä. 22, 1: 416, 24; Rä. 35, 3. Niþerwearþ, 413, 15; Rä. 32, 6.

niþer-weardes, -weard; adv. *Downwards, in a downward direction*:—Niþerweardes *per praeceps*, Mt. Kmbl. Rush. 8, 32. Nyþerwerd *deorsum versum*, Ælfc. Gr. 38; Som. 41, 63. Niþerwurd *in praeceps, deorsum*, Hpt. Gl. 499, 66.

niþe-weard. v. neoþe-weard.

niþ-full; adj. *Envious, malicious*:—Ǽfre biþ se niþfulla (*envious*) man on gedréfednysse, forðan ðe se anda his mód ǽlcere gástlícere blisse benǽmþ, Homl. Th. i. 606, 2. Se niþfulla wer ... se gesibsuma wer *the malicious man ... the man that loves peace*, Basil admn. 6; Norm. 46, 20, 22, 30. Feóndes niþfulles fácne *hostis invidi dolum*, Hymn. Surt. 3, 21. Ðeós costung is of ðam niþfullan deófle, Boutr. Scrd. 23, 10. Niþfullum *rancida, amara*, Hpt. Gl. 475, 73. Ða geseah se hálga wer ðæs árleásan preóstes niþfullan ehtnysse, Homl. Th. ii. 162, 34. Níþfulra *lividorum*, Hpt. Gl. 519, 69.

níþful-líce; adv. *Maliciously, enviously*:—Gé ðone rihtwísan Crist níþfullíce ácwealdon, Homl. Th. i. 46, 25.

níþ-gǽst, es; m. *A malicious, malignant guest*:—Hé ðone níþgǽst (*the dragon*) slóh, Beo. Th. 5391; B. 2699. Under níþgysta (*the devils who persecuted Guthlac*) nearwum clommum, Exon. Th. 134, 21; Gú. 511.

níþ-geteón, es; n. *Injurious malice*:—Sigor áhwearf of norþmanna níþgeteóne (*when Abraham defeated those who carried away Lot*), Cd. Th. 124, 26; Gen. 2068.

níþ-geweorc, es; n. *Malicious, evil work*:—Ðeáh hé (*Grendel*) róf síe níþgeweorca, Beo. Th. 1370; B. 683.

níþ-grim[m]; adj. *Savage, cruel*:—Nýdwracu níþgrim, nihtbealwa mǽst, Beo. Th. 388; B. 193. Mé beþeahton þeóstru níþgrim, Ps. Th. 54, 5. Cf. hete-grim.

níþ-gripe (P), es; m. *A hostile grasp*:—Hyne (*Grendel*) sár hafaþ in níþgripe (MS. mid gripe, nídgripe [?]) nearwe befongen, Beo. Th. 1956; B. 976.

níþ-heard; adj. *Bold in battle, audacious*:—Níþheard cyning (*Constantine*), Elen. Kmbl. 389; El. 195: (*Beowulf*), Beo. Th. 4826; B. 2417. Ðá wearþ sum tó ðæs árod ðara beadorinca, ðæt hé in ðæt búrgeteld níþheard neðde, Judth. Thw. 25, 25; Jud. 277. [Cf. *O. H. Ger*. Nídhart (*proper name*).]

níþ-hete, es; m. I. *rancorous hate, enmity*:—Áwehte ðone wælníþ Nabochodonossor þurh níþhete, Cd. Th. 219, 2; Dan. 48. II. *affliction, grievous trouble*:—Ðe hié generede wið ðam níþhete (*the fiery furnace*), 233, 22; Dan. 279. III. *malice, wickedness*:—Æfter níþhete widmétednyssa heora *secundum nequitiam adinventionum ipsorum*, Ps. Spl. T. 27, 5. Cf. hete-níþ.

níþ-hete, es; m. *A malignant foe*:—Léton ðone hálgan bídan burhwealle néh, his níþhetum, nihtlangne fyrst, Andr. Kmbl. 1667; An. 836. Cf. scyld-hete.

níþ-hycgende *having hatred* or *malice in the heart*:—Slógon eornoste Assiria oretmæcgas níþhycgende nänne ne sparedon *with hate in their hearts Assyria's warriors they (the Hebrews) hewed, not one did they spare*, Judth. Thw. 24, 40; Jud. 233. Him (*Christ*) mid næglum þurhdrifan níþhycgende ða hwítan honda, Exon. Th. 68, 28; Cri. 1110.

níþ-hygdig; adj. *Having the mind disposed to strife, bold*:—Hyrsta swylce on horde ǽr níþhýdige men genumen hæfdon, Beo. Th. 6311; B. 3166. [*O. Sax*. níð-hugdig *maliciously disposed* (*applied to Herod and to the devil*).] Cf. níþ-heard.

níþing, es; m. *A villain, one who commits a vile action*:—Walreáf is níþinges dǽde, L. Ath. iv. 7; Th. i. 228, 3. Se cing and eall here cwǽdon Swegen for níþing (*Swegen had treacherously put Beorn to death*), Chr. 1049; Erl. 174, 31. [*Icel*. níðingr *a villain*. v. Cl. & Vig. Dict.] v. un-níþing.

níþ-líce; adv. *Cowardly, meanly*:—Earhlíce t níþlíce *muliebriter*, Hpt. Gl. 424, 1.

níþ-loca, an; m. *A place where one is shut up in misery*:—Under níþloc[an] gebunden, under bealuclommum, Exon. Th. 463, 3; Hö. 64.

níþor. v. neoþor.

níþ-plega, an; m. *Battle*, Andr. Kmbl. 827; An. 414.

niþre. v. niþera.

níþ-sceaða, an; m. *A malignant foe*, Exon. Th. 397, 23; Rä. 16, 24.

níþ-sele, es; m. *A hall where one is exposed to the hatred of a foe*:—[Grein reads níþ-sele *a hall low down, beneath the water*.] Hé [in] níþsele náthwylcum wæs, ðær him nǽnig wæter wihte ne sceþede, Beo. Th. 3030; B. 1513.

níþ-syn. v. níd-syn.

níþþas, niþas; pl. m. (a poetical word used only in the plural) *Men*:—Niþþas findaþ gold, gumþeóda bearn, Cd. Th. 14, 27; Gen. 225. Niþþa bearna ǽrest ealra, 69, 14; Gen. 1135: 77, 33; Gen. 1284: Beo. Th. 2015; B. 1005: Exon. Th. 167, 34; Gú. 1070. Niþþa nergend, 140, 18; Gú. 612. Niþþa gehwylcum, 360, 15; Wal. 6. Geneósian niþa bearna ealra þeóda *ad visitandas omnes gentes*, Ps. Th. 58, 5: 65, 3: 71, 17. Niþa náthwilc, Beo. Th. 4436; B. 2215. Niþa gehwam un-ásecgendlíc, Elen. Kmbl. 928; El. 465. Hé from sceolde niþþum hweorfan *he must die*, Cd. Th. 74, 16; Gen. 1223: 75, 5; Gen. 1235. Hé is niþum swǽs, is ðín milde mód ofer manna bearn, Ps. Th. 99, 4. Neáh is Drihten niþum eallum ðe hine mid sóþe hige sécaþ *prope est Dominus omnibus invocantibus eum in veritate*, 144, 19. Ðú eart mihtum swíð niþas tó nergenne, Cd. Th. 234, 1; Dan. 285. [*Goth*. nithjis *a kinsman*: *Icel*. niðr; *pl*. niðjar *a son, kinsman*.]

níþ-weorc, es; n. *Battle, conflict*:—Níþweorca heard *brave in battle*, Chr. 973; Erl. 124, 26.

níþ-wracu; gen. -wræce; f. *Severe punishment*:—Ða fǽmnan hét þurh níþwræce nacode þennan, and mid sweopum swingan, Exon. Th. 253, 28; Jul. 187. Cwom Nabochodonossor of níþwracum (*his exile among the beasts of the field*), Cd. Th. 257, 28; Dan. 664.

níþ-wundor, es; n. *A wonder that bodes evil, a portent*:—Ðǽr mæg nihta gehwǽm níþwundor (niþ-, Grein) seón, fýr on flóde, Beo. Th. 2735; B. 1365.

nîwan, neówan, neón; *adv. Recently, lately, newly:*—Nîwan, neówan, *nuper*, Ælfc. Gr. 38; Som. 39, 58. Secgeaþ hî ðæt sume dæge ðider nîwan (*nuper*) côme cýpemen, Bd. 2, 1; S. 501, 4. Ða þing ðe ûs nîwan bodade syndon *ea quae nunc nobis nova praedicantur*, 2, 13; S. 516, 11. Ðonne man nîwan wíf nymþ *cum acceperit homo nuper uxorem*, Deut. 24, 5. Ðone consul ðe hié ðâ nîwan geset hæfdon, Ors. 2, 6; Swt. 86, 32. Gif hwelc man biþ witeþeów nîwan geþeówad, L. In. 48; Th. i. 132, 7. Ic eom se ðe nû nîwan com. Swilce hê swâ cwæde: Ic wæs geswutelod nû nîwan, Glostr. Fragm. 10, 2-4. For ðære swidlícan ehtnysse ðe ðâ nîwan âsprang æfter Carines slege *on account of the fierce persecution that just then had sprung up after the murder of Carinus*, Homl. Skt. 5, 326. Gif hwâ nîwan tô mynstres drohtnunge gecyrran wyle *if a man's wish to turn to a monastic life is but newly formed*, R. Ben. 96, 3. Gelamp nýwan *it happened lately*, Nicod. Thw. 8, 27; 19, 37. Seó nûgyt neówan is becumen and gelǽded tô Godes geleáfan *quae (ecclesia Anglorum) nuper adhuc ad fidem adducta est*, Bd. 1, 27; S. 489, 12. Hê eów neón gesceód *lately he harmed you*, Andr. Kmbl. 2354; An. 1178. [Cf. *O. H. Ger.* nîwanes *nuper*.] v..nîwane.

nîwan-âcenned *new-born:*—Ðâ wæs broht tô fulwihte nîwanâcenned cild, Shrn. 130, 7. Cf. nîw-cenned.

nîwan-cumen *recently come (to a particular belief), a neophyte:*—Nîwancumen *neofitus*, Wrt. Voc. ii. 59, 68. Cf. nîw-cumen.

nîwane; *adv. Lately, recently:*—Wênaþ ðæt ðæt ne síe eald gesceaft, ac síe geworden nîwane, Bt. 39, 3; Fox 216, 4. v. nîwan.

nîw-bacen; *adj. New-baked:*—Wê mid ûs nâmon nîgbacene hlâfas *panes calidos sumpsimus*, Jos. 9, 12.

nîw-cealct, -cilct *newly white-washed:*—On ânum nîcealtan (nîw-cilctan, MS. C.) hûse *in cubiculo nuper calce illito*, Ors. 6, 32; Swt. 286, 30.

nîw-cenned, -cend *new-born:*—Mid hyre nîcendum cilde *cum recens nato parvulo*, Bd. 2, 16; S. 520, 1.

nîw-cumen *newly come (to a particular belief), a neophyte:*—Nîcumen *neophytus*, Hpt. Gl. 480, 12. Se sylfa nîgecumena (nícumena, nîg-cumene) brôðor, R. Ben. 101, 15. Be nîgcumenra (nícumenra, Wells, Frag.) gebrôðra andfenge, 97, 2. Tǽce him mon siððan tô nîgcumenra manna hûse, 97, 11. v. nîwan-cumen.

nîwe, neówe; *adj.* I. *new, not yet used:*—Ne âsend nân scyp of nîwum reáfe on eald reáf; elles ðæt nîwe slît, and se nîwa scyp ne hylp ðam ealdan, Lk. Skt. 5, 36. Smyre mid nîre (MS. B. ânre) feþere, Lchdm. i. 234, 13. Hê léde hyne on hys nîwan byrgene, Mt. Kmbl. 27, 60. Gê ðær gemêtaþ nîwe byrgenne, Blickl. Homl. 147, 30. II. *new, recent, not of long standing, not long made:*—Nân man ne sent nîwe wín on ealde bytta; elles ðæt nîwe wín brycþ ða bytta ... Ac nîwe wín is tô sendenne on nîwe bytta ... And ne drincþ nân man eald wîn and wylle sôna ðæt nîwe, Lk. Skt. 5, 37-39. Ðâ hê (*the Roman name*) com ǽrest'tô Parþum, and wæs ðær swíðe nîwe, Bt. 18, 2; Fox 64, 13. Nîwe môna *neomenia*, Wrt. Voc. i. 16, 51. Se nîwa *neophytus*, i. 60, 64. Seó (*the English Church*) nûgyt is nîwe on geleáfan, Bd. 489, 41. Ðý læs se steall swâ nîwre cyricean tealtrian ongunne, 2, 4; S. 505, 11. Ðis gelimpþ seldon, and næfre bûton on nîwum mônan, Lchdm. iii. 242, 23. Nô on nîwan wylme, ac on lancsumere mynstres drohtnunge, R. Ben. 9, 6. Hê ðone winter mid ðý nîwan folce (*the newly converted Frisians*) wunode, Bd. 5, 19; S. 639, 26. Nîwan stefne *anew*, Cd. Th. 94, 1; Gen. 1555; Andr. Kmbl. 245; An. 123. Niówan, Beo. Th. 3582; B. 1789. On swâ niówan gefeán, Andr. Kmbl. 3336; An. 1672. Ðâ sceáwode Scyppend ûre his weorca wlite, nîwra gesceafta, Cd. Th. 13, 25; Gen. 208 : 55, 4; Gen. 889. On ðisum nîwum dagum *in these modern times*, Homl. Th. i. 608, 23 : Homl. Skt. 13, 177. Brembel ðe síen begen endas on eorþan; genim ðone neówran wyrttruman, Lchdm. ii. 292, I. III. *new (to anything), inexperienced:*—Swâ swíðe swâ ða nîwan Cristenan hit nîman mihte *in quantum rudes capere poterant*, Bd. 3, 22; S. 553, 10. Ðǽm nîwum *neotericis*, Wrt. Voc. ii. 60, 58. Ðæt is ðæt mon ða earce bere on ðǽm saglum ðætte ða gôdan lâreówas ða hâlgan gesomnunge lǽrende ða nîwan (niéwan, Cott. MSS.) and ða angeleáffullan môd mid hira lâre gelǽde tô ryhtum geleáfan *vectibus arcam portare est bonis doctoribus sanctam ecclesiam ad rudes infidelium mentes praedicando deducere*, Past. 22, 2; Swt. 171, 13. Nîwa lâre *rudimenta*, Rtl. 80, 3. IV. *new, novel, different from what has gone before:*—Árâs nîwe cing ðe nyste hwæt Iosep wæs, Ex. 1, 8. Hasterbal se nîwa cyning *Asdrubal novus imperator*, Ors. 4, 6; Swt. 176, 33. Hwæt is ðeós nîwe lâr, Mk. Skt. 1, 27. Nîwe circhâlgung (v. cyric-hâlgung) *encenia*, Wrt. Voc. i. 16, 52. Calic nîwe ǽ (*novi testamenti*), Mt. Kmbl. 26, 28. Nemde nîwan stefne; nama wæs gecyrred, Elen. Kmbl. 2119; El. 1061. Him ne wæs nǽnig earfoþe ðæt lîchomlíce gedâl on ðære neówan wyrde (*in their new condition*), Blickl. Homl. 135, 31. Ic eów sylle nîwe bebod, Jn. Skt. 13, 34. Singaþ Drihtne neówne sang, Ps. Th. 149, 1 : Ps. Surt. 32, 3. Hî hæfdon neówne gefeán gemêted, Elen. Kmbl. 1737; El. 870. Fægere word ðis synd ... ac forðon hî nîwe syndon and uncûþe, Bd. 1, 25; S. 487, 9. Lyt swîgode nîwra spella, se ðe næs geråd, Beo. Th. 5788; B. 2898. Hî sprecaþ nîwum tungum, Mk. Skt. 16, 17. Gelǽrdan biscepas swelce nîwe rǽdas swelce hié fol oft ǽr ealde gedydan,

Ors. 4, 7; Swt. 184, 2. Singaþ sangas neówe, Ps. Th. 95, 1. [*Goth.* niujis: *O. Sax.* nîwi: *O. Frs.* nîe : *O. H. Ger.* niuwi, nîwi *novus, recens, rudis, modernus.*] v. ed-nîwe.

nîwe, nîge; *adv. Newly, recently:*—Wê nîwe syndon tô ðissum geleáfan gedôn *we are newly turned to this faith*, Blickl. Homl. 247, 34. Syððan heó nîge cealfod hæfþ *after it (a cow) has recently calved*, L. R. S. 13; Th. i. 438, 19. Sceal mon lâcnian swilce âdle mid cû meolcum oððe gâte swâ nîge molcene drince (*or let him drink goat's milk as newly milked as possible*), Lchdm. ii. 218, 22 : 222, 13. v. nîwan.

niwel. v. neowol.

nîw-fara, an; *m. A new-comer, a stranger:*—Ic eom nîfara hider on eorþan beforan ðê and ælþeódig *incola ego sum apud te in terra, et peregrinus*, Ps. Th. 38, 15.

nîw-gecirred *newly converted:*—Nîgecerred *neophytus, novellus*, Hpt. 488, 4.

nîw-gehâlgod *newly consecrated:*—Hieu se nîgehâlgode (nîghâlgoda) cynincg, Homl. Skt. 18, 326.

nîw-gehwirfed *newly converted:*—Ðæt hê ða nîgehwyrfedan (nîghwurfedan, MS. C.; nîghwerfdan, MS. V.) mid fulluhte âþwôge, Homl. Skt. 5, 126.

nîw-hwirfed *newly converted:*—Nîhwurfed ł nîlǽred *neophytus*, Hpt. Gl. 480, 13. v. preceding word.

nîw-hworfen *newly converted:*—Betwux ðam nîghworfenum folce (*the recently converted people of Kent*), Homl. Th. ii. 130, 27.

nîwian; *p. ode To renew, renovate, restore:*—Nû mê Sethes bearn torn nîwiaþ, Cd. Th. 76, 16; Gen. 1258. Hê nîwade Cnutes lage (v. Freeman's Old English History, p. 241), Chr. 1064; Erl. 196, 2. Ne wrec ðû ða ǽrran yflu, bûton hî mon eft nîwige, Prov. Kmbl. 35. Swâ ðæt ðû ǽghwylce dæg ðone drenc nîwie (nîwige, MS. B.), Lchdm. i. 192, 15. Burh rǽran, and sele settan, salo nîwian, Cd. Th. 113, 3; Gen. 1881. Sâr nîwigan, Elen. Kmbl. 1878; El. 941. Eft nîwige *emendare*, Mt. Kmbl. p. 2, 12. Wǽren ǽrendracen gesend tô Ængla lande tô nîwianne ðone geleáfan, Chr. 785; Erl. 57, 17. Eorþan neówiende anseón *terrae novas faciem*, Hymn. Surt. 97, 34. [*Chauc.* newe: *Goth.* ana-niujan: *O. Sax.* nîwian: *O. H. Ger.* niuwôn, nîwôn *novare.*] v. ed-, ge-nîwian.

ni-wiht *nothing:*—Tô niwihte *ad nihilum*, Ps. Surt. 59, 14. v. nâ-wiht.

nîwinga, nîw-lǽred. v. nîwunga, nîw-hwyrfed.

nîw-lîc; *adj. New, fresh:*—Bearn ðíne swâ swâ nýwlícra elebergena *filii tui sicut nouellae oliuarum*, Ps. Lamb. 127, 3.

nîwlíce; *adv. Newly, recently:*—Nîwan *nuper*, nîwlícor *nuperius*, nîwlícost *nuperime*, Ælfc. Gr. 38; Som. 42, 11. Hér cumaþ tô eów nîwlíce twegen men, Homl. Th. ii. 494, 7. Hî hæfdon nîwlíce gesett *they had recently decreed*, Ors. 4, 10; Swt. 202, 26. Hió ðâ wæs nîwlíce cristen, 6, 4; Swt. 260, 12.

nîwnes, e; *f. Newness, novelty:*—Ne sceal him mon ânne mete gebeódan, ac missenlíce, ðæt seó niównes ðara metta mǽge him gôde beón, Lchdm. ii. 240, 15. Ðæs mônan nîwnys, Anglia viii. 310, 38. Ðâ wæs se dêma mid ða neównysse (*novitate*) swâ monigra heofonlícra wundra swýðe gedréfed, Bd. 1, 7; S. 478, 44. Mîne nîwnysse *juventutem meam*, Ps. Lamb. 42, 4.

niwol. v. neowol.

nîw-tirwed *new-tarred:*—Flotan nîwtyrwdne, Beo. Th. 595; B. 295.

nîwung, e; *f. A beginning, rudiment:*—Nîwunge *rudimenta*, Hpt. Gl. 428, 18. v. nîwe, III.

nîwunga; *adv. Anew:*—Nîwunga (niówunga, Rush.) *denuo*, Mk. Skt. Lind. 14, 40. Niúnge (niówunga, Rush.), Jn. Skt. 3, 3. Neówinga, Andr. Kmbl. 2787; An. 1396. v. ed-nîwinga.

nîwerne; *adj. Young, tender:*—Sum wíf mid hire nýwernan (MS. Bodl. niwernan, *glossed by* tenero) cilde, Homl. Th. i. 566, 5.

nixtnig. v. nihstnig.

nó, nôh, nôht, nolde, nom-, noma. v. nâ, ge-nôh, nâ-wiht, nyllan, nam-, nama.

nôn, es; *n.* I. *the ninth hour;* hora nona:—Prîm *prima;* undern *tertia;* middæg *sexta;* nôn *nona;* ǽfen *vesperum*, Wrt. Voc. i. 53, 10-15: R. Ben. 40, 13. Ða nigoþan tíde ðe wê nôn hâtaþ, Homl. Th. ii. 256, 35. Ðâ com nôn dæges, Beo. Th. 3204; B. 1600. Hî him tô gewunon nâman ðæt hî fæston tô nônes (*ad horam nonam*), Bd. 3, 5; S. 527, 9. Tô huil nônes *ad horam nonam*, Mt. Kmbl. Lind. 27, 45, 46. On tíde nônes, Mk. Skt. Rush. 15, 33, 34. Tô underne and tô nône ... and tô middæge, Lchdm. iii. 218-222, often. Fram middæge oþ nôn, H. R. 107, 9. Sele drincan on undern, on middæg, on nôn, Lchdm. ii. 140, 2. II. *the service held at the ninth hour, nones :*—Wê sungon nôn *cantavimus nonam*, Coll. Monast. Th. 33, 35. [*O. Sax.* nôn, nuon: *Icel.* nón: *n. nones, about three o'clock.*] v. ofer-nôn.

non, es; *m. The title given to the older by the younger monks:*—Ða yldran hyra gingran brôðor nemnen, and ða gingran hyra yldran nonnos (nonas, Wells, Frag.) nemnen, R. Ben. 115, 19. v. nunne.

nôn-gereord, es; *n. A repast after the service of nones:*—Siððan hý

ðone forman cnyl tô nône gehŷren, gangen hŷ ealle from hyra weorce and dôn hŷ gearuwe, ðæt hŷ mâgon tô cirican gân, ðonne mon eft cnylle. Ðonne eft æfter heora nôngereorde rædan hŷ eft heora bêc, R. Ben. 74, 8.

nôn-mete, es; m. *An afternoon meal:*—Nônmete *merenda*, Wrt. Voc. i. 38, 14: *annona*, 291, 2: ii. 8, 67. On xii mônþum ðû scealt sillan dînum þeówan men vii hund hlâfa and xx hlâfa, bûton morgenmetum and nônmetum, Salm. Kmbl. p. 192, 19. [*Prompt. Parv.* nunmete *merenda*, p. 360. v. note there.]

nôn-sang, es; m. *The service held at the ninth hour, nones:*—Ða seofon tîdsangas . . . nônsang, L. Ælfc. C. 19; Th. ii. 350, 7. *De officio nonae horae* (nônsang), Btwk. 216, 31: R. Ben. 39, 19: 40, 7. Nônsang wê singaþ *nonam psallimus*, Hymn. Surt. 60, 35.

nôn-tîd, e; f. *The ninth hour:*—On undern, on midne dæg, on nôntîde, Homl. Th. ii. 74, 9.

nôn-tîma, an; m. *The ninth hour:*—On nôntîman wê sculon God herian, forðam on ðone tîman Crist gebæd for ðam ðe him deredon, and siððan his gâst âsende, Btwk. 216, 31.

Normandîg, e; f. *Normandy:*—Willelm cyng fôr ofer sæ tô Normandîg, and Eádgâr cild com of Scotland tô Normandîge, Chr. 1074; Erl. 212, 3–4.

Nor-men. v. Norþ-mann.

Norren, Noren; adj. *Norse, Norwegian:*—Se Norrena cyng, Chr. 1066; Erl. 201, 12. Ôlaf ðæs Norna cynges sunu, 201, 34. [*Icel.* Norrœnn, Norœnn.]

[norþ]; adj. *In a northerly position:*—Ðæt folc ðe tô ðære norþerran byrig hiérde, Chr. 922; Erl. 108, 19. Hêt Eádweard cyning âtymbran ða norþran burg, 913; Erl. 100, 34. On ðæm dagum wæs ðæt norþmeste [rîce] micliende, Ors. 6, 1; Swt. 252, 12. Sciþþie ða norþmestan 1, 7; Swt. 40, 6. Ôþ ða norþmestan næssan on eorþan, Met. 9, 43. [*Icel.* nyrðri, norðari; nyrðstr, norðastr.] *See the compounds of which* norþ *forms the first part.*

norþ; adv. *In a northerly direction* or *position:*—Ðæt is norþ ehta hund mîla lang, Bd. 1, 1; S. 473, 11. Hié Baldred norþ ofer Temese âdrifon, Chr.¹ 823; Erl. 62, 20. Hié fôron norþ ymbûtan, 894; Erl. 91, 6. Symle swâ norþor swâ smælre *ever the further north, the narrower*, Ors. 1, 1; Swt. 18, 29. Hê ealra Norþmonna norþmest bûde, 17, 2. [*O. Sax.* norð: *O. Frs.* north: *Icel.* norðarr; *cpve.*; norðast; *super.*]

norþan; adv. *From the north:*—Se wind se ðe ær sûþan bleów, hine norþan âwearp, Bd. 2, 7; S. 509, 28. Gif hér wind cymþ westan oððe eástan, sûþan oððe norþan, Cd. Th. 50, 11; Gen. 807. See following words.

norþan-eástan; adv. *From the north-east:*—Ðonne se stearca wind cymþ norþan-eástan, Bt. 9; Fox 26, 19.

norþan-eástan *in be* norþan-eástan *to the north-east*, Ors. 1, 1; Swt. 24, 10: 16, 18.

norþan-eástan-wind *a north-east wind;* eurus, euroauster, circius, Wrt. Voc. i, 36, 13, 17. [Cf. *O. H. Ger.* norðôstir-wint *aquilo*.]

Norþan-hymbre; pl. *The Northumbrians, Northumbria, the people* or *province north of the Humber:*—Hér Ida fêng tô rîce, ðonon Norþanhymbra cynecyn onwôc, Chr. 547; Erl. 16, 7. Ðær wæs ungemetlîc wæl geslægen Norþanhymbra, 867; Erl. 72, 15. Norþanhymbra mægþ ðe Ceólwulf ofer is, Bd. 5, 24; S. 646, 28. Hér fôr se here of Eást-Englum on Norþanhymbre, Chr. 867; Erl. 73, 7. v. Norþ-hymbre.

norþan-weard; adj. *Northward:*—Sanctus Paulus wæs geseónde on norþanweardne ðisne middangeard, Blickl. Homl. 209, 30. Ða Pyhtas gefêrdon ðis land norþanweard *the Picts occupied the north of this land*, Chr. Erl. 3, 13.

norþan-westan; adv. *From the north-west;* a circio, Hpt. Gl. 512, 11: Wrt. Voc. ii. 3, 44: 98, 40.

norþan-westan *in be* norþan-westan *to the north-west*, Ors. 1, 1; Swt. 16, 5.

norþan-westan-wind *a wind from the north-west;* corus, aquilo *vel* boreas, Wrt. Voc. i. 36, 16, 18. [Cf. *O. H. Ger.* nortwesterwint *circius*.]

norþan-wind *a wind from the north:*—Norþanwind *septentrio*, Wrt. Voc. i. 36, 11. Ðæs norþanwindes ŷst, Bt. 9; Fox 26, 20. Stearc stormas and norþanwindas, 23; Fox 78, 27. v. norþ-wind.

norþ-dæl, es; m. I. *a northern part:*—Middaneardes norþdæl *Europa*, Hpt. Gl. 512, 20. Sió hæte hæfþ genumen ðæs sûþdæles mâre ðonne se cyle ðæs norþdæles hæbbe, Ors. 1, 1; Swt. 24, 29. Nû hæbbe wê âwriten ðære Asian sûþdæl; nû wille wê fôn tô hire norþdæle, Swt. 14, 6. Sittan on ðam norþdæle heofenan rîces, Homl. Th. i. 10, 25: Cd. Th. 3, 8; Gen. 32. Norþdæl *aquilonem*, Ps. Th. 89, 11. Hê wæs mid firde farende on Sciþþie on ða norþdælas, Ors. 1, 10; Swt. 44, 7. Peohtas ongunnon eardigan ða norþdælas ðysses eálondes, Bd. 1, 1; S. 474, 18. II. *the north:*—Breoton is geseted betwyh norþdæle and westdæle *Britannia inter septentrionem et occidentem locata est*, S. 473, 9. Ðonne âstîgeþ blôdig wolcen from norþdæle, Blickl. Homl. 91, 32.

Norþ-Dene; pl. *The North-Danes*, Beo. Th. 1571; B. 783: Ors. 1, 1; Swt. 16, 25, 27.

norþ-duru *a door on the north side of a building:*—Beforan ðære norþdura, Blickl. Homl. 203, 34. [*Icel.* norðr-dyrr; *n. pl.*]

norþ-eást; adv. *North-east.* v. following words.

norþeást-ende *the north-east end:*—Ôþ ðone norþeástende ðisses middangeardes, Ors. 1, 1; Swt. 14, 14.

norþeást-lang; adj. *Long in a north-easterly direction:*—Brittania is norþeástlang *Britannia per longum in boream extenditur*, Ors. 1, 1; Swt. 24, 12.

norþ-ende *the north end* or *part:*—Ðŷ þriddan dæge seó eorþe on ðæm norþende and on ðam eástende sprecaþ him betweónum, Blickl. Homl. 93, 11. Ðone norþende ðære eaxe (*the north-pole*), Met. 28, 14.

Norþ-Engle; pl. *The inhabitants of the north of England:*—On Norþ-Engla lage stent . . . be Norþ-Engla lage, L. Eth. vii. 13; Th. i. 332, 7–10.

norþerne; adj. I. *northern:*—Norþerne ŷst, Met. 6, 14. Norþerne wind *africum*, Ps. Lamb. 77, 26. Of Japhet com ðæt norþerne mennisc be ðære norþsæ . . . Europa on norþdæle [is gedæled] Japhetes ofspringe, Ælfc. T. Grn. 4, 37. Hine gelæhton sume ðæs norþernan folces *some of the Northumbrians seized him (after a battle between Northumbrians and Mercians)*, Homl. Th. ii. 356, 29. Ða norþerne men *the men from the north of England*, Chr. 1064; Erl. 196, 2. II. applied to the Scandinavians:—Guma norþerna (guman norþerne, other MSS.), 937; Erl. 112, 18. Godrum se norþerna cyning, 890; Erl. 86, 27.

norþe-weard; adj. *Northward, north:*—Norþeweard, ðær hit smalost wære, hit mihte beón þreora mîla brâd tô ðæm môre *the northern part of Norway, where it was narrowest, might be three miles broad to the mountains*, Ors. 1, 1; Swt. 18, 31. Ðonne is tôemnes ðæm lande sûþeweardum Sweóland, ôþ ðæt land norþeweard; and tôemnes ðæm lande norþeweardum Cwêna land *alongside the south of the country (Norway), up to its northern part, lies Sweden; and alongside its northern part the country of the Fins*, Swt. 19, 1–3. Ðæt Babylonicum wæs ðæt forme, and on eásteweardum; ðæt æfterre wæs ðæt Crêcisce, and on norþeweardum, 2, 1; Swt. 60, 3. Æt Baddanbyrg westeweardre and norþeweardre . . . of foxhylle norþeweardre, Cod. Dip. Kmbl. ii. 249, 26, 34. From easteweardan ðisses middangeardes ôþ westeweardne, and fram sûþeweardum ôþ norþeweardne, Bt. 18, 1; Fox 62, 4. v. norþ-weard.

norþ-folc, es; n. *The northern division of ? people;* (a) *the people of the north of England:*—Humbre tôsceádeþ sûþfolc Angelþeóde and norþfolc, Bd. 1, 25; S. 486, 18. (b) *the people of Norfolk, Norfolk:*—Hê wæs geboren on Norþfolce. Ðâ geaf se cyng his sunu ðone eorldôm on Norþfolc and Sûþfolc, Chr. 1075; Erl. 213, 4–5: 1085; Erl. 218, 21.

norþ-gemære *a boundary to the north:*—Ðara landa norþgemæro sindon æt ðæm beorgum Caucasus, Ors. 1, 1; Swt. 10, 26, 33.

Norþ-Gyrwas; pl. *The northern division of the Gyrwas:*—Norþ-Gyrwa syxhund hŷda, Cod. Dip. B. i. 414, 19.

Norþhâmtûn *Northampton:*—Ne innan Lægreceastre scîre, ne innan Norþhâmtûne, Chr. 1087; Erl. 224, 36.

norþ-healf, e; f. *The north-side, the north:*—Æt ðæs weofudes sîdan ðe ys on norþhealfe *ad latus altaris, quod respicit ad aquilonem*, Lev. 1, 11: Blickl. Homl. 209, 1: Ps. 47, 2: Ors. 1, 1; Swt. 12, 13: Swt. 22, 13. [*Icel.* norðr-hâlfa *northern region:* O. H. Ger. nord-halba *the north side*.]

norþ-here, es; m. *An army belonging to the north:*—Heó (*the English force*) gehergade swîðe micel on ðæm norþhere, ægðer ge on mannum ge on gehwelces cynnes yrfe, and manega men ofslôgon ðara Deniscena, Chr. 910; Erl. 100, 13.

Norþ-hymbre; pl. *The Northumbrians, Northumbria, the people* or *province north of the Humber:*—Norþhymera cyning, Homl. Th. ii. 356, 23. Norþhymbra cining, Chr. 761; Erl. 53, 15. Hér bræc se here on Norþhymbrum ðone friþ, 911; Erl. 100, 16. Hér fôr se here on Norþhymbre, 867; Erl. 72, 7: 873; Erl. 76, 18. v. Norþan-hymbre.

Norþ-hymbre; adj. *Northumbrian:*—Ða Norþhymbran leóde, Swt. A. S. Rdr. 95, 9.

Norþ-hymbrisc; adj. *Northumbrian:*—Tô Norþhymbriscum gereorde, Swt. A. S. Rdr. 97, 58.

norþ-land *a northern land:*—Hê fôr on Sciððie ða norþland, Ors. 1, 2; Swt. 30, 3.

norþ-lane *a north lane:*—Ôþ norþlanan tô stræte, Cod. Dip. Kmbl. i. 1, 15.

norþ-leóde; pl. *The north-folk of England, Angles:*—Norþleóda cynges gild (ðæs cyninges wergyld mid Engla cynne), L. Wg. 1; Th. i. 186, 2.

norþ-lîc; adj. *Northern:*—Ðære norþlîcan *boreali*, Wrt. Voc. ii. 12, 46.

Norþ-mann, es; m. *A man belonging to a northern country.* I. *a Norseman, Norwegian* or *Dane:*—Ealle ða ðe on Norþhymbrum bûgeaþ, ægþer ge Englisce ge Denisce ge Norþmen, Chr. 924; Erl. 110, 17. Gewitan him ðâ Norþmen nægledcnearrum, 937; Erl. 115, 2. Ða Cwênas hergiaþ hwîlum on ða Norþmen (*Norwegians*) ofer ðone môr, hwîlum

da Norþmen on hý, Ors. 1, 1; Swt. 19, 3–5: 16, 36. Ða Normen áhton sige, Chr. 1066; Erl. 199, 40: 200, 26. Hé sǽde ðæt Norþmanna land (*Norway*) wǽre swýðe lang and swýðe smæl, Swt. 18, 24. Hé (*Ohthere from Halgoland*) ealra Norþmonna norþmest búde, 17, 2. On his dagum cômon ǽrest iii scipu Norþmanna . . . Ðæt wǽron ða ǽrestan scipu Deniscra manna ðe Angelcynnes land gesôhton, Chr. 787; Erl. 57, 21–25. Ðǽr geflêmed wearþ Norþmanna bregu, 937; Erl. 112, 33. Wǽran ǽr under Norþmannum, 942; Erl. 116, 15. *The word occurs as a proper name :*—Norþman Leófwines sunu, 1017; Erl. 161, 6. II. *referring to other countries :*—Norþmen (*those who attacked Sodom*), Cd. Th. 120, 16; Gen. 1995. Norþmonna, 124, 25; Gen. 2068. Norþmonnum, 119, 9; Gen. 1977. [*Icel.* norð-maðr *a Norwegian: O. H. Ger.* nord-man.]

norþmest. v. norþ; *adj. adv.*

Norþ-Mirce; *pl. The North-Mercians :*—Wið Norþ-Myrcum, Bd. 3, 24; S. 557, 37.

norþ-portic *a north-porch :*—On ðǽre cyricean norþportice *in porticu aquilonali*, Bd. 2, 3; S. 557, 37.

Norþriga, an; *m. A Norwegian :*—Cnut cyningc ealles Engla landes cyningc, and Dena cyningc, and Norþrigena cyningc, L. C. E. pref.; Th. i. 358, 4.

norþ-rihte; *adv. Due north :*—Ðá fôr hê norþryhte be ðæm lande *he sailed due north along the coast*, Ors. 1, 1; Swt. 17, 9, 12.

norþ-rodor *the north part of the sky*, Exon. Th. 178, 33; Gú. 1253.

norþ-sǽ *a northern sea :*—Norþsǽ *mare arctoum*, Wrt. Voc. i. 41, 66. Of Japhet com ðæt norþerne mennisc be ðære norþsǽ, Ælfc. T. Grn. 4, 38. Án geweorc on Defnascíre be ðære norþsǽ, Chr. 894; Erl. 91, 8. [*Icel.* norðr-sjór.]

norþ-sceáta *a northern promontory*, Ors. 1, 1; Swt. 28, 3.

Norþ-Scottas *the Northern Scots*, Bd. 3, 3; S. 526, 12.

norþ-þeód *a northern people :*—Hergung ðara norþþeóda (*the peoples who harried Britain after the Romans went*), Bd. 1, 14; S. 482, 38.

Norþ-Walas, -wealas; *pl. The Welsh, Wales :*—Ða cyningas on Norþ-Wealum, Howel and Cledanc, Chr. 922; Erl. 108, 27. Se here . . . hergodon ǽgðer ge on Cornwealum and on Norþ-Wealum, 997; Erl. 134, 9. Ælfgár eorl gesôhte Griffines geheald on Norþ-Wealan, 1055; Erl. 190, 3. Ecgbryht lǽdde fierd on Norþ-Walas, 828; Erl. 64, 12: 853; Erl. 68, 10. Hí hergodon on Norþ-Wealas, 918; Erl. 102, 25.

Norþ-Wealh-cynn, es; *n. The Welsh*, Chr. 922; Erl. 108, 28. Sum dǽl ðæs Norþ-Wealcynnes, 894; Erl. 92, 21.

norþ-weard; *adj. North :*—Hê búde on ðæm lande norþweardum, Ors. 1, 1; Swt. 17, 3. v. norþe-weard.

norþ-weard; *adv. Northward :*—Hê éfste norþweard, Chr. 1016; Erl. 154, 28.

norþweardes; *adv. Northwards :*—Hié ða herehýþ woldon ferian norþweardes ofer Temese, Chr. 894; Erl. 90, 23.

norþ-weg *a way going to the north :*—Nearwe genýddon on norþwegas wiston him be sûþan Sigelwara land, Cd. Th. 181, 29; Exod. 68. [*Icel.* norðr-vegar; *pl.*]

norþ-west; *adv. North-west :*—Se þridda [gára liþ] norþwest, Ors. 1, 1; Swt. 24, 5.

norþwest-ende *the north-west end :*—Thyle is on ðam norþwestende ðisses middaneardes, Bt. 29, 3; Fox 106, 24: Ors. 5, 3; Swt. 220, 23.

norþwest-gemǽre *a north-west boundary :*—Ðære Affrica norþwestgemǽre is æt ðæm ilcan Wendelsǽ, Ors. 1, 1; Swt. 8, 31.

Norþ-wíc *Norwich :*—Hér com Swegen tô Norþwíc . . . Ða geræfde Ulfkytel wið ða witan on Eást-Englum, Chr. 1004; Erl. 139, 17. Ða geaf se cyng his sunu ðone eorldóm on Norþfolc and Súþfolc; ða lǽdde hê ðæt wíf tô Norþwíc, 1075; Erl. 213, 6.

norþ-wind *a north wind :*—Twegen norþwindas *circius et boreus*, Wrt. Voc. ii. 21, 55. [*O. H. Ger.* nord-wind *aquilo, boreas*.] v. norþan-wind.

Norweg, es; *m.* [*The plural seems the more usual form.*] *Norway :*—Sume férdon tô Norwæge, Chr. 1070; Erl. 209, 30. Hér fôr Cnut cyng tô Norwegum (Norwegon, Erl. 162, 37), 1028; Erl. 163, 13. Hér com Ólaf cyng eft intô Norwegum, 1030; Erl. 163, 16. Harold cyng of Norwegon, 1066; Erl. 199, 37. Com Harold of Norwegan, Erl. 200, 12. Harold cyng on Norwégan, 200, 18, 27, 34, 40. [*Icel.* Noregr, *occasionally* Norvegr (vegr = *way*).]

nose. v. nosu.

nos-, nosu-gristle *the gristle* or *cartilage of the nose :*—Nosgrisele *internasus*, Wrt. Voc. ii. 48, 31. Nosugrisle *cartilago*, i. 64, 49. v. *under* nosu.

nosle. v. nostle.

nos-þyrel, -þyrl, -terl, es; *n. A nostril :*—Dô on ðæt næsþyrl (nos-, MS. B.), Lchdm. i. 352, 4. Nosþyrla *nares*, Wrt. Voc. i. 43, 19. Úteweard nosterle *pinnulae*, 43, 22. Nosterla hǽr *vibrissae*, 21, 52. Se brǽþ on heora nosþyrlum, Homl. ii. 98, 9. Dô on ða næsþyrlu (nos-, MS. B.), Lchdm. i. 72, 21. [*Chauc.* nose-thirl, -þril: *Wick.* nose-nese-þril, -þril: *Prompt. Parv.* nese-thyrlys, *naris : O. Frs.* nosterle.] v. næs-þyrel.

nostle, nosle, an; *f. A fillet, band :*—Nostle *fascia*, Wrt. Voc. i. 26,

8: *ansa*, ii. 6, 34. Nosle *vel* særcláþ *fasciola*, i. 40, 62. Nostlena *vittarum*, ii. 87, 65. Mid nostlum (noslum, Hatt. MS.) gebunden, Past. 13, 2; Swt. 86, 10.

nosu, neosu; *gen.* a *and* e; *also* an; *f.* I. *the nose :*—Nosu *nasus*, Wrt. Voc. ii. 62, 4: i. 43, 17: 64, 48: *naris*, ii. 60, 37: *nasus vel naris*, i. 70, 29. Eal ufweard nosu *columna ;* foreweard nosu *pirula*, 43, 18, 21. Eal nosu *columna*, ii. 16, 49. Nose grystle *internasus vel interfinium*, i. 43, 20. *Odoratus* stænc on ðæra nosa, Homl. Skt. 1, 198. Wið ðæt hwam on nosa (nosan, MS. B.) wexe, Lchdm. i. 116, 11. Se ðe hæfþ miccle nosu *nasatus*, Ælfc. Gr. 43; Som. 45, 10: Past. 11, 1; Swt. 65, 3–4. Hé hæfþ medemlíce nosu, Homl. i. 456, 18: 568, 33. Hé hæfde midmycle nosu þynne, Bd. 2, 16; S. 519, 34. Wið blôdryne of nosum, Lchdm. i. 72, 17: 352, 3. Gif hwylcum weargbrǽde weaxe on ðam nosum, 86, 1. Wé gestincaþ mid úrum nosum, Past. 56; Swt. 433, 20: Bd. 5, 12; S. 628, 42. Ða telgran habbaþ ǽgðer ge eágon ge nosa (nosan, MS. B.), Lchdm. i. 318, 11. Nose hí habbaþ *nares habent*, Ps. Th. 134, 17: 113, 14. II. *a ness, a piece of land projecting into water :*—Of hliþes nosan, Beo. Th. 3789; B. 1892. Æt brimes nosan, 5599; B. 2803. Cf. næss, næssa. [*Laym.* neose, nose: *O. E. Homl.* nease: *A. R.* neose: *Havel.* nese: *Prompt. Parv.* nese, nose: *Chauc.* nose: *O. Frs.* nose: *Icel.* nös: *O. H. Ger.* nasa.] v. nasu.

not, es; *m. A mark, sign :*—Mé þingþ wynsumlíc ðæt ic ðæra preósta notas ðám bôcerum gekýðe ðe læs ðe hig witan ðæt ða rímcræftige weras sýn bûtan cræftigum getácnungum, Anglia viii. 333, 17–19. v. wæl-not.

-note. v. æ-note.

notere, es; *m. One who makes notes :*—Notera ł wrítera *notariorum*, Hpt. Gl. 473, 12. v. not-writere.

nôþ, e; *f.* I. *temerity, presumption, boldness, daring :*—Ðu sylfa meaht gecnáwan ðæt ic ðisse nôþe wæs nýde gebǽded ðæt ic ðé sôhte *thyself may'st know, that I was by need compelled to the presumption of visiting thee*, Exon. Th. 263, 1; Jul. 343. II. *an adventurous band (?) :*—Semninga on sealtne wǽg mid ða nôþe (*the sailors who have landed on the whale thinking it an island*) niþer gewíteþ gársecges gæst (*the whale*), 361, 31; Wal. 28. [*O. H. Ger.* nand *temeritas, praesumtio*.] v. nêþan.

nôþ *occurs often as a component of proper names.* v. Txts. 642.

nôðer. v. ná-hwæðer.

notian; *p.* ode. I. *to make use of, employ, enjoy ;* (a) *with gen. :*—Gif ðú his wel notast hwæt biþ wæstmbǽrre *if you make good use of it* (dung), *what is more productive?* Homl. Th. ii. 408, 34. Ða ðe ðisses middangeardes notigaþ swelce hí his nô ne notigen *qui utuntur hoc mundo, tanquam non utantur*, Past. 50, 2; Swt. 389, 1–2. Eall moncyn and ealle nêtenu ne notigaþ náwðer neáh feórþan dǽles ðisse eorþan, Bt. 18, 1; Fox 62, 8. Ðæt hý (*garments*) synd gemǽte ðám ðe hyra notiaþ, R. Ben. 89, 19. Nota ðæs wísdômes ðe ðú habbæ, Shrn. 189, 18. Gif hé þurh ða gebedu gehǽled ne biþ, notige ðonne se abbod cyrfes, R. Ben. 52, 19. Notian ðara (*the garments*), ðe for hwylcere neóde on ýtinge faraþ, 91, 12. Betǽce ðǽm ðe heora (*tools*) notian sceolan, 56, 6. Ic wille mid ðære gefêrrǽdene libban and ðære áre mid him notian (*enjoy with them the property given to them*), Cod. Dip. Kmbl. iii. 344, 26. (b) *with dat. :*—Hwilc eówer ne notaþ cræfte mínon *quis vestrum non utitur arte mea?* Coll. Monast. Th. 31, 9. Hý scylun lǽca þeáwe notian, R. Ben. 51, 2. (c) *with acc. :*—Gold and seolfor sind gôde, gif ðú hí wel notast: gif ðú sylf yfel bist, ne miht ðú hí wel notian, Homl. Th. ii. 410, 8–9. (d) *case undetermined :*—Man ða reáf nime, ðe hê ǽr notode, R. Ben. 101, 24. Nota ðenna neód sig *use the medicine when need be*, Lchdm. i. 378, 18. II. *to discharge an office :*—Bûton hé forworhte, ðæt hê ðære hádnote notian ne môste, L. R. 7; Th. i. 192, 16. [*A. R.* notien: *O. and N.* ich notie: *Orm.* þu notesst: *Ayenb.* noteþ: *Icel.* nota *to make use of*.] v. be-, ge-notian; nyttian.

notu, e; *f.* I. *use, profit, advantage :*—Nittung *vel* notu *usus*, Wrt. Voc. i. 21, 39. Hæbbe se abbod á mid him gewrit ealra ðæra ǽhta; ðonne seó notu (*the use of tools, etc.*) on gebrôðra gewrixle biþ, sý ðæt gewis á mid ðam abbode, ðæt hê wite, hwæt betǽht sý and hwæt underfangen, R. Ben. 56, 8. Hí tô ðínre note gelǽnde wǽron, Bt. 14, 2; Fox. 44, 2. Ðæt se man tô note (*to profit*) wyrcean wille, Btwk. 222, 8. Hit læg wéste and gé his náne note ne hæfdon *it lay waste and you got no good from it*, Ors. 1, 10; Swt. 48, 25. Gif Drihten tô lytele note and nytwyrðnesse on his heorde angyt, R. Ben. 11, 2. [Se ðe ðæren brôðren note gewanie, God gewani his dages hér on werlde, Cod. Dip. Kmbl. iv. 215, 21.] II. *an office, employment :*—Ne nán gehádod man ne sceal him tô geteón ðæt hê Crist spelige ofer his hálgan híred, bûton him seó notu fram Godes láreówum betǽht sý, Homl. Th. ii. 592, 30. Hé geset ðé tô ðære ylcan note (*to the office of butler*), Gen. 40, 13. Ða hwíle ðe hié tô nánre ôðerre note ne mǽgen *while they are fit for no other employment*, Past. pref.; Swt. 7, 12. Ne rǽden gebrôðru, ne ne singen ne nánre endebyrdnesse, ac ða sýn gecorene tô ðære note, ðe hit dôn cunnon, R. Ben. 63, 6: 49, 18. Cristes gespelia hé is and his note and spelinge on mynstre healt, 10, 12. Ealne dæg hí fleardiaþ and nænige note dreógaþ *they trifle all day, and exercise no useful employ-*

ment, L. I. P. 14 ; Th. ii. 322, 25. On eallum betæhtum notum, R. Ben. 29, 5. III. *the discharge of an office, conduct of business* :—Gif hit beón mæg swylc notu þurh decanonas on mynstre sý gefadod . . . ðæt ná nán æulípig ne módige ðonne mynstres notu manegum biþ betæht *if possible, let such a conducting of its business by deans be arranged in the monastery, that no single person grow proud, when the conduct of business is committed to many*, 125, 8–11. [*O. E. Homl.* note *profit, use* : *O. and N.* note *office* : *Chauc.* note *business* : *Prompt. Parv.* note *opus, occupacio* : *O. Frs.* note *usus* : cf. *Icel.* not ; *pl. use, utility.*] v. nytt.

not-wrítere, es ; *m. One who writes notes* :—Notwrítera *notariorum*, Wrt. Voc. ii. 59, 66. v. notere.

nō-wiht. v. nā-wiht.

nū. I. *adv. Now, at this time* :—Nú *nunc* vel *modo*, Wrt. Voc. i. 76, 70. Ǽr odðe nú *dudum*, ii. 27, 56. Nú *nunc*, Ælfc. Gr. 38 ; Som. 39, 59. Ic hæbbe sumne cnapan, ðe nú (*modo*) hás ys for hreáme, Coll. Monast. Th. 19, 29. Hú him ðá speów mid wísdóme . . . and hú man útanbordes wísdóm and láre hieder on lond sóhte, and hú wé hié nú sceoldon úte begietan gif wé hié habban sceoldon, Past. Pref. ; Swt. 3, 8–13. Understandaþ, ðæt deófol ðás þeóde nú fela geára dwelode, Wulfst. 156, 8. Babylonia, seó ðe mǽst wæs and ǽrest ealra burga, seó is nú læst and wéstast, Ors. 2, 4 ; Swt. 74, 23. Nǽron ðára góde ne ðá, ne nú, 2, 5 ; Swt. 86, 12. Wá eów ðe nú hlihaþ, forðon gé eft wépaþ, Blickl. Homl. 25, 23. Ðú meaht geseón nú gén (*still*) swátge wunde, Exon. Th. 89, 17 ; Cri. 1458 : Beo. Th. 5711 ; B. 2859. Micel is nú géna lád ofer lagustreám, Andr. Kmbl. 844 ; An. 422 : 950 ; An. 475. Nú gyt (*adhuc*) lytel fæc and ná biþ synfull, Ps. Spl. 36, 10. Nú giet, Ors. 2, 4 ; Swt. 76, 1. Ne þearft ðú ðé ondrǽdan feorhcwealm nú giet (*as yet*), Cd. Th. 63, 26 ; Gen. 1038. Ic wát manig nú gyt micel mǽre spell, Andr. Kmbl. 1628 ; An. 815. Ðás tácno ðe ic nú hwíle big sægde *the signs that I have just now spoken of*, Blickl. Homl. 109, 6. Hé nú hwonne (*quandoque*) biþ on wuldre árísende, Bd. 2, 1 ; S. 500, 16. Swá swá wé nú rihte (*straightway, directly*) secgaþ, Ælfc. Gr. 15 ; Som. 17, 53. *Futurum tempus* is tówerd tíd, *stabo* ic stande nú rihte, 20 ; Som. 23, 9. *Modo* nú ðá odðe hwíltídum, 38 ; Som. 41, 37. Nú ðá *nunc*, Ps. Spl. 11, 5. Ðér sitt nú ðá mid his hálgum, Homl. Th. i. 182, 30 : Beo. Th. 857 ; B. 426 : Cd. Th. 51, 24 ; Gen. 831. II. *conj. Now, since, when* :—Nú ðonne nú ælc gesceaft onscunaþ ðæt ðæt hire wiðerweard biþ *since, then, every creature shuns that which is contrary to it*, Bt. 16, 3 ; Fox 56, 4 : Ors. 2, 4 ; Swt. 74, 26. Ðeáh hí nú eall hiora líf áwriten hæfdon . . . hú ne forealldodon ða gewritu ðeáh *now though they had written all their life, yet would not the writings wax old?* Bt. 18, 3 ; Fox 64, 36. Forhwí ðé hátan dysige men wuldor, nú ðú nǽne eart, 30, 1 ; Fox 108, 2. Ond nú (*since*) ðeós hálige tíd englum tó blisse wearþ, ðonne . . . , Blickl. Homl. 123, 1. Wé mágon geþencean, nú (*since*) ða sint Godes bearn genemned ðe sibbe wyrcaþ, Past. 47 ; Swt. 359, 12. Hú mæg hé hira bión orsorg, nú (*when*) se hierde cwæþ, 54 ; Swt. 427, 5. Nú . . . nú (*in principal and dependent clauses*) :—Wé wyllaþ nú eów gereccan óðres mannes gesihþe nú se apostol Paulus his gesihþe mannum ámeldian ne móste, Homl. Th. ii. 332, 26 : Cd. Th. 26, 8–9 ; Gen. 403–404 : Beo. Th. 857–865 ; B. 426–430. Ðonne is nú tó geþencenne on ðás hálgan tíd, nú wé úrne líchoman clǽnsiaþ, Blickl. Homl. 39, 1. Nú ic sceal geendian earmlícum deáþe, nú wolde ic gebétan, Swt. A. S. Rdr. 101, 205. Nú ðonne nú ða líchomlícan lǽcas ðus scyldige gerehte sint, nú is tó ongietanne . . . , Past. 49 ; Swt. 377, 21. III. *interj.* :—Nú is seóc se ðe ðú lufast *ecce quem amas infirmatur*, Jn. Skt. 11, 3. Sume syndon *ortativa* . . . *heia* nú lá, *age* nú lá ; ðis is eác menigfealdlíce, *agite* nú gé lá, Ælfc. Gr. 38 ; Som. 40, 27–28. Nú lá *age jam*, Wülck. Gl. 252, 43. [*Goth. Icel. O. Sax. O. Frs. O. H. Ger.* nu, nú.]

-nugan. v. be-, ge-nugan.

-numa. v. irfe-numa.

Numantie, Numentie, Numentíne, Numentínas ; *pl. The Numantians* :—Se consul fór on Numentíne, Ispania folc, Ors. 5, 2 ; Swt. 218, 29. Numentie áhnescaden, 5, 3 ; Swt. 222, 15. On Numantie, 220, 22. Numantia duguþ, 222, 8. Numentia fæsten, 5, 2 ; Swt. 218, 32. On Numantium, 5, 3 ; Swt. 220, 19. On Numentínas, Ispania þeóde, 5, 2 ; Swt. 218, 13.

Numentisc ; *adj. Of Numantia* :—Se wæs Numentisc, Ors. 5, 3 ; Swt. 222, 14.

nume-stán, es ; *m. A pebble* :—Cealc, numestán *calculus*, Wrt. Voc. ii. 13, 6.

Numeðe ; *pl. The Numidians* :—Numeðe, Ors. 4, 10 ; Swt. 200, 9. Numeðia cyning *Numidarum rex*, 5, 7 ; Swt. 228, 6.

numol ; *adj. Able to take* or *contain much* :—Numol *capax*, Ælfc. Gr. 9, 60 ; Som. 13, 41. Numul, gripul *capax, qui multum capit*, Wrt. Voc. ii. 128, 29. v. scearp-, teart-numol.

nūna ; *adv. Now* :—Núna *nunc*, Wülck. Gl. 254, 24. [*Icel.* núna.]

nunne, an ; *f. A nun, a vestal* :—Árwurðe wudewe vel nunne *nonna*, Wrt. Voc. i. 42, 9. Nunna, 72, 3. Nunne *sanctimonialis*, 284, 68. Wæs on ðam sylfan mynstre sum hálig nunne *erat in ipso monasterio quaedam sanctimonialis femina*, Bd. 4, 23 ; S. 595, 36. Caperronis wæs

---

hátenu heora goda nunne (*virgo vestalis*), Ors. 4, 4 ; Swt. 162, 31. Se ðe mid nunnan hǽme, gehálgodre legerstówe ne sý hé wyrðe, L. Edm. E. 3 ; Th. ii. 246, 6. Ǽfre ne geweorðe, ðæt crísten man gewífige on gehálgodre ænigre nunnan, L. Eth. vi. 12 ; Th. 318, 17. Gif hwá wið nunnan forlicge, sí ægðer his weres scildig, ge hé ge heó, L. N. P. L. 63 ; Th. ii. 300, 20. Be nunnan hǽmede. Gif hwá nunnan of mynstre út álǽde bútan kyninges léfnesse geselle hundtwelftig scitt. . . . Gif heó leng libbe ðonne se ðe hié út álǽdde, náge hió his ierfes ówiht. Gif hió bearn gestriéne, næbbe ðæt ðæs ierfes, L. Alf. pol. 8 ; Th. i. 66, 14–20. Be nunnena onfenge. Gif hwá nunnan mid hæmedþinge, odðe on hire hrægl, odðe on hire breóst bútan hire leáfe gefó sý hit twýbéte, 18 ; Th. i. 72, 7–10. Nunnan regollíce libban *let nuns live according to their rule*, L. Eth. v. 4 ; Th. i. 304, 27. Sum fǽmne of ðæra nunnena ríme *quaedam de numero virginum*, Bd. 5, 3 ; S. 616, 3. Nunnena pól, Cod. Dip. Kmbl. iii. 313, 26. [*Icel.* nunna : *O. H. Ger.* nunna, *from Lat.* nonna.]

nunn-híréd, es ; *m. A nunnery* :—Ðe ǽr ðes nunhírédes wes, Chart. Th. 232, 6.

nunn-, nunnan-mynster, es ; *n. A nunnery* :—Ðæt nunmynster (*monasterium virginum*) ðæt mon nemneþ Coludesburhg, Bd. 4, 25 ; S. 599, 19 : 5, 3 ; S. 615, 41. In tó nunnanmynstre, Chart. Th. 231, 35. [Cf. *Icel.* nunnu-klaustr *a nunnery*.] v. neáh-nunnan-mynster.

nunn-scrúd, es ; *n. The habit of a nun* :—Finde Æþelflæd án hyre nunscrúde, lóce hwæt hió betsð mæge, Chart. Th. 538, 12.

nȳd, nýhst. v. níd, níhst.

nyhtness, e ; *f. Abundance* :—Of nyhtnisse *ex abundantia*, Mt. Kmbl. Rush. 12, 34. v. ge-nyhtsum.

nyllan =ne willan :—Nylle ic út wítan, Met. 24, 52. Ic nelle *nolo*, ðú nelt *non vis*, hé nele *non vult*, wé nellaþ *nolumus* . . . nelle ðu *noli*, nelle gé *nolite*, . . . nellan *nolle*, Ælfc. Gr. 32 ; Som. 36, 16–19. Ðú nelt, Exon. Th. 250, 12 ; Jul. 126. Nyle hé, Ps. Th. 74, 8. Nán eówer nele *nemo vestrum vult*, Coll. Monast. Th. 28, 1. Nellaþ *nolunt*, 29, 3. Nyllaþ, Past. 5 ; Swt. 45, 18. Nolde, Jn. Skt. 7, 1. Nalde, Ps. Surt. 35, 4. Noldon, Mt. Kmbl. 22, 3. Gif ðæt wíf nele *si noluerit mulier*, Gen. 24, 5. Sam wé willan, sam wé nyllan, Bt. 35, 12 ; Fox 154, 7. Nyllan gé *nolite*, Ps. Th. 61, 11. Nellaþ, 61, 10. Bútan nellendes andsware, R. Ben. 20, 19.

nymne, nymþe, nýr, nyrwan, nyrwian, nyrwet, nyt. v. nefne, nimþe, neáh (*adv.*), nirwan, nirwet, nett, nytt.

nytan =ne witan :—Ic nát *nescio*, Jn. Skt. 9, 25. Nát ic hwilc wundorlíc þing, Shrn. 36, 18. Gif ðú nást *if you do not know*, Ælfc. Gr. 50, 17 ; Som. 51, 34. Gyt nyton hwæt gyt biddaþ, Mk. Skt. 10, 38. Wé witon ðæt God spæc wið Moyses ; nyte wé hwanon ðes is, Jn. Skt. 9, 29. Wé nyton (nutu wé, Lind. : niton wé, Rush.), Mt. Kmbl. 21, 27. Ge neton, Exon. Th. 282, 9 ; Jul. 660. Ic wiste ðæt ðú út áfaren wǽre, ac ic nyste hú feor, Bt. 5, 1 ; Fox 8, 33. Ðæt ðæt ic ǽr sǽde ðæt ic nyste (Cott. MS. nesse) . . . Ðú sǽdest ðæt ðú nystest (Cott. MS. nesse), 34, 12 ; Fox 154, 12–13. Ðú nysstest (Cott. MS. nesse) . . . ic nyste (Cott. MS. nysse), 35, 2 ; Fox 156, 33–34. Ðú nestest, 5, 3 ; Fox 12, 34. Hé nyste, Past. 15 ; Swt. 91, 13. Hí nysðon (nyston, MS. A.), Mt. Kmbl. 24, 39. Wé neston, Blickl. Homl. 17, 12. Wénst ðú ðæt ic nyte, Bt. 5, 3 ; Fox 12, 17. Ðæt hé nán ryht andwyrde nyte, 35, 1 ; Fox 156, 8. Nyte ðín wynstre hwæt dó ðín swýðre, Mt. Kmbl. 6, 3. Nytende, Lchdm. i. 164, 5, Him nytendum, 228, 1. Nytendum weardmannum *clam custodibus*, Ælfc. Gr. 47 ; Som. 47, 58. v. nát-.

nyten ; *adj. Ignorant* :—Of ðás nytenan menn ðíne mihta oncnáwon, Homl. Th. i. 62, 14. v. nytenness.

nȳten. v. níten.

nytenness, e ; *f.* I. *ignorance* :—Hwæt getácnaþ seó midniht bútan seó deópe nytennys, Homl. Th. ii. 568, 5. Ðæt men for nytennysse misfaran ne sceolon, 314, 5. Ðú cniht ne cúðest manna Hǽlend . . . Nú ic for ðínre nytennysse geornlíce bæd, i. 66, 30. Se ðe tódrǽfde ealle nytennysse ðære ealdan nihte, 36, 29. Crist mæg ðíne nytennysse (MSS. C. V. nyte-) onlíhtan, Homl. Skt. 5, 200. Gif folces man syngaþ þurh nytennysse *per ignorantiam*, Lev. 4, 27. Ðæt hyra nán þurh nytennesse hine beládian ne mæge, R. Ben. 127, 10. II. *laziness, disgrace, ignominy* :—Netenes ðam se ðe forlét *ignominia ei qui deserit*, Kent. Gl. 454. On his netenesse *in ignominia sua*, 615. For módes mínes nytenysse *propter mentis meae ignaviam*, Coll. Monast. Th. 25, 7.

nyt-líc ; *adj. Useful, profitable, beneficial* :—Ẹghwæðer (*the male and female pennyroyal*) ys nytlíc (MS. H. netlíc) . . . and hí on him habbaþ wundorlíce mihte, Lchdm. i. 204, 11. Móna se feórþæ wercu onginnan nytlíc ys *the fourth day of the moon is advantageous for beginning works*, iii. 184, 28. Mǽden (*a girl born on the eighth day of the moon*) is nytlíce, 188, 6. Is eallum mannum nytlíc, ðæt hié heora fulwihthádas wel gehealdan, Blickl. Homl. 109, 26. Ic ne gýmde ðara nytlícra geþeahta mínra freónda *utilia consilia spreveram amicorum*, Nar. 6, 26. Monig nytlíco þing *multa utilia*, Bd. 5, 20 ; S. 642, 19. [*O. H. Ger.* nuz-líh *utilis*.] v. un-nytlíc.

nytlícness, e ; *f. Usefulness, utility, useful property* :—Ic bidde ðé

vica pervica manegum nytlícnyssum tô hæbenne *te precor vica pervica multis utilitatibus habenda*, Lchdm. i. 314, 8.

nytness, e; f. *Use, utility, advantage, profit :*—Hwylc nytnys on blôde mînum *quae utilitas in sanguine meo*, Ps. Spl. 29, 1ι. Nâwiht nytnesse (*nihil utilitatis*) hafeþ seó æfæstnys ðe wê ôþ ðis hæfdon, Bd. 2, 13; S. 516, 3. Mid micelre nytnysse (*magna utilitate*) æghwæðeres folces, 3, 24; S. 557, 13: 5, 10; S. 623, 38. Tô lîchoman nyttnesse *for the advantage of the body*, Blickl. Homl. 57, 8. Mid allum ðæm nytnessum ge on fixnoþum ge on mêdwum ðe ðærtô belympaþ, Cod. Dip. Kmbl. v. 186, 5, 9.

nyt[t], e; f. *I. use, advantage, profit :*—Nyt *commodum*,Wrt.Voc. ii. 24, 63. Hundteóntig hî him sylfum tô nytte dydon *centum in suos usus habebant*, Bd. 4, 13; S. 583, 3: Ors. 2, 4; Swt. 72, 6. Niþum tô nytte, Exon. Th. 409, 10; Rä. 27, 27. Nebb biþ hyre æt nytte *it has a face for use*, 416, 23; Rä. 35, 3. Tô nyttum *ad pensas*, Wrt. ii. 4, 40: *ad expensas*, 7, 30: *ad penses*, 99, 23. *II. office, duty :*—Ðegn nytte beheóld, se ðe on handa bær hroden ealowæge, Beo. Th. 993; B. 494: 6228; B. 3118. [Bruttes neoren noht to nuttes, Laym. 13428: *Icel.* nyt; *gen.* nytjar; f. *use, enjoyment : O. H. Ger.* nuzzi.] v. cyric-, sund-, sundor-nytt.

nyt[t]; adj. *Useful, profitable, advantageous, beneficial :*—Hê monegum nyt wæs *multis utilis fuit*, Bd. 3, 23; S, 555, 33. Ic nât, hû nyt ic ða hwîle beó, ðe ic ðás word sprece, bûtan ðæt ic mîn geswinc âmirre, Ors. 4, 13; Swt. 212, 25. Wê næfdon ða geselþa, ðæt seó scipfyrd nytt wære ðisum earde, Chr. 1009; Erl. 141, 26. Tô nâhte nyt, Blickl. Homl. 57, 5. Hû nyt biþ ðæm men, ðéh hê geornlíce gehýre ða word ðæs hâlgan godspelles, gif hê ða nel on his heortan habban, 55, 6: Bt. 38, 5; Fox 206, 10 note. Nyttre fôre, Exon. Th. 393, 4; Rä. 12, 5. Ðæt hî hæfdon nyt ærend and nytne intingan sumne (*aliquid legationis et causae utilis*), Bd. 5, 10; S. 624, 21. Ðysne nyttan cræft ðéh hê árlíc nære *hanc utilem magis quam nobilem victoriam*, Ors. 2, 8; Swt. 92, 2. Ne gehýrde nænig man on his mûþe ôht elles nefne nytte spræce, Blickl. Homl. 225, 1. Hê ðone gôdan cræft dô nytne ôðrum mannum, Ælfc. Gr. pref.; Som. 1, 29. Ðæt land hyre nytt gedôe, Chart. Th. 470, 8: 472, 10. Næron Metode wîd lond ne wegas nytte, Cd. Th. 10, 13; Gen. 156: Beo. Th. 1592; B. 794. Nis nænig mære mægen, ðisse menniscan tydernesse nyttre, Blickl. Homl. 31, 30. Hê cwæþ, ðæt nyttre wære ðæt hié man gesealde, 75, 22. Wê mâgon beón nyttran æt him, Past. 32; Swt. 211, 21. Ræd biþ nyttost, Exon. Th. 341, 1; Gn. Ex. 119. [Is þe man nut þe sæhtnesse wurcheþ, Laym. 9470: *Goth.* [un-]nutis *Icel.* nytr : *O. H. Ger.* nuzzi *utilis*.] v. un-nytt.

nytte. v. nette.

nyttian; p. ode; *with gen. To make use of, enjoy :*—Ic nyttige *fungor, utor, perago*, Wrt. Voc. ii. 152, 22. Wuda and wætres nyttaþ, Exon. Th. 340, 12; Gn. Ex. 110. Sume ðæs seáwes ânlípiges nyttiaþ *some make of the juice only*, Lchdm. ii. 30, 16. Nyttade Noe sîðan ríces, Cd. Th. 96, 21; Gen. 1598. Ðises dû nytta ge on æfenne ge on underne, Lchdm. ii. 184, 25: 28, 16: 32, 25. Nyttigen baþes, 240, 24. Gehwæðeres (*both methods of cure*) sceal mon nyttian, 22, 7. [*A. R. O. E. Homl. Marh.* nutten: *Orm. Havel.* nitten : *O. H. Ger.* nuzzan.]

nyttol; adj. *Useful, advantageous, beneficial :*—Ðæt ilce (*the same treatment*) biþ nyttol wið hundes slite, Lchdm. ii. 86, 2.

nyttung, e; f. *Profit, advantage :*—Nittung *usus*, Wrt. Voc. i. 21, 39.

nyt-weorð (-wirðe); adj. *Useful, advantageous, profitable :*—Eálâ dû mîn wyln beó nytwyrðe *O mea ancilla, esto utilis*, Ælfc. Gr. 15; Zup. 101, 4. Hû se lâreów sceal beón nytwierðe (MS. Hatt. -wyrðe) on his wordum *ut sit rector utilis in verbo*, Past. 15; Swt. 88, 3. Ðá stôd ðær sum nytwurðe hûs, Blickl. Homl. 221, 7. Se biþ on eallum þingum nytwurðe, Lchdm. iii. 158, 6. Nytwyrðe, 188, 14. Se nytwyrða brôðor, R. Ben. 24, 18. Fleóþ ðonne ða nytwierðan (nyttwyrðan, Hatt. MS.) hiérsumnesse ðære lâre, and nyllaþ ðæs þencean hû hié mægen nytwierðuste (nyttweorðuste, Hatt. MS.) bión hiera nîhstum, Past. 5, 3; Swt. 44, 17–19. Ic gehýrde fela nytwurðe (-wyrðe, -werðe, -wyrða) þing (*multa utilia*), Ælfc. Gr. 15; Zup. 95, 18. Seó wiðerwearde wyrd byþ ælcum men nytwyrðre ðonne seó orsorge *plus hominibus reor adversam, quam prosperam prodesse fortunam*, Bt. 20; Fox 70, 29. Ða scipu næron on Fresisc gescæpene, bûte swâ him selfum þûhte ðæt hié nytwyrðoste beón meahten, Chr. 897; Erl. 95, 16.

nytweorð (-wirð) -líc; adj. *Useful :*—Nytwurðlíc (-wyrð-, -weorð-) *utilis*, Ælfc. Gr. 9, 28; Zup. 55, 5. Tô ðæs munstres nitwurðlícre þearfe *for the useful requirements of the monastery*, Chart. Th. 369, 28.

nytweorð(-wirð) -líce; adv. *Usefully :*—Nytwurðlíce (-wyrð-,-wirð-) *utiliter*, Ælfc. Gr. 38; Zup. 238, 15. Ða ðonne sint tô manianne ðe nytwyrðlíce (nyttweorðlíce, Hatt. MS.) læran meahton (*qui praedicare utiliter possent*), Past. 49, 1; Swt. 374, 21. Nytwierðlecust (nyttwyrðlícost, Hatt. MS.), 15; Swt. 94, 22.

nytweorþ (-wirð) -ness, e; f. *Usefulness, utility :*—Nytweorðnes *commoditas* i. *utilitas*, Wrt. Voc. ii. 132, 5. Hwæt wyrcst ðû ûs nytwyrðnesse *quid operaris tu nobis utilitatis*, Coll. Monast. Th. 27, 25.

nywol, nýxt. v. neowol, neáh.

O

Ô *ever.* v. â.

ôb. v. ô-web.

ob, ober, obet. v. of, ofer, ofet.

oc, ôcusta. v. ac, ôhsta.

oden, e; f. *A threshing-floor :*—Frymþa odene ðínre *primitias areae tuae*, Scint. 29. Beóþ sume on bûre, sume on healle, sume on ôdene, sume on carcerne, and lybbaþ ðeáh ealle be ânes hlâfordes âre, Shrn. 187, 23. On odene cylne macian, Som.

of; prep. *with dat.*, or adv. *Of, from, out of, off.* **I.** with the idea of motion, (a) as the opposite of *in, into :*—Se wyll âstâh upp of ðære eorþan *fons ascendebat e terra*, Ælfc. Gr. 47; Som. 47, 61. Hê âstâh of ðam wætere *ascendit de aqua*, Mt. Kmbl. 3, 16. 'Drihten âsette on sunnan his hûs, and of ðæm ût eode swâ swâ brýdguma of his brýdbûre.' Ðæt wæs ðonne ðæt se wuldorcyning on middangeard cwom forþ of ðæm innoþe ðære â clænan fæmnan, Blickl. Homl. 9, 30–33. Faran of stôwe tô ôðerre, 19, 23: Gen. 12, 4. (β) as the opposite of *on :*—Moises eode nyðer of ðam munte tô ðam folce, Ex. 19, 14. Crist of heofona heánnessum on ðinne innoþ âstígeþ, Blickl. Homl. 5, 13. **II.** with the idea of direction from, but at the same time continuous connection with an object from which an act or thing proceeds :—Drihten lôcaþ of heofenum *Dominus de caelo prospexit*, Ps. Th. 13, 3. Of wealle geseah weard Scyldinga, Beo. Th. 463; B. 229. Of ðam leóma stôd *from it stretched a ray*, 5532; B. 2769. Ic geseah Drihten of ansíne tô ansíne, Gen. 32, 30. On ðam dæge plegedon hié of horsum, Ors. 3, 7; Swt. 118, 29: 3, 9; Swt. 132, 19. **III.** with the idea of origin or source :—Ða nítenu of eallum cinne and of eallum fugelcynne, Gen. 7, 8. Ðá feóllon ða ciningas ofslagene of Sodoman and Gomorran *rex Sodomorum et Gomorrhae ceciderunt ibi*, 14, 10. Sum wer of Sceotta þeóde, Bd. 4, 25; S. 599, 27. Ða ôðre seofan syndon *derivativa*, ðæt is ðæt hí cumiaþ of ðám ôðrum, Ælfc. Gr. 15; Som. 17, 44. Of Geáta fruman syndon Cantware . . . Of Seaxum cômon Eást-Seaxan and Sûþ-Seaxan and West-Seaxan, Bd. 1, 15; S. 483, 21–24. Ða men of Lundenbyrig, Chr. 896; Erl. 94, 17. Ðás woruldgeselþa of heora âgnum gecynde and heora âgnes gewealdes nâuht gôde ne sient, Bt. 16, 3; Fox 54, 17. Wæs sió bysen of him (*the example that had its origin with them*) ofer ealle world, Ors. 1, 5; Swt. 34, 31. Hié woldon of ælcerre byrig him self anwald habban (*imperare singulae cupiunt*) . . . Ðá bædan hié Philippus ærest of ânre byrig, ðonne of ôðerre, ðæt hê him on fultume wære, 3, 7; Swt. 112, 19–23. Mê of brýde bearn ne wôcon, Cd. Th. 131, 30; Gen. 2184: Exon. Th. 433, 26; Rä. 51, 2. Him stent ege of ðé *timebunt te*, Deut. 28, 10. Wendan on Englisc, hwílum word be worde, hwílum andgit of andgite, Past. Swt. 7, 20. Hwæðer ænig man wære ðe ænige mærþa of ðam Hælende hæfde, St. And. 36, 31. Sôðfæstnesse, ða ðe ic gehýrde of Gode, Jn. Skt. 8, 40. **IV.** denoting the agent from whom an action proceeds, *by :*—Æþelstân wæs gecoren tô cynge of Myrcum, Chr. 924; Erl. 111, 34. Hér wearþ Eádward cing gecoren tô hlâfuorde of Scotta cinge and of Scotton and of eallum Norðhumbrum, Erl. 111, 11. Hê wæs of cilda mûþe gecnâwen and weorþad, Blickl. Homl. 71, 33. **V.** denoting the instrument :—Weorþian wê ða clâþas his hâdes, of ðæm wæs ûre gecynd geedneówod, 11, 9. Hê of .v. hlâfon and of twâm fixum fíf þûsend manna gefylde, St. And. 28, 32. **VI.** denoting material or substance :—Reáf of olfenda hærum, Mt. Kmbl. 3, 4. Gyld of golde *au idol of gold*, Cd. Th. 226, 22; Dan. 175. Adam ðe wæs of eorþan geworht, 23, 26: Gen. 365. Hæfdon of ðæm hreóde on scipwîsan geworht *factis ex harundine naviculis*, Nar. 11, 18. Offrunga of nýtenum, Lev. 1, 2. Ne biþ on hlâfe ânum mannes líf, ac of eallum ðæm worde ðe gæþ of Godes mûþe, Blickl. Homl. 27, 9. **VII.** denoting removal, separation, or privation :—Of slæpe âwreht, Homl. Th. i. 60, 19. Ðæt ðû of deáþe âríse, 66, 30. Álys ûs of yfele, Mt. Kmbl. 6, 13. Beó of ðysum hâl, Mk. Skt. 5, 34. Hê gehælde manega of âdlum ge of wîtum and of yfelum gâstum, Lk. Skt. 7, 21. Sundor of ðæm weorode *apart from the multitude*, Blickl. Homl. 15, 7. Âsceofene of gefeán neorxna wanges, 17, 15. Wæs âdæled wæter of wætrum, Cd. Th. 10, 5; Gen. 152. Dyde him of healse hring gyldenne, Beo. Th. 5610; B. 2809. Ðone cynelícann naman of Rôme byrig âdydon, Bt. 16, 1; Fox 50, 9. Ne þincþ mê nâuht ôðres of (*nothing different from*) dínum spellum, 36, 4; Fox 178, 24. Fixas cwelaþ gyf hí of wætere beóþ, Lchdm. iii. 272, 25. **VIII.** *as regards, about :*—Fela spella him sædon ða Beormas ægðer ge of hiera âgnum lande ge of ðæm landum ðe ymb hié ûtan wæron, Ors. 1, 1; Swt. 17, 31. **IX.** partitive :—Ic nyme of dînum gâste, Num. 11, 17. Heó genam of ðæs treówes wæstme, Gen. 3, 6: Lk. Skt. 20, 10. Syllaþ ûs of eówrum ele, Mt. Kmbl. 25, 8. Ic ne drince of ðysum eorþlícan wîne, 26, 29. Se Peohta þeóde of myclum dæle (*in great part*) geeode, Bd. 2, 5; S. 506, 20. Swâ án of ðyson, Mt. Kmbl. 6, 29. Ân eá of ðám hâtte Fison, Gen. 2, 11. Ðû ne gesihst ænigne of Godes ðam hâlgum, St. And. 16, 8: Exon. Th. 154, 5; Gú. 838. **X.** marking time :—Of ðam dæge, Jn. Skt. 11, 53. Of sunnan upgange,

Swt. A. S. Rdr. 98, 96. Of ðyssan forþ áwa tô worulde, Ps. Th. 112, 2. Of cildhâde, Elen. Kmbl. 1826; El. 915. **XI.** adverbially (a) denoting separation, removal, privation :—Ic ðê ðíne teþ of âbeáte *I knock out your teeth for you*, Lchdm. i. 326, 15. Bûton hê him wille fǽhþe of âceápian *unless he will buy off the feud from himself*, L. In. 74; Th. i. 150, 2. Petrus âcearf him of ðæt swýðre eáre, Jn. Skt. 18, 10. Ðonne ân tweó of âdón biþ, Bt. 39, 4; Fox 216, 19. Gif man cealf of âdrîfe, L. Alf. pol. 16; Th. i. 72, 1. Hê âslôh of ânys ðæra sacerda ealdres þeówan eáre, Mt. Kmbl. 26, 51. Âtió of ða þornas, Bt. 23; Fox 78, 22. Gif ðara lima hwilc of biþ, 37, 3; Fox 190, 27. Ceorf of ðæt lim, Homl. Th. i. 516, 4. Ealles ðæs ðe ðenne on biþ, bûtan ðæt man scel for hyre sâulle of dôn, Chart. Th. 534, 7. Ða reáf ðe hê him of dyde, R. Ben. 103, 1. Seó eádmôdnys heáwþ of ðære môdinysse heáfod *humilitas amputat caput superbiae*, Gl. Prud. 36 b. Him mon slôg ða handa of, ðâ ðæt heáfod, Ors. 4, 5; Swt. 168, 5. Wring of ða wyrta, Lchdm. iii. 58, 30. (b) denoting motion :—Man sceolde mid sâre on ðâs world cuman, and mid sâre of gewîtan, Blickl. Homl. 5, 29. Ðonne hwâ on ða leásunga befêhþ, ðonne ne mæg hê of, Past. 35; Swt. 239, 17. (c) denoting direction :—Stôd se leóma him of swylce fýren þecelle, Bd. 5, 23; S. 645, 29. (d) denoting origin or source :—Ðære þeóde ðe hê of com, 5, 19; S. 639, 37. On ðære béc ðe wê ðás of âlesan, 4, 10; S. 578, 15. Hê ænne calic sealde his gingrum of tô sûpenne, Homl. Th. ii. 244, 13. [*Goth. Icel. O. Sax.* af: *O. Frs.* of: *O. H. Ger.* ab.]

of- as a prefix modifies the words to which it is attached in many ways. Amongst these may be noticed (1) its intensive force *in such words* as of-georn, of-langod, of-lysted, of-calen, of-hyngrod, of-þyrsted. (2) its unfavourable force *in* of-lícian, of-unnan, of-þyncan. (3) the idea of attainment which it gives to (a) *verbs of motion* as of-faran, of-fêran, of-irnan, of-rîdan : (b) *verbs of inquiring, calling, etc.*, as of-âxian, of-clypian, of-spyrian. (4) the force of (a) *killing* which it gives to *verbs of striking, throwing, falling, etc.*, as of-feallan, of-hnîtan, of-hreósan, of-sceótan, of-stician, of-stingan : (b) *injury* which it gives to *verbs denoting rest* as of-licgan, of-sittan, *or those denoting action as* of-settan, of-tredan.

ofæt. v. ofet.

of-âxian, -âcsian; *p.* ode *To find out by asking, to learn* :—Ðâ hê ofâxode (*didicisset*) hwæt his suna him dydon, Gen. 9, 24; Chart. Th. 340, 27. Hê his brôðor slege ofâxode, Homl. Th. ii. 358, 5. Hê ofâxode æt ðâm lâreówum, ðæt Cristes þeówdôm ne sceal tón geneádod, 130, 14. Hê ofâcsode (*suspicabatur*) ðæt hê hæfde ǽrendo, Bd. 4, 1; S. 564, 48 note. Hit wearþ gecweden, ðæt man ofâxode on eallum his rîce, gif ǽnig mǽden mihte beón âfunden swâ wlitiges hiwes, Anglia ix. 29, 71. Ic ðê bidde ðæt ðû ofâxie ða næglas, H. R. 15, 23. Ðâ sænde hê his môder tô Hierusalem, tô ðam ðæt hió ðǽr ofâxian scolde ða hâlgan rôde, 7, 4. Cf. of-spyrian.

of-beátan; *p.* -beót; *pp.* -beáten *To kill by beating, to beat to death, to beat to pieces* :—Wê hit uneáþe mid ísernum hamerum ofbeóton *quam ferreis vix comminuimus malleis*, Nar. 21, 6. Claudium mid saglum ofbeótan *they beat Claudius to death with clubs*, Ors. 2, 6; Swt. 88, 26. Ðæt hî ofbeátun *ut trucident*, Ps. Lamb. 36, 15. Hêt se câsere ðone cempan mid saglum ofbeátan, Homl. Skt. i. 5, 455. Mid billum ofbeátan, Met. 9, 30. Sume wǽron mid wǽpnum ofslagene ôðre mid swipum ofbeátene *some were slain with weapons, others scourged to death*, Homl. Th. i. 542, 27.

of-blindian *to make blind* :—Ofblindade ǽgo hiora *excaecavit oculos eorum*, Jn. Skt. Lind. Rush. 12, 40.

of-brædels. v. ofer-brædels.

of-brytsig (?); *adj. Very broken* :—Ofbyrtstigum (? ofbrytsigum) *praeruptis, fractis*, Hpt. Gl. 454, 44.

of-calen *very cold* :—Petrus stôd ofcalen on ðam cauertûne, Homl. Th. ii. 248, 27. v. calan.

of-clipian; *p.* ode *To obtain by calling* :—Ðâ wolde se hâlga habban gewitan ðære wunderlícan gesihþe and ofclypode his diácon him hrædlíce tô (*the deacon was called and came*), Homl. Th. ii. 184, 33. Heó mid hreáme hyre hrǽddinge ofclypode *she had obtained help by her cries*, Homl. Skt. i. 2, 219.

of-cumende *derivative* :—Eahta synd frumcennede, and seofan ofcumende, Ælfc. Gr. 15; Som. 17, 34.

of-cyrf, es; *m.* **I.** *a cutting off, amputation* :—Hwæt getâcnaþ ðæs fylmenes ofcyrf, Homl. Th. i. 94, 32. **II.** *that which is cut off* :—Hê tôcearf his basing on emtwâ, and sealde ôðerne dǽl ðam earman wǽdlan, and mid ðam ofcyrfe hine eft bewǽfde (*wrapped himself in the remaining portion of the cloak*), Homl. Th. ii. 500, 27. Heó (*the cross*) is wîde tôdǽled mid gelômlícum ofcyrfum (*by the bits often cut off it*), H. R. 105, 14.

of-dǽl; *adj. Tending downwards, inclined to anything inferior* :—Hit biþ âmerred mid ðam lǽnum gôdum forðam hit biþ ofdælre ðærtô *it is led astray by the transitory goods, because it is more inclined to them;* ad falsa devius error abducit, Bt. 24, 2; Fox 82, 2. v. next word.

of-dǽle, an (?); *n. A downward slope, descent, incline* :—Hié nyllaþ gepyndan hiera môd swelce mon deópne pôl gewerige ac hê lǽt his môd

tôflôwan on ðæt ofdæle (ofdele, Hatt. MS.) gîemeliéste and ungesceádwîsnesse *they will not dam up their minds, as one banks up a deep pool, but he lets his mind flow away to the downward slope of carelessness and folly ;* quia (anima) se ad superiora stringendo non dirigit, neglectam se inferius per desideria expandit, Past. 39, 1; Swt. 282, 15. Hî sîen on ðæt ofdæle âsigen tô yfele and ðider healde, Bt. 24, 4; Fox 84, 28. Sió sunne scýft on ofdæle *the sun descends,* Met. 13, 58. [Cf. *Goth.* at ibdaljin this fairgunjis *ad descensum montis,* Lk. 19, 37 : *O. Sax.* te dale : *O. H. Ger.* ze tale *downwards.*] v. preceding word.

of-drǽd[d] *terrified, afraid* :—Ic fêrde ofdrêd *timens abii,* Mt. Kmbl. 25, 25. Befrân se sceaþa hwæt hê manna wǽre, oððe wǽre ofdrǽd, Homl. Th. ii. 502, 28. Hê ofdrǽdd wæs for his morþdǽdum, Ælfc. T. Grn. 18, 38. Hié beóþ mid ðæm ymbeþonce ofdrǽdde, Past. 35, 2; Swt. 238, 7; Homl. Skt. i. 23, 300. [*Laym. A. R.* (swiþe, sore) of-dred: *Orm.* off-dredd: *O. and N.* of-drad.]

of-dûne; *adv. Down* :—Ofdûne stîgan, gestîgan *to descend,* Mt. Kmbl. Lind. 3, 16 : 11, 23 : Rtl. 28, 9. Hê gefeóll ofdûne on ða flôr, Bt. 1; Fox 4, 3. Nis hire êþre tô feallanne ofdûne ðonne up, 33, 4; Fox 130, 38. Ðeáh ðû teó hwelcne bôh ofdûne tô ðære eorþan, 25; Fox 88, 22. Hié lêton hiera hrægl ofdûne tô fôtum, Ors. 3, 5; Swt. 106, 19. 'Wendaþ mîn heáfod ofdûne, forðon ðe mîn Drihten of heofenum âdûne tô eorþan âstâg.' Ðâ fæstnedan hié ða fêt up and ðæt heáfod ofdûne, Blickl. Homl. 191, 2–9.

of-earmian; *p.* ode *To have pity or compassion* :—Rihtwîsa ofearmaþ *justus misereretur,* Ps. Spl. 36, 22. Ofearmian *misereri,* 76, 9.

of-earmung, e; *f. Pity, compassion* :—On ofearmunga *in miseratione,* Blickl. Gl.

ofen, ofn, es; *m. An oven, a furnace* :—Ofen *fornax* vel *clibanus,* Wrt. Voc. i. 83, 14. Ofn, 34, 40. Se ofn (*caminus*) ðære singalan costnunge, Bd. 4, 9; S. 576, 29. Ða fûlnessa ðæs þýstran ofnes (*fornacis*), 5, 12; S. 629, 21: Cd. Th. 245, 13; Dan. 462. Axan of ðam ofene (*camino*), Ex. 9, 8. Ðâs þrî cnihtas hêt se cyning âwurpan intô byrnendum ofne (*the fiery furnace*), Ælfc. T. Grn. 8, 26. Geond ðone ofen, Cd. Th. 238, 13; Dan. 354. On fýres ofen (ofn, Lind.) *in caminum ignis,* Mt. Kmbl. 13, 42. Gif hwylc wîf seteþ hire bearn on ofen (*in fornacem*), L. Ecg. C. 33; Th. ii. 156, 35. On ofon (*clibanum*) gisended, Lk. Skt. Rush. 12, 28. Hî gáþ on ðîne ofnas (*furnos*), Ex. 8, 3. Ðæt man ða ofnas ontende, Homl. Skt. i. 5, 294. [*Goth.* auhns : *Icel.* ofn *and* ogn : *Dan.* ovn : *Swed.* ugn : *O. Frs.* oven : *O. H. Ger.* ovan.] v. hlâf-ofen (-ofn).

ofen-bacen; *adj. Baked in an oven* :—Ofenbacen hlâf *formentum,* Wrt. Voc. ii. 38, 60 : *fermentum,* i. 27, 24 : *clibanius panis,* 41, 21. Genim ðone cruman of ofenbacenum hlâfe, Lchdm. i. 132, 19. Bring clǽne ofenbacene hlâfas *sacrificium coctum in clibano, panes,* Lev. 2, 4.

ofen-raca, an; *m. An oven-rake, an instrument for clearing out an oven or furnace* :—Ofenraca *rotabulum,* Wrt. Voc. i. 16, 34 : 27, 10.

of-eode. v. of-gân.

ofer, ofor; *prep. adv.* **I.** *with dat.* generally with the idea of rest; (1) *above, over* :—Wæs hâlig leóht ofer wêstenne, Cd. Th. 8, 16; Gen. 125. Beheóld ofer leódwerum byrnende beám, 184, 20; Exod. 110. Mǽst hlifade ofer Hrôþgâres hordgestreónum, Beo. Th. 3802; B. 1899. Wîglâf siteþ ofer Bíowulfe, 5806; B. 2907. (2) denoting contact with anything, *upon, on* :—Hê gesette ofer stâne fêt mîne, Ps. Lamb. 39, 3. Hwonne hié ofer streámstaðe stæppan môsten (*might set foot on shore*), Cd. Th. 86, 21; Gen. 1434. Wind ofer ýðum *the wind on the waves,* Beo. Th. 3819; B. 1907. Ânra gehwylc hæfde sweord ofer his hype, Blickl. Homl. 11, 18. Sittende ofor eoselan folan, 71, 5. Ûre Dryhten sæt ofer winda fiðerum, Salm. Kmbl. p. 198, 16. (3) denoting extension over, *throughout, in, on* :—Hê wolde ǽgðær ge ofer heofenum ge ofer eorþan ûs his miltse gecýðon, Blickl. Homl. 39, 22: Gen. 4, 11. (4) denoting a higher degree, *beyond, more than* :—Ofer snâwe scînende, Ps. C. 50, 75; Ps. Grn. ii. 278, 75. (5) denoting the cause of an emotion, *over* (as in to rejoice over, etc.) :—Byþ on heofone blis ofer ðæm flǽsce sitt and his wealdan sceolde *mens carni praesidens,* Past. 36, 7; Swt. 257, 3. Ofer deóflum wealdeþ, Cd. Th. 263, 20; Dan. 765. (7) with the idea of movement, where the accusative might be expected :—Hleó wand ofer wolcnum, Cd. Th. 182, 23; Exod. 80. Up gewât líg ofer leófum, 231, 18; Dan. 249. Ofer ðære Reádan Sǽ eode Israela folc, Salm. Kmbl. p. 198, 20. (8) marking time, *after, beyond* :—Ðâ undergeat heó ðæt se brôðer ne môste his lífes brûcan ofer ðam ânum geáre, Homl. Th. ii. 146, 17. Se dæg biþ ofer eástrum, H. R. 99, 15. **II.** *with acc.* generally with the idea of movement. (1) denoting motion in a definite direction across, to the other side of an object :—Ofer sǽ *citra pontum,* Wrt. Voc. ii. 18, 68. Ofer landgemǽru *extra terminum* . . . ofer ðone ford *trans vadum,* ofer sǽ *trans mare,* Ælfc. Gr. 47; Som. 47, 29, 38. Ðâ cômon hî ofer ðære sǽs mûþan, Mk. Skt. 5, 1. Hié ofer sǽ gewiton, Chr. 885; Erl. 82, 25. Hié eodon ofer land *they went across the country,* 896; Erl. 94, 14 : Andr.

Kmbl. 2460; An. 1231. Ofer eástreámas ís brycgade *the ice threw a bridge across the rivers*, 2523; An. 1263. Hí wurpon heora waru ofor bord *they cast their wares overboard*, Homl. Th. i. 246, 2, 9. Ofer clif *per praeceps* (v. Mt. 8, 32, where the swine go over the cliff's edge), Wrt. Voc. ii. 72, 35. Ic út gange ofer mínre burge weall *transgrediar murum*, Ps. Th. 17, 28: Cd. Th. 90, 12; Gen. 1494. Ic cume ofer langne weg, 35, 13; Gen. 554. Se eádega bewlát ofer exle, 177, 7; Gen. 2926. (2) denoting motion which is diffused over a surface:—Streám út áweóll, fleów ofer foldan, Andr. Kmbl. 3046; An. 1526. Wíde ofer woruld ealle geseón, Cd. Th. 36, 2; Gen. 565: 42, 17; Gen. 675. Hé ofer ealle þeóde eágum wlíteþ, Ps. Th. 65, 6. Álæd upp ða froxas ofer eall Egipta land, Ex. 8, 5. Wæron gewurden þýstru ofer ealle eorþan, Mt. 27, 45: Blickl. Homl. 93, 18. Bufan ðæm máran wealle ofer ealne ðone ymbgong hé is mid stænenum wíghúsum beworht, Ors. 2, 4; Swt. 74, 20. Mann ús ofer eall (cf. *Ger.* überall) sóhte, Homl. Skt. i. 23, 450. Ða weorcstánas lágon ofer eall *lay scattered in all directions*, 23, 490. (3) denoting extension through a space, *throughout, among* :—Se wæs mærost ofer werþeóde, Beo. Th. 1802; B. 899. Heó wæs seó eádgeste ofer eall wífa cynn *she was most blessed among all the race of women*, Blickl. Homl. 13, 15. He sceal beón gehered ofor ealle þeóda, 71, 16. Hét hé beódan ofer ealle ða fird ðæt hié fóron ealle út ætsomne, Chr. 905; Erl. 98, 22. Wilnung leáses gilpes ofer eall folc, Bt. 18, 1; Fox 60, 24. (4) denoting motion from below, *over, above* :—Hefe upp ðíne hand ofer eall ðæt flód, and ofer burna and ofer móras, Ex. 8, 5. Mín unriht mé hlýpþ ofer heáfod, Ps. Th. 37, 4. Hié him ásetton segen gyldenne ofer heáfod, Beo. Th. 95; B. 48. Man slóh án geteld ofer ða hálgan bán, Swt. A. S. Rdr. 100, 150. Iudas þá ðær róda twá oter ðæt fæge hús, Elen. Kmbl. 1759; El. 881. Se ðe ástáh ofer heofenas *qui ascendit super caelos*, ofer heálíce dúne ástíh ðú *super montem excelsum ascende tu*, Ælfc. Gr. 47; Som. 48, 23. (5) denoting motion from above, *upon, on* :—Se hys hús ofer stán getimbrode, Mt. Kmbl. 7, 24. Hé hine ásette ofer ðæs temples scylf, Blickl. Homl. 27, 11. Feallaþ ofor ús, 93, 33: Elen. Kmbl. 2267; El. 1135. (6) denoting the object upon which an action or feeling takes effect :—Andreas sette his hand ofer ðara wera eágan . . . And eft hé sette his hand ofer hiora heortan, St. And. 12, 34–35. Sleáþ synnigne ofer seolfes múþ (*smite him over the mouth*), Andr. Kmbl. 2602; An. 1302. Sý hys blód ofer ús and ofer úre bearn *his blood be upon us, and upon our children*, Mt. Kmbl. 27, 25. Mín hand byþ ofer ðíne æceras and ofslihþ ðíne hors mid hefegum cwealme, Ex. 9, 3. Ða tácna ðe hé worhte ofer ða untruman men *the miracles he wrought upon the sick*, Homl. Th. i. 182, 1. Eftwyrd cymþ ofer middangeard, Cd. Th. 212, 17; Exod. 540. Se tán gehwearf ofer (*the lot fell upon*) ænne ealdgesíþa, Andr. Kmbl. 2209; An. 1106. Gé onfóþ ðæm mægene Hálges Gástes, se cymeþ ofor eów, Blickl. Homl. 119, 12. Ðín mildheortnes is mycel ofor mé, 89, 27. (7) denoting the object over which power is exercised :—His mægen wealdeþ ofer eall manna cyn, Ps. Th. 65, 6. Forðam ðe ðú wære getrýwe ofer lytle þing, ic gesette ðé ofer mycle, Mt. Kmbl. 25, 21. Ðú byst andweald hæbbende ofer týn ceastra . . . Beó ðú ofer fíf ceastra, Lk. Skt. 19, 17–19. Se hæfde mægen ofer ealle gesceafta, Blickl. Homl. 9, 15. Ríce ofer heofenstólas, Cd. Th. 1, 15; Gen. 8. Cræft móda gehwylces ofer líchoman, Met. 26, 106. Deáþ rícsade ofer foldbúend, Exon. Th. 154, 17; Gú. 843. (8) denoting degree (*a*) *above, more than* ; supra, super :—Ioseph wæs gleáwra ofer hí ealle, Ors. 1, 5; Swt. 34, 1. Hé lufode Iosep ofer his suna *he loved Joseph more than all his children*, Gen. 37, 3: 44, 20. Ne lufige ic nánwiht ðisses andweardan lýfes ofer ðæt (*éce líf*), ne furðum ðam gelíce, Shrn. 177, 14. Ða stówe ofer ealle óðre ic geceás, Blickl. Homl. 201, 7. Nys se leorningcniht ofer his láreów, ne þeów ofer hys hláford, Mt. Kmbl. 10, 24: Exon. Th. 105, 35; Gú. 33. Hit is áwriten ðæt seó góde antswaru sý ouer ða sélestan selene, R. Ben. 55, 8. Ðæs biscepes weorc sceolon bión ofer óðra monna weorc *debet actionem populi actio transcendere praesulis*, Past. 12; Swt. 75, 3. Is án steorra ofer óðre beorht, Met. 29, 19. Moises wæs se bilewitusta mann ofer ealle men, Num. 12, 3. Fram twentig wintrum and ofer ðæt *a vigesimo anno et supra*, 1, 3. (β) *beyond, besides* ; ultra :—Ofer ðæt (*ultra*) gé ne lætaþ hine ænig þing dón, Mk. Skt. 7, 12. Ne ofer ðæt (*ultra*) sweltan ne mágon, Lk. Skt. 20, 36. Ne læteþ hé ús nó costian ofer gemet, Blickl. Homl. 13, 9. Ðú sprycst ofer mæþe úre *ultra etatem nostram*, Coll. Monast. Th. 32, 11. Ðæt hæhste gód [is] ðætte man ne þurfe nánes óðres gódes ne eác ne récce ofer ðæt siððan hé ðæt hæbbe *id est bonum, quo quis adepto, nihil ulterius desiderare queat*, Bt. 24, 1; Fox 80, 13. Siððan ðú hí gecnáwan miht ðonne wát ic ðæt ðú ne wilnast nánes óðres þinges ofer ða (*you will desire nothing further*), 23; Fox 80, 3. Hié lícettaþ ðæt him ne síe náwuhtes cearu ofer ða ryhtwísnesse, Past. 41; Swt. 302, 10. Se ðe godgeldum onsæcge ofer (*besides*) God ánne, L. Alf. 32; Th. i. 52, 12. (9) denoting the passing over moral bounds, *in violation of, in opposition to, contrary to, against* :—Ofer Godes æ hé deþ *extra legem Domini facit*, Ælfc. Gr. 47; Som. 47, 29. Se wæs ofslagen ofer áþas and treówa *contra fidem jusjurandi peremptus est*, Bd. 2, 20; S. 521, 17: Chr. 894; Erl. 90, 5. Ætsæton ða Centiscan beæftan ofer his

bebod, 905; Erl. 98, 24: Blickl. Homl. 91, 16 : Exon. Th. 244, 5; Jul. 23. Wíte hé ðæt hé hit dé ófer Godes ést, and ofer ealra his háligra, and eác ofer monna godcundra háda and woruldcundra, Chart. Th. 131, 36: Exon. Th. 226, 10; Ph. 403: Cd. Th. 76, 2; Gen. 1251. Hié ær ofer hiera willan him tó gecierdon, Ors. 2, 5; Swt. 82, 10. Gecuron Brettanie Maximianus him tó cásere ofer his willan, 6, 35; Swt. 292, 15. (10) with words implying rest :—Standende ofer hig, Lk. Skt. 4, 39. Ne biþ forlæten stán ofer stán, Blickl. Homl. 79, 1. Hé fyrgenbeámas ofer hárne stán hleonian funde, Beo. Th. 2834; B. 1415. Æþelingas ofer heánne hróf fand sceáwedon, 1970; B. 983. (11) denoting the subject of discourse (cf. to talk *over*) :—Hé ofer benne spræc, wunde wælbleáte, 5442; B. 2724. Ofer Ysmahel ic gehírde ðé, Gen. 17, 20. (12) denoting the cause of an emotion (cf. I. 5) :—Heó hæfþ genóh on ðis andweardan lífe, ac heó hit hæfþ eall forsewen ofer ðé ánne (*simply on your account* [?]), Bt. 10; Fox 28, 26. (13) *without* :—Gif hé gesécean dear wíg ofer wæpen, Beo. Th. 1374; B. 685. (14) with words expressing time, (*a*) *after :*—Ofer middæg *post meridiem*, Gen. 3, 8. Ofor undern, Blickl. Homl. 93, 15. Ofer ealle tíd tó sáwenne *ultra omne tempus serendi*, Bd. 4, 28; S. 605, 39, 8. Ofer hyre deg . . . ofer mínnæ dæg (cf. æfter hæora dæge, 12), Chart. Th. 520, 7–34. Ne onbirigdon ðæs bigleofan ofer ðæt (*ultra*), Jos. 5, 12. Hé ne oncnáweþ ofer ðæt stówe *non cognoscet amplius locum*, Ps. Lamb. 102, 16. Hé ofer ðæt (*ultra*) deófulgyldum ne þeówde, Bd. 2, 9; S. 512, 7 : Ors. 5, 7; Swt. 230, 7 : R. Ben. 53, 16. Longe ofer ðis, Exon. Th. 172, 15; Gú. 1144. Ofer ða niht, Beo. Th. 1476; B. 736. (β) expressing duration, *through, during* :—Ofer ealle ða niht ðe wé férdon *during the whole night that we marched*, Nar. 12, 2. Hé hié slóg ofer ealne ðone dæg, 13, 14; Swt. 200, 21. Ða steorran ðe ofer ealne winter scínaþ . . . Ofer ealne sumor hí gáþ on nihtlícre tíde under ðissere eorþan, Lchdm. iii. 270, 24–26. Hí wunodon mid ðæm biscope ofer geár, and siðan genendon tó Antiochia, Homl. Skt. i. 3, 81. Ða sylfan sealmas sýn dæghwamlíce geedlæhte ofer ealle wucan, R. Ben. 43, 1. III. adverbially, or not followed by a case :—Ðæt ðú ne mihtst nænne weg findan ofer, Bt. 34, 4; Fox 138, 28. Hé eode tó ðære burge wealle, and fleáh út ofer, Ors. 5, 12; Swt. 244, 3. Ðonne céþþ hé hwær se weall unhéhst sý, and ðær ofer scýt (oferscýt?) *he observes where the wall is lowest, and over there he rushes*, Homl. Th. i. 484, 11. Án fiscere uneáþe hiene ænne ofer brohte, 2, 5; Swt. 84, 10. Mid Angelþeóde ðe hé ofer cyning wæs, Bd. 3, 6; S. 528, 3. Sió giémen ðære ciricean síe ðæm beboden ðe hié wel ofer mæge, Past. 5; Swt. 45, 1. Wese ús beorhtnes ofer Drihtnes úres, Ps. Th. 89, 19. Se cwellere him ofer stód *illi instante carnifice*, Bd. 4, 16; S. 584, 36 : Homl. Th. ii. 494, 27. Eall ðæt ofer biþ tó láfe is tó syllanne, swá swá Crist lærde : '*Quod superest date eleemosynam :*' ðæt ofer sí and tó láfe sellaþ ælmessan, Bd. 1, 27; S. 489, 26–30. Wé nú gehýrdon ðis hálige godspel beforan ús rædan, and ðéh wé hit sceolan eft ofer cweþan (*we must say it over again*), ðæt wé ðé geornor witon ðæt hit ús tó bysene belimpeþ éces lífes, Blickl. Homl. 15, 31. Ealle ðe ðær ofer beóþ getealde wintra, ða beóþ gewinn and sár, 89, 11. Hú þicke se heofon wære oððe hwæt ðær ofer wære, Bt. 35, 4; Fox 162, 24. Ofer ufa *desuper*, Mt. Kmbl. Rush. 21, 7. Ofer uppan *up above*, Met. 24, 27. [*Goth.* ufar : *O. Sax.* oƀar : *O. Frs.* over : *Icel.* ofr-; *and cf.* yfir : *O. H. Ger.* ubar.]

ófer, ófor, es; *m.* I. *an edge, border, margin :*—Ófr *margo*, Wrt. Voc. ii. 113, 45. Ófor, 55, 6. Ofer, Ælfc. Gr. 6; Som. 5, 51. On ðære lifre ófrum, Lchdm. ii. 204, 24. Smire ða ófras (*the borders of a cancer*) ðær hit reádige, 108, 20. II. *the land bordering on water, a river-bank, sea-shore, over* in local names, e. g. Over in Cambridgeshire, Wendover :—Strand *litus*, brerd *vel* ófer *crepido*, Wrt. Voc. i. 54, 24–25. On ðone ófer; ondlong ófres ðæt on Stánford, Cod. Dip. Kmbl. iii. 378, 20. Ondlong stræte, ðæt on reádan ófer, iii. 52, 17. On ðære eá ófre, Nar. 10, 14 : Byrht. Th. 132, 39; By. 28. On ófre ðæs foresprecenan streámes, Bd. 2, 3; S. 504, 18. Of sæs ófre, 4, 13; S. 582, 32. On ðam sealtum ófre, Homl. Th. ii. 146, 6. On (meres) ófre, Beo. Th. 2746; B. 1371. Ófras heá, streámas stronge, Exon. Th. 404, 14; Rä. 23, 7. On wætum stówum and on ófrum, Lchdm. i. 222, 19 : Hpt. Gl. 516, 70. Óbras, ófras *oras, sæ marmora*, Wrt. Voc. ii. 91, 72 : 64, 42. [*Laym. Havel.* over : *M. H. Ger.* uover : *Ger.* ufer : *O. Du.* oever.] v. eá-ófer.

ofer-æt, es; *m.* I. *over-eating, gluttony, excess in eating :*—Oferæt *ingluvies*, Ælfc. Gr. 12; Som. 15, 54. Se oferæt wierþ gehwierfed tó fierenluste *edacitas usque ad luxuriam pertrahit*, Past. 43, 2; Swt. 309, 14. Behealdaþ eów ðæt gé ne gehefegien eówer heortan mid oferæte (*in crapula*), 18, 2; Swt. 129, 19. Ða teþ ðe nú on oferæte blissiaþ, Homl. Th. i. 530, 32. Hine wið oferæt beorge, L. E. I. 24; Th. ii. 422, 3. Þurh oferæt *per commessationem*, Confess. Peccat. II. *rioting, feasting, an entertainment where excessive eating takes place :*—Ða hús ða ðe on tó gebiddenne geworhte wæron ða syndon nú on hús gehwyrfed oferæta (*commessationum*), 24, 25; S. 601, 13. Oferætum *commessationibus*, Kent. Gl. 888. [Cf. ofer-etes = *comessationes* (in Rom. 13, 13), Rel. Ant. i. 131, 32: *O. L. Ger.* ovar-át : *O. H. Ger.* ubar-âzi, -âzzi; *f.* crapula, commessatio.]

ofer-ǽte; *adj. Given to excess in eating, gluttonous :*—Ne sceal mon beón druncengeorn, ne oferǽte, R. Ben. 17, 15.

ofer-bæc *the upper part of the back.* v. next word.

oferbæc-getéung, e; *f. Contraction of the muscles at the back of the neck, tetanus* (cf. Lchdm. iii. 110, 16 sqq. :—Ðisne lǽcecræft man sceal dón mannum ðe hyra swyran mid ðam sinum fortogen beóþ, ðæt hé hys nǽn geweald náh, ðæt Gréccas hátaþ *tetanicus*) :—Oferbæcgetéung *titanus*, Wrt. Voc. i. 19, 22.

ofer-bebeódan *to command, rule :*—Ic wealdige *vel* oferbebeóde *imperito*, Wrt. Voc. i. 54, 52.

ofer-becuman *to supervene :*—For ðí ðe oferbecymþ gedéfnes *quoniam supervenit mansuetudo*, Ps. Lamb. 89, 10.

ofer-bídan *to outlast, outlive, survive :*—Gif hwylces weres forme wíf biþ deád, ðæt hé be leáfe óðer wíf niman móte; and gif hé ða oferbýt (*si supersit ei*) wunige hé á syððan wífleás, L. Ecg. P. ii. 20; Th. ii. 190, 3. Yldo oferbídeþ stánas, Salm. Kmbl. 599; Sal. 299. Ða oferbád (*survived*) Ælféh his bróðor, Chart. Th. 272, 12. Gif ic hire ouerbíde . . . gif heó mé ouerbíde, 583, 5–10. Hé ða bysgu oferbiden hæfde, Exon. Th. 135, 3; Gú. 518.

ofer-biterness, e; *f. Excessive bitterness;* amaritudo, Ps. Spl. 13, 6.

ofer-blice (?), an; *f. A superficies, surface :*—Oferbliocan *superficiem*, Txts. 181, 44.

ofer-blíðe; *adj. Over-cheerful :*—Ðǽm oferblíðum (*laetis*) is tó cýðanne ða unrótnessa ðe ðæræfter cumaþ, and ðám unblíðum sint tó cýðanne ða gefeán ðe him gehátene sindon, Past. 27; Swt. 187, 15 : 189, 4 : 61; Swt. 455, 22.

ofer-bord. v. ofer, II. 1.

ofer-brédan. I. *to overspread, overshadow, act as a covering over :*—Ðæt land biþ eal unnyt swá se fiicbeám hit oferbræt, Past. 45; Swt. 337, 13–15. Oferbrædeþ, Met. 7, 13. Heofonlíc leóht com ofer hí ealle and hí swá swá mycel scýte hí ealle oferbrǽdde, Bd. 4, 7; S. 575, 7. Wolcen oferbrǽdde hié *nubis obumbrans eos*, Mk. Skt. Rush. 9, 7 : Lk. Skt. Rush. 9, 34. Sticmǽlum mid wuda oferwexen, sticmǽlum mid grénum felda oferbrǽded, Homl. Th. i. 508, 24. Mid ðý feó oferbrǽded and beþeaht, Blickl. Homl. 199, 3. Bewrigen and oferbrǽded mid baswe godwebbe, 207, 16. Apollonius mid rósan rude wæs eal oferbrǽded, Ap. Th. 22, 4. II. *to overspread, put a covering over :*—God oferbrǽdde byrnendne heofon nette, Cd. Th. 182, 9; Exod. 73. [*Laym.* mid palle overbræd.]

ofer-brǽdels, es; *m. A covering, veil, garment :*—Cyrtel *vel* oferbrǽdels *palla*, Wrt. Voc. i. 16, 56. Oferbrédels *operimentum*, Kent. Gl. 853. Swá swá oferbrǽdels (*opertorium*) ðú áwenst hyg, Ps. Lamb. 102, 27. On ðam oferbrǽdelse (*velamento*) fyðera ðínra, Ps. Spl. 62, 8. Hé þencþ on ðam oferbrǽdelse (*surface*) his módes ðæt hé sciele monig gód weorc wyrcan, and hé þencþ mid innewearde móde ðæt hé gierneþ for gilpe . . . on hiera módes rinde . . . ac on ðam piðan . . . , Past. 9, 1; Swt. 55, 18–23. Oferbrǽdels *superhumeralis*, 14, 3; Swt. 83, 21. Hí áhófon ðone oferbrǽdels (*the veil*) of ðære byrgene, Homl. Skt. i. 8, 227. vii. of[er]-brǽdelsas, Chart. Th. 429, 26. [*Icel.* yfir-breizl *a coverlet.*]

ofer-brǽw, -bráw, es; *m. An eye-brow :*—Hæfþ mǽden tácn on oferbráwe, Lchdm. iii. 188, 5. [*O. H. Ger.* uber-bráwa *supercilium.*] v. ofer-bráw.

ofer-brecan *to infringe, violate (an agreement) :*—Hé oferbræc heora gecwedrǽdenne, Ors. 3, 6; Swt. 108, 8 : 5, 12; Swt. 242, 8.

ofer-bregdan, -brédan. I. *v. trans. To overspread, cover, draw a covering over :*—Se ða burh oferbrægd blácan líge, Andr. Kmbl. 3080; An. 1543. Niht oferbrǽd beorgas steápe, 2613; An. 1308. II. *intrans. To break out over a surface :*—Scamoniam geceós ðus brec on tú dó hwón on ðíne tungan gif hió hwíte oferbregdeþ swá meluc ðonne hió biþ gód *choose scammony then; break it in two, put a bit on your tongue, if it breaks out all over white as milk, it is good*, Lchdm. ii. 272, 18.

ofer-brú; *gen.* -brúwe; *f. An eye-brow :*—Mǽden (hæfþ) tácn on oferbrúwe swíðran, Lchdm. iii. 186, 25 : 192, 28. Oferbrúa *supercilia*, Wrt. Voc. i. 42, 69. Oferbrúwa *supercilium*, 64, 33 : 70, 40 : 282, 47. Betwux oferbrúan and brǽwum *intercilium*, 43, 4. Oferbrúum *supercili[i]s*, Txts. 172, 33. v. ofer-brǽw.

ofer-brycgian *to overbridge, make a bridge over :*—Ða hét Maxentius oferbrycgian ða eá mid scipum, Homl. Th. ii. 304, 22.

ofer-búgan (?) *to avoid, shun :*—Hié sindon suá micle wǽrlícor tó oferbúganne [ferbúgonne, Cott. MSS.] suá mon ongiet ðæt hié on máran ungewitte beóþ *tanto caute declinandi sunt, quanto insane rapiuntur*, Past. 40, 5; Swt. 295, 21. [*Ofer* is probably a mistake for *fer*. v. note on this passage, and *for-búgan*.]

ofer-cæfed *covered with ornamental work :*—Ofercæfedu *innexa*, Germ. 394, 353. Cf. be-cæfed *falerata*, Wrt. Voc. ii. 34, 67; cæfing *discriminale* (*ornamentum capitis mulieris*, Wülck. Gl. 656, 13), 141, 1 : *and see* ymb-cæfed.

ofer-ceald; *adj. Excessively cold*, Runic pm. Kmbl. 341, 14; Rún. 11. [Cf. *Icel.* ofr-kuldi *excessive cold.*]

ofer-cídan *to censure, reprove :*—Ða ðe wyrceaþ Sunnandæge æt ðam forman cyrre Grécas hý ofercídaþ (*arguunt*), L. Ecg. C. 35; Th. ii. 160, 31. Ðú ofercíddest *increpasti*, Ps. Spl. T. 118, 21.

ofer-cirr, es; *m. A passing over :*—Ofercerr *transmigratio*, Mt. Kmbl. Lind. 1, 11.

ofer-climban, -climman *to ascend, climb upon :*—Alexander ðone weall oferclom *cum murum escendisset*, Ors. 3, 9; Swt. 134, 13.

ofer-cræft, es; *m. Craft, fraud :*—Gif hwá mid his ofercræfte (*per fraudem*) wíf nýdinga nimþ, L. Ecg. P. ii. 13; Th. ii. 186, 20.

ofer-cuman. I. *to overcome, vanquish, subdue :*—Ofercymeþ hé ælle his feónd, Lchdm. iii. 170, 19. Ofercymþ *deicit, confudit*, Wrt. Voc. ii. 133, 68. Ofercom *obpressit*, 65, 35. Æþelfriþ Scotta þeóde mid gefeohte ofercom (*praelio conterens*), Bd. 1, 34; S. 499, 17 : Cd. Th. 178, 33; Exod. 21. Hé ðone feónd ofercwom, Beo. Th. 2551; B. 1273. Hié feónd heora þurh ánes cræft ofercómon, 1403; B. 699. Ðæm wergan gáste wiðstondan and ofercuman, Blickl. Homl. 135, 11 : 119, 21. Beswicen and ofercumen, 179, 5. Ðonne hié hwelc folc mid gefeohte ofercumen hæfdon, Ors. 2, 4; Swt. 70, 23. Níða ofercumen, Beo. Th. 1694; B. 845. Ofercumen *obpressus*, Wrt. Voc. ii. 65, 34. Ofercymen wæs *obstipuit*, 63, 9. Ðú mé hæfst ofercumenne mid ðínre gesceádwísnesse, Bt. 22, 1; Fox 76, 12. Hé ongitt hine selfne ofercumenne (-cymenne, Hatt. MS.), Past. 34, 1; Swt. 228, 20. Ðás men geseóþ ðæt hié synt ofercumene, Blickl. Homl. 189, 5. Ofercymene *consternati*, Wrt. Voc. ii. 91, 10. Ofercumenum leahtrum *devictis vitiis*, Prud. 28 a. II. *to come upon, reach, obtain :*—Ofercuom *obtinuit*, Wrt. Voc. ii. 115, 29. Ofercom, 63, 26. Nänne ne sparedon cwicera manna ðe hié ofercuman mihton (*spared none that they could come up with*), Judth. Thw. 24, 41; Jud. 235. His geféran ðý ofercumendan wóle (*pestilentia superveniente*) fordílgode wǽron, Bd. 4, 1; S. 563, 26.

ofer-cyme, es; *m. A coming upon, arrival :*—Ǽr ðon ðe hé mid ofercyme semninga deáþes ealle tíd hreówe forlure *priusquam subito mortis superventu tempus omne poenitendi perderet*, Bd. 5, 13; S. 632, 12.

ofer-cýðan *to bring stronger testimony than another :*—Wé cwǽdon be mannum . . . gif áþ burste oððe ofercýðed wǽre (*if the oath were not supported by a sufficient number of compurgators, or were disproved by testimony more strongly supported by oath.* Cf. mid áþe cýðan, gecýðan), ðæt hý siððan áþwyrðe nǽron, L. Ed. 3; Th. i. 160, 20.

ofer-dón *to overdo, do to excess :*—Ðonne sceal his steór beón mid lufe gemetegod, ná mid wælhreáwnysse oferdón, Homl. Th. ii. 532, 13. Ealle oferdóne þing dæriaþ *omnia nimia nocent*, Homl. Skt. i. 1, 163.

ofer-drenc, es; *m. Excessive drinking, drunkenness :*—Ða heáfodleahtras sind . . . singal oferdrenc . . . , Homl. Th. ii. 592, 6. Ðú woldest mé laðian, ðá ðá ic wæs mid ðé, ðæt ic swíðor drunce swilce for blisse ofer míne gewunan . . . Úre Hǽlend on his hálgan godspelle forbeád ðone oferdrenc eallum gelýfendum mannum . . . and ða hálgan láreówas æfter ðam Hǽlende áledon ðone unþeáw . . . for ðan ðe se oferdrenc fordéþ ðæs mannes sáwle and his gesundfulnysse, Ælfc. T. Grn. 21, 29–37. v. ofer-drync.

ofer-drencan *to overdrench, give a person too much to drink, to inebriate, intoxicate :*—Se ðe þurh fácn oðerne oferdrencþ (*inebriaverit*), fæste .xl. daga, L. Ecg. P. iv. 37; Th. ii. 214, 20 : Past. 36; Swt. 261, 14. Ðú oferdrenctest hig *inebriasti eam*, Ps. Spl. 64, 9. Hié hié selfe mid ealoþ oferdrencton, Ors. 5, 3; Swt. 222, 6. Se ðe ne wirnþ ðæs wínes his láre ða mód mid tó oferdrencanne . . . hé biþ oferdrenced mid ðæm drence mislícra giefa, Past. 49; Swt. 381, 5–6 : Bt. 24, 4; Fox 84, 33. Hý beóþ oferdrencte (*inebriabuntur*) on ðære genihte ðínes húses, Ps. Th. 35, 8 : Judth. Thw. 21, 22; Jud. 31. [*O. H. Ger.* ubar-trenkjan *inebriare*.]

ofer-drettan (?) *to take with violence :*—Wé oferdryttan *praeoccupemus*, Ps. Spl. 94, 2. v. ge-drettan, -dreccan.

ofer-drífan. I. *to cover by drifting :*—Ðeáh hit wind oððe sǽs flód mid sonde oferdrífen *though the wind or sea cover it by driving the sand over it*, Ors. 1, 7; Swt. 40, 1. II. *to overcome, refute, repel, defeat :*—Ðú ðe þióstro giduoles oferdrífest (*depellis*), Rtl. 38, 17. Se Hǽlend ne geswutolode ná him his mihte ac oferdráf hine geþyldelíce mid hálgum gewritum *the Saviour did not display his power to him (the devil) but overcame him patiently by the holy scriptures*, Homl. Th. i. 176, 11. Marcellus folgode ðam sceandlícan drý óððæt Petrus ðone árleásan oferdráf, Homl. Skt. i. 10, 197. Onsage oferdrífan *to refute an accusation*, 2, 206. Wé syndon fram ðe oferswýðde, ac wé ácsiaþ : Hwæt eart ðú swá wunderlíc on ánes mannes hiwe ús tó oferdrýfenne, Nicod. Thw. 16, 20. Gif hig sacan stande ðæt hig .viii. secgaþ and ða ðe ðǽr oferdrífene beóþ gilde heora ǽlc .vi. healfmarc *if they (the twelve) disagree, that which eight of them say shall stand : and those that in such case are out-voted shall each pay six half-marks*, L. Eth. iii. 13; Th. i. 298, 4.

ofer-drinc. v. ofer-drync.

ofer-drincan *to overdrink (one's self) :*—Ne oferdrincaþ gé eów wínes, L. E. I. 40; Th. ii. 438, 19. Gif hwylc bisceop hine oferdrince (*se inebriet*), L. Ecg. P. iv. 33; Th. ii. 214, 12. Beón oferdruncen *inebriari*, Lk. 12, 45. Ðæt mód, ða hwíle ðe hit biþ oferdruncen ðæs ierres, Past. 40; Swt. 295, 3. Swá hwá swá óðerne drencþ, hé wirþ self oferdruncen, 49; Swt. 381, 4. Swá swá mihti oferdruncon (*crapulatus*) fram wíne, Ps. Spl. 77, 71. Swá swá oferdruncen man wát ðæt hé sceolde tó his húse, and ne mæg ðeáh ðider áredian, Bt. 24, 4; Fox 84, 30. [*O. H. Ger.* ubar-trinkan.]

ofer-druncen, es; *n. Drunkenness, inebriety :*—Ne gerîseþ ǽnig unnytt mid bisceopum, ne doll ne dysig, ne tô oferdruncen, L. I. P. 9; Th. ii. 314, 31. Ðæt preóstas beorgan wið oferdruncen, and hit beleán ôðrum mannum, L. Edg. C. 57; Th. ii. 256, 13. Gif preóst lufige oferdruncen, L. N. P. L. 41; Th. ii. 296, 11. [*O. H. Ger.* ubar-trunkani *ebrietas, crapula.*]

ofer-druncenness, e; *f. Drunkenness, intoxication, rioting :*—Oferdruncennys *ebrietas*, L. Ecg. P. iv. 64; Th. ii. 224, 30. Gif munuc for oferdruncennysse (*ex ebrietate*) spîwe, iv. 34–36; Th. ii. 214, 14–19. Ne gewunigen gê tô oferdruncennisse (*non in ebrietatibus*), Past. 43, 9; Swt. 317, 18. Ða ofordruncennessa ðe hê lufode, Blickl. Homl. 195, 15.

ofer-drync, es; *m.* I. *excessive drinking, drunkenness :*—Behealdaþ eów ðæt gê ne gehefgien eówre heortan mid oferdrynce (*ebrietate*), Past. 18, 2; Swt. 129, 19. Hî fêrdon tô sumre wydewan hâm and ðǽr wǽron ondrencte mid oferdrynce, Guthl. 14; Gdwin. 62, 20. II. *an entertainment where excessive drinking takes place* (cf. ge-drinc) :—Hê begæþ unǽtas and oferdrincas *comessationibus vacat atque conviviis*, Deut. 21, 20. [*O. H. Ger.* ubar-trunk *ebrietas*: cf. *O. L. Ger.* obardrank.]

ofer-dyre *a lintel; superliminare*, Wrt. Voc. i. 290, 17. [*O. H. Ger.* ubar-turi *superliminare.*] Cf. ofer-gedyre.

ofere; *adv. From above; desuper*, Ps. Spl. 77, 27.

ofer-eáca, an; *m.* I. *an over-plus, a surplus, what remains over when a part has been taken :*—Ðone mǽstan dǽl ðæs folces hî ofslôgon, and ðone ofereácan áweg gelǽddon, Homl. Th. ii. 66, 4. Ða seofon mynstru hê gegôdode, ðone ofereácan his ǽhta hê áspende on Godes þearfum, 1118, 31. Ofereácan, Chart. Th. 482, 17. Ofæreácan, 554, 32. Wê niman eall ðæt hê áge, and niman ǽrest ðæt ceápgyld of ðam yrfe, and dǽle man syððan ðone ofereácan on .ii., L. Ath. v. 1, 1; Th. i. 228, 16: v. 6, 1; Th. i. 232, 28: v. 6, 3; Th. i. 234, 6. Ðæs geáres ofereácan fæste *reliquum anni jejunet*, L. Ecg. P. iv. 29; Th. ii. 194, 13. II. *an addition, augmentation :*—Oferêce *augmentum*, Rtl. 85, 33.

ofer-eald; *adj. Exceedingly old :*—Ðeáh hit gecyndelîc sý on menniscum gewunan, ðæt man mildheortnesse cýðe ðâm oferealdum and ðâm cildgeongum, R. Ben. 61, 12.

ofer-ealdormann, es; *m. A chief officer :*—Hê wæs hyre þéna hire hûses· and hire gefêrscipes oferealdormann *erat primus ministrorum et princeps domus ejus*, Bd. 4, 3; S. 567, 21.

ofer-eode. v. ofer-gán.

ofer-etol, -ettol; *adj. Given to excess in eating, gluttonous :*—Oferetol *edax vel glutto*, Wrt. Voc. i. 86, 49. Se mynstres hordere sî . . . wîs, sýfre and ná oferettol (-etol, MS. T.), R. Ben. 54, 8. Ðes oferetola man *hic comedo*, Ælfc. Gr. 36; Som. 38, 47. Gehiéren ða oferetolan ða word ðe Krist cuæþ : Behealdaþ eów ðæt eówre heortan ne sîn gehefegode mid oferæte, Past. 43, 9; Swt. 317, 8, 16.

ofer-etolness, e; *f. Excess in eating, gluttony :*—Ne gewunigen gê tô oferetolnisse *non in comessationibus*, Past. 43, 9; Swt. 317, 18.

ofer-fǽr *a passing over; transmigratio*, Mt. Kmbl. p. 12, 13 : Lind. 1, 17.

ofer-fǽreld, es; *m. n. A going across, passage, transit :*—Galilea is gecweden oferfǽreld, Homl. Th. i. 224, 10. Pasca getâcnaþ oferfǽreld, Anglia viii. 322, 2. Crist gewât þurh oferfǽreld of deápe tô lîfe, 330, 9. Heore is ðæt scip and se ouerfǽreld ðare hæuene *eorum* (the monks of Christchurch) *est navicula et transfretatio portus*, Chart. Th. 317, 38. Æfter oferfǽrelde sǽ reádre *post transitum maris rubri*, Hymn. Surt. 82, 7.

ofer-fædman; *p. de To cover in an embrace, to overspread, to envelope :*—Swilce hê oferfædmed ealne middangeard *as if it* (the tree of Nebuchadnezzar's vision) *would cover with embracing boughs all the world*, Cd. Th. 247, 24; Dan. 502. Þýstre oferfædmed *enveloped in darkness*, Exon. Th. 470, 12; Hy. 11, 14.

ofer-fǽtt; *adj. Too fat, obese :*—Oferfǽt *obesus*, Wrt. Voc. i. 51, 10.

ofer-faran. I. *intrans. To pass, go off :*—Ælþeódiglîce ic oferfare *peregre transeo*, Ælfc. Gr. 38; Som. 41, 28. Oferfare on munt swâ swâ spearwa *transmigra in montem sicut passer*, Ps. Spl. 10. 1. II. *trans.* (a) *to pass, cross* (a river, boundary, etc.) :—Ic Iordane eft ongeán oferfare mid twâm floccon, Gen. 32, 10. Gyf ðû Iordanan oferfærst, Glostr. Frag. 108, 19. Moyses oferfôr ða Reádan Sǽ, Wulfst. 210, 12. Oferfôren *egrederentur*, Hpt. Gl. 464, 64. Ðâ gebeótode ân his þegna ðæt hê mid sunde ða eá oferfaran wolde, Ors. 2, 4; Swt. 72, 29 : Bd. 1, 7; S. 478, 9. Ne ða ebban foldes mearce oferfaran môton, Met. 11, 70. (β) *to pass through, traverse :*—Hî forþ oferfôran folcmǽro land, Cd. Th. 108, 4; Gen. 1801. Siððan ðû ðone up áhafast forþ oferfarenne, Met. 24, 26. (γ) *to pass through* (a danger) :—Ða hyssas fǽrgryre fýres oferfaren hæfdon, Cd. Th. 245, 15; Dan. 463. (δ) *to pass through, penetrate :*—Oferfarende *penetrans*, Hpt. Gl. 493, 30. (ε) *to come upon, come across, meet with :*—Se here . . . slôgon and bærndon swâ hwæt swâ hî oferfôron ðe Danes slew and burnt whatever they came across, Chr. 1016; Eil. 157, 2.

ofer-feallan *to fall upon, to attack :*—Hié oferfeóllan ða ðe ða yrmþo genǽson, Blickl. Homl. 203, 19. [*Ger.* über-fallen.]

ofer-feng, es; *m. A clasp, buckle, latchet of a shoe :*—Oferfeng *fibula*, Wrt. Voc. i. 40, 53. Oferfengc, 74, 60 : *ligulam, fibulam*, Hpt. Gl. 523, 2. v. ofer-fôn.

ofer-feohtan *to conquer, vanquish :*—Oferfehtaþ *debellant*, Ps. Surt. 55, 4. Hî feónd oferfeohtaþ, Exon. Th. 150, 7; Gú. 775. Oft hî ofyrfuhtun (*expugnaverunt*) mê, Ps. Spl. C. 128, 2. Hæfde Drihten feónd oferfohten, Cd. Th. 289, 29; Sat. 405. Sió burg biþ micle ðê iédre tô oferfeohtanne ðe hió self fieht wið hié selfe *tanto ille sine labore superat, quanto et ipsa, quae vincitur, contra semetipsam pugnat*, Past. 38, 6; Swt. 277, 25. [*O. H. Ger.* ubar-fehtan *expugnare, devincere.*]

ofer-fêran. I. *to pass over* or *through, to cross, traverse :*—Ic oferfêrde (*transivi*) Iordane, Gen. 32, 10. Seó sǽ ðe se Hǽlend oferfêrde, Homl. Th. i. 182, 25. Oberfoerde *emenso*, Wrt. Voc. ii. 107, 22. Oferfêrde, 29, 33. Mid ðý wit oferfêrdon (*transissemus*) ðâs wununesse ðara eádigra gâsta, Bd. 5, 12; S. 629, 31. Ðet hî ne oferfêrdan *ne transirent*, Kent. Gl. 275. Sê môr swâ brâd swâ man mæg on twâm wucum oferfêran, Ors. 1, 1; Swt. 18, 34. II. *to come upon* or *across, meet with :*—Se here fêrde intô Myrcean and fordydon eall ðæt hê oferfêrde, Chr. 1016; Erl. 157, 12. v. ofer-faran.

ofer-fêre. v. un-oferfêre and next word.

ofer-fêrness, e; *f. Possibility of being crossed :*—On twâm stôwum is oferfêrnes *duobus tantum in locis est transmeabilis*, Bd. 1, 25; S. 486, 21.

ofer-firr, e; *f. Too great distance :*—Hit is feáwum mannum cûð for ðære oferfyrre *insula Thule, quae per infinitum a ceteris separata, vix paucis nota habetur*, Ors. 1, 1; Swt. 24, 21.

ofer-flêdan *to overflood, overflow, inundate, cover with water :*—Seó eá Nilus oferflêtt (-flêd, MS. M.: -flêt, MSS. P. L.) eall ðæt Egiptisce land, and stent oferflêde hwîlon mônaþ hwîlon leng *the river Nile floods all the land of Egypt, and continues in a state of overflow sometimes a month, sometimes longer*, Lchdm. iii. 252, 23. [*Ger.* über-fluthen.]

ofer-flêde; *adj. Overflowing its banks.* v. preceding word. [Cf. *O. H. Ger.* ubar-fluatida *superfluitas* : *Ger.* über-fluth *overflowing* (of a river).]

ofer-fleón *to fly over :*—Ic oferfleó *supervolo*, Ælfc. Gr. 47; Som. 48, 46. [*In Beo.* Th. 5043; B. 2525 *it might be better to take* over *separate from* fleón :—Nelle ic beorges weard ofer fleón fôtes trem *I mean not to flee the dragon* [by retiring] *over even part of a foot's space.*]

ofer-flîtan *to overcome in a contest, to confute :*—Hê ðê æt sunde oferflât, hæfde mâre mægen, Beo. Th. 1039; B. 517. Ymb ðone tîman wæs gegaderad iii. hund biscepa and eahtatiéne hiene tô oferflîtanne (*to confute Arius*), Ors. 6, 30; Swt. 284, 1.

ofer-flôwan. I. *to overflow, cover with water :*—Seó eá ðæt land middeweard oferfleów mid fôtes þicce flôde, Ors. 1, 3; Swt. 32, 6. II. *to overflow, pass beyond bounds :*—Gôd gemet, geheápod and oferflôwende hig syllaþ on eówerne bearm, Lk. Skt. 6, 38.

ofer-flôwend; *adj. Superfluous :*—Îdel and oferflôwend byþ eal ðæt tôforan ðysum is, R. Ben. 91, 4.

ofer-flôwendlîce; *adv. Superfluously :*—Oferflôwenlîce *superflue*, Hpt. Gl. 527, 57.

ofer-flôwend-, flôwed-, flôwen-ness, e; *f. Superfluity, exuberance :*—Oferflôwenes *superfluitas*, Wrt. Voc. i. 17, 9. Oferflôwendnys *affluentia*, 41, 10. Eall hit byþ oferflôwednyss and idel tôforan ðisum, R. Ben. 90, 5. Mid heora ouerflôwednresse ne gedrîfen ða gebrôðru, 60, 17: 108, 5. Gif hit gelimpþ for oferflôwennysse metes (*ex superfluitate cibi*), L. Ecg. P. iii. 14; Th. ii. 200, 30. Hê ne dranc mid oferflôwendnysse, Homl. Th. i. 168, 12 : ii. 218, 30. Wê nellaþ habban ûs tô lîfes bricum, ac tô oferflôwednyssum, 540, 11.

ofer-flôwness, e; *f. Superfluity, overflowing :*—Oferflôwnes *superfluitas*, Wrt. Voc. ii. 149, 69. Oferflôuwnys (*superfluitas*) ðæs gecyndes, Bd. 1, 27; S. 494, 1. Of oferflôwnyss, S. 496, 37. His lîchoma mid oforoflôwnessum gefrætwod wæs, Blickl. Homl. 195, 12.

ofer-fôn *to seize :*—Oferfêng *obuncabat*, Wrt. Voc. ii. 62, 69. Þeódrîc ðone þegn oferfêng, hêht healdan ðone hererinc, Met. 1, 69. Ða genâman him ǽfest tô ða ealdormen ðara sacerda, and hine sylfne oferfêngon, Blickl. Homl. 177, 21. Hê hiene oferfôn hêt, and áhôn, Ors. 4, 4; Swt. 164, 32. Oferfangen *comprehensus*, Wrt. Voc. ii. 133, 8. [*O. H. Ger.* ubar-fâhan *rapere*.] v. ofer-feng.

ofer-froren *frozen over :*—Ðâ wæs Donua seó eá swîðe oferfroren, Ors. 4, 11; Swt. 208, 1: 1, 1; Swt. 21, 17.

ofer-full; *adj. Over-full :*—Oferfull *crapulatus*, Ps. Lamb. 77, 65. [*Goth.* ufar-fulls : *O. H. Ger.* ubar-foll *crapulatus*.]

ofer-fylgan, -fylgean; *p. de To pursue, persecute, attack :*—Gif ðæm môde mon tô ungemetlîce mid ðære þreápunga oferfylgþ *si mentem immoderata increpatio affligit*, Past. 21, 7; Swt. 167, 15. Ðonne ða iersigendan menn ôðrum monnum oferfylgeaþ tô ðon suîðe ðæt hit mon forberan ne mæg *cum ita iracundi alios impetunt, ut declinari non possint*, Past. 40, 5; Swt. 295, 10. Assael hine unwǽrlîce mid anwealde þreátode and him oferfylgde *hunc* (Abner) *cum Asael vi incautae praecipitationis impeteret*, Swt. 295, 14.

ofer-fyll, e; -fyllu (o); *indecl. f. Overfulness, repletion, surfeit, excess in eating* or *in drinking :*—Gýfernys *vel* oferfil *gastrimargia*, Wrt. Voc.

i. 27, 21. Oferfyl *aplestia*, ii. 10, 12. Ælc oferfyl fēt unhǣlo, Prov. Kmbl. 61. Nǣfre oferfyl ne fílige, forðī nis cristenum monnum nān þing swā wiðerweardlíc swā swā oferfyl, R. Ben. 63, 19-21. Seó oferfyll simle fēt unþeáwas, Bt. 31, 1 ; Fox 110, 27 : Blickl. Homl. 37, 14. Wið manegum ādlum ða ðe cumaþ of oferfyllo, Lchdm. ii. 178, 10 : 244, 4. Hit gelimpeþ of oferfylle . . . for oferfyllo (*ex crapula*), Bd. 1, 27 ; S. 496, 36-42. On oferfylle (oferfyllo, Lind. Rush.) *in crapula*, Lk. Skt. 21, 34 : Blickl. Homl. 159, 18. Ðú scealt druncen fleón, and ða oferfylle ealle forlǣtan, Dóm. L. 32, 75. Níwes wínes oferfelle *musti crapulam*, Hymn. Surt. 97, 18. Þurh oferfylla and mænigfealde synna heora eard hý forworhton, Wulfst. 166, 29. [*Goth.* ufar-fullei : *O. H. Ger.* ubarfullí *crapula.*]

ofer-fyllan, *to fill to overflowing*, (of eating) *to feed to excess :*—Oferfylled *crapulatus*, Wrt. Voc. ii. 136, 56. Hý beóþ oferfyllede óþ spíweþan, R. Ben. 136, 25. [*Goth.* ufar-fulljan.]

ofer-gǣgan *to transgress :*—Hwí ofergǣge gē Godes word *cur transgredimini verbum Domini?* Num. 14, 41. v. for-gǣgan *and next word.*

ofer-gǣgedness, e ; f. *Transgression :*—Wē sceolon mid geswince ús metes tilian for Adames ofergǣgednysse, Homl. Th. ii. 462, 12 : 486, 26 : Boutr. Scrd. 18, 13. v. for-gǣgedness *and preceding word.*

ofer-gán ; *p.* -eode ; *pp.* -gán. **I.** *to overspread :*—Seó lyft ofergǣþ ealne middaneard, Lchdm. iii. 272, 17. **II.** *to overrun (a country, as a victorious army does), to conquer :*—Se here fór tó Sandwíc, and swā ðanon tó Gipeswíc, and ðæt eall ofereode, Chr. 993 ; Erl. 132, 4. Wǣndon ðæt hē sceolde ðæt land ofergán, 1070 ; Erl. 207, 24. Hí hæfdon ða ofergán .i. Eást-Engle, and .ii. Eást-Sexe . . . , 1011 ; Erl. 144, 33. **III.** *to pass a point or limit :*—Ic ofergaa wall *transgrediar murum*, Ps. Surt. 17, 30. Hē ofergǣþ ðone súðran sunnstede, Lchdm. iii. 252, 14. Gemǣre ðú settest ðæt nā hí ofergáþ (*transgredientur*), Ps. Spl. 103, 10. Ofereode *excederit*, Wrt. Voc. ii. 30, 41. **III a.** *to pass a moral limit, to transgress :*—Forhwon leorneras ðíne ofergǣþ gesetnisse ðara ældra, Mt. Kmbl. Rush. 15, 2. **IV.** *to pass across, traverse, cross :*—Hē ofereode steáp stánhliðo, Beo. Th. 2820 ; B. 1408. Hí ða Reádan Sǣ ofereodon, Homl. Th. ii. 200, 27 : Beo. Th. 5911 ; B. 2959. **V.** *to pass, pass off or away, be over, come to an end :*—Hú hrædlíce se eorþlíca hlísa ofergǣþ, Past. 59, 1 ; Swt. 447, 30. Ðæt ílce yfel ofereode bútan geblóte *pestilentia sine ullis sacrificiorum satisfactionibus sedata est*, Ors. 5, 2 ; Swt. 218, 3. Ða geswinc ðe ofergán sculon *quod transeundo laboratur*, Past. 52, 5 ; Swt. 407, 31. **V a.** *impers. with gen. To be over (with anything) :*—Ðæs ofereode ðisses swā mæg *it is all over with that, so may it be with this, that trouble is over, so may this be*, Exon. Th. pp. 377-379 ; Deór. 7, etc. **VI.** *to come upon, attack* (of disease, sleep, etc.) :—Wæterseócnyss hine ofereode, Homl. Th. i. 86, 9. Hine slǣp ofereode, Andr. Kmbl. 1640 ; An. 821. v. ofer-gangan.

ofer-gangan. I. *to cross (a boundary) :*—Ic ofergange (*transgrediar*) weall, Ps. Spl. 17, 31. Heora ǣnig óðres ne dorste mearc ofergangan, Met. 20, 71. **II.** *to conquer :*—Gē feónda gehwone ofergangaþ, Cd. Th. 213, 33 ; Exod. 561 [cf. Orm. 10228 : To werenn hemm wiþþ wiþerrþeod þatt wollde hemm oferrganngenn]. **III.** *to pass, pass off, be over :*—Hié gebidon ðæt se ege ofergongen wæs, Ors. 4, 2 ; Swt. 160, 31. **IV.** *to come upon* (of sleep) :—Mec slǣp ofergongeþ, Exon. Th. 422, 23 ; Rä. 41, 10. [*Goth.* ufar-gaggan.] v. ofer-gán.

ofer-gapian *to neglect, disregard :*—Ne hē þurh ðone trúwan his sacerdhádes ofergapige (ofergumige, other MS.) his gehýrsumnysse *let not the priest through trust in his priesthood be careless of his obedience*, R. Ben. 112, 2. [Cf. *O. H. Ger.* geffida *consideratio.*]

ofer-geáre ; *adj. Old, superannuated :*—Gif wyrm ete ða teþ genim ofergeáre holenrinde, Lchdm. ii. 50, 14. [Cf. *Ger.* über-jährig *superannuated.*] Cf. þrí-geáre.

ofer-geatu, e ; f. *Oblivion :*—Ða his cwide weoldan on ofergeate hæbben (*would have it buried in oblivion*, cf. *O. H. Ger.* habe in âgezze *obliviscere*, Grff. iv. 279), Ps. Th. 128, 6. Cf. be-geatu.

ofer-gedrync, es ; *n. Excessive drinking* or *feasting :*—Hié hæfdon wiste and plegan and oforgedrync, Blickl. Homl. 99, 21. v. ofer-drync.

ofer-gedyre, es ; *n. A lintel :*—Smíton on ægðer gedyre and on ða ofergedyru *ponent super utrumque postem et in superliminaribus domorum*, Ex. 12, 7. v. ofer-dyre.

ofer-gemet, es ; *n. Excess :*—Suá oft suá wé úre hand dóþ tó úrum múþe for giéfernesse ofergemet (*per immoderatum usum*), Past. 43, 5 ; Swt. 313, 14. [Cf. *O. H. Ger.* ubar-gamez ; *adj. supervacuus.*]

ofer-genga, an ; *m. One who goes over* or *beyond :*—Gif hé biþ on .xi. nihta ealdne mónan se biþ landes ofergenga *if he is born on the eleventh of the month, he will be a traveller about the land*, Lchdm. iii. 158, 1 : 160, 30.

ofer-geong, es ; *m. A going across* ; transmigratio, Mt. Kmbl. p. 12, 13. Cf. forþ-geong.

ofer-geótan *to cover by pouring, to suffuse :*—Ðara deófla þeóstro hē oforgeát mid his dǣm scínendan leóhte *he overcame the darkness of the devils by pouring upon it his shining light*, Blickl. Homl. 85, 8. Dreórige hleór sealtum dropum ofergeótaþ *suffuse the mournful face with tears*,

Dóm. L. 4, 36. Ðæt scyp wearþ ofergoten (*operiretur*) mid ýðum, Mt. Kmbl. 8, 24. Mid swāte ofergoten, Glostr. Frag. 104, 17. Mid wōpe ofergoten, Ælfc. T. Grn. 18, 2.

ofer-geotol, -geotolian. v. ofer-gitol, -gitolian.

ofer-gesett *placed above (others) :*—On óðre wísan sint tó manianne ða underþióddan on óðre ða ofergesettan *aliter admonendi sunt subditi, atque aliter praelati*, Past. 28, 1 ; Swt. 189, 15, 23.

ofer-getimbran *to raise a building :*—On ðæm stāne hí ciricean ofergetimbredon *they raised a church on that rock*, Blickl. Homl. 205, 5.

ofer-geweorc, es ; *n.* **I.** *a superstructure :*—Ðæs heáhaltares ofergeweorc *cibborium*, Wrt. Voc. ii. 23, 15. **II.** *a tomb, mausoleum :*—Mētton ofergeweorke *depicto mausoleo*, Coll. Monast. Th. 32, 35. Gē sind gelíce gemēttum ofergeweorcum, Homl. Th. ii. 404, 17. v. ofer-weorc.

ofer-gewrit, es ; *n. A superscription, an inscription :*—Hwæs anlícnys, ys ðis and ofergewrit (*suprascribtio*), Mt. Kmbl. 22, 20 : Homl. Skt. i. 23, 475. Ofergewritum *epigrammatibus*, Wrt. Voc. ii. 33, 23.

ofer-gífre ; *adj. Over-greedy, gluttonous* ; gulae deditus, Past. 23 ; Swt. 177, 4 : 43 ; Swt. 308, 16.

ofer-gíman *to neglect, disregard :*—Gif hwā ðis ofergýme, R. Ben. 129, 9. Gif hē ǣdor dyde, oððe ofergímde, oððe forgeat, 71, 15. Ðæs git ofergýmdon Hælendes word, Cd. Th. 295, 14 ; Sat. 486. Cf. ofergumian.

ofer-gímness, e ; *f. Watching over, observation :*—Mið ofergēmnise *cum observatione*, Lk. Skt. Lind. 17, 20.

ofer-gitan *to forget, neglect :*—Ealle þeóda ða ðe ofergitaþ (*obliviscuntur*) God, Ps. Spl. 9, 18. Ic ofergeat (*oblitus sum*) etan, 101, 5. Sum wýf ofergeat hyre cyld slǣpende, Shrn. 150, 30. Hí ofergēton (-geáton, MS. A.) (*obliti sunt*) ðæt hí hláfas ne nāmon, Mk. Skt. 8, 14. Hié ofergeáton Godes dómas, Cd. Th. 155, 32 ; Gen. 2581. Spec . . . ðæt hié ofergieton (sýn ofergytende, MS. B.) ðisse sǣwe ege, St. And. 8. 15. Ne ofergit ðú þearfan, Ps. Spl. 9 second, 14. Ofergyt, 73, 24. Oferget, Ps. Surt. 73, 23. Nylle ðú ofergiten *noli oblivisci*, Ps. Spl. 102, 2. Ofergeotan, Nar. 45, 7. Wǣre ðú ofergeotende mínre bysne, Bd. 2, 6 ; S. 508, 17. Ān nis of ðām ofergyten, Lk. Skt. 12, 6.

ofer-gitness, e ; *f. Forgetfulness, oblivion :*—On ðam lande ðe ofergytnes on eardige (this seems to correspond to *in terra oblivionis*, v. 12), Ps. Th. 87, 11. On ofergetnisse *in oblivione*, Lk. Skt. Rush. 12, 6.

ofer-gitol, -geotol ; *adj. Forgetful, oblivious :*—Ne eom ic ofergitol (-gittul, Ps. Th.) *non sum oblitus*, Ps. Spl. 118, 61 : 9, 19 : 9 second, 13. Ofergiṭṭol, Ps. Th. 118, 41. Worda ðínra ofergittul, 118, 15. Ofergyttol, 118, 43. Ne sý ofergyttol ac gemyndig, R. Ben. 24, 1. Nā ofergeotol ðara gebeda his þearfena, Ps. Th. 9, 12. Ofergeottul, 102, 2. Ofergeatul *obliviosus*, Rtl. 29, 7. Ofergeotole wé ne sind *obliti non sumus*, Ps. Surt. 43, 18. Ofergeotulæ (-geotole, Ps. Th.), 43, 21. Ofergeotole, Mt. Kmbl. Lind. 16, 5.

ofer-gitolian ; *p.* ode *To forget, be forgetful of :*—Nó ofergeoteliu word ðín *non obliviscar sermones tuos*, Ps. Surt. 118, 16. Ofergeotulas ðú *oblivisceris*, 12, 1. Ofergeoteliaþ *obliviscimini*, 49, 22. Alle þeóde ða ðe ofergeoteliaþ Dryhten, 9, 18. Ne ofergeotela ðú, 9, 33. Ofergeotelien *obliviscantur*, 58, 12.

ofer-gitolness, e ; *f. Forgetfulness, oblivion :*—Ofergitolnys (-geotulnis, Ps. Surt.) *oblivio*, Ps. Spl. C. T. 9, 19. Wið ða ádle ðe man litargum hāteþ, ðæt ys on úre geþeóde ofergytulnys (-gittolnes, MS. H.), Lchdm. i. 200, 8. In eorþan ofergytolnysse *in terra oblivionis*, Ps. Spl. 87, 13. Ða unþeáwas oft ábisegien ðæt mód mid ofergiotulnesse, Bt. 35, 1 ; Fox 154, 32. Ic eom myd earmlícre ofergiotolnesse ofseten, Shrn. 198, 21. On ofergeotolnisse, Blickl. Homl. 103, 16. Ofergeottolnisse *oblivionem*, Rtl. 61, 14. Ofergiottulnisse *ignorantia*, 167, 31.

ofer-glenged ; *part. Over-ornamented, too much adorned :*—Ne mót nān preóst beón on his girlum tó ranc, ne mid golde oferglæncged, L. Ælfc. P. 49 ; Th. ii. 386, 10.

ofer-grǣdig ; *adj. Over-greedy, too covetous :*—Menn beóþ ofer grǣdige woruldgestreóna, Wulfst. 81, 13.

ofer-gumian ; *p.* ode *To neglect, be careless about :*—Ne hē ofergumige ða hýrsumnesse ðæs hálgan regoles, R. Ben. 113, 2. [Cf. *Icel.* guma at einu *to take heed to a thing : O. Sax.* far-gumón *to neglect.*] v. ofer-gíman.

ofer-gyldan *to cover* or *ornament with gold :*—Ic ofergylde *auro*, Ælfc. Gr. 36 ; Som. 38, 39. Ealle ða grǣftas gē ofergyldaþ mid crǣfte, Homl. Skt. i. 8, 61. On ofergildum hrægle *in vestitu deaurato*, Ps. Lamb. 44, 10 : Homl. Th. ii. 586, 16. ii. sylure candelsticcan and ii. ouergylde, Cod. Dip. Kmbl. vi. 101, 26. Ða ofergyldan saglas sceolden stician on ðǣm gyldnum hringum, Past. 22 ; Swt. 171, 22.

ofer-gylden ; *adj. Gilded, covered with gold :*—Gif hé begytaþ ðæt hē hæbbe byrne and helm and ofergyldene (cf. golde fæted, ll. 8-9) sweord, L. Wg. 10 ; Th. i. 188, 21.

ofer-gyrd *overgirt :*—Ofergyrdum *recincta*, Germ. 394, 236.

ofer-habban (?) *to command, govern :*—Hý mōstan ðām læppan friþ gebicgean ðe hý under cyngces hand oferhæfdon [geweald ofer hæfdon (?)], L. Eth. ii. 1 ; Th. i. 284, 14.

ofer-hacele, an; *f. A cope, hood;* cappa, L. Ecg. C. 10, note; Th. ii. 140, 22. [Cf. *Icel.* yfir-hökull *a surplice.*]

ofer-heáfod; *adv. Generally, in every case:*—Ælc man oferheáfod sceolde cennan his gebyrde and his áre on ðære byrig ðe hé tó gehýrde, Homl. Th. i. 30, 4. [*Ger.* über-haupt.]

ofer-heáh; *adj. Excessively high:*—Æsc byþ oferheáh, Runic pm. Kmbl. 344, 23; Rún. 26.

ofer-healdan *to hold over, delay to do, neglect:*—Gif se gereáfa ðis oferheald, gebéte .xxx. scitt., L. Ath. i. prm.; Th. i. 198, 11. Cf. ofer-hebban.

ofer-healfheáfod *the upper half of the head:*—Forheáfod *anciput,* æfteweard heáfod *occiput,* oferhealfheáfod *sinciput,* Wrt. Voc. i. 42, 42–44.

ofer-hebban *to pass by, neglect, omit:*—Gif hit (*the holding a gemót*) hwá oferhebbe (-habbe, MS. B.) béte swá wé ǽr cwǽdon, L. Ed. 11; Th. i. 164, 23. Gif hé áht ðæs oferhæbbe ðe on úrum gewritum stent, L. Ath. v. 8, 5; Th. i. 236, 33. Ic wát ðæt ic his sceal fela oferhebban *ego cogor fateri me praeterire plurima,* Ors. 1, 8; Swt. 42, 1. Hit þencþ fela gódra weorca tó wyrcanne, gif hé worldáre hæbbe, and wile hit oferhebban, siððan hé hié hæfþ, Past. 9; Swt. 55, 16. [For ever hem (*the poor*) thou overhaf, Mapes 341, 1: *O. H. Ger.* ubar-hevan *praeterire, transire.*] Cf. ofer-healdan.

ofer-helian *to cover over, conceal:*—Neahte þeóstru ðú oferhelast (*detegis*), Hymn. Surt. 12, 12. Oferhelaþ *contegit,* 23, 11. Se ceác oferhelede ða oxan, Past. 16, 5; Swt. 105, 4. Gif hwá pytt ádelfe and hine ne oferhelie (*operuerit*), Ex. 21, 33. Tó oferhelianne, Glostr. Frag. 102, 2. Beón oferheled *obtegi,* Germ. 389, 22. Nis nán þing oferheled (*opertum*) ðe ne beó unheled, Lk. Skt. 12, 2.

ofer-helmian *to overshadow:*—Wudu wæter oferhelmaþ, Beo. Th. 2733; B. 1364.

ofer-heortness, e; *f. Excessive feeling:*—Mid oferheortnesse hé him wæs wánigende ǽgðer ge hys ágene heardsǽlþa ge ealles ðæs folces *with bursting heart he was bewailing both his own and the people's hard fortune,* Ors. 4, 5; Swt. 166, 20.

ofer-hergian *to ravage:*—Ceólwulf oferhergeade (-ode, MS. E.) Cantware, Chr. 796; Erl. 58, 10: 865; Erl. 70, 34. Eádweard oferhergade eall hira land, 905; Erl. 98, 20: 933; Erl. 110, 28. Hǽþne men oferhergeadon (-odon, MS. E.) Sceápíge, 832; Erl. 64, 18. Ða Gotan eów hwón oferhergedon, Ors. 1, 10; Swt. 48, 20. Heora land tó bismere oferhergodan, Blickl. Homl. 201, 23.

ofer-hídig, -higd, -hige. v. ofer-hygdig, -hygd, -hyge.

ofer-higian *to overreach* (?):—Sinc eáþe mæg gold on grunde gumcynnes gehwone oferhígian hýde se ðe wylle *easily may treasure, gold in the ground, overreach every man* (i. e. *make the effort at concealment vain*), *hide it who will,* Beo. Th. 5525; B. 2766.

ofer-híran. I. *not to listen to, to disregard, disobey:*—Ðé ealle gesceafta heórsumiaþ . . . bútan men ánum, se ðe oferheórþ, Bt. 4; Fox 8, 10. Swá weorþlícne sige hæfde swá hé ǽr unweorþlíce ðara goda biscepum oferhírde (*he disregarded the prohibition of the augurs*), Ors. 3, 10; Swt. 140, 4. Hié þurh his láre oferhiérdon ðǽm godum, 4, 12; Swt. 210, 2. **II.** *to overhear, hear:*—Swá ic mid mínum eárum oferhýrde, L. O. 8; Th. i. 180, 29: L. C. S. 23; Th. i. 388, 24. Se opeling (*Phalaris*) ǽgðer hæfde, ge his plegan ge his gewill, ðonne hé ðara manna (*those shut up in the brazen bull*) tintrego oferhiérde, Ors. 1, 12; Swt. 54, 28. Gé sylfe swutele gesáwon, and eác oferhýrdan ða bletsunge, Wulfst. 176, 4.

ofer-híre; *adj. Disobedient, regardless:*—Gif preóst on his scriftscíre ǽnigne man wíte Gode oferhýre, oððe on heáfodleahtrum yfele befeallene, L. Edg. C. 6; Th. ii. 244, 22.

ofer-hírness, e; *f. Disobedience, disregard, neglect, contempt:*—Ungelimp mid oferhýrnysse Godes beboda geearnod, L. Edg. S. 1; Th. i. 270, 12. But it occurs chiefly as a legal term *the disregard of an authoritative enactment* or *the fine for such disregard,* amounting to 120 shillings. Some of the offences to which it applies may be seen from the following passages:—Gif hwá bútan porte ceápige, ðonne sý hé cyninges oferhýrnesse scyldig, L. Ed. 1; Th. i. 158, 14. Ðæt se wǽre, ðe rihtes wyrnde, scyldig æt þriddan cyrre cyninges oferhýrnesse ðæt is .cxx. scitt., 2; Th. i. 160, 16. Ne underfó nán man óðres mannes man bútan ðæs leáfe ðe hé ǽr fyligde. Gif hit hwá dó, béte míne oferhýrnesse, 10; Th. i. 164, 18. Gif hwá gemót forsitte þríwa, gilde ðæs cynges oferhýrnesse . . . Gif hé nylle ða oferhýrnesse syllan, ðonne rídan ða yldestan men . . . Gif hwá nylle rídan mid his geféran, gilde cynges oferhýrnesse, L. Ath. i. 20; Th. i. 208, 26–210, 1. Gif hwá hreám gehýre and hine forsitte, gylde ðæs cynges oferhýrnysse, L. C. S. 29; Th. i. 392, 18. Ne quis pecuniam puram et recte appendentem sonet, monetetur in quocunque portu monetetur, in regno meo, super overhyrnessam meam, L. Eth. iv. 6; Th. i. 302, 15. Gé (*geréfan*) híraþ, cwæþ se cyngc, hwæt gé gelæstan sculan be (*on pain of incurring*) mínre oferhýrnysse, L. Ath. i. prm.; Th. i. 196, 15. See Schmid. A. S. Gesetz. s. v. [Cf. *Goth.* ufar-hauseins *disregard, disobedience.*]

ofer-hlæstan *to overload:*—Mid ðære herehýþe Rómane oferhlæstan heora scipa, Ors. 4, 6; Swt. 176, 18, 27. Hié (*the ships*) mon ne mehte mid monnum oferhlæstan, 5, 13; Swt. 246, 11.

ofer-hleápan *to overleap, pass by jumping:*—Ic oferhleápe *transitio,* Wrt. Voc. i. 60, 40. *Saltus lunae,* ðæt is, ðæs mónan hlýp, for ðan ðe hé oferhlýþ ǽnne dæg, Lchdm. iii. 264, 24. Ðæt hors slóg on ðam wege oferhleóp, Bd. 5, 6; S. 619, 17. All eorþlíc þing wæs oferhleápende (*transiliens*), 2, 7; S. 509, 14. v. next word.

ofer-hleápend, es; *m. One who overleaps;* transilitor, Wrt. Voc. i. 60, 41.

ofer-hleóðor; *adj. Not hearing, inattentive to sound:*—Se ðe ǽrest eáran worhte hú se oferhleóður ǽfre wurde *qui plantavit aurem, non audiet?* Ps. Th. 93, 9.

ofer-hleóðrian. I. *to outsound, exceed in sound:*—Ðeáh ánra gehwylc hæbbe gyldene býman, and ealra býmena gehwylc hæbbe .xii. hleóðor, and hleóðra gehwylc sý heofone heárre and helle deópre, ðonne ðæs hálgan cantices se gyldena organ hé hý ealle oferhleóðraþ, and ealle ða óðre hé ádýfeþ, Salm. Kmbl. p. 152, 12. **II.** *to exceed* (?):—Ne frign ðú unc nóhtes má for ðon wit habbaþ oferhleóðred [-leóred (?)] ðæt gemǽre uncres leóhtes *cave ne nos ulterius scisciteris jam excede terminos luci nostri,* Nar. 32, 7.

ofer-hlifian. I. *to tower above, rise high above:*—Sóna swá seó sunne sealte streámas heá oferhlifaþ, Exon. Th. 206, 3; Ph. 121. **II.** *to exceed, surpass, excel:*—Ofer[h]lyfaþ *praecellat, superemineat,* Hpt. Gl. 413, 48. Hé óðre oferhlifaþ *ceteris praeeminet,* Past. 17, 3; Swt. 111, 1. Iohannes ealle heáhfæderas and Godes wítgan oferhlifaþ, Shrn. 95, 10. **III.** *to tower over in a threatening manner:*—Oferhlifode ege heora ofer hig *incubuit timor eorum super eos,* Ps. Spl. M. 104, 36. Ofer[h]lifiende *minaci,* Wrt. Voc. ii. 85, 47.

ofer-hlifung, e; *f. Eminence, sublimity, excellence:*—Oferhlifung *eminentia, sublimitas, celsitudo,* Wrt. Voc. ii. 143, 37. Oferhlifinge *excellentiae,* Germ. 393, 52.

ofer-hlúd; *adj. Over-loud, noisy, clamorous:*—Oferhlúd *clamosa,* Wrt. Voc. ii. 131, 61. v. next word.

ofer-hlýde; *adj. Over-loud, noisy:*—Hé ne sý oferhlýde on stefne, R. Ben. 30, 14. v. preceding word.

ofer-hlýp, es; *m. A leap across or over, a bound:*—Ðes *saltus,* ðæt is ðes mónan oferhlýp, Anglia viii. 308, 24. For ðæs mónan oferhlýpe *id est, propter saltum,* 316, 43. [Cf. *Icel.* yfir-hlaup.]

ofer-hlýttrian *to clarify, strain:*—Ic oferhlýttrige *eliquo,* Ælfc. Gr. 37; Som. 39, 42.

ofer-hoga, an; *m. One who despises, a contemptuous, proud person:*—Se biþ Godes oferhoga ðe Godes bodan oferhogiaþ, L. I. P. 5; Th. ii. 308, 31. Hér sýn on earde oferhogan godcundra rihtlaga, Wulfst. 164, 12. Oferhogan *superbi,* Ps. Surt. 118, 122: 139, 6. Oferhogum *superbis,* 122, 4. Oferhogan *superbos,* ii. p. 200, 16.

ofer-hogian *to despise, contemn, scorn, disdain:*—Moyses symle ða nyrugde ðe God oferhogodan. Se ðe Godes bebod oferhogaþ, hé biþ on hǽðenra onlícnesse, Blickl. Homl. 49, 12–13. Sum fearhrýðer ðæs óðres ceápes geférscipe oferhogode, 199, 4. Hé ǽlce unsíuernysse oferhogode Chr. 1067; Erl. 204, 36. Ða oferhogode hé ðæt hé áðer dyde, Ors. 6, 34; Swt. 290, 21: Beo. Th. 4679; B. 2345. Hié ealle worlde weán oforhogodan, Blickl. Homl. 119, 16, 20. Oferhoga hí, and ádríf hí fram ðé, Bt. 7, 2; Fox 18, 8. Warniaþ ðæt gé ne oferhogian ǽnne of ðysum lytlingum, Mt. Kmbl. 18, 10. Ða gýmeleásan and ða oferhogiendan hé sceal mid wordum þreágan, R. Ben. 13. 15. v. preceding and next words and ofer-hycgan.

ofer-hogiend, es; *m. A despiser, contemner:*—Gyf hwylc bróðor ongyten biþ his yldrena geboda oferhogiend, R. Ben. 48, 6.

ofer-holt *a forest of spears which rise over the heads of those who bear them* (?):—Hié gesáwon fyrd Faraonis forþ ongangan oferholt wegan eóred líxan *they* (*the Israelites*) *saw Pharaoh's host advance, saw a forest of spears move* (*or saw them bearing a forest of spears*), *saw the band glitter,* Cd. Th. 187, 27; Exod. 157.

ofer-hragan *to come in storms* (?):—Wǽtum hé oferhrægeþ, gebryceþ burga geatu *it* (*snow*) *comes in damp storms on cities' gates, and breaks them,* Salm. Kmbl. 612; Sal. 305. [Cf. *Icel.* hragla *to sleet;* hregg *a storm.*]

ofer-hréfan *to roof over, cover with a roof, cover:*—Ðé oferhréf ufan mid hwítle *cover yourself over from above with a cloak,* Lchdm. ii. 76, 22. Porticas ealle swíðe fægere oferhrýfde, Blickl. Homl. 125, 25.

ofer-hréran *to overthrow:*—Oferhrýred *dirute,* Wrt. Voc. ii. 26, 13. Oferhrérede *obrutos,* 62, 71.

ofer-hrops *voracity:*—Ic brúce ðisum mettum mid sýfernysse swá swá dafnaþ munuce næs mid oferhropse *vescor his cibis cum sobrietate, sicut decet monacho, non cum voracitate,* Coll. Monast. Th. 35, 5.

ofer-hrýfan, -hrýred. v. ofer-hréfan, -hréran.

ofer-hycgan *to despise, contemn, disdain, scorn:*—Gif hé ðis (*lying at the feet of his superior*) oferhicge and hit dón nelle, R. Ben. 131, 7. Ðonne se mon oferhygþ (Hatt. MS. oferhýþ) ðæt hé bió gelíc óðrum monnum *dum homo hominibus esse similis dedignatur,* Past. 17, 4; Swt. 112, 3. Wé hine mid swá micle máran unryhte oferhycgeaþ swá hé læs forhogaþ ðæt hé ús ðonne giet tó him spane, siððan wé hiene oferhycggeaþ *tanto graviori improbitate contemnitur, quanto contemtus adhuc vocare non dedignatur,* 52, 4; Swt. 407, 17–19. Ðeáh hí hine oferhogden, ne for-

hogde hé hí nô, Swt. 405, 31. Ða lytegan sint tô manianne ðæt hí oferhycggen (-hycgen, Cott. MSS.) ðæt hié wieton, 30, 1; Swt. 203, 7. Oferhige hí and ádríf hí fram ðé, Bt. 7, 2; Fox 18, 8 note. Utan oferhycgan helm ðone miclan, Cd. Th. 280, 7; Sat. 252: 283, 15; Sat. 305. [*O. H. Ger.* ubar-hugjan *contemnere, aspernere :* cf. *Goth.* ufar-hugjan *to be puffed up.*] v. ofer-hogian.

ofer-hýd, -hýdig. v. ofer-hygd, -hygdig.

ofer-hygd, -hýd, e; *f.: es; n.:* -hygdu, -hýdu (o); *indecl. f.* [the plural is used with singular meaning, cf. ofer-méde, -méttu]. I. in a bad sense, *pride, arrogance :*—Hæfde hig ofyrhigd (-hýd, MS. T.) *tenuit eos superbia,* Ps. Spl. 72, 6. Oferhigd *supercilio,* Wrt. Voc. ii. 76, 20. Oferhygd, Cd. Th. 21, 22; Gen. 328. Wlenco, oferhýd, 258, 21; Dan. 679. Ðæs oferhýdes ord, 272, 3; Sat. 114. Se is kyning ofer eall ða bearn oferhygde (-hýde, Cott. MSS.) *ipse est rex super universos filios superbiae,* Past. 17, 4; Swt. 111, 22. In oferhygde *in superbia,* Ps. Surt. 16, 10: 58, 13. Hû mycel yfel ðé gelamp for ðínre gítsunga and oferhýdo and for ðínum ídlan gilpe, Blickl. Homl. 31, 14. Hí druncennesse and oferhýdo wæron heora swiran underþeóddende *ebrietati, animositati, sua colla subdentes,* Bd. 1, 14; S. 482, 26. Sum on oferhygdo þrinteþ, Exon. Th. 314, 33; Mód. 23. Nô wé oferhygdu (*or pl.?*) ánes monnes máran fundon, 118, 15; Gú. 240. Ðæt heofenlíce ríce ðæt ða ærestan men forworhtan þurh heora gífernesse and oferhygde, 25, 1. Se dóeþ oferhygde *qui facit superbiam,* Ps. Surt. 100, 7. Ða dóeþ oferhygd, 30, 24. Ðás þing wé sculon forgan oferhýd gýtsunge . . . *ab his debemus nos abstinere, a superbia, et avaritia . . . ,* L. Ecg. P. iv. 64; Th. ii. 224, 28. Næfde hé on him nâðer ne yrre ne oferhýd, Bd. 3, 17; S. 545, 8. Him oninnan oferhygda dǽl weaxeþ, Beo. Th. 3485; B. 1740. Oferhýda, 3525; B. 1760. Ða setl ðe deófol for his ofer-hygdum of áworpen wæs, Blickl. Homl. 121, 35. For oferhygdum, 156, 13: Cd. Th. 268, 4; Sat. 50: 269, 6; Sat. 69. Ne gedafenaþ ðé ðæt ðú andsware mid oferhygdum séce, Andr. Kmbl. 638; An. 319. Hwæt is wuldor ðín ðe ðú oferhygdum upp áræredest, 2637; An. 1320. Þurh oferhygda, Exon. Th. 316, 23; Mód. 53. Hé ofer-hýda ágan wolde *he would give way to pride,* Cd. Th. 287, 20; Sat. 370. II. in a good sense, *honourable pride* (?), *high spirit :*—Gif ðú gesáwe sumne swíðe wísne man ðe hæfde swíðe góda oferhýda and wǽre ðeáh swíðe earm hwæðer ðú woldest cweþan ðæt hé wǽre un-wyrþe anwealdes and weorþscipes *si quem sapientia praeditum videres, num posses eum vel reverentia, vel ea, qua praeditus est, sapientia, non dignum putare?* Bt. 27, 2; Fox 96, 24. [*O. H. Ger.* ubar-huht, -hucti *superbia.*]

ofer-hygd; *adj. Proud :*—Oferhygdum égan *superbo oculo,* Ps. Surt. 100, 5. Oferhygde (sic MS.) *superbi,* Ps. Th. 139, 5. Ða oferhygdan *superbi,* Ps. Surt. 118, 78: 118, 21. Ágyld edleán oferhygdum *redde retributionem superbis,* Ps. Spl. C. 93, 2. Tóstrægd oferhygd *dispersit superbos,* Lk. Skt. Rush. 1, 51.

ofer-hygdig, -hýdig, es; *n. Pride :*—Ðonne hí oferhýdig up áhófan and him wôhgodu worhtan and grófun *in sculptilibus suis emulati sunt eum,* Ps. Th. 77, 58.

ofer-hygdig, -hýdig; *adj. Proud, arrogant, haughty :*—Hé eode tô reordum mid tôcumendum mannum. Ða tælde hine án oferhýdig bisceop for ðon, Shrn. 129, 28. Ðone oferhygdgan *superbum,* Ps. Surt. 88, 11. Ða wæron hí æfter æþelborennysse oferhýdega, Homl. Th. ii. 174, 8. Ða ofer-hýdegan, Ps. Th. 118, 78. Égan oferhygdigra *oculos superborum,* Ps. Surt. 17, 28: 118, 69. Ofyrhýdigra, Ps. Spl. 118, 69. Ðú eallum oferhýdigum eáþmódnesse forgifest, Blickl. Homl. 141, 12. Oferhýdegum eágum *superbo oculo,* Ps. Th. 100, 5. Fyll ða oferhýdigan, 73, 22. Ða oferhýgdego *superbos,* Lk. Skt. Lind. 1, 51. [*O. H. Ger.* ubar-huctig *superbus.*]

ofer-hygdigian *to be proud :*—Ðonne oferhygdgaþ se árleása *dum superbit impius,* Ps. Surt. 9, 23.

ofer-hyge (?), es; *m. Pride, arrogance :*—Ðú mé oferhige (*or ofer hige?* mé ofer corresponding to *super me* in the Latin) on ealle gelǽddest *omnes elationes tuas super me induxisti,* Ps. Th. 87, 7.

ofer-hylmend; *m. One who conceals, who does not act openly :*—Ic oferhylmend ealle getealde ða on eorþan yfele wǽron *praevaricantes reputavi omnes peccatores terrae,* Ps. Th. 118, 119. [Cf. *Icel.* hylma yfir *to hide, conceal* (as a law phrase); yfir-hylma *to hide, cloak;* yfir-hylming *a hiding, cloaking.*]

ofer-hýran, -hýre, -hýrness. v. ofer-híran, -híre, -hírness.

ofer-hyrned; *adj. Having horns above :*—Úr býþ oferhyrned, Runic pm. Kmbl. 339, 8; Rún. 2.

oferian; *p.* ode *To exalt :*—Geferode *sublimati, exaltati,* Hpt. Gl. 428, 47.

ofer-ild, e; *f. Very great age :*—Him se deáþ geneálǽcþ for ðære oferylde, Wulfst. 147, 27.

ofering; *f. Superfluity :*—Gif ðú ofer gemet itst oððe drincst oððe cláþa ðé má on hæfst ðonne ðú þurfe seó ofering ðé wurþ tô sáre *cujus satietatem si superfluis urgere velis, quod infuderes fiet noxium,* Bt. 14, 1; Fox 42, 16. Hé wilnigen mid oferinge hiora gítsunga gefyllan *qui abundantiam suam ambitus superfluitate metiantur,* 14, 2; Fox 44, 14.

ofer-irnan. I. *to pass by running, cross :*—Ða hwíle ðe se môna ðære sceade ord oferyrnþ *while the moon is crossing the point of the shadow,* Lchdm. iii. 240, 26. II. *to run over, go over a subject :*—Nû wille wé eft oferyrnan ða ylcan godspellícan endebyrdnysse, Homl. Th. i. 104, 7. Wé wyllaþ scortlíce oferyrnan ða dígelystan word, 202, 29. III. *to come upon with violence, overwhelm, to come upon with surprise :*—Seó sǽ oferarn Pharao and ealle his crǽtu, ii. 194, 27. Mé slǽp oferarn *cum mihi somnus obrepsisset,* Bd. 5, 9; S. 622, 33.

ofer-lád, e; *f. A carrying across, translation :*—Oferlád *translationem,* Rtl. 62, 19. v. lád, III.

ofer-lǽdan *to oppress :*—Ða wæs se munt mid myceluat brôgan eall oferlǽded; and unhiérlíc storm of ðæm munte ástág, Blickl. Homl. 203, 7. [Shal neither kynge ne knyȝte, constable ne meire ouerlede þe comune, Piers. P. 3, 314: *Prompt. Parv.* ovyrledyu *opprimo;* ovyrledare *oppressor ;* ovyrledynge *oppressio.*]

ofer-lǽg, es; *n. A cloak :*—Oberlagu *amfibula* (amfibulum *birrum villosum,* Isidore), Txts. 111, 1.

ofer-leóf; *adj. Exceedingly dear :*—Éðel býþ oferleóf ǽghwylcum men, Runic pm. Kmbl. 344, 3; Rún. 23.

ofer-leóran. I. *to pass, pass away, pass by :*—Hé oferlióræs (-liórap, Rush.) from deáþe in life *transiet a morte in vitam,* Jn. Skt. Lind. 5, 24. Oferleóraþ *transeant,* p. 4, 10. Oferhlióras *transibunt,* Mk. Skt. Lind. 13, 31. Ðætte oferleórade (*transiret*) ðió tíd, 14, 35. Tíd ðætte hé oferliórde of ðissum middengeorde, Jn. Skt. Rush. 13, 1. Oferleórdun *transierunt,* Ps. Surt. 118, 136. Oferleór *transfer,* Lk. Skt. Lind. 22, 42. Oferlióra *transire,* Mt. Kmbl. Lind. 26, 42. II. *to pass moral bounds, deviate from right, transgress :*—Hæftas heársume ðæs hálgan word lyt oferleórdun, Exon. Th. 145, 21; Gú. 698. Oferliórende *praevaricantes,* Ps. Surt. 118, 119. Ofyrleórynde, -liórende, Ps. Spl. C. T. 118, 119. v. next word.

ofer-leórness, e; *f. Deviation from right, transgression :*—Dônde oferleórnisse *praevaricationes,* Ps. Surt. 100, 3.

ofer-libban *to outlive, survive :*—Wes ðet lond becueden his brôðar, gif hé Cyneþrýðe oferlifde, Chart. Th. 465, 19. Láf oððe oferlibbende *superstes,* Gr. 9, 26; Som. 11, 7. [*O. H. Ger.* ubar-lebên.]

oferlíce; *adj. Excessively :*—Hí mid heora synnum swá oferlíce swýðe God gegræmedon, ðæt hé lét Engla here heora eard gewinnan, Wulfst. 166, 18: 83, 14. [Cf. *Icel.* ofr-ligr *excessive.*]

ofer-líhtan *to outshine :*—Seó sunne oferlíht ealle óðre steorran and geþiústraþ mid hire leóhte, Bt. 9, tit.; Fox xii. 2.

ofer-líðan *to cross* (water), *sail across :*—Ástígende on scipe oferlád (*transfretavit*) ðone sǽe, Mt. Kmbl. Rush. 9, 1: Shrn. 88, 28: Cd. Th. 200, 26; Exod. 362. Oferlíðan *transire, transfretare,* Hpt. Gl. 492, 50. [*Goth.* ufar-leiþan.]

ofer-lufu, e, an; *f. Excessive love :*—Seó oferlufu eorþan gestreóna, Wulfst. 149, 4: 263, 24.

ofer-mǽcga, an; *m. A man superior to others, an illustrious person :*—Ofermæcga spræc dýre Dryhtnes þegn (*the angel sent to save Guthlac*), Exon. Th. 143, 21; Gu. 664. [Cf. *Icel.* ofr-menni *a mighty champion.*]

ofer-mægen, es; *n. Superior or overwhelming force :*—Wið ofer-mægenes egsan, Cd. Th. 127, 27; Gen. 2117. Hé hæfde wígena tô lyt wið ofermægene, Elen. Kmbl. 128; El. 64. Hyne Hetware hilde ge-hnǽgdon mid ofermægene, Beo. Th. 5827; B. 2917. Forst and snáw mid ofermægene eorþan þeccaþ, Exon. Th. 215, 6; Ph. 249. Him on swaðe fylgeþ A ofermægene, Salm. Kmbl. 187; Sal. 93. [Cf. *O. H. Ger.* ubar-meginôn *praevalere: Ger.* über-macht: *Icel.* ofr-efli *overwhelming force.*]

ofer-mǽned (?) *made too common* (?), *trite :*—Ofermǽned *contrita,* Wrt. Voc. ii. 19, 43: 92, 37.

ofer-mǽstan *to over-fatten :*—Swá ðæt úre líchama ne wurþe ofermǽst tô ídelum lustum, Bd. Whelc. 228, 25.

ofer-mǽte; *adj. Beyond measure, excessive, immoderate, immense :*—Ofermǽte *insolens,* Hpt. Gl. 526, 10. Moyses behelede ða ofermǽtan bierhto his ondwlitan, Past. 63; Swt. 459, 19. God hyra ofermǽtan ofermétto genyðerode, Ors. 1, 7; Swt. 38, 27. Hé hét ða ofermǽtan brycge mid stáne ofer gewyrcan, 2, 5; Swt. 84, 3. Æt ðám ofermǽtum wæterum *de multitudine aquarum,* Ps. Th. 17, 17. Ýða ofermǽta, Exon. Th. 53, 23; Cri. 855. [Cf. *Icel.* ofr-máta *excessively: Ger.* über-mässig.]

ofermǽt-lic; *adj. Immense :*—Ðonne swá ofermǽtlícu rícu onstyrede wǽron *ubi tot et talia regna mutata sunt,* Ors. 1, 12; Swt. 52, 10.

ofer-mǽtu (o); *indecl. f. Excess, presumption :*—Ádríf fram mé dysig and ofermǽto and sile mé wísdom, Shrn. 169, 16.

ofer-mádum, es; *m. A very valuable treasure, a treasure of surpassing worth,* Beo. Th. 5979; B. 2993.

ofer-méde, es; *n.:* -médu; *f.* [the plural form is used with singular meaning, cf. ofer-hygd, -méttu] *Pride :*—His ofermédu is fruma úres for-lores, Past. 41; Swt. 301, 8. Ofermédes *elationis,* Hpt. Gl. 433, 31. His engyl ongan oferméde micel áhebban, Cd. Th. 19, 19; Gen. 293. Ðæt hié ne ástigan on ofermédu, Blickl. Homl. 185, 14. Se ðe on ofermédum leofaþ, Exon. Th. 317, 33; Mód. 75. [*O. H. Ger.* ubar-muoti *superbia, elatio, animositas.*] v. ofer-mód, -méttu.

ofer-méde; *adj. Proud, arrogant, presumptuous :*—Cyning gefeaht wið ðone ofermédan (-môdigan, MS. E.) aldorman, Chr. 750; Erl. 48, 10. [*O. H. Ger.* ubar-muoti *superbus.*]

ofer-médla, an ; *m.Pride:*—Eahta syndan heáfodlíce synna ... eahtoþa is ofermédla, L. E. I. 31 ; Th. ii. 428, 8. Sóna swá ic mínes ofermédlan geswíce, 36 ; Th. ii. 436, 1. Gif hé on ofermédlan and on óðrum unþeáwum his líf lyfaþ, 32 ; Th. ii. 428, 33 : Cd. Th. 257, 14 ; Dan. 657.

ofer-médu. v. ofer-méde.

ofer-mete, es ; *m. Food in excess, a feast where food is in excess:*—Se ofermete ne befæst ús næfre Gode *esca nos non commendat Deo,* Past. 43, 9 ; Swt. 316, 19. Ofermettas *commessationes,* Bd. 4, 25 ; S. 601, 13 note.

ofer-méttu (o) ; *indecl. in sing. ; but declined in pl., where it is used with singular meaning,* cf. ofer-hygd, -méde : *perhaps all the instances which follow may belong to the plural, since eáþmétto takes a verb in the plural ; f. Pride, arrogance, haughtiness:*—Hine his hyge gespeón and his ofermétto ealra swíðost, Cd. Th. 22, 35 ; Gen. 351. Þurh heora miclan mód, and þurh ofermétto, 22, 7 ; Gen. 337 : 21, 30 ; Gen. 332. Hé biþ on ofermettu (-métto, Cott. MSS.) áwended ... hé ástág on ofermétto *in elationem permutatur ... intumuit,* Past. 3, 2 ; Swt. 35, 13–16. On ofermétto *in superbiam,* 19, 3 ; Swt. 147, 3 : Bt. 6 ; Fox 14, 34. God hyra ofermétan ofermétto genyðerode, Ors. 1, 7 ; Swt. 38, 28. God ða mæstan ofermétto gewræc on ðam folce, 6, 2 ; Swt. 256, 5. Ðe ofermétto dóþ *qui faciunt superbiam,* Ps. Th. 30, 27. Ofermétto *fastu,* Wrt. Voc. ii. 33, 62. Ðis synt ða ídelnyssa ðisse worulde : ærest is ofermétta (*arrogantia*), L. Ecg. P. i. 8 ; Th. ii. 174, 32. Ne gerísaþ heom príta, ne micele ofermétta, L. I. P. 10 ; Th. ii. 318, 32. Heora eáþmétto ne mihton náuht forstandan, ne húru heora ofermétta, Bt. 29, 2 ; Fox 106, 1. Ðonne weaxaþ ða ofermétta (cf. ðonan mæst cymeþ yfla oferméta, Met. 25, 44), 37, 1 ; Fox 186, 19. Mid his ágnum wordum ðone swiran gebiége his ágenra ofermétta *suo judicio superbiae cervicem calcat,* Past. 26, 3 ; Swt. 185, 15. On heora oferméttum *in superbia,* Ps. Th. 30, 20. On oferméttum áþunden, Past. proem. ; Swt. 25, 6. Biscopum gebiraþ ealdlíce wísan búton oferméttum, L. I. P. 10 ; Th. ii. 318, 31. Lǽt befeallan on ðæt éce fýr ðe him gegearcod wæs for heora oferméttum, Homl. Th. i. 12, 4 : Met. 5, 32 : Bt. 16, 1 ; Fox 50, 9–11. Mid ofermétum *superbia,* Past. 42, 2 ; Swt. 307, 7. Ne mæg hé wið ofermétta, Bt. 12 ; Fox 36, 10. Múþ heora spræc ofermétta (*superbiam*), Ps. Lamb. 16, 10 : Met. 7, 8.

ofer-micel ; *adj. Over-much, excessive :*—On ðære tíde wæs sió ofermycelo hǽlo on ealre worulde, Ors. 1, 7 ; Swt. 40, 3. Bútan hý ouermicel geswinc habben, R. Ben. 65, 17. [*Prompt. Parv.* ovir-mikel *nimius : Icel.* ofr-mikill.]

ofer-micelness, e ; *f. Over-greatness, excess :*—Náht framaþ eallum dæge lang ádreógan fæsten gif æfter ðam metta oferfylle oððe ofermicelnysse (*nimietate*) sáwl byþ ofersýmed, Scint. 13.

ofer-mód, es ; *n.*   I. *pride, arrogance, over-confidence :*—Feala worda gespæc se engel ofermódes, Cd. Th. 18, 12 ; Gen. 272. Ðá se eorl ongan for his ofermóde álýfan landes tó fela laðere þeóde, Byrht. Th. 134, 25 ; By. 89. [Gif hwa nulle for his ouermóð, oðer for his prude ... his scrift ihalden, O. E. Homl. i. 9, 30.]   II. *a high style* (?) :—Ofermód *coturnus,* Wrt. Voc. i. 19, 5. [*O. H. Ger.* ubar-muot *superbia : Ger.* über-muth.] v. ofer-méde, -métto.

ofer-mód ; *adj. Proud, arrogant, presumptuous :*—Ne sceal mon beón ofermód, R. Ben. 17, 15. Cild ácenned ofermód him sylfum gelícigende *a child born on the thirteenth day of the moon will be arrogant, pleasing himself,* Lchdm. iii. 190, 14. Se ofermóda cyning (*Lucifer*), Cd. Th. 22, 9 ; Gen. 338. On Torcwines dagum ðæs ofermódan cyninges *in the days of Tarquinius Superbus,* Bt. 16, 1 ; Fox 50, 8. Ðú ne scealt næfre gelíce déman ðam eáðmódan and ðam ofermódan, L. de Cf. 3 ; Th. ii. 260, 25. Hig wǽron ófermóde ongén hig *superbe egerint contra illos,* Ex. 18, 11. Ða ðe wǽron ofermóde on heora heortan, Blickl. Homl. 159, 10. On óðre wísan ða ofermódan on óðre ða wácmódan *aliter protervi, aliter pusillanimes,* Past. 23 ; Swt. 175, 18 : 32 ; Swt. 209, 1. Ða eágan ðara ofermódena (*superborum*) ðú geeáðmétst, Ps. Th. 17, 26. Spell be ðam ofermódum cyningum, Bt. 37, 1 ; Fox 186, 1. Ofermódum *superbis,* Ps. Spl. 93, 2. Se Scyppend oft ða ofermódan geeáðmétte, Homl. Th. ii. 432, 20. [*O. Sax.* obar-mód.] v. ofer-méde.

ofer-módig ; *adj. Proud, arrogant, saucy, wanton :*—Mæden biþ ofermódig *a girl (born on the thirteenth day of the moon) will be saucy,* Lchdm. iii. 190, 16. Ofermódige *superbi,* Ps. Th. 118, 51. Ofermódigra *superborum,* Ps. Spl. 118, 69. Ofermódigum *superbis,* 122, 5. Ða ofermódegan *superbos,* Bd. 3, 17 ; S. 545, 12. Tarcuinius ðe hira eallra wæs ofermódgast *Tarquinius Superbus,* Ors. 2, 2 ; Swt. 66, 28. [*O. Sax.* obar-módig : *O. H. Ger.* ubar-muotig *contumax : Ger.* über-müthig.]

ofer-módigian, -módgian, -módigan *to be proud* or *haughty, to be puffed up with pride :*—Ðonne se unrihtwísa ofermódegaþ (-módgaþ, Ps. Spl.) *dum superbit impius,* Ps. Th. 9, 21. Hwí ofermódige gé ofer óðre men for eówrum gebyrdum, Bt. 30, 2 ; Fox 110, 15 : 42 ; Fox 258, 15. Hí ofermódigaþ for ðæm welan, 39, 11 ; Fox 230, 23. Ne ofermódgiaþ (*superbiunt*) ða scírmenn ná for ðý, Past. 17, 2 ; Swt. 100, 17. Hié wið Gode ofermódgiaþ *contra Deum superbiunt,* 29 ; Swt. 201, 16. Ðé læs ðe hira fýnd ofermódegodun *ne forte superbirent hostes eorum,* Deut. 32, 27. Hwý gé ofer óðre men ofermódigen, Met. 17, 16. [Cf. *O. H. Ger.* ubar-muotón *superbire.*]

ofer-módigness, e ; *f. Pride, arrogance :*—Ofermódignis *superbia,* Ps. Spl. 72, 6: 30, 22 : 16, 11 : 100, 8. Seó eáðmódnes of ácearf heáfod ðære ofermódignesse *humilitas amputat caput superbiae,* Gl. Prud. 36 a ; 37 a : 38 a. Ofermódinysse *arrogantiae, inflationis,* Hpt. Gl. 523, 52. Ofermódignysse *insolentiam,* 526, 8. Ofermódignessa *superbia,* Mk. Skt. 7, 22.

ofer-módigung, -módgung, e ; *f. The being proud, pride :*—Hit is ungecyndlícu ofermódgung *contra naturam superbire est,* Past. 17, 2 ; Swt. 109, 11.

ofermód-líc ; *adj. Proud, arrogant, presumptuous :*—Mid ofermódlícum gilpe, Bt. 18, 4 ; Fox 66, 31. Hé sceal ða ofermódlícan word mid eáðmódlícum wordum gemetgian *ut verba praemissae superbiae verbis subjectae humilitatis impugnet,* Past. 54, 5 ; Swt. 423, 36. [*O. H. Ger.* ubarmuot-líh *sublimis.*]

ofermódlíce ; *adv. Proudly, arrogantly, insolently :*—Hí sprecaþ swíðe ofermódlíce *os eorum locutum est superbiam,* Ps. Th. 16, 9. Hit ofermódlíce férde, Blickl. Homl. 199, 17 : 201, 24. [*O. H. Ger.* ubarmuotlího *superbe, proterve, elate, hyperbolice.*]

ofermódness, e ; *f. Pride :*—Ofermódnys *superbia,* Ps. Spl. 73, 24, 4. Ofermódnes eáðmódnes *superbia ... humilitas,* Gl. Prud. 31 a : 29 a : 30 a : 32 a : 33 a. Bebeorh ðé wið ofermódnysse *cave te a superbia,* L. Ecg. C. proem. ; Th. ii. 132, 10.

ofer-níd, -neód, e ; *f. Extreme need :*—Gif hit oferneód beó *si valde necesse sit,* L. Ecg. P. iii. 14 ; Th. ii. 200, 33.

ofer-niman.   I. *to take by violence, to violate :*—Be ðam men ðe wíf oððe mæden oferuimþ mid unrihtum þingum *de homine qui mulierem vel puellam per fraudem constuprat,* L. Ecg. P. ii. 13 tit. ; Th. ii. 180, 22. [*The section to which the title refers is as follows :*—Gif hwá mid his ofercræfte wíf oððe mæden nýdinga nimþ tó unrihthǽmede, Th. ii. 186, 20.] Gif ænig man ofernyme unbeweddod mǽden *si invenerit vir puellam virginem, et apprehendens concuberit cum illa,* Deut. 22, 28. Hé eode in tó mé ðæt hé mé ofernáme *ingressus est ad me, ut coiret mecum,* Gen. 39, 14.   II. *to take away, carry off :*—Sóna wæs ðæt ætter ofernumen *vidimus rasuram totam vim veneni absumisse,* Bd. 1, 1 ; S. 474, 39. [*O. H. Ger.* ubar-neman *to take away.*]

ofer-nón *the latter part of the day, afternoon :*—Middæg *sexta :* nón *nona :* ofernón oððe geloten dæg *suprema :* æfen *vesperum,* Wrt. Voc. i. 53, 12–15.

ofer-rǽdan.   I. *to read over* or *through :*—Ic oferrǽde *perlego,* Ælfc. Gr. 28, 6 ; Som. 32, 15. Oferrǽdan *perlegere,* Hpt. Gl. 439, 4 : Homl. Th. i. 166, 7. Ðá heó ða gewrita oferrǽd hæfde, Ap. Th. 20, 20 : 21, 12.   II. *to consider :*—Oferrǽdan ꝼ hycgean *coniici,* Hpt. Gl. 439, 4.

ofer-ranc ; *adj. Over-luxuriant, extravagant, sumptuous :*—God lǽteþ reáfian eówere dohtra heora gyrla and tó oferrancra heáfodgewǽda, Wulfst. 46, 1.

ofer-reccan *to convince, confute, convict :*—Gif hine mon oferricte ðæt hé ne móste londes wyrþe beón *if it should be proved against him that he was disqualified for holding land,* Chart. Th. 141, 11. Forðon hé ðus cwæþ ðæt hé ða lotwrencas oferwunne and oferreahte *quatenus et illos victrix ratio frangeret,* Past. 30 ; Swt. 205, 17. Ðú hæfst mé swíðe rihte oferreahte (-rehtne, MS. Bod.) *thou hast completely convinced me,* Bt. 34, 3 ; Fox 138, 11. Ðonne is betere ðæt hié mid ryhtre race weorðen oferreahte and mid ðære race gebundene and ofersuíðde *prodest, ut in suis allegationibus victi jaceant,* Past. 30 ; Swt. 205, 3. Ðý læs ðonne hié oferhyggaþ ðæt hié síen oferreahte útane mid óðerra manna ryhtum lárum hié ðonne síen innan gehǽfte mid ofermétum *ne dum rectis aliorum suasionibus foris superari despiciunt, intus a superbia captivi teneantur,* 42, 2 ; Swt. 307, 6. Cf. ofer-stǽlan.

ofer-renc[u], e ; *f. Over-luxuriance, extravagance :*—Manege ðe mid oferrence glengdan hý sylfe, Wulfst. 46, 2.

ofer-rícsian *to dominate, rule over :*—Hé him geþafode ðæt hit mid anwalde him móste oferrícsian, Past. 17, 8 ; Swt. 119, 19.

ofer-rídan *to cross on horseback :*—Sealde hé ðæt betste hors Aidane, ðæt hé on ðam mihte fordas oferrídan, Bd. 3, 14 ; S. 540, 18.

ofer-rówan *to cross by rowing :*—Ðá hét hé his leorningcnihtas faran tó scipe, and oferrówan ðone brym, Homl. Th. ii. 384, 19.

ofersǽ-líc ; *adj. Transmarine :*—On ðam ofersǽlícum dǽlum *in transmarinis partibus,* Bd. 3, 28 ; S. 560, 13. Cf. ofersǽwisc.

ofer-sǽlig ; *adj. Exceedingly fortunate, more than happy :*—Se biþ gesǽlig and ofersǽlig ðe swylce cwyldas mæg forbúgon, Dóm. L. 16, 246.

ofer-sǽlþ, e ; *f. Pleasure or happiness that exceeds due bounds :*—Gif ðú wilnast ðæt ðú wel mǽge ðæt sóðe leóht sweotole oncnáwan ðú forlǽtan scealt ídle ofersǽlþa unnytne gefeán (cf. gif ðú wilnige ðæt sóðe leóht oncnáwan áfyr fram ðé ða yfelan sǽlþa and ða unnettan, Bt. 6 ; Fox 14, 32) *tu si vis cernere verum, gaudia pelle,* Met. 5, 27.

ofersǽwisc ; *adj. From beyond the sea, transmarine :*—Ofersǽwisc rind *bark from beyond the sea* (*cinnamon*), Lchdm. ii. 52, 3. Landferþ se ofersǽwisca hit gesette on Léden, Glostr. Frag. 10, 21. Hé (*Benedict of Wearmouth*) ða ciricean gefretwade mid godcunde wísdóme and mid woroldlícum frætwum ofersǽwiscum, Shrn. 50, 32.

ofer-sáwan *to oversow :*—Ðá com his feónda sum and oferseów (*superseminavit*) hit mid coccele, Mt. Kmbl. 13, 25. [*O. H. Ger.* ubar-sâan.]

ofer-sceadwian *to cover with a shadow, overshadow :*—Ic ofersceade-wige *obumbro*, Wrt. Voc. i. 54, 59. Ðæs Heáhstan miht ðē ofersceadaþ, Lk. Skt. 1, 35. Genip ofersceadude hig, 9, 34. Seó lyft hí ofersceade-wude, Mk. Skt. 9, 7. Ðū oferscadudest (-sceaduwedest, Ps. Lamb.) *obumbrasti*, Ps. Spl. 139, 8. Ofersceadwa *obumbra*, Ps. Surt. 139, 8. [*Goth.* ufar-skadwjan.]

ofer-sceatt, es ; *m. Money in excess (of a loan), interest :*—Ic onfēnge ðæt ðe mín is mid ofersceatta (*cum usura*), Mt. Kmbl. Rush. 25, 27.

ofer-sceáwian *to overlook, superintend :*—Preóstum gedafenaþ, ðæt hí heora biscope beón eádmódlíce underþeódde, and hē hí ofersceáwige, and heora wísan begíme, swá swá his nama swēgeþ: his nama is gecweden *episcopus*, and ofersceáwigend on Englisc, ðæt hē ofersceáwige symle his underþeóddan, L. Ælfc. P. 37 ; Th. ii. 378, 25–30. [*Episcopus* . . . is on Englisc scawere, for he is iset to þon þet he scal ouerscawian mid his ȝeme þa lewedan, O. E. Homl. i. 117, 7.]

ofer-sceáwigend *a superintendent ;* episcopus. v. preceding word.

ofer-sceótan. v. ofer, III.

ofer-scínan *to cover with light, illumine :*—Næs nā ðæt ān ðæt leóht ða dūne āne oferscíneþ, ac eác swylce ða burh, Blickl. Homl. 129, 2. Beorht wolcn hig ofersceán *nubes lucida obumbravit eos*, Mt. Kmbl. 17, 5. Ðonne his (*the moon's*) leóma ealne middaneard oferscíne, Anglia viii. 323, 7.

ofer-scūwan, -scūan *to overshadow :*—Wolken oferscūade (-scýade, Lind.) hiæ, Mt. Kmbl. Rush. 17, 5. [*Icel.* yfir-skyggja.]

ofer-seám, es ; *m. A bag :*—Oferseámas *sacculos*, Lk. Skt. Lind. 12, 33.

ofer-sécan *to make too great demands upon, put to too severe a trial, press too hard :*—Wæs sió hond tō strong seó (MS. se) ðe mēca gehwane swenge ofersōhte *the hand was too strong, which with its stroke put every blade to too severe a trial,* i. e. Beowulf struck so hard that any sword would be broken, Beo. Th. 3655 ; B. 2686. [*O. H. Ger.* ubar-suochian, Grff. vi. 84.]

ofer-segl, es ; *m. A top-sail :*—Oversegl *artemon*, Wrt. Voc. ii. 100, 76.

ofer-seglian *to cross by sailing :*—Ðā āstāh hē on scyp and oferseglode (*transfretavit*), Mt. Kmbl. 9, 1.

ofer-sendan *to transmit :*—Ic ofersende *transmitto*, Ælfc. Gr. 28, 4 ; Som. 31, 40.

ofer-seócness, e ; *f. Extreme sickness :*—Unfæstende man húsles ne ābirige, būton hit for oferseócnesse sí, L. Edg. C. 36 ; Th. ii. 252, 2 : 30 ; Th. ii. 250, 20.

ofer-seolfrian *to cover with silver :*—Hié eall heora wǽpn ofersyle-fredan *deargentatis armis*, Ors. 3, 10 ; Swt. 138, 31. Eall heora wǽpn wǽron ofersylefreda, 3, 11 ; Swt. 146, 23. Ofersylfrede (-seolfrade, Ps. Lamb.) *deargentatae*, Ps. Spl. 67, 14.

ofer-seón. I. *to observe, survey, see :*—Ðū ðe ealle gesceafta ofer-sihst *thou that dost survey all creatures*, Bt. 4 ; Fox 8, 20. Æfter ðære wísan ðe ic hit oferseah *quemadmodum inspexi*, Nar. 2, 9. Swā ic mid mínum ēgum oferseah, and mínum eárum oferhýrde, L. O. 8 ; Th. i. 180, 29 : L. C. S. 23 ; Th. i. 388, 24. Ðū ealle míne fýnd eágum ofer-sāwe *super inimicos meos respexit oculus tuus*, Ps. Th. 53, 7. Ðæt hié heora sylfra eágon oforsēgon and heora eáron gehýrdon *what they had seen with their own eyes and heard with their ears*, Blickl. Homl. 121, 1. Oft wē ofersēgon þeóda þeáwas, Exon. Th. 118, 9 ; Gū. 237. Selfe ofersāwon ðā ic cwom, Beo. Th. 842 ; B. 419. Ofersewen *respectus*, Ps. Spl. 72, 4. [*O. H. Ger.* ubar-sehan *respicere, superspicere*.] II. *to overlook, neglect, despise :*—Ða ðe tō ðam þríste sýn, ðæt hig God oferseóþ and swā mæniges hāliges mannes dōm, Wulfst. 270, 23.

ofer-síman *to overload, oppress :*—Gif metta oferfylle sāwl byþ ofersýmed *si ciborum satietate anima obruatur*, Scint. 13. Warniaþ ðæt eówere heortan ne sýn ofersýmede mid oferfylle, R. Ben. 64, 1 : 138, 11. Ðæt ða unstrangan ofersýmede heora þeówdōm ne forfleón, 121, 23. [þe burden ðe hē haddeus mide ouersemd, O. E. Homl. ii. 65, 4.]

ofer-sittan. I. *to sit upon, occupy, take possession of :*—Ofer-sēton *obsederunt*, Ps. Surt. 21, 13. Ofersētun sāwle míne *occupaverunt animam meam*, 58, 4. Ðone mǽstan dǽl his hæfþ sǽ oferseten *the greatest part of the sea has occupied*, Bt. 18, 1 ; Fox 62, 11. [We maȝen ouersitten þis lond, Laym. 8035.] II. *to desist from, abstain from :*—Ic ofersitte *supersideo*, Ælfc. Gr. 47 ; Som. 48, 45. Ic gylp ofersitte *I abstain from boasting*, Beo. Th. 5050 ; B. 2528. Wit sculon secge ofersittan *we shall abstain from the sword, not make use of swords*, 1372 ; B. 684. [Cf. *Prompt. Parv.* ovyrsyttynge of dede or time *omissio.*]

ofer-slǽp, es ; *m. Excessive sleep :* — Wið overslǽpe, Lchdm. i. 342, 14.

ofer-sleán *to reduce, subdue :*—Ðæt ða munecas furþor restan ðonne healfe niht ðæt seó dæges þigen tōfered sý on ðære nihtlícam reste and seó hǽte ðære þigene oferslegen *that the monks may rest more than half the night, so that the food of the day may be distributed through the body in the nightly rest and the food subdued*, R. Ben. 32, 15.

ofer-slege, es ; *n. A lintel :*—Oferslege oððe þrexwold *limen*, Ælfc. Gr. 9, 12 ; Som. 9, 28. Oferslæge, Wrt. Voc. i. 85, 65. Sprengaþ on ðæt oferslege (*superliminare*) . . . ðonne hē gesihþ ðæt blōd on ðam oferslege, Ex. 12, 22–23. On hyra gedyrum and oferslegum, Homl. Th. i. 310, 29 : ii. 40, 12 : 264, 1 : 266, 8. [*Prompt. Parv.* ovyrslay of a dcore *superliminare.*] v. ofer-dyre, -gedyre.

ofer-slop, es ; *n. An over-garment, surplice :*—Oferslop hwīt habban,

blisse getācnaþ. Oferslop bleófāh habban ǽrende fullīc getācnaþ, Lchdm. iii. 200, 5–7. On oferslopum *in stolis*, Lk. Skt. Lind. 20, 46. [His (the canon's) oversloppe nis nat worth a myte, Chauc. Group G. 633 : *Icel.* yfir-sloppr.] v. next word.

ofer-slype, es ; *m. An over-garment, surplice :*—Ðæt mæssepreósta ǽnig ne cume binnan circan dyre būton his oferslipe (-slope), L. Edg. C. 46 ; Th. ii. 254, 10. Hē is ymbscrýd mid hwítum oferslype *he is clad in a white upper garment*, Homl. Th. i. 456, 19.

ofer-smeáung, e ; *f. Excessive consideration of a subject :*—Sió ofersmeáung mirþ ða unwísan, Past. 15, 6 ; Swt. 97, 17.

ofer-sprǽc, e ; *f. Excessive speaking, loquacity :*—Ne biþ nǽfre sió ofersprǽc būtan synne *in multiloquio non deerit peccatum*, Past. 38, 8 ; Swt. 279, 23. Āídlode on ofersprǽce *multiloquio vacantes*, 38, 1 ; Swt. 271, 10. On ídle ofersprǽce *supervacuis verbis*, 38, 6 ; Swt. 277, 11. Ðonne mon mid ungedafenlícre and unwǽrlícre ofersprǽce ða heortan gedweleþ ðara ðe ðǽrtō hlystaþ and eác se lāriów biþ gescinded mid ðære ofersprǽce *cum apud corda audientium loquacitatis incauta importunitate laevigatur, et auctorem suum haec eadem loquacitas inquinat*, 15, 5 ; Swt. 95, 19–21. Gelimpeþ ðæt his word beóþ gehwyrfedo tō unnyttre ofersprǽce *contingit, ut magistri lingua usque ad excessus verba pertrahatur*, 21, 7 ; Swt. 165, 18. [*O. H. Ger.* ubar-sprāhhi.]

ofer-sprǽce ; *adj.* I. *speaking too much, talkative, loquacious :*—Se ðe ofersprǽce biþ *multiloquio subditus*, Past. 15, 6 ; Swt. 97, 6. Se ofersprǽcea wer *vir linguosus*, 38, 8 ; Swt. 279, 21. Ne beó ðū tō ofersprǽce ac hlyst ǽlces monnes worda swíðe georne ' *give every man thy ear, but few thy voice,*' Prov. Kmbl. 58. Salamon cwæþ, ðæt sēlre wǽre tō wunigenne mid león and dracan ðonne mid yfelan wífe and ofersprǽcum, Homl. Th. i. 486, 33. Ða . . . ofersprǽcean *multiloquio vacantes*, Past. 38, 6 ; Swt. 277, 3 : 38, 1 ; Swt. 271, 14. [Cf. *O. H. Ger.* ubar-sprācha zungun *linguam magniloquum.*] II. *saying more than is just or true* (v. ofer-sprecan) :—Ða fācnes fullan weoloras and ða ofersprǽcan *labia dolosa*, Ps. Th. 11, 3. Ða ofersprēcan ðe mē yfel cweðaþ *qui maligne loquuntur adversum me*, 34, 24.

ofer-sprǽdan *to overspread, cover :*—Beón ðǽr (*in the house for strangers*) symble bedd genihtsumlíce ofersprǽdde, R. Ben. 84, 23.

ofer-sprecan. I. *to say too much, use too many words :*—Ne fiýt ðū wið ānwilne man ne wið ofersprecenne *don't dispute with an obstinate man, or with one using too many words*, Prov. Kmbl. 5. II. *to say too much, more than is just :*—Ofersprecendes *obloquentis*, Ps. Lamb. 43, 17. [*O. H. Ger.* ubar-sprehhan *blasphemare.*]

ofer-sprecol ; *adj.* I. *given to talk too much, talkative, loquacious :*—Se ðe ofersprecol biþ *multiloquio serviens*, Past. 38, 8 ; Swt. 279, 20. Se ofersprecola wer *vir linguosus*, R. Ben. 30, 5. II. *given to extravagant, inconsiderate speech :*—Ofersprecelum *procacibus, imprudentibus*, Hpt. Gl. 452, 14 : 507, 24.

ofer-sprecolness, e ; *f. Talkativeness, loquacity ;* superfluitas locutionis, Past. 43, 1 ; Swt. 308, 16.

ofer-stǽlan *to confute, convince, convict :*—Ic oferstǽle *confuto*, Wrt. Voc. i. 34, 15. Oberstaelid *confutat*, ii. 105, 32. Oferstǽleþ, 15, 31. Ic eom geþafa ðæt ic eom swíðe rihte ofersteled, and ic beó ealne weig micle gefegenra ðonne ðu mē myd þillícum ofærstǽlest, ðonne ic ǽfre wǽre ðonne ic ōðerne man oferstǽlde *I allow that I am very properly confuted, and I am always much more pleased when you confute me with such arguments, than I ever should be when I confuted another man*, Shrn. 197, 32–35. Ne beó ðū tō ānwille ; fordam ðe is gerisenlícre ðæt ðū sí mid rihte oferstǽled, ðonne ðū oferstǽle ōðerne man mid wōge, Prov. Kmbl. 8. Ðæt hí ðæs deófles leásunge mid Godes sōðfæstnysse oferstǽlan, Homl. Th. ii. 100, 9. Oberstaelende *convincens*, Wrt. Voc. ii. 104, 37. Oferstǽlende, 14, 60 : *confutans*, 23, 45 : Hpt. Gl. 436, 37. Oberstaeled *convicta*, Wrt. Voc. ii. 104, 45. Ðonne hē oferstǽled biþ *when he is convinced*, Past. 6 ; Swt. 46, 16. Hē biþ ðonne oferstǽled ðæt hē Godes feónd is *he will then be convicted of being God's foe*, Homl. Th. i. 612, 24. Gif hwā mǽne āþ on hāligdōme swerige, and hē oferstǽled weorðe *if a man commit perjury on a relic, and he be convicted*, L. C. S. 36 ; Th. i. 398, 5 : 37 ; Th. i. 398, 12. Oferstǣlede *confutati, superati, convicti, redarguti*, Hpt. Gl. 475, 19.

ofer-steall, es ; *m. Opposition :*—Gif ic ðísum dracan tō forswelgenne geseald eom hwī sceal ic elcunge þrowian for eówerum oferstealle (*the opposition which was offered by the prayers said at the speaker's bedside*), Homl. Th. i. 534, 20. Cf. wiðer-steall.

ofer-stealla, an ; *m. A survivor :*—Heó wýscte ðæt heó nānne æfter hyre ne forlēte, ðē lǽs gif hyra hwylc wǽre hyre oferstealla, ðæt se nemyhte on heofenum beón hyre efngemæcca, Shrn. 151, 13.

ofer-stellan *to cross :*—Hit sum slóg oferhleóp and oferstǽlde (*transiliret*), Bd. 5, 6 ; S. 619, 17.

ofer-steppan *to over-step, to cross, exceed :*—Ic ofersteppe weall *trānsgrediar murum*, Ps. Lamb. 17, 30. Ðū oferstōpe *du supergressa es*, Kent. Gl. 1151. Ne oferstepe ðū ealde gemǣro *ne transgrediaris terminos antiquos*, 854. Seó sǽ ne mōt ðone þeorcwold oferstæppan (-steppan, Met. 11, 69) ðære eorþan, Bt. 21 ; Fox 74, 26. [*O. H. Ger.* ubar-stephen *transgredi, excedere.*]

3 B

ofer-stígan.   I. *to mount, scale, surmount, rise above :*—Ic heofonas oferstíge, Exon. Th. 482, 24; Rä. 67, 6. Sume ða ýða hé becerþ mid ðý scipe sume hit oferstígþ *some of the waves the steersman avoids with the ship, some it surmounts,* Past. 56, 3 ; Swt. 433, 3. Heó ða þýstre ðysses andweardan middangeardes oferstáh *praesentis mundi tenebras transiens,* Bd. 3, 8 ; S. 532, 3. Hé on ánre diégelre stówe ðone munt oferstág, Ors. 4, 6 ; Swt. 172, 21. Breóst oferstág brim weallende eorlum oþ exle *the boiling sea rose above the breast up to men's shoulders,* Andr. Kmbl. 3146 ; An. 1576. Eles gecynd is ðæt hé wile oferstígan ælcne wætan, Homl. Th. ii. 564, 12.   II. *to transcend, surpass, excel, overcome, exceed :*—Ic oferstíge *excelleo,* Ælfc. Gr. 26 ; Som. 28, 45. Ðú ealle ðíne yldran on ríce feor oferstígest (*transcendas*), Bd. 2, 12 ; S. 514, 9. Hé ealra ðóðerra heáhfædera mægen oferstígeþ, Blickl. Homl. 167, 23. Yldo oferbídeþ stánas, heó oferstígeþ stýle, Salm. Kmbl. 600 ; Sal. 299. Oferstíhþ *excedit,* i. *superat,* Wrt. Voc. ii. 145, 71. Ðæs gebodes micelnes his mihta oferstíhþ, R. Ben. 128, 14. Hé ongeat ðæt hé oferstág hine selfne *semetipsum noverat transcendere,* Past. 16, 2 ; Swt. 101, 13. Oforstág, Blickl. Homl. 163, 28. Oferstáh, Homl. Th. i. 70, 11. Hí swiðra oferstág weard *a stronger guard overcame them,* Exon. Th. 116, 3 ; Gú. 201. Oferstíge *excedat,* Wrt. Voc. ii. 145, 72. Ǽr ðan ðe ðæs dæges lenge oferstíge ða niht, Lchdm. iii. 256, 13. Oferstigan *percellerent, supereminerent,* Hpt. Gl. 489, 27. Ða yldo mid þeáwum oferstígende *aetatem moribus transiens,* Bd. 5, 19 ; S. 637, 4. Ða oferstígendan lufe *the surpassing love,* Homl. Th. ii. 408, 22. [*Goth.* ufar-steigan : *O. H. Ger.* ubar-stígan *transcendere, transire, exsuperare : Icel.* yfir-stiginn *overcome.*]

ofer-stige, es ; *m. Astonishment, extasy :*—Hé cwæþ tó him sylfum : 'Nú ic wæs of ðam rihtan wege mínes ingeþances, ac betere hit biþ ðæt ic eft fare út of ðysum porte, ðý læs ðe ic tó swíðe dwelige . . . gewíslíce ic hér ongyten hæbbe ðæt mé hæfþ gelæht fæste mínes módes oferstige, ðæt ic nát hú forgeare hú ic hit ðus macige,' Homl. Skt. i. 23, 551–556. v. ofer-stigenness.

oferstígend-líc ; *adj. Superlative :*—Sume synd *superlativa,* ðæt is oferstígendlíce, Ælfc. Gr. 5 ; Som. 4, 63.

ofer-stigenness, e ; *f. A passing over :*—Geleórednysse ł oferstigenysse *extaseos, transgressionis,* Hpt. Gl. 413, 9. v. ofer-stige.

ofer-swimman *to cross by swimming :*—Oferswam ða sioleþa bigong sunu Ecgþeówes, Beo. Th. 4723; B. 2367. [*O. H. Ger.* ubar-swimman *tranare.*]

ofer-swíðan ; *p.* -swíðde, *but also* -swáð *To prove stronger than or superior to another, to overcome, overpower, conquer, surpass :*—Obersuíðo *vinco,* Wrt. Voc. ii. 123, 69. Ic oferswíðe *vinco,* Ælfc. Gr. 28 ; Som. 32, 17. Ic nardes stenc oferswíðe mid mínre swétnesse, Exon. Th. 423, 29; Rä. 41, 29. Ðú ðe úre wiðerwinnan oferswíðst, Homl. Skt. i. 11, 233. Ðú oferswíðest deáþ, Blickl. Homl. 141, 13. Hé on his mægenes weorþunga oferswíð ealra ðóðerra Godes martira wuldor, 167, 25. Gyf strengra hine oferswýð (-swíð, MSS. B. C.) *si fortior vicerit eum,* Lk. Skt. MS. A. 11, 22. Oferswýðeþ, Beo. Th. 564; B. 279. Ðú oferswíðdest ðone deófol, Homl. Skt. i. 3, 436 : Blickl. Homl. 157, 4. Hé ðone ealdan gedwolan oforswíðde, 7, 13. Gaius Julius se cásere Brettas oferswíðde, Chr. Erl. 4, 24 : Ors. 1, 2 ; Swt. 30, 22. Hé þurh Godes mihte ðone cwelmbǽran drenc oferswíðde, Homl. Th. i. 72, 12. Heó þurh martyrdóm ðisne middaneard oferswáð, Homl. Skt. i. 2, 4. Hé ðone feónd oferswáð, Shrn. 13, 30. Ða ðe mid sygefæstum deáþe middangeard oferswíðdon, Homl. Th. i. 84, 31. Oferswíð ðás cristenan þurh tearte wíta, Homl. Skt. i. 11, 137. Ðonne hé ðone áwyrgdan gást oferswíðe, Blickl. Homl. 31, 31. Deófol oferswíðan, 29, 1 : Elen. Kmbl. 2354; El. 1178. Mid swinglan oferswíðan, Bd. 1, 7 ; S. 478, 1 : 1, 25 ; S. 487, 1. Mid gedwylde láre oferswíðan, Homl. Th. i. 44, 26. Ðæt geþyld oferswíðeneð *patientia victrix,* Prud. 25 a. Hí habbaþ deófol oferswíðed, Blickl. Homl. 35, 4. Oferswíðod, Homl. Skt. i. 1, 8 : 4, 57. Is betre ðæt hié weorðen gebundene and oferswíðede, Past. 30; Swt. 204, 4 ; Blickl. Homl. 145, 13. Wit sýn oferswíðede, 181, 30. Oferswíðdum leahtrum *devictis vitiis,* Prud. 28 b. v. un-oferswíðende.

ofer-swíðe ; *adv. Over-much, too much :*—Sý hé snotor and ná oferswíðe ne þreáge, R. Ben. 121, 3. Ða heáfodmen lufedon swíðe and oferswíðe gítsunge on golde and on seolfre, Chr. 1086; Erl. 220, 5. [*Ouermuchel* and ouerswuðe ivonded, A. R. 178, 9.]

ofer-swíðestre, an ; *f. A conqueror ; victrix,* Wrt. Voc. ii. 141, 68.

ofer-swíðness, e ; *f. Oppression, distress :*—Oferswíðnisse *pressura,* Lk. Skt. Rush. 21, 25 : Jn. Skt. Rush. 16, 33.

ofer-swíðrian *to prevail, conquer :*—Ic oforswíðrode ongén hine *praevalui adversum eum,* Ps. Lamb. 12, 5. Wé oferswíðredon (-swíðdon MS. F.) on ðysum eallum þurh ðone ðe ús lufode '*in all these things we are more than conquerors through him that loved us*' (Rom. 8, 37), R. Ben. 27, 12.

ofer-swíðung, e ; *f. Oppression, distress ;* pressura, Jn. Skt. Lind. 16, 21, 33 : p. 7, 17.

ofer-swógan *to cover thickly :*—Mid þýstro genipum ðæs muntes cnoll eall oferswógen wæs, Blickl. Homl. 203, 9. v. á-swógan, ge-swógen.

ofer-tæl, es ; *n. An odd number :*—Ðæra pipercorna sý ofertæl, ðæt ys ðý forman dæge án and þrittig, and ðý óðrum dæge seofontýne, and ðý þriddan dæge þreótýne, Lchdm. i. 288, 8.

ofer-teldan *to cover with an awning :*—Segle ofertolden, Cd. Th. 182, 26; Exod. 81. [Al þe cure ouertild, Jul. 9, 8.]

ofer-teón.   I. *to draw one thing over another, to cover by drawing one thing over another :*—Ðonne ic oferteó heofenan mid wolcnum *cum obduxero nubibus coelum,* Gen. 9, 14 : Homl. Th. i. 22, 11. Woruld miste oferteáh, þýstrum biþeahte, Exon. Th. 178, 35 ; Gú. 1254. Hé nǽfre eft nolde ealne middaneard mid nánum flóde oferteón, Scrd. 21, 21. Ðonne se fulla móna wyrð ofertogen mid þýstrum, Bt. 39, 3 ; Fox 214, 29 : Met. 9, 16.   II. *to bring to an end, finish :*—Ofertogen *finitum,* Wrt. Voc. ii. 134, 4.

ofer-þearf, e ; *f. Extreme need :*—Gif ðæs oferþearf síe ǽr mete, ðæt hé spíwan mæge, Lchdm. ii. 226, 9 : Wulfst. 134, 21. Ágan ða yldran ðæs oferþearfe, ðæt hí heora gingran Gode gestrýnan, 38, 23. For oferþearfe ilda cynnes, Elen. Kmbl. 1039; El. 521. Ða unþeáwas habbaþ oferþearfe hreówsunga, Bt. 31, 1 ; Fox 110, 27.

ofer-þearfa, an ; *m. One in extreme need :*—Is seó bót gelong æt ðe ánum oferþearfum *on thee alone depends the remedy for those in dire need,* Exon. Th. 10, 17 ; Cri. 153.

ofer-þeccan *to cover :*—Blódig wolcen oforþecþ ealne ðysne heofon, Blickl. Homl. 91, 33 : 93, 2. Eall eorþe biþ þeóstrum oforþeaht, 93, 6. Mid forste oferþeaht *covered with ice,* Homl. Skt. i. 11, 143. Þicce vel oferþeaht *condensa,* i. *spissa, secreta,* Wrt. Voc. ii. 135, 65. Hé onwreáh ða eorþan ðe ǽr wæs oferþeaht mid feóndum *revelabit condensa,* Ps. 28, 7. Mid þeóstrum oferþeht, Homl. Th. ii. 350, 17. Scip mid ýðum oferþeht, 378, 15 : Hexam. 6 ; Norm. 10, 18 : Exon. Th. 353, 10 ; Reim. 10. [*Ger.* über-decken.]

ofer-þeón ; *p.* -þáh, -þeáh ; *pl.* -þugon, -þungon ; *pp.* -þogen, -þungen *To thrive beyond others, to excel, surpass :*—Ic oferþeó *excello,* Ælfc. Gr. 37; Som. 39, 28. Oft on lǽwedum háde mid gódum weorcum man oferþíhþ ðone munuchád, Past. 52, 10 ; Swt. 411, 36. Oferþýhþ, R. Ben. 12, 16. Hé oferþeáh biscopes, Shrn. 17, 11. Bútan hwylc óðerne mid geearnunge oferþeó, R. Ben. 12, 21. Bonan mǽndon ðæt hý monnes bearn oferþunge, Exon. Th. 128, 10 ; Gú. 402. Oferþuge *praestaret, superaret, superexcelleret,* Hpt. Gl. 480, 1. Oferþeón *praestare, antecellere,* 417, 62 : *melior esse,* 418, 67. Ic hæbbe ðé oferþogen, Homl. Th. i. 448, 34 : Homl. Skt. i. 3, 209. Seó hæfþ ealle óðru wíf oferþungen mid clǽnnesse, Bt. 10 ; Fox 28, 21 : 33, 4 ; Fox 132, 7 : Met. 20, 194 : Past. 32, 2 ; Swt. 213, 11. Hé wénþ ðæt hé hæbbe hié oferþungne on his lífes geearnunge *transcendisse se vitae meritis credit,* 17, 3 ; Swt. 111, 15. [*Goth.* ufar-þeihan.]

ofer-þrymm, es ; *m. Exceeding power :*—Ǽr ðon se wlonca dæg bodige þurh býman brynehátne lég egsan oferþrym *ere that august day (doom's day) announce by the trumpet fire burning-hot, over-powering terror,* Exon. Th. 448, 10 ; Dóm. 52.

ofer-þungen. v. ofer-þeón.

ofer-togenness, e ; *f. The condition of being covered :*—Wið eágena ofertogennysse *ad albuginem oculorum,* Lchdm. i. 176, 16. v. ofer-teón.

ofer-trahtnian *to comment upon, expound :*—Langsum hit biþ ðæt wé ealne ðisne lofsang ofertrahtnian, Homl. Th. i. 202, 28.

ofer-tredan *to trample upon, tread under foot :*—Se geleáfa ofertret ðæt deófolgyld *fides conculcat idolatriam,* Prud. 9 a. Seó gýtsung manega ofertret *avaritia multos sternit,* 25 a. Seó rúmgyfolnes ða gýtsunge mid cneówum and mid fótum ofertræd *largitas avaritiam genibus et calcibus perfodit,* 68 a. [Þe Laferrd oferrcomm ⁊ oferrtradd te deofell, Orm. 12493.]

ofer-trúwa, an ; *m. Over-confidence :*—For ðam ofertrúwan on ðam friþe *from over-confidence in the truce,* L. Ath. v. 8, 7 ; Th. i. 238, 5.

ofer-trúwod *possessed by over-confidence, over-confident :*—Ðæm lár-eówe is swíðe smeálíce tó underséceanne be ðǽm weorcum ðara ofertrú-wudena *subtiliter ab arguente discutienda sunt opera protervorum,* Past. 32, 1 ; Swt. 208, 13.

ofer-wacian *to keep watch over, act as a guard :*—Julianus wýcode wið ða eá Eufraten, and him oferwacedon syfanfealde weardes, Homl. Skt. i. 3, 271.

ofer-wadan *to cross by wading :*—Ðá gebeótode Cirus ðæt hé his þegn on hire swá gewrecan wolde ðæt hié mehte wífmon be hiere cneówe oferwadan *rex iratus ulcisci in amnem statuit, contestans eum feminis vix genua tingentibus permeabilem relinquendum,* Ors. 2, 4 ; Swt. 72, 33. [*O. H. Ger.* ubar-watan *pertransire.*]

ofer-wealdend, es ; *m. One who rules over others, ruler, governor :*—Ealles oferwealdend, Elen. Kmbl. 2469; El. 1236.

ofer-weaxan *to cover by growing, over-grow :*—Hǽlend wæs sprecende tó Abrahame and wæs cweðende ðæt his sǽd oferweóxe ealle ðás woruld, Blickl. Homl. 159, 26. Mid wuda oferwexen, 207, 27 : Homl. Th. i. 508, 23.

ofer-weder, es ; *n. Storm, tempest :*—Heora scipu sume þurh oferweder wurdon tóbrocene, Chr. 794; Erl. 59, 22.

ofer-wenian *to become insolent :*—Oberwenide *insolesceret,* Wrt. Voc. ii. 111, 24. Oberwænidae, Ep. Gl. 12 d, 20.

ofer-weorc, es ; *n. A superstructure, a tomb :*—Oferwurces *sarcophagi, tumba,* Hpt. Gl. 488, 51. [Oferr þatt arrke wuss An oferwerrc wel timmbredd (*the mercy-seat*), Orm. 1035.] v. ofer-geweorc.

ofer-weorpan. I. *to overthrow, throw down :*—Nim eorþan, oferweorp mid dînre swiðran handa under dînum swiðran fêt, Lchdm. i. 384, 19. Gif hê hié oferweorpe, mid x sciłł. gebête, L. Alf. pol. 11; Th. i. 68, 15. Ðŷ gewunelîcan þeáwe horsa æfter wêrinysse hit (*the horse*) ongan walwian and on gehwæðære sîdan hit oferweorpan *consueto equorum more, quasi post lassitudinem in diversum latus vicissim sere volvere coepit*, Bd. 3, 9; S. 533, 40. Mid ðŷ storme onwend and oferworpen *tempestate convulsa*, Past. 26; Swt. 181, 11. II. *to throw* (*water, etc.*) *upon, to sprinkle :*—Oferwurpe ðû mid ðŷ wætere ealle burgwaran, Exon. Th. 467, 3; Hö. 133. Se ðe mid wætere oferwearp wuldres cynebearn, Menol. Fox 315; Men. 159. III. *intrans. To fall down :*—Oferwearp ðâ wêrigmôd, wígena strengest, ðæt hê on fylle weard, Beo. Th. 3090; B. 1543. [Uorte holden þet schip, þet uðen ne stormes hit ne ouerworpen, A. R. 142, 11. He oferrwarp þeȝȝre bordess, Orm. 15567 note.]

ofer-wîgan *to overcome in fight, conquer :*—Yldo oferwîgeþ wulf, heó oferbîdeþ stânas, Salm. Kmbl. 598; Sal. 299.

ofer-willan. I. *to boil so that a liquid is reduced in quantity :*—Oferwylle ôþ ðone þriddan dæle, Lchdm. ii. 216, 3, 4 : 228, 18 : 238, 10. II. *to overboil, boil too much :*—Nim ðæt wæter ðe pyosan wæran on gesodene oferwilleda, 286, 29.

ofer-winnan *to overcome, conquer, vanquish, subdue :*—Se ðe his môd gewylt is betera ðonne se ðe burh oferwinþ, Homl. Th. ii. 544, 10. Oferwinnaþ *debellant*, Blickl. Gl. Gif ûre fŷnd ûs oferwinnaþ *expugnatis nobis*, Ex. 1, 10. Hê Soroastrem oferwann and ofslôh *Zoroastrem pugna oppressum interfecit*, Ors. 1, 2; Swt. 30, 11. Oferwan, 1, 6; Swt. 36, 17. Iudith seó wuduwe ðe oferwann Holofernem, Ælfc. T. Grn. 11, 15, 44. Hî oferwunnon mê *expugnaverunt me*, Ps. Spl. 128, 1. Oferwin onwinnende *expugna inpugnantes*, Blickl. Gl. Hê ðus cwæð ðæt hê ða lotwrenceas oferwunne, Past. 30, 2; Swt. 205, 17. Gif ðû wille ænige buruh oferwinnan (*expugnare*), Deut. 20, 10 : Jos. 10, 4. Ðæt hî mihton heora fŷnd oferwinnan, Bd. 1, 12; S. 480, 28. Seó ylce þeód wæs oferwunnen fram Eald-Seaxum, 5, 11; S. 626, 10, On ðǣm xxv. wintrum ðe hê winnende wæs hê nâ oferwunnen ne weard, Ors. 3, 7; Swt. 114, 6. Se mon hafaþ weán oferwunnen, Exon. Th. 475, 5; Bo. 43. Synd ða fŷnd oferwunnene, Gen. 14, 20. Oferwunnenum feóndum *devictis hostibus*, Prud. 4 a. [*O. H. Ger.* ubar-winnan *expugnare, superare, devincere : Icel.* yfir-vinna.]

ofer-wintran *to winter, pass the winter :*—Nân eówer nele oferwintran (*hiemare*) bûton mînum (*the shoemaker*) cræfte, Coll. Monast. Th. 28, 1. [*Ger.* über-wintern.]

ofer-wist, e; f. Excess in eating :—Sint tô manianne ða ofergîfran ðeáh hié ne mægen ðone unþeáw forlætan ðære gîfernesse and ðære oferwiste ðæt hê hûru hine selfne ne þurhstinge mid ðŷ sweorde unryhthǣmedes, ac ongiete hû micel ofersprǣc cymeþ of ðære oferwiste *admonendi sunt gulae dediti, ne in eo, quod escarum delectationi incubant, luxuriae se mucrone transfigant, et quanta sibi per esum loquacitas insidietur, aspiciant*, Past. 43, 5; Swt. 313, 6–10. Hî lufiaþ oferwiste and îdele bliese, L. I. P. 14; Th. ii. 322, 24.

ofer-wistlîc glosses *supersubstantialis*, Mt. Kmbl. Lind. 6, 11.

ofer-wlenced *possessed of superabundant means, very opulent :*—Hié andwyrdon ðæt hit gemâlîc wǣre ðæt swâ oferwlenced cyning sceolde winnan on swâ earm folc swâ hié wǣron *responderunt, stolide opulentissimum regem adversus inopes sumsisse bellum*, Ors. 1, 10; Swt. 44, 12.

ofer-wlencu (o); f. Ostentation, superabundant means :—Ða ðe hyra lîfes þurh lust brûcan îdelum ǣhtum and oferwlencum, gierelum gielplîcum, Exon. Th. 127, 21; Gû. 389. v. preceding word.

ofer-wrecan *to overwhelm :*—Oberurecan *obruere*, Wülck. Gl. 35, 14.

ofer-wreón; p. -wráh, -wreáh, pl. -wrugon, -wrugon; pp. -wrigen, -wrogen *To cover, cover over, veil, hide, conceal, overspread :*—Ic oferwreó *nubo*, Ælfc. Gr. 28; Som. 31, 19 : *cooperio*, 30; Som. 34, 43. Ðû ðe oferwrîhst mid wæterum ða uferan hire *qui tegis aquis superiora ejus*, Ps. Lamb. 103, 3. Geswinc welera heora oferwrîhþ (-wrîd, Ps. Surt. : -wrŷhþ, Ps. Spl.) hî *labor labiorum ipsorum operiet eos*, 139, 10. Oferwrîhþ (-wrîd, Ps. Surt. : -wrŷcþ, Ps. Spl.) *operit*, 146, 8. Oferwrîhþ *operit*, Kent. Gl. 323. Seó sôðe lufu Godes and manna oferwrŷhþ ða mengo synna *charity covereth a multitude of sins*, L. E. I. 36; Th. ii. 434, 39, 37. Mycel mægen ðone heofon oforþecþ and oforwrŷhþ, Blickl. Homl. 93, 3. God ælce stôwe gefylþ and ufan oforwrŷhþ, 19, 27. Seó sunne scînþ geond ealle eorþan gelîce, and ealle eorþan brâdnysse endemes oferwrŷhþ, Lchdm. iii. 236, 13. Unrehtwîsnesse mîne ic ne oferwráh (-wreáh, Ps. Spl. C. T.), Ps. Surt. 31, 5. Ðû oferwrige *operuisti*, 84, 3. Oferwráh (-wreáh, Ps. Spl.) *operuit*, 43, 17 : 68, 8. Seó sǣ ealle his crætu and riddan mid ŷðan oferwreáh, Homl. Th. ii. 194, 28. Þicce genip oferwrêh ðone munt, Ex. 19, 16. Oferwreogan (*contexerunt*) mê þŷstru, Ps. Lamb. 54, 6. Ne ne beóþ gecyrred oferwreón (tô oferwreónne, Ps. Lamb. : oferwreón, Ps. Surt.) eorþan *neque convertentur operire terram*, Ps. Spl. 103, 10. Sume âgunnon oferwreón (*velare*) his ansŷne, Mk. Skt. 14, 65. Nacode wê sceolan oferwreón, L. E. I. 32; Th. ii. 428, 25. Næs Salomon oferwrigen (*coopertus*) swâ swâ ân of ðyson, Mt. Kmbl. 6, 29. Ðæt dysig ðæt hit ǣr mid oferwrigen wæs, Bt. 39, 3; Fox 216, 6.

Sŷn oferwrigene *operiantur*, Ps. Lamb. 70, 13. Synna beóþ oferwrigenne for dædbôte, L. E. I. 36; Th. ii. 434, 22. Ðǣr stôd ân æmtig cŷf oferwrogen, Homl. Th. ii. 178, 34. Se ðe wæs hwîlon gescrîd mid golde, hê læg ðâ oferwrogen mid moldan, Chr. 1086; Erl. 221, 3. Hwîtum gegyrlan oferwrohne (-wrogenne, MS. A.), Mk. Skt. 16, 5. Mid hwam beó wê oferwrogene? Mt. Kmbl. 6, 31. Oferwrogne *contecta*, Hpt. Gl. 417, 48.

ofer-wrîgels, es; n. A covering :—Ofyrwrîgyls *opertorium*, Ps. Spl. C. 101, 28. Oferwrîgelsum *operculis*, Wrt. Voc. ii. 62, 55.

ofer-writ *a writing upon a subject, a letter :*—In oferwurit his *in epistola sua*, Mt. Kmbl. p. 8, 20.

ofer-wundenness, e; f. Experiment, proof :—Oferwundennyssum *experimentis*, Hpt. Gl. 419, 38.

ofer-wyrcan *to cover by working, to work a covering over something, to overlay :*—Hê hit him eft hâm bebeád on ânum brede âwriten and siððan hit âwriten wæs hê hit oferworhte mid weaxe *qui omnia civibus suis per tabellas scriptas, et post cera superlitas, enunciabat*, Ors. 4, 5; Swt. 168, 14. Næfre nænig man ða lǣstas sylfe ufan oferwyrcean ne mihte, ne mid golde, ne mid seolfre, Blickl. Homl. 125, 35. Hûs (*the temple*) oferworht mid golde and mid hwîtan seolfre, Ælfc. T. Grn. 7, 35 : Blickl. Homl. 125, 25. Hê wæs bebyrged and oferworht syððan (*a tomb was erected*), Homl. Skt. i. 21, 19.

ofer-ŷð, e; f. An excessive wave, wave of a tempestuous sea :—Nâ selleþ on ēcnysse oferŷðe rihtwîse *non dabit in aeternum fluctuationem justo*, Ps. Spl. 54, 25.

ofesc, e; f. A border (?) :—Ðis syndon ðæs landes gemǣru ... Ǣrest of Seferne be hîgna gemǣre ... and swâ be ðære alra ofesce (*along the border of elders ?*) on ða neówan dîc, Cod. Dip. Kmbl. iii. 393, 11. v. owisc.

ôfest. v. ôfost.

ofet, es; n. Fruit, pulse :—Obet *fraga*, Wrt. Voc. ii. 109, 20. Ofet *fraga* (cf. streówberge *fraga*, 59), 36, 9 : 150, 28. Ofet, wæstm *fruges, frumenta*, 151, 31. Ofet *legumen*, i. 38, 54. Ðis ofet *the fruit of the tree of knowledge*, Cd. Th. 46, 12; Gen. 655. Ofæt, 33, 11; Gen. 518. Ofett, Exon. Th. 202, 29; Ph. 77. Ofetes wôs *ydromellum*, Wrt. Voc. i. 27, 43. Hit ofetes noman âgan sceolde, Cd. Th. 44, 34; Gen. 719. Ofætes, 30, 4; Gen. 461. Bergena oððe ofeta *bacciniorum*, Wrt. Voc. ii. 87, 29. Ofætum ł wurtum *leguminibus*, Hpt. Gl. 444, 71 : *holusculis, leguminibus*, 494, 47. [*Ayenb.* þet ovet of þine wombe : *O. H. Ger.* obaz, obez *pomum, grosa : Ger.* obst.]

of-faran *to come up with those who are pursued, to overtake, to get near enough to attack, to reach and attack :*—Ðâ Philippus wæs cirrende and forneán offeóll ða ðe hit ǣr forcurfan *revertenti Philippo Triballi bello obviunt*, Ors. 3, 7; Swt. 118, 1. Ðâ cômon Tarentine tô heora âgnum scipum, and ðâ ôðre hindan offôran, and hié ealle him tô gewildum gedydan bûton v *Tarentini Romanam classem praetereuntem hostiliter invaserunt, quinque tantum navibus per fugam elapsis*, 4, 1; Swt. 154, 6. Ðâ offôron hié ðone here hindan æt Buttingtûne and hine ðǣr besǣton *they pursued and came up with the Danes at Buttington, and there besieged them*, Chr. 894; Erl. 92, 22. Hié offôron ðone here hindan, ðâ hê hâmweard wæs, and him ðâ wið gefuhton, 911; Erl. 100, 26. Ðâ ne mehte seó fird hié nâ hindan offaran, ǣr hié wǣron inne on ðan geweorce, 894; Erl. 93, 7. Se cyng ofslôh heora swâ feala swâ hê offaran mihte, 1016; Erl. 157, 8. Ic tô fare *adeo*, ic eom offaren *adeor*, Ælfc. Gr. 37; Som. 39, 1. v. of-fêran, -fylgan, -irnan, -rîdan.

of-feallan *to fall upon, kill by falling, destroy :*—Hit hreás underbæc and forneán offeóll ða ðe hit ǣr forcurfan *the tree fell backwards, and by its fall very nearly killed those who before were cutting it down*, Homl. Th. ii. 510, 2. Sigferþ cyning hine offeóll and his lîc ligþ æt Wimburnan *King Sigferth laid violent hands on himself, and his body lies buried at Wimborne*, Chr. 962; Erl. 124, 2. Hê geslôg xxv dracena and hine ðâ [of] deáþ offeóll *he slew 25 dragons and then death fell upon him*, Salm. Kmbl. 430; Sal. 216. Seó môdinys wyle offeallan ða eádmôdnysse *superbia inruere vult super humilitatem*, Prud. 32 b. Æfter his fielle wearþ ðara cásera mǣgþ offeallen *caesarum familia consumta est*, Ors. 6, 6; Swt. 262, 6. v. of-fillan.

of-fellan. v. of-fillan.

of-fêran *to overtake (an enemy) :*—Pharao tengde æfter mid eallum his here and offêrde hî æt ðære Reádan Sǣ *Pharaoh pressed after with all his host, and overtook the Israelites at the Red Sea*, Homl. Th. ii. 194, 16 : Chr. 948; Erl. 118, 19. Se cyng fêrde him (*the Danes*) æt hindan, and offêrde hî innan Eást-Seaxan, and ðær tôgædere heardlîce fêngon, 1016; Erl. 158, 1. v. of-faran.

of-ferian *to bear off :*—Hê fræt fîftŷne men, and ôðer swylc ût offerede lâdlîcu lâc, Beo. Th. 3171; B. 1583.

of-fillan *to kill by felling, to kill by causing to fall, to destroy :*—Gif mon ôðerne æt gemǣnum weorce offelle (-fealle, MSS. B. H.), L. Alf. pol. 13; Th. i. 70, 9. Hî woldon heó sylfe offyllan oððe ǣdrencan *ruina perituri aut fluctibus absorbendi*, Bd. 4, 13; S. 582, 33. Ðâ hêt se dêma ðæt wîf weorpan on seáþ and ðǣr mid stânum offellan, Shrn. 89, 29. Cf. of-feallan.

offrian; p. ode, ede *To offer, bring a sacrifice* or *gift in honour of another :*—Ic offrige onsǣgednyssa *immolavi hostiam*, Ps. Spl. 26, 10.

Onsegdnisse ic offriu ðé *holocausta offeram tibi*, Ps. Surt. 65, 15. Ðú offrast án celf, Ex. 29, 10, 18, 20. Offrede *litarat, sacrificabat*, Hpt. Gl. 415, 13. Hió offreðe hiore ansegednesse *immolavit victimas suas*, Kent. Gl. 285. Hé offrude lác Gode his fæder *mactatis victimis Deo patris sui*, Gen. 46, 1. Offrode, Homl. Th. ii. 456, 34. [Hé offrede hit (*the body of St. Florentine*) Crist and sce Peter, Chr. 1013; Erl. 149, 21.] Hí offrodon (*immolaverunt*) twelf cealfas, Ex. 24, 5. Mesiane noldon ðæt Læcedemonia mægdenmenn mid heora ofreden, Ors. 1, 14; Swt. 56, 16. Ðé ofreden (*offerent*) cyningas gefe, Ps. Surt. 67, 30. Lǽtaþ ús faran and offrian (*sacrificemus*) úrum Gode, Ex. 5, 17. Tó offrienne *litaturas*, Hpt. Gl. 522, 25. [*O. Frs.* offria: *O. L. Ger.* offrôn: *Icel.* offra: *O. H. Ger.* opfarôn: *from Lat.* offerre.] v. ge-offrian.

offrung, ofrung, e; *f.* I. *the offering of a sacrifice* or *gift :*—Hit wæs gewunelíc on ealdum dagum, ðæt man Gode ðyllíce lác offrode on cucan orfe; ac seó offrung is nú unálýfedlíc, Homl. Th. ii. 456, 35. II. *an offering, sacrifice :*—Ic áxige hwǽr seó offrung (*victima*) sig . . . God foresceáwaþ ða offrunge, Gen. 22, 7–8. Hwæðer is máre, ðe offrung (ofrung, MS. A.), ðe ðæt weofud ðe gehálgaþ ða offrunge (ofrunge, MS. A.)? Mt. Kmbl. 23, 19. Melu oððe offrung *ador*, Ælfc. Gr. 9, 21; Som. 10, 32. Offrung *sacrificium*, Wrt. Voc. i. 28, 49. Offrung *oblatio*, 28, 43. Wylt ðú ús syllan offrunge *hostias quoque et holocausta da nobis*, Ex. 10, 25. Ofrunga *libamina, sacrificia*, Hpt. Gl. 487, 72. Offrunga *holocausta*, 509, 61: *holocaustomata*, 521, 71. [*O. L. Ger.* offrunga: *O. H. Ger.* opfarunga.]

offrung-disc, es; *m. A paten* [? v. húsel-disc] :—Ânnæ offringdisc intó Nunnamynstær (*she gives*) *one paten to the Nuns' monastery*, Chart. Th. 553, 17.

offrung-hláf, es; *m. Sacrificial bread, the shew-bread :*—Hé æt ða offringhláfas *panes propositionis comedit*, Mt. Kmbl. 12, 4.

offrung-sang, es; *m. A hymn sung when an offering is made :*—Nú sceole wé healdan úrne palm, ðþ ðæt se sangere onginne ðone offringsang, and geoffrian ðonne Gode ðone palm, Homl. Th. i. 218, 9.

of-fylgan, -fyligan *to come up with, overtake by pursuit;* assequi, Lk. Skt. Lind. Rush. 1, 3.

of-fyllan. v. of-fillan.

of-gán. I. *to demand what is due, seek satisfaction for, require, exact :*—Ic ofgá his blódes gyte æt ðínum handum *I will require the shedding of his blood at thy hands*, Homl. Th. i. 340, 24. God ofgæþ his feoh æt eów, 554, 19. Ic wille ofgan æt ðé his blód, i. 6, 27. Ic wille ofgán ða scép æt eówrum handum, 242, 11. Ic wolde mín ágen ofgán mid ðam gafole, ii. 554, 9. Ofgán *exigere*, Wülck. Gl. 257, 29. II. *to require what is not due, to exact with violence, extort :*—Ic wille mid tintregum æt ðé ofgán ðises þinges insiht *I will extort from thee with torments an account of this thing*, Homl. Th. i. 590, 22. Mid ðám tintregum hé wolde his æhta æt him ofgán, ii. 180, 18. III. *to require what is not one's due but is granted as a favour or for a fair equivalent, to obtain, hold by allowance of another :*—His bróðer wearþ his yrfenuma swá swá hé hit æt ðam cynge ofeode *his brother was his heir, according to the concession he had obtained of the king*, Chr. 1098; Erl. 235, 8. Gif óðres mynstres ár on óðres mynstres rýmette lége ðæt ðes mynstres ealdor ðe tó ðam rýmette fénge ofeode ðæs óðres mynstres áre mid swilcum þingum swylce ðam híréde ðæ ða áre áhte gecwéme wǽre *if one monastery's property lay in the space allotted to another, that the chief of the monastery that accepted the space should hold the other monastery's property on such conditions as should be agreeable to the society that owned that property*, Chart. Th. 231, 10–18. Hé beád ǽlcon his þegna ðe ǽnig land on ðan lande hafde ðæt hí hit ofeodon be ðes biscopes gemédon oððe hit ágéfon *that they should hold it in accordance with the bishop's pleasure, or give it up*, 295, 11. [Ich wille ðæt hit cume in ongeáen, ôðer ðæt man hit ofgô on hise gemôd, 387, 22.] Eádmund æþeling bæd ðone híréd ðæt hé móste ofgán (*have, hold*) ðæt land . . . Ðá cwæþ se cing ðæt hé nolde ðæt ðæt land mid ealle út áseald wǽre, ac ðæt ðæt land eft intó ðære hálgan stówe ágéfen wǽre, 300, 13–33. Ofgán tó rihtan gafole *to hold at a fair rent*, 355, 23: 478, 21. Ofgán land wið gersuman, 587, 7. God wile ðæt wé mid gemáglícum bénum his mildheortnesse ofgán *God wishes us to seek for his mercy by importunate prayers*, Homl. Th. ii. 126, 5. Wé sceolon mid hálgum mægnum ðone eard ofgán ðe wé þurh leahtras forluron *with holy virtues must we obtain the country, that we lost through vices*, i. 118, 33. [Ich hit wulle uorto ofgon (*gain*) þine heorte, A. R. 390, 13. To ofgon her lyflode, Piers P. 9, 106.] IV. *to start off, make a beginning of anything :*—Se ðe hine belecge ofgá his sprǽce mid foráðe *let him start his suit with a preliminary oath*, L. O. Ð. 6; Th. i. 354, 30: L. Ath. i. 23; Th i. 212, 4: L. C. S. 22; Th. i. 388, 14, 17: 30; Th. i. 394, 4. v. next word.

of-gangan. I. *to require :*—Ic ofgange *exigo*, Ælfc. Gr. 28, 6; Som. 32, 13. Eówer blôd ic ofgange (*requiram*) æt eallum wilddeórum and eác æt ðam men; of ðæs weres handa ic ofgange ðæs mannes líf, Gen. 9, 5. II. *to extort, exact what is not due :*—Ofgang ða mádmas *extort* (from St. Lawrence) *the treasures* (of the church, about which he would say nothing), Homl. i. 420, 26. III. *to acquire, obtain :*—Syle mé ðínne wíneard . . . ic ðé (*Naboth*) ôðerne finde oððe mid feó ofgange *give me thy vineyard . . . I will find thee another or will*

acquire it by purchase, Homl. Skt. i. 18, 175. Ne sý nán man ðe ðyses landes ǽniges dǽles brúke, bútan hé hit ofgange æt ðam híwum mid rihtum landrihte, Cod. Dip. Kmbl. iii. 435, 34. v. preceding word.

of-gangende *derivative :*—*Dirivativum*, ðæt is ofgangende, Ælfc. Gr. 18; Som. 20, 58. Sume (*pronouns*) synd *derivativa*, ðæt synd ofgangende, 15; Som. 17, 33. Hwæt sí betwux ðám genitvum ðæra frumcennedra pronomina and ðæra ofgangendra, Som. 19, 41.

ofgangend-líc *derivative :*—*Dirivativa*, ðæt is ofgangendlíc, Ælfc. Gr. 14; Som. 17, 4: 17; Som. 20, 35.

of-georn; *adj. Too eager, elated :*—Ofgeorn[um] *subnixis*, Hpt. Gl. 485, 45.

of-geótan. I. *to moisten by pouring, souse, soak :*—Ofgeót mid ealaþ *moisten the plants by pouring ale on them*, Lchdm. ii. 140, 15: iii. 28, 16. Ofgeót mid wætere, 48, 5. Ofgeót hý áne niht mid wýne ðanne on morgen nim ða leáf cnuca hý . . . and ofgeót hý mid ðan ylcan wíne ðe hý ǽr ofgotene wǽron *soak them a night with wine, then in the morning take the leaves, pound them . . . and soak them with the same wine that they were soaked with before*, 130, 30–132, 2. Sele wernióð on wearmum wætere ofgotenne, ii. 182, 6. II. *to put out a fire by pouring water on it :*—Hit biþ gelíc, ðæt man mid wætere ðone weallendan welm (líg, MS. D.) ofgeóte, ðæt hé leng ne môt ríxian, swá man mid ælmessan synna ealle álýseþ, Wulfst. 257, 21. Hí woldon ðæt fýr mid wætere ofgeótan, Homl. Th. ii. 166, 7.

of-geráð; *adj. Appropriate :*—Ðeáh ic hwílum gecoplíce funde, ic nú wépende ofgerádra worda misfô, Bt. 2; Fox 2, 9.

of-gesleán *to slay :*—Hí ðǽr ofgeslógan (ofslógon, MS. E.) micel wæl, Chr. 992; Erl. 130, 35. v. of-sleán.

of-gestígness, e; *f. Descent :*—Ofgestígnisse *descensionis*, Mt. Kmbl. p. 8, 4.

of-gifan *to give up, leave, abandon :*—Obgibeht (= ofgifeþ) *destituit*, Wrt. Voc. ii. 105, 77. Hé Dena land ofgeaf *he left the Danes' land*, Beo. Th. 3813; B. 1904. Hé ðás woruld ofgeaf *he died*, Cd. Th. 72, 30; Gen. 1194. Hé ðone beám ofgeaf *he* (Christ) *left the cross* (*when he was taken to be buried*), Exon. Th. 45, 35; Cri. 729. Hí flet ofgeáfon *they gave up their halls* (*when they died*), 290, 7; Wand. 61. Mec deádne ofgeáfun fæder and môder, ne wæs mé feorh ðágen in innan, 391, 7; Rä. 10; 1. Hí heora land ofgeáfan *patria profugi*, Ors. 1, 4; Swt. 32, 20. Ofgéfon, Cd. Th. 6, 8; Gen. 85. Ne ofgif ðú mé *ne elonges a me*, Ps. Th. 70, 11. Hé ða goldburg ofgifan (*leave*) wolde, Andr. Kmbl. 3309; An. 1657. Ofgefen *distitutum*, Wrt. Voc. ii. 106, 58. Ofgifene, 25, 59: Bd. 4, 9; S. 577, 3. [*O. H. Ger.* aba-geban *destituere: O. Sax.* af-geban.]

of-habban *to keep from, hold back, restrain :*—Gif ðú ðæt (*letting the people go*) git dôn nelt and ðæt folc ofhæfst (*retines*), Ex. 9, 2. [*Goth.* af-haban.] Cf. of-healdan.

of-hagian *to be inconvenient :*—Gif his scrifte ofhagie, séce man tó ðam leódbiscope, Wulfst. 275, 5. Cf. on-hagian.

of-healdan *to withhold, keep back, retain :*—Hé lét niman of hyre ealle ða betstan gærsuma ðe heó ofhealdan ne mihte *he had all the best valuables, that she could not keep back, taken from her*, Chr. 1035; Erl. 164, 23. Mí gauil hé hauiþ ofhealden *my rent he has withheld*, Chart. Th. 427, 30. [*Ayenb.* to ofhealde *to retain*.] v. of-habban.

of-hearmian; *v. impers. To cause grief :*—Ðá ofhearmode (ofearmode ? v. of-earmian) Gode heora yrmða *God was grieved at their miseries*, Jud. 11, 1. [Cf. *Icel.* harmr *grief, sorrow;* harmar einum *it vexes one*.]

of-hende; *adj. Out of one's hand, taken away, lost to one :*—Gif him ǽnig ðara of hende wyrþ *if any one of them is lost to him*, Met. 25, 34. [*Icel.* af-hendr.] v. on-hende.

of-hnítan *to kill by butting, to gore to death :*—Gif se oxa wer oððe wíf ofhnít *if an ox gore a man or woman, that they die*, Ex. 21, 29: L. Alf. 21; Th. i. 48, 27.

of-hreósan. I. *to overwhelm, cover, bury;* obruere :—Oft eorþstyrung fela burhga ofhreás, Homl. Th. i. 608, 26. Ðæt ne ða sleacgiendan hé (*sompnolentia*) ofhreóse (*obruat*), Hymn. Surt. 18, 15. Swylce hé sý mid moldhýpan ofhroren, Homl. Th. i. 492, 33. Sume (*martyrs*) mid stánum ofhrorene, 542, 30. Mid sande ofhrorene *operti humo* Num. 16, 33. Ofhrorenne *obrutum*, Hpt. Gl. 487, 25. Ofrorene *obruti*, 506, 7: *obrutos*, 478, 77. II. *to fall down :*—Heofonas berstaþ, tungol of hreósaþ, Exon. Th. 58, 12; Cri. 934.

of-hreówan. I. *to cause grief or pity* (a) *impers. with dat. of pers. and gen. of the cause :*—Mé ofhríwþ mé *miseret*, Ælfc. Gr. 33; Som. 37, 24. Ðæs sceápes untrumnesse him ofhreáw (-hreów, MS. F.), R. Ben. 51, 20. Him ofhreów ðæs mannes *he was sorry for man*, Homl. Th. i. 192, 16. (b) *with dat. of pers. and nom. of cause, or a clause introduced by* ðæt :—Ðá ofhreów ðam munece ðæs hreóflian mægenleást *the powerlessness of the leper excited the pity of the monk*, 336, 11. Mé ofhreów ðæt hí né cúðon ða godspellícan láre, 2, 22. II. *to feel pity :*—Se mæssepreóst ðæs mannes of hreów, Swt. A. S. Rdr. 102, 216. Iohannes of hreów ðære méder dreórignysse, Homl. Th. i. 66, 21.

of-hyngrod *very hungry :*—Eádige beóþ ða ðe sind of hingrode rihtwísnysse, Homl. Th. i. 204. Ofhingrod æfter rihtwísnysse, 550, 34–

35. [Gif þu ert ofhungred efter þe swete, A. R. 376, 18: Laym. offingred: Piers P. afingred.]

of-irnan. I. *to overtake (by running)*:—Færþ hê (*the evening star*) æfter ðære sunnan, óþ hê ofirnþ ða sunnan hindan, Bt. 39, 13; Fox 234, 2. Ðæs wîtegan cnapa, Gyezi, ofarn Naaman, Homl. Th. i. 400, 17. [þe abbed æfter Uortiger rad & sone gon ofærne Uortigerne, Laym. 13149.] II. *to tire with running*:—Hê wæs swîðe ofurnen *he was very tired with running*, Jud. 4. 19.

of-lǽtan. I. *to give up, relinquish*:—Gif ðû ǽr ðonne hê worold oflǽtest *if you die before him*, Beo. Th. 2371; B. 1183. Ða se ellorgâst oflêt lîfdagas and ðâs lǽnan gesceaft, 3248; B. 1622. Lîf oflǽtan, Cd. Th. 65, 28; Gen. 1073. [Goth. af-lêtan ahman *to give up the ghost*.] II. *to let off, cause (blood) to flow*:—Ðæt him (hine, Cott. MS.) mon ofiête blôdes on ðam earme, Bt. 29, 2; Fox 104, 23. [Goth. af-lêtan *to let off, forgive, dismiss*: Ger. ab-lassen.]

oflǽte, -lâte, -lête, an; f. I. *an oblation, offering*:—Oflǽtan oblationem, Ps. Spl. C. 39, 9. Oflâtan oblationes, Ps. Surt. 50, 21. II. *a sacramental wafer*:—Eal ðæt tô húsle gebirige, ðæt is, clǽne oflête, clǽne wîn, and clǽne wæter, L. Edg. C. 39; Th. ii. 252, 13. Behealde hê ðæt his oflêtan ne beón ealdbacene, L. Ælfc. C. 36; Th. ii. 360, 26. Benedictus âsende âne ofelêtan, and hêt mid ðære mæssian, Homl. Th. ii. 174, 26. III. *a wafer like the sacramental wafer*:—Man sceal niman .vii. lytle oflǽtan swylce man mid ofraþ, Lchdm. iii. 42, 3. [Erest þat husel beð ouelete and win, O. E. Homl. ii. 97, 33. Icel. oblâta, oblât *a sacramental wafer*: O. H. Ger. oblâta oblatio: Ger. oblate *wafer*. From Mid. Lat. oblâta.]

of-langod; *part. Seized with an excessive longing* or *desire*:—For ðære sibbe he wearþ oflangod ungemetlîce *he was seized with an immense longing on account of the love he bore his father and mother*, Homl. Th. ii. 176, 1. Oflongad, Exon. Th. 443, 13; Kl. 29. [Laym.: O. and N. of-longed.]

oflâte. v. oflǽte.

of-lecgan *to lay down*:—Eom ic on lâme oflegd *infixus sum in limo*, Ps. Th. 68, 1. [Goth. af-lagjan *to lay down, put away*.]

of-leógan *to lie, be false*:—Ða ælþeódgan bearn mê oflugon *filii alieni mentiti sunt mihi*, Ps. Th. 17, 43.

of-licgan *to oppress, to hurt by lying upon*:—Gif hwâ on slǽpe his bearn oflicge ðæt hit deád wurðe *si quis in somno infantem suum oppresserit, et mortuus sit*, L. M. I. P. 41; Th. ii. 276, 10.

of-lícian *to displease, be displeasing*:—Gode swŷðe oflícaþ heora ceorung, Homl. Skt. i. 21, 240. Swâ hwæt swâ him oflícaþ ðeáh hit hâlig sŷ hié hit lǽtaþ unâlŷfed *whatever they do not like, though it be holy, they profess that it is not permitted*, R. Ben. 9, 19. Ða oflícode mê þearle ðæt ic eft tô ðam lîchaman sceolde, Homl. Th. ii. 354, 10. Gif hwam seó lâr oflícige, 216, 23.

of-linnan *to cease, leave off*:—Ðæt wæter oflan and mâ of heora mûþe hit ne eode *the water stopped, and it no longer came out of the mouth of the image*, Blickl. Homl. 247, 8. [Goth. af-linnan *discedere*.]

of-lysted, -lyst; *part. Possessed with a very strong desire, very desirous for* (with gen. of object):—Eubolus wearð swâ mycclum oflyst Basilies lâre, ðæt him ne lyste nânes metes, Homl. Skt. i. 3, 42: Bt. 35, 6; Fox 168, 23. Ða wæs ðes man swîðe oflyst ðæs Hǽlendes tôcymes, Homl. Th. i. 136, 6. Þeód wæs oflysted metes, Andr. Kmbl. 2226; An. 1114: 2454; An. 1228. Ða wæs hê swŷðe oflysted ðæt hê ðæs eádigan weres blôd âgute *he was possessed with a very strong desire to shed the holy man's blood*, Guthl. 7; Gdwin. 44, 22. Ða ðe sind oflyste rihtwîsnysse, Homl. Th. i. 204, 1: Exon. Th. 464, 3; Hö. 81. [Oflust æfter deores flæsce, Laym. 30554.]

of-manian *to exact a fine or due*:—Ofmanige se bisceop ða bôte tô ðæs cynges handa, L. Edg. ii. 3; Th. i. 266, 19, note.

of-munan *to recall to mind, recollect*:—Ðonne hê hit eft ofman æfter lytlum fæce *cum post paululum haec ipsa ad memoriam revocat*, Past. 33, 7; Swt. 225, 19. Ne ofman hê nǽfre nânwuht forðæm nǽfre nâuht hê ne forgeat *he never recalls anything, for he never forgot anything*, Bt. 42; Fox 356, 30. Ic wât æác ðæt ic hyt hæfde swâ clêne forgeten ðæt ic hyt nǽfre eft ne ofmunde *I know too that I should have so clean forgotten it, that I should have never again recalled it*, Shrn. 198, 4.

of-myrðrian *to murder*:—Men hine ofmyrðrodon, Chr. 979; Erl. 129, 7.

ofn. v. ofen.

of-neádian *to obtain by force, extort*:—Nû cýdde man mê ðet Æþelwold and ic sceoldon ofneádian ða bôc æt Leófríce *I have been informed that Athelwold and I must have obtained the charter from Leofric by force*, Chart. Th. 295, 32. [O. Frs. of-nêda: cf. Ger. ab-nöthigen.]

ofnet *a closed vessel*:—Geseóþ ofnete *seethe in a closed vessel* (vasculo clauso vel operto), Lchdm. ii. 30, 24.

of-niman *to fail*:—Be ðam ðe him his sprǽc ofnimþ *de eo cui sermo deficit*, L. Ecg. P. i. tit. 3; Th. ii. 170, 6. [Ger. ab-nehmen *to decrease, wane*.]

ôfost, ôfest, ôfst, e; f. Haste, speed:—Ôfost is sêlost tô gecŷðanne hwanon eówer cyme sŷ *the quicker you make known where you come*

from the better, Beo. Th. 518; B. 256: 6007; B. 3007. Ôfest, Cd. Th. 196, 18; Exod. 293. Ôfst and hradung, R. Ben. 3, 11. Swâ hwylc preóst swâ wyrne (*refuses to baptize a man*) for ôfste his fôre *quicunque presbyter festinandi itineris sui causa deneget*, L. Ecg. C. 6; Th. ii. 138, 21. Se cnapa hit mid ôfste gegearcode *puer festinavit et coxit illum*, Gen. 18, 7. Mid ôfste (oefeste, Lind.) *cum festinatione*, Mk. Skt. 6, 15: Lk. Skt. 1, 39: Jn. Skt. 11, 31. Wê secgaþ nû mid ôfste ðâs endebirdnisse, for ðan ðe wê oft habbaþ ymbe ðis âwriten, Ælfc. T. Grn. 3, 30. Ðeós woruld is on ôfste *this world is hurrying on* (*to its end*), Wulfst. 156, 5: Cd. Th. 191, 32; Exod. 223. Beó ðu on ôfeste *hasten*, Beo. Th. 777; B. 386. On ôfoste, 5487; B. 2747. Þorh ôfst *per anticipationem*, Wrt. Voc. ii. 116, 77. Ôfestum *hastily*, Cd. Th. 140, 32; Gen. 2343. Ôfestum miclum, 177, 15; Gen. 2930. Ôfstum, 153, 8; Gen. 2535: 161, 29; Gen. 2672. [Laym. ovest.]

ôfostlíce; *adv. Hastily, speedily, in haste*:—Hê stôp ôfostlíce tôforan ðam biscope and feól tô his fôtum *festinus accedens ante pedes episcopi conruit*, Bd. 3, 14; S. 540, 36. Ongan ôfostlíce ðæt hof wyrcan, Cd. Th. 79, 24; Gen. 1316. Gewît ðu ôfostlíce fêran, 172, 24; Gen. 2849. Mê ôfestlíce gehŷr *velociter exaudi me*, Ps. Th. 101, 2. Ôfstlíce *cursim, velociter*, Hpt. Gl. 446, 48. Éfstende wê sceolon etan úre eásterlícan blisse, and ôfstlíce wê sceolon Godes bebodu healdan, Anglia viii. 323, 36: Cd. Th. 150, 6; Gen. 2487. Ôfostlícor, Exon. Th. 17, 18; Cri. 272. [O. Sax. ôbast-, ôfst-líko.]

of-rídan *to overtake by riding, overtake*:—Éfstaþ ardlíce and gê hig ofrídaþ *persequimini cito, et comprehenditis eos*, Jos. 2, 5. Abram éfste wið ðæs heres óþ ðæt hê hig ofrâd, Gen. 14, 14. Se cyng hêt rídan æfter, and ne mehte hine mon ofrídan, Chr. 901; Erl. 98, 1: 877; Erl. 78, 21. v. of-faran.

of-sacan *to deny a charge*:—Gif hwâ ofsacan wille, dô ðæt mid eahta and feówertig fulborenra þegena, L. Ath. iv. 7; Th. i. 228, 3. [I ne mai hit noзt ofsake, P. L. S. 15, 60. Cf. Icel. af-saka *to exculpate*.]

of-sceacan *to shudder, shake with fear*:—Ofscôc *exhorrui*, Hpt. Gl. 504, 10.

of-sceamian *to put to shame*:—Gif ðû ðe ofsceamian (onsceamian, MS. Cott.) wilt ðînes gedwolan, Bt. 3, 4; Fox 6, 16. Hê wearþ ofsceamod, Homl. Skt. i. 12, 214. Se drŷ stôd eádmôd and ofsceamod *the sorcerer stood humble and ashamed*, Homl. ii. 416, 30. Hí gecyrdon him hâm hearde ofsceamode, 518, 31. [O. and N. of-schomed: R. Glouc. of-ssamed.]

of-sceótan. I. *to wound or kill by shooting an arrow* or *by hurling a weapon*:—Wulfstân ðone forman man mid his francan ofsceát, Byrht. Th. 134, 1; By. 77. Hæþcyn his mǽg ofscet blôdigan gâre, Beo. Th. 4870; B. 2439. Hê hiene ne meahte ofsceótan mid ðæm bismere *quem commovere in ipsa contumeliarum jaculatione non potuit*, Past. 33, 7; Swt. 227, 9. Hê mid geættrode flâne hine ofsceótan wolde, Homl. Th. i. 502, 18. Hê weard ofscoten mid ânre flâne *sagitta ictus interiit*, Ors. 1, 2; Swt. 30, 13. Ðǽr weard Leostenas mid ânre flân ofscoten *ibi Leosthenes telo e muris jacto perfossus occiditur*, 3, 11; Swt. 144, 27. Mid fŷrenum flânum ofscotene (ofsceotene, 7), Homl. Th. i. 506, 1. II. Ofscoten *elf-shot, diseased from an elf's shot*, Lchdm. ii. 156, 25: 290, 21. The disease consists in an over-distension of an animal's stomach from the swelling up of clover and grass, when eaten with the morning dew on it. See the Glossary and Jamieson's Scottish Dictionary *elf-shot*. v. next word.

of-scotian *to shoot, wound* or *kill with an arrow, spear*, etc.:—Hê hêt hine mid strǽlum ofscotian, ðæt hê wæs ðara swâ full swâ igl biþ byrsta, Shrn. 55, 8. Ne ofsleá hine nân man mid his handa ac sî hê mid stânum oftirfod oððe mid flânum ofscotod *manus non tanget eum, sed lapidibus opprimetur aut confodietur jaculis*, Ex. 19, 13. Mid flânum ofscotod and mid stânum oftorfod *sagittis, saxis contriti*, Ors. 4, 11; Swt. 206, 14. v. preceding word.

of-sendan *to reach by sending, send for, summon*:—Ofsǽnd se cyng Godwine eorl *the king sent for earl Godwin*, Chr. 1048; Erl. 178, 7. Ða sende se cyng æfter ðâm scypon ðe hê ofsendan mihte *the king sent after the ships that his summons could reach*, 1049; Erl. 172, 39. [Heo him radden ðat he ofsende magan, Laym. 15748. Þis kyng ys knyзtes let ofsende, R. Glouc. 122, 21.] v. of-faran.

of-seón *to see, observe*:—Ða ofseah hig Godes engel *cum invenisset eam angelus Domini*, Gen. 16, 7. Ða ofseah hê ǽnne geongne man, Homl. Skt. i. 23, 545. Se hâlga wer oferseah ealne middaneard, and ofseah lǽdan ânes biscopes sâwle tô heofenum, Homl. Th. ii. 184, 30. Se apostol ofseah hwǽr sum ûþwita lǽdde twegen gebrôðru, i. 60, 22. Ofsión, Met. 21, 38.

of-setenness, e; f. Siege:—Ofsetenesse *obsidione*, Wrt. Voc. ii. 63, 13. v. of-sittan.

of-settan *to beset, press hard, oppress*:—Hê hig ofsette and geswencte *he oppressed and afflicted them*; servierunt ei, Jud. 3, 8. Fearras fǽtt ofsetton ł ymbsǽton (obsederunt) mê, Ps. Lamb. 21, 13. Feónd ûrne ofsete (comprime), Hymn. Surt. 11, 33. Mid untrumnesse oððe bysegum ofset, R. Ben. 58, 15: 59, 3. Mid weorces geswince ofsette, 63, 17: Homl. Th. ii. 120, 8. Se ðe on wræcsîð gesihþ mid micelum gyltum

heom ofsett getácnaþ *if a man dreams of being in exile, it betokens that he will be weighed down with great crimes,* Lchdm. iii. 212, 23. Swéte etan on manegum leahtrum biþ ofsett hit getácnaþ *to dream of eating sweets betokens a man will be sunk in many faults,* 202, 25. Ofsettum *obsessis,* Wülck. Gl. 251, 5. v. next word.

of-sittan.   I. *to sit upon, press down by sitting :*—Heó ofsæt ðone selegyst *she (Grendel's mother) pressed down the hall-guest (Beowulf, who had fallen),* Beo. Th. 3094 ; B. 1545. Nú sceal se ðe wile sittan æt Godes gereorde ðæt gærs ofsittan, ðæt is, ðæt hé sceal ða flǽsclícan lustas gewyldan, Homl. Th. i. 188, 26.   II. *to sit upon, oppress :*—Gif hé (*a king*) his folc ofsit, ðon biþ hé *tyrannus,* Ælfc. Gr. 50, 20 ; Som. 51, 47 : Homl. Th. i. 242, 4. Swongornes hí ofsit, and hí mid slǽwþe ofercymþ, Bt. 36, 6 ; Fox 180, 33. Godes fýnd ðe ða earman ofsittaþ, Jud. Thw. 156, 5. Ðú wilt cweþan ðæt ungemetfæstnes hí ofsitte, Bt. 36, 6 ; Fox 182, 2. Ete ælþeódig folc ðíne tilunga and ðé mid bismore ofsittan *sis calumnian sustinens,* Deut. 28, 33. Ofseten mid ðǽm ðístrum ðisses andweardan lífes *praesentis vitae tenebris pressus,* Past. 11, 1 ; 65, 7. Ic eom mid earmlícre ofergiotolnesse ofseten, Shrn. 198, 21. Ðæt mód sǽde ðæt hit wǽre ofseten (cf. ofþrycced, Fox 24, 14) mid ðæs láðes sáre, Bt. 8, tit. ; Fox x. 19.   III. *to sit upon, occupy, take possession of* (with idea of force or wrong) :—Ðæt sió oferflównes ðæra geþohta ne meahte ofsittan ðæs sacerdes heortan *quatenus sacerdotale cor nequaquam cogitationes fluxae possideant,* Past. 13, 1 ; Swt. 77, 11. Eall ðæt seó sǽ his ofseten hæfþ *quantum maria premunt,* Bt. 18, 1 ; Fox 62, 12.   IV. *to sit about, besiege :*—Fearas mé ofsǽton (*obsederunt*), Ps. Th. 21, 10, 14.   V. *to repress, check, prevent motion ;* cf. of-standan :—Ðus ðú scealt ða yfelan ofsetenan wǽtan út ádón *thus shalt thou remove the evil, repressed humours,* Lchdm. ii. 24, 7. v. preceding word.

of-sleán *to kill* or *wound by a blow, to kill, slay :*—Ic ofsleah wildeór *ego jugulo feras,* Coll. Monast. Th. 21, 19. Ic on morgenne ofsleá mánes wyrhtan *in matutino interficiebam omnes peccatores,* Ps. Th. 100, 8. Gif man mannan ofslæhþ, L. Eth. 21 ; Th. i. 8, 3. Ofsleahþ, 6 ; Th. i. 4, 6. Ofslehþ, 7 ; Th. i. 4, 9. Gif hwá his cild ofslihþ tó deáþe, L. M. I. P. 8 ; Th. ii. 268, 1. Se geréfa nyste hwæt hié wǽron, and hiene mon ofslóg, Chr. 787 ; Erl. 56, 15. Hé (*the elephant*) ofslóg micel ðæs folces, Ors. 4, 1 ; Swt. 156, 12. Ofslóh, Cd. Th. 60, 18 ; Gen. 983. Ðú ofslóge (*percussisti*) ealle ða ðe mé wiðerwearde wǽron, Ps. Th. 3, 6. Hí willaþ mé ofsleán *interficient me,* Gen. 20, 11. Eall his weorod ofslegen wæs, Bd. 1, 34 ; S. 499, 33. Wurdon begen ofslægene ða aldormen, Chr. 800 ; Erl. 60, 8. Fela þúsenda ofslægenra, 871 ; Erl. 74, 24. [*Goth.* af-slahan *to slay.*]

of-slegenness ; *f. Killing, destruction :*—Sceáp ofslegennysse *oves occisionis,* Ps. Spl. 43, 25.

of-slítan *to wound by the bite* (of a snake, dog, etc.) :—Ða ðe ofslitene wǽron (*the Israelites who were bitten by the serpents*), Num. 21, 9.

of-smorian *to choke, strangle, suffocate :*—Hiene ofsmorode his ealdormon *dolo comitis sui strangulatus,* Ors. 6, 36 ; Swt. 294, 9. Mid ðǽm brǽþe ofsmorod (*suffocatus*), 6, 32 ; Swt. 288, 2.

of-sníðan *to kill by cutting, to slaughter* (an animal) :—Ðæt lamb ðe se ealda Israhel ofsnáð, Homl. Th. ii. 264, 28 : Gen. 22, 13. Ða námon hig án biccen and ofsnidon hit, 37, 31. Swilce se sunu wǽre geoffrod and se ramm ofsniden, Homl. Th. ii. 62, 27.

of-sprǽc, e ; *f. An outspeaking, utterance :*—Gydde ł ofsprǽce *elogio, textu, locutione,* Hpt. Gl. 460, 65.

of-spring, es ; *m. Offspring, progeny, posterity, descendants :*—Ofspring styrps, Ælfc. Gr. 9, 58 ; Som. 13, 36. Ofsprincg *progenies,* 12 ; Som. 15, 53. Eall heora ofspring ðe him of com, Ælfc. T. Grn. 3, 11. Eall heora ofspринc boren and unboren, Cod. Dip. Kmbl. iv. 263, 29. Gif his sunu and sunu-sunu swá micel landes habban, siððan biþ se ofsprinc (cf. hiora æftergengas, 24) gesíðcundes cynnes, L. Wg. 11 ; Th. i. 188, 11. Ic sette feóndrǽdene betweox ðínum ofspringe and hire ofspringe, Gen. 3, 15. Ðínum ofspringe (*semini tuo*) ic forgife ðis land, 12, 7 : 13, 15. Ic dó ðínne ofspring swá menigfealdne swá ðære eorþan dust, 13, 16. Hí gesworen habbaþ ge for hý sylfe ge for heora ofspryng (gingran, MS. B.), L. A. G. tit. ; Th. i. 152, 17. Ðis sý gedón for Síferþ and for his ofsprincg tó hyra sáwle þearfe, Cod. Dip. Kmbl. ii. 300, 15. [*Icel.* af-springr.]

of-spyrian *to find out by inquiry* or *search :*—Se ðe hit ofspyraþ, hé áh ðæt meldfeoh, L. In. 17 ; Th. i. 114, 4. Cf. of-áxian.

ófst. v. ófost.

of-standan *to remain standing, keep* (trans. or intrans.) *in the same place* or *condition, stop in a place :*—Swá raðe swá ðæt scrín in biþ geboren, swá ofstint (oft stint, Thw.) se stream *aquae in una mole consistent,* Jos. 3, 13. Gif him ofstondeþ on innan ǽnigu ceald wǽte *if any cold humour stops in them,* Lchdm. ii. 194, 15. Sele him on hátum wǽtre gewlecedum ða wyrta drincan ðý læs ðæt pic ofstande mid ðý dýr duste *give him the herbs to drink in hot water made lukewarm, lest the pitch be left sticking with the other dust,* 252, 4. Ðæt ofstandene þicce horh *the thick foulness that has refused to move,* 194, 21. Wág ofstonden unde stormum *a wall unmoved amid storms,* Exon. Th. 476, 21 ; Ruin.

11. Ofstondene beóþ *sive* ofstonden feoh *integri restitutione,* Wrt. Voc. ii. 49, 34–35. [Cf. *O. Sax.* is (*of the temple*) afstandan ni skal stên oðar óðrumu.]

of-steppan *to trample upon :*—Ofstæppaþ heora swuran swíðe mid fótum. Ða dydon ða ealdormen swá, swá him dihte Josue, and ðæra cynega swuran forcúðlíce trǽdon, Jos. 10, 24. v. of-tredan.

of-stician *to wound* or *kill by a thrust, to stab, pierce, transfix :*—Ofsticoþ *configet,* Kent. Gl. 844. Ic ofsticode hyne *jugulavi aprum,* Coll. Monast. Th. 22, 17. Antonius hiene selfne ofsticade *Antonius sese ferro transverberavit,* Ors. 5, 13 ; Swt. 246, 30. Se kásere álýfde ðám cnihtum ðæt hí hyne (*St. Casianus*) ofslógon mid heora writbredum, and hine ofsticodon mid hira writýrenum, Shrn. 117, 29. Ðá hét hé ðone pápan (*Alexander*) ofstician, 79, 8. Ofstikian bár *jugulare aprum,* Coll. Monast. Th. 22, 13. Hé swealt ofsticod fram him sylfum, Chr. 2 ; Erl. 5, 19. [*Ger.* ab-stechen.] v. of-stingan.

ófstig ; *adj. Hasty, swift :*—Ófstige *percita, velocissima,* Germ. 392, 73.

of-stingan *to wound* or *kill by a thrust, to stab, pierce :*—His ealdgefána sum hiene ofstang *a Pausania occisus est,* Ors. 3, 7 ; Swt. 118, 34. Hé hiene (*the elephant*) on ðone nafelan ofstang, 4, 1 ; Swt. 156, 11. Hé (*Pilate*) hiene selfne ofstong *sua se transverberans manu,* 6, 3 ; Swt. 258, 10 : Shrn. 33, 5. Hé wolde ofstingan Eádwine, ac hé ofstang Lillan his þegn, Chr. 626 ; Erl. 23, 29 : 755 ; Erl. 48, 23. Ne ofstong Æfner hiene nó mid dý speres orde ac mid hindewerdum ðam sceafte *Abner non eum recta, sed aversa hasta transforavit,* Past. 40, 6 ; Swt. 297, 10. Sunu gif hé (*an ox*) ofstinge (*gore*), L. Alf. 21 ; Th. i. 50, 2. Hét Fabianus ðæt hí man begen ofstunge, Homl. Skt. i. 5, 405. Ðæt ic ðé ne dyrre ofstingan *ne compellar confodere te in terram,* Past. 40, 5 ; Swt. 295, 16. Hér Ædmund cyning wearð ofstungen, Chr. 948 ; Erl. 117, 8. v. of-stician.

ófstlíce. v. ófostlíce.

of-swelgan *to swallow up, devour :*—Deáþ forsiehþ ða æþelo, and ðone tícan gelíce and ðone heánan ofswelgþ, Bt. 19 ; Fox 68, 33. Cf. for-swelgan.

of-swingan *to scourge to death :*—Sume hié ofslógon sume ofswungon sume wið feó gesealdon *omnes bello utiles caesi, reliqui pretio venditi sunt,* Ors. 4, 1 ; Swt. 154, 8.

oft ; *adv. Oft, often :*—Oft (*saepe*) hé fylþ on fýr, and gelómlíce (*crebro*) on wæter, Mt. Kmbl. 17, 15. Oft (oftust, Lind. Rush.) *sepe,* Mk. Skt. 5, 4 : *interdum,* Wrt. Voc. ii. 48, 38. Hú oft *quotiens,* Lk. Skt. 13, 34. Swá oft swá ða óðre hergas út fóron, ðonne fóron hié, Chr. 894 ; Erl. 90, 5. Hí beóþ ðæs ðe lator ðe hí oftor ymbþeahtiaþ, Past. 56 ; Swt. 435, 2. Hwílum hé wæs on horse sittende, ac oftor on his fótum gongende, Bd. 4, 27 ; S. 604, 12. Hé oftust on gebedum áwunode, 3, 12 ; S. 537, 22. Oftost, Met. 4, 28. [*Goth.* ufta : *O. Sax.* oft, ofto : *O. Frs.* ofta : *Icel.* oft, opt : *O. H. Ger.* oft, ofto.] v. for-oft.

of-talu, e ; *f. The successful defence made against a claim :*—Seó sprǽc wearð ðam cynge cúð. Ða ðá him seó talu cúð wæs, ðá sende hé gewrit tó ðam arcebisceope, and beád him ðæt hé and hys þegenas hý on riht gesémdon be ontale and be oftale *the suit became known to the king. When the claim was known to him, he sent a writ to the archbishop, and commanded him that he and his thanes should settle it rightfully according as the claim was to be allowed or rejected, according as the verdict was for or against the claim* (cf. *Icel.* bera kvið á einn, af einum *to give a verdict for, against a person*), Chart. Th. 302, 14–22.

of-teón ; *pp.* -togen *and* -tigen.   I. *to withdraw :*—Hé hine ofteáh ðære fóre *subtraxit se illi profectioni,* Bd. 5, 9 ; S. 623, 23.   II. *to take away what a person has, deprive a person of anything* (with dat. or acc. of person, gen. of thing, or dat. of person and acc. of thing) :—Ic ofteó *derogo,* Wrt. Voc. i. 39, 27. Gif mon him oftíhþ ðara þenunga and ðæs anwealdes *detrahat si quis superbis vani tegmina cultus,* Bt. 37, 1 ; Fox 186, 10. Oft ic sýne ofteáh, ábiende beorna unrím, Exon. Th. 270, 21 ; Jul. 468. Wé oftugon ðé londes wynna, 130, 15 ; Gú. 438. Búton seó syncopa ðone i (*of the gen. pl.*) ofteó, Ælfc. Gr. 10 ; Som. 14, 55. Nó Ælmihtig ealra wolde Adam and Eve árna ofteón, Cd. Th. 58, 29 ; Gen. 953. Gif him gebyreþ ðæt him wyrþ ðara þenunga oftohen (oftogen, Met. 25, 31), Bt. 37, 1 ; Fox 186, 14. Ðú biþ seó bodung oftogen, Homl. Th. ii. 530, 30. Oftigen biþ him torhtre gesihþe *he shall be deprived of clear sight,* Exon. Th. 335, 29 ; Gn. Ex. 41.   III. *to withhold, keep back, deny a person anything :*—Ic ðé ofteó mínne fultum . . . Ic ofteó míne rénscúras *I will withhold from thee my help . . . I will withhold my rain-showers,* Homl. Th. ii. 102, 32–33. Gehelp ðú earmra manna mid ðam dǽle ðe ðú ðé sylfum oftíhst, i. 180, 12. For synnum oftíhþ se Ælmihtiga Wealdend hwílon mannum bigleofan, ii. 462, 21. Gif wé Godes láre eów ofteóþ, 554, 18. Hond feorhsweng ne ofteáh *the hand refused not to strike a fatal stroke,* Beo. Th. 4972 ; B. 2489 : 3045 ; B. 1520. Gé him ǽghwæs oftugon hrægles nacedum móses meteleásum *ye withheld from them everything, raiment from the naked, food from the hungry,* Exon. Th. 92, 8 ; Cri. 1505. [And wó só míne cwyde ofte God him ofte heuenríches *and whoso refuses to carry out my testament, may God refuse him the kingdom of heaven,* Chart. Th. 515, 30.]

Hé hēt hire ofteón ǽtes and wǽtes, Homl. Skt. i. 8, 129 : Blickl. Homl. 37, 28. Ðæt ðám gódum, ðe hit (doctrine) gehealdan willaþ, ne sý oftogen seó gástlíce deópnyss, Homl. Th. ii. 96, 4 : i. 370, 8. Swelce snytro swylce óðrum ieldran gewittum oftogen is negatas senibus dignitates, Bt. 8 ; Fox 24, 29. [Goth. af-tiuhan : O. H. Ger. aba-ziohan abstrahere.]

of-þænnan to moisten :—Gegnid on wín, ofþæne wel, Lchdm. iii. 90, 7. Ofþæne mid ecede, 184, 15. Obðaenit madidum, Wrt. Voc. ii. 113, 72. Ofðæned madefactus, 80, 48. Ofþænda and gesodena on ecede, Lchdm. ii. 180, 15. Mid hláfes cruman ofþendum mid cealdwǽtre, 82, 7.

of-þecgan to consume, destroy :—Æþelinga bearn ecgum ofþegde consumed by the sword, Cd. Th. 120, 30 ; Gen. 2002.

of-þencan to remember :—Gif ðú ofþence hwæthwugu ðæs ðe ðín niéhsta ðē wiðerweardes gedón hæbbe si recordatus fueris, quia frater tuus habet aliquid adversum te, Past. 46, 4 ; Swt. 349, 10.

of-þinen too moist (?) :—Hig wǽron gemyndige ðæs tōweardan hungres ðý læs ða ofþinenan corn in brord gehwyrfden and hig forcurfon ða sǽd they (the ants) were mindful of future hunger, and lest the grains that were too moist should sprout, they bit them, Shrn. 41, 5. Cf. ofþænnan.

of-þístrian to darken, obscure :—Ða ðe ofþýstrode synd qui obscurati sunt, Ps. Spl. 73, 21.

of-þringan to throng, crowd, press upon :—Ðeós menigu ðe ofþrincþ 'the multitude throng thee and press thee' (A. V.), Homl. Th. ii. 394, 15. Ðæt hí hine ne ofþrungon 'lest they should throng him' (A. V.), Mk. Skt. 3, 9.

of-þriton, Jud. 4, 24. v. next word.

of-þryccan to press, oppress, repress, cumber, occupy forcibly :—Ic ofþricce premo, Ælfc. Gr. 28, 4 ; Som. 31, 14. Hwí ofþricþ hé ðæt land quid terram occupat ficulnea, Lk. Skt. 13, 7. Ofþrect comprimit, Kent. Gl. 654. Ofþrecþ exprimit, 1120. Se draca mē þearle ofþryhþ, Homl. Th. i. 534, 25. Ymbhídignyssa ofþriccaþ ðæt mód, ii. 92, 15. Ofþrihte compressit, Hpt. Gl. 465, 21. Míne sáwle feóndas míne ofþryhtum (occupaverunt), Ps. Th. 58, 3. Ofþrihton, Blickl. Gl. Hig hine ofþriton (·þrihton?) opprimebant eum, Jud. 4, 24. Ofþriccetan mē grynu deáþes praeoccupaverunt me laquei mortis, Ps. Lamb. 17, 6. Feónd úrne ofþrece (comprime), Hymn. Surt. 11, 33. Ofþrice reprime, 13, 7. Ofþreccan comprimant, 17, 32. Wē ofþriccan praeoccupemus, Ps. Spl. M. 94, 4. Ofþriccende deprimentes, 88, 41. Mid unrōtnessum ofþrycced, Ps. Th. 38, arg. : Bt. 8 ; Fox 24, 14. Biþ ofþrecced opprimitur, Kent. Gl. 974. Beón ofþryht deprimi, Rtl. 66, 25. Ofþrihte compressa, Hpt. Gl. 490, 13.

of-þryccedness, e ; f. Distress, trouble :—Biþ mycel ofþriccednys (pressura) ofer eorþan, Lk. Skt. 21, 23 : Homl. Th. i. 608, 24. Fram ofþriccednysse (a refuge) from trouble, Blickl. Gl.

of-þrycness, e ; f. Oppression, repression :—Swá þrycce se magister ða belde on ðæm oferblíðum ðæt ðær ne weaxe on him sió ofþrycnes ðæs eges ðe cymþ of ðæs yflan blódes flównesse sic in illo reprimatur repente oborta praecipitatio, ut non convalescat impressa ex conspersione formido, Past. 61, 1 ; Swt. 455, 22. Seó Sūþ-Seaxna mægþ for ðære grimman feónda ofþrycnesse ágenne biscop habban ne mihte the people of Sussex on account of the cruel oppression of their foes could not have a bishop of their own, Bd. 4, 13 tit. ; S. 581, 38.

of-þryscan to beat down, repress, suppress :—Concutit i. turbat, terreat tóscæcþ, ofþrysceþ percutit, Wrt. Voc. ii. 136, 48. Ða ðe ofþryscaþ ða styringe ðæs flǽsclícan lustes qui compressis motibus carnis, Past. 52, 6 ; Swt. 409, 1. Ða hié suíðe stíðlíce árásigeaþ and mid ealle ofþrysceaþ hos asperitate rigidos semper invectionis premunt, Past. 19, 2 ; Swt. 145, 1. Ðæt hé on him selfum ofþrysce ða lustas his unþeáwa in semetipso suggestiones vitiorum reprimat, 14, 5 ; Swt. 85, 12. [Cf. O. H. Ger. druski excute, Grff. 5, 265.] v. ge-þryscan.

of-þrysman, -þrysmian to destroy by choking :—Gewilnunga ðæt word ofþrysmaþ (-þrysmiaþ, MS. A.) concupiscentiae verbum suffocant, Mk. Skt. 4, 19.

of-þyncan with dat. of pers. and (a) gen. of cause, (b) nom. of cause, (c) cause given by a clause. I. to cause regret or sorrow :—Mē ofþincþ penetet, Ælfc. Gr. 33 ; Som. 37, 21. Mē ofþincþ ðæt ic hig worhte poenitet me fecisse eos, Gen. 6, 7. Hit mē ofþincþ, Lk. Skt. 17, 4. Ofþinceþ ðe ealles ðe ðú tó yfele hæfst geworht? L. Ecg. C. proem. ; Th. ii. 130, 43. Ðá ofþúhte Pharao ðæt hē ðæt folc swá freólíce forlēt, Homl. Th. ii. 194, 15. II. to cause displeasure or offence :—Ðonne him hira scylda ná ne ofþyncþ si minus contra culpas accenditur, Past. 21, 5 ; Swt. 161, 2. Hine drehton nihtlíce gedwimer swá ðæt him ðæs slǽpes ofþúhte so that sleep was displeasing to him, Homl. Th. i. 86, 19. Ðá ofþúhte heora ceorlum ðæt mon ða þeówas freóde, Ors. 4, 3 ; Swt. 162, 15. Ðá ofþúhte ðæt ánum ðæs cyninges geféran, Lchdm. iii. 424, 16. Ðá ðæs ofþúhte, ðæt se þeóden wæs strang, Cd. Th. 279, 32 ; Swt. 247. Ðá sceolde ðám gigantum ofþincan þæt hē hæfde hiera ríce it is said that the giants were displeased at his having their kingdom, Bt. 35, 4 ; Fox 162, 11. Mæg ðæs ofþincan þegna gehwam, Beo. Th. 4070 ; B. 2032. Hit wæs swíðe ofþyncende ðám óðrum consulum it gave great offence to the other consuls, Ors. 5, 9 ; Swt. 232, 21. Mid ðon ðe hé geweóx, him

ða ofþyncendum and ðǽm Perseum ðæt hié on his eámes anwalde wǽron, Ors. 1, 12 ; Swt. 52, 18. Him ða ofþyncendum ðæt his folc swá forslagen wæs, 2, 5 ; Swt. 80, 23.

of-þyrsted, -þyrst ; part. Possessed with exceeding thirst, very thirsty, athirst :—Hé wearþ swíðe ofþyrst sitiens valde, Jud. 15, 18. Eádige beóþ ða ðe sind ofhingrode and ofþyrste æfter rihtwísnysse, Homl. Th. i. 550, 34. Níþes ofþyrsted thirsting for strife, Cd. Th. 3, 7 ; Gen. 32. Ofþyrsted gástes dryncnes, Soul Kmbl. 80 ; Seel. 40. [Cf. Goth. afþaursiþs thirsty.]

of-tige. v. of-tyge.

of-torfian to stone, to kill by casting stones or similar missiles :—Hí ús oftorfiaþ mid stánum lapidibus nos obruent, Ex. 8, 26. Hí hine oftorfodon mid bánum and mid hrýðera heáfdum, Chr. 1012 ; Erl. 146, 18. Hig wyllaþ mē oftorfian populus lapidabit me, Ex. 17, 4. Mid stánum oftorfian lapidibus opprimere, Num. 14, 10 : lapidare, Jn. Skt. 8, 5 : Homl. Th. i. 48, 2 : 196, 12. Fela mid stánum oftorfod saxis contriti, Ors. 4, 11 ; Swt. 206, 15. Hér wæs scs Stephanus oftorfod, Chr. 34 ; Erl. 6, 15 : Ælfc. T. Grn. 9, 31. v. of-tyrfan.

oft-rǽde ; adj. I. frequent :—Hæglas and snáwas and se oftrǽda rēn leccaþ ða eorþan on wintra, Bt. 39, 13 ; Fox 234, 16. II. ready at many times :—Gafolswán sceal beón swá ic ǽr be beócere cwæþ (cf. l. 3, beóceorl sceal hwíltídum geara beón on manegum weorcum tó hláfordes willan) oftrǽde tó gehwilcon weorce the swain must be, as I said before of the beekeeper, generally ready for any work, L. R. S. 6 ; Th. i. 436, 18. Bydele gebyraþ ðæt hē sý weorces frigra ðonne óðer man forðan hē sceal beón oftrǽde he must be always ready, 18 ; Th. i. 440, 7.

oftrǽd-líc ; adj. Frequent :—Ðis syndon ða wǽpena ðe deófol mid oferswíðed biþ, oftrǽdlíce rǽdinga háligra bóca and gelómlíce gebedu, L. E. I. 2 ; Th. ii. 404, 2. Hié Alexander uneáðe oferwonn ǽgðer ge for ðære sumorhǽte ge eac for ðǽm oftrǽdlícan gefeohtum, Ors. 3, 9 ; Swt. 132, 32. Oftrǽdlíca gefeoht crebra bella, 6, 30 ; Swt. 282, 31.

oftrǽdlíce ; adv. Frequently, often, habitually :—Oftrǽdlíce crebro, frequenter, Wrt. Voc. ii. 136, 81. Se Hǽlend oftrǽdlíce (frequenter) com ðyder, Jn. Skt. 18, 2. Gif man hine oftrǽdlíce (ex consuetudine) ofer drince, L. Ecg. P. iv. 33 ; Th. ii. 214, 12. Ǽghwæðer óðerne oftrǽdlíce út drǽfde, Chr. 887 ; Erl. 86, 12. Hē oftrǽdlíce fór mid miclum gefeohtum on Sciððie, Ors. 1, 2 ; Swt. 30, 2 : 4, 12 ; Swt. 208, 33.

of-tredan to tread down, trample upon, injure or destroy by treading :—Ða ýða árison ac Drihten hí oftrǽd . . . Ðeáh ðe árleáse woruldmenn árison ongeán ús swá ðeáh Crist oftret heora heáfod, Homl. Th. ii. 388, 18–22. hit oftrǽd and hié tó loman gerēnode duos et .l. calcatos inutiles fecit, Nar. 15, 25. Ða hors hí (Jezabel) oftrǽdan huxlíce under fótum, Homl. Skt. i. 18, 347. Oftredan ðæt gærs and ofsittan, Homl. Th. i. 188, 25. Swá hwæt swá ðæs gódan sǽdes on swylcum wege befylþ, biþ mid yfelum geþohtum oftreden, ii. 90, 19. Ðǽr wǽron xxx M ofslagen and æt ðǽm geate oftreden triginta millia caede prostrata et compressione suffocata, Ors. 6, 4 ; Swt. 260, 18. [Orm. off-tredenn (gluterrnesse).]

oft-síþ, es ; m. A time that often occurs :—Hwæt hē hæfde Godes þeówum on oftsíþas tó láðe gedón what he had ofttimes done to hurt God's servants, Ors. 6, 34 ; Swt. 290, 29. [A. R. ofte-siðen : Chauc. ofte-siþes : Ayenb. ofte-ziþes : Icel. opt-sinnis, -sinnum ofttimes.]

of-tyge, es ; m. A holding back, withholding (v. of-teón, III) :—Ungelimp mid synnum geearnod, swíðost mid ðam oftige ðæs neádgafoles ðe Cristene men Gode gelǽstan sceoldon on heora teóþingsceattum misfortune merited through sins, especially through the keeping back of the tax that Christian men ought to pay to God in their tithes, L. Edg. S. 1 ; Th. i. 270, 13. Mid ǽnigum oftige Godes gerihta, 270, 30.

of-tyrfan to stone :—Hiene oftyrfdon his ágene geféran ab exercitu suo lapidibus coopertus interiit, Ors. 4, 6 ; Swt. 172, 28. v. of-torfian.

of-unnan. I. in a bad sense, to begrudge a person (dat.) anything (gen.), wish to deprive a person of anything :—Se biþ ðæm ísene gelíc se ðe ofan his níhstan his lífes ferro utitur, qui vitae proximi insidiatur, Past. 37, 3 ; Swt. 269, 7. Se ðe (the devil) him (hermits) lífes ofonn, Exon. Th. 107, 10 ; Gú. 56 : 265, 7 ; Jul. 377. II. to refuse to grant :—Ðám ðe gē forgifenysse ofunnon him biþ oftogen seó forgyfenys to whom ye refuse forgiveness, from them shall forgiveness be withheld, Homl. Th. i. 370, 8. [O. Sax. af-unnan : cf. O. H. Ger. ab-unst invidia, livor : Ger. ab-gunst : Icel. af-und (öfund).]

of-weorpan to kill by casting (a stone, etc.), to knock down and kill by a missile :—David nam fíf stánas and ðeáhhweðere mid ánum hē ðone gigant ofwearp, Blickl. Homl. 31, 18. David mid his liðeran ofwearp ðone geleáfleásan ent, Ælfc. T. Grn. 7, 18 : Homl. Skt. i. 18, 18, 23. Hē weard mid áne stáne ofworpen saxo ictus occubuit, Ors. 4, 1 ; Swt. 158, 32. Gif oxa wer ofslóge, síe hē mid stánum ofworpen, L. Alf. 21 ; Th. i. 48, 32 note : 50, 5 note. [Goth. af-wairpan stainam lapidare : Ger. ab-werfen.]

of-worpian to kill by casting (stones, etc.) :—Mid stánum ofworpod, L. Alf. 21 ; Th. i. 48, 28, 32 : 50, 5. v. preceding word.

of-wundrian to be astonished :—Ðætte ofwundradon alle ut ammirarentur omnes, Mt. Skt. Rush. 2, 12. v. next word.

of-wundrod astonished :—Sarra cwæþ ðā ofwundrod, Gen. 21, 6. Ic eom swīðe ungemetlīce ofwundred hwī eów þince ... *vehementer admiror*, Bt. 13; Fox 40, 5. Seó cwēn wæs tó ðan swīðe ofwundrod, ðæt heó næfde furþor nænne gāst, Homl. Th. ii. 584, 18. Maria and Ioseph wæron ofwundrode ðæra worda, i. 144, 15. [Wurþen men swīðe ofwundred and ofdrēd, Chr. 1135; Erl. 261, 1.]

ōga, an ; m. **I.** the feeling which is excited in a person, *terror, dread, horror, great fear* :—Óga *horror*, Ælfc. Gr. 9, 21; Som. 10, 26 : *metus*, 11; Som. 15, 12 : *pavor*, Hymn. Surt. 3, 23. Micel ōga (*horror*) him becom, Gen. 15, 12. Būtan ōgan (*absque terrore*) hē hine gerest, Ælfc. Gr. 47; Som. 48, 4. Ðā clypode hē mid micclum ōgan, Homl. Th. ii. 98, 3. **II.** the exciting cause of such a feeling :—Beó eówer ege and ōga ofer ealle nītenu *terror vester ac tremor sit super cuncta animalia*, Gen. 9, 2. On līgette is ōga, Homl. Th. i. 222, 32. For hellewītes ōgan (*on account of the terror which hell-torment causes*), oððe for ðæs ēcan līfes wuldre, R. Ben. 19, 17. **III.** *an object which excites fear*, a terrible, horrible thing :—Hē hēt Ðeódolum standan æt ðam mūþe (*of the fiery furnace*) ðæt hē for ðam ōgan (*on account of the terrible spectacle*) him ābūgan sceolde, Homl. Th. ii. 310, 33. Ógan (egsan, Lk. Skt. 21, 11) of heofenum and micele tācna *terrores de caelo et signa magna*, 538, 32. Ógana *terribilium*, Blickl. Gl. God him sende swiðlīce ōgan (*the ten plagues*), Ælfc. T. Grn. 5, 18. [Cf. *Goth.* ōgan *to fear*; ōgian *to terrify*: *Icel.* ōgn *dread, terror*; œgja *to frighten*; œgi-ligr *terrible*.] Cf. ege.

ō-gengel *a bar*, *bolt* : obex, Wrt. Voc. ii. 63, 28 : 115, 32. [Ō=on ? v. next word.]

ō-heald, -hilde; *adj. Sloping, inclined* :—Óhǽldi (ōhaelði, Ep. Gl. 21 d, 16) *pendulus*, Wrt. Voc. ii. 117, 19. Óhylde, 68, 10. Clifig ᵻ ōhyld (not tōhyld) *clivosus*, i. 19, 4. Óheal[d ?] *clivosa, tortuosa*, Germ. 392, 53. Hōhyldo *prona*, 400, 118. [*O. H. Ger.* uo-hald, -haldi *proclivus*. For the prefix ō-, cf. on-, ā-hildan.]

ōhsta, an ; *m. The arm-pit, oxter* (in northern dialects, e. g. Yorkshire, Cumberland, Scotland) :—Óhsta *ascella*, Wrt. Voc. ii. 10, 5. Ócusta, Ep. Gl. 2 b, 19. Óxtan *ascilla*, Wrt. Voc. i. 283, 9. Cf. ōxn.

ōht. v. āht.

ōht, e ; *f. Fear, terror* (? cf. ōga and *Icel.* ōtti *fear*), or *hostile pursuit, persecution, active enmity* (? cf. ēhtan and *O. H. Ger.* āhta *persecutio*, āhtunga *persecutio*) :—Wæron ðā gesōme ða ðe swegl būan wrōht wæs āsprungen ōht mid englum and orlegnīð *then were at peace the dwellers in heaven, discord was at an end among angels, and enmity* (*or fear ?*) *and war*, Cd. Th. 6, 5; Gen. 84. Ðær on fyrd hyra fǽrspell becwom, ōht inlende (*the pursuit by the Egyptians, or the terror which their coming caused*); egsan stōdan wælgryre weroda, 186, 9; Exod. 136.

oht-rip (?) *harvest*; messis, Mt. Kmbl. Lind, 9, 38 : Lk. Skt. Lind. 10, 2.

ō-hwǽr, -hwanon, -hylde. v. ā-hwǽr, -hwanon, ō-heald.

ō-leccan, -liccan, -lǽcan; *p.* -lecte, -lehte, -lǽhte. **I.** *to treat gently, to soothe, caress* :—Ic ōlǽce *blandior*, Ælfc. Gr. 31; Som. 35, 51. Ólecceþ *favet*, Wrt. Voc. ii. 147, 19. Ólehte *delinuit*, 138, 50. Hē him ōlecte ðā hē cuæþ *cui blandiens dicit*, Past. 26, 1; Swt. 181, 10. God hwīlon geōlǽhþ, and hwīlon beswingþ. Nǽre nān tihting, gif hē ūs ne ōlǽhte, Homl. Th. ii. 330, 3. Ólecce *demulceat, blanditur*, Wrt. Voc. ii. 138, 68. Óleccende *blandiens*, 127, 8. Óleccendra *palpantum*, 116, 51. **II.** *to be obsequious, pay court to, fawn upon, flatter, to try to gain a person's good will by unworthy means* :—Óleccaþ *adolatur*, Wrt. Voc. ii. 127, 7. Þeófum ðū ne ōlæce, ne yfeldǣdum ne geþwǣrlæce, Homl. Skt. i. 21, 361. Hē nolde ōlǽcan ǽnigum rīcan mid geswǣsum wordum, Homl. Th. ii. 514, 13. Gif ðū wille ðæt ðē monige ōlǽcan ðonne ōlǽce ðū ānum swīðe georne *if you wish many to pay court to you, do you sedulously pay court to one*, Prov. Kmbl. 79 : 80. Mē riht ne þinceþ ðæt ic ōleccan þurfe Gode æfter gōde ǽnegum, Cd. Th. 19, 12; Gen. 290. **III.** *to gain good will by worthy means, to propitiate, be submissive* :—Ðæm (God) ōleccaþ ealle gesceafte ðe ðæra ambehtes āwuht cunnon (cf. ðam þeówiaþ ealle ... ða ðe cunnon, Bt. 21; Fox 72, 30), Met. 11, 8. Ólǽce Gode ānum *try to please God only*, Prov. Kmbl. 80. Hē wolde onginnan him ōleccan mid his hearepan *he (Orpheus) would attempt to propitiate them (the gods of Hell) with his harp*, Bt. 35, 6; Fox 168, 14 : Cd. Th. 118, 3; Gen. 1959. Uton wē Gode ōliccan, Exon. Th. 366, 15; Reb. 12. **IV.** *of things to gratify, charm, give pleasure* :—Ealle ða ōðru gōd ōleccaþ ðam mōde and hit rēt se lust āna ōlecþ ðam līchoman ānum swīðost *cetera omnia jucunditatem animo videantur afferre*, Bt. 24, 3; Fox 84, 23–25. Swilce hȳ wǣron rihte ðā hī ðē mǣst geōleccan swilce hī nū sindon ðeáh ðe hȳ ðē ōleccan on ða leásan sǽlþa *talis erat, cum blandiebatur, cum tibi falsae illecebris felicitatis alluderet*, 7, 2; Fox 18, 2.

ōleccere, es ; *m. A flatterer* :—Leás ōlecere *parasitus*, Wrt. Voc. i. 74, 36. Hē geliéfþ ðæt hē suelc sīe suelce hē gehiérþ ðæt his ōlicceras secgaþ ðæt hē sīe, Past. 17, 3; Swt. 111, 11.

ōleccung, e ; *f.* **I.** *soothing, caressing, gentleness of treatment* :—Ólǽcung *delinimentum*, Wrt. Voc. i. 54, 69. Ólæcunge *blandimentorum*, Hpt. Gl. 485, 48. Hū gesceádwís se reecere sceal bión on his þreáunga

and on his ōleccunga *quae esse debet rectoris discretio correptionis et dissimulationis*, Past. 21, tit. ; Swt. 151, 6. Ðā āhsode heó hine georne mid hire ōlǽcunge, on hwam his miht wǽre, Jud. 16, 6. Ólǽcunge *blanditiae*, Ælfc. Gr. 13; Som. 16, 17. **II.** *flattery, fawning, adulation* :—Ólǽcung *adulatio*, Hpt. Gl. 527, 40. Wyrð ðæt mōd besuicen mid ðæra ōlicunga (ōliccunga, Cott. MSS.) ðe him underþiédde beóþ, Past. 17, 3; Swt. 111, 7. Ne wilna nānes monnes ōlǽcunga, Prov. Kmbl. 80. Hē nǽfre nǽnigum woruldrīcum men þurh leáse ōlǽcunga onbūgan nolde, Blickl. Homl. 223, 28. Ðonne him hæfþ gewunnen ðæs folces ōlecunga (*favor popularis*), Bt. 24, 3; Fox 82, 23. **III.** *of things, charm, allurement* :—Óliccung *jocunda*, Wrt. Voc. ii. 127, 2. Ne hine ne geloccige nān ōliccung (ōlicung, Cott. MSS.) tō hiere willan *non blanda usque ad voluptatem demulceant*, Past. 14, 3; Swt. 83, 18. Forsió hē ǽlce ōlicunge (ōliccunge, Cott. MSS.) ðisses middangeardes *blandimenta mundi despiciat*, 14, 2; Swt. 83, 6. Hit gewarenaþ ǽgðer ge wið heora þreáunga ge wið ōlecunga *nec formidandas fortunae minas, nec exoptandas facit esse blanditias*, Bt. 7, 2; Fox 18, 24. v. leás-ōlecung.

ōleht-word, es ; *n. A flattering speech* :—Hwǽr syndon ða ðe hié heredan, and him ōlyhtword sprēcan ?, Blickl. Homl. 99, 26.

ōl-fæt (= āl-fæt. v. Wrt. Voc. ii. 135, 39), es ; *n. A cooking vessel ; coculum*, Wrt. Voc. i. 24, 41.

olfend, es : offenda, an ; *m. A camel* :—Olfend *camelus vel dromeda*, Wrt. Voc. i. 22, 58 : *camelus*, 78, 8. Āfyred olbenda *dromidus*, ii. 106, 66. Ófyrit olfenda, 25, 68. Under ānes olfendes (*cameli*) sæle, Gen. 31, 34. Gescryd mid oluendes hǣrum, Mk. Skt. 1, 6. Gē drincaþ ðone olfend (olbendu, Rush.), Mt. Kmbl. 23, 24. Of olfenda hǣrum āwunden, Blickl. Homl. 169, 2. Hē nam tȳn olfendas (*camelos*), Gen. 24, 10. [*Orm.* olfent *a camel*: *Goth.* ulbandus ; *m.*: *O. Sax.* olbundeo ; *m.*: *O. H. Ger.* olpenta ; *f.*: *Icel.* ūlfaldi ; *m.*]

olfend-mere, an ; *f. A she-camel* :—Þrītig gefolra olfendmyrena mid heora coltum *camelos foetas cum pullis suis triginta*, Gen. 32, 15.

ōliccan. v. ōleccan.

oll *contempt, insult, contumely* (in the phrase mid olle) :—Se deófol cwæþ mid olle ðæt hē wolde æt ðam weorce gecuman, Homl. Th. i. 166, 15. Hē āxode ðā mid olle (*contemptuously*) : Eart ðū lā God ? Homl. Skt. i. 9, 72. Man tǽleþ and mid olle gegrēteþ (*insults*) ealles tō gelōme ða ðe riht lufiaþ, Wulfst. 164, 19.

ōl-þwang (*better ?* āl-, cf. ōl-fæt), es ; *m. A strap* :—Ólþwongas *corrigie*, Wrt. Voc. ii. 22, 47. [*Icel.* āl (mod. ōl) *a strap*.]

ōm *rust* :—Oom *rubigo*, Wrt. Voc. ii. 119, 34. Ōm *erugo, vitium ferri*, 144, 3 : Ælfc. Gr. 9, 3; Som. 8, 58. Ðær ōm (*aerugo*) hit fornimþ, Mt. Kmbl. 6, 19, 20. Ómm, Homl. Th. ii. 104, 29. Yldo ābīteþ īren mid ōme, Salm. Kmbl. 601 : Sal. 300. v. brand-(brond-)ōm.

om-. v. am-.

ōman ; *pl. f. Erysipelas, erysipelatous inflammations* :—Óman *ignisacrum*, Wrt. Voc. ii. 45, 34 : 110, 52. Lǽcedōmas wið ǽlces cynnes ōmum, Lchdm. ii. 98, 21. In the section of which this is the heading the word frequently occurs. Of hōmena stiéme cymþ eágna mist, 26, 26. Wið hōmum, nim gāte horn ... dō on ða hōman, i. 350, 17–20. Wið hōmum (ōman, MS. O.), bāres scearn ... ða hōman hyt bēteþ, 360, 10–11. [*Icel.* āma ; *f.* and āmu-sótt *erysipelas.*] v. next word.

ōm-cynn, es ; *n. Corrupt humour* :—Ðū meaht clǽnsian ðæt ōmcyn, Lchdm. ii. 82, 18. v. ōmig.

omer *a bird's name, hammer* (in yellow-*hammer*) :—Omer (emer, Ep. Gl. 23 e, 31) *scorelus*, Wrt. Voc. ii. 120, 6. Amore *scorellus*, i. 281, 18. In Cd. Dip. Kmbl. iii. 118 omerlond occurs. [*O. H. Ger.* amero : *Ger.* ammer.] v. clod-hamer.

ō-middan. v. on-middan.

ōmig ; *adj.* **I.** *rusty* (v. ōm), *rust-coloured* :—Ðær wæs helm monig eald and ōmig, Beo. Th. 5519 ; B. 2763. Dȳre swyrd ōmige þurhetene, 6090 ; B. 3049. Dȳ lǽs ðæt ōmige fæt mid ealle tōberste, gif hē mid ungemete scǽfþ, R. Ben. 121, 3. Anfiltes hōmiges *incudis*, Hpt. Gl. 417, 64. Ómigum *vel īsengrǽgum ferrugineo*, Wrt. Voc. ii. 147, 66. **II.** *inflammatory* (v. ōman) :—Wyrð gegaderodu ōmig wǽte on ðǽre wambe, Lchdm. ii. 218, 16. On ðam magan ōmigre wǽtan gefylled, 178, 9. v. next word.

ōmiht ; *adj. Full of inflammation* :—Þis sint tācn ðæs hātan magan ōmihtan ... Ðæs hātan magan tācn sindon ðonne hē biþ mid ōmum geswenced, Lchdm. ii. 192, 24. Ða ōmihtan þing *the inflammatory symptoms*, 82, 21.

on, an ; *prep. adv.* **A.** *with dat. or inst.* **I.** expressing local relations, (1) rest upon and contact with an object, *on* :—Hig stōdon on nyðewerdum ðam munte, Ex. 19, 17. Hē on dōmsetle sittende wæs, Bd. 5, 19 ; S. 639, 43. Him on bearme lǽg mādma mænigo, Beo. Th. 80 ; B. 40. On him byrne scān, 815 ; B. 405. Se on foldan læg, Byrht. Th. 138, 29 ; By. 227. Hē on meare rād, 138, 53 ; By. 239. *And metaphorically* :—On eów scyld siteþ, Exon. Th. 131, 2 ; Gū. 449. (2) dependence upon an object :—Hié hine on rōde āhēngon, Blickl. Homl. 7, 11. Ðæs on ðam beáme geweóx, Cd. Th. 31, 11 ; Gen. 483 : Exon. Th. 202, 27 ; Ph. 76. (3) extension over a surface :—Deófles rīce on ðyssum middangearde, Blickl. Homl. 7, 13. Ðæt mycel hǽto wǽre Cristes

geleáfan on Norþanhymbra þeóde, Bd. 2, 14; S. 518, 5 : Exon. Th. 201, 2 ; Ph. 50. (4) nearness :—Hí námon him wintersetl on Temesan, Chr. 1009; Erl. 143, 4. (5) *in* or *at* a place, or *with* a person, cf. æt :—Ða ðe wǽron on ðam mynstre Æbbercurnig, ðæt is geseted on Engla-lande, Bd. 4, 26; S. 602, 35. Hé on sinoþe sittende wæs, 5, 19; Bd. 639, 43. Gewundad on gefeohte, 4, 26; S. 603, 14. Hí ðone líchoman on cneówum bégde, 4, 11; S. 580, 10. On ðam dóme standeþ, Exon. Th. 95, 22; Cri. 1561. On beóre *at a feast*, 330, 14; Vy. 51. Hé ánne cnapan gesette on hyra middele, Mk. Skt. 9, 36: Lk. Skt. 21, 21. Ða clǽnan heortan God geseóþ. On ðære gesihþe wesaþ ealle geleáffulle, Blickl. Homl. 13, 27. Ða sǽton on portum *qui sedebant in porta*, Ps. Th. 68, 12. Ic mundbyrd on ðé hæfde, 70, 5. Is mildheortnisse miht on (*apud*) Drihtne, 129, 7. ¶ *Like Icelandic* á *it occurs in names of places* :— On his mynstre ðe is cweden on Hripum, Bd. 5, 18; S. 636, 45. (6) with verbs of motion :—Se ðe on heofenum cuman ne mót, Homl. Th. ii. 452, 4. (7) rest where one object is contained in another, or is surrounded by others, *within*, *among*; and metaphorically *in* (the power of, etc.) :— Drihten wæs uppan him on fýre, Ex. 19, 18. Drihten is on ðínre heortan, and on ðínum innoþe, Blickl. Homl. 5, 11. Sum mon scínende on hwítum gegyrelan, Bd. 5, 19; S. 640, 39. Twegen weras on hwítum reáfe, Lk. Skt. 24. 4. Ic wát ðæt ic (ðæt Mód) on libbendum men eom, and ðeáh on deádlícum, Bt. 5, 3; Fox 12, 27. Gif hit on heora anwealde wǽre, 11, 1; Fox 32, 2. Ic hí on lufan mínre hæfde *quae dilexi*, Ps. Th. 118, 47. On þeóstre, Exon. 94, 27; Cri. 1546. On Juda ealdrum *among the princes of Judah*, Mt. Kmbl. 2, 6. Ðú eart gebletsud on wífum, Lk. Skt. 1, 28. (8) marking the seat of feeling, thought, etc., *in*, *within*, *at* :—Ða ongan hé smeágan on him selfum, Bt. 1, 1; Fox 2, 18. Ða fóre ðe hé on his móde gelufad hæfde, Bd. 5, 19; S. 637, 27. Yr on móde, Cd. Th. 4, 33; Gen. 63. Murnan on móde, 45, 31; Gen. 735. Se unrihtwísa cwyþ on his móde, Ps. Th. 13, 1 : 54, 6. II. expressing temporal relations, (1) marking a point of time, *on*, *at*, *in* :— On ðære tíde Drihten cwæþ tó mé, Deut. 10, 1. Ðonne cymþ ðæs weles hláford on ðam dæge ðe hé ná ne wénþ, and on ðære tíde ðe hé nát, Mt. Kmbl. 24, 50. Swá byþ on worulde endunge, 13, 40. On anginne *in principio*, Gen. 1, 1. On mergen, St. And. 18, 28. On æfenne, 20, 14. (2) marking a period, past or future, *within*, *in the course of*, *in*, *during* :— Ðes tówyrpþ Godes templ, and on þrím dagum (*in triduo*) hyt eft getimbraþ, Mt. Kmbl. 27, 40: Cd. Th. 266, 1; Sat. 15 : Lchdm. iii. 262, 23. Gé sweltaþ on litelre hwíle, Deut. 30, 18. Hé nówiht elles dyde on eallum ðam dagum, Bd. 2, 14; S. 518, 8. On ðæs biscopes tíde, 4, 12; S. 580, 34. On ðissum geáre, Chr. 889; Erl. 86, 22. On ðý ylcan gére, 896; Erl. 93, 34. On geóguþe, Exon. Th. 288, 22; Wand. 35. On ðissum lífe, 448, 12; Dóm. 53. On dís andweardan lífe, Bt. 10; Fox 28, 25 : 11; Fox 30, 23. On sumera sunne scíneþ, Cd. Th. 233, 15; Dan. 276. On geárdagum, 287, 16; Sat. 368. On fyrndagum, Exon. Th. 313, 17; Mód. 11. Ðæt feoh ðe mon ðam ferdmannum on geáre sellan sceolde, Bt. 27, 4; Fox 100, 14. Þriwa on gére *tribus vicibus per singulos annos*, Ex. 23, 14. Ic fæste tuwa on wucan, Lk. Skt. 18, 12. Ðæt hridder tóbærst on ðære lǽne *the sieve broke during the loan*, Homl. Th. ii. 154, 16. Heó cwæþ ðæt heó wǽre wydewe on ðam geáre *she said that she had been a widow during the last year*, Homl. Skt. i. 2, 154. On ðam ðe Godwine eorl and Beorn eorl lágon on Pevenseá, Chr. 1050; Erl. 175, 14. III. expressing other relations, (1) *on*, *a-* (as in *a-foot*) :—Heó on hire fótum gesund hám hwearf, Bd. 3, 9; S. 534, 14. Sceal on ánum fét féran, Exon. Th. 415, 5; Rä. 33, 6. Ðú gǽst on ðínum breóste, Gen. 3, 14. (2) with verbs of taking, depriving, etc., *from* (cf. æt) :—Ðone mǽstan dǽl ðæra ǽhta ða ðe on ðé genumene wǽron, Bd. 5, 19; S. 640, 46 note : Bt. 7, 3; Fox 20, 29. Se ðe gold on óðrum reáfaþ, 13; Fox 38, 13. Ðæt (*what*) hé on him gereáfade (bereáfode, MS. C.), Ors. 3, 11; Swt. 146, 30. Ða ǽrestan cyningas ðe West-Seaxna lond on Wealum geeodon, Chr. Erl. 2, 10: Exon. Th. 118, 20; Gú. 242. Náðer ne mehte on óðrum sige geræcan, Ors. 3, 1; Swt. 96, 33. Hwæðer heora sceolde on óðrum sige habban, 4, 1; Swt. 156, 1. (3) marking the object of thought, feeling, etc., *on*, *in*, *at* :—Manege wundrodon on his láre (*or acc.*?), Mk. Skt. 6, 2. Ic on ðínum bebodum móte gemetegian, Ps. Th. 118, 47. (4) marking the means or instrument, *by*, *with* :—Ic hæfde gemynt ðé tó árwurðienne on æhtum and on feó, Num. 24, 11. On tympanis, Ps. Th. 67, 24. Ic on mínum múþe mihta Drihtnes andette, 108, 29. Hé nówiht fremian mihte on his láre ðære þeóde, Bd. 3, 5; S. 527, 24. Heó geleornod hæfde on onwrihgennysse, 3, 8; S. 531, 35. On ðæs engles wordum wæs gehýred ðæt þurh hire beorþor sceolde beón gehǽled eall wífa cynn, Blickl. Homl. 5, 22. On ðǽm upstige ðære róde eall úre líf Drihten getremede, 9, 35. Hí wurdon on fleáme generede, Chr. 894; Erl. 92, 33. Se deófol wæs oferswíðde on ðam ylcum gemetum ðe hé ǽr Adam oferswíðde, Homl. Th. i. 178, 1. (5) marking the material or components of which a thing is made, *of*, *consisting of* or *in* :—Mycelne aad on beámum and on ræftrum and on wágum and on watelum and on þacum, Bd. 3, 16; S. 542, 22. Lác on mæssereáfum and on bócum, Homl. Th. ii. 132, 7 : Gen. 21, 27. Ðæt gafol biþ on deóra fellum, and on fugela feðerum, Ors. 1, 1; Swt. 18, 17. Unrím getæl on horsum and on múlum and on olfendum and on elpendum.

Nar. 9, 14. Swá micel ungewiss, ǽgðer ge on sǽs fyrhto, ge on wéstennum wildeóra, ge on þeóda gereordum, Ors. 3, 9; Swt. 136, 24. (6) marking that *in* which a quality or property resides, *in respect to*, *in the matter of*, *in* :—Se wæs in bóccræftum and on woruldþeáwum se rihtwísesta, Bt. 1, 1; Fox 2, 13. Hé áxode gif hé cúðe áht on lǽcecræfte. Apollinaris him cwæþ tó : ' Ne cann ic náht on lácnunge,' Homl. Skt. i. 22, 40–41. Æþele on geðyrdum, 11, 1; Fox 30, 31. On dǽde unæþele, Bd. 2, 15; S. 518, 37. On ríce gestrangod, 4, 26; S. 603, 19. On wísdóme þeónde, Homl. Th. ii. 154, 11. Foremihtig on feþe, Beo. Th. 1944; B. 970. Spédig on ðam æhtum ðe heora spéda on beóþ, Ors. 1, 1; Swt. 18, 8. Wæstmberende on ælces cynnes blǽdum, 1, 3; Swt. 32, 13. Beorht on blǽdum, Cd. Th. 247, 20; Dan. 500. (7) marking state, condition, occupation, *in*, *of* :—Heora líc bíþ on marmorstánes hwítnysse, Nar. 38, 9. Ðú forþfærst on sybbe, ðonne se tíma cymþ, on gódre ylde, Gen. 15, 15 : Bd. 2, 15; S. 519, 14 : 5, 19; S. 641, 14. Ða welegan hé forlǽteþ on ídelnesse . . . Drihten is on ðínum fultume, Blickl. Homl. 5, 9–12. Þurhwunian on rihtum geleáfan and on fulfremedlícum weorcum, 77, 19. On sorhgum beón, 5, 29. On stilnesse, Bt. 7, 2; Fox 18, 11. Gif hé wyrþ on ungeþylde, 11, 1; Fox 32, 33. Ða wæs cyning on hreón móde, Beo. Th. 2619; B. 1307. On ungearwe *at unawares*, Ors. 1, 10; Swt. 46, 34. Eal ðæt folc wæs on blǽdran, 1, 7; Swt. 38, 6. Job sæt on ánre wunde, Homl. Th. ii. 452, 27. Ða wearþ hé on slǽpe, Glostr. Frag. 6, 26. Hí wǽron on ðan mǽstan hungre, Ors. 1, 5; Swt. 32, 26. Hé wénde ðæt hié wolden Hannibale on fultume beón, 4, 10; Swt. 196, 7. On feáwum stówum wíciaþ Finnas, on huntoþe on wintra, and on sumera on fiscaþe, 1, 1; Swt. 17, 5. Sum man wæs betogen ðæt hé wǽre on stale, Homl. Skt. i. 21, 265. (8) marking measure, *at* (*a distance*), *of* (*the weight of*), etc. :—Weðeras on oxna micelnesse, Nar. 33, 16: Lchdm. i. 314, 21. Six wæterfatu . . . ǽlc wæs on twegra sestra gemete (*capientes singuli metretas binas*), Jn. Skt. 2, 6. Ic geseah sumne gildenne dalc on fíftigum entsum, Jos. 7, 21. Án æstel on fíftegum mancessa, Past. pref.; Swt. 9, 1. Hí ðæt feoh gesetton on þrittig scillingum *they fixed the money at thirty shillings*, Homl. Th. ii. 242, 18. Se gewát on wésten ðá hé wæs on twentigum geára, and on ðam hé wunode óþ ðæt hé wæs on fíf and hundteóntig geára, Shrn. 52, 16–18. Hig flugon on twegra elna heáhnisse bufan eorþan, Num. 11, 31. Wæs seó stów hwæthwugu on healfre míle fram ðære ceastre wealle, Bd. 1, 7; S. 478, 31. Is ðæt eálond fram ðære ylcan cyricean feor út on gársecge seted húhugu on nygan mílum, 4, 27; S. 603, 30 : 2, 3; S. 504, 26: Shrn. 29, 31. Álecgaþ hit on ánre míle ðone mǽstan dǽl fram ðæm túne, ðonne óðerne ðæne þriddan, óþ ðe hyt eall álëd biþ on ðære ánre míle (*within the one mile*); and sceall beón se lǽsta dǽl nýhst ðæm túne, Ors. 1, 1; Swt. 20, 30–33. Hí hine bebyrgdon on ðære æfteran míle fram ðære ceastre, Shrn. 115, 16: Blickl. Homl. 193, 19. (9) marking degree :—On swíðe lytlon hiera hæfþ seó gecynd genóg, on swá myclum heó hæfþ genóg swá wé ǽr sprǽcon, Bt. 14, 1; Fox 42, 10–11. (10) marking manner, *with*, *in* :—Hláfas on lilian beorhtnysse scínende, and on hrósan brǽde stýmende, Homl. Th. ii. 136, 28. His gewǽda scinon on snáwes hwítnysse, 242, 7. Hí cómon on þrím floccum, Homl. Th. ii. 450, 13. Se on hrǽdnesse mycele menigo fornom, Bd. 1, 14; S. 482, 30. (11) denoting end, purpose :—Ásceacaþ ðæt dust of eówrum fótum on gewitnysse, Mk. Skt. 6, 11. ' Mín blód, ðæt ðe biþ ágoten on synna forgifennysse ' . . . Hí hálgodon hláf and wín on his gemynde, Homl. Th. ii. 268, 1–3. (12) *in accordance with* :—Ðæt hé irne on his willan, Bt. 11, 1; Fox 32, 21. (13) *of* (*such and such a name*) :—Wæs sum man on naman Zacheus, Lk. Skt. 19, 2 : 23, 50. Castel on naman Emaus, 24, 13. (14) *in* (*the name of*) :—Hé him geswór on his goda noman, Ors. 4, 6; Swt. 178, 11. (15) *without* a case following :—Deófol ðé sticaþ on, Jn. Skt. 7, 20. Seó wyrd (hit) ðé on geniman ne mihte, Bt. 11, 2; Fox 34, 14. Swelce him nǽfre gelíc yfel an ne becóme, Ors. 3, 10; Swt. 140, 10. For ðæm ungemetlícan feóndscipe ðe úre éhtende on sindon, 2, 5; Swt. 80, 36. Ðæt him mon sceolde an má healfa on feohtan ðonne on áne, Swt. 80, 27. Him man on líhþ, Bt. 30, 1; Fox 108, 8 : Prov. 70. On secgan *to bring a charge against*, Deut. 19, 16 : Mt. 26, 62. B. with acc. I. expressing local relations. (1) motion, actual or figurative, which is external to the object expressed by the word which *on* governs, *upon*, *on*, *on to*, *to* :—Hé ástáh on ðone munt, Mt. Kmbl. 5, 1. Se deófol lǽdde hine on swíðe heáhne munt, 4, 8 : Cd. Th. 220, 11 ; Dan. 69. Speún hine on ða dimman dǽd, 43, 3; Gen. 685. Gewát Abraham on ða wígróde, 125, 24; Gen. 2084. Se wuldorcyning on middangeard cwom, Blickl. Homl. 9, 32. Áhón on heáfene beám, Exon. Th. 261, 3; Jul. 309. Com hungur on Bryttas, Bd. 1, 14; S. 482, 15. Hé wæs ádrifen ðæt hé com up on Frysena land *pulsus est Fresiam*, 5, 19; S. 639, 20. Hé his ágene tungan wearp hine on ðæt neb foran *linguam in os tyranni abjecit*, Bt. 16, 2; Fox 52, 25. Hí spætton on hyne, Mt. Kmbl. 27, 30. On ðone andwlitan men slógun, Exon. Th. 69, 19; Cri. 1123. Wé wyllaþ fón on ðone traht ðissere rǽdinge, Homl. Th. i. 206, 21. (2) marking motion from without to the inside, *into*, *among* :—Sume feóllon on þornas, Mt. Kmbl. 13, 7. Sume feóllon on góde eorþan, 13, 8. Ða cómon híg on ða stówe ðe ys genemned Golgotha, 27, 33. Crist of

heofona heánessum on đĭnne innoþ ástĭgeþ, Blickl. Homl. 5, 14. On đás world cuman, 5, 28. Geléded on his đæt ǽrre mynster, Bd. 5, 19 ; S. 641, 17. Hé on scip eode, S. 639, 19. Gód geár com on Breotone land, I, 14; S. 482, 21. Gǽstas hweorfaþ on écne eard, Exon. Th. 64, 31 ; Cri. 1046. On đæt micle morþ men forweorpan, Cd. Th. 43, 15 ; Gen. 691. Sum man becom on đa sceaþan, Lk. Skt. 10, 30, 36. On ealle þeóda *among all nations*, 24, 47. Hé on đa duru eode, Chr. 755 ; Erl. 48, 32. (3) *marking position or direction* :—Ic stande on đás healfe *ego in hac parte sto*, Ælfc. Gr. 47 ; Som. 47, 50. On đa swýđran healfe, Mt. Kmbl. 26, 64. Án on đa swýđran healfe, and óđer on đa wynstran, 27, 38. On healfa gehwone, Beo. Th. 1604; B. 800. On đæt steórbord ... on đæt bæcbord *on the starboard ... on the larboard*, Ors. 1, 1 ; Swt. 17, 10–11. (4) *denoting conjunction* (*in the phrase* on án), *continuously, together, anon, at once* :—Feówertig daga and feówertig nihta on án, Gen. 7, 12 : Homl. Th. i. 178, 5. On đa gesworene *conjurati*, Wrt. Voc. ii. 20, 22. Ealle deáde men sculon đone myclan dôme gesécan, and đa synfullan sculon đanon on án tó helle faran, Wulfst. 126, 20. In đone ealdan secgmôr ; of đam on án betwénan ácwudu and wulleleáh ; and swá ǽfre betwyx đám twám wudan, Cod. Dip. Kmbl. vi. 218, 23. (5) *with verbs denoting division or separation* :—Úre ieldran ealne đisne ymbhwyrft đises middangeardes on þreó tódǽldon ; and hié đa þrié dǽlas on þreó tónemdon, Ors. 1, 1 ; Swt. 8, 1–4. Tósliten on twegen dǽlas, Mt. Kmbl. 27, 51 : Chr. 894 ; Erl. 90, 16. Hit tóbærst on emtwá, Homl. Th. ii. 154, 16. **II.** *expressing temporal relations.* (1) *marking a point of time*, on, in, at :—On đone dæg (*in die*) đe God gesceóp man, Gen. 5, 1. On midne dæg *in meridie*, Deut. 28, 29. On ǽlcne tíman *omni tempore*, 11, 1 : 14, 23. On ealle tíd, Ex. 18, 22, 26. On dægrêd *diluculo*, 8, 20. On đa tíd *tunc*, Bd. 4, 26 ; S. 603, 3 : *eo tempore*, 2, 16 ; S. 519, 38. On đone forman Eásterdæg, 5, 23, tit. ; S. 645, 3. Đá hit đá on morgendæg wæs, Nar. 16, 21 : 22, 1. Heora wíse on nænne sæl (*on no occasion*) wel ne gefór, Ors. 4, 4 ; Swt. 164, 13. Eten đa gebrôđru on twá mǽl, R. Ben. 65, 14. (2) *marking a period of time* :—On écnesse *for ever*, Blickl. Homl. 13, 30. On đás lǽnan tíd *in this life*, Exon. Th. 364, 1 ; Wal. 64. On sumeres tíd, 212, 12 ; Ph. 209. On nánes cynges dæg, Chr. 1009 ; Erl. 22, 22. **III.** *in metaphorical expressions.* (1) *into* (one's power, etc.) :—Gif hig on hand gáþ *if they submit*, Deut. 20, 11 : Bd. 1, 14 ; S. 482, 16. Hine sylfne on þeówdôm gesealde đara muneca, 5, 19 ; S. 637, 12. Hé ealle Assirie on Persa anwald gedyde, Ors. 2, 1 ; Swt. 62, 3. (2) *expressing hostile action, against* :—Hé wonn on Scíđđie, 2, 5 ; Swt. 78, 8. Đá gelédde hé here on Peohtas, Bd. 4, 26 ; S. 602, 19. Ne dó ús swá swá wé dydon on đisne ælþeódigan, St. And. 22, 21. Óđer biþ tó ungemetlíce átyht on đæt đe hió mid ryhte irsian sceal, óđer on đæt đe hió ne sceal biþ tó swíđe onbærned, Past. 40, 4 ; Swt. 293, 12–14. Hié ealle on đone cyning wǽrun feohtende, Chr. 755 ; Erl. 48, 35. (3) *expressing agreement, in accordance with* :—Hé hét sumne biscep secgan on his gewill hwá his fæder wǽre, Ors. 3, 9, tit. ; Swt. 3, 13. Heó on his willan spræc, Cd. Th. 44, 1 ; Gen. 701. On riht *a-right* :—Đæt hié healdan Godes ǽwe on riht, Blickl. Homl. 45, 9 : 47, 35. (4) *denoting change from one state to another* :—Đá wendon hié hié on hiora ágan gepióde, Past. pref. ; Swt. 7, 2. Wendan on Englisc, Swt. 7, 18. Wêsten hé geworhte on wídne mere, Ps. Th. 106, 34. Hé wendeþ stán on wídne mere, 113, 8. Hí on heora ágen dust hweorfaþ, 103, 27. (5) *marking the object of thought, emotion, speech, trust, sight* :—Ne gladige on đæt nôđer ne cyning ne woruldríca, Lchdm. iii. 442, 35. Hé getrúwode on ídel gylp, 51, 6. Đa đe on Drihten getreówaþ, 124, 1 : 70, 13. On hine gelýfende, Past. 130, 18. Gód ys on Drihten tó þenceanne, Ps. Th. 117, 8. Sete on Drihten đín gehygd, 54, 22. Hycgan on ellen, Cd. Th. 191, 22 ; Exod. 218. On đæt wundor seón, 261, 25 ; Dan. 731. On đæt bearn starian, Exon. Th. 21, 27 ; Cri. 341. Hé on đone æþeling lôcude, Chr. 755 ; Erl. 48, 33. Se deófol đe andode on đæs munuces sôđan lufe, Homl. Th. ii. 156, 8. Hwæt gôdes mágan wé secgan on đa flǽsclícan unþeáwas, Bt. 31, 1 ; Fox 110, 24. Higeteónan spræc on fæmnan, Cd. Th. 136, 22 ; Gen. 2262. Ne sceal nán mann secgan on hine sylfne đæs đe hé wyrcende næs, Homl. Skt. i. 12, 177, 195. (6) *marking the object in relation to which an action takes place* :—Đæt đa woruldsǽlþa on đé (*erga te*) onwenda sint, Bt. 7, 2 ; Fox 16, 29. (7) *marking end or purpose* :—Ic wylle gán on fiscaþ, Jn. Skt. 21, 3. Hí him on fultum cýgdon đa godcundan árfæstnesse, Bd. 4, 26 ; S. 602, 9. (8) *marking price* :—On gold bebycgean, 2, 12 ; Bd. 514, 39. Hé bebohte bearn Wealdendes on seolfres sinc, Cd. Th. 301, 7 ; Sat. 578. (9) *marking manner* :—Nemned on Læden *Pastoralis*, and on Englisc *Hierdebôc*, Past. pref. ; Swt. 7, 19. Hér sýn on earde on mistlíce wísan hláfordswican manege, Wulfst. 160, 7. On ýdel *necquicquam*, Wülck. Gl. 256, 14. (10) *in* (*the name of*), *by* (in adjuration) :—Ic eów hálsige on đone Drihten, đe gescóp heofenas and eorþan, and on đa Hálgan þrynnesse and on đa twelf apostolas and on ealle Godes hálgan and on đa cyrcan, đe gé tó gelýfaþ, and on đæt hálige fulluht, Wulfst. 232, 12–16. (11) *not followed by a case, or as adverb* :—Hí gegearwodon wægen and on ásetton đa fæmnan, Bd. 3, 9 ; S. 534, 9. Fæht hine on Penda, 3, 14 ; S. 539, 18. On đæm dǽle đe Decius on ofslagen wæs, Ors. 3, 10 ; Swt. 138, 15. Hé on đa sunnan mæg on lócian, Met. 22, 20. Đá fêrde hé tó heofonum, him on lócigendum *while they looked on*, Homl. Th. i. 294, 1. Đeþ hé wyrplas on, Exon. Th. 332, 19 ; Vy. 87. Hine on cymeþ wracu, Cd. Th. 63, 33 ; Gen. 1041. Hine Abraham on his ágene hand sette, 167, 17 ; Gen. 2767. Ræsdon on sôna, Andr. Kmbl. 2670 ; An. 1336. *The word is often used in translating Latin words with the prefix* in-, *thus on* beládan *inferre, on* gebringan, *on* heápian *ingerere, on* gehreósan *ingruere.* (12) *with other adverbs* :—Đér stôd disc on, Bd. 3, 6 ; S. 528, 14. Ne ic đér nán þing on ne cann, St. And. 28, 24 : 40, 3. Đér wæs on Leo (*at the synod*), Chr. 1046 ; Erl. 171, 12. Ealles đæs đe đér đenne on biþ, Chart. Th. 534, 5. Swá swá Drihten cwæþ on ǽr, Jos. 11, 23 : Gen. 6, 6. Hé cýđde his forþsíþ on ǽr *he foretold his death*, Homl. Th. ii. 186, 23. Hió þyrstende wæs on symbel (*for ever*) mannes blôdes, Ors. 1, 2 ; Swt. 30, 27. Hé nyste bútan hí sungon đone lofsang forþ on, Homl. Skt. i. 21, 236. [*Goth.* ana : *O. Sax. O. Frs. O. H. Ger.* an : *Icel.* á.]

on-. The prefix, when used with verbs, for the most part corresponds with the *O. H. Ger.* int-, *Ger.* ent-; e. g. on-lísan, -lúcan, -týnan, -wreón.

on-ǽht, e; f. *Possession* :—Ic sellu đé þeóde erfeweardnisse đíne and onǽhte đíne gemǽru eorþan *dabo tibi gentes hereditatem tuam et possessionem tuam terminos terrae*, Ps. Surt. 2, 8.

on-ǽlan; p. de. **I.** *to set fire to, to ignite, kindle* (lit. and figurative) :—Hú ne onǽlþ (*accendit*) heó hyre leóhtfæt? Lk. Skt. 15, 8. Hé hiene onǽlþ mid đam tapure đæs godcundan liéges, Past. 36 ; Swt. 259, 12. Ne byrnþ on đé đæt đæt đú on lífe ne onǽldest þurh leahtras, Homl. Th. ii. 338, 16 : 344, 26. Ne onǽl đú đé sylfum đæt ǽce fýr, i. 594, 27. Ne onǽle gé nán fýr on đam dæge *non succendetis ignem per diem sabbati*, Ex. 35, 3. Đá hét se ealdorman onǽlan ormǽte ád, Homl. Th. ii. 484, 7 : Exon. Th. 277, 13 ; Jul. 580. Drihtenes fýr wearþ onǽled (*accensus*), Num. 11, 1, 3. Cola onǽlde synd *carbones succensi sunt*, Ps. Spl. 17, 10. **II.** *to burn* (cf. *anneal*), *consume by burning* :—Đás fýr onǽlþ manna sáwla ... Đis fýr onǽlþ ǽlcne be his gewyrhtum, Homl. Th. ii. 338, 6–17. Đá námon Nadab and Abiud híra stôrcillan and onǽldon đǽron ungehálgod fýr, Lev. 10, 1. Nim fela tunnan and dô hí đǽr on innan, onǽl hí siđđan ealle, Homl. Skt. i. 4, 260. **III.** *to make hot with fire* :—Hé hét onǽlan đone ofen swíđe þearle, Homl. Th. ii. 20, 1. **IV.** *to make hot* (in a metaphorical sense), *to inflame, to excite intense feeling, to kindle passions* :—Ic (*the devil*) hine đæs swíđe synnum onǽle, đæt hé byrnende from gebede swíceþ, Exon. Th. 264, 31 ; Jul. 372. Onǽled *incensum*, Wrt. Voc. ii. 133, 18. Mid đære lufe onǽled đara worda, Ap. Th. 18, 27. Onǽled mid đæm andan his hiéremonna unþeáwa, Past. 21 ; Swt. 159, 8. v. in-ǽlan.

on-ǽlet, es ; n. *Lightning* :—Onǽletu ł lígetu *fulgura*, Ps. Lamb. 143, 6.

on-ǽdele; adj. *Natural, in accordance with the nature of a thing* :—Eallum treówum, đe him onǽdele biþ, đæt hit on holte hýhst geweaxe (cf. đám treówum đe him gecynde biþ up heáh tó standanne, Bt. 25 ; Fox 88, 21), Met. 13, 51.

on-ál, es ; n. *A burning, kindling ; also what is burnt* :—Hé nemde đære stówe naman ' onál ' (*incensio*), for đam đe Drihtenes fýr wæs đær onǽled, Num. 11, 3. Mid onále ramma *cum incenso arietum*, Ps. Lamb. 65, 15. Onál *incensum*, 140, 2. Onál *incensa*, 79, 17 : *incendia*, Hpt. Gl. 510, 18. Cf. ál- (aal-) geweorc *ignarium*, *and see* on-ǽlan.

on-áscunung, e ; f. *Abomination, detestation* :—Fram onáscununga *abominationem*, Ps. Spl. 87, 8.

on-ásendedness glosses *immissio*, Ps. Lamb. 77, 49.

on-bæc ; adv. *A-back, backward, behind* :—Gang đú sceocca onbæc, Mt. Kmbl. 4, 10. Ná gewát onbæc heorte úre *non recessit retro cor nostrum*, Ps. Spl. 43, 21. v. next word.

on-bæcling ; adv. *Back, backward, behind* ; *retrorsum* :—Gán onbæcling *to go back, retire*, Blickl. Homl. 27, 20 : 31, 12 : Ps. Th. 43, 19. Cer đé onbæcling *get thee behind*, Cd. Th. 308, 26 ; Sat. 698. Forhwí gengdest đú onbæcling *quare conversus es retrorsum?* Ps. Th. 113, 5. Đú hæfst ús gehwyrfde onbæclincg *avertisti nos retrorsum*, 43, 12. Đá feól hé fǽringa onbæcling, Blickl. Homl. 223, 11.

on-bærnan ; p. de. **I.** *to set fire to, to light* (a fire), *to kindle* (a) *literal* :—Hió hié mid flexe bewundon and onbærndon hit *they wrapped them round with flax, and set fire to it*, Ors. 4, 1 ; Swt. 158, 6. Đá héton đa dêman micel fýr onbærnan, Shrn. 53, 15 : Exon. Th. 277, 11 ; Jul. 579. Lyft biþ onbærned, 64, 26 ; Cri. 1043. (b) *figurative* :—Hé on monigra geleáfsumra heortan đæs gástlícan leóhtes gyfe onbærnde, Bd. 2, 2 ; S. 502, 30. Đæt fýr đæt đú sylfa on đé onbærndest, Guthl. 5 ; Gdwin. 38, 18. Is onbærned đín yrre, Ps. Th. 78, 5. **II.** *to burn, consume by burning* :—Fýr onbærneþ, 79, 15. **III.** *to heat, inflame* :—Mid đisse pannan hierstinge wæs Paulus onbærned *Paulus hujus sartaginis urebatur frixura*, Past. 21, 6 ; Swt. 165, 3. Óđer on đæt đe hió ne sceal irsian biþ tó swíđe onbærned (*inflammatur*), 40, 4 ; Swt. 293, 14. Mid hátheortnesse onbærnedne, 40, 6 ; Swt. 295, 25. Hié beóþ onbærnde mid ǽfeste, Blickl. Homl. 25, 7. **IV.** *to kindle desire for anything, to incite* :—Monigra monna môd tó worulde forhogenesse onbærnde (*accensi*) wæron, Bd. 4, 24 ; S. 596, 37.

on-bærnness, e; f. *Incense:*—Mid onbærnysse ramma *cum incenso arietum*, Ps. Spl. 65, 14. v. an-, in-bærness.

on-bǽru; f. [on = un?] *Wrong behaviour, vexation, anger:*—Hé ðæs onbǽru habban ne meahte ac hé háte lét teáras geótan *he could not be vexed at it (Guthlac's death), but he shed hot tears*, Exon. Th. 165, 12; Gú. 1827. [Cf. (?) O. H. Ger. un-gipárida *fastidium, ira, rabies*.]

on-básnung, e; f. *Awaiting, expectation;* expectatio, Rtl. 4, 34.

on-bégness. v. on-bígness.

on-bén, e; f. *A prayer asking for something (evil) to come upon a person, an imprecation:*—Hí ús mid heora wiðerwordum onbénum and wyrinessum éhtaþ *qui adversis nos imprecationibus persequuntur*, Bd. 2, 2; S. 504, 4.

on-beódan; p. -beád; pl. -budon; pp. -boden. I. *to bid, order:*—Ðú onbude hǽlu *qui mandas salutem*, Ps. Surt. 43, 5: 118, 138. Hé onbeád *ipse mandavit*, 148, 5. Ðá onbeád Basilla and cwæþ, Shrn. 86, 17. Ðá onbeád heó him ðæt hé hire tó onsænde all ða gesíðwíf, 87, 20. Hé onbeád ðæt hé of Róme cóme, Bd. 1, 25; S. 486, 25. II. *to announce, tell, proclaim, send word:*—Hé hit him hám bebeád (onbeád, MS. C.) *he sent them home word of it*, Ors. 4, 5; Swt. 168, 13. Him Pilatus onbeád ymbe Cristes tácnunga *Pilatus ad Tiberium repulit de Christi virtutibus*, 6, 2; Swt. 254, 23. Word unreht onbudun (*mandaverunt*) wið mé, Ps. Surt. 40, 9. Eác beámas onbudon, hwá hý sceóp, Exon. Th. 72, 9. Agustinus hét him onbeódan ðæt hér wǽre mycel riip, Bd. 1, 29; S. 498, 4. [*Goth.* ana-biudan : O. Sax. an-beodan : cf. O. H. Ger. in-piotan.]

on-beornan, -brinnan; pp. -burnen. I. *to set fire to, to kindle:*—Abraham ádfýr onbran, Cd. Th. 203, 4; Exod. 398. II. *to inflame:*—Se innoþ wyrþ onburnen, Lchdm. ii. 278, 9. Wǽte unwineu, 218, 14.

on-beran; pp. -boren *To diminish, enfeeble, impair, destroy:*—Onboren *inminutus*, Wrt. Voc. ii. 49, 60. Wæs ðam báncofan neáh geþrungen breósthord onboren wæs se blíða gǽst fús on forþweg *disease pressed the body hard, the mind was enfeebled, the glad spirit was eager for departure*, Exon. Th. 158, 29; Gú. 917. Ðá wæs hord rásod onboren beága hord *the hoard was explored, the treasure of rings rifled*, Beo. Th. 4557; B. 2284. Ǽghwylc gecwæþ ðæt him heardra nán hrínan wolde íren ǽrgód ðæt ðæs aglǽcan blódge beadufolme onberan wolde *everyone agreed that no weapon would wound Grendel's claws, no sword would destroy (or harm) the monster's hand*, Beo. Th. 1985; B.990.

on-bíd (-bid?), es; n. *Awaiting, expectation:*—Næs ðæt onbíd long ðæt . . . *it was not long to wait, before . . .*, Exon. Th. 156, 18; Gú. 876. Long is ðis onbíd worulde lífes *long in this life is this waiting for the next*, 164, 30; Gú. 1019. Hé on tweógendlícan onbíde wæs hwæðer hé wið Rómánum winnan dorste *he was waiting in doubt (cunctans) whether he durst fight with the Romans*, Ors. 4, 11; Swt. 204, 29. v. an-bíd, onbíd-stów.

on-bídan; p. -bád; pl. -bidon; pp. -biden. I. *to abide, wait, remain:*—Onbád ôþ ðæt ǽfen cwom, Beo. Th. 4594; B. 2302. Hé onbád ðæt feówertig wintra hweðer hié gecyrran woldan *he waited the forty years to see whether they would change*, Blickl. Homl. 79, 4. Onbíd hér seofon and twentig nihta, 231, 5: 237, 33. Hér sceolon hí onbídan, Soul Kmbl. 121; Seel. 61. II. *to wait for, expect, with gen.:*—Ic uncres gedáles onbád, 75; Seel. 37. Ic ðín onbád, Ps. Th. 118, 116. Gif wífes wer sig on hæftnýde geláded, onbýde (*expectet*) heo his .vi. winter, L. Ecg. C. 26; Th. ii. 152, 4. Wé sculon óðres onbídan, Lk. Skt. 7, 20. Willaþ gé mín onbídan? Blickl. Homl. 233, 30, 27. III. *to wait on, attend upon:*—Onbídendum *prestulanti*, Wrt. Voc. ii. 65, 58.

on-bidian *to wait:*—Onbidedon, Chr. 1006; Erl. 140 note 8. v. an-bidian.

onbíd-stów, e; f. *A place in which to wait:*—On hwylcere anbídstówe ðín sáwl bídan móte dómes dæges *in quo commorationis loco animae tuae expectare liceat diem judicii*, L. Ecg. P. iv. 65; Th. ii. 226, 8.

on-bídung. v. an-bídung.

on-bígan; p. de *To cause to bend, to subdue:*—Heó míne sáwle onbígdon *incurvaverunt animam meam*, Ps. Th. 56, 7. Heora módes heánesse ealle eorþcyningas onbégan mihton *their loftiness of soul could make all the kings of the earth to bend*, Blickl. Homl. 119, 21. v. onbúgan.

on-bígness, e; f. *Bending, curvature:*—Onbégnes *curvatura*, Wrt. Voc. ii. 64, 25: 79, 2.

on-bindan; p. -band; pp. -bunden *To unbind, set free, disclose:*—Seó wiðerwearde wyrd onbint and gefreóþ . . . mid ðam hió geopenaþ hú tiedre ðæs andweardan gesǽlþa sint, Bt. 20; Fox 72, 2. Hé onband beadurúne, Beo. Th. 1006; B. 501. Æfter ðon onbind, Lchdm. ii. 250, 20. Wæs onbunden *enodaretur*, Hpt. Gl. 490, 73. Onbund[en?]-um *exertis*, Wrt. Voc. ii. 31, 50. v. un-bindan.

on-birgan; p. de (*with gen. and acc.*) *To taste of, taste, take (food):*—Gif hé bitres onbyrgþ, Met. 12, 11: 13, 23. Onbirigþ, Bt. 23; Fox 78, 26: 25; Fox 88, 11. Sume ðe deáþ ne onbyrigeaþ (-byrgeaþ, MS. A.: -byrigaþ, MS. B.), Mt. Kmbl. 16, 28. Onbyrigeaþ (-byrgaþ

MS. A.), Mk. Skt. 9, 1. Nán ðara manna ne onbyrigeaþ (-byriaþ, MS. A.) mínre feorme, Lk. Skt. 14, 24. Hé his (*the water*) onbergde, Shrn. 64, 9. Onbyrigde (-byrgde, MS. A.), Mt. Kmbl. 27, 34: Jn. Skt. 2, 9: Homl. Th. i. 136, 8. Onberede, Bt. 23, tit. ; Fox xiv, 9. Ne hí siððan ne onbirigdon ðæs bigleofan, Jos. 5, 12: Homl. Th. ii. 168, 2-3. Onbyrigdon, i. 18, 1: Blickl. Homl. 209, 8. Onbyrgaþ *gustate*, Ps. Spl. 33, 8. Ic hæbbe bóca onbyrged, Salm. Kmbl. 3; Sal. 2.

on-birging, e; f. *Tasting, taking (food):*—Wið ǽttres onbyrgingce, Lchdm. i. 136, 12.

on-birgness, e; f. *Taste, tasting:*—Seó wæs wynsumu on ðære onbyrignesse . . . Manige men þurh ðyses wǽtan onbyrignesse wurdan gehǽlde, Blickl. Homl. 209, 9-12.

on-birhtan *to illumine:*—God hié onbyrhte mid andgite, Blickl. Homl. 105, 31.

on-bítan; p. -bát; pp. -biten (*with gen.*) *To taste of, partake of:*—Se ðæs wǽstmes onbát, Cd. Th. 30, 21; Gen. 470: 42, 22; Gen. 677. Gif wulf ǽniges cynnes orf tóslíte, and hit forðon deád beó, ne onbíte (*gustet*) his nán Cristen man, L. Ecg. P. iv. 29; Th. ii. 212, 26. Anbíte, iv. 28; Th. ii. 212, 23. Gecýðan ðæt heó ðæs forstolenan ne onbite, L. In. 57; Th. i. 138, 10. Ne sceal hé huniges onbítan, Lchdm. ii. 222, 20. Ne hit se mon drincan meahte, ne his ǽnig neát onbítan ne meahte, Nar. 8, 32. Nǽnigre wǽtan onbítan, Guthl. 2; Gdwin. 16, 24. [O. Sax. an-bítan: O. H. Ger. an-bízan.]

on-blǽstan *glosses* inrumpere, Wrt. Voc. ii. 48, 49.

on-blandan *to intermingle, to infect (with moral evil):*—Hé lungre áhóf wóðe wiðerhýd g weán onblonden *he raised at once his voice, hostile and harmful (cf. the use of geblanden in similar phrases)*, Andr. Kmbl. 1350; An. 675.

on-bláwan *to breathe into, inspire, inflate:*—Onblǽwþ litrat (?), Wrt. Voc. ii. 53, 60. Onblá[wende] *inspirans, inflans*, Hpt. Gl. 442, 29. Mid elreordre dysignesse onbláwne *barbara inflati stultitia*, Bd. 2, 5; S. 507, 13.

on-bláwness, e; f. *Inspiration:*—Seó onbláwnes ðære heofonlícan onfæþmnesse, Blickl. Homl. 7, 26.

on-blótan *to offer, sacrifice:*—Abraham onbleót ðæt lác Gode, Cd. Th. 177, 21; Gen. 2933.

on-bregdan, -brédan; p. brægd, -bréd, pl. brugdon, -brudon. I. *with dat. acc.* (?), *To move quickly:*—Heáfde onbrygdeþ þriwa áscæceþ *the Phenix thrice moves its head (bowing to the sun; igniferum caput ter venerata)*, Exon. Th. 207, 18; Ph. 143. Onbrǽd recedes múþan raþe æfter ðon on flór treddode *Grendel opened the door violently and stepped on to the floor of the hall*, Beo. Th. 1450; B. 723. II. *intrans. To move (oneself) quickly, to start (from sleep):*—Ðá on morgne mid ðý hit dagode ðá onbréd ic *postero die matutino expergefactus diluculo*, Nar. 30, 30: Bd. 3, 27; S. 559, 16. Ðá onbréd Gúþlác of ðam slǽpe, Guthl. 6; Gdwin. 42, 13. Hé of slǽpe onbrægd, Elen. Kmbl. 150; El. 75. Swá hé of hefigum slǽpe onbrude, Bd. 5, 19; S. 640, 27.

on-bring, es; m. *Instigation:*—Se man ðe hine sylfne ofslihþ mid wǽpne oððe mid (for, MS. X.) hwylcum mislícum deófles onbringe (*instigatione*), L. Ecg. P. ii. 5; Th. ii. 184, 5.

on-brinnan. v. on-beornan.

on-brucol; adj. *Rugged:*—Anbrucolne *preruptam*, Germ. 402, 85.

on-bryce, es; m. *An irruption, attack:*—Onrǽs, onbryce *irruptionem, ingressionem*, Hpt. Gl. 464, 66.

on-bryrdan; p. de. I. *to instigate, stimulate, incite, inspire, animate:*—Onbryrde *instigavit*, Wrt. Voc. ii. 44, 82. Hé hí tó geleáfan onbryrde, Blickl. Homl. 107, 2. Hí se héhsta Déma mid elne onbryrde *inspired her with courage*, Judth. Thw. 22, 37; Jud. 95. Git mid fullwihte onbryrdon ealne ðisne middangeard, Exon. Th. 467, 10; Hó. 136. Onbryrdan beorman míne *to leaven with my leaven*, 266, 10; Jul. 396. Hit nis git se tíma ðæt ic þé heálícor mæge onbryrdan *firmioribus remediis nondum tempus est*, Bt. 5, 3; Fox 14, 14. Onbryrdendum (*instigante*) feónde ealra góda, Bd. 2, 22; S. 553, 14. Onbryrd (*compunctus*) mid lufan ðæs upplícan ríces, 4, 12; S. 580, 36. Wearþ Johannes swá onbryrd þurh ðæt tácen, ðæt hé his brýde on mægþháde forlét, Homl. Th. i. 58, 16. Him wearþ onbryrded breóstsefa, Exon. Th. 122, 15; Gú. 306. Æfter heora láre ða ðe wǽron godcundlíce onbryrde *juxta divinitus inspiratam doctrinam*, Bd. 4, 17; S. 585, 34. Sceolan wé beón áwehte and onbryrde tó godcundre láre, Blickl. Homl. 33, 23. II. *to excite to a feeling of compunction:*—Hé wæs onbryrded (*compunctus*) mid gemynde his synna and weóp, Bd. 3, 27; S. 559, 2. Ðǽr mon ðæt godspel sægþ maniges mannes heorte biþ onbryrded, Blickl. Homl. 47, 32. v. in-bryrdan.

on-bryrding, e; f. *An exciting, a stimulus:*—Onbryrdinge *instinctu*, Wrt. Voc. ii. 44, 34.

on-bryrdness, e; f. *Instigation, stimulus, inspiration, compunction:*—Mid wíne onbryrdnysse *vino compunctionis*, Ps. Spl. 59, 3. Mid onbryrdnysse ðæs upplícan éðles *with the stimulus that is given by the land on high*, Homl. Th. ii. 550, 19. Mid godcundre onbryrdnysse monad *divino admonitus instinctu*, Bd. 1, 23; S. 485, 24: 4, 32; S. 611, 39. Mid ða godcundan onbryrdnesse monad, 5, 6; S. 620, 1. Þurh ðæs

sôþan Godes onbryrdnysse *inspirante Deo vero*, 2, 13 ; S. 517, 17 : Blickl. Homl. 119, 18. v. in-bryrdnes.

on-búgan; *p.* -beáh. **I.** *to bend :*—Ðonne ic onbúge *when I (a bow) bend*, Exon. Th. 405, 16 ; Rä. 24, 3. **II.** *to bend in reverence* or *submission, to bow :*—Heó tô hyre môdor cneówum onbeáh, Lchdm. iii. 428, 13. His gebrôðru onbugon tô him (*proni adorantes*), Gen. 50, 18. Ða ðe nolden tô his libbendum lîchaman onbúgan, ða nú eádmôdlîce on cneówum âbúgaþ tô his deádum bânum, Chr. 979; Erl. 129, 20 : Ors. 6, 9; Swt. 264, 9. **III.** *to submit, yield :*—Ðú eart rihtwîs and nânum ne onbîhst, Homl. Th. ii. 298, 33. Hê nænigum woruldrîcum men þurh leáse ôlecunga swîðor onbúgan nolde, ðonne hit riht wære, Blickl. Homl. 223, 28. Beó ðú onbûgende ðínum wiðerwinnan, Mt. Kmbl. 5, 25. **IV.** *to bend aside, deviate :*—Ic onbúgan ne môt of ðæs gewealde ðe mê wegas tæcneþ, Exon. Th. 383, 24; Rä. 4, 15. v. an-búgan, on-bígan.

on-bútan; *prep.* (*adv.*) *with dat. acc. About.* **I.** *of place :*—Gewrîðe onbútan (MS. H. âbútan) ðæs mannes swyran, Lchdm. i. 160, 23. Feówer circulas onbútan ðære sunnan, Chr. 1104; Erl. 239, 18. Se here sceolde bión getrymed onbútan Hierusalem, Past. 21; Swt. 161, 25. Seó eá gæþ onbútan ðæt land, Gen. 2, 11. Ðæt folc him sæh eall onbútan, Homl. Skt. i. 23, 651. **II.** *of time :*—Onbútan Martines mæssan and gyt lator, Chr. 1089; Erl. 226, 19. **III.** *with ðær :*—Æt Hocneratûne and ðær onbútan, 917; Erl. 102, 14. Ofer eall ðær onbútan, Homl. Skt. i. 23, 490, 660.

on-býgan, -byrgan. v. on-bígan, -birgan.

on-cennan; *p.* de To bear (*a child*), *bring forth :*—Mæden sceal oncennan sunu, Ælfc. T. Grn. 9, 11. Sî oncenned *nascatur*, Kent. Gl. 984. v. â-cennan.

on-cîgung, e; *f. Invocation, invoking :*—Ðerh onceigunge *per invocationem*, Rtl. 114, 3 : 122, 3. Onceigince, 147, 27.

on-cirran, -cerran, -cyrran; *p.* de. **A.** *in a physical sense.* **I.** (a) *to turn* (*trans.*) *make a change in position or direction :*—Hê oncyrde hine tô Paule *he turned to Paul*, Blickl. Homl. 183, 30 : 185, 36. Ðæt hié hine môston on ôðre sîdan oncyrran, 227, 19. Andwlitan út oncyrran *faciem avertere*, Ps. Th. 131, 10. Wênst ðú ðæt ðú ðæt hwerfende hweól ðonne hit on ryne wyrþ mæge oncyrran *tu vero volventis rotae impetum retinere conaris?* Bt. 7, 2; Fox 18, 36. Oncerran, Met. 10, 39. Oncirredre *preposterò*, Wrt. Voc. ii. 67, 23. (b) *to turn* (*into another form*) :—Ða lâstas on ôðerne mægwlite oncyrran, Blickl. Homl. 127, 19. **II.** *to turn* (*intrans.*) *to go :*—Ða oncerde se wind from ðære byrig, Bd. 3, 16; S. 543, 7. Hê þyder oncirde, Beo. Th. 5933; B. 2970 : 5895; B. 2951. Ýða ongin eft oncyrde, Andr. Kmbl. 932; An. 466. **B.** *in a metaphorical sense.* **I.** (a) *to turn, make a person adopt a line of conduct,* etc. :—Se nýdde Clementem ðæt hê Cryste wiðsôce, ðâ ne mihte hê hyne oncyrran *he could not turn him*, Shrn. 150, 18. Angan þencean hú hê þider meahte Crêcas oncerran, Met. 1, 61. (b) *to turn, change :*—Ðú ða wyrde oncyrrest *fata mutabis*, Nar. 31, 24. Hí mê ðæt on edwît eft oncyrdan *factum est mihi in opprobrium*, Ps. Th. 68, 10. Nergend him naman oncyrde, Elen. Kmbl. 1004; El. 503. Ne meahte hê ðæs wealdendes willan oncirran, Beo. Th. 5707; B. 2857. Hí woldon his môd oncyrran, Andr. Kmbl. 2921; An. 1463. Hê ne meahte hire môd oncyrran *he could not make her change her mind*, Exon. Th. 256, 4; Jul. 226. (c) *to turn from good to bad, to pervert :*—Ðus ic sôðfæstum môd oncyrre, 264, 13 ; Jul. 363. Ðæt wê þurh misgedwield môd oncyrren, 262, 2 ; Jul. 268. Hyge wæs oncyrred (*by a magical drink*), Andr. Kmbl. 72; An. 36. Ðú miht ongiton hú se mîn weorþscipe for worulde is oncerred *quantum decus ornamentis nostris decesserit, vides*, Bt. 10 ; Fox 30, 15. (d) *to turn aside, avert :*—Ðîn yrre fram ûs oncyrre, Ps. Th. 84, 4. Oncyrran, 78, 5. Oncyrran môd from his Meotude, Exon. Th. 124, 8; Gú. 336. (e) *to turn back, reverse* (*a sentence*), *revoke :*—Ðú yrre ðîn eft oncyrdest, Ps. Th. 70, 19. Hê ða yrmðu eft oncyrde æt his upstige, Exon. Th. 38, 30; Cri. 614. Ða word oncyr *retract the words*, 251, 13 ; Jul. 144. Wæs se dôm oncyrred Euan ungesælignesse, Blickl. Homl. 3, 8. Wearþ se sârlîca cwide eft oncerred, 123, 7. **II.** *to turn* (*intrans.*) :—Hié fram heora unrihtum oncyrron, Blickl. Homl. 109, 20.

on-clifiende; *adj. Sticking to, persistent :*—On forhæfednysse and on clænnysse fæsthafule and onclyfiende *in abstinence and purity constant and persistent*, Cod. Dip. B. i. 154, 37. v. clifian.

on-clipian *to invoke :*—Enos ongan ærest onclypian (*invocare*) Drihtnes naman, Gen. 4, 26.

on-cnǽwe; *adj. Known, recognised :*—Oncnǽwe *cognitum*, Ps. Spl. T. 33, 5. Cf. ge-cnǽwe.

on-cnáwan; *p.* -cneów; *pp.* -cnáwen *To know;* noscere, cognoscere, agnoscere :—Oncnáwen *nosco, cognosco,* ic ancnáwe *agnosco,* Ælfc. Gr. 28, 1 ; Som. 30, 31–32. Tô angitanne and tô oncnâwenne *animadverti,* Wrt. Voc. ii. 2, 44. Beón oncnáwen *conici* (cf. 23, 50), 23, 78. **I.** *to know, recognise,* (1) *to identify an object through being acquainted with its characteristics,* (a) *of persons :*—Se oxa oncneów his hláford, Homl. Th. i. 42, 25. Hí hine (*Jesus*) on ðam gereorde oncneówon, ðone ðe hî ne mihton on onwrigenysse hâliges gewrites oncnáwan, 284, 33–34. Ðæt is êce lif, ðæt hi ðê oncnáwon sôðne God, and ðone

de ðú âsendest, 42, 14 : ii. 362, 22. Frán hwæðer hit oncneówe his fôstermôdor, Bt. 3, 1 ; Fox 4, 28. Ðeáh ðe hê wundra fela gecýðde, synnige ne mihton oncnáwan ðæt cynebearn, Andr. Kmbl. 1131 ; An. 566. Hê is ancnáwen *dinoscitur, agnoscitur,* Hpt. Gl. 440, 32. Ðú wære æfre fæmne oncnáwen, Glostr. Frag. 106, 8. Biþ oncnáwen Drihten dômas wyrcende *cognoscetur Dominus judicia faciens,* Ps. Spl. 9, 17. (b) *of things :*—Heáh feorran hê oncnáwaþ *alta a longe cognoscit,* 137, 7. Ná ic hit swâ oncneów swâ hit ðîn æ hafaþ *I did not recognise it* (*what was said*) *as what is in thy law,* Ps. Th. 118, 85. Ic his word oncneów, ðêh hê his mægwlite bemiðen hæfde, Andr. Kmbl. 1710 ; An. 857 : Beo. Th. 5102 ; B. 2554. Se assa oncneów his hláfordes binne, Homl. Th. i. 42, 25. Ðæt ðú oncnáwe (*cognoscas*) ðara worda sôðfæstnesse, Lk. Skt. 1, 4. Ðæt ðâs nytenan menn ðîne mihta oncnáwon, Homl. Th. i. 62, 14. Hié ðæt ongeotan ne cúðan, ðæt hié gehýrdon, ne ðæt oncnáwan ne mihton ðæt hié gesáwon, Blickl. Homl. 105, 29: 95, 10. Ðú meaht sôða gesælþa sôna oncnáwan, Met. 12, 30 : Elen. Kmbl. 790 ; El. 395. Oncnáwan hwǽr wê sǽlan sceolon sundhengestas, Exon. Th. 54, 1 ; Cri. 862. Ðîne fôtswaða nǽron oncnáwene, Ps. Lamb. 76, 20. (2) *to recognise a fact* (which is generally stated in a clause beginning with *ðæt* :—Wundra weorc ðîne and sáwle mîn oncnáweþ (*knows that thy works are wonderful*), Ps. Spl. 138, 13. Be ðam oncnáwaþ ealle men, ðæt gê synt mîne leorningcnihtas, Jn. Skt. 13, 35. Ic oncneów (*cognovi*) ðæt ðú ondrǽtst swýðe God, Gen. 22, 12. Ðá se dêma ðæt oncneów and ongeat (*persensit*), ðæt hê hine oferswîðan ne mihte, Bd. 1, 7 ; S. 478, 1 : 5, 9; S. 623, 21. Ðá cwæþ eal folc ðæt hê Godes sunu wǽre, and ðæt fulfremedlîce oncneówan, Blickl. Homl. 177, 20. Hig oncneówon, ðæt hig nacode wǽron, Gen. 3, 7 : Mk. Skt. 12, 12. Oncnáw ðæt mîne welan syndon gewitene, Blickl. Homl. 113, 24. Tô ðam earde ðe flêwþ meolce and hunige, swâ swâ gê of ðissum wæstmum oncnáwan mágon, Num. 13, 28. Geseón and oncnáwan and swîðe gearelîce ongeotan ðæt ðisses middangeardes ende neáh is, Blickl. Homl. 107, 22 : 115, 5. Be ðam man mihte oncnáwan ðæt se cniht nolde wácian ðæt ðam wîge, Byrht. Th. 131, 16 ; By. 9. **II.** *to know, understand, attain to a knowledge of :*—Gyt gê ne oncnáwaþ ne ne ongitaþ, Mk. Skt. 8, 17. Ðú mîne geþohtas oncneówe *intellexisti cogitationes meas,* Ps. Th. 138, 2, 3. Ðá oncneówon hig be ðam worde *cognoverunt de verbo,* Lk. Skt. 2, 17: Homl. Th. i. 30, 32. Hê ða yldestan lǽrde, ðæt heó wîsdômes word oncnáwan, Ps. Th. 104, 18. Ða mægnu tweónedon be ðære gýtsunge, ðæt hió fullîce hió ne oncneówon, Gl. Prud. 64 a. Oncnáwaþ ða ðing ðe eówre bearn nyton, Deut. 11, 2. Dysige ðæt oncnáwan *stulti sapite,* Ps. Th. 93, 8. Ic ðînra worda ne mæg wuht oncnáwan, Cd. Th. 34, 8 ; Gen. 534. Oncnáwan, hú hine lýgneðon leáse, Exon. Th. 69, 12 ; Cri. 1119. **III.** *to know, learn by observation, observe, perceive :*—Gif ic mê unrihtes oncneów áwiht on heortan *iniquitatem si conspexi in corde meo,* Ps. Th. 65, 16. Oncnáw onsýne Cristes ðînes *respice in faciem christi tui,* 83, 9. Oncnáw paþas mîne *cognoscite semitas meas,* Ps. Spl. 138, 22. Ða deóflu æteówiaþ ðære synfullan sáwle hyre mânfullan dǽda, ðæt heó oncnáwe mid hwilcum feóndum heó ymbset biþ, Homl. Th. i. 410, 9. **IV.** *to acknowledge,* (1) *make acknowledgment of a fault :*—Wê oncnáwaþ eal ðæt wê geworhton on worldrîce, ne mágon we hit dyrnan, Hy. Grn. 7, 90. Ðæt hê mihte oncnáwan his mânfullan dǽda on ðam hæftneðe, Ælfc. T. Grn. 8, 21. (2) *to acknowledge a greeting :*—Iosep hig oncneów árfullîce *clementer resalutatis eis,* Gen. 43, 27. (3) *to acknowledge the power of another* (?) :—Elias eorl ðe ða Mannie of ðam cynge geheóld and oncneów (-cweow, MS.), Chr. 1110 ; Erl. 243, 11.

on-cnáwenness, e; *f.* **I.** *recognition, knowledge* (*that an object is what it really is*) :—Wê habbaþ ðæt êce lîf þurh geleáfan, and on-cnáwennysse ðære Hálgan Þrynnesse, gif wê ða oncnáwennysse mid ár-wurþnysse healdaþ. Witodlîce gif Godes oncnáwennys ûs gearcaþ ðæt êce lîf, swâ miccle swîðor wê êfstaþ tô lybbenne swâ micclum swâ wê swîðor on ðissere oncnáwennysse þeónde beóþ. Sôðlîce ne swelte wê on ðam êcan lîfe, ðonne biþ ûs Godes oncnáwennys fulfremed.... Ac wê sceolon on andwerdum lîfe leornian Godes oncnáwennysse ... ðæt wê môton becuman tô his fulfremedan oncnáwennysse, Homl. Th. ii. 362, 2–364, 9. Hê nolde him æteówian his oncnáwennysse *he would not let them recognise him* (cf. l. 16, hê him ne geswutelode hwæt hê wæs), 284, 12. Ða deóflu æteówiaþ ðære synfullan sáwle hyre mânfullan dǽda ... Tô eorþan heó biþ ástreht þurh hire scylda oncnáwennysse (*on recognising her guilt*), i. 410, 12. **II.** *acknowledgment, recognition of a claim :*—Ðonne ys ðis seó oncnáwennis ðe hê hæfþ God mid gecnáwen ... on circlicum mâðmum, Chart. Th. 429, 7.

on-cnáwness, e ; *f. Knowledge, conception :*—Hê hiene bedǽlþ ðære oncnáwnesse ðæs uplecan leóhtes *a luce se supernae cognitionis excludit,* Past. 11, 4; Swt. 69, 24.

on-cnyssan *to cast down :*—Ðú mê yfela feala oft oncnyssedest *thou didst strike me down with many evils,* Ps. Th. 70, 19. Oncnyssyde *depulsae,* Ps. Spl. C. 61, 3.

on-cunnan; *-cúðe; pp. -cunnen To accuse a person* (acc.) *of something* (gen., clause beginning with *ðæt,* or with prep. *be, for*), *to blame, charge, lay to a person's charge :*—Ðonne oncann hê hiene selfne for ðære hræðhýdignesse ðe hê ǽr tô fela sealde *occasionem contra se im-*

patientiae enquirit, Past. 44, 4 ; Swt. 325, 16. Ic him mīn wedd beád, ðæt ic hyra nǽfre nǽnne ne oncúðe, for ðon ðe hý on riht sprǽcon, Chart. Th. 486, 21. Mē mīne ágen word sócon swýðe oncúðan *verba mea execrabantur*, Ps. Th. 55, 5. Ðonne oncúðon (*impugnabant*) hié mē būtan scylde, Past. 46, 7 ; Swt. 355, 15. Hié sylfe be ðon oncúðon, ðæt hié swá ne dydon, Blickl. Homl. 215, 12. Gif hwá óðerne godborges oncunne, L. Alf. pol. 33 ; Th. i. 82, 5. Ðý lǽs ðec Meotud oncunne, ðæt ðū sý wommes gewita, Exon. Th. 301, 13 ; Fä. 18. Ðæt ūs God ne þurfa oncunnan for ðǽræ waniungæ *nec nobis Deus debeat imputare hanc imminutionem*, Chart. Th. 163, 25. Oncunnen *notatus*, Wrt. Voc. ii. 114, 81. Hē wæs oncunnen (*accusatus*) fram ðam ylcan cyninge, Bd. 5, 19 ; S. 640, 9. Tó oncunnyne oncunnysse *ad excusandas excusationes*, Ps. Spl. M. C. 140, 4. v. next word.

on-cunness, e ; *f. An accusation* (?), *excuse* (?) :—Tó ácunnenne oncunnisse *ad excusandas excusationes*, Ps. Surt. 140, 4. v. on-cunnan.

on-cunning, e ; *f. An accusation* :—Mid gelómlícum oncunningum *crebris accusationibus*, Bd. 3, 19 ; S. 548, 3.

on-cweðan. I. of animate beings, *to reply, respond* :—Oft mec slǽpwērigne secg grētan eode, ic him oncweðe, Exon. Th. 387, 18. Him Andreas oncwæð, Andr. Kmbl. 540 ; An. 470 : 1109 ; An. 555. Him Babilone weard andswarode and oncwæð, Cd. Th. 229, 3 ; Dan. 211 : 53, 23 ; Gen. 865. Judas cwæð . . . him oncwæð cáseres mǽg, Elen. Kmbl. 1334 ; El. 669. Stormas stánclifu beótan, him stearn oncwæð, Exon. Th. 307, 14 ; Seef. 23. Swilce ealle ða anlícnyssa ðe ðære byrig tó godon gesette wǽron, ðæt hí ealle ætgædere oncwǽdon and ánre stemne clypedon, ðæt hí áweg ðanon woldon . . . and swilce ða strǽta ealle eác oncwǽdon, Homl. Skt. i. 23, 93–98. Ne sculon mæssepreóstas būtan óðrum mannum mæssan syngan, ðæt hē wíte hwone hē grēte, and hwá him oncwæðe, L. E. I. 7 ; Th. ii. 406, 23. Ðæt hió ðære cwēne oncweðan meahton . . . swá hió him tó sóhte, Elen. Kmbl. 648 ; El. 324. II. of inanimate things, *to echo back, give back a sound, reply* :—Oncwyð *remugiet*, Hpt. Gl. 513, 12. Scyld scefte oncwyð, Fins. Th. 12 ; Fin. 7. Ðæt him se weald oncwyð . . . wudu eallum oncwyð, Met. 13, 46–50 : Bt. 25 ; Fox 88, 20. Oft oncwæð ýð óðerre, Andr. Kmbl. 884 ; An. 442.

on-cýð[ð], e ; *f. Grief, distress* :—Denum eallum wæs weorce on móde, oncýð eorla gehwǽm, syððan Æscheres hafelan mētton, Beo. Th. 2844 ; B. 1420. Hæfde Eást-Denum gilp gelǽsted, swylce oncýððe ealle gebētte, 1664 ; B. 830.

on-cýðan *to make known, announce* :—Ðá ðá ic on eard com ic oncýðde ealle folce hwæt ic on Rôme gedôn hæfde, Chart. Th. 117, 1.

oncýð-dǽd, e ; *f. A deed causing distress, an injury* :—Oncýðdǽda wrecan, Andr. Kmbl. 2360 ; An. 1181.

on-cýðig ; *adj. Suffering from the want of something* (?), *not acquainted with, a stranger to anything* (?) ; cf. un-cýðig :—Elnes oncýðig *suffering from weakness* (?) or *a stranger to strength* (?), Elen. Kmbl. 1446 ; El. 725. *The term is used of Judas, to whom the previous lines 1392–3 refer* :—Mēðe and meteleás, mægen wæs geswiðrod.

ond, ond—. v. and, and—.

on-dǽlan *to impart, infuse* ; *infundere*, Rtl. 17, 11 : 85, 39 : Lk. Skt. Lind. 10, 34.

on-dǽlend, es ; *m. One who imparts* :—Mægna sellend and bloedsunga ondǽlend *virtutum dator et benedictionum infusor*, Rtl. 103, 38.

ond-efen, on-derslíc, -deslíc, -desn. v. and-efn, on-dryslíc, -drysnu.

on-dôn *to undo, open* :—Ondêst *solvat*, Wrt. Voc. ii. 120, 79. Ðonne andydan ða duru, Ors. 3, 5 ; Swt. 106, 14. Siont ondône *aperientur*, Kent. Gl. 232.

on-dôung, e ; *f. A putting in, injection* :—Mid ondôunge wyrtdrences þurh horn sió wamb biþ tó clǽnsianne, Lchdm. ii. 260, 11.

on-drǽdan ; *p.* -drêd, -drǽd, -dreard, -dreord ; *pp.* -drêd *to dread, fear* ; *timeo.* I. with construction undetermined :—Ondrêt *obstupuit*, Hpt. Gl. 510, 23. Ondreard *timuit*, Mt. Kmbl. Lind. 2, 22. Ic ondreord *timui*, Ps. Surt. 118, 120. Ondreord *timuit*, 63, 10. Ondreordun, 63, 6. Ondreardon *timuerunt*, Mt. Kmbl. Lind. 9, 8. Ondreardon (-dreordun, Rush.), Mk. Skt. 10, 32 : 11, 18. II. with acc. or gen. of object, and (a) with a reflexive dative :—Ic ondrǽde mē God *Deum timeo*, Gen. 42, 18. Ic mē ondrǽde *timeo, metuo*. Se ðe him ondrǽt, sumes þinges hē him ondrǽt, *timeo Deum* ic mē ondrǽde God ; *timeor* ic eom ondrǽd, ðæt is, ðæt sumum menn stent ege fram mē, Ælfc. Gr. 19 ; Som. 22, 62–64. Ne ðū ðē nihtegsan ondrǽdest, Ps. Th. 90, 5. Se ðe him ǽlc wolcn ondrǽdt, ne rípþ se nǽfre, Past. 39 ; Swt. 285, 18. Hē him ondrǽt his deáþes, Homl. Skt. i. 12, 87. Hwá him ne ondrêde ðæs cyninges irre? Ap. Th. 2, 18. (b) without the reflexive dative :—Ic hine swíðe ondrǽde, Gen. 32, 11. Ðū ondrǽtst swýðe God, 22, 12. Se ðe ǽgðer ondrǽt, ge ðone ðe hine ondrǽt, ge ðone ðe hine nā ne ondrǽt, Bt. 29, 1 ; Fox 104, 5–6. Herodes ondrêd (-dreard, Lind. : -dreord, Rush.) Johannem, Mk. Skt. 6, 20. Ðæt hig hine ondrêdon, swá swá hig ondrêdon Moysen, Jos. 4, 14. III. with the prep. *from* :—Swá egefull wæs Alexander ðá ðá hē wæs on eásteweardum ðissum middangearde, ðætte ða from him ondrêdan ðe wǽron on westeweardum, Ors. 3, 9 ; Swt. 136, 7. Hié alle from him ondrêdon,

ðæt hí hié mid gefeohten, 1, 10 ; Swt. 48, 16. IV. without an object, and with reflexive dative, *to be afraid* :—Hié word Drihtnes gehýrdon and ondrêdon him, Cd. Th. 53, 15 ; Gen. 861. Ða weras ðá ðæt gesáwon hié him swíðe ondrǽdon, and cwǽdon, Blickl. Homl. 247, 16. Ne ondrǽd ðū ðē, Elen. Kmbl. 162 ; El. 81. Ne wilt ðū ðē ondrǽdan Zacharias, Blickl. Homl. 165, 7. Him ðá ondrǽdendum ðǽm gebrôðrum, Ors. 1, 5 ; Swt. 34, 1.

on-drǽdendlíc ; *adj. To be feared, terrible* :—Hē wæs swíðe strang and swíðe ondrǽdendlíc he (*William Rufus*) *was very severe, and very terrible*, Chr. 1100 ; Erl. 235, 39. Gif ðes bealdwyrda biscop ácweald ne biþ, siððan ne biþ úre ege ondrǽdendlíc, Homl. Th. i. 420, 3. Ðises godspelles geendung is swíðe ondrǽdendlíc : 'Fela sind gelaðode, and feáwa gecorene,' ii. 82, 3.

on-drǽding, e ; *f. Dread, terror* :—Hié selfe wǽron on ðære ondrǽdinge hwonne hié on ða eorþan besuncene wurden, Ors. 2, 6 ; Swt. 88, 14. Hē sume hwíle wênde ðæt hine mon gefôn sceolde, and hē for ðære ondrǽdinge ðæs ðe swíðor on ðæt weorod þrong, 5, 12 ; Swt. 244, 12.

on-drencan *to inebriate* :—Hí wǽron ondrencte mid oferdrynce, Guthl. 14 ; Gdwin. 62, 20. v. in-drencan.

on-drincan *to drink of* (with gen.) :—Ða ðe on wege weorðaþ wætres æt hlimman deópes ondrincaþ *de torrente in via bebet*, Ps. Th. 109, 8. Ðá ondranc se ðæs wætres, and sealde hit ðǽm brêðer . . . and se ondranc eác ðæs wætres, Shrn. 64, 11–12. Bæd ðæt hē him onsende wínes ondrincan, Bd. 5, 5 ; S. 618, 11. Sioððan hié hæfdon ondruncen ðæs wætres *potata aqua*, Nar. 13, 28.

on-drislíc. v. on-dryslíc.

on-druncnian *to get drunk* :—Beóþ ondruncniende *inebriabuntur*, Ps. Spl. T. 35, 9.

on-dryslíc, -drystlíc, -dyrstlíc, -deslíc ; *adj. Terrible, dreadful* :—Ús is tó geþencanne hú onþrislíc (-dryslíc : egeslíc, other MS.) hit on bôcum gecweden is, L. Ath. i. prm. ; Th. i. 196, 4. Cwæð ðæt se mon wǽre ondrysenlíc (onderslíc, MS. T. : ondrislíc, MS. B.) on tó seónne (*terribilis aspectu*), Bd. 2, 16 ; S. 519, 35. Ondeslíc *terribilis*, Rtl. 69, 4 : *orror* (?), 162, 28. Ácwellan ondryslícum wítum, Shrn. 111, 10. Þreágan mid ondrystlícum wítum, 104, 16. Gif hwilc mon síe on ondyrstlecum wísum, and hē sý mínes naman gemyndig, Drihten, gefriða ðū hine from ðǽm brôgan, 101, 30. Sum sume swíðe ondryslícu (*tremenda*) secgende wæs, Bd. 5, 12 tit. ; S. 627, 3. v. following words.

on-drysne ; *adj.* I. applied to that which is evil, *terrible, dreadful, awful* :—Firen ondrysne *terrible crime*, Beo. Th. 3869 ; B. 1932. II. applied to that which is good, *awful, exciting awe or reverence, venerable* :—Him wæs freán engla word ondrysne, Cd. Th. 173, 14 ; Gen. 2861. Wæs hē for his árfæstum dǽdum eallum his gefêrum leóf and weorð and ondrysne *he was beloved, honoured and reverenced by all his companions for his pious deeds*, Blickl. Homl. 213, 12. Ðæt hý messan singan and ða andrysnan þênunge mid árwyrþnesse gefyllen, R. Ben. 140, 5. *See other examples under* an-drysne.

on-drysness, -desness, e ; *f. Fear* :—Ondesnisse *timoris*, Rtl. 3, 24.

on-drysnlíc, -drysenlíc ; *adj. Terrible* :—Mē ætýwde ondrysnlíco gesihþ *visio mihi tremenda apparuit*, Bd. 5, 19 ; S. 640, 36. Ondrysenlíc *terribilis*, 2, 16 ; S. 519, 35. Ðá ætýwde hire micel mon and ondrysnlíc, Shrn. 106, 9. Hē wæs of líchoman álǽded, and hē geseah má ondrysnlíces and eác wundorlíces ðonne hē mihte ásecgan, 51, 31. v. on-dryslíc, *and see other examples under* an-drysenlíc.

on-drysnu, -desnu ; *f.* I. fear :—Fore ondesne (ondesnum, Rush.) *propter metum*, Jn. Skt. Lind. 19, 38 : 20, 19. Ðætte sió forsewennes him ege and ondrysnu on gebringe *ut ostensa desperatio formidinem incutiat*, Past. 37, 2 ; Swt. 265, 19. Hē wolde ðǽm fortrúwodum monnum andrysnan hálwendes eges on gebrengean *ut praecipitatis vim saluberrimi timoris, infunderet*, 49, 5 ; Swt. 385, 16. Ðonne esne ondrysnum his hláforde cwemeþ, Th. Th. 122, 2. II. *reverence* :—Hié hæfdan miccle lufan and geleáfan tó ðære ciricean, ond eác heálíco ondrysnu (*profound reverence for the church*), Blickl. Homl. 205, 9. v. an-drysno.

on-dryslíc, -dyrstlíc. v. on-dryslíc.

on-dwǽscan *to extinguish* :—Se môna ðe byþ andwǽsced oððe áteorod, Anglia viii. 316, 38. v. á-dwǽscan.

on-ealdian *to grow old* :—Onealdodon bán míne *inveteraverunt ossa mea*, Ps. Spl. 31, 6.

on-eardian *to inhabit* :—Oneardiaþ on ðam *inhabitabunt in ea*, Blickl. Gl. Rihtwíse oneardiaþ (*inhabitabunt*), Ps. Spl. 36, 31. Onearda *inhabita*, 36, 28. Ealle oneardigende ymbhwyrft *omnes inhabitantes orbem*, 32, 8.

on-eardiend, es ; *m. An inhabitant* :—Ne on heora êðele ne sy þinc oneardiendes *et in tabernaculis eorum non sit qui inhabitet*, Ps. Th. 68, 26.

on-efn, -emn, -em *by, near* :—Hí gemêtton fýr, and hláf onem *they found a fire, and bread close by*, Homl. Th. ii. 262, 5. Onefen ðone hagan . . . norþ onefen ðæt gelád, Cod. Dip. Kmbl. ii. 150, 10–13.

Onemn ðæm *at the same time*, Ors. 3, 9; Swt. 128, 33. *See* efn, emn *for other examples.*

on-égan; *p.* de *To fear*:—Sǽton him at wíne, wealle belocene, ne onégdon ná orlegra níð, Cd. Th. 259, 25; Dan. 697. Ic mē onégan (onagen, MS.) mæg, ðæt mē wráðra sum wǽpnes ecge feore beneóte, 109, 28; Gen. 1829. Ni anoegun (anoegu ná?) ic mē aerigfaerae egsan brógum, Txts. 151, 13. Cf. óga.

on-erian *to plough up*:—Wē má lufiaþ ðone æcer ðe ǽr wæs mid þornum áswógen and æfter ðæm ðe ða þornas beóþ áheáwene and se æker biþ onered bringþ gódne wæsðm *plus terram diligimus, quae post spinas exarata fructus uberes producit*, Past. 52, 9; Swt. 411, 18.

on-éðung, e; *f. In-breathing, inspiration*:—From onoeðunge gástes, Ps. Surt, 17, 16.

onettan; *p.* te.   **I.** *to hasten, move rapidly*:—Ǽlc wlite tô ende ēfsteþ and onetteþ, Blickl. Homl. 57, 28. Tô ðam onet Egeas unforwandodlíce, Homl. Th. i. 592, 17. Deáþ eów ǽlce dæg tôweardes onet, Bt. 39, 1; Fox 210, 28. Eall moncynn irnaþ and onettaþ, 37, 2; Fox 188, 14. Hé onette on ðære byrig him tô fultume, Jos. 10, 33. Hé wið mín onette, Homl. Th. ii. 352, 4. Wið ðæs fæstengeates folc onette, Judth.Thw.23,39; Jud.162. Ðá onette Abrahames mǽg tô ðam fæstenne, Cd. Th. 153, 3; Gen. 2533. Ēfste ðá swíðe and onette forþ foldwege, 174, 3; Gen. 2872. Hié swíðe on ðá úre wíc onetton and in ða feóllon *ad castra confluxere*, Nar. 13, 14. Onettad *agitate*, Wrt. Voc. ii. 99, 56. Onettendum (*festinantibus*) cretum, 147, 80.   **II.** *to make a quick movement, to anticipate*:—Onette *occupavit*, 63, 30. Hé gebrægd his swurd and wolde mē ofsleán ðér ic him ne onette and ic ðæt wíf gegripe be hire earme and mē tôforan ábræd and ðér ðis nǽre ðonne wǽre mín blód instæpe ágoten *he drew his sword and would have slain me, if I had not anticipated him, and had seized the woman by her arm, and drew her before me; and if it had not been for this, my blood would have been straightway shed*, Shrn. 39, 16.   **III.** *to be quick in one's movements* or *actions, be active, quick* or *busy*:—Byrig fægriaþ wongas wlitigaþ woruld onetteþ *fair grow the towns, beauteous the plains, the world is quickened* (*in the spring*), Exon. Th. 308, 34; Seef. 49. Lēg onetteþ *busy shall the flame be* (*at the day of judgment*), 448, 17; Dóm. 55: 212, 29; Ph. 217. Sceal onettan se ðe ágan wile líf æt Meotude ðenden him leóht and gǽst somod fæst seón *diligent must he be, while light and spirit hold fast together, who life will receive at the hands of the Lord*, 96, 24; Cri. 1529. Rǽd sceal mon secgan, dæges onettan (cf. *the night cometh, when no man can work*), 342, 11; Gn. Ex. 141. [Cf. *O. H. Ger.* anazzan *sollicitare, excitare, inflammare, hortari, instigare.*]

onettung, e; *f. Hastening, haste, precipitation*:—Oft ða oferblíðan weorðaþ gedréfde for ungemetlícre onettunga *gravatur usu immoderatae praecipitationis*, Past. 61, 1; Swt. 455, 15.

on-fægnian *to shew gladness*:—Ðære helle hund ongan onfægnian mid his steorte *Cerberus shewed his gladness by wagging his tail*, Bt. 35, 6; Fox 168, 17 note.

on-færeld, es; *n. An in-going, entrance*:—Gesáwon onfæreldu *viderunt ingressus*, Ps. Spl. 67, 26. v. an-, in-færeld.

on-fæstnian *to transfix, pierce*:—Hig geseóþ on hwæne hig onfæstnodon *videbunt in quem transfixerunt*, Jn. Skt. 19, 37. Onfæstna (*configo*) ege ðínum flǽsc míne, Ps. Spl. 118, 120.

on-fæðmness, e; *f. Embrace*:—Seó onbláwnes ðære heofonlícan onfæðmnesse sý gewindwod on ðē (*the Virgin Mary*), Blickl. Homl. 7, 26.

on-fangenness, e; *f. Receiving, reception, acceptance*:—Mid Gode nis anfangenness (onfangenes, MS. T.) nánra háda bútan geearnunge ánre (cf. *God is no respecter of persons, but he that worketh righteousness is accepted with him*, Acts 10, 34–35), R. Ben. 13, 4. Seó onfangenes ðæs ríces is Godes gódnysse, Homl. Th. ii. 80, 23. Nán ásolcen man nis orsorh be onfangennysse Godes feós, 556, 24, 33. Mid onfangennesse (*perceptione*) ðæs Drihtenlícan líchoman, Bd. 4, 3; S. 568, 39. For onfangenysse (*susceptionem*) gesta, 1, 27; S. 489, 8. v. on-fengness.

on-fealdan; *p.* -feóld *To unfold, unwrap*:—Hé onfeóld hys hrægl æt hys sceoldrum, Shrn. 98, 17. v. un-fealdan.

on-feall *a swelling, fellon*:—Wið onfealle, gefóh fox, ásleah of cucum ðone tuxl, læt hleápan áweg, bind on næsce, hafa ðē on, Lchdm. ii. 104, 12. Drenc wið onfealle, 102, 27: 104, 1, 3, 4, 6. Lǽcedomas wið ǽlces cynnes ómum ond onfeallum and báncoþum, 98, 21: 102, 20. Wið innanonfealle, 106, 9. Onfelle, 106, 10.

on-feallende; *part. On-rushing*:—From ðære onfeallendan *ab ingruenti*, Wrt. Voc. ii. 3, 34. Ða unstillnesse ðara onfeallendra menigeo *tumultus irruentium turbarum*, Bd. 3, 19; S. 549, 32.

on-feng, es; *m.* [v. fón (on)].   **I.** *laying hold of, seizing*:—Be cirliscre fǽmnan onfenge. Gif mon on cirliscre fǽmnan breóst gefó, L. Alf. pol. 11; Th. i. 68, 13. Be nunnena onfenge (andfencgum, MS. B.: anfenge, MS. H.), 18; Th. i. 72, 7. Be þeófes onfenge æt þiéfþe, L. In. 28; Th. i. 120, 4. Secg wundaþ beorna gehwylcne ðe him ǽnigne onfeng gedéþ *sedge cuts every one that lays hold of it*, Runic pm.

15; Kmbl. 342, 14.   **II.** *taking*, with the idea of wrongful taking:—Be wuda onfenge (andfenge, MS. H.: anfenge, MS. B.) bútan leáfe, L. In. 44; Th. i. 130, 1. Be unáliéfedes mæstennes onfenge, 49; Th. i. 132, 11.   **III.** *defence, protection* (cf. and-fenga):—Wǽron ða hálgan on onfenge manna sáulum, Blickl. Homl. 209, 29.   **IV.** *attack, onset, assault*:—Wurdon hié on ðam onfenge forhte, and on fleám numen, Andr. Kmbl. 2679; An. 1341. Hé hine scilde wið onfengom earmra gǽsta, Exon. Th. 126, 24; Gú. 376; 133, 15; Gú. 490. v. an-feng.

on-fenge, es; *m. A receptacle*:—Anfengce *receptaculum*, Hpt. Gl. 498, 32. Anfencgas *receptacula*, 408, 51.

on-fenge; *adj. Taken, accepted*:—Onfenge *adsumtus*, Mk. Skt. Lind. 16, 19. Mið ðý onfenge woeron *assumtis*, Lk. Skt. Lind. 9, 10: *acceptis*, 9, 16. Ān geonfenge (onfenge, Rush.) biþ *una assumetur*, 17, 35. Onfengo *suscepta*, Rtl. 9, 7. v. and-fenge.

on-fengness, e; *f. Reception, acceptance*:—Seó onfengnes Cristes geleáfan, Bd. 2, 9; S. 510, 12. Be ðære onfengnysse Cristes geleáfan *de percipienda fide Cristi*, 2, 13 tit; S. 515, 33. Ymb xl nihta ðæs sǽdes onfengnesse *xl dies post semen receptum*, L. Ecg. C. 30, note; Th. ii. 154, 36. Þurh ða onfengnesse ðæs Hálgan Gástes, Blickl. Homl. 135, 35. v. and-fengness.

on-findan; *p.* -fand, -funde.   **I.** *to find out, discover, detect*:—Ic anfinde *deprehendo*, Wrt. Voc. ii. 25, 32. Gif mec onfindeþ wíga, ðǽr ic búge, Exon. Th. 396, 20; Rä. 16, 7. Ic mē sylf onfand ðæt . . . *I discovered that . . .*, Blickl. Homl. 177, 6. Ic hine onfand, and hine onbændan hét, Salm. Kmbl. 550; Sal. 274. Ne ic culpan in ðē ǽfre onfunde, Exon. Th. 11, 30; Cri. 178. Hú Boetius hí wolde berǽdan, and þeódríc ðæt anfunde, Bt. 1, tit; Fox x. 2. Onfundan *deprenderint*, Wrt. Voc. ii. 25, 33. Gif hé wæccende weard onfunde búan on beorge, Beo. Th. 5675; B. 2841. Gif hwylc bróðor on lytlum gyltum byþ onfunden, R. Ben. 49, 2.   **II.** *to find out from experience, become aware of, perceive, be sensible of*:—Ic onfinde *experiar*, Wrt. Voc. ii. 32, 7. Hé ðæt ðonne onfindeþ, ðonne se fǽr cymeþ *he will find it out, when the peril comes*, Exon. Th. 449, 18; Dóm. 73. Ðá se gist onfand ðæt se beadoleóma bítan nolde, Beo. Th. 3049; B. 1522. Landweard onfand (*became aware of*) eftsíð eorla, 3785; B. 1890. Onfunde, 1504; B. 750: 1622; B. 809. Ðá hé ðá onfunde, ðæt hé deád beón sceolde, Bt. 29, 2: Fox 104, 20. Onfunde *comperit*, i. *intellexit, cognovit, invenit*, Wrt. Voc. ii. 132, 63. Ǽr hine ða men onfunden ðe mid ðam cyninge wǽrun, Chr. 755; Erl. 48, 31. On ðæs wífes gebǽrum onfundon ðæs cyninges þegnas ða unstilnesse, Erl. 50, 2. Hú fela onfundun (*were sensible of*), ða gefélan ne mágun, Dryhtnes þrowinga, Exon. Th. 72, 27; Cri. 1179. Onfindaþ ðæt and ongeotaþ *intelligite*, Ps. Th. 93, 8. Onfinden *sapiant*, Germ. 389, 16. Onfinden *experiamur*, Wrt. Voc. ii. 31, 42. Onfindende *expertur*, 31, 62. Onfunden, ongeten *expertus, cognitus*, i. *probatus, inventus*, 145, 47. Heó onfunden wæs men were aware of her presence, Beo. Th. 2591; B. 1293.   **III.** *to meet with, experience, suffer*:—Hé weán oft onfond, Exon. Th. 377, 16; Deor. 4.

on-findend, es; *m. One who finds out*; *inventor*, Germ. 391, 1.

on-flǽscness, e; *f. Incarnation*:—On ðære sôþan onflǽscnesse, Blickl. Homl. 81, 29.

on-fligen, es; *n. Infectious disease*:—Nú mágon ðás .viiii. wyrta wið .viiii. áttrum and wið nygon onflýgnum, Lchdm. iii. 36, 16. v. next word.

on-flyge, es; *m. Infectious disease, disease which, as it were, flies at people*:—Ðú miht wið áttre and wið onflyge, Lchdm. iii. 32, 2, 16, 30. v. *preceding word and* ongeflogen; *and cf. Icel.* á-flog, *flying at a person, fighting.*

on-fón; *p.* -féng; *pp.* -fangen (*with gen. dat. acc.*).   **I.** *to take*:—Calic hǽlu ic onfóu, Ps. Surt. 115, 13. Hé mycelne dǽl ðæs landes on anweald onféng, Bd. 1, 3; S. 475, 12. Mód Bryttas onféngon *they took courage*, 1, 16; S. 484, 19. Se Ælmihtiga onféng ðæt hiw úre tyddran gecynde. Geþencean wē, gif óðer nýten wǽre tô háligienne, ðonne onfénge hé heora hiwe, ac hé wolde úrum hiwe onfón, Blickl. Homl. 29, 2–6. Ðá nýddon hine hys yldran tô ðæm ðæt hé sceolde woroldlícum wǽpnum onfón, 213, 2. Se hálga hêht his heorþwerod wǽpna onfón, Cd. Th. 123, 5; Gen. 2040. Wífe onfón *uxorem ducere*, L. Ecg. C. 26; Th. ii. 152, 3.   **II.** *to take what another appoints* or *grants, to receive, have given* (a) *of material things*:—Seó sául onféhþ hire líchoman, Blickl. Homl. 57, 16. Adames cynn onféhþ flǽsce, Exon. Th. 63, 33; Cri. 1029. Ðá onféngon hig syndrige penegas, Mt. Kmbl. 20, 10. (b) *of non-material things*:—Ic ne onfó gewitnesse fram menn, Jn. Skt. 5, 34. Se ðe Godes word mid blisse onféhþ, Mt. Kmbl. 13, 20. Gē onfóþ ðæm mægene Hálges Gástes, Blickl. Homl. 119, 11. Hé onféng for worlde mycelne noman, 43, 34. Hí lêfnysse onféngon, Bd. 1, 26; S. 488, 5. Hé ðonne mid lǽwedum mannum onfó ðæs heardestan þeówdómes *let the hardest service be assigned to him among laymen*, Blickl. Homl. 49, 5. Onfón synna forgifnesse, 45, 7. Méde onfón 83, 15. Freódóm onfón, Ap. Th. 5, 19. Hé ðam upplícan ríce gehyhte tô anfónne, Bd. 3, 6; S. 528, 5.   **III.** *to take what another offers, receive favourably, accept*:—Gif ðú on God gelýfan wilt, ic ðæs drences onfó, Homl.

Th. i. 72, 17. Se yfela déma onféhþ feó, Blickl. Homl. 61, 30. Dryhten onféhþ eallum dǽm gódum ðe ǽnig man gedéþ his dǽm néhstan of árfæstre heortan, 37, 25. Onfóh ðissum fulle, Beo. Th. 2342 ; B. 1169. Ic bidde ðé ðæt ðú onfó ðissa láca, Gen. 33, 10. Gif hí sibbe mid Godes mannum onfón ne woldan, ðæt hí wǽron unsibbe fram heora feóndum onfónde, Bd. 2, 2 ; S. 503, 30. **IV.** *to receive a person* (a) *for entertainment, assistance or protection* :—Swá hwylc swá ánne lytling onféhþ, se onféhþ mé, Mt. Kmbl. 18, 5. Israhel onféhþ eallum his cnihtum *suscepit Israhel*, Blickl. Homl. 159, 20. Martha onféng Crist on hire hús, 73, 9. Onfóh ús on ðæt scip, 233, 7. Onfóþ mínre méder on neorxna wonge, 157, 32. Onfóh ðú ðínum esne, Ps. Th. 118, 122. Ðæt hé onfénge ðære eádigan Marian sáwle, Blickl. Homl. 155, 12. His ðá ða nolde onfón, ðe hiene mon tó brohte, Ors. 5, 2 ; Swt. 218, 34. Conon mid micle gefeán onfangen wæs, 3, 1 ; Swt. 98, 25. (b) *in a special sense of receiving at the baptismal font, or at confirmation, to stand sponsor to a person* :—His (*Godrum*) se cyning onféng æt fulwihte, Chr. 878 ; Erl. 80, 22. Æt ðam fulwihte hyre onféng sum Godes þeów, Shrn. 140, 22. Hine onféng æt fulluhtbæþe him tó godsuna Æþelwald, Bd. 3, 22 ; S. 553, 44. Ne hé náh mid rihte óðres mannes tó onfónne æt fulluhte ne æt bisceopes handa, L. C. E. 22 ; Th. i. 374, 2. Ic his hæfde ǽr onfongen æt biscopes handa, Ch. Th. 169, 27. **V.** *to undergo a rite, undertake a duty* :—Hié fulwihte onféngon, Blickl. Homl. 203, 24. Ðonne wile hé onfón rihtre ondetnesse, 155, 1. **VI.** *to conceive* :—Gif heó bearn onféhþ *si infantem conceperit*, L. Ecg. 6, 19 ; Th. ii. 146, 29. Seó unwæstmfæstnes fram him fleáh, and seó clǽnnes onféng, Blickl. Homl. 163, 19. Ic wæs mid unrihtwísnesse onfangen *in iniquitatibus conceptus sum*, Ps. 50, 6. **VII.** *to take to, to begin ; incipere* [cf. *O. H. Ger.* ana-fáhan : *Ger.* an-fangen] :—Ðonne ðæt vers geendaþ on ðam naman ðe hit eft onféhþ, Anglia viii. 331, 24. Ǽrest on cattes stán ... eft on cates stán ðǽr hit (*the boundary*) ǽr onféng, Cod. Dip. Kmbl. iii. 313, 33. v. á-fón.

on-fónd, es ; *m. One who undertakes* or *supports* :—Onfónd mínre hǽle *susceptor salutis meae*, Ps. Lamb. 88, 27. v. on-fón, **IV.**

on-foran. **I.** *prep. Before, afore* :—Onforan winter, Chr. 895 ; Erl. 93, 30. **II.** *adv. Before, in front* :—Beóþ onforan eágan, Ps. Th. 113, 13.

on-fordón ; *part. Destroyed* :—Bearn onfordónra *filios interemptorum*, Ps. Lamb. 101, 21. Cf. on-forwyrd.

on-foreweardan ; *prep. adv. In front, in the front of, in the earlier part of* :—Onforeweardan ðysre race *in the earlier part of this narrative*, Homl. Skt. i. 23, 790. Malchus eode onforeweardan (*led the way*) in tó his ðam hálgan geféran, 23, 752.

on-forht. v. an-forht.

on-forhtian *to fear, be afraid* :—Ne ondrǽdaþ gé eów ne gé ne onforhtion *nolite timere ne paveatis*, Deut. 31, 6. Onforgtigan *timere*, Germ. 388, 40. v. á-forhtian.

on-forwyrd, es ; *n. Destruction* :—Fornam hine eofor (onforwyrd, MS. T.) of wuda *exterminavit eam aper de sylva*, Ps. Spl. 79, 14. God gelédeþ hí on pitt onforwyrdes *in puteo interitus*, 54, 26.

on-fundelness, e ; *f. Experience, proof* :—Ðysse wyrte onfundelnysse manega ealdras geséðaþ *many authorities testify to the efficacy of this plant from experience*, Lchdm. i. 140, 9. Hyt déþ onfundelnysse ðæs sylfan þinges *it will give proof of the same thing, the second method will prove as efficacious as the first*, 162, 1.

on-fundenness, e ; *f.* **I.** *experience, experiment* :—Onfundenness *experimentum*, i. *testamentum*, Wrt. Voc. ii. 145, 49 : *experientia*, 145, 52. **II.** *finding out, discovery* :—Ðú ásettest rǽdels gehýr ðú ða onfundennesse ymbe ðæt ðú cwǽde *you have set a riddle, hear the meaning discovered of what you have said*, Ap. Th. 4, 22.

onga [*should have been given under* anga], an ; *m. A sting* :—Onga *aquilium*, Wrt. Voc. ii. 100, 59 : 7, 12. Mé of bósme fareþ ǽttren onga (*an arrow*), Exon. Th. 405, 18 ; Rä. 24, 4. [*O. H. Ger.* ango *aculeus : Icel.* angi *a spine, prickle.*]

on-gǽgum (?) *towards* :—Ongǽgum west *towards the west*, Ch. Th. 70, 18.

on-galan *to charm* :—Stefne ongalendra *vocem incantantium*, Ps. Spl. 57, 5 : Blickl. Gl.

on-gang, es ; *m.* **I.** *an entrance* :—Ongongas *ingressus*, Ps. Spl. C. 67, 26. **II.** *an irruption, attack, a going with violence* :—Ongong *incursus*, Wrt. Voc. ii. 111, 44 : *irruptio*, 111, 47. Ongeong (-gong, Rush.), *impetus*, Mk. Skt. Lind. 5, 13.

on-geador ; *adv. Together* :—Ongeador sprǽcon, Beo. Th. 3195 ; B. 1595.

on-geagn, -gegen, -gǽgn, -gegn, -geán, -gán, -geán, -gén. **A.** *prep. often following a case.* **I.** *with dat.* (1) *marking position, opposite, over against, against* :—Breoton ... ðam mǽstum dǽlum Európe nyccle fæce ongegen (-gén, MS. C. : -gǽgn, MS. B.) *Britannia ... maximis Europae partibus multo intervallo adversa*, Bd. 1, 1 ; S. 473, 10. Nebo on ðam lande Moab ongeán (*over against*) Iericho, Deut. 32, 49. Gangaþ on ðǽs ceasterwíc ðe inc ongeán standeþ, Blickl. Homl. 69, 35. Ðá arn hé and gestód

ongeán (*opposite*) ðam lége, 221, 11. Wæs ongeán ðyssum wæterscipe glæsen fæt *a glass vessel was placed so that the water ran into it*, 209, 4. Mín syn biþ symble ongeán mé *my sin is ever before me*, L. E. I. 30 ; Th. ii. 426, 40. (2) *marking motion, towards, in the direction of, to meet, in the way of* :—Héht his þegnas hine beran ongeán ðæm fýre *jussit se obviam ignium globis efferi*, Bd. 2, 7 ; S. 509, 24. Bǽd ðæt him mon brohte ðone triumphan ongeán, Ors. 5, 12 ; Swt. 240, 2 : Shrn. 129, 21. Him com seó menio ongeán (-gægn, Lind. Rush.), Jn. Skt. 12, 18. Férdon ongeán ðæm héðnum *they marched against the heathens*, Blickl. Homl. 203, 2. (3) *marking opposition, hostility in action or feeling, against* :—Swá se wind swíðor slóg on ðone lég, swá bræc hé swíðor ongeán ðæm winde, 221, 13. Hé hié lǽrde, ðæt hié hié forþ trymedan ongeán heora feóndum, 201, 36. Ic niste ðæt ðú stóde ongeán mé *I knew not that you opposed me*, Num. 22, 34. Ne hit on ðæm bryne wandode ðæs hátan léges ðe him wæs ongeán, Nar. 15, 21. Ðonne storm cyme mínum gǽste ongegn, Exon. Th. 455, 33 ; Hy. 4, 59. (4) *denoting waiting for what is coming, against, for the reception of, to receive* :—Ongeán gramum gearowe stódon *stood ready for the attack of the foes*, Byrht. Th. 134, 46 ; By. 100. Biþ súsla hús open ongeán áðlogum *open against the coming of the perjurers*, Exon. Th. 98, 10 ; Cri. 1605. Him biþ fýr ongeán *fire awaits them*, 446, 7 ; Dóm. 18. (5) *marking direction where no actual motion takes place* :—Seó eádge biseah ongeán gramum, 280, 12 ; Jul. 628. (6) *in reply to* :—Hig cwǽdon mé ongeán, St. And. 40, 14. (7) *denoting contrast* :—Ongeán ðam *e contra*. Ongeán ðyssum spelle, Bd. 5, 13 ; S. 632, 2–4. Swá wé oftor hig (*our sins*) gemunaþ, swá forgyt God hyra hraðor ... Ðonne ongeán ðon (*on the contrary*) swá wé oftor misdǽda forgytaþ, swá gemon hig God geornor, L. E. I. 30 ; Th. ii. 426, 36. (8) *in return for, as an equivalent for* :—Hé hine on eorþan streccan ongan, ongeán ðam heó eác hí ástrehte, Glostr. Frag. 102, 6. Ongeán ðam andgyte se deófol forgifþ stuntnysse, Wulfst. 59, 6–19. Cf. **II.** 7. **II.** *with the acc.* (1) *marking position, opposite, over against* (v. foran) :—Án ðǽra gárena liþ ongeán ðæt ígland ðe Gades hátte, óðer ongeán ðæt land Narbonense, se þridda ... ongeán ðæne múðan, Ors. 1, 1 ; Swt. 24, 3–6. Hé sæt ðǽr ihm getæht wæs ongeán ðone cyngc, Ap. Th. 14, 13. Ðá sæt se Hǽlend ongén (-geán, MS. A. : -gægn, Lind. : -gegn (*with dat.*), Rush.) ðone tollsceamol, Mk. Skt. 12, 41. (1 a) *to meet that which is moving* :—Gif ðú ðínes scipes segl ongeán ðone wind tóbrǽdst, Bt. 7, 2 ; Fox 18, 32. (2) *marking motion, towards, in the direction of, to meet* :—Se man ðe ongeán ús (*dat.*?) gǽþ, Gen. 24, 65. Seó eá gǽþ ongeán ða Assiriscan, 2, 14. Moises hig út álǽdde ongeán Drihten, Ex. 19, 17. Ðá gegaderode ðæt folc tógædere ongén Moises, 16, 2. Hí férdon ongén (-geaen, Lind. : -gægn, Rush.) ðone brýdguman, Mt. Kmbl. 25, 1, 6. Woldon ferian ða herehýð ongeán ða scipu, Chr. 894 ; Erl. 90, 24. Ðá flugon ða légetu ongeán ða héðnan leóde, Blickl. Homl. 203, 10. (2 a) *against, in a direction opposite to* :—Ongeán stréam *in a direction opposite to that in which the stream flows*, Cod. Dip. B. i. 502, 3 : ii. 374, 10. (3) *denoting hostility, resistance, or opposition in action or feeling, against, with, contrary to, in opposition to* :—Se lég ongan sleán and brecan ongeán ðone wind, Blickl. Homl. 221, 12. Æfter hǽðenum gewunan, ongeán heora cristendóm, Homl. Th. i. 100, 20. Ongén þúsendfealde deriende cræftas *contra mille nocendi artes*, Wrt. Voc. ii. 135, 29. Ongién allo ús wiðerwordnisse swíðre giræc *contra cuncta nobis adversaria dexteram extende*, Rtl. 14, 38. Him láð wǽre ðæt hí ongeán heora cynehláford standan sceoldan, Chr. 1048 ; Erl. 178, 31. Wearþ swíðe gestired se here ongeán ðone biscop, 1012 ; Erl. 146, 13. Ðæt heó yrsige ongeán leahtras (-es, MS.), Homl. Skt. i. 1, 104. Hé gewát yrre ongén hig, Num. 12, 9. Ðæt folc ... ceorodon ongeán God ... Wé sprǽcon ongeán God, 21, 5–7. Hwylce wróhte bringe gé ongeán ðysne man, Jn. Skt. 18, 29. Næfst ðú náne mihte ongeán (*adversum*) mé, 19, 11. Ic ne mæg nó wiðcweþan ne furþum ongeán ðæt geþencan *I cannot contradict, I cannot even have a conception contrary to it*, Bt. 34, 1 ; Fox 134, 29. (4) *marking direction where no actual motion takes place* :—Hí elciaþ ongeán ðone deáþ, and mid ealle ne forfleóþ ... Úre Álýsend ne elcode ná ongeán ðone deáþ ac hé hine oferswíðde *Enoch and Elias delay to meet death, and do not at all avoid it ... Our Redeemer did not delay to meet death, but he overcame it*, Homl. Th. i. 308, 2–8. Hé ne dorste beseón ongén God, Ex. 3, 6. Hé fægnaþ ongeagn (-geán, Cott. MSS.) ðara óðerra word *he rejoices at the words of the others*, Past. 17, 3 ; Swt. 111, 10. Ðæt cild ongeán his Hláford hyhte and hine hälette *the hope of the child went out to meet his Lord, and he hailed him*, Blickl. Homl. 165, 29. (5) *in reply to* :—Ne andwyrtst ðú nán þing ongén ða ðe ðiss ðé onsecgeaþ, Mt. Kmbl. 26, 62. (6) *denoting contrast or comparison* :—Seó næddre is geset on ðam godspelle ongeán ðone fisc *in the gospel the serpent is put in contrast to the fish*, Homl. Th. i. 252, 1. Feáwa ongeán getel ðæra wiðercorena *few in comparison with the number of the reprobate*, 536, 32. (7) *against* as in to set one thing *against* another, *as an equivalent for, in return for, in exchange for* :—Þolige cyle ongeán (*in atonement for*) ða hlíwþe, L. Pen. 16 ; Th. ii. 284, 5. Hé gesealde twá gegrynd ongén ðes mynstres mylne, Ch. Th. 231, 24 : 232, 3. Ælfríc sealde ðæt land æt Hacceburnan ongeán ðæt land æt Deccet, 288, 12. Hig of ðam Iúdeum for

ánum penige xxx gesealdon, ongeán ðæt ðæt ða Iúdeas úrne Hǽlend mid xxx penegum gebohton, St. Aud. 36, 26. (8) marking readiness for a coming event, *against, ready for:*—Híg lédon forþ hira lác ongeán ðætte Iosep in eode *they made ready their presents against Joseph came*, Gen. 43, 25. Ðonne sceolde fyrd út ongeán ðæt hí up woldon, Chr. 1010; Erl. 144, 4. (9) marking time, *towards:*—Fela ongeán winter hám tugon, Chr. 1096; Erl. 233, 22. **B. as an adverb.** (1) marking position, *opposite:*—Ic stande on ðás healfe and ðú ongeán *ego in hac parte stó, tu contra*, hér is se *contra* adverbium, Ælfc. Gr. 47; Som. 47, 50. Is Gotland on óðre healfe ongeán, Ors. 1, 1; Swt. 19, 20. Se hundredman ðe ðár stód ágén (ongeán, MS. A.: ongægn, Lind. Rush.) *ex adverso stabat*, Mk. Skt. 15, 39. (2) marking motion:—Ðá com mycel windes blǽd foran ongeán (*in the opposite direction*), Blickl. Homl. 199, 21. Ætstód se stréam and ongan tó þindenne ongeán (*in the direction opposite to that in which had come*), Jos. 3, 16. (3) denoting return, reversal of a previous action, *again, back; Lat.* re-:—Ða bodan ongeán cómon tó Jacobe, Gen. 32, 6. Ic fare eft ongeán, Num. 22, 34. Hé gewende ongeán tó ðam cynge, Chr. 1048; Erl. 178, 5. Ongeán cirran *reverti*, Gen. 8, 7. Ongeán fleón *refugere*, Ælfc. Gr. 28, 6; Som. 32, 47. Ongén sceát, ongeán hwyrfde *retrorsit*, Hpt. Gl. 505, 59. (4) with verbs of speaking, *in reply:*—Sóhte gylpword ongeán, Cd. Th. 17, 23; Gen. 264. Ða wergendan ne sceal mou ná ongeán werian, R. Ben. 17, 13. Brimmanna boda, ábeód ongeán, Byrht. Th. 133, 13; By. 49. (5) marking direction without actual motion, *towards:*—Ðonne hé síþ ongán *cum viderit*, Ps. Th. 57, 9. Hié ongeán lócian ne mihton, Blickl. Homl. 203, 11. (6) denoting opposition or resistance:—Ðá stód Grantabrycgscír fæstlíce ongeán, Chr. 1010; Erl. 143, 20. Nolde seó burhwaru ábúgan, ac heóldan mid fullan wíge ongeán, 1013; Erl. 148, 12. Ealle ða yldestan menn on West-Seaxon lágon ongeán swá hí lengost mihton ac hí ne mihton nán þing ongeán wealcan *all the chief men of Wessex resisted as long as ever they could, but they could not offer any effectual opposition*, 1036; Erl. 165, 1–3. Ongén sette *objecte*, Wrt. Voc. ii. 115, 25. (7) marking contrast, *on the other hand:*—God sette beforan eów líf and gód, and ðær ongén deáþ and yfel, Deut. 30, 15. (8) marking repetition, *again:*—Drihten cwæþ: Dó ðíne hand on ðíne bósum . . . Ðá cwæþ hé: Teóh eft ðíne hand on ðínne bósum. Ðá teáh hé hig ongeán, Ex. 4, 6–7. [*O. Sax.* an-gegin: *O. H. Ger.* in-gagan *and* in-gegin, -gegini: *Icel.* í-gegn *and cf.* gagn-.]

ongeán-cirrendlíc; *adj. Relative:*—Relativum ðæt is ongeáncyr-rendlíc, Ælfc. Gr. 38; Som. 40, 62.

ongeán-cyme, es; *m. A return:*—Útfæreld his fram fæder, ongeáncyme (*regressus*) his tó fæder, and utrene tó helle, ongeáncyme (*recursus*) tó setle Godes, Hymn. Surt. 44, 17, 23.

ongeán-flówende *refluent:*—Ongénflówende ýða *reciproca*, Hpt. Gl. 418, 41. Ongéndflówendum wæterum *reciprocis fluentis*, 462, 1.

ongeán-weard; *adj. Going against* or *towards:*—Hé him ongeánweard wæs *he was on his way to meet him*, Ors. 6, 31; Swt. 284, 32. Ongeánwurde *obvia*, Hpt. Gl. 499, 65.

ongeánweard-líc; *adj. Adversative:*—At (*the conjunction*) is ongeánweardlíc, Ælfc. Gr. 44; Som. 45, 40.

ongeánweardlíce; *adv. Adversatively*, Ælfc. Gr. 44; Som. 45, 50.

on-geboren; *adj. In-born:*—Ongeborene *ingenitam*, Hpt. Gl. 514, 2.

on-gebroht; *adj. Imposed:*—Be ongeb[r]ohtum *de inrogata*, Hpt. Gl. 514, 62.

on-gecígung, e; *f. Invocation:*—Þerh ongiceiging *per invocationem*, Rtl. 99, 28.

on-gefeoht, es; *n. Attack, assault:*—From ǽlcum ongifeht *ab omni impugnatione*, Rtl. 98, 26: 122, 5.

on-geflogen; *part. Attacked with disease:*—Gif men his leoþu acen oððe [hé] ongeflogen sý, Lchdm. i. 86, 21. Cf. on-flyge.

on-gefremming, e; *f. Imperfection:* — Ongefremminge míne (*imperfectum meum*) gesáwon eágan ðíne, Ps. Spl. 138, 15.

on-gegen, -gegn. v. on-geagn.

Ongel. v. Angel.

on-gemang. I. *prep. with dat. Among:*—Ongemong óðrum mannum, Bt. 35, 6; Fox 168, 6. Ðá ongan ic ongemang óðrum mislícum and manigfealdum bisgum ða bóc wendan on Englisc, Past. pref.; Swt. 7, 17. Eác ðæm golde and ðæm líne wæs ongemang *purpura*, 14; Swt. 85, 9. Ongemang ðæm ðe *whilst*, 45; Swt. 339, 24. Ongemang ðæm *meanwhile*, Jn. Skt. 4, 31. **II.** *adv.:*—Gif wé Sanctus Paulus láre sume ongemong secgaþ *if we introduce some of St. Paul's teaching*, 40; Swt. 291, 13. Gif wé Æfneres dǽda sume hér ongemong secgaþ, Swt. 295, 13. [*O. Sax.* an-gemong (*as adv.*).] v. gemang.

on-gemet; *adj. Immense:*—Ongemetum *immensis*, Wülck. Gl. 250, 23. v. un-gemet.

ongemet-hát; *adj. Exceedingly hot:*—Wyl on wætere, beþe hine mid ongemethátum *boil in water, foment him with it exceedingly hot*, Lchdm. ii. 338, 22.

ongend=(?) ongén (cf. *the form of the word under* ongeán-flówende), Exon. Th. 323, 28; Víd. 85.

---

on-geótung, e; *f. Pouring in:*—Clǽnsa ǽrest ða wambe mid drences ánfealddre ongeótunge, Lchdm. ii. 234, 26.

on-geþwǽre. v. un-geþwǽre.

on-gewiss; *adj. Uncertain:*—Ongewissu *incerta*, Ps. Spl. 50, 7. v. un-gewiss.

on-gifan. I. *to give back:*—Nime man ðínne assan and hine ná ne ongife *asinus tuus rapiatur, et non reddatur tibi*, Deut. 28, 31. **II.** *to forgive, pardon:*—Ðú ðe ongǽfest *qui ignoscis*, Rtl. 40, 33. v. á-gifan.

on-gildan. I. *to pay* (*a penalty for*), *to be punished for* (with gen. acc. of crime or clause):—Banan heardlíce grimme ongildaþ, ðæs hié oft gilp brecaþ, Salm. Kmbl. 265; Sal. 132. Hé ðæs wraðe ongeald, Cd. Th. 111, 26; Gen. 1861: 253, 20; Dan. 598. Hú eall moncyn angeald ðæs ǽrestan monnes synna mid miclum teónum and wítum *ab initio et peccare homines et puniri propter peccata*, Ors. 5, 15; Swt. 250, 27. Hú swíðe hí his anguldon from heora ágnum cǽsere *ut Caesare punirentur*, 6, 2; Swt. 256, 6. Weorces onguldon deópra firena þurh deáþes cwealm, Exon. Th. 153, 22; Gú. 829: 226, 23; Ph. 410. Ðæs ða byre siððan grimme onguldon gafulrǽdenne, 161, 15; Gú. 959. Sceal wearh ongildan, ðæt hé ǽr fácen dyde *he shall pay the penalty for previous wrong-doing*, Menol. Fox 573; Gn. C. 56. Sceolde hé ða dǽd ongyldan, Cd. Th. 19, 23; Gen. 295. Monig sceal ongieldan sáwel súsles *shall be tormented*, Exon. Th. 304, 17; Fä. 71. **II.** *to pay:*—Hwylc hira óðrum sceolde tó fóddurþege feores ongildan *which should pay for the others' food with his life*, Andr. Kmbl. 2204; An. 1103. **III.** *to give an offering, to offer:*—Ðér hǽðene men deóflum onguldon, Blickl. Homl. 221, 3. [Cf. *O. Sax.* a-, ant-(an-)geldan: *O. H. Ger.* ant-(en-, in-)geltan: *Ger.* ent-gelten.] v. á-, an-gildan.

on-gin[n], es; *n.* **I.** *a beginning:*—Ðæs weges ongin, ðe tó Criste lǽt, ne meg beón begunnen on fruman bútan sumre ancsumnysse, R. Ben. 5, 16. Næs his frymþ ǽfre, eádes ongyn, Exon. Th. 240, 13; Ph. 638. His ríces ongin (*original condition*) nǽfre gewonaþ, Blickl. Homl. 9, 16. **II.** *an attempt, undertaking, enterprise:*—Micel is ðæt ongin ðínre gelícan ðæt ðú forhycge hláford úrne *it is a great undertaking for the like of thee to despise our lord*, Exon. Th. 250, 15; Jul. 127. Gif ðú gewítest ána from éþele, nis ðæt ongin wiht, 119, 2; Gú. 248. Ongin, 123, 22; Gú. 326. Be ðam onginne ðe hé ongan, ðæt wésten swá ðína eardigan, Guthl. 4; Gdwin. 28, 7. Ðú miht æt Gode ábiddan ðæt ðú wilt wið ðæs drýg onginne, Blickl. Homl. 187, 19. Onginnum *incoeptis*, Hpt. Gl. 515, 15. **III.** *action, proceeding:*—Gesticulatio angin *jocus* ł *actus*, 473, 61. Wrætlíc þúhte stánes ongin (*the stone spoke*), Andr. Kmbl. 1482; An. 742. Ýða ongin *the violent action of the waves*, 931; An. 466. **IV.** *action, activity, active life, actions, endeavours:*—Ðér wæs wuldres wynn, wígendra þrym, ædelíc onginn, næs ðær ǽnigum gewinn, 1775; An. 890. Ðæt se ǽresta dǽl his onginnes and lífes wére tó geleáfan gecyrred, Blickl. Homl. 211, 30. Drihtne úres anginnes nán þing dígle ne biþ . . . 'Beforan ðé is eall mín gewilnung,' R. Ben. 25, 9. [*O. Sax.* ana-, an-gin: *O. H. Ger.* ana-gin, -ginni.] v. an-gin.

on-ginnan; *p.* -gan[n]; *pl.* -gunnon; *pp.* -gunnen. **I.** *to begin, set about, set to work:*—Ic onginne *inchoo*, Ælfc. Gr. 24; Som. 25, 39: *incipio*, 28, 6; Som. 32, 42: *ineo*, 37; Som. 39, 1. Wæs ongunnen *ordiretur*, Hpt. Gl. 494, 11. (a) where the action begun is given by the verb in the infin. or in the gerund.:—Ic onginne tó wearmigenne *calesco*, 35; Som. 38, 4. Hé onginþ (*incipiet*) tó álýsenne his folc of þeówte, Jud. 13, 5. Ðá ongan ic ða bóc wendan on Englisc, Past. pref.; Swt. 7, 17. Ðú ðe ongunne (*coepisti*) ætýwan ðíne mǽrþe, Deut. 3, 24. Se ongan ǽrest onclypian Drihtnes naman, Gen. 4, 26. Ongan se Hǽlend bodian, Mt. Kmbl. 4, 17. Ðá ongan hine langian on his cýþþe, Blickl. Homl. 113, 14. Ongan se Hǽlend him andswarigende tó cwepan, Mk. Skt. 13, 5. Hí ongunnon ða eár pluccigean, 2, 23. (b) where a case follows:—Se mon ðe gód onginneþ and ðonne áblinneþ . . . Se ðe gód onginneþ and on ðon þurhwunaþ, Blickl. Homl. 21, 34–36. Freme ðæt ðú ongunne, 189, 3. Raðe ðæs hié óðer ongunnon wið Macedonie *cui Macedonicum bellum continuo successit*, Ors. 4, 11; Swt. 202, 32. Ongin ðæt ðú onginnest, Blickl. Homl. 187, 22. Ðæt fæsten ongunnen wæs instepes ðæs ðe hé of ðæm fulwihte ástág, 35, 5. (c) where the verb is used intransitively:—Ða six onginnaþ of ðam stæfe e, and geendiaþ on him sylfum; x ána onginþ on ðam stæfe i, Ælfc. Gr. 2; Som. 2, 57–58. **II.** *to attempt, endeavour* (with infin.):—Ic onginne *conabor*, Wrt. Voc. ii. 24, 77: *nitar*, 60, 3. Ðæt ic geseó ða mé onginnaþ dón ða werrestan tintrega *that I may see those who are trying to inflict on me the worst tortures*, Blickl. Homl. 229, 24. Hiene Hannibal áspón ðæt hé ðæt gewin leng [ne] ongan *Hannibal induced him to carry on the struggle longer*, Ors. 4, 11; Swt. 204, 31. Se náht freomlíces ongan on ðǽre cynewísan *nihil omnino in re militari ausus est*, Bd. 1, 3; S. 475, 20. Ða ongunnon (*tentabant*) ða scypmenn ða ancras upp teón, 3, 15; S. 541, 40. Oþ ðæt ongite ðæt hé mǽge ábiddan æt Gode ðæt hé onginne (-ginne, MSS. Cot.) *until he finds that he can obtain by prayer from God what he endeavours to get*, Past. 10; Swt. 61, 22. Ðæt ic dorste ðis

weorc ongynnan *ut hoc opus adgredi auderem*, Bd. pref.; S. 472, 12. Hé wolde onginnan him ôleccan, Bt. 35, 6; Fox 168, 13. **III.** *to act strenuously:*—Hí on ðam gewinne werlíce ongunnon, Homl. Th. ii. 502, 5. Onginnaþ werlíce, i. 188, 31. Onginnaþ esnlíce *viriliter agite*, Deut. 31, 6. **IV.** *to make an attempt upon, to attack:*—Gramhýdige mé mid unrihte oft onginnaþ *injusti insurrexerunt in me*, Ps. Th. 85, 13. Ðonne ûs mânfulle menn onginnaþ (*insurgerent*), 123, 2. Ðonne yfle unmǽgas onginnaþ, mécum gemétaþ, swá gé mé dydon, Vald. 2, 23. Mé strange ongunnon *irruerunt in me fortes*, Ps. Th. 58, 3 : 61, 3. Gif hí sceoldon eoïor onginnan, Exon. Th. 344, 20 ; Gn. Ex. 176. [*O. H. Ger. in-ginnan inchoare, incipere, conari, moliri, niti.*]

on-ginnendlíc; *adj. Inchoative:*—Ôðer hiw is gehâten *inchoativa*, ðæt is onginnendlíc, forðan ðe hit getâcnaþ weorces anginn, Ælfc. Gr. 35; Som. 38, 2.

on-ginness, e; *f. A beginning, undertaking:*—Onginnissum *inceptis*, Wrt. Voc. ii. 86, 26. Hine hêt ðæt hé ðam hâlwendan ongynnessum georne befulge *eum coeptis insistere salutaribus jussit*, Bd. 5, 19; S. 637, 11.

on-girwan; *p.* -girede *To divest, strip:*—Hé hine middangeardes þingum ongyrede and genacodade *se mundi rebus exuit*, Bd. 4, 3 ; S. 567, 24. Ongyrede hine ða geong hæleþ . . . gestâh hé on gealgan heánne, Rood Kmbl. 77 ; Kr. 39. Hé wæs líchoman ongyrwed *corpore exutus*, Bd. 3, 19; S. 547, 34 : 5, 12; S. 631, 5. Ongered *exuta*, Ps. Surt. ii. p. 202, 17. Ongirede *exutas*, Wrt. Voc. ii. 33, 18.

on-git, es; *n. Understanding:*—Ongit (ondgit, Cott. MSS.) wîsdômes, Past. 14; Swt. 85, 3. Ongyt *intellectum*, Ps. Spl. 31, 10. v. and-git.

on-gitan, -gietan, -giotan, -geotan; *p.* -geat, -get; *pl.* -geáton, -géton; *pp.* -giten, -gieten *To perceive:*—Ic ongite *comperio*, Ælfc. Gr. 30; Som. 34, 46. Ongiotaþ *animadvertite*, Kent. Gl. 230. Ðá hé ongitende wæs *animadverterit*, Wrt. Voc. ii. 3, 9. Ongeten, onfunden *expertus, cognitus*, i. *probatus, inventus*, 145, 47. **I.** *to perceive, see:*—Gif ðú gesihst hwylcne ungesǽligne mon and ongitst hwæðhwegu gôdes on him, Bt. 38, 3 ; Fox 200, 15. Hí ðær hwîlum synne ongytaþ ðǽr ðe syn ne biþ, Bd. 1, 27 ; S. 494, 26. Gif hí hwilcne mon on ðám landum ongytaþ oððe geseóþ ðonne feorriaþ hí and fleóþ *sed hominem cum viderint longe fugiunt*, Nar. 36, 21. Ðín wuldor ongitaþ woruldcyningas, Ps. Th. 101, 13. Siððan hé beácen (*the miracle of the fiery furnace*) onget, Cd. Th. 246, 33; Dan. 488. Ðæt ic ǽrwelan ongite, gearo sceáwige, Beo. Th. 5489 ; B. 2748. Ðæt hié Geáta clifu ongitan meahton, 3827; B. 1911. Gefeán mon mihte on his andwleotan ongytan, Blickl. Homl. 223, 35. **II.** *to perceive by hearing:*—Ic ðæs þeódnes word ongeat, Exon. Th. 175, 11; Gû. 1193. Gif ðú sanges stæfne gehýrdest and ðú heofonlíc weorud ongeáte ofer ûs cuman, Bd. 4, 3; S. 568, 31. Hié horn galan ongeáton, Beo. Th. 5880; B. 2944. **III.** *to perceive, feel (pain, etc.):*—Ðonne ne ongitest ðú ǽnig sâr, Lchdm. i. 368, 26. Ðonne ne ongyt hé ná mycel tô geswynce ðæs sîðes, 102, 6. Ongæt gumena aldor hwæt him Waldend wræc wîteswingum, Cd. Th. 111, 29; Gen. 1863. Swá ðæt se seóca ðone stenc ne ongite, Lchdm. i. 304, 23. **IV.** *to feel, be of opinion, judge:*—Ðeáh ðe be ðyssum willan misenlíce cynn monna missenlíce ongite *quamvis de hac re diversae hominum nationes diversa sentiant*, Bd. 1, 27 ; S. 495, 14. **V.** *to know, hear of, find out:*—Wé witon manige foremǽre weras forþgewitene ðe swîðe feáwa manna â ongit *that very few men ever hear of*, Bt. 19; Fox 70, 13. Wé oft ongytaþ ðæt ârîseþ þeód wið þeóde *we often hear of nation rising against nation*, Blickl. Homl. 107, 27. Eall ðæt hé oððe on gewritum oððe on ealdra manna sægenum ongeat (*cognoverat*), Bd. prep.; S. 471, 24. Sumu ða ðe ic sylf ongitan (*cognoscere*) mihte þurh gesægene, S. 472, 30. Ne mæg ic nâne cwica wuht ongitan . . . ðe ungenéd lyste forweorpan *si animalia considerem . . . nihil invenio, quod, nullis extra cogentibus, ad interitum sponte festinet*, Bt. 34, 10; Fox 148, 13. Miht ðú ongitan hwæðer ðú âuht ðe deórwyrþre habbe ðonne ðé sylfne *do you know whether you have anything more precious to you than yourself?* 11, 2; Fox 34, 9. **VI.** *to perceive, understand:*—Ælc ðæra ðe Godes word gehýrþ and ne ongitt (*intelligit*), Mt. Kmbl. (MS. A.) 13, 19. Ongyte (ongete, Lind.) gé ealle ðâs þing ? Ðâ cwǽdon hig : Wé hit ongytaþ, 13, 51. Ne ongyte gé gyt *nondum intelligitis ?* Mk. Skt. 8, 21. Onfindaþ ðæt and ongeotaþ *intelligite*, Ps. Th. 93, 8. Ðý læs hig mid heortan ongyton (ongeotan, Rush.) *ne corde intelligant*, Mt. Kmbl. 13, 15. Hié hiora (*books*) nânwuht ongiotan (ongietan, Cott. MSS.) ne meahton, Past. pref.; Swt. 5, 12. Ðæt wé ðý geornor ongietan meahton tâcen, ðæt se fugel þurh bryne beácnaþ, Exon. Th. 236, 13; Ph. 573. Ongeotan, Blickl. Homl. 15, 13 : 131, 23 : 105, 28. **VII.** *to recognise, know*, (a) *to take a person or thing to be what it really is:*—Gif ðú sôðne God lufast and ongitest gǽsta hleó, 245, 23; Jul. 49. Wið ðæs ðú wilt higian ðon ǽr ðe ðú hine ongitest *towards that thou wilt strive as soon as thou dost recognise it*, Bt. 11, 2; Fox 34, 8. Se man ðe swereþ mân and eft his gilt onget, Lev. 5, 4 : Met. 22, 16. Ða neát ongitaþ hira gôddénd *the brutes know their benefactors*, Elen. Kmbl. 717; El. 359. Hé Godes good on his dǽde ongeat *he recognised the goodness of God in that deed of his*, Blickl. Homl. 215, 33. Witon wé ðæt ûre Drihten mid ûs wæs on ðæm scipe, and wé hine ne ongeáton, 235, 22. Ongytaþ Godes mildheortnesse seó is nû mid ûs geworden

recognise in this the mercy of God that has been now shewn to us, 235, 20. Ne mé ǽnig ongitan wolde *non erat qui agnosceret me*, Ps. Th. 141, 4. Ðéh ic engla þeóden ongitan ne cûðe, Andr. Kmbl. 1802 ; An. 903. Nû wé mâgon ougytan hwæt ðæt gerýne getâcnaþ *now we know what the mystery means*, Blickl. Homl. 17, 13. Wé mâgon ongytan on ðæm ûre tydran gecynd *we may see in that* (the temptation of Christ by the devil) *our weak nature*, 33, 35 : 95, 11. Ne mæg ic fullíce ongitan æfter hwǽm ðú spyrast *I don't quite know what you are asking for*, Bt. 34, 9; Fox 148, 1. Nû ðú hæfst ongyten ða wanclan treówa ðæs blindan lustes, 7, 2 ; Fox 18, 2. Heó (*a woman dressed in man's clothes*) wæs fram hire fæder ongitenu *she was recognised by her father*, Shrn. 31, 15. (b) *to recognise a fact or circumstance*, (1) the fact stated in a clause :—Ðú ongitst ðætte ðú git hæfst ðone mǽstan dǽl ðínra gesǽlþa, Bt. 10 ; Fox 28, 6. Ðonne ongit hé, hû lytel hé biþ, 12, 1 ; Fox 60, 28. Ðá se déma oncneów and ongæt, ðæt hé hine mid swinglan oferswîðan ne mihte, Bd. 1, 7; S. 478, 1. Oferswîðan ða men ðe hié ongeáton ðæt wiðerwearde wǽron, Blickl. Homl. 135, 12. Leóde ongéton, ðæt ðǽr Drihten cwom, Cd. Th. 183, 12; Exod. 90. Hé wolde ðæt hé on ðon ongeáte, ðæt ðon mon ne wæs, se ðe him ætýwde, Bd. 2, 12; S. 514, 25. Ic wundrige hwí ðú ne mǽge ongitan, ðæt ðú eart nû git swîðe gesélig, Bt. 10 ; Fox 28, 34. Ðæt is tô ongytanne ðæt âcennede wǽron wæstmas gôdra dǽda, Bd. 3, 23 ; S. 554, 23. Ðú hæfdest ongiten, ðæt mé selfum þúhte, ðæt ic hæfde forloren ðæt gecyndelíce gôd, Bt. 35, 2; Fox 156, 17. Heó ongieten hæfde, ðæt heó eácen wæs, Exon. Th. 378, 3 ; Deór. 10. (2) the fact referred to by the pronoun ðæt :—Fýren wolc[en] âstâh of heofonum, and hit ymbsealde ealle ða ceastre. Mid ðý ðæt (*the circumstance just related*) ongeat Andreas, Blickl. Homl. 245, 32. Hié ðæt ongeáton, ðæt hé leng mid him wunian nolde, 135, 22. (3) the fact given by accus., (a) with infin. :—Ðæt hié ongieton mîn mægen on ðé wesan, 241, 14. (b) without infin. :—Hé ongeat Titum hwéne monþwǽrran ðonne hé sceolde, and Timotheus hé ongeat hâtheortran ðonne hé sceolde, Past. 40; Swt. 291, 21–23 : Blickl. Homl. 219, 5. Hit ongeat his lâre swîðe tôtorenne, Bt. 3, 1 ; Fox 4, 31. Ðæt Môd swîðe ðæt hit hit ǽghwonan ongeáte scyldig (cf. Ic mé ongite ǽghwonan scyldigne, 8 ; Fox 24, 13), tit.; Fox 10, 19. (4) with the passive :—Hé wæs tô cinge ongyten *he was recognised as king*, Blickl. Homl. 71, 32. Ðonne hé biþ ongieten æfstig, Past. 13; Swt. 79, 12. **VIII.** *to know (of sexual intercourse):*—Ic nǽnigne wer ne ongeat, Blickl. Homl. 7, 22. v. an-gitan.

on-gitenness, e; *f.* **I.** *understanding, knowledge:*—Hé wæs gefeónde ðære ongytenesse (*agnitione*) ðæs sôðan Godes biganges, Bd. 2, 13; S. 517, 13. Tô ongytenysse (*ad agnitionem*) ðæs sôðan Godes, 2, 9; S. 511, 3. **II.** *meaning, purport* (cf. and-git, III) :—Ðeós ongitenys (þes ongitenysse, MS.) mínre untrumnysse ys ðæt of ðisum líchaman sceal beón se gást álǽded *the meaning of my illness is, that the spirit shall be taken away from the body*, Guthl. 20 ; Gdwin. 80, 22.

on-gitness, e; *f. The understanding, intellect:*—Of alre ongetnisse *ex tcto intellectu*, Mk. Skt. Rush. 12, 33.

ongnere, es ; ongnora (?), an; *m. The corner of the eye* (?):—Eághyll from ðam ongnoran *glebenus*, Wrt. Voc. ii. 42, 7. Ongneras *irqui*, 46, 30.

on-gripe, es; *m. Attack, assault :*—Ðæra wyrma ongrype and ðæra sorhwîta mǽst, Wulfst. 187, 2. [*O. H. Ger. ana-griffe (dat.) tactu*, Grff. iv. 318: *and cf. Icel. â-grip in the phrase lítill âgripum*].

on-grisla, an; *m. Dread, horror :*—Wæs se munt mid mycelum brôgan and mid ongryslan eall oferlǽded, Blickl. Homl. 203, 7.

on-grislíc; *adj. Horrible, dreadful :*—Ðá becwôm sum ongrislíc wîse (*horrenda res*) on hié, Nar. 10, 32. Ðæt ongrislíce gemôt *the last day*, Wulfst. 186, 15. Angryslíc, Dôm. L. 14, 225. Ongrislíces andwlitan *horrido vultu* . . . ongrislícre ansîne *horrendae visionis*, Bd. 5, 13 ; S. 633, 1–5. On ðære angrislícan gesihþe *horridae visionis*, 5, 12 ; S. 628, 19. Ongrislíco hǽr *horridi crines*, 5, 2 ; S. 615, 1. Ongristlíce on stefne, Guthl. 5 ; Gdwin. 34, 26.

on-gryrelíc; *adj. Horrible :*—Hí hine lǽddon on ðam ongryrlícan (-gryslícan ?) fiðerum, Guthl. 5 ; Gdwin. 36, 24.

on-gunnenness; e; *f. An undertaking :*—Hé bæd ðæt hé ða ârsæstan ongunnenysse gefylde *patiit eum pia coepta complere*, Bd. 3, 23 ; S. 554, 40.

on-gyldan, -gynness, -gyrede, -gytan. v. on-gildan, -ginness, -girwan, -gitan.

on-hâdian *to degrade from holy orders :*—Gif preóst ôðerne man ofsleá . . . hine biscep onhâdige, L. Alf. pol. 21 ; Th. i. 76, 1.

on-hǽle; *adj. Whole, entire :*—Gemengde beóþ onhǽlo gelâc engla and deófla *the entire hosts of angels and devils shall be joined together*, Exon. Th. 56, 5 ; Cri. 896.

on-hǽle; *adj. Secret, hidden :*—Ne lǽt ðú ðinne ferþ onhǽlne, ðegol ðæt ðú deópost cunne, onhǽle ic ðé mîn dyrne gesecgan, gif ðú mé ðinne hygecræft hylest, Exon. Th. 333, 9; Gn. Ex. 1. Gif mec onhǽle ân onfindeþ, ðǽr ic wîc bûge (cf. gif ic mǽgburge môt mîne gelǽdan on dégolne weg, 397, 15–17), 396, 19; Rä. 16, 7. Wîd is ðes wésten, wræcsetla fela, eardas onhǽle earmra gǽsta, 121, 7; Gû. 268 : 123, 13;

Gū. 322. Wið onhǽlum ealdorgewinnum *against secret and deadly foes*, 134, 9; Gū. 505.

on-hǽled *infirm, ill* :—Đa ðe on unhǽle (onhǽlede, MS. C.) wǽron, Ors. 4, 4; Bos. 80, 40.

on-hǽtan. I. *to heat* :—Hēt hē ðone stān onhǽtan, Ors. 4, 8; Swt. 186, 19. Blôd onhǽtan, Salm. Kmbl. 88; Sal. 43. Ofn onhǽtan, Cd. Th. 229, 31; Dan. 225. Onhǽted, 231, 7; Dan. 243. Đá ðæt (*the brazen bull*) onhǽt wæs, Ors. 1, 12; Swt. 54, 28. II. of violent emotion, *to inflame* :—Hira môd ne beóþ onhǽt mid nānre manunge, Past. 52; Swt. 411, 7. Heorte is onhǽted, Judth. Thw. 22, 30; Jud. 87.

on-hagian; *p.* ode; *v. impers. with dat. or acc. of pers. To be within a person's power or means, to be in accordance with a person's will or convenience* :—Eádig byþ se ðe ðam þearfan gefultumaþ, gif hine tô onhagaþ (*if it be in his power*); gif hine ne onhagaþ, ðonne ne lícaþ him his earfoþu, Ps. Th. 40, 1. Mē ne onhagaþ nū ða bôc ealle tô asmǽáganne, Shrn. 200, 22. Đonne hit (*the mind*) onhagaþ tô ðǽm úteran *si facultas exterior suppetat*, Past. 53, 6; Swt. 17, 13. Ne anhagode heora cyninge ðæt hē wið hié mehte būton fæstenne gefeohtan, Ors. 4, 5; Swt. 168, 21. Hié hergodon ǽghwǽr be ðam sǽ ðǽr hié onhagode (*wherever it suited them*), Chr. 918; Erl. 102, 25. Đá seó fyrd gesomnod wæs ðá ne onhagode heom ðártô būton ðæt wǽre ðæt se cyng ðǽr mid wǽre *they would not be satisfied unless the king were there too*, 1016; Erl. 153, 27. Đæra hálgena þrowunga ðe mē tô onhagode on Englisc tô áwendene *that I have had the opportunity of translating into English*, Homl. Skt. i. pref. 37. Gelǽste binnan twelf mônþum būton hire ǽr tô onhagige *unless it be convenient to pay earlier*, L. C. S. 74; Th. i. 416, 17. Gif hine tô swá mycelum ne onhagige *si tantum facultatis ei non suppetat*, L. Ecg. P. iv. 60; Th. ii. 222, 3. Gif hine onhagige (*si facultatem habeat*), gefreóge ǽnne man, ii. 24; Th. ii. 192, 12: L. Pen. 14; Th. ii. 282, 9–12: Homl. Th. i. 180, 12. Đone dǽl ðe him tô onhagie, 398, 17. Gif ðē onhagige, ðæt ðú hit (*the law*) healdan mǽge, far ðú in; gif ðē ne onhagige, far ðē freoh ðider ðú wille, R. Ben. 97, 23. Đa hýrsumnesse beginne ðeáh hine hwôn onhagige *though he have little power* (or *inclination*), 128, 19. Gif mon tô gôdum weorcum ne onhagie habban gôdne willan *if people have not the means for good works, let them have good will*, Bt. 41, 2; Fox 246, 10.

on-hátan *to promise* :—Hyre nales frǽtwe onhēht, Exon. Th. 249, 28; Jul. 118. Đæt ic deófolgieldum gaful onháte, 251, 27; Jul. 151. [Cf. *O. H. Ger.* ant-heizan *vovere, spondere, polliceri*: *O. Sax.* ant-hētan.]

on-hátian *to grow hot* :—Onhátode *incanduisset*, Wrt. Voc. ii. 47, 4.

on-heáw, es; *m. A block to hew on* :—Onheáwas *codices*, Wrt. Voc. ii. 104, 38: 135, 60: 14, 62. [*M. H. Ger.* ane-hou *incus*.]

on-hebban; *p.* -hôf (*the weak form* -hefde *also occurs*); *pp.* -hafen. I. *to lift up, raise* (the eyes, voice) :—Đonne ic mec onhæbbe, and hí onhnígaþ tô mē, Exon. Th. 412, 28; Rä. 31. 7. Đá onhôf Laurentius his ēgan up, Shrn. 116, 4. Petrus onhôf his stefne, Blickl. Homl. 149, 21. II. *to raise* (as barm does), *to leaven* :—Ne ete gē nān þing onhafenes, Ex. 12, 19. III. *to take up, begin* (cf. Icel. hefja *to begin*) :—Ic ðás unhýrlícan fers onhefde mid sange, Dôm. L. 2, 11. IV. *to take away* :—Óþ ðæt onhafen biþ (*auferatur*) se môna, Ps. Spl. 71, 7. V. metaph. *to lift up, exalt* (generally in a bad sense) :—Ǽlc ðæra ðe hine onhefþ, hē sceal beón geeádmēt, Homl. Th. i. 202, 33. Ǽlc ðe hine anhefþ, hē biþ geneoþerad, and ǽlc ðe hine geneoþeraþ, hē biþ mid weorþmynte onhafen. . . . Ac hwæt gif ic mín môd on môdignesse anhôfe? R. Ben. 22, 11–19. Đa de God ondrǽdaþ, and hý þurh heora gôdan dǽda ne anhebbaþ, 4, 2. Hē eðelþrym onhôf, rýmde and rǽrde, Cd. Th. 98, 2; Gen. 1634. Tô tēhte ðām rícan, ðæt hí ne onhôfon hí, Homl. Th. i. 378, 18. Ne onhebbe hine nān man on his weorcum, ii. 80, 29. v. an-, á-hebban, -hefan, *and next word*.

on-hefednes, e; *f. Exaltation* :—Gif wē ðone hrôf ðǽre heálícan eáðmôdnesse getellan willaþ and tô ðǽre heofonlícan anhefednesse cuman þencaþ, R. Ben. 23, 2.

on-heldan. -heldedness. v. on-hildan, -hildedness.

on-hende; *adj. On hand, demanding attention* :—Hié forgeátan ðara útera gefeohta ðe him anhende wǽron, Ors. 2, 6; Swt. 88, 24. [Cf. *Icel.* á-hendr *within reach.*] v. of-hende.

on-herian -hering. v. on-hyrian, -hyring.

on-hetting, e; *f. Persecution* :—Onhettincga *persecutiones*, Hpt. Gl. 476, 17. v. hettend.

on-hildan, -hieldan, -heldan, -hyldan. I. *trans.* (1) of actual motion, *to lean, incline, recline, bend down* :—Onheldeþ hine and falleþ *inclinabit se et cadet*, Ps. Surt. 9, 31. Se biscop hine onhylde tô ānre ðære studa, Bd. 3, 17; S. 543, 37: 4, 9; S. 577, 7. Hē his heáfod onhylde swá swá hē slápan wolde, 3, 11; S. 536, 30: 4, 24; S. 599, 6: Exon. Th. 178, 14; Gū. 1244. Walle onhældum *parieti inclinato*, Ps. Surt. 61, 4. Onhylded *reclinem*, Wrt. Voc. ii. 78, 80. (2) metaphorically (a) with the idea of favourable disposition towards a person or thing, *to incline* :—Tô mē ðín eáre onhyld, Ps. Th. 101, 2. His

---

breóst sīen onhielde tô forgiefnesse, Past. 10, 1; Swt. 61, 12. Onhelded wið ðæs gecyndes, Met. 13, 11. (b) with the idea of subjection, *to bow, bend* :—Mid hwelcum monnum mágon gē onheldon eówra feónda swyrbán, Shrn. 86, 22. (c) *to turn from the right course* :—Hié onhældon in ðē yfel *declinaverunt in te mala*, Ps. Surt. 20, 12. (d) *to cause to sink* :—Onhælde sind ríce *inclinata sunt regna*, 45, 7. II. *intrans. To decline, deviate, incline, sink* :—Heofones gym west onhylde, Exon. Th. 174, 32; Gū. 1186. Onhylde (onhældeþ, Ps. Surt.) of ðysum on ðys (*inclinavit*), Ps. Spl. 74, 8. Alle onhældon *omnes declinaverunt*, Ps. Surt. 13, 3. Onheldan *declinare*, 16, 11. Onhældende *declinantes*, 100, 4.

on-hildedness, e; *f. Declining* :—Onheldednis *declinatio*, Ps. Surt. 72, 4.

on-hindan; *adv. Behind* :—Womb wæs onhindan áþrunten, Exon. Th. 419, 6; Rä. 38, 1. Ǽtterne tægel hafaþ onhindan, Fragm. Kmbl. 38; Leas. 21.

on-hinder. v. hinder.

on-hinderling; *adv. Back* :—Onhinderling hweorfaþ míne feóndas *convertentur inimici mei retrorsum*, Ps. Th. 55, 8: 69, 3. v. on-bæcling.

on-hirdan *to comfort, strengthen, encourage* :—Manegum wearþ hige onhyrded þurh his hálig word, Apstls. Kmbl. 105; Ap. 53: Elen. Kmbl. 1678; El. 841.

on-hiscan. v. on-hyscan.

on-hlídan; *p.* -hlád. I. *trans. To open, unclose* :—Onhlídest (*aperis*) ðū ðíne handa, Ps. Th. 144, 17. Undôþ eówre geatu, and onhlídaþ ða ēcan geata, 23, 7, 9. Deáþræced heolstorcofan onhliden weorþaþ, Exon. Th. 201, 1; Ph. 49. Đǽr biþ open eádgum tôgeánes, onhliden hleóðra wyn, heofonríces dura, 198, 18; Ph. 12. Carcernes duru opene fundon, onhliden hamera geweorc, Andr. Kmbl. 2155; An. 1079. II. *intrans. To be disclosed, to appear* :—Óþ ðæt wuldres gim onhlád *until the sun shewed itself*, 2539; An. 1271.

on-hnígan. I. *trans. To bend down, bow, press down* :—Onhnígaþ *incumbunt*, Wülck. Gl. 255, 11. Onhnígendre *grassante*, Hpt. Gl. 421, 19. Biþ wuhta gehwilc onhnigen tô hrúsan, Met. 31, 13. Onhnigenum heáfde simle his gesyhþa ádúna on eorþan besette, R. Ben. 31, 8. II. *intrans. To bend down, bow* :—Hí onhnígaþ tô mē, Exon. Th. 412, 29; Rä. 31, 7. Ealle eáðmôdlíce tô Criste sylfum onhnigan, Blickl. Homl. 203, 23: Cd. Th. 227, 3; Dan. 181. Man mæg tô ðǽm lástum onhnígan, and ða cyssan, Blickl. Homl. 127, 10.

on-hnyscan. v. on-hyscan.

on-hohsnian (?) *to abominate, detest* :—Đæt onhohsnode (MS. onhohsnod, *the* s *has been afterwards inserted between the* h *and* n) Hemninges mǽg, Beo. Th. 3892; B. 1944. Cf. on-hyscan.

on-hrægel; *n. A covering, sheet* :—Wæfelsum, onhræglum *sabanis*, Hpt. Gl. 490, 43.

on-hrǽs. v. on-rǽs.

on-hreódan *to adorn* (?). v. on-reódan.

on-hrēran. I. of actual movement, *to stir, agitate, move violently* :—Đonne hí (*the waves*) wind onhrēreþ, Ps. Th. 88, 3: Met. 7, 27. Đonne micla ýsta onhrēraþ hronmere, 5, 10. Fiscas ðe onhrēraþ hreó wǽgas, Exon. Th. 194, 19; Az. 141. Eorþan ðú onhrērdest *commovisti terram*, Ps. Th. 59, 2. Onhrērdan, 76, 15. Đonne hine mon drincan welle, onhrēre eft, Lchdm. ii. 270, 13. Ne mæg him se flǽschoma hond onhrēran, Exon. Th. 311, 22; Seef. 96. Eorþe biþ onhrēred of hire stôwe, Blickl. Homl. 91, 36. Lyft wæs onhrēred, Cd. Th. 208, 13; Exod. 482. Grund is onhrēred, deópe gedrēfed, Andr. Kmbl. 786; An. 393. II. metaph. *to move, disturb, agitate* :—Đone ungeþyldegan swíðe lytel scúr ðære costunga mæg onhrēran (-hrēran, Hatt., Past. 33; Swt. 224, 5. Ne mæg hine ǽnig onhrēran (*non commovebitur*), ðe eardfæst byþ on Hierusalem, Ps. Th. 124, 1. Eall heofena mægen biþ onwended and onhrēred, Blickl. Homl. 91, 28. III. of emotions, *to stir up, arouse, excite* :—Mægen wæs onhrēred, Cd. Th. 192, 4; Exod. 226. Wæs merefixa môd onhrēred, Beo. Th. 1103; B. 549. Hete wæs onhrēred, 5101; B. 2554: Andr. Kmbl. 2606; An. 1304: 2788; An. 1396.

on-hrínan; *with gen. dat. To touch* :—Sió sunne ne onhrínþ nô ðæs dǽles ðæs heofenes ðe se môna on irnþ, ne se môna nô ne onhrínþ ðæs dǽles ðe sió sunne on irnþ, Bt. 39, 13; Fox 232, 27–29. Đa hundas . . . his nāne onhrinon, Shrn. 145, 5. Đá ne onhrán ðæt fýr him, 53, 24. Onhrín ðissum muntum *tange montes*, Ps. Th. 143, 6.

on-hrine (?), es; *m. Touch* :—Ne ðe ǽniges yfeles onhrine (onryne, MS. B.) dereþ, Lchdm. i. 328, 1. Cf. æt-hrine.

on-hrôp, es; *m.* I. *importunate clamour, importunity* :—For his onhrôpe hē árist and sylþ him his neóde *propter improbitatem surget, et dabit illi quod habet necessarios*, Lk. Skt. 11, 8: Homl. Th. i. 248, 32. Se brôðor ðe hine synderlíce gebiddan wile, ne sý gelet mid (þurh, W.F.) ǽniges ôðres onhrôpe, R. Ben. 81, 9. II. *abusive language, reproach* :—Hosp ł onhrôp *improperium*, Ps. Lamb. 68, 20. [Cf. *O. H. Ger.* ana-ruof *appetitio*.] v. hrôp.

on-hupian *to draw back, recoil* :—Đonne ðæt môd ongit hine selfne on swelcne spild forlǽd ðonne wiðtremþ hē and onhupaþ and ondrǽt

him ðæt ðæt hé ǽr lufode *dum mens sese in praecipitium pervenisse deprehendit, gressum post terga revocet, pertimescens quae amaverat*, Past. 58, 2 ; Swt. 441, 28. [Cf. *Icel.* hopa aptr, ā hæl, undan *to draw back.*]

on-hwelan *to resound :*—Onhwileþ *reboat*, Wrt. Voc. ii. 94, 74. v. hwelan.

on-hweorfan ; *p.* -hwearf. **I.** *trans.* *To change, reverse :*—Metod onhweorfeþ heortan ðíne (*of Nebuchadnezzar's transformation*), Cd. Th. 251, 27 ; Dan. 570. Hé cwide (*the curse pronounced against Adam*) eft onhwearf, Exon. Th. 39, 7 ; Cri. 618. Eft is ðæt onhworfen, is nú swá hit nó wǽre freóndscipe uncer, 443, 2 ; Kl. 23. Hwý is ðis gold ádeorcad and ðæt ædeleste hiew hwý wearþ þit onhworfen *quomodo obscuratum est aurum, mutatus est color optimus*, Past. 18, 3 ; Swt. 133, 11. **II.** *intrans. To change, turn, revert :*—Manegum cyninge onhwearf se anweald and se wela óþ ðæt hé eft wearþ wædla *qui reges felicitatem calamitate mutaverint*, Bt. 29, 1 ; Fox 102, 13. Hé (*Nebuchadnezzar*) eft onhwearf wódan gewittes, Cd. Th. 255, 21 ; Dan. 627. v. next word.

on-hwirfan ; *p.* de. **I.** *to turn* (of actual motion), (a) *trans. :*—Ic mé wille nú onhwyrfan tó ðisse bǽre, Blickl. Homl. 151, 14. (b) *intrans :*—Swá swá hweól onhwerfþ, Bt. 33, 4 ; Fox 132, 13. **II.** *to invert, transpose :*—Agof (boga) is mín noma eft onhwyrfed, Exon. Th. 405, 13 ; Rä. 24, 1. **III.** *to change, turn :*—Mé onhwyrfdon ða mé grome wurdon of ðære gecynde ðe ic ǽr cwic beheóld, 485, 24 ; Rä. 72, 2. Ðú geómrast forðam ðe seó woruldsǽlþ onhwyrfed is, Bt. 7, 1 ; Fox 16, 9. v. preceding word.

**on-hwirfedness, e ; *f.* Change, mutation :*—Sóð God búton ælcere onhwerfednesse, Shrn. 167, 34.

on-hyldan. v. on-hildan.

**on-hyreness, e ; *f.* Imitation :*—Ðone weg ðære onhyrenesse *viam imitationis*, Past. 16, 4 ; Swt. 103, 14. Tó onhyrenesse (*ad imitationem*) ðæra eádigra apostola, Bd. 4, 28 ; S. 606, 26 : 1, 27 ; S. 492, 23.

**on-hyrian ; *p.* ede *To imitate, emulate* (with dat. acc.) :*—Hwílum ic onhyrge gúþfugles hleóþor, Exon. Th. 406, 20 ; Rä. 25, 4 : 391, 2 ; Rä. 9, 10. Mon onhyreþ dysegum neátum *homo comparatus est jumentis insipientibus*, Ps. Th. 48, 11. Se ðe hit gehýreþ hé onhyreþ ðam *ad imitandum bonum auditor sollicitus instigatur*, Bd. pref. ; S. 471, 15 : Bt. 41, 5 ; Fox 254, 5. Ða cild onhyriaþ ealdum monnum, 36, 5 ; Fox 180, 10. Ðonne wé onhyrigaþ Criste, Past. 51 ; Swt. 397, 1. Ðæt hý ne onhyredon ðǽm yfelwillendum, Ps. Th. 36, arg. Ne onhyre (*emulatus fueris*) ðam þe byþ orsorh, 36, 1. Onhyriaþ, 36, 1. Ne ðú ne onhere *ne emuleris*, Kent. Gl. 58. Ne onherie *ne emuletur*, 885. Ðæt wé onhyrigen ðǽm þeawum, Past. 34 ; Swt. 231, 3 : Swt. 229, 15. Onhyrgean wé ðone blindan, Blickl. Homl. 21, 9. Wé sceolan onherian Marian ðære ðe smerede Hǽlendes fét, 75, 11. Onhyrian (-hirian, Cott. MS.), Bt. 40, 4 ; Fox 240, 4 : Bd. 1, 7 ; S. 477, 2. Ðæt onhyrian woldan, 4, 3 ; S. 569, 43. Heora líf onhyrian wolde, 4, 13 ; S. 582, 24 : 5, 9 ; S. 622, 12. Onhyrigean, 1, 26 ; S. 487, 32. Onhyrgan, 3, 18 ; S. 545, 43. Hé ðære frymþelícan cyrican líf wæs onhyrigende, 1, 26 ; S. 487, 28 : 4, 23 ; S. 593, 15. Onhyrgende, L. Ecg. P. iv. 68, 8 ; Th. ii. 228, 29.

**on-hyriend, es ; *m. One who imitates* or *emulates :*—Onhyrgend *emulatores*, Wrt. Voc. ii. 31, 28. Onhyrgend[r]as, 85, 25.

**on-hyring, e ; *f.* Imitation, emulation :*—Anhering *emulatio*, Wrt. Voc. ii. 143, 48. Gód anda and anhering áscyreþ fram synna leahtrum, and lǽt tó Gode, R. Ben. 131, 13. Ðonne wé onhyrigaþ Criste and eác ða onhyringe gefyllaþ *tunc legem Christi imitando complemus*, Past. 51, 3 ; Swt. 397, 2.

on-hyscan ; *p.* te. **I.** *to mock, make a jest of :*—Drihten on-hnyscþ (? -hyscþ, MS. T.) hine *Dominus irridebit eum*, Ps. Spl. 36, 13. Ic wénde ðæt hí mec onhyscte *illudi me a senibus existimavi*, Nar. 25, 22. **II.** *to reproach, abuse, speak ill of :*—Ðonne men eów onhiscaþ (*exprobaverint*), Lk. Skt. 6, 22. Gebiddaþ for ða ðe eów onhyscaþ (-hisceaþ) *pro calumniantibus vos*, 6, 28. Ðæt man ða onhisce swýðe for worulde and hý unweorðige, Wulfst. 168, 6 : 70, 12. **III.** *to detest :*—Ic unrihta gehwylc onhyscte *iniquitatem abominatus sum*, Th. 118, 163. Hí onhysctan æghwylcne mete *omnem escam abominata est anima eorum*, 106, 17.

on-in *within :*—Onin mé *intra me*, Ps. Spl. 38, 4.

on-innan. v. innan, V.

on-irnan *to yield, give way :*—Duru sóna onarn fýrbendum fæst, syððan hé hire folmum [hrán], Beo. Th. 1447 ; B. 721. Duru sóna onarn þurh handhrine háliges gástes, Andr. Kmbl. 1998 ; An. 1000. [*O.H.Ger.* int-rinnan *evadere, abire, profugere.*]

**on-irning, e ; *f.* Attack, assault :*—Diówlíca onerninge tósliten beón *diabolica incursione lacerari*, Rtl. 36, 1.

on-íwan *to shew :*—Drihten ús lífes wegas anýweþ, R. Ben. 3, 5. Ic ðé bidde ðæt ðú mé ðíne wísan on, St. And. 10, 14. Seó hlǽdder ðe Jacobe on swefne wearþ anýwed, R. Ben. 23, 5.

on-lǽnan ; *p.* de ; *with gen.* or *acc. of the loan. **I.** *to lend, grant :*—Ic eów onlǽne ðæs gewítendan, and ic eów geselle ða þurhwun-

iendan, Past. 46, 5 ; Swt. 351, 13. Se cræft ðe him Crist onlǽnþ, **Met.** 10, 37. Hé úre ðé onlǽnde æfter his bebodum tó brúcanne, Bt. 7, 5 ; Fox 24, 9. Gif hwá his wǽpnes óðrum onlǽne, L. Alf. pol. 19 ; Th. i. 74, 3 : L. In. 29 ; Th. i. 120, 10, 12, 14. Hí ðé onlǽnde wǽron, Bt. 7, 3 ; Fox 20, 6. **II.** *to lease, let :*—Denewulf and ða hýwan on Wintanceastre ænlǽnaþ Ælfréde his deg XL. hída landes, Chart. Th. 147, 27. Cf. on-león.

on-lǽtan. **I.** *to release, relax :*—Ðonne forstes bend Fæder onlǽteþ, Beo. Th. 3223 ; B. 1609. Ðonne him sigera weard his gewealdleðer wille onlǽtan, Met. 11, 28, 75. **II.** *to let a thing go on, to continue :*—Tó anlǽtenne *continuanda*, Wrt. Voc. ii. 135, 19. Fæstan twegen dagas on ðære wucan, bútan hý ouermicel geswinc habben. Gif hý út an æcere wurc habben, ðæs middæges gereord is singallíce tó anlǽtenne (on-, MS. T.), R. Ben. 66, 1.

**on-lang ; *prep.* Along :*—Onlong Mǽse, Chr. 882 ; Erl. 82, 7. v. and-lang.

**on-léc, es ; *m. On-look, regard :*—Onléce *respectu, intuitu*, Hpt. Gl. 487, 50. v. léc.

on-leccan (?) *to reproach, blame :*—Onlehton (bysmrydon, MS. C.) *irritaverunt*, Ps. Spl. T. 105, 8. Cf. hosp (on leccungæ, MS. T. : tó bysmre, MS. C.) *irrita*, 88, 34, *and see* lǽcing.

on-lecgende *on-lying :*—Wyrc him onlecgende sealfe, Lchdm. ii. 200, 8.

**on-legen, e ; *f. An on-laying,* (*medicinal*) *application :*—Onlegen (ἐπίθεμα) tó trymmanne ðone magan, Lchdm. ii. 180, 24. Mid onlegene swá swá mon of swelcum þingum wyrcþ . . . Lácna mid onlegena beres, 82, 14-24. Gesodene wuduæpla and hlafes cruman and swilce onlegena, 190, 15.

on-leóhtan. v. on-líhtan.

**on-leóhtness, e ; *f. Illumination ; illuminatio*, Ps. Lamb. 138, 11. v. on-líhtness.

on-león ; *p.* -láh, -leáh ; *pl.* -ligon ; *pp.* -ligen. **I.** *to grant the loan of something* (gen. of loan) :*—Gielde se ðæs wǽpnes onláh, L. Alf. pol. 19 ; Th. i. 74, 6 : Beo. Th. 2939 ; B. 1467. Onligenre *inpactae*, Wrt. Voc. ii. 111, 31. **II.** *to grant, bestow :*—Sum ǽhta onlíhþ, sum biþ wonspédig, Exon. Th. 295, 9 ; Crä. 30. Metod onláh Médum aldordómes, Cd. Th. 258, 25 ; Dan. 681. Ungelíc ðam óðrum stede ðe mé mín hearra onlág, 23, 12 ; Gen. 358. Hé mé láre onlág, Elen. Kmbl. 2489 ; El. 1246. God hyre sigores onleáh, Judth. Thw. 23, 16 ; Jud. 124.

on-lésan, -lésness. v. on-lísan, -lísness.

**on-líc ; *adj. Like, similar :*—Heáp synnigra híge onlíc, Ps. Th. 91, 6. Gelamp óðer wundor ðissum onlíc, Blickl. Homl. 219, 7 : 223, 14. Eal hé ǽr on onlíc weorc áteáh, 215, 5. Manigfeald onlíc wundor ðysum ðær wǽrom æteówed, 209, 14. Monige sindon mé suíðe onlíce on ungelǽrednesse, Past. proem. ; Swt. 25, 7. Se is lyfte onlícusð on hiwe, 14 ; Swt. 85, 5. Se fugel is onlícost peán, Exon. Th. 219, 25 ; Ph. 312. Onlícust, 189, 20 : Az. 62. v. an-líc.

**on-líce ; *adv. Like, in like manner :*—Ealle ða ríca forheregian . . . swíðe onlíce ðam micelan flóde, Bt. 16, 1 ; Fox 50, 6 : Met. 8, 47 : Elen. Kmbl. 197 ; El. 99. Onlícost dydon swelce him nǽfre ǽr ðæm gelíc yfel an ne becóme, Ors. 3, 10 ; Swt. 140, 10 : Past. 17 ; Swt. 123, 7. v. an-líce.

**on-lícness, e ; *f. Likeness, image :*—Idese onlícnes *the form of a woman*, Beo. Th. 2706 ; B. 1351 : Andr. Kmbl. 1461 ; An. 731. Hé hæfþ mon geworhtne æfter his onlícnesse, Cd. Th. 25, 19 ; Gen. 396 : Exon. Th. 424, 10 ; Rä. 41, 37. v. an-lícness.

on-líesan. v. on-lísan.

on-líhtan, -leóhtan. **I.** *of places or things, to illumine, make bright, cause to shine,* (a) literally :*—Mycel leóht onleóhte ðæt carcern, St. And. 4, 4. Óþ ðæt ðære sunnan leóman hine (*the moon*) eft onlíhton, Lchdm. iii. 240, 27. Onleóhtende *inluminans*, Hymn. Surt. 15, 22. Ealle steorran weorþaþ onlíhte and gebirhte of ðære sunnan, Bt. 34, 5 ; Fox 140, 5. (b) metaph. :*—God onlýhteþ (*illuminet*) andwlitan his ofer ús, Ps. Spl. 66, 1. Onlýht (*inlustra*) usic Drihten. Beów ðinne, 30, 20. Tó hwon yldestú middangeard tó onlýhtenne, Blickl. Homl. 7, 33. **II.** *of persons,* (a) *to give sight to, make the sight clear :*—Drihten blinde on heora eágum onleóhteþ *Dominus illuminat caecos*, Ps. Th. 145, 7. Heó gegódaþ and onlíht ðæra eágena scearpnysse, Lchdm. i. 72, 15. Ðæt se ylca ða dohter ðæs ealdormannes blinde onlíhte, Bd. 1, 18 ; S. 484, 30. Ðæt wundor worhte, ðæt hé ðone blindan onlýhte, Blickl. Homl. 19, 19. Mon geseah hine blinde onlýhtende, 177, 15. Ðá geseah hé sóna gesundfullum eágum, þurh ðone ylcan onlíht ðe hine ǽr áblende, H. R. 107, 28 : Homl. Skt. i. 21, 275. (b) *to clear the mental vision, to enlighten :*—Sóð leóht ðæt onlýht ælcne man, Jn. Skt. 1, 9. Se hálga Gást ealle ða englas onlíht, Ælfc. T. Grn. 2, 14. Worda mé heora wíse onleóhteþ, Ps. Th. 118, 130. Ne onlíhtaþ hí náuht ðæs módes eágan, Bt. 34, 8 ; Fox 144, 32. Ðú simle míne sáwle onlíhtest, Homl. Th. i. 74, 31. Hé hió onlýhte mid ðæs Hálgan Gástes gife, Blickl. Homl. 145, 6. Onliht ða eágan úres módes mid ðínum leóhte, Bt. 33, 4 ; Fox 132, 33 : Ps. Th. 12, 4. Onleóht heorte manna, Hymn. Surt. 23, 1. Crist mæg ðíne nytennysse onlíhtan, Homl. Skt. i. 5, 200. Manegum wearþ mód onlíhted, Apstls. Kmbl. 104 ; Ap. 52. Wé wurdon onlíhte þurh geleáfan, Homl. Th. i. 154, 21. Onlýhte, Blickl.

Homl. 161, 14.   **III.** *to give light* (with dat.) :—Ðæt hit onlíhte eallum ðe on ðam húse synt, Mt. Kmbl. 5, 15. Onlíhtan ðám ðe on þýstrum sittaþ, Lk. Skt. 1, 79.   **IV.** *intrans.* *To shine* :—Ic onlíhte oððe scýne *luceo*, Ælfc. Gr. 35 ; Som. 38, 8. Sæterdæg onlýhte (*inlucescebat*), Lk. Skt. 23, 54. Onlíhton (*illuxerunt*) lígrascas ðíne, Ps. Spl. 76, 18 : 96, 4. Heora wegas onlíhton, Blickl. Homl. 137, 2. Swá onlíhte (*luceat*) eówer leóht, Mt. Kmbl. 5, 16. Onlióhte *inlucescat*, Kent. Gl. 206.

on-líhting, e ; *f. Illumination, enlightening* :—Onlýhtinga *illuminatio*, Ps. Spl. 43, 5. Onlíhtinge, Ps. Lamb. 26, 1. Onlíhting, 138, 11. On onlíhtinge fýres, 77, 14.

on-líhtness, e ; *f. Illumination* :—Onlýhtnes (-líhtnes, Ps. Lamb.) *illuminatio*, Ps. Spl. 26, 1 : 43, 5. Seó onlýhtnes Cristes godspelles, Bd. 2, 9 ; S. 511, 10 : 2, 2 ; S. 502, 29. v. on-leóhtness.

on-lísan.   **I.** *to unloose* (real or metaphorical bonds) :—Ðæt bearn benda onlýseþ, Exon. Th. 5, 12 ; Cri. 68. Hé ða tungan onlýsde, Blickl. Homl. 167, 10. Hire bendas wæron onlýsede, 89, 25. Onlýsde, 87, 36.   **II.** *to release, deliver, liberate* :—Mín líf of ðære écean forwyrde ðú onlýsdest, 89, 4. Cyning onlésde (*solvit*) hine, Ps. Surt. 104, 20. Tó onliésanne ða gehæftan on helle, Past. 58 ; Swt. 443, 10. Siððan seó sáwl of ðam carcerne ðæs líchoman onliésed biþ, Bt. 18, 4 ; Fox 68, 15. Onlésed, unsǽled *desolutus*, i. *liberatus*, Wrt. Voc. ii. 139, 29 : 138, 50. Swá hwylcne swá hé on eorþan álýsde, ðæt se wǽre on heofonum onlýsed, Blickl. Homl. 49, 18. Fram swá myclum cwylmnessum onlýsed beón, Bd. 4, 9 ; S. 577, 10 : 5, 19 ; S. 639, 42. Onlýsed ðý líchaman *solutus corpore*, 3, 19 ; S. 548, 29.

on-lísness, e ; *f. Deliverance, redemption* :—Ða ðe on helle synt biddaþ ðínre onlésnesse *ask for deliverance by thee*, Blickl. Homl. 81, 23 ; 67, 3. Onlésnisse *redemptio*, Lk. Skt. Lind. Rush. 21, 28.

on-líðian *to become pliant, to yield* :—Sceal hira ánra gehwylc onlútan and onlíðigan ðe hafaþ læsse mægen, Salm. Kmbl. 713 ; Sal. 356.

on-lóciend, es ; *m. An on-looker, spectator* :—Heó wæs swíðe lufigendlíc eallum onlóciendum, Anglia ix. 30, 97.

on-lúcan.   **I.** literally, *to unlock, open* :—Ðæs ceasterhlides onlúcan . . . ða fæstan locu nænig óðer eft onlúceþ, Exon. Th. 20, 7-20. Onlæc *reserat*, Wrt. Voc. ii. 119, 2. Suelce ic gesáwe sume duru onlocene, Past. 21, 3 ; wt. 155, 6.   **II.** metaph. *to open, disclose, reveal* :—Ðæt word ðære þreáunge is cǽg forðam hit oft onlýcþ (anlýcþ, Hatt. MS.) and geopenaþ ða scylde *clavis est sermo correptionis* ; *quia culpam detegit*, Past. 15, 2 ; Swt. 90, 11. Hié ne ongietaþ ná hú suíðe hié onlúcaþ hiera mód mid ðæm unþeáwe ofermétta *quantum se vitiis superbiendo aperiat, non agnoscit*, 38, 1 ; Swt. 271, 22. Módhord onleác and ðus wordum cwæþ, Andr. Kmbl. 344 ; An. 172. Leóþucræft onleác, Elen. Kmbl. 2499 ; El. 1251. Wordhord onleác, Beo. Th. 524 ; B. 259. Engla helm tuddorspéd onleác (*revoked the sentence of barrenness*), Cd. Th. 166, 24 ; Gen. 2752. Hwylc ðæs hordgates, cǽgan cræfte, ða clamme onleác, Exon. Th. 429, 30 ; Rä. 43, 12. Wærc in gewód líchord onleác *pain hath invaded me, hath opened to itself a way within my body*, 163, 31 ; Gú. 1002 : 170, 26 ; Gú. 1117. Ðæt mon onlúce ða heardan heortan *duritiam cordis aperire*, Past. 21, 3 ; Swt. 155, 2. Onlúcan gǽstes cǽgon, Cd. Th. 211, 6 ; Exod. 522. Ic hæbbe lárcræftas onlocen, Salm. Kmbl. 5 ; Sal. 3.

on-lútan *to lout, bend down, bow* :—Hé onlýtt tó ðissum eorþlícum, suá ðæt neát for gífernesse onlýt tó ðære eorþan, Past. 21, 3 ; Swt. 157, 2-4. Ælc gesceaft ealle mægene symle onlýt wið his gecyndes, Met. 13, 66. Hié him tó onluton and hine weorþodan swá cinige geríseþ, Blickl. Homl. 69, 31 : 87, 7. Anlúte him eaðmódlíce tó mid ðam heáfde, R. Ben. 83, 11. Hira sceal ánra gehwylc onlútan, ðe hafaþ læsse mægen, Salm. Kmbl. 713 ; Sal. 356. Ælc gesceaft biþ heald onloten (-locen, Fox) wið hire gecynde, Bt. 25 ; Fox 88, 7.

on-lýhtan, -lýhting, -lýhtness, -lýsan. v. on-líhtan, -líhting, -líhtness, -lísan.

on-mǽdla. v. on-médla.

on-mǽlan *to address* :—Him Babilone weard yrre andswarode, eorlum onmælde, grimme ðám gingum oncwæþ, Cd. Th. 229, 1 ; Dan. 210.

on-mang, *prep. with dat. Among* :—Onmang folce, Lev. 24, 10 : Homl. Skt. i. 23, 92. Onmang óðrum mannum, 23, 478. Onmang ðam ðe hí on wópe wæron *whilst they were weeping*, 23, 246.

on-mearc. v. mearc.

on-mearcung, e ; *f. An inscription* :—Onmercunge *inscribtionem*, Lk. Skt. Rush. 12, 24.

on-médan (?) *to take upon one's self, to presume* (*the following passage should be given under* médan) :—Ondsware ýwe se hine on méde wordum secgan hú se wudu hátte *let him give answer, who will take upon himself to say in words, what the name of that wood is*, Exon. Th. 437, 30 ; Rä. 56, 15. v. next word.

on-médla (-medla, Grimm, Grein), an ; *m.*   **I.** *pride, glory, magnificence* :—Ald onmédla is gecyrred *the glory of earlier times is changed*, Elen. Kmbl. 2529 ; El. 1266. Ðæt geó guman heóldan, ðenden him on eorþan onmédla wæs, Exon. Th. 51, 13 ; Cri. 815. Dagas sind gewitene, ealle onmédlan eorþan ríces, 310, 27 ; Seef. 81.   **II.**

pride, arrogance, presumption :—For onmédlan, Beo. Th. 5844 ; B. 2926. Him for onmædlan eorre geworden, Cd. Th. 291, 11 ; Sat. 429.   **III.** *courage, boldness* :—For hwam ne móton wé ealle mid onmédlan (*boldly*) gangan in Godes ríce, Salm. Kmbl. 704 ; Sal. 351. v. an-medla.

on-métan *to paint, cover as with colour* :—Ðú mid sárlícre sceame onméttest (-meltest, Th.) *perfudisti eum confusione*, Ps. Th. 88, 38. v. á-métan.

on-middan ; *prep. Amid, in the middle of* :—Onmiddan ðæm hwǽte *in medio tritici*, Mt. Kmbl. 13, 25. Onmiddan ðám þýstrum, Bd. 5, 12 ; S. 628, 19. Onmiddan ðære byrig, Homl. Skt. i. 23, 609. Onmiddan eówrum sceáfum, Gen. 37, 7. v. á-middan, midde.

on-mirran *to hinder, obstruct, disturb* :—Ic bebeóde ðæt ðisne freódóm nænig mínra æfterfylgendra eft ne onwende, ne on nænigum dǽlum hyne ne onmyrdon, Chart. Th. 390, 31. v. á-myrran.

on-mitta, an ; *m. A measure* ; *exagium*, Wrt. Voc. ii. 30, 49 : 144, 45. v. an-mitta.

on-mód ; *adj. Bold, courageous*, Exon. Th. 146, 29 ; Gú. 717. v. an-mód.

on-munan *to esteem, consider* (*worthy*), *think* (*highly of*). (a) with acc. of person and adj. denoting worth :—Búton ic openlíce gecýþe ðæt ic God sylfa sý, ne onmun ðú mé nánre áre wyrþne, Blickl. Homl. 181, 36. Ælc ðara ðe sie under ðæm gioke hláfordscipes hé sceal his hláford ǽghwelcre áre wierþne onmunan *quicumque sunt sub jugo servi, dominos suos omni honore dignos arbitrentur*, Past. 29 ; Swt. 201, 23. (b) with acc. of person alone :—Hé úsic on herge geceás tó ðyssum síðfæte, onmunde úsic mǽrþa *he thought us fit for great deeds*, Beo. Th. 5273. Ðá cuǽðon hié ðæt hié ðæt hié ne onmunden ðon má ðe eówre geféran *then they said, that they did not consider themselves entitled to accept the offer*, ' *any more than your comrades did*,' Chr. 755 ; Erl. 50, 24. Miclum geblissod ðæt hié God wolde onmunan swá micles ofer menn ealle *Andrew was greatly rejoiced that God deemed his disciples worthy of such high regard beyond all men* (*in granting them the vision they had seen*), Andr. Kmbl. 1789 ; An. 897.

on-nytt *useless* :—Onnitte *inutiles*, Ps. Spl. 13, 4. v. un-nytt.

ono *if* :—Ono nú ðæt wíf wel gedyrstgade *si igitur bene praesumsit*, Bd. 1, 27 ; S. 494, 19. Ono hé wiste hine on wónyssum geeácnode, hé ðá geómrade hine fram scylde ácennede *qui enim in iniquitatibus conceptum se noverat, a delicto se natum gemebat*, S. 495, 24. Ono (ond ?) gif (*si autem*) hé gehæfted wæs, hwæt hé ðonne ne feaht, S. 497, 37 : 3, 24 ; S. 557, 29. [Cf. (?) Goth. an : O. H. Ger. inu, enu.]

on-orettan *to perform with effort, to accomplish* (*a difficult undertaking*) :—Nó hé ofer Offan eorlscype fremede (*he did not excel Offa*), ac Offa geslóg cyneríca mǽst ; nænig efeneald him eorlscipe máran onorette áne sweorde *no one of equal age had done such heroic deeds*, Exon. Th. 321, 4 ; Víd. 41. Iudiscféða án onorette uncúð gelád *the tribe of Judah by itself performed the difficult and unknown course* (*the passage of the Red Sea*), Cd. Th. 197, 25 ; Exod. 313.

on-orþung, e ; *f. A breathing in or* :—Fram onorþunge (*inspiratione*) gástes yrres ðínes, Ps. Lamb. 17, 16.

onoþa, an ; *f. Fear* :—Onoþa *formido*, Wrt. Voc. ii. 35, 75. Anoþa, 109, 3. [Cf. anathe *sollicitudine, cura* in Papias, quoted by Graff. i. 267.]

on-pennian *to unpen, open* :—Ðæt wæter, ðonne hit biþ gepynd, hit miclaþ . . . Ac gif sió pynding wierþ onpennad, ðonne tóflewþ hit eall, Past. 38, 6 ; Swt. 277, 8.

on-rád, e ; *f. Riding on horseback* :—Sécen hié him broc on onráde and on wǽne *let them seek for themselves fatigue in riding on horseback and in a carriage*, Lchdm. ii. 184, 13.

on-ræfniendlíc ; *adj. Intolerable* ; intolerabilis, Ps. Spl. 123, 4.

on-rǽs, es ; *m. On-rush, attack, assault, violent motion* :—Onrǽs *impetus*, Ælfc. Gr. 11 ; Som. 15, 12. Flódes onrǽs *fluminis impetus*, Ps. Spl. 45, 4 : Ps. Surt. ii. 189, 40. Hyne þurhþýdde mid egeslícum onhrǽse, Homl. Skt. i. 3, 274. Onrǽs *irruptionem*, Hpt. Gl. 464, 66. Ðone onrǽs his hátheortnesse *fervoris sui impetu*, Past. 40, 5 ; Swt. 297, 20.

on-rǽsa (?), an ; *m. Attack, irruption* :—Onrǽsan *inruptiones*, Wrt. Voc. ii. 45, 1.

on-rǽsan. v. rǽsan.

on-reádan ; *p.* -reód *To redden, stain* :—Onreód *inbuit*, Wrt. Voc. ii. 111, 65. v. on-reódan.

onred, es ; *m.* (?) *The name of some plant* :—Onred, hámwyrt . . . onredes einfela, Lchdm. ii. 104, 14-15. Genim onred, 270, 26.

on-reódan, *p.* -reád *To redden* :—Brynegield onreád (-hread, MS.) rommes blóde, Cd. Th. 177, 18 ; Gen. 2931. v. reódan.

onrettan. v. orrettan.

on-riht ; *adj. Right, proper* :—Se wuldorcyning gesette ýðum heora onrihtne ryne, Cd. Th. 10, 35 ; Gen. 167. v. on-rihtlíce *and next word*.

on-riht, es ; *n. A right* (?) :—Hálige þeóde, Israéla cyn, onriht Godes *God's peculiar people*, Cd. Th. 200, 18 ; Exod. 358. [Cf. Icel. eiga rétt á einum *to have rights over a person*.]

on-rihtlíce ; *adv. Rightly, duly* :—Ða láreówas sceolan synfullum mannum tǽcan, ðæt hié heora synna cunnon onrihtlíce geandettan, Blickl. Homl. 43, 16.

on-rihtwísness, e; f. *Unrighteousness, iniquity:*—Onrihtwísnyssum *iniquitatibus,* Ps. Spl. 52, 2.

on-rísan *to arise:*—Mín yrre onríst ongén hig *iraseetur furor meus contra eum,* Deut. 31, 17.

on-ryne, es; m. **I.** *a running on, course:*—Onryne tíde *cursu temporis,* Hymn. Surt. 36, 8. **II.** *a running on or against, an attack:*—Ne ðé æniges yfeles onryne (anryne, MS. H.) dereþ, Lchdm. i. 328, 1, MS. B.

on-sacan. **I.** *to attack, strive against:*—Ne biþ cwénlíc þeáw ðætte freoþuwebbe feores onsæce leófne mannan (*to strive with a man for his life*), Beo. Th. 3889 ; B. 1942. **II.** *to resist, refuse to comply with a demand:*—Ðeáh ðú onsóce ðæt ðú sóþ godu lufian wolde *though you have refused to love the true gods,* Exon. Th. 254, 8 ; Jul. 194. Hé ne trúwode ðæt hé sæmannan onsacan mihte, hord forstandan bearn and brýde, Beo. Th. 5901 ; B. 2954. **III.** *to deny,* (a) *of persons, to declare that one has no knowledge of a person:*—Ne ðé onsæco (-sæcco, Lind.) ic *non te negabo,* Mk. Skt. Rush. 14, 31. Ðú mé onsæces *me negabis,* 14, 72. Se ðe mé onsaekeþ (-sæccas, Lind.) beforan monnum, onsaece ic ðone beforan fæder mínum, Mt. Kmbl. 10, 33. (b) *to refuse a person what he wants:*—Gif huá wil æfter meh gecyme onsæce (andsæce, Rush.) hine seolfne *abneget semetipsum,* 16, 24. (c) *to refuse to acknowledge a claim, not to allow the truth of a statement,* in a legal sense *to deny a charge:*—Ne onsace ic náuht, ðæt seó eádignes síe ðæt héhste gód ðises andweardan lífes, Bt. 24, 3 ; Fox 84, 14. Ða onsóc se óðer, and cwæþ hé him nán feoh ne sealde, Shrn. 127, 26. Hé onsóc (andsóc, Rush.) mid aað, ðæt ic ne conn ðone monno, Mt. Kmbl. Lind. 26, 72, 70. Forðam hié his cræftas onsócon (*they would not acknowledge his powers, would not bow down to the golden image*), Cd. Th. 230, 1 ; Dan. 226. Se ðæs onsóce, ðætte sóþ wære Waldend, se hié álýsde, 244, 20 ; Dan. 451. Ðonne sceal hé be .lx. hýda onsacan ðære þíefþe gif hé áðwyrðe biþ. Gif Englisc onstal gá forþ, onsace be twýfealdum, L. In. 46 ; Th. i. 130, 13–15 : Th. i. 132, 1 : 28 ; Th. i. 120, 8. **IV.** *to make excuse:*—Ongunnun alle onsaca (-sacca, Lind.) *coeperunt omnes excusare,* Lk. Skt. Rush. 14, 18. Cf. of-sacan. **V.** *to sacrifice* (v. on-secgan):—Onsacende *litaturus,* Wrt. Voc. ii. 53, 57.

on-sæc; adj. **I.** *freed from a charge, excused* (cf. Icel. sekr):—Hæfe mec onsæcne *habe me excusatum,* Lk. Skt. Lind. Rush. 14, 18. **II.** *denying:*—Mec ðú bist onsæcc (-sæcen, Rush.) *me es negaturus,* Mk. Skt. Lind. 14, 30. Mec ðú bist onsæc *me negabis,* 14, 72.

on-ségan *to cause to sink down, to prostrate:*—Ærðon hine deáþ onségde, Exon. Th. 171, 32 ; Gú. 1135. Hú hí (*hell*) bútan ende éce stondeþ, ðæm ðe ðær for his synnum onségd weorþeþ, 446, 27 ; Dóm. 28. Selegesceotu synd onségd (?), Ps. Th. 82, 6.

on-sége; adj. *Falling upon, assailing, attacking:*—Wé ær ðysan oftor brǽcan, ðonne wé béttan, and ðý is ðisse þeóde fela onsæge. Ne dohte hit nú lange inne ne úte, ac wæs here and hunger, bryne and blódgyte on gewelhwylcon ende, Wulfst. 159, 7 : 128, 14 : 243, 2. Hæðcynne wearþ gúþ onsæge *war had come upon Hæðcyn,* Beo. Th. 4960 ; B. 2483 : 4159 ; B. 2076. [*O. H. Ger.* ana-seigi *infestus.*]

on-sægedness, e; f. **I.** *the rite, act of sacrifice or offering:*—Onsægednys lofes ǽrwurþaþ *sacrificium laudis honorificabit me,* Ps. Spl. 49, 24. Ðonne sceal hé hine áhabban fram onsægdnysse (*immolatione*) ðæs hálgan gerýnes, Bd. 1, 27 ; S. 497, 4. Hé rícels bærnde in Godes ansægdnesse, Shrn. 133, 29. In onsægdnesse (-sægedenesse, MS. A.), Mt. Kmbl. 9, 13. Aarone tó fylste tó ðam ǽlicum onsægednyssum, Num. 18, 2. Onsægdnyssum and offrungum, Mk. Skt. 12, 33. Onsægdnessa *cerimonias,* Wrt. Voc. ii. 15, 81. **II.** *what is offered at the rite, a sacrifice, oblation:*—Nis ná tó onsónne seó hálige onsægdnes (*eucharistiam accipere*), L. Ecg. C. 35 ; Th. ii. 160, 37. Gif seó onsægednys on eorþan fealle, L. Ecg. P. iv. 15 ; Th. ii. 216, 15. Ne offra ðú ðínre onsægdnysse (*victimae*) blód uppan beornman, Ex. 23, 18. Þurh lác ðære hálwendan onsægednesse (*hostiae*), Bd. 4, 22 ; S. 592, 22. Gode onsægednesse beran, S. 592, 25 : 5, 10 ; S. 624, 32. Ða onsægdnysse ða ðe fram eów deóflum wǽron ágoldene, 1, 7 ; S. 477, 36. Onsægdnisse onsǽcgan *victimis placare,* Nar. 20, 5. Ic ðé onsegednesse brohte, Ps. Grn. ii. 279, 120. Onsegednesse *victimas,* Wülck. Gl. 61, 8 : 71, 40.

on-sægness, e; f. *A sacrifice:*—Onsægnessa *holocausta,* Blickl. Gl.

on-sægung, e; f. *An offering in sacrifice;* immolatio:—Onsægung, Wrt. Voc. i. 28, 48. Onsǽcgiung, ii. 49, 45.

on-sǽlan *to untie, unfasten:*—Onsǽl meoto sigehréð secgum, Beo. Th. 983 ; B. 489. Ðonne geméte gyt eoselan gesǽlede and hire folan ; onsǽlaþ hié, Blickl. Homl. 69, 36. Onsælid *desolutus,* Wrt. Voc. ii. 105, 80. Hæft wæs onsǽled, Cd. Th. 215, 15 ; Exod. 583.

on-sagu, e; f. *A charge brought against a person, accusation:*—Á biþ andsæc swíðere ðonne onsagu, i. e. *in a case where a charge is brought against a person, and it is met with a denial attested by the proper legal formalities, the case against him fails,* L. Eth. ii. 9 ; Th. i. 290, 17. [Cf. the somewhat similar principle which follows :—Ágnung biþ nér ðam ðe hæfþ ðonne ðam ðe æfter sprecþ. See also Grmm. R. A. 856.] Ðá cwæþ Eugenia ðæt heó eáþe mihte Melantian onsage oferdrífan

(*refute the charge*), Homl. Skt. i. 2, 206. Manega mid leásum onsagum geneálæhton *multi falsi testes accessissent,* Mt. Kmbl. 26, 60. v. onsecgan.

on-sand, e; f. *A sending to another:*—Onsande *immissiones,* Ps. Spl. 77, 54. Onsanda, Blickl. Gl.

on-sáwen *sown:*—Sǽd onsáwen, Exon. Th. 215, 14 ; Ph. 253.

on-scǽgan (?) *to mock, deride, reproach:*—Hí tǽldon ł onscǽgdon (onsǽgdon ?) ł hig hyspton mé *subsannaverunt me,* Ps. Lamb. 34. 16.

on-sceacan. **I.** *to shake:*—Heó feðera onsceóe, Cd. Th. 88, 26 ; Gen. 1471. Onsacene *concusa,* Wrt. Voc. ii. 15, 80. **II.** *to shake off, remove:*—Onscacan (-seacan, Wrt.) *detestare,* Wrt. Voc. ii. 106, 28. Onsceæcannæ onsceacnessum (ob-, MS.) *excusandas excusationes,* Ps. Spl. T. 140, 4. Cf. á-sceacan.

on-sceacness. v. preceding word.

on-sceamian, -sceoniendlíc, -sceonung. v. of-sceamian, on-scuniendlíc, -scunung.

on-sceortian *to grow short:*—Swá ða dagas forþ onsceortiaþ *as the days go on shortening,* Shrn. 96, 3. Cf. á-sceortian.

on-scunian, -scynian, -sceonian. **I.** *to regard with loathing, to abhor, detest, execrate:*—Ic onscunige (-sceonige) *abhominor, detestor,* Ælfc. Gr. 25 ; Som. 26, 63. Drihten onscunaþ (*abominatur*) ealle ðás þing, Deut. 18, 12. Ealle Egiptisce onscuniaþ (*detestantur*) scéphyrdas, Gen. 46, 34. Onscuniaþ *abhominentur,* Wrt. Voc. ii. 3, 59. Ic hit swíðe onscunode *multum detestans,* Bd. 3, 17 ; S. 545, 3. Word mín onscunedon (*execrabantur*) wið mé, Ps. Surt. 55, 6. Ða anscunedon hiene his ágene leóde, and monige from him cirdon, Ors. 3, 11 ; Swt. 152, 12. Ðonne hé biþ æfstig wið óðra manna yfelu anscunige eác his ágenu *cum contra aliena vitia aemulatur ostenditur, quae sua sunt, exequatur,* Past. 13, 2 ; Swt. 79, 12. Ða wæs ic ðæt swíðe onscuniende, and mé láþ wæs, Bd. 5, 12 ; S. 630, 32. Onscunigende gefeoht *exosus bella,* Ælfc. Gr. 41 ; Som. 44, 12. Onscunede *exosam,* Wülck. Gl. 55, 18. **II.** *to regard with disfavour, to refuse, reject, shun:*—Ǽlc gesceaft onscunaþ ðæt ðæt hire wiðerweard biþ *quae sunt adversa, depellit,* Bt. 16, 3 ; Fox 56, 4. Se ðe ðis gewrit gehýreþ hé flýhþ ðæt yfel and onscunaþ *devitando quod noxium est,* Bd. pref. ; S. 471, 16. Ðá onscunode hé ðæt and cwæþ *qui renuens ait,* Gen. 48, 19. Hé onscunede unrihthǽmed *ille recusabat stuprum,* 39, 10. Gé onscunedon (*rejected*) ðone Scippend, and gedwolan fylgdon, Elen. Kmbl. 739 : El. 370. ,Onscuna ðú leásunga (cf. fleóh leasunga, Ex. 23, 7), L. Alf. 44 ; Th. i. 54, 14. Godes willan onscunian *Dei voluntate resistere,* Gen. 50, 19. Heora ealde þeáwas onscunian and forlǽtan *priscis abdizare moribus,* Bd. 2 ; S. 502, 35. Nis ná tó onsceoniene seó sóðe gecyrrednys *non est rejicienda vera conversio,* L. Ecg. P. i. 2 ; Th. ii. 172, 10. **III.** *to regard with fear:*—Ondrǽdeþ ł onscynaþ *formidet,* Jn. Skt. Lind. 14, 27 : *metuit,* Rtl. 125, 25. Onscyniaþ *opriant* (= *aporiant*), Wrt. Voc. ii. 65, 16. Onscunede *exorruit,* 33, 14. Ðeáh hí men oððe hundas wið eodan, hí hí ná ne onscunedon . . . and nán heort ne onscunode nænne león *though men or dogs went against them (wild beasts), they were not afraid of them . . . and the hart was not afraid of the lion,* Bt. 35, 6 ; Fox 168, 2–9. Gif ðú heora untreówa onscunige *si perfidam perhorrescis,* 7, 2 ; Fox 18, 8. Onscunien (*revereantur*) feónd míne, Ps. Surt. 34, 4, 26. Se onscunienda þystel *carduus orrens,* Wrt. Voc. ii. 22, 43. Onscuniende *aporians,* 2, 23 : 4, 74. Onscunigende, 78, 30. Anscunendi *aporiens,* 100, 41. **IV.** *to irritate:*—Ábealg *vel* onscunede *exacerbavit, i. provocavit, adflixit,* 144, 56. Onscynedun *exacervaverunt,* Ps. Surt. 106, 11.

on-scuniendlíc, -scunigendlíc, -sceoniendlíc; adj. *Abominable, detestable, execrable:*—Onscunigendlíc *þerosus,* Ælfc. Gr. 33 ; Som. 36, 62 : *detestabilis,* Bd. 3, 9 ; S. 533, 9. Cristendóm wæs ðær onscunigendlíc, Homl. Skt. i. 2, 330. Onscuniendlíc *execrandum,* Wrt. Voc. ii. 33, 20. Ða onscuniendlecan *execranda,* 33, 5. Onscuniendlícan *probrosas,* 66, 31. Onscuniendlíce (*abominabiles*) gewordene synd, Ps. Spl. 13, 2 : Ps. Surt. 52, 2. Onsceoniendlíce, Ps. Th. 52, 1. Ða anlícnessa ealra onscuniendlícra niétena *omnis animalium abominatio,* Past. 21, 3 ; Swt. 155, 14. Anscunigendlícra (anscunigendra, Cott. MSS.), Swt. 153, 22.

on-scunung, -sceonung, e; f. **I.** *abomination, execration:*—Of onscununge *execratione,* Ps. Spl. C. 58, 14. Hí setton mé on onscununge (*abominationem*) him, 87, 8. Ðonne gé geseóþ ða onsceonunge (*abominationem*) ðære tóworpennysse, Mt. Kmbl. 24, 15. **II.** *irritation, exasperation:*—In onscununge *in exacervatione,* Ps. Surt. 94, 9.

on-scynian. v. on-scunian.

on-scyte, es; m. **I.** *an attack, assault:*—Salomon ðæt mǽre hús Gode betǽhte, him and his folce tó gescyldnysse wið ælces yfeles onscyte *as a protection against the assault of every evil,* Homl. Th. ii. 578, 23. **II.** *an attack in words, a calumny, backbiting:*—Mǽst ælc óðrum derede wordes and dǽde ; and hūru unrihtlíce mǽst ælc óðerne æftan heáweþ mid scandlícan onscytan [and mid wróhtlácan, MS. E.], Wulfst. 160, 5. For ídelan onscytan hý scamaþ, ðæt hý bétan heora misdǽda, 165, 7.

on-sécan *to require something* (gen.) *of a person* (acc.) :—Ne onsêcþ *non quaeret*, Ps. Spl. T. 9 second, 4. Đǽr .xxx. wæs and feówere eác feores onsôhte þurh wǽges wylm *then was life required of thirty-four by the rage of the wave* (cf. *under* ā-sécan, Ps. 118, 95), Exon. Th. 283, 13; Jul. 679.

on-secgan. I. *to sacrifice, offer* :—Ic onsecge *sacrificabo*, Ps. Surt. 53, 8. Ic ðé tifer onsecge, Ps. Th. 65, 12. Gif man medmycles hwæthwega deóflum onsægþ (*immolaverit*), L. Ecg. C. 32; Th. ii. 156, 15. Hé lác onsægde, Cd. Th. 107, 21; Gen. 1792. Hé gild onsægde, 172, 11; Gen. 2842. Hé lác onsægde (*of Christian service*), Exon. Th. 168, 28; Gú. 1084. Mesiane noldon ðæt Læcedemonia mægdenmenn mid heora oífreden and heora godum onsægden, Ors. 1, 14; Swt. 56, 16. Hié Gode eáðmódlíce lác onsægdon, Blickl. Homl. 201, 14. Onsecgaþ gé him mid sôðfæstnesse wæstmum, 41, 10. Ne yld ðú ðæt ðú ðám myclan godum mid ús onsecge *diis magnis sacrificare ne differas*, Bd. 1, 7; S. 477, 36. Se ðe godgeldum onsæcge ofer God ánne, L. Alf. 32; Th. i. 52, 12. Gif ðú onsecgan nelt sôðum godum, Exon. Th. 253, 3; Jul. 174. Ðu scealt Isaac mé onsecgan, Cd. Th. 172, 30; Gen. 2852. Ðǽm godum onsægdnisse onsæcgan *victimis placare*, Nar. 20, 5. Ongunnan heora bearn blôtan feóndum, sceuccum onsæcgean *immolaverunt filios suos, et filias suas daemoniis*, Ps. Th. 105, 27. Onsægd síe *turificatur*, Wrt. Voc. ii. 88, 51. [Cf. *O. H. Ger.* insaket *litat*; insaket *delibatus*; insaget þim *delibor, sacrificio*.] II. *to deny, renounce, abjure* (*O. H. Ger.* antsagén *renunciare, abjurare, excusare* : *Ger.* ent-sagen) :—Gif mon síe dumb oððe deáf geboren ðæt he ne mǽge his synna onsecggan (-sæcgan, MS. H.; ætsacan, MS. B.) ne andettan, bête se fæder his misdǽda, L. Alf. pol. 14; Th. i. 70, 15.

on-segedness. v. on-sægedness.

on-sendan. I. *to send off, despatch* (*an emissary*) :—Onsende *direxit*, Wrt. Voc. ii. 27, 19. Him his sunu hám onsende *filium remisit*, Ors. 4, 11; Swt. 206, 2. Hine God ús onsende, Beo. Th. 770; B. 382. Se ǽrest ár hider onsende, Andr. Kmbl. 3207; An. 1606. Ðá onbeád heó him ðæt hé hire tó onsænde all ða gesídwíf, Shrn. 87, 21. Ðæt hé Angelþeóde onsende láreówas, Bd. 2, 1; S. 501, 29. Hwylcne Arcebiscop hé onsendan mihte to Angolþeódes cyricum, 4, 1; S. 563, 29. Tó ǽlcum biscepstôle on mínum ríce ic wille áne (*a copy of the translation*) onsendan, Past. pref.; Swt. 9, 1. Onsended *distinatus*, Wrt. Voc. ii. 26, 64: 28, 15. Ic wæs hider onsended, Blickl. Homl. 9, 20. Hé wæs of heofenum onsended, 131, 13: Chr. 430; Erl. 10, 18. Onsendum gewritum *missis literis*, Bd. 2, 10; S. 512, 17. II. *to send forth* or *out*, (a) literal :—Ðǽr wǽron on carcerne ccxlviii wera and xlix wífa, ða Andreas ðanon onsende, Blickl. Homl. 239, 15. (b) metaph. *to emit* (*an odour*, etc.) :—Of ðǽre stôwe mycel swêtnes onsended wæs, Bd. 5, 12; S. 629, 35. Seó beorhtnes ðæs onsendan leóhtes, 4, 7; S. 575, 9. (c) *to send forth* (*the spirit*), *to give up* (*the ghost*) :—Sôna swá hé ðás word gecwæþ, hé his gást onsende, Blickl. Homl. 191, 29. Heó hire gást onsænde, and hire líchoma resteþ on Ðæssalonica ðære ceastre, Shrn. 70, 28. Heó onsænde hire gást tó Gode, 107, 31. Hé sceal þurh gáres gripe gást onsendan, Andr. Kmbl. 374; An. 187.

on-seón. v. an-sýn.

on-seón *to regard, look on* :—Wliteseón wrǽtlíc weras onsáwon, Beo. Th. 3305; B. 1650. Freónd onsigon (feónd onségon?) láðum eágan landmanna cyme, Cd. Th. 189, 2; Exod. 178. [*O. H. Ger.* ana-sehan *intueri* : *Ger.* ansehen.]

on-setenness, e; *f.* *Laying on, imposition* :—Ðæm gáste ǽghwelc gefullwad man onféhþ þurh biscopa handa onsetenesse, Shrn. 85, 19.

on-setness, e; *f.* I. (cf. settan) *constitution, appointment* :—From onsetnisse middangeardes *a constitutione mundi*, Lk. Skt. Lind. Rush. 11, 50. II. (cf. sittan) *ambush, artifice, plot* :—Allo onsetnisse fióndes *omnes insidias inimici*, Rtl. 121, 40. v. next word.

on-setnung, e; *f.* *Plot, wile* :—Onsettnungo diúbles *insidias diaboli*, Rtl. 147, 13.

on-settan *to oppress, bear down* :—Hé hig yfele onsette *vehementer oppresserat eos*, Jud. 4, 3. Ða Cristenan him mid heora wǽpnum hýndon and onsetton, Blickl. Homl. 203, 17. v. settan, on-sittan.

on-sícan; *p.* -sác *To sigh, groan* :—Ðá onsác se wísdóm and cwæð; Eálá, Bt. 26, 2; Fox 92, 24 : 40, 3; Fox 238, 7.

on-sién. v. an-sýn.

on-sígan. I. *of gentle, gradual movement, to sink, decline, descend* :—Ðonne se dæg gewit, and seó niht onsíhþ tó wérium mankynne, Anglia viii, 320, 2. Simbel onsáh dæg *sollempnis urgebat* (*vergebat?*) *dies*, Hymn. Surt. 96, 1. Ðeáh seó sunne ofer midne dæg onsíge and lúte tó ðære eorþan, Bt. 25; Fox 88, 25. Wǽre onsigen *vergeretur*, Wrt. Voc. ii. 81, 27. Onsígendum (*vergente*) ǽfene, Hymn. Surt. 34, 28. Fornumen mid onsígendre ylde *with declining years*, Basil admn. 8; Norm. 50, 20. II. *of violent movement* :—Gif hí oncneówon ða geníðerunge ðe him onsíhþ, Homl. Th. i. 408, 8. Swearte gástas mid micclum þreáte him onsigon, ðæt hí his sáwle gegripon, 414, 10. Hé bodode ðæt him wæs Godes grama onsígende, 246, 17. Mé wæs onsígende se stranga wynd, St. And. 28, 13. Onsígendum *ingruenti*, Hpt. Gl. 503, 32.

on-sín, -sién, -sýn, e; *f.* *Lack, want* :—Ðæt eów nǽfre ne biþ þurh gife míne gódes onsién, Exon. Th. 30, 16; Cri. 480. Him nǽnges wæs willan onsýn, ne welan brosnung, 151, 24; Gú. 800. Nis on ðæm londe ne sár wracu ne wǽdle gewin ne welan onsýn *luctus acerbus abest, et egestas obsita pannis*, 201, 13; Ph. 55. Ðǽr him nǽnges wæs eádes onsýn, 225, 32; Ph. 398.

on-síne, -sýne; *adj.* *Visible* :—Hé mé fore eágum onsýne wearþ, Exon. Th. 177, 17; Gú. 1228: Andr. Kmbl. 1820; An. 912. Cf. gesýne.

on-sinscipe (?), es; *m.* *Wedlock* :—Ðyssum mánfullum onsinscype (gesinscipum, MS. T.) wǽron sǽde gemengde *huic nefando conjugio dicuntur admixti*, Bd. 1, 27; S. 491, 22. [*Perhaps* on *should be written before* ðyssum *instead of being a prefix.* v. sinscipe.]

on-sittan. I. *to occupy* :—Ic onsitte *insideo*, Ælfc. Gr. 26; Som. 29, 6. Ðô þré acres ðe hé onsit, Cod. Dip. Kmbl. iv. 259, 20. Ðone hagan ðe hé sylf onsæt, 39, 13. Hit wæs his lǽn ðæt hé onsǽte, Chart. Th. 173, 5. Onsíte sǽnacan, Exon. Th. 474, 7; Bo. 26. II. *to oppress* (cf. colloquial *to sit on a person*) :—For ðám heardum weorcum ðe him onsæt, Ex. 6, 9. Gehreás ł onsæt egsa heora ofer hig *incubuit timor eorum super eos*, Ps. Lamb. 104, 38. Hé álýseþ þearfan ðæt him se welega ne mæg wiht onsittan *liberavit pauperem a potente*, Ps. Th. 71, 12. III. (with a different prefix, cf. *O. H. Ger.* int-sizzen *metuere* : *Goth.* and-sitan *to regard*) *to fear* (*taking like* ondrǽdan *a reflexive dative*) :—Nô ic mé onsitte *non vereor*, Wrt. Voc. ii. 61, 46. Ne ic mé herehlôþe helle þegna swíðe onsitte, Exon. Th. 166, 15; Gú. 1043. Ðæt is ðæt án ðæs ic eallan dæg mé onsitte, Homl. Skt. i. 23, 730. Hí onsǽton and ondrédon ðæt wé heom grame beón woldon, 23, 273. Godes him ondrédon hete, heofoncyninges níþ swíðe onsitte, Cd. Th. 48, 1; Gen. 769. Ðonne ðú ðé selfum swíðost onsitte, Met. 5, 38. Ðú ðé láðra ne þearft hæleþa hildþræce onsittan, Cd. Th. 130, 10; Gen. 2157: Beo. Th. 1198; B. 597: Exon. Th. 397, 22; Rä. 16, 23. Hé wæs him onsittende ðæt hine sum man gecneówe, Homl. Skt. i. 23, 494.

on-slǽge. v. on-slege.

on-slǽpan, -slépan; *p.* te *To sleep, fall asleep* ; obdormire :—Wérig gesette his leomu tó restenne and hwæthwugo onslépte (slép, MS. B.), Bd. 2, 6; S. 508, 11. Onslǽpte (slép, MS. B.), 4, 11; S. 579, 33 : S. 580, 2 : 4, 24; S. 597, 11 : S. 599, 7 : 4, 31; S. 610, 31. v. next word.

on-slápan; *p.* -slép *To sleep, fall asleep* :—Heó hwôn onslép, forðon ðe heó wæs on ðære sǽ swíðe geswenced, Shrn. 60, 17. Andreas ásette his heáfod ofer ænne his discipula and hé onslép, Blickl. Homl. 235, 13. [*Goth.* ana-slépan : *O. H. Ger.* int-slâfan : *Ger.* ent-schlafen.] v. ā-slápan *and preceding word.*

on-slege, es; *m.* *A blow struck on something* :—Onslægiun *inflictis*, Wrt. Voc. ii. 92, 53. Onslægum, 47, 27.

on-spannan. I. literally, *to unfasten, unclasp* :—Þegn winedryhten his wǽtere gelafede, and his helm onspeón, Beo. Th. 5440; B. 2723. II. metaph. *to open the mind, to speak, disclose the thoughts* :—Ongan reordigan, wordlocan onspeónn, Andr. Kmbl. 940; An. 471. Onspeón, 1342; An. 671 : Elen. 172; El. 86 : Exon. Th. 247, 16; Jul. 79.

on-sprǽc, e; *f.* *A suit involving a claim* or *accusation, claim, charge* :—Se môste his hláford áspelian, and his onspǽce gerǽcan, L. R. 3; Th. i. 192, 3. Bǽdon ðæt heó môsten gesyllan hire morgengyfe wið ðan ðe se cing ða egeslícan onspǽce álête (*the charge is previously stated* : ðæt hé wǽre on ðám unrǽde, ðæt man sceolde on Eást-Sexon Swegen underfôn), Chart. Th. 540, 21.

on-spreca, an; *m.* *One who brings a claim* or *charge* :—Ðone áþ funde ðe se onspreca (*claimant*) on gehealden wǽre, L. Ed. 1; Th. i. 158, 20. v. sprecan.

on-spreccan *to enliven, to make sprack* (?) [Sprack *lively, active,* Halliw. Dict.: *Icel.* sprækr *active.*] :—Ðá wæs wæstnum áweaht world onspreaht (-spreht, MS.), Exon. Th. 353, 8; Reim. 9. [Cf. (?) ich sprechi in ham sprekes of lustes swa luðere ðæt ha forberneþ, Marh. 15, 21.]

on-sprecend, es; *m.* *An accuser, plaintiff* :—Ðá ongon Higa him specan on mid ôðran onspecendan, Chart. Th. 169, 22.

on-springan. I. *to burst asunder* :—Seonowe onsprungon, burston bánlocan, Beo. Th. 1639; B. 817. II. *to spring* or *burst forth,* (of streams), *to rise* :—Ðǽr lagustreámas, wyllan onspringaþ, Exon. Th. 202, 2; Ph. 63. Ðǽr se flôd onsprang, Andr. Kmbl. 3269; An. 1637. Ealle eorþan ǽddre onsprungon ongeán ðam heofonlícan flôde, Wulfst. 206, 18. [*Ger.* ent-springen.]

on-sprungenness, e; *f.* *Defect, want* :—Onsprungennes *eclipsis, i. solis vel lunae defectio*, Wrt. Voc. ii. 142, 22. v. ā-sprungenness.

on-stæl, es; *m.* *Arrangement, disposition* :—Ðá (*at the creation of man*) wæs fruma níwe ælda tudres, onstæl wynlíc, fæger and gefeálíc fæder wæs ácenned Adam ǽrest, Exon. Th. 151, 17; Gú. 796. v. on-steall.

on-stæpe, es; *m.* *Entrance, ingress* :—Onstæpas *ingressus*, Ps. Spl. 67, 26.

on-stál, es; *m.* (?) *A charge, accusation:*—Gif Englisc onstál gá forþ . . . Gif hit biþ Wilisc onstál, L. In. 46; Th. i. 130, 15–16. Onstáles *invectionis, illationis*, Hpt. Gl. 448, 53. v. stǽlan.

on-steall, es; *m. Institution, provision:*—Gode ælmiehtigum sî þonc ðætte wê nû ǽnigne onstal habbaþ láreówa, Past. pref.; Swt. 4, 1. v. on-stæl *and* on-stellan.

on-stedfullness, e; *f. Instability:*—Onstydfullnisse *instabilitas*, Rtl. 192, 19.

on-stellan *to institute, give rise to, set on foot, bring in, be the author of, set (an example):*—Ðú scealt greót etan swá ðú wróhte onstealdest *thou (the serpent) hast brought sin into the world*, Cd. Th. 56, 12; Gen. 911: 57, 22; Gen. 932. Hé in wuldre wróhte onstalde, 287, 19; Sat. 369. Ða onstealdon ða heretogan ǽrest ðone fleám *the leaders were the first to fly*, Chr. 993; Erl. 132, 15. Swá hit (*persecution*) Nero onstealde, Ors. 6, 6; Swt. 262, 12. Créca gewinn ðe of Læcedemonia ǽrest onsteled (stæled, MS. C.) wæs *dominandi Lacedaemoniorum cupiditas, quantas causas certaminum suscitavit*, 3, 1; Swt. 100, 11. Hé wuldres gehwæs ord onstealde *omnium miraculorum auctor exstitit*, Bd. 4, 24; S. 597, 21. Se ðæs orleges or onstealde, Beo. Th. 4806; B. 2407. Ðe ðæs oferhýdes ord onstaldon, Cd. Th. 272, 4; Sat. 114. Abraham bysene onstealde geleáffullum *Abram exemplum credentium fuit*, Gl. Prud. 1: Blickl. Homl. 7, 9: 23, 16. Ða láreówas sceolan gódes lífes bysene onstellan ðǽm ðe him æfter fylgeon, 81, 6. Wolde ic eów on ðon bysne onstellan, Andr. Kmbl. 1942; An. 973. Ða godcundan leán mínre sáule mid gerêce, swǽ hit míne ærfenuman onstellen (*appoint*), Chart. Th. 477, 12. Onstaelde (ox-, Wrt.) *idoneus*, Wrt. Voc. ii. 110, 51.

on-stépan *to raise:*—Onstép mínne hige in gearone rǽd, Exon. Th. 454, 25; Hy. 4, 38.

on-steppan *to walk, go:*—Ðú onstæpst *gradieris*, Ps. Spl. 31, 10. Lege on lange hwíle óþ ðæt hé onstæppe, Lchdm. ii. 126, 17. v. steppan.

on-stígend, es; *m. One who ascends or mounts:*—Hors and onstígend áwearp in sǽ *equum et ascensorem projecit in mare*, Ps. Surt. ii. 187, 4.

on-sting, es; *m. Authority:*—Icc nelle geþafian ðæt ǽnig mann ǽnigne onstingc habbe on ǽnigum þingum oððe on ǽnigum tíman bútan se abbod, Chart. Th. 362, 3. Ǽnige onsting, 369, 24. Ic nelle geþafian ðæt ǽni man ǽnine onstyngc hæbbe *nolo permittere ut quis jus habeat*, Cod. Dip. Kmbl. iv. 202, 17. Ðæt gé nán onsting ne hauuen of ðat mynstre bútan swá micel swá ðone abbot wille *ut nec tu nec quisquam successorum episcoporum quicquam hujus aecclesiae usurpet praeter abbatis uoluntatem*, v. 29, 20.

on-stíran *to govern:*—Rícsiendum on ēcnysse and onstýrendum his cyrīcean ðám ilcan Drihtne *regnante in perpetuum et gubernante suam ecclesiam eodem Domino*, Bd. 4, 5; S. 572, 4.

on-stíðian *to make hard:*—Onstíðade (*induravit*) hiora hearta, Jn. Skt. Lind. Rush. 12, 40.

on-stregdan *to sprinkle:*—Ðú onstrigdes (*asparges*) mec mid ysopan, Ps. Surt. 50, 9.

on-styreness, e; *f. Movement:*—Nalæs ðæt án óðra lima ac swylce eác ðære tungan onstyrenesse *non solum caeterorum membrorum, sed et linguae motu*, Bd. 4, 9; S. 577, 17.

on-styrian I. *to move, stir (of physical motion):*—Se líchoma ná ne onstyreþ siððan seó sáwl him of biþ, Blickl. Homl. 21, 27. Onstyredan, drifan *agitabant*, Wrt. Voc. ii. 3, 39. Heó nǽnig lim onstyrian mihte, Bd. 4, 9; S. 577, 4. Onstyrgan (*commoveri*) foet míne, Ps. Surt. 65, 9. II. *to move, stir up, excite:*—Hí mycle fyrhto onstyredon ðám monnum ðe hí sceáwodon, Bd. 5, 23; S. 645, 23. Unsibbe onstyrian, Cd. Th. 281, 14; Sat. 271. III. *to move, disturb, agitate* (a person):—Seó gedréfednes mæg ðæt mód onstyrian, Bt. 5, 3; Fox 12, 24. Se ðe mæg eorþware onstyrian, Ps. Th. 98, 1. Hý fægniaþ gif ic onstyred beó *exultabunt si motus fuero*, 12, 5: 32, 7. Hwæt arun gē onstyred *quid turbamini*, Mk. Skt. Rush. 5, 39. Onstyred and onǽled mid andan, Past. 21; Swt. 159, 7: Blickl. Homl. 199, 16. Onstyred mid heora wordum, 225, 23. Eal seó burh wæs onstyred, 71, 13. Hié beóþ on heora móde mid mislícum geþohtum onstyrede, 19, 9.

on-sund; *adj.* I. *of persons, sound, whole, uninjured:*—Sum cild wearþ tó deáþe tócwýsed. Seó móder bær ðæs cildes líc tó ðám gemynde ðæs hálgan Stephanes, and hit sóna geedcucode and ansund æteówode, Homl. Th. ii. 26, 28. Onsund, Exon. Th. 278, 5; Jul. 593. Heó árás ansund of ðam bedde, Homl. Skt. i. 22, 52. Ábeád ðæt hié hine ealles onsundne eft gebrohten of ðære folcsceare *the king ordered that Abraham should be brought again out of Egypt safe and sound*, Cd. Th. 112, 15; Gen. 1871. Hé áléde his tunecan uppon ðám deádum, and hí ansunde árison, Homl. Th. i. 74, 3: Andr. Kmbl. 2023; An. 1014: 3244; An. 1625. II. *of things, sound, entire, perfect, without flaw or injury:*—Ne wearm weder ne winterscúr wihte gewyrdan, ac se wong seómaþ onsund, Exon. Th. 199, 3; Ph. 20: 200, 21; Ph. 44. Nán cyneríce ne stent náne, hwíle ansund, gif hí gesóme ne beóþ, Homl. Skt. i. 13, 238. For ðære clǽnnysse his ansundan mægþhádes. Hé on ēcnysse on ungewemmedum mægþháde þurhwunode, Homl. Th. i. 58, 7. Hine getác-

node God tó ansundre hǽle, ii. 512, 13. Ansundre *integro*, Hpt. Gl. 525, 61. God hine (*Enoch*) genam mid ansundum líchaman of ðissum lífe, Ælfc. T. Grn. 3, 42. Ðenden gǽst and líc geador síþedan onsund on earde, Exon. Th. 285, 16; Jul. 715. Ðá wurdon ða gymstánas swá ansunde, ðæt nán tácen ðære ǽrran tócwýsednesse næs gesewen, Homl. Th. i. 62, 16. v. an-sund.

on-sundness, e; *f. Soundness, freedom from physical or moral flaw:*—Andsumnysse (ansundnysse?) *integritatis, virginitatis*, Hpt. Gl. 444, 53. Gefēg ðás bricas tó ansundnysse *join these broken gems together so that they may again be whole*, Homl. Th. i. 62, 8. v. ansundness.

on-sundrian *to separate, take apart:*—Nǽnig heora, of ðám ðe hí áhton, ówiht his beón onsundrad cwæþ *none of them said that anything they owned was his separate property*, Bd. 1, 27; S. 489, 15 note. v. á-sundrian.

on-sundrum, -sundran, -sundron; *adv.* I. *separately, severally, separated one from the other, apart:*—Onsundron *separatim*, onsundron hé sit *singillatim sedet*, Ælfc. Gr. 38; Som. 40, 39. Onsundran *altrinsecus, hinc et inde*, Hpt. Gl. 410, 2. Uton biddan onsundron æt Gode, ic æt mínum Gode . . . and gé eác swá dón *let each pray severally to his own God*, Homl. Skt. i. 18, 107. Ðá ná gestód hé ná ǽlcne onsundran (*each separately*), 23, 177. Hié sǽton onsundran *they (Adam and Eve) did not sit together*, Cd. Th. 52, 11; Gen. 842. Stande hé ealra ýtemest, oððe on ðam stede, ðe se abbod swá gémeleásum monnum tó stealle onsundrum betǽht hæfþ . . . Wé forðí tǽhton ðæt hý on úteweardan oððe onsundrum standen, ðæt . . ., R. Ben. 68, 10–17. Nǽnig heora, of ðám ðe hí áhton, ówiht his beón onsundran cwæþ, Bd. 1, 27; S. 489, 15. Deáh bútú on ánum men síen, ðeáh biþ ǽgðer him onsundron, Bt. 16, 3; Fox 54, 35. Ǽlc ðæra gesceafta hæfþ his ágenne eard onsundron, 33, 4; Fox 130, 24. For ǽghwylc onsundran riht ágieldan, Exon. Th. 372, 24; Seel. 97. II. *in retirement from others, apart:*—Ðá fērde hé onsundron *secessit*, Mt. Kmbl. 14, 13. Hé lǽdde hig onsundron (*seorsum*), 17, 1. Hé nam his leorningcnihtas onsundron *assumsit discipulos secreto*, 20, 17. Uton gán onsundron (*seorsum*) . . . Hí fóron onsundran, Mk. Skt. 6, 31–32. III. *making distinction from others, especially:*—Ic onsundrum ða stówe lufige, and ofer ealle óðre ic hié geceás, Blickl. Homl. 201, 6.

on-swǽtende, Wrt. Voc. ii. 47, 31. v. sprecan.

on-swápan. v. swápan.

on-swebban *to put to sleep (but generally of the sleep of death), lay to rest (in the grave):*—Onsuebbaþ *sepeliant*, Wrt. Voc. ii. 120, 44. Onsuebdum *sopitis*, 120, 73. [O. *Sax.* an-swebian: O. H. Ger. in-, intsueppen *sopire*.] v. á-swebban.

on-swégan. v. swégan.

on-swífan. I. *to swing, turn:*—Bordrand onswáf wið ðam gryregieste Geáta dryhten *Beowulf turned his shield against the approaching fire-drake*, Beo. Th. 5112; B. 2559. II. *to turn aside, divert:*—Ne mæg mon ǽfre ðý ēð ǽnne his cræftes beniman, ðe mon oncerran mæg sunnan onswífan and ðisne swiftne rodor of his rihtryne, Met. 10, 40.

on-swógan. v. swógan.

on-symbelness, e; *f. A solemn festival:*—Mid ðý heó gesǽgon ðone biscop mæssan (mæssena, MS. B.) onsymbelnesse mǽrsian (*celebratis missarum sollemniis*), Bd. 2, 5; S. 507, 12.

on-sýn. v. on-sîn.

on-talu, e; *f. A successful claim, a charge that is established.* v. of-talu.

on-ténan. v. on-týnan.

on-tendan; *p.* -tende; *pp.* -tended, -tend. I. *to kindle, set fire to, to fire:*—Gif fýr síe ontended . . . gebēte ðone æfwerdelsan se ðæt fýr ontent, L. Alf. pol. 27; Th. i. 50, 27–28. Ontend þreó candela, Lchdm. iii. 286, 6. Ðe ðæt fýr ontende *qui ignem succenderit*, Ex. 22, 6. Ða hálgan tihton ðæt man ða ofnas ontende (-tænde, MSS. C. V.), Homl. Skt. i. 5, 294. Hí on ða burh feohtende wǽron, and eác hí mid fýre ontendan woldon, Chr. 994; Erl. 133, 12. Ðoane hé (*the moon*) of hyre (*the sun*) ontend byþ, Lchdm. iii. 242, 12. Ontend *succensus*, Ps. Spl. 507, 17. Antend, 471, 22. Antendne *succensam, ardentem*, 464, 36. II. *to kindle emotion or passion, to excite, inflame:*—Sume se deófol ontent tó gýtsunge, Homl. Th. i. 240, 25. Ðíne gebedu geancsumiaþ mē and ontendaþ, 458, 4. [O. E. Homl. A. R. Jul. Marh. ontenden: Goth. tandjan *to kindle, light;* in-tandjan *to consume with fire*.]

on-tendness, e; *f.* I. *a burning, fire:*—Hé hēt gearcian ða tunnan tó heora bærnette . . . Hí wurdon gebrohte tó ðám tunnum and tó ðære ontendnysse, Homl. Skt. i. 4, 307. Ontendnyssa *incendia*, Hpt. Gl. 499, 42. Antendnyssum *globis*, 489, 68. Ontyndnissum *incendiis*, 440, 4. II. *metaph. fire, that which kindles passion:*—Eugenia cwæþ ðæt heó wǽre gálnysse ontendnyss, Homl. Skt. 2, 173. III. *passion, vehement desire:*—Ðære forīgerlícere ontendnysse *adulterinae titillationis*, Hpt. Gl. 505, 68. (yndnysse, 520, 33. Hé uneáðe ðære líchamlícan ontendnysse wiðstandan mihte, Homl. Th. ii. 156, 26.

Geangsumod mid ðæra ormǽtan ontendnysse and hrýmende . . . 'For-gif mē ðam men ðe mín mód mē tō spenþ, Homl. Skt. i. 3, 387 : 3, 397. Ic on ðǽ ádwesce ealle ontendnysse, 4, 171. **IV.** *burning sensation, inflammation :*—Hē unscrýdde hine ealne, and wylode hine sylfne on ðǽm þiccum bremlum and þornum and netelum . . . and swá þurh ðære hýde wunda ádwæscte his módes wunda; for ðan ðe hē áwende ðone unlust tó sárnysse, and þurh ða ýttran ontendnysse ácwencte ða inran, Homl. Th. ii. 156, 27–33.

on-þanc, Wrt. Voc. ii. 132, 40. v. or-þanc.

on-þenian *to extend, stretch, bend (a bow) :*—Hí onþeneden (*intende-runt*) boga, Ps. Spl. T. 63, 3. v. á-þenian.

on-þeón. **I.** *to prosper :*—Se wæs wreccena wíde mǽrost ofer werþeóde wígendra hleó ellendǽdum ; hē ðæs ǽr onþáh (*so at first he prospered*), Beo. Th. 1805; B. 900. **II.** *to be successful in one's efforts, to prove serviceable :*—Gamele ne móston hilde onþeón *the aged might not be of service in battle* (in the preceding lines it is mentioned that the very young were excluded from the army), Cd. Th. 193, 5 ; Exod. 241. Oft ic secga seledreáme sceal fægre onþeón ðǽr guman drincaþ *oft must I prove of excellent service to festivity in hall, where men drink*, Exon. Th. 480, 14; Rä. 64, 2.

on-þracian (-þrácian ?); *p. ode To fear, dread :*—Ic anþracige (and-, MS. F. : á-, MS. O.) *vereor*, Ælfc. Gr. 27 ; Zup. 162, 1. Ic andþracige (onþracie, MS. T.) *horreo*, ic onginne tó onþracigenne (and-) *horresco*, 35 ; Zup. 212, 3–4. Sum dēma wæs se God ne ondréd ne nánne man ne onþracude (*reverebatur*) . . . Ðá cwæþ hē : Ðeáh ic God ne ondrǽde ne ic man ne onþracige (*revereor*), Lk. Skt. 18, 2–4. Ðú ne onþraced-est (*horruisti*) mǽdenes innoþ, Hymn. L. 16. Ða ðe middaneard anþracode (*inhorruit*) Hymn. Surt. 132, 10. Ðá wearþ hē mycclum áfyrht and anþracode ðæt his ríce feallan sceolde, Homl. Th. i. 82, 5. Anþracian *revereantur*, Ps. Spl. 69, 2.

on-þræc (-þrǽce ?); *adj. Horrible, dreadful :*—Iulianus mid anþrǽcum hreáme forswealt, Homl. Th. i. 452, 16. Mid unásecgendlícum wítum áfyllede and mid anþrǽcum stencum, 68, 6. Ðá cwæþ ðæt wíf betwux ðám anþrǽcum wítum, Homl. Skt. i. 12, 191.

on-þringan (?) :—Beofode ðæt eálond foldwong onþrong [onsprong *the earth cracked with the shaking* (?)], Exon. Th. 181, 28 ; Gú. 1300.

on-þunian (?) *to swell out, exceed due bounds :*—Ic eom ufor ealra gesceafta ðára ðe worhte Waldend úser ; se mec ána mæg geþeón þrymme, ðæt ic onrinnan (onþunian *is suggested by Grein*) ne sceal, Exon. Th. 427, 15; Rä. 41, 91.

on-þweán *to wash, cleanse by washing :*—Wē nǽron mid fulwihte hér on eorþan onþwægen, Shrn. 53, 21. Gif gē willaþ onþwegene beón *si vultis ablui*, Bd. 2, 5 ; S. 507, 16.

on-þyncan *to seem, appear :*—Ðý læs ðæt eów seó sægen monigfeald-lícor biþ onþúhte tó wrítanne *ne sim scribendi multiplex*, Nar. 3, 29.

on-tígan *to untie, set free :*—Seó sáwl færþ swíðe fredlíce tó heofonum siððan heó ontíged biþ and of ðam carcerne ðæs líchoman onlíesed biþ, Bt. 18, 4; Fox 68, 14.

on-tige. v. on-tyge.

on-timber, es; *n.* **I.** *material :*—Ðæt óðer antimber *materia*, Wrt. Voc. ii. 57, 42. Ðæt antimber ðe hē of gesceóp gesceafta, Hexam. 4; Norm. 6, 22. Nis hit nán wundor ðeáh mon swilc ontimber gewirce, Shrn. 164, 1. **II.** *metaph. reason, occasion ; materia :*—Swilce him gerýmed sý and antimber geseald, ðæt hē God bereáfige, Lchdm. iii. 444, 1. For ðisum antimbre ic gedyrstlǽhte ðæt ic ðás gesetnysse undergann, Homl. Th. i. 2, 26. v. and-timber.

on-timberness, e ; *f. Instruction :*—Tó ontimberness ðæra æfterfyl-igendra *ad instructionem sequentium*, Bd. 4, 17; S. 585, 16. v. next word.

on-timbran *to instruct, edify :*—Hē monig þúsendo heora mid sóð-fæstnesse worde wæs ontimbrende (*instituens*), Bd. 5, 19 ; S. 639, 23. Æþellíce ontimbred and gelǽred *nobiliter instructus*, 5, 23 ; S. 646, 19 : 5, 19 ; S. 637, 36 : 5, 22 ; S. 644, 18. Hē wolde mid his láre and mid his lífes bysene beón ontimbred, Blickl. Homl. 217, 14.

on-tíned (-tímed ?) *well-supplied :*—Gif .vii. dæge sunne scíneþ, mycele wæstmas on treówum beóþ . . . Gif ðí .x. dæge sunne scýneþ, ðonne byþ sé and ealle ǽá mid fixum ontíned, Lchdm. iii. 166, 13.

ontre, an ; *f. Radish,* Lchdm. ii. 78, 26 : 76, 5. v. antre.

on-tydran *to nourish, support :*—Hú þyncþ eów hú seó sibb gefæstnad wǽre, hwæðer hió síe ðæm gelícost ðe mon nime ǽnne eles dropan, and drýpe on án micel fýr, and þence hit mid ðæm ádwæscan ? ðonne is wén, swá micle swíðor swá hē þencþ ðæt hē hit ádwæsce, ðæt hē hit swá micle swíðor ontydre *pax ista an incentivum malorum fuit ? stillicidium illud olei, in medium magnae flammae cadens exstinxit fomitem tanti ignis, an aluit ?* Ors. 4, 7 ; Swt. 182, 22–26.

on-tydre ; *adj. Weakened, debilitated, effete :*—Ontudri *effetum*, Wrt. Voc. ii. 106, 82. Ontydre *effeto, i. sine foetu, ebetato, debilitato, evacuato, exinanito,* 142, 46.

on-tyge, es; *m. What one takes upon one's self, an undertaking :*—Gif hwylc abbod geþafaþ ðæt mæssepreóst oððe diácon in tó mynstre gange tó ðý ðæt hý messan singan . . . hý nán þing gedyrstlǽcen ne

nǽnne ontige on ðam mynstre bútan ðære mæssan ánre *if any abbot per-mit a mass-priest or deacon to enter a monastery for the purpose of celebrating mass . . . they shall not presume to do anything or take any-thing upon themselves except only the mass*, R. Ben. 140, 3–10. v. teón (on), tyge.

on-tyhtan *to incite, instigate, impel :*—Wæs ðæt gifeðe tó swíð ðe ðone ðyder ontyhte, Beo. Th. 6164; B. 3086.

on-týnan *to open.* **I.** of places or things, (a) *to make an opening in :*—Seó eorþe hié ontýnde and hió forswealh ðæt wæter, Blickl. Homl. 247, 15. (b) *to open, allow to burst forth :*—Hē ús ontýneþ heofones þeótan, 39, 31. (c) *to open so as to admit of ingress or egress :*—Him se áwyrgda ongeán helle ontýneþ, Exon. Th. 364, 10 ; Wal. 68. Ðæt hí Godes cyricean ontýndon (*aperirent*), Bd. 3, 30 ; S. 562, 16. Ðē is ñeorxna wang ontýned, Andr. Kmbl. 209 ; An. 105. (d) *to open (a door) :*—Geatu ontýnaþ, Exon. Th. 36, 15 ; Cri. 576. Gif sió duru ontýned biþ . . . búton ðú ða duru ontýne, Past. 21 ; Swt. 157, 15–19. Síe manna gehwam behliden helle duru, heofones ontýned, éce geopenad engla ríce, Elen. Kmbl. 2458; El. 1230. Heofonríces duru sceal þurh ðē onténed beón, Blickl. Homl. 9, 3. (e) *to open* the mouth, lips, *to speak :*—Ic antýne (ontýne, MS. C. T.) on bigspellum múþ mínne, Ps. Spl. 77, 2. Ðá ontýnde Hǽlend his múþ, and wæs sprecende, Blickl. Homl. 159, 25. Ontýn weoloras míne, Ps. Grn. ii. 279, 116. (f) *to open* the eyes (one's own), *to look,* (another's), *to give sight to :*—Ðá ontýnde ic míne eágan, lócade on hine, Bd. 5, 6 ; S. 619, 39. Hē his eágan ontýnde, biseah tó heofones ríce, Exon. Th. 180, 6 ; Gú. 1275. Þweah ða eágan and ontýne, Lchdm. ii. 26, 25. Blindra manna eágan ontýnan, Jn. Skt. 10, 21. Hyra eágan wǽrun ontýnede, Mt. Kmbl. 9, 30. (g) *to open* the ears, *to listen to a person :*—Hē him mildheort-nesse eáron ontýnde, Blickl. Homl. 107, 1. Ontýn eárna hleóþor, Ps. Grn. ii. 278, 77. **II.** *to disclose, reveal, display :*—Se ðe líf on-týneþ, Exon. Th. 2, 15 ; Cri. 19. Forðæm wæs gecweden tó ðæm lyteg-an feónde ðe ðæs ǽrestan monnes mód ontýnde on ðæs æples gewil-nunge *unde hosti callido, qui primi hominis sensum in concupiscentia pomi aperuit*, Past. 43, 2 ; Swt. 309, 17. Þín tunge ontýnde fácn, Ps. Th. 49, 20. Ðú mē ðíure snetera hord selfa onténdes, Ps. Grn. ii. 278, 71. David his synna hord selfa onténde, 277, 28. Hwonne ús líffreá leóht ontýne, Exon. Th. 2, 31 ; Cri. 27. Ðæt ic móte ðis gealdor tóþum ontýnan *that I may utter this incantation*, Lchdm. i. 400, 5. Ús is wuldres leóht ontýned, Cd. Th. 299, 28 ; Sat. 557 : Exon. Th. 102, 17; Cri. 1674; Andr. Kmbl. 3222 ; An. 1614. Ðǽr is wuldres bléd ontýned, Cd. Th. 302, 5 ; Sat. 594.

on-týnan (?) *to cover :*—Ða stówe wæs ontýnende *ea loca operiens* (did the translator read *aperiens* ?), Bd. 4, 7 ; S. 575, 12.

on-tyndness. v. ontendness.

on-týnness, e ; *f.* **I.** *an opening, aperture :*—Se heofon tóbyrst and eall engla cynn lóciaþ þurh þa ontýnnesse on manna cynn, Blickl. Homl. 93, 24. **II.** *discovery :*—Be cierlisces monnes ontýnnesse (betogenesse, MSS. B. H. Schmid takes *ontýnesse = ontigenesse*, and Thorpe translates ' of accusing a "ceorlish" man ;' but the section deals with the discovery of the theft. Cf. too, L. In. 18 ; Th. i. 114, 5, which is a section to the same effect as the present one : Be cirliscum þeófe gefongenum) æt þíefþe. Se cierlisca mon se ðe oft betygen wǽre þíefþe, and ðonne æt sídestan synnigne man gefó, L. In. 37 ; Th. i. 124, 20.

on-ufan ; *prep. with dat. adv.* **I.** *of place, upon, on :*—Ðæt preóst ne mæssige búton onufan gehálgedon weofode, L. Edg. C. 31 ; Th. ii. 250, 22. Ða forwurdon ðe him (*the elephant*) onufan wǽron, Ors. 4, 1 ; Swt. 156, 13. Ða men ðe him onufan gáþ, Lk. Skt. 11, 44. Hí ðone Hǽlend onufan setton, 19, 35. **II.** of time, *beyond, after :*—Fór Eádweard cyning onufan hærfest, Chr. 923; Erl. 110, 1. v. ufan.

on-unspéd (?). e ; *f. Indigence, poverty :*—For hwý ofergytest onun-spéde (*inopiae*) úre, Ps. Spl. 43, 27.

on-unwis (?); *adj. Foolish, ignorant :*—Wiðmeten is nítenu[m] onunwísum (*insipientibus*), Ps. Spl. 48, 12. v. next word.

on-unwísdóm, es ; *m. Folly, ignorance :*—Ic wæs unwísum nētenum gelíc geworden. Ac ðú Drihten onunwísdómes ne wes ðú gemyndig, Blickl. Homl. 89, 10. v. preceding word.

on-uppan ; *prep. with dat. adv.* **I.** *upon, on :*—Se Hǽlend rád onuppan ðam assan, Jn. Skt. 12, 14. Stód ǽren ceác onuppan twelf ǽrenum oxum, Past. 16 ; Swt. 105, 2. Hē wearþ bebyrged, and him læg onuppan fela byrdena eorþan, Homl. Skt. i. 12, 56 : 14, 114. Hē sæt ðǽr onuppan, 13, 25. Ðonne man bringe offrunge nime smedeman and geóte ele onuppan, Lev. 2, 1. Hí lédon hyra reáf uppan hig, and setton hyne onuppan, Mt. Kmbl. 21, 7. Hí gemétton fýr and fisc on-uppan, Homl. Th. ii. 292, 4. **II.** *besides, over and above :*—Hē hét ácwellan ða rícostan witan, and onuppan ágenne bróðor and his módor ofbeátan, Met. 9, 28. v. uppan.

on-wacan. **I.** *to awake, cease to sleep :*—Sóna ðæs ðe heó on-wóc *ubi vigilavit*, Bd. 3, 9 ; S. 534, 11 : 4, 31 ; S. 610, 37. Ðá of slǽpe onwóc, swefn wæs æt ende, eorþlíc æðeling, Cd. Th. 249, 2 ; Dan.

524. Se wyrm onwóc, Beo. Th. 4563; B. 2287. Đa men onwócan, and ût urnon, Ors. 4, 2; Swt. 160, 22. II. *to arise, spring, be derived, be born :*—Đû wâst đæt đû of mînre (*the speaker is Eve*) dehter, Drihten, onwóce, Blickl. Homl. 89, 20: Cd. Th. 292, 12; Sat. 439. Hêr Ida fêng tô rîce, đonon Norþanhymbra cynecyn onwóc, Chr. 547; Erl. 16, 8. Him onwóc heáh Healfdene, Beo. Th. 112; B. 56. Beornas onwócan, cynn æfter cynne cende wæron, Ps. Th. 104, 11. Hwǽr ûs hearmstafas onwócan, Cd. Th. 58, 2; Gen. 940. Hié begeton feówertig bearna đæt đonon menio onweócon, 294, 25; Sat. 476.

on-wacan, -waccan, e; *f. An awakening, arousing, incentive :*—Onwaccano mægna *incitamenta virtutum*, Rtl. 74, 24. v. wacan.

on-wacnian; *p.* ode *To wake up, rouse one's self :*—Onwacnigeaþ nû, wîgend mîne, Fins. Th. 28; Fin. 10. v. on-wæcnan.

on-wadan. I. *to make one's way into, to penetrate :*—Oft hira mód onwód under dimscûan deófles lârum, Andr. Kmbl. 280; An. 140. II. *to enter with irresistible force, to make one's self master of, take possession of :*—Wîfa wlite onwód folcdriht wera *the beauty of the women made its way to the hearts of the men*, Cd. Th. 76, 20; Gen. 1260. Hié wlenco onwód, 155, 27; Gen. 2579. v. an-wadan.

on-wǽcan *to soften, mollify, cause relaxation of severity :*—Đæt wé mihtiges Godes mód onwǽcen, Cd. Th. 26, 7; Gen. 403.

on-wǽcnan; *p.* ede. I. *to awake :*—Hit ne onwæcneþ tô đon đæt hit eft on ierne mid hreówsunga. Ac hit wilnaþ đæt hit tô đon onwæcne, đæt hit mǽge eft weorþan oferdruncen, Past. 56; Swt. 431, 22–25. Đonne onwǽcneþ eft winleás guma, Exon. Th. 289, 8; Wand. 45. Đá hí onwǽcnedun *vigilantes*, Lk. Skt. 9, 32. 'Nû us is tîma đæt wé onwǽcnen of slǽpe.' Đon eft hé cwiþ: 'Onwǽcnaþ, gé ryhtwîsan,' Past. 63; Swt. 459, 33–461, 1. Fordytte đæt eáre mid đǽre wulle đonne đû slápan wille, and dó eft of đonne đû onwǽcne, Lchdm. ii. 42, 26. II. *to rise, spring, be derived :*—Đonne hé (*the Phenix*) of ascan onwǽcneþ, Exon. Th. 240, 34; Ph. 648. Monig sceal siđđan wyrt onwǽcnan, 191, 4; Az. 83. Đanon ǽtorcyn ǽrest gewurdon onwǽcned, Salm. Kmbl. 439; Sal. 220. v. next word.

on-wǽcnian, -wecnian; *p.* ode *To awake, arise, be roused, be raised :*—Of mistlícum dryncum onwǽcnaþ sió wóde þrág đære wrǽnnysse, Bt. 37, 1; Fox 186, 17. Đonne (*at the sound of the archangel's trumpet*) of đisse moldan men onwecniaþ, deáđe of duste árîsaþ, Cd. Th. 302, 23; Sat. 604. v. on-wacnian, -wæcnan.

on-wǽmme. v. un-wemme.

on-wǽre (?) *unripe :*—Genim onwǽre sláh đæt seáw, and wring þurh cláþ on đæt eáge, sóna gǽþ of (*the white spot will go off*) gif sió sláh biþ grêne, Lchdm. ii. 32, 18.

on-wǽstm *increase, increment :*—Onwǽstem *incrementum*, Rtl. 69, 19.

on-wǽterig. v. un-wæterig.

on-wald, -walh, -wealh. v. on-weald, -wealh.

on-wealcan; *v. trans. To roll :*—Dryhtnes bibod geofonflóda gehwylc georne behealdeþ đonne merestreámas wæter onwealcaþ *each ocean flood carefully observes the Lord's command, when the sea-streams make the water roll*, Exon. Th. 193, 25; Az. 127. Cf. Sóna swá đû geseó đæt đû hyre (*the mandrake*) geweald hæbbe, genim hý sóna on hand, swá andwealc hí, and gewring đæt wós of hyre leáfon, Th. An. 116, 22.

on-weald, es; *m. Power :*—Sý him ár and onwald *to him be honour and power*, Exon. Th. 241, 28; Ph. 663. Hié hiere onwaldes hié (*Rome*) beniman woldon; and heó hwæđere onwealg on hiere onwalde æfter þurhwunade, Ors. 2, 1; Swt. 62, 22–24. Se geeode đæt eálond and Rómāna onwealde underþeódde, Bd. 1, 3; S. 475, 18. Ne lǽt áwyrgde ofer ûs onwald ágan, Exon. Th. 10, 28; Cri. 159. Đa kyningas đe đone onwald hæfdon đæs folces . . . hié heora onweald gehióldon, Past. pref.; Swt. 3, 5–7. Đû áhtest alra onwald (*power over all*), Cd. Th. 268, 24; Sat. 60. Đæt gé min onweald ágan mósten, Exon. Th. 131, 9; Gú. 453. Ûs álêfan êcne onwald, Cd. Th. 272, 11; Sat. 118. Wé hine oferswýđdon and ûs in onweald geslógon eal his londrîce *regi superato acceptaque in conditiones omni ejus regione*, Nar. 3, 22. Wé ealle his þeóde on onwald onfêngon, 4, 6. v. an-, and-weald, on-wealda.

on-weald (?); *adj. Powerful :*—Đá Dryhten of deáþe árás onweald (-wealh?) of eorþan, Exon. Th. 168, 9; Gú. 1075.

on-wealda, an; *m. One who has power, a ruler :*—Ic gelýfe in êcne onwealdan ealra gesceafta, Exon. Th. 140, 14; Gú. 610. [*O. H. Ger.* ana-walto: *Ger.* an-walt. In *O. H. Ger.* ana-walton (*gen. dat. pl.*) also translates *potestatum, potestatibus*, v. Grff. i. 813. Cf. on-weald.] v. an-wealda.

on-wealg. v. next word.

on-wealh, -walh; *adj. Whole, entire :*—Onwalh *integer*, Wrt. Voc. ii. 44, 25. Of anwealh *integro*, Hpt. Gl. 525, 61. I. *literal, sound, uninjured, uncorrupted :*—Ealne his lîchoman gemêtton onwealhne and gesundne (*integrum*), Bd. 4, 30; S. 608, 37. Ealle đa scýtan đe se lîchoma mid bewunden wæs onwealge ætýwdon *lineamina omnia quibus involutum erat corpus integra apparuerunt*, 4, 19; S. 589, 21. Đa lástas á onwalge beóþ and on đære ilcan onsýne đe hié on forman on đa eorþan bestapene wæron, Blickl. Homl. 127, 20. II. *metaph. :*—Heó onwealg on hiere onwalde æfter þurhwunade *regnat incolumis*, Ors.

2, 1; Swt. 62, 23. Wæs hyre mægdenhád onwalg, Exon. Th. 87, 6; Cri. 1421. Đæt gecyndelíce gewitt biþ anwalg untósliten, Past. 52, 2; Swt. 405, 5. Đa óđre stondaþ on anwalgre hǽlo, Swt. 403, 23. Andswarede đæt hé on đyssum hæfde fæstne geleáfan and onwalhne *integram se in hoc habere fidem respondebat*, Bd. 3, 13; S. 539, 4. Geleáfan onwealhne and unwemmede heóldan, 1, 4; S. 475, 33: 4, 10; S. 578, 27. III. *of time, whole, entire :*—Onwalhge wican *ebdomade integra*, 4, 27; S. 604, 31. Geár onwealh *anno integro*, 3, 1; S. 523, 28. Onwalhge niht *noctes integras*, 4, 25; S. 599, 30. [*O. H. Ger.* anawalg *absolutus*.]

on-wealhlíce; *adv. Entirely :*—Đa mægenu đæs gódes weorces đe hé Gode ûtan anwealglíce forgeaf *tantae virtutis sacrificium, quod integrum foras immolant*, Past. 33, 5; Swt. 220, 22.

on-wealhnesse, e; *f. Wholeness, soundness, integrity.* I. *literal :*—þurh đa heora onwalhnesse gecýded is *it is made evident by the unchanged condition of the footsteps*, Blickl. Homl. 127, 27. II. *metaph. purity, chastity, integrity :*—Andwealhnys *integritas*, *religio sanctitas*, Hpt. Gl. 433, 50. Andhwælhnysse *integritatis*, 414, 74. Andwealcnysse, 432, 47. Andwealhnysse, 452, 32. Anhwealhnysse, 461, 46. Andwealhnysse *integritatis, castitatis*, 465, 71. Onwealhnesse *integritatis*, Wrt. Voc. ii. 44, 24. Mid êcre onwalhnesse (*integritate*) mægþhádes, Bd. 4, 19; S. 587, 25. Anwalhnysse, Homl. Th. ii. 564, 6. Andwælhnysse *integritatem, pudicitiam*, Hpt. Gl. 463, 57.

on-weard; *adj. Proceeding against, taking action against :*—Warnige se abbod đæt hé þurh andan ne sý onweard đam profaste *let the abbot take heed that he be not acting against the provost from hatred*, R. Ben. 126, 11.

on-wecnian. v. on-wæcnian.

on-weg; *adv. Away, off.* I. *with verbs of motion :*—Óđer þing wiston đa wîfmenn đa hý onweg cyrdon *when they went away (from the sepulchre)*, Exon. Th. 460, 13; Hö. 16. Gif đû onweg cymest *if you come away (alive from the fight)*, Beo. Th. 2769; B. 1382. Fêran onweg, Exon. Th. 373, 4; Seel. 103. Onweg (áweg) fleón, Ors. 4, 2; Bos. 79, 15: Bd. 4, 22; S. 591, 11. Onweg gewîtan, Blickl. Homl. 117, 1. Onweg hweorfan, Beo. Th. 534; B. 264. Hé onweg đanon feorhlástas bær, 1693; B. 844. II. *with verbs of taking, removing, separating, etc.* :—Onweg áceorfan *amputare*, Ps. Spl. T. 118, 39. Onweg ádón *to put away*, Bd. 3, 5; S. 524, 3. Onweg ádrîfan *to drive away, expel*, 2, 5; S. 507, 28. Onweg áhebban *to remove*, 1, 27; S. 493, 7. Onweg álǽdan, 5, 3; S. 616, 36. Onweg ániman, Blickl. Homl. 55, 9. Onweg áteón *to withdraw, subtract*, Bd. 4, 17; S. 586, 9. v. á-weg.

onweg-ácirrednes, e; *f. A turning away (from right belief), apostasy :*—Seó onwegácerrednes fram Cristes geleáfan Angelcyninga *apostasia regum Anglorum*, Bd. 3, 9; S. 533, 8.

onweg-álǽdnes, e; *f. A taking away, removal :*—Ond for đære gelómlícum onwegálǽdnesse (*frequenti ablatione*) đære hálgan moldan wæs mycel seáþ geworden, Bd. 5, 18; S. 635, 31.

onweg-gewite, es; *m. A going forth :*—In onweggewite *in excessu*, Ps. Surt. 115, 11.

onweg-gewitenness, e; *f. Going forth, departure :*—Æfter his onweggewitenesse (*abscessum*) of Breotene, Bd. 3, 7; S. 530, 12. v. áweggewiteness.

on-wendan. I. *to turn, change :*—Đû hí onwendest *mutabis ea*, Ps. Th. 101, 23. Hé onwendeþ his hiw, Lchdm. ii. 204, 9. Werþióde his (*the morning-star*) noman onwendaþ, hátaþ hine æfenstiorra, Met. 29, 29. Mé onhwyrfdon of đære gecynde đe ic ǽr beheold, onwendan mîne wîsan, Exon. Th. 485, 29; Rä. 72, 5. Onwend đec in gewitte *think differently*, 251, 12; Jul. 144. On đæs bisceopes anwealde đæt biþ hwæđer hé hit onwende đe ná *utrum mutet an non*, L. Ecg. C. 33; Th. ii. 158, 13. Đa menn đeáh wisston đæt hió mid đam drýcræfte ne mihte đara manna mód onwendan đeáh hió đa lîchoman onwende *nec potentia gramina, membra quae valeant licet, corda vertere non valent*, Bt. 38, 1; Fox 196, 8–10: Met. 26, 101–104. Đû ne meaht hiora sidu and heora gecynd onwendan, 7, 2; Fox 18, 31. Nis mé tîd mîn lîf tô onwendenne *non est mihi tempus vitam mutandi*, Bd. 5, 14; S. 634, 32. Onwended ne biþ ǽfre tó ealdre, Exon. Th. 203, 11; Ph. 82. Nán gewuna ne mæg nánum men beón onwended, Bt. 7, 1; Fox 16, 23. Gif đû wênst đæt hit on đê gelong sê đæt đa woruldsǽlþa on đê swá onwenda sint đonne eart đû on gedwolan *tu, fortunam putas erga te esse mutatam? erras*, 7, 2; Fox 16, 30. II. *to change one thing for another, to exchange :*—Heó wæs genumen of middanearde and eall đæt sár and đone deáþ mid êcre hǽlo and lîfe onwende, Bd. 4, 19; S. 589, 7. III. *to turn, change a direction, to avert, divert, turn aside :*—Nǽfre gê mec of đissum wordum onwendaþ đendan mec mîn gewit gelǽsteþ, Exon. Th. 124, 33; Gú. 347. 'Onwend đê tô đê sylfum'... Hé hine đá onwende from đisse worlde begangum, Blickl. Homl. 113, 26–30. Onwende hê his mód áweg, Lchdm. ii. 284, 15. Nǽfre đû đæs swíđlíc sár gegearwast đæt đû mec onwende đissa worda, Exon. Th. 246, 5; Jul. 57. Wênst đú đæt đû đæt hwerfende hweól, đonne hit on ryne wyrþ, mǽge oncyrran? Ne miht đû đon má đara woruldsǽlþa hwearfunga onwendan, Bt. 7, 2; Fox 18, 37. Ne mihte

hæleþ weán onwendan, Beo. Th. 384; B. 191: Exon. Th. 130, 19; Gú. 440. Ðara unstillena gesceafta styring ne mæg nó weorþan onwend of ðam ryne ðe him geset is, Bt. 21; Fox 74, 4. Bróc biþ onwended of his rihtryne, Met. 5, 19. Biþ him se wela onwended, and wyrþ him wíte gegearwod, Cd. Th. 28, 5; Gen. 431: Blickl. Homl. 195, 28. Sýn hié from heora wónessum onwende, 109, 20. **IV.** *to change the position of a thing, to invert, turn upside down,* (a) literal:—Sceal mín ród onwended beón; mín heáfod sceal beón on eorþan gecyrred, and míne fét tó heofenum gereahte, Blickl. Homl. 191, 5. Onwendedre endebyrdnysse *ordine prepostero,* Wrt. Voc. ii. 64, 33. (b) figurative, *to subvert, disturb, upset:*—Hond synfulra ne onwendeþ (*moveat*) mec, Ps. Surt. 35, 12. Ðis is ðæt mennisc ðe ealle míne dǽda mid heora wordum (*destroyed by their words the effect that my actions should produce*), ðæt hié mé ne gelýfdon, Blickl. Homl. 175, 25. Næ síe tó ðon gedurstig, ne cyning, næ bisceop, ðæt ðæs mínæ gife onwǽndæ (*commoveat*), Cod. Dip. Kmbl. v. 218, 28. Nǽfre ic ne míne lástweardas geþrístlǽcen ðæt heó hit (*the grant of certain dues*) onwenden, Ch. Th. 29, 14: Cd. Th. 26, 11; Gen. 405. Hwæt miht ðú his onwendan? Nú hé hafaþ ealle ðíne þeóstro geflémed, Blickl. Homl. 85, 21. Sibb ǽfre ne mæg wiht onwendan ðam ðe wel þenceþ *nothing can destroy the ties of kindred in the case of a right-minded man,* Beo. Th. 5195; B. 2601. Hé (*Julian*) wolde ðone Cristendóm onwendan, Ors. 6, 31; Swt. 286, 3. Hié ealle ða worold on hiora ágen gewill onwendende (*evertendo*) wǽron, 1, 10; Swt. 48, 10. Eall heofona mægen biþ onwended and onhréred, Blickl. Homl. 91, 27: 93, 13. Biþ se maga onwent and tóbrocen, Lchdm. ii. 218, 18. **V.** *to cause to change for the worse, to give a wrong direction, pervert:*—Se yfela déma onwendeþ ðone rihtan dóm for ðæs feós lufon, Blickl. Homl. 61, 31. Hié (*bribes*) wísra monna word onwendaþ, L. Alf. 46; Th. i. 54, 18. Beorht wǽron burgrǽced . . . meodoheall monig mandreáma full óþ ðæt ðæt onwende wyrd seó swíðe until fate wrought disastrous change, Exon. Th. 477, 15; Ruin. 25. Drync unheórne, se onwende gewit wera, Andr. Kmbl. 69; An. 35. Mid ðý ðe hié ðone drenc druncon, hraþe heora mód wæs onwended, Blickl. Homl. 229, 14. Is mín flǽsc frécne onwended *caro mea immutata est,* Ps. Th. 108, 24. **VI.** *intrans.* *To return:*—Heora gást gangeþ onwendeþ on ða eorþan ðe hí of cómon *exiet spiritus ejus, et revertetur in terram suam,* 145, 3. v. á-wendan.

on-wendedlíc; *adj. Changeable:*—Gyf se midwinter byþ on Frigendæge, ðonne byþ onwendedlíc winter, Lchdm. iii. 164, 8. v. á-wendedlíc.

on-wendedness, e; *f. Change, alteration:*—Nis him onwendednes *non est illis commutatio,* Ps. Th. 54, 20. Onwendednis *inmutatio,* Ps. Surt. 76, 11. In onwendednissum *in commutationibus,* 43, 13. v. next word.

on-wendness, e; *f.* **I.** *change:*—Ðǽre godcundnesse nǽnig onwendnesse on carcerne wæs of ðǽre menniscan gecynde, Blickl. Homl. 19, 24. **II.** *turning, movement* (v. onwend, **IV**)—Onwendnisse heáfdes *commotionem capitis,* Ps. Surt. 43, 15. v. preceding word.

on-weorpan *to throw aside, turn aside:*—Hine se wind onwearp fram ðære byrig *mutati ab urbe venti,* Bd. 3, 16; S. 543, 8. [Cf. *O. H. Ger.* int-werfan *dissociare.*]

on-weorpness, e; *f. A throwing on:*—Ðæt lég swíðe weóx and him nǽnig mon mid wǽtra onweorpnesse (*injectu*) wiðstondan meahte, Bd. 2, 7; S. 509, 20.

on-wéstan *to lay waste, desolate:*—Sý wunung heora onwést (*deserta*), Ps. Spl. 68, 30. v. á-westan.

on-wícan *to yield, retreat:*—Onwícan *cessere,* Wrt. Voc. ii. 14, 23. [Cf. *O. H. Ger.* int-wíchan *cedere, recedere: Ger.* ent-weichen.]

on-willan *to cause to boil;* fig. *to cause passion or emotion to be violent:*—Ðá wæs eft swá ǽr ealdfeónda níþ onwylled *then again as before hot waxed the hate of former foes,* Exon. Th. 125, 30; Gú. 362. v. á-wellan, -wyllan.

on-wille; *adj. Desired:*—'Ac gé hine gesundne ásettaþ ðǽr gé hine sylfne genóman' . . . Ongon ðá leófne síð dragan Dryhtnes cempa tó ðam onwillan eorþan dǽle tó *the hermitage to which he* (*Guthlac*) *desired to go, and from which the fiends had removed him,* Exon. Th. 145, 25; Gú. 700.

on-windan. **I.** *to unwind, unfasten, loosen:*—Ðonne forstes bend Fæder onlǽteþ, onwindeþ wǽlrápas, Beo. Th. 3224; B. 1610. Báncofan onband, breóstlocan onwand, Elen. Kmbl. 2498; El. 1250. **II.** *to retire, retreat:*—Hærn eft onwand . . . wǽdu swǽðorodon, Andr. Kmbl. 1062; An. 531.

on-winnende *assailing, attacking:*—Se onwinnenda here *the attacking army,* Homl. Th. ii. 432, 4. Míne geféran mé betǽhton ðam onwinnendum feóndum, Homl. Skt. i. 7, 351. v. winnan.

on-wist, e; *f. The being in a place, dwelling, habitation:*—Gesealde sigora waldend onwist ǽðles Abrahames sunum *God granted to Abraham's descendants to live in a country,* Cd. Th. 178, 27; Exod. 18. Cf. onwunung.

on-wlát (?) *form, appearance:*—Anwláten (-es?) *formae,* Hpt. Gl. 523, 61.

on-wlite, es; *m. Face:*—Onwlite *patham,* Txts. 172, 17. v. and-wlite.

on-wóh. v. wóh.

on-worpenness, e; *f. An injection;* fig. of a feeling which has been inspired:—Ðá ic getíhtode bi ðære gítsunge onworpennesse and ðá wæs ic gesprecende ðone man and sécende wæs ðæs þinges cúðnesse æt him, Shrn. 36, 19.

on-wrecan *to avenge:*—Ðý læs on him gesewen sí ðás þing onwrecen beón ne in eis illa ulcisci videantur, Bd. 1, 27; S. 491, 28. v. á-wrecan.

on-wreón *to uncover, disclose:*—Ne onwríh ðú *ne reveles,* Kent. Gl. 960. Onwreónde *discooperiens,* Wrt. Voc. ii. 25, 73. **I.** literal, *to uncover, open, remove a covering:*—Hé his hrægl onwrág *retecto vestimento,* Bd. 2, 6; S. 508, 23. Ðá onwrigon hý hire ondwliton *discooperto vultus indumento,* 4, 19; S. 589, 16. Onwreóh (-wríh, Ps. Surt.) eágan míne, Ps. Spl. T. 118, 18. Onwreón ða duru ðæs geteldes, Bd. 4, 19; S. 589, 14. **II.** figurative, *to make known, shew forth, reveal, discover:*—Heó onwríhþ hire ǽwelm, donne heó geopenaþ hiore þeáwas, Bt. 20; Fox 70, 25. Hé his miltse onwreáh, Blickl. Homl. 107, 20. Ðá com yrnan sum olbenda, and se cwæþ . . . 'Ne tódǽlaþ gé ðara háligra líchoman' . . . ðá dydon hý swá him ðæt dumbe neát onwreáh, Shrn. 136, 2. Ic ðé háte ðæt ðú ðás gesyhþe secge mannum, onwreóh wordum ðæt hit is wuldres beám, Rood Kmbl. 191; Kr. 97. Bæd ðæt hé him on spellum gecýðde, onwrige worda gongum, hú . . . , Exon. Th. 171, 29; Gú. 1134. Iudas ðé mæg sóð gecýðan, onwreón wyrda gerýno, Elen. Kmbl. 1174; El. 589. Ðú scealt biddan ðæt móte beón open and onwrigen hwæt hé sý, Blickl. Homl. 185, 4: Bd. 2, 12; S. 512, 32. Is onwrigen wyrda bigang, Elen. Kmbl. 2245; El. 1124. **III.** *to shew the* (*hidden*) *meaning of anything, to explain:*—Ic wéne ðæt ðás word ne sind eów fullcúðe, gif wé hí openlícor eów ne onwreóþ, Homl. Th. i. 580, 27. Augustinus ús onwreáh ðissere rǽdinge andgit, ii. 384, 21. Ðás word sind sceortlíce gesǽde, and we is neód ðæt wé hí swutelícor eów onwreón, i. 278, 14. Onwríon *explicare,* Kent. Gl. 1152. **IV.** *to shew, display so as to avoid concealment:*—Ðá seó fǽmne onwráh ryhtgerýno, Exon. Th. 12, 34; Cri. 195. Onwreóh (-wríh, Ps. Surt.) Gode ðíne wegas, Ps. Th. 36, 5. Gif his sáule gyltas óðerum monnum dígle beóþ and him sylfum cúðe, mid his andetnesse onwreó ða his abbode, R. Ben. 72, 5. **V.** *to display what is bad, to expose:*—God hine (*the sorcerer*) onwrýhþ gyt, ðeáh ðe wit hine ne geopenian, Blickl. Homl. 187, 17. Seó hálige ǽ forbeódeþ ða sceondlícnysse onwreón mǽgsibba . . . Ne onwreóh ðú sceondlícnysse ðínes fæder . . . se ðe gedyrstigaþ onwreón ða sceondlícnysse his steópméder . . . se onwríhþ his fæder sceondlícnysse, Bd. 1, 27; S. 491, 6-16. Womdǽda onwreón, Exon. Th. 270, 18; Jul. 467. **VI.** *of the operations of the Deity, to reveal:*—Dryhten ðú ðe ðás þing onwrige lytlingum, Mt. Kmbl. 11, 25. Hit ðé ne onwreáh flǽsc ne blód, ac mín Fæder, 16, 17: Bd. 2, 12; S. 512, 24. Áne bóc on his wítegunge, ðe him God sylf onwreáh, Ælfc. T. Grn. 9, 44. Deáh ðe him God onwríge wísdómes gǽst, Exon. Th. 273, 14; Jul. 516. Ðam ðe se sunu wyle ðone Fæder onwreón, Mt. Kmbl. 11, 27. Drihten hire forþfore wæs geeáþmódad tó onwreónne, Bd. 4, 23; S. 595, 36. Sum gesihþ ðe God ðysan menn onwreogan hæfþ, Homl. Skt. i. 23, 747. [*O. H. Ger.* int-ríhan *revelare.*]

on-wrigenness, e; *f. An uncovering, discovery:*—Onwrigenys *apocalypsis,* Hpt. Gl. 435, 43. **I.** *a removal of that which obscures* or *conceals:*—Leóht tó onwrigennysse þeóda *lumen ad revelationem gentium;* a light to lighten the Gentiles, Homl. Th. i. 136, 22. **II.** *an explanation, exposition* (v. on-wreón, **III**):—Circlícere anwrigennesse *ecclesiasticae traditionis, expositionis,* Hpt. Gl. 410, 36. Hæbben ða ungelǽredan inlendisce ðæs hálgan regules cýðe þurh ágenes gereordes anwrigennesse *by means of an explanation* (*translation*) *in their own tongue,* Lchdm. iii. 442, 9. **III.** *an exposure of a person's real character* (v. on-wreón, **V**):—Nú neálǽceþ ǽgðer ge ðín onwrigennes ge uncer gecýðnes *the vanity of your pretensions will be exposed, and the reality of our claims will be made manifest,* Blickl. Homl. 187, 23. **IV.** *a revelation, manifestation made to the eye* or *to the ear by divine power* (v. on-wreón, **VI**):—Heó sægde ðæt heó geleornod hæfde on onwriðgennysse (MS. T. onwrignesse) ðæt hire forþfore wǽre swíðe neáh. Sǽde heó him ðæt seó onwrihgnes ðyslíc wǽre. Cwæþ ðæt heó gesáwe micelne þreát, Bd. 3, 8; S. 531, 35-38. Se Hǽlend geswutelode him ða tóweardan onwrigenysse (*a revelation of the future*), ðe ðære hé áwrát ða bóc ðe is geháten Apocalipsis, Homl. Th. i. 60, 1. Helena hí (*the cross*) áfunde þurh Cristes onwrigenesse *through a revelation made by Christ,* H. R. 99, 8. Him ða upplícan onwrigenesse wiðstódon *superna illi oracula restiterunt,* Bd. 5, 9; S. 622, 21.

on-wrigness, -wrihnes, e; *f. Revelation:*—Of onwrihnesse geendad *revelatione saturatus,* Mt. Kmbl. p. 9, 6. v. on-wrigenness, **IV.**

on-wríðan *to unwrap, to release from a covering:*—Seó hét heáfod onwríðan *she bade unwrap the head* (*of Holofernes*) *from the bag in which it had been put,* Judth. Thw. 24, 5; Jud. 173.

on-wríðung, e; *f. A band:*—Onwríðung (-wrítung, Wrt.) *ligamentum,* Wrt. Voc. ii. 53, 76.

on-writing; e; *f. An inscription;* inscriptio Lk. Skt. Rush. 20, 24.

on-wunian *to dwell, inhabit:*—Ic onwunige (*inhabitabo*) on ðínum getelde, Ps. Lamb. 60, 5. Ðæt ic onwunige (*inhabitem*) on húse Drihtnes

Ps. Spl. 26, 7. Đū onwunast (*habitabis*) on heom, 5, 13. Hī onwuniaþ *inhabitabunt*, 55, 6. Hig onwuniaþ on worlde *inhabitabunt in saeculum*, Ps. Lamb. 36, 29. Onwuna on gelaðunge *inhabita terram*, 36, 3.

on-wunung, e; *f.* **I.** *a habitation, dwelling*:—Gewurðe him wēste eall his onwunung *fiat habitatio ejus deserta*, Ps. Th. 108, 7. Gewȳt fram mē, and far ūt of mȳnre onwununge, Nicod. 27; Thw. 15, 11. Đonne forlǣt se hālga gāst ða onwununge, and ðǣr sōna wyrþ deófol inne, Wulfst. 280, 9. **II.** *persistence, perseverance*:—Mid singalre ānrǣdnesse ł onwununge *assidua (perpetua) instantia*, Hpt. Gl. 407, 66.

on-wyllan, -ȳwan. v. on-willan, -īwan.

on-ȳdan *to pour in*:—Tunge witan swylce lagoflōd onȳðaþ *lingua sapientis quasi diluvium inabundabit*, Scint. 65.

oo-. v. ō-.

open; *adj. Open.* **I.** *not shut*, (a) *allowing ingress or egress*:—Heofen biþ open on sumum ende ... and mycel mægen forþ cymeþ þurh ðone openan dǣl, Blickl. Homl. 93, 1. Open scrǣf, Cd. Th. 212, 10; Exod. 537. Open wæs ðæt eorþærn (*the sepulchre*), Exon. Th. 460, 18; Hö. 19. Đīn carcern open wē gemētton, Blickl. Homl. 239, 27. Gē geseóþ opene heofenas (*caelum apertum*), Jn. Skt. 1, 51. (b) *of a door*:—Đonne andydan hié ða duru ðe on ða healfe open wæs (*the door that opened on that side*) ... and mið ðæm ðe hié ðara dura hwelce opene gesāwon, Ors. 3, 5; Swt. 106, 14–16. Biþ oft open eádgum tōgeánes onhliden heofonrīces duru, Exon. Th. 198, 16; Ph. 11. Hié gemētton ðæs carcernes duru opene, Blickl. Homl. 239, 24. (c) *of the eyes*:—Mid openum eágum gesión, Met. 20, 257. (d) *of wounds, not closed up*:—Đa openan dolg, Exon. Th. 68, 24; Cri. 1108: Rood Kmbl. 93; Kr. 47. **II.** *not covered, not protected*:—Seó cirice is ufan open and unoferhrēfed, Blickl. Homl. 125, 26, 30. Open burh *urbs patens*, Kent. Gl. 975. **III.** *declared, public*:—Đa bēc (*of the Old Testament*) synd gehātene seó ealde gecȳðnyss and seó ealde ǣ, ðæt is, open lagu ðe God gesette Israhēla folce, Hexam. 1; Norm. 2, 19. **IV.** *not secret, not concealed, discovered, brought to light* (in reference to things where concealment is desired):—Hwanon ys ðis word open geworden (*palam factum*), Ex. 2, 14. Ne dēþ nān man nān þing on ðiglum ac sēcþ ðæt hit open sȳ (*in palam esse*), Jn. Skt. 7, 4. Đæt mōte beón open and onwrigen hwæt hé sȳ, Blickl. Homl. 185, 4. Se ðe mānaþ swerige and hit him on open wurðe *he that commits perjury, and the crime is clearly proved against him*, L. Ath. i. 25; Th. i. 212, 18. Gif open morþ weorðe āgife man māgum ðone banan *if in a case of murder the murderer be discovered, let him be given up to the kinsmen of the murdered man*, L. C. S. 57; Th. i. 406, 25. Æt openre þȳfþe *in case of discovered theft*, 26; Th. i. 392, 3. Opene weordaþ monna dǣde *the deeds of men shall be brought to light (at the last day)*, Exon. Th. 64, 32; Cri. 1046. **V.** *without attempt at concealment*:—Antonius him (*Octavianus*) onbeád gewin and openne feóndscipe, Ors. 5, 13; Swt. 246, 1. Blisse on openum, Lchdm. iii. 200, 8. On ða openan tīd *the last day when nothing is concealed*, Exon. Th. 96, 9; Cri. 1571. **VI.** *manifest, clear, plain, evident*:—Đā cwæþ hē: 'Genōg sweotol ðæt is ðætte for ðȳ sint gōde men gōde ðe hī gōd gemētaþ.' Đā cwæþ ic: 'Genōg open hit is' *certum est, adeptione boni, bonos fieri. Certum*, Bt. 36, 3; Fox 176, 29. Se ðe unwīslīce leofaþ biþ open sott, ðeáh him swā ne þince, Homl. Skt. i. 13, 132. Is seó wyrd mid eów open orgete, Andr. Kmbl. 1517; An. 760. Đā āgann Landfranc atȳwian mid openum gesceáde (*with manifest reason*), ðæt hē mid rihte crafede ðǣs ða hē crafede, Chr. 1070; Erl. 208, 17. [*O. Sax.* opan: *O. Frs.* epen: *Icel.* opinn: *O. H. Ger.* offan.]

open-ears, -ærs, es; *m. A medlar*; mespila, Wrt. Voc. i. 32, 50. (v. *Halliw. Dict.* openers.)

openere, es; *m. One who opens*:—Aprilis quasi aperilis ... swylce hē sȳ openere. On his tīman beóþgeopenade trȳw tō blōwanne, Anglia viii. 326, 5.

openian; *p.* ode. **I.** *intrans.* (a) *to open, to become open*:—Openaþ *patebit*, Kent. Gl. 401. Byrgenu openodon, Homl. Th. ii. 258, 5. Openige nū ðīn fæðm, Blickl. Homl. 7, 24. Byrgen opnigende (*patens*) is race heora, Ps. Spl. 5, 10. Openiendum heofonum caelis patentibus, Bd. 4, 9; S. 576, 37. Opniendum, Hpt. Gl. 514, 55. (b) *to become manifest*:—Đæs līf mid heálīcum tācnum heofonlīcra wundra openode *cujus vita sublimis crebris miraculorum patebat indiciis*, 4, 30; S. 608, 26. **II.** *trans.* (a) *to open, unclose*:—Openast (*aperis*) ðū hand ðīne, Ps. Spl. 144, 17. Seáþ hē openode *lacum aperuit*, 7, 16. Opnyaþ mē gatu rihtwīsnysse, 117, 19. (b) *to disclose, manifest*:—Gefeohtu gesiþþ blisse hit openaþ *if he sees fights, it is a sure sign of joy*, Lchdm. iii. 200, 8. Hē cȳðde and openade ðæt hē Cristen wǣre *se Christianum esse prodiderat*, Bd. 1, 7; S. 477, 22. Đæt hē nǣnigum mā openade ne cȳðde hit openaþ (*patefaceret*), 5, 9; S. 623, 15. Hord openian *to discover the treasure*, Beo. Th. 6105; B. 3056. Openiende *propalat*, Wrt. Voc. ii. 66, 17. [*O. Sax.* oponōn: *O. H. Ger.* offanōn: *O. Frs.* epenia: *Icel.* opna.] v. ge-openian.

open-līc; *adj. Open, public*:—Openlīc *publicum*, Germ. 398, 45. Openlecre *puplica*, Wrt. Voc. ii. 66, 55. Openlecum (opauletet, Wrt. ii. 3, 61) *a puplicis*, Wülck. Gl. 343, 28. [*O. H. Ger.* offan-līh *publicus*.]

open-līce; *adv. Openly.* **I.** *publicly, in a way by which not a*

few only are affected:—Eft cymþ God swīðe openlīce (*in a way to be seen by all*), Ps. Th. 49, 3. Hié openlīce ðæt gesetton (*they publicly decreed*) ðæt hē swungen wǣre ōþ ðæt hē swylte, Blickl. Homl. 193, 3. Wæs ðis ðara wundra ǣrest ðe ðes eádiga wer openlīce beforan ōðrum mannum geworhte, 219, 3 : Homl. Th. i. 58, 15. Hē funde āne tabulan eall āwritene and hī openlīce rǣdde (*read it out to the by-standers*), Homl. Skt. i. 23, 767. **II.** *without concealment, without reserve, freely*:—Hē spræc openlīce (*palam*), Mk. Skt. 8, 32. Nān man spæce openlīce be him for ðæra Iudēa ege, Jn. Skt. 7, 13. Đā for hē næs nā openlīce ac dȳgollīce, 7, 10. Monige scylda openlīce wietena (*aperte cognita*), Past. 21, 2; Swt. 152, 1. Đa dīglan gyltas man sceal dīgelīce bētan, and ða openan openlīce, Homl. Th. i. 498, 10. **III.** *plainly, evidently, clearly, manifestly*:—Swelce hē open, līce cuæðe *si aperte dicat*, Past. 21, 2; Swt. 153, 11: Blickl. Homl. 81; 19. Būton ic openlīce gecȳþe ðæt ic God sylfa sȳ *unless I make it evidently appear that I am God himself*, 181, 36. Se wæs openlīce ūþwita, Bt. 19: Fox 70, 8: Met. 13, 72. Hū ne is ðē genōg openlīce geeówad, Bt. 24, 3; Fox 84, 19: 32, 2; Fox 116, 33. Ic ongite openlīce ... Ic wolde ðeáh hit fullīcor and openlīcor of ðē ongitan *video ... sed ex te cognoscere malim apertius*, 33, 1; Fox 120, 2–9: 39, 2-Fox 212, 11. Openlīce *manifeste*, Hpt. Gl. 460, 59. Sume sindon openlīce forgitene *some plainly are forgotten*, Met. 10, 60. **IV.** *without obstruction, at large*:—Wolde openlīcor (*latius*) ætȳwan seó god-cunde ārīsætnyss on hū myclum wuldre Cūþbyrht æfter his deáþe lifede, Bd. 4, 30; S. 608, 24. Ic wēne ðæt ðū nyte hwæt ðis gemǣne, būton wē of ōðrum bōcum ðis openlīcor secgan (*give a fuller account*), Boutr. Scrd. 18, 27. Đæs þing wē willaþ openlīcor gecȳðan ðonne ðæt lȳden dō, Anglia viii. 298, 25 : Chr. 1106; Erl. 240, 35. [*O. Sax.* opan-līko: *O. H. Ger.* offanlīhho *palam, publice, evidenter.*]

openness, e; *f. Openness, publicity*:—Gend openysse *per publicum*, Hpt. Gl. 524, 5. [*O. H. Ger.* offannussi *apocalypsis*.]

openung, e; *f. Manifestation, revelation*:—Seó openung ðæs dæges (*the day of judgment*) is swīðe egesfull eallum gesceaftum, Blickl. Homl. 91, 19. [*O. H. Ger.* offenunga *manifestatio, declaratio.*]

ōr. **I.** *beginning, origin*:—Ōr ł fruma *initium*, Mk. Skt. Lind. 13, 8. Dæges ōr onwōc geleáfan *the day-spring of belief awoke*, Apstls. Kmbl. 130; Ap. 65. Næs him fruma ǣfre ōr geworden, Cd. Th. 1, 11; Gen. 6. Đǣr wæs yfles ōr, Andr. Kmbl. 2763; An. 1384. On ðæm wæs ōr writen fyrngewinnes, Beo. Th. 3381; B. 1688. Ōr and ende, Exon. Th. 492, 6; Rä. 81, 10. Cwealmes on ōre *at the beginning of the destruction*, Cd. Th. 153, 32; Gen. 2547. Gif ðū his ne meaht ōr āreccan *if you cannot tell even the beginning of your dream*, 224, 9; Dan. 133. Secgan ōr and ende *to tell from first to last*, Andr. Kmbl. 1297; An. 649. Ic ðē yfla gehwylces ōr gecȳðe ōþ ende forþ, Exon. Th. 263, 21; Jul. 353. Suē hit wundra gihuaes ōr āstelidæ (cf. ord onstealde, Bd. 4, 24; S. 597, 21) *quomodo ille omnium miraculorum auctor exstitit*, Txts. 149, 4. Orleges ōr onstellan, Beo. Th. 4806; B. 2407: Exon. Th. 386, 10; Rä. 4, 59. Ne can ic Abeles ōr ne fōre hleómǣges sīð *I know not Abel's life from its beginning or its later course*, Cd. Th. 61, 33; Gen. 1006. **II.** *front, van*:—Wæs on ōre heard handplega, 198, 22; Exod. 326: Beo. Th. 2087; B. 1041. Heriges on ōre, Andr. Kmbl. 2213; An. 1108. Cf. ord.

or. This form occurs in A. Sax. only as a prefix, but in Goth. *us*, in Icel. *or*, *ur*, in O. H. Ger. *ur* it is found also as a preposition. It has the meaning *without*, e. g. or-mōd ; also that of *original, early*, e. g. or-eald. v. ōre.

ōra, an; *m. A border, edge, margin, bank* (mostly in place names, *-or* in Windsor, Bognor. v. Cod. Dip. Kmbl. iii. xxxv: Leo, A. S. Names, p. 92):—In ðone stede ðe is gecueden Cerdices ōra, Chr. 495; Erl. 14, 10: 514; Erl. 14, 21. Æt Cerdices ōran, Erl. 2, 3. Đonan on ðone ōran foran wið-eástan Ecgulfes setl west be ðæm ōran eft tōweard setle, Cod. Dip. Kmbl. ii. 216, 2–3. Siððan ðū gehȳrde on hlīþes ōran galan geác on bearwe, Exon. Th. 473, 28; Bo. 21. On ōra[n] his hrægles *in oram vestimenti ejus*, Ps. Spl. 132, 3.

ōra, an; *m. Ore, metal in an unreduced state*:—Ælces kynnes wecg vel ōra *metallum*, Wrt. Voc. i. 34, 67. Seolfor ðe byþ seofon sīðon āmered syððan se ōra ādolfen byþ, Ps. Th. 11, 7. Gedolfene ōran *effossa rudera*, Germ. 396, 190. Hit is eác berende on wecga ōrum āres and īsernes leádes and seolfres *quae etiam venis metallorum, aeris, ferri, et plumbi et argenti faecunda*, Bd. 1, 1; S. 473, 23. Seó eorþe is cennende wecga ōran *terra parens metallorum*, Nar. 2, 15. Gold-ōrum ł-wecgum *auri metallum*, Hpt. Gl. 449, 14. [Cf. golt, seluer, stel, irn, copper, mestling breas: al is icleopet or, A. R. 284, note b.] v. ōre.

ōra, an; *m. A species of money introduced by the Danes* (cf. Icel. *eyrir*, the eighth part of a mark):—Þolie twelf ōrena mid Denum .xxx. sciłł. mid Englum, L. E. G. 7; Th. i. 170, 16. Bēte man ðæt æt deádum menn mid .vi. healfmarce, and æt cwicon mid .xii. ōran, L. Eth. iii. 1; Th. i. 292, 11. Ita quod xv. (xvi ?) ore libram faciant, iv. 9; Th. i. 303, 9. In the Law of the Northumbrian Priests, Th. ii. 290 sqq., this money is often mentioned. Ōro *mnas*, Lk. Skt. Lind. Rush. 19, 13: Rush. 19, 16.

oraþ. v. oroþ.

or-bléde; *adj. Bloodless :*—Orbléde *exsangues*, Wrt. Voc. ii. 32, 39.

orc, es; *m. A cup, can, tankard, flagon :*—Orc *orca* (cf. orca *a tankard*, Wülck. Gl. 599, 16: *a cane*, 771, 29), Wrt. Voc. i. 25, 4. Blót-(blód-)orc *uas in quo sacrificabant res impias*, Germ. 307, 514. Orce *calice*, Hpt. Gl. 435, 39. Bollan steápe, swylce eác orcas, Judth. Thw. 21, 15; Jud. 18: Beo. Th. 6087; B. 3047. Hé geseah orcas standan, fyrnmanna fatu, 5514; B. 2760. Orcas *crateras*, Ex. 24, 6. [*Goth.* aurkeis *a cup* : *O. Sax.* ork.]

orc, es; *m. The infernal regions* (orcus) :—Orc *orcus*, Ep. Gl. 16 f, 36: Wrt. Voc. ii. 115, 61. Orcþyrs oððe heldeófol *Orcus* (*the god of the infernal regions*), 63, 49.

or-ceápe, -ceápes, -ceápunga, -ceápungum; *adv. Without payment, without cause, for nothing, gratis, gratuitously :*—Ne þurfon gé wénan ðæt gé ðæt orceápe sellon, ðæt gé under Drihtnes borh syllaþ, þeh gé sóna ðære méde ne ne onfón, Blickl. Homl. 41, 12. Orceápes *gratis*, Hpt. Gl. 478, 42. Beó hé frióh orceápunga, L. Alf. 11 ; Th. i. 46, 3. Hí onwunnon mé orceápunga (*gratis*), Ps. Spl. M. 119, 6. Orceápungum, Ps. Lamb. 108, 3.

orc-eard, -geard. v. ort-geard.

or-ceás; *adj. Free from complaint, not chargeable (with a fault) :*—Orceás *inmunis*, Wrt. Voc. ii. 91, 50; *inmunes*, 111, 14. Orceásne *immunem, immaculatum, castum*, Hpt. Gl. 474, 72. Orceáse í unwemme *immunes, incontaminati, inviolatas*, 447, 43. v. ceás, ceást, *and next word.*

or-ceásness, e; *f. Immunity, freedom from fault :*—Orceásnes *inmunitas*, Wrt. Voc. ii. 46, 59 : 77, 34. Seó orceásnys, Hpt. Gl. 433, 57. Orceásnysse í uniwemnysse *immunitatis*, 434, 27. Orceásnysse *immunitatem, castitatem*, 461, 41.

orcen (?) *a sea-monster :*—Ðanon untydras ealle onwócon, eotenas and ylfe and orcneas [orcenas (?). *Grein reads* orc-néas, *with which compare* orc-þyrs *under* orc] swylce gigantas, Beo. Th. 225; B. 112. [Cf. (?) *Icel.* orkn (örkn) *a kind of seal.*]

or-cnáwe, -cnǽwe; *adj. Recognisable, evident :*—Ðær orcnáwe (wearþ) þurh teóncwide tweógende mód, Andr. Kmbl. 1540 ; An. 771. Ðá wæs orcnǽwe (on-, Kmbl.) idese síðfæt, Elen. Kmbl. 457; El. 229. v. ge-, on-cnáwe.

orc-þyrs. v. orc.

ord, es; *m.*   I. *a point,* (a) of a weapon :—Ælces wǽpnes ord *mucro*, Wrt. Voc. i. 35, 35. Se ord (ðæs speres), L. Alf. pol. 36; Th. i. 84, 17. Seaxes ord, Exon. Th. 472, 6; Rä. 61, 12. Wordes ord breósthord þurhbræc, Beo. Th. 5576; B. 2791. Ne ofstong hé hiene mid ðý speres orde. Ðæt is ðonne swelc mon mid forewearde orde stinge ... suá suá Assael wæs deád bútan orde *non cum recta, sed aversa hasta transforavit ... quasi sine ferro moriuntur*, Past. 40, 5 ; Swt. 297, 10–23. Mid gáres orde, Cd. Th. 92, 2; Gen. 1522. Hé sette his swurdes ord tógeánes his innoþe, Homl. Th. ii. 480, 14. Ðæt gebearh feore wið ord and wið ecge (cf. *Icel.* með oddi ok eggju) *it protected life from thrust and cut*, Beo. Th. 3102; B. 1549. (b) *putting a part for the whole, a spear, pointed weapon :*—Mé sceal wǽpen niman, ord and íren (*spear and sword*), Byrht. Th. 139, 12 ; By. 253. Hwá ðær mid orde mihte on fǽgean men feorh gewinnan, wígan mid wǽpnum, 135, 31 ; By. 124. Hit is mycel nédþearf ðæt hié man forspille, and mid írenum þislum and ordum hié man sleá, Blickl. Homl. 189, 30. Hildesercum, bordum and ordum, Elen. Kmbl. 469; El. 235. (c) *of other point-shaped, conical things :*—Ord *apicem*, Wrt. Voc. ii. 73, 64. Ða hwíle ðe se móna ðære sceade ord (*the shadow of the earth*) ofer yrnþ, Lchdm. iii. 240, 26. Hafaþ tungena gehwylc xx orda, hafaþ orda gehwylc engles snytro, Salm. Kmbl. 461–464; Sal. 231–232. (d) *of persons,* (1) *one who is at the topmost point, a head, chief, prince :*—Æþelinga ord *Christ*, Exon. Th. 32, 19; Cri. 515 : 46, 22; Cri. 741 : 53, 5; Cri. 846 : Elen. Kmbl. 785; El. 393. Burgwarena ord, 462, 22 ; Hö. 56. (2) *of position, head, front :*—Se ðe on orde geóng *he who went at the head of the band*, Beo. Th. 6242; B. 3125.   II. *line of battle, forefront :*—Se ord on here acies, Ælfc. Gr. 5; Som. 4, 14. Hí Pantan streám bestódon, East-Seaxena ord and se æschere, Byrht. Th. 133, 52; By. 69. Elamitarna ordes wísa, Cd. Th. 121, 3; Gen. 2004. On orde stód Eádweard *Edward stood in the forefront of the battle*, Byrht. Th. 139, 52; By. 273.   III. *the beginning, origin, source* (applied to persons and things) :—Se ðe (*the devil*) is ord ælcere leásunge and yfelnysse, Homl. Th. i. 4, 29. Se leahter (*pride*) is ord and ende ælces yfeles, ii. 220, 34. Ord moncynnes (*Adam*), Cd. Th. 68, 2; Gen. 1111. Dæges ord *day-break*, 174, 10 ; Gen. 2876. Sume úre þéninghbec onginnaþ on Aduentum Domini ; nis ðeáh ðær forðý ðæs geáres ord, Homl. Th. i. 98, 27. From orde óþ ende forþ, Elen. Kmbl. 1176; El. 590. Hé folc-mægþa fruman áweahte, æþelinga ord, ðá hé Adam sceóp, 77, 20 ; Gen. 1278. Sóna ongeat cyning ord and ende ðæs ðe him ýwed wæs, 225, 30; Dan. 162. Ord onstellan *to make a beginning, be the source of*, 272, 4; Sat. 114 : Bd. 4, 24; S. 597, 21. Ðæt ðín sprǽc hæbbe ægðer ge ord ge ende, Past. 49; Swt. 385, 13. [*Laym., A. R., O. and N.* ord : *Orm.* ord and ende : *O. Sax. O. L. Ger. O. Frs.* ord : *O. H. Ger.* ort *angulus, aculeus, acies, initium* : *Icel.* oddr *the point of a weapon, head of a troop, leader.*]

or-dǽle; *adj. Not hiving* or *taking part in a thing, not participat-*

ing :—Ordǽle *expers*, Wrt. Voc. ii. 31, 48 : 90, 67. Ordǽla *expers*, i. *ignarus, alienus, sine parte, imperitus, inscius, privatus*, Wülck. Gl. 232, 23.

or-dǽl, -dél; *generally neuter, but an apparently fem. acc. pl.* ordéla *occurs*, L. Edg. C. 24; Th. i. 248, 28. (Cf. *O. H. Ger.* which has fem. and neut. forms.) In the sense of *judicial decision, judgment* the word is used by *O. Frs. O. Sax. O. H. Ger.* (v. Richthofen, the Heliand and Graff), but in *A. Sax.* it is found only in the special sense, which belongs also to the *O. Frs.*, of a decision which follows an appeal to the Deity. The ordeal was thus connected with religion, and attended by religious ceremonies. In L. Ath. i. 23; Th. i. 210, 26, it is said with respect to the person who is to undergo the ordeal ' féde hine sylfne mid hláfe and mid wætere and sealte and wyrtum ǽr hé tó gán scyle, and gestande him mæssan ðæra þreora daga (*the three days preceding the ordeal*) ælcne, and geoffrige tó, and gá tó húsle ðý dæge ðe hé tó ðam ordále gán scyle, and swerige ðonne ðane áþ, ðæt hé sý unscyldig ðære tihtlan ǽr hé tó ðam ordále gá.' Before taking the Eucharist and going to the ordeal a solemn form of adjuration was addressed to the person concerned, that unless he was conscious of innocence he should desist. v. Rtl. 114, 13–23. The further proceedings in connection with the ordeal by hot water or by hot iron are detailed in L. Ath. iv. 7; Th. i. 226, 8. After the fire to be used in heating was carried into the church, none were to enter but the priest and the accused. When the iron was hot or the water boiled, two men for the accused, two for the accuser, were admitted, to see that the proceedings were fairly conducted. When hot water was employed, if it were a case of *ánfeald tihtle*, the hand was plunged in up to the wrist, if of threefold, up to the elbow. When the hot iron was used, a weight of one pound or of three pounds, according to the case, had to be carried nine feet. The hand was then sealed up, and its condition, when un- wrapped at the end of three days, determined the guilt or innocence of the accused. See also L. Ath. i. 23; Th. i. 212, 2–10. Further refer- ence to the difference in degree is made in Ath. iv. 6; Th. i. 224, 13 : L. Edg. H. 9; Th. i. 260, 18. Among those who were to be subjected to this form of trial are mentioned convicted perjurors, who after con- viction are not ' áþwyrðe ac ordáles wyrðe,' L. Ed. 3; Th. i. 160, 18–21 : the man who was charged with plotting against his lord, or with being guilty of ' cyricbryce,' or with practising witchcraft and similar illicit arts underwent the threefold ordeal, L. Ath. i. 4–6; Th. i. 202, 1–17 ; and the same trial was appointed in the case of incendiaries, L. Ath. iv. 6; Th. i. 224, 11–19, and of coiners, L. Ath. i. 14; Th. i. 206, 17–25 : L. Eth. iii. 8; Th. i. 296, 12–16. The ordeal is also mentioned as being the only method of meeting an accusation in a case between English and Welsh, ' ne stent nán óðer lád æt tihtlan búte ordál betweox Wealan and Englan,' L. O. D. 2; Th. i. 354, 1–2. The ordeal must take place in a king's burgh, ' Ælc ordál beó on ðæs kyninges byrig, L. Eth. iii. 6; Th. i. 296, 4, and upon fastdays and festivals could not be used, ' ordél and áþas syndan tócwedene freólsdagum and rihtfæstendagum,' L. E. G. 9; Th. i. 172, 10: L. Eth. v. 18; Th. i. 308, 24–27: vi. 25; Th. i. 320, 24–27: L. Edg. C. 24; Th. ii. 248, 27. Wé forbeódaþ ordál and áþas freólsdagum and ymbrendagum and lenctendagum and rihtfæsten- dagum and fram *aduentum domini* óþ *octabas epiphanie* and fram *septua- gesima* óþ fifténe niht ofer eástran, Wulfst. 117, 14. See Schmid. A. S. Gesetz., Grmm. R. A. pp. 863 sqq., 908 sqq., and cf. cor-snæd. As an instance of the occurrence of the word elsewhere than in the Laws, see Chart. Th. 432, where the phrase *áþ and ordél* occurs several times. [*O. Frs.* or-, ur-dél : *O. Sax.* ur-deili : *O. H. Ger.* ur-teil, -teila, -teili *judicium, sententia.*] v. ísen-, wæter-ordál.

ordál-ísen, es ; *n. The iron used in the ordeal*, L. Ath. iv. 6; Th. i. 224, 14.

ord-bana, an; *m. One who slays with* (*the point of*) *a weapon* (ord, cf. ecg-bana), *a murderer :*—Ic fylde mid folmum ordbanan Abeles (*Cain*), Cd. Th. 67, 7; Gen. 1097.

ord-ceard. v. ort-geard.

ord-fruma, an; *m.*   I. *of things, source, origin :*—Ordfruma *origo*, Ælfc. Gr. 9, 3; Som. 8, 58. Ós byþ ordfruma ælcere sprǽce, Runic pm. Kmbl. 340, 5; Rún. 4.   II. *of persons,* (1) *author, source,* (a) applied to the Deity :—Crist, ordfruma ælcere gife, Homl. Th. ii. 526, 7. Ordfruma ealre clǽnnesse, Blickl. Homl. 13, 21. Drihten is ordfruma (*auctor*) ealra eádignesse, Bd. 4, 30; S. 609, 16. God, lífes ordfruma, Exon. Th. 14, 30; Cri. 227. Ordfruma ealra gescafta, Cd. Th. 292, 17; Sat. 442. (b) applied to others :—Se wæs ordfruma (*auctor*) ðæs gefeohtes, Bd. 3, 24; S. 556, 32. Danaus ðæs yfeles ord- fruma *scelerum fabricator Danaus*, Ors. 1, 8; Swt. 40, 16 : Nicod. 6; Thw. 3, 14 : 29; Thw. 17, 4. (2) *chief, head, prince :*—Wæs mín fæder æþele ordfruma, Beo. Th. 531; B. 263. Daniel wæs ordfruma earmre láfe, Cd. Th. 225, 10; Dan. 152. Ðonne beóþ ða synfullan genýderade mid heora ordfruman, swá hé genýderad wearþ, Blickl. Homl. 33, 1. [*O. Sax.* ord-frumo : *O. H. Ger.* ort-frumo *auctor.*]

ord-stapu; *gen.* -stæpe; *f. A step of a pointed instrument, the prick* or *wound made by a sharp point :*—Oft mec ísern scód sáre on sídan; ic nǽfre meldade monna ængum, gif mé ordstæpe egle wǽron, Exon. Th. 485, 19; Rä. 71, 16.

ord-wíga, an; *m. A warrior who fights with a pointed weapon* (? cf. gár-wíga), or *one who fights in the van* (? v. ord, II) :—Ordwýga! ne lǽt ðín ellen gedreósan tó dæge, Wald. 9; Vald. 1, 6.

óre, an; *f. A mine, place in which ore is dug* :—Ísern óre *ferri fodina, in quo loco ferrum foditur*, Wrt. Voc. ii. 148, 11. v. óra ore.

or-eald; *adj. Of great age* :—Caron wæs swíðe oreald, Bt. 35, 6; Fox 168, 20. [*O. H. Ger.* ur-alt *valde senex, grandaevus, veteranus, decrepitus*.]

or-eldo. v. or-ildu.

orel, es; *n.: orl*, es; *m. A garment, veil, mantle* :—Orel, ryft cycla[s], Wrt. Voc. ii. 131, 38. Orelu *oraria*, 65, 5. Winpel *vel* orl *ricinum;* orl *orarium vel ciclas*, i. 17, 1–3: *stola vel ricinum*, 40, 34. Orlas *ciclas vel oraria*, 59, 40. Hé geglǽngde mé mid orle (*the monastic veil?*), Homl. Skt. i. 7, 36. Wimplum ł orlum *cycladibus*, Hpt. Gl. 486, 47: *velaminibus*, 526, 54. [*Goth.* aurali *a napkin: O. H. Ger.* oral *strophium, peplum, flammeolum*. From *Lat. orale*.]

orenum, Nar. 24, 2. v. or-wéne.

oreþ, oreþian. v. oroþ, orþian.

oret, es; *n.* (?) *Struggle, labour* :—Ðonne ðú ðínes gewinnes wæstme byrgest etest oretes *labores fructuum tuorum manducabis*, Ps. Th. 127, 2. v. following words.

oreta. v. oretta.

oret-mæcg, es; *m. A combatant, warrior, champion* :—Hí (*the Jews*) slógon eornoste Assiria oretmæcgas (*the army of Holofernes*), Judth. Thw. 24, 39; Jud. 232. Oretmecgas (*Beowulf and his band*), Beo. Th. 669; B. 332: 732; B. 363: (*Hrothgar's men*), 967; B. 481. Orettmæcgas (*the disciples*), Andr. Kmbl. 1328; An. 664. Weóld Walum and Scottum and Bryttum eác byre Æðelrédes, Englum and Sexum, oretmæcgum, Chr. 1065; Erl. 196, 30. v. next word.

oret-mæcga, an; *m. A combatant, athlete* :—Oretmæcga *agonista*, Wrt. Voc. ii. 1, 2. Oretmæcgan *anthletae*, 3, 3.

oret-stów, e; *f. A place where a struggle is carried on, a place for wrestling* :—Oretstówe ł winstówe ł plegstówe *scammatis*, Hpt. Gl. 405, 39. Oredstówe *in scammate*, 478, 48.

oretta, an; *m. One who strives, a combatant, warrior, champion* :—Wearp ðá wunden mǽl yrre oretta (*Beowulf*), Beo. Th. 3068; B. 1532: 5070; B. 2538. David, eádig oretta, Andr. Kmbl. 1757; An. 881. Beorn beaduwe heard ... ánrǽd oretta ... Cristes cempa (*St. Andrew*), 1965; An. 985. Þegnas lǽrde eádig oreta (*St. Andrew*), eorlas trymede, 925; An. 463. Eádig oretta andwíges heard (*Guthlac*), Exon. Th. 112, 21; Gú. 147. Swá sceal oretta compian, 122, 33; Gú. 315. Godes orettan swencan, 136, 15; Gú. 541.

orettan. v. on-orettan.

orf, es; *n. Cattle, live stock* :—Ælce geáre byþ orf ácenned, and mennisce menn tó mannum ácennede, ða ðe God gewyrcþ swá swá he geworhte ða ǽrran, Hexam. 12; Norm. 20, 20. Cuce orf, L. Edg. S. 8; Th. i. 274. 25. Swá mycel órfes wæs ðæs geáres forfaren, swá nán man ǽr ne gemunde, Chr. 1041; Erl. 169, 7. Hé nam him on orfe and on mannum and on ǽhtum swá him gewearþ, 1052; Erl. 183, 22. Hé hæfde on orfe micele ǽhte *fuerunt ei oves et boves*, Gen. 12, 16. Ælces cynnes orf *animantia diversi generis*, Ex. 12, 38. Habbaþ ðæt orf eów gemǽne *omnia animantia diripiens vobis*, Jos. 8, 2. Hí námon eall ðet orf ðe hí mihton tó cuman, ðæt wæs fela þúsend, Chr. 1064; Erl. 196, 5. Drífaþ hider eówre orf *adducite pecora vestra*, Gen. 47, 16.

orf-cynn, es; *n. Cattle* :—Næs orfcynnes nán máre búton vii. hrúðeru, Chart. Th. 429, 5. Of eallum orfcinne *de jumentis in genere suo*, Gen. 6, 20.

orf-cwealm, es; *m. Pestilence among cattle, murrain* :—On ðisum geáre wæs swá mycel orfcwealm swá man ne gemunde fela wintrum ǽr, Chr. 1054; Erl. 188, 5. Ús stalu and cwalu, stric and steorfa, orfcwealm and uncoþu ... derede swýðe þearle, Wulfst. 159, 10.

or-feorm; *adj. Unprovided, destitute, worthless* :—Ðæt biþ feóndes bearn, hafaþ grundfúsne gǽst Gode orfeormne (of feormne, MS.) wuldor-cyninge (*a godless spirit*), Exon. Th. 316, 16; Mód. 49. Ða (*the heathen gods*) sind geásne góda gehwylces, idle, orfeorme, unbiþyrfe, 255, 20; Jul. 217. Hwider hweorfaþ wé hláforðleáse, góde orfeorme, synnum wunde (cf. gif wé gewítaþ fram ðé, ðonne beó wé fremde from eallum ðǽm gódum ðe ðú ús gegearwodest, Blickl. Homl. 233, 31–33), Andr. Kmbl. 812; An. 406. Gǽstas góde orfeorme, wuldre bescyrede, 3233; An. 1619: Judth. Thw. 25, 21; Jud. 271.

or-feormness, e; *f. Want of cleanliness* (v. feormian *to cleanse*), *squalor* :—Orfeormnisse *squalores*, Wrt. Voc. ii. 121, 8. v. or-firme.

orf-gebitt, es; *n. Grazing;* herbitum, Wrt. Voc. i. 39, 24.

or-firme; *adj. Uncleanly, squalid* :—Hí wǽron fúlíce and orfyrme on heora beardum, Guthl. 5; Gdwin. 34, 22. v. or-feormness *and next word*.

or-firm[u]; *f. Squalor* :—Orfiermae, orfermae *squalores*, Txts. 96, 933. v. preceding word.

or-gálscipe (?), es; *m. Wantonness* :—Orgálscype (on gálscype (?), orgelscipe (?)), wrǽnscipe *petulantia*, Hpt. Gl. 525, 74.

organ, es; *m. A song* :—Se organ *the Pater Noster* (cf. v. 47, where it is called *cantic*), Salm. Kmbl. 107; Sal. 53. Gif hé ðæs organes ówiht cúðe, 65; Sal. 33. Organa swéz ðe from englum biþ sungen, L. E. I. pref.; Th. ii. 400, 11. v. organian.

organe (organa (?); cf. *O. H. Ger.* organa; *f.*), an; *f.: organon; pl.* organa; *n. A musical instrument* :—Organon, Exon. Th. 207, 4; Ph. 136. Ða organa wǽron getogene, and ða bíman gebláwene, Th. Ap. 25, 15. Organan *organo*, Ps. Surt. 150, 4. On salig wé úre organan up áhengan *in salicibus suspendimus organa nostra*, Ps. Th. (Spl. T., Surt.), 136, 2. Iubal wæs fæder herpera and ðæra ðe organan macodun *Iuba, fuit pater canentium cithara et organo*, Gen. 4, 21. [*Icel.* organ; *n.*]

organe, an; *f. Marjoram;* origanum vulgare :—Organe. Deós wyrt ðe man *origanum* and óðrum naman ðam gelíce organan nemneþ, Lchdm. i. 236, 9–11: 282, 23.

organian, orgnian *to sing to the accompaniment of a musical instrument* :—Ic orgnige (organige, MS. H.), Ælfc. Gr. 28, 7; Som. 32, 62.

orgel *pride* :—Hwǽr is heora prass and orgol búton on moldan beþeaht and on wítum gecyrred? Wulfst. 148, 32. [Woreldes richeise wecheð orgel on mannes heorte, O. E. Homl. ii. 43, 17. *The form* orguil *occurs*, p. 63. Heó leapeð into horel (orhel, MS. T.: orȝel, MS. C.), A. R. 224, 2. Cf. *French* orgeuil (*to which Brachet assigns a German origin*): *Ital.* orgoglio.] v. orgel-líc.

orgel-dreám, es; *m. The sound of a musical instrument* :—Orgeldreáme *organo*, Blickl. Gl.

orgele (? cf. *O. H. Ger.* orgela: *Ger.* orgel; *f.*: orgles, Alis. 191) *an organ, a musical instrument*. v. preceding word.

orgel-líc; *adj.* I. *proud, arrogant, disdainful* (v. next word). II. *deserving scorn* or *disdain* :—Hwý sceal ǽnigum menn þyncean tó orgellíc ðæt hé onbúge tó óðres monnes willan *qua conscientia dedignatur homo alienae voluntati acquiescere?* Past. 42, 2; Swt. 307, 15.

orgel-líce; *adv. Proudly, arrogantly, haughtily, insolently* :—Hé hine swá orgellíce up áhóf and bodode ðæs ðæt hé úþwita wǽre ne cýðde hé hit mid nánum cræftum ac mid leásum and ofermódlícum gilpe *hominem, qui non ad verae virtutis usum, sed ad superbam gloriam falsum sibi philosophi nomen induerat*, Bt. 18, 4; Fox 66, 29. Ða áxode Pilatus hine orgollíce, Homl. Th. ii. 250, 29. Orgellíce, Homl. Skt. i. 9, 76. Hé cwæþ orgællíce, 5, 449. Hé forþ stæpþ wel orglíce swylce hwylc cyng of his giðfenre stæppe geglenged, Anglia viii. 298, 34.

orgelness, e; *f. Pride, elation* :—Orgelnysse *elationis*, Hpt. Gl. 432, 54.

orgel-word, es; *n. An arrogant, insolent speech* :—Ða cwæþ se ealdorbiscop mid orgelworde, Homl. Th. ii. 248, 21.

or-gete, -gyte, -geate; *adj. To be perceived, manifest* :—Ðæt tácn núgyt is orgyte (*pervidetur*), Ors. 1, 7; Swt. 38, 34. Orgeate, Exon. Th. 76, 12; Cri. 1238: 347, 6; Sch. 8. Tácen orgeatu, 75, 3; Cri. 1216. Is gesýne sóþ orgete cúð oncnáwen, ðæt ðú cyninges eart þegen geþungen, Andr. Kmbl. 1052; An. 526. Is seó wyrd mid eów orgete geþungen, Andr. Kmbl. 1052; An. 526. Is seó wyrd mid eów orgete, 1517; An. 760. Andrea orgete wearþ folces gebǽro, 3137; An. 1571. Ic eów secgan mæg sóþ orgete, 1702; An. 853. Ðú meaht geseón orgete on mínre sídan swátge wunde, Exon. Th. 89, 17; Cri. 1458.

or-gilde; *adj. Unpaid for*, applied to one for whom the wergild is not paid :—Gif hine (*the man who has broken his pledge, and will not submit to the penalty*) mon ofsleá, licgge hé orgilde, L. Alf. pol. 1; Th. i. 60, 15. v. ǽ-gilde.

orglíce, or-gyte. v. orgellíce, or-gete.

orgol. v. orgel.

or-hlyte; *adj. Having no share of, not participating in, free from, without* :—Orhlyte oððe bedǽled *expers*, Ælfc. Gr. 9, 43; Som. 13, 1: 47; Som. 48, 44. Orhlita *exsors*, Wrt. Voc. i. 51, 48. Wá ðære sáwle ðe orhlyte hyre líf ádríhþ ðæra háligra mihta, Homl. Th. i. 346, 25. Orhlyte ýdeles gylpes, ii. 286, 28. Ne bist ðú orhlyte eallunge ðæra wítena *you shall not altogether escape those torments*, 310, 27. Ðæt gé eallunge ðæs andgites orhlyte ne sýn, 188, 28. Eádiges orhlytte, Andr. Kmbl. 1359; An. 680. [*O. H. Ger.* ur-hlozi, -hlozzo *exsors*.] Cf. wan-hlyte.

orige (?) *in the following passage* :—Se ðe þeóf geféhþ hé áh .x. scitt. ... Gif hé ðonne óþierne and orige (orrige, MS. H.) weorðe ðonne biþ hé wítes scyldig *he who catches a thief shall have ten shillings ... If he (the thief) run away, and gets clear off* (?), *then shall he* (the captor. For the responsibility of one who lets a thief escape, see L. In. §§ 36, 72) *be liable to fine*, L. In. 28; Th. i. 120, 7.

or-ildu (o); *f. Extreme old age* :—Hine (*death*) gelettan ðæt hé ðý lator cymþ, ge furþum óþ oreldo hí hine hwílum lettaþ (*put off death until extreme old age*), Bt. 41, 2; Fox 246, 9. Ǽ ic wundor ðín weorðlíc sægde and ic ðæt swá oryldu áwa fremme *usque nunc pronuntiabo mirabilia tua, et usque in senectam et senium*, Ps. Th. 70, 16. v. or-eald.

orl. v. orel.

or-læg, -leg, es; *n.* (?) *Fate* :—Nó ic (*Daniel*) wið feohsceattum ofer folc bere Drihtnes dómas, ac ðé (*Belshazzar*) unceápunga orlæg secge, worda gerýnu *I will tell thee thy fate (by explaining the writing on the wall*), Cd. Th. 262, 19; Dan. 746. Hé ðonne á tó ealdre orleg dreógeþ *he then for ever and ever undergoes his fate in hell* (cf. *Icel.* drýgja örlög, to 'dree' one's 'weird'), Exon. Th. 446, 29; Dóm. 29. [*O. H. Ger.* ur-lag; *m. fatum: Icel.* ör-lög; *n. pl. fate;* also *war.*] v. or-lege.

orlæg-gífre; *adj. Eager to cause death* (?) :—Ismahel biþ unhýre orlæggífre wiðerbreca wera cneórissum, Cd. Th. 138, 6; Gen. 2287.

or-leahter = *dis-crimen :*—Orleahter *discrimen,* i. *periculum, damnum,* Wrt. Voc. ii. 140, 82. Orhlættras *discrimina,* Hpt. Gl. 450, 43.

or-leahter; *adj. Blameless, faultless :*—Ðæt wæs ān cyning æghwæs orleahtre, Beo. Th. 3776; B. 1886. Æghwylc mennisc leahter on ðæm eádigan Sancte Iohanne cennendum gestilled wæs, and hié on eallum heora lífe orleahtre gestódan, Blickl. Homl. 163, 17.

orleg-ceáp, es; *m. Battle-bargain, fighting* (?) :—Ðær wæs eáðfynde eorle orlegceáp se ðe ǽr ne wæs níðes genihtsum *there might fighting be easily found for the man that before had not had enough of war,* Cd. Th. 120, 13; Gen. 1994.

or-lege, es; *n.* I. *war, strife, hostility :*—Ðá wæs orlege eft onhréred, níð upp árás, Andr. Kmbl. 2605; An. 1304. Ic ðæs orleges or anstelle *(speaking of the strife of the elements),* Exon. Th. 386, 9; Rä. 4, 59. Se ðæs orleges or onstealde, Beo. Th. 4805; B. 2407. Ðonne wē on orlege hafelan weredon, ðonne hniton fēþan, 2657; B. 1326. Nalæs late wǽron eorre æscberend tó ðam orlege, Andr. Kmbl. 94; An. 47: 2411; An. 1207. Hēt wǽpen on ðam orlege formeltan, 2293; An. 1148. Hȳ hine brēgdon, budon orlege, egsan and ondan, Exon. Th. 136, 5; Gú. 536. Ðú hafast þurh ðín orlegu ofer witena dóm wísan gefongen, widsæcest tó swíðe ðínum brýdguman *thou hast by thy hostile proceedings acted contrary to the judgement of wise men, dost reject too violently thy suitor,* 248, 17; Jul. 97. II. *a place where hostility is shewn :*—Cwæ̂don ðæt hē on ðam beorge byrnan sceolde . . . gif hē monna dreám o ðam orlege eft ne wolde sylfa gesēcan, 114, 3; Gú. 167. Ðá ðú heán and earm on ðis orlege ǽrest cwóme, 129, 24; Gú. 426. (In both passages the word seems to mean the place which Guthlac had selected for his dwelling, and from which the evil spirits, that before occupied it, wished to drive him.) Hafaþ nú se hálga helle bireáfod ealles ðæs gafoles ðe hí geárdagum in ðæt orlege swealg, 35, 18; Cri. 560. [Cf. *O. Sax.* orlegas (-lages, -lagies) word *battle-cry :* O. Frs.* or-loch *war :* O. H. Ger.* or-loge, -liugi *bellum,* Grff. ii. 137: *Icel.* or-lygi *fate, battle :* Dan.* or-log *warfare at sea :* Du.* or-log *war.* v. Grmm. D. M. 381, 817.] v. or-læg *and next word.*

or-lege; *adj. Hostile :*—Wēpaþ and heówaþ eall orlegu folc, for ðam úre God eów hæfþ ofercumen . . . orlega þeóda hē âlēde under úre fēt, Ps. Th. 46, 1–3. Ne onēgdon ná orlegra níð, ðeáh ðe feónda folc fēran cwóme, Cd. Th. 259, 26; Dan. 697.

orleg-from; *adj. Stout in battle :*—Oft ic gǽstberend cwelle compwǽpnum; cyning mec . . . hwílum lǽteþ sceacan orlegfromne, Exon. Th. 401, 21; Rä. 21, 15.

orleg-hwíl, e; *f. Battle-time, time of war :*—Nú is leódum wēn orleghwíle, Beo. Th. 5814; B. 2911. Fela ic gúþræsa genæs, orleghwíla, 4845; B. 2427. [Cf. *O. Sax.* orlag-hwíla *the hour of death.*]

orleg-níð, es; *m. Hostility, strife,* Cd. Th. 6, 6; Gen. 84: 56, 20; Gen. 915.

orleg-stund, e; *f. A time of trouble, time when the unfavourable decree of fate is carried out :*—Dreóge þ earfoþu orlegstunde, Salm. Kmbl. 750; Sal. 374.

orleg-weorc, es; *n. War-work, action :*—Se ðæt orlegweorc *(the defeat of the people of Sodom)* gecȳdde, Cd. Th. 122, 2; Gen. 2020.

or-mǽte; *adj. Immense, excessive :*—Ormǽte gigas, Hymn. Surt. 44, 13. Ormǽde, 112, 23. Ðǽr læg sum ormǽta stán, Homl. Th. ii. 164, 29. Duru ormǽte, Exon. Th. 19, 32; Cri. 309. Þreát ormǽte, 270, 14; Jul. 465. Þreá ormǽte, Andr. Kmbl. 2333; An. 1168. Hē mid ormǽtre angsumnysse wæs gecwylmed, Homl. Th. i. 88, 5. Bifigende mid ormǽtre cwacunge, 504, 28. For ðære ormǽtan fretnysse, ii. 542, 20. Hié woldon ormǽte feoh gegaderian, Bt. 24, 2; Fox 82, 17. Ðá gesomnode man ormǽte fyrde, Chr. 1001; Erl. 137, 10. Ða ormǽtan *minacem,* Wrt. Voc. ii. 55, 1. Ormǽte buccan *magnicaper,* i. 23, 58. Lecgan him onuppan ormǽte *(ingentia)* weorcstánas, Jos. 10, 27. Ic dreág yfel ormǽtu, Exon. Th. 280, 10; Jul. 627. Þurh ða ormǽtan ēhtnyssa, Homl. Th. i. 6, 2. [*Orm.* orr-mæte.]

or-mǽte; *adv. Excessively, exceedingly, without measure :*—Mē ðínes húses heard ellenwód æt ormǽte *(or adj.?),* Ps. 68, 9.

[**ormǽt-líc**; *adj. Excessive :*—Ðises geáres wurdon ormǽtlíca wæ̂dera, Chr. 1117; Erl. 246, 14.]

ormǽtness, e; *f. Excess, immensity :*—Hátheorte láreówas þurh wódnysse hátheortnysse lāre gemet tó ormǽtnysse wælhreównysse gecyrraþ *iracundi doctores per rabiam furoris disciplinae modum ad inmanitatem crudelitatis convertunt,* Scint. 32. Þurh ormǽtnysse ðæs godcundlícan leóhtes, Homl. Th. ii. 186, 15. Micelre ormǽt[nysse] *mirae magnitudinis,* Hpt. Gl. 454, 77. Nǽht elles gestincan búton unstenca ormǽtnysse, Wulfst. 139, 8.

or-met (?) *a very great mass, something immense :*—Ormetum *molibus,* Wrt. Voc. ii. 55, 75 : 114, 20. Cf. ge-met.

or-met[t], -mete; *adj. Excessive, without measure :*—Ymbhogena ormete rēn *(cf.* se rēn ungemetlícs ymbhogan, Bt. 12; Fox 36, 19), Met. 7, 36. Hē mid ormettum mynum mē gefretewode *he decked me with priceless jewels,* Homl. Skt. i. 7, 37.

or-mód; *adj. Without courage, hopeless, despairing :*—Ðis folc is geirged and ormód ongēn eów *elanguerunt omnes habitatores terrae,* Jos.

2, 9. Se ðe hine forþencþ, se biþ ormód, Bt. 8; Fox 24, 18. Wæs ðá ormód eorl, âre ne wēnde, ne on ðam fæstene frófre gemund, Met. 1, 78 : 5, 30. Mín sylfes gást wæs ormód worden *defecit spiritus meus,* Th. 76, 4. Ðȳ læs hē ormód sȳ ealra þinga, Exon. Th. 294, 12; Crä. 14. Ne beó ðú tó ormód ðeáh ðe sí on unriht gedēmed *be not too much discouraged, though judgement be given wrongfully against thee,* Prov. Kmbl. 34. Ða lytelmódan ðonne hié ongietaþ hiera unbǽldo, hié weorðaþ oft ormóde *(in desperationem cadunt),* Past. 32, 1; Swt. 209, 8: Homl. Th. i. 536, 6: Nar. 32, 23. Hig ormóde (orwēne, MS. D.) ne gedó, L. de Cf. 1; Th. ii. 260, 14. [*O. H. Ger.* ur-mót *disperatus.*]

or-módness, e; *f. Desperation, despair :*—Ormódnes *disperatio,* Wrt. Voc. ii. 140, 72. Mid ðȳ hē ús geseah on ormódnesse *(in desperatione)* gesette, Bd. 5, 1; S. 614, 5. Ðá se earma man ðus mid ormódnesse sprecende wæs *sic loquebatur miser desperans,* 5, 13; S. 633, 21. Tó ormódnesse *ad desperationem,* Past. 14, 3; Swt. 83, 19: 21, 7; Swt. 165, 19. Hǽfde hine seó deófollíce strǽl mid ormódnysse gewundode : wæs se eádiga wer Gúðlác mid ðære ormódnysse þrí dagas gewundod, ðæt hē sylfa nyste hwider hē wolde mid his móde gecyrran, Guthl. 4; Gdwin. 28, 13–17. Ic habbe ongiten ðíne ormódnesse . . . ðú sǽdest ðæt ðú wǽre bereáfod ǽlces gódes, Bt. 5, 3; Fox 12, 31. Ic eom geunrótsod fulneáh óþ ormódnesse, 41, 2; Fox 246, 14.

orne; *adj. Unhealthy, harmful :*—Mid Godes fultume ne wyrð him nán orne *with God's help no harm will be done him,* Lchdm. iii. 16, 5. Wið ornum útgange, 70, 25. v. un-orne.

ornest, es; *n. Trial by battle :*—Gif Englisc man beclypaþ ǽnigne Frænciscne mann tó orneste for þeófte . . . oððe for ǽnigan þingan ðe gebyrige ornest for tó beónne . . . hæbbe hē fulle leáfe swá tó dónne. And gif se Englisca forsæcþ ðæt ornest, W. ii. 1; Th. i. 489, 5–9 : ii. 2–3; Th. i. 489, 11–25. v. eornost, orrest.

oroþ, orþ, es; *n. Breath, breathing :*—Oroþ oððe gást *flamen,* Wrt. Voc. ii. 37, 11 : *flatus, spiritus,* 149, 32 : anhela, Rtl. 192, 21. Hē oroþ stundum teáh . . . swá wæs ðæt áfen oroþ up hlæden, Exon. Th. 178, 17–30; Gú. 1245–1252. Heora oruþ biþ swylce líg *ignem et flammam flantes,* Nar. 34, 32 : Beo. Th. 5107; B. 2557. Orþ *spiraculum (cf.* lífes orþung *spiraculum vitae,* Gen. 2, 7), Kent. Gl. 757. Orþas ł hfæstes (= orþes ł fnæstes) *spiritus,* Hpt. Gl. 464, 24 : 454, 66. Oreþe *aura,* Wrt. Voc. ii. 6, 56 : *flatu,* 38, 9. Wið ǽttorsceaþan *(dragon)* oreþe, Beo. Th. 5671; B. 2839. Eallinga gewǽced and ðam orþe belocen, Glostr. Frag. 102, 13. Hē mid langre swóretunge ðæt orþ of ðam breóstum teáh, Guthl. 20; Gdwin. 84, 20. Þurh áttres oraþ, Salm. Kmbl. 441; Sal. 221. Ðú him on ðvdest oruþ and sáwul, Hy. Grn. 9, 55. Oroþo anhelae, Rtl. 192, 25.

orped; *adj. Grown up, of full strength, stout, active, bold :*—Lá orpeda cleric, gif ðú wylle witan ða terminos ðe wē ymbe sprǽcon, wite hwylc gēr hyt sȳ ðæs mónan ðæt man hǽt *lunaris,* Anglia viii. 325, 5. Swá gedafenaþ esnum ðam orpedan, ðonne hē gód weorc ongynþ, ðæt he georhlíce beswynce, 324, 17. Orped[n]e, snelne *adultum,* Hpt. Gl. 485, 25. [Orpud *audax, bellipotens,* Promp. Parv. 371, v. *note for other examples of the word.* Þe guode kniȝt and orped, þet heþ guod herte and hardi, Ayenb. 183, 6. *Jamieson gives* orpit *proud.*] v. next word.

orpedlíce; *adv. Boldly, in full force :*—Wē willaþ âmearkian ðás epactas and eác ða regulares lunares, ðæt hig openlíc[r]e and orpedlíce standun beforan ðæs preóstes gesyhþe *that they may stand out clearly and boldly in sight of the priest,* Anglia viii. 301, 31. [Cf. But for þe emperour hadde out of his companye þe orped man *(virum strenuum)* Bonefacius, þe emperour dede noþing orpedliche *(nihil strenue egit),* Trev. v. 231, 13–15. He orpedly strydeȝ, Bremly broþe, Gaw. 2232. Þenne orppedly in to his hous he hyȝed to Sare, Allit. Pms. 56, 623.]

orrest *battle :*—Hē hine on orreste ofercom, Chr. 1096; Erl. 233, 4. [A Danish form, *Icel.* orrosta *battle.* *Orm.* he wass Inn orresst ȝæn þe deofell.]

orretscipe, es; *m. Infamy, disgrace :*—Ðæs unhlíseádgan orretscipe *infamis,* Wrt. Voc. ii. 44, 49. Orretscipe *infamis,* 85, 11.

orrettan *to disgrace, put to shame, cover with confusion :*—Orretteþ *turpabat,* Wrt. Voc. ii. 91, 18 : *subfundit,* 78, 19. Onretteþ (or-?) *deturpans,* 26, 56 : 82, 56. Cf. georrettan *infamare,* 47, 26 : 92, 33. v. ge-orettan.

or-sáwle; *adj. Without soul, lifeless :*—Orsáule *exsangue,* Wrt. Voc. ii. 33, 28. Nǽs ðá deád ðá gyt, ealles orsáwle, Judth. Thw. 23, 6; Jud. 108. Saga ðæt heó láme bilúce líc orsáwle in þeóstorcofan, Exon. Th. 173, 28; Gú. 1167.

or-sceattinga; *adv. Gratuitously, free of charge :*—Hí láreówas orsceattinga sealdon *magisterium gratuitum praebere curabant,* Bd. 3, 27; S. 558, 27. Cf. or-ceápe, -ceápunga.

or-sorg, -sorh; *adj. with gen.* I. *free from care, without anxiety, secure, prosperous :*—Orsorh *securus,* Kent. Gl. 365 : Wrt. Voc. i. 83, 59. Orsorg *lentus,* ii. 96, 62 : *consors,* 15, 23 : 105, 18. Orsorh wæpna *securus armorum,* Ælfc. Gr. 41; Som. 44, 9. Se tó ánra ðara burga *(the cities of refuge)* geflíehþ ðonne mæg hē beón orsorg ðæs monnsliehtes *he may be without anxiety as to the manslaughter he has committed; reus perpetrati homicidii non tenetur,* Past. 21, 7; Swt. 167, 20. Ne þorítest ðú ðe nánwuht ondrǽdan . . . Ðonne ðú ðonne orsorg wǽre, Bt. 14, 3;

Fox 46, 30. Næs ic næfre swá emnes módes ðæt ic eallunga wære orsorg, ðæt ic swá orsorg wære ðæt ic náne gedréfednesse næfde, 26, 1 ; Fox 90, 26. Seó wiðerwearde wyrd byþ ǽlcum men nytwyrðre ðonne seó orsorge (prospera), 20 ; Fox 70, 30. Orsorg líf lǽdaþ woruldmen wíse, ðonne hé forsihþ eorþlícu gód and ðara yfela orsorh wunaþ, Met. 7, 43. Hé furþon orsorh ne brícþ his genihtsumnysse he does not enjoy even his abundance without anxiety, Homl. Th. i. 64, 34. Uton lǽtan bión ðás sprǽce and bión unc ðæs orsorge secure concludere licet, Bt. 34, 7 ; Fox 144, 18. Tó upáhafen for orsorgum woruldgesǽlþum (cf. on ðínre orsorgnesse, Fox 14, 35) too much uplifted on account of untroubled earthly felicity, Met. 5, 33. II. secure from danger, safe :—Orsorg tuta, Wrt. Voc. ii. 123, 2. Samson eode him swá orsorh of heora gesihþum, Jud. 16, 3. Hé ús sealde orsorh wuldor (glory secure from the assaults of men), Blickl. Homl. 151, 12. Ða hálgan martyras orsorge becómon tó wulderbeáge ðæs écan lífes, Homl. Th. i. 416, 9. Wit begrá ǽr wǽron orsorge we before were safe from both (hunger and thirst), Cd. Th. 50, 5 ; Gen. 804. Wé beóþ for eów and eów orsorge gedóþ (cf. wé gedóþ eów sorhléase securos vos faciemus ; we will secure you, Mt. Kmbl. 28, 14), Nicod. 17 ; Thw. 8, 23. [O. H. Ger. ur-sorg securus.]

orsorg-líc ; adj. with gen. Secure :—Ðæt líf ðara gesinhíwena mæg bión orsorglíc ǽlcra wíta conjugalis vita a suppliciis secura est, Past. 51, 6 ; Swt. 399, 22.

orsorglíce ; adv. I. without anxiety :—Geoffra Gode ðone te ðú getuge, ðæt ðú ðý orsorglícor becume tó ðam ǽðelan wulderbeáge offer to God him whom thou hast brought up, that with the less anxiety thou mayest come to the noble crown of glory, Homl. Th. i. 418, 5. II. carelessly, rashly :—Ðæt hiera nán ne durre grípan suá orsorglíce on ðæt ríce, Past. 4, 2 ; Swt. 41, 5. III. securely, safely :—Forðam ðe hit swá earfoðe is ǽnegum menn tó witanne hwonne hé geclǽnsod síe, hé mæg ðý orsorglícor (tutius) forbúgan ða þegnunga, 7, 2 ; Swt. 51, 6. Hí woldon ðý máran anweald habban, ðæt hý mihton ðý orsorglícor ðissa woruldlusta brúcan, Bt. 24, 2 ; Fox 82, 15. Sió nafu fǽrþ micle fæstlícor and orsorglícor ðonne ða felgan, 39, 7 ; Fox 220, 30.

orsorgness, e ; f. I. freedom from care or anxiety, tranquillity :—Caru cura, orsorhnys securitas, Wrt. Voc. i. 83, 60–61. Sibb and orsorhnes pax et securitas, Bd. 4, 25 ; S. 601, 29. II. prosperity :—Dysigra monna orsorgness (prosperitas) hí fordéþ, Past. 50, 2 ; Swt. 387, 34. Ða míne sǽlþa and seó orsorgnes prosperitas mea, Bt. 10 ; Fox 26, 26. Seó orsorhnes... seó wiðerweardnes prospera fortuna... adversa 20 ; Fox 72, 4. Ðæt ðú ðé ne anhebbe on ðínre gesundfulnesse and on ðínre orsorgnesse, 6 ; Fox 14, 35. Cuman tó ræste and tó orsorgnesse, 25 ; Fox 88, 31. Hú forht hé sceal beón for ǽlcre orsorgnesse prospera formidanda, Past. 3 ; Swt. 33, 5 : Swt. 35, 1, 2, 8. Ðe ðisses middangeardes orsorgnesse ne gímþ, ne him náne wiðerweardnesse ne andrǽt qui prospera mundi postposuit, qui nulla adversa pertimescit, 10, 1 ; Swt. 61, 8. Orsorgnesse prosperitatem, 50, 1 ; Swt. 387, 22.

ort-geard, es ; m. An orchard, garden :—Orcyrd hortus, Wrt. Voc. i. 84, 51. Orceard, orcird, orcyrd, orcgyrd, ordceard, Ælfc. Gr. 8 ; Zup. 28, 11. Se ordceard, Cod. Dip. Kmbl. iv. 72, 5. Of ǽlcum treówe ðises orcerdes, Gen. 2, 16. Ðú ðe eardast on fríondes ortgearde (orcgearde, MS. Hat.), Past. 49, 2 ; Swt. 380, 14. Suá se ceorl déþ his ortgeard, 40, 3 ; Swt. 293, 4. God áplantode wynsumnisse orcerd (the garden of Eden), Gen. 2, 8. Beóþ hyra orcerdas mid ǽpplum áfyllede, Lchdm. iii. 252, 22. Seó eorþe stód mid holtum ágrówen ... mid ǽppelbǽrum treówum and mid orcgeardum, Hexam. 6 ; Norm. 12, 6. [Goth. aurti-gards.]

ortgeard-weard, es ; m. A gardener :—Orcerdweard ortulanus, Wrt. Voc. i. 84, 52.

orþ. v. oroþ.

or-þanc, es ; m. n. Original, inborn thought. I. mind, genius, wit, understanding ; ingenium :—Orþanc ingenium, cræftica artifex, Wrt. Voc. i. 47, 8–9. Líflíces orþa[nces] vivacis ingenii, Hpt. Gl. 407, 40–43. Hé genam þurh heora láre on his orþance ða egeslícan dǽda, Ælfc. T. Grn. 17, 21. Nú wolde ic ðæt ða ǽdela[n] clericas ásceócon fram heora andgites orþance ǽlce sleacnysse, Anglia viii. 301, 4. Gif ðonne [man] mid orþonce (skilfully) ðisses þinges fundian wille, Lchdm. i. 100, 6. Yfele orþance malo ingenio, Wrt. Voc. ii. 56, 8. Orþancas ingenia, Germ. 397, 423. Orþancum ingeniis, Wülck. Gl. 250, 5. II. a skilful contrivance or work, artifice, device, design :—Orþanc molimen (cf. searo molimen, Wrt. Voc. ii. 54, 29), Ælfc. Gr. 9, 12 ; Som. 9, 32. His ofermédu is fruma úres forlores and se orþonc (argumentum, cf. searwe argumenta, Wrt. Voc. ii. 84, 69) ðe we mid áliésde siendon is Godes eáðmódnes, Past. 41, 1 ; Swt. 301, 9. Mid orþance argumento, Hpt. Gl. 439, 3 : Wrt. Voc. ii. 9, 12. Orþonce, gleáwnysse argumento, 2, 11. Hwá is ðæt ðe cunne orþonc clǽne (the creation) nymðe éce God ? Cd. Th. 266, 6 ; Sat. 18. Orþancas argumenta, commenta, Hpt. Gl. 479, 68. Orþanc commenta, i. machinamenta, excogitata, astutia, argumenta, machinationes, ficta, fraudes, sarwa dicta, mendacia, Wülck. Gl. 206, 42–46. Orþonc machinamenta, Wrt. Voc. ii. 113, 74. Orþoncum, searwum commentis, 14, 82. Orþancum machinamentis, Hpt. Gl. 477, 9 : argumentis, 486, 19. Stán in goldfate smiþa orþoncum biseted, Exon. Th. 219, 8 ; Ph. 304 : Beo. Th. 817 ; B. 406. Ealle ða orþancas

tóslíteþ, Salm. Kmbl. 145 ; Sal. 72. ¶ Orþoncum skilfully, cunningly, ingeniously, with art :—Orþanc[um ?] subtiliter, sagaciter, Hpt. Gl. 407, 21. Is se sweora orþancum geworht (cunningly wrought), Exon. Th. 483, 15 ; Rä. 69, 3 : Beo. Th. 4180 ; B. 2087. Ðæt orþancum ealde reccaþ, Cd. Th. 200, 19 ; Exod. 359. [O. H. Ger. ur-dank argumentum, commentum.]

or-þanc, es ; m. [or without] Thoughtlessness, want of thought :—Nǽnig man scile ott orþances (heedlessly) út ábrédan wǽpnes ecgge, Salm. Kmbl. 329 ; Sal. 164.

or-þanc ; adj. Cunning, skilful :—Ceastra beóþ feorran gesýne, orþanc enta geweorc, wrǽtlíc weallstána geweorc, Menol. Fox 463 ; Gn. C. 2. Orþonc ǽrsceaft, Exon. Th. 477, 1 ; Ruin. 16. Mé þurh hrycg wrecen hongaþ under án orþonc píl, óðer on heáfde, Exon. Th. 403, 23 ; Rä. 22, 12. Hwǽr com heora snyttro and seó orþonce gláunes, and se ðe gebregdnan dómas démde ? Blickl. Homl. 99, 31.

orþanc-bend ; m. f. A skilfully contrived band :—Bewrigene orþoncbendum, Exon. Th. 429, 35 ; Rä. 43, 15.

orþancscipe, es ; m. Art, mechanical art, mechanics :—Orþancscipe mechanica (the word occurs at the end of a list of the arts. Cf. in a similar list searocræft mechanica, Wrt. Voc. ii. 81, 61), Hpt. Gl. 479, 61. Orþancscipe mechanicam, i. peritiam ł fabricam rerum, 528, 65. Searwa, orþanscipes (-as ?) molimina, ingenia, 502, 54.

orþian ; p. ode To breathe, pant :—Ic orþige spiro, Ælfc. Gr. 19 ; Som. 22, 42. Ic on orþige inspiro, 47 ; Som. 48, 44. Animal is ǽlc þing ðæt orþaþ, 5 ; Som. 4, 41. Ðonne se sacerd cristnaþ, ðonne orþaþ hé on ðone man, Wulfst. 33, 18. Gást oreþaþ spiritus spirat, Jn. Skt. 3, 8. Þurh ðæt lyft wé orþiaþ and eác ða nýtenu, Hexam. 4 ; Norm. 8, 18. Ælc þing ðe orþode omne quod spirare poterat, Jos. 10, 40. Orþode palpitavit (palpavit, MS.), Germ. 402, 73. Orþige palpitet, 398, 116. Hé ne gedyrstlǽceþ ðæt hé furþon orþige he dare not even breathe, Homl. Th. i. 456, 10. Hé earfoþlíce orþian mihte, 86, 8. Ðá ongann hé tó édele ðæs upplícan lífes mid eallum gewilnungum orþian then began he to pant for the country of the life above with all his desires, ii. 118, 26. Orþiende swétnyssa spirans balsama, Hymn. Surt. 98, 19.

orþung, e ; f. I. breathing, breath :—Ðæra dracena orþung ácwealde ðæt earme mennisc, Homl. Th. ii. 474, 6. Syllaba is stæfgefég on ánre orþunge geendod, Ælfc. Gr. 3 ; Som. 3, 13. Of orþunga gástes graman ðínes ab inspiratione spiritus irae tuae, Ps. Spl. 17, 18. Nán mann ne nýten nǽsþ náne orþunge búton þurh ða lyfte, Lchdm. iii. 272, 22. Oþ ða nýhstan orþuncge until his latest breath, L. Ælf. E. 4 ; Th. ii. 392, 10. God on áblewo on hys ansíne lífes orþunge (spiraculum vitae), Gen. 2, 7. II. a breathing-hole (? cf. preceding passage), a pore :—Orþung spiramentum vel porus, Wrt. Voc. i. 54, 67. v. on-orþung.

or-treówe, -triéwe, -trýwe ; adj. I. despairing, hopeless :—Ðá him eorla mód ortrýwe wearþ, Cd. Th. 187, 21 ; Exod. 154. Wé tó wáce hýraþ úrum Drihtne, and wé tó ortreówe (-trýwe, MS. A.: -trúwe, MS. C.) syndan Godes mihta and mildheortnesse, Wulfst. 91, 14. Hié æt níhstan wǽron ortriéwe (-treówe, MS. C.) hwæðer him ǽnig moneáca cuman sceolde, Ors. 4, 1 ; Swt. 158, 19. II. faithless, perfidious :—Ortrúes cyuesdómes perfidi pellicatus, Hpt. Gl. 521, 33. Ortreówra cempena perfidorum militum, 415, 48.

or-treównes, e ; f. Want of faith or confidence, mistrust :—Ortreównes diffidentia, desperantia, Wrt. Voc. ii. 140, 18. Hé æteówde ða wunda ðǽm ungeleáffullum mannum, forðon ðe hé nolde ðæt ǽnig ortrýwnes wǽre embe his ǽriste, Blickl. Homl. 91, 3.

or-trúwian ; p. ode To be without hope of, to despair of :—Hé ortrúwode his Drihtnys mildheortnysse he despaired of his Lord's mercy, Ælfc. T. Grn. 17, 25. Tó ortrúwienne desperandum, Wülck. Gl. 250, 36. v. next two words and ge-ortrúwian.

or-trúwung, e ; f. Despair :—Se ðe forgyfenysse be synne ortrúwaþ swýðor be ortrúwunge ðænne be synne áfealþ. Ortrúwung geýcþ synne qui veniam de peccato desperat plus de desperatione quam de peccato cadit. Desperatio auget peccatum, Scint. 34.

or-trýwan to despair of :—Ne ortrýwan hig Godes mildheortnysse ne desperent illi de misericordia Dei, L. Ecg. P. ii. 20 ; Th. ii. 190, 7.

or-trýwe. v. ortreówe.

or-tydre ; adj. Without offspring, barren :—Ontydre (Wülcker reads ortydre) effeto, sine foetu, Wrt. Voc. ii. 142, 46. v. on-tydre.

oruþ. v. oroþ.

or-wearde ; adv. Without guard, in an unprotected condition :—Syððan orwearde ǽnigne dǽl secgas geségon on sele wunian, lǽne licgan after men saw any part (of the dragon's hoard) lying there without its warder, Beo. Th. 6245 ; B. 3127.

or-wegnes, e ; f. Inaccessibility, remoteness :—Orwegnes devia, s. loca secreta et abdita, quasi extra via, vel invia, sine via, Wrt. Voc. ii. 139, 55.

orweg-stíg, e ; f. A path difficult of access :—Orwegstíg devia callis (-us, MS.), Wrt. Voc. ii. 139, 57. Horwegstíg (but cf. horu-weg), 25, 25.

or-wéna ; adj. with gen. Hopeless, despairing :—Ðá wearþ his ágen sunu yfele geuntrumed, and orwéna lífes læg æt forþsíðe, Homl. Skt. i. 3, 301 : Beo. Th. 2008 ; B. 1002 : 3134 ; B. 1565 : Exon. Th. 329, 27 ; Vy. 40. Friþes orwéna, 261, 25 ; Jul. 320. Ic eom orwéna, ðæt . . . ,

Cd. Th. 134, 10 ; Gen. 2222. Wǽron orwénan éðelrihtes, 191, 7 ; Exod. 211. Sindon gé firenum bifongne, feores orwénan, Exon. Th. 139, 27 ; Gú. 599. v. next word.

or-wéne ; *adj. with gen.*   I. *not having ground for hope, without hope, despairing :*—Biþ orwéne ðæt hé ne mǽge ða bóte áberan *desperet posse se emendationem perferre*, L. Ecg. P. i. 4 ; Th. ii. 172, 23. Hé wearþ his lífes orwéne, Homl. Th. i. 86, 28. Hé læg his lífes orwéne, Homl. Skt. i. 21, 301 : Glostr. Frag. 6, 18 : Chart. Th. 339, 22. Hié ðæs écan lífes orwéne wǽron, Blickl. Homl. 85, 27. Hié wǽron orwéne hwæðer..., Ors. 4, 9 ; Swt. 192, 4.   II. *not giving ground for hope, desperate, despaired of :*—Wénstú ðæt ic sceole sprecan tó ðissum treówleásan men and tó ðissum orwénan drý (*this desperate sorcerer*), Blickl. Homl. 183, 32. Æt orwénum lífe *when life is despaired of ; in extremitate vitae*, L. Ecg. P. i. tit. x ; Th. ii. 170, 18. Wé ða bútan orenum (orwénum ?) þingum mete þigdon *ab securis nobis epule capiuntur*, Nar. 24, 2. See preceding word.

or-wénness, e ; *f. Despair, hopelessness :*—Ðonne biþ him seó or-wénnys (*desperatio illa*) tó máran synne geteald, L. Ecg. P. i. 4 ; Th. ii. 172, 24. Hwí sprecst ðú mid swá micelre orwénnysse ? Homl. Th. i. 534, 22. On orwénnysse his ágenre hǽle *in despair of his own salvation*, Ælfc. T. Grn. 17, 24. Woldon hý geteón in orwénnysse Meotudes cempan, Exon. Th. 136, 27 ; Gú. 547.

or-weorð, -wurð, es ; *n. Ignominy, shame :*—Gefyl ansýne heora of orwurðe (*ignominia*), Ps. Spl. C. 82, 15. v. or-wirðu.

or-wíge ; *adj.*   I. *defenceless, without power of fighting :*—Or-wíge *inbellem*, Wrt. Voc. ii. 45, 69 : 111, 81. Ofsleán mé míne fýnd orwíge *decidam merito ab inimicis meis inanis*, Ps. Th. 7, 4. Saga hú ðú wurde ðus wígþríst ðæt ðú mec ðus fæste gebunde æghwæs orwíge (*without any power of resisting*), Exon. Th. 268, 18 ; Jul. 434.   II. *not liable to a charge of homicide*, said of one who, under the circumstances mentioned in the following passages, caused a person's death, but was not exposed on that account to the consequences which usually followed homicide (cf. *Icel. víg homicide*):—Wé cweþaþ ðæt mon móte mid his hláforde feohtan orwíge (onwíge, MS. H.), gif mon on ðone hláford fiohte ; swá mót se hláford mid ðý men feohtan (cf. Unicuique licet domino suo sine wita subvenire, L. H. I. 82, 3 ; Th. i. 590, 2). Æfter ðære ilcan wísan mon mót feohtan mid his geborene mǽge, gif him mon on wóh on feohteþ. And mon mót feohtan orwíge, gif hé geméteþ óðerne æt his ǽwum wífe, betýnedum durum oððe under ánre reón, oððe æt his dehter ǽwum-borenre, oððe æt his swister, oððe æt his médder ðe wǽre tó ǽwum wífe forgifen his fæder, L. Alf. pol. 42 ; Th. i. 90, 20–30. (Cf. L. H. I. 82, 4–8 ; Th. i. 590, 5–22.)

or-wirðed *disgraced*, cf. ge-oruuierdid *traductus*, Txts. 100, 990. Georuuyrde, 103, 2042. Georwyrðed *traducta*, Wrt. Voc. ii. 85, 14.

or-wirðlic ; *adj. Ignominious, shameful :*—God hine forlét in ðisse nyþerlícan worulde swá orwyrþlíche déþ þrowian, ðæt hé hine wolde in ðære heán worulde gelædan, Shrn. 123, 10.

or-wirðu, *indecl.* ; -wirð, e ; *f. Ignominy, shame, dishonour :*—Gefyl onsíene heora mid orwyrðe *imple facies eorum ignominia*, Ps. Surt. 82, 17. Mé ðín dohtor hafaþ geýwed orwyrðu *thy daughter hath shewn me dishonour*, Exon. Th. 246, 29 ; Jul. 69.

or-yldu. v. or-ildu.

ós *a divinity, god*, the Anglo-Saxon form of a word whose existence in Gothic is inferred from a passage in Jornandes, 'Gothi proceres suos quasi qui fortuna vincebant non pares homines sed semideos, id est, *Anses* vocavere.' The Icelandic, which throws out *n* before *s*, as the Anglo-Saxon does (cf. *Icel.* gâs : *A. S.* gôs), has *áss* ; *pl. æsir*, a term which has an application in the opening chapters of the Yngling Saga very similar to that attributed to *anses* among the Goths : Odin, Thor, and other personages of the Scandinavian mythology are the Æsir. Particularly apparently did the term refer to Thor, so that the proper name Ás-björn is used as the equivalent of Þor-björn. As the first part of Scandinavian proper names it occurs frequently, and it is in the same dependent character that it mostly, if not exclusively, is found in Anglo-Saxon and O. H. German. Thus Ós-beorn, Ós-lác, Ós-wine, Ós-weald preserve the word which is found in Ás-björn, Ás-lákr, Ás-mundr, and this is certainly the independent *áss*. The *O. H. Ger.* Ans-gár shews the same word. Whether *ós* in the sense of *god* occurs as an independent word is doubtful. It is the name of the Rune ᚩ, which in the Runic poem is accompanied by the following verse :—

'Ós byþ ordfruma ǽlcre sprǽce
Wísdómes wraðu and witena frófur
And eorla gehwam eádnis and tóhyht.'
           Runic pm. Kmbl. 340, 5–10 ; Rún. 4.

Kemble translates ós by mouth (as if the Latin word had been taken ?), but if the verse is old, the reference might be to Woden. Cf. the account of Óðinn in the Ynglinga Saga : þar þóttust Óðins menn eiga alt traust, er hann var, c. 2. Óðinn var göfgastr af öllum, ok af honum námu þeir allir íþróttirnar : því at hann kunni fyrst allar ok þó flestar. ... Hann ok hofgoðar hans heita ljóðasmiðir, því at sú íþrótt hófst af þeim í norðr-löndum, c. 6. See also c. 7, and Salm. Kmbl. p. 192 : Saga mé hwá

ǽrost bócstafas sette ? Ic ðé secge Mercurius (= *Woden*) se gygand. Further in Lchdm. iii. 54, in a charm, occurs a genitive pl. *ésa :*—Gif hit wǽre ésa gescot, oððe hit wǽre ylfa gescot, oððe hit wǽre hægtessan gescot, nú ic wille ðín helpan. Ðis ðé tó bóte ésa gescotes, &c. ... But though on the comparison of other forms, a nom. pl. ǽs might be inferred for *ós*, the change of vowel would not occur in the genitive, which should be *ósa*. *Ésa* would point to a singular és (cf. ést ; *Goth.* ansts). The meaning however of the word is that given to *ós*. See Grmm. D. M. p. 22.

ósle, an ; *f. An ousel, blackbird :*—Óslae *merula*, Txts. 78, 665 : Wrt. Voc. ii. 114, 1. Ósle, i. 281, 17. [*O. H. Ger.* amsala, amisala : *Ger.* amsel.]

osogen = á-sogen (?):—Osogen wǽre *sugillaretur* [cf. wǽre forsocen (*in margin* forgnegen), *sugillaretur*, Hpt. Gl. 484, 68], Wrt. Voc. ii. 82, 23.

óst, es ; *m.* (?) *A knot, knob :*—Ost *nodus*, Txts. 80, 688 : Wrt. Voc. ii. 60, 66. Copses, óstes *cippi*, Hpt. Gl. 482, 61. Yfele treówes on óste yfel nægel oððe wecg on tó fæstnigenne ys *male arboris nodo malus clavus aut cuneus infingendus est*, Scint. 27. Of ðǽm óstum ðæs treówes flóweþ út swétes stences wǽte, Shrn. 67, 29.

oster-hláf, es ; *m. An oyster-patty :*—Osterhláfas sint tó forbeódanne, Lchdm. ii. 210, 28. See Lchdm. iii. Glossary.

oster-scill, e ; *f. An oyster-shell :*—Mid ostorscyllum gecnucud and gemenged, Lchdm. i. 338, 16.

Óst-Gotan ; *pl. The Ostrogoths :*—Þeódríc Óstgotona cyning, Shrn. 85, 26.

óstig ; *adj. Knotty, rough, scaly :*—Óstig gyrd *scorpio*, Wrt. Voc. i. 21, 17. Óstig *nodosus*, óstigre *nodosa*, óstigum *nodosis*, Hpt. Gl. 483, 66, 65, 57. Ósties, rúches *nodosi*, 482, 60. Óstie *squamigeros, scabrosos*, 464, 45. Þý óstihan *nodosa*, Wrt. Voc. ii. 93, 37.

óstiht ; *adj. Knotty, rough :*—Óstihtum *nodosi*, Wrt. Voc. ii. 82, 2 : 60, 65.

ostre, an ; *f. An oyster :*—Ostre *ostrea*, Wrt. Voc. ii. 63, 71 : i. 65, 67 : *ostrea vel ostreum*, 77, 70. Ðonne cumaþ ða oftost of mettum and of cealdum drincan swá swá sindon cealde ostran and æpla, Lchdm. ii. 244, 2 : Coll. Monast. Th. 24, 9. [From Latin.]

Óst-sǽ *the Baltic with the Cattegat*, the water east of Denmark and of the Scandinavian peninsula as that on the western coast is called Westsǽ, Ors. 1, 1 ; Swt. 17, 3 :—Be norþan Súþdenum is ðæs gársecges earm ðe mon hǽt Óstsǽ ... Norþdene habbaþ be norþan him ðone ilcan sǽs earm ðe mon hǽt Óstsǽ, Ors. 1, 1 ; Swt. 16, 23–28. [*Ger.* Ost-see *the Baltic* : cf. *Icel.* fara â Austrveg, a phrase used of trading or piratical expeditions in the Baltic.]

ót-. v. óþ.

oter, otr, es ; *m. An otter :*—Otr *lutrus*, Txts. 74, 585. Otor, Wrt. Voc. ii. 51, 18 : *lutria*, i. 22, 49. Ottor *sullus*, 121, 51. Oter *lutrius*, i. 78, 15. Of oteres hole, Cod. Dip. Kmbl. iii. 418, 17. [*Icel.* otr : *O. H. Ger.* ottar, oter.]

oter-hola, an ; *m. An otter's hole :*—Of ðam oterholan, Cod. Dipl. Kmbl. iii. 23, 30.

óþ ; *prep.*   I. *with dat.* (1) local, marking a point reached, *to*, *unto, as far as :*—Fram eástdǽle óþ westdǽle, and fram súþdǽle óþ norþdǽle, Gen. 28, 14. (2) referring to time, *until :*—Fram Davide óþ Daniele ðam wítegan, Ælfc. T. Grn. 7, 13. (3) marking extent, degree, *so much as :*—Nis se ðe dó gód, nis óþ ánum (*usque ad unum*), Ps. Spl. 13, 2 : 52, 4.   II. *with acc.* (1) local, marking a point reached, *to, up to, as far as :*—Óþ eorþan endas, Deut. 28, 64 : Ps. Th. 71, 8. Ðú nyðer fǽrst óþ helle, Mt. 11, 23. Hé him æfter rád óþ ðæt geweorc, Chr. 878 ; Erl. 80, 15. Hé him æfterfylgende wæs óþ v míla tó ðære byrig Cartanense *ad quintum lapidem a Carthagine statuit*, Ors. 4, 5 ; Swt. 168, 32 : 3, 4 ; Swt. 104, 2 : 4, 10 ; Swt. 194, 7. Óþ Eufraten, Cd. Th. 133, 6 ; Gen. 2206. (1 a) in phrases marking extent, degree or measure :—Óþ ðæt *eatenus vel eotenus*, Wrt. Voc. i. 61, 30. Óþ hielt *capulo tenus*, Wrt. Voc. ii. 19, 7. Óþ ða hylta, Ælfc. Gr. 47 ; Som. 48, 7. Hí druncan óþ ða drosna *usque ad feces biberunt*, Som. 47, 45. Óþ mannes breóst heáh *as high as a man's breast*, Blickl. Homl. 127, 6 : 245, 33. Hié ðæt gild gefyldan eal óþ grund, 221, 33. Ðæt hí óþ forwyrd fordíligade ne wǽron, Bd. 1, 16 ; S. 484, 17. Se Ægipta slóh frumbearn æghwylc ealra óþ ða nýtenu (*down to the very beasts*), Ps. Th. 134, 8. Seóð ðonne óþ huniges þicnesse, Lchdm. ii. 30, 7. (2) temporal, *until, to, unto :*—Óþ ðisne dæg *usque in praesentem diem*, Gen. 32, 4. Óþ ðás dagas, Ex. 9, 18. Nú óþ ðis *hactenus*, Bd. 4, 22 ; S. 591, 15 : Blickl. Homl. 175, 12. On ðære hwíle óþ ðæt *up to the present time*, Homl. Skt. i. 4, 265. Óþ ǽfen *usque ad vesperum*, Ælfc. Gr. 47 ; Som. 47, 46. Óþ ende his lífes, Blickl. Homl. 21, 36. Óþ ðone deáþ, 59, 30. Óþ ðæt *until :*—Óþ ðæt (*donec*) hé forgite ða þing ðe ðú him dydest, Gen. 27, 45 : Beo. Th. 4084 ; B. 2039 : Andr. Kmbl. 535 ; An. 268. Óþ ðæt hiene án swán ofstang, Chr. 755 ; Erl. 48, 22. Ót ðet *donec*, Ps. Surt. 70, 18. Óþ ðe *until :*—Fóron forþ óþ ðe hié cómon tó Lundenbyrig, Chr. 894 ; Erl. 91, 13. Óþ ðe hé eall forweorðeþ, Ps. Th. 139, 11 : Beo. Th. 1302 ; B. 649. (2 a) with other prepositions :—Óþ in ældu *usque in senecta*, Ps. Surt. 70, 18. Ða gestód hé æt ánum éþþyrle óþ forþ nihtes, Homl. Th. ii. 184, 27. Óþ tó dæge *usque hodie*, Bd. 1, 15 ; S. 483, 27. Óþ gyt tó

dæge, 4, 4; S. 571, 16. [Cf. *Goth.* und; *prep.*; unte *conj.*: *O. Sax.* unt; *prep.*; und; *conj.*: *Icel.* unz *conj.*: *O. H. Ger.* unz, v. Grff. i. 363–366.] v. next word.

ōþ, *conj. Until*:—Wuna mid him, ōþ ðínes bróður yrre geswíce, Gen. 27, 44: Mt. Kmbl. 10, 11. Hé hæfde ða, oþ hé ofslóg ðone aldorman, Chr. 755; Erl. 48, 20. Ðæt mód glít niðor and niðor, oþ hit mid ealle áfielþ, Past. 38; Swt. 279, 3: Cd. Th. 22, 14; Gen. 340. v. preceding word.

ōþ- as a prefix of verbs, *from, away*. Cf. *æt* for similar meaning. [Cf. *Goth.* *untha*-thliuhan *to escape*.]

ōþ-beran *to bear forth, bear away* (cf. æt-beran):—Nó ic eów sweord ongeán oþberan þence, Exon. Th. 120, 20; Gú. 274. Mec sǽ oþbær on Finna land *the sea bore me forth to the land of the Fins*, Beo. Th. 1163; B. 579: Exon. Th. 404, 20; Rä. 23, 10. Sumne fugel oþbær (*bore off*) ofer heánne holm, 291, 13; Wand. 81.

ōþ-berstan *to break away, escape*:—Hé oþbærst tó wuda, Cod. Dip. Kmbl. iii. 291, 17. [Rannulf út of ðam túre on Lunden nihtes oþbærst, ðær hé on hæftneþe wæs, Chr. 1101; Erl. 237, 40.] Gif se bana oþbyrste, L. H. E. 2; Th. i. 28, 1. Cf. æt-berstan.

ōþ-bregdan, -brédan *to take away, carry off*:—Ða burgleóde oþbrúdon ða snore mid hiere suna, and hí sendon on óðer fæstre fæsten, Ors. 3, 11; Swt. 148, 21. Hæbbe hé Godes unmiltse, se ðe ðis áwende and ðer stówe oþbréde, Cod. Dip. Kmbl. ii. 4, 3. Hé (*Nero when Rome was burning*) bebeád his ágnum monnum ðæt hié gegripen ðæs licgendan feós swá hié mæst mehten, and tó him brohten, ðonne hit mon út oþbrúde, Ors. 6, 5; Swt. 260, 32. Siððan wearþ Adame eardríca cyst oþbróden, Exon. Th. 153, 15; Gú. 826. Oþbrog[d]en *ademptam*, Wrt. Voc. ii. 9, 18. Oþbródenum hwelpum *raptis fetibus*, Kent. Gl. 607. v. æt-brédan.

ōþ-cirran *to turn away, be perverted*:—Gif sóðfæstra þurh myrrelsan mód ne oþcyrreþ (neod cyrreþ, MS.) *if by seduction the mind of the righteous is not perverted*, Exon. Th. 262, 26; Jul. 338. Cf. on-cirran; *intrans.*

ōþ-clifan *to cleave to, adhere*:—Him sár oþclífeþ, Exon. Th. 78, 1; Cri. 1267.

ōþ-cwelan *to die*:—Gif sió hond sié oþcwolen *if the person be dead*, L. In. 53; Th. i. 134, 17.

ōþ-dón *to put out*:—Gif hwá óðrum his eáge oþdó (of dó, MS. H.), L. Alf. 19; Th. i. 48, 20. Cf. æt-dón.

ōþ-eáwan. v. oþ-íwan.

ōþ-éhtian *to drive away*:—Se ðe ðis feoh oþfergean þence, oððe ðis orf oþéhtian þence *he that thinks of carrying off this cattle, or of driving it away*, Lchdm. i. 384, 15.

ōðel, es; *m. Home, native country*:—Abraham ferede æðelinga bearn óðle niór, mægeþ heora mágum, Cd. Th. 126, 7; Gen. 2091. v. éðel.

Ōðen, es; *m. Odin* (the Scandinavian form of the word which appears in Anglo-Saxon as *Wóden*):—Ðes gedwolgod (*Mercurius*) wæs árwurðe betwux eallum hæðenum on ðam dagum, and hé is Óðon geháten óðrum naman on Denisce wísan. Nú secgaþ sume ða Denisce men on heora gedwylde, ðæt se Iouis wǽre, ðe hý Þór hátaþ, Mercuries sunu, ðe hí Óðon namiaþ, Wulfst. 107, 6–11. Þór and Ówðen (Open, MS. F.), 197, 19.

ōþ-eode, -eówan. v. oþ-gán, -íwan.

ōðer, *indef. prn. and ordinal, used as adj. and as subst., always of strong declension.* I. *when two definite objects are referred to*, (1) *one of two*:—Him wearþ óðer eáge mid ánre flán út áscoten *ictu sagittae oculum perdidit*, Ors. 3, 7; Swt. 112, 15. Hé hyne onsende myd twám mæssepreóstum . . . ðá forþférde ðæra mæssepreósta óðer, Shrn. 98, 28. Him bærst micel wund on óðrum þeó *in one of his thighs*, 109, 14. Ðá gewearþ him ðæt hí twegen tó ánwíge eodon . . . calle gecwǽdon, ðæt gif ǽnig man wolde heora óðrum (*either of them*) fylstan, ðæt man hine sóna gefénge, H. R. 101, 21. Ðær wearþ Pirrus wund on óðran earme (*transfixo brachio*) . . . Hí námon treówu, and slógon on óðerne ende ísene næglas, Ors. 4, 1; Swt. 158, 2–5. Þurhscoten underneoðan óðer breóst, 3, 9; Swt. 134, 23. Wund þurh ðæt óðer cneów, 4, 6; Swt. 180, 6. Án strǽl hyne gewundode on hys óðer gewenge, Shrn. 97, 14. Se ðe hæbbe twá tunecan, selle óðre ðam ðe náne næbbe, Blickl. Homl. 169, 13. Óðer twega, *or without these forms, one of two alternatives*:—For ðam óðer twega, oððe hié nǽfre tó nánum men ne becumaþ, oððe hié nǽfre fæstlíce ne þurhwuniaþ, Bt. 11, 1; Fox 30, 26. Ða wilniaþ óðer twega, oððe . . . , oððe, 24, 2; Fox 82, 8. Wite hé ðæt óðer ðara, oððe hé sceal ðæs hádes þolian, oððe hit gebétan, L. E. I. 14; Th. ii. 412, 1: 9; Th. ii. 408, 11. Ðæt hió óðer ðara dydon, oððe . . . oððe . . . , Chart. Th. 167, 22: Ors. 3, 7; Swt. 114, 23. Him sǽdon ðæt hié óðer dyden, oððe hám cómen oððe hié him woldon óðerra wera ceósan, 1, 10; Swt. 44, 21. (2) *the second of two, other*:—Se óðer consul Duilius *Duilius, alter consul*, 4, 6; Swt. 172, 8. Hé for ðære geómrunga ðæs óðres deáþes leng on ðam lande gewunian ne mihte, Blickl. Homl. 113, 11. Án mann hæfde twegen suna. Ðá cwæþ hé ealswá tó ðam óðrum, Mt. 21, 30. Ðæt mon ierne from geate tó óðrum, Past. 49; Swt. 383, 8. Fram ende oþ óðerne *from one end of the church to the other*, Glostr. Frag. 12, 17. Hafa ðé (*Leah*) tó gemæccan, and ic gife ðé ða óðre (*Rachel*), Gen. 29, 27. Hé sette his ǽnne sunu tó ealdormen, and óðerne tó cyninge, Homl. Th. ii.

480, 21. (3) *when óðer is applied to each of two*:—Ðara óðer bewiste his byrlas, óðer his bæcestran, Gen. 40, 2. Óðer is se ǽresta apostol, óðer se néhsta, Blickl. Homl. 171, 8. Ðæt se óðer beó árǽred from ðæm óðrum *ut alter regatur ab altero*, Past. 17, 1; Swt. 107, 23. On twǽm gefylcum, on óðrum wǽron ða hǽðnan cyningas, on óðrum ða eorlas, Chr. 871; Erl. 74, 16–18. Ægðer óðerne ofslóg, Ors. 2, 3; Swt. 68, 18. Uncer láþette ægðer óðer ðeáh ðe hé hit óðrum ne sǽde, Shrn. 39, 22. II. *when the reference is not limited to two objects.* (1) *marking a sequence, other, second in a series, next, following* an object already mentioned:—Se forma . . . se óðer, and se þrydda oþ ðone seofoþan, Mt. 22, 25, 26: Ælfc. Gr. 49; Som. 49, 55. Wæs ðis ðara wundra ǽrest . . . Eft gelamp óðer wundor, Blickl. Homl. 219, 6: 221, 18: 223, 13. Ðære óðre eá naman *nomen fluvii secundi*, Gen. 2, 13. Fram dæge tó óðrum *from day to day*, Blickl. Homl. 107, 25. Faran of ´stówe tó óðerre, 19, 23. Án æfter óðrum, Cd. Th. 266, 22; Sat. 26. Hé sette hine on his óðer cræt (*currum suum secundum*), Gen. 41, 43. Ǽrest . . . óðre síþe . . . þriddan síþe, Blickl. Homl. 47, 16. Óþre dæge *next day*, 175, 18. Ðá fór hé swá feor swá hé meahte on ðæm óðrum þrím dagum (*in the next three days*) gesiglan, Ors. 1, 1; Swt. 17, 13. Wearþ syfan geár se ungemetlíca eorþwela, and hí æfter ðæm wǽron on ðan mǽstan hungre óðre syfan geár, 1, 5; Swt. 32, 26: Gen. 29, 27. (1 a) *with swilc, another such*, a repetition of what has preceded:—Ðá com ungemetlíc rén . . . eft wearþ óðer swelc rén, Ors. 4, 10; Swt. 194, 20. Medmicel pipores, óðer swilc cymenes, Lchdm. ii. 256, 5. His mágas hine wið óðær swilc (*contra simile quid*) gescyldan, L. Ecg. P. addit. 29; Th. ii. 236, 31. (2) *marking difference from the subject, or from something already referred to, other, different, somebody else, something else*:—Ðú nimst wíf and óðer man líþ mid hire, Deut. 28, 30. Ne þearf nán mon wénan ðæt hine óðer mon mǽge álésan, Blickl. Homl. 101, 13. Gif útancymene oxa óðres oxan gewundaþ, Ex. 21, 35. Eart ðú ðe tó cumenne eart, oððe wé óðres sceolon ábídan, Mt. Kmbl. 11, 3. Gif wé willaþ on óðres góde beón gefeónde, Blickl. Homl. 62, 30. Leófre mé ys, ðæt ic hig sylle ðé ðonne óðrum men, Gen. 29, 19. Ðæt man tó óðrum lǽdðe hæbbe, Blickl. Homl. 63, 36. Ðæt ǽlc stán ne sý fram óðrum ádón, 79, 1. Heora ongon ǽlc cweðan tó óðrum, 149, 29. Ða lǽstas on óðerne mægwlite oncyrran, 127, 19. On óðre wisan, 205, 21. Nú hæbbe wé broht óðer sylfor (*aliud argentum*), Gen. 43, 22. Seó wyrd oft oncyrreþ and on óðer hworfeþ, Nar. 7, 28. Mid hire syndan Godes apostola and óðre, Blickl. Homl. 143, 10. Petrus and óðre Cristes þegnas, 145, 27. Ða ðe wóhhæmed begangaþ mid óðerra ceorla wífum, 61, 14. Sceattas ge on lande, ge on óðrum þingum, 51, 7. Hé gesett hys wíngerd myd óðrum tilion, Mt. Kmbl. 21, 41. Hé him tó genymþ seofun óðre gástas, 12, 45. (2 a) óðer . . . óðer *other . . . than, different from*:—Nú is *participium* of worde and of worde cymþ, biþ swá ðeáh óðer dǽl and óðer þing óðer his ealdor biþ, Ælfc. Gr. 41; Som. 43, 14. Ðonne gá heó in óðer hús óðer heo út ofeode, Lchdm. iii. 68, 21. Gif ðú wilnast ðæt heó óðre þeáwas nimen óðre (óðer, Cott. MSS.) heora willa and heora gewuna is, Bt. 7, 2; Fox 18, 28. (2 b) *with the indefinites sum, ǽnig,* etc.:—Helias afterde óðer witega, Homl. Th. i. 364, 18. Wæs his néhmága sum ðæt hine swýðor lufode ðonne ǽnig óðor man, Blickl. Homl. 113, 10. Máran wræce ðonne ǽfre ǽr ǽnigu óðru gelumpe, 79, 10. Wæs ðæt wæter biterre ðonne ic ǽfre ǽnig óðer bergde, Nar. 8, 30. Hé nǽnigum óðrum ǽrne sceþþan ne mihte, Blickl. Homl. 221, 16. Ne bideþ hé æt ús nǽnig óðor edleán búton . . . , 103, 21. Nǽnige óðre búton ða ǽne, 185, 9. Ðara óðerra manna nán árian wolde, 215, 1. Mid manegum óðrum gástlícum mægenum, 73, 28. Ðæt geleáfulle folc Iudéa, and eác óðor manig ða ðe beóþ Gode underþeódde, 79, 31. Ðás wundor and manig óðer, 219, 22. Óðre wundro manega, 177, 18. Augustinum and óðre monige munecas, Bd. 1, 23; S. 485, 27. Lufian wé úrne Drihten ofer ealle óðer þing, Blickl. Homl. 11, 33. Sanctus Iohannes gǽþ beforan eallum óðrum wítgan and ealra óðerra heáhfædera mægen hé oferstígeþ, 167, 22. (3) *denoting that part of a whole which is not yet mentioned, other, the, rest, remaining*:—Micel ðæs folces hié ofer sǽ ádrǽfdon, and ðæs óðres ðone mǽstan dǽl hié geridon, Chr. 878; Erl. 78, 31. Sum fearhryðer ðæs óðræs ceápes geférscipe oferhogode, Blickl. Homl. 199, 4. Seó hand wæs gelíc ðam óðrum flǽsce *erat similis carni reliquae*, Ex. 4, 7. Ða óðre nigon *consonantes* synd gecwedene *mutae*, Ælfc. Gr. 2; Som. 3, 1. Ða óðre (*ceteri*) cwǽdon, Mt. 27, 49. Petrus and ða óðre apostolas, Blickl. Homl. 149, 5. Wæs heora sum rédra ðonne ða óðre, 223, 7. Wyrtruma ealra óðerra synna, 65, 3. Ðæt deófol cwæþ tó ðam óðrum deóflum, 243, 10. Hig cýðdon eall ðis ðam endlufenum and eallum óðrum (*ceteris omnibus*), Lk. Skt. 24, 9. [*Goth.* anþar: *O. Sax.* óðar: *O. Frs.* óther: *O. H. Ger.* andar: *Icel.* annarr.]

ōðerlíce; *adv. Otherwise, differently*:—Se ðe óðerlícor gedyrstlæce underhníge ðære regullícan þreále *que autem aliter presumpserit, discipline regulari subjaceat*, R. Ben. 87, 19. [*Goth.* anþarleiko *otherwise*: cf. *O. Sax.* óðar-lík: *O. H. Ger.* andar-líh: *Icel.* annar-ligr.]

ōþ-éwan. v. oþ-íwan.

ōþ-fæstan. I. *to entrust, commit to the charge of another*:—Oþ ðæt ic mé gebidde tó him and mín gást oþfæste *I commit my spirit*

into his hands, Nar. 46, 34. Heó hyre mægþhâd Gode ôþfæste, 40, 16. Gif hwâ ôþfæste his friénd feoh, L. Alf. 28; Th. i. 50, 29: L. Alf. pol. 20; Th. i. 74, 15. Gif hwâ ôðrum his unmagan ôþfæste, and hé hine on ðære fæstinge forferie, 17; Th. i. 72, 4. Se ðe wile hwilc sæd ôþfæstan ðâm drîum furum, Bt. 5, 2; Fox 10, 30. Þæt hié sîen tô liornunga ôþfæste, Past. pref.; Swt. 7, 12. **II.** *to inflict, impose* (pain, punishment. Cf. æt-fæstan) :—Ne meahton hié deáþ *(Kemble has deáþe, in which case the verb belongs to* I) *ôþfæstan they could not inflict death (on Christ)*, Elen. Kmbl. 952; El. 477. Drihten hæfde wîte clomma[s] feóndum ôþfæsted *the Lord had imposed penal chains on the fiends*, Cd. Th. 292, 23; Sat. 445.

ôþ-faran *to escape* :—Siððan hié feóndum ôþfaren hæfdon, Cd. Th. 181, 21; Exod. 64.

ôþ-feallan. I. *to fall away, cease to have connection with* :—Ôþfealle se wer *(in the case of a man who, upon a charge of theft, being forsaken by his kinsmen, forfeits his freedom)* ðâm mâgum *the kinsmen shall have no further concern in the* 'wer,' L. Ed. 9; Th. i. 164, 13. Cf. æt-feallan. **II.** *to fall away, fail, decay* :—Gif hwam seó spræc ôþfylþ *if speech fail a man*, Lchdm. ii. 288, 18. Æfter his fielle wearþ ðara câsera mægþ offeallen (ôþ-, MS. C.) *Caesarum familia consumta est*, Ors. 6, 5; Swt. 262, 6. Swâ clǽne hió *(learning)* wæs ôþfeallenu on Angelcyn *so utterly was learning decayed in England*, Past. pref.; Swt. 3, 13.

ôþ-feolan *to cleave, stick* :—Ôþfealh *heresceret*, Wrt. Voc. ii. 42, 46. Cf. æt-feolan, -felgan.

ôþ-ferian *to bear off* :—Ic unsôfte ðonan feorh ôþferede næs ic fǽge ðágyt *not easily thence (the conflict with Grendel's mother) did I bring away life, but not then had my hour come* (cf. *last passage under* ôþ-lǽdan), Beo. Th. 4288; B. 2141. Ðæt hé nǽfre nabbe foldan ðæt hit ôþferie . . . Se ðe ðis feoh ôþfergean *(carry off, steal)* þence, Lchdm. i. 384, 9–15. Hî willaþ ôþfergan, ðæt ic friþian sceal; ic him ðæt forstonde, Exon. Th. 398, 13; Rä. 17, 7. Cf. æt-ferian.

ôþ-fleógan *to fly away* :—Se ânhoga ôþfleógeþ feðerum snel, Exon. 222, 11; Ph. 347.

ôþ-fleón *to flee away, escape* :—Favius heánlîce hâmweard ôþfleáh, Ors. 3, 10; Swt. 140, 14. Ða ðe tô him mid scypum ôþflugon tô ðǽm beorgum *ad se ratibus confugientes*, 1, 6; Swt. 36, 11 : 2, 8; Swt. 94, 8. Sume binnan ðæt fæsten ôþflugon, Swt. 92, 23. Ða ðe him *(Joshua)* ôþflugon, ðám feóllon stânas on uppan, and hî fordydon, Homl. Th. ii. 214, 2. Ðám monnum ðe ôþflugon ofer ðone weall, Chr. 921; Erl. 107, 12. Uneáðe mehte ǽnig ðǽm Gallium ôþfleón, Ors. 2, 8; Swt. 94, 11. Wilniende ðæt hî ǽlcum gewinne ôþflogen hæfdon, 1, 4; Swt. 32, 21.

ôþ-flîtan *to get from another by litigation* :—Ðá ongon Higa him specan sôna on, and wolde him ôþflîtan ðæt lond *then Higa at once began the case against him, and wanted to get the land from him by the litigation*, Chart. Th. 169, 23.

ôþ-gân *to go away, escape* :—Ôþeodon, Beo. Th. 5860; B. 2934.

ôþ-glîdan *to glide away*, Salm. Kmbl. 804; Sal. 401.

ôþ-grîpan *to snatch away* :—Gif wên wǽre ðæt hé ðær hwylce mihte deófle ôþgrîpan of Criste gecyrran *si quos forte ex illis ereptos Satanae ad Christum transferre valeret*, Bd. 5, 9; S. 622, 19.

ôþ-healdan *to withhold, keep back* :—Gif hwelc folc biþ mid hungre geswenced, and hwâ his hwǽte gehýt and ôþhielt hû ne wilt hé ðonne hiera deáþes *si populus fames atteret, et occulta frumenta ipsi servarent, auctores proculdubio mortis existerent*, Past. 49, 1; Swt. 377, 9. Ðæt hé nǽfre nabbe hûsa ðæt hé hit *(stolen property)* ôþhealde, Lch. i. 384, 10.

ôþ-hebban *to elevate, exalt, lift up* :—Ða welan ðe ǽlcne ofermôdne ôþhebbaþ *abundantia, quae sublevat*, Past. 26, 2; Swt. 183, 18. Hé hine ôþhôf (ot-, Cott. MSS.) innan his geþohte eallum ôðrum monnum *cunctis in cogitatione se praetulit*, 4, 2; Swt. 39, 15. Ða ofersettan mon sceal swâ manian ðæt se hiera folgoþ hî ne ôþhebbe *admonendi sunt praelati, ne eos locus superior extollat*, 28, 1; Swt. 189, 17.

ôþ-hilde; *adj. Content* :—Ânum were ôþhylde heó ne biþ *she will not be content with one man*, Lchdm. iii. 188, 6. Ôþhelde (cf. êþhylde, l. 1), 194, 14. v. eáþ-, êþ-hylde.

ôþ-hleápan *to run away, escape* :—Gif hé ût ôþhleápe, L. Eth. i. 1; Th. i. 282, 11. Cf. æt-hleápan.

ôþ-hýdan *to hide away* :—Uneáðe mehte ǽnig ðám Gallium ôþfleón oððe ôþhýdan *hardly could any one escape or hide from the Gauls*, Ors. 2, 8; Swt. 94, 11.

ôþ-ican *to add to* :—Ôtêctun *addiderunt*, Ps. Surt. 68, 27. Cf. æt-ýcan.

ôþ-irnan *to run away, escape* :—Hé ðære eorþan ǽfre ne ôþrineþ, Met. 20, 138. Gif hé ôþierne, L. In. 28; Th. i. 120, 7. v. æt-irnan.

ôþ-îwan, -êwan, -eáwan, -eówan, -íewan, -ýwan. **I.** *to shew* :—Ic ôþeówe *ostendam*, Ps. Spl. 49, 24. Ne ðû mê ôþíewest ǽnig tâcen, Cd. Th. 34, 19; Gen. 540. Ôteáweþ *ostendit*, Ps. Surt. 4, 6. Hé ôþiewde openlîce ðæt hé ǽr gehýd hæfde, Ors. 6, 34; Swt. 288, 32. Ôþiwde, Ps. Spl. 77, 14. Ôþíewde, Cd. Th. 44, 24; Gen. 714. Hér cometa hiene ôþiewde, Chr. 729; Erl. 46, 5. Ðær ðû mê ôþêwe, Bt. 22, 2; Fox 78, 11. Wearþ ôþíewed ân îgland, Ors. 6, 4; Swt. 260, 14. Ôþewed, Met. 29, 34. Open and ôþeáwed, Exon. Th. 98, 9; Cri. 1605. Ôþýwed,

52, 25; Cri. 839. **II.** *to shew one's self, to appear* :—Ic ôteáwu *apparebo*, Ps. Surt. 16, 15. Sió sunne eldum ôþêweþ, Met. 13, 60. Ôþýweþ, Exon. Th. 56, 24; Cri. 905. Ic ôþeówde *apparui*, Ps. Spl. 62, 3 Met. 28, 74. Hér ôþiewde cometa se steorra, Chr. 678; Erl. 40, 5 : 773 ; Erl. 52, 23. Ôþýwde, Elen. Kmbl. 325; El. 163. Ôteáwdon *apparuerunt*, Ps. Surt. 17, 16. Ôþeówdun, Exon. Th. 28, 17; Cri. 448. In bôcum ne cwiþ ðæt hý in hwîtum hræglum ôþýwden, 28, 30; Cri. 454. Cf. æt-ýwan.

ôþ-lǽdan *to lead away, carry off* :—Hé Israhêlas ealle ôþlǽdde *eduxit Israel*, Ps. Th. 135, 11. Âlýs mê and ôþlǽd lâðum wætrum *eripe me et libera me de aquis*, 143, 12. Ic þence ðis feoh tô lufianne, næs tô ôþlǽdanne . . . hé nǽfre nabbe landes ðæt hé hit ôþlǽde, Lchdm. i. 384, 4–9. Ic eom ôþlǽded gôdum *excussus sum*, Ps. Th. 108, 23. Hié ôþlǽded hæfdon feorh of feónda dôme *life had they withdrawn from the foes' power* (cf. Beo. Th. 4288 *under* ôþ-ferian), Cd. Th. 214, 15; Exod. 569. Cf. æt-lǽdan.

Ôðon. v. Ôðen.

ôþ-rîdan *to ride away* :—Cyning in ôþrâd forþ onette *the king (Christ after the doors of Hell had opened) rode away into Hell, hastened on*, Exon. Th. 461, 24; Hö. 40.

ôþ-rôwan *to row off* :—Hié ût ôþreówon *they rowed out and away*, Chr. 897; Erl. 96, 7.

ôþ-sacan (*with gen.*). **I.** *to deny* (a statement) :—Hwâ ôþsæcþ ðæs? Bt. 26, 2; Fox 92, 21. Ne mæg ic ðæs ôþsacan, forðam ðe ic his wæs ǽr geþafa, 34, 3; Fox 138, 15 : 33, 1; Fox 122, 2 : 34, 9; Fox 146, 34. Nân mon ne mæg ôþsacan ðæt sum gôd ne sîe ðæt hêhste, 34, 1; Fox 134, 9. **II.** *to deny* (an obligation, a charge, etc.) :—Gange feówra sum tô and ôþsace *(deny a charge of robbing)*, L. Eth. ii. 4; Th. i. 286, 18. Borges mon môt ôþsacan gif hé wât ðæt hé ryht deþ, L. In. 41; Th. i. 128, 2. Cf. æt-sacan.

ôþ-sceacan *to run away, escape* :—Gif hé ôþsceóce (-seoce, MS.), L. Ath. v. 6; Th. i. 234, 11.

ôþ-sceótan *to shoot away, escape, turn aside, hurry off* :—Swâ hwâ swâ ôþscýt fram ânnysse ðæs geleáfan *whoever turns aside from the unity of the faith*, Homl. Th. i. 370, 17. Man gehylt ðæt hé hæfþ gif hé him ondrǽt ðæt hit him ôþsceóte *a man guards what he has, if he is afraid that it will escape from him*, Prov. Kmbl. 18.

ôþ-scûfan *to push* (intrans.) *away, move away* :—Hé gesêceþ (-aþ, MS.) Syrwara lond corðra mǽste. Him se clǽna ðǽr ôþscûfeþ scearplîce *(the Phenix moves off quickly from the attendant birds)* ðæt hé in scade weardaþ on wudubearwe wêste stôwe biholene and bihýdde hæleþa monegum *dirigit in Syriam celeres longaeva volatus, secretosque petit deserta per avia lucos, hic ubi per saltus silva remota latet*, Exon. Th. 209, 9; Ph. 168.

ôþ-seóce. v. ôþ-sceacan.

ôþ-spurnan, -spornan *to strike against, stumble* :—Hió ôþsper[n]þ *impingetur*, Kent. Gl. 769. Ðê læs ðîn fôt ôþsporne, Blickl. Homl. 27, 14. Næs gecweden ðæt his fôt æt stâne ôþspurne, 29, 31. Cf. æt-spurnan.

ôþ-spyrning, e; *f. An offence, a stumbling-block* :—Bûto ôtspurnince *absque offendiculo*, Kent. Gl. 528. Cf. æt-spyrning.

ôþ-standan. I. *to stop in one's course, to come to a standstill* :—Ðonne ôþstandeþ se blôdgyte sôna, Lchdm. i. 88, 10. Sôna ðæt blôd ôþstænt, 180, 3. Ðæt unstille hweól ôþstôd, Bt. 35, 6; Fox 168, 32. **I a.** metaphorically, *to cease to act* :—Gif se hlyst ôþstande, ðæt hé ne mǽge gehiéran, L. Alf. pol. 46; Th. i. 92, 23. **II.** *to remain standing, remain* :—Uneáþe ǽnig grot staþoles ôþstôd, Ors. 6, 1; Swt. 252, 23. Ðæt is lang tô sæcganne, hû ða wurdon generede in ðǽre Noes earce, ða ðe ðær tô lâfe ôþstôdon, Wulfst. 206, 30. **III.** *to remain standing and so prove an obstacle* :—Ðæt swefn swîðe ôþstôd manegum mînra leóda *(the dream interpreted by Daniel)*, Cd. Th. 246, 23; Dan. 483. Cf. æt-standan.

ôþ-stillan *to put a stop to, to stop* :—Ðonne biþ hit *(hæmorrhage)* sôna ôþstilled, Lchdm. i. 82, 5. Cf. æt-stillan.

ôþ-swerian *to abjure, deny on oath* :—Ðá ôþswôran hié mid ðam bismerlîcestan âðe ðæt hit mid hiera on fultume nǽre ðêh ðe ða âðas wǽren neár mâne ðonne sôðe *turpissimam rupti foederis labem adcumulavere perjurio*, Ors. 4, 3; Swt. 162, 10. Gif hlôþ ðis gedô and eft ôþswerian (æt-, MS. B.) wille, L. Alf. pol. 31; Th. i. 80, 16. Gif mon tô ðam men feoh getême ðe hé ǽr ôþswaren (ætsworen, MS. B.) hæfde, and æft ôþswerian wille, ôþswerige (æt-, MS. B.) be ðam wîte . . . Gif hé ôþswerian nylle . . ., L. In. 35; Th. i. 124, 10–12.

ôþ-swîgan *to stop speaking, become silent* :—Hé spræc tô his liornæra sumum, and ðá fǽringa ôþswîgde hé suæ hé hwæshwegu hercnade, Shrn. 72, 24.

ôþ-swimman *to swim off* :—Ða âne ðe ût ôþswymman mihton (ætswummon, MS. A.) tô ðam scipum, Chr. 915; Erl. 105, 11.

ôþ-teón *to take away* :—Him biþ slǽp ôþtogen *sleep deserts them*, Lchdm. ii. 232, 14.

oððe; *conj.* **I.** *or* :—Gif seó offrung beó of sceápon oððe of gâtum, Lev. 1, 10. Geeácnode ic hig ealle oððe âcende ic hig, Num. 11, 12. Hwâ geworhte mannes mûþ oððe hwâ geworhte dumne oððe deáfne and blindne oððe geseóndne? Ex. 4, 11. **I a.** *in conjunction with*

óðer :—Hí woldon óðer twega, líf forlǽtan oððe leófne gewrecan, Byrht. Th. 137, 59; By. 207: Wald. I, 16. **II.** oððe ... oððe ... *either ... or* (a) :—Oðða (oððe, MS. B.) mid freóndscipe oðða mid gefeohte *vel amicitia vel ferro*, Bd. 1, 1; S. 474, 26. Ðonne fôron hié oððe mid oððe on heora healfe, Chr. 894; Erl. 90, 6. Ða scipu eall oððe tôbrǽcon oððe forbærndon oððe tô Lundenbyrig brohton oððe to Hrôfesceastre, Erl. 91, 25. (b) *with* óðer, *aðer* :—Hé sǽde ðæt hé wolde óðer, oððe ðæt libban oððe ðǽr licgan, 901; Erl. 96, 32. Hét ðæt hié óðer sceoldon, oððe ðæt lond æt him ǽlesan, oððe hé hié wolde fordôn, Ors. 1, 10; Swt. 44, 9: 44, 21. Hié óðer forleósan woldon, oððe hira ágen líf oððe Porsennes, 2, 3; Swt. 68, 28. Nú ðonne óðer twega, oððe ðara nán nis, oððe hí nánne weorþscipe nabbaþ, Bt. 27, 4; Fox 100, 16. Gif onfunden biþ ðæt hé aðer oððe ... oððe ..., L. E. I. 16; Th. ii. 412, 11.

óþ-þeódan *to disjoin, dismember* :—Ðú ðæt gehéte ðæt ús heteófra hild ne gesceóde, ne líces dǽl oþþeóded, ne sinu ne bân on swaðe lágon, ne loc of heáfde tô forlore wurde, Andr. Kmbl. 2842; An. 1423.

óþ-þicgan *to take from* :—Him frumbearnes riht freóbrôðor oþþah, Cd. Th. 199, 14; Exod. 338.

óþ-þingian *to get from another on unfair conditions* :—Gif hwylc mæssepreóst onfunden biþ ðæt hé ... ǽnige mêdsceat selþ oððe sealde, for ðí ðe hé wilnige óðres preóstes cyrcean oþþingian, L. E. I. 16; Th. ii. 412, 13.

oððon ; *conj. Or* :—On cyriclícum þingum oððon on earmra manna hýððum oððon on hernumena bygenum oððon on sumum þingum, L. I. P. 19; Th. ii. 328, 10–12. Swá oft swá man fullaþ oððon húsel hálgaþ, 328, 21.

óþ-þringan *to force away from one* (*oftenest in phrases* líf, feorh, etc., oþþringan *to take a person's life*) :—Ðá geleornedon hú híe him mehten ðæt líf oþþringan, and him gesealdon ätor drincan, Ors. 3, 9; Swt. 136, 15. Se ðe mid gáres orde óðrum aldor oþþringeþ, Cd. Th. 92, 3; Gen. 1523: Exon. Th. 330, 11; Vy. 49. Ecghete fǽgum feorh oþþringeþ, 310, 8; Seef. 71. Ðám ic ealdor oþþrong, 272, 17; Jul. 500: Judth. Thw. 24, 12; Jud. 185. Hú hé Israélum eáþost meahte guman oþþringan *how he might most easily force away men from Israel* (*carry the Israelites captive*), Cd. Th. 219, 8; Dan. 51. Unc mágas uncre sculon eard oþþringan *our kinsmen shall take our home from us*, Exon. Th. 496, 9; Rä. 85, 11. Cf. æt-þringan.

óþ-wendan *to turn away, divert* :—Uton oþwendan hit (*the kingdom of heaven*) monna bearnum, Cd. Th. 26, 8; Gen. 403.

óþ-windan *to get away, escape* :—Ân scip oþwand, Chr. 897; Erl. 95, 27. Cf. æt-windan.

óþ-wítan *to reproach with a fault, lay to a person's charge, to taunt* :—Oþwíteþ *improperabit*, Ps. Spl. M. 73, 11. Hwý óðwíte gé wyrde eówre, ðæt hió geweald nafaþ? Met. 27, 4. Wé sindon cumen tô ðæm gódan tîdun ðe ús Rômáne oþwítaþ *we are come to the good times that the Romans taunt us with*, Ors. 4, 7; Swt. 182, 15. Oþwát *improperavit*, Ps. Spl. M. 73, 19. Oþwiton *exprobaverunt*, 88, 11. Dryhten him swelc oþwát *the Lord charged them with such a fault*, Past. 1, 2; Swt. 27, 13 : 15, 1 ; Swt. 89, 16. Ðæt wé him sume opene scylde oþwíeten, 32, 1; Swt. 209, 22. Ðæt hé mé ðæt ne ôtwíte *ut non hoc nobis imputet*, Bd. pref.; S. 472, 32. Uton gangan ðæt wé bysmrigen bendum fæstne, oþwíton him his wræcsîð *let us go and insult the captive, taunt him with his misery*, Andr. Kmbl. 2715; An. 1360. Ne meaht ðú ðínre wyrde náuht oþwítan ne ðín líf nó getǽlan, Bt. 10; Fox 30, 3 : Beo. Th. 5983; B. 2995. Cf. æt-wítan.

óþ-wyrcan *to do harm to* (?) :—Ic þence ðis feoh tô witanne næs tô oþwyrceanne *I intend to keep this cattle not to harm it* (?), Lchdm. i. 384, 5.

óþ-yrnan, -ýwan, otor. v. óþ-irnan, -íwan, oter.

otor *for* ofer (?), Cd. Th. 220, 19; Dan. 73.

ó-wæstm, es ; *m. A shoot, sprout, branch* :—Ówestem *propago*, Ps. Surt. ii. p. 195, 13. Ówæstm *surculus*, Wrt. Voc. ii. 121, 48. Ówæstmas *antes*, 9, 21. Ówæstmum *stirpidum*, 75, 70. Ówæsmum *stirpis*, 89, 20. Ða ówæstmas beóþ swá mycle, and swá fægere swá swá ðæs deóres bearn ðe unicornus hätte, Ps. Th. 28, 5. v. on-wæstm.

ó-web, -wef, es ; *n. Woof* :—Óweb *vel* áb (ób, Wülck. Gl. 188, 12) *trama vel subtemen*, Wrt. Voc. i. 59, 50 : *cladicla*, ii. 139, 59. Ówef *cladica*, 104, 13 : 14, 43. [Cf. *trama* ... *est filum inter stamen discurrens, abbe*, Wülck. 617, 13.]

ó-wér = ó-hwǽr.

ó-wérn ; *adv. Anywhere*, Th. An. 101, 16. (*Smith's Bede*, 595, 3, has ówhwǽr.)

ó-wiht. v. á-wiht.

owisc, e ; *f. A margin* (?) :—Ðanon tô gráfes owisce, andlang owisce tô wege, Cod. Dip. Kmbl. iii. 388, 25.

Ówðen. v. Óðen.

oxa, an ; *m. An ox* :—Oxa *bos* ... oxa on ðam forman teáme *unus*, on ðam æfteran teáme *binus*, Wrt. Voc. i. 23, 39, 47–48 : ii. 48, 36. Oxa *bova*, i. 287, 54. Wilde oxa *bubalus*, 22, 46. Oexen *boves*, Ps. Surt. 49, 10 : ii. p. 191, 11. Ân getýme oxena, Lk. Skt. 14, 19. Oxna hyrde *aubobulcus*, Wrt. Voc. i. 287, 63. iiii oxnum gers mid cyninges

oxnum, Cod. Dip. Kmbl. ii. 64, 29. Ðá genam Abimelech oxan and scép, Gen. 20, 14. ¶ The value of an ox as given in the Laws was 30 pence :—Oxan mon sceal gyldan mid .xxx. p̃., L. O. D. 7 ; Th. i. 356, 4. Oxan tô mancuse, L. Ath. v. 6, 2 ; Th. i. 234, 1. .xxx. pǽn scyldig oððe ánes oxan, v. 8, 5 ; Th. i. 236, 31. [*Goth.* auhsa: *Icel.* uxi : *O. H. Ger.* ohso.] v. feld-, steór-oxa. The word is found in many place-names; see e. g. Cod. Dip. Kmbl. vi. 320.

oxan-slyppe, an ; *f.* Oxlip ; primula veris elatior, Lchdm. ii. 32, 26 : iii. 30, 8.

óxn, e ; *f. The arm-pit* :—Ôxn *ascella*, Wrt. Voc. i. 43, 65 : 64, 70. Under his óxne *sub ascella sua*, Kent. Gl. 992. Heó ðone fúlan stenc ðæra óxna áfyrreþ, Lchdm. i. 284, 7. [*O. H. Ger.* uohsana *ascella*.] Cf. óhsta.

Oxna-ford *Oxford* :—Tô Oxnaforda, Chr. 912 ; Erl. 100, 31. On Oxnaforda, 1015 ; Erl. 151, 17.

oxna-lyb *ox-heal* ; helleborus foetidus and h. viridis, Lchdm. iii. Glossary.

óxta. v. óhsta.

P.

For the Runic P, see *peorð*.

pád, e ; *f. An outer garment, coat, cloak* :—Paad *pretersorium*, Wrt. Voc. ii. 118, 34, 15: 68, 40–41. [*Goth.* paida : *O. Sax.* pêda : *O. H. Ger.* pheit *camisa, indusium*.] v. here-pád, hôp-páda; hasu-, salu-, salowig-pád, -páda.

pǽca, an ; *m. A deceiver* :—Se ðe sægþ ðæt hé lufie God, and his beboda ne healdeþ, hé biþ ðonne him sylf leás, and biþ his ágen pǽca, Basil admn. 4 ; Norm. 40, 21.

pǽcan ; *p.* pǽhte ; *pp.* pǽht *To deceive* :—Swylce hié mid sceare and munuces hiwe God pǽcen (pǽcean, MS. T.) *as if deceiving God with the tonsure and the appearance of a monk*, R. Ben. 9, 15. Hý óðer specaþ, óðer hý þencaþ, and lǽtaþ ðæt tô wǽrscype, ðæt hý óðre mágan swá swicollíce pǽcan, Wulfst. 55, 3. Pǽcht *decepta, seducta*, Hpt. Gl. 449, 42. v. á-, be-pǽcan.

pægel *a wine-vessel, a pail* :—Pægel (Wright gives *wægel*, but see Anglia viii. 450) *gillo*, Wrt. Voc. i. 25, 26. [Cf. *Dan.* pægel *half a pint*.]

pæll, pell, es ; *m.* **I.** *a pall, covering, cloak, costly robe* :—Pǽl (pell) *pallium*, mid pælle (pelle) gescrýd *palliatus*, Ælfc. Gl. Zup. 257, 3–4. Pǽl *pallium*, Blickl. Gl. Weofod mid reádum pælle gescrýd (*the altar was in the church dedicated to St. Michael.* v. next passage), Homl. Th. i. 508, 16. Mid háligdôme of ðæs Hǽlendes rôde and of Marian reáfe and of Michaheles pelle, Homl. Skt. i. 6, 73. Volosianus ðone pæll ástrehte ðe Dryhtnes andwlytan on wæs befealden, St. And. 46, 13. iiii. pellas, and iiii. cuppan, Chart. Th. 519, 23. Mycel ðǽr wæs gegaderod on golde and on seolfre and on faton and on pællan, Chr. 1086 ; Erl. 223, 30. **II.** *purple, a purple garment* :—Of ðam biþ geweorht se weolocreáda pæl *quibus tinctura coccinei coloris conficitur*, Bd. 1, 1 ; S. 473, 20 note. Pællas *purpuram*, Coll. Monast. Th. 27, 7. [*Icel.* pell *costly stuff. From Lat.* pallium.] v. next word.

pællen, pellen ; *adj. Purple, rich or costly* (*of garments*) :—Hé hyne on pællenre scýtan befeóld, St. And. 42, 13. V. pællene weofodsceátas, Chart. Th. 429, 25. Bicgaþ eów pællene cyrtlas, ðæt gé tô lytelre hwíle scínon swá swá rôse, Homl. Th. i. 64, 13. Se cyning gesýmde gold and seolfor and deórwurðe gymmas and pællene gyrlan uppon olfendas, 458, 24. Se ríca on his pællenum gyrlum cwyþ : 'Nis se loddere mid his tættecon mín gelíca,' 256, 8. Se cásere dyde of his purpuran and his pellenan gyrlan, H. R. 103, 18. [*Laym.* pallen (curtel).]

pælme, pǽran. v. palm, á-, for-pǽran.

pærl (?) The word, which occurs in a list of terms connected with writing, is glossed by *enula*, which elsewhere glosses *horselene* :—Pærl *enula, bôcfel pergamentum*, Ælfc. Gr. Zup. 304, 7.

pæþ, pæþ, es ; *m.* : e ; *f.* (?) *A path, track* :—Pæþ, paþ *semita*, Ælfc. Gr. 7 ; Zup. 25, 3. Manna paþ *semita*, deóra paþ *callis*, Wrt. Voc. i. 37, 41–42. Pæþ *semita*, 80, 37. Wegleás pæþ *invium*, 53, 61. Pæþ *callis, iter pecudum*, Wrt. Voc. ii. 127, 58. Paþ *callis*, 14, 10. Paat, 103, 48. Andlang oxna pæþes, Cod. Dip. Kmbl. v. 215, 10. Ðone kyng gerihtan of ðam dweliandan pæþe (*from the path of error*), Chr. 1067 ; Erl. 204, 30. Ne mihton forhabban helpendra paþ merestreámes môd (*they could not stop the course of the rushing water*), Cd. Th. 208, 23 ; Exod. 487. Gerece mé on rihtne pæþ (*semitam*), Ps. Th. 26, 13. Lǽr mé ðíne paþas (*semitas*), 24, 3 : Ps. Spl. 8, 8 : Homl. Th. i. 360, 32 : 362, 16. Ðeáh willniaþ ealle þurh mistlíce paþas cuman tô anum ende, Bt. 24, 1 ; Fox 80, 8. Ic ondrǽde ðæt ic ðé lǽde hidres ðidres on ða paþas of ðínum wege, 40, 5 ; Fox 240, 21. On paþum (*semita*) beboda ðínra, Ps. Spl. 118, 35. *The word seems feminine in the following* :—Andlang paþæ ... ǽc ðæ standaþ in ðære paþæ, Cod. Dip. Kmbl. iii. 175, 36–176, 6. *In the Northern Gospels* pæþ *is an alternative gloss with* dene :—Pæþ t̃ dene *uallis*, Lk. Skt. Lind. Rush. 3, 5 : *chaos*, 16, 26. [*O. Frs.* path, paed: *O. H. Ger.* pfad *callis, semita*.] v. ân-, flet-, gegn-, here-, mearc-, míl-pæþ.

pæþþan; *p.* de *To tread* (*a path*), *to traverse* :—Tungol gársecges grundas pæþeþ *the sun* (*after it has set*) *treads ocean's depths as its path*, Exon. 350, 29; Sch. 71. Eorþgræf pæþeþ *it makes its way along a trench*, 439, 26; Rä. 59, 9. Sume fótum twám foldan peþþaþ, sume fiérfête, Met. 31. 10. Ic mearcpaþas træd, móras pæþde, Exon. 485, 8; Rä. 71, 10. [Cf. *O. H. Ger.* pfadôn *to go along a path*, Grff. 3, 326.]

pætig. v. prættig.

pál, es; *m.* I. *a pale, pole, stake* :—Pál *palus*, Wrt. Voc. i. 84, 69. II. *a kind of hoe* or *spade* :—Delfîsen *vel* spadu *vel* pál *fossorium*, 16, 14. [*Icel.* páll *a kind of hoe* or *spade*; *a pale* : *O. H. Ger.* pfâl *palus*. From Latin.]

palent, es; *m.*: palente, palendse, an; *f. A palace* :—On ðam mǽran palente ðǽr ðǽr se cyning wæs oftost wunigende, Anglia ix. 28, 31. Ðæt seó cwén ne cume nǽfre heononforþ intô ðînum pallente, 29, 64. On strǽte oððe on palentan, Lchdm. iii. 206, 6. Æt ðæs cáseres palendsan (palentsan, Bos.), Ors. 6, 21; Swt. 272, 23. Hé bræc ðæne palant (ða palentan, MS. D.), Chr. 1049; Erl. 172, 21. [*O. Frs.* palense : *O. Sax.* palencea : *O. H. Ger.* pfalanza, pfalinza *basilica, praetorium, aula, palatium.* From a Mid. Lat. form *palantium.* v. Kluge Dict. s. v. pfalz.]

palent-líc; *adj. Relating to a palace* :—Tô ðǽm palentlícum *ad palatinas*, Wrt. Voc. ii. 2, 67. [*O. H. Ger.* pfalenz-lîh *palatinus.*]

palm, es; palma, an (?); *m.*: pælme, an; *f. A palm* :—Palm *palma*, Wrt. Voc. i. 32, 61. Se palm is sigebeácen, Homl. Th. ii. 402, 10: i. 218, 10. Swé swé palma *ut palma*, Ps. Surt. 91, 13. Swælce pælme *quasi palma*, Rtl. 65, 33. Pælmana *palmarum*, 95, 8. Palmana, Jn. Skt. Lind. Rush. 12, 13. [*O. Sax. O. H. Ger.* palma : *Icel.* pálmr *a palm-tree.*]

palm-æppel *the fruit of the palm, a date* :—Palmæppel *dactulus*, Wrt. Voc. ii. 26, 63 : 89, 33. Palmæppla *nicolaos*, 83, 55. Palmæpla, 60, 67.

palm-bearu *a palm-grove* :—Palmbearwes *palmeti*, Wrt. Voc. ii. 75, 77.

Palm-sunnandæg *Palm Sunday:* —Gyf se terminus becymþ on ðone Sunnandæg ðonne byþ se dæg Palmsunnandæg, Lchdm. iii. 244, 16. On Palmsunnandæg, Rub. Lk. Skt. 19, 29. [*Icel.* pálmsunnudagr.]

palm-treów *a palm-tree* :—Palmtreów *palma*, Ps. Lamb. 91, 13. Palmtreó *palmes*, Jn. Skt. Lind. Rush. 15, 4. Ðǽr wǽron hundseofontig palmtreówa (*palmae*), Ex. 15, 27. Palmtreówa (-trýwa) twigu *ramos palmarum*, Jn. Skt. 12, 13.

palm-twig *a palm-branch:* —Palmtwig *palma*, Wrt. Voc. i. 32, 61; Blickl. Gl. Onfôh ðissum palmtwige, Blickl. Homl. 137, 25. Heó álegde ðæt palmtwig ðe heó ǽr onféng, 139, 4. Se gewuna stent ðæt gehwǽr on Godes gelaþunge se sacerd bletsian sceole palmtwigu on ðisum dæge (*Palm Sunday*), Homl. Th. i. 218, 3.

palmung *glosses* palmes, Jn. Skt. Lind. Rush. 15, 2.

palm-wicu *the week which begins with Palm Sunday:* —On ðære palmwucan, Rub. Lk. Skt. 22, 1 : Rub. Jn. Skt. 12, 1, 24.

palstr *a spike* or *something with a point:* —Palester, plaster, palstr *cospis*, Txts. 50, 225. Palstre *cuspite*, Wrt. Voc. ii. 21, 58.

pan-mete *cooked food:* —Ælces cynnes panmete *ferculum*, Wrt. Voc. ii. 38, 59. Ponmete *vivertitum*, i. 290, 42.

pang, Wrt. Voc. i. 289, 52, *an error for* þung (?).

panic, es; *n.*(?) *A sort of millet*; panicum :—Panecis fíf scillinga gewyht, Lchdm. iii. 124, 8. Nym panic, 118, 28. [*O. L. Ger.* penik : *M. H. Ger.* pfenich.]

panne, an; *f. A pan:* —Panne *patella*, Wrt. Voc. i. 24, 51. Mid ðisse pannan hiersinge wæs Paulus onbærned, Past. 21; Swt. 165, 3. Of brádre pannan *de sartagine*, Wrt. Voc. ii. 26, 11. Wyl on pannan, Lchdm. ii. 308, 28. Ðǽr wǽron inne geseted hweras and pannan, and hé clypte ða hweras and cyste ða pannan, ðæt hé wæs eall sweart, Shrn. 69, 27-29. [*O. H. Ger.* pfanna : *O. Frs.* panne : *Icel.* panna.] v. brád-, bráde-, brǽding-, bræg-, côcer-, fýr-, heáfod-, hearste-, holo-, hyrsting-, îsen-panne.

Pante, an; *f. The river Blackwater in Essex:* —Hí Pantan streám bestôdon, Eástseaxena ord and se æschere, Byrht. Th. 133, 50; By. 68. Wôdon wælwulfas ofer Pantan, 134, 41; By. 97. Seó ǽreste stôw is on Pante staþe ðære eá *prior locus est in ripa Pentae amnis*, Bd. 3, 22 ; S. 553, 8.

pápa, an; *m. A pope:* —Ðá wæs on ða tíd Vitalianus pápa ðæs apostolican sétles ealdorbiscop *sede apostolicae tempore illo Vitalianus praeerat*, Bd. 4, 1; S. 563, 23. Gregorius se hálga pápa, Homl. Th. ii. 116, 24. Æfter ðæs pápan geendunge, 122, 18. Tô pápan gecoren, 122, 31. Tô pápan gehálgod, 124, 1. [*Icel.* páfi. *From Latin* papa.]

pápan hád, es; *m. The papal dignity:* —Gregorius pápanhád onféng, Homl. Th. ii. 126, 24.

páp-dóm, es; *m. The papacy:* —Gregorius féng tô pápdóme, Chr. 592; Erl. 19, 33. [*Icel.* páfa-dómr.]

paper, es; *m.*(?) *Papyrus:* —Paper *papirus*, Wrt. Voc. ii. 92, 12.

papig. v. popig.

papol-stán, es; *m. A pebble-stone, pebble:* —Gæþ tô ðære sǽstrande and feccaþ mé papolstánas, Homl. Th. i. 64, 3. Popelstánas *lapillulos*, Hpt. Gl. 449, 18. [*Wick.* pibbil-ston.]

páp-seld, es; *n. The papal see:* —Hé hié lǽrede ðæt hié raðost tô

Rôme sendon tô ðæm pápan, and ðone pápan and ðæt pápseld ðæt hié beáhsodan hwæt him ðæs tô rǽde þúhte, Blickl. Homl. 205, 20.

páp-setl, es; *n. The papal throne:* —Hé sæt on ðam pápsetle ændlefen geár, Shrn. 49, 17.

part, es; *m. A part:* —Ðes part oððe ðes dǽl, Ælfc. Gr. 41; Som. 43, 2. Ðisses partes, 16; Som. 20, 11. On ðisum parte, 17; Som. 20, 32.

Parthe; *pl. The Parthians:* —Parthe forhergodon Mesopotamian, Ors. 6, 24; Swt. 276, 6. Partha cyning, 5, 11; Swt. 236, 3. Partha gewin, Swt. 236, 26. Hié hæfdon gewin wið Parthe, 6, 13; Swt. 268, 6, 8. Hé com ǽrest tô Parþum, Bt. 18, 2; Fox 64, 12.

paþ. v. pæþ.

páwa, peá, an; *m.*: páwe, an; *f. Peacock, peahen:* —Páwa *pavo*, Ælfc. Gr. 9, 3; Som. 8, 34. Pauua, Txts. 90, 826. Pawa, Wrt. Voc. i. 77, 24. Páwe, *pavo, pavus*, 29, 4. Fuglas ða ðe heard flǽsc habbaþ, páwa, swan, æned, Lchdm. ii. 196, 19. On ðære ylcan stôwe byþ óðer fugelcynn fenix hátte ða habbaþ cambas on heáfde swá páwan *in eo monte est avis fenix que habet cristas quasi orbes pavonis*, Nar. 39, 4. Se fugel (*the phenix*) is onlícost peán, Exon. Th. 219, 25; Ph. 312. [A pruest proud as a *po*, Pol. Songs, 159, 15 : *Wick.* poos; *pl.* : *O. H. Ger.* pfáwo : *Icel.* pá or pái (*as a nickname*). From Latin.]

Peác-land *the Peak of Derbyshire:* —Eádweard cyning fôr ðonan (*from Nottingham*) on Peácland tô Badecan wiellon (*Bakewell*), Chr. 924; Erl. 110, 11. v. next word.

Peác-, Péc-sǽtan; *pl. The occupiers of the Peak:* —Pécsǽtna [land is] twelf hund hýda, Cod. Dip. B. i. 414, 17.

pearroc, es; *m. An enclosure:* —Pearroc, pearuc *clatrum*, Txts. 50, 224. Pearruc, Wrt. Voc. i. 34, 7. Pearruc *cauea*, Germ. 400, 62. On ðisum lytlum pearroce búgiaþ swiðe manega þeóda *hoc ipsum brevis habitaculi septum plures incolunt nationes*, Bt. 18, 2; Fox 62, 27. Ðis sindon ða landgemǽro. Ǽrest ... on Bogeles pearruc; of Boceles pearruce, Cod. Dip. Kmbl. v. 277, 11. Hié (*the English*) bedrifon hié (*the Danes*) on ánne pearruc, and besǽton hié ðǽr útan, Chr. 918; Erl. 102, 35. Pearruca *clatrorum*, Hpt. Gl. 489, 75. Pearroca, Wrt. Voc. ii. 18, 63. Of pearrocum *de clatris*, 26, 52 : 18, 62. Of pearrucum, Hpt. Gl. 484, 44 : 508, 29. Ðæs gemǽre is on eásthealfe spachrycg, on súðan plumwearding pearrocas, Cod. Dip. Kmbl. i. 258, 12. [*O. H. Ger.* pferrih, pfarrih. From Celtic : *Welsh* parwg.]

Pedrida, Pedreda (e ?) *the river Parret:* —Æt Pedridan (Pedredan, MS. E.) múþan, Chr. 845; Erl. 66, 23 : 658; Erl. 34, 2.

pell, pellen. v. pæll, pyllen.

pellican, es; *m. A pelican:* —Ic geworden eom pellicane gelíc se on wéstene wunaþ, Ps. Th. 101, 5.

pending. v. pening.

Péne; *pl. The Carthaginians;* Poeni :—Ðæt hié wið Péna folce mehte ... Ðá flugon Péne ... Hanna, Péna cyning, Ors. 4, 6; Swt. 170, 21-25.

pening, penning, pending, penig, pennig, es; *m. A penny* (1) referring to other than English coinage:—Ðes peningc (pening, penig) *hic as*, Ælfc. Gr. 9, 25; Zup. 50, 14. Fals pening *paracaraximus*, Wrt. Voc. i. 57, 34. Penninge *hymenis* (?), ii. 96, 71. Peninge, 43, 27. Bringaþ mé ðone pening (*denarium*), Mk. Skt. 12, 15. Ðá brohton hí him ánne peninc (penig, MS. A. : penning, Lind.), Mt. Kmbl. 22, 19. Hé sealde ǽlcon ǽnne penig (penning, Lind.) ... Ðá onféngon hig ǽlc his pening (suindrigo penningas, Lind.) ... syndrige penegas *singulos denarios*, 20, 2-10. Pening, Homl. Th. ii. 78, 27. Mé sind wana penegas *desunt mihi nummi*, Ælfc. Gr. 32; Som. 36, 37. Wé eác wiernaþ úrum cildum úrra peninga mid tô plegianne *pueris nummos subtrahimus*, Past. 50, 4; Swt. 391, 27. Hig sealdon hine wið þrítigum penegum, Gen. 37, 28. (2) of English coinage, a silver coin, the 240th part of a pound :—Fíf penegas gemaciaþ ǽnne scillingc, and xxx. penega ǽnne mancs, Ælfc. Gr. 50; Som. 52, 8. Gá seó wǽge wulle tô .cxx. p̄. (tô healfan punde, MS. G.), L. Edg. ii. 8; Th. i. 270, 3. Tén hund (pund ?) peñd ... Gedǽle hé ǽlcum Godes þíowe peñd ... preóténe hund (pund ?) pending, Chart. Th. 471, 5-26. xiii. pund pendingæ, 474, 9. Mid .v. pundum mǽrra pæninga (*denarii meri*), L. Alf. pol. 3; Th. i. 62, 10. Gif mon men eáge of ásleá, geselle him mon .lx. scill. and .vi. pæningas and þriddan dǽl pæningas (peniges, MS. H.) tô bôte, 47; Th. i. 94, 3-5. Hire mægþhádes wurð, ðæt synd twelf scillingas be twelf penigon (cf. Se rihtscylling. byþ á be .xii. penegum *legitimus solidus semper est .xii. denariorum*, L. Ecg. P. iv. 60 : Th. ii. 222, 7), Ex. 21, 10. (3) as a weight, *pennyweight:* —Án *uncia* stent on feówer and twentig penegum. Twelf sídon twelf penegas beóþ on ánum punde, Anglia viii. 335, 17. Pund ealoþ gewihþ .vi. penegum mǽre ðonne pund wætres, and .i. pund wínes gewihþ .xv. penegum mǽre ðonne .i. pund wætres, etc., Lchdm. ii. 298, 16-26. Gegníd on mortere ðætte pening gewege, 18, 3 : 134, 25. Swylce swá .iii. penegas gewegen, 52, 13 : 110, 17. Wið lúsum; cwic seolfor, án pening seolfres, 124, 24. Drenc biþ on peninge *the dose will be a pennyweight*, 272, 24. Ceorf nygan penegas *cut up nine pennyweights*, iii. 8, 2. Man ðysses wyrttruman genime týn penega gewihte, i. 260, 17. Hý man wegeþ, swá man déþ gold wið penegas, and gif ða penegas teóþ swiðor ðonne ðæt gold, ðonne miswyrþ ðam men hraðe,

Wulfst. 240, 2-4. [*O. L. Ger.* penning : *O. Frs.* panning : *O. H. Ger.* pfenning, pfenting : *Icel.* penningr.] v. ælmes-, healf-, heorþ-, hundred-, Rôm-, seám-pening.

pening-hwirfere, es; *m. A money-changer :* — Pennighwyrfere *mensularius*, Wrt. Voc. i. 57, 31.

pening-mangere, es; *m. A money-dealer :*—Pennigmangere *collybista*, Wrt. Voc. i. 57, 32. Peningmongere, ii. 22, 36.

pening-sliht, es; *m. The striking of money :*—Gæfil ł penningslæht *tributum vel censum*, Mt. Kmbl. Lind. 17, 25.

pening-wǽg, e; *f. A penny-weight :*—Wið lûsum; cwic seolfor and eald butere; ân pening seolfres, and tû peningwǽge buteran, Lchdm. ii. 124, 24.

pening-weorþ, -wurþ, es; *n. A penny-worth :*—Hafa ân penigweorþ swefles, Lchdm. iii. 38, 28. Æt ǽlcon gegyldan ǽnne peningc oððe ân peningcwurþ weaxes, Chart. Th. 605, 26. Twâ hund peningweorþ hlâfes, Homl. Th. i. 182, 9.

penn, es; *m. A pen, fold :*—On penn; of ðam penne, Cod. Dip. Kmbl. iii. 456, 3-4: 25, 21. On hacapenn foreweardne, 412, 13.

penn *a disease of the eye, pin, a kind of cataract :*—Ðis is seó sēleste eáhsalf wið ēhwærce and wið miste and wið penne, Lchdm. i. 374, 2.

pennian. v. on-pennad. [Cf. þe pit tineþ his muð ouer þe man þe lið on fule synnen . . . gif ure ani is þus penned, O. E. Homl. ii. 43, 27.]

Pentecosten; *m.* (?) *Pentecost, the fiftieth day after the resurrection, Whitsuntide :*—On Pentecostenes dæg com se Hâlga Gâst ofer ða apostolas, Btwk. 214, 29. On ðǽre Pentecostenes wucan, Rubc. Lk. Skt. 5, 17 : 8, 40. On ôðerne Pentecostenes mæssedæg, Rubc. Jn. Skt. 3, 16. On Pentecostenes mæsseǽfen, 14, 15.

Penwiht-steort, es; *m. The Land's End in Cornwall :*—Se here . . . wendon eft âbûtan Penwiht-steort (Penwið-, MS. C. : Penwæd-, MS. D.) on ða sûþhealfe, and wendon in tô Tamermûþan, Chr. 997; Erl. 135, 10. [The Welsh form is *Pengwayd*, v. Earle's note.]

Peohtas; *pl. The Picts :*—Ðâ fērdon Peohtas in Breotone . . . Mid ð¯y Peohtas wîf næfdon . . . ðæt is mid Peohtum healden . . . Ðridde cynn Breotone onfēng on Pehta dǽle, Bd. 1, 1; S. 474, 17-25. On Peohta gereorde, S. 474, 4. Pehta cynn, 5, 24; S. 646, 33. Hî sceoldon feohton wið Pyhtas (Pihtas, MS. A.). Heó ðâ fuhton wið Pyhtas, Chr. 449; Erl. 13, 6.

peonia, an; *m.* (?) *Peony :*—Peonia *peonia*, Wrt. Voc. i. 69, 22. Ðeós wyrt ðe man peonian nemneþ, Lchdm. i. 168, 14. [The Latin form of the accusative, *peoniam*, occurs, 170, 4.]

peorð *the name of the Runic p.* Its meaning is doubtful. Grimm notices the name for *f* in the old Sclavonic alphabet, *fert*, and the Persian name for one of the figures on the chess-board, *ferz.* Kemble seems to take the latter, translating the word by *chess-man :* but it is doubtful whether the knowledge of chess was early enough among the Teutons to allow of this interpretation. v. Zacher Das Runenalphabet, pp. 7-9. The verse which accompanies the Rune in the Runic poem is the following :—Peorð byþ symble plega and hlehter wlancum ðǽr wîgan sittaþ on beórsele blîðe ætsomne, Runic pm. Kmbl. 341, 1-6; Rûn. 14.

pere(u), an; *f. A pear :*—Seó peru *hoc pirum*, Ælfc. Gr. 6; Som. 5, 59. Pere, Wrt. Voc. i. 285, 59. Healfreáde peran *crustumie vel volemis vel insana vel melimendrum*, 39, 25. (Cf. *hec volemus a° permayn-tre*, 191, col. 2 : *hoc volemum a° permayne*, 192, col. 2.) Peran, Lchdm. ii. 176, 18. [*Icel.* pera : *O. H. Ger.* bira.]

pere-wôs, es; *n. Perry, a drink made from pears :*—Perewôs *sapa*, Wrt. Voc. i. 27, 50. (The word occurs in a list of drinks.)

persa. v. medema.

Persc-ware; *pl. The Persians :*—Of Perscwara mǽgþe, Shrn. 55, 32.

Perse, Persēas; *pl. The Persians :*—Ðâ wǽron ða Perse geegsade, Ors. 2, 5; Swt. 78, 13 : 3, 1; Swt. 98, 30. Persa cyning, 2, 4; Swt. 74, 29. Persa rîce . . . Persēa rîce, 2, 5; Swt. 78, 2, 31. Wið Persum, Swt. 82, 23. On Persēum, 78, 30. Hié sendon on Perse, 3, 1; Swt. 98, 19.

Persida *Persia :*—Tô ðam earde ðe is gehâten Persida, Homl. Th. ii. 482, 2.

Persisc; *adj. Persian :*—Seó reáfung ðæs Persiscan feós, Ors. 2, 5; Swt. 84, 21 : Jud. Thw. 162, 23.

persoc, es; *m. A peach;* malum persicum :—Genim persoces leáf, Lchdm. iii. 58, 27. Æppla and peran and persucas, ii. 176, 18. [*M. H. Ger.* pfersich.]

persoc-treów, es; *n. A peach-tree :*—Persoctreów *persicarius*, Wrt. Voc. i. 32, 52.

peru. v. pere.

pervince, an; *f. Periwinkle* (plant) :—Pervincæ *vinca*, Wrt. Voc. i. 31, 65. Pervince, 79, 34.

petersilige, an; *f. Parsley :*—Petersilie. Ðás wyrte man *petroselinum* nemneþ, Lchdm. i. 240, 6. Petresilige, iii. 24, 9. Petorsilian sǽd, ii. 314, 29 : 228, 26. Ða wyrt petersilian, 206, 27 : 234, 8. [*O. H. Ger.* petarsile : *Ger.* petersilie.]

peþþan, peþþan; v. pæþþan, prættig.

Petrus; *gen.* Petres; *m. The apostle Peter :*—Ðâ genam Petrus hyne . . . Ðâ beseah hē hyne and cwæþ tô Petre, Mt. Kmbl. 16, 22-23. Se Hǽlend com on Petres hûse, 8, 14. Hē sceare (Petres mearce, MS. B.) onfēng, Bd. 3, 18; S. 546, 10. Be Peteres mæssan, Wulfst. 272, 9.

---

philosoph, es; *m. A philosopher :*—Paminunde ðæm strongan cyninge and ðæm gelǽredestan philosophe, Ors. 3, 7; Swt. 110, 22. Hié sealdon Demostanase ðæm philosophe licgende feoh, 3, 9; Swt. 124, 1.

pic, es; *n. Pitch :*—Ðis pic *haec pix*, Ælfc. Gr. 9, 63; Som. 13, 54 : Wrt. Voc. ii. 117, 39. Hlûttor pic *resin*, Lchdm. ii. 44, 24 : 72, 25. Genim pices lytel, 96, 12. Weallendes pices, 252, 1 : Dôm. L. 14, 199. Heó smirode hine mid tyrwan and mid pice, Ex. 2, 3. Ðâ hēt se câsere meltan on hwere leád and scipteoran and pic, Shrn. 91, 7 : Lchdm. ii. 318, 4. [*O. L. Ger.* pik : *O. H. Ger.* peh : *Icel.* bik.]

pîc, es; *m. A point, pointed instrument, pike :*—Piic *acisculum*, Wrt. Voc. ii. 98, 39. Pîc, 4, 23 : i. 17, 31. [Cf. his pic he nom on honden & helede hine under capen . . . þene pic he bilæfde, Laym. 30849. A Celtic word.] v. horn-pîc.

pîcan *to use a pîc, to remove by means of a pîc, to pick :*—Lēt him pŷcan ût his eágan, and ceorfan of his handa, Chr. 796; Erl. 58, 33. [Pykyn *purgo*, Prompt. Parv. 397 : to piken and to weden, Piers P. 16, 17.]

pic-bred (?) *glosses* glans, Wrt. Voc. i. 33, 58 (at the end of a list of names of trees).

picen; *adj. Pitchy, of pitch :*—Picen hell *piceus Tartarus*, Hymn. Surt. 142, 30. On ðære picenan eá, Blickl. Homl. 43, 28.

pician; *p.* ode *To pitch, cover with pitch :*—Crocca gepicod ûtan, Lchdm. ii. 26, 23.

pîcung; *f. A pricking :*—Pîcung *stigmata*, Wrt. Voc. ii. 121, 39. v. pîc.

pîe; *f. An insect :*—Hundes pîe (pēo, Ps. Spl. C.) *cynomia*, Ps. Surt. 104, 31. Lûs *peducla*, hnitu *ascarida*, pîe *ladasca*, Wrt. Voc. j. 287, 45-47. *Ladasca* pîae, *briensis* hondwyrm, Wrt. Voc. ii. 112, 48.

pihment *a pigment, drug :*—Of ôþrum pyhmentum, Lchdm. iii. 136, 29. Cf. next word.

pihten *part of a loom :*—Pihten, Anglia ix. 263, 12. Pihtine *pectine*, Hpt. Gl. 494, 26.

pîl, es; *m. A stick with a point, something pointed :*—Dægmǽles pîl *gnomon*, Wrt. Voc. i. 86, 42. Ða Walas âdrifon sumre eá ford ealne mid scearpum pîlum (stængum, MS. D.) greátum innan ðam wetere (cf. Cassobellannus ripam fluminis ac pene totum sub aqua vadum acutissimis sudibus praestruxerat, Bd. 1, 2), Chr. Erl. 5, 10. Heó (sea-holly) hafaþ stelan hwîtne, on ðæs heáhnysse ufeweardre beóþ âcennede scearpe and þyrnyhte pîlas (sharp and thorny prickles), Lchdm. i. 304, 1. Hē gehæfte hî on ânum micclum stocce, and mid îsenum pîlum heora îlas gefæstnode, Homl. Skt. i. 5, 388. [*O. H. Ger.* pfîl *pilum, arundo.* From Lat. *pilum.*] v. hilde-, orþanc-, searo-, wæl-pîl; and dægmǽls-pîlu.

pîle, an; *f. A stake.* v. temes-pîle.

pîle, an; *f. A mortar :*—Ðeáh ðu portige ðone dysegan on pîlan swâ mon corn dēþ mid pîilstæfe ne meaht ðu his dysig him from âdrîfan *si contuderis stultum in pila, quasi ptisanas feriente desuper pilo, non auferetur ab eo stultitia ejus*, Past. 37, 2; Swt. 267, 1. Swilce hit on pîlan gepîlod wǽre *quasi pilo tusum*, Ex. 16, 14. [From Latin *pila.*]

pîle *a pillow.* v. pyle.

pilece, an; *f. A robe of skin, pelisse :*—Pylece *pellicie*, Wrt. Voc. i. 81, 68. Hwî worhte God pylcan Adame and Eve æfter ðam gylte? Ðæt hē geswutelode mid ðâm deádum fellum ðæt hî wǽron ðâ deádlîce, Boutr. Scrd. 20, 28. [He to-rended þe olde pilche of his deadliche uelle, A. R. 362, 29. Pylche *pellicium, pellicia*, Prompt. Parv. 397; see the note, where many instances of the word are given. *O. H. Ger.* pelliz : *Icel.* piliza, pilla *a fur coat.* From Latin.]

pîlere, es; *m. One who pounds in a mortar :*—Pîlere *pilurius*, Wrt. Voc. i. 34, 52. v. next word.

pîlian; *p.* ode *To pound in a mortar :*—Se ðe pîlaþ *vel* tribulaþ *pilurus vel pistor*, Wrt. Voc. i. 20, 26. v. preceding word and pîle, pîlstampe, -stoc.

pillan (?) *to peel* (of skin) :—Ðis lâcecræft sceal tô ðan handan ðe ðæt fell of pyleþ, Lchdm. iii. 114, 13.

pill-sâpe, an; *f. Silotrum* (?), Wrt. Voc. i. 27, 32.

pîl-stæf. v. pîle.

pîl-stampe, an; *f. A pestle;* pilum, Wrt. Voc. i. 34, 51.

pîl-stocc, es; *m. A pestle;* pila, Wrt. Voc. i. 86, 6.

pîlstre, an; *f. A pestle,* pila, Wrt. Voc. i. 34, 50.

pîn-beám, es; *m. A pine-tree :*—Se hâlga wolde âheáwan ǽnne pînbeam, Homl. Th. ii. 508, 24.

pinca. v. pynca.

pînere, es; *m. One who torments :*—Hlâferd his gesalde hine ðæm pînerum (*tortoribus*), Mt. Kmbl. Lind. 18, 34 : Germ. 399, 265.

pinewincle. v. winewincle.

pîn-hnutu; *gen. dat.* -hnyte; *pl.* -hnyte; *f. A pine-nut, fir-cone :*—Seó eorþe stent on gelîcnesse ânre pînnhnyte, Lchdm. iii. 258, 6. Genim of pînhnyte .xx. geclǽnsodra cyrnela, ii. 180, 19. [*Prompt. Parv.* pynote *pinum.*]

pînian; *p.* ode *To torment, torture :*—Ðâ pîneden hié hiene mid ðæm ðæt hié his hand forbærndon, anne finger and ânne, Ors. 2, 3; Swt. 68, 22. Pînedon *excruciabant*, 6, 11; Swt. 266, 15. Ðæt hē his heortan and his môd mid hreówsunga suîðe pinige *ut per afflictionem poenitentiae cor prematur*, Past. 28, 6; Swt. 199, 25. Ðâ hēt hē hî pînian (pînigan, MS. C.), Homl. Skt. i. 5, 371. Ðonne onginþ hē hý tô pînianne on

mistlícre wîsan, Wulfst. 195, 1. Gnættas ægðer ge ða men ge ða nýtenu pîniende wæron, Ors. 1, 7; Swt. 36, 31. Pîniendum *cruciante*, Hpt. Gl. 503, 36. [*O. H. Ger.* pînôn : *Icel.* pína. From Latin.]

pinn. I. *a pin, peg* :—Ne sceolde hê nân þing forgýman ðe æfre tô note mehte; ne mûsfellan, ne ðæt git læsse is, tô hæpsan pinn, Anglia ix. 265, 10. [From Latin *pinna*.] II. *an instrument for writing, a pen* :—Mid pinn ł urittsæx *calami*, Mt. Kmbl. p. 2, 17. [From *penna?* or *pinna?*]

pinne (?), an; *f. A flask, bottle* :—Ic (*sutor*) wyrce of him (*cutes et pelles*) flaxan (pinnan) *facio ex iis flascones*, Coll. Monast. Th. 27, 35.

pinness, e; *f. Torment, pain* :—Tô ðare helware stîðe pînnesse, Chart. Th. 369, 34.

pinsian; *p.* ode *To weigh, judge, estimate, consider, examine* :—Geþænce ælc man hû swíðe man pinsaþ ða sâwle on dômes dæg, ðonne man sett ða synne and ða sâwle on ða wæge and hý man wegaþ, swâ man dêþ gold wið penegas, Wulfst. 239, 24. Hê holrede ł pinsode *pensavit, cogitavit*, Hpt. Gl. 443, 76. Hê sceáwode hine selfne and pinsode *he observed and weighed himself*, Past. 7, 2; Swt. 51, 15. Pinsige ælc mon hiene selfne georne, 10, 2; Swt. 63, 18. Pinsiende *inquirendo, scrutando*, Hpt. Gl. 411, 26. [*Lat.* pensare.] v. â-pinsian.

pinsung. v. â-pinsung, Hpt. Gl. 447, 73.

pintel *virilitas, membrum virile*, Wrt. Voc. i. 65, 29. [Pyntyl *veratrum, tentigo, priapus*, 184, 11. Pyntylle *veretrum*, 186, col. 2. Pyntyle, 208, col. 1. Also see Cath. Angl. p. 281. s. v. *pyntelle*, and the note. Leo 200, 41 gives a Platt-deutsch *pint* with the same meaning.]

pín-treów, es; *n. A pine-tree* :—Pîntreów *pinus*, Wrt. Voc. i. 32, 54; 79, 80: 285, 60. Þûfbæres pîntreówes *frondentis pini*, Hpt. Gl. 68; Lchdm. ii. 216, 5. Ðæt man pîntreów bærne tô glêdum and ðonne ða glêda sette tôforan ðam seócum men, 284, 12.

pín-treówen, -tríwen; *adj. Belonging to a pine-tree* :—Cyrnlu of pîntrýwenum (-treów-, MS. O.) hnutum, Lchdm. i. 250, 9.

pínung, e; *f. Torment, torture, pain* :—Rôd[e] pînung *crucis tormentum*, Rtl. 24, 11. Tô pînunge *ad poenam*, 103, 17. For his gylta pînunga *in criminum suorum cruciatum*, L. Ecg. P. ii. 5; Th. ii. 184, 8. Pînunga, L. Edg. C. 13; Th. ii. 268, 19. Mid ungemetlícre pînunge hê (*Phalaris*) wæs ðæt folc cwielmende, Ors. 1, 12; Swt. 54, 18. Pînunge *tormento*, Hpt. Gl. 503, 20. Pînungum *cruciatibus*, 502, 70.

pínung-tól, es; *n. An instrument of torture* :—Decius hêt gearcian eall ðæt pînungtól, Homl. Th. i. 428, 18. Mid eallum ðisum pînungtôlum getintregod, 424, 22.

pipat (?) *glosses* accipiter, Wrt. Voc. ii. 10, 36.

píp-dreám, es; *m. The sound of the pipe* :—Pîpdrâm singan gehýreþ gehende blisse *to hear (in a dream) the sound of the pipe shews joy at hand*, Lchdm. iii. 208, 22.

pipe, an; *f.* [superscript 4] *pipe*. (1) *as a musical instrument* :—Pîpe odðe hwistle *musa*, Wrt. Voc. i. 73, 60. Hearpe and pîpe drêmaþ eów on beorsele, Wulfst. 46, 16. i. silfren pîpe, Chart. Th. 429, 20. (2) *of other tubes* :—In pîpan; of pîpan in widi brôc, Cod. Dip. Kmbl. iii. 380, 2. Dô mid pîpan on, Lchdm. ii. 126, 3. Mid ondônge wyrtdrences þurh horn odðe pîpan, 260, 11 : 224, 28. [*O. L. Ger.* pîpa : *Icel.* pîpa : *O. H. Ger.* pfîfa *fistula, calamus, camena.*] v. sang-pîpe.

pípere, es; *m. A piper, player on the flute* :—Pîpere *tibicen*, Wrt. Voc. i. 73, 59 : 289, 55. Reódpípere *auledus*, 60, 46. Se Hælend geseah hwistleras (pîperas, Rush.), Mt. Kmbl. 9, 23. [*Icel.* pîpari : *O. H. Ger.* pfîfari *tibicen*.]

pípfan; *p.* te *To breathe, blow* :—Pîpfendes *spirantis, sufflantis*, Hpt. Gl. 450, 76. Ût â-pýfhte (-pîpfte?) *exhalavit, exspiravit*, 472, 42.

piplian *to grow pimply* :—Wið teter and pypylgende (pipligende, MS. B.) lîc, Lchdm. i. 234, 10. Wið pypelgende (pipligende, MS. B.) lîc ðæt Grêcas erpinam (ἑρπης) nemnaþ, 266, 20. [*Lat.* papula *a pimple*.]

pipor, es; *m. Pepper* :—Piper (*other MSS.* pipor) *piper*, Ælfc. Gr. 9, 18; Zup. 44, 2. On ðâm londum biþ pipores genihtsumnys . . . Ðone pipor mon swâ nimeþ, Nar. 34, 21–23. Genim langes pipores .x. corn, Lchdm. ii. 186, 8. Of blacum pipore, 234, 2. Genim gebeátenne pipor, 186, 4. [*Icel.* pipar : *O. H. Ger.* pfeffar. From Latin.]

pipor-corn, es; *n. A pepper-corn* :—Genim .xvii. piporcorn, Lchdm. i. 74, 4. Ðæra pipercorna sý ofertæl, 288, 8.

pipor-horn, es; *m. A horn for holding pepper* :—Man sceal habban . . . sealtfæt . . . piperhorn, Anglia ix. 264, 19.

piporian; *p.* ode *To pepper* :—Pipra hit sýðan swâ swâ man wille, Lchdm. iii. 76, 9. Cf. Gepipera mid .xx. corna, ii. 182, 21. Gepiporod wyrtdrenc, 182, 7. [*Icel.* pipra.]

pir-gráf, es; *m. An orchard of pear-trees* :—On pirgráf, Cod. Dip. Kmbl. v. 284, 23.

pirige, an; *f. A pear-tree* :—Ðeós pirige *haec pirus*, Ælfc. Gr. 6; Som. 5, 59 : Wrt. Voc. i. 32, 51 : 80, 9. Pirge, ii. 117, 35. On gerihte tô ðære pirigan, Chart. Th. 145, 28. Ðis sindon ða londgemæra . . . ærest of Piriforda on ða dîc ; andlang dîc on ða pyrigan ; of ðære pyrigan, Cod. Dip. Kmbl. iii. 76, 27–30. Æt ðære pirian, 52, 18. On ða pyrian, ii. 205, 15. *The word, as in* Piriford, *is found in local names, e. g.* Pirig-fliát, Pyrihom, Pirigtûn, vi. 322, col. 2. [*Chauc. Piers P.* pirie.]

pís; *adj. Heavy, weighty* :—Byrðenna hefiga ł pîsa *onera gravia*, Mt. Kmbl. Lind. 23, 4. [From Latin *pensus*.] v. pinsian, pís-líc, pîsian.

pise, an; *f. A pea* :—Pise *lenticula*, Wrt. Voc. ii. 50, 75. Piose, 112, 63. Pysan *lentis*, 51, 50. Pisan hosa *siliqua*, 120, 58 : Lk. Skt. Lind. 15, 16. Heó hafaþ sæd on ðære mycele ðe pysan, Lchdm. i. 316, 10. Beán, pisan *cicer*, Wrt. Voc. ii. 14, 37. Pisan gesodena on ecede, Lchdm. ii. 180, 15. Geseáwe pysan *juicy peas*, 254, 15. Nim ðæt wæter ðe pyosan wæran on gesodene, 286, 29. Ðonne sceal man ða langnysse (*of the root*) tôceorfan on pysena gelícnysse, i. 260, 15. On pysena wôse, 260, 25. Pysena seáw, ii. 220, 10. Pysena broþ, 278, 18. Healde hî hine wið pisan and wið ða þing ðe windigne æþm on men wyrcen, 214, 2. v. mûse-pise.

pise-cynn, es; *n. A kind of pea* :—Sum pysecynn hâtte lenticulas, Lchdm. ii. 190, 16.

pîsian (?) *to weigh* :—Geþænce ælc man hû swíðe man pinsaþ (pysæþ, MS. H.) ða sâwle, Wulfst. 239, 26. v. pís.

pisle, an; *f.* (?) *A warm* (?) *chamber* :—*Scriptorium* pisle, fer-(fýr-?) hûs (or? *pis(a)le* fýrhûs), Wrt. Voc. i. 58, 58. [Cf. *O. Frs.* pisel *a chamber* : 'pisel, pesel ist in Niedersachsen, Dietmarschen, Nordfriesland und Süddänemark, phiesel in Baiern für verschiedene arten von gemächern noch gangbar,' Richthofen. *O. H. Ger.* pfisel *pisalis, pisale, pirale*, Grff. 3, 352. 'Pisalis videtur fuisse vestiarium seu vestiaria theca,' Du Cange.]

pís-líc; *adj. Heavy* :—Woeron ego hiora píslíco ł hefigo (*ingravati*), Mk. Skt. Lind. Rush. 14, 40. v. *next word and* pís.

píslíce; *adv. Heavily* :—Píslíce ł hefiglíce, *graviter*, Mt. Kmbl. Lind. 13, 15 : Lk. Skt. Lind. Rush. 11, 53. v. preceding word.

pistol, es; *m. An epistle, letter* :—Be ðam spræc se pistol æt ðyssere mæssan, Homl. Th. ii. 330, 13. Ðone pistol ðe Hieronimus sette be forþsîðe Marian, 438, 3. Se apostol Iacob âwrît on his pistole, Boutr. Scrd. 22, 47. Iacob se rihtwîsa âwrât ânne pistol, Ælfc. T. Grn. 14, 9, 13. Petrus âwrât twegen pistolas, 14, 7, 12, 16, 19. [*Icel.* pistil. From Latin.]

pistol-bóc; *f. A book containing the Epistles* :—Hê (*the priest*) sceal habban ða wæpna tô ðam gástlícum weorce . . . ðæt synd ða hálgan bêc, saltere and pistolbóc, godspellbóc and mæssebóc, L. Ælfc. C. 21; Th. ii. 350, 11–13. Hê (*bishop Leofric*) hæfþ ðiderynn (*St. Peter's minster at Exeter*) gedôn . . . i. pistelbêc . . . Hê ne funde on ðam mynstre ðâ hê tô fêng bôca nâ mâ bûton . . . . i. pistelbóc . . . , Chart. Th. 430, 8–29. [Cf. *Icel.* pistla-bók.]

pistol-rædere, es; *m. He who reads the epistle in church*, R. Conc. 5.

pistol-ræding, e; *f. A lesson in the church-service* :—Lucas ûs manode on ðisre pistol-rædinge, Homl. Th. i. 294, 13 : ii. 380, 23. (Both passages refer to the Acts of the Apostles.)

pistol-roce, es; *m. The vestment worn when reading the epistle* :—v. fulle mæssereáf, ii. dalmatica, iii. pistolroccas, Chart. Th. 429, 22.

pipa, an; *m. Pith, the soft inner part of the stem of a plant* :—Eall se dæl se ðe ðæs treówes on twelf mónþum geweaxeþ, hê onginþ of ðam wyrt-rumum, and swâ upweardes grêwþ ôþ ðone stemn, and siððan andlang ðæs pîþan and andlang ðære rinde ôþ ðone helm, Bt. 34, 10; Fox 150, 2. Þeahtigaþ on hiera môdes rinde monig gôd weorc tô wyrcanne, ac on ðam pîþan biþ ôðer gehýded, Past. 9; Swt. 55, 23. Nim ellenes pîþan, Lchdm. iii. 90, 2.

plæce, plæse, an; plæts, e; *f. A place, an open space, a street* :—In huommum ðara plæcena *in angulis platearum*, Mt. Kmbl. Lind. 6, 5. On plæcum (on plætsa, Rush.) *in plateas*, Lk. Skt. Lind. 10, 10 : 14, 21. In plaecum *in plateis*, 13, 26. In plæcum (plæsum, Rush.), Mk. Skt. Lind. 6, 56. In plægiword ł on plæcum *in plateis*, Rtl. 36, 7. In plæcum, 65, 37. [*Prompt. Parv.* plecke *or* plotte *porciuncula*. 'Pleck is given by Cole, Ray, and Grose as a North-country word, signifying a place;' note, p. 405. *Icel.* plâz : *M. H. Ger.* platz, *m.* : (*both introductions of the end of thirteenth century*). From Latin.]

plægan, plægi-word (plæce-worþ), plæts. v. plegan, plæce.

plætt *a sounding blow, a smack* : in the compound *eár-plætt* :—Drihten ûs sealde hælu þurh ðâm eárplættum, Homl. Th. ii. 248, 25. [Plat *a blow with the fist*, Jamieson's Dict.] v. next word.

plættan; *p.* te *To give a sounding blow, to smack* :—Hî plætton hyne mid hyra handum *dabunt ei alapas*, Jn. Skt. 19, 3. [He come plattinde (*tramping, making a noise with the feet*), Havel. 2282. Plette; *pl.* hurried, 2613. His heued of he plette (*struck*), 2626. Plat, 2755. Platch *to make a heavy noise in walking, with quick short steps*, Jamieson's Dict. *O. Du.* platten, pletten : *M. H. Ger.* blatren, platren *to strike noisily* : *Ger.* platzen. Of onomatopoetic origin ; cf. *smack*.] v. eár-plættan *and preceding word*.

plagian. v. plegan.

plante, an; *f. A plant, shoot* :—Swê swê niówe plant[e] *sicut novella*, Ps. Surt. 43, 12. Gesæwena plantan *plantaria*, Wrt. Voc. i. 39, 13. Ðæt is sió hálige gesomnung ðæt eardaþ in æppeltûnum ðonne hié wel begáþ hira plantan and hiera impan ôþ hié fulweaxne beóþ *ecclesia quippe in hortis habitat, quae de viridatem intimam exculta plantaria virtutum servat*, Past. 49, 2; Swt. 381, 17. [*Icel.* planta : *O. H. Ger.* pflanza. From Latin.] v. mixen-plante.

plantian; *p.* ode *To plant* :—Ðû plantast (*plantes*) wîneard and ne brîcst his, Deut. 28, 30. Gê plantiaþ, 28, 39. Gê plantigeaþ, Lev. 19, 23. Hî heora heortan wyrtruman on ðisum andwerdum lîfe plantiaþ,

Homl. Th. ii. 132, 7. Abraham plantode ǽnne holt, Gen. 21, 33: Mt. Kmbl. 15, 13. Hwæðer se anweald hæbbe ðone þeáw ðæt hé unþeáwas áwyrtwalige of rícra manna móde, and plantige ðær cræftas on? Bt. 27, 1; Fox 94, 24. Sanctus Paulus underféng ða hálgan gesomnunga tó plantianne, suâ se ceorl déþ his ortgeard, Past. 40; Swt. 293, 3. [Icel. planta: O. H. Ger. pflanzôn.] v. ã-, ge-plantian.

plant-sticca, an; m. A gardening-tool, a dibble (?):—Plantsticca pastinatum, Wrt. Voc. i. 16, 13. [Cf. Ital. pastinare to dig.]

plantung, e; f. I. planting:—Wîntwiga plantung propaginatio, Wrt. Voc. i. 39, 5. II. what is planted, a plant:—Ǽlc plantung (plantatio) ðe mîn heofenlíca fæder ne plantode byþ áwurtwalod, Mt. Kmbl. 15, 13. Plontung rósæs plantatio rosae, Rtl. 65, 35. Ðara bearn swá swá ǽdele plantunga, Ps. Spl. 143, 14: Blickl. Gl. Plantunga seten plantaria, Wrt. Voc. ii. 65, 76. [O. H. Ger. pflanzunga propagatio, plantarium, plantatio.]

plaster, es; n. (?) A plaster:—Tó plastre gewyrc, Lchdm. i. 272, 23: 304, 20. Hwí ne bidst ðú ðé beþunga and plaster lîfes lǽcedómes æt lîfes freán cur tibi non oras placidae fomenta medelae? Dóm. L. 6, 80. [O.H. Ger. pflastar; n. cataplasma, cementum. From Latin [em]plastrum.]

platian; p. ode To cover with plates: in the compound á-platian:—Áplatad obryzum, nitidum, Hpt. Gl. 417, 18. Áplatedum obryzo, 456, 47. v. next word.

platung, e; f. A plate, thin piece of metal:—Platung (? platum, Btwk.), smǽte gold obrizum, Hpt. Gl. 489, 34. Platungum brateolis, laminis, Wrt. Voc. ii. 127, 17. v. preceding word.

plega, an; m. I. play, quick movement:—Plega gesticulatio, Wrt. Voc. ii. 41, 36. P.legan gestum, Hpt. Gl. 474, 10. II. play, (athletic) sport, game; often in poetry applied to fighting, see the compounds:—Plega ludus, Ælfc. Gr. 8; Som. 7, 30. Ðes plega hic jocus, 13; Som. 16, 27: Wrt. Voc. i. 85, 30. Plaega palestra, ii. 116, 5. Mid ðám þíowum wæs on symbel mín plega hunc continuum ludum ludimus, Bt. 7, 3; Fox 20, 34: Exon. Th. 46, 27; Cri. 743. Ealle ða hwíle ðe ðæt líc bíþ inne, ðær sceal beón gedrync and plega, Ors. 1, 1; Swt. 20, 26. Ðǽr wæs heard plega wælgára wrixl (the battle between the four kings and the five), Cd. Th. 120, 4; Gen. 1989. Plǽges saltationis, Mk. Skt. p. 3, 11. Ic mé tó ðam plegan gemengde ludentibus me miscui, Bd. 5, 6; S. 619, 11. Bebudon Rómána godas ðǽm senatum ðæt mon theatrum worhte him tó plegan, Ors. 4, 12; Swt. 208, 33. Ðá hié æt hiora theatrum wǽron mid heora plegan . . . heora plegan begán, 6, 2; Swt. 256, 10-14. Ðá cild rídaþ on heora stafum, and manigfealde plegan plegiaþ, Bt. 36, 5; Fox 180, 9. Wé forbeódaþ ǽgðer ge plegan, ge unnytta word, ge gehwylce unnyttnesse in ðám hálgan stówum tó dónne, L. E. I. 10; Th. ii. 408, 22. Hié wǽron welige . . . and heora plegan wǽron genihtsume . . . Hié hæfdon wiste and plegan and oforgedrync, Blickl. Homl. 99, 17-21. Plegan allusiones, Wrt. Voc. ii. 9, 44: colludia, 20, 71. Plegena ludorum, 50, 25. III. clapping with the hands, applause (v. plegan, IV):—Ðæm plegan plausu, Wrt. Voc. ii. 67, 26. v. æsc-, ecg-, gilp-, gúþ-, hand-, hearm-, hyht-, lind-, níþ-, secg-, stæf-, sund-, sweord-, wíg-plega, next word, and the compounds with pleg-.

plegan, plǽgan, plegian, plagian, plǽgian; p. de, ede, ode To play; ludere:—Ic plege ludo, Ælfc. Gr. 28, 4; Som. 31, 23: Wrt. Voc. ii. 53, 29. Plegade lusit, 53, 28. Plegende ludens, Kent. Gl. 279: 995. I. to play, move about sportively, frolic, dance:—Hornfisc plegode, glǽd geond gársecg, Andr. Kmbl. 740; An. 370. Hlóh ðá and plegode boda bitre gehugaþ, Cd. Th. 45, 10; Gen. 724. Plǽgede saltasset, Mk. Skt. Lind. Rush. 6, 22. Pleagade saltavit, Mt. Kmbl. Rush. 14, 6. Ne plǽgde gé, Lind., ge ne plagadun, Rush. non saltastis, 11, 17. Ðæt folc sæt and æt and dranc, and árison and plegedon, Ex. 32, 6. Ðæt folc . . . eodon him plegean, Past. 43; Swt. 309, 14. Men willaþ binnan Godes húse bysmorlíce plegian, L. Ælfc. C. 35; Th. ii. 357, 2 note. Gesíon sǽmearh plegean, Elen. Kmbl. 490; El. 245. Ðæt wíf geseah Ismael plegan, Cd. Th. 168, 6; Gen. 2778. Ðá geseah hé plegan micel cnihta weorod be ðæs sǽs waroþe, Shrn. 78, 27. Án plegende cild arn under wǽnes hweowol, 32, 11. Swá plegende lamp quasi agnus lasciviens, Kent. Gl. 214. Seofon nacode wîmmen urnon plegende on heora gesihþum, Homl. Th. ii. 162, 32. II. to play, to divert or amuse one's self:—Ða ðe dwollíce plegaþ æt deádra manna líce, and ǽlce fúlnysse ðǽr forþteþ mid plegan, Homl. Skt. i. 21, 308. Tarentíne ðæt folc plegedon binnan heora byrg æt heora þeatra the Tarentines were taking their amusement at the theatre, Ors. 4, 1; Swt. 154, 2. Wé lǽraþ ðæt preóst ne beó hunta ne hafecere ne tæflere ac plege on his bócum we enjoin that a priest be neither a hunter nor a hawker nor a gamester, but let him find his amusement in his books, L. Edg. C. 64; Th. ii. 258, 8. II a. to play (a game), exercise one's self in any way for the sake of amusement:—Ða cild rídaþ on heora stafum, and manigfealdne plegan plegiaþ, Bt. 36, 5; Fox 180, 9. Samson plegode him ætforan ludens Samson, Jud. 16, 27. On ðæm dæge plegedon hié of horsum, Ors. 3, 7; Swt. 118, 29. II b. to play (with anything):—Hé mid bǽm handum upweard plegade he waved both hands aloft, Elen. Kmbl. 1609; El. 805. Ðá pleogede hé mid his

wordum, Bd. 2, 1; S. 501, 25. Wé wiernaþ úrum cildum úrra peninga mid tó plegianne, Past. 50; S. 361, 27. II c. to play with a person, toy; in a bad sense, to make sport of:—Sarra beheóld, hú Agares sunu wið Isaac plegode, Gen. 21, 9. Ðære helle hund ongan fægenian mid his steorte and plegian wið hine (Orpheus), Bt. 35, 6; Fox 168, 17. Plegan, Exon. Th. 429, 10; Rä. 43, 2. II d. to play (for something), strive after:—Ðis is se ilca ðe ðú longe for his deáþe plegodest this is the same for whose death thou hast long played, Blickl. Homl. 85, 19. III. to play on an instrument:—Plǽgiendra (plegiyndra, Ps. Spl. C.) timpanan tympanistriarum, Ps. Surt. 67, 26. IV. to clap the hands in applause (v. plega, III):—Flódas plǽgiaþ (plegiaþ, Ps. Spl. C.) flumina plaudent, Ps. Surt. 97, 8. Plagiaþ (plegaþ, Ps. Spl. C.) plaudite, 46, 2. v. plega.

plegere, es; m. A player, athlete, wrestler:—Nacod plegere gimnosophista (the glosser seems to have misunderstood the word, which is rendered by heáhláreów, Wrt. Voc. ii. 40, 40, and by weoroldsnottor, 81, 52), Wrt. Voc. i. 17, 10. v. pleg-mann.

pleg-hús, es; n. A play-house, theatre:—Ðæs heofenlícan pleghúses coelestis theatri, Hpt. Gl. 447, 62.

plegian. v. plegan.

pleg-líc; adj. Relating to play of any kind:—Ðæs pleglícan olimpiaci, Wrt. Voc. ii. 64, 20. Pleglícum scenico, Hpt. Gl. 474, 6: palaestrico, 489, 60. Ðý pleglícan plegan scenica ludicra, Wrt. Voc. ii. 90, 54. Ða pleglícan theatrales, 75, 17. Pleglícum palaestricis, gymnicis, Hpt. Gl. 405, 6, 9.

pleg-mann, es; m. A player, athlete, wrestler:—Plegmanna gymnicorum, Hpt. Gl. 407, 39. þurh plegemen ł gligmen ł gleáwe per gymnosophistas, 406, 72. Swilce wittige ł gleáwe leorneras ł plegmen velut sagaces gymnosophistas, 404, 78. Plegmen gymnosophistas, ðǽm wǽrstlícum palestricis, Wrt. Voc. ii. 74, 53-54. v. plegere.

plegol; adj. Playful, sportive, jocose:—Hwílon wacodon menn ofer án deád líc, and ðǽr wæs sum dysig mann plegol ungemetlíce, and tó ðám mannum cwæþ swylce for plegan, ðæt hé Swýðun wǽre, Homl. Skt. i. 21, 292.

pleg-scild, es; m. A small shield:—Plegscylde pelta, Wrt. Voc. ii. 65, 69. [Cf. lytel scyld pelta, ða lǽssan scyldas peltae, i. 35, 28, 59.] Truman pleigscelde tuta pelta, Hpt. Gl. 424, 38.

pleg-scip, es; n. A small ship, a yacht (?); parunculus, Wrt. Voc. i. 56, 35. v. next word.

pleg-stów, es; f. A place for play, a gymnasium, wrestling-place, amphitheatre:—Oretstówe ł winstówe ł plegstówe scammatis, Hpt. Gl. 405, 41. Plegstówe amphitheatri, Wrt. Voc. ii. 3, 13. On plegstówe (bleg-, MS.) oððe on wafungstówe andbidian hine gesihþ styrunge sume getácnaþ if a man in a dream sees himself waiting in an amphitheatre or theatre it betokens some disturbance, Lchdm. iii. 206, 15. Plegstów[a] ł winstówe palaestrarum, Hpt. Gl. 478, 50. Plegstówa palaestrarum, Wrt. Voc. ii. 66, 50. On plegestówum in gymnasio, 40, 20.

pleoh; gen. pleós; n. Danger, hurt, peril, risk:—Nys ðæt nǽnig pleoh nullum ei est periculum, L. Ecg. C. 40; Th. ii. 166, 5. Swylce hit nán pleoh ne sý, ðæt se preóst libbe swá swá ceorl, L. Ælfc. C. 6; Th. ii. 344, 18: Wulfst. 269, 28. Lǽsse pleoh byþ ðam men, ðæt hé flǽsces brúce on Lenctenfæstene, ðonne hé wífes brúce, 286, 3: Homl. Th. i. 178, 34. Ðæt wæs swíðe micel pleoh ðæt ðú swá wénan sceoldest, Bt. 5, 3; Fox 14, 5. Hit biþ his pleoh ná mín, Ælfc. Gr. pref.; Som. 2, 2. Wénaþ sume menn ðæt nán pleoh ne sý on deórwurðum gyrlum, Homl. Th. i. 328, 25. Hé búton pleó tó his fixnoþe gecyrde, ii. 288, 26. Pleó periculo, Hpt. Gl. 457, 40. Gif hié síen gelíc ord and hindeweard sceaft ðæt síe bútan pleó (cf. si cuspis et acies lancee pari sustentacione respondeant, sine culpa sit, L. H. I. 88, 3; Th. i. 595, 12-14), L. Alf. pol. 36; Th. i. 84, 19. Philippus Mæcedonia ríce ealle hwíle on mician pleó and on miclan earfeþan hæfde, Ors. 3, 7; Swt. 110, 28. Gif ðú ofer gemet itst . . . seó ofering ðé wurþ oððe tó sáre . . . oððe tó plió cujus satietatem si superfluis urgere velis, aut injucundum, quod infuderis, fiet, aut noxium, Bt. 14, 1; Fox 42, 17. Hwá mæg ǽhta wilnian bútan plió nú se swelc plioh ðæron gefór se ðe his nó ne wilnode quis opes quaerat innoxie, si et illi extiterunt noxia, qui haec habuit non quaesita, Past. 50, 4; Swt. 393, 9. Hwelc mágon beón máran gehát ðonne mon geháte for his freónd ðæt hé underfoo his sáule on his pleoh spondere pro amico est alienam animam in periculo suae conversationis accipere, 28, 3; Swt. 193, 7. [O. Frs. plê, plí danger.] v. pliht.

pleó-líc; adj. Dangerous, perilous, hurtful, hazardous:—Hit swýðe pleólíc is, ðæt man on ðám hálgum stówum áðer oððe ðæt dó oððe ðæt sprece ðæt ðǽm stówum ne gedafenaþ, L. E. I. 10; Th. ii. 408, 27. Mé þincþ ðæt ðæt weorc (translating Genesis) is swíðe pleólíc (dangerous, because a foolish person might misapply what he read), Ælfc. T. Grn. 22, 8. Ne becymst ðú nǽfre tó ðam pleólícum leahtre, Homl. Th. ii. 208, 31. Gif hié (seó menigo ðínra monna) yfele sint ðonne sint hié ðé pleólícran and geswincfulran gehæfd ðonne genǽfd si vitiosi moribus sunt, perniciosa domus sarcina, Bt. 14, 1; Fox 42, 22. Hiora ingewinn him wǽron forneáh ða mǽstan and ða pleólecestan, Ors. 2, 6; Swt. 88, 29. v. un-pleólíc.

pleón; *p.* pleah; *with gen.* To risk, expose to danger :—Se ilca David miclum his ágenes herges pleah (pleh, Cott. MSS.) *the same David exposed his host to great danger*, Past. 3, 2 ; Swt. 37, 7. Se ðe on ðæm gefeohte ðisses andweardan lifes nile suincan ne his selfes plión, 34, 1 ; Swt. 229, 20. v. pleoh, pliht.

plett, e ; f. (?) *A fold* :—Oðre scíp ic hafo ða ðe ne sindun of ðisse pletta (from ðissum plette, Lind.) . . . biþ ánn pletta (án plette, Lind.), Jn. Skt. Rush. 10, 16. In scípa plett ł locc *in ouile ouium*, Lind. 10, 1. [From Latin *plecta* a hurdle. Cf. hyrdle ł bige *plecta*, Hpt. Gl. 497, 71.]

plicettan (?) *to expose to danger* :—Plicet *adludit* (*adlidit?*), Germ. 397, 20. Cf. pliht.

plicgan *to scrape, scratch* :—Plicged (plicgeð ?) *scalpit*, Germ. 396, 255. [Cf. (?) *Chauc. p.* plighte ; *pp.* plight *plucked.*]

pliht, es ; *m.* : e ; f. *Danger, damage* :—Mid micclan plihte *cum magno periculo*, Coll. Monast. Th. 26, 37. Ne biþ ǽnig gewemmed líchama tó plihte (*dangerously, harmfully*), gif hit ne lícaþ ðam móde, Homl. Skt. i. 9, 85. Gyf hit (*stolen property*) on hýdelse funden sý, ðonne mæg ðæt forfangfeoh leóhtre beón, forðam [hit] biþ on læsse plihte (*with less danger than when taken from the thief*) begytan, L. Ath. iv. 6 ; Th. i. 226, 6. Plihtas *pericula*, Ps. Surt. 114, 3. [*Laym.* pliht *harm, danger*; e. g. him muchel plihte ilomp (*he was murdered*), 4003 : *O. Frs.* plicht *periculum* : *O. H. Ger.* pfligida *periculum*.] v. next word.

plihtan ; *p.* te *To bring danger upon* an object (*dat.*), *to compromise* [*To plight* has later the meaning of to promise under peril of forfeiture, to make a solemn engagement for which one has to answer] :—Gif hwá bútan leáfe of fyrde gewende ðe se cyng sylf on sý plihte him sylfum and ealre his áre *it shall be at the peril of life and property*, L. Eth. v. 28 ; Th. i. 310, 29 : vi. 35 ; Th. i. 324, 10. Gif ǽnig ámánsumad man . . . on ðæs cynges neáweste gewunige, ǽr ðam ðe hé hæbbe godcunde bóte georne gebogene, ðonne plihte him sylfum and eallan his ǽhtan, v. 29 ; Th. i. 312, 3. Plihte hí heora áre and eallon heora ǽhton, vi. 36 ; Th. i. 324, 14. Gif hwá útlahne hæbbe and healde plihte him sylfum and ealre his áre, L. C. S. 67 ; Th. i. 410, 18. Plihte tó him sylfum and ealre his áre, L. Eth. ix. 42 ; Th. i. 350, 2.

plihtere (?) *one that watches in the prow of a ship* :— Pliclitere (plihtere ?) ł ancremen *proreta*, Hpt. Gl. 406, 55. [Cf. *O. H. Ger.* pfliht *prora*, Grff. 3, 360.]

pliht-líc, adj. *Dangerous* :— Plyhtlíc þingc hit ys gefón hwæl *periculosa res est capere cetum*, Coll. Monast. Th. 24, 21. Ðrý dagas syndon on geáre ðe wé *egiptiaci* hátaþ, ðæt is on úre geþeóde plihtlíce dagas ; on ðám ná tó ðæs hwón for nánre neóde ne mannes ne neátes blód sý tó wanienne, Lchdm. iii. 76, 11–14.

plóg, es ; *m. A plough* ; with this meaning the word occurs in *Icel.* and *O. H. Ger.*, but in *A. S.* it seems to mean *land, a plough of land* (cf. Cath. Angl. p. 284 :—a ploghe of land *carrucata*. In the *Tale of Gamelyn*, the knight, bequeathing his estate says :—

'Johan myn eldeste sone shall have *plowes* fyve,
And my myddeleste sone fyf *plowes* of lond.'

Plowlond *carrucata*, þat a plow may tylle on a day, Prompt. Parv. 405. In Ælfric's Colloquy the ploughman says : Ælce dæg ic sceal erian fulne æcer oððe máre. *Pleuch* a quantity of land for caring for which one plough suffices, Jamieson's Dict.), the word *sulh* being used to denote the implement :—Ic hit (*property*) ágnian wille tó ágenre æhte, ðæt ic hæbbe, and næfre ðé myntan ne plot ne plóh, ne turf ne toft, ne furh ne fótmǽl, L. O. ; Th. i. 184, 6. [*Icel.* plógr ; *m. a plough* ; plógs-land *an acre* : *O. H. Ger.* pfluoc *aratrum*.]

plot *a plot of ground.* v. preceding word. [*Prompt. Parv.* plotte *porciuncula*.]

pluccian, ploccan ; *p.* ode *To pluck, pull away, tear* :—Ic tótere oððe pluccige oððe tǽse *carpo*, ic of ápluccige *excerpo*, Ælfc. Gr. 28, 4 ; Som. 31, 21. Plucciaþ *carpunt, vellint*, Wrt. Voc. ii. 128, 77. Ploccaþ *disceptant, lacerant*, 140, 59. Pluc[ciaþ] *decerpint*, Hpt. Gl. 408, 37. Ða ðe ðæra treówa bógas heówon . . . sind ða láreówas on Godes cyrcan, ðe plucciaþ ða cwydas ðæra apostola, Homl. Th. i. 212, 35. His leorningcnihtas ða eár pluccedon (*uellebant*), Lk. Skt. 6, 1. Pluccian *plumemus* (cf. scecele scecen wé *plectro plumemus*, Wrt. Voc. ii. 66, 79–80 : 83, 77–78), Hpt. Gl. 497, 73. Pluccian (*later MS.* plockien) *vellere*, Mt. Kmbl. 12, 1. Pluccigean, Mk. Skt. 2, 23. Ic wolde gadrian (pluccian, MS. M.) sum gehwæde andgyt of ðære béc ðe Beda se snotera láreów gesette, Lchdm. iii. 232, 2. [*Icel.* plokka, plukka : *M. H. Ger.* pflücken : *Du.* plukken.]

plúm-bléd, e ; f. *Fruit of the plum-tree* :—Plúmbléda ete neahtnestig *let him eat plums after his night's fasting*, Lchdm. ii. 230, 13.

plúme, an ; f. *A plum* (fruit or tree) :—Seó plúme *hoc prunum*, Ælfc. Gr. 6 ; Som. 5, 60. Plumae *prunus*, Txts. 88, 822 : *plumum*, 87, 1600. [*Prompt. Parv.* plowme *prunum* : *Icel.* plóma : *M. H. Ger.* pflúme. From Latin.] v. plýme, plúm-treów.

plúm-feðer, e ; f. *Down* :—Plúmfeðera hnescnyss geonglíce lima ná gehlýwe *plumarum mollities iuuenilia membra non foveat*, Scint. 43.

plúm-seáw, es ; *n. Plum-juice* :— Nim plúmséwes ánes scyllinges gewyht, Lchdm. iii. 114, 21.

plúm-slá *a sloe, wild-plum* ; *pruniculus*, Wrt. Voc. i. 33, 28.

plúm-treów, es ; *n. A plum-tree* :—Ðis plúmtreów *haec prunus*, Ælfc. Gr. 6 ; Som. 5, 60 : Wrt. Voc. i. 32, 55 : 33, 33 : 80, 10 : *plummus*, 285, 56. Plúmtréu *plunas*, ii. 117, 44. Nim plúmtreówes leáf, Lchdm. ii. 310, 19.

plýme, an ; f. *A plum* (fruit or tree) :—Plýme *prunum*, Wrt. Voc. i. 285, 57 : *prunus*, ii. 68, 45. v. plúme.

poc-ádl. v. next word.

pocc, es ; *m. A pock, pustule, ulcer* :—Poccas *ulcera*, Wrt. Voc. ii. 90, 73. Gif poc sý on eágan, Lchdm. iii. 4, 1 : 14, 31. Wið ómena geberste . . . sleah feówer scearpan ymb ða poccas útan, and lǽt yrnan ða hwíle ðe hé wille, 44, 1 : ii. 100, 4. Wið pocádle . . . Mid. hunige smire ðær hit út sleá on ðone poc . . . Sealf wið pocádl . . . Drenc wið poccum . . . Wið poccum swíðe sceal mon blód lǽtan . . . gif hié út sleán ælcne man sceall áweg ádelfan mid þorne, and ðonne win oððe alordrenc drýpe on innan, ðonne ne beóþ hý gesýne, 104, 14–106, 6. See the note on this section. [*Prompt. Parv.* pokke, sekenesse *porrigo, variolus* : *Piers P.* 20, 97 : Kynde come after wiþ many kene sores, As pokkes and pestilence.]

pohha, poha, pohcha, pocca, an ; *m. A poke, pouch, bag* ; as a medical term *sinus* :—Pohha (poha, Lind.) *pera*, Mk. Skt. Rush. 6, 8. Pohha (pocca, Lind.), Lk. Skt. 9, 3. Ðý læs ðider in yfel pohha (*sinus*) gesíge, Lchdm. ii. 208, 18. Sift ðonne, dó on pohhan (*bag*), lege under weofod, 138, 27. Dó on ænne pohchan, iii. 48, 5. 'Se ðe mêdsceattas gaderaþ, hé legeþ hié on þyrelne pohchan (*sacculum*).' An þyrelne pohchan se legþ . . . , Past. 45, 4 ; Swt. 343, 20. [*Prompt. Parv.* pooke *sacculus*: *Chauc. Piers P.* poke : *Icel.* poki : *O. Du.* poke. A Celtic word, Irish poc, *Gaelic* poca *a bag*.] v. nest-pohha *and next word*.

pohhed ; adj. *Baggy, loose* :—Hý gelyst ælces (ealces, MS.) ýdeles habbaþ síde earmellan and pohhede hosa stípe reáf hý anscuniaþ *they take pleasure in every vanity, they have wide sleeves and loose hose, close-fitting garments they avoid*, R. Ben. 136, 23.

pól, es ; *m. A pool* :—Salamon sǽde ðætte swíðe deóp pól wǽre gewered on ðæs wísan monnes mód *aqua profunda verba ex ore viri*, Past. 38, 7 ; Swt. 279, 15. Hié nellaþ gepyndan hiora mód, swelce mon deópne pool gewerige, 39, 1 ; Swt. 283, 14. Maurus þurh Godes mihte eode uppon yrnendum wætere, on ánum wídgyllan póle, Homl. Skt. i. 6, 12. Tó ðæm póle *ad natatoriam*, Jn. Skt. 9, 11. In tó póle, Cod. Dip. Kmbl. iii. 424, 17. On pól ; of póle út on Auene, 456, 1–2. In póll, 399, 14. Út on hreódpól, ii. 29, 10. [*O. H. Ger.* pfuol *palus*.] v. fisc-, hwirf-, mylen-pól, *and* pull.

polente (?), an ; f. *Parched corn* :—Hig ǽton polentan (*polentam*), Jos. 5, 11.

pollegie, polleie, an ; f. *Pennyroyal* ; mentha pulegium :—Polleie, Lchdm. ii. 296, 23 : 350, 26. Pollege, ðæt on englis dwyrcge dwosle, i. 380, 10. Genim polleian, 118, 4 : ii. 318, 7. Genim pollegian, 138, 26 : iii. 4, 9 : 16, 10. Pollegan, 28, 26 : 48, 9. [*O. H. Ger.* polei, pulei : *Ger.* polei. From Latin.]

pollup, es ; *m. A scourge* (?) :—Mistlíce þreála gebyriaþ for synnum, bendas oððe dyntas oððe pollupas oððe carcernþýstra, lobban oððe bælcan, L. Pen. 3, note ; Th. ii. 278, 26.

popig *poppy* :—Papig *papaver*, Ælfc. Gr. 9, 18 ; Som. 9, 62. Popig, Wrt. Voc. i. 31, 7 : 68, 56. Popei, ii. 116, 48. Baso popig *astula regia*, i. 66, 65. Popaeg, Txts. 90, 824. Popeg *cucumis*, 52, 253. Popig, Wrt. Voc. ii. 15, 54. Popi *cucumus*, 17, 27. Wilde popig *saliunca*, i. 31, 8. Popig . . . ðe Grécas *moecorias* and Rómáne *papauer album* nemnaþ and Engle hwít popig hátaþ, Lchdm. i. 156, 17–20. Him is tó sellanne lactucas and súþerne popig inneweard, ii. 212, 12.

popul *a poplar* (? ; but cf. popylle *lolium*, Wrt. Voc. i. 234, 2), in popul-finig :—Of ðam ellene tó populfinige ; of populfinige tó Lamb-hyrste, Cod. Dip. Kmbl. iii. 219, 8. The second part of the compound occurs again v. 194, 2–3 : 195, 10. [*Prompt. Parv.* popul-tre.]

por-leác, es ; *n. A leek* :—Porleác *porrus*, Wrt. Voc. i. 31, 2. Wé hæfdon cucumeres and pepones and porleác *in mentem nobis veniunt cucumeres et pepones porrique*, Num. 11, 5. v. next word.

porr, es ; *n.* (?) *A leek* :—Por *porrum*, Wrt. Voc. i. 286, 12. Nim merwes porres leáf, Lchdm. ii. 84, 31. Heáfdehtes porres, 230, 10. Dó sealt and merce tó, and porr, 284, 2. Por, 186, 19 : 278, 19. [*O. H. Ger.* pforro : *Ger.* pors.]

port, es ; *m. n.* I. *a port, haven* :—Wið ðone gársecg is se port ðe mon hǽt Caligardamana, and be súþaneástan ðæm porte is ðæt ígland Deprobane, and be norþan ðæm Gandes múþan . . . is se port Samera. Be norþan ðæm porte is se múþa ðære íe Ottorogorre, Ors. 1, 1 ; Swt. 10, 8–13. Ðonne is án port on súþeweardum ðæm lande, ðone man hǽt Sciringes heal . . . Of Sciringes heale hé seglode on fíf dagan tó ðæm porte ðe mon hǽt æt Hæþum, Swt. 19, 10–23. Hé hine gelædde tó ðam porte (*ad portum*) ðe is nemned Cwentowíc, Bd. 4, 1 ; S. 564, 44. II. *a town* :—Port *castellum*, Wrt. Voc. i. 36, 28. Wíc oððe lytel port *castellum*, 84, 42. Hwæt fremaþ ðære burhware ðeáh ðe ðæt port (*the town*) beó trumlíce on ælce healfe getimbrod, gif ðær biþ án hwæm open forlǽten, ðæt se onwinnenda here þurh ðam infær hæbbe ? Homl. Th. ii. 432, 3. On ælche healfe ðæs portes, Chart. Th. 226, 25. Hwá rít intó

ðam port *quis equitat in civitatem?* Ælfc. Gr. 5 ; Som. 3, 52. In burug ł in port *in civitate*, Mt. Kmbl. p. 15, 19. Gif ðú hér on porte (*Ephesus*) geboren wære, hwær synt ðíne mágas ðe ðé áséddon, Homl. Skt. i. 23, 679. Ic wille ðæt nán man ne ceápige bútan porte, ac hæbbe ðæs portgeréfan gewitnesse oððe óðera manna ðe man gelýfan mæge. And gif hwá bútan porte ceápige, ðonne sý hé cyninges oferhýrnesse scyldig, L. Ed. 1 ; Th. i. 158, 10–14. Wé cwædon ðæt man nænne ceáp ne ceápige bútan porte ofer .xx. penega, ac ceápige ðǽr binnan on ðæs portgeréfan gewitnesse, L. Ath. i. 12 ; Th. i. 206, 8–10. Ælc ceáping sý binnan porte, i. 13 ; Th. i. 206, 16. Nán man ne mynetege bútan on porte, i. 14 ; Th. i. 206, 19. Lecge án .c. tó wedde, healf landrícan and healf cinges geréfan binnan port, L. Eth. iii. 7 ; Th. i. 296, 8. Ðá com se here tó Hamtúne (*Northampton*) and ðone port forbærndon, Chr. 1010 ; Erl. 144, 14. Burgas ł portas *civitates*, Mt. Kmbl. p. 16, 10. Portas *castella*, Mk. Skt. Lind. Rush. 6, 6. [Latin *portus*. ' *Portus* est conclusus locus quo importantur merces et inde exportantur. Est et statio conclusa et munita,' Du Cange. Cf. Port- *in place-names*, e.g. Port-stræt, Cod. Dip. Kmbl. vi. 323.]

port, es ; *m. A gate, entrance :*—Port ł dure ł gæt *portam*, Mt. Kmbl. 7, 13. Eode ðe Hǽlend in tempel in ðone port (*in porticu*) Salamonnes, Jn. Skt. Rush. 10, 23. Fíf portas *quinque porticos*, Lk. Skt. Rush. 5, 2. Ða him sǽton sundor on portum *qui sedebant in porta*, Ps. Th. 68, 12. [*O. Frs.* porte : *O. Sax.* porta : *O. H. Ger.* pforta ; *f.* : *Icel.* port ; *n.* From Latin *porta*.]

Port, es ; *m. The name attributed to one of the Saxon invaders of Britain, apparently an inference from a place-name :*—Hér cuom Port on Bretene . . . on ðære stówe ðe is gecueden Portesmúþa, Chr. 501 ; Erl. 14, 12.

port-cwén, e ; *f. A harlot, woman of the town :*—Portcuoene ł synnful *peccatrix*, Lk. Skt. Lind. 7, 37, 39. Mið portcuoenum *meretricibus*, 15, 30. Portcuoenes *meretricis*, Rtl. 106, 28. Portcuoene *meretrici*, 106, 30. Portcuoeno *meretrices*, Mt. Kmbl. Lind. 21, 31, 32. [Cf. *Icel.* port-kona *a harlot* ; port-hús *a brothel* ; port-lífi *prostitution*.]

Portes-múþa. v. Port.

port-geat, es ; *n. The gate of a town :*—Portgeat *porta*, Wrt. Voc. i. 36, 37 : 84, 38. Fare ðæt wíf tó ðam portgate *perget mulier ad portam civitatis*, Deut. 25, 7. Ðá ðá hé geneálǽhte ðam portgeate (cf. ðære ceastre gate, Lk. Skt. 7, 12), Homl. Th. i. 490, 30. Ðæt portgeat getácnaþ sum líchamlíc andgit ðe menn þurh syngiaþ, 492, 13. Hé ða portgeatu ealle beeode, Homl. Skt. i. 23, 507.

port-geréfa, an ; *m. A port-reeve* (v. port, II) :—Portgeréfa oððe burhwita *municeps*, Wrt. Voc. i. 18, 41. Ðes portgeréfa *hic prefectus urbis*, Ælfc. Gr. 14 ; Som. 16, 56. Man cýðde ðam portgeréfan (*the case is one of buying in the market at Ephesus*), Homl. Skt. i. 23, 643. Port-reeves of London, Canterbury, Bodmin, and Bath are mentioned in the charters, and from the Laws (v. under *port*, II) it is seen that one of the duties of such officials was to witness all transactions by bargain and sale effected within the *port*. See Kemble's Saxons in England, ii. c. 5. [Robert of Gloucester mentions two portreeves of Oxford, ' William the Spicer and Geffray of Hencsei that tho were Portreven,' p. 540.] [*Icel.* port-greifi.]

port-geriht, es ; *n. A town-due, due paid by a town :*—Ðæs túnes cýping and seó innung ðara portgerihta *uillae mercimonium censusque omnis civilis*, Cod. Dip. Kmbl. iii. 138, 10.

portian ; *p.* ode *To pound, bray in a mortar :*—Ðeáh ðú portige ðone dysegan on pílan swá mon corn déþ mid piilstæfe ne meaht ðú his dysi him from ádrífan *si contuderis stultum in pila, quasi ptisanas feriente desuper pilo, non auferetur ab eo stultitia ejus,* Past. 37, 2 ; Swt. 265, 25. v. pyrtan.

portic, es ; *m.* **I.** *a porch, covered entrance, portico :*—Portic *porticus*, Ælfc. Gr. 11 ; Som. 15, 22 : Wrt. Voc. i. 58, 2. Se mere hæfþ fíf porticas. On ðám porticon læg mycel menigeo geádludra, Jn. Skt. 5, 2–3. **II.** *an enclosed place, a place roofed in :*—Sinewealt cleofa *vel* portic *absida*, lytle porticas *cancelli*, Wrt. Voc. i. 58, 34, 37. Ic Eádwine munek læi innan mínan portice (*cell*) anbútan nóntíde, Chart. Th. 321, 31. Portic *abscidam* (*absidam*), Wrt. Voc. ii. 9, 45. **III.** *part of a church, porch, vestibule ; also an arched recess.* ' *Porticus* aedis sacrae propylaeum in porticus formam exstructum, in quo consistebant Catechumeni et Poenitentes: improprie pro sanctuarium, seu orientalis ecclesiae pars in qua majus altare erigi solet,' Du Cange :—Hálig portic *sanctuarium*, Ps. Surt. 72, 17 : 73, 7 : 82, 13. Of ðæs portices dura þærscwolde wæs gesýne ðæt ða swaðo wǽron ǽrest útwearde ongunnen . . . Ðeós circe mid ðýs portice mihte húhwego fíf hund manna befón, Blickl. Homl. 207, 10–14. His líchaman Eorcenwald on portice (*in porticu*) his cyrcan sumre geheóld . . . Ðá dydon hí his líchaman up of ðam portice and on cyrcan neáh weofode byrgan wolde, Bd. 3, 19 ; S. 550, 5–10. Wæs hé bebyriged on Scē Paules portice (*porticu*), se is on Scē Andreas cyrícean, 5, 23 ; S. 645, 18. His líchoma on ðære cyrícean norþportice (*porticu aquilonali*) wæs bebyriged ; in ðam eác swylce ealra ðæra æfterfylgendra ærcebiscopa líchoman syndon bebyrged bútan twegra ; heora líchaman sindon on ðære cyrícean sylfre gesette, forðan ðe on ðone forecwedenan portic má ne mihte, 2, 3 ; S. 504, 34–38. Ðæt hé wíbedas

sette and porticas worhte and tódǽlde binnan ðære cyrícean weallum *ut poneret altaria, distinctis porticibus intra muros ecclesiae,* 5, 20 ; S. 641, 42. Synd þrý porticas emb ða ciricean útan geworhte, and ða ealle fægere ufan oferworhte and oferhrýfde, Blickl. Homl. 125, 23. [*O. H. Ger.* pforzih *porticus, vestibulum, peribolus, atrium.*] v. húsel-, norþ-, súþportic.

port-mann, es ; *m. A towns-man, citizen :*—Portman *civis*, Wrt. Voc. i. 84, 39. Eádgár æþeling com mid eallum Norþhymbram tó Eoferwíc, and ða portmenn wið hine grídedon, Chr. 1068 ; Erl. 207, 2. Se portgeréfa and ða yldostan portmenn (*of Ephesus*), Homl. Skt. i. 23, 749.

port-stræt, e ; *f. A town-road, public way :*—In ðære portstræt ; and swá æfter ðære stræte, Cod. Dip. Kmbl. iii. 36, 22. Of ðære portstræte, 52, 20. Portstreet occurs as a proper name, vi. 323, col. 2.

port-wara, an ; *m. A citizen :*—Lulla gebohte ðis lond miþ ealra ðeassa portweorona gewitnesse, Cod. Dip. Kmbl. ii. 3, 11.

port-weall, es ; *m. A town-wall :*—Man geþeáh ábútan ðone portweall, Homl. Skt. i. 23, 267. Ða heáfodleásan man héngc on ða portweallas, and man sette heora heáfda búton ðám portweallon on ðám heáfodstoccum, and ðǽr flugon hrócas and hremmas intó ðære byrig geond ða portweallas, and tósliton ða hálgan Godes dyrlingas, 23, 73–80.

port-wer, es ; *m. A citizen ;* civis, Rtl. 187, 23.

posa. v. pusa.

posel *a small lump, a pill :*—Gǽten smeoro geþýd tó poslum swelge let him swallow goat's grease squeezed to pills, Lchdm. i. 354, 9. v. next word.

posling, es ; *m. A pill :*—Wyrc lytle poslingas feówer *make four little pills*, Lchdm. i. 76, 23. v. preceding word.

post, es ; *m. A post, pedestal :*—Post *basis*, Wrt. Voc. i. 47, 20 : *postis*, 86, 29 : Ælfc. Gr. 9, 28 ; Som. 11, 45. Under ðám sylfum postum *sub ipsos postes*, 47 ; Som. 48, 17. Hé áhéng ðæt dust on ænne heáhne post . . . Ðæt hús weard ða forburnen búton ðam ánum poste, Swt. A. S. Rdr. 101, 186–191. [*O. H. Ger.* pfosto. From Latin.]

postol, es ; *m. An apostle :*—Ðara postolra *apostolorum*, Lk. Skt. p. 2, 2. Ða ðe cwědun ðás tó ðǽm postolum *quae dicebant apostolas haec,* Rush. 24, 10. [*Icel.* postuli : *O. H. Ger.* postul.] v. apostol.

potian *to push, thrust, strike, butt :*—Hwæt wǽron hí, búton fearra gelícan, ðá ðá hí, mid leáfe ðære ealdan ǽ, heora fýnd mid horne líchamlícere mihte potedon? Homl. Th. i. 522, 25. Ða deóflu hý potedon and þoddetton ða earman sáwle and héton hý út faran raðe of ðam líchaman swíðe heardlíce, Wulfst. 235, 15. [From Celtic, *Gael.* put *to push, thrust : Welsh* pwtio *to push, poke.*]

pott, es ; *m. A pot :*—Dó on ænne neówna pott, Lchdm. i. 378, 21. [From Celtic, Welsh *pot.*]

prætt, es ; *m. Craft, art, wile, trick :*—Præt, prætt *astu*, Ælfc. Gr. 43 ; Zup. 257, 8. Wó dómas and prættas, Anglia viii. 336, 40 : Wulfst. 245, 2. Prættum *artibus*, Hpt. Gl. 459, 23. Ongeán þúsendfealde derigende prattas *contra mille nocendi artes*, 424, 46. [Prat, pratt *a trick, wicked action*, Jamieson's Dict.: cf. *Laym.* mid pretwrenche, 81 : mid prætwrenchen (2nd MS. felle wrenches), 5302 : *Icel.* prettr *a trick.*] v. next word.

prættig, pættig ; *adj. Wily, crafty, astute :*—Præt *astu*, pættig *astutus*, Ælfc. Gr. 43 ; Zup. 257, 8. Ic beó pættig *callidus fio*, 26, 2 ; Zup. 154, 11. Pætig *callida*, Germ. 389, 21 : *astutus*, Wrt. Voc. i. 76, 14. Petig *sagax vel gnarus vel astutus vel callidus*, 47, 36. Næddre seó pætige *serpens ille callidus*, Hymn. Surt. 61, 32. Wille gē wesan prættige (*versipelles*), Coll. Monast. Th. 32, 27. Prættigustan deóre *callidissime bestiole*, Wrt. Voc. ii. 127, 50. [*Scot.* pratty and ill-pretty *tricky :* cf. *Orm.* nis he nohht hinnderrȝæp ne pratt. In *Prompt. Parv.* praty *elegans, formosus. Icel.* prettugr, prettóttr *deceitful, tricky ;* pretta *to deceive.* Perhaps of Celtic origin. Cf. Cornish *prat* an act or deed, a cunning trick.] v. preceding word.

práfost, prófost, es ; *m.* **I.** *an officer :*—Geréfa oððe práfost *prepositus*, Wrt. Voc. i. 72, 67. Valerianus Decies práfest ðæs cáseres *Valerian, officer of the emperor Decius*, Shrn. 117, 12. Valerianus se práuost, 117, 16. Pharaones þénas swungon ða ðe bewiston Israéla folces . . . Ðá cómon Israéla folces práfostas (*praepositi*) *the officers of the children of Israel* (A. V.), Ex. 5, 14–15. **II.** *an officer of a monastery ;* praepositus : v. Smith's Dict. of Christian Antiquities, ' *praepositus* the second in command under the abbot in a monastery, the *prior claustralis ;*' also ' that member of a chapter who takes charge of the administration of the capitular estates :'—Be mynstres práfaste. For oft hit getímaþ, ðæt swýðe hefigtýme ungeþwærnessa on mynstre áspringaþ þurh ðæs geendebyrdan prófostes misfadunge . . . him þincþ, ðæt hé sý óðer abbod . . . ðis gelimpþ swíðust on ðám stówum, ðǽr se prófost on gýmenne biþ geset fram ðám ylcan biscopum oððe abbodum, ðe ðone abbod . . . on ðam weordmente settan . . . Him þincþ, ðæt hé ðam abbode ne þyrfe hýran . . . Wé forði foresceáwiaþ . . . ðæt eal mynstres fadung on ðæs abbodes dóme and tæcinge simle stande . . . Gif . . . hit ðam abbode rǽd þince, swá hwilcne swá se abbod geceóse mid geþeahte ðara bróðra ðe God ondrǽdaþ, sette ðæne him tó práuoste. Se sylfa práuost dó mid árweorðnesse eal ðæt him fram ðam abbode getǽht biþ

. . . forðí swá miclan swá hé furður on weorðmynte forlǽten biþ, swá miclan hé sceal geornlícor healdan regules beboda, R. Ben. pp. 124–125: 46, 21. Ðá æteówde Benedictus on swefne hine sylfne ðam munece ðe hé tó ealdre geset hæfde ofer ðam mynstre, and his prófoste samod, Homl. Th. ii. 172, 15. Ðá hé ðá monig geár biscophád þegnode and swylce eác ðysses mynstres gémyne dyde, and ðǽr práuast and ealdormen gesette *qui cum annis multis episcopatum administraret, et hujus quoque monasterii statutis propostis curam gereret*, Bd. 3, 23; S. 555, 7. [*Icel*. prófastr: *O. H. Ger*. prôbist *praepositus, economus*. From Latin forms *praepostus, propostus*.] v. mynster-práfost.

práfost-folgoþ, es; *m. The office of provost*:—Gif se práfast þurh þreále nele gerihtan, hé sý áworpen of ðam práfastfolgoþe (*de ordine prepositure*), R. Ben. 126, 5.

práfost-scír, e; *f. Provostship*:—Ða sylfan him (*the provost*) práfost-scíre (prófost-, MSS. O. F.) betǽhtan, ðe ðǽne abbod tó abbodháde gecuran, R. Ben. 124, 16.

pranga, Wrt. Voc. i. 56, 50, *read* wranga.

prass *pomp, array, parade*:—Hwǽr syndon dêmra dômstôwa? hwǽr ys heora rícetere and heora prass and orgol, búton on moldan beþeaht and on wítum gecyrred? Wulfst. 148, 32. Se cásere fôr intô Efese mid ðrymme and mid prasse, Homl. Skt. i. 23, 26. Hí Pantan streám mid prasse bestôdon, Eást-Seaxena ord and se æschere *they stood by Panta's stream in proud array, the East-Saxon line and the host of the ashen boats*, Byrht. Th. 133, 51; By. 68.

predicere, es; *m. One who announces, a preacher*:—Praedico ic bodige oððe foresecge, *praedicator* prydecere (predicere, MSS. C. U.), Ælfc. Gr. 47; Zup. 276, 1. [*O. H. Ger*. predigari: *Icel*. prēdikari.]

predician; *p*. ode *To preach*:—Hé fêrde Godes ríce prediciende (*euangelizans*), Lk. Skt. 8, 1. [*O. L. Ger*. predikôn: *O. H. Ger*. predigôn: *Icel*. prēdika. From Latin *praedicare*.]

prénan. v. be-prénan.

preón, es; *m. A pin, brooch, fastening*:—Preón *vel* oferfeng *vel* dalc *fibula*, Wrt. Voc. i. 40, 53. Dolc oððe preón *spinther*, 74, 59. Hió becwiþ hyre ealdan gewíredan preón is an .vi. mancussum, Chart. Th. 537, 35. Ic geann mínre yldran dehter . . . ánes bendes and twegea preóna[s] and ánes wífscrúdes ealles, 530, 21. Menum ł preónum *monilibus*, Hpt. Gl. 434, 71. Mynas, preánas *lunulas*, 458, 30. [Þe vikelare ablent þene mon and put him preon in eien, A. R. 84, 2. Gol prenes and ringes, Gen. and Ex. 1872. Scot. preyne, prene, prin *a pin made of wire*: Icel. prjónn (Vigfusson compares with Gael. prine) *a pin, knitting-pin*: M. H. Ger. pfrieme: Ger. pfriem: Du. priem. Cf. also M. English prene to stick with a pin: Yorkshire Dialect prin-cod *a pincushion*: Scot. prein to pin; prein-cod, -head *pin-cushion, -head*: Icel. prjóna *to knit*.] v. eár-, feax-, mentel-preón.

preóst, es; *m. A priest*:—Preóst *clericus*, Wrt. Voc. i. 42, 24: 71, 77. Hé wæs tó preóste bescoren fram him *attonsus est ab eo*, Bd. 5, 19; S. 638, 21. (v. be-sceran.) Riht is ðæt preóstas regollíce libban, L. I. P. 16; Th. ii. 324, 2. Wé lǽraþ ðæt preóstas geóguþe geornlíce lǽran, L. Edg. C. 51; Th. ii. 254, 25. Wé lǽraþ ðæt preósta gehwilc, tóeácan láre, leornige handcræft georne, 11; Th. ii. 246, 16. [*O. L. Ger*. prêstar: *O. Frs*. prêstere: *O. H. Ger*. priestar, prêstar: *Icel*. prestr. From Latin *presbyter*.] v. hand-, híréd-, mæsse-, mynster-preóst.

preóst-hád, es; *m. Priest-hood*:—Sumne Godes mann preósthádes *clericum quendam*, Bd. 1, 7; S. 476, 36. Gé sint ácoren kynn Gode and kynelíces preósthádes *vos autem genus electum regale sacerdotium*, Past. 14, 5; Swt. 85, 19. Iulianus nolde gehealdan his preósthád on riht, Homl. Skt. i. 3, 290.

preóst-heáp, es; *m. A band of priests, the clergy*:—On preóstheápe *in clero*, Wrt. Voc. ii. 45, 22.

preóst-lagu, e; *f. Law affecting priests*:—Norþhymbra preósta lagu . . . Ælc preóst ðeáh on riht þe .xii. festermen ðæt hé preóstlage wille healdan mid rihte, L. N. P. L. 2; Th. ii. 290, 1–16.

preóst-scír, e; *f. The district in which a priest exercises his duties, a parish*:—Ne spane nán mæssepreóst nánne mon of óðre cyrcean hýrnysse tó his cyrcan, ne of óðre preóstscýre lǽre, ðæt mon hys cyrcan geséce, L. E. I. 14; Th. ii. 410, 31.

preówt-hwíl, e; *f. The time taken to close and open the eye, the twinkling of an eye*:—Preówthwíle, beorht (bearhtme?) *atomo* (ἐν ἀτόμῳ *in an instant*. See also Anglia viii. 318, 43:—564 *atomi* wyrcaþ án *momentum*, 4 *momenta* gefyllaþ *minutum*, 2½ *minuta*, gewyrcaþ ánne prican, 4 prica gewyrceaþ áne tíd), Hpt. Gl. 462, 9. On ánre preówthwíle on ðære endenéxtan býman *in ictu oculi, in nouissima tuba*, Homl. Th. ii. 568, 23. Cf. be-príwan *to wink with the eye*, Wulfst. 148, 13.

press, e; *f. A press* (in a list of requisites for spinning), Anglia ix. 263, 12. Cf. *Pannicipium* a presse, Wülck. 600, 14: *vestiplicium*, 619, 10.

prica, an; *m*. pricu (e), *a point, spot, dot*:—; *f.* I. *a point, prick, dot*:—Prica *punctus*, Ælfc. Gr. 28, 7; Som. 32, 57. Se forma prica on ðam ferse is geháten *media distinctio*, ðæt is, onmiddan tôdál, 50; Som. 51, 15. Mǽltanges prica *centrum*, Wrt. Voc. i. 39, 62. Án i oððe án prica ne gewît fram ðǽre ǽ *iota unum aut unus apex non praeteribit a lege*, Mt. Kmbl. 5, 18. Ðonne miht ðú ongitan ðæt eorþan ymbhwyrft is eall

wið ðone heofon tó mettanne swylce án lytel pricu (lytlu price, Cott. MS.) on brádan brede *omnem terrae ambitum ad coeli spatium puncti constat obtinere rationem*, Bt. 18, 1; Fox 62, 4. Swilce án prica (price, Cott. MS.), Fox 62, 20. Hé sǽde ðæt eal ðes middaneard nǽre ðe mǽre dríges landes ofer ðone mycelan gársecg, ðonne man ǽnne prican ápricce on ánum brádum brede, Wulfst. 146, 21. Heó hæfþ *æghwylcum* leáfe twá endebyrdnyssa fægerra pricena, and ða scínaþ swá gold, Lchdm. i. 188, 14. II. *a very small portion* (cf. *Fr*. ne point) (a) *of space*:—Ne gǽþ heora náðer ǽnne prican ofer ðam ðe him gesette is, Lchdm. iii. 252, 17. (b) *of time, the fourth or fifth part of an hour*:—Feówer *puncti*, ðæt synt prican, wyrcaþ áne tíd on ðære sunnan ryne, and forðan ys se prica gecweden forðan seó sunne ástíhþ pricmǽlum on ðam dægmǽle . . . Syx and hundnigontig prican beóþ on ðam dæge, and ða prican habbaþ *minuta* twá hund and feówertig, Anglia viii. 317, 16–24. Se án dæg hæfþ syx and hundnigontig prica (?) . . . feówer prica (?) gewyrceaþ áne tíd, 318, 10, 46: 320, 12 (cf. prican, l. 20). In Lchdm. iii. 222 the *prica* is a fifth of an hour:—On ánre nihta eald môna, and on .xxix. scínþ .iiii. pricena lengce. On twegra nihta eald môna, and on .xxviii. scínþ áne tíd and iii pricena, etc.: cf. with the calculations on this page the statement at 242, 7:—Dæghwamlíce ðæs mônan leóht byþ weaxende oððe waniende feówer prican. See also Homl. Th. i. 102, 30.

pricel, es; *n.* (?) *A prickle, sharp point*:—Seó rǽding pingþ ðæne scoliere mid scearpum pricele, Anglia viii. 308, 1. Wið priclom *contra stimulos*, Lk. Skt. p. 3, 6. [*Prompt. Parv*. prykyl *stimulus, aculeus: Du*. prikkel.] v. pricels.

pricele (a?), an; *f. m.* (?) *A point, very small thing*:—Foruord ł pricle *iota*, pricle ł stæfes heáfod *apex*, Mt. Kmbl. 5, 18. Ðone hlætmesto pricclu (pricla, Rush.) *nouissimum minutum*, Lk. Skt. Lind. 12, 59.

pricels, es; *m.* (?) *A sharp point*:—Pricelsum *stimulus*, Hpt. Gl. 514, 13. v. pricel.

prician, priccan *to prick*:—Ic pricige *pungo*, Ælfc. Gr. 28, 5; Som. 31, 59: 28, 7; Som. 32, 57. *Punctus a pungendo dicitur*, forðan ys se prica gecweden, forðan hé pricaþ, Anglia viii. 317, 18. Ðornas priciaþ, Homl. Th. ii. 88, 20. Hé hêt ðæs pápan lima gelôme prician, 312, 11. Ðonne man ǽnne prican ápricce on ánum brádum brede, Wulfst. 146, 21.

pric-mǽlum; *adv. By points*. v. prica, II b.

pricung, e; *f. Pricking*:—Ðornas priciaþ and ða welan gelustfulliaþ. Hí sind þornas ðonne hí ða sáwla tôteraþ mid pricungum mislicra geþohta, Homl. Th. ii. 88, 22.

prím *prime, the first hour, six o'clock*; also *the service held at that hour*, v. prím-sang:—Prím *prima*, undern *tertia*, middæg *sexta*, Wrt. Voc. i. 53, 10–12. Onginnaþ heáfudcwido tó prím (*ad primam*), Rtl. 166, 17. Gibedd tó prím, 171, 27. On ðysum tídum wé herien úrne scyppend . . . on dægréd, on prím, on undern, on middæg, on nôn, on ǽfen, on nihtsange, R. Ben. 40, 13. Ic sang prím and seofon seolmas, Coll. Monast. Th. 33, 27. [*Icel*. prími; *m.*: príma; *f.*: prím; *n.*]

prím-sang, es; *m. Prime-song, the service at the first hour*:—Ða seofon tídsangas . . . prímsang . . ., L. Ælfc. C. 19; Th. ii. 350, 6: R. Ben. 40, 6. Ælce Sunnanniht bútan Lenctene . . . dægrédsang, prímsang . . . mid alleluian sýn gesungden, 39, 18.

princ (?) *a prick*:—On prince *in ictu, in puncto*, Hpt. Gl. 462, 8. [Jamieson gives prink *to prick*.]

prior, es; *m. A prior*:—Hine God geuferade ðæt hé weard prior, Chart. Th. 445, 34.

prít. v. prýt.

príwan. v. be-príwan, preówt-hwíl.

prod-bor (?):—On prodbore *in foro*, Mt. Kmbl. Rush. 11, 16. On protbore, 20, 3. [Cf. (?) bor *and* prod *a pointed instrument*; to prod *to prick*, Jamieson, and common in many parts of England, as if *foro* were connected with *forare*.]

prófast. v. práfost.

prófian; *p*. ode *To esteem* or *regard as*:—Gif feorrancumen man búton wege gange, and hé ðonne náwðer ne hrýme ne hé horn ne bláwe, for þeóf hé is tó prófianne ðe is tó be *regarded as a thief*, L. Wih. 28; Th. i. 42, 25: L. In. 20; Th. i. 116, 2. [Cf. *Icel*. prófaðr *convicted of: nema þeir fengi af sér prófat unless they can clear themselves of it*.]

prút; *adj. Proud, arrogant*:—Mægen prútes unnytt Gode *virtus superbi inutilis Deo*, Scint. 17. Sáwl prútes (*superbi*) byþ forlǽten fram Gode, 17. Wiðerwyrdnyss prúte (*sublimes*) geniþerude, 46. Ðǽr mihton geseón Winceastre leódan rancne (prútne, MS. F., v. note, p. 336) here and unearhne, ðæt hí be hyra gate tó sǽ eodon, Chr. 1006; Erl. 140, 26. [Þa iward þe king on mode prut, Laym. 8828. Prud (*the opposite of* edmod), A. R. 176, 17. *Icel*. prúðr *gallant, brave, magnificent*.] v. prút-líce, -scipe, prútung, prýt.

prutene, an; *f. A plant-name, artemisia abrotanon*:—Ðone súþernan wermôd, ðæt is prutene, Lchdm. ii. 236, 20.

prútlíce; *adv. Proudly, in a stately manner, magnificently*:—Wel gelôme hig áspyriaþ ðæs solecismus unþeáwas . . . and eác hig prútlíce gýmaþ ðæs miotacismus gefleard, Anglia viii. 313, 25. Wé prútlíce (*in splendid fashion*) gecýðaþ uplendiscum preóstum ðæt wé be ðissum circul

gerǽdd habbaþ, 325, 40. [*Icel.* prúðliga *stately, magnificently.* Cf. prúð-leikr *show, ornament.*]

prút-scipe, es; *m. Pride, arrogance:*—Prútscipes *arrogantiae, superbiae*, Hpt. Gl. 432, 50. Múþ heora sprǽc prútscipe (*superbiam*), Ps. Lamb. 16, 10.

prútung, e; *f. Pride, haughtiness:*—Prútunge *fastu, elatione, superbio*, Hpt. Gl. 434, 13.

prýd, prýde, prydecere. v. prýt, prýte, predicere.

prýt, e; *f. Pride, pomp:*—Môd ofermôdignysse mid prýte byþ gewemmed *animus superbiae tumore corrumpitur*, Scint. 13. Mid nánre prýte ðú ofermôdiga *nulla elatione superbias*, 46. Ne gerîsaþ heom príta ne ídele rænca ne micele ofermétta, L. I. P. 10, note; Th. ii. 318, 31. Riht is ðæt abbodas nǽfre ymbe woruldcara ne ídele prýda ne carian tô swýðe, 13; Th. ii. 320, 35. [We ne beoð iboren for to habbene nane prudu ne nane oðre rencas, O. E. Homl. i. 7, 27.] v. woruld-prýt *and next word*.

prýte, an; *f. Pride, haughtiness:*—Of ýdelum gylpe biþ ácenned prýte, Homl. Th. ii. 220, 32. Prýte heáge út áwyr[p]þ *elatio excelsos deiecit*, Scint. 46. Gelíce ðám dwǽsan ðe for heora prýtan léwe (sáre, MS. C.), (*on account of the infirmity of pride in them*) nellaþ beorgan, Wulfst. 165, 9. Se ðe for his prýdan Gode nele hýran ... hê sceal misfaran, 178, 19. Sume men for heora prýtan forhogiaþ ðæt hí hýran godcundan ealdran, L. Eth. vii. 21; Th. i. 332, 33. [þat ece fer þe ham ȝearcod was for hare prede, O. E. Homl. i. 221, 1. *Laym.* prude, prute : *R. Glouc.* prute : *Ayenb.* prede : *Icel.* prýði ; *f. an ornament*; also *pride, pomp, bravery.*] v. preceding word.

prýtian *to be* or *make proud, shew pride:*—Prítigeaþ *pipant*, Wrt. Voc. ii. 88, 80. [þe luttele mon ... bute he mote himseluen *pruden*, he wole maken fule luden, Salm. Kmbl. p. 247, 25. Ofte onder þe uayre robes is þe zaule dyad be zenne, and mameliche ine þan þet ham *predeþ*. Yef þe pokoc him *prette* uor his uayre tayle, and þe coc uor his kombe hit ne is no wonder. ... Ac man ... ne ssel him naȝt *prede*, Ayenb. 258, 20–27. Prydyñ or wax prowde *superbio*, Prompt. Parv. 413.]

psalm, psealm. v. sealm.

púcel, es; *m. A goblin, demon:*—Wudewásan *faunos*, púcelas *priapos*, Germ. 394, 242. [Halliwell gives *puckle* a spirit or ghost. Cf. He wurþ bitauht þe *puke*, Misc. 76, 120. He shulde putten of so þe pouke (*the devil*), Piers P. 14, 190. *Icel.* púki *a devil, imp: Dan.* pokker *devil, deuce : Welsh* pwca : *Irish* púca *sprite, hobgoblin*, hence Puck in Shakspere.]

pucian *to poke* (?), *strike:*—Pucigende *repens*, Germ. 397, 493.

pudd, es; *m. A ditch, furrow :*—Puddas *sulcos*, Germ. 399, 338 [cf. puddle.]

puduc, es; *m. A wen:*—Puducas *strumas*, Germ. 396, 258.

puerisc; *adj. Boyish:*—Dý pueriscan *puerio*, Wrt. Voc. ii. 94, 48.

Pulgare; *pl. The Bulgarians:*—Hiliricos ðe wê Pulgare hátaþ, Ors. 3, 7; Swt. 110, 33.

pull, es; *m.: e; f. A pool, creek* :—Ondlong ðære strǽt tô máwpul; andlang pulles, Cod. Dip. Kmbl. iii. 79, 30–31. Of seges mere in ðæs pulles heáfod ... of þornbrycge in ðone pull, and æfter ðam pulle in baka brycge ... in dodhǽma pull, of ðam pulle in stream, 386, 12–19. Tô crampulle, 79, 12. Andlang Osríces pulle, 18, 18 : 19, 3. On æstege pul, 444, 7. [*Icel.* pollr. A Celtic word: *Welsh* pwll : *Irish* poll, pull.] v. pyll.

pullian; *p.* ode *To pull, pluck :*—Ða hreáþemýs on úre ondwlitan sper[n]don and ús pulledon *vespertiliones in ora uultusque nostros ferebantur*, Nar. 15, 7. Gif him þince ðæt hê sceáp pullige, ne biþ ðæt gôd, Lchdm. iii. 176, 7. v. á-pullian.

pull-spere, es; *n. A pool-spear, a reed* (*such as grows by pools*, cf. hreód-pól *under* pól) :—Pulsper *harundinem*, Mt. Kmbl. Lind. 11, 7.

pumic, es; *m.* (?) *Pumice:*—Of felle áscafen mid pumice, Lchdm. ii. 100, 15. [*O. H. Ger.* pumiz *pumex*.] v. next word.

pumic-stán, es; *m. Pumice-stone:* pumex, Wrt. Voc. i. 38, 26.

pund, es; *n. A pound.* **I.** as a weight *without reference to money :*—Án *uncia* stent on feówer and twentig penegum ; twelf sîðon twelf penegas beóþ on ánum punde, Anglia viii. 335, 18. *Libra is* pund on Englisc, Ælfc. Gr. 50, 30; Som. 52, 8. Pund *praesorium* (*pressorium*), Wrt. Voc. ii. 118, 25. Maria nam án pund (*libram*) deórwyrðre sealfe, Jn. Skt. 12, 3. Ðæt ísen ðe biþ tô prímfealdum ordále, ðæt wege .iii. pund, and tô ánfaldum án pund, L. Edg. H. 9; Th. i. 260, 13. **II.** as a money-denomination, (a) *of English money ; a pound*, 240 *pence :*—xx. scillinga beóþ on ánum punde, and twelf sîðon twentig penega byþ án pund, Anglia viii. 306, 35. Gá seó wǽge wulle tô .cxx. p̄. (tô healfan punde, MS. G.), L. Edg. ii. 8; Th. i. 270, 3. (b) *of other money :*—Ánum hê sealde fíf pund (*talenta*), Mt. Kmbl. 25, 15, 16, 20, 22. Hê sealde týn pund (*mnas*), Lk. 19, 13. Týn þúsend punda *decem millia talenta*, Mt. Kmbl. 18, 24. Pundes *libelli*, Wrt. Voc. ii. 52, 53 : 91, 44. **III.** as a measure (cf. wæter-pund *norma*, Wrt. Voc. i. 39, 60) *a pint* *,* that is, *a pound of water is a pint of water, and a pint of water is a pint for all liquids*, Lchdm. ii. 402 :—Pund eles gewihþ .xii. penegum læsse ðonne pund wætres, and pund ealoþ gewihþ .vi. penegum

---

máre ðonne pund wætres, etc., Lchdm. ii. 298, 16–26. [*O. L. Ger.* punt : *O. Frs.* pund : *O. H. Ger.* pfunt : *Goth. Icel.* pund. From Latin *pondo*.]

pund *a pound, an enclosure.* Cf. Si pundbreche, i. infractura parci, fiat, L. H. I. 40; Th. i. 540, 5. See also pyndan.

pundar, pundur *a plumb-line:*—Pundar *perpendiculum, modica petra de plumbo, quam ligant in filo quando aedificant parietes*, Txts. 112, 36. [Cf. punder *librilla*, 'librilla est baculus cum corrigia plumbata, ad librandum carnes,' Prompt. Parv. 416. Halliwell gives *punder*, to balance evenly, as an East-country word. *Icel.* pundari *a steel-yard*.]

pundere, es; *m. One who weighs :*—From boecerum ł punderum *a librariis*, Mt. Kmbl. p. 2, 2.

pundern. v. wǽg-pundern.

pundern-georn (?) ; *adj. Desirous of weighing* or *considering* (?) :—Punderngeð *ponderator*, Kent. Gl. 545.

pund-mǽte ; *adj. Weighing a pound:*—Gif hý on twá mǽl etaþ, sý gehealden ðæs pundmǽtan hláfes se þridda dǽl tô ðam ǽfengife, R. Ben. 63, 16.

pund-wǽg, e; *f. A pound-weight, a pound:*—Mon sceal simle tô beregafole ágifan æt ánum wyrhtan six pundwǽga, L. In. 59; Th. i. 140, 6. .xx. pundwǽga (-wêga, MS. B.) fôðres, 70; Th. i. 146, 19.

pung, es; *m. A small bag, purse :*—Pung *cassidele*, Txts. 54, 297 : Wrt. Voc. i. 291, 19 : ii. 13, 39. [*Goth.* puggs *a purse: Icel.* pungr : *O. H. Ger.* scaz-pfung *marsupium*.]

pungetung, e; *f. A pricking:*—Sió wamb gefêlþ sár ðonne se mon mete þigeþ and pungetunga and unlust metes, Lchdm. ii. 216, 21. v. pyngan.

punian ; *p.* ode *To pound, beat, bray :*—Puna *pound* (*the roots*), Lchdm. iii. 292, 19. [Cf. Wicklif, Mt. 21, 44 : it shal to gidre poune (*conteret*) hym. Halliwell gives *pun* as a West-country word, and quotes Florio : 'to stampe or punne in a morter.'] v. ge-punian.

Púnice ; *pl. The Carthaginians :*—Him cômon ongeán Púnice mid swá fela scipa *eo Carthaginienses cum pari classe venerunt*, Ors. 4, 6 ; Swt. 176, 11 : 172, 25 : 180, 5. Wæs geendad Púnica ðæt æfterre gewinn *bellum Punicum secundum finitum est*, 4, 11 ; Swt. 202, 31. Ðiss gewearþ Púnicum on ðæm teóþan geáre heora gewinnes, 4, 6 ; Swt. 176, 5. Claudius fôr eft an Púnice, Swt. 178, 31.

punt *a punt, flat-bottomed boat :*—Punt *pontonium*, Wrt. Voc. i. 47, 63 : *caudex*, 56, 26 : *trabaria vel caudex*, 63, 36. [From Latin *ponto*.]

pur *a bittern* (?) :—Hæferblǽte *vel* pur *bicoca*, Wrt. Voc. i. 21, 42. Ráradumbla, ðæt is pur *onocrotalus*, 62, 21. [*Purre* two sea-birds, the tern and the black-headed gull; pirre-, pyr-maw *a sea-bird*, E. D. S. Publ. Antrim and Down Glossary.]

pur-lamb, es; *n. A pur-lamb* (pur-lamb *a wether-lamb*, West of England, E. D. S. Publ. Old Farming Words, No. 6) :—Ðæt lamb sceal beón ánwintre purlamb clǽne and unwemme *erit agnus absque macula, masculus, anniculus*, Ex. 12, 5.

purpure, an; *f. A purple garment :*—Constantinus hiene benǽmde ǽgðer ge ðæs onwaldes ge ðære purpuran ðe hê werede, Ors. 6, 31 ; Swt. 284, 23. Hí scrýddon hine mid purpuran *induunt eum purpura*, Mk. Skt. 15, 17. Hê gemétte his ágenne sunu mid purpurum gegieredne (*purpuratus*) ... hit næs þeáw ðæt ǽnig ôðer purpuran werede búton cyningum, Ors. 4, 4; Swt. 164, 30–35. Hiene hêt iernan on his ágenum purpurum, 6, 30 ; Swt. 280, 12. Hié woldon ða onwaldas forlǽtan, and ða purpuran álecgan ða hié weredon *purpuram imperiumque deponerent*, Swt. 280, 21. [*Goth.* paurpaura : *Icel.* purpuri. From Latin.]

purpuren ; *adj. Purple:*—Purpuren hrægl *clavus vel purpura*, Wrt. Voc. i. 40, 13.

pusa, posa, an ; *m. A bag, scrip :*—'Nolite portare sacculum ne peram : ' 'Ne bere gê mid eów pusan oðde codd ' ... Hwæt mǽnþ se pusa búton woruldlíce byrþene, Homl. Th. ii. 532, 19–24. Se ríca berþ máre ðonne hê behôfige tô his fôrmettum, se þearfa berþ ǽmtigne pusan, i. 254, 31. Áwurp stánas in tô ðam pusum, and besenc hý on sǽlicum ýþum, ii. 418, 6. Posa *peram*, Mk. Skt. Lind. 6, 8 : Lk. Skt. Lind. 9, 3 : 10, 4. [*O. H. Ger.* pfosa *marsupium, bursa : Icel.* posi *a bag;* cf. púss.]

puslian *to pick out the best bits:*—Wyl on meolcum ôþ ðæt hié sýn wel mearuwe, pusla snǽdmǽlum *pick them out by a bit at a time*, Lchdm. ii. 356, 13. 'Peuselen *summis digitis varia cibarria carpere*,' Kilian.

pýcan, Pyhtas, pylece. v. pícan, Peohtas, pilece.

pyle, es; *m.* (?) *A pillow:*—Pyle *cervical*, Ælfc. Gr. 9, 5 ; Som. 9, 2 : *pulvillus*, Wrt. Voc. i. 284, 60 : *pulvinar*, 81, 60. Lytel pile *pulvillus*, 25, 48. Wá ðǽm ðe willaþ under ǽlcne elnbogan lecggean pyle and bolster under ǽlcne hnecean. ... Se legeþ pyle under ǽlces monnes elnbogan .. *vae his qui consuunt pulvillos sub omni cubito manus, et faciunt cervicalia sub capite universae aetatis. ... Pulvillos sub omni cubito manus ponere est* ..., Past. 19, 1 ; Swt. 143, 13–15. Hit wæs þeáw mid him ðæt hie mon ymbe .xii. mônaþ dyde ǽlces consules seti áne pyle hiérre ðonne hit ǽr wæs, Ors. 5, 11 ; Swt. 236, 7. [*Prompt. Parv.* pyliwe: *O. H. Ger.* pfuliwi ; *n.* From Latin *pulvinus.*]

pyll, es; *m. A pool, pill* ('*Pill*, a small creek, *Hereford*. The channels through which the drainings of the marshes enter the river are termed *pills*,' Halliwell. *Pill*, a pool, a creek, E. D. S. Publ. Cornish Gloss. See also Seebohm's English Village Community, pp. 149–150) :—

Andlang díce west on pull; of pylle on ford . . . eft on gerihte innan mycela pyll; of mycela pylle on smala pyll; andlang pylles . . . on ða díc innan holapyll; andlang holapylles, Cod. Dip. Kmbl. iii. 449, 11–22. v. pól, pull.

pynca, an; *m. A point :*—On pincan *in puncto*, Hpt. Gl. 492, 77. Cf. pyngan.

pyndan *to shut up, dam.* [Moni punt hire word uorte leten mo ut, as me deþ water and ter mulne cluse, A. R. 72, 10. To pynde *includere*, Cath. Angl. 280.] v. for-, ge-pyndan; pynding.

pynding, e; *f. A, dam :*—Ðæt wæter, ðonne hit biþ gepynd, hit miclaþ . . . ac gif sió pynding wierð onpennad, ðonne tóflêwþ hit eall, Past. 38, 6; Swt. 277, 8.

pyngan; *p.* de To *prick :*—*Punctus a pungendo dicitur* ; forðan ys se prica gecweden, forðan hé pingþ oððe pricaþ, Anglia viii. 317, 18. Seó ræding pingþ ðæne scoliere mid scearpum pricele, 308, 1. Hé wærlíce hine pynge mid sumum wordum *animum pungant*, Past. 40, 5; Swt. 297, 8. [Arthur up mid his spere . . . and pungde uppen Frolle þar he was on grunde, Laym. 23933. From Latin *pungere*.]

pyretre, an; *f. Bertram ;* pyrethrum parthenium, Lchdm. iii. 12, 19.

pyrige. v. pirige.

pyrtan; *p.* te To *strike, beat :*—Wæs sceacen *vibratur*, pyrte *ferit*, Germ. 401, 47. v. portian.

pyse. v. pise.

pytt, es; *m.*   I. *a pit, hole in the ground, a grave :*—Pyt *puteus*, Wrt. Voc. i. 84, 58. *Scrobs* ys pytt oððe díc, Ælfc. Gr. 9, 51; Zup. 66, 10. Heora mód ys swâ deóp swâ grundleás pytt *sepulcrum patens est guttur eorum*, Ps. Th. 5, 10. Gif hwâ pytt (*cisternam*) âdelfe and hine ne oferhelie and ðær fealle on oxa oððe assa, gilde ðæs pyttes hláford ðæra nýtena wurð, Ex. 21, 33–34. Pytte *baratrum*, Wrt. Voc. ii. 11, 68. On fúlan pyt; of ðam pytte on dene, Cod. Dip. Kmbl. iii. 77, 20. On wulfputt; of ðam pytte on ða wôgan æc, 449, 31–32. Tô wulfpytte, Cod. Dip. B. i. 280, 20. Gelæste man â ðone sâwelsceat æt openum pytte (cf. æt openum græfe, L. Eth. v. 12; Th. i. 308, 5), Wulfst. 118, 7. Uton dôn hine on ðone ealdan pytt (*cisternam*), Gen. 37, 20. Ic wæs on pytt beworpen *in lacum missus sum*, 40, 15. Hé âdylfþ ðone pytt *lacum aperuit*, Ps. Th. 7, 15. Hwylces eówres assa befealþ on ânne pytt (*puteum*), Lk. Skt. 14, 5. Hé hêt âdelfan ænne gehwædne pytt, Homl. Th. ii. 162, 2. On hiere bryne gemulton ealle ða onlícnessa tôgædere and on pyttas besuncan, Ors. 5, 2; Swt. 216, 3.   II. *a pit* (as in *pitted* with small-pox) :—Pyt ful wyrmses *serpedo* (cf. *serpedo* a mesylle, 224, 9 : a tetere, 267, 48), Wrt. Voc. i. 20, 4. [*O. H. Ger.* pfuzzi, pfuzza *puteus, cisterna* : *Ger.* pfütze : *Icel.* pyttr. From Latin *puteus*.] v. gang-, hor-, lâm-, môr-, rysc-, wæter- pytt, *and next word*.

pytted *pitted* (v. pytt, II), *marked with hollows :*—Ic gean mínon brêðer ðæs swurdes mid ðam pyttedan hiltan, Chart. Th. 559, 23.

Q

This letter occurs but seldom in Anglo-Saxon; in those native words where *qu* is now found, e. g. *quick, quoth, cw* or *cu* was written, *cwic, cuic, cwæþ, cuæþ*. In the glossary (belonging to the eighth century) given in Wrt. Voc. ii. 98 sqq. are six instances of words beginning with *qu*, and four others occur in the same volume; in the Blickling Gloss the form *quêmde* glosses *complacebam*, and the foreign word *reliquias* retained its original form.

R

râ, râha; *gen.* rân; *m. A roebuck, a roe :*—Râha *capria*, Wrt. Voc. ii. 103, 19. Raa *capriolus*, 129, 58 : *capia* (= *caprea*), 128, 47. Râ *capria*, 16, 79 : i. 288, 15. Gyf man on huntuþe rân oððe rægean mid flâne gewæceþ, Lchdm. i. 166, 24. Mære on huntunge heorta and râna *cervorum caprearumque venatu insignis*, Bd. 1, 1; S. 474, 41. Ic gefeó heortas and rânn *capio cervos et damas*, Coll. Monast. Th. 21, 31. Râ ł gæt *capras*, Rtl. 119, 16. *The word is found in names of localities*, e. g. On râhweg: ðæt ondloung râhweges on râhdene, Cod. Dip. Kmbl. iii. 378, 22. Ðonan wið heortsolwe; ðonne wið râhgelega, 391, 32. [*Prompt. Parv.* roo *capreus, capreolus* : *O. H. Ger.* rêho *capreolus* : *Icel.* râ a *roe*.] v. râh-deór, ræge.

rabbian; *p.* ode To *rage :*—God læt ðone deófol Antecrist rabbian and wêdan sume hwíle, Wulfst. 84, 11. [*From Latin.*]

raca, an; *m. A rake :*—Raca *rastrum* vel *rastellum*, Wrt. Voc. i. 15, 10. [*O. H. Ger.* rehho *rastellum.*] v. ofen-raca, racu, ræce.

-raca. v. ærend-raca.

racca, an; *m. A cord, which forms part of the rigging of a ship;* cf. *Icel.* rakki *the ring by which the sailyard moves round the mast :*—Racca *anguina* (cf: cops *anguina*, 56, 56 : bogen streng *anguina*, 35, 26. The word occurs among a list of names for ropes under the heading *de nave et partibus ejus*), Wrt. Voc. i. 63, 63.

racente, an; *f. A chain, fetter :*—Licgaþ mê ymbe îrenbendas, rídeþ acerntan sâl, Cd. Th. 24, 3; Gen. 372. Gebunden mid gyldenre racentan

vinctum compedibus aureis, Ors. 3, 9; Swt. 128, 12. Geræped mid his racentan, Met. 13, 8 : 25, 37 : 26, 78. Racentan slítan, 13, 29. Sleán on ða raccentan and on copsas, Bt. 38, 1; Fox 194, 32. Geseah hé his sylfes ungesælige stôwe and carcern (racetan, MS. B.) *videt suum infelix carcerem*, Bd. 5, 14; S. 635, 3. Hié hine hæfdon geþreátodne mid fýrenum racentum, Blickl. Homl. 43, 31. Ðonne hié mon on racentum beforan hiera triumphan drifon *regibus catenatis ante currum actis*, Ors. 5, 1; Swt. 214, 16. Restan on racentum, Cd. Th. 28, 11; Gen. 434. [*O. H. Ger.* rahhinza *baga* : *Icel.* rekendr; *pl. f. a chain.*] v. next two words.

racent-teáge, an; *f. A chain :*—Se ðe tôbræc ða raceteágan ymbûtan eówrum swuran *qui confregi catenas cervicum vestrarum*, Lev. 26, 13. v. next word.

racent-teáh; *gen.* -teáge; *f. A chain, fetter :*—Racenteáh *catena*, Wrt. Voc. i. 86, 30. Glæsen fæt on seolfenre racenteáge âhangen, Blickl. Homl. 209, 5. Unforedlícre racentâgæ *inextricabili collari*, Hpt. Gl. 455, 10. Mid rûmre racenteáge, Salm. Kmbl. 587; Sal. 293. Fæste mid îsenum racenteágum gewriðen, Homl. Th. i. 456, 9. Hé wæs mid racenteágum (raccentêgum, Lind.) gebunden *vinciebatur catenis*, Lk. Skt. 8, 29. Hine nân man mid racenteágum (raceteágum, MS. A.: racantêgum, Rush.) ne mihte gebindan. For ðam hé oft mid racenteágum (racontêgum, Rush.) gebunden tôslât ða raceteága (racontêge, Rush.), Mk. Skt. 5, 3–4. Gebundene on heardum racenteágum *vinctum catenis*, Jud. 16, 21. [*Laym.* raketeȝe: *O. E. Homl.* raketehe.] v. sweor- racentteáh *and preceding words.*

racete, raceteáh. v. racente, racent-teáh.

racian; *p.* ode.   I. *to direct, rule* (cf. reccan) :—Ðæt is ðæt hêhste gôd ðæt eallum swâ gereclíce racaþ and swâ eáðelíce hit eall set *est summum bonum, quod regit cuncta fortiter, suaviterque disponit*, Bt. 35, 4; Fox 162, 1. Gif hí næfdon ænne God ðe him eallum stiórde and racode and rædde, 34, 12; Fox 154, 5. Hé sceal rædan and racian (reccean, MS. T.) ôðra manna sâulum, R. Ben. 14, 6. Hé þeódum sceal racian (rædan, Kmbl.) mid rihte, Andr. Kmbl. 1041; An. 521.   II. *to take a course or direction, to run* (cf. racu a '*rake*') :—Hé his tungan gehealde ðæt hió ne racige on unnytte spræca *ne lingua per verba inutiliter defluat*, Past. 38, 5; Swt. 275, 19. Ne biþ nâ gebeorhlíc, ðam ðe wið God hæfþ forworht hine sylfne, ðæt hé tô hrædlíce intô Godes hûse æfter ðam racige, ac stande ðær ûte, Wulfst. 155, 21. [Cf. (?) *Scott.* raik *to move expeditiously* ; rack *a swift pace* : *Chauc.* rakel *hasty* : *Icel.* rakr *straight* ; rak-leið, -leiðis *straightway* : *Swedish* raka *to run hastily.*]

racsan, raxan *to stretch one's self after sleep :*—Swâ hé of hefegum slæpe raxende âwôce, Guthl. 12; Gdwin. 60, 6. [Cf. *Après dormir* il ço espreche *raskyt hym*, Wrt. Voc. i. 152, 25. He (*sloth*) his brest knocked and roxed (raxed, MS. W.: roskid, MS. B.) and rored, Piers P. 5, 398. *Scott.* rax *to stretch the limbs.*]

racsian (?) :—Racsode *libet*, Wrt. Voc. ii. 53, 62.

racu, e; *f.*   I. *an exposition, explanation, orderly account, narrative :*—Racu *historia*, Wrt. Voc. ii. 42, 56. Geþeahtung, gesceád *vel* racu *conlatio*, i. *conductio, comparatio, conciliatio*, i. *datio, contentio*, 134, 44. Gesytnys ł racu *textus*, Hpt. Gl. 505, 61. Ús ne segþ nâ seó racu (*the narrative*), tô hwam hé hine sette, Ælfc. T. Grn. 19, 3 : Jud. Thw. 156, 10. Ðætte on Arones breóstum sceolde beón âwriten sió racu ðæs dômes *ut in Aaron pectore rationale judicii imprimatur*, Past. 13, 1; Swt. 77, 9. Ðære býcnendlícan race *allegoricae expositionis*, Bd. 5, 24; S. 647, 42. Race *historiae*, Hpt. Gl. 459, 68: *prosae*, 528, 1. Of racu *relatione*, 480, 24. Ic eom geþafa ðæs ðe ðû segst forþam ðe ðû hit hæfst geseþed mid gesceádwíslícre race *assentior, cuncta enim firmissimis nexa rationibus constant*, Bt. 34, 9; Fox 146, 8. Ðû spenst mé on ða mæstan spræce and on ða earfoþestan tô gereccenne. Ða race (*the explanation*) sôhton ealle úþwitan, 39, 4; Fox 216, 15. Ic wolde reccan sume race, 41, 4; Fox 252, 14. Race *narrationem*, Hpt. Gl. 522, 54: Lk. Skt. 1, 1. Raca *conlationes*, Wrt. Voc. ii. 134, 47. Racum *relatibus*, Hpt. Gl. 529, 39. Hit is geræd on gewyrdelicum racum *in historical narratives*, Homl. Th. i. 58, 10. Ræde him mon ða raca oððe lif ðæra heáhfædera, R. Ben. 66, 17.   II. *comedy :*—Racu, tûnlíc spæc *comedia*, Wrt. Voc. i. 27, 13: 82, 63.   III. *the art of exposition, rhetoric :*—Swâ gedêþ se dreámcræft ðæt se mon biþ dreámere and seó racu deþ ðæt hé biþ reccere *sic musica musicos, rhetorica rhetores facit*, Bt. 16, 3; Fox 54, 32.   IV. *an account, reckoning :*—Ðær wæs uneten racu unc gemæne; ic onfêng ðin sâr ðæt ðû môste gesælig mínes êðelríces eádig neótan, Exon. Th. 89, 24; Cri. 1460. [*O. H. Ger.* rahha *res, causa, ratio, fabula, circumlocutio.*] v. martyr-, riht-, swefn- racu; reccan.

racu, e; *f. A rake :*—Hé sceal habban race (cf. man sceal habban ofurace, 265, 2), Anglia ix. 263, 7. v. raca, ræce.

racu, e; *f. A 'rake'* (*rake* a mountain track across a steep, Cumberland Gloss. e. g. the Lord's rake on Sca-fell), *a hollow path, bed of a stream :*—Cf. Andlang bróces; ðanon . . . on ða ealdan eárace, Cod. Dip. Kmbl. v. 122, 15; *and see* stream-racu. [Ryde doun þis ilk *rake*, bi ȝou rokke syde, Gaw. 2144. Out of the *rake* of riȝtwisnes renne suld he nevire, Alex. 3384.]

racu, e; f. *Rack* (?), *cloud, storm*:—Ic wille ǽhta and ágend eall ácwellan ða beútan beóþ earce bordum ðonne sweart racu (*the black clouds that overspread the sky at the Deluge*) stígan onginneþ, Cd. Th. 81, 34; Gen. 1355. [Cf. (?) In rede rudede upon *rak* rises þe sunne, Gaw. 1695. A *rak* and a royde wynde rose in hor saile, A myst and a merkenes was meruell to se, Destr. Tr. 1984. Or cf. (?) *Icel.* raki *wet*, rakr; *adj. wet*.]

Raculf, Ræculf, Reaculf, Reculf, Raculf-ceaster *Reculver* in Kent; Regulbium:—In ðam mynstre ðe is Reaculf nemned, Bd. 5, 8; S. 621, 33. Abbot on Raculfe, Chr. 692; Erl. 43, 13. Reculf, 669; Erl. 34, 26. See Cod. Dip. Kmbl. vi. 324.

rád, e; f. I. *riding, going on horseback* or *in a carriage*. v. rǽd-wægen:—Þeáw wæs ðam ylcan biscope ðæt hé ðæt weorc ðæs godspelles má þurh his fóta gange fremede ðonne on his horsa ráde *moris erat eidem antistiti opus evangelii magis ambulando per loca quam equitando perficere*, Bd. 4, 3; S. 566, 32. Nán mon for ðý ne rít ðe hine rídan lyste, ac rít for ðý ðe hé mid ðære ráde earnaþ sume earnunga. Sume mid ðære ráde earniaþ ðæt hié síen ðý hálran, Bt. 34, 7; Fox 144, 5–8. Ðá wearð his hors gesíclod, and feóll wealwigende geond ða eorþan... Hé begann ða tó gereccenne hú him on ráde getímode, Swt. A. S. Rdr. 101, 178. Gif mon on mycelre ráde oððe on miclum gangum weorðe geteorad, Lchdm. i. 76, 4. Ðæt man funde ǽnne man tó ráde oððe tó gange, L. Ath. v. 4; Th. i. 232, 15. Rynestrong on ráde, Exon. Th. 400, 9; Rä. 20, 7. II. *a going in a ship*:—Sió cwén bebeád áras fýsan tó ráde, sceoldon Rómwarena ofer heánne holm hláford sécean, Elen. Kmbl. 1960; El. 982. II. *an expedition on horseback*; in a hostile sense *a raid*:—Ðonne rídan ða yldestan men tó... and nimon eall ðæt hé áge, and fó se cyning tó healfum, tó healfum ða men ðe on ðære ráde beón, L. Ath. i. 20; Th. i. 210, 7. Gif áðor oððe mǽg oððe fremde ða ráde forsace, L. C. S. 25; Th. i. 390, 24. Cyninges þegnas oft ráde onridon, Chr. 871; Erl. 76, 11. III. *a road*; in the compounds brim-, hran-, hweogol-, segl-, streám-, swan-, wíg-rád. IV. *the name of the Runic R*. v. Exon. Th. 440, 10; Rä. 59, 15. *See also next word.* [*Icel.* reið *riding*; *a raid.*] v. mid-, on-, setl-, swegl-, þunor-rád.

rád, e; f. *Furniture (of a house), harness (of a horse)*:—Rád byþ on recyde rinca gehwylcum sëfte and swíðhwæt ðam ðe sitteþ onufan meare mǽgenheardum ofer mílpaþas *in the house is for each man furniture soft, and (the furniture for the horse, the harness) very strong for him that sits on the stout steed, traversing the roads*, Runic pm. Kmbl. 340, 11; Rún. 5. [Cf. *Icel.* reiða *implements, outfit*; reiði; *n. harness*: reiði; *m. tackle, harness.*] v. brand-rád, ge-rǽde, rǽde-sceamol.

rád, L. Wih. 10; Th. i. 38, 21. v. rǽd.

-rád. v. ge-, sam-rád.

rád-cniht, es; m. *A title equivalent to that of* sixhynde man:—Si autem talis occiditur qualem supra nominavimus rádcniht, et quidam Angli vocant sixhændeman, Text. Roff. p. 38. In domo hominis, quem Angli vocant rádcniht, alii vero sexhendeman, Schmid. A. S. Gesetz. 93, note 6.

rád-here, es. v. rǽde-here.

rád-hors, es; n. *A horse for riding, a saddle-horse*:—Man sceal lǽtan hine rídan on ðæs cyninges rádhorse, Anglia ix. 35, 235. [*Wick.* rood-hors *a horse for a chariot*: *O. H. Ger.* reit-hros *currilis equus*.]

-rádian, rador, v. ge-rádian; rodor.

radre glosses bovistra, Wrt. Voc. ii. 11, 26: 102, 10.

rád-stefn, e; f. *A term of service performed by a mounted person* (?):—Gif þegen geþeáh ðæt hé þénode cyning, and his rádstefne rád on his hírëde, L. R. 3; Th. i. 190, 19. v. stefn.

rád-wægen. v. rǽd-wægen.

rád-wérig; adj. *Weary with riding* or *journeying*, Exon. Th. 401, 19; Rä. 21, 14.

rǽcan; p. rǽhte. I. *intrans. To reach, extend, stretch forth*:—Ic wíde rǽce ofer engla eard, Exon. Th. 482, 26; Rä. 67, 7. Yldo rǽceþ wíde, Salm. Kmbl. 588; Sal. 294. Heó rǽhte mid handum tó heofoncyninge, Cd. Th. 292, 7; Sat. 437: Beo. Th. 1499; B. 747. Rǽhton wíde geond werþeóda wróhtes telgan, Cd. Th. 61, 1; Gen. 990. Ne hé sóðfæste lǽteþ ðæt hí tó unrihte willen handum rǽcean *ut non extendant justi ad iniquitatem manus suas*, Ps. Th. 124, 4. II. *trans. To reach, hold forth, offer, present*:—Ic rǽce porrigo vel porgo, Ælfc. Gr. 28, 5; Som. 31, 46. Hé ys se ðe ic rǽce (*porrexero*) hláf, Jn. Skt. 13, 26. For hwon ne rǽcst (*porrigis*) ðú ús ðone hwítan hláf? Bd. 2, 5; S. 507, 14. Rǽcþ (*porrigit*) hé him scorpionem? Lk. Skt. 11, 12. Ðǽr (*in hell*) hý leomu rǽcaþ (*stretch forth*) tó bindenne, Exon. Th. 99, 8; Cri. 1621. Eall ða weoruldgód hé gefeónde þearfum rǽhte and sealde *cuncta pauperibus erogare gaudebat*, Bd. 3, 5; S. 526, 26. Hé hláf brǽc and him rǽhte, Lk. Skt. 24, 30. Se óðer rǽhte forþ (*protulit*) his hand, Gen. 38, 28. Heó rǽhte hire handa him tó, Th. Ap. 27, 1: Past. 36, 1; Swt. 247, 21. Ðara ánra ðe for neóde him þénunge æt ðæs mynstres ingange rǽcan scylon, R. Ben. 139, 11. Se gebúr sceal erian healfne æcer and rǽcan (cf. on bærene gebringan, Chart. Th. 145, 1) ðæt sǽd on hláfordes berne, Cod. Dip. Kmbl. iii. 450, 35. [*O. Frs.* rëka: *O. H. Ger.* reihhen.] v. á-, ge-, mis-rǽcan.

---

rǽcc, es; m. *A dog that hunts by scent*:—Rǽcc bruccus, Wrt. Voc. i. 288, 29. [*Rache* a dog that pursues and discovers his prey by the scent, Jamieson. Rihht alls an hunnte takeþþ der wiþþ hise ȝæpe racchess, Orm. 13505. See other passages in Halliwell's Dict. *Icel.* rakki a dog.]

rǽce, an; f. *A rake*:—Rǽce rastrum, Wrt. Voc. ii. 98, 28. v. raca.

rǽced, reced, es; m. n. *A house, hall, palace*:—Reced sélesta (*Hrothgar's hall*), Beo. Th. 828; B. 412: 1545; B. 770. Rǽced, 3603; B. 1799. Wið ðæs recedes weal, 658; B. 326: 1452; B. 724. His (*Lot*) recedes hleów, Cd. Th. 147, 18; Gen. 1441. Se beorn (*Noah*) reste on recede, 95, 25; Gen. 1584. In rǽcede, Exon. Th. 314, 21; Mód. 17: 413, 11; Rä. 32, 3. Recyde, Runic pm. Kmbl. 340, 11; Rún. 5. Ic seah rǽplingas in rǽced fergan, Exon. Th. 435, 2; Rä. 53, 1. Con hé sídne rǽced fæste gefégan, 296, 7; Crä. 47. In ðæt dimme rǽced (*a prison*), Andr. Kmbl. 2618; An. 1310. Reced, Beo. Th. 2479; B. 1237. Hwearf geond ðæt síde reced, 3966; B. 1981. Ðæt (*Hrothgar's hall*) wæs foremǽrost receda, 625; B. 310. Receda wuldor, Salomones templ, Cd. Th. 219, 23; Dan. 59. Hié on Sodoman wlítan meahton, gesáwon ofer since salo hlífian, reced ofer reádum golde, 145, 11; Gen. 2404. Rǽced, Exon. Th. 381, 4; Rä. 2, 6. [*O. Sax.* rakud *used of the Temple.*] v. burg-, deáþ-, eorþ-, gim-, heáh-, heal-, hlín-, horn-, sund-, wín-rǽced, *and next word.*

rǽced-líc; adj. *Pertaining to a palace, palatine*:—Rǽcedlíce *palatina*, Wrt. Voc. ii. 116, 7.

rǽcing, e; f. *Reaching, holding out, offering, presenting, extending*:—Hláfes mid rǽcing *panis porrectione*, Jn. Skt. p. 7, 3. Mið rǽcing honda *extensione manuum*, 8, 11.

rǽd, es; m. I. *counsel, advice*:—Rǽd consilium, Wrt. Voc. i. 73, 23. Ðæt hit nǽfre næs náðer ne his gewile, ne his geweald, ne his rǽd, L. C. S. 76; Th. i. 418, 12. Is micel þearf ðisse þeóde helpes and rǽdes, Wulfst. 243, 5: Elen. Kmbl. 1103; El. 553. Sum woruldwita wæs swýðe wís on rǽde Acitofel gehâten... Ðá wæs se Acitofel mid Absalone on rǽde and rǽdde him hú hé mihte beswícan his fæder, Homl. Skt. i. 19, 196–203. Ðíne heortan tó rǽde gecyr *turn thine heart to listen to good advice*, Blickl. Homl. 113, 27. On ðone Drihten næs ic æt rǽde ne æt dǽde, ðǽr man mid unrihte N. orf ætferede, L. O. 3; Th. i. 178, 17. Gyf mon ðone hláford teó ðæt hé (*the accused person*) be his rǽde út hleópe, L. Eth. i. 1; Th. i. 282, 5, 12. Gif þeów ete his sylfes rǽde if *a slave eat flesh during a fast of his own accord* (i. e. *when his master does not give the meat*). v. the paragraph which precedes), L. Wih. 15; Th. i. 40, 11. Rǽde, 10; Th. i. 38, 21. Ic ða féng on mínne ágenne rǽd, Chart. Th. 322, 10. Gehýr míne word and mínne rǽd, Ex. 18, 19. Ðæt hí ðæs cynges rǽd hæfdon and his fultum and ealra witena, Chr. 1048; Erl. 178, 22. Rǽd gelǽran *to give good advice* (cf. sellan hǽlwende geþeahte, Bd. 1, 1; S. 474, 14), pref.; Erl. 3, 10. Rǽd sóhton consulunt, Wrt. Voc. ii. 21, 3. II. *counsel, prudence, intelligence*:—Nis nán wísdóm ne nán rǽd náht ongeán God, Homl. Th. i. 82, 14. Ongeán ðam wíslícan rǽde ðe of Godes ágenre gyfe cymþ, se wiðerrǽda deófol sǽwþ réceleásnesse, Wulfst. 53, 6. Se man ána hæfþ gesceád and rǽd and andgit, Homl. Skt. i. 1, 99. III. *counsel, course of action that results from deliberation, plan, a resolution taken after deliberation, ordinance, decree*:—Sý ðes rǽd gemǽne eallum leódscipe, L. Edg. S. 2; Th. i. 272, 33. Se rǽd wæs ǽrre on his rǽdfæstum geþance, ðæt hé wircan wolde ða wunderlícan gesceafta, Ælfc. T. Grn. 2, 4. Hé him tó rǽde genom ðæt... *cui rei consilium utile ratus est, ut*... Ors. 4, 5; Swt. 166, 27. Hé Rómanum tó rǽde gelǽrde, ðæt hié fôren on Hannibales land, 4, 10; Swt. 200, 1. On ðisum rǽde (*the conspiracy against William Rufus*) wæs Oda, Chr. 1087; Erl. 224, 5. Ðæne rǽd (*paying the Danes*) gerǽdde Síric, 991; Erl. 131, 19. Rǽd geþencean, Cd. Th. 19, 4; Gen. 286: 35, 28; Gen. 561. Rǽd áhicgan, 122, 24; Gen. 2031: 131, 24; Gen. 2181. Rǽdas consulta, Wrt. Voc. ii. 19, 54: consulta, consilia, 133, 80: Hpt. Gl. 504, 75: decreta, judicia, edicta, 433, 19. Gelǽrdan biscepas swelce níwe rǽdas swelce hié fol oft ǽr ealde gedydan, Ors. 4, 7; Swt. 184, 2. Manna wísdóm and heore rǽdas syndon náhtlíce ongeán Godes geþeaht, Chr. 979; Erl. 129, 26. IV. *what is advisable, benefit, advantage*:—Rǽd opere pretium, Wrt. Voc. ii. 64, 31. Rǽd biþ gif hé nimþ mealwan *it will be worth his while to take mallow*, Lchdm. ii. 238, 13. Biþ nú micel rǽd, ðæt he him gebycge ðæt éce líf, Homl. Skt. i. 12, 122. Ðonne biþ hire rǽd ðæt frýnd ða forword habban, L. Edm. B. 7; Th. i. 256, 2. Ðæt heó ús sý þingere ond-weardes rǽdes and éces wuldres *that she be for us an advocate for present profit and eternal glory*, Blickl. Homl. 159, 34. Hí him tó rǽde and tó frófre fundon *aliquid commodi adlaturum putabant*, Bd. 1, 12; S. 481, 7: 2, 5; S. 507, 31. Ðis him tó rǽde gecuron *hoc esse tutius decernebant*, 3, 5; S. 485, 34. Tó rǽde Angelcynne *to the advantage of the English*, 2, 1; S. 501, 39: Blickl. Homl. 199, 30: 205, 12. Tó hǽle and tó rǽde, 227, 4. Ðam þeódscype tó langsuman rǽde *to the lasting benefit of the nation*, L. I. P. 4; Th. ii. 308, 6. Eów sylfum tó rǽde, Ælfc. T. Grn. 12, 2. Rǽd árædian *to determine what is advisable*, L. Eth. vi. 40; Th. i. 324, 28. Ða ðe heora sylfra rǽd forlǽtaþ *those who forsake their own advantage*, Blickl. Homl. 103, 16. Rǽda fyrmest ðæt manna gehwylc ofer ealle óðre þinc ǽnne God lufige, L. I. P. 24; Th. ii.

338, 1. **V.** *a council :*—Hé eode tó ðæra Judéiscra ræde and befrân, hwæt hí him feós geúðon, Homl. Th. ii. 242, 16. Se cyng beád heom ðæt hí cómon mid. xii. mannum intó ðæs cynges ræde, Chr. 1048; Erl. 180, 11. **VI.** *as a part of proper names, generally under the form* rēd (red ?). For a list of such names v. Txts, 603 sqq., and for similar *O. H. Ger.* names v. Grff. ii. 463. [*O. Sax.* rád : *O. Frs.* rēd : *O. H. Ger.* rât : *Icel.* ráð.] v. feorh-, folc-, mis-, un-ræd.

ræd[e] *in composition of adjectives*, v. ân-, fæst-, heard-, hwæt-, læt-, wiðer-ræd[e].

rædan. *Two verbs originally distinct seem to coalesce under this form, the strong* rædan ; *p.* reórd, rēd ; *pp.* ræden : *Goth.* ga-rēdan : *O. Sax.* rādan ; *p.* rēd, ried : *O. Frs.* rēda ; *p.* rēd : *O. H. Ger.* rātan ; *p.* riet, riat : *Icel.* ráða ; *p.* rēð, *and the weak* rædan ; *p.* rædde : *Goth.* ga-raidjan : *O. H. Ger.* ant-reitjan *ordinare* : *Icel.* g-reiða. *The strong forms are rare.* **I.** *to counsel, give advice* :—Ic ræde ðé *consulo tibi*, Wrt. Voc. i. 49, 37. Girwan Godes tempel, swá hire gâsta weard reórd, Elen. Kmbl. 2043 ; El. 1023. Hé rád and rædde, rincum tæhte hú hí sceoldon standan, Byrht. Th. 132, 18 ; By. 18. Ðæt folc eall hrýmde, swá swá Josue him rædde, Jos. 6, 5 (20, Grn.). Rædende *consulentes, consilium dantes*, Hpt. Gl. 491, 20. **II.** *to ask advice, consult a person* :— Ic frîne *vel* ic ræde *consulo*, i. *inquiro* (cf. ic frîne ðé *consulo te*, i. 49, 38),Wrt. Voc. ii. 133, 79. **II a.** *to consult, deliberate, take counsel upon a matter* (acc.) *with* (wið) *a person* :—Justinus rædde wið ða cristennan, hwæne hí tó bisceope ceósan wolde, Homl. Th. i. 434, 28. Wið ðone rædde Chromatius, and be hiu ræde underféng ealle ða cristenan, Homl. Skt. i. 5, 323. Him þúhte and ðæm ðe he hit wið rædde, L. Ath. v. 12 ; Th. i. 240, 27. Hí gamenlíce ræddon *callide cogitantes*, Jos. 9, 4. Ðá gesomnedon hí gemót and þeahtedon and ræddon hwæt him tó dónne wære *initum est consilium quid agendum*, Bd. 1, 14 ; S. 482, 36. Ðá rēdon (ræddan, MSS. C.) hí him betweónum *consultatione habita*, Ors. 1, 14 ; Swt. 56, 20. Ðá ongunnon ða Pharisēi rædan *consilium inierunt*, Mt. Kmbl. 22, 15. Bisceopum gebyreþ ðæt he mid his rӕde mӕgan wunian wel gebungene witan,ðæt hí wið rædan mӕgan, L. I. P. 10 ; Th. ii. 316, 23. Man rædan sceolde hú man ðisne eard werian sceolde, Chr. 1010; Erl. 144, 7. **II b.** *to debate, speak in council* (or (?) *to read.* v. **VI b**) :—Rādaþ (rædaþ) ł maðeliaþ *concionantur, sermocinantur, loquuntur*, Hpt. Gl. 461, 1. Rædende ł wordiende *concionandi, loquentes*, 461, 35. **II c.** *to deliberate for the good of any one, look to, provide for* :—Mínre sáwle ræd on ēcnysse *animae meae in aeternum consules*, L. Ecg. P. iv. 67 ; Th. ii. 228, 3. Rædende *consulens*, i. *consilium tenens, providens*, Wrt. Voc. ii. 133, 77. Rædende *consulentes, succurrentes*, Hpt. Gl. 491, 20. **III.** *to resolve after deliberation, to determine, decide* :—Ðæt folc ræde be him ðæt hí woldon hine áhebban tó cyninge ... Ðá ðá Crist ongeat ðæs folces willan, Homl. Th. i. 162, 3–6. Ac ðeáh man hwæt rædde, ðæt ne stód furðon ænne mónaþ, Chr. 1010; Erl. 144, 9. Hí ræddon ðæt hí woldon ðone cyng gesettan út of Englalandes cynedóme, 1075; Erl. 213, 10. Hí ealle ánmódlíce ræddon ðæt ealle his gesetnyssa áýdlode wæron, Homl. Th. i. 60, 4. Rædan *decernere*, Wrt. Voc. ii. 27, 67. **III a.** rædan on (cf. *Icel.* ráða â *einn to attack one*) *to proceed against, take action against a person* :— Wæs ðam eorle Godwine and his sunan gecýdd, ðæt se cyng and ða menn ðe mid him wæron woldon rædon on hí, Chr. 1048; Erl. 178, 30. **IV.** *to rule, govern, direct* (with dat. or inst.) :—Ðú ðe Israéla éðelum cynne reccest and rædest *qui regis Israel*, Ps. Th. 79, 1. Hé ræt ûs and recþ *ipse reget nos*, 47, 12. Drihten mē ræt (*regit*), 22, 1. God ðe ræt and gewissaþ eallum gesceaftum, Chart. Th. 239, 34. Hé reht and ræt eallum gesceaftum, swá swá gód steóra ánum scipe, Bt. 35, 3 ; Fox 158, 25. God ðe him stiórde and racode and rædde, 34, 12 ; Fox 154, 6. Ðætte God rædde and weólde ealles middaneardes, 35, 2 ; Fox 156, 31. Ðæt hē (*the abbot*) sceal rædan and racian óðra manna sáulum, R. Ben. 14, 6. Hwā meahte iéð monnum rædan bútan scylde *quis principari hominibus tam sine culpa potuisset*? Past. 3, 1 ; Swt. 33, 16. Ðam ðe hié (*the Church*) wel ofer mæge and hiere wel rædan cunne *ei qui hanc bene regere praevalet*, 5, 2 ; Swt. 45, 1. Ic mæg rædan on ðis ríce, Cd. Th. 19, 10; Gen. 289. Ða ðe ðý ríce rædan sceoldon, 259, 4; Dan. 686. Wolde dóm Godes dædum rædan gumena gehwylcum *the decree of God would govern the deeds of every man*, Beo. Th. 5709; B. 2858. **V.** *to have the disposal of, have possession of* :— Ðone máððum ðe ðú mid rihte rædan sceoldest, 4119; B. 2056. Ðenden hié ðý ríce rædan móston, Cd. Th. 216, 18; Dan. 8. Bútan hý ðý reáfe rædan mótan, Exon. Th. 110, 6; Gú. 103. **VI.** *to read* (a) *as in to read* a riddle, *to explain* ; conjicere :—Ic ræde swefn *conicio*, Ælfc. Gr. 28, 6; Som. 32, 40. Módor ne rædoþ (-aþ, MS.) ðonne heó magan cenneþ, hú him weorðe geond woruld sceapen *a mother cannot read a boy's fate at his birth*, Salm. Kmbl. 741; Sal. 370. Ræde se ðe wille hú wunda cwæden, Exon. Th. 441, 11; Rä. 60, 16. Ræd hwæt ic mæne, 479, 18; Rä. 62, 9. Ðá ongan hē mid gleáwe móde rædan *coepit sagaci animo conjicere*, Bd. 3, 10; S. 534, 21, MS. B. (b) *to read a book*; legere :—Ic ræde *lego*, ðú rætst *legis*, Ælfc. Gr. 22; Som. 24, 1. Ræt *legit*, 44; Som. 45, 49. On hwylcum dæge man ræt .ix. kl'. apr. swá fela beóþ *concurrentes* ... gif man ræt ðæne datu-

rum on Sunnandæg ðænne byþ ân, Anglia viii. 302, 19–20. Se ðe ræt (rædæ, Rush.), Mt. Kmbl. 24, 15. Hé rædde his bóc ðam folce, Ex. 24, 7. Hé him gebæd and his béc rædde, Bd. 4, 3 ; S. 567, 4. Ne rædde gē (gē hreórdeþ, Rush.) hwæt Dauid dyde, Mt. Kmbl. 12, 3. Ne rædde, gē (gē ne reórdade, Rush.), 19, 4. Ræddon (reórdadun, Rush.), 21, 16. Rædde (reórdun, Rush.), 21, 42. Mē lyst rædan *lecturio*, Ælfc. Gr. 34; Som. 37, 56. Rēða *to read*, Mt. Kmbl. p. l, 8. Hé árás tó rēdanne, Lk. Skt. Rush. Lind. 4, 16 ; Rtl. 195, 16. Hē mē sealde bóc tó rædanne, Bd. 5, 13 ; S. 632, 37. Ðæt gewrit wæs ræded beforan ðam cyninge, 5, 21 ; S. 643, 11. Ðá ðæt godspel rædd wæs, Blickl. Homl. 161, 9. Wē gehýrdon ðá ðá Esaias se wítga rædan wæs, 167, 28. **VII.** *to prepare* (?) :—Hé sceal ælcre wucan erian .i. æcer and rædan sylf ðæt sæd on hlafordes berne, L. R. S. 4 ; Th. i. 434, 15. (Cf. *last passage under* rǽcan.) v. á-, be-, for-, ge-, mis-, ofer-rǽdan.

rǽd-bana, an ; *m. One who contrives a person's death, but is not the actual perpetrator* :—Gif man secge ðæt hē wǽre dǽdbana oððe rǽdbana *if he be said to be the actual perpetrator of homicide, or the deviser of it*, L. Eth. ix. 23 ; Th. i. 344, 26. Cf. Qui ad occidendum innoxium redbana vel dedbana fuerit, L. H. I. 85, 3 ; Th. i. 592, 13. [*Icel.* ráð-bani : *see also* bana-ráð *the planning a person's death* ; ráða einum bana *to plot a person's death*. v. Grmm. R. A. 626.]

rǽd-bora, an ; *m. A counsellor* ; *also translates consul* :—Rǽdbora *consiliarius*, Wrt. Voc. i. 73, 22. Hé (*the Messiah*) biþ geháten wundorlíc, rǽdbora, strang God, Homl. Th. ii. 16, 5 ? Dóm. L. 42, 38. Aðelwold ðe is mín rǽdbora *a secretis noster Athelwoldus*, Chart. Th. 241, 27 : Beo. Th. 2655 ; B. 1325. God næfþ nænne rǽdboran, Ælfc. T. Grn. 24, 24. Hí hæfdon him Consulas ðæt wē cweðaþ Rǽdboran, Jud. Thw. p. 161, 22. Seó gerædnes ðe Angelcynnes witan and Wealhþeóde rǽdboran gesetton, L. O. D. tit. ; Th. i. 352, 2. Rǽdboran *jurisperiti*, Wrt. Voc. ii. 46, 41. Rēdboran, 112, 13. Rǽdborena *juris peritorum*, Hpt. Gl. 524, 68. Cf. ræd-gifa.

rǽde (?), an ; *f. A reading, lesson* :—Ðiós rēdo *haec lectio*, Lk. Skt. p. 11, 16. Ðió rēdo *quae lectio*, 11, 5. Rēdes *lectionis*, Mt. Kmbl. p. 10, 16. Ðara rēda *lectionum*, 13, 13. Tó rēde *ad lectionem*, Rtl. 126, 1. Hálige rædan (rædincge, MS. T.) hē sceal lustlíce gehýran, R. Ben. 18, 9.

-rǽde. v. ge-ræde ; *n.*

rǽde ; *adj. Ready, prompt* :—On hwan mæg se iunga on gódne weg riht[r]an ne (ðe ?) rædran ræd gemittan ðonne hē ðíne wîsan word gehealde *in quo corrigit junior viam suam* ? *in custodiendo sermones tuos*, Ps. Th. 118, 9. Rædan (?) biionges *exercitationis*, Wrt. Voc. ii. 29, 59. v. ge-ræde, ræde-gafol, ræde-sceamol, rædness.

rǽde ; *adj. Mounted* :—Rædum here *equitatu*, Hpt. Gl. 525, 25. v. ræde-cempa, -here, -mann.

rǽde-cempa, an ; *m. A mounted soldier* :—Rædewíga *vel* [ræde]-cempa *equester, qui equitat*, Wrt. Voc. ii. 143, 66. v. ræde-here.

rǽde-gafol ; *n. Rent that can be paid all at once, as opposed to rent that is discharged by service rendered, and consequently takes time for its payment* :—Gif mon geþingaþ gyrde landes oððe máre tó rædegafole and geereþ gif se hláford him wile ðæt land árǽran tó weorce and tó gafole ne þearf hé him onfón gif hē him nán botl ne selþ *if a man takes a yard of land or more at a fixed rent and ploughs it, if the lord wants to get service as well as rent, the tenant need not take the land, if the lord does not give him a dwelling*, L. In. 67 ; Th. i. 146, 3. [Cf. *Icel.* reiðupenningar *ready money*.]

rǽde-here, es ; *m. A mounted force, cavalry* :—Rædehere *cerethi*, Wrt. Voc. ii. 15, 76 : *cerethei*, 130, 15. Of rādehere *equitatu*, Hpt. Gl. 525, 25. Alexandres næs ná má geslægen ðonne hundtwelftig on ðæm rædehere *in exercitu Alexandri centum et viginti equites defuere*, Ors. 3, 9; Swt. 124, 21. Ægðer ge an gangehere ge on rædehere (rád-, MS. C.), 4, 1; Swt. 154, 24. Earnulf gefeaht wið ðæm rēdehere (ræde-, MS. B.: rád-, MS. D.), Chr. 891 ; Erl. 88, 2.

rǽdelle. v. next word.

rǽdels, es ; *m.* : e ; *f.* : rædelse, rædelle (?), an ; *f.* **I.** *counsel, consideration* :—Seó rédelse and ðæt geþeaht úrra feónda geteorode, Ps. Th. 9, 6. **II.** *debate, speech in council* (v. rædan, II b) :—Rædelse *concionis, locutionis*, Hpt. Gl. 461, 4. **III.** *conjecture, imagination, interpretation* (v. rædan, VI a) :—Ræswung *vel* rædels *conjectura*, i. *opinatio, estimatio, interpretatio*, Wrt. Voc. ii. 133, 53. Ðeáh se leása wéna and sió rædelse ðara dysigena monna tiohhie ðæt se anweald síe ðæt hēhste gód (*hominum fallax opinio*), Bt. 27, 3 ; Fox 98, 32. Eall ðis ðú gerehtest tó sóðe swíðe gesceádwíslíce búton ælcre leásre rædelsan *haec nullis extrinsecus sumtis, sed altero ex altero fidem trahente, insitis domesticisque probationibus explicabas*, 35, 5 ; Fox 164, 31. Hrædelse *conjectura, argumentatione*, Hpt. Gl. 443, 19. Of rædelse *conjectura*, 460, 11. **III a.** *the imaginative faculty* :—Hē (*man*) hine ongit þurh his rædelsan (*imaginatio*) synderlíce, þurh his gesceádwísnesse (*ratio*) synderlíce, Bt. 41, 5 ; Fox 252, 19. **IV.** *a dark saying, enigma, riddle* :—Rædels *aenigma*, Ælfc. Gr. 9, 1 ; Som. 8, 23. Hē ásette rædels ðus cweðende. Swá hwilc man swá mínne rædels riht árǽde onfó se mýnre dohtor tó wífe, and se ðe hine misráede, sý hē beheáfdod, Ap. Th. 3, 8–11. *The riddle is given on* p. 4. Ða clamme ðe ða rædellan

(rǽdelsan?) wið rýnemenn heóld, Exon. Th. 423, 31; Rä. 43, 13. Ic sprece tó him openlíce næs þurh rédelsas (*per aenigmata;* dark speeches, A. V.), Num. 12, 8. [*Wick.* redels: *Piers P.* redel, ridel: *Prompt. Parv.* rydyl or probleme *enigma: M.H. Ger.* rātsal: *M.L. Ger.* rēdelse.] v. rǽsele.

rǽdelse. v. preceding word.

rǽde-mann, es; *m.* *A horseman:*—Nâwðer ne ðam horse ne ðæm rédemen ne wyrð geborgen of his ágnum cræfte, Ps. Th. 32, 15. [*Icel.* reið-maðr.]

rǽden[n], e; *f.* I. *a condition, stipulation:*—Rǽden *conditio,* Hpt. Gl. 436, 1. Rédin *condicio,* Wrt. Voc. i. 288, 44: ii. 17, 10. Ǽlc gebúr sylle .vi. hláfas ðam inswáne ðonne hé his heorde tó mǽstene drífe, on ðam sylfum lande ðe ðeós rǽden on stænt, L. R. S. 4; Th. i. 434, 22. Rǽdenne *condicione,* Wrt. Voc. ii. 104, 59. Ðan (on ða?) rǽdenne *ea conditione,* Hpt. Gl. 492, 8. On ða rǽdenne ðe hé him gá tó honda, L. In. 62; Th. i. 142, 3. Ðú bist Godes bearn þurh ða rǽdenne ðæt ðú ðinne feónd lufige, Homl. Th. i. 56, 7. Raedinnae *condiciones,* Ep. Gl. 7 f, 13. II. *rule, direction* (v. rǽdan, IV):—Hǽfdon sume mid áþum gefǽstnod ðæt hié on hire rǽdenne (rǽdinge, 193, col. 2) beón woldan *would be under her rule,* Chr. 918; Th. i. 192, 12. III. *a reckoning, estimating:*—Raedinne *taxatione,* Wrt. Voc. ii. 122, 1. The word occurs as the second part of many nouns, when its force is much the same as that of the suffixes *-ship, -hood, -red,* denoting *a state, condition.* v. bed-, bróðor-, burh-, camp-, feónd-, folc-, freónd-, gafol-, gebed-, gecwid-, gefér-, heord-, híw-, hús-, land-, mǽg-, mann-, meodo-, nám-, teón-, þing-, treów-, un-, weorc-, wíg-, worold-rǽden[n].

rǽdend, es; *m.* *A ruler, one who possesses control over anything* (v. rǽdan, IV):—Rodera rǽdend *the Deity,* Chr. 975; Erl. 126, 17: Beo. Th. 3114; B. 1555: Andr. Kmbl. 1253; An. 627. Dreáma rǽdend, Exon. Th. 358, 34; Pa. 55. [*O. Sax.* rādand (*Christ*).] v. mago-, sele-rǽdend.

rǽdend-líc; *adj. Pertaining to a decree* or *statute* (v. rǽdan, III):—Ðǽm rǽdendlícum *decretalibus,* Wrt. Voc. ii. 26, 45.

rǽden-gewrit, es; *n.* *A writing containing a condition* or *stipulation, a written agreement, a note of hand:*—Ic him sealde úre ágen rǽdengewrit, ðæt wǽre him tó ðam geráde ðæt land tó lǽten, ðe mon ǽlce gére gesylle fíftene scillingas clǽnes feós ðam bisceope, Chart. Th. 168, 12. Rǽdinggewrit (rǽden-?) *cirographum,* Wrt. Voc. i. 20, 51.

rǽdere, es; *m.* I. *a reader, one who reads:*—Rǽdere *lector,* Wrt. Voc. i. 72, 6; Ælfc. Gr. 9, 21; Som. 10, 40. Ðe ðǽre wucan rǽdere (rédere, 7, 23). Gebrôðra gerecorde ne sceal beón bútan háligre rǽdinge. Ne nán ne gedyrstlǽce, ðæt hé fǽrlíce bóc gelǽcce and ðǽr bútan foresceáwunge onginne tó rǽdenne, ac ðǽre wucan rǽdere on ðone Sunnandæg mid bletsunge hit beginne ... Nánes mannes stefn gehýred ne sý bútan ðæs rǽderes ánes, R. Ben. 62, 2–15. II. *a reader, scholar:*—Swá swá ða geleáfullan rǽderas hit gesetton, Lchdm. iii. 256, 21. III. *a reader, lector, the second of the seven orders:*—Seofon hádas syndon gesette on cyrcan ... ôðer is lector ... Lector is rǽdere, ðe rǽd on Godes cyrcan, and biþ ðǽrtó gehádod ðæt hé bodige Godes word, L. Ælfc. C. 10–12; Th. ii. 346, 25–32. Rédere réderes forlonge foreboderes ł ceigeras fruma from wítgum ðǽm is gecuoedin ceig *lector; lectores dudum praecones vel clamatores, initium es prophetis, quibus dicitur, Clama,* Rtl. 194, 1–4. IV. *a reader of riddles, a diviner* (v. rǽdan, VI a):—Wiccum, fram rǽderum *pythonibus,* Hpt. Gl. 504, 67. v. bóc-rǽdere.

rǽde-sceamol, es; *m.* *A reading-bench* (?); *a 'ready,' prepared bench, bench with furniture, a couch,* cf. *Icel.* reiðu-stóll, *and see* rád:—On rǽdescamole *in pulpito,* Wrt. Voc. ii. 45, 3. Rǽdescamelas *fulchra* (cf. *fulcra* eal bedreáf, Wrt. Voc. i. 59, 33 : *fulcris, thoris, lectis,* Wülck. Gl. 245, 28), 36, 36.

rǽdes-mann, es; *m.* I. *a counsellor, adviser, councillor:*—Ealle ðæs cynges rǽdesmen, Chart. Th. 330, 8 : Chr. 1039; Erl. 167, 19. II. *a steward, manager:*—Ǽt Steorran ðe ðá wæs ðæs kinges rǽdesman, Chart. Th. 339, 12. [*Icel.* ráðs-maðr *a manager, counsellor, steward.*]

rǽdesn (?), e; *f.* *A cluster of grapes;* bacido [cf. clyster *bacido, botrus,* Wrt. Voc. i. 33, 31] :—Rédisn *vacedo* (in a list *de lignis*), Wrt. Voc. i. 285, 43. Rédisnae *bacidones* (cf. raedinne *bacidones,* 43, 260: rǽdenne, Wrt. Voc. ii. 10, 59), Txts. 44, 1.

rǽdestre, an; *f.* *A female reader:*—Rédestre, Ælfc. Gr. 9, 21; Som. 10, 40. Rǽdystre 9, 64; Som. 13, 63. Rǽdistre, Wrt. Voc. i. 72, 7.

rǽde-wíga. v. rǽde-cempa.

rǽd-fæst; *adj. Wise, prudent:*—Se deófol gemacaþ ðæt se man þurh leáse hiwunge déþ swylce hé rǽdfæst sý ðe rǽdes ne gýmeþ *the devil causes the man by a false show to act as if he were wise, who cares not for wisdom,* Wulfst. 53, 9. Ðæt ic on ðínum rihte rǽdfæst lifige, Ps. Th. 142, 11. Ðínes ríces rǽdfæst wulder *gloriam magnificentiae regni tui,* 144, 12. Him in gást becwom rǽdfæst sefa, Cd. Th. 257, 3; Dan. 652: Exon. 468, 23; Hy. 5, 4. Se ðe symle byþ rǽdfest, Wald. 108 ; Vald. 2, 26 : Cd. Th. 90, 20 ; Gen. 1498. Árís and gereorda ðé mid

rǽdfæstum móde, Homl. Skt. i. 18, 185. Se rǽd wæs ǽfre on his rǽdfæstum geþance, Ælfc. T. Grn. 2, 5. Ðæt hig mâgon árîsan, gif hig rǽdfæste beóþ, 19, 5. Rincas rǽdfæste, Exon. Th. 347, 15; Sch. 13. Cf. rǽd-leás.

rǽdfæstlíce. v. un-rǽdfæstlíce.

rǽdfæstness, e; *f. Readiness to follow good counsel, adherence to right courses:*—Eahta sweras syndon ðe rihtlícne cynedóm up wegaþ ... rǽdfæstnes (*persuabilitas*), L. I. P. 3 ; Th. ii. 306, 20.

rǽd-findende *furnishing counsel, advising;*—Rǽdfindende *consulentes,* Wrt. Voc. ii. 24, 29. Cf. rǽd-hycgende.

raedgasram glosses hyadas, Txts. 69, 1035.

rǽd-geþeaht, es; *n.* *Counsel:* — *Consilium,* ðæt is rǽdgeþeht on Englisc, Wulfst. 51, 6. Elene héht Eusebium on rǽdgeþeaht gefetian, Elen. Kmbl. 2101; El. 1052. Héht gefetigean tó rúne ðone ðe rǽdgeþeaht þurh gleáwe miht georne cúðe, 2322; El. 1162.

rǽd-gifa, an; *m. One who gives counsel, a counsellor, councillor, adviser;* mostly of the king's advisers; *it also translates consul:*—Rǽdgifa *consiliator,* Wrt. Voc. i. 50, 1. Stígand ðe wæs ðæs cinges rǽdgifa and his handprést, Chr. 1051 ; Th. i. 317, col. 2. Rǽdgifan *consulem,* Germ. 397, 560. Ðis sindon ða gerǽdnessa ðe Engla rǽdgifan gecuran and gecwǽdan, L. Eth. vi. 1; Th. i. 314, 3. Ealle ðæs kyninges rǽdgyfan (*conciliarii*), Chart. Th. 326, 7. Ðone rǽd ðe ic mid mínum rǽdgyfum gerǽdd hæbbe, 307, 10. Rǽdgifena *juris peritorum,* Hpt. Gl. 524, 69. [Cleope nu to rǽde þine rǽdgiuen gode, Laym. 11615. *O. Sax.* rād-gebo: *O. Frs.* rēd-jeva: *O. H. Ger.* rāt-gebo: *Icel.* rāð-gjafi.] Cf. rǽd-bora *and next word.*

rǽd-gift glosses consulatus, senatus *in the following instances:*—Rǽdgiftes *consulatus,* Hpt. Gl. 412, 64. Rǽdgyft *senatu,* Hymn. Surt. 105, 34. Rǽdgifte *senatum,* Germ. 398, 108.

rǽd-hycgende; *part. Having wise counsel in the mind, prudent, sagacious:*—Ðú ðé ánne genim tó gesprecan symle rǽdhycgende, Exon. Th. 301, 28; Fä. 5.

rǽdic (rǽdic?), es; *m.* *A radish:*—Rǽdic *raphanum vel radix,* Wrt. Voc. i. 31, 37: *vermenaca,* 68, 65; *hierobotanim,* ii. 43, 52. Rédic, Lchdm. ii. 276, 10. Syle ðane rǽdic tó þicganne ... se rǽdic, 286, 10–14. Hrǽdic, iii. 20, 26. Genim hrǽdic nyðeweardne, 46, 1. [*O. H. Ger.* rātih, retih : *M. H. Ger.* retich : *Ger.* rettich. *From Lat.* radic-em.]

rǽding, e; *f.* I. *reading:*—Bisceopes dægweorc. Ðæt biþ mid rihte his gebedu ǽrest, and ðonne his bócweorc, rǽding, L. I. P. 8 ; Th. ii. 314, 19. Ǽmtigaþ eów tó rǽdinge *vacate lectioni,* hé begæþ his rǽdinge *vacuus est lectionibus,* Ælfc. Gr. 33 ; Som. 37, 14. Ðæs ðe ic on rǽdinge ne mihte fullíce ásmeágan, Wulfst. 65, 22. Beó ðú ábisgad ymbe rǽdinge *attende lectioni,* Past. 22, 1; Swt. 169, 17. I a. *a reading, a single act of reading:*—Ofthrǽdlíce rǽdinga háligra bóca, L. E. I. 2 ; Th. ii. 404, 2. Capitula rǽdinga, R. Ben. 43, 2. II. *what is read, reading, a passage in a book, a lesson:*—Ðis Englisc ǽtýwþ hwæt seó foresette rǽding (*passage*) mǽnþ, Anglia viii. 298, 9. Seó rǽding cwyþ ðæt ðǽr ys gyt on ǽlcum tácne healftíd, 298, 31 : 300, 32 : 309, 1. Sý án rǽdincg gerǽd of ðǽre ealdan cýðnesse *let one lesson from the Old Testament be read,* R. Ben. 34, 12. Agustinus ús onwreáh ðissere rǽdinge (*the lesson for the day*) andgit, Homl. Th. ii. 384, 21. Swá swá gé gehýrdon on ðissere rǽdinge (*the homily which precedes*), Homl. Skt. i. 11, 284. Hé lufode hálige rǽdinge ... ealle his gefǽran sceolde sealmas leornian oððe sume rǽdinge, Swt. A. S. Rdr. 97, 62–65. Gé sculon singan sunnanúhtan ǽfre nigon ræpsas mid nigon rǽdingum, L. Ælfc. P. 44; Th. ii. 384, 5. Man þreó rǽdinga rǽde, R. Ben. 33, 14. Wé willaþ on ðisre stówe ða seofon rǽdinga (*passages*) áwrítan ðe ymbe ða seofon geár synd gedihte ... Ðás rǽdinga syndon wíde cúðe, Anglia viii. 314, 18–22. III. *rule, government* (v. rǽdan, IV):—Hǽfdon sume mid áþum gefǽstnod ðæt hí on hire rǽdinge (rǽdenne, other MSS.) beón woldon, Chr. 918; Erl. 105, 30. v. béc-, bóc-, pistol-rǽding.

rǽding-bóc; *f. A book containing the lessons, a lectionary:*—Se mæssepreóst sceal habban ða wǽpna tó ðam gástlícum weorce ... ðæt synd ða hálgan béc ... rǽdingbóc, L. Ælfc. C. 21 ; Th. ii. 350, 14. ii. forealdode rǽdingbéc swíðe wáke (cf. ii. sumerrǽdingbéc and i. winter-rǽdingbóc, 16), Chart. Th. 430, 30. v. Maskell's Monumenta, vol. i. c. 3.

rǽding-gewrit, rǽdistre. v. rǽden-gewrit, rǽdestre.

rǽd-leás; *adj.* I. *without counsel, unwise, inconsiderate, rash, ill-advised:*—Rédeleás *preceps,* Ælfc. Gr. 9, 55 ; Som. 13, 27. Gleáw ne wæs gumríces weard, réðe and rǽdleás, Cd. Th. 226, 26 ; Dan. 177. II. *without wise direction, in confusion:*—Ðá ðis (*the destruction of certain ships*) cúð wæs tó ðam óðrum scipon ... wæs ðá swilc hit eall rǽdleás wǽre *it was as if there were no counsel anywhere, as if everything was in confusion,* Chr. 1009; Erl. 142, 9. III. *lacking what is advantageous or beneficial, miserable, desolate* (v. rǽd, IV):—Gé Godes cræfta nán þing ne gýmaþ, ðý is folces forfaren mǽre ðonne scolde oððe þearf wǽre, and for ðam hit weard swá rǽdleás ðe hit Godes beboda forgýmde *the people is become so miserable, because it neglected God's commandments,* Wulfst. 46, 20. Ðæt rǽdleáse hof (*hell*), Cd. Th. 3, 32 ; Gen. 44. [Nabbich in me wisdom ... and am redleas ... Drihten ase þu ert redlease (*gen. pl.*) red, red me þet am redles

O. E. Homl. i. 211, 32-213, 1. Nis nevre mon redles Ar his heorte beo witles, O. and N. 691.] [*O. H. Ger.* ráti-lô *sabsque consilio : Ger.* rat-los : *Icel.* ráđ-lauss *shiftless, confused. foolish.*]

ræd-líc ; *adj. Advisable :*—Him đā rǽdlecre geþūhte đæt hē wiđ ōđerne here friþ genāme đæt hē đone ōđerne đē iéđ ofercuman mehte *proviso ad tempus consilio, unum denuntiato bello adpetit, alterum pacta pace suspendit,* Ors. 3, 1 ; Swt. 96, 15 : 4, 13 ; Swt. 212, 16. Tō smeágenne wiđ his witan hwet heom eallum rǽdlícost þūhte, Chr. 1006 ; Erl. 141, 4. [*Icel.* ráđ-ligr.]

ræd-líce ; *adv.* I. *wisely, skilfully, cleverly :*—Hē rǽdlíce slôh swā hē hine (*the ball*) nǽfre feallan ne lēt, Ap. Th. 13, 5. Đæt hē meahte đæt folc đȳ wíslícor and đȳ rǽdlícor lǽran, Past. 18, 2 ; Swt. 131, 18. II. *advisedly, deliberately, designedly, on purpose :*—Rǽdlíce *consulto,* Ælfc. Gr. 38 ; Som. 41, 35. On đam gemôtan đeáh rǽdlíce wurđan on namcūđan stôwan, L. Eth. ix. 37 ; Th. i. 348, 17. [*Icel.* rádliga *cleverly.*] v. ān-, fæst-, un-rǽdlíce.

ræd-mægen ; *es ; n. Beneficial force(?), force that is productive of good* or *abundant good* (?), cf. lof-mægen (v. rǽd, IV):—Đā wæs wǽstmum āweaht world onspreht . . . rǽdmægne oferþeaht *the world was aroused to fruitful life, and overspread by productive force,* Exon. Th. 353, 10 ; Reim. 10.

rædness, *es ; f.* I. *readiness, promptness :*—Rǽdnis (hrǽdnis ?) *pernicitas,* Txts. 182, 75. On rǽdnysse *in maturitate,* Blickl. Gl. Đone þôþor mid swiftre rǽdnesse geslegene ongeán gesǽnde tô đam plegendan cynge, Ap. Th. 13, 4. Rǽdnisse (hrǽdnisse ?) *concursionibus,* Wrt. Voc. ii. 105, 24. Rǽdnessum, 15, 26. II. *an arrangement, agreement, condition :*—Đæt đeós gerǽdnis stondon môte in ēcnesse, and đis syndon đara manna naman đe æt đǽre rédnisse wǽron, Chart. Th. 168, 30. v. ge-rǽdness, rǽde.

ræd-rípe (hrǽd-?) ; *adj. Soon ripe, premature :*—Rǽdrípe wæstm *praecoquus fructus,* Wrt. Voc. i. 39, 22. Rǽdrípe wínberige *praecoquae,* 38, 61.

ræd-snotor ; *adj. Wise in counsel, prudent, sagacious :*—Nǽfre ic sǽlidan sélran métte rǽdsnotteran, Andr. Kmbl. 946 ; An. 473. [*Icel.* ráđ-snotr *sagacious.*]

ræd-þeahtende ; *part. Consulting, deliberating :*—Gesǽton sigerófe rǽdþeahtende ymb đa rôda þreó, Elen. Kmbl. 1734 ; El. 869 : 895 ; El. 449. Cf. rǽd-hycgende *and next word.*

ræd-þeahtere, *es ; m. A counsellor, adviser :*—Đa (*the senators*) wǽron simbel binnan Rômebyrg wuniende, tô đon đæt hié heora rǽdþeahteras wǽron, Ors. 2, 4 ; Swt. 72, 3. Đara twentigra monna đe hē him tô fultume hæfde ācoren, đæt his rǽdþeahteras wǽron *viginti viros sibi consilii causa legerat,* 6, 2 ; Swt. 256, 3.

ræd-þeahtung, *e ; f. Counsel :*—Hē wæs gemǽrsad ofer ealle ōđere cyningas ǽgđer ge mid his miclan fultume ge mid his rǽdþeahtunge ge mid his wígcræfte *ob magnitudinem virium consiliorumque summam belli nomenque traduxit,* Ors. 4, 1 ; Swt. 154, 27.

ræd-wægn, *es ; m. A vehicle, chariot :*—Hē hiene hēt iernan beforan his rǽdwǽne *ante vehiculum ejus,* Ors. 6, 30 ; Swt. 280, 13. Cf. *Icel.* reiđ-skjótr, *but see also* hræd-wægn.

ræd-wita, *an ; m. A counsellor, one wise in counsel :*—Ríce rǽdwitan, Dom. L. 18, 298.

ræfan (?) ; *p.* te *To involve, wrap :*—Hí weorþaþ gerǽfte (gerǽpte (?) cf. gereáped (-rǽped ?) Met. 25, 48) mid đǽre unrôtnesse, Bt. 37, 1 ; Fox 186, 21. [*Icel.* reifa *to swaddle.*]

ræfnan ; *p.* de. I. *to endure, suffer, undergo :*—Rǽfnde *perpetitur,* Wrt. Voc. i. 66, 66. Đeáh hē deáþes cwealm rǽfnan sceolde, Exon. Th. 240, 24 ; Ph. 643. II. *to do, perform, accomplish, carry out :*—Đa đe rǽfnaþ hēr wordum and weorcum wuldorcyninges lāre, 149, 20 ; Gū. 764 : 139, 17 ; Gū. 594. Neáh is Drihten eallum đe his willan hēr wyrceaþ georne and his hyge swylce elne rǽfnaþ, Ps. Th. 144, 19. Hié đæt ôfstum miclum rǽfndon, Judth. Thw. 21, 9 ; Jud. 11. Rǽfn elne đis, đæt đū nǽfre fǽcne weorđ freónde đínum, Exon. Th. 302, 3 ; Fä. 30. v. ā-rǽfnan *and cf.* dreógan *for the same two meanings.*

ræfnendlic, rǽfniendlíc, rǽfsan. v. un-ārǽfnendlíc, ā-rǽfnian, on-rǽfniendlíc, rǽpsan.

ræfter, *es ; m. A rafter, beam :*—Rǽfter *tignum,* Wrt. Voc. i. 26, 40 : 82, 14 : 290, 5. Reftras *amites,* Txts. 36, 11. Rǽftras, Wrt. Voc. ii. 6, 58 : *anses,* 10, 58. Reafteres *vel* latta *asseres,* i. 58, 35. Mycelne aad on beámum and on rǽftrum and on wāgum and on watelum and on þacum *congeriem trabium, tignorum, parietum, virgeorum & tecti fenei,* Bd. 3, 16 ; S. 542, 22. Ǽrest man āsmeáþ đæs hūses stede, and eác man đæt timber beheáwþ, and đa syllan man fǽgere gefēgþ, and đa beámas gelegþ, and đa rǽftras tô đǽre fyrste gefæstnaþ, Anglia viii. 324, 7-9.

ræge, *an ; f. A roe, a wild she-goat :*—Rǽge *caprea,* Wrt. Voc. i. 78, 31 : *capriole,* ii. 129, 59. Hrǽge *damula vel* đa, 22, 65. Rǽge, ii. 16, 80. Mynster đe is nemned an Hrǽge heáfde (*ad Caprae caput*), Bd. 3, 21 ; S. 551, 18. Ic gefeó rǽgan *capio capreas,* Coll. Monast. Th. 21, 31. Rǽgean (rǽgan, MS. B.), Lchdm. i. 166, 24. [*O. H. Ger.* reia *caprea.*] v. rā.

ræge-reósa, *an ; m. : -*reóse (?) ; *f. A ridge of muscles at the side of*

the spine running up the back :—Lǽcedômas wiđ rægereósan sāre, Lchdm. ii. 14, 26. Wiđ rægereósan, rūdan swā grēne, seóþ on ele and on weaxe, smire mid đone rægereósan. Eft nim gāte hǽr, smēc under đa brēc wiđ đǽs rægereósan, 146, 1-3. Be đam nafolan and đam rægereósan and bæcþearme, 230, 26. Biþ đæt sār fram đam nafolan ôþ đone milte and on đa winestran rægereósan, and gecymþ æt đam bæcþearme, 232, 3-6. v. Lchdm. ii. Glossary.

ræg-hár ; *adj. Grey like the goat* (v. rǽge) :—Oft đæs wāg gebād rǽghár and reádfáh ríce æfter ōđrum *oft did its wall, grey and redstained, see change of rule,* Exon. Th. 476, 19 ; Ruin. 10.

rægiming (?) :—*A clapping of the wings* (?) :—*Pullorum* cocca, *plausu* blisse *laetitiae* fiđerslehte (*in margin*) rægiminge, Hpt. Gl. 518, 51-54.

rægu. v. ragu.

réman. v. ā-rǽman.

réming (?) :—Heofenlícre rǽminge *celibea Tempe,* Wrt. Voc. ii. 130, 54.

rænc. v. renc.

ræpan ; *p.* te *To bind* (*with a rope*), *make captive :*—Hí férdon ǽghwider and þa earme folc rǽpton (rýpton, MS. C.) and slôgan (cf. rǽpling), Chr. 1011 ; Erl. 145, 6. Cyspan and mid racentan rǽpan, Met. 26, 78. [*Icel.* reipa *to fasten with a rope.*] v. ge-rǽpan.

ræping. v. next two words.

rǽpling, rǽping, *es ; m. One bound, a captive, prisoner, criminal :*—Wæs đā rǽpling se đe ǽr wæs Angelcynnes heáfod (*of archbishop Ælfheah taken captive by the Danes*), Chr. 1011 ; Erl. 145, 19. Hē (*St. Paul*) wæs đýder (*to Rome*) rǽpling gelǽded, Blickl. Homl. 173, 7. Rǽplinga *damnatorum,* Wrt. Voc. ii. 26, 54. Se wæs gebunden mid đǽm rǽplingum *qui cum seditiosis erat vinctus,* Mk. Skt. 15, 7. On cweartern đǽr man đæs cyninges rǽplingas heóld, Gen. 39, 20. Rǽplingas his *vinctos suos,* Ps. Spl. 68, 32. Rǽplingas unbindan, Dôm. L. 4, 48. Ic geseah rǽpingas in rǽced fergan . . . đa wǽron genamne nearwum bendum, gefeterade fæste tôgædre (*two buckets of a draw-well*), Exon. Th. 435, 1 ; Rä. 53, 1.

rǽpling-weard, *es ; m. A keeper of prisoners :* — Rǽplingcweardes *collegiati,* Wrt. Voc. i. 18, 45. Rǽpingweardas *collegiati,* ii. 134, 52.

ræps, reps, *es ; m. A response* (*in the service of the church*) :—Ǽfengebed *vespertinum officium, reps responsorium,* rǽding *lectio,* Wrt. Voc. i. 28, 32. Sý ān rǽding gerǽd, and ān swýđe scort ræps æfterfylige, R. Ben. 34, 13. Ān rǽding, æfter đam reps (ræps, MSS. O. F. : ryps, MS. T.), ymen, fers and lofsang, 36, 21. Ǽfter đǽm glorian đæs feórþan repses (rǽpses, MS. O.), 35, 18. Man þreó rǽdinga rǽde and þrý rǽpsas. Æt đam þriddan repse singe se sangere 'Gloria Patri,' 33, 14-16 : 35, 8-10. On đisum dagum wē forlǽtaþ on ūrum repsum 'Gloria Patri,' Homl. Th. ii. 224, 26. Gē sculon singan sunnanūhtan ǽfre nigon ræpsas mid nigon rǽdingum, L. Ælfc. P. 44 ; Th. ii. 384, 5.

ræpsan ; *p.* te *To seize* (?), *to reprove* (?) :—Raebsid uuaes, repset uaes, rǽpsit wæs *interceptum est* (cf. ā-raepsid, -repsit *interceptum,* 511), Txts. 68, 523. Raefsed, refset, raefsit *interpellari,* 70, 526. Refsede *intercepit,* 69, 1082. Cf. Fornoom *intercepit,* 71, 1083. Ārǽsed *interceptum,* 69, 1067. Ārǽsed wæs *interceptus est,* Wrt. Voc. ii. 46, 31. [*O. H. Ger.* refsan ; *p.* rafsta *corripere, increpare, arguere, reprehendere.*]

ræpsung, *e ; f.* I. *seizing* (?), *reproving* (?) :—Raepsung *interceptio* (v. *preceding word, and cf. O. H. Ger.* rafsunga *correptio, invectio, increpatio),* Txts. 69, 1068. II. *an interval :* — Seó niht hæfþ seofan dǽlas . . . Ōđer is *uesperum,* đæt is ǽfen, đonne se ǽfensteorra betwux đǽre repsunge æteówaþ, Lchdm. iii. 244, 1. *Vesperum* đæt ys ǽfen ođđe hrepsung, Anglia viii. 319, 28.

ræran ; *p.* de *To cause to rise, to rear, raise.* I. *to lift up, move from a lower to a higher position :*—Hē ūs tô roderum up hlǽdre rǽrde, Exon. Th. 437, 11 ; Rä. 56, 6. Hí tô heofenum up hlǽdræ rǽrdon, Cd. Th. 101, 1 ; Gen. 1675. Hié tô gūþe gārwudu rǽrdon, 198, 20 ; Exod. 325. Rǽre up đín heáfod and geseoh đis đæt Simon dyde, Blickl. Homl. 187, 35. II. *to raise* (*a building*) :—Đū rǽrst hūs *domum aedifices,* Deut. 28, 30. Hí wíbed setton neáh đam đe Abraham ǽror rǽrde, Cd. Th. 113, 7 ; Gen. 1883. Đæt beácen (*the tower of Babel*) đe rǽran ongunnon Adames eaforan, 101, 13 ; Gen. 1681. Ongunnon him bytlian and heora burh rǽran, 113, 1 ; Gen. 1880. III. *to set up, establish* (*a law, institution, etc.*) :—God sibbe rǽreþ ēce tô ealdre engla and monna, Exon. Th. 43, 16 ; Cri. 689. Hē Cristes cyricean on his ríce geornlíce timbrede and rǽrde *ecclesiam Christi in regno suo multum diligenter aedificare ac dilatare curavit,* Bd. 3, 3 ; S. 525, 37. Man unriht rǽrde and unlaga manege, Wulfst. 156, 12. Đonne rǽre man cyninges munde, đæt is đæt hý ealle đam sémende syllan đæt cyninges mund stande, L. E. G. 13 ; Th. i. 174, 20. Se đe unlage rǽre ođđe undôm gedéme, L. C. S. 15 ; Th. i. 384, 9. Ys his handgeweorc ryhte dômas đa hē rǽran wyle *opera manuum ejus judicium,* Ps. Th. 110. 5. IV. *to raise, offer* (*a prayer*) :—Hyra þeódnes dôm đæt hié tô đam beácne (*the golden image*) gebedu rǽrde, Th. 227, 24 ; Dan. 191. V. *to raise, begin, give rise to, excite* (*ill-feeling*) :—Rǽrde *exagitabat,* Wrt. Voc. ii. 30, 22. Oft hí þrǽce rǽrdon . . . feóndscype rǽrdon *oft were their violence and enmity roused,* Exon. Th. 243, 18-22 ; Jul. 12-14. Hǽteþ þrǽce rǽran . . . đæt hí fūsic binden and in bǽlwylme

swingen, 262, 16; Jul. 333. Fǽhþe rǽran, 113, 14; Gū. 157. Ne cūðon fírena fremman . . . elles ne ongunnon rǽran on roderum nymþe riht and sóð, Cd. Th. 2, 18; Gen. 21. Geflitu rǽran, Elen. Kmbl. 884; El. 443. Sǽce rǽran, 1879; El. 941. **VI.** *to rouse, excite* :—Saga hwā mec rǽre ðonne ic restan ne mót, oððe hwā mec stæððe ðonne ic stille beom, Exon. Th. 387, 2; Rä. 4, 73. **VII.** *to raise, elevate, exalt, promote* :—Gif ðū sóðne God lufast and his lof rǽrest, 245, 22; Jul. 48 : 103, 17; Cri. 1681 : 111, 23; Gū. 131. Se ǽrest ǽdelinga ēðelþrym rŷmde and rǽrde, Cd. Th. 98, 24; Gen. 1635. Uton beón ā ūrum hláforde holde, and ǽfre eallum mihtum his wurðscipe rǽran, L. C. E. 20; Th. i. 372, 9; Wulfst. 119, 14. Hū neáh ðære tíde wǽre ðætte ða bróðru árisan sceolden and Godes lof rǽran and heora ūhtsang singan *quam prope esset hora qua fratres ad dicendas Domino laudes nocturnas excitari deberent*, Bd. 4, 24; S. 599, 4. [*Goth.* raisjan : *Icel.* reisa.] v. ā-rǽran.

rǽrend, rǽrness. v. ā-rǽrend, -rǽrness.

rǽs, es; *m.* **I.** *a race, swift* or *violent running, rush* :—Wæs se þridda hlŷp, rodorcyninges rǽs ðā hē on róde āstāg, Exon. Th. 45, 30; Cri. 727. Micle rǽse (*magno impetu*) worn tódrifen wæs on sǽ, Mk. Skt. Rush. 5, 13. Mycelum rǽse, Lk. Skt. 8, 33. Ðæt hors sum slóg on ðam wege mid swiðran rǽse (*valentiore impetu*) oferhleóp, Bd. 5, 6; S. 619, 17. Ongeán ðam rǽse ðæs forþgotenan streámes *contra impetum fluvii decurrentis*, 5, 10; S. 625, 7. Hē hēt hwílon ða hundas ætstandan ðe urnon on ðam rǽse deórum getenge *he sometimes ordered the dogs to stop that were running at full speed close upon the game*, Homl. Th. ii. 514, 25. **II.** *an onset, attack* :—Beadumægnes rǽs, Cd. Th. 198, 28; Exod. 329. Hit ofslóh mínra þegna xxvi. āne rǽse (*in one onslaught*), Nar. 15, 25. Ðā weard líg tólŷsed, leád wíde sprong, hæleþ wurdon acle for ðŷ rǽse, Exon. Th. 277, 27; Jul. 587. Hē gūðe rǽs mid his freádryhtne fremman sceolde, Beo. Th. 5246; B. 2626. Gūðe rǽsum, 4702; B. 2356. [*Laym.* ræs, res, reas *an attack, onslaught : Allit. Pms.* to run in on a res *to rush in : Icel.* rás; *f. a race, running*.] v. beadu-, deáþ-, feónd-, gār-, gūþ-, hand-, heaðo-, hilde-, mǽg-, mægen-, on-, scyte-, sweord-, syn-, wæl-rǽs.

rǽs (?), -we; *f. Counsel, deliberation* :—Ðonne merestreámas meotudes rǽswum (*or from* rǽswa?) onwealcaþ, Exon. Th. 193, 24; Az. 126. *And see* rǽs-bora, rǽswian.

rǽsan; *p.* de *To rush, move violently* or *impetuously; inruere* :—Rǽsde *inruit*, Wrt. Voc. ii. 111, 56. **I.** *of actual movement* :—Seó hǽtu rǽsde on ða ðe ðæt fŷr ǽlde, Bd. 3, 16; S. 543, 9. Hē, getogene ðŷ wǽpne, rǽsde on ðone cyning, 2, 9; S. 511, 21. Hē ūt rǽsde on ðone ǽpeling, Chr. 755; Erl. 48, 34. Ðe stranga wind ðǽr on rǽsde, Shrn. 81, 32. Hit on ūs and on ūre wícstówe rǽsde, Nar. 15, 20 : Beo. Th. 5373; B. 2690. Hiá rǽsdon (*inruerunt*) on hine, Mk. Skt. Lind. Rush. 3, 10. Hundas rǽsdon on ðone apostol, Blickl. Homl. 181, 21. Hié rǽsdon on gífrum grápum, Andr. Kmbl. 2670; An. 1336. Wǽron hŷ reówe tó rǽsanne gífrum grápum, Exon. Th. 126, 27; Gū. 377. Rǽsed eode impetu abiit, Mt. Kmbl. Rush. 8, 32. **II.** *of violent action, to proceed against with violence, to assault, attack* :—Se hláford ne scríþ freónde ne feónde, ac hē rēðigmód rǽst on gehwilcne wēdehunde gelícost (cf. se ne murnþ náuþer ne friénd ne fiénd ðe mā ðe wēdende hund, Bt. 37, 1; Fox 186, 7–8), Met. 25, 17. Hine (deáþ) rǽscþ on gífrum grápum, Exon. Th. 161, 34; Gū. 968. Hū longe on rǽsaþ (*inruitis*) gē on men, Ps. Surt. 61, 4. On rǽsdun (*inruerunt*) in mē stronge, 58, 4. Ðæs burhgerēfan sunu wolde rǽsan on hī on ðæm scandhūse and hí bysmrian, Shrn. 56, 11. **III.** *of precipitate action, to rush* (*into anything*) :—Oft mon biþ suíðe rempende, and rǽsð suíðe dollíce on ǽlc weorc and hrǽdlíce, Past. 20; Swt. 149, 12. Geþence se láríow ðæt hē unwǽrlíce forþ ne rǽse on ða sprǽce, 15; Swt. 95, 9. v. be-, ge-, in-, þurh-rǽsan; fǽr-rǽsende.

rǽs-bora, an; *m. A counsellor, one who takes thought for the public good, a leader, guide* :—Rǽsbora (*Abraham*), Cd. Th. 108, 24; Gen. 1811. Andreas þanc gesǽgde rícum rǽsboran (*the Deity who in disguise had guided Andrew's ship*), Andr. Kmbl. 769; An. 385. Rēðe rǽsboran (*the chiefs of the Mermedonians*), 277; An. 139. Rǽsborena [rǽs- (?), rǽd- (?)] *jurisperitorum*, Wrt. Voc. ii. 87, 38. v. rǽs, *and cf.* rǽd-bora.

rǽsc. v. líg-rǽsc.

rǽscan; *p.* te *To move quickly* (cf. *rash*), *to quiver* (*of light*), *to glitter* :—Fēr rǽscendum leóhte *ignis vibrante lumine*, Hymn. Surt. 94, 1. v. rǽsc, rǽscettan.

rǽscettan; *p.* te *To crackle, make a crackling noise as fire does, to sparkle* :—Fŷren líg braslaþ, rǽsct and ēfesteþ, Dóm. L. 10, 152. Ðæt rēðe flód rǽscet fŷre, 12, 165. Rǽscetteþ *crepitat*, Wrt. Voc. ii. 94, 61. Rǽscettaþ, cyrmþ, scylþ *crepitat*, i. *resonat*, 136, 73. Rǽscettan *crepitarent*, rǽscettende *crepitantes*, 18, 9–10. Rǽscetende *crepitantes*, 78, 10. [*O. H. Ger.* raskezzan *scintillare, singultare*.] v. *preceding word*.

rǽsc[t]ung, e; *f. Sparkling, gleaming, coruscation* :—Hrǽscetunga *coruscationes*, Hpt. Gl. 509, 31. v. líg-rǽscetung *and preceding word*.

rǽsele, an; *f. A conjecture, solution of a riddle* :—Gif ðū mǽge rǽselan gesecgan, Saga hwæt hió hātte, Exon. Th. 421, 34; Rä. 40, 28. v. rǽswan, *and cf.* rǽdels, III.

rǽsn, es; *n. A plank, a ceiling* :—Rǽsn *asser*, Ælfc. Gr. 9, 18; Som. 9, 59. Rǽfter *tignus*, beám *trabs*, wāh *paries*, rǽsn *laquear*, Wrt. Voc. i. 290, 5–8 : ii. 52, 4. [*Goth.* razn *a house* : *Icel.* rann.]

rǽst, rǽstan. v. rest, restan.

rǽswa, an; *m.* (a word used only in poetry). **I.** *a counsellor* :—Cwæð ðā se ðe wæs cyninges rǽswa (cf. 'the king spake unto his counsellors . . . They answered and said unto the king,' Dan. 3, 14), wís and wordgleáw, Cd. Th. 242, 11; Dan. 417. **II.** *one who takes thought* (*for the public good*), (a) *a prince, king* :—Se rǽswa (*Nebuchadnezzar*), 256, 14; Dan. 640. Werodes rǽswa, Babilone weard, 246, 31; Dan. 487. Folca rǽswa, Caldea cyning, 257, 34; Dan. 667. Ealwealdan Gode, þeóda rǽswan, Andr. Kmbl. 3243; An. 1624. Folccyningas, leóde rǽswan, Cd. Th. 125, 6; Gen. 2075. (b) *a leading man, chief person, leader* :—Ðā weard forht manig folces rǽswa *many a chief man among the Mermedonians*, Andr. Kmbl. 2174; An. 1088. Gesetton Sennar leóda rǽswan leófum mannum heora, 99, 34; Gen. 1656 : 100, 25; Gen. 1669. Folces rǽswan (*the chief men with Holofernes*), Judth. Thw. 21, 10; Jud. 12. Leóda rǽswan (*the chief men of Bethulia*), 24, 8; Jud. 178. Hæleþa rǽswan, dugoþ dómgeorne (*the high priest and his fellows*), Andr. Kmbl. 1384; An. 692. Módgleáwe men, middangeardes rǽswan, Salm. Kmbl. 362; Sal. 180. Rǽswan herges, *the leaders of the host*, Cd. Th. 192, 20; Exod. 234. Hē beforan fremede folces rǽswum (*the chief men among the Jews*), Andr. Kmbl. 1238; An. 619. [*Icel.* ræsir *chief, captain, king*.] v. ge-, here-rǽswa, rǽs-bora, *and next word*.

rǽswan, rǽswian, rǽsian, rēsian; *p.* ede, ode *To think, suppose, suspect, consider, conjecture* :—Tó ðǽm sóðum gesǽlþum ðe ðín mód oft ymbe rǽsweþ *ad veram felicitatem, quam tuus somniat animus*, Bt. 22, 2; Fox 78, 7. Rēsiaþ *comminiscimus*, Wrt. Voc. ii. 18, 7. Rēsiat, 77, 24. Hié eallneg rǽswaþ and ondrǽdaþ ðæt hí mon tǽlan wille *they are always suspecting and dreading that people want to blame them*, Past. 35, 2; Swt. 239, 6. Ðū rǽswedest (*existimasti*) swíðe unryhte ðæt ic wǽre ðín gelíca, Ps. Th. 49, 22. Hē rēsade (*suspicabatur*) ðæt hē hæfde ðæs Cǽseres ǽrendo sum tó Breotone cyningum . . . Ðā hē ongeat ðæt hit swā ne wæs swā hē rēsade, Bd. 4, 1; S. 564, 48–565, 3. Rǽswodan, spǽcan, wǽron gemunende *comminiscuntur*, Wrt. Voc. ii. 24, 1. Ne rēccaþ hwæt him mon ymbe rǽswe *mala de se opinari permittunt*, Past. 59, 1; Swt. 447, 28. Ðā ongan hē mid gleáwe móde þencan and rēsian (rēsian, MS. C.) *coepit sagaci animo conjicere*, Bd. 3, 10; S. 534, 21 note. Rēsigan *opinare*, Wrt. Voc. ii. 115, 55. Rǽswian *conici, conari*, 131, 79. Rēsenðe ic eom *suspicatus sum*, Ps. Surt. 118, 39. v. next word.

rǽswung, rēsung, e; *f. Supposition, conjecture* :—Rǽswung *vel* rǽdels *conjectura*, i. *opinatio, estimatio, interpretatio*, Wrt. Voc. ii. 133, 53. Rēsung *conjectura*, 104, 35. Rēsong, 77, 72. Rēsunge *ratiunculus*, 119, 14.

rǽt, *a rat* :—Rǽt *raturus* (in a list of animals), Wrt. Voc. i. 22, 48. [*O. Du.* ratta : *O. H. Ger.* rato; *m.* ratta; *f.*: *Icel.* rotta; *f.*]

rǽw, rāw, e; *f. A row, line* :—Ðonon on ða rǽwe (*hedge-row*); of ðære reáwe on Temese, Cod. Dip. Kmbl. v. 275, 20. Sele ðonne drincan sume on (on sume, MS.) rǽwe nigon dagon *nine days in succession*, Lchdm. ii. 238, 10. Cf. He sende hem so muche honger and luþer geres a-rewe, R. Glouc. 252, 2. Is seid of euerich on a-rewe, A. R. 90, 10. For þre niȝtes a-rowe he seiȝ þat same siȝt, Chron. Vilod. 68 [in Stratmann]. *The word also occurs in* hæsel-, hege-, hlinc-, stān-, widig-rǽw, Cod. Dip. Kmbl. iii. xxxv. Cf. *also* gerǽwud fēða *acies*, Wrt. Voc. i. 18, 26. Standaþ on gerǽwe, Cod. Dip. Kmbl. iii. 424, 8. Hí on gerǽwe sǽton, Homl. Skt. i. 23, 779.

rafan. v. be-rafan.

rāfian *to involve, wrap up* (?) cf. *Icel.* reifa *to swaddle* ; or *to unloose, disclose*, cf. *Icel.* reifa *to rip up, disclose*. v. ā-rāfian, rǽfan.

rǽge. v. rǣge.

raggig; *adj. Shaggy, bristly, ragged* as applied to the rough coat of a horse :—Raggie *setosa*, Hpt. Gl. 524, 16. [*Icel.* rögg *shagginess*; *a* tuft (cf. *rug-headed kernes* in Macbeth); raggaðr *tufted* : *Swed. Dan. dial.* ragg *rough hair* : *Swed. dial.* raggig *shaggy* : *Dan. dial.* raggad *shaggy*.]

ragu, e; *f. Lichen* :—Ragu *mosilicum*, Wrt. Voc. i. 67, 63 : *mosiclum*, 287, 33 : ii. 114, 18 : mossiclum, 55, 74 : *mosicum*, 114, 26. Ragu *sedulium*, 120, 46. Hæseles ragu *the lichen of hazel*, Lchdm. ii. 96, 2. Cristes-mǽl-ragu *lichen off a crucifix*, 346, 23. Ragu and meós (*rubigo*) fornymþ ealle eówre landes wæstmas, Deut. 28, 42. v. berc-, slāhþorn-ragu.

ragu-finc, es; *m. The name of some bird* :—Ragofinc *scutatis, scutatus*, Wrt. Voc. i. 62, 44 : 281, 15 : *barrulus*, ii. 10, 79. Reagufinc *bariulus*, 101, 62.

rāha. v. rā *and next word*.

rāh-deór, es; *n. A roe-buck* :—Rāhdeór *capreus*, Wrt. Voc. i. 22, 66 : *capreolus*, 78, 30. Rāhdeóres mearh, Lchdm. iii. 2, 25.

ram. v. ramm.

ram-gealla, an; *m. Ram-gall* (a plant name); *menyanthes trifoliata* :—Ramgeallan ðone fāgan, Lchdm. ii. 124, 13. Hramgeallo, 140, 13.

3 E

ram-hund (?) :—De canibus quos ramhundt vocant, L. C. F. 32 ; Th. i. 430, 7.

ramm, es (*a wk. gen. pl. occurs*) ; *m.* **I.** *a ram* :—Ramm *aries*, Wrt. Voc. i. 23, 52. Ram, 78, 46. Rom *berbex*, ii. 12, 71 : 126, 3. Rommes blôd, Cd. Th. 177, 20 ; Gen. 2932. Geoffra mē ǽnne þríwintre ramm, Gen. 15, 9 : 22, 13. Beorgas wǽron blîðe swá rammas, Ps. Th. 113, 6. Bringaþ him eówra ramma bearn, 28, 1 : Ps. Spl. 65, 14. Twentig rammena *arietes viginti*, Gen. 32, 14. Rammum gelîce, Ps. Th. 113, 4. **II.** *an instrument for pounding* or *battering* :—Aries biþ ram betwux sceápum and ram tô wealgeweorce, Ælfc. Gr. 5 ; Som. 4, 15. Ram tô wurce *aries*, Wrt. Voc. i. 34, 57. Ram *aries*, andweorc tô wealle *cimentum*, wealwyrhta *cimentarius*, 85, 26–28. Þerscaþ ðone weall mid rammum, Past. 21 ; Swt. 161, 6. 'Gáþ tô mid rammum' ... Hē bierþ rammas ymbûtan ðæt môd his hiéremonna, ðonne hē him gecýð mid hû scearplícum costungum wē sint ǽghwonon ûtan behringcde, and se weall ûres mægenes þurhþyrelað mid ðan scearpan ramman (ðǽm scearpan rammum, Cott. MSS.) ðara costunga, Swt. 163, 10–18. [*O. H. Ger.* ramm *aries, vervex.*]

rán, es ; *n. Unlawful seizure of property, robbing* :—Rán quod dicunt apertam rapinam, L. W. iii. 12 ; Th. i. 493, 6. [*Icel.* rán.]

rán *rained.* v. rînan.

ranc ; *adj.* **I.** *proud, haughty, arrogant, insolent* : the word remains with a somewhat different meaning in *rank,* used of coarse but fertile growth :—Gif ǽnig man hæbbe môdigne sunu and rancne (*protervum*) ðe nelle hîran his fæder and his mēder, Deut. 21, 18. Ne beón gē tô rance ne tô gylpgeorne, Wulfst. 40, 19 : 81, 15. Sume munecas synd tô wlance and ealles tô rance, L. I. P. 14 ; Th. ii. 322, 12. Hî taliaþ ðē wyrsan for heánan gebyrdan ða ðe heora yldran on worolde ne wurdan welige ne wlance ne on lǽnan liffæce rance ne rîce *they account the worse for humble birth, those whose forefathers were not of great wealth or of high estate in the world, nor in this poor life-space proud or rich,* L. Eth. vii. 21 ; Th. i. 334, 4. [Forr þatt te33 shollden Crist forseon þurrh þe33re modignesse, þatt follc, þatt haffde beon til þa heh follc and rannc on eorþe, Orm. 9622. So were theih daungerouse for wlaunke ; And siththen bicom ful reulich, that thanne weren so ranke, Pol. Songs 341, 390.] **II.** *applied to dress, showy* (cf. *brave* in Shakspere) :—Witaþ ðæt ne môt mid rihte nân preóst beón ne on his girlum tô ranc ne mid golde oferglæncged, L. Ælfc. P. 49 ; Th. i. 386, 10. Ne gē ne sceolon beón rance mid hringgum geglengede, L. Ælfc. C. 35 ; Th. ii. 358, 5. v. ofer-ranc. **III.** *bold, valiant* (*Icel.* rakkr *courageous, bold*) :—Ðǽr mihton geseón Winceastre leódan rancne here and unearhne *a host bold and fearless,* Chr. 1006 ; Erl. 140, 26.

ranclîce ; *adv.* **I.** *showily* (v. ranc, **II**) :—Ne eówer reáf ne beó tô ranclîce gemacod, L. Ælfc. C. 35 ; Th. ii. 358, 6. **II.** *boldly* (v. ranc, **III**) :—Ymbe ða feówer tîman wē wyllaþ cýðan iungum preóstum mâ þinga ðæt hig mâgon ðē ranclîcor ðæs þing heora clericum geswutelian, Anglia viii. 312, 18. [*Icel.* rakk-liga *boldly, valiantly.*]

ranc-strǽt, e ; *f. A road in which bravery is displayed* (?) :—God ðē wǽpnum lǽt rancstrǽte forþ rûme wyrcan *God let thee with weapons work an ample road where thy bravery was shewn* (of *Abraham's rescuing Lot*), Cd. Th. 127, 17 ; Gen. 2112. v. ranc.

rand, es ; *m.* **I.** *a brink, edge, margin, shore* :—Áræs ða bî ronde rôf oretta (cf. gesæt ða on næsse nîðheard cyning, Beo. Th. 4825 : hlǽw holmwylme neáh, 4814), Beo. Th. 5069 ; B. 2538. Of ðam fúlan brôce wið westan randes æsc *to the west of the ash tree on the bank* (?), Cod. Dip. B. ii. 259, 8. [Cf. *later English* rand *border, strip,* slice :—Rawe3 and rande3, Allit. Pms. 4, 105. Randes of bakun, Piers P. Crede 763. Rand *a narrow stripe,* Jameson. Rand *the edge of the upper leather, a seam of a shoe,* Bailey. *Icel.* rönd *a stripe* : *Ger.* rand *border, edge, margin.*] **II.** the word however is used generally of a shield, denoting the whole or part of it. (1) Denoting a part, *the boss of a round shield,* cf. rand-beáh and *O. H. Ger.* rant *umbo.* The word seems to have a different meaning in Icelandic : 'á fornum skjöldum var títt at skrifa rönd þá er baugr var kallaðr, ok er við þann baug skildir kenndir.' v. Cl. and Vig. s.v. baugr. Grein gives *margo clypei* as the meaning in the following passages, but *umbo* suits the sense : see too Worsaae's Primeval Antiquities of Denmark, pp. 31–2 : 51–3, where instances of early shields are given :—Rand sceal on scylde fæst fingra gebeorh *a boss must be on a shield, a sure protection for fingers* (*which grasped the shield just behind the boss*), Menol. Fox 534 ; Gn. C. 37. Lîgýðum forborn bord wið rond *the buckler against the boss burned with the flames,* Beo. Th. 5339 ; B. 2673. (2) Denoting the whole, *a shield, buckler* [*Icel.* rönd *a shield*] :—Rand dynede, campwudu clynede, Elen. Kmbl. 100 ; El. 50. Ðonne rond and hand on herefelda helm ealgodon, Andr. Kmbl. 18 ; An. 9 : 824 ; An. 412. Hē under rande gecranc *slain he sank under his shield,* Beo. Th. 2423 ; B. 1209. Ðæt hē mē ongeán sleá, rand geheáwe, 1368 ; B. 682. Siððan ic hond and rond hebban mihte *since I could bear arms,* 1316 ; B. 656. Hond rond gefeng geolwe linde, 5212 ; B. 2609. Scyldes rond fæste gefêgan wið flyge gâres *to join together firmly the shield's disk against the flight of javelin,* Exon. Th. 297, 11 ; Crä. 65. Beorhte randas, Beo. Th. 468 ; B. 231.

Rondas regnhearde, 657 ; B. 326. Ða hî on ðone Reádan Sǽ randas bǽron, Ps. Th. 105, 8 ; Cd. Th. 199, 2 ; Exod. 332. Rincas randas wǽgon, 123, 22 ; Gen. 2049. Bæd ðæt hyra randan (randas ?) rihte heóldon, Byrht. Th. 132, 22 ; By. 20. v. bord-, calc-, gafol-, geolo-, hilde-, sîd-rand.

rand-beáh, -beág, es ; *m. The boss of a shield* or *the shield itself* ; *buculus, bucula* (cf. *bucula the boss of a shield,* Isidore), *bucularis, umbo, testudo* (cf. *scyld testudo, clipeus,* Wrt. Voc. i. 35, 57) :—Randbeáh umb[r]o, Wrt. Voc. i. 84, 33 : Ælfc. Gr. 9, 3 ; Som. 8, 34. Îsen randbeág *ferreus umbo,* ii. 147, 79. *Umbo* randbēh *vel bucula,* i. 35, 29. Randbeáh *buculus,* 288, 13. Rondbeág, ii. 11, 37. Rondbaeg, 102, 29. Randbeág *buculus vel bucularis,* 126, 65. Randbeáh *testudo,* Ælfc. Gr. 9, 3 ; Som. 8, 60. Swilce lytel pricu on brâdan brede oððe rondbeáh on scilde, Bt. 18, 1 ; Fox 62, 5. Randbeáges *umbonis,* Wrt. Voc. ii. 86, 83 ; Hpt. Gl. 521, 8. Hrandbeága *testudine,* 495, 47. Under þiccum randbeáge *subter densa testudine,* Ælfc. Gr. 47 ; Som. 48, 29. Randbeág *testudinem,* Hpt. Gl. 423, 58. Randbeágum *umbonibus,* 424, 6 : Wrt. Voc. ii. 76, 45. [*O. H. Ger.* rant-pouc.] v. rand.

rand-burh *a town that acts as a shield* (?), *a fortified town, a frontier town* (?) :—Rîce gerêfa rondburgum weóld, eard weardade, Exon. Th. 243, 32 ; Jul. 19. Randbyrig (*the walls formed by the waters of the Red Sea when the Israelites passed through it*) wǽron rofene *were riven* (*when the Egyptians attempted to cross*), Cd. Th. 207, 7 ; Exod. 463. Or are the walls formed by the water compared to the arrangement of the line of battle when the shields overlapped, called *scild-burh* q. v.? v. next word.

rand-gebeorh *a protection such as that afforded by a shield* :—Se âgend up árǽrde reáde streámas in randgebeorh *the Lord hath raised the Red Sea's waters as a protecting shield* (cf. the waters were a wall unto them, Ex. 14, 29), Cd. Th. 196, 24 ; Exod. 296.

rand-hæbbend es ; *m. One who has a shield, a warrior* :—Ôðer nǽnig sēlra nǽre rondhæbbendra, Beo. Th. 1726 ; B. 861.

rand-wîga, an ; *m. A warrior with a shield, a warrior* :—Rîce randwîga (*Æschere*), Beo. Th. 2600 ; B. 1298. Rôfne randwîgan, 3590 ; B. 1793. Randwîgena rǽst (*the camping of the Israelites on their march*), Cd. Th. 186, 5 ; Exod. 134. Randwîgum frætwa dǽlan, 171, 14 ; Gen. 2828.

rand-wîgend, -wîggend (-wiggend ?), es ; *m. A warrior with a shield, a warrior* :—Rondwîggende (*the men of Holofernes*), Judth. Thw. 21, 9 ; Jud. 11 : 21, 15 ; Jud. 20. Nû ic gumena gehwǽne ðyssa burhleóda biddan wylle randwîggendra (*the people of Bethulia*), 24, 14 ; Jud. 188 : (*the descendants of Abraham*), Cd. Th. 205, 13 ; Exod. 435.

ráp, es ; *m. A rope, cord, cable* :—Ráp *funiculus vel funis,* Wrt. Voc. i. 15, 19 : 75, 4. Ráp *vel* strenc *funiculus, modicus funis,* ii. 151, 66. Ráp *rudens,* i. 285, 18. Heó lêt hig ût mid ânum langum rápe (*per funem*), Jos. 2, 15. Rápas *funes vel restes,* Wrt. Voc. i. 56, 58 : *lora,* ii. 51, 40 : *restes,* 93, 4 : *funes,* Ps. Th. 118, 61. Hig hine gebundon mid twâm bǽstenum rápum (*novis funibus*) ... Ða rápas tôburston, Jud. 15, 13–14 : 16, 9. Hwæt beóþ ða feówere fǽges rápas ? Gewurdene wyrda, ða beóþ ða feówere fǽges rápas, Salm. Kmbl. 661–668 ; Sal. 331–333. Rápa *nodorum,* Wrt. Voc. ii. 61, 68. Rápum *rudentibus, funibus,* Hpt. Gl. 529, 27. Dū gedydest ðæt wē mǽtan ûre land mid rápum, Ps. Th. 15, 6. Swá swá hē mid gildenum rápum âhafen wǽre, Bd. 4, 9 ; S. 576, 36. Ânra gehwilc manna is gewriðen mid rápum his synna, Homl. Th. i. 208, 4. Hē worhte âne swipe of rápum (of strengum *of small cords,* Jn. Skt. 2, 15), 406, 7. [*Goth.* raip ; *n.* : *Icel.* reip ; *n.* : *O. H. Ger.* reif ; *m.*] v. ancor-, bealu-, helpend-, mǽrels-, mæst-, met-, net-, scip-, stig-, sund-, wǽl-ráp.

ráp-gang (?), es ; *m. Rope-dancing* :— Rápgong (MS. -gon. Cf. l. 33, where gegon is written for gegong [v. p. 33, 65]) *funambulus,* Wrt. Voc. ii. 38, 36. The meaning seems to require rap-gonga (-genga ?).

ráp-gewealc (?), es ; *n. A coil of rope* (?), *a cord* :— Rápe gewælc *funiculum,* Ps. Spl. T. 104, 10.

rápincel, es ; *n. A cord, string, rope* :—Rápincel *funiculus,* Cant. M. ad fil. 9. On rápincle tôdâles *in funiculo distributionis,* Blickl. Gl. : Ps. Spl. 77, 60. Rápincel *funiculum,* 104, 10. Mîn rápincel ðû âsmeádest *funiculum meum investigastis,* Ps. Lamb. 138, 2.

ráp-lîc ; *adj. Of rope* :—Ráplîc *funale.* Germ. 399, 469.

ráre-dumla, -dumle, an ; *m. f. A bittern* :—Ráredumlæ *onocrotalum, avis quae sonitum facit in aqua,* Shrn. 29, 6. Ráradumbla *onocrotalus,* Wrt. Voc. i. 62, 21. Ráredumle, 280, 26 : 63, 70 : buban, 126, 61. [*M. H. Ger.* rôr-tumel *a bittern* : *Ger.* rohr-dommel : *M. Du.* roes-domel. *O. H. Ger.* has horo-tûbil, -tumil.] [Cf. for the second part of the word, *dumble-dore,* the name given in some places to the bee.]

rárian ; *p.* ode. **I.** of human beings, *to wail, lament loudly* :— Seó dreórige môdor samod mid ðam lîcmannum rárigende hî âstrehte æt ðæs apostoles fôtum, Homl. Th. i. 66, 18. **II.** of other than human beings, *to roar, bellow* :—Hwîlum dióflu him ráredon on swá hrýðro, Shrn. 141, 10. Rárende ł bellende *rugiens,* Mt. Kmbl. p. 9, 14. Ðære rárigendan *bombosa,* Wrt. Voc. ii. 89, 8.

rárung, e; *f.* *Roaring, loud cry*; barritus, Wrt. Voc. ii. 10, 68 : 79, 29 : 125, 18.

rásettan; *p.* te *To move impetuously, to rage* (of fire) :—Hé (*Nero*) wolde fandian, gif ðæt fýr (*at the burning of Rome*) meahte swá longe reád rásettan, swá hé secgan gehérde, ðæt Troia burg ofertogen hæfde léga leóhtost, Met. 9, 14. Blác rásetteþ reáda líg, réðe scríþeþ, Exon. Th. 51, 1 ; Cri. 809. Cf. ræs.

rásian; *p.* ode *To explore* :—Ðá wæs hord rásod, onboren beága hord, Beo. Th. 4556 ; B. 2283. v. á-rásian.

raðe (*aspirated and unaspirated forms occur, and each can alliterate ; the two forms are given separately.* v. hraðe); *adv. Quickly, soon, at once, directly, without hesitation* :—Raðe ilico, Wrt. Voc. ii. 44, 68 : ocius, 64, 47. Ræðe *ultro*, 90, 8. Heó nam raðe (*cito*) hyre wæfels, Gen. 24, 65. Cwelle hy man raðe (*statim*), L. Ecg. C. 39 ; Th. ii. 164, 1. Ðæt hine mon slóge swá raðe swá mon hiora fiénd wolde *that they should kill him as soon* (*with as little compunction*) *as they would their enemies*, Ors. 1, 12 ; Swt. 52, 35. Ða men wæron swá raðe deáde swá ðæt yfel him an becom, 4, 5 ; Swt. 166, 7. Raðe ðæs *directly afterwards*, 3, 10 ; Swt. 138, 33. Hêht lífes brytta leóht forþcuman ofer rúmne grund ; raðe wæs gefylled heáhcininges hæs, Cd. Th. 8, 13 ; Gen. 123 : 95, 26 ; Gen. 1584 : Exon. Th. 93, 15 ; Cri. 1526 : Beo. Th. 1453 ; B. 724. Ðæs cymþ raðor *iste egredietur prius*, Gen. 38, 28. Ne þincþ eów nó ðý raðor (*none the sooner*) heora genóh, Bt. 13 ; Fox 38, 31 : 30, 1 ; Fox 108, 9. Nán man hit náh tó geáhnianne raðost þinga (*at the earliest*) ǽr syx mónþum æfter ðam he forstolen wæs, L. C. S. 24 ; Th. i. 390, 12. Swá ðæt cild raðost ǽnig þing specan mǽge *as soon as ever the child can speak*, Wulfst. 39, 8. Ðonne mágon wé hí swá raðost (*in the quickest manner possible*) tó ryhte gecierran, Past. 32 ; Swt. 209, 21.

ráw. v. ræw.

ráwan (?) *to cut in strips* (?) (v. ræw.) Cf. Geráwende *infindens*, Wrt. Voc. ii. 91, 24. Geráwende slítende and ceorfende *infindens*, 47, 22. Geráwen hrægel *segmentata vestis*, i. 40, 10.

rawe, Wrt. Voc. ii. 128, 39. v. rupe.

reád; *adj. Red* :—Reád deáh *coccus*, Wrt. Voc. i. 40, 40. Reád teafor *minium*, 46, 74. Se reáda telg, Exon. Th. 408, 21 ; Rä. 27, 15. Reád *ruber*, Wrt. Voc. ii. 119, 35 : *flavum, fulvum*, 108, 70 : *roseus, vel rubeus, vel pheniceus*, i. 46, 50 : *croceus*, Hpt. Gl. 524, 37. Reádde lǽmene fatu *alsierina*, Wrt. Voc. i. 41, 47. (a) Of plants or fruit :—Reáde wínberige *ceraunis*, 38, 62. Reáde clefre *calta*, 67, 72. Rǽde clæfer, 288, 49. Reád clæfré, Lchdm. ii. 312, 19. Rósena reáde heápas, Dóm. L. 18, 286. Mid reádum rósum *cum purpureis rosis*, Hpt. Gl. 511, 4. (b) of gold :—Reád gold *aurum obrizum*, Wrt. Voc. i. 38, 33 : Met. 19, 6 : Cd. Th. 219, 24 ; Dan. 59. (c) of fire, sky :—Reád líg, Dóm. L. 10, 149, 152 : Exon. Th. 51, 2 ; Cri. 810. Ðes heofon ys reád (*rubicundum*), Mt. Kmbl. 16, 2. (d) of blood :—Sió reáde ród *the bloodstained cross*, Exon. 68, 11 ; Cri. 1102. [*Goth.* rauds : *O. Sax.* ród : *O. Frs.* rád : *O. H. Ger.* rót : *Icel.* rauðr.] v. bleó-, geolu-, weolc-reád ; reód.

reáda glosses *tolia vel porunula* :—Smǽle þearmas *ilia*, reáda *tolia vel porunula*, bǽcþearm *entales*, Wrt. Voc. i. 44, 46–48.

reádan. v. on-reádan.

reád-basu; *adj. Reddish purple* :—Ðér synt ða reádbeswean blóstman grówende, L. E. I. prm. ; Th. ii. 400, 5.

reádda, an ; *m. The robin redbreast* :—Raedda *rubisca*, Wrt. Voc. ii. 119, 38. Cf. rudduc.

reáde; *adv. Redly, in red* :—Hire andwlita biþ reáde wan *livid with a red tinge*, Lchdm. ii. 348, 19. Ða wearþ beám monig blódigum teárum birunnen reáde and þicce, Exon. Th. 72, 22 ; Cri. 1176. Ic eom reáde bewæfed *I am clothed in red*, 484, 2 ; Rä. 70, 1.

reád-fáh; *adj. Red-stained, having patches of red colour* :—Wág reádfáh, Exon. Th. 476, 19 ; Ruin. 10.

reádian; *p.* ode *To be or become red* :—Ic reádige *rubeo*, Ælfc. Gr. 26, 2 ; Som. 28, 42. Reádaþ þe heofun *rutilat coelum*, Mt. Kmbl. Rush. 16, 3. Reádode *purpurescit*, Wrt. Voc. ii. 67, 8. Reádede, Hpt. Gl. 503, 51. Reádode *rubescunt*, Hymn. Surt. 52, 31. Æppel ðe ðonne gyt ne reádige, Lchdm. i. 330, 22. Smire ða ófras ðǽr hit reádige, ii. 108, 20. Reádian *rubescere*, Hymn. Surt. 49, 17. Reádiendum *rubente*, 91, 33. [*O. H. Ger.* róten *rutilare, rubere, erubescere*.] v. reódian.

reád-leáf(?); *adj. Having red leaves* :—On ða hreádleáfan ǽc, Cod. Dip. Kmbl. v. 179, 26.

reádness, e ; *f. Redness* :—Reádnyss *rubor*, Ælfc. Gr. 9, 21 ; Som. 10, 28. Seó reádnes ðæs rósan, Blickl. Homl. 7, 29. Seó reádnes ðæs swyles *rubor tumoris*, Bd. 4, 19 ; S. 589, 31. Reádnysse *ostro, purpura, vermiculo*, Hpt. Gl. 503, 49 : 522, 6.

reád-staled; *adj. Having a red stale or stalk* :—Reádstalede hárhuna, Lchdm. i. 378, 19.

reád-stán(?), es ; *m. Ruddle, red ochre* :—Rédestán *sinopide*, Wrt. Voc. ii. 120, 63. [*O. H. Ger.* rót-stein *sinopis*, Grff. 6, 688.]

reáf, es ; *n.* **I.** *spoil, booty* :—Reáf *exuviae, spolia*, Wrt. Voc. ii. 146, 33 : *exuvias*, 31, 56 : 93, 1. Weorðlíc reáf *spolia*, Ps. Th. 67, 12. Se ðe beorna reáf manige (*spolia multa*) mêteþ, 118, 162. Seó gýtsung

hyre reáf (*spolia*) on ðære wynstran sídan scylt, Gl. Prud. 56 a. Hý ðý reáfe rædan mótan, Exon. Th. 110, 5 ; Gü. 103. **II.** *raiment, a garment, robe, vestment* :—Reáf *vestis vel vestimentum vel indumentum*, Wrt. Voc. i. 81, 40 : *cultus*, 39, 70. Heó æthrán his reáfes (*vestimenti*) fnæd. Heó cwæþ sóðlíce : Ic beó hál gyf ic hys reáfes æthríne, Mt. Kmbl. 9, 20–21. Tó hwí sint gé ymbhýdige be reáfe ? 6, 28. Twegen weras on hwítum reáfe *in veste fulgenti*, Lk. Skt. 24, 4. Ne scríde nán wíf hig mid wǽpmannes reáfe ne wǽpman mid wífmannes reáfe, Deut. 22, 5. Hé scrýdde hine mid línenum reáfe *cum stola byssina*, Gen. 41, 42. Hláf tó etenne and reáf tó werigenne, 28, 20. Ðæt hálige reáf ðæt Aaron wereþ, Ex. 29, 29. Johannes hæfde reáf of olfenda hǽrum, Mt. Kmbl. 3, 4. His reáf (*vestimenta*) wǽron swá hwíte swá snáw, 17, 2: Hí sǽton on blacum reáfum weán on wênum, Cd. Th. 191, 10 ; Exod. 212. Ðá dyde heó of hire wydewan reáf *depositis viduitatis vestibus*, Gen. 38, 14. [*Laym.* reaf, ræf *a robe* : *O. Sax.* nód-róf *rapine* : *O. Frs.* ráf *robbery, booty* ; *also a pledge* : *O. Du.* roof : *O. H. Ger.* roub *spolia, praeda* : *Icel.* val-rauf *spoils taken from the slain.*] v. bed-, búr-, deáþ-, gúþ-, heaðo-, here-, lenden-, síd-, wæl-reáf.

reáfere, es ; *m. A reaver, robber, spoiler* :—Reáfere *raptor vel praedo, vel spoliator*, Wrt. Voc. i. 47, 49 : *raptor*, 76, 8 : *agressor*, 19, 7. Hreáfere *praedo, raptor*, Hpt. Gl. 501, 34. Gif hwilc þeóf oððe reáfere gesóhte ðone cyning, ðæt hé hæbbe nigon nihta fyrst, L. Ath. iv. 4 ; Th. i. 222, 26. Ueriathus wæs micel þeófmon and on ðære stalunge hé weard reáfere *Viriathus latro, primum infestando vias, deinde vastando provincias*, Ors. 5, 2 ; Swt. 216, 8. Gif ðú on hwilcum men ongitst ðæt hé byþ gítsere and reáfere, ne scealt ðú hine ná hátan man, ac wulf, Bt. 37, 4 ; Fox 192, 15. Ne sǽde ðæt godspel ðæt se ríca (*Dives in the parable*) reáfere wǽre, ac wæs uncystig, Homl. Th. i. 328, 18. Scyld sceal cempan, sceaft reáfere, Exon. Th. 341, 23 ; Gn. Ex. 130. Ic ne eom swylce óðre men, reáferas (*raptores*), Lk. Skt. 18, 11. Rýperas and reáferas and ðás woruldstrúderas, L. I. P. 2 ; Th. ii. 304, 19 : Wulfst. 165, 35 : L. C. S. 7 ; Th. i. 380, 5. [*O. E. Homl.* reaferes, reveres ; *pl.* : *A. R.* reavares : *Laym.* ræveres ; *Piers P.* reveres : *O. H. Ger.* roubari *raptor, predo* : *Ger.* räuber : *Icel.* raufari, reyfari.]

reáfian; *p.* ode. **I.** *to plunder, rifle, spoil, waste, rob* (1) a person :—Úte hí reáfaþ (*vastabit*) swurd, Deut. 32, 25. Se ðe reáfaþ man leóhtan dæge *he who robs a man by daylight*, L. Eth. iii. 15 ; Th. i. 298, 11. Æghwá mec reáfaþ, Exon. Th. 482, 4 ; Rä. 66, 2. Gé reáfiaþ (*spoliabitis*) Egipte, Ex. 3, 22. Ðenden reáfode rinc óðerne, Beo. Th. 5962 ; B. 2985. Wígfrecan wæl reáfedon, 2429 ; B. 1212. Reáfodon (*diripuerunt*) hine ealle oferfarende wæg, Ps. Spl. 88, 40. Reáfa *vastes*, Kent. Gl. 936. Gif hwylc man reáfige (*spoliaverit*) óðerne æt his dehter, L. Ecg. P. iv. 13 ; Th. ii. 208, 7. Swiðor ðonne hié reáfian earme and unscyldige, Blickl. Homl. 63, 17. (2) a place :—Ic folcsalo bærne, ræced reáfige, Exon. Th. 381, 4 ; Rä. 2, 6. Ic lond reáfige, 394, 7 ; Rä. 13, 14. Se snáw gebryceþ burga geatu, reáfaþ swiðor mycle ðonne se swiðra níð, Salm. Kmbl. 65 ; Sal. 307. Reáfiaþ hine (*the vineyard*) ealle ða farende, Ps. Spl. 79, 13 : Blickl. Gl. Hý hergiaþ and heáwaþ, rýpaþ and reáfiaþ and tó scipe lǽdaþ, Wulfst. 163, 12. Rib reáfiaþ ræðe wyrmas, Soul. Kmbl. 220 ; Seel. 113. Ic reáfode beám and ða blǽda æt, Cd. Th. 55, 28 ; Gen. 901. Ðonne man his hús reáfige (*diripiet*), Mk. Skt. 3, 27. Hord reáfian, Beo. Th. 5540 ; B. 2773. Helle weallas forbrecan, ðære burge þrym reáfian, Exon. Th. 461, 15 ; Hö. 36. **II.** *to seize, take as a robber takes* :—Reáfiaþ *rapiunt*, Kent. Gl. 4. Ic forþ ágef ða ðe ic ne reáfude ǽr *quae non rapui tunc exolvebam*, Ps. Th. 68, 5. [*Goth.* bi-raubon : *Icel.* raufa : *O. Frs.* rávia : *O. Du.* róven : *O. H. Ger.* roubón.] v. á-, be-, gereáfian.

reáfigend, es ; *m. A spoiler, a plunderer* :—Ic bidde míne æftergengan, ciningas and þeóde wealdendras, ðæt gé ne sýn cyrcean reáfgendras, ac ðæt gé sýn geornfulle bewerigendras Cristes ágenre landáre, Cod. Dip. Kmbl. iii. 350, 26.

reáfigende; *adj. Ravening, rapacious* ; rapax, Ælfc. Gr. 9, 60 ; Som. 13, 42.

reáf-lác, es ; *n. m.* **I.** *rapine, robbery, spoliation, plundering* :—Ðis synt ða ídelnyssa ðysse worlde . . . gýtsung and reáflác (*rapina*) and manslihtas, L. Ecg. P. i. 8 ; Th. ii. 174, 34. Heáfodleahtras sind . . . reáflác, gítsung . . . , Homl. Th. i. 6, 29. Ús rýpera reáflác derede swíðe, Wulfst. 159, 11. Gé synt innan fulle reáfláces *pleni rapina*, Mt. Kmbl. 23, 25. Full reáflace and unrihtwísnesse, Lk. Skt. 11, 39. Nellaþ gé tó reáflác ræda þencean *in rapinis nolite concupiscere*, Ps. Th. 61, 10. Hé weard reáflíce, and on ðæm reáflác (*in the course of his plundering*) hé him geteáh tó micelne monfultum, and monege túnas oferhergeade, Ors. 5, 2 ; Swt. 216, 8. On reáflác *in rapinam*, Wrt. Voc. ii. 45, 20. Be reáflace. Gif hwá binnan ðám gemǽrum úres ríces reáflác deó, L. In. 10 ; Th. i. 108, 8 (*where see note*). Gif ciricgrið ábrocen beó, sí hit þurh feohtlác, sí hit þurh reáflác, L. Eth. ix. 4 ; Th. i. 340, 22 : L. C. S. 48 ; Th. i. 402, 30. Gif hwá reáflác gewyrce, ágife and forgylde, 64 ; Th. i. 410, 2. Ðæt hé begange nán reáflác, Homl. Th. ii. 46, 4. Þurh rícra reáflác, Wulfst. 166, 23. [*Unwrenches, stele oðer refloc oðer drunkenesse, O. E. Homl.* ii. 79, 29. Þe vox of giscunge

haueð þeos hweolpes . . . þeof þe, reflac . . . , A. R. 202, 19. Þe king his ræflac makede (his lond al forverde, 2nd MS.), Laym. 9939. Ðeft and reflac ðhugte him no same, Gen. and Ex. 436.] II. *what is taken, spoil, booty, plunder* :—Reáflác *preda*, Wrt. Voc. i. 35, 39 : ii. 146, 33. Ælc bit ðæs reáflác es ðe him on genumen biþ, oððe eft oðres gítsaþ, Bt. 26, 2 ; Fox 92, 17. Man wolde biddan ðæs reáflác es ðæt hé hit sciolde ágyfan and forgyldan, Chart. Th. 289, 27. Ágife hé ðone reáflác *he shall restore what has been seized*, L. In. 10 ; Th. i. 108, 9. Tódǽlan reáflac *dividere spolia*, Ps. Spl. T. 67, 13.

reáfol ; *adj. Rapacious* :—Reáfol *captator*, Germ. 397, 19. Cild ácenned þriste reáful ofermód him sylfum gelícigende *a child born on the thirteenth day of the moon will be bold, rapacious, arrogant, pleasing himself*, Lchdm. iii. 190, 14. v. next word.

reáfolness, e ; *f. Rapacity* :—Reáfulnesse *rapacitatis*, Hpt. Gl. 508, 44.

reáfung, e ; *f. Plundering, spoiling* :—Atheniensum se sige and seó reáfung ðæs Persiscan feós tó máran sconde wurde forðon siððan hié welcgran wǽron hié eác bleáðran gewurdon *castra regiis opibus referta ceperunt, non parvo quidem antique industriae damno. Nam post hujus praedae divisionem, aurum Persicum prima Graeciae corruptio fuit*, Ors. 2, 5 ; Swt. 84, 21.

reám, es ; *m. Cream* :—Wið ðon ðe mon blóde hrǽce and spíwe ; genim god beren mela and hwít sealt, dó on reám oððe góde fléte, Lchdm. ii. 314, 2. [Cristened we weore In red rem Whon his bodi bledde on þe Beem, H. R. 146, 144. Ream (*subst. and verb*) *cream*, Jamieson. *See also Halliwell's Dict. where instances of milkes rem are given under* ream. *Du.* room : *M. H. Ger.* roum : *Icel.* rjómi.]

reáma. v. reóma.

réc, es ; *m. Reek, smoke* :—Réc *fumus*, Wrt. Voc. i. 284, 18 : 66, 45 : ii. 36, 54. Of ðære stówe steám up árás swylce réc, Elen. Kmbl. 1604 ; El. 804. Réce hí gelícast geteoriaþ *sicut deficit fumus, deficiant*, Ps. Th. 67, 2. Geondfolen fýre, réce and reáde lége, Cd. Th. 3, 31 ; Gen. 44. In onlícnesse uppástígendra yselena mid réce *instar favillarum cum fumo ascendentium*, Bd. 5, 12 ; S. 628, 23. Hé geseah ðone líg ðæs fýres and ðone réc ofer ðære burge wallas áhefenne, 3, 16 ; S. 543, 2 : Cd. Th. 155, 26 ; Gen. 578. Bráde lígas, swilce eác ða biteran récas, 21, 17 ; Gen. 325. Ic folcsalo bærne, récas stígaþ haswe ofer hrófum, Exon. Th. 381, 5 ; Rä. 2, 6. [*O. Frs.* rék : *O. Sax.* wíh-rók : *O. Du.* rook : *O. H. Ger.* rouh : *Icel.* reykr.] v. swefel-, wæl-, wudu-réc.

récan ; *p.* réhte *To smoke* (trans.), *steam* :—Réhte (Wrt. reþte ; Wülck. reohte) *fumarat*, Wrt. Voc. ii. 151, 55. Ðám mannum ðe fram ðære teóþan tíde ne geseóþ, ðæs ylcan drinces smýc heora eágan onfón and mid ðam broþe récen, and ða lifre wǽten, and gníden and mid smyrgen, Lchdm. i. 346, 22. [*O. H. Ger.* rouhan ; *p.* ta *thurificare, suffire, vaporare* : *Icel.* reykja ; *p.* ta *to smoke* (trans.).] v. reócan.

récan, réccan (reccan ?) ; *p.* róhte *To care, reck,* (1) with gen. :—Ne can ic eów ne ic eówer récce *I know you not and I care not for you*, L. Ælfc. P. 40 ; Th. ii. 380, 3. Ðú ǽfre ne récst ǽniges þinges (cf. ðú ne wilnast nánes óðres þinges, Bt. 23 ; Fox 80, 2) ofer ða áne, Met. 12, 31. Biþ micel rǽd ðam ðe his sylfes récþ, Homl. Skt. i. 12, 122, 132. Se deáþ swelces ne récþ, Bt. 19 ; Fox 68, 32. Ne wǽpna ne récceþ, Beo. Th. 873 ; B. 434. Ne récceþ hí ðara metta, Bt. 25 ; Fox 88, 19. Hí habbaþ cornes swá fela swá hí mǽst récceaþ (réccaþ, MSS. P. S.) *they have as much corn as ever they care for*, Lchdm. iii. 254, 5 : Wulfst. 132, 21. Wé willaþ nú on Englisceum gereorde secgean ðám ðe his (*the book*) récceaþ, Basil prm. ; Norm. 32, 14. Hwæt róhte ic hwæðer ic wǽre gyf ic ne lyfde, oððe hwæt róhte ic ðæs lýfes gyf ic náwiht nyste, Shrn. 194, 2. Hé lǽrde ðæt ða þearfan ne wénden ðæt God heora ne róhte, Ps. Th. arg. 48. Ðǽr læig ðæt reáf beæftan, forðon ðe hé ne róhte ðæs eorþlícan reáfes, syððan hé of deáþe árás, Homl. Th. i. 224, 4. Feores hí ne róhton, Byrht. Th. 139, 27 ; By. 260. Hié ðæs ne róhton, Cd. Th. 79, 31 ; Gen. 1319 : 228, 13 ; Dan. 201 : Exon. Th. 88, 17 ; Cri. 1441. Gif ðú ðínes feores récce, 119, 30 ; Gú. 262. Gif ðú aldres récce, Cd. Th. 160, 27 ; Gen. 2656. Gif hwelc wíf forlǽt hiere ceorl and nimþ hire wénestú récce hé hire ǽfre má (*numquid revertetur ad eam ultra?*) Past. 52, 3 ; Swt. 405, 12 : L. Alf. 12 ; Th. i. 46, 15 : L. A. G. prm. ; Th. i. 152, 6. (1 a) used impersonally, with acc. of person :—Hí ðæs metes ne récþ (cf. above, Bt. 25 ; Fox 88, 19), Met. 13, 45. [Cf. me ne reccheð (naut I ne recche, MS. C.) *non requiro*, A. R. 104, 21.] (2) with a preposition :—Ðú eart sóðfæst and ðú ne récst be ǽnegum menn (*non curas quemquam*), Mk. Skt. 12, 14. (3) with a clause :—Ne récþ God, ðeáh ic ðus dó *non requiret Deus*, Ps. Th. 9, 33. Hwæt réce wé hwæt wé sprecan *quid curamus quid loquamur?* Coll. Monast. Th. 18, 14. Gé ne récceþ hweðer gé áuht tó góde dón, Bt. 18, 4 ; Fox 66, 20. Hié ne récceaþ hwæðer, Past. 19, 2 ; Swt. 145. 21. Se cyng ne róhte ná hú swíðe synlíce ða geréfan hit þbegeátan, ne hú manige unlaga hí dydon, Chr. 1086 ; Erl. 220, 12. Hí woldon on elþeódignesse beón, hí ne róhton hwǽr, 891 ; Erl. 88, 8. Men ne róhton hwæt hý worhtan, Wulfst. 163, 16. [*Laym.* rehchen, recchen (*with gen.*): *O. and N.* recche ; recþ (*3rd pers.*) : *Piers P. Chauc.* recche, rekke : *Havel.* recke : *O. Sax.* rókian : *O. H. Ger.* ruohian : *Icel.* rœkja.]

reccan ; *p.* reahte, rehte. I. *to stretch, extend* :—Wið hǽr-scearde . . . onsníð mid seaxse, seówa mid seolce fæste . . . gif tósomne teó réce mid handa *for harelip . . . cut with a knife, sew fast with silk . . . if there be contraction (where the stitches are) smooth out with the hand*, Lchdm. ii. 56, 9. II. *to hold out to another, to give* ; porrigere :—Hærfest tó honda hérbúendum rípa receþ (cf. hærfest bryngþ rípa bléda, Bt. 39, 13 ; Fox 234, 15), Met. 29, 63. Eall ðæt ofer biþ tó láfe on heora weoruldspédum árfæstum and gódum is tó recceanne and tó syllanne *omne quod superest, in causis piis ac religiosis erogandum est*, Bd. 1, 27 ; S. 489, 27. III. *to stretch one's steps, to tend, to go, stray* :—Hé nát hwider hé recþ mid ðam stæpum his weorca *quo gressus operis porrigat, nescit*, Past. 11, 1 ; Swt. 65, 9. Gif hé (*a close*) biþ untýned, and recþ (receþ, MS. H.) his neáhgebúres ceáp in on his ágen geat, L. In. 40 ; Th. i. 126, 14. [Swa sone swa heo mihten ut of scipe heo rehten, Laym. 25646.] IV. *to unfold a tale, to narrate, recite, tell, say* :—Recceo *alligeo* (*allego*), Txts. 39, 139. Ic recce (*narrabo*) ealle wundra ðíne, Ps. Spl. 9, 1. Ic ðe má be Gode recce, Bt. 35, 3 ; Fox 158, 9. Ðonne hé eall ðis recþ and sægþ, Blickl. Homl. 91, 14. Hwæt synt ða spǽca ðe gyt recceaþ (*confertis*) inc betwýnan, Lk. Skt. 24, 17. Hé rehte him óðer bigspel *aliam parabolam proposuit illis*, Mt. Kmbl. 13, 24, 31. Ðá reahte heora ǽgðer his spell *each of them told his tale*, Chart. Th. 170, 14. Hé him his earfoþa rehte, Guthl. 19 ; Gdwin. 76, 19. Ymb ðæt reahte Paulus swíðe wel *quod bene Paulus exprimit*, Past. 51, 1 ; Swt. 395, 11. Rehte, 51, 2 ; Swt. 395, 26. Hé him rehte hú myccle scipbrocu hé gebád, Blickl. Homl. 173, 6. Spell ðæt ús reahte Platon, Met. 22, 53. Rehte, Beo. Th. 4226 ; B. 2110. Hé Dryhtnes lof reahte, Exon. Th. 111, 23 ; Gú. 131. Ðam wit rehton (*narravimus*) uncer swefen, Gen. 41, 12. Ne nán ne dyrstlǽce ðæt hé óðrum recce, oððe mid wordum gecýðe, hwæt hé bútan mynstre geseah, R. Ben. 128, 4. Reccan *expedire*, Wrt. Voc. ii. 31, 26. Ic ðe mæg reccan sum spell, Bt. 38, 1 ; Fox 194, 1. Reccan race, 38, 6 ; Fox 208, 4. Bigspell reccan *in parabolis loqui*, Mk. Skt. 12, 1. Reccean and secggan, Blickl. Homl. 55, 28. Godes béc reccean and rǽdan, and godspell secggean, 111, 17. Reccean ymbe Dauides dǽda sume, Past. 28 ; Swt. 196, 10. Sint tó recceanne ða godcundan cwidas *divinae sententiae proferendae sunt*, 37, 2 ; Swt. 265, 22. Tó lang ys tó reccenne *too long to tell*, Beo. Th. 4192 ; B. 2093. Hé his intingan wæs reccende *causam dicturus*, Bd. 5, 19 ; S. 639, 19. Reccendes *loquentis, narrantis*, Hpt. Gl. 460, 68. [Ic þe wulle ræcchen (telle of, 2nd MS.) deorne runen, Laym. 14679. V. *to unfold the meaning of anything, to explain, interpret, expound* :—Eall hé his leorningcnihtum ásundron rehte (*disserebat*), Mk. Skt. 4, 34. Rehte *interpraetabatur*, Lk. Skt. 24, 27. Ðá wæs ic ungleáw ðæs geþeódes . . . ðá rehte hit me bisceop and sægde, Nar. 29, 16. Hú gleáwlíce hé ðæt swefen rehte *quod prudenter somnium dissolvisset*, Gen. 40, 16. Rece, wísworda gleáw, hwæt sió wiht síe, Exon. Th. 415, 19 ; Rä. 33, 13. Hé begann se deófol tó reccanne hálige gewritu and hé leáh mid ðære race *here the devil began to expound holy writ, and he was false in his exposition*, Homl. Th. i. 170, 4. [Ðe king him bad ben harde and bold, If he can rechen ðis dremes wold ; He told him quat him drempte o niht, And Josep recheðe his drem wel rigt, Gen. and Ex. 2121–4.] VI. *to unravel a difficult case, give a solution of a difficult question* :—Wé sǽdan hú wé hit reahtan and be hwý wé hit reahtan *we said what decision we had come to in the case, and on what grounds we had come to it*, Chart. Th. 171, 5–7. VII. *to rule, direct, guide* :—Eal ic under heofones hwearíte recce, Exon. Th. 424, 3 ; Rä. 41, 33. Ðú recest (*reges*) hí, Ps. Spl. 2, 9. Ðú ðe reccest and rǽdest *qui regis*, Ps. Th. 79, 1. Hé rǽt ús and recþ *reget nos*, 47, 12 : Mt. Kmbl. 2, 6. Receþ *regit*, Bd. 5, 18 ; S. 635, 34. Ðes cásere framlíce rehte ða cynewísan *fortissime rempublicam rexit*, 1, 5 ; S. 476, 7. Hé Ispania heóld and rehte *Hispaniam regebat*, 1, 8 ; S. 479, 29 : 2, 2 ; S. 500, 10 : 4, 27 ; S. 603, 35. Justus reahte ða gesomnunge æt Hrofes ceastre, 2, 7 ; S. 509, 10. Steóran and reccan ðone anweald ðe mé befæst wæs, Bt. 17 ; Fox 58, 27. Sealde hé ðæt mynster tó reccanne his bréðer, Bd. 3, 23 ; S. 555, 15. Tó healdanne and tó reccanne micelne dǽl ríces, 5, 19 ; S. 638, 3. Hé ða cyricean wæs reccende and stýrende, S. 639, 12. VII a. *to correct* :—Seó cyrice sum þing þurh wælm receþ (*corrigit*), 1, 27 ; S. 491, 30. [Goth. uf-rakjan *to stretch out* : O. Sax. rekkian : O. H. Ger. reckian *tendere, extendere, expandere, porrigere, narrare, explicare, disserere* : Icel. rekja *to unwind, spread out, unfold*.] v. á-, and-, be-, ge-, ofer-reccan.

-recce. v. earfoþ-recce.

récce-leás. v. réce-leás.

reccend, es ; *m. A ruler, governor.* (1) applied to the Deity :—God eálá ðú micele reccend (*rector*), Hymn. Surt. 72, 1 : Exon. Th. 2, 12 ; Cri. 18. Þeóda reccend, Ps. Th. 101, 1. God is ealra þinga reccend, Bt. 35, 5 ; Fox 166, 9. Dryhten úre reccend is hé ðara læssena ríca reccend is, Ors. 2, 1 ; Swt. 58, 22–25. God is scyppend and reccend ealra his gesceafta, Blickl. Homl. 185, 27 : Met. 4, 30. Ealra gesceafta reccend and stýrend, Wulfst. 255, 17. Án metod, reccend and ríce, Cd. Th. 252, 17 ; Dan. 580 : Exon. Th. 422, 8 : Rä. 41, 3. (2) used of earthly rulers :—Ðæt folc biþ gesǽlig þurh snoterne cyning, sigefæst and

gesundful þurh gesceádwîsne reccend, Homl. Th. ii. 320, 2. Nis ðeós þeód wyrðe ðæt hî swylcne reccend and cyning (*as Oswine*) habban, Bd. 3, 14; S. 541, 8.

reccend-dóm, es; *m. Ruling, directing, governance :*—Reccendóm (recen-, other MSS.) *regimen*, Ælfc. Gr. 9, 12; Som. 9, 30. Ús (*priests*) befæst is seó gýming Godes folces and se recenddóm heora sáwla, L. E. I. 1; Th. ii. 402, 10. Be ðære byrðenne ðæs reccendómes (reccen-, Cott. MSS.) *de pondere regiminis*, Past. 3, tit.; Swt. 33, 4. Recendómes, 17, 7; Swt. 119, 4. Se underféng sáula reccendómes *animas suscepit regendas*, R. Ben. 14, 11. *Rex* kyning is gecweden *a regendo*, ðæt is, fram reccendóme, Ælfc. Gr. 50; Som. 51, 40. Cyninge is nama gesett of sóðum reccendóme, Homl. Th. ii. 318, 33. v. rece-dóm.

recceness, e; *f. An interpretation, explanation :*—Sóð reccenise *vera interpretatio*, Mt. Kmbl. p. 2, 6.

reccere, es; *m.* **I.** *speaker, rhetorician.* v. racu, III. **II.** *an interpreter.* v. swefn-reccere. **III.** *a ruler, director :*—Hú se láreów (*rector*) sceal bión clæne on his móde. Se reccere (*rector*) sceal bión simle clæne on his geþohte, Past. 13, 1; Swt. 75, 18–19. Se reccere, se ealdormonn, 17, 1; 107, 5, 8. Ðone ealdordóm ðe se reccere for monigra monna þearfe underféhþ, 17, 7; Swt. 119, 6. Offa Mercene reccere, Cod. Dip. B. i. 340, 10. Recceras *presbiteri*, Wrt. Voc. ii. 67, 14. [O. H. Ger. *rechari executor, doctor, assertor*.] v. freá-reccere.

reced. v. ræced.

rece-dóm, es; *m. Ruling, governance, guiding :*—Recedóm (recendóm?) *regimen, dominium*, Hpt. Gl. 412, 69.

réce-leás; *adj. Careless, reckless :*—Réccileás *prefaricator*, Wrt. Voc. ii. 118, 8. Ymb ða gýmene his écre hælo hé wæs tó sæne and tó réceleás *erga curam perpetuae suae salvationis nihil omnino studii & industriae gerens*, Bd. 3, 13; S. 538, 19. Tó hwam wurde ðú swá réceleás ðæra gyfena ðe ic ðé geaf? Wulfst. 258, 13 note. Hié ne wéndon ðætt æfre menn sceolden swæ réceleáse (récce-, Cott. MSS.) weorðan, Past. pref.; Swt. 5, 23. Ðú wéndest ðæt steórleáse men and réceleáse wæron gesælige *nequam homines atque nefarios felices arbitraris*, Bt. 5, 3; Fox 14, 1. Se démþ stíðne dóm ðám réceleásum æt ðam æfterran tócyme, Homl. Th. i. 320, 18. Gif hé hwíltídum ðám réceleásum stýrþ, ðonne sceal his steór beón mid lufe gemetegod, ii. 532, 12. [ʒiff þatt he wære reckelæs to ringenn hise belles, Orm. 932. Ðe unwreste herde (*iners pastor*) syneʒeþ on gemeleste alse he þat is recheles, O. E. Homl. ii. 39, 19. Alle beoð untohene and rechelese hinen, bute ʒef he ham rihte, i. 245, 27. O. H. Ger. *ruahha-lôs negligens : Ger.* ruch-los.]

réceleásian; *p. ode To be negligent or careless :*—Tó hwon réceleásedest ðú ðære gife ðe ic ðé geaf? Wulfst. 258, 15. [O. H. Ger. *ruahha-lôsôn negligere*.]

réceleáslíce; *adv. Negligently, carelessly, without attention :*—Nis ús náwht récceleáslíce tó gehíranne *neque negligenter audiendum est*, Past. 57, 4; Swt. 439, 31.

réceleásness, e; *f. Carelessness, negligence :*—*Improvidentia*, ðæt is réceleásnys, Wulfst. 52, 18. Ongeán ðam wíslícan ræde se wiðerræda deófol sæwþ réceleásnesse, and eác gemacaþ ðæt se man þurh leáse hiwunge déþ swylce hé rædfæst sý, 53, 7. Ic andette mínes módes réceleásnessa Godes beboda, L. Edg. C. 8; Th. ii. 262, 32.

réceleást, e; *f. Carelessness, negligence, heedlessness :*—Swá hwæs swá his irsung willaþ, ðonne gehét him ðæs his réccelést, Bt. 37, 1; Fox 186, 24. Réccelést, Met. 25, 53. Hí for heora slæwþe and for gímeléste and for récceléste forléton unwriten ðara monna þeáwas ðe on heora dagum foremæroste wæron, Bt. 18, 3; Fox 64, 34. Ðæt hé swá stíere ðæm ungeþyldegum irsunga swá hé ðone hnescan þafettere on récceléste ne gebrenge *sic ab impatientibus extinguatur ira, ut remissis ac lenibus non crescat negligentia*, Past. 60; Swt. 453, 25. [Þurh mannes gémeléste and þurh mannes recheleste, O. E. Homl. ii. 45, 4.]

récels, es; *n. Incense :*—Him lác brohton ðæt wæs gold récels (récils, Rush.) and myrre (récels, Lind.) *obtulerunt ei munera, aurum, tus, et murram*, Mt. Kmbl. 2, 11: Homl. Th. i. 78, 28. Récels *thymiama, odoramentum incensi*, Hpt. Gl. 442, 1: *incensum*, Rtl. 88, 30: Bt. 38, 2; Fox 196, 32. Roecels, Lk. Skt. Lind. I, 9. Ic eom on stence strengre ðonne rícels, Exon. Th. 423, 19; Rä. 41, 24. Rícels *incensum*, Ps. Th. 140, 2. Récilc *balsamum, myrra*, Rtl. 65, 39, 41: 68, 30. Rǽcelc (?) *thuribulum*, 70, 27. Récelces *myrrae*, 4, 13. Réceles, Jn. Skt. Lind. 19, 39. Genim ðás ylcan wyrte for rýcels (récels, MS. O.), Lchdm. i. 302, 6. [O. E. Homl. recheles: A. R. rechles: Orm. recless: Prompt. Parv. rychellys, richelle *thus, incensum* : Icel. reykels.]

récels-búc, es; *m. A vessel for holding incense :*—Rýcelsbúce acerrâ (cf. fæte oððe glédfæte accerrâ, 5, 66: *hec acerra* a schyp for censse, i. 230, col. 2), Wrt. Voc. ii. 9, 36. [*Perhaps the following should be put here :*—Of ðam æscene ðe is óðre namon hrýgilebúc gecleopad, Chart. Th. 439, 26.]

récels-fæt, es; *n. A censer :*—Þriéféste rícelsfæt *cythropodes*, Wrt. Voc. ii. 15, 60. Nim ðín récelsfæt *tolle thuribulum*, Num. 16, 46. Fýr ofslóh ða óðre ðe offrodon ðone stór ðær hig heóldon ða récelsfatu, 16, 35. [O. E. Homl. rechel-fat: Orm. recle-fatt: Gen. and Ex. recle-fat.]

---

récelsian; *p. ode To cense with incense :*—Récelsa hine and séna gelóme, Lchdm. ii. 344, 18. [Cf. Zacharie gede in þe temple mid his rechelfat to rechelende þe alter, O. E. Homl. ii. 133, 36.]

récels-reóce (?), ap; *f. Burning incense :*—On ðone tíman man offrode on ðære ealdan ǽ, and mid récelsreócan on ðam temple ðæt weofod georne weorðode, Btwk. 218, 8.

recen; *adj.* **I.** *ready, prompt.* v. recenian :—Mæg sige syllan se ðe symle byþ recon and rædfæst, Wald. 108; Vald. 2, 26. [Cumeð her forð, and beð alle reken, And lereð wel quat he sal speken, Gen. and Ex. 3485. Louerd, ic (*Moses*) am unreken of wurdes, 2817. My rankor refrayne for þy reken (*apt*) wordes, Allit. Pms. 60, 756. (*See the glossary for other instances. See also* rekenli *in the same work, and in Sir Gawayne.*) O. Frs. rekon (*of a road which is clear*): L. Ger. reken. v. Richthofen. Cf. O. Sax. rekôn *to make ready, set in order*.] **II.** *swift, quick* (cf. recene) :—Blâc ræsetteþ recen reáda líg réðe scríþeþ geond woruld *bright and swift rushes the red flame, fierce strides through the world*, Exon. Th. 51, 2; Cri. 810. v. full-recen. **III.** *coming swiftly and so causing terror* (? cf. fǽr *and its compounds*) :—God ðe on Ægyptum æðele wundur worhte and recene wundar on ðam Reádan Sǽ *Deus qui fecit magnalia in Aegypto, terribilia in Mari Rubro*, Ps. Th. 105, 18.

recendóm. v. recend-dóm.

recene; *adv. Quickly, straightway, at once :*—Recene (recone, Lind.) *protinus*, Mk. Skt. Rush. 1, 29. Hét him recene tó his sunu gangan, Cd. Th. 53, 20; Gen. 864: 134, 41; Gen. 2228. Ðú nú recene beheald *intende*, Ps. Th. 29, 1. Recone ł sóna *confestim*, Mk. Skt. Rush. 5, 29: *cito*, 9, 39: *statim*, Lind. 14, 45. Recune (recone, Rush.) *continuo*, Jn. Skt. 4, 27. Yrn ricene forðan ðe se streám berþ áweg Placidum, Homl. Th. 160, 7: Cd. Th. 309, 12; Sat. 708. Saga ricene mé hwǽr seó ród wunige, Elen. Kembl. 1243; El. 623: 1211; El. 607. Ic ðonne ricene reste syððan, Ps. Th. 54, 6. Ricone, Beo. Th. 5958; B. 2983. Rycene, Ps. Th. 108, 11. Ne sceal næfre his torn tó rycene beorn of his breóstum ácýðan, Exon. Th. 293, 7; Wand. 112. Ðæt hé recenust tó þrowunge becóme *ad martyrium ocius pervenire*, Bd. 1, 7; S. 478, 11. v. recen, recenlíce.

recenian; *p. ode To arrange, dispose, reckon.* [Cristess kinn o modere hallfe be weppmann shollde reccnedd ben, Orm. 2055. Alle sunnen sunderliche ne muhte no mon rikenen, A. R. 210, 7. Him ne poruayþ of his receninge, and wel wot þet rekeni him behoueþ, Ayenb. 19, 6. Reknyñ or cowntyn, rekkyn, rekene, *computo*, Prompt. Parv. 428. O. Frs. rekenia *to reckon* : O. H. Ger. rehhanôn *parare, rationem ponere, disponere* : cf. Goth. rahnjan.] v. ge-recenian, recen.

recenlíce; *adv. Quickly, immediately, at once, straightway :*—Eodun hreconlíce (*cito*) from byrgenne, Mt. Kmbl. Lind. 28, 8. Hreconlíce (ricenlíce, Rush.) *protinus*, Mk. Skt. Lind. 1, 18. Reconlíce (ricenlíce, Rush.) *continuo*, 1, 31. Reconlíce (recunlíce, Rush.) *protinus*, 6, 25. v. recen.

recenness, e; *f. A narrative, history :*—Recennysse *historiae*, Hpt. Gl. 474, 30. v. ge-recedness.

recettung, e; *f. Eructation :*—Recetunge *eructantia*, Ps. Spl. C. 143, 16: Ps. Surt. 143, 13. v. rocettan.

recu, e; *f. Guidance, direction, correction* (v. reccan, VII, VII a) :—Seó (*Hilda*) gódre rece and híhd intingan þegnade *occasionem salutis et correctionis ministravit*, Bd. 4, 23; S. 594, 42.

(-)réd(-), rede, réde-stân. v. (-)rǽd(-), reðe, reád-stân.

rédian (?) *to furnish, provide :*—Noe ongan Nergende lác rædfæst rédian (MS. reðran. *Bouterwek suggests* rénian), Cd. Th. 90, 20; Gen. 1498. v. á-rédian.

réfa, an; *m. A prefect :*—Ðá hét Ualerianus se réfa hí ácwellan, Shrn. 121, 26. v. ge-réfa.

réf-land. v. sundor-geréfland.

regen. v. regn-.

regn, rén, es; *m. Rain :*—Blódig regn and fýren fundiaþ ðás eorþan tó forswylgenne and tó forbærnenne, Blickl. Homl. 93, 3: 91, 34. Nǽnig reng on ðám stówum ne com, Bd. 4, 13; S. 582, 28. Rén *pluvia*, Wrt. Voc. i. 52, 43. Fǽrlíc rén *imber*, 52, 63. Se rén weard forboden, Gen. 8, 2. Ðá com rén (regn, Lind.: rægn, Rush.) *descendit pluvia*, Mt. Kmbl. 7, 25. Nǽnig dǽl regnes ne ungewidres in cuman ne mæg, Blickl. Homl. 125, 33. Hé áríman mæg rægnas scúran dropena gehwelcne, Cd. Th. 265, 22; Sat. 11. Þurh dropunge deáwes and rénes, Ps. Th. 64, 11. Líget hé tó regne wyrceþ *fulgura in pluviam fecit*, 134, 7. Mid heofonlícon réne, Bt. 7, 3; Fox 22, 13: Met. 7, 23. Wilsumne regn wolcen bringeþ, Ps. Th. 67, 10: 146, 8: Cd. Th. 82, 34; Gen. 1372. Rén, Gen. 7, 4: Met. 7, 14, 21. Ðás windas and ðás regnas syndon ealle his, Blickl. Homl. 51, 20. Regna scúr, Cd. Th. 252, 10: Dan. 576. Nalles wolcnu regnas bæron, 14, 4; Exon. Th. 213. Regnas (rénas, Ps. Spl.), Ps. Th. 104, 28. Ðá ábæd se wítega æt Gode ðæt hé sceolde him rénas forgyfan, Lchdm. iii. 276, 21. [Goth. rign; *n.*: Icel. regn; *n.*: O. Sax. regin, regan: O. Frs. rein : O. H. Ger. regan.] v. morgen-, wæl-regn.

regn-, *in the compounds* regn-heard, -meld, -þeóf, -weard *has an*

intensive force, implies greatness, might. The word occurs as part of many proper names, e. g. Rǽdwoldes sunu wæs Regenhere geháten, Bd. 2, 12 ; S. 515, 10. Some of these e. g. *Reginald* are still used. [Cf. *Goth.* raginôn *to rule ;* ragineis *a ruler, counsellor ;* ragin *ordinance, counsel : Icel.* regin ; *pl. n.* (in ancient poems) *the gods, the rulers of* the universe ; forming part of compounds, *mighty, great ;* ragn-, rôgn-*in proper names : so O. Sax.* regin- : *O. H. Ger.* ragin-, regin- *in proper names,* v. Grff. ii. 384.]

regnan. v. rignan.

regn-boga, an ; *m. A rainbow :*—Rênboga *iris,* Wrt. Voc. i. 52, 42 : *yris vel arcus,* 76, 33. Hwî wæs se rênboga tô wedde gesette mancynne ? God gesette ðone rênbogan tô wedde tô ðam beháte ðæt hé nǽfre eft nolde ealne middaneard mid nánum flôde oferteón . . . Se rênboga cymþ of ðam sunbeáme and of wǽtum wolcne, Boutr. Scrd. 21, 19–26. Ic sette mînne rênbogan (*arcum*) on wolcnum, Gen. 9, 13. [*O. H. Ger.* regan-bogo : *Icel.* regn-bogi.]

regn-dropa, an ; *m. A raindrop :*—Hagol cymþ of ðam rêndropum ðonne hî beóþ gefrorene up on ðǽre lyfte, Lchdm. iii. 278, 19. [*O. H. Ger.* regan-tropfo.]

regn-heard ; *adj. Exceedingly hard, wondrous hard :*—Rondas regnhearde, Beo. Th. 657 ; B. 326. [Cf. *Icel.* regin-djúpr *very deep ;* regindjúp *the mighty deep : O. H. Ger.* Regin-hart.] v. regn-.

regnian ; *p.* ode *To set in order, arrange, dispose, regulate :*—Tungelcræftum *Chaldaeorum,* scincræfta *hierophantorum,* ða ðæt womfreht rêniaþ *ariolorum,* wyrmgalera *marsorum,* Wrt. Voc. ii. 82, 6–9. Gemyne ðu, mucgwyrt, hwæt ðu ámeldodest, hwæt ðu rênadest, Lchdm. iii. 30, 29. Hû geworhte ic ðæt ðæt ðú mê ðus swîðe searo rênodest *how have I deserved that you should lay such a snare for me?* Cd. Th. 162, 9 ; Gen. 2678. Inwitnet ôðrum bregdan, dyrnum cræfte deáþ rê[nian], Beo. Th. 4343 ; B. 2168. Sum biþ searocræftig goldes and gimma ðonne him gumena weard háteþ máððum rênian *one is a cunning workman in gold and gems, when a prince of men bids him set a jewel,* Exon. Th. 296, 33 ; Crä. 60. Wrôhtas tô webgenne, ne searo tô rênigenne *to set a trap,* Blickl. Homl. 109, 30. Hé geseh twegen ôðre gebrôðru remigende (rênigende (?) : *the later MSS. have* reniende, renigende ; *the Lindisfarne MS. glosses* reficientes *by* geboeton ł gestricedon) hyra nett, Mt. Kmbl. 4, 21. [*Goth.* raginôn *to rule.*] v. be-, ge-regnian (-rênian).

regniend, es ; *m. One who arranges :*—Rihtes rêniend, Elen. Kmbl. 1756 ; El. 880. v. *preceding word.*

regnig ; *adj. Rainy :*—Hit wæs rênig weder, Exon. Th. 380, 18 ; Rä. 1, 10. Rênig sumer, Lchdm. iii. 162, 33.

regn-líc ; *adj. Rainy :*—Rênlíc *pluvialis,* Ælfc. Gr. 9, 28 ; Som. 11, 36. Rînlícum (rên-) *pluvio, pluviali,* Germ. 401, 14. Ða regenlícan weter *pluviales aquas,* Ps. Surt. 77, 44. [*Icel.* regn-ligr.]

regn-meld, e ; *f. A mighty, solemn announcement :*—Gemyne ðu, mucgwyrt, hwæt ðu ámeldodest æt regenmelde, Lchdm. iii. 30, 30. v. regn-.

regn-scúr, es ; *m. A shower of rain, a shower :*—God sende byrnende rênscúr, Gen. 19, 24. On Ægipta lande ne cymþ nǽfre nán winter ne rênscúras, Lchdm. iii. 252, 20 : Homl. Th. i. 64, 30. Ic ofteó mîne rênscúras, ii. 102, 33. God sylþ rênscúras ðam rihtwísum and ðam unrihtwísum, 216, 19. Rênscúras *imbres,* Ps. Spl. 77, 49. [*Icel.* regn-skúr ; *f.*]

regn-þeóf, es ; *m. An arch-thief :*—Regnþeóf ne lǽt [mê] on sceade sceððan, Exon. Th. 453, 14 ; Hy. 4, 14. Swá nú regnþeófas ríce dǽlaþ (cf. regintheobos farstelad (Mat. vi. 19). Hel. 1646), Cd. Th. 212, 12 ; Exod. 538. [Cf. *also O. Sax.* regin-skaðo.] v. regn-.

regn-wæter, es ; *n. Rain-water :*—Gefulle mid rênwætere, Lchdm. ii. 26, 24. Baþu of rênwætere, 222, 12. [*O. H. Ger.* regan-wazar : *Icel.* regn-vatn.]

regn-weard, es ; *m. A mighty guard :*—Yrre wǽron begen rêðe rênweardas (*Beowulf and Grendel*), Beo. Th. 1544 ; B. 770. [Cf. *O. H. Ger.* Ragin-wart.] v. regn-.

regn-wyrm, es ; *m. An earth-worm :*—Regnwyrm *lumbricus,* Wrt. Voc. ii. 113, 26 : 71, 13. Rênwyrm, 51, 23. Rênwyrm *vel* angeltwicce, i. 24, 31. [*O. H. Ger.* regan-wurm *lumbricus : Ger.* regenwurm.]

regol, es ; *m.* **I.** *a rule :*—Se gewuna is strǽngra on ælcum worde ðon his regol sý, Ælfc. Gr. 30, 4 ; Som. 34, 67. Sume gáþ of ðam regole, forðan ðe se gewuna is strengra, *eruo* ic nerige, *erutus* generod. Nú wolde se regol ðæs cræftes habban of ðam *eruturus,* ac se gewuna hylt *eruturus,* 41 ; Som. 44, 24–26. Ðis is lǽwedra regol æfter bôclícere gesetnysse, Homl. Th. ii. 94, 8. Se Hǽlend him tǽhte ðone regol, ðæt hí sceoldon yfel mid gôde forgyldan, i. 372, 31. Ðone eásterlícan regol *the rule for determining Easter,* Lchdm. iii. 264, 16. Ðonne byþ hé getealed tô ðam mônþe and be his regolum ácunnod, 250, 6. On mynsteres reogolum gelǽred *monasterii regulis erudita,* Bd. 1, 27 ; S. 489, 10. Hé symle rihte regolas Godes cyricean (*catholicas ecclesiae regulas*) lufode, 5, 19 ; S. 638, 33. Rûme regulas geongra monna *the lax rules of young men,* Exon. Th. 131, 23 ; Gú. 460. **II.** *a rule, pattern,*

standard, norm :—Ða leásan wîtegan wǽron gedwolmen, and woldon áwendan ðone sôðan geleáfan of ðam rihtan regole tô heora gedwyldum, Homl. Th. ii. 404, 9. Regol *normam vite,* regol *normam,* Wrt. Voc. ii. 59, 53–54. **III.** *as an ecclesiastical term,* (a) *a single rule or prescript, a canon :*—Ðæs regles *canonis,* Jn. Skt. p. 1, 12. Reglas *canones,* Mt. Kmbl. p. 2, 18. Bôc ðara reogola *librum canonum,* Bd. 4, 5 ; S. 572, 25. (b) *the body of rules which guide a particular order of ecclesiastics, a rule,* e. g. the Benedictine *rule :*—Hér beginþ seó foresprǽc muneca regules, R. Ben. L, 1. Wite se abbod, eal ðæt hé dô, ðæt hé hit dô mid gehealdsumnesse ðæs regoles, 16, 6. Munecas ðe under regole (*sub regula*) lifigeaþ, Bd. 4, 4 ; S. 571, 21. Intô Sanctus Benedictus regole, Chart. Th. 548, 4. Ðæt forme muneca cyn is mynstermonna, ðæt is ðara ðe under regule and abbodes tǽcinge wuniaþ, R. Ben. 9, 4. On ælcum þingum hié sceoloþ habban ðone regol tô láreówe, 15, 20. Benedictus nam ðone hálgan regol ðe hé mid his handum áwrát, Homl. Skt. i. 6, 66. [*O. H. Ger.* regula *regula, canon : Icel.* regla. From *Lat.* regula.] v. munuc-, riht-regol.

regol-bryce, es ; *m. A breach of rule,* v. regol, **III** :—þurh gelǽredra regolbryce and þurh lǽwedra lahbryce *through the breach of their rule by clerks and the breach of the law by laymen,* Wulfst. 166, 22.

regol-fæst ; *adj. Observing a rule, regular* (of ecclesiastics) :—Rincas rægolfæste, Menol. Fox 88 ; Men. 44.

regol-lagu, e ; *f. Monastic law, the law to which the member of a monastic body is subject :*—Mynstermunuc gǽþ of his mǽgþlage ðonne hé gebýhþ tô regollage, L. C. E. 5 ; Th. i. 362, 28.

regol-líc ; *adj. As an ecclesiastical term* (v. regol, **III**). **I.** *regular, in accordance with monastic rules ;* regularis :—Regollíces *regularis,* Hpt. Gl. 526, 17. Fram ðam hé ðæt gemet leornode regollíces þeódscipes *a quibus norman disciplinae regularis didicerat,* Bd. 3, 23 ; S. 554, 35. On rihtum lífe and on reogollícum *recte vivendo et regulariter,* 4, 6 ; S. 574, 19. Libbaþ regollícan life, sécaþ eówre cyrican, and gefyllaþ eówre tíde aa on gesette tíman, L. I. P. 20 ; Th. ii. 330, 19. On reogollícne þeódscipe, Bd. 3, 3 ; S. 526, 9. **II.** *in accordance with the canons of the church, canonical :*—Bútan sealmsange reogollícre tíde *praeter canonici temporis psalmodiam,* 3, 27 ; S. 559, 10. Tô reogollícum þeáwe rihtra Eástrena *ad ritum Paschae canonicum,* 5, 22 ; S. 643, 38. Æfter regollícre wísan, Lchdm. iii. 428, 15. Regulícra *canonicorum,* Hpt. Gl. 512, 36. Ðǽm regolecum *canonicis,* Wrt. Voc. ii. 24, 19. Ða reogollícan gesettnysse háligra fædera *canonica patrum statuta,* Bd. 4, 5 ; S. 571, 40. [*Icel.* reglu-ligr.]

regollíce ; *adv. Regularly, in accordance with rule* (v. *preceding word*) :—Ða þing ðe regollíce gedêmed wǽron *quaeque erant regulariter decreta,* Bd. 2, 4 ; S. 505, 36. Ðæt biscopas and abbudas, munecas and mynecena, preóstas and nunnan tô rihte gebúgan and regollíce libban, L. Eth. v. 4 ; Th. i. 304, 27. Sacerd ðe regollíce libbe, L. C. E. 5 ; Th. i. 362, 8. Riht is ðæt mynecena mynsterlíce macian . . . and ǽ regollíce libban, L. I. P. 15 ; Th. ii. 322, 35. Ðæt gehádode menn regollíce libban, and lǽwede lahlíce heora líf fadian, 18 ; Th. ii. 324, 26 : Wulfst. 160, 1. Ðæt abbodas and munecas regollícor libban, L. Eth. ix. 31 ; Th. i. 346, 27.

regol-líf, es ; *m. A life according to ecclesiastical rules :*—Ðá gestaþelode hé ðǽr mynster and ðæt tô reogollífe gesette *fundavit ibi monasterium, ac regulari vita instituit,* Bd. 4, 13 ; S. 283, 12. Gif man folciscne mæssepreóst mid tihtlan belecge ðe regollíf næbbe, ládige hine swá swá diácon ðe regollífe libbe, L. Eth. ix. 21 ; Th. i. 344, 20 : L. C. E. 5 ; Th. i. 362, 17. [*Icel.* reglu-líf.]

regol-sticca, an ; *m. A ruler :*—Reogolsticca *regola,* Wrt. Voc. i. 81, 29. þwyrnyssa beóþ gerihte ðonne þwyrlícra manna heortan þurh regolsticcan ðære sôðan rihtwísnysse beóþ geemnode, Homl. Th. i. 362, 28. [*Icel.* reglu-stika.]

regol-weard, es ; *m. The guardian of a rule,* (1) *an authority in the matter of the observance of a rule* (v. regol, **I**) :—Se circul ðe ys gecíged *none aprilis,* hé sceal mid his ealdorscipe ealle ða ôðre gerihtan and gereccan, ðæs ðe ða regolweardas (*those who state with authority what the rule or rules on the point may be*) ús hêton secgan, Anglia viii. 329, 8. (2) *One who sees that a rule* (v. regol, **III**) *is observed, a provost,* v. prâfost :—Ðæs mynstres prâfost and reogolweard wæs in ða tíd Boisel . . . Æfter ðon . . . wæs Cúþberht ðæs ylcan mynstres regolweard geworden *cui tempore illo praepositus Boisil fuit . . . Postquam Cudberct eidem monasterio factus praepositus,* Bd. 4, 27 ; S. 603, 37–43. Se ylca Bosel wæs reogolweard ðæs mynstres on Mailros under Eatan ðam Abbude *idem Boisil praepositus monasterii Mailrosensis sub Abbate Eata,* 5, 9 ; S. 622, 29. Ond ðás forecuædenan suæsenda all ágefe mon ðæm reogolwarde, and hé brytnige swæ hígum mæst rêd síe, Chart. Th. 460, 37. Se reogolweord, 460, 16. (3) *a ruler :*—Sum reiglword (regoloword, Rush.) *quidam regulus,* Jn. Skt. Lind. 4, 46. Se reglword *regulus,* 4, 49.

regul, reht. v. regol, riht.

relic-gang, es ; *m. A going to visit relics :*—Seó tíd is nemned *laetania majora* . . . on ðǽm dæge eall Godes folc mid eáðmódlíce

relicgonge sceál God biddan ꝺæt hē him forgefe siblíce tíd, Shrn. 74, 10. Letanias, ꝺæt is ꝺonne bēne and relicgongas, 79, 29.

reliquias; *pl. m. Relics of saints :*—Ꝺisra reliquia dǽl hæfde sum mæssepreóst . . . Hē ꝺa cyste ontýnde ꝺara reliquia, Bd. 4, 32; S. 611, 30–34. Æt his reliquium wæs sum man gehǽled, S. 611, 9. Mon byrþ his heáfod tô reliquium, Shrn. 57, 26: Blickl. Homl. 127, 12, 16. Mid háligdôme of ꝺæs Hǽlendes rôde . . . and of Martines reliquium, Homl. Skt. i. 6, 74. Ofer his reliquias ꝺæt heofonleóht wæs scínende and deófolseóce æt his reliquium wǽron gelácnode, Bd. 3, 11; S. 535, 6–8. Ꝺæt þurh his reliquias geworden wæs, 4, 32; S. 611, 12. Hē sette ꝺa reliquias on heora cyste, S. 612, 1. Swā hwylc mann swā hrîneþ ꝺîne reliquias oꝺꝺe ꝺîne bân, Nar. 49, 4.

remigende, Mt. Kmbl. 4, 21. v. regnian.

rempan *to go headlong* (like an animal butting with its horns (?), cf. gerumpenu nǽdre *coluber cerastes,* Wrt. Voc. ii. 15, 68), *be precipitate :*—Oft mon biþ suîꝺe rempende and rǽsþ suîꝺe dollîce on ǽlc weorc and hrǽdlîce and ꝺeáh suîꝺe ꝺæt hit sîe for arodscipe and for hwætscipe *saepe praecipitata actio velocitatis efficacia creditur,* Past. 20, 1; Swt. 149, 12. v. note. [Cf. þei rempede þem to reste, Mand. (quoted by Stratmann).]

rên. v. regn.

renc, e; *f. Pride, pomp, vanity, bravery, display :*—Bisceopum gebyreþ ꝺæt hî ne hēdan ne woruldwlence ne îdelre rence, L. I. P. 10; Th. ii. 316, 30. Ægwhylce wlence and îdele rence forhogian swā gebyreþ munecum, 14; Th. ii. 322, 9. Ne gerísaþ biscopum prîta ne îdele rænca ne micele ofermētta, 11; Th. ii. 318, 32. Be îdelum rencum. *Pro eo, quod eleuate sunt filie Sion, etc.* For ofermēttan, hē cwæꝺ, and îdelan rencan eówra leóda, Wulfst. 45, 21–23. [We ne beoꝺ iboren for to habbene nane prudu ne forꝺe nane oꝺre rencas, O. E. Homl. i. 7, 27.] v. ofer-renc, ranc.

renge, rynge, ringe (?), an; *f. A spider* or *a spider's web :*—Renge *aranea,* Blickl. Gl. Úre gǽr swā swā lobbe ꝶ rynge beóþ ásmeáde *anni nostri sicut aranea meditabuntur,* Ps. Lamb. 89, 9. Áýdlian ꝺu dydest swā swā ǽtterloppan ꝶ ryngan sáwle his *tabescere fecisti sicut araneam animam ejus,* 38, 13.

reng-wyrm, es; *m. A maw-worm, a worm in the intestines :*—Wiꝺ ꝺæt rængcwyrmas (rengc-, MS. B.; rýnwyrmas with a gloss *lumbrici,* MS. H.) dergen ymb nafolan, Lchdm. i. 168, 9. Wiꝺ ꝺæt rengwyrmas ymb ꝺone nafolan wexen, 218, 14. [Cf. *O. H. Ger.* pouh-wurm *lumbricus.*]

rendan; *p.* de *To rend, tear, cut :*—Óꝺre ꝺa twigu gibēgdun ꝶ rendun (rindon, Lind.) ꝺa telge of ꝺǽm trēum *alii frondes caedebant ab arboribus,* Mk. Skt. Rush. 11, 8. Ceorfas ꝶ rendas (hrendas, Lind.) *succidite,* Lk. Skt. Rush. 13, 7. [Scipen gunnen helden bosmes þer rendden, Laym. 7849. Heo hauede bipiled mine figer, irend of al þe rinde *decorticauit ficum meam,* A. R. 148, 23. þe reue rende his claꝺes, Jul. 70, 7. *O. Frs.* renda *to tear* ; rend a *rent.*] v. tô-rendan.

rendrian. v. Lchdm. ii. Gloss.

rēnian, rēnig, rēn-lîc. v. regnian, regnig, regn-lîc.

reó. v. reôwe.

reóc; *adj. Fierce, savage :*—[Grendel] grim and grǽdig, reóc and rēꝺe, Beo. Th. 244; B. 122.

reócan; *p.* reác *To reek, send forth smoke* or *steam :*—Ꝺonne hē (*helleborus albus*) tôbrocen byþ, hē rýcþ eal swylce hē smîc of him ásænde, Lchdm. i. 260, 8. Muntas reócaþ *montes fumigant,* Ps. Th. 103, 30: 143, 6. Reác *exalabat,* Wrt. Voc. ii. 32, 46: *fumarat,* 151, 55. Wel on wætere, ꝶ reócan on ꝺa eágan ꝺonne hit hát sîe, Lchdm. ii. 18, 24: 32, 7. Reócan *fumare,* Germ. 395, 70. Reócende *anhelans,* 400, 92. Rēcende *fumigans,* Mt. Kmbl. Lind. 12, 20. Heó ꝺæra máꝺma ne rôhte ꝺe má ꝺe reócendes meoxes, Homl. Skt. i. 7, 20. Bútan rēnscûrum and reócendum deáwe, 18, 57. Ꝺæs hreóflian lîc mid reócendum stence, Homl. Th. i. 336, 33. Æt hreócendum heorꝺe, Wulfst. 170, 21. Reócendne (reccendne, MS.) weg, Cd. Th. 177, 19; Gen. 2932. Reócende hrǽw, Judth. Thw. 26, 7; Jud. 314. Hreócendum *fumigabundis,* Hpt. Gl. 516, 29. [*O. H. Ger.* riuhhan *fumigare : Icel.* rjúka.] v. rēcan.

reóce. v. rēcels-reóce.

reód; *adj. Red :*—Se ꝺe ǽror com, se wæs reód (*rufus*) and eall rûh, Gen. 25, 25. Ꝺonne ꝺû (*the body*) wǽre glæd and reód and gôdes hiwes, ꝺonne wæs ic (*the soul*) blác and swýꝺe unrôt, Wulfst. 140, 27: L. E. I. prm.; Th. ii. 398, 14. Ꝺâ Moises hæfde gefaren ofer ꝺa Reódan Sǽ, Ex. 15, 1. Hié wǽron sume reóde sume blace sume hwîte *quaedam rubentibus scamis erant quaedam nigri et candidi coloris,* Nar. 13, 17. [*Icel.* rjóꝺr.] v. bleó-reód, reáð.

reód, es; *n. Red, red colouring :*—Reóde gnídan *fucare,* Wrt. Voc. ii. 37, 49.

reódan; *p.* reáð. **I.** *to redden, stain with blood :*—Deáþwang rudon, Andr. Kmbl. 2006; An. 1005. **II.** *to redden a person by causing blood to flow from a wound, to wound, kill :*—Næs ꝺeós eorþe besmiten beornes blôde ꝺe hine bil ruꝺe (cf. ne seó eorþe besmiten mid ofslegenes monnes blôde, Bt. 15; Fox 48, 15), Met. 8, 34. Se eorl wolde

sleán eaferan sînne, ecgum (MS. eagum) reódan magan mid mēce, Cd. Th. 204, 2; Exod. 412. [*Icel.* rjôꝺa *to redden* (*with blood*); rjóꝺa kiꝺr eins *to slay a person.*] v. on-reódan.

reodian (?) :—Ic þragum þreodude and geþanc reodode, Elen. Kmbl. 2476; El. 1239.

reódian; *p.* ode *To be* or *become red :*—Ic reádige (reódige, MS. O.) *rubeo,* Ælfc. Gr. Zup. 154, 7. v. á-reódian.

reód-mûþa, an; *m. The name of a bird :*—Reódmûþa *faseacus, nomen avis,* Wrt. Voc. ii. 146, 56.

reód-naesc glosses *partica,* Wrt. Voc. ii. 116, 61.

reófan; *p.* reáf, *pl.* rufon; *pp.* rofen. *To break, rend, rive :*—Randbyrig wǽron rofene, Cd. Th. 207, 7; Exod. 463. [*Icel.* rjúfa *to break, rip up.*] v. be-reófan.

reohhe, an; *f. The name of a fish :*—Reohhe *fannus,* Wrt. Voc. i. 56, 5. Reohche, 77, 66. [*Laym.* rehꝫe (rohꝫe, 2nd MS.): *Du.* rog *a ray : Dan.* rokke *a ray.*]

reóma, an; *m. A membrane, ligament :*—Se reóma ꝺæs brægenes *cartilago,* Wrt. Voc. ii. 22, 58. Biþ ꝺæt brægen ûtan mid reáman bewefen on ꝺære syxtan wucan, Lchdm. iii. 146, 4. [A rym (*other MS.* reme) þat es ful wlatsome Es his (a man's) garment when he forth sal com, þat es noght bot a blody skyn þat he byfor was lapped in, Pricke of Conscience, 520. See Nares' Glossary, s.v. rim. *O. Sax.* reomo *the latchet of a shoe : O. H. Ger.* riumo *corrigia, lorea, balteus, habena : Ger.* riemen.]

reomig-môd. v. reónig-môd.

reón *mourning, lament :*—Woldan wērigu wîf wôpe bimǽnan æþelinges deáþ, reóne bereótan, Exon. Th. 459. 27; Hö. 6. Cf. rýn.

reónian; *p.* ode *To whisper, mutter :*—Reónigende *mussitantes,* Hpt. Gl. 472, 5. v. ge-reónian, rûnian.

reónig; *adj. Mournful, sad, gloomy, weary :*—Â mîn hige sorgaþ reónig reóteþ and geresteþ nô *ever hath my heart care, mournful laments and hath not rest,* Elen. Kmbl. 2163; El. 1083. Hē ꝺǽr þreó mētte in ꝺam reónian hofe (*in the hole in which they were buried*) rôda ætsomne greóte begrauene, 1664; El. 834. In ꝺam reóngan hâm *in that gloomy dwelling* (*hell*), Exon. Th. 274, 8; Jul. 530. v. preceding word.

reónig-môd; *adj. Sad at heart, weary :*—Wæs him ræste neód reónigmôdum *need of rest was there for him weary-hearted,* Exon. Th. 167, 32; Gû. 1069. Ꝺonne gewîciaþ wērigferꝺe . . . hæleþ beóþ on wynnum reónigmôde ræste geliste *the weary seafarers are eager for rest,* 361, 21; Wal. 23. Fēꝺan sǽton reónigmôde (reomigmôde, MS. *Grimm suggests a comparison with Gothic* rimis *quiet*) reste gefēgon wērige æfter wæꝺe, Andr. Kmbl. 1183; An. 592.

reónung, e; *f. : es; n. Whispering, muttering :*—Nânes mannes stefn oꝺꝺe reónung gehýred ne sý, bútan ꝺæs rǽderes ânes *nullius musitatio uel vox audiatur nisi solius legentis,* R. Ben. 62, 14. v. ge-reónung.

reopa, reopan. v. ripa, repan.

reord, e; *f. : es; n. Speech, tongue, language, voice :*—Reord ꝺîn ꝺæc gecýꝺeþ *loquela tua manifestum te facit,* Mt. Kmbl. Rush. 26, 73. Reord wæs eorþbûendum ân gemǽne '*and the whole earth was of one language, and of one speech,*' Cd. Th. 98, 25; Gen. 1635. Reord up âstâg *voices rose high,* Exon. Th. 246, 16; Jul. 62. Æt ealra manna gehwæs mûþes reorde *from the voice of each man's mouth,* Soul Kmbl. 186; Seel. Verc. 93. Herian God hâligum reorde, Hy. 3, 58. Heofonrîces weard sprǽc hâlgan reorde, Cd. Th. 89, 22; Gen. 1484: 248, 10; Dan. 511. Wit scîran reorde song âhôfan, Exon. Th. 324, 32; Vîd. 103. Geác monaþ geómran reorde, 309, 7; Seef. 53. Ꝺæt him ꝺa swā cûþe wǽron swā his âgene reorde ꝺe hē on âcenned wæs *ut tam notas ac familiares sibi eas* (Latin and Greek), *quam nativitatis suae loquelam haberet,* Bd. 5, 23; S. 645, 17. Stefn in becom under hârne stân . . . hordweard oncneów mannes reorde, Beo. Th. 5103; B. 2555. Hî gehýrdon hlûde reorde, ꝺînes mûþes ꝺa mǽran word, Ps. Th. 137, 5. Hē reorde gesette eorþbûendum ungelîce, Cd. Th. 101, 19; Gen. 1684. Se hâlga wer hergende wæs Meotudes miltse, and his môdsefan rehte þurh reorde, Exon. Th. 188, 25; Az. 51: 111, 24; Gû. 131. Ic glidan reorde mûþe gemǽne, 406, 23; Rä. 25, 5. Se ꝺe reorda gehwæs ryne gemiclaþ ꝺara ꝺe noman Scyppendes hergan willaþ, 4, 3; Cri. 47. Reordana *locutionem,* Jn. Skt. p. 7, 10. Hý mislîce mongum reordum wôꝺe hôfun, Exon. Th. 156, 6; Gû. 870. Fugla cynn songe lofiaþ, mǽraþ môdigne meaglum reordum, 221, 22; Ph. 338. Hē ûs syleþ missenlîcu môd, monge reorde, 334, 9; Gn. Ex. 13. [*Orm.* reord, rerd *sound, voice : Ps.* rorde *sonus : Ayenb.* ecko, þet is þe rearde þet ine þe heꝫe helles comþ aꝫen. and acordeþ to al þet me him zayþ : *Goth.* razda *speech, tongue : O. H. Ger.* rarta *modulatio : Icel.* rödd *voice.*] v. ge-reord.

reord, e; *f. A meal, refection, food :*—Reorde mîn *refectio mea,* Mk. Skt. Rush. 14, 14. Fēd feora wôcre ôꝺ ic ꝺære lâfe lagosîꝺa eft reorde rýman wille, Cd. Th. 81, 12; Gen. 1344. Hē wæs swā gistlîþe, ꝺæt hē for Godes lufon eode tô reordum mid ꝺâm tôcumendum mannum, Shrn. 129, 27. v. ge-reord.

-reord. v. el-reord.

reordan; *p.* de *To take food, eat :*—Reordendum *cenantibus* i. *vescentibus,* Wrt. Voc. ii. 130, 72. v. ge-reordan.

reord-berend, es; *m. One gifted with speech, a man :*—Tô midre

nihte syđđan reordberend reste wunedon, Rood Kmbl. 5; Kr. 3: Cd. Th. 223, 21; Dan. 123. Ealle reordberend, hæleþ geond foldan, Exon. Th. 18, 4; Cri. 278. Reordberende, earme eorþware, 24, 8; Cri. 381: 63, 26; Cri. 1025. Sceall æghwylc reordberendra riht gehýran, Elen. Kmbl. 2561; El. 1282. Đǽr leán cumaþ reordberendum, Exon. Th. 84, 5; Cri. 1369. Hé reordberend lǽrde under lyfte, Andr. Kmbl. 838; An. 419.

reord-hús, es; *n.* *A house* or *room where meals are taken* :—Reordhús *cenaculum*, Mk. Skt. Lind. 14, 15.

reordian; *p.* ode. **I.** *to speak, say, talk* :—Sleáþ synnigne ofer múþ, tó feala reordaþ, Andr. Kmbl. 2604; An. 1303. Đus reordiaþ ryhtfremmende, Exon. Th. 240, 1; Ph. 632. Đá reordDe Waldend and worde cwæþ, Cd. Th. 76, 6; Gen. 1253. Reordode, 161, 30; Gen. 2673. Heáhcyning him tó reordode, 130, 28; Gen. 2166. Sceal se wonna hrefn fela reordian, earne secgan hú him æt ǽte speów, Beo. Th. 6043; B. 3025. Ongan reordigan rǽdum snottor, wordlocan onspeónn, Andr. Kmbl. 637; An. 469. Wolde reordigean ríces hyrde hálgan stefne, Cd. Th. 194, 5; Exod. 256. Him biþ reordiende éce Drihten, ofer ealle gecwyþ, 304, 7; Sat. 626. Se Hǽlend his gingrum tó sprǽc ymbe Godes ríce, samod mid him reordigende, Homl. Th. i. 294, 18. **II.** *to read* :—Ne reordaþ *non legistis*? Mt. Kmbl. Rush. 12, 5. Gé ne reordade *non legistis*, 19, 4. Reordadun, 21, 16. Seó bysen đǽs rihtan geleáfan fram eallum đe hine gehýrdon ođđe reordedon þancwurþlíces wæs onfangen *exemplum catholicae fidei ab omnibus qui audiere vel legere gratan-tissime susceptum*, Bd. 4, 18; S. 587, 13. [He reordien gan, and þas word sæide, Laym. 22174.]

-reordig. v. el-reordig.

reordung, e; *f.* *Taking food, refection* :—Riordung mín *refectio mea*, Mk. Skt. Lind. 14, 14. v. ge-reordung.

reosan glosses *pissli* (in a list of plant names), Wrt. Voc. i. 68, 44.

reóst *a rest* (*rest the wood on which the coulter of a plough is fixed,* Halliw. Dict.) :—Sules reóst *dentale, s. est aratri pars prima in qua vomer inducitur quasi dens*, Wrt. Voc. ii. 138, 72: *dentalia*, 106, 20: 25, 28. [Cf. *O. H. Ger.* riostar *stiva, dentile.*]

reót (?) (*joyous*) *sound* (?), *gladness* (?) :—Gesyhþ sorhcearig on his suna búre wínsele wéstne, reóte berofene . . . nis đǽr hearpan swég, gomen in geardum, Beo. Th. 4905; B. 2457. v. (?) reótan.

reótan; *p.* reát. **I.** *to make a noise* :—Reótaþ (wreotaþ, MS.) *crep-ita*[*n*]*t*, Wrt. Voc. ii. 21, 44. Reát (hreát ?) *desteruit* (*stertuit* ?), *somniavit*, 139, 17. **II.** *to make a noise in grief, to lament, wail* :—Reóteþ meówle, seó đe hyre bearn gesihþ brondas þeccan, Exon. Th. 330, 5; Vy. 46. Cerge reótaþ fore onsýne éces déman, 52, 20; Cri. 836. Hý (*sinners*) reótaþ and beofiaþ fore freán forhte, 75, 32; Cri. 1230. **III.** *to weep, shed tears* :—Lyft drysmaþ, roderas reótaþ, Beo. Th. 2756; B. 1376. [*O. H. Ger.* riuzan; *p.* róz *flere, plangere, stridere: Icel.* rjóta *to roar, rattle.*] v. be-, wið-reótan.

reótig; *adj.* *Sad, mournful, tearful* :—Đonne hit wæs rénig weder, and ic reótugu sæt, Exon. Th. 380, 19; Rä. 1, 10.

reów; *adj.* *Fierce, cruel* :—Sume wurdon bisencte under reóne streám, sume ic róde bifealh, Exon. Th. 271, 12; Jul. 481. Wǽron hý reówe tó rǽsanne gífrum grápum, 126, 26; Gú. 377: Andr. Kmbl. 2669; An. 1336. v. blóđ-, deáþ-, flyge-, gúþ-, wæl-reów.

reówe, an; *f.* *A rug, mantle, covering* :—Reówu *tapeta*, Wrt. Voc. i. 289, 50. Reówe *lena*, línen reówe *lena linea*, ii. 53, 71–72. Mon mót feohtan orwíge gif hé geméteþ óđerne æt his ǽwum wífe betýnedum durum ođđe under ánre reón, L. Alf. pol. 42; Th. i. 90, 27. Reówan and hwítlas wacsan *lenas sive saga lavare*, Bd. 4, 31; S. 610, 11. v. rúwa, rýhe.

repan (?); *p.* ræp, *pl.* rǽpon *To reap* :—Hié reopaþ *metent*, Ps. Surt. 125, 5. Manig men rǽpon heora corn onbútan Martines mæssan and gyt lator, Chr. 1089; Erl. 226, 19. [I gaf hem red þat ropen To seise to me with her sykel þat I ne sewe neure, Piers P. 13, 374.] v. wín-repan, rípan.

reps, repsan, repsung, rǽsele, résian. v. ræps, rǽpsan, rǽpsung, rǽsele, rǽswan.

rest, e; *f.* **I.** *rest, quiet, freedom from toil* :—Sǽterndæges rest (*requies sabbati*) ys Drihtne gehálgod, Ex. 16, 23. Nis nán gesceaft gesceapen đara đe ne wilnige đæt hit đider cuman mǽge đonan đe hit ǽr com, đæt is tó rǽste and tó orsorgnesse. Seó rǽst is mid Gode, Bt. 25; Fox 88, 29–32: Met. 13, 71. Ne đǽr biþ hungor ne þurst . . . ac đǽr biþ seó éce rǽste, Blickl. Homl. 65, 20. Heó reste stówe funde, Cd. Th. 88, 17; Gen. 1466. Wæs him rǽste neód, Exon. Th. 167, 31; Gú. 1068. Ic sylle đe reste *requiem dabo tibi*, Ex. 33, 14: Ps. Th. 114, 7. Hé gǽþ sécende reste, Mt. Kmbl. 12, 43. Wé rǽste habbaþ, forđon đe đú sylest úrum leomum rǽste, Blickl. Homl. 141, 10–11: 41, 33. Mid gódum dǽdum man geearnige him đa écean rǽste, 101, 26. Hwonne him Freá reste ágeáfe, Cd. Th. 86, 9; Gen. 1428. **II.** *rest, repose, sleep* :—Rest *dormitatio*, Kent. Gl. 894. Hé his limo on reste gesette and onslǽpte *membra dedisset sopori*, Bd. 4, 24; S. 597, 10. Đá hé đá tó reste eode *dum iret cubitum*, 3; S. 525, 12. Be muneca reste. Ænlýp-ige munecas geond ænlýpige bed restan, R. Ben. 47, 2. **III.** *a*

place of rest, resting-place :—Đú eart seó séfte rǽst sóđfæstra, Bt. 33, 4; Fox 132, 34. Đæt is sió án rǽst eallra úrra geswinca, sió án hýþ byþ simle smyltu, 34, 8; Fox 144, 27. Đis is mín rest đe ic on worulda woruld wunian þence, Ps. Th. 131, 15. Đé is eđelstól gerýmed, rest fæger on foldan, Cd. Th. 89, 26; Gen. 1486. Wíc, randwígena rǽst, 186, 5; Exod. 134. **IV.** *a bed, couch* :—Đǽr biþ rest of elpenda báne geworht *lectus eburneus*, Nar. 38, 32. Wæs his seó æþeleste rest on nacodre eorþan, Blickl. Homl. 227, 10. Salomones reste wæs mid weardum ymbseted.—Hwæt wæs seó Salomones rǽste . . .? Ac hwæt mǽnde đæt syxtig wera stondende wǽron ymb đa reste? 11, 16–23. Rǽst a sepulchre, Exon. Th. 459, 28; Hö. 6. On mínre reste *per stratum meum*, Ps. Th. 62, 6. Míne cnihtas synt on reste (*in cubili*) mid mé, Lk. Skt. 11, 7. Wæs án gesittende beforan his reste (*ante lectulum ejus*), Bd. 4, 11; S. 579, 38. Swá swá oferdruncen man wát đæt hé sceolde tó his húse and tó his rǽste, Bt. 24, 4; Fox 84, 31. Heó ásette đa hand æt hire heáfdum on hire rǽste, Shrn. 60, 1. Hé on his reste gestáh *lectulum conscendens*, Bd. 3, 27; S. 559, 15: Cd. Th. 134, 22; Gen. 2228. Seó wlitignes heora rǽsta and setla, Blickl. Homl. 99, 33. Rǽst *recubitos*, Mt. Kmbl. Lind. 23, 6. [*O. Sax.* resta, rasta *a couch: O. H. Ger.* resti *requies, quies, dormitio, pulpitum; rasta a stage in a journey: Goth.* rasta *a mile: Icel.* röst *a stage.*] v. æfen-, bed-, flet-, fold-, land-, niht-, sele-, wæl-rest.

restan; *p.* te To rest. **I.** *intrans.* (a) *of persons* (1) *to cease from toil, be at rest* :—Ic đonne reste *requiescam*, Ps. Th. 54, 6. Eádige beoþ þearfena gástas, and hié restaþ on heofena ríce, Blickl. Homl. 159, 29. Hý bídinge móstun æfter tintergum tídum brúcan, restan ryne-þrágum, Exon. Th. 115, 3; Gú. 184. Đa restendan *pausantis*, Wrt. Voc. ii. 66, 21. Restendum *fereatis*, i. *quietis, securis*, 147, 59. (2) *to rest on a couch, to sleep* :—Đonne hié restaþ đonne restaþ hié búton bedde and bolstre ac on wildeóra fellum heora bedding biþ *homines accubantes et quiescentes sine ullis cervicalibus stratisque, tantum pellibus ferarum*, Nar. 31, 10. On đǽre tíde đe óđre men slépon and reston *caeteris quiescentibus et alto sopore pressis*, Bd. 2, 12; S. 513, 37. On niht ǽr hé rǽste, Blickl. Homl. 47, 18. Ænlýpige munecas geond ænlýpige bed restan, R. Ben. 47, 3. Đá bæd hé his þeng on ǽfenne . . . đæt hé him stówe gegearwode, đæt hé restan mihte, Bd. 4, 24; S. 598, 31. Đá hé gesette his leomu tó restenne *cum ad quiescendum membra sua posuisset*, 2, 6; S. 508, 11. Đá wæs heo restende on sweostra slǽperne, 4, 23; S. 595, 39. (3) *to rest in death, lie dead, lie in the grave* :—Augustinus on Brytene rest on Cantwarum, Menol. Fox 206; Men. 104. Gerusalem is gereht 'sibbe gesyhþ,' forđon đe hálige sáula đǽr restaþ, Blickl. Homl. 81, 2. Reste hé đǽr *Christ lay in the sepulchre*, Rood Kmbl. 138; Kr. 69. Hæfdon eđelweardas ealdhettende swyrdum áswefede, hié on swađe reston (*of the Assyrians slain in battle*), Judth. Thw. 26, 12; Jud. 322. (b) *of things, to remain unmoved* or *undisturbed, be still* :—Flǽsc mín resteþ (*requiescet*) on hyhte, Ps. Sel. 15, 9. Reste đǽr eówer sib, Lk. Skt. 10, 6. Đám folcum sceal sacu restan, Beo. Th. 3719; B. 1857. Se æđeling hét streámfare stillan, stormas restan, Andr. Kmbl. 3151; An. 1578. Đín ríce restende biþ oþ đæt đú eft cymst, Cd. Th. 252, 26; Dan. 584. **II.** *trans. with reflex. acc.* *To rest one's self* (1) *of cessation from toil* :—Đú rest đe nú on eorþan, and ic mid sáre tó helle sceal beón lǽded, L. E. I. prm.; Th. ii. 398, 16. Đǽr hí æđelingas inne restaþ, Runic pm. Kmbl. 340, 22; Rún. 6. Reste đæt folc hit *sabbatizavit populus*, Ex. 16, 30. On six dagon God geworhte heofon and eorþan and on đam seofoþan hé hine reste (*ab opere cessavit*), 31, 17. Ic mé mæg restan on đissum racentum, Cd. Th. 28, 11; Gen. 434. (2) *of rest on a couch or in sleep* :—Đá reste hine se biscop đá-giet and mid wildeóra fellum wæs bewrigen . . . Đá áwehte ic đone bisceop, Nar. 31, 1. On đæt hús đe heó hié inne reste, Blickl. Homl. 147, 2. Đonne hé reste hine, đonne wæs his seó æþeleste rǽst on nacodre eorþan, 227, 10. Hú se beorn (*Noah in his drunken sleep*) hine reste on recede, Cd. Th. 95, 25; Gen. 1584. [*O. Sax.* restian: *O. Frs.* resta: *O. H. Ger.* restan *requiescere, dormire, cubare.*] v. ge-restan.

rèstan (?) *to exult* :—Hæfdon beorgas blíđe sǽle and rammum đá rèstan *gelíce montes exultaverunt ut arietes*, Ps. Th. 113, 4. [Grein compares the word with *O. H. Ger.* hlút-reisti, -reisig *clamosus, canorus.*]

rest-bedd, es; *n.* *A bed, couch* :—Đeáh ic on mín restbedd gestíge *si ascendero in lectum stratus mei*, Ps. Th. 131, 3.

reste-dæg, es; *m.* *A day of rest, a day when no work is to be done, a Sabbath* :—Restedæg *feriatus*, Wrt. Voc. i. 22, 20. Restedagas *feriati dies*, ii. 148, 6. Gehálga đone restedæg . . . Se seofoþa dæg ys Drihtnes restedæg: ne wirc đú nán weorc on đam dæge, Ex. 20, 8–10. Mannes sunu ys restedæges hláford, Mt. Kmbl. 12, 8. On ánum đara restedaga se nú Sunnandæg is nemned, Bd. 3, 17; S. 545, 30.

resten-dæg, es; *m.* *A day of rest, Sabbath* :—Đæt þridde bebod is 'Beó đú gemyndig đæt đú đone restendæg gehálgige' . . . Se Sæternes-dæg wæs gehäten restendæg . . . on đam dæge læg Cristes líc on byrigene, and hé árás of deáþe on đam Sunnandæge, and se dæg is cristenra manna restendæg, Homl. Th. ii. 206, 3–33. Se seofoþa dæg is mín se hälga

restendæg ... healdaþ gē mínne restendæg, Wulfst. 210, 17–21. Gedaf-enaþ ǽlcum men tô habbenne restendæg, 227, 22.

resten-geár, es; n. A year in which work is not done:—Ne sáw ðu ðonne (in the seventh year) ne rip ne ðínne wíneard ne wirce, forðam ðe hit biþ restengēr, Lev. 25, 4–5.

rest-gemána, an; m. Conjugal intercourse; concubitus:—Hié noldan leng heora hláforda ne heora wera ræstgemánan sécean, Blickl. Homl. 173, 16. Restgemanan, Lchdm. i. 350, 10.

rest-hús, es; n. A sleeping-chamber:—Hē ðǽr hæfde án resthús (cubiculum), Bd. 3, 17; S. 543, 23.

rest-leás; adj. Restless, disturbed:—Á biþ ungestillod and restleás ðe mid ðám unþeáwum beléd biþ, R. Ben. 121, 14.

rēsung. v. rǽswung.

rētan; p. te To cheer, gladden, comfort:—Geseóþ hū blíþe ða earman beóþ, ðonne hî mon mid mete and mid hrægle rētaþ, Blickl. Homl. 41, 29. Ealle ða óðru gōd óleccaþ ðam mōde and hit rētaþ, Bt. 24, 3; Fox 84, 24 note. Hū se wísdóm hine eft rēte and rihte mid his andsworum, tit. 5; Fox x, 10. Ðæt dolh-rēt mid ferscre buteran, Lchdm. ii. 354, 5. Wudewan and steópcild hȳ (eorlas and heretogan) sculon rētan and þearfena helpan, L. I. P. 11; Th. ii. 318, 26. Se hālga ongann wígendra þreát wordum rētan, Andr. Kmbl. 3215; An. 1610. Ða wǽdlan sint tô frēfranne and tô rētanne (offerre consolationis solatium), Past. 26, 1; Swt. 181, 6. v. ā-, un-rētan; rōt.

rētend, es; m. One who cheers or comforts:—Wǽron wē oft gemyngode ðæt wē sceoldan beón wudewena helpend and steópcilda árigend and earmra rētend and wēpendra frēfriend, Wulfst. 257, 4.

reðe; adj. Right, just:—Ðū (God) eart hālig lǽce, rede and rihtwís, rūmheort hláford, Hy. Grn. 7, 63. Mē ðín se gōda gást lǽdde ðæt ic on rihte weg reðne fērde spiritus tuus bonus deducet me in viam rectam, Ps. Th. 142, 11. Ic on wísne weg worda dínra, reðne rinne, 118, 32. Hí cȳðan dínes mægenþrymmes mǽre wuldur, riht and reðe, ríces dínes, 144, 11. Ic ðæt ongeat dōmas díne reðe rihtwíse cognovi quia aequitas judicia tua, 118, 75. Synd his dōmas reðe mid rǽde rihte gecȳðde rectum judicium tuum, 118, 137.

reðe; adj. Fierce, cruel, savage. It glosses the following Latin words, efferus, Ælfc. Gr. 14; Som. 16, 57: ferus, 38; Som. 41, 45: trux, 9, 67; Som. 14, 10: ferox, 9, 66; Som. 14, 6: Wrt. Voc. ii. 108, 37: funestus, 34, 12: infestus, 45, 26: durus, crudelis, asper, 142, 19: severus, immansuetus, 142, 44: austerus, 1, 20: furibundus, valde iracundus, Hpt. 450, 1: truculentus, 518, 34. Roeðe asper, Lk. Skt. Lind. 3, 5. I. applied to persons (a) in a bad sense:—Ðes (Ishmael) byþ reðe (ferus) man and winþ wið ealle and ealle wið hyne, Gen. 16, 12. Ealle his ǽhta ríce reðe mann gedǽle may a rich and cruel man divide all his possessions; scrutetur foenerator omnem substantiam ejus (Grein takes reðemann and compares Gothic raþjo), Ps. Th. 108, 11. Gif hē (a king) his folc ofsit, ðon biþ hē tyrannus, ðæt is reðe, Ælfc. Gr. 50, 20; Som. 51, 47. Grim and grǽdig, reóc and reðe (Grendel), Beo. Th. 244; B. 122. Ðá wæs ellenwōd, yrre and reðe, frēcne and ferþgrim fæder wið dehter, Exon. Th. 251, 5; Jul. 140. Sum árleás hine wolde sleán on his heáfde, ac ðæt wæpen wand áweg of ðæs reðan handum, Homl. Th. ii. 510, 23. Burhrūnan, reðe furie, Wrt. Voc. ii. 151, 77. Twegen ðe hæfdon deófolseócnesse wǽron swíðe reðe (saevi nimis), Mt. Bos. 8, 28. Ealle swíðe erre wǽron. Ðá wæs heora sum reðra and hátheortra ðonne ðá óðre, Blickl. Homl. 223, 6. (b) of justifiable severity, severe, stern, austere, zealous:— Strang wæs and reðe se ðe wǽtrum weóld (the Deity at the time of the flood), Cd. Th. 83, 8; Gen. 1376. Biþ ðonne (at the day of judgment) ríces weard reðe and meahtig, yrre and egesful, Exon. Th. 93, 19; 1528. Reðe biþ Dryhten æt ðam dóme, Soul Kmbl. 196; Seel. Verc. 98. Óðer biscop, reðes mōdes mon austerioris animi vir, Bd. 3, 5; S. 527, 20. Reðe and strǽce for ryhtwísnesse justitiae severitate districti, Past. 5, 1; Swt. 41, 19. Ðá wæs se bysceop mycle ðig reðra on gódum weorcum ðe hē ymbe ða cúðlícan mēde gehȳrde, Shrn. 98, 19. God sylfa ðonne ne gȳmeþ nænges mannes hreówe ... ac biþ ðonne reðra and þearlwísra ðonne ǽnig wilde deór, oððe ǽfre ǽnig mōd gewurde, Blickl. Homl. 95, 30. Wolde heofona helm helle weallas forbrecan ... reðust ealra cyninga (Christ at the harrowing of hell), Exon. Th. 461, 16; Hö. 36. II. applied to animals, wild, savage, fierce:—Reðe deór bellua, Wrt. Voc. i. 22, 40. Reðe nȳten feralis bestia, ii. 147, 53. Rib reáfiaþ reðe wyrmas, Soul Kmbl. 221; Seel. 113. III. applied to things (punishment, calamity, etc.), severe, cruel, fierce, dire:—Reðe wyrd fortuna aspera, Bt. 40, 1; Fox 236, 6–7. Wæs þreálíc þing þeódum tōweard, reðe wíte, Cd. Th. 79, 30; Gen. 1319. Weard him on slǽpe gecȳðed ðætte ríces gehwæs reðe sceolde gelimpan eorðan dreámas ende wurðan on sleep was made known to him that of every kingdom a terrible end should befall, an end be of the joy of earth, 223, 4; Dan. 114. Líg reád and reáðe, Dóm. L. 152. Reðe, Exon. Th. 51, 3; Cri. 810. Sprecan reðe word (of the judgment passed on the wicked), 50, 11; Cri. 798. Hē him sylfum reðne dóm and heardne geearnaþ, Blickl. Homl. 95, 34. Regnas reðe, háte of heofenum, Ps. Th. 104, 28: Met. 7, 27. Reðum wítum ferocibus cruciatibus, Hpt. Gl. 487, 10. Mid ðȳ hí cwǽdon ðæt ðæt is wundor

ðæt ðū swá reðe forhæfednesse and swá hearde habban wylt andswarede hē: 'Heardran and reðran ic geseah' cum dicerent: 'Mirum quod tam austeram tenere continentiam velis,' respondebat: 'Austeriora ego vidi,' Bd. 5, 12; S. 631, 34. v. un-reðe.

reðe-hygdig; adj. Right-minded:—Wel biþ ðam eorle ðe him oninnan hafaþ reðehygdig wer rūme heortan well will it be for that man who, being a mortal right-minded, hath a liberal heart within him, Exon. Th. 467, 15. v. reðe.

reðe-mód; adj. I. in a bad sense, of fierce or savage mind, Cd. Th. 4, 2; Gen. 47. Ábrecan ne meahton reðemóde (the people of Sodom who were trying to break into Lot's house) reced æfter gistum, 150, 15; Gen. 2492. II. of justifiable severity or anger, of stern or severe mind, wroth :—God reðemód reorde gesette eorþbúendum ungelíce, 101, 18; Gen. 1684: 218, 2; Dan. 33.

reðen (?); adj. Wild:—Ðæt hē hine gereordode mid ðám reðenum (MS. U. reðum) nȳtenum, Homl. Skt. i. 10, 102.

reðig; adj. Fierce, savage, cruel:—Reðig ferox, Wrt. Voc. ii. 35, 17.

reðigian; p. ode To rage, be furious:—Godes yrre ys ofer hig and his wíte reðegaþ egressa est ira a Domino, et plaga desaevit, Num. 16, 46. Reðegadon furuerunt, insanierunt, Wrt. Voc. ii. 151, 71.

reðig-mód; adj. Of fierce or savage mind:—Hē reðigmód rǽst on gehwilcne wēdehunde (reðe, MS., but cf. wēdende hund, Fox 186, 8) wuhta gelícost, Met. 25, 17.

reð-líc; adj. Cruel, deadly:—Reðlíc scinhiw ferale monstrum, Wrt. Voc. ii. 147, 53. Deriendlícan, reðlícan feralia, i. lugubria, tristia, noxia, luctuosa, mortifera, mortalia, 147, 50. Mid reðlícum feralibus, 34, 20.

reðlíce; adv. Fiercely, furiously:—Roeðlíce violenter, Wrt. Voc. ii. 123, 47.

reðness, e; f. Fierceness, rage, cruelty, severity. It glosses the following Latin words, ferocitas, Ælfc. Gr. 9, 25; Som. 10, 65; Wrt. Voc. ii. 34, 11: austeritas, 1, 19: feritas, i. crudelitas, inclementia, duritia, 148, 2: furor, 151, 69: furia, insania, 151, 73. Reðnyssa efferata, Germ. 399, 380. I. applied to persons, (a) in a bad sense:—Ðone lǽddon feówer áwyrgde englas mid mycelre reðnesse and hine besencton on ða fȳrenan eá, Blickl. Homl. 43, 29. (b) of justifiable severity:—Ðæt hē his hieremonna yfelu tô hnesclíce forberan ne sceal ac mid miclum andan and reðnesse him stiére subditorum mala, quae tolerare leniter non debent, cum magna zeli asperitate corrigere, Past. 21, 5; Swt. 16, 1, I. II. applied to animals, savageness, fierceness, ferocity:—Hē ealle mid wildeórlícre reðnysse (ferocitate ferina) deáþe gesealde, Bd. 2, 20; S. 521, 26. Wið hunda reðnysse and wiðerrǽdnysse: se ðe hafaþ hundes heortan mid him, ne beóþ ongeán hine hundas cēne, Lchdm. i. 372, 3. III. applied to things (reproof, calamity, etc.), harshness, severity:—Seó reðnes ðæs stormes saevitia tempestatis, Bd. 5, 1; S. 614, 9. Reðnes cyles frigoris asperitas, 5, 12; S. 631, 30. Se ðe wunde lácnigean wille gióte wín on ðæt sió reðnes ðæs wínes ða wunde clǽnsige ... Swá eác ðam láreówe is tô monianne ða liéðnesse wið ða reðnesse quisquis sanandis vulneribus praeest, in vino morsum doloris adhibeat ... Miscenda ergo est lenitas, cum severitate, Past. 17, 11; Swt. 125, 10–13. Síe ðǽr eác reðnes ræs ðeáh tô stíþ sit vigor, sed non exasperans, Swt. 127, 2. Ðonne sió lár wint on reðnesse suíður ðonne mon niéde scyle cum sese increpatio, plus quam necesse est in asperitatem pertrahit, 21, 7; Swt. 167, 8.

reðra, an; m. An oarsman, sailor, rower:—Reðra nauta, Ælfc. Gr. 7; Som. 6, 43. Reðra remex, 9, 61; Som. 13, 47: Wrt. Voc. i. 48, 9: 63, 77. Roeðra, ii. 119, 3. Roedra, Ep. Gl. 22 d, 25. v. ge-reðra, rōðer.

reðran, -reðre, -reðru. v. reðian, þri-reðre, ge-reðru.

reðscipe, es; m. Rage, fierceness, fury:—Reðscipas vel hátheortnessa furias, iras, Wrt. Voc. ii. 151, 77.

rēwet[t], es; m. n. (?) I. rowing:—Forhwí ne fixast ðu on sǽ? Hwilon ic dó, ac seldon, forðan micel rēwyt mē ys tô sǽ quia magnum navigium mihi est ad mare, Coll. Monast. Th. 24, 3. On rēwette swincende laborantes in remigando, Mk. Skt. 6, 48. Hí wǽron on rēwette, Homl. Th. i. 162, 10. On ðǽre sǽ swuncon on nihtlícum rēwette, ii. 384, 25. Gif hwá hreóhnysse on rēwytte þolige, Lchdm. i. 302, 5. II. a ship; navigium:—Lǽtaþ ðæt nett on ða swíðran healfe ðæs rēwettes (nauigii), Jn. Skt. 21, 6: Homl. Th. ii. 290, 11.

ribb, es; n. A rib:—Ribb costa, Wrt. Voc. i. 65, 17. Rib, ii. 105, 29. Ðá genam hē án ribb of his sídan and gefilde mid flǽsce ðǽr ðǽr ðæt ribb wæs. And geworhte ðæt ribb tô ánum wífmen, Gen. 2, 21–22. Hæfde fela ribba, Exon. Th. 415, 9; Rä. 33, 8. Rib reáfiaþ reðe wyrmas, 373, 21; Scel. 113. Hwílum cnysseþ ðæt sár on ða rib, Lchdm. ii. 258, 4. [O. H. Ger. rippa; f. rippi; n.: Icel. rif; n.] v. hrycg-ribb.

ribbe, an; f. The herb hounds-tongue; cynoglossum officinale:—Ribbe cinoglosa, Wrt. Voc. i. 286, 23: ii. 104, 2: canes linga, 102, 51: quinquenerbia, i. 68, 33. Ribbe. Ðás wyrte ðe man cynoglossam and

ôðrum naman ribbe nemneþ, and hý eác sume men *linguam canis* hāteþ, Lchdm. i. 210, 16–19. Ribban seáw, ii. 40, 29. Genim ribban, 36, 23.

ribb-spácan; *pl. n.* 'Rib-spokes,' *the brisket* (?) [cf. *Icel.* bring-spelir 'breast-tails,' *the brisket* or *part where the lower ribs are joined with the cartilago ensiformis*] :—Ribbspácan *radiolus*, Wrt. i. 283, 47.

Ríc-, -ríc=ríce, q. v. are found in English, as in other dialects, helping to form proper names. For a list of such names see Txts. pp. 629–630, and for *O. H. Ger.* Graff ii. 389.

ríca, an ; *m. A powerful person, ruler* :—Feórðan dǽles ríca *tetrarcha*, Lk. Skt. 3, 1 : 9, 7. Nán ðara cyninga ðe cumaþ æfter mē, oððe ealdorman oððe óðer ríca, Chart. Th. 243, 13. Wulf biþ se unrihtwísa ríca ðe bereáfaþ ða eádmódan, Homl. Th. i. 242, 3. Hē nolde ólǽcan ǽnigum rícan mid geswǽsum wordum, ii. 514, 13. Ðonne gesihst ðū ða unrihtwísan cyningas and ða ofermódan rícan bión swíðe unmihtige, Bt. 36, 2 ; Fox 174, 27. v. fyðer-, land-ríca, ríce.

rícceter. v. ríceter.

ríce ; *adj.* **I.** of persons, (a) *powerful, mighty, great, possessed of power* :—Oft gebyreþ ðæm monþwǽran ðonne hē wierð riéce (ríce, Cott. MSS.) ofer óðre menn *nonnunquam mansueti, cum praesunt*, Past. 40, 1 ; Swt. 287, 23. Freá ælmihtig biþ á ríce ofer heofonstólas heágum þrymmum, Cd. Th. 1, 14 ; Gen. 7. Ríce þeóden (*God*), 53, 21 ; Gen. 864 : (*Hygelac*). Beo. Th. 2422 ; B. 1209. Ríce randwíga (*Æschere*), 2600; B. 1298. Ríce Drihten *Dominus*, Ps. Th. 96, 1. Wite se ríce man (*vir potens*) ðe him God hæfþ micelne welan and æhta ðyses lífes tó forlǽten, L. Ecg. C. 2 ; Th. ii. 136, 3. Hū mæg ðǽr ánes ríces monnes naman cuman ðonne ðǽr mon furþum ðære burge naman ne geheórþ ðe hē on hámfæst biþ, Bt. 18, 2 ; Fox 64, 2. Ríccræ wífe hrægl *regillum vel peplum vel pella vel amiculum*, Wrt. Voc. i. 40, 32. Hē nǽnigum rícum men ǽfre ǽnig feoh syllan wolde *nullam potentibus saeculi pecuniam umquam dare solebat*, Bd. 3, 5 ; S. 527, 12. Drihten ne wandaþ for rícum ne for heánum *Dominus personam non accipit*, Deut. 10, 17. Ne dēm nán unriht ne árwurða ðone rícan *non injuste judicabis nec honores vultum potentis*, Lev. 19, 15. Áhóf ic rícne (riicnæ, Ruth. Cross) cyning (*Christ*), Rood Kmbl. 88 ; Kr. 44. Se cyning and se bisceop and monige óðre ǽfæste weras and ríce *rex cum antistite et aliis religiosis ac potentibus viris*, Bd. 4, 28 ; S. 606, 12. Guman ríce and heáne *men, great and small*, Exon. Th. 415, 18 ; Rä. 33, 13. Ríccera *potentum*, Wülck. 253, 29. Ríccra gesetnes *senatus consultum*, Wrt. Voc. i. 20, 66. Ríccra manna need *vis potestatis*, 21, 28. Him mon þyngode tó ðám rícum (*the judges*), Bt. 38, 7 ; Fox 208, 29. Hē áwearp ða rícan (*potentes*) of setle, Lk. Skt. 1, 52. Hē (*God*) hæfþ nænne rícran, ne furþum nænne gelícan, Bt. 42 ; Fox 258, 5. Gyf ðū ðæt gerǽdest ðe hér rícost eart *if you decide on this who are here in command*, Byrht. Th. 132, 55 ; By. 36. Wæs Alexandreas ealra rícost monna cynnes, Exon. Th. 319, 21 ; Vid. 15. Monege óðre ðe of Macedonian rícoste wǽron *multi Macedoniae principes*, Ors. 3, 9 ; Swt. 130, 24. Hē hēt ácwellan ða rícostan witan and ða æðelestan, Met. 9, 25. (b) *rich, possessed of wealth* :—Eáðere ys olfende tó farenne þurh nǽdle þyrel ðonne se ríca and se welega on Godes ríce gá *facilius est camelum per foramen acus transire quam divitem intrare regnum Dei*, Mk. Skt. 10, 25. Ðá ðá se Hǽlend spræc be ðam rícan, ðá cwæþ hē : 'Sum ríce man wæs'... Cúð is eów ðæt se ríca biþ namcúðre on his leóde ðonne se þearfa, Homl. Th. i. 330, 3–6. 'Hē forlét ða rícan ídele.' Ðæt sind ða rícan ða ðe mid módignysse ða eorþlican welan lufiaþ swíðor ðonne ða heofonlícan. Fela ríccra manna geþeóþ Gode, ðæra ðe swá dóþ. Swá swá hit áwriten is : 'Ðæs rícan mannes welan sind his sáwle álýsednyss,' 204, 3–7. Be ríccera (rícra manna, W. F.) and þearfena (bearna) andfenge. Gif hwylc ríce mon and æþelboren his bearn Gode on mynstre geoffrian wile, R. Ben. 103, 9–11. Rícra grundleás gítsung æhta, Met. 7, 14. Rícum mannum *divitibus*, Bd. 3, 5 ; S. 527, 9. **II.** of things, *strong, powerful, mighty, potent* :—Wǽron hyra rǽdas ríce, siððan hié rodera waldend wið ðone hearm gescylde, Cd. Th. 245, 3 ; Dan. 457. God rícum mihtum wolde ðæt him eorþe geseted wurde, 6, 34 ; Gen. 98. Gegníd swefl tó duste ... meng wið ealde sápan, and sie swefl rícra *let the sulphur be the stronger ingredient*, Lchdm. ii. 108, 16. Sió (*jaundice*) biþ ealra ádla rícust, 106, 20. [*Goth.* reiks *mighty, powerful, having authority, great* : *O. Sax.* ríki : *O. Frs.* ríke : *O. H. Ger.* ríhhe *magnus, potens, magnificus, dives* : *Icel.* ríkr. The word passed into the Romance tongues. *Fr.* riche : *Ital.* ricco : *Span.* rico ; ricos omes *the grandees.*] v. med-, sige-, woruld-ríce.

ríce, es ; *n.* **I.** *power, authority, dominion, rule, empire, reign*, (a) *referring to sovereigns or nations* :—Tó becume ðín ríce *adveniat regnum tuum*, Mt. Kmbl. 6, 10. Biornwulfes ríce Mercna cyninges *the reign of Biornwulf king of Mercia*, Chart. Th. 70, 8. Ealle stærwríteras secgaþ ðæt Asiria ríce æt Ninuse begunne ... From ðæm ǽrestan geáre Ninuses ríces óþðæt Babylonia burg getimbred wæs wǽron lxiiii wintra ... ðý ilcan geáre gefeóll Babylonia and eall Asiria ríce and hiora anwald, Ors. 2, 1 ; Swt. 60, 25–32. Wæs Maximianes ríce brád, Exon. Th. 243, 10 ; Jul. 8. Ríces *imperii*, Wrt. Voc. ii. 44, 42. Wihtrǽde ríxigendum ðē fíftan wintra his ríces, L. Wih. prm.; Th. i. 36, 5. Under

fíftiga cyninga ríce, Ors. 1, 8 ; Swt. 42, 4. Tó ríce fón *to become king, assume the royal authority*, 4, 6 ; Swt. 178, 19 : Chr. 675 ; Erl. 36, 10 : 754; Erl. 48, 17. Hér Certic and Kynríc onféngon West-Seaxna ríce ... and siððan ríxadon West-Seaxna cynebarn of ðam dæge, 519 ; Erl. 15, 24. Hér Ceadwalla ongan æfter ríce winnan, 685 ; Erl. 40, 16. On ðæs cyninges ríce foreweardum *cujus regni principio*, Bd. 5, 2 ; S. 614, 24. Dū ealle cyninga ða ðe on Breotene wǽron ǽr ðē in mihte and on ríce (*potestate*) oferstígest, 2, 12 ; S. 514, 9. Ymb xxxi wintra ðæs ðe hē ríce hæfde *after he had reigned thirty-one years*, Chr. 755 ; Erl. 48, 26. Wē witon ðæt onwealdas from him sindon wē witon eác ðæt ealle ríca sint from him forðon ealle onwealdas of ríce sindon. Nū hē ðara lǽssena ríca reccend is hū micle swíðor wēne wē ðæt hē ofer ða máran síe *omnem potestatem a Deo esse (omnes) recognoscunt. Quod si potestates a Deo sunt, quanto magis regna, a quibus reliquae potestates progrediuntur? Si autem regna diversa, quanto aequius regnum aliquod maximum*, Ors. 2, 1 ; Swt. 58, 23–26. (b) *referring to others in authority* (bishops, consuls, etc.) :—Biscepes burgbryce mon sceal bētan, ðǽr his ríce biþ *where he has jurisdiction*, L. In. 45 ; Th. i. 130, 8. Brihtwold biscop fēng tó ðam ríce (biscopstóle, MS. F.) on Wiltúnscíre *Brihtwold became bishop of Wiltshire*, Chr. 1006 ; Erl. 140, 2. Ðæt is ðæt hí (*men*) swíðost wilniaþ tó begitanne, wela and weordscipe and ríce, Bt. 24, 4 ; Fox 86, 28. Nán man for his ríce ne cymþ tó cræftum ac for his cræftum hē cymþ tó ríce and tó anwealde, 16, 1 ; Fox 50, 20–22. Hwí ðū (*Boethius*) swá manigfeald yfel hæfdest on ðam ríce ðe hwíle ðe ðū hit hæfdest, 27, 2 ; Fox 96, 13. Ne forsáwe hē (*Catullus*) nó ðone óðerne (*Nonium in curuli sedentem*) swá swíðe, gif hē nán ríce ne nǽnne anweald næfde, 27, 1 ; Fox 96, 7. Biþ ǽlc dysig mon ðý unweorðra ðe hē máre ríce hæfþ, 27, 2 ; Fox 98, 11. **II.** *the district in which power is exercised, a kingdom, realm, a diocese* :—Biscop *episcopus* ; bisceopscír *vel* biscopríc *dioecesis* ; cyncg *rex* ; ríce *regnum*, Wrt. Voc. i. 42, 2–6. Eal ðæt ríce widgeondan Jordanem *omnis regio circum Jordanen*, Mt. Kmbl. 3, 5. Gif him ðæt ríce losaþ *if heaven be lost to them*, Cd. Th. 28, 12 ; Gen. 434. Hū mihtest ðū sittan onmiddum gemǽnum ríce (*intra commune omnium regnum*) ðæt ðū ne sceoldest ðæt ilce geþolian ðæt óðre men, Bt. 7, 3 ; Fox 22, 17. Danaus, of his ríce ádrǽfed *regno pulsus*, Ors. 1, 8 ; Swt. 40, 17. Eall Italia ríce ðæt is betwux ðám muntum and Sicilia ðam eálonde, Bt. 1 ; Fox 2, 4. Ðá fērde ðeós spǽc embe eall ðæt ríce (*regionem*), Lk. Skt. 7, 17. Hē wealdeþ sídum rícum, Ps. Th. 71, 8. Of rícum (*regionibus*) hē gaderode hig, Ps. Spl. 106, 2. Se deófol æteówde him ealle middangeardes rícu, Mt. Kmbl. 4, 8. Ða heofonlícan ríco, Bd. 5, 19 ; S. 641, 15. Ða ēcan ríceo, 2, 5 ; S. 507, 7. **II** a. *the people inhabiting a district, a nation* :—Cumaþ folc feorran tógædere and rícu eác, Ps. Th. 101, 20. [*Goth.* reiki *power, authority* : *O. Sax.* ríki : *O. Frs.* ríke : *O. H. Ger.* ríhhi *regnum, imperium, regio* : *Icel.* ríki.] v. abbod-, bisceop-, brego-, bryten-, cyne-, eard-, eást-, eorþ-, ēðel-, fæder-, gum-, heofon-, þrym-, west-, woruld-ríce.

ríce-dóm, es ; *n. Power, rule, dominion* :—Ðín rícedóm ofer ús ríxie 'thy kingdom come,' Wulfst. 125, 9. [*O. Sax.* ríki-dóm *power* : *O. Frs.* ríke-dóm : *O. H. Ger.* ríhhi-tuom *imperium* ; *divitiae* : *Icel.* rík-dómr *power* ; *wealth.*]

rícels, ricene, ricenlíce. v. rēcels, recene, recenlíce.

rícen[n], e ; *f. A female endowed with power, a goddess* :—Rícenne *Diane*, Wrt. Voc. ii. 26, 76 : 86, 63.

ríceter[e], es ; *n.* **I.** *power, dominion, rule, greatness, glory* :—Ríceter *gloria*, Germ. 389, 41. Wē ne sceolon ða rícan for heora rícetere wurðian *we are not to honour the great ones for their greatness*, Homl. Th. i. 128, 22. Ðam láreówe gedafenaþ ðæt hē hogie hū manegra manna sáwle hē mǽge Gode gestrýnan ... ná hū micel hē mǽge mid his rícetere him tó geteón *it behoves the teacher to strive how many men's souls he can gain for God, not how much he can draw to himself by his power*, ii. 532, 30. Gyf kyng mid his rícetere his folc ofsit, ðon biþ hē *tyrannus*, Ælfc. Gr. 50, 20 ; Som. 51, 47 : Homl. Th. i. 242, 4. Wite se abbod, ðæt he gýmenne ðara untrumra sáula tó rihtre lácnunge underfēng, and ná for rícetere ðe hē ofer ða hæbbe ðe hále syndon, R. Ben. 51, 12. Ne ongyte wē ðæt ðǽr ǽnigra háda andfencg wǽre, ðæt is ðæt ǽnig líues rícetere, ac ǽlc be his neóðe and untrumnesse ancnáwen wǽre *we do not understand that in this case there was any acceptance of persons, that is that recognition was made of any one in proportion to the greatness of his position in life, but of each according to his need and weakness*, 57, 21. Smeáge se abbod hū hē swíðor ðam sáwlum fremian mǽge, ðonne hē hogige embe rícitere his andwealdes, 118, 21. Hríceter *monarchium, principatum, regnum*, Hpt. Gl. 414, 17. Ríciter, 422, 26 : 511, 11. Rícetere ł ealdordóm, 453, 41 : 465, 26. Ríceter *potentiam*, Blickl. Gl. Ðone ealdordóm and ðæt ríceter ðe se reccere for monigra monna þearfe underfēhþ hē hine sceal eówian útan, Past. 17, 7 ; Swt. 119, 61. Hē dyde him ðæt ríceter tó sida and tó gewunan *ministerium regiminis vertit in usum dominationis*, 17, 9 ; Swt. 121, 19. **II.** *power improperly used, violence, force* :—Hē (*Lucifer*) wolde mid rícetere him ríce gewinnan, Ælfc. T. Grn. 2, 42. Ðæt nán ðara cyninga ðe cumaþ æfter mē oððe ealdorman oððe óðer ríca mid ǽnigum ríccetere oððe unrihte ðiss ne áwende, Chart. Th. 243, 13:

Homl. Th. i. 82, 21. Hú mæg, oððe hú dear ǽnig lǽwede man him tó geteón þurh ríccetere Cristes wícan ? ii. 592, 27.

rícettan (?) *to rule :*—Rírciten (riccetan?) *gubernare,* Hpt. Gl. 414, 20.

ríc-líc ; *adj. Great, splendid, magnificent :* — Ungemetlíce ríclíc lýf *excessively splendid mode of life,* Shrn. 184, 8. [O. H. Ger. ríh-líh *splendidus : Icel.* ríku-ligr.] v. next word.

ríclíce ; *adv.* I. *powerfully, with power, as one possessing power :*—Gē budon swíðe ríclíce and swíðe ágendlíce *vos cum austertate imperabatis eis et cum potentia,* Past. 19, 2 ; Swt. 145, 5. On ðám dagum ríxode Æþelbyrht on Cantwarebyrig ríclíce, and his ríce wæs ástreht fram ðære micclan eá Humbre óð súðsǽ, Homl. Th. ii. 128, 18. Ríclíce ꝼ stranglíce ꝼ ríclícost *potentissime,* Ps. Lamb. 44, 4. II. *splendidly, sumptuously :*—Sum welig man dæghwamlíce ríclíce (*splendide*) gewistfullude, Lk. Skt. 16, 19. [O. H. Ger. ríhlího *splendide, festive, mirifice : Icel.* ríkuliga *magnificently ; strictly* (of observance).]

rícs. v. rysc.

rícsere, es ; *m. A ruler :*—Rícsares aldormen *dominationes principatum,* Rtl. 113, 12.

rícsian ; *p.* ode. I. *to exercise or have power, to rule, govern, reign :*—Eálá ðú scippend heofones and eorþan ! ðú ðe on ðam ēcan setle rícsast ! Bt. 4 ; Fox 6, 30. Hē rícsaþ (*regnabit*) on ēcnesse, and hys ríces ende ne byþ, Lk. Skt. 1, 33. Ríxaþ, Ps. Th. 9, 36. Ríhcsaþ, Ps. Spl. 95, 9. ' Hī rícsodon (-edon, Hatt. MS.) næs ðeáh mínes þonces ' . . . Ða ðe swá rícsiaþ (-ieaþ, Hatt. MS.) hī rícsiaþ of hira ágnum dóme *ipsi regnaverunt, et non ex me* '. . . *Ex se regnant, qui* . . . , Past. 1, 2 ; Swt. 26, 14–16. On ðám dagum ríxode Æþelbyrht cyning on Cantwarebyrig, Homl. Th. ii. 128, 17. Circe rícsode on ðam íglonde, Met. 26, 57. Ríhcsode *regnavit,* Ps. Spl. 92, 1 : 96, 1. Gif ðín willa síe ðæt rícsie se ðe on róde wæs, Elen. Kmbl. 1544 ; El. 774. Se mǽra wyrhta ðe ríhsigende wylt eal ðæt hē geworhte, Lchdm. iii. 432, 15. Wihtrǽde ríxigendum, L. Wih. prm. ; Th. i. 36, 4. II. *with the idea of supremacy secured by, or exercised with, force or violence, to domineer, dominate, tyrannize, exercise violence :*—Swá nū ríxiaþ gromhýdge guman, Exon. Th. 445, 26 ; Dóm. 13. Deáþ rícsade ofer foldbūend, 154, 16 ; Gū. 843. Ríxade, 154, 2 ; Gū. 836. Se þeódsceaða (*famine*) rícsode, Andr. Kmbl. 2233 ; An. 1118. Swá ríxode and wið rihte wan ána wið eallum (*of Grendel's successful raids on Hrothgar's hall*), Beo. Th. 290 ; B. 144. Án ongan deorcum nihtum draca rícsian, 4429 ; B. 2211. Gif wē áslaciaþ ðæs weddes ðe wē seald habbaþ, ðonne mǽge wē wēnan ðæt ðás þeófas willaþ ríxian gyta swíðor ðonne hig ǽr dydon *these thieves will get the upper hand yet more than they did before,* L. Ath. v. 8, 9 ; Th. i. 238, 23. Ðæt hē mǽge ríxian and wealdan ealra his feónda and dón him tó yfele ðæt ðæt hē wylle *omnium inimicorum suorum dominabitur,* Ps. Th. 9, 25. Deáþ him furðor ne biþ rícsend *mors illi ultra non dominabitur,* Rtl. 26, 33. Drihten rícsandra *Dominus dominantium,* 101, 10. II a. *of things, to prevail :*—Ða yfelan wǽtan weorþaþ gegaderode on ðone magan, and ðǽr ríxiaþ mid scearfunga innan, Lchdm. ii. 176, 7. On ðisse þeóde ríxode unrihta fela *in this nation many a wrong has prevailed,* Wulfst. 128, 3. Gif preóst forhele hwæt on his scriftscíre betweox mannum tó unrihte ríxigen (ríxige? ríxigende?) gebête hē *if a priest conceal anything in his district between men that may have force to cause injustice, let him make amends,* L. N. P. L. 42 ; Th. ii. 296, 14. [O. H. Ger. ríhhison *regnare.*] v. ofer-rícsian.

rícsiend, es ; *m. A ruler :*—Rícsand *rector,* Rtl. 102, 15.

rícsung, e ; *f. Rule, dominion :*—Rícsunges *dominationis,* Rtl. 174, 19.

rid *a swinging, swaying ; in* sand-rid *a quicksand.* v. rídan. [Icel. rið *sway, swing.*]

rídan ; *p.* rád, *pl.* ridon. I. *to ride on horseback ; equitare :*—Hwílum ic on wloncum wicge ríde, Exon. Th. 489, 14 ; Rä. 78, 7. Hwá rít intó ðam porte *quis equitat in civitatem ?* Ælfc. Gr. 5 ; Som. 3, 52. Ðín cyning rít uppan tamre assene, Mt. Kmbl. 21, 5. Hú ne wást ðú ðæt nán mon for ðý ne rít ðe hine rídan lyste, ac rít for ðý ðe hē mid ðære ráde earnaþ sume earnunga, Bt. 34, 7 ; Fox 144, 5–7. Ðonne rídeþ ǽlc hys weges, Ors. 1, 1 ; Swt. 21, 4. Sum mon rád be ðære stówe, Bd. 3, 9 ; S. 533, 30. Him (*the Danes*) Ælfréd and cyninges þegnas oft ráde on ridon, Chr. 871 ; Erl. 76, 11. Ofer ðý cræte curran, ofer ðý cwēne reodan, Lchdm. iii. 32, 10. Ymbe hlǽw riodan hildedeóre, Beo. Th. 6319 ; B. 3170. Rídan ða yldestan men tó ðære byrig, L. Ath. i. 20 ; Th. i. 208, 29. Deáh ðe hē gewuna wǽre ðæt hē má eode ðonne ride, Bd. 3, 14 ; S. 540, 17. Ne wæs álýfed ðæt hē mōste bútan on myran rídan, 2, 13 ; S. 517, 7 : 4, 3 ; S. 566, 33. Nalæs rídende on horse, ac on his fótum gangende, 3, 28 ; S. 560, 33. Ðes rídenda here *hic equester exercitus,* Ælfc. Gr. 9, 18 ; Som. 10, 2. Hē ásent rídendne here, Wulfst. 200, 21. Rídende men *equites,* Gen. 50, 9. II. *to ride* (of other modes of transport as a vessel *rides* on the waves) :—Wíde rád ðæt scip ofer holmes hrincg, Cd. Th. 84, 3 ; Gen. 1392. Fana up rád *the ensign* (*the fiery pillar*) *moved aloft,* 193, 18 ; Exod. 248. Ðæt hē (*a vessel*) scyle fǽmig rídan ýða hrycgum, Exon. Th. 384, 24 ; Rä. 4, 32. III. *without the idea of progress, to ride* (as in *to ride at anchor*), *to swing, rock :*—Licgaþ mē ymbe írenbendas rídeþ racentan sál

the chain swings (or *presses?*) *on me,* Cd. Th. 24, 3 ; Gen. 372. Swá biþ geómorlíc gomelum eorle tó gebídanne ðæt his byre ríde giong on galgan *that his son swing on the gallows,* Beo. Th. 4882 ; B. 2445. Sum sceal on galgan rídan, seomian æt swylte, óþðæt báncofa blódig ábrocen weorðeþ, Exon. 329, 13 ; Vy. 33. [O. *Frs.* rída : O. H. Ger. rítan *to ride* (*on horseback or in a carriage*) : Icel. ríða *to ride, to swing, sway.*] v. á-, æfter-, be-, for-, ge-, of-, ofer-, óþ-rídan.

ridda, an ; *m.* I. *a horseman, rider :*—Ridda oððe rídende *eques,* Ælfc. Gr. 9, 26 ; Som. 11, 8. Ridda *homo equo portatus,* Wrt. Voc. ii. 143, 65. Se ridda (cf. sum wegfarende mann, l. 168) fērde forþ on his weg, Swt. A. S. Rdr. 100, 175. II. *a mounted soldier :*—Hors and ðone riddan hē áwearp on sǽ, Cant. Moys. Feówer hund and þúsend cræta hē hæfde and twelf þúsend riddena, Homl. Th. ii. 578, 3. Pharao him filigde mid his crætum and gilplícum riddum. . . . Seó sǽ ealle his crætu and riddan oferwreáh, 194, 22–27. Ða gemētte Martinus ǽnne nacodne þearfan, and his nán ne gýmde, ðeáh ðe hē ða riddan ðæs bǽde . . . Ða hlógon ða cempan sume, 500, 19–28.

rídel. v. for-rídel.

rídend, es ; *m. A horseman, knight :*—Rídend swefaþ, hæleþ *knights and warriors sleep the sleep of death,* Beo. Th. 4906 ; B. 2457.

[**rídere, es ;** *m. A knight :*—Hē begeat ðone castel æt Albemare and ðárinne hē sette his cnihtas . . . Æfter ðisum hē begeat má castelas and ðǽrinne his ríderas gelógode, Chr. 1090 ; Erl. 226, 30. [M. H. Ger. rítare : Icel. ríðari.]]

ride-soht. v. hrid-suht.

rídusende (?) *swaying, swinging* (?) :—Rídusende (-aendi, -endi) *pendulus,* Txts. 87, 1562. Cf. rídan (?).

rid-wiga, an ; *m. A mounted soldier :*—þrittig rídwígena *turma,* Wrt. Voc. i. 18, 24.

rif (?) ; *adj. Fierce :*—Ic wiste ðæt úre fór wæs þurh ða lond and stówe ðe missenlícra cynna eardung in wæs rifra wildeóra *ego sciebam per bestiosa loca nobis iter esse,* Nar. 10, 5. Ða rifsta *ferociora,* Rtl. 125, 31.

rífe ; *adj. Rife, abundant :*—Ðere .vii. niht gyf wind byþ, fír byþ swíðe rýfe ðý geáre, Lchdm. iii. 164, 21. [Baluwe þer wes riue, Laym. 631 : 4544 : 20079 : 20672. þenne scullen blissen wurðen riue, 32107. þa hæðene weoren swa riue & auere heo comen, 14542. Alle worldes wele ham is inoh riue, H. M. 29, 22. Lauerd, mi hele so rife, Ps. 26, 1 note. Of him cam kinde mikil and rif, Gen. and Ex. 1252. *Icel.* rífr *munificent, abundant ;* ríf-ligr *large, munificent.*]

rifeling, es ; *m. A kind of shoe or sandal :*—Rifelingas *obstrigilli* (*obstrigilli calcei, qui per plantas consuti, ex superiore parte corrigia constringuntur,* Isidore), Wrt. Voc. i. 26, 25. [Rewelyns, rivlins *shoes* or *sandals of raw hide,* Jamieson's Dict. *See also* riveling in Halliwell's Dict.]

Riffeng *Riphaei montes :*—Of Riffeng ðam beorgum, Ors. 1, 1 ; Swt. 8, 15.

rift, rifte, es ; *n. A veil, curtain, cloak :*—Rift *laena,* Wrt. Voc. ii. 112, 42 : *palla,* 116, 35 : *biuligo, niger velamen,* 126, 38 : *cicla,* 131, 28. Hwítel ꝼ ryft *sagum,* i. 284, 62 : *pallium,* Ps. Surt. 103, 6. Ðý áwundenan ryfte *plumario,* Wrt. Voc. ii. 77, 15. Hē nywolnessa swá swá ryfte (*pallium*) him tó gewǽde woruhte, Ps. Th. 103, 7. Sprenge se sacerd seofon síðon on ðæt ryft (*velum*), Lev. 4, 17. [O. H. Ger. pein-refta *tibarii : Icel.* ript ; *f.* ; ripti ; *n. a kind of cloth* or *linen jerkin.*] v. bán-, cneó-, fleóg-, hálig-, wáh-rift.

rifter, riftr, es ; *m. An instrument for reaping, a sickle, scythe :*—Riftr *falx,* Txts. 62, 430. Rifter, Wrt. Voc. ii. 35, 1. Wíngeardseax, rifte[r] *vel sicul falx,* 146, 76. Riftre *falce,* 79, 69. Riftras *falcis,* 108, 19.

riftere, es ; *m. A reaper :*—Riftre *messor,* Wrt. Voc. ii. 56, 55 : 71, 30. Riptere, i. 74, 68. Ðæt geríp is micel and ða rifteras feáwa, Homl. Th. ii. 520, 16. Riftra[s] *messores,* Mt. Kmbl. Rush. 13, 39. Ic cweþe tó riftrum mínum *dicam messoribus,* 13, 30. Se bær his ryfterum mete tó æcere, Homl. i. 570, 33.

rige, rigen. v. ryge, rygen.

rignan, rínan ; *p.* rínde. [A *strong preterite occurs in the Blickling Gloss,* rán *pluit.* Cf. In Elyes tyme heuene was yclosed þat no reyne ne rone (roon, MS. W. : roen, MS. R. : ron, MS. B. : raynade, MS. C.), Piers P. 14, 62.] I. *to rain, to cause rain to fall,* (a) *with the agent expressed :*—Ic ríne *pluo,* Ælfc. Gr. 28 ; Som. 30, 53. Hē rýnde ofer synfullan grin, Ps. Spl. 10, 7. Hit ágan rínan xl. daga and xl. nihta tósomne ðæm mǽstan rēne, and seó eorþe rínde ealswá swíðe of hire eásprencgum angēn ðam heofenlícan flóde, Wulfst. 217, 1. Hēt hē ða wídan duru wolcen ontýnan heá of heofenum and hider rignan manna *mandavit nubibus desuper, et januas coeli aperuit ; et pluit illis manna manducare,* Ps. Th. 77, 25. (b) *with the agent not expressed :*—Rínþ *pluit,* Ælfc. Gr. 22 ; Som. 24, 6. Hit rínde feówertig daga, Gen. 7, 12 : Mt. Kmbl. 7, 27. Hyt rínde fýr and swefl of heofone *pluit ignem et sulphur de coelo,* Lk. Skt. 17, 29. Swá gelíc swá . . . sý fýr onǽled and ðín heall gewyrmed and hit ríne and sníwe and styrme úte, Bd. 2, 13 ; S. 516, 17. Drihten lēt rínan hagol *pluit Dominus grandinem,* Ex. 9, 24. Hē lǽt rínan (regneþ,

and ða unrihtwísan, Mt. Kmbl. 5, 45. **II.** *to rain, to fall* (of rain) :—Rîneþ blôdig regn æt æfen *a bloody rain shall fall at even,* Blickl. Homl. 91, 34. Mon geseah weallan blôd of eorþan and rînan meolc of heofonum *sanguine e terra, lac visum est manare de coelo,* Ors. 4, 3 ; Swt. 162, 7. [Goth. rignjan : Icel. rigna, regna : O. H. Ger. reganôn.]

riht, es ; *n.* **I.** *that which is straight* or *erect, a plumb line* :— Reht *perpendiculo,* Wrt. Voc. ii. 81, 26. **II.** *that which is straight* in a metaphorical sense, *right, law, canon, rule* :—Mennisc riht *jus ; gecynde riht jus naturale ;* ânre burge riht *jus civile ;* ealra þeóda riht *jus gentium ;* cempena riht *jus militare ;* ealdormanna riht *jus publicum,* Wrt. Voc. ii. 49, 5–10. Riht ł Godes riht *fas,* 38, 71. Scipmanna riht *rodia lex,* i. 20, 50. Reht Rômwara *jus Quiritum,* Rtl. 189, 13. Ryhtes wyrðe *entitled to call in the aid of the law,* Chart. Th. 170, 3. Godes rihtes wiðerbreca, Blickl. Homl. 175, 8. Wiþerwearde Godes beboda and ðæs gástlícan rihtes *opponents of God's commands and of the spiritual law,* 135, 13. Lufige man Godes riht georne, L. Eth. vi. 30; Th. i. 322, 23. Æfter þeáwe árwurðra rihta *juxta morem canonum venerabilium,* Bd. 4, 5 ; S. 572, 5. **III.** *what is in accordance with law, human or divine, what is just or proper, right, justice, equity* :— Ðá cwædon ealle ða weotan ðæt mon uðe ðære cyrcan rihtes swá swel swá ôðerre ... And Eþelwald cwæð ðæt hé ælcre circan aa his ðæla rihtes uðe, Chart. Th. 140, 7–16. Hí rihtes ne gýmdon, Andr. Kmbl. 278 ; An. 139. Gif mon ne mihte hí tô rihte gecyrron, ðæt hí heora wôhdæda geswícan woldan, Blickl. Homl. 45, 27. Godes lof mid rihte begán, 43, 4. Mid rihte Gode þeówian, 45, 29. Hí mê habbaþ benumen mînes naman ðe ic mid rihte habban sceolde, Bt. 7, 3; Fox 20, 28. On ðínum rihte *in aequitate tua,* Ps. Th. 142, 11. Filige rihtlíce ðam rihte *juste quod justum est persequeris,* Deut. 16, 20. Dêmaþ ælcon men riht *quod justum est judicate,* i, 16. Gif wê sôþ and riht on úrum lífe dôn willaþ, Blickl. Homl. 129, 32. Hê á tô æghwylcum sôþ and riht sprecende wæs and dônde, 223, 30. On riht *a-right, by rights, according to what is just or proper* :—Healdan Godes æwe on riht, 45, 9, 22. Ðære cyrican on riht þeówian, 49, 4. Nis eów forboden ðætte æhta habban, gif gé ða on riht strênaþ, 53, 28. Hê férde mid ðam þingum ðe his on riht wæron *quae juris sui erant,* Gen. 31, 21. **III a.** *what is just in the case of a criminal, just punishment, justice* :—Dô ðam þeófe his riht, swá hit ær Eádmundes cwide wæs, L. Edg. H. 2 ; Th. i. 258, 9. **IV.** *what properly belongs to a person, what may justly be claimed, a right, due* :— Ðá sôna wæs Eþelwald ðæs wordes, ðæt hé nô ðes rihtes (*the right to certain woodland*) wiðsacan wolde, Chart. Th. 140, 12. Nelle ic ða rincas rihte benæman *I will not deprive the men of what rightly belongs to them,* Cd. Th. 129, 32; Gen. 2152. Hê hafaþ mec bereáfod rihta gewylces, feohgestreóna, Elen. Kmbl. 1817; El. 910. **V.** *what is due from a person, duty* :—Ðæt biþ ðæs recceres ryht ðæt hê þurh ða stemne his láriówdômes ætiéwe ðæt wuldor ðæs uplícan êðles *debitum rectoris est supernae patriae gloriam per vocem predicationis ostendere,* Past. 21, 5; Swt. 159, 22. Ús is riht micel ðæt wê rodera weard wordum herigen (cf. nú wê sceolan (*debemus*) herigean heofonríces weard, Bd. 4, 24 ; S. 597, 20), Cd. Th. 1, 1 ; Gen. 1. **VI.** *what agrees with a proper standard, what is correct or exact, the rights of a case, the truth* :—h and k geendiaþ on a æfter rihte, Ælfc. Gr. 2 ; Som. 3, 4. Hê ne mæg beón æfter rihte gecweden, bûton ðæt andgit beó ær foresæd, 15 ; Som. 18, 43. Se ðe secgan wile sôþ æfter rihte, Beo. Th. 2103 ; B. 1049. Ðæt wíf sæde him eall ðæt riht *dixit ei omnem veritatem,* Mk. Skt. 5, 33. Ðæt hê be ðære rôde riht getæhte, Elen. Kmbl. 1199 ; El. 601. Hê fram Scē Pauline ðæt riht (*rationem*) leornade ðæs hálgan geleáfan, Bd. 2, 9 ; S. 512, 9. On riht *a-right, correctly, properly* :—Ne eart ðú fullfremedlíce ne on riht gefullad *non es perfecte baptizatus,* 5, 6 ; S. 620, 6. Gif ic ðíne unrótnesse on riht ongiten hæbbe, Bt. 7, 1 ; Fox 16, 7. **VII.** *an account, a reckoning ; ratio, mostly in such phrases as* riht ágildan *to render an account* :—Hió ágeofaþ be ðæm reht *reddent rationem de eo,* Mt. Kmbl. Rush. 12, 36: Lk. Skt. Lind. 16, 2. Hê sceal ealra his dôma riht ágyldan beforan ðæm rihtwísan dêman on dômesdæge, R. Ben. 16, 7. Hê sceal mid his sáwle ânre Gode riht ágyldan ealles ðæs ðe hê on worlde tô wommum gefremede, Blickl. Homl. 113, 3. Wê sceolan riht ágyldan for ealles úres lífes dædum, 63, 31. Reht setta *rationem ponere,* Mt. Kmbl. Lind. 18, 23. Hyra lífes riht, Exon. Th. 84, 18 ; Cri. 1375. Ænige rihte áræsnan *ulla ratione tolerare,* Bd. 5, 12 ; S. 631, 30. [O. Sax. reht : O. Frs. riucht : O. H. Ger. reht *jus, justitia, judicium, aequitas, rectitudo, ratio* : Icel. rêttr ; *m. right, law ; due, claim.*] v. æ-, eald-, êðel-, folc-, ge-, land-, leód-, níd-, on-, sundor-, swæn-, un-, word-, woruld-riht ; *and* á-riht.

riht; *adj.* **I.** *of direction,* (a) literally, *straight, erect, direct* :— Seó heá rôd ryht áræred *raised erect,* Exon. Th. 66, 3 ; Cri. 1066. Rihtes síþfætes *directi callis,* Wrt. Voc. ii. 140, 55. Rihtre *directo,* 27, 69. Rihtre stíge *recto tramite,* Bd. 1, 12 ; S. 481, 8. Faran be rihtum wege (*via publica*), Num. 20, 17. Deós wyrt hafaþ rihte stelan, Lchdm. i. 316, 8. Wæs ðæt ilce hús hwemdragen, nalas æfter gewunan mennisces weorces ðæt ða wægas wæron rihte, ac git swíðor on scræfes onlícnesse ðæt wæs æteówed, and gelômlíce ða stánas swá of ôðrum clife út sceoredon,

Blickl. Homl. 207, 17–20. Ðá áxode ic hwylc se weg tô ðære eá ealra ríhtost wære, Glostr. Frg. 108, 28. Hê ðan rihtestan wege ðyder tô geférde, Guthl. 3 ; Gdwin. 20, 12. (b) metaphorically, *right, straight* :— Ða men ðe bearn habban, tæcean him lífes weg and rihtne gang tô heofonum, Blickl. Homl. 109, 18. Ic him lífes weg rihtne gerýmde, Rood Kmbl. 175 ; Kr. 89. Hí ðá gelædde lífes ealdər ðær hí on rihne weg (*in viam rectam*) eodan, Ps. Th. 106, 6. Wærun Godes mînes gangas rihte, 67, 23. Dôþ hys síðas rihte (ræhta, Lind.) *rectas facite semitas ejus,* Mt. Kmbl. 3, 3. **I a.** *right ; dexter.* v. rihthand. **II.** *agreeable to the spirit of law, human or divine, just, equitable* :—Hwí ne dême gê ðæt riht (*justum*) is? Lk. Skt. 12, 57. Dêmað rihtne dôm *justum judicium judicate,* Jn. Skt. 7, 24. Rihte syndon ðîne dômas, Blickl. Homl. 89, 6. Beóþ rúmmôde ryhta gestreóna *liberal of gains justly acquired,* Exon. Th. 106, 31 ; Gú. 49. **III.** *satisfying the requirements of a law or regulation, legitimate, lawful, regular* :—Riht canonicus *a regular canon,* L. Ælfc. C. 5 ; Th. ii. 344, 12. Heora riht cyning *legitimus rex,* Bd. 4, 26 ; S. 603, 18. On rihte æwe *in lawful marriage,* Wulfst. 304, 21. Ða ðe on rihtum hæmede beóþ *qui in legitimo matrimonio sunt,* L. Ecg. C. 25 ; Th. ii. 150, 22. Ða men ðe bearn habban, læran hié ðám rihtne þeódscipe (*regular discipline*), Blickl. Homl. 109, 17. On rihtne tîman (cf. on gesetne tîman, Th. ii. 296, 3) tîda ringan, L. Edg. C. 45 ; Th. ii. 254, 5. **IV.** *satisfying the demands of conduct, right, proper, fitting* :—Ys hit riht ðæt man ðam câsere gafol sylle *licet nobis dare tributum caesari,* Lk. Skt. 20, 22. Riht ðæt is ðæt ealle geleáffulle men ðis feówertig daga on forhæfdnesse lifgean, Blickl. Homl. 35, 8. Hê næfre nænigum woruldrícum men swíðor onbúgan nolde, ðonne hit riht wære, 223, 29. Ðæt me riht ne þinceþ, ðæt ic óleccan áwiht þurfe Gode, Cd. Th. 19, 11 ; Gen. 289. Reáfode, swá hit riht ne wæs, beám on bearwe, 55, 29 ; Gen. 901 : Byrht. Th. 137, 23 ; By. 190. Gif hire forþsíð getímige ær him, ðonne is hit rihtast ðæt hé ðanon forþ bûton ælcum wífe wunige, Wulfst. 304, 23. Saulscat is rihtast ðæt man gelæste aa æt openum græfe, 311, 12. **V.** *satisfying the requirements of a standard, right, correct, true, orthodox* :— Riht gewrit *orthography,* Ælfc. Gr. 50, 16 ; Som. 51, 22. Hwæt rêce wê hwæt wê sprecan bûton hit riht spræc (*recta locutio*) sý, Coll. Monast. Th. 18, 14. Eálá ðætte ðis moncyn wære gesælig gif heora môd wære swá riht and swá geendebyrd swá swá ða ôðre gesceafta sindon *O felix hominum genus, si vestros animos amor, quo coelum regitur, regat,* Bt. 21 ; Fox 76, 1. Hit is swíðe ryht spell ðæt Plato sæde ... Ða cwæþ ic : 'Ic eom geþafa ðæt ðæt wæs sôþ spell ðæt Plato sæde,' 35, 1–2 ; Fox 156, 8–14. Hê wæs riht cyning *he* (Constantine) *was a true king,* Elen. Kmbl. 26 ; El. 13. Ðæt is se rihta geleáfa, Blickl. Homl. 21, 17 : Bd. 1, 21 ; S. 485, 9. Ðonne wile hê onfón rihtre ondetnesse (*true confession*), Blickl. Homl. 155, 1. Mid rihtum ondgite, 63, 29. Hê ongon hí læran ðæt hí rihte sibbe betwih him hæfdon ... Hí ne woldan rihte Eástran healdan on heora tíd *coepit eis suadere ut pace Catholica secum habita ... Non Paschae Dominicum diem suo tempore observabant,* Bd. 2, 2 ; S. 502, 8–11 : 5, 19; S. 638, 33. Sum mæg godcunde reccan ryhte æ *one can expound the law divine and true,* Exon. Th. 42, 11 ; Cri. 671. Hæbbe ælc man rihtne anmittan and rihte wægan and rihte gemetu ðæt hig náðer ne sín ne máran ðonne hit riht sig *pondus habebis justum et verum et modius aequalis et verus erit tibi,* Deut. 25, 15. [Goth. raihts : O. Sax. reht : O. Frs. riucht : O. H. Ger. reht *rectus, justus, aequus* : Icel. rêttr.] v. forþ-, ge-, on-, un-riht, *and the compounds of which* riht *is the first part.*

rihtæþel-cwén, e ; *f. A legitimate wife* :— Ðæt syndon Godes wiðersacan ... unrihthæmeras ... and ða ðe habbaþ má ðonne heora rihtæþelcwéne, Wulfst. 298, 18.

riht-æw, e ; *f.* **I.** *legitimate matrimony* :—Gehádedum mannum is beboden, ðæt hí cýþan sceolan folce hwæt on hálgum bôcum áwriten is, and hí wísian, hú hí rihtæwe healdan sceolan, Wulfst. 304, 18. **II.** *a legitimate wife* :—Be ðam men ðe hæfþ his rihtæwe (*legitimam suam uxorem*), L. Ecg. P. ii. tit. x ; Th. ii. 180, 16. Se man ðe his rihtæwe forlæt and ôðer wíf nimþ, ii. 8 ; Th. ii. 184, 21. Ðonne hê his rihtæwe ærest hám bringþ, ii. 21 ; Th. ii. 190, 11. Gif hwylc man wið ôðres rihtæwe hæmþ, ii. 10 ; Th. ii. 186, 6.

rihtan ; *p.* te. **I.** *to right, to restore to a proper position that which is displaced, erect, direct* :—Hê mid handum eft on heofonríce rihte rodorstôlas he (*God*) *with his hands again in the heavenly kingdom restored the celestial seats (after the expulsion of the rebellious angels),* Cd. Th. 46, 24 ; Gen. 749. Tô rehtanne foet úsra in woege sibbe *ad dirigendos pedes nostros in via pacis,* Lk. Skt. Rush. 1, 79. **II.** *to right a person, to replace a person in the rights of which he is wrongfully deprived* :— Heó smeádan hú heó mehton monige men ryhtan, ge godcundra háda ge weorldcundra, ge on londum ge on má ðara þinga ðe heó on forhaldne wæran, Chart. Th. 139, 25. **III.** *to make right that which is faulty, set right, rectify, correct, amend* :— Ne sêce ic nô ða bêc, ac ðæt ðæt ða bêc forstent, ðæt ic ðín gewit swíðe rihte *that I may set thy mind thoroughly right,* Bt. 5, 1 ; Fox 10, 20. Sume dêman myccle swíðor rihtaþ Godes folc ðonne hié reáfian earme. Ða dêman beóþ æghwær ge ðæt hié him selfum heora synna bebeorgaþ ge eác ôðre syngiende

rihtaþ, Blickl. Homl. 63, 16-25. Wē boetas t wē hrihð *corrigimus*, Mt. Kmbl. p. 2, 2. Se wīsdōm hine rēte and rihte mid his andsworum, Bt. tit. 5 ; Fox x, 9. Gemeta and gewihta rihte man georne *let weights and measures be made correct with all diligence*, L. C. S. 9 ; Th. i. 380, 24. Ælþeódige mæn, gif hió hiora hæmed rihtan (amend) nyllaþ, of lande gewīten, L. Wih. 4 ; Th. i. 38, 1. Tō rihtanne t tō boetanne *emendasse*, Mt. Kmbl. p. 2, 12.   IV. *to keep right, direct, rule* :—Angelþeóde ðe hē rihte *gens Anglorum quam regebat*, Bd. 3, 3 ; S. 525, 29. Hē ða circan heóld and rihte *rexit ecclesiam*, 3, 20 ; S. 550, 32 : 1, 23 ; S. 485, 23. Ða sylfan stōwe ðe Eata mid abbudes onwalde heóld and rihte (*regebat*), 4, 27 ; S. 604, 41. Ðys eówde stýran and rihtan, Blickl. Homl. 191, 28. [*Goth.* ga-raihtjan : *O. Sax.* rihtian *to erect, to rule* : *O. Frs.* riuchta : *O. H. Ger.* rihtan *erigere, corrigere, dirigere, ordinare, regere* : *Icel.* rētta *to right.*] v. ge-rihtan.

riht-andswaru, e ; *f. An answer that corrects, a reproof, rebuke* :—Se mann ðe on his mūþe næþ nāne rihtandsware *homo non habens in ore suo increpationes*, Ps. Th. 37, 14.

riht-aþelu(o) ; *pl. True nobility* :—Ealle sint emnæþele, gif wē willaþ ðone fruman sceaft geþencan . . . and sidðan eówer ælces ācennednesse. Ac ða ryhtæþelo biþ on ðam mōde, næs on ðam flǣsce, Bt. 30, 2 ; Fox 110, 19 : Met. 17, 20.

riht-cynn, es ; *n. A genuine stock, a race really derived from a particular source* :—Moyses wæs ðæs rihtcynnes *Moses was of the true stock of Abraham*, Wulfst. 13, 6.

riht-cynecynn, es ; *n. A legitimate royal family* :—Antigones him ondrēd Ercoles ðæt ðæt folc hiene wolde tō hlāforde geceósan for ðon ðe hē ryhtcynecynnes wæs *timens ne Herculem Macedones quasi legitimum regem praeoptarent*, Ors. 3, 11 ; Swt. 150, 10. Dauides cynnes, ðæs rihtcynecynnes, Blickl. Homl. 23, 29. [Cf. Se cyng (*Henry*) genam Mahalde him tō wīfe . . . of ðan rihtan Ænglalandes kynekynne, Chr. 1100 ; Erl. 236, 36-39.]

riht-dōnde *right-doing* :—Gif wē beóþ rihtdōnde, Blickl. Homl. 51, 14. Seó duru ðæs heofonlīcan rīces biþ ontýned ðæm rihtgelýfendum monnum and ðǣm rihtdōndum, 61, 10.

rihte ; *adv.* I. *of direction, right, due, directly, straight* :—Swā oft æspringc ūt āwealleþ of clife hārum, and gerēclīce, rihte flōweþ, irneþ wið his eardes (*runs straight on in its course*), Met. 5, 14. Ryhte beeástan him *due east of them*, Ors. 1, 1 ; Swt. 16, 3. [v. eást-, norþ-rihte.] II. *of time, directly, straightway* :—Send nū rihte *mitte jam nunc*, Ex. 9, 19. Nū rihte ðū gesihst *jam nunc videbis*, Num. 11, 23. Gif ic on helle gedō hwyrft ænigne, ðū mē æt byst efne rihte, Ps. Th. 138, 6. [v. ðær-rihte.] III. *in accordance with justice or equity, justly* :—Hē ymbhwyrft eorþan dēmeþ sōðe and rihte *judicabit orbem terrae in aequitate*, 95, 13. IV. *rightly, well, in a manner suited to the circumstances of a case, fittingly, properly, duly* :—Rihte ys hē genemned Jacob ; nu hē beswāc mē, Gen. 27, 36 : Exon. Th. 9, 7 ; Cri. 139. Wæs swīðe ryhte (*recte*) tō ðæm witgan gecweden, Past. 21 ; Swt. 153, 16. Ne ǣnig wið ōðerne getrýwlīce ne þohte swā rihte swā hē scolde, Wulfst. 160, 2 : Cd. Th. 127, 32 ; Gen. 2119. Hū gōd is God ðam ðe mid heortan rihte hycgeaþ *quam bonus Deus his, qui recto sunt corde*, Ps. Th. 72, 1 : 62, 6. Scylan eard niman on ðīnre ansýne ða mid rǣde hēr rihte lifigeaþ *habitabunt recti cum vultu tuo*, 139, 13. V. *correctly, in the proper manner, exactly, accurately, truly* :—Wē biddaþ ðē, Lāreów, ðæt ðū tǣce ūs sprecan rihte (*to speak Latin correctly*), Coll. Monast. Th. 18, 8. Bæd ðæt hē hyra randan rihte heóldon, Byrht. Th. 132, 23 ; By. 20. Swā wæs on ðæm scennum þurh rūnstafas rihte (*correctly*) gemearcod, hwam ðæt sweord geworht ǣrest wǣre, Beo. Th. 3395 ; B. 1695. Swylce hý wǣron rihte . . . swilce hī nū sindon *they were exactly such then as they are now*, Bt. 7, 2 ; Fox 18, 1. Heó is swīðe ryhte feówerscýte *it* (*Babylon*) *is very accurately quadrangular*, Ors. 2, 4 ; Swt. 74, 13. Ryhtor cweþan *to say with greater accuracy*, 5, 1 ; Swt. 214, 9.

rihte-bred, es ; *n. An instrument for measuring, a square* :—Rihtebred *norma*, Wrt. Voc. ii. 60, 20 : 114, 83 : *linea*, 54, 16.

rihtend, es ; *m. A ruler* :—Eálā ðū ælmihtiga scippend and rihtend (*rector*) eallra gesceafta, Bt. 4 ; Fox 8, 10. Hē ðǣhyreþ cyning mǣdlan, rodera ryhtend sprecan, Exon. Th. 50, 10 ; Cri. 798.

rihtere, es ; *m. A ruler, director* :—Ic wāt ðætte God rihtere is his āgnes weorces . . . Gesege mē nū ðū cwist ðæt ðū nāht ne tweóge ðætte God ðisse worulde rihtere sīe . . . *operi suo conditorem praesidere Deum scio* . . . *Dic mihi, quoniam a Deo mundum regi non ambigis* . . ., Bt. 5, 3 ; Fox 12, 5-14. [*O. H. Ger.* rihtari *rector, regulus, rex, judex* : *Icel.* rēttari *a justiciary.*]

rihtes ; *adv. Right, straight* :—Foran rehtes in ða rōde *straight on to the cross*, Cod. Dip. Kmbl. iii. 392, 6.

riht-fæderencynn, es ; *n. Lineal descent* or *descendants on the father's side* :—Hiera ryhtfæderencyn gǣþ tō Cerdice *they are lineally descended on the father's side from Cerdic*, Chr. 755 ; Erl. 50, 33 : 784 ; Erl. 56, 5. Gif heó bearn næbbe, feó ðonne an hire rehtfæderen[cynnes] sió nēste hond, Chart. Th. 481, 21. v. riht-mēdrencynn.

riht·fæstendæg, es ; *m. A regularly appointed fastday* :—Ǣlc ðara

mauna ðe yt oððe drincþ on ðam hālgan lenctene oððe on rihtfæstendagum, Homl. Skt. i. 12, 76 : Wulfst. 117, 15.

riht-fremmend, es ; *m. One acting rightly* :—Ðus reordiaþ ryhtfremmende, Exon. Th. 240, 2 ; Ph. 632. Geát hǣden hildfruma hāligra blōd, ryhtfremmendra, 243, 9 ; Jul. 8. Hǣlu būtan sāre ryhtfremmendum, 101, 9 ; Cri. 1656.

riht-gefremed ; *adj. Rightly constituted, orthodox* :—Forðon ðe hē Wilfriþ rihtgefremedne gemētte *quia catholicum Vilfridum comperit*, Bd. 5, 19 ; S. 638, 34.

riht-gegilda, an ; *m. One who is legally a member of a guild* :—Æt ǣlcon rihtgegyldan, Chart. Th. 606, 14.

riht-geleáffull ; *adj. Holding a true belief, orthodox* :—Fram ðam rihtgeleáffullum bisceope *ab episcopo orthodoxo*, L. Ecg. P. addit. 5 ; Th. ii. 232, 19. Mid ðǣre hālgan and mid ðǣre rihtgeleáffullan gesomnunge *cum sancta ecclesia*, Bd. 3, 17 ; S. 545, 31. For rihtgeleáffulra sibbe *pro pace Catholica*, 2, 2 ; S. 502, 2. Ðæt rīce ðam unrihtwīsan cyninge āferran and on ryhtgeleáffulra and on rihtwīsra anwald gebringan, Bt. 1 ; Fox 2, 19. Rihtgeleáffulum *orthodoxis*, Wrt. Voc. ii. 62, 66.

riht-gelífed ; *adj. Possessed of a true belief, orthodox, catholic* :—Eal rihtgelýfed folc sceal gefeón on ðone his tōcyme, Blickl. Homl. 167, 14. Ðæs rihtgelýfdan geleáfan *orthodoxiae*, Wrt. Voc. ii. 65, 13. Ða hālgan gelaþunge rihtgelýfdan *sanctam aecclesiam catholicam*, Apstls. Crd. Rihtgeléfedan, Blickl. Homl. 111, 9. Of rihtgeléfedum lārum *orthodoxis dogmatibus*, Hpt. Gl. 468, 12. Drihten, ðū ðe cwǣde on ðīnum godspelle tō eallum rihtgelýfedum mannum *omnibus fidelibus hominibus*, L. Ecg. P. iv. 67 ; Th. ii. 226, 39. Rihtgelýfdum, Blickl. Homl. 171, 14. [cf. *Icel.* rētt-trúaðr *orthodox.*]

riht-gelífende *having a true belief, faithful* :—Seó duru rīces biþ ontýned ðæm rihtgelýfendum monnum, Blickl. Homl. 61, 9. Ic beó līfes gǣst on eallum rihtgelýfendum on mē, 185, 34. [Cf. *Icel.* rētt-trúandi *orthodox.*]

riht-geþancod ; *adj. Right-minded, having right thoughts* :—Ða rihtgeþancodon *rectos corde*, Ps. Lamb. 7, 10. Rihtgeþancedon, 10, 3.

riht-gewitt, es ; *n. Right mind* :—Ðā wæs heó of hyre ryhtgewitte *she was out of her mind*, Shrn. 141, 18.

riht-hǣmed, es ; *n. Legitimate matrimony* :—Cirraþ tō eówrum ryhthǣmede, Past. 16, 1 ; Swt. 99, 17. Æfter ðon wǣre on rihthǣmed (riht hǣmed ?) geþeódod *postea in matrimonio jungatur*, L. Ecg. C. 19 ; Th. ii. 146, 2. v. unriht-hǣmed.

riht-hand, a ; *f. The right hand* :—Se Hǣlend be ðǣre ryhthanda mē genam, Nicod. 21 ; Thw. 11, 5. Se Hǣlend Adam be ðǣre rihthand genam, 30 ; Thw. 17, 24.

riht-handdǣda, an ; *m. The actual perpetrator of a crime* :—Gif hwā wrace dō on ǣnigum ōðrum būtan on ðam rihthanddǣdan, L. Edm. S. 1 ; Th. i. 248, 12.

riht-heort ; *adj. Upright in heart* :—Mid rihtheortum *qui recto sunt corde*, Ps. Th. 93, 14. Ðǣm rehtheortum *rectis corde*, Ps. Surt. 111, 4. [*O. H. Ger.* reht-herzi.]

riht-hīwa, an ; *m. A legitimate consort* :—Monige beóþ ðara ðe hié gehealdaþ wið unryhthǣmed and swāðeáh his āgenra ryhthīwena ne brýcþ swā swā hē mid ryhte sceolde *multi sunt, qui scelera quidem carnis deserunt, nec tamen in conjugio positi usus solummodo debiti jura conservant*, Past. 51, 6 ; Swt. 399, 8.

riht-hláford, es ; *m. A rightful lord* :—Gif wīf ofer hire rihthláford ōðerne man hæbbe *si mulier, praeter dominum suum legitimum, alium habet virum*, L. Ecg. P. ii. 7 ; Th. ii. 184, 19.

riht-hláfordhyldu ; *indecl.* : -hyld, e ; *f. Fidelity justly due to a lord* :—Uton beón ā ūrum hláforde holde and getreówe . . . forðam eall ðæt wē ǣfre for rihthláfordhelde dōþ, eal wē hit dōþ ūs sylfum tō mycelre þearfe, Wulfst. 119, 15 : 299, 27. v. hláford-hyldo.

rihting. v. rihtung.

rihtlǣcan ; *p.* -lǣhte *To make right, rectify, correct, amend* :—Gif hē ðonne (*after punishment*) swā ne bēte and rihtlǣce, hē sý of ðam ealdorscype āworpen, R. Ben. 46, 19. Se ðe ǣr ðysum misdyde, ðæt hē hit georne gebēte and rihtlǣce hine sylfne, Wulfst. 277, 2. Uton wē nū ǣlces yfeles geswīcan and rihtlǣcan ūs sylfe on eallan þingan, 174, 30. Æfter ðam ðe hē sylf geriht wearþ hē began georne mynstera wīde geond his cynerīce tō rihtlǣcynne *after his own life was ordered aright, he began to set the monasteries in order*, Lchdm. iii. 440, 2. v. ge-rihtlǣcan.

riht-lǣce, es ; *m. A genuine physician, one who is really a doctor* :—Se ðe his broces bōte sēcþ būton tō Gode sylfum and tō his hālgum and tō rihtlǣcum hē drýhþ deófles wyllan *he that seeks a remedy for his malady except from God and from his saints and from regular doctors, he does the devil's will*, Wulfst. 12, 12.

rihtlǣcung, e ; *f. Correction, making right* :—Ða underþióddan sint tō maniane ðæt hié ðara unþeáwas ðe him ofergesette bióþ tō swīðe and tō þrislīce ne eahtigen . . . ðý lǣs hié for ðære ryhtlǣcinge weorþen upāhæfene, Past. 28 ; Swt. 197, 2. Tō ðǣm dōmbōcum ðe se heofonlīca Wealdend his folce gesette tō rihtlǣcunge ealra forgǣgednyssa, Homl. Th. ii. 198, 20. Hrihtlǣcinge *ratiocinationis*, Hpt. Gl. 481, 78.

riht-laga, an; _m. Right_ or _just law, equity_ :—Rihtlaga is, ðæt man ôðran gebeóde, ðæt hé wylle ðæt man him gebeóde, Wulfst. 274, 11. v. next word.

riht-lagu, e; _f. Right_ or _just law_ :—Oferhogan godcundra rihtlaga, Wulfst. 164, 12. Ða ðe godcunde láre and woruldcunde rihtlage wyrdan on ǽnige wísan, 168, 8.

riht-líc; _adj._   I. _right, just_ :—Rihtlíc _fas, justum_, Hpt. Gl. 460, 16. Ic tôcwýse eówer deófolgyld, and biþ ðonne rihtlíc geþúht ðæt gé geswýcon eówres gedwyldes, Homl. Th. i. 70, 33. Gif hiora hwilc swá heardheort wǽre, ðæt hé nâne hreówsunge ne dyde, ðæt hé ðonne hǽfde rihtlíc wíte, Bt. 41, 3; Fox 248, 16.   II. _right, fitting, adapted to due requirements_ :—Hú wolde ðe lícian gif hwylc swíðe ríce cyning wǽre and næfde nǽnne frýne mon on eallon his ríce, ac wǽron ealle þeówe. Ða cwæþ ic : ' Ne þúhte hit mé nâuht rihtlíc, ne eác gerisenlíc, gif him sceoldan þeówe men þénigan,' 41, 2 ; Fox 244, 26. Rihtlíc ðæt wæs ðæt se blinda be ðæm wege sǽte wǽdliende ; forðon ðe Drihten sylfa cwæþ : ' Ic eom weg sôðfæstnesse,' Blickl. Homl. 17, 30 : 29, 17.   II a. _adapted, fitted, entitled_ :—Ðeáh beóþ ða foremǽrran and rihtlícran tô herigenne ða ðe beóþ mid cræftum gewyrðode, Bt. 30, 1 ; Fox 108, 24.   III. _right, in accordance with reason_ :—Ðæt wǽre rihtlíc tô ongytenne (_merito intelligendum_) ðæt ealle ða ðe Godes willan worhton, fram ðam ðe hí gesceapene wǽron, ðæt hí ðonne wǽron fram him éce méde tô onfônne, Bd. 3, 22 ; S. 552, 21.   IV. _right_ as regards conduct, _righteous_ :—Ðæt biþ rihtlíc líf, ðæt cniht þurhwunige on his cnihthâde, oþ ðæt hé on rihtre mǽdensǽwe gewífige, L. I. P. 22 ; Th. ii. 332, 28. Ðonne mon hwæt ryhtlíces and gerisenlíces geþencþ _quando et si qua jam justa, si qua honesta cogitantur_, Past. 21, 3 ; Swt. 155, 24. Ðonne hé ðæm ryhtlícum inngeþonce his híeremonna foresægþ ða diéglan sǽtenga ðæs lytegan feóndes _quando rectae intentioni audientium hostis callidi circumspectas insidias praedicit_, 21, 5 ; Swt. 163, 13. Eahta sweras syndon ðe rihtlícne cynedôm trumlíce up wegaþ _there are eight pillars that firmly sustain a rightly conducted royal authority_, L. I. P. 3 ; Th. ii. 306, 19. [_O. H. Ger._ reht-líh _fas, jus, justus, regularis, canonicus : O. L. Ger._ reht-, riht-lík : _O. Frs._ riucht-lik : _Icel._ rétt-legr _just, due, meet._]

rihtlíce; _adv._   I. _rightly, justly, with justice_ or _equity_ :—Rihtlíce _juste_, rihtlícor _justius_, rihtlícost _justissime_, Ælfc. Gr. 38 ; Som. 40, 50. Him getîmode swíðe rihtlíce ðæt hí mid hiora ârleásan hláforde ealle forwurdon, Homl. Th. i. 88, 30. Ðú rihtlíce dǽlest mete ðínum mannum, Hy. 7, 70.   II. _rightly, in a manner which suits the circumstances of a case_ :—Swíðe ryhtlíce hit wæs âwriten ǽfter ðæm nítenum ðæt ða heargas wǽron âtiéfrede _recte post animalia idola describuntur_, Past. 21, 3 ; Swt. 157, 6 : 21, 5 ; Swt. 163, 21. Æfter ðon wé singaþ rihtlíce on his lof : ' Hǽl ûs on ðǽm hêhstan,' Blickl. Homl. 81, 27. Hú ne belimpþ se weorþscipe tô ðam ðe hine geweorþaþ? ðæt is tô herianne hwǽne rihtlícor, Bt. 14, 3 ; Fox 46, 13.   III. _rightly, in accordance with rules_ or _regulations, regularly_ :—Gewunelíce ī rihtlíce _rite_, Ælfc. Gr. 38 ; Som. 41, 44. Rihtlíce gehâlgad _canonice ordinatus_, Bd. 3, 28 ; S. 560, 28. Ða þénunge hé rihtlíce gefyllan ne mihte _ministerium regulariter implere nequibat_, 5, 6 ; S. 620, 9. Gif hé rihtlíce (_in such a way as to observe the rules imposed by Christianity_) Cristen beón wille, 4, 5 ; S. 573, 18. Ða munecas beádon hine (_the abbot_) ðæt hé sceolde healdan hí rihtlíce, Chr. 1083 ; Erl. 217, 5. Ða witan cwǽdon ðæt him nán leófre hláford nǽre ðonne heora gecynde hláford gif hé hí rihtlícor healdan wolde (_if he would rule with better observance of the laws_) ðonne hé ǽr dyde, 1014 ; Erl. 150, 7.   IV. _rightly_ as regards conduct :—Wé sceolan gôd weorc wyricean and rihtlíce libban, Blickl. Homl. 75, 13 ; 109, 13. Riht is ðæt gehâdode men ðam lǽwedum wísian hú hí heora ǽwe rihtlícost sculon healdan, L. I. P. 22 ; Th. i. 332, 28. v. on-rihtlíce.

riht-lícettere, es; _m. A thorough hypocrite_ :—Fela manna wyrð þurh deófol forlǽred swá ðæt hé eal ôðer specaþ and ôðer hiwiaþ, ôðer hý þencaþ ; and ða beóþ rihtlíceteras, Wulfst. 54, 14.

riht-líf, es; _n. A right life_ :—Wýf tô onfônne tô rihtlífe, Lchdm. iii. 176, 22.

riht-liþlíc; _adj. Articulate_ :—Rihtliþlícu _articulata_, Wrt. Voc. ii. 9, 46.

riht-médrencynn. v. médren-cynn, _and_ cf. riht-fæderencynn.

riht-meterfers, es; _n. Correct hexameter verse_ :—Ðæt rihtmetervers sceal habban feówer and twentig tîman, Anglia viii. 314, 10.

riht-munuc, es; _m. A true monk_ :—Beóþ rihtmunecas, gif hý libbaþ be ðam geswince, heora âgenra handa, R. Ben. 73, 19.

rihtnes, e; _f._   I. _rightness, straightness, perpendicularity_ :—Perpendicula walþrǽd, ðæt is rihtnesse [þrǽd], Wrt. Voc. ii. 91, 67. Cf. rihtung-þrǽd.   II. _rightness, justice, equity_ :—On rihtnesse _in aequitate_, Ps. Th. 97, 9 : 110, 5. Rehtnise, Rtl. 102, 17.   III. _in the following passage_ rehtnis _glosses_ ratio, Rtl. 113, 32 : 32, 32 : Mt. Kmbl. Lind. 12, 36 : 18, 24 : 25, 19. [_O. L. Ger._ reht-, riht-nussi _justitia : O. H. Ger._ rehtnissa _justitia, aequitas._]

riht-norþanwind, es; _m. A due north wind_ :—Ða sceolde hé ðær bídan ryhtnorþanwindes, Ors. 1, 1 ; Swt. 17, 17.

riht-racu, e; _f. A correct account_ :—Ða lýfde hé ðæt hé môste beón ryhtes wyrðe for mí[n]re forspǽce and ryhtrace, Chart. Th. 170, 4.

riht-regol, es; _m. A correct rule, a canon_ :—Rihtregula _canonum_, Hpt. Gl. 526, 16.

riht-ryne, es; _m. A right course_ :—Se brôc ðeáh hé swíðe of his rihtryne ðonne ðǽr micel stân of ðam heáhan munte oninnan fealþ and hine tôdǽlþ and him his rihtrynes wiðstent, Bt. 6 ; Fox 14, 27-30 : Met. 5, 20. Oncerran ðisne swiftan rodor of his rihtryne, 10, 41.

riht-scilling, es; _m. A lawful shilling_ :—Se rihtscylling byþ â be .xii. penegum _legitimus solidus semper est_ .xii. _denariorum_, L. Ecg. P. iv. 60; Th. ii. 222, 7.

riht-scrífend, es; _m. One who declares the sentence of the law, a lawyer_ :—Rihtscrífend ī dômsettend _jurisconsultus, jurisperitus_, Wrt. Voc. ii. 49, 17 : i. 20, 69.

riht-scytte; _adj. Sure of aim_ :—Sum biþ ryhtscytte, sum leóþa gleáw, sum on londe snel, Exon. Th. 296, 15 ; Crä. 51.

riht-smeáung, e; _f. Right reasoning, argument_ :—Rehtsmeáwung _argumentum_, Mt. Kmbl. p. 11, 10.

riht-tíd, e; _f. A proper time_ :—Hé ða Eástran on heora rihttíde ne heóld _Pascha suo tempore non observabat_, Bd. 3, 17 ; S. 545, 18. v. next word.

riht-tíma, an; _m. A right, proper time_ :—Ælc wuht from Gode wiste his rihttîman, Bt. 5, 3 ; Fox 12, 8.

rihtung, e; _f._   I. _direction, guidance_ :—Bisceope gebyreþ ǽlc rihting . . . Hé sceall gehâdode men gewissian, ðæt heora ǽlc wite hwæt him gebyrige tô dônne, L. I. P. 7 ; Th. ii. 312, 9. Gyrd rihtingce _virga directionis_, Ps. Spl. 44, 8. Him God hâlige ǽ sette tô heora lífes rihtinge, Homl. Th. i. 558, 21. Ðonne mann wísdôm sprecþ manegum tô þearfe and tô rihtinge, Ælfc. T. Grn. 21, 28. Rihtinga _directiones_, Ps. Lamb. 98, 4.   II. _correction, setting right_ :—Rihtingc _correctio_, 96, 2. Bisceopes dægweorc . . . his gebedu ǽrest, and ðonne his bôcweorc, rǽding oðð on rihting (_correcting manuscripts?_), L. I. P. 8 ; Th. ii. 314, 19. On ða gerád ðæt seó bôc heam sý geara, gyf hý hyre beþurfan tô ǽnire rihtinge _on the condition that the charter be ready for them, if they need it for any correction_, Chart. Th. 588, 17.   III. _correction, reproof_ :—For ðære geornfulnesse ðære ryhtinge ne síe hé tô stíð tô ðære wrace _ne correptionis studia privatus dolor exasperet_, Past. 13, 2 ; Swt. 79, 11.   IV. _a direction, rule_ :—Ne scylen hý beón bútan regole, ðæt is lífes rihtinge, R. Ben. 61, 14. Ðisne regul, ðæt is lífes rihtunge, wé âwriton tô ðý ðæt wé hine on mynstre healden, 132, 14.   V. _a translation of the technical term regularis_ [_Regulares apud compotistas, seu computi ecclesiastici conditores, alii sunt solares, alii lunares. Regularis solis sunt numerus invariabilis datus mensi, qui, adjunctus concurrenti, declarat qua feria septimanae quilibet mensis iniret, cujus fuerit regularis. Dicitur regularis a regula quia invariabilis est. Regularis lunaris est numerus invariabilis, datus mensi ad inveniendum lunam in kalendis mensium singulorum, Ducange_] :—De regularibus feriarum dicamus . . . Januarius and October habbaþ twâ rihtinga, and Februarius and Martius and November gladiaþ on fífum, and Aprilis and Julius habbaþ âne rihtinge, and Maius hæfþ þrý, and Agustus mid feówrum glitnaþ, Junius âna hæfþ syx rihtinga, and September and December mid heora seofon gefêrum gladiaþ, Anglia viii. 302, 1-4. Cf. Aprilis hæfþ ânne regularem, 303, 40. De regularibus lunae. Gyf ðú wille witan ðæra rihtinga gesceád ðe geþungene preóstas cweþaþ _lunares_, 305, 8. The word occurs often in the treatise from which these passages are taken. [_O. H. Ger._ rihtunga _regimen, reformatio, emendatio, dispositio._]

rihtung-þrǽd, es; _m. A directing thread, a plumb-line_ :—Wealles rihtungþrǽd _perpendiculum_, Wrt. Voc. i. 39, 64.

riht-weg, es; _m. A right way_ :—Se ðe secge ðæt hé on Crist gelýfe fare se ðæs riht-weges ðe Crist sylf fêrde _qui se dicit in Cristum credere debet ambulare sicut et ipse ambulavit_, Wulfst. 65, 25. Gebringan on rihtwege ða ðe ǽr dweledan, 75, 2 : 49, 19.

riht-wer, es; _m. A legitimate husband_ :—Gif wíf hire rihtwer (_virum suum legitimum_) forlǽt, L. Ecg. P. ii. 8 ; Th. ii. 184, 25.

riht-westende, es; _m. The extreme western limit_ :—Hire ryhtwestende _ultimus finis ad occidentem_, Ors. 1, 1 ; Swt. 8, 32.

riht-willend, es; _m. One whose desires are right_ :—Ðú eart ân ðara rihtwillendra, Bt. 15, 1 ; Fox 10, 6.

riht-wís; _adj. Righteous, just_ :—Rihtwís _justus_, Wrt. Voc. i. 75, 69. Rihtwís _justus_, rihtwísre _justior_, ealra rihtwísost _justissimus_, Ælfc. Gr. 5 ; Som. 4, 65 ; 9, 21 ; Som. 10, 20. Rihtwís dêma, Hy. 6, 7. Se ðe underfêhþ rihtwísne on rihtwíses naman, hé onfêhþ rihtwíses mêde, Mt. Kmbl. 10, 41. Unscyldig ic eom fram ðyses rihtwísan blôde (_a sanguine justi hujus_), 27, 24. Ðé ic geseah rihtwísne ætforan mé, Gen. 7, 1. Dômas ðíne rihtwíse _aequita judicia tua_, Ps. Th. 118, 75, 172. Rihtwíse _non errantes_, Wrt. Voc. ii. 61, 52. Fíftig rihtwísra manna _quinquaginta justi_, Gen. 18, 24. Heó is rihtwísre (_justior_) ðonne ic, 38, 26. Boetius wæs on woruldþeáwum se rihtwísesta, Bt. 1 ; Fox 2, 4. [_O. H. Ger._ reht-wís : _Icel._ rétt-víss.] v. un-rihtwís.

riht-wís (?), e; -wíse (?), an; _f. Righteousness, justice_ :—Rihtwíse and sybbe hý cyston _justitia et pax osculatae sunt_, Ps. Spl. 84, 11. [Cf. _Icel._ rétt-vísa, -vísi.]

riht-wísend, es; *m.* *A Sadducee:*—Ðá hé geseh manega ðæra sunderhálgena and ðæra rihtwísendra tó his fulluhte cumende, Mt. Kmbl. 3, 7.

rihtwísian; *p.* ode *To justify:*—Gé rihtwísiaþ eów ætforan mannum and God cann eówere heortan *vos justificatis vos coram hominibus, Deus autem novit corda vestra* (Lk. 16, 15), Homl. Th. ii. 404, 15. *v.* ge-rihtwísian, riht-wís.

riht-wísian; *p.* ode *To direct aright, rule:*—Ðú cwist ðæt ðú náht ne tweóge ðætte God ðisse worulde rihtere síe (rihtwísige, Cott. MS.) *a Deo mundum regi non ambigis*, Bt. 5, 3; Fox 12, 14. *v.* wísian.

rihtwís-líc; *adj. Righteous, just, rational:*—For ryhtwíslícum andan *per zelum justitiae*, Past. 17, tit.; Swt. 107, 7 : 21, 6; Swt. 163, 20. [*O. H. Ger.* rehtwís-líh *rationabilis.*] *v.* next word.

rihtwíslíce; *adv. Rationally, justly:*—Hú mæg ǽnig man ryhtwíslíce and gesceádwíslíce ácsigan, gif hé nán grot rihtwísnesse on him næfþ, Bt. 35, 1; Fox 156, 5 : Met. 22, 45. *v.* preceding word.

rihtwísness, e; *f.* **I.** *righteousness, justice:*—Óðer mægen (ðære sáwle) is *justitia*, ðæt is rihtwísnys; þurh ða heó sceal God wurðigan and rihtlíce libban, Homl. Skt. i. 1, 159. On rihtwísnesse wege *in via justitiae*, Mt. Kmbl. 21, 32. Abram gelífde Gode and hit wæs him geteald tó rihtwísnisse (*ad justitiam*), Gen. 15, 6. Ealle rihtwísnesse gefyllan, Mt. Kmbl. 3, 15. Rihtwísnysse sprecan, Ps. Spl. 57, 1. Geléd mé on rightwísnysse ðíne, 5, 9. Rechtwísnisse, Ps. Surt. 44, 5. Gif hí míne rihtwísnessa (*justificationes*) gewemmaþ, Ps. Th. 88, 28. **II.** *rightness, reasonableness, reason:*—Ða sceare onfón sculon ðe wé gehýraþ fulle beón ealre rihtwísnesse *hanc accipere tonsuram quam plenam esse rationis audimus*, Bd. 5, 21; S. 643, 23. Hú mæg ǽnig man ryhtwíslíce and gesceádwíslíce ácsigan, gif hé nán grot rihtwísnesse on him næfþ? Nis nán swá swíðe bedǽled ryhtwísnesse, ðæt hé nán ryht andwyrde nyte, gif men ácsaþ. Plato cwæþ : 'Swá hwá swá ungemyndig síe rihtwísnesse, gecerre hine tó his gemynde, ðonne fint hé ðǽr ða ryhtwísnesse gehýdde mid ðæs líchoman hæfignesse, Bt. 35, 1 ; Fox 156, 5–12 : Met. 22, 43–60. *v.* un-rihtwísness.

riht-wrítere, es; *m. One who writes correctly:*—Rihtwrítera *orthographorum, rectorum scriptorum*, Hpt. Gl. 410, 72. Rihtwríterum *ortagraphorum*, Wrt. Voc. ii. 64, 22 : 75, 41.

riht-wuldriende *orthodox:*—Wé wǽron smeágende rihtne geleáfan and rihtwuldriende. Ðás wé syndon árfæstlíce fyligende and rihtwuldriende *tractantes fidem, rectam et orthodoxam . . . Hos sequentes nos pie atque orthodoxe*, Bd. 4, 17 ; S. 585, 28–34.

rím, es; *n. Number:*—Rím miclade monna mǽgþe geond middangeard, Cd. Th. 75, 21; Gen. 1243. His dógora wæs rím áurnen, 98, 6; Gen. 1626. Seofon geteled rímes, 80, 30; Gen. 1336. Ic feówertig folce ðyssum wintra rímes wunade neáh *forty years in number I dwelt near this folk*, Ps. Th. 94, 10. Æfter ríme fíf Moyses bóca *juxta numerum librorum*, Bd. 1, 1; S. 474, 1. Weaxendum ðam ríme geleáfsumra *crescente numero fidelium*, 4, 5; S. 573, 12. Gecuron hí of heora ríme gemetfæstne man *elegerunt ex suo numero virum modestum*, 5, 11; S. 625, 43. On ríme *in catalogo*, Wrt. Voc. ii. 84, 31. Hundtwelftig geteled ríme wintra, Cd. Th. 76, 27; Gen. 1263. On wera ríme gewurðod, 127, 8; Gen. 2107. Rím dæga mínra *numerum dierum meorum*, Ps. Surt. 38, 5. Is nú worn wintra sceacen twá hund oððe má geteled ríme, ic ne mæg áreccan nú ic ðæt rím ne can, Elen. Kmbl. 1267; El. 635. Meotod wolde manna rím, fela þúsenda, forþ gelǽdan, Cd. Th. 289, 22; Sat. 401. [*O. Sax.* un-rím : *O. Frs.* rím : *O. H. Ger.* rím *numerus* : *Icel.* rím.] *v.* cneó-, dæg-, dógor-, ende-, fæðm-, ge-, geár-, getæl-, mann-, niht-, scilling-, un-, winter-rím.

rima, an; *m. A rim, border, bank, coast:*—Rima *crepido*, Wrt. Voc. ii. 15, 45. Rimo, Txts. 55, 601. [Cf. *Icel.* rim *a rail*; rimi *a strip of land*.] *v.* bord-, dæg-, sǽ-, súþ-, tóþ-rima.

ríman; *p.* de. **I.** *to count, number:*—Ducentesimus se ðe biþ on ðam twám hundredum æftemyst, ðon hí man rímaþ, Ælfc. Gr. 49; Som. 50, 5. Næs þeáw ðæt mon ǽnig wæl on ða healfe rímde ðe ðonne wieldre wæs *mos est, ex ea parte quae viceret occisorum non commemorare numerum*, Ors. 4, 1; Swt. 156, 22. Cyninges þegnas oft ráde in ridon ðe mon ná ne rímde, Chr. 871; Erl. 76, 12. Gif ic hí ríman onginne *dinumerabo eos*, Ps. Th. 138, 16. Hé mæg ríman steorran *qui numerat multitudinem stellarum*, 146, 4. **II.** *to enumerate, recount, describe in succession:*—On ðam is godcundnesse wén ðe manna ingehygd wát and can and heora heortena deágol ealle smeáþ and rímeþ *divinity is to be looked for in him that knows the minds of men, and scrutinizes and tells one by one the secrets of their hearts*, Blickl. Homl. 179, 27. Hú nytt rehton wé and rímdon ða cǽga búton wé eác feáwum wordum ætiéwen hwæt hié healden *quid utilitatis est, quod cuncta haec collecta numeratione transcurrimus, si non etiam admonitionis modos per singula pandamus?* Past. 23; Swt. 179, 11. Hwæt sceal ic má ríman yfel endeleás? Exon. Th. 272, 27; Jul. 505. Háligra manna naman rímende and gebedo singende *laetanias canentes*, Bd. 1. 25; S. 487, 4. **III.** *to calculate, compute, count up:*—Ða reáferas geþenceaþ swíðe oft hú micel hié sellaþ swelce hié ða métsceattas rímen (*quasi mercedem numerant*),

Past. 45, 4 ; Swt. 343, 16. For ege ðínum graman ðínne tó rímanne (*dinumerare*), Ps. Spl. 89, 13. [Beón] rímed *computari, numerari*, Hpt. Gl. 482, 24. **IV.** *to account, esteem as :*—Gé beóþ mé talade and rímde on bearna stæl, Exon. Th. 366, 11 ; Reb. 10. [*O. H. Ger.* ge-ríman.] *v.* á-, ge-ríman.

rím-áþ, es; *m. An oath taken by a person and by the number of persons he brings with him as compurgators* (cf. the expressions in Norse law *tylptar-, séttar-eiðr*, oaths in which twelve, six persons respectively took part], L. Ath. i. 9 ; Th. i. 204, 15. *v.* cyre-áþ.

rím-cræft, es; *m. The science of numbers, arithmetic:*—Ða seofon cræftas on ðam beóþ geméted ealle weoruldwýsdómas, ðæt ys ǽrest *arythmetica*, ðæt ys rýmcræft, Shrn. 152, 13. Rímcræft *arithmetica*, Hpt. Gl. 479, 56 : Wrt. Voc. ii. 81, 58 : 3, 7. Uton witan hwæt *saltus lunae* sý tó sóðe . . . oððe hwá hine ðæs wurðscipes cúðe ðæt hé sceolde gestandan on ðam rímcræfte *that he should have a place in the science of computation*, Anglia viii. 308, 22. Ða ðe ǽr wǽran on rímcræfte rihte getogene *those who were correctly instructed in the art of computing*, Chr. 975; Erl. 126, 1. Hæfdon hié on rímcræfte áwriten wera endestæf hwænne hié tó móse meteþearfendum weorðan sceoldon *they (the cannibal Mermedonians) had numbered the days of their captives who were to be food to satisfy their hunger*, Andr. Kmbl. 268 ; An. 134. *v.* gerím-cræft.

rím-cræftig; *adj. Skilful in computation :*—Tó þám rihtungum ðe rímcræftige preóstas cweþaþ *lunares*, Anglia viii. 300, 27. On ðam eahta geárum ðe rímcræftige weras on Grécisc hátaþ *ocdoade*, 315, 23 : 327, 34–36. Rýmcræftige, Menol. Fox 89 ; Men. 44. *v.* next word.

rím-cræftiga, an ; *m. One skilful in computation :*—Béda se árwurða rímcræftiga, Anglia viii. 301, 33.

-ríme. *v.* earfoþ-ríme.

rímere, es ; *m. A computer, reckoner, calculator :*—Betwux ðisre sprǽce sceal se rímre geþencean, ðæt hé gedó ðæt Februarius mónþ ðý geáre hæbbe þrittig nihta ealdne mónan, Anglia viii. 307, 34.

rím-getæl, es ; *n. A number :*—Rímgetæl daga *the appointed number of days*, Cd. Th. 85, 25; Gen. 1420. Drihten lét weaxan eft heora rímgetel, 166, 29; Gen. 2755.

rímian. *v.* ge-rímian.

rimpan (hrimpan ?) *to wrinkle, rumple.* [Gerumpenu nædre *coluber cerastis*, Wrt. Voc. ii. 15, 68. Ðære gehrumpnan *rugosa*, 91, 15. Cf. also hry[m]pellum *rugis*, 95, 73. *O. H. Ger.* [h]rimpfan (hrimfit *terit*): rampf *caperrabat*; girumpfan *rugosus, contractus.* *v.* Grff. ii. 512 : cf. *Ger.* rümpfen.]

rimpel (? hrympel. *v.* preceding word). [*Prompt. Parv.* rympyl *ruga*; rymplyd *rugatus*: *M. H. Ger. O. Du.* rimpel.]

rím-talu, e ; *f. A number, tale :*—Læt mec, mihta God, on rímtale ríces ðínes wunigan, Elen. Kmbl. 1636 ; El. 820.

Rín; *m.*; *f. The Rhine :*—Sió eá ðe man hǽt Rín, Ors. 1, 1 ; Swt. 22, 23. Neáh Rínes ófre ðære ié, Swt. 14, 32. Beeástan Ríne, Swt. 14, 36. On ðæm londe beeástan Rín, Chr. 887; Erl. 86, 7. On cyrican Colonie ðære ceastre bí Ríne, Bd. 5, 10; S. 625, 22. Ða wurpon hí heora líchoman út on Ríne ða eá, S. 624, 42. [*O. H. Ger.* Rín; *m.* : *Icel.* Rín ; *f.*]

rínan. *v.* rignan.

rinc, es; *m. A man* (a poetical term) :—Se rinc (*Enoch*) on líchoman lisse sóhte, Cd. Th. 73, 12; Gen. 1203 : (*Abraham*), 107, 17; Gen. 1790. Com ðá tó recede rinc (*Grendel*) síðian, Beo. Th. 1445 ; B. 720. Árás ðá se ríca (*Hrothgar*), ymb hine rinc manig, þegna heáp, 804; B. 399. Ðá wæs rinc manig, gúðfrec guma, ymb ðæs geongan feorh breóstum onbryrded, Andr. Kmbl. 2234; An. 1118. Ðæt wæs rihtwís rinc (*Boethius*), Met. 1, 49. Ðæs rinces (*Abraham*) se ríca ongan cyning (*God*) costigan, Cd. Th. 172, 16; Gen. 2845. Junge rince ł hysse *ephebo robusto*, Hpt. Gl. 488, 1. Rófe rincas (*the fallen angels*), Cd. Th. 19, 4; Gen. 286 : (*those who occupied Shinar*), 99, 24; Gen. 1651. [Heo smiten togædere, helmes þere gullen . . . , rinkas feollen (mani m[en] þer fulle, 2nd MS.), Laym. 5188. *Piers P.* renke : *O. Sax.* rink : *Icel.* rekkr (frequent in poetry, but in prose it occurs only in old law phrases].] *v.* beadu-, fyrd-, gum-, gúþ-, heaðo-, here-, hilde-, magu-, sǽ-rinc.

rinc-getæl, es ; *n. A number of men, a host :*—Ðæt wæs wíglíc werod; wác ne grétton in ðæt rincgetæl rǽswan herges, Cd. Th. 192, 19; Exod. 234.

rind, e ; **rinde**, an ; *f. Rind.* **I.** *of a tree, the bark :*—Rind *cortix*, Wrt. Voc. i. 285, 78. Rinde *cortex*, 79, 68. Súðerne rind *cinnamonum, resina*, 131, 9. Ofersǽwisc rind, Lchdm. ii. 52, 3. Rómánisc rind, i. 376, 5. Andlang ðæs piþan and andlang ðære rinde óþ ðone helm, Bt. 34, 10; Fox 150, 3. Of corntreówes rinde *de cortice corni*, Wrt. Voc. ii. 27, 6. Gif hé beget and yf rinde sió ðe cymþ of neorxna wonge ne dereþ him nán átter ; ðonne cwæþ se ðe ðás bóc wrát ðæt hió wǽre torbegete, Lchdm. ii. 114, 3 : 92, 29. Wé ne mágon geseón on ðam cyrnele náðor ne wyrtruman, ne rinde, ne leáf, Homl. Th. i. 236, 18. Rinda *cortices* (*codices*, MS.), Wrt. Voc. ii. 135, 60. Rinda *librorum*, Hpt. Gl. 417, 46. Of corntreówes rindum *de cortice corni*,

Wrt. Voc. ii. 138, 7. Rindum *corticibus* (*codicibus*, MS.), 75, 46. Rinde *libros*, 53, 18. Dó of ða rinda, Lchdm. ii. 98, 11. **I a.** metaphorically :—Þeahtigaþ on hiera módes rinde monig gód weorc tô wyrcanne, ac on ðam piþan biþ óðer gehýded, Past. 9, 1 ; Swt. 55, 22. *The word occurs in combination with names of trees*, e. g. apuldor-, æsc-, âc-, elm-, holen-, sealh-, slâhþorn-, wipi-grind. **II.** *of other things, crust, rind :*—Rinde *crustula*, Wrt. Voc. ii. 137, 22. Rindan *crustulae*, Hpt. Gl. 462, 77. Wê hêdaþ ðæra crumena ðæs hlâfes, and ða Judéiscan gnagaþ ða rinde, Homl. Th. ii. 114, 34. Rinda *crusta* (this is omitted from) Wrt. Voc. i. 41, 23. Rindum *crustulis*, Hpt. Gl. 496, 23 : 497, 15. [*O. Du.* rinde : *O. H. Ger.* rinta *cortex, liber.*]

-rindan, rinde. v. be-rindan, rind.

rinde-clifer (?) *a wood-pecker* (?), *a bird that sticks to, or scratches the bark of trees* (?) [cf. clifer, clifrian, clifian] :—Rindeclifre *ibin*, Wrt. Voc. ii. 48, 34.

rinden ; *adj. Of bark :*—Of rindenum *corticeo*, Germ. 390, 43.

rind-leás ; *adj. Without bark ;* decorticatus, Wrt. Voc. i. 61, 14.

rine, rinel, ring, ringan. v. ryne, rynel, hring, hringan. [*Add under the last :*—Ðæt man on rihtne tíman tída ringe, L. Edg. C. 45 ; Th. ii. 254, 5 : L. N. P. L. 36 ; Th. ii. 296, 3.]

rinnan ; *p.* rann, *pl.* runnon ; *pp.* runnen *To run :*—Ic on wîsne weg worda ðínra rinne *viam mandatorum tuorum cucurri*, Ps. Th. 118, 32. Satan seolua ran and on susle feóll, Cd. Th. 309, 20 ; Sat. 712. Wæn æfter ran, Runic pm. Kmbl. 343, 32 ; Rún. 22. Gif lioþole út rynne, Lchdm. ii. 12, 24. Blôd and wæter út bicwôman rinnan fore rincum, Exon. Th. 69, 3 ; Cri. 1115. [*Goth.* rinnan : *O. Frs.* rinna : *O. Sax. O. H. Ger.* rinnan : *Icel.* renna, rinna.] v. â-, bi-, ge-, óþ-rinnan, *and* irnan.

rinelle, an ; *f. A brook, stream :*—Rinnellan *rivos*, Ps. Surt. 64, 11. Cf. rynel.

rio-. v. reo-.

ríp, es ; *n.* **I.** *reaping, harvest :*—Ðæt ríp (*messis*) is worulde endung, Mt. Kmbl. 13, 39. Micel ríp (*messis*) ys, and feáwa wyrhtyna. Biddaþ ðæs rýpes Hláford ðæt hê sende wyrhtan tô his rípe, 9, 37–38 ; Lk. Skt. 10, 2. Ðæt ríp (rípes tíd, Lind.), Mk. Skt. 4, 29. Ðæt hêr wære mycel ríp, Bd. 1, 29 ; S. 498, 4. On hærfeste wícode se cyng on neáweste ðare byrig, ða hwíle ðe hié hira corn gerypon, ðæt ða Deniscan him ne mehton ðæs rípes forwiernan, Chr. 896 ; Erl. 94, 7. Ær wintres cyme on rýpes tíman, Exon. Th. 214, 28 ; Ph. 246. Twuga on geáre æne tô mæþe and óðre tô rípe *twice a-year, once at hay-time and the other at harvest*, Cod. Dip. Kmbl. ii. 400, 30. His men beón gearuwe ge tô rípe ge tô huntoþe, v. 162, 28. Huíto sint tô hrippe (*ad messem*), Jn. Skt. Lind. 4, 35. **II.** *what is reaped* or *gathered in*, *a sheaf of corn* (cf. Whan thou repist corn in the feeld, and forȝetist and leeuest a repe, Wickl. Deut. 24, 19. See also Halliw. Dict. *reepe* a sheaf) :—Rípu gaderian blisse getácnaþ, Lchdm. iii. 208, 15. **II a.** of other products [cf. wín-reopad *vendemiant*, Ps. Surt. 79, 13] :—Wíngeardas (-es, MS.) rípe fulle gesihþ blisse getácnaþ *if he sees vineyards full of fruit ready to gather, it betokens joy*, 210, 32. v. ge-, oht-ríp, *and next word*.

rípa, an ; *m. A sheaf :*—Berende rypan (Ps. Surt. reopan) heora *portantes manipulos suos*, Ps. Spl. 125, 8 : 128, 5. v. ríp, II.

rípan ; *p.* ráp, *pl.* ripon *To reap, cut corn ;* metaph. *to derive advantage :*—Ic rípe *meto*, Ælfc. Gr. 28, 3 ; Som. 30, 63. Ðú rípst ðine æceras *tui agros metis*, 15 ; Som. 19, 46. Hláford ðú rípst ðær ðú ne seówe. . . . Ðú wistest ðæt ic rípe (hrippo, Lind.) ðær ic ne sâwe, Mt. Kmbl. 25, 24-26. Hrippes, Lk. Skt. Lind. 19, 21. Hú ne secge gê ðæt nú gyt synt feówur mónþas ær man rípan mæge . . . geseóþ ðás eardas ðæt hig synt scíre tô rípene (rýpanne, MS. A). And se ðe ripþ (hrioppaþ, Lind.) nimþ mêde, Jn. Skt. 4, 35-36. Heofonan fuglas ne sâwaþ ne hig ne rípaþ (rioppas, Lind.), Mt. Kmbl. 6, 26. Eal manna bearn sorgum sâwaþ, swâ eft rípaþ, Exon. Th. 6, 19 ; Cri. 86. Ða hié heora corn ripon, Ors. 4, 8 ; Swt. 188, 27. Gif wê eów ða gâstlícan sæd sâwaþ, hwónlíc biþ ðæt wê eówere flæslícan þing rípon, Homl. Th. ii. 534, 27. On ðam man ne mæg nâðer ne erian ne rípan, Gen. 45, 6. [*O. E. Homl.* repen ; *p. pl.* reopen : *Laym.* repen ; *p. pl.* : *Ayenb.* ripe : *Wick.* repe : *Piers P.* ropen, repen ; *p. pl.* : *Chauc.* ropen ; *p. part.*] v. ge-rípan, repan.

rípan, rýpan ; *p.* te *To spoil, plunder :*—Ða syndon rýperas ðe scoldan beón hyrdas folces. Hý rýpaþ ða earman bútan ælcere scylde, L. I. P. 12 ; Th. ii. 320, 16. Hý hergiaþ and heáwaþ, rýpaþ and reáfiaþ and tô scipe lædaþ, Wulfst. 163, 12. Ðér þeáfas ofdelfes t hrypes *ubi fures effodiunt*, Mt. Kmbl. Lind. 6, 19. Hí fêrdon æghweder flocmælum and heregodon úre earme folc, and hí rýpton (ræpton, MS. E.) and slôgon, Chr. 1011 ; Erl. 145, 26. Fram rýpendum t bereáfiendum *a diripientibus*, Ps. Lamb. 34, 10. v. be-rýpan, *and* cf. reáfian.

rípe ; *adj. Ripe, mature :*—Rípe deáþ *matura mors*, Wrt. Voc. i. 39, 19. Swíðe rípe *matura satis*, ii. 58, 36. Swâ swâ rípe yrþ *quasi maturam segetem*, Bd. 1, 12 ; S. 480, 35. Se westmbæra hærfest bringþ rípa blêda, Bt. 39, 13 ; Fox 234, 15 : Met. 29, 63. [*O. Sax.* rípi : *O. H. Ger.* rîfi.] v. sæd-, un-rípe.

rípe (?) es ; *n.* or (?) rípu ; *indcl.:* ríp, e ; *f.* (cf. *O.H.G.* rîfi ; *f. maturitas: Ger.* reife) *Ripeness, maturity :*—On rípe *in maturitate*, Ps. Th. 118, 147.

rípere, es ; *m. A reaper :*—Ða ríperas (hrípemenn, Lind.) *messores*, Mt. Kmbl. 13, 39. On ðam ríptíman ic secge ðám ríperum (hrippemornum, Lind.), 13, 30. Cf. riftere.

rípere, es ; *m. A robber, plunderer, spoiler :*—Rýperas and reáferas Godes graman habban, bûton hig geswícan, L. C. S. 7 ; Th. i. 380, 5. Má is ðæra rýpera ðonne rihtwísra, and is earmlíc þing, ðæt ða syndon rýperas ðe scoldan beón hyrdas folces, L. I. P. 12 ; Th. ii. 320, 14–16. Cyning sceal rýperas and reáferas and ðás woruldstrúderas hatian and hýnan, 2 ; Th. ii. 304, 19 : Wulfst. 266, 28 : 165, 35. Ús stalu and cwalu . . . and rýpera reáfiác derede swíðe þearle, 159, 11. Cf. reáfere.

rípian ; *p.* ode *To grow ripe, to mature :*—On hærfest wæstmas rípiaþ, Anglia viii. 312, 23. Dô ðæt sunne scíne ðæt ðíne æceras rípion *cause the sun to shine, that thy fields may ripen*, Homl. Th. ii. 104, 3. Rípian *maturescere*, Wrt. Voc. ii. 3, 27 : Hpt. Gl. 419, 64. [*O. Sax.* rípón : *O. H. Ger.* rîfôn.] v. ge-rípian, ful-rípod, un-gerípod.

ríp-ísern, es ; *n. A sickle, an instrument for reaping :*—Rípísern *falcem*, Mk. Skt. Lind. Rush. 4, 29.

ríp-mann. v. rípere.

ripness, e ; *f. Ripeness, maturity, season of ripeness, harvest :*—Hrípnes *messis*, Mt. Kmbl. Lind. 13, 29. On rípnysse *in maturitate*, Ps. Lamb. 118, 147. Cf. rípung.

ríptere. v. riftere.

ríp-tíma, an ; *m. Harvest-time :*—Lætaþ ægðer weaxan óþ ríptíman, and on ðam ríptíman ic secge ðám ríperum, Mt. Kmbl. 13, 30.

rípu. v. rípe ; *n.*

rípung, e ; *f.* **I.** *ripening :*—Seó sunne tempraþ ða eorþlícan wæstmas ægðer ge on wæstme ge on rípunge, Lchdm. iii. 250, 19. **II.** *ripeness, maturity :*—Tô ðæra mynstres geate sý geatweard geset, eald and wís . . . seó rípung his gestæþþignesse sý swylc, ðæt hine ne worian ne scríðan ne lyste *ad portam monasterii ponatur senex sapiens . . . cujus maturitas non sinat eum vagari*, R. Ben. 126, 17. On rípunga *in maturitate*, Ps. Spl. 118, 147. Se þridda tíma ys autumnus. . . . Bôceras getrahtniaþ ðæne naman for ðære rípunge oððe for ðære gaderunge. Hig cweþaþ autumnus *propter autumationem vel propter maturitatem*, Anglia viii. 312, 27.

rípung, e ; *f. Spoliation, plundering :*—Fordêminge and rýpincge *proscriptionem, fraudationem*, Hpt. Gl. 480, 38.

-rís. v. ge-rís *rabies*, Wrt. Voc. ii. 118, 67. Cf. rísan *to seize*.

rísan ; *p.* rás, *pl.* rison ; *pp.* risen. **I.** *to rise :*—Âlýs mê from lâðum ðe mê lungre on rísan (onrísan ?) willaþ *ab insurgentibus in me libera me*, Ps. Th. 58, 1. **II.** *to be fitting, becoming* (*the most usual form is* ge-rísan, *q.* v. cf. come *and* become, venire *and* convenire, *Ger.* fallen *and* ge-fallen *for similar development of meaning*) :—Ne rí-eþ *non decent*, Kent. Gl. 681. Ðér ne ríseþ *ubi non debet*, Mk. Skt. Rush. 13, 14. [*Goth.* ur-reisan : *O. Sax.* rîsan : *O. Frs. Icel.* rísa : *O. H. Ger.* rîsan *cadere* (cf. stígan *which can be used of upward or downward motion*).] v. â-, on-rísan, *and next word*.

rísan ; *p.* rás ; *pp.* risen (*different word from preceding ?*) *To seize, snatch away, carry off :*—Rísende is rísende wulf *lupus rapax*, Bd. 1, 34 ; S. 499, 27. Se rísenda *rabula*, Wrt. Voc. ii. 88, 68. Ðære rísendan *rapaci*, 79, 83. Wulfas rísænde t woedende *lupi rapaces*, Mt. Kmbl. Rush. 7, 15. v. ge-rísan, -rís, rǽs (?).

risc, risel, rísende. v. rysc, rysel, rísan.

risn (?) *a pair of compasses :*—Risn *cercinum* [? risl (hrisel q.v.) κερκις], Wrt. Voc. ii. 130, 30.

risne ; *adj. Fitting, becoming, suitable :*—Hé sôna ðám risne andsware (congrua responsa) onsende, Bd. 1, 27 ; S. 488, 35, MS. B. v. ge-risene *and next word*.

risne (?), es ; *n. What is fit* or *suitable ;* congruum :—Habbaþ eów swylc massereáf and swylce béc and swylce húselfata swylce gê mid risnum (*decently*) eów ða befæstan þênunga þênian mágon, L. E. I. 4 ; Th. ii. 404, 27. v. ge-risene ; *n. and preceding word*.

risoda (?) *rheum :*—Ða yfelan wætan on ðam seócum men ðe biþ swâ swâ horh oððe risoda oððe gillistre, Lchdm. ii. 282, 11.

ríþ, es ; *m.* (v. eá-ríþ) : e ; *f.* : ríþe, an ; *f. A rithe* (v. Halliw. Dict. and Leo A. S. Names of Places, p. 86 : the word is still to be found in North Frisian in the form ride, rie, to denote the bed of running water), *a small stream :*—Ríþ *rivus* . . . lytel ríþ *rivulus*, Wrt. Voc. i. 54, 20-27 : *rivus*, 80, 62. Burne t rípe *latex*, Hpt. Gl. 447, 4. Norþ tô blacan rîþe, andlang rîþe, Cod. Dip. B. i. 296, 33. On fúlan rîþe, andlang rîþe, Cod. Dip. Kmbl. i. 257, 32. On âne rîþe, andlang rîþes (cf. oꝺ ðære rîþe, 24), iii. 385, 28-29 : 386, 5. Hinc ad ælrithe, ab ipso rivo ad fraxinum unum, 373, 19. Ðér fleów of ðam flinte wæter . . . ðæt hí druncon of ðære ânre rîþe, Num. 20, 11. Rîþe rivo, Hpt. Gl. 490, 30. On ða rîþe, Cod. Dip. Kmbl. iii. 10, 25. Óþ ða litlan rîþe, andlang rîþe . . . tô ða rîþe, ðon andlang rîþe, 12, 15-21. Swâ swâ sum mical æwelm, and irnon manige brôcas and rîþan (rîþa, Cott. MS.) of, Bt. 34, 1 ; Fox 134, 10. Rîþa *torrentum, rivulorum*, Hpt. Gl. 499, 54. Rîþum *rivulis*, 448, 61. Hrípum, 477, 37. Eorþan rîþum *terrae rivulis*, Hymn. Surt. 17, 12. Ic geseah ða wlitegan swilce culfran ástígende ofer streámlicum rîþum, Homl. Th. i. 444, 10. Swelce hit eall

lytlum ríþum tórinne, Past. 38 ; Swt. 277, 12 : 65 ; Swt. 469, 5 : Met. 5, 20. Tô ðam lande ðe flêwþ on ríþum meolce and hunies, Num. 16, 14. v. wæter-ríþe *and next word.*

ríþig, es ; *n. : e ; f.* (?) *A stream :*—Hit cymeþ on ðæt lytle ríþig, of ðæm ríþige, Cod. Dip. Kmbl. iii. 33, 1. On ðæt ríþig, ondlong ríþiges, 378, 15. Swá on ða ealdan díc, andlang díces on âne ríþige, of ðære ríþe on âne ealde díc, 385, 24. On hweólríþig, 381, 8.

riðða, rixe, ríxian. ryðða, rysc, rícsian.

rocc, es ; *m. An upper garment :*—Rocc *callicula,* Wrt. Voc. i. 26, 11. Deórfellen roc *mastruga,* roc *toral,* 82, 3–4. Rooc (rocc ?) *toral,* 25, 64. Gǽten vel broccen rooc (rocc ?) *melotes* vel *pera,* 40, 27. Mid rocce beón gescríd, orsorhnysse getâcnaþ, Lchdm. iii. 200, 12. [*O. Frs.* rok : *O. H. Ger.* rocch *tunica, melotes* : *Ger.* rock : *Du.* rok : *Icel.* rokkr.] v. biscop-, breóst-, pistol-rocc.

rocc *what is chewed* (?), *a cud* (?) :—Edreced roc *rumen* (cf. edrecéþ, ceóweþ *ruminet,* l. 15), Wrt. Voc. ii. 97, 18. v. ed-roc.

rocc *a rock.* v. stân-rocc.

roccettan, roccettan ; *p.* te *To eructate, utter ;* eructare : — Roketto ł bilketto forþ ða ðe âhýded wǽrun *eructabo abscondita,* Mt. Kmbl. Rush. 13, 35. Roccetteþ *eructuat,* Ps. Surt. 18, 3 : *eructuavit,* 44, 2. Roccetaþ *eructabunt,* 118, 171. Bylcetteþ *eructuat,* i. *a corde emittit,* Wrt. Voc. ii. 144, 13. Bleów ł roccette *ructabat,* 96, 1.

rôd, e ; *f.* I. *a rod, pole.* v. segl-rôd. II. *a measure of land :*—Se haga is fram ðære eástwardes .xxviii. rôda lang and sûþwardes .xxiiii. rôda brâd and eft ðanon westwardes on sæferne .xix. rôda long, Cod. Dip. Kmbl. ii. 150, 6–9. III. *a cross, rood* (as in Holyrood) :—Ðeós rôd *haec crux,* Ælfc. Gr. 9, 67 ; Som. 14, 8. Rôd *crux* vel *staurus,* Wrt. Voc. i. 26, 52. Wîtestengces, rôde *eculei,* rôde *gabuli,* Hpt. 478, 70–74. Ic bidde ðé for ðære hâlegan rôde tâcne, Bt. 42 ; Fox 260, 3. Hê hine gesênade mid Cristes rôde tâcne *signans se signo sanctae crucis,* Bd. 4, 24 ; S. 599, 6. Hî mearcodon mid blôde Tau, ðæt is rôde tâcen, Homl. Th. ii. 266, 8. Se Hǽlend rôde tâcen ofer Adam geworhte, Nicod. 32 ; Thw. 17, 29. Ðæt gê sceolan þurh ðæt treów mýnre rôde oferswýdan ðone deáþ, Thw. 17, 21. Sige forgeaf cyning ælmihtig þurh his rôde treó, Elen. Kmbl. 294 ; El. 147. Ðá gefæstnodon Judéi hine rôde gealgan . . . Mancynna ealdor ðære rôde gealgan underfêng, Homl. Th. i. 588, 16–19. Hæt Pharao ðé âhôn on rôde (*in cruce*), Gen. 40, 19. Gâ nyþer of ðære rôde, Mt. Kmbl. 27, 40. Ðone hig nýddon ðæt hé ðære hys rôde, 27, 32 ; Jn. Skt. 19, 17. Hê ðær þreó mette rôde ætsomne, Elen. Kmbl. 1665 ; El. 834. III. *a crucifix.* v. sweor-rôd. [*O. Sax.* rôda *a cross:* *O. Frs.* rôde *patibulum:* *O. L. Ger.* ruoda *virga; rood* (*a measure*) : *Icel.* rôða *a rood, crucifix:* *O. H. Ger.* ruota *virga.*] v. wearh-, wyn-rôd.

rôd-begenga, an ; *m. One who worships a cross :*—Rôdbigenga *crucicola,* Wrt. Voc. ii. 137, 23. v. rôd-weorðiend.

rôd-bora, an ; *m. One who bears a cross :*—Rôdbora *crucifer,* Germ. 389, 1.

rôde-hengen[n], e ; *f. A cross, crucifixion :*—Hwæt hæfþ ðes man gefremod, ðæt hê rôdehengene wyrðe sý, Homl. Th. i. 596, 2. Hêt hine âhôn on rôdehengcene, 594, 29. Ðá ðá hê on rôdehengene mancynn âlýsde, 58, 20. On rôdehengene genæglod, 82, 25. Hê (*the penitent thief*) geandette his synna on ðære rôdehengene, ii. 78, 22. Úre Hǽlend rôdehengene underbeáh, 600, 6.

rôde-wirðe ; *adj. Deserving crucifixion :*—Gangaþ ût git godwrecan and gongaþ ût git rôdewyrðan, Shrn. 43, 8.

rôd-fæstnian ; *p.* ode *To crucify :*—Gerôdfæstnad *crucifixus,* Apstls. Crd. 24.

rodor, rador, es ; *m.* I. as a technical term, *the firmament, the heaven of the fixed stars :*—Sunne *sol,* môna *luna,* roder *firmamentum,* Wrt. Voc. i. 41, 55–57 : 70, 8. Lyft *aer,* hroder *aether,* 52, 56. Se rodor ymbfêhþ ûtan eall ðás niþerlícan gescæfte, Shrn. 63, 9. Sió eorþe is nioþor ðonne ǽnig ôðru gesceaft bûton ðam rodore, forðam se rodor hine hæfþ ǽlce dæg ûtane . . . on ǽlcere stôwe hê is hire emnneáh, Bt. 33, 4 ; Fox 130, 20, 23. Siððan wæs rodor ârǽred and ryne tungla, folde gefæstnad, Exon. Th. 272, 12 ; Jul. 498. Radores *aethrae* (MS. *uetre*), Wrt. Voc. ii. 92, 43. Hwá unlǽredra ne wundraþ þæs roderes færeldes, hû hê ǽlce dæge ûton ymbhwyrfþ ealne ðisne middaneard, Bt. 39, 3 ; Fox 214, 10. Rodres, Met. 28, 3. Ðû mihtest ðé fleógan ofer ðam fýre ðe is betwux ðam rodore and ðære lyfte, and mihtest ðé fêran mid ðære sunnan betwyx ðâm tunglum and ðonne weorþan on þam rodore, Bt. 36, 2 ; Fox 174, 9–12 : 33, 4 ; Fox 130, 15. Ofer rodere ryneswiftum, Met. 24, 28. Micel swêg gǽþ of ðam scînendan rodore, ðeáh wê for ðam mycclan fyrlene hit gefrêdan ne mâgon, Boutr. Scrd. 18, 43. Se godcunda foreþonc stýreþ ðone rodor and ðá tunglu, Bt. 39, 8 ; Fox 224, 7. Ðás twelf tâcna (*the signs of the Zodiac*) synd swâ gehiwode on ðam heofenlícum roderum (rodere, MSS. R. L. P.), Lchdm. iii. 246, 6. II. mostly as a poetical term, *the heavens, sky, upper regions :*—Rodores candel *the sun,* Beo. Th. 3148 ; B. 1572. Hroderes *aethera,* Hpt. Gl. 521, 23. Roderes *Olimpi,* Wrt. Voc. ii. 64, 61. Ðæs heálícan roderes *celsi Olymphi,* Hymn. Surt. 55, 3. Under radores ryne, Elen. Kmbl. 1586 ; El. 795. Fram rodere Crist scînþ *ab ethere Christus promicat,* Hymn. Surt. 37, 8. Wunigende on rodore *manens*

Olimpho, 91, 19. Sende him of heán rodore God gâst ðone hâlgan, Cd. Th. 230, 21 ; Dan. 236. Roderas *aethera,* Kent. Gl. 273. Lyft drysmaþ, roderas reótaþ, Beo. Th. 2756 ; B. 1376. Dryhten, rodera rǽdend, Andr. Kmbl. 1253 ; An. 627. Rodra weard, Exon. Th. 394, 23 ; Rä. 14, 7. Rodera weard *God,* Cd. Th. 1, 2 ; Gen. 1. Rodora rîce *heaven,* 308, 5 ; Sat. 688. Under roderum, 7, 21 ; Gen. 109. Steám up ârâs swylce rêc under radorum, Elen. Kmbl. 1604 ; El. 804. Alwalda worhte rûme roderas, Exon. Th. 341, 30 ; Gen. Ex. 134. [*O. Sax.* radur.] v. beorht-, eást-, gim-, heáh-, norþ-, sûþ-, up-, west-rodor.

rodor-beorht ; *adj. Heavenly bright :*—Rodorbeorhtan tunglu, Cd. Th. 239, 12 ; Dan. 369.

rodor-cyning, es ; *m. The king of heaven, Christ :*—Þurh ðæs hýhstan meaht, rodorcyninges giefe, se ðe on rôde treó geþrowade, Exon. Th. 269, 8 ; Jul. 447 : 45, 30 ; Cri. 727 : Elen. Kmbl. 1771 ; El. 887. Radorcyninges rôd, 1245 ; El. 624.

rodor-lîc ; *adj.* I. *of the firmament* (v. rodor, I) :—Se roderlíca *ethereus,* Wrt. Voc. ii. 144, 25. *Firmamentum* is ðeós roderlíce heofen mid manegum steorrum âmétt, Lchdm. iii. 254, 8. Hî (*Enoch and Elias*) sind genumene tô lyftenre heofenan, nâ tô rodorlícere, Homl. Th. i. 308, 3. Godes rîce on rodorlícere heofonan, ii. 330, 27. II. *celestial, heavenly* (v. rodor, II) :—Cǽgbora se roderlíca (*aethereus*) mid ôðrum apostolum, Hymn. Surt. 118, 11. Cæstergewaran rodorlíce *cives aetherei,* 57, 4. Hî faraþ tô heofonum and rodorlíce wununga underfôþ, Homl. Skt. i. 5, 83.

rodor-lîhtung, e ; *f. The illumination of the heavens, the dawn :*—Roderlîhtinge *auroram,* Ps. Lamb. 73, 16.

rodor-stôl, es ; *m. A celestial throne :*—Hê mid handum his on heofonríce rihte rodorstôlas, Cd. Th. 46, 24 ; Gen. 749.

rodor-torht ; *adj. Heavenly bright :* — (Rodor)torht ryne regen gestilled, Cd. Th. 85, 17 ; Gen. 1416.

rodor-tungol, es ; *n. A star of heaven :*—Torr ârǽrde tô rodortunglum, Cd. Th. 100, 21 ; Gen. 1667.

rôd-weorðiend, es ; *m. A worshiper of the cross :*—Rôdwurþiend *crucicola, crucis adorator,* Hpt. Gl. 403, 30. v. rôd-begenga.

rôf ; *adj. Valiant, stout, strong* (used only in poetry) :—Rôf oretta, heard under helme (*Beowulf*), Beo. Th. 5070 ; B. 2538. Rôf rûnwita (*Guthlac*), Exon. Th. 167, 30 ; Gû. 1068. Wîs hǽleþ, maga môde rôf, Andr. Kmbl. 1249 ; An. 625. Ânrǽd oretta, maga môde rôf, 1967 ; An. 986. Ârâs ðá mægene rôf, 2936 ; An. 1471 : 3348 ; An. 1678. Dǽdum rôf, æþeling ânhýdig, Beo. Th. 5326 ; B. 2666. Ðeáh hê (*Grendel*) rôf sîe niþgeweorca, 1369 ; B. 682. Rôfne randwígan restan lyste, 3590 ; B. 1793. Fýrdraca rǽsde on ðone rôfan, 5373 ; B. 2690. Hǽleþas heardmôde, rôfe rincas (*the fallen angels*), Cd. Th. 19, 4 ; Gen. 286. Ðæt wǽron mǽre men (*the apostles*), frome folctogan and fyrdhwate, rôfe rincas, Andr. Kmbl. 17 ; An. 9. Rincas wǽron rôfe, randas wǽgon forþ fromlíce, Cd. Th. 19, 4 ; Gen. 2049. Ic on morgen gefrægn môdes rôfan hebban herebýman, 183, 28 ; Exod. 98. [*O. Sax.* rôf (ruob).] v. æsc-, beadu-, cwyld-, cyne-, dǽd-, ellen-, gûþ-, hand-, heaþo-, hete-, hyge-, mægen-, môd-, mund-, sǽ-, sige-, un-camp-, þræc-rôf. -rôf. v. secg-rôf.

rogian (?) :—Heán sceal gehnîgan, âdl gesîgan, ryht rogian, Exon. Th. 340, 30 ; Gn. Ex. 119.

Rôm, e ; *f. Rome :*—Ðá wæs âbrocen burga cyst, beadurincum wæs Rôm gerýmed, Met. 1, 19. Hêr onfêng Ecgbriht pallium æt Rôme, Chr. 735 ; Erl. 47, 19. Petrus gesæt biscepsetl on Rôme, 45 ; Erl. 6, 20. Hêr sendon Brytwalas tô Rôme, 443 ; Erl. 10, 21 : 721 ; Erl. 44, 25. Ðæt hê of Rôme côme, Bd. 1, 25 ; S. 486, 25. Hê mid ealre his firde wið Rôme weard farende wæs, Ors. 5, 11 ; Swt. 236, 9. ¶ *The combination* Rôme-, Rôma-burh *is also frequent :*—Wearþ Rômeburg getimbred fram twâm gebrôðrum, Ors. 2, 2 ; Swt. 64, 21. Swâ mildelíce wæs Rômeburg on fruman gehâlgod, Swt. 66, 4. Twâm geárum ǽr Rômaburh âbrocen wǽre . . . wæs Rômaburh âbrocen fram Gotum, Bd. 1, 11 ; S. 480, 10–12. On Rômebyrig, Apstls. Kmbl. 22 ; Ap. 11. Hêr Gotan âbrǽcon Rômeburg, Chr. 409 ; Erl. 10, 7.

Rômáne (Rôme ?), Rômánan ; *pl. The Romans :*—Nǽfre siþan Rômáne ne rícsodon on Bretone, Chr. 409 ; Erl. 10, 9 : 418 ; Erl. 10, 13. Rômánan gesâwon fîren cleáwen feallan of heofenum, Shrn. 30, 5. Claudius ôðer Rômâna cyninga, Chr. 47 ; Erl. 6, 23. Hê onfêng pallium from Rômâna biscope, 736 ; Erl. 46, 21. Rômâna burh, 409 ; Erl. 11, 10. Rômâna rîce, Ors. 2, 2 ; Swt. 64, 11. Ealra ðara Rômâna wíf, Swt. 66, 29. Wǽron ealle Italie Rômânum on fultume, 4, 11 ; Swt. 208, 7.

Rômánisc ; *adj. Roman :*—Se Rômánisca câsere Octavianus, Homl. Th. i. 30, 1. Se Rômánisca here, Bd. 1, 12 ; S. 480, 33. Man Rômánisces cynnes, i. 16 ; S. 484, 18. On ðære hâlgan Rômániscan cyricean, i. 27 ; S. 489, 33. Fram ðam Rômániscan Pâpan, 2, 20 ; S. 522, 19. Ealde Rômánisce weorce geworhte, i. 33 ; S. 498, 31. Gûþlac ys on Rômánisc *belli munus,* Guthl. 2 ; Gdwin. 10, 24. Ðá yfel ðe þeódríc wið ðam Rômániscum witum dyde, Bt. 1 ; Fox 2, 15. Him leófre wæs ðæt hié Rômánisce cyningas hæfden ðonne of heora âgnum cynne, Ors. 3, 5 ; Swt. 106, 25. Ealle ða Rômániscan men þe Hannibal geseald hæfde, 4, 11 ; Swt. 204, 7.

3 F

Rôme-burh, -scot. v. Rôm, Rôm-gesceot.

Rôm-feoh; *gen.* -feós; *n. Peter's pence.* [William of Malmesbury attributes to Ethelwulf the institution of this tax: 'Ethelwulf went to Rome (v. Chron. 855) and there offered to St. Peter that tribute which England pays to this day,' bk. 2, c. 2; but in the earlier and similar payment by Offa, established in 787, may probably be seen the origin of the *Rômfeoh* in England, v. Stubbs, Const. Hist. i. 230. The Chronicle several times during Alfred's reign contains the notice that 'Wesseaxna ælmessan' were sent to Rome, but the first notice in the laws of Rômfeoh occurs in the agreement between English and Danes, to which his son Edward was a party: 'Gif hwá Rômfeoh forhealde gylde lahslit mid Denum, wíte mid Englum,' Th. i. 170, 2. The penalty, which is not here stated, was a heavy one, as will be seen from the passages given below. There is no mention in these of any being exempted from the contribution on the score of insufficient means, but in the laws of Edward the Confessor, in that which treats 'de denario Sancti Petri qui Anglice dicitur Rômescot,' it is said: 'Omnis qui habuerit .xxx. denariatas vive pecunie de suo proprio in domo suo, lege Anglorum dabit denarium Sancti Petri.' Further with regard to the time of payment it is enacted: 'Iste (denarius) summoniri debet in festivitate sanctorum Apostolorum Petri et Pauli, et ultra festum Sancti Petri ad Vincula non detineatur,' Th. i. 446. So too in the laws of William I: 'Cil ki ad aueir champestre xxx. deñ vaillant deit duner le deñ sein Piere,' Th. i. 474. And see note on p. 170. See too the laws of Henry I: 'Romfech in festo Sancti Petri ad Vincula debet reddi,' Th. i. 520. v. Ducange s.v. Denarius S. Petri.]:—Wê bebeódaþ ælcum cristenum men ... Rômfeoh ... Gif hit hwá dón nelle, sý hê ámánsumod, L. Edm. E. 2; Th. i. 244, 17. Rômfeoh gelæste man æghwilce geáre be Petres mæssan; and se ðe ðæt nelle gelæstan, sylle ðártðeácan .xxx. peninga, and gilde ðam cyninge .cxx. sciłł., L. Éth. ix. 10; Th. i. 342, 24. Rômfeoh gelæste man be Petres mæssan; and se ðe ofer ðæne dæg hit healde, ágyfe ðam bisceope ðæne penig, and ðærtó .xxx. penega and ðam cingce .cxx. sciłł., L. C. E. 9; Th. i. 366, 15. Rômfeoh gelæste man æghwilce geáre be Peteres mæssan; and se ðe ðæt ne gelæste, sylle ðærtðeácan .xxx. peninga tó Rôme and gylde ðam cynge on Engla lage .cxx. scillinga, Wulfst. 272, 9. [Cf. *Icel.* Rôma-skattr.] v. Rômpening *and next word.*

Rôm-gesceot, es; *m. n. Peter's pence:*—Man syddan ðæt Rômgesceot be him sende, swá man manegan geáran æror ne dyde, Chr. 1095; Erl. 232, 33. [Hê com æfter þe Rômescot, 1123; Erl. 250, 39.] v. preceding word.

rômian; *p.* ode; *with gen. To strive after:*—Is ðes ænga stede (*hell*) ungelíc swíðe ðam ôðrum ðe wê ær cúðon on heofonríce ... ðeáh wê hine for ðam Alwealdan ágan ne môston rômigan úres ríces *though we are prevented by the Almighty from possessing our former place and from striving after our former power* (cf. Ic eom ríces leás as *marking the inability for further striving on the part of Lucifer*, 24, 3; Gen. 372), Cd. Th. 23, 15; Gen. 350. [*The word seems to be the O. Sax.* rômôn, *to aim at, strive after;* cf. rômôd gî rehtoro things, Hel. 1690. *O. H. Ger.* râmen (*with gen.*) *intendere.*]

rômig (?); *adj. Blackened, sooty:*—Rômei *catabatus* (cf. hrúmig *caccabatus*, 13, 17), Wrt. Voc. ii. 102, 56. [Cf. (?) *O. H. Ger.* raamac, hrâmac *furva.*] v. hrúmig.

Rôm-pening, es; *m. A penny paid to Rome.* v. Rôm-feoh:—Sig ælc Rômpenig ágifen be Petres mæssedæge æiþer ge uppon lande ge on ælcan porte, Shrm. 208, 32. Rômpenegas (cf. seó ælc heorþpenig ágifen be Petres mæssedæg, 116, 4), Wulfst. 113, 11. Wê willaþ ðæt ælc Rômpænig beó gelæst be Petres mæssan tó ðam bisceopstôle, and wê willaþ ðæt man namige on ælcon wæpengetæce .ii. trýwe þegnas and ænne mæssepreóst, ðæt hí hit gegaderian. Gif cyninges þegn oððe ænig landríca hit forhæbbe, gilde .x. healfmearc, healf Criste, healf cynge. Gif hwilc túnes-man ænigne pænig forhæbbe, gilde se landríca ðone pænig, and nime ænne oxan (cf. *the fine of 30 pence in the passages given under* Rôm-feoh, *and the value of an ox, v.* oxa) æt ðam men, L. N. P. L. 57–59; Th. ii. 298, 29–300, 7.

Rôm-waran, -ware; *pl. The people of Rome, the Romans:*—Hú ungemetlíce gê Rômware bemurciaþ, Ors. 1, 10; Swt. 48, 17. Rômwara sundorriht *jus Quiritum*, Wrt. Voc. ii. 49, 11. Se ærra Rômwara cásere Julius, Bd. 1, 2; S. 475, 2. Rômwara ríce, 1, 3; S. 475, 13. Rômwarena hláford, Elen. Kmbl. 1961; El. 982. Micel sído mid Rômwarum, Bt. 27, 1; Fox 96, 2. [*Icel.* Rôm-, Rûm-verjar.]

Rôm-, Rûm-wealh; *gen.* weales; *m. A Roman* (cf. Bret-walas *the Britons*):—Reht Rômwala *jus Quiritum*, Rtl. 189, 13. Ic wæs mid Rúmwalum, Exon. Th. 322, 27; Víd. 69. v. wealh.

rop *the colon.* v. ropp.

rop (?) *broth:*—Rop (broþ?) *jus* (*in a list* de suibus), Wrt. Voc. i. 286, 55.

róp; *adj. Liberal, bountiful:*—Ðeós lyft byreþ lytle wihte, ða sind sanges wlite *they* (*the birds*) *are bountiful of song*, Exon. Th. 439, 2; Rä. 58, 3. v. next word.

rópness, e; *f. Liberality:*—Roopnis *liberalitas*, Wrt. Voc. ii. 113, 2. Rópnes, 51, 10.

ropp, es; *m. An intestine, the colon:*—Rop *colum* vel *intestinum*, Wrt. Voc. ii. 134, 60: *extale*, 145, 29. Roop *colus* (in a list of parts of the body), i. 45, 20. Hrop *colum*, 19, 55. Be wambe coþum and tácnum on roppe and on smælþearmum, Lchdm. ii. 230, 16–18. Tíhþ innan ðone rop and on ðæt smælþearme, 232, 15. Roppum *extalibus*, Wrt. Voc. ii. 32, 11. [He naȝt ne his roppes bote wynd, Ayenb. 62, 32. v. *Halliw. Dict.* ropes: *O. Du.* rop.]

rop-wærc, es; *m. Colic:*—Ropwærc *colica*, Wrt. Voc. ii. 134, 68. Hropwyrc, i. 19, 56.

rórend. v. rôwend.

róscian *to dry by fire.* v. ge-róscian, Wrt. Voc. i. 288, 60: ii. 116, 31. v. róstian.

róse, an; *f. A rose:*—Rôse *rosa*, Wrt. Voc. i. 30, 13: 79, 60. Rôsa, 69, 24. Ðære rôsan wlite, Bt. 9; Fox, 26, 20. Ðæra rôsena blóstman getácniaþ mid heora reádnysse martyrdóm, Homl. Th. i. 444, 13. [*Icel.* rôs: *O. H. Ger.* rôsa. From Latin.]

rôsen; *adj. Of roses;* roseus, rosatus:—Mid wlite rôsenum *decore roseo*, Hymn. Surt. 105, 20. Mid rôsenan ele gemencged, Lchdm. i. 302, 3. On rôsenne *in rosatum*, Hpt. Gl. 483, 25.

rôsig; *adj. Rosy:*—Mid rôseum hiwe ofergoten, Homl. Th. ii. 334, 30.

rôstian; *p.* ode *To roast, dry by a fire:*—Gerôstode *passos*, Wrt. Voc. ii. 67, 60. [*O. H. Ger.* rôsten *torreri, frigere.*] v. róscian.

rot *scum*, Lchdm. ii. 204, 1: 286, 4. v. hrot.

rót; *adj.* I. *glad, cheerful:*—Ðær moncyn môt for Meotude rót sóðne God geseón and aa in sibbe gefeón, Exon. 355, 33; Reim. 86. v. un-rót, rêtan, rót-hwíl, rótlíce, rótness. II. *noble, excellent:*—Se gôda man swá hê swíðor áfandod biþ, swá hê rótra biþ, and neár Gode, óþ ðæt hê mid fulre geþincþe færþ of ðisum lífe tó ðam ecan lífe. Se yfela swá hê oftor on ðære fandunge ábrýd, swá hê forcúðra biþ, and deófle neár, óþ ðæt hê færþ of ðisum lífe tó ðam ecan wíte, Homl. Th. i. 268, 26–31. Drihten cwæþ, ðæt wê sind miccle rôttran ðonne ða fugelas (cf. Besceáwiaþ ða hrefnas ... gê synt hyra sélran, Lk. Skt. 12, 24); forðan ðe se man is ðe Gode geþíhþ ealra gesceafta rôtost, and Gode leófost, buton ðam heofenlícum englum ðe næfre ne syngodon, ii. 462, 31–34. On ðam ilcan geáre forbarn ðæt hálige mynster on Lundene ... and ðæt mæste dæl and ðæt rótteste ealle ðære burh, Chr. 1086; Erl. 220, 20.

róðer, es; róðra, an; *m. A rower, sailor:*—Rôðer *nauta*, Wrt. Voc. i. 48, 8. Rôðra, 63, 28. v. rêðra.

róðer, es; *n. An oar, a rudder* (i. e. an oar for steering):—Rôðr *tonsa*, Wrt. Voc. ii. 122, 48. Rôthor, Ep. Gl. 26 d, 29. Rôðer *remus*, Wrt. Voc. i. 73, 77. Rôðres blæd *palmula*, 48, 15. Ne mæg scip nó stille gestondan, búton hit ankor gehæbbe, oððe mon mid rôðrum ongeán tió (*pull against the stream with oars*), Past. 58; Swt. 445, 13. [*O. H. Ger.* ruodar *remus, palmula, clavus, gubernaculum.*] v. scip-, steór-rôðer, ge-rêðru.

roð-hund, es; *m. A large dog;* molossus. [In later English vocabularies *molossus* is translated by *blood-hound* and *band-dog.* v. Wrt. Voc. i. 177, 15: 187, col. 2.]:—Roðhund *molossus*, Wrt. Voc. ii. 114, 24: 56, 41: i. 288, 27. Rothundas *molosos*, ii. 91, 9. Hroðhund *inutilis canis*, i. 23, 36. [Cf. *O. H. Ger.* rudo *molossus* (v. Grff. ii. 490): *Ger.* rüde.] v. ryðða.

róðra. v. róðer.

rót-hwíl, e; *f. A time of refreshing:*—Ælc rihtwís man, ðonne hê ðysne sealm singþ, wilnaþ him sumere rôthwíle on ðissere worulde, and éc reste æfter ðisum, Ps. Th. 14, arg. Forlæt mê nú tó sumre rôthwíle on ðisse weorulde ær ic hire of gewíte *remitte mihi ut refrigerer prius quam eam*, 38, 16.

rotian; *p.* ode *To rot, get corrupt, ulcerate, putrify:*—Ðonne se læce on untíman lácnaþ wunde, hió wyrmseþ and rotaþ *secta immature vulnera deterius infervescunt*, Past. 21, 2; Swt. 153, 3. Hit ne rotode *non computruit*, Ex. 16, 24. Míne wunda rotedan and fúledon *computruerunt et deterioraverunt cicatrices meae*, Ps. Th. 37, 5. Gif sió wund swíðe rotige óþ ðæt hê ðæt wursm of múþe hræce, Lchdm. ii. 202, 25. Ær se seoloc (*silk thread*) rotige, 56, 8. Mid ðam (*myrrh*) man smyraþ rícra manna líc ðæt hig rotian ne mágon, Anglia viii. 299, 48. [Cf. *Icel.* rotinn *rotten:* rotna *to putrefy, rot.*] v. for-rotian, rotung.

rót-líc. v. un-rótlíc.

rót-líce; *adv. Cheerfully:*—Nú ðú ðus rótlíce and ðus glædlíce tó us sprecende eart *qui tam hilariter nobiscum velut sospes loqueris*, Bd. 4, 24; S. 598, 37. v. un-rótlíce.

rótness, e; *f.* I. *gladness, cheerfulness:*—Of rótnise (un-r.?) *de merore*, Rtl. 41, 5. From rótnise *a tristitia*, 69, 34. v. un-rótness. II. *comfort, protection:*—Rótnys (gebeorh, Ps. Th.: frófr, Ps. Spl. T.) *refugium*, Blickl. Gl. Rótnes ł ner (rótsung, Ps. Spl. T.) ðam þearfan *refugium pauperi*, Ps. Lamb. 9, 10. On húse rótnysse *in domum refugii*, Ps. Spl. 30, 3.

-rotigendlíc, -rotodness. v. un-forrotigendlíc, for-rotodness.

rótsian. v. ge-, un-rótsian, *and next word.*

rótsung, e; *f. Comfort, protection, cheering:*—Rôtsung *refugium*, Ps. Spl. T. 9, 9.

rotung, e; f. I. *corruption, putrefaction* :—Mín rotung on byrgenne *dum descendo in corruptionem*, Ps. Th. 29, 8. II. *a sore accompanied with putrefaction, an ulcer* :—Rotung *ulcus*, Wrt. Voc. i. 20, 15.

rów; *adj. Quiet, calm, mild* :—Se cleweþa (*itch*) biþ 'swíðe rów, and ðeáhhwæðere gif him mon tó longe fylgþ, hé wundaþ and sió wund sáraþ, Past. 11, 6; Swt. 71, 19. [*Icel.* rór, *quiet, calm.*] v. next word.

rów, e; f. *Quiet, rest* :—Ðær hý bídinge móstun æfter tintergum tídum brúcan, ðonne hý of waþum wérge cwóman restan ryneþrágum, rówe gefégon, Exon. Th. 115, 4; Gú. 184. [Biteache mi gast and mi bodi baðen to ro and to reste, Marh. 20, 5. Cristess reste and Cristess ro, Orm. 7042. *O. H. Ger.* ruowa *quies, requies* : *Icel.* ró.]

rówan; *p.* reów *To go by water, to row* or *sail* :—Ic rówe *navigo*, Ælfc. Gr. 24; Som. 25, 40. Ic ástíge mín scyp and rówe (*navigo*) ofer sǽlíce dǽlas, Coll. Monast. Th. 26, 31. Wérig sceal se wið winde róweþ, Exon. Th. 345, 12; Gn. Ex. 187. Drihten tó ðam lande reów, Homl. Th. ii. 378, 31. 'Utun seglian ofer ðisne mere.' And hig seglydan ða. Ða hig reówun ða slép hé (*navigantibus illis obdormivit*), Lk. Skt. 8, 23, 26. Ða óðre leorningcnihtas reówon *navigio venerunt*, Jn. Skt. 21, 8. Hí geféngon hine and wurpon hine on ðone bát and reówan tó scipe, Chr. 1046; Erl. 174, 18. Ða git on sund reón, ðǽr git eagorstreám earmum þehton, mǽton merestrǽta, mundum brugdon, Beo. Th. 1029; B. 512. Ða wit on sund reón, 1083; B. 539. Ðonne mót hé swá rídan, swá rówan, swá swilce fǽrelde faran swylce tó his wege gebyrige, L. E. I. 24; Th. ii. 420, 24. Seó sǽ is hwíltídum smylte and myrige on tó rówenne, Homl. Th. i. 182, 32. [*Icel.* róa *to row.*] v. be-, ofer-, óþ-rówan.

rówan (?) :—On hlíor rówuit *adplaudat*, Wrt. Voc. ii. 99, 37.

rówend, es; *m. A rower, sailor* :—Nǽfre ic sǽlidan sélran mętte ... rówend (rórend, MS.) rófran, Andr. Kmbl. 945; An. 473. Ðæt scip wile hwílum stígan ongeán ðone stream, ac hit ne mæg, búton ða rówend hit teón, Past. 58; Swt. 445, 11. v. scip-rówend.

rówet[t] *glosses* remigium :—Rówette *remigio*, Hpt. Gl. 529, 14. v. réwet[t].

równess, e; f. *Rowing* :—Wé ne mid seglinge ne mid równesse (*neque velo neque remigio*) ówiht fremian mihte, Bd. 5, 1; S. 613, 25.

rówung, e; f. *Rowing* :—Winnende in rówinge *laborantes in remigando*, Mk. Skt. Rush. 6, 48. On scip ł on róuing *nauigio*, Jn. Skt. Lind. 21, 8.

rudduc *a ruddock* (v. Halliw. Dict.), *a robin red-breast* :—Rudduc *rubisca*, Wrt. Voc. i. 29, 20 : 62, 36.

rúde (?) *roughness of the skin, scab* :—Seó rúde *or* se rúda (se rude, MS.) on ðam men *scamma in homine*, Wrt. Voc. i. 45, 30. [*O. L. Ger.* rútha *scabies* : *O. H. Ger.* rúda, rúdo *scabies, impetigo* : *Ger.* räude. Cf. (?) *Icel.* hrúðr *crust* or *scab on a sore*. This form seems to point to *hrúde* as the earlier form in English.]

rúde, an; f. *Rue* :—Rúde *ruta*, Wrt. Voc. i. 30, 40 : 69, 1 : 79, 18. Wildre rúdan seáw, Lchdm. ii. 26, 10. Mintan and rúdan *mentam et rutam*, Lk. Skt. 11, 42. Rútan, Wrt. Voc. ii. 73, 46. [*O. H. Ger.* rúta : *Ger.* raute. From Latin?]

rudig; *adj. Ruddy* :—Rudi *purpureus, rubicundus*, Hpt. Gl. 475, 8. [Rudi scheome, A. R. 330, 20. Þi rudi neb schal as gres grenen, H. M. 35, 22.]

rud-molin (?) *redshanks* or *water pepper*; polygonum hydropiper, Lchdm. ii. 342, 12. v. note and glossary.

rudu, e; f. *Red, redness, redness of the cheeks, the countenance* (?) :—Anwlita *vel* rudu *vultus*, Wrt. Voc. i. 42, 52. Mid rude *rubore*, Hpt. Gl. 507, 63. Ða geseah se cyngc ðæt Apollonius mid rósan rude wæs eal oferbrǽded, Ap. Th. 22, 4. Gezabel gehiwode hire eágan and hire neb mid rude, Homl. Skt. i. 18, 342. [The rude of monnes nebbe þet seið ariht his sunnen, A. R. 330, 29. Þe rose mid hire rude, O. and N. 443. Cf. *Icel.* roði *redness.*]

rúg. v. rúh.

Rug-ern *rye-harvest, the name of a month* :—Sextan dæge Rugernes, L. Wih. proem.; Th. i. 36, 6. [Cf. *O. Frs.* arn : *O. H. Ger.* aran, arn *messis*, and see Grmm. Gesch. D. S. 58.]

rúh; *adj.* I. *rough, hairy, shaggy* :—Rúh *hispidus, hirsutus*, Wrt. Voc. ii. 43, 15–16 : 90, 17 : i. 51, 20. Rúh hrægel *amphibalum*, 25, 65. Óxn *vel* rúh óxn *ascella vel subhircos*, 43, 65. Rúh scó *pero*, ii. 68, 6. Se wæs reód and eall rúh *totus in morem pellis hirsutus*, Gen. 25, 25. Mín bróðer ys rúh and ic eom sméðe, 27, 11. Gif him þince ðæt hé habbe rúh líc, Lchdm. iii. 170, 24 : Exon. Th. 407, 14; Rä. 26, 5. Rúwes nát hwæt, 479, 17; Rä. 62, 9. Rúhne wǽfels *yrcum tegimen*, Hymn. Surt. 103, 31. Rúhne (rihne, MS.) hine gesíhþ gewordenne, Lchdm. iii. 208, 29. Leáf beóþ rúge and bráde, ii. 254, 13. Ða gesáwe wé rúge (*pilosos*) wífmen and wǽpnedmen, wǽron hié swá rúwe and swá gehǽre swá wildeór, Nar. 20, 3–5. Ða rúwan (*pilosae*) handa wǽron swilce ðæs yldran bróður, Gen. 27, 23. Seó clǽne beó blósman gegrét swá lange ðæt hyre ða rúwan þeóh wurþaþ swýðe gehefegode, Anglia viii. 324, 13. Rúwe *hirta*, Germ. 398, 258. Hrúhe wulla *hirsutas lanas*, Hpt. 524, 13. II. *rough, untrimmed, uncultivated* :—Rúg *frondosa*, Wrt. Voc. ii. 151, 16. Ne turf ne toft, ne land ne lǽse, ne fersc ne mersc, ne rúh ne rúm, Lchdm. iii. 286, 24. Tó ðære rúwan hecgan, Cod. Dip. Kmbl. ii. 172, 32. On ðone rúwan hlync; andlang ðæs rówan linces, v. 297, 22. On rúwan beorg; of rúwan beorge, 277, 18. On ða rúgan þyrnan; of ðære þyrnan, iii. 419, 12. Ða férdon begen þurh ða rúgan fennas, Guthl. 3; Gdwin. 20, 25. III. *rough, knotty* :—Rúches *nodosi*, Hpt. Gl. 482, 60. IV. *rough, undressed* :—.xxx. ombra rúes cornes, iv. ambru meolwes, Chart. Th. 40, 9. [Þet ruwe vel, A. R. 120, 23. Nis þet iren acursed þet iwurðeð þe swarture and þe ruhure so hit is ofture iviled? 284, 17. Margareet sette hire fot uppon his ruhe necke, Marh. 12, 12. Sharrp and ruhh and gatelæs þurrh þorrness and þurrh breress, Orm. 9211. Mid ruȝe felle, O. and N. 1013. Sridde ȝhe Jacob and made him ru, Gen. and Ex. 1539. *O. H. Ger.* rúh *hirtus, hirsutus, hispidus, villosus, scaber, asper* : *O. Du.* rouw, rúgh, rú.]

rúm, es; *m.* I. *local, room, space* :—Under rodera rúm, Cd. Th. 71, 5; Gen. 1166. Hig næfdon rúm on cumena húse *non erat eis locus in diversorio*, Lk. Skt. 2, 7. II. *temporal, space of time* :—Næhtes rúme *noctis spatio*, Rtl. 36, 35. Þerh alle tído rúmo *per omnium horarum spatia*, 171, 41. III. *time which allows unhindered* or *unhurried action, opportunity* :—Rúm wæs tó nimanne londbúendum on hyra ealdfeóndum herereáf *the men of the land had ample opportunity of taking the spoil from their ancient foes*, Judth. Thw. 26, 7; Jud. 314. Hig ne móston rúm habban ðæt hig hit on riht gebócon (*Aegyptiis nullam facere sinentibus moram*), Ex. 12, 39. Fýrdraca rǽsde on ðone rófan ða hirn rúm ágeald (*when the opportunity was given him*), Beo. Th. 5374; B. 2690. Deáþ ðæs ne scrífeþ ðonne him rúm forlǽt rodora Waldend, Met. 10, 30. [*Goth.* rúms : *O. Sax.* rúm : *O. H. Ger.* rúm : *Icel.* rúm ; *n.*] v. ge-rúm.

rúm; *adj.* I. *local, roomy, spacious, ample, extensive* :—Se weg is swíðe rúm (cf. *Goth.* rúms wigs) ðe tó forspillednesse gelǽt *spatiosa via quae ducit ad perditionem*, Mt. Kmbl. 7, 13. Ðeós sǽ micel and rúm (*spatiosum*), Ps. Spl. 103, 26. Behealde hé hú wídgille ðæs heofones hwealfa biþ, and hú neara ðære eorþan stede is, ðeáh heó ús rúm þince, Bt. 19; Fox 68, 23. Rúma rodor *the spacious firmament on high*, Met. 28, 16. Ðære sunnan ryne is swíðe rúm, and ðæs mónan ryne is swíðe nearo, Lchdm. iii. 248, 7. Rúme ríce *a realm far-reaching*, Cd. Th. 254, 13; Dan. 611. Rúmes *spatiosae, ampli*, Hpt. Gl. 434, 45 : 493, 29. Ðú gesettes in stówne rúmre (*loco spacioso* ; in roume stede, E. E. Psalt.) foet míne, Ps. Surt. 30, 9. On sumne smeðne feld and rúmne (*amplam*), Bd. 5, 6; S. 618, 40. Ðis rúme land *the earth*, Cd. Th. 7, 31; Gen. 114. Ða rúman *patula*, Wrt. Voc. ii. 94, 61. Hié úte wilniaþ ðara rúmena wega ðisse worulde *causarum secularium foras lata itinera expetunt*, Past. 18, 4; Swt. 135, 6. Sóhton rúmre land, Cd. Th. 99, 25; Gen. 1651. Geseah ic ðone rúmestan (*latissimus*) feld, Bd. 5, 12; S. 629, 19. I a. *roomy, open, unencumbered.* v. rúmian :—Ne fersc ne mersc, ne rúh ne rúm *neither uncleared nor cleared* (?) *land*, Lchdm. iii. 286, 24. Þurh ða rúman *per patentes*, Wrt. Voc. ii. 69, 7. II. *temporal, long, extended* :—Bútan him se cyng rúmran fyrstes geunnan wolde, L. Eth. vii. 4; Th. i. 330, 12. III. *of mental qualities, ample, great, liberal* :—Ic mæg þurh rúmne sefan rǽd gelǽran, Beo. Th. 561; B. 278. Rúmran geþeaht, Elen. Kmbl. 2480; El. 1241. IV. *unrestricted, clear, free from conditions* :—Ðæt hé hit hæbbe swá rúm tó bóclonde swá hé ǽr hæfde tó lǽnlonde, Cod. Dip. Kmbl. iii. 258, 29. Ðé weorð on ðínum breóstum rúm *your mind will be freed from the trammels hitherto restricting it*, Cd. Th. 33, 13; Gen. 510. V. *not restrained within due limits, lax* :—Rúme regulas, Exon. Th. 131, 23; Gú. 460. VI. *ample, far-reaching* :—Ðíne dómas synd rihte and rúme, Hy. 7, 15. VII. *liberal.* v. rúm-gifa, -gifol, -mód :—Wel biþ ðam eorle ðe him oninnan hæfþ rúme heortan (*liberal in giving alms*), Exon. Th. 467, 16. v. rúm-heort. VIII. *great, noble, august* :—Ðære rúman a[u]guste, Wrt. Voc. ii. 5, 22. Rúmum *augusto, regali*, Hpt. Gl. 487, 29. Þurh ðæs rúme *per augustam*, Wrt. Voc. ii. 65, 59. Ðæs æþelan golde rúme *fausta*, 33, 76. [He wollde ȝifenn uss heoffness rume riche, Orm. 3689. Mi nest is holȝ and rum, O. and N. 643. He made ys wey roume ynou, R. Glouc. 303, 28. Make this place rom, Chauc. Reeves T. 206. *Goth.* rúms *spatiosus* : *O. Frs.* rúm *spacious, open* : *O. H. Ger.* rúmi *spatiosus, amplus* : *Icel.* rúmr.] v. ge-rúme.

rúma, an; *m. Separation* :—Rúma *discidium*, i. *separatio, divisio*, Wülck. 223, 25. [Cf. He gede on rum *he went apart*, Gen. and Ex. 400. On, a roume *at a distance*, Strat. Dict.]

rúmaþ. v. rýman.

rúme; *adv.* I. *local, widely, far and wide, so as to extend over a wide space* :—Cyning rúme rícsaþ *a king* (*the Deity*) *rules far and wide*, Met. 24, 32. Rúme geondwlítan ymb healfa gehwone, Exon. Th. 4, 30; Cri. 60. Heó wíde hire willan sóhte and rúme fleáh, Cd. Th. 87, 29; Gen. 1456 : 86, 10; Gen. 1428. Gehýran mæg ic rúme and swíðe geseón, 42, 14; Gen. 673 : 132, 9; Gen. 2190. Hié ne meahton leng somed blǽdes brúcan ... ac sceoldon ða rincas rúmor sécan, ellor éðelseld, 113, 31; Gen. 1895 : 115, 1; Gen. 1913. II. *liberally, extensively, amply, abundantly, in a high degree* :—Hyt rúme ða wyrmas forþ gelǽdeþ *it plentifully brings out the worms*, Lchdm. i. 282, 23.

Drihten rûme lḗt willeburnan on woruld þringan, Cd. Th. 82, 35; Gen. 1372: 75, 20; Gen. 1243. Ðû meaht his rûme rǣd geþencan *for this in ample measure may'st thou devise means*, 35, 27; Gen. 561. Ne willaþ rûmor unc landriht heora, 114, 27; Gen. 1910. Wes ðissum leódum árfæst gif ðē Alwalda scirian wille ðæt ðû rúmor (*more liberally than now is in your power* (?)) môte on ðisse folcsceare frætwa dǣlan, 171, 15; Gen. 2828.   **III.** *without restriction* or *encumbrance, without the pressure of care.* v. rûm-heort, II:—Ðá (*after Judith's prayer was answered*) wearþ hyre rûme (cf. *Ger.* aufgeräumt *of good cheer*) on môde, Judth. Thw. 22, 39; Jud. 97.   **IV.** *without obstruction, plainly, clearly*:—Emmanuhel, ðæt is gereht rúme : Nû is God sylfa mid ûs, Exon. Th. 9, 13; Cri. 134.   **V.** *without contraction, in full:*—Ðē ic ásecgan ne mæg rûme áreccan (*relate at length*), ne gerîm witan heardra heteþonca, 261, 12; Jul. 314.   [*O. Sax.* rûmur; *cpve. further*: *O. H. Ger.* rûmo procul, longe.]

rûmed (rûm-mḗd ?)-líc ; *adj. Ample, large, liberal*:—Hwæt rûmed-líces oððe micellíces oððe weordfullíces hæfþ se eówer gilp? Bt. 18, 1; Fox 62, 21. Mid rûmedlícum ælmessum, Shrn. 80, 10. v. *next word and* rûmmôd-líc.

rûmedlíce ; *adv.*   **I.** *liberally*:—Hē swá gifol is and swá rûmedlíce gifþ, Bt. 38, 3; Fox 202, 14. Ða ic rûmodlíce (rûmmôdlíce ?) gescarode, Cod. Dip. Kmbl. v. 331, 2. Ne ôðerra monna ne reáfaþ, ne hiera rûmedlíce dǣlaþ, Past. 23; Swt. 177, 7. Ðonne hwá ǣgðer ge mete ge hrægl þearfeudum rûmedlíce (rûmodlíce, Hatt. MS.) selþ, 44; Swt. 326, 20.   **II.** *at large, fully*:—Ðis ðæt wē nû feám wordum árîmdon wē willaþ hwḗne rûmedlícor (*paulo latius*) áreccean, 12; Swt. 75, 17. Rûmerlícor [rûmed-?] *latius, multiplicius*, Hpt. Gl. 420, 30. v. preceding word.

rûm-gál ; *adj. Rejoicing in ample space in which to move* (applied to the dove when sent from the ark) :—Seó culufre wîde fleáh ôþ ðæt heó rûmgál reste stôwe funde *far the dove flew, in flight unconfined rejoicing, until a place of rest she found* (cf. heó rûme fleáh, 87, 29; Gen. 1456), Cd. Th. 88, 16; Gen. 1466.

rûm-gifa, an; *m. A liberal giver*:—Hē wæs eallum rûmgifa *manu omnibus largus*, Bd. 3, 14; S. 540, 8. v. next word.

rûm-gifol ; *adj. Liberal, bountiful, munificent*:—Rûmgifol, cystig *prodiga, larga*, Germ. 395, 18. Monig biþ ágiéta his gôda and wilnaþ mid ðý geearnigan ðone hlísan ðæt hē síe rûmgiful *se effusio sub apellatione largitatis occultat*, Past. 20; Swt. 149, 7. Ic Oswald þurh ða rûmgiflan Godes cyste tô biscope gehádod, Cod. Dip. Kmbl. ii. 400, 25. Hē gewende ðe Rôme mid ðam rûmgyfolan (-geofolan, MS. V.) þegne, Homl. Skt. i. 5, 330.

rûmgifolness, e; *f. Liberality, bounty, munificence*:—Seó rûmgifolnes (*largitas*) winþ ongeán ða gýtsunge, Gl. Prud. 65. Rûmgyfolnes, 67: 68-70. Hwá áwent gîtsunge mid rûmgifulnysse bûtan strece ? Homl. Th. i. 360, 6.

rûm-heort ; *adj.*   **I.** *of liberal heart, liberal, munificent* :—Rûmheort *dapsilis*, Wrt. Voc. ii. 27, 12. Rûmheort hláford (*the Deity*), Hy. 7, 63. Mē wine Scyldinga fela leánode . . . rûmheort cyning, Beo. Th. 4227; B. 2110: 3602; B. 1799. Rûmheort beón mearum and máþmum *to be liberal of gifts*, Exon. Th. 3391; Gn. Ex. 87.   **II.** *with mind free from oppression, untroubled.* v. rûme, III :—Se weg ðe tô life lǣt is ûs tô gefarenne mid rûmheortum môde and mid gôdum and glædum geþance *dilatato corde curritur via mandatorum Dei*, R. Ben. 5, 22.

rûmheortness, e; *f. Liberality, munificence*:—Syndon eahta heálíce mægenu . . . ðæt is rûmheortnys (*largitas*) . . . , Wulfst. 68, 19. Eahta sweras syndon ðe rihtlícne cynedôm trumlíce up wegaþ . . . rûmheortnes (*largitas*), L. I. P. 3; Th. ii. 306, 20. Rûmheortnesse *liberalitatis*, Wrt. Voc. ii. 52, 32; 79, 52. Rûmheortnesse *liberalitatem*, 86, 52.

rûmian ; *p.* ode *To get free from encumbrance*:—Ðonne rûmaþ him sôna se innaþ, Lchdm. i. 76, 13. v. rûm, I a.

rûm-líc ; *adj.*   **I.** *gracious, liberal, benign*:—Rûmlíc *benignus* (*Deus*), Rtl. 104, 32. Rûmlícum helpe *benigno favore*, 17, 35.   **II.** *liberal, abundant, plentiful*:—Nû wille wē ðis águnnene weorc mid rûmlícum wæstme begán, Anglia viii. 300, 6. Se ðe mid fôdan ðære upplícan lufe biþ gefylled, hē biþ swilce hē sý mid rûmlícum mettum gemæst, Homl. Th. i. 522, 32. v. next word.

rûmlíce ; *adv.*   **I.** *largely, fully, at large, at length*:—Ðæt hî rûmlíce roccettaþ swîðe, Ps. Th. 143, 16. Ðás þing rûmlíce gecýðan, Anglia viii. 303, 48. Ymbe ðás þing rûmlícor sprecan, 321, 36. Tô-dǣledlícor *vel* rûm[lícor ?] *differentius*, Wrt. Voc. ii. 140, 15.   **II.** *liberally*:—Gif wē lustlíce and rûmlíce ða welan dǣlaþ earmum monnum ðe God ûs ǣr sealde, Blickl. Homl. 49, 32.   **III.** *graciously, kindly, benignly*:—Rûmlíce *clementer*, Rtl. 89, 38: Mt. Kmbl. p. 16, 7. [Heó rumliche hit (*silver and gold*) ʒef þon kempan, Laym. 2452. *O. H. Ger.* rûmlího *large*.]

rûm-môd ; *adj.*   **I.** *of liberal mind, liberal in giving*:—Hē þearfum rûmmôd (*largus*) wæs, Bd. 3, 6; S. 528, 11. Sýn wē rûm-môde þearfendum mannum and earmum ælmesgeorne, Blickl. Homl. 109, 14. Sellaþ ælmessan, beóþ rûmmôde ryhtra gestreóna, Exon. Th. 106, 30; Gû. 49.   **I a.** *too liberal, profuse*:—Swá ða rûmmôdan fæsthaf-

olnesse lǣren swá hî ða uncystigan on yfelre hneáwnesse ne gebrengen *sic prodigis praedicetur parcitas, ut tamen tenacibus periturarum rerum custodia non augeatur*, Past. 60; Swt. 453, 28.   **II.** *benignant, gracious, kind*:—Rûmmôd and mildheart is God *benignus et misericors est Deus*, Rtl. 5, 8: Bt. 42; Fox 258, 22: Lk. Skt. Lind. 6, 35. Rûmmôd *clemens*, Rtl. 74, 10. *The word translates* paracletus, Rtl. 120, 1: Jn. Skt. Lind. 14, 16, 26: 15, 26.

rûmmôdlíc. v. next word.

rûmmôdlíce ; *adv.*   **I.** *liberally*:—Gif wē blíþe and rûmmôdlíce hî (*the tenth part of our goods*) dǣlan willaþ earmum mannum, Blickl. Homl. 51, 10.   **II.** *graciously, favourably*:—Rûmmôdlíce *propitius*, Rtl. 2, 5: 22, 38: *clementer*, 14, 36: *clementissime*, 98, 16.

rûmmôdness, e; *f.*   **I.** *liberality*:—Ðýlæs ða rûmmôdnessa sió unrôtnes gewemme *ne largitatem tristitia corrumpat*, Past. 44, 3; Swt. 323, 10.   **II.** *favour, grace, kindness*:—Snotor rûmmôdnise *sapiens benignitas*, Rtl. 105, 1. Rûmmôdnise *clementiam*, 41, 5: *propitiationem*, 17, 25.

rûmness, e; *f.*   **I.** *breadth, a broad space*:—Ða rûmnisse Jericho feldes *latitudinem campi Jericho*, Deut. 34, 3.   **II.** *breadth, amplitude, abundance*:—Wæs swá mycel rûmnes on him ðæs hálgan geleáfan and swá mycele hē tô ðære Godes lufan hæfde *there was in him so great abundance of the holy belief, and he had besides so great love for God*, Guthl. 20; Gdwin. 82, 8.

rûm-well (= -full ?) *spacious*:—Rûmwelle weg *spatiosa via*, Mt. Kmbl. Lind. 7, 13.

rûn, e; *f.*   **I.** *a whisper* (v. rûnian), hence *speech not intended to be overheard, confidence, counsel, consultation* [cf. *Goth.* rûna niman *to take counsel*]:—On hyne nǣnig monna cynnes mihte wlítan nymþe se môdiga hwæne neár hête rinca tô rûne gegangan (cf. gangan te rûnu, an rûna, Hel. 1273, 5064), Judth. Thw. 22, 7; Jud. 54. Gesittan tô rûne *to sit in consultation*, Beo. Th. 346; B. 172. Gesittan sundor tô rûne, Andr. Kmbl. 2324; An. 1163. Swá cwæþ snottor on môde gesæt him sundor æt rûne *sat apart communing with himself* (cf. nim thû ina sundar te thî an rûne, Hel. 3227), Exon. Th. 293, 5; Wand. 111. Gefetigan tô rûne (cf. *Icel.* heita rûn at rûnum *to consult*), 246, 15; Jul. 61: Elen. Kmbl. 2319; El. 1162. Eodon fram rûne, 821; El. 412. Rûne besittan, Andr. Kmbl. 1254; An. 627. Ic Sîward cinges þegen æt rǣde and æt rûnan (cf. þegno betst (*Peter*) te is herron sprak an rûnun, Hel. 3096), Cod. Dip. Kmbl. iii. 351, 17. Hē (*Christ*) feówertig daga folgeras sîne rûnum (cf. Jesus . . . being seen of them forty days, and speaking of the things pertaining to the kingdom of God, Acts 1, 3) ārētte, Hy. 10, 36.   **II.** *a mystery, cf. geryne*:—Rûn biþ gerecenod, Cd. Th. 211, 12; Exod. 525. Bæd him áreccan hwæt seó rûn (*the dream*) bude, 250, 6; Dan. 542. Healdaþ æt heortan hálige rûne, Exon. Th. 282, 1; Jul. 656: Elen. Kmbl. 666; El. 333. Dryhtnes word, hálige rûne, 2336; El. 1169. Déglum rûnum *mystice*, Jn. Skt. p. 4, 4.   **III.** *a secret*:—Rûne healdan *to keep one's counsel*, Exon. Th. 338, 31; Gn. Ex. 87.   **IV.** *of that which is written, with the idea of mystery or magic*:—Ðæt hē him bôcstafas árǣdde and árehte hwæt seó rûn (*the writing on the wall of Belshazzar's palace*) bude, Cd. Th. 262, 9; Dan. 741. Hæfdon hié on rûne and on rîmcræfte áwriten wera endestæf, Andr. Kmbl. 267; An. 134.   **V.** *a rune, a letter.* v. rûn-stæf:—Enge rûne (*referring to* ᛏ = nîd), Elen. Kmbl. 2521; El. 1262. Rǣd sceal mon secgan, rûne wrítan, leóþ gesingan, Exon. Th. 342, 7; Gn. Ex. 139. Hē hine ácsade hwæðer hē ða álýsendlícan rûne cûþe and ða stafas mid hine áwritene hæfde þe swylcum menn leásspell secgaþ ðæt hine mon forðon gebindan ne mihte *interrogare coepit, an forte literas solutorias de qualibus fabulae ferunt, apud se haberet, propter quas ligari non posset*, Bd. 4, 22; S. 591, 25. [Ofte heo eoden to ræde ofte heo heolden rune (ʒeode to roune, 2nd MS.), Laym. 25332. Þan kaisere heo radden þat he write runen (writes makede, 2nd MS.), 25340. Godess dærne rǣd and run, Orm. 18719. Godes derne runes and his derne domes, A. R. 96, 4. [*Goth.* rûna *counsel, a mystery*: *O. Sax.* rûna *counsel, conference*: *O. H. Ger.* rûna *susurrio, mysterium, litera*, v. Grff. ii. 523: *Icel.* rûn *counsel, mystery, a letter*.] v. beadu-, hete-, hyge-, inwit-, leóþu-, searo-, wæl-rûn.

-rûn *in* burh-rûn:—Burgrûne *Parcas*, Wrt. Voc. ii. 116, 10. [Cf. -rûn *in proper names in Icel.* e. g. Sig-, Öl-rûn: and see Grmm. D. M. 376.] v. -rûne.

-rûna. v. ge-rûna, hell-rûna (-rune ?). [Cf. *Icel.* rûni *a counsellor*] v. -rûne.

rûn-cofa, an; *m. The chamber of secret counsel, the mind, breast*:—Hē mæg on his rûncofan rihtwîsnesse findan on ferhþe fæste gehýdde (cf. ðonne fint hē ðær (on his gemynde) ða ryhtwîsnesse gehýdde, Bt. 35, 1; Fox 156, 11), Met. 22, 59.

rûn-cræftig ; *adj. Skilled in explaining mysteries*: — Ne mihton árǣdan rûncræftige men (cf. *the astrologers, Chaldeans, and the soothsayers*, Dan. 5, 7) eagles ǣrendbēc (*the writing on the wall of Belshazzar's palace*), Cd. Th. 261, 31; Dan. 734.

-rûne. v. helle-, leód-rûne, burh-rûnan, *and* -rûn. [Cf. *Icel.* rûna *a counsellor*.]

rūnere, es; *m. A whisperer :*—Ðes rūnere *hic susurro,* Ælfc. Gr. 36; Som. 38, 51. [*O. H. Ger.* rūnari *susurro, musitator.*] v. next word.

rūnian; *p.* ode *To talk low, whisper, mutter :*—Ic rūnige *susuro,* Ælfc. Gr. 36; Som. 38, 53. Tôgeánes mē rūnedon (*susurrabant*) ealle fŷnd mīne, Ps. Spl. C. 40, 8. Ðeáh ðē mon hwylces hlihge, and ðū ðē unscyldigne wite, ne rēhst ðū hwæt hȳ rǣdon oððe rūnion, Prov. Kmbl. 12. Ða rūniendan *musitantes,* Wrt. Voc. ii. 54, 72. Rūnigendum stefnum, Guthl. 5; Gdwin. 36, 1 note. [His egen to sen, his muð to runien, O. E. Homl. ii. 107, 19. Ræden and runan (rouni, 2nd MS.), Laym. 2331. *Chauc.* Piers P. roune *to whisper :* Prompt. Parv. rounin *susurrare : O. L. Ger.* rūnan *susurrare : O. H. Ger.* rūnēn *susurrare, musare, musitare : O. Du.* rūnen.] v. reónian, rȳnan.

rūn-lic; *adj. Mystical :*—Færme his rūnlīce ł deóplīce *cenae ejus misticae,* Mk. Skt. p. 5, 11. Cf. rȳne-līc.

runol (*for* hrunol, cf. *Icel.* hrunull *foul-smelling*); *adj. Foul, stinking* (?) :—Wið dȳ (ða, MS.) runlan ǣttre, Lchdm. iii. 36, 17.

rūn-stæf, es; *m. A* (*runic*) *letter, a rune.* Cf. rūn, V :—Drȳ sind in naman rūnstafas, Exon. Th. 440, 9; Rä. 59, 15. Ic mæg þurh rūnstafas rincum secgan, ðam ðe bēc witan, 429, 17; Rä. 43, 6. Wæs on ðām scennum þurh rūnstafas rihte gemearcod, hwam ðæt sweord geworht wǣre, Beo. Th. 3394; B. 1695. Ðā áxode se ealdorman ðone hæftling hwæðer hē þurh drȳcræft oððe þurh rūnstafas his bendas tôbrǣce, Homl. Th. ii. 358, 11. On the subject of Runes see Kemble's paper in Archaeologia, vol. xxviii; the Preface to Dr. George Stephens' Handbook of Runic Monuments; Dr. Isaac Taylor's Greeks and Goths, and the same writer's work 'The Alphabet.'

rūnung, e; *f. Whispering, soft speech :*—Seó sôðfæste fǣmne hyre lāca ne rôhte ne hyre rūnunga, Homl. Skt. i. 2, 149.

rūn-wita, an; *m.* I. *a privy councillor, one acquainted with a person's secrets :*—Deád is Æschere mīn rūnwita and mīn rǣdbora, Beo. Th. 2654; B. 1325. II. *one acquainted with mysteries, a sage :*—Rôf rūnwita (*Guthlac*), Exon. Th. 167, 30; Gū. 1068.

rupe (?) :—Rupe (rūwe (?), cf. rūh) oððe drisne *capillamenta* (cf. rawe, drisne *capillamenta,* ii. 128, 39), Wrt. Voc. i. 28, 73.

rusce, an; *f. Rushy ground* (?) :—Tô ðære wulfruscan, Cod. Dip. Kmbl. iii. 131, 7. v. rysc.

rust, es; *m. n.* (?) *Rust erugo,* Wrt. Voc. ii. 107, 37 : 29, 46. *Erugo* rust, ôm, *vel tinea .i. vitium frumenti* vel *ferri,* 144, 3 : Mt. Kmbl. Lind. 6, 19. Rost, Txts. 60, 397. Of ruste vel ôme *erugine .i. rubigine,* Wrt. Voc. ii. 144, 5. Ðǣr wæs suîðe suîðlīc gesuinc and ðeáh ne meahte monn him of ániman ðone miclan rust. . . Hē wolde from ūs ádôn ðone rust ūrra unþeáwa, Past. 37; Swt. 269, 11–15. Ǣrest ic wille beón gefremed in litlum weorce, ðæt ic mǣge sum rust (sinnrust (?) v. syn-rust) on weg ádrīfan of mīnre tungan, Shrn. 35, 20. [*O. Sax.* rost : *O. H. Ger.* rost.] v. syn-rust.

rustig; *adj. Rusty :*—Ða wurdon Janes dura fæste betȳned and his loca rustega *Jani portas ipse clausit. Quas obseratas otio ipsa etiam rubigo signavit,* Ors. 5, 15; Swt. 251, 21. [*O. H. Ger.* rostag *scabrosus.*]

rūte *rue.* v. rūde.

rūwa, an; *m. A rug, covering, tapestry :*—Hió becwið Eádgyfe līnnenne rūwan, Chart. Th. 537, 27. Ðeáh ðe ða rícestan hātan him reste gewyrcan of marmanstāne and mid goldfrætwum and mid gimcynnum eal āstǣned and mid seolfrenum rūwum and godwebbe eall oferwrigen, Wulfst. 263, 4. v. reówe, rȳhe.

ruxlan = hruxlan *to make a noise :*—Ruxlende *tumultuantes,* Mt. Kmbl. Rush. 9, 23. v. ge-hruxl.

rȳan (?), rȳn (cf. *for similar form of infinitive* þȳn); *p.* rȳde *To roar, rage :*—Hwȳ rȳð (rȳnþ?) ǣlc folc *quare fremuerunt gentes ?* Ps. Th. 2, 1. Seó leó gif heó blôdes onbirigþ heó gemonþ ðæs wildan gewunan hire eldrana onginþ ðonne rȳn and hire racentan slītan (cf. *the corresponding passage in the Metres :* Onginþ racentan slītan, rȳn, grymetigan, Met. 13, 29) *si cruor horrida tinxerit ora, resides olim redeunt animi, fremituque gravi meminere sui, laxant nodis colla solutis,* Bt. 25; Fox 88, 13. [Cf. *O. H. Ger.* rohôn *rugire,* Grff. ii. 431.] v. rȳung.

rȳcels. v. rēcels.

ryddan (hryddan? v. hryding) *to strip :*—Árȳdid *expilatam,* Wrt. Voc. ii. 108, 4.

ryden, es; *n. The name of some plant :*—Wirc beþinge, nim ðæt reáde ryden, Lchdm. ii. 340, 5.

rȳe, rȳfe, ryft. v. rȳhe, rīfe, rift.

ryge, es; *m. Rye :*—Ryge *sicalia,* Wrt. Voc. ii. 120, 53 : *singula,* i. 287, 18. Riges seofoþa, Lchdm. ii. 48, 20. [*Icel.* rugr; *m.* : cf. *O.L. Ger.* roggo : *O. H. Ger.* rokko.]

rygen; *adj. Rye, of rye :*—Of rigenum melwe, Lchdm. ii. 236, 9. Of sûrre rigene grût, 342, 17. Genim rigen healm and beren, 148, 11. Genim rigen mela, 148, 22.

rȳhe, rȳe, an; *f. A rug, rough covering, blanket :*—Rȳhae, rȳe *villosa,* Txts. 106, 1080. Hrȳhae, rȳae, rȳe *tapeta,* 102, 1020. Līnin rȳhae, rȳee *villa,* 106, 1081. Rīhum *tapetibus,* 114, 120. v. reówe, rūwa.

ryht. v. riht.

rȳman; *p.* de. I. *to make roomy, extend, spread, enlarge, amplify :*—Ðū rȳmdest *dilatasti,* Ps. Lamb. 4, 2. Hē ēðelþrym rȳmde and rǣrde, Cd. Th. 98, 24; Gen. 1635. Sôð metod rȳmde, wîde wǣðde *spread and drove the waters widely,* 208, 7; Exod. 479. Ðæt se gîtsere his land mid unryhte rȳme, Past. 44, 8; Swt. 329, 21. Hū feor wolde gē rȳman eówer land *quousque vos extenditis?* Swt. 331, 1. Ic eft reorde under roderum rȳman wille *I will multiply food again under heaven* (*after the deluge*), Cd. Th. 81, 13; Gen. 1344. Hira mearce tô rȳmanne *ad dilatandum terminum suum,* Past. 48, 2; Swt. 367, 15. Heora hūs tô rȳmende, Chart. Th. 436, 18. II. *to make clear by removing obstructions, to clear a way* (lit. and metaph.) :—Hē sáwlum rȳmeþ lîfwegas, Exon. Th. 148, 4; Gū. 739 : 436, 6; Rä. 54, 10. Ðonne rȳmeþ hē ðam deádan tô ðam áþe ðæt hine môton his mǣgas unsyngian *by such conduct he clears the way for an oath on behalf of the dead man, so that his* (*the dead man's*) *kinsmen may exculpate him,* L. In. 21; Th. i. 116, 7. Gif getrȳwe gewitnes him tô ágenunge rȳmþ *make the way to possession clear for him,* L. Eth. ii. 9; Th. i. 290, 20. Ðæt syndan Antecristes þrǣlas ðe his weg rȳmaþ, Wulfst. 55, 9. Ða ðe ingang rȳmaþ, Salm. Kmbl. 442; Sal. 221. Se engel āwylte ðæt hlid; nā ðæt hē Criste ûtganges rȳmde, Homl. Th. i. 222, 9. Se engel rȳmde him weg þurh ðæt fȳr, ii. 344, 13. Ic wille rȳman mînne bertūn and þeáno geeácnian (*I will pull down my barns and build greater,* Lk. 12, 18), 104, 1 : Wulfst. 286, 19. Seó sealf wile ǣrest ða dolh rȳman, and ðæt deáde flǣsc of etan, Lchdm. ii. 332, 24. III. *to make room by removing one's self, yield, give place :*—Ic fare āweg oððe ic rȳme (rume, MS. W. : hryme, hrime, other MSS.) *caedo* (cf. Wot no mon þe time wanne he sal henne rimen, O. E. Misc. 113, 170), Ælfc. Gr. 28, 4; Zup. 171, 9. Se ôðer rȳmþ him setl, Homl. Th. i. 248, 17. Rūmaþ, steppaþ *cedunt,* Wrt. Voc. ii. 19, 19; 87, 64. Rȳmde *cessit,* 81, 75. Ā man rȳmde (*retreated*) fram ðære sǣ, and hî fērdon ǣfre forþ æfter, Chr. 999; Erl. 135, 35. Hî rîmdon heora feóndum *they left the field clear for their foes,* 1015; Erl. 152, 16. Rȳm ðysum men setl *da huic locum,* Lk. Skt. 14, 9. Rȳmaþ him (*cease to oppose him*) ðæt hē mē leng ne swence, Homl. Th. i. 534, 17. [*Laym.* rumen *to clear* (*a way*), *to yield, give place :* R. Glouc. rume *to clear* (*a way*) : Piers P. roume *to keep clear of :* O. Sax. rūmian *to clear :* O. Frs. rēma : O. H. Ger.* rūmman *cedere, abire, laxare :* Icel. rȳma *to make room, clear, to quit, leave.*] v. ge-rȳman.

rȳmet[t], es; *n.* I. *space, extent :*—Seó cyrce mid hire portice mihte fîf hund manna eáðelīce befôn on hire rȳmette, Homl. Th. i. 508, 14. Nā swylce on eástdǣle synderlīce sȳ his (*God's*) wunung . . . se ðe ǣghwǣr is andweard nā þurh rȳmyt ðære stôwe ac þurh his mægenþrymnes andweardnysse *he who is everywhere present, not through the extent of the place in which he dwells, but through the presence of his glory,* 262, 9. Eall ðæt rȳmet ðe eówer fôtswaþu on bestæpþ ic eów forgife *omnem locum, quem calcaverit vestigium pedis vestri, vobis tradam,* Jos. 1, 3. II. *clear space, room* (v. rȳmetleást) :—Ðær næs nān rȳmet on ðam gesthūse, Homl. Th. i. 30, 14. Hit is gedôn swā ðū hēte, and hēr gyt is rȳmet æmtig, ii. 376, 9. III. *extension, clearance :*—Eádgár mid rȳmette (*by extending the limits of their property and so removing the claims which interfered with the monasteries standing within a ring fence*) gedîhligean hēt ða mynstra on Wintanceastre . . . and ðæt āsmeágan hēt, ðæt nān ðæra mynstera ðǣr binnan þurh þet rȳmet wið ôðrum sace næfde, ac gif ôðres mynstres ār on ôðres rȳmette lēge (*if the property of one monastery should lie within the part given by the extension to another*) ðæt ðes mynstres ealdor, ðe tô ðam rȳmette fēnge, ofeode ðæs ôðres mynstres āre mid swilcum þingum swylce ðam hîrede, ðæ ða āre āhte, gecwēme wǣre, Chart. Th. 231, 2–18. v. Lchdm. iii. 417 on this charter. IV. *extension of a person's well-being :*—Ða (*certain property*) ic gescarode mē sylfum and mînum foregengum and eftyrgengum tô ēcum rȳmete *to the furtherance of the eternal well-being of myself and of my predecessors and successors,* Cod. Dip. Kmbl. v. 331, 3.

rȳmetleást, e; *f. Want of room :*—Maria hire sunu for rȳmetleáste (v. rȳmet, II) on ānre binne gelēde, Homl. Th. i. 34, 22.

rymg. v. rȳung.

rȳmþ, e; *f. Amplitude;* amplitudo (cited by Lye). [Heo bigunnen arumðe (*in large numbers*) ræsen to somne, Laym. 27492. *Prompt. Parv.* rymthe *spacium; oportunitas vel spacium temporis.*]

rȳn. v. rȳan.

rȳnan; *p.* de *To roar :*—Sume hî sǣdon ðæt hió sceolde forsceoppan tô león, and ðonne seó sceolde sprecan, ðonne rȳnde hió, Bt. 38, 1; Fox 194, 34. Ða ðe león wǣron ongunnon lāðlīce yrrenga rȳna (rȳnan (?), rȳan (?)), Met. 26, 84. v. rȳan.

ryne, es; *m. A course, run, running,* both in the sense of *motion* and in that of *the path in which motion takes place.* I. *of a ship :*—Ānes ceóles ryne on London *free entrance of one ship into the port of London* (cf. ego indico me dedisse unius navis incessum in portu Lundoniae, 220, 18–22), Cod. Dip. B. i. 221, 21. II. *of other things, of the heavenly bodies, an orbit :*—Nǣron nô swā gewîslīce ne

swā endebyrdlíce hiora (*the various members of the created world*) stede and hiora ryne funden on hiora stówum and on hiora tídum gif ān unāwendendlíc God nǽre *non tam certus naturae ordo procederet, nec tam dispositos motus, locis, temporibus explicaret, nisi unus esset qui has mutationum varietates manens ipse disponeret*, Bt. 35, 2; Fox 158, 3. Roder *firmamentum*, ryne *cursus*, middaneard *mundus*, Wrt. Voc. i. 41, 57-59. Dǽre sunnan ryne is swíðe rúm, and ðæs mónan ryne is swíðe nearo, Lchdm. iii. 248, 7-8. Siððan wæs rodor ārǽred and ryne tungla gefæstnad, Exon. Th. 272, 13; Jul. 198. Ryne *curriculo, cursu*, Hpt. Gl. 457, 18. Ealle gesceafta symle sculon ðone ilcan ryne eft gecyrran, Met. 11, 37. Ða mǽran tungl āwðer ððres rene ā ne gehríneþ, 29, 10. Tunglu ða ðe ryne healdaþ, Cd. Th. 239, 13; Dan. 369. **II a.** metaph. *course, uninterrupted progress* (cf. that the word of the Lord may have free course, 2 Thes. 3, 1) :—Se ðe reorda gehwæs ryne gemiclaþ, ðara ðe noman Scyppendes þurh horscne hād hergan willaþ, Exon. Th. 4, 4; Cri. 47. **III.** of fluids, *a course, water-course, a flow, flux* of blood :—Ða ætstód ðæs blódes ryne *fluxus sanguinis*, Lk. Skt. 8, 44: Mk. Skt. 5, 29. Seó eá ætstent on hire ryne, Jos. 3, 13. Hí nāmon twelf stānas on ðæs streámes ryne *de medio Iordanis alveo*, 4, 8. Plantud nēh ryne (rynum, Ps. Th.) wetæra *secus decursus aquarum*, Ps. Spl. 1, 3. Wæter ða nū under roderum heora ryne healdaþ, Cd. Th. 10, 20; Gen. 159. Wið rynas wætera, Ps. Lamb. 1, 3. **IV.** of time, *course, cycle, lustre* :—Geár *annus*, tíd *tempus*, ryne *cursus*, Wrt. Voc. i. 52, 38-40. Ryne *cyclus*, rynum *cyclis*, ii. 20, 64-65 : 137, 73. Ða se ryne ðissa geára gefylled wæs *quo completo annorum curriculo*, Bd. 3, 9; S. 533, 9. Ryne *lustro*, Wrt. Voc. ii. 50, 42. **V.** *course* of life :—Honorius æfter ðon ðe hē ða gemǽro his rynes gefyllde of ðissum leóhte leórde (*postquam metas sui cursus implevit*), Bd. 3, 20; S. 550, 25. Gif ðū hine lufast on ðínes lífes ryne, ðe ðē is ungewiss, Basil admn. 8; Norm. 52, 8. **VI.** *currus* is translated by *ryne* in Ps. Spl. T. 67, 18 *and* Cant. Moys. Thw. 29, 10. [Bi his blodi rune þet ron inne monie studen, O. E. Homl. i. 207, 10. Þe stronge rune of þat blodi stream, Marh. 7, 12. Þer is mest neod hold hwon þe tunge is o rune, A. R. 74, 21. **Goth.** runs: O. *Frs.* blôd-rene: O. H. Ger. run *meatus*: cf. Icel. runi *a flux, stream*.] v. blôd-, eft-, forþ-, gegn-, on-, riht-, streám-, up-, út-ryne.

-ryne; *adj.* v. dæg-, hider-, hwider-ryne.

rȳne, es; *n. A mystery, mysterious saying* :—In rȳne *in misterio*, Lk. Skt. p. 3, 3. Tó wuttanne clǽne rȳne ł āsægdnise (*mysterium*) ríces Godes, Lind. 8, 10. Rȳne ongietan reádan goldes guman galdorcwide gleáwe beþuncan *let men understand the mysterious speech of the red gold* (*a ring which is represented as speaking*), *wisely consider its charm*, Exon. Th. 432, 26; Rä. 49, 6. Clǽno rȳno ł gesægdnise ł diópnise *mysteria*, Mt. Kmbl. Lind. 13, 11. v. ge-rȳne.

ryne-gǽst, es; *m. A guest or foe that comes swiftly* (?), a term used for lightning :—Feá ðæt gedýgaþ ðara ðe gerǽcaþ rynegiestes wǽpen *few escape whom the lightning strikes*, Exon. Th. 386, 8; Rä. 4, 58.

rȳnegu in hel-rȳnegu *pythonissa*, Wrt. Voc. ii. 50, 9.

rynel, es; *m. A runner, messenger, courier* :—Rynel *cursor*, Wrt. Voc. i. 76, 24 : Ælfc. Gr. 36; Som. 38, 24. Renel, Kent. Gl. 949. Pilatus hēt geclypian his ænne rynel and hym tó cwæþ : Yrn and clypa tó mē ðone ðe ys Jesus genemned. Se rynel swā dyde and myd mycelum ófste wæs forþyrnende . . . Hí clypodon tó Pilate : Hēte ðū ðynne bydel and ðynne rynel hym swā ongeán cuman ? Nicod. 3; Thw. 2, 5-16 : 4; Thw. 2, 19-36. Renula *cursorum*, Hpt. Gl. 406, 8. Rynela *concurrentium*, Anglia viii. 302, 33 (v. samod-rynel). v. for-rynel.

rynel, es; *m. A stream* :—Rynelas *rivos*, Ps. Spl. 64, 11 : Blickl. Gl. cf. rinnelle.

rȳne-líc; *adj. Mystical*; *mysticus*, Hymn. Surt. 48, 25 : 87, 15. v. ge-rȳnelíc, rún-líc.

rȳnelíce; *adv. Mystically*; *mystice*, Hymn. Surt. 68, 13. v. ge-rȳnelíce.

rȳne-mann, es; *m. One skilled in explaining mysteries* :—Ða clamme ðe ða rǽdellan wið rȳnemenn heóld, Exon. Th. 429, 32; Rä. 43, 13.

ryne-strang; *adj. Strong for the course*, Exon. Th. 400, 9; Rä. 20, 7.

ryne-swift; *adj. Swift in its course* :—Ofer uppan rodere ryneswiftum, Met. 24, 28.

ryne-þrág; *f. A space of time* :—Hý bídinge mōstun tídum brúcan . . . restan ryneþrágum, Exon. Th. 115, 3; Gū. 184.

ryne-wægn, -wǽn, es; *m. A swift vehicle, a chariot* :—On rynewǽnum *in curribus*, Ps. Th. 19, 7.

rynge. v. renge.

rynig; *adj. Good at running* :—Sum biþ rynig, sum ryhtscytte, Exon. Th. 296, 14; Crä. 51. [Cf. (?) He gon to rusien swa þe runie (wode, 2nd MS.) wulf þenne he cumeð of holte, Laym. 20123.] v. wíd-rynig.

ryniga (?), an; *m. Liquid that runs off* (?) :—Wel mintan on sealtes rynian, Lchdm. ii. 76, 2. Genim rynian sealt[es], gehæt, þweah mid ðý, 156, 16.

rynning, e; *f. Rennet*; *coagulum*, Wrt. Voc. i. 27, 70. [*Gloucestershire* running *rennet*, E. D. S. Gloss. B. 4. 'Earning, yearning, cheese-rennet, or that which curdles milk,' Brockett. 'Runnet, called in Derby-shire *erning*; it runs the milk together,' Pegge. E. D. S. Gloss. C. 3.] v. ge-runnen.

rȳpe, rȳpan, rȳpere. v. rípe, rípan, rípere.

ryplen (?); *adj. Made of broom* :—Ryplen (þýfflen ? v. þýfel) *sparteus*, Germ. 399, 457.

rysc; *m. f.* (?) : rysce, an; *f. A rush* :—Risc *juncus*, Wrt. Voc. i. 31, 30 : ii. 112, 18. Risce, i. 68, 35. Resce *juncus vel scyrpus*, 79, 66. Spyrte biþ of rixum gebróden. Rixe weaxst gewunelíce on wæter-igum stówum, Homl. Th. ii. 402, 8-10. Risce *papyro, junco*, Hpt. Gl. 483, 69. Grównys hreódes and ricsa *viror calami et junci*, Bd. 3, 23 ; S. 554, 23. Ricsa wyrttruman, Lchdm. ii. 234, 8. Rixum *juncis*, Wrt. Voc. ii. 97, 21. Ða heó geseah ðone windel on ðām rixum (*in papyrione*), Ex. 2, 5. [*Ayenb.* resse : *Piers P.* rische, reshe, rusche : *Chauc.* rishe : *Prompt. Parv.* rische, rusche : *M. H. Ger.* rusch ; *f. a rush* : *Du.* rusch ; *n.* From Latin *ruscus*.] v. eá-(ǽ-, eó-)risc, -rixe.

rysc-bedd, es; *n. A bed of rushes* :—On ðæt riscbed, Cod. Dip. Kmbl. iii. 428, 31.

rysce. v. rysc.

ryscen; *adj. Of rushes, rush* :—Riscene weocan *fila scirpea* (*juncea*), Germ. 391, 15.

rysc-leác; *n. Rush leek, rush garlick*; allium scharnoprassum :—Riscleác *allans* (*allium*?), Wrt. Voc. ii. 10, 40.

rysc-pytt, es; *m. A pit or pool in which rushes grow* :—In hriscpyt ; of hriscpytte intó ðere díc, Cod. Dip. Kmbl. 385, 2-3.

rysc-steort, es; *m. A promontory where rushes grow* :—Æt riscsteorte ; of ðam hriscsteorte, Cod. Dip. Kmbl. v. 217, 12-13.

rysc-þýfel, es; *m. A rush-bed, bed of rushes* :—Riscþýfel *juncetum*, Wrt. Voc. i. 63, 73 : *juvencibus*, 287, 261. Risc *juncus* ; riscþýfel *jungetum* ; riscþýfel *juvencibus*, ii. 45, 75-77. Risc-, ry[s]c-thýfel *jungetum*, Txts. 68, 517. Andlang ðære díc on riscþýfel, Cod. Dip. Kmbl. v. 215, 4.

rysel, rysele, es; *m. Fat* :—Rysel *adeps*, Wrt. Voc. i. 71, 10: *axungia*, ii. 101, 37. Rysle *arvina*, 2, 61 : 92, 15. Rysele, 80, 44. Rysle *ilium*, 48, 33. Genim hænne rysele . . . gôse rysele, Lchdm. ii. 40, 10-12. Swínes rysl, Homl. Th. ii. 144, 29. Ðū nimst ðone rysel, Ex. 29, 13. Ðū nymst ðone rysle of ðam ramme, 29, 22. Ðone risel, Lev. 3, 9. Ryslas ealra eáfisca, Lchdm. ii. 30, 1. [O. L. Ger. rusli, hrusli *arvina*.]

ryðða, an; *m. A large dog, mastiff, blood-hound* :—Ryðða *molossus*, Wrt. Voc. i. 23, 35: 78, 52. Riðða, ii. 56, 41. Hē getígde ænne or-mǽtne ryððan innan ðam geate ðær Petrus inn hæfde, ðæt hē hine ābítan sceolde, Homl. Th. i. 372, 34. v. roð-hund.

rytran. v. a-ritrid *expilatam*, Txts. 58, 372.

rȳung (?), e ; *f. Roaring, groaning, grunting* :—Ic wiste ðæt swín wǽron ðǽm elpendum láðe and hiora rymg (rȳung ? v. rȳan) hié meahte āfyrhton *quorum grunnitas timere bestias noveram*, Nar. 21, 26. Hríung (?) *suspirium*, Wrt. Voc. i. 19, 34.

S

For the Runic S see Sigel.

sā; *gen.* sān; *m. A tub, pail, vessel* :—Saa *libitorium*, Txts. 35, 17. [*Prompt. Parv.* soo *or* cowl, vessel *tina*. He kam to þe welle, water updrow, And filde þer a michel *so*, Havel. 933. *So, soa* with two ears, to carry on a stang, Ray's North-country words. *Soa, soe* a tub ; commonly used for a brewing-tub only, but sometimes for a large tub in which clothes are steeped before washing, E. D. S. Pub. Lincolnshire. 'In Bedfordshire, what we call a *coal* and a *coal-staff*, they call a *sow* and a *sow-staff*,' Kennett. *Icel.* sār a *cask* : *Dan.* saa : *Swed.* sâ.]

saban, es; *m.* (?) *A sheet* :—On sabanum, *id est* scéte *in sabanis* (cf. on scétum *in sabanis*, 48, 47), Wrt. Voc. ii. 82, 57. [In Mt. 27, 59 the Gothic version translates σινδόνι by *sabana*. O. H. Ger. saban, sapon ; *m.* sabanum, sindon, teristrum, linteum : *Gk.* σάββανον : *Mid. Lat.* sabanum : *Span.* sabana *a sheet.* Diefenbach il. 770 cites an Arabic word *sabaniyat* fine stuff for girdles, veils, etc., with the derivation of it from the name of the town Sabano near Bagdad.]

Sabat, es; *m.* (?) *The Sabbath* :—Sabates *sabbati*, Mt. Kmbl. p. 20, 5. [Cf. *Goth.* Sabbato, Sabbatus.]

Sabíne, a ; *pl. The Sabines* :—Hú Rómāne and Sabíne him betweónum wunnon, Ors. 2, 4 tit.; Swt. 2, 19. Tó ānwíge gangan wið swā fela Sabína, 2, 4; Swt. 72, 16.

Sabínisc; *adj. Sabine* :—Ðæt Sabínisce gewinn, Ors. 2, 4; Swt. 68, 32 : Swt. 72, 8.

sac. v. sacu.

sac (sæc ?); *adj. Accused, charged, guilty* :—Swerian ðæt hig nellan nænne sacleásan man forsecgean ne nænne sacne forhelan *let them swear that they will not bring a charge against an innocent man, nor conceal one who is justly charged*, L. Eth. iii. 3; Th. i. 294, 5. v. un-sac, sǽc.

-saca. v. and-, ge-, wiðer-saca. [O. Sax. -sako: O. Frs. -seka: O. H. Ger. -sahho. Cf. Goth. ni sakjis ἄμαχος.]

sacan; p. sôc, pl. sôcon; pp. sacen. I. to fight, strive, contend:—Þeódscypas winnaþ and sacaþ heom betweónan, Wulfst. 86, 8. Hé geseh twegen Ebreísce him betwýnan sacan conspexit duos Hebraeos rixantes, Ex. 2, 13. Ic (Beowulf) sceal fôn wið feónde and ymb feorh sacan, Beo. Th. 883; B. 439. Gód sceal wið yfele, líf sceal wið deáþe, leóht sceal wið þýstrum, fyrd wið fyrde, feónd wið ôðrum, lâð wið lâðe ymb land sacan, Menol. Fox 568; Gn. C. 53. Sceal fǽge sweltan and dôgra gehwam ymb gedâl sacan middangeardes, Exon. Th. 335, 4; Gn. Ex. 28. Ðû tǽlnissum wið ða sêlestan sacan ongunne thou didst attempt to strive with the best (the gods) with insults, 254, 23; Jul. 206. Wǽran sacende emulabantur, Wrt. Voc. ii. 33, 10. II. to disagree, act in opposition, not to be, or not to act, in unison, to wrangle:—Ðonne se abbod and se prâfost ungerâde beóþ and him betwyx sacaþ dum contraria sibi inuicem sentiunt, R. Ben. 124, 19. Ne ða ôðre ongeán ðæt ne sacan (wiðcweðon, Wells Frag.) the others shall offer no opposition to the decision, 119, 2. Dôm stande ðâr þegenas sammǽle beón; gif hig sacan (disagree), stande ðæt hig .viii. secgaþ, L. Eth. iii. 13; Th. i. 289, 3. Ðæt hê sôce altercaretur, sermocinaretur, Hpt. Gl. 476, 67. III. of litigation, to bring a suit:—Ðâ sôc Wulfstán on sum ðæt land Wulfstan brought a suit laying claim to some of the land, Chart. Th. 376, 7. IV. to bring a charge against one, to accuse, blame:—Hû micla wið ðec sacas cýðnessa quanta adversum te dicant testimonia, Mt. Kmbl. Lind. 27, 13. Mê mîne âgen word sylfne sôcon verba mea execrabantur, Ps. Th. 55, 5. Monige cýðnisse leóse hiæ gicwēdun tô sacanne wið him multi testimonium falsum dicebant aduersus eum, Mk. Skt. Rush. 14, 56. Swâ hwæt þwyr and gebolgen môd ... sacendes hâtheortnys hit is nâ lufu þreáginge quicquid proteruus et indignus animus protulerit, objurgantis furor est, non dilectio correctionis, Scint. 36. V. to refuse, deny. v. on-sacan:—Sæccendum sedlum negatis sedibus, Mt. Kmbl. p. 18, 14. [Goth. sakan to strive, rebuke: O. Sax. sakan to rebuke, blame: O. H. Ger. sahhan litigare, increpare, objurgare: Icel. saka; wk. to fight, blame, accuse.] v. æt-, be-, for-, fore-, ge-, of-, on-, ôþ-, wið-, wiðer-sacan; sacian.

sacc, es; m. A sack, bag:—Ne bere gê sacc ne codd sacculum neque peram, Lk. Skt. 10, 4. Sæc sacculum, Kent. Gl. 208. Hig fyldon hira saccas (saccos) and lêdon hira ælces feoh on his sacc ... Ðâ undyde hira ân his sacc ... hê ðæt feoh geseah on his sacces (saculi) mûþe, Gen. 42, 25, 28. Ðâ guton hig hira hwǽte of hira saccon, 42, 35. Fylle hira saccas and lege hira ælces feoh on his âgenne sacc, 44, 1. [Goth. sakkus: O. H. Ger. sac: Icel. sekkr.], Cf. bí-sæc, sæcc.

sacerd, es; m. A priest (the term is not confined to the Christian priesthood):—Sacerd vel cyrcþingere sacerdos, Wrt. Voc. i. 42, 23: Rtl. 125, 1. Hæfde se sacerd (sacerdos) on Madian seofon dohtra, Ex. 2, 16. Moises heóld his mǽges sceáp ðæs sæcerdes on Madian, 3, 1. Putifares dohtor ðæs sacerdes of ðære byryg, Gen. 41, 45. Hê slôh ðæs sacerdes (hêhsacerdas, Lind. Rush.) her, Mk. Skt. 14, 47. Ða word ðæs sacerdes vox praedicatoris, Past. 21, 5; Swt. 163, 1. Ðone clǽnan sacerd (Christ), Exon. Th. 9, 19; Cri. 137. Suîðe ryhte ða sacerdas (sacerdotes) sint gehâtene sacerdas, ðæt is on Englisc clǽnseras, forðæm hié sculon lâtteówdôm gearwian ðâm geleáffulum, Past. 18, 7; Swt. 139, 14. Ða sacerdas of Leuies cynne, Deut. 27, 1, 14: Ps. Th. 77, 64. Moyses and Aaron sôðe sacerdas, 98, 6: Andr. Kmbl. 1483; An. 743. Ða mæssepreóstas wǽron ðus gehâtene ... Ðâ ða gemynegodan sacerdos (-as ?) côman erant presbyteri ... Venientes memorati sacerdotes, Bd. 3, 21; S. 551, 19. Ðæra sacerda ealdor princeps sacerdotum, Mt. Kmbl. 26, 51: Blickl. Homl. 77, 8: 239, 28. Hýrde wê ðæt Jacob fore sacerdum swilt þrowade, Apstls. Kmbl. 141; Ap. 71. [From Latin. Anglo-Saxon alone seems to have borrowed this word.] v. ealdor-, heáh-sacerd.

sacerd-bana, an; m. One who slays a priest:—Hér syndan sacerd-banan, Wulfst. 163, 27: 266, 27.

sacerd-gerisne, adj. Befitting a priest:—Hê hæfde sacerdgerisene ealdorlícnysse auctoritatem sacerdote dignam, Bd. 3, 17; S. 545, 11.

sacerd-hâd, es; m. Priest-hood:—Ðâ Zacharias his sacerdes hâdes (sacerdhâdes, MSS. A. B. C.) breác cum sacerdotio fungeretur, Lk. Skt. 1, 8. Æfter gewunan ðæs sacerdhâdes hlotes, 1, 9. Ðæt hê gesette on sacerdhâd Judas ðam folce tô bisceope that he might ordain Judas bishop of the people, Elen. Kmbl. 2108; El. 1055. Bisceophâdas vel sacerd-[hâdas] flaminea, i. episcopali gradu, Wülck. 239, 23.

sacerd-land, es; n. Land assigned to priests:—Bûtan ðam sacerdlande absque terra sacerdotali, Gen. 47, 26.

sacerd-líc, adj. Priestly, sacerdotal:—Sacerdlíc sacerdotium, Rtl. 25, 31: sacerdotalis, 195, 4. Sacerdlíce þênunge dôn officium sacerdotale agere, Bd. 4, 5; S. 573, 4. Be sacerdlícum hræglum de vestibus sacerdotum, Bd. 5, 24; S. 647, 38.

sac-full; adj. I. contentious, quarrelsome:—Hê biþ swîðe sacful and micele ungeþwǽrnesse and mǽnigfealde saca on ðære geferǽdenne wyrcþ scandala nutriunt et dissensiones in congregatione faciunt, R. Ben. 124, 8. Ne ænig man ne sý tô sacfull ne ealles tô geflitgeorn, Wulfst.

70, 19: Lchdm. iii. 428, 34. Sacful wíf litigosa mulier, Kent. Gl. 690. Mid secfullan (rixosa) wífe, 790. [Ȝif þe cristene mon bið sacful, O. E. Homl. i. 109, 1.] II. given to accusation (v. sacan, IV):—Ne beó ðû sacfull non eris criminator, Lev. 19, 16.

sacian; p. ode To strive, brawl:—Gif men saciaþ si rixati fuerint viri, Ex. 21, 22. Fela sind ðe wyllaþ fracodlíce him betwýnan sacian many there are that will shamefully brawl among themselves, Homl. Th. ii. 294, 1. v. and-sacian; sacan.

sac-leás; adj. I. free from charge or accusation, innocent:—Swerian hig ðæt hig nellan nǽnne sacleásan man forsecgean ne nǽnne sacne forhelan, L. Eth. iii. 3; Th. i. 294, 5. Fiónge mec habbaþ sacleósne (sacleás, Lind., cf. Icel. saklaust without cause) odio me habuerunt gratis, Jn. Skt. Rush. 15, 25. II. free from charge or contention, unmolested, secure:—On ðæt gerâd ðæt ðes cynges men sacleás beón môston on ðâm castelan ðe hí ǽr þes eorles unþances begiten hæfdon, Chr. 1091; Erl. 227, 9. Eádgâr æþeling wæs gefangen; ðone lêt se cyng syððan sacleás faran, 1106; Erl. 241, 20. Sacleáso iwih wê gedôeþ securos vos faciemus, Mt. Kmbl. Lind. 28, 14. [Ðo þe hadden on þesse liue alle here sunnes forleten and bet ... alle he quað hem saclese, O. E. Homl. ii. 171, 35. Wass Crist sacclæs o rode naȝȝledd, Orm. 1900. Sacles (without strife, freely) he let hin welden it so, Gen. and Ex. 916. Icel. sak-lauss innocent, not guilty. Sackless still remains in Northern dialects, but seems to have got a meaning, with which innocent also is used, that of silly, simple. v. Jamieson, Halliwell, and E. D. S. Publications.]

sacu, e; f. I. strife, contention, dissension, sedition, dispute:—Sacu seditio, Wrt. Voc. i. 21, 30. Seó sacu (seditio) ârâs, Num. 16, 42. Wearð sacu (rixa) betwux Abrames hyrdemannum and Lothes ... Abram cwæð tô Lothe: 'Ic bidde ðæt nân sacu (jurgium) ne sig betwux mê and ðê,' Gen. 13, 7, 8. Ðanun mæg âspringan seó mǽste sacu and se mǽsta swice ealra ungeþwǽrnessa exinde grauissima occasio scandalorum oriri potest, R. Ben. 129, 8. Drihten cwæð: 'Ðonne gê gehýraþ on middangearde gefeoht and sace ne beó gê âfyrhte.' Gefeoht belimpþ tô feóndum and sacu tô ceastergewarum. Mid ðâm wordum hê gebícnode ðæt wê sceolon þolian wiðútan gewinn fram ûrum feóndum and eác wiðinnan fram ûrum nêhgebûrum lâðlíce ungeþwǽrnyssa, Homl. Th. ii. 538, 12–17. Hê (Caligula) mǽnde ðæt ðǽr ðâ næs swelc sacu swelc ðǽr oft ǽr wæs, and hé self fôr oft on ôðra lond, and wolde gewin findan, ac hê ne mehte bûton sibbe, Ors. 6, 3; Swt. 256, 28. Sceal Geáta leódum and Gâr-Denum sib gemǽnum, and sacu restan, Beo. Th. 3719; B. 1857. Ðære ðe wæs for sumere sace (propter seditionem) on cwerterne, Lk. Skt. 23, 25. Moises genemde ða stôwe Costung for Israhêla bearna sace propter jurgium filiorum Israel, Ex. 17, 7. Ðæt hié under ðære sibbe tô ðære mǽstan sace becôme, Ors. 4, 7; Swt. 182, 28. Sace militiam, Hpt. Gl. 494, 70. Grendel wan wið Hróðgâr, wæg singale sæce, sibbe ne wolde, Beo. Th. 310; B. 154. Lǽt sace restan, lâð leódgewin, Exon. Th. 254, 21; Jul. 200. Saca lites, Kent. Gl. 575. Of sacum rixis, 635. Ne mæg ic âna âcuman eówre saca (jurgia), Deut. 1, 12. Mænigfealde saca on ðære geferǽdenne wyrcþ dissensiones in congregatione faciunt, R. Ben. 124, 9. II. distress, trouble, affliction, persecution:—Ðǽr eów is sacu bûtan ende grim gǽstcwalu in hell is trouble without end for you devils, fierce torment of spirit, Exon. Th. 142, 27; Gû. 650. Ðǽr biþ â gearu wraðu wannhâlum wîta gehwylces sæce and sorge there shall be ever ready for the wretched support against every infliction, against distress and care, Elen. Kmbl. 2059; El. 1031. Ne þearft ðû sâr níwigan and sæce rǽran (cf. Gi werðat ôk sô sâlige thes iu saka biodat liudi blessed are ye when men shall persecute you, Hel. 1336), 1879; El. 941. Ðǽr hê hæfþ eal sâr and sace, hungor and þurst, wôp and hreám, and weána mâ ðonne æniges mannes gemet sý ðæt hié âríman mǽge, Blickl. Homl. 61, 36. Seó sunsciéne slege þrowade, sace singrimme, Exon. Th. 256, 11; Jul. 230. III. crime, guilt:—Nis ðǽr on ðam londe synn ne sacu non huc adit scelus infandum (cf. O þæt an bukk he leȝȝde All þeȝȝre sake and sinne, Orm. 1335. He alâtan mag saka endi sundea he can forgive sins, Hel. 1009), 201, 10; Ph. 54. Ðâ wæs synn and sacu Sweóna and Geáta, wrôht gemǽne, Beo. Th. 4935; B. 2472. IV. a contention at law, a suit, cause, action:—Nân sacu ðe betweox preóstan sî ne beó gescoten tô world-manna sôme no suit that there may be between priests shall be referred to the adjustment of secular men, L. Edg. C. 7; Th. ii. 246, 3. Gif man ôðerne sace tihte if one man bring a suit against another (cf. ef man hwemu saka sôkea, Hel. 1522), L. H. E. 8; Th. i. 30, 11. Hit betere wǽre ðæt heora seht tôgædere wurde ðonne hý ænige sace hym betweónan heóldan it would be better that they should come to an agreement than that they should carry on any suit between them, Chart. Th. 377, 3. V. jurisdiction in litigious suits. For the first time apparently in charters of Edward the Confessor the phrase sac and sôc or sôcn occurs, and in them it is frequent. It is thus explained in the Latin version of an Anglo-Saxon charter where it is found:—Ic an heom ealswâ ðæt hý habben ðǽrofer saca and sôcna iis (sanctus Petrus et fratres Westmonasterienses) etiam concedens ut insuper habeant priuilegium tenendi curiam ad causas cognoscendas et dirimendas

lites inter uasallos et colonos suos ortas, cum potestate transgressores et calumniae reos mulctis efficiendi easque leuandi, Cod. Dip. Kmbl. iv. 202, 7, v. Stubbs, Const. H. i. 184, Cod. Dip. Kmbl. i. xliii sqq., Grmm. R. A. 854 sq. [*Laym,* sake *strife: O. and N.* cheste *and* sake: *Goth.* sakjô *strife: O. Sax.* saka: *O. L. Ger.* saca *res, causa: O. Frs.* sake, seke *causa, res: O. H. Ger.* sahha *lis, causa, occasio, negotium, res: Icel.* sök *a charge, a crime, a suit, cause, sake.*] v. sæcc.

-sacung. v. wið-, wiðer-, yfel-sacung.

sáda, an; *m. A cord, halter, snare :*—Swelce sâdo (sâde, Rush.) *tamquam laqueus,* Lk. Skt. Lind. 21, 35. Grin biþ on sâdan tôrænded *laqueus contritus est,* Ps. Th. 123, 7. Mid sâde (*laqueo*) hine âwrigde, Mt. Kmbl. Lind. 27, 5. [*O. H. Ger.* seito *laqueus, pedica, tendicula.*] v. wealh-sâda.

Sadducéas; *pl. The Sadducees :*—Eodun tô him Fariséas and Sadducéas, Mt. Kmbl. Rush. 16, 1. Sadducéa *Sadducaeorum,* 16, 6.

Sadducéisc; *adj. Sadducean :*—Hé hét ða Saducéiscan stylle beón, Mt. Kmbl. 22, 34.

sadian; *p.* ode. **I.** *to satisfy, satiate.* [*O. H. Ger.* satôn *saturare.* Cf. *Icel.* seðja *to satisfy.*] v. ge-sadian. **II.** *to become satisfied, to get satiated* or *tired :*—Mê þincþ ðæt ðú sadige hwæt hwegunges and ðé þincen tô ælenge ðás langan spell *methinks thou art getting somewhat wearied and these long discourses seem to thee too protracted,* Bt. 39, 4 ; Fox 218, 5.

sadol (-el, -ul), es; *m. A saddle :*—Sadol *sella,* Wrt. Voc. ii. 120, 33 : i. 83, 70. Sadul, 23, 19. Hé hêht eahta mearas on flet teón, ðara ânum stôd sadol, ðæt wæs hildesetl heáhcyninges, Beo. Th. 2080; B. 1038. [*O. H. Ger.* satal, satul; *m.: Icel.* söðull; *m.*] v. seám-sadol.

sadol-beorht; *adj. Having a splendid saddle :*—Þrió wicg sadolbeorhte (cf. sadol searwum fâh, since gewurðad, 2080; B. 1038), Beo. Th. 4356; B. 2175.

sadol-boga, an; *m. A saddle-bow :*—Sadolboga *carpella,* Wrt. Voc. i. 291, 16 : ii. 128, 71. Sadulboga, 103, 4. Sadelboga, 17, 34 : *corbus,* 22, 46. Sadulboga, i. 23, 18. [*Icel.* söðul-bogi: *O. H. Ger.* satalbogo.]

sadol-felg, e; -felge, an; *f. The pommel of a saddle ;* pella (cf. Spanish *pella* a ball, anything made in a round form) :—Sadulfelgae, -felge *pella,* Txts. 88, 818. Sadolfelg, Wrt. Voc. ii. 68, 9. Sadolfelg (? Wrt. radolfelt), i. 291, 15.

sadolian; *p.* ode *To saddle :*—Ic sadelige hors *sterno,* Ælfc. Gr. 28, 1 ; Som. 30, 34. [*Icel.* söðla: *O. H. Ger.* satalôn.] v. ge-sadelod.

sǽ; *m. f.;* gen. sǽs, sǽes, sǽ, sǽwe, seó; nom. *pl.* sǽs, sǽ; *dat.* sǽm, sǽum, sǽwum. Sea. The word is found in the following glosses :—Sǽ *mare* vel *aequor,* Wrt. Voc. i. 41, 62 : 70, 13. Brym, sǽ *aequor,* 53, 50. Sǽ *lutex* (*latex*?), ii. 53, 17, Ðæs ýþiendan sǽs *fluctivagi ponti,* 149, 61. And sǽ *et salis,* 32, 28. Mid sǽ *cum pelago,* 21, 27. Ofer sǽ *citra pontum,* 18, 68. Ða héwnan sǽs *marmora glauca,* 57, 7. Sǽ *marmora,* 91, 73. **I.** *sea* (water as opposed to air and earth) :—On ðæm dæge gewíteþ heofon and eorþe and sǽ, and ealle ða þing ðe on ðæm syndon, Blickl. Homl. 91, 21. God gescóp ðone rodor betweoh heofone and eorþan and betweoh ðæm twǽm sǽum, ðæm uplícan and ðæm niðerlícan. Se uplíca sǽ . . . céleþ ðære tungla hǽto, and se rodor ymbféhþ útan eall ðás niðerlícan gesceafte, sǽ and eorþan, Shrn. 63, 5-10. On syx dagum Crist geworhte heofenas and eorþan, sǽs and ealle gesceafta, L. Alf. 3; Th. i. 44, 13. **II.** *sea* (as opposed to land) :—Ðonne ðú wyte ðæt sǽ sí ful *at high water,* Lchdm. iii. 176, 18. Ús drîsaþ ða ællreordan tô sǽ, widscúfeþ ús seó sǽ ðám ællreordum, Bd. 1, 13 ; S. 481, 44. Ðæs sǽes flôdes weaxnes, 5, 3 ; S. 616, 16. On sǽs (sǽes, Lind.: sǽas, Rush.) grund *in profundum maris,* Mt. Kmbl. 18, 6. For gedréfednesse sǽs swéges, Lk. Skt. 21, 25. Sǽs earm, Ors. i, 1 ; Swt. 24, 6, 14. Gang tô ðære sǽs waroþe . . . Hé eode tô ðære sǽ, Blickl. Homl. 231, 29-36. Gán ofer sǽs ýþa, 177, 18. Geswencede of ðisse sǽwe hreónesse, 233, 26 : 235, 1. Hreónesse ðære sǽwe, 235, 5. Monigra ceápstôw of lande and of sǽ cumendra, Bd. 2, 3 ; S. 504, 19. Bât on sǽwe, Exon. Th. 458, 12 ; Hy. 4, 99 : Andr. Kmbl. 1029; An. 515. Æt fulre seó, Lchdm. iii. 178, 18. On siewe (?sǽwe), Cant. Moys. Thw. 29, 4. Ða mætte hié micel ýst on sǽ, Chr. 877; Erl. 78, 18. Hié micel ðæs folces ofer sǽ ádrǽfdon, 878 ; Erl. 78, 30 : Bd. 1, 15 ; S. 484, 7. Ofer ðone sǽ, 1, 12 ; S. 481, 2. Gif hwá his ágenne geleód bebycgge ofer sǽ, L. In. 11 ; Th. i. 110, 4. God gecîgde ða drîgnesse eorþan and ðæra wætera gegaderunga hé hét sǽs, Gen. 1, 10. Sǽs up stigon, Cd. Th. 83, 6 ; Gen. 1375. Ðæt ðás deópan sǽ drî geworðaþ, Ps. Th. 65, 5. Beútan eallum sǽwum, 138, 7. **III.** *sea* (as opposed to water inland) :—For hwî ne fixast ðú on sǽ? (cf. ic wyrpe max mîne on eá, 23, 9). Hwîlon ic dô, ac seldon, for ðam micel réwyt mé ys tô sǽ, Coll. Monast. Th. 24, 1-5. Sǽs tôslúpan, eal sealt wæter, Lchdm. iii. 36, 27. **IV.** *a sea :*—Him is be-eástan se sǽ ðe man Arfatium hêt, and westan and be-norþan Creticum se sǽ, Ors. 1, 1 ; Swt. 26, 32 : 28, 1. Néh ðæm clife ðære Reádan sǽs, Swt. 12, 20. Be ðære reódan sǽ, Ex. 14, 9. Betwih ðære sǽ seó is nemned Adriaticus, Blickl. Homl. 197, 21. **V.** *of inland water, a sea, lake :*—Sume men secgaþ seó eá ðæt wyrcþ micelne sǽ *aliqui auctores ferunt fluvium vastissimo lacu*

---

exundare, Ors. 1, 1 ; Swt. 12, 24. On ðære sǽ *in the sea* (*of Galilee*), Mt. Kmbl. 8, 24. [*Goth.* saiws: *O. Sax.* séo, séu: *O. Frs.* sê: *O. H. Ger.* séo: *Icel.* sær, sjór, sjár; *gen.* sævar; *dat.* sævi, sæ.] v. eást-, heáh-, norþ-, Ost-, Wendel-, west-, wíd-sǽ.

sǽ-æbbung. v. æbbung.

sǽ-ǽl, es; *m. A sea-eel :*—Sǽǽl *murenula* (cf. *hec murenula* a lamprun, i. 222, col. 2), Wrt. Voc. ii. 57, 74.

sǽ-ælfen[n], e; *f. A sea-elf, sea-nymph :*—Sǽælfenne *Naiades,* Wrt. Voc. ii. 62, 32 : 59, 12. Sǽelfen, i. 60, 18.

sǽ-bât, e; *f. A sea-boat :*—On sǽbâte, Andr. Kmbl. 876 ; An. 438 : 980 ; An. 490. Ic on holm gestâh, sǽbât gesæt, Beo. Th. 1270; B. 633 : 1795 ; B. 895.

sǽ-beorh *a sea-hill, a hill* or *cliff against the sea :*—Ealle gerîman stânas on eorþan, steorran on heofonum, sǽbeorga sand (MS. sund ; but cf. Ic ðínne ofspring gemenigfylde swâ swâ steorran on heofonum and swâ swâ sandceosol on sǽ, Gen. 22, 17), Cd. Th. 205, 25 ; Exod. 441. Hú geweard ðé ðæt ðú sǽbeorgas sécan woldes, merestreáma gemet, ofer cald cleofu ceóles neósan, Andr. Kmbl. 615; An. 308.

sǽ-burh *a maritime town :*—Hé gewunade in *Capharnaum ðæt is sǽburug (-caestrae, Rush.) habitavit in Capharnaum maritima (*note on Capharnaum : in ðær byrig Capharnaum is genemned and maritimam cuoeð, forðon ðyú burg is on sǽ), Mt. Kmbl. Lind. 4, 13. [*Icel.* sæborg *a sea-side town.*]

sǽc; *adj.* **I.** *hostile, offensive, hateful :*—Tô áscamelícum *ad detestabilem, ad odiosum,* sǽcum *invisum, exosum,* meltestran húse *lupanar,* Hpt. Gl. 500, 58-62. v. next word. **II.** *guilty, charged with guilt.* v. on-sæc, sac ; *and cf. Icel.* sekr *guilty, convicted.*

-sæc. v. and-, eoful-, wiðer-sæc.

sæcc, es; *m. Sacking, sack-cloth :*—Hé árás of ðam wâcan sæcce ðe hé lange onuppan dreórig wæs sittende, Homl. Skt. i. 23, 802. Ðú slite hǽran (sæcc, MS. C.) mîne *concidisti saccum meum,* Ps. Spl. 29, 13. v. sacc, seccing.

sæc[c], e; *f. Strife, contest, conflict :*—Â wæs sæc, Elen. Kmbl. 2512 ; El. 1257. Ðær biþ ceóle wén slîðre sæcce *there* (*at the rocky shore*) *the vessel may expect fierce conflict,* Exon. Th. 384, 17 ; Rä. 4, 29. Hé sæcce ne wéneþ tô Gâr-Denum, Beo. Th. 1205 ; B. 600. Se æt sæcce gebâd wîghyre wrâðra, 3241 ; B. 1618 : 1910 ; B. 953. Ðam æt sæcce wearð Weohstân bana méces ecgum *Weohstan felled him in fight with the edge of the falchion,* 5218 ; B. 2612. Nægling geswâc æt sæcce (*in fight with the fire-drake*), 5355 ; B. 2681. Tír geslôgon æt sæcce *gained glory in battle,* Chr. 937; Erl. 112, 4: Erl. 114, 8. Æt sæcce forweorþan *to perish in battle,* Judth. Thw. 25, 32 ; Jud. 289. Æt wîgge spéd, sigor æt sæcce, Elen. Kmbl. 2363 ; El. 1183. Hé feorg gesealde æt sæcce, Apstls. Kmbl. 117 ; Ap. 59. Ic oflsóh æt ðære sæcce (*the battle with Grendel's mother*) húses hyrdas, Beo. Th. 3334 ; B. 1665. Hé tô sæcce bær wǽpen wundrum heard *he to battle bore a weapon wondrous hard,* 5366 ; B. 2686. Se ðe sæcce genæs *he who came safe from conflict* (*Beowulf*), 3959 ; 1977. Sæcce sécean, 3982 ; B. 1989. Nô hé him ðam sæcce ondréd, ne him ðæs wyrmes wîg for wiht dyde, 4687 ; B. 2347. Sæcce fremman *to fight,* 4991 ; B. 2499 : Exon. Th. 496, 28 ; Rä. 85, 21. Hí hæfdon sæcce gesôhte, sceolde sweordes ecg feorh ácsigan, Andr. Kmbl. 2265 ; An. 1134. Hé wælfǽhþa dǽl sæcca gesette *he composed many a deadly feud and quarrel,* Beo. Th. 4062 ; B. 2029. Cf. sacu.

sæccan (?) *to fight, contend :*—Oft ic sceal wið wǽge winnan and wið winde feohtan, somod wið ðám sæcce (? sæcce fremman *or* sécan, v. *preceding word ; but cf. also* sacian, sacan), Exon. 398, 3 ; Rä. 17, 2.

sæccing, es; *m. Sacking, a bed made of sacking :*—Hí on sæccingum (*in grabatis*) bǽron ða untruman, Mk. Skt. 6, 55. v. sæcc.

sǽc-dôm, sǽc-ceaster. v. sceac-dôm, sǽ-burh.

sǽ-ceosol *sand* or *gravel on the sea-shore :*—Sǽceosol *arena maris,* Gen. 32, 12. Sǽcysul *calculus,* Wrt. Voc. i. 38, 23.

sǽcg, sǽcgan, sǽcgend. v. secg, secgan, secgend.

sǽ-cir[r] *the retreat of the sea* (*when the waves drew back and left a passage for the Israelites*), Cd. Th. 196, 13 ; Exod. 291.

sǽclian. v. sîclian.

sǽ-clif *a cliff by the sea :*—Swâ fela welena swâ ðara sondcorna beóþ be ðisum sǽclífum, Bt. 7, 4 ; Fox 22, 27.

sǽ-coco, es; *m. A cockle :*—Hwæt féhst ðú on sǽ? Crabban muslan sǽcoccas *cancros, musculos, neptigallos,* Coll. Monast. Th. 24, 11. [Cf. a farthing-worth of muscles were a feste for suche folke, oþer so fele Cockes (cokkys, MS. G.: cokeles, MS. I.), Piers P. C text x. 95. Welsh cocs *cockles.*]

sǽ-col, es; *n. Jet ;* gagates, Wrt. Voc. ii. 42, 25.

sǽ-cyning, es; *m. A sea-king, a king who was powerful on the sea :*—Helm Scylfinga, ðone sélestan sǽcyninga ðara ðe in Swiórîce sinc brytnade, Beo. Th. 4754 ; B. 2382. [*Icel.* sæ-konungr.]

sǽd; *adj. with gen. Sated, weary, filled, having had one's fill* (the word is not used in the sense of modern *sad*) :—Sǽd *effetus,* i. *plenus,* Germ. 396, 215. Ðær læg secg mænig . . . wérig wîges sǽd *many a warrior lay dead there . . . : of war had had his fill,* Chr. 937; Erl. 112,

20. Beadoweorca sǽd, Exon. Th. 388, 4; Rä. 6, 2. Wiste wlonc and wînes sǽd, 369, 11; Seel. 39. Swíðe ǽtan and sade wurdan *manducaverunt et saturati sunt nimis*, Ps. Th. 77, 29. Hí sǽde wǽron *saturavit eos*, 80, 15. [Goth. saþs: O. Sax. sad: O. L. Ger. sad: O. H. Ger. sat satur: Icel. saðr (saddr).] v. hilde-, un-, wín-sǽd; sadian.

sǽd, es; *n.* **I.** *seed, what is sown, that part of a plant which propagates*:—Senepes sǽd *granum sinapis*, Mk. Skt. 4, 31. Ðæt treów sceolde sǽde eft onfón *the tree should again bear seed*, Cd. Th. 251, 12; Dan. 562: 252, 24; Dan. 583. Ealle treówu ðe habbaþ sǽd on him silfon heora ágenes cynnes *universa ligna quae habent in semetipsis sementem generis sui*, Gen. 1, 29. Ðam men ðe seów gód sǽd on his æcyre, Mt. Kmbl. 13, 24. Ût eode se sǽdere his sǽd tó sáwenne, Mk. Skt. 4, 3. Swylce man wurpe gód sǽd (*sementem*) on his land, 4, 26. **I** *a.* fig. *seed, that from which anything springs*:—Ðæt hálige sǽd gewát, ðæt him ǽr of ðæs láreówes múþe bodad wæs, Blickl. Homl. 55, 29. Ðeáh biþ sum corn sǽdes gehealden symle on ðære sáwle sóðfæstnesse: ðæs sǽdes corn biþ simle áweaht mid áscunga, Met. 22, 37–41. Gif wé eów ða gástlícan sǽd sáwaþ, Homl. Th. ii. 534, 26. **II.** *the ripe fruit, that from which the seed is taken*:—Hí heora sylfra sǽd sníþaþ *they shall reap their crops*, Ps. Th. 125, 5. Se háta sumor giereþ and drígeþ sǽd and bléda, Met. 29, 61. **III.** *fruit, growth*:—Of wlite wendaþ wæstma gecyndu, biþ seó síðre tíd sǽda gehwylces mǽtræ in mægne, Exon. Th. 105, 1; Gú. 16. **IV.** *sowing, sǽd-tima*:—Sǽd and geríp sumor and winter ne geswícaþ *sementis et messis, aestas et hiems non requiescent*, Gen. 8, 22. **V.** applied to animals, *seed, progeny, posterity*:—Sǽd *crementum* (in a list 'de homine et de partibus ejus'), Wrt. Voc. i. 282, 26: ii. 16, 39. Weres sǽd, 44, 55. Mín sǽd him þeówaþ, Ps. Th. 21, 29. Ðæt sǽd ðara unrihtwísra forwyrð, 36, 28. Tó Abrahame wæs cweþende ðæt his sǽd oferweóxe ealle ðás woruld, Blickl. Homl. 159, 26. Swá hé spræc tó Abrahame and hys sǽde, Lk. Skt. 1, 55. Ðæt his bróðor nime his wíf and his bróðor sǽd wecce, Mk. Skt. 12, 19. [Goth. mana-seþs: O. H. Ger. sât: Icel. sáð *seed, crop*.] v. god-, lín-, un-, wǽd-sǽd.

sǽd-berende *seed-bearing*:—Eorþe swealh sǽdberendes (v. sǽd, **V**) Sethes líce, Cd. Th. 69, 33; Gen. 1145. Grówende wirte and sǽdberende *herbam viventem et facientem semen*, Gen. 1, 29.

sǽd-cynn, es; *n.* *A kind of seed*:—Ǽghwilc sǽdcyn *omne genus seminarum*, Wrt. Voc. i. 55, 30. Sǽdere gebyreþ ðæt hé hæbbe ǽlces sǽdcynnes ǽnne leáp fulne, ðonne hé ǽlc sǽd wel gesáwen hæbbe ofer geáres fyrst, L. R. S. 11; Th. i. 438, 9.

Sǽ-Dene; *pl.* *The sea-Danes, Danes of the islands* (?), or *Danes skilled in sea-faring* (?):—Sigehere lengest Sǽ-Denum weóld, Exon. Th. 320, 13; Víd. 31. Cf. Sǽ-Geátas.

sǽ-deór, es; *m.* *A sea-beast* (cf. Milton's 'that sea-beast Leviathan'):—Hine swencte on sunde sǽdeór monig, Beo. Th. 3025; B. 1510. Hé hét his ágene men hine sǽndan on ðone sǽ, and ða sǽdeór hine sóna forswulgon, Shrn. 54, 27. Hý mon wearp in sǽdeóra seáþ, 133, 11. Gif hit on Frigedæig þunrige, ðæt tácnaþ sǽdeóra cwealm, Lchdm. iii. 180, 17. [Icel. sjó-dýr.]

sǽdere, es; *m.* *A sower*:—Sǽdere *sator, seminator*, Hpt. Gl. 461, 73. Sum sǽdere férde tó sáwenne his sǽd, Homl. Th. ii. 88, 12: Mk. Skt. 4, 3. Be sǽdere, L. R. S. 11; Th. i. 438, 8. v. next word.

sǽdian; *p.* ode *To sow, provide seed for land*:—Folgere gebyreþ ðæt hé on twelf mónþum .iii. æceras geearnige, óðerne gesáwene and óðerne unsáwene; sǽdige sylf ðæne *he must provide the seed for the latter himself*, L. R. S. 10; Th. i. 438, 5.

sǽd-leáp, es; *m.* *A basket* or *o[t]her vessel of wood carried on one arm of the husbandman, to bear the seed which he sows with the other, a seed-leap* (Essex), *seed-lip* (Oxford). V. E. D. S. Pub. B. 18; also *seed-lop*, v. Old Country and Farming words, iii. Hopur or a seed lepe *satorium, saticulum*, Prompt. Parv. 246. A sedlepe *saticulum*, Wülck. Gl. 609, 28: *semilio*, 611, 11:—Sǽdleáp, Anglia ix. 264, 13. [Ðæt acersæd hwǽte, ðæt is twegen sédlǽpes, and ðæt bærlíc, ðæt is þrǽ sédlǽpas, and ðæt acersǽd áten, ðæt is feówer sédlǽpas, Chr. 1124; Erl. 252, 34–36. In the note on this passage *seed-lip* is said to be still used in Somersetshire.] v. leáp.

sǽd-líc; *adj.* *Seminal*:—Sǽd sǽdlíc *semen seminalem*, Rtl. 146, 17.

sǽdnaþ, es; *m.* *Sowing*:—Sǽdnaþ *satio, seminatio*, Wrt. Voc. i. 37, 50.

sǽdness, e; *f.* *Satiety, repletion*:—Óþ sǽdnesse *ad congeriem, congestionem, nauseam, satietatem*, Germ. 391, 30.

sǽ-draca, an; *m.* *A sea-dragon, sea-serpent*:—Sǽdracan *leviathan .i. serpens aquaticus*, Hpt. Gl. 424, 55. Gesáwon æfter wætere wyrmcynnes fela, sellíce sǽdracan, sund cunnian, Beo. Th. 2856; B. 1426.

sǽd-tíma, an; *m.* *Seed-time, time for sowing*:—Sǽdtíma and hærfest, sumor and winter ne geswícaþ nǽfre, Hexam. 7; Norm. 12, 28. [Icel. sáð-tími *the sowing season*.] v. sǽd, **IV**.

sǽ-earm, es; *m.* *An arm of the sea*:—Scýt se sǽearm up of ðæm sǽ westrihte, Ors. 1, 1; Swt. 22, 4.

sǽ-ebbung, -elfen. v. sǽ-æbbung, -ælfenn.

sǽ-fǽreld *a sea-passage*, used in reference to the attempt made by

the Egyptians to pass the Red Sea:—Ðá hí (*the Egyptians*) oninnan ðæm sǽfǽrelde wǽron, Ors. 1, 7; Swt. 38, 33.

sǽ-fæsten *the fastness* or *stronghold which the sea constitutes*:—Óþ ðæt sǽfæsten landes æt ende leódmægne forstód *the sea was a stronghold which blocked the further passage of the Israelites*, Cd. Th. 185, 24; Exod. 127.

sǽ-faroþ *the sea-shore*:—Ceólas léton æt sǽfearoþe sande bewrecene, Elen. Kmbl. 501; El. 251. Sǽfaroþa sand, Cd. Th. 236, 18; Dan. 323. v. sǽ-waroþ.

Sæfern, e; *also indecl. f. The river Severn*:—Hié gedydon innan Sæferne múþan, Chr. 918; Erl. 102, 24. On Sæferne staþe, 894; Erl. 92, 23. Hié gedydon æt Sæferne, ðá fóron be Sæferne . . . be westan Sæfern, 92, 14–20. Be Sæfern, 896; Erl. 94, 15. Be westan Sæferne, Bd. 5, 23; S. 646, 21. Of Seferne, Cod. Dip. Kmbl. iii. 393, 10. Of Sæfern, 405, 29. Westweardes on Sæferne, ii. 150, 9, 14. *Latin forms in the charters are* Saberna, i. 64, 11: Sabrina, 84, 2: Saebrina, ii. 59, 18.

Sæfern-múþa, an; *m. The mouth of the Severn*:—On súþhealfe Sæfernmúþan, Chr. 918; Erl. 104, 4: 997; Erl. 134, 8.

sǽ-fisc, es; *m. A sea-fish, fish that lives in the sea*:—Fleógende fuglas and sǽfiscas *volucres coeli et pisces maris*, Ps. Th, 8, 8. Swelaþ sǽfiscas, wǽgdeóra gehwylc swelteþ, Exon. Th. 61, 19; Cri. 987. Óðre sǽfisca cynn, 363, 19; Wal. 56. [Ifulled mid gode sæfisce, Laym. 22550. Icel. sǽ-fiskr.]

sǽ-flód, es; *m. n.* **I.** *an incoming tide, flood* (as opposed to ebb):—Grécas hátaþ *malina* sǽflód ðonne hyt wixst, and *ledon* ðonne hyt wanaþ (cf. *ledona* nepflód vel ebba, *malina* heáhflód, Wrt. Voc. i. 57, 11–12), Anglia viii. 327, 29. Wæs án burg sió wæs néh ðæm sǽ óþ án sǽflód com and hié áwéste *civitas repentino maris impetu abscissa, atque desolata est*, Ors. 2, 7; Swt. 90, 20. On ðissum geáre com ðæt mycele sǽflód, and ærn swá feor up swá nǽfre ǽr ne dyde, and ádrencte feala túna, Chr. 1014; Erl. 151, 14. Ðises geáres ásprang up tó ðan swíðe sǽflód, and swá mycel tó hearme dyde swá nán man ne gemunet ðæt hit ǽfre ǽror dyde, 1099; Erl. 235, 24. Sǽflóde *indruto*, Wrt. Voc. ii. 48, 27. **II.** *the sea, the water of the sea*:—Ðá fandode forþweard scipes (*Noah*) hwæðer sincende sǽflód wǽre, Cd. Th. 86, 28; Gen. 1437. Heofen and eorþe síde sǽflódas *coeli et terra, mare*, Ps. Th. 68, 35. [He lætte bi sæflode ȝearkien scipen gode, Laym. 2630.]

sǽ-flota, an; *m. A ship*:—Næs him cúð hwá ðam sǽflotan sund wísode, Andr. Kmbl. 761; An. 381. [Cf. He makede muchul sæflot, Laym. 4530.]

sǽ-fór, e; *f. A journey by sea, a voyage*:—Nis ðæs módwlonc mon ofer eorþan . . . ðæt hé á his sǽfóre sorge næbbe, Exon. Th. 308, 19; Seef. 42.

sǽ-fugol *a sea-fowl*. Sǽfugl, as a proper name, occurs in the genealogy of Ælle of Northumbria, Chr. 560; Erl. 16, 29. [Icel. sjó-fugl.]

sǽgan; *p.* de *To cause to sink*:—Óþ ðæt seó sunne on súþrodor sǽged weorðeþ (cf. Só gisēgid wurð sedle náhor hēðra sunna, Hel. 5715), Exon. 207, 15; Ph. 142. v. on-sǽgan; sígan.

-sǽge. v. on-sǽge.

sǽ-geáp; *adj. Roomy enough for sea voyages* (of a ship):—Sǽgeáp naca, Beo. Th. 3797; B. 1896.

Sǽ-Geátas; *pl. The seafaring* (?) *Geats*:— Ða Sǽ-Geátas sélran næbben tó geceósenne cyning ǽnigne, Beo. Th. 3704; B. 1850. Sǽ-Geáta (*Beowulf and his companions*) síðas, 3976; B. 1986. Cf. Sǽ-Dene.

sǽgedness *a sacrifice*, Mk. Skt. Lind. Rush. 12, 33. v. on-sægedness.

sǽ-gemǽre, es; *n. A sea-border, coast*:—Sǽgemǽro *maritima*, Lk. Skt. 6, 17. On ðam sǽgemǽrum, Mt. Kmbl. 4, 13.

sǽgen, sæcgen, segen, e; *f.* **I.** *a saying, statement, assertion*:— Ðá sægde se Clitus ðæt Philippus máre hæfde gedón ðonne hé. Hé ðá Alexander áhleóp, and hiene for ðære sægene oíslóg, Ors. 3, 9; Swt. 130, 30. Heora biscopas from hiora godum sǽden ðæt hié ðæt gefeoht forbuden. Ac Papirius ða biscepas for ðære sægene swíðe bismrade, 3, 10; Swt. 140, 2. Se Hǽlend cwæð: 'Ic sittende beó æt mínes Fæder swiðran.' Ðá cwæð se ealdorbiscop: 'Hwæt þincþ eów be ðissere segene, Homl. Th. ii. 248, 22: 320, 31: 484, 1. Gyf hé ðé segþ ðæt he hwethwugu gesáwe . . . hweðer ðé áwuht æt his segene tweóge, Shrn. 196, 17. Ðú ne tweódast ymbe Honorius segene, hwí tweóst ðú ymbe hera þegena sǽcgena, 197, 21–23. Hié sǽdon ðæt sió sibb of his mihte wǽre ac hé fleáh ða sægene *he would not admit what they said*, Ors. 3, 5; Swt. 106, 33. Sægenum *assertionibus*, Wrt. Voc. ii. 3, 62. Hié wiston ðe ðæs engles sægenum, ge be heora sige ge be ðara hǽðenra manna fleáme, Blickl. Homl. 203, 3. Sæcgenum, Ps. Th. 144, 7. **II.** *what is said generally, tradition, report, story*:—Ðæt is fyrn sægen (fyrn-sægen? cf. fyrn-gewrit, -gid) *it is an old story*, Andr. Kmbl. 2977; An. 1491. Ic wolde gewitan hweðer sió segen sóð wǽre ðe mé mon be ðon sægde *I wanted to know whether the story I had been told about it was true*, Nar. 24, 15. Of ealdra manna gewritum oððe sægene *ex scriptis vel traditione priorum*, Bed. pref.; S. 472, 19. Se hlísa ðe þurh yldra manna segene tó ús becom *opinio quae traditione majorum ad nos perlata est*, 2,

1 ; S. 501, 2. On gewritum oððe on ealdra manna sægenum *munimentis literarum vel seniorum traditione*, pref. ; S. 471, 27. Sægenum *scriptis*, 472, 5. III. *a narration, relation* (whether spoken or written) :— Ð̣y̌ læs ðæt eów seó sægen monigfealdlícor biþ onþúhte tó wrítanne ic ða wille lǽton ðe ðǽr gewurdon *ne sim scribendi multiplex, priora facta praecognita praetereo*, Nar. 3, 29. [*Icel.* sögn *a tale, report.*] v. ge-, sóþ-sægn, eald-gesegen.

sæ̌-genga, an ; *m.* I. *a sea-goer, a mariner :*—Ða gleáwe sǽgenga (gleáwan sǽgehgan ?) wel hig understandaþ ðæt eorþlíce líchamlíce beóþ fulran on weaxendum mónan ðonne on wanigendum *the skilful mariners well understand that earthly, corporeal things are fuller with a waxing than with a waning moon*, Anglia viii. 327, 21. II. *a vessel, ship :*—Sǽgenga forð, fleát fámigheals forþ ofer ýðe, bundenstefna ofer brimstreámas, Beo. Th. 3821 ; B. 1908 : 3769 ; B. 1882.

sæ̌-geset, es ; *n. A maritime district :*—Saegesetu (-seotu) *promaritima*, Txts. 82, 728. Sǽgesetu, Wrt. Voc. ii. 68, 33.

sæ̌gl, -sægness, sægnian. v. sigel, on-sægness, segnian.

sæ̌-grund (or sǽ (*gen.*) grund), es ; *m. The depth of the sea, the bottom of the sea :*—Ne mé forswelge sǽgrundes deóp *neque obsorbeat me profundum*, Ps. Th. 68, 15. Paulus áwrát be him sylfum, ðæt hé ǽnne dæg and áne niht on sǽgrunde ádruge, Homl. Th. ii. 574, 14. Sǽgrunde neáh (cf. ðis fis (*the whale*) wuneð wið ðe se grund, Misc. 16, 517), Beo. Th. 1133 ; B. 564. Þurh ðone sǽgrund (*profundum maris*, cf. tó sǽs grunde, l. 18, *and* on sǽs grund, Mt. Kmbl. 18, 6) is getácnod hira ende, Past. 2 ; Swt. 31, 20. Fán Gode besenctun on sǽgrund sigefæstne wer, Menol. Fox 421 ; Men. 212. Ic styrge wíde sǽgrundas, Exon. Th. 382, 12 ; Rä. 3, 10 : Cd. Th. 196, 9 ; Exod. 289.

-sægung. v. on-sægung.

sæ̌-hengest, es ; *m.* I. *a sea-horse, hippopotamus :*—Sǽhengest *ipotamus*, Wrt. Voc. ii. 48, 30. II. *a sea-steed, ship :*—Hú ðú wǽgflotan, sǽhengeste, sund wísige, Andr. Kmbl. 975 ; An. 488. Cf. sǽ-mearh.

sæ̌-hete (or sǽ (*gen.*) hete), es ; *m. Raging of the sea :*—Mid ðý wé wið ðam winde and wið ðam sǽ (sǽhete, MS. Ca.) campodan *cum vento pelagoque certantes*, Bd. 5, 1 ; S. 613, 27.

sæ̌-holm, es ; *m. Sea :*—Sǽholm oncneów, gársecges begang, ðæt ðú gife hæfdes, Andr. Kmbl. 1058 ; An. 529.

sæ̌ht, sæhtlian. v. seht, sahtlian.

sæ̌l, sel, es ; *n. A hall :*—Ic seah ræplingas in ræced fergan under hróf sales, Exon. Th. 435, 3 ; Rä. 53, 2. Gæst yrre cwom, ðǽr wé sǽl weardodon, Beo. Th. 4157 ; B. 2075. Ne gód hafoc geond sǽl swingeþ, 4520 ; B. 2264. Hý sǽl timbred (æltimbred, MS., the alliteration requires s) ongytan mihton ; ðæt wæs foremǽrost receda, 620 ; B. 307. Heorot (*Hrothgar's hall*), sincfáge sel, 336 ; B. 167. Geond ðæt síde sel, Andr. Kmbl. 1523 ; An. 763. Wuna salu sinchroden *halls splendidly decorated*, 3342 ; An. 1675. Salo, Cd. Th. 113, 3 ; Gen. 1881. Gesáwon ofer since salo hlífian, reced ofer reádum golde, 145, 10 ; Gen. 2403. [Wyn for to schenche, after mete in sale, Horn. 1107. Þyse renkeȝ schal neuer sitte in my sale my soper to fele, Allit. Pms. 41, 107. Such a freke watȝ neuer in þat sale er þat tyme, Gaw. 197. *O. H. Ger.* sal *exsolium, coenaculum ;* daz sal *templum : Icel.* salr *a hall.*] v. beág-sel, burg-, folc-, horn-sæl ; sele, salor.

sæ̌l, es ; *m. :* e ; *f.* I. *time, occasion :*—Ðá becom se apostol æt sumum sǽle (*on one occasion*) tó ðære byrig Pergamum, Homl. Th. i. 62, 24 : 70, 23. On sumne sǽl *quandoque*, Ælfc. Gr. 38 ; Som. 40, 66. Heora wíse on nǽnne sǽl wel ne gefór, Ors, 4, 4 ; Swt. 164, 13. Ðás wyrte man mæg niman on ǽlcne sǽl *this plant may be gathered at any time*, Lchdm. i. 112, 3. II. *a fit time, season, opportunity, the definite time at which an event should take place :*—Ðéh ðe sel síe *etiamsi oportuerit*, Mt. Kmbl. Lind. 26, 35. Ðá Godan sǽl þúhte ðá gesóhte hé ðone kynincg *when it appeared to Goda a favourable opportunity, he visited the king*, Chart. Th. 202, 30. Hí wundiaþ, ðonne se sǽl cymeþ, Fragm. Kmbl. 43 ; Leás. 23. Ðá wæs sǽl and mǽl, ðæt tó healle gang Healfdenes sunu *it was the proper time for Hrothgar to go to the banquet-hall*, Beo. Th. 2021 ; B. 1008. Oþ ðæt sǽl álamp (cf. Ðá seó tíd gelamp, ðæt . . . , Met. 26, 17) ðæt hió Beówulfe medoful ætbær *till the proper time arrived for her to present the mead cup to Beowulf*, 1249 ; B. 622 : 4123 ; B. 2058. Ic ofslóh æt ðære sæcce ðá mé sǽl ágeald (*when opportunity was offered me :* cf. ðá him rúm ágeald 5374 ; B. 2690) húses hyrdas, 3335 ; B. 1665 : Cd. Th. 121, 11 ; Gen. 2008. Seó sǽl gewearð (cf. seó tíd gewearð, ðæt se eorl ongan ǽðele cennan, 74, 25 ; Gen. 1227), ðæt his wíf sunu on woruld brohte, 72, 14 : Gen. 1186. Se sǽl cymeþ, ðæt heó dómes dæges dyn gehýre, Salm. Kmbl. 648 ; Sal. 323. Ne mihte ná lengc manna ǽnig hine sylfne bedyrnan ac gehwá tó sǽles (*at once*) móste clipian, Homl. Skt. i. 23, 115. Wit þencaþ sǽles bídan siððan sunne Metod up forlǽt *we intend to wait till after sunrise*, Cd. Th. 147, 10 ; Gen. 2437. Sǽles bídeþ hwonne heó cræft hyre cýþan móte, Exon. Th. 413, 28 ; Rä. 32, 12. Hé sóhte ða seel (sél, Rush.) ðætte hine salde *quaerebat opportunitatem ut eum traderet*, Mt. Kmbl. Lind. 26, 16. III. *time* as in *bad* or *good times, circumstance, condition.* v. IV :—Nú ís sǽl (*a time of misery*)

cumen, þreá ormǽte, Andr. Kmbl. 2332 : An. 1167. Storm oft holm gebringeþ in grimmum sǽlum *storm oft brings ocean into a furious condition*, Exon. Th. 336, 20 ; Gn. Ex. 52. Jacob byþ on glædum sǽlum *exultabit Jacob*, Ps. Th. 52, 8. Hæfdan beorgas blíðe sǽle *montes exultaverunt*, 113, 14. Sael gewynsumie roede *casus secundet asperos*, Ps. Surt. ii. 201, 11. IV. *happiness, good fortune, good time, prosperity* (often in pl.) :—On ðære stówe wé gesunde mágon sǽles bídan, Cd. Th. 152, 21 ; Gen. 2523. Mæg snottor guma sǽle brúcan, gódra tída, Exon. Th. 104, 12 ; Gú. 6. Sǽlum geblissad *gladdened with all joys*, 207, 12 ; Ph. 140. Siteþ sorgcearig, sǽlum bidǽled, 379, 5 ; Deór. 28. Syngum tó sǽlum (cf. After liked him ful wele for al was turned him to sele, C. M. 4432) *for the happiness of sinners*, 84, 21 ; Cri. 1377. Ne trín ðú æfter sǽlum, sorh is geníwod, Beo. Th. 2648 ; B. 1322. ¶ On sǽlum, sǽlum *in a state of happiness, happy* [cf. þu ware a sele gief ich was wroð, O. E. Homl. ii. 183, 17. Heora færð wes on sæle *was prosperous*, Laym. 1310. Selden sal he ben on sele (selde wurþ he blyþe and gled, Jes. MS.), Misc. 121, 301] :—Þá wæs þeód on sǽlum (*joyous*), Beo. Th. 1291 ; B. 643. On sálum, 1218 ; B. 607. Ðú on sǽlum wes *be fortunate*, 2345 ; B. 1170. On sǽlum *in times of prosperity*, Met. 2, 2, 7. Folc wæs on sálum, Cd. Th. 184, 13 ; Exod. 164 : 214, 5 ; Exod. 564 : Elen. Kmbl. 387 ; El. 194. [All middellǽrdess sceþe and sel, Orm. 14304. For quoso suffer cowþe syt (*trouble*), sele wolde folȝe, Allit. Pms. 92, 5. *Goth.* sélei *goodness : Icel.* sǽla *bliss, joy, happiness.*] v. gyte-, heáh-sǽl ; sǽlþ.

sæ̌-lác *a gift* or *present* or *offering that comes from the sea* or *from a lake :*—Beówulf maþelode : Hwæt wé ðé ðás sǽlác (*what B. had brought to Hrothgar from Grendel's lake-dwelling*) brohton tíres tó tácne, Beo. Th. 3308 ; B. 1652 : 3253 ; B. 1624.

sæ̌-lád *a course* or *way on the sea :*—Wé on sǽláde (*in our course*) brecaþ ofer bæðweg, Andr. Kmbl. 1022 ; An. 511. Hie on sǽláde wíf tó Denum feredon *they on the watery way took the woman to Denmark*, Beo. Th. 2319 ; B. 1157. Hé tó gyrnwræce swíðor þohte ðonne tó sǽláde *his thoughts were turned rather to vengeance effected by wiles than to taking his way over the sea*, 2283 ; B. 1139. [Cf. *Icel.* sjó-leiði *a seaway ;* sjó-leiðis *by sea.*]

sæ̌-láf *what is left by the sea*, applied to the spoils of the Egyptians drowned in the Red Sea :—Ongunnon sǽláfe dǽlan, ealde mádmas, reáf and randas, Cd. Th. 215, 16 ; Exod. 584.

sæ̌lan ; *p.* de *To happen, betide, fortune* (e. g. in Spenser) :—Gif hié ǽrfeweard ne gestriónen, oðða him sylfum ǽlles hwæt sǽle . . . Gif him elles hwæt sǽleþ, Chart. Th. 471, 30–472, 1. Sǽlde unc on þám brocum swá unc gesǽlde (sǽlde, Kmbl.) *happen what might to us in those troubles*, 485, 23. Hú ðé sǽle *how it may happen to thee, what your success may be*, Andr. Kmbl. 2710 ; An. 1357. v. ge-, tó-sǽlan.

sæ̌lan ; *p.* de. I. *to fasten with a cord :*—Hé sǽlde tó sande síðfætdmed scip oncerbendum fæst, Beo. Th. 3838 ; B. 1917. Wedera leóde sǽwudu sǽldon, 457 ; B. 226. Hwær wé sǽlan sceolon sǽhengestas ancrum fæste, Exon. Th. 54, 3 ; Cri. 863. Ymb geofenes stæþ gearwe stódon sǽlde sǽmearas, Elen. Kmbl. 455 ; El. 228. II. fig. *to restrain, repress, confine :*—Dómgeorne dreórigne hyge oft in heora breóstcofan bindaþ fæste. Swá ic módsefan mínne sceolde oft feterum sǽlan, Exon. Th. 287, 29 ; Wand. 21. Sǽlde sǽgrundas *the bound seadepths* (in contrast with the relaxing of the bonds which held the sea, when a passage was made through it for the Israelites), Cd. Th. 196, 9 ; Exod. 289. [*Goth.* in-sailjan.] v. á-, ge-, on-, un-sǽlan ; sál.

sæ̌-land *a maritime district :*—Mín gafolfisc ðe mé árist be sǽlande *maritimos pisces qui mihi contingere debent annualiter per thelonei lucrum*, Chart. Th. 308, 1. [Cf. *Icel.* Sjó-land (*a local name*).]

sæ̌ld. v. seld.

sæ̌len ; *adj. Of sallow :*—Sǽlenum *salignis*, Wrt. Voc. ii. 89, 50. [*O. H. Ger.* salahin *salignus.*] v. sealh.

sæ̌-leóda. v. sǽ-lida.

sæ̌-leóþ *a sea-song, song sung by the sailors in rowing, to keep stroke :*—Sǽleóþes *celeumatis* (κέλευμα), Wrt. Voc. ii. 22, 24.

sæ̌-líc ; *adj. Of the sea :*—On sǽlícum strande *on the sea-shore*, Homl. Th. ii. 62, 10. Of sǽlícum grunde, 138, 11. On sǽlícere ýðe *in the water of the sea*, 138, 8. Hí fixodon on sǽlícum ýðum, i. 576, 21. Gedréfed on ðám sǽlícum ýðum ðyssere worulde, ii. 388, 7. On sǽlícum *in glarigeris*, Hpt. Gl. 465, 3 : *in marinis*, 473, 71. Ðæt hí Seaxna þeóde ofer ðám sǽlícum (? of ðám ofersǽlícum) dǽlum him on fultum gecýgdon *ut Saxonum gentem de transmarinis partibus in auxilium vocarent*, Bd. 1, 14 ; S. 482, 39. Ic rówe ofer sǽlíce dǽlas *navigo ultra marinas partes*, Coll. Monast. Th. 26, 33. Drihten gegaderode ða sǽlícan ýða fram ðære eorþan brádnysse, Hexam. 6 ; Norm. 10, 16. Ða sǽlícan nýtenu (*two seals*), Homl. Th. ii. 138, 15. v. ofersǽ-líc.

sæ̌-lida, -leoda, an ; *m. A sea-goer, sailor :*—Snottor sǽleoda (*Noah*), Cd. Th. 201, 18 ; Exod. 374. Gehýrst ðú, sǽlida ! . . . brimmanna boda ! Byrht. Th. 133, 4 ; By. 45. Ic ǽfre ne geseah ǽnigne mann ðe gelícne steóran ofer stæfnan . . . Ic georne wát ðæt ic ǽfre ne geseah on sǽleodan syllícran cræft *I have never seen in a seaman more wondrous skill*, Andr. Kmbl. 999 ; An. 500. Næfre ic sǽlidan sélran mǽtte, 941 ;

An. 471. Offa đone sǽlidan slôh, Byrht. Th. 140, 10; By. 286. Cf. sǽ-liđend.

sǽlig *blessed, fortunate.* [*O. Sax.* sâlig : *O. L. Ger.* sâlig, sêlig : *O. H. Ger.* sâlig *beatus, felix.*] v. earfoþ-, ge-, gewif-, heard-, ofer-, un-, wansǽlig, *and next word.*

sǽliglîce ; *adv. Happily :*—Sêliglîce *feliciter*, Rtl. 79, 30. [*O. Sax.* sâliglîko : *O. H. Ger.* sâliglîhho *feliciter.*] v. ge-sǽliglîce.

sǽligness. v. ge-sǽligness.

sǽ-liđend, es ; *m. A seaman, sailor, seafarer ; also a ship, cf.* sǽgenga :—Secgaþ sǽliđend, Beo. Th. 826 ; B. 411 : 3640 ; B. 1818 : 5604 ; B. 2806. Sǽgdon sǽliđende, 760 ; B. 377. Se đe bisenceþ sǽliđende, eorlas and ŷđmearas, Exon. 363, 4 ; Wal. 48. [*O. Sax.* sêo-liđandi.]

sǽ-liđende ; *adj. Seafaring :*—Se mǽra wæs hâten sǽliđende weallende Wulf, Salm. Kmbl. 422 ; Sal. 211. [Sǽ-liđende men, Laym. 7821.]

sǽlmerige, an ; *f. Brine :*—Sǽlmerige (sǽll-, sel- ; -mærige) *salsamentum*, Ælf. Gr. 30 ; Zup. 192, 18. [Cf. *Span.* salmuera *brine : Ital.* salamoja : *Fr.* saumure : *Lat.* sal-muria ; cf. *Gk.* ἅλμυρος *briny.*]

sǽlþ, e ; *f. A dwelling, abode :*—Bare hié gesáwon heora lîchaman næsdon on đam lande đâ giet sǽlþa gesetena *bare they (Adam and Eve after the fall) saw their bodies, they had not yet in the land dwellings appointed*, Cd. Th. 48, 33 ; Gen. 785. [*O. Sax.* seliđa ; *f. a dwelling : O. L. Ger.* salitha, selitha *tabernaculum, habitaculum : Goth.* salithwa ; *f. a mansion, lodging, guest-chamber : O. H. Ger.* salida, selida ; *f. mansio, domicilium, habitaculum.*]

sǽlþ, e ; *f. Happiness, joy, felicity, good fortune, prosperity* (the word is generally in the plural) :—Ic nû hæbbe ongiten đæt đa mîne sǽlþa and seó orsorgnes đe ic ǽr wênde đæt gesǽlþa beón sceoldan nâne sǽlþa ne sint *I have now seen that my prosperity and security, that I supposed were certainly happiness, are none* ; non infitiari possum prosperitatis meae velocissimum cursum, Bt. 10 ; Fox 26, 25–27. Hâtan đæt sǽlþa đe nâne ne beóþ, 16, 3 ; Fox 56, 25. Âfyr fram đê đa yfelan sǽlþa and unnettan *gaudia pelle*, 6 ; Fox 14, 32. Đæm men þincþ đeáh hé sê godcundlîce gesceádwîs đæt hé on him selfum næbbe sǽlþa genôge bûton hé mâre gegaderige đara ungesceádwîsena gescefta đonne hê beþurfe *divinum merito rationis animal, non aliter sibi splendere, nisi inanimatae supellectilis possessione videatur*, 14, 2 ; Fox 44, 19. Ys micel niédþearf đæt mon hiene wiđ đa ungemetlîcan sǽlþa warenige, Past. 27 ; Swt. 189, 6. Hŷ weorđgeornra sǽlþa tôslîtaþ *they destroy the fortunes of the ambitious*, Salm. Kmbl. 697 ; Sal. 348. Heofenas blissiaþ sealte sǽstreámas sǽlþe habbaþ, Ps. Th. 95, 11. [*O. Sax.* sâlda : *O. L. Ger.* sâlda *salus, salutare : O. H. Ger.* sâlida *felicitas, beatitas, bona fortuna* ; v. Grmm. D. M. pp. 822 sqq. on *Sælde = Fortuna : Icel.* sælđ *bliss.*] v. ge-, ofer-, un-, woruld-sǽlþ.

sǽltna (?) *a bird's name* :—Saeltna, Wrt. Voc. ii. 119, 37 : seltra, i. 281, 8 : salthaga, 62, 36 gloss *rubisca* which is in the last case also glossed by *rudduc* the robin redbreast. v. rudduc.

sǽl-wang, es ; *m. A fertile plain, plain :*—Hê be wealle geseah wundrum fæste under sǽlwange sweras unlytle *by the wall he saw huge pillars with their bases wondrous fast underground*, Andr. Kmbl. 2984 ; An. 1495. Hwîlum mec mîn freá fæste genearwaþ, sendeþ đonne under sǽlwonge (MS. sal-), Exon. Th. 382, 27 ; Rä. 4, 2. Ic geseah hors ofer sǽlwong þrægan, 400, 3 ; Rä. 20, 3. Hê geseah sîde sǽlwongas synnum gehladene, Cd. Th. 78, 14 ; Gen. 1293.

sǽ-mann, es ; *m.* **I.** *a seaman, one who journeys by sea :*—Sǽmen æfter fôron flôdwege, Cd. Th. 184, 11 ; Exod. 105. Sǽmanna sîđ, 208, 4 ; Exod. 478. Gâras, sǽmanna searo, Beo. Th. 663 ; B. 329. Hê sǽmannum onsacan mihte, 5900 ; B. 2954. Sigel sǽmannum symble byþ on hihte, Runic pm. Kmbl. 342, 15 ; Rún. 16. **II.** *when English affairs are referred to the word is used of the Scandinavians :*—Wâlâ đære woruldscame đe nû habbaþ Engle. Oft twegen sǽmen ođđe þrŷ drîfaþ đa drâfe cristenra manna fram sǽ tô sǽ, Wulfst. 163, 5. Mê sendon tô đê sǽmen snelle, Byrht. Th. 132, 41 ; By. 29. Gif đu wille syllan sǽmannum feoh, 132, 58 ; By. 38. Hê his sincgyfan on đam sǽmannum wrec, 139, 63 ; By. 278. [*Icel.* sjó-mađr *a seaman, mariner.*]

sǽ-mearh *a sea-horse, a ship :*—Ús bær heáhstefn naca, snellíc sǽmearh, Andr. Kmbl. 533 ; An. 267. Meahte gesión brimwudu myrgan, sǽmearh plegan, Elen. Kmbl. 490 ; El. 245. Fearoþhengestas, sǽmearas, 455 ; El. 228. Heáhstefn scipu, sǽmearas, Exon. Th. 361, 5 ; Wal. 15. [For similar terms in Icelandic v. Corpus Poeticum Boreale, vol. ii. p. 458.] Cf. sǽ-hengest.

sǽmend, sǽmest, sǽmestre. v. sêmend, sǽmra, seámestre.

sǽ-mêđe ; *adj. Weary with being on the sea :*—Sǽmêđe (*Beowulf and his companions on their arrival at Hrothgar's palace*), Beo. Th. 655 ; B. 325.

sǽ-minte, an ; *f. Sea-mint :*—Sǽminte *nereta* (cf. sea-minte *nereta*, Lchdm. iii. 304, col. 1), Wrt. Voc. i. 68, 39 : *althea*, 68, 79.

sǽmninga. v. semninga.

sǽmotu (?) glosses *fustrum* (*frustum?*), Wrt. Voc. ii. 152, 10.

sǽmra ; *adj.* (without positive) *Inferior, worse :*—Symle wæs đŷ sǽmra đonne ic sweorde drep ferhþgenîđlan *ever was the deadly foe the worse when I struck him with the sword*, Beo. Th. 5752 ; B. 2880. Hit is sǽmre nû *it is worse now (than in the golden age)*, Met. 8, 42. Ic lǽre đæt hê gŷme ǽgđer ge đæs sêlran ge đæs sǽmran *I advise him to take care both of the more and of the less important matters*, Anglia ix. 260, 10. Hnâhran rince, sǽmran æt sæcce, Beo. Th. 1910 ; B. 953. Gif đû sôđne God lufast. . . Gif đû tô sǽmran gode hætsþ hǽđen feoh, Exon. Th. 245, 28 ; Jul. 51 : 264, 9 ; Jul. 361. Đa sǽmran *deteriora*, Wrt. Voc. ii. 139, 38. Đû byst se ilca se đû ǽr wǽre, ne beóþ đîn winter wiht đê sǽmran (*anni tui non deficient*), Ps. Th. 101, 24. Hî dweligende sêcaþ đæt hêhste gôd on đa sâmran (sǽmran, Cott. MS.) gesceafta *id* (good) *error humanus a vero atque perfecto ad falsum imperfectumque traducit*, Bt. 33, 1 ; Fox 120, 12. Sǽmust *vel* wyrst *pessima*, Blickl. Gl. Ne wǽron đæt gesîþa đa sǽmestan, Exon. Th. 326, 8 ; Wîd. 1. Cf. sâm-.

sǽm-tinges. v. sam-tinges.

sǽ-naca, an ; *m. A sea-going vessel*, Exon. Th. 474, 7 ; Bo. 26.

sǽne ; *adj. Slow, dull, sluggish, inactive :*—Ymb đa gŷmene his êcre hǽlo hê wæs tô sǽne *erga curam perpetuae suae salvationis nihil omnino studii gerens*, Bd. 3, 13 ; S. 538, 19. Ne sceal se tô sǽne beón, ne đissa lârna tô læt, Exon. Th. 450, 16 ; Dôm. 88. Sǽne môd *a sluggish mind*, 122, 32 ; Gû, 314. Wæs đæt sǽne cyning, 322, 23 ; Wîd. 67. Eálâ đæt đû woldest đæs sîđfætes sǽne weorđan (*slow to undertake the journey*), Andr. Kmbl. 408 ; An. 204 : 422 ; An. 211 : Elm. Kmbl. 440 ; El. 220. Næs his brôđor læt, sîđes sǽne, Apstls. Kmbl. 67 ; Ap. 34. Nǽron đa twegen tohtan sǽne, lindgelâces, 150 ; Ap. 75. Đone sǽnan đe biþ tô slâw đû scealt hâtan assa mâ đonne man *segnis ac stupidus torpet? asinum vivit*, Bt. 37, 4 ; Fox 192, 19. Mægencræft môda gehwilces ofer lîchoman lǽnne and sǽnne *might of the mind over the body weak and dull*, Met. 26, 106. Hê (*a sea serpent*) on holme wæs sundes đê sǽnra (*the slower in swimming*), đâ hyne swylt fornam, Beo. Th. 2876 ; B. 1436. Ic sceal sêcan ôđerne ellenleásran cempan sǽnran *I must seek another warrior less courageous and active*, Exon. Th. 266, 9 ; Jul. 395. [*O. H. Ger.* seine : *Icel.* seinn : *Dan.* seen : *Swed.* sen. Cf. *Goth.* sainjan *to be slow, to tarry.*] v. â-sânian.

sǽ-næss, es ; *m. A ness or promontory stretching into the sea, a cape :*—Sǽnesse *promontorio*, Hpt. Gl. 420, 7. Đa lîđende land gesáwon brimclifu blîcan, beorgas steápe, sîde sǽnæssas, Beo. Th. 451 ; B. 223. Sǽnæssas geseón, windige weallas, 1146 ; B. 571.

sǽ-nett *a net for fishing in the sea :*—Sǽnet *sagene*, Wrt. Voc. i. 68, 14.

sǽp, es ; *n. Sap :*—Sǽp *succus*, Hpt. Gl. 450, 12. Cederbeám *cedrus*, his sǽp *cedria*, Wrt. Voc. i. 33, 39. Đâ weard beám monig blôdigum teárum birunnen, þæt weard tô swâte, Exon. Th. 72, 23 ; Cri. 1177. Đæs swêtestan sæpes *suavissime succi*, Hpt. Gl. 411, 58. Seó drîge gyrd đe næs mid sæpe âcucod, Homl. Th. ii. 8, 17. Sep *sucum*, Germ. 391, 18. [*Ayenb.* þet zep ; *O. H. Ger.* saf : *Icel.* safi ; m.] v. stôr-sæp.

sǽpig ; *adj. Full of sap, succulent :*—Sǽpig stela *succulentus cauliculus*, Hpt. Gl. 419, 45. [*Prompt. Parv.* sapy or fulle of sap *cariosus*.] v. unsæpig.

sǽppe, an ; *f. The spruce fir :*—Sǽppe *abies*, Wrt. Voc. i. 285, 40. [Cf. *Lat.* sappinus *from which Fr.* sapin.]

sǽp-spôn *a chip or shaving with sap in it :*—Genim geongre âcrinde hand fulle . . . sceafe đæt grêne, wylle đa sæpspône on cûmeolce, Lchdm. ii. 292, 27.

sǽr (= rǽr ?), Ps. Th. 7, 6.

sǽ-rima, an ; *m. The sea-shore, coast :*—Hî mycel yfel gedydon ǽgđer ge on Defenum ge wel hwǽr be đæm sǽriman, Chr. 897 ; Erl. 95, 20 : 994 ; Erl. 133, 19. [Bî đa sǽrime âhwǽr in Engelande *in littore marino alicubi in Anglia*, Chart. Th. 422, 2.] [Bî þisse sǽrime, Laym. 6216.]

sǽ-rinc, es ; *m. A sea-man, one who journeys by sea* (used of the Scandinavians, cf. sǽ-mann) :—Hine ymb monig snellic sǽrinc (*of Beowulf and his companions*), Beo. Th. 1384 ; B. 690. Sende se sǽrinc (*one of the Danes attacking Byrhtnoth*) sûþerne gâr, Byrht. Th. 135, 46 ; By. 134.

sǽ-rîric *a reed-bed in the sea* (?), *an ait :*—Swylce wôrie bî ôfre sondbeorgum ymbseald sǽrŷrica mǽst, swâ đæt wênaþ wǽglîþende đæt hý on eálond sum eágum wlîten (*the reference is to the whale, which mariners mistake for an island*), Exon. Th. 360, 24 ; Wal. 10. [Cf. *O. H. Ger.* rôrahi *arundinetum.*]

sǽ-rôf ; *adj. Active on the sea, strong in rowing :*—Đonne sǽrôfe snelle mægne ârum bregdaþ, Exon. Th. 296, 25 ; Crä. 56.

sǽs *a seat.* v. sess.

sǽ-sceaþa, an ; *m. A sea-robber, pirate :*—Sǽsceaþan *piratici*, Wrt. Voc. ii. 68, 12.

sǽ-sîđ *a sea-journey, voyage*, Beo. Th. 2302 ; B. 1149.

sǽ-snægl, es ; *m. A sea-snail :*—Sǽsnæl *chelio, testudo vel marina gagalia*, Wrt. Voc. i. 24, 32. Sǽsnæglas *conchae vel cochleae*, 56, 7 : ii. 136, 14.

sǽ-strand, es; _m. Sea-shore :_—Sǽstrand _litus_, Wrt. Voc. i. 80, 59. Swā mænigfealde swā swā sandceosol on sǽstrande, Jos. 11, 4: Wulfst. 198, 22. Beraþ ða stānas tō sǽstrande, Homl. Th. i. 68, 29. [Heo stepen up a sæstrond, Laym. 9235. _Icel._ sævar-strönd.]

sǽ-streám, es; _m. Sea-stream, water of the sea :_—Ðonne sæstreámas flōwaþ _elationes maris_, Ps. Th. 92, 5. Sǽstreámas sealte, 79, 11: Andr. Kmbl. 391; An. 196: 1497; An. 750. Swearte sǽstreámas, Cd. Th. 80, 9; Gen. 1326. Sǽstreámum neáh, 193, 22; Exod. 250. Ic his swīðran hand settan þence ðæt hē sǽstreámum syððan wealde _ponam in mari manum ejus, et in fluminibus dexteram ejus_, Ps. Th. 88, 22. Sicilia sǽstreámum in, Met. 1, 15. [He iwende ouer sea-streames, Laym. 326. Þu steorest te sea stream þ hit fleden ne mot fir þan þu markedest, Marh. 9, 34. _O. Sax._ seó-strōm.]

sǽt, e; _f. An ambush, a place where one lies in wait :_—Hȳ sǽtiaþ mīn and siittaþ swā gearwe swā seó leó dēþ tō ðam ðe hẽ gefōn wyle and swā swā his hwelp byþ gehȳd æt ðære sǽte _susceperunt me sicut leo paratus ad praedam, et sicut catulus leonis habitans in abditis_, Ps. Th. 16, 11. Deórhege heáwan and sǽte haldan _to maintain the places from which the deer might be shot_ (?), L. R. S. 2; Th. i. 432, 15. The Latin version has _stabilitatem observare_; Leo takes _sǽte_ = hedges, and Schmid translates 'in ordnung erhalten.' [_Icel._ sāt; _f._ ambush.] v. sǽtian.

sǽta _a resident, inhabitant._ The form occurs only in compounds, and these are for the most part in the plural. There is also beside the weak -_sǽtan_ a strong -_sǽte._ v. Dorn- (Dor-), Dūn-, Peác-, Sumor-, Wil-sǽte (-sǽtan). Other instances of the suffix are given in Bd. 4, 12; S. 581, 34, where _Hrypensis ecclesia_ is translated _Hrypsǽtna cyrice:_ Hiisētena munecas _Hiiensæ monachi_, 5, 22; S. 644, 24: and in Cod. Dip. B. i. 414. It also forms part of common nouns, v. burh-, ende-, land-sǽta: with which may be compared _O. L. Ger._ land-sētio: _O. H. Ger._ himil-sāzo: _Ger._ land-saſs. See too the compounds of sittend[e].

sǽtan, -sǽte; _subst._, -sǽte; _adj._, sæten, Sæter-dæg. v. sǽtian, sǽta, and-sǽte, seten, Sætern-dæg.

sǽtere, es; _m. One that lies in wait, one that waylays._ **I.** _a robber;_ latro:—Þeáf and settere _fur et latro_, Jn. Skt. Lind. 10, 1. Þeáfas and setteras _fures et latrones_, 10, 8. **II.** fig. _one who acts insidiously;_ insidiator, seductor:—Se sǽtere (_insidiator_), ðæt is se dióful, hē hine spænþ on wōh, Past. 53, 7; Swt. 417, 23. Ðonne cymþ se lytega sǽtere (_seductor_) tō ðæm slāwan mōde, and āteleþ him eall ðæt hē ǽr tō gōde gedyde, 65, 2; Swt. 463, 12. Hī sendon sēteras (_insidiatores_) ðætte genōmo hine on word, Lk. Skt. Lind. 20, 20. v. sǽt, sǽtian.

Sǽtern-dæg, Sæternes-, Sæter-, Sæteres-dæg, es; _m. Saturday; dies Saturni_:—Sæterndæges rest _requies sabbati_, Ex. 16, 23. On Sæterndæg, Mk. Skt. 9, 2, Rbc. Sæterndæg (sæter-, MS. A.), Lk. Skt. 23, 56. Sæterdæg (sæternes-, MS. A.), 23, 54. Sæternesdæg, Mt. Kmbl. 16, 28, Rbc.: 20, 29, Rbc. On ðone Sæternesdæg, Chr. 1012; Erl. 146, 12: Shrn. 70, 7. Sæternesdæg of Saturno Iovis fæder, Anglia viii. 321, 11. Se seofoða dæg is se Sæternesdæg, Homl. Th. ii. 206, 6. Ǽghwylce Sæternesdæge _per omne sabbatum_, Bd. 2, 3; S. 504, 40. Seternesdæg _Sabbatum_, Mt. Kmbl. Lind. 12, 8. Sæstresdæg (Sæternes-, MS. T.), R. Ben. 37, 23: 38, 8. On ðæm Sæteresdæge, Blickl. Homl. 71, 30. [Saturnus heo (_the forefathers of the English_) ȝiven Sætterdæi (Sateres-dai, 2nd MS.), Laym. 13933. _Orm._ Saterrdaȝȝ. High German and Scandinavian take a different form, but Frisian and Dutch agree with English. v. Grmm. D. M. pp. 114-5; 226-7.]

Sǽtern-, Sæter-niht, e; _f. Friday night, the night between Friday and Saturday:_—His (_Christ_) līc læg on byrgene ða Sæterniht and Sunnanniht _his body lay in the sepulchre on the nights of Friday and Saturday_, Homl. Th. i. 216, 27. [_R. Glouc._ Sater-niȝt.]

sǽ-þeóf, es; _m. A sea-thief, a pirate:_—Heáh-sǽþeóf _archipiratta_, Wrt. Voc. ii. 5, 28.

sæþerige, an; _f. Savory;_ satureia hortensis:—Sæþerian sǽd, Lchdm. ii. 314, 19: iii. 72, 8. v. saturege.

sǽtian, sǽtan; _p._ ode _To lie in wait for, waylay_ (with gen.):—Forðam hē hine ne meahte mid openlīcum gefeohte ofersuīdan sǽtaþ ðonne diógollīce and sēcþ hū hē hine mǽge gefōn _quia enim publico bello perdidit, ad exercendas occulte insidias exardescit_, Past. 33, 7; Swt. 227, 13. Hē sǽtaþ (_insidiatur_) ðæt hē bereáfige ðone earman, Ps. Th. 9, 30. Se synfulla sǽtaþ ðæs rihtwīsan _observabit peccator justum_, 36, 12. Hȳ sǽtiaþ mīn _susceperunt me_, 16, 11. Ðū scealt fiersna sǽtan, Cd. Th. 56, 18; Gen. 913. Hū ǽghwelc syn biþ sǽtigende ðæs þióndan monnes _quomodo unumquodque peccatum proficientibus insidietur_, Past. 21, 5; Swt. 161, 24. Feóndas and sǽtendan sāwle mīnre _inimici et qui custodiebant animam meam_, Ps. Th. 70, 9. Sǽtendum _insidiantibus_, Lk. Skt. p. 10, 5. [_Icel._ sæta _to lie in wait for_ (with dat.): _M. H. Ger._ sāzen.] v. sǽtnian, sǽt, sǽtere.

sǽtilcas:—Ne ymbe sciphergas sǽtilcas ne hērdon ne furþum fira nān ymb gefeoht sprecan, Met. 8, 31. _Grein_ suggests scealcas, cf. næs scealca nān _in_ v. 21; _the corresponding prose is:_—Ne gehẽrde nōn mon

ða get nānne sciphere, ne furþon ymbe nān gefeoht sprecan, Bt. 15; Fox 48, 14-16.

sǽtnere, es; _m. One who lies in wait._ v. sǽtnian, sǽtere; but used in the following case to gloss _seditiosus :_—Mid sētnerum _cum seditiosis_, Mk. Skt. Lind. 15, 7. v. sǽtnung.

sǽtnian; _p._ ode _To lie in wait for_ (with gen.):—Ðā wǽron ðǽr Sarocine gesamnode, ðæt hig sǽtnodan manna, Shrn. 37, 34. v. sǽtian.

sǽtnung, e; _f._ **I.** _a lying in wait, plot, snare._ v. sǽtung:—Hē hine bæd ðæt hē his līf gescylde wið swā mycles ēhteres sǽtnungum _obsecrans ut vitam suam a tanti persecutoris insidiis tutando servaret_, Bd. 2, 12; S. 513, 5. Hē him ða sǽtnunge (_insidias_) gewearnode ðæs unholdan cyninges, S. 515, 11: 5, 23; S. 646, 37 note. Sētnungum _insidiantes_, Lk. Skt. Lind. Rush. 11, 54. **II.** in the following passages the word glosses _seditio_. v. sǽtnere :—On sǽtnuncge (setnong, Lind.) _in seditione_, Mk. Skt. Rush. 15, 7. Fore sǽtnunge _propter seditionem_, Lk. Skt. Rush. 23, 19, 25.

sǽtung, e; _f. A lying in wait, plot, snare :_—Sǽtunge _aucupatione_, Wrt. Voc. ii. 7, 43. Sǽtunge, 101, 25. Gif him þince ðæt hē feala earna ætsomne geseó, ðæt biþ yfel nīð and manna sǽtunga and seara, Lchdm. iii. 168, 11. Ðonne hē foresægþ ða dīeglan sǽtenga ðæs lytegan feóndes _quando hostis callidi circumspectas et quasi incomprehensibiles insidias praedicit_, Past. 21, 5; Swt. 163, 14. Scottas ne sǽtincge ne gestrodu wið Angelþeóde syrwaþ _Scotti nil contra gentem Anglorum insidiarum moliuntur aut fraudium_, Bd. 5, 23; S. 646, 37.

sǽ-upwearp _what is thrown up on land by the sea, jetsum :_—Ic habbe gegeofen Ælfwine abbod . . . ða sǽupwearp on eallen þingen æt Bramcæstre, Chart. Th. 421, 33.

sǽ-wǽg _a wave of the sea :_—Sealte sǽwǽgas, Cd. Th. 240, 9; Dan. 384.

sǽ-wǽter, es; _n. Sea-water :_—Genim celeþonian seáw and sǽwæter, Lchdm. ii. 28, 12.

sǽ-wang, es; _m. The plain by the sea, the shore :_—Gewāt se hearda æfter sande sǽwong tredan, wīde waroþas, Beo. Th. 3933; B. 1964.

sǽ-wār _sea-weed :_—Sǽwaar _alga_, Wrt. Voc. i. 31, 35. Cf. waar _alga_, ii. 99, 29. See E. D. S. Pub. Plant Names, s. v. _waur_.

sǽ-waroþ _the sea-shore :_—Be sǽwaroþe and be æáðrum, Bt. 32, 3; Fox 118, 17: Met. 19, 21.

sǽ-weall, es; _m._ **I.** _a sea-wall, a cliff by the sea :_—Higelāc wunode sǽwealle neáh, Beo. Th. 3853; B. 1924: Exon. Th. 471, 15; Rä. 61, 1. **II.** _a wall formed by the sea :_—Sǽweall āstāh (cf. Ðæt wæter (_of the Red Sea_) stōd swilce twegen hẽge weallas, Ex. 14, 22), Cd. Th. 197, 6; Exod. 302.

sǽ-weard _sea-ward, keeping watch and ward on the sea-coast ;_ it was a duty that might be required in some cases of the thane and of the 'cotsetla':—Of manegum landum māre landriht ārīst tō cyniges gebanne . . . sǽweard (_the section refers to the_ 'thegen'), L. R. S. 1; Th. i. 432, 8. Werige his (_the_ 'cotsetla') hlāfordes inland, gif him man beóde, æt sǽwearde, 3; Th. i. 432, 28. Cf. the description of Beowulf's landing :—Ðā of wealle geseah weard Scyldinga, se ðe holmclifu healdan scolde, etc., Beo. Th. 463 sqq.

sǽ-weg _a sea-way, a path through the sea :_—Sǽfiscas ða faraþ geond ða sǽwegas _pisces maris qui perambulant semitas maris_, Ps. Th. 8, 8. [_Icel._ sjó-vegr.]

sǽ-wērig; _adj. Weary with being on the sea :_—Sǽwērige slǽp ofereode, Andr. Kmbl. 1651; An. 817: 1723; An. 864. [We beoþ sæwerie men, Laym. 4619.]

sǽwet, es; _n. Sowing :_—Ofer ða tīd ðæs sǽwetes _ultra tempus serendi_, Bd. 4, 28, tit.; S. 605, 8.

sǽ-wīcing, es; _m. A viking :_—Randas bǽron sǽwīcingas (_the tribe of Reuben_) ofer sealtne mersc, Cd. Th. 199, 3; Exod. 333.

sǽ-wiht, e; _f. A sea-animal :_—Ðeós eorþe is berende missenlīcra fugela and sǽwihta _this land is productive of divers fowls and sea-animals_ (_the Latin has_ insula . . . avium ferax terra marique diversi generis), Bd. 1, 1; S. 473, 15.

sǽ-wilm, es; _m. A billow :_—Gẽ him syndon ofer sǽwylmas hider wilcuman, Beo. Th. 792; B. 393.

-sǽwisc. v. ofer-sǽwisc.

sǽ-wudu _a ship :_—Hī sǽwudu sǽldon _they fastened their ship to the shore_, Beo. Th. 457; B. 228.

sæx. v. seax.

sǽ-ȳþ, e; _f. A wave of the sea :_—Sǽȳþa _vel_ holmas _equomaria_, Wrt. Voc. ii. 143, 74. Hī sǽȳþa swīðe brēgaþ, Runic pm. Kmbl. 343, 23; Rūn. 21. [_O. Sax._ seo-ūðia.]

safine, an; _f. Savine;_ juniperus savina :—Sauine. Genim ðās wyrte, ðe man _sabinam_, and ōðrum naman wel ðam gelīc, _sauinam_ hāteþ, Lchdm. i. 190, 13: iii. 16, 8: 58, 20. Safine, 22, 31. Lytel sauinan, 30, 15. Safinan dust, ii. 250, 27. Genim safinan, 100, 10: 294, 24: iii. 44, 5. Safenan, 46, 3: ii. 312, 11. Sauinan, iii. 38, 26.

saftriende _rheumatic:_—Saftriende _reumaticus_, Wrt. Voc. i. 45, 48. Cf. sǽp.

sāg (?) :—Ic heáfod hæbbe and heáne steort, eágan and eáran and ǽnne foot, hrycg and heard nebb, hneccan steápne and sīdan twā, sāg on middum, eard ofer ældum, Exon. Th. 490, 3 ; Rä. 79, 5.

saga, an ; *m. A saw :*—Saga *serula,* Wrt. Voc. i. 16, 17 : *serra,* 39, 67. v. sagu.

saga, an ; *m. A saying, story, statement :*—Ðīn saga biþ geswutelod, gif ðū ðone sylfan encgel bitst, ðæt hē mīnne sunu ansundne árǽre, Homl. Skt. i. 7, 193. v. sagu.

sagian. v. secgan.

-sagol. v. leás-, sōþ-, unsōþ-, wǽr-sagol.

sāgol (v. sowel *fustis,* Wrt. Voc. i. 94, 22, soþsawel *veridicus,* 90, 19), es ; *m. A staff, cudgel, club :*—Sāgol oððe stæf *fustis,* Ælfc. Gr. 9, 28 ; Som. 11, 44 : *fustis,* Wrt. Voc. i. 84, 28. Ða sāglas (*vectes*) sticiaþ inn on ðām hringum ða earce mid tō beranne . . . Ðæt is ðonne ðæt mon ða earce bere on ðǽm sāglum, Past. 22, 1 ; Swt. 171, 5–12. Hié Claudium mid sāglum ofbeótan, Ors. 2, 6 ; Swt. 88, 26. Hēt ða cwelleras mid stearcum sāglum hine beóton, Homl. Th. i. 424, 32. Mid stīðum sāglum beátaþ, 432, 12 : 468, 33. Hēt his cwelleras ðone hālgan beátan mid heardum sāglum. Ðā bǽrst sum sāgol intō ānes beáteres eágan, Homl. Skt. i. 4, 142. Mid swurdum and sāhlum *cum gladiis et fustibus,* Mt. Kmbl. 26, 47, 55. Hē stafas ł sāhlas īsenne tōbræc *vectes ferreos confregit,* Ps. Lamb. 106, 16. [Ælc bær an honde ænne saȝel (staf, 2ND MS.) stronge, Laym. 12280.]

sagu, e ; *f. A saw :*—Sage *serram,* Germ. 400, 531. Hē sceal habban æcse, adsan, sage, Anglia ix. 263, 2. [O. H. Ger. saga, sega ; *f. serra, lima : Icel.* sög ; *f. a saw.*] v. saga.

sagu, e ; *and indecl.? f.* I. *a saw, say* (to say one's *say*), *saying, statement, story, tale :*—Racu, sagu *sermo,* Hpt. Gl. 433, 12. Nis ðis nān gedwimor ne nān dwollīc sagu, Jud. Thw. p. 159, 27. Ic hāte healdan hī ōþ ðæt heora sagu āfandod sȳ, Homl. Th. ii. 484, 3. Teónan ðū wyrcst ūs mid ðisse sage *haec dicens nobis contumeliam facis,* Lk. Skt. 11, 45. Sagu *dictu* (cf. gesæʒene *dictu,* 28, 47), Wrt. Voc. ii. 140, 7. Hī sǽdon ðam kinge ðæt hē hæfde swȳðe āgylt wið Crist. . . . Ðā læg se king and āsweartode eall mid ðare sage, Chart. Th. 340, 1. Gehȳr ðū ðās race nū swilce leáse sagu ac geworden þing *audi fabulam, non fabulam sed rem gestam,* Ælfc. T. Grn. 16, 12. Geendebrednege ða sago þinga *ordinare narrationem rerum,* Mt. Kmbl. p. 7, 2, 9. *Fabulae* synd ða saga ðe menn secgaþ ongeán gecynde, Ælfc. Gr. 50, 29 ; Zup. 296, 5. Spellenga, sagena *sermonum,* Hpt. Gl. 505, 77. Ic ðīnra bysna ne mæg, worda ne wīsna wuht oncnāwan, sīðes ne sagona, Cd. Th. 34, 9 ; Gen. 535. Sagum *fabulis,* Lk. Skt. p. 2, 10, 11. II. *saying, narration, telling, report :*—Se hlīsa ðe þurh yldra manna segene (sage, MS. B.) tō ūs becom *opinio quae traditione majorum ad nos perlata est,* Bd. 2, 1 ; S. 501, 2. III. *statement of a witness, testimony :*—Tō hwī wilnige wē ǽnigre ōðre sage *quid adhuc egemus testibus,* Mt. Kmbl. 26, 65. Ne gehȳrst ðū hū fela sagena (*quanta testimonia*) hig ongēn ðē secgeaþ, 27, 13. Hī sōhton leáse saga (*falsum testimonium*) ongēn ðone Hǽlend, 26, 59. IV. *a saying before-hand, foretelling :*—Of sage *fatidicum,* Wrt. Voc. ii. 147, 22. Saga presagia, 67, 46. Sagum praesagminibus, vaticinationibus, divina-tionibus, Hpt. Gl. 448, 64. [Ælc his saȝe sæide, Laym. 26345. Heo wenden þat his sawen (2ND MS. sawes) soðe weren, 749. A. R. saȝe, sawe, sahe : Chauc. Piers P. sawe : O. H. Ger. saga *assertio, narratio, sermo, enuntiatio : Icel.* saga *story, tale.*] v. on-, sōþ-sagu ; saga.

saht, sales. v. seht, sæl.

sāl, es ; *m. : e ; f.* (?) I. *a rope, cord, line, bond :*—Licgaþ mē ymbe īrenbendas, rīdeþ racentan sāl, Cd. Th. 24, 3 ; Gen. 378. Ðā wæs he mæste segl sāle (cf. O. H. Ger. segil-seil *rudens*) fæst, Beo. Th. 3816 ; B. 1906. Sālum *nexibus,* Wrt. Voc. ii. 60, 74. II. *a rein :*—Sālas [*h*]*abenas,* 4, 58 : 6, 22. Sālum ł gewealdeþerum *habenis,* 42, 60. III. *the loop which forms the handle of a vessel* (?) :—Sāl *ansa* (cf. hringe *ansa,* 284, 7, *and see* nostle. *The word occurs under the heading* nomina vasorum), Wrt. Voc. i. 25, 11. IV. *the fastening of a door :*—Repagulum sāl [-panra ?], Wrt. Voc. i. 10, 3. Sāle *repagula,* ii. 119, 4. V. *a necklace, collar :*—Sweorclāþ *vel* [sweor]tēg *vel* [sweor ?]sāl *collarium,* 134, 49. Sāle *collario,* 18, 17. Saule *callario* (saale *collario*), 78, 71. [Soole, beestys teyynge *ligaculum ; restis* a sole to tie beasts, Prompt. Parv. 463. Hi drayeþ myd such sol, Misc. 51, 162. O. H. Ger. seil ; *n. funis, rudens, lorum, habena, restis : Icel.* seil ; *f. a line.*]

sala, an ; *m. A sale :*—Ceáp *distractio,* sala *venditio,* Wrt. Voc. i. 55, 55. [O. H. Ger. sala ; *f. traditio : Icel.* sala ; *f. a sale.*]

salf. v. sealf.

salfige, an ; *f. Sage :*—Saluige *salvia,* Wrt. Voc. i. 79, 49. Salfige, Lchdm. iii. 22, 31. Saluie. Genim ðās wyrte ðe man saluian nemneþ. . . . Genim ðās ylcan wyrte salfian, i. 218, 6–11. Saluian sǽd, iii. 72, 7 : ii. 358, 18. Nim salfian, iii. 48, 3. Wyl sealuian, 44, 17. [O. H. Ger. salbeia, salveia : Ger. salbei. *From Latin.*]

salh *a sallow.* v. sealh.

sallettan *to play on the harp, sing to the harp, sing psalms :*—Singaþ him and salletaþ *cantate ei et psallite ei,* Ps. Th. 104, 2.

salm. v. sealm.

salness, e ; *f. Darkness, duskiness :*—Conticinium, ðæt ys swītīma oððe salnyssa tīma, Anglia viii. 319, 29. v. salu.

salor *a hall, palace :*—Eów ðeós cwēn laðaþ tō salore (cf. tō hofe, 1111 ; El. 557), Elen. Kmbl. 1100 ; El. 552 : 764 ; El. 382. v. sæl, sele.

salo, salowig, salpanra, salt, salt-haga. v. salu, saluwig, sāl IV, sealt, sæltna.

saltere, es ; *m.* I. *a stringed musical instrument, a psaltery :*—Saltere *sambucus,* Wrt. Voc. i. 289, 26 : *psalterium,* Ps. Spl. 80, 2 : 107, 2. On saltere syngaþ him *in psalterio psallite illi,* 32, 2 : 91, 3 : 143, 11 : 150, 3. Cimbalan oððe psalteras oððe strengas ætrīnan, Lchdm. 202, 14. II a. *the book of Psalms :*—Se saltere ys ān bōc, ðe hē (*David*) gesette þurh God betwux ōðrum bōcum on ðǽre bibliothecan, Ælfc. T. Grn. 7, 26. II b. *a psalter, a service-book containing the book of Psalms divided into certain portions for Matins, and the Hours, so as to be gone through in the course of the week :*—Hē (*the mass-priest*) sceal habban ða wǽpna tō ðam gāstlīcum weorce . . . ðæt synd ða hālgan bēc, saltere and pistolbōc, godspellbōc and mæssebōc, L. Ælf. C. 21 ; Th. ii. 350, 12 : L. Ælfc. P. 44 ; Th. ii. 384, 1. ii. salteras and se þridda[n] saltere swā man singþ on Rōme, Chart. Th. 430, 11. ¶ Saltere singan *to sing psalms taken from the psalter :*—Hē gehāt gehēt . . . ðæt hē ǽghwylce dæge ealne saltere āsunge *vovit votum quia quotidie psalterium totum decantaret,* Bd. 3, 27 ; S. 599, 11. Hē āsong ǽlce dæge tuwa his saltere and his mæssan, Shrn. 134, 17. Singe eal gefērrǽden ætgædere heora saltere ða þrȳ dagas,Wulfst. 181, 21. Ǽlc brōður singe twegen salteras sealma . . . vi. mæssan oððe .vi. salteras sealma *each brother shall sing two portions of psalms from the psalter,* Chart. Th. 614, 7, 11. [O. H. Ger. saltari, psaltari *psalterium ;* salzara *sambucus : Icel.* saltari *a psalm-book.*]

saltian ; *p.* ode *To dance :*—Gē ne saltudun (sealtedon, MS. A.) *non saltastis,* Lk. Skt. 7, 32. [O. H. Ger. salzōn. *From Latin.*]

salu ; *adj. Dusky, dark :*—Ic sylfa [eom] salo, Exon. Th. 489, 21 ; Rä. 48, 11. [O. H. Ger. salo *fuscus, furvus, ater, niger : Icel.* sölr *yellow.*] v. following words.

salu-brūn ; *adj. Dark-brown :*—Hrefn sweart and sealobrūn, Fins. Th. 70 ; Fin. 35.

salu-neb ; *adj. Dark-faced :*—Se wonna þegn, sweart and saloneb, Exon. Th. 433, 9 ; Rä. 50, 9.

salu-pād ; *adj. Dark-coated :*—Ða sind blace swīðe, swearte, salopāde, Exon. Th. 439, 1 ; Rä. 58, 3. Cf. saluwig-pād.

saluwig-feðera ; *adj. Of dusky plumage :*—[Hrefn] salwigfeðera, Cd. Th. 87, 13 ; Gen. 1448.

saluwig-pād ; *adj. Dark-coated, having dark plumage :*—Hrefn sal-wigpād, Exon. Th. 329, 20 ; Vy. 37. Earn salowigpāda, Judth. 24, 28 ; Jud. 211. Lētan hrǽ bryttian saluwigpādan ðone sweartan hræfn, Chr. 937 ; Erl. 115, 10.

sāl-wang, sal-warp. v. sǽl-wang, sealt-wearp.

salwian *to make dark, to blacken :*—Heó (*the dove*) nolde ǽfre under salwed bord (*in the ark, which was dark-coloured from the pitch that had been smeared over it*) syððan ætȳwan, Cd. Th. 89, 15 ; Gen. 1481. [Cf. O. H. Ger. gi-salwian *decolorare ;* salwet *obscuratum ;* salawi *fuscatio.* v. Grff. vi. 183.] v. salu.

sam ; *conj. Whether, or* (cf. swā . . . swā = *whether . . . or*) :—Sam hī þyrfon, sam hī ne þurfon, hī willaþ ðeáh, Bt. 26, 2 ; Fox 92, 29. Sam wē willan, sam wē nyllan, 34, 12 ; Fox 154, 7 : 40, 1 ; Fox 234, 34. Hȳ gedōþ ðæt ǽgðer fætels biþ oferfroren sam hit sȳ sumor, sam winter, Ors. 1, 1 ; Swt. 21, 17. Sam hȳ fæsten sam hȳ ne fæsten *omni tempore siue jejunii siue prandii,* R. Ben. 66, 14. Sam hē hine miclum lufige, sam hē hine lytlum lufige, sam hē hine mydlinga lufige, Shrn. 194, 13. Wið wunda som hȳ sȳn of īserne, som hȳ sȳn of stence, oððe fram nædran, Lchdm. i. 166, 9. [Sam . . . sam *whether . . . or,* O. E. Homl. ii. 107, 8.]

sam- as a prefix denotes *agreement, combination.* v. sam-mǽle, -rǽd, -winnende, -wist. [*Icel.* sam-.]

sām- *half-* ; the prefix denotes *imperfection.* Cf. sǽmra. [O. Sax. sām- : O. H. Ger. sāmi- : Lat. semi- : Gk. ἡμι-.]

Samaringas, Samaritane, Samaritanisce ; *pl. The Samaritans :*—Innan Samaritana ceastre (in burgum ðæra Samaritanesca, Lind. ; in cæstra Samaringa, Rush.) *in civitates Samaritanorum,* Mt. Kmbl. 10, 5. Tō Samaritaniscum, Jn. Skt. Rush. Lind. 4, 9. v. next word.

Samaritanisc ; *adj. Samaritan, of Samaria :*—Ðā fērde sum Samaritanisc man wið hine, Lk. Skt. 10, 33. Ðes wæs Samaritanisc, 17, 16 ; Jn. Skt. 8, 48. Ðā cwæþ ðæt Samaritanisce wīf. . . . 'Ic eom Samaritanisc wīf ; ne brūcaþ Judéas and Samaritanisce metes ætgædere,' Jn. Skt. 4, 9. [O. H. Ger. Samaritanisc.]

sām-bærned ; *adj. Half-burnt :*—Sāmbærnd *semiustus,* Hpt. Gl. 508, 56.

sām-boren ; *adj. Born out of due time :*—Sāmboren *abortus,* Wrt. Voc. ii. 10, 6. Cf. ful-boren.

sām-bryce *a violation only partially effected :*—Tō hādbōte, ðar

sámbryce wurðe, bête man georne be ðam ðe seó dǽd sý, L. E. B. 9; Th. ii. 242, 9. *The term is in contrast with* ful-bryce *in the preceding sections.* v. sám-wyrcan.

sám-cwic, -cucu; *adj. Half-dead:*—Sum mǽden hê gehǽlde, ðæt ðe læg on legerbedde seóc, sámcucu geþúht, Homl. Th. ii. 510, 25. Hê sámcucu læg, Homl. Skt. i. 6, 164: L. Ælfc. C. 31; Th. ii. 354, 10. Hê (*Anthony*) bebeád ðæt hiene mon on ða ilcan byrgenne tó hiere (*Cleopatra*) swá sómcucre álegde, Ors. 5, 13; Swt. 246, 31. Hí forlêton hine sámcucene *semiuiuo relicto*, Lk. Skt. 10, 30. Sum môder bær hire sámcuce cild, Homl. Th. ii. 150, 16. [*O. Sax.* sám-quik: *O. H. Ger.* sámi-quek.]

same (*always in combination with* swá); *adv. Similarly, in the same way.* (1) Swá same:—And eft Lǽdenware swá same wendon ealla on hiora ágen geþeóde *and again the Romans in the same way translated all into their own language,* Past. pref.; Swt. 6, 3. Deós wyrt is swýðe scearpnumul wunda tó gehǽlenne, swá ðæt ða wunda hrǽdlíce tógædere gáþ; and eác swá some hió gedéþ ðæt flǽsc tógædere clifaþ, Lchdm. i. 134, 12: Elen. Kmbl. 2553; El. 1278. Ðæt hié lufan Dryhtnes and sybbe swá same sylfra betweónum gelǽston, 2411; El. 1207: 2565; El. 1284. On Adame and on his eafrum swá some, Cd. Th. 25, 24; Gen. 399. Is ðæt fýr swá same on ðam wætre and on stánum eác, Met. 20, 150: 24, 33. Deór efne swá some faraþ, Exon. Th. 358, 30; Pa. 53. (2) Swá same swá:—Hú ne forealldodon ða gewritu swá some swá ða wríteras dydon, Bt. 18, 3; Fox 66, 1. Twá ðara gecyndu habbaþ nêtenu swá same swá men, 33, 4; Fox 132, 5. Ðǽr wífmenn feohtaþ swá same swá wǽpnedmen, Ors. 2, 4; Swt. 76, 27. [*O. Sax.* so sama, só sama só: *O. H. Ger.* sama, só sama, só sama só.]

samen; *adv. Together:*—Giurnan tuoege somen (*simul*), Jn. Skt. Rush. 20, 4. Wêrun somen Simon Petrus and Didimus, 21, 2. [*Baþe* samenn, Orm. 377. Sitte samen, R. Brun. *Goth.* samana: *O. Sax.* saman: *O. Frs.* samin, semin: *O. L. Ger.* samen, samon: *O. H. Ger.* saman *simul:* Icel. saman.]

sám-grêne; *adj. Half-green, backward* (of a plant):—Spelt sámgrêne *far serotina,* Wrt. Voc. ii. 36, 41.

sám-hál; *adj. Not in perfect health, weak:*—Nú ne beóþ náht fela manna ætsamne, ðæt heora sum ne sí seóc and sámhál, Wulfst. 273, 10. [*O. H. Ger.* sámi-hail *debilis.*]

sam-heort; *adj. Of one heart, of the same disposition;* concors:—Singaþ samheorte sangas Dryhtne, Ps. Th. 149, 1.

sam-híwan; *pl. Members of the same household or family:*—Samhíwna yrfebêc *jus liberorum,* Wrt. Voc. i. 20, 46. Somhíwena yrfebêc, ii. 49, 14.

sam-hwilc; *pron. Some:*—Þeówne .lx. Somhwelcne fíftegum (mid fíftig, MSS. B. H.) the '*wer*' *for the* '*þeów*' *is* 60 *shillings. For one kind it is* 50 (?), L. In. 23; Th. i. 118, 4. Swá hwæt swá ús God sylle mǽre ðonne wê nêde brúcan sceolan . . . , ne sylþ hê hit ús tó ðon ðæt wê hit hýdon, oððe tó gylpe syllan samhwylcum mannum ðe náht swíðe God ne lufiaþ, Blickl. Homl. 53, 17. Cf. swá hwilc.

sám-lǽred; *adj. Imperfectly taught:*—Wê lǽraþ ðæt ǽnig gelǽred preóst ne scænde ðone sámlǽredan, ac gebête hine gif hê bet cunne, L. Edg. C. 12; Th. ii. 246, 19. Hieronimus ádwæscte ða dwollícan gesetnysse ðe sámlǽrede men sǽdon be hire forþsíðe, Homl. Th. ii. 438, 6. *Barbarismus* and *solocismus* bêcumaþ of ðam sámlǽredum leáslíce geclypode oððe áwritene, Ælfc. Gr. 50, 22; Som. 51, 52.

sam-mǽle; *adj. Agreed, come to an agreement:*—Gif hý ðonne ǽlces þinges sammǽle beón *if they then be agreed in everything,* L. Edm. B. 6; Th. i. 254, 19. Ðæt dóm stande ðǽr þegenas sammǽle beón, L. Eth. iii. 13; Th. i. 298, 3. Hêr swutelaþ on ðisum gewrite hú Wulfríc and Ealdrêd wǽron sammǽle ymbe ðæt land æt Clife, Cod. Dip. Kmbl. ii. 300, 5. Dene and Engle wurdon sammǽle æt Oxnaforda, Chr. 1018; Erl. 161, 16. [Cf. *Icel.* sam-mǽli *an agreement;* sam-mǽlask á eitt *to agree in a thing.*] Cf. mǽlan, mǽl.

sám-milt, -melt; *adj. Half-digested:*—Se geþigeda mete hefegaþ ðone magan, and hê ðone sámmeltan (*the half-digested food*) þurh ða wambe út sent, Lchdm. ii. 186, 22. v. miltan.

-samne. v. æt-, tó-samne.

samnian; *p.* ode. I. *v. trans.* (1) *to collect, assemble, bring together, gather:*—Ða swêtestan somnaþ and gædraþ wyrta wynsume and wudublêda *colligit succos et odores divite silva,* Exon. Th. 211, 6; Ph. 193. Somnas his huǽte *congregabit triticum suum,* Mt. Kmbl. Lind. 3, 12. Nát hwam hit gaderaþ ł somnaþ ða, Ps. Spl. 38, 10. Hê ðyder folc samnode, Cd. Th. 230, 5; Dan. 228. Hié here samnodon, Andr. Kmbl. 2250; An. 1126. Wê somnadon ł geadredon ða *colligimus ea,* Mt. Kmbl. Lind. 13, 28. Sommas (somnigas, Rush.) ða ðe hiá gelǽfdon, Jn. Skt. Lind. 6, 12. Swylce man fyrde trymme and samnige, Blickl. Homl. 91, 32. Fyrde somnian, Chr. 1016; Erl. 154, 2. Folc somnigean, Cd. Th. 191, 19; Exod. 217. (2) *to draw together, join, unite:*—Ðonne samnaþ hió ða wunde and hǽlþ, Lchdm. ii. 22, 11. (3) *to get materials together for a poem to compose:*—Ic ðysne sang fand samnode wíde *I was author of this poem, gathered its matter far*

and wide, Apstls. Kmbl. 4; Ap. 2. Ne wêne ðæs ǽnig ælda cynnes, ðæt ic lygewordum leóþ somnige (*that I compose my lay of lying words*), write wôðcræfte, Exon. Th. 234, 29; Ph. 547. II. *intrans.* (1) *to collect, assemble, come together:*—Sellendum ðê him hí somniaþ *dante te illis, colligent,* Ps. Spl. 103, 29. Somnode *conglobatur,* Wrt. Voc. ii. 19, 34: 91, 20. Duguþ samnade, Andr. Kmbl. 250; An. 125. Mægen samnode, Elen. Kmbl. 110; El. 55: 120; El. 60. Hí gederedon ł somnodon tógeánes mê *convenerunt adversum me,* Ps. Spl. 30, 17. (2) *to draw together, join, unite:*—Ðá weóxon ða fýr swýðe and hí tógædere þeóddon and samnedon óþ ðæt ða hí wǽron on ǽnne unmǽtne lêge geánede and gesomnade *crescentes vero ignes usque ad invicem sese extenderunt, atque in inmensam adunati sunt flammam,* Bd. 3, 19; S. 548, 21. (3) *to glean:*—Hê mid his sceáfe ne mæg sceát áfyllan ðeáh ðe hê samnige swíðe georne *non implevit sinum suum qui manipulos colligit,* Ps. Th. 128, 5. [*Laym.* somnien, sumnien: *Orm.* sammnenn: *O. Sax.* samnôn: *O. Frs.* samena, somnia: *O. H. Ger.* samanôn: *Icel.* samna.] v. ge-samnian.

samnung, e; *f. An assembly, council:*—Somnung *synagoga,* Mk. Skt. Lind. Rush. 1, 23: Lk. Skt. Lind. Rush. 4, 15, 16: *concilium,* Mk. Skt. Lind. Rush. 14, 55: Lind. 15, 1: Mt. Kmbl. Lind. 26, 59: *congregatio,* Rtl. 173, 3. v. ge-samnung.

samnunga, sæmninga, semninga; *adv. All at once, on a sudden, suddenly, forthwith, immediately;* continuo, subito, repente:—And ða hig ðæt sprêcon samninga (samnunga, MSS. A. B.) se hana creów *et continuo athuc illo loquente cantauit gallus,* Lk. Skt. 22, 60. Hí hine samnuncga (*subito*) scearpum strêlum on scotiaþ, Ps. Th. 63, 4. Ða ásceán samninga mycel leóht, Blickl. Homl. 145, 12. Somnunga, 239, 31. Hié sume semninga sweltaþ, Lchdm. ii. 176, 9. Sæmninga, Blickl. Homl. 141, 27. Ðis is feáwra manna dǽd, ðæt hí ealle eorþlíce þing sæmninga forlǽtan mágon, Homl. Th. ii. 398, 33. Hí semninga sneóme forwurdon *subito defecerunt et perierunt,* Ps. Th. 72, 15: Bd. 1, 7; S. 477, 1. Ða ástód hê semninga *exsurrexit repente,* 2, 9; S. 511, 20. Ða geseah hê semninga (*subito*) mon wið his gangan, 2, 12; S. 513, 34. Hit semninga (*subito*) on ús rǽsde, Nar. 15, 19, 11. Ða cômon semninga twegen englas, Blickl. Homl. 221, 27: Exon. Th. 257, 5; Jul. 242: Beo. Th. 3284; B. 1640. Oþ ðæt semninga sunu Healfdenes sêcean wolde ǽfenreste, 1293; B. 644. Hê (*the whale*) semninga on sealtne wǽg niþer gewíteþ, Exon. Th. 361, 29; Wal. 27. Ðá wæs semninga geworden mycel þunorrád, Blickl. Homl. 145, 28: Exon. Th. 31, 5; Cri. 491. Mec semninga slǽp ofergongeþ, 422, 22; Rä. 41, 10: Andr. Kmbl. 927; An. 464: 1639; An. 821.

samnung-cwide, es; *m. A collect:*—Somnungcwido *collecta,* Rtl. 2, 1.

samod; *adv. Together.* I. marking association in joint action:—Ealle hí áhyldon samod onnitte gewordene sint *omnes declinaverunt, simul inutiles facti sunt,* Ps. Spl. 13, 4. Ða unrihtwísan forweorðaþ samod (*simul*), 36, 40. Cumaþ út samod Ilfing and Wisle (*the two rivers have a common channel*), Ors. 1, 1; Swt. 20, 10. Stôd his handgeweorc (*Adam and Eve*) somod on sande, nyston sorga wiht tó begnornianne, Cd. Th. 16, 12; Gen. 242. Ne beóþ wê leng somed, 168, 20; Gen. 2785. Somod eardedon Meotudes bearn and se monnes sunu, Exon. Th. 8, 30; Cri. 125. Tó gebede feóllon sinhíwan somed, Cd. Th. 48, 19; Gen. 778. Samed síþian, Exon. Th. 434, 17; Rä. 52, 2. I a. of mutual or reciprocal action:—Hié fela sprǽcon sorhworda somed, Cd. Th. 49, 8; Gen. 789. Cf. samod-geflit. I b. marking union or junction. v. samod-cumende. II. with numerals or with *eall:*—Him wæs bám samod lond gecynde, Beo. Th. 4399; B. 2196. Ðendan bú somod, líc and sáwle, lifgan môte, Exon. Th. 81, 20; Cri. 1326. Þreó tácen somod, 76, 7; Cri. 1236. Seofon winter samod *seven years in unbroken succession,* Cd. Th. 256, 11; Dan. 639. Ic eów bidde ðæt gê mê secgan hwylce gemete gê cóman ealle samod tó mê, Blickl. Homl. 143, 20. Hê eal innan samod forswǽled wæs *within he was one mass of inflammation,* Homl. Th. i. 86, 5. III. marking association of similar objects or circumstances, with nearly the force of *and, both . . . and, also, too:*—Somod *jamque,* Wrt. Voc. ii. 45, 31. Weras wíf samod *men and women,* Andr. Kmbl. 3330; An. 1668. Weras, heora wíf somed, Cd. Th. 146, 7; Gen. 2418. Hê ðone healsbeáh gesealde, þrió wicg somod, Beo. Th. 4355; B. 2174. Ðú geworhtest heofon and eorþan, sǽs sídne fæðm, samod ealle gesceaft, Elen. Kmbl. 1455; El. 729. Ongan his feax teran and his hrægl somod, Judth. Thw. 28; Jud. 282. Somod for his hǽlo ðæs cyninges and ðǽre þeóde ðe hê fore wæs *pro salute illius, simul et gentis cui praeerat,* Bd. 2, 12; S. 512, 29. Niht somod and dæg, Cd. Th. 239, 25; Dan. 375. Swylce ic his willan wylle sêcean, samed (*also, likewise, at the same time*) andettan . . . , Ps. Th. 110, 2. Ðú ðínra bearna bearn sceáwige; geseó samed gangan sibb ofer Israhêl, 127, 7: Exon. Th. 69, 16; Cri. 1122. IV. in combination with *ætgædere, mid:*—Sende mihtig God his milde gehigd and his sóðfæst môd samod ætgædere, Ps. Th. 56, 4: 88, 21. Ðǽr wæs sang and wópad samod ætgedere, Beo. Th. 2131; B. 1063. Gáras stódon samod ætgædere, 662; B. 329. Ðú ðe samod mid mê swête gripe metas *qui simul mecum dulces capiebas cibos,*

Ps. Spl. 54, 15. Graton samod mid ðǽm cnihtum feóll tó Johannes fótum, Homl. Th. i. 62, 17. Cwom samod mid ðǽm swylce Assur *etenim Assur simul venit cum illis*, Ps. Th. 82, 7. Ða ðe someð mið hine āstigun *quae simul cum eo ascenderant*, Mk. Skt. Rush. 15, 41. Hē gesette ðone mónan fulne on eástdǽle mid scīnendum steorrum samod, Lchdm. iii. 238, 28. **IV a. with *anlíce* :—**Hí me ymbsealdon samod anlíce swā beón *circumdederunt me sicut apes*, Ps. Th. 117, 12 : 142, 4 : 147, 5. Samod anlíce . . . swā swā *sicut*, 123, 6. Samod anlícast swā *velut*, 78, 2 : *ut*, 91, 11 : *sicut*, 127, 4. **V.** translating the prefix *con-* in Latin words :—Ic samod awende *converto*, Ælfc. Gr. 37 ; Som. 39, 14. Ic samod cume *convenio*, Som. 39, 5. Ic samod fealde *complico*, 24 ; Som. 25, 52. Ic samod fealle *concido*, 28 ; Som. 32, 62. Ic samod fleó *confugio*, Som. 32, 49. Ic samod wurpe *conicio*, Som. 32, 40. Somud mengaþ wē *comminiscimur*, Wrt. Voc. ii. 18, 7. Somod geþwǽrende *concordantes*, 24, 8. [*Laym. A. R.* somed : *Goth.* samath : *O. Sax.* samad, samod.] v. next word.

samod ; *prep. with dat.* With, at :—Samod ǽrdæge (*with the coming of the dawn*) eode æþele cempa self mid gesíðum, Beo. Th. 2627 ; B. 1311. Frófor eft gelamp sārigmódum somod ǽrdæge *with day came comfort to the sadhearted*, 5877 ; B. 2942. Cf. mid ǽrdæge.

samod-cumende *flocking together :*—Samadcumendum folcum *populis confluentibus*, Hpt. Gl. 455, 71 : 518, 45.

samode (?) :—Tala . . swylce ic nǽfre on eallum ðǽm fyrngewritum findan ne mihte sóðe samode [samnode (?) *collected* or (?) *composed*, v. samnian, I. 3. Or cf. (?) *Icel.* semja (kvǽði, bók) *to compose* (*a poem, book*)], Salm. Kmbl. 17 ; Sal. 9.

samod-eard, es ; *m. A common country :*—Git (*Guthlac and his sister*) ā mósten in ðam ēcan gefeán mid ða sibgedryht somudeard niman, Exon. Th. 184, 19 ; Gū. 1346.

samod-fæst ; *adj. Fast joined together :*—Sceal onettan, se ðe āgan wile líf æt Meotude, ðenden him leoht and gǽst somodfæst seón, Exon. Th. 96, 28 ; Cri. 1581.

samod-geflit, es ; *n. Strife, conflict :*—Somodgeflit *concertatio*, Wrt. Voc. ii. 24, 18.

samod-gesíþ, es ; *m. A companion, comrade :*—Samodgesíþ *coheres*, Germ. 400, 575.

samod-herung, e ; *f. A praising :*—Samodhering *conlaudatio*, Blickl. Gl.

[**samodlíce ;** *adv. Together, unitedly in a body :*—Iedon ealle samodlíce tó ðone kyng, Chr. 1123 ; Erl. 250, 10.]

samod-rynelas ; *pl.* translates the technical term *concurrentes :*—Ða *concurrentes* synt samodrynelas genemned, Anglia viii. 302, 10.

samod-swégende *translates the Latin* consonantes :—Ða óðre stafas syndon gehātene *consonantes*, ðæt is, samodswégende, forðan ðe hí swégaþ mid ðǽm fíf clypiendlícum, Ælfc. Gr. 2 ; Som. 2, 49.

samod-þyrlíc ; *adj. Concordant :*—Somodðyrlíce *concordi*, Wrt. Voc. ii. 22, 13. Cf. (?) ge-þweran.

samod-willung, e ; *f. A boiling together, condensing :*—Somodwellunge *concretione*, Wrt. Voc. ii. 23, 19.

samod-wunung, e ; *f. A living together :*—Him is tó forbeódenne ǽghwilc gemāna . . ge ǽt, ge drinc, ge samodwunung on hūsum, L. E. I. 26 ; Th. ii. 422, 31.

samod-wyrcende *co-operating :*—Somodwyrcendum *cooperante*, Wrt. Voc. ii. 24, 76.

sam-rád ; *adj. Harmonious, united :*—Se cræftga gefērscipas fæste gesamnaþ ðæt hí hiora freóndscipe forþ on symbel untweófealde treówa gehealdaþ sibbe samráde *the mighty one unites societies firmly, so that for ever they continue to maintain their friendship, faith sincere, peace unbroken*, Met. 11, 96. Cf. ge-rád.

sámran, Bt. 33, 1 ; Fox 120, 12. v. sǽmra.

sám-soden ; *adj. Half-cooked :*—Gif man āwiht blódiges þicge on healfsodenum (sámsodenum, MSS. X. Y.) mete *si quis cruentum quid comederit in semicocto cibo*, L. Ecg. C. 40 ; Th. ii. 166, 2.

sám-swéled ; *adj. Half-burnt :*—Sámswǽlede *semiustus*, sámswǽled *semiustus*, Hpt. Gl. 508, 55–57.

sam-tinges (sǽm-, sem-) ; *adv. In close connection* (as regards time), *immediately, forthwith,* continuo :—Meahtest ðē full recen on ðǽm rodere upan siððan weorþan, and ðonne samtenges æt ðǽm ǽlcealdan steorran, Met. 24, 18. Swā hraðe swā ðæt wolcn styrode, swā síðode samtinges eal seó fyrd after ðam wolcne, Homl. Th. ii. 196, 11. Ðā nolde hē hí sǽmtinges ācwellan ac lēt him fyrst *he would not kill them immediately, but allowed them time*, 424, 14. Ðā āwurpon ða hǽðenan sóna heora gedwyld, and tó heora Scyppende sǽmtinges gebugon, 510, 3 : 230, 18. Ðæt man hí ofslóge sǽmtinges ealle, Anglia ix. 32, 165. Snáw cymþ of ðam þynnum wǽtan ðe byþ gefroren ǽr ðan hē tó dropum geurnen sy, and swā semtinges (sǽm-, MS. P.) fylþ, Lchdm. iii. 278, 25. [Cf. *Icel.* sam-tengja *to join, consent* ; sam-tenging *a connection.*] v. tengan, ge-tenge.

sam-winnende *struggling together :*—Ða samwinnendan *conluctantia, depugnantia*, Wrt. Voc. ii. 134, 62. [Cf. *Icel.* sam-vinnandi *working together.*]

sám-wís ; *adj. Dull, foolish :*—Wēnaþ sámwíse (cf. ða dysegan men, Bt. 32, 3 ; Fox 118, 22) ðæt hí on ðis lǽnan mǽgen lífe findan sóþa gesǽlþa, Met. 19, 34. Ða sámwísan (*hebetes*) sint tó manianne ðæt hié wilnien tó wiotonne ðæt ðæt hié nyton, Past. 30, 1 ; Swt. 201, 7. Cf. med-wís.

sam-wist, e ; *f. A living together, cohabitation, matrimony :*—Samwist *jugalitas*, Hpt. Gl. 438, 63. Samwiste *matrimonii*, 481, 36 : *copulae, connubii*, 485, 57 : *copulae*, 508, 75. Samwiste *contubernium*, 511, 76. Ne ceara ðū (*Hagar*) fleáme dǽlan somwist incre, Cd. Th. 137, 27 ; Gen. 2280. þeáh his líc and gǽst hyra somwiste, sinhíwan tú, gedǽled (-de ?), Exon. Th. 160, 9 ; Gū. 941. Somwist, 172, 28 ; Gū. 1150. Samwista *contubernia*, Hpt. Gl. 416, 27 : 520, 54. [*O. H. Ger.* samwist : *Icel.* sam-vist.]

sám-worht. v. sám-wyrcan.

sam-wrǽdnes, e ; *f. Combination, union :*—Eall ðæt ðætte ānnesse hæfþ þæt wē secgaþ ðætte síe ða hwíle ðe hit ætsomne biþ and ða samwrǽdnesse wē hātaþ gód *everything that has unity, that, we say, exists, while it maintains its unity, and the union of its parts we call good ; omne, quod est, unum esse, ipsumque unum bonum esse didicisti,* Bt. 37, 3 ; Fox 190, 23. Cf. wrǽd, wrǽd-mǽlum.

sám-wyrcan *to do a thing incompletely :*—Gif hwā on fyrde griðbryce fulwyrce . . . Gif hē sámwyrce . . ., L. C. S. 62 ; Th. i. 408, 23. [Cf. sám-bryce.] Fæsten wæs sámworht *the fort was not finished*, Chr. 892 ; Erl. 88, 34. Stāntorr (*the tower of Babel*) sámworht stód, Cd. Th. 102, 16 ; Gen. 1701.

sanct ; *m. A saint :*—Hē wæs on lífe eorþlíc cing, hē is nū æfter deáþe heofonlíc sanct, Chr. 979 ; Erl. 129, 10. Ða mynstermenn noldon ðone sanct underfón, Swt. A. S. Rdr. 100, 149. Hē gesóhte ðone sanct, Glostr. Frag. 6, 8 : 8, 10. Ðǽr habbaþ englas eádigne dreám, sanctas singaþ, Cd. Th. 286, 20 ; Sat. 355 : 279, 18 ; Sat. 240. Ðý ylcan dæge ealra wē healdaþ sancta symbel, Menol. Fox 367 ; Men. 200. *The Latin forms* sanctus, sancta (*also* sancte) *are used before proper names :*—Sanctus Johannes, se mon Sancte Johannes, Sanctus Johannes líf, Blickl. Homl. 163. Sancta Maria, 5, 30. Sancta Marian (*gen.*), 165, 27.

sand, es ; *m.* [? *or should the passages that follow be put under* sand ; *f. ? cf. the later application of* witness *to a person*] *A messenger, envoy :*—Ðā wæs Lýfing ᵬ. mid ðam kincge . . . Ðā com Xᵖes cyrc sand tó ðam ᵬ. and hē forð (fór ?) ðā tó ðam kincge *bishop Lyfing was then with the king . . . Then came a messenger* (or *message* ?) *from Christchurch to the bishop, and he* (the bishop) *went then to the king*, Chart. Th. 339, 26. Dæg byþ Drihtnes sond deóre mannum mǽre Metodes leóht *day is the Lord's messenger* (or *message* ?) *dear to men, God's glorious light*, Runic pm. Kmbl. 344, 9 ; Rūn. 24. On ðís ylcan geáre com ðæs Pāpan sande (sand ?) hider tó lande ; ðæt wæs Waltear bisceop *in the same year came the Pope's legate to this country ; that was bishop Walter*, Chr. 1095 ; Erl. 232, 28. [Here sandes feórðen betwyx heom and hí togædere cómen and wurðe sǽhte *their envoys went between them, and they came together and were reconciled*, 1135 ; Erl. 261, 20. Sonden commen betwenen ðe soðe word me seiden, Laym. 4651. Euerich wo is Godes sonde. Heie monnes messager, me schal heiliche underuongen, A. R. 190, 15. In alle our neoden sendeð þeos sonden (*prayers*) touward heouene, 246, 22.]

sand, e ; *f.* **I.** *a sending, mission, message :*—Paulus cwæð : ' Ðā ðā ðǽra tída gefyllednys com, ðā sende God Fæder his sunu tó mancynnes ālýsednysse.' Seó wurðfulle sand wearð on ðisum dæge gefylled, Homl. Th. i. 194, 17. Gregorius is rihtlíce Engliscre þeóde apostol, forðan ðe hē þurh his rǽd and sande ūs fram deófles biggengum ætbrǽd, ii. 116, 28. Nū com ic tó eów þurh ðæs Almihtigan sande, 296, 20. Ðes ylca apostol becom þurh Godes sande tó Ethiopian, 472, 11. [*Laym.* sande, sonde *a message* ; sondes mon *a messenger* : *Orm.* sanderr-man : sander-men, Chr. 1135 ; Erl. 249, 28 : *C. M.* sandir-men : sander-bodes, *O. E. Homl.* ii. 89, 22 : *Prompt. Parv.* sond or sendynge *missio* : sond or ȝyfte sent *eccenium* : *O. H. Ger.* -santa, santi- *missio*, Grff. vi. 239.] v. onsand. **II.** *a mess* (from Latin *mitto*), *a dish of food, victuals :*—Wista *vel* sand *dapes* vel *fercula*, Wrt. Voc. i. 26, 63. Sand *daps*, 82, 64 ; Ælfc. Gr. 9, 54 ; Som. 13, 20. Godes engel cwæð : ' Abacuc, bær ðone mete tó Babilone ' . . . Ðā clypode se Abacuc : ' Ðū Godes þeówa, nim ðās lāc ðe ðē God sende ' . . . And hē ðā ðǽre sande breác, Homl. Th. i. 572, 8. Ðā genemnode se hālga wer ðæt wíf ðe hí gelaðode, and ða sanda tealde ðe heó him gebær, ii. 168, 5. Sanda *obsonia*, Germ. 394, 297. Sandae, sondae *commeatos*, Txts. 46, 188. Sanda *ferculorum, epularum*, Hpt. Gl. 444, 57. [Of everilc sonde . . . most and best he gaf Benjamin, Gen. and Ex. 2295.] v. preceding word.

sand, es ; *n.* **I.** *sand, gravel :*—Sand *glarea, glitis*, vel *samia*, Wrt. Voc. i. 22, 8 : *arena*, 37, 32. Sande *sablo*, ii. 89, 36. Hē behídde hyne on ðam sande (*sabulo*), Ex. 2, 12. Sume men secgan ðæt seó eá síe eást irnende on ðæt sond, and ðonne besince eft on ðæt sand, and ðǽr nēh síe eft flówende up of ðam sande, Ors. 1, 1 ; Swt. 12, 20–23. Ða tódǽlaþ ðæt wæsmbǽre land and ðæt deádwylle sand ðe syððan līþ

súþ on ðone gársecg *qui dividit inter vivam terram et arenas jacentes usque ad oceanum*, Swt. 26, 19. **II.** *sand by the sea, sands, sea-shore :*—Sand sǽ *arena maris*, Ps. Spl. 77, 31. Sǽfaroþa sand, Cd. Th. 236, 18; Dan. 323. On sande *on the shore of the Red Sea*, 315, 5; Excd. 302. Nacan on sande, Beo. Th. 596; B. 295 : 3796; B. 1896. Gewát him se hearda æfter sande sǽwong tredan, 3932; B. 1964. Ic wæs be sande sǽwealle neáh, Exon. Th. 471, 14; Rä. 61, 1. Swá swá hradu ýst windes scip tóbrycþ on ðam sandum neáh ðære byrig ðe Tarsit hátte, Ps. Th. 47, 6. [*O. Sax. O. Frs.* sand : *O. H. Ger.* sant *arena, sabulum: Icel.* sandr.] γ. eolh-sand.

sand-beorh *a sand-hill, sand-bank :*—Ondlong weges tó sondbeorge, Cod. Dip. Kmbl. iii. 402, 11. Sondbeorgum ymbseald, Exon. Th. 360, 23; Wal. 10. Se ðe wille fæst hús timbrian ne sceall hé hit nó settan up on ðone héhstan cnol and eft se ðe wille fæst hús timbrian ne sette hé hit on sondbeorhas *quisquis volet perennem cautus ponere sedem, montis cacumen alti, bibulas vitet arenas*, Bt. 12; Fox 36, 11. Sondbeorgas, Met. 7, 10.

sand-ceosol, es; *m. Sand, gravel :*—Sandceosel *arena*, Wrt. Voc. i. 80, 64. Sandcesel, 54, 32. Sandceosol on sǽ *arenam in litore maris*, Gen. 22, 17. Sandceosol on sǽstrande, Jos. 11, 4. Sandceosol on sǽlicum strande, Homl. Th. ii. 62, 9. Sandcysel, Wulfst. 198, 22. Hé getimbrode hys hús ofer sandceosol *supra arenam*, Mt. Kmbl. 7, 26. Hí beóþ gemenigfylde ofer ðære sǽ sandceosol *they shall be multiplied above the sand of the sea*, Homl. Th. ii. 524, 21. [Cf. *Ger.* kiesel-sand *gravel.*]

sand-corn *a grain of sand :*—Gif míne synna and mín yrmþ wǽron áwegene on ánre wǽgan, ðonne wǽron hí swǽrran gesewene ðonne sandcorn on sǽ, Homl. Th. ii. 454, 24. Swá fela welena swá ðara sondcorna beóþ be ðisum sǽclifum *quantas pontus versat arenas*, Bt. 7, 4; Fox 22, 27. Hí beóþ ofer sandcorn manige *super arenam multiplicabuntur*, Ps. Th. 138, 16. [*Icel.* sand-korn.]

sand-geweorp, es; *n. A sand-bank, quicksand :*—Sandgewurp *syrtis*, Wrt. Voc. i. 63, 72. On sandgeweorp *in sirtim*, ii. 45, 66. [Cf. *O. H. Ger.* sant-wurfi *syrtis*.] v. next word and sand-hrycg.

sand-gewyrpe, es; *n. A sand-heap :*—Tó sandgewyrpe, of sandgewyrpe út an Temese, Cod. Dip. Kmbl. vi. 228, 25.

sand-grot *a grain of sand :*— Geríman sǽs sondgrotu, Exon. Th. 466, 6; Hö. 117.

sand-hliþ *a sand-hill by the sea :*—Gewát him ofer sandhleoþu tó sǽs faruþe, Andr. Kmbl. 471; An. 236.

sand-hof *a house in the sand, the grave :*—Líc orsáwle sceal in sondhofe wunian, Exon. Th. 173, 31; Gú. 1169.

sand-hrycg *a sand-bank :*—Ðes sandhrycg *haec syrtis*, Ælfc. Gr. 9, 78; Som. 14, 34.

sand-hyll *a sand-hill :*—Sondhyllas *alga* (cf. waar *alga*, 99, 69, wára *sablonum*, strand *sablo*, Hpt. Gl. 502, 76), Wrt. Voc. ii. 99, 73.

sandig ; *adj. Sandy :*—Sandig *arenosa*, sandegum *arenosis*, Hpt. Gl. 502, 73, 75. Sandigum, 449, 25. Deós wyrt wihst on sandigum landum, Lchdm. i. 94, 7 : 100, 16.

sandiht ; *adj. Sandy, dusty :*—Hiora gemitting wæs on sondihtre dúne, ðæt hié for duste ne mehton geseón, hú hí hí behealdan sceolden, Ors. 5, 7; Swt. 230, 15. Of ðam stáne on ðone sandihtan hærepoþ, Cod. Dip. Kmbl. iii. 453, 22.

sand-land *the sea-shore :*—Se hærnflota (*the ship*) æfter sundplegan sondlond gespearn, grond wið greóte, Exon. 182, 11; Gú. 1308.

sand-rid *a quick-sand :*—Sandrid *syrtes*, Wrt. Voc. i. 57, 19. v. rid, *and* cf. sand-geweorp.

sand-seáþ *a sand-pit :*—Ofer ðene héþ inn on ðam sandseápe; of ðam sandseápe, Cod. Dip. Kmbl. iii. 384, 26. Of ðære sǽc on ða sandseápas, 80, 2 : 169, 4.

sang, es; *m.* **I.** *song, singing, (a)* of human or angelic beings :—Sárlíc sang *trenos* (θρῆνος), Wrt. Voc. i. 28, 18. Twegra sang *bicinium*, 25. Ungeswége sang *diaphonia*, 34. Geþwǽre sang *armonia*, 39. Auswége sang *simphonia*, 40. Wuldres weard wordum herigaþ þegnas . . . þǽr is sang æt selde, Cd. Th. 306, 12; Sat. 663. Ðǽr wæs sang and swég samod ætgædere . . . gomenwudu gréted, gid oft wrecen, Beo. Th. 2130; B. 1063 : 180; B. 90. Ðǽr wæs singal sang and swegles gong, wlitig weoroda heáp, Andr. Kmbl. 1737; An. 871. Ðǽr is engla song, eádigra blis, Exon. Th. 100, 31; Cri. 1650. Magister cyriclíces sanges *magister ecclesiasticae cantionis*, Bd. 2, 20; S. 522, 27. Songes magister *cantandi magister*, 4, 2; S. 565, 38. Ðá hé ðá ðis leóþ ásungen hæfde, ðá forlét hé ðone sang, Bt. 24, 1; Fox 80, 5. Ðǽr (*in heaven*) wé hálgan Gode sang ymb seld secgan sceoldon, Cd. Th. 279, 9; Sat. 235. Gesǽton sigerófe sang áhófon *lifted up their voices in song*, Elen. Kmbl. 1733; El. 868. (*b*) of birds or animals :—Winsum sanc (*of birds*), Met. 13, 50. Fugla cynn songe lofiaþ módigne, Exon. Th. 221, 20; Ph. 337. Mǽwes song, 406, 25; Rä. 25, 6. Earn sang áhóf, Elen. Kmbl. 58; El. 29. Wulf sang áhóf, 224; El. 112. (*c*) of sound caused by inanimate things; v. býme-sangere, sang-cræft, singan :—Ealle hearpan strengas se hearpere grét mid ánre honda, ðý hé wile ðæt hí ánne song singen, ðeáh hé hié ungelíce styrige *idcirco chordae consonam*

modulationem reddunt ; quia uno quidem plectro, sed non uno impulsu feriuntur, Past. 23; Swt. 175, 9. **II.** *a singing, chanting :*—Se biscop and se mæssepreóst sceolan mæssan gesingan . . . and ða ðe on heofenum syndon, hí þingiaþ for ða ðe ðyssum sange fylgeaþ, Blickl. Homl. 45, 36. **III.** *song, poetry.* v. sang-cræft. **IV.** *a song, a poem to be sung* or *recited :*—Se hálga song gehýred wæs, Exon. Th. 181, 23; Gú. 1297. Ðá hæfde hé mé gebunden mid ðære wynnsunnesse his sanges *me carminis mulcedo defixerat*, Bt. 22, 1; Fox 76, 6. Mé Gúðhere forgeaf máþþum songes tó leáne, Exon. Th. 322, 22; Víd. 67. Galan sigeleásne sang, Beo. Th. 1578; B. 787. Ðonne hé gyd wrece, sárigne sang, 4885; B. 2447. Ic ðysne sang (*the poem which follows*) fand, Apstls. Kmbl. 1; Ap. 1. Word sanga *verba cantionum*, Ps. Spl. 136, 3. Singaþ ús ymnum ealdra sanga ðe ge on Sione sungan *hymnum cantate nobis de canticis Sion*, Ps. Th. 136, 4. Sangum *carminibus*, Hpt. Gl. 519, 50. Singaþ sangas Drihtne and him neówne sang singaþ *cantate Domino canticum novum*, Ps. Th. 149, 1 : 95, 1. [*Goth.* saggws: *O. Sax.* sang : *O. Frs.* song: *O. H. Ger.* sang : *Icel.* söngr. v. ǽlen-, brýd-, byrig-, cyric-, dæg-, dægréd-, foranniht-, galdor-, heáf-, hearp-, líc-, lof-, mæsse-, middæg-, niht-, nón-, offrung-, prím-, sealm-, tíd-, úht-, undern-, wóþ-, yfel-sang.]

sang, *song a bed :*—Song t bedd *stratum*, Mk. Skt. Lind. 14, 15 : Lk. Skt. Lind. 22, 12. [*Icel.* sæing, sæng: *Dan.* sæng: *Swed.* säng *a bed.*]

sang-bóc ; *f.* **I.** *a music-book, a book with the notes marked for singing :*—Nota ðæt is mearcung. Ðæra mearcunga sind manega and mislíce gesceapene, ægðer ge on sangbócum ge on leóþcræfte, Ælfc. Gr. 50, 15; Som. 51, 20. **II.** *one of the service books, containing* '*besides the canticles, the hymns which were used in the Anglo-Saxon churches.*' v. Maskell's Monumenta Ritualia, i. cii :—Ðæt synd ða hálgan béc . . . , L. Ælfc. C. 21; Th. ii. 350, 13. Mæssepreóst sceal habban . . . sang-béc . . . , L. Ælfc. P. 44; Th. ii. 384, 1. Nú sindon ðǽr (*in the church at Exeter*) ii. fulle sangbéc, Chart. Th. 430, 8. [*Icel.* söng-bók.]

sang-cræft, es; *m.* **I.** *the art of singing, music* (*vocal or instrumental*) :—Sangcræft *musica* (in a list of the arts), Hpt. Gl. 479, 46. Wæs he swýðest on cyricean sangcræft getýd Rómánisce þeáwe *maxime modulandi in ecclesia more Romanorum peritum*, Bd. 4, 2; S. 566, 19. On sangcræft geléred *cantandi sonos edoctus*, 5, 20; S. 646, 6. **I** *a. an art of singing :*—Biþ ðæs hleóþres swég (*the voice of the Phenix when singing*) eallum songcræftum swétra and wlitigra, and wynsumra wrenca gehwylcum, Exon. Th. 206, 25; Ph. 132. **II.** *the art of composing poetry :*—Hé (*Cædmon*) þurh Godes gife ðone sangcræft onféng *gratis canendi donum accepit*, Bd. 4, 24; S. 596, 41.

sangere, es; *m.* **I.** *a singer :*—Sangere *cantor*, Wrt. Voc. i. 28, 17 : 72, 6. Idel sangere *temelici* (θυμελικὸς *a musician, singer*), 39, 40. Wé witan ðæt þurh Godes gyfe ceorl wearþ tó eorle, sangere tó sacerde, and bócere tó biscope, L. Eth. vii. 11 : Th. i. 334, 8. Bútan Jacobe ðam sangere, Bd. 4, 2; S. 565, 37. Se bisceop ðǽr gesette ðǽr sangeras and mæssepreóstas and manigfealdlíce circicean þegnas, Blickl. Homl. 207, 31. **II.** *a poet :*—David wæs sangere sóðfæstest, swíðe geþancol tó þingienne þeódum sínum wið ðane Sceppend, Ps. C. 50, 6. [Alse þe holi sangere seið on his loft songe, O. E. Homl. ii. 117, 22. *O. H. Ger.* sangari *cantor, psalmista : Icel.* söngvari.] v. býme-, cyric-sangere.

sangestre, an; *f. A female singer, songstress :*—Sangestre (-ystre) *cantrix*, Ælfc. Gr. 9, 64; Som. 13, 63. Sangystre, Wrt. Voc. i. 72, 5.

sang-pípe, an; *f. A musical pipe :*—Sangpípe *camena*, Germ. 389, 26. -sánian. v. á-sánian, sǽne.

sáp, e; *f.* (?) *Amber, resin, pomade :*—Sáp, smelting (cf. smulting *electrum*, 94, 61) *succinum* vel *electrum*, Wrt. Voc. i. 38, 31. Reádre deáge (*in margin*, also) *rubro stibio* (*the word occurs in a passage treating of dressing the hair*, cf. *the passage in Pliny describing the use and invention of* '*sapo :*' Gallorum hoc inventum rutilandis capillis; fit ex sebo et cinere optimus fagino et caprino, duobus modis, spissus ac liquidus: uterque apud Germanos majore in usu viris quam feminis), Hpt. Gl. 435, 17. v. sápe *and next word.*

sáp-box *a box for resin :*—Man sceal habban leóhtfæt, blácern, cyllan, sápbox, Anglia ix. 264, 22.

sápe, an; *f. Soap, salve* (? v. sáp) :—Sápe *sapo* (*sopo*, MS.), Wrt. Voc. i. 86, 12: *lumentum*, ii. 54, 4. Hé biþ ðonne áþwogen fram his synnum þurh ða untrumnysse, swá swá horig hrægl þurh sápan, Homl. Th. i. 472, 6. [Monie of þas wimmen smurieð heom mid blanchet, þet is þes deofles sápe (*unguent?*), O. E. Homl. i. 53, 24. Þe wreche peoddare more noise he makeð to ʒeien his sope, þen a riche mercer al his deorewurðe ware, A. R. 66, 18. *O. H. Ger.* seifa *sabona, smigma;* also *resina.*] v. ár-, pill-sápe.

sár, es; *n.* **I.** referring to the body, (1) *pain, suffering, soreness :*— Mé sár gehrán, wærc in gewód, Exon. Th. 163, 28; Gú. 1000. Sár gewód ymb ðæs beornes breóst, Andr. Kmbl. 2494; An. 1245. Mid sáre geswenced, mid mislícum ecum and tyddernessum, Blickl. Homl. 59, 7. On sáre his líchoma sceal hér wunian, 61, 1. Hǽlu bútan sáre,

Exon. Th. 101, 8; Cri. 1655. Ða ðe on sáre seóce lágun, 83, 14; Cri. 1356. Hé sár ne wiste *he did not feel pain*, Cd. Th. 12, 3; Gen. 179. (2) *a pain, pang, sore, wound* :—Nis ðær ǽnig sár geméted, ne ádl, ne ece, Blickl. Homl. 25, 30. Hé byð ðæs sáres hál, Lchdm. i. 352, 2. Wið eágena, eárena, sídan, wambe, &c. sáre, i. 2, sqq. On his módor sáre hé biþ ácenned, Blickl. Homl. 57, 35. Ðýlæs hwelc ðara niéhstena ðæs ofslægenan for ðæm sáre (*the mortal wound caused by the slipping of an axe*) hine ofsleá, Past. 21; Swt. 167, 3. Mugcwyrt ðæt sár ðara fóta of genimþ, Lchdm. i. 102, 16. Gif sió wamb biþ windes full, ðonne cymþ ðæt of wlacre wǽtan; sió cealde wǽte wyrcþ sár an . . . ðonne déþ ðæt ðæt sár áweg, Lchdm. ii. 224, 24. Nǽfre ðú ðæs suíðlíc sár gegearwast heardra wíta, ðæt ðú mec onwende worda ðissa, Exon. Th. 246, 2; Jul. 55. Ðú ðæs sár (*stripes and blows*) áber, Andr. Kmbl. 1912; An. 958. Ðæt gé him sára gehwylc gehǽlde *that you should heal every wound for him*, Exon. Th. 144, 11; Gú. 676. Leomu hefegodon sárum gesóhte *his limbs waxed heavy, visited by pains*, 159, 21; Gú. 930. Ádle gebysgad, sárum geswenced, 170, 11; Gú. 1110. Ðá wæs heó eft hefigod mid ðǽm ǽrran sárum *prioribus adgravata doloribus*, Bd. 4, 19; S. 589, 5. Se Hǽlend his þegnum sǽde ða sár ðe hé ádreógan wolde, Blickl. Homl. 15, 33. Hié ealle líchomlícu sár oforhogodan, 119, 20. II. *of the mind*, (1) *grief, pain, trouble, sorrow* :—Ne biþ ðǽr sár ne gewinn, ne nǽnig unéþnes, Blickl. Homl. 103, 35. Wépende sár, Exon. Th. 79, 14; Cri. 1290. Is sáwl mín sáres and yfeles gefylled *repleta est malis anima mea*, Ps. Th. 87, 3. Tó tácnunge sorges and ánfealdes sáres, Bt. 7, 2; Fox 18, 21. Hí hí forlǽtaþ on ðam mǽstan sáre, 7, 1; Fox 16, 13. Hé heora helpend wæs on heora sáre, Bd. 3, 9; S. 533, 26. (2) *a grief, sorrow, pain, wound* :—Hit wæs swá gewunelíc on ealdum dagum, ðæt gif hwam sum fǽrlíc sár (*affliction*) becóme, ðæt hé his reáf tótǽre, Homl. Th. ii. 454, 14. Ðeáh him mon hwæt wiðerweardes doo, oððe hé hwelce scande gehiére be him selfum, hé æt ðæm cierre ne biþ onstyred . . . ac æfter lytlum fæce hé biþ onǽled mid ðý fýre ðæs sáres, Past. 33; Swt. 225, 20. Ðá ðæt mód ðillíc sár cwéþende wæs, Bt. 5, 1; Fox 8, 24. Lufu him sára gehwylc symle forswíðede, Exon. Th. 160, 4; Gú. 938: 176, 31; Gú. 1218. Æfter ðære menigeo mínra sára ðe me on ferhþe gestódan *secundum multitudinem dolorum meorum in corde meo*, Ps. Th. 93, 18. Ða angunnenan sár *conceptos dolores*, Wrt. Voc. ii. 136, 12. [*Goth. sair : O. Sax. O. L. Ger. O. Frs. O. H. Ger. sér dolor, supplicium, amaritudo, ulcus : Icel. sár a sore; a wound.*] v. líc-sár, *and next word*.

sár; *adj. Sore, painful, grievous, distressing*, (1) *of physical pain* :—Se lǽca ðe sceal sáre (*yfela*, MS. Y.) wunda wel gehǽlan, hé mót habban góde sealfe ðǽrtó, L. Pen. 4; Th. ii. 278, 15. Ne wæs hyra ǽnigum síde ðý sárra, ðeáh hý swá sceoldan reáfe birofene slítan haswe bléde, Exon. Th. 394, 20; Rä. 14, 6. Wé wieton ðæt sió diégle wund biþ sárre ðonne sió opene, Past. 38; Swt. 273, 22. (2) *of mental pain* :—Ðá hé ðæs mannes deáþ swá earmlícne gehýrde ðá wæs him ðæt swíðe sár *when he heard the man's death was so miserable, it was very grievous to him*, Blickl. Homl. 219, 14. Ne wæs hyre bróðra deáþ on sefan swá sár, Exon. Th. 377, 25; Deór. 9. Ðæt ðam hálgan wæs sár on móde, Cd. Th. 96, 11; Gen. 1593: 27, 30; Gen. 425. Ðæt wæs Satane sár tó geþolienne, Andr. Kmbl. 3375; An. 1691. Ðonne hí sáres hwæt siófian scioldon (*cf. ðonne hí sceoldan heora sár siófian*, Bt. 38, 1; Fox 194, 35), Met. 26, 82. Bídan sáran sorge, Cd. Th. 266, 26; Sat. 28. Forlǽt sáre sorgceare, Exon. Th. 13, 27; Cri. 209. Hearm, sáre swyltcwale, Andr. Kmbl. 2735; An. 1370. Morþra, sárra sorga, Judth. Thw. 24, 10; Jud. 182: Rood Kmbl. 157; Kr. 80. Manaþ sárum wordum *prompts with words that wound*, Beo. Th. 4122; B. 2058. Ealle ða sáran edwíta ðe hé ádreág, Blickl. Homl. 97, 15. Uncúðne eard cunnian, sáre síþas *to make trial of a land unknown, of travails sore*, Exon. Th. 87, 2; Cri. 1419. Cwæð ðæt hym wurde on móde, sorga sárost, Cd. Th. 122, 19; Gen. 2029. [*O. Sax. O. L. Ger. O. Frs. O. H. Ger. sér tristis : Icel. sárr sore; wounded.*] v. un-sár.

Saracene, Sarocine, Sarcine; *pl.* Sarracene *Sarasene*, Ors. 1, 1; Swt. 12, 5. Wǽron ðǽr Sarocine gesamnode ðæt hig sǽtnodan manna, Shrn. 37, 34. Wit urnon for Sarcina hergunge, 42, 9. Se hefegosta wól Sarcina þeóde Gallia ríce forhergedon *gravissima Sarracenorum lues Gallias vastabat*, Bd. 5, 23; S. 645, 31. On India Saracena *in India Saracenorum*, Rtl. 196, 35. [*Cf. Icel. Serkir : O. H. Ger. Sarci, Serzi Arabes.*]

Saracenisc; *adj. Saracen* :—Hé gegaderode of ðám Saraceniscum swíðe micele fyrde, Jud. Thw. p. 162, 25. [*Cf. Icel. Serkneskr : O. H. Ger. Sarcisc, Sarzisc Arabicus.*]

Saracen-, Sarcin-ware; *pl. The Saracens* :—Ðá hergodon ða hǽþnan Sarcinware on þa stówe (*Sardinia*), Shrn. 122, 25.

sár-benn, e; *f. A painful wound* :—Wæs ðæs hálgan líc sárbennum soden, swáte bestémed, bánhús ábrocen, blód ýþum weóll, Andr. Kmbl. 2479; An. 1241. Sárbennum gesóht, Exon. Th. 163, 11; Gú. 992.

sár-bót, e; *f. Compensation paid for inflicting a wound*, L. W. I.; Th. i. 470, 21. [*Icel. sár-bætr ; pl.*]

Sarcine, Sarcin-ware. v. Saracene, Saracen-ware.

sár-cláþ, es; *m. A bandage for a wound* :—Sárcláþ *ligatura*, Wrt. Voc. i. 20, 18: ii. 53, 77: *fasciola*, i. 40, 62: ii. 39, 75.

sárcren (?) *disposed to soreness* :—On ðám monnum ðe habbaþ swíðe gefélne and sárcrenne magan *a very sensitive stomach and one easily made sore*, Lchdm. ii. 176, 9.

sár-cwide, es; *m.* I. *a speech that is intended to give pain, injurious* or *affronting speech, reproach, bitter words* :—Ne gedafenaþ ðé ðæt ðú andsware mid olerhygdum séce sárcwide *it befits thee not to seek an answer with arrogance and bitter words*, Andr. Kmbl. 693; An. 320. Synnige ne mihton þurh sárcwide sóð gecýðan, 1929; An. 967. Ðú ús ásettest on sárcwide úrum neáhmannum *posuisti nos in contradictionem vicinis nostris*, Ps. Th. 79, 6. Hé ðæt eal þolaþ, sárcwide secga, Exon. Th. 458, 2; Hy. 4, 94. Ic worn for ðé hæbbe sídra sorga and sárcwida, hearmes gehýred, and mé hosp sprecaþ, tornworda fela, 11, 14; Cri. 170. II. *a speech in which grief is expressed, a lament* :—Ic nyste ǽr ðú ðé self hit mé gerehtest mid dínum sárcwidum *I did not know until you yourself told it me with your lamentations*, Bt. 5, 1; Fox 8, 34. Nú sceal ic siófigende wreccea giómor singan sárcwidas *flebilis moestos cogor inire modos*, Met. 2, 4.

Sardinie; *pl. The Sardinians, the people* or *the island of Sardinia* :—Hú Sardinie wunnon on Rómane, Ors. 4, 7, tit.; Swt. 4, 16. On Sicilium and on Sardinium ðǽm íglondum, 4, 7; Swt. 164, 23.

sáre; *adv. Sorely, grievously, bitterly* :—Wǽron earme men sáre beswicene (*sorely deceived*) and hreówlíce besyrwde, Wulfst. 158, 11 note. Hrinon hearmtánas hearde and sáre drihta bearnum, Cd. Th. 61, 5; Gen. 992. Mé ðæt cynn hafaþ sáre ábolgen *that race hath angered me sore*, 76, 14; Gen. 1257. Forgrípan gumcynne grimme and sáre heardum mihtum, 77, 15; Gen. 1275. Sum sáre angeald ǽfenreste *one paid a heavy price for his night's rest*, Beo. Th. 2507; B. 1251. Hé cenþ unriht and hit cymþ him sáre *it shall trouble him sorely*, Ps. Th. 7, 14. Hí sáre sprecaþ *they speak bitterly*, 63, 4. Wé sittaþ and sáre wépaþ (*cf. Icel. gráta sáran : Scot. to greet sair*), 136, 1. Wæs se hálga wer sáre geswungen, Andr. Kmbl. 2791; An. 1398. [*O. Frs. sére : O. Sax. O. H. Ger. séro dolenter : Ger. sehr.*] v. emn-sáre.

sárettan; *p. te To lament, complain* :—Hé sárette ðætte ða synfullan sceoldan bytlan onuppan his hrycge *supra dorsum suum fabricasse peccatores queritur*, Past. 21, 2; Swt. 153, 9. Ðæt ilce sárette se wítga contra hos propheta conqueritur, 37, 2; Swt. 267, 2. [*O. H. Ger. sérazzan dolere.*]

sár-ferhþ; *adj. Sore at heart, wounded in spirit* :—Ðæt wíf (*Sarah complaining to Abraham about Hagar*) módes sorge, sárferhþ sægde : 'Ne fremest ðú riht wið mé,' Cd. Th. 135, 17; Gen. 2244. Cf. sárig-ferhþ.

sárga, an; *m. Some kind of trumpet* :—Trúðhorn oððe sárga *lituus*, Wrt. Voc. i. 73, 67. Sárgana *salpicum, tubarum*, Hpt. Gl. 445, 11.

sárgian; *p. ode.* I. *to make sad* (*sárig*), *to grieve* (*trans.*), *afflict, wound* :—Hí sárgiaþ fremdne flǽschoman, Salm. Kmbl. 220; Sal. 109. II. *to be* or *become sad, to grieve* (*intrans.*), *languish* :—Hé sárgaþ ðæs *he is grieved at it*, Past. 33; Swt. 227, 21. Se bisceop hefiglíce sárgode be ðam fylle and mínre forwyrde *episcopus gravissime de casu et interitu meo dolebat*, Bd. 5, 6; S. 619, 32. Eágan míne sárgodon *oculi mei languerunt*, Ps. Spl. 87, 9. Ðá ongan hé forhtian and sárgian *et coepit pavere et taedere*, Mk. Skt. 14, 33. Sárgiende I sorhful *dolens*, Ps. Lamb. 68, 30. Sárgiendne fréfrian *dolentem consolari*, R. Ben. 17, 3 MS. O. [*O. H. Ger. sérágón to sadden, pain, wound.*] v. be-, ge-sárgian.

sárgung, e; *f. Lamentation, grief* :—Ðǽr is sorgung and sárgung and á singal heóf, Wulfst. 114, 5. Beó ðú forþloten tó sárgungum *esto pronus ad lamenta*, Scint. 6. v. be-sárgung.

sárian; *p. ode.* I. *to feel pain for, feel sorry for* :—Heó is má tó sárianne *magis dolendum*, Bd. 1, 27; S. 496, 40. II. *to be sore* (v. sár; *adj.*), (1) *of physical pain* :—Hé (*the disease*) wundaþ and sió wund sáraþ *the wound gets painful*, Past. 11; Swt. 71, 20. Ða liran ðara lendena sáriaþ, Lchdm. ii. 216, 24. (2) *of mental pain, to grieve, be sad* :—Ic sárige on mínum wítum *I grow sad in my punishments*, 94, 7. Wé sáriaþ ealle, forðon þe wé seóþ ðíne líchaman beón cwylmed, 42, 2. Ðín fæder and ic sárigende (*dolentes*) ðé sóhton, Lk. Skt. 2, 48. Sáriendne (sáriende, MS. T.) fréfrian, R. Ben. 17, 3. [*O. Sax. gi-sérid afflicted : O. Frs. sérd : O. H. Ger. séren, sérón vulnerare, dolere.*]

sárig; *adj.* I. *feeling grief, sorry, sorrowful, sad* :—Ðá wæs Petrus sárig *contristatus est Petrus*, Jn. Skt. 21, 17: Homl. Th. ii. 248, 11. Ic mé sylfa eam sárig þearfa *pauper et dolens ego sum*, Ps. Th. 68, 30. Ðá sceolde se hearpere weorþan swá sárig ðæt hé ne mihte on gemong óðrum monnum beón *the harper (Orpheus) is said to have become so afflicted with grief, that he could not live among other men*, Bt. 35, 6; Fox 166, 6. Se is swíðe sárig for ðínum earfoþum and for ðínum wræcsíþe, 10; Fox 28, 18. Ðæs ðe he swá geómor weard, sárig for his synnum, Exon. Th. 450, 15; Dóm. 88. Hé weard swíðe sári *graviter accepit*, Gen. 48, 17. Ne forseoh sáriges béne, Ps. Th. 54, 1. On salig wé sárige úre organan áhéngan, 136, 2. Ðá wurdon hiora wíf swá

sârige on hiora môde, and swâ swîðlíce gedréfed, Ors. 1, 10; Swt. 44, 29. Híg wæron sârie (*dolentes*) for hira geswince, Num. 11, 1. Monge ðe hine sârge gesôhtun, freórigmôde, Exon. Th. 155, 12; Gû. 859. Sôhton sârigu tú (*the two women at the sepulchre*) sigebearn Godes, 460, 2; Hö. 11. Sârge gê ne sôhton, ne him swǽslíc word frôfre gespræcon, 92, 19; Cri. 1511. II. *expressing grief, mournful, sad, bitter:*—Hê ðâ wêpende wêregum teárum his sigedryhten sârgan reorde grětte, Andr. Kmbl. 120; An. 60. Ðonne hê wrece sârigne sang, Beo. Th. 4885; B. 2447. Sârige teáras, Ps. Th. 55, 7. [*O. Sax. O. H. Ger.* sêrag *dolens, amarus.*] v. efen- (em-) sârig.

sârig-ferhþ; *adj. Sad in soul:*—Geseoþ sorga mǽste synfâ men sârigferþe, Exon. Th. 67, 4; Cri. 1083. Cf. sâr-ferhþ.

sârig-môd; *adj. Sad-hearted, of mournful mood:*—Ðonne fêhþ seó weâlâf sorhful and sârigmôd geómrigendum môde synne bemǽnan, Wulfst. 133, 13. Geneósige ða ðe beóþ sârigmôde and seóce, L. Pen: 16; Th. ii. 282, 28. Frôfor eft gelamp sârigmôdum, Beo. Th. 5876; B. 2942. [Þa wes he sarimod and sorhful an heorten, Laym. 29791. Sorimod and wroþ, O. and N. 1218. Forfrigted folc and sorimod, Gen. and Ex. 3520. *O. Sax.* sêrag-môd.]

sârigness, e; *f. Sadness:*—Hwæt mæg beón wôp oððe sârignys, gyf ðæt næs se mǽsta ǽgdres, Homl. Skt. i. 23, 102. [Hê hig funde slǽpende for unrôtnesse (*later MS.* sârignesse) *dormientes prae tristitiam*, Lk. Skt. 22, 45. Tristicia þet is þissere worlde sarinesse, O. E. Homl. i. 103, 22. Þer wes sarinesse (wowe, 2nd MS.), sorreʒen inoʒe, Laym. 27560. In eche sorinesse, O. E. Misc. 76, 125. Wiþ muchel sorinesse, Horn. 922]

sâr-líc; *adj.* I. *giving occasion for sorrow, sad, mournful, lamentable, grievous:*—Wâ lâ wâ! ðæt is sârlíc ðæt swâ leóhtes andwlitan men sceolan âgan þýstra ealdor *heu, proh dolor! quod tam lucidi vultus homines tenebrarum auctor possidet*, Bd. 2, 1; S. 501, 15. Sârlíc tô cweðene *dolendum dictu*, Hpt. Gl. 447, 25. Nô his lífgedâl sârlíc þúhte secga ǽnigum *to no man did his death seem occasion for sorrow*, Beo. Th. 1688; B. 842. Sârlíc symbel (*the eating of the forbidden fruit*), Exon. Th. 226, 15; Ph. 406. Sârlíc síþfæt (*the journey to hell*), 446, 20; Dôm. 25. Se sârlíca cwide: 'Terra es et in terram ibis' *that sad sentence,* 'Dust thou art and to dust thou shalt return,'* Blickl. Homl. 123, 7. Mid sârlícre sceame *confusione*, Ps. Th. 88, 38. I a. *causing pain, grievous:*—Éþung biþ sârlíc *the breathing is painful*, Lchdm. ii. 258, 17. Wê witon unrîm ðara monna ðe ða ēcan gesǽlþa sôhtan nallas þurh ðæt ân ðæt hí wilnodon ðæs lichomlîcan deáþes ac eác manegra sârlícra wîta hié gewilnodon *multos scimus beatitudinis fructum non morte solum, verum etiam doloribus suppliciisque quaesisse*, Bt. 11, 2; Fox 36, 4. II. *expressing sorrow or grief, sad, mournful:*—Sârlíc sang *trenos*, Wrt. Voc. i. 28, 18. Sârlíc blis *cantilena*, ii. 128, 13. Hê sit mid sârlícum andwlitan, nât ic hwæt hê besorgaþ, Ap. Th. 15, 10. Hê cwæð mid sârlícre stemne, Swt. A. S. Rdr. 101, 205. Sârlíc leóþ *tragoediam*, Wrt. Voc. ii. 82, 37. Hwîlum gyd âwræc sârlíc, Beo. Th. 4224; B. 2109. [Næs heo næuere swa sarlic, þ wes Wenhauer þa quene, sarʒest wimmone, Laym. 28457. *O. H. Ger.* sêr-lîh *grievous.*]

sârlíce; *adv.* I. *in a manner that causes or is attended by physical pain, sorely, painfully:*—Job sæt sârlíce eal on ânre wunde, Homl. Th. ii. 452, 27. Blôd ðæt wæs sârlíce âgoten, Ps. Th. 78, 11. Ðē sculon slîtan sârlíce swearte wihta, Soul Kmbl. 145; Seel. 73. Hê sôhte hû hê sârlícast, þurh ða wyrrestan wîtu, meahte feorhcwale findan, Exon. Th. 276, 25; Jul. 571. II. *in a manner that causes mental pain, sorely, grievously, lamentably:*—Ðæt mín fôt ful sârlíce âsliden wǽre, Ps. Th. 93, 17. Hî mê on dígle deorce stôwe settan sârlíce, 142, 4. Hit oft swîðe sârlíce gebyrede ðæt wríteras forlēton unwritene ðara monna dǽda ðe on hiora dagum foremǽroste wǽron *it has often happened most lamentably, that writers have left unwritten those men's deeds that in their days were most distinguished*, Bt. 18, 3; Fox 64, 32. III. *in a manner that expresses sorrow or grief, sorely, bitterly, heavily:*—Apollonius sârlíce sæt, Ap. Th. 14, 21. Sârlíce wêpende *weeping bitterly*, Gen. 21, 16. Ða onsác se Wîsdôm ðen *Wisdom sighed heavily*, Bt. 26, 2; Fox 92, 24: 40, 3; Fox 238, 7: Wulfst. 133, 14. Ðâ wǽron hié ealle sôna unrôte, and sârlíce gebǽrdon, Blickl. Homl. 225, 14. [*O. Frs.* sêrlîke.]

Sarmondisc; *adj. Sarmatian:*—Nêh ðæm gârsecge ðe mon hâteþ Sarmondisc *Sarmatico aversi oceano*, Ors. 1, 1; Swt. 8, 16.

sârness, e; *f.* I. *bodily pain:*—On sârnysse ðú âcenst cild *in dolore paries filios*, Gen. 3, 16. Freoh fram deápes sârnysse, Homl. Th. i. 76, 14. II. *mental pain, affliction, grief:*—Geopenige úre sârnys (*the trouble arising from a pestilence*) ûs infær sôðre gecyrrednysse, ii. 124, 7. Gehrepod mid heortan sârnisse *tactus dolore cordis*, Gen. 6, 6. Hê ðis eal mid sârnesse beheóld, Ap. Th. 14, 19. Âfirsa fram him his sârnesse, 16, 14. *Heu* geswutelaþ môdes sârnesse, Ælfc. Gr. 5; Som. 4, 1. Helle sârnyssa mē beeodon, and ic on mínre gedréfednysse Drihten clypode, Homl. Th. ii. 86, 17. Ðæt beóþ ða angin, hê cwæð, ðara sârnessa . . . ða sorga and ða sârnessa ðe on woruld becumaþ, Wulfst. 89,11–14.

Sarocine, Sarracene. v. Saracene.

sâr-seófung, e; *f. Complaint:*—Sârseófunge *querulosis* quiðungum *questibus*, Wrt. Voc. ii. 76, 18–19.

sâr-slege, es; *m. A painful blow, a blow that wounds or pains:*—Wê ða heardestan wîtu geþoliaþ þurh sârslege, Exon. Th. 262, 31; Jul. 341: 275, 8; Jul. 547. Ne môstun hý Gûþláces gæste sceþþan, ne þurh sârslege sâwle gedǽlan wið líchoman, 115, 31; Gû. 198. Ðâ wæs hê swungen särslegum, swât ýðum weóll, Andr. Kmbl. 2551; Ann. 1277.

✝ sâr-spell, es; *n. A sorrowful speech, a lament:*—Ic secge ðis sârspell and ymb síþ spræce, Exon. Th. 458, 6; Hy. 4, 96.

sâr-stæf, es; *m. A term intended to pain, an insult, a reproach:*—Godes andsacan sægdon sârstafum swíðe gehēton ðæt hê deápa gedâl dreógan sceolde *God's adversaries said with bitter words, vehemently vowed, that he should suffer death*, Exon. Th. 116, 10; Gû. 205.

sârung, e; *f. Mourning, lamentation:*—Ðǽr is sorgung and sârgung (sâruncg, MS. K.) and ā singal heóf, Wulfst. 114, 5.

sâr-wilm, es; *m. A painful burning, a feverish heat:*—Soden sârwylmum (cf. âdle gebysgad, sârum geswenced, 170, 10-11), Exon. Th. 171, 7; Gû. 1123.

sâr-wîs (?) *dull:*—Ða sârwîsan (Cott. MS. sâmwîsan), Past. 30, 1; Swt. 203, 7. v. sâm-wîs.

sâr-wracu; *gen.* -wræce; *f. Sore tribulation:*—Nis ðǽr synn ne sacu ne sârwracu (sâr wracu?), Exon. Th. 201, 11; Ph. 54. Swâ ðæt ēce líf eádigra gehwylc æfter sârwræce sylf geceóseþ, 224, 27; Ph. 382: 274, 2; Jul. 527.

Satan, es; *m. Satan:*—God cwæð ðæt se hêhsta hâtan sceolde Satan, Cd. Th. 22, 23: Gen. 345: 22, 27; Gen. 347. Hê wæs fram Satane gecostnod, Mk. Skt. 1, 13: Exon. Th. 93, 6; Cri. 1522: Andr. Kmbl. 3374; An. 1691. *The Greek form* Satanas *with acc.* Satanan *also occurs,* Mk. Skt. 3, 23: Lk. Skt. 10, 18; *and* Satanus, Cd. Th. 287, 22; Sat. 371: 292, 27; Sat. 447.

saturege, an; *f. Savory;* satureia hortensis, Lchdm. iii. 24, 4. [*M. H. Ger.* satereie: *Ger.* saturei.] v. sæþerige.

Saturnus; *gen.* Saturnes; *m.* I. *Saturn the god:*—Ðæs (*Jove's father*) nama wæs Saturnus, Bt. 38, 1; Fox 194, 17: Met. 26, 48. Tô ðam cealdan stiorran ðe wê hâtaþ Saturnes steorra (cf. Met. 24, 31, *where the star is called Saturn:* ðone steorran Saturnus londbûende hâtaþ), Bt. 36, 2; Fox 174, 13. II. *the name occurs often in the* Dialogue of Salomon and Saturn.

sauine. v. safine.

sâwan; *p.* seów, sêw; *pp.* sâwen. I. lit. (a) *to sow* (seed in a field):—Túncersan ðe mon ne sǽwþ, Lchdm. ii. 22, 13. Weard sâweþ on swæd mín, Exon. Th. 403, 11; Rä. 22, 6. Hig ne sâwaþ *non seminant*, Lk. Skt. 12, 24. Hlâford hû ne seów (seówe, MS. A.) ðú gôd sǽd on ðínum æcere *Domine, nonne bonum semen seminasti in agro tuo?* Mt. Kmbl. 13, 27. Út eode se sǽdere hys sǽd tô sâwenne [sēde ł sēdege, Lind.]. And ðâ ðâ hê seów, 13, 3-4. Ðâ hê sēw (seów, MS. A.) Mk. Skt. 4, 4. Hê wîngeard sette, seów sǽda fela, Cd. Th. 94, 9; Gen. 1559. Be ðæm âworpnan engle is âwriten ðæt hê sêwe ðæt weód on ða gôdan æceras *cum bonae messi inserta fuissent zizania*, Past. 47, 1; Swt. 357, 17. Gehýre gê ðæs sâwendan (*seminantis*) bigspell, Mt. Kmbl. 13, 18. Sâwondum *seminanti*, Kent. Gl. 370. (b) *to sow* (a field with seed):—Hí seówon æceras *seminaverunt agros*, Ps. Spl. 106, 37. Ne sâw ðú ðínne æcyr mid gemengedum sǽde *agrum tuum non seres diverso semine*, Lev. 19, 19. Six geár ðú scealt sâwan sex annis seres agrum tuum, 25, 3. II. fig. *to sow the seeds of anything, to originate, do an action which produces a result, implant:*—Se eorþlica anweald ne sǽwþ (*inserit*) ða cræftas ac lisþ uuþeáwas, Bt. 27, 1; Fox 94, 25. Âworpen man on ǽlce tíd sâweþ wrôhte *homo apostata omni tempore jurgia seminat*, Past. 47, 1; Swt. 357, 22. Se ealda inwit sâweþ, Fragm. Kmbl. 67; Leás. 35. Ða hēr on teárum sâwaþ hí eft fægerum gefeán sníðaþ *qui seminant in lacrymis, in gaudio metent*, Ps. Th. 125, 5: Exon. Th. 6, 18; Cri. 86. Hê monigfealde môdes snyttru seów and sette geond sefan monna, 41, 29; Cri. 663. Sibbe sâwaþ on sefan manna, 30, 31; Cri. 487. [*Goth.* saian: *p.* saisô: *O. Sax.* sâian: *p.* sâida, sêu: *O. Frs.* sêa: *O. H. Ger.* sâjan; *p.* sâta : *Icel.* sâ; *p.* seri, *later* sâði.] v. â-, be-, ge-, geond-, ofer-, on-, tô-sâwan.

sâwel (ol, ul), sâwl, sâul, sôwhul, e; *f. The soul:*—Sâwul *anima*, Wrt. Voc. i. 76, 30. Sâwl, 42, 32: Sâul, 282, 23: ii. 7, 75. I. *the soul, the animal life:*—Ic secge mínre sâwle: 'Eálâ sâwel, ðú hæfst mycele gôd . . . gerest ðē, et, drinc, and gewista.' Ðâ cwæð God tô him: 'Lâ dysega, on ðisse nihte hig feccaþ ðíne sâwle fram ðē' . . . Ic eów secge: 'Ne beó gê ymbehýdige eówre sâwle, hwæt gê etan . . . Seó sâwul ys mâ ðonne se líchama, Lk. Skt. 12, 16–23. Mannes Sunu com ðæt hê sealde his sâwle líf (ferh, Rush.) tô âlýsednesse for manegum, Mt. Kmbl. 20, 28. Gif hwâ eácniend wíf gewerde . . . gif hió deád síe, selle sâwle wið sâwle, L. Alf. 18; Th. i. 48, 19. Se ðe gemēt hys sâwle (sáule ł ferh, Rush.), se forspilþ hig; and se ðe forspilþ his sâwle for mē, hê gemēt hí, Mt. Kmbl. 10, 39: 16, 25; Jn. Skt. 12, 25. Genera sâwle míne fram ârleásum, Ps. Spl. 16, 14. Sâwle sêcan *to try to kill*, Beo. Th. 1606; B. 801. Ðæt hê gefriðie heora sâwla fram deápe, and hí fêde on hungres tíde, Ps. Th. 32, 16. II. *the soul, the intellectual and immortal principle in man:*—Hwæt gelýfeþ se líchoma bûtan þurh ða sâwle? Geþencean ða men ðæt hié heora sylfra sâwla geseón ne mâgon; ac eal

swâ hwæt swâ se gesênelîca lîchama dêþ, eal đæt dêþ seó ungesýnelîce sâwl þurh đone lîchoman; and đonne seó sâwl hié gedǽleþ wið đone lîchoman, hwylc biþ hê đonne bûton swylce stân, oððe treów? Ne hê hine nâ ne onstyreþ, siðđan seó ungesýnelîce sâwl him of biþ, Blickl. Homl. 21, 21–28. Se êcea dǽl, đæt is seó sâwl, III, 32. Seó sâul mid gâstlicum þingum on êcnesse leofaþ, 57, 15. Ealle nien-lîchomlîce sweltaþ, and đeáh seó sâwl biþ lîbbende. Æt seó sâwl færþ swîðe freólîce tô heofonum, siðđan heó of đam carcerne đæs lîchoman onliésed biþ, Bt. 18, 4; Fox 68, 13. Sâwl and lîcchoma wyrcaþ ânne mon .. tô đære sâwle and tô đam lîchoman belimpaþ ealle đâs đæs monnes good, ge gâstlîce ge lîchomlîce ... Đonne is đære sâwle gôd wærscipe and gemetgung and geþyld and rihtwîsnes and wîsdôm and manege swelce cræftas, 34, 6; Fox 140, 28–35 : 34, 10; Fox 148, 3–4. Nû tô đam sôþan gefeán sâwel fundaþ, Exon. Th. 178, 3; Gû. 1238: 233, 12; Ps. 523. Gewât sâwol sêcean sôðfæstra dôm, Beo. Th. 5633; B. 2820. Sâwul, Byrht. Th. 136, 64; By. 177. Seó ýdelnes is đære sâwle feónd, L. E. I. 3; Th. ii. 404, 11. Hwæt is đæt đæm men sý mâre þearf tô þencenne đonne embe his sâuwle þearfe? Blickl. Homl. 97, 20. Nýtenu and deór, fixas and fugelas hê gesceóp on flǽsce bûtan sâwle, Homl. Th. i. 276, 4. On hwilcum dǽle hæþ se man Godes anlîcnysse on him? on đære sâwle ... Đæs mannes sâwl hæfþ on hire þreó þing, đæt is, gemynd and andgit and willa ... Ân sâwul is, and ân lîf and ân edwist seó đe hæfþ đâs þreó þing ... Đeáhhwæðere nis nân đæra þreora seó sâwul, ac seó sâwul þurh đæt gemynd gemanþ, þurh đæt andgit heó understent, þurh đone willan heó wile swâ hwæt swâ hire lîcaþ, 288, 15–30. Se man is êce on ânum dǽle, đæt is, on đære sâwle; heó ne geendaþ nǽfre, 16, 16. Ne mâgon hig đa sâwle ofsleán, Mt. Kmbl. 10, 28. Sâuwle, Blickl. Homl. 43, 23. Monna sâwla sint undeáþlîce and êce, Bt. 11, 2; Fox 34, 33. Gebid heó sînna sôwhula, Txts. 124, 5. Gemyndige ûre sâula þearfe, Blickl. Homl. 101, 16. Đæt hê ûre sâula gelǽde on gefeán, 211, 8.   **III.** *a soul, a human creature (after death)* :—Đa hâlgan sâwla cleopodan tô Drihtne : ' Âstîg nû đû hafast helle bereáfod,' 87, 20. Hâlige sâula đêr (*in Jerusalem*) restaþ, 81, 2. Hê geseah đæt on đæm clife hangodan manige swearta sâula be heora handum gebundne ... Đis wǽron đa sâula đa đe hér on worlde mid unrihte gefyrenode wǽron, and đæs noldan geswîcan ǽr heora lîfes ende, 209, 34–211, 7. Seó menigo hâligra sâula đe ǽr gehæftnede wǽron (*those who were released when Christ descended to Hell*), 87, 7. Heora (*the angels'*) êþel sceolde geseted weorþan mid hâlgum sâwlum ... mid đære menniscan gecynde, 121, 34. Mid eallum đǽm sâulum đe hé on worlde mid rihte tô Gode gecyrraþ, 57, 25 : 89, 29 : 95. 22. Drihten đa hâlgan sâuwla đonon (*from Hell*) âlǽdde, 67, 19. [*Goth.* saiwala : *O. Sax.* sêola : *O. Frs.* sêle : *O. L. Ger.* sêla, sîla : *O. H. Ger.* sêla, sêula : *Icel.* sâla.]   v. or-sâwle.

sâwel-berend *a being with a soul* :—Sâwlberendra, niðđa bearna, grundbûendra, Beo. Th. 2013; B. 1004.

sâwel-cund; *adj. Spiritual* :—Sâwelcund hyrde, Exon. Th. 121, 14; Gû. 288.

sâwel-dreór *life-blood* :—Hê geblôdegod wearð sâwuldriôre, Beo. Th. 5379; B. 2693. Besmiten mid sâwldreóre, Cd. Th. 91, 31; Gen. 1520.

sâwel-gedâl *the parting of soul and body, death* :—Ne biþ đæs lengra swice sâwelgedâles đonne seofon niht fyrstgemearces, đæt mîn feorh heonan on đisse eahteþan ende gesêceþ, Exon. Th. 164, 7; Gû. 1008. Cf. lîf-gedâl.

sâwel-gescot *soul-scot* :—Đat sâwulgesceot sceolon đa canonicas habban, Chart. Th. 609, 14, 29. v. sâwel-sceatt.

sâwel-hord *the treasure of life, life guarded as a treasure in the body, the body full of life* :—Oþ đæt sâwlhord, bâncofa blôdig, âbrocen weorþeþ, Exon. Th. 329, 15 ; Vy. 34. Oþ sâwlhord *to the very soul*, Ps. Th. 77, 49.

sâwel-hûs *the body* :—Đis sâwelhûs, fǽge flǽschoma, Exon. Th. 163, 34; Gû. 1003. Deáþ sôhte sâwelhûs, 170, 19; Gû. 1114.

sâwel-leás; *adj.*   **I.** *without life* (v. sâwel, **I**) :—Sâwulleás (sâwl-, MS. F.) *exanimis*, Ælfc. Gl. 9, 28; Zup. 56, 16. Hê feóll geswôgen swylce hê sâwllîce wǽre, Homl. Skt. i. 21, 299. Hî þwôgon đone sâwlleásan lîchaman, 20, 97. Magoþegna bær đone sêlestan sâwolleásne, Beo. Th. 2817; B. 1406. Sâwulleásne, 6059; B. 3033. Sâwelleásne, Exon. Th. 329, 21; Vy. 37. Hêht đâ âsettan sâwlleásne, life belidenes lîc on eorþan, Elen. Kmbl. 1751; El. 877.   **II.** *without soul* (v. sâwel, **II**) :—On đæs mannes sâwle is Godes anlîcnyss, for đam is se mann sêlra đonne đa sâwulleásan nýtenu, đe nân andgit nabbaþ embe heora âgenne Scýppend, Hexam. 11; Norm. 18, 22.

sâwel-sceatt, es ; *m. An ecclesiastical due, to be paid for every deceased person to the clergy of the church to which he belonged, in consideration of the services performed by them in his behalf.* It was to be paid before the funeral rites were completed, though the regulation would hardly be carried out in cases where grants of land were made. It appears to have been one of the objects of the early gilds, to provide for the payment of this fee :—Sâwlsceat *vel* syndrig Godes lâc *dano* (*dona ?*), Wrt. Voc. i. 28, 44. The passages dealing with the subject in the Laws are the following :—Ic wille đæt mîne gerêfan gedôn đæt man âgife đa

ciriscsceattas and đa sâwlsceattas tô đam stôwum đe hit mid riht tô gebirige, L. Ath. I. prm.; Th. i. 196, 9. Gelǽste man sâwlsceat (sâul-, MS. A.) æt ǽlcan cristenan men tô đam mynstre đe hit tô gebyrige, L. Edg. I, 5; Th. i. 264, 24. And sâulsceat is rihtast đæt man symle gelǽste æt openum græfe ; and gif man ǽnig lîc of rihtscriftscîre elles hwâr lecge, gelǽste man sâulsceat swâ đeáh intô đam mynstre đe hit tô hýrde, L. Eth. v. 12; Th. i. 308, 4–7: vi. 20–21; Th. i. 320, 4–8: ix. 13; Th. i. 342, 33 : L. C. E. 13 ; Th. i. 368, 5–8. To the same effect it is said in Wulfstan's Homilies :—Eác wê lǽraþ đæt cristenra manna gehwylc understande, đæt hê æfter forþsîðe bûtan sâwulsceatte ne licge on mynstre, ac gelǽste man â đone sâwelsceat æt openum pytte, 118, 4–7. Sâulsceat is rihtast đæt man gelǽste aa æt openum græfe, 311, 12. The *sâwelsceat* is sometimes determined in amount by the will of the deceased :—Ic gean intô Élig ... đêr mînes hlâfordes lîchoma rest, đara þreó landa đe wit geheótan Gode ... and đes beáhges gemacan, đe man sæalde mînum hlâforde, tô sâwlescæatte, Chart. Th. 524, 14–30. See too Shrn. 159, and Turner's Anglo-Saxons, bk. vii. c. xiv. Kemble, Cod. Dip. i. lxii, remarks that in lands leased by the Church, and exclusively in such, there is frequently a stipulation for the payment of sâwelsceat. For the practice in the case of gilds, see Chart. Th. 609, 10–18 :—Æt ǽlcum forðfarenum gildan æt ǽlcum heorþe ǽnne penig tô sâwulsceote, sê hit bonda, sê hit wîf, đe on đam gildscipe sindon ; and đat sâwulgesceot sceulon đa canonicas habban, and swilce þênisce dôn for hig swilce hig âgon tô dône.

sâwel-scot.   v. preceding word (at last passage).

sâwel-þearf, e ; *f. What is necessary or beneficial for the soul* :—Ic wes smeágende ymb mîne sâulþearfe, Chart. Th. 474, 18.

sâwend, es ; *m. A sower* :—De sêdere ł sâwend *seminans*, Mk. Skt. Rush. 4, 3. Se sâwena (sâwend ?) *qui seminat*, Mt. Kmbl. Rush. 13, 3. Gehêraþ gelîcnisse đæs sâwendes *audite parabolam seminantis*, 13, 18. Cf. leóhtsâwend *lucisator*, Germ. 389, 2.

sâwere, es ; *m. A sower* :—Ût eode se sâwere his sǽd tô sâwenne, Mt. Kmbl. 13, 3. v. word-, wrôht-sâwere.

sâwlian; *p.* ode *To give up the ghost, expire* :—Hê ne geswâc his gebeda oþ đæt hê sâwlode, Homl. Th. ii. 518, 1. Flaccus hêt đone preóst beswingan oþ đæt hê sâwlode, Homl. Skt. i. 10, 291. Sôna swâ hê đyder com swâ sâwlode đæt mǽden, 22, 101 : Homl. As. 59, 202. v. next word.

sâwlung, e ; *f. The giving up the ghost, expiring* :—Cwæđ sum hâlig biscop đâ hê wæs on sâwlenga đe đeossum fæder : Arsenius đû wǽre eádig forðon đû hæfdest â đâs tîd beforan đînum eágum *a certain holy bishop, when he was expiring, said of this father* : 'Arsenius, blessed wert thou, for ever hadst thou this hour (*the hour of death*) before thine eyes,' Shrn. 106, 26.

.sca-, scâ-, scǽ-; scæ-. v. scea-; sceá-; scea-, sce-.

scæd, Wrt. ii. 120, 8. v. sceabb.

scǽnan; *p.* de *To break* :—Đa cômon đa cempan, and sôna đæra sceaðena sceancan tôbrǽcon. Hî gemêtton Crist deádne, and his hâlgan sceancan scǽnan ne dorston, Homl. Th. ii. 260, 10. Đa gemettan ne môston đæs lambes bân scǽnan, ne đa cempan ne môston tôbrecan his (*Christ's*) hâlgan sceancan, 282, 7. [Helmes gullen ... sceldes gunnen scenen, Laym. 31234. Breken brade sperren, bordes scænden, 5186. Cf. (?) *Icel.* skeina *to scratch, wound slightly*.] v. ge-, tô-scǽnan.

-scǽre. v. ǽ-scǽre.

Scald *the Schelde* :—Hêr fôr se here up on Scald, Chr. 883 ; Erl. 82, 15.

Scariothisc ; *adj. Of Scariot* :—Judas se Scariothisca ; forðon hê com of đǽm tûne đe Scariot hâtte, Blickl. Homl. 69, 5 : Mk. Skt. Lind. Rush. 14, 43.

scaþel, Dôm. L. 30, 58. v. staþel.

sceáb, sceaba. v. sceáf, sceafa.

sceabb, scæb, sceb, es ; *m. Scab, a scab* :—Sced (scaeb ?) *scara* (*scara* vulneris crusta, Du Cange. Cf. *Span.* escara *the scurf* or *scar of a sore*, Wrt. Voc. ii. 120, 8. Đone leahtor đe Grêcas achoras (ἀχῶρας) nemnaþ, đæt ys sceb (scæb, MS. B.), Lchdm. i. 322, 17. Wið sceb (scæb, MSS. H. B.), 150, 5 : 316, 22. Wið sceab, 66, 21. Se hæfþ singalne sceabb se đe nǽfre ne blinþ ungestæđđignesse. Đonne bî đæm sceabbe swîðe ryhte sió hreófl getâcnaþ đæt wôhhǽmed *jugem habet scabiem, cui carnis petulantia sine cessatione dominatur. Per scabiem recte luxuria designatur*, Past. 11, 5 ; Swt. 70, 3–4. Gif hê hæfde singale sceabbas *si jugem scabiem habens fuerit*, 11, 1 ; Swt. 65, 6. [*Ger.* schabe *scab, itch* : *Dan.* skab : *Swed.* skabb.]

sceabbed; *adj. Having scabs* or *sores* :—Sceabbede, ǽttren *purulentus*, Hpt. Gl. 519, 32.

sceacan, scacan ; *p.* sceóc, scôc ; *pp.* sceacen, scacen, scæcen.   **I.** *to shake* (intrans.), *quiver* :—Gerd from uinde styrende ł sceæcende, Mt. Kmbl. Lind. 11, 7.   **II.** but generally used of rapid movement, (1) of living creatures, *to flee, hurry off, go forth* (cf. (?) colloquial *shack* to *rove about*) :—Đâ sceóc hê on niht fram đære fyrde him sylfum tô myclum bysmore *he fled at night from the English army to his great disgrace*, Chr. 992; Erl. 130, 32. Hê sceóc dîgellîce of đære byrig *he hurried off secretly from the town*, Homl. Th. ii. 154, 12. Sceócon môdige maguþegnas morþres on luste *they hurried on lusting for murder*, Andr. Kmbl. 2280; An. 1141. Hê behêt đæt hê nǽfre siðđan of đam

mynstre sceacan nolde *he promised that he would not leave the monastery in a hurry again*, Homl. Th. ii. 176, 28. Hwî woldest ðú sceacan bútan mínre gewitnisse *cur ignorante me fugere voluisti?* Gen. 31, 27. Deófol ongon on fleám sceacan, Exon. Th. 280, 17; Jul. 630: Judth. Thw. 25, 34; Jud. 292. Hí gewiton in forwyrd sceacan *they hurried to perdition*, Andr. Kmbl. 3187; An. 1596. On gerúm sceacan, Exon. Th. 401, 20; Rä. 21, 14. On lyft scacan, fleógan ofer foldan, Cd. Th. 280, 32; Sat. 263: Beo. Th. 3610; B. 1803. [Nes þer nan biscop·þ forð on his wæi ne scoc, na munec ne nan abbed þ he an his wæi ne rad, Laym. 13246.] (2) of material things, *to move quickly, to be flung, be displaced by shaking* :—Hwîlum hâra scôc forst of feaxe *at times the hoar frost was thrown from my hair*, Exon. 498, 26; Rä. 88, 7. Strǽla storm, strengum gebǽded, scôc ofer scyldweall, Beo. Th. 6227; B. 3118. (3) of immaterial things (time, life, thought, etc.), *to pass, proceed, depart* :—Ðonne mín sceaceþ líf of líce *when my life takes flight from the flesh*, Beo. Th. 5478; B. 2742: Exon. Th. 327, 4; Wîd. 141. Swǽ giémeleáslíce oft sceacaþ úre geþohtas from ûs ðæt wé his furðum ne gefrēdaþ *curae vitae ex sensu negligenti quasi nobis non sentientibus procedunt*, Past. 18, 7; Swt. 138, 20. Seó tíd gewât sceacan *time passed on*, Cd. Th. 9, 2; Gen. 135. Is nû worn wintra sceacen, Elen. Kmbl. 1263; El. 633. Ðá wæs dæg sceacen, Beo. Th. 4602; B. 2306: 5448; B. 2727. Ðá wæs winter scacen, 2277; B. 1136. Wæs hira blǽd scacen *their glory had departed*, 2253; B. 1124. Biþ se wēn scǽcen, Exon. Th. 50, 23; Cri. 805. Biþ his líf scǽcen, 329, 25; Vy. 39. Biþ týr scecen, 447, 27; Dôm. 45. **III.** *to shake* (trans.) :—Ic sceace (scace, scace) *concutio*, Ælfc. Gr. 28, 4; Zup. 169, 7. Gûðweard gumena wælhlencan sceóc, Cd. Th. 188, 31; Exod. 176. Sceacas (scæcas, Rush.) ðæt asca of fôtum iúrum *excubite te pulverem de pedibus vestris*, Mk. Skt. Lind. 6, 11. Wæs sceacen *vibratur*, Germ. 401, 47. **IV.** *to weave* (cf. bregdan) :— Scecen wē *plumemus* (cf. windan *plumemus*, 83, 78 : *plumarium opus dicitur quod ad modum plumarum texitur*, Du Cange), Wrt. Voc. ii. 66, 80. [O. Sax. skakan *to depart* ; ellior skôk *he died* : cf. O. H. Ger. untscachondes *flutivagi*, Grff. vi. 412 : *Icel.* skaka *to shake* (trans.).] v. â-, of-, on-, ôþ-, tô-sceacan.

sceacdôm (?), es; *m. Flight, hurried departure* :—Nolde nâ Iacob cýðan his scæcdôm (sæcdôm, Thw.) his sweore *noluit Jacob confiteri socero suo, quod fugeret*, Gen. 31, 20. v. preceding word.

sceacel, es; *m.* **I.** *a shackle* :—Sceacul *vel* bend *columbar*, Wrt. Voc. i. 16, 44. **II.** the word also glosses *plectrum* :—Secele oððe slegele scecen wē *plectro plumemus*, ii. 66, 78–80. Sceacelas *plectra*, 89, 10. [*Prompt. Parv.* schakkyl *numella*. Ancren schulen ine so wide scheakeles pleien ine heuene . . . Þet tet bódi schal beon hwar so euer þe gost wule in one hondhwule, A. R. 94, 25. O. Du. schakel *the link* or *ring of a chain* : Icel. skökull *the pole of a carriage* : Swed. skakel *the loose shaft of a carriage* : Dan. skagle *a trace for a carriage*.] v. sweor-sceacel; sceacan.

sceácere, es; *m. A robber* :—Þeáf and sceácere *fur et latro*, Jn. Skt. Lind. 10, 1. Þeáfas and sceácaras *fures et latrones*, Mt. Kmbl. p. 8, 1. Mið sceácerum (sceácrum, Rush.) í mið sétnerur. *cum seditiosis*, Mk. Skt. Lind. 15, 7. [O. H. Ger. scâhhâre *latro*; scâh *latrocinium, praeda*: O. Frs. skâk *booty* ; skêka *to rob* : Du. schaak *abduction*.] v. next word.

sceácerian. v. tô-sceácerian.

sceacga, an; *m. The hair of the head* ; cf. shaggy :—Feax, sceacga coma, Wrt. Voc. ii. 22, 56. [Cf. Icel. skegg *the beard* : Dan. skæg : Swed. skägg.] v. next word.

sceacged; *adj. Having hair on the head, shagged* :—Sceacgede *comosus*, Wrt. Voc. ii. 22, 71. [Cf. Icel. skeggjaðr *bearded*.] v. preceding word.

sceac-líne, sceacness, sceacul. v. sceát-líne, on-sceacness, sceacel.

scæd, es; *m. ?* :—Siblingchyrst and Trowincsceadas and Rocisfald, Cod. Dip. Kmbl. iii. 123, 8.

scæd, scæd, scad, sced, es; *n. Shade;* fig. *shelter, protection* :—Æfter sceades sciman, Salm. Kmbl. 233; Sal. 116. Scedes, Cd. Th. 271, 15; Sat. 106. On sceade (scade, MS. B.) âhôn, Lchdm. i. 284, 21. On ðam sceade his geteldes *in abscondito tabernaculi sui*, Ps. Th. 26, 6. Manna bearn hopiaþ tô ðæm sceade ðínra fiðera *filii hominum in protectione alarum tuarum sperabunt*, 35, 8. Ðonne on sceade weaxeþ, Exon. Th. 214, 5; Ph. 234. Hé in scade weardaþ, on wudubearwe, wêste stôwe, 209, 10; Ph. 168. Ðæt gē mec mid searocræftum under scæd scúfan môtan, 142, 20; Gú. 647. Sceadu beóþ bidyrned, ðǽr se leóhta beám leódum byrhteþ, 67, 16; Cri. 1089. Sceadu sweðerodon, Andr. Kmbl. 1672; An. 838. Sceado (sceaðo, MS.), Cd. Th. 184, 27; Exod. 113. Scadu, Exon. Th. 179, 16; Gú. 1262. Deorc deáþes sceadu dreógan, 8, 15; Cri. 118. Sunne ofer sceadu scíneþ, 212, 14; Ph. 210. Under sceadu bregdan *to kill*, Beo. Th. 1419; B. 707. Dæg ǽresta geseah deorc sceado sweart swiðrian, Cd. Th. 8, 33; Gen. 133. v. leáf-scead, sceadu.

sceád, scâd, es; *n. Shed* (in water-*shed*), *a division, distinction, reason, reckoning* :—Ðú scealt gyldan scád wordum ðínum *thou shalt give an account* (*of thine actions*) *in words*, Dôm. L. 73. [Haueð wit and schad bituhhe god and uuel, O. E. Homl. i. 255, 30. Shæd and skill, Orm. 5534. Niss bitwenen junnc and hemm nan shæd i manness kinde, 6229. Schead ba

of god and of uvel, Kath. 240.   O. L. Ger. scēth *discrimen* : O. H. Ger. sceit *discissio*.] v. ge-, tô-, unge-sceád.

sceáda (sceáde; *f.* (?)), an ; *m. The top of the head, parting of the hair* :—Hē tôfylleþ feaxes scádan *conquassabit verticem capilli*, Ps. Th. 67, 21. [Crulle was his heer, and as the gold it schon . . . Ful streyt and evene lay his joly schood. Miller's Tale, 130. The nayl y-dryven in the schode a-nyght, Knight's Tale, 1149. v. *Halliwell's Dict.* shed, *and* E. D. S. Pub. Lincolnshire, shed *the parting of the hair*. Cf. *Prompt. Parv.* schodynge of the heede *discrimen* : O. L. Ger. scēthlo, sceithlo *vertex* (*capilli*) : O. H. Ger. sceitila *vertex* ; fahs-sceitila *cervix capilli*.] v. preceding word.

sceada. v. niht-scada.

sceádan, scâdan ; *p.* scēd, sceád (v. tô-sceádan) ; *pp.* sceáden. **I.** *trans.* (1) *to separate, divide, make a line of separation between* :— Eádmund Myrce geeode swâ Dor scâdeþ, hwîtan wylles geat and Humbra eá brâda brimstreám *Edmund conquered Mercia, which Dor, Whitewell's gate, the river Humber, the broad estuary, divides* (*from Northumbria*), Chr. 942; Erl. 116, 9. From Egypta eðelmearce swâ Nilus sceádeþ, Cd. Th. 133, 10; Gen. 2208. Ðonne sceádene beóþ ða synfullan and ða sôðfæstan on ðam mǽran dæge, Exon. Th. 375, 33; Seel. 147. (2) *to distinguish, decide* :—Scâdeþ *discriminet*, Wrt. Voc. ii. 27, 20. Scâdet, 93, 34. Ðonne biþ gǽsta dóm sceáden swâ hí geworhtun ǽr *then shall the spirits' doom be decided, according to their deserts*, Exon. Th. 76, 2; Cri. 1233. Sceáden mǽl *the appointed time* (?), Beo. Th. 3882; B. 1939. (3) *to scatter, shed* :—Nim beolonan sǽd sceád on glēda *take seed of henbane, scatter it on gledes*, Lchdm. ii 38, 1 : 52, 2. Sceád (scâd, MS. B.), i. 82, 7. Gnîd tôgædere and scâd on, ii. 134, 3. Ðæt mela biþ gôd on tô sceádenne, 94, 3. [*See also the compounds* (omitted *in their proper places*) :—Besceád, 54, 21. Ofersceáde, 182, 2.] Tô scēdende blôd *ad effundendum sanguinem*, Ps. Spl. T. 13, 6. **II.** *intrans.* (1) *to separate, divide, part* :—Tigelum sceádeþ hróstbeáges hrôf (rôf, MS.) *the woodwork of the roof parts from the tiles*, Exon. Th. 477, 29; Ruin. 31. Ðonne dæg and niht scâde *when day and night separate* (*at morning twilight*), Lchdm. ii. 116, 19. Ðonne dæg and niht furþum scâde, 346, 14: 356, 6: iii. 6, 7. Ðonne dæg scáde and niht, ii. 138, 16. (2) *to be distinguished, to differ* :—Scâdaþ *discrepent*, Wrt. Voc. ii. 27, 1: 88, 39. (3) *to scatter, shed* :—Ðonne sceádaþ ða wyrmas on ðæt wæter, Lchdm. ii. 38, 4. [He shodeð þe gode fro þe iuele, O. E. Homl. ii. 67, 24. Eiðer of þisse teres schedde þe apostel, i. 157, 33. Þe halwe men schedden teres, 157, 15. Redde blod scede (sadde, 2nd MS), Laym. 5187. He shadde him fra menn, Orm. 3200. Shædenn hemm fra Criste, 1209. Tobrekeð hore uetles and schedeð hore clennesse, A. R. 166, 7. His blode þet he shedde for us, 312, 19. Scheaden þet chef urom þe clene cornes, 270, 27. Blod isched, 402, 21. So wurð ligt fro ðisternesse o sunder sad, Gen. and Ex. 58. On sunder shad, 148. Goth. skaidan *to divide, separate* : O. Sax. skēdan, skēthan *trans. and intrans.) to separate* : O. L. Ger. scēthan, sceithan : O. Frs. skēda, skētha *to separate, to decide* : O. H. Ger. sceidan *separare, segregare, discernere, distinguere, discriminare, judicare*.] v. â-, for-, ge-, tô- (be-, ofer-, v. I. 3 above) sceádan.

sceadd *a shad* :—Ic geann Ælfhelme and Wulfâge ðæra landa betwux Ribbel and Mærse and on Wirhalum . . . on ðæt gerád ðonne sceaddgenge sý ðæt heora ǽgðer sylle .iii. þúsend sceadda intô ðære stôwe æt Byrtûne *I grant to Ælfhelm and Wulfeah the lands between the Ribble and the Mersey, and in Wirral . . . on the condition that, when shad are in season, each of them give .iii. thousand shad to the convent at Burton*, Chart. Th. 544, 21–31.

sceadd-genge ; *adj. Seasonable for shad.* v. preceding word.

sceáde-sealf, e ; *f. A salve that may be shed on a place* (? v. sceádan, I. 3), *a medicinal powder* :—Sceádesealf tô eágum, Lchdm. ii. 300, 6. Wyrc gôde drîge scâdesealfe : nim gebærned sealt and piper and hwîtewudu, gegnîd tô duste âsift þurh clâð, dô lytlum on, 308, 22.

sceadiht ; *adj. Shady* :—Of munte scedehtum *de monte umbroso*, Ps. Surt. ii. p. 189, 16.

sceádlíce ; *adv. Reasonably, rationally* :—Gif hē gesceádlíce (sceádelíce, Wells Frag.) mid eáðmódnesse and mid sôþre lufe hwilcu þing on mynstre tǽle *si qua rationabiliter et cum humilitate caritatis reprehenderit*, R. Ben. 109, 8. v. ge-, un-sceádlíce.

sceadu; *gen.* sceaduwe, sceadwe, sceade; *f. Shadow, shade* :—Sceadu *umbra*, Wrt. Voc. i. 77, 8. **I.** *a shadow* (cast by an object) :— Seó sceadu byþ tô underne seofon and twentigoþan healfes fôtes *the shadow* (*of the dial-gnomon*) *will be twenty-six and a half foot long at nine o'clock* (*on Christmas day*), Lchdm. iii. 218, 4 (and often on this and following pages). Nis ðeós woruldlíce niht nân þing bûton ðære eorþan sceadu betweox ðære sunnan and mankynne . . . Seó sceadu ástíhþ up ôþ ðæt heó becymþ tô ðære lyfte ufeweardan, and ðonne beyrnþ se môna hwîltídum, ðonne hé full byþ, on ðære sceáde ufeweardre and fǽggeteþ oððe mid ealle âsweartaþ, 240, 18–24. On India lande wendaþ heora scada (sceada, MSS. R. P.) on sumera sûðweard and on wintra norðweard. Eft on Alexandria on ðam sumerlícan sunn-

stede on middæge ne byþ nân sceadu on nânre healfe, 258, 12-16. His sceadu gehælde ða untruman, Homl. Skt. i. 10, 19. Dagas mîne swâ swâ scadu âhyldon, Ps. Spl. 101, 12 : 143, 5. Swâ ðû on scimiendre sceade lôcige *sicut umbra*, Ps. Th. 143, 5. Dagas mîne swâ swâ sceaduwa âhyldon, Ps. Lamb. 101, 12. **II.** *shade* as opposed to light, *shadow* (lit. and fig.), *darkness* :—Ða ðe nân sceadu (scadu, Cott. MSS.) ne geþiéstraþ ðære twiéfealdnesse *quos nulla umbra duplicitatis obscurat*, Past. 35, 4; Swt. 243, 23. Þýstro hæfdon bewrigen mid wolcnum wealdendes hræw, sceadu forþeode wann under wolcnum, Rood Kmbl. 108; Kr. 54. Oferwreâh ûs scadu deâþes, Ps. Spl. 43, 22. On midlunge sceaduwe dæþes, 22, 4. On scade (sceaduwe, Ps. Lamb.) deâþes, 106, 10. Ðis andwearde lif is swîðe anlîc sceade, and on ðære sceade nân mon ne mæg begitan ða sôðan gesǽlþa, Bt. 27, 3; Fox 98, 19. On midde ða sceade deâþes, Ps. Th. 22, 4. Ðâ gesundrode sigora Waldend leóht wið þeóstrum, sceade wið scîman, Cd. Th. 8, 22; Gen. 128. For hwon sêcest ðû sceade, 54, 8; Gen. 874. **III.** *shadow, protection* :— Under scaduwe fiðera ðînra gescyld mê, Ps. Spl. 16, 10. Hî slêpon ûte on triówa sceadum *umbras dabat altissima pinus*, Bt. 15; Fox 48, 12. **IV.** *a shady place, shade, arbour* :—Scadu *scena* (cf. geteld *scena vel tabernaculum*, i. 37, 15), Wrt. Voc. ii. 119, 80. Sceadwe *scenam*, 80, 1. **V.** *shadow* as opposed to substance, *an obscure image* :—Seó ealde ǽ wæs swilce sceadu, and seó nîwe gecýðnis is sôðfæstnys, Homl. Th. i. 356, 1. Genôg ic ðê hæbbe nû gereht ymbe ða anlîcnessa and ymbe ða sceadwa ðære sôðan gesǽlþe *hactenus mendacis formam felicitatis ostendisse suffecerit*, Bt. 33, 1; Fox 118, 34. [*O. E. Homl.* sceadewe, shadewe : *A. R.* scheadewe : *Goth.* skadus : *O. Sax.* skado : *O. H. Ger.* scato.] v. beám-, heolstor-, niht-, scûr-sceadu; scead.

sceadu-geard, es; *m. A shady enclosure* :—Sceadugeardas *Tempe*, Wrt. Voc. ii. 122, 17.

sceadu-genga, an; *m. One who walks in darkness* (v. sceadu, II) :— Com on wanre niht scrîðan sceadugenga (*Grendel*), Beo. Th. 1410; B. 703. Cf. niht-genga.

sceadu-helm, es; *m. The cover of night, darkness* :—Niht, scaduhelma gesceapu, Beo. Th. 1304; B. 650.

sceadwian, sceadewian; *p.* ode *To cover with shadow* :—Hê scadewode (scaduaþ, Ps. Lamb. : sceadewede, Blickl. Gl.) *obumbrabit*, Ps. Spl. 90, 4. [*Goth.* ufar-skadwjan: *O. Sax.* skadowan, scadoian : *O. L. Ger.* scedeuuan : *O. H. Ger.* scatewen.] v. ofer-sceadwian ; sceadwung.

sceádwíslíc. v. ge-, un-sceádwíslíc, *and next word*.

sceádwíslíce; *adv. With discretion, rationally* :—Gif ðû him sceádwíslíce æfter spyrast, Bt. 13; Fox 38, 3. v. ge-sceádwíslíce.

sceádwísness, e; *f. Reason* :—Ðâ cwæþ seó Gesceádwísnes (Sceádwîsnes, Cott. MS.), Bt. 5, 3; Fox 12, 1. Ic wêne ðæt hyt mîn sceádwîsnes (*reason*) wêre, Shrn. 164, 29. Sceádwîsnyssum *ratiociniis*, R. Ben. Interl. 17, 6.

sceadwung, e; *f. An overshadowing* :—On sumum earde dagas beóþ lengran, on sumon scyrtran for ðære eorþan sceadewunge (sceadwunge, MS. R.) *in one land days are longer, in another shorter, because of the way in which the shadow falls on the earth*, Lchdm. iii. 258, 4. Se fulla môna fǽrlîce fâgettaþ ðonne hê ðæs sunlícan leóhtes bedǽled biþ þurh ðære eorþan sceadwunge (*by the casting of the earth's shadow*), Homl. Th. i. 610, 1. v. be-sceadwung.

sceáf, es; *m. A sheaf, bundle.* **I.** in the following glosses :— Sceáfes *fascis*, sceáfe *fasculo* (*fasciculo*), Wrt. Voc. ii. 34, 62-63. Sceáfas *areoli*, 7, 16 : garbas, 40, 60 : garbas, manipulas, 89, 19. Sceabas, scêbas *areoli*, Txts. 38, 30 : garbas, 66, 468. Sceáfum *fasciculis*, Hpt. Gl. 520, 19. **II.** *a sheaf* (*of corn*) :—Mê þûhte ðæt wê bundon sceáfas (*manipulos*) on æcere and ðæt mîn sceáf ârise ômiddan eówrum sceáfum and eówre gilmas âbugon tô mînum sceáfe, Gen. 37, 7. Gýme hê ðæt nâðor ne misfare ne corn ne sceáf, Anglia ix. 260, 12. Mid his sceáfe sceát âfyllan, Ps. Th. 128, 5. Hê nǽnne sceáf (*manipulum*) ne rîpþ, Past. 39, 2; Swt. 287, 3. Heora sceáfas (*manipulos*) beraþ, Ps. Th. 125, 6. **II a.** *a bundle* (*of herbs*) :—Dippaþ ysopan sceáf (sceaft, Thw.) on ðam blôde *fasciculum hyssopi tingite in sanguine*, Ex. 12, 22. Syndrige sceáfas *separate bundles* (*of rue, dill, mint, and marche*), Lchdm. ii. 188, 24. Rûdan sceáfas þrý, 216, 2. [*O. H. Ger.* scoub: *Ger.* schaub: *Du.* schoof: *Icel.* skauf *a fox's brush.*]

sceafa, an; *m. A plane* :—Sceaba *runcina*, Txts. 92, 853. Scafa *olatrum*, Wrt. Voc. i. 287, 11 : ii. 64, 13. Hê sceal habban æcse, adsan, scafan, sage, Anglia ix. 263, 2. [*Prompt. Parv.* schave or schavynge knyfe *scalpellum, scalprum*: *O. H. Ger.* scaba *plana, asperella*: *Ger.* schabe: *Du.* schaaf *a plane*: *Icel.* skafa *a scraper*.] v. mǽl-sceafa, sceafan.

Sceáfa, an; *m. The name of a king of the Lombards* :—Sceáfa weóld Longbeardum, Exon. Th. 320, 21; Vîd. 33. *See also* Scyld Scêfing, Beo. Th. 7; B. 4.

sceafan, scafan; *p.* scôf; *pp.* sceafen, scafen *To shave, scrape, shred, polish* :—Scaebe *poleo*, Wrt. Voc. ii. 117, 63. Gif hê ðæt ômige fæt mid ungemete scæfþ *dum nimis cupit eradere eruginem*, R. Ben. 121, 4. Hê scôf on halig wæter of ðam hâlgan treówe, Swt. A. S. Rdr. 102, 216. Man scôf ðæra bôca leâf and ða sceafþan dyde on wæter *rasa folia*

codicum, et ipsam rasuram aquae immissam, Bd. 1, 1; S. 474, 37. Monige men sprytlan âcurfon and on wæter scôfan, 3, 17; S. 544, 45, col. 1. Sceaf (scaf, MS. B.) gâte horn on þrý scenceas, Lchdm. i. 352, 11 : 344, 13. Sceafe ðæt grêne, ii. 292, 26. Ðû scealt hine scafan on wæter . . . and ðære reádan eorþan dǽl scafe ðærtô, ii. 290, 11-13. [*Goth.* skaban: *O. L. Ger.* scavan *scalpere*: *O. H. Ger.* scaban, scapan *scabere, scalpere, radere*: *Icel.* skafa.] v. â-, be-, ge-sceafan (-scafan).

sceáf-fôt; *adj. Splay-footed* :—Scâbfoot, scaabfôt, scâffo[o]t *pansa*, Txts. 90, 832. Scâffôt, Wrt. Voc. i. 288, 78. [Cf. *Icel.* skeift *askew, oblique*; skeifa *a horse-shoe.*]

sceáf-mǽlum; *adv. In sheaves* or *bundles* :—Gadriaþ ǽrest ðone coccel, and bindaþ sceáfmǽlum, Mt. Kmbl. 13, 30.

sceafoþa, sceafþa, scæfþa, an; *m.* (*or -e; f.* ?) *A shaving, chip, what is shaved, scraped, or rubbed off* :—Ðâ gehâlgode ic wæter and scæfþan dyde on ðæs foresprecenan treówes *tunc benedixi aquam, et astulam roboris praefati inmittens*, Bd. 2, 13; S. 539, 5. Ða scæfþan ðe ðæron genumene wǽron lǽcedôm bǽron *astulae de illo abscissae solent adferre medelam*, 4, 6; S. 574, 9. Man scôf ðara bôca leáf and ða sceafþan (*ipsam rasuram*) dyde on wæter, 1, 1; S. 474, 38. Monige spônas and sceafþan (*astulas*) nimaþ, 3, 2; S. 524, 31 : 3, 17; S. 544, 44, col. 2. Genim heorotes sceafoþan of ðam horne, Lchdm. ii. 72, 13. Genim heorotes sceafoþan of felle âscafen mid pumice, 100, 14.

sceaft, es; *m. A smooth, round, straight stick or pole, a shaft.* **I.** generally (1) *the shaft of a spear* (cf. *Icel.* skaft *the shaft*, spjót *the point*) :—Spereleás sceaft *contus*, Wrt. Voc. i. 35, 42. Gif se ord sîe þreó fingre ufor ðonne hindeweard sceaft, L. Alf. pol. 36; Th. i. 84, 17, 18. His sceaft ætstôd ætforan him, and ðæt hors hine bær forþ, swâ ðæt ðæt spere him eode þurh ût, Homl. Skt. i. 12, 53. Hê sceáf mid his scylde, ðæt se sceaft tôbærst, and ðæt spere sprengde, Byrht. Th. 135, 52; By. 136. Gâr sceal on sceafte, ecg on sweorde, Exon. Th. 346, 12; Gn. Ex. 202. [He igrap his spere stronge . . . þe scæft al tobrac, Laym. 6494.] Or (2) *a spear* :—Sceaft *asta, quiris*, Wrt. Voc. i. 35, 18 : 84, 24. Ðes sceft (scæft, sceaft) *cuspis*, Ælfc. Gr. 9, 28; Zup. 56, 4. Scyld sceal cempan, sceaft reáfere, Exon. Th. 341, 23; Gn. Ex. 130. Sceaftes t speres ðînes *hastae tuae*, Cant. Ab. 11. Ðæt yrre ðæt geþyld mid ðam sceafte (mid his spere, B.) slihþ *ira patientiam conto percutit*, Glos. Prud. A. 18. Scyld sceft oncwyð, Fins. Th. 12; Fin. 7. Hlyn wearð on wîcum scylda and sceafta, Cd. Th. 124, 13; Gen. 2062. Deáwig sceaftum, 199, 25; Exod. 344. Hig bǽron lange sceaftas, and ne côman hig nâ tô feohtanne, ac ðæt hig woldan mid hlôpe geniman, Shrn. 38, 9. **II.** *the shaft of an arrow* :—Sceaft feðergearwum fûs, Beo. Th. 6228; B. 3118. [Þe ssaft (*the arrow that killed William Rufus*), þat was wyþoute, gryslych he tobrec, R. Glouc. 419, 2.] **III.** *a pole* :—Fana hwearfode scîr on sceafte, Met. 1, 11. Ic gegaderode mê stuþan sceaftas . . . Ic lǽre ǽlcne ðara ðe manigne wǽn hæbbe, ðæt hê menige tô ðam ilcan wuda ðâr ic ðâs stuþan sceaftas cearf, Shrn. 163, 5-14. [Moyses made a wirme of bras, And henget hege up on a saft, Gen. and Ex. 3899.] **III a.** *something shaped like a shaft, a taper* :—Swâ swâ eles gecynd biþ ðæt hê beorhtor scîneþ ðonne wex on sceafte (*wax in the form of a taper* or (?) *a wax candle in a candlestick*, cf. candelstæf), Blickl. Homl. 129, 1. **IV.** The word occurs in the passage that defines the distance to which the king's ' grið ' extended, but the origin of the phrase, of which it forms part, is not evident :— Ðus feor sceal beón ðæs cinges grið fram his burhgeate ðær hê is sittende on feówer healfe his, ðæt is, .iii. mîla, and .iii. furlang, and .iii. æcera brǽde, and ix. fôta, and .ix. scæfta munda, and .ix. berecorna, L. Ath. iv. 5; Th. i. 224, 7-10. [Cf. Tria miliaria, et .iii. quarantene, et .ix. acre latitudine, et .ix. pedes, et .ix. palme, et .ix. grana hordei, L. H. i. 16; Th. i. 526, 15. As the name of a measure of about six inches the phrase continued to exist. Stratmann gives *schaftmonde*, Nares cites a passage from Harrington's Ariosto in which *shaftman* occurs ; in Ray's Collection (1691) *shafman, shafmet, shaftment* is explained 'the measure of the fist with the thumb set up.' v. also Halliwell's Dict., and Jamieson's, s.v. *schaftmon, shathmont.* For the latter form see Sir W. Scott's Antiquary, c. 8 (at the end). [*O. Sax.* skaft *a spear* : *O. H. Ger.* scaft *hastile, hasta, jaculum, telum, arundo* : *Icel.* skapt, skaft *a shaft, haft* (*of an axe*).] v. deoreþ-, here-, lôh, wæl-sceaft.

sceaft, es; *m. : f.* **I.** *creation, origin* :—Ealle sint emnæðele gif wê willaþ þone fruman sceaft geþencan and ðone Scippend . . . Ac ǽlc mon ðe allunga underþeóded biþ unþeáwum forlǽt his Sceppend and his fruman sceaft *si primordia vestra auctoremque Deum spectes, nullis degener exstat, ni vitiis pejora proprium deserat ortum*, Bt. 30, 2; Fox 110, 17-21. **II.** *a creation, what is created, a creature* :—Ealre sceafte fæder *omniparens*, Germ. 389, 2. Fram fruman gesceafte (scæftes, Lind.) *ab initio creaturae*. Mk. Skt. 10, 6. Of frymmðe ðære gesceafte (ðæs sceæftes, Lind.) ðe God gesceóp *ab initio creaturae quam condidit Deus*, 13, 19. Bodiaþ godspell ealre gesceafte (êghwelcum sceafte, Lind.) *praedicate euangelium omni creaturae*, 16, 15. Gif God næfde on eallum his rîce nâne frige sceaft (gesceaft, Cott. MS.), Bt. 41, 2; Fox 244, 29. Forðæm sint ðâs sceafta (gesceafta, Cott. MS.), 41, 5; Fox 252, 30. Alra þinga t sceafta *omnium rerum*, Mt. Kmbl. p. 12, 16. [Our schaft

wele knawes he *ipse scit figmentum nostrum*, Ps. 102, 14. Godd þatt alle shaffte wrohhte, Orm. pref. 58. Swilc safte (*the tabernacle*) was ear neuere on werlde brogt, Gen. and Ex. 3628. For be a man faire or foule it falleth nouȝte for to lakke þe shappe ne þe shafte · þat God shope hymselue, Piers P. B. 11, 387. *O. Sax.* -skaft: *O. H. Ger.* -scaft.] v. ær-, ed-, frum-, ge-, geó-, hyge-, meotud-, nafel-, orleg-, self-, un-, wan-sceaft.

-sceaft; *adj.* v. feá-sceaft.

Sceaftes-burh *Shaftesbury* in Dorset :—Æt Sceaftesbyrig, Chr. 1036; Erl. 164, 9. Tó Scæftesbyrig, 980; Erl. 129, 34. See also Cod. Dip. Kmbl. vi. 329, col. 1.

sceaft-lóha, an ; *m.* (or -e ; *f.?*) *The strap attached to the shaft of a missile* :—Scaeptlóan *hastilia telorum*, Txts. 66, 489. Sceptlóum *amentis*, 42, 106. v. lóh-sceaft, mæst-lón, sceaft-tog.

sceafþa. v. sceafoþa.

sceaft-tog (?) *the strap attached to the shaft of a missile* :—Sceptog *ammentum*, Wrt. Voc. ii. 100, 11. v. sceaft-lóha.

sceaga, an ; *m. A shaw, small wood, copse, thicket.* The word is found in many local names, and was preserved in various dialects, e.g. *shaw* a small shady wood in a valley, E. D. S. Pub. B. 7 (West Riding) : a wood that encompasses a close, B. 16 (Sussex). *Shaws* broad belts of underwood, two, three, and even four rods wide, around every field, Farming words, 4 (Sussex). *Shaw* a natural copse of wood, Cumberland. The word occurs in the following passages of charters :— Juxta silvam quam dicunt Toccansceaga, Cod. Dip. Kmbl. i. 121, 24. Mariscum uocabulo Scaga, quam etiam circumfluit Iaegnlaad, 190, 6 : 160, 28. On brēmeles sceagan eásteweardne, ii. 172, 28. On ðone langan sceagan westeweardne ; of langan sceagan on ðæt hǣðene byrgils, iii. 85, 19-20. Onbūtan færsscagan, 229, 29. Rihte ũt þurh ðone sceagan óþ ða lēge, 406, 27. Of ðære byrig þwyres ofer ðane sceagan, 460, 2. þurh Beaddes scagan, v. 166, 10. [At a schaȝe syde, Gaw. 2161. In a schaȝe (*the reference is to the gourd under which Jonah sat*) þat schade ful cole, Allit. Pms. 105, 452. Wodschaweȝ, 9, 284. For love of hym thou lovedst in the shawe, I mene Adon, Tr. and Cr. 3, 671. Thane schotte owtte of þe schawe schiltrounis many, Mort. A. 1765. In ȝone dyme schawes, 1723. See also Halliwell's Dict. and Nares' Glossary. Cf. (?) *Icel.* skaga *to project.*]

sceagod. v. sceacgod.

sceal *shall.* v. sculan.

sceál, scál (?) *a shoal, troop, band* :—Ic be hondum mót hǣðenre (-ra?) sceál grípan tó grunde, Godes andsacan, Cd. Th. 281, 8 ; Sat. 268. Cf. Mid his handscále, Beo. Th. 2638; B. 1317.

sceale, es ; *m.* **I.** *a servant* :—Eálá ic eom ðín ágen esne Dryhten and ðín swylce eom sceale ombehte (cf. ambeht-sceale) and ðínre þeówan suna *O Domine, quia ego servus tuus, ego servus tuus, et filius ancillae tuae*, Ps. Th. 115, 6. Ic eom ðín hold sceale *tuus sum ego*, 118, 94. Dô ðínes scealces (*servi*) sáwle blíðe, 85, 3. Tó scealce *in servum*, 104, 15. Hǣl ðínne sceale *salvum fac servum tuum*, 85, 2 : 88, 17. Hê Moyses sende his sylfes sceale *misit Moysen servum suum*, 104, 22. Beseoh on ðíne scealcas *respice in servos tuos*, 89, 18. Babilone weard hêt his scealcas scúfan ða hyssas in bǣlblyse, Cd. Th. 230, 10 ; Dan. 231. **II.** *as a term of reproach* :—Ða hine heówon hǣðene scealcas, Byrht. Th. 137, 5 ; By. 181. Hwîlum ic gehêre helle scealcas, gnorniende cynn, Cd. Th. 273, 8 ; Sat. 133. **III.** *a man, soldier, sailor* :—Scealc (*Beowulf*) hafaþ dǣde gefremede, ðe wê ealle ǣr ne meahton, Beo. Th. 1883; B. 939. Eode scealc monig swíð-nicgende tó sele searowundor seón, 1841 ; B. 918. Hú mæg ðæt gesceádwîs scealc (cf. gesceádwîs mon, Bt. 28 ; Fox 100, 30) gereccan, ðæt hê him ðý sêlra þince, Met. 15, 14. Brugdon scealcas (*the Jews who defeated the Assyrians*) of sceáðum scîrmǣled swyrd, Judth. Thw. 24, 38; Jud. 230. Næs scealca nán *there was no one*, Met. 8, 21. Scipu mid scealcum *ships with their crews*, Exon. Th. 362, 3 ; Wal. 31. [Þer wes moni bald scalc (cniht, 2nd MS.), Laym. 19126. Heo wenden bi þen scelden þat hit heore scalkes (men, 2nd MS.) weoren, 4219. Schalk *a knight*, Gaw. 160. *Goth.* skalks ðoûlos : *O. Sax.* skalk *servus* : *O. Frs.* skalk *a servant, slave* : *O. H. Ger.* scalch *servus, famulus, manceps* : *Icel.* skálkr *a rogue.* v. Grmm. R. A. 302, and Grff. vi. 480 sqq. for compounds.] v. ambeht-, beór-, freoðo-scealc.

sceald. v. dæg-sceald.

sceald-húlas glosses *paupilius*, Wrt. Voc. ii. 116, 21. v. next word.

sceald-þýfel (-hýfel), es ; *m. A thicket* :—Scaldhýflas, scald[t]hýblas alga, algè ; scaldhýflas *vel* sondhyllas *alga*, Txts. 38, 58. ‘Scaldhýflas alga, scaldhúlas *paupilius*, are errors. Scealdþýfelas, *fruteta, thickets*, occurs in Greg. Dial.’ Lchdm. iii. 343, col. 2. [Cf. (?) *O. H. Ger.* scald *sacer ;* scald-eiche *ilex :* and see Grmm. D. M. 615.]

scealfor, e ; *f.* : scealfra, an ; *m. A diver* (bird) :—Scalfr, scalfur *mergulus*, Txts. 78, 647. Scealfr *mergus*, Wrt. Voc. i. 29, 13. Scealfor *turdella, mergula*, 63, 15, 16 : *mergulus*, 280, 11 : ii. 56, 18 : 89, 54. Scealfra *mergus vel mergulus*, i. 77, 27. Grǣdigre scelfre *voracis mergulae*, Hpt. Gl. 418, 70. Ða geseah hê swymman scealfran on flóde, and gelóme doppettan ádúne tó grunde ēhtende þearle ðære eá fixa . . . Ða

hêt Martinus ða fugelas ðæs fixnoðes geswícan, and tô wêstene síðian ; and ða scealfran gewiton áweg tô holte, Homl. Th. ii. 516, 6-12.

scealga, scylga, an ; *m. The name of a fish* :—Scealga *rocea*, Wrt. Voc. i. 77, 67. Scylga, 55, 77.

scealian. v. á-scealian.

sceallan, scallan ; *pl. Testiculi*, Lchdm. i. 330, 13 : 336, 15 : 358, 21.

scealu, e ; *f.* **I.** *a shell, husk* :—Scealu *glumula*, Wrt. Voc. ii. 40, 23. Scalu, scala, Wrt. Voc. 66, 462. Scale † hule *glumula*, Hpt. Gl. 439, 50. v. æpel-, beán-, stán-scealu. **II.** *a platter, dish, cup* :— .VI. mæsene sceala, Chart. Th. 429, 30. **III.** *the scale of a balance* :—Ðeós wǣge † scalu *haec lanx*, Ælfc. Gr. 9, 73 ; Som. 14, 18. Scale lanx, twá scale *balances*, Wrt. Voc. i. 38, 39-40. v. wǣg-scealu. [*O. Sax.* skala *a drinking-vessel* : *O. L. Ger.* scala *concha* : *O. H. Ger.* scala *patera, cratera, concha, gluma* : *Icel.* skál *a bowl, a scale* (of a balance).]

sceám, es ; *m. A white horse* (?) :—Etsomne cwom .LX. monna wicgum rídan, hæfdon .XI. eoredmacgas frídhengestas, IIII. sceámas (cf. (?) hyra bloncan, 405, 5 ; Rä. 23, 18), Exon. Th. 404, 8 ; Rä. 23, 4.

sceamel. v. sceamol.

sceam-fæst ; *adj. Shamefast* (corrupted later into *shamefaced.* v. 1 Tim. 2, 9 where Wicklif has *schamefastnesse*, the modern copies of the A. V. *shamefacedness*; the Revised Version has restored *shamefastness*), *modest, bashful* :—Scamfæst *verecundus vel pudens*, Wrt. Voc. i. 51, 31. Sceamfæst *verecundus*, 86, 56. Seó scamfæste nǣcednys *pudibunda* (*pudica* .i. *erubescens*) *nuditas*, Hpt. Gl. 492, 53. Mǣden is sceamfæst, Lchdm. iii. 188, 6. Scamfæst, 192, 2. On óðre wîsan sint tó lǣranne ða scamleásan, on óðre ða scamfæstan (*verecundi*), Past. 31 ; Swt. 205, 21. [Sannte Marȝe wass shammfæst, Orm. 2175. Wyfmen þet byeþ ssamuest, Ayenb. 222, 20. Schamefast chastite, Chauc. Kn. T. 1197. Schamefast *verecundus, pudorosus*, Prompt. Parv. 443.] v. un-sceamfæst.

sceam-full ; *adj. Modest, chaste* :—Sceomfull *pudica*, Rtl. 108, 25. Sceomfullre *verecundia*, 110, 3. [Schrift schal beon . . . edmod, scheomeful, dredful, A. R. 302, 23. *Dan.* skam-fuld *shamefaced, ashamed.* Chaucer uses the word in its modern sense *ignominious*, As shamful deeth as herte may deusye Come to these Juges, C. T. Group C. 290.]

sceamfullness ; *e; f. Modesty ; pudicitia.* v. un-sceamfullness.

sceamian ; *p. ode.* **I.** *to feel shame, be ashamed* (with gen. of cause) :—Ic ðæs nǣfre ne sceamige *non erubescam*, Ps. Th. 24, 1. Ne ic ne scamige *nec confundar*, Ps. Spl. 30, 20. Gif wê scomiaþ ðæt wê tó uncúðum monnum suelc sprecen *si homo apud hominem, de quo minime praesumit, fieri intercessor erubescit*, Past. 10, 2 ; Swt. 63, 5. Weorðaþ gescende and hiora scamiaþ ða tó Sione hete hæfdon *confundantur et revereantur, qui oderunt Sion*, Ps. Th. 128, 3. Ná ic ne scamode *non confundebar*, Ps. Spl. 118, 46. Ðiós sǣ cwið ðæt ðú ðín scamige Sidon *erubesce Sidon, ait mare*, Past. 52, 8 ; Swt. 409, 33. Hit is cyn ðæt wê úre scomigen, 52, 4 ; Swt. 407, 15. Sceamian heora ealle mîne fýnd *erubescant omnes inimici mei*, Ps. Th. 6, 8. Scamien, 69, 3. Scamien (*confundantur*) heora ealle ða unrihtwîsan, 24, 3. Heora æfstu ealle sceamien, 69, 4. For hwî hî ne mágan heora má sceamigan ðonne fægnian? Bt. 30, 1 ; Fox 108, 7. Nô hê ðære feohgyfte scamigan þorfte, Beo. Th. 2057; B. 1026. Ðú ne þearft sceamian, Soul Kmbl. 286 ; Seel. 147. For hwon sêcest ðú sceade sceomiende? Cd. Th. 54, 8 ; Gen. 874. Sceomiande man sceal in sceade hweorfan, Exon. Th. 337, 19 ; Gn. Ex. 67. Ða deóflu wendon sceamigende áweg, Wulfst. 236, 26. Hý (*Beowulf's followers who had failed him in his need*) scamiende scyldas bǣron, ðǣr se gomela læg, Beo. Th. 5692; B. 2850. **II.** *to cause shame* (used impersonally with dat. or acc. of person, gen. of cause, or with *for*, or the cause given in a clause) :—Mê sceamaþ *pudet*, Ælfc. Gr. 33 ; Som. 37, 22. Oft ðone geþyldegestan scamaþ ðæs siges ðe hî ofer ðone dióful hæfde, Past. 33, 7 ; Swt. 227, 19. Menn scamaþ for gódan dǣdan swýðor ðonne for misdǣdan, Wulfst. 164, 16. Ðæs ús ne scamaþ ná, ac ðæs ús scamaþ swýðe, ðæt wê bóte áginnan, 165, 39. Hý scamaþ, ðæt hý bêtan heora misdǣda, 165, 8. Ða woroldlecan lǣcas scomaþ, ðæt . . . , Past. 1, 1 ; Swt. 25, 20. Mê sceamaþ ðæt ic wædlige *mendicare erubesco*, Lk. Skt. 16, 3. Gehwam sceamaþ, ðæt he wáclíce gescrýd cume, Homl. Th. i. 528, 21. Him ðæs sceamode, 18, 12 : Gen. 2, 25. Ða sceamode ealle his wiðerwinnan, Lk. Skt. 13, 17. Hwá biþ gescended, ðæt mê for ðæm ne scamige? Past. 21, 6 ; Swt. 165, 5. Forgif ús úre synna, ðæt ús ne scamige eft, Hy. 7, 84. Ne sceamige nánum men, ðæt hê ánum láreow his gyltas cýðe . . . him sceal sceamian ætforan Gode, Homl. Th. ii. 602, 30. Ðæt mê ne sceamie *non erubescam*, Ps. Th. 24, 18. Hú ne scolde hire sceamian *nonne debuerat rubore suffundi?* Num. 12, 14. Ðonne fægniaþ hî ðæs ðe hî sceamian sceolde, Bt. 30, tit. ; Fox xvi, 6. Ðonne mæg hine scamian ðære brǣdinge his hlîsan, 19 ; Fox 68, 24 : Met. 10, 13. Ne þearf ðê ðæs eaforan sceomigan, Cd. Th. 140, 14 ; Gen. 2327. [*Goth.* skaman (*reflex.* *with gen.*) : *O. L. Ger.* scamón : *O. H. Ger.* scamón, scamēn : *Icel.* skamma *to shame ;* skammask *to be ashamed.*] v. á-, for-, e-, of-, on-sceamian.

sceamig. v. un-sceamig.

sceamisc ; *adj. Of which one is to be ashamed ; pudendus* :—Scamescan lim *veretrum*, Wrt. Voc. ii. 96, 54.

sceam-leás; *adj. Shameless, bold, impudent, wanton :*—Scamleás *impudens,* Wrt. Voc. i. 47, 45. Scamleás *frontosa,* Hpt. Gl. 506, 77. Scamleáse *procax,* 525, 57. Scomleás *impudens,* Wrt. Voc. ii. 44, 38. Se lǽce biþ micles tô beald and tô scomleás (*praesumtione percussus*) ðe gǽþ lǽcnigende, and hæf þ on his ágnum nebbe opene wunde unlácnode, Past. 9, 2; Swt. 61, 3. Of ðysse scamleásan scylde gecǽnsa mê *a delicto meo munda me,* Ps. Th. 50, 3. On ôðre wîsan sint tô lǽranne ða scamleásan (*impudentes*), on ôðre ða scamfæstan . . . Ðone scamleásan mon mæg ðý bet gebêtan ðe hine mon suíður þreáþ, Past. 31, 1; Swt. 205, 21–207, 5. Ðú hine ongeáte swíðe sceamleásne bûton ǽlcum gôdum þeáwe, Bt. 27, 2; Fox 96, 18. God ða sceamleásan (*the people of Sodom*) fordyde, Gen. 19, 24. [*O. H. Ger.* scama-lôs *impudens, procax : Icel.* skamnlauss *without disgrace.*]

sceamleás-lîc; *adj. Shameless, wanton :*—Dauit wæs mid ofermêttum gewundad, and ðæt gecýðde on Urias slæge, for ðære scamleáslecan gewilnunge his wîfes, Past. 3, 2; Swt. 35, 24.

sceamleáslîce; *adv. Shamelessly, impudently :*—Be ðâm Sodomitiscum ðe ongeán gecynd sceamleáslîce syngodon, Boutr. Scrd. 22, 38. Hî swíðe grǽdilîce eorþcundum lustum filigaþ and oft swíðe sceamleáslîce on manna gesyhþe, R. Ben. 139, 28. Hié sceamleáslîce gielpaþ ðisses hwîlendlícan onwaldes *improbe de temporali potestate gloriantur,* Past. 19, 2; Swt. 145, 9. Swâ hê sceamleáslîcor his yfel cýð (*impudenter innotescit*), 55, 1; Swt. 427, 25.

sceamleást, e; *f. Shamelessness, want of modesty, impudence, lasciviousness :*—Sceamleást *impudicitia,* Mk. Skt. 7, 22. Scamlêstan (-lêste?) *impudentiam,* Hpt. Gl. 526, 7.

sceam-lîc; *adj.* I. *shamefast, bashful :*—Scæmlîc, seó scamfæste *pudibunda, pudica, erubescens,* Hpt. Gl. 492, 53. II. *shameful, base, disgraceful, ignominious :*—Ðâ ongan hê him secgan hû lytel and hû scomlîc ðæs monnes lîf biþ hêr on worolde . . . and hû wuldorlîc seó êce eádignes biþ, Shrn. 92, 16. Sceomlîc *corruptibilis,* Rtl. 6, 1. Scildige scamlîcre forgǽgednysse *praevaricationis rei,* Jos. 6, 18. Nys ûs nâ tô secgenne ðone sceamlîcan morþ (*the disgraceful events at the siege of Jerusalem*) ðe ðǽr gedôn wæs, Ælfc. T. Grn. 21, 15. Ðæt hê ða sceamlîcan þing and ða mânfullan begǽþ *se res turpes et scelestas committere,* L. Ecg. P. ii. 6; Th. ii. 184, 11. Wæs ðæt feorþe wîte ðæt ealra scamlîcost wæs ðæt hundes fleógan cômon *post muscas caninas inferentes tam gravia tormenta quam turpia,* Ors. 1, 7; Swt. 38, 1. [Þenne were his cun iscend mid scomeliche witen, Laym. 20462. Eni velunge bitweone mon and ancre is so scheomelich and so naked sunne, A. R. 116, 3. *O. H. Ger.* scama-lîh *verecundus, pudibundus ; turpis, foedus.*] v. á-, un-sceamlîc.

sceamlîce; *adv. Shamefully, disgracefully :*—Ða ðe ǽwbryce ne wyrceaþ wôlîce and sceamlîce, Homl. As. 19, 140. Hê sceandlîce (scamelîce, MS. N.) sâwlode, 59, 202.

sceam-lim, es; *n. The private member :*—Sceamlim, gecyndlim *dedecus,* Germ. 390, 120.

sceamol, es; *m. A bench, stool.* The word remains in the form *shambles,* properly stalls or benches on which butchers expose meat for sale :—Sceamul *scabellum,* Wrt. Voc. i. 81, 24. Scamol *subsellium,* 289, 24. Scamel, sceamul, sceamol *scabellum,* Ælfc. Gr. 8; Zup. 31, 7. Scamul, scæmol, Ps. Spl. 98, 5. Ðara mynetera sceamelas *mensas nummulariorum,* Mt. Kmbl. 21, 12. Sceomolas, Blickl. Homl. 71, 18. Swâ forþ be efise tô lippan hamme; ðæt tô ðâm scamelan; swâ forþ tô stapole, Cod. Dip. Kmbl. v. 184, 14. [Þe halewen makeden of al þe worlde ase ane stol (scheomel, MS. C.: schamel, MS. T.) to hore uet, A. R. 166, 16. I sal set þe faas of þe schamel of þi fete to þe, Ps. 109, 1. *O. Sax.* fôtskamel : *O. H. Ger.* scamal *scabellum, subsellium : Ger.* schemel *a stool : Dan.* skammel. *From Lat.* scamellum.] v. fôt-, rǽde-, rǽding-sceamol.

sceamu, e; *f.* I. *the emotion caused by consciousness of unworthiness or of disgrace,* in a good sense (v. sceam-fæst, -full, -leás, -lîc), *modesty, bashfulness ;* in a bad sense, *shame, confusion :*—Sceamu *pudor* . . . reádnyss oððe sceamu *rubor,* Ælfc. Gr. 9, 21; Som. 10, 17–18. Scamu, scoma, scomo *pudor,* Txts. 84, 732. Scame *pallor,* Hpt. Gl. 474, 77. Scamu *rubor,* 475, 9. Se ðe nû ne mæg his gyltas for sceame ânum men geandettan, him sceal sceamian ðonne ætforan heofenwarum, and seó scamu him biþ endeleás, Homl. Th. ii. 604, 3–6. Ðú mid sceame (sceoma, Lind.: scomo, Rush.) nyme ðæt ýtemeste setl *incipias cum rubore nouissimum locum tenere,* Lk. Skt. 14, 9. Ðonne biþ hê self geládod wiþ hine selfne mid his ágenre scame and mid his geþylde, Past. 21, 1; Swt. 151, 18. Ðonne árâs hê for sceome *he got up because he was ashamed of his inability to play the harp,* Bd. 4, 24; S. 597, 7. II. *what causes a feeling of shame, disgrace, shame :*—Scoma *obprobrium,* Rtl. 190, 29. Micel hýnþ and sceamu (*verecundia*) hyt ys men neîle wesan ðæt ðæt he ys, and ðæt ðe he wesan sceal, Coll. Monast. Th. 32, 3. Ælce dæge byþ mîn sceamu (*verecundia*) beforan mê, Ps. Th. 43, 17. Byþ ðâm scand and sceamu *operiantur confusione et pudore,* 70, 12. Hû mæg mâre scamu mannum gelimpan, Wulfst. 162, 3. Sceome gihênedo *confusione contempnata,* Rtl. 27, 31. Sceame, Ps. Th. 88, 38. Ic his feóndas gegyrwe mid scame *inimicos ejus induam confusione,* 131, 19. Ðeós woruld scyldwyrcende in scome byrneþ, Exon.

Th. 232, 6; Ph. 502. Ne scomu dôaþ *neque calumniam faciatis,* Lk. Skt. Rush. 3, 14 : *contumiliam.* 11, 45. Sceame dreógan, habban, þrowian *to be put to shame, be disgraced :*—Beóþ gescende and sceame dreógaþ mîne fýnd *confundantur et revereantur inimici mei,* Ps. Th. 69, 2. Habban sceame *confundantur,* 85, 16. Ne sceolon æt mê ǽnige habban sceame *non erubescant in me,* 68, 7. Sume mǽgon habban ælles wouruldwelan genôg ac hî habbaþ ðeáh sceame ðæs welan gif hî ne beóþ swâ æðele on gebyrdum swâ hî woldon *huic census exuberat, sed est pudori degener sanguis,* Bt. 11, 1; Fox 30, 31. Ðæs ealdfeôndes scyldigra scolu scome þrowedon, Exon. Th. 114, 20; Gû. 175: 269, 5; Jul. 445: 369, 31; Seel. 49. Hî scoma mǽste dreógaþ, 78, 15; Cri. 1274. Mið scomum (sceofmum, Lind.) miclum tô giworhtun *contumeliis affecerunt,* Mk. Skt. Rush. 12, 4: Exon. Th. 153, 19; Gû. 828. III. *the private part* (v. sceam-lim):—Him sî âbrogden swâ of brêchrægle hiora sylfra sceamu, Ps. Th. 108, 28. Forhwon wrîhst ðû sceome? Cd. Th. 54, 13; Gen. 876: 58, 7; Gen. 942: 95, 3; Gen. 1573. Scama, ða wǽpenlîcan limo *preputia,* Wrt. Voc. ii. 69, 16. Scamu, 68, 60. [*O. Sax.* skama *shame, disgrace : O. L. Ger.* scama *confusio, reverentia : O. H. Ger.* scama *verecundia, reverentia, pudor, rubor, confusio, ignominia, turpitudo : Icel.* skömm *a shame, outrage.*] v. âr-, hleór-, woruld-sceamu.

sceamung, e; *f. Shaming, disgrace :*—Ðú canst gescændnysse ł sceamunga mîne *tu scis confusionem meam,* Ps. Lamb. 68, 20. v. forsceamung.

sceanca, an; *m.* I. *a shank, shin, the leg from the knee to the foot :*—Sceanca *crus,* Ælfc. Gr. 9, 33; Som. 12, 22: Wrt. Vóc. ii. 137, 21: i. 71, 56. Scance(-a?) *crus,* sceanca[n] *crura,* 44, 68. Gif se sconca biþ þyrel beneoðan cneówe, L. Alf. pol. 63; Th. i. 96, 16. Gif monnes sconca biþ of áslagen wið ðæt cneóu, 72; Th. i. 98, 19. Him blæces hundes deádes ðone swýðran fôtes sceancan (fôtscancan, MS. B.), Lchdm. i. 362, 27. Sconcan *crura,* Wrt. Voc. i. 65, 41. Scancan, ii. 17, 43. Sceancan *crura,* scancan *tibiae,* i. 283, 69–70. Lǽcedômas wið scancena sâre, and gif scancan forade synd, Lchdm. ii. 6, 10. Sindon ða scancan (*of the Phenix*) scyllum biweaxen *crura tegunt squamae,* Exon. Th. 219, 20; Ph. 310. Scancan *tibias,* Hpt. Gl. 482, 64: Kent. 982. Sconca[n?] *suras,* Wrt. Voc. ii. 93, 5. Ðæt man forbrǽce hyra sceancan (*crura*), Jn. Skt. 19, 31, 32, 33. Se sceocca gewrâð his sceancan, Homl. Skt. i. 11, 223. Sconcan, Salm. Kmbl. 203; Sal. 101. II. *the upper part of the leg* (= þeóhsceanca) :—Ic wille ðæt gê fêðaþ ân earm Engliscmon . . . Âgyfe mon hine . . . ân sconc spices oððe ân ram weorðe iiii. peningas, L. Ath. i. prm.; Th. i. 198, 7. [*Dan. Swed.* skank *a shank :* cf. *Germ.* schenkel.] v. earm-, fôt-, hôh-, þeóh-sceanca.

sceanc-bend, es; *m. A band for the leg, a garter :*—Scangbendas *periscelides,* Wrt. Voc. i. 40, 55.

sceanc-forod; *adj. Broken-legged :*—Ðæt sceáp ðæt sceoncforad (scanc-, Cott. MSS.) wæs, Past. 17, 9; Swt. 123, 9. Scancforedum men, Lchdm. ii. 66, 21.

sceanc-gebeorg, es; *n. A protection for the leg, a greave :*—Bânberge, scan[c]gebeorg *ocreas,* Wrt. Voc. ii. 97, 35.

sceanc-gegirela, an; *m. Clothing for the leg, a garter :*—Scancgegirelan *periscelides,* Wrt. Voc. ii. 67, 38.

sceanc-lira, an; *m. The fleshy, brawny part of the shank, the calf of the leg :*—Scanclira *surra,* Wrt. Voc. i. 283, 71.

sceand, es; *m. An infamous person, a buffoon, charlatan :*—Scond *scurra,* Wrt. Voc. ii. 120, 5. Ðonne sægde Petrus, ðæt hê wǽre leás drý and sceand and scyldig æswica *then Peter said that he (Simon the sorcerer) was a false sorcerer and a shameless impostor and a guilty deceiver,* Blickl. Homl. 175, 7. Sume hî wyrcaþ heora wôgerum drencas, ðæt hî hî tô wîfe habbon; ac ðyllíce sceandas sceolan sîðian tô helle, Homl. Skt. i. 17, 159.

sceand, e; *f.* I. *shame, disgrace, infamy, ignominy :*—Byþ ðâm scand and sceamu *operiantur confusione et pudore,* Ps. Th. 70, 12. *Ignominium* sconde hlêwung (cf. (?) ge-lêwan) *sive* fraceþu, *idem et infamium,* Wrt. Voc. ii. 49, 30. Sume wurdon getawod tô scande *some were shamefully entreated,* Chr. 1076; Erl. 214, 39. Is him ôðer earfeþu scyldgum tô sconde, Exon. Th. 78, 14; Cri. 1274. Sylfum tô sconde *to thine own disgrace,* 90, 27; Cri. 1480. Ðú sceonde æt mê [ne] anfenge ac gefeán eallum *thou gottest not disgrace from me, but gladness ever,* Cd. Th. 54, 9; Gen. 874. Ne þurfun gê wênan ðæt gê mec mid searocræftum under scæd sconde (*with ignominy*) scúfan môtan, Exon. Th. 142, 20; Gû. 647. Unwlite oððe sconde *dedecus,* Wrt. Voc. ii. 27, 35. Hî sceande âgon *confundantur,* Ps. Th. 108, 27. Sceonde fremman ylda bearnum *to bring disgrace on men,* Cd. Th. 149, 3; Gen. 2469. II. *a shameful, infamous, or abominable thing, that which brings disgrace :*—Ðonne is suîðe micel scand *ignominiosum valde est,* Past. 22, 2; Swt. 173, 1. Hê ne wolde ða sceonde (*the drunkenness of Noah*) hleómâgum helan, Cd. Th. 95, 20; Gen. 1581. Flǽsc scandum þurhwaden, Exon. Th. 78, 32; Cri. 1283. Ðú ðone lîchoman scondum gewemdest, 91, 5; Cri. 1487. Âscamode, scondum gedreahte, 79, 32; Cri. 1299. Geseoh ða scande and ða wierrestan þing ðe ðâs menn hêr dôþ *vide abominationes pessimas, quas isti*

faciunt hic, Past. 21, 3 ; Swt. 153, 20 : Swt. 155, 9. Sconde, Swt. 155, 8. [Þatt wass hiss aʒhenn shame ʒ shande, Orm. 11956. He makede to sconde *he disgraced*, Laym. 7032. Unk schal itide harm and schonde, O. and N. 1733. Þu schalt haue schonde, Horn. 714. To spouse þe emperours doȝter yt ner hym no schonde, R. Glouc. 65, 12. *Goth.* skanda αἰσχύνη: *O. H. Ger.* scanta *ignominia, confusio.*]

sceand-full ; *adj. Shameful, infamous, vile :*—Hé (*John the Baptist*) wæs heáfde becorfen for scandfulra wífa béne, and for scondfulles gebeór-scypes hleahtre, Shrn. 123, 6-8. [Him wule þunche swiðe strong and swiðe scondful þet he scal al aȝeuen and seoddan bisechen milce et þan ilke monne þe he haueð er istolen, O. E. Homl. i. 31, 2.]

sceand-hús, es ; *m. A house of ill fame, a brothel :*—Ðá heó ðæt nolde, ðá hét hé hí nacode lǽdan tó sumum scandhúse . . . Ðæs burhgeréfan sunu wolde rǽsan on hí on ðæm scandhúse, Shrn. 56, 7-11.

sceand-líc ; *adj.* I. *of persons, that acts in a disgraceful way, infamous, base, vile :*—On ánre tíde twá mǽdencild cumaþ, and biþ ðæt án sydeful and ðæt óðer sceandlíc, Homl. Skt. i. 5, 280. Hierusalem winþ for rihtwísnysse, and Babilonia winþ ongeán for unrihtwísnysse . . . Ðære heofonlícan Hierusalem cyning is Crist, ðære scandlícan Babilonian cyning is deófol, Homl. Th. ii. 66, 32. Ðá com ðæs geréfan suna mid his sceandlícum gegadum, Homl. Skt. i. 7, 164. God sende tó ðám sceandlícum mannum (*the people of Sodom*) twegen englas, 13, 207. II. *of things,* (a) *that is vile in its nature* or *circumstances, disgraceful, foul, shameful, obscene :*—Scandlíc hosp *ridiculosum opprobrium,* Hpt. Gl. 524, 73. Gif hit ǽr sceondlíc wæs, ne biþ hit nó ðÿ fægerre, Bt. 14, 3 ; Fox 46, 16. Seó gesceádwísnes ; nis ðæt scandlíc cræft, fordæm hit nænig hafaþ neát búton monnum, Met. 20, 188. Scandlícre fúlnesse *spurcae obscoenitatis,* Hpt. Gl. 447, 19. Of scondlícum geþohte *ex turpi cogitatione,* Bd. 1, 27 ; S. 497, 5. Mid sceandlícum willan *with foul lust,* Homl. Skt. i. 7, 170. Ðín módor gewíteþ of weorulde þurh scondlícne deáð and unárlícne *miserando turpissimoque exitu,* Nar. 31, 29. Ǽlc óðerne æftan heáweþ mid scandlícan onscytan, Wulfst. 160, 5. Hé sang scandlícan leóþ, and plegode scandlíce plegan, Shrn. 121, 10. Sceondlícum *corruptibilibus,* Rtl. 24, 36. Ic wille geswigian Tontolis and Pilopes ðara scondlícestena spella *nec mihi nunc enumerare opus est Tantali et Pelopis facta turpia, fabulas turpiores,* Ors. 1, 8 ; Swt. 42, 8. (b) *that causes shame, disgraceful :*—Hit is scondlíc ymb swelc tó sprecanne hwelc hit ðá wæs *pudet erroris humani,* 1, 10 ; Swt. 48, 4. [Wið scondliche deaðe, Laym. 2274. *O. H. Ger.* scant-líh *turpis, probrosus, ignominiosus, teter, lugubris.*]

sceandlíce ; *adv.* I. *in a disgraceful manner, disgracefully, shamefully, infamously :*—Heó lyfde sceandlíce, swá swá swín on meoxe, Homl. Skt. i. 3, 528. Nán cristen man ne sceal sceandlíce flítan, 13, 122. Him wand út his innoþ æt his setle, and hé sceandlíce sáwlode, Homl. As. 59, 202. II. *opprobriously, reproachfully, insultingly :*—Hiera wíf [sægdon] ðæt hié óðer gener næfden, búton hié on heora wífa hríf gewiton. Hí ðá, æfter ðæm ðe ða wíf hié swá scondlíce geræht hæfdon, gewendan eft ongeán ðone cyning, Ors. 1, 12 ; Swt. 54, 5. Gif man mannan bismærwordum scandlíce gréte *if one man insult another by abusive words,* L. H. E. 11 ; Th. i. 32, 5. Ne sceolon æt me ǽnige habban sceame sceandlíce ðe ðínes sídes biddaþ (bídaþ ?) *non erubescant in me, qui expectant te,* Ps. Th. 68, 7.

sceandlícness, e ; *f. Shame, disgrace, dishonour :*—Seó hálige ǽ forbeódeþ ða sceondlícnysse (*turpitudinem*) onwreón mǽgsibba, Bd. 1, 27 ; S. 491, 6, 12. Hé [ne] mæg mid weorce begán ða sceondlícnesse (scond-, MS. Hatt.) *qui turpitudinem non exercet opere,* Past. 11, 7 ; Swt. 72, 5.

sceandness. v. ge-sceandness.

sceand-word, es ; *n. A vile, foul word,* or *an opprobrious, abusive word :*—Ðæt ic (*the devil*) wolde, ðæt hý (*wicked men*) ðé (*God*) áfremdedon and ðíne circean forgeáton and æt mé leornedan sceandword, Wulfst. 255, 15.

sceán-feld. v. scín-feld.

sceap, es ; *n. A private part :*—Hé getælde his fæder Noe, ðær he on his sceape lócode, Anglia xi. 2, 53. Wið gicþan ðæra sceapa, Lchdm. i. 38, 15. v. for-, ge-, land-sceap.

sceáp, scép, scíp, es ; *n. A sheep :*—Scép *ovis,* Wrt. Voc. i. 23, 54. Ðæt dysige scép, Ps. Th. 118, 176. Sceáp sceal gongan mid his fliése óþ midne sumor, L. In. 69 ; Th. i. 146, 10. Emban ceápgild . . . sceáp tó sciťť., L. Ath. v. 6, 2 ; Th. i. 234, 2. Man healde .iii. niht hÿde and heáfod (*of a slain ox*), and sceápes eall swá, L. Eth. iii. 9 ; Th. i. 296, 19. Nán scyldwyrhta ne lecge nán scépes fell on scyld, L. Ath. i. 15 ; Th. i. 208, 10. Eówu biþ mid hire giunge sceápe sciťť. weorð óþ ðæt .xiii. niht ofer Eástron, L. In. 55 ; Th. i. 138, 7. Sceáp mon sceal gildan mid sciťť., L. O. D. 7 ; Th. i. 356, 6. Hwylc man ys ðe hæbbe án sceáp (scép, Rush. : scíp, Lind.), Mt. Kmbl. 12, 11. Sceáp (scép, Rush. : scíp, Lind.) ðe hyrde nabbaþ, 9, 36. Scípo *oves,* Rtl. 19, 37. Sceápa hús *ovile,* Wrt. Voc. i. 15, 21. Sceápa locu *caule,* 16, 6 : ii. 23, 11. Lambra sceápa *agni ovium,* Ps. Spl. 113, 6. Þreó heorda sceápa *tres greges ovium,* Gen. 29, 2. Heald míne sceáp (scíp, Rush. : scípo, Lind.) *pasce oves meas,* Jn. Skt. 21, 17. Ic drífe sceáp míne tó heora leáse, Coll. Monast. Th. 20, 11. [*O. Frs.* sképp, schép : *O. L. Ger.* skáp :

O. H. Ger. scâf.] v. snǽding-sceáp. The word occurs in local names, v. Cod. Dip. Kmbl. vi. 328, 329.

sceáp-ǽtere, es ; *m. The carcase of a sheep* (?) :—Ánan esne gebyreþ tó metsunge .xii. pund gódes cornes, and .ii. scípǽteras, and i. gód metecú, L. R. S. 8 ; Th. i. 436, 27.

sceapen. v. earm-sceapen.

sceápen ; *adj. Of a sheep :*—Sceápen smera (cf. on sceápes smerwe, l. 9), Lchdm. ii. 128, 16. Ete sceápen flǽsc and nán óþer, 358, 22. [*O. H. Ger.* scâfin *ovinus.*]

sceáp-heord, e ; *f. A flock of sheep :*—Nimaþ eówre hrýðerheorda and eówer sceápheorda and eówer orf *oves vestras et armenta assumite,* Ex. 12, 32.

sceáp-heorden, es ; *n. A hovel, shed :*—Býre *vel* sceápheorden *magalia vel mappalia vel capanna,* Wrt. Voc. i. 58, 31.

sceáp-hirde, es ; *m. A shepherd :*—Abel wæs sceáphyrde *fuit Abel pastor ovium,* Gen. 4, 2. Hwílum weard geworden sceáphyrde tó cynge, L. Eth. vii. 22 ; Th. i. 334, 10. Scéphyrde *oppilius,* Wrt. Voc. ii. 65, 10. Scýphyred (-hyrde ? cf. gáta hierde *titurus,* 288, 21) *titirus,* Wrt. Voc. i. 18, 57. Swá swá sceáphyrde tósceát sceáp fram gátum, Wulfst. 288, 2. Scéphyrdas *opiliones,* Coll. Monast. Th. 19, 3. Godes engel ætíwde sceáphirdon, Shrn. 29, 31. Be sceáphyrdan. Sceáphyrdes riht ætíwde . . . , L. R. S. 14 ; Th. i. 438, 21.

Sceáp-íg, e ; *f. Sheppy* (= *Sheep-island,* cf. Far-oe, *Icel.* fær *a sheep*) :—Hér hæþne men ǽrest on Sceápíge (-ege, MS. E.) ofer winter sǽtun, Chr. 855 ; Erl. 68, 23. Hér hæþne men oferhergeadon Sceápíge, 832 ; Erl. 64, 18.

sceáp-scearu, e ; *f. Sheep-shearing :*—Ðá fór hé tó his scépscere, Gen. 38, 12.

sceapung. v. for-sceapung.

sceáp-wæsce, an ; *f. A place for washing sheep,* the word remains as a place-name in *Sheepwash,* in Worcestershire :—Of ðam stáne on sceáp-wæscan ; andlang sceápwæscan, Cod. Dip. v. 48, 6. Andlang sceápwæscan tó sceápwæscan forda, 174, 11. Tó ðære sceápwæscan, 298, 4. Juxta fluvium qui dicitur Stúr, ad uadum nomine Scépesuuasce, i. 155, 23.

sceáp-wíc, es ; *n. A sheep-fold :*—Tó sceápwícan, Cod. Dip. Kmbl. iii. 405, 5.

scear, es ; *m.* (?) *A plough-share :*—Scer, scær, scear *uomis,* Ælfc. Gr. 9, 28 ; Zup. 55, 16. Scear *vomer,* Txts. 35, 32. Scear *vomer vel vomis,* Wrt. Voc. i. 15, 1 : 74, 72. Scer, 287, 6. Hwanon ðam yrþlinge sylan scear oððe culter, Coll. Monast. Th. 30, 29. Gefæstnodon sceare and cultre mid ðære syl *confirmato vomere et cultro aratro,* 19, 19. Hé sceal habban scear, culter and eác gádíren, Anglia ix. 263, 4. [*Chauc. Piers P. Prompt. Parv.* schare : *O. Frs.* skere, schere : *O. H. Ger.* scar, scaro *vomer.*]

sceár, e ; *f. A pair of shears* or *scissors ;* but the word is generally used in the plural (dual ?) as the modern *shears, scissors :*—Scér *forfex,* Wrt. Voc. ii. 36, 65. Scéroro, scéoro *forfices,* Txts. 60, 401. Ísernscéruru *forfex,* 65, 903. Sceára *forfex,* Wrt. Voc. i. 86, 21. Sceára *forficis,* ii. 1, 15. Tange *forcipis,* tang *forceps,* sceára *forficis,* 33, 35-37. Tangan, tange *forcipis,* sceáre[n] *forficis,* Hpt. Gl. 417, 75. Hí ne sceoldon hira loccas lǽtan weaxan ac hié sceoldon hié efsigean mid sceárum *non comam nutrient, sed tondentes attondent capita sua,* Past. 18, 7 ; Swt. 139, 14. Ne hé his loccas mid sceárum wanode, Shrn. 93, 9. Hé sceal habban horscamb and sceára (*shears*) . . . sceára (*scissors*), nǽdle, Anglia ix. 263, 8-15. Cf. Rægelsceára *forfices,* fexsceára *forfices,* Wrt. Voc. ii. 150, 21, 22. [My berd, myn heer . . . That nevere yit ne felte offensioun of rasour ne of schere, Chauc. Kn. T. 1559. A shepster (*sutrix*) shere, Piers P. 13, 331. Schere (scherys) to clyppe wythe *forfex,* Prompt. Parv. 445, col. 2. *O. Frs.* skére, schére ; *f.: O. H. Ger.* scári ; *pl. forpices :* scára *forfex : M. H. Ger.* schære : *Ger.* schere : *Icel.* skæri ; *n. pl. shears.*] v. secg-gescére.

sceára. v. secg-sceára.

scear-beám, es ; *m. The wood to which the ploughshare is fixed* (?) :—Scearbeám *brigacus,* Wrt. Voc. ii. 127, 21.

sceard, es ; *n. A shard, sherd, pot-sherd, tile :*—Scearda *testarum,* Germ. 398, 257. [Gower uses *sherd* for the scale of a dragon, ' a dragon whose scherdes schinen as the sonne,' iii. 68, 5 : and in Shakspere *shard* denotes a beetle's hard wing-case, v. Nares' Glossary. *M. H. Ger.* scharte *a sherd : Ger.* scharte.] v. croc-sceard ; scirden.

sceard ; *n. A gap, notch :*—Dó of ðam feórþan deále eall ðæt seó sǽ his ofseten hæfþ and eall ða sceard ðe heó him on genumen hæfþ *subtract from this fourth part (of the earth) all of it that the sea has covered, and all the gaps (bays and creeks) it has taken ;* huic quartae, si quantum maria premunt subtraxeris, Bt. 18, 1 ; Fox 62, 13. [*Shard* a gap remains long in some dialects. v. E. D. S. Pub. Gloss. B. 15, 19 (Wiltshire). *O. Frs.* skerd *a notch, cut, gash : M. H. Ger. Ger.* scharte : *Icel.* skarð *a notch, chink, gap.*] v. díc-, hær-sceard, *and next word.*

sceard ; *adj.* I. *notched, hacked, having gaps* or *rifts :*—Ic geann Ælmǽre ðæs sceardan swurdes *the hacked sword* (cf. *Icel.* með skarða skjöldu *with hacked shields,* Chart. Th. 561, 1, 23. Tó ðam sceardan beorge (cf. ðone tóbrocenan beorg ðe is tóclofen, Cod. Dip. Kmbl. ii. 251, 5), of ðam sceardan beorge tó ðam rúgan hlǽwe, Cod.

Dip. B. iii. 170, 2. On sceard hweogl (?), Cod. Dip. Kmbl. iii. 419, 11. Hrôfas sind gehrorene . . . scearde scûrbeorge, Exon. Th. 476, 9; Ruin. 5. **II.** *gashed, mutilated :*—Gif eáre sceard weorðe, L. Eth. 42; Th. i. 14, 7 : 48; Th. i. 14, 13. **III.** *deprived :*—Hê wæs his mæga sceard, freónda gefylled on folcstede, beslagen æt sæcce, and his sunu forlêt on wælstôwe, Chr. 937; Erl. 114, 6. (Cf. *Icel.* hafa, bera skarðan hlut *to get worsted.*) [*O. Sax.* skard : *O. Frs.* skerde *cut, gashed :* *O. H. Ger.* scart ; lid-scart *murcus;* lid-scartî *mutilation;* scartsam *scabrosus :* *M. H. Ger.* schart : *Ger.* schartig : *Icel.* skarðr.] v. scirdan, *and previous word.*

scearfian ; *p.* ode *To scrape, cut into shreds :*—Genim ða ylcan wyrte, scearfa hŷ ðonne, and gnîd swŷðe smale tô duste, Lchdm. i. 70, 14 : 80, 16 : 344, 13 note. Scearfa smæle, ii. 322, 25. Scearfaþ *succidite* . . . gescearfa ðû *succides,* Lk. Skt. Lind. 13, 7, 9. Scearfige ealle ðæs rinda tôgædere, Lchdm. iii. 14, 4. [*O. H. Ger.* scarbôn *concidere.*] v. sceorfan, *and next two words.*

scearflian ; *p.* ode *To scrape :*—Scearfla on wæter, Lchdm. i. 184, 18.

scearfung, e ; *f. Scraping, scarifying :*—Ða wætan ða yfelan weorðaþ gegaderode on ðone magan, and ðǽr rîxiaþ mid scearfunga innan, Lchdm. ii. 176, 7. Áberan ða strangan scearfunga ðæra wǽtena, 176, 10.

scearian *to grant.* v. ge-scearian.

scearn, es ; *n.* Sharn (v. E. D. S. Pub. Gloss. B. 17), *dung, filth :*—Scearn, scern *fimus,* Ælfc. Gr. 13 ; Zup. 83, 13. Gor, scear[n] *letamen,* Wrt. Voc. ii. 50, 38. Swê swê scearn (*stercus*) eorþan, Ps. Surt. 82, 11. Gôse scearn, ðonne hió ne ete, Lchdm. ii. 92, 15. Scearnes *fimi,* Wrt. Voc. ii. 95, 75. Scearn (oxena) *fimum,* Coll. Monast. Th. 20, 1. [*O. Frs.* skern : *Icel.* skarn ; *n. dung :* *Dan.* skarn *dung, muck, filth.*]

scearn-fîfel. v. scearn-wifel.

scearn-wibba, an ; *m. A dung-beetle :*—Scærnwibba *scarabeus,* Wrt. Voc. i. 77, 52. v. next word.

scearn-wifel, es ; *m. A dung-beetle :*—Scearnwifel (-fîfel, MS.) *scarabeus,* Wrt. Voc. i. 23, 69. [Halliwell gives *sharn-bug,* a cockchafer, as a Sussex word. Cf. Ssarnboddes (*beetles*) þet louieþ þet dong, Ayenb. 61, 32. *Icel.* tord-yfill *a beetle.*]

scearp ; *adj.* **I.** *sharp, having a fine edge* or *point :*—Seaxes ecg scearp, Exon. Th. 70, 21 ; Cri. 1142. Ic eom heard and scearp, ingonges strong, 479, 19 ; Rä. 63, 1. Genim ðæs scearpan þistles moran, Lchdm. ii. 314, 11. Scearpe gâras, Cd. Th. 124, 18 ; Gen. 2064. Ða Walas ádrifon sumre eá ford ealne mid scearpum pîlum, Chr. Erl. 5, 10. Scearpre ðonne ǽni sweord, Ps. Th. 44, 4. Nǽdle scearpran, Exon. Th. 373, 33 ; Seel. 119. Scearpeste stânas *cautes* vel *murices,* Wrt. Voc. i. 38, 22. **II.** *sharp* to the taste, *pungent, acid :*—Sió scearpe docce *oxylapatium,* Wrt. Voc. ii. 65, 50 : Lchdm. iii. 304, col. 2. Meng wið scearpum ecede, i. 354, 22 : ii. 72, 16. On wîne wel scearpum, 180, 16. Mettas ge drincan ða ðe habban hât mægen and scearp, 184, 10. Ðæs scearpestan wînes .v. sestras, 252, 8. **II a.** *acrid :*—Ða yfelan wǽtan sceorfendan and scearpan, Lchdm. ii. 176, 20. **III.** *sharp* of speech (cf. *sharp*-tongued) :—Hê biþ scarp and biter and scearp on his wordum, Lchdm. iii. 162, 13. ʻWǽron hyra tungan tô yfele gehwam ungemet scearpe, Ps. Th. 56, 5. **IV.** *sharp, keen, severe,* of pain or of that which causes pain :—Syððan com se scearpa hungor and ádyde hî mid ealle, Chr. 1086; Erl. 219, 37. Biþ ðæt sár scearpre ðonne wel welmes sár, Lchdm. ii. 206, 3. **V.** *sharp, rough* (v. scearpness, III) :—Ðǽr sint swîðe scearpe wegas and stânihte *situ terrarum montoso et aspero,* Ors. 1, 1 ; Swt. 10, 25. **VI.** *sharp, keen, active, strenuous :*—Ða ásende hê him tô ðone scearpan here of Rômâna rîce mid rêðum wǽpnum, Homl. Th. ii. 302, 18 : Homl. As. 61, 244. Ða geceás hê him geféran ða ðe ǽgðer ge on heora dǽdum ge on heora gelǽrednesse frome and scearpe wǽron Godes word tô bodienne and tô lǽranne *electis sociis strenuissimis et ad praedicandum verbum idoneis, utpote actione simul et eruditione praeclaris,* Bd. 5, 9 ; S. 622, 25. **VI a.** of things, *effectual, penetrating,* cf. scearplîce :—Hyre (*black horehound*ʼ miht ys scearp, Lchdm. i. 310, 7. Seó sunne scînþ mid hyre scearpan leóman, Homl. As. 43, 484. **VII.** *sharp, keen,* of sight :—Scearp gesihþ *acies,* Ælfc. Gr. 5 ; Som. 4, 14. Sió sŷn biþ ðŷ scearpre, Lchdm. ii. 30, 21. **VIII.** *sharp, keen, acute,* of understanding :—Scearp angyte *acre ingenium,* Ælfc. Gr. 9, 18 ; Som. 9, 66. Bûton hê hæbbe swâ scearp andget swâ ðæt fŷf, Bt. 39, 4 ; Fox 216, 28. Hû ðû eart gleáw and scearp, Exon. Th. 463, 27 ; Hö. 76. Sceal scearp scyldwîga gescâd witan worda and worca, se ðe wel þenceþ, Beo. Th. 581 ; B. 288. Scearpe *arguto,* Wrt. Voc. ii. 9, 64. Tôsceád simle scearpe môde in sefan ðînum, Exon. Th. 303, 1 ; Fä. 46. Ða ongeat hê mid scearpre gleáwnysse *ille, ut vir sagacis ingenii, intellexit,* Bd. 3, 9 ; S. 533, 42. [*O. Sax.* skarp : *O. Frs.* skerp : *O. H. Ger.* scarf : *Icel.* skarpr.] v. beadu-, efen-, heoru-, mylen-, un-scearp.

scearpe ; *adv. Sharply, keenly.* **I.** literal :—Ða fugelas ðe be flǽsce lybbaþ syndon scearpe gebilode *the birds that live on flesh are sharp-billed,* Hexam. 8 ; Norm. 14, 19. **II.** referring to seeing, observing :—Scearpe gesceáwian, Ps. Th. 93, 9. Se ðe ealra scearpost lôcianne mæg, Shrn. 187, 1.

scearpe, an ; *f. A scarification :*—Âsleah âne scearpan on ðam dolge,

---

Lchdm. ii. 142, 21 : 144, 6. Stande on heáfde, âsleá him mon fela scearpena on ðam scancan, ðonne gewît ût ðæt âtter þurh ða scearpan, 154, 2–4. Wið onfealle : genim hæslenne sticcan oððe ellenne, wrît ðinne naman on, âsleah þrŷ scearpan on, gefylle mid ðŷ blôde ðone naman, weorp ofer eaxle oððe betweoh þeóh on yrnende wæter . . . Ða scearpan âsleá, and ðæt eall swîgende gedô, 104, 6–11 : 84, 4 : 100, 4 : 126, 21 : 130, 10.

scearp-ecged ; *adj. Sharp-edged :*—God hêt ðæt hê nâme scearpecgedne flint, Homl. Th. i. 92, 33.

scearpian ; *p.* ode *To scarify, make an incision in the skin :*—Scearpa him ða scancan, Lchdm. ii. 46, 24 : 76, 13 : 126, 20. Scearpige and smire mid hâtan ele, 130, 7 : 284, 8. Dû scealt ymb .iii. niht scearpian, 264, 1. Scearpigean, iii. 132, 31.

scearp-lîc ; *adj. Sharp, keen, searching, effectual :*— Hwæt is sió þyrelung ðæs wâges bûton scearplîcu and smeálîcu fandung ðæs môdes ðæt mon onlûce ða heardan heortan *quid est parietem fodere, nisi acutis inquisitionibus duritiam cordis aperire* ? Past. 21, 3 ; Swt. 155, 1. Ðonne hê him gecŷd mid hû scearplîcum costungum wê sint æghwonon ûtan behrincgde *cum tentationum aculeos nos undique circumdantes innotescit,* 21, 5 ; Swt. 163, 16. Hû ne gesceóp ðê se scaþa scearplîce bysne *nonne exempla tibi dabat latro* ? Dôm. L. 53.

scearplîce ; *adv.* **I.** *sharply, keenly, smartly, effectually, quickly :*—Scearplîce *efficaciter, velociter,* Wrt. Voc. ii. 142, 56. Hyt ys gelŷfed ðæt heó scearplîce gehǽle, Lchdm. i. 154, 9. Heó gehǽlþ ðæt sár tô ðam scearplîce, ðæt hê eác gân dyrre *it heals the pain (gout) so smartly, that he may even venture to walk,* 176, 8 : 210, 9 : Exon. Th. 209, 9 ; Ph. 168. **II.** *sharply, keenly* (of the mind) :—Ða ðe meahton smeálîce and scearplîce mid hiera andgite ryht geseón *qui videre recta subtiliter per ingenium poterant,* Past. 11, 4; Swt. 69, 6. **III.** *sharply, painfully :*—Scearplîce *acerbatim,* Txts. 181, 47. Stingaþ hine scearplîce on ðone mûð, Wulfst. 141, 7.

scearpness, e ; *f. Sharpness.* **I.** referring to the sight :—Scearpnes *acies,* Wrt. Voc. ii. 2, 19. Sió scearpnes ðæs æpples *acies pupillae,* Past. 11, 4 ; Swt. 69, 3. Seó scearpnes mînra eágena nis nû mid mê *lumen oculorum meorum non est mecum,* Ps. Th. 37, 10. Heó (*betony*) gegôdaþ ðæra eágena scearpnesse, Lchdm. i. 72, 16. Hî ðæs môdes eágena scearpnesse nâuht gebêtaþ tô ðære sceáwunga ðære sôðan gesǽlþe, Bt. 34, 8 ; Fox 144, 32 : Met. 21, 24. **II.** referring to the mind :—On his môdes scearpnesse *aciem mentis,* Past. 16, 1 ; Swt. 99, 9. Wæs hê nâwiht hefig . . . ne hê cnîhtlîce gâlnysse næs begangende . . . ac on his scearpnysse hê weóx, Guthl. 2; Gdwin. 12, 13–20. **III.** *roughness* of surface (v. scearp, V) :—Ealle wôhnyssa beóþ gerihte and scearpnyssa gesmêðode, Homl. Th. i. 360, 34. **IV.** *acidity, pungency :*—Sió scearpnes *the acidity of the humours,* Lchdm. ii. 28, 1. Ðæs ecedes afre scearpnes, 224, 22. Sê lîchama gefêlþ ðæs sealtes scearpnesse, Wulfst. 35, 6. **V.** *efficacy :*—For ðære sealfe scearpnesse (*to make the salve effectual*) genim wîfes meoluc, ii. 28, 7. v. un-scearpness.

scearp-numol ; *adj. Efficacious :*—Ðeós wyrt ys swŷðe scearpnumul (-el, MS. B.) nîwe wunda and wîde tô gehǽlenne, swâ ðæt ða wunda hrædlîce tôgædere gâþ, Lchdm. i. 134, 10. Ðeós wyrt is swîðe scearpnumul wið ðæt âttor, 152, 3. Swâ se lǽcedôm yldra byþ, swâ hê scearpnumulra and hâlwendra byþ, 242, 5.

scearp-sîne, -siéne, -sŷne ; *adj. Sharp-sighted :*—Gif hwâ biþ swâ scearpsêne (-siéne, Cott. MS.) . . . swâ swâ Aristoteles sǽde ðæt deór wǽre, ðæt mihte stânas þurhseón . . . gif ðonne hwâ wǽre swâ scearpsiéne, Bt. 32, 2 ; Fox 116, 19–23. v. un-scearpsîne.

scearp-smeáðung, e ; *f. A sharp, strict examination, argument :*—Scearpsmeáðung *argumentum,* Mt. Kmbl. p. 12, 7. Scearpsmeáwunges *argumenti,* 13, 9.

scearpþanclîce ; *adv. Acutely, effectually :*—Scearpþanclîce *efficaciter,* Scint. 32.

scearp-þancol ; *adj. Acute, subtle :*—Ða scearpþanclan witan ðe ðone twŷdǽledan wîsdôm tôcnâwaþ, Lchdm. iii. 440, 28.

scearpung, e ; *f. Scarifying :*—Lâcna mid scearpinge, Lchdm. iii. 82, 23. Mid gelômlîcre scearpunge, hwîlum mid miclum, hwîlum mid feáwum, 84, 2. Lǽcedômas and scearpunga wið sîdan sâre, 262, 24.

scear-seax, es ; *n. A razor :*—Scearsex *rasorium,* Wrt. Voc. i. 35, 21. Scersaex *novacula,* Ps. Surt. 51, 4. Scirseax, Wrt. Voc. ii. 70, 17. Scyrseax, 60, 44 : *culter,* 15, 58. Scyrseax scearp *machera acuta,* Blickl. Gl. Ða sacerdas ne sceoldon nô hiera heáfdu scieran mid scearseaxum (scier-, Cott. MS.) *sacerdotes caput suum non radent,* Past. 18, 7 ; Swt. 138, 14. [*O. L. Ger.* scar-, scer-sahs *novacula :* *O. H. Ger.* scar-, schersahs *novacula, rasorium :* cf. *Icel.* skar-öx *a carpenter's adze.*]

scearu, scyru, e ; *f.* **I.** *a cutting, shaving :*—Scaro *tonsura,* Wrt. Voc. ii. 70, 18. Gif preóst sceare misgŷme beardes oððe feaxes, L. N. P. L. 34; Th. ii. 294, 27. **II.** *a shearing* of sheep :—Fêrde Laban tô his sceápa sceare *ad tondendas oves,* Gen. 31, 19. **III.** *the ecclesiastical tonsure.* v. L. Ecg. E. 152–154 ; Th. ii. 124, 9–24 :—Tô sceares gefe *ad tondendi gratiam* (in ' oratio ad capilaturam '), Rtl. 97, 4 : 95, 31. Ða wǽron scorene ealle munecas and sacerdas on ðone beh Sceͤ

Petres sceare, Bd. 5, 21; S. 643, 29. Tô reogollícum þeáwe rihtra Eástrena and scyre *ad ritum Paschae ac tonsurae canonicum*, 5, 22; S. 643, 38. Tô scare, 5, 22; S. 643, 38, note. Hêr Eádberht Norþhymbra cining fêng tô scære, Chr. 757; Erl. 53, 6. Ðæt hié heóldon ða ciriclecan scare, 716; Erl. 44, 19. Hê sceare onfêng, Bd. 3, 18; S. 546, 10: 5, 19; S. 636, 26. Ða sceare onfôn, 5, 21; S. 643, 22. Hê onfêng preósthâdes scare, Shrn. 50, 27. Ða ðe beóþ gehâdode fram Scyttiscum bisceopum oððe fram Bryttiscum, ða ðe sceare nabbaþ swâ ôðre cyriclíce preóstas, L. Ecg. P. Addit. 5; Th. ii. 232, 17. Wê lǽraþ ðæt ǽnig gehâdod man his sceare ne helige, L. Edg. C. 47; Th. ii. 254, 12. **IV.** *a share.* v. folc-, hearm-, land-, leód-, sceáp-scearu.

scearu, e; *f. The share*; pubes :—Mannes scaru *alvus*, Wrt. Voc. ii. 10, 26. Scare *ilium*, i. 44, 45. Biþ ðæt sâr on ða swiðran healfe on ða scare, Lchdm. ii. 232, 4: 232, 23. [Heo þuruh stihten Isboset adun into schere. Her seið seint Gregorie: ' In inguinem ferire est etc.' þe ueond þuruh stihð þet scher, A. R. 272, 12-14. Schare *pubes*, Wrt. Voc. i. 183, 29. The shore *le penul*, 148, 17. Schere *pubes*, 246, col. 2. Schore, *privy part of a man pubes*, Prompt. Parv. 448. v. Lchdm. ii. Glossary.]

sceat. v. sceatt.

sceát, es; *m.* **I.** *a corner, an angle* (v. -scîte); applied to the earth or heaven, *corner, quarter* (cf. the Edda: Þeir görðu þar af himinn ok settu hann yfir jörðina með fjórum skautum. Hence himin-skaut *the four quarters of the heavens*; heims-skaut *the poles*):—Ðâ wæs heora lâr sâwen and strogden betuh feówer sceátum middangeardes, Blickl. Homl. 133, 33. From feówerum foldan sceátum ðâm ýtemestum eorþan ríces englas blâwaþ býman, Exon. Th. 55, 6; Cri. 879. Lege on ða feówer sceáttas ðæs ærnes *lay at the four corners of the house*, Lchdm. ii. 142, 11. **II.** *a projection, promontory* (cf. sceáta) :—Bætweónæ ða twægen brômfeldas andlang ðæs alarsceátæs (*along the alder-covered piece of land which thrusts itself out into the fields*) on ðonæ fûlan brôc, Cod. Dip. Kmbl. v. 84, 12. **III.** *a nook, corner, region* (*in the phrases* eorþan, foldan sceát) :—Is feor heonan eástdǽlum on æþelast londa . . . nis se foldan sceát mongum gefére *est locus in primo felix oriente remotus*, Exon. Th. 198, 1; Ph. 3. Sceal fromcynne folde ðíne sîd land manig geseted wurðan eorþan sceátas *with thine offspring shall earth be settled, many a wide land, earth's regions*, Cd. Th. 133, 5; Gen. 2206. Foldan sceátas (sceattas, MS.), 204, 33; Exod. 428. Ic ne wât hwǽr mín brôþor on wera æhtum eorþan sceáta eardian sceal *I know not in what corner of earth my brother must dwell*, Exon. Th. 496, 25; Rä. 85, 19. Hê ne mette middangeardes, eorþan sceáta (sceatta, MS.) mundgripe mâran, Beo. Th. 1508; B. 752. Fyllaþ eówre fromcynne foldan sceátas, Cd. Th. 92, 26; Gen. 1534: 247, 25; Dan. 502. Drihten hâteþ hêhenglas bêman blâwan ofer burga sceotu geond foldan sceátas, 302, 21; Sat. 603: Exon. Th. 445, 20; Dôm. 10. Faraþ geond ealle eorþan sceátas, Andr. Kmbl. 664; An. 332: Exon. Th. 309, 22; Seef. 61. Hê ne mæg ðone (hlísan) tôbrêdan ofer ðæs nearowan eorþan sceátas (cf. tôbrêdan ofer ða nearwan eorþan âne, Bt. 19; Fox 68, 25), Met. 10, 17. **IV.** *a lap, bosom :*—Gif ðæs môdes forhæfdnes mid ungeþylðe ne âscôke ða sibbe of ðæm sceáte ðære smyltnesse *nisi mentes abstinentium impatientia a sinu tranquillitatis excuteret*, Past. 43, 3; Swt. 311, 15. Of midum sceáte (*sinu*) ðínum, Ps. Surt. 73, 11. Of his ðæm fæderlícan sceáte, Blickl. Homl. 5, 15. Gyld gramhýdigum on sceát hiora (*in sinu eorum*), Ps. Th. 78, 13. Ne mæg hê sceát âfyllan *non implevit sinum suum*, 128, 5. Gripon unfægre under sceát werum scearpe gâras *sharp spears fixed cruel fangs within the breasts of men*, Cd. Th. 124, 17; Gen. 2064. In sceát âlegd t bewedded t befest *desponsata* (cf. gesceátwyrpe *despondi*, Wrt. Voc. ii. 25, 72, *and* Icel. bera, leiða â skaut *of the ceremony which was a recognition of a child's legitimacy or of a person's adoption*. v. Cl. and Vig. Dict. skaut, 3, *and* Grmm. R. A. p. 160), Mt. Kmbl. Rush. 1, 18. Gif hió ôðrum mæn in sceát bewyddod sî *if she be betrothed to another man*, L. Ethb. 83; Th. i. 24, 5. **IV a.** *the bosom, surface of the earth :*—On ðone sêlestan foldan sceátes (*Thorpe would read sceáta*, cf. **III**) ðone fira bearn nemnaþ neorxna wong *in the fairest part of earth's surface, which the children of men call Paradise*, Exon. Th. 225, 28; Ph. 396. Geond eorþan sceát *over earth's surface*, 331, 8; Vy. 65. Ic wât ðætte wile woruldmen tweógan geond foldan sceát bûton feá âne (cf. went fulneáh eall moncyn on tweónunga, Bt. 4; Fox 8, 18), Met. 4, 52. Sió forme eld geónd eorþan sceát (cf. seó forme eld ðises middangeardes, Bt. 15; Fox 48, 3), 8, 5. Ofer foldan sceát, Exon. Th. 428, 22; Rä. 42, 5. Ofer ealne foldan sceát, 5, 21; Cri. 72. Deófol gefeallaþ in sweartne lêg under foldan sceát, 94, 2; Cri. 1534. **V.** *a bay; sinus :*—Wæs hê besenced on sumne sæs sceát *demersus est in sinu maris*, Bd. 1, 33; S. 499, 6. **VI.** *a garment :*—Sceát *vel* heortes hýd *nebris*, Wrt. Voc. i. 26, 26. Ðâ âstôd hê semninga and getogene ðý wǽpne under his sceáte ræsde on ðone cyning (cf. *Icel.* hann hafði und skauti sér leyniliga handöxi) *exsurrexit repente, et evaginata sub veste sica, impetum fecit in regem*, Bd. 2, 9; S. 511, 21 : Exon. Th. 431, 3; Rä. 45, 2 : 391, 18; Rä. 10, 7. **VII.** *a cloth, napkin :*—Sceát *manuterium vel mantele*, Wrt. Voc. i. 82, 38 : ma[n]tile, 290, 72 : ii. 56, 48 : *gausape*, 41, 13. Ealle neádbehêfe þing, ðæt is . . . nǽdl sceát weaxbreda *omnia necessaria, id est . . . acus, mappula, tabule*, R. Ben. 92, 3. Ðæt hê Godes gifa ne becnytte

on ðæm sceáte his slǽwþe, Past. 9; Swt. 59, 16. Nam ðære moldan sumne dǽl, gebond on his sceáte (*inligans in linteo*) . . . Âhêng hê ðone sceát (*linteolum*) on âne studu, Bd. 3, 10; S. 534, 24-29. Seóþ eft mid sceáte ôðres godwebbes, Lchdm. i. 332, 5. **VII a.** with the idea of concealment, *cloak, fold :*—Ne mágon gê ða word gesêðan ðe gê hwíle nû on unriht wrigon under womma sceátum, Elen. Kmbl. 1162; El. 583. [*Goth.* skauts; *m. the hem of a garment, skirt: O. Frs.* skât, *skirt : O. H. Ger.* scôz; *m. f. gremium, sinus;* scôza; *f. gremium, sinus, lacinia : Icel.* skaut; *n.*] v. beód-, feder-, grund-, weofod-sceát, sceáta, scîte.

sceáta, an; *m.* **I.** *a corner, angle :*—Sicilia is þrýscýte (*tria habet promontoria*) on ælces sceátan ende siindon beorgas. Ðone norþsceátan man hæt Polores . . . and se sûþsceáta hâtte Bachinum . . . and ðone westsceátan man hæt Libeum . . . se þridda sceáta is ân hund and syfan and hund syfantig míla westlang, Ors. I, 1; Swt. 28, 2-9. **II.** *the lower corner of a sail* (cf. sheet *the rope fastened to the lower corner of a sail: Icel.* skaut, skaut-reip *the sheet of a sail*) :—Sceáta *pes veli*, Wrt. Voc. i. 63, 59. **III.** *bosom, lap :*—Geond ealne ymbhwyrft eorþan sceátan, Exon. Th. 359, 26; Pa. 68. **IV.** *a cloth, napkin :*—Hê geseah Godes engel drýgan mid sceátan sci Laurentius limu, Shrn. 115, 23. [*O. H. Ger.* scôzo; *m. gremium, sinus : Icel.* skauti *a kerchief* used as a purse by knitting all four corners together so as to make a bag.] v. preceding word.

sceát-codd, es; *m. A bag, wallet, sack :*—Metefætels vel sceátcod *sitarchia*, Wrt. Voc. i. 16, 39. [Cf. *Icel.* skauti (*given under the preceding word*.)]

sceáþ, scǽþ, e; *f. A sheath :*—Sceáþ *vagina*, Wrt. Voc. i. 35, 19 : 84, 25. Sweord of sceáþe âtugon ða synfullan *gladium evaginaverunt peccatores*, Ps. Spl. 36, 14 : Judth. Thw. 22, 26; Jud. 79. Of scêþe, Byrht. Th. 136, 37; By. 162. Ða sweord on heora sceáðum behýdde wǽron *gladii reconduntur in vaginas*, Prud. 72 a. Brugdon scealcas of sceáþum scîrmǽled swyrd, Judth. Thw. 24, 38; Jud. 230. Sceáþum, Cd. Th. 120, 9; Gen. 1992. Hê âwende his sword intô ðære sceáþe, Homl. Th. i. 482, 32. On scæáþe (scêþe, MSS. A. B. C.), Jn. Skt. 18, 11. On hys scǽþe, Mt. Kmbl. 26, 52. [*O. Sax.* skêðia : *O. H. Ger.* sceida *theca, vagina : Icel.* skeiðir; *pl. a sheath.*]

sceaþa, an; *m.* **I.** *one who does harm, a criminal, wretch, miscreant, an enemy :*—Sceaþa, deógol dǽdhata (*Grendel*), Beo. Th. 554; B. 274. Nû earttû (*Satan*) earm sceaþa in fýrlocan feste gebunden, Cd. Th. 268, 19; Sat. 57. His feónd âfyllan ðe ðone scaþan (*the assassin Eomer*) sende, Chr. 626; Erl. 23, 34. Fýnd t sceaþan *inimici*, Ps. Lamb. 9, 7. Gewîtaþ, âwirgede woruldsorga, of mínes þegenes môde, forðam gê sind ða mǽstan sceaþan, Bt. 3; Fox 4, 24. Scyppeод sceaþan onfêngon syngum hondum, Exon. Th. 70, 2; Cri. 1132. Beraþ linde forþ in sceaþena gemong *bear the linden shields forth into the press of the foe*, Judth. Thw. 24, 17; Jud. 193. Wælstreámas (*the waters of the Deluge*) werodum swelgaþ, sceaþum scyldfullum, Cd. Th. 78, 32; Gen. 1302. **I a.** *a spiritual enemy, fiend, devil :*—Se sceaþa (*the devil who tempted Eve*), 38, 14; Gen. 606. Sceaþa, Satanes þegn, Salm. Kmbl. 234; Sal. 116. Ðæt hê ûs gescilde wið sceaþan wǽpnum, lâþra lygesearwum, Exon. Th. 48, 22; Cri. 775 : Andr. Kmbl. 2584; An. 1293. Fǽcnum feónde hýrdes, sceþþendum sceaþan, Exon. Th. 85, 24; Cri. 1396. Helle hæftling, scyldigne sceaþan, Salm. Kmbl. 257; Sal. 128. Sceaþan (*the fallen angels*) hwearfdon earme æglêcan geond ðæt atole scref, Cd. Th. 269, 13; Sat. 72. In ðæt sceaþena scræf *hell*, 304, 20; Sat. 633. Scyldwyrcende sceaþan (*the fallen angels*), Elen. Kmbl. 1521; El. 762. **II.** *a spoiler, robber :*—Sceaþa *predo*, Wrt. Voc. ii. 88, 66. Hê is þeóf and sceaþa *ille fur est et latro*, Jn. Skt. 10, 1 : Exon. Th. 54, 20; Cri. 871. Se sceaþa *the thief* (on the cross), Homl. Th. ii. 78, 18. ' Hwæt eart ðû ðe ðýn ansýn ys swylce ânes sceaþan.' Hê (*the penitent thief*) hym andswarode : ' Sôð gê secgaþ ðæt ic sceaþa wæs and ealle yfelu on eorþan wyrcende,' Nicod. 32; Thw. 18, 19-22. Hê (*Judas*) wæs gîtsere and se wyresta sceaþa, Blickl. Homl. 69, 11. Swâ swâ tô ânum sceaþan (*ad latronem*) gê fêrdon, Mk. Skt. 14, 48 : Lk. Skt. 22, 52. Sceaþena scip *paro*, Wrt. Voc. i. 56, 27. Hí habbaþ dǽmena and sceaþena dǽda, Blickl. Homl. 63, 9. Oðer hine scyhte ðæt hê sceaþena gemôt nihtes sôhte (cf. hê (*Guthlac*) menigfeald wæl felde and slôh and of mannum heora æhta nam, Guthl. 2; Gdwin. 14, 5-6), Exon. Th. 109, 31; Gû. 98. Gê hit dôþ sceaþum tô scrafum ' *ye have mad it a den of thieves*,' Blickl. Homl. 71, 20. Hê wæs on mycelre freccednysse on wêstene betwux sceaþum, Homl. Th. i. 392, 7. Sum man becom on ða sceaþan ða hine bereáfodon *homo quidam incidit in latrones qui etiam despoliauerunt*, Lk. Skt. 10, 30. **III.** with a favourable meaning, *a warrior :*—Scaþan onetton, wǽron æþelingas eft tô leódum fûse tô farenne, Beo. Th. 3610; B. 1803. Scaþan scîrhame tô scipe fôron, 3794; B. 1895. [*O. Sax.* skaðo *a robber, evildoer*.] v. âtor-, dol-, fǽr-, feónd-, folc-, fyrn-, gilp-, gûþ-, hell-, helle-, hearm-, leód-, lyft-, mân-, môr-, níþ-, sǽ-, syn-, þeód-, þeóf-, ûht-, wam-, wícing-sceaþa, *and next word*.

sceaþa, an; *m. Scathe, harm, injury :*—Cwæð ðæt sceaþena mǽst eallum heora eaforum æfter siððan wurde on worulde, Cd. Th. 85, 4; Gen. 549. [*O. H. Ger.* scado *damnum, noxia, detrimentum : Icel.* skaði *scathe, harm, damage.* Cf. *Goth.* skaþis *wrong.*] v. scepþ[u].

sceaþa (?), sceáþ (?) *a nail :*—Tácon đara sceađana (sceođona, Rush.) . . . styd đara sceađana *figuram clauorum . . . locum clauorum,* Jn. Skt. Lind. 20, 25. v. horn-sceaþa.

sceaþan; *p.* scód, sceód; *pp.* sceaþen. [*This strong form seems almost confined to the poetry, the prose making use of* sceþþan, q. v.] *To scathe, hurt, harm, injure,* (a) with dat. :—Đe ne sceaþeþ ǽnig, Ps. Th. 90, 7. Oft ic óđrum scód, Exon. Th. 401, 22; Rä. 21, 15. Hê tóswengde líges leóman, swá hyra líce ne scód, 189, 16; Az. 60: 197, 9; Az. 187. Se đe nǽngum scód, 90, 1; Cri. 1467. Đæt éce níþ ældum scód, 346, 5; Gn. Ex. 200. Us hearde sceód freólecu fǽmne (*Eve*), Cd. Th. 61, 15; Gen. 997: 245, 17; Dan. 464. Sió hǽleþum sceód (*punished ?*), Elen. Kmbl. 1415; El. 709. Him đa cwyđe frêcne scódon, Cd. Th. 96, 20; Gen. 1597. Scódun, Exon. Th. 134, 30; Gû. 516. Đæt him feóndes hond æt đam ýtemestan ende ne scóde, 129, 1; Gû. 414. Sceaþen is mê sáre, frêcne on ferhþe, Cd. Th. 53, 31; Gen. 869. (b) with acc. :—Oft mec ísern scód sáre on sídan, Exon. Th. 485, 14; Rä. 71, 13. (c) without a case :—Ne ic ne scaþe (scaþeđ, MS.) *neque nocebo,* Ps. Spl. 88, 33. Đý læs scyldhatan sceaþan mihton, Andr. Kmbl. 2296; An. 1149. [*Goth.* skaþjan; *p.* skôþ.] v. sceþþan, sceaþian.

sceaþ-dǽd, e; *f. A misdeed, crime* :—Scæþdǽd *facinus,* Wrt. Voc. i. 21, 27. Scepđǽd, ii. 39, 33. [Þat he hine awreke a þan awarriede uolke, þa hine isend hafden mid heore scađededen, Laym. 29578.]

sceaþel, e; *f. A shuttle (?)* :—Hê sceal habban fela tôwtôla . . . cranc-stæf, sceaþele, seámsticcan, Anglia ix. 263, 14.

sceaþenness, e; *f. Injury, damage* :—An wíf mihte gegán bútan ælcere sceaþenysse fram sǽ tô sǽ ofer eall đis eálond *ut etiam si mulier vellet totam perambulare insulam a mari ad mare, nullo se laedente valeret,* Bd. 2, 16; S. 520, 2. Hê oft stormas fram his sylfes sceþenisse and his gesérena scylde and wiđsceáf *tempestates a sua suorumque laesione repellere consueverat,* 2, 7; S. 509, 32.

sceaþfullíce, sceaþfulness. v. un-sceaþfullíce, un-sceaþfulness.

sceaþian; *p.* ode *To hurt, harm, spoil, rob* :—Ne sceaþa đu *thou shalt not steal,* Wulfst. 66, 18. Đæt deófol tô swyđe ne sceaþige, L. I. P. 7; Th. ii. 312, 26. Gif hwylc þeódsceaþa sceaþian onginneþ, Th. ii. 310, 24: L. C. E. 26; Th. i. 374, 29. Scađian, Wulfst. 191, 19. Se đe wǽre sceaþigende (scaþiende), weorđe se tiligende on rihtlícre tilþe, 72, 12. [*O. L. Ger.* scathan; *pp.* ge-scathot: *O. Frs.* skathia : *O. H. Ger.* scadôn *nocere* : *Icel.* skađa; *p.* skađađi.] v. ge-sceaþian; sceaþan, sceþþan.

sceaþung, e; *f. Injury, damage* :—Ge landfeoh ge fihtewíte ge stale ge wôhceápung ge burhwealles sceatung (sceaþinge ?) ge ælc đæra wônessa đe tô ænigre bôte gebyrie, đæt hit áge healf đære cyrcean hláford, Chart. Th. 138, 18.

sceát-líne, an; *f. The sheet of a sail, the rope fastened to the lower end of a sail* :—Sceátlíne (sceac-, MS.) *propes,* Wrt. Voc. i. 56, 62 : 63, 58. Cf. fôtrâp *propes,* 48, 25, *and* Icel. skaut-reip.

sceatt, es; *m.* I. *property, goods, wealth, treasure* :—Scaet *bona,* Txts. 44, 157. Scet *bona,* scettas bon[i], Wrt. Voc. ii. 11, 22–23. Scættas bo[n]i, 126, 45. Hê cwæđ đæt đê æniges sceates þearf ne worде on worulde, Cd. Th. 32, 15; Gen. 503. Nys unc sceattes wiht tô mete gemearcod, 50, 24; Gen. 813. Nǽron hî bescyrede sceattes willan *non sunt fraudati a desiderio suo,* Ps. Th. 77, 29. [Swá manega gersumas on sceat and on scrúd and on bôkes swá nán man ne mæi tællen, Chr. 1070; Erl. 209, 14.] Hî námon ealle his wépna and gold and seolfor and ealle his sceattas đe hî mihton geáxian, 1064; Erl. 194, 17 : 1069; Erl. 207, 14 : 1071; Erl. 210, 23. On geweald woroldcyninga đæm sélestan đara đe sceattas dǽlde, Beo. Th. 3377; B. 1686. I a. *of property which is paid as a price or contribution, price, gift, bribe, tax, tribute, money, goods* :—Anweald on sibbe smyltnesse gehealdan mid gefeohte ođđe mid scette (*by fighting or by paying tribute*), Lchdm. iii. 436, 15. Ne wanda đu for nánum scette for đam mêdsceattas âblendaþ wísra manna geþancas *non accipies munera, quia munera excoecant oculos sapientum,* Deut. 16, 19. Æt đam lande đe arcebisceop gebohte mid his ágenan sceatte (*with his own money*), Cod. Dip. Kmbl. iv. 86, 10. Godwine geann Leófwine đæs dænnes . . . æt đon sceatte (*at the price*) đe Leófsunu him geldan scolde, đæt is, feówertig penega and twâ pund and eahta âmbra cornes, vi. 178, 11 : Cod. Dip. B. i. 544, 4. Hê begeat swíđe mycele sceatt of his mannan . . . fêrde syđđan intô Normandîge *he (William) levied a large sum of money from his men . . . and afterwards went into Normandy,* Chr. 1085; Erl. 219, 10. Mænige gefôþ hwælas and micelne sceat đanon begytaþ *multi capiunt cetos, et magnum pretium inde acquirunt,* Coll. Monast. Th. 25, 3 : Ps. Spl. 61, 4. Mænig welig man is đe wolde mycelne scet and ungerîm feós syllan, gif hê hit gebicgan mihte, Homl. Skt. i. 12, 101. Gif hit fâcne is him man his scæt âgefe *if the marriage-contract be fraudulent, what he has paid shall be returned to him,* L. Eth. 77; Th. i. 22, 3 : 78; Th. i. 22, 4. Gif man mannan ofsleá, ágene sцætte and unfâcne feó gehwilce gelde, 30; Th. i. 10, 4 : 31; Th. i. 10, 7. Abram underfêng fela sceatta for hire hê hæfde đâ on orfe and on þeówum and on olfendum and on assum micele ǽhte *Abram bene usi sunt propter illam, fueruntque ei oves et boves et*

asini *et servi et cameli,* Gen. 12, 16. Đa bodan cômon mid sceattum habentes *divinationis pretium in manibus,* Num. 22, 7. Gif đû đæt gerǽdest, đæt đu wille syllan sǽmannum feoh . . . wê willaþ mid đâm sceattum ûs tô scype gangan, Byrht. Th. 132, 62; By. 40. Hêr fôr se cyng ofer sǽ and hæfde mid him gíslas and sceattas (*the contributions he had levied*), Chr. 1067; Erl. 203, 34. ¶ Teóþa sceatt *a tithe* :—Đæs hereteámes ealles teóþan sceat sealde '*he gave him tithes of all*' (Gen. 14, 20), Cd. Th. 128, 5; Gen. 2122. Bringaþ gê on mîn beren eówerne teóþan sceat (Malachi 3, 10), Blickl. Homl. 39, 26: 53, 11. Đonne lǽre ic eów, đæt gê syllon eówre teóþan sceattas earmum mannum, 49, 19: 43, 3. Abram his teóþan sceattas (*decimas*) offrede, Prud. 5 a: L. Alf. 38; Th. i. 52, 31. II. *a piece of money, a coin* :—Sceat *obulum,* Wrt. Voc. ii. 64, 78. Nis woruldfeoh đe ic mê ágan wille, sceat ne scilling (cf. *O. Frs.* mit schat ende mit schillinge : *O. H. Ger.* scaz unde schilinch), Cd. Th. 129, 13; Gen. 2143. Ne þearf ic N. sceatt ne scilling, ne pænig ne pæniges weorđ, L. O. 11; Th. i. 182, 9. Se mê beág forgeaf on đam siex hund wæs smætes goldes gescyred sceatta scillingríme, Exon. Th. 324, 9; Víd. 92. Hî behêton hire sceattas *dabimus tibi singuli mille et centum argenteos,* Jud. 16, 5. Wê đê mid ûs willaþ ferigan . . . siđđan gê eówre gafulrǽdenne âgifen habbaþ, sceattas gescrifene, Andr. Kmbl. 593; An. 297. II a. *as the name of an English coin the word is found in the form scætt in the laws of Ethelbert of Kent.* It is inferred from a comparison of passages in these that the value of the scætt in Kent was one-twentieth of a shilling, v. Thorpe's Glossary. The *sceatt* is also mentioned in the Mercian law, Th. i. 190, 5, where '30,000 sceatta' is equivalent to '120 punda.' This would give 250 sceatts to the pound. In the Northern Gospels *dragmas decem* is glossed by 'fíf sceattas teásîđum,' while the West-Saxon version has 'týn scyllingas.' If the sums here given may be regarded as equal, the *sceatt* would be worth a West-Saxon penny, the value which it appears to have in the Mercian law. The coin then seems to be of different values in Kent and in the more northern parts of England. [*Goth.* skatts ἀργύριον, δηνάριον, μνᾶ : *O. Sax.* skatt *money, property, piece of money* : *O. Frs.* skett : *O. H. Ger.* scaz *substantia, mobilia, pretium, lucrum, pecunia, aes, denarius, quadrans, obolus* : *Icel.* skattr *tribute.*] v. feoh-, fere-, freó-, geþing-, gif-, mán-, mêd-, ofer-, teóþing-, wæstm-sceatt; scír-gesceatt.

-sceatte, -sceattinga, sceát-weorpan. v. twí-sceatte, or-sceattinga, sceát, IV.

sceáwend-sprǽc, e; *f. Buffoonery, the speech of the theatre* :—Sceáwendsprǽc *scurrilitas (scarilitas,* MS.), Wrt. Voc. ii. 96, 65. v. sceáwere, V.

sceáwend-wíse, an; *f. A jesting song, song of a jester* :—Ic sceáwendwísan hlúde onhyrge, Exon. Th. 391, 1; Rä. 9, 9. v. preceding word.

sceáwere, es; *m.* I. *an observer, one who examines into a matter* :—Wê willaþ đæt se sceáwre wite mid fullum geráde, đe đis gewrit âspyraþ, Anglia viii. 331, 1. Đone dôm đæs sceáweres *spectatoris judicium,* Past. 15, 3; Swt. 93, 6. II. *a spy* :—Hê sende sceáwere (sceáware, Lind.) *misso speculatore,* Mk. Skt. Rush. 6, 27. Gê synd sceáweras *exploratores estis,* Gen. 42, 9, 14. Leáse sceáweras, Beo. Th. 511; B. 253. Moises sende twelf sceáweras, Num. 13, 4: Jos. 2, 1. III. *a watch-tower (?)* :—Sceáwere *specula* (the word occurs in a list of military terms), Wrt. Voc. i. 36, 4. IV. *a mirror* :—Sceáwere *speculea* (in a list of words connected with dress. Cf. Alse hit bi þe wimman and bi sheawere . . hie bihalt hire sheawere . and cumeđ hire shadewe þaronne, O. E. Homl. ii. 29, 10. Godes word is ase a uayr ssewere, ine huam me yziʒt alle þe lakkes of þe herte, Ayenb. 202, 21. Sheweres glasses. (A. V.), Wick. Isaiah 3, 23), 40, 54. V. *a buffoon, an actor* (v. sceáwend-sprǽc) :—Sceáwera *scurra rum,* ii. 90, 13. [*O. H. Ger.* scouwari *spectator, contemplator, scrutator.*] v. be-, fore-, steór-sceáwere.

sceáwian; *p.* ode. I. *to look* :—Ic sceáwode tô swiđran considerabam *ad dexteram,* Ps. Spl. 141, 5 : Ps. Th. 141, 4. II. *to look at, observe, behold, see* :—Đonne hê đæs fâcnes fintan sceáwaþ, Exon. Th. 315, 17; Mód. 32. Dryhten sceáwaþ hwǽr đa eardien đe his ǽ healden, 105, 19; Gú. 25. Đǽr hî sceáwiaþ Scyppendes giefe, 220, 28; Ph. 327. Đǽr hit eágum folc sceáwiaþ *in conspectu omnis populi,* Ps. Th. 115, 8. Đú đæs eágan eall sceáwadest gesêge fyrenfulra wíte *oculis tuis considerabis, et retributionem peccatorum videbis,* 90, 8. Sceáwode *conspicatur,* Wrt. Voc. ii. 18, 26: 80, 71. Đâ sceáwode Scyppend úre his weorca wlite, Cd. Th. 13, 21; Gen. 206. Hî sceáwodon Scyppend engla, 298, 18; Sat. 535. Đê wæter sceáwedon *viderunt te aquae,* Ps. Th. 76, 13: Beo. Th. 265; B. 132: 1971; B. 983. Sceáwa heofon, Cd. Th. 132, 6; Gen. 2189. Đæt ic đîn wuldur sceáwige *ut viderem gloriam tuam,* Ps. Th. 62, 2. Đú đînra bearna bearn sceáwige (*videas*), 127, 7. Đa mon mæg sceáwian gehealdene on Cantwara cyricean *quae in ecclesiae Cantiae conservata monstrantur,* Bd. 2, 20; S. 522, 10: Beo. Th. 1685; B. 840. Onwreóh đú míne eágan, đæt ic wel mǽge on đînre ǽ sceáwian wundur, Ps. Th. 118, 18. Đæt hê môste God sceáwian, Cd. Th. 297, 29; Sat. 524. Andgiettâcen (*the rainbow*)

sceáwigan, 93, 4; Gen. 1540. Þæt mæg mon on bôcum sceáwigean, hû monega gewin hê dreógende wæs, Ors. 1, 11; Swt. 50, 25. Hwylce ða nû synd tô sceáwigenne *quales illi nunc appareant*, L. Ecg. P. iv. 66; Th. ii. 226, 21. Tô sceáwianne, Exon. Th. 57, 7; Cri. 915. Sceáwiendum *contemplantibus, intuentibus*, Wrt. Voc. ii. 134, 83.　　III. *to look at, look on with favour, to regard, have respect to* :—Ic sceáwiu wegas ðíne '*I will have respect unto thy ways*' (A. V.), Ps. Surt. 118, 15. Hê hyra dǽde sceáwaþ *God will regard the deeds of the charitable*, Exon. Th. 106, 35; Gû. 51. Hê sceáwode ða eáþmôdnesse his þeówene *respexit humilitatem ancillae suae*, Blickl. Homl. 7, 3. Sceáwa (*respice*) ðis folc, Ex. 33, 13. Cyning eallwihta Caines ne wolde tiber sceáwian '*to Cain and to his offering the Lord had not respect*' (A.V. Gen. 4, 5), Cd. Th. 60, 9; Gen. 979.　　IV. *to look at with care, consider, inspect, examine, scrutinize, reconnoitre* :—Sceáwaþ *speculatur*, Wülck. Gl. 250, 8. Ðá ðæt eall gedôn wæs swá se geótere ðæm æðelinge ǽr behêt se æðeling ðæt ðá sceáwode *when all that was done as the founder (Perillus) promised the prince (Phalaris), the prince then inspected it*, Ors. 1, 12; Swt. 54, 29. Se cyng sceáwode ðæt mâdmehûs and ða gersuman ðe his fæder ǽr gegaderode, Chr. 1086; Erl. 223, 27. Ðonne seó ádl cume ǽrest on ðone mannan, ðonne sceáwa his tungan, Lchdm. ii. 280, 8. Sceáwiaþ ða lilian hû hí wexaþ *considerate lilia quomodo crescunt*, Lk. Skt. 12, 27. Ic eów bidde ðæt ânra manna gehwylc sceáwige hine sylfne on his heortan, Blickl. Homl. 57, 33: 107, 13. Moyses sende and hêt sceáwian Azer *misit Moyses, qui explorarent Jazer*, Num. 21, 32. Iosue ásende twegen sceáweras ðígellíce and hêt sceáwian ðæt land, Jos. 2, 1. Him ða fêran gewât land sceáwian, Cd. Th. 106, 33; Gen. 1780: Beo. Th. 2831; B. 1413. Hord sceáwian, 5481; B. 2744. Land sceáwigan, Cd. Th. 115, 16; Gen. 1920. Ðá ongon ic geornlícor ða stôwe sceáwigan and geond ða bearwas gongan *igitur perambulare totum nemus incipio*, Nar. 27, 20. Ceós ðê menn ðæt mágon sceáwigean ðone eard *mitte viros, qui considerent terram*, Num. 13, 3. Gê cômon ðis land tô sceáwienne, Gen. 42, 12.　　V. *to look out, seek for, select, choose, provide* :—Ðá sceáwode man þreó þegnas of ðam gemôte *three thanes were chosen from the moot* (*to go on a certain business*), Chart. Th. 337, 12. Gyf ðú ǽnigne gôdne heorde hæbbe . . . sceáwa hyne mê; gyf ðú ðonne nænne swá gerâdne næbbe, sêc hýne oð ðú hyne finde, Shrn. 164, 31. Se ðe ðás gemôt forbûge, ðonne sceáwige (scifte, MS. D.) man of ðam gemôte ða ðe him tô rídan, L. Edg. ii. 7; Th. i. 268, 15 : L, C. S. 25 ; Th. i. 390, 18. Him Loth gewât wíc sceáwian ôþ ðæt hié eorþscræf fundon *Lot went seeking a dwelling, until they found a cave*, Cd. Th. 156, 24; Gen. 2593. Drihtnes earc fôr beforan him þrí dagas sceáwiende ða wícstôwa *providens castrorum locum*, Num. 10, 33.　　VI. *to shew (favour, respect,* etc.*), to grant*, v. ge-sceáwian, I :—Ðá geornde se eorl gríðes and gísla . . . Ðá wyrnde him mann ðera gísla and sceáwode him mann .v. nihta grið út of lande tô farenne *then the earl asked for safe-conduct and hostages. . . . The hostages were refused him, and safe-conduct during five days was granted him to go out of the country*, Chr. 1048; Erl. 180, 11–14. [O. Sax. skawôn *to see, observe* : O. L. Ger. scauwôn, scouwôn *respicere, despicere* : O. Frs. skawia, skowia *to see, inspect* : O. H. Ger. scawôn, scauwôn, scouwôn *videre, conspicere, intendere, considerare, contemplari, scrutari, speculari, perpensare, censere*.] v. be-, ge-, geond-, ofer-sceáwian.

sceáwigend. v. leóht-, ofer-sceáwigend.

sceáwung, e; f.　　I. *a looking at, contemplation, consideration* :— Embeþonc *vel* sceáwung *circumspectio*, Wrt. Voc. ii. 131, 27. Tô dígolnesse and tô stilnesse becom ðære godcundan sceáwunge ancorlífes *ad anachoreticae contemplationis silentia secreta pervenit*, Bd. 4, 28; S. 605, 11. Se biþ eallenga blind se ðe nôht ne ongiet þe ðam leóhte ðære uplecan sceáwunge *caecus quippe est, qui supernae contemplationis lumen ignorat*, Past. 11, 1; Swt. 65, 7. Sceáwunga, 16, 1; Swt. 99, 2. For ðære sceáwungge ðara ungesewenlícra þinga *invisibilium contemplatione*, Swt. 99, 8. Tô ðære sceáwunga ðære sôþan gesǽlþe, Bt. 34, 8; Fox 144, 33 : Met. 21, 24. Sceáwunge *intuitu*, Wülck. Gl. 250, 7. Sceáunge *aspectu*, Rtl. 74, 7. Ǽrest ic hyt leornode myd ðam eágum, syððan myd ðam ingeþance . . . ac syððan ic hyt ongyten hæfde ðá forlæt ic ða sceáwunga mid ðám eágum, Shrn. 175, 8.　　II. *respect, regard* :— Nis scáwung heora deápes *non est respectus morti eorum*, Ps. Lamb. 72, 4.　　III. *reconnoitring, surveying, examination* :—Swíðost hê fôr ðider, tôeácan ðæs landes sceáwunge, for ðæm horschwælom, Ors. 1, 1; Swt. 17, 35.　　IV. *a spectacle, show* :—Al ðe here hiora ða ðe tôgedre cômun tô sceáwunga ðæt *ad spectaculum istud*, Lk. Skt. Rush. 23, 48. Ðá hêt Neron gewyrcean mycelne tor, and beád ðæt eall ðæt folc côme tô ðisse sceáwunga (*the spectacle of Simon flying from the tower*), Blickl. Homl. 187, 13.　　V. *a show, appearance, pretence* :—Under sceáwunge longes gibedes *sub obtentu prolixae orationis*, Mk. Skt. Rush. 12, 40.　　VI. *as a technical term, the same as* ostensio, *which occurs* L. Eth: iv. 2; Th. i. 300, 20, *and is explained in Du Cange*: Tributum a mercatóribus exigi solitum pro facultate ostendendi et exponendi merces in nundinis. *Seáwing, sceáwing is mentioned as being granted to the church at Westminster by Edward the Confessor in English charters*, Cod.

Dip. Kmbl. iv. 213, 11 : 215, 7: *and the form* sceáwing *occurs in Latin charters*, Chart. Th. 359, 4 : 411, 29. [O. H. Ger. scouwunga *consideratio, contemplatio, tuitio, providentia, spectaculum, speculum*.] v. blôd-, for-, fore-sceáwung.

sceb, scecel. v. sceabb, sceacel.

scecgan (?); p. scægde *To jut out, project, be distinguished*. [Cf. Icel. skaga; p. skagði *to project*.] v. tô-scecgan.

sced, scedan, scedeht, scefe, Scefing. v. scead, sceádan, sceadiht, scyfe, Sceáfa.

Scede-land, Sceden-íg. *The latter, occurring* Beo. 3376 ; B. 1686, *is the same as the Icel.* Skân-ey, *in Wulfstan's narrative,* Scôn-êg (q. v.): *the former* (in pl.) *seems to denote all the Danish or Scandinavian lands* :—Blæd wíde sprang Scyldes eaferan Scedelandum in, Beo. 38; B. 19.

scegð, scæð, es; m.: e; f. *A light, swift vessel* :—Scægð *trieris*, Wrt. Voc. i. 64, 1. Sceið, 56, 13. Litel scip *vel* sceigð *scapha vel trieris*, 47, 61. Ic gean mínre scǽðe for mínre sáwle intô Hramsêge healfe ðam abbode and healfe ðam hírêde, Chart. Th. 598, 9. Syððan hê tô lande cymþ, ðonne forlǽt hê ðæt scyp standan ; for ðam him þincþ syððan ðæt hê mǽge ǽð bútan faran ðonne mid. Eáðre mê þincþ ðeáh myd scêðþe on lande tô farande, ðonne mê þynce mid ðám eágum bútan ðære gesceádwísnesse ǽnigne creft tô geleornianne, Shrn. 175, 11–15. Scehð *liburnam, navim*, Hpt. Gl. 406, 51. Hêr bebeád se cyng ðæt man sceolde ofer eall Angelcynn scipu wircean ; ðæt is ðonne of þrým hund hídum and of x hídon ænne scegð (scægð, MS. D.), Chr. 1008; Erl. 141, 18. See note. Scêðhas *curuanas* (?), Wrt. Voc. ii. 137, 52. [A word taken from the Danes. *Icel.* skeið; *f. a swift-sailing ship of war*.] v. next word.

scegð-mann, es; m. *A member of the crew of a* scegð, *a Dane, a pirate* (cf. wícing, sǽ-man, flot-man, scip-here *and similar terms applied to the Danes*) :—Wícing *vel* scegðman *pirata vel piraticus vel cilix*, Wrt. Voc. i. 18, 59. Wícing oððe scegðman (scægð-, scǽð-, sceigð-) *pirata*, Ælfc. Gr. 7; Zup. 24, 9. Gif man secge on landes mann ðæt hê orf stǽle oððon man slôge, and hit secge án sceiðman and án landes mann (*a Dane and a native Englishman*), L. Eth. ii. 7; Th. i. 288, 8. Ægelsig þe Reáda and Winsig Scægðman, Chart. Th. 337, 17. v. preceding word.

scehdun, Exon. Th. 61, 6; Cri. 980. v. scildan.

-scel. v. wæl-scel.

scel, sceld (*a shield*), sceld (*a fault*), sceldig, scel-êge. v. scill, scild, scyld, scyldig, sceolh-íge.

scelfan; p. scealf, pl. sculfon *To shake, quiver, totter* :—On ðyssum stapelum sceall ǽlc cynestôl standan mid rihte on cristenre þeóde, and áwácie heora ænig, sôna se stôl scylfþ . . . áwácie se cristendôm, sôna scylfþ se cynedôm, L. I. P. 4; Th. ii. 308, 1–7: Wulfst. 267, 18. Ne hrisil scelfaeð, Txts. 151, 7. [*Icel.* skjálfa; *p.* skalf *to shiver, shake, quiver.*]

scell. v. scill.

scellan; p. sceall; pl. scullon *To sound, make a noise* :—Scylþ, cirmþ *crepitat, resonat*, Wrt. Voc. ii. 136, 72. [Cum qð þe culure wið schillinde stefne, Marh. 19, 19. O. L. Ger. ir-scal *increpuit*: O. H. Ger. scellan; p. scal, pl. scullon *sonare, clangere, tinnire, crepitare*: *Icel.* skjalla; p. skall, pl. skullu *to clash, clatter.*] v. scillan.

scelle *glosses* concisium, Wrt. Voc. ii. 105, 10: 15, 15: Wülck. Gl. 214, 7. [Cf. M. H. Ger. zer-schellen *to shatter*: *Icel.* skellr *a loud splash*; *a smiting, beating*. Or (?) cf. Goth. skilja *a butcher*: *Icel.* skilja *to divide.*] v. scellan, wæl-scel.

scelliht. v. scilliht.

Sceltifêre (?); pl. *The Celtiberians* :—Se mǽsta ege from Sceltiuêrin ingens Celtiberorum metus, Ors. 4, 12; Swt. 208, 24.

scenc, e; m. *A draught, cup* :—Scenc ðú sylst ûs *potum dabis nobis*, Ps. Spl. C. 79, 6. Cælc t scenc wætres caldes *calicem aquae frigidae*, Mt. Kmbl. Lind. 10, 42. Drince scenc fulne, Lchdm. ii. 116, 21. Genim ðysse ylcan wyrte seáw ânne scenc (scænc, MS. H.), i. 110, 21. Nim þrý scenceas (scæncas, MS. B.) gôdes wínes, 90, 19: 110, 10. [He lette heom bringen schenches of feole cunne drenches, Laym. 13461. M. H. Ger. shanc *a cup*.] v. medu-scenc.

scencan; p. te *To skink* (v. Nares' Glossary for instances of the use of this word), *to pour out liquor for drinking, to give to drink* (lit. and fig.) :—Ðú scæncst *potabis*, Ps. Lamb. 35, 9. Ðæt gôde wín ðæt hê scencþ nû geond his geladunge, Homl. Th. ii. 70, 11. Ðonne scencþ hê ða scylde mid ðære bisene ǽlcum ðæra ðe him ænges yfles tô wênþ *cunctis mala credentibus per exemplum culpa propinatur*, Past. 59, 5; Swt. 451, 24. Heó bær drincan and ûs eallum þênade and scencte ôð ðæt ðæt gereorde gefylled wæs *obtulit poculum, coeptumque ministerium nobis omnibus propinandi usque ad prandium completum non omisit*, Bd. 5, 4; S. 617, 26. Þegn, se ðe on handa bær hroden ealowǽge, scencte scír wered, Beo. Th. 996; B. 496. Feónd byrlade ðære idese, and heó (*Eve*) hyre were scencte, Exon. Th. 161, 12; Gû. 957. Mê þyrste, and gê mê scencton (cf. drincan sealdon, l. 21) . . . Hwænne gesáwe wê ðê þurstigne, and wê ðê scencton? Homl. Th. ii. 108, 4–11 : i. 336, 3 :

Wulfst. 288, 15. Ðá hí him betwih beadowíg scencton ðæs heofonlíces lífes *dum sese alterutrum coelestis vitae poculis debriarent*, Bd. 4, 29; S. 607, 17. Scencean *propinare*, Engl. Stud. ix. 40. Deáþes scencende drenc *mortis propinans poculum*, Hymn. Surt. 31, 15. [Nom heo (*Rowena*) ane bolle of ræde golde & heo gon scenchen, Laym. 14962. And tu . . . ne shennkesst nohht tatt wise, ne birrlesst tu þin hird, Orm. 15403. þe drynke for to schenche, R. Glouc. 118, 12. Schenkyñ drynke *propino*, Prompt. Parv. 445 (v. note). O. Frs. skenka : O. H. Ger. scenchen *fundere, propinare, ministrare, porrigere* : Icel. skenkja *to serve drink, fill one's cup*: cf. O. Sax. skenkio *a skinker, cupbearer*: O. L. Ger. skenki-vaz *cyathus*.] v. bi-, forþ-scencan.

scencel, scencen, gloss *acrum*, Wrt. Voc, ii. 10, 44 : i. 16, 4.

scencing-cuppe, an ; *f. A cup in which drink is served* :—Heó bit ðæt hí findon betweox him twá smicere scencingcuppan intó beódern for hí, Chart. Th. 536, 7. [Cf. O. L. Ger. skenki-vaz *cyathus* : O. H. Ger. scenche-bechar *calix* ; scenche-uaz *poculum*.]

scendan ; *p. de To put to shame, to abuse, insult, harm* :—Ic scendo *confundam, qui invehendo reprehendit*, Rtl. 1, 25. Ðone scamléasan mon mæg ðý bet gebétan ðe hine mon suíður þreáþ and sciend (scent, Cott. MSS.) *impudentes melius corrigit, qui invehendo reprehendit*, Past. 31, 1 ; Swt. 207, 6. Grendel nǽnegum áraþ leóde Deniga ac swefeþ ond scendeþ (? MS. sendeþ. Leo, Heyne, Grein refer to *sand*, q. v., and would translate by *feasts*) Grendel *spares no man of the Danes, but slays and puts to shame*, Beo. Th. 1204; B. 600. Ealne ðæne bysmor wé gyldaþ mid weorðscype ðám ðe ús scendaþ *all the disgrace we repay with honour to those who bring shame on us*, Wulfst. 163, 10. Hwilcan geþance mæg ǽnig man ǽfre geþencan on his móde ðæt hé tó sacerdan heáfod áhylde . . . and sóna ðæræfter hí scyrde oððe scynde mid worde oððe weorce *injure or abuse them with word or deed*, L. Eth. vii. 27 ; Th. i. 334, 36. Wé lǽraþ, ðæt ǽnig gelǽred preóst ne scænde ðone sámlǽredan, ac gebéte hine, gif hé bet cunne, L. Edg. C. 12 ; Th. ii. 246, 18. Biscopas ná sceótan ná tó lǽwedum mannum ne ne scendan ná hý sylfe *bishops shall not refer (their disputes) to laymen, nor bring disgrace upon themselves*, L. I. P. 10 ; Th. ii. 316, 36. Giþyll scendende *aura corrumpens*, Rtl. 121, 40. Scend ł forhogod *confunditur, spernitur*, Hpt. Gl. 419, 4. Scende (*con-fusi*) wǽron ealle ðe mé yfel tó ǽr gesóhton, Ps. Th. 70, 22. ¶ *With dat.* :—Se deópa seáþ mid wíta fela folcum scendeþ, Exon. Th. 94, 33; Cri. 1549. [Also ase þu wult schenden þene schucke, A. R. 316, 11. Men ne wolden scenden, Laym. 14167. Shennd and shamedd, Orm. 1985. Uor to ssende and to destrue, Ayenb. 28, 22. Schendyñ *confundo, culpo* ; scheut *culpatus, vituperatus, confusus, destructus*, Prompt. Parv. 445, col. 1. O. L. Ger. scendan *confundere* : O. H. Ger. scenten.] v. ge-scendan.

-scende. v. un-scende.

scendele (?), an ; *f. Abuse, reproach :*— Fore scendla ł scending *propter improbitatem*, Lk. Skt. Lind. Rush. 11, 8.

scendness, scendþ(u). v. ge-scendness, ge-scendþ(u) (Ps. Surt. 108, 29).

scendung, e ; *f. Abuse, harm :*—Scendung *afflictio*, Rtl. 86, 16. Fore scending *propter improbitatem*, Lk. Skt. Lind. 11, 8. v. for-scendung.

scéne, scén-feld. v. scíne, scín-feld.

scenn, e ; *f.* (?) *A plate of metal on the handle of a sword* (?) (Worsaae, Primeval Antiquities, pp. 29, 49, notes that the handles of some of the early swords were covered with plates of gold. v. hilt) :—Wæs on ðæm sweord scíran goldes þurh rúnstafas rihte gemearcod, hwam ðæt sweord geworht ǽrest wǽre, Beo. Th. 3392 ; B. 1694.

sceó *a cloud* (?) :—Scearp cymeþ sceó wið óðrum, ecg wið ecge (*of the coming together of clouds charged with electricity*), Exon. Th. 385, 8 ; Rä. 4, 41. [O. Sax. skio : Icel. ský *a cloud*.]

sceó *a shoe*, sceocca, -sceód, sceófan, sceofl, sceógan, sceóh. v. scóh, scucca, scógan, scúfan, scofl, scógan, scóh.

sceóh; *adj.* I. *shy, timid, fearful* :—Nú mín hreðer is hreóh, heówsíþum sceóh, Exon. Th. 354, 10 ; Reim. 43. II. *wanton* (?) :—Ðæs sción *petulantis* (*peculantis*, Wrt.), Wrt. Voc. ii. 89, 24. [Loked þet ȝe ne beon nout iliche þe horse þet is scheouh, and blencheð uor one scheadewe . . . To scheowe heo beoð mid alle, þet fleoð uor ane peinture, þet þuncheð ham grislich uorto biholden, A. R. 242, 8–12. Schey or skey as hors, Prompt. Parv. 444, col. 2. M. H. Ger. schiech *fugax, pavidus*. Cf. O. H. Ger. sciuhen *expavescere, terrere* : Ger. scheuchen *to scare* : scheuche *a bugbear* : Dan. sky *fear*.] v. next word.

sceóh-mód ; *adj. Fearful (wanton?) of heart* :—Se synsceaþa tó scipe sceóhmód ðhstreám sóhte, Exon. Th. 282, 32 ; Jul. 672. v. preceding word.

sceolh, sceol ; *adj. Oblique, wry* :—Of ðæm sceolan *de scevo*, Wrt. Voc. ii. 26, 67 : 85, 10. Sceolan *scevi*, 91, 47. [O. H. Ger. scelah *strabo, strabus, obliquus* : Ger. scheel : Icel. skjálgr *oblique, squinting*. Cf. skelly *to squint* (Yorks.).] v. next word.

sceolh-eágede ; *adj. Cross-eyed, squinting* :—Scelgégede *strabo*, Wrt. Voc. i. 75, 42. Sceolégede (scyl-, MSS. D. H. J. : -eágede, MS. J.) *strabo*, Ælfc. Gr. 9, 3 ; Zup. 36, 12. Scyleágede *strabus*, Wrt. Voc. i.

45, 56. Scylégede *luscus*, 43, 8. [Sculeiȝede, 89, 64. *Dan.* skel-öjed.] v. preceding and following words.

sceolh-íge ; *adj. Cross-eyed, squinting* :—Sceolhégi, sceolégi, scelége *scevus, strabus, torbus*, Txts. 98, 981. Sceolíge *strabos*, Wrt. Voc. ii. 92, 64. [*Icel.* skjól-eygr *squinting*.] v. preceding word.

sceolu, sceom-, sceón *to shoe*. v. scolu, sceam-, scógan.

sceón, scýan (?), scýn (?) ; *p. de To go quickly, fly* :—Ðonne ic forþ sció *when I depart (die)*, Cd. Th. 67, 20 ; Gen. 1103. Ðæt fýr scýde (scynde ?) tó ðám ðe ða scylde worhton, 232, 26 ; Dan. 266. [Cf. (?) Goth. skéwjan *to go* : O. H. Ger. scehanto *vagendo*, Grff. vi. 417 ; skihtig *fugax*, 418.] v. sceóh, and next word.

sceón ; *p. de To fall to a person's lot* :—Gif unc bán fordsíð scæt on Rómeweȝe *if death be the lot of both of us on the journey to Rome*, Chart. Th. 583, 29. Heom (heo, MS.) on riht sceóde (sceo, MS.) gold and godweb Iosepes gestreón *gold and purple, Joseph's treasure rightly fell to the share of the Egyptians in the Red Sea* (*after the destruction of the Egyptians in the Red Sea*), Cd. Th. 215, 21 ; Exod. 586. v. ge-sceón, and pre-ceding word.

sceonca, sceond, sceóne. v. sceanca, sceand, scíne.

sceóness, sciéness, scinness, scýness, scynness, e ; *f. Suggestion, per-suasion, incitement* :—Seó scynnes biþ þurh deófol *suggestio fit per diabolum*, Bd. 1, 27 ; S. 497, 13. On scynnesse, S. 497, 24. Mid scýnesse, S. 497, 10. Deófol mid hire (*the serpent's*) ðære yfelan sceónesse and fácne beswác ðone ǽrestan wífmon, Blickl. Homl. 5, 1. Sió scyld ðe hiene þurh sciénesse (scinnesse, Cott. MSS.) costaþ *vitium, quod per suggestionem tentat*, Past. 13, 2 ; Swt. 79, 22. Þurh scynuysse, Bd. 1, 27 ; S. 497, 12, 17. Hié swíður fylgaþ deófles lárum and his sceónessum, Blickl. Homl. 25, 11. Uncysta cumaþ oft þurh deófles sceónessa, 19, 7. v. scýan.

sceop, sceoppa, sceoppend, sceór, sceorf. v. scop, scoppa, scippend, scúr, scorf.

sceorfan ; *p.* scearf, *pl.* scurfon ; *pp.* scorfen *To gnaw, bite, scarify* :—Se (*hiccup*) cymþ of yfelum wǽtan slítendum and sceorfendum ðone magan. Gif se seóca man áspíwþ ðone yfelan bítendan wǽtan ou weg, ðonne forstent se geohsa. Spíwe ða deah ðám monnum ðe gihsa hié innan scyrfþ, Lchdm. ii. 60, 18–25 : 176, 20. Gif hé geféle ðæt se geohsa hine innan sceorfe on ðone magan, 62, 10. Gærstapan frǽton ealle ða gærscídas ðe bufan ðære eorþan wǽron ge furðon ða wyrttruman sceorfende wǽron *locustarum nubes, exhaustis omnibus, ipsas quoque radices seminum persequentes*, Ors. 1, 7 ; Swt. 38, 12. v. for-sceorfan ; scearfian ; ge-sceorf.

sceorian. v. scorian.

sceorp, es ; *n. Dress, apparel* :—Gemǽtte Macheus his ágenne sunu mid purpurum gegierdne. Hé hiene ðá for ðæm girelan gebealg . . . and wénde ðæt hé for his forsewennesse swelc sceorp werede, Ors. 4, 4 ; Swt. 164, 33. Somnite áwendan on óðre wísan heora sceorp *Samnites novum habitum sumentes*, 3, 10 ; Swt. 138, 30. Of manegum landum mǽre landriht áríst tó cyniges gebanne . . . scorp tó friðscipe (*apparel for those on board?*), L. R. S. 1 ; Th. i. 432, 8. v. fyrd-, gúþ-, heoru-, hilde-, hleó-, sige-sceorp ; ge-scirpla, scirpan.

sceorpan ; *p.* scearp *To scrape, to irritate* :—Gif man [hwæt?] sceorpe on ðone innaþ *if anything irritate a man in the insides*, Lchdm. iii. 44, 27. v. ge-sceorpan, and cf. sceorfan.

sceort, sceot. v. scort, scot.

sceót ; *adj. Quick, ready* :—Hweðer hé carful sý and sceót (gesceót, W. F.) tó godcundum weorce and tó hýrsunnesse *si sollicitus est ad opus Dei, ad obedientiam*, R. Ben. 97, 16. [*Icel.* skjótr *swift*.]

sceóta, an ; *m. A kind of trout, a shoate, shot* [' Carew makes a dis-tinction between the trout and *shot*. "The latter," he says, "is in a manner peculiar to Devon and Cornwall. In shape and colour he re-sembleth the Trowts : howbeit in biggnesse commeth farre behind him." The *shoates* with which is Tavy fraught.—Browne's Brit. Past.,' E. D. S. Pub. E. Cornwall Gloss. *Shote*, a small kind of trout, W. Corn-wall] :—Hwilce fixas geféhst ðú ? . . . sceótan (*tructos*), Coll. Monast. Th. 23, 33.

sceótan ; *p.* sceát, *pl.* scuton, sceoton ; *pp.* scoten. I. *to shoot*, (a) *cast a missile*, with acc. of missile :—Ðæt yrre scýt his spere ongeán ðæt geþyld *ira lanceam suam jacit contra patientiam*, Gl. Prud. 20 b. Ða wǽpna ðe ðæt yrre scét (*miserat*), 21 b. Hig sceoton hyra strǽlas tó ðære hynde, Shrn. 148, 6. (b) *to shoot* (intrans.) :—Ic torfige oððe sceóte *jacio*, Ælfc. Gr. 28, 6 ; Som. 32, 38. Se ðe of flánbogan fyrenum sceóteþ, Beo. Th. 3493 ; B. 1744. Hé hygegár léteþ, scúrum sceóteþ, Exon. Th. 315, 22 ; Mód. 35. Hé on bord sceát, Byrht. Th. 139, 46 ; By. 270. Hé mid geǽttredum strǽle ongan sceótan wið ðæs ðe hé geseah ðæt hrýþer stondan, Blickl. Homl. 199, 19. II. *to shoot* an object, *hit* an object with a missile :—Wyrd gǽst scýt, heó gǽr bireþ, Salm. Kmbl. 875 ; Sal. 437. Ðonne hié (*the serpent*) mon slóg oððe sceát, Ors. 4, 6 ; Swt. 174, 7. Hé óðerne sceát, Byrht. Th. 135, 67 ; By. 143. Tó ðam ðæt hí mágon sceótan ða unscyldigheortan *ut sagittent rectos corde*, Ps. Th. 10, 2. Ðær læg secg mænig ofer scild scoten, Chr. 937 ; Erl. 112, 19. Gif ðú wǽre on fell scoten, Lchdm. iii. 54, 4–7. II a. where

the weapon is the subject :—Ðā đone ilcan welegan mon se strǽl sceát, đæt hē sóna deád wæs, Blickl. Homl. 199, 23.   **III.** *to shoot, make an object move rapidly, push* (cf. *to shoot a bolt*) :—Đonne man đa sulh forþ drīfe, and đa forman furh on sceóte, Lchdm. i. 404, 2. Belūcaþ đa ǽrenan gatu and tōforan on sceótaþ đa ýsenan scyttelsas, Nicod. 27 ; Thw. 15, 15. Hē lēt dragan up đæne deádan Harald and hine on fen sceótan, Chr. 1040 ; Erl. 166, 24.   **III a.** *to give a person help* in escaping (cf. *Icel.* skjóta einum brott, undan *to let a person escape*) :—Gyf hine man teó đæt hē hine (*the criminal*) ūt sceóte, L. Edg. H. 6 ; Th. i. 260, 9.   **IV.** *to shoot, move rapidly, dart, run, plunge, rush,* (a) of living things :—Swā swā dēþ se đe his feóndum ofer sumne weall ætfleón wile, đonne cēpþ hē hwǽr se weall unhēhst sý, and đǽr ofer scýt, Homl. Th. i. 484, 11. Hē scēt innan sǽ *misit se in mare,* Jn. Skt. 21, 7. Hē unscrýdde hine sylfne and scǽt intó đam mere, Homl. Skt. i. 11, 211. Ān culfre scǽt (sceát, MS. V.) of đam fýre intó đǽre eá, 3, 73. Hī ānmódlīce hine tō scuton *they ran upon him with one accord* (Acts vii. 57), Homl. Th. i. 46, 34 : 404, 4 : ii. 496, 19. Seó dene wæs āfylled nīd manna sāwlum đa scuton of đam fýre intó đam cyle (*utrumque latus erat animabus hominum plenum, quae vicissim hinc inde videbantur quasi tempestatis impetu jactari,* Bd. 5, 12), 350, 10. Gif đū Godes sunu sý, sceót ādūn (*mitte te deorsum,* Mt. 4, 6) . . . Đæt wǽre swīđe gilplīc dǽd, gif Crist scute đā ādūn, i. 170, 1, 21. (b) of inanimate things :—On đa burnan đe of đam munt scýt *in torrentem, qui de monte descendit,* Deut. 9, 21. Đǽr sciét se Wendelsǽ up of đæm gārsecge *Tyrrheni maris faucibus oceani aestus immittitur,* Ors. 1, 1 ; Swt. 8, 25. Scýt, Swt. 8, 32. Seó eá scýt ūt on đone gārsecg, Swt. 14, 14. Đǽr ocærburna ūtt scýt on sǽ, Cod. Dip. Kmbl. iii. 175, 31 : 424, 4. Seó lacu ūt scyt, 422, 14, 26. Đonne đa wolcnan sceótaþ betweón hyre (*the sun*) and đē, Shrn. 201, 25. Him on gafol forlēt feówer wellan scīre sceótan, Exon. Th. 420, 1 ; Rä. 39, 4. (c) of speech :—Hē đæs geanwyrde wes ætforan eallum đam mannum đe đær gegaderode wǽron, đeáh him đæt word of scute his unnþances *though the remark burst from him involuntarily,* Chr. 1055 ; Erl. 189, 6.   **V.** *to run* (of a road, etc.) :—On đam wege đe scýtt tō đam pytte *per viam, quae ducit ad puteum,* Gen. 24, 62. Tō đǽre fyrh đe scýt sūþrihte tō đǽre miclan strǽt . . . Oþ đone weg đe scýt tō fealuwes leá . . . tō đam wege đe scýt tō đam hricgge, Cod. Dip. Kmbl. iii. 422, 4–19, 20, 25. Oþ đæt se weall eást sciát, ii. 86, 20.   **VI.** *to refer* a case to a person or court :—Ðus wrāt Hieronimus. Gif hwā elles secge, wē sceótaþ tō him, Homl. Th. ii. 306, 19. Đā nolde hē, būtan hit man sceóte tō scīregemōte, Chart. Th. 288, 19. Gif preóst dōm tō lǽwedum sceóte, đe hē tō gehādedum scolde, L. N. P. L. 5 ; Th. ii. 290, 22. Wē lǽraþ, đæt nān sacu đe betweox preóstan sī ne beó gescoten tō woruldmanna sōme, ac sēman heora āgene gefēran, ođđe sceótan tō đam biscope, L. Edg. C. 7 ; Th. ii. 246, 5 : L. I. P. 10 ; Th. ii. 316, 36. Se engel andwyrde : Uton sceótan tō Godes dōme, Homl. Th. ii. 338, 33.   **VII.** *to advance* money, *contribute, pay* (cf. scot) :—Hē forgeaf Middel-Sexon đæt feoh đæt hē heom fore sceát, Chart. Th. 551, 12. Sceóte ǽlc gegylda ǽnne gyldsester fulne clǽnes hwǽtes, 6. Sceóte man ælmessan, Wulfst. 170, 18. Sceóte man æt ǽghwilcre hīde pænig ođđe pæniges weorđ, 181, 4. Wē cwǽdon đæt ūre ǽlc scute .iiii. pæng tō ūre gemǽne þearfe . . . and forgyldon đæt yrfe đe syđđan genumen wǽre đe wē đæt feoh scuton, L. Ath. V. 2 ; Th. i. 230, 15–17.   **VIII.** *to shoot* (of sharp pain) :—Wiđ sceótendum wenne, Lchdm. ii. 324, 25 : iii. 30, 3. [*O. Frs.* skiata *to shoot* : *O. H. Ger.* sciozan *jaculari, sagittare, ferire : Icel.* skjóta *to shoot* with a weapon (dat.) ; *to push quickly* ; *to refer a case* to (til) *another ; to pay.*] v. ā-, be-, for-, ge-, of-, óþ-, þurh-, un-, under-sceótan ; scotian.

sceótend, es ; *m.* One who shoots, *a warrior* :—Sceótend wǽron gūþe gegremede, Judth. Thw. 26, 2 ; Jud. 305 : Beo. Th. 1411 ; B. 703 : Met. 1, 11. Sceótend sendaþ flāngeweorc, Exon. Th. 42, 20 ; Cri. 675. Hlyn scylda and sceafta, sceótendra fyll, Cd. Th. 124, 14 ; Gen. 2062. Ofer sceótendum, 184, 24 ; Exod. 112 : 129, 14 ; Gen. 2143.

sceoþa, sceó-þwang, sceotian, Sceottas, sceotung, sceóung, sceó-wyrhta. v. sceaþa, scōh-þwang, scotian, Scottas, scotung, scóung, scóh-wyrhta.

sceþ, scepen, sceppan, scer, scēr. v. sceáþ, scypen, scippan, scear, sceár.

sceran, sciran, sceoran ; *p.* scær, scear ; *pl.* scǽron, sceáron ; *pp.* scoren.   **I.** *to cut, shear* :—Đonne sweord swīn ofer helme scireþ, Bco. Th. 2579 ; B. 1287. Hǽleþ higerófe linde heówon, scildburh scǽron, Judth. Thw. 26, 2 ; Jud. 305. Lǽtaþ īren ecgheard ealdorgeard sceáran, Andr. Kmbl. 2364 ; An. 1183. Đæt hi hlīpen unwillende on đæt scorene clif unþeáwa *quia per multa etiam, quae non appetunt, iniquitatum abrupta rapiuntur,* Past. 33, 1 ; Swt. 215, 8. Scearde scūrbeorge, scorene, gedrorene, Exon. Th. 476, 10 ; Ruin. 5. Scorenum *rassis,* Wrt. Voc. ii. 84, 77.   **II.** *to shave hair* :—Ic scere *tondeo,* Ælfc. Gr. 26, 6 ; Som. 29, 9 : *rado,* 28, 4 ; Som. 31, 24. Ǽghwā mīn heáfod scireþ, Exon. Th. 482, 6 ; Rä. 66, 3. Ne hē his loccas mid sceárum wanode, ne his beard mid seaxe scear, Shrn. 93, 9. Ne gē eów ne efesion ne beard ne sciron *neque in rotundum attondebitis comam, nec radetis barbam,* Lev. 19, 27. Ne eówre hǽr ne sciron *nec facietis calvitium super mortuos,* Deut. 14, 1. Suiđe ryhte wæs đǽm sacerde forboden đæt hē his heáfod sceáre (*caput*

radere), Past. 18, 7 ; Swt. 139, 25. Heáfdu scieran mid scierseaxum *caput radere,* Swt. 139, 12. Se ylca preóst com tō Gūđlāce, đæt hē hine wolde scyran, Guthl. 7 ; Gdwin. 44, 20. Tō scearanne (beard) *ad tondendum,* Rtl. 97, 16.   **III.** *to cut* the hair of the head :—Heó scear hyre feax swā swā weras, and gegyrede hý mid weres hrægle, Shrn. 133, 13.   **IV.** *to shear* sheep :—Hī sculan waxan sceáp and sciran on hiora āgenre hwīle *they shall wash and shear sheep in their own time,* Chart. Th. 145, 13. Sceáp scyran, Anglia ix. 261, 10. Hē fōr scēp tō sciranne *ad tondendas oves,* Gen. 38, 13. [*O. Frs.* skera : *O. H. Ger.* sceran *tondere : Icel.* skera *to cut.*] v. ā-, be-, ge-sceran.

scerden, scerian. v. scirden, scirian.

scericge, an ; *f. An actress* :—Sca Pilagia wæs ǽryst mima in Antiochia đǽre ceastre, đæt is scericge (scēwicge (?) cf. sceáwere *scurra ;* or scernicge (?), cf. *O. H. Ger.* scern *scurrilitas, spectaculum ;* scernari *scurra, histrio*) on ūrum geþeóde, Shrn. 140, 11.

scern, scerpan. v. scearn, scirpan.

scerran (?) *to harness* an animal to something [:—Se yrþlincge unscenþ (-scerþ?) đa oxan *arator disjungit boves,* Coll. Monast. Th. 20, 27. Cf. *Ger.* an-, aus-schirren *to harness, un-harness.*]

scerwen, scerpen (?) *a scattering* (?), *sharing* (?), *giving* (?) (cf. bescerwan *to deprive*) :—Denum eallum weard cēnra gehwylcum eorum ealuscerwen *there was a fine feast for all the Danes* (?) (the reference is to the disturbance caused by the fight between Beowulf and Grendel), Beo. Th. 1542 ; B. 769. Myclade mereflōd meoduscerwen (scerpen, MSS.) weard æfter symbeldæge *the flood increased ; a fine feast was there after the banquet* (the reference is to the flood which came from the stone pillar, and swept away some of the Mermedonians. Cf. Đæt wæs biter beórþegu : byrlas ne gǽldon . . . đǽr wæs ǽlcum genóg drync sōna gearu, 3063–3069 ; An. 1534–1537), Andr. Kmbl. 3051 ; An. 1528. v. Grmm. A. and E. pp. xxxvi, 133, and note to Wülcker's ed. of Grein.

scéte, scēþ, sceþ-dǽd, sceþenes. v. scīte, sceáþ, sceaþ-dǽd, sceaþenness.

sceþness, e ; *f. Hurt, harm* :—Hē eft fērde būtan sceþnysse ǽniges sāres, Guthl. 16 ; Gdwin. 68, 27.

sceppan ; *p.* sceþede *To scathe, hurt, harm, injure* (a) with dat. :—Ic næ ngum sceþþe, Exon. Th. 407, 9 ; Rä. 26, 2. Nǽfre him deáþ sceþeþ, 203, 23 ; Ph. 88. Đonne þunorrād biþ, ne sceþeþ đam men đe đone stān (*agate*) mid him hæfþ, Lchdm. ii. 296, 30 : 162, 19. Ne sceþ đē nān wiht, iii. 178, 25. Eów seó wergþu sceþþeþ scyldfullum, Elen. Kmbl. 619 ; El. 310. Ūs seó wyrd scyþeþ, Andr. Kmbl. 3121 ; An. 1563. Nǽnig geweald deáþes him sceþþaþ *leti nil jura nocebunt,* Bd. 2, 1 ; S. 500, 21. Đás þing sceþþaþ đam eágum, Lchdm. ii. 26, 21. Him đæt ne sceþede, Shrn. 84, 29 : 131, 1 : Beo. Th. 3033 ; B. 1514 : Blickl. Homl. 161, 32 : 169, 6. Đa sǽdeór hyre ne sceþedon, Shrn. 133, 11. Hū đū sōdfæstum swīđast sceþþe, 263, 14 ; Jul. 349. Đý lǽs him gielp sceþþe, Exon. Th. 43, 6 ; Cri. 684 : 299, 11 ; Crä. 100. Đam mon sceal sellan đa mettas đa đe wambe nearwian and đam magan ne sceþþan, Lchdm. ii. 278, 18. Đý lǽs hī him and his freóndum sceþeden *ne sibi suisque nocerent,* Bd. 2, 7 ; S. 509, 35. Se līg ne mæg nā sceþþan đisse fǽmnan, Shrn. 130, 32 : Blickl. Homl. 129, 15 : 221, 17 : Ps. Surt. 104, 14 : Cd. Th. 273, 33 ; Sat. 146. Scyþþan, Andr. Kmbl. 2096 ; An. 1049. Đæt Scottas him nōht sceþþende ne āfuhton, Bd. 4, 26 ; S. 602, 25. (b) with acc. :—Se līg đa 'stuþo sceþþan ne meahte *flamma destinam laedere nullatenus sinebatur,* 3, 17 ; S. 544, 33. Ne mæg him bryne sceþþan wlitigne wuldorhoman, Exon. Th. 196, 23 ; Az. 178. (c) without a case :—Ic sceþþu *nocebo,* Ps. Surt. 88, 34. Regnþeóf ne lēt on sceade sceþþan, Exon. Th. 453, 15 ; Hy. 4, 15 : Beo. Th. 492 ; B. 243. Sceþþende *nocens,* Wrt. Voc. ii. 130, 12. Seó sceþþende wǽta, Bd. 4, 19 ; S. 589, 1. Đū hýrdes sceþþendum sceaþan, Exon. Th. 85, 24 ; Cri. 1396. [*Icel.* skeðja ; *p.* skaddi.] v. ge-sceþþan ; sceaþan, sceaþian.

sceþþend, es ; *m. One who harms, a foe, adversary* :—His āras ūs gescildaþ wiđ sceþþendra earhfarum, Exon. 47, 27 ; Cri. 761 : 126, 23 ; Gū. 375. Sceþþendum *adversaris,* Rtl. 113, 40.

sceþþig, scæþþig ; *adj. Hurtful, noxious* :—Scyldig ođđe scæđđig (sceaþþig, MS. U.) *sons,* Ælfc. Gr. 9, 29 ; Zup. 63, 15. v. un-sceþþig.

sceþþigness. v. un-sceþþigness.

sceþþ[u], e ; *f. Hurt, injury* :—Wiđ fótswylum and sceþþum (scæþþum, MS. H. B.), Lchdm. i. 342, 18. v. sceaþa.

sceþwrǽc ; *adj. Hurtful, noxious, hostile* :—Đæm (*St. John*) ne sceþede nǽnig scyld đisse sceþwracan worlde, Blickl. Homl. 161, 33.

sceucca. v. scucca.

scía, an ; *m. The shin ; crus* :—Scía *crus,* Txts. 54, 299. Scīu (scīa, sciæ, Rush.) *crura,* Jn. Skt. Lind. 19, 31–33.

sciccels, sciccel, es ; *m. A cloak, mantle* :—Scicilse *melote, mantile, veste,* Hpt. Gl. 440, 72. Hē hine unscrīdde đam heálfan scicelse đe hē on hæfde, Th. Ap. 12, 22. Hē wæs mid horhgum sicelse bewǽfed, 13, 26. Hī scrýddon hyne mid weolcenreádum scyccelse (*clamys*), Mt. Kmbl. 27, 28, 31. Geteáh his seax and genam his sciccels đe hē him on hæfde, tōsnāđ hine on twā, Blickl. Homl. 215, 6. Đā tōcearf hē his scyccel on

twã, and hyne gesealde healfne ðam þearfendum men, Shrn. 146, 36. [Cf. *Icel.* skikkja *a cloak.*] v. next word.

sciccing *a cloak, cape :*—Scicing, scinccing, scicgiug *cappa,* Txts. 50, 245. Sciccing, Wrt. Voc. i. 284, 64 : ii. 13, 24. v. preceding word.

scîd (?) *a course* (?) :—Scîd (ryne, MS. T.) *currus* (*cursus* ?), Ps. Spl. C. 67, 18. [Cf. (?) *Icel.* skeið *a race, course.*]

scîd, es ; *n.* A *shide* or *shingle, a piece of wood split thin, a billet :*—Scîd *scindula* (in a list *de igne*), Wrt. Voc. i. 284, 15 : 66, 41. Scîdum *scindulis,* ii. 120, 12 : 80, 21. [Stickes kan ich breken . . . and kindlen ful wel a fyr . . . ful wel kan ich cleuen shides, Havel. 917. Schyyd or astelle *teda, asula, astula,* Prompt. Parv. 446, col. 1. Go shape a shippe of shides and of bordes, Piers P. 9, 131. O. *Frs.* skîd : O. H. *Ger.* scît : M. H. *Ger.* schît : *Ger.* scheit ; *n.* : *Icel.* skîð ; *n.* : *a billet, firewood.*]

scîd-hreác, es ; *m.* A *heap of shingles* or *billets :*—.iiii. foðera âclofenas gauolwyda tô scîdhræce on hiora âgenre hwîle, Chart. Th. 145, 6. [Cf. *Icel.* skîða-hlaði *a pile of firewood :* *Ger.* scheiter-haufen *a funeral pile.*]

scîd-weall, es ; *m.* A *wooden fence, palings :*—Scîdwealles eorþbyri *vallum,* Wrt. Voc. i. 37, 34. [From sæ to sæ eode þæ dich (*the wall of Severus*) . . . þer ufenen he maakede scidwal, Laym. 10354. Cf. *Icel.* skîð-garðr *wooden palings, a wooden fence.*]

sciell, sciéne, sciéness, scieppend, scier-, scierpan, scife. v. scill, scîne, sceóness, scippend, scear-, scyrpan, scyfe.

sciftan ; *p.* te. I. *to divide, separate into shares :*—Fôn ða yrfenuman tô lande and tô æhtan, and scyftan hit swîðe rihte *the heirs shall succeed to the land and property, and shall divide it with perfect justice,* L. C. S. 79 ; Th. i. 420, 17. [*Shift* to divide, *Sussex.* A division of land among co-heirs is called a *shifting,* Halliwell Dict. Cf. *Icel.* skipta arfi, landi.] II. *to appoint, ordain, arrange.* Cf. *shift* used of a set of men which succeeds another in work that is carried on continuously, e.g. in a mine :—Ða scyfte man Beorn (Harold, MS., but cf. l. 21 : Ða læg Godwine eorl and Beorn eorl on Pefensæ) up ðæs cynges scipe ðe Harold eorl ær steórde, Chr. 1046 ; Erl. 174, 4. Moyses be Godes âgenum dihte rihte lage scyfte, Wulfst. 176, 8. Scifte man of ðam gemôte ða ðe him tô rîdan *those who may go to him shall be appointed from the meeting,* L. Edg. ii. 7 ; Th. i. 268, 15. [Schyftyñ or part a-sundyr *sepero, disgrego ;* Schyftyñ or partyñ or delyñ *divido, partior,* Prompt. Parv. 446, col. 1. Eter gate me his scyft, and þer me hi togesceodeð, O. E. Homl. i. 237, 30. Prestess and dæcness shifftedenn (*arranged*) hemm betwenenn whillc here shollde serrfenn first, Orm. 470.] v. ge-, tô-sciftan (-scyftan).

-scîgan *in* ge-scîgan :—Heora ælc sceal ân .c. þearfendra manna fêdan and ealle ða gescŷgean (*provide them all with shoes*), Chart. Th. 616, 26.

scilbrong. v. scilfrung.

scilcen[n], e ; *f.* A *female servant* or *slave, a woman of bad character :*—Hê gemacode ðæt seofon nacode wîmmen urnon plegende on heora gesihþum, ðæt heora môd wurde ontend tô gâlnysse þurh ðæra scylcena plegan, Hom. Th. ii. 162, 33. [Al nis bute ase a schelchine to seruien þe leafdi, A. R. 12, 24. M. H. *Ger.* schelkin *serva.*] v. scealc.

scild, sceld, scyld, es ; *m.* I. *a shield, a piece of defensive armour :*—Scyld *scutum* vel *clipeus* vel *parma,* lytel scyld *pelta* . . . scyld *clipeus, testudo,* lytel scyld *ancile,* ða læssan scyldas *peltae* vel *parmae,* Wrt. Voc. i. 35, 27–28, 57–59. Scyld *cetra,* ii. 20, 9 : *pelta,* 68, 4. Sceld *scutum,* i. 289, 30 : Ps. Th. 75, 3. Scyld sceal gebunden, leóht linden bord, Exon. Th. 339, 15 ; Gn. Ex. 94. Scyld sceal cempan, 341, 22 ; Gn. Ex. 130. Rand sceal on scylde, Menol. Fox 534 ; Gn. C. 37. Næfde hê scyld (*scutum*) æt handa, ðæt hê ðone cyning mid gescyldan mihte, Bd. 2, 9 ; S. 511, 22. Nân scyldwyrhta ne lecge nân scêpes fell on scyld, L. Ath. i. 15 ; Th. i. 208, 11. Ðær læg secg mænig ofer scild scoten, Chr. 937 ; Erl. 112, 19. Dynedan scildas, Judth. Thw. 24, 24 ; Jud. 204. Scylda *parmarum,* Wrt. Voc. ii. 96, 30. Eorles heregeata . . . ehta spera and eall swâ feala scylda, L. C. S. 72 ; Th. i. 414, 7. Hlyn weard on wîcum scylda and sceafta, Cd. Th. 124, 13 ; Gen. 2062. Sîde scyldas, randas regnhearde, Beo. Th. 656 ; B. 325. II. fig. *a shield, protection :*—Scild mîn bist ðû *refugium meum es tu,* Ps. Spl. T. 70, 4. Ðam biþ Dryhten scyld, Exon. Th. 229, 31 ; Ph. 463. III. *scyld* in the following passage is used of a bird's back (as being shield-shaped ? or can *scyld* here be connected with *sculdor* ? cf. (?) *shield-bone* = shoulder-blade quoted by Halliwell. *Icel.* skjöldr is used of shield-shaped things) :—Is se scyld ufan frætwum gefêged ofer ðæs fugles bæc, 219, 17 ; Ph. 308. [*Goth.* skildus : O. *Sax.* skild : O. *Frs.* skeld : O. H. *Ger.* scilt : *Icel.* skjöldur.] v. bôc-, ge-, pleg-scild.

Scild, es ; *m.* The *name of the ancestor of the Danish kings.* His story is given in the opening canto of Beowulf. According to the Ynglinga Saga, c. 5, one of Odin's sons is Skjöldr. v. Scildingas.

scildan, scyldan, sceldan, sceoldan ; *p.* de. I. *to shield, protect, guard, defend :*—Ic hine scylde *protegam eum,* Ps. Th. 90, 14. Hwâ forstandeþ hié, gif ðû hié ne scyldest ? Blickl. Homl. 225, 19. Ne þearf him ondrædan ænig, gif hine God scyldeþ, Exon. Th. 49, 6 ; Cri. 781.

Se godcunda anweald hî scilde, Bt. 39, 10 ; Fox 228, 12 : Exon. Th. 195, 33 ; Az. 165. Mê nama Dryhtnes scylde, Ps. Th. 117, 12 : Cd. Th. 247, 31 ; Dan. 505. Ðara gâsta ðe hine scildon *defensiones spirituum bonorum,* Bd. 3, 19 ; S. 548, 36. Scild ûsig *tuere nos,* Rtl. 79, 16 : 84, 15. Him wæs lŷfnesse seald, ðæt hê him môste scyldan aud besecgan, Bd. 5, 19 ; S. 640, 11. Scildende *protegente,* Rtl. 103, 34. Ðætte wê sîe scildad *defendi,* 75, 5. Scylded beón *tueri,* Wrt. Voc. ii. 88, 59. ¶ Scyldan wið *to shield from, guard against :*—Ic ðe wið weána gehwam scylde, Cd. Th. 131, 3 ; Gen. 2170. Wið ða speru hié scildaþ, Past. 35, 4 ; Swt. 245, 10. Mê sôðfæstnes mîn scylde wið feóndum *scuto circumdabit te veritas ejus,* Ps. Th. 90, 5. Scilde, Lchdm. ii. 238, 5 : Exon. Th. 126, 22 ; Gû. 375. Ða englas hine scildon wið ðæs fŷres frêcennesse, Bd. 3, 19 ; S. 548, 32 : Exon. Th. 496, 4 ; Râ. 85, 9. Wê ûs wið hâm sceldan ðæs ðe wê mihton, Nar. 14, 29. Ða wið flôdum foldan sceldun (scehdun, MS.), 61, 6 ; Cri. 980. Hû hî hî sylfe scyldan sceolan wið deóflu, Blickl. Homl. 47, 22. Utan scyldan ûs wið ðone hâtan bryne ðe wealleþ on helle, L. C. S. 85 ; Th. i. 424, 15. *Without an object :*—God, se ðe wið ofermægnes egsan sceolde, Cd. Th. 127, 28 ; Gen. 2117. Wê lêraþ ðæt man wið heálíce synna scylde georne, L. C. E. 23 ; Th. i. 374, 7. II. *to make a defence :*—Siððan hê his hyspinge gehêred hæfde ða scylde hê ongeán swîðe ungeþyldelíce *after he had heard his abuse then he made a defence in reply very impatiently,* Bt. 18, 4 ; Fox 66, 35. v. ge-scildan.

scild-burh ; *f.* I. *a battle-array in which men stood shield to shield* [cf. the account of the battle of Stamford-bridge : 'Siðan fylkti Haraldr Konungr liði sínu, lêt fylkingina langa ok ekki þykka ; þá teygði hann armana aptr â bak, svâ at saman tôku, var þat þâ viðr hringr, ok þykkr ok jafn öllum megin ûtan, skjöldr við skjöld.' Saga Haralds Harðráða, c. 92. When this arrangement is abandoned, they are said 'bregða skjaldborginni,' c. 95] :—Weard scyldburh tôbrocen, Byrht. Th. 138, 56 ; By. 242. Hæleþ higerôfe scildburh scæron, Judth. Thw. 26, 2 ; Jud. 305. II. *a city which affords protection, a city of refuge.* v. scild, II :—Sôðfæste men in heora fæder rîce scînaþ in sceldbyrig (*heaven*), Cd. Th. 283, 23 ; Sat. 309. Grimm would translate the word here by 'aula clypeis tecta,' and compares it with the description of Valhalla in the Edda, 'skjöldum þökt, lagt gyltum skjöldum, svâ sem spánþak,' D. M. 662. [O. H. *Ger.* scilt-burg *testudo.*]

scilden, es ; *m.* A *protector, guardian, defender :*—Scylend *protector,* Ps. Spl. T. 17, 21 : Ps. Spl. 58, 12 : Ps. Th. 26, 2 : 83, 9 : Blickl. Homl. 141, 14. v. ge-scildend.

scilden[n], e ; *f.* Protection :—Scildenne, scildiunae *tutellam,* Txts. 103, 2073.

scildere, es ; *m.* A *shielder, protector :*—Ðû eart mîn scyldere *protector meus,* Ps. Th. 17, 3.

scild-freca, an ; *m.* A *warrior with a shield :*—Ðonne scyldfreca ongeán gramum gangan scolde, Beo. Th. 2071 ; B. 1033.

scild-hreáda. v. next word.

scild-hreóða, -hrêða, an ; *m.* Shield-covering, (1) *a shield, buckler :*—Scinon scyldhreóðan, Cd. Th. 184, 26 ; Exod. 113. (2) *the arrangement of shields as in the* scild-burh, q. v. :—Scyldrêðan *testudine,* Wrt. Voc. ii. 96, 31. Sumum wîges spêd giefeþ æt gûþe, ðonne gârgetrum ofer scildhreádan (-hreóðan?) sceótend sendaþ (cf. *the passage under* scild-weall), Exon. Th. 42, 19 ; Cri. 675. v. bord-hreóða, and cf. hrêðan *melote,* Wrt. Voc. ii. 56, 63.

scildig. v. scyldig.

Scildingas ; *pl.* The *descendants of Scild,* or more generally *the Danes.* The word occurs often in Beowulf, and is also found in the compounds Âr-, Here-, Sige-, þeód-Scildingas. [*Icel.* Skjöldungar.]

scildness, e ; *f.* A *protection, defence :*—Scildnisse *defensionis,* Rtl. 41, 13 : *protectionis,* 97, 18. v. ge-scildness.

scild-rêða. v. scild-hreóða.

scild-rída (= hreóða?), an ; *m.* A *phalanx :*—Ðeáh hî wyrcen getruman and scyldrîdan wið mê *si consistant adversum me castra,* Ps. Th. 26, 4.

scild-truma, an ; *m.* A *phalanx ; testudo :*—Under þiccum scyldtruman *subter densa testudine,* Ælfc. Gr. 47 ; Som. 48, 29. Of sceltruman *testudine,* Hpt. Gl. 475, 66. [He makede his sceldtrume swulc hit weoren an hær wude, Laym. 16371. A scheltrone *hec acies,* Wrt. Voc. i. 240, 9.]

scildung, e ; *f.* Shielding, protection :—Ða deófellícan flân wurdon ealle âdwæscte þurh ðæs gewæpnodan engles scyldunge, Homl. Th. ii. 336, 10. Scilding *tutum,* Rtl. 100, 3.

scild-weall, es ; *m.* A *shield-wall, the shields held by a line of soldiers :*—Ðonne stræla storm scôc ofer scyldweall, Beo. Th. 6227 ; B. 3118. Cf. scild-burh.

scild-wîga, an ; *m.* A *warrior who bears a shield :*—Scearp scyldwîga, Beo. Th. 581 ; B. 288.

scild-wyrhta, an ; *m.* A *shield-maker :*—Sceldwyrhta *scutarius,* Wrt. Voc. i. 289, 31. Be scyldwyrhtum. Nân scyldwyrhta ne lecge nân scêpes fell on scyld ; and gif hê hit dô, gilde .xxx. scitt., L. Ath. i. 15 ; Th. i. 208, 9–11. Andlang flæscmangara strête ðet it cymþ tô scyld-

wyrhtana stræte; andlang scyldwyrhtana stræte eást eft ꝺæt hit cymþ tô Leófan hagan, Cod. Dip. Kmbl. vi. 135, 18–20.

scilfe, an; f. A shelf, ledge, floor :—Gescype scylfan on scipes bôsme (cf. With lower, second, and third stories shalt thou make the ark, Gen. 6, 16), Cd. Th. 79, 4; Gen. 1306. [Cf. Icel. Hliꝺ-skjálf; f. Odin's seat whence he looked out on all the world.]

Scilfingas; pl. A Swedish royal family, the Swedes :—Helm Scylfinga ꝺone sêlestan sæcyninga ꝺara ꝺe in Swióríce sinc brytnade, Beo. Th. 4752; B. 2381: 5200; B. 2603. The compounds Gûþ-, Heaꝺo-Scilfingas also occur, and the singular Scylfing, Beo. Th. 4968; B. 2487. Scilfing, 5928; B. 2968. [Icel. Skilfingar; pl. the name of a mythical royal family; skilfingr a prince (poet.). v. Grmm. D. M. 343.]

scil-fisc, es; m. A shell-fish :—Monige sint cwucera gesceafta unstyriende, swâ swâ scylfiscas sint, Bt. 41, 5; Fox 252, 21. Mettas ꝺe gôd blôd wyrceaþ, swâ swâ sint scilfixas, Lchdm. ii. 244, 24. [Icel. skel-fiskr.]

scilfor; adj. Yellow, of the colour of gold :—Of scylfrum hiwe flava specie, Wrt. Voc. ii. 149, 21. Of scilfrum flava auri specie, Hpt. Gl. 419, 23.

scilfrung, e; f. Shaking, balancing, swinging :—Hwær com seó wlitignes heora ræsta and setla . . . and seó scylfring heora leóhtfata ꝺe him beforan burnon the swinging (?) of the lamps that burnt before them, Blickl. Homl. 99, 34. Scilbronge libramine, Wrt. Voc. ii. 88, 72. Cf. skelfan, and Icel. skjálfra to shake.

scilian; p. ode To separate, part, remove :—Eádwerd cing scylode ix scypa of mâle and hî fôron mid scypon mid eallon anweg King Edward put nine ships out of commission, and they went away ships and all, Chr. 1049; Erl. 174, 38. Cf. (?) âscelede (-scerede?) dividuntur, Hpt. Gl. 438, 50. [He wass skiledd ut fra þe follc þurrh halig lif, Orm. 16860. Our king, That wic men fra god sal schille, Met. Homl. 152, 9. Schyllyn owte segrego, Prompt. Parv. 446. Icel. skilja to separate, part, divide.] v. â-scilian.

scilig; adj. Shaly. v. stân-scilig.

scill, scell, scyll, e; f.   **I.** a shell, shell-fish :—Musclan scil conca, Wrt. Voc. ii. 15, 35. Scel, 105, 37. Scel echinus, i. piscis, cancer, 142, 24: 106, 75. Musclan ł scille de concha, Hpt. Gl. 417, 10. Scille vel sæsnæglas conchae vel cochleae, Wrt. Voc. i. 56, 7. Scellum concis, ii. 15, 18.   **II.** the shell of an egg :—Se rodor ymbfêhþ ûtan eall þâs niꝺerlícan gescæfte, swâ seó scell ymbfêhþ ꝺæt æg, Shrn. 63, 10 : Met. 20, 174. Fæger swylce hê of ægerum ût âlæde, scîr of scylle, Exon. Th. 214, 4; Ph. 234.   **III.** a scale of a fish, serpent, etc. :— Hió dyde sciella tô bisene his heor neohtum and ꝺus cwæꝺ : Ælces fisces sciell biþ tô ôꝺerre gefêgeꝺ sub squamarum specie de ejus satellitibus perhibetur : Una uni conjungitur, Past. 47, 3; Swt. 361, 17. Sumum (serpents) scinan ꝺa scilla swylce hié wæron gyldene, Nar. 13, 19. Ꝺonne hió (the serpent) mon slôg oꝺꝺe sceát, ꝺonne glâd hit on ꝺæm scyllum, swelce hit wære smeꝺe ísen, Ors. 4, 6; Swt. 174, 8. Sindon ꝺa scancan scyllum biweaxen crura tegunt squamae, Exon. Th. 219, 21; Ph. 21. Ne ete gê nânne fisc bûton ꝺa ꝺe habbaþ finnas and scilla, Lev. 11, 9.   **IV.** a shell-shaped dish (?) or simply a shell :—Nim león gelynde, mylt on scylle (a dish or a shell?), Lchdm. i. 364, 24. Wyrme on scille, ii. 42, 16 : 310, 6. [Goth. skalja a tile : Icel. skel a shell.] v. æg-, oster-, sæ-, weolc-scill.

scill; adj. Sonorous, sounding :—Scyl wæs hearpe, Exon. Th. 353, 44; Reim. 27. [Cf. Heo song so lude and so scharpe Riht so me grulde schille harpe, O. and N. 142. With a shil vois, Parten. 1997. Schylle and sharpe acutus, sonorus. Schylly and scharply acute, aspere, sonore, Prompt. Parv. 446. Cf. O. H. Ger. scall sonus, sonitus; scella tintinnabulum : Icel. skillr a loud splash; skella a rattle.] v. next word.

scillan to cause to sound :—Scyllendre concrepante, scyllende concrepans, Hpt. Gl. 518, 48. [O. H. Ger. scellan; p. scalta to cause to sound : Icel. skella.] v. scellan.

scilliht; adj. Shell (of fish) :—Ꝺû scealt sellan scellihte fiscas, Lchdm. ii. 196, 21 : 254, 19. Scellehte, 227, 17.

scilling, es; m.   **I.** as a denomination of English money (uncoined), a shilling. The shilling appears to have been of different values in different parts of the country; in Wessex five pennies make a shilling : Fíf penegas gemacigaþ æune scillingc, Ælfc. Gr. 50; Som. 52, 8 : and with this statement agree several passages of Henry I.'s Laws, e. g. c. 93, §§ 3, 19, where unus solidus = v denarii, duo solidi = x denarii. In Mercia four pennies go to the shilling. According to Mercian law (Th. i. 190) the ceorl's wergild was 200s., the thane's six times as much, 1200s., the king's, which is six times the thane's, is £120; so that 7200s. = 120 × 240d., i.e. the shilling is four pennies. With this agrees L. W. i. 11; Th. i. 473, where it is said : Solidum Anglicum quatuor denarii constituunt. In the Norman time the shilling is twelve pennies. This reckoning seems to be taken in earlier times. v. riht-scilling and Ex. 21, 10. The word is of constant occurrence in the Laws and Charters; from the latter the following passage may illustrate the point that the shilling was a denomination of value, not a coin : Biscop gesalde six hund scillinga on golde, Chart. Th. 90, 21. It also occurs as

a weight : Genim of ꝺysse wyrte petroselini swýꝺe smæl dust ânes scillinges gewihte, Lchdm. i. 240, 11.   **II.** as denoting foreign money the word is used to translate various words :—Scylling numisma, Wrt. Voc. i. 57, 30. Scilling obelus, ii. 63, 68 : stater, Mt. Kmbl. Rush. 17, 27. Scylling (scilling, Lind., Rush.) dragmam, Lk. Skt. 15, 9. Nis woruldfeoh ꝺe ic mê âgan wille, sceat ne scilling, Cd. Th. 129, 13 ; Gen. 2143. Hundraþ scillinga centum denarios, Mt. Kmbl. Lind. 18, 18. Þriim peninga ł scillinga, Jn. Skt. Lind. 12, 5. Þrítig scillinga triginta argenteos, Mt. Kmbl. 26, 15. Þûsend scyllinga on seolfre mille argenteos, Gen. 20, 16. Feówerhund scillinga (siclos), 23, 16. Hê hêt heora ælcum fiftig scyllinga tô sceatte syllan, Homl. Th. i. 88, 4. [Goth. skilliggs : O. Frs. skilling : O. L. Ger., O. H. Ger. scilling solidus, aureus : Icel. skillingr.] v. mene-, riht-, wægn-scilling.

Scilling, es; m. The name of a poet :—Wit Scilling for uncrum sigedryhtne song âhôfan, Exon. Th. 324, 31 ; Vîd. 103.

scilling-rîm, es; n. A reckoning by shillings :—Se mê beág forgeaf, on ꝺam siex hund wæs smætes goldes sceatta scillingrîme a ring containing gold to the value of six hundred shillings, Exon. Th. 324, 10; Vîd. 92.

scíma, an; m. Shadow, gloom :—Ne hêr (in hell) dæg lýhteþ for scedes sciman, Cd. Th. 271, 15; Sat. 106. Hýdeþ hine æghwylc æfter sceades sciman, Salm. Kmbl. 233; Sal. 116. [Cf. Uualdandes craft scal thi scadouuan mid skimon virtus altissimi obumbrabit te, Hel. 279. M. H. Ger. scheme a shadow, mask; Icel. scheme a shadow; Ger. schemen.] v. scimian.

scíma, an; m. Splendour, brightness, light :—Ꝺonne ꝺære sunnan scíma hâtast scînþ, Bt. 5, 2 ; Fox 10, 28 : Cd. Th. 232, 23 ; Dan. 264. Ꝺæs leóhtes scíma wæs swâ mycel cujus radius lucis tantus exstitit, Bd. 4, 7 ; S. 575, 17 : 5, 10 ; S. 625, 9. Se scíma gâstlícre beorhtnysse, Guthl. 2 ; Gdwin. 12, 22 : Exon. Th. 44, 4 ; Cri. 697. Wuldres scíma scân, 179, 12 ; Gû. 1260. Mîn se swêtesta sunnan scíma, Iuliana, 252, 21 ; Jul. 166. Ne nænig ꝺæl leóhtes scíman geseón mihte ne minimam quidem lucis alicujus posset particulam videre, Bd. 4, 10 ; S. 578, 20. Sió beorhtnes ꝺære sunnan scíman, Bt. 34, 8 ; Fox 146, 4 : 39, 3 ; Fox 216, 1 : 4 ; Fox 6, 33. Metod æfter sceáf scírum scíman æfen, Cd. Th. 9, 5 ; Gen. 137. Ꝺâ gesundrode Waldend sceade wiꝺ scíman, 8, 22 ; Gen. 128. Se môna gehrân mid his scíman (splendore) ꝺæm treówum ufeweardum, Nar. 30, 7. God hira môd onlíeht mid ꝺæm scíman (radio) his geþe, Past. 35, 4 ; Swt. 243, 21 : 48; Swt. 369, 16. Fore scíman prae fulgure, Ps. Surt. 17, 13. Seó sunne scíman ne hæfde the sun was eclipsed, Bd. 3, 27 ; S. 558, 11. Swâ ꝺæt ic mihte geseón swîꝺe lytellne scíman leóhtes, Bt. 35, 3 ; Fox 158, 29. Niht ne genîpþ ꝺæs heofenlícan leóhtes scíman nox nulla rapit splendorem lucis amoenae, Dôm. L. 16, 254. Þýstro hæfdon bewrigen Wealdendes hræw, scírne scíman, Rood Kmbl. 107; Kr. 54. [Goth. skeima φανή : O. L. Ger. scímo splendor, fulgor, nitor : O. Sax. dag-skímo : O. H. Ger. skímo splendor, fulgor, effulgentia, radius, fax : Icel. skími a gleam of light.] v. æfen-scíma.

scímian; p. ode To grow dark, (of the eyes) to be dazzled, bleared :— Mîne eágan scímiaþ lippio, Ælfc. Gr. 30, 5 ; Som. 34, 59. Swâ ꝺæt nân man ne mihte for ꝺam mycclum leóhte hire on beseón . . . and swâ hí hî geornlícor sceáwodon, swâ scimodon heora eágon swîꝺor, Homl. Skt. i. 7, 153. Beóþ his dagas dêmde gelíce swâ ꝺû on scimiendre sceade lôcige, Ps. Th. 143, 5. v. scíma.

scímian; p. ode To shine, glisten :—Ic scímige (scíne, MS. W.) mico, Ælfc. Gr. 24; Zup. 138, 1. Scímande (scínende, Rush.) coruscans, Lk. Skt. Lind. 17, 24. Cf. Be hiora hiwe . . . beóþ æblæce and eal se líchoma âscímod (shiny), Lchdm. ii. 232, 2. [Þat hus schineꝺ ant schimmeꝺ, O. E. Homl. i. 257, 35. Schan (schimede ant schan, MS. B.), Marh. 2, 34. Wiꝺ schimmende sweord, 19, 30. Schiminde (schiminde, other MS.) hire nebscheaft, Jul. 55, 4. O. H. Ger. scímit micat.] v. scíma.

scímrian to shine, glisten :—Scymriendes wætes cerulei gurgitis, Germ. 401, 10. [Þat hus schineꝺ ond schimmeꝺ (schimereꝺ, MS.), O. E. Homl. i. 257, 35. Hit schemereꝺ and schon, Gaw. 772. Þat eadi trume of schimerinde meidenes, H. M. 21, 34. Du. schemeren : Ger. schimmern : Swed. skimra. Cf. scimeringe crepusculum, Grff. vi. 512.]

scín, scinn, es; n. An extraordinary appearance, a deceptive appearance, a spectre, evil spirit, phantom :—Scín portentum, Txts. 87, 1611. Scíu fantasma, i. nebulum (-am?), Wrt. Voc. ii. 37, 43 : 95, 65 : prestigiis, 79, 5. Bôcstafa brego bregdeþ scína feónd be ꝺam feaxe, læteþ flint brecan scínes sconcan, Salm. Kmbl. 203; Sal. 101. Egsa âstígeþ monna cynne ꝺonne blâce (blace?) scotiaþ scríþende scín (the spirits of the storm) scearpum wæpnum, Exon. Th. 385, 29; Rä. 4, 52. Swâ biþ scinna þeáw, deófla wîse, 360, 4; Wal. 31. Scinnum scenis (cf. scína gríma, 94, 904), Txts. 97, 1831. Ꝺam deófle wiꝺstandan ꝺonne hê his wôd scinn (wôde scín, MS. H.) tôbrædeþ to oppose the devil, when he spreads abroad his mad spirits (?), Wulfst. 80, 4. Cf. Ꝺa hæþenan deófle offrodon . . . and ꝺa brædas ꝺæs flæsces stigon upp on ælce healfe eall swilc hit mist wære . . . ꝺa hæþenan on swilcon deófolscíne (altered to -scinne) blissedon, Homl. Skt. i. 23, 39. Deófulscinnu þurh gebed beóþ oferswýþede demonia per orationem uincuntur, Scint. 7. [Cf. O. H. Ger. gi-scín

fantasma. v. Grmm. D. M. 450, 867.] v. scinna *and the compounds with* scín-.

scín (?) *brightness, shine.* [O. Sax. (sunnon) skín: O. Frs. (sunna) skín: O. H. Ger. scín *jubar*: M. H. Ger. schín: Ger. schein: Icel. sól-, tungl-skin.] v. sun-scín.

scínan; *p.* scán, *pl.* scán, sceán *To shine.* I. lit.:—Ic scíne *splendeo*, Ælfc. Gr. 26, 2; Som 28, 42. Scínþ *candescit*, Past. 14, 6; Swt. 89, 1. Swá se lígræsc scínþ (*fulget*), Lk. Skt. 17, 24: Bt. 5, 2; Fox 10, 29. Ðonne seó sunne on heofone beorhtost scíneþ. 9; Fox 26, 15. Scýneþ ðes móna, Fins. Th. 13; Fin. 7. Ða steorran scínaþ beforan ðam mónan, and ne scínaþ beforan ðære sunnan, Bt. 39, 3; Fox 214, 30. Scaan *ardebat*, Wrt. Voc. ii. 101, 3. Scán, 7, 29. Se steorra (*comet*) scán .iii. mónþas, Chr. 678; Erl. 41, 4. His ansýn sceán (*resplenduit*) swá swá sunne, Mt. Kmbl. 17, 2: Bd. 5, 12; S. 628, 13: Cd. Th. 185, 19; Exod. 125. Seó ród sceán swá heofenes tungol, Shrn. 149, 11. His ansýn eal sceán swá swá sunne, and his gewæda scinon on snáwes hwítnysse, Homl. Th. ii. 242, 7. Hwær is seó eorðe ðe næfre sunne on ne sceán? In ðære reádan sæ, Salm. Kmbl. 198, 14. Wígbord scinon, Cd. Th. 207, 14; Exod. 466. Eoforlíc scionon, Beo. Th. 612; B. 303. Ân cyn ys *olocryseis*, ðæt is on úre geþeóde gecweden, ðæt heó eall golde scíne, Lchdm. i. 242, 13. Hig scínon (*luceant*) on ðære heofenan fæstnysse, Gen. 1, 15. Sunnan leóma cymeþ scýnan, Exon. Th. 56, 18; Cri. 902. Scínende *refulgens*, Lk. Skt. 9, 29. Beorhtnes scínendes steorran *fulgor stellae*, Bd. 5, 12; S. 629, 5. Scínendes léges, 4, 13; S. 581, 15. Scínendum *limpidis*, Wrt. Voc. ii. 50, 34. II. fig.:—Ðonne scínaþ ða rihtwísan, Mt. Kmbl. 13, 43. Se nama se ðe mid him swá lange sceán and bryhte *nomen quod apud eos tam diu claruerat*, Bd. 1, 12; S. 480, 39: 3, 13; S. 538, 39. Seó stów on ðære ðe ðu ðæt fægereste weorud on geóguþhádnesse gesáwe scínan and wynsumian *locus iste in quo pulcherrimam hanc juventutem jocundari ac fulgere conspicis*, 5, 12; S. 630, 15. Ðæt mód swá beorhte ne mót blícan and scínan, Met. 22, 35: Bt. 35, 1; Fox 156, 2. Ðæt ðú móste hálig scínan, eádig on ðam écan lífe, Exon. Th. 87, 19; Cri. 1427. On wordum and on dædum beorht and scínende *verbo et actibus clarus*, Bd. 3, 19; S. 547, 4. On scínendre *praepollenti*, Hpt. Gl. 491, 1. [*Goth.* skeinan: O. Sax. skínan: O. Frs. skína: O. H. Ger. scínan: Icel. skína.] v. á-, be-, ge-, geond-, ofer-, ymb-scínan.

scín-bán, es; *n. A shin, shin-bone:*—Scina *vel* scinbán *tibiae*, Wrt. Voc. i. 44, 72: 71, 58. [A schynbone *sura*, 247, col. 2. Oc (*cervus*) leigeþ his skinbon on oðres lendbon, Misc. 12, 359. M. H. Ger. schine-bein: Ger. schien-bein: Du. sheen-been.]

scín-, scinn-cræft; *m.* I. *the art by which deceptive appearances are produced, magic:*—Ðis synt ða ídelnyssa ðisse worlde ... scíncræft *hae sunt vanitates hujus mundi ... ars magica*, L. Ecg. P. 1, 8; Th. ii. 174, 34. Hié ne angeátan mid hwelcum scinncræfte and mid hwelcum lotwrence hit deófla dydon, Ors. 3, 3; Swt. 102, 17. Hé behét ánum drýmen sceattes, gif hé mid his scýncræfte (scín-, MS. O.) him ðæt mæden mihte gemacian tó wífe, Homl. Skt. i. 3, 365. Beó ic scyldig, gif ic his scýncræft ne mæg ádwæscan mid mínum drýcræfte, 14, 57. Hý wæron tó sáre beswicene þurh ðæs sweartan deófles scíncræft, Wulfst. 198, 18. II. *a magic art or trick:*—Scíncræfte *praestigia*, Wrt. Voc. ii. 66, 59. Wiccan beóþ tó helle bescofene for heora scíncræftum, Homl. Th. ii. 330, 29. Hí mid mislícum scýncræfton ðæt folc dwelodon, 482, 4. Hé wolde ðære fæmnan mód on his scíncræftum onwendan tó hæðendóme, Shrn. 135, 1. Ða cwædon hí, ðæt hit scinn-cræftas ne cúþan, 90, 10. Se sceocca eów lærþ ðyllíce scíncræftas, Homl. Skt. i. 17, 106. ¶ *In the following the word is glossed as if it were* scíncræftiga:—Scíncræfta *hierophantorum*, Wrt. Voc. ii. 43, 25: 82, 7: Hpt. Gl. 483, 7.

scín-cræftiga, an; *m. A magician, sorcerer:*—Gif wiccan oððe wigleras, scíncræftigan oððe hórcwénan on earde wurðan ágitene, fýse hí man georne út of ðysan earde, L. Eth. vi. 7; Th. i. 316, 20.

scíne, scéne, sceóne, scióne, scýne; *adj. Beautiful, fair, bright:*—Is se forrynel fæger and sciéne, Met. 29, 25: Cd. Th. 41, 14; Gen. 656. Cwæð ðæt his líc wære leóht and scéne, 17, 26; Gen. 265. Wæstm wlitig and scéne, 30, 16; Gen. 467. Deór wundrum scýne (*the panther*), Exon. Th. 356, 20; Pa. 19. Is seó womb wundrum fæger, scír and scýne, 219, 16; Ph. 308. Mægþ scýne, Beo. Th. 6025; B. 3016. Se scýna stán, Andr. Kmbl. 1532; An. 767. On stede scýnum, Exon. Th. 70, 33; Cri. 1148. Ic ðe swá sciénne gesceapen hæfde, 85, 6; Cri. 1387. Hé forlærde idese sciéne, Cd. Th. 43, 34; Gen. 700. Hé geseah Euan stondan sceóne gesceapene, 35, 3; Gen. 549. Tó sceáwianne ðone scýnan wlite, Exon. Th. 57, 8; Cri. 915. Forhwon forléte ðú líf ðæt scýne, 90, 7; Cri. 1470. Sceóne lambru, Ps. Th. 113, 4, 6. Gimmas swá scýne, Exon. Th. 43, 27; Cri. 695: 219, 1; Ph. 300. Fuglas scýne, 237, 17; Ph. 591. Þurh ða scénan scínendan rícu ðæs Fæder *per Patris fulgenti regna paratu*, Dóm. L. 18, 294. Him wíf curon scýne and fægere, Cd. Th. 76, 5; Gen. 1252. Hyrsta scýne, Judth. Thw. 26, 9; Jud. 317. Hiwbeorhta and scýnra, Exon. Th. 357, 10; Pa. 26. Wurdon ðín gesceapu scénran, Cd. Th. 32, 14; Gen. 503. Eue idesa sciénost, 51, 4; Gen. 821. Scénost, 39, 17; Gen. 626. Sceónost,

44, 5; Gen. 704. Engla scýnost, 22, 10; Gen. 338. [Feier and sceone (scene, 2nd MS.), Laym. 2299. Regan þ sceone (scene, 2nd MS.), 3098. A steorrne ... brihht and shene, Orm. 3431. Scone and faȝȝerr, 15665. A þusent fold schenre þen þe sunne, A. R. 100, 4. Heo as schene as schininde sunne wende up aloft, Marh. 19, 14. Emelye hire yonge suster schene, Chauc. Kn. T. 114. Æfter sharpe shoures moste shene is þe sonne, Piers P. 18, 409. *Goth.* skauns: O. Sax. skóni: O. Frs. skéne: O. L. Ger. scóni *lucidus*: O. H. Ger. scóni *splendidus, splendens, formosus, venustus, pulcher, speciosus*: Ger. schön.] v. ælf-, sun-, þurh-, wlite-scíne.

scínefrian *to glitter:*—And scínefrian *ac micare*, Wrt. Voc. ii. 6, 33.

scínendlíc; *adj. Clear, bright:*—Beorht ł scínendlíc ł leóht *lucidum*, Ps. Lamb. 18, 9. v. þurh-scínendlíc.

scínere, scinnere; *m. One who produces deceptive appearances* (v. scín), *a magician:*—Scinneras *emaones*, Txts. 59, 746. Scíneras, scinneras *scienicis*, 98, 952.

scín-feld; *dat.a*; *m. The beautiful, Elysian field,* applied to Tempe:—Hwæt synt ða twegen men on neorxna wange? Enoch and Helias. Hwær wuniaþ hý? Malifica and Intimphonis (in Tempis?), ðæt is on sunfelda and on sceánfelda (sceón-?), Salm. Kmbl. 202, 2. On scénfeldum *in Tempis*, Wrt. Voc. ii. 47, 16: 89, 72.

scín-gedwola, an; *m. A delusion produced by magic, delusive appearance, phantom:*—Scíngedwolan *nebulam*, Wrt. Voc. ii. 61, 30.

scín-gelác, es; *n. A magical practice:*—Hí ongunnon secgan ðæt hit drýcræftum gedón wære scíngelácum ðæt se stán mælde *they said that it was done by the sorcerer's arts, by magical practices, that the stone spoke*, Andr. Kmbl. 1531; An. 767.

scín-, scinn-hiw, es; *n. A form produced by magic, phantom, spectre:*—Scínhiw *prestigium*, Wrt. Voc. i. 21, 61: *fantasma*, ii. 33, 82. Scínlác *vel* [scín]hiw *fantasia, i. imaginatio, delusio mentis*, 147, 42. Réþlíc scinnhiw *ferale monstrum*, 147, 53: Hymn. Surt. 142, 12. Ne eom ic ná scinnhiw (*phantasma*), swá swá gé wénaþ, Homl. Th. ii. 388, 26. Scínhiowes *fantasiae*, Ps. Surt. ii. p. 190, 11. Scínhiwe[s] *phantasmate*, Wrt. Voc. ii. 67, 5. Scínhiwe, 34, 1. Wiccecræftas, scínhiw *prestigias*, 66, 25.

scín-hosu, e; *f. A shin-hose, a covering for the lower part of the leg, a greave:*—Scínhose *ocreis*, Hpt. Gl. 521, 5.

scín-lác, es; *n.* I. *magic, necromancy, sorcery:*—Scýnláce *necromantia*, Hpt. Gl. 482, 74. Se mec gescyldeþ wið ðíuum scínláce, Exon. Th. 255, 15; Jul. 214. Hí sædon ðæt hió sceolde mid hire scínláce (cf. mid hire drýcræft, Bt. 38, 1; Fox 194, 30) beornas forbrédan and mid balocræftum weorpan on wildra líc, Met. 26, 74. Twegen drýas ða worhton micel scínlác mid twám dracum, Shrn. 131, 29. II. *a particular act of magic, a sorcery, delusion produced by magic:*—Hí ðæt hæfdon gedón mid yflum scínlácum, Shrn. 90, 10: 75, 18. Ða ðe galdorcræftas begangaþ, and mid ðæm unwære men beswícaþ, and hí áweniaþ from Godes gemynde mid heora scínlácum, Blickl. Homl. 61, 25. Scíndlácum, Shrn. 141, 27. III. *delusion, superstition, frenzy, rage:*—Scínlác *fantasia, i. imaginatio, delusio mentis*, Wrt. Voc. ii. 147, 42. Ðætte gifearria from ðære stówe ælc scínlác and ymbcerro diúbles fácnes *ut discedat ab eo loco omnis fantasia vel versutia diabolicae fraudis*, Rtl. 120, 33. Næs his scínlác ne his hergiung on ða fremdan áne ac hé gelíce slóg and hiénde ða ðe him wæron mid farende *nec minor ejus* (Alexander) *in suos crudelitas, quam in hostem rabies fuit*, Ors. 3, 9; Swt. 130, 19. Máufulles scínlác *fanaticae superstitionis*, Hpt. Gl. 488, 41: 509, 39. Scínlác[e] *superstitione*, 500, 70. Sume Rómâna wíf on swelcum scínláce wurdon and on swelcum wôdan dreáme *incredibili rabie et amore scelerum Romanae matronae exarserunt*, Ors. 3, 6; Swt. 108, 25. IV. *a delusive appearance, a spectre, apparition, phantom:*—Hí cwædon: Hyt ys scínlác *dicentes: Quia phantasma est*, Mt. Kmbl. 14, 26. Scínlác *nebulo*, Hpt. Gl. 501, 16. Scýnláce *praestigia*, 482, 74. Tó fleánne ælc scínlác díuoles *ad effugandum omne fantasma diaboli*, Rtl. 100, 33. Deós wyrt (ἀστέριον) scíneþ on nihte swilce steorra on heofone, and se ðe hý nytende gesihþ, hé sægþ ðæt hé scínlác geseó, Lchdm. i. 164, 6. Scínlác *monstra*, Wrt. Voc. ii. 56, 15: *nebulones*, Hpt. Gl. 501, 73. Wið deófulseócnysse and wið yfelre gesihþe, wulfes flæsc gesoden ... ða scínlác ðe him ær ætýwdon ne geunstillaþ hý hine, Lchdm. i. 360, 13-16. Ðý læs cild sý hreósende, oððe scínlác méte, 350, 13. Ða ðe scínlác þrowien etan león flæsc; ne þrowiaþ hý ofer ðæt ænig scínlác, 364, 22. Scínláca *praestigiarum*, Hpt. Gl. 501, 68. Galdrás *praestigias*, scínlác *fantasias*, 459, 16. Scínlácu gesihþ, gestreón of ungewéndum hit getácnaþ, Lchdm. iii. 204, 18. [O. H. Ger. scín-leih *monstrum*.] v. Grmm. D. M. 450.

scín-læca, -láca, an; *m. A magician, necromancer, sorcerer:*—Scínlæcan (-læcean, -lécan) *nebulonis*, Txts. 81, 1372: *nebulis* (*nebulonis?*), Wrt. Voc. ii. 60, 60: 79. 4. Ðæs leásan scínlæcan *falsi nebulo*, 147, 2. Sabastianus ongon hine (*St. Victor*) nédan tó deófolgelde; ðá hé ðæt ne geþafede, ðá hét hé sumne scínlæcan him sellan etan ðæt flæsc ðæt wæs geæÞtred, Shrn. 84, 27. Hí gefetton Escolafius ðone scínlácan mid ðære ungemetlícan nædran ðe mon Epithaurus hét *horrendum illum Epidaurium colubrum, cum ipso Aesculapii lapide advenerint*, Ors. 3, 10; Swt. 140,

9 : 3, 10, tit.; Swt. 3, 19. Scínlǽcan *magi*, Wrt. Voc. ii. 39, 11. On helle beóþ ða scínlǽcan, ða ðe galdorcræftas begangaþ, Blickl. Homl. 61, 23. Ða fǽmnan ðe gewuniaþ onfón gealdorcræftigan and scínlǽcan (-lácan, MS. H.) and wiccan, ne lǽt ðú ða libban, L. Alf. 30; Th. i. 50, 10. v. two following words.

scín-lǽce, an ; f. *A woman who practises magic, a sorceress :*—Ða cwǽdon Rómware ðæt heó wǽre drýegge and scínlǽce, Shrn. 56, 13.

scín-lǽc[e], -lác ; adj. *Magical, phantasmal :* — Hí him héton gefeccean tô Escolapius ðone scínlácan mid ðære scínlǽcan (-lácan, MS. L.) nǽdran, Ors. 3, 10, tit.; Swt. 3, 19. Álêsedo from ælcum ongifeht scínelácum *libera ab omni inpugnatione fantasmatica*, Rtl. 98, 26. v. preceding words.

scín-líc ; adj. *Of the nature of an apparition, phantasmal :*—Suoefno and næhta scínelíco *sompnia et noxia fantasmata* (the glosser seems to have read *noctes fantasmaticae?*), Rtl. 180, 16.

scinn, scinnere. v. scín, scínere.

scinna, an ; m. *An evil spirit, spectre :*—Blace hworfon scinnan (*the fallen angels*) forscepene, sceaþan hwearfdon geond ðæt atole scref (*hell*), Cd. Th. 269, 12 ; Sat. 72. Ðæt hié leóda landgeweorc láþum beweredon scuccum and scinnum, Beo. Th. 1882 ; B. 939. v. scín.

scínness, e ; f. *Brightness, splendour :*—Ðe môna ne seleþ scínisse (*splendorem*) his, Mk. Skt. Lind. 13, 24.

scín-seóc ; adj. *Haunted by apparitions :*—Scínseócum men wyrc drenc of hwítes hundes þoste, Lchdm. i. 364, 4.

scinu, e ; f. *A shin :*—Scinu *cruscula*, Wrt. Voc. ii. 137, 20. Scinu vel scinbán *tibiae*, i. 44, 72. Scyne oððe scinbán *tibia* (ae?), 71, 58. Scina, 65, 42. Scancan, scina *tibias*, Hpt. Gl. 482, 64. [O. H. Ger. scina *tibia*.] v. scin-bán.

sció, scioppa. v. sceón, scoppa.

scip, es ; m. *A patch, clout :*—Ne ásend nán man scyp (scep *altered to* scyp, MS. A.: ðæt ésceapa *commisuram*, Lind.) of níwum reáfe on eald reáf ; elles ðæt níwe slít, and se níwa scyp (*as before in* MS. A. *and* Lind.) ne hylpþ ðam ealdan, Lk. Skt. 5, 36 ; Mt. Kmbl. 9, 16. Scyp (*also* scep, MS. A.: later MSS. scep, scyp) *assumentum*, Mk. Skt. 2, 21.

scip, es ; n. *A ship :*—Scip *navis* vel *faselus*, scipu *rates*, sceort scip *naviscella* vel *cimba*, vel *campolus* vel *musculus*, litel scip *scapha*, Wrt. Voc. i. 47, 55–61. Scip *ratis*, horsa scip *ypogavus*, swift scip *archiromacus*, sceaþena scip *paro*, ânbýme scip *trabaria*, 56, 11–28. Scip barca, ii. 12, 19 : *caraba*, 22, 34. Foreweard scip *prorostris*, 68, 48. Scipes botm *carina*, scipes hláford *nauclerus*, i. 48, 3-4. Scipes flór *fori* vel *tabulata navium*, 63, 40. Lytlum scipe *cimbula*, ii. 22, 34. Scipe *cercilo*, 17, 72 : 76, 30 (cf. aesc *cercilus*, 103. 56). Ðá wende hé on scype (scipp, Lind.) *ascendens nauem*, Lk. Skt. 8, 37. Scyp ástígan, Lchdm. iii. 184, 13. Swá eode hé on scyp, Bd. 4, 1 ; S. 564, 47. Scipu *classes*, Wrt. Voc. ii. 14, 46. Scypu (sciopu, Rush.: scioppu, Lind.) *naues*, Jn. Skt. 6, 23. Scipu (sciopu, Lind.), Lk. Skt. 5, 2 : *nauiculas*, 5, 7. Scypo (scioppo, Lind.) *naues*, 5, 11. Sceopu, Ps. Surt. 47, 8 : 103, 26. [Goth. O. Sax. .O. L. Ger. O. Frs. Icel. skip : O. H. Ger. scif.] v. ǽrend-, ceáp-, fird-, flot-, for-, horn-, hýð-, lang-, pleg-, troh-, unfriþ-scip.

scip-bíme, e ; f. *A ship-trumpet :*—Scypbýman *classicam tubam*, Germ. 391, 48.

scip-broc, es ; n. *Trouble, hardship, or labour when journeying in a ship :*—Paulus him rehte hú myccle scipbrocu hé gebád on ðæm síþe St. Paul related to them the hardships he had undergone on his voyage to Rome, Blickl. Homl. 173, 6.

scip-brucol ; adj. *Causing shipwreck :*—Scypbrucules wæles *nauifragi gurgitis*, Germ. 401, 9.

scip-bryce, es ; m. *Ship-wreck, what comes ashore from wrecks :*—Ic habbe gegeofen Ælfwine abbod intô Ramesége . . . scipbryce and ða sǽupwarp on eallen þingen swá wel swá ic hit mê seolf betst habbe bí ða sǽrime áhwǽr in Engelande, Chart. Th. 421, 33. (Cf. L. H. i. 10, 1 ; Th. i. 519, 4 *where among the rights* (jura) *belonging to the king* naufragium *is mentioned*.) [Cf. Icel. skip-brot *wreck drifted ashore*.]

scip-cræft, es ; m. *Naval power, strength in ships :*—Swegen sende hider and bæd him fylstes ongeán Magnus, ðæt man sceolde sendan .L. scypa him tô fultume. Ac hit þúhte unrǽd eallum folce, and hit wearð gelet þurh ðæt ðe Magnus hæfde micelne scypcræft, Chr. 1048 ; Erl. 173, 7.

scip-drincende (-drencende? *see* uêre gidruncen *mergeretur*, l. 31) *making shipwreck :*—Paulum scipdrincende gifriáde *Paulum naufragantem liberavit*, Rtl. 61, 33.

scipe, es ; m.   **I.** *pay, stipend :*—Scipe vel bigleofa *stipendium*, Wrt. Voc. i. 20, 33. [Hi nolleþ paye þet hi ssofle, and hi ofhealdeþ þe ssepes of ham þet doþ hare niedes, Ayenb. 39, 5 (the word occurs several times in this work). Withholdyng or abrigging of the schipe or the hyre or the wages of servauntes, Chauc. Persones T. (De Ira). And cf. Ne mihte ic of þan kinge habben sciping ; ich spende mine ahte þa wile þa heo ilaste, Laym. 13656.]   **II.** *state, condition, dignity, office :*—Hæbbe ic mínes cynescipes gerihta swá mín fæder hæfde, and míne þegnas hæbben heora scipe (cf. se dêma ðe ôðrum wôh dême . . . þolige á his þegenscipes, L. Edg. ii. 3 ; Th. i. 266, 15–18) on mínum tíman swá hý

hæfdon on mínes fæder, L. Edg. S. 2 ; Th. i. 272, 28.   ¶ -scipe -*ship*, helps to form many nouns.   [O. Frs. -skipe, -skip : O. Sax. -skepi.]

scipen. v. scypen.

scipere, es ; m. *A sailor :*—Hé tealde ðæt his sciperes woldon wændon fram him, bûton hé ðê raðor côme . . . His sciperes gefêngon hine and wurpon hine on ðone bát, Chr. 1046 ; Erl. 174, 13–18. [From Scandinavian (?). Icel. skipari *a mariner*.]

scip-fæt, es ; n. *A vessel in the form of a ship :*—Húseldisc *patena*, scipfæt *cimbia* (the word occurs under the heading *nomina vasorum*), Wrt. Voc. i. 25, 32. Cf. Hec *acerra* a schyp for censse, 230, col. 2. Wright has the following note on this entry : The *nef*, a vessel in the form of a ship, used in the church from an early period to hold the incense, as well as other articles.

scip-farend, es ; m. *A ship-farer, sailor :*—Aidan ðám scypfarendum (*nautis*) ðone storm tôwardne foresægde, Bd. 3, 15 ; S. 541, 16. v. next word.

scip-férend, es ; m. *A sailor :*—Wǽron hié on gescirplan scipfêrendum onlíce, eálíðendum, Andr. Kmbl. 500 ; An. 250. v. preceding word.

scip-fird, e ; f. *A naval force* or *expedition, a fleet :*—Ðá ðeós scipfyrd (*the naval expedition described in the preceding paragraph*) ðus geendod wæs, Chr. 1009 ; Erl. 142, 15. Wé næfdon ða gesélþa ðæt seó scipfyrd nytt wǽre ðisum earde, 1009 ; Erl. 141, 26. Ðá cýdde man in tô ðære scipfyrde, ðæt hí mann eáðe befaran mihte, Erl. 141, 33. *See* land-fird *for other passages.* [Humber King & al his fleote & his muchele scipferde comen on Albanaces londe, Laym. 2156.]

scip-firdung, e ; f. *A naval force* or *armament :*—Æt ðam ende ne beheóld hit nánþing seó scypfyrding ne seó landfyrding, Chr. 999 ; Erl. 134, 36. Burhbóta and bricbóta áginne man georne on æghwilcon ende, and fyrdunga eác, and scipfyrdunga ealswá, L. Eth. vi. 32 ; Th. i. 322, 32.

scip-flota, an ; m. *A sailor :*—Hettend crungun Sceotta leóda and scipflotan (*the Danes*), Chr. 937 ; Erl. 112, 11.

scip-forðung, -fyrðung, e ; f. *Preparation of ships :*—Burhbóta and bricgbóta and scipforðunga (-fyrðunga, MS. B.) áginne man georne (cf. wærlíc biþ ðæt man æghwilce geáre sóna æfter Eástron fyrdscipa gearwige, L. Eth. vi. 33 ; Th. i. 324, 3), L. C. S. 10 ; Th. i. 380, 27   v. scipfyrðung.

scip-fylleþ *the private jurisdiction exercised over a group of three hundreds.* The word occurs in a charter of Edgar granting to Bishop Oswald certain privileges connected with three hundreds, and in reciting the request that had been made to the king it is said : ' quatinus posset ipse (Oswald) cum monachis suis unam naucupletionem, quod Anglice scypfylleð dicitur, per se habere.' The grant of the request is then stated : ' Ego Eadgarus Oswaldo episcopo annuo et dono huius libertatis priuilegium . . . ut ipse episcopus cum monachis suis de istis tribus centuriatibus . . . construant (constituant, Chart. Th. 214) unam naucupletionem, quod Anglice dicitur scypfylleð oððe scypsôcne, in loco quem ob eius memoriam Oswaldeslaw deinceps appellari placuit, ubi querelarum causae secundum morem patriae et legum iura iure discernantur ; habeatque ipse episcopus debita transgressionum . . . et omnia quaecunque rex in suis hundredis habet,' Cod. Dip. Kmbl. vi. 240. The connection between the sense in which the word seems to be used in the charter and the meanings of the two parts of the compound may perhaps be found in the entry under the year 1008 in the Chronicle. It there apparently states, that from every three hundred hides one ship should be furnished to the national fleet. v. Stubbs' Const. Hist. i. 105, and cf. Kemble's Saxons in England, i. 255. The word *fylleþ* occurs in the compound *winter-fylleþ*, q. v. ; cf. also Icel. skip-sókn *a ship's crew.*

scip-fyrðrung, e ; f. *Fitting out of ships :*—Ymbe scypfyrðrunga, ðæt æghwylc geset sý sóna ofer Eástran, L. Eth. v. 27 ; Th. i. 310, 26. v. scip-forðung.

scip-gebroc, es ; n. *Shipwreck :*—Ðæt hié æfter ðæm scipgebroce him ða sǽ ondrǽden *ut mare post naufragium metuant*, Past. 52, 1 ; Swt. 403, 12. Ic ðê bidde ðæt ðú mê on ðæm scipgebroce ðisses andweardan lífes sum bred geræce ðínra gebeda *in hujus quaeso vitae naufragio orationis tuae me tabula sustine*, 65, 7 ; Swt. 467, 24. Hwelce tibernessa hié dreógende wǽron on hungre ge on scipgebroce, Ors. 1, 11 ; Swt. 50, 19.

scip-gefeoht, es ; n. *A naval battle* or *war :*—Scypgefeoht *bellum classicum*, Germ. 389, 42.

scip-gefére (?), es ; n. *A going by ship, navigation, sailing :*—Hé on his scipgefêre hwearf eft tô Cent *rediit Cantiam navigio*, Bd. 2, 20 ; S. 521, 41.

scip-getawu *furniture of a ship :*—Gerêþru vel scipgetawu *aplustre*, Wrt. Voc. i. 56, 19.

scip-gild, es ; n. *A ship-tax, a tax to supply funds for the maintenance of a fleet :*—Swá fela sýðe swá menn gyldaþ heregyld oððe tô scipgylde *quotiens populus universus persolvit censum Danis, vel ad naves seu ad arma*, Chart. Th. 307, 24.

scip-hamer, es ; m. *A hammer carried in the hand, by which a signal is given to the rowers :*—Sciphamor *portisculus* vel *hortator remigum*, Wrt. Voc. i. 48, 20.   Ꝩ. hamer.

scip-here, es; m.   I. *a collection of ships of war, a naval force, a fleet of war :*—Sciphere *classis*, Ælfc. Gr. 9, 28; Som. 11, 56: Wrt. Voc. i. 73, 75: *classica*, ii. 131, 62. Flota, sciphere *clasis*, 14, 45. Sciphere eów nymþ *reducet te Dominus classibus in Aegyptum*, Deut. 28, 68. On ðæs sǽs waroþe tô sûþdǽle ðanon ðe hí sciphere on becom *in litore oceani ad meridiem quo naves eorum habebantur*, Bd. 1, 12; S. 481, 11. Ðý ilcan geáre gegadrode micel sciphere on Ald-Seaxum, and ðǽr weard micel gefeoht, Chr. 885; Erl. 84, 6. Gif ǽnig sciphere on Engla lande hergie, L. Eth. ii. 1; Th. i. 284, 15. Ðý sumera fôr Ælfrêd cyning ût on sǽ mid sciphere and gefeaht wið .vii. sciphlæstas, 875; Erl. 78, 6. Persa cyning sende Conon mid scipehere (scip-, MS. C.), Ors. 3, 1; Swt. 96, 25. Ðâ côman hí sôna mid sciphere *mox advecti navibus*, Bd. 1, 12; S. 480, 34. Ðæt on land Dena lâðra nǽnig mid scipherge sceðþan ne meahte, Beo. Th. 491; B. 243. Ne gehêrde nôn mon ðǽget nánne sciphere, ne furþon ymbe nán gefeoht sprecan, Bt. 15; Fox 48, 14. Se cyng wæs west on Defnum wið ðone sciphere (*acting against the Danish fleet*), Chr. 894; Erl. 92, 26. On ðysum geáre wæs micel unfriþ on Angelcynnes londe þurh sciphere, 1001; Erl. 136, 2. Sciphergas, Met. 8, 31.   II. *the men of a ship of war :*— Ælfrêd cyning gefeaht wið feówer sciphlæstas Deniscra monna and ðara scipa tû genam . . . and tuegen scipheras him on hond eodon, and ða wǽron miclum forslægene, ǽr hié on hond eodon (cf. ðara scipa twá genâmon . . . and twá him on hand eodon, and ða men wǽron myclum ofslagene, ǽr hî on hand eodon, MS. E.), Chr. 882; Erl. 82, 12.

sciphere-líc; adj. *Relating to a fleet, naval :*— Scipherelícum *classicis*, Hpt. Gl. 406, 40.

scip-hlǽder, e; f. *A ship's ladder, a ladder for passing from a ship to the shore :*—Sciphlǽder *pons*, Wrt. Voc. i. 63, 53. Sciphlædder *ponsis*, 56, 47.

scip-hlǽst, es; m.   I. *the body of (fighting) men on a ship :*— Claudius se consul fôr an Pûnice and him Hannibal ût on sǽ ongeán com and ealle ofslôg bûton .xxx. sciphlæsta ða ôþflugon tô Libeum ðæm íglande *Claudius consul contra hostem profectus superatus est. Et ipse quidem cum triginta navibus Lilybaeum confugit*, Ors. 4, 6; Swt. 178, 32. Hêr gefeaht Ecgbryht cyning wið .xxxv. sciphlæsta, Chr. 833; Erl. 64, 19: 837; Erl. 66, 5: 840; Erl. 66, 19. Ælfrêd cyning gefeaht wið .vii. sciphlæstas and hiera ân gefêng and ða ôðru gefliémde, 875; Erl. 78, 6. Ælfrêd cyning gefeaht wið feówer sciphlæstas Deniscra monna, and ðara scipa tû genam, 882; Erl. 82, 10.   II. *a ship of burden, a transport :*—Sciplæst *oneraria*, Wrt. Voc. i. 63, 71. Scyphlæst *honeraria*, ii. 43, 10 (cf. hlaestscip *honeraria*, 110, 46).

scip-hláford, es; m. *A ship-master :*—Sciphláford *nauclerus*, Wrt. Voc. i. 56, 16.

scipian *to take shape :*—Ðonne gelimpþ ðǽræ ('*he mother*) manigfeald sâr ðonne ðæs byrþres líc on hire innoþe scypigende biþ, Lchdm. iii. 146, 15. v. scyppan.

scipian; p. ode *To put in order, equip, man* a ship :—Ðá lǽt Eádweard cyng scypian xl snacca, Chr. 1052; Erl. 183, 33. [*From* (?) *Icel.* skipa *to give order or arrangement to things, to man* a ship.]

scipian; p. ode *To take ship :*—Se eorl on Wiht scipode and intô Normandíg fôr, Chr. 1091; Erl. 228, 12. v. ge-scipian.

scipincel, es; n. *A small ship :*—Scipincel *carabus*, Wrt. Voc. i. 48, 1: 64, 3. Scippincel *navicula*, 56, 12.

scip-lâd, e; f. *Sailing, navigating :*—Hê wolde on scyplâde mid ða fǽmnan hâm hweorfan *navigio cum virgine redire disponebat*, Bd. 3, 15; S. 541, 27.

scip-líc; adj. *Relating to a fleet, naval :*—Ða men ða ðe beóþ winnende in sciplícum gewinne, Shrn. 35, 12. Ðǽm sciplícum *classicis*, Wrt. Voc. ii. 75, 7. Flotlícum, sciplícum *classicis*, 131, 63. Sciplícum herium *classicis cohortibus*, Hpt. Gl. 406, 39. [*O. H. Ger.* scef-lîh *nauticus, navalis.*]

scip-liðend, es; m. *One who goes in a ship :*—Hê cwæð tô ðǽm scipliðende . . . ða scipliðende ðæt gehêrende mearcedon ðone dæg, Shrn. 85, 30–86, 2. Ealla ða þing ða ðe scypliðendum (*navigantibus*) nydþearflícu gesewen wǽron, Bd. 5, 9; S. 622, 26. v. next word.

scip-liðende; adj. *Going in a ship, sailing :*—Hê sǽde scipliðendum monnum, Shrn. 85, 28: Homl. As. 117, 17. Ða scipliðendan *navigeros*, Wrt. Voc. ii. 61, 35.

scip-mǽrels *a ship-rope :*—Scipmǽrls *tonsilla*, Wrt. Voc. i. 57, 4. v. mǽrels.

scip-mann, es; m. *A mariner, sailor;* nauta, navarchus :—Scypman *nauta*, Hymn. Surt. 6, 26. Scipmen *navarcas*, Wrt. Voc. ii. 62, 15. (1) *a sailor, one of a ship's crew :*—Ðá ongunnon ða nýdlingas and ða scypmen ða ancras on ðone sǽ sendan woldon ðæt scyp mid gefæstnian *tentabant nautae anchoris in mare missis navem retinere*, Bd. 3, 15; S. 541, 40. Volosianus hêt hys scypmen swiðe forþ rôwan, St. And. 44, 4. (2) *one who goes on trading voyages :*—Scipmanna (-e, MS.) myrt *teloneum*, Wrt. Voc. i. 37, 10. Ðǽm scipmannum is beboden gelíce and ðǽm landbûendum, ðæt ealles ðæs ðe him on heora ceápe geweaxe hig Gode ðone teóþan dǽl âgyfen, L. E. I. 35; Th. ii. 432, 27. [*Arður* him ot scipe fusde and hehte þat his scipmen brohten hine to Romerel,

Laym. 28308. Agrayþed ase byeþ þe ssipmen ine ssipe, þet ase zone ase he yhyerþ þane smite of þe lodesmanne hi yerneþ, Ayenb. 140, 22. See Chaucer's Prologue, vv. 388–410. *Icel.* skip-maðr *one of a crew.*]

scippan, scieppan, sceppan; p. scôp, sceóp; pp. sceapen, scepen.   I. *to shape, form :*—Ic hiwige oððe scyppe *fingo*, Ælfc. Gr. 28, 5; Som. 31, 61.   II. *to create* (of the act of the Deity) :—Ðû scyppest eorþan ansýne *renovabis faciem terrae*, Ps. Th. 103, 28. Ælmihtig fæder ðe ða sciran gesceaft sceópe and worhtest, Hy. 10, 2. Waldend scóp wudige môras, Exon. Th. 193, 11; Az. 120: 132, 1; Gû. 456. Ðâ hê Adam sceóp, Cd. Th. 77, 21; Gen. 1278. Swâ gôd Sceoppend rihtlíce sceóp eall ðæt hê sceóp, Bt. 39, 2; Fox 214, 12. Heortan clǽne scyp (*crea*) on mê, Ps. Lamb. 50, 12. God gesceóp ealle gesceafta, and deófol ðáne gesceafta scyppan ne mæg, Homl. Th. i. 102, 1. Hê (*God*) selcúðe syððan scyppan nolde, Hexam. 12; Norm. 20, 15. Ic scyppendum wuldorcyninge hýrde, rícum dryhtne, Exon. Th. 453, 16; Hy. 4, 15. Hê bebeád and sceapene synd *ipse mandavit, et creata sunt*, Ps. Spl. 32, 9.   III. *to shape for one* (*dat.*) *as his fate* (*acc.*), *to assign as a person's lot.* v. ge-sceap :—Scôp *censebat*, Wrt. Voc. ii. 19, 28: 91, 1. Unc Dryhten scôp síþ ætsomne, Exon. Th. 494, 3; Rä. 82, 2. God monna crǽftas sceóp and scyrede ǽghwylcum on eorþan eormencynnes, 332, 34; Vy. 95. Ðâ sceóp freá ælmihtig fágum wyrme wíde síþas, Cd. Th. 55, 32; Gen. 903: 110, 21; Gen. 1841. Hû him weorðe geond woruld wîdsîþ sceapen, Salm. Kmbl. 744; Sal. 371. Ðǽr eów is hâm sceapen, Exon. Th. 142, 25; Cri. 649. Wæs sió wrôht scepen wið Hugas, Beo. Th. 5819; B. 2913.   III a. *to destine, adjudge* a person (*acc.*) *to anything :*—Sceóp and scyrede Scyppend ûre oferhîdig cyn engla of heofnum *our Creator adjudged the presumptuous race of angels to banishment from heaven*, Cd. Th. 5, 1; Gen. 65. Ic eom wiht on gewin sceapen *I am a creature destined to strife*, Exon. Th. 400, 15; Rä. 21, 1: 405, 14: Rä. 24, 2.   III b. *in the phrases* naman *or* tô naman scyppan *to give a name :*—Him se pâpa Petrus tô noman scôp *cui papa Petri nomen imposuerat*, Bd. 5, 7; S. 620, 43. Scôp him Heort naman, Beo. Th. 157; B. 78. Se apostol sceóp ðære cyrcan naman ' resurrectio,' Homl. Th. ii. 474, 33. Ríce menn sceópon heora bearnum naman be him sylfum, i. 478, 9. Sceópan, Shrn. 47, 26. Gê fægniaþ ðæt gê môton sceppan ðone naman, Bt. 16, 3; Fox 56, 24. [*Goth.* skapjan: *O. Sax.* skeppian: *O. Frs.* skeppa: *O. H. Ger.* scepfen, skeffen: *Icel.* skepja. Cf. also *O. H. Ger.* scaffan: *Icel.* skapa.] v. â-, for-. ge-scippan, -sceppan.

Scippend, es; m. *The Creator :*—Ðû Scippend heofones and eorþan, Bt. 4; Fox 6, 30: Past. 7; Swt. 49, 17: Cd. Th. 234, 15; Dan. 292: Andr. Kmbl. 556; An. 278. Scieppend *Creator*, Rtl. 145, 24. Scæppend, 166, 29. Scæpend, 180, 8. Sceppend, Bt. 34, 10; Fox 150, 12: Cd. Th. 283, 24; Sat. 309. Sceoppend, Bt. 39, 13; Fox 234, 21: 14, 2; Fox 44, 27: 33, 4; Fox 132, 13. Scyppend, Hexam. 13; Norm. 22, 2: Cd. Th. 5, 2; Gen. 65. [*O. E. Homl.* sceppende, scuppend: *A. R.* schuppinde: *Orm.* shippennd.]

scip-râp, es; m. *A cable :*—Sciprâpas *rudentes*, Wrt. Voc. i. 48, 24: 57, 1. Hiora (*walruses*) hýd biþ swiðe gôd tô sciprâpum . . . Ðæt gafol biþ on ðǽm sciprâpum, ðe beóþ of hwæles hýde geworht and of seoles . . . Se byrdesta sceall gyldan . . . twegen sciprâpas; ǽgðer sý syxtig elna lang, ôðer sý of hwæles hýde geworht, ôðer of sioles, Ors. 1, 1; Swt. 18, 1–23.

scip-rêðra, an; m. *A sailor :*—Scyprêðra *nauita*, Germ. 389, 39. Hê on scyp eode, and myd hys scyprêðrum hys segl up âhôf, and forþ seglode, St. And. 38, 32.

scip-rôðer, es; n. *An oar or a rudder for a ship :*—Sciprôðor *navalia*, Wrt. Voc. i. 61, 37.

scip-rôwend, es; m. *One who rows in a ship, a sailor, one of a crew :*—Sciprôwend *nauta*, Wrt. Voc. ii. 61, 33.

scip-ryne, es; m. *A course or channel for ships :*—Hê lêt delfon ân mycel gedelf and wolde ðæt scipryne sceolde ðǽrinne licgean eall swâ hig dydon on Sandwíc *he had a great trench dug and intended that in it ships could run, just as they did at Sandwich*, Chart. Th. 341, 16.

scip-setl, es; n. *A seat or bench for rowers :*—Scipsetl *transtra*, Wrt. Voc. i. 48, 14: 64, 8.

scip-sôn v. scip-fylleþ.

scip-steall, es; m. *A place for a ship :*—Andlang streámes on scypsteal, God. Dip. B. iii. 316, 16.

scip-steóra, -stýra, an; m. *A steersman, pilot :*—Swíðe eáðe mæg on smyltre sǽ ungelǽred scipstiéra (-stióra, Cott. MSS.) genôh ryhte stiéran *quieto mari recte navem imperitus nauta dirigit*, Past. 9, 2; Swt. 59, 1. Swâ swâ gôd scipstýra (-stioera, Cott. MS.) ongit micelne wind ǽr hit weorþe, Bt. 41, 3; Fox 250, 13. [*Icel.* skip-stjóri *a skipper.*]

scip-steorra, an; m. *The Pole-star :*—Twegen steorran standaþ stille . . . ðone norðran wê geseóþ; ðone hâtaþ menn scipsteorra, Lchdm. iii. 270, 20.

scip-teóꞂa, -teara, -tara, -tera, an; m.: -ter, -teoro (u), -tearo; gen. -tearos; n. *Pitch :*—Scipter *bitumen*, Wrt. Voc. ii. 126, 36. Sciptearo, Lchdm. ii. 66, 8. Sciptearos lǽst, 126, 8. Sciptaran *bituminis*, Wrt. Voc. ii. 11, 77. Scipteran, 82, 40. Scipteran *bitumine*, 84, 41. Dô

3 H 2

gódne sciptaran tô, Lchdm. ii. 326, 14. Đã hēt se cāsere meltan on hwere leád and scipteoran and pic, Shrn. 91, 7. Dô sciptearo tô, Lchdm. ii. 122, 17 : 124, 10.

scip-toll, es ; *m. Passage money :*—Sciptol *naulum* (cf. a schyppes tolle *hoc naulum*, 274, col. 2), Wrt. Voc. i. 56, 49. [Icel. skip-tollr.]

scip-wealh ; *gen.* -weales ; *m. A servant whose service is connected with ships :*—Đæt land is sum inland, sum hit is đãn scipwealan tô gafole gesett (*the land in question lies by the Severn*), Cod. Dip. Kmbl. iii. 450, 19.

scip-weard, es ; *m. One who has charge of a ship :*—Scipweardas, Andr. Kmbl. 596 ; An. 297.

scip-weorod, es ; *n. The crew of a vessel :*—Scipweredes (-weardes ?) *naucleri*, Wrt. Voc. ii. 59, 48.

scip-wíse, an ; *f. The fashion or form of a ship :*—On scipwísan geworht *made in the fashion of a ship*, Nar. 11, 20. Đã nam heó ánne riscenne windel on scipwísan gesceapenne *sumpsit fiscellam scirpeam*, Ex. 2, 3.

scip-wyrhta, an ; *m. A shipwright :*—Scipwyrhta *navicularius*, Wrt. Voc. i. 19, 13.

scír, e ; *f.* I. *office, charge, business, administration, government :*—Scír *procuratio*, Wrt. Voc. i. 57, 36 : 288, 58. Sciir, ii. 117, 71. Scír *dispensatio*, 106, 51 : 25, 55 : 140, 65 : *negotium*, 59, 65. Đonne se mōna biþ .xx. niht, and .i. and .xx. niht, đæt biþ scír odde ceáp in đem swefne tôweard, Lchdm. iii. 160, 8. Scír *prefecturae*, Wrt. Voc. ii. 66, 2. Ne gewanige se reccere nã đone ymbhogan đære inneran scíre for đære ábisgunge đære úterran *sit rector internorum curam in exteriorum occupatione non minuens*, Past. 18, 1 ; Swt. 127, 13. Persa cyning đeonne đone ealdorman his scíre, Ors. 3, 1 ; Swt. 96, 22. Scíre *negotio*, Ps. Surt. 90, 6. Hē wið đa scíre (*the office of bishop*) ne winne, Past. 10, 1 ; Swt. 61, 11. Hū dear se grípan on đa scíre đæt hē ærendige ōðrum monnum tô Gode *qua mente apud Deum intercessionis locum pro populo arripit ?* 10, 2 ; Swt. 63, 7. Se đe đone sacerdhád onfēhþ, hē onfēhþ friccan scíre *praeconis officium suscipit, quisquis ad sacerdotium accedit*, 15, 2 ; Swt. 91, 21 : 45, 1 ; Swt. 337, 15. Ágyf đíne scíre *give up thine office (of steward)*, Lk. Skt. 16, 2. Paulinus đære cyrican scíre (*curam*) onfēng, Bd. 2, 20 ; S. 522, 15. Hē forlēt đa scíre đæs mynstres his brēđer *reliquit monasterii et animarum curam fratri suo*, 3, 19 ; S. 549, 39. For intingan đære cynelícra scýra *negotiorum regalium causa*, 3, 23 ; S. 551, 1. Him leófre wæs se cristendóm tô begānne đonne his scíra tô habbanne *omnes officium quam fidem deserere maluerunt*, Ors. 6, 31 ; Swt. 286, 8. Ic ne oncneów scíre *non cognovi negotiationes*, Ps. Surt. 70, 15. I a. *where the term refers to an English official :*—Se đe þeóf gefēhþ . . . and hē hine đonne álǣte . . . gif hē ealdorman sié, þolie his scíre, L. In. 36 ; Th. i. 124, 19. II. *a district, province*, as an ecclesiastical term *diocese, parish :*—Scír *provincia*, Wrt. Voc. i. 54, 3. Sió scír hátte Hálgoland đe hē (*Ohthere*) on búde, Ors. 1, 1 ; Swt. 19, 9. Hē áxode hū đære þeóde nama wǣre đe hí of cōmon . . . Gyt đã Gregorius befrān hū đære scíre nama wǣre đe đa cnapan of álǣdde wǣron, Homl. Th. ii. 120, 27–33. Scíre biscopas *vicari episcopi*, Rtl. 194, 33. Hí feórdon fram đære scíre bisceope, and God him foresceáwode on sumere ōðre scíre on Francena ríce fulgóde wununge, Homl. Skt. i. 6, 122. On Alexandiscre scýre, 2, 29. Tô Cappadoniscre scýre, 3, 88. Đã gemunde se ealdorman (*Pilate*) đæt Herodes wæs on đære scíre, Homl. Th. ii. 250, 31. Đæt mynster gesett on Angel-seaxna scíre and eác ōðer mynster on đære ylcan scíre *monasterium situm in provincia Saxonia, atque aliud monasterium in eadem provincia*, Cod. Dip. B. i. 154, 25 : Swt. A. S. Rdr. 100, 154. Scíre *parochiam*, Hpt. Gl. 427, 38. Liódbiscopas in scírum and londum gesettedo, Rtl. 194, 35. Scíre *provincias, regiones*, Hpt. Gl. 451, 17. Scíra *provincias*, 512, 12. Đis wundor ásprang geond đa gehendan scíra, Homl. Th. i. 562, 20. II a. *the people of a district, a tribe :*—Hē is swýđe rihtwýs wer, đæt wāt eall đeós scýr, Homl. Skt. i. 10, 120. Twā scíra, đæt ys, Iude and Benjamin, Ps. Th. 45, arg. Đis sind đe wǣron đa ǣdelestan ealdras geond đa scíra *hi nobilissimi principes multitudinis per tribus et cognationes suas*, Num. 1, 16. III. *as a technical English term, a shire :*—Hæbbe man scírgemót, and đær beó on đære scíre bisceop and se ealdorman, L. Edg. ii. 5 ; Th. i. 268, 4 : ii. 3 ; Th. i. 266, 19. Đære scíre bisceop *episcopus provinciae*, L. Ecg. P. iii. 11 ; Th. ii. 200, 4. Him man sealde gíslas of ǣlcere scíre, Chr. 1013 ; Erl. 148, 1. Gif man wille of boldgetale in ōðer boldgetæl hláford sēcan, dô đæt mid đæs ealdormannes gewitnesse đe hē ǣr in his scíre folgode, L. Alf. pol. 37 ; Th. i. 86, 4. Gif man spor gespirige of scýre in ōðre . . . drífan hí đæt spor oþ hit man đam gerēfan gecýđe, fô hē syđđan tô and ádrífe đæt spor út of his scíre, L. Ath. v. 8, 4 ; Th. i. 236, 20–23. Đæt ǣlc gerēfa nǣme đæt wedd on his ágenre scíre, v. 10 ; Th. i. 240, 1. Ne nime nán man nāne nǣme ne innan scíre ne útan scíre, L. C. S. 19 ; Th. i. 386, 12. Gif hwā fare unáliéfed fram his hláforde oððe on ōðre scíre hine bestele, L. In. 39 ; Th. i. 126, 10. Hēde se đe scíre healde, L. R. S. 4 ; Th. i. 434, 33. III a. *The people of a shire, the community inhabiting a shire :*—Nān scír nolde ōðre gelǣstan æt nýxtan *at last no shire would help another*, Chr. 1010 ; Erl. 144, 11. Đã sealde Leófwine ealdorman . . . and eal seó scír his land clǣne, Chart. Th. 376, 14 : L. C. S.

19 ; Th. i. 386, 15. Se đe land gewerod hæbbe be scíre gewitnesse, 80 ; Th. i. 420, 20. Wæs se cyng đã điderweardes mid đære scíre đe mid him fierdedon, Chr. 894 ; Erl. 90, 32. Hí lifedon of Eást-Seaxum and of đām scírum đe đǣr nýxt wǣron, 1002 ; Erl. 143, 5. IV. as an ecclesiastical term, the district in charge of an ecclesiastic (bishop, etc.), *a diocese, parish :*—Swā biscop him tǣce đe hit on his scýre sý, L. Edm. S. 4 ; Th. i. 250, 2. Gif man ǣnig líc of rihtscíre lecge, L. Eth. vi. 21 ; Th. i. 320, 6. Gif preóst on unriht út of scíre hád begite, gilde .xii. ór, and þolie his hādes, búton scíre biscop him hādes geunne, L. N. P. L. 12 ; Th. ii. 292, 13. Đises ys ealles wana .xxxiii. hída of đām hídun đe ōðre bisceopas ǣr hæfdon intó hyra scýre, Cod. Dip. Kmbl. iii. 327, 12. Nǣnigum heora álýfed sí ǣnige sacerdlíce þēnunge dōn búton đæs bisceopes leáfe đe hí on his scíre (*parochia*) gefeormade sín, Bd. 4, 5 ; S. 573, 5. [O. H. Ger. scíra *procuratio, negotium*.] v. biscop-, burh-, gerēf-, hām-, mæssepreóst-, mynster-, prāfost-, preóst-, riht-, scrift-, toll-, tūn-scír ; and see Stubbs' Const. Hist. i. 109 sqq. ; Kemble's Saxons in England, bk. i. c. 3.

scír ; *adj. Clear, bright :*—Scír *limpidus*, Wrt. Voc. i. 46, 54. Sciir *sublustris*, Txts. 96, 941. I. *of living creatures, bright, brilliant, splendid, resplendent :*—Scír Metod (*God*), Beo. Th. 1962 ; B. 979. Scír cyning (*Christ*), Exon. Th. 71, 9 ; Cri. 1153. Is seó womb (*the phenix*) wundrum fæger, scír and scýne, 219, 16 ; Ph. 308 : 214, 4 ; Ph. 234. Ic eom ásceáden from đære scíran driht (*the heavenly host*), Cd. Th. 275, 26 ; Sat. 177. Đone scíran Scippend, Elen. Kmbl. 740 ; El. 370. I a. *of a quality :*—Gē đa scíran miht (*the power of Christ*) dēman ongunnon, Elen. Kmbl. 620 ; El. 310. II. *of inanimate things*, (a) *of vegetation, bright, brilliant, white :*—Ofer hine scír cymeþ mínre segnunga sōđfæst blóstma *super ipsum florebit sanctificatio mea*, Ps. Th. 131, 19. Geseóþ đās eardas đæt hig synt scíre (*albae*) tô rípene, Jn. Skt. 4, 35. (b) *of metals, stones, etc., bright, lustrous, glittering, brilliant :*—Sceán scír werod (*the band with glittering armour*), Cd. Th. 185, 19 ; Exod. 125. Hringíren scír, Beo. Th. 650 ; B. 322. Scíran goldes, 3393 ; B. 1694. Hē gewyrceþ scírne mēce, Exon. Th. 297, 8 ; Crä. 65. Hyrste beorhte, reáde and scíre, 392, 25 ; Rä. 12, 2. Scíre burstan múras and stānas, 70, 22 ; Cri. 1142. Scíre helmas, Judth. Thw. 24, 17 ; Jud. 193. (c) *of glass, clear, transparent :*—Swā đæt scíre glæs đæt mon ýþæst mæg eall þurhwlítan, Exon. Th. 78, 33. (d) *of water, clear, limpid :*—Ofter Pantan, ofer scír wæter, Byrht. Th. 132, 42 ; By. 98. Đa hlútran and đa scíran wæter *liquidas lymphas*, Wrt. Voc. ii. 50, 10. [Þurh ān scýr wæter Brādan ǣ hātte, Chr. 656 ; Erl. 31, 16.] (e) *of wine, bright, clear, pure, neat :*—Wínes scíres *vini meri*, Ps. Surt. 74, 9. Syle drincan on scíran wíne, Lchdm. i. 342, 23. Nalles scír wín hí ne druncan, Bt. 15 ; Fox 48, 9 : Met. 8, 21. Scír wered, Beo. Th. 996 ; B. 496. (f) *of light and light-giving things, bright, clear, brilliant :*—Heofontorht swegl (*the sun*) scír, Exon. Th. 52, 2 ; Sch. 74 : 486, 18 ; Rä. 72, 17. Sunne scír and beorht, Met. 30, 9. Sió scíre scell (*the firmament*), 20, 174. Metod æfter sceáf scírum scíman ǣfen ǣrest, Cd. Th. 9, 5 ; Gen. 137. Scírne scíman, Rood Kmbl. 107 ; Kr. 54. Fleógan þurh scírne dæg, Exon. Th. 439, 15 ; Rä. 59, 4. Þurh đa scíran neaht, Met. 20, 229. Blāce stódon ofer sceótendum scíre leóman, Cd. Th. 184, 25 ; Exod. 112. On sumera đonne đa hātostan weder synd and đa scíran dagas hwítan, Lchdm. iii. 256, 2. Scippend scírra tungla, Met. 4, 1 ; 20, 8. Hwí hí (*stars*) ne scínen scírum wederum, 28, 45. (g) *of the world :*—Þurh đa scíran gesceaft, Exon. Th. 286. 7 ; Jul. 728. (h) *of a banner :*—Fana hwearfode scír on sceafte *the flag fluttered gleaming bright on its staff*, Met. 1, 11. (i) *of the voice, clear :*—Wit Scilling scíran reorde song áhófan, Exon. Th. 324, 32 ; Víd. 103. [Iss all þeȝȝre spell shir atter and shir galle, Orm. 15383. Clene off grediȝnesse and off galnesse skir and fre, 8015 : Prompt. Parv. schyre, as water and oþer lycure *perspicuus, clarus*. Þe mihte of schir and of clene bone, A. R. 246, 26. Đat skie scir, Gen. and Ex. 3848. Goth. skeirs *clear, evident :* O. Sax. skír, skíri (wín, watar) : O. L. Ger. scíri : O. Frs. skíre : M. H. Ger. schír : Icel. skírr *clear, bright, pure.*]

scíran *to cut.* v. sceran.

scíran *to discharge an office.* v. ge-scíran.

scíran ; *p.* de. I. *to make clear what is hidden or obscure, declare, tell, make known :*—Drihten đæt ongeat and geseah, đæt se deófol đone Iudas lǣrde, đæt hē hine belǣwde. Ac đæt hē đeáhhwæđere gedyldelíce ábær and gemetfæstlíce scírde (*did not declare it in terms of strong reprobation*), Homl. As. 154, 68. Gif hié eallunga forberan ne mǣgen đæt hié hit ne scíren, đonne sprecen hié ymbe his unþeáwas, Past. 28 ; Swt. 198, 9. Đæt hit sceáden mǣl scýran mōste, cwealmbealu cýđan, Beo. Th. 3883 ; B. 1939. [God đe soðe shire, Gen. and Ex. 2036.] II. *to make clear by distinguishing between things, to distinguish, decide :*—Scíro *disceptavero*, Txts. 57, 688. Is gehāten đæt hé wille cueđan, ' Gewítaþ from mē áwiergde.' Ne scírþ hé nô hwæđer hié reáfodeu oððe hwelc oðer yfel fremeden (*no distinction is made in the sentence between various kinds of evil*), Past. 44 ; Swt. 329, 7. Đæt gé ne scíraþ *you do not bring out that (the difference between a man in his youth and in mature age) clearly*, Exon. Th. 132, 21 ; Gū. 476. Hē hēt wurpan ac hē ne scýrde on hwæđere healfe hí đæt net wurpan

sceoldon *he bade throw, but he did not decide on which side they were to throw the net*, Homl. Th. ii. 290, 9. **III.** *to bring a charge* against a person :—Scîrde *actionabatur* [*or is the verb here connected with* scîr *an office?* cf. gescîra *uilicare*, Lk. Skt. Lind. 16, 2, folcgerêfa *actionator*, Wrt. Voc. i. 17, 30], Wrt. Voc. ii. 99, 11 : 3, 55. Ealdarmon swŷđe spræcon and wiđ mê wrâđum wordum scîrdan *principes adversum me loquebantur*, Ps. Th. 118, 23. [Cf. Nes nan mon þat durste word sciren, Laym. 16822.] **IV.** *to get clear* of obligation, trouble, etc., *get exemption* :—Gif hwylc man đone ândagan forgêmeleásige, æt forman cyrre :iii. messan, æt óđerum cyrre .v., æt þriddan cyrre ne scîre his nân man (*no man shall be exempt from the obligation*), bûtun hit sié for mettrumnesse ođđe for hlâfordes neóde, Chart. Th. 614, 18. Đæm folce wæs ægđres waa ge đæt hié đæt mæste yfel forberan sceoldon ge eác đæt hié his scîran ne dorstan *there was trouble to the people on both accounts, that they had to bear a very great evil, and that they durst not get rid of it*, Ors. 3, 7 ; Swt. 114, 32. [*A. R.* schiren *to make pure*: Goth. gaskeirjan *to interpret*: Icel. skîra *to purify, clear* from a charge ; skŷra *to explain, solve, decide.*]

scîr-basu; *adj. Bright purple* :—Scîrbasu benetum (venetus *caeruleus*, Ducange), Wrt. Voc. ii. 125, 30.

scîr-biscop, es ; *m. The bishop of a shire* or *diocese* (v. scîr, **III.** 2) :—Bête đæt, swâ se scîrbisceop and eal scîrwitan dêman, Wulfst. 173, 30. [Đe scŷrbiscop *episcopus dioceseos*, Cod. Dip. Kmbl. v. 28, 32.]

scirdan; *p.* de *To hurt, injure* :—Hwilcan geþance mæg ænig man geþencan on his môde, đæt hê tô sacerdan heáfod âhylde, . . . and hî hrædlîce siđđan scyrde ođđe scynde mid worde ođđe weorce, L. Eth. vii. 27; Th. i. 334, 35. Đa đe godcunde lâre and woruldcunde rihtlage wyrdan and scyrdan on ænige wîsan, Wulfst. 168, 9. [*Icel.* skerđa *to diminish: O. H. Ger.* giscartit uuerd *dolet.*] v. sceard ; *adj.*.

scirden; *adj. Of tiles* or *sherds* :—Scerden *testeum*, Germ. 400, 553. v. sceard *a sherd.*

scîre (?), an ; *f. An enclosure, precinct* :—Portic *porticus*, scîre *peribolum*, heall *aula*, Wrt. Voc. i. 58, 3. [Cf. (?) Andlang scîre on hweđels heal, Cod. Dip. Kmbl. v. 358, 15.]

scîre; *adv.* **I.** of light, *clearly, brightly* :—Scîre scînan, Exon. Th. 67, 15 ; Cri. 1089 : Andr. Kmbl. 1671 ; An. 838 : Salm. Kmbl. 679 ; Sal. 339. **II.** of the voice, *clearly* :—Saga hwætt ic hätte đa (đe ?) swâ scîre nige (cîge ?), sceáwendwîsan hlûde onhyrge, hæleþum bodige wilcumena fela wôþe mînre, Exon. Th. 390, 29 ; Rä. 9, 9.

scîr- (scir-?)ecg ; *adj. Having a bright* (*cutting?* cf. sceran) *edge* :—Swurd scearp and scîrecg, Lchdm. i. 390, 7. Cf. brún-ecg.

scirfe-mûs. v. scyrfe-mûs.

scîr-gemôt, es ; *n. A shire-mote, a meeting of the duly qualified men of a shire* :—Hêr swutelaþ on þissum gewrite đæt ân scîrgemôt sæt æt Ægelnôþes stâne be Cnutes đæge cinges. Đær sæton Æđelstân biscop and Ranig ealdorman . . . and đær wæs Bryning scîrgerêfa . . . and ealle đa þegnas on Herefordscîre, Chart. Th. 336, 22. Gif hê æt đam þriddan cyrre nân riht næbbe, đonne fare hê feórþan sîđe tô scîrgemôte, L. C. S. 19 ; Th. i. 386, 14. Hæbbe man tuwa on geáre scîrgemôt, L. Edg. ii. 5; Th. i. 268, 3. Habbe man twâ scîrgemôt on geáre, L. C. S. 18 ; Th. i. 386, 5. See Stubbs' Const. Hist. s. v. shiremoot.

scîr-gerêfa, an ; *m. A shire-reeve, sheriff*, '*the judicial president of a shire.*' v. Stubbs' Const. Hist. i. 113 ; Kemble's Saxons in England, bk. ii. c. v. The word glosses *preses* in Wrt. Voc. i. 18, 11. (1) of a secular official, v. scîr, **IV.** :—Ælfnôþ scîrgerêfa, Chr. 1056; Erl. 190, 29. Ân scîregemôt sæt æt Ægelnôþes stâne . . . đær wæs Bryning scîrgerêfa, Cod. Dip. Kmbl. iv. 54, 14. On Æđelwines scîregerêfan gewitnesse, 10, 27. (2) of an ecclesiastic, v. scîr, **IV.** :—Đonne sceall Cristes scîrgerêfan (*the bishop*) đæt witan, and ymbe đæt dihtan and dêman, swâ swâ bêc tæcan, L. I. P. 25 ; Th. ii. 340, 8.

scîr-gesceatt, es ; *n. The property of a see* :—Æđelríc bisceop grêt freóndlîce Æđelmær : and ic cŷđe đæt mê is wana æt đam scŷrgesceatte đus micelys đe mîne foregengan hæfdon . . . Đises ys ealles wana .xxxiii. hîda of đam hîdun đe óđre bisceopas ær hæfdon intô hyra scŷre, Cod. Dip. Kmbl. iii. 327, 4.

scîr-ham; *adj. Having bright armour* :—Scacan scîrhame (*Beowulf and his followers*) tô scipe fôron, Beo. Th. 3794 ; B. 1895.

scirian; *p.* ede ; *pp.* scired, scirred (v. â-scirred) *To separate, divide* (v. scirung, â-, tô-scirian), but used only metaphorically of setting apart something as a person's lot, *to ordain, assign, allot, dispense* :—Swâ missenlîce meahtig Dryhten geond eorþan sceát eallum dæleþ, scyreþ and scrîfeþ, Exon. Th. 331, 10 ; Vy. 66. God geond middangeard monna cræftas sceóp and scyrede, 332, 34 ; Vy. 95. Đara gifena đe him tô duguþe Drihten scyrede, Cd. Th. 221, 13 ; Dan. 87. Sceóp đa and scyrede Scyppend ûre oferhîdig cyn engla of heofnum *then did our Creator adjudge and ordain the presumptuous race of angels to banishment from heaven*, Cd. Th. 5, 1 ; Gen. 65. Gif đê Alwalda scirian wille đæt đû môte *if the All-ruler be pleased to grant thee opportunity*, 171, 12 ; Gen. 2827. Sceolde him beón deáþ scyred *should death be the lot doomed him*, 31, 15 ; Gen. 485. Sié hira dæl scired mid Marian *may their part be assigned with Mary*, Elen. Kmbl. 2462 ; El. 1232. Đær womsceaþan on đone

wyrsan dæl scyrede weorþaþ, hâteþ Scyppend him gewîtan on đa winstran hond, Exon. Th. 75, 26 ; Cri. 1227. [*O. Sax.* skerian : *O. H. Ger.* scerian.] v. â-, be-, ge-scerian, -scirian.

scîriendlîc; *adj. Derivative* :—Scyriendlîc *dirivativum, deductum*, Wrt. Voc. ii. 140, 44. v. â-scirgendlîc.

scîrig-mann, es ; *m. Apparently the same as* scîr-mann, q. v. The form occurs only in one (Kentish) charter, where 'Wulfsige preóst se scîrigmann' is twice mentioned, Cod. Dip. Kmbl. vi. 127, 128. In a later Latin version of this charter the term is rendered *scîrman* and explained by *judex comitatus, judex provinciae*, Chart. Th. 275, 276, and in this sense it is taken by Kemble, v. Saxons in England, ii. 168 sq.. In another charter the same person is mentioned, but without the title : a grant of land is made by Ethelred to Winchester ' ofer Wulfsiges dæg preóstes,' Cod. Dip. Kmbl. vi. 135. This document is dated 996 ; somewhat later, in the time of Cnut, Wulfsige preóst is mentioned in connection with Kent, but then Æđelwine is scîregerêfa, Cod. Dip. Kmbl. iv. 10. In another charter (before 1011) Leófrîc is scîresman in Kent. For the form scîrig-, cf. (?) hŷrig-mann.

scîr-mæled; *adj. Brightly marked, bright with inlaid ornaments* :—Scîrmæled swyrd, Judth. Thw. 24, 38 ; Jud. 230. v. mâl-sweord.

scîr-mann (scîre-, scîres-), es ; *m.* **I.** *an official, officer, ruler, one who discharges the duties of a* scîr (v. scîr, **I**) :—Scîrman *procurator*, Wrt. Voc. i. 57, 37. Wæs scîremonn (Pontius Pilatus) *procurante Pontio Pilato*, Lk. Skt. Lind. 3, 1. Scîremon (sgiiremonn, Lind.) *dispensator*, Lk. Skt. Rush. 12, 42. Swâ sceal gôd scŷrman (*a reeve* or *bailiff*) his hlâfordes healdan, dô ymbe his âgen swâ swâ hê wylle, Anglia ix. 260, 16. Ne ofermôdgiaþ đa scîrmenn nâ for đŷ *nequaquam praepositi ex hoc superbiunt*, Past. 17, 2 ; Swt. 109, 18. Hwæt elles meahte beón getâcnod þurh Ezechiel bûton đa scîrmenn *per Ezechielem praepositorum persona signatur*, 21, 3 ; Swt. 153, 24. **II.** *an inhabitant of a district* (v. scîr, **II**) :—Gregorius befrân, hû þære scîre nama wære, đe đa cnapan of âlædde wæron. Him man sæde, đæt đa scîrmen wæron Dere gehâtene, Homl. Th. ii. 120, 33. **III.** *as a technical English term* = scîr-gerêfa. v. Stubbs' Const. Hist. i. 113, Kemble's Saxons in England, ii. 158 :—Æđelwine scîrman (*in the next charter he is called* scîregerêfan, iv. 10, 27), Cod. Dip. Kmbl. iv. 9, 29. Ufegeat scîreman, 304, 17. Đâ com đider se scŷresman Leófrîc, 266, 24 : 267, 11. Gif hwâ him ryhtes bidde beforan hwelcum scîrmen ođđe óđrum dêman, L. In. 8 ; Th. i. 106, 21. v. scîrig-mann.

scîrness, e ; *f. An explanation, declaration* (?) :—Scîrnis *ypoteseo bassio*, Wrt. Voc. i. 289, 73.

scirpan; *p.* te. **I.** *to sharpen, whet* :—Scyrpþ *acuit*, Engl. Stud. ix. 40. Hî hwetton (scyrptun, MS. C.) tungan heora *acuerunt linguam suam*, Ps. Spl. 139, 3. Scerptun, Ps. Surt. 139, 4. **II.** metaph. *to make active, arouse* :—Symle hê sceal his hŷrmen scyrpan mid manunge tô hlâfordes neóde and him eác leánian be đam đe hŷ earnian, Anglia ix. 260, 23. v. â-, ge-scirpan.

scirpan; *p.* te ; *pp.* ed *To clothe* :—Engel hine scirpeþ (scierpeþ) on cwicum wædum, Salm. Kmbl. 278 ; Sal. 138. v. ge-scerpan, sceorp.

-scirpla, scir-seax. v. ge-scirpla, scear-seax.

scîr-þegen, es ; *m. The thane of a shire* :—Đises is tô gewitnesse . . . Godwine eorl . . . Ælfwine abbod . . . and ealle scîrþegenas on Hâmtûn-scîre, Cod. Dip. B. i. 544, 8.

scirung, e ; *f. Separation, dismission, rejection* :—Gif hê swâ biþ đæt hê ne sŷ wyrđe đære scyrunge (scirunge, MS. T.: đæt hê wurđe ne beó, đæt hê beó đanon âscyred, Wells Frag.) *si non fuerit talis qui mereatur proici*, R. Ben. 109, 21.

scîr-wered; *adj. Bright, clear* :—Wuldres scîma æđele ymb æđelne andlonge niht scân scîrwered, Exon. Th. 179, 15 ; Gû. 1262. Cf. swegl-wered.

scîr-wita, an ; *m. A chief man* (wita, q.v.) *of a shire* :—Bête đæt, swâ se scîrbisceop and eal scîrwitan dêman, Wulfst. 173, 30. Gebête đæt, swâ scîrewitan geceósan, 172, 4.

scîtan, scŷtan (?) *to shoot* (of a plant), *flourish* :—Næfre on his weorþige weá âspringe mearce mâ scŷte (sprŷte ?) mân inwides *may ill never fail in his place, rather may guile flourish in his borders*; non defecit de plateis ejus usura et dolus, Ps. Th. 54, 10. [*Or does* scŷte *belong to* sceótan? cf. *for change of vowel in subjunctive* hlîpen, Past. Swt. 215, 7.]

scîtan; *p.* scât, *pl.* sciton ; *pp.* sciten *Cacare.* [He sched out his bowels and his lyf wiþ þe dritt þat he schoote (shote) *effudit viscera et vitam cum ipsis stercoribus*, Trev. 5, 153. Prompt. Parv. schytyn *merdo, stercoro : O. H. Ger. scîzan : Icel.* skîta.] v. be-scîtan.

scîte, scête, scŷte, an ; *f. A sheet, piece of linen cloth* :—Scête, lođa *sandalium*, Wrt. Voc. ii. 119, 55. Scŷte sindo, i. 25, 47 : 81, 61 : 284, 58. Wæfelses t scŷtan *sindonis*, Hpt. Gl. 494, 13. Mid scîtan begîrd, Ap. Th. 12, 17. Heó hire feax gerædde and hî mid scŷtan besweóp *crines composuit, caput linteo cooperuit*, Bd. 3, 9 ; S. 534, 13. Sum iungling mid ânre scŷtan bewæfed (*amictus sindone*), Mt. Skt. 14, 51, 52. Josep bewand đone lîchoman mid clænre scŷtan (scêtan, Rush.), Mt. Kmbl. 27, 59 : Nicod. 11 ; Thw. 6, 11 : 13 ; Thw. 6, 31 : Guthl.

20; Gdwin. 84, 8. Hí bewundon his líc mid línenre scýtan, Homl. Th. ii. 260, 35. Hé ðone andwlytan (*the face on S. Veronica's handkerchief*) genam, and hyne on pællenre scýtan befeóld, and eác heó wæs gewefen myd golde. And ða scýtan hé dyde ðá on án gylden fæt, St. And. 42, 11-15. Scétan *sindonem*, Kent. Gl. 1148. On scétum *in sabanis*, Wrt. Voc. ii. 48, 47. v. beód-, hop-, -scíte (-scýte).

-scíte *-cornered*. v. feówer-, feðer-, þrí-scíte (-scýte).

scitel, scytel *dung* (?) :—Nim heortes scytel and cnuca tó duste, Lchdm. i. 336, 18. Nim fearres scytel, cnuca and gníd swíðe smale, 368, 12. v. scítan.

Sciððeas, Sciððie, Sciððige, a; *pl. The Scythians* or (using the name of the people where now the name of the country would be used) *Scythia* :—Ða Sciððeas, Bt. 18, 2; Fox 64, 10. Uesoges wolde him tó geteón ... ðone norþdæl, ðæt sint Sciþþie; and hú ii ædelingas wurdon áfliémed of Sciððium, Ors. 1, 10, tit.; Swt. 1, 25. Hé wonn on Sciððe ... His heres wæs seofon hund þúsenda, ðá hé on Sciððie fór. Huæðere ða Sciððie noldon hiene gesécan tó folcgefeohte, 2, 5; Swt. 78, 8-11. Eall Sciððia lond, 1, 1; Swt. 14, 22. Hé wæs mid firde farende on Sciððie on ða norþdælas, 1, 10; Swt. 44, 7 : 2, 4; Swt. 76, 4. On Sciððie (Sciððige, Bos. 43, 42), Swt. 72, 24.

Sciððia, Sciððiu; *indecl.*: Sciððie, an; *f. Scythia* :—Gotan of Sciððiu mægþe, Bt. 1; Fox 1, 1. Of Sciððia, Met. 1, 2. Wurdon twegen ædelingas áfliémde of Sciððian, Ors. 1, 10; Swt. 44, 25. Ðæt lond mon hætt þa ealdan Sciððian, 1, 1; Swt. 14, 17. v. preceding word.

scitol; *adj. Purgative* :—Mettas ðe late melten and swá ðeáh ne synd scitole, Lchdm. ii. 178, 1.

scittan. v. scyttan.

scitte, an; *f. Looseness of the bowels, diarrhœa* :—Wið ðon ðe men mete untela melte and gecirre on yfele wætan and scittan, Lchdm. ii. 226, 6. [*Prompt. Parv.* skytte or flux *fluxus, lienteria, dissenteria, dyaria* : *Icel.* skita *diarrhœa.*]

Scittisc. v. Scyttisc.

scl-. v. sl-.

scó, scobl, scocca. v. scóh, scofl, scucca.

scocere ? :—Innan scocera wege, Cod. Dip. Kmbl. v. 107, 9.

-scód. v. drýg-, ge-, un-scód; scógan.

scóere. v. scóhere.

scofettan; *p.* te *To drive* hither and thither :—For ðam hit is openlíce cúð ðætte sió úterre ábisgung ðissa woruldþinga ðæs monnes mód gedréfþ and hine scofett (scofed, Cott. MSS.: cf. scied, *3rd pers. sing.* of sceótan, Swt. 70, 7) hidres ðædres óþ þæt hé áfielþ of his ágnum willan *cum indubitanter constet, quod cor externis occupationibus tumultibus impulsum a semetipso corruat*, Past. 22, 1; Swt. 169, 13. Cf. scúfan.

scofl, e; *f. A shovel* :—Scofl *trulla*, Wrt. Voc. i. 289, 19 : ii. 122, 67. Ísern scobl *vatilla*, 123, 12. Scofle, spadu *capella, tuba*, 128, 36. Scoble *palas*, 116, 13. Hé sceal habban spade, scofle, Anglia ix. 263, 6. [*Du.* schoffel; *f.* Cf. *O. H. Ger.* scúvala *pala, vanga* : *Ger.* schaufel.] v. fýr-, gléd-, meox-, steór-, wind-scofl.

scógan, scógean, sceógan, scóan (? v. scóung), **sceón**; *p.* scóde; *pp.* scód. sceód *To shoe, put on* (one's) *shoes, furnish with shoes* :—Ic scóge (sceóge) mé *calceo vel calcio*, Ælfc. Gr. 26, 6; Zup. 158, 8. Se engel cwæð : Begyrd ðé, and sceó (gisceó ðec, Rtl. 58, 11) ðé, and fylig mé, Homl. Th. ii. 382, 9. Sceógiaþ *calciate*, Engl. Stud. ix. 40. Sceógeaþ eówre fétt, Past. 5, 2; Swt. 44, 10. Cf. Ða mægas hine anscógen óðre fét ðæt mon mæge siddan hátan his tún ðæs anscódan tún *unum ei pedem propinquus discalciet, ejusque habitaculum domum discalceati vocet*, Swt. 43, 16. Se biþ mid ryhte óðre fét anscód (on-, Cott. MSS.), and hine mon scyle on bismer hátan se anscóda (*discalceatus*), Swt. 45, 8. [Scheoinde ou & cloðinde *putting on your shoes and clothes*, A. R. 16, 4. Heo scoiden (soide hire stedes, 2nd MS.), Laym. 22291. Ræftres mid irene iscod, 7831. *O. H. Ger.* scuohón; *p.* scuohta : *Icel.* skóa, skúa *to shoe.*] v. -scígan, -scód.

scóh, scó, sceó : *gen.* scós, sceós; *n. pl.* scós, sceós; *gen.* sceóna; *dat.* scóm, scóum; *the Ancren Riwle has the weak plural* scheon; *m. A shoe* :—Scóh *caliga*, Wrt. Voc. ii. 103, 11 : 127, 67 : 13, 43 : *calcarium*, i. 291, 29. Scó *fico*, 26, 17. Rúh scós óþ *pero*, ii. 78, 6. Tríwen sceó *coturnus*, i. 26, 21. Gif se innera dæl ðæs sceós (scós, MS. B.) byþ fixen hýd, Lchdm. i. 342, 11. Þuong scóes (giscóes, Rush.) *corrigiam calciamenti*, Jn. Skt. Lind. 1, 27. Dó on ðínne winstran scóh, Lchdm. i. 396, 3. Scó *calciamentum*, Ps. Spl. T. 59, 9. Scós *galliculae*, Wrt. Voc. ii. 41, 53. Wífes sceós *baxeae*, unhége sceós *talares*, i. 26, 20-23. Næron his scós forwerode, Homl. Th. i. 456, 21. Wíde sceós hangodan on hira (*the Saracens*) fótum, Shrn. 38, 8. His sceóna þwanga, Mk. Skt. 1, 7. Sceóea, Lk. Skt. Lind. 3, 16. Hí brohton swínes rysl his scón tó gedreóge, Homl. Th. ii. 144, 29. On ðínum sceón (scón, MS. B.), Lchdm. i. 330, 5. Sceón (scóeum, Lind.) *calciamentis*, Lk. Skt Rush. 22, 35. Sceówum, p. 4, 7. Ic wyrce sceós *facio ficones*, Coll. Monast. Th. 27, 33. N lt ðú habban yfele sceós, and wylt swá ðeáh habban yfel líf. Ic bidde ðé ðæt ðú læte ðé ðín líf deórre ðonne ðíne sceós, Homl. Th. ii. 410, 15-18. Habbaþ eówre scós on eówrum fótum, Anglia viii. 322, 19. Scóas (Lind. scóea), Mt. Kmbl. Rush. 10, 10. Scóeas,

Lk. Skt. Lind. 15, 22. [*Goth.* skóhs : *O. Sax.* skóh : *O. Frs. O. L. Ger.* scó : *O. H. Ger.* scuoh : *Icel.* skór.] v. slífe-, slýpe-, steppe-scóh; hand-sció; ge-scý.

scóhere, scóere, es; *m. A shoemaker* :—Scoehere *sutrinator*, Txts. 115, 122. Scóere, 101, 1962. [*Icel.* skóari.]

scóh-nægel, es; *m. A shoe-nail* :—Scóhnegl *clavus caligaris*, Wrt. Voc. ii. 104, 15. Scóhnægl *clavus caliculoris*, 131, 54.

scóh-þegn, es; *m. A servant who attends to shoes* :—Be sceóhþénum *de calciariis*, R. Ben. Interl. 91, 9.

scóh-þwang, es; *m. The thong* or *latchet of a shoe* :—Ic ne eom wyrðe ðæt ic hys sceóþwancg (shoþuong, O. E. Homl. ii. 137, 33. Shoþwang, Orm. 10387) uncnytte *non sum dignus soluere corrigiam calciamentorum ejus*, Lk. Skt. 3, 16. Sceóþwang, Jn. Skt. 1, 27. Gisceó ðec sceóhþongum ðínum *calcia te caligas tuas*, Rtl. 58, 11. [*Icel.* skóþvengr.]

scóh-wyrhta, an; *m. A shoemaker.* From the description of his work given by the sceówyrhta (*sutor*) in Ælfric's Colloquy, Thorpe, p. 27, he seems to have been a general worker-in leather. Besides boots and shoes he makes harness, leather bags and bottles :—*Facio calceamenta diversi generis, subtalares et ficones, caligas et utres, frenos et phaleras et flascones et calidilia, calcaria et chamos, peras et marsupia.* [*M. H. Ger.* schuoch-wurhte.] v. sútere.

scól. v. scolu.

scola *a debtor* :—Gescolan *condebitores*, Wrt. Voc. ii. 105, 23. [*Goth.* skula : *O. Sax.* skolo : *O. H. Ger.* scolo *debitor.*]

scola (scóla ? v. scolu) *a learner* :—Gescola *condiscipulus, conscolaris*, Hpt. Gl. 459, 66.

scolere (scólere ?), es; *m. A scholar, learner* :—Nim ðú lá geornfulla scoliere, Anglia viii. 304, 16. Seó ræding þingþ ðæne scoliere, 308, 1. Ða scolieras witon ðe synt getýdde on bóclícum cræfte, 314, 9 : 335, 42. Ðám scolierum ðás þing gecýdan, 303, 48. Ðæt ænig preóst ne underfó óðres scolere, L. Edg. C. 10; Th. ii. 246, 14. [*O. H. Ger.* scuolari *scholaris, discipulus.*] v. non-scólere.

scol-(scól-)mann, es; *m.* **I.** *one who attends a school, a scholar* :—Scól *scola*, scólman *scholasticus*, Wrt. Voc. i. 75, 27-28 : 46, 62. **II.** *one who belongs to a band* (v. scolu, II), *a follower, client* :—Scolman *cliens*, 46, 62.

scolu, scól (*these two forms may give the later* shoal, school *as* col, cól *give* coal, cool), e; *f.* **I.** *a school* :—Scól *scola*, Wrt. Voc. i. 75, 27. Scól *scola*, se ðe on scóle (sceóle, MS. U.) ys *scholasticus*, Ælfc. Gr. 5; Zup. 11, 13-15. Ðý ilcan geáre forborn Ongolcynnes scolu, Chr. 816; Erl. 62, 7. Constantinus hiene benæmde ðære scole ðe hé on leornode, Ors. 6, 31; Swt. 284, 24. His líc líþ on Angelcynnes scole, Chr. 874; Erl. 76, 26. Of scole *ex scole*, Wrt. Voc. ii. 31, 64 : 95, 14. Hú ne eart ðú se mon ðe on mínre scole wære áféd and geláered, Bt. 3, 1; Fox 4, 19. Eubolus underféng ðone cnapan tó lárlícre scole ... On ðære ylcan scole wæs Iulianus, Homl. Skt. i. 3, 14-16. Ic becom tó Cristes scole, 2, 244. Maria wunode on ealra ðæra apostola gýmene on ðære heofonlícan scole embe Godes æ smeágende, Homl. Th. i. 440, 8. Sum leorningman on scole *scholasticus quidam*, Bd. 3, 13; S. 538, 18. Ic (Ethelwulf) on Róme Englisce scole gesette, Chart. Th. 116, 33. Se (Marinus) gefreóde Ongelcynnes scole be Ælfrédes béne West-Seaxna cyninges, Chr. 885; Erl. 84, 19. Cildru on scole betæcan, Lchdm. iii. 184, 27 : 188, 18. **II.** *a band or troop of people, a shoal, school* (in *school* of fishes) :—Him on healfa gehwone heofonengla þreát ymbútan faraþ, ælbeorhtra scolu, Exon. Th. 58, 2; Cri. 929. Synfulra here, womfulra scolu, 94, 5; Cri. 1535 : 98, 15; Cri. 1608 : 114, 19; G. 175. Seó deóre scolu (*the heavenly host*, 235, 21; Ph. 235. Árleásra sceolu, Elen. Kmbl. 2600; El. 1301 : 1523; El. 763. Éce fýr wæs Satane and his gesíðum mid, deófle, gegearwad, and ðære deorcan scole, Exon. Th. 93, 9; Cri. 1523. Ðæt gesælige weorud gesihþ ðæt fordóne, ... byrnendra scole, 77, 6; Cri. 1252. Hé gesomnode miccle scole and wered his geþoftena, Guthl. 2; Gdwin. 14, 2. Ðá wearð stearc storma gelác ... út feor ádráf on Wendelsæ wígendra scola, Met. 26, 31. [*O. Sax.* skola *a band, troop* : *O. H. Ger.* scuola *schola* : *Icel.* skóli *a school. From Latin.*] v. geneát-, hand-, þegn-, þeóf-scolu.

scom-. v. sceam-.

scóm-hylt, e; *f. A shady wood, thicket, shrubbery* :—Scoomhylti *frutices*, Wrt. Voc. ii. 39, 60. [Cf. (?) *Icel.* skúmi *shade, dusk.*] Cf. holt.

scon-. v. scean-.

Sceón-ég *Skaane, a district forming the southernmost part of the Scandinavian peninsula, formerly belonging to Denmark, but since 1658 to Sweden : the Icelandic form is* Skáney. The name occurs in Ors. 1, 1; Swt. 19, 35.

scop, sceop, es; *m. A poet* :—Scop *liricus*, unwurð scop *tragicus vel comicus*, Wrt. Voc. i. 60, 5, 9. Scop *comicus*, 291, 25 : ii. 17, 38. *Comicus, s. est qui comedia scribit, cantator, vel artifex canticorum seculorum, idem satyricus*, i. scop, *joculator, poeta*, 132, 16. Se hæðena scop *Pompeius historicus*, Ors. 1, 5; Swt. 32, 28. Terentius se mæra

Cartaina scop *Terentius comicus*, 4, 10 ; Swt. 202, 26. Gerîseþ gód scop gumum, Exon. Th. 341, 18 ; Gn. Ex. 128. Scop hwîlum sang hâdor on Heorote, Beo. Th. 997 ; B. 496. Hrôdgares scop, 2137 ; B. 1066 : Exon. Th. 379, 21 ; Deór. 36. Scop oðde leóðwyrhta *poeta*, Wrt. Voc. i. 73, 68. Ðes sceop *hic poeta*, ðises scopes *huius poetae*, Ælfc. Gr. 7 ; Zup. 24, 6 : 36 ; Zup. 215, 8. Wîtega oððe sceop *vates*, 10 ; Zup. 77, 3. Be ðam wæs singende sum sceop *unde tragicus exclamat*, Bt. 30, 1 ; Fox 106, 31. Swâ Parmenides se sceop geddode, 35, 5 ; Fox 166, 8. Omerus se gôda sceop on his leóþum swîðe herede ðære sunnan gecynd, 41, 1 ; Fox 244, 4. Ðær wæs hearpan swêg, swutol sang scopes, Beo. Th. 180 ; B. 90. Omerus wæs ðæm mæran scope (*Virgil*) magistra betst, Met. 30, 4. Gecuron him ânne scop tô cyninge . . . se heora cyning ongan singan and giddian, Ors. 1, 14 ; Swt. 56, 29. Unweorþe scopas *tragedi* vel *comedi*, Wrt. Voc. i. 39, 39. Scopas *lyrici*, i. 54, 9 : *vates*, Hymn. Surt. 119, 18. Fram ðisum sceopum ic gehýrde leóþ, Ælfc. Gr. 7 ; Zup. 24, 2. [Scopes þer sungen, Laym. 30615. *O. H. Ger.* scof *poeta, vates.* Cf. (?) *Icel.* skop *railing, mocking.*] v. æfen-, ealu-, sealm-scop.

-scop, -sceop. v. wîd-scop.

scop-cræft, es ; *m. The poet's art, poetry :*—Sceop *poeta*, ic leornige sceopcræft (scop-) *poetor*, Ælfc. Gr. 36 ; Zup. 215, 9.

scop-gereord, es ; *n. Poetic diction, the language of poetry :*—Swâ hwæt swâ hê of godcundum stafum þurh bôceras geleornode, ðæt hê in sceopgereorde (*verbis poeticis*) geglencde, Bd. 4, 24 ; S. 594, 34.

scop-leóþ, es ; *n. A poem :*—Se heora cyning ongan singan and giddian and mid ðæm scopleóþe heora môd swîðe getrymede *Tyrtaei ducis composito carmine et pro concione recitato accensi*, Ors. 1, 14 ; Swt. 56, 32. Hê (*Nero*) ongon wyrcan scopleóþ be ðæm bryne *Iliadem decantabat*, 6, 5 ; Swt. 262, 1. Swâ hit an scopleóþum sungen is *quod poeta descripsit*, 2, 4 ; Swt. 72, 20. [*O. H. Ger.* scof-leod.]

scop-lîc ; *adj. Poetic :*—Mid meterlîcum fôtum ł scoplîcum *pedibus poeticis*, Hpt. Gl. 411, 4. [*O. H. Ger.* scof-lîh *poeticus :* cf. *O. L. Ger.* scop-lîco *poetice.*]

scoppa, an ; *m. A shop, a booth* or *shed for trade or work* (cf. work-*shop*):—Hê geseh ða welegan hyra lâc sendan on ðone sceoppan (*in gazophilacium*), Lk. Skt. 21, 1. [The bowiares ssope hii breke, & the bowes nome echon, R. Glouc. 541, 16. Euerych soutere þt halt shoppe, English Gilds, 358, 22. Marchantz beshetten hym in here shope, Piers P. 2, 213. Schoppe *opella, propala*, Prompt. Parv. A shoppe or a werkehous *operarium*, Wülck. Gl. 599, 10. A schope *opella*, a hordhows *gazafilacium*, 730, 3-6. A schoppe *opella*, a treserhouse *gazafilacium*, 804, 28, 29. Cf. *O. H. Ger.* schof *a building without walls ;* also *a vestibule :* Ger. schuppen *a shed.*] v. scypen.

scora, an ; *m. A hairy garment :*—Bânrift *tibialis*, scora *tricilo*, Wrt. Voc. i. 289, 16.

scorf, sceorf, scurf, scruf, es ; *m.* (?) *Scurf :*—Hyt âfeormaþ ðone leahtor ðe Grêcas hostopyturas hâtaþ, ðæt ys scurf ðæs heáfdes, Lchdm. i. 322, 16. Wið scurfe and nebcorne, 68, 10. Wið heáfodsâr, ðæt ys wið scurf, 116, 23. Wið scruf (scurf, MSS. H. B.) and wið sceb, 316, 22. Wið scurfum, 356, 23. Swâ mycel hreófla and sceorfa on his heáfde hæfde ðæt him næfre nænig feax on ðam uferan dæle ðæs heáfdes âcenned beón mihte *scabiem tantam ac furfures habebat in capite, ut nil unquam capillorum ei in superiore parte capitis nasci valeret*, Bd. 5, 2 ; S. 614, 44. [Scrofe or scalle *glabra*, Wrt. Voc. 179, 9. Scurf of scabbys *squama*, scurfe of metel *scorium*, Prompt. Parv. 451. *O. H. Ger.* scorf *scabies :* Ger. schorf ; *m. :* Icel. skurfur ; *f. pl.*]

scorfed, sceorfed, scurfed ; *adj. Rough, scabbed :*—Wið scurfedum nægle (*unguium scabritiem*); nim gecyrnadne sticcan, sete on ðone nægl wið ða wearta, Lchdm. ii. 150, 4. [þé ssouede (*leprous*), þe scallede, Ayenb. 224, 6.] v. next word.

scorfende, sceorfende, scurfende ; *part. Getting rough* or *scabby :*—Wið scurfendum næglum ad *scabiem unguium*, Lchdm. i. 370, 9. v. preceding word.

scorian ; *p.* ode *To refuse, reject an offer, repudiate :*—Ða ðe ne gelýfaþ þurh âgenne cyre hî scoriaþ nâ þurh gewyrd *those who do not believe refuse by their own choice, not by fate*, Homl. Th. i. 114, 12. Ða sceorede ðâ gyt se yldesta hæðengylda mid mycelre þwyrnysse *the chief idolater still refused (Christianity) with much perversity*, 72, 9. [Cf. *O. L. Ger.* scurgan *avertere, expellere :* *O. H. Ger.* scurgan *trudere, impellere, propellere ;* fer-scurgan *repellere.*] v. wið-scorian *and next word.*

scorian ; *p.* ode *To project, jut :*—Ða stânas swâ of ôðrum clife ût sceoredon, Blickl. Homl. 207, 20. [Cf. *O. H. Ger.* scorrên *prodire*, fram-, furi-scorrên *pro-, e-minere*, Grff. vi. 539.] v. preceding word.

scort ; *adj. Short.* **I.** marking the length of an object :—Scort sinewealt stân *cilindrus*, Wrt. Voc. i. 41, 35. Sceort beó wið eorþan *cama*, 41, 31. Ðæt ic ðê môste getæcan swâ sceortne (scortne, Cott. MS.) weg swâ ic scyrtestne findan meahte, Bt. 40, 5 ; Fox 240, 17. Hê hæfþ scyrtran (sceortran, MS. R.) sceade ðonne seó sunne, Lchdm. iii. 252, 13. On lxv and þreó hundræd scy[r]tran and lengran ða ædron beóþ tôdælede, 146, 6. **II.**

marking height, *not tall :*—Hê (*Zacchaeus*) wæs scort on wæstme, Homl. Th. i. 580, 30. **III.** of time, (1) of a period of time :—Tô scortre hwîle *for a short time*, Past. 36, 6 ; Swt. 255, 11. Ðæt wê sceolan on ðisse sceortan tîde geearnian êce ræste, Blickl. Homl. 83, 2. Ðû ðam winterdagum selest scorte tîda, Bt. 4 ; Fox 8, 4. Sceorta, Met. 4, 20. Nis nænig mon ðe wite hwæðer ðis þûsend sceole beón scyrtre ðe lengre, Blickl. Homl. 119, 6. Dagas ne synd nâðor ne længran ne scyrtran ðonne hî æt fruman wæran, Lchdm iii. 252, 19. Se mônaþ (*February*) is ealra scyrtost (scyrtst, MSS. P. M. : scirtst, MS. L.), 264, 8. Scyrtest, Anglia viii. 306, 8. (2) marking duration, (a) *short-lived, brief :*—Deáh se hlîsa ðara foremærena monna hwîlum lang sié, hê biþ ðeáh swîðe scort tô metanne wið ðone ðe næfre ne geendaþ, Bt. 18, 3 ; Fox 66, 18. Hû ne biþ simle ðæt lange yfel wyrse ðonne ðæt scorte, 38, 2 ; Fox 198, 12. Ðæt wuldor ðysses middangeardes is sceort and gewîtende, Blickl. Homl. 65, 15. (b) *not occupying much time :*—Hwâ ne wundraþ ðætte sume tunglu habbaþ scyrtran hwyrft (*an orbit that requires less time to complete*) ðonne sume habban, Bt. 39, 3 ; Fox 214, 18. Wê hit sæcgaþ eów on ða scortostan wîsan *we will tell it you in the briefest fashion*, Homl. Skt. i. 4, 140. (c) as a grammatical term :—Seó forme geendung is on scortne a, Ælfc. Gr. 9, 1 ; Zup. 32, 17. Mid fiffêtedum ł scertrum *brachycatalectico*, Hpt. Gl. 409, 27. [*O. H. Ger.* scurz. Cf. *Icel.* skortr *want.*] v. next word.

scortian ; *p.* ode. **I.** *to get short, shorten* (intrans.) :—Se dæg ðonne sceortaþ, Lchdm. iii. 250, 23. Se sceortigenda (scort-, MS. L.) dæg . . . se langienda dæg, 252, 8. **II.** *to make short* (? cf. þenne cumeð þe deofol and him scorted his dazes, O. E. Homl. i. 25, 14. To schorte oure weie, Chauc. Prol. 791). **III.** *to run short, fail :*—Ðætte ne scortige (sceortiga, Lind.) gileófa ðîn *ut non deficiat fides tua*, Lk. Skt. Rush. 22, 32. [Cf. *Icel.* skorta *to run short.*] v. a-, ge-, on-sceortian ; scyrtan.

scortlîc ; *adj. Short, of time, not lasting :*—Sceortlîc ł hwîlendlîc *momentaneum*, Scint. 214, 10.

scortlîce ; *adv.* **I.** of time, *shortly, before long, soon :*—Nû gyt scortlîce ł lytel fæc and ne byþ se synfulla *adhuc pusillum et non erit peccator*, Ps. Lamb. 36, 10. Scortlîcor *maturius, citius, velocius*, Hpt. Gl. 527, 14. **II.** of speech, narrative, etc., *shortly, briefly, compendiously :*—Scortlîce *strictim, breviter*, 492, 27. Scortlîce (*breviter*) ic hæbbe nû gesæd ymbe ða þrié dælas, Ors. 1, 1 ; Swt. 10, 3 ; 1, 14 ; Swt. 58, 7 : Ælfc. Gr. 10 ; Zup. 76, 3. Nû wylle wê sum þing scortlîce eów be him gereccan, Homl. Th. ii. 118, 3. Sceortlîce *summatim, breviter*, vel *commatice*, Wrt. Voc. i. 55, 15 : *strictim*, ii. 82, 74. Nû is ôðer cwyde be gôdum mannum sceortlîce gecweden, Homl. Th. i. 484, 20. Wê willaþ furðor swîðor sprecan, and wê secgaþ nû sceortlîce, Lchdm. iii. 240, 2.

scortnesse, e ; *f.* **I.** *shortness* (of time) :—Ðonne byrneþ on scortnisse gramen hys *cum exarsent in brevi ira ejus*, Ps. Spl. 2, 13. Ða scortnesse ðysse worulde and ða êcnesse ðæs tôweardan lîfes, Homl. As. 168, 117. **II.** *a short account, an epitome* (cf. a brief, *and* v. scortlîce, II) :—Manega synd ðæt coniunctiones, ðe wê ne mâgon nû secgan on ðissere sceortnysse, Ælfc. Gr. 44 ; Zup. 266, 8. Wê habbaþ gesæd on ðisre sceortnysse, hû God geswutelode ða sôðfæstan godspelleras, Homl. Skt. i. 15, 219.

scort-wyrplîc ; *adj. Of early fulfilment, coming to pass shortly :*—On .xv. nihta sceortwyrplîc ðæt bid. On .xvi. nihta æfter langre tîde hit âgæþ *a dream on the fifteenth night of the month will be of early fulfilment. On the sixteenth it will come to pass after a long time*, Lchdm. iii. 156, 2.

scot, es ; *n.* **I.** *a shot, a shooting :*—Hié his siððan wæran swîðe êhtende ge mid scotum (gesceotum, MS. C.) ge mid stâna torfungum ge mid eallum heora wîgcræftum, Ors. 3, 9 ; Swt. 134, 15. [**II.** *a shot, missile.* v. ge-sceot, *and cf.* No man . . . Nò maner schot, ne pollax, ne schort knyf Into the listes sende, Ch. K. T. 1686. See also the cognate words.] **III.** *a rapid movement* (v. sceótan, **IV, III**, ge-sceót (*read* -sceot), **II**), *a rush, dart :*—Leax sceal on wæle mid sceotte scrîðan, Menol. Fox 539 ; Gn. C. 40. **IV.** *a scot* (as in *scot and lot, scot-free*), *a shot* (as in to pay one's *shot*), *a contribution, tax.* v. sâwel-scot, sceótan, **VII.** **V.** *a building.* v. sele-scot, ge-sceot, **III.** [*O. Frs.* scot *a missile ; a contribution, tribute : O. H. Ger.* scoz ; *n. telum, jaculum : M. H. Ger.* schoz ; *m. tribute, tax : Icel.* skot ; *n. a shot, shooting ; a missile ; a contribution.*] v. ge-sceot (-scot) ; scyte.

scota, an ; *m. One who shoots* or *hurls, a soldier :*—Gescota *communipularius*, Wrt. Voc. ii. 104, 82 : 132, 49. [Icel. and-skoti *an adversary.*] Cf. scytta.

scotere (?) *one who shoots* or *hurls, a warrior :*—Nô hê ðære feohgyfte for scoterum (? scotenum, MS.) scamigan þorfte, Beo. Th. 2056 ; B. 1026.

scot-freó ; *adj. Scot-free, exempt from imposts :*—Scotfrê and gafolfrê, Cod. Dip. Kmbl. iv. 215, 32 : 191, 18.

scotian, sceotian ; *p.* ode. **I.** (1) *to shoot a person with a weapon :*—Hwâ sceotaþ ðæt deofol mid weallendum strælum ? Se Pater Noster sceotaþ ðæt deofol, Salm. Kmbl. p. 148, 1-3. Hî scotiaþ hine

sagittabunt eum, Ps. Lamb. 63, 6. Wḗ mid strǣlum hié scotodon, Nar. 22, 18. Ðæt hý scotien rihtheortan, Ps. Spl. 10, 2. Hí unscyldige mid bogan scotian þenceaþ *ut sagittent immaculatum*, Ps. Th. 63, 3. Ðū scealt mid hálgum Godes wordum ðínne feónd scotian, Basil admn. 2; Norm. 36, 7. Hý wǣron mid strǣlum scotode, Shrn. 135, 29. (2) *to shoot* a weapon at a person, *to hurl* :—Ðæt yrre hys spere scotaþ ongeán ðæt geþyld *ira lanceam suam iacit contra patientiam*, Gl. Prud. 20 a. Drihten lígeas sceotaþ *Dominus jaculatur fulgura*, Bd. 4, 3; S. 569, 22. Hē sceotaþ his flán and his scearpe spere ongeán his wiðerwinnan, Basil admn. 2; Norm. 36, 5. Of heofene dōm scotad is, Ps. Surt. 75, 9. (3) *to shoot* (intrans.) :—Hí hine scearpum strḗlum on scotiaþ, Ps. Th. 63, 4. Gif ðḗ man scotaþ tō, Homl. Th. ii. 538, 10. Scotiaþ scrḗfende scín scearpum wǣpnum, Exon. Th. 385, 28; Rä. 4, 51. Mid ðám strǣlum ðæs hálgan sealmsanges hē wið ðám ǣwerigedum gástum sceotode, Guthl. 3; Gdwin. 24, 12. Sume scotedon mid arewan tōweard ðam háligdōme. . . . Hí scotedon swíðe, Chr. 1083; Erl. 217, 19–25. II. *to shoot, move rapidly* :—Steorran fóran swýðe scotienda [cf. *O. H. Ger.* diu scozonten fiur (*a shooting star*)], 744; Erl. 49, 2. [*Laym.* scotien (mid flan).] v. of-scotian.

Scot-land, es; *n.* I. *Ireland*, where the Scottas lived before migrating to the country now called Scotland :—On westende (*of Europe*) is Scotland, Ors. 1, 1; Swt. 8, 27. Ấn diácon wearð forþféred on Sceotlande (cf. an Scotta eálonde, 215, 21), and ðæs diácones nama wæs Njál háten, Wulfst. 205, 16. Hí cômon on Scotland (*Hiberniam*) upp, Bd. 1, 1; S. 474, 10. II. *Scotland* :—Hḗr fór Æþelstán cyning on Scotland (tō Scotlande *in Scotiam*, MS. F.), Chr. 934; Erl. 111, 9. Hē (*Cnut*) fôr tō Scotlande, and Scotta cyng him tō beáh, Mælcolm, 1031; Erl. 163, 20. Hē (*Furseus*) fērde geond eal Ýrrland and Scotland, Homl. Th. ii. 346, 29. v. Scottas.

scot-lira, an; *m. The fleshy part of the leg, the calf of the leg* :—Scotliran *suras*, Lchdm. i. lxxiv, 19. Cf. spear-lira.

scot-spere, es; *n. A spear for hurling, a javelin* :—Scotsper[a], gára *jaculorum*, Hpt. Gl. 405, 52.

Scottas; *pl. The Scots*, a race found first in Ireland, whence a part migrated to North Britain, which from them got the name Scotland. (1) *Scots of Ireland* :—Þrié Scottas cuômon tō Ælfrḗde cyninge on ánum báte būtan ǣlcum gerēþrum of Hibernia, Chr. 891; Erl. 88, 5. Ðá forþgongenre tíde æfter Bryttum and Peohtum þridde cynn Scotta Breotone onféng . . . Ða wǣron cumene of Hibernia Scotta eálonde . . . Hibernia is ágendlíce Scotta ēþel, heonan cōman seó þeód Scotta, Bd. 1, 1; S. 474, 24–42. Com of Hibernia Scotta eálande Fursius . . . Wæs Furseus of ðam æþelestan cynne Scotta, 3, 19; S. 547, 2–25. In Hibernia mǣgþe, ðæt is on Scotta lande, Shrn. 51, 30. On Sceotta land, Wulfst. 205, 7. Scotta land, eálond, 215, 17, 21. Gif næddre sleá man, ðone blacan snegl ǣwæsc on háligwætre, sele drincan oððe hwæthwega ðæs ðe fram Scottum cốme *a little water that has come from Ireland* (because of its peculiar efficacy (?). Cf. Bede's statement of the cures worked on those who were bitten by snakes through the application of water in which scrapings from the leaves of Irish books were put, Bd. 1, 1; S. 474, 36–39), Lchdm. ii. 110, 15. (2) *Scots of Scotland* :—Eádrēd gerád eal Norþhymbra land him tō gewealde, and Scottas him áþas sealdan, Chr. 946; Erl. 118, 1. Hine geces tō hláforde Scotta cyning and eall Scotta þeód, 924; Erl. 110, 14. Crungun Sceotta leóda, 937; Erl. 112, 11, 32. Férde bodiende betwux Ýrum and Scottum and sidðan ofer eal Angelcynn, Homl. Th. ii. 346, 35. Mid Scottum ic wæs and mid Peohtum (*or under* (1) Cf. Scotta cynn Breotone onféng on Pehta dǣle, Bd. 1, 1; S. 474, 24), Exon. Th. 323, 15; Víd. 79.

scottettan (?) *to move about quickly* (? cf. sceotan, III, IV; scotian, II), *to dance, leap* :—Sceottet (or = (?) sceóteþ: t *for* þ *occurs in verb inflexions in the same glossary*, e.g. geþwǣrat, 397, 439) *saltat*, Germ. 394, 222.

scotung, e; *f.* I. *shooting* :—Wunda ðe ða wælhreówan hǣðenan mid gelômre scotunge on his líce macodon, Th. An. 123, 33. II. *what is shot, a missile* :—Hí synt scotunga oððe flána *ipsi sunt jacula*, Ps. Lamb. 54, 22. Sceotunga, Ps. Spl. C. 54, 24. Scotunge ðíne *jacula tua*, Ps. Surt. ii. p. 190, 15. Hē wæs biset mid heora scotungum swylce ýles byrsta, Th. An. 122, 17. Wið ðám scotungum ðara werigra gásta hē hine mid gástlícum wǣpnum gescylde, Guthl. 3; Gdwin. 24, 5. For ðæs fýres sceotungum *on account of the flashes of lightning*, Lchdm. iii. 280, 15. v. scotian.

scóung, e; *f. A provision of shoes* :—His mete and scóung and glófung him gebyreþ *he is to have his food and shoes and gloves provided for him*, L. R. S. 10; Th. i. 438, 6.

scrád, *a moving body* (? v. scríðan), *a vessel* (?), *a body of travellers* (? cf. *Icel.* skreið *a shoal, flock*) :—Scrifen scrád glád þurh gescád in brád, wæs on lagustreáme lád, Exon. Th. 353, 15; Reim. 13.

scrádung. v. screádung.

scrǣf, es; *m. Some kind of bird, a cormorant* (?) :—Scraeb *merga*, Wrt. Voc. ii. 114, 6. Screb *ibinem* (ἴβιν, cf. ibin avis in Affrica habens longum rostrum, 4), Shrn. 29, 19. [Cf. (?) *Icel.* skarfr *the green cormorant*.]

scræf, screaf, scref, es; *n.* I. *a cave, cavern, hollow place in the earth* :—Scrǣf *spelunca*, Wrt. Voc. i. 38, 21. Ðǣr (*hell*) biþ fýr and wyrm, open ēce scræf, Cd. Th. 212, 10; Exod. 537. Cirice on scræfes onlícnesse, Blickl. Homl. 197, 18. Hē fērde tō ðam munte and on ánum scræfe (*in spelunca*) wunode, Gen. 19, 30: 23, 11. Hē hēt wilian tō ðam scræfe (*ad os speluncae*) micele weorcstánas, Jos. 10, 18. Scræfe *crypta*, Wrt. Voc. ii. 24, 59. Scrafe *antro*, Hpt. Gl. 483, 76. Tô ánum micclum screafe under ánre dūne, Homl. Th. ii. 424, 21. Tō screfe ł scrife *ad cloacum*, Hpt. Gl. 515, 72. Hí ne mihton ofer ðæt scræf, Blickl. Homl. 201, 16. Cwôman wyrmas of ðǣm neáhdūnum and scrafum *ex vicinis montium speluncis*, Nar. 14, 6. On wēstenum and on scrǣfum, Bd. 1, 8; S. 479, 21. Scrǣfu *speluncas, concavas petras*, Wrt. Voc. ii. 129, 66. Screafu *cavernas*, 21, 64. II. *a miserable dwelling, a den* :—Neara scræf *gurgustulum* vel *gurgustium*, i. 58, 29. Nihthrefne gelíc ðe on scræfe eardaþ *sicut nycticorax in domicilio*, Ps. Th. 101, 5. Gē mín hūs habbaþ gedôn sceaðum tô screafe (cf. gē worhtun ðæt tô þeófa cote, Mt. Kmbl. 21, 13), Homl. Th. i. 406, 3. Se hæfde on byrgenum scræf (*domicilium*), Mk. Skt. 5, 3. Geond ðæt atole scræf (*hell*), Cd. Th. 272, 33; Sat. 129: 290, 22; Sat. 419. Scref, 266, 23; Sat. 26: 269, 15; Sat. 73. Gē mín hūs dōþ sceaþum tô scrafum, Blickl. Homl. 71, 20. Ðē is leófre on ðisum wácum scræfum ðonne ðū on healle heálíc biscop sitte (cf. ðá wolde se hálga sum hūs timbrian, 144, 31), Homl. Th. ii. 146, 28. On wáclícum screafum oððe hulcum lutigende, i. 544, 30. v. dūn-, eorþ-, wīte-, wráþ-scræf.

scrǣtte, an; *f. An adulteress, a harlot* :—Scrættena *moecharum, meretricum*, Hpt. Gl. 507, 2. Scrættena (scræftena, MS.) *scortarum*, 524, 1. In fifteenth century vocabularies *skratt, skrate* translates *armifrodita*, Wrt. Voc. i. 217, 23: 268, 64; see also Cath. Angl. 325; and in this sense Halliwell gives *scrat* as a word in dialects of the North. *Scritta* is the form glossing *hermaphroditus* in Ælfric's Glossary, Wrt. Voc. i. 45, 28. Corresponding forms but with different meanings are found in *O. H. Ger.* scraz; *pl.* scrazza *pilosi, incubi*; screzza *larvae*; scratun; *pl. pilosi, larvae*: *Icel.* skratti; *m. a wizard, warlock; goblin, monster.* Cf. Old Scratch. v. Grmm. D. M. 447 sqq.

scrallettan *to make a loud sound* :—Ðonne wín hweteþ beornes breóstsefan stígeþ cirm on corþre cwide scralletaþ missenlíce *when wine excites a man's mind, clamour arises in the company, they cry out with speech diverse*, Exon. Th. 314, 27; Mōd. 20. Sum sceal mid hearpan æt his hláfordes fôtum sittan snere wrǣstan lǣtan scralletan *one shall sit with a harp at his lord's feet, bend the strings, make them send forth loud sound*, 332. 10; Vy. 83. [Cf. *Dan.* skralde *to sound loud*; and see *shrill* in Skeat's Etym. Dict.]

screáde, an; screád, e; *f. A piece cut off, a shred, a screed, paring* :—Screáde *sceda*, Wrt. Voc. i. 46, 70. Screádan *praesegmina, praecisiones*, 40, 9. Æppelscreáda *quisquiliae*, 22, 13. [Gif heo mei sparien eni poure schreaden (schiue, MS. T.: schraden, MS. C.), A. R. 416, 2. Hauede he non so god brede, Ne on his bord non so god shrede, þat he ne wolde þorwit fede Poure, Havel. 99. Schrede or clyppynge of clothe or oþer þynge *scissura, presegmen*, Prompt. Parv. 448. *O. Du.* schroode: *O. H. Ger.* scrôt: *Ger.* schrot. Cf. *Icel.* skrjóðr *a shred, strip*.]

screádian; *p.* ode *To shred, cut up* or *off, pare*, (of trees) *to prune* :—Būton ða lǣreówas screádian ða leahtras þurh heora láre áweg, ne biþ ðæt lǣwede folc wæstmbǣre, Homl. Th. ii. 74, 16. Ðá hēt hē (*Herod*) him his seax árǣcan tô screádigenne (cf. æppelscreáda *quisquiliae*) ǣnne æppel, i. 88, 9. [He (*Herod*) badd himm briungenn ænne cnif An appell forr to shrædenn, Orm. 8118. Sceldeð eower sceldes al of þe smal enden, Laym. 5866. Wortes or othere herbes . . . she shredde and seeth, Chauc. Cl. T. 227. Cf. He shred (*concidit*) the wild gourds into the pot of pottage, 2 Kings iv. 39. Schredyn or schragge trees *sarculo, sarmento*; schredyn wortys or oþer herbys *detirso*, Prompt. Parv. 448. *O. Du.* schrooden: *O. H. Ger.* scrôtan; *p.* screot *demere, tondere*: *Ger.* schroten *to cut, gnaw*.] v. á-, ge-screádian, *and next word.*

screádung, e; *f.* I. *pruning, trimming* :—Screádung *putatio*, Wrt. Voc. i. 39, 3. [Schredynge of trees and oþer lyke *sarmentacio, sarculacio*, Prompt. Parv. 448.] v. next word. II. *what is cut off, a shred, cutting, fragment, paring, leaving* of food :—Screádunga *fragmentorum*, Mt. Kmbl. Lind. 14, 20: Jn. Skt. Lind. Rush. 6, 12, 13: Mk. Skt. Lind. 6, 43. Scrádunga, Rush. 6, 43. Of screádungum *de micis*, 7, 28. Screádungo *reliquias*, Lk. Skt. Lind. Rush. 24, 43.

screádung-ísen, es; *n. An instrument for pruning* or *trimming* :—Wíngeardes screádungísen *sarculus*, Wrt. Voc. i. 16, 11.

screáf. v. scræf.

screáwa, an; *m. A shrew-mouse* :—Screáwa *mus araneus*, Wrt. Voc. i. 24, 29: *musiranus*, ii. 55, 80: *massiranus*, 71, 24. Screuua, screáuua, scraeua *musiranus*, Txts. 78, 649. [Cf. *Chauc. Piers P. Prompt. Parv.* schrewe, shrewe *pravus*.]

screb, scréc, scref. v. scræf; *m.* scríc, scræf.

scremman; *p.* de *To make a person stumble, put a stumbling-block in a person's way* :—Ne wirige ðū deáfe ne scremme ðū blinde *non maledices surdo, nec coram coeco pones offendiculum*, Lev. 19, 14. [*The word, like* scrimman, q. v., *seems to suggest comparison with forms in*

mp. Cf. *Icel.* skreppa; *þ.* skrapp *to slip; the causative of this verb might appear in English as* scrempan, *whose meaning would be that given to* scremman. Scrincan, screncan *are parallel, as regards meaning, to* scrimman, scremman.]

screncan; *p.* te *To lay a stumbling-block in a person's way, trip up, ensnare:*—Eft hē cwæd: 'Ne screnc ðū ðone blindan' . . . Se screncþ ðone blindan ðe ðone ungesceádwísan mirþ *protinus adjunxit:* '*Nec coram coeco pones offendiculum*' . . . *Coram coeco offendiculum ponere est . . . ei, qui lumen discretionis non habet, scandali occasionem praebere* Past. 59, 6; Swt. 453, 1–4. Cf. Hē þurh ealle uncysta ða mōd gescrencþ *per universa vitia animum supplantat*, 11, 6; Swt. 73, 2. Healden hié ðæt hié ða ne screncen ða ðe gáþ on ryhtne weg tōweard ðæs hefonríces *ne ad ingressum regni tendentibus obstaculum fiunt*, 9; Swt. 59, 19. [Ute we bidden God ðæt he us shilde þerwið þat he (*the devil*) us ne shrenche and seien: *Custodi me a laqueo*, O. E. Homl. ii. 209, 18. Þe deouel þat weneð me to schrenchen ant schunchen of þe weie þat leadeþ to eche lif, Jul. 34, 1. He wile scrennkenn hemm Full hefig fall to fallenn, Orm. 11861. To scrennkenn ure sawless, 2618.] v. ā-, for-, ge-screncan.

scrence, screncedness, screncend. v. ge-, mis-scrence, ge-screncedness, for-screncend.

screón (?); *p.* scráh *To cry out, proclaim; dicare:*—Forscráh *abdicavit* (in Lye). [O. H. Ger. scrían *clamare*.]

screpan; *p.* scræp, *pl.* scræpon; *pp.* screpen *To scrape, scratch:*—Scriopu *scalpio*, Txts. 97, 1828. Scripið, scripith, scribid *scarpinat*, 95, 1805. Screpes *scratches* (? the word glosses *arescit.* v. scrípan), Mk. Skt. Rush. 9, 18. Screp ðæt blōd of, Lchdm. ii. 262, 6: 38, 20. [Þet he screpe zennes of al of oure herten, Ayenb. 98, 19. *But generally later English forms seem to represent a verb* scrapian. Cf. *Icel.* skrapa: *O. Du.* schrapen:—Heo schulden schreapien þe eorðe up of hore putte, A. R. 116, 15. Al þet scrift ne schreapeð nout of, 344, 13. Shame shrapeth his clothes, Piers P. 11, 423. Scrapyñ (shrapyn) awey *abrado*; scrapyñ (schrapyn) *scalpo, scalpito*, Prompt. Parv. 450.] v. ā-, be-screpan.

scréþe, scroepe; *adj. Suitable, adapted, convenient:*—Hit (*Britain*) is gescræpe (scroepe, MS. C.) on læswe sceápa and neáta *alendis apta pecoribus ac jumentis*, Bd. 1, 1; S. 473, 13. v. ge-scræpe.

screpu (?), e; *f. A curry-comb:*—Æren screop[u?] *strigillus*, Txts. 99, 1935. Scroepe *strigillum*, 99, 1906.

scríc, scréc *a kind of thrush, screech, skrike* [v. E. D. S. Pub. Provincial names of birds, where *screech, skrike* are given as names of the missel-thrush, p. 1, and *screech bird, screech thrush* as those of the fieldfare (*turdus pilaris*), p. 6]:—Scríc, scréc, scruc (scriic ?) *turdus*, Txts. 103, 2069. Scríc *turdus*, Wrt. Voc. i. 29, 30 : 281, 20.

scrid, es; *n. A carriage, chariot, litter:*—Scrid *basterna*, Wrt. Voc. ii. 101, 49 : *carracutium, vehiculum*, 121, 81. Scrid (*currus*) Godes, Ps. Surt. 67, 18 : ii. p. 187, 14. Scrides *basterne*, Wrt. Voc. ii. 11, 80. Scrides, Hpt. Gl. 504, 15. On scride l on cræte *in carruca*, Wrt. Voc. ii. 47, 42. Ðá hēht se cásere gesponnan fiówer wildo hors tō scride and hine gebundenne in ðæt scrid āsetton . . . Hió geleáddon ðæt scrid on heá dūne, Shrn. 71, 34. Heó wæs on gyldenum scryd, 156, 11. Screoda siex hun[dred]a *six hundred chariots* (cf. Exod. 14, 7), Exon. Th. 468, 9; Phar. 5. Lígbærum scridum *vel* crætum *flammigeris quadrigis*, Wrt. Voc. ii. 149, 14. v. scriða.

-scrid *in* ful-scrid. v. scrýdan, IV.

Scride-finnas; *pl. m. A people who, according to Jornandes and Procopius, seem to have inhabited the present Russian Lapland and other tracts thereabouts, and even to have extended into the present Swedish Finnland.* [Procopius, σκριθίφινοι; Adam of Bremen, *Scritefinni*; Paulus Diaconus, *Scritobini*; Saxo Grammaticus, *Scricfinni*.]:—Scridefinnas, Ors. 1, 1; Swt. 16, 36. Ic wæs mid Scridefinnum, Exon. Th. 323, 16; Víd. 79. The distinguishing prefix seems to refer to the use of snowshoes or skates, cf. *Icel.* skriðr *a sliding motion*, skríða *to glide, slide* in snowshoes.

scrid-wægn, -wǽn, es; *m.* **I.** *a chariot:*—Hú seó gesceádwísnes bæd ðæt mōd ðæt hit sǽte on hire scridwǽne (cf. on hrædwǽne, 36, 1; Fox 174, 1), Bt. 36, tit.; Fox xviii, 4. **II.** *sella curulis:*—Sittan on gerénedum scridwǽne *in curuli sedere*, 27, 1; Fox 96, 1 : 27, tit.; Fox xiv, 22.

scrid-wísa, an; *m. A charioteer:*—Scridwísa *auriga*, Wrt. Voc. i. 39, 38.

scrif. v. scræf, ge-scrif.

scrifan; *p.* scráf, *pl.* scrifon; *pp.* scrifen *To decree, appoint* (cf. ge-scrif):—Scribun *promulgarunt*, Wrt. Voc. ii. 117, 74. Scriben *decerni*, 106, 22. **I.** *to decree to a person as his lot, to allot, assign:*—Swā missenlíce Dryhten eallum dǽleþ, scyreþ and scrífeþ, Exon. Th. 331, 10, Vy. 66. Him (*God*) þonc ǽghwā secge ðæs ðe hē for his miltsum monnum scrífeþ, 333, 7; Vy. 98. **II.** *to fix as his lot* for a person :— Ic sceal sēcan ða hámas ðe ðū mē ǽr scrife *I must visit the abodes that you (the body) have made my (the soul's) portion*, 371, 4; Seel. 70. Brūcan swylcra yrmþa swā ðū unc ǽr scrife, 373, 2; Seel. 102. Ðæt

wyt gesáwon heofona wuldor swylc swā ðū mē ǽr scrife, 375, 25; Seel. 143. **III.** *to decree after judgment, to adjudge, doom, inflict, impose, pass as a sentence* upon a person :—Folca gehwylcum Scyppend scrífeþ bí gewyrhtum eall æfter ryhte, 75, 12; Cri. 1220. Scrífeþ bí gewyrhtum meorde monna gehwam, 286, 8; Jul. 728. Gif hē bétan mōte sylle wið his lífe swā hwæt swā man him scrífe *si pretium ei fuerit impositum, dabit pro anima sua, quidquid fuerit postulatus*, Ex. 21, 30. Þrowige hē (*a pledgebreaker*) swā biscep him scrífe, L. Alf. pol. 1; Th. i. 60, 10. Þolige hē (*a criminal priest*) ǽgðres ge hádes ge eardes, and wrǽcnige swā wíde swā pápa him scrífe, L. Eth. ix. 26; Th. i. 346, 6. Ðǽr ábidan sceal maga máne fáh, hú him Metod scrífan wille, Beo. Th. 1963; B. 979. Heó woldan ðíne dōmas gehýran, and hú ðū ðám forworhtum scrífan woldest, Wulfst. 254, 17. **IV.** as an ecclesiastical term, *to shrive, to impose penance after confession, to hear confession and then impose penance:*—Ðonne sacerd mannum fæsten scrífeþ *quum sacerdos jejunium hominibus injungit*, L. Ecg. C. 1; Th. ii. 132, 25. Ðæt hē hit swā gebétt hæbbe, swā him his scrift scrífe, L. Ath. i. 25; Th. i. 212, 22. Wē lǽraþ ðæt ǽlc preósta scrífe and dǽdbóte tǽce ðam ðe him andette, L. Edg. C. 65; Th. ii. 258, 9. Ofer ealle ða scíre ðe hē (*scrift*) on scrífe, I. I. P. 7; Th. ii. 314, 5. Man sceal ðam unstrangan men líðelícor déman and scrífan ðonne ðam strangan . . . Man sceal on godcundan scriftan ge on woruldcundan dōman ðás þingc tōsceádan, L. C. S. 69; Th. i. 412, 5. Hire nán preóst scrífan ne mōt *neque ulli presbytero confessionem ejus accipere licebit*, L. Ecg. P. ii. 16; Th. ii. 188, 6. His scrift him sceal swā scrífan, swā hē on his dǽdum gehýreþ, ðæt him tō dōnne biþ . . . Gyf hwā tō ðam (*making peace*) cyrran nylle, ðonne ne mæg hē ðam scrífan, L. E. i. 36; Th. ii. 432, 37–434, 4. **V.** *to care for, regard* [cf. *O. Sax.* bi-skríban (*with gen.* or *prep.*) *to care about*], (a) *with gen.* :—Deáþ ðæs ne scrífeþ (cf. se deáþ swelces ne rēcþ *mors spernit altam gloriam*, Bt. 19; Fox 68, 32), Met. 10, 29. Rihtes ne scrífe, 25, 53. Ne scrífe hē ðæs hlísan būton hú hē ryhtosð wyrce *opus rectitudinis appetitio ignoret favoris*, Past. 44, 3; Swt. 323, 16. (b) *with dat.* :—Se hláford ne scrífþ, se ðam here waldeþ, freónde ne feónde, feore ne æfstum (cf. se ðe hiora welt ne murnþ náuðer ne friénd ne fiénd, Bt. 37, 1; Fox 186, 7), Met. 25, 15. (c) *with a clause* :—Hí (*the people of Sodom*) forléton eallinga ðone brídele ðæs eges, ðá hí ne scrion hwæðer hit wǽre ðe dæg ðe niht, ðonne ðonne hí syngodon, Past. 55, 1; Swt. 427, 31. [He (*the pope*) ne scal scriuen of þine weoridlifen, þat þine sunen alle scullen þe from falle, Laym. 32074. Þe preost shall shrifen þe and huslenn, Orm. 6128. Him for to hoslon au for to shriue, Havel. 361. Schryvyñ or here schryftys *audire confessiones*, Prompt. Parv. 449. *Also* schriven (*reflex.*) *means to confess* :—Ich chulle schriuen me *confitebor*, A. R. 344, 6. Mede shroue (shrof, shroof) hire of hire shrewednesse . . . Thanne he assoilled hir sone, Piers P. 3, 44. Schryvyñ or ben aknowe synnys yn schryfte *confiteor*, Prompt. Parv. 449. *O. Frs.* scríva *to impose a punishment.*] v. for-, ge-scrífan; riht-scrífend; scrift.

scrifen *painted* (?):—Scrifen scrád glád, Exon. Th. 353, 15; Reim. 13. [Cf. (?) *Icel.* skrifa *to paint*; scrifan *a picture.*]

scrífend. v. riht-scrífend.

scrift, es; m. I. *what is prescribed as a punishment, a penalty* (cf. scrífan, III) :—Ic (*bishop Werferth*) him (*Eadnoth*) sealde ðæt lond and ða béc . . . and ūre ágen rǽdengewrit ðæt wǽre him tō ðam geráde ðæt land tōlǽten ðe mon ǽlce gére gesylle fíftene scillingas clǽnes feós ðam bisceope and him eác ðone ne scrift (scrift ne?) healde *our agreement that the land was resigned to him on the condition that fifteen shillings a year be paid to the bishop, and also that the penalty* (the land had before been subject to the condition that if it were not held by a person in orders it must pass to the church at Worcester; this condition was now removed) *be not maintained in respect to him* (*Eadnoth*), Chart. Th. 168, 18. **I a.** as an ecclesiastical term, (1) *penance imposed after confession* :—Gif feorhlyre wurþe, tóeácan ðam rihtwere, twā pund tō bóte mid godcundan scrifte, L. E. B. 2; Th. ii. 240, 16; also in five following paragraphs. Ǽgðer man sceal ge on godcundan scriftan ge on woruldcundan dōman ðás þingc tōsceádan *these things* (the various circumstances of persons) *are to be discriminated in the penances of the church and in the sentences of the law*, L. C. S. 69; Th. i. 412, 11. (2) *confession* which is followed by penance, *shrift* :—Ðæt hē scriftes gyrnde and húsles *quod confessionem et eucharistiam desideravisset*, L. Ecg. P. i. 3; Th. ii. 172, 19 : 9; Th. ii. 176, 7. Gif preóst fulluhtes oððe scriftes forwyrne, L. N. P. L. 8; Th. ii. 292, 1. Ǽghwylc cristen man . . . gewunige gelómlíce tō scrifte; and unforwandodlíce his synna gecýþe, L. Eth. v. 22; Th. i. 310, 5. Gá man tō scrifte (*ad confessionem*), Wulfst. 181, 3. [Scrift ihalden *to carry out the penance imposed*, O. E. Homl. i. 9, 31. Nimen scrift *to accept penance*, Laym. 18395. Takenn shriffte, Orm. 6613. Schrift (*confession*) and penitence, A. R. 8, 6. Þe holy ssrifte (*confession, one of the seven sacraments*), Ayenb. 14, 8. Schryfte *confessio*, Prompt. Parv. 449. *Icel.* skript, skrift, *confession, penance.*] **II.** *one who passes sentence, inflicts punishment, a judge* (v. scrífan, III):—Wá is worulde scriftum, būtan heó mid rihte dōmas reccan, Wulfst. 263, 18. **II a.** as an ecclesiastical term, *one*

who hears confession and imposes penance, a confessor :—Ða bóte áberan ðe his scrift (*confessarius*) him tǽcþ, L. Ecg. P. i. 4; Th. ii. 172. 24. Ðonne sceal se scrift hine áhsjan be ðǽm ðe he him andettaþ, hú ða þing gedón wǽron, L. E. I. 31; Th. ii. 428, 10. His scrift him sceal swá scrífan, swá he ðonne on his dǽdum gehýreþ, ðæt him tó dónne biþ, 36; Th. ii. 432, 37; Homl. Th. ii. 94, 9. Libban ðam life, ðe scrift us wísige, Wulfst. 112, 18. Ne mæg þurh ðæt flǽsc se scrift geseón on ðære sáwle, Exon. Th. 80, 13; Cri. 1307. Bête he be his scriftes geþeahte, L. C. E. 23; Th. i. 374, 8 : L. P. M. 1; Th. ii. 286, 15. Hé ondette ælce costunge ðam móde his scriftes *tentationes suas menti pastoris indicet*, Past. 16, 5; Swt. 105, 16. Cweðe his andetnessa tó his scrifte, and ðus cweðe : Ic andette Ælmihtigum Gode and mínum scrifte ðam gástlícan lǽce ealle synna, L. de Cf. 6; Th. ii. 262, 18–21 : Blickl. Homl. 43, 20. Gif him þince ðæt he wið his scrift sprece, ðæt tácnaþ his synna forgyfennysse, Lchdm. iii. 174, 14. Úre mísdǽde bétan, swá úre scriftas ús tǽcon, Wulfst. 142, 12. Andettan úre synna úrum scriftan, 115, 12 : Blickl. Homl. 193, 22. [Hit ibeten swa þin scrifte þe techet, O. E. Homl. i. 19, 3. Widuten schriftes leaue, A. R. 418, 24.]

scrift-bóc ; *f.*   **I.** *a penitential, a book stating the penances to be enjoined after confession for various sins :*—i. scriftbóc on Englisc, Chart. Th. 430, 20; cf. L. Ælfc. P. 44; Th. ii. 384, 2. Swá hwylc swá ðæs scriftbóce tilige tó ábrecanne *quicunque Confessionale hoc violare conatus fuerit*, L. Ecg. P. Addit.; Th. ii. 238, 8. Ða mæssepreóstas sceolan heora scriftbéc mid rihte tǽcan and lǽran, swá swá hie úre fæderas ær démdon, Blickl. Homl. 43, 8.   **II.** *a discourse referring to penance :*—Lárspel and scriftbóc (*the title of the homily*), Wulfst. 242, 22.

scrift-scír, e ; *f. The district in which a confessor exercises his functions :*—Gif preóst on his scriftscíre ænigne man wite Gode oferhýre, L. Edg. C. 6; Th. ii. 244, 22 : 9; Th. ii. 246, 12 : 15; Th. ii. 246, 26. Sacerda gehwylc on his scriftscíre, Wulfst. 79, 17. Sacerdum gebyreþ on heora scriftscírum, L. I. P. 7; Th. ii. 312, 38 : 19; Th. ii. 326, 2. v. riht-scriftscír.

scrift-sprǽc, e ; *f. Confession :*—Gif deáþscyldig man scriftsprǽce gyrne, ne him man nǽfre ne wyrne, L. E. G. 5; Th. i. 168, 24 : L. C. S. 44; Th. i. 402, 4.

scrimman ; *p.* scramm *To shrink, draw up, contract :*—Gif monnes fót tó hommum scrimme and scrince (cf. monegum men gescrincaþ his fét tó his homme, 68, 3), Lchdm. ii. 6, 15. [Cf. scram *distorted ; benumbed with cold :* scrambed *deprived of the use of a limb by a nervous contraction of the muscles :* scrimed *shrivelled up :* shrammed, shrimmid *benumbed with cold :* scrimp *to spare, pinch :* shrump-shouldered *humpbacked*, all from Halliwell's Dictionary : scrimp, scrimpit *scanty, contracted*, Jamieson's Dict. *M. H. Ger.* schrimpfen : *Ger.* schrumpfen *to wrinkle : Dan.* skrumpe *to shrink, shrivel ;* skrumpen *shrunk, shrivelled.* See also Skeat's Dictionary s. v. *shrimp.*] v. screnman.

scrín, es ; *n.*   **I.** *a chest, coffer, casket, box in which precious things are kept :*—Scrín *arca vel scrinium*, Wrt. Voc. i. 26, 49 : *capsella*, 33, 62 : *arca*, 80, 79. Ðæt hálige scrín *the ark of the covenant*, Homl. Th. ii. 214, 35 : Jos. 4, 7. Godes scrín, 7, 6 : Num. 14, 10. Ðæt scrín, Jos. 3, 8, 13. Hé (*Judas*) hæfde scrín (*loculos*) and bær ða þing ðe man sende, Jn. Skt. 12, 6 : 13, 29. Hire scrín mid hiræ háligdómæ, Chart. Th. 553, 12.   **II.** *a receptacle for the relics of a saint, a shrine :*—Se earm weard geled on scríne of seolfre ásmiðod on Sancte Petres mynstre, Swt. A. S. Rdr. 99, 143. Ðá þwóh man ða hálgan bán, and bær intó ðære cyrcan on scríne, 100, 158. Ic genam ða reliquia ðære hálgan fǽmnan and hí gesætte on scrín ðæt ic sylf ær of stáne geworhte *ego tuli reliquias beatae Margaretae et reposui in scrinio, quod feci de lapide*, Nar. 49, 7. Ðá gebrohte se bisceop ealle ða hálgan bán on gelimplícum scrýnum, and gelógodon hí up on cyrcan, Homl. Skt. i. 11, 275. Ðá wolde se cásere wyrcan him eallum (*the seven sleepers*) gyldene scrýn, Homl. Th. ii. 426, 22. [Hí námen ðære (*in the minster*) twá gildene scrínes and .lx. seolferne, Chr. 1070; Erl. 209, 11.]   **III.** *a cage* in which a criminal is confined :—Hig Pilatum on ánum ýsenum scrýne gebrohton on ðære byrig Damascum, and hyne myd scrýne myd eallum on feastum cwearterne beclýsdon, St. And. 38, 8 : 44, 19. [*O. H. Ger.* scríni *scrinium, loculus : Icel.* skrín *a shrine.* From Latin.]

scrincan ; *p.* scranc, *pl.* scruncon ; *pp.* scruncen.   **I.** *of a plant, to wither away, dry up, shrivel :*—Mid ðam mǽstan bleó hý (*the male and female pennyroyal*) blówaþ ðonne neálíce óðre wyrta scrincaþ and weorniaþ, Lchdm. i. 204, 13. Scrincan *marcescere*, Hpt. Gl. 419, 74.   **II.** *of a living being, to pine away, become weak :*—Hé scrinceþ *arescit ;* he pineth away (A. V.), Mk. Skt. Lind. 9, 18. Ðá weard se cyning (*Belshazzar*) tó ðan swíðe áfyrht, ðæt he eal scranc (cf. Then the king's countenance was changed, and his thoughts troubled him, so that the joints of his loins were loosed, and his knees smote one against another, Dan. 5, 6), Homl. Th. ii. 436, 2. [Þu scalt scrinchin (deʒe, 2nd MS.), Laym. 2278. And hy scrynketh for shome, P.S. 158, 7.]   **III.** *to contract, shrink :*—Ða tán scrincaþ (-eð, MS.) up, Lchdm. iii. 48, 28. Gif sino scrince . . . oððe gif monnes fót tó hommum scrimme and scrince, ii. 6, 13–15. v. á-, for-, ge-scrincan.

scrind *swiftness* (?) :—Ofer ðæne (sǽ) mægene oft scipu scríþende

scrinde fleótaþ *over the sea oft sail the ships strongly and swiftly*, Ps. Th. 103, 24. [*Grein compares Lith.* skrindus *flying, running swiftly.*]

scrípan (?) ; *p.* te *To waste away, wither* :—Scrépes arescit, Mk. Skt. Rush. 9, 18. [Cf. *Icel.* skrjupr *frail : Norweg.* skrypa *to waste : Swed. dial.* skryyp *to shorten ;* skryp *weak.*]

scripel. v. eár-scrypel.

scripen, scripende gloss *austerus*, Lk. Skt. Lind. Rush. 19, 21.

scripp, es ; *n.* (?) *A scrip, bag :*—Petrus forlét lytle þing, scripp and net, ac he forlét ealle þing, ðá ðá he for Godes lufon nán þing habban nolde, Homl. Th. i 394, 7. [Horn tok burdon and scrippe, Horn. 1061. Palmere with pike ne with scrippe, Piers P. 5, 542. Scrippe *pera*, Prompt. Parv. 450. *Icel.* skreppa.]

scrippa, an ; *m.* ? :—Of ðære díc on ðone midmestan scrippan, Cod. Dip. Kmbl. v. 78, 27.

scriptor *occurs in the compound* tíd-scriptor *chronographus*, Wrt. Voc. ii. 131, 8. [Cf. *O. H. Ger.* scriptora ; *n. pl.* scriptores.]

scrida or **scriðu,** an ; *m.* or *f. A chariot :*—Scriðena t cræta *bigarum, curruum*, Hpt. Gl. 457, 77. v. scrid.

scríðan ; *p.* scráð, *pl.* scridon ; *pp.* scriden, scriðen.   **I.** *to go, take one's way to a place :*—Drihten gecwyð : 'Ástígaþ nú áwyrgde in ðæt wítehús.' Sóna æfter ðæm wordum werige gástas hwyrftum scríðaþ in ðæt sceaðena scræf, Cd. Th. 304, 17; Sat. 631. Men ne cunnon hwyder helrúnan hwyrftum scríðaþ, Beo. Th. 329; B. 163. Com on wanre niht scríðan sceadugenga, 1410; B. 703.   **II.** *to go hither and thither, go about, wander :*—Líg scríðeþ geond woruld wíde *fire shall spread itself far and wide through the world*, Exon. Th. 51, 3; Cri. 810. Fífte cyn is wídscriþelra hleápera, ðe under muneces gegyrlan æghwider scríðaþ, R. Ben. 135, 21. Bana wíde scráð (*of the destroying angel that smote the firstborn of Egypt*), Cd. Th. 180, 3; Exod. 39. Ðæt hine ne worian ne scríðan (*uagari*) ne lyste, R. Ben. 126, 18. Swá scríðende hweorfaþ gleómen, Exon. Th. 326, 27; Víd. 135.   **III.** *of the gliding motion of a ship, cloud, etc., or of the motion of a heavenly body in its orbit :*—Ne æt me hrisil scríðeþ, Exon. Th. 417, 20; Rä. 36, 7. Sió scíre scell scríðeþ ymbútan dógora gehwylce *the heavens make one revolution each day*, Met. 20, 174 : 28, 16. Sunne tungl scríðaþ leng útan ymb eall ðis, 28, 8. Wolcnu scríðaþ *clouds sail along*, Menol. Fox 486; Gn. C. 13. Leax sceal on wǽle mid sceote scríðan, 539; Gn. C. 40. Sægl (*the sun*) gewát under scríðan, Andr. Kmbl. 2913; An. 1459. Léton scríðan bronte brimþisan, Elen. Kmbl. 474; El. 237. Scríðende (*revolving*) færþ hweóle gelícost, Met. 20, 216. Scríðende oft scipu scríðende fleótaþ (*the storm-clouds*), Exon. Th. 385, 29; Rä. 4, 52. Ofer ðæne (sǽ) oft scipu scríðende fleótaþ, Ps. Th. 103, 24.   **IV.** *of the increase or decrease of light :*—Heó ðæt leóht geseah ellor scríðan, Cd. Th. 48, 9; Gen. 773. Niht ofer ealle scríðan cwóme, Beo. Th. 1305; B. 650.   **V.** *of the coming of times or seasons, of the passage of time :*—Ðæs scríð ymb seofon niht Weódmónaþ on tún, Menol. Fox 270; Men. 136. Dagas forþ scridun, Exon. Th. 160, 12; Gú. 942. Ofer nidða bearn nihtrím scridon, 167, 35; Gú. 1070. Cymeþ wlítig scríðan on tún Maius, Menol. Fox 152; Men. 77. Ðenden him ðeós woruld scríðende scínan móte, Exon. Th. 97, 3; Cri. 1585. Mín feorh ende geséceþ dæg scríðende, ðonne dógor beóþ mín forþ scriden, 164, 10–16; Gú. 1011. [Þa com Scottene king scriðen to hirede, Laym. 10799. He scráð (com, 2nd MS.) to þisse londe, 4109. Tweien scalkes scriden under bordes & skirmden, 8405. None of þe Normandes fro þam might skrith, Min. v. 68. To scrythe *labi*, Cath. Angl. 326. *O. Sax.* scríðan, scríðan *to go, pass* (of time, light): *O. H. Ger.* scríttan *gradi : Icel.* skríða *to creep, crawl ; to glide, slide.*] v. geond-, tó-, þurh-, ymb-scríðan.

scriðe, es ; *m. A course :*—Ða habbaþ scyrtran scriðe and færeld, ymbhwerft lǽssan ðonne óðru tungl, Met. 28, 11. [*O. H. Ger.* scrit *gradus, passus : Icel.* skriðr *a creeping* or *sliding motion.*]

scriðol, scritta. v. wíd-scriðol, scrætte.

Scrobbes-burh ; *f. Shrewsbury :*—Ðá férdon hí intó Stæffordscíre and intó Scrobbesbyrig, Chr. 1016; Erl. 154, 4. Cf. Civitas Scrobbensis, Cod. Dip. Kmbl. ii. 137, 24.

Scrobbesbyrig-scír, *and later* Scrob-scír ; *f. Shropshire :*—Ðá wæs se cyng gewend ofer Temese intó Scrobbesbyrigscíre, Chr. 1006; Erl. 140, 29. Hugo eorl of Scrobscíre, 1094; Erl. 230, 37.

Scrob-sǽte, -sǽtan ; *pl. The men of Shropshire ;* also used where now the name of their district would be used, *Shropshire :*—Ðá fyrdedon hí intó Stæffordscíre and intó Scrobsǽton, Chr. 1016; Erl. 154, 22.

scroepe. v. scrépe.

scrofell, es ; *n.* (?) *Scrofula :*—Cyrneles and scrofelles and æghwylces yfles, Lchdm. iii. 62, 22.

scrúd, es ; *n.*   **I.** *dress, clothing, attire :*—Hræglung *vestitus*, scruud *habitus*, Wrt. Voc. i. 39, 69. Hwæt begytst ðú of ðínum cræfte? Bigleofan and scrúd (*vestitum*) and feoh, Coll. Monast. Th. 23, 5. Hé sylþ him andlyfene and scrúd *dat ei victum et vestitum*, Deut. 10, 18 : L. Pen. 15; Th. ii. 282, 26.   **II.** *an article of dress, a garment :*—Scrúd *vestis, clamis*, Wrt. Voc. i. 25, 50. Cildes scrúd *praetexta*, 25, 56. Sléfleás scrúd *colobium*, sléfleás ancra scrúd *levitonarium*, 40, 20, 21.

Scrúde *melote, veste*, Hpt. Gl. 492, 52. Hé sealde hira ǽlcum twá scrúd (*stolas*), and hé sealde Beniamine fíf scrúd, Gen. 45, 22. [*Laym.* scrud: *Orm.* shrud: *A. R.* schrud: *Ayenb.* ssroud: *Piers P.* shroud *dress, garment*: *Icel.* skrúð *shrouds of a ship, tackle*.] v. beadu-, byrdu-, gúþ-, munuc-, nun-, ofer-, wíf-scrúd.

scrúd-fultum, es; *m. Assistance in providing clothing*; the word occurs in grants made to religious houses of funds for the provision of clothing :—Ealle ða sócna ofer ðæt fennland him (*the monks of Ely*) tó scrúdfultume (cf. stent causas seculares emendandae fratrum loco manentium victui vel vestitui necessaria ministrantes, p. 238), Chart. Th. 242, 18. Twelf hída tó scrúdfultume ðam hírēde (*Winchester*), 499, 13. Ic habbe gifen ðæt land intô Sanctes Petres mynstre intô Baðan ðám munecan tô scrúdfultume, Cod. Dip. Kmbl. iv. 171, 15. v. next word.

scrúd-land, es; *n. Land given to provide means for buying clothing, land given as scrúdfultum*, q. v. :—Hé geunn(-ann?) ðæs landes æt Orpedingtúne for his sáwle intô Cristes cyrican ðám Godes þeówum tô scrúdland, Chart. Th. 329, 19.

scrudnian, scrutnian; *p.* ode *To examine carefully, consider, investigate* :—Ic scrudnige ł ic smeáge bebodu Godes mínes *scrutabor mandata Dei mei*, Ps. Lamb. 118, 115. And Drihten on micelre folces menige smeáþ and scrutnoþ (scrudnaþ, MS. T.) hwæt ða feáwa syndan ðe his willan wyrcean willen *et querens Dominus in multitudine populi*, R. Ben. 2, 16. Míne gebróðra, scrutniaþ mid hú wáclícum wurðe Godes ríce biþ geboht, Homl. Th. i. 582, 25. Twá þing sind ðe wé sceolon carfullíce scrutnian, ii. 82, 25. Scrutniende *scrutando, investigando*, Hpt. Gl. 410, 12. Tô ásmeáganne mid scrutniendre scrutnunge, Anglia viii. 302, 36. Scrudinend (scrudniend?) *scrutantes*, Ps. Spl. 63, 6. [Cf. *O. H. Ger.* scrodôn, scrutôn *scrutari*.] v. á-scrudnian.

scrudnung, scrutnung; *f. Examination, investigation, enquiry* :—Hé began mid geornfulre scrudnunge smeágan and áhsian be ðám gebodum ðæs hálgan regules, Lchdm. iii. 440, 20. Tô ásmeáganne mid scrutniendre scrutnunge, Anglia viii. 302, 36.

scrúd-waru, e; *f. Habit, dress* :—Ðæt hé (*a monk*) healdan wille his clǽnnisse and munuclíce scrúdware, L. Eth. v. 6; Th. i. 306, 9: vi. 3; Th. i. 314, 27.

scruf, scrutnian, scrutnung. v. scorf, scrudnian, scrudnung.

scrybb, e; *f. Scrub, underwood, shrubbery* :—Of ðare stánstrǽte andlang scrybbe, Chart. Th. 525, 21.

scrýdan; *p.* de. **I.** *to put clothes on a person, to clothe* a person with (*mid*) a garment, *to dress* :—Ic mé scrýde *induo*, Ælfc. Gr. 28, 2; Zup. 167, 2. Ic [mé] scrýdde mid hǽran *induebar cilicio*, Ps. Spl. 34, 15. Heó scrýdde Iacob mid ðam deórwurðustan reáfe *vestibus valde bonis induit eum*, Gen. 27, 15. Hine man efosode and scrýdde hine and brohte hine tô ðam cynge *Joseph totonderunt, ac veste mutata ob'ulerunt ei*, 41, 14. Hé scrídde (*vestivit*) ðone bisceop mid línenum reáfe, Lev. 8, 7. Ic wæs nacud and gé mé scrýddon (*operuistis*), Mt. Kmbl. 25, 36, 38. Módor, scrýd (*vesti*) ðínne sunu, Ælfc. Gr. 18; Zup. 111, 3 Ne scríde nán wíf hig mid wǽpmannes reáfe, Deut. 22, 5. **II.** *to clothe, to furnish with clothes, provide with clothes* :—Hé scrýt mé wel and fêtt, Coll. Monast. Th. 22, 33. Gif ǽcyres weód God scrýt, Mt. Kmbl. 6, 30. Scrýtt, Lk. Skt. 12, 28. Hé hí fédan scolde and scrýdan, Chr. 1012; Erl. 147, 11. Hingrigendum mete syllan and nacode scrýdan, Blickl. Homl. 213, 18. **III.** *to put on a garment* :—Wlite ðú scrýddest *decorem induisti*, Ps. Spl. 103, 2. Línen reáf scrédan sume seócnysse ge'ácnaþ (*in a dream*) *to put on a linen garment betokens some sickness*, Lchdm. iii. 206, 30. **IV.** *to rig a ship* (cf. *shrouds of a ship*: *Icel.* skrúð *the shrouds of a ship, standing rigging; tackle, gear*) :—Is ðeós bát fulscríd, Andr. Kmbl. 992; An. 496. [He hine lette ueden, he hine lette scruden, Laym. 8945. Nolde þe neodfule ueden ne schruden, A. R. 214, 17. He wollde shridenn uss wiþþ heofennlíke wæde, Orm. 3676. He ne hauede nouth to shride but a kowel, Havel. 963. Ssrede þe poure, Ayenb. 90, 25. *Icel.* skrýða *to clothe, dress*.] v. ge-, mis-, un-, ymb-scrýdan; wan-scrýd.

scrynce; *adj. Withered* :—Menigo ðara unhálra blindena haltra scryngcara (giscrungenra, Rush.: forscruncenra, W. S.) *multitudo languentium, caecorum, claudorum, aridorum*, Jn. Skt. Lind. 5, 3. Cf. scrence, *and for the inflexion* tuoegara, 8, 17.

scúa. v. scúwa.

scucca, sceucca, sceocca, scocca, an; *m. A devil, demon*; in sing. generally *the devil, Satan, Beelzebub* :—Wæs se scucca (*Satan*) him betwux. Tô ðæm cwæð Drihten : 'Hwanon côme ðú?' Se sceocca andwyrde: 'Ic férde geond ðás eorþan,' Homl. Th. ii. 446, 25-27. Se scucca, 452, 13, 17. Se sceocca, 448, 4. Gang ðú sceocca (sceucca, MS. A.) on bæc *vade Satanas*, Mt. Kmbl. 4, 10. Æfter ðæs sceoccan (scoccan, Thw.) êhtnysse, Homl. Th. ii. 450, 3. Sceoccan *Belzebulis*, Germ. 399, 267. Sceoccan betǽht tô flǽsces forwyrde, R. Ben. 50, 1. Deóful ł scuccan *Zabulun*, Hymn. Surt. 115, 15. Ða áwyrigedan sceoccan (scuccan), Homl. Th. i. 68, 1 : Wulfst. 249, 1. Þurh ðara scuccena lotwrencas, Bt. 39, 6; Fox 220, 14. Scucna englas, Blickl. Homl. 189, 7. Ðæt hié leóda landgeweorc láþum beweredon scuccum

and scinnum, Beo. Th. 1882; B. 939. Ongunnan heora bearn blôtan feóndum, sceuccum onsæcgean *immolaverunt filios suos et filias suas daemoniis*, Ps. Th. 105, 27. *The word is found in the name of a place*, Scuccanhláu, Cod. Dip. Kmbl. i. 196, 1. [Þu scheomelese schucke (*the reeve that condemned St. Margaret*), Marh. 7, 26. Þe laðe unwiht, þe hellene schucke, H. M. 41, 35. Schenden þene sckucke (schucke), A. R. 316, 11. Þe scucke wes bitweonen, Laym. 276. Þu (*the reeve before whom Juliana was brought*) þat schucke art schucken (shuken, Bod. MS.) herien, Jul. 56, 2.]

scucc-gild, es; *n. An idol* :—Hí sceuccgyldum guldan *servierunt sculptilibus eorum*, Ps. Th. 105, 26.

scúdan *to shake, tremble, shiver, shudder* :—Hý (*Adam and Eve*) on uncýððu scomum scúdende scofene wurdon on gewinworuld *they shivering with shame into a strange land were thrust, into a world of struggle*, Exon. Th. 153, 19; Gú. 828. [Cf. *O. Sax.* skuddian: *O. Frs.* skedda: *O. H. Ger.* scuten, scutten *to shake* (trans.): *O. L. Ger.* scuddinga *excussus*.]

scúfan, sceúfan, sceófan; *p.* sceáf, *pl.* scufon, sceufon, sceofon, sceofon; *pp.* scofen, sceofen *To shove, push, thrust*; trudere, praecipitare :—Ic sceúfe (sceófe, scúfe) *praecipito*, Ælfc. Gr. 24; Zup. 137, 11: trudo, 28, 4; Zup. 171, 1. Scífþ *trudit*, Hpt. Gl. 406, 71. Scúfaþ *praecipitate*, Wrt. Voc. ii. 68, 78. **I.** *to shove, push, try to move something* :—Hé sceáf mid ðam scylde, ðæt se sceaft tôbærst, Byrht. Th. 135, 50; By. 136. Sume sceufon, sume tugon, and seó Godes fǽmne hwæðre stôd, Shrn. 154, 26. **II.** *to shove, thrust, cause to move with violence*. (1) literal :—Ðá ne gelífde Apollonius ðæt heó his gemæcca wǽre ac sceáf hí fram him, Ap. Th. 25, 6. Hé sceáf reáf of líce, Cd. Th. 94, 20; Gen. 1564. Hí dracan scufon, wyrm ofer weallclif, Beo. Th. 6254; B 3131. 'Uton hine underbæc sceófan' . . . Hí ðá hine underbæc scufon . . . ac hé næs ácweald þurh ðam heálícan fylle, Homl. Th. ii. 300, 14-20. Hét his scealcas scúfan ða hyssas in bǽlblýse, Cd. Th. 230, 11; Dan. 231: Exon. Th. 142, 21; Gú. 647. Leahtra leáse in ðæs leádes wylm scúfan, 277, 21; Jul. 584. Scúfan scyldigne in seáþ, Elen. Kmbl. 1380; El. 692. Ús ys miht geseald ðé tô sceófanne on ðás wítu ðisse deópnysse, Guthl. 5; Gdwin. 38, 17. (2) of proceedings which imply violence, *to thrust into prison, out of a place, etc.* :—Drihten heó (*the fallen angels*) furðor sceáf in ðæt neowle genip, Cd. Th. 292, 24; Sat. 445. Hig scufon (*ejecerunt*) hine of ðære ceastre, Lk. Skt. 4, 29. Sume scufon heora mágas forþ tô heofenan ríce, and férdon him sylfe tô helle wíte, Homl. Th. ii. 542, 22. Búton man ágeáfe Eustatius and his men heom tô hand sceofe *unless Eustace were given up and his men were handed over to them*, Chr. 1052; Erl. 179, 22. Se cyning wæs yrre wið mé and hét sceófan mé on cweartern *me retrudi jussit in carcerem*, Gen. 41, 10. Gé (*devils*) scofene wurdon fore oferhygdum in êce fýr, Exon. Th. 140, 5; Gú. 605. Hý (*Adam and Eve*) scofene wurdon on gewinworuld, 153, 20; Gú. 828. **III.** *to shove, push, cause to move* (*without notion of violence*) :—Hí scufon út heora scipu and gewendon heom begeondan sǽ, Chr. 1048; Erl. 180, 15 : Beo. Th. 436; B. 215. **IV.** *of the production of natural phenomena* :—Metod æfter sceáf ǽfen, Cd. Th. 9, 4; Gen. 136. Ðá wæs morgenleóht scofen and scynded, Beo. Th. 1840; B. 918. [Cf. Grmm. D. M. 706.] **V.** *to push* a person's cause, *advance, forward*, cf. scyfe, II :—Scúfeþ Freá forþwegas folmum sínum, Cd. Th. 170, 13; Gen. 2812. **VI.** *to urge, prompt* a thought or action, cf. scyfe, III :—Mid ðý se weriga gást ða synne scýfþ on môde *cum malignus spiritus peccatum suggerit in mente*, Bd. 1, 27; S. 497, 19 note. **VII.** *to push on or forward, to move* (intrans.) :—Merecondel (*the sun*) scýft on ofdæle, Met. 13, 58. Werige gástas scúfaþ tô grunde in ðæt nearwe níþ, Cd. Th. 304, 21; Sat. 633. [*Goth.* skiuban: *O Frs.* skúva: *O. H. Ger.* sciuban: *Icel.* skýfa (*wk.*) *to shove, drive, push*.] v. á-, æt-, be-, for-, ôþ-, tô-, wið-scúfan.

[sculan]; ic, hé sceal, scal, ðú scealt, *pl.* wé sculon, sceolon; *p.* sceolde, scolde, scealde, scalde; *subj. prs.* scyle, scile, sciele, scule. **I.** *to owe*; debere :—Án him sceolde (scalde, Rush.: áhte tô geldanne, Lind.) týn þúsend punda. Se hláford forgeaf him ðone gylt. Se þeówa gemétte hys efenþeówan, se him sceolde (sculde,.Rush.) án hund penega, and hé cwæð; 'Ágyf ðæt ðú mé scealt,' Mt. Kmbl. 18, 24, 28. Hú mycel scealt ðú (áht ðú tô geldanne, Lind.) mínum hláforde? Lk. Skt. 16, 5, 7. Gif hwá ôðrum scyle (scule) borh oððon bôte, gelǽste hit georne, L. Eth. v. 20; Th. i. 308, 31. [Cf. Uoryef me þet ich þe ssel, Ayenb. 115, 29. By the feith I shal Priam, Tr. and Cr. iii. 472.] **II.** *denoting obligation or constraint of various kinds, shall, must, ought,* (*I*) *have or am* (with infin.), *am bound*, with an infinitive expressed or that may be inferred from a preceding clause. (1) denoting a duty, moral obligation :—Ðú scealt on ǽghwylce tíd Godes willan wercan, Blickl. Homl. 67, 33. Nǽnig mon ne sceal lufian ne ne gēman his gesibbes, gif . . . (*it is a man's duty not to love*), 23, 16. Swá sceal oretta á in his môde Gode compian, Exon. Th. 122, 33; Gú. 315. God sceal mon ǽrest hergan, 333, 15; Gn. Ex. 4. Swá hire eaforan sculon æfter lybban, ðonne hie lǽd gedôþ, hié sculon lufe wyrcean, Cd. Th. 39, 12; Gen. 624. Næs fela manna, ðe hogade ymbe ða bôte swá georne, swá man scolde (sceolde, MS. B.), Wulfst. 156, 12. Hé (*the bishop*) ne cúðe dôn his gerihte swá wel swá

hé sceolde, Chr. 1047; Erl. 177, 9. Ðá andswarede se cyning ðæt hé
ǽgðer ge wolde ge scolde ðam geleáfan onfón *rex suscipere se fidem et
velle et debere respondebat*, Bd. 2, 13; S. 515, 35. Hwider hyra gehwylc
faran scolde, Blickl. Homl. 229, 5. Seó lufu ðe wé tó úrum Hǽlende
habban sceoldan, 109, 4. Forðæm ne scyle nán wís man nǽnne mannan
hatian, Bt. 38, 7; Fox 210, 15. (2) *shall, ought* as being fit, right,
proper, in accordance with reason :—Ic mid grápe sceal fón wið feónde,
Beo. Th. 881; B. 438. Hwý sceal ic æfter his hyldo þeówian . . . ic
mæg wesan god swá hé, Cd. Th. 18, 33; Gen. 282. Se ðe tó reccendóme
cuman sceal *qui ad regimen venire debeat*, Past. 11; Swt. 61, 5. Forðan
sceal gehycgan hæleða ǽghwylc, ðæt hé ne ábǽlige bearn Waldendes,
Cd. Th. 276, 25; Sat. 194. Ðonne gé geseóþ ðære tówurpednysse
ásceonunge standan ðǽr heó ne sceal (rīseþ, Rush.), Mk. Skt. 13, 14.
Be úre ǽ hé sceal (gedaefnaþ, Lind.) sweltan *debet mori*, Jn. Skt. 19, 7.
Seó cyrice sceal fédan ða ðe æt hire eardiaþ, Blickl. Homl. 41, 27: 47, 21.
Hwæt sculon wé nú dón tó ðam ðæt wé mǽgon cuman tó ðám sóþum
gesǽlþum *quid nunc faciendum, ut illius summi boni sedem reperire
mereamur?* Bt. 33, 3; Fox 126, 32. Dēmaþ ūs hwylcum deáðe wé
sweltan sceolon, for ðam ðe wé ðone Hǽlend tó deáðe gesealden, St. And.
36, 16. Oncnáwan hwǽr wé sǽlan sceolon sundhengestas, Exon. Th.
54, 3; Cri. 863. Ne sceole gé swá sófte sinc gegangan, Byrht. Th.
133, 32; By. 59. Ne sceolon unc betweónan teónan weaxan, Cd. Th.
114, 10; Gen. 1902. His weorc sceolon beón ðæs weorðe, ðæt him
óðre menn onhyrien, Past. 11, 1; Swt. 61, 17. Ic worda gespræc má
ðonne ic sceolde, Andr. Kmbl. 1848; An. 926: Hy. 3, 43. Ðone
máððum ðe ðú mid rihte rǽdan sceoldest, Beo. Th. 4119; B. 2056.
Swylc sceolde secg wesan æt þearfe, 5410; B. 2708. Gúþbill geswác,
swá hit nó sceolde, 5164; B. 2585. Oft mon forlǽt ðone ege ðe hé
mid ryhte on him innan habban scolde, Past. 4, 1; Swt. 37, 18. Hé ús
lǽrde, hú wé ús gebiddan sceoldan, Blickl. Homl. 19, 36. Hí cuædon,
ðæt hié ðæt tó his honda healdan sceoldon, forðæm hira nán næs on fædrenhealfe tó geboren, Chr. 887; Erl. 86, 4. Hú hié libban sceoldon, Cd. Th.
52, 30; Gen. 851. Hié níþ áhófon, swá hié nó sceoldon, Elen. Kmbl.
1673; El. 838. Gif ic scile *etsi oportuerit me*, Mk. Skt. Lind. 14, 31.
Hú hé scyle (scile, Cott. MSS.) eall earfoðu forsión *quod adversa quaeque
despicienda sunt*, Past. 3; Swt. 33, 4. Ne scyle nán mon blǽcern ǽlan
under mittan, 5, 1; Swt. 43, 2. Hú gehiérsum ðǽm ðe hé mid ryhte
hiéran sciele, 9; Swt. 56, 14: 10; Swt. 60, 6. (3) denoting obligation
to perform an engagement, to do appointed work, to carry out the terms
of an agreement :—Wísdómes beþearf se ðære æðelan sceal andwyrde
ágifan *he will need wisdom to whom the task of giving an answer is
assigned*, Elen. Kmbl. 1085; El. 545. Sume sceolon (*it will be the task
of some*) hweorfan geond hæleþa land, Cd. Th. 281, 11; Sat. 270. Næs
ðæt forma síð ðæt hit (*the sword*) ellenweorc æfnan scolde, Beo. Th.
2933; B. 1464. Ðonne scyldfreca ongeán gramum gangan scolde, 2073;
B. 1034. Ðone ende ðe Æðereð healdan sceolde, Chr. 894; Erl. 92, 2.
Hí woldon ðisne eard healdan, and hé hí fédan scolde and scrýdan, 1012;
Erl. 147, 10. Būtan ðǽm monnum ðe ða burga healdan scolden, Erl.
90, 19. Sceótend swǽfon, ða ðæt hornreced healdan sceoldon, Beo. Th.
1413; B. 704. His scipu sceoldan cumon ongeán, ac hí ne mihton, Chr.
1000; Erl. 137, 3. Gnornian hú oft hé feohtan scule (scyle, Cott. MS.),
Bt. 40, 3; Fox 238, 10. (4) denoting bidding, commanding :—'Hwæt
sceal ic singan?' Cwæð hé : 'Sing mē frumsceaft,' Bd. 4, 24; S. 597, 16.
Hǽlend him cwæð : 'Ðú scealt fylgean mē,' Blickl. Homl. 23, 14: Cd. Th.
139, 15; Gen. 2310: 172, 29; Gen. 2851. Scealtú mid ǽrdæge ceól
gestígan, Andr. Kmbl. 439; An. 220. Ic secge ðæt hé sceal wesan
Ismahel háten, Cd. Th. 138, 2; Gen. 2285. Ne sceolon gé míne ða
hálgan hrínan, Ps. Th. 104, 13. Ne scule gé hit þurhteón, 4, 5. Sægþ
on ðissum bócum, ðæt Drihten cwǽde, ðæt ðis mennisce cyn ne sceolde
áginneleásian, ðæt hié sealdon heora wǽstma fruman for Gode, Blickl.
Homl. 41, 4: Exon. Th. 15, 9; Cri. 233. Se (*God*) ús ðás láde sceóp, ðæt
wé on Egiptum sceolde ús fremu sēcan, Cd. Th. 110, 23; Gen. 1842. Hé
ús gesette ðæt wé hine biddan sceoldan *he made this ordinance for us, that
we should pray to him*, Blickl. Homl. 21, 3. Ða þing ðe ic eów foresægde,
ðæt gé dón sceoldon, 131, 34. Landfranc bebéad ðan munecan, ðæt hí
scoldan hí unscrýdan, Chr. 1070; Erl. 208, 8. Hé oncwæð, ðæt hié gyldan
sceolde, Cd. Th. 229, 5; Dan. 212. 'On ðæt fýr gē (*the wicked on the
day of judgment*) hreósan sceolan.' Ne mágon hí gehýnan heofoncyninges
bibod, Exon. Th. 93, 11; Cri. 1524. (5) where the obligation results from
a law, statute, regulation :—Se byrdesta sceall gyldan fíftýne meardes fell,
Ors. 1, 1; Swt. 18, 19. Nū sceal beón ǽfre on lí abbod, and ná
biscop, and ðan sculon beón underþeódde ealle Scotta biscopas, Chr.
565; Erl. 18, 6. Sceolde sweordes ecg feorh ácsigan, Andr. Kmbl.
2266; An. 1134. Se ðe scyle (*since the regulations of the Penitential
require it*) āne wucan dǽdbóte dón, L. Ecg. C. 2; Th. ii. 134, 13.
(6) denoting the necessity of fate, of the order of providence, *shall,
must* as being decreed by fate or providence :—Ðú scealt greót etan ðíne
lífdagas, Cd. Th. 56, 9; Gen. 909. Ðú eart eorþe, and þú scealt eft tó
eorþan weorðan, Blickl. Homl. 123, 9. Gyt scyl (sceal, MS. A.) beón
gefylled ðæt be mē áwriten is, Lk. 22, 37. Sceal hine wulf etan *his fate
will be to be eaten by a wolf* (cf. swá missenlíce Dryhten eallum dǽleþ,

331, 6; Vy. 64), Exon. Th. 328, 5; Vy. 12 (*and often*). Mon sceal on
eorþan geong ealdian, 333, 21; Gn. Ex. 7. Gǽþ á wyrd swá hió scel,
Beo. Th. 915; B. 455. Hié (*the Jews*) God sylfne áhéngon; ðæs hié
sculon wergðu dreógan, Elen. Kmbl. 420; El. 210: Exon. Th. 455, 28;
Hy. 4, 56. Hí ðǽr geférdon máran hearm ðonne hí ǽfre wéndon ðæt
him ǽnig burhwaru gedón sceolde *more than they ever expected it would
be the fate of any citizens to do them*, Chr. 994; Erl. 132, 22. Ðá hé
from sceolde niþþum hweorfan, Cd. Th. 74, 15; Gen. 1222. Nǽnig
heora þohte ðæt hé ðanon scolde eft gesēcean folc *every one of them
thought himself fated not to visit his people again*, Beo. Th. 1387;
B. 691. Ðonne ðú forþ scyle metodsceaft seón, 2363; B. 1179:
Cd. Th. 63, 27; Gen. 1038. Se dæg ðe hé sceole wið ðæm líchomon
hine gedǽlon, Blickl. Homl. 97, 20. Hwæðer ðis þúsend sceole beón
scyrtre þe lengre, 119, 6. Scile, Beo. Th. 6335; B. 3177. Ðeáh gé
wénen ðæt gé lange libban scylan, Bt. 19; Fox 70, 15. Nele se
Waldend ðæt forweorðan scylen sáula ússe *it is not God's will, that our
souls be destined to destruction* (but cf. hé nyle ðæt ða sáula forweorðan,
Bt. 34, 8; Fox 144, 37), Met. 21, 34. (7) *to be forced, must* because
there is no possible alternative, because one cannot help one's self :—
Nū sceal ic (*Hagar*) on wēstenne witodes bídan, Cd. Th. 137, 16;
Gen. 2274. Ic (*Satan*) sceal bídan in bendum, 268, 1; Sat. 48. Ic
teáras sceal geótan, Exon. Th. 11, 18; Cri. 172. Ne sceal ic míne
onsýn fore eówere mengu míþan, 144, 16; Gú. 679. Ðú scealt
furþor síþfæt secgan, 261, 18; Jul. 317. Blind sceal his eágna þolian,
335, 27; Gn. Ex. 39. Ðe ðam gesuincum hé sceal hine selfne
geþencean ðeáh hé nylle *in adversis ad sui memoriam nolens etiam
coactusque revocatur*, Past. 3, 1; Swt. 35, 7. Sculon hié ðás helle sēcan,
Cd. Th. 26, 14; Gen. 406. Ðú neorxna wonges wlite nýde sceoldes
ágiefan, Exon. Th. 86, 11; Cri. 1406. Ordfruma earmre láfe ðære ðe
ðǽm hǽðenan hýran sceolde, Cd. Th. 225, 13; Dan. 153. Scolde,
Beo. Th. 20; B. 10: 1935; B. 965. Hyne Hetware gehnǽgdon mid
ofermægene, ðæt se byrnwíga būgan sceolde, 5829; B. 2918. Sceoldon
wrǽcmæcgas ofgiefan grēne beorgas, Exon. Th. 116, 5; Gú. 202.
Ðonne hí siófian scioldon *when they could not help sighing*, Met. 26, 82.
Ðý læs ic scyle leng þrowian, Andr. Kmbl. 154; An. 77. Hé tó foo gif
hé niéde sciele *coactus ad regimen veniat*, Past. 9; Swt. 59, 9. (8) *to
be obliged, must, shall* because from the conditions or nature of a case
no alternative is admissible, because a conclusion is inevitable :—Gif ðæt
wíf nele hider tó lande mid mē, sceal ic lǽdan ðínne sunu eft tó ðam
lande ðe ðú of férdest? Gen. 24, 5. Nú ic eówer sceal frumcyn witan,
ǽr gé furþur féran, Beo. Th. 508; B. 251. Ic forworht hæbbe hyldo
ðíne, forðon ic lástas sceal weán on wēnum wíde lecgan, Cd. Th. 63, 3;
Gen. 1026. Ðú meaht be sumum tácnum ongietan, hwæs ðú wēnan
scealt *what with certainty you may expect*, Past. 21, 3; Swt. 157, 20.
Se ðe wille Drihtne bringan gecwēme lácfæsten, ðonne sceal hé ðæt mid
ælmessan fullian, Blickl. Homl. 37, 18. Nú sceal hē sylf faran, ne mæg
his ǽrende his boda beódan, Cd. Th. 35, 18; Gen. 556. Ðonne hé æt
hilde sceall lífes tiligan, Salm. Kmbl. 320; Sal. 159. Eart ðú ðe tó
cumenne eart, hwæðer ðe wé óðres scylon (sceolon, MS. A.: sculon,
MSS. B. C.) onbýdan (*expectamus*), Lk. Skt. 7, 19, 20. Sceolon, Mt.
Kmbl. 11, 3. Forðon wit sculon unc staðolwangas rúmor sēcan, Cd. Th.
114, 29; Gen. 1911. Ðǽr hig ǽnne sculan eard weardian, Ps. Th.
132, 1. Sculun, Runic pm. Menbl. 343, 21; Rún. 21. Ðý sceolon
gelýfan eorlas, hwæt mín æðelo sién, Andr. Kmbl. 1466; An. 734. Ne
sceolon mē þegenas ætwítan *men shall not reproach me* (*because there
will not be the slightest grounds for reproach*), Byrht. Th. 138, 14; By.
220. Wēnde ic ðæt ðú ðý wærra weorðan sceolde *I expected that you
must have got more cautious*, Exon. Th. 268, 1; Jul. 425. Ðá sceolde
hé ðǽr bídan ryhtnorþanwindes, Ors. 1, 1; Swt. 17, 17. Scolde herebyrne sund cunnian, Beo. Th. 2890; B. 1443. Hít ofetes noman ágan
sceolde, Cd. Th. 44, 35; Gen. 719. Ne meahton leng somed heora
begra ðǽr ǽhte habban, ac sceoldon ða rincas ðý sēcan ellor ēðelseld,
113, 29; Gen. 1894. Mē þincþ wundor tó hwon ðú sceole for ówiht
ðysne man habban ungelēredne fiscere *what reason obliges you to hold
this man, an ignorant fisherman, as of any account?* Blickl. Homl. 179, 13.
Gif hine mon tó genēdan scyle, and hé elles nylle *if there is no other
course open but to compel him*, L. Alf. pol. 1; Th. i. 60, 13. Seó orsorge
wyrd simle líhþ, ðæt mon scyle wēnan, ðæt heó seó sió sóþe gesǽlþ,
Bt. 20; Fox 70, 30. (9) denoting need, *shall, must*, where an end is
to be attained or a task to be completed or a purpose to be served :—
Hwæt sceal ic mā secgean fram Sancte Iohanne *what more need I say of
St. John?* Blickl. Homl. 169, 24. Ðæt scell ǽgleáwra mann ðonne ic
mē tælige findan on ferðe *a more learned man than I reckon myself is
necessary to perform the task*, Andr. Kmbl. 2965; An. 1485. Sculan
wē gyt martíra gemynd mā áreccan, Menol. Fox 136; Men. 68. 'Satan
ic ðǽr (*in hell*) sēcan wille.' . . . Sceolde hē ða brádan lígas sēcan, Cd. Th.
47, 20; 763. Nihtweard (*the fiery pillar*) sceolde wícian ofer weredum,
185, 2; Exod. 116. Tó hwon sceolde ðeós smyrenes ðus beón tó lose
gedón *what end was to be served by thus wasting this ointment?* Blickl.
Homl. 69, 6. Hwý gé ǽfre scylen unrihtfióungum eówer mód drēfan
quid tantos juvat excitare motus? Met. 27, 1. (10) denoting the

certainty of a future event, that results from a settled purpose or decision :—Ic gefremman sceal eorlíc ellen oððe endedæg mínne gebídan *I am determined to do or die*, Beo. Th. 1277; B. 636. Mid eárum ne sceal ic (*it is settled that I shall not*) gehéran ðære béman stefne, Cd. Th. 275, 13; Sat. 171. Ðú scealt deáðe sweltan *thou shalt surely die*, Gen. 2, 17; Ps. Th. 118, 39. Ðæt ðú sunu Dryhtnes cennan sceolde, Exon. Th. 19, 10; Cri. 298. Hé (*Christ*) wiste, ðæt seó burh (*Jerusalem*) sceolde ábrocen weorþan, Blickl. Homl. 77, 29. On ðære nihte ðe hý on ðone dæig tógædere fón sceoldan, Chr. 992; Erl. 130, 32. Hæfdon hié on rúne áwriten wera endestæf, hwænne hié tó móse weorðan sceoldon, Andr. Kmbl. 274; An. 137. (10 a) denoting the certainty of a result under proper conditions :—Ðú him fæste hel sóðan spræce, swá ðú mínum scealt feore geborgan *you are then certain to save my life*, Cd. Th. 110, 113; Gen. 1837. Forðan ðe (*on account of his previous conduct*) hé sceal éce wíte þrowian, Homl. Th. i. 66, 14: Blickl. Homl. 41, 32. Hú sceal mín cuman gǽst tó geóce? Exon. Th. 124, 10; Gú. 337. Se hláew sceal tó gemyndum mínum leódum heáh hlífian on Hronesnæsse, Beo. Th. 5600; B. 2804. Wé cwǽdon ǽr, ðæt se sceolde lytel sáwan, se ðe him ðone wind ondréde, Past. 39; Swt. 285, 23. Wéndun gé ðæt gé Scyppende sceoldan gelíce wesan, Exon. Th. 141, 33; Gú. 636. Ðá héht se cásere sceolde tó gemyndum mínum leódum heáh hlífian on Hronesnæsse . . . ðæt ða wildan hors scealden iornan on hearde wegas and him ða limo all tóbrecan, Shrn. 72, 1. Hé fægenaþ ðæs, hú hiene mon sciele (scyle, Hatt. MS.) herigean, Past. 8; Swt. 54, 7. Scile (sciele, Hatt. MS.), 9; Swt. 54, 19. Hé wéneþ ðæt hé sceole tó heofenum áhafen weorþan, Blickl. Homl. 185, 5. Gif wé ǽnige bóte gebídan sculan (scylen, MS. B.) *if improvement in our condition is certainly to take place*, Wulfst. 157, 2. (11) denoting probability :—Neron cwæð tó Paule : 'Forhwon ne sprecst ðú, Paulus?' Ðá andswarede him Sanctus Paulus : 'Wénstú ðæt ic sceole sprecan tó ðissum treówleásan men' *do you think it likely that I shall speak to this false man?* Blickl. Homl. 183, 32. (12) as an auxiliary :—Ic sceal rǽdan tó merigen *lecturus sum cras*, ðú scealt rǽdan *lecturus es*, hí sceolon (sceolan, sculon) rǽdan *lecturi sunt*, Ælfc. Gr. 24; Zup. 136, 10–12. Óðer *participium* is tówerdre tíde se ðe rǽdan sceal *lecturus . . . ðæt ðe sceal beón geræd legendus*, 41; Zup. 246, 10–15. Se ðe wyle oððe sceal sprecan *loquuturus*, Zup. 247, 15, 11: 248, 6. Se ðe sceal beón gecyssed *osculandus*, 248, 7. Sceal habba ł hæfis *habebit*, Mt. Kmbl. Lind. 1, 23. Hæfeþ ł hé scile habba, 6, 24. Wé stíges ł wé scilon stíge *ascendimus*, 20, 18. Gé sciolon geseá ł gé geseás *videbitis*, 13, 14. Ne héras hiá ł ne sciolon gehéra *non audiunt*, 13, 13. Ðonne ðú ǽfre on moldan man gewurde oððe ǽfre fulwihte onfón sceolde, Soul Kmbl. 172; Seel. 86. On ðæs engles wordum wæs gehýred, ðæt þurh hire beorþor sceolde beón gehǽled eall wífa cynn and wera, Blickl. Homl. 5, 23. Ðá bæd Swegen hine ðet hé sceolde faran mid him, Chr. 1046; Erl. 174, 12. Wéndon ðæt hig sceoldon máre onfón *plus essent accepturi*, Mt. Kmbl. 20, 10. Ða ðonne ðe sió godcundde stefn þreáde and cuæð ðæt hié scolden leásunga wítgian *quos divinus sermo falsa videre redarguit*, Past. 15, 2; Swt. 91, 8. Hyra þeáw wæs ðæt hí ða untruman in leásan sceoldan, Bd. 4, 24; S. 598, 28. Deáh hé micel áge, and him mon erigan scyle æghwelce dæg æcera þúsend, Met. 14, 4. (13) denoting an assertion not made by the speaker, when a statement is matter of report [cf. *Ger.* sollen, and the use of *should* in the following passage :—There was something said about ane Campbell, that suld hae been concerned in the robbery, and that he suld hae had a warrant frae the Duke of Argyle, Rob Roy 1, 219] :—Be ðære frécnan coþe ðe se mon his útgang þurh ðone múþ sceal (*is said*) áspíwan. Hé sceal oft bealcettan, Lchdm. ii. 236, 13. Ys sǽd, ðæt Diana ðás wyrta findan scolde, i. 106, 5, 23: 120, 4. Ðú gehérdest reccan on ealdum leásum spellum, ðætte Iob sceolde beón se héhsta god, Bt. 35, 4; Fox 162, 6. Ðá sǽdon hí, ðæt ðæs hearperes wíf sceolde ácwelan, and hire sáwle mon sceolde lǽdon tó helle. Ðá sceolde se hearpere weorðan swá sárig. . . Ðá hé ðider com, ða sceolde cuman ðære helle hund ongeán hine . . . se sceolde habban þrió heáfdu, 35, 6; Fox 168, 3–17: 38, 1; Fox 194, 30–34. Deáh hé Cristen beón sceolde *though he was said to be a Christian*, Bd. 2, 20; S. 521, 29. Fundon ðá leáse gewitan ðe forlugon Naboð ðæt hé sceolde wyrigan God (*they brought reports of his blasphemy*), Homl. Skt. i. 18, 197. Ulf biscop com and forneáh man sceolde tóbrecan his stef *the report was that they were very near breaking his staff*, Chr. 1047; Erl. 177, 7. Swá swá manige men sǽdon þe hit geseón sceoldan *who were said to have seen it*, 1098; Erl. 235, 5: 1100; Erl. 235, 33. **III.** *without an infinite* (1) denoting constraint, necessity, need, fixed purpose :—Ealle wyrd forsweóp míne mágas, ic him æfter sceal *I must after them*, Beo. Th. 5625; B. 2816. Hé sceal néde tó ðara hláforda dóme ðe hé hine ǽr underþeódde *non facit, quod optat, ipse dominis pressus iniquis*, Bt. 37, 1; Fox 186, 28. Sió manbót ðe ðam hláforde sceal *the fine that must be to the lord*, L. In. 76; Th. i. 150, 16. Tó myclan bryce sceal micel bót nýde, and tó miclum bryne wæter unlytel, Wulfst. 157, 8. Earc sceal dý máre ðe ark must be the bigger, Cd. Th. 79, 19; Gen. 1313. Hié tó helle sculon, Cd. Th. 45, 26; Gen. 732. Xersis áscade hwæt sceolde æt swá lytlum weorode mára fultum búton ða áne ðe him ǽr ábolgen wæs *Xerxes demanded what*

a greater force was needed for in dealing with so small a band, than those only with whom he had before been angry, Ors. 2, 5; Swt. 80, 16. Eall swá hí sceoldon tó Sandwíc *as if they had or purposed to go to Sandwich*, Chr. 1049; Erl. 174, 26. Ðæt hé of disse worlde sceolde, Blickl. Homl. 225, 5. Ðonne seó eorþe him on ufan scealde *when the earth came to be put upon them*, Shrn. 81, 2. Ær hé onweg scyle *before he die*, Exon. Th. 310, 14; Seef. 74. (2) denoting obligation, fitness, propriety, use (cf. *Ger.* wozu soll dies?) :—Ðys sceal on twelftan dæg *this is the proper gospel for twelfth-day* (cf. ðys godspel gebyraþ, Rubc. 1, 18), Mt. Kmbl. Rubc. 2, 1 (*and often*). Hwæt scal ðé swá láðlíc stríð *what good will the strife do you?* Cd. Th. 41, 28; Gen. 663. Rǽd sceal mid snyttro . . . til sceal mid tilum, Exon. Th. 334, 26; Gn. Ex. 22. Wíta sceal geþyldig, ne sceal nó tó hátheort, 290, 15; Wand. 65. Hige sceal þe heardra, mód sceal ðe mára, ðe úre mægen lytlaþ, Byrht. Th. 140, 62; By. 312. Hí gecnáwan ne cunnan ne ða medtrymnesse ne eác ða wyrta ðe ðærwið sculon *the herbs that are proper for the disease*, Past. 1, 1; Swt. 25, 22. Óðre wyrtdrencas sculon (*are proper*), Lchdm. ii. 208, 3. Ðás wyrte sculon tó lungensealfe, iii. 16, 6. Hwæt sceolon (sculon, MS. H.) hí gesǽde nú wé swerian ne móton *what good would they (adverbia jurativa) do stated, now we may not swear?* Ælfc. Gr. 38; Zup. 227, 10. Hé áxode ðone cásere hú hé embe hí sceolde *how he was to deal with them*, Homl. Skt. i. 5, 370. Ne meahte geþencan hú ymb ðæt sceolde *what ought to be done about it, how the matter ought to be dealt with*, Exon. Th. 378, 7; Deor. 12. Hwæt sceoldon (*deberent*) hig mé búton ic cúþe temian hig *what good would they (hawks) be to me unless I knew how to tame them?* Coll. Monast. Th. 25, 23. Hié be ðæm wiston hwider hié sceoldon *they knew by that in which direction they had to go*, Ors. 3, 5; Swt. 106, 15. Hié wiston hú hié tó ðæm elpendon sceoldon *they knew the proper way of attacking the elephants*, 4, 1; Swt. 156, 17. Warnige man ðone stiwerd tó hwylcere stówe ðæt líc sceole, Chart. Th. 607, 15. Hwæt sceoldon ðé úre ælmessan? Wulfst. 240, 15. [**Goth.** [skulan]; *prs.* skal, *pl.* skulum; *p.* skulda : *O. Sax.* [skulan]; *prs.* skal, *pl.* skulun; *p.* skolda : *O. Frs.* skila; *prs.* skal, skel, skil, *pl.* skilun ; *p.* skolde: *O. H. Ger.* scolan ; *prs.* scal, *pl.* sculumes ; *p.* scolta : *Icel.* skulu ; *prs.* skal, *pl.* skulum ; *p.* skyldi.]

sculdor; *pl.* (*dual?*) sculdru (-o), sculdra ; *m. A shoulder* :—Sculdur *scalpula*, Wrt. Voc. ii. 120, 18 : *scapulus*, i. 64, 68. Sculdor, 283, 6. Sculdor *scapula*, 44, 27. His sculdor and his hleór wurdon ontende mid ðam fýre, Homl. Th. ii. 344, 16. Wæs ðæt bærnet on his sculdre ǽfre gesewen, 346, 26. On his sculdre *in humero*, Bd. 3, 19; S. 549, 15. Oþ ðone swíðran sculdor, Lchdm. ii. 198, 19. Duru ðæt mannes heáfod ge ða sculdro mágan in, Blickl. Homl. 127, 9. Sculdra *scapula* (-*ae?*), Wrt. Voc. i. 71, 19. On bǽm sculdrum *in utroque humero*, Past. 14, 3; Swt. 83, 9, 21. Hé onfeóld hys hrægl æt hys sceoldrum, Shrn. 98, 17. In scyldrum ł bæccum *in humeros*, Mt. Kmbl. Lind. 23, 4 : Lk. Skt. Lind. Rush. 15, 5. Hí dydon ánne hwítel on hira sculdra *pallium imposuerunt humeris suis*, Gen. 9, 23 : Bd. 3, 19 ; S. 549, 1. Sculdru (sculdra, MS. X.), L. Ecg. C. 9; Th. ii. 140, 10. Gif mon óðrum ða sculdor forsleá, L. Alf. pol. 73; Th. i. 98, 21. Se sacerd smyreþ breóst and sculdru (sculdran, MS. E.), Wulfst. 35, 16 : Lchdm. ii. 260, 17. [Schuldren ; *pl.* Jul. 49, 18. He let smyte of ys hede by þe ssoldren, R. Glouc. 313, 7. *O. Frs.* sculder : *O. H. Ger.* scultarra *humerus, scapula, spadula*.]

sculdor-hrægel, es ; *n. A garment to cover the shoulders* :—Sculdorhrægl *superhumerale*, Wrt. Voc. i. 81, 44.

sculdor-wærc, es ; *m. Pain in the shoulders* :—Wið sculdorwærce and earma, Lchdm. ii. 340, 12 : 6, 2.

scult-héta. v. scyld-héta.

scunian, sceonian ; *p.* ode. **I.** *to shun, fear, avoid a thing from fear* :—Hé his hatunge fleáh and scunode, Guthl. 19 ; Gdwin. 76, 16. **II.** *to be afraid* :—Scunian *revereantur*, Ps. Spl. T. 69, 2. **III.** *to detest, abhor* :—Mid ǽne wurð hé gescunned *uni animo detestetur*, Chart. Th. 318, 37. [Mi uader scunede (sonede, 2nd MS.) þene cristindom & þa hæðene laʒen luuede to swiðe, þa we sculleð seonien (hatie, 2nd MS.), Laym. 14868. Birrþ þe shunenn (*avoid from fear*) to follʒhenn ohht tærinne, Orm. 4502. Ancren owen to hatien ham, and schunien, þ heo ham ne iheren, A. R. 82, 23. Þu ahtest þis werc ouer alle þing to schunien (*avoid with abhorrence, abhor*), H. M. 35, 11. Al hit him uleh and scunede, þet him er luuede, O. E. Homl. i. 79, 29. ʒif him wrattheth, þe ywar and hi weye shonye (*avoid from fear*), Piers P. prol. 174.] v. á-, on-scunian.

scunung, e ; *f. An abomination* :—On scunungum *in abominationibus*, Cant. M. ad fil. 16. v. á-, on-scunung.

scúr, sceór, scyúr, es ; *m.* : e ; *f.* (?) **I.** *a shower, storm of rain, snow, hail, etc.* :—Scúr *nimbus*, Wrt. Voc. i. 52, 60 : 76, 42. Scyúr (scúr, Rush.), Lk. Skt. Lind. 12, 54. Ðes scúr *hic imber*, Ælfc. Gr. 9, 18 ; Zup. 43, 7. Swylce scúr ofer gærs *quasi imber super herbam*, Cant. M. ad fil. 2. Rénes scúr, Exon. Th. 215, 1 ; Ph. 246. Regna scúr, Cd. Th. 252, 10 ; Dan. 576. Hægles scúr, 50, 13 ; Gen. 808. Syððan (*after the overflow of the Nile*) tó twelf móndum ne cymþ ðær nán óðer scúr, óð ðæt seó eá eft up ábrece, Lchdm. iii. 254, 2. Ðonne

sceór cymeþ, Andr. Kmbl. 1024; An. 512. Đā wæs geblissod seó Godes-burh for đam cyme đæs scūres đe hý geclǣnsode *fluminis impetus laetificat civitatem Dei*, Ps. Th. 45, 4. Scūre *nimbo*, Wrt. Voc. ii. 61, 54: *inserenae*, Hpt. Gl. 514, 15. Scūras *imbres*, Ps. Lamb. 77, 44. Geþēn-sume scūras *coloni nimbi*, Wrt. Voc. ii. 134, 28. Wealcaþ hit (*hail*) windes scūras (? MS. scūra), Runic pm. Kmbl. 341, 6; Rūn. 9. Scūra *procellarum*, Hpt. Gl. 509, 20 [H]reósendlícum scūrum *ruituris imbribus*, 499, 64: 501, 6: Wrt. Voc. ii. 47, 15. Seó lyft liccaþ đone wǣtan of ealre eorþan and of đære sǣ and gegaderaþ tô scūrum, Lchdm. iii. 276, 13. Weal sceal wiđstondan storma scūrum, Exon. Th. 281, 25; Jul. 651. I a. metaph. *a shower* of missiles:—Flāna scūras, Judth. Thw. 24, 34; Jud. 221: Elen. Kmbl. 234; El. 117. Hygegār lēteþ, scūrum sceóteþ, Exon. Th. 315, 22; Mód. 35. I b. *a shower* of blows of a hammer falling on a weapon (?):—Scearpne mēce scūram heardne, Judth. Thw. 22, 26; Jud. 79. Cf. scūr-heard. II. metaph. *a storm, trouble, disquiet*:—Swā đeós woruld fareþ scūrum (cf. scȳr-mǣlum) scyndeþ *hurries on stormily*, Exon. Th. 469, 24; Hy. 11, 7. [*Goth.* skūra (*windis*) λαῖλαψ, *procella: O. Sax.* skūr *a missile, weapon:*—That man ina wītnôdi wāpnes eggiun, skarpun skūrun, Hel. 5138. *O. H. Ger.* scūr *tempestas, grando; also of weapons:*—Dō lēttun sē askim scrītan, scarpēn scūrim, Hildebrandslied 66. *Icel.* skūr *a shower; a shower* of missiles; vāpna, hjālma skūr *id.*] v. hægl-, hagal-, hilde-, regn-, winter-scūr; scȳr-mǣlum, *and next word.*

scūra (-e; *f.?*), an; *m. A shower:*—Hē āriman mæg rægnas scūran dropena gehwelcne, Cd. Th. 265, 22; Sat. 11.

scūr-beorh, gen. -beorge; *f. A shelter against storm:*—Hrôfas sind gehrorene . . . scearde scūrbeorge, Exon. Th. 476, 9; Ruin. 5.

scūr-boga, an; *m. A rain-bow:*—Đonne ic scūrbogan mînne iéwe, Cd. Th. 93, 5; Gen 1541.

scurf, scurfed, scurfende. v. scorf, scorfed, scorfende.

scūr-heard; *adj. Made hard by blows* (v. scūr, I b; *and* cf. heoru hamere geþuren, Beo. Th. 2575; B. 1285):—Sweordes ecg, scerp and scūrheard, Andr. Kmbl. 2267; An. 1135. Đæt him fēla lāf (*the sword*) ne meahte scūrheard sceþþan, Beo. Th. 2070; B. 1033.

scūr-sceadu (*or* -scead; *n.*); *f. A protection against storms* (cf. *umbrella*):—Nys unc wuht beforan tô scūrsceade, Cd. Th. 50, 23; Gen. 813.

-scuta in an-scuta *falarica*, Hpt. Gl. 425, 14. [Cf. (?) He þa fla lette gliden bi Corineus siden Corineus bleinte & þene scute biberh, Laym. 1461.]

scutel *a dish:*—Scutel *catinus*, Wrt. Voc. i. 290, 21 : ii. 17, 17. [Scotylle *scutella*, 257, 15. *O. H. Ger.* scuzzila *scutula, scutella, discus, catinus, lanx: Icel.* skutill *a dish.* From Latin (?) *scutella.*]

scutel *and* scytel, es; *m.* I. *a dart, missile, arrow:*—Sciutil *jaculum, sagitta*, Txts. 110, 1177, 1179. Scytelum cilda *sagittis parvulorum*, Ps. Th. 63, 7. [*Icel.* skutill *an instrument shot forth, a harpoon.* Cf. scytyl *a shuttle*; schytle, chyldys game *sagitella*, Prompt. Parv. 447. Schetylle *navecula*, Wrt. Voc. i. 235, 3.] II. *the tongue of a balance* (?):—Scytel *momentum*, 76, 632. Scutil, Wrt. Voc. ii. 71, 20. Scutel, 56, 52. [*M. Lat.* momentum *languette de bilance.* Cf. schytylle, schityl, onstabyl *preceps*, Prompt. Parv. 447.]

scūwa, scūa, an; *m. The shadow* thrown by an object:—Oferwrāh muntas scūa his *operuit montes umbra ejus*, Ps. Surt. 79, 11. Dægas mîne swē swē scūa (*umbra*) onhældun, 101, 12 : 143, 4. Ic eom scūan gelíc swýþe āhylded *sicut umbra cum declinat*, Ps. Th. 108, 23. II. *shade, darkness*:—Mid đý wit forþgongende wǣron under đam scūwan đære þýstran nihte *cum progrederemur sola sub nocte per umbras*, Bd. 5, 12; S. 628, 14. II a. fig. *shadow:*—Scūa déaþes umbra *mortis*, Ps. Surt. 43, 20. In midle scūan deáþes, 22, 4. Ālǣd mē ūt of đyses carcernes hūse and of deáþes scūan, Blickl. Homl. 87, 35. Scūia (scūa, Rush.), Mt. Kmbl. Lind. 4, 16. In scýa, Rtl. 168, 9. Sealde him deorcne deáþes scūwan, Ch. Th. 293, 15; Sat. 455. III. *shadow, protection:*—In on fægerum scūan fiđera đînra gewîcie in umbra *alarum tuarum spero*, Ps. Th. 56, 1. Under scūan fiđra đînra gesild mē *sub umbra alarum tuarum protege me*, Ps. Surt. 16, 8. IV. *shadow* as opposed to substance:—Scūan ł leásunge *fallacis*, Hpt. Gl. 459, 14. [Screne *or* scu *or* spere *scrinium, ventifuga*, Prompt. Parv. 450. Spere *or* scuw (schuu), 468. Þe skuues of the scowtes, Gaw. 2167. Cf. *Goth.* thairh skuggwan δι᾽ ἐσόπτρου, 1 Cor. 13, 12. *O. H. Ger.* scūwo *umbra: Icel.* skuggi *a shadow; a spectre.*] v. dǣd-, deáþ-, dim-, heolstor-, hlin-, niht-scūwa (-scūa).

scūwan, scūan (?) *to shade*. [*O. H. Ger.* scūit *adumbrat: Icel.* skyggva *to overshadow.*] v. ofer-scūwan.

scȳan (*for* scȳhan), scȳn (?); *p. de To prompt, urge, persuade, suggest:*—Đa ǣrestan synne se weriga gāst scȳde . . . Forđon mid đý se weriga gāst đa synne scȳþ (scȳ̄þ, MS. C.: scȳþ, MS. T.) on môde *primam culpam serpens suggessit . . . Cum enim malignus spiritus peccatum suggerit in mente*, Bd. 1, 27; S. 497, 14–20. Wē getǣceþ ł scȳaþ him *nos suadebimus ei*, Mt. Kmbl. Rush. 28, 14. Cf. scȳhend, scȳend *maulistis*, Txts. 78, 654. Scȳhend *malistis*, Wrt. Voc. ii. 55,

52. [Cf. *O. H. Ger.* scūhenti *exhortans*, Grff. vi. 417.] v. sceóness, scyhtan.

scyccels. v. scíccels.

scydd, es; *m. Alluvial ground* (?):—Đis synt đa denbæra . . . hudelinga scydd, Cod. Dip Kmbl. ii. 195, 19. On timbersǣd in stæpa cnolles scydd on hanslǣdes heáfdan, iii. 380, 26. Haec sunt pascua porcorum . . . in communi silua pascuale quod dicitur Palinga Schittas (scyddas ?), ii. 303, 19. [Cf. *M. H. Ger.* schüt: *Ger.* schutt.]

scȳde, Cd. Th. 232, 26; Dan. 266. v. sceón.

scȳend. v. scȳan.

scȳe-uange:—Scōe ł scȳeuange (-þwange ?) *calciamentum*, Ps. Spl. T. 59, 9.

scyfe, es; *m.* I. of rapid motion caused by a push (metaph.), *precipitation*, v. scūfan:—Word scyfes *verba praecipitationis*, Ps. Lamb. 51, 6. Hié weorđaþ oft āscrencte on đæm scyfe đære styringe hira môdes đæt hî hira selfra ne āgon đý māre geweald đe ôđerra monna *motionis impulsu praecipites quaedam velut alienati peragunt*, Past. 33, 1; Swt. 215, 12, 17. I a. glossing *preceps*:—Seó ūs on scefe gedwelde teáh mid wegleásum *quae nos in preceps errore traxit devio*, Hymn. Surt. 24, 11. II. *furtherance* of a project, *the pushing* of a matter, *prompting, instigation* in a good sense, cf. scūfan, V:—Ælc burghemet beó be his dihte ge scife swîđe rihte, L. I. P. 7; Th. ii. 312, 21. III. *prompting, instigation* in a bad sense, cf. scūfan, VI:—Se đe þurh deófles scyfe on synna befealle, L. C. E. 23; Th. i. 374, 9. Befeallen þurh deófles scyfe on heálíce misdǣde, Wulfst. 102, 13. v. niđer-scyfe.

scyfel, e: scyfele, an; *f. [Shovel in shovel-hat ?] A covering for a woman's head;* mafors (mafors *operimentum capitis maxime feminarum,,*Ducange):—Hacele *capsula*, cōp *ependiten*, scyfele *mafors*, nunne *sanctimonialis*, Wrt. Voc. i. 284, 67. Scyfla, scybla *maforte* (-ae), Txts. 77, 1267: Wrt. Voc ii. 55, 38. Scyfelum *mafortibus*, 55, 39 : 87, 63. [Cf. (?) *scuffle*, a linen garment worn by children to keep their clothes clean, a pinafore, an apron (Sussex). *Icel.* skupla; *f.*; skypill; *m. a woman's hood hiding or shading her face.*]

scyftan, -scȳgean, scȳhend. v. sciftan, -scígan, scȳan.

scyhtan; *p.* te *To instigate, prompt, urge:*—Mē nædre beswāc and mē neódlíce tô forsceape scyhte and tô scyldfrece, Cd. Th. 55, 22; Gen. 898. Ôđer him đas eorþan ealle sægde lǣne . . . Ôđer hine scyhte đæt hē sceađena gemôt nihtes sôhte, Exon. Th. 109, 30; Gū. 98. [We schuchted hine ueor awei hwon we dođ deadliche sunne, A. R. 312, 10, MS. C.] v. scȳan.

scyl, scylcen, scyld *a shield*. v. scill, scilcen, scild.

scyld, e: scyldu (o); *indecl. f.* I. *guilt, sin, crime, fault:*—Hē sume māndǣd gefremede đā seó scyld đā tô his heortan hwearf đā onscunode hē hî hefelíce *sceleris aliquid commiserat, quod commissum, ubi ad cor suum rediit, gravissime exhorruit*, Bd. 4, 25; S. 599, 34. Sitte sió scyld (*the killing of a slave*) on him, L. Alf. 17; Th. i. 48, 15. On eów scyld siteþ, Exon. Th. 131, 2; Gū. 449. Is Euan scyld eal for-pynded, 7, 6; Cri. 97. Hē his scylde forgyfenysse bæd *veniam reatus postulans*, Bd. 3, 22; S. 553, 33. Đa byrđenne suā nicelre scylde *tanti reatus pondera*, Past. 2, 2; Swt. 31, 14. Būtan scylde *sine culpa*, 3, 1; Swt. 33, 16: L. H. E. 12; Th. i. 32, 9. Hî būton ǣlcere scylde (*without being guilty of any crime*) wurdon fordône, Bt. 29, 2; Fox 104, 30. Æt openre scylde *flagrante delicto*, L. In. 37; Th. i. 124, 23. Hafaþ đæt môd hwylcehugu scyldo *habet animus aliquem reatum*, Bd. 1, 27; S. 496, 42. Synna, scylda *piacula*, Wrt. Voc. ii. 66, 78. Mîne scylde *delicta mea*, Ps. Th. 68, 6: Ps. Surt. 58, 13. Scylđa, Past. 32, 2; Swt. 211, 20. Scelda, Ps. C. 45. Brôđres scyldo *fratris vitia*, Mt. Kmbl. p. 15, 5. Āscyred scylda gehwylcre, deópra firena, Elen. Kmbl. 2624; El. 1313: 937; El. 470. Hwîlum biþ gôd wærlíce tô miđanne his hiéremonna scylda (*vitia*), Past. 21, 1; Swt. 151, 9. II. *a debt, due:*—Ryhtlícor cweđan đæt wē him gielden scylde đonne wē him mildheortnesse dôn *justitiae debitum potius ·olvimus, quam misericordiae opera implemus*, 45, 1; Swt. 335, 19. Āgefnæ beón đa scylde *reddi debitum*, Mt. Kmbl. Rush. 28, 25. Hē đa scyld forlēt wiđ hine *debitum dimisit ei*, 27 : 30. [Sculd *scelus*, Wrt. Voc. i. 95, 74. *O. Sax.* skuld *a crime; a due: O. L. Ger.* sculd: *O. Frs.* skelde, schuld, schild: *O. H. Ger.* sculd, sculda *causa, facinus, noxa, injuria, crimen, debitum, meritum: Icel.* skuld, skyld *a due, tax; sake.*] v. deáþ-, frum-, ge-, god-, mān-, nîd-scyld.

scyldan, scyldend. v. scildan, scildend.

scyldan, scyldian *to charge, accuse:*—Hý gān .xii. sume and gescyld-igen (gescylden, *other MS.*) hine, L. Ath. i. 11; Th. i. 206, 3.

scyld-frecu, e; *f Guilty greed:*—Mē (*Eve*) scyhte tô scyldfrece fāh wyrm þurh fægir v ord, Cd. Th. 55, 23; Gen. 898.

scyld-full; *adj. Guilty, criminal, sinful, wicked:*—Ic (*Adam*) wreó mē scyldfull, Cd. Th. 53, 30; Gen. 869. Bearn Godes on wergum folce wíf curon, scyldfulra mægđ, scȳne and fægere, 76, 4; Gen. 1252. Đonne sweart wæter swelgaþ sceađum scyldfulium *when the deluge swallows the wicked*, Gen. 1302: Elen. Kmbl. 619; El. 310.

scyldgian, scyldgung. v. scyldigian, scyldigung.

scyld-hǣta, an; *m. One who demands a due* or *debt, a bailiff:*—

Sculthēta *exactor*, Wrt. Voc. ii. 107, 70. Scyldlǽta (-hǽta?) *exactor*, i. *postulator*, 144, 54. [*O. Frs.* skeltata, skelta (der stellvertreter des grafen, v. Richthofen, pp. 1023 sqq.): *O. L. Ger.* sculd-hēto (quicunque villicus est abbatis quod nos vulgo dicimus *sculthētho*): *O. H. Ger.* scultheizo *vilicare, tribunus, procurator, exactor populi*: *Ger.* schult-heisz. Cf. *Goth.* dulga-haitja *a creditor*.]

scyld-hata, an; *m. One who hates wrongfully, an enemy*:—Scyldhatan, ealdgenīðlan, Andr. Kmbl. 2095; An. 1049. Scyldhatan, egle ondsacan, 2295; An. 1149. v. next word.

scyld-hete, es; *m. An enemy, a foe*:—Mid scyldhetum, werigum wrōhtsmiðum, Andr. Kmbl. 170; An. 85. v. preceding word.

scyldian *to commit a fault*:—Gesette God ǽ scyldiendum *legem statuit delinquentibus*, Ps. Th. 24, 7. v. scyldigian, scyldan.

scyldig; *adj.* I. *guilty, sinful, criminal*:—Scyldig *reus*, Wrt. Voc. i. 49, 1: 86, 61: *sons*, Ælfc. Gr. 9, 39; Zup. 63, 14. Gif man wāt, ðæt ōðer mān sweraþ, hē biþ scildig (*portabit iniquitatem suam*), gif hē hit forhilþ, Lev. 5, 1. Wæs gecueden tō ðæm scyldegan folce *delinquenti populo dicitur*, Past. 15, 1; Swt. 91, 2. Ðæt hē hine scyldigne ongete *reum se cognoscat*, Bd. 1, 27; S. 496, 33: Elen. Kmbl. 1380; El. 692. Hū hē ðæt scyldige werud forscrifen hefde, Cd. Th. 267, 4; Sat. 33. Scyldge men, Exon. Th. 71, 10; Cri. 1153. Scyldigra scolu, 98, 15; Cri. 1608: 132, 22; Gū. 476. Hendum scyldigra *manibus nocentium*, Rtl. 24, 11. Heó nāuht ne þreāþ ðām scildigum, Bt. 4; Fox 8, 13. Earfeþu scyldgum tō sconde, Exon. Th. 78, 14; Cri. 1274. Hié ða scyldigan þearlwīslīce dēmaþ, Blickl. Homl. 63, 20. Stræc wið ða unryhtwīsan and wið ða scyldgan, Past. 17, 5; Swt. 113, 23. I a. *guilty* of committing a crime, (1) with gen. of crime:—Se biþ ēces gyltes scyldig *reus erit aeterni delicti*, Mk. Skt. 3, 29. Morðres scyldig, Beo. Th. 3370; B. 1683. Deáðes scyldig *guilty of causing death*, L. In. 5; Th. i. 104, 13. Mansleges scyldig, Blickl. Homl. 189, 34. Morðres scyldige . . . deáþes scyldige, 65, 10–11: H. R. 107, 1. (2) with inst. of crime:—Synnum scyldig, Beo. Th. 6135; B. 3071. Dǽdum scyldige, Cd. Th. 76, 35; Gen. 1267. Lehtrum scyldige, Andr. Kmbl. 2434; An. 1218. I b. *guilty* against (*wið*) a person:—Ǽlc man ðe yfel dēþ mid yfelum willan is scyldig wið God, H. R. 105, 33: Cd. Th. 250, 20; Dan. 549. Menn wǽron deádlīce and wið heora Drihten scyldige, Hexam. 17; Norm. 24, 26: Blickl. Homl. 47, 21. II. *responsible* for, *liable* for, *chargeable* with an ill result, (1) with gen.:—Gif hwylc mæssepreóst untruman men sprǽce forwyrne, and hē on ðære tyddernesse swelte, sȳ hē on dōmes dæg ðære sāwle scyldig (*ejus animae reus*), L. Ecg. P. i. 2; Th. ii. 172, 29. Hē sceal mid rēðnesse him stiére ðȳlæs hē sié scyldig ealra hira scylda *ne culparum omnium reus ipse teneatur*, Past. 21, 5; Swt. 161, 1. Gif hwelc gōd lǽce gesihþ, ðæt his hwam þearf biþ, and ðonne for his slǽwþe āgiémeleāsaþ ðæt hē his helpe, ðonne wille wē cweðan ðæt hē sié genōg ryhtlīce his brōðor deāþes scyldig, 49, 1; Swt. 377, 21. Swā feala earmra manna swā on ðæs rīcan neáweste sweltaþ, and hē him nele syllan his teóþungsceatta dǽl, ðonne biþ hē ealra ðara manna deāþes sceldig, Blickl. Homl. 53, 7. (2) with inst.:—Gif God him ne āraþ, ðonne beóþ hié suā monegum scyldum scyldige suā hié manegra unþeáwa gestīran meahton mid hiora lārum, gif hī ongemong monnum beón wolden *ex tantis rei sunt, quantis venientes ad publicum prodesse potuerunt*, Past. 5, 3; Swt. 45, 22. III. *liable* for a debt, *bound* by an obligation:—Swā hwā swā swereþ on ðæs temples golde se ys scyldig *qui juraverit in auro templi, debet*, Mt. Kmbl. 23, 16. Suǽ uoe forgefon scyldgum ūsum (*debitoribus nostris*), Mt. Kmbl. Lind. 6, 12. Syndrigum scyldgum *singulis debitoribus*, Lk. Skt. Lind. 16, 5. IV. *liable* to forfeiture, *forfeiting* (1) with gen. of forfeit:—Gif hwā ymb cyninges feorh sierwie, sié hē his feores scyldig and ealles ðæs ðe hē āge, L. Alf. pol. 4; Th. i. 64, 1: L. Ath. v. 1, 4; Th. i. 230, 6, 12. Hē æt wīge gecrang, ealdres scyldig, Beo. Th. 2680; B. 1338: 4128; B. 2061. Feores sceldig, Ps. C. 20. Sȳ hē scyldig his sylfes and ealles ðæs ðe hē āge, L. Ath. iv. prm.; Th. i. 220, 12. Beó hē .cxx. scill. scildig wið ðone cing, L. Ath. v. 1, 5; Th. i. 230, 11: L. In. 4; Th. i. 104, 10. Beó hē wið ðone cyninge scyldig ealles ðæs ðe hē āge, Wulfst. 271, 26. Se ðe ðæt gecwēme ne dēþ, beó hē his inganges scyldig, Ch. Th. 606, 21. Sȳ hē his tungan scyldig, L. Edg. ii. 4; Th. i. 266, 25. (2) with inst.:—Ðū, ealdre scyldig, deāþe sweltest, Exon. Th. 250, 9; Jul. 124. Gebeád ðæt se wǽre aldre scyldig, se ðæs onsōce, Cd. Th. 244, 19; Dan. 450. V. *liable* to punishment, *deserving* of punishment:—Scyldig *obnoxius*, Wrt. Voc. ii. 115, 31. (1) with gen. of punishment:—Hē is deāþes scyldig *reus est mortis*, Mt. Kmbl. 26, 66: Mk. Skt. 14, 64. Ðū eart wið mē deāþes scyldig *dignus es morte*, Bd. 4, 22; S. 591, 41. (2) with dat. (?):—Se ðe ofslihþ se byþ dōme (dōmes, MS. A.) scyldig *qui occiderit, reus erit judicio*, Mt. Kmbl. 5, 21. [Þe bið al swa sculdig þe þet uuel iþeuað swa þe þe hit deð, O. E. Homl. i. 113, 2. *A. R.* schuldi: *O. Sax.* skuldig *guilty, liable* to a payment or penalty: *O. Frs.* skeldech: *O. H. Ger.* sculdig *reus, culpabilis, meritus, debitus, debitor, obnoxius*.] v. feorh-, for-, god-, hand-, mān-, morþor-, þeóf-, þurh-, twī-, un-, wam-scyldig.

scyldigian, scyldgian; *p.* ode *To sin*:—Wið ða scyldgiendan (scyldgigendan, Hatt. MSS.) *contra peccantem*, Past. 21, 1; Swt. 151, 23. v. for-, ge-scyldigian; scyldian.

scyldiglīc. v. un-scyldiglīc.

scyldigness, e; *f. Guiltiness*:—Synnignise ɫ scyldignise *reatum*, Rtl. 42, 33: 103, 17.

scyldigung, scyldgung, e; *f. A criminal charge*:—Be ðon ðe scyldgunge bǽde æt ofslegenum. Wē cwǽdon, se ðe scyldunga (be ðon ðe scyldgunga, *other* MS.) bǽde æt ofslagenum þeófe ðæt hē eode þreora sum tō . . . and ðone āþ syllen ðæt hȳ on heora mǽge nāne þȳfþe nyston . . . and hȳ gān siþþan .xii. sume and gescyldigen hine *of him who asks for the charge* (*in order to refute it*) *in the case of a slain thief. We ordained, he that should ask for the charge in the case of a slain thief, that he should go with two others . . . and they shall make oath that they knew of no theft on the part of their kinsman . . . and afterwards twelve on the other side shall go and bring the charge against him* (*the thief*) (cf. Qui culpam exigit de fure occiso, L. H. i. 74, 2; Th. i. 578), L. Ath. i. 11; Th. i. 204, 26.

scyld-lǽta, scyldung. v. scyld-hǽta, scyldigung.

scyld-leás; *adj. Guiltless*:—Scyldlǽs *insons*, Lchdm. i. lxiii, 2.

scyld-wreccende *punishing guilt*:—Hell scyldwreccende, Exon. Th. 71, 25; Cri. 1161.

scyld-wyrcende *committing sin* or *guilt*:—Ðū (*the soul*) ðone līchoman scyldwyrcende gewemdest, Exon. Th. 91, 4; Cri. 1487. Ðonne ðeós woruld scyldwyrcende byrneþ, 232, 5; Ph. 502: 269, 4; Jul. 445. Ðū womfulle, scyldwyrcende sceaþan āwurpe, Elen. Kmbl. 1520; El. 762.

scyl-ēgede. v. sceolh-eágede.

scylf, scylp, es; *m.* I. *a peak, crag, tor* (in local names):—Ðonon ofer ealne ðone hǽþfeld tō Hnæfes scylfe, Cod. Dip. Kmbl. iii. 130, 37. Tō byrnan scylfe, 38, 36. Sticule scylpas *scabri murices*, Germ. 399, 446. Scylfa *scopulorum*, Hpt. Gl. 421, 43. II. *a turret, tower, pinnacle*:—Se deófol gesette hine uppan ðam scylfe ðæs heágan temples, Homl. Th. i. 166, 17: 170, 1. Wē biddaþ ðæt ðū āstīge tō ðam sticelan scylfe . . . Hwæt ða bōceras hine gebrohton tō ðæs temples scylfe, ii. 300, 1–3. Hē hine āsette ofer ðæs temples scylf, Blickl. Homl. 27, 11. Scylfas *maciones*, Wrt. Voc. ii. 59, 29: *pinnas*, Blickl. Gl. Ða torras and ða scylfas on him bǽron ða elpendas *elephanti superpositas turres gestaverunt*, Nar. 4, 16. [Cf. (?) *O. H. Ger.* sculpa *gleba*.] v. stān-scylf; scylfig.

scylfe. v. scilfe.

scylfig, scylpig; *adj. Craggy, rocky*:—Scylpige *scopulosas*, Hpt. Gl. 529, 29. v. scylf.

Scylfingas, scylfor, scylfring, scylga, scylian, scylig, scyll, scylp, scylpig, scymrian. v. Skilfingas, scilfor, scilfrung, scealga, scilian, scilig, scill, scylf, scylfig, scimrian.

scyltumend (? fultumend), es; *m. A helper*:—Drihten is mīn scyltumend and mīn gescyldend *Dominus adjutor meus et protector meus*, Ps. Th. 27, 8.

scyndan, scendan; *p.* de. I. *intrans. To hurry, hasten*:—Swā ðeós woruld fareþ, scūrum scyndeþ, Exon. Th. 469, 24; Hy. 11, 7. Brimwudu scynde, 182, 5; Gū. 1305. Scynde Gregorius in Godes wære, Menol. Fox 77; Men. 38. Scynde beaduþreáta mǽst tō hilde, Elen. Kmbl. 60; El. 30. Fǽge scyndan (*of death by violence*), Exon. Th. 271, 29; Jul. 489. Hī ǽghwonon tō him ēfston and scyndon, Guthl. 15; Gdwin. 66, 10: Bd. 4, 27; S. 604, 8. Ðā ongunnan monige ēfstan and scyndan tō gehȳranne Godes word *coepere plures ad audiendum verbum confluere*, 1, 26; S. 488, 11: Guthl. 2; Gdwin. 14, 25. Hē gewāt scrīðan, tō gesceape scyndan, Beo. Th. 5133; B. 2570. Manna freóndscipe biþ swīðe hwīlwendlīc and swīðe scendende (cf. gnornscendende), Blickl. Homl. 195, 26. II. *trans.* (1) *To cause to hasten, to hurry*:—Ðā wæs morgenleóht scofen and scynded, Beo. Th. 1840; B. 918. (2) *to urge, incite, exhort*:—Se feónd his (*Judas*) heortan tō ðan lǽrde and scynde, ðæt hē Drihten tō deáðe belǽwde, Homl. As. 153, 55. Hū mon monige scyndan scyle (*de exhortatione multis exhibenda*) tō ðæm ðætte his gōdan dǽda ne weorðen tō yflum dǽdum, Past. 60; Swt. 453, 6. [*O. Sax.* far-skundian *to incite, egg on*: *O. H. Ger.* scuntan *sollicitare, suggerere, urgere*: *Icel.* skynda *to hasten*: *Dan.* skynde.] v. ā-, ge-scyndan.

-scynde. v. un-scende.

scyndel?:—Tīwesdæges nama wæs of Martie, Iovis sunu ðæs scyndles (cf. (?) scyndan, II. (1); scyndel *one who causes swift movement*, referring to the lightning (?). Or cf. (?) scendan, scendele; scyndel, scendel *a shameful person* (?)), Anglia viii. 321, 16.

scyndendlīce; *adv. Hurriedly, hastily*; consummatim, Wrt. Voc. ii. 18, 41: 82, 75.

scȳne, scȳ-nes, scyp. v. scīne, sceó-ness, scip.

scypen, e; *f. A shippen* [in some northern dialects; also pronounced *shup'm* (Cumberland)], *a cow-house, stall*:—Scypen *bovile*, Wrt. Voc. ii. 12, 72. Scipen, 126, 59: *bostar* vel *boviale*, i. 58, 25. Scepen, steal, *vel* fald *bovile, stabulum*, 15, 23. Ða þing tō begānne ðe tō scipene belimpaþ, Anglia ix. 260, 4. Ūt wæs gongende tō neáta scypene (*ad*

stabula jumentorum), Bd. 4, 24; S. 597, 9. Nǽnig mann scypene his neátum ne timbreþ, I, I; S. 474, 32. Andlang dîces on dæs cinges scypena; of dan scypenum on dæt riscbed, Cod. Dip. Kmbl. vi. 62, 27. Scipena behweorfan, Anglia ix. 261, 18. [Schepyn *boscar* (*-tar?*), Wrt. Voc. i. 178, 10. Schyppune *boster*, 204. col. 2. The schepne brennyng with the blake smoke, Chauc. Kn. T. 1142.] Cf. scoppa.

scyppan, scyppend, scŷr, scyran, scŷran, scyrdan. v. scippan, scippend, scîr, sceran, scîran, scirdan.

scyrfe-mûs, e; *f. A shrew-mouse:*—Scirfemûs *sorex*, Wrt. Voc. ii. 71, 27. Cf. sceorfan.

scyrft *a scraping* (?); scansio, Wrt. Voc. ii. 119, 78. Cf. sceorfan.

scyrian, scyriendlîc. v. scirian, sciriendlîc.

scŷr-mǽlum; *adv. Stormily:*—Seó orsorhnes gǽþ scŷrmǽlum swâ dæs windes þys *prosperam fortunam videas ventosam*, Bt. 20: Fox 72, 4. v. scûr.

scyrpan, scyr-seax. v. scirpan, scear-seax.

scyrtan; *p.* te *To make short, to shorten:*—Gif God his hwîle ne scyrte (gescyrte, MSS. B. C.), Wulfst. 19, 9. v. ge-scyrtan; scortian.

scyrte (-a; *m.?*), an; *f. A short garment, skirt, kirtle:*—Scyrtan *pretexta*, tunecan *togae*, Germ. 393, 143. [He ches stiue here to shurte and gret sac to curtle, O. E. Homl. ii. 139, 16. Arður warp an his rugge a ræf swiðe deore, ænne cheisil scurte & ænne pallene curtel, Laym. 23761. He broucte bred in his shirte or in his couel, Havel. 768. He yaf ofte his kertel and his sserte to þe poure, Ayenb. 191, 9. *M. H. Ger.* schurz: *Ger.* schurz *an apron*: *Icel.* skyrta *a kind of kirtle*.]

scyrting, e; *f. A shortening, an abridgement:*—Gif hwilc gelǽred man ðâs race (*the homily on Job*) oferrǽde, ðonne bidde ic ðæt hê ðâs scyrtinge ne tǽle, Homl. Th. ii. 460, 6.

scyrtra, scyrtest, scyru, scŷtan. v. scort, scearu, scîtan.

scyte, es; *m.* I. *shooting:*—Hié fortendun ðæt swîðre breóst foran ðæt hit weaxan ne sceolde ðæt hié hæfden ðŷ strengran scyte (*ne sagittarum jactus inpedirentur*), Ors. I, 10; Swt. 46, 13. Dryhten dǽleþ sumum wyrp odðe scyte, Exon. Th. 331, 17; Vy. 69. II. *a shot, blow:*—Scytum *ictibus*, Hpt. Gl. 478, 76. III. *what is shot or thrown, a javelin, dart:*—Scytas *iacula*, Lchdm. i. lxix, 9. [He þene scute biberh, Laym. 1461. Mid scute of eien, A. R. 60, 16. Wið þe schute wite heo hire, 62, 1. An carpenter that sset the ssute, R. Glouc. 537, 4. *O. Frs.* sket: *O. H. Ger.* scuz *jactus*.] v. on-, under-, ût-, wæter-scyte.

scyte. v. scîte.

scyte-finger, es; *m. The forefinger;* digitus secundus quo sagittatur:—Scytefinger *index* vel *salutaris*, Wrt. Voc. i. 44. 5. Bêcnend, scytefinger *index*, ii. 46, 35. Gif se scytefinger biþ ofâslegen, sió bôt biþ .xv. scill., L. Alf. pol. 57; Th. i. 96, 1. *In Ethelbert's Laws the fine is only eight shillings*, L. Eth. 54; Th. i. 16, 10. Scytefingres, Anglia viii. 326, 28. Euenmicel swâ ðu mǽge mid ðînan scitefingre tô ðînum þuman befôn, Lchdm. iii. 6, 21. Mid scetefingre ðu gebêcnest *indice prodis*, Hymn. Surt. 104, 5. [Cf. *O. Frs.* skot-finger.] v. scytel-finger.

scyte-heald, -healden; *adj.* I. *bent so as to shoot downwards* (cf. scyte-rǽs), *sloping steeply:*—Scyteheald *preceps*, Wrt. Voc. ii. 68, 77. II. *oblique, inclined:*—Scytehald *obliquum*, 115, 13. Sió scytehealde onbêgnes *obliqua curvatura*, 64, 24. Sió scytehealde *obliqua*, 79, 1. Scytehealden, 62, 61.

scytel *dung.* v. scitel.

scytel *a dart.* v. scutel: *a bolt,* v. scyttel.

scytel-finger *the arrow-finger, the forefinger:*—Scytelfinger (scyte-? v. scyte-finger) *index*, Wrt. Voc. i. 71, 31.

scytels. v. scyttels.

scyte-rǽs, es; *m. A headlong rush:*—On scyterǽs odðe on fǽrfyll, unforesceáwadlîc *in preceps*, Wrt. Voc. ii. 47, 43.

scytere, es; *m.* I. *a shooter, an archer.* v. scyteres (sciteres) *clif, flôde, streám,* Cod. Dip. Kmbl. vi. 330. [*Icel.* skytari *a shooter.*] II. *one that moves swiftly* (?):—Ad rivulum qui scitere dicitur, Cod. Dip. Kmbl. v. 102, 29. Cf. scytta.

scytling. v. ût-scytling.

scytta, an; *m. A shooter, an archer:*—Sagittarius ðæt is scytta, Lchdm. iii. 246, 2. Strǽlbora and scytta *arcister*, Wrt. Voc. ii. 7, 32. Ðâ gebende ân scytta his bogan, Homl. Skt. i. 18, 219. On scyttan fǽn, Cod. Dip. Kmbl. iii. 132, 22. On scyttan mere; ðæt on scyttan dûne, 381, 11. Wulfsiges môdor scyttan, vi. 212, 5. Ðâ gegaderade Regulus ealle ðâ scyttan ðe on ðǽm fǽrelte wǽron, ðæt hié (*the serpent*) mon mid flânum ofercôme, Ors. 4, 6; Swt. 174, 5. Wǽron on his fyrdinge twelf þûsenda scyttena, Homl. As. 104, 55. ' [Alle þe scutten, Laym. 27046, *O. H. Ger.* scuzzo *sagittarius: Icel.* skyti *one who shoots or hurls.*]

scyttan; *p.* te. I. *to cause rapid movement, to shoot* a bolt, *to shut:*—Ic scytte sum loc *sero*, Ælfc. Gr. 37; Zup. 220, 2. II. *to discharge* a debt:—Ic wille ðæt man selle ðæt land et Fersafeld ... and recna man iungere Brûn ân marc gol and mid ðan lâue scytte man mîna

borgas (*my loans shall be paid off*), Chart. Th. 568, 19. [Schutted þet þurl to, A. R. 96, 10. *Ayenb.* ssette: *Piers P.* shutte, shette: *Wick.* schitte: *O. Frs.* sketta *to stop, close.*] v. for-scyttan.

-scytte. v. riht-scytte.

scyttel, scytel, es; *m. A bar, bolt:*—Ealle ða îsenan scyttelas helle loca wurdan tôbrocene, Blickl. Homl. 87, 5: 8ӡ, 7: Nicod. 27; Thw. 15, 24. Scyttelas *vectes*, Ps. Spl. 106, 16. Scetelas, Kent. Gl. 658. [A gardin besset myd tuo ssetteles, Ayenb. 94, 30. Schyttyl *pessulum* vel *pessellum*, Prompt. Parv. 447: ondoynge of schettellys *apercio*, 365.] v. scutel *and next word.*

scyttels, scytels, es; *m. A bar, bolt:*—Ða scytelses (scittelsas, MS. O.) tôburston, Homl. Skt. i. 3, 348. Openiaþ ðâs gatu and ða fæstan scytelsas, Wulfst. 230, 31. Scytelsas *seras*, Ps. Spl. 147, 2. Scettelsas, Hymn. Surt. 122, 28. Scyttylsum *vectibus*, Germ. 399, 349. [Þet (*the cross*) is þet scutles þe ðe deofol ne mei nefre tocysan, O. E. Homl. i. 127, 35.] v. fore-scyttels, *and preceding word.*

Scyttisc; *adj. Scottish, Scotch* (v. Scottas):—Ðær læg secg mænig ... guma norþerna ... swilce Scittisc eác, Chr. 937; Erl. 112,' 19. Scyttisc gecost gealdor wið ǽlcum ǽttre, Lchdm. ii. 10, 23. Scyttysces cynnes *natione Scottus*, Bd. 3, 21; S. 551, 16. Gif hê hæfþ Scyttisc weax, Lchdm. ii. 114, 11: iii. 46, 17. Scittisc, ii. 156, 26. ¶ Of speech:—Sind on ðis îglande fîf geþeóde ... Scyttisc, Chr. Erl. 3, 3. Se cyning Scyttysc (*linguam Scottorum*) geleornad hæfde, Bd. 3, 3; S. 525, 39. On ðam mynstre ðe on Scyttisc is nemned Rathmelsigi, 3, 27; S. 558, 35.

se, sió, Lchdm. ii. 260, 1; *m.:* seó, ðeó, Blickl. Homl. 65, 13; se, Lchdm. ii. 228, 8; *f.:* ðæt; *n.* I. a demonstrative adjective, *the, that.* (I) marking an object as before-mentioned or already well-known (a) with substantives:—Se Hǽlend, Mt. Kmbl. 3, 13. Se steorra stód ofer ðǽr ðæt cild wæs, 2, 9. Wæs se engel sprecende tô ûres Drihtnes mêder, Blickl. Homl. 5, 2. Seó heofon biþ gefeallen æt ðæm feówer endum middangeardes, 93, 4. Seó eorþe, Lchdm. iii. 254, 15. Seó sǽ and se môna geþwǽrlǽcaþ him betweónan, 268, 12. Seó lyft *the air*, 272, 20. On ðone gemânan ðæs brýdguman and ðære brýde, Blickl. Homl. 11, 5. Hê fægnode ðæs miclan weorces ðærre ceastre, Past. 4; Swt. 39, 15. Ðæt mon ða earce bere on ðæm saglum, 22; Swt. 171, 12. Mid ðŷ selflîce se Dêma biþ geniéded tô ðæm ierre, 4; Swt. 39, 10. Ðŷ þearlan dôme (*by the severe sentence just mentioned*) hê forleás his mennisce, Swt. 39, 23. Ðæt mǽsten is gemǽne tô ðâm (*those mentioned in the charter*) ân and twentigum hîdum, Cod. Dip. Kmbl. v. 319, 29. On hâte ða ahsan, Lchdm. ii. 32, 13. (b) with adjectives:—Se dumba spræc, Mt. Kmbl. 9, 33. Ðâ æthrân hê ðæs blindan hand, Mk. Skt. 8, 23. (c) with numerals:—Ða þrý cômon, Cd. Th. 221, 24; Dan. 93. Ðîna âgna treówa and seó godcunde lufu and se tôhopa, ða þreó ðê ne lǽtaþ geortrêwan be ðam êcan lîfe, Blk. 10; Fox 32, 8. (d) with proper names:—Se Iohannes *the same John* (A.V.), Mt Kmbl. 3, 4. Se (*the one in question*) Cynewulf oft feaht wið Bretwalum ... Hê wolde âdrǽfan ænne æþeling, se wæs Cyneheard hâten, and se Cyneheard wæs ðæs (*the one previously mentioned*) Sigebryhtes brôður, Chr. 755; Erl. 48, 24–28. Fêng Carl tô ðam westrîce ... se Carl wæs Hlôþwîges sunu, se Hlôþwîg wæs Carles brôður, se wæs Iuþyttan fæder, ... and hié wǽron Hlôþwîges suna. Se Hlôþwîg wæs ðæs aldan Carles sunu; se Carl wæs Pippenes sunu, 885; Erl. 84, 10–17. Seó Asia (*Asia Minor*), Ors. 1, 1; Swt. 12, 11. Him Iosep gehealp. From ðæm Iosepe ..., 1, 5; Swt. 32, 28. (2) marking an object which is further described (a) by an adjective:—Se heofonlîca cyning, Blickl. Homl. 5, 18. Mîn se heofonlîca Fæder, Mt. Kmbl. 18, 35. Se earma upâhafena, Past. 26; Swt. 183, 13. Se dysega ungeþyldega, 33; Swt. 220, 9. Ðeó deáþberende uncyst, Blickl. Homl. 65, 13. Mid hire ðære yfelan sceónesse beswâc ðone ǽrestan wîfmon, 5, 1. Ðone yfelan fæsðrǽdan willan fulneáh nân wind ne mæg âwecgan, Past. 33; Swt. 225, 6. ¶ The weak declension usually occurs with the demonstrative, but in the following instances strong forms are found:—On ðam seócum men, Lchdm. ii. 282, 11. Snâw cymþ of ðam þynnun wǽtan, iii. 278, 23. Of ðam hâtum bæðe, Homl. Th. i. 58, 29. Ða gleáwe sǽgenga[n] hig understandaþ, Anglia viii. 327, 21. Ða anbestungne saglas, Past. 22; Swt. 171, 11. For ðære sceáwunge ðara ungesewenlîcra þinga, 16; Swt. 99, 8. Ðara eádigra apostola, Bd. 5, 19; S. 637, 31. Orhlyte ðæra hâligra mihta, Homl. Th. i. 346, 26. (b) by a pronoun:—Mon sceal suâ manian ðæt se hiera folgoþ hîe ne ôþhebbe, Past. 28, 1; Swt. 189, 17. Ða mîne sǽlþa and se mîn weorðscipe, Bt. 10; Fox 30, 14–15. Ǽnigne dǽl ðara ðînra gesǽlþa, 11, 1; Fox 32, 26. (c) by a numeral:—Ðæt þridde gebed, Homl. i. 264, 16. Hyt eall âlêd biþ on ðære ânre mîle, Ors. 1, 1; Swt. 20, 32. Ðâ âxode se câsere ðone ænne preóst, Homl. Th. ii. 310, 14. Ðis synt ðæra twelf Apostola naman, Mt. Kmbl. 10, 2. Hû mon scule blôdlǽse on ðara six fîfa ǽlcum on mônþe forgân, Lchdm. ii. 146, 19: 148, 2. (d) by a genitive:—Ðâ wæs gesended ðæt goldhord ðæs mægenþrymmes on ðone bend ðæs clǽnan innoþes, Blickl. Homl. 9, 28. Se emnihtes dæg, Lchdm. iii. 256, 26. Nêh ðæm clife ðære Reádan Sǽs, Ors. 1, 1; Swt. 12, 19. Ða diógolnesse ðæs þriddan hefones, Past. 16; Swt. 99, 8. (e) by a phrase:—Ðara twentiges hîda landgemǽra tô

Burhtûne. . . . Đara .vii. hîda landgemǣra æt mǣđelgâres byrig, Cod. Dip. Kmbl. iii. 429, 25–32. (f) by an appositive :—Saul se cyning, Past. 3 ; Swt. 35, 14. Membrađ se ent . . . Ninus se cyning, Ors. 2, 4 ; Swt. 74, 9–10. Đæt land Cilia . . . seó sǣ Euxinus . . . se hêhsta beorg Olimpus . . . Nilus seó eá ; . . . nêh đam beorge Athlans, 1, 1 ; Swt. 12, 11–21. (g) by a clause, v. IV :—Eart đû se Beowulf, se đe wiđ Brecan wunne, Beo. Th. 1016 ; B. 506. Seó Ægyptus đe ûs neár is, Ors. 1, 1 ; Swt. 12, 16. Seó menigo đe beforan fêrde, Blickl. Homl. 71, 9. Sý đæs cynnes orf đe hit sý, L. Ff. ; Th. i. 226, 3. Gif esne eorlcundne mannan ofslæhþ đane đe sió (whoever it be), L. H. E. 1 ; Th. i. 26, 8 : 3 ; Th. i. 28, 4. Đa hwîle đe hié tô nânre ôđerre note ne mǣgen, Past. pref. ; Swt. 7, 12. Oft mon forlǣt đone ege and đa fæstrǣdnesse đe hê mid ryhte on him innan habban scolde, 4, 1 ; Swt. 37, 17. Đa twelfe đe mid him wǣron, Mk. Skt. 4, 10. Đa fîf hlâfas đe se cnapa bær getâcniaþ đa fîf bêc đe Moyses sette, Homl. Th. i. 186, 13. Hwæt đa sume dreógaþ, đe đa wrǣclâstas wîdost lecgaþ, Exon. Th. 309, 13 ; Seef. 56. (h) by a clause in apposition :—Ne sceal hê đæt ân dôn, đæt hê âna wacie, Past. 28 ; Swt. 193, 21. (i) by relation to other objects mentioned :—Se đe ne gǣþ æt đam gete intô sceápa falde, Jn. Skt. 10, 1. Hig gefyldon đa[fatu] ôþ đone brerde, 2, 7. Irnende on đæt sond, and đonne besince eft on đæt sand, Ors. 1, 1 ; Swt. 12, 22. (3) with adjectives used as epithets :—Salomonn se snottra, Past. 4, 1 ; Swt. 37, 16. Hit is Hǣlend se Nazarenisca, Blickl. Homl. 15, 19. Sidroc eorl se alda and Sidroc eorl se gioncga, Chr. 871 ; Erl. 74, 22. Eádweard se langa, Byrht. Th. 139, 53 ; By. 273. (4) marking an object as the representative of a class :—Ys seó æx tô đæra treówa wyrtruman âsett, Mt. Kmbl. 3, 10. Hû nys seó sâwl sêlre đonne mete, 6, 25. Đa lîchamlîcan gôd bióþ forcúþran đonne đære sâwle cræftas . . . Seó fægernes đæs lîchoman geblissaþ đone mon, Bt. 24, 3 ; Fox 84, 5–8. Ǣr đan đe đæs dæges lenge oferstîge đa niht, Lchdm. iii. 256, 13. Bere is swîđe earfoþe tô gearcigenne, and đeáhhwæđere fêt đone mann, đonne hê gearo biþ, Homl. Th. i. 188, 5. (4 a) marking genus :—Se mon homo, Bd. 1, 27 ; S. 497, 40. Se mann âna gǣþ uprihte, Bt. 41, 6 ; Fox 254, 29. (5) marking a definite whole or a class of objects :—Hié hâtaþ đa landmen (the natives) Nuchul, Ors. 1, 1 ; Swt. 12, 24. On ôđre wîsan mon sceal manian đa blîđan, on ôđre đa unrôtan . . . đa underþiéddan . . . đa ofer ôđre gesettan . . . đa woroldwîsan . . . đa dysegan, Past. 23 ; Swt. 175, 14–17. Đæra Persiscra cyning . . . gegaderode of đâm Saraceniscum micele fyrde, Jud. Thw. 162, 23. Hê clypode đa geladodan tô đam gyftum. . . . Hê sǣde đâm geladedon, Mt. Kmbl. 22, 3–4. (6) with abstract nouns where modern English would not use the article :—Sió hǣlu đone mon gedêþ lûstbǣrne, Bt. 24, 3 ; Fox 84, 9. Gif se weorđscipe and se anweald gôd wǣre, 16, 3 ; Fox 54, 8. Hê đa geþyld đe is môdur ealra mægena for đǣm unwrence đære ungeþylde forlǣt, Past. 33, 1 ; Swt. 215, 20. Þurh đa wilnunga đære woroldâre, 3 ; Swt. 33, 8. On đǣm gesundfulnessum đæt môd wierđ upâhafen ; and on đǣm earfeđum hit biþ geeáđmêdd. On đære gesundfulnesse mon forgiett his selfes ; on đǣm gesuincum hê sceal hine selfne geþencean. On đære orsorgnesse . . . on đǣm earfođum. . . . Suîđe oft monn biþ đære earfođnesse lâreówđôme underþiéded, Swt. 35, 4–10. (6 a) where an abstraction is personified :—Se Wisdóm and seó Gesceádwîsnes, Bt. 3 ; Fox 6, 13 (and often). II. as a demonstrative pronoun, he, she, it, that, (1) referring to a person or thing :—Se wæs betera đonne ic, Beo. Th. 943 ; B. 469. On đâm ys sǣd, and đæt sweart, Lchdm. i. 278, 1. Heó hafaþ leáf sinewealte and đa bitere, 290, 18. Đonne hî eów êhtaþ on đysse byrig, fleóþ on ôđre ; and đonne hî on đære eów êhtaþ, fleóþ on đa þryddan, Mt. Kmbl. 10, 23. Hê se câsere hine lǣdan tô his deófolgelde, đæt hê đǣm gulde, Shrn. 88, 22. Heó hafaþ stelan and đone on bôgum geþûsne, Lchdm. i. 298, 20. Đa swungon hî đæne, Mk. Skt. 12, 3. Hê sorgaþ ymb đa and biþ đara suîđe gemyndig, Past. 4, 1 ; Swt. 37, 19. Đa wǣron ealle đa wîf geladede ; đara wæs iii hund and hundeahtatig, Ors. 3, 6 ; Swt. 108, 32. Đæt hê nânes þinges bûton đǣm þurfe, Bt. 24, 4 ; Fox 86, 6. Gesyllan .xv. leaxas and đa góde, Cod. Dip. Kmbl. iii. 296, 1. (2) referring to the subject dealt with in a clause that, it :—Gif gê gesâwen hwelce mûs đæt wǣre hlâford ofer ôđre mýs . . . hû wunderlîc wolde eów đæt þincan, hwelce cehhettunge gê woldon đæs habban, Bt. 16, 2 ; Fox 52, 1–4. Hwylc đæs cyninges geleáfa wǣre, đæt æfter his deáþe mid wundrum wæs gecýþed, Bd. 3, 19 ; S. 533, 15. Đæt hê đæs (for praying in a certain place) hæfde mêde wiđ God, Shrn. 88, 32. Hê đæs (for beheading a saint) dyde hreówsunga, 89, 18 : Ps. Th. 28, 7 : 30, 1. Ic hit scortlîce secgan scyle, hwâ đæs (the stirring up of strife) ordfruman wǣron, Ors. 5, 9 ; Swt. 232, 18. Hû his gesceafta weaxaþ and eft waniaþ, đonne đæs tîma cymþ, Bt. 34, 10 ; Fox 150, 13. Heora æfterfyligendas wǣron deófolgylde folgiende, for đam Mellitus and Iustus of Breotene gewiton, Bd. 2, 5 ; S. 506, 3. Nis hit lang (feor) tô đon, 4, 24 ; S. 599, 5. Gif eárun sýn innan sâre, and đǣr wyrms sý, on dô đa ylcan sealfe, heó ys swýđe gód tô đam, Lchdm. i. 358, 17. Se hearpere suîđe ungelîce đa strengas styreþ, and mid đý gedêþ đæt . . . , Past. 23 ; Swt. 175, 7. (2 a) in apposition with a clause : see also V :—Wê nyston đæt hê đæs girnan wolde, đæt wê ûrne brôđur đyder lǣddon, Gen. 43, 7. Hié wǣron gebrocede mid

ceápes cwilde, ealles swîđost mid đæm, đæt manige đara sêlestena cynges þêna forþferdon, Chr. 897 ; Erl. 94, 32. Đa næs long tô đon, đæt wê tô sumre eá cwôman, Nar. 8, 19. Næs lang tô đý đæt his brôđor đyses lǣnan lîfes tîman geendode, Lchdm. iii. 434, 25. Se scamfæsta hæfþ genôh on đæm tô his bettrunge, đæt his lâreów hine suîđe lythwôn gemyndgige his unþeáwa, Past. 31 ; Swt. 207, 3. Wê leornedon æt him đæt wê flugen đa ôliccunga đisses middangeardes, and eác đæt, đæt wê his ege ûs ne ondrêden, 3 ; Swt. 33, 23. (3) đæt referring to an object of any gender or number :—Đæt (se ǣwelm ealra góda) eart đû, Bt. 33, 4 ; Fox 132, 30. Hê đæt is, se đa gebundenan ût âlǣdde, St. And. 14, 43. Đæt is mid Estum þeáw, đæt . . . , Ors. 1, 1 ; Swt. 21, 11. Đæt is Iohannes gewitnes, Jn. Skt. 1, 19. Godes bearn, đæt wǣron góde men, Gen. 6, 2. Đa eágan, đæt beóþ đa lâreówas, and se hrycg, đæt sint đa hiéremenn, Past. 1 ; Swt. 29, 12 : Nar. 34, 2, 7. Đæt wǣron eall Finnas, Ors. 1, 1 ; Swt. 17, 26. Đæt wǣron fiéftiéne hund þûsend monna, 3, 9 ; Swt. 128, 22. (3 a) đæt is = there is :—Đæt nis nân man đætte sumes eácan ne þurfe, Bt. 24, 4 ; Fox 86, 6. (4) one in contrast with another :—For hwî se góda lǣce selle đam hâlum men sêftne drenc, and ôđrum hâlum strangne, Bt. 39, 9 ; Fox 226, 10. Đonne lufaþ sum đæt sum elles hwæt one man likes one thing, another something else, Bt. 33, 2 ; Fox 122, 24. III. as a relative :—Sum hîredes ealdor wæs, se (qui) plantode wîngerd, Mt. Kmbl. 21, 33. Nys nân þing dýhle, đæt ne wurđe geswutelod, 10, 26. Đonne tôdǣlaþ hî his feoh, đæt tô lâfe biþ, Ors. 1, 1 ; Swt. 20, 28. Gif đû sý his discipul, se is cweden Crist, St. And. 8, 13. Ondrǣd đê Drihten and his rôdtâcn, beforan đǣm forhtigaþ heofon and eorþe, 20, 25. Đæt ic eów secge on þýstrum, secgaþ hyt on leóhte, Mt. Kmbl. 10, 27. Manige synt on đisse ceastre, đa sculon geleófan on mînne naman, St. And. 12, 7. III a. where relative and antecedent are included in the same word :—Môste on êcnisse æfter lybban se đæs wæstmes onbât, Cd. Th. 30, 21 ; Gen. 470 : 63, 8 ; Gen. 1029. Đæt gê on eare gehýraþ bodiaþ uppan hrôfum, Mt. Kmbl. 10, 27. Đonne đû hæfdest đæt đû noldest, ođđe næfdest đæt đû woldest, Bt. 26, 1 ; Fox 90, 31. Đæt hî tôweorpen đæt God geteohhad hæfþ tô wyrcanne, Ps. Th. 10, 3. Cum and geseoh đæt hié mê dôþ, St. And. 16, 34. ¶ where the construction is incomplete :—Eác sculon wiotan đa ofergesettan đæt đæt hié unâliéfedes þurhteóþ, swǣ manigra wîta hié beóþ wyrđe, swǣ swǣ hié manna on wôn gebrohten, Past. 28 ; Swt. 190, 6. IV. (see also I. 2 g) in correlative sentences where antecedent and relative are represented (1) by se . . . đe :—Gif him gebyrige đæt hê on đæs hwæt befoo đe wiđ his willan sié, Past. 28 ; Swt. 198, 23. Ne þearf hê nânes þinges bûton đæs đe hê on him selfum hæfþ, Bt. 24, 4 ; Fox 86, 8. Đa gife ic wylle tô đon đon đe ic heóld I will put the gift to the use for which I kept it, Guthl. 20 ; Gdwin. 84, 12. (1 a) by se . . . đe hê :—Forđon mæg gehycgan se đe his heorte deáh, Cd. Th. 282, 8 ; Sat. 283. Đæs bihofaþ se đe his hâlig gǣst wîsaþ, Exon. Th. 123, 34 ; Gû. 332. (2) by se . . . se :—Se ilca se monegum yfelum geárode, Past. 3 ; Swt. 35, 24. Đonne cymeþ se man se đæt swiftoste hors hafaþ, Ors. 1, 1 ; Swt. 20, 36. Se þurhwunaþ ôþ ende, se byþ hâl, Mt. Kmbl. 10, 22. Gif đæt wæs đæt seldon gelomp, Bd. 3, 5 ; S. 527, 2. Beó đæt þinga, đæt hit beó, đæt se man tô note wyrcean wille, Btwk. 222, 8. Hê for Godes ege dêþ đæt đæt hê dêþ, Past. 22 ; Swt. 169, 4. Herigan đæt đæt hê fæstrǣdes wiste, 32, 2 ; Swt. 213, 7. (2 a) by se . . . se hê :—Đæt is se Abraham, se him (= đe him to whom) engla God naman âsceóp, Cd. Th. 201, 30 ; Exod. 380. (3) by se . . . se đe :—Đys ys se be đam đe gecweden ys, Mt. Kmbl. 3, 3. Se đe brýde hæfþ, se is brýdguma, Jn. Skt. 3, 29. Seó ilce burg, seó đe mǣst wæs, seó is nú lǣst, Ors. 2, 4 ; Swt. 74, 22. Đæt mon ne wæs, se đe him ætýwde, Bd. 2, 12 ; S. 514, 25. Đæt đe âcenned is of flǣsce, đæt is flǣsc ; and đæt đe of gâste âcenned is, đæt is gâst, Jn. Skt. 3, 6. Đa, đa đe hi, qui, Rtl. 5, 33. Đæs monnes nama wæs, se đe hî beheáfdade, Dorotheos, Shrn. 89, 17. Đa eallreordan þeóde, đara đe hî đa gereorde ne cûþan, gesêcan, Bd. 1, 23 ; S. 485, 33. Đætte tǣlwyrđes sié, đæt hié đæt tǣlen, Past. 28 ; Swt. 195, 24. Hî nâmon him wîf of eallum đâm, đa đe hig gecuron, Gen. 6, 2. (3 a) irregular constructions :—Se, seđe ǣr worolde rîcsode on hefenum, hit is âwriten, Iudeas woldon hine dôn tô cyninge, Past. 3 ; Swt. 33, 12. Se ilca, seđe wênde đæt hê wǣre ofer ealle ôđere menn, him gebyrede . . . , 4 ; Swt. 39, 24. Se hondwyrm, se đe secgas seaxe delfaþ, Exon. Th. 427, 24 ; Rä. 41, 96. Se biþ leófast, se đe hym God syleþ gumena rîce tô gehealdenne, 326, 21 ; Víd. 132. Cf. Hê weorđeþ eádig se đe hine God geceóseþ beatus quem elegisti, Ps. Th. 64, 4. (4) by se . . . se se :—Swâ þyncþ him, đæt hê him niédscylde sceolde, se hê him sealde, Past. 9 ; Swt. 57, 6. Đæt đæt lator biþ, đæt hæfþ angin, Homl. Th. i. 284, 7. Ǣlc mon tiohhaþ him đæt tô sêlestum goode đæt đæt hê swîđost lufaþ, Bt. 33, 2 ; Fox 122, 23. (4 a) irregular :—Se Drihten, se đæs (= đe his) setl ys on heofenum, Ps. Th. 10, 4. V. in adverbial or conjunctional forms. In phrases such as for đam đe the pronominal element was represented later by that, as in Shakspere, and is now usually omitted altogether. (1) Nô (nalæs, nallas nó) đæt ân đæt . . . ac not only . . . but also, Bt. 21 ; Fox 74, 17 : 22, 1 ; Fox 76, 13 : 37, 3 ; Fox 190, 18 : Guthl. 5 ; Gdwin. 30, 23. (2) Đæs (a) in reference to time, or sequence of events, marking the point

from which measurement is made, *after* :—Sume men secgen ðæt hire ǽwielme sié on westende Affrica, and ðonne folraðe ðæs (*very soon after*) sié eást irnende on ðæt sond, Ors. I, I ; Swt. 12, 21. Fulraðe ðæs ic clipode tó him, Bt. 22, I ; Fox 76, 8. Ðæs on morgen *the next morning*, Ors. 3, 4 ; Swt. 104, 5. Ðæs on ðæm æfterran geáre *anno ab hoc proximo*, 4, 6 ; Swt. 172, 17. Ðæs ymb iii geár *tertio anno*, Swt. 176, 24. Ðæs ymb iii niht, Chr. 871 ; Erl. 74, 6, 14, 25. Wífes wer gif hé forþfærþ ymbe .xii mónaþ ðæs heó mót niman óðerne *mortuo viro, post annum licet mulieri alium accipere*, L. Ecg. C. 19 ; Th. ii. 146, 10. ¶ Ðæs ðe :—Ðæs ðe ðá seó costung gestilled wæs, ðá wǽron forþgongende ða geleáfsuman, Bd. I, 8 ; S. 479, 19 : 3, 22 ; S. 552, 39. Sóna ðæs ðe hí on ðis eálond cómon, ðá compedon hí, I, 12 ; S. 480, 29. Ðæs ðe ... ðá sóna, 5, 6 ; S. 620, II. Æfter siextegum daga ðæs ðe ðæt timber ácorfen wæs *intra sexagesimum diem quam arbores caesae erant*, Ors. 4, 6 ; Swt. 172, 4. Hé ðá gyt lifde æfter ændlefan geárum ðæs ðe [hé] wæs bebyrged, Shrn. 82, 15. Ðæt wæs ymb twelf mónaþ ðæs ðe hié ǽr hider cómon, Chr. 894 ; Erl. 93, 14 : 895 ; Erl. 93, 32 : 896 ; Erl. 94, 23. (b) marking degree, proportion, *so* (cf. colloquial use of *that* = so, with adjectives) :—Nǽre flód ðæs deóp, merestreám ðæs micel, ðæt his mín mód getweóde, Cd. Th. 51, 26, 27 ; Gen. 832, 833. Nó ðæs fród leofaþ gumena bearna ðæt ðone grund wite (*so wise as to know*), Beo. Th. 2737 ; B. 1366. Wurde ðú ðæs gewitleás, ðæt ðú þonc ne wisses, Exon. Th. 90, 12 ; Cri. 1473. Nis ǽnig ðæs horsc ne ðæs hygecræftig ðe ðín fromcyn mǽge gesêþan, 15, 24 ; Cri. 241. Wé ús wið him sceldan ðæs ðe wé mihton *we protected ourselves against them as far as we could*, Nar. 14, 29 ; Ps. Th. 10, 3 : Homl. Th. ii. 550, 20 : L. Eth. v. 23 ; Th. i. 310, II : vi. I ; Th. i. 314, 6 : Lchdm. ii. 86, 23. Næs ic nǽfre git náne hwíle swá emnes módes, ðæs ðe ic gemúnan mǽge (*from what*, or *as far as, I can remember*), Bt. 26, I ; Fox 90, 25. ¶ with comparatives :—Ðá clypodon hig ðæs ðe má (*so much the more*), Mt. Kmbl. 20, 31 : Mk. Skt. 10, 26. Sió wund biþ ðæs ðe wierse and ðý máre, Past. 17 ; Swt. 123, 18 : 18 ; Swt. 131, 16. Ðæt hié wéndon ðæt hié ðæs ðe (*tanto*) untǽlwyrðran wǽren ðe (*quanto*) hié wéndon ðæt hé nyste hira leóhtmódnesse, 32, 2 ; Swt. 215, I. (b I) with *tó* :—Tó ðæs mycel ðæt ... so great that ..., Bd. I, I ; S. 474, 13. Wæs seó eorþe tó ðæs heard and tó ðæs stánihte ðæt ..., 4, 28 ; S. 605, 27. Nis nán tó ðæs lytel ǽwelm, ðæt hé ða asǽ ne gesêce, Bt. 24, I ; Fox 80, 24. Hé him ðæs leán forgeald tó ðæs ðe hé in ræste geseah Grendel lícgan *he gave him reward for that so*, or *to such a degree, that he saw Grendel lie dead*, Beo. Th. 3175 ; B. 1585. (c) marking agreement, *according to what, as* :—Wé him answæredon ðæs ðe hé ús áxode *respondimus juxta id quod fuerat sciscitatus*, Gen. 43, 7. Hú hé him ondwyrdan sceolde ðæs hé hiene áscade *quid sibi tamquam consulenti responderi velit*, Ors. 3, 9 ; Swt. 126, 30. Ðæs ðe (*ut*) hié gesawen is, Bd. I, 25 ; S. 487, 12 : Bt. 24, 3 ; Fox 84, 10. Swá efne ðæs ðe *ita ut*, Bd. I, 34 ; S. 499, 20. And se mon biþ ðæs ðe swá tó cweþanne sí ǽghwæðer ge gehæfted ge freó *itaque homo est, ut ita dixerim, captivus et liber*, I, 27 ; S. 497, 40. Ðæs ðe béc secgaþ *as books say*, St. And. 26, 6. (d) *because, since* :—Waa mé ðæs ic swigode *vae mihi quia tacui*, Past. 49, 2 ; Swt. 379, 24. (3) Ðæm, ðam, ðan, ðon (ðe). (a) with a comparative :—Gif hé ne biþ ðon raþor gelácnod, Lchdm. ii. 200, 20. (b) *with prepositions* :—Æfter ðæm ðe Rómeburg getimbred wæs *urbe condita*, Ors. 4, 6 ; Swt. 170, 19 (*and often*). Æfter ðæm ðe Cartainiense gefliémde wǽron hié wilnedon friþes *Carthaginenses, fracti bellis, pacem poposcerunt*, Swt. 174, 23. Ǽr ðæm ðe Rómeburh getimbred wǽre, I, 3 ; Swt. 32, I (*and often*). Ǽr ðam ðe *donec*, Mt. Kmbl. 12, 20. Ǽr ðon, Past. 33, I ; Swt. 215, 15. Wurdon viiii folcgefeoht gefohten ... and bútan ðam ðe him cyninges þegnas oft ráde onridon ðe mon ná ne rímde *there were nine pitched battles ... and besides king's thanes often made raids upon them, that were not counted*, Chr. 871 ; Erl. 76, 10. For ðæm ðe (I) *for, because* :—Eádige synt ða líðan ; for ðam ðe (*quoniam*) hí eorþan águn, Mt. Kmbl. 5, 4. For ðam, 5, 3. For ðon ðe *quia*, 7, 13. Ða Deniscan sǽton ðǽr behindan, for ðæm hiora cyning wæs gewundod, Chr. 894 ; Erl. 91, 2. For ðæm ðe, 91, 28. For ðam, Ps. Th. 9, 13. Ðý ... for ðam *therefore ... because*, Bt. 36, 7 ; Fox 184, 15. (2) *therefore* :—Hé for ðæm nolde, ðý hé mid his folce getíúwode ðæt hé hiene beswícan mehte, Ors. 2, 4 ; Swt. 76, 8 : Bt. 38, 2 ; Fox 188, 16. For ðon (*therefore*) ic ðé bebióde, Past. pref. ; Swt. 5, I. (3) *for the purpose, in order* :—Geþence gé hwæt gé sién ; for ðam ðæt gé eówer mód gemetgien *pensa, quod es ; ut se spiritus temperet*, Past. 21, 4 ; Swt. 159, 14. Mid ðæm ðe *whilst, when, as*, of simultaneous events :—Mid ðæm ðe ða burgware swá geómorlíc angin hæfdon ðá com se cyning self mid his scipe *inter haec procedit ipse de navi sua imperator*, Ors. 4 ; Swt. 166, 14. Mid ðam ðe se apostol stóp intó ðære byrig, ðá bær man him tógeánes ánre wydewan líc, Homl. Th. i. 60, II. Ongemang, onmang ðam (ðe) *whilst, meanwhile* :—Ongemang ðæm ðe hié wilniaþ ðæt hié gifule þyncen, Past. 45, 3 ; Swt. 339, 24. Seó sunne sáh tó setle onmang ðam ðe hí on wópe wǽron, Homl. Skt. i. 23, 246 : Chr. 1105 ; Erl. 240, 4. Ongemang ðam (*interea*) his leorningcnihtas hine bǽdon, Jn. Skt. 4, 31. Tó ðam (I) marking degree *so, to such a degree* :—Ðá wǽron hié tó ðæm gesárgode, ðæt hié ne mehton Súð-Seaxna lond útan berówan, Chr. 897 ; Erl. 96, 8. Men tó

ðam dyrstige ðæt hí ðæt gold nimen *men so bold as to take the gold*, Nar. 35, 9 : Bt. 11, I ; Fox 32, 32. Nis nán tó ðam ungelýfedlíc spel ... ðæt ic hym ne gelífe, Shrn. 196, 18. Tó ðam ðú mé hæfst gerétne ðæt ..., Bt. 22, I ; Fox 76, II. (2) marking purpose, *to the end (that)* :—Tó ðæm ðæt (*ut*) hé forleóse heora gemynd, Ps. Th. 33, 16 : 10, 2. Ne com hé ná tó ðam on eorþan ðæt him mon þénade, Past. 17 ; Swt. 121, 8. Tó ðæm ðæt, pref. ; Swt. 5, 3. Ða cwóman tó ðon ðæt hié woldan ús wundigan *nos adlacessere temptabant*, Nar. 22, 17. Wið ðam ðe *in return for, on condition (that)*, connecting two clauses containing mutual concessions, v. wið :—Se cyng and his witan him (*the Danes*) gafol and metsunga behétan wið ðam ðe hí heora hergunga geswicon, Chr. 1011 ; Erl. 144, 22 : Past. 36, 6 ; Swt. 255, 3, 9. (4) Ðæt *in oþ ðæt*. v. oþ, II. 2. (5) Ðý, ðí, ðig (1) *therefore, so* :—Ðý him is micel þearf, ðonne hé tela lǽrþ, ðæt hé eác tela doo, Past. 28 ; Swt. 193, 12 ; Bt. 36, 7 ; Fox 184, 14. Ðý *ideo*, L. Ecg. P. i. 15 ; Th. ii. 178, 29. Ðig *itaque*, Th. ii. 176, 15. (2) *because* :—Wénst ðú, ðæt ealle ða þing ðe góde sint, for ðý góde sint, ðý hí habbaþ hwæthwegu gódes on him, 34, 9 ; Fox 146, 30 : Ors. 2, 4 ; Swt. 76, 8. (3) with comparatives, *the, any* :—Búton ðú mé ðý gesceádlícor óðer gerecce, Bt. 39, 2 ; Fox 214, 7. Hió ne biþ ðý néar ðære sǽ ðe hió biþ on midne dæg, 39, 3 ; Fox 214, 28. Ðæt hié hira selfra ne ágon ðý máre geweald ðe óðerra monna, Past. 33, I ; Swt. 215, 13. Hié woldon ðæt hér ðý mára wísdóm on londe wǽre ðý wé má geþeóda cúðon, Past. pref. ; Swt. 5, 24. (2) *with prepositions*. For ðý (ðe), (a) *therefore* :—For ðý ... ðý *therefore ... because*, Bt. 34, 9 ; Fox 146, 30. For ðý ... for ðæm *therefore ... because*, Past. 21 ; Swt. 157, 10. (b) *because* :—Ðæt wæs for ðý ðe hié wǽron benumene ðæs ceápes, Chr. 895 ; Erl. 93, 17. Mid ðý (ðe) (a) *of time, when, as* :—Mid ðý ðe hé ðis gebed gecweden hæfde, Blickl. Homl. 229, 27 : 231, 7. Sumre tíde mid ðý ðe wé wǽron mid úrum Drihtne, 235, 2. Mid ðí ðe, 237, 17. Mid ðí hé ðis cwæð, hé ástáh on heofonas, 237, 15. Mid ðý *cum*, Bd. 4, 24 ; S. 598, 33. Mid ðý *cum*, Mt. Kmbl. Lind. 24, 15 (*and often*). (b) denoting a cause or consequence, *when, as, since* :—Mid ðý Peohtas wíf næfdon, hí bǽdon him wífa fram Scottum, Bd. I, I ; S. 474, 19. (c) *though* :—Gif hé eów forhogige and eów ne wylle árísan tógeánes mid ðý eówer má is *sin autem vos spreverit, nec coram vobis adsurgere voluerit, cum sitis numero plures*, 2, 2 ; S. 503, 13. Tó ðý ... ðæt *to the end that* :—Ne com hé for ðý ðæt hé wolde his eorþlíce ríce mid riccetere him tó geteón ; ac tó ðí hé com ðæt hé wolde his heofenlíce ríce geleáffullum mannum gyfan, Homl. Th. i. 82, 20-24 : ii. 226, 9. Ne dó ná se Godes þeówa Godes þénunge for sceattum, ac tó ðý ðæt hé geearnige ðæt éce wuldor þurh ðæt, L. Ælf. C. 27 ; Th. ii. 352, 23. [Gothic and Icelandic have forms corresponding with the *nom. m. f. se, seó*, and O. Sax. also has a masculine *se ;* in other dialects the dental forms prevail throughout. In the Lindisfarne Gospels ðe (= ipse, Mt. Kmbl. 15, 24), ðiú (= quae, 24, 15) are used, but also *se ðe* (= qui, 6, 4). In later English þe, þeo replace *se, seó*.]

se co :—Se ðeáh *yet, still*, Exon. Th. 13, 31 : Cri. 211 : 159, 30 ; Gú. 934 : 328, 24, Vy. 22 : 454, 6 ; Hy. 4, 28 : 455, 12, 18 ; Hy. 4, 48, 51 : 495, 13 ; Rä. 84, 7. Hwæðre se ðeáh, 417, 27 ; Rä. 36, II. Efne se ðeáh, 421, 33 ; Rä. 40, 27 : 482, 2 ; Rä. 66, I. Se ðeána, 127, 3 ; Gú. 380. Sete hí samod anlíce swá se wægnes hweól *þone illos ut rotam*, Ps. Th. 82, 10. [Hi rihtleceden þat folc swa se hi mihten, O. E. Homl. i. 235, 32. *Se* in combinations *hwat se, alse* is frequent in later English.] v. swá, nese (?).

seað, seáða, seáfian, seaht, seal, sealcan, seald. v. seód, seáða, seófian, seht, sealh, á-sealcan, solcen, sellan.

sealdness, e ; *f. Giving* :—Sealdnesse *dandi*, Wrt. Voc. ii. 28, 7. v. ge-saldniss.

sealf, e : sealfe, an (?) ; *f. Salve, ointment* :—Salf, salb *malagma*, Txts. 77, 127. Sealf, Wrt. Voc. i. 68, 6. Smyrels *vel* sealf *unguina vel unguenta*, 49, 29. Fota, i. *confortata vel* sealf, ii. 149, 76. Smyrels oððe sealfe *unguentum*, i. 74, 8. Sealfe *nardi*, Hpt. Gl. 517, 28. For hwí wæs ðisse sealfe forspilledness ? Deós sealf mihte beón geseald, Mk. Skt. 14, 4, 5 : Jn. Skt. 12, 3, 5. Wyrc tó salfe (sealfe, MSS. H. B.), Lchdm. i. 110, 18. Sealfe *fotu*, Wrt. Voc. ii. 90, 74. Lǽcedómnessa oððe sealfe *cataplasma*, 18, 31. [O. Sax. salba : O. H. Ger. salb and salba (gen. -a and -un) *unctio, unguentum, malagma, cataplasma*.] v. bæþ-, bán-, ciper-, cú-, dolh-, eág-, eár-, ele-, múþ-, sceáde-, smeoru-, tóþ-, weax-, weax-hláf-, wen-sealf.

sealf-box, es ; *m.? n.? A box for ointment* :—Ǽn wíf hæfde hyre sealfbox deórwyrþes nardes, and tóbrocenum sealfboxe ofer his heáfod ágét, Mk. Skt. 14, 3 : Lk. Skt. 7, 37.

sealf-cynn, es ; *n. An ointment* :—Sealfcyn (seals-, Wrt.) *amaracium* (cf. *Span.* unguento amaracino *a sort of ointment made of marjoram*), Wrt. Voc. ii. 7, 74.

sealfian ; *p.* ode *To salve, anoint* :—Sealfode *fotam*, Wrt. Voc. ii. 37, 16 : 85, 22. Gisalbot *delibutus*, Txts. 56, 325. [Salue me mine wunden, Marh. 5, 30. Eзhesallfe to sallfenn þe follkes herrtess eзhe, Orm. 9427. Þatt mann þatt smeredd iss and sallfedd, 13243. Buten зif heorte wunden beon isalued, A. R. 274, 30. *Goth.* salbón *to anoint* : O. Sax. salbón : O. H. Ger. salbón *ungere, fovere, impinguare*.]

sealf-læcnung, e; *f. Curing by means of salves or ointments :—Far-macida in Latinum medicamina sonat, id est* sealflǽcnung, Wrt. Voc. ii. 39, 19. v. next word.

sealf-lǽcnung, e; *f. Pharmacy;* pharmacia, Wrt. Voc. i. 20, 27. v. preceding word.

sealh, salig, es; *m. A sallow, sally, selly* (v. E. D. S. Pub. Plant Names, p. 607) :—Salch, salh *salix*, Txts. 94, 89. Sealh *amera*, Wrt. Voc. i. 285, 61. Seal, ii. 8, 41. Seales rinde, Lchdm. iii. 14, 2. Reádes seales leáf (*red sally* lythrum salicaria, Plant Names, p. 413), 58, 28. Ge-nim sealh, ii. 18, 26 : 86, 7. On salig (saligum, Ps. Lamb.: salum, Ps. Surt. : sealum, Ps. Spl.) wē úre organan up áhēngan *in salicibus suspen-dimus organa nostra*, Ps. Th. 136, 2. Salhas *salices*, Txts. 113, 58. Selas *saliunculas*, Hpt. Gl. 408, 56. [In selihes (salyhes, MS. H.), Ps. 136, 2. *Chauc.* salwes: *Prompt. Parv.* salwhe: *O. H. Ger.* salaha; *gen.* -un; *f. salix, saliuncula: Ger.* sahl-weide: *Icel.* selga (cf. selly, *Yorks.*); *f.*] v. following words.

sealh-beorh *a hill where sallows grow :*—Tó sahlborge, Cod. Dip. Kmbl. iii. 451, 17.

sealh-hangra *a meadow where sallows grow :*—On sealhangran, Cod. Dip. Kmbl. vi. 234, 18.

sealh-hyrst *a sallow-copse :*—Tó sealhyrstæ foreweardræ, Cod. Dip. Kmbl. v. 256, 1.

sealh-rind *the bark of sallow :*—Nim sealhrinde, Lchdm. ii. 98, 9. Grēne sealhrinde, 318, 9.

Sealh-wudu *Selwood :*—Be eástan Sealwyda, Chr. 878 ; Erl. 80, 9. Sealwuda, 894 ; Erl. 92, 19.

sealm, p세alm, psalm, es; *m. A psalm, song* (a) in a general sense :—*Psalmus, propie* hearpsang; *canticum* psalm, æfter hearpan sang; *psalmus* ǽr hearpan sang, Wrt. Voc. i. 28, 36–38. On fatum sealmes *in vasis psalmi*, Ps. Spl. 70, 24. Syngaþ Gode sealm, 67, 4. Salma *psalmorum*, Ps. Surt. 70, 21. On sealmum wē drýman him *in psalmis jubilemus ei*, Ps. Spl. 94, 2. Ðæt ic Gode sealmas singe, Ps. Th. 56, 9, 11. (b) the psalms of David :—David wītegode fela ymbe Crist, swā swā ús cýðaþ ða sealmas ðe hē gesang, Ælfc. T. Grn. 7, 25. On Moyses ǽ, on wītegum and on sealmum, Lk. Skt. 24, 44. On psalmum (salmum, Cott. MSS.) Past. 48; Swt. 375, 1. (c) with special reference to the services of the church :—Hú fela psealma on nihtlīcum tīdum tó sing-enne synt, R. Ben. 6, 15. Sealma, 33, 5 : 6, 22. Nǽnig mon ne dorste for hine sǽalmas ne mæssan singan, Bd. 5, 14; S. 634, 35. Seofon seolmas, Coll. Monast. Th. 33, 29. [*O. L. Ger.* salm: *O. H. Ger.* salmo, psalmo : *Icel.* sálm.] v. bletsing-, gebed-, lof-sealm.

sealma, selma, an; *m. A couch :*—Selma, benc *sponda*, Txts. 98, 955. Gewíteþ ðonne on sealman, Beo. Th. 4911; B. 2460. (Cf. Lazarus answēbit ist an selmon, Hel. 4008.) [*O. Frs.* bed-selma *bedstead.*]

sealm-cwide, es; *m. A psalm :*—On stefne sealmcwides *uoce psalmi*, Ps. Lamb. 97, 5.

sealm-fæt :—On sealmfatum *translates in vasis psalmorum*, Ps. Th. 70, 20.

sealm-getæl, es; *n. A tale* or *number of psalms :*—Ðæs sealmgetæles is elles tó lyt, R. Ben. 43, 19.

sealm-glig, -gliw, es; *n. Psalmody :*—On sealmglige *in psalterio*, Blickl. Gl. Sealmglywe, Ps. Lamb. 143, 9.

sealmian; *p.* ode *To play on the harp* (*and sing*) :—Ic singe and sealmige *cantabo et psallam* (I sal sing and salme, Ps.), Ps. Spl. M. 107, 1.

sealm-leóþ, es; *n. A psalm :*—Sealmleóþ and hearpswēg *psalterium et cythara*, Blickl. Gl.

sealm-lof, es; *n. A psalm :*—Sealmlof *psalmus*, Ps. Lamb. 146, 1 : 17, 50 : *psalterium*, 107, 3. Sealmlof cweðaþ *psallite*, 97, 4.

sealm-loflan *to sing psalms :*—Singaþ him and sealmlofiaþ him *cantate ei et psallite ei*, Ps. Lamb. 104, 2.

sealm-sang, es; *m.* **I.** *a psalm :*—Sealmsang *psalmus*, Ps. Lamb. 146, 1. Salmsang, 60, 9. On sealmsangum *in psalmis*, Hymn. Surt. 7, 34. **II.** *psalm singing, psalmody :*—Ðá ðā se sealmsang gefylled wæs *expletis psalmodiis*, Bd. 4, 7; S. 575, 2. Ðæs dæglīcan sealmsanges *diurne psalmodie*, Wrt. Voc. ii. 141, 61 ; R. Ben. 34, 9. On fæstenne and on sealmsange, Blickl. Homl. 199, 34. Gif se man sealm-sang ne cunne *si homo psalmos cantare nesciat*, L. Ecg. P. iv. 61 ; Th. ii. 222, 16. Sealmsang *melodiam*, Wrt. Voc. ii. 56, 77. **III.** *the making and reciting of psalms :*—Ða twegen fixas getācnodon sealmsang and ðæra wītegena cwydas. Ân ðæra bodode Cristes tócyme mid sealmsange and óðer mid wītegunge. Nú sind ða twā gesetnyssa, ðæt is sealmsang and wītegung, Homl. Th. i. 188, 16–19. [*O. H. Ger.* salm-sang *psalmus, psalterium, psalmodia.*]

sealm-scop, es; *m. A writer* or *maker of psalms, a psalmist* (generally *the psalmist* David) :—Se sealmscop (salm-, Cott. MSS.), Past. 1 ; Swt. 29, 8. Salmscop, 14; Swt. 85, 23. Psalmscop (-scop, Cott. MSS.), 37; Swt. 273, 13 : 275, 21. Se sealmscop, Blickl. Homl. 55, 12 : 57, 1 : L. Ecg. P. i. 9; Th. ii. 176, 14: Homl. Th. ii. 82, 30. Sealmscopes *psalmigraphi*, Hpt. Gl. 430, 40. Heáhfæderas, wītigan, sealmscopas, Blickl. Homl. 105, 10 : Wulfst. 250, 18.

sealm-traht, es; *m. A commentary on the psalms* or *on a psalm :*—

Swā swā Hieronimus se wîsa trahtnere áwrât on sumum sealmtrahte, Homl. As. 36, 297.

sealm-wyrhta, an; *m. A psalmist :*—Se psalmwyrhta (*David*), Homl. Th. ii. 82, 32. Sealmwyrhta, Ælfc. T. Grn. 1, 24.

sealo-brún. v. salu-brún.

sealt, es; *n. Salt* (lit. and fig.) :—Sealt *sal*, Wrt. Voc. i. 82. 89. Gē synt eorþan sealt (salt, Lind., Rush.) : gyf ðæt sealt áwyrþ, Mt. Kmbl. 5, 13 : Mk. Skt. 9, 49, 50. Hwylc manna werodum þurhbrýcþ mettum búton swæcce sealtes *quis hominum dulcibus perfruitur cibis sine sapore salis?* Coll. Monast. Th. 28, 17. Nim ácorfenes sealtes (*rock salt*) ðæt wæter ðe ðǽrof gǽþ, Lchdm. ii. 246, 18. Hwítes sealtes, iii. 20, 26. Greát sealt *rock salt*, 40, 20, 10 : i. 158, 34. [*Goth.* salt : *O. Sax. O. L. Ger.* salt : *O. H. Ger.* salz : *Icel.* salt.] See following words, Cod. Dip. Kmbl. vi. 331, col. 2, and Leo on Anglo-Saxon Names, p. 27.

sealt, salt; *adj. Salt,* (1) of that which is naturally salt :—For hwam wæs seó sǽ sealt geworden? Moises áwearp ða .x. word in ða sǽ, and his teáras ágeát in ða sǽ ; for ðam weard seó sǽ sealt, Salm. Kmbl. 188, 15–19. Sealt wæter *the sea*, Ps. Th. 68, 2 : Cd. Th. 13, 6 ; Gen. 198. Brim sceal sealt weallan, Menol. Fox 552 ; Gn. C. 45. On sealtum mersce *in salsuginem*, Ps. Spl. 106, 34. Oþ ðone sealtan mere *usque ad lacum Salinarum*, Ors. 1, 1 ; Swt. 26, 8. Ofer sealtne (saltne, Cott. MSS.) sǽ, Past. pref.; Swt. 9, 8. Sió onlícnes sendde mycel wæter þurh hiora múþ swā sealt (*very salt*), Blickl. Homl. 245, 25. Eahtoðe wæs sealtes pund, ðanon him wǽron ða teáras sealte, Salm. Kmbl. 180, 16. Sealte ýða, Cd. Th. 205, 26 ; Exod. 441. Sealte sǽwǽgas, 240, 9 ; Dan. 384. Sealte streámas, Exon. Th. 206, 2 ; Ph. 120. Sealte flódas, Ps. Th. 68, 14. Swēg sealtera wætera, 76, 13. Salte sǽstreámas, Andr. Kmbl. 1497 ; An. 750. (2) of that which is artificially salt, *salt* (meat) :—Tú hrieðeru, óðer sealt, óðer fersc, Ch. Th. 158, 27. Forgá sealtes gehwæt, Lchdm. ii. 56, 23. Ete sealtne mete and nówiht fersces, iii. 28, 24. Sele ðú him sealte mettas, 182, 13 : 184, 8. [*O. Frs.* salt : *Icel.* saltr.] v. un-sealt.

sealt-ærn, -ern, es; *n. A salt-house, a place where salt is prepared :*—.i. sealtern, Cod. Dip. Kmbl. ii. 64, 28. Bútan ðem sealtern and bútan ðem wioda ðe tó ðem sealtern limpþ, 66, 22. Sealtearn, iii. 426, 19.

sealt-bróc, es; *m. A brook that runs from salt works* (?) :—Of saltere-wellan eástriht on saltbróc ; and swā ondlong saltbróces, Cod. Dip. Kmbl. iii. 206, 32.

sealten; *adj. Salt, salted :*—In ðǽm ðe biþ salten *in quo salietur*, Mt. Kmbl. Rush. 5, 13.

sealtere, es; *m. A salt-worker :*—Sealtere *salinator*, Wrt. Voc. i. 74, 10. Sealtere, saltere, Ælfc. Gr. 9, 21 ; Zup. 47, 2. Sealtere, hwæt ús fremaþ cræft ðīn ? . . . Nân eówer blisse brýcþ on gereorduncge oððe mete, búton cræft mín gistlīþe him beó, Coll. Monast. Th. 28, 5–11. On ðone saltherpaþ ; and swā ondlong ðæs herpaþes ðæt on salteredene . . . on salterewellan ; of salterewellan eástriht on saltbróc, Cod. Dip. Kmbl. lii. 206, 28–32. Sealtera cumb, 412, 24. In saltera weg ; of sealtera wege, 80, 16. [*Prompt. Parv.* saltare or wellare of salt *salinator*.]

sealt-fæt, es; *n. A vessel for salt, a salt-cellar :*—Sealtfæt *salinare* vel *salinum*, Wrt. Voc. i. 26, 59 : *vas salis*, 290, 23 : Anglia ix. 264, 18. Se Hǽlend bestang ðone hláf on ðæt sealtfæt ðe him beforan stód, Homl. As. 163, 254. [*O. H. Ger.* salz-faz *salinum : Icel.* salt-fat.]

sealt-hálgung, e ; *f. Salt-hallowing;* benedictio salis :—Salthálguncge tó acrum ł in húsum, Rtl. 117, 33.

sealt-herepaþ *a road to salt-works.* v. sealtere, *and cf.* sealt-strǽt.

sealt-hús, es; *n. A house where salt is prepared* (?) or *sold* (?); *sali-narium*, Wrt. Voc. i. 56, 49. [*O. H. Ger.* salz-hús *salsamentarium.*]

sealtian *to dance.* v. saltian *and next word.*

sealticge, an ; *f. A dancer :*—Hēt Herodes ðæt heáfod beran on disce and sellan ânre sealticgan (*the daughter of Herodias who danced before Herod*) hire plegan tó mēde, Shrn. 123, 2.

sealt-leáf glosses mozicia, Wrt. Voc. ii. 59, 35.

sealt-leáh ; *gen.* -leáge ; *f. A salt lea ; hence Saltley :*—Of ðan swínhagan ðæt on sealtleáge ; and of sealtleáge in ðone hyrstgeard, Cod. Dip. Kmbl. iii. 400, 1.

sealt-mere, m : *A salt mere* or *marsh; hence Saltmere.*—Tó sealtmere ; of sealtmere, Cod. Dip. Kmbl. iii. 82, 3.

sealtness, e ; *f. Saltness :*—Eorþan wæstmbēre sealtsæleðan ł tó sealt-nesse *terram fructiferam in salsuginem*, Ps. Lamb. 106, 34. In saltnisse *in salsilaginem*, Ps. Surt. 106, 34.

sealt-seáþ, es ; *m. A salt-pit, salt-spring :*—Hafaþ eác ðis land sealt-seáþas *habet fontes salinarum*, Bd. 1, 1 ; S. 473, 22. [Cf. *O. H. Ger.* salz-suti *salina.*]

sealt-stán, es ; *m.* **I.** *rock salt :*—Ðis mæg tó eáhsalfe : genim geoluwne stán (*ochre*) and saltstán, Lchdm. i. 374, 14. **II.** *a stone formed of salt, a pillar of salt :*—Heó on sealtstánes sóna wurde anlícnesse ǽfre siððan, Cd. Th. 154, 31 ; Gen. 2564. Lothes wíf weard áwende tó ánum sealtstáne (*in statuam salis*), Gen. 19, 26 : Anglia vii. 48, 472. [*O. H. Ger.* salz-stein : *Icel.* salt-steinn. In English *salt-stone*

somewhat later means a rock in the sea, translating *cautes*, Wrt. Voc. i. 256, col. 1.]

sealt-strǽt, e; *f. A road to salt-works* (?); hence *Saltstreet:*— Andlang sealtstrǽte, Cod. Dip. Kmbl. iii. 38, 20. Ondlong ðære sealt-strǽt, 160, 13. Tô ðære sealtstrǽte, 263, 24. Cf. sealt-herepaþ.

sealt-wîc, es; *n. A place where salt is sold*; hence *Saltwych:*—In unico emptorio salis quem nos Saltuuic uocamus, Cod. Dip. Kmbl. i. 81, 9. Æt Saltwîc, v. 143, 21.

sealt-wille, -welle, an; *f. A salt spring* or *well*; hence *Saltwell:*— In saltwyllan; of saltwyllan, Cod. Dip. Kmbl. iv. 70, 24. Ða saltwǽlla ł of sæltwǽlla *a saliua* (translator seems to have read *salina*), Mt. Kmbl. p. 1, 5.

sealt-ŷþ, e; *f. A salt wave, sea-wave:*—Ðæt ic sealtŷþa gelác cunnige, Exon. Th. 308, 5; Seef. 35. Sealtŷþa geswing, 356, 7; Pa. 8.

seám, es; *m. A seam:*—Heáfodpanne *capitale*, heánnes ðære heáfod-pannan *cacumen capitalis*, seám ðære heáfodpannan *cerebrum*, brægen *cervellum*, Wrt. Voc. ii. 22, 51-55. Seám *panicenũ*, 116, 8. His tunece wæs eal bûton seáme (*inconsutilis*, Jn. 19, 23), Homl. Th. ii. 254, 32. Geclǽm ealle ða seámas mid tyrwan, i. 20, 33. [*O. Frs.* sâm: *O. H. Ger.* saum *ora, lacinia, limbus: Icel.* saumr.]

seám, es; *m.*   I. *a seam, a load, burden* [a *seam* of corn is a quarter, eight bushels ; a *seam* of wood is a horse-load ; a *seam* of dung 3 cwts. (Devon), v. E. D. S. Pub. Reprinted Glossaries, and Farming Words 1, 3, 7. Bailey gives a *seam* of glass as 120 lbs.] :—Seám *vel* berþen *sarcina*, Wrt. Voc. i. 16, 27 : Ælfc. Gr. 9, 32 ; Zup. 59, 3. Seáme *sarcina*, Hpt. Gl. 528, 35. Gê sŷmaþ men mid byrþenum (seámum, Lind.: seómum, Rush.) . . . and gê ne ãhrînaþ ða seámas mid eówrum ánum fingre, Lk. Skt. 11, 46. Wæs þridde healf þúsend mûla ðe ða seámas (*sarcinas*) wǽgon, Nar. 9, 10 : 23, 1-2. II. *the furniture of a beast of burden :*—Rachel hig hæfde gehŷdd under ánes olfendes seáme (*subter stramenta cameli*), Gen. 31, 34. III. *that in which a burden may be carried, a bag :*—Bûta seáme (seóme, Rush.) *sine sacculo*, Lk. Skt. Lind. 22, 35. Nællaþ gié gebeara seám (seóm, Rush.) *nolite portare sacculum*, 10, 4. IV. as a technical term, *a service which consisted in supplying the lord with beasts of burden ;* summagium, sagmegium :—Hê sceal beón gehorsad, ðæt hê mǽge tô hláfordes seáme ðæt (*the horse*) syllan oððe sylf lǽdan, swǽðer him man tǽce, L. R. S. 5 ; Th. i. 436, 6. [I shal assoille þe myselue for a seme of whete, Piers P. 3, 40. Seem of corne *quarterium*, Prompt. Parv. 452. *O. H. Ger.* soum *sagma, sella, sarcina.* From *Lat.* (*Gk.*) sagma, later *salma* ; cf. *Ital.* salma ; *Fr.* somme.] v. ofer-seám ; sîman.

seámere, es ; *m. A tailor* :—Seámere *sartor*, Wrt. Voc. i. 74, 12. Seámere, seamyre, Ælfc. Gr. 30, 2 ; Zup. 190, 6 note. Seámere *burdus* (*burdus sutor vestiarius*), Wrt. Voc. i. 21, 47. Se smiþ secgþ . . . Hwanon seámere (*sartori*) nǽdl ? nis hit of mînon geweorce ? Coll. Monast. Th. 30, 33.

seámere, es ; *m. A beast of burden, a mule :*—Hors *equus*, hengest *caballus*, seámere *burdus* (= *burdo* ; hic burdo, i. genitum inter equum et asinam, 219, col. 1), Wrt. Voc. i. 287, 42-44. Seámere *burdus*, oxa *bova*, ii. 11, 61-62. [*O. H. Ger.* soumari *burdo, saumarius, dromedarius: Ger.* säumer.]

seámestre, an ; *f. One who sews, a tailor, sempstress* (though the noun is feminine it seems not confined to females, cf. bæcestre) :— Seámestre *sartrix*, Wrt. Voc. i. 74, 13. *Sarcio* . . . of ðam is *sartor* seámystre (-estre, *other MSS.* seámere) *sartrix* heó, Ælfc. Gr. 30, 2 ; Zup. 190, 6. Hió becweð Eádgyfe áne crencestræn and áne sêmestran, ôðer hátte Eádgyfu, ôðer hátte Æðelyfu, Cod. Dip. Kmbl. vi. 131, 32. Fíf pund Ælffæhe mín sǽmestres, Chart. Th. 568, 10. [Sadlers, souters, semsteris fyn, Destr. Tr. 1585. Good semsters be sowing . . . good hus-wifes be mending, Tusser 176, 7.]

seám-hors, es ; *n. A pack-horse* ; sagmarius equus, Wrt. Voc. i. 23, 13. [*Ger.* saum-ross.]

seám-penig, -pending, es ; *m. A toll of a penny on a load* (of salt) :— Se wægnscilling and se seámpending gonge tô ðæs cyninges handa swâ hê ealning dyde æt Saltwîc, Cod. Dip. Kmbl. v. 143, 20. Cf. *statio sive inoneratio plaustrorum* mentioned in connection with *salis coctiones*, 125, 31. v. Kemble's Saxons in England, ii. 329.

seám-sadol, es ; *m. A pack-saddle* ; sagma, Wrt. Voc. i. 23, 12. [*O. H. Ger.* soum-satol *sagma : Ger.* saum-sattel.]

seám-sticca, an ; *m. Some part of a weaver's apparatus :*—Hê sceal fela tôwtôla habban . . . seámsticcan, scearra, nǽdle, Anglia ix. 263, 14.

seár and **siére** ; *adj. Sear, dry, withered, barren :*—Hit stent on ðam siéran bôchagan ; andlang ðes siéran bôchagan, Cod. Dip. Kmbl. v. 70, 32. Seáre *steriles*, Germ. 402, 69. [His body wex alle seere, R. Brun. 18, 25. With seere braunches, blossoms ungrene, Chauc. R. R. 4752. Seere or dry, as treys or herbys *aridus*, Prompt. Parv. 453. *O. Du.* sore *dry* ; zoor *dry, withered*, or seare (Hexham) : *L. Ger.* soor *dry*.] v. seárian.

Sear-burh. v. Searo-burh.

seárian ; *p.* ode *To grow sear, wither, pine away :*—Eorþan indryhto ealdaþ and searaþ, Exon. Th. 311, 9 ; Seef. 89. His leáf and his blǽda

ne fealwiaþ ne ne seáriaþ *folium ejus non decidet*, Ps. Th. 1, 4. Grênu leáf wexaþ . . . hŷ eft onginnaþ seárian, Shrn. 168, 22. Hé (*Regulus*) slâpan ne mehte, ôþ hê swâ seárigende his lîf forlêt, Ors. 4, 6 ; Swt. 178, 24. [*Prompt. Parv.* seeryñ or dryyñ or welkyñ, dryyn up *areo, aresco: O. H. Ger.* ar-sôrên *emarcescere* ; un-saorentlîh *immarcescibilis*.] v. â-, for-seárian.

searo. v. searu.

Searo-burh *Salisbury :*—In ðære stôwe ðe is genemned [æt] Searobyrg (-byrig, Searobuurh, Sælesberi), Chr. 552 ; Th. pp. 28, 29. Tô Searebyrig, 1086 ; Th. 353, 18. Tô Searbyrig, 1003 ; Th. pp. 252, 253. [Seres-byrig (Særes-), 1123 ; Th. 374, 5, 20, 24, 34.]

searu, searo, [w]e ; *f.:* [w]es ; *n. Device, design, contrivance, art.*   I. in the following glosses it is uncertain whether the word is used with a good or with a bad meaning :—Sarwo *adventio*, Wrt. Voc. ii. 99, 38. Searo *molimen*, 54, 29. Searwe *molimine*, 89, 64. Searwe *argumenta*, 84, 69. Searwum *commentis*, 14, 82 : 80, 76. Seorwum, 104, 75. Seara *machinas*, Hpt. Gl. 510, 21.   II. in a bad sense, *craft, artifice, wile, deceit, stratagem, ambush, treachery, plot :*—Searu *factio* (cf. fâcn *factiones*, 64 ; bepǽcunga *factione*, Hpt. Gl. 474, 26), Wrt. Voc. ii. 33, 81. Gleáwnisse and seare (sceare, Wrt. *alias*) *astu*, Wrt. Voc. ii. 9, 27. Mid searwe on gewald gedôn *per proditionem tradere*, Ors. 1, 12 ; Swt. 52, 27. Swîðor beswicen for Alexandres searewe ðonne for hy gefeohte *non minus arte Alexandri superata, quam virtute Macedonum*, 3, 9 ; Swt. 124, 19. Mid searuwe ácwellan *morti tradere*, Ps. Th. 108, 16. Ðara feónda searo beswîcan and ofercuman, Blickl. Homl. 201, 29. Searo rênian *to lay a snare*, 109, 30 : Cd. Th. 162, 9 ; Gen. 2678. Þurh ðæs deófles searo dôm forlǽtan, 39, 27 ; Gen. 632 : Exon. Th. 153, 7 ; Gû. 822 : 227, 6 ; Ph. 419. Þurh îdel searu, Ps. Th. 138, 17 : Elen. Kmbl. 1438 ; El. 721. Swilt þurh searwe *death by treachery*, Andr. Kmbl. 2695 ; An. 1350. Searwa *molimina* (*magorum*), Hpt. Gl. 502, 53. Sarwa *mendacia*, Wrt. Vôc. ii. 132, 41. Full fácnes and searuwa *plenum dolo*, Ps. Th. 9, 27 : Met. 9, 27. In searwum *in insidiis*, Ps. Surt. 9, 29. Searwum *factionibus*, Wrt. Voc. ii. 34, 9. Mid sibbe wê cômon næs mid searwum *pacifice venimus nec quidquam machinamur mali*, Gen. 42, 11. Beswicen mid deófles searwum *daemonica fraude seductus*, Bd. 5, 13 ; S. 632, 26. Mid searewan (his searum, MS. C.) *consiliis*, Ors. 3, 7 ; Swt. 112, 18. Searowum beswicene, Andr. Kmbl. 1489 ; An. 745. Hié þurh seara (*per insidias*) ofslægene wurdon, Ors. 1, 10 ; Swt. 44, 28. Ðâ funde hê swîðe yfel geþeaht and searwa ymb hira lîf *contra eorum vitam consilium praebuit*, Past. 54, 4 ; Swt. 423, 15. Gif hwâ ofsleá his ðone nêhstan þurh searwa, L. Alf. 13 ; Th. i. 48, 1 : Blickl. Homl. 83, 33. Hwylce searwa se drŷ ârefnde *what artifices the sorcerer practised*, 173, 8. Nyston ða searwe ðe him sǽton bæftan *ignorans quod post tergum laterent insidiae*, Jos. 8, 14.   III. in a good sense, *art, skill, contrivance,* (in the adverbial inst. searwum *skilfully, ingeniously, with art*) :—Searwum ásǽled, Cd. Th. 207, 21 ; Exod. 470. Salem stôd searwum (or IV ?) áfæstnod, weallum geweorðod, 218, 17 ; Dan. 40. Sadol searwum fáh (cf. searu-fáh), Beo. Th. 2080 ; B. 1038. Earmbeága fela searwum gesǽled (cf. searu-sǽled), 5521 ; B. 2764 : Exon. Th. 438, 10 ; Rä. 57, 5 (cf. searu-bunden) : 216, 17 ; Ph. 269. Bûr átimbran, searwum ásettan, 411, 27 ; Rä. 30, 6.   IV. *that which is contrived with art, a machine, engine, fabric :*—Stæflidere oððe searu *ballista, machina belli*, Wrt. Voc. ii. 10, 62. Searu *ballista, catapulta, vel machina belli*, 125, 9. Middanearedes wyrhta seares *mundi factor machinae*, Hymn. Surt. 29, 9. Ic seah searo hweorfan, grindan wið greóte, giellende faran, Exon. Th. 414, 29 ; Rä. 33, 3.   IV a. *armour, equipment, arms :*—Byrnan, gûðsearo gumena, gâras . . . sǽ-manna searo, Beo. Th. 663 ; B. 329. Beran beorht searo, Cd. Th. 191, 23 ; Exod. 219. Licgeþ lonnum fæst . . . swîðe swingeþ and his searo hringeþ, Salm. Kmbl. 534 ; Sal. 266. Hringîren song in searwum (*coats of mail*), Beo. Th. 651 ; B. 323 : 5053 ; B. 2530. Secg on searwum, 503 ; B. 249 : 5392 ; B. 2700. Geseah on searwum (*among the arms*) sigeeádig lîð, 3118 ; B. 1557. Searwum gearwe *equipped*, 3631 ; B. 1813. [*Goth.* sarwa ; *n. pl.* τὰ ὅπλα, πανοπλία : *O. H. Ger.* saro ; gi-sarwi, -sarwa *lorica, armatura, arma : Icel.* sörvi *a necklace ; armour.*] v. beadu-, bealu-, fácen-, fǽr-, fyrd-, gûþ-, hláford-, inwit-, lâþ-, lyge-, nearu-searu ; siru ; and cf. or-þanc.

searu-bend ; *m. f. A cunning, curious clasp* or *fastening :*—Glôf searobendum fæst, sió wæs orþoncum eall gegyrwed diófles cræftum, Beo. Th. 4179 ; B. 2086. Cf. orþanc-bend.

searu-bunden ; *adj. Cunningly fastened, bound with art :*—Wunden gold, sinc searobunden, Exon. Th. 437, 7 ; Rä. 56, 4.

searu-cǽg, e ; *f. An insidious key :*—Flânþracu feorh onleác searo-cǽgum gesôht (*of the insidious attacks of disease*), Exon. Th. 170, 27 ; Gû. 1118.

searu-ceáp, es ; *n. An ingenious piece of goods, a curious implement :*— Næfde sellîcu wiht folme, exle ne earmas, sceal on ánum fêt searoceáp (cf. searo, IV) swîfan, Exon. Th. 415, 6 ; Rä. 33, 7.

searu-céne ; *adj. Bold in arms* or *skilfully daring :*—Wæs Dauid æt wîge sôð sigecempa, searocŷne man, câsere creaftig, Ps. C. 10. Cf. searu-grim.

searu-cræft, es; *m.* I. *a treacherous art, wile, stratagem, an artifice, a machination, plot*:—Searecræft *molimen*, Hpt. Gl. 502, 56. Searocræft *machinam*, Wrt. Voc. ii. 54, 28. Þurh diófles searucræft, Cod. Dip. Kmbl. ii. 304, 26. Þurh searocræft, Andr. Kmbl. 217; An. 109. Searecræftum *argumentis*, Hpt. Gl. 471, 27: *machinamentis*, 478, 54. Bepæht mid dæs deófles searocræftum, Homl. Th. i. 192, 17: Exon. Th. 136, 13; Gú. 540: 142, 19; Gú. 646. Ealdfeónda níþ searocræftum swíð, 110, 25; Gú. 113. Searecræfts *machinas* (*fraudulentas*), Hpt. Gl. 474, 15. Ðe hé ne beswíce þurh his searucræftas (searo-, searæ-), Wulfst. 97, 8. Uton forsleón mân and morþor and searacræftas, 115, 9. Swíðe forsyngod þurh swicdómas and þurh searacræftas, 164, 3. II. *art, skill, cunning, a cunning art* (in a good sense, v. next word):—Wuldres ealdor gesweotula þurh searocræft ðín sylfes weorc, Exon. Th. 1, 16; Crä. 9. Ða róde mid ðám ædelestum eorcnanstánum besetton searocræftum (*cunningly, skilfully,* cf. searu, III), Elen. Kmbl. 2049; El. 1026. Ne hí searocræftum godweb giredon, Met. 8, 24. III. *an engine, machine* (cf. searo, IV):—Stæfliðera *ballista*, searecræftes *machinae*, Hpt. Gl. 487, 22.

searu-cræftig; *adj.* I. *skilful, skilled in* (with gen.), *cunning* (in a good sense):—Snottor, searocræftig sáwle rædes, Frag. Kmbl. 80; Leás. 42. Sum biþ searocræftig goldes and gimma, Exon. Th. 296, 29; Crä. 58. II. *wily, cunning* (in a bad sense), 416, 7; Rä. 34, 7.

searu-fáh; *adj. Curiously, cunningly coloured* (cf. gold-fáh):—Herebyrne síd and searofáh, Beo. Th. 2892; B. 1444.

searu-geþræc, es; *n. A store of things in which art is displayed*:—Seón and sécean searogeþræc (*the dragon's hoard*), wundur under wealle, Beo. Th. 6196; B. 3102 [cf. geþræce *apparatu*, Wrt. Voc. ii. 85, 72].

searu-gim[m], es; *m. A curious gem, precious stone*:—Seærogim *topazion*, Ps. Spl. T. 118, 127. His égan scinan swá searagym, Nar. 43, 15. Searogemme *unio*, Wrt. Voc. ii. 89, 34. Meregrota oððe gymmas (saragimmas, MS. V.) *margaritae*, Nar. 37, 29. Stán, searogimma nán (ælces cynnes gimmas ne ..., Bt. 34, 8; Fox 144. 31), Met. 21, 21: Beo. Th. 2318; B. 1157. Ðæt ic ærwelan, goldæht ongite, gearo scéawige sigel, searogimmas (*the dragon's hoard*), 5491; B. 2749: Exon. Th. 478, 5; Ruin. 36.

searu-grim; *adj. Fierce in arms* or *skilfully fierce, having fierceness accompanied by skill*:—Gif ðín hige wære swá searogrim swá ðú self talast *if thy spirit had been as cunningly fierce* (?) *as thyself reckons*, Beo. Th. 1192; B. 594. Cf. searu-céne.

searu-hæbbend[e] [*one*] *having armour, armed*:—Slæpe tóbrugdon searuhæbbende *the warriors started from sleep*, Andr. Kmbl. 3054; An. 1350. Searohæbbendra, 2934; An. 1470: Beo. Th. 480; B. 237: Exon. Th. 468, 12; Phar. 6.

searu-líc; *adj. Ingenious, cunning, clever, displaying art* or *skill*:—Ðæt (*writing being able to convey a message*) is wundres dæl, on sefan searolíc ðam ðe swylc ne conn, Exon. Th. 472, 4; Rä. 61, 11. Sum hafaþ searolíc gomen gleódæda, 298, 9; Crä. 82. v. next word.

searulíce; *adv. Ingeniously, cunningly, cleverly, with art* or *skill*:—Sum mæg searolíce wordcwide wrítan, Exon. Th. 42, 14; Cri. 672. Is se finta sum splottum searolíce beseted, 218, 19; Ph. 297. Ne hí gimreced setton searolíce, Met. 8, 26.

searu-net[t], es; *n.* I. *an armour-net,* or *a net ingeniously wrought, a coat of mail*:—On him byrne scán, searonet seowed smiþes orþancum, Beo. Th. 816; B. 406. II. *a net of treachery* or *guile, a net* (metaph.), *a snare, wile*:—Mé elþeódige inwitwrásne, searonet seóþaþ, Andr. Kmbl. 127; An. 64. Searonettum beseted *beset with snares,* 1885; An. 945.

searu-níþ, es; *m.* I. *hostility to which effect is given by treachery, crafty enmity*:—Ic ne sóhte searoníþas ne ne swór fela áþa on unriht *I did not have recourse to the arts of the treacherous foe, nor swore many oaths wrongfully,* Beo. Th. 5469; B. 2738: 2405; B. 1200. Swá wæs Biówulfe, ðá hé biorges weard sóhte, searoníþas (*the wily hostilities of the dragon, who used poison to destroy his foe,* cf. ǽttorsceaþa, 5670, and *is called* inwitgest, 5333. Cf. *too* inwit-níþ), 6126; B. 3067. II. *armour-hate* (v. searu, IV a), *martial strife, the strife of armed men, battle*:—Nó ic wiht fram ðé swylcra searuníþa secgan hýrde, billa brógan, 1168; B. 582.

searu-píl, es; *m. An implement with a point*:—Mín heáfod is homere geþuren, searopíla wund, sworfen feóle, Exon. Th. 497, 17; Rä. 87, 2.

searu-rún, e; *f. A cunning mystery*:—Searorúna gespon, Exon. Th. 347, 20; Sch. 15.

searu-sǽled; *adj. Cunningly tied*:—Nelle ic unbunden ǽnigum hýran, nymþe searosǽled (cf. searu, III, *and* searu-bunden), Exon. Th. 406, 12; Rä. 24, 16.

searu-þanc, es; *m.* I. *a cunning* (in a bad sense) *thought, device, artifice, wile*:—Geþeóddum searaþancum *adhibitis argumentis,* Hpt. Gl. 502, 16. Eác ic gelærde Simon searoþoncum, ðæt hé sacan ongon, Exon. Th. 260, 16; Jul. 298. Sume ic mínum hondum searoþoncum (*cunningly, craftily*) slóg, 272, 4; Jul. 494. Searoþancum beseted *beset with snares* (v. searu-net), Andr. Kmbl. 2511; An. 1257. II

a cunning (in a good sense) *thought, skilful device*:—þurh sefan snyttro, searoþonca hord, Past. pref.; Swt. 9, 10. Saga sóðcwidum, searoþoncum, gleáwwordum wísfæst, hwæt ðis gewǽdu sý, Exon. Th. 418, 3; Rä. 36, 13. Se wítga, snottor searuþancum, Elen. Kmbl. 2377; El. 1190. Georne smeádon, sóhton searoþancum (*sagaciously, shrewdly*) hwæt sió syn wǽre, 827; El. 414. Se wínsele fæste wæs írenbendum searoþoncum (*skilfully, cunningly*) besmiþod, Beo. Th. 1554; B. 775. Cf. or-þanc.

searu-þancol; *adj. Of cunning thought, cunning, sagacious, wise*:—Searoþoncol mægþ (*Judith*), Judth. Thw. 23, 28; Jud. 145. Nis ǽnig secg searoþoncol tó ðæs swíðe gleáw, Exon. Th. 14, 16; Cri. 220. Ðe (*which*) secgas searoþoncle seaxe delfaþ, 427, 26; Rä. 41, 97. Gesǽton searuþancle sundor tó rúne, Andr. Kmbl. 2323; An. 1163. Mon ǽnig searoþoncelra, Judth. Thw. 26, 17; Jud. 331.

searu-wrenc, es; *m. A crafty trick, treacherous device*:—Hé hié biddende wæs ðæt hié mid sume searawrence from Xerse áwende, Ors. 2, 5; Swt. 82, 21. v. siru-wrenc.

searu-wundor, es; *n. A wonderful thing in implements* or *engines* (v. searu, IV, *and* cf. searu-píl. The term is applied to Grendel's arm, which had been torn away by Beowulf):—Eode scealc monig searowundor seón, Beo. Th. 1844; B. 920.

searwaþ, L. N. P. L. 40; Th. ii. 296, 10. v. next word.

searwian; *p.* ode *To act with craft* or *treachery, to feign*:—Hé searwaþ *fingitur*, Wrt. Voc. ii. 132, 13. Hió searwaþ *insidiatur*, Kent. Gl. 191. Gif preóst ordál misfadige, gebéte ðæt. Gif preóst searwaþ be winde, gebéte ðæt *if a priest do not conduct an ordeal rightly, let him make 'bót.'* If a priest uses deceit in respect to the wrapping up of the hand or arm exposed to the ordeal, let him make 'bót,' L. N. P. L. 39, 40; Th. ii. 296, 9–10. Sinc searwade *treasure played the traitor* (left its possessor (?)), Exon. Th. 353, 62; Reim. 37. Searw[a] ð[ú] *insidieris,* Kent. Gl. 935. Searwiende *machinans*, 151. Hé cwæð him tó særwigendum móde (*insidiously*), Homl. Th. ii. 308, 6. v. sirwan *and next word.*

searwung, e; *f. Treachery, artifice, plot, snare*:—Hé sit mid searwungum *sedet in insidiis,* Ps. Lamb. 9 second, 8. v. sirwung.

seáþ, es; *m. A pit, hole, well, reservoir, lake*:—Seáþ *lacus,* Ælfc. Gr. 11; Zup. 79, 10: Ps. Spl. 7, 16: 27, 1: Mk. Skt. 12, 1: *lacus, lacuna,* Wrt. Voc. i. 54, 31: *fovea,* ii. 150, 10: Ps. Spl. 7, 16: 56, 9: *puteus,* Bd. 5, 12; S. 628, 16: *cisterna,* Wrt. Voc. ii. 24, 4: Kent. Gl. 102: *barathrum,* Hpt. Gl. 422, 50: *cloaca,* 484, 19: 508, 70. Ðær is se seáþ ðæs singalan susles ... *After ðam ðe ðú deád bist, ðonne cymst ðú tó helle* ... and ðín seáþ biþ twegea cubita wíd and feówra lang, Nar. 50, 23–29. On hú grundleásum seáþe *on how bottomless a pit,* Bt. 3, 2; Fox 6, 8. Ðá wæs ðǽr on óðre sídan ðæs hláwes gedolfen swylce mycel wæterseáþ wǽre. On ðam seáþe ufan Gúþlác him hús getimbrode, Guthl. 4; Gdwin. 26, 8. Danihel læg betwux seofan leónum on ánum seáþe, Homl. Th. i. 488, 5. Héht scúfan scyldigne on drígan seáþ, Elen. Kmbl. 1382; El. 693. In synna seáþ, Exon. Th. 267, 10; Jul. 413. Ðǽr syndon twegen seáþas (*lakes*) ... heora wíde is .cc. míla ðæs læssan mílgetales, Nar. 36, 25. [Inne deope seaðen setten þa deade, Laym. 841. O. Frs. sáth: M. H. Ger. sôt *puteus.*] v. adel-, cealc-, fǽr-, helle-, horu-, lám-, sand-, sealt-, wæter-, wulf-seáþ.

seáða, an; *m. 'A feeling as if the cavity of the body were full of water swaying about,'* Cockayne. The word glosses *tendiculum,* Wrt. Voc. ii. 77, 3:—Wið seáðan (seádan, 4, 18), Lchdm. ii. 56, 10.

seáw, es; *n. Juice, moisture, humour*:—Genim túncersan ... dó in ða nosu ðæt se stenc mǽge on ðæt heáfod and ðæt seáw, Lchdm. ii. 22, 14. Genim cileþoniam seáwes cucler fulne, 28, 2. Ys sǽd ðæt se earn wylle mid ðam seáwe (*of wood lettuce*) his eágan hreppan and wǽtan, i. 128, 12. Seáw *ius,* 80, 13: 128, 18. Ðæt seáw sele on cuclere súpan, ii. 120, 19. Gemeng wið huniges seáw *mix with pure honey,* 30, 7. Feallan lǽtaþ seáw of bósme, wǽtan of wombe, Exon. Th. 385, 20; Rä. 4, 47. Seá *sucum,* Txts. 182, 83. Cumaþ ða ádla on [of ?] yflum seáwum, Lchdm. ii. 176, 5. [*Used later of food.* With diverse spieces The flesh ... She taketh and maketh thereof a sewe, Gow. ii. 325, 4, Seew, Wick. Gen. 27, 4. I wol nat tellen of her strange sewes, Chauc. Sq. T. 67. Sew *cepulatum,* Wülck. Gl. 572, 9: Prompt. Parv. 454. O. H. Ger. sou; *n. succus, venenum, alimentum:* cf. Icel. söggr *dank, wet:* saggi; *m. moistness.*] v. lip-, plúm-seáw; ge-seáw; *adj.*

seax, es; *n.* I. *a knife, an instrument for cutting*:—Seax *cultellus,* Wrt. Voc. i. 287, 3. Seax oððe scyrseax *culter,* ii. 15, 58. Saex, 105, 69. Ðæt stǽnene sex ðe ðæt cild ymbsnáþ, Homl. Th. i. 98, 10. Seaxes ord, Exon. Th. 472, 6; Rä. 61, 12. Seaxes ecg, 70, 20; Cri. 1141. Sníþ mid seaxse, Lchdm. ii. 56, 7. Ða hét hé him his seax árǽcan tó screádigenne ǽnne æppel, Homl. Th. i. 88, 9. Nim ðæt seax ðe ðæt hæfte sié fealo hrýðeres horn and sién .iii. ǽrene næglas on, Lchdm. ii. 290, 22. Sting ðín seax on ða wyrte, 346, 12. Hý begyrde resten and náne sex (seax, MSS. T. F.) be heora sídan næbben *cultellos ad latus non habeant,* R. Ben. 47, 10. Wirc ðé stǽnene sex *fac tibi cultros lapideos,* Jos. 5, 2. II. *as a weapon, a short sword, dagger*:—Ðǽr gebrǽgd ðara hǽðenra manna sum his seaxe; ða hé hine

ðǽ stingan mynte, ðá nyste hé fǽringa hwǽr ðæt seax com, Blickl. Homl. 223, 16. Heó hyre seaxe geteáh, brád, brúnecg, Beo. Th. 3095; B. 1545. Hé (*St. Martin*) tócearf his basing on emtwá mid sexe, Homl. Th. ii. 500, 26. Geteáh his seax, Blickl. Homl. 215, 6. [*O. L. Ger.* sahs: *O. Frs.* sax: *O. H. Ger.* sahs *cultrum, semispathium: Icel.* sax *a short sword.*] v. blód-, ceorf-, hand-, hup-, lǽce-, nægel-, scear-, þeóh-, wæl-seax; *and* cf. sagu.

Seax- *in proper names:*—Sigeferþ Seaxing, Seaxa Sledding (*in a list of East Saxon kings*), Txts. 179, 23. Cf. *Icel.* Járn-Saxa = *iron-chopper,* the name of an ogress in the Edda. Ðá féng tó Eást-Seaxna ríce Swíþhelm Seaxbaldes suna, Bd. 3, 22; S. 553, 42. Ðæs cyninges (*Anna of East Anglia*) dohter Sexburh, 3, 8; S. 531, 24: Chr. 639; Erl. 27, 6. Hér forþférde Cénwalh (*of Wessex*), and Seaxburg án geár rícsode his cuén æfter him, 672; Erl. 34, 34. Gesecg Seaxnêting (*East Saxon*), Txts. 179, 16. Cf. Saxnôt *in the formula of renunciation.* v. Grmm. D. M. 184. Seaxrêd (*East Saxon*), 179, 19. Seaxulf biscop (*of Lichfield*), Bd. 4, 6; S. 573, 40. Saxulf (Sǽx-), Chr. 656; Erl. 30, 2, 10.

seax-ben[n]. v. six-ben[n].

Seaxe; Seaxan; *pl. The Saxons,* (1) in connection with England:—Cômon hí of þrim folcum ðám strangestan Germanie, ðæt [is] of Seaxum and of Angle and of Geátum ... Of Seaxum, ðæt is of ðam lande ðe mon hâteþ Eald-Seaxan, cóman Eást-Seaxan (-Seaxa, -Sexa, Chron. 449) and Sûþ-Seaxan (-Sexa, Chron.) and West-Seaxan (-Sexa, Chron.), Bd. 1, 15; S. 483, 20–24. Ðá wǽron Seaxan sécende intingan, S. 483, 36. On Germanie ðanon Engle and Seaxan cumene wǽron, 5, 9; S. 622, 14. Engle and Seaxe, Chr. 937; Erl. 115, 19: Menol. Fox 368; Men. 185. Sexna kyning, 459; Men. 231. Æt Seaxena handa forwurþan, Chr. 605; Erl. 21, 28. Englum and Sexum (Sæxum), 1065; Erl. 196, 30. Ðæt spell ðæt ic áwrât be Angelþeóde and Seaxum, Bd. pref.; S. 471, 10. (2) continental Saxons:—Ðý ilcan geáre gegadrode micel sciphere on Ald-Seaxum, and ðǽr wearþ micel gefeoht ... and ða Seaxan hæfdun sige, Chr. 885; Erl. 84, 8. Ic wæs mid Seaxum, Exon. Th. 322, 12; Vîd. 62. [*O. H. Ger.* Sahsun: *Icel.* Saxar. For the connection of Seaxe(-an) with Seax, v. Grmm. Gesch. D. S. c. xxiii.] v. Eald- (Ald-), Eást-, Sûþ-, West-Seaxe.

Seax-land, es; *n. England:*—Com Gûðrum on eástdǽle Sexlandes, Shrn. 16, 4.

sécan, sêcean; *p.* sôhte; *pp.* sôht *To seek.*   I. (1) *to try to find, to look for, make search for:*—Ic séce míne gebrôðru *fratres meos quaero,* Gen. 37, 16. Hwæne sécst ðú? Jn. Skt. 20, 15. Se ðe sécþ, hê hyt fint, Mt. Kmbl. 7, 8. Hwæðer gê willen on wuda sécan gold ðæt reáde? ... Hit witena nán ðider ne séceþ (cf. gê hit ðær ne sécaþ, ne finde gê hit nô, Bt. 32, 3; Fox 118, 9), Met. 19, 8. Ðonne gê Drihten sécaþ, ðonne gemête gê hine, gif gê hine mid inweardre heortan séceaþ, Deut. 4, 29. Gê séceaþ (soecas, Lind.) ðone Hǽlynd, Mt. Kmbl. 28, 5. Ic áxode hine, hwæt hê sôhte, Gen. 37, 15. Ðín fæder and ic sárigende ðé sôhton, Lk. Skt. 2, 48. Hí sôhton hyne, Mt. Kmbl. 21, 46: Blickl. Homl. 241, 12. Mannes sunu com sécean (tó soecanne, Lind.) and hâl dôn ðæt forweard, Lk. Skt. 19, 10. Sécende God *requirens Deum,* Ps. Spl. 13, 3. (2) *to try to get* (the source from which a thing is sought marked by tô):—Ic monnes feorh tó slagan séce (MS. sêðe) *I will require man's life of the slayer,* Cd. Th. 92, 7; Gen. 1525. Ic tó Drihtne séce ðæt ic gôd æt him begitan môte *quaesivi bona tibi,* Ps. Th. 121, 9. Gif ðú ðé tó swá mildum mundbyrd sécest, Exon. Th. 252, 29; Jul. 170. Heó úrne fultum sêhþ, Homl. Th. ii. 112, 18. Gumena gehwylcum ðara ðe geóce tó him séceþ, Andr. Kmbl. 2307; An. 1155. Ðǽr is help gearu manna gehwylcum ðam ðe séceþ tó him, 1818; An. 911. Gê hí sécaþ tó fremdum gesceaftum, Bt. 14, 2; Fox 44, 17, 29. Sûþ-Seaxna mǽgþ him biscopþéninge séceaþ tó West-Seaxna biscope, Bd. 5, 23; S. 646, 24. Ðæt se án ne ætburste ðe hê sôhte, Homl. Th. i. 82, 13. Hwílum man ceás ða men ðe noldan swician ... and syððan hit man sôhte be ðam ðe nearwlícast cûðan swician *at one time the men were chosen that would not deceive ... and since they have been looked for among those that could most oppressively deceive,* L. I. P. 12; Th. ii. 320, 24. Ús is nêðþearf ðæt wê sécan ðone lǽcedôm úre sáuwle, Blickl. Homl. 97, 31. Biddon wê Drihten ðæs leóhtes ðe nǽfre ne geendaþ ... ðæt leóht wê sceolan sécan, ðæt wê môtan habban mid englum gemǽne, 21, 14. Bearn Godes brýda ongunnon on Caines cynne sécan, Cd. Th. 75, 33; Gen. 1249. Woldon tó dûnscræfum drohtoþ sécan, Andr. Kmbl. 3077; An. 1541. Uton sibbe tó him sécan, Exon. Th. 365, 11; Wal. 87. Seócan, Ps. C. 109. Hwæt elles is tó sécanne wið ðam hungre nymþe andlyfen, Bd. 1, 27; S. 494, 16. Hê gǽþ sécende reste, Mt. Kmbl. 12, 43. Sió ǽ sceal beón sôht on ðæs sacerdes mûþe, Past. 15; Swt. 91, 17. (3) *to try to attain an end, strive to effect a purpose, aim at, strive after, make something the object of endeavour:*—Ic ne séce mínne willan ac ðæs ðe mê sende, Jn. Skt. 5, 30: 8, 50. Hwæt sécst ðú? 4, 27. León hwelpas sécaþ, ðæt hî God gedême, Ps. Th. 103, 20. Gif hê ðone dôm ofer hine sôhte *if the other tried to get judgment upon him,* L. Alf. 49; Th. i. 56, 33. Ðá hálgan ðe on ðyssum life náht ne sôhton ne ne gyrndon tó hæbbene, Blickl. Homl. 53, 25. Hí sôhton

hine him tó hláforde and tó mundboran *they tried to get him to be their lord and protector,* Chr. 921; Erl. 107, 29: 922; Erl. 108, 20, 28. Gif ðæt riht tó hefig sý, séce siþþan ða líhtinge tó ðam cynge, L. Edg. ii. 2; Th. i. 266, 11. (4) *to try to find out* by investigation or examination:—Hwylc séceþ ðæt ðe sôðfæst byþ *veritatem quis requiret?* Ps. Th. 60, 6. Sôhte synnum fáh, hú hê sárlícast meahte feorhcwale findan ... Feónd hine gelǽrde, Exon. Th. 276, 24; Jul. 571. Georne smeádon, sôhton searoþancum, hwæt sió syn wǽre, Elen. Kmbl. 827; El. 414. Ongan on sefan sécean sôðfæstnesse weg tó wuldre, 2295; El. 1149. Ic ðíne gewitnesse wylle sécan *testimonia tua exquisivi,* Ps. Th. 118, 22. Lǽcedôm sécan *medicamentum explorare,* Bd. 1, 27; S. 494, 18. Hwílum beóþ ða wǽtan on ðære wambe filmenum, ðonne sceal mon ðæt wíslíce sécean, Lchdm. ii. 222, 24. (5) *to try to learn* by asking, *to ask:*—Ða mê cunnon andsware cýðan tácna gehwylces ðe ic him tó séce, Elen. Kmbl. 638; El. 319. Ðá cwæð Maria tó ðǽm engle: Hwæt is ðín nama? Ðá cwæð se engel tó hire: Hwæt sécestû mínne naman? Blickl. Homl. 137, 29. Hê ðá Drihtnes willan sôhte *he tried to learn what was the will of the Lord,* 225, 30. Wíslíce gê dyde, ðætte mannum bedîgled wæs on eorþan, ðæt gê ðæt on heofenas tó Gode sôhtan, 201, 2. Tó sécenne, 205, 27. Ic wât ðæt hió wile sécan (*ask.* Cf. Ðá seó cwên ongan friccgan, 1116; El. 560) be ðam sigebeáme, Elen. Kmbl. 840; El. 420.   II. *to go* or *come to:*—Oft sécende *frequentantem,* Wrt. Voc. ii. 34, 18. (1) *to seek* a person, *to visit* (cf. *Ger.* be-suchen):—Ðér beóþ gegearwoda Godes mildheortnessa ðǽm mannum ðe ða líchoman séceaþ þurh heora gebedo, Blickl. Homl. 193, 21. Ða ðe æfter deáþe Dryhten sécaþ, Andr. Kmbl. 1200; An. 600. Ðá hê ðone cyningc sôhte *when he visited the king,* Ors. 1, 1; Swt. 18, 10. Sárge gê ne sôhton *ye did not visit the afflicted,* Exon. Th. 92, 19; Cri. 1511. Hig ðæs wyrðe wǽron ðæt Godes englas hig sôhton, L. E. I. 25; Th. ii. 422, 15. Séc nú ðínne þeów, Blickl. Homl. 87, 31. Hider ic wille ðæt wê sêcan Sce Petre, Chr. 656; Erl. 31, 32. Satan ic sêcan wille, Cd. Th. 47, 15; Gen. 761. Gewît ðú ðínne eft waldend sécan *go back again to your master,* 138, 17; Gen. 2293: Andr. Kmbl. 1886; An. 945. (1 a) *to seek* a person for protection, *to take refuge with* a person. v. sôcn, VI. 2:—Gif hwilc þeóf oððe reáfere gesôhte ðone cing ... hê hæbbe nigon nihta fyrst. And gif hê ealderman oððe abbud oððe þegen séce, hæbbe þreora nihta fyrst, L. Ath. iv. 4; Th. i. 222, 28. (2) *to seek* a place, *to visit, resort to:*—Hê (*the phenix*) sunbeorht gesetu sêceþ, Exon. Th. 217, 11; Ph. 278. Ða men ðe ðyder cóman and ða hálgan stôwe sôhton, Blickl. Homl. 125, 28: 201, 11. Hí syððan gewunelíce ðider sôhton *they afterwards resorted thither,* Homl. Th. i. 504, 6. Sêce man hundredgemôt, L. Edg. ii. 5; Th. i. 268, 2. Ðæt ðeós onlícnes eorþan sêce *fall to earth,* Andr. Kmbl. 1462; An. 731. Ðeáh heorot holtwudu séce, Beo. Th. 2743; B. 1369. Ðæt hí secggan ðæm folce ðæt hí sunnandagum Godes cyrican georne sêcan, Blickl. Homl. 47, 28: L. C. E. 2; Th. i. 358, 14. Gif hié ǽnigne feld sécan wolden *if they should attempt to come into the open country,* Chr. 894; Erl. 90, 11. Gewitan him Norþmen Difelin sêcan, 937; Erl. 115, 4. Ðonne sculon hié ðás helle sêcan, Cd. Th. 26, 14; Gen. 406: 136, 30; Gen. 2266. Óðerne êðel sêcan, Blickl. Homl. 23, 6. Mere sêcan *to go to sea,* Exon. Th. 474, 5; Bo. 25. (3) *to seek* immaterial things, *to go to war, resort to* artifice, etc.:—Ic ne sôhte searoníþas, ne ne swór fela áþa on unriht, Beo. Th. 5469; B. 2738. Se wuldres dǽl sigorleán sôhte *the soul has gone to its reward,* Exon. Th. 184, 14; Gû. 1344. Se rinc sôhte óðer líf, Cd. Th. 98, 9; Gen. 1627. Hí clǽnsunge bæþes sôhton, Bd. 1, 27; S. 495, 16. Hié noldan leng heora hláforda ne heora wera rǽstgemânan sêcean, Blickl. Homl. 173, 16. Ðá ðú gehogodest sæcce sêcean, Beo. Th. 3982; B. 1989: 5117; B. 2562. Fǽhþe sêcan, 5020; B. 2513.   III. *to seek with hostile intent* (as in *to seek* a person's life), *to try to get at, to go to attack:*—Mê fyrenfulle fǽcne sêcaþ, wyllaþ mê lifes ásêcean *me expectaverunt peccatores, ut perderent me,* Ps. Th. 118, 95. Him (hié, hí *other MSS.*) mon mid óðrum floccum sôhte, Chr. 894; Erl. 90, 14. Hié micle fierd gegadrodon and ðone here sôhton æt Eoforwícceastre, 867; Erl. 72, 13. Ða ðe míne fýnd wǽron, and míne sáwle sôhton mid níðe, Ps. Th. 69, 2: 85, 13: Mt. Kmbl. 2, 20. Hié alle from him ondrêdon, ðæt hí hié mid ecgsêhte sôhte, Ors. 1, 10; Swt. 48, 17. Sêcan míne fýnd míne sáwle *persequatur inimicus animam meam,* Ps. Th. 7, 5. Ðá hié gewin drugon, and on healfa gehwone heáwan þohton, sáwle sêcan, Beo. Th. 1606; B. 801. Sêcean sáwle hord, sundur gedǽlan líf wið líce, 4835; B. 2422. [*Goth.* sôkjan: *O. Frs.* sêka: *O. Sax.* sôkian: *O. L. Ger.* suocan: *O. H. Ger.* suohhan *quaerere, petere, exquirere, arcessire, appetere, invisere: Icel.* sœkja *to seek, fetch; to visit, frequent; to prosecute* (*a suit*); *to attack.*] v. á-, for-, ge-, geond-, ofer-, on-, under-sêcan.

secg, es; *m. n. Sedge;* carex, gladiolus, lisca :—Ðis secg (segc) *haec carex,* Ælfc. Gr. 9, 61; Zup. 69, 16. Segg, secg, saecg *gladiolum,* Txts. 66, 463. Sech *carex,* 50, 251. Seic, 115, 151. Secg, Wrt. Voc. ii. 13, 28. Segc, i. 79, 65. Segg, 67, 3. Secg *gladiolum,* ii. 40, 70. Segc, 70, 29. Secgg, i. 67, 55. Secg *lisca,* ii. 53, 45: *carex* vel *sabium* vel *lisca,* i. 31, 28. Endlefan snǽda reádes secges, Lchdm. ii. 102, 17. Handfulle secges, 356, 1. Wyl neoþoweardne secg, 52, 16: 66, 5.

[Eolug-secg *papyrus*, Wrt. Voc. ii. 67, 58. Ilug-segg, Txts. 86, 781. *See also* eolhx, hamer-, môr-secg. Grein cites risc-seccas *carices*.] Cf. secg *a sword*.

secg, es; *m. A man* (used only in poetry):—Secg oððe meówle *man or maid*, Exon. Th. 387, 15; Rä. 5, 5. Nis ænig eorl under lyfte, secg searoþoncol, 14, 16; Cri. 220. Se beorn, sêfteádig secg, 309, 12; Seef. 56. Secg, lagucræftig mon, Beo. Th. 422; B. 208. Swylc sceolde secg wesan, þegen æt þearfe, 5410; B. 2708. Beówulf, sigoreádig secg, 2626; B. 1311. Ðær læg secg mænig, guma norþerna, Chr. 937; Erl. 112, 17. Secgas and gesíþas fôron tô gefeohte, Judth. Thw. 24, 22; Jud. 201. Seccas, Chr. Th. 124, 23; Gen. 2067. Wæron æscwýgan, secggas ymb sigecwên sîðes gefýsde, Elen. Kmbl. 519; El. 260. Rôm-ware, secgas sigerôfe, 93; El. 47. Ðá ic sæbât gesæt mid mînra secga gedriht, Beo. Th. 1271; B. 633. [Laym. seg, sæg; *pl.* segges: Piers P. segge: O. Sax. segg: Icel. seggr (*poet.*).] v. ambyht-, ærend-, sele-secg.

secg, es; *m. The sea*:—*Salum* seeg (secg?) *vel mare*, Txts. 95, 1786. Segg, seg *salum*, 98, 966. Segc, Wrt. Voc. i. 289, 37. v. gâr-secg.

secg, e; *f. A sword*:—Wit sculon secge ofersettan, gif hê gesêcean dear wîg ofer wæpen, Beo. Th. 1372; B. 684. Secgum ofslegene, Cd. Th. 120, 27; Gen. 2001. [Cf. Icel. ben-sægr *as a name for the sword*.] Cf. secg *sedge, and* sagu; *and* secg-hwæt, -plega.

secga, an; *m. One who says or tells, an informant*:—Ne ic nân sôðre wât, bûte swâ mîn secga mê sæde, L. O. 4; Th. i. 180, 12. [Þer weore segge (*or from* seg *a man* (?). *The other MS. has* gleomenne) songe, Laym. 5109. Cf. O. Sax. O. H. Ger. sago: O. Frs. sega, *in compounds*.]

secgan, secgean, secggan, secggean, sæcgan; *p.* sægde, sæde; *pp.* sægd, sæd [*Forms as from an infin.* sagian—sagast, sagaþ; *p.* sagode; *imp.* saga, *are given here*.] *To say* (of written or spoken words). **I.** *to say* certain words, the words used being given:—Hê segþ: Gê ne mágon cuman ðyder ic fare, Jn. Skt. 8, 22. Gif hwâ segþ, corban, Mk. Skt. 7, 11. Sege folce: Ðis sind ða dagas, Lev. 23, 2. Secgaþ ðæs hûses hlâforde: Úre lâreów secgþ: Hwâr is mîn gysthûs, Mk. Skt. 14, 14. Hwæðer is êðre tô secgenne tô ðam laman: 'Ðê synd ðîne synna forgyfene,' hwæðer ðe cweðan: 'Árîs, nim ðîn bed, and gâ, Mk. Skt. 2, 9. Wê gehýrdon hine secgan: Ic tôwurpe ðis tempel, 14, 58. **I a.** of words, *to mean*:—*Cantica canticorum*, ðæt segþ on Englisc ealra sanga fyrmest, Ælfc. T. Grn. 7, 42. **II.** with acc. (1) where the object denotes a collection of words, a story, poem, regulation, etc., *to tell* a tale, *recite* a poem, *pronounce*, *deliver*:—Ic bî mê secge ðis sârspell, Exon. Th. 458, 6; Hy. 4, 96. Ðonne ic ðê æfenlâc secge, Ps. Th. 140, 3. Ðus word ðe ðú mê sagast, Exon. Th. 247, 26; Jul. 84. Ðú worn fela ymb Brecan spræce, sægdest from his sîðe, Beo. Th. 1068; B. 532. Ðá sæde hê him sum bigspel, Lk. Skt. 12, 16. Se magorǽswa mægþe sînre dômas sægde (cf. O. Sax. ēo-sago: O. Frs. ā-sega: Icel. segja lög; lögsögu-maðr), Cd. Th. 98, 4; Gen. 1625. Éce rǽdas Moyses sægde, 210, 17; Exod. 516. Sægde eorlum Abimeleh waldendes word, 161, 19; Gen. 2667. Wordum sægde Lameh unârlíc spel, 66, 27; Gen. 1090. Wê lofsonga word sǽdon, 274, 18; Sat. 156. Abeód eft ongeán, sege ðînum leódum miccle lâþre spell, Byrht. Th. 133, 14; By. 50. Nâne gewitnesse æfter him ne saga ðú, L. Alf. 40; Th. i. 54, 5. His naman secgeaþ mid sealmum, Ps. Th. 65, 1. Secgan spell, Bt. 13; Fox 36, 31: 30, 1; Fox 106, 30. Andsware secgan *to return answer*, Elen. Kmbl. 752; El. 376: 1131; El. 567. Sang secgan *to sing a song*, Cd. Th. 279, 10; Sat. 235. Naman sæcgean, Ps. Th. 141, 8. Ðonne wê gehýron Godes bêc reccean and rǽdan, and godspell seccgean, Blickl. Homl. 111, 17. Hié forgytaþ ðæt hié hwêne ǽr gehýrdon reccean and secggan, 55, 28. Hwæt sceal ic mâ secgean fram Sancte Iohanne? 169, 24. Ðæt him ǽr of ðæs lâreówes mûþe wæs bodad and sægd, 55, 31: 69, 19. Býþ sægd nama Drihtnes *ut annuntient nomen Domini*, Ps. Th. 101, 19. ¶ where the object is included in a genitive:—Ðæs ðú mê wylle wordum secgean *from what you tell me*, Cd. Th. 162, 2; Gen. 2675. (1 a) where the written form of a word is referred to:—Ic mæg þurh rûnstafas secgan naman ðâra wihta, Exon. Th. 429, 18; Rä. 43, 6. (2) where the object denotes that which is spoken about, *to speak of, tell, relate, narrate, declare, announce, give an account of* something:—Ic ðê orlæg secge *I will tell thee thy fate*, Cd. Th. 262, 19; Dan. 746. Ic Gode líf mîn secge *vitam meam nuntiavi tibi*, Ps. Th. 55, 7. Ic mîne earfeþu sæcge *tribulationem meam pronuntio*, 141, 2: 54, 17. Ðú sagast lífceare, Cd. Th. 54, 17; Gen. 878. Ðis gewrit oððe hit gód sagaþ be gôdum mannum, oððe hit yfel sagaþ be yfelum mannum *sive historia de bonis bona referat . . . seu mala commemoret de pravis*, Bd. pref.; S. 471, 14. Mîn mûþ sægeþ (*pronuntiabit*) ðîne mægenspêde, Ps. Th. 70, 14. Hî secgeaþ (*narrabunt*) eall ðîn wundur, 144, 5. Gé scyldigra synne secgaþ, Exon. Th. 132, 23; Gû. 477. Nêh ðæm clife ðe ic ǽr sǽde *that I spoke of before*, Ors. 1, 1; Swt. 12, 30. Heó sǽde him eall ðæt riht, Mk. Skt. 5, 33. Hǽlend his þegnum sǽde his þrowunga, Blickl. Homl. 15, 33. Sagode *refert*, Germ. 396, 10. Hê sîðfæt sægde, Cd. Th. 256, 31; Dan. 649. Hit forhæfed geweard, ðætte hié sǽdon swefn cyninges, 225, 2; Dan. 148. Bodan þurh hleóþorcwide hyrdum cýððon,

sægdon sôðne gefeán, Exon. Th. 28, 23; Cri. 451. Ic ðê hâte, ðæt ðú ðâs gesyhþe secge mannum, Rood Kmbl. 190; Kr. 96. Ne wê wîtegan habbaþ, ðæt ús andgytes mâ secgen, Ps. Th. 73, 9. Hî ðîne mihte sæcgeon *potentiam tuam pronuntiabunt*, 144, 4. Ic ðê secgan wille or and ende, Andr. Knibl. 1296; An. 648. Hê secgan ongan swefnes wôman, Cd. Th. 249, 32; Dan. 539. Ðæt ðú hellwarum hyht ne âbeóde, ah ðú him secgan miht sorga mǽste, 308, 21; Sat. 696. Nô ic wiht fram ðê swylcra searuníþa secgan hýrde, billa brôgan, Beo. Th. 1169; B. 582. Ðara ârfæstra dǽda sume gehýran sæcgan, Blickl. Homl. 213, 26. Wê gehýraþ oft secggan worldrícra manna deáþ, 107, 29. Ne his snytru mæg secgean ǽnig. Ps. Th. 146, 5. Hî sculon his weorc sæcgean *annuntient opera ejus*, 106, 21. (3) *to express in words* feelings of gratitude, admiration, etc., *to give* thanks, glory, etc., to a person (cf. Ger. Dank sagen):—Ic ðara frætwa þanc wuldurcyninge wordum secge, Beo. Th. 5583; B. 2795. Wê ðé wuldur sæcgeaþ, Ps. Th. 78, 14. Hê sægde him ðæs leánes þanc, Beo. Th. 3623; B. 1809. Secggan wê him þanc ealra his miltsa, Blickl. Homl. 103, 25. Þancas secggan, 115, 22. Ðæm Scyppende lof and wuldor secgean ðara ára,'123, 4. Lof secgan Dryhtne, Andr. Kmbl. 2011; An. 1008: Exon. Th. 138, 34; Gû. 586. Ðæs wê ealles sculon secgan þonc and lof, 38, 25; Cri. 612. Hê for his hǽlo Drihtne þanc secgende wæs *pro sua sanitate Domino gratias referens*, Bd. 4, 31; S. 610, 38. (4) where the object is a pronoun referring to a clause:—'Eart ðú Iudéa cining?' Ðá andswarude hê: 'Ðú hit segst,' Lk. Skt. 23, 3. Saga mê ðæt, for hwon sécest ðú sceade, Cd. Th. 54, 6; Gen. 873. Gif ðú wille mildheortnesse ús dôn, sæge ús ðæt hrædlíce, Blickl. Homl. 233, 19. Dryhten micellíce dyde; seggaþ ðis in alre eorþan, Ps. Surt. ii. p. 184, 15. Ic ðæt londbûenden secgan hýrde, ðæt hié gesâwon . . ., Beo. Th. 2697; B. 1346. *(all that had been seen and heard*) mancynne bodian and secgan, Blickl. Homl. 121, 4. Is ðæt sægd, ðæt . . ., Bd. 3, 2; S. 524, 16. (5) where the verb is of incomplete predication, *to declare* a person or thing so and so:—Ic secge hine mǽran ðonne ǽnigne wîtgan, Blickl. Homl. 165, 3. Se hæfde mægen ofer calle gesceafta ðe hê tôwearde sægde, 9, 16. Oðer him ðâs eorþan ealle sægde lǽne, Exon. Th. 109, 15; Gû. 90. Hî ðone clǽnan sacerd sægdon tôweard, 9, 20; Cri. 137. Ða hâlgan hine tôweardne sægdon, Blickl. Homl. 81, 31. Hié hine scyldigne sægdon, 173, 33. Hié sægdon hine sundorwisne, Elen. Kmbl. 1172; El. 588. **III.** with gen.:—Swâ se secghwata secggende wæs lâðra spella, Beo. Th. 6049; B. 3028. **IV.** where the object is a clause, *to say, tell*:—Ic secge ðé, ðæt ðú eart Petrus, Mt. Kmbl. 16, 18. Nû segþ ús seó bôc, ðæt God âfcdde ðone here, Ælfc. T. Grn. 5, 32. Seó bôc segþ, hû hê fêrde, 6, 5. Heó mê sagaþ, ðæt . . ., Exon. Th. 246, 30; Jul. 69. Swâ Arculfus sagaþ, ðæt hê gesâwe . . ., Shrn. 95, 31. Ðæs is tô tâcne, sæcgeaþ men, ðæt oft .XL. manna . . . ðæt hî hî be handum nôman and of sǽs ôfre ût feóllan, Bd. 4, 13; S. 582, 30. Ic wordum sægde, ðæt Sarra mîn sweostor wǽre, Cd. Th. 163, 25; Gen. 2703. Sæge Adame, hwilce ðú gesihþe hæfst, 38, 35; Gen. 617. Saga mê, hwylces cynnes ðú sí, Bd. 1, 7; S. 477, 26. Secgaþ mê, hwæt git gesâwon, Gen. 40, 8. Secgge Petrus, hwæt ic þence, Blickl. Homl. 181, 8. Ic eów bidde, ðæt gê mê secgan, hwylce gemete gê côman ealle samod tô mê, 143, 20. Ðæt hî secggan, ðæt . . . 47, 26. Secgan, hû him æt ǽte speów, Beo. Th. 6044; B. 3026: Exon. Th. 437, 31; Rä. 56, 16. Be songe secgan, hwǽr ic sǽlast wisse goldhrodene cwên, 324, 26; Víd. 100. Seggan, ðæt ic gesǽlig mon wǽre, Bt. 2; Fox 4, 13. Secgian hwæðer wǽre twegra strengra, Salm. Kmbl. 851. Micel is tô secgan, ðæt hê âdreág, Exon. Th. 134, 4; Gû. 502. Long is tô secganne, hû . . ., 421, 23; Rä. 40, 22: Andr. Kmbl. 2961; An. 1483. Swâ hit is nú hrædost tô secganne be eallum ðæm woruldgesǽlþum . . . ðæt ðǽr nân wuht on nis ðæs tô wilnianne seó *postremo idem de tota concludere fortuna licet, in qua nihil expetendum*, Bt. 16, 3; Fox 56, 29. Ðæt is nú hrædost tô secganne, ðæt ic wilnode weorþfullíce tô libbanne ða hwíle ðe ic lifede, 17; Fox 60, 14. Sægd is, ðæt . . ., Blickl. Homl. 61, 16. Se wæs sǽd ðæt his brôðor wǽre Oswíes sunu *qui frater ejus et filius Oswiu esse dicebatur*, Bd. 4, 26; S. 603, 7. **V.** where the verb is used impersonally (cf. Icel. segir *it is told*):—Hit segþ on bôcum, ðæt . . ., Wulfst. 146, 16. Swâ hit hýrefter segeþ, L. Wih. pref.; Th. i. 36, 13. Hî ecton ða ǽ ðyssum dômum ðe hýrefter sægeþ, L. H. E. pref.; Th. i. 26, 7. Hér segþ, hú se æþela wæs sprecende, Blickl. Homl. 55, 3. Gehýraþ hwæt hér segþ on ðissum bôcum be Sancta Marian, 137, 20. Segeþ ðǽron, ðæt sum ríce man wǽre on ðære burh, 197, 27. Sægþ on ðissum bôcum, ðæt . . ., 41, 3. Hér sægþ be ðisse tîde ârwyrþnesse, hú Drihten hine selfne geeáþmédde, 65, 29. [Hér] sagaþ, ðæt Idpartus ðam câsere hǽlo bodade, Lchdm. i. 326, 1. **VI.** where the verb is used absolutely (secgan be, fram, ymbe *to speak of*):—Swâ swâ ic nú æt feáwum wordum secge, Bd. 3, 17; S. 545, 14. Swâ swâ seó bôc sagaþ, 3, 19; S. 547, 32. Swâ wê eft secgeaþ, 3, 21; S. 551, 31. Tô ðæm gesǽlþum, ðe wê secgaþ ymb, Met. 21, 4. Swâ ic ǽr sǽde, Chr. 894; Erl. 92, 6. Mê lyste bet, ðæt ðú mê sǽdest sume hwíle ymbe ðæt, Bt. 34, 6; Fox 142, 12. Gehêraþ hú Lucas sægde be ðisse tîde, Blickl. Homl. 15, 4. Heáhfæderas sægdon and cýððon, sealmsceopas sungon and sægdon, 105, 9-10. 'Ic hæbbe ðê tô secgenne sum þing.' Ðá cwæð hê: 'Lâreów sege ðænne,' Lk. Skt. 7,

40. Saga mē from đam lande, Salm. Kmbl. 418; Sal. 209. Đū đone mǣngengan mē helan woldest, swȳđor đonne mīnum þegnum secgean, Bd. 1, 7; S. 477, 20. Hwylcumhwego wordum secgan be đære ārwyrþnesse đisse hālgan tīde, Blickl. Homl. 115, 29. Secggean, 211, 12. Wē nū gehȳrdon of hwylcumhugu dǣle secggan be đǣm eádmōdnessum, 103, 18. **VII.** secgan on (with acc., dat.) to ascribe to a person, lay to the charge of, accuse of, attribute to :—Ne mæg se scrift geseón on đare sāwle, hwæđer him mon sōđ đe lyge sagaþ on hine sylfne, Exon. Th. 80, 16; Cri. 1308. Đæs hē sceal fægnian, đæt hī him sōđ on secggaþ, Bt. 30, 1; Fox 108, 10. Ne andwyrtst đū nān þing ongēn đa đe điss đē on secgeaþ nihil respondes ad ea, quae isti adversum te testificantur ? Mt. Kmbl. 26, 62. Hī wrōhta and yfel on sægdon, Bd. 3, 19; S. 548, 35. Wæs kȳđed đæt his wrēgend leáse wiđ hine syredon and on sægdon probatum est accusatores ejus falsas contra eum machinasse calumnias, 5, 19; S. 640, 14. Gif ǣnig mann ōđerne wrēge and him hwilcne gilt on secge si steterit testis mendax contra hominem, accusans eum praevaricationis, Deut. 19, 16. Gif đē mon sōđ on secge, Prov. Kmbl. 70. Gif man secge on landesmann, đæt hē orf stǣle, L. Eth. ii. 7; Th. i. 288, 7. Đæm gielpnan biþ leófre đæt hē secge on hine selfne gif hē hwæt gōdes wāt ge þeáh hē nyte hwæt hē sōđes secge him is leófre đæt hē leóge eligit arrogans bona de se vel falsa jactari, Past. 33, 2; Swt. 217, 14. Hwæt gōdes māgan wē secgan on đa flǣsclícan unþeáwas quid de corporis voluptatibus loquar ? Bt. 31, 1; Fox 110, 24. Geunsōđian đæt him man on secgan wolde to disprove what a man would charge him with, L. Edg. ii. 4; Th. i. 266, 4. Ne mōt nān mann secgan on hine sylfne đæs đe hē wyrcende næs, Homl. Skt. i. 12, 177. Ic nelle secgan unsōđ on mē sylfe, 195. [O. Frs. sega, sedsa : O. Sax. seggian : O. H. Ger. sagēn : Icel. segja, seggja.] v. ā-, be-, for-, ge-, on-, sōþ-secgan.

secge, an; f. Speaking, speech :—Mē nāwđer deág secge ne swīge neither speech nor silence will avail me, Exon. Th. 12, 23; Cri. 190. Cf. secga.

secgend, es; m. A speaker, relater, narrator :—Nǣnig tweógende secgend mē đis sǣde non quilibet dubius relator hoc mihi narravit, Bd. 3, 15; S. 542, 7. Sió leásung simle deret đǣm secggendum, Past. 35, 1; Swt. 237, 10. [Icel. segendr, seggendr ; pl. sayers, reporters.]

secg-gescēre (?) sedge-shears (?), a name of the grasshopper :—Secggescēre vel hāman cicad[ae], Txts. 51, 464. v. sceár.

secg-hwæt; adj. Vigorous or bold in using the sword :—Se secghwata, Beo. Th. 6048; B. 3028.

secgihtig; adj. Sedgy, full of sedge or reeds :—Secgihtig vel hreódihtig carecta, loca caricis plena, spinacurium, Wrt. Voc. ii. 129, 14.

secg-leác, es ; n. Chive garlic, rush garlic, rush leek (v. E. D. S. Pub. Plant Names) ; allium schoenoprasum, L'chdm. ii. 128, 11 : iii. 28, 11.

secg-plega, an ; m. Sword-play, battle :—Æt đam secgplegan, Andr. Kmbl. 2705; An. 1355. Cf. sweord-plega.

secg-rōf a host of men (?) :—Cwōman wōldagas swylt eall fornom secgrōf wera death carried off the host of men, Exon. Th. 477, 20; Ruin. 27. [Cf. O. H. Ger. ruaba ; f. numerus : Icel. segg-fjöld a host of men ; and rinc-getæl, folc-getæl.]

secg-sceára, -scāra (-scara ?), an ; m. A corn-crake or a quail :—Secgscāra ortigometra (cf. erschen ortigomera, ii. 63, 53 : edischen, 115, 67), Wrt. Voc. i. 63, 21. v. E. D. S. Pub. Names of Birds, p. 177, where bean crake, grass drake, meadow drake, gorse duck are given as names of the corn-crake. [Cf. (?) Icel. skári a sea-mew.]

sēcness, e ; f. Seeking, visiting, visitation :—Tīde soecnisse (sōcnises, Lind.) tempus visitationis, Lk. Skt. Rush. 19, 44.

sēdan to satisfy [:—Āsoedan satiare, Wrt. Voc. ii. 119, 68. Gesēdeþ (-sedeþ? v. next word : but cf. Goth. ga-sōþjan) satiavit, Ps. Th. 106, 4.]

seddan to satisfy. v. un-āsedd ; sadian.

sēde, sēdege to sow, Mt. Kmbl. Lind. 13, 3. v. sǣdian.

seding-līne, sedl. v. steding-līne, setl.

Sedlingas (?) Ethiopians :—Sedlingum (Rēdlingum ?) Aethiopia, Ps. Spl. T. 67, 34.

see, seeg. v. seón, secg the sea.

sefa, an ; m. Understanding, mind, heart :—Sefa sensus (cf. gewit sensus, 42, 35), Wrt. Voc. i. 64, 17 : 282, 27. Sefa nearwode (of Noah when drunk), Cd. Th. 94, 32 ; Gen. 1570. Him (Nebuchadnezzar on recovery from his madness) in gāst becwom rǣdfæst sefa, 257, 2 ; Dan. 652. Næs him hreó sefa, Beo. Th. 4367 ; B. 2180. Gif đīn hige wǣre, sefa swā searogrim, swā đū self talast, 1192 ; B. 594. Him wæs leóht sefa, hyge untyddre, Andr. Kmbl. 2504 ; An. 1253 : Exon. Th. 164, 33 ; Gū. 1021. Geómor sefa, mōd morgenseóc, 458, 3 ; Hy. 4, 94 : Beo. Th. 98 ; B. 49. Leóht sefa, ferhþ gefeónde, Elen. Kmbl. 346 ; El. 173. Weá biþ in mōde, siofa synnum fāh, Frag. Kmbl. 28 ; Leás. 16. Mōdcræfte sēc þurh sefan snyttro, Exon. Th. 28, 5 ; Cri. 442. Sēcan sefan gehygdum, Cd. Th. 219, 4 ; Dan. 49. Sefan sīdne geþanc, 249, 26 ; Dan. 536. Sefan (seofan, MS. A.) snytro, Salm. Kmbl. 133 ; Sal. 66. On sefan (ondgete, Ps. Surt. 77, 72) in sensu, Blickl. Gl. Hié đam Hālgan Gāste onfengon on heora sefan, Blickl. Homl. 137, 6. On wērigum sefan, Exon. Th. 74, 18 ; Cri. 1208. On mildum sefan, 83, 6 ; Cri. 1352. On sīdum sefan, 169, 17 ; Gū. 1096. On sārgum sefan, 183, 20 ;

Gū. 1330. Tō ontȳnenne mīne sefan, Nar. 40, 30. Ic heom ābleonde hera sefan, 45, 7. Þurh rūmne sefan rǣd gelǣran, Beo. Th. 561 ; B. 278. Begēm ūrum sefum intende nostris sensibus, Hymn. Surt. 22, 3. Urum sefum leóht gearce nostris sensibus lumen prebe, 53, 22. v. breóst-, ferhþ- (firhþ-, fyrhþ-), mōd-, wīs-sefa.

sēferlíce, sēfian, sēfre. v. sȳferlíce, seófian, sȳfre.

sēfte ; adj. Soft :—Delicatus, i. tenerus, querulus, amoenus unbrocheard vel sēfta, Wrt. Voc. ii. 139, 40. **I.** of persons, gentle, mild, not stern :—Drihten is swȳđe sēfte suavis est Dominus, Ps. Th. 33, 8. Weorđ ūrum synnum sēfte and milde propitius esto peccatis nostris, 78, 9. **II.** of medicine, mild, not strong :—Đæt is, for hwī se gōda lǣce selle đam hālum men sēftne drenc and swētne, and ōđrum hālum biterne and strangne, Bt. 39, 9 ; Fox 226, 11. **III.** of rest, sleep, undisturbed, untroubled :—Đū eart seó sēfte ræst sōđfæstra, Bt. 33, 4 ; Fox 132, 34. **IV.** easy, comfortable, pleasant, without pain or discomfort :—Rād byþ on recyde rinca gehwylcum sēfte, Runic pm. Kmbl. 340, 13 ; Rūn. 5. Dōþ sīđfæt sēftne and rihtne, Ps. Th. 67, 4. Ful sēfte seld, đæt hī sǣton on, 88, 3. Hē his līchoman forwyrnde sēftra setla and symbeldaga, Exon. Th. 111, 33 ; Gū. 136. Sēlre mē wæs and sēftre, Ps. Th. 118, 71. Đone deáþ hē him gedēþ sēftran đonne ōđrum monnum, Bt. 39, 10 ; Fox 228, 10. **IV a.** in a bad sense, luxurious, voluptuous, effeminate :—Đȳ ne sceolde nān wīs man wilnian sēftes līfes gif hē ǣnigra cræfta rēcþ neque enim vos in provectu positi virtutis, diffluere deliciis, et emarcescere voluptate venistis, Bt. 40, 3 ; Fox 238, 13. [O. H. Ger. semfti.] v. ge-sēfte ; sōfte.

sēft-eádig (?) ; adj. In easy circumstances, free from hardships :—Se beorn ne wāt, eft eádig (sēfteádig, Grein) secg, hwæt đa sume dreógaþ, đe đa wræclāstas wīdost lecgaþ, Exon. Th. 309, 12 ; Seef. 56.

sēftness, e ; f. Quiet, repose, freedom from disturbance :—Hié woldon hiera dagas on sēftnesse geendian ut in privato otio consenescerent, Ors. 6, 30 ; Swt. 280, 22. Hī gewurdon on đære sēftnysse (of the seven sleepers), Homl. Skt. i. 23, 261.

sego, segel, -segel a seal, segen a saying, segen a sign, segl sun. v. secg, segl, in-segel, sægen, segn, sigel.

segl, swegel, segel, es ; m. n. **I.** a sail :—Segl artemon, Wrt. Voc. ii. 7, 24. Segl velum, se mǣsta segl acateon, se mēdemesta segl epidromas, se lesta segl dalum, i. 56, 48-53. Segel velum, lytel segel dalum, 48, 22, 23. Đes segl hic carbasus, đās seglu haec carbasa, Ælfc. Gr. 13 ; Zup. 86, 3. Đā wæs se mǣste merehrægla sum, segl sāle fæst, Beo. Th. 3816 ; B. 1906. Đæt scip wæs ealne weg yrnende under segle, Ors. 1, 1 ; Swt. 19, 34. Nefne hē under segle yrne, Exon. Th. 345, 11 ; Gn. Ex. 186 : Andr. Kmbl. 1009 ; An. 505. Be đæs scipes segele, Bt. tit. 7 ; Fox x. 16. Gif đū đīnes scipes segl ongeán đone wind tōbrǣdst, đū lǣtst eal eówer færeld tō đæs windes dōme, 7, 2 ; Fox 18, 32. Fealdan đæt segl to furl the sail, 41, 3 ; Fox 250, 15. Eówre seglas sendon geseted your sails are set, Shrn. 60, 11. Seglu vela, Wrt. Voc. i. 63, 54. **I a.** used metaphorically of the fiery and cloudy pillars :—Swegl sīđe weóld the pillar governed their journey, Cd. Th. 184, 10 ; Exod. 105. Hæfde God sunnan sīđfæt swegle ofertolden, swā đa mæstrāpas men ne cūđon, ne đa seglrōde geseón meahton, 182, 26 ; Exod. 81. Fyrd geseah, hū đær hlifedon hālige seglas, 183, 10 ; Exod. 89. **II.** a veil, curtain :—Đæs temples segl, Exon. Th. 70, 16 ; Cri. 1139. **III.** a flag, banner (?) :—Segl larbanum (labarum (?). Labarum signum militare Romanorum, pensile, ex panno aut serico contectum, et transversario antennae specie ligno affixum, a suprema conti parte pendens. v. segl-gird, II), Wrt. Voc. ii. 52, 8. [O. Sax. segel: O. H. Ger. segal velum, artemon, carbasus: Icel. segl; n.] v. ofer-segl.

seglan, siglan, seglian ; p. de, ede, ode To sail :—Đā hē hāmweard seglde, Ors. 4, 10 ; Swt. 202, 1. Hē siglde đā eást be lande, 1, 1 ; Swt. 17, 16. Se sciphere sigelede (seglode, MS. E.) west ymbūtan, Chr. 877 ; Erl. 78, 17. Hē hys segl up āhóf, and swȳđe forđ seglode, St. And. 38, 33. Ūt on sǣ tō seglanne, Prov. Kmbl. 64. [O. H. Ger. segelen: Icel. sigla.] v. ge-seglian.

segl-bōsm, es ; m. The swelling out of a sail, sail swelled out by the wind :—Seglbōsm carbasus, Wrt. Voc. ii. 13, 57 : 103, 28 : carbasus, tumor veli, 128, 53. Seglbōsmas carbasa, vela navium, 54 : carbasa, 88, 24.

segl-gerēde, es ; n. Sail-furniture, tackle :—Hē becwæđ his lāford his beste scip and đa segelgerēda đārtō domino suo meliorem suarum navium unam cum sibi pertinentibus armamentis contulit, Chart. Th. 549, 18. [Cf. Icel. segl-reiði sail-rigging.]

segl-gird, es ; m. : e ; f. **I.** a sail-yard, yard of a ship :—Seglgærd antenna, Wrt. Voc. ii. 100, 30. Segelgyrd antenna, i. 48, 17 : antenna vel temo, 56, 39. Mæst sceal on ceóle, segelgyrd (Grein takes this = sail-gird, and as applying to the mast) seomian, Menol. Fox 509 ; Gn. C. 25. Đa twegen endas đære seglgyrde cornua, Wrt. Voc. i. 56, 40 : 48, 18. Segelgyrda antennarum, ii. 5, 41 : 88, 25. Segelgyrdena, mǣsta antennarum, Hpt. Gl. 529, 18. Segelgyrdas antennas, 97, 29. **II.** the cross rod from which a banner hangs (? v. segel, **III**) :—Segelgyrd labara, Wrt. Voc. ii. 78, 24. [Prompt. Parv. seyl-

ȝerd *antenna*. Cf. *O. H. Ger.* segal-poum *antenna;* also *malus: Icel.* segl-viðr *a yard.*] Cf. segl-ród.

seglian. v. seglan.

segling, e; *f. Sailing :*—Ðæt wē ne mid seglinge ne mid rōwnesse ōwiht fremian mihte *ut neque velo neque remigio quicquam proficere valeremus.* Bd. 5, 1; S. 613, 25. Hē mid seglunge binnon ānum dæge com tō Antiochian, Ap. Th. 6, 27.

segl-rád, e; *f. The sail-road, the sea :*—Sīð on seglráde, Beo. Th. 2863; B. 1429.

segl-ród, e; *f. A sail-yard,* Cd. Th. 182, 29; Exod. 83. (v. segl, I a.) [*O. H. Ger.* segal-ruota *antenna.*] Cf. segel-gird.

segn, segen, es; *m. n. A sign :*—Segn *signum,* Wrt. Voc. ii. 120, 61. I. *a sign, mark, token :*—Abraham sette friðotācn (*circumcision*) on his selfes sunu, hēht ðæt segn wesan (wegan?) heáh gehwilcne, ðe his hīna wæs wǣpnedcynnes, Cd. Th. 142, 32; Gen. 2370. II. *a military standard, banner, an ensign :*—Segn ban[*dum*], Txts. 45, 278. Segn, seng, segin *labarum* (v. segl, III), *vixilla,* 73, 1167. Seign (segin?) *vexilla,* 105, 2093. His segen se wæs mid golde and mid godewæbbe gefrætewod and ofer his byrigenne geseted *vexillum ejus super tumbam auro et purpura compositum adposuerunt,* Bd. 3, 11; S. 535, 31. Segn, Beo. Th. 5909; B. 2958. Ðā wæs þúf hafen, segen for sweótum, Elen. Kmbl. 247; El. 124. Sió býman stefen and se beorhta segn, Exon. Th. 65, 30; Cri. 1062. Segnes gúþfana *labara,* Wrt. Voc. ii. 49, 74. Segne *pendiculo* (cf. *labarum,* signum *pensile*), 66, 48. Hæfdon him tō segne beácen ārǣred, gyldenne león *the tribe of Judah had a golden lion for their standard,* Cd. Th. 198, 7; Exod. 319. Hē under segne sinc ealgode *fighting under his flag he defended his treasure,* Beo. Th. 2412; B. 1204. Hié him āsetton segen gyldenne heáh ofer heáfod, 94; B. 47: 2046; B. 1021. Hē siomian geseah segn eallgylden, gelocen leóþocræftum, 5528; B. 2767: 5546; B. 2776. Ðæt nalæs ðæt ān ðæt hī segen fore him bǣron æt gefeohte ac swylce eác on sibbe tíde . . . him mon symble ðæt tācen beforan weg *ut non solum in pugna ante illum vexilla gestarentur, sed et tempore pacis . . . semper antecedere signifer consuesset,* Bd. 2, 16; S. 520, 9. Segn and síde byman, Salm. Kmbl. 907; Sal. 453. Wið ðone segn foran þengel rād, Cd. Th. 188, 23; Exod. 172. Segnas stódon *standards were stationary,* 214, 7; Exod. 565: 197, 4; Exod. 302. Eall mín weorod . . . herebeácen and segnas beforan mē lǣddon *totum agmen me . . . sequebatur cum signis et uexillis,* Nar. 7, 16. II a. used· metaphorically :—Wynrōd segn sōðfæstra *the cross, the standard of the righteous,* Salm. Kmbl. 471; Sal. 236. Gesáwon randwīgan segn (*the pillar of fire*) ofer sweóton, Cd. Th. 185, 23; Exod. 127. [From Latin.] v. eafor-heáfod-segn; segnian.

segn-berend, es; *m. One bearing a standard* (or *crest?*), *a warrior :*—Ne mæg meo oferswīðan segnberendra ǣnig ofer eorþan, nymþe se āna God, Exon. Th. 423, 13; Rä. 41, 20. v. next word.

segn-bora, an; *m. A standard-bearer :*—Hē (*John*) wæs segnbora ðæs ufancundan Kyninges, Blickl. Homl. 163, 22. Segnbora *draconarius* (draconarius *vexillifer, qui fert vexillum ubi est draco depictus*), i. *vexillarius, signifer,* Wrt. Voc. ii. 142, 5. Segnboran, tácnboran *draconarii vel vexillarii vel signiferi,* i. 21, 66.

segn-cyning, es; *m. A king before whom a banner is borne :*—Him ðǣr segncyning (*Grein would read* sigecyning; *but cf.* (?) *the passages from Bede under* segn) wið ðone segn foran rād, Cd. Th. 188, 22; Exod. 172.

segne, an; *f. A seine, sean, a drag-net :*—Næs ðiú segni tōsliten *non est scissum rete,* Jn. Skt. Lind. 21, 11. Of suegna fiscum *de saginae piscibus,* Mt. Kmbl. p. 17, 6. Ongelíc segne *simile saginae,* Lind. 13, 47. Sendas ðæt nett ì segna *mittite rete,* Jn. Skt. Lind. 21, 6. Segni, 8. Hí ongunnon sǣlafe segnum dǣlan, Cd. Th. 215, 17; Exod. 584. [(Pecher) de nase *wit a seyne,* Wrt. Voc. i. 159, 7. *O. Sax.* segina: *O. Frs.* seine: *O. H. Ger.* segina *sagena.* From Latin; cf. *Fr.* seine.]

segnian, sēnian; *p.* ode. I. *to make the sign of the cross upon anything in token of blessing or consecration, to bless, consecrate :*—Se biscop nam hláf and sēnode *essent manus ad panem benedicendum missuri,* Bd. 3, 6; S. 528, 15 note. Ðā sang hē orationem ofer hine and hine bletsode and sēnode *dixit orationem, ac benedixit eum,* 5, 5; S. 618, 8. Sēnade, 5, 6; S. 619, 42. Hē mid his handum húsel sēnode, Homl. Skt. i. 3, 114. Ðā hē sēnade ðæt fæt ðe ðæt āttor on wæs, ðā tōbærst hit, Shrn. 65, 11. Sǣnade, 52, 32. Ðonne ðú hláf brece, sēna ðú ða cruman, 53, 18. Ðeáh· ðe man wafige wundorlíce mid handa, ne biþ hit ðeáh bletsung, búta hē wyrce tácn ðǣre hálgan róde . . . Mid þrým fingrum man sceall sēnian and bletsian, H. R. 105, 22. Hine sylfne sēniende *signando sese,* Bd. 4, 24; S. 599, 13. II. *without reference to the sign of the cross :*—Segnade earce innan āgenum spēdum Nergend, Cd. Th. 82, 21; Gen. 1365: 83, 35; Gen. 1390. III. *of speech* (?) :—Uē sægnade *bene dicimus,* Jn. Skt. Lind. 8, 48. [We sculen ure forheafod mid þere halie rode tacne seinian, O. E. Homl. i. 127, 25. Godd feder ant his sune iseinet (*blessed*), Marh. 23, 18. Þanne sat sleuthe up and seyned hym swithe, Piers P. 5, 456. Swa sal I saine þe, Ps. 62, 5. *O. Sax.* seginón: *O. H. Ger.* seganôn *benedicere: Icel.* signa *to sign,*

consecrate, in heathen times, with Thor's hammer, in Christian times, with the cross; *to bless.*] v. ge-segnian.

segnung, sēnung, e; *f. Blessing, consecration :*—Ofer hine cymeþ mínre segnunga blōstma *super ipsum florebit sanctificatio mea,* Ps. Th. 131, 19. Wæs hē lǣded tō Brytta biscopum and hē nǣnige hǣle ne frōfre þurh heora segnunge (þegnunge?) onfēng *qui cum oblatus Brittonum sacerdotibus, nil curationis vel sanationis horum ministerio perciperet,* Bd. 2, 2; S. 502, 26. v. hláf-sēnung.

seht, es; *m. : f.* I. *a settlement, an agreement, terms arranged between two parties by an umpire, a peace between two powers :*—Se seht ðe Godwine eorl worhte betweónan ðam arcebisceop and ðam hírēde æt Scē Augustine, and Leófwine preóste, Chart. Th. 349, 19. Spǣcon ðá Leófríces freónd and Wulfstānes freónd, ðæt hit betere wǣre, ðæt heora seht tōgædere wurde, ðonne hý ǣnige sace hym betweónan heóldan; sōhtan ðá hyra seht. (*The terms are then given.*) Ðis wæs úre ealra seht, 377, 1–13. Syððan ðæs cáseres seht wæs and Baldwines, Chr. 1050; Erl. 173, 33. Hí tōhwurfon mid ðisum sehte (*the agreement between Edmund and Cnut*), 1016; Erl. 159, 6. Ðá fērdon betwux Rōðbeard eorl and Eádgār æðeling and þæra cinga sehte ᷄swā gemacedon. (*The terms are then given.*) On ðisum sehte weard Eádgār eþeling wið ðone cyng gesæhtlad, 1091; Erl. 228, 1–8. [Fērden þe ærcebiscop and te wíse men betwux heom and makede ðæt sahte ðæt . . ., 1140; Erl. 265, 30.] II. *peace, friendship :*—Syððan seaht and sibb mycelre tíde betwyh ða ylcan cyningas and heora ríce áwunode, Bd. 4, 21; S. 590, 25 note. Ðæt ða cyningas seht nāmon (cf. friþ niman) heom betweónan, Chr. 1016; Erl. 159, 1. Hí mōston mid ealle ðæs cynges wille folgian, gif hí woldon land habban oððe wel his sehta, 1086; Erl. 222, 35. [Sib and sæhte sculde bēn betwyx heom and on al Engleland, 1140; Erl. 265, 32. Betere weore sæhte þene swilc unisibbe, Laym. 9844. God lihte to eorðe uorte makien þreouold seihte, A. R. 250, 2. *Taken from the Danes* (?) cf. *Icel.* sátt *a settlement, agreement; peace.*] v. un-seht *and following words.*

seht; *adj. In agreement* about the terms of a settlement, *agreed :*—Hí wurdon sehte ðæt ða gebrōðra ealle geeodon of ðam lande bútan ánum, Cod. Dip. Kmbl. vi. 195, 25. Hí him ðæs gætíðodon wið swylcon gersumen swylce hí ðá sehte wǣron *such as they were then agreed upon,* 198, 16. Hí wurdon sehte on ða gerād ðæt . . ., Chr. 1093; Erl. 229, 25. Weard se cyng and his brōðor sehte . . . and eall Normandíg æt him mid feó álísde, swā swā hí ðá sehte wǣron, 1096; Erl. 233, 17. Sæhte, 1077; Erl. 215, 10. [Sehte, 1120; Erl. 248, 1. Sæhte, 1135; Erl. 261, 21. Þus iwerað Brennes saht (isehte, 2nd MS.) whit his broðer, Laym. 5114. Hiss bodiȝ wiþþ hiss gast sammtale & sahhte wurrþe, Orm. 5731. Cf. *Icel.* sáttr verða á eitt *to agree on.*] v. un-seht *and next word.*

sehtan; *p.* te *To bring about agreement between* people, *to settle a dispute :*—Cristenum cyninge gebyreþ ðæt hē eall cristen folc sibbie and sehte mid rihtre lage, L. I. P. 2; Th. ii. 304, 12: Wulfst. 266, 17. Ðæt wē habban ús gemǣne sibbe and sōme, and ǣlce sace sehtan, 272, 23. Bisceop sceal beón symle ymbe sōme and ymbe sibbe . . . Hē sceal georne saca sehtan and friþ wyrcan, L. I. P. 7; Th. ii. 312, 14. [A porueance . . . thut lond uor to seyte, R. Glouc. 533, 15. We schul saughte sone (cf. we schulle ben at oon, 156), Chauc. Tale of Gamelyn, 150. Ȝe schulle sauȝte (*agree*), Piers P. A-Text, MS. T. 4, 2. *Icel.* sætta *to bring about agreement.*] v. ge-sehtian.

sehtlian (?); *p.* ode. I. *to settle, bring to an agreement, settle a dispute between* people (the word seems to occur only in the later part of the Chronicle) [:—Ðá eodon gōde men heom betwēnan and sahtloden heom, Chr. 1066; Erl. 203, 27. Ða twegen kyngas wurðon sæhtlod, 1070; Erl. 209, 26. II. *to come to an agreement :*—Ðá feórden ðe wíse men betwyx þe kinges freónd & te eorles freónd & sahtlede suá ðæt . . . Sithen sahtleden þe king and Randolf eorl, 1140; Erl. 264, 31–35. Þe eorles sæhtleden wyd þemperice, Erl. 265, 6.] [Forr to sahhtlenn hemm towarrd hiss Faderr, Orm. 351. When a sawele is saȝtled to dryȝtyn, Allit. Pms. 72, 1139. Ȝe schulle saghtlyn, Piers P. A-Text, MS. U. 4, 2.] v. ge-sæhtlian.

sehtness, e; *f. Agreement, accord, concord, peace :*—Ðám dómbócum ðe se heofonlíca Wealdend his folce gesette tō sōme and tō sehtnesse, Homl. Th. ii. 198, 19. [Geaf ðone cyng .xl. marc goldes tō sahtnysse, Chr. 1066; Erl. 203, 29.] [Crist wass borenn her sahhtnesse & griþþ to settenn, Orm. 3515. He sahtnesse wrohte, Laym. 2809. Sæhtnesse underfon *to accept terms,* 8262. Næfde þa sehtnesse ilast buten seouen ȝere urist, 30137. 'Pax vobis.' Seihtnesse beo bitweonen ou, A. R. 250, 5.] v. ge-sehtness.

[seim [*from earlier* segem (?)] *fat, lard :*—Seime ì fetnesse *adipe,* Ps. Spl. T. 62, 6. [Ge ne schulen eten ulesche ne seim, A. R. 412, 26. *See Halliw. Dict.* saim, seam, *and cf. Fr.* sain: *Ital.* saime. *From late Lat.* sagimen.]]

sel *a hall,* **sel** *a season.* v. sæl, sél.

sél (*the positive form does not occur, but is found in Layamon*); *cpve.* sēlra, sēlla; *spve.* sēlest, sēlost; *adj. Good.* I. *of health :*—Sóna seó blǣdder tō sēiran (*to a healthier condition*) gehwyrfeþ, Lchdm. i.

206, 15. **II.** *good, worthy, having excellent qualities* or *proper-ties* :—Sancte Iohannes wæs māra and sélra eallum ōðrum mannum, Blickl. Homl. 163, 20. Sýlra, 161, 24. Ðeáh hine se dysiga dō tō cyninge, hú mæg gesceádwís scealc gereccan, ðæt hē him ðý sélra sié oððe þince, Met. 15, 15. Nænig sélra nære rondhæbbendra ríces wyrðra *no warrior was worthier, more deserving of rule*, Beo. Th. 1725; B. 860. Næs mid Rōmwarum sincgeofa sélla *among the Romans was not a prince of nobler character*, Met. 1, 50. Bōþ his sylfes swíðor micle ðonne se sélla mon, Exon. Th. 315, 11; Mōd. 29. Him wearþ sélle líf bihýded, 227, 3; Ph. 417. Wē sculon ídle lustas forseón and ðæs sélran gefeón, 47, 19; Cri. 757. Ðæt hē fēre him tō ðam sélran ríce (*heaven*), 352, 24; Sch. 102. On sýllan mon, 377, 20; Deór. 6. Uton wē georne teolian ðæt wē ðe beteran sýn & ðe sélran for ðære lāre ðe wē gehýrdon, Blickl. Homl. 111, 19. Gē sōhtun ða sǽmran and ða séllan nō dēmdan æfter dǽdum, Exon. Th. 131, 30; Gú. 463. Ðú se sélusta Theophilus *optime Theophile*, Lk. Skt. 1, 3. Hláford mín and brōðor ðín se sélesta, Exon. Th. 183, 26; Gú. 1333. On gōdre and on sélestre heortan *in corde bono et optimo*, Lk. Skt. 8, 15. Nymaþ of eówrum sélustan wæstmum, Gen. 43, 11. **III.** *good of its kind,* (a) *of persons, possessing the excellences of a class, excellent, well-qualified, skilful, efficient* :—Hē ðæs wǽpnes onláh sélran sweordfrecan, Beo. Th. 2940; B. 1468. Nǽfre ic sǽlidan sélran mētte, Andr. Kmbl. 942; An. 471. Ic fæste binde swearte wealas, hwilum séllan men, Exon. Th. 393, 23; Rä. 13, 4. Omerus se gōda sceop ðe mid Crēcum sélest wæs . . . Firgilius wæs mid Lǽdenwarum sélest, Bt. 41, 1; Fox 244, 4-6. Cwēna sélost, Drihtnes mōdor, Menol. Fox 334; Men. 168. Ealra sigebearna ðæt séleste and ǽþeleste, Exon. Th. 33, 4; Cri. 520. Twegen wǽron biscopas and twegen mǽssepreóstas ealle ða sélestan *omnes sacerdotes fuere praeclari*, Bd. 3, 23; S. 555, 19. Manige ðara sélestena cynges þéna forþférdon, Chr. 897; Erl. 94, 32. (b) *of things* :—Næs sincmáðþum sélra on sweordes hád *there was no greater treasure in the shape of a sword*, Beo. Th. 4392; B. 2193. Hí nǽfre song séllan ne hýrdon, Exon. Th. 325, 8; Víd. 108. Ídel stód húsa sélest, Beo. Th. 294; B. 146. Hof séleste (*the ark*), Cd. Th. 84, 6; Gen. 1393. Éce líf, sélust sigeleána, Elen. Kmbl. 1051; El. 527. Blícan swá ðæt séloste gold, H. R. 15, 35. Seó séleste gesælþ, Bt. 24, 2; Fox 82, 3. Biþ Drihten úre se sélosta scyld *the Lord will be our most effectual shield*, Blickl. Homl. 13, 10. Heó hié gegyrede mid ðon sélestan hrægle, 139, 7. **III a.** *marking the rank or class of a person* :—Ðone sélestan (*of the highest class*) . . . ðane óðerne . . . ðane þriddan, L. Ethb. 26; Th. i. 8, 12. **IV.** *good, advantageous, to one's interest, advisable* :—Is hit micle sélre ðæt wē hine álýsan, Andr. Kmbl. 3124; An. 1565. Sélle, Exon. Th. 371, 15; Seel. 76. 'Him sylfum sélle þynceþ leahtras tō fremman, 266, 33; Jul. 407. Ne mæg ðec séllan rǽd mon gelǽran, 119, 4; Gú. 249. Wē ðē mágon sélre gelǽran, Andr. Kmbl. 2706; An. 1355. Ða forléton wē ða frécnan wegas and ðǽm sélran wē férdon, Nar. 17, 13. Ðæt him soelest wǽre ðæt hié friþes wilnaden *nullam esse residuam spem, nisi in petenda pace*, Ors. 4, 10; Swt. 202, 18. Hē brytniæ swǽ hígum maest rǽd sié and ðaem sáwlum soelest, Chart. Th. 461, 2: 465, 33. Ófest is sélost, Cd. Th. 196, 18; Exod. 293: Andr. Kmbl. 3129; An. 1567: Beo. Th. 518; B. 256. Hwæt sélest wǽre tō gefremmanne, 351; B. 173: Elen. Kmbl. 2328; El. 1165. Ellen biþ sélast ðam ðe sceal dreógan dryhtenbealu, Exon. Th. 183, 4; Gú. 1322. Biþ andgit ǽghwǽr sélest, Beo. Th. 2123; B. 1059. Is hit ealles sélest tō sécenne hwæt ðæs willa sié, Blickl. Homl. 205, 27. **V.** *good, honourable, noble, proper* :—Deáþ biþ sélla eorla gehwylcum ðonne edwítlíf, Beo. Th. 5773; B. 2890. Sélre biþ ǽghwæm ðæt hē his freónd wrece, ðonne hē fela murne, 2773; B. 1384: Andr. Kmbl. 640; An. 320. Ðé ðæt sélre geceós, ēce rǽdas, Beo. Th. 3523; B. 1759. Hé smeáde hwæt him sélest (or under **III**) tō dónne wǽre *quid sibi esset faciendum tractabat*, Bd. 2, 9; S. 512, 15. Maria geceás ðone sélestan dǽl, Lk. Skt. 10, 42. **VI.** *of value, precious* :—Ðú golde eart, sincgife sýlla, Andr. Kmbl. 3016; An. 1511. Hú nys seó sáwl sélre ðonne mete *nonne anima plus est quam esca?* Mt. Kmbl. 6, 25. Ne hýrde ic guman ǽnigne bringan ofer sealtne mere sélran lāre, Menol. Fox 204; Men. 103. Gē synt sélran ðonne manega spearuan, Mt. Kmbl. 10, 31. Gif hē nele ðone sélestan dǽl Gode gedǽlan, Blickl. Homl. 195, 7. **VII.** *good, happy, pleasant* :—On ðǽm sélran þingum *in secundis rebus*, Nar. 7, 26. Wē dreámas hefdon sélrum tídum, Cd. Th. 267, 29; Sat. 45. [Þu scalt uurþan sæl *thou shalt prosper*, Laym. 1234. Cloten hauede enne sune þe sel (bold, 2nd MS.) wes, 4071. Mid selere strengðe wurðliche, 21654. Seoue þusend selere (boldere, 2nd MS.) þeinen, 18011. Ich wulle sende to selen mine þeinen, 25162. Ne isæh na man selere cniht nenne, 21166. Þat us is selest (best, 2nd MS.) to don, 918. In al þat sel is, H. M. 47, 34. *Goth.* sēls *good, kind: Icel.* sǽll *blest, happy.*] v. next word.

sél, soel; *also* sélor; *adv.* (*cpve.*) *Better.* **I.** *of health* :—Cwæð ðæt heó gelýfde ðæt hire sóna sél wǽre *quia crederet eam mox melius habituram*, Bd. 5, 3; S. 616, 11. Sóna ic wæs wyrpende and mé sél wæs *statim melius habere incipio*, S. 616, 34: 5, 5; S. 618, 4. Sóna him biþ sél, Lchdm. iii. 288, 19. Him biþ soel *bene habebunt*, Mk. Skt. Lind. 16, 18. **I a.** *of moral or spiritual well-being* :—Ne mæg ic gehycg-

an, hwý him on hige þorfte á ðý sæl wesan, Met. 15, 10. **II.** *of knowledge* :—Gé sind searowum beswicene oððe sél nyton, mōde gemyrde, Andr. Kmbl. 1490; An. 746. Findaþ ða ðe fyrngewritu sélost cunnen, Elen. Kmbl. 748; El. 374. **III.** *of the operation of the senses* :—Hē biþ suá micle sél gehiéred, suá hē ufor gestent, Past. 14, 1; Swt. 81, 17. **IV.** *denoting excellence in act or in conduct* :—Nō ðý sél dyde, ac ðam æðelinge oferhygd gesceód, Cd. Th. 246, 35; Dan. 489. Ne gefrægn ic nǽfre wurðlícor æt hilde sixtig sigebeorna sél gebǽran, Fins. Th. 77; Fins. 38: Beo. Th. 2029; B. 1012. Hwylc hira sélast simle gelǽste hláforde æt hilde, Andr. Kmbl. 821; An. 411. Bet gē rǽdaþ *melius legitis*, sélost (sǽlost, MS. T.) hí rǽdaþ *optime legunt*, Ælfc. Gr. 5; Zup. 9, 17. Hwǽr ic sélast wisse cwén giefe bryttian, Exon. Th. 324, 28; Víd. 101. **V.** *denoting advantage or profit* :—Hwæt byþ ús tō mǽde (ús ðý soel, Lind.), Mt. Kmbl. 19, 27. Tō hwan hió ða næglas sélost and deórlícost gedōn meahte, Elen. Kmbl. 2315; El. 1158. **VI.** *denoting success or good result, with (more) success, (more) effectually, to (more) purpose* :—Ic gelýfe ðe sél and ðý fæstlícor ferhþ staþelige, Elen. Kmbl. 1589; El. 796. Ne gefrægn ic nǽfre sixtig sigebeorna medu sél forgyldan, Fins. Th. 79; Fins. 39. For ðý ðe mon ðæs feorme ðý soel gelǽste, Chart. Th. 474, 12. Næs him wihte ðe sél *he did not succeed any the better*, Beo. Th. 5368; B. 2687. Sél æfter wælrǽse wunde gedýgan *to be more successful in escaping wounds*, 5054; B. 2530. Se æcer syððan gegreów .c. síða sélor ðonne hē ǽr dyde, Shrn. 137, 25. Hú man sélost mæg synna forbúgan *how sins may most effectually be avoided*, Ælfc. T. Grn. 7, 38. Hú ic ðíne sóðfæstnesse sélest heólde, Ps. Th. 118, 54, 26. Hié hígon gefeormien swǽ hié soelest þurhtión mægen, Chart. Th. 476, 31. **VII.** *with verbs of liking or pleasing* :—Hē nǽnum menn sél ne úðe ðonne mē *there was no one he would sooner give it to than to me*, Chart. Th. 485, 17. Ða men ðe ic mínes erfes seólest onn, 480, 20. Se getreówa man sceal syllan his gōd on ða tíd ðe hine sylfne sélest lyste his brúcan, Blickl. Homl. 101, 20. Hí genáman ðæs folces ðe ðǽr tō láfe wæs and him sélost lícodan, 79, 21.

seld, es; *n.* **I.** *a seat, that on which one sits, a throne; sedes* :—In heofene seld his *his throne is in heaven* (A. V.), Ps. Surt. 10, 5: 44, 7. Dōm gegearwung seldes ðínes, 88, 15: 96, 2. Of dúne sette maehtge of selde, ii. p. 200, 20: Cd. Th. 275, 17; Sat. 173: 276, 12; Sat. 187. Ðǽr is sang æt selde (*the throne of God*), 306, 12; Sat. 662. Sang ymb seld secgan, 279, 9; Sat. 235. Siteþ him on heofnum, hafaþ wuldres bearn his seolfes seld, 301, 27; Sat. 588. God siteþ ofer seld hálig his, Ps. Surt. 46, 9: 9, 8. Ealdormenn sǽton on seldum, Ps. Th. 118, 23. Hí on seldon sǽton æt dōmum, 121, 5. **II.** *a seat, residence, mansion, hall* :—Scyppendes seld, Salm. Kmbl. 160; Sal. 79. Ðá hē ða mænego (*the rebellious angels*) ádráf of ðæm heán selde (*heaven*), Cd. Th. 277, 10; Sat. 202. Cwom Daniel in ðæt seld gaugan, 225, 9; Dan. 151: 262, 1; Dan. 737. Engel lēt his hand cuman in ðæt heá seld (*Belshazzar's hall*), 261, 7; Dan. 722. Hié tempel strudon, Salomanes seld, 260, 19; Dan. 712. Com tō Heorot, ðǽr Hring-Dene geond ðæt sǽld swǽfon, Beo. Th. 2564; B. 1280. Wǽron on ðyssum felda unríme gesomnunge manna and monig seld (*or to I?*) gefeóndra weorada *erant in hoc campo innumera hominum conventicula, sedesque plurimae agminum laetantium*, Bd. 5, 12; S. 629, 25. Ða heallícan seld *palatias zetas*, Wrt. Voc. ii. 81, 23. Hú hé eft gesette swegeltorhtan seld, Cd. Th. 6, 27; Gen. 95. Heáhgetimbru, seld on swegle, Exon. Th. 137, 10; Gú. 557. [Ǽr he arise of selde, Laym. 25988. *Cf. Goth.* salithwa; *f. a mansion, chamber: O. Sax.* selida, selda: *O. H. Ger.* selida; *f. domicilium, mansio, habitaculum, tabernaculum.*] v. ān-, biscop-, cear-, ēðel-, heáh-, medu-, pāp-, sundor-, þrym-, weard-seld; selde.

-selda. v. ge-selda.

seldan (-on, -un, -um); *cpve.* seldnor; *adv.* *Seldom, rarely* :—Seldan (-on) *raro*, Ælfc. Gr. 38; Zup. 240, 12: Bt. 16, 1; Fox 50, 14. Oft nalæs seldan, Ps. Th. 74, 4. Tō seldan hit biþ, beó hit seldor on dæg ðonne seofon síðum, Btwk. 194, 11. Oft (*of? cf. Icel.* of- *too, and* v. of-) seldan hwǽr æfter leódhryre lytle hwíle bongár búgeþ *too rare are the cases in which after the fall of men the deadly weapon retires*, or *often after slaughter the spear is seldom at rest*, i. e. in most cases frequent strife follows (cf. the first passage under *seld-hwanne, and* seldum hwonne), Beo. Th. 4063; B. 2029. Him seldon teola gespeów, Ors. 4, 5; Swt. 168, 19: Bd. 1, 1; S. 474, 31: Met. 28, 71. Seldon wē ǽnig seolfor fundon, Nar. 5, 15. Hwílon ic dó ac seldon *aliquando facio, sed raro*, Coll. Monast. Th. 24, 3. Se ðe him ealneg wind ondrǽt, hē sǽwþ tō seldon, Past. 39, 2; Swt. 285, 18. Seldun, 9; Swt. 57, 16. Seldum ǽfre, Salm. Kmbl. 540; Sal. 269. Ac ðeáh hí seldum hwonne (cf. seld-hwanne) beswemede weorþon ðonne sleáþ hé eft on ða solu *but though on rare occasions they (swine) get washed, at such times they return to the mire*, Bt. 37, 4; Fox 192, 28. Ðæt dysie folc ðæs hit seldnor gesihþ swíðor wundriaþ, Met. 28, 66. [*O. Frs.* sielden: *O. H. Ger.* seltan; *cpve.* seltanor: *Icel.* sjaldan; *cpve.* sjaldnor; *spve.* sjaldnast.] v. un-seldan, seldor.

seld-, sel-cúþ; *adj.* *Little known, strange, wonderful, unfamiliar* :—Se seldcúþa tungel gebícnode ðæs sóðan cyninges ácennednysse, Homl,

Th. i. 106, 27. Hē wæs oflyst đæs seldcūþan sōnes (*the sound of Orpheus' harp*), Bt. 35, 6 ; Fox 168, 23. Hī willaþ simle hwæthwegu nīwes and seldcūþes eówian, 34, 4 ; Fox 138, 29. Đū hwerfest ymbūton sume wunderlīce and seldcūþe spræce, 35, 5 ; Fox 164, 17. Dīglu þing tæcan and seldcūþe, 39, 4 ; Fox 216, 13. Selcūþe reáf *varias vestes*, Coll. Monast. Th. 27, 9. [Þeo wimon was mid ane sune þat wes a selcuđ bearn (wonderful to telle, 2nd MS.), Laym. 280. Þatt wass sellcuþ mecleȝȝc, Orm. 19217. Gif him þuncheđ wunder & selkuđ of swuch onswere, A. R. 8, 26. Gret outrage we se . . . in selcouthe maners, Pr. C. 1518.]

seld-cyme, es; *m. A rare visit :*—Wēna mē đīne seóce gedydon, đīne seldcymas, Exon. Th. 380, 27 ; Rä. 1, 14. [Cf. *Icel.* sjald-kvæmr *seldom coming.*]

selde, an; *f. A porch :*—Selde *proaula* (*porticus* a porche, *proaula idem est*, 204, col. 2), i. *domus coram aula*, Wrt. Voc. i. 57, 46. v. sumor-, winter-selde ; seld.

seld-guma, an ; *m. A hall-man, one who has a place in a lord's hall, a retainer :*—Næfre ic māran geseah eorl ofer eorþan đonne is eówer sum . . . nis đæt seldguma (*he is no mere retainer*). Grein translates ʻvir qui semper in domo manet.' Heyne says ʻ*seldguma* ist hier offenbar der gemeine Mann, der nur ein *seld* besitzt, im Gegensatz zu dem edeln, der einen *hof* zu eigen hat.' But *seld* is used of royal residences, so that Bugge's explanation seems better, ʻen mand som holder til en hövdings sal, en mand som er traadt i en hövdings tjeneste '), Beo. Th. 504; B. 249. Cf. sele-secg.

seld-hwanne ; *adv. Seldom, rarely :*—Oft đonne đæt mōd đæs fæstendan biþ mid đȳ irre ofseten, đonne cymþ sió blis seldhwanne, swelce hió sié elþeódig, Past. 43, 6; Swt. 313, 24. Đeáh seldhwænne leáf geseald sié tō sprecenne *quamvis rara loquendi concedatur licentia*, R. Ben. 21, 16. Heó wolde seldhwænne hire līc bađian, Homl. Skt. i. 20, 44. Seldhwonne biþ đætte āuht manegum monnum ānes hwæt līcige,∙Bt. 18, 3; Fox 64, 29. [Swuch ouh wummone lore to beon liđe and seldhwonne sturne, A. R. 428, 25. Cf. *Icel.* sjald-stundum *rarely*.]

seld-, sel-, syl-līc ; *adj.* I. *strange, extraordinary, wonderful :*—Đis godspel þincþ dysegum mannum sellīc, Homl. Th. ii. 466, 9. Nū þincþ eów đis syllīc tō gehȳrenne, L. Ælfc. C. 6; Th. i. 344, 16 : Wulfst. 269, 26. Is đæt sellīc þincg, đæt hī ne wundriaþ hū . . . , Met. 28, 53. Næfde sellīcu wiht sȳne ne folme, Exon. Th. 415, 2 ; Rä. 33, 5. Glōf sīd and syllīc searobendum fæst, Beo. Th. 4178; B. 2086. Ic seah sellīc þing singan, Exon. Th. 413, 9 ; Rä. 32, 3. Đa rēđan león and đa sellīcan (syl-) pardes and đa egeslīcan beran, Hexam. 9 ; Norm. 14, 33. Sellīce sædracan, Beo. Th. 2856; B. 1426. Syllīce tācn, Blickl. Homl. 91, 29. Syllīce stānas *monstrous stones*, 189, 15. Seldlīcra fela *many wonderful creatures*, Exon. Th. 193, 34; Az. 131. Hit is sellīcre đæt hiora ænig ne mæg būtan ōđrum bión, Met. 11, 50. Hī đær gesēgon syllīcran wiht, Beo. Th. 6069; B. 3038. II. *having unusual good qualities, excellent, admirable :*—Þeódnes cynegold sōđfæstra gehwone sellīc glengeþ, Exon. Th. 238, 19 ; Ph. 606 : 341, 16 ; Gn. Ex. 127. Is đes middangeard missenlīcum wīsum gewlitegad, wrættum gefrætwad, sīþum sellīc, 414, 28 ; Rä. 33, 3. Freólīc sellīc, 492, 29 ; Rä. 81, 23. Wundor syllīc (*the pillar of fire*), Cd. Th. 184, 17 ; Exod. 109: Rood Kmbl. 25 ; Kr. 13. Hē wundur worhte seldlīc, Ps. Th. 125, 3. Ænlīcra and fægerra, symle sellīcra, Exon. Th. 357, 17 ; Pa. 30. Him (*the phenix*) sette sōđ cyning sellīcran gecynd ofer fugla cyn, 221, 4 ; Ph. 329. Ic æfre ne geseah syllīcran cræft, Andr. Kmbl. 1000; An. 500: Rood Kmbl. 8 ; Kr. 4. [*Laym.* sel-, sil-, seol-, sul-lich : *O. E. Homl.* sullic : *Jul.* sul-lich : *O. and N.* sel-, seol-lich : *Goth.* silda-leiks : *O. Sax.* seld-līk.]

seld-, sel-, syl-līce ; *adv.* I. *strangely, wonderfully :*—Næfre hié đæs sellīce bleóum bregdaþ, Salm. Kmbl. 300; Sal. 149. Singeþ syllīce, 539; Sal. 269. II. *wonderfully well, excellently, admirably :*—Iericho wæs sellīce getimbrod, mid seofon weallas beworht and wel wiđinnan geset, Homl. Th. ii. 212, 25. Syllīce hyt đæt āttor tōsceádeþ, Lchdm. i. 352, 13.

seldnor, seldon. v. seldan.

seldor ; *cpve.* : seldost ; *spve.* (the positive seems expressed by seldan, which however has a comparative seldnor) ; *adv. More seldom, less frequently :*—Seldan *raro*, seldor *rarius*, ealra seldost (-ast, MS. H.) *rarissime*, Ælfc. Gr. 38 ; Zup. 240, 13. Tō seldan hit biþ, beó hit seldor on dæg đæt wē God herian đonne seofon sīđum, Btwk. 194, 11. Bæþ đām untrumum, swā oft swā hit framige, sȳ geboden ; hālum sȳ seldor getīđod, R. Ben. 61, 1. Đæt ungestæđđige folc wundraþ đæs đe hit seldost gesihþ, Bt. 39, 3 ; Fox 216, 2. [Gon seldere þene he sholde to his chirche, O. E. Homl. ii. 207, 26. *Icel.* sjaldar.]

seld-sīne, -sȳnde ; *adj. Seldom seen, uncommon, unfamiliar :*—Cirus geāhsade đæt đæm folce seldsiéne and uncūđe wæron wīnes dryncas, Ors. 2, 4 ; Swt. 76, 12. Ælc seldsȳnde fisc đe weorđlīc biþ, Cod. Dip. Kmbl. iii. 450, 27. [Hit is seltsene on eorđe, H. M. 27, 22. Our speche schal beon seldcene, A. R. 80, 19. *Icel.* sjald-sénn.]

sele, es ; *m. A hall, house, dwelling :*—Cwom bytla (*Guthlac*) tō đam beorge . . . wæs sele (*his hermitage*) nīwe, Exon. Th. 146, 24 ; Gū. 714.

Sele sceal stondan, sylf ealdian, 343, 16 ; Gn. Ex. 158. Sele (*Heorot, Hrothgar's hall*) hlifade, heáh and horngeáp, Beo. Th. 163 ; B. 81. Đes sele, receda sēlest, 827 ; B. 411. Đes windiga sele (*hell*), Cd. Th. 273, 14; Sat. 136. Hē on temple gestōd . . . Hē anlīcnesse geseh on seles (*or from sæl*, cf. 1523 ; An. 763) wāge, Andr. Kmbl. 1428 ; An. 714: Exon. Th. 394, 17 ; Rä. 14, 4. Þegen đe on cinges sele his hlāforde þēnode, L. R. 3 ; Th. i. 192, 1. Hē (*Pharaoh*) lǣdan hēht wīf tō his selfes sele, Cd. Th. 111, 17 ; Gen. 1857. Geseah hē engles hand in sele (*Belshazzar's hall*) wrītan, 261, 16 ; Dan. 727. Hié tō sele (*the Danish king's hall*), gangan cwōmon, Beo. Th. 652 ; B. 323. In sele đam heán, 1431 ; B. 713 : (*Hygelac's hall*), 3973 ; B. 1984. On sele *in the dragon's cave*, 6248; B. 3128. Tō sele *to the prison*, Andr. Kmbl. 2624 ; An. 1313. Cyning mec on sele weorþaþ, Exon. Th. 401, 12 ; Rä. 21, 10. Ic sōhte sele sinces bryttan, hwær ic findan meahte đone đe in meoduhealle mec frēfran wolde, 288, 2 ; Wand. 25 : Beo. Th. 1657 ; B. 826 : 4694 ; B. 2352. Sele āsettan, sīdne ræced fæste gefēgan, Exon. Th. 296, 6 ; Crä. 47. Brūcan đæs boldes đe ūs gearwaþ gǣsta ealdor ; đæt is sigedryhten đe đone sele frætweþ, 450, 24 ; Dôm. 92. Innan on đone ealdan sele, Cod. Dip. Kmbl. iii. 406, 13. Đone werigan sele (*hell*), Cd. Th. 285, 4 ; Sat. 332. Ongunnon heora burh rǣran and sele settan, salo nīwian, 113, 2 ; Gen. 1881. [*O. Sax.* sele ; *m.* : *Icel.* salr ; *pl.* salir : cf. *O. H. Ger.* seli-hûs : *Goth.* saljan *to dwell, abide.*] v. bān-, beág-, beór-, burg-, burn-, deáþ-, dreór-, dryht-, eorþ-, gæst-, gold-, grund-, gūþ-, heáh-, horn-, hring-, hrōf-, nīþ-, will-, wīn-, wind-, wyrm-sele ; sæl.

sele ? :—Winter ȳþe beleác īsgebinde ōþ đæt ōđer com geár in geardas swā nū gyt dēþ đa đe sele (=sǣle?) bewitiaþ wuldortorhtan weder *winter shut up the waves with bonds of ice, until another year came to men's dwellings ; so still the new year comes, and brilliant weather (as is apparent to those) who keep constant watch on the seasons*, Beo. Th. 2275; B. 1135. But see Heyne's Beowulf, or Paul and Braune, Beiträge, 12, 31.

sele-dreám, es ; *m. Mirth of the hall, joyous life of the hall, festive pleasure :*—Beorgas wæron blīđe gebǣrdon swā rammas wurdan gesweoru swā on seledreám swā on sceápum beóþ scēóne lambru *montes, quare exultastis ut arietes, et colles velut agni ovium*, Ps. Th. 113, 6. Oft ic secga seledreám sceal onþeon, Exon. Th. 480, 13 ; Rä. 64. 1. Goldburg ofgifan, secga seledreám, beorht beágselu, Andr. Kmbl. 3310; An. 1658: Beo. Th. 4496; B. 2252. Swǣfon seledreámas, Cd. Th. 179, 29 ; Exod. 36: Exon. Th. 290, 3 ; Wand. 93.

sele-ful[1], es ; *n. A cup used in a hall :*—Hē geþah symbel and seleful Beo. Th. 1242; B. 619.

sele-gescot, -gesceot, es ; *n. A tabernacle :*—In selegescote đīnum *in tabernaculo tuo*, Ps. Surt. 14, 1. Selegesceote, Ps. Th. 60, 3. Đeáh đe ic on mīnes hūses hyld gegange ođđe selegesceot *si introiero in tabernaculum domus meae*, 131, 3, 5, 7. Đæt selegescot, hūs tō wynne (*the body*), Exon. Th. 90, 28 ; Cri. 1481. Selegescotu *tabernacula*, Ps. Th. 77, 28. Selegesceotu, 82, 6 : 107, 6. On đīnum selegescotum, 146, 11. v. sele-scot, ge-sceot.

sele-gist, es ; *m. A guest in a hall :*—Heó ofsæt đone selegyst (*Beowulf who was in Hrothgar's hall*), Beo. Th. 3094; B. 1545.

selen, sellen, sylen, e ; *f.* I. *a gift :*—Ic đē nū āfyrre fram mīnre selene đe ic đē forgeaf, Wulfst. 258, 14. Seó gōde antswaru sȳ ouer đa sēlestan selene *sermo bonus super datum optimum*, R. Ben. 55, 9. Sylena *donaria*, Germ. 394, 343. Gāstlicra sellena ꝉ gifa *sanctorum donorum*, Hpt. Gl. 414, 37. Syllena, 473, 50. Mid selenum hē gewelgie *donis maneret*, Hymn. Surt. 4, 32. Đū onfēnge selena *accepisti dona*, Ps. Spl. 67, 19. Gōde sylena syllan, Mt. Kmbl. 7, 11. Sylene, Lk. Skt. 11, 13. II. *a giving, donation, grant :*—His handseten and sælen, Cod. Dip. Kmbl. ii. 89, 12. Ic geeácnode tō đare ǣrran sylene tȳn þūsenda ælfixa, Chart. Th. 242, 11. Ic đās sīre selene trymme, 106, 10. Þurh his sylene and gyfe *ipso largiente*, Bd. 2, 12; S. 515, 24. Þurh ælmyssan sylene *per erogationem eleemosynae*, L. Ecg. P. iv. 63 ; Th. ii. 222, 32. Mid gebedum and mid wæccum and mid ælmessa sylenum, Wulfst. 228, 20. III. *the habit of giving, liberality, munificence :*—Sylen *liberalitas*, mid sylene *munificentia*, Hpt. Gl. 466, 52, 49. Cystigre sylene *prodiga liberalitate*, 517, 36. v. ælmes-, hand-, mann-selen, -silen.

seleness, selnes *tradition* ; *traditio*, Mt. Kmbl. Lind. 15, 2, 3 : Mk. Skt. Lind. Rush. 7, 3, 9.

sele-rǣdend, es ; *m. One who takes part in the councils held in a hall, a counsellor of a prince :*—Manige cōmon snottere selerǣdend, symble gefēgon beornas burhweardes cyme, Andr. Kmbl. 1317; An. 659. Men ne cunnon secgan tō sōđe, selerǣdende (-rǣdenne, MS.), hæleþ under heofenum, hwā đæm hlæste onfēng, Beo. Th. 102; B. 51. Ic đæt leóde mīne, selerǣdende, secgan hȳrde, 2696; B. 1346.

sele-rest, e ; *f. A bed in a hall :*—Hine ymb monig sǣrinc selereste gebeáh (*of Beowulf and his men when sleeping in Hrothgar's hall*), Beo. Th. 1384; B. 690.

sele-scot, es ; *n. A tabernacle, dwelling :*—Gewyrce wē þreó selescotu (*tabernacula*), Mt. Kmbl. Rush. 17, 4. Fuglas heofunas habbaþ selescota (*nidos*), 8, 20. v. sele-gescot.

sele-secg, es; *m. A hall-man, a retainer who has a place in his lord's hall :*—Gemon hē selesecgas and sincþege, hū hine his goldwine wenede tō wiste, Exon. Th. 288, 20; Wand. 34. Cf. seld-guma.

sele-þegn, es; *m. A hall-thane, chamberlain* :—Him (*Beowulf*) seleþegn sīdes wērgum forþ wīsade, se ealle beweotede þegnes þearfe (*the chamberlain who saw after everything Beowulf needed*), Beo. Th. 3592; B. 1794.

sele-weard, es; *m. A hall-warder, guard of a hall :*—Hæfde hē Grendle tōgeánes seleweard āseted, Beo. Th. 1338; B. 667.

self, seolf, silf, sylf; *pron.*   **A.** *self, very, own.*   **I.** with a noun (α) which it immediately follows:—Đam ðe se þeóden self seceþ niht naman, Cd. Th. 9, 10; Gen. 139. Drihten sylf, Blickl. Homl. 41, 4 : 51, 6. God selfa cuman wille, 163, 31. Hē, Drihten selfa, cwæð, 165, 2. Drihten sylfa, 39, 25. God seolfa, Cd. Th. 286, 11; Sat. 350. Nǽniges gebyrd ciricean ne mǽrsiaþ nemþe Cristes sylfes and ðyses Iohannes, Blickl. Homl. 161, 11. From ðære dura selfre ðisse béc *ab ipso libri hujus exordio*, Past. proem.; Swt. 25, 11. Gode sylfum underþeódde, Blickl. Homl. 109, 22 : 73, 12. Gearo mód ge eác swylce deáþ sylfne tō þrowienne *paratum vel etiam ad moriendum animum*, Bd. 1, 26; S. 487, 38. On ðæt dægrēd sylf, Judth. Thw. 24, 24; Jud. 204. Rômāne selfe sǽdon, Ors. 5, 3; Swt. 220, 20. Nǽnig man ða lárias sylfe ufan oferwyrcean ne mihte, Blickl. Homl. 125, 35. Hē ða deádan sylfe āwehte, 173, 29. (β) which it follows, but not immediately :—Nergend com nihtes self, Cd. Th. 159, 12; Gen. 2633. Đeáh ðe ðæt hús ufan open sȳ sylf, Blickl. Hom. 125, 30. Mē sægde ðæt wíf hire wordum selfa, Cd. Th. 160, 11; Gen. 2648. Hē mid hondum Hǽlend genom sylfne be sīdan, 299, 5; Sat. 545. (γ) along with a personal pronoun in the dative :—Pilatus on hys dômerne hym sylf āwrāt ealle ða þyng, Nicod. 34; Thw. 19, 33. Oðra gesceafta weordaþ him selfe tō náuhte, Met. 11, 87. (δ) which it immediately precedes:—On ðā sylf cyning wrāt, wuldres God, Andr. Kmbl. 3017; An. 1511. Hēht sylf cyning him Abraham tō, Cd. Th. 161, 27; Gen. 2671. Hit is se seolfa sunu Waldendes, 289, 11; Sat. 396. Se sylfa cyning lȳsde (hié) of firenum, Exon. Th. 74, 20; Cri. 1209. Sylfes ðæs folces, 481, 20; Rä. 65, 6. Under ðam sylfum norþdǽle middangeardes *sub ipso septentrionali vertice mundi*, Bd. 1, 1; S. 473, 29. Ic tō sylfum Drihtne cleopode, Ps. Th. 54, 16. Đæt ða sylfan ȳþa wǽron āhofene ofer ðæt scip, Blickl. Homl. 235, 6. Đæt ða sylfan his lāreówas æt his mūþe leornodan *that his very teachers learned from his mouth*, Bd. 4, 24; S. 598, 8. (ε) which it precedes, but not immediately :—Bidon þegnas ... swā him sylf bebeád swegles āgend, Exon. Th. 34, 16; Cri. 543. Bīdan selfes gesceapu heofoncyninges, Cd. Th. 52, 12; Gen. 842 : 36, 4; Gen. 566. Weard sylfum ætȳwed ðam cásere swefnes wôma, Elen. Kmbl. 138; El. 69.   **II.** with a pronoun. (1) in agreement with a personal pronoun denoting the subject of the sentence and (α) following it immediately:—Ic sylf (seolf, Lind. : solfa, Rush.) hit eom *ipse ego sum*, Lk. Skt. 24, 39. Heó sylf hié þeówen nemde, Blickl. Homl. 13, 13. Đæt hé sylfa cwæð, 13, 26 : 95, 5. Beó hē sylfa syxta, L. C. S. 30; Th. i. 394, 5, MS. G. Sȳ hē scyldig his sylfes, L. Ath. iv. prm.; Th. i. 220, 12. Hē swīðor mīnes feores wilnade ðonne his selfes, Nar. 8, 6. Mid his sylfes willum, willan *ultro*, Bd. 1, 7; S. 477, 15, 22. Gif þeów ete his sylfes rǽde, L. Wih. 15; Th. i. 40, 11. Hwæt segst ðú be ðē sylfum (seolfum, Lind. : fore ðec solfne, Rush.), Jn. Skt. 1, 22. Heó hæfde hire sylfre geworht ðæt mǽste wīte, Blickl. Homl. 5, 26. Gif his rīce on him sylfum biþ tōdǽled ... Gif ðæt hús ofer hit sylf ys tōdǽled ... Gif Satanas winþ ongén hine sylfne, Mk. Skt. 3, 24–26. Mē siolfne, Chart. Th. 476, 19. Ic swerige þurh mē sylfne *per memetipsum juravi*, Blickl. Homl. 9, 23. Nū mæg sôð hit sylf gecȳðan, 187, 16. Đone anwald ūre selfra, Past. 33; Swt. 220, 7. Suā micle giéman ūrra niéhstena suā suā ūre selfra, 5; Swt. 45, 12. Hī hiora selfra sibbe manwald nabbaþ, Bt. 16, 3; Fox 54, 18. Hiora seolfra hǽlo, Nar. 30, 18. Heó hié selfe āweredon ... him leófre wæs ðæt hié hié seolfe fornēðdon, Ors. 5, 3; Swt. 220, 23–26. Eáþmôdgiaþ eów sylfe, Blickl. Homl. 99, 3. (α 1) with irregular construction:—Đeáh ðe hí synd of miclum dǽle heora sylfes anwealdes *quamvis ex parte sui sint juris*, Bd. 5, 23; S. 647, 3. (β) following the pronoun, but not immediately :—Hē eác self biþ gecostod, Past. 16; Swt. 104, 20. Hwæt hē mē self bebeád *what he himself bade me*, Cd. Th. 34, 10; Gen. 535. Hē his brȳde ofslôh self mid sweorde, Met. 9, 31. Gif hē wille sylf Godes dômas gedēgan, Blickl. Homl. 43, 11. Hē wæs þridda sylf, Elen. Kmbl. 1707; El. 855 : Andr. Kmbl. 1330; An. 665. Ne wēn ðú ðæt ic tō ānwillíce winne wið ða wyrd, fordam ic hit nó selfe ne ondrǽde, Bt. 20; Fox 70, 21. Đā dā wē hit nôhwæðer ne selfe ne lufodon, Past. pref.; Swt. 5, 6. Sylfe, Blickl. Homl. 53, 1 : 223, 20. Hié wēnaþ ðæt hié wísran sién selfe ðonne ôðre, Past. 48, 1; Swt. 365, 20. (γ) along with a pronoun in the dative :—Đū meaht nū ðē self geseón, Cd. Th. 38, 23; Gen. 611. Hē feóll him silf *quem percussit Josue ad internecionem*, Jos. 10, 33. Hī weorþaþ him selfe tō náuhte, Bt. 21; Fox 74, 36. (δ) preceding the pronoun :—Đǽr syndon dǽlas on sylfre hire *cujus participatio ejus in idipsum*, Ps. Th. 121, 3. (2) in agreement with a demonstrative :—Þurh ðæs sylfes hand ðe ic ǽr onsended wæs, Soul Kmbl. 111; Seel. 56. (3)

with a possessive :—Be mīnre seolfre nídþearfe *de propio meo periculo*, Nar. 9, 24. On ðīnes silfes hand, Hy. 7, 83. Đīn rīce and ðīnes sylfes feorh, Blickl. Homl. 185, 1. Mīnes sylfes mūþ *os meum*, Ps. Th. 77, 2. Đīnre sylfre sunu, Exon. Th. 21, 23; Cri. 339. Wē sceoldon ūrra selfra waldan, Past. 33; Swt. 220, 5.   **II a.** where the pronoun with which *self* agrees is not the subject of the sentence:—Hē (*Claudius Marcellus*) fôr on ðone ende Hannibales folces ðe hē self (*Hannibal*) on wæs, and hiene selfne (*Hannibal*) gefiémde, Ors. 4, 9; Swt. 192, 11–13. Antonius forlēt Octauianuses swostor and him selfum onbeád gewin, 5, 13; Swt. 244, 32. Đæt man tō ôðrum lǽþþe hæbbe and hine hatige and tǽle behindan him sylfum, Blickl. Homl. 65, 1. Neoptolomus com tō Antigone ... Đā sende Antigones hiene selfne (*Neoptolomus*), Ors. 3, 11; Swt. 146, 9. Æðelstān wið Anláf gefeaht and his firde ofslôh and āflímde hine sylfne, Jud. Thw. p. 163, 10. Đā gelȳfde ic him ... beswang hine and tō heora sylfra dôme āgeaf, Blickl. Homl. 177, 24.   **III.** standing alone :—Oft gebyreþ, ðonne se scrift ongit ðæs costunga, ðe hē him ondetteþ, ðæt eác self biþ mid ðǽm ilcum gecostod, Past. 16; Swt. 105, 20. Hit Scipia hām onbeád, ... and eác self sǽde, ðā hē hām com, Ors. 4, 12; Swt. 208, 34. Seolf, Cd. Th. 143, 5; Gen. 2374. Nime fíf and beó sylf sixta, L. C. S. 44; Th. i. 402, 7, MSS. A. G. For hwon wrīhst ðū sceome, and ðīn sylf þecest líc, Cd. Th. 54, 15; Gen. 877. Is ðīn āgen sprǽc innan fȳren, sylf swīðe hāt *ignotum eloquium tuum vehementer*, Ps. Th. 118, 140. Đǽr habbaþ englas dreám, sanctas singaþ, ðæt is seolfa for God, Cd. Th. 286, 21; Sat. 355. Đā onfēng hē gǽste ... and sylfa his wunda āwrāþ, Bd. 4, 22; S. 590, 36. Hē his torn gewræc selfes mihtum, Cd. Th. 4, 26; Gen. 59 : Beo. Th. 1404; B. 700. Hē beáhhordes brūcan môste selfes dôme, 1794; B. 895. Sleáþ synnigne ofer seolfes mūþ, Andr. Kmbl. 2602; An. 1302 : Cd. Th. 248, 17; Dan. 514. Gest hine clǽnsie sylfes āþe, L. Wih. 20; Th. i. 40, 19. Sylfæs, 18; Th. i. 40, 14. Sylfum tō sconde *to thine own shame*, Exon. Th. 90, 27; Cri. 1480. Se swôre for sylfne æfter his rihte, L. R. 4; Th. i. 192, 6. Se cásere hēht eft gearwian sylfe tō sīðe, Elen. Kmbl. 1998; El. 1001.   **III a.** along with a pronoun in dative :—Biþ him self sunu and fæder *ipsa sibi proles, suus est pater*, Exon. Th. 224, 12; Ph. 374. Đæt ðū ūs sunnan onsende, and ðē sylf cyme, 8, 8; Cri. 114. Nime fíf and beó him sylf sixta, L. C. S. 44; Th. i. 402, 7. Him sylfa, 30; Th. i. 394, 5. Eall ðis māgon him sylfe geseón, Exon. Th. 69, 6; Cri. 1116.   **IV.** denoting voluntary or independent action (not inflected?). Cf. *Goth.* Silbô airtha akran bairith αὐτομάτη ἡ γῆ καρποφορεῖ, Mk. 4, 28; *and see* self-dēma, -líc, -sceaft, -will, -wille, -willende :—Genim tūncersan, sió ðe self weaxeþ, and mon ne sǽwþ, Lchdm. ii. 22, 12. Gif hē wíf self hæbbe gange hió of mid him. Gif se hlāford him wíf sealde sié hió ðæs hláfordes *if he have a wife that he got himself, let her go out with him. If the lord gave him a wife, she shall be the lord's*, L. Alf. 11; Th. i. 46, 4. Gif hit cucu feoh wǽre and hē secgge ðæt hit self ācwǽle *died a natural death*, L. Alf. 28; Th. i. 52, 2. Marius and Silla gefôran him self, and Cinna wæs ofslagen, Ors. 5, 11; Swt. 236, 24. Hié woldon of ǽlcerre byrig him self anwald habban *imperare singulae cupiunt*, 3, 7; Swt. 112, 20. Hē ne mihte hine handum self mid hrægle wryón, Cd. Th. 95, 1; Gen. 1572. Đonne wearp seó eorþe hit sôna sylf (*of its own accord*) of hire, Blickl. Homl. 127, 2. Đone sylf ne mæg man āspyrigean *man left to himself cannot investigate it*, Elen. Kmbl. 930; El. 466.   **B.** (*the*) *same*, (α) with a demonstrative :—Đū eart se sylfa God ðe ūs ādrife fram dôme, Ps. Th. 107, 10. Đæt ilce geþanc and seó sylfe carfulnyss ðe heom amang ðam nihtslǽpe wæs on heora heortan, eall ðā hí āwacodon hí ðæt sylfe geþohton, Homl. Skt. i. 23, 441. Đæt selfe wæter þegnunge gearwode beforan his fôtum, Blickl. Homl. 247, 10. Weorðeþ sunne sweart gewended ... Môna ðæt sylfe, Exon. Th. 58, 19; Cri. 938 : 387, 25; Rä. 5, 10. Ic ðē sǽde ǽr on ðisse selfan béc (cf. on ðisse ilcan béc, Bt. 37, 1; Fox 186, 25), Met. 25, 54. On ðære sylfan nihte ... On ðam sylfan mynstre, Bd. 4, 23 ; S. 595, 33, 36. On ðam sylfan leóhte, S. 596, 3. On ðam sylfan stede ðe ðū him settest, Ps. Th. 83, 6. Đȳ sylfan dæge, Exon. Th. 71, 12; Cri. 1154: Menol. Fox 94; Men. 47. Dôn ðæt sylfe, Past. 44, 3; Swt. 323, 21. Đæt seolfe, L. E. G. proem.; Th. i. 166, 9. Hí cumaþ tōgeánes Antecriste ... and beóþ ofslegen þurh ðone sylfan feónd, Ælfc. T. Grn. 3, 45. On ða sylfan tíde, Blickl. Homl. 171, 19. Heó tôfereþ ðæt sār; ðæt sylfe heó dēþ mid wíne gecnucud, Lchdm. i. 190, 18 : Ps. Th. 81, 3 : 83, 6 : 128, 1. His freónda forspðec forstent him eal ðæt sylfe, swylce hit sylf spǽce, Wulfst. 38, 17. (β) alone :—Ic mē on mūþe mægene hæbbe, and ic sôðfæst word on sylfan healde, Ps. Th. 118. 43. On selfe wísan *in the same fashion*, Lchdm. ii. 72, 17. [*Goth*. selba: O. *Frs*. O. *Sax*. O. *L. Ger*. self : O. *H. Ger*. selp : *Icel*. sjálfr.] v. selfe.

self-ǽta, an; *m. An eater of those belonging to its own species,* (applied to man) *a cannibal, anthropophagus* :—Đū scealt fēran ... ðǽr sylfǽtan eard weardigaþ ... swā is ðære menigo þeáw, ðæt hié uncūðra ǽngum ne willaþ feores geunnan, Andr. Kmbl. 350; An. 175.

self-ǽte, an; *f.* A plant name, *wild oat*(?) :—Selfǽte, eoforþrote, Lchdm. ii. 312, 15. Wyl on buteran selfǽtan, 80, 13. [Cockayne cites *O. H. Ger.* selbēza *senecion*, iii. 344, col. 1.]

self-bana, an; *m. One who kills himself, a suicide:*—Selfbona *bic-tonatus* (l. *biothanatus,* qui mortem sibi ipsi consciscit aut qui violenta morte perit), Wrt. Voc. ii. 126, 10. Selfbonan *biothanatas,* 11, 69. Seolfbonan (-boran, MS.), 101, 74. Selfbanan *biothanatos,* Hpt. Gl. 469, 26.

self-cwalu, e; *f. Self-slaughter, suicide:*—Sum sceal ful earmlíce ealdre linnan . . . and hine tó sylfcwale secgas nemnaþ *speak of: him a committing suicide,* Exon. Th. 330, 24; Vy. 56. [Cf. We scole witan, þet nan seolfcwale, þet is aȝenslaȝa, ne cumeð to godes riche, O. E. Homl. i. 103, 3.]

self-, selfe-déma, an; *m. One who depends upon his own judgment* [cf. ǽlc ídel mon liofaþ æfter his ágenum dóme, Past. 39; Swt. 283, 21], used of a certain kind of monks called *sarabaitae,* monachi qui nulla regula approbati . . . proprio arbitratu vivunt:—Þridde cyn is muneca ealra atelucost, sylfdémena (*sarabaitarum*), ðe nó on regules and lāreówa tǽcinge ne beóþ āfandode, swā swā gold on heorðe . . . Ðæt feórþe muneca cyn ðe is wīðscripuþ genæmned . . . hié synt wyrsan ðænne ða sylfedéman (sylf-, MS. T.), R. Ben. 9, 10–10, 2. Hý āscyriaþ hý sylfe fram mynsterlícum þeáwum and heora āgenum lustum filiaþ, hý sint Egyptiscan gereorde genemnede *sarabagite* oððe *renuite,* ðæt ys sylfedéman and wiðersacan, 136, 12. [Cf. Icel. sjálf-dæmi *judgment given in a case by one of the parties themselves.*]

selfe; *adv. In the same way,* in combination with *swā:* cf. gelíce, same:—Hé forlǽt lífes frumsceaft and his āgene ǽðelo swā selfe, Met. 17, 25. Hæfþ ða wilnunga welhwilc néten and ða yrsunga eác swā selfe, 20, 192, 199. [Cf. O. Sax. sō *self also, likewise:* O. H. Ger. sō selp (sō), selp sō *sic, sicut.*]

self-líc; *adj. Of one's own accord, spontaneous, voluntary:*—Selflíces *spontaneae, ultroneae, voluntariae,* Hpt. Gl. 436, 75.

self-líce, es; *n. Self-love, self-complacency, self-satisfaction, conceit, arrogance:*—Ðonne ðæt selflíce gegriépþ ðæt mód þæs recceres *amor proprius cum rectoris mentem ceperit,* Past. 19, 1; Swt. 143, 5. Ðonne āhefþ hé hine on his móde . . . mid ðý selflíce se Déma biþ genieded tó ðæm ierre, 4, 2; Swt. 39, 10. Ðæm lytegan is ǽresð tó beleánne hiera selflíce ðæt hié ne wénen ðæt hié sién wíese . . . hé biþ ǽr ūpāhæfen on selflíce for hys lotwrencium *in sapientibus hoc primum destruendum est, quod se sapientes arbitrantur,* 30, 1; Swt. 203, 9, 18. Ðýlæs hé sié āhafen on his móde and on oferméttum āþunden and þurh ðæt selflíce his gódan weorc forleóse *ne perfecta opera tumor elationis extinguat,* proem.; Swt. 25, 7. Hé hiene up āhefeþ on his móde on suelc gielp and on suelc selflíce *se apud se per arrogantiam exaltat,* 11, 4; Swt. 71, 1. Ðæt freódóm ne gewende on selflíce and on ofermétto *ut libertas in superbiam non erumpat,* 19, 3; Swt. 147, 3. Upāhafene þurh selflíce, Bt. 3, 4; Fox 6, 25.

self-líce; *adj. Self-satisfied, self-complacent, conceited, arrogant:*—Oft se welega and se wædla habbaþ suā gehweorfed hira þeáwum ðæt se welega biþ eáðmód and sorgfull and 'se wædla biþ upāhæfen and selflíce *plerumque personarum ordinem permutat qualitas morum, ut sit dives humilis, sit pauper elatus,* Past. 26, 2; Swt. 183, 11. Selflícne secg *the self-satisfied man,* Met. Introd. 7.

self-myrþe (?); *adj. Self-destructive:*—Betweónan sylfmyrþe *inter biothonatas,* Wrt. Voc. ii. 80, 2. v. next word.

self-myrþere (?), an; *m. One who destroys himself, a suicide:*—Betweónan selfmyrþras (-an? *but cf.* Icel. myrða *to murder:* O. H. Ger. murdit *jugulat:* Ger. morden) *inter biothanatas,* Wrt. Voc. ii. 46, 61. Cf. self-bana.

self-sceaft, es; *m. Self-shaping, spontaneous generation,* applied to Adam, who had not father and mother:—Adam maþelode ðǽr hé on eorþan stód selfsceafte guma *a man by spontaneous generation,* Cd. Th. 33, 20; Gen. 523.

self-will, es; *n.* (?) *Self-will, one's own will, free-will:*—Be ðám ðe beóþ hyra sylfwilles (*sua sponte*) gefullode, L. Ecg. C. 17, tit.; Th. ii. 128, 30. Be selfwille *ultro,* Wrt. Voc. ii. 73, 27. Getǽc mé sumne mann ðara ðe ðé gesǽlegost þince and on his selfwille sý swíðost gewiten *who most has had things his own way,* Bt. 11, 1; Fox 32, 16. Gif ðú ne wilt wirde steóran ac on selfwille sígan lǽtest *if thou wilt not guide fate, but lettest her go at her own will,* Met. 4, 50. [Cf. O. H. Ger. (pī) selpwillin *sponte, ultro;* Icel. með, at sjálfvilja *of one's own will.*] v. self-willes.

self-wille; *adj. Voluntary, spontaneous:*—Mid selfwilre *spontanea,* Hpt. Gl. 415, 11: *spontaneo, voluntario,* 439, 11. For ðan selfwillan *propter spontaneum,* 413, 33. [*Goth.* silba-wiljós *voluntarii,* 2 Cor. 8, 3.]

self-willende; *adj. Voluntary:*—Rēn sylfwillendne *pluviam voluntariam,* Ps. Lamb. 67, 10. [*Goth.* silba-wiljandi galaith *sua sponte profectus est,* 2 Cor. 8, 17: Icel. sjálf-viljandi.]

self-willes; *adv. Voluntarily, of one's own accord:*—Selfwilles *ultro,* Wrt. Voc. ii. 92, 74. Sylfwilles *sponte,* Ælfc. Gr. 38; Zup. 234, 19: *ultro,* Zup. 237, 2. (1) *of persons:*—Drihten on róde selfwilles þrowode, H. R. 17, 21. Ealle hyra unlustas hí sceolon gebētan sylfwilles on ðyssum lífe, oððe unþances æfter ðyssum lífe, Homl. Th. i. 148, 27.

Wrýt nū sylfwylles ðæt ðū wiðsace Criste, Homl. Skt. i. 3, 379. Hé sylfwilles menniscnesse underféng, Wulfst. 15, 12. Hé gǽþ sylfwilles twā míla tó ānre geneádod, R. Ben. 28, 6. (2) *of things:*—Ðonne his wæstmas weaxaþ sylfwilles *quae sponte gignet humus,* Lev. 25, 5. Sylfwilles (*ultro*) seó eorþe wæstm beraþ, Mk. Skt. 4, 28. v. self-will.

selan. v. sylian, sēl.

sella, an; *m. A giver:*—Ðone glædan syllan *hilarem datorem,* R. Ben. Interl. 25, 6.

sellan, sillan, syllan; *p.* salde, sealde; *pp.* sald, seald *To give something* (*acc.*) *to somebody* (*dat.*). **I.** *of voluntary giving, to put into the possession* of a person, *transfer ownership* from one to another:—Ic sello Werburge ðás lond, Chart. Th. 480, 30: 481, 5. Ðæt land ic sylle eów tó ágenne, Ex. 6, 8. Ealle ðás rícu ic sylle (sello, Lind.: selle, Rush.) ðé, Mt. Kmbl. 4, 9. Ðū sāwlum selest ginfæsta gifa, Met. 20, 226. Eówer Fæder syleþ (selleþ, Rush.) gód ðám ðe hyne biddaþ, Mt. Kmbl. 7, 11. Hig wǽron ðíne, and ðū hý sealdest mé, Jn. Skt. 17, 6. Salde *inpendebat* (cf. geben wæs *inpendebatur,* 21), Wrt. Voc. ii. 111, 24. Ecgbryht salde Basse mæssepríoste Reculf mynster on tó tymbranne, Chr. 669; Erl. 34, 25. Hié saldon hiera nefum Wiehte eálond, 534; Erl. 14, 33. Ðeáh Balac mé sille goldes ān hús full, Num. 22, 18. Hí ne mágon sellan ðæt hí gehātaþ, Bt. 16, 1; Fox 90, 16. Nelle gē syllan (sella, Lind.) ðæt hālige hundum, Mt. Kmbl. 7, 6. Gē cunnun góde sylena eówrum bearnum syllan (sellan, Rush.), 7, 11. Biþ sald *dabitur,* Kent. Gl. 338. **II.** *to give* what one is bound to give, *to pay* tribute, *offer, dedicate* to God:—Sylle mé ðín forme bearn. Dó eall swā of hríðerum . . . syle (*reddas*) hit mé on ðam ehtuþan dæge, Ex. 22, 29, 30. Norþmonnum níede sceoldon gombon gieldan and gafol sellan, Cd. Th. 119, 12; Gen. 1978. Gafol syllan, Chr. 1006; Erl. 141, 10. Hí willaþ eów tó gafole gāras syllan, Byrht. Th. 133, 7; By. 46. Hié nǽfdan for him lamb tó syllenne, Blickl. Homl. 23, 26. **III.** *to give, furnish* or *supply with* food, medicine, poison, etc.—Hwā sylþ ūs flǽsc? . . . Drihten eów silþ flǽsc and gē etaþ, Num. 11, 18. Byrelas sealdon wín of wunderfatum, Beo. Th. 2327; B. 1161. Hié him sealdon āttor drinccan, Blickl. Homl. 229, 16: Ealle ða mettas ge drincan ða ðe habban hāt mægen and scearp sele þicgean, Lchdm. ii. 184, 10 (often in Leechdoms). Him man metsunge syllan sceolde, Chr. 1006; Erl. 141, 10. Hé wolde syllan his assan fóddur, Gen. 42, 27. **III a.** *with infin.* instead of *acc.*—Hwílum ic deórum drincan selle, Exon. Th. 393, 25; Rä. 13, 5. Gehwylc mé drincan sealde, 484, 24; Rä. 71, 6. **IV.** *to give* one thing for another. (a) *to sell* for (*wið*) a price:—Ic sylle wið wirðe *vendo,* Ælfc. Gr. 28, 8; Zup. 181, 17. Hwí ne sealde heó ðás sealfe wið þrím hundred penegon, Jn. Skt. 12, 5. Sume man wið feó sealde, Chr. 1036; Erl. 164, 34: Blickl. Homl. 79, 22. (b) *to sell* at (*tó*) a price:—Sēlre ys ðæt wē hine syllon tó ceápe Ysmahēlitum, Gen. 37, 27. (c) *to sell:*—Hé sylþ (*vendit*) eall ðæt hé āh, Mt. Kmbl. 13, 44. Hé worhte his weorc tó sceoran nihtum, and sealde on ðone Sæternesdæg, Homl. Th. ii. 356, 6. Hí sealdon heora gymstānas, i. 62, 21. Ne eów ne ofþince ðæt gē mé sealdon (*vendidistis*) on ðis ríce, Gen. 45, 5. Syllaþ (*vendite*) ðæt gē āgon, Lk. Skt. 12, 33. Nān man hig nā undeóror ne sylle (sille, MS. D.); and gif hwā hí undeóror sylle, gilde ǽgðer .xl. scillinga, ge se ðe hí sylle ge se ðe hí bycge, L. Edg. ii. 8; Th. i. 270, 3–6. Móna se óðer on eallum þingum tó dónum nytlíc ys, bicgan, syllan, scip āstígan, Lchdm. iii. 184, 13. Ða syllendan *vendentes,* Lk. Skt. 19, 45. (d) *to give* in payment:—Hé sealde his ðone reádan gim, ðæt wæs his ðæt hālige blód, mid ðon hé ūs gedyde dǽlnimende ðæs heofonlícan ríces, Blickl. Homl. 9, 36. Eall ðæt feoh ðe hié wið ðæm weorce sellan woldon, Ors. 4, 12; Swt. 210, 4. Syllan feoh wið freóde, Byrht. Th. 132, 58; By. 39. **V.** (a) *to give* into the keeping of, *hand over, deliver, commit, entrust:*—Gif ðū mé sylst underwedd, Gen. 38, 17. Ic befæste ðé ðæt eówde ðæt ðū mé sealdest, Blickl. Homl. 191, 27. Hé hire sāule sealde Sarcte Michahele, 147, 13. Hé sealde his sweord ombihtþegne, Beo. Th. 1349; B. 672. Hié sealdon ānum unwísum þegne Miercna ríce tó haldanne, and hé him gíslas salde, Chr. 874; Erl. 76, 26–28. Hié sealdon hiera suna tó gíslum *they gave their sons as hostages,* Ors. 4, 11; Swt. 204, 4. Hí on wedde sealdon, hwæt hý hyre syllan woldon, Homl. As. 196, 24. Ðā wæs ic mid gýmenne mínra māga seald tó fédanne and tó lǽranne Abbude Benedicte, Bd. 5, 23; S. 647, 22. (b) *to give* a woman to be a man's wife:—Ðā wolde se fæder hí sellan sumum æþelon men tó brýde, Shrn. 31, 6. Nyme hé hig tó rihtwífe. Gif se fæder hig him syllan nelle, Ex. 22, 17. (c) *to give over* to a hostile power, *deliver up* to. (1) *with dat.:*—Ne syle (*tradas*) ðū unscyldigra sāwla deórum, Ps. Th. 73, 18. Ðý lǽs ðe ðín wiðerwinna ðé sylle ðam dēman, and se dēma ðé sylle ðam þéne, Mt. Kmbl. 5, 25. (2) *without dat.:*—Ne syle mé ne ne send mé mid ðám synfullan *ne tradas me cum peccatoribus,* Ps. Th. 27, 3. (3) *with prepositions:*—Hé sealde on edwít ðe mé ǽr trǽdan, Ps. Th. 56, 3. Hé sealde his folc sweordes under ecge, 77, 62. Ne syle mé tó ðara módes willan, 26, 14. Ne dó mé ne syle on ðone biterestan deáþ, Blickl. Homl. 229, 26. Ne syle ðū mé in wīta forwyrd, Frag. Kmbl. 14; Leás. 9. Mannes sunu ys tó syllenne on manna handa, Mt. Kmbl. 17, 22. (4) *with dat. and prep.:*—Drihten him sealde ða burh on his

handa, Jos. 10, 32. (d) with a bad sense, *to deliver wrongfully, to betray;* cf. colloquial *to sell* a person. Mannes sunu þū mid cosse sylst (seles, Rush.: selles, Lind.), Lk. Skt. 22, 48. Nū is gehende se ðe mē sylþ (seleþ, Rush.: selleþ, Lind.), Mk. Skt. 14, 42. Ðæt mon ne selle his weorðscipe fremdum menn, Past. 36; Swt. 249, 21. **VI.** *to give up, yield up:*—Hē feorh seleþ *he dies*, Beo. Th. 2745; B. 1370. **VII.** *to give forth, produce, be the source of:*—Ne seleþ ðē wæstmas eorþe, Cd. Th. 62, 17; Gen. 1015. Sume sealdon (saldun, Rush.: saldon, Lind.) wæstm, Mt. Kmbl. 13, 8. God lǣteþ hrusan syllan blǣda beornum, Runic pm. Kmbl. 341, 23; Rūn. 12. **VII a.** *to give light, emit* sound:—Sylle se friccea his stefne, Blickl. Homl. 163, 31. Leóht sellan, Bt. 6; Fox 14, 23. **VIII.** where the object is immaterial, (a) *to give an answer, a pledge, a promise,* etc.:—Ic eów treówa mīne selle, Cd. Th. 92, 29; Gen. 1536. Ic ðē wǣre mīne selle, 132, 35; Gen. 2203. For ðīnum gebode ðe ðū mē sealdest, Blickl. Homl. 241, 33. Se Hǣlend him ne sealde nāne andsware, Jn. Skt. 19, 9. Hī sealdon āþas, Met. 1, 24. Him lof syllaþ, Ps. Th. 65, 1. Heora ǣlc sylle ðone āþ, ðæt . . . , L. Edg. S. 6; Th. i. 274, 15. ¶ where the object is expressed by a clause:—Ic eów behāta and on hand selle, ðæt gē sculon finden reste eowre sāwlen, Homl. As. 171, 29. Ðæt hȳ ealle ðam sēmende syllan, ðæt cyninges mund stande, L. E. G. 12; Th. i. 174, 22. Slaga sceal his forspecan on hand syllan, and se forspeca māgum, ðæt se slaga wille bētan wið mǣᵹþe. Ðonne gebyreþ ðæt man sylle ðæs slagan forspecan on hand, ðæt se slaga môte mid griþe weddian, L. Edm. S. 7; Th. i. 250, 14–17. (b) *to give leave, consent, forgiveness,* etc.:—Ðyssum wordum ðder ealdormann geþafunge sealde (*tribuens assensum*), Bd. 2, 13; S. 516, 13. Hē him ne sealde leáfe, Homl. Th. ii. 380, 5. Nis nān tweó ðæt hē forgifnesse syllan nelle ðam ðe hié geearnian willaþ, Blickl. Homl. 65, 8. (c) *to give help, pain, peace, victory,* etc.:—Ic ðē mīne sylle sibbe, Andr. Kmbl. 194; An. 97. Ðū sylest ūrum leomum rǣste, Blickl. Homl. 141, 11. Se ðe sigor seleþ, Cd. Th. 170, 5; Gen. 2808. Sile ðīne āre ðīnum earminge, Hy. 2, 3. Ūs fultum sile, 7, 80. Gif Drihten him sige syllan wolde, Bd. 3, 24; S. 556, 18. Nū biþ ðæm seald Drihtnes mildheortnes, Blickl. Homl. 49, 24. ¶ where the object is expressed by a clause:—Ne svleþ hē sôðfæstum, ðæt him ȳþende môd innan hreðre, Ps. Th. 54, 22. Gūþlāce engel sealde, ðæt him sweðraden synna lustas, Exon. Th. 109, 1; Gū. 83. Syle mē, ðæt ðū mē generige nīða gehwylces, Ps. Th. 118, 169. (d) *to give* punishment, reward:—Sealde him wītes clom, Cd. Th. 193, 11; Sat. 453. Leán sellende eallum, 240, 34; Dan. 396. (e) *to give, endow with* a capacity, life, sight, understanding, etc.:—Ðū sylest andgit eallum eorþbūendum, Ps. Th. 118, 130. Ðū man geworhtest and him sealdest word and gewitt and wæstma gecynd, Hy. 9, 56. Ðū sealdest ǣlcre gecynde āgene wīsan, 7, 66. Sealde hē dumbum gesprec, Andr. Kmbl. 1153; An. 577. Syle mē heortan clǣne, Ps. Grn. 50, 11. Ǣghwylc ðe him eágna gesihþ cyning syllan wolde, Exon. 350, 22; Sch. 67. Ðē biþ ēce līf seald, Elen. Kmbl. 1052; El. 527. ¶ with the gerund:—Heáh geweorc furþor āspyrgen ðonne him freá sylle tô ongietanne, Exon. Th. 348, 17; Sch. 29. (f) *to give* one's heart to a person:—Nemne ic Gode sylle hȳrsumne hige, Exon. Th. 124, 12; Gū. 338. [*Goth.* saljan *to offer: O. Frs.* sella *to give, sell, pay: O. Sax.* sellian *to give: O. H. Ger.* sellan *tradere: Icel.* selja *to hand over, to sell.*] v. ā-, be-, for-, ge-, ymb-sellan; un-seald.

sellend, es; m. **I.** *a giver:*—God gôdra mægna sellend (*dator*), Rtl. 103, 36. Sigora sellend (*the Deity*), Exon. Th. 282, 24; Jul. 668: 359, 10; Pa. 64. Syllend, 284, 30; Jul. 705. Drihten se is ordfruma and syllend (*largitor*) ealra eádignesse, Bd. 4, 30; S. 609, 17. Hihton hī on God, ðæra gôda syllend, Homl. Th. ii. 328, 1. Hē lufaþ ðone glædan syllend, 212, 9. **II.** *a betrayer:*—Se sellend his *traditor ejus*, Mk. Skt. Lind. Rush. 14, 44. v. ǣ-sellend.

sel-līc, selma, selmerige, selnes, sēlost, sēlra, seltra. v. seld-līc, sealma, sǣlmerige, seleness, sell, sæltna.

sēma, an; m. *An arbitrator, umpire:*—Sēma (sȳma, sīma) *sequester*, Ælfc. Gr. 9, 18; Zup. 43, 16. v. sēman.

sēman *to load.* v. sīman.

sēman; p. de; pp. ed. **I.** with acc. of person, (1) *to bring to an agreement* those who have a dispute:—Ðā hēt hē hié sēman. Ðā wæs ic ðara monna sum ðe ðærtô genemned wǣran . . . Ðā wē hié sēmdan *then bade the king to bring them* (the parties in a dispute about some land) *to an agreement. Then was I one of the men who were nominated for the purpose . . . When we had brought them to an agreement*, Chart. Th. 170, 6–35. (2) *to satisfy* a person in a matter of doubt or difficulty:—Sēme ic ðē recene ymb ða wrætlīcan wiht, Salm. Kmbl. 504; Sal. 252. **II.** with acc. of thing, *to settle* a dispute:—Hī sace sēmaþ, sibbe gelǣraþ, Exon. Th. 334, 22; Gn. Ex. 20. **III.** used intransitively, *to arbitrate, bring about agreement:*—Nān sacu ðe betweox preóstan sī ne beó gescoten tô worldmanna sôme, ac sēman and sibbian heora āgene geféran, L. Edg. C. 7; Th. ii. 246, 6. Gif hē healt weorð, ðær môtan freónd sēman, L. Ethb. 65; Th. i. 18, 14. v. ge-sēman; sôm.

sēmend, sǣmend, es; m. *One who brings about agreement between parties in a dispute, an arbitrator, umpire:*—Ðæt hȳ ealle gemǣnum

handum of ǣᵹðere mǣᵹþe on ānum wǣpne ðam sēmende syllan, ðæt cyninges munde stande, L. E. G. 12; Th. i. 174, 22. Ymb .iii. niht gesēcan hiom sǣmend, L. H. E. 10; Th. i. 30, 18. v. preceding word.

sēmestre, semian, semle. v. seámestre, seomian, symble.

semnendlīce; adv. *By chance, fortuitously:*—Semnendlīce *fortuito*, Wrt. Voc. ii. 37, 10: 80, 40: *fortuis*, 84, 78.

semninga, senap. v. samnunga, senep.

senatus *the senate, senators.* The treatment of this word in the translation of Orosius is somewhat exceptional. The Latin form *senatus* occurs in the nom. and acc., but in the former *senatas, senatum,* and in the latter *senatum, senatos* are also used; in the gen. *senatuses, senatusa* are found, and in the dat. *senatum;* in every case but one (?) the word is plural. The Latin *senator* is also used, though the word *witan* is generally employed to denote the senators:—Sceoldon ealle hiera senatus (senatas, Bos. 43, 5) cuman . . . sceoldon hiera senatus (-as) rīdan, Ors. 2, 4; Swt. 70, 24, 28. Ealle heora senatus *senatores,* 4, 9; Swt. 190, 19. Ealle ða senatus *omnis senatus,* 5, 12; Swt. 240, 13. Ðā wolde ān (woldan, Bos. 70, 36) senatus hiene āweorpan . . . Ðā bæd his fæder ðæt ða senatum (*altered to* senatus *in other MSS.*) forgeáfen ðæm suna ðone gylt, 3, 10; Swt. 140, 14–16. Se consul bæd ðætte senatus him fultum sealdon, 4, 9; Swt. 192, 22. Ða senatus him hæfden ða dǣd forboden . . . Ne mehten ða senatus nǣnne consul under him findan, 4, 10; Swt. 196, 7–10. Būton his āgnum fultume and būton ðara senatuses, 5, 12; Swt. 242, 1. Hē forneáh nānne ðara senatusa ne lēt cucne *plurimos senatorum ad mortem coegit,* 6, 2; Swt. 256, 1. (Cf. Ðara senatorum xxxv *triginta quinque senatores,* 6, 4; Swt. 260, 23: 6, 14; Swt. 268, 28.) Hē sende tô ðǣm senatum ðæt hē ðæt irre gesette wið hié, 4, 11; Swt. 206, 26: 2, 6; Swt. 88, 12. Hē hit sǣde ðǣm senatum, ða wurdon hié alle wið hiene wiðerwearde *senatus indignatione motus,* 6, 2; Swt. 254, 25: 5, 12; Swt. 244, 16. Romulus gesette senatum, 2, 4; Swt. 70, 36. Ðēh hē hit wið ða senatus hǣle, 4, 10; Swt. 196, 16. Hē sette senatus, 5, 12; Swt. 242, 28. Ðæt hē sprǣce wið ða senatos (-us *other MSS.*), 4, 11; Swt. 206, 29: 4, 13; Swt. 210, 16: 5, 5; Swt. 226, 16.

sencan; p. te; pp. ed. **I.** *to sink* (trans.), *plunge, immerse:*—Wæs his gewuna ðæt hē hine on ðam streáme sencte *solebat in flumine supermeantibus undis immergi,* Bd. 5, 12; S. 631, 21. **II.** *to submerge, flood* with water:—Abraham wolde his sunu cwellan folmum sīnum fȳre(?)sencan mæges dreóre (*flood the pile with his son's blood*), Cd. Th. 176, 4; Gen. 2906. Ne biþ flôd tô sencende (tô stencende (?) *dissipans*) ða eorþan, Gen. 9, 11. [Forte reauin hire bodi and i þea sea senchen, Jul. 79, 1. *Goth.* saggkwjan: *O. Sax.* be-senkian: *O. H. Ger.* sencan *mergere: Icel.* sökkva.] v. ā-, be-, ge-sencan.

sendan; p. sende; pp. sended, send *To send, cause to go.* **I.** where the object is a living thing, (1) *to send* after (*æfter*), on an errand, for a purpose, *despatch:*—Ic sende ǣrendracan tô mīnum hlāforde, Gen. 32, 5. Ic ēow sende swā swā sceáp gemang wulfas, Mt. Kmbl. 10, 16. Hē sent ǣrendracan, Lk. Skt. 14, 32. God sendeþ his engla gāstas tô ǣrendwrecum, Blickl. Homl. 203, 14. Hē ūsic sendeþ ðæt wē sôðfæstra môd oncyrren, Exon. Th. 261, 34; Jul. 325. Ða twegen leorningcnihtas ðe Crist sende æfter ðam assan, Homl. Th. i. 206, 23. Ðā sendon hī hyra leorningcnihtas tô, Mt. Kmbl. 22, 16. Gif ðū wylt hine mid ūs sendan, Gen. 43, 4. Hē mē on ðisne sīð sendan wolde, Exon. Th. 460, 35; Hö. 27. Se ðe englas gehēt wið mē tô sendenne, Blickl. Homl. 181, 26. Fram Gode hē is send, 247, 19. Oþ ðæt ðū gefylle ðīne þegnunge tô ðære ðe ðū sended eart, 233, 28. Hē senden (?) wæs tô hādianne, and Wilfreð on Gallia rīce tô hādianne sended wæs, Bd. 4, 2; S. 566, 12, 13. Ðā wæs culufre sended, Cd. Th. 88, 13; Gen. 1464. Ealle Drihtnes apostolas beóþ sende ðē tô bebyrgenne, Blickl. Homl. 137, 27. (2) with a sense of compulsion or violence, *to send* to prison, into exile, etc.:—Se ðec on wræc sendeþ, Cd. Th. 251, 26; Dan. 569. Se ðec sended in ða sweartesten wītebrôgan, Elen. Kmbl. 1858; El. 931. Hē hine on fȳr and on wæter sende, Mk. Skt. 9, 22. Wē iii hæfdon cniehtas gebunden in fȳres leóman, nū ic ðær iiii men sende tô sīðe (cf. geseó tô sôðe *in the version given,* Cd. Th. 242, 8; Dan. 416), Exon. Th. 196, 16; Az. 175. Ðara ðe hē of hleó sende, Cd. Th. 7, 7; Gen. 102. Hié mē sendon on ðis carcern, Blickl. Homl. 237, 31. Hē wile ða sáula sendan on ēce wītu, 95, 4: 125, 2. Wē wǣron on ðysne wræcsīþ sende, 23, 6. **II.** where the object is not a living creature, *to send* a message, present, help, etc.:—Ðū senst ūrne hlāf dæghwamlīce, Hy. 7, 68. Dryhten sendeþ þurh monnes hond mīne þearfe, Exon. Th. 121, 22; Gū. 292. Meotud monnum dǣleþ, syleþ sundorgiefe, sendeþ wīde āgne spēde, 293 23; Crä. 5. Sende ic Wylfingum ealde mādmas, Beo. Th. 946; B. 471. Ðē sende God ðās helpe, Cd. Th. 33, 15; Gen. 520. Sende ðā his bēne fore bearn Godes, Andr. Kmbl. 3224; An. 1615. Sendon hira bēne fore bearn Godes, 2055; An. 1030. Þinga gehwylces ðara ðe ðīn weorc wylle tô cunnunge, Exon. Th. 453, 32; Hy. 4, 23. Ǣrendgewrit suelce hit from ūs send sié, Past. 32; Swt. 213, 18. **III.** *to send, move* to a place of rest, *put, lay:*—Ic sende mīne hond on ðās fǣmnan *I will lay my hand on this woman,* Shrn. 130, 27. Ðū sāwle sendest intô ðam flǣsce, Hy. 7, 4. Ðonne se wæstm hine forþbringþ, sôna hē sent his

sicol, Mk. Skt. 4, 29. Sumum wordlaþe sendeþ on his mōdes gemynd Exon. Th. 41, 32 ; Cri. 664. In eorþan fæþm sendaþ lîchoman, 231, 12 ; Ph. 488. Ælmihtig eácenne gāst in sefan sende, Cd. Th. 246, 28 ; Dan. 486 : Beo. Th. 3688 ; B. 1842. Hié sendon rāp on his sweoran, Blickl. Homl. 241, 24. Ðæt on ðone hālgan handa sendan fæderas ūsse, Elen. Kmbl. 912 ; El. 457. Uton sendon rāp on his swyran, Blickl. Homl. 241, 10. **IV.** with a stronger sense of motion, *to send* a missile, *cast* lots, *throw, hurl* :—(a) Ðonne sceótend sendaþ flāngeweorc, Exon. Th. 42, 20 ; Cri. 675. 'Nū, anlīcnes, sænd mycel wæter þurh þínne mūþ.' Sió onlīcnes sendde mycel wæter þurh hiora mūþ, Blickl. Homl. 245, 20–24. Gūþfrecan gātas sendon in heardra gemang, Judth. Thw. 24, 35 ; Jud. 224. Hié sendon hlot him betweónum, Blickl. Homl. 229, 5. Send ðe nyþer of ðisse heáhnesse, 27, 12. Ðā hēt ic feá stræla sendan in ða burh innan, Nar. 10, 22. Hē geseh ða welegan hyra lāc sendan on ðone sceoppan, Lk. Skt. 21, 1. Sendende hyra nett on ða sæ, Mt. Kmbl. 4, 18. Seó stræl wæs sended, Blickl. Homl. 199, 22. (b) of the operations of Nature, *to send* rain, *fire*, etc. :—Drihten sende regn, Cd. Th. 82, 33 ; Gen. 1371. Him brego engla wylmhātne līg tō wræce sende, 156, 6 ; Gen. 2584. God eástan sende leóhtne leóman, Judth. Thw. 24, 16 ; Jud. 190. Sceolde hē sendan þunras and lȳgetu, Bt. 35, 4 ; Fox 162, 13. Ic sendan gefrægn swegles aldor swefl of heofonum, Cd. Th. 153, 17 ; Gen. 2540. (c) *to send* punishment, pestilence, etc. :—Drihten sende on hié māran wræce, Blickl. Homl. 79, 9. Ðæt God wolde sendan hungor and ādla on manna ceáp, Wulfst. 209, 28. **V.** *to send forth, emit* a sound :—Heofenfuglas sendaþ stefne mycle *dabunt voces suas*, Ps. Th. 103, 11. **VI.** where the object is not expressed, *to send* a message or a messenger. (1) to or after (*tō, æfter*) a person or thing, *to send for* (æfter) :—Hēr sende se cyng tō ðam here, Chr. 1011 ; Erl. 144, 20 : 1048 ; Erl. 180, 9. Ðā sende se cing æfter ðām scypon, 1049 ; Erl. 172, 39. Ðā sende se cyng æfter eallon his witan, 1048 ; Erl. 178, 13. Hī sendon on Perse æfter Conone, Ors. 3, 1 ; Swt. 98, 19. Ðā sendon hié on Affrice tō Cartaginenses æfter fultume, 4, 1 ; Swt. 160, 2. Ðæt hié tō Rōme sendon tō ðæm pāpan, Blickl. Homl. 205, 19. (2) where the person or thing sent to or for is not stated :—Hī sendon geond eall ðæt land, and brohton tō him ealle untrume, Mt. Kmbl. 14, 35. [*Goth.* sandjan : *O. Frs.* senda : *O. Sax.* sendian : *O. H. Ger.* sentan : *Icel.* senda.] v. ā-, āgen-, for-, fore-, geond-, in-, of-, ofer-, on-, tō-sendan.

sendeþ, Beo. Th. 1204 ; B. 600. v. scendan.

sendlíc ; *adj. To be sent* :—Ða sendlícan gebrōðra on wege *dirigendi fratres in viam*, R. Ben. Interl. 113, 4.

sendness, e ; *f. A sending, dismission* :—Sendnessa *missarum* (*Low Latin missa dimissio*), Wrt. Voc. ii. 56, 71 : 80, 70.

senep (-ap, -op), es ; *m. Mustard* :—Senep *sinapis*, Wrt. Voc. i. 31, 47. Senap, 69, 20. On ða gelícnesse geworht ðe senop biþ getemprod tō inwīsan, Lchdm. ii. 184, 22. Gelíc senepes corne, Mt. Kmbl. 13, 31 : Lk. Skt. 13, 19. Senepes sǣd, Mk. Skt. 4, 31 : Lchdm. ii. 20, 11. Mid sinope gnīde, 186, 6. Gerēnodne senep, 184, 9 : 20, 22. [*Goth.* sinapis (*gen.*) : *O. H. Ger.* senaf : *Ger.* senf.]

senep-sǣd, es ; *n. Mustard-seed* :—Nim senepsǣd, Lchdm. iii. 88, 15.

sengan ; *p.* de ; *p.* ed *To singe, scorch* :—Gýme eác swān ðæt hé æfter sticunge his slyhtswýn wel behweorfe, sæncge, L. R. S. 6 ; Th. i. 436, 16. [*Chauc.* senge ; *pp.* seind : *Prompt. Parv.* sengiñ *ustulare* : *O. Frs.* senga : *M. H. Ger.* sengen : *Du.* zengen : cf. *Icel.* sangr *burnt, scorched* ; sengja *a singed taste*.] v. be-sengan, unbesenged ; singan.

sēnian, senn, senoþ, seó (*pron.*), seó (*verb*). v. segnian, synn, seonoþ, se, sî.

seó ; *gen.* seón, seó ; *acc.* seón, seó ; *f. m.* (?) *The pupil, apple of the eye* :—Seó *pupilla vel pupula*, Wrt. Voc. i. 43, 1 : 64, 40 : *papilla, papula*, 282, 53, 54. Seó sceal in eágan, Exon. Th. 341, 8 ; Gn. Ex. 123. Ðæs (ðære ?) seó hringc *circulus*, Wrt. Voc. i. 42, 72. Hē heóld hig swā his eágan seón (*quasi pupillam oculi sui*), Deut. 32, 10 : Ps. Spl. 16, 9. Seán, Ps. Surt. 16, 8. Sión, Kent. Gl. 177. Swylce hē hreppe ða seó mînes eágan, Homl. Th. i. 390, 15 : 516, 23. Seón *pupillae*, Wrt. Voc. i. 65, 8. Seóna *pupillarum*, Hpt. Gl. 404, 28. Sión *pupillis*, Lchdm. i. lxx, 6. Seóum, lxxiv, 7. [*O. H. Ger.* seha (*acc.* sehun, sehe, *n. pl.* seha, sehun, v. Grff. vi. 123) *pupilla, acies*.]

seóbgende. v. seófian.

seóc ; *adj. Sick, ill.* **I.** of bodily infirmity or disease :—Sum seóc man *quidam languens*, Jn. Skt. 11, 1. Se is seóc *infirmatur*, 3. Hé wæs *infirmabatur*, 6. Seóc hé biþ ðe tō seldan ieteþ, Exon. Th. 340, 16 ; Gn. Ex. 111. Seonobennum seóc, 328, 17 ; Vy. 19 : Beo. Th. 5473 ; B. 2740 : 5800 ; B. 2904. Gif mon sȳ ðære healfdǣdan ādle seóc, Lchdm. ii. 184, 31. Seó lange mettrumnes ðæs seócan mannes, Blickl. Homl. 59, 28. Swā swā lǣca gewuna is ðonne hió seócne (siócne, Cott. MS.) mon gesióþ, Bt. 36, 4 ; Fox 178, 26. Ða ðe on sāre seóce lāgun, Exon. Th. 83, 15 ; Cri. 1356. Feóllon wergend bennum seóce, Cd. Th. 118, 29 ; Gen. 1972. Seócra manna hūs *nosocomium*, Wrt. Voc. i. 58, 52. Ofer seóce (*aegrotos*) hí hyra handa settaþ and hí beóþ hāle, Mk. Skt. 16, 18. ¶ used as a noun :—þurh his hrepunge beóþ gestrangode ða unstrangan seócan, Homl. Skt. i. 7, 54. **II.** of moral disease :—Hǣðne wǣron begen, synnum seóce, Exon. Th. 246, 21 ; Jul.

65. Gif hé his seócum ðæt is synfullum dǣdum ealle lācnunge gegearwade *si morbidis eorum actionibus universa fuerit cura exibita*, R. Ben. 11, 5. **III.** of mental disquiet, *sick* at heart, *ill* at ease, *sad* :—Ne beó ðū on sefan tō seóc, Exon. Th. 166, 29 ; Gū. 1050. Seóc and sorhful, Cd. Th. 281, 20 ; Sat. 275. Ic ðysne sang sīðgeómor fand on seócum sefan, Apstls. Kmbl. 3 ; Ap. 2. Wēna mē ðîne seóce gedydon, Exon. Th. 380, 26 ; Rä. 1, 14. [*Laym.* seoc, seac, sec, sæc : *Orm.* seoc, sec : *A. R.* sec, sic : *Chauc.* sek, sik : *Wick.* seek, siik : *Ayenb.* zik : *Prompt. Parv.* seek : *Goth.* siuks : *O. Sax.* seok, siok, siak : *O. Frs.* siak, siek : *O. H. Ger.* sioh, siuh, sieh : *Ger.* siech : *Icel.* sjúkr.] v. bræc-, brægen-, deófol-, ellen-, fefer-, feónd-, feorh-, fylle-, gebræc-, gewit-, heaðu-, lifer-, lim-, milte-, mód-, món-, mónaþ-, morgen-, scín-, wæter-, wamb-, wan-, wit-seóc.

seócan *to seek.* v. sēcan.

seócen (?) ; *adj. Troubled with sickness* :—On ðæs seócnan (seócan ?) tíd *in this time of sickness*, Exon. Th. 166, 11 ; Gū. 1041.

seóclian. v. sîclian.

seócness, e ; *f. Sickness, illness, disease* :—Ðæt God wolde sendan ǣrest hungor and ādla on manna ceáp, ǣr ðæt fŷr cóme on heó, and heó mid mislícre seócnesse æt mannum genyman, Wulfst. 209, 30. v. deóful-, fylle-, lifer-, mód-, mónaþ-, ofer-, wæter-seócness.

seód, es ; *m. A money-bag, purse, pouch* :—Seód *marsupium vel marsippa* (cf. *marsupium* a purse, 197, 16), Wrt. Voc. i. 40, 65 : 83, 12. Kyninga seód *fiscus*, ii. 39, 80. Ðā ic eów sende būtan seóde (*sacculo*) and code . . . Se ðe hæfþ seód gelíce nime codd, Lk. Skt. 22, 35, 36. Seódas *marsupia*, Hpt. Gl. 500, 40 : Wrt. Voc. ii. 55, 9. Siódas, 84, 37. Ðæt feoh ðæt hí hæfdon on heora seódum, Homl. Skt. i. 23, 262. Seódas *loculos*, Wrt. Voc. ii. 52, 22 : 74, 18. Wyrcaþ seódas (seádas, Rush.) seádo, Lind. *sacculos*) ða ðe ne forealdigeaþ, Lk. Skt. 12, 33. Seádo *loculos*, Jn. Skt. Lind. 12, 6. Seódas, Blickl. Homl. 69, 11. [*Icel.* sjóðr *a money-bag* : cf. *O. H. Ger.* siut *sutura*.]

-seódan. v. ā-seódan.

seód-cist, e ; *f. A coffer* :—Seódcist (seód, cist ?) *loculum*, Wrt. Voc. ii. 74, 46. Seódcyst, 52, 23.

seodu, seofa, seofan, seofen. v. sidu, sefa, seofon.

seófian, sēfian, sȳfian ; *p.* ode. **I.** *trans. To lament, complain of* :—His sylfes earfoþu hé seófaþ tō Drihtne, Ps. Th. 3, arg. Gilleþ geómorlíce and his gyrn sēfaþ, Salm. Kmbl. 536 ; Sal. 267. Hē seófode his ungelimp tō Drihtne, Ps. Th. 7, arg. : 3, arg. Hleahtor ālegdon sorge seófedon *laughter they laid aside, woes they bewailed*, Exon. Th. 116, 2 ; Gū. 201. Ne forlæt hē nō ða seófunga ðæt he ne seófige his eormþa *humanum miseras haud ideo genus cesset flere querelas*, Bt. 7, 4 ; Bt. 22, 29. Sege hē hwæðer ðū mid rihte mæge seófian (siófian, Cott. MS.) dîna unsǣlþa *poterisne de infortunio jure caussari*? 10 ; Fox 28, 8. Ðonne hī sceoldan hiora sār siófian, ðonne grymetodan hí, 38, 1 ; Fox 194, 35 : Met. 26, 82. Ongan sîðfæt seófian, sār cwānian, Exon. Th. 274, 22 ; Jul. 537. Synna bemǣnan and sārlíce sȳfian (sîf-, seóf-), Wulfst. 133, 14. Hū Boetius his sār seófiende wæs, Bt. tit. 2 ; Fox x, 4. ¶ with cognate accusative :—Seó seófung ðe ðū siófodost *the complaint you made*, 41, 3 ; Fox 246, 26. **II.** *intrans. To lament, complain of* (be, ymbe) :—Hwæt (*why*) seófast ðū wið mē *quid igitur ingemiscis?* Bt. 7, 3 ; Fox 20, 14. Hí seófiaþ be heora feóndum, Ps. Th. 16, arg. Be Iudan Scarioth hē seófode tō Drihtne, 3, arg. Seófade, seáfade ł (seó)mænde *ingemescens*, Mk. Skt. Lind. Rush. 8, 12. Ceare seófedun ymb heortan, Exon. Th. 306, 20 ; Seef. 10. Ðā ongunnon ða hîwan seófian be ðære untrumnysse *cum familiares de infirmitate quererentur*, Bd. 3, 9 ; S. 534, 6. Ne sceal hē sȳfian (seófian, MS. T.) ne mǣnan ymb woruldspēda *ne causetur de minore substantia*, R. Ben. 14, 13. Seófende wæs *maerens erat*, Mk. Skt. Lind. Rush. 10, 22. Be ðæm Dryhten siófigende cwæð *unde Dominus queritur dicens*, Past. 48, 3 ; Swt. 369, 4. Sceal ic siófigende wōpe gewǣged wreccea giómor singan sārcwidas *flebilis moestos cogor inire modos*, Met. 2, 2. **III.** uncertain :—Sȳfaþ *causatur*, i. *querelatur, causam dicit*, Wrt. Voc. ii. 130, 10. Seófade *causavit*, 130, 11. Sȳfiende *cupide*, i. *avare*, 137, 36, 64. Seóbgendum *querulis*, 106, 9. [Cf. *O. H. Ger.* sûftôn *gemere, ingemiscere, suspirare* : *Ger.* seufzen.]

seofon, syfon ; *when used without a following noun it is declined, nom., acc. seofone* ; *g. seofona* ; *d. seofonum. Seven*, (1) as adjective :—Mid ūs wǣron seofun (-on, MS. A.) gebrōðru, Mt. Kmbl. 22, 25. Ða seofon gōdan geár, Gen. 41, 53. His heres wæs seofon hund þūsenda, Ors. 2, 5 ; Swt. 78, 10. Seofon nihta fyrst, Elen. Kmbl. 1385 ; El. 694. On ðǣm seofon wǣstmbǣron geárum, Gen. 41, 47. Hē ābād ōðre seofon dagas, 8, 10. Hē him tō genymþ seofun (-en, MS. A. : seofona, Lind. : siofun, Rush.) ōðre gāstas, Mt. Kmbl. 12, 45. (2) without a following noun :—Ðā nam se þridda hig, and swā ealle seofone (-ene, MS. A. : siofuno, Lind. : ða siofune, Rush.), Lk. Skt. 20, 31. Ealle seofon (-en, MS. A. : -an, MS. B. : ða seofona, Lind. : ða siofune, Rush.) hí hæfdon, Mk. Skt. 12, 22. Hwylces ðara seofona biþ ðæt wîf, 12, 23. Hwylces ðæra sufona (seofena, MS. A. : of ðæm seofonum, Lind. : ðara siofuna, Rush.), Mt. Kmbl. 22, 28. Ðā com seofona sum, Andr. Kmbl. 2623 ; An. 1313. Mid feáwum brōðrum, ðæt is seofonum oððe eahtum, Bd. 4, 3 ; S. 567.

4. Ðû seofone genim tudra gehwilces, Cd. Th. 80, 27; Gen. 1335. Geseh hê hyrdas standan seofone ætsomne, Andr. Kmbl. 1987; An. 996. Syfone, Beo. Th. 6235; B. 3122. [*Goth.* sibun: *O. Sax.* sibun: *O. Frs.* saven, sigun: *O. H. Ger.* sibun: *Icel.* sjau.]

seofon-feald; *adj. Sevenfold :*—Seofonfeald wracu biþ sealde for Cain and hundseofontig seofonfeald for Lamech, Gen. 4, 24. Hê onbryrt ûre môd mid seofonfealdre gife, Homl. Th. i. 326, 12. Gyld seofonfealde wrace, Ps. Th. 78, 13: Gen. 4, 15. Him ofer wacedon syfanfealde weardes, Homl. Skt. i. 3, 271.

seofonfealdlîce; *adv. Sevenfold, seven times :*—Geclâsnad seofenfaldlîce *purgatum septuplum*, Ps. Surt. 11, 7: 78, 12.

seofon-leáfe, an; *f. Seven-leaves, setfoil; potentilla tormentilla :*—Seofonleáfe. Ðeós wyrt ðe man *eptafilon* and ôðrum naman *septifolium* nemneþ and eác sume men seofonleáfe, Lchdm. i. 232, 1-3. [Cf. *O. H. Ger.* sibun-blat *heptaphyllon.*]

seofon-nihte; *adj. Seven days old :*—Se .vii. nihta môna is gôd on tô fixiane, Lchdm. iii. 178, 13. On .vii. nihtne mônan, 178, 9.

seofon-stirre, es; *n. The Pleiades :*—Sifunsterre (sibun-) *pliadas,* Txts. 86, 762. [Cf. *O. H. Ger.* sibun-stirni, -stirri, es; *n. pliades, orion : Ger.* sieben-gestirn : *Icel.* sjau-stirni ; *n. the Pleiades.*]

seofonteóþa, -teogoþa *seventeenth :*—Se wæs seofonteogeþa fram Agusto, Bd. 1, 5; S. 476, 6. Ðý seofonteóþan dæge, 3, 24; S. 557, 12. On ðone seofenteóþan dæge ðæs mônþes, Gen. 7, 11. Seofontegðan, Shrn. 91, 32.

-seofontig. v. hund-seofontig.

seofon-tîne *seventeen :*—Æfter seofentŷnum nihtgerîmes, Menol. Fox 50; Men. 25. Hê lyfode seofentŷne gêr, Gen. 47, 28. Seofontŷne, Bd. 1, 5; S. 476, 2, 15; S. 519, 13.

seofontine-nihte; *adj. Seventeen days old :*—On .xvii. nihte mône, Lchdm. iii. 180, 7.

seofon-wintre; *adj. Seven years old :*—Mid ðŷ ic wæs seofonwintre *cum essem annorum septem,* Bd. 5, 23; S. 647, 21. Ic wæs syfanwintre, Beo. Th. 4847; B. 2428. Ðâ ðâ hê syfonwintre wæs, Homl. Skt. i. 3, 5. [*Icel.* sjau-vetra *seven years old.*]

seofoþa *seventh :*—Tô ðære seofoþan (ðió seofunda, Lind.: ðŷ siofunda, Rush.) tîde, Jn. Skt. 4, 52. Oþ ðone seofoþan (tô ðæm seofunda, Lind.: siofund, Rush.), Mt. Kmbl. 22, 26. On ðone seofeþan dæg, Gen. 2, 2.

seofoþa *bran.* v. sifeþu.

seófung, e; *f. Lamenting, complaining, complaint :*—Hwî biþ elles swelc seófung and swelce dômas *unde forenses querimoniae?* Bt. 26, 2; Fox 92, 16. Ðis is seó ealde siófung ðe ðû longe siófodost (siófodes, Cott. MS.) *vetus haec est querela,* 41, 3; Fox 246, 25. Ne beó ðû tô ceástful; of irsunge wyxt seófung, Prov. Kmbl. 23. Ic ne mæg ádreóhan ðîne seófunga for ðam lytlan ðe ðû forlure, Bt. II, 1; Fox 30, 20. Forlætan ða seófunga his eormþa *miseras fugare querelas,* 29, 3; Fox 106, 20: Met. 16, 7. v. sâr-seófung; seófian.

seohhe, an; *f. A strainer :*—Seohhe *colatorium,* Wrt. Voc. i. 24, 52. Man sceal habban seohhan, Anglia ix. 264, 18. [A mylke syhe *colum,* Prompt. Parv. 79, note 1. A sigh-clout, Halliwell Dict. (*under* sie). Sye-dish *a milk-strainer,* E. D. S. Pub. country words, 6. Cf. *O. H. Ger.* sîha *colum, colatorium: Ger.* seche: *Icel.* sîa *a strainer.*] v. seón *to strain,* seohtre.

seoh-tor[r] (?), es; *m. A look-out place* (?) :—Ofer ðone cnol tô ðæn seohtore (-torre ?), Cod. Dipl. Kmbl. iii. 451, 14.

seohtre, sihtre, an; *f. A pipe through which a small stream is directed, a drain :*—Andlang seohtran, Cod. Dip. B. i. 295, 11. Tô ðare reádan sihtran, 296, 28. In wætan sihtran (cf. sîce, 382, 7); of ðam wætan sîce, Cod. Dip. Kmbl. iii. 386, 10. Ad locum qui dicitur hylsan seohtra, 373, 12.

seol, seolc, seolcan, seolcen, seolc-wyrm, seolf. v. seolh, seoluc, â-seolcan, seolucen, seoluc-wyrm, self.

seolfor, siolufr, silofr, sylfor (-er, -ur), es; *n. Silver :*—Seolfor *argentum,* Wrt. Voc. ii. 8, 52. Seolfer, i. 85, 7. Seolfur, Ps. Th. 134, 15. Feówer hund scillinga seolfres, Gen. 23, 16. Fîftig yntsena seolfres, Deut. 22, 29. Hwîtes seolfres, Jos. 7, 21. Silofres, Salm. Kmbl. 62, MS. B.; Sal. 31. Siolufres (siolofres, Cott. MSS.), Past. 37; Swt. 269, 4. Tô siolofre, Swt. 266, 20. Ic sealde siolfor (sylofr, Cott. MSS.), 48; Swt. 369, 6. Silofr, Swt. 368, 20. Hwîtan seolfre bêtan, Cd. Th. 165, 14; Gen. 2731. Sylfore, Exon. Th. 395, 4; Rä. 15, 2. Næbbe gê seolfer (sulfer, Lind.: sylfur, Rush.), Mt. Kmbl. 10, 9. Wênst ðû ðæt wê ðînes hlâfordes seolfor stælon, Gen. 44, 8. Sealde him tô bôte gangende feoh and glæd seolfor, Cd. Th. 164, 24; Gen. 2719. [*Goth.* silubr: *O. Frs.* selover, selver, silver : *O. Sax.* silubar, siloƀar : *O. H. Ger.* silabar, silbar : *Icel.* silfr.] v. cwic-seolfor.

seolfor-fæt, es; *n. A vessel of silver :*—Seolforfatum *argenteis vasis,* Bd. 4, 1; S. 563, 21. [*O. H. Ger.* silbar-faz. Cf. *Icel.* silfr-bolli, *and many similar cpds.*]

seolfor-gewiht, es; *m. Silver-weight, the scale of weight by which silver is weighed,* where the pound is of sixteen ounces :—Se sester sceal wegan twâ pund be sylfyrgewyht, Lchdm. iii. 92, 14. v. sester, **II.**

seolfor-hammen; *adj. Silver-coated :*—Ænne seolforhammenne blædhorn, Chart. Th. 559, 24.

seolfor-hilt; *adj. Silver-hilted :*—Ic geann mînon brêðer ânes seolferhiltes swurdes, Chart. Th. 560, 10. Ðæs sealferhiltan swurdes ðe Ulfcytel âhte, 559, 13.

seolfor-hilted; *adj. Silver-hilted :*—Twâ seolforhilted sweord, Chart. Th. 544, 4.

seolfor-smiþ, es; *m. A silver-smith, worker in silver :*—Seolforsmiþ *argentarius,* Wrt. Voc. i. 73, 31. Seolfersmiþ, 47, 13. Ic hæbbe smiþas . . . seolforsmiþ *habeo fabros . . . argentarium,* Coll. Monast. Th. 29, 35. [*O. H. Ger.* silbar-smid: *Icel.* silfr-smiðr.]

seolfor-stycce, es; *m. A piece of silver, a coin :*—Ðæt þrîtig seolforsticca *the thirty pieces of silver (given to Judas),* Anglia xi. 8, 3.

seolfren, seolofren, seolfern, silfren, sylofren, sylfren; *adj. Silvern, of silver :*—Sylofren sinc, Met. 21, 21. Glæsen fæt on seolfrenre racenteáge, Blickl. Homl. 209, 4. In seolfren fæt belûcan, Elen. Kmbl. 2050; El. 1026. Hafaþ silfren (seolofren, MS. B.) leáf, Salm. Kmbl. 129; Sal. 64. Nim mînne sylfrenan læfyl, Gen. 44, 2: Bd. 1, 25; S. 487, 3. Seolferne silfer coins, Mt. Kmbl. p. 20, 2. Sylfrenu (selfrenu, Cott. MS.) fatu, Bt. 36, 1; Fox 172, 19. Ða seolfrenan stânas, 34, 8; Fox 144, 31. Sweopum seolfrynum, Salm. Kmbl. 287; Sal. 143. Sylfrenum, Homl. Th. ii. 212, 30. Ne wyrce gê sylfrene godas, Ex. 20, 23 : 3, 22. [*Goth.* silubreins: *O. Frs.* selvirn: *O. Sax.* silubrin: *O. H. Ger.* silbarîn.]

seolfrian. v. be-, ofer-seolfrian.

seolh; *gen.* seoles; *m. A seal, sealgh, selcht* (v. Jamieson's Dict.), *sea-calf :*—Seolh *focca,* Wrt. Voc. ii. 149, 81 : *bromus marinus,* i. 22, 54: *focus,* 281, 58. Seol *foca,* 55, 79 : *focus,* ii. 38, 48. Ðæs wyrt onsænde seolh ofer sæs hrygc, Lchdm. iii. 34, 15. Of seoles hŷde, Ors. 1, 1 ; Swt. 18, 18. Sioles, 18, 23. Seolas *vituli marini,* Bd. 1, 1; S. 473, 16. [*O. H. Ger.* selah: *Icel.* selr.]

seolh-bæþ, es; *n. The seal's bath, the sea* (cf. fisces, ganotes bæþ) :—Mec wind wîde bær ofer seolhbaþo, Exon. 392, 21; Rä. 11, 11.

seolh-wæd (?), -pæð (?), es; *n. The seal's ford, path, the sea :*—Hié on ŷðum æðelinga wunn ofer seolhwaðu (-wadu ?, -paðu ?) geseón mihton, Andr. Kmbl. 3424; An. 1716. Cf. *preceding word and* mearc-pæð, -wæd.

seolh-ŷða (?); *pl. The waves where the seal swims :*—Oferswam ðâ sioleða (siolŷða ? cf. flóda, holma begang. Or (?) sioleþ *still water.* Cf. *Goth.* ana-silan : *Swed. dial.* sil *still water.* v. Heyne's note) bigong sunu Ecgþeówes, Beo. Th. 4723; B. 2367.

seolofren. v. seolfren.

seoluc (-oc), seolc, es; *m. Silk :*—Seolc *sericum,* Wrt. Voc. i. 40, 2. Gôd geolo seoluc, Lchdm. ii. 10, 16 : 106, 22. Seowa mid seolce fæste, smire mid ðære sealfe ǽr se seoloc rotige, 56, 7-8: 358, 25. Heora wǽda sioloce siowian, Met. 8, 24. Gyf man mǽte ðæt hê seoluc oððe godweb hæbbe, Lchdm. iii. 174, 29. [*Icel.* silki. *From Latin* sericum (?). But see Kluge, Etymol. Wörterb. *under* seide.]

seolucen, seolcen; *adj. Silken, of silk :*—Seolcen *bombicinum,* Wrt. Voc. i. 39, 72. Siolcen, ii. 11, 67 : 75, 74. Seolcen gegerla *bombicinium,* 126, 50. Seolcen âb *tramasericum,* i. 40, 4. Seolce[n] hnygele *platum,* 40, 38. Silcen *serica,* Hpt. Gl. 417, 37. Seolocenra hrægla, Bt. 15; Fox 48, 11. v. eal-seolcen.

seoluc-wyrm, es; *m. A silk-worm :*—Siolucwyrm *bombix,* Wrt. Voc. ii. 12, 22. Seolcwyrm, i. 24, 6 : 40, 1.

seomian, siomian, semian; *p.* ode *To rest.* (1) *to remain suspended, to hang, to lower* as a cloud :—Hit bærneþ boldgetimbru, seomaþ steáp, Salm. Kmbl. 827; Sal. 413. Deorc deáþscûa seomade *the dark shadow of death hung over them,* Beo. Th. 324; B. 161. Sum sceal on galgan rîdan, seomian æt swylte, Exon. 329, 14; Vy. 34. Mæst sceal on ceóle segelgyrd seomian *the mast shall be fixed in a boat and the yard hang from it,* Menol. Fox 509; Gn. C. 25. Hê siomian geseah sægn, Beo. Th. 5527; B. 2767. Geseah deorc gesweorc semian, Cd. Th. 7, 20; Gen. 109. (2) *to remain supported, to lie so as to press, lie heavily, lie securely :*—Se wong seomaþ eádig and onsund, Exon. Th. 199, 2; Ph. 19. Seomaþ (-ad, MS.) wîr ymbe ðone wælgim, 400, 19; Rä. 21, 3. Seomaþ sorgcearig *lies troubled,* 285, 4; Jul. 709. Hê siomode in sorgum seofon nihta fyrst, Elen. Kmbl. 1384; El. 694. Flota stille bâd, seomode on sole scip, Beo. Th. 609; B. 302. Heó on wrace seomodon, Cd. Th. 5, 15; Gen. 7. Him on healfa gehwam hettend seomedon mægen oððe merestreám *on each side of them lay foes pressing, the Egyptian force or the Red Sea,* 191, 4; Exod. 209. Ðǽr ic seomian wât ðinne sigebrôðor *I know thy brother lies in prison there,* Andr. Kmbl. 365; An. 183.

seó-mint, *plant name,* altea *vel* eviscus, Wrt. Voc. i. 32, 12. v. sǽminte.

seón *to be :*—See esse, Mt. Kmbl. p. 1, 11. v. eom, sî.

seón; *p.* seah, *pl.* sâwon, sǽgon, sêgon; *pp.* sewen, sawen. **I.** *to see* with the eyes, (1) *with acc. :*—Oft ic wîg seó, Exon. Th. 388, 6; Rä. 6. 3. Ic seah wundorlîce wiht, 495, 1; Rä. 84, 1. Ne seah ic medudreám mâran, Beo. Th. 4033; B. 2014. Hî wuldres þegn eágum

sáwon, Andr. Kmbl. 3355; An. 1681. Ðæs ðe (hió) ælda bearn eágum sáwe, Exon. Th. 493, 7; Rä. 81, 26. Eode scealc monig searowundor seón, Beo. Th. 1844; B. 920: 2735; B. 1365: Cd. Th. 125, 25; Gen. 2084. (2) *with acc. and infin.* :—Ic seah turf tredan .vi. gebrôðor, Exon. Th. 394, 10; Rä. 14, 1: 400, 1; Rä. 20, 1: 414, 29; Rä. 33, 3: 434, 15; Rä. 52, 1. (3) *with acc. and predicative adj. or participle*:— Hý grim helle fýr gearo tô wîte andweard seóþ, Exon. Th. 78, 8; Cri. 1271. Ne seah ic elþeódige môdiglîcran, Beo. Th. 678; B. 336. Hý God upstîgende eágum sêgun, Exon. Th. 34, 3; Cri. 536. (4) *with clause*:—Hý on ða clǽnan seóþ, hû hî blissiaþ, Exon. 79, 6; Cri. 1286: Beo. Th. 5428; B. 2717. II. *to see, to visit*:—Nǽnig cêpa ne seah (geseah, Bt. 15; Fox 48, 13) ellendne wearod *nondum nova litora viderat hospes*, Met. 8, 29. Hât in gân seón sibbegedriht, Beo. Th. 779; B. 387. Uton êfstan seón and sêcean searogeþræc, 6195; B. 3102. II a. metaph.:—Hê heán gewât deáþwîc seón, Beo. Th. 2555; B. 1275. III. *to see, perceive, discern, understand*:—Ic seó ðê, ðæt is, ðæt ic ongite ðînne willan bûtan tweón, Ps. Th. 5, 3. Sôðfæst blissaþ, ðonne hê sîþ hû ða ârleásan ealle forweorðaþ, 57, 9. Sioh nû sylfa ðê, hû ðec heofones cyning gesêceþ, Exon. Th. 4, 27; Cri. 59. Seh ðê *ecce*, Ps. Surt. 32, 18: 38, 6. Sih ðê, Mt. Kmbl. Rush. 19, 16, 27: 24, 25, 26. Wênaþ ða dysgan, ðæt ǽlc mon sié blind swâ hî sint, and ðæt nân mon ne mǽge seón (gesíon, Cott. MS.) ðæt hî gesíon ne mágon, Bt. 38, 5; Fox 206, 21. IV. *to see* (as in *to see death*), *to experience*:—Mec ongan hreówan ðæt moncynnes tuddor sceolde mâncwealm seón, Exon. Th. 86, 33; Cri. 1417. Morðorleán seón, 98, 24; Cri. 1612. Hê forþ gewât metodsceaft seón *he died*, Cd. Th. 104, 31; Gen. 1743: B. 2364; B. 1180. V. *with prepositions, to look* at, on :—On ðæt ða folc seóþ, Exon. Th. 80, 2; Cri. 1301. Seóþ on êce gewyrht, 448, 29; Dôm. 61. Ealle synd gedrêfede ðe hî on sióþ *conturbati sunt omnes qui videbant eos*, Ps. Th. 63, 8. Secg seah on unleófe, Beo. Th. 5719; B. 2863. Folc tô sǽgon, 2849; B. 1422: Elen. Kmbl. 2208; El. 1105. Ðǽr hî tô sêgon, Andr. Kmbl. 1422; An. 711: Exon. Th. 260, 3; Jul. 291. Sêgun, 31, 14; Cri. 495. Hî cômon on ðæt wundor seón, Cd. Th. 261, 25; Dan. 731. Fægre leomu on tô seónne, Blickl. Homl. 113, 22. [*Goth.* saihwan: *O. Sax.* sehan: *O. Frs.* sîa: *O. H. Ger.* sehan: *Icel.* sjá.] v. be- (bi-), for-, fore-, ge-, geond-, of-, ofer-, on-, þurh-, ymb-seón.

seón (*from* sîhan); *p.* sâh, *pl.* sigon; *pp.* sigen (cf. león), seowen (v. â-seowen, Lchdm. ii. 26, 11), siwen (v. â-siwen, Lchdm. ii. 124, 14), seón (v. bi-seón, Exon. Th. 67, 13; Cri. 1088). I. *trans. To strain, filter*:—Siid *excolat*, Wrt. Voc. ii. 107, 71. Seóh þurh claþ, Lchdm. ii. 24, 1: iii. 14, 18. II. *intrans. To run* as a sore, ooze, trickle :— Managum men lîþseáu sýhþ, Lchdm. ii. 132, 10. Ðæt se lǽce sceolde âsceótan ðæt geswell; ðâ dyde hê swâ, and ðǽr sâh ût wyrms, Homl. Skt. i. 20, 64. Wiþ seóndre exe, Lchdm. iii. 70, 20. Wiþ seóndum geallan, Lchdm. ii. 314, 7, 10. Wiþ seóndum ômum, 102, 9. Eal ðæt folc wæs on blǽdran and ða wǽron berstende and ða worms ût siónde (*ulcera manantia*), Ors. 1, 7; Swt. 38, 7. [Mid þornene crune his heaued was icruned, swa þet þet rede blod seh ut on iwulche half, O. E. Homl. i. 121, 12. Syynge or clensynge *colacio, colatura*, Prompt. Parv. 455. I sye mylke, Cath. Ang. 339, n. 3. Halliwell gives *sie* as a word still in use in Derbyshire. *O. H. Ger.* sîhan *colare, excolare, liquare*: *Ger.* seihen: *Icel.* sîa *to strain*.] v. â-, ge-seón; seohhe, sîgan.

seón; seondon, -seonod. v. sîn, wlite-, wundor-seón.

seonoþ, sionoþ, senoþ, sinoþ, synoþ (-aþ, -od), es ; *m. A synod, council, meeting*:—Sinoþ *sinodus*, Wrt. Voc. i. 72, 76. I. mostly used of the councils of the Christian Church:—Seonod (sinoþ) wæs æt Âcleá, Chr. 782; Erl. 57, 6. Wæs senoþ (sinoþ, MS. E.) æt Heorotforda, 673; Erl. 56, 2: 822; Erl. 62, 13. Hêr wæs gefiltfullîc senoþ æt Cealchýþe, 785; Erl. 56, 7. Se hâlga sinoþ, Bd. 4, 17; S. 585, 41. Æfter ealles sinoþes dôme, 3, 7; S. 530, 35. Be ðǽm sinoþe se wæs geworden on ðam felda se wæs genemned Hæþfeld, 4, 17; S. 585, 7. On ðam miclan synoþ æt Greátanleáge, L. Ath. i. 26; Th. i. 214, 7. Eádmund cyning gesomnode micelne sinoþ tô Lundenbyrig ǽgðer ge godcundra hâda ge woruldcunda, L. Edm. E. proem.; Th. i. 244, 2. Gif preóst sinoþ forbûge, gebête ðæt, L. N. P. L. 44; Th. ii. 296, 16. Monega þeóda Cristes geleáfan onfêngon; ðâ wurdon monega seonoþas gegaderode, L. Alf. 49; Th. i. 58, 2. Ðás feówer sinoþas (*the councils of Nice, Ephesus, Constantinople, and Chalcedon*), Wulfst. 270, 15. II. in other senses:—Bǽdon ðæt eft ôðer seonaþ wǽre (*of the meeting between Augustine and the British Christians*), Bd. 2, 2; S. 502, 36. Tô sionoþe (*the Council called by Constantine to enquire about the cross*), Elen. Kmbl. 307; El. 154. Hig tô ðæra Iudêa synoþe cômon, Nicod. 18; Thw. 8. 31. Wile fæder engla seonoþ gehêgan Exon. Th. 231, 23; Ph. 493. [The word is borrowed also by *O. Frs.* and *O. H. Ger.*] v. bisceop-seonoþ, *and following words.*

seonoþ-bôc; *f. A book containing the decrees of a synod* :—Hié on monegum seonoþum monegra menniscra misdǽda bôte gesetton, and on monega senoþbêc hý writon, hwǽr ânne dôm, hwǽr ôðerne, L. Alf. 49 ; Th. i. 58, 15.

seonoþ-dôm, es; *m. The decree of a synod* :—Seonoþdômas reccan, Elen. Kembl. 1101; El. 552.

seonoþlîc; *adj. Synodal, of a synod* or *meeting* :—Ðâ wæs sionoþlîc gemôt, Chart. Th. 70, 10. Ðǽre sinoþlîcan dǽde *synodicae actionis*, Bd. 4, 5; S. 572, 1. Mid sinoþlîcum stafum *synodalibus literis*, 4, 17; S. 585, 15.

seonoþ-stôw, e; *f. A place for a synod* or *meeting, a place of assembly*:—Sinaþstôw *conciliabulum, locus sinodalis*, Wrt. Voc. ii. 136, 19. Geseóþ gê ðæt hê ǽrest tô ðǽre sinoþstôwe (*ad locum synodi*) cymeþ, Bd. 2, 2; S. 503, 9.

seonu, sionu, senu, sinu, synu; *gen.* seonwe, sine; *weak forms also occur* ; *f. A sinew, nerve, tendon* :—Sionu *nervus*, Wrt. Voc. ii. 114, 67. Sinu, i. 71, 42. Gif sin[o] scrince . . . and gif sino clæppette, Lchdm. ii. 6, 13–15. Ðâ æthrân hê his sine on his þeó *tetigit nervum femoris ejus*, Gen. 32, 25. Healt for ðære sinwe (synewe, MS. B.) wunde, L. Alf. pol. 75; Th. i. 100, 5. Gif man on sinwe besleá æt blôdlǽtan, Lchdm. ii. 16, 8. Gif mon ða greátan sinwe (synewe, MS. B.) forsleá, L. Alf. pol. 75; Th. i. 100, 3. Gif ða smalan sinwe (synewan, MS. B.) mon forsleá, 76; Th. i. 100, 8. Nellaþ folc etan sine (*nervum*), Gen. 32, 32. Seonuwa [beóþ] fortogene, Lchdm. iii. 48, 28: 50, 5. Seonwe onsprungon, Beo. Th. 1639; B. 817. Seonwe (sina, Soul Kmbl. 217), Exon. Th. 373, 19; Seel. 111. Sionwe, Andr. Kmbl. 2849; An. 1427. Senwe *nerve*, Wrt. Voc. i. 283, 37. Sena, 65, 15. Sinu (-a ?), 44, 23. Gif sinwe sýn forcorfene, Lchdm. ii. 328, 5. Wiþ sina sâre, i. 84, 10. Wiþ ðara sina bifunge, 104, 27. Sina togung, 136, 9. Syna, 136, 19. Sina getog, 356, 3. Seonowum beslîtan, Exon. Th. 371, 9; Seel. 73. Seonwum (synum, Soul Kmbl. 123), 370, 20; Seel. 62. Mid râpum of sinum geworhte *nerviceis funibus*, Jud. 16, 7. Se lîchama wæs geboren mid blôde and mid bânum, mid felle and mid sinum, Homl. Th. ii. 270, 19. Ârǽran of duste flǽsc and bân, sina and fex, i. 236, 21. On ða sâran sinua, Lchdm. ii. 282, 6. For flǽsc and for bân and for sinnwan, L. Edg. C. 9; Th. ii. 264, 4. Sinuwa, Anglia xi. 101, 47. [*C. Frs.* sini(-e), sene, sin: *O. H. Ger.* senawa: *Icel.* sin; *pl.* sinar and sinur.] v. hôh-sinu.

seonu-ben[n], e; *f. A wound* or *injury of a sinew* :—Seonobennum seóc *crippled*, Exon. Th. 328, 17; Vy. 19. v. next word.

seonu-bend (?), e; *f. A bond made of sinews* (?) :—Siððan hine Nîðhâd on nêde legde swoncre seonobende (Grein would read -*benne*, which is more in accordance with the story in the Edda, that Völund had the sinews of the knees cut : v. Thorpe's note on this passage, and his Northern Mythology, i. 86. For confusion of *benne* and *bende* see *ben*), Exon. Th. 377, 19; Deór. 6. v. preceding word.

seonu-dolh, es; *n. A wound of a sinew* :—Benna weallaþ, seonodolg swâtige, Andr. Kmbl. 2811; An. 1408.

seonu-wealt (sionu-, sinu-, sino-, sin-, syne-); *adj. That may be always rolled, round* :—Sinewealt gesceap *volubile scema*, Wrt. Voc. i. 55, 18. Sionuualt *torosa* (*teres*?), ii. 122, 54. Sionewaltum *conteriti* (*cum teriti*?), 21, 56. I. *round, circular, cylindrical*:—Sinewealt cleofa *absida*, Wrt. Voc. i. 58, 34. Synewealg wafungstede *amphitheatrum*, 37, 1. Sineweald trendel *circulus*, Hpt. Gl. 418, 16. Se môna went his hricg tô ðære sunnan, ðæt is, se sinewealta ende ðe ðǽr onlýht biþ, Lchdm. iii. 242, 14. Ðæs sinewealtan hringes *teretes*(-*is* ?) *cycli*, Wrt. Voc. ii. 89, 60. Timbredon men seonewalte (cf. cyrice is sinhwyrfel . . . seó is unoferhrêfed, Blickl. Homl. 125, 21) cirican, Shrn. 80, 37. Hæfde ðæt deór seonowealt heáfod swelce môna, Nar. 20, 27. Heó is leáfun sinewealton, Lchdm. i. 290, 8, 18. II. *round, spherical, globular,* of a building, *having a concave roof* or *dome*:—Seó heofon is sinewealt, Lchdm. iii. 232, 20: Boutr. Scrd. 18, 24. Sineweald cliuen *rotundus, teres globus*, Hpt. Gl. 446, 67. Corn sonuuald (*the manna eaten in the desert*), Jn. Skt. Lind. 6, 31, rubc. Hyre wyrttruma ys synewealt *the root is a bulb*, Lchdm. i. 152, 16. Seó byrgen (*Christ's tomb*) is sinowalt hûs âcorfen of ânum stâne, Shrn. 68, 35. Of sinuwealtum cliwene *ex teriti glomere*, Wrt. Voc. ii. 31, 20: 83, 19. Sinewæltum, Hpt. Gl. 494, 17. Wyrc hit sineweald *make it into a ball*, Lchdm. i. 72, 21. Sinewealte swammas *volvi*, Lchdm. i. 30, 28. Heó hafaþ berian synewealte, Lchdm. i. 276, 24. v. sin-, sin-hweorfol.

seonuwealtian *to reel, not to stand firmly* :—Sinewealtigan (wine-, Wrt.) *vacillare*, Wrt. Voc. ii. 88, 48.

seonuwealtness, e; *f. Roundness, circularity, sphericity* : — Sinewealtnes *globositas*, Wrt. Voc. i. 55, 19. Ðǽre eorþan sinewealtnes *the sphericity of the earth*, Lchdm. iii. 258, 10. For ðǽre eorþan sinewealtynysse, 260, 11.

seonu-wind *an artery* :—Sinewind *arteriae*, Wrt. Voc. ii. 8, 29. [Cf. (?) *O. H. Ger.* sen-âdra *arteria, nervus*.] Cf. wind-ǽdre, sin-.

seordan, seordan (?); *p.* searð *To violate* :—Ne serð ðû ôðres monnes wif *non moechaberis*, Mt. Kmbl. Lind. 5, 27. [From (?) Scandinavian. Cf. *Icel.* serða *stuprare* : *M. H. Ger.* serte. See Altdeutsche Gespräche. Nachtsag vom W. Grimm, p. 18.]

seóslig; *adj. Afflicted, troubled, vexed* :—Se hâlga wer ælda gehwylces ðe hine seóslige sôhtun hǽlde lîc and sâwle *the holy man healed body and soul of all that in affliction sought him*, Exon. Th. 157, 29; Gû. 899. Cf. sûsl.

-seóþa. v. newe-seóþa.

seóðan (? cf. seód, á-seóðan?) *to put in a bag, wrap up* :—Bewind ðone æppel on weolcreádum godwebbe, and seóð eft mid sceáte óðres godwebbes, and beheald ðæt ðes lǽcedóm ne hríne ne wæteres ne eorþan, Lchdm. i. 332, 5.

seóþan; *p.* seáþ, *pl.* sudon; *pp.* soden. **I.** *to seethe, boil, cook in a liquid* :—Ic seóþe *coquo*, Ælfc. Gr. 28, 5; Zup. 175, 16. Gif ðu seóþest rúdan on ele, Lchdm. ii. 206, 23. Gif mon sý̂ gárleác on henne broþe, Lchdm. ii. 276, 15. Seóþ on wætere tô þriddan dǽle, i. 72, 2. Seóþ on wîne, 134, 4. Seóþaþ (*coquite*) eówerne mete beforan ðæs temples dura, Lev. 8, 31. Seóþe on strangum wîne, Lchdm. i. 142, 2. Seóþan ða þingc ðe tô seóþenne synd *coquere quae coquenda sunt*, Coll. Monast. Th. 29, 19. **II.** metaph. (1) with the idea of purification, *to subject to a fiery ordeal, to try as with fire* :—Seóþeþ swearta lêg synne on fordónum . . . ôþ ðæt hafaþ ældes leóma woruldwidles wom forbærned, Exon. Th. 62, 1; Cri. 995. Ðu mê sude mid ðam fýre monegra earfoþa swá swá gold *igne me examinasti*, Ps. Th. 16, 3. (2) *to subject to great pain, to afflict grievously* :—Mê elþeódige searonet seóþaþ *me barbarian snares afflict* (?), Andr. Kmbl. 127; An. 64. Mid ðý hê ða lange mid swîgendum nearonessum his môdes and mid ðý blindan fýre soden wæs *cum diu tacitis mentes angoribus et caeco carperetur igni*, Bd. 2, 13; S. 513, 34. Herebryht wæs mid singale untrumnesse soden and swenced, 4, 29; S. 607, 41. Sorgwylmum soden, Exon. Th. 166, 21; Gû. 1046: 171, 7; Gû. 1123: 177, 32; Gû. 1236. Sárbennum soden, Andr. Kmbl. 2479; An. 1241. (2 a) *to reduce by pain or disease* :—Heó swá swýðe mid ða untrumnysse soden wæs ðæt ða bán án tô láfe wǽron *in tantum ea infirmitate decocta est, ut vix ossibus haereret*, Bd. 4, 9; S. 577, 15. (3) *to prepare food for the mind, to make fear, hope, etc., subjects with which the mind may be occupied* ; cf. to feed a person with hopes :—Ic ðæs môdceare sorhwylmum seáþ *on account of your dangerous journey anxiety was that fear I prepared for my mind*, Beo. Th. 3990: 1993. Swá ða mǽlceare maga Healfdenes singala seáþ *Hrothgar had that care ever ready to feed his mind with*, 382; B. 190. [O. Frs. siatha : O. H. Ger. siodan : Icel. sjóða : cf. Goth. sauths *a burnt-offering*.] v. á-, be-, for-, ofer-, tô-seóþan; ge-, healf-, sám-, un-soden.

seoððan, seotl, seotol, seotu, seóung. v. siððan, setl, sweotol, set, eág-seóung.

seowian, seówan, siwian; *p.* ode; *pp.* od, ed, id *To sew* :—Sióuu *sarcio*, Wrt. Voc. ii. 119, 52. Ic siwige *sarcio*, Ælfc. Gr. 30, 2; Zup. 190, 6. Ic sywige (siwige, MS. R.), 28, 3; Zup. 167, 6. Heó siwaþ (seowaþ, MS. U.) *illa suit*, 15; Zup. 97, 6. Siwaþ (siuieþ, Lind.: siowes, Rush.) *assuit*, Mk. Skt. 2, 21. Sum sútere siwode (seowode, MS. C.) ðæs hálgan weres sceós, Homl. Skt. i. 15, 23. Hig siwodon ficleáf and worhton him wǽdbréc, Gen. 3, 7. Seowa mid seolce fære, Lchdm. ii. 56, 7. Wǽda sioloce siowian, Met. 8, 24. Byrne, searonet seowed smiþes orþancum, Beo. Th. 816; B. 406. Golde siowode *segmentata*, Wrt. Voc. ii. 95, 49. [Goth. siujan : O. Frs. sia : O. H. Ger. siuwan : Icel. sýja : Lat. suere.] v. á-, be-seowian (-siwian); ge-seówan; ge-siwed.

sépan (seppan?); *p.* te *To cause to perceive, to teach* :—Se stán sépte sacerdas sweotolum tácnum, Andr. Kmbl. 1483; An. 743. Ðus mê fæder mîn unweaxenne wordum lǽrde, sépte sóðcwidum, Elen. Kmbl. 1057; El. 530. Hyssas heredon Drihten for ðam hǽðenan folce, sépton (MS. stepton) hié sóðcwidum, and him sǽdon fela sóðra tácna, Cd. Th. 244, 10; Dan. 446. [Cf. (?) O. Sax. af-sebbian; *p.* -sóf *to perceive* : O. H. Ger. int-suab; *p.* : M. H. Ger. en-seben *to perceive, understand*. v. Grff. vi. 168.]

serc, syrc, syric, es; *m.* : serce, syrce, an; *f. A shirt, shift, smock, tunic, sark* (Scott.) :—Loða, serc *colobium*, Hpt. Gl. 493, 76. Smoc *vel* syrc, Wrt. Voc. i. 25, 60. Syric *colobium vel interula*, 81, 69. Syrc *suppar, interula*, 59, 24. Serc *armilausia*, 284, 61 : ii. 8, 16. Serce, 100, 77 : 7, 4. Swátfáh syrce, Beo. Th. 2226; B. 1111. Ðæt hê hæbbe syric (*tunicam*), R. Ben. 89, 10. Genôh is munuce ðæt hê hæbbe twegen syricas (*tunicas*), for ðære nihtware and for ðæs reáfes þweále, 91, 3. Syrcan, gûþgewǽdo *shirts of mail*, Beo. Th. 458; B. 226: 673; B. 334. [Icel. serkr *a shirt*; hring-, járn-serkr *a shirt of mail*.] v. beadu-, heoru-, here-, hilde-, leoþu-, lîc-, under-serc (-serce).

serede, serð, serwan. v. sirwan, seorðan, sirwan.

sess, es; *m. A seat, bench* :—Ses, sæs *transtrum*, Txts. 103, 2050. Hê gesæt on sesse, Beo. Th. 5427; B. 2717: 5506; B. 2756. [Icel. sess; *m.*]

sessian; *p.* ode *To subside* :—Sǽ sessade (sǽs essade, MS.), smylte wurdon merestreáma gemeotu, Andr. Kmbl. 905; An. 453.

sester, seoxter, es; *m.* **I.** *a vessel, jar, pitcher* :—Sester *amfora*, Wrt. Voc. i. 24, 36 : 83, 23. Hê hêt heora ǽlcne geniman ânne ǽmtigne sester . . . Hig slôgon tôgædere ða sestras (*lagenas*), Jud. 7, 16–19. Cristallisce dryncìatu and gyldne sestras wǽron forþborenne *crystallina vasa potatoria et sextariola aurea invenimus*, Nar. 5, 14. **II.** *a measure* for liquids or for dry things; its capacity is uncertain. (a) as an English measure :—Twegen sestres sápan and twege[n] hunies and þrê ecedes, and se sester sceal wegan twá pund be sylfyrgewyht, Lchdm. iii.

92, 14. Cf. Unum sextarium mellis triginta duarum unciarum, Cod. Dip. Kmbl. iv. 285, 1. Wæs swýðe mycel hungor, and corn swá dýre, swá nân mann ǽr ne gemunde, swá ðæt se sester (*Henry of Huntingdon renders this* : 'sextarius frumenti, qui equo uni solet esse oneri') hwǽtes eode tô .lx. penega and eác furðor, Chr. 1043; Erl. 169, 31. xv pund (yntsan? cf. 'sextarius medicinalis habet uncias decem,' note on this passage) wætres gáþ tô sestre, Lchdm. ii. 298, 26. Fîftêne sestras lîðes aloþ, Chart. Th. 105, 12. Twelf seoxtres beóras, 158, 22. (b) as a foreign measure :—Under sestre *sub modio*, Mt. Kmbl. Lind. 5, 15. Hund sestra (*cados*) eles, Lk. Skt. 16, 6. Ǽlc wæterfæt wæs on twegra sestra gemete oððe on þreora *capientes singuli metretas binas uel ternas*, Jn. Skt. 2, 6. Gecned þrí sestras (*sata*) smedeman, Gen. 18, 6. Habbaþ emne gemetu and sestras *sint justus modius aequusque sextarius*, Lev. 19, 36. [O. H. Ger. sehstári, sehtári : Ger. sester, sechter *a measure of grain, twelve bushels; measure of liquids, sixteen quarts. From Lat.* sextarius. Cf. Fr. sêtier (*for grain*) *twelve bushels; for liquids, two gallons* : Ital. sestiere *a pint-measure*.] v. wîn-sester.

set, es; *n. A seat.* **I.** of the sun, *the place where the sun sets* :—Midðý tô sete eode sunne *cum occidisset sol*, Mk. Skt. Rush. 1, 32. Gewât sunne tô sete glîdan, Andr. Kmbl. 2498; An. 1250: 2610; An. 1306. Tô sete sîgeþ, Menol. Fox 221; Men. 112. Cf. set-gang, setl. **II.** of men, *a place where people remain*, of an army, *a camp, entrenchment*, cf. to sit down before a place :—Ne com se here ofter eall úte of ðǽm setum ðonne tuwwa, ôðre siþe ða hié ǽrest tô londe cômon . . . ôðre siþe ða hié of ðǽm setum faran woldon (cf. Ða Deniscan sǽton ðǽr behindan, 91, 1), Chr. 894; Erl. 90, 19–22. **III.** of animals, *a place where animals are kept, a stall, fold*, or *where they feed, pastures* :—Seotu *bucitum* (cf. hrýðra fald *bucetum*, Wrt. Voc. i. 15, 22), Txts. 47, 339. Seto *stabula*, 99, 1903. Siota, Wrt. Voc. i. 289, 11. ['In sedibus quies imperturbata.' I þe sette is reste & eise bitocned, A. R. 358, 23. Þat folc hafden alle iзeten and arisen from heore seten, Laym. 30841. O. H. Ger. sez *sedes, suggestus* : Icel. set *the sitting-room*, v. Cl. & Vig. Dict.] v. ge-set, -sete (*read* -set), ymb-set.

séta (seta?), setel. v. sǽta, setl.

seten, [n]e; *f.* **I.** *a set, shoot, branch* :—Setene *propagines*, Ps. Surt. 79, 12. v. ymb-seten. **II.** *a nursery, plantation* :—Setin pla[n]taria, Wrt. Voc. ii. 117, 49. Plantunga seten *plantaria*, 65, 76. Ǽghwilc wæstma seten ða ðe ne sette fæder mîn *omnis plantatio quam non plantavit Pater meus*, Mt. Kmbl. Lind. 15, 13. **III.** *what is planted* or *set* :—Gif mon gesêdcundne monnan ádrîfe, fordrîfe ðý botle næs ðære setene (*the ejected tenant was not to be deprived of what he had planted* (?); or seten,V, *he was to be compensated for the cultivation of the land* (?)), L. In. 68; Th. i. 146, 8. **IV.** *a cultivated place.* v. land-seten, I, *and* feldsætennum *campo*, Ps. Lamb. 77, 12. **V.** *planting, cultivation.* v. land-seten, II. **VI.** *a setting, putting.* v. hand-seten. **VII.** *a stopping.* v. blôd-seten. *See also* in-seten.

-setenness, sétere. v. ge-setenes, on-setenness, sǽtere.

set-gang, es; *m. Setting* of the sun :—Ofer setgong *super occasum*, Ps. Surt. 67, 5 : 49, 2 : 103, 19. v. set, setl-gang.

séðan; *p.* de *To declare true, affirm, attest, prove* :—Ic séðe *testificor*, Ælfc. Gr. 25; Zup. 146, 3. Ealle hálige gewritu sóðlîce séðaþ, ðæt se is Hǽlend Crist, Homl. Th. ii. 414, 9. Hê árás on ðam þriddan dæge, swá swá gewritu séðaþ, 598, 4. Sum ôðer séðde and cwæþ *alius quidam affirmabat dicens*, Lk. Skt. 22, 59. Is séðende and cweðende *adstipulatur*, Wrt. Voc. ii. 2, 17. Sume (*adverbs*) syndon con- *vel* adfirmativa, ðæt synd fæstnigende oððe séðende, Ælfc. Gr. 38; Zup. 226, 11. Séðende ðæt Crist is Godes Sunu '*proving that this is very Christ*' (Acts 9, 22), Homl. Th. i. 388, 4. v. á-, ge-séðan; sôðian, séðend, séðung.

seðe, Cd. Th. 92, 7; Gen. 1525 : seðel. v. sécan, I (2), setl.

séðend, es; *m. One who affirms* or *asserts* :—Séðend *stipulatorem* (cf. trymmend *stipulatorem*, Wrt. Voc. ii. 88, 2), Hpt Gl. 527, 34. v. ge-séðend.

set-hrægl, es; *n. A cloth for covering a seat* :—Setrægl *tapeta*, Wrt. Voc. i. 82, 19. Ic gean tô Cristes weofede ânre lytlere goldenre rôde and ânes sethrægles (*an altar-cloth?*), Chart. Th. 564, 10, 18. Ân lang healwáhrift and þrió sethrægl, 538, 4. [Cf. Icel. set-klǽði.] v. setl-hrægl.

séðung, e; *f. Attestation, affirmàtion, proof* :—Séðunge *adstipulatione, adsertione, adfirmatione*, Hpt. Gl. 444, 41. Hwæne mǽrsiaþ ðás wundra mid heora séðunge, Homl. Th. ii. 34, 5. Hê ðæs árleásan eáre gehælde tô séðunge sôðre godcundnysse (*in proof of true divinity*), 248, 2. Hê heora goda geendunge mid swutelum séðungum gewissode, i. 558, 16. Séðingum *assertionibus*, Hpt. Gl. 525, 35. Séðincgum, 409, 53. v. ge-séðung; séðan.

setin. v. seten.

setl, sedl, seðl, seotl, sotl, seatl, sitl (-el, -ol, -ul), es; *pl.* setl, setlu, sotelas, setlas (*North.*); *n. m.* (?) **I.** *that on which one sits, a settle, seat, place to sit* :—Setl *sella*, Wrt. Voc. i. 83, 70. Sotol, 289, 23. Gá nû tô setle, symbelwynne dreóh' . . . Geât geóng sôna setles neósan, swá se snottra hêht, Beo. Th. 3576; B. 1786. Se wæs setles yldest (on

setle yldost, MS. B.) *he had the chief seat*, Bd. 5, 13; S. 633, 4. Sæt Agustinus on sotole, 2, 2; S. 503, 15. Hé hét him úte setl gewyrcean, 1, 25; S. 486, 38. Mé hé wið his sylfes sunu setl getæhte, Beo. Th. 4031; B. 2013. Ofer setol *super sellam*, Kent. Gl. 304. Sotelas *sella*, Germ. 393, 143. Seó wlitigness heora ræsta and setla, Blickl. Homl. 99, 33. Hé his líchoman forwyrnde séftra setla and symbeldaga, Exon. Th. 111, 33; Gú. 136. On dæm forþmestum seatlum (seotlum, Rush.) sitta in sunnungum and ða forþmesto setla æt farmum *in primis cathedris sedere in sinagogis et primos discubitos in cenis*, Mt. Skt. Lind. 12, 39. Hé út áwearp ða setl ðara mynetera, Blickl. Homl. 71, 19. Hyra setlu (ceatlas, Lind.: settlas, Rush. *cathedras*) ðara ðe culfran sealdon hé tóbræc, Mt. Kmbl. 21, 12. Lufigaþ ðæt æreste sætil (*recubitos*) æt éfengereordum and forþmestu setulas (seatlas, Lind. *cathedras*) on heora somnungum, Rush. 23, 6. Seotlas, Mk. Skt. Rush. 11, 15. Ða yldstan setl (seatlas, Lind., Rush.), Lk. Skt. 20, 46. I a. *an official seat* of a king, judge, etc., *a throne, judgment-seat*:—On swiðre sedles Godes, Rtl. 27, 33. Fore sedle *before the throne*, 47, 26. Ðú Scippend heofones ðú ðe on ðam écan setle rícsast, Bt. 4; Fox 6, 30. Setle *solio*, Wrt. Voc. ii. 142, 13. Ðonne sitt hé ofer his mægenþrymme setl (seðel, Lind.: on sedle, Rush.), Mt. Kmbl. 25, 31. Hit is swíðe gewunelic ðætte dómeras & ríce menn on setelum sitten, Past. 56; Swt. 435, 21. Gé sittaþ ofer twelf setl (seatla tuelfa, Lind.: on sedlum twelfe, Rush.) démende, Mt. Kmbl. 19, 28. I b. metaph. *seat, place, position*:—Hé áwearp ða rícan of setle (sedle, Lind., Rush.), Lk. Skt. 1, 52. Se sit on wôles setle, se ðe yfel wyrcþ mid geþeahte, Past. 56; Swt. 435, 19–22: Ps. Th. 1, 1. Ðú setst ús on ðæt setl ðínes Sceoppendes, Bt. 7, 5; Fox 24, 2. Ofer seatul (on setule, Rush.) Moyses, Mt. Kmbl. Lind. 23, 2. Him sylþ God his fæder Dauides setl (sedle, Lind.: seðel, Rush.), Lk. Skt. 1, 32. On sotelum sôðfæstra *in cathedra seniorum*, Ps. Th. 106, 31. In reference to the heavenly bodies, tô setle gân, etc. (cf. *Fr.* le coucher du soleil, le soleil se couche) *to set*:—Syððan sunne beó on setle *after sunset*, Lchdm. iii. 8, 19. Ðonne heó (*the sun*) tô setle gæþ, Bt. 39, 3; Fox 214, 27: Salm. Kmbl. 186, 6. Ðá ðá sunne eode tô setl *cum occubuisset sol*, Gen. 15, 17. Ær sunne tô setle eode *usque ad occasum solis*, Ex. 17, 12. Ðá sunne tô setle eode *cum occidisset sol*, Mk. Skt. 1, 32. Sunne sáh tô setle, Chr. 937; Erl. 112, 17. Ðonne heó (*the sun*) on setl eode, Bt. 5, 23; S. 645, 26. Ðonne hió on setl glídeþ, Met. 28, 39. Se æfenstiorra on setl glídeþ, 29, 27, 31. On setel, Salm. Kmbl. 202, 34. v. setl-gang. II. *a seat, place where one abides, an abode, a residence, dwelling*:—Him wæs geseald setl on swegle ðær hé symle môt eardfæst wesan, blíðe bídan, Exon. Th. 149, 5; Gú. 757: 125, 15; Gú. 354. Geswíc ðisses setles, 119, 3; Gú. 249. Ða stôwe his seþles *locum sedis illius solitariae*, Bd. 3, 16; S. 542, 36. Hé eft tô ðam fæderlícan setle eode, Blickl. Homl. 115, 33: 129, 12. Ðá næfde hé nán setl hwær hé sittan mihte, for ðan ðe nán heofon nolde hine áberan, Ælfc.¹ T. Grn. 2, 45: Ps. Th. 88, 37: Exon. Th. 116, 31; Gú. 215. On prestes setl (*a hermitage*?), Cod. Dip. Kmbl. iii. 416, 29. Ða hálgan setl sceoldon weorþan gefylde mid ðære menniscan gecynde, Blickl. Homl. 121, 34: Cd. Th. 6, 10; Gen. 86. Gumena ríce, secga sitlu, Met. 9, 42. ¶ *a stall for animals*:—On ðam (*in the ark*) ðú scealt gerýman rihte setl ælcum eorþan tudre, Cd. Th. 79, 1; Gen. 1304. II a. as an ecclesiastical term, *a see*:—Sanctus Gregorius ðæs Rômâniscan setles bisceop, Lchdm. iii. 432, 24. Ðæs Apostolícan setles, Bd. 1, 23; S. 485, 23: 4, 1; S. 563, 23. Hér Rômâne ðone pâpan of his setle áfliémde, Chr. 797; Erl. 58, 14. On setl biscopstôles *in sedem pontificatus*, 5, 23; S. 646, 32. II b. metaph. *seat* of a disorder, etc., *dwelling-place* of non-material things:—Ðý læs ingæ se fiónd in sáuelo hiora & seðel habba ne mægi, Rtl. 117, 31. III. *the part of the body on which one sits, the seat*:—Wið gicþan ðæs setles, Lchdm. i. 218, 10. Gif se uíc weorðe on mannes setle geseten, iii. 30, 16. Wríð under ðæt setl neoþan, i. 366, 17. Him wand út his innoþ æt his setle, Homl. As. 59, 201. IV. *a sitting, the being in*, or *assuming, a sitting position*; sessio:—Hé frægn for hwon hé âna swâ unrôt on stâne wæccende sæte . . . 'Ne tala ðú ðæt ic ne cunne ðone intingan ðínre unrôtnesse and ðínre wacone and ânlépnesse ðínes sedles' *ne me aestimes tuae moestitiae & insomniorum & solitariae sessionis causam nescire*, Bd. 2, 12; S. 513, 41 note. Ðú mín setl (*sessionem*) oncneówe and mínne ærist æfter gecýddest, Ps. Th. 138, 1. IV a. *stay, residence*:—On ðæm setle ðe hé ðær sæt *during the stay he made there*, Chr. 922; Erl. 108, 22. IV b. as a military term, *a siege*:—Him (*the besiegers*) ðæt setl (*obsidio*) swiðor derede ðonne ðám ðe ðærinne (*in Veii*) wéron, Ors. 2, 8; Swt. 90, 24. Porsenna ðæt setl forlét *Porsenna raised the siege*, 2, 3; Swt. 68, 30. Ðá forlét hé ðæt setl *ab obsidione discessit*, 3, 11; Swt. 146, 20. [Heo isetten Iacob on Cristes selt, O. E. Homl. i. 93, 9. Adam set on the setle of unhele, ii. 59, 25: Ps. 1, 1. Ich mai þe finde at þe rumhuse . . . þu sittest and singst behinde þe setle, O. and N. 594. Our loverd sal sitt . . . opon þe setil of his mageste, Pr. C. 6122. *Goth.* sitls; *m. a seat, throne, nest*: *O. H. Ger.* sez[z]al *cathedra, sponda, solium, tribunal*; sedal, sethal, sedhal sedes, thronus, triclinium, occasus (*solis*): *O. Frs. O. Sax.* sedel.] v. ân-, ancer-, ancor-, beór-, bisceop-, burhgeat-, cyne-, dôm-, eðel-, ge-, heáh-, hilde-, lâreów-, medu-, pâp-, scip-, sunder-, þrym-, út-, weard-, wræc-setl; beorg-seðel; set.

setla. v. ân-, cot-, ge-, wésten-setla.

setlan; *p.* [e]de. I. *trans. To settle, seat, put in a position of rest*:—Wæglíðende setlaþ sæmearas, and ðonne in ðæt églond up gewítaþ, Exon. Th. 361, 5; Wal. 15. II. *intrans. To settle, take a position of rest*, of the sun, *to set*. v. setlung *and the Mid. E. forms*. [Þatt allderrmann þatt heʒhesst wass Att tatt bridale settledd (ἀρχιτρίκλινος), Orm. 15285. Til þe sunne wæs setled to reste, Will. 2452. Him thoughte a goshauk . . . Setlith on his beryng, Alis. 484.]

setl-gang, es; *m. Setting* of the heavenly bodies, generally of the sun, (1) marking time:—Ðá bád se sacerd sunnan setlgonges, forðon sunnan trió âgefeþ ondsware æt ðæm upgonge & eft æt setlgonge, Nar. 27, 15–18. Sunne, setlgonges fús, Exon. Th. 174. 34; Gú. 1187. Æfter sunnan setlgonge, Chr. 773; Erl. 52, 24. Ær sunnan setlgange, Bd. 1, 27; S. 495. 7. Æt sunnan setlgange, Blickl. Homl. 93, 16. Sunne hire setlgang healdeþ *sol cognovit occasum suum*, Ps. Th. 103, 18. Ðá se æþela glæm setlgong sôhte, Exon Th. 178, 32; Gú. 1253. (2) marking place, *the west*:—Be ðam wege ðe liþ tô sunnan setlgauge *by the road that runs to the west*, Deut. 11, 30. Fram sunnan upgange oþ hire setlgang *from the east unto the west*, Ps. Th. 49, 2. [*Ps.* setl-gang. Cf. *O. H. Ger.* sedal-gang: *O. Sax.* gangan, sígan to sedle, werðan an sedle (*of the sun.* v. Grmm. D. M. 700, R. A. 817.] v. setl, I c, set-gang, *and next word.*

setl-gangende; *adj.* (*ptcp.*) *Setting*:—Setlgangendre sunnan *occidenti*, Bd. 5, 23; S. 645, 27.

setl-hrægl, es; *n. A covering for a seat*:—vii. setlhrægel, Chart. Th. 429, 28. v. set-hrægl.

setl-râd, e; *f. Setting* of the sun:—Æfter sunnan setlráde, Cd. Th. 184, 19; Exod. 109. Cf. setl-gang.

setlung, e; *f.* I. *a taking of a seat, a sitting down*:—Ðú understôde setlunge míne and æriste míne *tu cognovisti sessionem meam et resurrectionem meam*, Ps. Lamb. 138, 2. II. *setting* of the sun; occasus:—Seó niht hæfþ seofon dælas fram ðære sunnan settlunge (setlunge, MS. P.), Lchdm. iii. 242, 26. Æfter sunnan setlunge, 266, 5. Fram sunnan uprine oþ setlunge, Ps. Spl. 112, 3. Setellung, 49, 2. v. setlan.

sétnere, sétnung. v. sætnere, sætnung.

setness, e; *f.* I. *an ordinance, a regulation, an institution*:—Ðis is seó gerædnes ðe Eádulf hæfþ gerád tô setnesse, Cod. Dip. Kmbl. iii. 295, 32. Gé forlætaþ Godes bebod and healdaþ manna laga (setnesse, Lind.: setnisse, Rush. *traditionem*), Mk. Skt. 7, 8, 3, 13. Setnesa, Mt. Kmbl. Lind. 15, 2. The word glosses also *testimonium*, Mt. Kmbl. p. 1, 11: *testamentum*, p. 2, 5. II. *constitution, arrangement*:—From setnisse middangeardes *a constitutione mundi*, Mt. Kmbl. Rush. 25, 34. [Heo makeden ane sætnesse . . . þe ælc cheorl eæt his sulche hæfde grið, Laym. 4258. Godess laʒhe & hiss hallʒhe settnesse þeʒʒ didenn fallen dun, & hofenn affterr þeʒʒre wille settnessey, hu mann birrde Godess laʒhe follʒhenn, Orm. 16836–43.] v. â-, fore-, ge-, in-, on-, wið-setness.

setnung. [*Icel.* setning.] v. frum-setnung.

settan; *p.* sette, *pp.* seted, set[t] (*generally transitive, but see* XII). I. *to set, place, put, cause to take a certain position*:—Ic sette mínne rênbogan on wolcnum, Gen. 9, 13. Ic sette max on stôwe gehæppre, Coll. Monast. Th. 21, 13. Hwæðer gé settan eówer nett on ða hêhstan dúne, ðonne gé fiscian willaþ? Ic wât ðæt gé hit ðær ne settaþ. Hwæðer gé eówer net út on ða sæ lædon, ðonne gé huntian willaþ? Ic wêne ðæt gé hí ðonne seton up on dúnum, Bt. 33, 3; Fox 118, 11–15. Ne hí ne ælaþ hyra leóhtfæt and hit under cýfe settaþ, Mt. Kmbl. 5, 15. Heó (*the fallen angels*) God sette on ða sweartan helle, Cd. Th. 20, 20; Gen. 312. Hé sette his ða swiðran hand (cf. mid ða swiðran hand, 514, 21) him on ðæt heáfod, Bd. 2, 12; S. 515, 19. Hí ðá nô ða studu úton tô ðam wage tô fultume ne setton, ac hí heo on ða cyricean setton, 3, 17; S. 544, 37. Hié setton hié æt ðære byrgene dura, Blickl. Homl. 155, 8. Ðá hé bebyrged wæs, hié settan him hyrdas tô, 177, 26. Setton scyldas wið weal *they set their shields against the wall*, Beo. Th. 655; B. 325. Sete ðín hand under mín þeóh, Gen. 24, 2: 48, 18. Se ðe wille fæst hús timbrian ne sceall hé hit nô settan up on ðone hêhstan cnol (*must not take the top of a hill as a site for his house*) . . . and eft se ðe wille fæst hús timbrian, ne sette hé hit on sondbeorhas, Bt. 12; Fox 36, 7–11. Ðá lét hé hine on hæft settan *he had him put into prison*, Chr. 1036; Erl. 164, note 3. Hé gearwe hæfde reliquias in tô settenne, Bd. 5, 11; S. 625, 37. I a. *to set down*:—Ðá hét se apostol ða bære settan, Homl. Th. i. 60, 16. II. figurative, *to set to work, set before one a choice, set a mark*, a name, one's mind, *lay a charge, a curse*, etc., *upon one, put one in a position, put into one's power*, etc.:—Ic sette beforan eów bletsunga and wiriginssa, Deut. 11, 26. Ic hine wergþo on míne sette, Cd. Th. 105, 20; Gen. 1756. Swâ hit mé sealde se ðe ic hit nú on hande sette, L. O. 3; Th. i. 180, 4. Ðú setst (settes, Cott. MSS.) ús on ðæt setl ðínes Sceoppendes, Bt. 7, 5; Fox 24, 2. Swâ hwæt swâ ðú mé on settest and bebeódest tô dônne, Bd. 4, 25; S.

600, 4. God him sette naman, Homl. Th. i. 12, 31. Hé him naman on sette, Mk. Skt. 3, 17. Abraham sette friþotácn on his selfes sunu, Cd. Th. 142, 29; Gen. 2369. Hine Abraham on beácen sette, 167, 19; Gen. 2768. Gé setton mé in edwít ðæt . . . *you laid to my reproach, that* . . ., Exon. Th. 131, 21; Gú. 459: Cd. Th. 165, 8; Gen. 2728. Gé ða wintergerím on gewritu setton, Elen. Kmbl. 1305; El. 654. Sete heora ealdormenn, swá ðú Oreb dydest *make their nobles like Oreb* (A.V.), Ps. Th. 82, 9. Sete on Drihten ðín gehygd, 54, 22. Setton hí hine on borh *they shall make him give security*, L. Ath. i. 20; Th. i. 208, 30: 210, 7. Deáþ settan *to kill*, Elen. Kmbl. 955; El. 479. Wíte settan *to impose punishment*, Cd. Th. 76, 33; Gen. 1266. On gewrit settan *to put into writing*, L. Alf. 49; Th. i. 58, 22. Wutan ús tó symbeldæge settan, Ps. Th. 117, 25. **II a.** of travelling, cf. lecgan *and* Ger. zurücklegen :—On weg setteþ wíse gangas, Ps. Th. 84, 12. Sceal ic nú wreclástas settan, síðas wíde, Cd. Th. 276, 15; Sat. 189. Gesundne síð settan *to make a safe journey*, Elen. Kmbl. 2008; El. 1005. **III.** *to set, plant :*—Sette *pastinat*, Wrt. Voc. ii. 96, 52. Hé leác sette *he set vegetables*, Shrn. 61, 20. Hé wíngeard sette, seów sǽda fela, Cd. Th. 94, 8; Gen. 1558: 172, 7; Gen. 2840. Settan *pastinare*, Wrt. Voc. ii. 116, 6. Settende *pastinantem*, 66, 18. Settum beámum anlíce *sicut novellae plantationes*, Ps. Th. 143, 14. **IV.** *to set, fix, implant :*—Hé módes snyttru seów and sette geond sefan monna, Exon. Th. 41, 29; Cri. 663. Settaþ on eówerum heortum, ðæt gé ne þurfon ásmeágan, hú gé andwyrdan sceolon, Homl. Th. ii. 542, 3. Uton wé ðæs dæges fyrhto on úre mód settan, Blickl. Homl. 125, 6. **V.** *to set, fix, appoint* a limit, time, place (cf. *set* day, time in A.V.) :—In ðam frumstóle ðe him freá sette, Exon. Th. 349, 25; Sch. 51. Hí settan dæg tó ðæt man tó ðam lande scolde faran *they appointed a day for going to the land*, Chart. Th. 376, 16. Ðæt ic ðé symbledæg sette, Ps. Th. 75, 7. Settan gemǽro, Ex. 19, 23. Mearce settan, Cd. Th. 171, 19; Gen. 2830. **VI.** *to set* a task, *ordain, establish* a law, regulation, *appoint* a condition :—Wé settaþ æghwelcere cirican ðis frið, L. Alf. pol. 5; Th. i. 64, 8. Hé sette gecamp geleáffullum sáwlum, Homl. Th. i. 64, 18. Se ðe ða ealdan lé sette, 94, 4. Sylfa sette, ðæt ðú sunu wǽre efeneardigende, Exon. Th. 15, 14; Cri. 236. Ǽ ðú mé sete, Ps. Th. 118, 33. Gif gé nú gesáwen hwelce mús, ðæt wǽre hláford ofer óðre mýs and sette him dómas, Bt. 16, 2; Fox 52. 2. **VII.** *to build, erect :*—Hús settan and tún timbrian, Shrn. 163, 16. Ongunnon heora burh rǽran and sele settan . . . weras on wonge wíbed setton, Cd. Th. 113, 2–5; Gen. 1881–2. **VIII.** *to set up, institute, found, establish :*—Hwá ǽrost bócstafas sette? Salm. Kmbl. p. 192, 6. Hé sette scole *instituit scholam*, Bd. 3, 18; S. 545, 44. Ǽgðer ge cyninga rícu settan ge ceastra timbredon, Ors. 1, 10; Swt. 48, 9: Met. 1, 4. Ðæt wæs weallfæstenna ǽrest ealra ðara ðe ædelingas settan héton, Cd. Th. 65, 3; Gen. 1060. **VIII a.** of the operations of the Deity :—Ðú dæg settest and deorce niht, Ps. Th. 73, 16: 138, 11: Exon. Th. 258, 33; Jul. 274. Ðá hé ðisne ymbhwyrft ǽrest sette, 422, 17; Rä. 41, 7: Cd. Th. 265, 29; Sat. 15. Ðá ðú wǽre settende ðás sídan gesceaft, Exon. Th. 22, 23; Cri. 356. **IX.** *to set, base, found :*—Gif ðú wísdóm timbrian wille, ne sete ðú hine uppan ða gítsunga, Bt. 12; Fox 36, 11. **X.** *to appoint* an officer or a person to an office or duty :—Hine tó ealdormenn sette, Ps. Th. 104, 16. Hé sette hine on his húse tó hláfwearde *constituit eum dominus domus suae*, 104, 17. Sette hé getreówe borgas, L. Eth. i. 1; Th. i. 280, 19: L. C. S. 30; Th. i. 394, 8. **X a.** *to appoint* something for a purpose :—Bæd þrymcyning, ðæt hé him ða weádǽd tó wræce ne sette, Elen. Kmbl. 988; El. 495. **XI.** *to settle* a quarrel, *allay* animosity, *compose* a difference :—Witan scylon fǽhþe settan, L. Edm. S. 7; Th. i. 250, 13. **XII.** *intrans.* To *settle, abate, subside :*—Lege uppa þat geswollene and hyt sceal sóna settan, Lchdm. iii. 86, 19. Ðonne biþ ðæs innoþes sár settende, i. 74, 9. **XIII.** *to compose* a book, etc. :—Ic ðás bóc wrát and sette . . . ic sette feówer béc, Bd. 5, 23; S. 647, 32–37. For ðisum þingum ic ðás bóc sette, Guthl. prol.; Gdwin. 4, 26. Scē Isidorus ðe ðás bóc sette *qui hunc librum instituit*, L. Ecg. P. i. 6; Th. ii. 174, 16. Dauid ða sealmas sette, ðe wé æt Godes lofsangum singaþ, Homl. Th. ii. 576, 5. Se cyng hét ðone arcebisceop bóc settan *the king ordered the archbishop to draw up a charter*, Chart. Th. 376, 3. [*Goth.* satjan: *O. Sax.* settian: *O. Frs.* setta: *O. H. Ger.* sezzan: *Icel.* setja.] v. á-, an-, be- (bi-), for-, fore-, ge-, in-, of-, on-, tó-, un-, wið-, ymb-settan.

settaþ, Ps. Th. 9, 29 *for* sǽtaþ (?).

settend, es; *m.* An *ordainer, appointer :*—Ðæt ðú ána eart éce Drihten, weroða Waldend, sigora settend (sigeróf settend), Exon. Th. 188, 17; Az. 47), Cd. Th. 237, 5; Dan. 333. v. dóm-settend.

set-þorn *some kind of tree :*—Andlang fura on setþorn; of setþorne on fúlan ríþig, Cod. Dip. Kmbl. iii. 436, 14.

sétung. v. sǽtung.

sewte, Andr. Kmbl. 1483; An. 743. v. sépan.

sex. v. seax, six.

sí *be :*—Him sí ábrogden hiora sceamu, Ps. Th. 108, 28. Hwæt hér sí gedón, Blickl. Homl. 179, 34. Hwæðer hit sig ðe sóð ðe leás, Gen. 42,

16. Ðæs sig Metode þanc, Beo. Th. 3561; B. 1778. Ðæt gé witen hwæt hit sié, Past. 8; Swt. 53, 13. Gif ðú sié Godes sunu, Blickl. Homl. 27, 7. Him sió wuldor, Hy. 8, 4. Ðæt ðæt betst sý, ðæt mon seó foremǽre, Bt. 34, 2; Fox 82, 10. Gif heó leng sý ðonne hé *if she live longer than he*, L. Edm. B. 3; Th. i. 254, 13. Deáh ðe heora hundred seó, Ps. Th. 89, 10. Hé cwyð ðæt ic seó teónum georn, Cd. Th. 36, 34; Gen. 581: 309, 4; Sat. 704. Gyf ðú Godes sunu sý (sig, MS. A.: sié, Rush.), Mt. Kmbl. 4, 3. Sib sý (sig, MS. A.) eów, Lk. Skt. 24, 36. Ðæt gé ne sín (sié, Lind.) ymbhýdige, Mt. Kmbl. 6, 25. Sín (sién, Hatt. MS.) hira eágan áþístrode, Past. 1; Swt. 28, 9. Ðæt sién gewemmede ealle, Blickl. Homl. 245, 22. Ðæt mé æfter sié eaforan síne yrfeweardas, Cd. Th. 131, 28; Gen. 2183. Seón, Exon. Th. 96, 28; Cri. 1581. Sín (sé, Lind.: sié, Rush.), Mt. Kmbl. 6, 1. Ðæt hí sýn (sié, Lind.: sié, Rush.) án, Jn. Skt. 17, 11, 21, 22, 23. v. eom.

sib[b], e; *f.* **I.** *relationship :*—Sybbe *propinquitas*, Hpt. Gl. 469, 55. Gif hwá sibleger gewyrce, gebéte ðæt be sibbe mǽðe (*according to the degree of relationship*), L. C. S. 52; Th. i. 404, 25. Sameramis gesette ðæt nán forbyrd nǽre æt geligere betwuh nánre sibbe, Ors. 1, 2; Swt. 30, 35. On ðæs láfe ðe swá neáh wǽre on woruldcundre sibbe, L. Eth. vi. 12; Th. i. 318, 16. For ðære mǽglícan sibbe (*of Christ and John*), Homl. Th. i. 58, 6. Ðá com Swein eorl and bæd Beorn eorl, ðe wæs his eámes sunu, ðæt hé his geféra wǽre tó ðam cynge. Hé wende ðá for ðære sibbe mid him, Chr. 1050; Erl. 175, 18. Hréðel (*the grandfather of Beowulf*) sibbe gemunde, næs ic (*Beowulf*) him láðra beorn ðonne his bearna hwylc, Beo. Th. 4854; B. 2431. Hé biþ his móder twám sibbum (*in two relationships, in double relationship*) getǽht, ðæt hé biþ ǽgðer ge sunu ge bróðer, Wulfst. 193, 7. **I a.** in a spiritual sense, cf. *gossip :*—Se cyning him tó godsuna onféng and tó tácne ðære sibbe him twá mǽgþe forgeaf (*in signum adoptionis, duas illi provincias donavit*), Bd. 4, 13; S. 582, 9. **II.** *friendliness, kindness*, the opposite of hostility :—Sibbe cos *pacis osculum*, R. Ben. 82, 6. Ne gehýrde nǽnig man on his heortan óht elles búton mildheortnesse and sibbe, Blickl. Homl. 225, 2. Ne mihte hé mid ðone cyning sibbe habban, ac mycel ungeþwǽrnys betwih him árás, Bd. 3, 14; S. 539, 35. Feóndscype dwæscaþ, sibbe sáwaþ, Exon. Th. 30, 31; Cri. 487. Á ic sibbe wið ðé healdan wille *I will ever maintain my friendliness to thee*, 177, 33; Gú. 1236. Gé hý mid sibbum sóhtun *ye visited the sick with kind attentions*, 83, 22; Cri. 1360. **III.** *peace*, the opposite of war :—Ǽgðer ge on sibbe ge on gewinne, Bt. 24, 2; Fox 82, 11. Hé him gebeád wið his sibbe (*in pretium pacis*) unrím máþma, Bd. 3, 24; S. 556, 8. Gif hí sibbe mid Godes mannum onfón ne woldan ðæt hí wǽron gefeoht fram heora feóndum onfónde *si pacem cum fratribus accipere nollent, bellum ab hostibus forent accepturi*, 2, 2; S. 503, 29: Chr. 605; Erl. 21, 28. Se bisceop betweox ðám cyningum sibbe geworhte, Bd. 4, 21; S. 590, 11. Eall ðeós worold geceás Agustuses frið and his sibbe, Ors. 5, 15; Swt. 250, 17. On ða tíd (*in the golden age*) wæs sibba geníhtsumnes (*an utter absence of wars*), Blickl. Homl. 115, 9. **IV.** *peace, concord, unity, absence of dissension* or *variance :*—Suá dætte án sibb (sib, Cott. MSS.) Godes lufe bútan ǽlcum ungeráde ús gefége tósomne, Past. 36; Swt. 253, 22. Ongeán ðæt sint tó manienne ða ðe ða sibbe sáwaþ ðæt hié swá micel weorc tó unwærlíce ne dón and húru ðær ðær hié nyton hwæðer sió sibb (sib, Cott. MSS.) betre betwux gefæstnuþ biþ ðe ne biþ fordæm swá swíðe swá hit dereþ ðætte ǽnig wana sié ðære sibbe betwux ðǽm goodum swá swíðe hit eác dereþ ðæt hió ne sié gewanod betwux ðǽm yfelum. Fordæm gif ða unryhtwísan hiera yfel mid sibbe gefæstnigaþ and tósomne gemengaþ ðonne biþ geícced hiera mægen at contra admonendi sunt pacifici, ne tantae actionis pondus levigent, si, inter quos fundare pacem debeant, ignorent. Nam sicut multum nocet, si unitas desit bonis, ita valde est noxium, si non desit malis. Si ergo perversorum nequitia in pace jungitur, profecto eorum malis actibus robur augetur, 47, 3; Swt. 361, 5–12. Beó mannum sib and sóm gemǽne, and ǽlc sacu getwǽmed, L. Eth. v. 19; Th. i. 308, 29. Sibb, vi. 25; Th. i. 320, 28. Crist ðe ys ðære sibbe ealdor, Ælfc. T. Grn. 8, 1. Sibbe (sibbes, Lind., Rush.) bearn, Lk. Skt. 10, 6. Mid sibbe *cum consensu*, Ps. Spl. 54, 15. Ðá wiste hé sumne híréd ðe ungeþwǽre him betweónum wǽron . . . hé wolde ðæt hié ealle on sibbe wǽron, Blickl. Homl. 225, 9. God sylfa bebeád ðæt wé sóðe sibbe heóldan and geþwǽrnesse ús betweónum habban, 109, 15: Ps. Th. 33, 14. Ne wéne gé ðæt ic cóme sybbe on eorþan tó sendanne; ne com ic sybbe tó sendanne, ac swurd, Mt. Kmbl. 10, 34. Sybbe . . . tódǽl, Lk. Skt. 12, 51. Habbaþ sibbe betwux eów, Mk. Skt. 9, 50. Ðonne forlǽtaþ hí ða sibbe ðe hí nú healdaþ, and winþ heora ǽlc on óðer, and forlǽtaþ heora geférrædenne, Bt. 21; Fox 74, 34: Elen. Kmbl. 2411; El. 1207. **V.** *peace, freedom from disturbance* or *molestation, tranquillity :*—Gerusalem is gereht sibbe gesyhþ (cf. sib-gesihþ), fordon ðe hálige sáula ðær restaþ, Blickl. Homl. 81, 1. Nú is ǽghwonon hreám and wóp and sibbe tólésnes, 115, 16. Iethro cwæþ: 'Gá on sybbe,' Ex. 4, 18. Hí ða feówertig wintra wunodon on sibbe *quievit terra per quadraginta annos*, Jud. 5, 32. Beóþ on sibbe ðás þing ðe hé áh, Lk. Skt. 11, 21. Hú wéne gé hwelce sibbe ða weras hæfdon, ðonne heora wíf swá monigfeald yfel dónde wǽron? Ors. 1, 10; Swt. 50, 2. **V a.** *the peace* of a country, the

king's *peace* :—Ða kyningas ǽgðer ge hiora sibbe ge hiora onweald innanbordes gehíóldon, Past. pref. ; Swt. 3, 6. Hē (*Augustus*) bebeád ðæt eall moncynn áne sibbe hæfde, Ors. 5, 14 ; Swt. 248, 20. **VI.** *peace* of mind, *freedom from agitation, fear*, etc. :—Sib sí mid eów, ne ondrǽde gē eów, Gen. 43, 23. Sý sibb betwux eów ; ic hit eom, ne beó gē ná áfyrhte, Homl. Th. i. 220, 13 : Jn. Skt. 20, 19. Ús biþ gearu sóna sibb æfter sorge, Andr. Kmbl. 3134; An. 1570. Léton ðone hálgan swefan on sibbe, blíðne bídan, 1663; An. 834. Wē mótan his ða wuldorfæstan onsýne mid sibbe sceáwian, Blickl. Homl. 103, 29. [Sæhte and sibbe, Laym. 6096. Off Daviþess kin and sibbe, Orm. 3315. We ne muʒe grið ne sibbe macie, O. E. Homl. i. 243, 14 : O. and N. 1005. *Goth.* sibja *relationship, adoption : O. Sax.* sibbia *relationship : O. Frs.* sibbe : *O. H. Ger.* sippa, sibba *adfinitas, propinquitas, pax, foedus : Icel.* sifjar ; *pl. affinity ;* Sif *the wife of Thor ; she was the goddess of the sanctity of the family and wedlock.* v. Grmm. D. M. 286 and R. A. 467.] v. bróðor-, cneów-, dryht-, friðu-, mǽg-, mǽgþ-, neáh-, níð-, un-sib[b], *and next word.*

sib[b] ; *adj. Sib* (dial. e. g. Lancashire, Scottish), *related ;* also absolute, *one related, a relation* (In *god-sibbas* the word is inflected as a noun, cf. *Icel.* sifr *a near relation.* In the passage below, Lk. 14, 12, the form may be taken as a weak noun, cf. *Icel.* sifi *a relation by marriage,* guð-sifi *a god-sib : O. L. Ger.* sibbeo : *O. H. Ger.* sibbo *consanguineus*) :— Ne biþ ná gelíc ðæt man wið swustor gehǽme and hit wǽre feor sibb (or ? feorsibb ; cf. neáh-sibb), L. C. S. 52 ; Th. i. 404, 28. Ðær ne byþ sybbes lufu tó óðrum, Wulfst. 146, 13. Ðære sibban ob *cognate,* Wrt. Voc. ii. 64, 26. Hē biþ his móder on twám wísum tó sibbum getǽht, ðæt hē biþ ǽgðer ge sunu ge bróðer, Wulfst. 193, 7. Hát in gán seón sibbe gedriht samod ætgædere, Beo. Th. 779 ; B. 387 : 1462 ; B. 729. (*Grein* takes sibbe as *gen.* of sibb, *Thorpe and Heyne make it the first part of a compound.* Cf. sib-gedryht.) Ðonne se deáþ cymeþ ásundraþ ða sibbe ða ðe ǽr somud wǽron líc and sáwle *when death comes, it separates then relations, who before were together, body and soul,* Exon. Th. 367, 7 ; Seel. 4. Hē (*Augustus*) bebeád ðæt ǽlc mǽgþ tógædere cóme, ðæt ǽlc man ðý gearor wiste hwǽr hē gesibbe (sibbe, MS. C.) hæfde, Ors. 5, 14 ; Swt. 248, 17. Sibbo ł cúðo menn (gisibbe, Rush.) *cognatos,* Lk. Skt. Lind. 14, 12. [Hiss follc, þatt wass himm sibb o moderr hallfe, Orm. 19144. Sohhtenn himm betwenenn sibbe and cuþe (v. Lk. 2, 44), 8922. Bitwhwe sibbe, vlesliche oðer gostliche, A. R. 204, 20. Iosep bad sibbe (*his kinsmen*) cumen him biforen, Gen. and Ex. 2503. Who is sibbe to þis seuene . . . he is wonderliche welcome, Piers P. 5, 634. Sybbe or of kynne *consanguineus,* Prompt. Parv. 455. *Goth.* un-sibis *impious : O. Frs.* sibbe *related : O. H. Ger.* sippe.] v. ge-, neáh-sib[b].

sibæd *sifted* (?) ; arbatae, Txts. 43, 216. v. sife, sifeþa.

sib-ǽðeling, es ; *m. A prince and kinsman* :—Sibǽðelingas (*Beowulf and Wiglaf ;* a few lines before the former is spoken of as the *mǽg* of the latter), Beo. Th. 5409 ; B. 2708.

sibban (?) ; *p.* sifde (?) *To rejoice* :—Sifeþ *gaudet,* Bd. 5, 23 ; S. 646, 35 note. [*Goth.* sifan ; *p.* sifaida *to rejoice.*]

sibbian ; *p.* ode *To make people friends, make peace between disputants, reconcile* :—Ða seðe ða unryhtwísan tósomne sibbaþ (*pace sociat*), hē seleþ ðære unryhtwísnesse fultom, Past. 47, 3 ; Swt. 361, 22. Sipbade *paciscitur,* Lk. Skt. p. 11, 2. On .iiii. nyhta mónan mónan sibba ða cídenda[n] men, and ðú hie gesibbast, Lchdm. iii.\176, 25. Cyninge gebyreþ, ðæt hē eall cristen folc sibbie and sehte, L. I. P. 2 ; Th. ii. 304, 12. Sybbie, Wulfst. 266, 17. Wē lǽraþ, ðæt nán sacu, ðe betweox preóstan sí, ne beó gescoten tó worldmanna sóme, ac sǽman and sibbian heora ágene geféran, L. Edg. C. 7 ; Th. ii. 246, 4. Ða wǽron on ðam tíman ungeþwǽre preóstas, ða hē wolde sibbian, Homl. Th. ii. 516, 5. v. ge-, un-sibbian.

sib-cwide, es ; *m. A speech professing peace and friendliness, fair words* :—Ða leásan men, ða ðe mid tungan treówa gehátaþ fægerum wordum . . . hafaþ on gehátum hunigsmæccas, smeðne sybcwide, Fragm. Kmbl. 54 ; Leás. 29.

sib-fæc, es ; *n. A degree of relationship* :—Ǽfre ne geweorðe, ðæt cristen man gewífige in .vi. manna sibfæce on his ágenum cynne, ðæt is binnan ðam feórþan cneówe, L. Eth. vi. 12 ; Th. i. 318, 14 : L. C. E. 7 ; Th. i. 364, 22. Cf. Christiani ex propinquitate sui sanguinis usque ad septimum gradum connubia non ducunt, Th. i. 257, note b, and ii. 19, note 1. v. Grmm. R. A. 468.

sib-gebyrda ; *pl. f. Relationship* :—Ic (*Abraham*) eom fædera ðín (*Lot*) sibgebyrdum, Cd. Th. 114, 8 ; Gen. 1901.

sib-gedryht, e ; *f.* **I.** *a band of kinsmen* :—Bád eall seó sibgedriht (*the Israelites*) somod ætgædere, Cd. Th. 191, 13 ; Exod. 214. **II.** *a peaceful band* :—Swinsaþ sibgedryht (*the host of spirits who live in the peace and tranquillity of heaven*), Exon. Th. 239, 8 ; Ph. 618. In ðam écean geféan mid ða sibgedryht somud eard niman, 184, 18 ; Gú. 1346.

sib-gemágas ; *pl. m. Kinsmen* :—Heáhlond stigon sibgemágas (*Abraiam and Isaac*), Cd. Th. 202, 10 ; Exod. 386.

sib-georness, e ; *f. Eagerness for peace and kindness, love* :—Sybgeornes *caritas Dei et proximi,* Wulfst. 69, 2. Sibgeornes, 189, 21.

sib-gesihþ, e ; *f. A vision of peace* :—Sibgesyhþe *Hierosolymae* (v. sib, V), Hpt. Gl. 447, 56.

sibi. v. sife.

sib-lác, es ; *n. A peace-offering* :—Ic ðe wille gesyllan míne siblác (*hostias pacificas*), L. Ath. i. prm. ; Th. i. 196, 21.

sib-leger, es ; *m. An incestuous person* :—Be siblegerum. And æt siblegerum ða witan gerǽddan, ðæt cyng áh ðone uferan and bisceop ðone nyðeran, L. E. G. 4 ; Th. i. 168, 13–15. Cf. for-liger ; *m.*

sib-leger, es ; *n. Incest* :—Be siblegere. Gif hwá sibleger gewyrce gebéte ðæt be sibbe mǽðe, L. C. S. 52 ; Th. i. 404, 24. Wearþ ðes þeódscype swýðe forsyngod þurh sibblegeru and þurh mistlíce forligru, Wulfst. 164, 5 : 165, 31.

sib-líc ; *adj. Of peace* :—Mid siblícum cosse, Homl. Skt. i. 22, 31. God biddan ðæt hē forgefe siblíce tíd and smyltelíco gewidra, Shrn. 74, 11. Wē sceolan ús geearnian ða siblecan wǽra Godes and manna, Blickl. Homl. 111, 3.

sibling, es ; *m. A relation, kinsman* :—Sibling *affinis vel consanguineus,* Wrt. Voc. i. 72, 46 : Homl. Th. i. 516, 14. Hæfst ðú suna oððe dohtra oððe áðum oððe ǽnigne sibling? Gen. 19, 12. Gebróðru *vel* siblingas *fratres,* Wrt. Voc. i. 52, 3. Ofsleáþ ðás ealdras, ðonne beóþ heora siblingas tó heófunge geneádode, Homl. Th. i. 88, 1. Fæder and móder and flǽsclíce siblingas, 398, 8. Úre frýnd geseón and úre siblingas gegrétan, ii. 526, 33. Siblingum *contribulibus, propinquis, parentibus,* Hpt. Gl. 472, 23. v. ge-sibling.

sib-lufu, an ; *f. Kindly affection, kindness, love such as exists between kinsmen* :—Ic (God) tó eów mid siblufan gecyrre þurh milde mód, Exon. Th. 366, 6 ; Reb. 8 : 40, 7 ; Cri. 635. Hié (*the fallen angels*) of siblufan Godes áhwurfon, Cd. Th. 2, 25 ; Gen. 24. Git mē sibblufan and freóndscipe cýðaþ, 152, 3 ; Gen. 2514.

[sib-rǽden[n], e ; *f. Affinity, relationship* :—Þes ilce Willelm hæfde ǽror numen ðes eorles dohter of Angeow tó wífe oc hí wǽron sibðen tótweamde for sibrǽden, Chr. 1127; Erl. 255, 21. The king him let uor sibrede todele fram is wif, R. Glouc. 492, 9. A sybredyn *consanguinitas,* Cath. Ang. 338, where see note. See also sib-rit, sibbe-ridge, -red banns of marriage. E. D. S. Pub. B. 16.]

sib-sum ; *adj. Peaceable, pacific, friendly* :—Sibsum *pacificus,* Rtl. 39, 9. Eálá ðú sóða and ðú sibsuma, Crist ælmihtig, Exon. Th. 14, 5 ; Cri. 214. Ða Gotan lustlíce sibbsumes friðes æt eów biddende sindon *the Goths willingly ask for a friendly peace at your hands ;* Gothi societatem Romani foederis precibus sperant, Ors. 1, 10 ; Swt. 48, 22. Sibsume ł friðgeorne (ł friðsume, Rush.) *pacifici,* Mt. Kmbl. Lind. 5, 9. [*O. H. Ger.* sippi-sam *pacificus.*] v. ge-sibsum.

sibsumian. v. ge-sibsumian.

sibsumlíce ; *adv. Peaceably, in peace* :—Sibsumlíce gebunden mid dínum bebode, Bt. 33, 4 ; Fox 128, 31. v. ge-sibsumlíce.

sibsumness, e ; *f. Peaceableness, peace, tranquillity* :—Hē ðæt ríce heóld on gódre geþuærnesse and on micelre sibsumnesse, Chr. 860 ; Erl. 70, 24. Lufa sibsumnysse and geþwǽrnysse, Wulfst. 247, 1. [Þa weren alle mid sibsumnesse, O. E. Homl. i. 91, 17. Sibsumnesse eu beo among *pax vobiscum,* Misc. 54, 599.] v. ge-sibsumness.

sibsumung, sibun. v. ge-sibsumung, seofon.

síc, es ; *n. : but* síce, es ; *m. seems also to occur. A sike.* '*Sike* a watercourse ;' applied to a natural as well as to an artificial stream ; the latter usually constructed to receive the contents of field gutters, for discharge into the river.' Mid-Yorks. Gloss. See also E. D. S. Pub. 13, 15, and Old Farming Words, III :—*Sike* a quillet or furrow. Jamieson gives *sike* a rill. Cuddie Headrigg says 'I took up the syke a wee bit.':—Of ðam mere west . . . ðonne innan ánne síce, ðonne andlangc síces ðæt cymþ tó ðam horpytte, Cod. Dip. Kmbl. iii. 37, 20–22. Of ðæm beorge on ðæt síc ; ondlong síces ofer ðone bróc, 38, 28 : 35, 7. In wǽtan síce ; of ðæm wǽtan síce in ða bakas, 382, 7 : 386, 11. In ðæt wǽte sícc ; of ðam síce, 386, 16. On ðæt eástre síc, 438, 28. In ðæt síc, 31, 12. [Syke *rivus,* Wrt. Voc. i. 195, col. 2. *Icel.* sík a ditch, trench : *O. H. Ger.* gi-síh *stagnum, lacus, palus* (cf. *Scott.* sike *a marshy bottom with a small stream running through it*), Grff. vi. 58.] Cf. seohtra.

sícan, sýcan ; *p.* te *To cause to suck, to suckle, give suck* :—Ðú sýcst hálgum breóste *lactas sacrato ubere,* Hymn. Surt. 75, 43. Ða breóst ðe ne síctun (sýctun, MS. A.) *ubera quae non lactauerunt,* Lk. Skt. 23, 29. v. ge-sícan, á-sícyd ; súcan.

sícan ; *p.* sác. **I.** *to sike* (still in dial. e. g. Lancashire), *sigh, groan* :—On mínum bedde ic síce and wēpe, Ps. Th. 6, 5. **II.** *to sigh for, long* :—Ðæt wǽron ða tída ðe ǽfter sícaþ *en tempora . . . quibus recordatio suspirat,* Ors. 2, 8 ; Swt. 92, 35. [*Seoruh*fulnesse made him siken sore, A. R. 110, 13. Wepenn & sikenn sare & suhhʒhenn, Orm. 7924. Þe king gon siche (sike, 2nd MS.) sare, Laym. 12772. He sikede, Jul. 20, 9. Sike, Horn. 426 : Havel. 291. She neither weep ne syked, Clerkes Tale 545. Thanne syked Sathan, Piers P. 18, 263.] v. á-, on-sícan, *and following words.*

siccettan. v. sicettan.

síce, es ; *m. A sigh, groan ; gemitus* :—Ic mē on Godes helde bebeode

widð ðane sāra[n] sice, wið ðane sāra[n] slege, wið ðane grymma[n] gryre ... and wið eal ðæt lāð ðe intó land fare, Lchdm. i. 388, 12. [He weorp a sic as a wiht þat sare were iwundet, Jul. 21, 12. He ne fecheð noht þe sore siches on neðerward his heorte, O. E. Homl. ii. 83, 26. Mid seoruhfule sikes, A. R. 284, 3. Ðor sat his moder in sik and sor, Gen. and Ex. 1239. With a sik she seyde, Tr. and Cr. 3, 207. Amang his sobbes and his sikes sore, 4, 50.]

sicel. v. sicol.

sicerian; p. ode To ooze, of a fluid, to make way through a small opening :—Swîðe lytlum sicerað ðæt wæter and swîðe ðêgellîce on ðæt hlece scip and ðeáh hit wilnaþ ðæs sió hlûde ŷð ðêþ on ðære hreón sǽ bûton hit mon ǽr ût âweorpe by very small quantities and with very great secrecy does the water make its way into the leaky ship, and yet it has the same intention as the loud wave in the rough sea, unless it be cast out beforehand ; hoc agit sentina latenter excrescens, quod patenter procella saeviens, Past. 57, 1 ; Swt. 437, 14. [Ger. sickern to ooze, trickle.]

sicet[t], es ; n. A sigh, groan :—On siccetum in gemitibus, Ps. Lamb. 30, 10.

sicet[t]an, siccet[t]an ; p. te. I. to sigh, groan :—Sicetit singultat, Wrt. Voc. ii. 120, 50. Ðâ begann se ealda siccetan and mid wôpe wearþ ofergoten, Ælfc. T. Grn. 18, 1. II. as opposed to expressing grief by speech (?) :—Ða unryhtwîsan sicettaþ (siccettaþ, Cott. MSS.) on ðǽm þiéstrum impii in tenebris conticescent, Past. 11, 1 ; Swt. 65, 12. Siccitan conticiscent, silebant, Wrt. Voc. ii. 135, 15.

sicet[t]ung, siccet[t]ung, e ; f. A sigh, sob, heavy or short breathing, sighing :—Siccetung suspirium, Wrt. Voc. i. 19, 34. Siccitung singultus, 46, 19. Mê ðiós siccetung hafaþ âgǽled, ðes geocsa, Met. 2, 4. Mîn geár wǽron on sicetunga and on gestæne (in gemitibus), Ps. Th. 30, 11. Sicetunge singultu, Hpt. Gl. 514, 66. In sicettunge and geoxunge in singultum, Wrt. Voc. ii. 46, 8. Getogene sicetunge ducta suspiria, Hpt. Gl. 511, 41. Heófunga sicetungum lamentorum singultibus, 472, 57. Siccitungum, 504, 63. Hê angsumlîce siccetunga teáh swâ ðæt hê earfoðlîce orðian mihte he drew his breath painfully and heavily, so that he could hardly breathe, Homl. Th. i. 86, 8. Hê wearþ ðâ gesícelod and siccetunga teáh of niwellîcum breóste on bedde licgende he fell ill and drew sighs from the bottom of his heart, as he lay in his bed, Homl. Skt. i. 7, 65.

Sicilie; pl. The Sicilians, the people of Sicily, or (as in the older stage of the language the name of a people was used where now that of their country is put) Sicily. [In this sense the Latin form also occurs :—Sicilia, églond micel, Met. 1, 15. Sicilia ðæt îgland is þryscŷte, Ors. 1, 1 ; Swt. 28, 2. On Sicilia ðæm londe, 2, 6 ; Swt. 88, 31. Betwux ðâm muntum and Sicilia ðam eálonde, Bt. 1 ; Fox 2, 4] :—Sicilie ungeráðe wǽron him betweónum, Ors. 2, 7 ; Swt. 90, 6. Hit Sicilia fela ofslôg, 2, 6 ; Swt. 88, 32. Sicilia folc, burh, 4, 6 ; Swt. 170, 20, 30. Sicilia îglond insulas Siciliae, Swt. 172, 30. On Sicilium in Sicilia, 4, 4 ; Swt. 164, 23 : 5, 3 ; Swt. 222, 27. Of Sicilium ex Sicilia, 4, 6 ; Swt. 174, 20. Hî wunnon on Sicilie (adversus Siculos), 4, 5 ; Swt. 168, 19. Hê gefôr mid firde an Sicilie cum in Sicilia bellum gereret, Swt. 166, 6 : 4, 10 ; Swt. 194, 3.

Sicilisc; adj. Sicilian : — Sicili[s]c, Sicul inberdli(n)c † burhleód, Sicilisc inbyrdlincg siculus indigena, Sici[li]ensis incivis, Hpt. Gl. 499, 35–39. Se Sicilisca Siculus, Wrt. Voc. ii. 84, 26.

siclian, sícelian ; p. ode To sicken, be or fall sick :—Lange hê sîclaþ diu egrotat, Lchdm. iii. 151, 8. Sícclaþ (sîclaþ, MS. T.), 13. [Ðâ wæs Leófric abbot of Burh æt þ ilca feord, and sæclode ðǽr, and com hâm, and wæs dǽd sône ðǽr æfter, Chr. 1066 ; Erl. 203, 12. Þat ilce ðæi þat Martin abbot of Burch sculde þider faren, þa sæclede hê & ward dêd .iv. no. Jañ., 1154 ; Erl. 266, 10.] [Leste oure soule secli, A. R. 50, 20. O. H. Ger. siechelôn languere.] v. ge-síclian.

sicol (-el, -ul), es ; m. A sickle :—Ðes sicol haec falx, Ælfc. Gr. 9, 72 ; Zup. 73, 6 : Wrt. Voc. i. 85, 2 : falciola vel falcicula, 34, 63. Sicul falx, ii. 146, 77. Sicel baxus, 12, 53 : Wülck. Gl. 193, 9. Ne rîp ðû nâ mid sicele (falce), Deut. 23, 25. Hê sent his sicol mittit falcem, Mt. Skt. 4, 29. Hê sceal sicol habban, Anglia ix. 263, 5. [O. H. Ger. sihhila ; f. falx, falcicula : Ger. sichel. Probably from Latin secula.]

sicor; adj. with gen. Secure from, free from guilt and the punishment it brings, safe, free from danger or harm, sure, certain, free from doubt :—Swâ ûs biþ æt Gode ðonne wê wið hine gesyngiaþ ; ðeáh wê nǽfre eft swâ ne dôn, gif wê ðæt gedône mid nânum þingum ne bêtaþ ne we hreówsiaþ, ne bió wê nô ðæs sicore ; gif ûs ðæt ne mislîcaþ ðæt ûs ǽr lîcode, ðonne ne biþ hit nô ûs færgiefen. Ðeáh wê nû nâuht yfeles ne dôn on ðisse worulde, ne sculon wê ðeáh forðŷ bión tô orsorge, gif wê nâuht tô gôde ne dôþ ; forðæmðe swîðe fela unâléfedes wê oft geþenceaþ. Hû mæg se ðonne bión orsorg, se ðe him self wât, ðæt hê gesyngaþ ita et cum Deo delinquimus, nequaquam satisfacimus, si ab iniquitate cessamus, nisi voluptates quoque, quas dileximus, e contrario appositis lamentis insequamur. Si enim nulla nos in hac vita operum culpa maculasset, nequaquam nobis hic adhuc degentibus ipsa ad securitatem innocentia nostra sufficeret ; quia illicita animum multa pulsarent. Qua ergo

mente securus est, qui perpetratis iniquitatibus ipse sibi testis est, quia innocens non est? Past. 54, 5 ; Swt. 425, 3, 10. [Hi harm hadde, hii wende þat hii siker were, Laym. 9401 (2nd MS.). Dead is þe king, & siker þu miht hider comen, 15092. Wâ wes Brutten þere, þenne heo wenden beon sikere, 29289. Be þu sikerr þatt te shall þe ʒifenn eche blisse, Orm. 4844. Beoð ancren wise, þet habbeð wel bituned ham aʒein þe helle leun, uorte beon þe sikerure, A. R. 164, 12. Ne migten he siker ben, for magnie of ðo woren ouertaken, Gen. and Ex. 876. Þat ich mowe a siker bold arere, R. Glouc. 116, 1. Syker þou be Engelond ys nou þyn, 359, 9. Hit is sikerest in þi heeued (safest to sprinkle water on the head at baptism), Shoreham. Þai salle þare syker and certayne To have endeless joy, Pr. C. 8559. A man hath most honour To deyen ... whan he is siker of his goode name, Chauc. Kn. T. 2191. Her none sikerer þan other, Piers P. 12, 162 note. O. Frs. sikur (-er) free from guilt ; sure, trustworthy : O. Sax. (sundiono) sikur (-or) : O. H. Ger. sihhur securus, immunis, liber, tutus. From Latin securus.]

sîd; adj. I. wide, broad, spacious, ample, extensive. (a) applied to the world, universe, ocean, etc.:—Ðiós sîde gesceaft þênaþ and þíowaþ the wide world ministers and serves, Met. 29, 76. Eorþe and sîd wæter earth and ocean broad, Cd. Th. 7, 2 ; Gen. 100. Geseah sceado swiðrian geond sîdne grund, 8, 35 ; Gen. 134. Sǽs sîdne grund, Exon. Th. 349, 2 ; Sch. 40. Geond sîdne sǽ, 53, 19 ; Cri. 853. Sǽs sîdne fæðm, Elen. Kmbl. 1454 ; El. 729. Is ðæs fŷres frumstôl ofer eallum ôðrum gesceaftum geond ðisne sîdne grund, Met. 20, 127. (b) applied to a tract of land, to a kingdom, etc., v. sîd-land :—Sîde rîce a broad realm, Beo. Th. 4404 ; B. 2199. Nyttade Noe mid sunum sînum sîdan rîces, Cd. Th. 96, 24 ; Gen. 1599. Unlytel dǽl sîdre foldan (the district of Sodom and Gomorrah), 154, 5 ; Gen. 2551. Sennar sîdne and wîdne Shinar's plain broad and wide, 99, 33 ; Gen. 1655. Sîde sǽlwongas, 78, 14 ; Gen. 1293. Sîde sǽnæssas, Beo. Th. 451 ; B. 223. Hê wealdeþ sîdum rîcum he shall rule broad realms, Ps. Th. 71, 8. (c) applied to a comparatively small surface :—Ic bere sîdne scyld, Beo. Th. 879 ; B. 437. Setton sîde scyldas wið weal, 656 ; B. 325. Sîde weallas, Exon. Th. 1, 9 ; Cri. 5. (d) applied to a number of people who cover a wide space, v. sîd-folc :—Sêcan sîde herge, Exon. Th. 33, 12 ; Cri. 524. Weorode, sîde herge, Beo. Th. 4683 ; B. 2347. Sîde worude (? worulde, MS.), Cd. Th. 118, 11 ; Gen. 1963. Ofer sîd weorod, Elen. Kmbl. 316 ; El. 158. Sîde þeóde, Ps. Th. 117, 10. Sîde hergas, Cd. Th. 194, 14 ; Exod. 260 : Andr. Kmbl. 1304 ; An. 652. (e) figuratively, far-reaching, large :—Geþolode wine Scyldinga weána gehwylcne, sîdra sorga, Beo. Th. 300 ; B. 149. Ic worn hæbbe sîdra sorga gehŷred, Exon. Th. 11, 13 ; Cri. 170. Ne behwylfan mæg heofon and eorþe his wuldres word wîddra and sîddra ðonne befæðman mæge foldan sceátas (stretching too far and wide to be embraced), Cd. Th. 204, 31 ; Exod. 427. II. capacious, ample, spacious, large :—Glôf sîd, Beo. Th. 4178 ; B. 2086. In sîdum ceóle, Exon. Th. 345, 10 ; Gn. Ex. 186. On ðyssum sîdan sele, Cd. Th. 273, 3 ; Sat. 131. Geond ðæt sîde sel, Andr. Kmbl. 1523 ; An. 763. Con hê sîdne ræced fæste gefêgan, Exon. Th. 296, 7 ; Crä. 47. II a. figuratively of the capacity of the mind :—On sîdum sefan, Exon. Th. 169, 17 ; Gû. 1096. Þurh sîdne sefan, Beo. Th. 3456 ; B. 1726. Sefan sîdne geþanc and snytro cræft, Cd. Th. 249, 26 ; Dan. 536. III. long, hanging, of ample length, of clothes, hair, etc., v. sîd-feax :—Sîd reáf swilce mêteras wyrceþ on anlícnesse toga, Wrt. Voc. i. 41, 3. Iohannes geseah ûrne Drihten mid alban gescrîdne, and seó wæs sîd niðer ôþ ða andcleówa (it reached down to the ancles, cf. Icel. kné-, skó-sîðr reaching to the knee, the shoes (of dress), L. Ælfc. P. 15 ; Th. ii. 370, 3. Herebyrne sîd (cf. Icel. brynja rûm ok sîd), Beo. Th. 2892 ; B. 1444. Mid sîdum bearde (cf. Icel. sîtt skegg), Homl. Th. i. 466, 24. Sîde beardas, 456, 18. Se beard and ðæt feax him wǽron ôþ ða fêt sîde (cf. Icel. lokkar sîðir til jarðar), Shrn. 120, 25. Hî habbaþ beardas ôþ cneów sîde and feax ôþ helan barbas habentes usque ad genua, comas usque ad talos, Nar. 35, 2 : 38, 8. Wîf habbaþ beardas swâ sîde ôþ heora breóst, 38, 2. [Now wers men short and now syde, Pr. C. 1534. Syyd, as clothys talaris, Prompt. Parv. 455 where see note. See also Halliwell Dict. side. Icel. sîðr long, hanging.]

sîd-ádl, e ; f. Pleurisy :—On sîdan lama vel sîdádl pleuriticus, Wrt. Voc. i. 19, 31. Cf. sîd-wærc.

sîdan; adv. From a wide area :—Of gehwilcum stôwum wŷdan and sŷdan gegaderod, Cod. Dip. B. ii. 389, 23. Cf. next word.

sîde; adv. Widely, extensively, amply :—Sîde prolixius, Hpt. Gl. 526, 60. ¶ The word generally occurs along with wîde, far and wide :—Sîde and wîde longe lateque, Wrt. Voc. ii. 53, 59 : Cd. Th. 8, 3 ; Gen. 118: El. 554 ; El. 277. Hê Godes lof rǽrde wîde and sîde, Chr. 959 ; Erl. 119, 26 : Cd. Th. 1, 20 : Gen. 101. Is wuldur ðîn wîde and sîde ofer ðâs eorþan ealle in omnem terram gloria tua, Ps. Th. 56, 6, 13. Gesamnaðon weras wîde and sîde, Andr. Kmbl. 3273 ; An. 1639. Cyningas hine wîde worðodon wîde and sîde, Chr. 975 ; Erl. 125, 23. Ealra lêca ðæra ðe gewurde wîde oððe sîde, Hy. 1, 7. [Þis wes itald wide and side, Laym. 29902. Wide and side spelledd iss, Orm. 5900. Sidder (hanging) lower, Piers P. 5, 193.] Cf. preceding word.

sîde, an ; f. I. a side, flank, of living things :—Sîde latus, Wrt.

Voc. i. 44, 24 : ii. 51, 72 : *lumbus*, 113, 29. Wið ðære swiðran sídan sáre and ðære winestran, Lchdm. ii. 6, 3. On sídan lama *pleuriticus*, Wrt. Voc. i. 19, 31. Hé Hælend genom be sídan, Cd. Th. 299, 5 ; Sat. 545. Hit (*the horse*) ongan walwian and on gehwæðere sídan hit oferweorpan (*in diversum latus vicissim sese volvere*), Bd. 3, 9 ; S. 533, 40. Án ðæra cempena geopenode his sídan (sídu, Lind. : sído, Rush.) mid spere, Jn. Skt. 19, 34. Sídan (ða sídu ł ðæt sídu, Lind. : ða sído, Rush.) *latus*, 20, 20. II. *side* of a house, ship, etc. :—Duru ðú setst be ðære sídan (*the side of the ark*), Gen. 6, 16 : Past. 22 ; Swt. 169, 24. Ðæt scyp on sídan licgende, Bd. 5, 9 ; S. 623, 21. III. marking direction on this or that *side* :—Ðeós þridde India hæfþ on ánre sídan þeóstru, on óðere gársecg, Homl. Th. i. 454, 14. Æfre byþ on sumre sídan ðære eorþan dæg, and æfre on sumre sídan niht, Lchdm. iii. 234, 27 : Anglia viii. 319, 39. IV. of descent, cf. on the father's, mother's *side* :—Hig wæron ácennede of Constantines sídan, ðæt ys of gestreónde, Shrn. 97, 6. [O. Sax. síða : O. Frs. síde : O. H. Ger. síta : Icel. síða.]

síde, an ; f. *Silk* :—Sídan *sericum*, Coll. Monast. Th. 27, 7. [O. H. Ger. sída *sericum* : Ger. seide. From *Mid. Lat.* seta. Cf. *Span.* seda : *Ital.* seta : *Fr.* soie.] v. síd-wyrm, síden.

síd-ece, es ; m. *Side-ache* :—Drenc wið sídece, Lchdm. iii. 48, 9, 18.

síde-ful[1] ; adj. I. of good behaviour or manners, honest, modest, virtuous, sober :—Sídeful *pudicus*, Wrt. Voc. i. 51, 33. Se árfæst snoter eádmod sidefull séfre clæne wæs *qui pius, prudens, humilis, pudicus, sobrius, castus fuit*, Hymn. Surt. 137, 1. Sidefull mann and mid þeáwum gefrætwod, Homl. Th. i. 596, 31. Sideful *pudica, casta*, Hpt. Gl. 439, 16. On ánre tíde twá mædencild cumaþ, and biþ ðær án sydefull and ðæt óðer sceandlíc, Homl. Skt. i. 5, 280. Sidefulre *pudicae, castae*, Hpt. Gl. 428, 48. Ða heáhfæderas wæron sidefulle on þeáwum and syferlíce lybbende, Homl. As. 37, 327. Wé witon ðæt manega sydefulle clericas (*many honest clerks*) nyton hwæt byþ *quadrans*, Anglia viii. 306, 27. II. of dress, sober, modest, decorous :—Mid háligre drohtnunge and sidefullum gyrlan, Homl. Th. i. 546, 25. [Sannte Marȝe wass shammfasst & daffte & sedefull, Orm. 2175.] v. un-sidefull.

sidefullíce ; adv. *Virtuously, decorously* :—Sidefullíce *honeste*, Germ. 389, 33.

sidefulness, e ; f. *Honesty, modesty, sobriety* :—Clænnyss and sidefulnys eówres líchaman and sáule *castitas atque sobrietas corporis simul et spiritus vestri*, Cod. Dip. B. i. 155, 13. Sidefulnysse *pudicitiae*, Hpt. Gl. 433, 56. Mæg[þ]hádlícere sidefulnysse *pudicitiae* (*castitatis*) *virginalis*, 440, 65 : 447, 9. Wífmen ne beón bútan sidefulnysse, Homl. Skt. i. 13, 120. v. un-sidefulness.

síde-líc ; adj. *Sober, sedate, modest* :—Of sidelícre ansýne *serio*, Germ. 389, 36. [O. H. Ger. situ-líh *moralis, deliberatus* : Ger. sitt-lich : Icel. sið-ligr *well-bred*.]

sidelíce ; adv. *In a proper manner, suitably* :—Monige scylda openlíce witene beóþ tó forberanne ðonne ðæs þinges tíma ne biþ ðæt hit mon sidelíce gebétan mæge . . . Ac ðonne se láreów ieldende sécþ ðone tíman ðe he his hiéremenn sidelíce on þreátigean mæge . . . *nonnulla aperte cognita mature toleranda sunt, cum rerum minime opportunitas congruit, ut aperte corrigantur . . . Sed cum tempus subditis ad correptionem quaeritur . . .*, Past. 21, 2 ; Swt. 153, 1–6. [O. H. Ger. situlího *rite* : Icel. siðliga *nicely*.]

síden ; adj. *Silken, of silk* :—Síden *sericum*, Hpt. Gl. 417, 34. [O. H. Ger. sédín *sericeus* : Ger. seiden.]

-síden[n]. v. ælf-síden.

sidesa (?), sidsa, an ; m. *A charm* (?), *magical influence* (?) :—Wið ælfe and wið uncúþum sidsan, Lchdm. ii. 296, 10. [Cf. (?) *Icel.* síða *to work a charm* ; seiðr *a spell, charm, enchantment*.] Cf. ælf-síden.

sideware, an ; f. *Zedoary* :—Nim sidewaran, Lchdm. iii. 10, 30. [O. H. Ger. sitiwar *citawar* ; Ger. zitwer : *Low Lat.* zedoaria, zeduarium (v. hoc zeduarium zeduarye, Wrt. Voc. i. 227, col. 1) *from Arabic* zedwár. *From a French form* citoual *comes Mid. E.* zeduale, A. R. 370, 11, cetewale, Chauc. Group B 1951, see Skeat's note on the passage.]

síd-fæðme ; adj. *Broad of bosom, of a ship, broad in the beam* :—Hé sælde tó sande sídfæðme scip, Beo. Th. 3839 ; B. 1917. Cf. wíd-fæðme.

síd-fæðmed ; adj. *Broad-bosomed, broad-beamed* :—Seomode on sole sídfæðmed scip, Beo. Th. 610 ; B. 302.

síd-feax, -feaxe, -fexe ; adj. *With long hair* :—Absalon wæs sídfeaxe, Homl. Skt. i. 19, 221 MS. U. Sídfexe *capillatus*, Ælfc. Gr. 43 ; Zup. 256, 10 note. Hí lange tíd eodon ealle unscorene and sídfeaxe, Th. Ap. 6, 12. Sume gáþ sídfeaxe, ðæt hý þurh ðæt wiðmetene sýn Samuele and Elian and óðerum hálgum ðe sídfeaxe wæron, R. Ben. 135, 27–30. v. síd, III, *and next word*.

síd-feaxode, -fexede ; adj. *Long-haired* :—Absalon wæs sídfæxede (-feaxode, MS. D.), Homl. Skt. i. 19, 221. Sídfexede *capillatus*, Ælfc. Gr. 43 ; Zup. 256, 10. v. *preceding word*.

síd-folc, es ; n. *A people occupying an extensive space*, (1) *a multitude* :—Sídfolc micel (*the multitude that accompanied St. Juliana's body*),

Exon. Th. 284, 4 ; Jul. 692. (2) *a great people, great nation* :—God hí of sídfolcum gesamnade *Dominus de regionibus congregavit eos*, Ps. Th. 106, 2. v. síd, I d ; wíd-folc.

sídian ; p. ode *To make* or *to become wide, ample* (síd) :—Sídaþ, Exon. Th. 354, 53 ; Reim. 65. v. be-sídian ; sídung.

síd-land, es ; n. *A broad, spacious land* :—Sceal fromcynne folde ðíne, sídland manig, geseted wurðan, Cd. Th. 133, 3 ; Gen. 2205. Sæs and sídland, 148, 3 ; Gen. 2451. Cf. wíd-land.

sídling-weg, es ; m. *A road that runs obliquely* (?) :—Ofer feld on ða rihtlandgemære on ðone sídlingweg tó wuda, Cod. Dip. Kmbl. iii. 446, 19. Cf. Halliwell Dict. *sidelings aslant, sideways* : Jamieson *sideling, oblique* ; *sydlingis obliquely, not directly*.

sído. v. sidu.

síd-rand, es ; m. *A broad shield* :—Ðá wæs on healle heardecg togen sweord ofer setlum, sídrand manig hafen, Beo. Th. 2583 ; B. 1289.

sidsan, Lchdm. ii. 296, 10. v. sidesa.

sidu, seodu, siodu (o) ; *gen. dat.* a ; m. I. *a custom, use, manner, habit, practice* :—Ðæt heó cóme tó him mid hire cynehelme, swá swá heora seodu wæs, Anglia ix. 28, 31. Micel sido mid Rómwarum wæs, Bt. 27, 1 ; Fox 96, 2. Se sido ðe sume men secgaþ ðæt [hé] sié méde wyrðe, sume men secgaþ ðæt hé sié wýtes wyrðe, 39, 9 ; Fox 226, 4. Hé dyde him ðæt ríceter tó sida (sioda, Cott. MSS.) and tó gewunan *ministerium regiminis vertit in usum dominationis*, Past. 121, 9 ; Swt. 121, 19. Ðú ne meaht hiora sidu and heora gecynd onwendan, Bt. 7, 2 ; Fox 18, 30. God gesette unáwendendlícne sido and þeáwas his gesceaftum, 21 ; Fox 74, 1 : Met. 11, 12. Þeóda swíðe ungelíca ægðer ge on spræce ge on þeáwum ge on eallum sídum *nationes lingua, moribus, totius vitae ratione distantes*, Bt. 18, 2 ; Fox 62, 30. I a. *a religious practice, a rite* (cf. *Icel.* siðr *religion, faith*, Kristinn, heiðinn siðr *Christianity, heathenism*) :—Moyses wolde Obab ob ðæs hæðendómes siðum áládan *cum Hobab a gentilitatis conversatione vellet educere*, Past. 41, 5 ; Swt. 304, 9. II. *good conduct, morality, modesty* :—Hádlícere side (fæmnhádlícere sidefulnysse (?) v. sidefulness) *virginalis pudicitiae* (*castitatis*), Hpt. Gl. 449, 4. Side (? -fulnysse) *pudicitia, castitate*, 454, 53. Ðá kyningas (*of England*) ægðer ge hiora sibbe ge hiora siodo (sido, Cott. MSS.) ge hiora onweald gehióldon *the kings maintained peace, morality, and power*, Past. pref. ; Swt. 3, 7. Gif hé þurh cúscne siodo læst mína lára *if by modest conduct he carry out my instructions*, Cd. Th. 39, 2 ; Gen. 618. [*Goth.* sidus góds *boni mores* : O. Sax. sidu *a custom* : O. Frs. side : O. H. Ger. situ *mos, consuetudo, habitus, usus, ritus, indoles, moralitas* : Icel. siðr.] v. land-, un-sidu.

sídung, e ; f. *An extension, augmentation* :—Ymbe ðises bissextus gefyllednysse wé wyllaþ rúmlícor iungum cnihtum geopenian . . . ðæt hig syððan his sýdung óðrum gecýðon . . . De augmentatione bissexti, (*then follows the promised account*), Anglia viii. 306, 16.

síd-wærc, es ; m. *A pain in the side* :—Wið sídwærce, Lchdm. ii. 62, 24 : 256, 12 : iii. 20, 20.

síd-weg, es ; m. *A road that stretches far*; in the plural *distant parts* :—Ðá wæs gesamnod of sídwegum mægen unlytel, Elen. Kmbl. 564 ; El. 282. Fugla cynn on healfa gehwone heápum þringaþ sígaþ sídwegum contrahit in coetum sese genus omne volantum, Exon. Th. 221, 19 ; Ph. 337. Cf. wíd-weg.

síd-wyrm, es ; m. *A silk-worm* :—Siolucwyrm oððe sídwyrm *bombix*, Wrt. Voc. ii. 12, 21. Sýdwyrm, i. 24, 6. [O. H. Ger. sída-wurm.]

sió, siemle, sién (be), sién (*vision*), siendon, sient, siére, sierede, siex. v. sí, simle, sí, sín, sind, seár, sirwan, six.

sife, es ; n. *A sieve* :—Sibi *crebrum*, Wrt. Voc. ii. 105, 41. Sife *crebrum, cribellum*, 136, 62 : *cribrum*, i. 34, 41 : *cribra vel cribellum*, 83, 20. Lytel sife *cribellum*, 34, 42. Ásift smale þurh smæl sife *sift fine through a fine sieve*, Lchdm. ii. 94, 2 : 72, 28. Man sceal habban . . . syfa . . . hérsyfe, Anglia ix. 264, 13. [O. H. Ger. sib ; n. *cribrum, cribellum*.] v. hær-, windwig-sife.

sifer, sifeþ. v. sýfer, sibban.

sifeþa, seofoþa ; pl. f. : *but also* sifeþa, an ; m. I. *siftings, bran, chaff* :—Sifeþa *furfur*, Wrt. Voc. i. 67, 49 : *acus*, 83, 19. Sifiþan, siuida *furfures*, Txts. 65, 940. Syfeþa, Wrt. Voc. ii. 38, 75. Swá swá mon melo sift ; ðæt melo þurhcrýpþ ælc þyrel and ða siofoþa (syfeþa, Cott. MS.) weorþaþ ásyndred, Bt. 34, 11 ; Fox 152, 3. Genim ðæs wyrte sæd on ele gesodene and mid syfeþon gemencged, Lchdm. i. 282, 1. Dó seofoþa on sealt wæter, ii. 262, 13. Riges seofoþa, 48, 20. Oferwylle on ðam selfan ecede sifeþan, 250, 23. II. *useless seeds, tares* :—Áta ł sifþa ł unwæstm *zizania*, Mt. Kmbl. Lind. 13, 38. Sifþe, 13, 25. Sifþena *zizaniorum*, p. 17, 5. [Syvedys or brynne or palyys *furfur*, Prompt. Parv. 457.]

sífre. v. sýfre.

siftan ; p. te *To sift, pass through a sieve* :—Ic syfte *cribro*, Ælfc. Gr. 24 ; Zup. 137, 10. Siftiþ (-it, -id) *crebrat*, Txts. 55, 596. Syfteþ, Wrt. Voc. ii. 15, 44. Sift, 136, 61. Swá swá mon melo sift (seft, Cott. MS.), Bt. 34, 11 ; Fox 152, 2. Sifte, *cribraret*, Wrt. Voc. ii. 74, 3. Syfte, 15, 57. v. á-, be-, ge-siftan.

sifþa(e), sig. v. sifeþa, sí.

sig (?) *himself :*—Se ꝺe gebysmreþ sig *qui se polluerit*, L. Ecg. iv. 68, 16; Th. ii. 230, 14.

sígan ; *p.* sáh, *pl.* sigon; *pp.* sigen. **I.** *to pass from a higher to a lower position, to sink, descend, decline, fall down :*— Hē (*a man hung on a tree*) on wyrtruman sígeþ, fealleþ on foldan, Exon. Th. 328, 29; Vy. 25. Ðá hē on eorþan sáh *cadens in terram*, Bd. 3, 12; S. 537, 31. Hí áheówon ꝺæt treów ꝺæt hit brastliende sáh tó ꝺam hálgan were. Ðá worhte hē ongeán ꝺam hreósendum treówe róde tácn, Homl. Th. ii. 508, 33. Him sáh (*here, or from* seón (?), *but cf. Icel.* höfðu út sigit iðrin í þat sárit) se innoþ eall út, L. Ælfc. C. 3; Th. ii. 344, 6. Sitte gē sigewíf, sígaþ tó eorþan (*in a charm for bees*), Lchdm. i. 384, 24. Ðú gestaþoladest eorþan swíðe wundorlíce . . . náunwuht eorþlíces hí ne healt, ꝺæt hió ne síge, and nis hire ēꝺre tó feallanne ofdúne ꝺonne up, Bt. 33, 4; Fox 130, 37. Ne mæg hió hider ne ꝺider sígan, Met. 20, 165. Hit hreósan wile, sígan sond æfter rēne, 7, 23. Ic sígan lǣte wællregn ufan *I will cause to descend destructive rain from above*, Cd. Th. 81, 23 ; Gen. 1349. Gewāt se wilda fugel earce sēcan, wērig sígan tó handa hálgum rince, 88, 9; Gen. 1462. Sígende *præceps*, Germ. 399, 460. [Þe kinge sah to grunde (deide, 2nd MS.), Laym. 10255. Scal þi saule siꝨen to helle 14589.] **I a.** *to sink as the sun to its setting :*—Heó (*the sun*) síhþ tó ꝺam tácne (*Aries*) ōþ ǣfen, Anglia viii. 307, 20. Tungla torhtast tó sete sígeþ, Menol. Fox 221; Men. 112. Ealle stiorran sígaþ æfter sunnan under eorþan grund, Met. 29, 15. Sió æþele gesceaft (*the sun*) sáh tó setle, Chr. 937; Erl. 112, 17. [The sunne arist anes a dai and eft sigeꝺ, O. E. Homl. ii. 109, 22.] **I b.** *in a figurative sense :*— Ða men ꝺe sígaþ on ꝺisses middangeardes lufan ōþ ꝺæt hié áfeallaþ of hiera ryhtwísnessum *cadentes a sua rectitudine animas, atque in hujus mundi se delectatione reclinantes*, Past. 19, 1; Swt. 143, 16. Mē on sáh unrihtes feala *declinaverunt in me iniquitates*, Ps. Th. 54, 3. Swá swá wē sigon ǣr on ꝺæt unálíefede ōþ ꝺæt wē áfeóllon *qui per illicita defluendo cecidimus*, Past. 54, 5; Swt. 425, 15. Ðonne áginþ hē sylf sígan oꝺꝺe áfylþ *inclinavit se et cadet*, Ps. Th. 9, 30. Forlǣte heteníþa gehwone sígan, Exon. Th. 352, 23; Sch. 101. **II.** *to move towards a point* (cf. *to make a* descent *upon a place*) :—Fugla cynn on healfa gehwone heápum þringaþ sígaþ sídwegum *contrahit in coetum sese genus omne volantum*, Exon. Th. 221, 19; Ph. 337. Godwine sáh him ǣfre tówerd Lundenes mid his líþe ꝺæt hē com tó Súþgeweorce *Godwin kept moving towards London with his force until he came to Southwark*, Chr. 1052; Erl. 184, 19. Ðæt folc him sáh eall onbútan *the people pressed upon him on all sides*, Homl. Skt. i. 23, 650. Eall seó burhwaru sáh út ætgædere ongeán ꝺæs cáseres tócyme *the whole town moved out together in the direction of the emperor's approach*, 814. Guman sigon ætsomne, Beo. Th. 619; B. 307. Gif ꝺú ne wilt wirde steóran ac on selfwille sígan lǣtest (cf. *gif* seó wyrd *swá* hweorfan mót on yfelra manna gewill, and ꝺú heore nelt stýran, Bt. 4; Fox 8, 18), Met. 4, 50. Him englas tógeánes heápum cwóman sígan, Exon. Th. 34, 30; Cri. 550. [Engles siþen in heouene, Jul. 77, 7. Heo siꝨen to his hærme, Laym. 8682. Forð heo gunnen siꝨen, 29071.] **II a.** *of the movement of time :*—Iunius síhþ tó mancynne . . . Agustus síhþ tó mannum, Anglia viii. 311, 6–17. Solmónaþ sígeþ tó túne, Menol. Fox 32; Men. 16. **II b.** *figurative :*—Sigon tó slǣpe *they sank to sleep*, Beo. Th. 2506; B. 1251. Hine man þreáge mid teartran steóre ꝺæt is him síge on swingella wracu (*verberum vindicta in eum procedat*), R. Ben. 52, 7. [Wið þene sele brudgume þat siheꝺ alle selhꝺe of *from whom proceeds all happiness*, H. M. 47, 35.] **III.** *to ooze, run* as matter. v. seón :—Gif ꝺæt brægen út síge *if the brain protrude*, Lchdm. ii. 22, 19. Lǣt sígan út on sum fæt *let it drain out into a vessel*, iii. 48, 6. **IV.** *to strain, filter, act as a filter,* cf. (?) síɡere :—Sígende sond rēn swylgþ *bibulae arenae*, Bt. 12; Fox 36, 12, 16. [O. Sax. sígan *to sink* (of the sun); *to proceed:* O. Frs. síga : O. H. Ger. sígan *declinare, ruere:* Icel. síga *to sink down, slide.*] v. á-, ge-, on-sígan; sígend, *and* seón.

sigdi. v. síꝺe.

sige, es; *m. A fall, setting* of the sun :—Sió sunne ꝺonne hió on sige weorðeþ (cf. Bt. 25; Fox 88, 25), Met. 13, 111. v. niꝺer-sige.

sige, es; *m. Victory, triumph.* **I.** *success in war* :—Sige *victoria*, Wrt. Voc. i. 84, 19. Ic siges mihte eów sille, ꝺæt gē eów tó gamene feónda áfillaþ, Wulfst. 132, 19. Se cyng áhte siges geweald *victory remained with the king*, Chr. 1066; Erl. 201, 12. Hí mid mycele sige (*triumpho magno*) hám fóran, Bd. 1, 12; S. 480, 32. Palm getácnaþ syge, Homl. Th. i. 218, 11. Sige forgifan *to grant victory*, Bd. 2, 9; S. 511, 36: Elen. Kmbl. 288; El. 144. Sige syllan, Val. 2, 25. Sige habban *to conquer, be victorious*, Num. 31, 18. Hæfde sige *vincebat, superabat*, Ex. 17, 11. Ða Cretense hæfdon ðone grimlecan sige *cruentiorem victoriam Cretenses exercuerunt*, Ors. 1, 9; Swt. 42, 28. Sige gerēcan, gesleán, gewinnan *to gain the victory*, 3, 1; Swt. 96, 33: Bd. 1, 16; S. 484, 22: Num. 21, 1. Sige niman, onfón *to obtain the victory*, Chr. 800; Erl. 60, 9: 845; Erl. 66, 24: Bd. 1, 16; S. 484, 21. Hié ꝺæt an missenlícum sigum dreógende wǣron, Ors. 4, 7; Swt. 182, 3. Ðæt hié mec mid heán sigum (*cum sublimibus tropheis*) geweorðedon, Nar. 24, 24. **II.** *success in conflict :*—Siges *triumphi*, Hpt. Gl.

447, 76. Mid sigerlícum sige *triumphali tropheo*, 473, 41: Hymn. Surt. 44, 27. Sige onsendan *to make victorious*, Salm. Kmbl. 487; Sal. 244. Heó bád ꝺone ēcan sige, Bd. 4, 23; S. 593, 14. Sigas *triumphos*, Hymn. Surt. 47, 20: *victorias*, 129, 24: *trophea*, 131, 22. **II a.** *success in commerce :*—Oxan grasiende gesihþ sige ceápas (-es? *or* sigeceápas?) getácnaþ, oxan slápende gesihþ yfelnysse ceápes getácnaþ, Lchdm. iii. 200, 9. [The word occurs often as one of the components of proper names: e.g. see Txts. 512–513. SiꝨe (syꝨe, siꝨen) habben, Laym. 23896: 17409: 16199. SiꝨe winnenn, Orm. 5461. Sy *triumph*, Jul. 11, 16. **Goth.** sigis: *O. Sax.* sigi: *O. H. Ger.* sigi, sigu: *Icel.* sig.] v. weorc-, word-sige, *and* sigor.

sige-beác[e]n, es; *n.* **I.** *a sign* or *monument of victory gained, a trophy :*—Se palm is sigebeácen, Homl. Th. ii. 410. Ðǣr ꝺæt heofonlíce sigebeácen (*trophæum*) ārǣred beón sceolde, Bd. 3, 2; S. 524, 35. Æþelinges (*Christ*) ród, sigebeácen (cf. sige-beám, -bearn) sōꝺ, Elen. Kmbl. 1772; El. 888. Be ꝺam sigebeácne (*the cross*), 336; El. 168. Sēlest sigebeácna (*the cross*), 1946; El. 975. Sigebēcn, sigebeácn *tropea, signa*, Txts. 103, 2043. Ðis sigbēcn, 124, 2. **II.** *an ensign that is to lead to victory, a banner :*—Mid sigebeácne *vexillo, signo*, Hpt. Gl. 450, 35. Ārǣraþ eówer sigebēcn, and onginnaþ eówer gefeoht, Homl. Skt. i. 5, 59. v. sigor-beác[e]n.

sige-beáh ; *g.* -beáges ; *m. That which encircles the head of the victor, a crown :*—Hē onfēng sigebeáh (*coronam*) ēces lífes, Bd. 1, 7; S. 478, 34. Sigbēg, Jn. Skt. Lind. 19, 2: Rtl. 1, 15. Sigbēh, 6, 1.

sige-beám, es; *m. A tree on which a victory is gained,* generally *the cross :*—Se sigebeám *the cross*, Rood Kmbl. 25 ; Kr. 13 : 251; Kr. 127: Elen. Kmbl. 1927; El. 965. Be ꝺam sigebeáme, on ꝺam þrowode þeóda Waldend, 840; El. 420: 885; El. 444. Sēlest sigebeáma, 2053; El. 1028. Sigebeámas þrý (*the three crosses at the crucifixion*), 1691; El. 847. v. sige-beácen, -bearn.

sige-bearn, es; *n. A victorious child,* applied to Christ :—His gást onsende sigebearn Godes, Elen. Kmbl. 959; El. 481: Exon. Th. 460, 3; Hö. 11. Ðæt sygebearn, 461, 29; Hö. 43. Ealra sigebearna ꝺæt sēleste, 33, 3; Cri. 520.

sige-beorht ; *adj. Rendered illustrious by victory, triumphant :*—Hié swá sigebeorhte and swá gebǣgde mid mycelre blisse tó hám fóran, Blickl. Homl. 203, 30. Cf. *the proper name* Sigebryht, -berht, Chr. 755; Erl. 48, 18: Txts. 512. v. sigor-beorht.

sige-beorn, es; *m. A victorious warrior :*—Ne gefrægn ic æt wera hilde sixtig sigebeorna sēl gebǣran . . . Hig fuhton fíf dagas, swá hyra nán ne feól, Fins. Th. 76; Fin. 38. [Cf. *Icel.* Sig-björn (*proper name*).]

sige-bíme, an ; *f. A trumpet which is sounded after victory :*—Sungon sigebýman (*after the Israelites had escaped from the Egyptians*), Cd. Th. 214, 6; Exod. 565. [Cf. *Icel.* sigr-lúðr.]

sige-bróꝺor ; *m. A victorious brother,* used in speaking to St. Andrew of St. Matthew, who was not daunted by his heathen captors, Andr. Kmbl. 366; An. 183.

sige-cempa, an ; *m. A victorious warrior :*—Wæs Dauid æt wíge sōꝺ sigecempa, Ps. C. 50, 10.

sige-cwēn, e ; *f. A victorious queen,* applied to Elene, Elen. Kmbl. 519; El. 260: 1992; El. 998.

sige-dēma, an ; *m. A victorious, triumphant judge, the irresistible judge of the day of judgment :*—Se sigedēma, freá mihtig (*Christ*), Andr. Kmbl. 1322; An. 661. Ne beóþ ꝺǣr (*at the last judgment*) forþ borene sigele tó ꝺam sigedēman, Wulfst. 254, 1: Exon. Th. 65, 28; Cri. 1061.

sige-dryhten, es; *m. A victorious lord,* (1) as a complimentary epithet of an earthly chief :—Sigedrihten mín, aldor Eást-Dena, Beo. Th. 788; B. 391. Sigedryhten mín (*the departed Guthlac*), Exon. Th. 184, 24; Gú. 1349. Wit for uncrum sigedryhtne song áhófan, 324, 33; Víd. 104. (2) as an epithet of the Deity :—Þeóda Waldend, sigedryhten mín, Andr. Kmbl. 2905; An. 1455: Exon. Th. 176, 19; Gú. 1212: Ps. C. 50, 119. Þeóden engla, sōꝺ sigedrihten, Hy. 6, 34. Ðú eart selfa sigedrihten God, Met. 20, 260. Ðonc secgan sigedryhtne, ꝺæs ꝺe hē hine sylfne ús sendan wolde, Exon. Th. 9, 1; Cri. 128: Andr. Kmbl. 1753; An. 879. Sigedrihten, mihtigne God, Cd. Th. 33, 21; Gen. 523: 48, 20; Gen. 778. [O. Sax. sigi-drohtin (*applied to the Deity*).]

sige-eádig ; *adj. Blessed with victory, victorious :*—Sigeeádig bil, Beo. Th. 3119; B. 1557. [Cf. *Icel.* sigr-sæll.] v. sigor-eádig.

sige-fæst ; *adj. With victory secured, victorious, triumphant.* (1) applied to persons :—Sigefæst *victor*, Wrt. Voc. i. 84, 18. Sigfæst *triumphator*, Rtl. 122, 12. And hē sigefæst swá eft hám fērde *sicque victor in patriam reversus*, Bd. 2, 9; S. 512, 5: Exon. Th. 460, 26; Hö. 23. Þurh cyningces wísdóm folc wyrꝺ gesǣlig, gesundful and sigefæst, L. I. P. 2; Th. ii. 306, 5. Hē oflslóh mid ꝺam sigefæstan here eall ꝺæt mennisc, Jos. 10, 40. Hí sigefæste ofer sǣ férdon, Bd. 1, 12; S. 481, 15. Sigefæste *triumphabiles, triumpho plenos*, Hpt. Gl. 489, 33. Hý beóþ ꝺý gesundran and ꝺý sigefæstran, Exon. Th. 408, 29; Rä. 27, 19. Se sigefæstesta cyning *victoriosissimus rex*, Bd. 3, 7; S. 529, 16. (2) applied to things :—Sigefest wuldor, Hy. 8, 4. Sigefæstne

hám, Menol. Fox 298; Men. 150. Sigefæst tâcon *victricia signa*, Bd. 1, 8; S. 479, 24: H. R. 105, 21. Sigefæstan gûþfanan *victricia, victoriosa*, Hpt. Gl. 447, 54. v. sigor-fæst.

sigefæstan; *p. te To triumph :*—Sigefeston *triumphant*, Txts. 182, 77. v. ge-sigefæstan.

sigefæstness, e; *f. Victoriousness, triumph :*—Hê wîtgode be Cristes sigefæstnesse, ðâ ðâ hê on heofonas âstâh Ps. Th. 23, arg. Ðeáh ânra gehwylc wind hæbbe twelf sigefæstnissa, Salm. Kmbl. 152, 3. Sigefæstnissum *triumphis*, Rtl. 93, 7: 75, 19. v. sigorfæstness.

sige-folc, es; *n. A victorious* or *triumphant people :*—Heó (*Judith*) ðæt word âcwæþ tô ðam sigefolce (*the Jews who were about to destroy the Assyrians*), Judth. Thw. 23, 32; Jud. 152. Ðǽ wæs þeód on sǽlum, sigefolca swêg, Beo. Th. 1292; B. 644: Menol. Fox 593; Gn. C. 66.

sige-gealdor, es; *n. A charm that gives victory :*—Ic mê on ðisse gyrde belûce . . . wið eal ðæt lâð ðe intô land fare; sygegealdor ic begale, sigegyrd ic mê wege, Lchdm. i. 388, 14.

sige-gefeoht, es; *n. A victorious battle, a victory :*—On sigegefeohtum ellreordra cynna *in victories over foreign races; in expugnandis barbaris*, Bd. 3, 3; S. 525, 25.

sige-gird, e; *f. A rod that brings victory.* v. sige-gealdor.

sige-hrêmig; *adj. Exultant with victory, triumphant :*—Gesæt sigehrêmig on ða swiðran hand êce eádfruma (*Christ*) âgnum Fæder, Exon. Th. 33, 25; Cri. 531: Hy. 8, 30.

sige-hrêð *fame gained by victory :*—Onsǽl sigehrêð secgum *tell men of the fame you have won* (cf. the account of his deeds which Beowulf had given to Hrothgar), Beo. Th. 984; B. 490. Cf. gûþ-hrêð.

sige-hrêðig; *adj. Triumphant.* (1) applied to men :—Dômeádig cempa . . . sigehrêðig (*Guthlac*), Exon. Th. 146, 4; Gû. 704. Hig ne wêndon ðæt hê sigehrêðig sêcean côme þeóden *they did not expect that Beowulf would come triumphant* (*from his fight with Grendel's mother*) *and visit Hrothgar*, Beo. Th. 3198; B. 1597: 5505; B. 2756. (2) applied to the Deity :—Se Ælmihtiga . . . gesette sigehrêðig sunnan and mônan, 188; B. 94.

sige-hwîl, e; *f. A time of victory, the hour of victory :*—Wedra helm feónd gefylde . . . Ðæt ðam þeódne wæs sîðes sigehwîl, Beo. Th. 5413; B. 2710.

sigel, sægl, segl; *n.* (?) *The sun;* also *the name of the rune = S :*— ⅀ sǽmannum symble byþ on hihte (cf. Icelandic Runic poem—Sôl er landa ljómi), Runic pm. Kmbl. 342, 15; Rûn. 16. Woruldcandel scân, sigel sûþan fûs, Beo. Th. 3936; B. 1966. Wuldres tâcen swylce hâdre sægl, Andr. Kmbl. 178; An. 89. Hâdor sægl gewât under scrîðan, 2911; An. 1458. Heáfdes segl *the sun of the head, the eye* (cf. Icel. enni-máni, -tungl = *the eye*), 100; An. 50. [Goth. sauil; *n. the sun :* Icel. sôl; *f.*] v. heáðo-sigel; sigel-beorht, -hearwa, -hweorfa, -torht, -waras.

sigel, sigl; *n.* (?) *A clasp, brooch, jewel :*—Sigl, sigil *bulla*, Txts. 45, 331 : *fibula*, 63, 874: *sibba*, 97, 1856. Sigl *bulla, gemma*, Wrt. Voc. ii. 126, 70: *fibula*, 148, 57. Sigil *bulla*, i. 288, 7. Sigel, ii. 11, 34: *fibula*, 35, 42. [Cf. O. H. Ger. sigilla; *f. lunula. From Latin* (?) *sigillum.*] v. sigle.

sigel-beorht; *adj.* I. *sun-bright, bright with the sun, sunny :*—Wintres dæg sigelbeorhtne genimþ hærfest mid herige hrîmes and snáwes *winter's day takes captive sunny autumn with its army of frost and snow*, Menol. Fox 404; Men. 203. Bringþ sigelbeorhte dagas sumor tô tûne, 175; Men. 89. II. *bright as the sun :*—Sitt sigelbeorht swegles brytta on heáhsetle *ille sedens solio fulget sublimis in alto*, Dôm. L. 117. [*Icel.* sôl-bjartr.] Cf. sigel-torht.

sige-leán, es; *n. A reward of victory, prize, palm :*—Sigeleán ł edleán *palma*, Hpt. Gl. 482, 5: 432, 75 : *triumphus, palma*, 424, 53. Ðæt wê brûcan sigeleáne *ut perfruamur bravio*, Hymn. Surt. 129, 18. Simon and Thaddeus beornas beadorôfe sceoldon þurh wǽpenhete sigeleán sêcan, Apstls. Kmbl. 161; Ap. 81. Êce lîf, sélust sigeleána, Elen. Kmbl. 1051; El. 527. [Goth. sigis-laun *bravium*.] v. sigor-leán.

sige-leás; *adj.* I. *without victory, unsuccessful in conflict, defeated :*—Engle nû lange [wǽron] eal sigeleáse *the English now for a long time have been deserted by victory*, Wulfst. 162, 15. Hý sigeleáse (*defeated*) ðone grênan wong ofgiefan sceoldan, Exon. Th. 130, 33; Gû. 447 : 141, 6; Gû. 623 : Cd. Th. 20, 20; Gen. 312. I a. of an expedition, *unattended by victory :*—Sigeleásne sîð, Exon. Th. 120, 17; Gû. 273. I b. of a song, *that tells of defeat :*—Gehŷrdon galan Godes andsacan sigeleásne sang, Beo. Th. 1578; B. 787. [O. H. Ger. sigu-lôs.]

sige-leóþ, es; *n. A song of triumph :*—Ðâ wæs sigeleóþ (cf. Icel. sigr-ljóþ) galen on herefelda, Elen. Kmbl. 248; El. 124. Engla þreátas sigeleóþ sungon (*when Guthlac came to Heaven*), Exon. Th. 181, 6; Gû. 1289.

Sigel-hearwa (Sîl-), an; *m. An Ethiopian :*—Se deófol wearþ ǽteówod swylce ormǽte Sîlhearwa, Homl. Th. i. 466, 24. Hê him ǽtŷwde micelne Sigelhearwan, ðǽm wæs seó onsŷn sweartre ðonne hrûm, Shrn. 120, 24. Twegen blace Sîlhearwan, Homl. Th. ii. 496, 17 : Homl. Skt. i. 4, 285. Sigylhearwan (Sielhearwæn, MSS. T.) *Aethiopes*, Ps. Spl. 71, 9. Sigel-

hearwena (Sŷl-, Ps. Spl.) folc, Ps. Surt. 73, 14 : ii. p. 189, 36. Ethiopia, ðæt is ðæra Sîlhearwena rîce, Homl. Th. ii. 472, 13 : i. 454, 12. Ðæra Sîlhearwena land *terra Aethiopiae*, Gen. 2, 13. Sîllhearewena (Sîlhearwena, MSS. R. P.) land, Lchdm. iii. 258, 18. Ðû sealdest Sigelhearwan (-as, MS.) tô môse *dedisti in escam populo Aethiopum*, Ps. Th. 73, 14. Cf. Sigel-waras.

Sigelhearwen; *adj. Ethiopian :*—For his Sigelhearwenan wîfe *propter uxorem ejus Aethiopissam*, Num. 12, 1. Sŷlhearwenre, *Aethiopica*, Hpt. Gl. 514, 49.

sigel-hweorfa, an; *m. A plant-name, a word equivalent in meaning to the Greek heliotrope.* It is found as the representative of foreign words in the following :—Sigelhweorfa *heliotropus*, Wrt. Voc. i. 68, 5, 80: Lchdm. iii. 302, col. 1. Sigelhuerpha *eliotropia*, id. Sigelhueorua *nimphea*, 304, col. 1: *solsequia*, 305, col. 1. Sigelwearfa. Ðeós wyrt ðe Grêcas *heliotropus*, and Rômâne *uertamnum* nemnaþ, and eác Angle sigelhweorfa hâtaþ, Lchdm. i. 152, 21. Sigilhweorfa *eliotropus*, 254, 11. In the following no foreign equivalent is given :—Sigelhweorfa, ii. 94, 25 : iii. 24, 4. Nim nioþoweardne sigelhweorfan, 326, 17. See Lchdm. ii. 404, col. 2.

sigel-hweorfe, an; *f. A plant name :*—Sigelhwerfe *solsequium* vel *heliotropium*, Wrt. Voc. i. 30, 30. *Eleotropam*, Grece; Latine, *solsequium*, idem sigelhweorfe, ii. 32, 26. Nim sigelhweorfan ða smalan unwæscene, Lchdm. ii. 108, 23. v. preceding word.

sige-lîc; *adj. Victorious :*—Ða sigelîcan *victricia*, Wrt. Voc. ii. 78, 21: *victoria*, 92, 4. v. sigor-lîc.

sigel-torht; *adj. Bright with sunshine* or *bright as the sun*, cf. sigelbeorht :—Swâ wæs ealne dæg ôððæt ǽfen com sigeltorht (*epithet of ǽfen or of* Andrew?) swungen, Andr. Kmbl. 2493; An. 1248.

**Sigel-waras, -ware, -; *pl. The Ethiopians :*—Mannkynn sweartes hiwes . . . ða man hâteþ Sîlhearwan (Sigilwara, MS. V.), Nar. 38, 30. Hine Sigelwearas (*Aethiopes*) sêceaþ, Ps. Th. 71, 9. Folc Sigelwara *populus Aethiopum*, 86, 3. Sigelwara land, Cd. Th. 182, 2; Exod. 69. Hê (*St. Matthew*) gelǽrde Sigelwara mǽgþe, and of Sigelwarum hê flŷmde twegan drŷas, Shrn. 131, 27: Apstls. Kmbl. 127; Ap. 64. Cf. Sigelhearwa.

sige-mêce, es; *m. A victorious sword, a sword wielded by a victor's hand*, Exon. Th. 93, 24; Cri. 1531.

Sigen, e; *f. The Seine;* Sequana :—Andlang Sigene, Chr. 887; Erl. 84, 31. Be Sigene (Signe, MS. A.), 660; Erl. 35, 8: 897; Erl. 94, 28. [O. H. Ger. Sigana.]

sigend, es; *m. Movement of the sea, wave :*—Sîgend *flustra*, i. *undae*, Wrt. Voc. ii. 35, 62. Flôd *flustra*, sîgendum *flustris*, 33, 33 : 76, 63. v. sîgan.

sígere (?), es; *m. A glutton :*—Sîgiras (siras, Corpus Gl.) *lurcones, avidi*, Txts. 72, 568. v. sîgerian, *and cf.* (?) sîgan, IV.

sige-reáf, es; *n. A triumphal robe; toga palmata*, Wrt. Voc. i. 41, 4. **sigerian.** v. sigorian.

sígerian (?) *to act as a glutton :*—Sîgergendum *lurconibus*, Wrt. Voc. ii. 76, 34. v. sîgere.

sige-rîce; *adj. Victorious, triumphant.* (1) applied to the Deity :—Witig Drihten . . . sigerîce, Cd. Th. 179, 11; Exod. 27. (2) applied to men, cf. *prop.* name Sigerîc :—Gif gê (*the Israelites*) gehealdaþ hâlige lâre, gê gesittaþ sigerîce beórselas beorna, Cd. Th. 213, 34; Exod. 562. [*Ger.* sieg-reich.]

sige-rôf; *adj. Of victorious energy, triumphantly active.* (1) applied to a warrior or to a king :—Sigerôf kyning (*Hrothgar*), Beo. Th. 1243; B. 619 : (*Constantine*), Elen. Kmbl. 315; El. 158: 141; El. 70. Wǽron Rômware secgas sigerôfe, 93; El. 47 : Judth. Thw. 24, 8; Jud. 177. (2) without reference to battle :—Mîn yldra fæder sigerôf sægde, frôd fyrnwiota, Elen. Kmbl. 873; El. 437. Sigerôfne (*St. Andrew*), Andr. Kmbl. 2451; An. 1227. Gesǽton sigerôfe . . . rǽdþeahtende, Elen. Kmbl. 1732; El. 868. Sigerôfra (*the saints in glory*), Lchdm. i. 390, 4. (3) applied to the Deity :—Êce Dryhten, sigerôf settend, Exon. Th. 188, 17; Az. 47.

sige-sceorp, es; *n. Triumphal apparel*, Exon. Th. 341, 16; Gn. Ex. 127.

sige-sîþ, es; *m. A victorious expedition* or *journey :*—Oft dǽdlata dôme foreldit sigisîtha gahuem *generally the dilatory man is too late for glory, for every successful undertaking*, Txts. 152, 9.

sige-spêd, e; *f. Triumphant faculty, ability that gains its ends :*—Ðê God sealde sâwle sigespêd and snyttro cræft *God hath given thee effectual power of soul and wisdom's art*, Elen. Kmbl. 2341; El. 1172. Ic on ðê oncnâwe wîsdômes gewit, sigespêd geseald, Andr. Kmbl. 1291; An. 646. v. sigor-spêd.

sige-tâc[e]n, es; *n. A sign of victory :*—Ðæt hâlige sigetâcen (*the cross*), Blickl. Homl. 97, 13. Hê sigetâcen sende *misit signa*, Ps. Th. 134, 9. v. sigor-tâc[e]n, sige-beác[e]n.

sige-þeód, e; *f. A victorious people, a powerful people :*—Hyne g.sôhton on sigeþeóde hearde hildefrecan, Beo. Th. 4415; B. 2204 : Exon. 473, 23; Bo. 19. Sigeþeóda (*the victorious Goths*), Met. 1, 4. Secgeaþ his wuldor geond sigeþeóde (*inter gentes*), Ps. Th. 95, 3.

sige-þreát, es; *m. A triumphant band,* Exon. Th. 53, 2; Cri. 844.

sige-þúf (?), es; *m. A banner that conducts to victory, a victorious banner :*—Stópon secgas and gesíþas, bǽron þúfas (sigeþúfas?), Judth. Thw. 24, 22; Jud. 201. Cf. sige-beác[e]n.

sige-tiber, es; *n. A sacrifice for victory* (? cf. *Icel.* sigr-blót) :—Wolde líge gesyllan his swǽsne sunu tó sigetibre, Cd. Th. 203, 12; Exod. 402. v. sigor-tiber.

sige-torht; *adj. Splendid with victory, triumphant :*—Sigetorht árás éce Drihten, Cd. Th. 279, 19; Sat. 240. Cf. sige-beorht.

sige-tudor, es; *n. A victorious, triumphant progeny,* applied to the human race, Exon. Th. 154, 5; Gú. 838.

sige-wǽpen, es; *n. A weapon with which victory is won,* Beo. Th. 1612; B. 804.

sige-wang, es; *m. A plain where victory is won, a glorious plain.* (1) where actual fighting has taken place :—Se mǽsta dǽl ðæs heriges læg on ðam sigewonge, Judth. Thw. 25, 36. (2) where actual fighting is not referred to, *a place in which evil is overcome :*—Smeolt wæs se sigewang (*the place where St. Andrew's heathen enemies had been overwhelmed*), Andr.·Kmbl. 3160; An. 1583. Smylte is se sigewong (cf. *ðæt torhte lond,* l. 19, wlitig is se wong eall, 198, 8, *the dwelling-place of the Phenix*), Exon. Th. 199, 29; Ph. 33: 146, 23; Gú. 714. Mennisce áras on ðam sigewonge (*Guthlac's dwelling-place*) helpe gemétton, 157, 18; Gú. 893.

sige-wíf, es; *n. Grimm supposes this word may be a general denomination of wise women,* D. M. 402; *the passage in which it occurs is a charm, where it is addressed to bees when swarming :*—Sittaþ gē, sigewíf, sígaþ tó eorþan, Lchdm. i. 384, 24.

sigle, es; *n. A necklace, collar, band for the neck :*—Ne beóþ ðær forþ borene sigele ne beágas ne heora heáfodgold, Wulfst. 253, 23. Đá gemétte heó under hrægele gylden sigele (*monile*), Bd. 4, 23; S. 595, 5. Háma ætwæg Brósinga mene, sigle and sincfæt, Beo. Th. 2404; B. 1200. In mínum sweoran ic mē gemon beran ða ýdlan byrþenne gyldenra sigla *in collo me memini supervacua monitium pondera portare,* Bd. 4, 19; S. 589, 27: Beo. Th. 2318; B. 1157. Hí on beorg dydon bēg and siglu . . . hyrsta, 6308; B. 3164. [*Icel.* sigli *a necklace.*] v. máðum-sigle; sigel, in-sigle.

sigle, an; *f. Rye :*—Siglan dust, Lchdm. ii. 126, 7. [*Lat.* secale; *later* segale, sigalum, sigla: cf. *Ital.* segale: *Fr.* seigle.]

sigor, es; *m. Victory, triumph :*—Mē oferswíðde se wyrresta sigor, Shrn. 37, 24. Sigor eft áhwearf of norþmonna níðgeteóne, æsctír wera, Cd. Th. 124, 24; Gen. 2067. Sigores *palmam,* Wrt. Voc. ii. 67, 32. Mid sigores wuldre tó heofonum ástígan, Wulfst. 199, 13. Swegles ealdor hyre (*Judith*) sigores onleáh, Judth. Thw. 23, 16; Jud. 124. Sigere *tropheo,* Hpt. Gl. 508, 64. Elne gewurðod, dóme and sigore, Cd. Th. 129, 3; Gen. 2138. Hlísfulne sigor, *famosum tropheum,* Wrt. Voc. ii. 147, 29. Ic sceal his róde sigor (*the triumph of Christ's cross*) swíðor wíscan ðonne ondrǽdan, Homl. Th. i. 594, 20. Sigor æt sæcce, Elen. Kmbl. 2363; B. 1183. Folc ðe hē on deóflum genom þurh his sylfes sygor, Exon. Th. 36, 24; Cri. 581. Sigera *triumphorum,* Hpt. Gl. 425, 33. Ðyssum sigorum ðú Godes biscop blissian miht *hisque Dei consul factus laetare triumphis,* Bd. 2, 1; S. 500, 31. Mid ðǽm siogorum geweorðad *triumphans,* Nar. 28, 4. ¶ *The word occurs often in reference to the Deity* (cf. *in* Icel. Sig-föður *one of Odin's names,* sig-tívar *the gods of victory,* sigr-goð *a god of victory*) :—Swegles aldor se ðe sigor seleþ, Cd. Th. 170, 5; Gen. 2808. Ðæt hē sigora gehwæs ána weolde (cf. Hans (*Odin's*) menn trúðu því, at hann ætti heimilan sigr í hverri orrostu, Ynglinga Saga, c. 2), Exon. Th. 276, 5; Jul. 561. Sigores ágend, ealdor, freá, fruma, God, weard, Cd. Th. 307, 11; Sat. 678: Hy. 3, 20: Exon. Th. 25, 21; Cri. 404: 19, 2; Cri. 294: Andr. Kmbl. 1519; An. 761: Exon. Th. 15, 29; Cri. 243. Sigora dryhten, freá, God, sellend, settend, sóðcyning, waldend, weard, Cd. Th. 63, 23; Gen. 1036: Exon. Th. 242, 18; Ph. 675: Elen. Kmbl. 2613; El. 1308: Exon. Th. 359, 17; Pa. 64: Cd. Th. 237, 5; Dan. 333: Exon. Th. 75, 29; Cri. 1229: Cd. Th. 8, 19; Gen. 126: 106, 13; 1770. Bidde ic sigere (-a?, -es?) Godes miltse, Lchdm. i. 390, 10. [*Icel.* sigr.] v. hréð-, wíg-sigor; sige, *and following words.*

sigor-beác[e]n, es; *n. A symbol of victory,* applied to the cross, Elen. Kmbl. 1967; El. 985. v. sige-beác[e]n.

sigor-beorht; *adj. Triumphant,* epithet of Christ, Exon. Th. 1, 18; Cri. 10. v. sige-beorht.

sigor-cynn, es; *n. A triumphant, glorious race,* epithet of the Seraphim, Elen. Kmbl. 1506; El. 755.

sigor-eádig; *adj. Blessed with victory, victorious,* Beo. Th. 2626; B. 1311: 4693; B. 2352. v. sige-eádig.

sigor-fæst; *adj. Victorious, triumphant.* (1) as an epithet of the Deity :—Se Sunu (*Christ*) wæs sigorfæst on ðam síðfate, Rood Kmbl. 297; Kr. 150. God sigorfæst, Exon. Th. 217, 18; Ph. 282. (2) of an angel :—Meahtig Meotudes þegn, sigorfæst, 176, 30; Gú. 1218. (3) *of a passion :*—Brondhát lufu, sigorfæst in sefan, 160, 3; Gú. 938. v **sige-fæst.**

sigorfæstness, e; *f. Victoriousness :*—Be sigerfestnisse and swíðmódnisse úses Drihtnes mid ðǽm hē ða hǽþnan ofercom, Anglia xi. 173, 12. v. sigefæstness.

sigorian, sigerian, sigrian; *p.* ode *To vanquish, triumph over, triumph :* —Ic sigerie (sigerige, sigrige, sigrie) *triumpho,* Ælfc. Gr. 24; Zup. 137, 5. Ic sigrige be Cristes mádmum, and ic ðíne tintregu ne gefréde, Homl. Th. i. 424, 33. Fullfremed sóðlufu middaneardes sigoraþ ealdor *perfecta caritas mundi triumphat principem,* Hymn. Surt. 123, 38. Sigerode *triumphat,* 105, 32. Sigoriende *triumphans,* 85, 9. Sigriende, Germ. 395, 4. Sigerendes *triumphantis,* Hpt. Gl. 455, 64. [*O. H. Ger.* ubar-sigirôn *triumphare: Icel.* sigra : *Dan.* seire.]

sigoriend, sigriend, es; *m. A victor :*—Sigriend *victor,* Hymn. Surt. 38, 7.

sigor-leán, es; *n. A reward of victory, prize :*—Dryhten hyre (*Judith*) geaf sigorleán in swegles wuldre, Judth. Thw. 26, 26; Jud. 345. Sigorleán sécan, Exon. Th. 154, 29; Gú. 850: 184, 14; Gú. 1344. Sigorleán habban, Elen. Kmbl. 1246; El. 623. Sigorleánum onfón, Cd. Th. 176, 27; Gen. 2918. Tó sigorleánum sellan, Exon. Th.·97, 14; Cri. 1590. v. sige-leán.

sigor-líc; *adj. Triumphal :*—Ðæt sigorlíce leóþ *carmen triumphale,* Wrt. Voc. ii. 23, 48: Hpt. 438, 16. Mid sigerlícum sige *triumphali tropheo* (*victoria*), 473, 40. v. sige-líc.

sigor-spéd, e; *f. Abundant success :*—Is help gearu æt mǽrum, manna gehwylcum sigorspéd geseald, Andr. Kmbl. 1817; An. 911. Mē is miht ofer eall, sigorspéd geseald, 2868; An. 1437. v. sige-spéd.

sigor-tác[e]n, es; *n. A sign of victory, a convincing sign :*—Godspel bodian, secgan sigortácnum *to preach the gospel, tell it with convincing proofs* or *with marks shewing how it had prevailed,* Exon. Th. 169, 3; Gú. 1089. v. sige-tác[e]n.

sigor-tiber, es; *n. A sacrifice for victory* or *deliverance :*—Wes ðú on ófeste . . . ðæt ðú lác onsecge sigortifre *hasten to offer with a sacrifice, that may deliver you from your peril,* Exon. Th. 257, 30; Jul. 255. v. sige-tiber.

sigor-weorc, es; *A victorious work, a victory :*—Sigorworca hréð, Cd. Th. 198, 2; Exod. 316. [*Icel.* sigr-verk *a victory.*]

sigor-wuldor, es; *n. Triumphant glory, the glory of the victor :*—Háligra sáula gesittaþ in sigorwuldre Dryhtnes dreámas, Exon. Th. 109, 21; Gú. 93.

sigrian. v. sigorian.

sigsonte P *a plant name,* Lchdm. i. 74, 11 : 102, 24.

-siht, -sihte. v. ge-, in-, út-siht, blód-, út-sihte.

sihþ, e; *f. A vision :*—Bóc ðæra sigðana *apocalypsis,* Jn. Skt. p. 1, 11. Đa sihðe (gisihðe, Rush.) *quae vidissent,* Mk. Skt. Lind. 9, 9. v. æt-, ge-sihþ.

sihtre, silcen. v. seohtre, seolucen.

silf, silfren, Síl-hearwa, sillan. v. self, seolfren, Sigel-hearwa, sellan.

Sillende *Zealand,* Ors. 1, 1; Swt. 19, 20, 23.

sil-líc, silofor. v. seld-líc, seolfor.

siltan; *p.* te *To salt, season :*—Ic sylte *condio,* Ælfc. Gr. 30; Zup. 192, 13. Selt *condit,* Wrt. Voc. ii. 135, 55. On ðæm ðe gē hit syltaþ (*condistis*), Mk. Skt. 9, 50. Selte mon hiora mettas, Lchdm. ii. 234, 14. Láreówum gedafenaþ ðæt hí mid wísdómes sealte geleáffulra manna mód sylton, Homl. Th. ii. 536, 17. ge-, un-silt (-sylt).

síma, an; *m. A cord, rope :*—Satan læg símon gesǽled (cf. *Icel.* sím-bundinn), Cd. Th. 47, 23; Gen. 765. [*O. Sax.* símo *a cord: O. Frs.* sím: *Icel.* síma; *n.; cf. also* seimr *a string: Dan.* sime *a seton.*]

síman; *p.* de *To load, put a burden* (seám) *on :*—Gē sýmaþ (sēmaþ, Lind.) men mid ðám byrþenum . . . and gē ne áhrínaþ ða seámas mid eówrum ánum fingre *oneratis homines oneribus . . . et ipsi uno digito uestro non tangitis sarcinas,* Lk. Skt. 11, 46. Sýmaþ *onerant,* Engl. Stud. ix. 40. Hig sýmdon hira assan *oneratis asinis,* Gen. 44, 13. Sýmaþ eówre assan, 45, 17. v. ge-, ofer-síman (-sýman).

simbel, symbel, simel; *adj. Continual, perpetual.* [*The word occurs only in the adverbial forms* simbles, simble, on simbel (cf. on ídel), *and the compounds* simbel-farende, -geféra; *similarly O. H. Ger.* simpal *for the most part appears in adverbial forms, but Graff* vi. 26 *gives one instance of its adjective use,* simplêm *assiduus. Icel.* simul *ever, is preserved in only one or two passages*] :—On simbel *ever, always, continually* :—Hí hiora freóndscipe forþ on symbel gehealdaþ *they continue ever to maintain their friendship,* Met. 11, 94. Hió þyrstende wæs on symbel mannes blódes *she was continually thirsting for human blood;* haec, sanguinem sitiens, inter incessablia homicidia, Ors. 1, 2; Swt. 30, 27. Ðǽr se ríca hyne reste on symbel nihtes *where the ruler ever rested at night,* Judth. Thw. 22, 2; Jud. 44. v. following words.

simbel-farende; *adj.* (*ptcp.*) *Always travelling, wandering, roving :* —Đa simbelfarendan Æthiopes *Aethiopum gentes pervagantes,* Ors. 1, 1; Swt. 26, 16. v. next word.

simbel-geféra, an; *m. One who continually goes with another, a constant companion :*—Nis hit nó ðæt án ðæt swá eáðe mæg wiðerweard gesceaft wesan ætgædere symbelgeféran, ac hit is sellícre ðæt hiora ǽnig

ne mæg bûtan ôðrum biôn *it is not only that it is so easy for opposites to be able to be constant companions, but it is more extraordinary that no one can exist without another*, Met. 11, 50. v. preceding word.

simble, symble, simle, siemle, semle, symle; *adv. Ever, always.* **I.** *continually, continuously, without intermission.* (1) alone :—Symble mid ðê semper tecum, Ps. Th. 72, 18. Symble fŷr oððe gâr *ever fire or piercing cold*, Cd. Th. 20, 29; Gen. 316. Simle *diuturne*, Wrt. Voc. ii. 139, 23. Hié simle lôcigeaþ tô ðære eorþan *ad terram semper inclinantur*, Past. 21, 3; Swt. 155, 20. Hié wæron simle healfe æt hâm, healfe ûte, Chr. 894; Erl. 90, 17. In ðê sâule sôðfæstra simle gerestaþ, Exon. Th. 4, 16; Cri. 53: Met. 20, 238. Semle, 20, 198. Ic siemle mid ðê beô, Bt. 7, 3; Fox 22, 23. Hê symle Drihtne folgode, Homl. Th. i. 58, 17. Symle wesan on lustum, Cd. Th. 30, 26; Gen. 472. Ne swylteþ hê symle ac him eft cymeþ bôt *he does not die for ever, does not remain dead, but remedy comes again to him*, Exon. Th. 419, 13; Rä. 38, 5. ¶ in clauses with a comparative :—Symle biþ ðŷ heardra ðe hit sæstreámas swiðor beátaþ *it keeps getting harder the more the waves beat it*, Cd. Th. 80, 7; Gen. 1325; Beo. Th. 5752; B. 2880: Salm. Kmbl. 485; Sal. 243. (2) with words of similar meaning :—His sôðfæstnyss wunaþ symble êce *justitia ejus manet in seculum seculi*, Ps. Th. 110, 2. Symble on êcnesse *in aeternum*, 118, 142. Simle singales beclŷsed, Exon. Th. 20, 25; Cri. 323. Singallice simle, Met. 7, 46. Forþ simle, Exon. Th. 23, 30; Cri. 376. Symle âwo tô ealdre, 149, 6; Gû. 757. Â symle, 459, 10; Hy. 4, 114. ¶ with comparative :—Dê biþ â symble of dæge on dæg drohtaþ strengra, Andr. Kmbl. 2768; An. 1386. **II.** *on every occasion* or *opportunity, without missing, in unbroken succession :*—Faraþ six dagas simble (*without missing a day*) ymb ða burh, Jos. 6, 3. Symle biþ gemyndgad morna gehwylce, Beo. Th. 4891; B. 2450. Symble gefêgon burhweardes cyme *they rejoiced whenever he came*, Andr. Kmbl. 1318; An. 659. Ðú simle mænst, gif ðê ænies willan wana biþ, Bt. 11, 1; Fox 30, 21. Ðæt môd siemle biþ gebunden ðær ðissa twega yfela âuðer rícsaþ *whenever either of these two evils prevails, the mind is bound*, 6; Fox 16, 2. Næfre ic ða gebeahte sêcan wolde, ac ic symle mec âscêd ðara scylda, Elen. Kmbl. 936; El. 469. Ðæt hê symle oftost God weorþige, Exon. Th. 27, 17; Cri. 432: 243, 34; Jul. 20. Symle hŷ Gûðlác fromne fundon, ðonne hŷ neósan cwôman, 123, 7; Gû. 319: 205, 6; Ph. 108. ¶ where a series of times is mentioned :—Symble (symle, Exon. Th. 367, 19) ymbe seofon niht *every seven days*, Soul Kmbl. 19; Seel. 10: Andr. Kmbl. 313; An. 157. Simle ymb .xii. mônaþ, Chart. Th. 461, 9: 474, 5: 475, 3. [*O. Sax.* simbla, simla : *O. H. Ger.* simble *semper.*] v. simbel.

simbles, simles; *adv. Ever, always :*—Â ic simles wæs on wega gehwam willan ðínes georn on môde, Andr. Kmbl. 128; An. 64. [*O. H. Ger.* simles, simples *semper.*] v. simbel.

simblian, simlian *to frequent :*—Symligaþ ł oftginiósaþ *frequentant*, Rtl. 15, 17.

simblunga, simlunga; *adv. Always, continually :*—Symlinga *jugiter*, Rtl. 33, 17. Symlunge *continuo*, 59, 33. Symlinga *continua (-o?)*, 17, 5. **simel**, simering-wyrt. v. symbel, symering-wyrt.

sín, seón, sién, sŷn, e; *f.* **I.** *power of seeing, sight, vision :*—Smire on ða eágan, sió sŷn biþ ðŷ scearpre, Lchdm. ii. 30, 21. Se hwæl se ðe gârsecges grund bihealdeþ sweartan sŷne *the whale that beholds the depths of ocean with darkened sight*, Exon. Th. 427, 20; Rä. 41, 94. Ne wyrt ðæt ða seón *it does not injure the sight*, Lchdm. ii. 26, 14. Se ðe hire ða siéne onláh, ðæt heó swâ wîde wlîtan meahte, Exon. Th. 38, 16; Gen. 607. Oft ic sŷne ofteáh, âblende beorna unrîm, Exon. Th. 270, 21; Jul. 468. Næfde sellîcu wiht sŷne ne folme, 415, 3; Rä. 33, 5. **II.** *the instrument of sight, the eye :*—Sŷne *pupillam*, Hpt. Gl. 487, 54. [He feide þe sene to þe egen, þe hlust to þe earen, O. E. Hom. ii. 25, 12. 3iff þatt tin eʒhe iss unnhal o þe sêne, Orm. 9394. Ich (*the owl*) habbe gode sene, O. and N. 368. *Goth.* siuns *sight :* O. Sax. siun *sight; eye : Icel.* sjón, sŷn *sight; eye.*] v. an-, heáfod-sŷn (-sién).

sín; *possess. pron. His, her, its, their;* suus. This pronoun, which is regularly used in the cognate dialects, rarely occurs in English prose, where its place seems to have been early taken by the genitive of hê, heó, hit. **I.** referring to a sing. masc. :—Gif hæleþa hwilc eágum môdes sínes (cf. his môdes, Bt. 34, 8; Fox 146, 3; Met. 21, 38. Him Hrôðgâr gewât tô hofe sînum, Beo. Th. 2477; B. 1236. Harold hŷrde holdlíce hærran sínum, Chr. 1065; Erl. 198, 13. Man æt ðam âgende sînne willan æt gebicge, L. Ethb. 82; Th. i. 24, 4. Esne wið dryhten gebête sîne hŷd, L. Wih. 10; Th. i. 38, 22. **II.** referring to a sing. fem. :—Bær seó brimwylf hringa þengel tô hofe sînum, Beo. Th. 3019; B. 1507. Heáfod on hand âgeaf Iudith gingran sînre, Judth. Thw. 23, 21; Jud. 132. Ðæt wîf (wíf though neuter is represented by a fem. pron.) ðín heáfod tredeþ mid fôtum sínum, Cd. Th. 56, 16; Gen. 913. **III.** referring to a plural :—Ðec Israéla herigaþ, herran sînne (þinne, MS.), 240, 28; Dan. 393. Gebid sînna sôwhula, Txts. 124, 5. Âhealtedon fram stîgum sŷnum ł fram heora paðum *claudicaverunt a semitis suis*, Ps. Lamb. 17, 46. [*Goth.* seins : *O. Sax. O. Frs. O. H. Ger.* sîn : *Icel.* sînn (sinn).]

sín be. v. sî.

sin- (sine-, seonu-, v. cpds.). The form does not occur as an inde-

---

pendent word; as a prefix it has usually the force *ever, everlasting;* but in some cases it seems to denote *magnitude*, e.g. sin-here; cf. *O. H. Ger.* sin-vluot *the deluge.* [*O. Sax. O. Frs. O. H. Ger.* sin- : *Icel.* sî- (*but in the phrase* sî ok æ *the independent word is found*) : cf. *Goth.* sinteins *continual, daily.*]

sin-birnende *ever burning, continually burning :*—Hit (*the fire of Etna*) simle biþ sinbyrnende, Met. 8, 52.

sinc, es; *n.* (*used only in poetry*) *Treasure, gold, silver, jewels :*—Gold gerîseþ on guman sweorde, sinc on cwêne, Exon. Th. 341, 17; Gn. Ex. 127. Sinc, gold on grunde, Beo. Th. 5522; B. 2764. Ða ðe seolfres beoþ since gecoste *qui probati sunt argento*, Ps. 67, 27. Gesâwon ófer since salo hliðian, reced ofer reádum golde, Cd. Th. 145, 9; Gen. 2403. Bereáfodon receda wuldor (*the temple*) reádan golde, since and seolfre, 219, 25; Dan. 60. Sadol searwum fâh, since gewurþad, Beo. Th. 2081; B. 1038: 3234; B. 1615. Se wyrm ligeþ since (*the hoard which it guarded*) bereáfod, 5486; B. 2746. Cyning mec gyrweþ since and seolfre, Exon. Th. 401, 11; Rä. 21, 10. Seah on sync, on sylfor, on searogimmas, 478, 4; Ruin. 36. Tô heánlíc mê þinceþ, ðæt gê mid ûrum sceattum tô scype gangon unbefohtene . . ne sceole gê swâ sôfte sinc gegangan, Byrht. Th. 133, 33; By. 59. Leóda gôd, sûðmonna sinc, Cd. Th. 121, 28; Gen. 2017. Hê bebohte bearn Wealdendes on seolfres sinc, 301, 7; Sat. 578. Hê beágas dælde, sinc æt symle, Beo. Th. 162; B. 81. Ðone hring hæfde Higelác nŷhstan síðe, siððan hê under segne sinc ealgode, Beo. Th. 2413; B. 1204. Ðú ða mádmas Higeláce onsend; mæg ðonne on ðæm golde ongitan, ðonne hê on ðæt sinc stariaþ . . . , 2975; B. 1485. ¶ Sinces brytta, hyrde *a dispenser, guardian of treasure, a prince*, cf. sinc-gifa, *and* Sinca baldor . . . Hrêðel cyning geaf mê sinc and symbel, 4853; B. 2431. Ðone sêlestan sæcyninga ðara ðe sinc brytnade, 4756; B. 2383 :—Sinces brytta, goldwine gumena (*Hrothgar*), 2344; B. 1170. Sinces brytta, folces hyrde, 1219; B. 607: Exon. Th. 288, 3; Wand. 25: (*Holofernes*), Judth. Thw. 21, 22; Jud. 30. Sinces brytta, aðelinga helm (*Pharaoh*), Cd. Th. 111, 18; Gen. 1857. Sinces hyrde, Melchisedec, 126, 27; Gen. 2101. [*O. Sax.* sink.] v. fæted-sinc.

sincan; *p.* sanc, *pl.* suncon; *pp.* suncen. **I.** *to sink :*—Ða ingon sincan *cum coepisset mergi*, Mt. Kmbl. Rush. 14, 30. Hwæðer sincende sæflôd wære, Cd. Th. 86, 27; Gen. 1437. **II.** *to act as an aperient :*—Gif ðæt sié ômihte wæte innan, tyhte hié mon ût mid liþum mettum sincendum, and ne læt inne gesittan on ðam líchoman, Lchdm. ii. 218, 14. [*Goth.* siggkwan : *O. Sax.* sinkan : *O. H. Ger.* sinchan : *Icel.* sökkva.] v. â-, be-, ge-sincan.

sin-ceald; *adj. Ever-cold :*—Sincalda sæ, Cd. Th. 207, 25; Exod. 472.

sin-cealdu; *indecl. f. Continual cold :*—Ne mæg ðær rên ne snâw, ne sunnan hætu, ne sincaldu wihte gewyrdan, Exon. Th. 198, 29; Ph. 17.

sinc-fæt, es; *n.* **I.** *a costly vessel, a vessel of gold* or *of silver*, cf. mâðum-fæt :—Hordweard (*the dragon*) sincfæt (cf. fæted wæge, 4553; B. 2282, dryncfæt dŷre, 4601; B. 2306) sôhte, Beo. Th. 4589; B. 2300: B. 2231. Ides sincfato sealde . . . hió Beówulfe medoful ætbær, 1248; B. 622. Forsôc hê ðâm syncfatum, beága mænigo, Vald. 1, 28. **II.** *a receptacle for treasure, a casket*, cf. hord-fæt :—Hê ætwæg Brôsinga mene, sigle and sincfæt, Beo. Th. 2404; B. 1200.

sinc-fâg, -fâh; *adj. Variegated with costly ornament :*—Ic winde sceal sincfâg swelgan, Exon. Th. 395, 29; Rä. 15, 15. Heorot, sincfâge (cf. goldfâh *applied to Heorot*, 621; B. 308) sel, Beo. Th. 336; B. 167.

sinc-gestreón, es; *n. Treasure :*—Hê wolde ofgifan secga seledreám and sincgestreón, beorht beágselu, Andr. Kmbl. 3311; An. 1658. Ic ðê an tela sincgestreóna, Beo. Th. 2456; B. 1226. Hringum þênede, sincgestreónum fættan goldes, 2189; B. 1093.

sinc-gewæge, es; *n. A weight of treasure, abundance of treasure :*—Oft rinc gebâd ðæt hê in sele sæge sincgewæge *it was a frequent experience to see abundance of treasure in the hall*, Exon. Th. 353, 24; Reim. 17.

sinc-gifa, an; *m. A treasure-giver, a prince, chief* who was expected to be liberal in his gifts. Cf. other compounds of gifa :—Næs mid Rômwarum sincgeofa sélla (*of Boethius*), Met. 1, 50. Hŷ (*the disciples*) word ne gehyrwdon hyra sincgiefan (*Christ*), Exon. Th. 29, 9; Cri. 460. On hyra sincgifan (*Beowulf*), Beo. Th. 4611; B. 2311. Sincgyfan, 2688; B. 1342. Se ðe wât his sincgiefan holdne beheledne hê sceal heán hweorfan *he who knows his gracious lord buried shall wander downcast*, Exon. Th. 183, 13; Gû. 1326. Hê his sincgyfan (*Byrhtnoth*) wrec, Byrht. Th. 139, 62; By. 278. Cf. sinc-gim, -þegu.

sinc-gifu, e; *f. A gift of treasure, costly gift :*—Ðú golde eart, sinc-gife sylla, Andr. Kmbl. 3016; An. 1511.

sinc-gimm, es; *m. A precious gem, jewel :*—Fyrdrincas fôron . . . hyrstum gewerede. Ðær wæs gesŷne sincgim locen, hláfordes gifu, Elen. Kmbl. 528; El. 264.

sinc-hroden; *adj.* (*ptcp.*) *Treasure-laden, adorned with costly ornaments :*—Hec biddan hêt se ðisne beám âgrôf, ðæt ðú sinchroden gemunde

. . . , Exon. Th. 473, 11; Bo. 13. Salu sinchroden *halls richly adorned*, Andr. Kmbl. 3342; An. 1675.

sinc-máðum, es; *m. A treasure* :—Næs sincmáððum sélra on sweordes hád *there was no greater treasure in the shape of a sword*, Beo. Th. 4392; B. 2193.

sinc-stán, es; *m. A jewel* :—Gylden máðm, sylofren sincstán (cf. ða gyldenan stánas and ða seolfrenan, Bt. 34, 8; Fox 144, 30), Met. 21, 21.

sinc-þegu(o), e *or indecl.*; *f. Acceptance of treasure* the gift of a lord : —Sceal sincþego and sweordgifu eówrum cynne álicgean . . . syððan æðelingas gefricgean eówerne dómleásne dæd *for your kin shall receiving a lord's costly present and gift of sword be no more . . . after men learn your inglorious deed* (*the desertion of their lord, Beowulf, at his need*), Beo. Th. 5760; B. 2884. Gemon hé sincþege *he remembers receiving costly presents from his lord*, Exon. Th. 288, 21; Wand. 34. Cf. sinc-gifa, *and see other cpds. of* þegu.

sinc-weorðung, e; *f. A costly decoration, jewel* :—Ic ðé beága lyt, sincweorðunga, syllan meahte, Andr. Kmbl. 543; An. 272: 953; An. 477. Him Elene forgeaf sincweorðunga, Elen. Kmbl. 2435; El. 1212.

sind, synd, sint, sient, siont, synt, sindon, seondon, siendon, syndon *are* :—Hig sind strengran ðonne wé, Num. 13, 32: Met. 10, 33. Synd, Ps. Th. 21, 26: Cd. Th. 19, 7; Gen. 287. Sint, Num. 13, 17: Andr. Kmbl. 696; An. 348: Elen. Kmbl. 1484; El. 744. Sient (sint, Cott. MSS.), Past. 28; Swt. 197, 4: Bt. 11, 1; Fox 32, 32: 16, 3; Fox 54, 18. Siont, Kent. Gl. 232. Synt (synd, MS. A.), Mt. Kmbl. 6, 26, 28: Cd. Th. 114, 14; Gen. 1904. Sindon, Bt. 42; Fox 256, 10, 14. Sindan, 5, 3; Fox 14, 19: Met. 20, 149. Seondon, Cd. Th. 271, 12; Sat. 104: 309, 13; Sat. 709. Seondan (siendon, Cott. MS.), Bt. 3, 4; Fox 6, 24. Siendon (sindon, Cott. MSS.), Past. 6; Swt. 47, 8: Cd. Th. 235, 4; Dan. 301. Syndun, Ps. Th. 58, 10. v. eom.

sind (= síð?) :—Yfla ðara ðe ic gefremede nalæs feám sindon (cf. gylta ðara ðe ic gefremede nales feám síðum, Elen. Kmbl. 1633; El. 818; *also* Andr. Kmbl. 1210; An. 605: Hy. 4, 65), Exon. Th. 263, 24; Jul. 354.

sinder, es; *n.*: sindra (-e?), an; *m.* (*f.?*) *Dross, impurity of metal* :— Sinder *scoria*, Wrt. Voc. ii. 120, 4. Sindor *caries, putredo lignorum* vel *ferri*, 129, 11. Synder *scorium*, i. 86, 18. Syndran blæccan *scoriae atramento*, Hpt. Gl. 421, 50. Nim seolferun syndrun, Lchdm. iii. 112, 24. Ðiss folc is geworden nú mé tó sindrum *versa est mihi domus Israel in scoriam*, Past. 37, 3; Swt. 267, 17. Seaxes ecg sindrum begrunden (*with all impurities ground off*), Exon. Th. 408, 3; Rä. 27, 6. [*O. H. Ger.* sintar *scoria, purgamen: Icel.* sindr; *n. dross.*]

sinder-óm *rust* :—Sinderóme *ferrugine*, Wrt. Voc. ii. 35, 35.

sin-dolh *a lasting, very great wound* :—Him on eaxle weard syndolh sweotol, Beo. Th. 1638; B. 817.

sindon. v. sind.

sin-dreám, es; *m. Everlasting joy, joy of heaven* :—Wuldres áras . . . in sindreáme, Elen. Kmbl. 1478; El. 741. Tó heofonríces gefeán hweorfan móstan and ðær siððan á in sindreámum tó wídan feore wunian móstun, Exon. Th. 154, 20; Gú. 811: 164, 23; Gú. 1016: 225, 6; Ph. 385.

-síne. v. eág-, ge-, on-, scearp-síne, -sýne.

sineht; *adj. Sinewy* :—Mid sinehtum limum gehæfd, Lchdm. ii. 242, 19.

sin-éðe, sine-wealt, sine-wind. v. sin-íðe, seonu-wealt, seonu-wind.

sin-freá, an; *m. A perpetual lord, a husband* :—Nænig nefne sinfreá *none but her wedded lord*, Beo. Th. 3873; B. 1934. Cf. sin-híwan.

sin-fulle, an; *f. House-leek*; *sempervivum tectorum* :—Sinfulle *sempervivum*, Wrt. Voc. i. 68, 64; but the word also glosses *eptafolium*, ii. 106, 83: 107, 31: 30, 50: i. 286, 30: *parulus*, 286, 37 : *pariulus*, ii. 67, 64: *paliurus*, 116, 38. Genim ðás wyrte ðe man *sempervivum* and óðrum naman sinfulle nemneþ, Lchdm. i. 236, 20. Genim sinfullan, ii. 190, 2. Nim ða miclan sinfullan, 240, 12. See Lchdm. iii. 305, col. 1: ii. 405, col. 1.

sin-gal; *adj.* **I.** referring to things of the next life, *everlasting, perpetual* :—Dreám ys singal *canor est jugis*, Hymn. Surt. 58, 4. On ðam heofenlícum éðele is singal leóht, Lchdm. iii. 240, 12 : Homl. Th. i. 238, 5 : Rood Kmbl. 280; Kr. 141. Ðær (*in hell*) is á singal sorh, Wulfst. 26, 8. **II.** referring to things of time, *continual, constant, without intermission* :—Swá singal gebiórscipe *quasi juge convivium*, Kent. Gl. 521. Hine gedreht singal slæpleást, Homl. Th. i. 86, 16. Singal oferdrenc, ii. 592, 6. Ðis is singal sacu, Elen. Kmbl. 1808; El. 906. Singal gesíþ *a constant companion*, Exon. Th. 257, 4; Jul. 242. Se singala ege ne læt nænne mon gesælige beón *continuus timor non sinit esse felicem*, Bt. 11, 2; Fox 34, 28 : 12; Fox 36, 28. Mid ðæm singalum geþohte *ab hac cogitatione continua*, Past. 11, 7; Swt. 72, 6. Geleáfan singalum *fides jugis*, Hymn. Surt. 44, 39. Mid singalre éstfulnysse *sedula devotione*, 88, 15. Singalre *assidua*, Hpt. Gl. 407, 65. Men habbaþ singalne andan betwuh him, Bt. 39, 3; Fox 214, 33. Mid singalum bénum *sedulis questibus*, Hymn. Surt. 127, 14. Mid singalum gebedum *orationibus adsiduis*, Bd. 4, 28; S. 606, 29. **II a.** of the

regular succession of time, *daily* (cf. *Goth.* sinteins *daily* (*bread*)) :—Syle ús hláf úserne ðone singalan, Exon. Th. 469, 4; Hy. 5, 8. Singal tído *diurna tempora*, Rtl. 164, 36. **II b.** of an unbroken series, *in succession, continuous* :—Þurh syx singal geár *per sex continuos annos*, Bd. 4, 23; S. 595, 17: 5, 9; S. 623, 27. **III.** *of long continuance, lasting* :—Wæs seó éhtnys[se] singalre (*diuturnior*) eallum ðám ærgedónum, 1, 6; S. 476, 24. v. following words.

singale, singala; *adv. Ever, continually, constantly* :—Singale *olim*, Wrt. Voc. ii. 115, 48. Ðeáh hine se wind . . . swence, and hine singale (seó singale? cf. seó singale gémen, Bt. 12; Fox 36, 28) gémen gæle, Met. 7, 50. Singala, Beo. Th. 382; B. 190. v. next word.

singales; *adv. Ever, continually* :—Ic singales wæg módceare micle, Beo. Th. 3559; B. 1777: Exon. Th. 115, 15; Gú. 190. Simle singales, 20, 25; Cri. 323: 24, 31; Cri. 393. Syngales, Beo. Th. 2274; B. 1135.

singal-flówende; *adj.* (*ptcpl.*) *Continually flowing* :—Singalflówende eá *fluvius*, Wrt. Voc. i. 54, 18.

singallíce; *adv. Perpetually, continually, constantly* :—Hieremias wilnode singallíce (*sedulo*) hine geþiédan tó ðære lufan his Scippendes, Past. 7, 1; Swt. 49, 16 : Blickl. Homl. 101, 27. Wé him gyldaþ singallíce, and hý ús hýnaþ dæghwamlíce, Wulfst. 163, 10. Swíðe singallíce beswícþ monna mód, Bt. 18, 1; Fox 60, 20. Hí (*Cherubim*) singallíce singaþ '*they continually do cry,*' Elen. Kmbl. 1490; El. 747. Syle drincan singallíce nigon dagas, Lchdm. i. 230, 22. Hine æghwonan God singallíce simle gehealdeþ, Met. 7, 46 : Bt. 12; Fox 36, 27. [*O. H. Ger.* sincallíhho *jugiter.*]

singalness, e; *f. Constancy, perseverance, assiduity* :—Ánrædnys ł singalnys *perseverantia, assiduitas*, Hpt. Gl. 434, 18.

singal-ryne, es; *m. A continual running* of water :—Singalrenes ł swift[renes] *decursus*, Hpt. Gl. 418, 51.

singan; *p.* sang, song, *pl.* sungon; *pp.* sungen *To sing.* **I.** *used absolutely*; (1) of persons, (a) *to sing, recite, relate musically* or *in verse* :—Singan *modulare*, singe *modulabor*, Wrt. Voc. ii. 57, 2, 3. Ic Gode singe *gaudebo* Deo, Ps. Th. 74, 8. Ic Drihtne singe *cantabo Domino*, 103, 31. Hwæt is ðis folc ðe ðus hlúde singeþ? Blickl. Homl. 149, 30. Ðær habbaþ englas eádigne dreám, sanctas singaþ, Cd. Th. 286, 20; Sat. 355. Scop hwílum sang on Heorote, Beo. Th. 997; B. 496. Singende heáp *chorus*, Wrt. Voc. i. 28, 27. Hé geseah Matheus ænne sitton singende, Blickl. Homl. 237, 23. (b) *to compose verse, narrate* :—On ðé ic singge *in te decantatio mea*, Ps. Th. 70, 5. Song hé be middangeardes gesceape and be fruman moncynnes, Bd. 4, 24; S. 598, 9 : Exon. Th. 44, 33; Cri. 712. Be ðam Moyses sang, Elen. Kmbl. 674; El. 337. Swá se wítega sang, Menol. Fox 119; Men. 59. Wítgan sungon be Godes bearne, Elen. Kmbl. 1119; El. 561. (2) of other living creatures :—Se fugel singeþ, Exon. Th. 206, 9; Ph. 124: Salm. Kmbl. 539; Sal. 269. Fugelas singaþ, gylleþ græghama, Fins. Th. 9; Fin. 5. Se hana sóna hlúdswége sang *immediately the cock crew*, Homl. Th. ii. 248, 33 : Shrn. 30, 29. Sang se wanna fugel, Cd. Th. 119, 22; Gen. 1983. Mæw singende, Exon. Th. 307, 11; Seef. 22. (3) of inanimate resonant objects :—Ic þurh múþ sprece, wrencum singe, Exon. Th. 390, 15; Rä. 9, 2. Wiht is wrætlíc, singeþ þurh sídan, 483, 13; Rä. 69, 2. Se hearpere gedéþ, ðæt hearpan strengas náwuht ungelíce ðæm sone ne singaþ ðe hé wilnaþ, Past. 23; Swt. 175, 8. In ðæm dæge singaþ ða býman, Wulfst. 183, 10. Syngaþ, L. E. I. prm.; Th. ii. 396, 8. Hringíren scír song in searwum, Beo. Th. 651; B. 323. Ic seah sellíc þing singan, Exon. Th. 413, 10; Rä. 32, 3. **II.** with a cognate accusative, or followed by the words used or by a clause; (1) of persons (a) *to sing* a song, *recite* a poem, prayer, formula, etc., *read* aloud :—Wé singaþ on his lof : '*Hæl ús on ðæm hehstan,*' Blickl. Homl. 81, 27. Heáhgealdor ðæt snotre men singaþ *a charm that wise men recite*, Ps. Th. 57, 4. Hí singaþ Metude lof, Exon. Th. 239, 7; Ph. 617. Ðegnas singaþ, ðæt ðú sié hláfdige, 18, 14; Cri. 283. Ða lióþ ðe ic songe, Bt. 2; Fox 4, 7. Heó '*Magnificaþ*' sang, Blickl. Homl. 159, 1. Crist sylf sang Pater Noster ærest, L. C. E. 22; Th. i. 372, 26. Engla þreátas sigeleóþ sungon, Exon. Th. 181, 6; Gú. 1289. '*Sing mé hwæthwegu.*' Ða andswarede hé : '*Ne con ic nán þing singan,*' Bd. 4, 24; S. 597, 12. Sing ðás gebedsealmas, Lchdm. iii. 12, 6. Singan sangas *cantare canticum*, Ps. Th. 136, 4. Ða ongan hé singan ða fers and ða word ðe hé næfre ne gehýrde, Bd. 4, 24; S. 597, 17. Ðú singan *dicere carmen*, 597, 31. Cwide singan, Salm. Kmbl. 171; Sal. 85. Singan Pater Noster, 333; Sal. 166. Hé wæs ymen singende, Blickl. Homl. 147, 3. On ðære hálgan cyricean biþ sungen ðæt hálige gerýne, 77, 15. Wæs se wítedóm beforan sungen, Elen. Kmbl. 230; El. 1154. (b) *to narrate in verse, write* :—Se scop sang, ðæt má manna fægnodon . . . , Bt. 30, tit.; Fox xvi. 4. Sealmsceopas sungon and sægdon, ðæt se wolde cuman, Blickl. Homl. 105, 10. For hwam wolde gé secgan oððe singan, ðæt ic gesællíc mon wære, Met. 2, 17. (2) of other living creatures :—Earn sang hildeleóþ, Judth. Thw. 24, 28; Jud. 211. Wulfas sungon æfenleóþ, Cd. Th. 188, 7; Exod. 164. (3) of inanimate things :—Seó byrne sang gryreleóþa sum, Byrht. Th. 140, 7; By. 284. Horn song fúslíc leóþ, Beo. Th. 2851; B. 1423. Ealle hearpan strengas

hē grēt mid ānre honda, ðȳ ðe hē wile ðæt hī ānne song singen, Past. 23; Swt. 175, 9. **III.** where the subject of the song is the object of the verb, *to sing about, recite* or *compose a poem about* something :—Ic dīne strengðu singe, Ps. Th. 58, 16. Ic mildheortnesse and dōm Drihtnes singe and secge, 100, 1. Cwæþ hē : 'Hwæt sceal ic singan?' Cwæþ hē : 'Sing mē frumsceaft,' Bd. 4, 24; S. 597, 16. [*Goth.* siggwan *to sing, read aloud: O. Sax.* singan : *O. Frs.* singa : *O. H. Ger.* singan *canere, cantare, decantare, psallere, modulari, edere, jubilare: Icel.* syngva (-ja) *to sing; to ring* (of metals, etc.), *whistle* (of the wind).] v. ā-, be-, ge-singan.

-singe, -singend. v. ge-singe, æfter-, fore-, mid-singend.

singend-līc ; *adj. That may be sung :*—Singendlīce *cantabiles,* Ps. Spl. 118, 54.

singian *to sin.* v. syngian.

sin-grēne, an ; *f.* A plant name (lit. *ever-green*), sin-green (sen-, sim-), *house-leek ; sempervivum tectorum :* see E. D. S. Pub. Plant Names, s. v. sen-green (sin-, sim-), and Lchdm. ii. 405, col. 1. Besides *sempervivum* the word glosses several other names :—Singrēne *titemallos,* Wrt. Voc. i. 68, 33 : *temolus* ? *titemallos,* Lchdm. iii. 305, col. 1. Syngrēne. Ðeós wyrt ðe man *temolum* and ōðrum naman singrēne nemneþ, i. 152, 12. Singrēne *colatidis,* iii. 301, col. 2 : Wrt. Voc. i. 69, 4. Nim singrēnan, Lchdm. ii. 56, 22. Ða smalan singrēnan, 54, 2. [*Iovis barba jubarbe,* singrene, Rel. Ant. i. 37, col. 2. Howsleke or sengrene *barba Jovis, semperviva,* Prompt. Parv. 251, where see note. *Ger.* sin-grün *and Dan.* sin-grøn *is periwinkle.* Cf. *Icel.* sī-grænn ; *adj. evergreen.*]

sin-grim ; *adj. Ever-fierce, of unceasing fierceness :*—Sace singrimme, Exon. Th. 256, 11 ; Jul. 230.

sin-here ; *gen.* -her(i)ges ; *m. An immense army :*—Besæt sinherge sweorda lāfe *he besieged the fugitives with an immense army,* Beo. Th. 5864 ; B. 2936.

sin-hīgscipe. v. sin-hīwscipe.

sin-hīwan, -hīgan ; *pl. Members of a family united by the lasting bond of marriage, a married pair :*—Sinhīwan (*Adam and Eve*), Cd. Th. 48, 19 ; Gen. 778 : 49, 9 ; Gen. 789 : Exon. Th. 153, 9 ; Gū. 823. Hyra somwist sinhīwan (*body and soul*) gedælden, 160, 10 ; Gū. 941 : 284, 17 ; Jul. 698. [*O. Sax.* sin-hīwun (-iun) : *O. Frs.* sin-hīgen, sinnane, senne : *O. H. Ger.* sin-hīun *conjuges.*] v. ge-sinhīwan, *and following words.*

sin-hīwian *to marry :*—Ne sinīgaþ (synnīgaþ, Rush.) *neque nubunt,* Lk. Skt. Lind. 20, 35. v. ge-sinīgan.

sin-hīwscipe, es ; *m. The lasting family relation of marriage :*—God sinhīgscipas gesamnaþ mid clænlīcre lufe *conjugii sacrum castis nectit amoribus,* Bt. 21 ; Fox 74, 38. v. ge-sinīgscipe, *and cf.* sin-scipe.

sin-hweorfende, -hwurfende *ever-turning, round :*—Sintredende (-trendende ?) ? sinhwurfende *teretes, rotundos,* Hpt. Gl. 408, 73. v. *next word, and cf.* seonu-wealt.

sin-hwurfol, -hwyrfel ; *adj. Round, cylindrical :*—Sinuurbul, sinuulfor, siunhuurful (sinu- ?) *teres,* Txts. 104, 1047. Ðonne is swīðe mycel cyrice getimbred, and is sinhwyrfel on wilewīsan geworht (cf. *under* seonu-wealt *passage from* Shrn. 80, 37). Blickl. Homl. 125, 21. [*Cf. O. H. Ger.* sin-, sina-[h]werbal *teres, rotundus : Icel.* sī-valr.]

sinīgaþ. v. sin-hīwian.

sin-īðe, -ēðe ; *adj. Very gentle :*—Mid sinēðre ondōunge wyrtdrences þurh horn oððe pīpan sió wamb biþ tō clænsianne, Lchdm. ii. 260, 11.

sinlīce. v. ge-sinlīce.

sinnan ; *p.* sann, *pl.* sunnon ; *pp.* sunnen ; *with gen. To care for, mind, heed :*—Ne ic mē eorþwelan ōwiht sinne, ne mē mid mōde micles gyrne, Exon. Th. 121, 18 ; Gū. 290. Hē wæs swungen sārslegum . . . hrā weorces ne sann (*the body cared not for pain*), Andr. Kmbl. 2556 ; An. 1279. Hié fægerra (-o, MS.) lyt for ædelinge idesa (-e, MS.) sunnun ac hié Sarran swīðor micle wynsumne wlite heredon *they* (*Pharaoh's nobles*) *heeded little fair women before the prince, but much more did they praise the winsome beauty of Sarah,* Cd. Th. 111, 10 ; Gen. 1853. [*Cf. Icel.* sinna (*wk.*) *to care for, mind, give heed to.*]

sin-niht, e *and* es (v. niht) ; *f. Continual night, perpetual darkness :*—Ða ðe in þeóstrum sæton sinneahtes *those who sat in the shades of perpetual darkness,* Exon. Th. 8, 13 ; Cri. 117. Hām sweart sinnehte (*hell*), Exon. Th. 142, 26 ; Gū. 650. Hȳ ābīdan sceolon in sinnehte, 99, 29 ; Cri. 1632. Sinnihte, 94, 20 ; Cri. 1543 : Cd. Th. 3, 27 ; Gen. 42 : Salm. Kmbl. 138 ; Sal. 68. Grendel sinnihte heóld mistige mōras, Beo. Th. 325 ; B. 161 : (*of the darkness of chaos*), Cd. Th. 7, 20 ; Gen. 109. Synnihte, 8, 2 ; Gen. 118. [*Cf. O. Sax.* sin-nahti *the darkness of hell.*]

sin-nīþ, es ; *m. Continued enmity* or *trouble,* Exon. Th. 354, 27 ; Reim. 52.

sinoþ, sino-walt. v. seonoþ, seonu-wealt.

sin-rǣden[n], e ; *f. A perpetual, lasting condition, wedlock :*—On ōðre wīsan sint tō manienne ða ðe mid synnrǣdenne bióþ gebundene *aliter admonendi sunt conjugiis obligati,* Past. 51, 1 ; Swt. 393, 22. Ða ðe beóþ gebundne mid synrǣdenne *conjugati,* Swt. 393, 21. Cf. sinhīwan, -hīwscipe, -scipe.

sinscipe, es ; *m. Marriage, wedlock :*—Sinscipe *conjungium* vel ma-

trimonium, Wrt. Voc. i. 72, 11. Senscipe *consortium, matrimonium,* Hpt. Gl. 469, 44 : *jugalitas,* 416, 25 : 417, 5. Ðrȳ hādas . . . mæigðhād, wudewan hād, and riht sinscype, Homl. Th. i. 148, 7. Sinscipe, 604, 30. Mē nū ne lyst nānes synscipes ac ðæs Hælendes geþeódnysse mid gehealdenre clēnnisse, Homl. Skt. i. 43, 37. Heó wunode twelf geár on ðæs cynincges synscype, 20, 16. Hū miht ðū ðam Ælmihtigan his brȳde beniman and dīnum sinscipe geþeódan, Homl. Th. ii. 476, 33. Ða ðe on sinscipe wuniaþ *married people,* i. 448, 2. Ða ðe beóþ mid sinscipe (syn-, Hatt. MS.) gebundene *conjugati,* Past. 23 ; Swt. 176, 21. Gif hwā on swilcum mānfullum sinscipe (*conjugio*) þurhwunaþ, L. M. I. P. 20 ; Th. ii. 270, 20. Tō senscipum *ad commercia, connubia,* Hpt. Gl. 490, 54. Gesamnaþ sinscipas, clænelīce lufe *conjugii sacrum castis nectit amoribus,* Met. 11, 91. v. ge-, on-sinscipe, *and preceding word.*

sin-snǣd, e ; *f. A huge bit :*—Grendel slǣpendne rinc slāt . . . synsnǣdum swealh (*swallowed by huge bits,* or *by bits that followed each other continuously ?*), Beo. Th. 1490 ; B. 743.

sin-sorh(g), e ; *f. Continual trouble :*—Habban breostceare, sinsorgna gedreag, Exon. Th. 444, 10 ; Kl. 45.

sint. v. sind.

sin-pyrstende *ever thirsting :*—Alexander tōēcan ðæm ðe hē hiénende wæs ǣgðer ge his folc ge ōðerra cyninga hē wæs sinþyrstende monnes blōdes *Alexander humani sanguinis insaturabilis, sive hostium sive etiam sociorum, recentem tamen semper sitiebat cruorem,* Ors. 3, 9 ; Swt. 130, 31.

sin-tredende. v. sin-hweorfende.

sin-trendel, -tryndel ; *adj. Round, circular,* or *globular :*—Dō hyt syntrændel (sinetrundæl, MS. V.? sinetrum del, MS. H.) *make a ball of it,* Lchdm. i. 106, 17. Sintryndel lytel scyld *ancile,* Wrt. Voc. i. 35, 58. v. *next word, and cf.* sin-hwurfol, seonu-wealt.

sin-trendende (?) *ever-turning, round :*—Sintredende (-trendende ?) *teretes, rotundos,* Hpt. Gl. 408, 73. v. *preceding word.*

sinu, sin-wealt. v. seonu, seonu-wealt.

sin-wrǣnness, e ; *f. Continual wantonness :*—Synwrǣnnys vel gālscipe *saturiasis,* Wrt. Voc. i. 19, 51.

sio-, sió-. See generally seo-, seó-.

sió (be), siodo, siofa, siofoþa, sioleða, siolf, siota. v. sī, sidu, sefa, sifeþa, seolh-ȳða, self, set.

sipian, Siras(-e), sīras. v. sypian, Syras, sīgere.

siru ; *gen.* sirwe ; *f. An artifice, a snare, wile, crafty device;* as a military term, *an ambush :*—Gif hwā gewea[l]des ofsleá his ðone nēhstan þurh syrwa (*with guile,* Exod. 21, 14), L. Alf. 13 ; Th. i. 48, 1, note. Sette syrwa þone insidias, Jos. 8, 2. v. searu, sirwe.

siru-tūn (?) *a place for an ambush, lurking-place :*—Syretum (-tūn ?) *latibulum,* Wrt. Voc. ii. 54, 27. v. *preceding word.*

siru-wrenc, es ; *m. An artifice, crafty trick, wile :*—Hī ymbsǣton Cantwareburuh and hī in tō cōman þuruh syruwrencas (syre-, MS. E.), Chr. 1011 ; Erl. 145, 29. v. searu-wrenc.

sirwan, sirwian, sirewan ; *p.* sirwde, sirwede, sirede, sirewede, sir-wode. **I.** in a good sense, *to plan, devise, use art in doing* something :—Hē (*the Creator*) serede and sette eorþan dǣlas, Cd. Th. 265, 29 ; Sat. 15. **II.** in a bad sense, (1) *trans. To plan, contrive, devise, plot, attempt with craft :*—Hī ne sǣtincge ne gestrodu wið Angelþeóde syrwaþ *nil contra gentem Anglorum insidiarum moliuntur aut fraudium,* Bd. 5, 23 ; S. 646, 37. Syrwiaþ *concinnant (iniquitatem),* Blickl. Gl. Ðām ðe mē syrwedan yfel *qui quaerunt mala mihi,* Ps. Th. 70, 12. Hī fācen geswipere syredan *astute cogitaverunt consilium,* 82, 3 : Andr. Kmbl. 1220 ; An. 610. Beó serewede *moliretur, machinaretur,* Hpt. Gl. 487, 23. (2) *with a clause :*—Hē angan sierwan hū hē hiene beswīcan mehte, Ors. 1, 12 ; Swt. 52, 3. (3) *without a case* (a) *in the following glosses :*—Syrwaþ *moliuntur,* Wrt. Voc. ii. 54, 30. Serwede *machinaretur,* Hpt. Gl. 509, 73. Serwedon *machinabantur,* 520, 4. Serewedan, 506, 5. Seredon *concinnabant,* Wrt. Voc. ii. 20, 26. Seruende *convenientes,* 105, 26. Syrwende, 15, 28. (b) *to lie in wait, plot :*—Hē syrwþ (Ps. Lamb. syrwaþ) swā swā leó *insidiatur quasi leo,* Ps. Spl. second 9, 10. Se ðe nānþing ne syrwde *qui non est insidiatus,* Ex. 21, 13. Syrede, Beo. Th. 324 ; B. 161. Se syrwienda deóful, Wulfst. 107, 22. (c) *with prep. to lie in wait for, plot against :*—Ðū syrwst ongeán hyre hō, Gen. 3, 15. Deófol syrwþ ymbe Godes gelaðunge, Homl. Th. i. 240, 1. Mē manige ymb mægene syrwaþ, Ps. Th. 54, 18. Ða syrwde Herodias ymbe hine *Herodias insidiabatur illi,* Mk. Skt. 6, 19 : Homl. Th. i. 82, 20. Ða ðe ymbe ðæs cildes feorh syrwdon *those who sought the child's life,* 88, 18 : ii. 112, 33. Ða ðe emb his feorh syredon *quos in necem suam conspirasse didicerat,* Bd. 2, 9 ; S. 512, 4. Mē seredon ymb secgas monige, hū heó mē deáþes cwealm hrefnan mihten, Cd. Th. 296, 6 ; Sat. 498. Gif hwā ymb cyninges feorh sierwie (syrwie, MSS. B. H.), L. Alf. pol. 4 ; Th. i. 62, 15. Seó næddre wolde syrwan ongeán hire hó, Boutr. Scrd. 20, 12. v. be-, ge-sirwan (-serian, -syrewian, -syrian, -syrwan).

sirwe, an ; *f. An artifice, device, plot, wile :*—Syrwan (serwan) *insidiae,* Ælfc. Gr. 13 ; Zup. 84, 14. v. siru.

sirwian. v. sirwan.

sirwung, e; *f. Plotting, machination, contrivance:*—Beó áîđlod Amanes sirwung ongeán đǽm Judēiscum, Homl. As. 101, 308. Be hláfordes syrwunge. Gif hwá embe cyningc ođđe hláford syrwie *of plotting against a lord. If any man plot against king or lord,* L. C. S. 58; Th. i. 408, 1. Gif hwá ofsleá his đone nēhstan þurh syrwunge (*with guile,* Exod. 21, 14), L. Alf. 13; Th. i. 48, 1, note. Hé cýdde his fácenfulle syrewunge, Homl. Th. i. 82, 18. Mid syrewungum hē becom tô đǽre cynelícan geþincþe, 80, 34. God heóld hine wiđ đǽs deófles syrwungum, ii. 454, 3. Serewungum *machinamentis,* Hpt. Gl. 478, 54. Syrwunga *insidias,* Hymn. Surt. 47, 26. Samson heora syrwunga undergeat, Jud. 16, 3. v. searwung.

sise-mús *a dormouse:*—Sisemûs *glis,* Wrt. Voc. i. 22, 56: 78, 22. [*O. H. Ger.* sise-, zise-mûs: cf. (?) sise-sang *carmen lugubre,* sisegomo *pelicanus.*]

síþ, es; *m.* **I.** *going, journeying, travel:*—Síþes ámyrred *hindered from going,* Cd. Th. 24, 16; Gen. 378. Síþes wērig *weary of swimming,* Beo. Th. 1162; B. 579. Síþes sǽne *slow in travelling,* Apstls. Kmbl. 67; Ap. 34. Ne æt hâm ne on síþe ne on ǽnigre stôwe *neither at home, nor when travelling, nor in any place,* L. I. P. 9; Th. ii. 314, 33: Exon. Th. 339, 34; Gn. Ex. 104. Se đe of síþe cwom feorran gefēred, Salm. Kmbl. 356; Sal. 177. **I a.** *going from this world:*—Is nû fûs đider gǽst síþes georn, Exon. Th. 164, 27; Gú. 1018. Ic eom síþes fûs, 166, 30; Gú. 1050: 212, 10; Ph. 208. Líf biþ on síþe, 213, 6; Ph. 220: 328, 32; Vy. 26. Beó đú on síþ gearu, 172, 24; Gú. 1148. **II.** *a journey, voyage, course, expedition:*—Síþ wæs gedǽled *the course of the Israelites and Egyptians no longer a common one,* Cd. Th. 190, 31; Exod. 297. Lust leófes síþes (*the journey out of Egypt*), 180, 31; Exod. 53: Andr. Kmbl. 2084; An. 1043. Cwēn síþes (*her voyage to Palestine*) gefeah, Elen. Kmbl. 494; El. 247. Ne lǽt đú đec síþes getwǽfan, lâde gelettan, lifgendne mon ongin mere sēcan, Exon. Th. 474, 2; Bo. 23. Nó wǣgflotan wind síþes getwǽfde, sǣgenga fôr forþ ofer ýþe, Beo. Th. 3820; B. 1908. Flôwan môt ýþ ofer eal lond, ne wile heó áwa đæs síþes geswícan, Salm. Kmbl. 647; Sal. 323. Hú myccle scipbrocu hē gebâd on đǽm síþe đe hē (*St. Paul*) wæs đyder rǽpling gelǣded, Blickl. Homl. 173, 7. Æghwelc mon đe on đǽm síþe wǣre *every man that was on the expedition,* L. Alf. pol. 29; Th. i. 80, 8: Ps. Th. 76, 2: Andr. Kmbl. 1590; An. 796: Exon. Th. 451, 13; Dôm. 103. Ne gǽle gē mínne síþ, nû míne fēt gongaþ on heofenlícne weg, Blickl. Homl. 191, 21. Waldend sende here on langne síþ, Cd. Th. 5, 8; Gen. 68. Hēt mē on đysne síþ faran, 32, 7; Gen. 499. Heó on síþ gewât wēsten sēcan, 136, 29; Gen. 2265. Hí tugon longne síþ in hearmra hond, Exon. Th. 228, 19; Ph. 440. Gif đú hafast mid đē wulfes hrycghǽr on síþfǽte, bûtan fyrhtu đú đone síþ gefremest, ac se wulf sorgiaþ ymbe his síþ, Lchdm. i. 360, 22. Gegân sorhfulne síþ, Beo. Th. 2560; B. 1278. Síþ ásettan, Elen. Kmbl. 1990; El. 997. Hwílum ûs earfoþlíce gesǣleþ on sǣwe đēh wē síþ nesan frēcne gefēran *at times we have hard hap at sea, though we come safe from and perform our dangerous voyage,* Andr. Kmbl. 1030; An. 515. Đǽre sunnan síþ behealdan, Exon. Th. 203, 27; Ph. 90. Hwylce Sǽ-Geátas síþas wǣron: 'Hû lomp eów on lâde?' Beo. Th. 3977; B. 1986. Síþa rest *rest from journeys,* Cd. Th. 86, 8; Gen. 1427. Wíde síþas, 55, 36; Gen. 905: 276, 16; Sat. 189. **II a.** *the journey* of the spirit from this world, cf. forþ-síþ:—Ne mæg mon foryldan đone deóran síþ, Salm. Kmbl. 723; Sal. 361. Mín dohtor is on ýtemestum síþe (*in extremis*), Mk. Skt. 5, 23. **III.** *coming, arrival:*—Hió rícsode on đǽm íglonde đe Aulixes com tô líþan; cûđ wæs sôna æđelinges síþ, Met. 26, 62: Andr. Kmbl. 88; An. 44. Geseah Iohannes sigebearn cuman tô helle, ongeat Godes sylfes síþ, Exon. Th. 462, 15; Hö. 52: Beo. Th. 1007; B. 501: 3946; B. 1971. Sorgian for his síþe, Cd. Th. 49, 30; Gen. 800. **IV.** *a proceeding, course* of action, *way of doing, conduct:*—Hí deófle offredon, swâ him ǽfre se síþ hreówan mihte, Homl. Skt. i. 23, 64: Beo. Th. 6109; B. 3058. Hē hafaþ mec bereáfod rihta gehwylces; nis đæt fǽger síþ, Elen. Kmbl. 1819; El. 911. Ne biþ swylc earges síþ *such is not a coward's way,* Beo. Th. 5076; B. 2541: 5058; B. 2532: 5166; B. 2586. Ic ne mæg đînra worda ne wísna wuht oncnáwan síþes ne sagona *I cannot understand aught of thy words or of thy ways, of thy proceeding or of thy sayings,* Cd. Th. 34, 9; Gen. 535. Ne can ic Abeles fôre, hleómǽges síþ, 61, 34; Gen. 1007. Nû đú seolfa miht síþ úserne (*our course of action,* as described in the command of Christ given in the preceding lines, or *our journey,* cf. faraþ l. 663, fôre, 673) gehýran, Andr. Kmbl. 680; An. 340. [Þæt te schal bireowe þat sið, þat tu eauer dides te into swuch þeowdom, H. M. 9, 2. A nyđ đat weldeþ al his sið, Gen. and Ex. 274.] **V.** denoting that which occurs to a person, how a person fares, *the course of* events *in the case of a person, lot, condition, fate, experience:*—Secgan hwelc siđđan wearþ herewulfa síþ *to say what happened afterwards to the war-wolves,* Cd. Th. 121, 25; Gen. 2015. Hú đæs gǽstes síþ æfter swyltcwale geseted wurde *how it might be appointed that the spirit should fare after the death-pang,* Andr. Kmbl. 310; An. 155. Tô hwon đînre sáwle síþ (þing, Vercel.) wurde *what the lot of thy soul would come to be,* Exon. Th. 368, 11; Seel. 20. Đæt wæs hreówlíc síþ eallre đissere þeóde,

đæt hē swá rađe his líf geendade, Chr. 1057; Erl. 192, 20. Wá heom đæs síđes đe hí men wurdon *alas for them that it was their lot to be born men,* Wulfst. 27, 3. Hú lange wilt đú bewēpan Saules síþ, đonne ic hine áwearp, đæt hē leng ne ríxige? Homl. Th. ii. 64, 4: Cd. Th. 49, 14; Gen. 792. Wē đē gecýđaþ síþ úserne *we will tell thee what happened to us* (the incidents are then related), Andr. Kmbl. 1719; An. 862. [Iob minegede alle his wrecche síđes (*all the miseries he had experienced*), O. E. Homl. ii. 169, 9. Mi muchel unseli síđ (unselhđe, Bod. MS.), Jul. 46, 8.] **VI.** *a path, way:*—Brim, sǣmanna síþ, Cd. Th. 208, 4; Exod. 478. Hié tô helle sculon on đone sweartan síþ (cf. the account of Hermôđr going to Hell: Hann reið đökkva dala ok diúpa), 45, 27; Gen. 733. Dôþ hys síþas (*semitas*) rihte, Mt. Kmbl. 3, 3: Mk. Skt. 1, 3. **VII.** *a time* (cf. colloquial *go,* and *Dan.* gang), (1) with ordinals:—Eft ôđre síþe hē fērde *iterum secundo abiit,* Mt. Kmbl. 26, 42: Gen. 27, 36. Đæt deófol hine genam þriddan síþe, Blickl. Homl. 27, 16. (2) with cardinals:—Se hēt forbǽrnan ealle Rômeburh on ǽnne síþ (*all at once*), Bt. 16, 4; Fox 58, 4. Oftor đonne on ǽnne síþ *oftener than once,* Beo. Th. 3163; B. 1579. On þrý síþas drince *let him drink it at three times,* Lchdm. i. 352, 13. Ǽne síþa (síþe, MS. C.) *once,* Bd. 4, 5; S. 572, 44. Hig fērdon seofon síþon embe þa buruh, Jos. 6, 15: Gen. 33, 3: Lk. Skt. 17, 4. (2 a) used in multiplying numbers:—Feówer síþon seofon beóþ eahta and twentig &c., Anglia viii. 302, 47 sqq. Cweþ .xii. síþum twelf, 298, 22. Endleofan síþon hund þúsenda . . . eahtatýne sýþum hundteóntig þúsenda, Blickl. Homl. 79, 19, 22. (2 b) marking degree:—Heó hǽfde seofon síþum beorhtran sáule, 147, 16. [Spenser uses *sithe* in the sense of *time.* *Goth.* sinþ[s] *time : O. Sax.* síđ ; *m.* way, *journey ; a time : O. H. Ger.* sind ; *m. iter, trames ; vicis : Icel.* sinn ; *n.* (in adverbial phrases) *a time : Dan.* sind (in numeral forms, e. g. tre-*sinds-*tyve *three times twenty, sixty*).] v. bealu-, cear-, earfoþ-, eft-, ellor-, forþ-, from-, gryre-, hâm-, heonan-, hin-, lagu-, láþ-, neó-, oft-, sǽ-, sige-, spild-, un-, unrǽd-, ût-, wíd-, wíg-, wil-, wrǽc-síþ ; sind ; manig-síþes ; ge-síþ.

[**síþ**] ; *cpve.* síþra ; *spve.* síþest, síþ[e]mest ; *adj. Late:*—Biþ seó síþre tíd sǣda gehwylces mǣtræ in mægne, Exon. Th. 104, 31; Gú. 16. Se síþemesta dôm (síþemesđa demm, Hatt. MS.) *extrema damnatio,* Past. 2; Swt. 30, 21. Sardanopolus wæs se síþmesta cyning đe on đǽm londe rícsade *novissimus apud Assyrios regnavit Sardanapalus,* Ors. 1, 12; Swt. 50, 29. Đæt ǽreste . . . đæt síþmeste ríce *primum . . . novissimum regnum,* 2, 1; Swt. 60, 5. Him lásta wearþ síþast gesýne *the last trace of them was seen,* Exon. Th. 270, 34; Jul. 475. Đæt đǽm þeódne wæs síþas[t] sigehwíl (*his last hour of victory*), Beo. Th. 5413; B. 2710. On đǽm ǽrestan and on đǽm síþmestan (onwealdum), Ors. 2, 5; Swt. 86, 17. Síþmestan, 6, 1; Swt. 254, 1. Gesǣt tô symble síþestan (síd-, MS.) dæge cyning, Cd. Th. 259, 34; Dan. 701. Mǣssige man swâ fela mæssan . . . and æt đare síþmæstan dô man absolutionem, L. P. M. 3; Th. ii. 288, 10. ¶ *In the adverbial phrase* æt síþestan, síþ[e]mestan *at last, in the end:*—Gif hē æt síþestan (síþmestan, MS. H.) sié gefongen, L. In. 18; Th. i. 114, 7: Beo. Th. 6018; B. 3013: Cd. Th. 217, 31; Dan. 31. Æt síþemestan *novissime,* Mt. Kmbl. 22, 27. [*Icel.* síđari ; *cpve. later ;* síđastr ; *spve. last.*] v. next word.

síþ. **I.** *adv.* (1) *Late, after some time :*—Síþ *sero,* Wrt. Voc. ii. 88, 22. Him đá síþ oncwæþ, sôna ne meahte oroþ up geteón, Exon. Th. 163, 19; Gú. 996. Síþ and late, Judth. Thw. 25, 24; Jud. 275. Tô síþ, Exon. Th. 96, 3; Cri. 1568. ¶ *In phrases with* ǽr (cf. *O. Sax.* ni síđ noh ēr: *O. H. Ger.* ēr enti síd : *Icel.* ár ok síð, síð ok snemma) :—Ǽr and síþ *early and late, always,* Beo. Th. 4993; B. 2500. Síþ and ǽr, Cd. Th. 177, 24; Gen. 2934: Exon. Th. 38, 5; Cri. 602. Ǽr ođđe síþ, ǽfre *ever, at any time,* 56, 1; Cri. 894: 65, 12; Cri. 1053: 471, 28; Rä. 61, 8. Míne gyltas đe ic síþ ođđe ǽr ǽfre gefremode, L. de Cf. 11; Th. ii. 264, 24: Elen. Kmbl. 1947; Elen. 975. Sýþ ođđe ǽr, Menol. Fox 398; Men. 200. Ne síþ ne ǽr *never,* Elen. Kmbl. 480; El. 240. Ne ǽr ne síþ, 1140; El. 572. Sume ǽr, sume síþ, Exon. Th. 154, 25; Gú. 848. Hē síþor fôr on leófes lást, Cd. Th. 199, 10; Exod. 336. (2) *later, afterwards ; postmodum :*—Ǽrest hí sculon ongietan đæt hí fleón đæt đæt hí luniaþ đonne mágon hí síþ iéđelíce ongietan đæt đæt is tô lufianne đæt æft in flugon *prius videant fugienda, quae amant, et sine difficultate postmodum cognoscant amanda esse, quae fugiunt,* Past. 58, 1; Swt. 441, 14. **II.** *prep.* cf. siđđan, *After :*—Síþ đam *after that,* Exon. Th. 110, 14; Gú. 107. **III.** *conj. After :*—Síþ heora tuuege dæg ágân sint, Cod. Dip. Kmbl. ii. 47, 2. [*Goth.* seithu *sero ;* ni thanaseiths *no longer : O. Sax.* síđ ; *cpve. O. H. Ger.* síd (*adv., prep. with dat., conj.*) ; *cpve.* síđor : *Ger.* seit : *Icel.* síð ; *cpve.* síđarr ; *spve.* síđast.]

síþ-berend, es ; *m. A scythe-bearer, a mower :*—Síþberend *vel* mǽþre *falcarius* i. *falciferens,* Wrt. Voc. ii. 146, 80.

síþ-bôc ; *f. An itinerary :*—Síþbôc *itinerarium,* Hpt. Gl. 454, 19.

síþ-boda, an ; *m. One who announces that a journey or march is to begin,* applied to the pillar of cloud, Cd. Th. 193, 21; Exod. 250.

síþ-boren *late-born :*—Of đǽm síþborenum *de post fetantes,* Ps. Surt. 77, 70: Wrt. Voc. ii. 138, 84.

síþ-dagas ; *pl. Latter days, later times :*—On síþdagum ácenned *born in the latter days,* Elen. Kmbl. 1274; El. 639. Cf. ǽr-dagas.

sîþe (*from* sigþe), es; *m. A scythe, implement for mowing:*—Sigdi, sîþe *falcis*, Txts. 62, 430. Sîþe, Wrt. Voc. ii. 35, 1: *falx*, 38, 51: i. 34, 64: *falcastrum*, 16, 16: 85, 3: ii. 33, 74. Befeóll ân sîþe of ðam snæde intô ânum deópan sæþe, Homl. Th. ii. 162, 10. Hê sceal habban . . . sîþe, Anglia ix. 263, 5. [*Icel.* sigðr; *m.*: sigd (*in Norway*) *a sickle*.]

sîþemest. v. sîþ.

sîþ-fær, es; *n. A way, journey:*—Wið sýðfære *juxta iter*, Ps. Spl. 139, 6.

sîþ-fæt, es; *in sing. generally masc., in pl. neut.*   **I.** *a journey, expedition:*—'Se sîþfæt is ðyder tô lang, and ðone weg ic ne con.' Drihten him tô cwæþ: 'Andreas ic ðînne sîþfæt gestaþelode,' Blickl. Homl. 231, 26–8: Andr. Kmbl. 840; An. 420: Elen. Kmbl. 458; El. 229. Ðæt gewin ðæs sîþfætes *labor itineris*, Bd. 1, 23; S. 486, 1. Ðone intingan his sîþfætes *itineris sui causam*, 4, 1; S. 563, 24: Andr. Kmbl. 407; An. 204. Sîþfates, Elen. Kmbl. 439; El. 220. Ðæt folc wearþ þrît mid ðam sîþfæte *taedere coepit populum itineris*, Num. 21, 4. Ðû mê hafast on ðissum sîþfæte sibbe gecýðed, Andr. Kmbl. 715; An. 358. Hê byþ on sýþfæte and gysthûses beþearf, L. E. I. 32; Th. ii. 430, 25. Ðý ongunnenan sîþfate, Bd. 5, 19; S. 641, 2: Kent. Gl. 307: Cd. Th. 211, 4; Exod. 521: Judth. Thw. 26, 19; Jud. 336. Ðonne hwâ sîþfæt onginnan wille, ðonne genime hê ðás wyrte artemisiam, and hæbbe mid him, ðonne ne ongyt hê nâ mycel tô geswynce ðæs sîþes, Lchdm. i. 102, 4. Ðone sîþfæt him ceorlas lythwón lôgon, Beo. Th. 406; B. 202: Exon. Th. 274, 3; Jul. 527. Ongan sîþfæt (*his journey* or (?) *his fate*, cf. sîþ, V) seófian, wyrd wânian, 274, 22; Jul. 537.   **II.** *a path, course, way, road:*—Weg *via*, sîþfæt *iter*, Wrt. Voc. i. 53, 59. Rihtes sîþfætes *directi callis*, ii. 140, 55. Sîþfæte *tramite*, Hpt. Gl. 513, 26. Sunnan sîþfæt *the sun's path*, Cd. Th. 182, 25; Exod. 81. Ealne gôdne sîþfet *omnem semitam bonam*, Kent. Gl. 20. Ðâ oncierde ðæt scip on wônne sîðfæt *the ship took a wrong course*, Shrn. 60, 8. Sîþfatu *calles*, 27. Sîþfata *semitas, vias*, Hpt. Gl. 457, 9. Ðâ forlêton wê ða frêcnan wegas and sîþfato, Nar. 17, 13.   **III.** fig. *a way, path, course:*—Sîþfæt ârleásra losaþ *iter impiorum peribit*, Ps. Spl. 1, 7. Gerece mê on sîþfæte (*semita*) rihtum, 26, 17. Gesundfull (gesundne, Ps. Th. 67, 20) sîþfæt dô ûs, 67, 21. Sîþfæt sægde ðe hê mid wilddeórum âteah *told of his life with the wild beasts*, Cd. Th. 256, 31; Dan. 649. Nû ðû ædre const sîþfæt mînne. Ic sceal sârigferþ hweorfan . . . *now thou shalt speedily know my course. Mournful must I wander* . . . , Exon. Th. 184, 30; Gû. 1352. Hine geheald ôþ ðæt hê his sîþfæt secge ealne from orde (*the devil is then made to give an account of his proceedings*), 259, 20; Jul. 285: 261, 20; Jul. 318. Sîþfatu *semitas*, Ps. Spl. 24, 4.   **IV.** *course of time* (?):—Ðâ wæs æfter sîþfæte ðæt mægen on him weóx *in course of time it came to pass that strength grew in him*, Guthl. 2; Gdwin. 12, 25.

sîþ-from; *adj. Good at travelling, bold in journeying:*—Sîþfrome, searwum gearwe, wîgend (*Beowulf and his men when ready for their homeward voyage*), Beo. Th. 3630; B. 1813: Andr. Kmbl. 493; An. 247. Land Persêa sôhton sîþfrome Simon and Thaddeus, Apstls. Kmbl. 153; Ap. 77: Andr. Kmbl. 1281; An. 641: Exon. Th. 157, 17; Gû. 893.

sîþ-geómor; *adj. Sad and weary with travel:*—Ic ðysne sang sîþgeómor fand, on seócum sefan samnode wîde, hû ða æþelingas ellen cýðdon, Apstls. Kmbl. 2; Ap. 1.

sîðian; *p.* ode *To journey, go, travel:*—Hwider sîðast ðû bûtan ðînum bearne? Homl. Th. i. 416, 33. Ðær ic sîðade *juxta iter*, Ps. Th. 139, 5. Hê ðider sîðode, Homl. Th. ii. 516, 6. Sum undercyning hine bæd ðæt hê hâm mid him sîðode, i. 128, 6. Ðâ ðâ se Hælend sîðode, sum man him cwæþ tô: 'Ic wille sîðian mid ðê and ðê folgian,' Homl. Skt. i. 16, 154. Nænig wæs ðæt hê eft sîðade hyhta leás, Exon. Th. 157, 24; Gû. 896. Þurh ðê Freá on ðâs eorþan ût sîðade, 21, 4; Cri. 329. Hig intô helle cuce sîðodon *descenderunt vivi in infernum*, Num. 16, 33. Hig sîðodon ealle tô Egipta lande, Ælfc. T. Grn. 5, 3. Sîðedon, Cd. Th. 121, 13; Gen. 2009. Hine cneówmægas mid sîðedon, 104, 13; Gen. 1734. Ðæt ic hláfordleás hâm sîðie, wende fram wîge, Byrht. Th. 139, 9; By. 251. Ðær gê sîðien, Cd. Th. 195, 6; Exod. 272. Sîðien and færen *comitentur*, Wrt. Voc. ii. 22, 14. For ðê sceal ælc flæsc forþ sîðian *ad te omnis caro veniet*, Ps. Th. 64, 2. Ðâ com eorl sîðian on Egypte, Cd. Th. 110, 27; Gen. 1844. Gewât him hâm sîðian, 130, 18; Gen. 2161. Hêht hine twegen men mid sîðian, 173, 28; Gen. 2868. Samed sîðian, Exon. Th. 434, 17; Rä. 52, 2. Up sîðian, Hy. 3, 56. Sîðigean, Andr. Kmbl. 1657; An. 831. Ic eom engel Godes ufan sîðende, Exon. Th. 258, 7; Jul. 261.   ¶ *of the spirit's journey to another world:*—Æfter deáþe somod sîðiaþ sâwla mid lîce, 237, 2; Ph. 584. Scolde se ellorgâst on feónda geweald feor sîðian, Beo. Th. 1621; B. 808. Ðæt mîn sâwul tô ðê sîðian môte, Byrht. Th. 136, 65. [*O. E. Homl.* sîðian: *Laym.* siðen: *O. Sax.* sîðón: *O. H. Ger.* sindôn: *Icel.* sinna.]   v. for-, gemid-, mid-, wræc-sîðian.

sîþ-lædness, e; *f. A leading* or *taking away:*—Sîþlædnisse *abductione*, Ps. Surt. ii. p. 195, 39. Cf. onwegâlædness.

sîþlîce; *adv. Late* (?), *after some time, at last, in the end, lately:*—

Eft ðâ siððan ôðre twegen swearte hremmas sîþlîce cômon and his hûs tæron mid heardum bile *again afterwards two other black ravens came after some time, and tore his house with hard bill*, Homl. Th. ii. 144, 21. Næs Petrus gewunod tô nânre wæpnunge ac ðær wæron twâ swurd sîþlîce gebrohte *Peter was not accustomed to arms, but two swords had lately* (?) *been brought there*, 248, 4.

sîþmæst, sîþor. v. sîþ.

sîþ-stap[p]el *a track, footstep:*—Ðæt ne sýn âstyrode sîþstapla mîne ł wegas ł fôtswaþu *ut non moueantur uestigia mea*, Ps. Lamb. 16, 5. v. under-stapplian.

sîþ[þ], e; *f. Travel, journey:*—Bæm wæs on sîþþe hæbbendes hyht *to both when journeying was the possessor's joy*, Exon. Th. 481, 12; Rä. 65, 2. v. gesîþ[þ].

sîðða; *adv. Afterwards:*—Ðonne meaht ðû siðða sôðes leóhtes habban ðînne dæl, Bt. Met. Fox 24, 59. v. next word.

siððan, siððon, syððan, seoððan. [*From* sîþ ðam; cf. Ger. seitdem.]   **I.** *adv. Afterwards, since:*—Gê faraþ siððan *postea transibitis*, Gen. 18, 5. Siððon, Exon. Th. 131, 33; Gû. 465. Sioððan, Elen. Kmbl. 2292; El. 1147. Syððan (*exinde*) ongan se Hælend bodian, Mt. Kmbl. 4, 17. Ðâ ongan hyne syððan hingrian *postea esuriit*, 4, 2. Hê biþ ðonne seoððan ðæm englum gelîc, Blickl. Homl. 49, 7. Siððon, 59, 7. Ða ðe seoððan after Cristes cyme wæron tô Gode gecyrrede, 81, 15. Ðâ æfter ðisse dæde his noma wæs â seoððan mære geworden, 219, 4. Â syððan ðenden wunaþ hûsa sêlest, Beo. Th. 571; B. 283. Siððan â, Andr. Kmbl. 2387; An. 1195: 2757; An. 1381. Seoððan â, Cd. Th. 289, 16; Sat. 398. Siððan æfre, Elen. Kmbl. 1012; El. 507. Hî sunnan ne geseóþ syððan æfre, Ps. Th. 57, 7. Â forþ sioððan, Ps. C. 103. Hraðe seoððan, Beo. Th. 3879; B. 1937. Nænig efenlîc ðam ær ne siððan, Exon. Th. 3, 21; Cri. 39.   **II.** *conj.* (1) *where the tense of the verb in the clause introduced by siððan is past, in the other clause present, since:*—Ðê is ungelîc wlite, siððan ðû læstes mîne lâre, Cd. Th. 38, 28; Gen. 613: Exon. Th. 44, 13; Cri. 702. Wê ælþeódige wæron, siððon se æresta ealdor Godes bebodu âbræc *we have been exiles, since Adam broke God's commands*, Blickl. Homl. 23, 4. Hû lang tîd is, siððan hê ðis gebyrede? Mk. Skt. 9, 21. Deós syððan ic ineode ne geswâc ðæt heó mîne fêt ne cyste, Lk. Skt. 7, 45. Manige geár syndon âgân nû seoððan ûre bisceopas tô mê gewreoto sende, Blickl. Homl. 187, 3. (2) *where the tense is the same in each clause, after:*—Ðonne biþ his wela for nâuht, siððan hî ongitaþ . . . , Bt. 27, 3; Fox 100, 2. Ðû scealt Isaac onsecgan, siððan ðû gestîgest dûne, Cd. Th. 172, 32; Gen. 2853: 174, 22; Gen. 2882. Him eorla môd ortrýwe wearþ, siððan hié gesâwon fyrd Faraonis, 187, 22; Exod. 155. Wæron Adames dagas, siððan (*postquam*) hê gestrînde Seth, Gen. 5, 4. Syððan, 18, 12. Syððan Iohannes geseald wæs, com se Hælend, Mk. Skt. 1, 14. Hwæt biþ hûton flæsc, seoððan se êcea dæl of biþ? Blickl. Homl. iii. 31: Cd. Th. 309, 7; Sat. 706. [*Later forms are* sithenes, *which gives modern* since, sin, *still used in dialects, and* sithe, sith, *which latter is common in Elizabethan writers.*]

sîþ-weg, es; *m. A road to travel on, high-road* (?):—Hê gehælde hygegeómre ðe hine gesôhtun of sîðwegum (sîð-? v. sîd-weg) *he* (*Guthlac, who lived in the wilderness*) *healed the sad in heart that from the travelled ways sought him*, Exon. Th. 155, 13; Gû. 859.

sîþ-weorod, es; *n. A band out on an expedition:*—Ne meahton sîþwerod gûþe spôwan, Cd. Th. 127, 22; Gen. 2114.

sîþ-wîf, es; *n. A noble lady:*—On sumes sîþwîfes (gôdes wifes, 2nd MS.) hûse *in domum inclytae matronae*, Nar. 49, 9. v. gesîþ-wîf.

sitl. v. setl.

sittan; *p.* sæt, *pl.* sæton; *pp.* seten.   **I.** *to sit, be seated:*—Ðû sitst on ðam heán setle, Ps. Th. 9, 4. Sitest, Hy. 8, 30. Ðû ðe sittest ofer cherubin, Ps. Th. 79, 2. On ðam ðe ofer ðæt [þrymsetl] sitt, Mt. Kmbl. 23, 22. God sitt ofer setle his, Ps. Spl. 46, 8. Ðe sit on his cynesetle, Ex. 11, 5. Siteþ, Cd. Th. 17, 16; Gen. 260. Se ðe sitteþ ofer cherubim, Ps. Spl. 98, 1. Hê on bolcan sæt, Andr. Kmbl. 610; An. 305. Weard on wicge sæt, Beo. Th. 578; B. 286. Hê æt fôtum sæt freán Scyldinga, 1004; B. 500. Maria sæt be Hælendes fôtum, Blickl. Homl. 73, 30. Wê on geflitum sæton *we sat engaged in discussions*, Salm. Kmbl. 862; Sal. 430. Hié æt swæsendum sæton, Cd. Th. 1688; Gen. 2779. Hæleþ in sæton, Andr. Kmbl. 724; An. 362. Site nû tô symle, Beo. Th. 982; B. 489. Geseah twegen englas sittan, ânne æt ðam heáfdon, ôðerne æt ðam fôtum, Jn. Skt. 20, 12. Sittan ofer ða eorþan, Mk. Skt. 8, 6. Sittan on scridwærne, Bt. 27, 1; Fox 96, 1. Tô sittanne on mîne swîðran healfe, Mt. Kmbl. 20, 23. Sittende, Lk. Skt. 22, 69. Uppan assan folan sittende, Jn. Skt. 12, 15. Sittendum wîfe under geled, Lchdm. i. 266, 6.   **I a.** *with reflexive dative:*—Ðâ him sæton sundor on portum, Ps. Th. 68, 24. Sæton him æt wîne, Cd. Th. 259, 23; Dan. 696.   **I b.** *of kneeling:*—Hié for ðam cumble on cneówum sæton, 227, 2; Dan. 180.   **I c.** *applied to the position of a bird at rest:*—Ic (*picus*) glado sitte, Exon. Th. 406, 26; Rä. 25, 7. Hê (*the phenix*) siteþ sîþes fûs, 212, 10; Ph. 208. Næfuglas under beorhhleoþum sittaþ, Cd. Th. 130, 14; Gen. 2159.   **II.** *to stay, dwell, sojourn, abide, reside, remain in a place*, (a) *of persons:*—Wê in carcerne sittaþ sorgende,

Exon. Th. 2, 28; Cri. 26. Đa đe on þýstrum sittaþ, Lk. Skt. 1, 79. Ealle đa đe sittaþ ofer eorþan ansýne, 21, 35. Unc mōdige ymb mearce sittaþ *dwell on our borders*, Cd. Th. 114, 21; Gen. 1907. On đam setle đe hē đǣr sæt *during the stay he made there*, Chr. 922; Erl. 108, 22. Inne on đǣm fæstenne sǣton feáwa cirlisce men *a few common men were living in the fort*, 893; Erl. 88, 33. Wē on đam gōdan rīce sǣton, Cd. Th. 27, 1; Gen. 411. Hæleþ lāgon, on swaþe sǣton (*were left behind dead*), 125, 10; Gen. 2077. Gang tō ciricean and site đǣr and stille wuna and geseoh đæt đū ūt đanon ne gonge ǣr seó ádl from đē gewiten sý *ingredere ecclesiam & ibi reside, quietus manens; vide ne exeas inde, nec de loco movearis, donec hora recessionis febris transierit*, Bd. 3, 12; S. 537, 9. Sitte gē on ceastre oþ gē sýn ufene gescrýdde, Lk. Skt. 24, 49. Se đe sitte uncwydd on his āre on līfe, L. Eth. iii. 14; Th. i. 298, 9. (a 1) *referring to warlike or hostile operations*, as in *to sit down before a place* (cf. *siege*), *to encamp:*—Đū sǣte ongeán đīnne brōþor (cf. *Icel.* sitja ā svikum við einn *to plot against one*), Ps. Th. 49, 21. Hē him æfter rād oþ đæt geweorc and đǣr sæt .xiiii. niht, Chr. 878; Erl. 80, 15. (Often in the Chronicle.) (b) *of things:*—Sió hefige eorþe sit đǣr niþere be đæs cyninges gebode, Bt. 39, 13; Fox 234, 13. Flōd mycel on sæt *there was a great flood in the river*, Bd. 3, 24; S. 556, 35. **II a.** *to continue* in a state or condition, *live* (in hope, fear, etc.), *remain* (silent, etc.):—Ic ā on wēnum sæt *I lived in constant expectation*, Cd. Th. 163, 18; Gen. 2700. Mǣre þeóden unblīđe sæt, Beo. Th. 261; B. 130. Sæt secg monig sorgum gebunden, weán on wēnan, Exon. Th. 378, 30; Deór. 24. Sitte ǣlc wuduwe werleás twelf mōnaþ, L. C. S. 74; Th. i. 416, 6. **III.** *with the idea of oppression* (cf. colloquial *to sit on a person, Icel.* sitja ā sēr *to restrain one's self*), *to sit on* or *bear heavy on, weigh, press, rest :*—Ne mē wiht an siteþ egesan áwiht ǣniges mannes *non timebo quid faciat mihi homo*, Ps. Th. 55, 9. Seó hefige byrþen siteþ on đǣm deádan līchoman đǣre byrgenne *the heavy burden of the tomb presses on the dead body*, Blickl. Homl. 75, 7: Lchdm. iii. 110, 23, 26. On eów scyld siteþ, Exon. Th. 131, 2; Gū. 449. Ūs Godes yrre hetelīce on sit, Wulfst. 162, 2. Đa yrmþa đe ūs on sittaþ, 157, 5. Swā sæt seó byrþen synna on đissum cynne, Blickl. Homl. 75, 9. For đǣm earfoþum đe him on sǣton *for the miseries that lay heavy on them*, Met. 26, 97. Sitte sió scyld on him, L. Alf. 17; Th. i. 48, 15. Ǣr đon đe him se egesa onufan sǣte, Judth. Thw. 25, 10; Jud. 252. **IV.** *to sit* in authority, *preside :*—Đæt mōd đe ofer đæm flǣsce sitt *mens carni praesidens*, Past. 36, 7; Swt. 256, 3. **V.** *trans.* To occupy a seat:—Sæt hē đæt biscopsetl .xxxvii. wintra, Bd. 5, 23; S. 646, 9. [*Goth.* sitan : *O. Sax.* sittian : *O. Frs.* sitta : *O. H. Ger.* sizzan : *Icel.* sitja.] v. ā-, æt-, be-, eft-, for-, fore-, ge-, of-, ofer-, on-, tō-, under-, ymb-sittan; *and next word.*

-sittende -*sitting*, -*occupying*, -*inhabiting*. v. benc-, burh-, flet-, hām-, heal-, in-, land-, þrym-, ymb-sittende.

siun-huurful. v. sin-hwurful.

siwen-īge, -ēge ; *adj. Blear-eyed :*—Se biþ siwenīge (-igge, Cot. MSS.) se đe his andgit biþ tō đon beorhte scīnende đæt hē mæge ongietan sōđfæstnesse, gif hit đonne áþīstriaþ đa flǣsclīcan weorc. On đæs siwenīgean (-iggean, Cott. MSS.) eágum beóþ đa æpplas hāle . . . Se biþ eallinga siwenīge (-igge, Cott. MSS.) đonne his mōd and his andgit đæt gecynd āscirpþ and hē hit đonne self gescint mid his ungewunan *lippis vero est, cujus quidem ingenium ad cognitionem veritatis emicat, sed tamen carnalia opera obscurant. In lippis quippe oculis pupillae sanae sunt . . . Lippus itaque est, cujus sensum natura exacuit, sed conversationis pravitas confundit*, Past. 11, 4; Swt. 67, 24-69, 9. Siwenēge *lippos*, Germ. 396, 284.

siwian *to sew.* v. seowian.

six, siex, syx *six.* **I.** *as adj. indecl. :*—Wirc six dagas, Ex. 20, 9. On six dagum God geworhte ealle þing, 20, 11. Æfter six (sex, Lind., Rush.) dagum, Mt. Kmbl. 17, 1. Betweox đara sex fīfa ælcum, Lchdm. ii. 148, 2. Sex *bis terna*, Wrt. Voc. ii. 12, 10. On siex dagum, Exon. Th. 105, 13; Gū. 22. Đa siex stafas sweotule bēcnaþ, 407, 4; Rā. 3, 16. Syx (sex, Lind., Rush.) dagon ǣr, Jn. Skt. 12, 1. **I a.** in multiplication :—Ceorles wergild is .cc. sciłł. Đegnes wergild is syx swā micel, L. M. L.; Th. i. 190, 3: **II.** *as subst. declined :*—Đā hyra syxe wǣron ācwealde, Shrn. 111, 10. On đam mynstre wǣron fīf brōþra ođđe syxe, Bd. 4, 13; S. 582, 22. Hē sǣde đæt hē syxa sum ofslōge syxtig, Ors. 1, 1; Swt. 18, 7. Ymbsealde sint mid sixum, Elen. Kmbl. 1481; El. 472. [*Goth.* saihs : *O. Sax.* sehs : *O. Frs.* sex : *O. H. Ger.* sehs : *Icel.* sex.]

six-benn, e ; *f. A wound made by a 'seax':*—Ealdorgewinna (*the fire-drake*) siexbennum seóc (cf. cyning wælseaxe gebrǣd . . . forwrāt Wedra helm wyrm on middan, 5400; B. 2703), Beo. Th. 5800; B. 2904.

six-ecge ; *adj. Hexagonal :*—Sixecge *exagonum*, Wrt. Voc. i. 55, 3. Sixecge bere *exaticum*, ii. 144, 58.

six-feald ; *adj. Six-fold :*—Sixfeald *exagonum*, sixfealdum leóþcræfte *exametro heroico*, Wrt. Voc. ii. 144, 46, 47. Siexfealdre anlīcnesse *sena paradigmata*, 89, 39.

six-fēte ; *adj. Having six feet* (of verse) :—Đæt syxfēte vers, Anglia

viii. 335, 13. Mid getelferse ł sixfētum *catalectico versu*, Hpt. Gl. 409, 21.

six-gilde ; *adj. Requiring six-fold payment* or *fine :*—Diácones feoh .vi. gylde *a deacon's property (when stolen) shall be paid for with a six-fold fine*, L. Ethb. 1 ; Th. i. 2, 5. v. -gilde.

six-hynde ; *adj. Of a class whose wergild is six hundred shillings :*—Gif wealh hafaþ fīf hýda hē biþ sixhynde, L. In. 24; Th. i. 118, 10. Be syxhyndum men. Gif hit sié syxhynde mon, [gielde] ǣlc mon .lx. sciłł., L. Alf. pol. 30; Th. i. 80, 11. Gif hió sié syxhyndu, 18; Th. i. 72, 14. Syxhyndes monnes burhbryce .xv. sciłł., 40; Th. i. 88, 10. Gif syxhyndum đissa hwæđer gelimpe, gebēte be đæs syxhyndan bōte, 39; Th. i. 88, 2-5. Syxhyndum men .c. sciłł. gebēte, 10; Th. i. 68, 10. ¶ applied to the wergild :—Æt twyhyndum were mon sceal sellan tō monbōte .xxx. sciłł., æt syxhyndum .lxxx. sciłł., L. In. 70; Th. i. 146, 14. v. twelf-hynde, and see Stubbs' Const. Hist. i. 161, note 3.

six-hyrnede ; *adj. Having six corners* or *angles :*—Sixhernede *sexangulatum*, Wrt. Voc. i. 55, 4.

six-nihte ; *adj. Six days old :*—Se đe biþ ācenned on .vi. nihte mōnan, Lchdm. iii. 160, 23 : 178, 6.

sixta ; *ord. num. Sixth :*—Se sixta (sexta) *sextus*, Ælfc. Gr. 49; Zup. 282, 17. Siexta wæs Ōswald, Chr. 827; Erl. 64, 4. Đā wæs syxte geár, Elen. Kmbl. 14; El. 7. Wæs đā sihste tīd, Exon. Th. 171, 8; Gū. 1123. Seista (sesta, Rush.), Mk. Skt. Lind. 15, 33.

sixteóþa ; *ord. num. Sixteenth :*—Se syxteóþa (six-) *sextus decimus*, Ælfc. Gr. 49; Zup. 283, 3. Sextegđa, Shrn. 91, 20.

sixtig ; *used as subs.* or *adj. Sixty :*—Syxtig *sexaginta*, Ælfc. Gr. 49; Zup. 281, 18. Salomones reste wæs ymbseted mid syxtigum werum . . . Hwæt mǣnde đæt syxtig wera strongera? Blickl. Homl. 11, 16-22. Æfter siextegum daga *intra sexagesimum diem*, Ors. 4, 6; Swt. 172, 4. Mid iii hund scipa and LXgum, Swt. 176, 25. Sexđig (sextig, Rush.), Mk. Skt. Lind. 4, 8. Sexdig ł sextih, Mt. Kmbl. Lind. 13, 23. Sexdeih, 13, 8.

sixtigoþa *sixtieth :*—Se sixteogoþa *sexagesimus*, Ælfc. Gr. 49; Zup. 283, 12.

sixtig-feald *sixty-fold*, Mt. Kmbl. 13, 8, 23.

sixtig-wintre *sixty years old :*—Hē wæs fīf and sixtigwintre, Gen. 5, 15, 18, 20, 21, 23.

sixtine *sixteen :*—Syxtýne *sedecim*, Ælfc. Gr. 49; Zup. 281, 13.

sixtine-nihte ; *adj. Sixteen days old :*—On .xvi. nihte mōnan, Lchdm. iii. 180, 3.

sixtine-wintre ; *adj. Sixteen years old :*—Đǣr georn .xvi. wintre mǣden, Shrn. 140, 1 : 141, 9.

slā (*from* slāhe), *gen.* slān : *but also* slāh, slāg, e ; *f. A sloe :*—Slā *brumela, bellicum*, Wrt. Voc. ii. 127, 26. Slāg *bellicum*, Txts. 45, 289. Genim onwǣre slāh đæt seáw . . . gif sió slāh biþ grēne, Lchdm. ii. 32, 18-20. Gewring tōsomne swilce sió ǣn slāh, 54, 6. Slān *moros*, Wrt. Voc. i. 285, 33 : ii. 56, 32. [Cockayne quotes from a late MS.: 'Acasia est succus prunellarum [im]maturarum, grene slane wose :' and *pl.* slon occurs Alis. 4983. In Baker's Northants. Gloss. slacen-, slaun-bush are given as used of the blackthorn. *O. H. Ger.* slēha, slēa *prunella, agacia: Ger.* schlehe : *Dan.* slaaen.] v. plūm-slā; slāh-þorn.

slacian, slæcian, sleacian ; *p. ode To slacken, relax an effort :*—Gif hē lithwōn slacode . . . his handa ne slacedon *sin autem paululum remisisset . . . factum est, ut manus illius non lassarentur*, Ex. 17, 11, 12. Đæt ne đa sleacgiendan (*pigritantes*) hē ofhreóse, Hymn. Surt. 18, 15. [Nullich neuer slakien to drien herd widuten, A. R. 134, 22. Ne schaltu seon me slakien to leuen, Jul. 26, 1. He mōt slakie his bendes, Laym. 23345 (2nd MS.). Cf. *Icel.* slakna *to get slack.*] v. ā-, tō-slacian; slæccan.

slacigendlīc, slǣ. v. ā-slacigendlīc, sleahe.

slæc, sleac, slec (v. slæcness) ; *adj. Slack.* **I.** *of persons* (1) *inactive, slothful, lazy, not willing to make an effort :*—Slæc *reses*, Wrt. Voc. ii. 118, 77. Sleac *piger*, i. 74, 33 : *lentus* vel *piger*, 49, 35. Sleac vel slāw *pigrus* vel *lentus*, 16, 48. Đū yfela þeówa and sleac *thou wicked and slothful servant*, Homl. Th. ii. 554, 7. Sǣgdon đæt hē sleac wǣre, æđeling unfrom, Beo. Th. 4381 ; B. 2187. Đæt đæm sleacan preóste ne þince tō mycel geswinc, đæt hē undō his eágan, Anglia viii. 317, 4. Tō swilcum sleacum cweđ se hīredes ealdor : 'Tō hwī stande gē hēr ealne dæg ýdele?' Homl. Th. ii. 78, 10. (2) *careless, negligent, remiss, not strict in the performance of duty :*—Ne tō stræc on đǣre lāre ne tō slæc on đǣre mildheortnesse *ne aut districtio rigida, aut pietas remissa*, Past. 17, 10; Swt. 125, 1. Se đe sleac wǣre tō gōdnesse, Homl. Th. ii. 100, 22. Se biþ wacigende . . . se biþ sleac and slǣpende, Btwk. 220, 32. Sleaces *socordis*, Germ. 388, 34. Ne beón gē tō slāpole ne tō sleace, ac scyldaþ eów georne wiđ deófles dare, Wulfst. 40, 21. Sleace tō ǣnig wyrcenne gōd *pigre ad aliquod operandum bonum*, Anglia xi. 117, 36. (3) *languid, ill :*—Slǣc *egra*, Wrt. Voc. ii. 107, 8 : 29, 18. **II.** *of things*, (1) *of physical movement, slow, gentle :*—Sum munuc mid sleaccre stalcunge his fōtswađum filigde, Homl. Th. ii. 138, 6. (2) *that makes inactive, sluggish :*—Wē sceolon āsceacan đone sleacan slǣp ūs fram, i. 602, 15. (3) *not attended with effort :*—Hit is ealles tō sleac munuca þeówdōm (*nimis iners seruitium*), gif hié lǣsse singaþ, R. Ben. 44, 18.

(4) *lax* of conduct :—Gemetgie ðæt fȳr ða bilewitnysse, ðæt heó tô sleac ne sȳ, Homl. Th. ii. 46, 8. Þeówode hē druncennesse and monigum ôðrum unâlȳfednessum ðæs sleacran lifes (*vitae remissioris*), Bd. 5, 14; S. 634, 15. [*O. Sax.* slak : *O. H. Ger.* slah : *Icel.* slakr.] v. un-slæc.

slæccan, sleccan (?) ; *p.* slæcte, slæhte *To make slack* or *slow, to delay* :—' Ðū ūs oftrædlíce mid elcunge geswænctést.' . . . Ðâ cwæþ se cyngc, ' Ðe læs ðe ic eów â leng slæce (slæcce ?),' Th. Ap. 20, 6. v. â-slæccan, ge-sleccan ; slacian.

slæcfull ; *adj. Slothful* :—Slacfulran for belâdunge *propter somnolentorum excusationes*, R. Ben. Interl. 55, 8.

slæcian. v. slacian.

slæclíc ; *adj. Slow* :—Mid sleacilera (sleaclícere ?) *sera, tarda*, Hpt. Gl. 472, 49. v. next word.

slæclíce ; *adv. Lazily, slothfully, languidly* :—Sleaclíce *enervatius*, i. *debilius*, Wrt. Voc. ii. 143, 54. Sume sleaclíce (scleac-, MS. F.) lâgon and slêpon, R. Ben. 68, 21. v. un-slæclíce.

slæcness, e ; *f. Sloth, inertness, laziness* :—Slecnes *accidia*, Wrt. Voc. ii. 5, 73 : 97, 5. Scleacnes *pigredo*, Kent. Gl. 694. **I.** *slowness of physical movement* :—Swâ swâ ðære sunnan sleacnys âcenþ ænne dæg and âne niht . . . swâ eác ðæs mônan swiftnys âwyrpþ ūt ænne dæg and âne niht, Lchdm. iii. 264, 19. **II.** *slowness in action* :—Ðæs þeówes sleacnys (*he seemed long in doing his errand*), Shrn. 43, 15. Wæs beboden ðæt hī sceoldon caflíce etan, forðan ðe God onscunaþ ða sleacnysse on his þegnum, Homl. Th. ii. 282, 3. **III.** *mental inertness* :—Nū wolde ic ðæt ða æðela[n] clericas âsceócon fram heora andgites orþance ælce sleacnysse, Anglia viii. 301, 4. **IV.** *remissness, slowness* in performance of duty :—Oft eác sió gôdnes ðære monþwærnesse biþ diégellíce gemenged wið sleacnesse . . . Wē sculon manian ða manþwæran ðæt hió fleón ðæt ðær suíðe neáh liegeþ ðære monnþwærnesse, ðæt is sleacnes, Past. 40; Swt. 289, 18–22.

slæcorness, e ; *f. Slackness, laziness, remissness* :—Ic ondette sleacornesse and slâpornesse, Anglia xi. 98, 40.

slæd, slêd, es ; *n. A slade* (in local names, e.g. Waters*lade*, v. W. Somerset Words, E. D. S. Pub., and in some dialects. ' *Slade* a breadth of greensward in ploughed land ; a flat piece of grass ; but now most commonly applied to a broad strip of greensward between two woods, generally in a valley,' Baker's Northampt. Gloss. ' Narrow strips of boggy ground running into the hard land at Rockland are called " The Slades,"' E. Anglian Gloss. *Slade* a breadth of greensward in ploughed land, or in plantations, E. D. S. Publ. Gloss. B. 7 (West Riding). In Levin's Manip. Vocab. (1570) a slade, valley = *vallis*, and Drayton uses the word in this sense, v. Nares ; see also Halliwell's Dict.), *low, flat, marshy ground, with a broad bottom, a valley.* (1) The word occurs not unfrequently in the charters, e. g. :—On slêdes heáfad, Cod. Dip. Kmbl. v. 148, 3. Andlang slædes on pyt, iii. 48, 24 : 407, 12. Tô brocces slæde, 233, 34. On ðæt slæd, 385, 28. Oþ ðæt niéhste slæd, 416, 21. On slæd, 25, 24. *It occurs also in composition* :—Tô wulfslæde, 456, 6. On Fugelslêd ; of ðam slêde, 48, 21. In barfodslæd ; and swâ on timberslæd . . . on hamslædes heáfdan, 380, 25–6. On fearnslæd, 385, 30. On ðæt riscslæd, 437, 15. Ondlong slæðbróces, 405, 17. (2) In other connections it is not common, but occurs in the following passage :—Dameris beforan ðæm cyninge farende wæs swelce heó fleónde wære oþ hió hiene gelædde on ân micel slæd . . . Ðær wearþ Cirus ofslægen and twâ þūsend monna mid him *Tomyris simulat diffidentiam, paulatimque cedendo, hostem in insidias vocat. Ibi quippe, compositis inter montes insidiis, ducenta millia Persarum cum ipso rege delevit*, Ors. 2, 4; Swt. 76, 29. Cf. Iulius ferde ut of Doure in to ane muchele slæde & his folc hudde, Laym. 8585. Heo talden whar me heom kepen mihte in ane slade deopen, 26887. Geond slades & geon dunen, 28365. By slente oþer slade, Allit. Pms. 5, 141. Loke a littel on þe launde on þi lyfte honde & þou schal se in þat slade þe self chapel, Gaw. 2147.

slæge. v. slege.

slægu, e ; *f. Slag, dross* :—Slaegu, slægu, slegu *lihargum* (= *lithargyrum*), Txts. 75, 1230. Slægu *liliagrum*, Wrt. Voc. ii. 51, 6.

slæht, slæhtan ; v. sliht, slihtan.

slæp, slôp, sleáp, slâp, es ; *m. Sleep* :—Befeóll slæp (*sopor*) on Abram, Gen. 15, 12. Hrædlíce se slæp becymeþ, Lchdm. i. 246, 17. Slæp biþ deáþe gelícost, Salm. Kmbl. 624; Sal. 611. Hine slæp ofereode, Andr. Kmbl. 1640; An. 821. Mec slæp ofergongeþ, Exon. Th. 422, 23; Rä. 41, 10. Slêp, Prov. Kmbl. 1. Gif ic mínum eágum unne slæpes, Ps. Th. 131, 4. Slêpes *soporis*, Ps. Surt. ii. p. 201, 38: *somni*, 202, 15. Hī wêndon ðæt hē hyt sæde be swefnes slæpe (*mitione somnii*), Jn. Skt. 11, 13. Mid ðȳ heó ðȳ slæpe tôbræd *somno excussa*, Bd. 4, 23; S. 596, 5 : Andr. Kmbl. 3053; An. 1529: Cd. Th. 161, 15; Gen. 2655. Of slæpe onwóc æþeling, 249, 2; Dan. 524. Tô slæpe; gâte horn under heáfod gelæd weccan hē on slæpe gecyrreþ, Lchdm. i. 350, 21–2. Sigon tô slæpe, Beo. Th. 2506; B. 1251. Se ðe for sleápe âwêd *freneticus* (cf. slæpleást), Wrt. Voc. i. 45, 72. Mid slæpe (slêpe, Lind., Rush.) gehefegude, Lk. Skt. 9, 32. Ealle hefige slæpe swundon *omnes somno torpent inerti*, Bd. 4, 23; S. 601, 11. Ic sôftum slæpe mē reste, Homl. Th. i. 566, 22. Gif hē ðære hnappunge ne swîcþ

*donne hnappaþ hē ôþ hē wierþ on fæstum slæpe *dormitando oculus ad plenissimum somnum ducitur*, Past. 28, 4; Swt. 195, 12. Ðȳ swîðan slæpe, Blickl. Homl. 205, 4. Slâpe *somno*, Eng. Stud. ix. 40, col. 1. Ðæt dust ðysse wyrte ðone slêp on gelædeþ, Lchdm. i. 286, 6: 158, 2. Næfþ hē nânne slæp, ii. 198, 25. Slêp, i. 158, 2. Sió slæwþ him giét on ðone slæp, Past. 39; Swt. 283, 8. Âsceacan ðone sleacan slæp, Homl. Th. i. 602, 15. Slæpa sluman, Exon. Th. 122, 31; Gū. 314. ¶ *The sleep of death* :—' Ic wille âwreccan hyne of slæpe ' . . . Se Hælend hit cwæþ be his deáþe, Jn. Skt. 11, 11. Up âstandan of slæpe ðæm fæstan, Andr. Kmbl. 1589; An. 796: Exon. Th. 55, 27; Cri. 890. [*Goth.* slêps : *O. Sax.* slâp: *O. Frs.* slêp: *O. H. Ger.* slâf.] v. frum-, niht-, ofer-slæp.

slæp, es ; *m.* (?) *A slippery, miry place* (?):—Ðis sind ða landgemæro . . . Ærest of ðan ealdan slæpe . . . tô ðan ealdan slæpe ðær hit ær ongan, Cod. Dip. Kmbl. vi. 112, 30–113, 3. On occan slæw (slæp ?), iii. 48, 19. [Cf. *O. H. Ger.* sleifa *labina* (labina *a myre*, Wulck. Gl. 591, 11 : *a fenne*, 797, 10): *Icel.* sleipr *slippery*. Slape *soft, slippery* is given in Halliwell as a North-country word. See also E. D. S. Pub. Gloss. B. 2 (E. Yorks.), ' slape *slippery* as a dirty path,' and Gloss. B. 7 (W. Yorks.), B. 15 (Ray's North-country Words).] Cf. slipor.

slæp-ærn, -ern, es ; *n. A dormitory* :—Slæpern *dormitorium*, Wrt. Voc. i. 58, 10. Hwær slæpst (?) On slæperne (*dormitorio*) mid gebrôþrum, Coll. Monast. Th. 35, 25: Bd. 4, 23; S. 595, 39. Canonicas, ðær seó âr sî, ðæt hî beóddern and slæpern habban mâgan, healdan heora mynster mid rihte, L. Eth. v. 7; Th. i. 306, 12. Ic begeat ðæt stæinene slæpern and ðærtô ðæs landes be sūþan ðæn slæpern .xxiiii. gerda on lange, Chart. Th. 156, 20–27.

slæpan, slæpan ; *p.* te. [*The Northern Gospels also shew forms from* slêpian :—Gif hē slêpaþ, Jn. Skt. Lind. 11, 12. Slêpiaþ ł slêpeþ ł ârísaþ (slêpiaþ ꝛ ârísas, Rush.), Mk. Skt. Lind. 4, 27. Slêpade (geslêpedon, Lind.) *dormitaverunt*, Mt. Kmbl. Rush. 25, 5.] **I.** *to sleep* :—Ðū slêpes, Mk. Skt. Lind., Rush. 14, 37. Slêpes *dormit*, Mt. Kmbl. Lind., Rush. 9, 24. Hwær resteþ (-aþ, MS.) ðæs mannes sâwul ðonne se lîchama slêpþ ? Salm. Kmbl. 188, 12. Slæpeþ *dormitet*, Ps. Lamb. 120, 3. Slæpeþ (slêpeþ, Ps. Surt.) *obdormiet*, Ps. Th. 120, 4. Tô slæpe ; wulfes heáfod lege under pyle ; se unhâla slæpeþ, Lchdm. i. 360, 18. Gif gē slæpaþ (slâpaþ, Ps. Surt.), Ps. Th. 67, 13. Slêpes, Lk. Skt. Lind. 22, 46. Hē æt ðæm stâne slæpte, Past. 16; Swt. 101, 18. Hwæðer hē wacode ðe slêpte, Bd. 2, 12; S. 513, 39. Ðâ hié slêptun (geslêpdon, Lind.) *cum dormirent*, Mt. Kmbl. 13, 25. Slêptun (slêpdon, Lind.) *dormierant*, 27, 52. Hneapedun ł slȳpton (in a later hand, v. Txts. p. 293) *dormierunt*, Ps. Surt. 75, 6. Ðeáh ðæt môd slæpe gôdra weorca, Past. 56; Swt. 431, 25. Mê lyste slæpan *dormitio*, Ælfc. Gr. 34; Zup. 211, 12 note. Ongunnon slæpan *dormitaverunt*, Ps. Th. 75, 5. Wæs ic slæpende, 56, 4: 77, 65. Ðâ gemêtte hē his geþoftan slæpende, Bd. 3, 27 ; S. 559, 15: Beo. Th. 1486; B. 741. Hē hig funde slæpende (slêpende, Lind., Rush.), Lk. Skt. 22, 45. **II.** *to sleep, lie* with a person :—Gif hwâ fæmnan beswîce unbeweddode, and hire mid slæpe (slêpe, MS. G.), L. Alf. 29; Th. i. 52, 6. [*Laym. p.* slæpte, slepte : *A. R. p.* slepte : *Orm.* sleppte.] v. ge-, on-slæpan ; healf-slæpende ; slâpan, slâpian.

slæp-bære ; *adj. Somniferous, soporific* :—Hys gecynde is swîðe hât and slæpbære, Lchdm. i. 284, 22.

slæpere, es ; *m. A sleeper* :—Ðæra eádigra seofon slæpera þrowung, Homl. Skt. i. 23, 1. v. slâpere.

slæpig ; *adj. Sleepy.* [*O. H. Ger.* slâfag.] v. un-slæpig.

slæp-leás ; *adj. Sleepless* :—Slæpleás *insomne*, Germ. 399, 263. [*O.H. Ger.* slâf-lôs.]

slæp-leást, e ; *f. Sleeplessness* :—Hine gedrehte singal slæpleást, Homl. Th. i. 86, 16. Wið slæpleáste, genym ðysse ylcan wyrte (*poppy*) wôs, smyre ðone man mid; sôna ðū him ðone slêp on senst, Lchdm. i. 158, 1. [Þe þet þuruh slôplêste âwêt *frenetus*, Wrt. Voc. i. 89, 81.]

slæpness, e ; *f. Sleepiness, drowsiness* :—Deófol ūs læreþ slæpnesse and sent ūs on slæwde, Homl. As. 168, 106.

slæpor ; *adj. Addicted to sleep* :—Ne beó ðū tô slæpor, forðan ðe slêp fêt unhælo ðæs lîchoman, Prov. Kmbl. 1. v. slâpornes.

slæp-wêrig ; *adj. Weary and sleepy, sleepily weary, so tired as to sleep*, cf. deáþ-wêrig ; or (?) *weary of sleep*, cf. symbel-wêrig :—Oft mec (*a mill-stone*) slæpwêrigne secg oððe meówle grêtan eode, Exon. Th. 387, 14 ; Rä. 5, 5.

slætan ; *p.* te [*causative of* sîtan ; cf. *bait* an animal, and *bite*] *To slate* [Halliwell quotes from a book of 1697 ' to *slate* a beast is to hound a dog at him ;' and in Ray's North-country Words (1691), E. D. S. Pub. Gloss. B. 15, ' to *slete* a dog,' is to set him at anything, as swine, sheep, etc. In Gloss. B. 17 the form is *sleat*. Jamieson also gives ' to *slate* to let loose, applied to dogs in hunting "], *bait, set dogs on, hunt with dogs* :—Man slætte ænne fearr, and se fear arn him tôgeánes, Homl. Skt. i. 12, 72. [Heo leiden to him, sum wið stan, sum wið ban, and sleatten on him hundes (sletten him wið hundes), Jul. 53, 16. To slætenn affter sawless, Orm. 13485. Tho hede the wrecne (*the wolf*) fomen inowe, That weren egre him to slete Mid grete houndes, and to bete, Rel. Ant. ii. 278, 23. Cf. *O. H. Ger.* sleizan *scindere, vellicare*.] v. next word.

slæting, e ; *f. Hunting* :—Hē (*William Rufus*) geátte mannan heora

3 L

wudas and slætinge (cf. William of Malmesbury's statement that he gave the English free leave to hunt), Chr. 1087; Erl. 225, 7. [Toward þan kinge heo weoren beien þær he wes an slæting (an hontyng, 2nd MS.), Laym. 12304. Bole slating, Alis. 200.] v. preceding word.

slǽw; slǽwan. v. sláw; á-, for-slǽwan, sláwian.

slǽwþ, e; f. *Sloth, laziness, inertness, torpor;* accidia, inertia, pigredo, torpor:—Se sixta leahter is *accidia* geháten, ðæt is ásolcennyss oððe slǽwþ on Englisc, Homl. Skt. i. 16, 296. Sió slǽwþ giétt slǽp on ðone monnan *pigredo immittit soporem,* Past. 39, 1; Swt. 283, 6. Slǽwþ *torpor,* Hymn. Surt. 26, 28. Slěuþ *pigredo,* Kent. Gl. 694. On ðæm sceáte his slǽwþe *in sudario lenti torporis,* Past. 9; Swt. 59, 16. From ðære slǽwþe his synna *a peccati torpore,* 28, 4; Swt. 193, 23. Slǽwþe *inertia,* Engl. Stud. ix. 40, col. 1. Hí for heora slǽwþe and for gímeléste forlēton unwriten ðara monna dǽda, Bt. 18, 3; Fox 64, 33. Ic wát ðæt swongornes hí mid slǽwþe ofercymþ, 36, 6; Fox 180, 34. Gyf hē for slǽwþe his hláfordes forgýmþ, ne biþ his ágnum wel geborgen, L. R. S. 20; Th. i. 440, 16. Slǽwþum *torporibus,* Hymn. Surt. 4, 10. v. un-slǽwþ.

slág *a sloe.* v. slá.

slaga, an; m. *A slayer, homicide;* interfector, percussor, lanio:—Slaga *lanio,* Wrt. Voc. ii. 53, 36. Hú ne biþ hē swelce hē sié his slaga (*mortis auctor*), ðonne hē hine mæg gehǽlan and nyle? Past. 38, 4; Swt. 275, 9. Gif man þeóf geméte, and hē hús brece, and hine man gewundie, se slaga biþ unscildig, Ex. 22, 2. Se slaga (cf. ðæs sleges andetta, 29; Th. i. 80, 7). L. Alf. 30; Th. i. 80, 12. The procedure in cases of homicide is given L. E. G. 13; Th. i. 174, 15 sqq., and L. Edm. S. 7; Th. i. 250, 12 sqq. Ic monnes feorh tó slagan sēce, Cd. Th. 92, 7; Gen. 1525. Slagum *interfectoribus,* Engl. Stud. ix. 40, col. 1. Se Hǽlend miltsian wolde his ágenum slagum, H. R. 107, 5. [O. H. Ger. (man-)slago.] v. ágen-, bróðor-, fæder-, mǽg-, mann-, môdor-, morþ-, morþor-slaga.

slágian, slág(h)-þorn, slagu (?), sláh, slahae. v. sláwian, sláh-þorn, mán-, morþor-slagu, slá, sleahe.

sláh-hyll *a hill where sloes grow:*—On sláhhyll, Cod. Dip. Kmbl. iii. 367, 3.

sláh-þorn, es; m. *A sloe-thorn, blackthorn:*—Slághþorn, sláchthorn, -dorn *nigra spina,* Txts. 81, 1380. Slághþorn, slágh- salach-thorn, 99, 1898. Slághþorn, Wrt. Voc. ii. 60, 39. Slágþorn, i. 285, 32. Ádelf niþeweardne sláhþorn, Lchdm. ii. 92, 30. [Le fourder (*slothorne*) que la fourdine (*slon*) porte, Wrt. Voc. i. 163, 1. *Dan.* slaaentorn.]

sláhþorn-ragu *lichen from a blackthorn,* Lchdm. ii. 144, 1.

sláhþorn-rind *bark of a blackthorn,* Lchdm. ii. 98, 7: 108, 11: 132, 9: iii. 58, 8.

sláhþorn-weg *a road along which blackthorns grow,* Cod. Dip. Kmbl. iii. 130, 27.

sláp. v. slǽp.

slápan; p. slěp, sleáp; pp. slápen *To sleep.* **I.** of natural sleep:—Slǽpst ðú? Mk. Skt. 14, 37. Heó slǽpþ, Mt. Kmbl. 9, 24: Jn. Skt. 11, 12. Simle hē biþ lóciende, ne slǽpþ hē nǽfre, Bt. 42; Fox 258, 8. Ðonne wē s'ápaþ, 34, 11; Fox 152, 5. Hwí slápe gē? Lk. Skt. 22, 46. Ic slép (sleáp, Ps. Spl.), Ps. Lamb. 56, 5. Hē slép, Gen. 2, 21: 28, 11: Bd. 3, 9; S. 534, 11. Óðre men slépon, 2, 12; S. 513, 37: Bt. 15; Fox 48, 12. Ealle slépun, Mt. Kmbl. 25, 5. Slápaþ *dormite,* Mk. Skt. 14, 41. Ðeáh hē slápe, Ps. Th. 40, 9: Lchdm. ii. 36, 9. Swelce se stióra slépe, Past. 56; Swt. 431, 30. Mě lyste slápan *dormiturio,* Ælfc. Gr. 34; Zup. 211, 12: Ps. Th. 3, 4: Ors. 4, 6; Swt. 178, 24: Bd. 3, 11: S. 536, 30: Shrn. 106, 23. Ðonne mon wile slápan gán, Lchdm. ii. 228, 5. Hě wæs slápende, Mk. Skt. 4, 38: Homl. Th. i. 566, 17. **I a.** figurative, *to sleep, be inactive, be motionless:*—For hwí slǽpst ðú, Drihten? Ps. Th. 43, 24. Ðæt mód slǽpþ ðæs ðe hit wacian sceolde, and wacaþ ðæs ðe hit slápan sceolde, Past. 56; Swt. 431, 27. Ðonne wē slápaþ fæste, ðonne wē nôhwæðer ne hit witan nyllaþ, ne hit bětan nyllaþ . . . ne slǽpþ hē nó fæsðe, ac hnappaþ . . . , 28; Swt. 195, 4–5. Ðæt ic (*the creation*) ne slépe siððan ǽfre, Exon. Th. 422, 20; Rä. 41, 9. **I b.** of death:—Ic slápe on deáþe, Ps. Spl. 12, 4. Lazarus slǽpþ . . . Se Hǽlend hit cwæþ be his deáþe, Jn. Skt. 11, 11. Ðæt míne eágan nǽfre ne slápan on swylcum deáþe, Ps. Th. 12, 4. Be ðam slápendum, ðæt is, be ðám deádum. Hwí sind ða deádan slápende gecwedene? . . . Ealle môton slápan on ðam gemǽnelícum deáþe, Homl. Th. ii. 566, 30–34. **I c.** of numbness in the limbs, *to sleep, be paralyzed:*—Gif wē tó lange sittaþ ús slápaþ ða lima, i. 490, 1. Gif þeóh slápan . . . lǽt reócan on ðæt lim ðætte slápe, Lchdm. ii. 66, 5–6. Wið slápende (*paralyzed*) líce, i. 380, 18. Cf. Wið áslápenum líce, ii. 12, 17. **II.** *to sleep, lie* with a person:—His hlǽfdige cwæþ tó him: 'Slǽp mid mē,' Gen. 39, 7. [Strong preterites as well as weak, are found in Chaucer and Langland. *Goth.* slépan: O. Sax. slápan: O. Frs. slēpa: O. H. Ger. slâfan.] v. á-, on-slápan, be-slǽpan (-slápan); slǽpan, slápian.

slápere, es; m. *A sleeper:*—Ðæra seofon slápera gemynd, Homl. Th. ii. 424, 8. v. slǽpere.

sláp-ern. v. slǽp-ærn.

slápfulness, e; f. *Sleepiness, drowsiness:*—Ungelimplíce slápfulnys [slápful (? cf. slápor)] *lethargus,* Wrt. Voc. i. 46, 1.

slápian; p. ode *To cause to sleep,* used impersonally with acc.; cf. O. H. Ger. mih sláphôta *dormitavit anima mea:*—Ne geþafa ðú ðínum eágum ðæt hié slápige ne ne hnappigen ðíne brǽwas . . . Ne slápige nó ðin eáge (eágan, Cott. MSS.) . . . Ðæt is ðæt mon his eáge lǽte slápian (slápan, slápigen, Cott. MSS.) *ne dederis somnum oculis tuis, ne dormitent palpebrae tuae . . . Ne dederis somnum oculis tuis . . . Somnum oculis dare, est . . . ,* Past. 28, 4; Swt. 193, 18–25. v. slápan, slǽpan.

slápol; adj. *Addicted to sleep, somnolent:*—Ne sceal mon beón tó slápol (*somnolentus*), R. Ben. 17, 16. Se ðe wǽre slápol, weorðe se ful wacor, Wulfst. 72, 13. Ne beón gē tó slápole ne ealles tó sleace, 40, 21. Tó ðam Godes weorce árísende, heora ǽlc óðerne myngige, ðæt ða slápule (-an, MS. F.) náne láde nǽbben, R. Ben. 47, 17. Hana ða slápolan þreáþ, Hymn. Surt. 7, 1. [Unilimpliche slápel *letargicus,* Wrt. Voc. i. 90, 1.]

slápolnes, e; f. *Somnolence, sleepiness:*—Seó slápolnys byþ gescrýdd mid wácum tætticum *dormitatio vestietur pannis,* Homl. As. 9, 237. Ádrǽf slápolnyssa *expelle sompnolentiam,* Hymn. Surt. 18, 13. Ásolcennys ácenþ ídelnysse and slápolnysse, Homl. Th. ii. 220, 25. Ic syngede þurh slǽwþe and þurh slápelnesse *per accidiam et somnolentiam,* Confess. Peccat.

sláporness, e; f. *Somnolence:*—Ic ondette slápornesse, Anglia xi. 98, 40. v. preceding word, and slápor.

slarige, an; f. *Clary;* salvia sclarea:—Slarege *sclaregia,* Wrt. Voc. i. 79, 16. Slarige, Lchdm. iii. 6, 10. Slarian sǽd, 72, 8. Slarian gôdne dǽl, ii. 58, 11. [From Latin.]

sláw, slǽw, sleáw; adj. *Slow, inert, sluggish, slothful, torpid:*—Sleac vel sláw *pigrus* vel *lentus,* Wrt. Voc. i. 16, 48. Sláw *reses* vel *deses* vel *piger,* 49, 30. Se ðe wǽre full sláw, weorðe se unsláw, Wulfst. 72, 14. Ðone sǽnan ðe biþ tó sláw ðú scealt hátan assa má ðonne man *segnis ac stupidus torpet? asinum vivit,* Bt. 38, 4; Fox 192, 20. Sió sláwe *torpens,* Wrt. Voc. ii. 60, 2. Môd ðæt sláwe *mens torpida,* Hymn. Surt. 37, 10. Ðú yfela þeów and sláwa (*piger*), Mt. Kmbl. 25, 26. Ðú sláwa gá ðē tó ǽmethylle *vade ad formicam, o piger,* Past. 28, 3; Swt. 191, 25. On óðre wísan sceal man manian ða sláwan (cf. late, Swt. 281, 16), on óðre ða ðe beóþ tó hrade, Past. '23; Swt. 175, 25. Ða sláwan (*pigri*) sint tó manianne ðæt hié ne forielden ðone tíman ðe hié tiola on dôn mǽgen, 39, 1; Swt. 281, 19. Sláwera *desidiosorum,* Wrt. Voc. ii. 28, 12. [Slak (slēu, MS. C.) an môde, Hel. 4962. O. H. Ger. sléo *hebes: Icel.* slær, sljór *blunt, dull: Dan.* slöv.] v. un-sláw.

sláwian; p. ode *To be* or *become slow, sluggish, inactive:*—Hwæs wilnast ðú ðæt ðú ne sláwedest swá micel geswinc tó gefremmanne *what dost thou desire, that thou hast not been slow to perform so great a labour,* Homl. Skt. ii. 23 b, 224. Wacige and swince ðár ongeán ðæt hé oft ǽr beslêp and sláwode, L. Pen. 16; Th. ii. 284, 3. Slágige (slacige?) ł sláwige *pigeat,* Hpt. Gl. 479, 5. [O. H. Ger. slêwēn *hebere, torpere:* cf. Icel. sljófa *to blunt.*] v. á-, for-sláwian.

sláwlíce; adv. *Slowly, sluggishly;* pigre:—Ðæt hié tó sláwlíce ðara ne giémen ðe · him befæste sién *ut a commissorum custodia minime torpescant,* Past. 28, 3; Swt. 191, 23. Ic wêne ðæt hē hiene snide sláwlícor (sláulícor, Hatt. MS.) *pigrius fortasse incideret,* 26, 3; Swt. 186, 3. [Ne dyde hē þ náht sláulíce, Anglia x. 143, 87. Man slawliche arised, and late to chireche goð, O. E. Homl. ii. 11, 35. *Icel.* slæ-, sljó-liga *slowly, dully, carelessly.*] v. un-sláwlíce.

slá-wyrm, es; m. *A slow-worm, blind-worm* (cf. a slaworme *cecula,* Cath. Angl. 343), *a kind of snake:*—Sláwyrm *stellio,* Wrt. Voc. i. 24, 25: 78, 60: *spalangius,* 24, 27: Hpt. Gl. 450, 26: *regulus* (cf. regulus est serpens, avis, et rex parvulus omnis, Wrt. Voc. i. 221, 9), Kent. Gl. 913: Engl. Stud. x. 40. Efete ł sláwyrm *stellio,* Ælfc. Gr. 9, 3; Zup. 35, 7 note. [Cf. *Norweg.* slo, orm-slo *a blindworm: Swed.* slå, orm-slå.] Cf. sleán *to strike.*

sleac, sleacian. v. slæc, slacian.

sleahe, slǽ; f. *A slay* (or *sley*), *a weaver's reed, an instrument of a weaver's loom that has teeth like a comb:*—Slahae *pectica,* Wrt. Voc. ii. 117, 23. Slǽ *pe[c]tica,* i. 282, 6. [Purvu de une lame (*slay*), Wrt. Voc. i. 157, 26. Sley *lamia, pecten,* 217, col. 2. Slaye *lanea,* 234, col. 2. Slay *pecten, lania,* Cath. Angl. 342, col. 2, and see note. Slay, webstarys loome *lanarium, radius,* Prompt. Parv. 458, col. 1.]

sleán; p. slóh, slógh, pl. slógon; pp. slagen, slægen, slegen. **A.** trans. **I.** *to strike* an object, *smite:*—Gif ðú slehst *si percusseris,* Kent. Gl. 880. Gif man óðerne mid fyste in naso slæhþ, L. Ethb. 57; Th. i. 16, 17. Ðæt fell hlýt, ðonne hit mon sliehþ, Past. 46; Swt. 347, 5. Ðæt ár ðonne hit mon slihþ, 37; Swt. 267, 24. Ðam ðe ðe slihþ (slyhþ, MS. A.: slǽþ, Lind.) on ðin gewenge, Lk. Skt. 6, 29. Ic sylfa slóh grēne tácne gársecges deóp, Cd. Th. 195, 21; Exod. 280. Ðonne hié (*the serpent*) mon slóg oððe sceáf, Ors. 4, 6; Swt. 174, 7. Hé ðone nǐðgæst slóh, ðæt ðæt sweord gedeáf, Beo. Th. 5392; B. 2699. Slóh ðá wundenlocc ðone feóndsceaþan fágum mêce, Judth. Thw. 23, 3; Jud. 103. Sume hyne slógon (slógan, Lind., Rush.) on his ansýne mid hyra handum, and cwǽdon: 'Sege hwæt is ðe ðē slóh (slóg, Rush.),' Mt. Kmbl. 26, 67. Mě weras slógon and swungon, Andr. Kmbl. 1927; An. 966. Hí mě mid sweopum slógun, Exon. Th. 88, 18; Cri. 142.

Ne sleá gē nānne *neminem concutiatis*, Lk. Skt. 3, 14. Sleáþ synnigne ofer seolfes mūþ, Andr. Kmbl. 2601; An. 1302. Se ðe sleá (*percusserit*) his fæder oððe his mōder swelte hē deáþe, Ex. 21, 15. Gehýrde ic ðæt Eádweard ānne slōge swiðe mid his swurde, Byrht. Th. 135, 13; By. 117. Ða beáh hē sleánde his breóst, H. R. 107, 11. Áhsa hwæðer hē æfre wære slegen on ða sídan, Lchdm. ii. 258, 23. Biþ slaegen *percellitur*, Wrt. Voc. ii. 117, 3. An slægenre *in pacte*, 48, 77. **II.** of special kinds of striking, (a) *to strike* coin, *to stamp* money (cf. similar use in O. Frs. and Icel.), cf. mynet-slege :—Wæs ðæs feós ofergewrit ðæs ylcan mynetsleges ðe man ðæt feoh on slóh, sōna ðæs forman geáres ðā Decius fēng tō rīce, Homl. Skt. i. 23, 476. Ælc mynetere ðe man tīhþ ðæt fals feoh slōge, L. Eth. iii. 8; Th. i. 296, 12. Godes feoh biþ befæst myneterum tō sleánne, Homl. Th. ii. 554, 14. (b) *to forge* a weapon (cf. *Icel.*), .cf. slecg-hamer :—Sæt smiþ, slóh seax, Lchdm. iii. 52, 27. **III.** of a serpent, *to sting* :—Gif næddre sleá man, Lchdm. ii. 110, 14. **IV.** *to strike so as to kill, to slay* :—Slēs ðū *occideris*, Ps. Surt. 138, 19. Hē slēþ *occideret*, 77, 34. Mann slihþ ðīnne oxan *bos tuus immoletur*, Deut. 28, 31. Ic slōg niceras, Beo. Th. 847; B. 421: Exon. Th. 272, 4; Jul. 494. Ðonne God hié slóg (*occideret*), ðonne sōhton hié hine, Past. 36, 3; Swt. 251, 20: Beo. Th. 217; B. 108. Slōgh, Bd. 3, 9; S. 533, 14. Hē slóh and fylde feónd, Cd. Th.124, 32; Gen. 2071. Se hagol slóh ealle ða þing ðe ūte wæron, ægðer ge men ge nýtenu, Ex. 9, 25. Slōgon *obruerunt*, Wrt. Voc. ii. 65, 20. Abraham ne sleah ðīn bearn, Cd. Th. 176, 18; Gen. 2913. Sleh, 204, 12; Exod. 418. Sleá man ðone leásan wītegan *propheta ille interficietur*, Deut. 13, 5. Ðás folc sleán mid cwealmþreá, Cd. Th. 151, 10; Gen. 2506. Se eorl wolde sleán eaferan sīnne, 203, 30; Exod. 411. On deáþ sleán (cf. *Dan.* at slaa ihjel) scyldige, 76, 34; Gen. 1267. Hē biþ ... tō sleánne oððe tō álýsenne, L. Wih. 28; Th. i. 42, 25. Hié wæron ða wæpnedmen sleánde, Ors. 1, 10; Swt. 48, 6. Wæs Fin slægen, Beo. Th. 2309; B. 1152. Sacerdas wæron slægene, Bd. 1, 15; S. 484, 1. Ða hæþenan wæron slægne, 3, 24; S. 556, 29. **V.** *to make by striking, to strike* fire, *to make* a mark, sound, signal *by a stroke* :—Ðá arn sum þeng and slóh tácen æt ðam gæte *cucurrit minister, et pulsans ad ostium*, Bd. 3, 11; S. 536, 17. Hē tácen mid his handa slóh *sonitum manu faciens*, 4, 3; S. 568, 6. Men tácen slógon, Guthl. 11; Gdwin. 54, 24 : 12; Gdwin. 58, 23. Sleah feówer scearpan, Lchdm. ii. 100, 3 : 142, 18. Sleá him ānne spearcan, 290, 17. **Va.** *to strike* a bargain (cf. *Icel.* slá kaupi) :—Hig slógon heora wedd ægðer tō ōðrum, Gen. 21, 27. **VI.** *to strike, drive so as to cause impact* :—Hē slóh fýr on feóndas *he drove the fire on to the foes*, Cd. Th. 237, 28; Dan. 344. **VI a.** metaph. :—Ic wēne gif wit uncre word tōsomne sleáþ, ðæt ðær ásprunge sum spearca sōþfæstnesse, Bt. 35, 5; Fox 164, 2. **VI b.** *to pitch* a tent, *drive* a stake into the ground (cf. *Icel.* slá landtjöldum; *Ger.* ein Lager schlagen) :— Iacob slóh his geteld on ðære dūne, Gen. 31, 25. Sleah ænne stacan onmiddan ðam ymbhagan, Lchdm. i. 395, 4. Ðá hēt Moises sleán ān geteld būtan hira wīcstōwe, Ex. 33, 7: Homl. Th. ii. 242, 8. Ða stōwa ðe gē eówre geteld on sleán sceoldon, Deut. 1, 33. **VI c.** *to cast* into chains (cf. *O. Frs.* on the helda slein) :—Hió sceolde ða men weorpan an wildedeóra líc and siððan sleán on ða raccentan and on copsas, Bt. 38, 1; Fox 194, 32. **VII.** *to move by a stroke, to strike* off a limb, etc. :—Hī slógon him of ðæt heáfod, Th. An. 122, 23. Sleá mon hond of ðæ fōt, L. In. 18; Th. i. 114, 7 : 37; Th. i. 124, 23. **VIII.** metaph. *to strike* with disease, punishment, etc., cf. a paralytic, apoplectic *stroke* :— Ic ástrecce mīne hand and sleá Egipta land on eallum mīnum wundrum, Ex. 3, 20. Sliét *concidet* (*cervices peccatorum*), Blickl. Gl. Hī mid ðý wīte ðæs foresprecenan wræces slægene wæron *praefatae ultionis sunt poena multati*, Bd. 4, 25; S. 601, 31. **B.** *intrans.* **I.** *to strike, make a stroke* :—Hē yrringa slóh *in anger he struck*, Beo. Th. 3135; B. 1565: 5350; B. 2679. On ðone eádgan andwlitan men hondum slógun, Exon. Th. 69, 22; Cri. 1124. Ðæt hē mē ongeán sleá, Beo. Th. 1367; B. 681. **I a.** *to strike* as a smith does :—Hē sulh heóld and on īren slóh and corn ðærsc and windwode, Shrn. 61, 18. **II.** *to kill* (the object not being expressed) :—Ne sleah ðū, L. Alf. 5; Th. i. 44, 17. Slyh (sleh, MS. A.), Mk. Skt. 10, 19. Þeóf ne cymþ būton ðæt hē stele and sleá, Jn. Skt. 10, 10. Hié wæron ða burg hergende and sleánde, Ors. 2, 8; Swt. 92, 16. **III.** *to move rapidly* (v. **A. VI.**), *rush, dash, break, take* a certain direction; cf. *to strike* into a path, across a country (cf. *Icel.* slásk *to betake one's self*) :—Gesca slæet *singultat* (cf. *Icel.* impersonal use slō á hann hlátri *he was seized with a fit of laughter*), Wrt. Voc. ii. 120, 50. Ðæt seolesburna sliht on meóne, Cod. Dip. Kmbl. iii. 13, 31. Deáh swīn beswemde weorþon, ðonne sleáþ hē eft on ða solu, Bt. 37, 4; Fox 192, 28. Hē on scip ástah and slóh ūt on ða sæ *put to sea*, Ap. Th. 6, 6. Se lēg slóh tō I eofonum, Shrn. 73, 36. Ða slóh ðær micel mist *a great mist came on suddenly*, Gen. 15, 17. Seó sæ slóh tōgædere *occurrerunt aquae*, Ex. 14, 27. Hē ofdræd slóh ádūn ðærrihte *terrified he straightway fell down as if struck* (cf. *Icel.* slá sèr niðr *to throw one's self down on a bed*), Homl. Skt. i. 23, 718. Ða slóh ðær micel leóht ūt æfter ðam englum (cf. *Icel. impers. use*, e.g. loganum slō ūt), Homl. Th. ii. 342, 7 ; 350, 24.

on slógan *incursere*, Wrt. Voc. ii. 48, 1. Drenc wið deádum swile ðæt hē ūt sleá, Lchdm. ii. 74, 18 : 102, 20. Ðý læs hit in sleá, 324, 3. Gif hié ūt sleán *if they* (*pocks*) *break out*, 106, 4. [*Goth.* slahan : *O. Sax.* slahan : *O. Frs.* slá : *O. H. Ger.* slahan : *Icel.* slá.] v. â-, be-, for-, ful-, ge-, of-, ofer-, tō-, wið-sleán; fýst-slægen.

sleáw, slēbe-scōh, sleccan. v. sláw, slífe-scōh, slæccan.

slecg, e; f. A sledge-hammer, mallet; malleus :—Slecg, hamur *malleus*, Wrt. Voc. ii. 57, 78. Slegc, i. 86, 16. Hwæt sylst ðū ūs on smiþþan ðīnre būton īsene fýrspearcan and swēgincga beátendra slecgea (*malleorum*), Coll. Monast. Th. 31, 7. Wē hit uneáþe mid īsernum hamerum and slecgum gefyldon *quam ferreis uix comminuimus malleis*, Nar. 21, 5. [The gret slegges, Parten. 3000. *Icel.* sleggja *a sledge-hammer* : *O. H. Ger.* slaga *malleus*.]

slecgettan ; p. te *To palpitate, beat, throb* :—Seó wamb cloccet, swā swā hit slecgete, Lchdm. ii. 220, 18. [*O. H. Ger.* slagazen *palpitare, tremere*.]

slēd, slēf, slēfan, slēfe. v. slæd, slíf, slífan, slífe.

slege, slæge, es; m. **I.** *a stroke, blow* :—Mē and mīne geféran mid ānum slege (*ictu*) hē (*the whale*) mæg besencan, Coll. Monast. Th. 24, 33. Gif hine mon geyfligie mid slege oððe mid bende, L. Alf. pol. 2; Th. i. 62, 3. Geswell ðe wyrð of fylle oððe of slege, Lchdm. ii. 6, 28. His eáge wand ūt mid ðam slæge, Homl. Skt. i. 4, 143. Slægum *ictibus*, Wrt. Voc. ii. 47, 54. Of wundum oððe of snīþingum oððe of slegum, Lchdm. ii. 82, 23. · **II.** of a serpent's sting, cf. sleán, **III** :—Wið nædran slege, Lchdm. ii. 10, 21 : 110, 22. **III.** *a striking, beating*, (a) *scourging* :—Seó sunsciéne slege þrowade, Exon. Th. 256, 10; Jul. 229. Þēh ðū þolie synnigra slege *though thou suffer scourging at the hands of sinners*, Andr. Kmbl. 1911; An. 958. (b) *stamping, coining*, v. mynet-slege, sleán, **II a.** (c) *clashing, collision*, v. sleán, **VI a** :— Slæge *conlisio*, Wrt. Voc. ii. 105, 27. Slege, 15, 29. Slægum *contunsionibus*, 24, 43. Slegum, 20, 32. **IV.** *a crash, clap* of thunder, cf. *Ger.* donner-schlag :—Ðær com swylce þunres slege, Nicod. 24; Thw. 13, 4. Hreám swā hlūd swā þunres slege, 27; Thw. 15, 5. [Wæs swyðe mycel lihtinge and ungemetlice slæge ðæræfter, Chr. 1118; Erl. 246, 40.] **V.** *a fatal stroke, slaying, slaughter, death* (by violence. On the difference between *slege* and *morþor* see Grmm. R. A. 625) :—Ðæra cildra slege (*the murder of the innocents*), Homl. Th. i. 80, 28. Hū nyt is ðē mín slæge *quae utilitas in sanguine meo*, Ps. Th. 29, 8. Nū is æghwonon yfel and slege, Blickl. Homl. 115, 16. Gif mon twýhyndne mon mid hlōðe ofsleá, gielde se ðæs sleges andetta sié ..., L. Alf. pol. 29; Th. i. 80, 7. For gecláensunge his unrihtes slæges *ob castigationem necis ejus injustae*, Bd. 3, 24; S. 557, 25. Æfter Pendan slæge *post occisionem Pendan*, S. 557, 30. Æfter his slæge (*interfectionem*), 3, 9; S. 533, 30. On Urias slege (slæge, Hatt. MS.), Past. 3; Swt. 34, 23. Be elþiódies monnes slege. Gif mon elþeódigne ofsleá, L. In. 23; Th. i. 116, 13. Mid his brōðor slege *parricidio*, Ors. 2, 2; Swt. 64, 23. Hē tihte ðæt folc tō ðæs Hælendes slege, Homl. Th. i. 292, 6 : 216, 15. Hē is geláed tō slege swā swā scēp, ii. 16, 20. Hī heora swuran gearcodon sylfwylles tō slege *they voluntarily prefered their necks for the fatal stroke*, Homl. Skt. i. 5, 47. Mid micelre gnornunge ymb ðæs cyninges slege, Ors. 2, 4; Swt. 76, 23. Þurh ðæs hyrdes slege byþ seó heord tōdræfed, Mt. Kmbl. 26, 31. **VI.** *a defeat, loss* inflicted on an army; clades :—Ðæt tácen nūgiet cūþ is on ðære eá noman ðæs consules sleges Fauiuses *testatur hanc Fabii cladem Allia, sicut Cremera Fabiorum*, Ors. 2, 8; Swt. 92, 17. Crist him gefylste tō his feónda slege (cf. hī álēdon heora fýnd, 96, 22), A. S. Rdr. 95, 13. **VII.** metaph. *a stroke* of affliction, punishment, disease, etc. v. sleán, **VIII** :—Ær ðæn se sǽrlica slege (*the pestilence*) ūs ástrecce, Homl. Th. ii. 124, 21. **VIII.** *an instrument for striking* (or to be put with the next word?). (a) *a slay* :—Slege *percussorium* (the word occurs among terms connected with weaving), Wrt. Voc. i. 59, 44. v. sleahe. (b) *a plectrum* [v. Hearp-slege *plectro*, Engl. Stud. xi. 64]. [*Goth.* slahs *a stroke, blow* : *O. Sax.* slegi *slaying* : *O. Frs.* slei : *O. H. Ger.* slag *plaga, ictus, tusio, percussio* : *Icel.* slagr *a blow, defeat*; cf. also slag; *n. a blow*; *a defeat, slaughter, loss*; *a stroke* of apoplexy.] v. brōðor-, deáþ-, dolg-, eár-, gegn-, hearm-, hearp-, hleór-, morþor-, mynet-, on-, sár-, sweord-, þeóf-slege(-slæge).

slege, es; n. A beam, bar. v. heáfod-, ofer-slege (-slæge). [Cf. *Icel.* slá; f. a cross-beam.]

slege-bītel, es; m. A beetle, hammer, mallet :—Sleah ðonne on mid slegebýtle, Lchdm. ii. 342, 7.

slege-fǽge; adj. Doomed to slaughter, doomed to death by the sword :— Slegefǽge hæleþ (*the Assyrians before their defeat*), Judth. Thw. 25, 7; Jud. 247.

slegel, es; m. An instrument for striking a harp :—Slegele *plectro*, Wrt. Voc. ii. 66, 79. [*O. H. Ger.* slegil *percussorium, maza* : *Ger.* schlägel : *Du.* slegel *a hammer, mallet*.]

sleg-neát, es; n. A beast to be slaughtered :—Hē geselle ǽghwelce gēre tuā slegneát (slægnæt, Chr. 852; Erl. 67, 39), Ch. Th. 105, 4. [Cf. *Icel.* slag-á *a ewe to be slaughtered*.] Cf. sliht-swín.

sleht, sleów, slēpan (*to sleep*), slēpan (*to drag*), slī. v. sliht, slíw, slǽpan, slípan, slíw.

slíc (?); adj. I. sleek, smooth. v. slícian. II. cunning, crafty, using smooth words (v. words given under slícian):—Ic wæs ána slícera đonne ealle óđre drýas sapientior eram omnium sapientium magorum, Nar. 50, 19. [Prompt. Parv. slyke or smothe lenis. With browis smothe and slyke (rimes with chike), Chauc. R. R. 542. Thowe make hem slyke and fatte ynough, Pall. I, 689. Icel. slíkr sleek.]

slíc[c] (?), es; n. A hammer:—Sleánde slicc (slicc for slecg?) mallei percutientes, Kent. Gl. 723, see the note. Hé sceal habban . . . slic (in a list of weaver's implements; slíc (?) an implement for smoothing what is woven, a sleek-stone, cf. slykston amethon, Wülck. Gl. 563, 26: letatorium, 593, 19. Slekstone lacinatorium, Wrt. Voc. i. 218, 2. A slikestone lucchier, 172, 15. See also Prompt. Parv. 458, note 2), Anglia ix. 263, 15. v. sliccan and slícian.

sliccan (?) to strike, slap (cf. (?) colloquial lick = to beat. Halliwell gives slick as an Oxfordshire word for a blow, slap):—Se đe his wiel slicþ (slieþ(?), slihþ (?)) mid girde qui percusserit servum suum virga, Ex. 21, 20. Gif men cídaþ and hira óđer his néxtan mid fýste slicþ(?), and hé deád ne biþ . . . hé biþ unscildig, đe hine slóh, 21, 18-19. Gif hwilc slicþ eacniende wíf, 21, 22. v. slíc[c].

slícian; p. ode To make sleek, smooth, or glossy:—Heó glytenode swá scýnende sunne ođđe nígslýcod hrægel, Shrn. 149, 8. [v. Prompt. Parv. 458, note 2, where 'to sleek clothes' is quoted from Kennett, and a passage from Walter de Bibelesworth is given (v. also Wrt. Voc. i. 172, 13): la dame ge ta koyf luche (slike). Til sleuth and slepe slyken his sides, Piers P. 2, 98. The word is also applied to making a fair show in speech:—Alle þine wordes beoþ isliked, And so bisemed and biliked, O. and N. 841. Wordes afaited and ysliked, Ayenb. 212, 2. He can so well his wordes slike, Gower ii. 365, 22. See, too, Jamieson's Dictionary, sleekie fawning and deceitful; sleekit smooth, shining (of the face); but also, deceitful; sleekit-gabbit smooth-tongued.] v. slíc.

slídan; p. slád; pp. sliden To slide, slip, fall. I. of actual movement, to slide, glide:—Đá cómon twegen deóflu tó him of đære lyfte slídan, Guthl. 5; Gdwin. 30, 16. **II. fig. to make a mistake, to fail, err:**—Đonne hé geong fareþ, hafaþ wilde mód, slídeþ geneahhe (makes many a slip), Salm. Kmbl. 758; Sal. 378. **III. to fall into an unhappy condition:**—Gif seó sáwl slídan sceal in đa écan wíte, Wulfst. 187, 16. **IV. to pass away, be transitory or perishable:**—Đeós mennisce tyddernes biþ swá slídende swá glæs, đonne hit scínþ and đonne tóbersteþ; ac Godes wuldor nafaþ nǽnigne ende, Shrn. 119, 23. Floeg đú wesan ealdor slídendes plegan (labentis ludi), Lchdm. i. lviii, 2. [Ɖer on geđ him one in one sliddrie weie, he slit & falleþ sone; and ter monie gođ togederes, . . . ᵹif eni uođ on uorte sliden, þe ođer breideđ hine up er þen he allunge salle, A. R. 252, 10-12. Mony folk slod to helle, H. R. 136, 157. Huanne þe on uot slyt, þe oþer him helpþ, Ayenb. 149, 2. M. H. Ger. slíten.] v. á-, æt-slídan; útásliden.

slíde, es; m. A slip, fall; Ælfc. Gr. 11; Zup. 79, 9. **I. of an actual slip:**—Đá wearþ mé slide and ic him (the horse) of áfeóll lapsus decidi, Bd. 5, 6; S. 619, 18. **II. fig. a slip into misfortune or error:**—Forđǽm hit ǽr hit nolde behealdan wiđ uunyt word, hit sceal đonne niédinga áfeallan for đæm slide, Past. 38; Swt. 279, 5. Đú generedest fét míne fram slide (de lapsu), Ps. Spl. 55, 13: 114, 8. Forwyrd ᵹ slide lapsum, ruinam, Hpt. Gl. 440, 61. Þurh synna slide through falling into sin, Exon. Th. 263, 13; Jul. 349. Slidas lapsus, Hymn. Surt. 7, 17. v. fǽr-slíde.

slíding. v. á-slíding.

slidor; adj. Slippery:—Ýs byþ ungemetum slidor, Runic pm. Kmbl. 341, 15; Rún. 11. Slideres lubrici, Hpt. Gl. 405, 46. Sýn heora wegas þýstre and slidore fiant viae eorum tenebrae et lubricum, Ps. Th. 34, 7. [Prompt. Parv. slydyr lubricus. þu schalt falle, þe wei is slider, O. and N. 956. To a dronke man the wey is slider, Chauc. Kn. T. 406: Gower iii. 14, 8.]

slidor, es; n. (?) I. a slippery, miry place; lubricum:—Turf gleba, sliddor labina (cf. labina a myre, Wülck. Gl. 591, 11: a fenne, 797, 10), sol volutabrum, moor uligo, Wrt. Voc. i. 37, 20-24. Cf. slǽp. **II. In a list giving names of things connected with ships,** slidor glosses pulvini (pulvini machinae quibus naves deducuntur et subducuntur in portum, Du Cange), 56, 54.

slidorian, slidrian; p. ede To slither (in various dialects; Dryden uses sliddering), to slide, slip:—Đonne hié on monigfealdum wordum slidrigaþ dum per multiplicia verba dilabuntur, Past. 38, 6; Swt. 277, 5. Míne fét ne slideredon non sunt infirmata vestigia mea, Ps. Th. 17, 35. Gif hý geseón đæt míne fét slidrien dum commoverentur pedes mei, 37, 16. [Prompt. Parv. slyderyn labo vel labor: O. Du. slideren. Cf. Vondunge is sliddrunge, A. R. 252, 14.]

slidorness, e; f. Slipperiness, a slippery place:—Slidornis lubricum, Blickl. Gl. (Ps. 34, 6): Ps. Spl. T. 34, 8. [Prompt. Parv. slydyrnesse labilitas.]

slíf, sléf, slýf, e: slífe, an; f. A sleeve:—Slýf manica, Wrt. Voc. i. 81, 70. Be slífan gebunden submanicatus, 21, 64. Slýfa manicae vel brachila, 25, 63. Slýfan manice, ii. 55, 23: 87, 58: bracile, 127, 14: manicas, 87, 43. Ǽghwelcere wunde beforan feaxe and beforan sliéfan

---

(sléfan, MS. B.: slýfan, MS. H.) and beneođan cneowe sió bót biþ twýsceatte máre (cf. 45; Th. i. 92, 20 for this double compensation when a wound was not concealed by the hair), L. Alf. pol. 66; Th. i. 96, 30. Synd gesealde from đam abbode ealle neádbehéfe þing, đæt is cugele . . . slýfa (slýfan, MSS. O. T.), gyrdel, R. Ben. 92, 3. Hé one hláf tóbræc and bewand on his twám slýfum, Homl. Th. i. 376, 30. Hé đone hláf gedyde on his twá sléfan, Blickl. Homl. 181, 17. v. earm-slífe.

slífan; p. sláf; pp. slifen To slive ('Slive to cut, slip, or slice off . . . Palsgrave, "I slyve a gylowfloure or any other floure from his branche or stalke."' Baker, Northants Gloss.) [Slyvyn a-sundyr findo, effisso. Cf. also slyvynge, cuttynge a-wey avulsio, abscissio; slyvynge of a tre or oþer lyke fissula. He al hool or of hym slyvere (a slice, cutting), Chauc. T. and C. iii. 138. Sliver = slice still used in Scotland. v. Jamieson's Dict.] v. tó-slífan.

slífan, sléfan; p. de To slip or put a garment on a person:—Hé hine sylfne ungyrede, and đæt reáf đe hé on hine hæfde hé sléfde on đone foresprecenan man . . . Sóna swá hé mid đan hrægle swá miccles weres gegyred wæs, Guthl. 16; Gdwin. 68, 18. [Slíve to dress carelessly, Cumb. A garment rumpled up about any part of the person is said to be slived. Sliver a short slop worn by bankers or navigators, Linc. It was formerly called a sliving. The sliving was exceedingly capacious and wide. Halliwell's Dict.] Cf. slípan, slíf (?), slífe-scóh.

slífe. v. slíf.

slífe-scóh a loose shoe easily drawn on, a slipper:—Socc, slébescóh soccus, Wrt. Voc. i. 120, 69. Cf. slífan, slípe-scóh.

slíf-leás; adj. Sleeveless:—Sléfleás scrúd colobium, sléfleás ancra scrúd levitonarium, Wrt. Voc. i. 40, 20, 21. Hæbban hý scapulare, đæt is gehwǽde cugelan and slýfleáse, R. Ben. 89, 13.

slifor; adj. Slippery, deceitful (?):—Slideres ᵹ sliferes lubrici, Hpt. Gl. 405, 46. [Cf. sliverly cunning, deceitful, Linc. Halliwell's Dict.] Cf. slipor.

sliht, sleaht, sleht, slieht, sliét, slyht (see the cpds.), es; m. I. a striking of coin. v. pening-sliht. II. a stroke, flash of lightning. v. líget-sliht. III. slaughter, death by violence:—Ðes sliht haec caedes, Ælfc. Gr. 9, 27; Zup. 53, 4. Æt eallum slyht[e?] and æt ealre đære hergunge đe ǽr đam gedón wǽre, ǽr đæt friđ geset wǽre . . . nán man đæt ne wrǽce ne bóte ne bidde, L. Eth. ii. 6; Th. i. 288, 1. Hú hé mid forhergiunge and mid heora mǽga slihtum on his geweald geniédde, Ors. 2, 5; Swt. 82, 17: 5, 11; Swt. 238, 5. **III a. the deadly stroke of disease:**—Đis folc is mid swurde đæs heofonlícan graman ofslegen, and gehwilce sind mid fǽrlícum slihte áwéste, Homl. Th. ii. 124, 10. **IV. what is to be killed, animals for slaughter. v. sliht-swín (cf. Icel. slátr butcher's meat; slátra to slaughter cattle):**—Gafolswáne gebyreþ đæt hé sylle his slyht be đam đe on lande stent. On manegum landum stent đæt hé sylle ǽlce geáre .xv. swýn tó sticunge, L. R. S. 6; Th. i. 436, 11. [Kath. slaht· Laym. slaht, slæht, sclæht, slezht: R. Glouc. slaᵹt. O. Sax. man-slahta; f.: O. Frs. slachte a blow, mortal blow; stamp, coining: O. H. Ger. slahta strages, occisio: Icel. sláttr; m. mowing; striking of an instrument.] v. fiđer(-el?)-, for-, hand-, hlóþ-, líget-, mǽg-, mann-, morþ-, morþor-, pening-, þeóf-, wæl-sliht; cf. slege.

sliht (?); adj. Level, smooth; in the cpd. eorþ-slihtes level with the ground (?):—Swá swá oxa gewunaþ tó áwéstenne gærs óþ đa wirttruman eorþslihtes mid tóþum (eats the grass to the root, to the level of the ground), Num. 22, 4. [Goth. slaihts wigs a level road: O. H. Ger. sleht planus: Icel. sléttr plain, level.]

slihtan; p. te To smite, slay:—Gif đú fallas ᵹ slæhtas cadens (translator seems to have read caedens in the second case), Mt. Kmbl. Lind. 4, 9. [Cf. O. H. Ger. slahtón mactare: Ger. schlachten.]

sliht-swín, es; A swine to be killed:—Gýme eác swán đæt hé æfter sticunge his slyhtswýn wel sæncge, L. R. S. 6; Th. i. 436, 16. [Cf. Ger. schlacht-vieh cattle to be killed; Ger. schlacht-neát. Cf. Ger. schlacht-vieh cattle to be killed.] v. sleg-neát.

slím, es; m. (?) n. (?) Slime, mud, mire:—Slím limus, Wrt. Voc. ii. 54, 14: borbus, cena, 126, 53. Ásæstnod ic eom on líme (slíme? cf. Ps. 68, 3: I am festened in slime depenense) grundes infixus sum in limo profundi, Ps. Spl. 68, 2. [M. H. Ger. slím; m.: Ger. schleim: Du. slijm: Icel. slím; n.]

slincan; p. slanc, pl. sluncon. I. to crawl:—Eodon đa wyrmas and sluncon wundorlíce, wǽron him đa breóst up gewende, Nar. 14, 8. Slincendes reptantis, Hymn. Surt. 28, 17. Hé gescóp eall wyrmcynn and creópende and fleógende and swymmende and slincgende, Anglia viii. 310, 17. **II. fig. to slink away:**—Se earma flýhþ uncræftiga slǽp slincan on hinder, Dóm. L. 240. [Cf. O. H. Ger. slíhhan repere, reptare.] v. next word.

slincend, es; m. n. A crawling thing, a reptile:—Ealle slincendu (Ps. Spl. slincende) omnia reptilia, Ps. Lamb. 68, 35: 103, 25. Fram đam slincendum óþ đa fugelas, Gen. 6, 7.

slingan; p. slang, pl. slungon To wind, twist, worm, move as a serpent. Cf. sling to move quickly, Var. dial. It also has the same meaning as slinch (slink). Halliwell's Dict.:—Gif heó (the adder) wat heáfod innan đone man bestingþ đonne slingþ (= slincþ?) heó mid ealle inn if it strikes its head into the man, then it winds itself quite in, Boutr. Scrd. 20, 15. [O. H. Ger. slingan: Ger. schlingen to wind: Icel. slyngva to wind.]

slipa (slypa?), an; *m. A viscous, slimy substance :*—Genim sealh and ele dó ahsan (tó?) gewyrc ðonne tó slypan . . . dó ðonne on ðone slipan, Lchdm. ii. 18, 26–28. Wyrc slypan of wætere and of axsan, iii. 38, 1. v. slipig, slipor, *and* slyppe.

slípan (?); *p.* sláp, *pl.* slipon *To slip, glide.* [He with feigned chere him slipeth (rimes with wipeth) *he slips off,* Gower ii. 347, 30. *Slype* to move freely, as any weighty body which is dragged through a mire, Jamieson's Dict. *O. H. Ger.* slífan *labi.*] Cf. slipor, *and see* slúpan.

slipan, slépan; *p.* te *To slip, put* something on or off. Cf. *slipe* to take away the outside covering from anything, Halliwell's Dict. *Slype* to strip off the skin or bark of anything, Jamieson's Dict. :—Se hláford hefig gioc slépte on ða swyran sínra þegena, Me . 9, 55. Se cyning slýpte his beáh of *the king slipped his ring off ;* tuiit rex annulum de manu sua, Anglia ix. 32, 158. [*Goth.* af-slaupjan thana fairnjan mannan *to put off the old man :* O. Sax. slópian *to slip* one's self from a bond : M. H. Ger. sloufen, ana-sloufen *induere.*] v. be-slépan ; un-slíped, slípe-scóh, slúpan ; *and* cf. slífan.

slípe-scóh *a slip-shoe* (Halliwell gives the word from a work dated 1615. Cf. *slip-shod*), a shoe easily slipped on, a slipper :—Slýpescós *soccus,* Wrt. Voc. i. 289, 7. v. slípan ; slífe-scóh.

slipig; *adj. Slippy, slimy, viscid :*—Mid slipigre and þiccere wætan, Lchdm. ii. 280, 4. Ða þiccan and ða slipigan (slipinga, MS.) wætan on ðam magan and ðæt þicce slipige horh ðú scealt mid ðam lǽcedómum wyrman and þynnian, 194, 20–22. Wǽtan þicce and slipegran, 178, 15. Of þiccum wǽtum slipegrum . . . Wið slipegrum wǽtum ðæs miltes, 246, 17. [*M. H. Ger.* slipfic.] v. next word.

slipor; *adj.* I. *slippery, not easy to hold, moving easily :*—Deófol nǽddre ys slipor ðæs gif heáfde ná byþ wiðstanden eall on innemystum heortan ðænne ná byþ ongyten byþ ásliden *diabolus serpens est lubricus, cuius si capiti non resistitur, totus in interna cordis, dum non sentitur, inlabitur,* Scint. 210, 9. II. *slipping easily, easily moved :*—Ymhídignysse ofþriccaþ ðæt mód, and unlustas tólýsaþ ; þwyrlíce þing ðe heora hláfordas dóþ geswencte fram carum, and slipere þurh unstæðdignysse, Homl. Th. ii. 92, 16. III. *foul :*—Fúl ne sý oððe slipor *nec feda sit nec lubrica,* Hymn. Surt. 5, 9. Ælc þing slipores ï fúles *omne lubricum,* 30, 9. Bedǽled angdite sliporum ï fúlum *excita sensu lubrico,* 3, 17. Gilt sliporne ï fúlne *culpam lubricam,* 15, 38. Ne tunge leás ne eágan syngian slipere *ne lingua mendax occulive peccent lubrici,* 24, 27. [Sliper *lubricum,* Ps. 34, 6. Nares gives several instances of *slipper* in sixteenth century, and Shakspere uses the form : A *slipper* and a subtle knave, Oth. ii. 1. *O. H. Ger.* slefar, Grff. vi. 506: *M. H. Ger.* slepfer.] Cf. slíder ; slǽpe, slípan (?).

sliporness, e; *f. Foulness :*—Beón út ánýdde slipornesse *sint pulsa lubrica,* Hymn. Surt. 36, 16.

slip-ræsn *a sliding beam* (?) :—Slypræsn *ferna,* Wrt. Voc. ii. 147, 75.

slipung (?), e; *f. Viscidity :*—Wið slipunge (slipigre? *the text has* slípegrum wǽtum. v. slipig) wǽtan ðæs miltes, Lchdm. ii. 166, 24.

slit. v. ge-, lah-slit.

slítan ; *p.* slát, *pl.* sliton ; *pp.* sliten *To slit, tear, rend.* I. *in the following glosses :*—Sclát *carpebat,* Wrt. Voc. ii. 103, 51. Bítende and slítende *mordax,* 57, 52. Slítende *mordens,* Kent. Gl. 580 : *corrumpens,* Hpt. Gl. 454, 68. Ic þeó sliten *carpor,* Wrt. Voc. ii. 21, 40. Wǽran slitene *carpebantur,* 22, 22. II. *to tear* a garment, rend :—Ðæra sacerda ealdor slát (*scidit*) hys ágyn reáf, Mt. Kmbl. 26, 65. Ne slíte wē hý *non scindamus eam,* Jn. Skt. 19, 24. Se heáhsacerd his reáf slítende, Mk. Skt. 14, 63. III. *to tear, split, rend, cleave, divide :*—Hē slát sǽ *interrupit mare,* Ps. Spl. 77, 16. Hē slát stán *interrupit petram,* 77, 18. IV. *to tear, rend,* as an animal does with the teeth or feet, a bird with its beak, etc. v. slite II, slítung :—Fótum ic fēre, foldan slíte, Exon. Th. 393, 17 ; Rä. 13, 1. Hrefn hine slíteþ, 329, 20 ; Vy. 37. Hine se wulf slíteþ, 342, 27 ; Gn. Ex. 148. Hē (*the evil spirit*) bítes and slítes hine, Mk. Skt. Rush. 9, 18. Heora heortan wyrmas ceorfaþ and slítaþ, Dóm. L. 12, 168 : 14, 210 : Exon. 497, 5 ; Rä. 85, 24. Hē (*Grendel*) slǽpendne rinc slát, Beo. Th. 1487 ; B. 741. Ða wyrmas mid ðǽm scillum gelíce mid ðē múþe eorþan sliton and tǽron *oribus scamisque humum atterentes,* Nar. 14, 12. Gif hund slíte, Lchdm. ii. 92, 10. Hié (*lions and bears*) noldon slítan hý (*St. Tecla*), Shrn. 133, 10. Gesáwon fuglas slítan, Cd. Th. 126, 1 ; Gen. 2088. Ðē sculon moldwyrmas slítan, Soul Kmbl. 145 ; Seel. 73. Hió (*the lioness*) onginþ racentan slítan (cf. brecan, Bt. 25 ; Fox 88, 13), Met. 13, 29. Se unclǽna gást hine slítende (*discerpens*), Mk. Skt. 1, 26. Slítende wulfas *ravening wolves,* Blickl. Homl. 63, 10. Slítendum ï terendum tóþreómum *rabidis (voracibus) gingivis,* Hpt. Gl. 423, 42. IV a. *fig. applied to inanimate subjects :*—Nú slít mē hunger and þurst, Cd. Th. 50, 2 ; Gen. 302. Hungor innan slát merewērges mód, Exon. Th. 306, 22 ; Seef. 11. Hí beóþ mec slítende (*of the waves tearing at an anchor*), 398, 11 ; Rä. 17, 6. V. *to tear, bite* (of pungent things, cf. slitol), *irritate* (of physical or mental irritation) :—Slíto (suto, Wrt., cf. slítung) *lacesso,* Wrt. Voc. ii. 112, 29. Slíteþ *lacessat,* 95, 32. Ðæt wín slít ða wunda *per vinum mordentur vulnera,* Past. 17, 10 ; Swt. 125, 9. Sliten oððe gremeden *lacessant,* Wrt. Voc. ii. 52, 54. Of yfelre wætan slítendre, Lchdm. ii. 4, 30. Of yfelum

wǽtan slítendum and sceorfendum, 60, 21. VI. *to tear* (fig.), *to destroy, waste, consume.* v. slítendlíc, slítere, slítness II :—Nán cræft nis Gode deórwyrðra ðonne sió lufu ne eft ðæm deófle nán cræft leóftǽlra ðonne hié mon slíte *nihil pretiosius est Deo virtute dilectionis, nil est desiderabilius diabolo extinctione caritatis,* Past. 47, 2 ; Swt. 359, 24. Tó slítenne (breccanne, Rush.) ae *solvere legem,* Mt. Kmbl. Lind. 5, 17. VII. *to carp at, back-bite.* v. bæc-slitol :—Æt ǽrestum lyst ðone monn unnyt sprecan be óðrum monnum & ðonne æfter firste hine lyst tǽlan and slítan ðara líf bútan scylde *ut prius loqui aliena libeat, postmodum detractionibus eorum vitam mordeat,* Past. 38, 7 ; Swt. 279, 7. VIII. *to tear* (intrans.) :—Godwebba cyst (*the veil of the temple*) eall forbærst . . . ðæs temples segl . . . sylf slát on tú, swylce hit seaxes ecg þurhwóde, Exon. Th. 70, 19 ; Cri. 1141. [*Prompt. Paru.* slytyñ *attero : O. Sax.* slítan *to tear, split :* O. Frs. slíta *to tear, break :* O. H. Ger. slízan *scindere, lacerare, laniare, lacessere, saevire, delere : Icel.* slíta *to slit, tear, break.*] v. á-, be-, for-, ge-, of-, tó-slítan ; wæl-slítende, sliten, un-sliten.

slit-cwealm *death by the tearing of animals :*—Neát ðe slitcwealm begéte *animalia quae lacerationem mortiferam nacta sunt,* L. Ecg. C. 40 ; Th. ii. 166, 24.

slite, es ; *m.* I. *a slit, tear, rent* in cloth, etc.:—Se slite byþ wyrsa *pejor scissura fit,* Mt. Kmbl. 9, 16 : Mk. Skt. 2, 21. II. *a rent, tear* made by an animal, *a bite.* v. slítan, IV :—Wið hundes slite, Lchdm. i. 148, 7. Íces slite oððe hundes, ii. 86, 2. Be hundes slite. Gif hund mon tóslíte oððe ábíte, L. Alf. pol. 23 ; Th. i. 78, 1. Wið nædran slite, Lchdm. ii. 10, 21. Wyrma slite, Exon. Th. 77, 4 ; Cri. 1251. Slita *morsuum,* Germ. 392, 30. III. *a coil* of a snake (?) :—Nædre sprotum slitas (?) líces clyniende *vipera sarmentis laqueos corporis inplicans,* Germ. 401, 24. IV. *a breach, infraction* of a law. v. lah-slit. [*O. H. Ger.* sliz : *Ger.* schlisz ; *m.* : cf. *Icel.* slit ; *n.*] v. folc-, lah-, wyrm-slite.

slíte, an (?) ; *f. A plant name, cyclamen, sowbread :*—Slíte. Ðeós wyrt ðe man *orbicularis* and óðrum naman slíte nemneþ, Lchdm. i. 110, 11. Slíte *cyclaminos,* iii. 301, col. 2 : *cyclamen,* Wrt. Voc. i. 67, 53 : *ciclamina,* ii. 131, 37.

sliten *schismatic, heretic :*—Slítenum *haereticis,* Mt. Kmbl. p. 10, 9. Lye gives sliterum (slitenum ?) *sagum haereticis fabulis,* Josc. (?). v. slítan.

slítend-líc ; *adj. Consuming, devouring, wasting.* v. slítan, VI :—Slítendlícum *lurconibus,* Wrt. Voc. ii. 52, 71.

slitenness (?), e ; *f. Tearing, laceration :* — Sliten[nesse] *morsum, lacerationem,* Hpt. Gl. 490, 62.

slítere, es ; *m.* I. *a waster, destroyer :*—' Hwæt is seó ungesǽlige sáwel?' Ðá sǽde hē him, ðæt hē wǽre cyrican slítere, Wulfst. 235, 24. II. *a consumer* of food, *a glutton :*—Slíteras *lurcones,* Wrt. Voc. ii. 52, 26. v. slítan, VI.

slípan *to harm, hurt, damage, destroy :*—Heoro slípendne, Exon. Th. 346, 10 ; Gn. Ex. 202. [Cf. *Goth.* ga-sleithjan *to injure.*] v. next word.

slíþe ; *adj. Dire, hard, cruel, hurtful, dangerous :*—Biþ ceóle wēn slíþre sæcce *the ship may expect dire strife,* Exon. Th. 384, 17 ; Rä. 4, 29. On ða slíþan tíd (*the crucifixion*), Elen. Kmbl. 1710 ; El. 857. Þurh slíþne níþ sáwle bescúfan in fýres fæþm, Beo. Th. 370 ; B. 184. Hē níþa gehwane genesen hæfde, slíþra geslyhta, 4787 ; B. 2398. [*Goth.* sleithis *dangerous, perilous, fierce : O. Sax.* slíði *dangerous, destructive, cruel : cf. O. H. Ger.* slídíc, *saevus, malus : Icel.* slíðr *fearful, dire ;* slíðr-hugaðr *atrocious ;* slíðr-liga *savagely.*] v. slíþen.

slíþe ; *adv. Cruelly :*—Bearn ðara ðe ofslegene slíþe wǽran *filios interemtorum,* Ps. Th. 101, 18.

slíþe (?) ; *adj. Formed, moulded ;* fictus. I. *graven* (of images) :—Ealle ðe gebiddaþ ða slíþan *omnes qui adorant sculptilia,* Ps. Spl. T. 96, 7. Hí offrodon ðæ slíððan *sacrificaverunt sculptilibus,* 105, 35. II. *feigned, false :*—Hē oncneów slíþe mód úre *cognovit figmentum* (taken by the translator = *fictam mentem?*) *nostrum,* Ps. Spl. T. 102, 13. v. slíþness, *and next word.*

slíþelíc ; *adj. Graven :*—Gebǽdon ða slíþelecæn *adoraverunt sculptile,* Ps. Spl. T. 105, 19.

slíþen ; *adj. Cruel, hard, evil :*—Slíðen *infastum,* Wrt. Voc. ii. 111, 66. Ðú wēndest ðæt ðiós slíþne wyrd ðæs worulde wende bútan Godes þeahte, Bt. 5, 3 ; Fox 14, 4. Fin eft begeát sweordbealo slíþen *dire harm from the sword overwhelmed Fin,* Beo. Th. 2298 ; B. 1147. Hú slíþen biþ sorg tó gefēran *how cruel is care as a comrade,* Exon. Th. 288, 12 ; Wand. 30. Hē him feorgbona þurh slíþen searo weorþeþ *a destroyer of life through cruel craft to him he becomes,* 362, 25 ; Wal. 42. On ða slíþan tíd *at that dread hour* (*of death*), 161, 27 ; Gú. 965. In ða slíþnan tíd *in the evil days of the present life,* 316, 22 ; Mód. 52.

slíþ-heard ; *adj. Excessively hard.* I. of living things, *very fierce, savage :*—Slíþherde deór (*the boar and the bear*), Exon. Th. 344, 22 ; Gn. Ex. 177. II. of inanimate things, *very hard, cruel :*—Mē habbaþ hringa gespong slíþhearda sál síþes ámyrred *the cruel chain has hindered me from going,* Cd. Th. 24, 15 ; Gen. 378.

slíþness, e ; *f. A formation* (?), *a graven image :*—Hí þeówedon slíþnesse *servierunt sculptilibus,* Ps. Spl. T. 105, 33. v. slíþe (?).

slítness (slit- ?), e ; *f.* I. *a tearing, rending, laceration.* v.

slítan, IV :—Ða slítnysse gedígean *a laceratione* (by wolves or dogs) *convalescere*, L. Ecg. C. 40; Th. ii. 166, 25.   II. *a wasting, destroying, desolation*. v. slítan, VI :—Slítnese *desolationis*, Mt. Kmbl. Lind. 24, 15.   v. from-, tó-slítness.

slitol; *adj.*   I. *pungent, biting*. v. slítan, V :—Slitul léc *mordax allium*, Germ. 394, 260.   II. *carping, backbiting*. v. bæc-slitol, slítan, VII.

-slitt. v. lah-þrí-slitt.

slítung, e; *f.*   I. *tearing, rending, biting*. v. slítan, IV :—Slítinc ł geter *dilaceratio*, Hpt. Gl. 499, 21. Fugelas hig fretaþ mid ðære biterustan slítunge *devorabunt eos aves morsu amarissimo*, Deut. 32, 24. Sume men fram ðara wyrma slítunge sweltaþ, Lchdm. ii. 176, 14.   II. *wasting, spoiling*. v. slítan, VI :—Slítunge *arpagine* (or under I ?), Wrt. Voc. ii. 5, 38 : 87, 72 (*Wright has* sutunge). [*Prompt. Parv.* slytynge *consumpcio: O. H. Ger.* slízunga *saevitia*.]

slíw, sleów, slíu, slí, es; *m.* The name of a fish, *a tench* or *a mullet* :—Síw *tinca*, Wrt. Voc. i. 55, 73 : *tinctus*, 281, 52. Sliú *tincus*, 66, 1. Sleów *mugilis*, ii. 57, 75. Slí *tincti*, Txts. 101, 2020. Slíi, 116, 221. [*O. H. Ger.* slío; *m. tinca, tincus: Ger.* schleie *a tench*.]

slóh, slóg; *gen.* slóges, slós; *dat.* slóg, slí; *acc.* slóg, slóh, slí; *m. n. A slough, hollow place filled with mire, a pathless, miry place* :—Slóh *devium*, orwegnes *devia*, s. *loca secreta, quasi invia, sine via*, Wrt. Voc. ii. 139, 53–56. Tó ðam ealdan slí; of ðam slí tó ðam lytlan beorhe, Cod. Dip. Kmbl. iii. 38, 27. In reádan slóe, 391, 31. On ðæt reáde slóh; of ðam slóh, 376, 5. On ðæt fúle slóh; of ðam slí, 406, 32. In ðone fúlan slí, 381, 5. On horgan slóh, Cod. Dip. B. ii. 394, 30. On reádan slóh, 398, 38. Ðæt hors sum slóg on ðam wege oferhleóp *equus quoddam itineris concavum transiliret*, Bd. 5, 6; S. 619, 17. Ðeáh se man nime ænne stán and lecge on fúl slóh, Wulfst. 239, 10. [Heo arist up of þe slo, O. and N. 1394. He hath also to do more than ynough To kepe him and his capel out of slough, Chauc. Mancip. Prol. 64. Skeat takes this to be a word borrowed from Celtic. v. Etym. Dict.]

slop *a loose, upper garment*. 'Slop a smock-frock; any kind of outer garment made of linen,' Halliwell's Dict. [These cuttid sloppis or anslets, that thurgh her schortness ne covereth not the schamful membres of men, Chauc. Pers. T. Sloppe, garment *mutatorium*, Prompt. Parv. 460, col. 1. Icel. sloppr *a gown, a loose garment, esp. a priest's gown*.] v. ofer-slop, *and cf.* slípan, slype.

-sloppe. v. cú-slyppe.

slota, an; *m. A bit, morsel* :—Betere ys slota (cf. bite, Kent. Gl. 587) drýge mid blisse ðænne hús full mettum mid sace *melior est bucella sicca cum gaudio quam domus plena uictimis cum iurgio*, Scint. 153, 12. [Lye gives sloca *bucella*, with a reference to Past. 47, an error probably for Scint. 47. If this were the form the word might be compared with Ger. schlucken : but Halliwell gives *slot* a small piece.]

sluma, an; *m. Slumber* :—Sleac mid sluman, Dóm. L. 240. Ðæt hine elne binóman slæpa sluman oððe sæne mód, Exon. Th. 122, 31; Gú. 314. [Upon a sloumbe, A. P. 97, 186. Cf. *Laym.* slumen *to slumber*.]

slúpan; *p.* sleáp; *pl.* slupon; *pp.* slopen *To slip, glide* :—Sóna swá ús seó sáwl of ðam líchaman slýpþ *simul atque anima de corpore se subduxerit*, L. Ecg. P. iv. 66; Th. ii. 224, 23. Gársecg wédde on sleáp (*of the Red Sea coming upon the Egyptians*), Cd. Th. 208, 28; Exod. 490. Hwílum ic wægfatu wíde tóþringe ... hwílum léte eft slúpan tósomne *sometimes I (the storm) drive apart the clouds, sometimes make them again glide together*, Exon. Th. 385, 3; Rä. 4, 39. [*Goth.* Thaiei sliupand in gardins *they which creep into houses*, 2 Tim. 3, 6. *O. H. Ger.* sliufan *to slip, creep*.] v. á-, tó-slúpan.

slúping, slýcod, slýf, slypa, slýpan. v. tó-slúping, slícian, slíf, slipa, slípan.

slype *a garment, slip*. [Slyp or skyrte *lascinia*, Prompt. Parv. 459, col. 2. *Slip* a child's pinafore; an outside covering, as a pillow-*slip* (= -case): in earlier times, a sheath, Halliwell's Dict. *Slip* an upper petticoat, Jamieson.] v. ofer-slype, slop.

slýpe-scóh. v. slípe-scóh.

slyppe, an; *f. A viscous, slimy substance* :—Wyrc slypan of wætere and of axsan, genim finol, wyl on ðære slyppan, Lchdm. iii. 38, 2. [Cf. slyp, slype, slypp *limus*, Prompt. Parv. 459, col. 2.] v. cú-, oxan-slyppe, *and* slipa.

slyp-ræsn. v. slip-ræsn.

smacian; *p.* de *To smack, pat, caress* :—Ic smacige *demulceo*, Hpt. Gl. 476, 72. [Cf. *Du.* smak *a loud noise: Dan.* smække *to smack, slap: Swed.* smacka.] v. ge-smacian.

smæc[e], es; *m. Smack, taste, savour* :—*Dulcis sapor* swéte smæc, i. *dulcis odor*, Wrt. Voc. ii. 142, 6. Ðone swétan smæc *nectar*, 61, 31. [Witt iss þurrh salltes smacc bitacnedd, Orm. 1653. Smech muðes & neoses smel, A. R. 276, 15. Smeorðrinde smoke smecche forcuðest, Marh. 9, 6. Maken he uynt smak in ane zoure epple þanne in ane huetene lhoue, Ayenb. 82, 21. Smak or taste *gustus*, Prompt. Parv. 460. *O. Frs.* smek[k]: *O. H. Ger.* smac (*dat.* smacche) *gustus, sapor*. v. hunig-smæc.

smæccan, smecgan; smæhte *To taste* :—Ic smæcce (smæcge, MS. J.) *sapio*, Ælfc. Gr. 28; Zup. 166, 6. ['Cum gustasset acetum noluit bibere;' þet is, he smeihte þet bittre drunch & wiððrouh him anon, A. R. 238, 21. Summe þinge þ me hauað ismeiht oðer smelled, 92, 4. Al þet ich abbe mid muþ ismaht, O. E. Homl. i. 189, 5. Unlouely þei smauȝte, Piers P. 5, 363. *O. Frs.* smekka: *O. H. Ger.* smecchen *sapere*.] v. ge-smæccan, -smecgan, *and preceding word*.

smæl; *adj. Small*.   I. in the following glosses :—Smæl *gracilis*, smælre *gracilior*, ealra smælst *gracillimus*, Ælfc. Gr. 5; Zup. 16, 8. Smel, smael, smal, Txts. 67, 992. Smæl *gracilis vel exilis vel subtilis*, Wrt. Voc. i. 51, 18. Greát and smæl *grossas et graciles*, ii. 41, 68.   II. *small, little, not great* :—Smæl þistle *carduus*, Wrt. Voc. i. 66, 66. Smæl æl *anguilla*, 281, 69. Se smala ciið ... se greáta beám, Past. 33; Swt. 224, 3. Æt ælcon smalon orfe penig, L. Ff.; Th. i. 224, 22. Dó tó smale netelan, Lchdm. ii. 68, 4. Smæle þearmas *the small guts*; ilia, Wrt. Voc. i. 44, 46. Ða gnættas and ða smalan wyrmas ... ge þeós lyttle loppe, Bt. 16, 2; Fox 52, 11. Flésc smælra fugla, Lchdm. ii. 180, 13. Smealum bryt (brycum?) *minutatim*, Hpt. Gl. 443, 11. Hæfaþ seó læsse smæle (smale, MSS. H. B.) leáf and gehwæde ... seó óðer hafaþ máran leáf and fætte, Lchdm. i. 264, 18.   III. *narrow, not broad* :—Hé sæde ðæt Nordmanna land wære swýðe lang and swýðe smæl ... ðær býne land is eásteweard brádost, and symle swá norðor swá smælre ... and norðeweard, hé cwæð, ðær hit smalost wære, ðæt hit mihte beón þreora míla brád tó ðæm móre, Ors. 1, 1; Swt. 18, 24–33. Andlangan ðes smalan pæces, Cod. Dip. B. ii. 600, 9.   IV. *slender, thin, not thick* :—Swiora smæl *a slender neck*, Exon. Th. 486, 11; Rä. 72, 15. Him ne hangaþ nacod sweord ofer ðam heáfde be smalan þræde, Bt. 29, 1; Fox 102, 28. Wið ðam smalan wyrme *for hair worm*, Lchdm. ii. 122, 18.   V. *fine* (of a powder, texture, etc.), *not coarse* :—Smæl hláf *artocobus* [*artocopa* (also *-us*) quaevis placenta, panis quidem dulciarius et arte confectus], Wrt. Voc. ii. 10, 47. Tú hund greátes hláfes (*coarse bread*) and þridde smales (*fine*), Chart. Th. 158, 26. Swíðe lytle beóþ ða dropan ðæs smalan rénes, Past. 57; Swt. 437, 12. Cnuca tó swíðe smalan duste, Lchdm. i. 240, 4. Génim swýðe smæl dust, 240, 11. Smæl beren mela, ii. 86, 24. Ásifte smale þurh smæl sife *sift through a fine sieve*, 94, 1 : 72, 28. Hí smalo hrægel wefaþ and wyrceaþ *texendis subtilioribus indumentis operam dant*, Bd. 4, 25; S. 601, 16. Heortes hornes ðæs smælestan dustes, Lchdm. i. 334, 19. Gníd swíðe ðæt hit sý ðæt smælste, iii. 18, 15.   VI. *of the voice, not loud*. v. smale, II. [*Goth.* smals: *O. Sax.* smal: *O. Frs.* smel: *O. H. Ger.* smal *gracilis, exilis, subtilis, minutus, strictus: Icel.* smár; *cf. also* smali *a sheep, small cattle*.] v. æ-smæl.

smæle *finely*. v. smale.

smæll, es; *m. A smack, blow with the open hand* :—Dynt ł smæll mið honde uutearde *alapam*, Jn. Skt. Lind. 18, 22. [Cf. At þan uorme smællen Romanisce veollen, Laym. 27052. *Icel.* smellr *a smacking or cracking sound: Dan.* smæld *a crack, smack: Swed.* smäll.] v. hand-smæll, smellan.

smæl-þearmas, -þyrmas; *pl. m. The small guts, intestines* :—Smælþearmas *intestina*, Wrt. Voc. i. 44, 44 : *inguina*, ii. 44, 4 : *jejuna*, 49, 51. Wið smælþearma sáre, Lchdm. ii. 236, 18. Smælþearmum *ilibus*, Wrt. Voc. ii. 44, 1. Be wambe coþum and tácnum on roppe and on smælþearmum, Lchdm. ii. 230, 16. Hé clænsaþ ðone magan and ða smælþyrmas, i. 80, 21. [*Icel.* smá-þarmar *the small gut, also the lower abdomen*.] v. next word.

smæl-þearme, es; *n. The small gut, lower abdomen* :—Wyrð gegaderodu ðmig wæte on ðære wambe oððe on ðam smælþearme, Lchdm. ii. 218, 17. Síhþ innan ðone rop and on ðæt smælþearme, 232, 15 : 246, 21. Ðá þydde Æfner hine mid hindewerde sceafte on ðæt smælþearme *percussit eum Abner aversa hasta in inguine*, Past. 40, 5; Swt. 295, 18. v. preceding word.

smær[e?], es; *m. A lip* :—Smæras (? *printed* sinæres) *labra*, Hpt. Gl. 457, 39. Reádum smærum *roseis labris*, 481, 25. Smærum *buccis*, 422, 72. Smærum, Lchdm. i. lxx, 6. [Cf. For hire speche he smere loh, Laym. 14981. Tho he (*the fox*) wes inne, smere he lou, Rel. Ant. ii. 272, 23.] v. gál-smere (*where read* gál-smære), *and next word*.

smæran (?); *p.* de *To laugh at, deride* :—Gehlógun ł smérdon (besmerdon? cf. besmeradun *in Rush.*) hine *deridebant eum*, Mt. Kmbl. Lind. 9, 24. v. preceding word.

smæte; *adj. Refined, pure* (of gold) :—Smaete gold *obrizum*, Wrt. Voc. ii. 115, 11. Smæte *obrizum*, 75, 72. Hié wurdan sóna tó ðam golde ðe man háteþ *obritsum*, ðæt is smæte gold, Shrn. 32, 21. Smæte gold ðæt in wylme biþ þurh ofnes fýr eall geclænsod, Elen. Kmbl. 2616; El. 1309. Beág on ðam siex hund wæs smætes goldes gescyred sceatta, Exon. Th. 324, 8; Vid. 91: Salm. Kmbl. 29; Sal. 15. On smætum *in obrizum*, Hpt. Gl. 449, 10. Hé hét smiðian of smætum golde áne lytle róde, Homl. Th. ii. 304, 16: Homl. Skt. i. 2, 113. [Kynehelm of smeate gold, Chr. 1070; Erl. 209, 7. Guldene ȝerde alre gold smeatest, Marh. 11, 24.]

smæte-gylden; *adj. Of refined gold* :—Smætegyldne *obridzum*, Wrt. Voc. ii. 89, 25. Ða smætegyldenan cláþas *auri obriza lammina*, 2, 7.

smale, smæle; *adv.* **I.** *finely* (v. smæl, **V**) :—Hundes tux gebærned and smale gegniden, Lchdm. i. 372, 1. Gegníd tó dúste swýðe smale, 196, 12: 198, 1, 15. Genim wæterhæfern gebærnedne and ðonne gegniden smale, ii. 44, 20. Genim swefl, gebeát swíðe smale, 88, 17: i. 358, 9. Ásift smale þurh smæl sife, ii. 94, 1. Getrifula smale, 90, 27. Ðeáh ðú hié smale tódæle swá dust, Bt. 13; Fox 38, 33. Ic hí tódælde swá smæle and swá swá dust beforan winde *comminuam eos ut 'pulverem ante faciem venti*, Ps. Th. 17, 40. Ðæs dustes smæle gecnucudes, Lchdm. i. 286, 2. Gegníd smæle on mortere, ii. 60, 1. Gebeát smæle, 88, 5. Gegníd tó duste swá ðú smalost mæge, 108, 15. **II.** *of the voice, not loudly* :—Ðæs cocces þeáw is ðæt hé micle hlúdor singþ on úhtan ðonne on dægréd ac ðonne hit neálǽcþ dæge ðonne singþ hé smælor and smicror *gallus profundioribus horis noctis altos edere cantus solet; cum vero matutinum jam tempus in proximo est, minutas ac tenues voces format*, Past. 63; Swt. 461, 3.

smalian; *p.* ode *To become small, slender*, etc. :—Fram mettum smaligan *to get slender by diet*, Lchdm. ii. 282, 29. [*Prompt. Parv.* smaliñ *minoro.*] v. next word.

smalung, e; *f. Diminishing, lessening* :—Lǽcedómas ða ðe þynnunge mægen habben and smalunge *medecines that have the power of thinning and reducing*, Lchdm. ii. 260, 23.

smeágan, smeán; *p.* smeáde; *ppr.* smeágende, smeánde; *pp.* smeád. **I.** *in the following glosses* :—Ic smeáge *scrutor*, Ælfc. Gr. 25; Zup. 145, 3: *meditor*, Wrt. Voc. i. 50, 3. Smeáþ *investigabit*, Kent. Gl. 652. Smeáde *disputavit*, Wrt. Voc. ii. 25, 61. Smégan *investigare*, Kent. Gl. 953. Tó smyágenne *tractanda*, 749. **II.** *used absolutely, or with prepositions* (be, on, ymbe), *to consider, meditate, inquire, deliberate* :—Ic smégu *meditabor*, Ps. Surt. ii. p. 185, 3. Hé smeáþ on his móde ymb ðis eorþlíce líf, Bt. 39, 7; Fox 224, 4. Be ðam gé smeágeaþ *de hoc quaeritis*, Jn. Skt. 16, 19. Ða senatores dæghwamlíce smeádon on ánum sindrian húse embe ealles folces þearfe, Thw. p. 161, 33: Nicod. 19; Thw. 9, 10: Homl. Skt. i. 3, 44. Ðá hig mid him smeádon *dum secum quaererent*, Lk. Skt. 24, 15. Smeá (smeáge, Lind.: smeóge, Rush.) and geseoh ðæt . . . *scrutare et vide quia*, Jn. Skt. 7, 52. Ðeáh wé ofer úre mǽþ þencen and smeágean, Past. 16; Swt. 101, 11. Ic mid eallum mínum ealdormonnum wæs smeágende be ðære hǽlo úrra sáwla, L. In. prm.; Th. i. 102, 7. Smeágende ymbe heora sáwla ærǽd, L. Edm. S. prm.; Th. i. 244, 5. Ymb his ǽ hé byþ smeágende *in lege ejus meditabitur*, Ps. Th. 1, 2. On eallum ðínum weorcum ic wæs smeágende, 76, 10. **III.** *to consider, ponder, examine, inquire into, discuss, search*, (1) *with acc.* :—Ðenden ic Godes bebodu smeáge *scrutabor mandata Dei*, Ps. Th. 118, 115. Ne sécþ hé nánwuht, ne ne smeáþ, for ðam ðe hé hit wát eall, Bt. 42; Fox 258, 1. Hwí smeágaþ hí unnytt *quare populi meditati sunt inania*, Ps. Th. 2, 1. Hwæt smeáde gé be wege *quid in via tractabatis*, Mk. Skt. 9, 33. Drihten, smeá míne geþohtas, Ps. Th. 25, 2. Smeágeaþ (smeás gié, Lind.: smeógas gé, Rush.) hálige gewritu *scrutamini scripturas*, Jn. Skt. 5, 39. Ðæt hé his ágene dǽda georne smeáge, Blickl. Homl. 109, 12. Ðeáh wé fela smeán (smeágen, Cott. MS.), wé habbaþ litellne gearowitan búton tweón, Bt. 41, 5; Fox 254, 9. Ðæt ic smeáde (*meditarer*) sprǽce ðíne, Ps. Spl. 118, 148. Ðú woldest míne láre smeágean, 22, 1; Fox 76, 25. Ic ðé sende ðæt spell tó rǽdanne and tó smeágeanne (*ad legendum ac probandum*), Bd. pref.; S. 471, 10. Com Mellitus tó Róme be ðam nýdþearflícum intingum Angelcyricean and hé ða wæs smeágende mid ðone pápan *venit Mellitus Romam de necessariis ecclesiae Anglorum cum papa tractaturus*, 2, 4; S. 505, 30. Godes mærþa smeágende, H. R. 105, 8. Scmegende wes *scrutata est*, Ps. Surt. 118, 129. Smégende (smeánde, Ps. Spl.), 118, 70. Biþ smeád *meditabitur*, 36, 30. (2) *with a clause introduced by ðæt*, hú, hwilc, hwæt, etc. :—Smeádon men oft, and gyt gelóme smeágaþ, hú se hláf máge beón áwend, Homl. Th. ii. 268, 7 : L. Ed. 4; Th. i. 162, 1. Ic smeáde mid mínra witena geþeahte, hú ic mæhte cristendómes mǽst árǽran, L. Edm. S. prm.; Th. i. 246, 19. Hé sóhte and smeáde (*tractavit*), hwæt tó dónne wǽre, Bd. 2, 5; S. 507, 29 : Elen. Kmbl. 826; El. 413. Maria smeáde and þóhte, hwæt seó hǽlettung wǽre, Blickl. Homl. 7, 16. Smeáge man geornlíce, hwæðer hit sóþ sí, Deut. 19, 18. Hí águnnon smeágan, hwilc of him ðæt tó dónne wǽre, Lk. Skt. 22, 23. Dauid ongan smeágan and þencan, hwilce ðæs gódan mannes dǽda wǽron, Blickl. Homl. 55, 12. Ðonne mót man smeágan and geornlíce spyrian hwár ða mánfullan v.ununge habban, L. Eth. ix. 40; Th. i. 348, 26. Mid wæccere móde is tó smeágeanne and tó geþencenne (*pensandum est*), ðæt Drihten bebeád, ðæt hí heora hrægel clǽnsodon, Bd. 1, 27; S. 496, 3. Ús is tó smeágeanne, ðæt Drihten on ðære costunge nolde his ða myclan miht gecýþan, Blickl. Homl. 33, 17. **III a.** *to seek an opportunity* :—Ðá smeáde hé ðæt hé hine gesáwe *querebat videre eum*, Lk. Skt. 9, 9. Hé smeáde geornlíce ðæt hé hyne wolde belǽwan *quaerebat opportunitatem ut eum traderet*, Mt. Kmbl. 26, 16. **IV.** *to accept as the result of inquiry, to suppose* :—Be ðisum þingum ne cunne wé smeágean nán óðer þing búton hit sig on Godes dóme gelang *de his rebus nihil aliud conjicere possumus, nisi quod ad judicium Dei pertineat*, L. Ecg. P. i. 13; Th. ii. 178, 16. v. á-, fore-, tó-, þurh-smeágan; smeah *and* cpds. *with* smeá-; cf. smúgan.

---

smeágelegen, e; *f. A syllogism* :—Smeágelegena *syllogismos*, Hpt. Gl. 503, 57. Cf. riht-smeáung, *and preceding word.*

smeágend-líc; *adj. Meditative* :—Smeágendlíc *meditativa*, Ælfc. Gr. 34; Zup. 211, 6.

smeágung, smeáwung, smeáung, sméung, smeáng, e; *f.* **I.** *search, inquiry, investigation* where something is lost :—On swylcere smeágunge (*the search for stolen cattle*), L. Edg. S. 12; Th. i. 276, 21. Habban ðæs ylcan smeágunge on mínum cucum orfe and on mínra þegena, 13; Th. i. 276, 24 : 14; Th. i. 276, 32. **II.** *inquiry carried on by the mind, inquiry, consideration, meditation, discussion, deliberation* :—Smeágung *studium*, Wrt. Voc. i. 51, 27. Sió smeáung and sió gesceádwísnes *ratiocinatio*, Bt. 39, 8; Fox 224, 4. Smeáung (Ps. Surt. smeáng) *meditatio*, Ps. Spl. 118, 24, 97, 99. Smeágunge *scrutinio*, 63, 6. On smeáwunge and on leornunge háligra gewrita *meditationi scripturarum*, Bd. 4, 3; S. 567, 29. Smeáunge, 1, 1; S. 474, 5: Past. 11; Swt. 67, 5. Smeánge, Ps. Surt. 38, 4. Hí hæfdon on ðam gemóte micle smeáunge geþeaht hwæt him tó dónne wǽre *illi tractatum magnum in concilio quid esset agendum habere coeperunt*, Bd. 3, 5; S. 527, 26. Ðá geseah se árleása ðídlan his smeágunge *then the impious king saw all his deliberation was of no avail*, Homl. Skt. i. 4, 399. Smeáunga yfle *cogitationes malae*, Mt. Kmbl. Lind. 15, 19. Smeáwunga, 9, 4. Smeáwungas (smeóunge, Rush.), Lk. Skt. Lind. 11, 17. Smeáungas (sméunges, Rush.), 2, 35. v. á-, ofer-, riht-, scearp-smeágung, -smeáung.

smeáh, smeóh; *adj.* **I.** *creeping in, penetrating* :—Wið smeógan wyrme, Lchdm. iii. 10, 17. v. smeá-wyrm. **II.** *subtle, crafty.* [Ðe man is ȝiep toȝenes him seluen! þat is smegh oðer man to bicharren and to biswiken, O. E. Homl. ii. 195, 5. Cf. Two þing ben in þe manne, on his þat clene kinde þat God haueþ þeron broht þureh his smehnesse (*wisdom, skill*), 205, 19. Smeihliche bicharede, 71, 28. Cf. Icel. í-smeygiligr *insinuating*.] *See* smeá-wrenc *and other compounds with* smeá-, *and* smeágan, ge-smeáh.

smeá-líc; *adj.* **I.** *searching, penetrating* (of inquiry, trial, etc.) :—Hwæt is sió þyrelung ðæs wǽges búton scearplícu and smeálícu fandung ðæs módes ðæt mon mid ðære . . . onlúce ða heardan heortan *quid est parietem fodere, nisi acutis inquisitionibus duritiam cordis aperire?* Past. 21, 3; Swt. 155, 1. **II.** *that goes to the root or heart of a matter, profound* :—Hú ðú mé hæfst áfréfrodne ǽgðer ge mid ðínre smeálícan sprǽce ge mid ðære wynsumnesse ðínes sanges *quantum me vel sententiarum pondere vel canendi jucunditate refovisti*, Bt. 22, 1; Fox 76, 10: tit.; Fox xiv, 6. **III.** *exquisite, choice* (?) :—Smeálícran *exquisitiores*, Wrt. Voc. ii. 145, 15.

smeá-líce; *adv.* **I.** *of inquiry, investigation, etc., searchingly, carefully, narrowly, closely* :—Hí smeálíce sóhtan *perquirentes subtilius*, Bd. 3, 10; S. 534, 37. Hí smeálíce sóhton ðone behíddan mete, Ælfc. T. Grn. 21, 12. Hit is smeálíce and geornlíce tó séceanne *subtiliter perscrutanda*, Past. 21, 1; Swt. 150, 11. Wé sculon swíðe smeálíce ðissa ǽgðer underþencean *hoc in utrisque est subtiliter intuendum*, 7, 1; Swt. 49, 23. Gesceád ða wé smeálíce geþencan sculan *discretio, quae subtiliter pensari debeat*, Bd. 1, 27; S. 496, 35. Smeálícor, Past. 11, 2; Swt. 67, 6. **II.** *of reasoning, thinking, etc., closely, deeply, acutely, with penetration* :—Hé ongann smeálíce þencan on his módes ingeþance *velut in augustam suae mentis sedem recepta*, Bt. 24, 1; Fox 80, 5. Mé þincþ ðæt wé mægen smeálícor sprecan and diógolran wordum *validioribus rationibus utendum puto*, 13; Fox 36, 32 : 13, tit.; Fox xii, 16. Ðonne ic ymbe swelc smeálícost þence *when I think most deeply about such a matter*, 10; Fox 26, 29. **III.** *of knowing, seeing, etc., clearly, accurately, exactly* :—Ða ðe meahton smeálíce and scearplíce mid hiera andgite ryht geseón *qui videre recta subtiliter per ingenium poterant*, Past. 11, 4; Swt. 69, 5. Ðeáh se láreów ðis eall smeálíce and openlíce gecýðe *cuncta haec licet subtiliter rector insinuet*, 21, 6; Swt. 163, 18. Se ðe wile geornlíce ðone Godes cwide singan sóðlíce (smeálíce, MS. B.), Salm. Kmbl. 171; Sal. 85. **IV.** *closely* :—Án cliwen suíðe nearwe and suíðe smeálíce gefealden, Past. 35; Swt. 241, 24. v. smeáh.

smeá-mete, es; *pl.* -mettas; *m. A delicacy* :—On ðás tíd (*Lent*) sceal beón forhæfednes gehwylcra smeámetta, L. E. I. 40; Th. ii. 438, 9. Disc mid cynelícum mettum (smeámettum, MS. B.) gefylled *discus, regalibus epulis refertus*, Bd. 3, 6; S. 528, 15.

smeán. v. smeágan.

smearcian, 'smercian; *p.* ode *To smirk, smile* :—Ic smercige *subrideo*, Ælfc. Gr. 47; Zup. 268, 8. Ðonne ðú smercodest and hlóge, ðonne weóp ic biterlíce, Wulfst. 140, 28. Ðá smearcode hé, Bt. 34, 10; Fox 148, 17. Smercode (smearcode, Cott. MS.), 34, 12; Fox 154, 8: 35, 4; Fox 160, 31: 40, 2; Fox 236, 22. Smercode, Blickl. Homl. 189, 4: Homl. Skt. i. 14, 126: Ap. Th. 19, 23. Ðá ongan hé smearcian, Bt. 39, 4; Fox 216, 14. Smercigende *subridendo*, Scint. 172, 17. Gúþlác tó smerciende féng *Guthlac received it smiling*, Guthl. 11; Gdwin. 56, 6. Mid smercigendum múþe, Homl. Th. i. 430, 34.

smeart; *adj. Smart, painful* :—Ic wylle swingan eów mid ðam smeartestum swipum, ðæt is, ðæt ic wítnige eów mid ðam wyrstan wíte, Wulfst. 295, 10. [Gif þi sulf one smerte discipline & drauh þet swete likunge into smeortunge, A. R. 294, 12. Stede and twei sporen and ane smearte

ʒerd, O. E. Homl. i. 243, 23. Mid smerte smiten of smale longe ʒerden, ii. 207, 6. Me him smæt mid smærte ʒerden, Laym. 20318. If men smot it with a yerde smerte (*adv.*), Chauc. Prol. 149.] v. smeortan.

smeáþ, e; *f. Meditation:—Ǽ ðin smeáþ (meditatio)* mín is, Ps. Spl. 118, 77. Cf. smeáʒung.

smeáþanclíce; *adv. Exactly, at large*; subtiliter :—Swá wé hér bufan smeáþanclíce áwriten habbaþ, Anglia viii. 309, 22.

smeá-þancol; *adj. Acute, subtle :—*Mid smeáþancelre trahtnunge *tenaci memoriae textu*, Hpt. Gl. 410, 64.

smeáþancol-líc; *adj. Subtle, crafty :—*Smeáþancollíce wriþan ꝓ cnottan cræftelícum *sertaque mystica dactylico*, Germ. 389, 28.

smeáþancollíce; *adv. Exactly, in a searching manner, thoroughly*; subtiliter :—Smeáþancelíce *subtiliter, eleganter*, Hpt. Gl. 431, 49. Hí smeádon swíðe smeáþancollíce ymbe ðæt éce líf *they went into the question of eternal life in the most searching manner,* Homl. Skt. i. 3, 44. Hé hí gewissode swíðe smeáþancellíce ymbe ðæs mynstres gebytlungum *he gave them most exact directions about the buildings of the monastery,* Homl. Th. ii. 172, 16. Hé lǽrþ manna mód mid godcundre láre smeáþancellíce, i. 412, 32.

smeáþancolness, e; *f. Exactness, strictness :—*Ðeáh wé witon hú fela gód oððe hú micele wé gefremodon nyte wé ðeáh mid hwylcere smeáþancelnysse se upplíca Déma ða áfandaþ, Homl. Th. ii. 80, 34.

smeáung. v. smeáʒung.

smeá-wrenc, es; *m. A crafty device, sharp trick :—*Hé begeat mid his smáhwrencan and mid his golde and seolfre eall dyrnunga, ðæt him geweard se þridda pæníg of ðære tolne on Sandwíc, Chart. Th. 339, 8. v. smeáh.

smeá-wyrhta, an; *m. A skilled workman, an artisan :—*Gif hé smeáwyrhtan hæfþ ðám hé sceal tó tólan fylstan, Anglia ix. 263, 16.

smeá-wyrm, es; *m. A penetrating worm, worm that makes its way into the flesh :—*Wið smeáwyrme (cf. wið smégea-wyrme, 302, 12) smiring... seó sealf ðone wyrm deádne gedéþ oððe cwicne of drífþ, Lchdm. ii. 332, 3–26. Wið sméga-wyrme, 126, 1. Wið smoega-wyrmum, 12, 14. v. smeáh.

sméc, smécan, smecgan. v. smíc, smícan, smæccan.

smedema, smeodema, smidema, smedma, an; *m. Fine flour, meal :—*Smeoduma *polenta,* Wrt. Voc. ii. 117, 51. Melewes smedma *simila,* 83, 65. Smedma of melwe *pollis,* Ælfc. Gr. 9, 28; Zup. 55, 15. Smedma *simila* vel *pollis,* Wrt. Voc. i. 41, 24. Hwǽtes smedma, Lchdm. ii. 108, 10. Gecned þrí sestras smedeman (*similae*), Gen. 18, 6 : Ex. 29, 40. Smideman, Lev. 2, 2. Genim smedman six yntsena gewihte, Lchdm. i. 150, 17. Mid hwǽtes smedeman *with the fat of kidneys of wheat* (A.V.); *cum medulla tritici*: cf. óþ smedeman *ad medullam,* Hpt. Gl. 410, 28. Ðá hláfas wǽron bérene. Bere is swíðe earfoþe tó gearcigenne, and ðeáh-hwæðere fét ðone mann, ðonne hé gearo biþ. Swá wæs seó ealde ǽ swíðe earfoþe tó understandenne, ac ðeáhhwæðere ðonne wé cumaþ tó ðam smedman, ðæt is tó ðære getácnunge, ðonne gereordaþ heó úre mód, Homl. Th. i. 188, 7. Genim ácrinde, wirc tó smedman, Lchdm. ii. 132, 19. Of mealtes smedman geworht, 332, 20. Genim hwǽtenes meluwes smedman, 134, 4. v. hwǽte-smedeme (*read* -a; *m.*).

smedemen, smedmen; *adj. Of fine flour :—*Smedmen hláf *similagineus panis*, Scint. 154, 1.

sméga-wyrm, sméh-wrenc. v. smeá-wyrm, -wrenc.

smellan (?); *p.* smeall *To crack, make a noise.* [*Mod. Icel.* smella ; *p.* small *to crack as a whip.*] v. smillan, smæll.

smelt, smylt, es; *m. A smelt :—*Smelt *sardina,* Wrt. Voc. i. 281, 71. Smylt *sartate,* 66, 7. Smeltas *sardas,* ii. 119, 63.

smelt (?). v. dolh-smeltas ; smelte *serene.* v. smylte.

smelting, smilting, e; *f. Amber :—*Smelting *electrum,* Wrt. Voc. i. 38, 31. Smilting, 34, 66. Smyltinc, 85, 14. Anlícnyssa gyldena and sylfrena, sume of smyltinga, sume of crystallan, Homl. Skt. i. 4, 165. [Smulting, Wrt. Voc. i. 94, 61. Cf. *O. H. Ger.* smelzi *electrum,* smelzida *electrum : Icel.* smeltr *enamelled.*]

smeócan; *p.* smeác, *pl.* smucon; *pp.* smocen. I. *intrans. To smoke, emit smoke :—*Smeógoþ *fumigant,* Ps. Spl. 103, 33. Muntas smeócaþ *montes fumigabunt,* 143, 6 : Wülck. Gl. 244, 35. Eall Sinai munt smeác (*fumabat*), Ex. 19, 18. Smeóce *fumet,* Germ. 393, 187. Heortes mearh gebærned óþ ðæt hyf smeóce, Lchdm. i. 338, 13. Eall folc gesáwon ðone munt smeócan, Ex. 20, 18 : Engl. Stud. ix. 40. Smeócende (smécende, Lind. : smíkende, Rush.) flex *linum fumigans,* Mt. Kmbl. 12, 20. II. *trans. To smoke, fumigate :—*Smeóc ðone man mid gáte hǽrum, Lchdm. i. 352, 1. Smeóce mid hæþe, 354, 23. Heortes hær beóþ swíðe góde mid tó smeócanne, 338, 4. [*Prompt. Parv.* smekyñ *fumo, fumigo.*] v. smícan, smocian, smíc.

smeodma, smeóh, smeolt. v. smedema, smeáh, smolt.

smeortan; *p.* smeart, *pl.* smurton; *pp.* smorten *To smart :—*Gnættas cómon mid fýrsmeortendum bitum *ignitos ciniphes,* Ors. 1, 7 ; Swt. 36, 30. [Þenne akeþ his heorte and smerteð, O. E. Homl. ii. 207, 21. Hire ne oc, ne ne smeart, 21, 27. Ðenne wile his heorte aken and smerten, 207, 34. Me iveleð hit bitterliche smeorten, A. R. 238, 29. Smertyñ *uro,* Prompt. Parv. 460. *O. H. Ger.* smerzan ; *p.* smarz *dolere.*] I. in

smeoru, smeru (o, a), wes ; *n. Fat, grease, suet, tallow.* I. in

the following glosses :—Smeoru *unguentum,* Wrt. Voc. ii. 124, 9. Unsilt smeoro *saevo,* 119, 45. Smero *sevo* (in a list 'de igne'), i. 284, 27. Unámaelte smeoruwe *pice, saevo,* ii. 117, 28. Smerwe *sevo,* 80, 45. Smeruwe, Hpt. Gl. 503, 18. Smerewe *arvina,* 471, 4. II. in the following passages :—Wið útsihte, hunig and unsylt smeoru and wex, Lchdm. iii. 18, 5. Heortes smeoro (smeru, smero), i. 338, 15 : 354, 4. Sceápes smeru, ii. 66, 7. Foxes smero, iii. 2, 25. Heorotes smera oððe gáte oððe góse, 68, 26 : 80, 18. Ðæt smeru wand út, Jud. 3, 22. Smeoruwes, Ps. Th. 62, 5. Beran smeruwes (smerwes, MS. B.), Lchdm. i. 216, 15. Mid gáte smeorwe (smerwe, MS. B.), 354, 1. Mid smeorwe *adipe,* Ps. Surt. 62, 6. Of swínes smerwe, Lchdm. ii. 66, 7. Ofer smere (*unguento*), Rtl. 115, 34. Cnucige wið eald smeoru (smera, MS. B. : smeru, MS. O.), Lchdm. i. 74, 21 : 86, 7. Genim heortes smeoruw (smeruw, MS. H. : smeru, MS. B.). Genim góse smero, 76, 9. Sceápen smera, ii. 128, 16 : 148, 20. Eal ðæt smeru hig forbærndon, Lev. 8, 25. [Smeredd ꝗ sallfedd þurrh nan eorþliʒ smere, Orm. 13244. *O. H. Ger.* smero *adeps, arvina, unctura : Icel.* smjör *grease, fat ; butter.*] v. flot-, heorot-smeoru.

smeoru-mangestre, an; *f. A butter-woman, woman who deals in butter and cheese :—*Smeremangestre, que mangonant in caseo et butiro, L. Eth. iv. 2; Th. i. 301, 5.

smeoru-sealf, e; *f. A grease-salve :—*Gif ðú wǽtan dést tó oððe smerusealfe, ne meaht ðú hit gelácnian, Lchdm. ii. 148, 23.

smeoru-þearm, es; *m. An entrail :—*Smeruþearm *extale,* Wrt. Voc. ii. 145, 29. Smæreþerm *julium* (in a list 'de suibus'), i. 286, 61.

smeoruwig; *adj. Fatty, greasy, unctuous :—*Eal ða wǽtan þing and ða smerewigan sint tó forbeódanne, Lchdm. ii. 210, 27. [*Icel.* smjörugr *greasy.*] v. un-smeoruwig.

smeoru-wyrt, e; *f. Smer-wort.* 'Aristolochia rotunda, in allusion to its use in ointments.' E. D. S. Plant Names. Halliwell gives 'smereworth the round birthwort, or the herb mercury.' It is found in the following glosses :—Smeoruwyrt *veneria,* Wrt. Voc. ii. 123, 33. Smeroruwyrt nam (naþ?) *silvatica,* 62, 39. Smerewyrt *aristolochia,* i. 67, 17 : Lchdm. iii. 300, col. 1. It occurs also in the Leechdoms :—Smerowyrt. Ðeós wyrt ðe man aristolochiam and óðrum naman smerowyrt nemneþ, Lchdm. i. 114, 9–11. Smeruwyrt, ii. 338, 13. Smerewyrt, 128, 15.

smér[e], smera, smercian, smereness, smerewig, smerian, smering, smeru, smerwan. v. smǽr[e], smeoru, smearcian, smireness, smeoruwig, smirwan, smiring, smeoru, smerwan.

smédan; *p.* de *To make smooth, to soothe :—*Him is tó sellanne ðæt ðone innoþ stille and sméðe, Lchdm. ii. 210, 20. v. ge-smédan, smédian.

sméðe; *adj. Smooth.* I. in glosses :—Sméðe *lenis,* Wrt. Voc. ii. 51, 48. Smoeðum *politis,* 117, 55. Ðæs sméðestan *politissimis,* 66, 27. II. *smooth, without roughness* or *inequalities* of surface :—Sméðe ringce *tinius,* Wrt. Voc. i. 40, 56. Mín bróður ys rúh and ic eom sméðe, Gen. 27, 11. Ðonne glád hit on ðam scyllum swelce hit wǽre sméðe ísen, Ors. 4, 6 ; Swt. 174, 8. Wæs cyrtil unrúh ꝓ smoeðe, Jn. Skt. Lind. 19, 23. An dún ful sméðe, Homl. Skt. i. 19, 109. On sméðum felda *on a plain,* Ors. 3, 11 ; Swt. 142, 14. Wé becóman on sumne sméðne feld (*in viam planam*), Bd. 5, 6; S. 618, 40. Ðeós wyrt biþ cenned on sméðum landum, Lchdm. i. 90, 3 : 298, 3. On sméðe (smoeðum, Lind., Rush.) wegas *in vias planas,* Lk. Skt. 3, 5. Hé hæfþ ðe sméðran líchoman, Lchdm. ii. 298, 13. III. *smooth, without discomfort* or *annoyance :—*Wǽron hyra gongas under Godes egsan sméðe and geséfte, Exon. Th. 146, 3 ; Gú. 704. IV. *smooth, suave, avoiding offence :—*Hé biþ hwílum tó ungemetlíce sméðe, hwílum tó ungemetlíce réðe *amor proprius mentem aliquando inordinate ad mollitiem, aliquando ad asperitatem rapit,* Past. 19, 1; Swt. 143, 7. V. *smooth, not irritating* (of food, medicine, etc.) :—Ne se mete ne sié tó scearp ne tó súr, ac sméðe and fǽt, Lchdm. ii. 196, 8. Eáðmylte mettas and scír wín and sméðe, 220, 13. Ða wambe man sceal clǽsnian mid sméðe wyrtdrence, 262, 17. Wyrc sméþe eágsealfe, 308, 27. VI. *smooth* (of words) :—Sméðne sybcwide, Frag. Kmbl. 54; Leás. 29. Ðám ðe ful sméðe sprǽce habbaþ, 20; Leás. 12. Ðone éle, ðæt wǽron ða smédan lyffetunga, Homl. Th. ii. 572, 1. Bepǽcean mid sméðan wordan, Homl. Skt. i. 23, 602. Se Hǽlend lufaþ swíðor ða dǽde ðonne ða sméðan word, Ælfc. T. Grn. 14, 34. VII. *of the voice, not harsh, melodious, harmonious :—*Stefen smoeðu *vox canora,* Ps. Surt. ii. p. 202, 5. v. un-sméþe ; smóþ.

smédian; *p.* ode ; *pp.* od. I. *to become smooth :—*Ðonne sméðaþ ðæt neb and hálaþ, Lchdm. i. 86, 8. II. *to make smooth :—*Ic sméðie *polio,* Wrt. Voc. i. 28, 74. [He wile foxliche smeþien mid worde, O. E. Homl. i. 31, 8. Rihteð and smeðeð þe heorte, A. R. 4, 23.] v. ge-smédian ; smédan.

smédness, e; *f.* I. *smoothness :—*Hé forgeaf hreóflium smédnysse, Homl. Th. i. 26, 11. II. *a smooth, level surface :—*Feld *campus,* smédnys *planities,* Wrt. Voc. i. 53, 49.

smíc, sméc, smýc, es ; *m. Smoke, vapour, steam :—*Swelce se bitresta smíc, Ors. 3, 11 ; Swt. 142, 20. Smíc *fumus,* Ælfc. Gr. 8; Zup. 28, 12 : Ex. 19, 18 : Homl. Th. ii. 68, 20. Hí losiaþ swá swá sméc, Bt. 27, 3 ; Fox 98, 31 : Ps. Th. 36, 19. Smýc, Hpt. Gl. 501, 78 : Shrn. 52, 33

Ða þicnyssa smíces stigon upp on ǽlce healfe, Homl. Skt. i. 23, 36. Ða ýsla up flugon mid ðam smíce, Gen. 19, 18 : Homl. Th. i. 530, 34. Se wǽta gǽþ up swylce mid smíce oððe miste, Lchdm. iii. 278, 9. Sméce gelíce *sicut fumus*, Ps. Th. 101, 3. On ðam fýre and on ðam smýce, Homl. Th. ii. 202, 32. Se wind ðæt fýr and ðone smíc ofer ða wallas dráf, Bd. 3, 16 ; S. 543, 1. Genim spices snǽd, lege on hátne stán, drince ðonne smíc, Lchdm. ii. 58, 17. Ðonne hé (*the root*) tóbrocen byþ, hé rýcþ eal swylce hé smíc of him ásænde, i. 260, 9. Ðæs drinces smýc héora eágan onfón, 348, 22. Sméc *vaporem*, Ps. Surt. ii. p. 202, 15. [Ne michte ut seon for smike, O. E. Homl. i. 161, 16. Smeche, ii. 220, 18. Smiche, 258, 20. Smec off recless, Orm. 1088. Smeke or smoke, Prompt. Parv. 460. M. H. Ger. smouch : Ger. schmauch.] v. smoca.

smícan, smécan ; *p.* te. **I.** *to smoke, emit smoke* :—Muntas smícaþ *montes fumigabunt*, Ps. Surt. 103, 32 : 143, 5. **II.** *to smoke, fumigate* :—Sume mid pice smícaþ, Lchdm. ii. 236, 9. Nim gáte hǽr, sméc under ða bréc wið ðæs rægereósan, 146, 3. Smíce mid fearne swíðe ða heáfon, 64, 26. [*Wicklif has a wk. past* smekide.] v. *smeócan, smocian*.

smicer ; *adj. Fair, fine, beautiful, elegant* :—Smicre *elegans, loquax*, Txts. 59, 737 : *elegans*, Wrt. Voc. ii. 29, 22. Smicerre ansíne *eleganti forma*, 30, 26. Smicere leóþe *carmine rithmico*, 23, 24. Windan mán igne smicerne wǽn and manig ǽnlic húse settan and fegerne tún timbrian, Shrn. 163, 16. Hió bit ðæt hí findon twá smicere scencingcuppan intó beóðern *she asks them to provide two fair goblets for the refectory*, Ch. Th. 536, 7. Ðæs smicerestan *politissimis*, Wrt. Voc. ii. 66, 26. [He warrþ till atell defell off shene smikerr enngell, Orm. 13679. O. H. Ger. smechar *elegans, delicatus*.]

smicere ; *adv. Finely, fairly, elegantly* : — Cræftlíce *vel* smicere *affabre*, ic smicere geglengce *orno*, Wrt. Voc. i. 54, 55–58. Smicere geworhte *fabrefactum*, ii. 33, 68 ; Shrn. 165, 27 ; Ps. Th. 118, 164, 84. Sió lufu scínþ suíðe smicere (*fulgescit*), Past. 14, 6 ; Swt. 87, 9. In burh raðe smicere cymeþ wlitig scríðan þrymlíce on tún Maius, Menol. Fox 150; Men. 76. Ðonne singþ hé smælor and smicror *minutas ac tenues voces format*, Past. 63 ; Swt. 461, 3.

smicerness, e ; *f. Elegance, neatness* :—Þurh smicernesse and hiwunge *hironiam* (= per ironiam ; irony is explained as combining elegance and dissimulation), Wrt. Voc. ii. 42, 53.

smidema. v. smedema.

smillan ; *p.* de. **I.** *to cause to crack* as a whip, etc. **II.** *intrans. To crack* as a whip :—Under smyllendum gyrdum weóp *crepantibus flevit sub ferulis*, Germ. 388, 7. [*Icel.* smella ; *p.* small *to crack*, as a whip; smella (*wk.*) *to cause to crack*.] v. smæll, *and cf. trans. and intrans. forms of* miltan.

smilt, smilting. v. smylt, smelting.

smirels, es ; *m. An unguent, ointment, unction, salve* :—Smyrels *vel* sealf *unguina vel unguenta*, Wrt. Voc. i. 49, 29 : *unguentum*. Hé gehǽlde án mǽden mid hálwendum smyrelse gehálgodes eles, Homl. Th. ii. 508, 14. Wé lǽraþ ðæt preósta gehwilc ǽgðer hæbbe ge fulluhtele ge seócum smyrels, L. Edg. C. 66 ; Th. ii. 258, 15. [Nicodemus brouhte smuriles uorte smurien mide ure Louerd, A. R. 372, 18. Þat swote smirles þat is icleopet basme, H. M. 13, 21. Kepen þe lich wiðuten smerles, Gen. and Ex. 2454. Þe Magdalene smerede Cristes uet mid þe precious smerieles, Ayenb. 187, 32. *Dan.* smörelse *grease*.]

smireness, e ; *f. Ointment, unguent* :—Cwæþ se wrítere ðæt Maria genáme án pund deórwyrðre smyrenesse (smerenesse, 69, 1). . . . Ðeós smerenes wæs geworht of ehtaténe cynna wyrtum, Blickl. Homl. 73, 17–20. Smirinis (smerenisse, Rush.) *unguentum*, Mt. Kmbl. Lind. 26, 12. Smirenisse *unguenti*, Rtl. 115, 41. Smyrenisse, Lchdm. i. 346, 9. Mið smiriniss *unguento*, Lk. Skt. Lind. 7, 38 : *oleo*, 46. Smyrenesse *unctum*, Wrt. Voc. ii. 91, 35. Smerenessa and sealf, Lchdm. ii. 19, 19 ; 158, 9. Hié selfe mid smirenissum hié smerwan, 224, 1.

smirian, smiring. v. smirwan, smirwung.

smirwan, smerwan, smirewan, smeruwan, smirian, smerian, smyrian ; *p.* smirede, oðe *To smear, anoint* :—Ic smirie míne flán on blóde, Deut. 32, 42. Ðú smirest *unges*, Ex. 29, 36. Ðú smyrest *linies*, Wrt. Voc. ii. 51, 46. On ðam dæge ðe hig man smiraþ *in die unctionis suae*, Lev. 6, 20. 'Smirewaþ (smiriaþ, Hatt. MS.) eówre eágan mid sealfe.' Ðonne wé smirewaþ (smirewaþ, Hatt. MS.) úre heortan eáge mid sealfe, Past. 11 ; Swt. 68, 10–12. Smiriaþ, Ps. Surt. 140, 6. Smirede *linivit*, Wrt. Voc. ii. 74, 15. Smyrede, 51, 47. Smerede *unxit*, Ps. Spl. 44, 9 : Blickl. Homl. 69, 2. Smyrede, 73, 18. Hé worhte fenn and smyrede (smiride, Lind., Rush.) míne eágan, Jn. Skt. 6, 13. Mín heáfod ðú mid ele ne smyredest, ðeós smyrede mid sealfe míne fét, Lk. Skt. 7, 46. Smyredon (smiredon, Lind.), Mk. Skt. 6, 13. Smire mid, Lchdm. ii. 132, 1 (and often). Smyre, i. 216, 5 (and often). Smyra ðín heáfod *unge caput tuum*, Mt. Kmbl. 6, 17. Þweah ǽr ðú hit smeruwe, Lchdm. ii. 156, 2. Gníde and smerwe, 186, 7. Hý hine smyrigon . . . ǽr hé hyne smyrige . . . hine ne mót nán mann smyrigan, L. Ælfc. C. 32 ; Th. ii. 354, 21–31. Hié selfe mid smirenessum hié smerwan, Lchdm. ii. 224, 1. Ða menn ðú scealt smerwan mid ðý ele, 194, 18 : 156, 4. Smirewan, 184, 2 : 238, 26. Smyrian, 118, 16. Smerian, Blickl. Homl. 73, 24 : 75, 17. Tó smirwanne, Lchdm. ii.

244, 19. Tó smerwanne, 288, 16. Tó smergenne, iii. 4, 14. Heó com tó smyrianne (smiriane, Lind.: smiranne, Rush.) mínne líchaman, Mk. Skt. 14, 8. [O. H. Ger. þi-smeruuit *unctus : Icel.* smyrja, smyrwa *to anoint*.] v. á-, be-, ge-, geá-smirwan, -smirian.

smirwung, smiring (-ung), e ; *f.* **I.** *anointing, unction* :—Ðus cwæð se apostol be ðære smyrunge seócra manna, L. Ælfc. C. 32 ; Th. ii. 354, 27. Gif se seóca man girnþ ðæt man hine smerige, hé dó ðonne his andetnesse ǽr ðare smerunge, and gif hé æfter ðare smyrunge hál wurð, hé mót flǽsces brúcan. On ðare smyrunge biþ lǽcedóm, L. Ælfc. P. 47, 48 ; Th. ii. 384, 27–32. **II.** *an ointment* :—Smiring *cassia*, Ps. Surt. 32, 9. Smyring *unguentum*, Ps. Spl. 132, 2. Balzaman smiring wið eallum untrumnessum, Lchdm. ii. 174, 7. Smyring, 288, 12. Gif ðú myhtest ǽnig þing fyndan on smyrunge oððe on wyrtum, ðæt ðu myhtest mýne wunde myd gehǽlan, St. And. 28, 17. Smerwunga wyrce of ele and of wermóde, Lchdm. ii. 182, 16.

smirwung-, smiring-ele, es ; *m. Oil for anointing* :—Of ðam smiring ele *de oleo unctionis*, Ex. 29, 21.

smítan ; *p.* smát, *pl.* smiton ; *pp.* smiten. **I.** *to daub, smear, smudge* :—Ðú nymst his blód and smítst ofer úteweard Aarones swýðre eáre, Ex. 29, 20. Smát, gemaercode *inpingit* (cf. *inpingit* gemaercode *vel signat*, 45, 59), Wrt. Voc. ii. 111, 57. Genim gáte tord, gemeng wið eced, smít on, Lchdm. ii. 68, 2. Genim ðæs hornes melo, meng wið wætere, smít on, 72, 14. Mid feðere smít on, 102, 8. Smíte mon ða sealfe ǽrest on ðæt heáfod, iii. 14, 29. Smíte of ðam sylfan blóde on ðæs weofodes hyrnan, Lev. 4, 18. Nymon of his blóde and smíton on ǽgðer gedyre, Ex. 12, 7. Ðissa (*oil, grease, and tar*) ealra emfela and ðara dusta ealra emfela, gemeng eal ceald tósomne, ðæt hit fram ðám wósum eal wel smítende [sí] (*may be adapted for smearing*), smire mid, Lchdm. ii. 126, 11. [Ofersmít mid ele, 180, 28.] **II.** *to defile, pollute* :—Wiáþ áþ smíteþ, Exon. Th. 354, 52 ; Reim. 64. Smiton *funestavere*, Wrt. Voc. ii. 109, 43. [*Goth.* be-, ga-smeitan *to smear, anoint : O. Frs.* smíta *to cast : O. H. Ger.* smízan *linere*. Later English takes the word in the sense of *strike*. In Mt. 26, 68 the later MS. has Hwæt ys se þe ðe *smat*, where the earlier has *slóh*. Brutus heom smat on, Laym. 534. He hoff þe swerd to smitenn, Orm. 14677. Ase ofte ase eni hund bininneð þe þine mete, nultu aþe ofte smiten ? A. R. 324, 23. So in later works.] v. be-, ge-smítan ; smittian.

smite (?), es ; *m. Pollution* :—Mustfleógan *vel* wurma smite *bibiones vel mustiones*, Wrt. Voc. i. 23, 75. v. must-fleóge.

smíte, an ; *f. A foul, miry place* (?) :—Ego mansam in loco qui celebri a solicolis nuncupatur æt Smítan uocabulo ministro meo largitus sum . . . Ðis is ðæra on híde landgemǽru tó Smítan . . . of ðæm slo tó Smítan ; of ðære Smítan tó berge, Cod. Dip. Kmbl. iii. 166, 2–20. Of smítan on ðone stán . . . of ðære apoldran innan smítan, v. 105, 13–36.

smitenness. v. be-smitenness.

smiþ, es ; *m. A smith, a worker in metals* or *in wood* :—Cudo ic smiðige ; eft gyf ðú cwedst *hic cudo*, ðonne byþ hit nama, smiþ, Ælfc. Gr. 36; Zup. 216, 10. Se smiþ *ferrarius* . . . se treówyrhta *lignarius*, Coll. Monast. Th. 30, 29. Smiþ *faber vel cudo*, Wrt. Voc. i. 73, 26 : *faber*, 286, 74. Fýres god, heile smiþ *Vulcanus*, ii. 95, 7. Wæs sum bróðor syndrilíce on smiþcræfte well gelǽred ; þeówode hé swýðe druncennesse and monigum óðrum unálýfednessum ðæs sleacran lífes, and hé má gewunode on his smiþþan dæges and nihtes sittan and licgean, ðonne hé wolde on cyricean singan and gebiddan . . . wið ðon ðe smiþ ðæs þýstran modes and dǽde his deáþe neálǽhte . . . , Bd. 5, 14 ; S. 634, 13–42. Gif smiþ monnes andweorc onfó, hé hit gesund ágife swá hé hit ǽr onfénge, L. Alf. pol. 19 ; Th. i. 74, 9. Módcræftig smiþ, ðonne hé gewyrceþ helm oððe hupseax, Exon. Th. 297, 2; Cri. 62. Wǽpna smiþ, Beo. Th. 2908 ; B. 1452. Hú nys se smiþ (smiþ t wyrihte *faber*, Lind.) Marian sunu, Mk. Skt. 6, 3. Ðes ys smiþes sunu *hic est fabri filius*, Mt. Kmbl. 13, 55. Byrne, searonet seowed smiþes orþancum, Beo. Th. 817; B. 406. Gif gesíþcund man fare, ðonne mót hé habban his smiþ mid him, L. In. 63 ; Th. i. 144, 3. Weorc, handweorc smiþa, Exon. Th. 408, 18 ; Rä, 27, 14 : 388, 16; Rä. 6, 8 : 401, 6 : Rä. 21, 7. Ic hæbbe smiþas, ísene smiþas, goldsmiþ, seolforsmiþ, ærsmiþ, treówwyrhtan, Coll. Monast. Th. 29, 35. ¶ In poetical compounds the word is used figuratively. v. gryn-, hleahtor-, lár-, wig-, wíg-, wróht-smiþ. [*Goth.* aiza-smiþa : *O. Frs.* smeth, smid : *O. H. Ger.* smid *faber, cudo : Icel.* smiðr.] v. ambiht-, ár-, gold-, ísen-, seolfor-, wundor-smiþ.

smiþ-cræft, es ; *m. Smithcraft, the craft* or *art of the worker in metal* or *wood* :—Wæs sum bróðor syndrilíce on smiþcræfte well gelǽred *erat fabrili arte singularis*, Bd. 5, 14 ; S. 634, 14.

smiþ-cræftig ; *adj. Skilled as a smith.* v. next word.

smiþ-cræftiga, an ; *m. One skilled in the smith's art* :—Tubal Cain smiþcræftega wæs, Cd. Th. 66, 15 ; Gen. 1084.

smiðian ; *p.* ode *To make* out of metal *or* wood, *to fashion, forge* :—Ic smiðige *cudo*, ðú smiðast *cudis*, Ælfc. Gr. 36; Zup. 216, 8 : 28, 6; Zup. 178, 10. Smiðode oððe gescóp *cuderet*, Wrt. Voc. ii. 19, 36. Hé hét smiðian of smǽtum golde áne lytle róde, Homl. Th. ii. 304, 16. Smiðian on smǽtum golde ánre culfran anlícnysse, Homl. Skt. i. 3, 126. Smeoðed *fabricata*, Hpt. Gl. 418, 3.

[Brien enne smið funde þe wel cuðe smiðie . . . þe smið gon to smiðeʒe ane pic, Laym. 30742-9. Ofte a ful hawur smið smeoðið a ful woc knif, A. R. 52, 8. A smith that in his forge smithed plowharneis, Chauc. C. T. 3760. To smythye wepne into sikul or to sithe, Piers P. 3, 305. **Goth.** ga-smiþón: **O. H. Ger.** smidón *fabricare, cudere*: **Icel.** smiða.] v. ā-, be-, ge-smiðian.

smiþlíce; *adv. After the manner of a smith, with skill*:—Smiþlíce *fabrile*, Wrt. Voc. ii. 108, 33: 35, 14: 146, 59. [O. H. Ger. smidilího *fabriliter.*]

smiþþe, an; *f. A smithy, a smith's workshop*:—Smiðoe *officina*, Wrt. Voc. ii. 64, 12: i. 34, 55: 73, 27. Smiþþe, 286, 75. Smiðþe *vel* weorchús, 58, 23. On smiððan *in conflatorio*, Kent. Gl. 1033. Hwæt sylst ðú (*the smith*) ús on smiþþan ðínre búton ísene fýrspearcan, Coll. Monast. Th. 31, 5. Hé má gewunode on his smiþþan dæges and nihtes sittan and licgean, ðonne hé wolde on cyricean singan and gebiddan, Bd. 5, 14; S. 634, 16. Gáþ tó smiððan and fandiaþ ðises goldes and ðissera gymstána, Homl. Th. i. 64, 6. Ðæt wíde geat be-eástan Welandes smiððan, Cod. Dip. Kmbl. v. 332, 23. [O. Frs. smithe: O. H. Ger. smitta, smidda *officina, fabrica*: Icel. smiðja.]

smiþu. v. gold-smiþu.

smitta (-e; *f.?*), an; *m. A smear, blot, mark, spot*:—Bútan smittan *sine macula*, R. Ben. Interl. 4, 3. Smyttena *naevorum, notarum*, Hpt. Gl. 421, 56. v. next word.

smittian; *p.* ode *To smear, pollute, defile*:—Smittodan *funestavere, maculavere*, Wrt. Voc. ii. 151, 60. Smittud *cacabatus*, Hpt. Gl. 514. 47. [Smitted *contaminata*, Ps. 105, 39. As reignes shall ben flitted Fro folk to folk, or whan they shal ben smitted, Chauc. T. and C. v. 1544. Ismitted (*smeared*) wið smirles, H. M. 13, 23. Bismitted (-smuddet, MS. T.) and bismeoruwed, A. R. 214, 22. Besmetted ine herte mid kueade þoʒtes, Ayenb. 229, 20. O. H. Ger. pi-smizzit *illitus, unctus.*] v. be-smittian; smítan.

smoc[c], es; *m. A smock, shift*:—Smoc *vel* syrc *colobium*, Wrt. Voc. i. 25, 60. Loþa, hom *vel* smoc *colobium, dictum quia longum est et sine manicis*, ii. 134, 37. [Smokke *interula*, 182, 1. Smok, schyrt *camisia, interula*, Prompt. Parv. 461. O.H. Ger. smoccho *interula*: Icel. smokkr.]

smoca, an; *m. Smoke*:—Ástáh smoca on yrre his *ascendit fumus in ira ejus*, Ps. Lamb. 17, 9. Út æt his nosu eode micel smocca, Nar. 43, 16. Hé nele ðone wlacan smocan wáces flǽsces wætere gedwæscan *nec vult lini tepidos undis exstinguere fumos*, Dóm. L. 51. v. smíc.

smocian; *p.* ode. **I.** *intrans. To smoke, emit smoke*:—Muntas smociaþ, Ps. Lamb. 103, 32. Smeócaþ ł smociaþ *fumigabunt*, 143, 5. Swilce án ofen eall smociende, Gen. 15, 17. Smocigende, Homl. Th. ii. 202, 24. **II.** *trans. To smoke*:—Genim ðú ðás ylcan wyrte and smoca ðæt cild mid, Lchdm. i. 116, 9. Smeóce (smoca, MS. R,) mid hǽþe, 354, 23. [Þa iseʒen heo a fur smokien, Laym. 25734. Smekyñ or smokyñ *fumo, fumigo*, Prompt. Parv. 460.] v. smeócan, smícan.

smoega-wyrm, smoh. v. smeá-wyrm, -a, in-smoh.

smolt, smeolt; *adj. Serene, quiet, peaceful*:—Smolt wæs se sigewong, Exon. Th. 146, 23; Gú. 714. Smeolt, Andr. Kmbl. 3160; An. 1583. Smolt regn *imbres*, Rtl. 85, 9 : *torrens*, Blickl. Gl. (Ps. 125, 4). Smolt biþ *serenum erit*, Mt. Kmbl. Lind. 16, 2. Smolt dæg ł restdæg (smolte dæge, Rush.) *sero die*, Jn. Skt. Lind. 20, 19. Éfern ł smolt (efern ꞇ smolt, Rush, Mk. Skt. Lind. 6, 47. Wé hæfdon smolte niht *nox serena reddita est nobis*, Nar. 23, 52. [With smeþe smylyng and smolt, Gaw. 1763.] v. smylte, *and next word*.

smolte; *adv. Quietly, mildly*:—Ðonne smolte (cf. smylte, Bt. 9; Fox 26, 17) blǽwþ súþan and westan wind under wolcnum, Met. 6, 8. [Cf. O. Sax. smultro gibárean (*of the wind and waves*).]

smoltlíce; *adv. Gently, quietly*:—Flówæþ seó welle swá fægeðe and swá smoltlíce swá hunig, Engl. Stud. viii. 477, 10. v. smylt-líc.

smorian; *p.* ode *To choke, suffocate*:—Wyrgeþ *vel* smoraþ st[r]angulat, Wrt. Voc. ii. 121, 32. Se esne genimende smorede hine (*suffocabat eum*), Mt. Kmbl. Rush. 18, 28. Ða þornas smoradun (*suffocaverunt*) hiǽ, 13, 7. [Wend he smore þat sede, C. M. 5573. All suld be smored, Pr. C. 7601. Smore wythe smeke *fumigo*, smoryd *fumigatus*, smorynge *fumigacio*, Prompt. Parv, 461. Halliwell gives *smore* as a word in northern dialects, and quotes Hall's Chronicles; and *smoor* is given as a Lincolnshire word, E. D. S. Pub.] v. ā-, for-, of-smorian.

smóþ; *adj. Smooth, unruffled*:—Mid smóðestum andwlite *serenissimo vultu*, Engl. Stud. ix. 40. v. un-smóþ, *and* sméðe.

smúgan; *p.* smeág, *pl.* smugon; *pp.* smogen *To creep, crawl, move gradually*:—Ic smúge *serpo*, Ælfc. Gr. 28, 4; Zup. 170, 15 : *crepo* (*serpo?*), Wrt. Voc. ii. 136, 84. Smúgaþ *serpunt*, Wülck. Gl. 248, 19. Smúgen(-an?) *serpere*, Hpt. Gl. 527, 49. Hé (æwelm) biþ smúgende geond ða eorþan, Bt. 24, 1; Fox 80, 26. [Nedre smuʒed derneliche, O. E. Homl. i. 153, 22, 32. Smuʒð, smuhgð digeliche, ii. 191, 7, 15, 17. M. H. Ger. smiegen: Icel. smjúga *to creep through* a hole, narrow space, etc.] v. ā-, under-smúgan; smeáʒan, *and next word.*

smúgendlíc; *adj. Creeping, reptile*:—Ealle slincendu ł smúendlícu *omnia reptilia*, Ps. Lamb. 68, 35.

smygel, smygels, es; *m. A burrow, place to creep into*:—Smygels

cuniculus, Wrt. Voc. ii. 137, 34. Smygelas *cuniculos*, 15, 51. Smygilas, smigilas, smyglas, Txts. 48, 199. [Cf. Icel. smuga *a narrow cleft to creep through, a hole*; smogall, smugall *penetrating*.] v. smúgan.

smyllende, smyltan. v. smillan, ge-smyltan.

smylte; *adj. Quiet, tranquil, calm, serene.* **I.** of physical calmness :—Se mónaþ (*June*) is nemned on úre geþeóde se ǽrra líða, for ðon seó lyft biþ þonne smylte, Shrn. 87, 34. Swilce seó heofone ðonne heó smylte (*serenum*) byþ, Ex. 24, 10. Hyt byþ smylte weder, Mt. Kmbl. 16, 2. Smylte weder biþ ðý þancwyrþre, gif hit hwǽne ǽr biþ stearce stormas and micle rénas and snáwas, Bt. 23; Fox 78, 26. Smylte reng *pluvia serena*, Bd. 4, 13; S. 582, 34. Smelt hagol *imber serotinus* (v. smolt), Kent. Gl. 560. Swá biþ sǽ smilte, Exon. Th. 336, 26; Gn. Ex. 55. Sió án hýþ byþ simle smyltu æfter eallum ýstum *that haven is ever calm after all the storms*, Bt. 34, 8; Fox 144, 28. Smylte is se sigewong, Exon. Th. 199, 29; Ph. 23. Smeltre *intempestae, tranquillae, serenae*, Hpt. Gl. 495, 4. Swíðe eáðe mæg on smyltre sǽ ungelǽred scipstiéra genóh ryhte stiéran, Past. 9; Swt. 59, 1. Ðonne heó baðaþ hí on smyltum wætre, Shrn. 85, 21. Smylte wedere *aure tenuis*, Wrt. Voc. ii. 4, 56. Seó sǽ mót brúcan smyltra ýþa, Bt. 7, 3; Fox 20, 23. Ic become tó ðære smyltestan hýðe, Guthl. prol.; Gdwin. 4, 20. **I a.** *gentle, mild*, of the wind :—Þurh ðone smyltan súþan-westernan wind, Bt. 4; Fox 8, 8. Hé ýste mæg oncyrran ðæt him windes hweoðu weorðeþ smylte *statuit procellam in auram*, Ps. Th. 106, 28. **I b.** fig. *favourable, prosperous*:—Smyltum belimpum *successibus*, Anglia xiii. 32, 132. **II.** of mental calm, *placid, serene, tranquil, unruffled*:—Cild ácenned smylte *a child born on the ninth day of the moon will be placid*, Lchdm. iii. 188, 12. Hé smylte móde and blíþe (*placida mente*) him eall forlét, Bd. 3, 22; S. 553, 20. Ðá frægn hé hwæðer hí ealle smylte mód (*placidum animum*) tó him hæfdon, 4, 24; S. 598, 40. Mid smyltre willsumesse *tranquilla devotione*, S. 599, 9, 10. Smylte ł blíðelíce árfæstnisse *sinceram pietatem*, Rtl. 48, 28. Smyltum þohtum *sinceris mentibus*, 7, 21 : 16, 37. v. mere-smylte; smolt, smyltness.

smylte; *adv. Quietly, mildly, gently*:—Ðonne smylte blǽweþ súþan-westan wind, Bt. 9; Fox 26, 17. v. smolte.

smylte-líc, smylting. v. smylt-líc, smelting.

smylt-líc; *adj. Tranquil, serene*:—Smyltelíco gewidra, Shrn. 74, 11. Smyltlícum *tranquilla*, Rtl. 39, 9. Smyltlícum *seneris* (*serenis?*), 98, 8.

smyltness, e; *f. Quiet, calm, serenity, tranquillity.* **I.** of physical calm :—Ðá bebeád hé ðam winde and ðære sǽ, and ðǽr wearð geworden mycel smyltness, Mt. Kmbl. 8, 26. Smyltnes, Mk. Skt. 4, 39 : Blickl. Homl. 235, 9. On smyltnysse lyfta *serenitate aerum*, Bd. 1, 1; S. 474, 30. **I a.** *the quiet of evening, evening*:—Midðý éfern ł smyltnis, (*sero*) wére áwordæn, Mk. Skt. Lind. 4, 35. Smyltnise, Jn. Skt. Lind. 6, 16. Nǽhtes smyltnisse *noctis quiete*, Rtl. 37, 35. **I b.** *gentleness, quietness* in action :—Hig hine mid ealre smyltnesse swá gelǽddon and on heora fiðerum ðǽron, ðæt hé ne mihte ne on scipe fægeror gefered beón, Guthl. 5; Gdwin. 40, 16, 14. **II.** *quiet, silence*:—Smyltnisse gesette *silentium inposuisset*, Mt. Kmbl. Lind. 22, 34. **III.** *placidity, calmness*:—Cara *cura*, oferfǽt *obesus*, smyltnys *pinguedo* (placidity?), Wrt. Voc. i. 51, 11. **IV.** *peace, tranquillity, quiet*:—Smyltnes wæs ofor eorþan and sibba genihtsumnes, Blickl. Homl. 115, 9. Þurh ðæt wierð tóslieten sió stilnes hiera hiéremonna módes and biþ gedréfed sió smyltnes hiera lífes *eo subditorum vitam dissipata quietis tranquillitate confundunt*, Past. 40, 1; Swt. 289, 8. Anweald on sibbe smyltnesse gehealdan, Lchdm. iii. 436, 13. Swefn smyltnysse and glædnysse gehátaþ, 156, 14. Tídlíc smyltnisse giráece and líf gibrenga éce *temporalem tranquilitatem tribuat et vitam conferat sempiternam*, Rtl. 31, 28. **V.** *calmness, composure*:—Ðý lǽs ða smyltnesse ðæs dómes gewemme tó hræd ierre, Past. 13; Swt. 79, 13.

smyréls, smyrian, smyring, smytta. v. smirels, smirwan, smirwung, smitta.

snaca, an; *m.*: snacu (?), e; *f. A reptile, a snake*:—Snaca *coluber*, Wrt. Voc. i. 78, 56: 287, 30: ii. 16, 75: Ælfc. Gr. 8; Zup. 27, 7. Sý Dan snaca on wege *fiat Dan coluber in via*, Wulfst. 192, 20. Snace *colubro*, Hpt. Gl. 409, 72. Gif ðú gesihst snacan ongeán ðé cuman, ongeán yfele wýfmen ðé bewerian mynegaþ, Lchdm. iii. 214, 9. Snacan *colubros*, Wrt. Voc. ii. 21, 37: *scorpiones*, Lk. Skt. 10, 19. [O. Du. snake: Icel. snákr (*only in poetry*).] v. ban-snace.

snace, e; *f.* (?) *A swift-sailing vessel*:—Ðá lét Eádweard cyng scypian xl snacca, Chr. 1052; Erl. 182, 36. Hé fór tó Scotlande mid xii snaccum, 1066; Erl. 201, 8. [(Borrowed from?) Icel. snekkja *a swift-sailing vessel*, belonging to the kind of '*lang-skip:*' Dan. snekke *a bark, sailing vessel.*]

snǽd, es; *m.* '*A piece of land within defined limits, but without enclosures, a limited circumscribed woodland or pasturage*,' Leo, Anglo-Saxon Names of Places, pp. 68-9. Or (?) *a clearing* in a wood. Cf. snǽdan, **II**:—Ic hire lǽte tó ðæt ceorla gráf tósundran . . . and se alhmunding snǽd hére intó preosda byrig, Cod. Dip. Kmbl. ii. 100, 16. Be ðam gráue ðæt hit cymþ intó ðam snǽde; and of ðam snǽde, iii. 399, 34. Ðét firhðe bituihu longanleág and ðem suðtúne and ða snádas illuc pertinentia, i. 261, 10. Tó Óswaldingtúne hiérþ holenhyrst . . . cyrþring-

hyrst, triphyrst, and insnádis(-as?) intô Óswaldingtûne, ii. 228, 4. *Also* snǽdfeld occurs iii. 399, 20:—On ðone lytlan snǽdfeld; *and* snádhyrst, i. 273, 6.

snǽd, es; *m. The handle of a scythe.* Under the forms *snathe, sneath, snead, sned* the word occurs in the glossaries of many dialects, e. g. Wilts, Somerset, Northamptonshire. Jamieson also gives it. v. E. D. S. Pub. Gloss. B. 15, 16, 19, C. 4:—Hwîlon befeóll ân sîðe of ðam snǽde intô ânum deópan seáðe. Benedictus heóld ðone snǽd bufon ðam wætere ðǽr ðæt îsen âsanc, and ðǽrrihte hit becom swymmende tô ðam snǽde, Homl. Th. ii. 162, 10–14.

snǽd, e; *f. A cut, slice, morsel, bit:*—Snǽd *offa*, Wrt. Voc. i. 82, 73: *morsus*, ii. 58, 12. Spices snǽd *offella* vel *particula*, i. 27, 19. Seó snǽd ðæs hûsles ðe heó þicgan sceolde, Homl. Th. ii. 272, 26 : Salm. Kmbl. 809; Sal. 404. Hê began tô etenne; hê feóll ðâ æt ðære forman snǽde, Homl. Skt. i. 12, 62. Ða sweartan snǽd *atram offam*, Wrt. Voc. ii. 90, 23 : 63, 14. Genim spices snǽde þynne, lege on hâtne stân, Lchdm. ii. 58, 16. Heorotes horn gebærned tô ahsan ... and mid hunige gewealcen tô snǽdum, 238, 2. Genim þreó snǽda, 52, 23. Genim fǽttes flǽsces, sele twâ snǽda, 268, 31. Nim of ðam gehâlgedan hlâfe feówer snǽda, iii. 290, 27. Ðâs sweartan snǽda *atras offulas*, Wrt. Voc. ii. 84, 40. Swâ swâ snǽda *sicut buccellas*, Ps. Spl. 147, 6. Snǽda *offulas, partes*, Hpt. Gl. 500, 78. [*Icel.* sneið *a slice.*] v. sin-snǽd.

snǽdan ; *p.* de. **I.** *to slice, cut into slices:*—On hunig gesnǽd, Lchdm. ii. 294, 9. **II.** *to snathe* [given by Halliwell as a northern word = *to prune* trees, and occurs in Ray's collection, E. D. S. Pub. Gloss. B. 15. Jamieson gives *sned* to prune, lop off, *sned* a branch pruned off.] *to lop, prune, cut* branches off trees:—Snêdit *putat*, Txts. 117, 249. Sume snêddun telgran of treówum *alii caedebant ramos de arboribus*, Mt. Kmbl. Rush. 21, 8. Hit biþ unnyt ðæt mon hwelces yfles bôgas snǽde bûton mon wille ða wyrtruman forceorfan ðæs staðoles *incassum foras nequitia ex ramis inciditur, si surrectura multiplicius intus in radice servatur*, Past. 33, 5; Swt. 222, 15. **III.** *to hew* or *trim* stones. [In this sense Jamieson gives *sned* as a word of northern Scotland.]:—Ðara werhtena ðe ðanæ stân sneóddon and fêgdon, Anglia xi. 5, 7. [Þe moder mid sexe hine tosnæde & al todælde, Laym. 4015. Þa quene ich al tosnaðde mid mine sweorede, 28050. *O. H. Ger.* gi-sneitôn *putare: Icel.* sneiða *to cut into slices.*] v. be-snǽdan ; snîdan.

snǽdan ; *p.* de *To take food, take a meal:*—Ðâ hê com to Cantwarbyrig, ðâ snǽdde hê ðǽr and his menn, and tô Dofran gewende, Chr. 1048; Erl. 177, 31. [*Icel.* snæða *to take a meal;* snæði *a meal;* snáð *food, meat.*] v. snǽding.

snǽdel, (more generally) snǽdelþearm, es; *m. The great gut:*—Snaedil *vel* þearm, snaedilþearm, snêdildaerm *extale*, Txts. 58, 381. Snǽdel, Wrt. Voc. i. 286, 59. Snǽdel(-?) *vel bæc-þearm extales*, 44, 48. Snǽdelþearm *extale*, ii. 29, 74 : 145, 29 : *fither*, 149, 1 : *fiber*, 38, 54. Snǽdelþearm *fithrem*, Lchdm. i. lxxii, 5.

snǽding, e; *f. A* (slight ?) *meal:*—Seó wucaþen nime snǽdinge (*mixtum*, = déjeûner, consistant en un verre de vin et un peu de pain, Migne. Cf. the translation of the passage, R. Ben. 63, 1 :—Ðære wucan rǽdere gange tô hlâfe and drince) ǽr ðan ðe hê âginne rǽdan, R. Ben. Interl. 70, 4. [*Icel.* snæðing *a meal.*] v. snǽdan *to take food, and next two words.*

snǽding-hûs, es; *n. An eating-house, a place where cooked meat is sold* :—Snǽdinghûs *popina*, Wrt. Voc. i. 58, 21.

snǽding-sceáp, es; *n. A sheep to be killed for eating:*—Hý teohhiaþ ûs him tô snǽdincgsceápum *aestimati sumus ut oves occisionis*, Ps. Th. 43, 23.

snǽd-mǽlum; *adv. By bits, a bit at a time:*—Pusla snǽdmǽlum *pick them out by a bit at a time*, Lchdm. ii. 356, 13.

snægel, snǽs. v. snegel, snás.

snǽsan; *p.* de *To spit, run through with a pointed implement or weapon:*—Gif mon hafaþ spere oðer eaxle and hine mon on âsnâseþ (âsnǽseþ, MS. H., snǽseþ, MS. B.), gielde ðone wer bûtan wîte ; gif beforan eágum âsnáse (âsnǽse, MS. H.) gielde ðone wer, L. Alf. pol. 36; Th. i. 84, 13. [Þe deoflen schulen mid helle sweordes al snesien (snesen, MS. C.: sneasin, MS. T.) ham þuruhut, A. R. 212, 22, *Icel.* sneisa *to spit.*] v. snás.

snǽð-feld. v. snǽd ; *m.*

snás, snǽs, e; *f. A spit, skewer :*—Snaas *veru*, Txts. 115, 144. Ân snǽs fisca oðde ôðra þinga *una serta;* a number of fish or other things run on to a stick, Wrt. Voc. i. 64, 9. Snásum *veribus*, ii. 91, 37 : *feribus*, 148, 7. [*Icel.* sneis; *f. a skewer : Dan.* snes *a score.*] v. snǽsan.

snǽð, es; *m.* (?) *A killing:*—Snáðes *occisionis*, Hpt. Gl. 478, 45.

snáw, es; *m. Snow:*—Snâw *nix*, Wrt. Voc. i. 52, 47. Swâ hwîte swâ snâw (snâ, Lind.: snáu, Rush.), Mt. Kmbl. 28, 3. Snáuw, Shrn. 50, 15. Snâua *nix*, Mk. Skt. Lind. 9, 3. Snâw cymþ of ðam þynnum wǽtan ðe byþ up âtogen mid ðære lyfte, and byþ gefroren ǽr ðan hê tô dropum geurnen sý, and swâ semtinges fylþ, Lchdm. iii. 278, 23. Ðǽr (*in Ireland*) seldon snáu leng ligeþ ðonne þrý dagas, Bd. 1, 1 ; S. 474, 31. Micle rênas and snáwas, Bt. 23 ; Fox 78, 28. Hæglas and snáwas, 39, 13; Fox 234, 16. Forstas and snáwas, Cd. Th. 239, 31; Dan. 378. Snâwum *nivibus*, Wrt. Voc. ii. 61, 45. [*Goth.* snaiws : *O. Sax.* snêu : *O. H. Ger.* snêo : *Icel.* snjór.]

snáw-ceald ; *adj. Cold as snow :*—Ðæt sió fýrene (ne) môt sunne gesêcan snâwcealdes weg monna (*but read* (?) mônan. Cf. Bt. 39, 13 ; Fox 232, 28) gemǽro, Met. 29, 8.

snáw-gebland, es; *n. A snow-storm:*—Fôr Hanníbal ofer Bardan ðone beorg, þéh ðe ymb ðone tíeman wǽren swâ micel snâwgebland swâ ðætte ǽgðer ge ðara horsa fela forwurdon ge ða elpendas ealle bûton ânum ge ða men selfe uneáðe ðone ciele genǽson *Annibal, cum in Etruriam transiret, in summo Apennino tempestate correptus, nivibus conclusus obriguit ; ubi magnus hominum numerus, jumenta complurima, elephanti pene omnes frigoris acerbitate perierunt*, Ors. 4, 8 ; Swt. 186, 34.

snáw-hwît ; *adj. Snow-white:*—Snâwhwît *niveus*, Wrt. Voc. i. 52, 48. Snâwître clǽnnysse *nivei pudoris*, Hymn. Surt. 104, 17. Mid snâwhwîtum hreóflan beslagen, Homl. Th. i. 400, 29. Sittende on snâwhwîtum horse, ii. 134, 27. Snâwhwîtne hlâf, Homl. Skt. i. 2, 405 : 18, 164. [*Icel.* snjó-hwîtr.]

snáwig ; *adj. Snowy.* v. next word.

snáwlíc ; *adj. Snowy*—Snâwlíc *nivalis*, Wrt. Voc. i. 52, 49. Se feórða heáfodwind hátte *septemtrio:* se blǽwþ norðan and cealde and snâwlíc (snâwig, MS. L.), Lchdm. iii. 274, 23. [*O. H. Ger.* snê-lîh *ninguidus: Icel.* snjó-ligr.]

snearu, an ; *f. A snare, noose:*—Snearan *tendiculam, decipulam, laqueum quod tenditur leporibus* ł *avibus*, Hpt. Gl. 429, 17. [*Icel.* snara *a snare:* cf. *O. L. Ger.* snari ; *n. fidis, fidicula.*] v. snêr.

snegel, snægel, snegl, snél, snǽl, es ; *m. A snail :*—Snegl, snél *limax, Txts. 75, 1220. Snegel, Wrt. Voc. i. 78, 63. Snǽgl, 24, 4 : ii. 51, 4. Snegel se ðe hæfþ hûs *testudo*, i. 78, 64. Snegl, snægl, snægel, Ælfc. Gr. 9, 3 ; Zup. 37, 8. Gehûsed snægl, Wrt. Voc. i. 24, 5. Snegl, snægl *marruca*, Txts. 77, 1283 : Wrt. Voc. ii. 55, 50 : *coclea*, 22, 3. Snægl *cuniculus*, 137, 34. Lytle sneglas *cocleae*, 104, 61. Snæglas, 135, 45. Mê is snægl swiftra, Exon. Th. 426, 7 ; Rä. 41, 70. Ðone blacan snegl ǽwæsc on hâligwǽtre, sele drincan, Lchdm. ii. 110, 14. Blace sneglas on pannan gehyrste, 144, 2. [*Icel.* snigill: *Dan.* snegl.] v. sǽ-snægl.

snell, snel ; *adj. Quick, active, strong.* **I.** in following glosses :—Snel *alacris*, Wrt. Voc. ii. 99, 75 : 6, 50 : *expeditus, velox, fortis*, 30, 17 : *explicitus, liber, efficatus*, 145, 35. Snelle *adultum*, Hpt. Gl. 485, 25. **II.** of rapid movement, *quick, rapid, swift:*—Sum biþ on londe snel, feþe spêdig, Exon. Th. 296, 17 ; Crä. 52. Fareþ feþrum snell, 206, 7 ; Ph. 123. Snel, 208, 29 ; Ph. 163. Hê is snel and swift and swîðe leóht *est levis et velox*, 220, 8 ; Ph. 317. Wæterþissa snel, 182, 2 ; Gú. 1304: Andr. Kmbl. 1009; An. 505. Snelle *veloces*, Ps. Spl. T. 13, 6. Fêrend snelle *swift emissaries*, Exon. Th. 246, 12 ; Jul. 60. Mê is snægl swiftra, snelra regnwyrm, Exon. Th. 426, 8 ; Rä. 41, 70. **III.** *active, prompt, ready, quick in action, bold.* [*Snell* is given in Jamieson's Dictionary with the meanings, *keen, severe; sharp* (of the air) ; *acute* (of the mind) ; *firm, determined.* Also in Cumberland it is used of the wind]:—Se snella sunu Wonrêdes, Beo. Th. 5934; B. 2971. Mê sendon tô ðê sǽmen snelle, Byrht. Th. 132, 41; By. 29: Cd. Th. 191, 26 ; Exod. 220: Exon. Th. 296, 25 ; Crä. 56. Snellra werod, cênra *the band of the bold and the brave*, Judth. Thw. 24, 21 ; Jud. 199. [Snel (strong, 2nd MS.) cniht wes Carric, Laym. 28860. Icel. snell *bold, active* : O. H. Ger. snell *alacer, acer, agilis, strenuus, robustus, pernix: Icel.* snjallr *valiant, brave ; ready of speech, eloquent.*] v. swîð-snell.

snel-líc ; *adj.* **I.** *moving rapidly, swift:*—Snellíc sǽmearh, Andr. Kmbl. 533; An. 267. **II.** *quick in action, ready, bold* :—Monig snellíc sǽrinc, Beo. Th. 1384; B. 690. [*M. H. Ger.* snellec *strenuus.*]

snellíce ; *adv. Rapidly, quickly, with activity :*—Sum sceal snellíce snêre wrǽstan *one rapidly bends the harpstrings*, Exon. Th. 332, 9 ; Vy. 82. [*O. H. Ger.* snellícho *strenue.*]

snelness, e; *f. Quickness, readiness, activity, agility :*—Hê slôh swâ hê hine (*the ball*) nǽfre feallan ne lét. Se cyngc ða oncneów ðæs iungan snelnesse, ðæt hê wiste ðæt hê næfde his gelícan on ðam plegan, Ap. Th. 13, 7.

sneóme, snióme ; *adv.* **I.** *swiftly, rapidly:*—His word yrneþ wundrum snióme *velociter currit sermo ejus*, Ps. Th. 147, 4. **II.** *quickly, immediately, at once :*—Hêt ôfstlíce up âstandan . . . sneóme of slǽpe ðæm fæstan, Andr. Kmbl. 1589; An. 796: Exon. Th. 55, 27; Cri. 890. Hî semninga sneóme forwurdon *subito defecerunt*, Ps. Th. 72, 15 : 106, 13. Snióme, 74, 7 : 103, 33 ; 123, 2. Swâ heó sǽ geseah, hê hió snióme fleáh, 113, 3. Sniómor, Cd. Th. 51, 21 ; Gen. 830. [*O. Sax.* sniumo : *O. H. Ger.* sniumo *velociter, cito, subito, statim;* sniumor, citius : cf. *Goth.* sniumundô *quickly;* sniumjan *to hasten.*]

sneorcan; *p.* snearc *To shrivel :*—Ic gesnerc swê deád from heortan *excidi tamquam mortuus a corde*, Ps. Surt. 30, 13. [Cf. þte hude swartete as hit snarchte (*shrivelled with the heat*), Marh. 18, 14. Cf. (?) *Icel.* snerkja *to wrinkle the face in displeasure* (?): *Scott.* snirk *to draw up the nose in contempt* or *displeasure.*]

sneówan; *p.* sneáw (?), sneówde (?) *To proceed, go, come, hasten :*—On brim sneóweþ snel under segle, Andr. Kmbl. 1008; An. 504. Mid ǽrdæge eástan sneóweþ (snoweþ, MS.) wlitig and wynsum (*of the sun*), Exon. Th. 350, 12 ; Sch. 62. Ða com beácna bearhtost (*the sun*) ofer

breomo sneówan, Andr. Kmbl. 484; An. 242: 3333; An. 1670. [*Goth.*
sniwan; *p.* snau, *pl.* snêwun *to go, come:* cf. (?) Icel. snöggr *sudden.*]

snér, e; *f. The string of a musical instrument:*—Snér *fidis,* Txts. 115,
148. Gellende snér, Exon. Th. 353, 40; Reim. 25. Snellíce snére
wrǽstan, 332, 9; Vy. 82. [*O.H.Ger.* snuor; *f. filum, lineolus:* cf. *Icel.*
snœri; *n. a twisted rope: Goth.* snôrjô a (*twisted*) *basket.*] v. snearu.

snerian. v. snirian.

snícan; *p.* snác, *pl.* snicon *To crawl, creep* (1) of the motion of a reptile:—[Sume wuhta] creópaþ and snícaþ, eall líchoma eorþan getenge (cf.
sume licgaþ mid eallon líchaman on eorþan and snícende faraþ, Bt. 41, 6;
Fox 254, 26), Met. 31, 6. Wyrm com snícan, Lchdm. iii. 34, 21. On
ðínum wambe and on ðínum breóstum ðú scealt snícan *pectore et ventre
repes,* Past. 43, 2; Swt. 311, 1. Snícan *serpere,* Txts. 180, 5. Ðǽr (*in
Ireland*) monn ǽnigne snícendne wyrm ne gesihþ *nullum ibi reptile videri
soleat,* Bd. 1, 1; S. 474, 33. Snícende *reptilia,* Ps. Surt. 103, 25. Ða
creópendan and ða snícendan (snícendan, Hatt. MS,), Past. 21, 3; Swt.
154, 18. (2) fig. of imperceptible movement:—Ða wunde snícaþ (*irre-
punt*) in ða innoðas mínes líchoman, Bd. 5, 13; S. 633, 18. [Snikeð in
and ut neddren, O. E. Homl. i. 251, 16. *Dan.* snige *to sneak:* cf. *Icel.*
sníkja (*wk.*) *to hanker after.*]

snid, snide, es; *m. A saw:*—Saga *vel* snide *serula,* Wrt. Voc. i. 16, 17. Snid
serra, 85, 1. Hié wǽron snidene mid snide *secti sunt,* Past. 30; Swt. 205, 13.

snid, es; *n. A slice, cut:*—Ðæt snid *copus,* Wrt. Voc. ii. 21, 59. [*Icel.*
snið; *n. a slice:* cf. *O. H. Ger.* snita; *f. buccella.*] v. ge-snid.

snide; *m. I. a cut, incision:*—Ða wunde ðæs snides *vulnus
incisurae,* Bd. 4, 19; S. 589, 17. Gif ðú wille on snide blód forlǽtan *if
you wish to let blood at an incision,* Lchdm. ii. 148, 10: 16, 5.   **II.**
slaying. v. sníðan, IV:—Swá swá scép tó snide *tamquam ouis ad occisionem,* Engl. Stud. xiii. 27, 9. [*O.H.Ger.* snit *concisio, laceratio.*]

snid-ísen, es; *n. A lancet:*—Ðonne ðú ongite ðæt geswel hnescige
and swiþrige, ðonne hrín ðú him mid snidísene and þníd listum, Lchdm.
ii. 208, 16.

snirian, (snerian?), snyrian; *p.* ede *To go quickly, hasten:*—Brimwudu
scynde, lagumearg snyrede tó hýðe, Exon. Th. 182, 7; Gú. 1306. Snyredon ætsomne, Beo. Th. 809; B. 402. Gesíón brecan ofer bædweg brimwudu snyrgan, sǽmearh plegan, wadan wǽgflotan, Elen. Kmbl. 488; El.
244. [Cf. *Icel.* snarr *swift;* snara *to make a quick turn, step out quickly.*]

sniring *a sharp rock;*—Stánum oððe snyringum *cautibus,* Wrt. Voc.
ii. 18, 15.

sníte, an; *f. A snite, snipe.* [Halliwell quotes: 'A snipe or snite, a
bird lesse than a woodcocke,' Baret, 1580, and gives *snite* as a word still
in use. See also E. D. S. Pub. Bird Names, p. 192.]—Sníte *vel* wudecocc *aceta,* Wrt. Voc. i. 29, 52. Sníte *acegia,* 62, 23: ii. 4, 36: 99, 14.
[In later glossaries *snyte* glosses *ibis,* i. 177, 29: 253, 1. *Prompt. Parv.*
snype or snyte *ibex.*]

sníðan; *p.* snáð, *pl.* snidon; *pp.* sniden.   **I.** *to cut, make an incision in* anything:—Snáð ðæt ís ðara háligra líchoman, Shrn. 62, 1.
Mec snáð seaxes ecg, Exon. Th. 408, 2; Rä. 27, 6.   **II.** *to cut as
a surgeon does, to lance or to amputate:*—Mon sníð ða bearneácnan wíf
secuerunt praegnantes, Past. 48, 2; Swt. 367, 14. Gif ðonne ðæt worms
up stíhþ tó ðon ðæt ðe þince ðæt hit mon sníþan mǽge and út forlǽtan
... ðonne hrín ðú him mid ðý snidísene and þníd listum ... ðonne ðú
hit tóstinge oððe sníþe, Lchdm. ii. 208, 11–21. Sníð oððe ceorf on ðæt
hále and ðæt cwice líc, 84, 28: 52, 2. Gód lǽce ðe wel cann wunda
sníðan, Past. 49; Swt. 377, 18. Ic wéne ðæt hé hiene snide sláwlícor,
gif hé him ǽr sǽde ðæt hé hiene sníðan wolde ... se lǽce, ðonne hé
cymþ ðone untruman tó sníðanne, 26; Swt. 186, 2–7.   **II a.** metaphorically:—Ðæt mon mǽge sníþan and bærnan his unþeáwas *ut culpae
morbos supplicio resecarent,* Bt. 38, 7; Fox 210, 3.   **III.** *to cut up
or to pieces:*—Ðone ramm ðú snitst tó sticcon, Ex. 29, 17. Hié wǽron
snidene mid snide *secti sunt,* Past. 30; Swt. 205, 13.   **IV.** *to cut so
as to kill, to slay* an animal (v. of-sníðan, sníðung, II):—Ðæra éwena
meolc gé brucon and ða ðe fǽtte wǽron gé snidon (*mactavistis*), L. Ecg.
P. iii. 16; Th. ii. 202, 24. Ða ealdan sacerdas cealf snidon, Homl. Th.
ii. 210, 19. God hét niman ánes geáres lamb and sníðan on Eástertíde,
40, 11: 262, 29.   **V.** *to cut stone, to hew:*—Ðæra wyrhtena ðe
ðæne stán snidon and fégdon, Anglia xi. 4, 12.   **VI.** *to cut hair:*—
Wið heáfodece, hundes heáfod gebærn tó absan and sníð ðæt heáfod;
lege on, Lchdm. ii. 20, 2.   **VII.** *to cut corn, to reap:*—Ða on
teárum sáwaþ, hí eft gefeán sníðaþ *in gaudio metent,* Ps. Th. 125, 5.
[Tacc Ysaac þin wennchell & sniþ itt alls itt wære an shep, Orm. 14666.
Goth. sneiþan *to reap; O. Sax.* snídan *to cut: O. Frs.* snítha: *O. H. Ger.*
snídan *secare, resecare, caedere, putare, dolere, attondere: Icel.* sníða;
p. sneið (*but* sníddi *also occurs*) *to cut, prune.*] v. á-, be-, ge-, of-, tó-,
ymb-sníðan; snǽdan.

sníðing. v. sníðung.

sníð-streó[w] *carline thistle* (?):—Sníthstreó *gacila,* Txts. 35, 13.
Sníðstreó, snídstreú, snídstreú *sisca, sista,* 97, 1868. Cf. Eoforþrote *scisca,*
35, 27: *scasa* ł *scapa* ł *sisca,* Lchdm. iii. 305, col. 1. In Spanish *sisca* is
the cylindrical sugar-cane.

sníðung, e; *f.   I. a cutting, cut* (v. sníðan, I):—Gif ða ómihtan

---

þing sýn útan cumen of wundum oððe of sníþingum oððe of slegum, Lchdm.
ii. 82, 22.   **II.** *slaying, slaughtering* (v. sníðan, IV):—Offrung
sacrificium, sníþung *mactatio,* Wrt. Voc. i. 28, 50. Sníðing, ii. 59, 10.

sníwan; *p.* de *To snow:*—Ic sníwe *ninguo,* Ælfc. Gr. 28, 5; Zup.
174, 8. Hit sníwþ *ningit,* 22; Zup. 128, 17. Sníuuith, sníuidh *ninguit,*
Txts. 78, 669. Sníweþ, Wrt. Voc. ii. 60, 14. Ða cwom ðær micel snáw
and swá miclum sníwde swelce micel flýs feoll, Nar. 23, 13. Norþan
sníwde, Exon. Th. 307, 30; Seef. 31. Swá swá hit ríne and sníwe and
styrme úte, Bd. 2, 13; S. 516, 17. [*Chauc.* snewede; *p.: Mand.* snew;
p., and a similar form remains in dialects. *O. H. Ger.* sníwan: *Icel.* has
a strong form snivinn; *pp.*] v. be-sníwod.

snóca, an; *m. A bend, bay* (?):—Of ðære díc on færscmærus west-
snócan; of ðam snócan on fúlan mære eástweardnæ *from the dike to the
western bay of fresh mere; from the bay to the east side of the foul mere,*
Cod. Dip. Kmbl. v. 344, 33.   With some variations the same boundaries
are given in a later charter:—De Elmede dych usque ad solemeres west-
snok; de solemeres westnok usque ad Horehyrne, iii. 119, 29.   [Cf. (?)
O. H. Ger. snôh; *forestum, nomine* bracten snôh, Grff. vi. 839.]

snód, e; *f. A snood, fillet, head-dress:*—Snód *cappa,* Wrt. Voc. ii. 103,
8: 13, 42: *capsa (çappa?),* 128, 34: *cinthium, mitra,* 131, 10: *vitta,*
i, 16, 65: 26, 5. Ðá lǽrde hí sum man, ðæt heó náme ǽnne wernægel
of sumes oxan hricge, and becnytte tó ánum hringe mid hire snóde ...
Ðá geseah heó licgan ðone hring on ðam wege mid snóde mid ealle ...
Ðá wénde heó ðæt se hring tóburste, oððe seó snód tóslupe, ac ðá ðá
heó áfunde ... ða snóde mid eallum cnottum fæste gewriðen ..., Homl.
Th. ii. 28, 16–26. Snóda *vittarum,* Hpt. Gl. 526, 57. Wæs ðæm deóre
se hrycg ácæglod swelce micel snóda (sníde?) *belua serrato tergo,* Nar. 20, 27.

snofl *mucus, snivel:*—Wið langum sáre ðæs heáfdes þurh horh oððe
þurh snofl, Lchdm. ii. 24, 4. v. next word.

snoflig; *adj. Full of snivel, having a cold in the head:*—Hiemps ys
winter, hé byþ ceald and wǽt ... Swá byþ se ealda man ceald and snoflig;
flegmata, ðæt byþ hraca oððe geposu, deriaþ ðam ealdan and ðam unhálan, Anglia viii. 299, 36.

snoru, e; *f. A daughter-in-law:*—Snoro *nurus,* Wrt. Voc. ii. 115, 3:
83, 83. Snoru, 73, 52: 60, 49: i. 52, 10. Snoru, snora, Ælfc. Gr. 11;
Zup. 79, 18. Swegr on hyre snore and snoru on hyre swegere, Lk. Skt.
12, 53. Scá Maria is Godfæder snoru and Godes suna módur and háligra
sáuwla sweger, Shrn. 118, 6. Hió genom hiere snore, Alexandres láfe, Ors.
3, 11; Swt. 148, 18. [*O. Frs.* snore: *Q. H. Ger.* snura: *Icel.* snor.]

snot *mucus from the nose, snot* [*found in the compound* ge-snot:—Wið
gesnote and geposum, Lchdm. ii. 54, 17. *O. Frs.* snotte: *M. H. Ger.*
snuz: *Dan.* snot]. v. snýtan.

Snotinga-hám *Nottingham:*—Hér fór se ilca here innan Mierce tó
Snotengahám (Snotinghám, MS. E.), Chr. 868; Erl. 72, 21. Fór hé tó
Snotingahám and gefór ða burg and hét hié gebétan and gesettan ægðer
ge mid Engliscum mannum ge mid Deniscum, 922; Erl. 108, 30. Hér
Eádmund cyning Myrce geeode, burga fífe, ... Snotingahám ..., 942;
Erl. 116, 13.

Snotingahám-scír, e; *f. Nottinghamshire,* Chr. 1016; Erl. 154, 8.

snotor, snottor (-er, -ur); *adj. Prudent, wise, sagacious:*—Snotor
prudens, Wrt. Voc. i. 47, 35. Snoter, 76, 12. Cwom Daniel tó dóme, se
wæs snotor, Cd. Th. 225, 8; Dan. 151. Nis nǽnig swá snotor ... ne
ðæs swá gleáw, nymþe God seolfa, 286, 8; Sat. 349. A sceal snotor
hycgean ymb ðysse worulde gewinn, Menol. Fox 570; Gn. C. 54: Beo.
Th. 1656; B. 826. Snotur, Ps. Th. 118, 23. Ðæs snottor in sefan ðæt
hé ána mǽge ealle geríman stánas on eorðan, Cd. Th. 205, 19; Exod.
438. Fród wita, snotor ár, Exon. Th. 313, 18; Mód. 2. Swá cwæð
snottor on móde, gesæt him sundor æt rúne, 293, 4; Wand. 111. Rǽdum
snottor, wís on gewitte, Andr. Kmbl. 938; An. 469. Se wítga snottor
searuþancum, Elen. Kmbl. 2377; El. 1190. Se snotera, Beo. Th. 2631;
B. 1313. Snotra, 6231; B. 3120. Snottra, 3577; B. 1786. Salomon
se snottra, Past. 4; Swt. 37, 16. Seó snotere mægð, Judth. Thw. 23, 17;
Jud. 125. Snottrum men snǽd óðglídeþ, Salm. Kmbl. 803; Sal. 401.
Háligne wer and snotorne *virum sanctum et sapientem,* Bd. 3, 23; S. 554, 9.
Ðú mé snoterne gedýdest *prudentem me fecisti,* Ps. Th. 118, 98. Ðone
snoteran Salomon, Ælfc. T. Grn. 7, 28. Mín sóðfæste snotere bídaþ *me
expectaverunt justi,* Ps. Th. 141, 9. Snotre men, 57, 4. Snotre *urbana,*
Hpt. Gl. 481, 40. Snottere seleræðend, Andr. Kmbl. 1317; An. 659.
Snottre and unwíse, Blickl. Homl. 107, 11. Snottre ceorlas, Beo. Th.
3187; B. 1591. Hwylc is wísra, wel snotera, Ps. Th. 106, 42. Engla
werod snotra, Hy. 3, 16. Snoterra mon, Salm. Kmbl. 502; Sal. 251. Gomol
snoterost, fyrngeárum fród, Menol. Fox 482; Gn. C. 11. Ðú oferswíþ-
dest ðone snótrestan helwerena cyning, Exon. Th. 275, 1; Jul. 543.
Burgsittendum ðam snoterestum, Elen. Kmbl. 553; El. 277. Ða ðe hé
wíseste and snoterste wiste *quos sapientiores noverat,* Bd. 2, 9; S. 512,
11. [Þet folc biþ iseli þurh snoterne biscop, O. E. Homl. i. 117, 19.
Uþwitess unndersstodenn þurrh snoterr gyn, Orm. 7087. *Goth.* snutrs:
O. H. Ger. snot[t]ar *prudens: Icel.* snotr.] v. fore-, forþ-, gearo-, hyge-,
mód-, rǽd-, þanc-, un-, word-, woruld-snotor; snytre.

snotor-líc; *adj. Wise, prudent, philosophical:*—On snoterlícum lárum
in philosophicis dogmatibus, Hpt. Gl. 459, 63. [*Icel.* snotr-ligr.]

snotorlíce; adv. Wisely, prudently, philosophically:—Snotorlíce sapienter, Ps. Lamb. 46, 8. Snotorlíce (snotur-, Rush.) ł wíslíce sapienter, Mk. Skt. Lind. 12, 34. Uton ðás þing geþencean swíþe snotorlíce & wíslíce, Blickl. Homl. 97, 1. Snotorlíce academice, Wrt. Voc. i. 61, 27. Ne hýrde ic snotorlícor guman þingian, Beo. Th. 3689; B.1842. [O.H.Ger. snotarlíhho: Icel. snotr-liga.] v. un-snotorlíce.

snotorness, e; f. Prudence, wisdom, sagacity:—Prudentia, ðæt ys snoternys, Wulfst. 247, 15: Homl. Skt. i. 1, 157. Hí (the innocents) wǽron gehwǽde and ungewittige ácwealde, ác hí árísaþ on ðam gemǽnelícum dóme mid fullum wæstme and heofenlícere snoternysse, Homl. Th. i. 84, 23. Snotornesse ł wísdóm sapientiam, Ps. Lamb. 48, 4. Salomon gesette þreó béc þurh his snoternesse, Ælfc. T. Grn. 7, 36.

snotorung. v. word-snotorung.

snotor-wyrde; adj. Prudent or wise of speech:—Herodes wearð gewrēged tô ðam cásere ... hē wæs snotorwyrde tô ðan swíðe, ðæt se cásere hine mid máran wurðmynte ongeán ásende, Homl. Th. i. 80, 9. Sum man wæs geháten Mercurius on lífe, se wæs swýðe fácenfull and deáh full snotorwyrde, Wulfst. 107, 1.

snúd swiftness, quickness:—Ús bær naca, snellíc sǽmearh, snúde bewunden (possessed by swiftness), Andr. Kmbl. 534; An. 267.

snúd; adj. Coming at once, coming soon or suddenly:—Biþ ǽghwylcum synwyrcendra on ða snúdan tíd (the day of judgment, which was to come suddenly, cf. Matt. 24, 39; or to come soon?), Exon. Th. 52, 32; Cri. 842. v. next word.

snúde; adv. At once, quickly, directly:—Snúde denuo, Jn. Skt. Lind. Rush. 3, 3. Gangaþ snúde go directly, Elen. Kmbl. 625; El. 313: 307; El. 154. Hēt hine snúde eft cuman bade him quickly return, Beo. Th. 3743; B. 1869. Se wyrm gebeáh snúde tôsomne, 5129; B. 2568. Snúde forsended, 1812; B. 904: Exon. 231, 12; Ph. 488: Judth. Thw. 22, 8; Jud. 55: 23, 17; Jud. 125. Wearþ snellra werod snúde gegearewod, 24, 21; Jud. 199. Mec Dryhten hēt snúde gesecgan, Exon. Th. 144, 10; Gû. 676. Snúde cýðan, 19, 7; Cri. 297: Elen. Kmbl. 890; El. 446: 3947; B. 1971: 4639; B. 2325. Ic snúde gefrægn, 5497; B. 2752. [Cf. Icel. ganga snúðigt to walk fast.]

snyrian, snyring. v. snirian, suiring.

[snýtan to clear the nose. (Prompt. Parv. snytyñ a nese or a candyl emungo, mungo. Snite, snyte in this sense remains in several dialects. O.H.Ger. snûzan emungere, nasumpurgare: Icel. snýta.) v. snyting, snot.]

snyðian to go as a dog with its nose to the ground (?):—Neb is mín niþerweard ... ic snyþige forð (it is a plough that speaks), Exon. Th. 403, 12; Rä. 22, 6. [Icel. snýðja to go sniffing like a dog, but applied also to the going of ships, and other things.]

-snyðian. v. be-snyðian.

snýting, e; f. A clearing of the nose, sneezing:—Snýtingc vel fneósung sternutatio vel sternutamentum, Wrt. Voc. i. 46, 20. [Prompt. Parv. snytynge of a nose or candyl munctura, emunctura.] v. snýtan.

snytre; adj. Wise:—Se ðe sigor seleþ snytrum mihtum, and ðín mód trymeþ godcundum gifum, Cd. Th. 170, 6; Gen. 2808. v. snotor.

snytrian; p. ode To be or to become wise:—Hwæt is se dumba, se ðe swíðe snyttraþ, hafaþ seofon tungan, hafaþ tungena gehwylc .xx. orda, hafaþ orda gehwylc engles snytro, Salm. Kmbl. 459; Sal. 230. Snytrian philosophari, Hpt. Gl. 527, 63.

snytro, snyttro, snytero(u); indecl. in sing.; pl. is used with the same force as sing.; f. Prudence, wisdom, sagacity:—Snytru sapientia, Mk. Skt. Lind. Rush. 6, 2. Hwǽr com heora snytro what has become of their wisdom? Blickl. Homl. 99, 31. Wera snytero, Cd. Th. 295, 25; Sat. 492. Se þurh snytro spéd smiðcræftega wæs, 66, 14; Gen. 1084. Ic eom gewis ðínra mægena and snytro, Lchdm. i. 326, 4. Snyttro, Elen. Kmbl. 586; El. 293. Hié ðære snytro gelýfdon, Cd. Th. 217, 25; Dan. 28. Full mid snyttro (snytrum, Rush.) plenus sapientia, Lk. Skt. Lind. 2, 40. Ealle ðú mid snyteru worhtest omnia in sapientia fecisti, Ps. Th. 103, 23. Wísdóm ł snytro sapientiam, Ps. Spl. 18, 8. Ic ðē gelǽrde swelce snytro swylce manegum ieldran gewittum oftogen is, Bt. 8; Fox 24, 28. Snyttro, 7, 3; Fox 20, 11. Þurh his godcundemeht and þurh his ēcean snyttro, Blickl. Homl. 121, 16. Tô hēranne snytro (snyttro, Rush.) Salomones, Mt. Kmbl. 12, 42. Þurh sefan snyttru, Past. pref.; Swt. 9, 10: Exon. Th. 28, 5; Cri. 442. Beoran on breóstum sibbe and snytero, Cd. Th. 277, 19; Sat. 207. Ealle heora snytru beóþ forglendred omnis sapientia eorum devorata eۦt, Ps. Th. 106, 26. Spræc sunu Arones snytra gemyndig, Cd. Th. 148, 28; Gen. 2463. Snyttra, Exon. Th. 304, 30; Fä. 78. Þurh snyttra cræft, Andr. Kmbl. 1261; An. 631. Ðara ðe geóce tô him sēceþ mid snytrum, 2307; An. 1155. On snytrum in sapientia, Ps. Th. 89, 14. Mid módes snyttrum, Beo. Th. 3416; B. 1706. Snyttrum wisely, prudently, Andr. Kmbl. 1292; An. 646. Deáh ánra gehwylc hæbbe ða .xii. snyttro Habrahames and Isaces and Iacobes, Salm. Kmbl. 150, 2. Þurh ða snyttra (snyttro, MS. O.) ðe ic fram ðam sôþan Gode onféng per sapientiam mihi a Deo vero donatam, Bd. 2, 13; S. 517, 3. Paulus ðæt lof Gode betǽhte ðe him snytera (snytra, MS. F.) and wísdóm sealde, R. Ben. 4, 6. [Goth. snutrei.] v. ge-, un-snytro.

snytro-cræft (or snytro (gen.) cræft, cf. þurh snyttra cræft, Andr. Kmbl. 1261; An. 631), es; m. Prudent skill, prudence, wisdom:—

Wundra mǽst, ðæt swylc snyttrocræft ǽnges hæleþa hreþer weardade, Exon. Th. 169, 28; Gû. 1101. Se mæg eal secgan, ðam biþ snyttrucræft bifolen on ferhðe, 42, 4; Cri. 667: 239, 18; Ph. 622. Sefan sídne geþanc and snytrocræft, Cd. Th. 249, 27; Dan. 536. Daniel gespræc þurh snyttrocræft, 253, 14; Dan. 595. Ða ðe fyrngewritu þurh snyttrocræft sēlest cunnen, Elen. Kmbl. 747; El. 374. Ða ðe snyttrocræft þurh fyrngewrito gefrigen hæfdon, 308; El. 154. Ðē God sealde sigespéd and snyttrocræft, 2342; El. 1172. Snyttrucræft, Exon. Th. 113, 10; Gû. 155. Nǽnig ðæs swíþe þurh snyttrucræft, 294, 21; Crä. 18. Ælmihtig eácenne gǽst in sefan sende, snyttrocræftas, Cd. Th. 246, 29; Dan. 486.

snytro-hús, es; n. The house of wisdom:—Hē ðá swá gelôme wiðsôc snytruhúse repulit tabernaculum Silon, Ps. Th. 77, 60.

soc, es; n. Suck, sucking at the breast:—On ðone dæg ðe man ðæt cild fram soce áteáh in die ablactationis ejus, Gen. 21, 8. [Sese₃ childer of her sok, A. P. 103, 391. Taken awei fro sok, or wenyd, Wick. (Isaiah 11, 8).] v. ge-soc.

socc, es; m. A sock, kind of shoe:—Socc soccus, Wrt. Voc. ii. 120, 70. Soccas pedules (cf. meó), R. Ben. Interl. 92, 1. [O.H.Ger. soc soccus, caliga, calicula: Icel. sokkr a sock. From Latin.]

sóchtha. v. sôhþa.

socian; p. ode I. to soak (trans.), to steep in a liquid:—Socodon coquebant, Germ. 399, 378. II. to soak (intrans.), to lie in a liquid:—Glǽdenan rinde lytelra gedô þreó pund on glæsfæt, gedô ðonne ðæs scearpestan wínes tô .v. sestras, ásete ðonne on háte sunnan ... ðæt hit socige .iiii. dagas oþþe má, Lchdm. ii. 252, 11. Dweorge dwostlan weorp on weallende wæter, lǽt socian on lange, 240, 7: iii. 14, 17. v. súcan.

sócn, e; f. I. a seeking, search, exploring:—land-sôcn, sēcan, I. 1. II. a seeking, desiring, trying to get. v. mete-sôcn, sēcan, I. 2. III. a seeking to obtain an end. v. hláford-sócn, sēcan, I. 3. IV. a seeking for information, question, inquiry. v. sôcn, I. 5:—Be monigum sócnum and frignyssum ða ðe him nýdþearflíce gesewen wǽron de eis quae necessariae videbantur quaestionibus, Bd. 1, 27; S. 488, 33. V. a seeking, visiting of a place, attendance at a place, resort. v. cyric-sócn, sēcan, II. 2:—Wē ûre synna georne bētan mid fæstene and mid ælmessan and mid ciriclícere sócne (with going to church), Wulfst. 134, 17. Ðá tôwende se biscop ðæt weofod and ða dwollícan sócne mid ealle ádwæscte (put an end to the resorting to the place, which had been supposed erroneously to be holy), Homl. Th. ii. 508, 5. Ic cýþe ðæt ic nelle sócne habban tô mínum hírēde ðone ðe mannes blód geóte ǽr hē hæbbe godcunde bôte underfangen ... I declare that I will not that he who sheds man's blood have resort to my court before he have undertaken ecclesiastical 'bôt' ..., L. Edm. S. 4; Th. i. 248, 22. [Cf. From sôcne þes folkes free from the resort of the people, Laym. 2365. Sookne or custom of hauntynge frequentacio, concursus, Prompt. Parv. 463, col. 2. Gret soken hadde this meller With whete and malt of al the londe aboute, Chauc. Reeve's T. 67.] VI. a seeking for protection or a place so sought, refuge, sanctuary, asylum, (1) in a general sense:—Ic sēce sócne refugio, of ðam is refugium sócn, Ælfc. Gr. 28, 6; Zup. 179, 13-14. Ðǽr se freónd wunaþ on ðære sócne ðe ic ða sibbe wíð hine healdan wille, Exon. Th. 145, 1; Gû. 688. (2) as a technical term in reference to the protection afforded by a church or by the king's court, etc. v. ciric-, friþ-sócn:—Gif hwilc þeóf oþþe reáfere gesôhte ðone cing oþþe hwylce cyrican and ðone biscop, hē hæbbe nigon nihta fyrst. And gif hē ealderman oþþe abbud oþþe þegen sēce, hæbbe þreora nihta fyrst. And gif hine hwá lecge binnan ðæm fyrste, ðonne gebēte hē ðæs mundbyrde ðe hē ǽr sôhte, oþþe hē hine twelfa sum lädige, ðæt hē ða sócne nyste. And sēce hē swylce sócne swylce hē sēce, ðæt hē ne sý his feores wyrðe bútan swá feola nihta swá wē hēr cwǽdon, L. Ath. iv. 4; Th. i. 224, 2. Be ciricena sócnum. Gif hwá ðara mynsterháma hwelcne for hwelcre scylde gesēce ðe cyninges feorm tô belimpe oþþe ôðerne fríone híerēd ðe árwyrðe sié, áge hē þreora nihta fierst him tô gebeorganne, L. Alf. pol. 2; Th. i. 60, 22. Cf. Si fur qui furatus est postquam concilium fuit apud Dunresfeld, vel furetur, nullo modo vita dignus habeatur, non per socnam, non per pecuniam, si per verum reveletur in eo, L. Ath. iii. 6; Th. i. 218, 30. VII. a seeking with hostile intent, an attack. v. hám-sócn, sēcan, III:—Ðære sócne (the hostility of Grendel) singales wæg módceare micle, Beo. Th. 3558; B. 1777. VIII. as a legal term, frequently in connection with sacu. Kemble says: 'Sócn is inquisitio, the preliminary and initiative in Sacu, in other words the right of investigating, necessary to and a part of power of holding plea,' Cod. Dip. Kmbl. i. xlv. But from a Latin version of a charter it would seem that sôcn was the power of seeking or levying fines; the English 'Ic an heom ðæt hý habben saca and sócna' is rendered by 'cedens ut habeant privilegium tenendi curiam ad causas cognoscendas et dirimendas lites inter vassallos et colonos suos ortas, cum potestate transgressores et calumniae reos mulctis afficiendi easque levandi,' iv. 202, 7. Other instances of the occurrence of the word, whose Latin form is often soca, are the following:—Ic habbe gegeofen ... Ælfwine abbod saca and sócna (sacam et socam, Lat.) ... And ic wylle ðæt seó sócne (soca, Lat.) wiðinnen Bichámdíc licge intô Ramesēge on eallen þingen swá full swá ic heó mēseolf áhte ... and se abbod and ða gebróðra intô Ramsēge habben ða sócne (socam, Lat.) ofer heom

. . . And in ǽlcer[e] scíre ðǽr sanctus Benedictus hafþ land inne [habbe hē] his saca and his sócne . . . swā hwylc man swā ða sócne āhe, Sanctus Benedictus habbe his freódóm on eallen þingen, 208, 19–209, 14. Mór-tūn and eal seó sócna ðe ðǽrtó hēreþ, vi. 148, 36. Ne gyrne ic ðínes ne sace ne sócne *I desire nothing of yours, neither your privileges nor your rights*, L. O. 14; Th. i. 184, 16. Cyninges þegenes heregeata ðe his sócne hæbbe, L. C. S. 72; Th. i. 414, 16. Nān man nāge nāne sócne ofer cynges þegen būton cyng sylf, L. Eth. iii. 11; Th. i. 296, 23. [þe reue of Rotland sokene, Piers P. 2, 110. *Goth.* sôkns *quaestio: O. H. Ger.* sóhni *inquisitio: Icel.* sókn *an attack;* as a law-term, *an action, prosecution; an assemblage of people* at church, etc.; *a parish* (*Dan.* sogn).] v. cyric-, friþ-, hām-, hláford-, land-, mete-, scip-sócn.

sod. v. ge-sod.

Sodoma, Sodome, an; *or indecl. The town of Sodom:*—Ða cininges of Sodoman and Gomorran . . . on ðām burgum Sodoma and Gomorra, Gen. 14, 10, 11. Hē eardode on ðære byrig Sodoma, 13, 12. Hig eodon tó Sodoman weard, 18, 22. On ðære byrig Sodoman, 18, 26. On Sodoman weallsteápe burg, Cd. Th. 145, 6; Gen. 2401. Woldon Sodome burh werian, 119, 6; Gen. 1975.

Sodome; *pl. The people of Sodom:*—Hí lǽrdon hira synna swā swā Sodome dydon . . . Gif Sodome hira synna hǽlen, Past. 55; Swt. 427, 28. Sodoma lande (eorðe Sodominga, Rush.), Mt. Kmbl. 10, 15. On Sodomum (Sodomingum, Rush.), 11, 23.

Sodomingas. v. preceding word.

Sodomisc; *adj. Of Sodom:*—Sodomisc cynn, Cd. Th. 116, 12; Gen. 1935. *Used as a noun,* sodomita:—Sodomisce .vii. geár fæston *sodomitae .vii. annos jejunent,* L. Ecg. P. iv. 68, 5; Th. ii. 228, 16.

Sodomitisc; *adj. Of Sodom:*—Ða Sodomitiscan menn, Gen. 13, 13. Sodomitiscra cining, 14, 17: 18, 20.

Sodom-ware; *pl. The people of Sodom:*—Cômon Sodomware, Cd. Th. 148, 4; Gen. 2451: 120, 18; Gen. 1996. Būton Sodomwarum ánum, Blickl. Homl. 79, 10.

sófte (sóft?); *adj. Soft:*—Sófte *suavis,* Ælfc. Gr. 9, 28; Zup. 54, 5. **I.** *soft* (of sleep), *quiet, undisturbed:*—Ic sóftum slǽpe mē gereste, Homl. Th. i. 566, 22. **II.** *soft, luxurious:*—Ne hē ne cume on wearmum bǽðe ne on sóftum bedde, L. Ælfc. C. 11; Th. ii. 280, 22. On ðam sóftum baðe, Homl. Skt. i. 11, 231. **III.** *gentle, not harsh, not stern.* v. sófte, **III** [:—He wæs swiðe gód and sófte man and dyde mycel tó góde, Chr. 1114; Erl. 244, 38. Hē milde man was and sófte and gód, 1137; Erl. 261, 31.] v. sēfte.

sófte; *cpve.* sóftor, sēft; *adv. Softly, gently:*—Sófte *suaviter,* Ælfc. Gr. 38; Zup. 228, 6: *gradatim,* Wrt. Voc. ii. 41, 37: *pedetemtim,* 81, 39: *sensim,* 120, 41. Ðone sófte langan *morosam,* 32, 6. **I.** *of sleep, rest, etc., softly, quietly, without disturbance:*—Hē sófte swæf, Cd. Th. 12, 2; Gen. 179. Reste hē hine sófte, Lchdm. ii. 292, 7: Ps. Th. 77, 65. **II.** *calmly, at ease, without trouble:*—Ðǽr mē sófte byþ, ðǽr ic beó fægere beþeaht fiðerum ðínum, Ps. Th. 60, 3. Hié sófte ðǽs bidon, Exon. Th. 10, 3; Cri. 146. Hí willniaþ manifeald earfolþe tó þrowianne, for ðam ðe hí willniaþ mǽran are mid Gode tó habbanne, ðonne ða habbaþ ðe sóftor libbaþ, Bt. 39, 10; Fox 228, 17: Shrn. 163, 20. Ðæt ic ðý sēft mǽge mín ālǽtan líf and leódscipe *that with mind the more at ease I may relinquish life and people,* Beo. Th. 5492; B. 2749. **III.** *gently, not harshly:*—Ðū sófte wealdest gesceafta, Met. 20, 7. Ðū sófte gedēst, ðæt hí ðé selfne gesión móten, 20, 272. **IV.** *without discord:*—Gebunden gesiblíce sófte tógædere, Met. 20, 68. **V.** *easily, without opposition:*—Ne sceole gé swā sófte sinc gegangan, ūs sceal ord and ecg ǽr gesēman, Byrht. Th. 133, 32; By. 59. [*O. Sax.* sáfto: *O. H. Ger.* samfto *facile.*] v. un-sófte.

sóftnes, e; *f. Softness, ease;* in a bad sense, *luxury, effeminacy:*—Heora fela wǽron mid olfendes hǽrum tó líce gescrýdde, and ðǽr láðode sóftnys, Homl. Th. ii. 506, 24. Mid sóftnysse and mid yfelum lustum, i. 270, 5: Homl. As. 15, 59. Ða ðe ðǽr (*in heaven*) singaþ ne swincaþ on ðam sange, ac mid sóftnysse būtan geswince hí heriaþ ðone Hǽlend, 43, 470. Sóftnysse *luxuriam,* Germ. 401, 19.

sogoþa, an; *m.* **I.** *hiccough, heartburn* (?):—Gyf men sý sogoþa getenge oððe hwylc innan-gundbryne . . . ðonne wēne ic ðæt hyt him wel fremie ge wið sogoðan ge wið ǽghwylcum incundum earfoðnyssum Lchdm. i. 196, 16–21. Of hómena stíeme and of wlǽtan cymþ eágna mist and sió scearpnes and sogoþa ðæt dēþ wið ðon is ðis tó dónne *the acidity and heartburn* (?) *cause that against which this is to be done,* ii. 28, 1. Wið sogoþan and geohsan ðe of milte cymþ, 248, 1. Ne yrne he ðe lǽs hē mid ðæs rynes ēðgunge hwylcne wleattan and sogeðan on his heortan ne āstyrige *lest the running cause nausea or give him heartburn* (?): the Lat. version has 'ut non scurilitas inveniat fomitem,' R. Ben. 68, 3. **II.** *gastric juice* (?):—Lǽcedómas ðe gesóge sind ge heáfde ge heortan and wambe and blǽdran and sogeþan, Lchdm. ii. 166, 3. v. ælf-sogoþa, sūgan.

soht. v. suht.

sóhþa? Sochtha glosses *iota,* Wrt. Voc. ii. 112, 4. *The word is written* sohctha, 45, 72. *Somner suggests* ioctha.

sol *a sole* (?), 'a collar of wood, put round the neck of cattle to confine

---

them to the stelch. ''A bow about a beestes necke.'' Palsgrave.' Halliwell. '*Sole,* a rope or halter to tie cattle in the stall,' Kennett's Parochial Antiquities. Among 'husbandlie furniture' Tusser gives '*soles,* fetters, and shackles [cf. *however* sál.] :—Sol *orbita,* Wrt. Voc. ii. 65, 6.

sol, es; *n.: solu,* we, e; *f. Mire* or *a miry place* [Halliwell gives *soul, sole* = a dirty pond, as a Kentish word] :—Sol *volutabrum,* Wrt. Voc. i. 37, 22. On grǽgsole burnan; andlang burnan on grǽgsole hagan, Cod. Dip. Kmbl. v. 336, 24. Wið Heortsolwe, iii. 391, 32. Of ðam wylle on ðæt heorotsol; of ðam heorotsole, ii. 249, 37. Iu ða heortsole; of ðære sole, iii. 380, 6. On ðæt sol; of ðan sole on ða ealdan strǽte, Cod. Dip. B. i. 518, 40. Sole *volutabro,* Wrt. Voc. ii. 97, 17. Tó sole ł fýlþe *ad volutabrum,* Hpt. Gl. 477, 70. Seomode on sole sídfæðmed scip, Beo. Th. 609; B. 302. Sió sugu hī wile sylian on ðære sole æfter ðæm ðe hió āþwægen biþ, Past. 54; Swt. 419, 27. Gif swín eft silþ on ðæt sol, Swt. 421, 3. Þonon ðæt cume in ða reádan sole, Cod. Dip. Kmbl. iii. 375, 8. In reádan solo, Txts. 431, 6. Ad stagnum quendam cujus vocabulum est Ceabban solo, Cod. Dip. Kmbl. iii. 388, 2. Tó Higsolon; of Higsolon, 219, 3. Swin simle willnaþ licgan on fúlum solum . . . ðeáh hí beswemde weorþon, ðonne sleáþ hē eft on ða solu and bewealwiaþ þǽron, Bt. 37, 4; Fox 192, 26–29. [Cf. sol; *adj. filthy:*—Wule a sol cloð et one cherre beon hwit iwaschen? A. R. 324, 1. His (the priest's) alter cloð great and sole, and hire (the priest's concubine's) chemise smal and hwit; and te albe sol, and hire smoc hwit, Rel. Ant. i. 129. Solwy *dirty,* Wrt. Voc. i. 171, 41. *O. H. Ger.* sol *volutabrum.*] v. Sol-mónaþ, solian, sylu, sylian.

sól, e; *f.* (?) *The sun:*—Ne ðé sunne on dæge sól ne gebærne *per diem sol non uret te,* Ps. Th. 120, 6. [*Goth.* sauil; *n.: Icel.* sól; *f.*] v. sunne.

solate, solcen, solen, solere. v. sólsece, a-, be-solcen, solu, solor.

solian; *p.* ode *To make* or *to become foul:*—Searo hwít solaþ sumur hát cólaþ eorðmægen ealdaþ ellen cólaþ *the armour* or *implement that was bright grows rusty, summer that was hot grows cool, earthly might grows old, strength grows chill,* Exon. Th. 354, 57; Reim. 67. [Cf. Nis noht so hot þat hit na coleþ, ne noht so hwit þat hit ne soleþ, O. and N. 1276. *O. H. Ger.* bi-, gi-solót *made filthy.*] v. sol, sylian.

Sol-mónaþ, es; *m. The old name for February:*—Ðonne se Sol-mónaþ biþ geendod, ðonne biþ seó niht feówertýne tída lang and se ðæg týn tída, Shrn. 59, 2. Solmónaþ sígeþ tó túne, Februarius, Menol. Fox 31; Men. 16. [The first part of the compound is of doubtful meaning. Bede says, 'Solmónaþ dici potest mensis placentarum, quas in eo diis suis [Angli] offerebant;' but there is no word *sol* = placenta, unless it be found in the gloss *panibus sol,* Epinal Glossary, ed. Sweet, p. 21 a, 11. Kluge takes the word to be sól = sun, and observes 'die form des kuchens war für die benennung massgebend,' Engl. Stud. viii. 479. *Sol* = mire would give a name that suggests the later February fill-dyke, and would not be inappropriate. The form *sille, selle* is found in some L. G. dialects, and also *sporkel,* which may be connected with *spurcalia.* See Grimm, Gesch. D. S. c. vi.]

solor, soler[e?], es; *m. An upper chamber, a soler.* v. Halliwell's Dict.:—Ic wiluige ðætte ðeós sprǽc stigge on ðæt ingeþonc ðæs leorneres suǽ suǽ on sume hlǽdre óððæt hió fæstlíce gestonde on ðæm solore ðæs módes *until it stand firmly in the upper chamber of the mind,* Past. proem.; Swt. 23, 18. Se fugel ofer heánne beám hús getimbreþ, and gewícaþ ðǽr sylf in ðam solere *in that upper chamber* (*its nest*), Exon. Th. 212, 2; Ph. 204. [Soler *solarium,* Wrt. Voc. i. 178, 12. Solere, 273, col. 2. Solere or lofte *solarium, hectheca,* Prompt. Parv. 464 (see note). Garytte, hey solere *specula,* 187. Wiclif (Jos. 2, 6) uses the word for the flat roof of a house. *O. Sax.* soleri *an upper room* (Mk. 14, 14). *O. H. Ger.* soleri, solær *solarium, coenaculum: Ger.* söller. *From Lat.* solarium.]

sólsece, sólosece, an; *f. Heliotrope:*—Sólsece vel sigelhwerfe *solsequium* vel *heliotropium,* Wrt. Voc. i. 30, 30. Sólsæce *solsequium,* 79, 15. Ðás wyrte ðe man solate and óðrum naman sólosece nemneþ, Lchdm. i. 178, 21. Cf. sólesege *solata,* iii. 305, col. 1. Halliwell gives solsekille.

solu, an, e (?); *f. A sole, a sandal:*—Solen *soleae,* Wrt. Voc. i. 26, 18. [*Goth.* sulja *a sandal: O. H. Ger.* sola, *pl.* solun, sola *solea, sandallo, planta. From Lat.* solea.]

solu *mire.* v. sol.

som, som-. v. sam, sam-.

sóm, e; *f.* **I.** *agreement, concord:*—Beó eallum mannum sibb and sóm gemǽne, and ǽlc sacu tótwǽmed, L. Eth. vi. 25; Th. i. 320, 28: L. C. E. 17; Th. i. 370, 10: Wulfst. 118, 3. Ðám dómbócum ðe se heofonlíca Wealdend hto sóme and to sehtnesse, Homl. Th. ii. 198, 19. Tó sibbe and tó sóme, Chart. Th. 231, 35. Hē sceal beón symle ymbe sóme and ymbe sibbe *he shall ever be engaged in promoting concord and peace,* L. I. P. 7; Th. ii. 312, 13. Sibbe and sóme lufie man georne, Wulfst. 73, 16. **II.** *the bringing about of concord, reconciliation, adjustment of differences:*—Nān sacu ðe betweox preóstan sí ne beó gescoten tó woroldmanna sóme *no dispute between priests shall be referred to the adjustment of secular men,* L. Edg. C. 7; Th. ii. 246, 4. Bisceopum gebyraþ, gyf ǽnig ðrum ābelge, ðæt man geþyldige ðð geférena sóme, L. I. P. 10; Th. ii. 316, 35. **III.** *an agreement, arrangement* of a matter in dispute:—Ús eallan ðe æt ðære sóme wǽran, Chart. Th. 171, 1. v. un-sóm; sēman, ge-sóm.

són, es; *m. A musical sound, music* vocal or instrumental:—Nán neát nyste nænne andan tô ôþrum for ðære mergþe ðæs sônes ... Hê wæs oflyst ðæs seldcûþan sônes (*the music of Orpheus' harp*), Bt. 35, 6; Fox 168, 11, 23. Ða hearpan strengas se hearpere suîðe ungelîce tiéhþ and styreþ and mid ðý gedêþ ðæt hî nâwuht ungelîce ðæm sône ne singaþ ðe hê wilnaþ *chordas tangendi artifex, ut non sibimetipsi dissimile canticum faciat, dissimiliter pulsat*, Past. 23; Swt. 175, 8. Gif hit mycel geférǽden is sýn hý (*the psalms*) mid antefene gesungene, gif seó geférǽden lytel is, sýn hý forðrihte bûtan sône gesungene *si major congregatio fuerit cum antiphonis, si vero minor in directum psallantur*, R. Ben. 41, 9. In efnum sônum *in consonantibus*, Mk. Skt. p. 1, 13. Sônas tô singanne on cyricean *sonos cantandi in ecclesia*, Bd. 4, 2; S. 565, 35. [*Icel.* sônn. From Latin.]

sôna; *adv. Soon, immediately, directly, at once*:—Sôna *actutum*, Wrt. Voc. ii. 5, 2: 82, 70: *extemplo*, 31, 45. Hî wǽron sôna deáde *they died at once*, Bd. 1, 12; S. 481, 22. Lege ðǽrtô, ðonne biþ hit sôna gebêt, Lchdm. i. 116, 13: 118, 11. Ælc cræft biþ sôna forealdod, Bt. 17; Fox 60, 10. And sôna (*statim*) gâst hine on wêsten genýdde, Mk. Skt. 1, 12, 10. Hî ðá sôna forlêton hyra nett, Mt. Kmbl. 4, 22. Ðá sôna (*continuo*) forscranc ðæt fictreów, 21, 19: Cd. Th. 53, 16; Gen. 862. Ðá sôna and hræðe ac[t]utum, Wrt. Voc. ii. 9, 17. Forhwon ne woldest ðû sôna hraþe ða dígolnesse mê cýþan *quare non citius hoc compertum mihi revelare voluisti?* Bd. 4, 25; S. 601, 21. Se ðe wille wyrcan wæstmbǽre lond, âtió of ðǽm æcere ǽrest sôna (*first of all*; cf. ǽrest, Bt. Fox 78, 22) fearn and þornas, Met. 1, 2, 25. Eft sôna *again*, Soul Kmbl. 134; Seel. 67. Sôna æfter ðæra daga gedrêfednesse *statim post tribulationem dierum illorum*, Mt. Kmbl. 24, 29; Cd. Th. 304, 14; Sat. 630. Sôna ðæs forman geáres ðá Decius fêng tô rîce, Homl. Skt. i. 23, 476. Sôna ðæs ðe hê ðam biscopsetle onfêng *ubi sedem episcopalem accepit*, Bd. 1, 33; S. 498, 29. Ðá sôna ðæs ðe ðis fæsten geworht wæs *quo mox condito*, 1, 12; S. 481, 12. Sôna hraþe ðæs ðe hê biscop geworden wæs *mox ut ipse pontificatus officio functus est*, 2, 1; S. 501, 34. Sôna ðæt him bet wæs *nec mora, melius habere coepit*, 3, 13; S. 539, 6. Sôna swâ seó sunne sealte streámas oferhlîfaþ, swâ se fugel of beáme gewîteþ, Exon. Th. 206, 1; Ph. 120. Sôna swâ ..., ðá, Met. 8, 1. [*O. Sax.* sân, sâno: *O. Frs.* sân: cf. *Goth.* suns.]

sôn-cræft, es; *m. Music*:—Sôncræft *musicam*, Anglia xiii. 38, 306.

sond, song. v. sand, sang.

sopa, an; *m. A sup, draught*:—On wearmum wætre drince betonican týn sopan, Lchdm. ii. 134, 22. Sûpe cû buteran .viii. morgnas .iii. sopan, 294, 1. [Þer (in hell) is o wateres flod ... a þusen saulen beoþ bi sore ofþurst ... ne moten heo biden neuer o sope, Misc. 152, 169. þyse renkeȝ schul neuer suppe on sope of my seve, Allit. Pms. 41, 108. *Icel.* sopi *a sup, mouthful.*] v. sûpan.

sopp-cuppe, an; *f. A sop-cup*, a cup into which sops were put:—Ic ann mînæn cinæhláfordæ ânræ sopcuppan, Chart. Th. 553, 31: 554, 4. Ic ann Ælfwerdæ ânræ sopcuppan and Æþelwerde ânæs drincæhornæs, 555, 4. Ânæ soppcuppan an þrým pundum, 527, 7. Twâ sopcuppan, 522, 22. See Brand's Popular Antiquities, on Nuptial Usages, ii. 84–6, and next word.

soppe (?) *a sop*. [Soppe *offa*, soppe in wyne *vipa*, Prompt. Parv. 465. Cf. *vipa* a wynsope, *offa* a ale sope, Wrt. Voc. i. 242, col. 1. Ase is a zop of hot bryead huanne me hit poteþ in wyn, Ayenb. 107, 5. Wel loved he by the morwe a sop in wyn, Chauc. Prol. 334. If he soupeth, he ete but a soppe of *spera-in-deo*, Piers P. 15, 175. *Icel.* soppa.] v. preceding word.

soppian *to sop*:—Genim hlâf, geseóð on gâte meolce, soppige on sûþerne [drenc], Lchdm. ii. 228, 31.

sorg. v. sorh.

sorgian, sorhgian (*and* sorgan, v. *pres. part.* sorgende); *p.* ode. I. *to care, be anxious, feel anxiety or care*, (a) with a clause:—Hê nalles sorgode hwæðer siððan â Drihten âmetan wolde wrece be gewyrhtum *he felt no anxiety as to whether the Lord would ever mete out vengeance according to deserts*, Met. 9, 34. Hî lyt sorgodon hwylc him ðæt geleán æfter wurde, Andr. Kmbl. 2456; An. 1229. (b) *with preps.* ymbe, for:—Hê sorgaþ ymb ða (*useless works*) and biþ ðara suîðe gemyndig and forgiett his selfes *mens fit in exteriorum dispositione sollicita, et sui ignara*, Past. 4, 1; Swt. 37, 19. Geþenceaþ ðæt gê winnaþ and â embe ðæt sorgiaþ, ðæt wê ûrne lîchoman gefyllan, Blickl. Homl. 99, 6. Ða ðe for his lîfe lyt sorgedon, Exon. Th. 116, 19; Gû. 209. Nô ðû ymb mînes ne þearft lîces feorme leng sorgian, Beo. Th. 907; B. 451. (c) absolute:—Hê sceal winnan and sorgian, ðonne se dæg cume ðæt hê sceole ðæs ealles îdel hweorfan, Blickl. Homl. 97, 25. Sorgiende *anxius*, Wrt. Voc. i. 287, 67. Sorgende, ii. 6, 66. Hû him woruldmanna seó unclǽne gecynd cearum sorgende hearde ondrêde, Exon. Th. 63, 10; Cri. 1017. Sume dæge ðæt hê sorgiende (*sollicitus*) bâd hwonne seó ádl tô him côme, Bd. 3, 12; S. 537, 6. Ac hwæðere sorhgiende môde geornlîce þohte *sed multum sollicitus ac sedula mente cogitans*, 2, 12; S. 514, 28. II. *to sorrow, grieve, be sorry*, (a) *with preps.* ymbe, for, on:—Gif ðû hafast mid ðe wulfes hrycghǽr ... on siðfæte, bûtan fyrhtu ðû ðone sîð gefremest, ac se wulf sorgaþ ymbe his sîð *the wolf will be sorry for his journey*, Lchdm. i. 360, 22. Swîþe on ðon sorhgedon ðæt hî ðam láreówe onfôn ne

woldon ðe hî him tô sendon *de non recepto quem miserant predicatore dolentes*, Bd. 3, 5; S. 527, 29. Wit hreówige mâgon sorgian for his sîðe, Cd. Th. 49, 30; Gen. 800. Sorgiende for ðâm ermþum, Bt. 38, 1; Fox 196, 7. (b) absolute:—Sorgedon Adam and Eve, and him oft betuh gnornword gengdon, Cd. Th. 47, 24; Gen. 765. Ne sorga, snotor guma, sêlre biþ ǽghwæm ðæt hê his freónd wrece, ðonne hê fela murne, Beo. Th. 2772; B. 1384. Ða woruldâre ðe ðû nû sorgiende ânforlête, Bt. 7, 3; Fox 20, 12: Cd. Th. 22, 28; Gen. 347. Ðǽr mon mæg sorgende folc gehýran hygegeómor, Exon. Th. 55, 28; Cri. 890. Sume ofer sǽ sorgiende (*dolentes*) gewiton, Bd. 1, 15; S. 484, 7. Him sorgendum sâr ôðclîseþ, Exon. Th. 77, 35; Cri. 1267. [*Goth.* saurgan *to be anxious*; *to sorrow*: *O. Sax.* sorgôn: *O. H. Ger.* sorgên: *Icel.* sorga.] v. be-, for-sorgian.

sorgung, e; *f. Sorrowing, grieving, sorrow, grief*:—Ðær (in hell) is sorgung and sârgung and â singal heóf, Wulfst. 114, 4.

sorh, sorg, sorhg, e; *f.* I. *care, anxiety*:—Sorg *accidia, tedium vel anxietas*, Wrt. Voc. ii. 99, 17: *cura*, 19, 62. Mec sorg dreceþ on sefan, ic ne mæg rǽd âhycgan, Cd. Th. 131, 21; Gen. 2179. Nis mê ðæs deáþes sorg *death causes me no anxiety*, Exon. Th. 125, 7; Gû. 350. Frêfrigende gesihþe seó him ealle ða nearonesse ðære gemyugedan sorhge âfyrde *visionem consolatoriam, quae omnem ei anxietatem memoratae sollicitudinis auferret*, Bd. 4, 11; S. 579, 34. Ða ðe nǽfre nânne mon buton sorge (*securum*) ne forlǽtaþ, Bt. 7, 2; Fox 18, 14. Ðæt gê lybbon eówre lîf bûtan ǽlcre sorge *absque ullo pavore*, Lev. 25, 18. Ûs biþ sibb æfter sorge, Andr. Kmbl. 3134; An. 1570. Ne biþ him on ðâm wîcum wiht tô sorge *there shall be nothing in heaven to cause them anxiety*, Exon. Th. 238, 29; Ph. 211. Gê mê lyt sorge sealdon *ye caused me little care*, 121, 13; Gû. 288. Ne ic ðæs deáðes hafu sorge on môde, 166, 12; Gû. 1041: 308, 20; Seef. 42: 376, 33; Seel. 164. Sorgum *curis*, Wrt. Voc. ii. 19, 63. Heorte mid sorgum gedrêfed, Judth. Thw. 22, 31; Jud. 88. Ferhð sorgum âsǽled, Cd. Th. 132, 18; Gen. 2195. II. *sorrow, grief, affliction, trouble*:—Ne biþ ðǽr sorg ne wôp, Blickl. Homl. 103, 36. Wât se ðe cunnaþ, hû slîþen biþ sorg tô geféran, Exon. Th. 288, 13; Wand. 30: 288, 30; Wand. 39. Mec sorg bicwom ... ic bihlyhhan ne þearf sîðfæt ðisne, 273, 33; Jul. 525. Ðæt wæs Satane sâr tô geþolienne, micel môdes sorg, Andr. Kmbl. 3376; An. 1692. Ne frîn ðû æfter sǽlum, sorh is genîwod, Beo. Th. 2649; B. 1322: Ps. Th. 118, 28. Sorh is mê tô secganne, hwæt ... *it is a grief to me to tell, what* ..., Beo. Th. 950; B. 473. Se Hǽlend wiste ðæt his gingran wolde unrôte beón ... Wǽron swâ manigfealdlîce sorga Cristes þegnum ... Wæs him micel langung and sorh on heora heortan, ða hié ðæt ongeáton, ðæt hê ðeng mid him lîchomlîce wunian nolde; hê hié ... frêfrede for ðære gelômlîcan sorge, Blickl. Homl. 135, 14–23. Nû hý ðê willaþ on murnunga gebringan ðonne hié ðe fram hweorfaþ tô hwæm cumaþ hî ðonne elles bûtan tô tácnunge sorge[s] and anfealdes sâres *si calamitosos fugiens facit, quid est aliud fugax, quam futurae quoddam calamitatis indicium*, Bt. 7, 2; Fox 18, 21. Ne hié sorge wiht, weorces wiston; ac hié wel meahton libban, Cd. Th. 49, 1; Gen. 785. Wraðu wîta gehwylces, sæce and sorge, Elen. Kmbl. 2059; El. 1031. Seó hreówsung ne beoþ nâ bûtan sorge, Bt. 31, 1; Fox 110, 29. Wedera helm æfter Herebealde heortan sorge weallende wæg ... mid ðære sorge, ðâ him sió sâr belamp, gumdreám ofgeaf, Beo. Th. 4937; B. 2468. Sægde him tô sorge, ðæt hý ðone grênan wong ofgiefan sceoldan, Exon. Th. 130, 32; Gû. 447: 39, 11; Cri. 620. Gê ðæs nǽfdon nâne sorge (*luctum*), Past. 32, 1; Swt. 211, 10. Weán cûðon, sâr and sorge, Cd. Th. 5, 21; Gen. 75: Beo. Th. 239; B. 119. Ðû his (*for it*) sorge ne þearft beran on ðînum breóstum, Cd. Th. 45, 28; Gen. 733. Ic ða sorge gemon, hû ic bendum fæst bisga unrîm dreág, Exon. Th. 280, 5; Jul. 624. Hyge weard mongum blissad sâwlum, sorge tôglidene, 71, 31; Cri. 1164. Sorga sârost, 122, 19; Gen. 2029. Sorga mǽst, 308, 22; Sat. 696. Weána gehwylcne, sîdra sorga, Beo. Th. 300; B. 149. Holofernus ðe ûs monna mǽst morþra gefremede, sârra sorga, Judth. Thw. 24, 10. Sorgna hâtost, Exon. Th. 163, 12; Gû. 992. Manna bearn sorgum sâwaþ, 6, 18; Cri. 86. Ne biþ hyra yrmðu âu tô wîte, ac ðara ôþerra eád tô sorgum, 79, 22; Cri. 1294. On wîte mid swâte and mid sorgum libban, Cd. Th. 31, 8; Gen. 482. Mid sorgum geswenced, Andr. Kmbl. 231; An. 116. Æghwilc man sceolde mid sâre on ðæs world cuman, ond hêr on sorhgum beón and mid sâre of gewîtan, Blickl. Homl. 5, 29. [*Goth.* saurga *sorrow, care*: *O. Sax.* sorga: *O. H. Ger.* sorga *cura, solicitudo, angor, moeror, labor*: *Icel.* sorg *care, sorrow.*] v. bealo-, cear-, gnorn-, hyge-, inwit-, môd-, nearu-, sin-, torn-, þegn-sorh (-sorg); be-, or-, unbe-sorh; *adj.*

sorh-byrðen *a burden of sorrow, a grievous trouble*:—Ðæt (*the drowning of a number of people*) wæs sorgbyrðen, Andr. Kmbl. 3063; An. 1534.

sorh-cearig; *adj. Having grievous care, oppressed with anxiety or sorrow, anxious, sorrowful*:—Siteþ sorgcearig sǽlum bidǽled, Exon. Th. 379, 4; Deór. 28: 278, 25; Jul. 603: 285, 4; Jul. 709: Beo. Th. 6294; B. 3152. Gesyhþ sorhcearig wînsele wêstne, 4901; B. 2455. Wreclástas settan sorhgcearig, Cd. Th. 276, 15; Sat. 189.

sorh-cearu *grievous care, painful anxiety*:—Næs him sorgcearu ðeáh his lîc and gǽst hyra somwiste gedǽled(-de ?), Exon. Th. 160, 6; Gû. 939.

sorheriunge *infestatione*, Wrt. Voc. ii. 45, 43. *Read* forheriunge. v. for-hergung.

sorh-full; *adj.*   I. *full of care* or *anxiety, careful, anxious,* (a) *feeling anxiety:*—Seldan snottor guma sorgleás blissaþ swylce dol seldon drýmeþ sorgful ymbe his forðgesceaft nefne hé fǽhþe wite *seldom does the prudent man rejoice without anxiety about his future, just as the fool seldom rejoices with trembling, unless he know that hostility* (or *death?* fǽhþe *from* fǽge?) *is near,* Exon. Th. 303, 19; Fä. 55. Symble beó gé sorhfulle for eówre sáwle hǽlo *ever be ye solicitous for your soul's salvation,* L. E. I. prm.; Th. ii. 394, 14. Ðæt hé sorgfulra sié ymb hine selfne *ut circa se solicitius vivant,* Past. 28, 2; Swt. 191, 19. (b) *attended with anxiety, causing anxiety:*—Ðeós woruld is sorhful *the present time is full of anxieties,* Wulfst. 189, 6. Ne incǽnig mon beleán mihte sorhfulne síð (*the perilous swimming match of Beowulf and Breca*), Beo. Th. 1028; B. 512.   II. *sorrowful, mournful, sad,* (a) *feeling sorrow* or *grief:*—Oft se welega and se wædla habbaþ suá gehweorfed hira þeáwum ðæt se welega biþ eáðmód and sorgfull, and se wædla biþ upáhæfen and selflíce, Past. 26, 2; Swt. 183, 11. Ic eom þearfa and sorhful *ego sum pauper et dolens,* Ps. Lamb. 68, 30. Sorhfull, Ps. Th. 85, 1. Ic sceal gnornian seóc and sorhful, Cd. Th. 281, 20; Sat. 275. Módor síðode sorhfull, sunu deáþ fornam, Beo. Th. 4244; B. 2119. Hig heora synna andetton mid sorhfullum móde, Jud. 10, 10. Hé hafaþ wérige heortan, sefan sorhfulne, Salm. Kmbl. 757; Sal. 378. Ða sorgfullan *illi quos caminus paupertatis excoquit,* Past. 26, 1; Swt. 183, 4. (b) *attended with* or *causing sorrow, grievous:*—Módor gegán wolde sorhfulne síð, sunu wrecan, Beo Th. 2560; B. 1278. Ða sorhfullan sáule wunde, Ps. C. 50, 141. Adam and Eve in ðäs deáðdene drohtað sóhton, sorgfulran gesetu, Exon. Th. 227, 7; Ph. 417. [*O. H. Ger.* sorg-fol *sollicitus*: *Icel.* sorg-fullr.]

sorh-leás; *adj.*   I. *free from anxiety* or *care, secure:*—Sorgleás *secura,* Rtl. 63, 10: 8. 23: 40, 15. Ic hit ðé geháte, ðæt ðú móst sorhleás swefan, ðæt ðú ondrædan ne þearft aldorbealu eorlum, Beo. Th. 3348; B. 1672. Ne sculon wé nǽfre sorhleáse beón, ac symble úrne deádes dæg beforan úres líchoman eágum settan, L. E. I. prm.; Th. ii. 396, 22. Wé gedóþ eów sorhleáse *securos vos faciemus,* Mt. Kmbl. 28, 14. Þeóf, ðe on þýstre færeþ, sorgleáse hæleð forféhþ, Exon. Th. 54, 24; Cri. 873. Cyning wæs þe sorgleásra (cf. módsorge wæg cyning, 122; El. 61), Elen. Kmbl. 193; El. 97.   II. *free from sorrow:*—Wé sorgleáse mótan wunigan in wuldre, Exon. Th. 22, 3; Cri. 346. [*Icel.* sorg-lauss.]

sorhleást, e; *f. Security:*—Gif ðú gesihst ðæt ðú on wætere fægere in gá oððe ofer gá, sorhleáste getácnaþ. Gif ðú gesihst ðæt ðú mid swurde bist begyrd, sorhleáste hit getácnaþ, Lchdm. iii. 212, 30–33.

sorh-leóþ, es; *n. A sorrowful song, a lay of grief:*—Gesyhþ sorhcearig on his suna búre winsele wéstne ... nis ðær hearpan swég, gomen in geardum swylce ðær iú wǽron. Gewíteþ ðonne, sorhleóð gǽleþ, Beo. Th. 4912; B. 2460. Ongunnon ðá (*after putting Jesus in the grave*) sorhleóð galan, Rood Kmbl. 134; Kr. 67.

sorh-líc; *adj. Sorrowful, causing sorrow, grievous, sorry, miserable:*—Hit is earmlíc and sorhlíc mannum tó gehýranne, eall ðæt man ús foresægþ, Wulfst. 241, 21. Ðonne biþ sorhlíc sár and earmlíc gedál líces and sáwle, 187, 14. Ðonne wyrd ehtnes grimlíc and sorhlíc *there shall be persecution cruel and grievous,* 89, 16. Stingaþ hine mid sorhlícum sáre on his heortan, 141, 9. Setl his ðú gesettest oppe on eorðan *sedem ejus in terra collisisti,* Ps. Th. 88, 37. [*O. H. Ger.* sorg-líh: *Icel.* sorg-ligr.] v. next word.

sorhlíce; *adv. Miserably, grievously:*—Herodes hys spere genam, and hyne sylfne ofstang; and hé swá sorhlíce hys lýf geendode, St. And. 34, 7. Ðær synd sorhlíce (cf. tó sorge, Dóm. L. 190) tósomne gemencged se þrosmiga lig and se þrece gycela, Wulfst. 138, 25. [Þonne biþ þ soule hus seoruhliche bereaued, Frägm. Phlps. 5, 39. Sorhliche heo gunnen clupien (hii gonne grede, 2nd MS.), Laym. 21883. *O. H. Ger.* sorglícho: *Icel.* sorgliga.]

sorh-lufu, e, an; *f. Love that is attended with anxiety* or *sorrow, hapless love:*—Him seó sorglufu slǽp ealle binom, Exon. Th. 378, 14; Deór. 16.

sorh-stæf, es; *m. Trouble, care, affliction:*—Æfter sorgstafum, Exon. Th. 282, 8; Jul. 660. Cf. sár-, hearm-stæf.

sorh-wilm, es; *m. Violent emotion of anxiety* or *sorrow:*—Soden sorgwælmum, Exon. Th. 177, 32; Gú. 1236. Sorgwylmum, 166, 21; Gú. 1046. Ic ðæs módceare sorhwylmum seáð, Beo. Th. 3990; B. 1993.

sorh-wíte, es; *n. A grievous punishment, torment:*—Ðara sorhwíta mǽst, Wulfst. 187, 2.

sorh-word, es; *m. A word expressive of care* or *sorrow:*—Hié (*Adam and Eve*) fela sprǽcon sorhworda, Cd. Th. 49, 8; Gen. 789.

sorig; *adj. Sorry, grieved:*—Hé biþ suíðe sorig (sárig, Cott. MSS.) *dolet,* Past. 33; Swt. 227, 8.

soru (?), e; *f. A particle of dust, bit of straw:*—Sore (stréu, Rush.) *festucam,* Mt. Kmbl. Lind. 7, 3, 4. v. seár, and cf. (?) *Icel.* sori *dross.* **sot.** v. sott.

sót, es; *n. Soot:*—Sót *fuligo,* Ælfc. Gr. 9, 3; Zup. 37, 4. Sót *fuligo,* deorces sótes *furvae fuliginis,* Hpt. Gl. 504, 6–8. Soote *fuligine,* Wrt. Voc. ii. 36, 28. Sooth, 109, 46. Meng wið sóte, Lchdm. ii. 76, 8. Meng ðærtó sót and sealt and sand, i. 356, 24. [*Icel.* sót; *n.*]

sotel. v. sotol.

sóþ, es; *n. Sooth.*   I. *truth in a general sense, conformity with an absolute standard:*—Ðæt is fruma worda ðínra ðæt ðér byþ sóð symble méted *principium verborum tuorum veritas,* Ps. Th. 118, 160. Ðæt his sóð fore ús genge weorðe, Exon. 147, 33; Gú. 736. Swá ic geornlícor ðæt sylfe sóþ sóhte swá ic hit læs métte. Nú ðonne ic ondette ðæt on ðysse láre ðæt sylfe sóþ scíneþ, ðæt ús mæg syllan éces lífes hǽlo, Bd. 2, 13; S. 516, 29–32. Ic on ðínum sóðe gancge *ambulabo in veritate tua,* Ps. Th. 85, 10.   I a. *truth, that which conforms to an absolute standard:*—Mid Sigelwarum sóð yppe weard, dryhtlíc dóm Godes, Apstls. Kmbl. 128; Ap. 64. Ða ðe Godes lage healdaþ and sóþes gelýfaþ, Wulfst. 4, 8. Of eorðan cwom ǽþelast sóða, Ps. Th. 84, 10.   I b. *truth, what is true in general:*—Se ðe lýhþ oððe ðæs sóðes ansaceþ, Salm. Kmbl. 365; Sal. 182. Hé can him gesceád betweox sóðe and unsóðe, Wulfst. 51, 29. Ic tó sóðe (*as a general truth*) wát ðæt biþ in eorle indryhten þeáw, ðæt hé his ferðlocan fæste binde, Exon. Th. 287, 9; Wand. 11.   II. *truth in regard to a particular circumstance, exact conformity with the facts of a case:*—Ðære gesyhþe sóþ (*its agreement with what actually occurred*) wæs hraþe gecýþed on ðære fǽmnan deáþe, Bd. 4, 8; S. 576, 10. Ðæs gehátes and ðæs wítedómes sóþ se æfterfyligenda becyme ðara wísena geséþde, 4, 29; S. 607, 35. Ne meaht ðú nó mid sóþe getælan ðíne wyrd ... hit is leásung ðæt ðú wénst ðæt ðú seó ungesǽlig, Bt. 10; Fox 28, 1. Is tó ðære tíde tælmet hwíle emne mid sóðe seofon and twentig, Andr. Kmbl. 227; An. 114.   II a. *truth, fidelity to a promise:*—Hé him gehét his ǽriste, swá hé mid sóðe (*in exact accordance with his promise*) gefylde, Blickl. Homl. 17, 4. Deópne áð Drihten áswór and ðone mid sóðe getrymede, Ps. Th. 131, 11. Gif hé on sóþe tóweard cyneríce geháteþ, Bd. 2, 12; S. 514, 7.   II b. *truth, reality, certainty, real condition of things, what really is:*—Nú mæg sóð hit sylf gecýþan *now can the truth declare itself,* Blickl. Homl. 187, 16. Ðæt andgyt biþ on ðæs mannes heortan ðe nele sóðes gelýfan, ðeh hé sylf his ágenum eágum eal ne gesáwe, Wulfst. 3, 20: 93, 22. Gé mengan ongunnon lyge wið sóðe, Elen. Kmbl. 613; El. 307. Gé wiðsócon sóðe, ðæt in Bethleme bearn cenned wǽre, 780; El. 390. Gif hit man tó sóðe ongite *if it is known as a fact,* Deut. 17, 4. Ic tó sóþe wát, Exon. Th. 275, 9; Jul. 547. Men ne cunnon secgan tó sóðe *men cannot certainly say,* Beo. Th. 101; B. 51. Secge ic ðé tó sóðe, ðæt . . ., 1184; B. 590. Gif ðú him tó sóðe sægst *if you tell him it as a fact,* Cd. Th. 36, 11; Gen. 570. Ic feówer men geseó tó sóðe *I really see four men,* 242, 8; Dan. 416. Syle mé ða tó sóðe *give me it really,* Ps. Th. 118, 144. Ic wát ðæt ðú sóþ segst, Bt. 26, 1; Fox 92, 8: Jn. Skt. 19, 15. Hé á tó ǽghwylcum sóð sprecende wæs, Blickl. Homl. 223, 29. Ðeáh gé ða ǽ cúðon, gé ne woldon sóð oncnáwan, Elen. Kmbl. 790; El. 395. Bútan ðú forlǽte ða leásunga and mé sweotollíce sóð gecýðe, 1377; El. 690. Ðeáh ic ðæt sóð tó láte gecneówe, 1412; El. 708. Hwæðer mon sóð ðe lyge sagaþ, Exon. Th. 80, 15; Cri. 1307. Wite ðú for sóð *be certain of this,* Bt. 7, 3; Fox 20, 17. Nǽni eft cymeþ ðe ðæt for sóð mannum secge, hwylc sý Meotodes gesceaft, Menol. Fox 590; Gn. C. 64. Ic eów fela wille sóða gesecgan, Exon. Th. 116, 30; Gú. 215.   II c. *affirmation of truth, asseveration:*—Preóst hine clǽnsie sylfæs sóðe ðus cweðende: Veritatem dico in Xþo, non mentior, L. Wih. 18; Th. i. 40, 14.   III. *truth, conformity with right, righteousness, equity, justice:*—Hú ic míne heortan heólde mid sóþe *justificavi cor meum,* Ps. Th. 72, 11. Ic sóð déme *ego justitiam judicabo,* 74, 2. Ðæt méste sóð *justitiam tuam,* 70, 18. Suna cyningces syle ðæt hé sóð healde *justitiam tuam da filio regis,* 71, 1. Se ðe his sóþ and riht symble healdeþ *justus,* 111, 6. Gif wé sóþ and riht on úrum lífe dón willaþ, Blickl. Homl. 129, 32. Se ðe sóð and riht fremeþ on folce, Beo. Th. 3405; B. 1700. Snyttra brúceþ ðe warnaþ him wommas worda and dǽda and sóþ fremeþ, Exon. Th. 304, 25; Fä. 80. Hié firendǽda tó frece wurdon sóð ofergeáton, Drihtnes dómas, Cd. Th. 155, 32; Gen. 2581. [*O. Sax.* sóð.] v. un-sóþ, *and next word.*

sóþ; *adj. Sooth, very, true.*   I. *the opposite of that which is false,* or *merely pretends,* or *has the appearance of, genuine, real:*—Ðæt hí oncnáwon ðæt ðú eart án sóþ God *ut cognoscant te solum Deum verum,* Jn. Skt. 17, 3. Tó sóþ God man, ðý hine dorste deófol costian, swylce hé wæs sóþ God, ðý him englas þegnedon, Blickl. Homl. 33, 33. Ðes is sóð wítega, Jn. Skt. 7, 40. Sóð leóht wæs *erat lux vera,* I, 9: 15, 1. Ælc sóþ wela and sóþ weorþscipe sindon míne ágne þeówas, Bt. 7, 3; Fox 20, 15. Ðis is sóð lǽcecræft, Lchdm. i. 376, 8. Se sóþa boda ðæs heán leóhtes Agustinus, Bd. 2, 2; S. 502, 31. Ðæt hig geleornigen ðæs gewinnes onlícnesse ðæt hig hiom eft nánwiht ondrædon in ðæs sóðan gewinnes gefiohte, Shrn. 35, 17. Ðæt wé úre synna béton mid sóþre hreówe, Blickl. Homl. 25, 17: 171, 12. Tó ðon sóþan angide gecyrran, 107, 15. Ðæt wé sóþe sibbe heóldan, 109, 15. Se ðe his godcundnesse mid sóþum wísum gerýmeþ, 179, 24.   II. *true, in conformity with the actual state of things:*—Mín gewitnes is sóþ, Jn. Skt. 8, 14: 19, 35. Ic eom geþafa ðæt ðæt is sóþ ðæt ðú ǽr sǽdest, Bt. 38, 2; Fox 196, 16. Sóþ is ðæt ic eów secgge, Blickl. Homl. 53, 2. Hé þohte on him sylfum hwæt his sóðes wǽre *he thought in himself what there was of it true,* Homl. Skt. i. 23, 545. Fela spella him sǽdon ða

Beormas . . . ac hē nyste hwæt ðæs sôþes wæs, Ors. 1, 1; Swt. 17, 33. Ealle ða word sind sôþe ðe Paulus sægþ, Blickl. Homl. 187, 2. Wite gē tô sôðum þingum *scito ergo*, Deut. 9, 6. Ic secge eów tô sôðum, 8, 19: Mt. Kmbl. 5, 32. Nis nān þing sôþre ðonne ðæt ðū segst, Bt. 26, 1; Fox 92, 12: 34, 4; Fox 138, 25. Hig biddan God ðæt hē ðæt sôðeste geswytelie, L. Ath. iv. 7; Th. i. 226, 30. III. *true, righteous, just* :—Ðæt ic sôðne dôm healde *custodire judicia justitiae tuae*, Ps. Th. 118, 106.ʌ Sî ðīn seó swiðre hand ofer sôðne wer *fiat manus tua super virum dexterae tuae*, 79, 16. Ða ðe wyllaþ sôðe dômas efnan *qui custodiunt judicium*, 105, 3. [*O. Sax.* sôð: *Icel.* sannr, saðr. Cf. Lat. -sent in prae-sent-.] v. un-sôþ, *and preceding word.*

sôþ *occurs in the Northern specimens apparently corresponding to Latin* pro *in compounds* :—Sôð wē cliopiaþ *provocamus*, Rtl. 42, 15. In sôð (sôðe, Rush.) cneóreso *in progenies*, Lk. Skt. Lind. 1, 50. Sôð cymes *procedit*, Jn. Skt. Lind. Rush. 15, 26. Sôðcuom *processit*, Rtl. 2, 37: *procedit*, 57, 6. Sôðfylga *prosequere*, 29, 36. Sôð gistrŷnd *progeniem*, 29, 28. Sôðlæde *producere*, 108, 36. *Also* (?) sôð-cwide *proverbium*: sôþ-secgan *pronunciare*: sôþ-tácen *prodigium*, q.v.

sôþ-bora (?), an; *m. A truth-bearer, one who has exact knowledge* :—Ðone hæleð higegleáwe hâtaþ wīde cométa be naman, cræftgleáwe men, wīse sôðboran (*other MSS. have* wôþboran, *which suits better the alliteration*), Chr. 975; Erl. 126, 27. v. wôþ-, ræd-bora.

sôþ-cwēde (?); *adj. Veracious* :—Sôðcuoed (sôðcweden, Rush.) *verax*, Jn. Skt. Lind. 7, 18: 8, 26. Sôðcuêd, 3, 33.

sôþ-cweden; *adj. True-spoken* (cf. fair-spoken), *speaking truly, veracious.* v. *preceding word.*

sôþ-cwide, es; *m.* I. *a true saying, a truth* :—Ic fela sette sôðcwida, Met. 2, 7: 7, 3. Ðæt ðeós onlícnes word sprece, secge sôðcwidum; ðỹ sceolon gelŷfan eorlas, hwæt mín ædelo sién, Andr. Kmbl. 1465; An. 733: Cd. Th. 294, 14; Sat. 471: 244, 10; Dan. 446: Elen. Kmbl. 1057; El. 530: Exon. Th. 418, 2; Rä. 36, 13. Hit is ælces môdes wīse ðæt sôna swā hit forlæt sôþcwidas swā folgaþ hit leásspellunga *eam mentium constat esse naturam, ut quoties abjecerint veras, falsis opinionibus induantur*, Bt. 5, 3; Fox 14, 16: Met. 6, 2: 8, 3. I a. *a proverb* :—Ðās sôðcwide (-cuido, Lind.) ȝ gedd cwæð he Hælend *hoc proverbium dixit illis Jesus*, Jn. Skt. Rush. 10, 6. In sôðcwidum *in proverbis*, 16, 25. II. *a righteous saying.* v. sôþ, III :—On ðīne sôðcwidas *in tuis justificationibus*, Ps. Th. 118, 48.

sôþ-cyning, es; *m. The king of truth or justice, the Deity* :—Ic wât geare, ðæt ðam līchryre (*the murder of Cain*) on lâst cymeþ sôðcyninges seofonfeald wracu, Cd. Th. 67, 13; Gen. 1100. [Sôþ *and* cyning *often occur together in the nominative,* ʰᵘᵗ *it is doubtful whether they form a compound.*] Cf. sôþ-fæder.

sôþe; *adv.* I. *truly, genuinely, really* :—Ic mē sôðe sāwle mīne tô Gode hæfde georne geþeóded, Ps. Th. 61, 5. Mīn sāwl on ðē sôðe getreóweþ, 62, 7: 118, 15. Is on sibbe his stôw sôþe behealden, 75, 2. II. *truly, in accordance with the facts of a case* :—Ic eów sôðe secgan wille, ðæt . . . , Andr. Kmbl. 915; An. 458. Word sôðe gebunden (*the facts were truly told in the poem*), Beo. Th. 1746; B. 871. Hī sôðe ne ongeáton *they did not rightly understand*, Ps. Th. 73, 5. III. *truly, in accordance with a promise, agreement, or forecast* :—Ic ða wære forð sôðe gelæste, ðe ic ðē sealde, Cd. Th. 139, 11; Gen. 2308: 142, 22; Gen. 2365. Beót eal wið ðē hē sôðe gelæste, Beo. Th. 1053; B. 524. Sceolde wîtedôm in him sylfum beón sôðe gefylled, Exon. Th. 14, 3; Cri. 213.

sôþes; *adv. Of a truth, verily, indeed, really* :—Sôþes ic secge ðē *amen dico tibi*, Mt. Kmbl. 5, 26. Sôðes ðū eart Godes sunu, Lk. Skt. 4, 41. Ðeh hē sylf his āgenum eágum eal ne gesāwe, ðæt sôðes is geworden, Wulfst. 3, 21.

sôþ-fæder *the father of truth or justice, the Deity* :—Ā tô worulde forð in engla dreáme mid sôðfæder symble wunian, Exon. Th. 7, 18; Cri. 103. Cf. sôþ-cyning.

sôþ-fæst; *adj.* I. *true, without deception* :—Ðes man is sôþfæst, ac git sindon bigswicon, Blickl. Homl. 187, 29. II. *true in deed, just, righteous, pious, without wickedness* :—Sôðfæst *justus*, Ps. Th. 114, 5: 57, 9: Mt. Kmbl. Lind. 1, 19: Rtl. 102, 15. Sôðfæst sunu, ðam wæs Seth noma, Cd. Th. 67, 25; Gen. 1106. Sôþfæst eart ðū, Drihten, and rihte syndon ðīne dômas, Blickl. Homl. 89, 6. Ânra gehwylc, sôðfæst ge synnig, Exon. Th. 233, 11; Ph. 523. Hwylc sêceþ ðæt ðe sôðfæst biþ *misericordiam et veritatem quis requiret ?* Ps. Th. 60, 6. Sîe ðe onfôes ðone sôðfæst (*justum*) in noma sôðfæstes (*justi*), Mt. Kmbl. Lind. 10, 41. Hī on ðīn sôðfæst weorc (*in justitiam tuam*) ne gangan, Ps. Th. 68, 28: 70, 14, 20, 22. Ðīn sôðfæst word *justificationes tuas*, 118, 20. Hē gecŷpde ðæt sôþfæsten mēn habbaþ mid him þeófas and synfulle men, Blickl. Homl. 75, 27. Hit (*the law*) sôðfæste siððan heóldon godfyrhte guman, Andr. Kmbl. 3026; An. 1516. Ðū eart seó sêfte ræst sôþfæstra *tu requies tranquilla piis*, Bt. 33, 4; Fox 132, 34: Blickl. Homl. 131, 23. Yfele geréfan ða ðe rihte dômas sôþfæstra manna onwendaþ, 61, 27. On ða swîþran healfe Drihtnes mid sôþfæstum sāwlum, 95, 22. Mon mid gôdum and sôþfæstum dædum geearnige him ða êcean ræste, 101, 26. Þurh sôþfæste dæda and þurh mildheortnesse weorc, 97, 2. Ne cwom

ic tô ceigenne sôðfæsto (*justos*) ah synfullo, Mk. Skt. Lind. 2, 17. III. *true in speech, veracious* :—Sôðfæst *verax*, Wrt. Voc. i. 76, 17: Mt. Kmbl. 22, 16: Mk. Skt. 12, 14: Jn. Skt. Lind. 3, 33: Ps. Th. 85, 14. Gefyrn sôðfæst sægde sum wôdbora, Esaias, Exon. Th. 19, 17; Cri. 302. Cwom Daniel snotor and sôðfæst, Cd. Th. 225, 8; Dan. 151. Sôðfæst word *verbum veritatis*, Ps. Th. 118, 43. Sangere hē (*David*) wæs sôðfæstest, Ps. C. 50, 6. [*O. Sax.* sôð-fast.] v. un-sôþfæst.

sôþfæstian *to justify* :—Gié sôðfæstigeþ *justificatis*, Lk. Skt. Lind. 16, 15.

sôþfæst-líc; *adj. True, sincere* :—Hī (*patriarchs and martyrs*) sungon sigedryhtne sôðfæstlíc lof *praise unfeigned*, Andr. Kmbl. 1754; An. 879. [Cf. uss birrþ soþfasstlike trowwenn þatt Godess Gast iss soþfasst Godd, Orm. 2995.]

sôþfæstness, e; *f.* I. *truth, faithfulness, good faith, sincerity* :—On worulda woruld wunaþ ðīn sôðfæstnes *thy faithfulness is unto all generations* (A. V.), Ps. Th. 118, 90: 56, 12. Ūs is wyrse ðæt wē ūrne ceáp teóþian gif wē willaþ syllan ūre ðæt wyrste Gode. Cwæþ se æþela lāreów: 'Onsecggaþ gē Drihtne mid sôþfæstnesse wæstmum,' Blickl. Homl. 41, 10. Ongan ðā geornlíce gástgerŷnum on sefan sêcean sôðfæstnesse (*in sincerity*; *or* (?) *gen. with weg*) weg tô wuldre, Elen. Kmbl. 2296; El. 1149. Sete ðīne hand under mīn þeóh and cŷð mē ðīne sôðfæstnysse, and swera mē, ðæt ðū mē næfre ne bebirge on Egipta lande, Gen. 47, 29. II. *truth, righteousness, justice* :—Beseah sôðfæstnes (*justitia*) of heofonum, Ps. Th. 84, 10, 12: 71, 7. His sôðfæstnyss wunaþ symble, 111, 8. Æ wæs geseald þurh Moysen, and gyfu and sôþfæstnes is geworden þurh Hælend Crist, Jn. i, 17. Cwæþ Pilatus: 'Nys nān sôþfæstnys on eorþan.' Se Hælend hym andswarode and cwæþ: 'Begŷn hū ryhte dômas ða dêmon ðe on eorðan syndon and anweald habbaþ, Nicod. 9; Thw. 5, 5. Se ðe wæs sôþfæstnesse bysen and cining ealre clænnesse forlêt mid him beón ðone godwracan þeóf, Blickl. Homl. 75, 25. Ic eom weg sôðfæstnesse, 17, 32. For sôðfæstnesse ðæt wē lufigen gesuinc, Past. 3; Swt. 35, 1. Se ðe hylt sôðfæstnysse on worulde, hē ðêþ dôm on teónan þoliendum, Ps.Spl. 145, 5. III. *truth of speech or thought* :—Deófol ne wunode on sôðfæstnesse, forðam ðe sôðfæstnes nis on him. Ðonne hē sprycþ leásunga, hē sprycþ of him sylfum, forðam ðe hē is leás. Gé ne gelŷfaþ mē forðam ðe ic secge eów sôðfæstnysse, Jn. Skt. 8, 44, 45. Ðū settest on mīnum mūðe ðīnre sôðfæstnysse word, Homl. Th. i. 74, 33. Hē mid ðære sôðfæstnysse stefne geweorþod wæs, Blickl. Homl. 165, 1. Se mon se ða sôþfæstnesse mid his mūþe sprecþ and hié on his heortan geþencþ, 55, 14. Ðonne ðære sôþfæstnysse gást cymþ hē lærþ eów calle sôþfæstnysse; ne sprycþ hē of him sylfum, ac hē sprycþ ða þing ðe hē gehŷrþ, and cŷð eów ða þing ðe tôwearde synt, Jn. Skt. 16, 13. v. un-sôþfæstness.

sôþ-gid *a true tale* :—Sôðgied wrecan, Exon. Th. 306, 2; Seef. 1: 314, 17; Môd. 15.

sôþ-hwæðere; *conj. However, yet, nevertheless* :—Sôðhueðre ic cueðo *verumtamen dico*, Mt. Kmbl. Lind. 26, 64: Jn. Skt. Lind. Rush. 12, 42.

sôðian; *p.* ode *To prove true* :—Sôðeþ *probat*, Mt. Kmbl. .p. 9, 9. Sôðadon *probarunt*, Jn. Skt. p. 7, 2. [Ich hit wulle sođien, Laym. 8491. *Icel.* sanna *to prove*, 'make good.] v. ge-, un-sôðian; sêðan.

sôþ-líc; *adj.* I. *true, genuine, unfeigned* :—Nænig ôðerne freóþ in fyrhðe, ðæt hē sôðlíce (*or adv. ?*) sybbe healde, gāstlíce lufe, Fragm. Kmbl. 72; Leás. 38. II. *true, right* :—Ne þincþ mē næfre nānwuht swā sôþlíc swā mē þincþ ðīn spell ðæm tímum ðe ic ða gehiére *cum tuas rationes considero, nihil dici verius puto*, Bt. 38, 5; Fox 204, 22. [*O. Sax.* sôð-lîk: *Icel.* sann-ligr *probable; just; fit.*] v. *next word.*

sôþlíce. I. *as adv. Truly, really, certainly, verily* :—Sôðlíce ðū eart Godes sunu *vere filius Dei es*, Mt. Kmbl. 14, 33: 27, 54. Sôðlíce ic secge eów *amen dico vobis*, 6, 16 (*and often*). Ðām ðe sôðlíce sêcaþ Dryhten, Ps. Th. 104, 3. Ðis wæs sôðlíce eádig wer *vere beatus vir*, Blickl. Homl. 223, 31. Ðū bist sôþlíce ær þrim dagum genumen of ðīnum līchoman *certainly before three days thou wilt be taken from the body*, 137, 25. Is sôðlíce se cwide gefylled, 139, 27. Swŷþe sôþlíce (*with great truth*) wē māgon geþencan, ðæt hit biþ deáþes ylding swîðor ðonne lîfes, 59, 31. Ic sôðlíce meahte ongitan, Exon. Th. 313, 24; Môd. 5. Se ðe ðē ðyslíce gife and swā mycle sôþlíce (-re, MS.) ðē tôwearde forecwyþ *is qui tanta taliaque dona veraciter adventura praedixerit*, Bd. 2, 12; S. 514, 13: Exon. Th. 9, 19; Cri. 137. Weras ða ðe eówre æ on ferhðsefan fyrmest hæbben, ða mē sôðlíce secgan cunnon, Elen. Kmbl. 633; El. 317: Beo. Th. 284; B. 141. Hī ðŷ sôðlícor ongeáton ðæt hit wæs sôðlíce his āgen līchoma, Shrn. 68, 33. Ic sôðlícost wēne, 164, 28. II. *as conj. Now, then, for;* representing Latin autem, ecce, enim, ergo, nam, vero :—Sôðlíce Iosep hyre wer *Joseph autem vir ejus*, Mt. Kmbl. 1, 19 (*and often*). Sôðlíce seó fæmne hæfþ on innoðe *ecce virgo in utero habebit*, 1, 23: 2, 9: 3, 17. Sôðlíce wē gesāwon hys steorran, *vidimus enim stellam ejus*, 2, 2: 3, 1: 4, 18 (*and often*). Gehŷre gē sôðlíce ðæs sāwendan bigspell *vos ergo audite parabolam seminantis*, 13, 18. Sôðlíce ic eom man under anwealde gesett *nam et ego homo sum sub potestate*, 8, 9. Sôðlíce ðæt ðe āsāwen wæs on ðæt gôde land *qui vero in terra bona seminatus est*, 13, 23, 29. [*O. Sax.* sôðlíko: *Icel.* sannliga.]

sôþ-sægen, -segen, e; *f. A true statement, statement of the truth,*

statement of the facts of a case:—Se Hǽlend nolde hine betellan mid
nánre sóðsegene ðeáh ðe hé unscyldig wǽre *the Saviour would not clear
himself by any statement of the truth, though he was innocent,* Homl. Th.
ii. 250, 11. Hí sceolon forsuwian heora gefêrena unþeáwas, ðý lǽs ðe hí
þurh heora sóðsegene ungeðyldige beón, 230, 17.

sôþ-sagol; *adj. Veracious* :—Sóðsagol *veridicus*, Wrt. Voc. i. 76, 18 :
verax, Ps. Lamb. 85, 13. Swá swá sôþsagol stǽrwrítere *quasi verax
historicus*, Bd. 3, 17; S. 545, 4. Se ðe wǽre leássagol, weorðe se sóþ-
sagol, Wulfst. 72, 16. [*Icel.* sann-sögull.] v. un-sóþsagol.

sôþ-sagu, e; *f.* I. *true speech, truth* :—On manna gehwylces
móde and múðe sóðsagu stande, Wulfst. 74, 16. II. *a true saying,
a history* :—Sóðsaga *historia*, Mt. Kmbl. p. 9, 4 : *historiae*, 7, 9. [þilke
soþsaȝe (*saw*), þat man schal erien and sowe þar he wenþ after sum god
mowe, O. and N. 1038. *Icel.* sann-saga *a true tale.*]

sôþ-secgan *to say truly, declare* :—Sóðsǽges *pronuntiat*, Jn. Skt. p. 4,
11 : 6, 15 (*see* sóþ =*pro*). Ðes man is sôþsecgende, Blickl. Homl. 187, 29.

sôþ-spell *a true story, history* :—Sóðspell *historia*, Mt. Kmbl. p. 9, 4.
[*O. Sax.* sóð-spell.]

sôþ-sprǽc *a true saying* :—Sóðsprǽco *eloquia*, Rtl. 171, 35.

sôþ-tácen *a true sign, prodigy* :—Sóðtáceno *prodigia*, Rtl. 43, 35
(*see* sóþ =*pro*).

sôþ-word *a true word* :—Ic Gode sealmas singe, sóðword sprece, Ps.
Th. 56, 9 : 118, 93.

sotol *a seat.* v. setl.

sotscipe, es; *m. Folly, stupidity* :—Sotscipe *hebetudo*, Wrt. Voc. i. 50,
60. [Sǽide se abbot of Clunni, þ hi heafdon foloron S. Ioħes mynstre
þurh hi and þurh his mycele sotscipe, Chr. 1131; Erl. 260, 8. Nolde þe
leodking his sothscipe (folie, 2nd MS.) bilǽuen, Laym. 3024. Muchel
sotschipe hit is uorto uorleosen uor one deie tene oðer tweolue, A. R.
422, 24.]

sott; *adj. Foolish, stupid*; substantively, *a fool* :—Sot *sottus*, Wrt. Voc.
i. 76, 16. Sott *hebes*, 50, 59. Se ðe his ágene sprǽce áwyrt, hé wyrcþ
barbarismus. Swylce hé cweðe ðú sôt ðǽr hé sceolde cweðan ðú sott,
Anglia viii. 313, 21. Ne biþ se ná wita, ðe unwíslíce leófaþ, ac biþ open
sott, ðeáh ðe him swá ne þince, Homl. Skt. i. 13, 132. [Þu ebure sot
(fol, 2nd MS.), Laym. 2271. Þa weoren Scottes ihalden for sottes, 21806.
Seide þ heo weoren sotten iueren, 17309. Nout to ȝunge preostes, ne to
sot olde, A. R. 336, 12. Lat sottes chide, O. and N. 297. The word is
of doubtful origin, v. Skeat's Etym. Dict. *sot.*]

spáca, an; *m.* I. *the spoke* of a wheel :—Ða sêlestan men faran
nêhst Gode, swá swá sió nafu ferþ nêhst ðære eaxe, and ða midmestan
swá swá spácan; for ðam ðe ælces spácan biþ óþer ende fæst on ðære nafe,
óþer on ðære felge . . . Ða felga hangiaþ on ðám spácan, Bt. 39, 7; Fox
222, 1–13. Spácan *radii*, Wrt. Voc. i. 16, 23 : 284, 47 : 66, 54. II.
part of the body [= ribb-spácan] :—Spácan *radioli*, 65, 21. [*O. L. Ger.*
spêca *radius* : *O. H. Ger.* speicha *radius, lignum in rota* : *Ger.* speiche.]

spad, spada. v. spadu.

spade *eunuchus* :—Eviratus, i. *effeminatus, eunuchus, enervus* spade,
Wrt. Voc. ii. 144, 37.

spádl. v. spátl.

spadu, an, e; *f.* : spada (?); *m. A spade* :—Spadu *fossorium*; spada
vanga; spad[u?] *scudicia vel fossorium*, Wrt. Voc. i. 16, 14, 8, 29.
Spadu, spædu *uanga vel fossorium*, Ælfc. Gl. Zup. 318, 17. Ic nát mid
hwî ic delfe, nû mê wana is ǽgðer ge spadu ge mattuc, Homl. Skt. ii.
23 b, 765. Sum underdealf ða duru mid spade, A. S. Prim. 87, 174.
Ðá genam hê áne spada[n?] and dealf ða eorþan, H. R. 13, 13. Spadan
vangas, Wrt. Voc. ii. 123, 10. [*O. L. Ger.* spado *sarculum, rastrum* :
Gk. σπάθη.]

spæc, es; *m.* (?) *n.* (?) *A thin twig, tendril, runner* :—Twig *ramus*,
spæc *framen* (cf. *framen* streáberie-wísan, 31, 70), Wrt. Voc. i. 285, 81 :
ii. 36, 57 : *cremium* (*cremia* ligna tenuia et arida), 151, 2. Ðara spaca
speldra *malleoli* (malleolus manipulus sparteus pice contectus quem incensum in muros jaciebant), 54, 57. [*O. H. Ger.* spah, spahha(o) *sarmentum, cremium, fasciculus ex siccis lignis, malleolus, ramus.*] v. spræc.

spêc, spædu. v. sprǽc, spadu.

spær; *adj. Spare, frugal* :—Spær mete *parcus cibus*, Scint. 52, 6.
[*O. H. Ger.* spar *parcus* : *Icel.* sparr.] v. spær-hende, -líc, -ness.

spærca. v. spearca.

spæren; *adj. Of plaster, of mortar* :—Spaeren, sparaen, sparen *gipsus*,
Txts. 67, 968. Spæren, Wrt. Voc. ii. 40, 67. v. spær-stán.

spær-habuc. v. spear-hafoc.

spær-hende; *adj. Of sparing hand, frugal, sparing* :—Spærhende
frugi vel parcus, Wrt. Voc. i. 76, 6. Uncystig oþþe spærhynde (-hende)
frugi, Ælfc. Gr. 9, 78; Zup. 74, 12. Spærhynde *parcus*, Germ. 392,
66. [*O. H. Ger.* un-sparahenti *prodigus.*]

spær-líc; *adj. Sparing, frugal* :—Swá sperlíc *tam frugalis*, Hpt. Gl.
494, 43.

spærlíce; *adv. Sparingly, sparely* :—Spærlíce *parce*, Scint. 156, 9.
Ðý mon dǽlþ spærlíce ðe mon nele hit forberste *sparingly people spend,
because they do not want to run short,* Prov. Kmbl. 19. Ic sperlícor mid
wordum sægde ðonne hié dǽdum gedón wǽrun *solere me parcius loqui*

quam gesta sint omnia, Nar. 2, 24. [*O. H. Ger.* sparalíhho *parce, fru-
galiter* : *Icel.* sparliga *sparingly.*]

spær-lira. v. spear-lira.

spærness, e; *f. Sparingness, frugality, parsimony* :—Spærnes *frugali-
tas*, i. *temperantia, parcitas*, Wrt. Voc. ii. 151, 29. Drences and metes
spearness *potus cibique parcitas*, Hymn. Surt. 9, 24. Spærnisse, Rtl. 163, 7.
Spærnesse *frugalitatis*, Hpt. Gl. 456, 56 : *frugalitatis, temperantiae,
moderationis*, 425, 64 : *frugalitatis, abstinentiae*, 496, 22 : 513, 61 :
parsimonia, penuria, temperantia, 454, 59.

spær-stán, es; *m. Gypsum, chalk* :—Spærstán *gipsum*, Wrt. Voc. i.
85, 22 : *creta argentea*, 37, 30.

spǽtan; *p.* te. I. *to spit* (a) intrans. :—Ic hrǽce oððe spǽte
screo, Ælfc. Gr. 26, 6; Zup. 158, 6. Ic spǽte *spuo*, 28, 3; Zup. 167,
10. Hé spǽtte on his eágan *expuens in oculos ejus*, Mk. Skt. 8, 23 : Jn.
Skt. 9, 6. Hí spǽtton on hine, Mt. Kmbl. 17, 30 : 26, 67. Hig spǽtton
him on *conspuebant eum*, Mk. Skt. 15, 19. Spǽte ðæt wíf on his nebb,
Deut. 25, 9. Suelce hié him on ðæt nebb spǽten, Past. 5; Swt. 45, 4.
Sume águnnon him on spǽtan (*conspuere eum*), Mk. Skt. 14, 65. Hé
spǽtende (*expuens*) his tungan onhrán, 7, 33. Spátende *expuentes*, Mt.
Kmbl. Lind. 27, 30. Hé byþ on spǽt *conspuetur*, Lk. Skt. 18, 32.
(b) trans. :—Ic spǽte áttor, Exon. Th. 405, 26; Rä. 24, 8 : 398, 27;
Rä. 18, 4. II. *to syringe, squirt* :—Gespǽt ða wunde, Lchdm. ii.
22, 22. v. geond-spǽtan.] v. be-spǽtan.

spǽtl. v. spátl.

spǽtlan, spǽtlian; *p.* ede. I. *to emit saliva, to foam* :—Spǽtleþ
spumat, Wrt. Voc. ii. 73, 37. II. *to spit* on anything :—Hié hine
bindaþ and spǽtliaþ on his onsýne, Blickl. Homl. 15, 11. Hié spǽtledon
on his onsýne, 23, 32. Spǽtlædon, 237, 11. Spǽtledon (-odon), Anglia
xii. 505, 14. v. spátlian.]

spǽtung, e; *f. Spitting, expectoration* :—Gelóme spǽtunga oððe hrǽcunga, Lchdm. ii. 174, 20.

spala, an; *m. A representative, substitute* :—Gif hé untrum byþ, begyte
him lahlícne spalan, L. Wil. ii. 2; Th. i. 489, 16. Cf. ge-spelia.

Spalda *a tribe name left in Spalding* (?). In a list giving the extent
of territory belonging to various districts in England it is said :—Spalda
syx hund hýda, Cod. Dip. B. i. 414, 20. Cf. Spaldyng, Cod. Dip. Kmbl.
vi. 333, col. 2.

spaldur *aspalt* ; aspaltum, Txts. 43, 228.

-span *allurement.* v. ge-span.

spanan; *p.* spôn, speón; *pp.* spanen *To allure, entice, lure, decoy,
attract, urge* :—Spenst *illicias*, Hpt. Gl. 524, 9. Spones *inlicias*, Wrt.
Voc. ii. 47, 7 : 87, 26. I. in a good sense, (a) with a preposition
marking the direction of aim :—Ðú spenst (spænst, Cott. MS.) mê on ða
mǽstan sprǽce and on ða earfoþestan tô gereccenne *ad rem me omnium
quaesitu maximam vocas*, Bt. 39, 4; Fox 216, 14. Swá earn his briddas
spænþ tô flihte *sicut aquila provocans ad volandum pullos suos*, Deut. 32,
11. Ðá hé spôn his hiéremen tô ðære geðylde *cum patientiam discipulis
suaderet*, Past. 33, 5; Swt. 222, 8. Ðá ðá hé his apostolas spôn of
ðissum andweardan tô ðæm êcan *cum ad venturam discipulos ex praesenti
provocaret*, 46, 5; Swt. 351, 11. Speón (spôn, Cott. MSS.), 17, 8; Swt.
121, 2. Speón, Andr. Kmbl. 1194; An. 597. Ælcne man spane hê of
synnum *let him draw every man from sins*, L. Edg. C. 16; Th. ii. 284,
14. Ða spone (spane, Cott. MSS.) ðe his ðeáwa giemaþ tô ryhte *spectatores suos ad sublimia invitet*, Past. 14, 2; Swt. 83, 2. Hé sende his
englas ûs hám tô spananne tô him *exhortantes angelos misit*, 52, 4; Swt.
405, 34. (b) with a clause :—God hine spænþ ðæt hé tô him gecierre
Deus ad se redire persuadet, Swt. 407, 10. Hí hine speónnan and lǽrdon
ðæt hé ða fôre ðurhtuge *eum id perficere suadebant*, Bd. 5, 19; S. 637,
26. Span ðú hine georne ðæt hé ðíne láre lǽste, Cd. Th. 36, 22; Gen.
575. Cwæð, ðæt hine his hige speóne, ðæt hé trymede getimbro, 18, 17;
Gen. 274. II. in a bad sense, (a) with a preposition :—Hine spænþ
his môd tô unnyttum weorce, Past. 4; Swt. 37, 18. Deófol hine on wôh
spaneþ, Salm. Kmbl. 1002 : Sal. 502 : 990 : Sal. 496. Hí spanaþ ðê tô
ðínre unþearfe, Bt. 7, 2; Fox 18, 9. On ðæm weorce ðe hine ǽr nán
willa tô ne spôn *quo non trahit desiderium*, Past. 33, 1; Swt. 215, 10.
Hió speón hine on ða dimman dǽd, Cd. Th. 43, 2; Gen. 684. Hé mid
listum speón idese on ðæt unriht, 37, 12; Gen. 588. Hí (*the conspirators
against William*) speónan ða Bryttas heom tô, Chr. 1075; Erl. 213, 14.
(b) with a clause :—Hé hiene spôn ðæt hé on Umenis unmyndlenga mid
here becóme *quem, ut Eumenem de insperato opprimat, perurget*, Ors. 3,
11; Swt. 146, 7. [*O. Sax.* spanan : *p.* spôn: *O. Frs.* spona: *O. H. Ger.*
spanan; *p.* spuon *suggerere, suadere, persuadere* : cf. *Icel.* spenja; *p.*
spandi *to allure*.] v. á-, be-, for-, ge-spanan.

Spáneas, *pl. The Spaniards or Spain* :—Betux Galleum and Spáneum,
Ors. 4, 8; Swt. 186, 15. [Cf. *Icel.* Spána-land *Spain.*] v. Spêne.

spanere, es; *m. One who entices* :—Sponera *lenonum*, Wrt. Voc. ii. 52,
42 : 84, 39. [*O. H. Ger.* spanari *hortator, suasor, persuasor, illex.*]

spang, e; *f. A clasp, fastening* :—Hæleðhelm on heáfod ásette and
ðone full hearde geband spênn mid spangum *drew the helmet firmly on
with its clasps*, Cd. Th. 29, 4; Gen. 445. [*O. H. Ger.* spanga; *f. seracula,
prena* : *Ger.* spange *a clasp* : *Icel.* spöng; *f. a clasp.*]

spann, e ; *f.* A *span* :—Span *vel* handbred *palmus*, Wrt. Voc. i. 43, 52. Wæs se líchoma sponne lengra ðære ðrýh *invenerunt corpus mensura palmi longius esse sarcofago*, Bd. 4, 11 ; S. 580, 5. [*O. H. Ger.* spanna ; *f. cubitus* : Icel. spönn ; *f. a span.*] Cf. ge-spann.

spannan ; *p.* spénn, speónn ; *pp.* spannen. **I.** *to join* one thing to another, *to attach, fasten, clasp*, (a) literal :—Hé helm spénn mid spangum (cf. *Dan.* spænde ved spænder, *Swed.* spänna med spänne *to buckle*) *he buckled on his helmet*, Cd. Th. 29, 4 ; Gen. 445. (b) figurative :—Wá eów ðe gadriaþ hús tó húse and spannaþ æcer tó ðæm óðrum *vae, qui conjungitis domum ad domum, et agrum agro copulatis*, Past. 44, 8 ; Swt. 329, 23. **II.** *to span, clasp.* v. ymb-spannan, spanning. [*O. H. Ger.* spannan ; *p.* spien *nectere ; intendere, contendere* : cf. *Icel.* spenna ; *p.* ta *to clasp ; to span.*] v. ge-, on-, ymb-spannan.

spanning, e ; *f. Spanning, bend, span* :—Eln *vel* spanning betwiox þuman and scitefingre *ulna*, Wrt. Voc. i. 43, 53.

spanu, e, an ; *f.* A *teat* :—Tittas *mammille*, spana *ubera*, Wrt. Voc. i. 283, 30. Tittas oððe sponan *mammillas*, Lchdm. i. lxxiv, 24. [Speen, spene *a cow's pap*, E. D. S. Gloss. B. 16 : C. 3. Speans *the teats of a cow*, C. 4. *Icel.* speni ; *m. a teat, dug : Norweg.* spæne : *Swed.* spene.]

spanung. v. for-, leás-spanung(-ing).

sparian ; *p.* ode. **I.** *to spare, to show mercy to, to refrain from injuring* or *destroying* :—Ic sparige oððe árige *parco*, Ælfc. Gr. 28, 7 ; Zup. 180, 12. Dætte hé spærio *parcere*, Rtl. 40, 19. (a) with acc. :— Ic geswerge ðæt ic ne sparige, ac on spild giefe, Exon. Th. 247, 27 ; Jul. 85. Hé áraþ (sparaþ, MS.C. : spearaþ, Ps. Surt.) *parcet*, Ps. Spl. 71, 13. Hié ne sparodan ða synfullan, ac slógon, Past. 46 ; Swt. 353, 16. Hí nánne ne sparedon ðæs herefolces, Jud. Thw. 24, 40 ; Jud. 233. Spara mé ðinne ðeów *parce servo tuo*, Ps. Th. 18, 11. (b) with dat. :— [Ne spareþ se fæder ðan sune ne' nán mann óðren ; ac ælc man winþ ongeán óðren, Shrn. 17, 27.] Swá ðæt ne cyricum ne mynstrum seó herehand ne sparode ne árode *ita ut ne ecclesiis quidem aut monasteriis manus parcerit hostilis*, Bd. 4, 26 ; S. 602, 8. God ne sparode his ágenum bearne, Homl. Th. ii. 62, 20. Ná hé sparode (spearede, Ps. Surt. v. 50) sáulum heora *non pepercit animabus eorum*, Ps. Spl. C. 77, 55. Spær esne ðínum *parce servo tuo*, Rtl. 168, 19 : 39, 38. **II.** *to spare, preserve, not to use, to leave alone, abstain from* :—Hé sparode ðæt góde wín óð his ágenum tócyme, Homl. Th. ii. 70, 10. Féðe ne sparode eorl, Cd. Th. 153, 6 ; Gen. 2534. Sindon ða loccas tó sparianne (-enne, Hatt. MS.) ðæm sacerde ðæt hí ða hýd behelien *capilli in capite sacerdotis servantur, ut cutem cooperiant*, Past. 18, 7 ; Swt. 141, 9. [*O. H. Ger.* sparôn *parcere, fovere : Icel.* spara *to spare.*] v. ge-sparian.

sparrian *to bar, shut.* [Sparren, sperren is not uncommon in later English. v. Stratmann's Dict. Cf. *O. H. Ger.* sperren *claudere.*] v. be-, ge-sparrad.

spátl, es ; *n. Spittle, saliva* :—Spátl *sputum*, Wrt. Voc, ii. 70, 21. Hé worhte fenn of his spátle *he made clay of the spotle* (Wick), Jn. Skt. 9, 6. Dín spátl spíw on, Lchdm. ii. 322, 7 : 24, 8 : 36, 17. Se ná ne forbeág mid hys nebbe ðara triówleásena monna spátl, Past. 36 ; Swt. 261, 9 : Exon. Th. 88, 7 ; Cri. 1436. Spádl, Elen. Kmbl. 600 ; El. 300. Spádl, Mt. Kmbl. Rush. 27, 30. Ða spætlu áþwógon úre sweartan gyltas, Homl. Th. ii. 248, 26. Spátlum *salivis*, Germ. 396, 283. Dæne ðe hý heora spátlum on spiwon, Wulfst. 183, 21. Spátlu *sputa*, Hymn. Surt. 80, 1. [Heo bispeteð hire mid hire blake spotle, A. R. 288, 10. Spotle *sputum*, screa, *saliva*, Prompt. Parv. 469, col. 2.]

spátlian ; *p.* ode *To spit out* :—Ic spátlige *pitisso*, Wrt. Voc. i. 46, 16. [I (*the old man*) spitte, I spatle, Rel. Ant. ii. 211, 34.] v. *next word,* and spætlan.

spátlung, e ; *f. Spitting out, spittle* :—Pituita, i. minuta saliva horas *vel* hræcunga *vel* spátlung, Wrt.Voc.i. 46, 15. [I (*Christ*) þolede schomeliche spateling of unwurdi ribauz, O. E. Homl. i. 279, 34. Þenched þet te worldes weldinde wolde þolien buffetes, spotlunge, blindfellunge, A. R. 188, 10.]

spearca, an ; *m.* A *spark.* **I.** literal :—Spærca *scintella*, Wrt. Voc. ii. 120, 21. Spearca *scintilla*, i. 66, 39 : 284, 14. Ne biþ ðær leóhtes án lytel spearca, Wulfst. 139, 11. Sleá hé him ánne spearcan, Lchdm. ii. 290, 17. Hí ásprungon up swá swá spearcan, Homl. Th. ii. 350, 23 : Bd. 3, 10 ; S. 534, 31. Dæt manega menn geseóþ feallan of ðære heofene, swylce hit sýn steorran, hit beóþ spearcan of ðam rodere, Anglia viii. 320, 33. His eágan wæron fýrene spearcan sprengende, Homl. Th. i. 466, 26. **II.** metaphorical :—Se spearca ðara gódra weorca, Past. 14 ; Swt. 87, 6. Sum spearca sóþfæstnesse, Bt. 35, 5 ; Fox 164, 2. Ne furðum án spearca mínes cynrenes nis mé forlæten, Homl. Skt. ii. 30, 206. Gif ða scyldigan ænigne spearcan wísdómes hæfdon, 38, 7 ; Fox 210, 9. Word spearcum fleáh áttre gelícost, Cd. Th. 274, 32 ; Sat. 162. v. fýr-spearca.

spearcian, spearcan (?) *To sparkle, emit sparks* :—Hé sweartade (spearcade ?) ðonne hé spreocan ongan fýre and áttre, Cd. Th. 269, 24 ; Sat. 78. Sparcendum *scintillante*, Hpt. Gl. 501, 5. [*Prompt. Parv.* sparkyn *scintillo.* It sparkede and full brith shon, Havel. 2144.] v. spircan.

spear-hafoc, es ; *m.* A *sparrow-hawk* :—Spaerhabuc *alietum* (alietus au hobey, Wülck. Gl. 562, 48), Wrt. Voc. ii. 99, 67. Spearhafuc, 7, 65 :

i. 280, 20. Spearhafoc *hetum*, 62, 16 : *accipiter vel raptor*, 29, 58 : *ismarus* (=*ismerlus*? cf. *French* émerillon *a merlin*), 63, 25. [Sparowhawke *nisus*, Wrt. Voc. i. 177, 14. *Icel.* sparr-haukr. Cf. *O. H. Ger.* sparwári *nisus: Ger.* sperber ; and the borrowed Romance forms, *Fr.* épervier ; *Ital.* sparviere.]

spear-lira, an ; *m. The calf of the leg* :—Spærlira *sura,'* Wrt. Voc. i. 44, 71 : 71, 55. Sperlira, 65, 43. Spærlirena *surarum*, Hpt. Gl. 478, 56. Spærlirum *suris*, 483, 37. Spærliran *suras*, 482, 65. On spearlirum *in suris*, Deut. 28, 35. Speoruliran *suras*, Lchdm. i. lxxi, 10. [Hose . . . þat spenet on his sparlyr & clene spures under, Gaw. 158. Sparluris, Wick. Deut. 28, 35.] v. spearwa.

spearlirede *having a large calf* :—Spærlirede *surosus*, Wrt. Voc. i. 45, 42.

spearnlian ; *p.* ode *To spurn, strike out with the feet, kick* :—Ðæt ðú ne spear[n]last *ut non calcitres*, Hpt. Gl. 463, 77. Se sticca him eode út þurh ðæt heáfod in tó ðære eorðan and hé ætforan hire spearnlode mid fótum *the nail went through his head into the earth, and he* (Sisera) *struck out with his feet before her*, Jud. 4, 21. Cf. spurnan.

spearwa, an ; *m.* A *sparrow* :—Spearuua, spearua, sperua *fenus*, Txts. 62, 435. Spearwa, Wrt. Voc. ii. 35, 22 : *passer*, i. 77, 29 : 281, 27 : Bd. 2, 13 ; S. 516, 17 : Ps. Spl. 83, 3. Spearewua, Wrt. Voc. i. 63, 7. Spearuua, Ps. Th. 10, arg. : 83, 3. Speara, Ps. Surt. 83, 4. Ðá geseah heó spearwan nest, Homl. As. 120, 116. Ic spearuwan gelíce gewearð, Ps. Th. 101, 6. Spearwan nystlaþ *passeres nidificabunt*, 103, 16. Spearwan (hrond-sparuas, Lind. : spearwas, Rush.), Mt. Kmbl. 10, 29, 31 : Lk. Skt. 12, 6. Beteran manegum spearwum, 12,'7. [*Goth.* sparwa : *O.H. Ger.* sparo : *Icel.* spörr.] v. neód-spearuwa.

spearwa, an ; *m. The calf of the leg* :—Sparuua, sparua, spearua *surum*, Txts. 94, 897. v. spear-lira.

spec, spéc, speca, specan. v. spic, spréc, spreca, sprecan.

specca, an ; *m.* A *speck, spot, blot* :—Ðone sweartan speccan *maculam pullam*, Wrt. Voc. ii. 57, 11 : 92, 84. Speccan *notae*, 114, 80 : 60, 18 : *scoriae*, Hpt. Gl. 421, 59. Smire ða speccan (*in a case of shingles*) mid ðære sealfe, Lchdm. ii. 88, 19. v. next two words.

specel (?) *adj. Speckled.* v. haran-specel, *and see* Lchdm. ii. 390, col. 2.

spec-fáh, *adj. Speckled, spotted, full of spots* :—Specfaag *maculosus*, Wrt. Voc. ii. 98, 25.

specol. v. sprecol.

sped *phlegm, rheum* :—Sped *petuita*, Wrt. Voc. ii. 117, 22 : 68, 18. Sped *glaucoma* (cf. spade the congealed gum of the eye, Halliwell's Dict.), Hpt. Gl. 447, 22. v. spediende.

spéd, e ; *f. Speed, success, means.* The word is found in the following glosses :—Spoed *proventus, praeventus*, Txts. 88, 815 : *successus*, 96, 940 : *praesidium*, 89, 1648. Spéd *proventus*, Wrt. Voc. i. 61, 25 : ii. 68, 44. Ðeós spéd *haec ops*, Ælfc. Gr. 9, 56 ; Zup. 67, 18. Spéde *facultatem*, Hpt. Gl. 437, 40. Spédum *successibus*, Wrt. Voc. ii. 76, 56. **I.** *speed, quickness ;* spédum *speedily, quickly* :—Gewiton him ædre æfter ðære spræce spédum féran, Cd. Th. 144, 32 ; Gen. 2398. Spédum sægde eorlum Abimeleh egesan geðreád Waldendes word, 161, 19 ; Gen. 2667. Him ða bróðor þrý spédum miclum (*very speedily*) hældon hygesorge heardum wordum, 122, 30 ; Gen. 2034. [Waterrstræm erneþþ towarrd te sæ wiþþ mikell sped ʒiff þatt itt nohht ne letteþþ, Orm. 18094.] **II.** *speed* (as in good *speed*), *success, prosperous issue* :—Ðæt mínre spræce spéd folgie *that success may attend my speech*, Ps. Th. 55, 4. Heó (*Sarah*) ne gelýfde ðæt ðære spræce spéd folgode *she did not believe that any happy result would follow those words, did not believe that she should have a son*, Cd. Th. 144, 4 ; Gen. 2384. Hit ne becymþ eów ná tó nánre spéde *vobis non cedet in prosperum*, Num. 14, 41. Ðær ríexaþ sib mid spéde *peace and happiness reign there*, Dóm. L. 267. On spéd *successfully, to purpose, with effect*, Beo. Th. 1750 ; B. 873 : Exon. Th. 387, 28 ; Rä. 5, 12. Swá wit him an spéd sprecaþ *we shall speak so as to convince him*, Cd. Th. 36, 21 ; Gen. 575. Ic on ðínre hælo hyldo sóhte and on ðínre spræce spéd sóðfæste *in salutari tuo, et in eloquio justitiae tuae*, Ps. Th. 118, 123. Wíges spéd *success in war*, Exon. Th. 42, 16 ; Cri. 673. Æt wigge spéd, sigor æt sæcce, Elen. Kmbl. 2362 ; El. 1182. Hié ðære spræce spéd ne áhton ða *the people at Babel had no advantage from speech*, Cd. Th. 101, 23 ; Gen. 1686. Se ðe him dóm forgeaf, spówende spéd (*good speed*), 246, 14 ; Dan. 479. Æt ðam spereníðe spéde lænan, 124, 8 ; Gen. 2059 : 187, 19 ; Exod. 153. **III.** *means, substance, abundance, wealth* :—Spéd ł dæl mín ðú eart *portio mea es*, Ps. Lamb. 118, 57 : 141, 6. Spéd *substantia*, Ps. Spl. 38, 7, 11 : 68, 2 : Ps. Th. 88, 40. His meahta spéd *the abundance of his powers*, Exon. Th. 240, 18 ; Ph. 640. Hé is mægna spéd, Cd. Th. 1, 6 ; Gen. 3. Wilna gehwilces weaxende spéd *a growing abundance of every thing to be desired*, 100, 7 ; Gen. 1660. Ic on mínre heortan hýdde georne ðæt ic ðínre spræce spéd gehealde *in corde meo abscondi eloquia tua*, Ps. Th. 118, 11 : 38. Tubal Cain þurh snytro spéd smiðcræftega wæs *Tubal Cain was a workman cunning through wealth of wisdom*, Cd. Th. 66, 14 ; Gen. 1084. Metod tóbræd þurh his mihta spéd monna spræce, 102, 6 ; Gen. 1696 : 306, 23 ; Sat. 668 : Exon. Th. 225, 25 ; Ph. 394. Hé ús giefeþ æhta spéd, welan ofer wíd lond, 38, 10 ; Cri. 604. Hwær sind spéða rícera

ubi sunt opes potentum, Wülck. Gl. 253, 38: Ors. 1, 1; Swt. 18, 8. Eorðan spéda, Soul Kmbl. 154; Seel. 77. Ðínre spréce spéde *eloquia tua*, Ps. Th. 118, 172. Ða ðe ðære mycelnesse hiora spéda gylpaþ *qui multitudine abundantiarum suarum gloriabuntur*, 48, 6. Ðú on ðínes mægenes mihte spédum sǽ gesettest *tu confirmasti in virtute tua mare*, 73, 13. Oðre him of hyra spédum (*de facultatibus suis*) þenedon, Lk. Skt. 8, 3. Mid eallum hira spédum ðe hig hæfdon *cum universa substantia eorum quam habebant*, Deut. 11, 6. 'Redemptio animae propriae divitiae' . . . wé sceoldon mid úrum spédum úrum sáulum ða écan gesǽlinesse begitan, Chart. Th. 124, 27. Mé ðín spréc spédum (*richly, abundantly*) cwycade *eloquium tuum vivificavit me*, Ps. Th. 118, 50. Ealle mynstres fata and spéde hé sceal beséon *omnia uasa monasterii cunctamque substantiam conspitiat*, R. Ben. 55, 1. Ðín sunu ðe hys spéde (*substantiam*) ámyrde, Lk. Skt. 15, 30. Gemicla ðú heora wín and heora worldlíce spéde, Shrn. 104, 26: Ps. Th. 51, 6. Hé nǽsþ rihtwísnysse spéda and wísdómes goldhordas ðe sind sóðe welan, Homl. Th. ii. 88, 28. **IV.** *power, faculty :*—Ðǽr wæs gesýne his seó sóðe spéd *videbitur in majestate sua*, Ps. Th. 101, 14. Þurh ðínra dǽda spéd dagas hér gewuniaþ *ordinatione tua perseverat dies*, 118, 91. Hafast ðu heáh mægen ðínes earmes spéd and ealle fýnd *in virtute brachii tui dispersisti inimicos tuos*, 88, 9. Þurh his ægne spéd witan, Exon. Th. 351, 9; Sch. 77. Syndon on ðissum Simone twá spéda, mannes and deófles, Blickl. Homl. 179, 10. Ðú eart mægena God, nis ðé gelíc on spédum, Ps. Th. 88, 7. Wæs heofonweardes gást ofer holm boren miclum spédum, Cd. Th. 8, 8; Gen. 121. Meotud monnum syleþ sundorgiefe, sendeþ wíde ágne spéde (*faculties peculiar to each*), Exon. Th. 293, 24; Crä. 6. **V.** *opportunity,* or *means* of doing anything :—Ðæt hé him spéde sealde ðæt hé ðær wunian móste for intingan his gebeda *ut sibi facultatem et licentiam ibidem orationis causa demorandi concederet*, Bd. 3, 23; S. 554, 29. Se ealdormon him spéde and lýfnesse sealde tó farene swá hwider swá hí woldan *major domus regiae copiam pergendi quoquo uellent, tribuit eis*, 4, 1; S. 564, 34. **VI.** *progeny* (?) :—On cederbeámum mid heora spédum spearwan nystlaþ, Ps. Th. 103, 16. [Huand iu thiu spôt cumid, helpe fon himile (cf. thurgh helpe and spede of prayer, Pr. C. 2882), Hel. 1901. *O. H. Ger.* spuot *celeritas, successus, provectus, prosperitas, substantia.*] v. æht-, freónd-, freoþo-, here-, land-, mægen-, sige-, sigor-, tuddor-, un-, wíg-, woruld-, wuldor-spéd.

spédan ; *p.* de *To have success, succeed* in doing something :—Eów betere is ðæt gé ðisne gárræs mid gafole forgyldon . . . ne þurfe wé ús spillan gif gé spédaþ tó ðam (cf. Gif hé ne geþeó búton tó healfre híde *if he succeeds in obtaining no more than a half hide*, Ll. Th. i. 188, 1) *for you is it better to buy off this attack . . . We need not destroy one another, if you succeed in doing this*, Byrht. Th. 132, 51; By. 34. [Swá hé spédde, swá hím Crist húðe, swá þet in féuna geáre wæs þ mynstre gare, Chr. 656; Erl. 30, 18. Hé spédde litel, and be gode rihte, for hé wæs án yuel man, 1140; Erl. 265, 17. His broþer heo him wolde binimen, ah he ne mihte speden, Laym. 403. He wollde winnenn Crist alls he wann Eve and Adam ȝiff þatt he mihhte spedenn, Orm. 12317. *O. H. Ger.* gi-spuotôn *accelerare*.] v. á-, ge-spédan.

sped-dropa (spéd-?), an; *m. A rheumy* (?) *drop* :—Mec (*a book*) fugles wyn (*a pen*) geond speddropum (*ink*) spyrede, Exon. Th. 408, 6; Rä. 27, 8. v. spéde.

spediende *suffering from rheum* or *phlegm* (?) :—Spediende (swed-, Wrt.) *molaricus* (the preceding words are *podagricus, flegmaticus, reumaticus*), Wrt. Vocab. i. 45, 49: ii. 58, 2. v. sped.

spédig ; *adj.* **I.** *having good speed, prosperous :*—Him féran gewát Abraham wíde óð ðæt hé tó Siem com síðe spédig (*prosperous in travel*), Cd. Th. 107, 3; Gen. 1783. **II.** *having means, wealthy, opulent, rich in material wealth :*—Hé wæs swýðe spédig man on ðǽm ǽhtum ðe heora spéda on beóþ, ðæt is, on wildrum, Ors. 1, 1; Swt. 18, 8. Ic ne eom swá spédig (*dives*) ðæt ic mǽge bicgean mé wín, Coll. Monast. Th. 35, 17. Of spé[digre] (of gestreónfulre *sumptuosa, copiosa*, Hpt. Gl. 491, 4. **III.** *rich, abounding in, abundant, copious :*—Mundbora meahtum spédig *a protector abundant in power* (*God*), Exon. Th. 143, 27; Gú. 667: 198, 14; Ph. 10: 305, 2; Fä. 82. Wæstmum spédig, Cd. Th. 169, 19; Gen. 2802. Mihtum spédge, 101, 25; Gen. 1687. Spédige, Ps. Th. 59, 5. **IV.** *powerful* (cf. ríce) :—Spédig *potens*, Ps. Lamb. 77, 65. Hé on eorðan byþ eádig and spédig *potens in terra erit*, Ps. Th. 111, 2. Se sunu wæs sigorfæst, mihtig and spédig, Rood Kmbl. 299; Kr. 151. Mægena God, milde and spédig *Deus virtutum*, Ps. Th. 79, 14. Dǽdum spédig, 67, 18: 104, 7. [*O. H. Ger.* spuotig *uber, efficax, brevis*.] v. æht-, féðe-, freónd-, gód-, gold-, heán-, land-, med-, þurh-, un-, wan-, wuldor-spédig.

spédlíce v. ge-spédiglíce.

spédigness, e ; *f. Wealth, opulence :*—Welan, spédignesse *opulentia*, Hpt. Gl. 491, 9.

spédlíce ; *adv. Successfully, efficaciously, powerfully, in a manner which produces a result :*—Him spédlíce spearwan hús begyteþ *the sparrow succeeds in finding a house for itself*, Ps. Th. 83, 3: 105, 2. Dó mé spédlíce cuicne *quicken me effectually*, 118, 154. Syle mé spédlíce ðæt ðú mé nerige *grant me effectual release*, 169: 170. Ðonne ic him spéd-

---

líce tó spræc and hí lǽrde *when I spoke to them with power and taught them*, 119, 6. [Cf. *O. H. Ger.* spuot-líh *prosper*.]

spédsumian, spel. v. ge-spédsumian, spell.

spelc, spilc *a splint :*—Monegum men gescrincaþ his fét tó his homme . . . dô spelc tó, Lchdm. ii. 68, 7. Wið foredum lime . . . dô spilc tó *apply a splint*, 66, 23. [Spelke *fissula*, Prompt. Parv. 468, col. 1. *Spelk* a splinter or narrow strip of wood. 'To spelk in Yorkshire, to set a broken bone; whence the splints used in binding up of broken bones are called spelks,' Kennett MS., Halliwell's Dict. *Icel.* spelkur, spjalkir; *pl. f. splints* for binding up broken bones.] v. spilcan.

spelcean. v. spilcan.

speld, es ; *n.; pl.* speld and speldru (? or speldra (*see below*) *from* speldr. Cf. 'Spelder of woode *esclat*, Palsgrave. The schafte to *spildurs* spronge, Avow. of Arthur,' Halliwell's Dict.) : speld, *f. A splinter, a thin piece of wood used as a torch, a torch :*—On spelde *in favillam*, Anglia xiii. 35, 213. Speldum *favillis*, 36, 234. Ðara spaca speldra *malleoli* (v. spæc), Wrt. Voc. ii. 54, 56. Biernende speld *tedas*, 95, 26. Spelde *tedas*, 82, 29. [*Will.* speldes (*splinters*) of a broken spear. *Mod. E.* spell, spill. *M. H. Ger.* spelte *splinter of a lance : Icel.* speld, spjald ; *n. a tablet :* spilda *a flake : Goth.* spilda *a tablet*.]

spelian ; *p.* ode *To act as the representative* of another, *to represent, to take,* or *stand in, the place of* another :—Pronomen spelaþ ðone naman . . . Gif ðú cwest : 'Hwá lǽrde ðé ?' ðonne cweðe ic : 'Dúnstán.' 'Hwá hádode ðé ?' 'Hé mé hádode :' ðonne set hé on his naman stede and spelaþ hine, Ælfc. Gr. 5; Zup. 8, 11–16. Se abbod, for ðig ðe hé Godes gespelia is (*quia uices Christi in monasterio creditur agere*), sig hláford geháten . . . for ðæs lufe ðe he spelaþ *for the love of him whom he represents*, R. Ben. 114, 24. Næs Isaac ofslegen ac se ramm hine spelode, Homl. Th. ii. 62, 25. Hé God spellode (spelode ?) *he* (*Nebuchadnezzar*) *put himself in the place of God*, Cd. Th. 257, 16; Dan. 658. Gif hé wrítan ne cunne bidde óðerne ðe cunne ðæt hine spelige *si non scit literas, alter ab eo rogatus scribat*, R. Ben. 100, 5. Nán gehádod man ne sceal him tó geteón, ðæt hé Crist spelige ofer his hálgan híred, búton him seó notu fram Godes láreówum betǽht sý, Homl. Th. ii. 592, 29. v. á-spelian ; ge-spelia, *and next two words.*

speliend, speligend, es ; *m. A representative, vicar :*—Pronomen is ðæs naman speliend, se spelaþ ðone naman, Ælfc. Gr. 5; Zup. 8, 12. Se cyning is Cristes sylfes speligend under him sylfum, Bd. Whelc. 151, 39. v. preceding word.

speling, e ; *f. The taking the place* of another, *the acting as the representative* of another :—Cristes gespelia hé (*the abbot*) is and his note and spelinge on mynstre heald *Christi uices agere in monasterio creditur*, R. Ben. 10, 12.

spell, es ; *n.* **I.** *a story, narrative, account, relation :*—Ðæt is mǽre spell (*the story of Lot's wife*), Cd. Th. 155, 2; Gen. 2566. Spelli *relatu*, Txts. 93, 1720. Ða rehton hí sum hálig spel *exponebant illi quendam sacrae historiae sermonem*, Bd. 4, 24; S. 597, 34. Se man sǽde fram helle síðfæte swylc sár spell (sárspell ?) swylce nǽfre ǽr on men ne becom ne náht oft siððan *the man told such a dismal story of the journey to hell as never before had come to men, and not often since*, Shrn. 49, 10: Cd. Th. 66, 31; Gen. 1092. Spel wrecan *to tell the story* (*of Beowulf's exploit*), Beo. Th. 1751; B. 873. Hwílum gyd áwræc, hwílum spell rehte, 4225; B. 2109. Lyt swigode níwra spella ac hé sóðlíce sægde *little of the story of what had just happened did he leave unsaid, but told truly*, 5788; B. 2898: 6050; B. 3029. Fela spella him sǽdon ða Beormas of hiera águm lande, Ors. 1, 1; Swt. 17, 31. Ðás níwan spel ic ðé ealle in cartan áwríte *has nouas explicaturas historias omnia cartis commendabo*, Nar. 3, 17. Ic mæg singan and secgan spell in meoduhealle, hú mé cynegóde cystum dohten, Exon. Th. 321, 31; Víd. 54. **I a.** *a historical narrative, history :*—Ic sette hálig spelle (*de historiis sanctorum*) áne bóc . . . Ðara abbuda stær and spell ðysses mynstres on twám bócum ic áwrát *historiam abbatum monasterii hujus in libellis duobus descripsi*, Bd. 5, 24; S. 648, 20, 28. Ic ðé sende ðæt spell ðæt ic áwrát be Angelðeóde and Seaxum *historiam gentis Anglorum quam edideram tibi transmisi*, Bd. pref.; S. 471, 9. Ic cýþe hwanan mé ðás spell (*the narratives contained in the history*) cóman, S. 471, 20. Hé spell martyra ðrowunge gesomnade *historias passionis martyrum congregans*, 5, 20; S. 641, 43. Ic longe spell hæbbe tó secgenne *uber dicendi materia est*, Ors. 2, 8; Swt. 94, 16. **I b.** *a false or foolish story, a fable :*—Ealdra cwéna spell *anilis fabula*, Wrt. Voc. i. 55, 24. Spel *vel* unnyt sprǽc *fabula, i. bella*, ii. 146, 64. Mé mánwyrhtan manige on spellum sægdon *narraverunt mihi iniqui fabulationes*, Ps. Th. 118, 85. Ðú gehérdest reccan on ealdum leásum spellum, ðætte Iob sceolde beón se héhsta god, Bt. 35, 4; Fox 162, 5: Met. 26, 2. Ða ongunnon leáse men wyrcan spell, and sǽdon ðæt hió sceolde mid hire drýcræft men forbrédan, Bt. 38, 1; Fox 194, 30. **II.** *an instructive talk, discourse, a philosophical argument,* as a theological term *a sermon, homily* (v. spell-bóc) :—Sunnandæges spell . . . Se diácon sæde fram ðysum fýre emne swá wé rǽdaþ on Sunnandæges spelle, Wulfst. 205, 4–206, 1. Ðæt nis tó spelle ac elles tó rǽdenne *it is not to be taken as a sermon, but to be read otherwise*, Lchdm. iii. 232, 6. Se wísdóm écte ðæt spell mid leoþe *wisdom, added verse to his argument*,

Bt. 12; Fox 36, 7. Secgan spell *to discourse*, 13; Fox 36, 31. Gehér nū ān spell be ðām ofermódum cyningum, 37, 1; Fox 186, 1: Met. 25, 1. Ongan Waldend wið Abraham sprecan sægde him unlytel spell *held with him long discourse*, Cd. Th. 145, 14; Fä. 25. Spella and lára rǣd-hycgende, Exon. Th. 301, 27; Fä. 25. Ða twā béc on hundeahtatigum spellum (*homilies*), Ælfc. Gr. pref.; Zup. 2, 15. Bæd ðæt [hē] him on spellum gecýðde, onwrige worda gongum, hū . . . , Exon. Th. 171, 28; Gú. 1133: Cd. Th. 33, 7; Gen. 516. Gif ðú gesihst gimmas deórwyrða findan, spellu (*parabolas*) getácnað, Lchdm. iii. 214, 1. **III.** *a say-ing, remark, sentence, statement of a single point, dictum*, cf. the later *spell* :—Hit is swíþe ryht spell ðæt Plato sæde (*the saying is then given*). Ðā cwæþ ic : 'Ic eom geþafa ðæt ðæt was sóð spell, ðæt Plato sæde, Bt. 35, 1, 2; Fox 156, 8–14: 38, 3; Fox 202, 19. Ic ðē mæg eáþe geand-wyrdan ðæs spelles *I can easily give you an answer on the point you have mentioned*, 41, 2; Fox 244, 24. **III a.** *a saying* that is to be re-peated to another, *a message, an announcement*. v. spell-boda, I, god-spell :—Brimmanna boda ábeód eft ongeán, sege ðínum leódum miccle láþre spell *give them a much less pleasant message*, Byrht. Th. 133, 15; By. 50. Drihten dóm forgeaf ðám ðe his spel beraþ *the Lord gave glory to those that bear his messages*, Cd. Th. 246, 15; Dan. 479. **IV.** *speech, language* of prose :—Ðū ðás bóc of Lǣdenum tó Engliscum spelle hæfde gewende, ðā geworhte hē hí eft tó leóþe, Bt. pref.; Fox viii, 9. [*Goth.* spill *a fable, tale*: *O. Sax.* spell: *O. L. Ger.* spell *fabulatio, para-bola*: *O. H. Ger.* spell *sermo, narratio, parabola, fabula, mythus*: *Icel.* spjall *a saying*.] v. bealu-, bí-, eald-, fǽr-, forþ-, god-, gúþ-, hilde-, inwit-, lár-, láþ-, leás-, leóf-, lyge-, morgen-, riht-, sár-, sóþ-, weá-, wil-spell.

spell-bóc *a book of homilies* :—.i. full spelbóc wintres and sumeres, Chart. Th. 430, 21.

spell-boda, an; *m.* **I.** *one who delivers a message*, or *brings intelligence, a messenger, an ambassador* :—Sancte Iohannes wæs gelíc Godes englum & hē wæs béme, Cristes fricca on ðysne middangeard, & wæs Godes Suna spellboda, Blickl. Homl. 163, 22. Hú ðæt wæs weallende spelboda, se ðe ðone Hǣlend on ðysne middangeard cumendne gesecgean wolde, 165, 33. Heora feóndas flód ádrencte ðæt ðæra ǽfre ne com ān spelboda *there was never a one left to tell the tale*, Ps. Th. 105, 10. Him andswarode Godes spelboda (*the prophet Daniel*) : 'Nó ic wið feohsceattum ofer folc bere Drihtnes dómas,' Cd. Th. 262, 12; Dan. 743: 249, 20; Dan. 533: *the angel Gabriel*, Exon. Th. 21, 17; Cri. 336. Ðus gieddade Godes spelboda (*Job*), 236, 9; Ph. 571. Godes spelbodan *the prophets*, 104, 22; Gú. 11. Godes spellbodan (*the angels who came to Lot*), Cd. Th. 150, 19; Gen. 2494. Spelbodan (*those who should have brought the news of Pharaoh's overthrow*), 210, 10; Exod. 513. Spelbodan *oratores*, Wrt. Voc. ii. 115, 68. **II.** *one who delivers a discourse, a public speaker* :—Spelboda *causidicus*, Wrt. Voc. ii. 130, 14.

spell-cwide, es; *m. Historical narrative* :—Ic wolde gesecgan and mid spellcwidum gemearcian, Ors. 3, 1; Swt. 100, 12.

spellian; *p.* ode. **I.** *intrans. To talk, converse, discourse* :—Ic spellige *fabulor*, Ælfc. Gr. 25; Zup. 145, 13. Hí ealne dæg fleardiaþ and spelliaþ, L. I. P. 14; Th. ii. 322, 25. Ðā hig spelledon (woeron spellendo, Lind. : spellende, Rush.) *dum fabularentur*, Lk. Skt. 24, 15. Mid deádum spellian, gestrión hit getácnaþ, Lchdm. iii. 202, 5. Man ne mót spellian ne sprǽce drífan binnan Godes cyrcan, Homl. Skt. i. 13, 69 : L. Ælfc. C. 35; Th. ii. 356, 28. Ðā se Wísdóm ðis teló āsungen hæfde, ðá ongan hē spellian, Bt. 37, 2; Fox 186, 34. Spelligan, 32, 1; Fox 114, 2. Spellien (spillian, Cott. MS.), 20; Fox 70, 20. **II.** *trans. To announce, proclaim, tell, utter* :—Hig spelliaþ ł tógǽnaþ and sprǣcaþ unrihtwísnesse *effabuntur et loquentur iniquitatem*, Ps. Lamb. 93, 4. Him wæs lust ðæt hē ðiossum leódum leóð spellode, Met. Introd. 4. Hié (*the prophets*) ðære sóðfæstnesse tácen spellodan and secgende wǣron, Blickl. Homl. 161, 20. [Ðat folc gan to spelien (vsi, 2nd MS.) Irlondes gan Laym. 10068. Speken heom togadere & speleden, 4051. Þe posstless forenn . . . till hæþenn follc to spellenn, Orm. 8528. Mardocheus speleð *amare conterens impudentem*, A. R. 170, 19. *Goth.* spillôn *to tell, an-nounce*: *Icel.* spjalla *to talk*.] v. ge-spellian.

spell-stów, e; *f. A place where announcements are made* (?) :—And-lang dene tó ðære spelstówe, Cd. Dip. Kmbl. iii. 429, 28.

spellung, e; *f.* **I.** *talking, conversation, discourse, narration* :—Ðý lǣs on mē mǣge ídel spellung oþþe scondlíc leásung beón gestǣled *ne aut fabulae aut turpi mendacio dignus efficiar*, Nar. 2, 20. Forbúgaþ ídele spellunge and dyslíce blissa *avoid idle conversation and foolish pleasures*, Homl. Th. i. 180, 13 : 148, 2 : ii. 336, 19 : Cd. Th. 304, 31; Sat. 638. Spellung *fabulositas*, Wrt. Voc. ii. 55, 23. **II.** *a tale, conversation, discourse, narrative* :—Fabulae, ðæt synd ídele spellunga, Ælfc. Gr. 50, 29; Zup. 296, 5. Spellenga *sermonum*, Hpt. Gl. 505, 77. Spellunga ł saga *fabulas*, 410, 54. Ídele spellunga *otiosas fabulas*, Con-féss. Peccat. Hí cýð[d]on mē spellunga *narraverunt mihi fabulationes*, Ps. Spl. 118, 85. [Spellunge and smecchunge (*talking and tasting*) beoð ine muðe boðe . . . we schulen speken nu of spellunge, and ter efter of herrunge, A. R. 64, 11. *O. L. Ger.* spellunga *tragoediae*.] v. eft-, leás-spellung.

spelt, es; *m.* (?) *spelt, corn* :—Spelt *planta*, Wrt. Voc. i. 75, 11 : 46, 66 :

faar, 287, 19 : ii. 34, 38. Spelt sámgréne *far serotina*, 36, 39. Hwǣtes, speltes *farris*, 34, 37. [*O. H. Ger.* spelza *spelta, far.* From Latin *spelta*.]

spén (?), es; *m. A fibre* :—Spénas *fibras*, Wrt. Voc. ii. 35, 52. Cf. spón.

spendan *to spend*. [*O. H. Ger.* spentôn *consumere, impendere, expen-dere*. From Latin.] v. ā-, for-spendan, *and next word.*

spendung, e; *f. Spending* :—Sum underféhþ eordlíce ǣhta and se sceal ðæs pundes spendunge Gode āgifan of his ǣhtum *one receives earthly possessions, and he must repay the spending of the pound to God out of his possessions*, Homl. Th. ii. 556, 29. [*O. H. Ger.* spentunga *dispensatio, impensa*.]

Spéne (Spene?); *pl. The Spaniards* :—Amilcor weard from Spénum ofslagen, Ors. 4, 7; Swt. 182, 31. v. Spáneas.

spennan *to allure.* v. for-spennen, -spennend[e], -spennestre, -spenning. [*O. H. Ger.* spennen *allicere, illicere, sollicitare, seducere* : *Icel.* spenja.] Cf. spanan.

spennels, es; *m. A clasp* :—Fibula .s. dicta quod ligat cnæp, sigl, spennels, Wrt. Voc. ii. 148, 58. [Cf. *O. H. Ger.* spenula *fibula* : *Icel.* spennill *a clasp.*] v. spannan.

speoftan (?); *p.* speaft *To spit* :—Speaft (speoft, Rush.; cf. ā-speaft, -speoft, Jn. Skt. Lind, Rush. 9, 6), Mk. Skt. Lind. 8, 23. Speufton *expue-runt*, Mt. Kmbl. Lind. 26, 67. Speofton, 27, 30. Speafton (speoftun, Rush.), Mk. Skt. 15, 19. [Gespeoftad biþ *conspuetur*, Lk. Skt. Lind. 18, 32.]

speówan; *p. de To spit* :—Hí on his leófor hyra spátl speówdon, Exon. Th. 69, 17; Cri. 1122. Gé mid horu speówdon on ðæs andwlitan, Elen. Kmbl. 594; El. 297. Hí ǣtre spiówdon, Exon. Th. 156, 34; Gú. 884. [Cf. *Icel.* spýja (*strong*).] Cf. spiwian, spíwan.

speówung, e; *f. Spewing, vomiting* :—Speówung *evomatio*, Wrt. Voc. ii. 144, 40. v. spíwing.

speowþa. v. spiweþa.

spere, es; *n. A spear, lance, pike, javelin* :—Spere *lancea, falarica*, Wrt. Voc. i. 35, 11 : 84, 17 : falarica, ii. 86, 82 : hasta, i. 287, 4 : ii. 43, 19. Getridwet spere *hasta*, i. 35, 40. His sceaft ǣtstód ætforan him, swā ðæt ðæt spere him eode þurh út, Homl. Skt. i. 12, 55 : Byrht. Th. 135, 53; By. 137. Nægle oððe spere *cuspide*, Wrt. Voc. ii. 21, 24. Ecg sceal on sweorde, ord spere, Exon. Th. 346, 14; Gn. Ex. 204. Mid spere *lancea*, Jn. Skt. 19, 34. Hé nam him spere on hand *accepit lanceam in manu*, Bd. 2, 13; S. 517, 8. Ða speru sóðfæsdnesse *veritatis jacula*, Past. 35, 5; Swt. 245, 9 : 38, 6; Swt. 277, 22. Spera *sparorum*, Wrt. Voc. ii. 96, 33. Mid sperum tósticad *confossum vulneribus*, Ors. 3, 9; Swt. 128, 14. Spiorum (swiorum, Wrt.) *contis*, Wrt. Voc. ii. 21, 57. Speoru *contos*, 104, 58. Speru, 14, 72 : 20, 15 : ansatas (*ansatas* ǣtgāras, 3, 68), 5, 44 : 88, 16. Speru, boltas *catapultas*, 18, 58 : 85, 16. Hí léton of folman feólhearde speru fleógan, Byrht. Th. 134, 63; By. 108. ¶ In the following the word refers to a shooting pain or stitch :—Út lytel spere gif hér inne sié, Lchdm. iii. 52, 18. [*O. Sax.* sper; *n.* : *O. Frs. O. H. Ger.* sper; *m. hasta, lancea, sparus, catapulta* : *Icel.* spjör; *n. pl.* (poetical).] v. ātor-, bār-, deáþ-, huntig-, pull-, scot-, wǣl-, wíg-spere.

spere-bróga, an; *m. Terror caused by the casting of spears or darts* :—Ic spǣte sperebrógan . . . mé of hrífe fleógaþ hyldepílas, Exon. Th. 398, 27; Rä. 18, 4.

spere-healf, e; *The male side or line* (in speaking of inheritance). Cf. swert-, gér-mǣge, Grmm. R. A. 470) :—Mín yldra fæder hæfde gecweden his land on ða sperehealfe, næs on ða spinlhealfe, Chart. Th. 491, 20. [Cf. spera-hand *in Richthofen O. Frs. Dict.*] Cf. wǣpned-healf, -hand.

spere-leás; *adj. Without a point* or *head* :—Spereleás sceaft *contus*, Wrt. Voc. i. 35, 42.

spere-mann. v. spyre-mann.

spere-níþ, es; *m. Spear-strife, battle* :—Him Drihten mihte æt ðam spereníðe spéde lǣnaþ, Cd. Th. 124, 7; Gen. 2059.

spere-wyrt, e; *f. A plant name*; the word translates *innule(-a) cam-pane(-a)*, Wrt. Voc. i. 68, 17 : Lchdm. i. 210, 7 : *nap silvatica*, Wrt. Voc. i. 31, 27.

speriend, sper-lira, sperlíce. v. spyriend, spear-lira, spærlíce.

sperran, spirran, spyrran; *p.* de *To strike, spar* :—Ðǣr eác cwóman hreáþemýs . . . and ða on úre ondwlitan sperdon and ús pulledon *et uespertilionum uis ingens . . . in ora uultusque nostros ferebantur* (the translator has read *feriebant?*), Nar. 15, 6. Spyrrynde *verberans*, Germ. 399, 411. [Cf. *Icel.* sperrask *to struggle* : *Ger.* sich sperren *to struggle, resist.*] v. next word.

sperring, spirring, spyrring, e; *f. Striking* :—Clifra spyrringe *ungula-rum arpagine* (cf. slítunge *arpagine*, Wrt. Voc. ii. 5, 38), Hpt. Gl. 526, 67. Spyrrince *arpagine*, Anglia xiii. 37, 297.

sperte. v. spyrte.

spic, es; *n. Bacon, lard, the fat flesh of swine* :—Hí lares ðás hús; ðanon ys gecweden *lardum* spic, forðan ðe hit on húsum hangaþ lange, Ælfc. Gr. 9, 17; Zup. 42, 17. Spic *lardum*, Wrt. Voc. i. 82, 15 : *larda*, 286, 52 : ii. 52, 1 : *tanea*, i. 26, 47. Spices snǣd *offella* vel *particula*, 27, 19 : ii. 65, 7 : Homl. Skt. ii. 25, 87. Man nime āne cuppan huniges and healfe cuppan clǣnes gemyltes spices, and mǣngc on gemang ðæt hunig and ðæt spic tógædere, Lchdm. iii. 76, 5. Ān sconc spices *a ham*,

L. Ath. i, prm.; Th. i. 198, 7. Hē ǽlce gēre āgefe ðǽm hīgum .iii. wēga spices, Chart. Th. 471, 14: 473, 28. Speces, 468, 24. Mid ealdan spice oþþe mid ferscre buteran, Lchdm. ii. 354, 5. Gemelte eald spic, 52, 20. Nim clǽne spic, iii. 40, 26. Ðonne hē spic behworfen hæfþ *when he has attended to the bacon,* L. R. S. 7; Th. i. 436, 23. Etan spicc, Homl. Skt. ii. 25, 111. ¶ *Spic* occurs in names of places where swine were fed, e. g. Holan-spic, Cod. Dip. Kmbl. i. pp. 115, 137, 184, but its meaning here is not evident. Kemble suggests that it may refer to the mast on which the swine were fed. [Per com spic (fleas, 2nd MS.), Laym. 24437. Spyk or fet flesche *popa,* Prompt. Parv. 469, col. 1. *O. L. Ger.* spec[-suīn]: *O. H. Ger.* spech *lardum: Ger.* speck: *Icel.* spik *fat of seals, whales, etc., blubber: Dan.* spek *blubber, lard: Swed.* späk *lard.*] v. offrung-spic.

spíca, an; *m. Spikenard; any aromatic herb* (?):—Ðeós smerenes wæs geworht of ehtatēne cynna wyrtum; ðǽr wǽron þreó ða betstan—ele, & nardus, & spíca (*or is this merely the Latin word?*), Blickl. Homl. 73, 21. Lǽcedōm . . . spícan wiþ ūtsihtan, and dracontjan wiþ fūle horas, . . . and balzaman smiring wiþ eallum untrumnessum, Lchdm. ii. 174, 4.

spic-hús, es; *n. A larder:*—Spichūs *lardarium,* Wrt. Voc. i. 58, 16: *lar* (kitchen?), Lchdm. i. lxiii. 3. [*O. H. Ger.* spech-hús *lardarium.*]

spícing, es; *m. A spike* (? Halliwell gives *spiking* a large nail, as a northern word):—Spícyngas gadirian oððe wyrcean, geswinc hit getácnaþ, Lchdm. iii. 200, 24.

spic-máse, an; *f. A titmouse:*—Spicmáse (*Wright prints* swic-) *parrula,* Wrt. Voc. i. 62, 40. [In E. D. S. Pub. Bird Names, p. 33, *blue spick* is given as the name of the blue titmouse in North Devon. Cf. *Icel.* spiki *a tit.*]

spíder *a spider* (?):—Hēr com in gangan in spíder wiht, hæfde him his haman on handa, Lchdm. iii. 42, 11. The passage is the beginning of a charm.

spigettan; *p.* te *To spit:*—Gif hire fæder spigette (*spuisset*) on hire nebb, Num. 12, 14. Ða ongan se Catulus him spigettan on, Bt. 27, 1; Fox 96, 5.

spilæg:—Spilæg se ǽtterne *spilagius,* Rtl. 125, 29.

spilc. v. þpelc.

spilcan, spelcean; *p.* te *To bind with splints:*—Ðæt sceáp ðæt sceoncforad wæs ne spilcte gē ðæt *quod fractum est, non alligastis,* Past. 17, 9; Swt. 123, 10. Gif scancan forade synd . . . hū mon spelcean scyle, Lchdm. ii. 6, 12. v. spelc.

spild, es; *m. Destruction, ruin:*—Spildes *internicionis,* Wrt. Voc. ii. 76, 65. Spilde geblonden, Exon. Th. 405, 27; Rä. 24, 8. Ic hī ne sparige, ac on spild giefe, 247, 28; Jul. 85. Spilth *pessum,* Wrt. Voc. ii. 116, 75. Ðætte hié ðone spild ðæs hryres him ondrǽden *ut praecipitem ruinam metuant,* Past. 52, 5; Swt. 407, 20. Ðæt mód . . . ongit hine selfne on swelcne spild forlǽd *mens . . . sese in praecipitium pervenisse deprehendit,* 58, 2; Swt. 441, 27. Ðurh deófles spild *through the ruin caused by the devil,* Elen. Kmbl. 2235; El. 1119. [Cf. *O. H. Ger.* spildi; *f. desperatio; effusio.*] v. for-spild.

spildan; *p.* de *To waste, destroy, make away with:*—Ðeáf ne cymes būta ðætte [hē] spildeþ (*perdat*), Jn. Skt. Lind. 10, 10. Seðe lufaþ sáuel his spildeþ (*perdet*) hiá, 12, 25. Ðū wilnast, ðæt ðū ðīne feore spilde, Andr. Kmbl. 568; An. 284. [*O. H. Ger.* spildan *effundere, expendere.*] v. for-spildan, *and* spillan.

spilde. v. an-spilde.

spild-síþ, es; *m. A journey undertaken with the object of causing destruction,* Cd. Th. 187, 18; Exod. 153.

spilian; *p.* ode *To play, sport, wanton:*—Hī lufiaþ ídele blisse . . . and ealne dæg fleardiaþ, spelliaþ and·spiliaþ, and nǽnige note dreógaþ, L. I. P. 14; Th. ii. 322, 25. Eówra leóda ðe spiliaþ and plegaþ and rǽdes ne hēdaþ, Wulfst. 45, 24. [Uortigerne mid his hirede hæhliche spilede, Laym. 13816. In blisse spilen, Gen. and Ex. 2532. *O. Sax.* spilón *to play, dance: O. H. Ger.* spilón *ludere, ludificare, lascivire.*]

spillan; *p.* de *To destroy:*—Suā huelc soecaþ sáuel his hāl gewyrca spilleþ hiá (*perdet illam*), Lk. Skt. Lind. Rush. 17, 33; Jn. Skt. Rush. 12, 25. Ðeóf ne cymeþ būta ðætte [hē] spilleþ (*perdat*), 10, 10. Ne spildic ł ne losade *non perdidi,* Lind. 18, 9. Eal ðæt God spilde *God destroyed it all,* Cd. Th. 154, 22; Gen. 2559. Sumne man tō Lundene lǽdde, and ðǽr spilde, Chr. 1096; Erl. 233, 9. Ðætte ne ic losige ł ic ne spillo *ut non perdam,* Jn. Skt. Lind. 6, 39. Ðætte ðū spilda *ut dissipes,* Rtl. 55, 22. Ne þurfe wē ūs spillan *we need not destroy one another,* Byrht. Th. 132, 50; By. 34. Sōhton hine tō spillanne *quaerebant eum perdere,* Jn. Skt. Lind. Rush. 10, 39. Swil[g]ra, gliw[e]ra [*in margin* spillendra (spiliendra?); *but see* onspillendra *parasitorum,* Anglia xiii. 28, 29] *parasitorum,* Hpt. Gl. 422, 37. [Wǽron six men spilde of here ægon, Chr. 1124; Erl. 253, 14. Ȝif ȝe hit willed ich hine uulle spillen, Laym. 880. Unleoden spilden al his þeoden, 22863. Speche þu maht spillen, ant ne speden nawiht, Jul. 24, 14. Late ye nouth mi bodi spille, Havel. 2422. To spille hem þat ben gulty, Piers P. 19, 298. Spyllyñ or destroyyñ *confundo,* Prompt. Parv. 469. *Icel.* spilla *to destroy, spoil.*] v. for-, ge-spillan; spildan.

---

spilling, e; *f. Destruction, waste:*—Nān þing . . . būton folces geswinc and feós spylling and heora feónda forðbylding, Chr. 999; Erl. 134, 37. [*Prompt. Parv.* spyllinge or lesynge or schendynge *confusio, deperdicio.*] v. feoh-spilling.

spind *fat:*—Spind *arbina,* Wrt. Voc. ii. 5, 54. Hrysel *vel* gelend *vel* spind (swind, Wrt.) *vel* swínes smere *arvina vel adeps,* i. 44, 20. [*O. L. Ger.* spind *arvina: O. H. Ger.* spint *adeps, arvina, pinguedo.*] v. hago-spind.

spindel. v. sprindel.

spinel, spinl, e; *f. A spindle:*—Spinil (spinel), *stilium vel fusa,* Txts. 98, 967: *nitorium,* 81, 1377. Spinel *fusum,* 65, 933. Spinl, Wrt. Voc. ii. 34, 30: *fusu,* 152, 12: *nitorium,* 60, 12: *fusus,* i. 26, 15: 82, 10: *fussum,* 281, 74. Spinle *fusi,* Wülck. Gl. 245, 23. Spinele *fuso,* Wrt. Voc. ii. 83, 21. Spinle, 34, 29: Hpt. 494, 20. Spinle *fussum,* Kent. Gl. 1142. Hē sceal . . . spinle habban, Anglia ix. 263, 10. [*O. L. Ger.* spinnila: *O. H. Ger.* spinnala, spinala *fusus.*] v. eár-, þráwing-, wealc-spinel(-spinl).

spinel-healf, e; *f. The female side* or *line:*—Mín yldra fæder hæfde gecweden his land on ða sperehealfe, næs on ða spinlhealfe, Chart. Th. 491, 21. [Cf. *O. Frs.* spindel-sída. v. Richthofen, O. Frs. Dict.] Cf. wífhand, *and see* spere-healf.

spinnan; *p.* spann, *pl.* spunnon; *pp.* spunnen. I. *to spin:*—Neo ic spinne, neui ic spann, neuisti vel nesti ðū spunne, neuistis vel nestis gē spunnon, *neuerunt vel nerunt* hī spunnon, Ælfc. Gr. 25; Zup. 147, 2–4. Ic spinne *neo,* Wrt. Voc. ii. 60, 13. Spinnaþ *neunt,* 19: Mt. Kmbl. 6, 28: Lk. Skt. 12, 27. Hig spinnaþ wulle *illae nent lanam,* Ælfc. Gr. 15; Zup. 97, 9. Nim ðone hweorfan ðe wíf mid spinnaþ, Lchdm. ii. 310, 22. Spunnun *neverant,* Wrt. Voc. ii. 119, 10. Ða of his leáfum and of his flýse ðæs treówes spuunon and swā eác tō godewebbe wǽron and worhtan *foliis arborum ex siluestri uellere uestes detexunt,* Nar. 6, 18. II. of the action of the tide on the sand :—Sand sǽcir span (Grein would read spān) *the ebb hath knit the sand together* (?), Cd. Th. 196, 13; Exod. 291. III. of convulsive movement (?), *to writhe, twist:*—Sum ungesceádwís man hine sylfne áhēng ðæt hē fótum span (*for* sparn? v. spornan) and his feorh forlēt *a certain foolish man hung himself, so that he moved his feet convulsively* (could not rest them on the ground?), *and gave up the ghost,* Homl. Th. ii. 504, 34. Heó hī sylfe on grine áhēng, ðæt heó fótum span, 30, 23. [*Goth. O. H. Ger.* spinnan: *Icel.* spinna.] v. á-, ge-spinnan; twí-spunnen.

spíówan, spiowian. v. speówan, spiwian.

spír *a spire* [v. E. D. S. Pub. Plant Names, where *spire* is given as the name of the reed and of various spiked grasses. The word is also used of tapering trees, v. Baker's Northampt. Gl.]:—Hreódes spír, Lchdm. ii. 266, 10. [*Prompt. Parv.* spyre of corne or herbe *hastula,* spyryñ as corne and oþer lyke *spico.* Imeind mid spire and grene segge, O. and N. 18. The word occurs in Chaucer and Piers Plowman. v. Skeat's note on the latter, 13, 180 (C text). Cf. *Icel.* spíra *a spar: Dan.* spire *a sprout;* spir *a spar: Swed.* spira *a spar; a sceptre; a pistil.*]

spircan. I. *to sparkle:*—Spircendre *scintillante,* Hpt. Gl. 429, 42. Spyrcendum *scintillantibus,* 499, 43. II. *to fall in drops.* v. spircing:—Hē hēt mycel ád ontendan on ymbhwyrfte ðæs mǽdenes and mid pice hī besprencgan and mid spyrcendum ele (*with oil that bespattered her*), Homl. Skt. i. 9, 118. v. for-spyrcan; spearcian.

spircing, e; *f. A sprinkling, dropping:*—Spyrcinge *aspergine,* Germ. 398, 225. v. previous word.

spirian, spirte. v. spyrian, spyrte.

spitel *a kind of spade, a spud, a spittle* ['*spittle* a spade, used for light digging, which is *spittling.* The square board, with a short flat handle, used in putting cakes into an oven, is a baking-*spittle,*' Mid-York. Gl. '*Spittle* a spade with a curved edge, used for grip-digging,' Holderness Gl. See also E. D. S. Pub. Gl. B, 2, 12, and Halliwell's Dict. In A. R. 384, 18, where one MS. has *spade,* another has *spitelstaf.*] v. hand-, wād-spitel, *and* spittan.

-spitel. v. wrōht-spitel.

spittan; *p.* te *To dig with a spittle:*—In Agusto and Septembri and Octobri man mæg māwan, wād spittan, fela tilða hām gæderian, Anglia ix. 261, 16. Cf. '*Spittle* to cut weeds with a spittle-staff,' E. D. S. Pub. Linc. Gl. '*Spittle ower* to dig over a piece of ground with a spade,' Holderness Gl. '*Spitter* a small tool with a long handle for cutting up weeds,' Halliwell's Dict. v. spitel.

spittan *to spit.* v. spyttan.

spitu, e; *f. A spit:*—Spitu *veru,* Wrt. Voc. i. 27, 9: 82, 66: Ælfc. Gr. 11; Zup. 80, 10. Ueru spitu, *ueribus* spitum, 14; Zup. 89, 13. [*O. H. Ger.* spiz *veru.*]

spíwan; *p.* spáw, *pl.* spiwon. I. *to spew, vomit, spit up* (a) with acc.:—Ðonne spíwaþ hié ðæt horh, Lchdm. ii. 194, 16. Hē spáw blôd, Homl. Skt. i. 12, 63. Hē spáw his innoð ūt þurh his mūð, Shrn. 66, 33. Ðonne man ða cild cwalde, ðonne spiwon hí ða meoloc, 33, 1. Hit eft spíwende, Blickl. Homl. 57, 7. (b) with dat.:—On ða ádle ðe mon wormse spíweþ (cf. worms spíwende, 208, 9), Lchdm. ii. 200, 22. Ic blóde spáu *vomebam sanguinem,* Bd. 5, 6; S. 619, 30. Holm heolfre

spáw, Cd. Th. 206, 9; Exod. 249. (c) without a case:—Stinge him gelóme on ða hracan, ðæt hé mǽge spíwan, Lchdm. ii. 62, 12. Gelóme tó spíwanne, 174, 21: 286, 20. Ða gebrǽd hé hine seócne, and ongan hine brecan tó spíwenne, Chr. 1003; Erl. 139, 9. **II.** *to spit:*— Geót ðæt blód on yrnende wæter, spíw þríwa æfter, Lchdm. ii. 76, 15. Ðonne is cynn, ðæt him spíwe ðæt wíf on ðæt nebb, Past. 5; Swt. 45, 2. [*Goth.* speiwan *to spit: O. Sax.* spíwan: *O. Frs.* spîga, spîa: *O. H. Ger.* spíwan vomere, spuere: *Icel.* spýja.] v. á-spíwan.

spiw-drenc, -drinc (spiwe-), es; *m. An emetic:*—Spiwedrenc, Lchdm. ii. 136, 25: 270, 19: 272, 4, 6. Se ðe hæfþ þearfe spiwdrinces, 60, 26. Tó spiwdrence, 268, 21. Wyrc spiwdrenc, 270, 27: 302, 17. Se man þurh spiwedrenc áspíwþ ðone wǽtan, 60, 22: 336, 1. Spiwedrencas, 170, 6.

spiwe, es; *m. A vomiting, vomit:*—Spiwe deah ðám monnum ðe for fylle gihsa slihþ, Lchdm. ii. 60, 23.

spiwe-drenc, spiwel. v. spiw-drenc, spiwol.

spíwere, es; *m. One who vomits:*—Spíwere vomex vel vomens, Wrt. Voc. i. 17, 6.

spiweþa, aṅ; *m.* **I.** *vomiting:*—Gif hié (*diseases*) cumen of oferfyllo, mid spiweþan hý mon sceal lytlian, Lchdm. ii. 178, 11. Wið miclan spiweþan, and hé ne mǽge nánne mete gehabban, 190, 8. Wið spiwþan, 190, 1. Ðurh spiweþan, i. 274, 21. Spiweþan dón tó vomit, iii. 214, 23. Hí beóþ oferfyllede óþ spiweþan, R. Ben. 136, 25. Drincan óð speowðan, Homl. Th. ii. 292, 35. **II.** *vomit, what is vomited:*— Lǽt spíwan . . . gesceáwa hwæðer ðe spiwða sý swá micel swá hé ǽr gedranc, Lchdm. ii. 286, 22. Gif hund ðone spiweðan frete *si canis vomitum illum devoraverit,* L. Ecg. P. iv. 47; Th. ii. 218, 5. Hund eft hwyrfde tó his spiwðan, Shrn. 37, 16.

spiwian; *p.* ode *To spit up, vomit* (with dat.):—Him bánlocan blóde spiowedan *their carcases spouted forth blood,* Exon. Th. 271, 3; Jul. 476. v. spíwan, speówan.

spíwing, e; *f. Spewing, vomiting:*—Spíwingc *evomitio,* Wrt. Voc. i. 46, 17. Spíwing, ii. 32, 57. v. blód-spíwing; speówung.

spiwol; *adj. Emetic, causing vomiting:*—Drince hé spiwles drences, Lchdm. ii. 264, 24. Drince se man spiwolne drenc, 216, 11. Speowolne drenc, 216, 16. Mid wyrtdrencum útyrnendum oþþe spiwlum oþþe migolum, 82, 17. v. líg-, un-spiwol.

spjungean. v. sponge.

splott, es; *m.* **I.** *a plot of land:*—Mann ðe áhte geweald ealles ðæs splottes æt Celian dúne, ðǽr scræf wæs tómiddes, ðe ða seofon hálgan lágon inne slápan, Homl. Skt. i. 23, 415. On clǽnan splott súðe-weardne, Cod. Dip. B. iii. 336, 23. **II.** *a spot:*—Is se finta fægre gedǽled sum brún sum basu sum blácum splottum searolíce beseted *cauda porrigitur fulvo distenta metallo, in cujus maculis purpura mista rubet,* Exon. Th. 218, 18; Ph. 296. [Cf. Hyre treówenan gesplottude cuppan, Chart. Th. 537, 33. Wicklif uses *splotti* = spotted in Gen. 30, 35; and Halliwell gives *splotch* as an East-country word for a splash of dirt.] v. æcer-, friþ-, land-, mǽd-splott.

splottian *to spot, blot.* v. preceding word.

spón, es; *m.: e; f.* (? v. sæp-spón) *A chip, shaving:*—Spón *astula,* Wrt. Voc. ii. 5, 63: *gingria,* 109, 71. *Fomes* spoon; idem *astula,* 39, 70. Geswǽled spoon *vel* tynder *fomes,* i. 39, 21. Monige of ðam treówe ðæs hálgan Cristes mǽles spónas and sceafþan nimaþ *multi de ipso ligno sacrosanctae crucis astulas excidere solent,* Bd. 3, 2; S. 524, 31: 3, 17; S. 544, 44. Genim ðone wyrttruman . . . þwít nigon spónas, Lchdm. ii. 292, 2. [*O. Frs.* spón: *O. H. Ger.* spán *hastula, carpenta: Icel.* spánn, spónn *a chip, splinter.*]

sponan *teats,* sponere. v. spanu, spanere.

spong, e; *f. A spongy excrescence* (?):—Gif on eágan weaxen reáde sponge drýpe on hát culfran blód . . . óþ ðæt ða sponge áweg synd, Lchdm. ii. 308, 17: 300, 5. v. next word.

sponge, an; *f. A sponge:*—Án heora genam áne spongean, Mt. Kmbl. 27, 48. Genim spjungean, gedó on scearp eced, Lchdm. ii. 192, 18. [*O. Sax.* spunsia: *O. H. Ger.* spunga.] v. spynge.

sponn, spoon. v. spann, spón.

spor, es; *n.* **I.** *a trace, track, spoor:*—Ne biþ ðǽr eþe ðín spor on tó findanne *vestigia tua non cognoscentur,* Ps. Th. 76, 16. Stande ðæt spor for ðone foreáð, L. Ath. iv. 2; Th. i. 222, 16. Wé noldon tó ðæm spore onlútan, Past. pref.; Swt. 5, 18: Exon. Th. 497, 8; Rä. 85, 26. Hwæt mæg bión dyslícre ðonne hwá lufige hwelcre wuhte spor on ðæm duste and ne lufige ðæt ðætte ðæt spor worhte *quid esse dementius potest, quam vestigia in pulvere impressa diligere, sed ipsum, a quo impressa sunt, non amare?* Past. 46, 5; Swt. 351, 1–2. Gif man spor gespirige of scýre on óðre, fón ða menn tó ðe ðǽr nýcst syndon, and drífan ðæt spor óð hit man ðám geréfan gecýðe; fó hé syþþan tó and ádrífe ðæt spor út of his lande, L. Ath. v. 8, 4; Th. i. 236, 20–23. Hé ús spor tǽce, v. 8, 7; Th. i. 238, 3. Gif ðú gesyxt wulfes spor, Lchdm. i. 360, 19. **II.** *a trace, vestige, mark left by anything* (of the marks made by weapons; cf. *Icel.* sverða, eggja spor, dólg-spor *a wound*):— Lǽtaþ hý láþra leána hleótan þurh wǽpnes spor (*by a wound*), Exon. Th. 280, 2; Jul. 623; Andr. Kmbl. 2362; An. 1182. Bealubenne, lícwunde spor, Cd. Th. 193, 1; Exod. 239. **III.** *tracing, tracking:*—Ðú

teohhast ðæt ðú spyrige æfter mé, and swíþor swincst on ðam spore ðonne hí dón, Bt. 38, 5; Fox 206, 14. Ðæt ǽlc man wǽre óðrum gelástfull ge æt spore ge æt midráde, L. Ath. v. 4; Th. i. 232, 11. Befæste mon ðæt spor landes mannum, L. O. D. 1; Th. i. 352, 5. [*O. H. Ger.* spor *vestigium, indago; Icel.* spor.] v. fót-, hóh-spor.

spora, spura, an; *m. A spur:*—Spora *calcar,* Txts. 47, 361: 110, 1164. Spura, Wrt. Voc. ii. 17, 3, 63: i. 84, 3: 288, 22: Hpt. Gl. 505, 70: Ælfc. Gr. 9, 16; Zup. 42, 10. *Calcaria* spuran *dicta, quia in calce hominis ligantur,* Wrt. Voc. ii. 127, 44. Spurum *calcaribus,* 17, 62. Hé heów ðæt hors mid ðam spuran (cf. *Icel.* höggva hest sporum), Ælfc. T. Grn. 18, 22. .ii. spuran on .iii. pundan, Chart. Th. 503, 8. [*O. H. Ger.* sporo: *Icel.* spori.] v. hún-, táh-spora, -spura; hand-spor(a?); sporu.

sporettan (?); *p.* te *To kick:*—Sporetteþ (spornetteþ?) *recalcitravit,* Ps. Surt. ii. p. 193, 7. *next word.*

sporettung (?), e; *f. Kicking:*—Sportengæ *calcaneum,* Ps. Spl. T. 55, 6. v. *previous word.*

spor-leþer, es; *n. A spur-leather:*—Spurleþera *calcaria* (amongst things made by the shoemaker), Coll. Monast. Th. 27, 35. [*O. H. Ger.* spor-leder *calcarium.*]

spornan, spurnan; *p.* spearn, *pl.* spurnon; *pp.* spornen. **I.** *to strike with the foot, spurn:*—Ðe lǽs ðú on stán fóte spurne *ne offendas ad lapidem pedem tuam,* Ps. Th. 90, 12. On ·spurnan *inpingere,* Wrt. Voc. ii. 44, 72. On spornendum fét *in offenso pede,* Scint. 187, 8. (See (?) passages under spinnan, spannan.) **II.** *to spurn, reject:*— Æfter ðæs mǽdenes sprǽce, ðe hine spearn mid wordum, Homl. Skt. i. 7, 64. [Makede he þe spurnen (*stumble*) ine wredðe, A. R. 188, 2. *O. Sax.* spurnan *to strike with the feet, tread: O. H. Ger.* spurnan (*also wk.*): *Icel.* sperna.] v. æt-, ge-, óþ-spornan, -spurnan.

spornere, es; *m. One who treads or strikes with the feet, a fuller:*— Spornere, spurnere *fullo,* Ælfc. Gr. 9, 3; Zup. 35, 2.

spornettan; *p.* te *To strike with the feet, kick:*—Ne spornette ðú *non calcitres,* Wrt. Voc. ii. 60, 61: 80, 10.

sporning, e; *f. A stumbling, stumbling-block:*—þurh sporningcge *per offendiculum,* Scint. 134, 5. [Cf. *O. H. Ger.* spurnida *offensio, scandalum.*] Cf. spyrning.

spor-plætt, es; *m. A kick* (?):—Spátlu spurplættas (eárplættas?) bendas ðú þrowodest *tu sputa, colaphos, vincula passus,* Hymn. Surt. 80, 1. v. plætt, *and next word.*

sporu (?), an; *f. A heel:*—Spuran míne *calcanei mei,* Ps. Spl. T. 48, 5. v. hél-spure.

spor-wrecel (?), es; *m. What is tracked after being driven off* (?):— Ðá forstæl hé ða unlǽdan oxan æt Funtial, and dráf tó cytlid, and hine mon ðǽræt áparade, and his speremon áhredde ða sporwreclas *the man who tracked him rescued the cattle that had been driven off* (?), Chart. Th. 172, 26.

spówan; *p.* speów *To succeed.* **I.** used personally with instrumental of that in which the person succeeds, *to be successful:*—Hú mæg hé ǽnige gewinne wið mé spówan *how can he succeed in any struggle with me?* Nar. 16, 20. Ne mót ic ǽnige rihte spówan, Elen. Kmbl. 1830; El. 917: Andr. Kmbl. 3087; An. 1546: Cd. Th. 127, 23; Gen. 2115: Exon. 35, 27; Cri. 564. Spówende spéd, 117, 16; Gú. 225: 139, 14; Gú. 593: Cd. Th. 246, 14; Dan. 479. **II.** used impersonally, *it succeeds* with a person (dat.) (1) absolute:—Him spéwþ ðe bet, Btwk. 222, 9. Ðá hié ongeáton, ðæt him ne speów, L. Alf. 49; Th. i. 56, 8. Him wiht ne speów, Judth. Thw. 25, 23; Jud. 274: Beo. Th. 5701; B. 2854. Gesǽh Pilatus ðæt him náuwiht speów (spéua, Lind.) *videns Pilatus quia nihil proficeret,* Mt. Kmbl. Rush. 27, 24. Him speów hwónlíce, Homl. Skt. i. 7, 94. Hú swýþe him speówe *quantum profecerit,* Bd. 2, 4; S. 505, 27. (2) with gen. of that in which a person succeeds:—Ðá ðá him ðæs (*the attempt to raise the dead*) ne speów, Homl. Th. ii. 474, 11. Ðé speów ðæs ðú wið freónd oðóe feónd fremman ongunne, Cd. Th. 170, 9; Gen. 2810. (3) the object of success governed by a preposition:—Ða ðe on eordlícum weorcum hwónlíce speówþ, Homl. Th. i. 526, 16. Hú him speów ǽgðer ge mid wíge ge mid wísdóme, Past. pref.; Swt. 3, 8. Hú him æt ǽte speów, Beo. Th. 6045; B. 3026. [*O. H. Ger.* spuon, spuoan (*wk.*).] v. ge-, mis-spówan.

spówendlíce; *adv. Thrivingly, prosperously, abundantly:*—Mé ofer cume hǽlu æfter ðínre sprǽce spówendlíce *veniat super me salutare tuum secundum eloquium tuum,* Ps. Th. 118, 41, 58: 147, 4.

spówness. v. forþ-spówness.

spracen; *n.* '*The berry-bearing alder; rhamnus frangula. Germ.* Spreckenholz: *Dan.* spregner: *Swed. dial.* sprakved,' Lchdm. ii. 406. The word glosses *apeletum* in Wrt. Voc. i. 285, 83: ii. 8, 43, for *alnetum* (Cockayne):—Genim spracen berindred, Lchdm. ii. 58, 8: 66, 3.

sprǽc *a shoot:*—Spraec *sarmentum,* Wrt. Voc. ii. 119, 48. [*Icel.* sprek *a stick.* Cf. *O. H. Ger.* sprachila *siliqua. Graff* also cites spraioh *sarmenta,* vi. 391.] v. spæc; sprǽte (?).

sprǽc, spréc, sprǽc, e; *f. Speech.* **I.** in the following glosses:— Sprǽce *disputationis,* Wrt. Voc. i. 28, 49. Godcundra spréca *divinorum eloquiorum,* Hpt. Gl. 442, 37. Sprǽce *faminem,* Wrt. Voc. ii. 37, 28: 95, 38. Sprǽce *fatu,* 38, 6. Spéce wíse *scema locutionis,* i. 55, 22.

Spréc *loquela*, 88, 7. Spréce *omelias*, 288, 53 : ii. 64, 16. Spéc *oracu-lum*, spréca *oraculorum*, 62, 59, 60 : Hpt. Gl. 503, 10. Spǽcum *oraculis*, 518, 33. Spréce *procacitate*, 506, 2. Spréc *sermo*, Wrt. Voc. ii. 120, 45. Gesmeád spréc *sermo commentitius*, i. 55, 25. **II.** *speech, talking :*—Ne sý ðér nán óðer spéc inne, búton ðæt hig biddan God . . ., L. Ath. iv. 7 ; Th. i. 226, 29. Ðæt hí sín gehýrede on hyra menig-fealdan spéce (spréce, MS. A. : spréc, Lind. Rush.) *in multiloquio suo*, Mt. Kmbl. 6, 7. **III.** *speech, the faculty of speaking :*—Gif spréc áwyrd weorð, L. Ethb. 52 ; Th. i. 16, 5. Be ðam ðe him his spréc ofnimþ *de eo cui sermo deficit*, L. Ecg. P. i, tit. 3 ; Th. ii. 170, 6. Gif hwam seó spréc ðþfylþ, Lchdm. ii. 288, 18. Strong on spréce, Exon. Th. 410, 9 ; Rä. 28, 13. **IV.** *skilful speech, speaking with art, eloquence :*—Spréc *eloquentia*, Hpt. Gl. 529, 57. Sumum men hé forgifþ wísdóm and spréce, Homl. Th. i. 322, 25. **V.** *what is said, a speech, saying, collection of words :*—Heard is ðeós spréc *durus est hic sermo*, Jn. Skt. 6, 60. Spéc, Kent. Gl. 503. Ic áhsige eów ánre spréce, gif gé mé ða spréce secgeaþ *interrogabo vos ego unum sermonem, quem si dixeritis mihi*, Mt. Kmbl. 21, 24. God geopenude Abrahame, hwæt hé mid ðére spréce mǽnde, Gen. 18, 20. For ðére spréce ðe ic tó eów spréc, Jn. Skt. 15, 3. 'Ðín sunu leofaþ.' Ðá gelýfde hé ðére spréce, 4, 50 : Lk. Skt. 1, 20. Hé ásende hí, ðus cwéðende : 'Faraþ . . .' Hí férdon æfter ðæs cyninges spréce, Homl. Th. i. 78, 22 : Cd. Th. 144, 3 ; Gen. 2384. Iudas him andwyrde and cwæð . . . Æfter ðyssere spréce, Homl. Skt. ii. 86, 317. Engla sum Abraham cýgde, hé stille gebád áres spréce, Cd. Th. 176, 11 ; Gen. 2910. Wiste spréca fela, wóra worda, 29, 5 ; Geu. 445. Ðá se Hélend geendode ðás spréca, Mt. Kmbl. 19, 1 : 26, 1. Spécce, Kent. Gl. 873. **VI.** *speech, language, talk, discourse, words :*—Þreó þing syndon ðe gebringaþ ðone geséligan tó heofenan ríce ; ðæt is, hálig geþanc and gód spéc (cf. ídele word, 9) and fullfremed worc, Wulfst. 299, 12. Mé ðín spréc cwycade *eloquium tuum vivificavit me*, Ps. Th. 118, 50 ; 140. Ne gelýfe wé ná for ðínre spréce (spréc, Lind. : spréce, Rush.) *propter tuam loquelam*, Jn. Skt. 4, 42. Þeáwlícre spéce *tropologium*, Hpt. Gl. 410, 44. Ðú him hel sóðan spréce *conceal the truth from him*, Cd. Th. 110, 12 ; Gen. 1837. Ic on ðisse byrig (*Sodom*) gehýre yfele spréce werod habban, 145, 20 ; Gen. 2408. Hí habbaþ on múðe milde spréce, Ps. Th. 58, 7. Ídele spréce, Hy. 7, 108. **VI a.** *of written words :*—For ðére gelícnisse his gelógodan spréce *from the likeness to his style*, Ælfc. T. Grn. 8, 43. **VII.** *a speech, language :*— Ðeóda ungelíca ǽgþer ge on spréce ge on ðeáwum . . . heora spréc is tódǽled on twá and hundseofontig, and ǽlc ðara spréca is tódǽled on manega ðeóda, Bt. 18, 2 ; Fox 62, 28–34. Hé reorde gesette eorðbúendum ungelíce, ðæt hié ðére spréce gesǽl ge ne áhton, Cd. Th. 101, 22 ; Gen. 1686. On Engliscre spréce, Ælfc. T. Grn. 1, 26. Hé sealde heora ǽlcum synderlíce spréce, ðæt heora ǽlcum wæs uncúð, hwæt óðer séde, 4, 11. Ealle men óðer spréce áne spréce, Gen. 11, 1. Ða apostolas cúðan ealle ða spréca ðe syndon swá wíde swá middaneard is, Wulfst. 294, 8 : 296, 1. Mid sprécum hiá sprecas níuum *linguis loquentur nouis*, Mk. Skt. Lind. 16, 17. **VIII.** *speech, e. g. to have speech of or with a person, conversation, consultation, conference, discussion :*—Nis ðæt lytelu spréc tó gehéganne (*of the day of judgment*), Exon. Th. 445, 17 ; Dóm. 8. Folc biþ gebonnen tó spréce, 451, 10 ; Dóm. 101. Se déma æfter langsumre spréce lét ða módor tó ðam suna. . . 'Béde ðú forðí ðínre módor spréce, ðæt ðú hí gebígdest fram mé,' Homl. Skt. i. 4, 341–357. Hé hét Agustinum tó his spréce cuman *jussit Augustinum ad suum advenire colloquium*, Bd. 1, 25 ; S. 486, 39 : Guthl. 9 ; Gdwin. 48, 21 : 11 ; Gdwin. 54, 4 : Cd. Th. 33, 6 ; Gen. 516. Æt spréce ðére *at that consultation*, 122, 29 ; Gen. 2034 : Bd. 2, 13 ; S. 516, 13. Æfter heora spréce, Jud. 3, 19. Gisomnadun ða biscopas tó spréce *colligerunt pontifices concilium*, Jn. Skt. Rush. 11, 47. Spréce and geþeahte habban *to treat, consult ; agere*, Bd. 1, 27 ; S. 492, 16. Cwæþ ðæt hé wolde mid his freóndum spréce and geþæht habban *cum amicis suis sese de hoc collaturum esse dicebat*, 2, 13 ; S. 515, 37. Hæfdon betwih him spréce and geþeahte *habito inter se consilio*, 3, 29 ; S. 561, 6. Ðá hí hæfdon lange spréce and geflit *longa disputatione habita*, 2, 2 ; S. 502, 13. Gif hwylc mæssepreóst untruman men spréce forwyrne (*colloquium denegaverit*), L. Ecg. P. i. 2 ; Th. ii. 172, 27. **VIII a.** *a question, case that requires explanation :*—Ungelíc ðære spréce ðe wé æfter spyriaþ, Bt. 38, 2 ; Fox 198, 25. Ðæt folc ðe hæfde ǽnige spréce eode. ut tó ðam getelde *omnis populus, qui habebat aliquam quaestionem, egrediebatur ad tabernaculum*, Ex. 33, 7. Ðú spenst mé on ða méstan spréce and on ða earfoþestan tó gereccenne . . . and uneáþe ǽnig cóm tó ende ðére spréce ; forðam hit is þeáw ðére spréce and ðære áscunge, ðætte simle ðonne ðér án tweó of áðón biþ, ðonne biþ ðér unrím ástyred . . . Swá is ðisse spréce ðe ðú mé æfter ácsast *ad rem me omnium quae-situ maximam vocas, cui vix exhausti quidquam satis sit ; talis namque materia est, ut una dubitatione succissa innumerabiles aliae succrescant*, Bt. 39, 4 ; Fox 216, 14–26. **IX.** *a sentence, decision, agree-ment, terms :*—Ðá com Putrael tó Bora and bed his forespéce tó Ælfríce. Ðá sette Bora ðás spéce wið Ælfríce : ðæt wes, ðæt Putrael sealde Ælfríce .viii. oxan, and gef Bora sixtig penga for ðere forespéce, and dide hine sylfne sacclés wið Ælfríce, Chart. Th. 628, 17. **X.**

a case, cause, suit, claim, (a) in a general sense :—Wið ðon ðe heó his spéce underfénge *in consideration of her receiving his suit* (Godwine asked for the lady in marriage), Chart. Th. 312, 14. Ðeáh hié ryhte spréce hæbban hiera yfel on him tó télanne *mala recte redarguunt*, Past. 28, 5 ; Swt. 197, 2. Ðú démst míne spréce *fecisti causam meam*, Ps. Th. 9, 4. (b) as a legal term :—Ðæt ðis ǽfre gesett spréc wǽre *that this for ever should be a settled suit*, Chart. Th. 203, 4 : 172, 2. Ongan ðá tó specenne on ðæt land . . . óð ðæt seó spréc weard ðam cynge cúð, 302, 15. Be dóme and spréce. . . . Gehwilc spréc hæbbe ándagan hwænne heó gelæst sý, L. Ed. proem. ; Th. i. 158, 3–7 : 11 ; Th. i. 164, 22. Ǽge-hwilcre spréce ðe mǽre sý ðonne .iiii. mancussas, L. A. G. 3 ; Th. i. 154, 9. Gif man mæssepreóst tihtlige ánfealdre spréce . . . æt þrimfealdre spréce, L. Eth. ix. 19 ; Th. i. 344, 11–13, 15–17. Fultum æt swá micelere spréce, L. Ath. v. 8, 3 ; Th. i. 236, 16. Gif ús feoh áríse æt úrum ge-mǽnum spréce, v. 3 ; Th. i. 232, 6. Æt cynges spéce, lecge man .vi. healfmarc wedd, L. Eth. iii. 12 ; Th. i. 296, 25. Clǽne ǽlcere spéce, L. C. S. 28 ; Th. i. 392, 12. Swá fela manna . . . tó gewitnesse gehwylc-ere spréce, L. Ath. v. i ; Th. i. 222, 11. Ǽlcne wítefæstne man ðe on spréce áhte (*gained at law, as the result of a suit*), Chart. Th. 557, 22. Hé dráf his spréce *he prosecuted his suit*, 376, 11. Ic spéce drífe mid fullan folcrihte, L. O. 2 ; Th. i. 178, 13. Habban ða geréfscypas begen ða fullan spéce gemǽne, L. Ath. v. 8, 4 ; Th. i. 236, 25. Man ne mót spréca drífan binnan Godes cyrican, L. Ælf. C. 35 ; Th. ii. 356, 29. **XI.** *talk* about a person or thing, *report, fame :*—Ðæs ðe má seó spréc be him férde, Lk. Skt. 5, 15. Ðá férde ðeós spréc be him, 7, 17. Hé ongan bodian and wídmǽrsian ða spéce, Mk. Skt. 1, 45. **XII.** in the Northern Gospels spréc translates words denoting places where there is speaking :—In spréce (spréc, Lind.) in synagoga, Mk. Skt. Rush. 6, 2. On spréce (spréc, Lind.) *in foro*, 12, 38 : Lk. Skt. Rush. 20, 46 : Lind. 7, 32. [O. Sax. spráka : O. Frs. spréke : O. H. Ger. sprâhha *lingua, loquela, sermo, sermocinatio, colloquium, eloquium, ratio, judicium, consilium, senatus*.] v. éfen-, æfter-, ærend-, burh-, bysmor-, dol-, eald-, edwít-, ellen-, for-, fore-, frécnen-, frum-, gedwol-, gegaf-, gilp-, hete-, Léden-, morgen-, of-, ofer-, on-, sceáwend-, scrift-, sód-, stunt-, teosu-, tó-, twí-, untíd-, wíder-, woruld-, ymbe-spréc (-spéc) ; -spréce, -sprec.

spréc-ærn, -ern, es ; *n. A place for speaking, court-house :*—In sprécern *in praetorium*, Jn. Skt. Lind. Rush. 18, 28 : 19, 9. Cf. spréc-hús.

spréc-cynn, es ; *n. A mode of speaking :*—Bóc be gesetnessum and gemetum spræccynna *libellum de figuris modisque locutionum*, Bd. 5, 24 ; S. 648, 42.

-spréce. [O. L. Ger. bi-spráki : O. H. Ger. ga-sprähhi.] v. ge-, god-spréce.

-spréce, -spéce ; *adj.* [O. Sax. -spráki : O. H. Ger. -sprähhi.] v. án-, fela-, ge-, gegaf-, ídel-, ofer-, stunt-, twí-, yfel-, ymb-spréce.

sprécelíc. v. ge-sprécelíc.

sprécful ; *adj. Talkative, loquacious :*—Wer sprécful *vir linguosus*, Ps. Lamb. 139, 12.

spréc-hús, es ; *n. A house for speaking :*—Spréchús *auditorium*, Wrt. Voc. i. 58, 11. Úþwitena spréchús *curia vel senatus*, 13. [O. L. Ger. sprác-hús *curia* : O. H. Ger. sprâh-hús *curia, consistorium, praetorium*.] Cf. spréc-ærn.

sprécleás ; *adj. Speechless, without the power of speech :*—Spécléase ł dume *elinguia*, Germ. 398, 72. [O. H. Ger. sprâhhalôs *elinguis*.]

-sprécness. v. twi-sprécness.

sprédan ; *p. de To spread, expand.* [O. L. Ger. te-spreidan *disper-gere* : O. H. Ger. spreiten *pandere, expandere, diffundere*.] v. ge-, ofer-, tó-sprédan, á-spreádan ; sprédung.

sprédung, e ; *f. Spreading, diffusion, propagation :*—Sprédung men-nisces cynnes *propagatio humani generis*, Rtl. 109, 4.

spréngan. v. sprengan.

spréte (?), spræt (?), es ; *n. A sprout, shoot :*—Sprétu (sprǽcu ? v. spréc) *labruscas*, Hpt. Gl. 454, 16. [Cf. (?) spreat, sprat, sprett *the jointed-leaved rush*, Jamieson's Dict. Sprat-barley *barley with very long beards* ; sprats *small wood*, Halliwell's Dict.]

spranc (?), es : spranca, an ; *m. A shoot, twig, sprig :*—Spranca (sprauta, Wrt.) *sirculus vel virgultum*, Wrt. Voc. i. 32, 44. Styb *vel* spranca (sprauta, Wrt.) *stirps*, 33, 57. Treówes sprancan *plante*, 39, 14. Deádbǽre sprancan *letiferas labruscas*, Hpt. Gl. 454, 17. Spranca *sarmen-torum*, 468, 22.

sprauta. v. preceding word.

spreáwlian ; *p. ode To sprawl, move convulsively :*—Spreáwlige *pal-pitet*, Germ. 392, 10. [Sprawlyñ *palpito* ; sprawlynge *palpitacio*, Prompt. Parv. 470 (and see note). Leyen and sprauleden in the blod, Havel. 475. Sprauleind with her winges twey, Gow. ii. 5, 11.]

-sprec, spréc. v. ge-, god-sprec, spréc.

spreca, speca, an ; *m. A speaker, one who speaks in council* (cf. spréc, **VIII**), *a councillor :*—Forht folces weard héht him fetigean sprecan síne, Cd. Th. 161, 18 ; Gen. 2667. [O. Frs. for-spreka : O. H. Ger. sprehho.] v. edwít-, for-, fore-, ge-, mid-, on-spreca (-speca).

sprecan, specan ; *p.* spræc, spæc ; *pl.* sprǽcon, spǽcon ; *pp.* sprecen, specen *To speak.* **I.** *to exercise the faculty of speech :*—Se dumba spræc,

Mt. Kmbl. 9, 33. Dumbe sprǽcon, Mk. Skt. 7, 37. Đú byst suwiende, and đú sprecan ne miht, Lk. Skt. 1, 20. Ǽnne lícþrowere . . . unsprecende forneán. . . . Basilius gelǽdde hine forð wel sprecande, Homl. Skt. i. 3, 489. Wæs eall weoruld sprecende on án gereord, Wulfst. 211, 19. Geseónde dumbe specende (sprecende, MS. A.), Mt. Kmbl. 15, 31. **II.** *to use words* in conversation, discourse, etc. :—Ic ne sprece tó đǽm, ac ic sprece tó đé, Bt. 38, 5; Fox 206, 12. Ic secge đis sárspell and ymb síþ sprǽce, Exon. Th. 458, 7; Hy. 4, 96. Hwæþer ic be mé sylfum spece. Se đe be him sylfum sprycþ, Jn. Skt. 7, 17, 18. Nú đú sprycst openlíce, 16, 29. Eorl óðerne tǽleþ behindan, spreceþ fægere beforan, Frag. Kmbl. 8; Leás. 5. Đá spræc se ofermóda cyning, Cd. Th. 22, 9; Gen. 338. Hió spræc him þicce tó, 43, 1; Gen. 684. Drihten wið Abrahame spræc, 139, 2; Gen. 2303. Hig spǽcon (sprǽcon, MS. A.) him betwýnan, Lk. Skt. 24, 14. Đæt đú ne belge wið mé, gif ic sprǽce. . . . Nú ic ǽne begann tó sprecanne tó mínum drihtne, ic wylle sprecan git, Gen. 18, 30-31. Ic eom ásend wið đé sprecan, Lk. Skt. 1, 19. Đonne hé spreocan ongan, Cd. Th. 269, 25; Sat. 78. **III.** with acc. (a) *where the object of the verb is* word *or a similar form* :—Ic đás word sprǽce, Exon. Th. 457, 12; Hy. 4, 82. Đú đa word spricest, 12, 2; Cri. 179. Se đe God sende sprycþ Godes word, Jn. Skt. 3, 34. Đú worn fela ymb Brecan sprǽce, Beo. Th. 1067; B. 531. Him ellenróf andswarode, word æfter spræc, 688; B. 341. Đæt gé on eárum sprǽcon, Lk. Skt. 12, 3. Hié fela sprǽcon sorhworda somed, Cd. Th. 49, 7; Gen. 788. Spǽcon, Ps. Th. 57, 3. Gilde ǽlc đe hit (*the exculpation on oath*) æt sprece .cxx. scill., L. Ath. i. 13; Th. i. 206, 6. Warna đæt đú nán þing elles ne sprece, búton đæt ic đé bebeóde, Num. 22, 35. Đis synd đa word đe đú scealt sprecan tó folce, Ex. 19, 6. Ongan hospword sprecan, Andr. Kmbl. 2632; An. 1317. Đæt ǽrende wæs sprecen, 3242; An. 1623; Beo. Th. 1290; B. 643. (b) *where the object of the verb is a word denoting the matter expressed in the words spoken* :—Ic rǽd sprece *I give counsel in my words*, Cd. Th. 115, 2; Gen. 1913. Đú bysmor spycst *blasphemas*, Jn. Skt. 10, 36. Tunga his sprecþ dóm, Ps. Spl. 36, 32. Se đe sóð spriceþ, Exon. Th. 3, 9; Cri. 33. Hé beót spriceþ, 290, 25; Wand. 70. Heó mé wom spreceþ, 402, 22; Rä. 21, 23. Đa đe sprecaþ sybbe, Ps. Spl. 27, 4. Hié sprecaþ fácen and inwit, Cd. Th. 145, 30; Gen. 2413. Fela hé mé láðes spræc, 39, 9; Gen. 622. Đam đe sár sprece sáwle mínre, Ps. Th. 108, 20. (c) *where the object is that which is spoken about, to mention* :—On swelcum cræftum swelce wé ǽr sprǽcon, Past. 9; Swt. 59, 12. Of đǽm beorgum đe wé ǽr sprǽcon (sǽdon, MS. L.), Ors. 1, 1; Bos. 17, 44. Wé gehýrdon hí sprecan Godes mǽrða mid úrum gereordum, Homl. Th. i. 314, 19. **III a.** *with a clause, to say* :—Hié sprǽcon, đæt hit betere wǽre, Ors. 2, 3; Swt. 68, 8. Đá gehýrde hé sumne đara bróþra sprecan, đæt hé wolde féran, Bd. 3, 2; S. 525, 5. **III b.** *with the words that are spoken* :—Hí sáre sprecaþ : ' Hwá gesyhþ úsic?' *dixerunt, Quis videbit eos?* Ps. Th. 63, 4. **IV.** with a gen. :—Míne fýnd sprǽcon mé yfeles, Ps. Th. 40, 8. **V.** with inst., *to speak in a language, with words* :—Ic sprece mongum reordum, Exon. Th. 390, 13; Rä. 9, 1. Beówulf beótwordum spræc, Beo. Th. 5014; B. 2510: Exon. Th. 253, 24; Jul. 185. Hé spræc him wordum tó, Ps. Th. 98, 7. Hé wordum wið his Waldend spræc, Cd. Th. 155, 22; Gen. 2576. Hé tó Noe spræc hálgan reorde, 89, 19; Gen. 1483. Hí sprǽcon úrum gereordum, Homl. Th. i. 314, 18. Tó Geátum spræc mildum wordum, Beo. Th. 2347; B. 1171. **VI.** with prep. :—Hé mid heardre đreá hí on spræc and hí gebétte *aspera illos invectione corrigebat*, Bd. 3, 5; S. 527, 11. Wé sind an specende *dicturi*, Wrt. Voc. ii. 28, 66. On specende *inspirans*, 93, 40. On spǽcende (swǽtende, Wrt.), 47, 31. Ongeán sprecendes *obloquentis*, Ps. Spl. 43, 18. ¶ *In technical terms*, v. sprǽc, **X**, sprecan æfter, on, ymb *to sue for, make a claim against, lay claim to* :—Đæt orf đæt ic on spece *the cattle that I lay claim to*, L. O. 2; Th. i. 178, 15. Ágnung biþ nér đam đe hæfþ đonne đam đe æfter sprecþ, L. Eth. ii. 9; Th. i. 290, 21. Đa fíf hída đe Ǽđelm Higa ymb spycþ. . . . Ongon Híga him specan on mid óðran onspecendan and wolde me óðflítan đát lond *the five hides about which Æthelm Higa has a suit . . . Higa along with other claimants began to make a claim against him (Helmstan), and wanted to get the land from him by litigation*, Chart. Th. 169, 17-24. Đá spræc ic on đa mágas then I made a claim against the kinsmen, 167, 18. Hé spræc on his ágene módor æfter sumon dǽle landes, 337, 4. Đá gemǽtæ hé đa swutelunga and đǽrmid on đæt land spæc, ongan đá tó specenne on đat land, 302, 12. Hine man tó rihte gelǽde đam đe him on sprǽcon (*those that bring charges against him*), L. Eth. i. 4; Th. i. 284, 1. Đone đǽ đe se gelýfan mihte đe on sprece, L. Ed. 1; Th. i. 158, 18. Đæt nán man on his yrfenuman ne spece *that no man bring an action against his heir*, L. Eth. iii. 14; Th. i. 298, 10. Đoñe ið syllan, đæt hé mid folcrihte on đæt land sprece, L. O. D. 1; Th. i. 352, 13. [*O. Frs.* spreka : *O. Sax.* sprekan : *O. H. Ger.* sprehhan.] v. á-, be-, for-, forþ-, ge-, mis-, ofer-sprecan ; un-sprecende, for-, fore-sprecen.

sprecan. v. on-sprecan.

sprecel *a spot* (?). v. haran-specel. [Cf. Spreckled *speckled*, Halliwell's Dict. : spreckly, spreckled, Jamieson's Dict. *O. H. Ger.* sprehhiloht *maculosus* : *Icel.* spreklóttr *speckled*.]

sprecend, sprecende. v. on-sprecend, un-sprecende.

sprecol, specol ; *adj. Talkative, loquacious* :—Wer sprecul *vir linguosus*, Ps. Spl. 139, 12. v. fela-, ofer-, swíð-sprecol.

sprecolness, e ; *f. Talkativeness, loquacity* :—Genihtsumian on gebeórscypum specolnyss gewunaþ *abundare in conuiuiis loquacitas solet*, Scint. 170, 15. v. ofer-sprecolness.

sprengan ; *p.* de *To cause to spring.* **I.** *to scatter* :—Đú gaderast đǽr đú ne sprengdest (*sparsisti*), Mt. Kmbl. 25, 24. His eágan wǽron spearcan sprengende, Homl. Th. i. 466, 26. **II.** *to sprinkle*, (a) *an object with something* :—Đú spren[g]st Aaron and his reáf, Ex. 29, 21. Hé nam đæt blód and sprengde đæt folc, 28, 8. (b) *something on to an object* :—Sprǽnge se mæssepreóst háligwæter ofer hig ealle, L. Ath. iv. 7; Th. i. 226, 23. Genim đás ylcan wyrte gesodene, sprengc intó đam húse, Lchdm. i. 264, 15. Nime se sacerd his blód and dyppe his finger đǽron and sprenge on đæt ryft, Lev. 4, 17, 6. (c) *government uncertain* :—Đá đá hé sprencde *dum rorat*, Germ. 402, 43. **III.** *to burst, crack* (cf. *to spring* a leak, *sprung*, applied to a bat) :—Hé sceáf mid đam scylde, đæt se sceaft tóbærst, and đæt spere sprengde (*shivered the spear-head*), đæt hit sprang ongeán, Byrht. Th. 135, 52; By. 137. **IV.** as a medical term, *to apply a clyster*. v. spring, IV (3) :—Đæt mon on morgen on sprenge, Lchdm. ii. 48, 24. [Sprenged on mid hali water, A. R. 16, 9. *O. H. Ger.* sprengen *quassare, rorare*: *Ger.* sprengen *to burst, scatter, sprinkle* : *Icel.* sprengja *to burst* : *Dan.* sprænge : *Swed.* spränga.] v. á-, be-, geond-sprengan.

spreót, es ; *m. A pole, sprit* (in bow-*sprit*) :—Spreót *contus*, Wrt. Voc. i. 33, 61. Ánes mannes lenge đe healt ánne spreót on his hand and strecþ hine swá feor swá he mæg árǽcan intó đere sǽ *statura unius hominis tenentis lignum quod Angle nominant* spreot, *et tendentis ante se quantum potest*, Chart. Th. 318, 10. Spreótas *trudes* vel *amites*, Wrt. Voc. i. 35, 43 : *trudes* (in a list of things connected with ships), 48, 13 : 57, 16 : 64, 7 : *ansatas*, ii. 3, 68 : *contos*, 14, 72. Spreótum, spreútum *contis*, Txts. 48, 211. [*Prompt. Parv.* sprete *contus* : *Du.* spriet *sprit* : *Dan.* sprød : *Swed.* sprót.] v. eofor-spreót.

spreótan. v. sprútan.

sprinca *glosses* circopythicos, Wrt. Voc. ii. 131, 29.

sprincan. v. springan.

sprincel, es ; *m. A wicker-basket* :—Sprinclum *fiscillis*, Wrt. Voc. ii. 108, 58 : 35, 43. [Cf. *Dan.* sprinkel, sprinkel-værk *trellis, lattice*.] Cf. tǽnel, windel.

sprincting, sprind. v. springung, springd.

sprindel *a tenter-hook* :—Sprindel (-il) *tenticum*, Txts. 101, 2003. Spindel, Wrt. Voc. i. 289, 18. v. next word.

-sprindlian. v. á-sprindlad.

spring, spryng, es ; *m.* (*but* eá-spring ; *n.*) **I.** *a source* of water :—Spring *casta* (*castalia*?), Wrt. Voc. ii. 129, 31. Ǽt đæs wæteres sprynge, Cod. Dip. Kmbl. iii. 389, 7. [*Prompt. Parv.* sprynge *scaturigo, scatebra* : *O. L. Ger.* gi-spring *fons* : *O. Sax.* aho-spring : *O. H. Ger.* ur-spring *fons*.] v. ǽ-, eá-, ge-, will-spring. **II.** *a springing, rising, spring* in day-*spring*. v. up-spring **III.** *what springs* up or from. [Sprynge of a tre or plante, springe or yonge tre *planta, plantula*, Prompt. Parv. 470.] v. of-spring. **IV.** as a medical term, (1) *an ulcer, a sore, pustule* :—Spryng *carbunculus*, Wrt. Voc. ii. 102, 46 : 13, 11 : *papula*, 116, 22. Carbunculus spring *vel* angset *vel* pustula, i. 19, 19. Tó sealfe wið springe, Lchdm. ii. 80, 8. Wið đæt man wille spring on gesittan, i. 2, 19. Láðlíc biþ đæs hreóflian líc mid menigfealdum springum and geswelle, Homl. Th. i. 122, 22 : 336, 33. Wið uncúðe springas đe on líchoman ácennede beóþ, Lchdm. i. 150, 14. Springas (sprincas, MS. B.), 262, 10. [Cf. *O. H. Ger.* gesprinc *pustula*.] v. fǽr-, wen-, wund-spring. (2) *a flux* :—Wið đæs magan sprige, ii. 190, 16 (where see note) : 192, 12. (3) *a squirting, sprinkling* :—Mon sceal ǽr mid wearmum springum and háte wætre beþian and þweán đa stówe, 202, 21. Mid spryngum, 206, 17 : 208, 14.

springan ; *p.* sprang, *pl.* sprungon ; *pp.* sprungen *To spring.* **I.** *to leap, bound* :—Đæt cild on sprang *the babe leaped in her womb* (Lk. 1, 41), Blickl. Homl. 165, 29. Hrá wíde sprong, syþðan hé drepe þrowade, Beo. Th. 3181; B. 1588. **II.** *to burst forth*, of a fluid *to spirt*, of sparks, etc., *to fly* :—Đæt spere sprang ongeán *the spear-head sprang out again* (*under the pressure of the shield*), Byrht. Th. 135, 53; By. 137. Leád wíde sprong *the drops of boiling lead flew far*, Exon. Th. 277, 24; Jul. 585. Swát ǽdrum sprong *the blood spirted from the veins*, Beo. Th. 5925; B. 2966. Wíde sprungon hildeleóman, 5158; B. 2582. Sprungon spearcan of đam múðe, Shrn. 120, 26. **III.** *to grow* as a plant :—Swá swá of ánum treówe springaþ manega bogas, swá gáþ of ánre lufe manega óðre mihta, Homl. Th. ii. 314, 22. Hig hrædlíce up sprungon, for đam đe hig næfdon đære eorðan dýpan, Mt. Kmbl. 13, 5. **IV.** *to rise* as the sun, cf. spring, II :—Up sprungenre sunnan *sole orto*, Mt. Kmbl. 13, 6. **V.** *to move* as a spring moves :—Þeáh đú teó hweiene boh ofdúne tó đære eorþan, swá đú hine álǽtst, swá sprincþ hé up, Bt. 25 ; Fox 88, 24. **VI.** *to spread, be diffused* :—Đa wíde springaþ *crebrescunt*, Hpt. Gl. 517, 4. Wíde springaþ, wídmérsiaþ, 471, 16. Đes hlísa sprang (spranc, Lind.)

ofer eall ðæt land *exiit fama haec in universam terram illam*, Mt. Kmbl. 9, 26. Sprang ł foerde *processit*, Mk. Skt. Rush. 1, 28 : Beo. Th. 36 ; B. 18 : Apstls. Kmbl. 12 ; Ap. 6. Ða sprang ðæt word *the report spread*, Homl. Th. i. 384, 8 : Ap. Th. 25, 13. Wíde springende *crebrescens*, Hpt. Gl. 519, 37 : 513, 21. [*O. Sax.* springan *to spring* as blood from a wound : *O. Frs.* springa : *O. H. Ger.* springan *to spring* as watet : *Icel.* springa *to burst, crack.*] v. á-, æt-, ge-, geond-, on-, tó-springan.

springd, sprind ; *adj. Active, vigorous :*—Snelne, sprindne *adultum, juvenem*, Hpt. Gl. 485, 26. Geþogenne ł sprindne *adultum, maturum*, 491, 13. Sprindne *adultum*, Anglia xiii. 34, 186. His geðoht is springdra and swiftra ðonne xii. ðúsendu háligra gásta, Salm. Kmbl. p. 150, 34. v. next word.

springdlíce, sprindlíce ; *adv. Actively, vigorously :*—Sprindlíce ł cáflíce *naviter, alacriter, agiliter, velociter*, Hpt. Gl. 405, 22. Fromlíce ł sprin-líce *naviter, velociter, viriliter* ł *fortiter*, 423, 71.

springe. v. æ-springe.

springung (?), e ; *f. Growth :*—Mǽda ł sprinctinge (sprincunge ?) ł grènnessa *prata, viriditates*, Hpt. Gl. 409, 38. v. á-springung.

spring-wyrt, e ; *f. Wild caper, caper-bush, -plant, -spurge ;* Euphorbia lathyris, Lchdm. ii. 104, 2 : 106, 1. [*O. H. Ger.* spring-wurz *actureda, lactaridia ;* springa *actureda, lactarida.*]

sprot, es ; *n. A sprout, shoot, twig, small branch :*—Sprote *with a rod* (?), Coll. Monast. Th. 23, 35. Sprota *sarmentorum, ramorum, qui de vinea exciduntur*, Hpt. Gl. 445, 32 : 489, 10 : *palmitum*, Germ. 401, 16. Sprotum *sarmentis*, 401, 24. [Halliwell gives *sprote*-wood as a word still in use for small wood or sticks for firing. Jamieson gives *sprōt* (1) the withered stump of any plant, broken and lying on the ground ; (2) the end of a branch blown off a growing tree ; (3) a chip of wood, flying from the tool of a carpenter. *O. L. Ger.* gi-sprot *surculum : Du.* sprot *a sprout, twig* (Hexham).] v. sprútan, *and next word.*

sprota, an ; *m.* I. *a sprout, shoot :*—Sprotena *sarmentorum*, Hpt. Gl. 478, 64. II. *a peg :*—Nægl oððe sprota *clavus*, Wrt. Voc. ii. 22, 10. [I ne have stikke, i ne have sprote, Havel. 1142. *O. H. Ger.* sprozzo *rung of a ladder : Ger.* sprosse : *Icel.* sproti *a shoot, twig ; a rod.*] v. preceding word.

sprott, es ; *m. A sprat :*—Ða myclan hwælas and ða lytlan sprottas and eall fisckynn, Anglia viii. 310, 18. [A sprott *hec epimera*, Wrt. Voc. i. 222, col. 2. *Du.* sprot : *Ger.* sprotte.]

-sprungenness. v. á-, on-sprungenness.

sprútan ; *p.* spreát, *pl.* spruton ; *pp.* sproten *To sprout.* [Blosme, þat beo ha eanes fulliche forcoruen, ne spruteð ha neauer eft, H. M. 11, 20. *Egredietur virga de radice iesse* an gerd sal spruten of iesse more, O. E. Homl. ii. 217, 25. In a night sua did it sprute, C. M. 11216. Sproutyū *pululo*, Prompt. Parv. 471. Faine sal he sproutand ai *laetabitur germi-nans*, Ps. 64, 11. *O. Frs.* sprúta ; *pp.* spruten.] v. á-spreótan (*read* -sprútan), geond-spreót.

sprýtan, sprítan (?) ; *p.* te *To sprout, spring* as a plant :—Of ðam blado bealwa gehwilces sprýtan (spryttan ?) ongunnon, Cd. Th. 61, 10 ; Gen. 995. v. spryttan.

sprytele, sprítele (?), an ; *f. A twig, chip.* (v. quotation from Jamieson's Dict. under *sprot*) :—Men of ðære ylcan styde spryttan ácurfon *astulis ex ipsa destina excisis*, Bd. 3, 17 ; S. 544, 43. [Halliwell gives *sprittel* a sprout or twig. Cf. *O. H. Ger.* spruzil : *M. H. Ger.* sprüzzel *rung of a ladder.* Or (?) cf. *M. H. Ger.* sprízel *a splinter.*] v. sprot *and* spreót.

spryttan ; *p.* te I. *intrans. To sprout, spring, germinate :*—Ðonne sprit his gird *germinabit virga ejus*, Num. 17, 5. Up spryt riht-wísnys *orietur justitia*, Ps. Lamb. 71, 7. Tó ðý hé sprytt, ðæt hé mid cwyldum fornyme swá hwæt swá hé ǽr sprytte, Homl. Th. i. 614, 9. Ðonne treówa spryttaþ, ðonne wite gé ðæt hit sumorlǽhþ, 614, 4. Ðonne treów and wyrta ǽrest up spryttaþ, Lchdm. ii. 148, 6 : Met. 29, 68. Up spryttende *pululantes*, Wrt. Voc. ii. 66, 4. Folc weóx swilce hig of eorðan spryttende wǽron *creverunt et quasi germinantes multiplicati sunt*, Ex. 1, 7. Ealle spryttende þingc *universa germinantia*, Hymn. T. P. 76. Eft spryttendum ðám twigum *renascentibus virgultis*, Bd. 1, 21 ; S. 485, 5. [He is þæt he wiði þet sprutted ut þe betere þ me hine ofte cropped, A. R. 86, 15.] II. *trans.* (a) *To put forth* a shoot, *bring forth* fruit :—Seó eorðe spryt hyre wæstmas eów, Homl. Skt. i. 13, 159. Ðes wíngeard sprytte Godes gecorenan, Homl. Th. ii. 74, 4 : i. 614, 10. Spritte seó eorðe grówende gærs *germinet terra herbam virentem*, Gen. 1, 11. God hét ða eorðan spryttan grówende gærs, Hexam. 6 ; Norm. 10, 33. Nǽnne wæstm tó spryttanne, Homl. Th. ii. 90, 18. (b) *to in-cite* (cf. þurh þes (*Ranulf's*) macunge and tóspryttinge se eorl þis land mid unfriðe gesóhte, Chr. 1101 ; Erl. 238, 1) :—Sprytte *instigavit*, Anglia xiii. 36, 245. Ða sprytte se deófol ðæt folc tó his (*Christ's*) slege, Homl. Th. i. 216, 14. Ðæt hé ðisne freóls ǽfre gefyrðrian wolde, and his bearn tó ðam ylcan sprittan wolde, Chart. Th. 116, 22. v. á-spryttan.

sprytting, e ; *f.* (*but pl. in* -as in Ps. Lamb. 79, 12) *A sprig, shoot, sprout, plant :*—Ne biþ spryttinge on wíngeardum *non erit germen in vineis*, Cant. Abac. 17. Spryttinc *incrementum*, spryttincgum ł eácnungum *incrementis, fructibus*, Hpt. Gl. 491, 56-59. Háligre spryttinge *almo*

germine, Hymn. Surt. 76, 3. Sprettinge forð bringende *germen proferens*, 19, 35. Sprittincga *plantaria, plantationes*, Hpt. Gl. 433, 34. Gescóp se ælmihtiga God eorðan and ealle eorðlíce spryttinga, Lchdm. iii. 234, 3. Hé ástrehte óþ flód his spryttingas *extendit usque ad flumen propagines ejus*, Ps. Lamb. 79, 12.

spura, spurnan, spurnere. v. spora, spornan, spornere.

spurul *glosses calcatiosus*, Txts. 110, 1162.

spynge, an ; *f. A sponge :*—Elpendes hýd wile drincan wǽtan gelíce and spynge déþ (*tanquam spongia*), Ors. 5, 7 ; Swt. 230, 27. Hí be-wundon áne spyngan (spingan, MS. B.) mid ysopo, Jn. Skt. 19, 29 MS. A. Spingan, Mk. Skt. 15, 36. Spincgan, Homl. Th. ii. 256, 32. Spync ł spynga, Lind. : spynge, Rush. Mt. Kmbl. 27, 46. v. sponge.

spyrcan, spyrcing. v. spircan, spircing.

spyrd, es ; *m.* The word glosses *stadium* (1) with the meaning *a course :*—Ða ðe in spyrde iornaþ *qui in stadio currunt*, Rtl. 5, 33. (2) with the meaning *a measure of distance :*—Swelce spyrdas fíftêne (spyrdum fíftênum, Lind.) *quasi stadiis quindecim*, Jn. Skt. Rush. 11, 18. Swelce spyrdo fífe and twoegentig *quasi stadia .xxv.*, 6, 19. Ðara spyrda *stadio-rum*, Lk. Skt. Lind. Rush. 24, 13. *In all these passages the West-Saxon uses* furlang. [*Goth.* spaurds (1) *a course ;* (2) *a distance : O. H. Ger.* spurt *stadium.*]

spyre-mann, es ; *m. One who tracks :*—His speremon *the man who tracked him*, Chart. Th. 172, 25. v. spor-wrecel.

spyrian ; *p.* ede, ode I. *to track, go in a track* (v. spor, spyre-mann), *follow, make a journey in search of something :*—Deáð spyraþ (spyreþ, Met. 27, 9) ǽlce dæge æfter fuglum and æfter diórum and æfter monnum, and ne forlǽt nán swæþ, ǽr hé geféhþ ðæt, ðæt hé æfter spyreþ, Bt. 39, 1 ; Fox 210, 28-212, 1. Nyle deáð ǽnig swæð forlǽtan, ǽr hé gehende ðæt hé hwíle ǽr æfter spyrede, Met. 27, 16. Mon mæg giet gesión hiora swæð ac wé him ne cunnon æfter spyrigean *we can still see their track, but we do not know how to follow the track after them*, Past. pref. ; Swt. 5, 16. II. *to make a track, go :*—Mec fugles wyn geond speddropum spyrede geneahhe . . . beámtelge swealg stóp eft on mec síþade sweartlǽst *me (a book) throughout the bird's joy (the pen) with drops made frequent tracks, . . . swallowed the tree's dye (ink), stepped on to me, journeyed with footprints black*, Exon. Th. 408, 7 ; Rä. 27, 8. Syndan onhrêrede anlícast hú druncen hwylc spyrige *as any drunken man makes his way*, Ps. Th. 106, 26. III. *to enquire, investigate, examine :*—Ða cwæþ se wísdóm : 'Hwí . . .?' Ða andswarode ic : 'Genóh ryhte ðú spyrast, swá hit is swá ðú segst, Bt. 26, 2 ; Fox 92, 18. Hí spyredan hwæt and hwonan hé wæs *investigantes unde vel quis esset*, Bd. 1, 33 ; S. 499, 11. Geléfe hé ðæt wit on riht spirien (spyrigen, Cott. MS.) *let him believe that we conduct the enquiry aright*, Bt. 38, 2 ; Fox 198, 27. Uton spirian (spyrian, MSS. G. I.) be bócan, hwæt ða geféran, ða ðe God lufedon, Wulfst. 130, 11. Tó ongann ðíne gerece spyrian georne *ut meditarer eloquia tua*, Ps. Th. 118, 148. Spirian *enucleare*, Hpt. Gl. 498, 16. Spiriende *indagando, inquirendo*, 410, 52 : *scrutando, investigando, meditando*, 479, 20. III a. *with* æfter, (1) *to enquire after* or *into, seek to know about :*—Ðære sprǽce ðe wit æfter spyriaþ *the subject into which we are enquiring*, Bt. 38, 2 ; Fox 198, 26. Se ðe wile wíslíce æfter ðam hlísan spyrian, ðonne ongit hé, hú lytel hé biþ, 18, 1 ; Fox 60, 28. Wé sceoldon eallon mægne spirian æfter Gode, 42 ; Fox 256, 1. (2) *to search after, seek to attain :*—Ealle men spyriaþ æfter ðam hêhstan góde. Ac ne mágon ða yfelan cuman tó ðam hrófe eallra góda, forðam hí ne spyriaþ on riht æfter, 39, 9 ; Fox 224, 24-27. Hwý nyllaþ hí spyrigan æfter cræftum and æfter wísdóme, 36, 6 ; Fox 180, 32. Spirigan, 35, 1 ; Fox 154, 19. Hí ǽfre ne lyst æfter spyrian, sécan ða gesǽlþa, Met. 19, 33. [Speer, speir *to ask* in Scot. and North-E. : *O. H. Ger.* spuren, spurien *investigare, indagare, sciscitari : Icel.* spyrja *to track ; to investigate ; to ask.*] v. á-, ge-, of-spyrian.

spyrigend, spyrgend, spyriend, es ; *m. An enquirer, investigator :*—Speriend *investigator*, Kent. Gl. 384. Godes spyrigendes *of an enquirer after God*, Salm. Kmbl. 281 ; Sal. 140. v. á-spyrigend.

spyrigness. v. á-spyrigness.

spyrigung, spyrgung, spyriung, e ; *f. Enquiry, investigation :*—Spir-iungum ł áxungum *argumentis*, Hpt. Gl. 524, 50. [*O. H. Ger.* spurunga *indagatio, investigatio.*] v. á-spyrgung.

spyrnung, e ; *f. Spurning.* v. æt-, óþ-spyrning.

spyrran, spyrring. v. sperran, sperring.

spyrte, an ; *f. A basket :*—Spyrte *fiscella*, Germ. 400, 492. Spirte *cistula*, Wrt. Voc. i. 288, 33 : ii. 17, 7. Of ðære láfe wǽron gefyllede seofon spyrtan. . . . Spyrte biþ, swá swá gé sylfe witon, of rixum gebroden, oððe of palmtwygum, Homl. Th. ii. 402, 6-9 : 396, 8. Siofun sperta *septem sportas*, Mt. Kmbl. Rush. 15, 37 : 16, 10. [*Lat.* sporta.]

spyttan *to spit :*—Spittas (-es, Lind.) *conspuent*, Mk. Skt. Rush. 10, 34. Spittadun *expuerunt*, Mt. Kmbl. Rush. 26, 67. Spittende *expuentes*, 27, 30. [Blod to spitten ant te speowen, Jul. 48, 18. Þenne spit leccherie meidenhad oþe nebbe, H. M. 17, 13. Spit him amidde þe bearde, A. R. 290, 20. Cf. *Ger.* speutzen, spützen : *Icel.* spýta.] v. ge-spittan.

staca, an ; *m. A stake :*—Nygon fêt of ðam stacan tó ðære mearce, L. Ath. iv. 7 ; Th. i. 226, 12. Ðær his bróðor heáfod stód on stacan

gefæstnod, Homl. Skt. ii. 26, 166. Wrît ðysne circul on ânum mealan stâne and sleah ǽnne stacan on middan ðam ymbhagan, and lege ðone stân on uppan ðam stacan, Lchdm. i. 395, 3-5. Mon hæfde ða burg mid stacum gemearcod, wulfas âtugan ða stacan up, Ors. 5, 5; Swt. 226, 17-19. Âlege ðone man upweard, drîf .ii. stacan æt ðâm eaxlum, Lchdm. ii. 342, 5. ¶ In the following passages there seems to be a reference to the method of witchcraft, that consisted in thrusting a pin or the like into the figure of a person, whom it was desired to injure. On this practice, see, *inter alia*, Brand's Antiquities, ed. Hazlitt, vol. iii. p. 65, Grmm. D. M. 1045, and the Glossary to Thorpe's edition of the Early Laws, s. v. stacung :—Gif hwâ drîfe stacan on ǽnigne man. . . . And gif se man for ðǽre stacunge deád biþ *si quis acus in homine aliquo defixerit.* . . . *Et si homo ex illa punctura mortuus sit,* L. Ecg. P. iv. 17; Th. ii. 208, 26-29 : L. Edg. C. 38; Th. ii. 274, 26-28. (In each case the section occurs amongst regulations dealing with witchcraft.) Ân wyduwe and hire sune drifon îserne stacan on Alsie, Wulfstânes feder . . . Man têh ðæt morð forð of hire inclîfan. Ða nam man ðæt wîf and âdrencte hî æt Lundenebrigce, Chart. Th. 230, 12-19. [*O. Frs.* stac[e].]

stacga (?), an; m. *A stag* :—Regalem feram, quam Angli staggon appellant, L. C. F. 24; Th. i. 429, 5. [Cf. *Icel.* steggi, steggr *a he-bird;* in modern usage also *a tom-cat.*]

stacung, e; f. *Staking, piercing with a stake.* v. *passages under* staca.

stæf, es; m. **I.** *a staff, stick* :—Staeb *olastrum*, Wrt. Voc. ii. 115, 49. Stæf, 63, 41: *baculus*, i. 80, 2 : *fustis*, Ælfc. Gr. 9, 28; Zup. 55, 9. Ðîn gyrd and ðîn stæf (*baculus*) mê âfrêfredon, Ps. Th. 22, 5. Mid gierde mon biþ beswungen, and mid stæfe hê biþ âwreðed. Gif ðǽr ðonne sié gierd mid tô ðreágeanne, sié ðǽr eác stæf mid tô wreðianne, Past. 17; Swt. 126, 2. Gangan bî stafe *to walk with the aid of a staff*, L. Alf. 16; Th. i. 48, 10 : Ex. 21, 19. Mid ylpenbânenon stæfe ða eorðan delfan, Lchdm. i. 244, 24. 'Hafa ðê mînne stæf on handa.' Se drý ðâ nam ðone stæf, Homl. Th. ii. 418, 1-2. Ða cild rîdaþ on heora stafum, and manigfealdne plegan plegiaþ, Bt. 36, 5; Fox 180, 9. Stafas *vectes*, Ps. Lamb. 106, 16. **II.** *a written character, a letter,* the old letters having been carved on staves. Cf. *Germ.* buch-stabe :—*Littera* is stæf on Englisc, and is se lǽsta dǽl on bôcum . . . Wê tôdǽlaþ ða bôc tô cwydum, and syððan ða cwydas tô dǽlum, eft ða dǽlas tô stæfgefêgum, and syððan ða stæfgefêgu tô stafum; ðonne beóþ ða stafas untôdǽledlîce; forðan ðe nân stæf ne byþ nâht, gif hê gǽþ on twâ. Ælc stæf hæfþ þreó ðing, *nomen, figura, potestas,* Ælfc. Gr. 2; Zup. 4, 18-5, 5. S, wuldres stæf, Salm. Kmbl. 225; Sal. 112 : 250; Sal. 124. Ic hâten eom, swâ ða siex stafas sweotule bêcnaþ, Exon. Th. 407, 4; Rä. 25, 10. Âwriten Grêciscum stafum, Lk. Skt. 23, 38. Gemêtte ic sweartum stafum âwritene eall ða mân ðe ic ǽfre gefremede, Bd. 5, 13; S. 633, 8. Oft gehwâ gesihþ fægre stafas âwritene, ðonne heraþ hê ðone writere and ða stafas, and nât hwæt hî mǽnaþ, Homl. Th. i. 186, 1-3 : Lchdm. iii. 290, 13. Ne cûðe hê bôclîce stafas . . . hê nænne stæf ne cûðe, Homl. Th. ii. 96, 24-30. **II a.** *a mark in writing* :—Stafum *apicibus*, Wrt. Voc. ii. 5, 29. **II b.** *a letter* as representing a minute detail :—Ân strica oððe ân stæf ðære ealdan ǽ ne biþ forgǽged, Homl. Th. ii. 200, 1. **III.** in pl. *a collection of written symbols, a letter, writing* :—Hê mê ealle on stafum âwrât, Bd. pref.; S. 472, 3. Ðysne geleáfan hê gýmde gefæstnian sinoþlîcum stafum . . . Ðara stafas is ðes fruma, 4, 17; S. 585, 14-17 : 41. Swâ hwæt swâ hê of godcundum stafum geleornode *whatever he learnt from the sacred writings*, 4, 24; S. 596, 33. Bæd hê ðone Abbud ðæt hê him sende trymmendlîce stafas and gewrito (*exhortatorias litteras*), 5, 21; S. 642, 38 : Chr. 167; Erl. 8, 15. Nim ðîne stafas and wrît hundeahtatig, Lk. Skt. 16, 7. **IV.** *letters, book-learning, literature* :—Bôclîcum stafum *litteris liberalibus*, Hpt. Gl. 503, 55. Hûmeta cann ðes stafas, ðonne hê ne leornode? Jn. Skt. 7, 15. Hê ðâ wæs in stafas and on leornunge getogen, Guthl. 2; Gdwin. 18, 6. [*Goth.* stabs *an element, a rudiment: O. Sax. O. L. Ger.* [bôk-]staf : *O. Frs.* stef : *O. H. Ger.* stap *baculus, virga, regula: Icel.* stafr *a staff, post ; a letter;* in pl. *learning.*] v. âr-, bôc-, candel-, cranc-, di[s]-, ende-, fâcen-, gebregd-, gleó-, gyrn-, hearm-, heg-, inwit-, leád-, pîl-, rûn-, sâr-, sorh-, wrôht-, wyrd-stæf (-stafas); stafa.

stæf-cræft, es; m. **I.** *the art of letters, grammar* :—Ic Ælfric wolde ðâs lytlan bôc âwendan tô Engliscum gereorde of ðam stæfcræfte, ðe is gehâten *grammatica* . . . forðan ðe stæfcræft is seó cǽg ðe ðæra bôca andgit unlîcþ, Ælfc. Gr. pref.; Zup. 2, 13-17. *Gramma* is on Englisc stæf, and *grammatica* is stæfcræft, 50; Zup. 289, 10. *Litteratus* se ðe can stæfcræft, 43; Zup. 257, 7. **II.** *skill in letters* (v. stæf, **IV**), *learning, study* :—Ðeodorus mid hâlgum gewritum and stæfcræftum hî (*the English*) georne hêt beón lǽrende *literarum sanctarum coeperint* (*Angli*) *studiis imbuî*, Bd. 4, 2; S. 565, 12. [Crist sceolde don us mid his mihte þat stefcreft ne mihte, O. E. Homl. i. 235, 35.] v. next two words.

stæfcræftig; adj. *Skilled in letters* :—Stæfcræftigra *grammaticorum*, Hpt. Gl. 410, 69. Stæfcræftira, 473, 16. Stæfcræftigera, 529, 34. Stæfcræftiera *grammaticorum, litteratorum*, 459, 58.

stæf-cyst, e; f. *Excellence in letters* or *learning, book-learning* :—

'Leornodest ðû ǽfre sealmas oþþe ôþre hâlige gewritu?' 'Ic stæfcyste ne leornode ne ðæra manna nânum ne hlyste ðe ða smeádon and rǽddon' '*didst thou ever learn psalms, or other holy writings?*' '*I never learned anything from books, nor have I listened to any of those men that have studied and read them,*' Homl. Skt. ii. 23 b, 593. Cf. stæf-cræft, **II**.

stæf-ford *Stafford* :—Æt Stæfforda, Chr. 913; Th. i. 186, col. 2.

stæfford-scîr *Staffordshire* :—Ðâ fêrdon hî intô Stæffordscîre, Chr. 1016; Erl. 154, 3.

stæf-gefêg, es; n. **I.** *a combination of letters* (a) that forms a syllable :—*Syllaba* is stæfgefêg on ânre orðunge geendod. *A domo fram hûse*, hêr is se *a* for ânum stæfgefêge; *ab homine*; hêr is se *ab* ân stæfgefêg. Hwîlon byþ ðæt stæfgefêg on ânum stæfe, hwîlon on twâm, etc., Ælfc. Gr. 3; Zup. 7, 4-11. Wê tôdǽlaþ . . . ða dǽlas tô stæfgefêgum and syððan ða stæfgefêgu tô stafum, 2; Zup. 5, 1-2. (b) that forms a diphthong :—*Dyptongus* is twyfeald swêg oððe twyfeald stæfgefêg, 4; Zup. 7, 13. *Diptongon*, ðæt ys twyfeald stæfgefêg, Anglia viii. 326, 4, **II.** *a forming of letters in writing* :—Stæfgefêg *literaturam*, Ps. Spl. 70, 17,

-stæf-lǽred. v. ge-stæflǽred.

stæf-leornere, es; m. *A learner of letters, a scholar* :—Stæfleornera *stoicorum*, Hpt. Gl. 479, 64. v. stær-leornere.

stæflîc; adj. **I.** *literal* :—Wê understandaþ ðæt gâstlîce andgit ðæra bôca, and hî rǽdaþ ða stæflîcan gereccednesse, Homl. Th. ii. 114, 35. Hî nellaþ understandan bûtan ðæt steaflîce (stæf-, MSS. C. D.) *the literal meaning*, Homl. Skt. ii. 25, 73. **II.** *literate* :—Stæflecum *liberalitatis* (*literatis?*), Wrt. Voc. ii. 53, 55. [*Icel.* stafligr *pertaining to letters.*]

stæf-liðere, an; f.: -liðera, an; m. *An engine for casting stones, a kind of sling* :—Staeblidrae, steblidrae, staeflidre *ballista*, Txts. 44, 136. Stæflidere, Wrt. Voc. ii. 10, 62 : *fundabulum*, i. 35, 31 : *fundibalum*, 84, 36 : *balista*, Hpt. Gl. 423, 63. Stæflidera *ballista*, 487, 21. Stæfliðera[n] *fundibulo*, 521, 12.

stæfn, stæfnan. v. stefn, stefnan.

stæf-plega, an; m. *A letter-game* or *a literary game* :—Staebplegan, staefplagan *ludi litterari, ludi litterali*, Txts. 72, 577. Stæfplegan, Wrt. Voc. ii. 51, 17.

stæf-rǽw, e; f. *A letter-row, an alphabet* :—Mid stæfrǽwe endebyrdnesse tôsceádene *alphabeti ordine distinctum*, Bd. 5, 24; S. 648, 40. [Cf. *Icel.* staf-róf *an alphabet.*]

stæf-róf glosses *elimentum*, Wrt. Voc. ii. 32, 24.

stæf-sweord, es; n. *A sword-stick* :—Stæfsweord *dolones*, Wrt. Voc. i. 35, 55. [*O. H. Ger.* stapa-swert *framea*.]

stæf-wîs; adj. *Skilled in letters, literate* :—Gelǽred, stefwîs, Lchdm. iii. 186, 24. v. un-stæfwîs.

stæf-writere, es; m. *A writer about letters* or *grammar* :—Stæfwrîterum *grammaticorum*, Wrt. Voc. ii. 41, 33 : 75, 40. The word glosses *historiographus*, 42, 45, but perhaps *stær-* should be read for *stæf-* : and 18, 67 stæfwrîterum glosses *caracteribus*, which seems an error.

stæg; es; n. *A stay, a rope supporting a mast* :—Stæg *safo* (in a list of nautical words), Wrt. Voc. i. 63, 60. [*Icel.* stag; n. *a stay: Dan.* stag. Cf. *O. French* estay (from German).] v. stæp.

stægel; adj. *Steep, abrupt* :—Staegilrae, stêgelræ, staegilre *praerupta*, Txts. 84, 747. Heánne beám stǽlgne (= stǽglne?) gestîgan, Exon. Th. 42, 27; Cri. 679. [Jamieson gives *stell* steep. *O. H. Ger.* steigal *abruptus. O. C. L. Ger.* stêgil *crepido.*]

stæger, e; f. *A staircase* :—Stæger *ascensorium*, Wrt. Voc. i. 26, 37. Hê âstâh up tô ðære stægre ðe stôd wið ðæs câseres botl, Homl. Skt. i. 5, 438. Hê feóll of ânre stægere, 18, 232. [On þe steire of fiftene stoples, O. E. Homl. ii. 165, 34. Þolemodnesse haueð þreo steiren, A. R. 282, 7.]

stæger; adj. *Steep.* [Þise twelve degres wern brode & stayre, A. P. 31, 1021. A cliffe so staire and so stepe, ib. 196, col. 1.] v. wiðer-stæger; stægel.

stæl, es; n. **I.** *a place* :—Stalu tô fuglum *umbrellas*, Txts. 107, 2153. **II.** *place, stead* :—Cristenum cyninge gebyraþ ðæt hê sý on fæder stæle cristenre þeóde, L. I. P. 2; Th. ii. 304, 23 : Beo. Th. 2963; B. 1479. Ic eom gesceádwîsnes and ic eom ǽlcum manniscum môde on ðam stale ðe seó hâwung byþ ðam eágum, Shrn. 178, 9. Gê beóþ mê talade on bearna stæl, Exon. Th. 366, 13; Reb. 11 : Cd. Th. 68, 7; Gen. 1113. **III.** *stead* (as in the phrase to stand a person in good *stead*. Cf. stælwirðe) :—Hié ðæm âdrǽfdan on nânum stale beón ne mehton *they could not be of any assistance to the exile*, Ors. 5, 9; Swt. 232, 23. **IV.** *situation, condition* :—Mê lyste witan be ðam gewitte, hweðer hyt æfter ðæs lîchaman gedâle and ðære sâwle weóxe ðe wanede, ðe hyt swâ on stæle stôde, ðe hyt swâ dyde, swâ hyt ǽr dæd on ðisse weorulde, ôðre hwîle weóxe ôðre hwîle wanode (cf. 200, 17-19), Shrn. 199, 26-30. v. æt-, on-stæl; stæl.

stælan; p. de *To* impute *a crime to* (*on, ongeán*) *a person, to charge, declare* something against a person :—Ic ðê þreáge and stæle beforan ðê and ðê cýðe eal ðâs yflu *arguam te, et statuam contra faciem tuam*, Ps. Th. 49, 23. Se deófol ða syndæda stæleþ on ða gâstas *the devil charges*

the spirits with their sinful deeds, Wulfst. 256, 7 : Exon. Th. 84, 16 ; Cri. 1374. Stǽleþ fǽhðe *declares enmity,* Cd. Th. 305, 2 ; Sat. 640. Hē būtan leahtrum wæs clǽne gemēted ðara ðinga ðe hine mon fore-wrēgde and on stǽlde *absque crimine accusatus fuisse inventus est,* Bd. 5, 19 ; S. 639, 31. Wē ðec sōð on stældun *we brought a true charge against thee,* Exon. Th. 130, 17 ; Gū. 439. Wið mē ārison léase gewitan and stǽldon on mē ðæt íc nāwþer ne nyste ne ne worhte *exurgentes testes iniqui quae ignorabam interrogabant me,* Ps. Th. 34, 12. Ic wolde an-dettan and stǽlan ongeán mē sylfne mīne scylda *pronuntiabo adversum me injustitias meas,* 31, 6. Synne stǽlan, Menol. Fox 569 ; Gn. C. 54. Fǽhðe ic wille on weras stǽlan (*of the threatened deluge*), Cd. Th. 81, 27 ; Gen. 1352. Ic gefrægn mǽg ōðerne billes ecgum on bonan stǽlan *I heard that one kinsman with the edge of the sword brought home to the slayer the death of the other* (? Eofor killed Ongentheow, who had slain his brother), Beo. Th. 4964 ; B. 2485. v. be-, ge-, ofer-stǽlan ; -stǽl.

stǽl-giest, es ; *m.* *A thievish guest* (of an insect eating a book) :—Þeóf in þýstro . . . stælgiest ne wæs wihte ðý gleáwra ðe hē ðám wordum swealg, Exon. Th. 432, 13 ; Rä. 48, 5.

stǽlgne. v. stǽgel.

stǽl-here ; *g.* -her(i)ges ; *m.* *A marauding band, predatory army* :—Hié fōron ūt mid stælherge nihtes . . . and genōmon unlytel ǽgðer ge on mannum ge on ierfe, Chr. 921 ; Erl. 106, 13. Drehton ða hergas West-Seaxna lond mid stælhergum, 897 ; Erl. 95, 9. Ðæt hié ða burga hira mōdes wið stælherigas behealden, Past. 33 ; Swt. 229, 5.

stǽl-hrán, es ; *m.* *A decoy-reindeer* :—Ða deór hí hātaþ hránas ; ðara wǽron syx stælhránas ; ða beóþ swýðe dýre mid Finnum, for ðæm hý fōþ ða wildan hránas mid, Ors. 1, 1 ; Swt. 18, 11.

stǽli *steel,* stǽllan *to put in a stall,* stǽllo. v. stēle, ge-stǽllan, steall.

stǽl-tihtle, an ; *f.* *A charge of theft* :—Be stǽltyhtlan (staltihtlan, MS. B.). Ðonne mon monnan betýhþ ðæt hē ceáp forstele, L. In. 46 ; Th. i. 130, 11 : L. O. D. 4 ; Th. i. 354, 14. Gif hwā þurh stæltihtlan freót forwyrce, L. Ed. 9 ; Th. i. 164, 10.

stǽl-wirðe ; *adj.* *Able to stand a person in good stead* (v. stæl, II), *serviceable* :—Se ðe geornlíce conn ongietan ðæt hē gadrige ðæt him stælwierðe sié *qui sollicite noverit sumere, quod adjuvat,* Past. 17, 5 ; Swt. 115, 3. Ða scipu ðe stælwyrðe wǽron binnan Lundenbyrig gebrohton *the ships that could be of service they brought into London,* Chr. 896 ; Erl. 94, 19. Hē gyfþ gooda gifa on ðissa wurlda ; þeáh hí éca ne sién, hí beóþ þeáh stælwyrða ða hwíle ðe wē on ðisse wurlde beóþ, Shrn. 192, 6. [In later English the word seems used more in the sense of the modern *stalwart*=strong :—Ic em hal and fere and strong and stelewurðe, ȝet ic mei longe libben, O. E. Homl. i. 25, 12. Þeo þat beoð stalewurðe and warpeð mid strencðe ut of hare heorte hare unwreste wil, Jul. 44, 7. Þeo þ stalewurðe beoð ant starke to ȝein me, Marh. 15, 32. Þou hart on stale-worþe (hende, 1st MS.) gome, Laym. 3812. Gurguont, stalworþe mon and hardy, R. Glouc. 39, 4. A man þat es yhung and light, Be he never swa stalworth and wyght, Pr. C. 689. Cf. stanndenn stallwurrþlig ȝæn þe deofless wille, Orm. 1194. Louerd mi stalwurnesse (stalworthhede, other MSS.) *Domine, virtus mea,* Ps. 17, 2.] Cf. nyt-wirðe.

stǽl-wyrt, e ; *f.* *Water starwort* :—Stǽlwyrt *callitriche* (cf. wæter-wyrt *callitriche,* 67, 18), Wrt. Voc. i. 68, 15.

stǽna (or -e ; *f.*), an ; *m.* *A stean, a pot of stone* or *earth* :—Stǽnan *gillone* (gillo *lagena, vas vinarium*), Wrt. Voc. ii. 42, 3. [Sete adun þine stene (*waterpot,* Jn. 4, 28), Misc. 85, 29. Stene (*cruse,* 1 Kings 17, 12), Wick. Into a stene lette hem be pressed, Pall. 4, 666. See Halliwell's Dict., and Spenser's F. Q. vii, stanza 42 : Upon an huge great earth-pot steane he stood. *O. H. Ger.* steinna *olla, cacabus.*]

stǽnan ; *p.* de   I. *to stone, cast stones at* :—Ðū stǽnæst (stǽnas, Lind.) ða ðe tō ðē sende wǽran, Mt. Kmbl. Rush. 23, 37. Ne stǽnas uē ðec *non lapidamus te,* Jn. Skt. Lind. 10, 32. Heó wæs stǽned oþ ðæt heó hire gāst onsænde ; ða com þunerrád and ofslōh ðone mǽstan dǽl ðæs folces ðe hí stǽnde, Shrn. 57, 34-36. Hig hine stǽndon, Jos. 7, 25. Stǽne hine man mid stānum, Lev. 20, 2. Ðā hē se dēma hine stǽnan, Shrn. 48, 28. Tō stǽnenna, Mt. Skt. Rush. 11, 8. Hí hine gelǽddon tō stǽnenne, Homl. Th. i. 46, 35. Hē for ðǽm stǽnendum gebǽd, 52, 19. Hý wǽron stǽned, and ða stānas wǽron on bæc gecyrred, Shrn. 135, 27. [Goth. stainjan : O. H. Ger. steinōn.] v. ge-, of-stǽnan.   II. *to adorn with* (*precious*) *stones.* [O. H. Ger. gi-steinen.] v. ā-stǽned.

stǽnen (in the oblique cases the -en is sometimes contracted or absorbed ; see below, and for other instances see under stapol) ; *adj.*   I. *stony.* v. next word :—Se āfeól of his horse ofer stǽnene eorþan, and him wǽron ða limo gecnyssed, Shrn. 126, 18. Of sandigum ꝉ stǽnenum *de arenosis,* Hpt. Gl. 449, 26.   II. *metaph. of stone, stony, hard as stone,* (1) *in a good sense* :—Ic ðē secge, ðæt ðū (*Peter*) eart stǽnen, and ofer ðysne stān ic timbrige mīne cyrcan, Homl. Th. i. 364, 23. (2) *in a bad sense* :—Hié wǽron stǽnenre heortan' and blindre, Blickl. Homl. 105, 27. Hí hæfdon stǽnene heardnysse on heora heortan, Homl. Th. ii. 236, 21. Hæfdon heortan stǽn[e]ne, Exon. Th. 40, 20 ; Cri. 641.   III. *stone, made of stone, built of stone* :—Stǽnen elefæt *alabastrum,* Wrt. Voc. i. 24, 40. Stǽnen cyrice *ecclesia de lapide facta,* Bd. 3, 23 ; S. 555, 12. Stǽnen bedd, Shrn. 69, 4. Ðæt stǽnna

fæt *alabastrum,* Mk. Skt. Rush. Lind. 14, 3. Be ðǽre stǽnenan strǽte *the paved way,* Blickl. Homl. 189, 13. Stǽnen weofod *altare lapideum,* Ex. 20, 25. Weall stǽnenne, Cd. Th. 101, 33 ; Gen. 1691. Wíf hæb-bende stǽnna (stǽna, Rush.) fulle smirinisse *mulier habens alabastrum unguenti,* Mt. Kmbl. Lind. 26, 7. Ða stǽnenan bredu *the tables of stone,* Past. 17 ; Swt. 125, 18 : Ex. 31, 18. Stǽnene (stǽnine, Lind.) wæter-fatu, Jn. Skt. 2, 6. Geond ealle ðás strǽt and stǽnene wegas, Homl. Skt. i. 14, 156. [Goth. staineins : O. Frs. stēnen : O. H. Ger. steinîn.]

stǽner (? v. stǽnen, I) *stony ground* :—In stǽrer (stǽnen ?) *in petrosa,* Mt. Kmbl. Lind. 13, 5. Stǽner, 20. Ofer stǽnere *super petrosa,* Mt. Skt. Rush. 4, 5, 16. Stǽnero, Lind. 4, 16.

stæng, stǽnig. v. steng, stānig.

stǽniglíc ; *adj. Stony* :—On stǽnilícum stōwum, Lchdm. i. 216, 20. **stǽniht.** v. stāniht.

stǽning, e ; *f.*   I. *stoning, casting of stones* :—Saulus heora mōd tō ðære stǽninge geornlíce tihte, Homl. Th. i. 50, 30 : ii. 236, 29 : Shrn. 32, 1.   II. *ornamenting with stones.* v. bleó-stǽning.

stǽpe, stepe, an ; *pl.* stæpas, stapas, stæpe ; *m.*   I. *a step, pace* (lit. and fig.) :—Stǽpe, stepe *passus,* Ælfc. Gr. 11 ; Zup. 79, 8. Ne mágon becuman ða stæpas ðæs weorces ðieder ðe hí wilnaþ, Past. 11 ; Swt. 65, 17. Ágotene synt mīne stapas (stǽpas, Spl.), Ps. Lamb. 72, 2. Ǽlc ðæra stæpa and fótlǽsta ðe wē tō cyricean weard gestæppaþ, Wulfst. 302, 26. Mid heora þeáwa stæpum Drihtne filiaþ, Homl. Th. i. 120, 28. Se ðe beforan ðǽm stapum his weorca ne lōcaþ, Past. 39 ; Swt. 287, 18. His weg and his stæpas tō sceáwianne, 18 ; Swt. 131, 21. Geriht mīne stapas on ðīne wegas, Ps. Th. 16, 5. Stapas, Ps. Lamb. 84, 14 : 118, 133 : Wulfst. 247, 2. Gelǽd mē on stige ðǽr ic stæpe mīne on ðīnum bebodum brýce hæbbe *deduc me in semitam mandatorum tuorum,* Ps. Th. 118, 33.   I a. *a step, pace* as a measure of distance :—Stǽpe *passus,* furlang *stadium,* Wrt. Voc. i. 38, 8. Nis ān stæpe ðæt seó eá wille ofer-yrnan, Wulfst. 211, 14. Ne gang ðū, mōna, ānne stæpe furðor, Jos. 10, 12. Swā hwā swā ðe genýt þūsend stapa, Mt. Kmbl. 5, 41.   II. *stepping, going* :—Germanus ðam healtan geongan his stǽpe geednīwode and ðam Godes folce geednīwode ðone stæpe rihtes geleáfan *Germanus claudo juveni incessum et populo Dei gressum recuperarti fidei,* Bd. i. 21 ; S. 485, 5-9. Strong on stæpe, Exon. Th. 498, 23 ; Rä. 88, 6.   III. *a step, that on which the foot may be placed* :—Ðā āstāh Isachar up on ðone ýtemestan stæpe *the topmost of the steps leading to the temple,* Homl. Ass. 129, 431. Stapas *vel* stīrápas *scansilia,* Wrt. Voc. i. 41, 34. On ðǽre hlǽddra is twā and sixti stapa, Anglia xi. 5, 22. Stæpena, 4, 11. Ne gā ðū on stapum tō mīnum weofode, Ex. 20, 26. Hē stīhþ be ðǽre hlǽddre stapum, Homl. Skt. i. 1, 22.   III a. *that on which the lower part of any thing rests, the step* of a mast, *a pedestal* :—Stepe *basis,* Wrt. Voc. ii. 12, 50. Hig fæstniaþ ðone stepe þurh ða þilinge, Shrn. 35, 14. Tredelas *vel* stæpas *bases,* Wrt. Voc. i. 21, 48. Hearpan stapas *cerimingius* (? v. stalu), Wrt. Voc. ii. 130, 40.   IV. *a degree* :—Hād oððe stæpe (stepe) *gradus,* Ælfc. Gr. 11 ; Zup. 79, 9. *Positivus* is se forma stæpe (stepe), *comparativus* is se ōðer stæpe (stepe), *superlativus* is se ðridda stæpe (stepe), 5 ; Zup. 15, 20. Synd þrý stæpas gecorenra manna. Se nýðemysta stæpe . . . Se ōðer stæpe is on wydewan hāde . . . Se hēhsta stæpe is on mægðhādes mannum, Homl. Th. ii. 70, 17-23 : 94, 15. Be ðam twelf stæpum eáðmōdnesse. Ðære forman eáðmōd-nysse stæpe is, R. Ben. 23, 16. Seofon stapas sindon hāligra hāda . . . Ðone forman stæpe bēte man mid āne punde . . . Æt ðam ōðrum stæpe twā pund tō bōte . . . Æt ðam þriddan stæpe, etc., L. E. B. 1-8 ; Th. ii. 240, 242. [O. Frs. stape : O. H. Ger. stapfo *passus, gradus, incessus, vestigium.*] v. in-, on-stæpe ; ord-stapu (*read* -stæpe) : in-stæpe, -stæpes.

stǽpe-gang, es ; *m. A step* :—Ic stepegongum weóld *I had control of my steps,* Exon. Th. 353, 34 ; Reim. 22.

stǽp-mǽlum, *adv.*   I. *step by step* :—Wæs gesewen micel cyrce tō ðǽre hí stæpmǽlum āstigon (cf. Blickl. Homl. 207, 11), Homl. Th. i. 508, 12.   II. *step by step* (fig.), *gradually, by degrees* :—Stæpmǽlum *gradatim, per singulos gradus,* Hpt. Gl. 497, 54 : Scint. 101, 13. Ðæt mōd glít niðor and niðor stæpmǽlum, Past. 38 ; Swt. 279, 3. Suǽ suǽ on sume hlǽdre, stæpmǽlum, proem. ; Swt. 23, 17 : Shrn. 188, 12. [O. H. Ger. stapf-mālum *gradatim.*]

stǽppa (or -e), an ; *m.* (or *f.*) *A step* :—Þūsend stǽppan *mille passus,* Mt. Kmbl. Rush. 5, 41.

stǽppan, steppan ; *p.* stōp ; *pp.* stapen *To step, go, proceed* :—Ic stæppe *gradior,* Ælfc. Gr. 29 ; Zup. 185, 18. Gange se wífman tō birgenne, and stæppe ofer ða byrgenne . . . Ðonne heó tō hyre hlāforde on reste gā, ðonne cweþe heó : ' Up ic gange, ofer ðē stæppe,' Lchdm. iii. 66, 18-26. Ic steppe on grēnne græs, Exon. Th. 396, 16 ; Rä. 16, 5. Ðonne stæpþ se sacerd tǽlleaslíce on ðone weg *tunc sacerdos irreprehensi-biliter graditur,* Past. 13, 1 ; Swt. 77, 18 : Homl. Th. i. 374, 21. Hē stæpþ beforan ðison folce *praecedet populum istum,* Deut. 3, 28. Stepeþ, Exon. Th. 264, 34 ; Jul. 374. Steppeþ, 499, 33 ; Rä. 88, 25. Rūmaþ, steppaþ *cedunt,* Wrt. Voc. ii. 19, 19 ; 87, 64. Stōp forð (*prodiit*) se deád wæs, Jn. Skt. 11, 44. Deáð neálǽcte, stōp stalgongum, sōhte sāwelhūs, Exon. Th. 170, 17 ; Gū. 1113. Se cyning stōp tōforan ðam biscope, Bd. 3, 14 ; S. 540, 36. Seó wífman stōp inn *ingressa,* Jud. 4, 21 : Cd

Th. 69, 16; Gen. 1136. Se apostol stóp intó đære byrig, Homl. Th. i. 60, 11: Ryrht. Th. 134, 3; By. 78. Hé wið đæs beornes stóp, 135, 41; By. 131. Hié stópon tó đam gysterne, Judth. Thw. 21, 29; Jud. 39: 24, 36; Jud. 227: Cd. Th. 95, 26; Gen. 1584. Stæppaþ ryhte, ne healtigeaþ leng, Past. 11; Swt. 65, 18. Ðýlæs hé ofer đone đerscold stæppe, 13; Swt. 77, 22: Lchdm. ii. 124, 6. Đæt hié stæppen on ryhtne weg, Past. 18; Swt. 131, 25. Đær ic stæppan scyle, Ps. Th. 16, 5: Cd. Th. 86, 22; Gen. 1434: Wulfst. 303, 10: Homl. Th. i. 118, 32. Steppan, Ps. Th. 31, 9: Wulfst. 239, 11: Cd. Th. 88, 2; Gen. 1459: 279, 35; Sat. 248. Com stæppende sum cempa, Homl. Th. i. 452, 14. [O. Sax. stóp, p.: O. Frs. steppa; p. stóp; pp. stapen: cf. O. H. Ger. stepfen, stapfôn.] v. æt-, be-, for-, fore-, forþ-, ge-, in-, of-, ofer-, on-, wið-stæppan (-steppan, -stapan. In the compounds, instead of stapan read stæppan).

stæppend. v. fore-stæppend.

stæppe-scóh a slipper :—Stæppescôs subtalaris, Wrt. Voc. i. 289, 8. Steppescôh, ii. 121, 73.

stæppung. v. fore-stæppung.

stær, es; m. A starling, a stare (the latter is the name used in some dialects. v. E. D. S. Pub. Bird Names, and Halliwell's Dict.) :—Staer sturnus, Wrt. Voc. ii. 121, 17. Stær, i. 63, 6: turdus, 77, 30. Stær turdus, se mære stær turdella, 29, 40, 41. Etan gebrǽdne stær, Lchdm. ii. 320, 4. Staras l hrondsparuas passeres, Mt. Kmbl. Lind. 10, 29. Staras (stearas, Rush.), Lk. Skt. Lind. 12, 6. [O. H. Ger. stara sturnus, turdus: Ger. staar: Icel. stari: Dan. stær a starling: Swed. stare.]

stǽr, stěr, steór, es; n. A history; historia :—Tó eallum đe đis ylce stǽr becyme úres cynnes tó rǽdanne omnes ad quos haec eadem historia pervenire poterit nostrae nationis legentes, Bd. pref.; S. 472, 33. Đæt getæl đæs hálgan stǽres and spelles ... Song hé eall đæt stǽr Genesis illum seriem sacrae historiae ... Canebat de tota Genesis historia, 4, 24; S. 598, 5–10. On đyssum úrum stǽre, 4, 23; S. 609, 33. Be stære Angelþeódes cyricean, 5, 24; S. 647, 16: 4, 22; S. 592, 31. On Ongelcynnes steóre, đæt is, on historia Anglorum, Shrn. 87, 7. Đara Abbuda stǽr and spell đysses mynstres on twám bócum ic áwrát, Bd. 5, 24; S. 648, 28. On đis úre cyriclíce stěr, 4, 7; S. 574, 28.

stær-blind; adj. Blind from giddiness, purblind, quite blind :—Stæ[r]-blind scotomaticus (cf. scotomaticorum, cecorum, 78, 20), Wrt. Voc. ii. 119, 81. Næfþ nán man tó đæs unhále æágan, đæt hé ne mǽge lybban be đare sunnan and hire nyttian, gyf hé ényg wiht geseón mæg, búton hé stareblind sí, Shrn. 187, 5. Sume unæáđe áwiht geseóþ; sume beóþ stæreblinde and nyttiaþ þeáh đare sunnan, 27. Stærbli[nde] scotomaticus, Hpt. Gl. 478, 20. [Bi daie þu (the owl) art stareblind, O. and N. 241. O. Frs. staru-, stare-, star-blind: O. H. Ger. stara-plint: Ger. staar-blind suffering from cataract: Dan. stær-blind purblind: Swed. starr-blind quite blind: cf. Icel. star-blinda blindness.]

stærced-, sterced-ferhþ; adj. I. having the mind strengthened, stouthearted, courageous :—Deaređlácende (the Huns) on Danúbie stærcedfyrhđe stæđe wícedon, Elen. Kmbl. 75; El. 38. Stercedferhþe hæleþ, Judth. Thw. 22, 9; Jud. 55. II. of hard or cruel mind :—Drôgon hine (St. Andrew) ymb stánhleođo stærcedferđe cruelhearted ones dragged him about the stony slopes, Andr. Kmbl. 2468; An. 1235.

stǽr-leornere (?), es; m. One who learns history, a historical scholar :—Stǽrleornera (? stæf-, v. stæf-leornere), leornera stoicorum (storicorum?), Hpt. Gl. 503, 64.

stǽrn. v. stearn.

stǽr-trahtere, -tractere, es; m. One who treats of history :—Stǽrtractere commentarius, Wrt. Voc. ii. 132, 42.

stǽr-writere, es; m. A writer of history, a historian :—Swá swá sôþsagol stǽrwrítere (verax historicus) đa þing đe be hnin oþþe đurh hine gewordene wǽron ic áwrát, Bd. 3, 17; S. 545, 5. Stæfwrítere (stǽr-?) historiographus, Wrt. Voc. ii. 42, 45. Swá swá Trogus and Iustinianus sédon heora stǽrwríteras, Ors. 4, 4; Swt. 164, 12: 2, 1; Swt. 60, 25.

stæþ (?) a stay; safon (in a list of nautical words. In a similar list stæg, q. v., occurs as the gloss), Wrt. Voc. i. 56, 63. Cf. stæđđan, stæđđig.

stæþ, es; n. A bank, shore, the land bordering on water :—Stǽđ ripa, Wrt. Voc. i. 54, 19. Stæđ vel brerd labram, margo, vel crepido, 57, 25. Of đæm mere đe Truso standeþ in stađe, Ors. 1, 1; Swt. 20, 9. On đam staþe đe is genemned Ypwines fleót, 449; Erl. 12, 2. Treówlícre hit is be stađe tó [swim]manne, đonne út on sǽ tó seglanne, Prov. Kmbl. 64. On geofenes stađe, Cd. Th. 215, 8; Exod. 582: Exon. Th. 361, 11; Wal. 18. On Sǽferne staþe, Chr. 894; Erl. 92, 23. Hé befeól ofer đam stæđe intó đam streáme, Homl. Th. ii. 160, 5: Elen. Kmbl. 76; El. 38. Of stæđe on óđer from one bank of the boundary stream to the other, L. O. D. 1; Th. i. 352, 4, 11: 2; Th. i. 354, 3: 6; Th. i. 354, 25. Sume cuce tó đam stæđe cômon, and đa man sôna ofslôh æt đare eá múđan, Chr. 794; Erl. 59, 23: Byrht. Th. 132, 32; By. 32. Æt Wendelsǽ on stæđe, Elen. Kmbl. 463; El. 232. Stæđ marginem, Hpt. Gl. 492, 72. Be wætera stađum, Ps. Th. 22, 2. Stæđum marginis, Wrt. Voc. ii. 58, 25. Betweox stæđum between those living on the two sides of the boundary stream, L. O. D. 2; Th. i. 352, 16. Oft stille wæter stađo brecaþ, Prov. Kmbl. 63. Streámas staþu beátaþ, Exon. Th.

382, 4; Rä. 3, 6: Met. 6, 15. Staþu ástígan geswinc getácnaþ. Of staþe niþer stígan gôdne tíman getácnaþ, Lchdm. iii. 210, 16. ¶ In the following passage the word seems to be masculine :—Wægn brohte beornas ofer burnan from stæde heáum, đæt hý stôpan up on óđerne of wǽge, Exon. Th. 405, 6; Rä. 23, 19. [Uppen Seuarne staþe, Laym. 7. Stathe a wharf, Halliwell's Dict. Goth. staths a shore: O. Sax. stað a bank, shore: O. H. Ger. stad, stado ripa, litus, margo.] v. bord-, eá-, streám-, súþ-, wǽg-stæþ.

stæþ-fæst; adj. Firm on the shore (? epithet of sea-cliffs), stable :—Heáhcleofu stíđ and stæđfæst, staþelas wið wǽge, Exon. Th. 61, 7; Cri. 981.

stæþ-hlípe; adj. Running to the shore (?), steeply sloping, precipitous :—Stæþhlépe divexum, i. inclinatum, pronum, Wrt. Voc. ii. 141, 52. Hí ne mihton ofer đæt scræf swá swæđhlýpe (stæþ- ?) [wæs] đér hí gongan [sceoldon] ǽr đon hié gerýmdon đone upgang and geworhtan they could not pass the cave, so steep was it where they had to go, before they had cleared and constructed the ascent, Blickl. Homl. 201, 16. v. next word.

stæþhlíplíce; adv. At a steep inclination :—Wæs đæt hús ... on scræfes onlícnesse ...; and gelômlíce đa stánas swá of óđrum clife stæđhlýplíce út sceoredon, Blickl. Homl. 207, 20. v. preceding word.

stæđig. v. stæđđig.

stæþ-swealwe, an; f. A sand-martin :—Staeđsuualwe ripariolus, Wrt. Voc. ii. 119, 22. Gif mon fundige wið his feónd tó gefeohtanne, stæþswealwan briddas geseóþe on wíne, ete đonne ǽr, Lchdm. ii. 154, 5.

stæđđan to make staid, to stay :—Saga hwá mec rére, đonne ic restan ne môt, oþþe hwá mec stæđþe, đonne ic stille beóm, Exon. Th. 387, 4; Rä. 4, 74.

stæđđig; adj. Staid, sober, sedate, grave :—Đæt cild Cúđberhtes dyslícan plegan mid stæđđigum wordum þreáde, Homl. Th. ii. 134, 7. v. ge-, un-stæđđig, and next word.

stæđđignys, e; f. Staidness, sedateness, gravity, seriousness :—Đǽr is stæđignyss ióguđe, Wulfst. 265, 8. Môderlícere stæđđinysse materna gravitate, Hpt. Gl. 469, 37. Hé on heálícere stæđđignysse symle þurhwunode he ever continued deeply serious, Homl. Th. ii. 134, 22. Gif wé đa ungesceádwíslícan styrunga on stæđđignysse áwendaþ, 210, 31. Fore stilnesse stæđđinesse propter taciturnitatis gravitatem, R. Ben. Inter. 26, 6.

stæþ-weall, es; m. The wall formed by the shore :—Sǽs up stigon ofer stæđweallas, Cd. Th. 83, 7; Gen. 1376.

stæþ-wyrt, e; f. A plant name, Cockayne suggests statice, Lchdm. ii. 78, 3.

stafa (?), an; m. A letter :—Stafana litterarum, Hpt. Gl. 460, 54. v. stæf.

stafian; p. ode To direct, dictate :—Se geréfa đone áđ him swôr swá hé hyne sylf stafode the steward swore the oath to Abraham, as Abraham himself dictated it, Gen. 24, 9. Abraham đurh wítegunge stafode đone áđ, Homl. Th. ii. 234, 34. [O. H. Ger. stabén dirigere: Icel. stafa eið to dictate an oath to a person.]

staggon. v. stacga.

-stál. v. ge-, on-, wiđer-stál; stælan.

stala one who steals. v. ge-stala.

stál-ærn, es; n. A place where charges are heard (? v. stǽlan. Or stál = staþel; cf. stálian = staþelian) :—Stálern consistorium, Wrt. Voc. ii. 133, 70.

stalaþ(-eþ), stalcung, staled. v. staþel, stealcung, reád-staled.

stal-gang, es; m. A stealthy step :—Deáđ neálǽcte, stôp stalgongum, sôhte sǽwelhús, Exon. Th. 170, 17; Gú. 1113. v. stalian, II.

stalian; p. ode I. to steal :—Se đe stalaþ on Sunnanniht, L. Alf. pol. 5; Th. i. 64, 22. Đæra þeófa đe staledon, L. Ath. i. 3; Th. i. 200, 24. Ne stala đú, L. Alf. 7; Th. i. 44, 19: Homl. Th. ii. 208, 24. Be stale. Gif hwá stalie (stalige) ... Gif hé stalie (stalige) on gewitnesse ealles his hírédes, gongen hié ealle on þeówot, L. In. 7; Th. i. 106, 14–17: 22; Th. i. 116, 9–10. Be þeófum. Gif þeóf ofer đæt stalige, L. Ath. i. 1; Th. i. 198, 25. II. to proceed stealthily, steal upon a person :—Hé oftrǽdlíce on Rômane stalade Marianum exercitum creberrimis incursionibus fatigavit, Ors. 5, 7; Swt. 230, 9. Læcedemonie hæfdon máran unstillnessa đonne hié mægenes hæfden and hlôđum on hié (the Thebans) staledon Lacedaemonii, inquieti magis quam strenui, tentant furta bellorum, Ors. 3, 1; Swt. 100, 2. v. for-, ge-stalian.

stálian (= staþelian) to confirm :—Stálige (staþelige, L. I. P. 4; Th. ii. 308, 3) man and strangie and trymme hí georne mid wíslícre Godes lage, Wulfst. 267, 21. Cf. (?) efenstáledan conficiebantur, Wrt. Voc. ii. 133, 31.

stalla, stallere, stal-tihtle. v. stealla, steallere, stæl-tihtle.

stalu, e; f. I. theft, stealing :—Stalu ne lufaþ náne yldinge stealing loves not any delay, Homl. Th. i. 220, 9. Be stale. Gif hwá stalie, L. In. 7; Th. i. 106, 14. Gif hwá Godes cyricean brece for stale, L. Ecg. P. iv. 24; Th. ii. 210, 30: Blickl. Homl. 75, 31. Sum wer wæs betogen đæt hé wǽre on stale, Homl. Skt. i. 21, 265. Se đe cyricean ǽhte mid stale áfyrde, Bd. 2, 5; S. 506, 30. Sume stale fremmaþ, 1, 27; S. 490 9, 5. Of đære heortan cumaþ stale (stala, MS. A.), Mt. Kmbl. 15, 19: Mk. Skt. 7, 22. Đa heáfodleahtras sind ... leásgewitnyssa, stala, Homl. Th. ii. 592, 5. Stala furtum, Wrt.

Voc. ii. 38, 31. Mōna se syxteóđa nānum þingum nytlíc nymþe stalum, Lchdm. iii. 192, 7. **II.** *what is stolen :*—Stalu biþ funden, 186, 14 : 188, 2. Gif hē næbbe, hwæt hē wiđ đære stale sylle, sylle man hine wiđ feó, Ex. 22, 3. Gif preóst mycele stale forstele, L. Ecg. C. 11 ; Th. ii. 140, 14. Gif hwylc man medeme þing stele, āgyfe đa stale đam đe hig āhte, L. Ecg. P. ii. 25 ; Th. ii. 192, 20 : iv. 24 ; Th. ii. 212, 1. **III.** *a fine payable for theft,* Chart. Th. 138, 17. See Kemble's Saxons in England, ii. 329. **IV.** *anything done by stealth :*—Đæt sćs Petrus on dæge folce be Criste sǣde, đonne wrāt sćs Marcus đæt on niht, and hē đæt hæl sće Petre ; for đon his godspell is swā cweden, *furtum laudabile,* hergendlíco stalo, Shrn. 74, 22. [To cumen bi stale ferliche, O. E. Homl. i. 249, 20. O. H. Ger. stala *furtum.*] v. ge-stalu.

stalu, e ; *f. A stale :*—Hearpan stala *the pieces of wood into which the strings are fixed* (?) : ceminigi, Wrt. Voc. ii. 130, 66 (cf. 40). [Scheome and pine beođ þe two leddre stalen þet beođ upriht to þe heouene, and bitweonen þeos stalen beođ þe tindes ivestned, A. R. 354, 18–20.]

stalung, e ; *f. Stealing, robbery :*—Ān hirde, se wæs Veriatus hāten, wæs micel þeófmon and on đære stalunge hē weard reáfere *Viriathus, homo pastoralis et latro, primum infestando vias, deinde vastando provincias,* Ors. 5, 2 ; Swt. 216, 7.

stam, stamm ; *adj. Stammering :*—Stom, wlisp *balbutus,* Txts. 45, 277 : *blessus,* 308. Stam *battulus* (*balbutus* ?), 109, 1150. Stomm *blessus,* stom, wlisp *balbutus,* Wrt. Voc. ii. 10, 72, 75. [Goth. stamms : O. H. Ger. stam, stamm : Icel. stamr, stammr.]

stamer (-or, -ur), *adj. Stammering :*—Stomer *balbutus,* Wrt. Voc. ii. 125, 12. Stamer *balbus,* i. 45, 51. Stamur, 75, 37. Stamor *blessus,* 288, 9. Stamerum *balbis,* ii. 81, 41 : Hpt. Gl. 478, 14 : 507, 45. [Cf. O. H. Ger. stamel *balbus.*]

stamerian, p. ode *To stammer :*—Stamaraþ *balbutit,* Germ. 392, 12. Mē þinceþ đæt mē sió tunge stomrige, Shrn. 42, 33. [Cf. O. H. Ger. stam[m]elón *balbutire.*]

stammettan, p. te *To stammer :*—Stommeteþ *mutulat,* Wrt. Voc. ii. 57, 68.

stampe *a pestle.* [O. H. Ger. stampf *pilum.*] v. píl-stampe ; stempan.

stān, es ; *m.* **I.** *stone as a material :*—Hig hæfdon tygelan for stān, Gen. 11, 3. Genim geoluwne stān *take ochre,* Lchdm. i. 374, 14. Se đe ofer đone stān (*supra petrosa*) āsāwen is, Mt. Kmbl. 13, 20. **II.** *a stone, a piece of stone :*—Se pitt wæs geheled mid ānum stāne . . . Hig āwylton đone stān of đam pitte, Gen. 29, 2–3. Hē nam stānas and lēde under his heáfod, 28, 11. **II a.** *a stone for building, wrought stone :*—Ne biþ lǣfed stān uppan stāne, Mt. Kmbl. 24, 2. Lóca hwylce stānas hēr synt, Mk. Skt. 13, 1. Holum stānum *fornicibus,* Wrt. Voc. ii. 40, 5. Ne tymbra đū đæt of gesnidenum stānum, Ex. 20, 25. **II b.** *a stone* (in its natural state or wrought) *that serves as a mark :*—Andlang herepađes west on đone þyrla[n] stān ; of đam stāne on đone hāran stān, Cod. Dip. Kmbl. iii. 406, 12 (*and often*). Hē nam đone stān and ārǣrde hine tō mearce, Gen. 28, 18. **II c.** *an image of stone :*—Se stān mǣlde for mannum (cf. ic bebeóde đæt đeós onlícnes word sprece, 1460 ; An. 731), Andr. Kmbl. 1532 ; An. 767. **II d.** *a stone* to which worship is paid. v. stān-weorþung :—Gehātaþ hý ælmessan þurh deófles lāre ođþon tō wylle ođđon tō stāne, Wulfst. 12, 3. Gif hwylc man his ælmessan gehāte ođđe bringe tō hwylcon wylle ođđe tō stāne, L. Ecg. P. ii. 22 ; Th. ii. 190, 24. Gif fridgeard sí on hwæs lande ābúton stān ođđe wille, L. N. P. L. 54 ; Th. ii. 298, 16. Đa gemearr đe man drífþ . . . on stānum, L. Edg. C. 16 ; Th. ii. 248, 6. Cf. Si quis ad fontes vel ad lapides votum voverit, L. Th. P. 27, 18 ; Th. ii. 34, 6–8. Gē þeówiaþ fremdum godum, stoccum and stānum, Deut. 28, 36. Hǣđenscipe biþ đæt man weorđige wæterwyllas ođđe stānas, L. C. S. 5 ; Th. i. 378, 20. **II e.** *a stone-that contains metal :*—Đa gyldenan stānas and đa seolfrenan *aureae arenae,* Bt. 34, 8 ; Fox 144, 30. **II f.** *a precious stone :*—Gerēnod mid golde and mid đæm stāne iacinta, Past. 14 ; Swt. 83, 24. Stāne gelícast gladum gimme, Exon. Th. 219, 5 ; Ph. 302. **II g.** *a stone in the bladder :*—On đære blǣdran stānas weaxaþ, Lchdm. ii. 238, 18 : i. 212, 22. **III.** *rock, a rock* (lit. and fig.) :—Đæt hig sucon hunig of stāne and ele of đam heardustan stāne, Deut. 32, 13. Hē lǣdde wæter of stāne (*de petra*), Ps. Spl. 77, 19. Gē tō đam lifgendan stāne stapol fæstniaþ, Exon. Th. 281, 30 ; Jul. 654. Ic stande beforan đē uppan Oreb stāne (*supra petram Horeb*), Ex. 17, 6. Đū eart Petrus and ofer đisne stān (*petram*) ic timbrige míne cyricean, Mt. Kmbl. 16, 18 : 7, 24. Stearcheort styrmde, stefn in becom under hārne stān, Beo. Th. 5100 ; B. 2553. Stānum *cautibus,* Wrt. Voc. ii. 18, 15. [Goth. stains : O. Sax. O. Frs. stēn : O. H. Ger. stein : Icel. steinn.] v. beácen-, ceosel-, clif-, cweorn- (cwyrn-), eá-, earcnan- (eorcnan-, eorcan-, eorclan-), flór-, gefóg-, gim-, grund-, hiéwe-, hwet-, hyrn-, loc-, mægen-, mǣr-, marma-, marman-, marmor-, mylen-, nume-, papol-, pumic-, sealt-, tæfl-, tigel-, weall-, weorc-stān.

stān-æx, e ; *f. A stone axe* ; or (?) *an implement for working stone* [Halliwell gives *stone-ax* a stone-worker's axe] :—Stānæx *bipennis,* Wrt. Voc. i. 34, 60. Stānex, 84, 68. For an account of stone axes found in England, see Wright's The Celt, the Roman, and the Saxon, pp. 69 sqq. : see also Nilsson's Stone Age, pp. 60 sqq. v. stān-bill.

stān-bæþ, es ; *n. A vapour bath made by the help of heated stones on to which water was poured :*—Dō on troh hāte stānas wel gehǣtte, gebeþe đa hamma mid đam stānbađe ; đonne hié sién geswāte, recce hē đa bān, Lchdm. ii. 68, 4–7. Stānbæþ, 10, 13 : 60, 9. Tō stānbæþe, 106, 16. Sele him stānbađu gelóme, 106, 25.

stān-beorh, -beorges ; *m. A stony elevation, rocky hill :*—On gerihte wiđ đæs lytlan stānbeorges up on hæslhille ; of đam stānbeorge ofer đa dene . . . tō đon lytlan stānbeorge ; of đam stānbeorge tō đon ođerum lytlan stānbeorge, Cod. Dip. Kmbl. v. 194, 15–18. Of riscmere on stānbeorg, iii. 453, 23. Stānbeorh, 381, 1. Stānbeorh steápne, Beo. Th. 4432 ; B. 2213. Sunt termini ab occasu stānbergas, Cod. Dip. Kmbl. i. 159, 14.

stān-berende *stone-bearing, stony :*—In đǣm stānberendum *in glanigeris* (*glarigeris* ?), Wrt. Voc. ii. 48, 51.

stān-bill, es ; *n. An implement of stone,* or *one used in working stone :*—Bill *marra,* stānbill *mastellas,* Wrt. Voc. ii. 57, 71. v. stān-æx.

stān-boga, an ; *m. A natural stone arch :*—Hē geseah stondan stānbogan, streám út đonan brecan of beorge, Beo. Th. 5083 ; B. 2545. Seah on enta geweorc hū đa stānbogan (*of the cave within the rock*) stapulum fæste ēce eorđreced innan healde, 5429 ; B. 2718. [Icel. steinbogi, steina-brú *a stone arch* or *bridge* (a natural one).]

stān-brycg, e ; *f. A stone bridge :*—Andlang brōces ōđ stānbrycge, sūđ from stānbrycge, Cod. Dip. Kmbl. iii. 429, 9–10. Andlang burnan ōđ hit cymþ đēr Blíđe út scýt ; đæt andlang Blíđan ōđ đa stānbriccge ; đæt eást of đære briccge, 421, 34. Đis sint đa landgemǣre . . . Ǣrest on stānbriccge ; of stānbriccge eást onlang Temese, v. 395, 29–31. Stānbricge (?) *lithostratos,* Wrt. Voc. i. 22, 6.

stān-bucca, an ; *m. A mountain goat :*—Đes stānbucca *hic cynyps,* Ælfc. Gr. 9, 57 ; Zup. 68, 5. [O. H. Ger. stein-boch *caper, Capricornus.*]

stān-burh *a town built with stone, a walled town* (?) :—Steápe stānbyrig, Cd. Th. 133, 17 ; Gen. 2212.

stanc *a sprinkling :*—Stanc *pluvicinatio,* Wrt. Voc. i. 46, 25. v. stancrian, stencan.

stān-carr *rock, stone :*—Stāncarr heard *petram durissimam,* Rtl. 19, 21.

stān-ceastel, -cistel, es ; *m. A chestnut-tree :*—Đonon sūđrihte wiđ đara stānceastla, and đonne of đæm stānceastlum, Cod. Dip. Kmbl. ii. 172, 16. On đane stāncistel, iii. 434, 33. v. cystel, stān-cist.

stān-ceosel, es ; *m. Sand :*—Hē getimbrode hys hús ofer stānceosel, Mt. Kmbl. 7, 24 MS. A. Stāncislas *glareas,* Hpt. Gl. 449, 16.

stān-cist, -cisten *a chestnut-tree :*—Of đane þorne on đo stāncysten on holencumbe ; of đane stāncyste on blacmanne bergh, Cod. Dip. Kmbl. iv. 8, 22. v. cisten-, cyst-beám, stān-ceastel.

stān-clif, es ; *n. A rocky cliff, a rock :*—Hē of stānclife burnan lǣdde *qui eduxit aquam de petra rupis,* Ps. Th. 135, 17. Beorgas đǣr ne muntas steápe ne stondaþ, ne stānclifu heáh hlífiaþ, Exon. Th. 199, 8 ; Ph. 22. Sume flugon æfter stānclifum, Elen. Kmbl. 269 ; El. 135. Stormas stānclifu beótan, Exon. Th. 307, 13 ; Seef. 23. Stāncleofu, Beo. Th. 5073 ; B. 2540.

stān-clúd, es ; *m. A rock :*—Haec Caribdis ān stānclúd on sǣ, Ælfc. Gr. 9, 78 ; Zup. 75, 7. Đā āhēng se munuc āne lytle bellan on đam stānclúde, Homl. Th. ii. 156, 5. God him (*the Israelites*) forgeaf wæter of heardum stānclúde, 264, 22 : Homl. Skt. i. 6, 279. Stānclúd *rupem,* Ps. Lamb. 113, 8. Swelce hit sié ungemong miclum and monigum stānclúdum tóbrocen *quasi per obviantia saxa frangatur,* Past. 9 ; Swt. 59, 7.

stān-cræftiga, an ; *m. A skilled worker in stone :*—Đæt wǣron .iiii. stāncræftigan . . . and nǣron nāne ōđre him gelíce ; hý gesenodon ælce morgen heora íseríngelóman, and đonne nǣron hý nā tóbrocene, Shrn. 146, 13.

stancrian *to sprinkle :*—Ic stancrige *pluvicino,* Wrt. Voc. i. 46, 26. Cf. stanc.

stān-cropp, es ; *m. Stone-crop* ; *sedum acre :*—Nim stāncroppes sǣd, Lchdm. iii. 72, 10.

stān-crundel *a tumulus of stones* (?) :—Tō đam stāncrundle, Cod. Dip. Kmbl. iii. 408, 33.

stand, es ; *m. A stand, stay, pause, delay :*—Midđý stondas (stando, Lind.) monige wērun *cum mora multa fieret,* Mk. Skt. Rush. 6, 35.

standan, p. stód, pl. stódon ; pp. standen *To stand.* **I.** of attitude, (1) of persons :—Đonne gē standaþ eów tō gebiddenne, Mk. Skt. 11, 25. Stand ofer đone man, Lchdm. ii. 104, 10. Stande on hæfde, 154, 2. Đa đe beóþ mid hiora āgnum byrđennum ofđrycte, đæt hié ne māgon standan, Past. 7 ; S. 50, 25. (2) of things :—Segnas stódon *the banners were raised,* Cd. Th. 214, 7 ; Exod. 565. Hē đǣr geseah swer standan *there was an upright column,* Blickl. Homl. 239, 21. **II.** of situation or position, (1) of persons :—Ic stande beforan đē uppan Oreb stāne, Ex. 17, 6. Ic niste, đæt đū stóde ongeán mē, Num. 22, 34. Ǣđelm self stód đǣrinne mid, Chart. Th. 171, 8. Đā stódan him twegen weras big, Blickl. Homl. 121, 23. Lǣde hig tō đære eardungstówe dura, đæt hig standon đǣr mid đē, Num. 11, 16. Pharao mætte, đæt hē stóde be ānre eá, Gen. 41, 1. Geseah hē deófol đǣr unfeor standan, Blickl. Homl. 227, 24. (2) of things :—Se port stent betuh Winedum and Seaxum and

Anglum, Ors. I, I; Swt. 19, 23. Se steorra āna stent, Met. 29, 16. Nis ꝺæt feor heonon, ꝺæt se mere standeþ, Beo. Th. 2729; B. 1362. Se tōꝺ se ꝺe bī ꝺam standeþ, L. Ethb. 51; Th. i. 16, 4. Ꝺa wīc ꝺe beforan inc stondeþ, Blickl. Homl. 77, 22. Sió burg stód bī ꝺære sǽ, Past. 52; Swt. 409, 33. Ꝺā com genip and stód æt ꝺære dura, Ex. 33, 9. Him æt heortan stód ord, Byrht. Th. 136, 3; By. 145: Beo. Th. 5352; B. 2679. Gemearca hū ꝺa tyrf ǽr stódon, Lchdm. i. 398, 5. Ꝺa stānas on ꝺæm mǽran temple . . . ǽr hié mon tô ꝺæm stede brohte ꝺe hié on standan sceolde, Past. 36; Swt. 253, 15. (3) of time :—Ꝺæt se dæg swīꝺe neáh stóde his forþfōre, Bd. 4, 3; S. 568, 16. **II a.** of situation or position in a figurative sense, denoting resistance, assistance, representation, degree, etc.:—Stande ꝺæs cyreáꝺ ofer .xx. peninga *let his oath be valid in matters above xx pence*, L. Ath. i. 15; Th. i. 204, 15. Stande ꝺæt spor for ꝺone foráꝺ, iv. 2; Th. i. 222, 16. Gif hwā on leásre gewitnesse stande, and hē oferstǽled weorꝺe, ne stande his gewitnesse syþþan for āht, L. C. S. 37; Th. i. 398, 12. Hū hē sceal swerigean ꝺe mid óꝺre on gewitnesse standaþ (-eþ?), L. O. 8; Th. i. 180, 26: 6; Th. i. 180, 17. Ælc man ꝺara ꝺe ꝺǽr mid stande, L. Ath. I, I; Th. i. 200, 3. Ꝺone wísdóm ꝺe on hālgum bócum stent, Homl. Th. i. 258, 14: L. Ath. v. 3; Th. i. 232, 9. On Gode standeþ mīn hǽle, Ps. Th. 61, 7. Stande hit on his āgenan gewealde, Chart. Th. 329, 35. Se ꝺe unriht gestreón on his handa stóde, L. Eth. ii. 9; Th. i. 290, 5. Wulfgeat wæs se forma man and Wulfmǽr is ꝺe óꝺer ꝺe hit nū on honda stant, Cod. Dip. Kmbl. iii. 260, 28. For ꝺare neóde, ꝺe ūs nū on handa stent *that we now have on our hands*, Wulfst. 181, 25. Gyf neód on handa stande *if the need present itself*, L. Edg. H. 2; Th. i. 258, 7. Ꝺonne stent se hē on his naman stede and spelaþ hine, Ælfc. Gr. 5; Zup. 8, 15. Ꝺa ꝺe stódon ongeán ūs *insurgentes in nos*, Ps. Th. 58, 11, 7. Ne manna getrýwꝺa tô āhte ne standaþ, Wulfst. 82, 11. Godu ꝺe ꝺissum folce tô freme stondaþ, Exon. Th. 250, 7; Jul. 123. Swā hī ufor stondaþ ꝺonne ꝺa óꝺre, Past. 52; Swt. 407, 21. Stond heó wiꝺ ǽttre, Lchdm. iii. 32, 21: 36, 6. Se wiꝺ mongum stód, wuldres cempa, Exon. Th. 121, 26; Gū. 294. **III.** of condition :—Heó grówende standeþ, Blickl. Homl. 197, 25: 109, 22. Be ꝺam cūþ standeþ ꝺæt hē fram deáþe gescylded wæs *quem a morte constat esse servatum*, Bd. 3, 23; S. 555, 27. Hē gearu standeþ, Ps. Th. 117, 2. Heó gewuldrad stondeþ, Blickl. Homl. 197, 10. Hē stent þeófscyldig, L. Eth. ii. 9; Th. i. 290, 16. Ꝺǽr geworht stondaþ Adam and Eue, Cd. Th. 27, 16; Gen. 418. Ꝺus hit stód on ꝺam dagum mid Englum *such was the condition of things among the English*, L. Eth. vii. 3; Th. i. 330, 9. Stód bewrigen folde mid flóde, Cd. Th. 10, 14; Gen. 156. Ꝺa ciricean giond eall Angelcynn stódon máꝺma and bóca gefyldæ, Past. pref.; Swt. 5, 10. Hié mōston stondan on fríðum anwalde *they might be in a condition of freedom*, 52; Swt. 405, 28. **IV.** of constitution :—Ic ongite ꝺæt sió sôþe gesǽlþ stent on gódra monna geearnunga and sió unsǽlþ stent on yfelra monna geearnungum *video quae sit vel felicitas, vel miseria in ipsis proborum atque improborum meritis constituta*, Bt. 39, 2; Fox 212, 12. Seó geladung ꝺe stent on mǽdenum and on cnapum, Homl. Th. ii. 566, 11. **V.** of occupation or action :—Petrus stód on gebedum *Peter was praying*, Blickl. Homl. 181, 21. Gif mæsse-preóst stande on leásre gewitnesse, L. Eth. ix. 27; Th. i. 346, 8: L. C. S. 37; Th. i. 398, 11. **VI.** to be fixed as a law or regulation :—Griꝺlagu ꝺus stent *the regulations are as follows*, L. Eth. vii. 9; Th. i. 330, 22. Ne stent nān óꝺer lād, L. O. D. 2; Th. i. 354, 1. Geneátriht is mistlīc be ꝺam ꝺe on lande stænt, L. R. S. 2; Th. i. 432, 12. Hwílon stód ꝺæt . . . *at one time the law was that* . . ., L. Ff.; Th. i. 226, 1: L. Eth. ii. 9; Th. i. 288, 29. Ꝺæt his griꝺ stande swā forꝺ swā hit fyrmest stód on his yldrena dagum *that the regulations be as full as ever they were*, iii. 1; Th. i. 292, 3. Stande betwux burgum ān lagu æt lādunge, L. C. S. 34; Th. i. 396, 22. **VII.** to remain undisturbed :—Hit fela wintra siþþan on ꝺæm stód *regnum Assyriorum diu inconcussa potentia stetit*, Ors. 2, 1; Swt. 60, 15. Stande þridda[n] dǽl ꝺære bôte inne *let a third part of the fine remain unpaid*, L. Alf. pol. 47; Th. i. 94, 6. Lǽt standan neáhterne, Lchdm. ii. 24, 21: 32, 11. Ꝺa ꝺe unne ꝺæt ꝺeós gerǽdnis stondon móte, Chart. Th. 168, 28. **VIII.** to stand still, cease to move, remain without motion, stop :—Gedón ꝺæt se Hǽlend stent, se ꝺe ǽr eode, Homl. Th. i. 156, 26. Hē clypode: 'Hǽlend, gemiltsa mín.' Ꝺā stód se Hǽlend, 152, 19. Hē sérde þurh his menniscnysse, and hē stód þurh ꝺa godcundnysse, 156, 34. Hwæt stondaþ gē hēr? Blickl. Homl. 123, 21. Ꝺa eá stódon, Bt. 35, 6; Fox 168, 8. **IX.** to reside, abide :—Ꝺa standendan munecas ꝺǽr *consistentes ibi monachi*, Bd. 4, 4; S. 571, 12. **X.** to continue, remain :—Ꝺenden standeþ woruld, Cd. Th. 56, 21; Gen. 915. Stande hē on þeówete, L. Ath. v. 12, 2; Th. i. 242, 5. Ꝺes middangeard eów ne mæg ealneg standan, Past. 51; Swt. 395, 29. **XI.** to stand, not to fall, to be upheld :—Ic getrymed fæste stande, Blickl. Homl. 225, 34. Seó godcunde meht staþolfæstlīce stondeþ, 19, 21. Hī on ꝺam geleáfan fæstlīce stódan, Bd. 2, 17; S. 520, 21. Ꝺæt dóm stande ꝺær þegenas sammǽle beón; gif hig sacan, stande ꝺæt hig .viii. secgaþ, L. Eth. iii. 13; Th. i. 298, 2-4. Hū mæg his rīce standan, Mt. Kmbl. 12, 26. Stondan, Blickl. Homl. 175, 15. **XII.** of direction (lit. and fig.) :—Him stent ege of ꝺe *timebunt te*, Deut. 28, 10. Ꝺēron stent ꝺam bisceope eahta marca goldes *eight marks are due*

to the bishop, Chart. Th. 595, 2. Swā micel ege stód deóflum fram eów *the devils stood in such awe of you*, Homl. Th. i. 64, 25: Ps. Th. 104, 33: Cd. Th. 249, 5; Dan. 525. Him stód stincende steám of ꝺam mūꝺe, Homl. Th. i. 86, 13. Him of eágum stód leóht unfæger, Beo. Th. 1457; B. 726. Fýrleóma stód geond ꝺæt atole scræf, Cd. Th. 272, 32; Sat. 128. Ic wille ꝺat se freóls stonde intó ꝺat minstre, Cod. Dip. Kmbl. iv. 219, 19. [Goth. standan: O. Sax. standan: O. Frs. standa: O. H. Ger. stantan: Icel. standa.] v. ā-, æt-, āgēn-, and-, be-, for-, fore-, ge-, of-, ôþ-, tô-, under-, wiþ-, ymb-standan; ān-standende.

standend, es; *m.* One who stands :—Ꝺrífaldo stondendo *ternos statores*, Rtl. 193, 35.

standendness. v. ā-standendness.

stán-fæt, es; *n.* A stone vessel :—On stánfate gehīded, Wald. 62. Com wíf hæbbende stánfæt (*alabastrum*), Mk. Skt. Rush. 14, 3. Midꝺẏ gebrocen wæs ꝺæt stánfæt, Lind. 14, 3. [O. Sax. stēn-fat.]

stán-fáh; *adj.* Many-coloured with stones, epithet of a road, Beo. Th. 645; B. 320: Andr. Kmbl. 2473; An. 1238.

stán-gaderung, e; *f.* A collection of stones, a wall :—Stángaderunge *maceriae*, Ps. Spl. T. 61, 3. Cf. stán-lesung.

stán-geat, es; *n.* An opening to pass through between stones :—On stángeat; of stángeate, Cod. Dip. Kmbl. iii. 81, 16.

stán-gedelf, es; *n.* A stone quarry :—Tô ꝺan stángedelfe; of ꝺam stángedelfe, Cod. Dip. Kmbl. iii. 77, 23. Æt ꝺæm stángedelfe, 366, 18. On ꝺæt stángedelf, v. 304, 21: vi. 144, 9.

stán-gefeall, es; *n.* A mass of fallen stones :—Twā wíf āhýddon ꝺone lýchaman under myclum stángefealle, Shrn. 152, 4.

stán-gefóg, es; *n.* A joining of stones in building :—Ꝺa ꝺe wyrcan cūꝺon stángefógum *those that could work at putting stones together*, Elen. Kmbl. 2039; El. 1021. v. gefóg-stán.

stán-gella, -gilla, an; *m.* A stone-yeller, a bird whose cry is heard among the rocks (gellan is used of the cry of the hawk, Rä. 25, 3), a pelican :—Stángella *pellicanus*, Wrt. Voc. i. 63, 20. Gelīc geworden ic eom ꝺam stángillan (-gyllan, MS. C.: stánegellan, Ps. Surt.) wēstene *similis factus sum pellicano solitudinis*, Ps. Spl. T. 101, 7.

stán-getimbre, es; *n.* A stone building :—Stángetimbru *moenia*, Wrt. Voc. ii. 54, 67.

stán-geweorc, es; *n.* Working in stone, stone-work :—Bæd hē ꝺæt hē him onsende sumne heáhcræftigan stángeweorces *architectos sibi mitti petiit*, Bd. 5, 21; S. 643, 1. On hire wurꝺmynte is ārǽred mǽre cyrce mid wundorlīcum stángeweorce, Homl. Th. i. 440, 18. Cf. stán-weorc.

stán-gripe, es; *m.* A seizing of stones, stones seized :—Ꝺeáh hē stángreopum (-greótum, Kmbl.) worpod wǽre *though he was stoned with the stones that they seized*, Elen. Kmbl. 1645; El. 824.

stán-hege, es; *m.* A stone fence, a wall :—Tô hwý tôwurpe ꝺū his stánhege *quid destruxisti maceriam ejus*, Ps. Lamb. 79, 13.

stán-hifete. v. stán-hīwet.

stán-hīpe, an; *f.* A stone-heap :—Andlang burhweges tô ꝺære stánhýpan, Cod. Dip. Kmbl. iii. 431, 10.

stán-hīwet, es; *n.* A stone-quarry :—Stánhýwet *lapidicina vel lapidicedum*, Wrt. Voc. i. 19, 17. Tô ꝺam stánhifete (-hīwete?); of ꝺam stánhifete (-hīwete?) tô ꝺam heꝺe, Cod. Dip. Kmbl. vi. 60, 24.

stán-hliþ, es; *n.* A rocky slope, a rock :—Mīn freónd siteþ under stánhliꝺe, Exon. Th. 444, 16; Kl. 48. Bīdaþ stánleoþu streámgewinnes, 384, 11; Rä. 4, 26. Under stánhliꝺum, Cd. Th. 219, 28; Dan. 61. Stánhleoꝺum, Elen. Kmbl. 1302; El. 653. Ꝺás stánhleoþu stormas cnyssaþ, Exon. Th. 292, 18; Wand. 101. Se æꝺeling hēt stormas restan ymb stánhleoꝺu, Andr. Kmbl. 3152; An. 1579. Æfter dūnscræfum ymb stánhleoꝺu, 2467; An. 1235. Ofereode æþelinga bearn steáp stánhliꝺo, stige nearwe, Beo. Th. 2822; B. 1409.

stán-hof, es; *n.* A house of stone :—Stánhofu stódan, Exon. Th. 478, 10; Ruin. 39.

stán-hol, es; *n.* A hole in rocks :—Hié (*serpents and wild beasts*) in stánholum hié selfe dīgliaþ *saxorum latebris occulta*, Nar. 6, 1. Ꝺā flugon hié in ꝺa wæter and hié ꝺǽr in ꝺam stánholum hýddon, 22, 13. [O. H. Ger. stein-hol *spelunca*.]

stán-hrycg, es; *m.* A ridge of rock :—Swilce betwux stánhrícgum *quasi inter Scyllam*, Hpt. Gl. 529, 22.

stán-hýwet. v. stán-hīwet.

stánig, stænig; *adj.* Stony, rocky :—Of ꝺan hǽꝺenan byrgelse on ꝺone stánigan beorh; of ꝺan stánigan beorge óꝺ ꝺa heáfda, Cod. Dip. Kmbl. iii. 454, 2-4. On ꝺone stánigan weg, vi. 186, 19. On stǽnig lond *in petrosa* . . . on ꝺa stánige lond *supra petrosa*, Mt. Kmbl. Rush. 13, 5, 20. Ꝺǽr synd swýꝺe scearpe wegas and stánige (stánihte, Laud. MS.), Ors. 1, 1; Bos. 16, 32. Ꝺeós wyrt biþ cenned on stánigum stówum, Lchdm. i. 102, 3. Stǽnigum, 212, 9 note: 216, 20 note: 256, 22. [O. H. Ger. steinag (-ig) *saxosus, petrosus*.]

stániht, stæniht; *adj.* Stony, rocky :—Wæs seó eorþe tô ꝺæs heard and tô ꝺæs stánihte *erat tellus durissima et saxosa*, Bd. 4, 28; S. 605, 27. Tô ꝺære stánehtan dæne, Cod. Dip. Kmbl. v. 179, 24. On stánehtan ford, iii. 389, 1. On ꝺone stánihtan ford, 168, 31. On ꝺone stánihtan weg, 409, 11. On stænihtum stówum, Lchdm. i. 212, 9. Sume feóllon

on stǽnihte *alia ceciderunt in petrosa*, Mt. Kmbl. 13, 5.　[*O. H. Ger.* steinaht : *Ger.* steinicht.]

stáninccl, es ; *n. A little stone :*—Stáninclu *lapillulos*, Anglia xiii. 31, 86.

stán-lesung, e ; *f. A gathering of stones, building with stones and without cement :*—Stánlesung *lithologia* (λιθολογέω to gather stones ; *to build with stones and without cement*), Wrt. Voc. i. 22, 5.　Cf. stángaderung.

stán-lím, es ; *m. Mortar :*—Stánlím *cimentum*, Wrt. Voc. ii. 131, 45.

stán-merece, -merce *parsley :*—Stánmerce *sigsonte*, Wrt. Voc. i. 68, 36.　[*Prompt. Parv.* stanmarche, herbe Macedonia, *Alexandria.*]

stán-rocc, es ; *m. A high rock, a peak ; an obelisk :*—Stánrocces *obolisci* (cf. *obolisci*, genus lapidis, 78, 17. *Obolisci* ðæs stánes, 82, 43), Wrt. Voc. ii. 62, 57.　Stánrocca, torra *scopulorum*, Hpt. Gl. 449, 15.　Stánrocca ł torra *scopulorum*, *saxorum eminentium*, 454, 47.　Cf. scylf.

stán-scealu, -scalu, e ; *f. Shale :*—Of Stúre on ða stánscale, Cod. Dip. Kmbl. iii. 378, 12.　v. next word.

stán-scilig ; *adj. Shaly, stony :*—Sum feóll ofer stánscyligean . . . ofer ða stánscylian *super petrosa*, Mk. Skt. 4, 5, 16.　v. preceding word.

stán-scræf, es ; *n. A cave in the rocks :*—Sča Maria hine ácende on ánum holum stánscræfe, Shrn. 29, 28 : 107, 28.　Gongaþ on ðis stánscræf, and git ðǽr métaþ weal, se is mid ifige bewrigen, 139, 26.

stán-scylf, es ; *m. A peak, rock :*—Stánscylfa *scrupearum* (*scrupea*, i. *aspera saxa*). . . . Of sandigum stánscilfum *de arenosis sablonibus*, Hpt. Gl. 449, 20, 25.

stán-strǽt, e ; *f. A road made with stones, a paved road :*—Ðonne forð ðæt hit cymþ tó ðare stánstrǽte ; of ðare stánstrǽte, Chart. Th. 525, 20.　Cf. stán-weg.

stán-stycce, es ; *n. A bit of stone :*—Stánsticcum *crustis* (*frustis ?*), Wrt. Voc. ii. 20, 61.

stán-torr, es ; *m.*　**I.** *a stone tower :*—Stántorr *the tower of Babel*, Cd. Th. 102, 14 ; Gen. 1700.　**II.** *a rock, crag, tor* (cf. stánrocc, -scylf) :—Ad locum qui stántor dicitur, Cod. Dip. Kmbl. v. 104, 2.

stán-wang, es ; *m. A stony plain :*—Stánwongas gróf, Exon. Th. 498, 24 ; Rä. 88, 6.

stán-weall, es ; *m. A wall of stone :*—Stánweal[les] *maceriae, muri*, Hpt. Gl. 409, 77.　Stánwealle (-walle, Ps. Surt.) *maceriae*, Ps. Spl. T. 61, 3.　Ða hwíle ðe mon worhte ða burg mid stánwealle, Chr. 921 ; Erl. 107, 27.　Ðæt wæter (*of the Red Sea*) him stód swilce stánweallas bufan heora heáfdum, Ælfc. T. Grn. 5, 27 : Homl. Ass. 105, 104.　Stánweallas tófeóllan, Shrn. 67, 19.

stán-weg, es ; *m. A road made with stones :*—On ealdan stánwege ; of stánwege, Cod. Dip. B. i. 417, 15.　[*O. Sax.* stēn-weg.]　Cf. stán-strǽt.

stán-weorc, es ; *n. Stone-work, stone-building :*—Hé worhte of seolfre ǽnne heáhne stýpel on stánweorces gelícnysse, Homl. Skt. ii. 27, 29. [*O. Sax.* stēn-werk.]　Cf. stán-geweorc.

stán-weorþung, e ; *f. Worship of stones :*—Wé læraþ ðæt preósta gehwilc forbeóde stánwurþunga, L. Edg. C. 16 ; Th. ii. 248, note 2.　v. stán, II d, and Grmm. D. M. 611.

stán-wurma, an ; *m. Colour got from a stone :*—Stánwurman *vermiculo, tinctura*, Hpt. Gl. 431, 34.

stán-wyrht (?), e ; *f. A stone building :*—Stánwyrhte *mationes* (cf. scylfas *maciones*, Wrt. Voc. ii. 59, 29), Wrt. Voc. i. 39, 55.

stán-wyrhta, an ; *m. A stone-wright, worker in stone, a mason :*—Stánwyrhta *latomi*, Wrt. Voc. i. 19, 16.　Stánwyrhtan *cementario*, Hpt. Gl. 459, 38.　From ðám stánwyrhtum *a cimentario*, Wrt. Voc. ii. 2, 40.

stapa, an ; *m. One who steps.*　**I.** a name given to the grasshopper or locust :—Stapan *locuste*, Wrt. Voc. ii. 52, 20 : 71, 62.　v. gærsstapa.　[*O. H. Ger.* houui-staffo *locusta*.]　**II.** in cpds. án-, eard-, hǽþ-, hild-, mearc-, mór-stapa.

stapol (-el, -ul), es ; *m.*　**I.** *a post, pillar, column :*—Stapul *batis* (*basis ?*), Wrt. Voc. ii. 12, 49 : *patronus* (in a list giving parts of a house), i. 26, 36.　Stapole *cione* (κίων a column, pillar), ii. 131, 41.　Of ðam beorge on ðone stapol ; of ðam stapole, Cod. Dip. Kmbl. iii. 14, 11 : 378, 15.　Tó ðam stǽnenan stapole ; ðonne andlang ðæs weges óð ðone stǽnan stapol ; of ðam stapole, 418, 28.　Æt stēnan steaple, Txts. Mm. no. 25.　Stapul ǽrenne, Andr. Kmbl. 2126 ; An. 1064.　Ælc riht cynestól stent on þrým stapelum, L. I. P. 4 ; Th. ii. 306, 31 : Wulfst. 267, 9.　Stánbogan stapulum fæste, Beo. Th. 5430 ; B. 2718.　Hé hēt stapulas ásettan *erectis stipitibus*, Bd. 2, 16 ; S. 520, 6.　Sweras unlytle, stapulas, Andr. Kmbl. 2986 ; An. 1496.　Staplas *columbas* (*l. columnas*), Mt. Kmbl. Lind. 21, 12.　**II.** *a step, threshold* (?) :—Hé tó healle gong, stód on stapole, geseah Grendles hond (*the hand had been laid in the hall*), Beo. Th. 1856 ; B. 926.　[*O. Frs.* stapul (-el) *a block* : *O. H. Ger.* stafol (-el) *basis* : *Icel.* stöpull *a pillar* : *Dan.* stabel *a boundary-stone, post.*]　v. stapola.

stapol ; *adj.* v. fore-stapol.

stapola, an ; *m. A post, stock, piece of wood standing upright in the ground :*—Licge ðæt íren uppan ðám glédan . . . lecge hit man syþþan uppan ðam stapelan (cf. stacan, l. 12), L. Ath. iv. 7 ; Th. i. 226, 28.

stapol-weg, es ; *m. A road marked out by posts* (?) :—Fram túnweges

---

ende forð be efise tó stapolwege ufeweardan, Cod. Dip. Kmbl. v. 281, 23.

stappa, stapplian. v. stoppa, under-stapplian.

stappel (?) *a step.*　[*O. H. Ger.* staffalun *passibus* : *Ger.* staffel *step, degree.*]　v. siþ-stap[p]el.

stare-blind. v. stær-blind.

starian ; *p. ode To stare, look fixedly, gaze (with* on, tó) :—Ðæt ic on ðone hafelan eágum starige, Beo. Th. 3567 ; B. 1781.　Starie, 5585 ; B. 2796.　Secga gehwylcum ðara ðe on swylc staraþ, 1997 ; B. 996 : 2975 ; B. 1485.　Wē on ðæt bearn foran breóstum stariaþ, Exon. Th. 21, 28 ; Cri. 341.　Ðe gē hēr on stariaþ, 33, 6 ; Cri. 521 : 36, 3 ; Cri. 570.　Him ðæt tácen wearð, ðǽr hē tó starude, Cd. Th. 260, 32 ; Dan. 718.　Ðe hire an eágum starede, Beo. Th. 3875 ; B. 1935.　Hí on mere staredon, 3211 ; B. 1603.　On ða beorhtan gescæft ne mót ic ǽfre má eágum starian, Cd. Th. 273, 22 ; Sat. 140 : Judth. Thw. 24, 9 ; Jud. 179.　Se earn mæg starian on ðære sunnan leóman, Homl. Skt. i. 15, 199.　Hí stóden æt ðæra dura stariende on ðæt leóht, 3, 133.　[Staryū wythe brode eyne *patentibus oculis respicere*, Prompt. Parv. 472.　*O. H. Ger.* starēn : *Ger.* starren : *Icel.* stara.]　v. ge-starian.

stapol (-el, -ul), es ; *m.*　**I.** *a foundation* (lit. or fig.) (cf. *staddle* the bottom of a hay-stack, E. D. S. Pub. Gloss. 15, 19) :—Stapol *fundamen*, Wrt. Voc. ii. 152, 15.　Se fruma se ðe stapol eallra góda ðe of him cumaþ, Bt. 34, 5 ; Fox 140, 4.　Biþ Drihten úre se trumesta stapol, Blickl. Homl. 13, 10.　On ðissum cwydum is se staðol ealles geleáfan, L. E. I. 22 ; Th. ii. 418, 29 : 29 ; Th. ii. 426, 1.　Hié oft gebidon on lytlum stapole and on unwēnlícum, Ors. 4, 9 ; Swt. 192, 34.　Wera gied sumes, þrymfæstne cwide and ðæs strangan stapol *a glorious saying and the strong man's firm support*, Exon. Th. 432, 12 ; Rä. 48, 5.　Staðol *fundum*, Hpt. Gl. 488, 6.　Dúna staðelas *montium fundamenta*, Deut. 32, 22.　Eorðan staþelas, Ps. Th. 81, 5 : 103, 6.　Steaðelas, Ps. Surt. 17, 8 : ii. p. 194, 9.　Staðulas, Cd. Th. 207, 28 ; Exod. 473.　Ða staþolas ðære cyrican, Bd. 2, 4 ; S. 505, 16.　Hé ða staþelas gesette ðæs mynstres, 3, 23 ; S. 554, 28.　On stapelum healdan, Exon. Th. 312, 14 ; Seef. 109.　**I a.** *the lower, firmer part, base* of a pillar, *trunk* of a tree :—Se is stemn and staðol ealra góda and of ðæm cumaþ eall gód, Bt. 34, 5 ; Fox 140, 2.　Hit biþ unnyt ðæt mon hwelces yfles bógas snǽde, búton mon wille ða wyrtruman forceorfan ðæs staðoles, Past. 33 ; Swt. 222, 16.　Gehēr ðú marmanstán . . . Lǽt nú of ðínum staðole streámas weallan, Andr. Kmbl. 3004 ; An. 1505.　Genim feówer tyrf on feówer healfa ðæs landes . . . Nim ele etc., and dó háligwæter ðæron, and drýpe on ðone staðol ðara turfa (*the lower side of the sods*), Lchdm. i. 398, 11.　**I b.** that on which a thing depends :—Staðul *cardo*, Wrt. Voc. ii. 20, 60.　**II.** *fixed condition, state, position :*—Hwylc se stapol is Angelcynnes ðeóde *qui sit status gentis Anglorum*, Bd. 5, 23 ; S. 645, 4.　Ic wæs smeágende be ðære hǽlo úrra sáwla and be ðam staðole úres ríces, L. In. pref. ; Th. i. 102, 8.　Hé hit nyle up árǽran tó ðam staðole fulfremedes weorces *ad virtutis statum consuetudo non erigitur*, Past. 11 ; Swt. 65, 16.　Hiera geðohtes staðol *cogitationis statum*, Swt. 67, 17.　Stede ł stalað (*l.* staðal) *statum, stabilitatem*, Hpt. Gl. 469, 12.　Hé geþyld lufige and ne áwácige ná ne his stapel ne lǽtende fram Gode búge, R. Ben. 27, 2.　Sette heora staðol sceápum anlíce *posuit sicut oves familias*, Ps. Th. 106, 40.　**III.** *a fixed position, station, place, site :*—Stapol wæs wyrta wlitetorhta (*the plain*) *was the site of beauteous plants*, Exon. Th. 484, 4 ; Rä. 72, 2.　Be ðære stówe staðole *secundum positionem loci*, R. Ben. 59, 1.　Staðelc, 88, 4.　Se wyrtruma stille wæs on staðole, Cd. Th. 252, 21 ; Dan. 582 : 251, 9 ; Dan. 561.　Wē stódon on staðole, Rood Kmbl. 141 ; Kr. 71.　Æsc byþ stíð staðule, stede rihte hylt, Runic pm. Kmbl. 344, 25 ; Rūn. 26.　Ic sceal bordes on ende stapol weardian (*keep my station* ; cf. Wulches cunnes þinges under þissen stane staðel habbeoð inumen (under þis ston uonieþ, 2nd MSS.), Laym. 15911), sto[n]dan fæste, Exon. Th. 496, 19 ; Rä. 85, 17.　Hē ús sealde mid englum ēce stapelas, 41, 26 ; Cri. 661.　Ðú álǽtan scealt lǽne staþelas, eard and plēol, Dóm. L. 30, 58.　**IV.** *the firmament, the heavens :*—Weard ætýwed steorra on staðole, Chr. 975 ; Erl. 126, 24 ; Edg. 50.　[*O. H. Ger.* stadal *scuria, horreum* : *Ger.* stadel : *Icel.* stöðull *a milking shed.*]　v. burh-, ēðel-, frum-, mód-, wēsten-stapol.

stapol-ǽht, e ; *f. An estate, landed possession*, Exon. Th. 353, 33 ; Reim. 22.

stapol-fæst ; *adj. Steadfast, stable, firm* ; stabilis, Ælfc. Gr. 9, 28 ; Zup. 55, 3.　**I.** in a physical sense :—On ðam feórþan mónþe hé (*the foetus*) biþ on limum staþolfæst, Lchdm. iii. 146, 11.　Staðolfæst stán (*glosses* Petrus), Mt. Kmbl. Lind. 16, 18.　Beðearf seó sáwel staðolfæstre brycge ofer ðone glideran weg hellewítes, Wulfst. 239, 14.　Staleðfæste (*l.* staðel-) tremmincge *firmo fulcimento*, Hpt. Gl. 439, 63.　**II.** *stationary, keeping in one place :*—Staþolfæst ne mæg gewunian in gebedstōwe, Exon. Th. 265, 1 ; Jul. 374.　Faraþ hý geond missenlíce þeóda, nǽfre staþolfeste, nǽfre wuniende, náhwǽr sittende, R. Ben. 135, 23.　**III.** *firm* in a moral sense, *unwavering, unyielding, constant :*—God is ána staþolfæst wealdend, Bt. 35, 3 ; Fox 158, 24 : Andr. Kmbl. 241 ; An. 121.　Staðulfæst, 2673 ; An. 1338.　Swíðe geþungen on his ðeáwum and staðolfæst on his wordum *not to be moved from what he had said*,

Blickl. Homl. 217, 7. Staðolfæst on hire heortan wið deófles costnungum, Wulfst. 237, 12. Beó strang and staðulfæst *confortare et esto robustus* Deut. 31, 7. Ic eów friðe healde strengðu staþolfæstre, Exon. Th. 31, 3; Cri. 490. Mid steaðulfestum aldum *cum stabilito sene*, Ps. Surt. ii. p. 194, 27. Sele mē staðolfæste heortan, Anglia xi. 114, 71. Staðolfæstne geðoht, Salm. Kmbl. 478; Sal. 239. 70 manna of folces ealdrum ðe ðū wite ðæt sîn staðulfæste and láreówas, Num. 11, 16. Onginnaþ esnlîce and beóþ staðulfæste *viriliter agite et confortamini*, Deut. 31, 6. v. un-, under-staþolfæst.

staþolfæstan. v. ge-staþolfæstan.

staþolfæst-lîc; adj. *Steadfast, firm*:—Mē sum staþolfæstlîc smyltnyss tô becom, Homl. Skt. ii. 23 b, 551.

staþolfæstlîce; adv. I. *in a physical sense, firmly*:—On ðam eahtoþan mônþe hē (*the foetus*) biþ eall staþolfæstlîce geseted, Lchdm. iii. 146, 19. II. *steadfastly, constantly, firmly*:—Seó godcunde meht ā staþolfæstlîce stondeþ, Blickl. Homl. 19, 21. Symble in Godes lofe wē sceolon staþolfæstlîce gewunigan, L. E. I. 42; Th. ii. 438, 32. Ðæt ðiós ûre sylene staðulfæstlîce ðurhwunian môte, Cod. Dip. Kmbl. v. 186, 12.

staþolfæstness, e; f. *Steadfastness, stability*:—Staðolfæstnys *stabilitas*, R. Ben. Interl. 23, 3: Ps. Lamb. 103, 5: *firmamentum*, 18, 2: *status*, Rtl. 108, 38. Steaðulfestnisse *stabilitatem*, Ps. Surt. 103, 5. v. ge-, un-staþolfæstness.

staþolfæstnian. v. ge-staþolfæstnian.

staþolfæstnung, e; f. *A foundation*:—Tô staðolfæstnunga *ad fundamentum*, Ps. Lamb. 136, 7.

staþolian; p. ode. I. *to establish, found, settle, fix*:—Ic tô ânum ðē môd staðolige *to thee alone do I keep my mind constant*, Andr. Kmbl. 164; An. 82. Staþelige, Exon. Th. 255, 30; Jul. 222. Ðū in God getreówdes ic in mînne fæder hyht staþelie *thou didst trust in God, I found my hope on my father*, 268, 25; Jul. 437. Ic ðy fæstlîcor ferhð staðelige, hyht untweóndne, on Crist, Elen. Kmbl. 1591; El. 797. Ðe ðæs hûses hróf staðeliaþ *qui aedificant domum*, Ps. Th. 126, 1. Ic on heofonum hâm staðelode, Cd. Th. 281, 23; Sat. 276. Staðelodest *fundasti*, Ps. Spl. 101, 26: 103, 6, 9. Se steaðelade eorðan ofer steaðulfestnisse his. Ps. Surt. 103, 5. Hē woruld staþelode, Exon. Th. 206, 22; Ph. 130. Ðær hē hungrium hâm staðelude *collocavit illic esurientes*, Ps. Th. 106, 35. Ðā hē æt Rôme Cristes cyricean staþelode *fundata Romae ecclesia Christi*, Bd. 2, 4; S. 505, 13. Se wealdend ðe ðæt weorc staðolade, Andr. Kmbl. 1598; An. 800: Met. 29, 87. Ðā heó in helle hâm staðelodon, Cd. Th. 266, 21; Sat. 25. Staðelodon, 286, 1; Sat. 345. Staðola ðū ða óðra on hira hâmon, Gen. 48, 6. Geleáfan fæste staðelian on ûrum heortum, Blickl. Homl. 111, 4. Staþelian, 115, 1. II. *to make steadfast, confirm, endow with steadfastness*:—Ne mîð ðū for menigo, ah ðinne môdsefan staðola wið strangum . . . herd hyge ðinne, heortan staðola, Andr. Kmbl. 2419–2428; An. 1212–1215. Staþelige man and strangie hî georne, L. I. P. 4; Th. ii. 308, 3. Se hâlga ongan hyge staðolian, Elen. Kmbl. 2186; El. 1094. Môd staþelian geleáfan, Exon. Th. 168, 26; Gú. 1083: 264, 15; Jul. 364. Ûre heortan rihtan and staðelion, Wulfst. 253, 18. v. ge-, gegrund-staðolian (-elian).

staþoliend, es; m. *A founder*; fundator, Ps. Lamb. 47, 2.

staþolness. v. môd-staþolness.

staþolung, e; f. *Founding, foundation, settling*:—Steaðelinge *plantationis*, Ps. Surt. 143, 12. Tô staþolungæ *ad fundamentum*, Ps. Spl. T. 136, 10. Staleðunga (*l.* staðelunga) *fundamina*, Hpt. Gl. 502, 71.

staþol-wang, es; m. *A plain to establish one's self in*. v. staþol, III:—Lǣteþ hió ða wlitigan wyrtum fæste stille stondan on staþolwonge (*in the field they occupy*), Exon. Th. 417, 4; Rä. 35, 8. Teón wē of ðisse stôwe and unc staþolwangas (*places where we may establish ourselves*) sēcan, Cd. Th. 114, 31; Gen. 1912.

stealc; adj. *Steep*:—Bîdaþ stille stealc stânhleoþu streámgewinnes, Exon. Th. 384, 11; Rä. 4, 26. On stealc hleoþa, 382, 6; Rä. 3, 7. Stealc hliþo stîgan, 498, 17; Rä. 88, 3.

stealcian. v. be-stealcian, *and next word*.

stealcung, e; f. *Stalking* (cf. deer-*stalking*), *cautious walking*:—On sumere nihte hlosnode sum óðer munuc his færeldes and mid sleaccre stealcunge his fôtswaðum filigde, Homl. Th. ii. 138, 6. [Cf. stalkyn or gon softe *serpo*, Prompt. Parv. 472. Though I wolde stalke and crepe, Gow. ii. 351, 18. With dredful fot than stalketh Palamon . . . in that grove he wolde him hyde, Chauc. Kn. T. 621.]

-steald. v. ge-, hæg-, hago-steald.

stealdan; p. steóld *To possess*:—Ic staðolæhtum steóld, Exon. Th. 353, 33; Reim. 22. [*Goth.* ga-staldan *to possess, gain*.]

steall, es; m. I. *a standing position*:—Setl gedafenaþ dēman, and steall fylstendum . . . Stephanus hine (*Christ*) geseah standende, forðan ðe hē wæs his gefylsta, Homl. Th. i. 48, 29. Syle hât drincan in stalle stonde gôde hwîle *give him the medicine hot to drink in a standing position; let him stand a good while*, Lchdm. iii. 28, 5. II. *the way matters stand, position of affairs, state, condition*:—Se steall cyricean *status ecclesiae*, Bd. 2, 4; S. 505, 10. On fræcenesse heora stealles in *periculo sui status*, 4, 25; S. 601, 18. Be ðisses biscopes lîfes stealle *de cujus statu vitae*, 5, 19; S. 637, 2. Be ðam stalle cyrican, 3, 19; S.

561, 7. On ðone ǣrran steall *priscum in statum*, 5, 20; S. 642, 10: 5, 24; S. 646, 38. Ðone stal ðæs rîces *regni statum*, 4, 26; S. 603, 8. III. *position, place*:—Horsa steal *carceres* (the starting-place in the circus), scridwîsa *auriga*, Wrt. Voc. i. 39, 37. On brǣdo his stealles *latitudine sui status*, Bd. 1, 1; S. 474, 29. Ðæt se sý furþor forlǣten on stealle and on setle (cf. on stede and on setle, 13, 1), se ðe furðor on geearnunge sý, R. Ben. 12, 19. Stande hē ealra ýtemest, oðþe on ðam stede ðe se abbod swā gemeleásum monnum tô stealle on sundrum betǣht hæfþ *ultimus omnium stet aut in loco quem talibus negligentibus seorsum constituerit abbas*, 68, 11. Ðæt hî nǣfre ne beón on stede ne on stealle, ðǣr ǣfre undón worðe ðæt ûre forgengles geúðen, Chart. Th. 348, 30. IV. *place, stead*:—Brihtwald gehálgode Tobian on his steall, Chr. 693; Erl. 43, 19. Steal, 780; Erl. 57, 1: 803; Erl. 61, 23. Stall, 779; Erl. 55, 38. Stal, 678; Erl. 41, 7: 727; Erl. 47, 2: 796; Erl. 59, 39. V. *a place for cattle, a stall*:—Stal *stabulum*, Wrt. Voc. i. 121, 11. Steal, i. 15, 23. Ðæra tamra nýtena steall, Boutr. Scrd. 21, 9. VI. *a place for catching fish*:—Lēt ða netto on stællo *laxa retia in capturam* (captura *locus piscosus, ubi capiuntur pisces*), Lk. Skt. Lind. 5, 4. (Cf. stell, a deep pool, in a river, where nets for catching salmon are placed, Jamieson.) [O. Frs. stall *standing*; place; stall: O. H. Ger. stall *stabulum, caula, praesepe*; *locus, statio, status*: Icel. stallr *a stall*; *shelf on which another thing is placed*.] v. æt-, bíd-, burg-, fore-, ge-, geard-, hege-, mylen-, ofer-, on-, scip-, treów-, wæter-, weal-, weard-, weofod-, weoh-, wîc-, wîg-, wið-, wiðer-steall (-steal); fæst-steall; adj. Cf. stæl, stede.

stealla, an; m. *A crab* (?):—Stalla *cancer* (carcer? cf. steall, III), Wrt. Voc. i. 291, 30.

-stealla. v. ge-, ofer-stealla.

steallere, stallere, es; m. *A marshall*. [The word occurs only in late documents; the passages given belong to Edward the Confessor's reign]:—On Esgēres stealres and on Roulfes steallres and on Lîfinges stealres gewitnesse, Cod. Dip. Kmbl. iv. 291, 13–14. Esgār stallere and Roberd stallere, 191, 11–12: 221, 13: Chr. 1047; Erl. 171, 31. [*Icel.* stallari.]

steallet. v. ān-steallet.

steallian *to take place*. v. forþ-steallian.

steám, stēm, stiém, es; m. I. *steam, hot exhalation, hot breath*:—Him (*Herod*) stôd stincende steám of ðam mûðe, Homl. Th. i. 86, 14. Forlǣt wynsumne rēc ástîgan . . . Ðā of ðære stôwe steám up árās swylce rēc, Elen. Kmbl. 1603; El. 803. Stenc ût cymeþ of ðam wongstede, wynsumra steám swæcca gehwylcum, Exon. Th. 358, 14; Pa. 45. Man pîntreów bærne tô glēdum . . . wende his neb tô and onfó ðam steáme (*the heat proceeding from the embers*), Lchdm. ii. 284, 16. Of hômena æþme and stiéme cymþ eágna mist; 26, 26. Fleó ða mettas ða ðe him stiém on innan wyrcen, 226, 10. II. *that which emits hot vapour, blood*:—Forlēton mē standan steáme bedrifenne *they left me* (*the cross*) *standing bespattered with blood*, Rood Kmbl. 123; Kr. 62. [A stem als it were a sunnebem, Havel. 591. Steem or lowe of fyre *flamma*, steem of hothe lycure *vapor*, Prompt. Parv. 473.]

steáp, es; m. *A stoup, drinking vessel, cup, flagon*:—Steáp *ciatum*, Wrt. Voc. i. 290, 78: ii. 17, 28. Micel steáp ful, Lchdm. ii. 294, 19. Se wînes steáp fægere gefylled is *calix vini meri plenus est*, Ps. Th. 74, 7. Steápes *poculi*, Hpt. Gl. 450, 6. Nalles wîn druncon scîr of steápe, Met. 8, 21. Dô steáp fulne wînes tô wôse, Lchdm. ii. 18, 4. Gif man ôðrum steóp ásette ðǣr mæn drincen āgelde .vi. scill. ðam ðe man ðone steáp áset *if a man remove* (?) *a cup from another when men are drinking, let .vi. s. be paid to the man from whom the cup was taken*, L. H. E. 12; Th. i. 32, 8–10. Steápas *fialas*, Wrt. Voc. ii. 149, 4. [A stope *hec cupa*, Wrt. Voc. 235, 16. O. H. Ger. stouf *calix, cyathus*: Icel. staup; *n. a cup, beaker*.]

steáp; adj. I. *lofty, high, towering*, of buildings, hills, etc.:—Se streám ætstôd swā steáp swā munt *the stream* (*Jordan*) *stood as high as a hill*, Homl. Th. ii. 212, 23. Wāg steáp gedreás, Exon. Th. 476, 22; Ruin. 11. Seó steápe burh on Sennar stôd, Cd. Th. 102, 15; Gen. 1700. Fýr steápes and geápes swôgende forswealh eall *fire everything lofty and spacious devoured roaring*, Cd. Th. 154, 16; Gen. 2556. On ðisum steápum munte, Homl. Skt. i. 13, 9. Worhton mid stânum ānne steápe beorh him ofer *congregaverunt super eum acervum magnum lapidum*, Jos. 7, 26. Steápne hróf, Beo. Th. 1857; B. 926. Þurh steápne beorg strǣte wyrcan, Exon. Th. 397, 11; Rä. 16, 18. Steápe dûne, Cd. Th. 172, 33; Gen. 2853. Steápe stânbyrig, 133, 17; Gen. 2212. Weallas steápe, Exon. Th. 383, 13; Rä. 4, 10. Beorgas ðǣr ne muntas steápe ne stondaþ, 199, 7; Ph. 22: Beo. Th. 450; B. 222: Andr. Kmbl. 1680; An. 842. I a. *of smaller objects*:—Heard and steáp (*the pillar into which Lot's wife was turned*), Cd. Th. 155, 8; Gen. 2569. Wið steápne rond *by the tall shield*, Beo. Th. 5126; B. 2566. Ic hæbbe hneccan steápne, Exon. Th. 490, 1; Rä. 79, 4. Bollan steápe *tall flagons*, Judth. Thw. 142, 6; Jud. 17. Hî habbaþ on heáfde helmas steápe (cf. O. Frs. with thene stápa helm. *Icel.* steypðir hjálmar), Wulfst. 200, 12. I b. *of fire, mounting high* (see also I c):—Hit ðurh hróf wadeþ, bærneþ boldgetimbru, seomaþ steáp and geáp, Salm. Kmbl. 827; Sal. 413. I c. *standing out, or up, prominent* [or *bright?* In later English *steap* applied to the eyes or to gems seems to have this meaning. 'Twa ehnen steappre þene steorren ant þene ʒimstanes,' Marh.

9, 4. In the note on this passage Cockayne gives other instances of this use, e. g. Schinende and schenre þen eni ʒimstanes, steapre þen is steorre. In Chaucer's line, Prol. 201, the meaning might be *prominent*. In the passage quoted below from Ælfric the Latin from which the description is taken has *oculi grandes*.] Gim sceal on hringe standan steáp and geáp, Menol. Fox 505; Gn. C. 23. Se steápa gim, Salm. Kmbl. 570; Sal. 284. Hé hæfþ steápe eágan, Homl. Th. i. 456, 17. II. *lofty, high, placed high :*—Oð ða steápan heofenan *to high heaven*, Homl. Th. i. 3, 500. [Þer wes moni steap (bold, 2nd MS.) mon, Laym. 1532. An lawe swiþe stæp and heh, Orm. 11379. O. Frs. stáp. v. I a above.] v. heaðu-, weall-steáp; stípel, stípan.

steápan. v. á-steápan, stípan.

stearc; *adj.* I. *stiff, rigid, not soft, not bending :*—Is seó eág-gebyrd stearc and hiwe stáne gelícast, Exon. Th. 219, 4; Ph. 302. Hláf and stán, streac and hnesce, Elen. Kmbl. 1226; El. 615. Stánas and ðæt starce ísen, Homl. Skt. i. 8, 29. Beátan mid stearcum stengum, Homl. Th. i. 428, 6. I a. fig. *unyielding, stiff-necked, obstinate :*—Heó wæron stearce, stáne heardran, Elen. Kmbl. 1126; El. 565. II. *hard, rough, strong,* of wind or weather :—Stearc winter *aspera hyems,* Coll. Monast. Th. 19, 17. Se stearca wind norþan-eástan, Bt. 9; Fox 26, 18. Se stearca storm, Met. 6, 11. Stearc storma gelác, 26, 29. Þurh ðone stearcan wind norþan and eástan, Bt. 4; Fox 8, 5. Stearce stormas, 23; Fox 78, 27. Gescyrped mid rinde wið ða stearcan stormas, 150, 8. III. *rough, attended with hardship, hard,* of living, discipline, etc. :—Hé ða stiðnyssa his stearcan bigleofan betwux læwedum folce geheóld, Homl. Th. ii. 148, 31. Se ðe mec lǽreþ from ðé on stearcne weg, Exon. Th. 259, 14; Jul. 282. Hú hé mihte swá stearce forhæfednysse (*rigid abstinence*) healdan, Homl. Th. ii. 354, 23. IV. *stern, severe :*—Hé (*William*) wæs milde ðám gódum mannum and ofer eall gemett stearc ðám mannum ðe wiðcwǽdon his willan . . . Hé wæs swýðe stearc man swá ðæt man ne dorste nán þing ongeán his willan dón, Chr. 1086; Erl. 221, 17, 32 : Erl. 222, 21. Hé ða heardheortan ðeóde mid stearcre ðreále and stíðre myngunge tó lífes wege gebígde, Homl. Th. i. 362, 34. V. *strong, impetuous, violent, vehement,* (a) lit. :—Hé of stánclife stearce burnan lædde, Ps. Th. 135, 17. (b) fig. v. stearc-heard :—Nán stefn búton stearce and heard wóp for wóhdǽdum, Wulfst. 139, 3. [O. Sax. stark : O. Frs. sterk : O. H. Ger. starc, starah *fortis, validus :* Icel. sterkr *strong.*]

stearc-ferhþ; *adj.* *Of harsh or stern soul :*—Hí stearcferþe cwellan þohtun, Exon. Th. 280, 29; Jul. 636.

stearc-heard; *adj.* *Violent, unrestrained :*—Stearcheard wóp *durus fletus,* Dóm. L. 200. v. stearc, V b.

stearc-heort; *adj.* *Stout-hearted :*—Stearcheort (*the fire-drake*), Beo. Th. 4566; B. 2288: (*Beowulf*), 5097; B. 2552. [Cf. O. Sax. stark-mód *valiant.*]

stearcian; *p.* ode *To grow stiff or hard :*—Stearcode *riget, durescit,* Germ. 402, 56. [His skyn shall starken, Rel. Ant. i. 65, 3. O. H. Ger. starcēn *solidari.*]

stearclíce; *adv.* *Strongly, vigorously, vehemently, fiercely :*—Ðá gewende se here tó Lundene and ða buruh útan embsæt and hyre stearclíce (cf. stranglíce, MS. E.) on feaht ægðer ge be wætere ge be lande *made a vigorous assault upon it by land and water,* Chr. 1016; Erl. 156, 32.

stearn, es; *m.* *Some kind of bird.* [*Starn* is a name for the starling in the Shetland Isles; the same bird is called a *starnel* in Northants. v. E.D.S. Pub., Bird Names, p. 73. *Starn* is used in Norfolk for the common tern: and *stern* is a name for the black tern, ib. pp. 202, 204] :—Stearn, stearno, stern *beacita* (according to Migne *beacita* is a woodcock or snipe), Txts. 45, 284. Stearn, Wrt. Voc. i. 281, 3: ii. 11, 1: *beatica,* i. 62, 32 : *beacita* vel *sturnus,* 29, 6: *fida,* ii. 108, 52. Stern, 35, 28. Stern *avis qui dicitur gavia,* Txts. 108, 1116. Stærn *stronus* (= *sturnus*), Wrt. Voc. i. 29, 39. Him stearn (*the tern*) oncwæð ísigfeþera, Exon. Th. 307, 14; Seef. 23.

steartlian; *p.* ede *To kick with the foot, stumble :*—Ðæt ðú ne spear[n]-last ł steartlest, steartlige *ut non calcitres,* Hpt. Gl. 464, 1. [In later English *startle* is used of quick movement :—A courser, startling as the fyr, Chauc. Leg. G.W. 1204. Thouʒ ne havest frend that ne wolde fle, come thouʒ stertlinde in the strete, Mapes 335, 24. See also Halliwell's Dict. *stertle.*]

steb. v. stybb.

stéda, an; *m.* *A stallion, an entire horse ;* the word is also used of a camel :—Hors *equus,* stéda *emisarius,* Wrt. Voc. ii. 30, 55 : *misarius,* 56, 39 : i. 287, 40. Stéda *faussarius,* hengst *canterius,* 23, 9. Hé hleóp on ðæs cyninges stédan *ascendens emissarium regis,* Bd. 2, 13 ; S. 517, 9 : Chart. Th. 501, 12. Ne hét Crist him tó lǽdan módigne stédan, Homl. Th. i. 210, 14 : Homl. Skt. ii. 27, 97. Ðonne lǽdaþ hý mid him olfenda myran mid hyra folan and stédan . . . ða stédan hý forlǽtaþ . . . ða æmettan ymbe ða stédan ábisgode beóþ *tollent camelos masculos et feminas illas quae habent foetas . . . masculi remanent . . . formicae masculos comedunt,* Nar. 35, 10–15.

stede, es; *m.* I. *a place, spot, locality :*—Mid wæter ymbtyrnd **stede** *circumlutus locus,* Wrt. Voc. i. 59, 15. Se stede ys hálig ðe ðú on

stenst *locus, in quo stas, sanctus est,* Jos. 5, 16. Ðes ænga stede (*Hell*), Cd. Th. 23, 9; Gen. 356. Hí cóman tó Brytene on ðam stede Heopwines fleót, Chr. 449; Erl. 13, 4. In ðone stede ðe is gecueden Cerdices óra, 495; Erl. 14, 10. Ðone stede healdan, Byrht. Th. 132, 21 ; By. 19. Tó hwí hremþ hit ðisne stede (*quid terram occupat?* Lk. 13, 7), Homl. Th. ii. 408, 5. Eode on woestigum styd (steyde, Rush.) *abiit in desertum locum,* Mk. Skt. Lind. 1, 35. Stydd, Lk. Skt. Lind. 10, 1. Hí sǽton tú winter on ðam twám stedum, Chr. 887 ; Erl. 84, 33. II. *of fixed position, a place which a person or thing occupies, an appointed place, station, site :*—Hú neara ðære eorþan stede is *arctum terrarum situm,* Bt. 19; Fox 68, 23. Ðæs fýres ágen stede is ofer eallum woruldgesceaftum gesewenlicum, 33, 4; Fox 130, 16. Heáfudponnes styd *calvariae locus,* Mt. Kmbl. 27, 33. Ǽr mon ða stánas tó ðæm stede brohte ðe hié on standan scoldon, Past. 36; Swt. 253, 15. Of hiora stede styrede, Met. 7, 25. On his ágenum stede, Ps. Th. 102, 21. Ne stande hé on his stede and endebyrdnesse, ac stande hé ealra ýtemest, R. Ben. 68, 10. Sig him geþafod, ðæt hé stede æfter ðam abbode healde, 106, 2. Æsc stede rihte hylt, Runic pm. Kmbl. 344, 26; Rún. 26. Nǽfþ náðer ne sǽ ne eá nǽnne stede búton on eorðan, Lchdm. iii. 256, 2. Gecerr suord ðín in styd his, Mt. Kmbl. Lind. 26, 52. II a. *place, standing, position, status :*—Ðes dǽl (*the participle*) næfþ nán angin ne nǽnne stede of him sylfum, ac byþ of worde ácenned and becymþ syþþan tó his ágenre geþingðe, Ælfc. Gr. 41; Zup. 244, 17. II b. *place, sphere of action :*—Gif ealle men on worulde ríce wǽron, ðonne næfde seó mildheortnys nǽnne stede, Wulfst. 287, 9. III. *of position in the case of a moving body :*—Ne stira ðú, sunne, of ðam stede furðor ongeán Gabaon . . . Ðá stód seó sunne on ðam stede, Jos. 10, 12, 13. IV. *standing as opposed to moving, stopping, standing still.* v. sunn-stede :—Hwæt is ðæs Hǽlendes stede oððe hwæt is his fær? Homl. Th. i. 156, 33. IV a. fig. *stability, unchanging condition, fixity :*—Nán stede nis úres líchaman ; cildhád gewít tó cnihtháde and cnihthád tó geðungenum wæstme, 490, 2. Stede ł staþal *statum, stabilitatem,* Hpt. Gl. 469, 12. IV b. *state, condition :*—Stede *status,* Wülck. 254, 31. On stede *statu,* Hpt. Gl. 458, 10. Swá hwæt swá stede (*statum*) módes áhwyrfþ, Scint. 106, 7. IV c. as a technical medical term *strangury :*—Wið stede and wið blǽddran sáre, Lchdm. i. 360, 4 : 338, 3. [*Goth.* staþs : O. Sax. stedi : O. Frs. sted, stid, steith : O. H. Ger. stat; *f. locus :* Icel. staðr.] v. æsc-, æl-, bæþ-, beorg-, burg-, camp-, deáþ-, ealh-, eard-, eolh-, eorþ-, folc-, gemót-, gener-, gléd-, heáfod-, heáh-, hleóðor-, hús-, land-, mearc-, meðel-, mylen-, sunn-, þing-, wang-, wíc-stede; cf. steall.

stede-fæst; *adj.* *Steadfast, constant, holding one's ground :*—Wíslíc wǽrscipe and stedefæst (styde-, MS. G.) módstaðol biþ witena gehwilcum weorðlícre micle, ðonne hé his wísan for ǽuigum þingum fǽgige tó swíðe, L. I. P. 10; Th. ii. 318, 38. Stódon stædefæste *they stood unyielding,* Byrht. Th. 133, 33 ; By. 127. Ne þurfon mé stedefæste hæleð wordum ætwítan, 139, 5 ; By. 249. [Icel. stað-fastr.]

stedefæstness, e ; *f.* *Steadfastness, constancy :*—Stydfæstnise *constantiae,* Rtl. 50, 4.

stedefulness. v. on-stedefulness.

stede-heard; *adj.* *Of enduring hardness (?), very hard :*—Strǽlas stedehearde, Judth. Thw. 24, 34 ; Jud. 223.

stede-leás; *adj.* *Without stability, unsteady, without power to retain one's place :*—Ðonne biþ hé ðam men gelíc, ðe árǽrþ sume heáge hlæddre and stíhþ be ðære hlæddre stapum, oð ðæt hé tó ðæm ænde becume, and wylle ðonne git stígan ufor; ástíhþ ðonne búton stapum, oð ðæt hé stedeleás fylþ, Homl. Skt. i. 1, 24. Stedeleáse steorran hreósaþ, Dóm. L. 107. [Icel. stað-lauss *unsteady.*]

stede-wang, es ; *m.* *A plain, open place :*—On ðam stedewange, Elen. Kmbl. 2040; El. 1021 : 1346; El. 675 : Andr. Kmbl. 1548; An. 775. Stedewangas, 667; An. 334. Æfter stedewonga stówum, Exon. Th. 154, 23; Gú. 847.

stede-wist, e ; *f.* *Stability, steadiness, constancy :*—Stedewist *subsistentia, perseverantia,* Hpt. Gl. 530, 4.

stedig ; *adj.* *Sterile, barren :*—Se ðe eardian déþ stedigne *qui habitare facit sterilem,* Ps. Lamb. 112, 9. Nǽron ðíne heorda stedige (*steriles*), Gen. 31, 38. Cf. (?) stede, II c, and see next word.

stedigness, e ; *f.* *Sterility, barrenness :*—Stedignysse sáwle mínre *sterilitatem animae meae,* Ps. Spl. 34, 14.

steding-líne, an ; *f.* *A rope that supports a mast, a stay :*—Stedinglíne *opisfera,* Wrt. Voc. i. 63, 61. S[t]edinglíne, 57, 2.

stefn, stemn, es ; *m.* I. *a stem of a tree :*—Hwæt wénst ðú for hwí ǽlc sǽd grówe innon ða eorþan and tó wyrtrumum weorþe on ðære eorþan, búton for ðý ðe hí tiohhiaþ ðæt se stemn and se helm móte ðý fæstor standon . . . Eal se dǽl, ðe ðæs treówes on twelf mónþum geweaxeþ, hé onginþ of ðám wyrtrumum and swá upweardes grěwþ óþ ðone stemn, Bt. 34, 10 ; Fox 148, 31–150, 2. Íc (*the cross*) wæs áheáwen holtes on ende, ástyred of stefne (*swefne,* Kemble) mínum, Rood Kmbl. 59 ; Kr. 30. Beám yldo ábreóteþ and bebriceþ telgum, ástyreþ stefn on síðe, áfylleþ hine on foldan, Salm. Kmbl. 594 ; Sal. 296. I a. fig. :—God is se stemn and staðol ealra góda, Bt. 34, 5 ; Fox 140, 2. Se ðorn

ðære gītsunga ne wyrð forsearod on ðæm helme gif se wyrttruma ne biþ færcorfen oððe forbærned æt ðæm stemne *si radix culpae non exuritur, nunquam per ramos avaritiae spina siccatur,* Past. 45, 3; Swt. 341, 11. **I b.** *a stem, stock, race.* v. leód-, þeód-stefn. **II.** *prow or stern* of a vessel:—Se æftera stemn *puppis,* Wrt. Voc. i. 63, 37. Tō lides stefne, Chr. 937; Erl. 112, 34. Of nacan stefne, Andr. Kmbl. 582; An. 291. Beornas on stefn stigon, Beo. Th. 429; B. 212. [*O. Sax.* stamn (*of a vessel*): *O. Frs.* stevne: *O. H. Ger.* stamm *stips, truncus, caudex* : *Icel.* stafn, stamn *prow* or *stern* of a vessel.] v. forþ-, frum-, steór-stefn; stefna, *and next word.*

-stefn, -stæfn; *adj.* v. brond-, heáh-stefn (-stæfn).

stefn, stemn, es; *m.* **I.** *a turn, time:*—Ðá besæt sió fierd hié (*the Danes*) ðær ða hwíle ðe hié ðær lengest mete hæfdon, að hié hæfdon heora stemn gesetenne *the English force had sat out its turn of service,* Chr. 894; Erl. 90, 31. But the word occurs mostly in phrases:—Ðá Noe ongan nīwan stefne (*anew, a second time*) hām staðelian, Cd. Th. 94, 2; Gen. 1555; Beo. Th. 5181; B. 2594. Eft . . . niówan stefne, 3582; B. 1789; Andr. Kmbl. 2607; An. 1305; Cd. Th. 113, 12; Gen. 1886. Hē hine Cyriacus syððan nemde nīwan stefne *he afterwards named him afresh Cyriacus,* Elen. Kmbl. 2119; El. 1061. Emb stemn *uicissim,* Germ. 388, 77. Emb stemn, Scint. 140, 17. **II.** *a body of persons who take their turn at any work* (v. fird-stemn), *the English military force* (?):—On stemnes peð (cf. here-paþ), Cod. Dip. Kmbl. v. 121, 33. v. stefnan, stefning.

stefn, stæfn, stemn, e; *f.* **I.** *a voice, sound uttered by the mouth* (lit. or fig.):—Stemn is geslagen lyft gefrēdendlīc on hlyste . . . Ælc stemn byþ geworden of ðæs mūðes clypunge and of ðære lyfte cnyssunge; se mūð drīfþ ūt ða clypunge, and seó lyft byþ geslagen mid ðære clypunge and gewyrð tō stemne. Ælc stemn is oððe andgytfullīc oððe gemenged; andgytfullīc stemn is, ðe mid andgyte biþ geclypod . . .; gemenged stemn is, ðe biþ būtan andgyte, swylc swā is hrýðera gehlōw and horsa hnægung, hunda gebeorc, treówa brastlung *et cetera,* Ælfc. Gr. 1; Zup. 4, 5–16. Stebn *vox,* Wrt. Voc. ii. 124, 18. Stefn of heofenum ðus cwæð, Mt. Kmbl. 3, 17: Mk. Skt. 1, 11. Seó ārleáse helwarena stefn wæs gehýred and heora gnornung, Blickl. Homl. 87, 3. Seó stemn ðære heortan biþ gedrēfed, 19, 9. Seó stemn ys Iacobes stefn, Gen. 27, 22. Seó stefen heom andswarode, Nicod. 24; Thw. 13, 5. Swā him seó stefen beád, Gl. Prud. 1 a. Sió býman stefne, Exon. Th. 65, 29; Cri. 1062. Heó clypode micelre stefne, Lk. Skt. 1, 42. Hē cūþre stæfne wæs hē sprecende, Bd. 4, 25; S. 600, 43. Gif ðū sanges stæfne gehýrdest, 4, 3; S. 568, 30. Hig gecnāwaþ his stefne, Jn. Skt. 10, 4. Hig mycelre stefne bǽdon ðæt hē wǽre āhangen; and hyra stefna swīðredon, Lk. Skt. 23, 23. Stæfna, Ps. Spl. 18, 3. Lǽðe cyrmdon fǽgum stæfnum, Cd. Th. 207, 5; Exod. 462. **II.** as a grammatical term, *form to mark relation*:—Se forma hād and se ōðer hād habbaþ ǽnlīpige stemna, forðan ðe hī beóþ ǽfre ætgædere and him betwýnan sprecaþ. Ðonne ic cwepe *ego* ic, and ðū cwest tō mē *tu* ðū, ðonne beó wyt ætgædere and for ði ne behōfaþ naðor ðissera *pronomina* nā mā stemna būton twegra. Se ðridda hād hæfþ syx clypunga, forðan ðe hē ys hwīlon mid, hwīlon on ōðre stōwe, Ælfc. Gr. 15; Zup. 93, 2–8. [*Chauc.* steven: the word is used by Gawin Douglas. *Goth.* .stibna: *O. Sax.* stemna: *O. Frs.* stemme: *O. H. Ger.* stimna, stimma, stemna, stemma.] v. þunorrād-, wæter-stefn.

stefn, e; *f. A summons, citation* (in\rād-stefn *a summons carried by a mounted person.* v. rād-stefn, where this meaning may be substituted for the one there given). [*Icel.* stefna *a summons, citation.*] v. stefnian.

stefna, an; *m. The prow or stern* of a vessel:—Lǽðe lides stefnan, Andr. Kmbl. 806; An. 403: 3411; An. 1709. Æt nacan stefnan, Exon. 306, 14; Seef. 7. Sum wǽg stefnan steórep, 296, 20; Crä. 54. Steóran ofer stæfnan, Andr. Kmbl. 989; An. 495. v. -stefn, *and next word.*

stefnan; *p.* de. **I.** *to regulate, direct, fix, institute :*—Hē stefnde Godes cyrican and Godes gesomnunga on ðære byrig eahta and twentig geára *he had the direction of God's church and God's congregations in that town eight-and-twenty years,* Shrn. 108, 6. Ongann timbrian ða stōwe ðæs mynstres ðe hē from ðam cyninge onfēng and mid regollícum ðeódscipum stæfnde *curavit locum monasterii, quem a rege acceperat, construere ac regularibus instituere disciplinis,* Bd. 3, 19; S. 547, 21 note. **II.** *to alternate :*—Staefnendra *alternantium,* Wrt. Voc. ii. 99, 74. Stefnendra, 6, 49. v. ge-stefnan; stefn; *m. a turn.*

stefnan, stefnian *to provide with a hem* or *börder, to fringe* [:—Bebyrde (cf. gebyrded *clabatum,* 104, 18. *Clavatum, sutum vel gebyrd,* 131, 57) oððe bestefnde *clavatae,* Wrt. Voc. ii. 20, 42. Gestefnode *clavate,* Anglia xiii. 37, 288.] v. stefning, II.

stefn-byrd, e; *f. Regulation, direction :*—Sceoldon eal beran stīþe stefnbyrd swā him se steóra bibeád missenlíce gemetu *all creatures had to submit to firm direction, as the guide ordered them, various modes,* Exon. Th. 349, 12; Sch. 45. v. stefnan *to regulate.*

-stefne; *adj. -voiced.* v. hlūd-stefne.

stefnettan, stemnettan; *p.* te *To stand firm* (?) :—Swā stemnetton stiðhugende hysas æt hilde, Byrht. Th. 135, 22; By. 122. [Hwi studgi ȝe nu and steuentið se stille, Kath. 59, 1265.]

stefnian; *p.* ode *To cite, summon* (with dat.) :—Stefnode man God-

wine eorle and Harolde eorle tō ðon gemōte . . . Ðá hí ðider cōmon, ðá stefnede heom man tō gemōte, Chr. 1048; Erl. 180, 3–6. Se cing him steofnode tō Glōweceastre, 1093; Erl. 228, 33. [Taken from Scandinavian (?); cf. *Icel.* stefna *to cite, summon* a person (*dat.*).]

stefnian. v. stefnan.

stefning, stemning, e; *f.* **I.** *a turn,* used of service where one set of persons replaces another. (In E. Cornwall Glossary *stemming* is given as 'a turn in succession, as when in dry seasons people have to take their regular turn for water at the common pump') :—Hié (seó fyrd) hæfdan heora stemninge (steminge, *another MS.*) gesetene, Chr. 894; Th. i. 166, col. 2, l. 14. v. stefn; *m.;* stefnan, II. **II.** *a border, hem :*—Stemning *vel* hem *limbus,* Wrt. Voc. i. 26, 6. v. limb-stefning; stefnan *to fringe.*

stela, steola, stæla; *m.* **I.** *the stalk* of a plant :—Steola *caulem,* Wrt. Voc. ii. 102, 53 : *cauliculus,* 103, 50 : 129, 84. Stela *caulem,* 13, 14 : *cauliculus,* 76, 11 : i. 33, 10. Healm *vel* stela *culmus,* i. *stramen spicarum,* ii. 137, 58. Sǽpig stela *succulentus cauliculus (ramusculus),* Hpt. Gl. 419, 45. Hyre (*leechwort*) stela byþ mid geþūfum bōgum, Lchdm. i. 248, 18. Genim ðysse wyrte wōs oððe ðone stelan mid ðam wæstme, 156, 21 : 160, 11 : 184, 20. Eleleáfes stelan, ii. 272, 23. Heó hafaþ nigon wyrttruman and swā fela stelena, i. 238, 17. Mid feówer reádum stǽlum (stelum, MS. B.), 154, 15. Genim nigon stelan, 230, 20. **II.** fig. :—Witan sceoldon smeágan hwilc ðæra stelenna ðæs cinestōles wǽre tōbrocen, and bētan ðone sōna. Se cinestōl stynt on ðisum þrīm stelum: *laboratores, bellatores, oratores,* Ælfc. T. Grn. 20, 15–19. [*O. H. Ger.* stil *thyrsus herbae* : *Ger.* stiel.] v. cawel-stela; -steled.

stelan; *p.* stæl, *pl.* stǽlon; *pp.* stolen *To steal* (with dat. of person from whom) :—Stilith *conpilat,* Wrt. Voc. ii. 105, 33. Stiled, 15, 32. Gif frigman frēum stelþ, L. Ethb. 9; Th. i. 6, 2. Se ðeo stelaþ on ðone dæg, ne geáhsaþ hit manna, Lchdm. iii. 178, 5. Stæl *conpilabat,* Wrt. Voc. ii. 22, 32. Wēnst ðū, ðæt wē ðīnes hlāfordes gold stǽlon, Gen. 44, 8. Ne stel ðū, Ex. 20, 15 : Mt. Kmbl. 19, 18. Ic stele *furer,* Kent. Gl. 1081. Þeóf ne cymþ būton ðæt hē stele *fur non uenit nisi ut furetur,* Jn. Skt. 10, 10. Gif frigman cyninge stele, L. Ethb. 4; Th. i. 4, 3. [*Goth.* stilan : *O. Sax. O. H. Ger.* stelan : *O. Frs. Icel.* stela.] v. be-, for-, ge-stelan; þeóf-stolen; stalian.

stēle *steel,* -steled. v. stíle, ān-steled, staled.

stellan, stillan; *p.* stealde; *pp.* steald. **I.** *to give a place to, set, place :*—Hē ōðrum yfele bisene stelep, Past. 28; Swt. 191, 12. Hwelce bisena hē ðær stellende wæs, Ors. 2, 2; Swt. 64, 24. **II.** *to take a place* (?), *to stand :*—Ðonne cumaþ upplíce eoredheápas stīþmǽgen āstyred styllaþ embūtan eal engla werod ēcne behlǽnaþ ðone mǽran Metod (cf. ðonne cumaþ ealle engla þreátas stīðe āstyrode standaþ ābūtan eall engla werod ēcne ymbtrynmaþ ðone mǽran kyning, Wulfst. 137, 14) *tum superum subito veniet commota potestas, coetibus angelicis regem stipata supernum,* Dōm. L. 114. [*Laym.* stalde; *p.:* A. R. stolde; *p.:* O. Sax. stellian : *O. H. Ger.* stellen.] v. ā-, on-stellan.

stellan; *p.* stealde, and stillan, styllan, stiellan; *p.* de *To leap, rush :*—Ðus hēr on grundum Godes ēce bearn ofer heáh hleoþu hlýpum stylde; swā wē men sculon heortan gehygdum hlýpum styllan of mægne in mægen, Exon. Th. 46, 28–36; Cri. 744–748. Ðonne hī ðæt mægen ðære unmētan hǽto ārǽfnan ne mihton ðonne stealdon hī eft on middan ðæs unmētan cyles mid ðý hí ðǽr nǽnige reste gemētan mihton stelldon (stældon, MS. T.) hī eft on middel ðæs unāðwæscendlícan líges *cum vim fervoris immensi tolerare non possent, prosiliebant in medium frigoris infesti; et cum neque ibi requiei invenire valerent, resiliebant rursus in medium flammarum inex̃inguibilium,* Bd. 5, 12; S. 627, 40–628, 1. Seó ofermōdnes stellan wile ofer eáðmōdnesse *superbia inruere vult super humilitatem,* Gl. Prud. 32 a. v. ā-, ge-, ofer-stellan (-styllan) ; still.

stel-mēle, es; *m. A vessel with a stem* or *handle :*—Stelmēlas, Anglia ix. 264, 11.

stel-scofl (?), e; *f.* The word apparently should mean *a shovel with a long handle* (v. stela), but it glosses *faselus :*—Steolscofle *faselo,* Germ. 400, 498.

stēm, stēman, stēming, steming, stemn, stemnettan, stemning. v. steám, stíman, stíming, steming, stemn, stefnettan, stefning.

stempan; *p.* te; *pp.* ed *To stamp, bray :*—Nim ysopo and stemp, Lchdm. i. 378, 20. [Cf. *O. H. Ger.* stampfōn *comminuere :* Icel. stappa *to stamp, bray.*] v. ā-stempan, *and next word.*

stemping-ísern, es; *n. A stamping-iron :*—Āgrafen, āstemped *cela-tum,* i. *pictum;* stempingísern *celon;* stempingísern *cilion, celox,* Wrt. Voc. ii. 130, 57–61.

stēnan; *p.* de. **I.** *to groan :*—Ic grymetige and stēne mid ealle mōde *rugiebam a gemitu cordis mei,* Ps. Th. 37, 8. [*Du.* stenen *to groan.*] **II.** *to cause to sound* (?) :—Com ðá wígena hleó þegna þreáte þrýðbord stēnan beaduróf cyning burga neósan (*came with clang of shields*), Elen. Kmbl. 302; El. 151. v. stinan.

stenc, es; *m.* **I.** *a smell, scent, odour :*—Ic eom on stence strengre ðonne rícels, Exon. Th. 423, 18; Rä. 41, 23. Stencas *sapores,* Kent. Gl. 1178. Mid ðære nose wē tōsceádaþ ða stencas, Past. 11, 2;

Swt. 65, 21. Gôde stencas and yfele, 56; Swt. 433, 22. **I a.** *a pleasant smell, fragrance, perfume:*—Ys mînes suna stenc swilce đæs landes stenc, đe Drihten bletsode, Gen. 27, 27. Swêtnys swâ đæra wynsumestra blôstmena stenc, Guthl. 20; Gdwin. 86, 19: Exon. Th. 363, 16; Wal. 54. Swîđe swête stenc, Blickl. Homl. 145, 29. Mycel swêtnysse stencg, Bd. 3, 8; S. 532, 18. Balzamum đæs betstan stences, Nar. 27, 22. Tô wynsumum stence, Lev. 1, 9. Đa swêtan stencas đara wuduwyrta, Blickl. Homl. 59, 3. Mid đâm fægrestum foldan stencum, Exon. Th. 198, 11; Ph. 8. **I b.** *an unpleasant smell, stench, stink:*—Se wôlberenda stenc đære lyfte, Bd. 1, 13; S. 482, 8. Đær slôh ût of đære niwelnysse ormæte stenc, Homl. Th. ii. 350, 25. Eall forwearđ for đæm stence, Ors. 5, 4; Swt. 226, 13. Stænce, 2, 6; Swt. 90, 1. Se lîchoma on đone heardestan stenc and on đone fûlostan biþ gecyrred, Blickl. Homl. 59, 12. Micgan stencgum *urinae foetoribus,* Hpt. Gl. 483, 3. Stencum, 516, 32: Homl. Th. i. 68, 7: ii. 374, 6. **II.** *the sense of smell:*—Stengc odor, odoratus, olfactus, Wrt. Voc. i. 42, 57, 61: Hpt. Gl. 488, 23. Đa fîf andgitu ûses lîchaman . . . stenc, Homl. Th. ii. 372, 26: i. 138, 27. [O. *Sax.* stank: O. *H. Ger.* stanc odor, odoratus, foetor.] v. ædel-, un-, wyrt-stenc; ge-, swôt-stence; *adj.*

stencan; *p.* te *To pant, emit breath with effort:*—Stenecendra renula *anhelantium cursorum,* Hpt. Gl. 406, 8. [Jamieson gives stank to gasp for breath. Cf. *Swed.* stânka *to pant.*]

stencan; *p.* te *To scatter:*—Se đe ne somnaþ se stenceþ *qui non congregat, spargit,* Mt. Kmbl. Rush. 12, 30. Đû somnast đær đû ne strenctæs (stenctæs ?, sprenctæs ?) *congregas ubi non sparsisti,* 25, 24. Đû stenctest (swenctest ?) đa eldeôdgan folc and hý âwurpe *afflixisti populos et expulisti eos,* Ps. Th. 43, 3. [*Goth.* ga-staggkwan *to dash: Icel.* stökkva *to cause to spring, sprinkle: Dan.* stænke *to sprinkle: Swed.* stänka *to sprinkle, scatter.*] v. tô-stencan; stincan *to spring.*

stenc-brengende; *adj. (ptcpl.).* Odoriferous:*—Stengcbrengendra blôstmana sigbêgo *odoriferas florum coronas,* Rtl. 77, 39.

stencednes, stencend, stencness, stencende. v. tô-stencedness, -stencend, -stencness, swôt-stencende.

stencness, e; *f. Scent, odour:*—Salde stencgnisse *dedit odorem,* Rtl. 4, 13.

steng, es; *m. A stang* (v. Halliwell's Dict.), *pole, stake, staff, cudgel, bar:*—Steng (stencg, stengc) *vectis,* Ælfc. Gr. 9, 28; Zup. 55, 10: Wrt. Voc. i. 26, 44. Stengc, 81, 29. Steng *clava,* ii. 104, 11: 14, 41: *claumentia,* 131, 55. Styng *clava,* i. 33, 60. Wiđ slege îsernes ođđe stenges (stænges, MS. H.), Lchdm. i. 132, 4. Wiđ wunda som hý sýn of îserne, som hý sýn of stence (stængce, MS. H.), 166, 10. Đâ hêt se dêma hî nacode gebindan tô ânum stænge, Shrn. 115, 13. Heáfod on steng (*stipitem*) âsettan, Bd. 3, 12; S. 537, 34. Stengcum *fustibus,* Hpt. Gl. 487, 48. Stencgum (stængum, Rush.), Mt. Kmbl. Lind. 26, 47. Stengum *sudibus,* Wrt. Voc. ii. 85, 53. Mid stengum đyrscan, Shrn. 55, 10. Mid stearcum stengum beátan, Homl. Th. i. 428, 6. Hât wyrcean twegen stengeas (stengas, Hatt. MS.) of đæm treówe đe is haten sethim *facies vectes de lignis sethim,* Past. 22, 1; Swt. 168, 22. [O. *H. Ger.* stanga; *f. fustis, vectis, contus: Icel.* stöng; *f. a pole.*] v. wîte-steng.

steola, steol-scofl. v. stela, stel-scofl.

steóp *a cup.* v. steáp.

steóp- *deprived of* a relative. The form seems to have been used in the first instance in combination with words denoting children, to mark loss of parents, and then to have been combined with father, mother to express the relation of one who married the mother or father of an orphan. It is a common Teutonic word. [O. *Frs.* stiap-, stiep-: *Du.* stief-: O. *H. Ger.* stiuf-: *Ger.* stief-: *Icel.* stjúp-: *Dan.* stif-: *Swed.* stjuf-, styf-.] v. stîpan, *and following words.*

steóp-bearn, es; *n. An orphan:*—Steópbearn *pupillus,* Ps. Vos. 81, 3. Đam steópbearne ic geheólp, Homl. Th. ii. 448, 14, 20. Đæt mann wydewan geneósige and steópbearnum gehelpe, Homl. Skt. i. 9, 63. [He scal biwerian widewan and steopbern, O. E. Homl. i. 115, 20. *Icel.* stjúp-barn.]

steóp-cild, es; *n.* **I.** *an orphan, one who has lost a parent:*—Steópcild *privignus,* Wrt. Voc. i. 50, 47: *pupillus,* 285, 1. Steópcilde *orphano,* Ps. Spl. 9 second, 17. Eówer bearn beóþ steópcild (*pupilli*), Ex. 22, 24. Steópcild *orphani,* Ps. Th. 108, 9. Heó wæs wuduwena and steópcilda ârigend, Lchdm. iii. 430, 1. Stêpcilda, Ps. Surt. 67, 6. Ne deriaþ wudewum and steópcildum, Ex. 22, 22: Blickl. Homl. 45, 1: Ps. Th. 108, 12. Đæt hî widuwan and steópcild gladian, L. Eth. vi. 47; Th. i. 326, 25. Steápcildo *pupillos,* Rtl. 29, 13. **II.** fig. *one deprived of protection:*—Wê wæron steópcild gewordene, forđan đe wê wæron âstýpede đæs heofenlîcan rîces, Wulfst. 252, 10. Ne læte ic eów steópcild, Jn. Skt. 14, 18.

steóp-dohtor; *f. A step-daughter:*—Steópdohter *filiaster,* Wrt. Voc. i. 72, 34. Stêpdohter, 51, 69. [Ic and Algîf mîn stêpdouter, Chart. Th. 583, 23.] [O. *H. Ger.* stief-tohter *filiastra: Icel.* stjúp-dôttir.]

steóp-fæder; *m. A step-father:*—Steópfaeder *bitricius,* Txts. 45, 300. Steópfaeder, steúpfaedcer, staupfoter, steúffeder *vitricius,* 107, 2124. Steópfæder *vitricus vel patraster,* Wrt. i. 52, 11: 72, 31: 284, 75: ii. 11, 10. Hê ofslôh ge his âgenne fæder ge his steópfæder (*vitricum suum*), Ors. 1, 8; Swt. 42, 22. [O. *Frs.* stiap-fader: O. *H. Ger.* stiuf-fater *vitricus: Icel.* stjúp-faðir.]

steóp-môdor; *f. A step-mother:*—Steópmôder *noverca,* Wrt. Voc. i. 72, 32: 284, 76. Steópmôdur, ii. 60, 33. Heó wæs Philippuses steópmôdor, Ors. 3, 7; Swt. 110, 26. Đæt mon hine menge mid his steópmêder, Bd. 1, 27; S. 491, 11. Steópmôdrum, S. 490, 35. Gê sume hæfdon eówre steópmôdur, Past. 32; Swt. 211 9. [O. *Frs.* stiap-môder: O. *H. Ger.* stiaf-môter: *Icel.* stiup-môðir.]

steóp-sunu, a; *m. A step-son:*—Steópsunu *filiaster,* Wrt. Voc. ii. 108, 69: 35, 61: *privignus,* i. 52, 12: 72, 33. Hê ofslôh his steópsunu, Ors. i. 8; Swt. 42, 22. [O. *H. Ger.* stiuf-sun *privignus: Icel.* stjúp-sonr.]

steór, es; *m. A steer, young bull, or cow:*—Đríuintri steór, steúr *prifeta,* Txts. 89, 1655: Wrt. Voc. ii. 68, 42. Steór *anniculus,* 10, 41: *juvencus, vel vitula,* i. 23, 43: *laudaris,* 287, 61: *ludares,* ii. 51, 22: *ludarius,* 113, 24. [*Goth.* stiurs *a calf: O. L. Ger.* stier *taurus: O. H. Ger.* stior *juvencus: Icel.* stjórr.] v. steór-oxa.

steór and stýr, e; *f.* **I.** *guidance, direction:*—Lâr vel steór *disciplina,* Wrt. Voc. i. 46, 57: 75, 31. Gyrd steóre *virga directionis,* Ps. Lamb. 44, 7. Đæt hê đoncfull sî stýre him đæs bebodenan folces *contentus sit gubernatione creditae sibi plebis,* Bd. 4, 5; S. 572, 33. God sette æ đam folce tô steóre, Ælfc. T. Grn. 5, 36: L. Eth. ix. 36; Th. i. 348, 14: L. Ælfc. P. 8; Th. ii. 366, 18: Boutr. Scrd. 18, 4. Gegrîpaþ stýre *adprehendite disciplinam,* Ps. Surt. 2, 12. **II.** *that which guides, a rule, regulation:*—Seó æ, đæt is se[ó] rihtwíse steór, ne gegrêt đone rihtwísan mid nânum yfele, Homl. Skt. i. 17, 19. Ælc mînra þegna đe đa steóre swâ healdan nelle swâ ic beboden habbe, L. Ath. v. 11; Th. i. 240, 21. **III.** *correction, discipline, reproof:*—Gif hê đâm receleásum stýrþ, đonne sceal his steór beón mid lufe gemetegod, Homl. Th. ii. 532, 12. Eallum him sceal beón ân steór and ân lâr æfter heora geearnunga anddyfene *una prebeatur in omnibus secundum merita disciplina,* R. Ben. 13, 7. Steór *correptio,* Scint. 117, 8. Đæt man cýde bûton steóre intingan, Homl. Th. ii. 590, 23. Wrænnes mid stîđre steóre lâre sî geweld *lascivia duro disciplinae paedagogio refrenetur,* Hpt. Gl. 432, 34: Homl. Th. i. 360, 18. Đæt wíse men sceolon settan steóre dysigum mannum, swâ đæt hî đæt dysig and đa undeáwas âlecgan, 268, 2. On steórum *in increpationibus,* Ps. Spl. 38, 14. [See O. E. Homl. i. 117, 21–35.] **IV.** *restraint, check:*—Đæt môd hæfþ fulfremedne willan tô đære wrænnesse bûtan ælcre steóre and wearne *animus voluptate luxuriae sine ullo repugnationis obstaculo delectatur,* Past. 11, 7; Swt. 73, 8. Đæs unrædes stîđferhđ cyning steóre gefremede (*checked that evil plan* (building the tower of Babel)), đâ hê reorde gesette eordbûendum ungelîce, Cd. Th. 101, 17; Gen. 1683. **V.** *punishment, penalty:*—Ic hæbbe gecoren hwæt seó steór beón mæge gif ænig man andbyrdnysse beginþ, L. Edg. S. 14; Th. i. 276, 31. Oft gé in gestalum stondaþ, đæs cymeþ steór of heofonum, Exon. Th. 132, 32; Gû. 481. Ægđer wære unnyt ge mildheortnes ge steór, gif hié ânlîpe wæron . . . Fordæm scel bión on đæm reccere đæt hê sié mildheortlîce wîtniende, Past. 17; Swt. 125, 3. Æfter đæs gyltes gemete sceal beón gelengen đære steóre gemet (*disciplinae mensura*), R. Ben. 48, 16. Đæt hî stýran (*punish*) ælcum đara đe đis ne gelæste . . . and on đære steóre ne sý nân forgifnes, L. Edg. S. 1; Th. i. 272, 8. Mid woruldcundre steóre *with punishment inflicted by the secular power,* L. Eth. vi. 50; Th. i. 328, 3. Mid worldlîcre steóre, ix. 15; Th. i. 344, 4: L. I. P. 2; Th. ii. 304, 18: Wulfst. 169, 8: 311, 16. Gif feohbôt ârîseþ swâ swâ woroldwitan tô steóre gesettan (*fixed as penalty*), L. Eth. vi. 51; Th. i. 328, 5. Đæt gehwilc man his teóđunge rihtlîce gelæste be đære steóre đe Eádgâr gelagede *under pain of the punishment that Edgar fixed by law,* Wulfst. 272, 8. Ceóse Dene be lagum hwylce steóre hý be đan healdan willaþ, L. Edg. S. 13; Th. i. 276, 28. Đâ âsende him God tô swýđlîce steóre (*he was carried away captive*), Homl. Skt. i. 18, 437. Tôscâdan ge on godcundan scriftan ge on woroldcundan steóran, L. Eth. vi. 52; Th. i. 328, 19. Hig gesetton woruldlîce steóra . . . and đa woruldbôte hig gesetton gemæne Criste and cyngé, L. E. G. prm.; Th. i. 166, 13. Geræde man fridlîce steóra and ne forspille for lytlum Godes handgeweorc, L. Eth. v. 3; Th. i. 304, 20: vi. 10; Th. i. 318, 3. Đonne wurđ seó heardnes stîđmôdre heortan gehnexad þurh grimlîce steóra and heardlîce đreála, Wulfst. 133, 19. Se rihtwísa ne þearf him ondrædan đa stîđan steóra đe Godes æ tæcþ, Homl. Skt. i. 17, 22. **V a.** *where the punishment is stated to be a money one, a fine, penalty:*—Đone feórđan pening on folclícre steóre, Chart. Th. 242, 30. Gif se landríca nelle tô steóre filstan *will not assist to levy the fine,* L. N. P. L. 54; Th. ii. 298, 19. [O. *H. Ger.* stiura *gubernaculum, clavus, stipendium.* v. Grmm. R. A. 298.] v. woruld-steór; steóran.

steór, es; *n. A rudder, helm.* [Itt iss sett att te ster to sterenn, Orm. 15258. Hys sterisman . . . the stere smote overe borde, Chauc. H. of F. i. 437. 3if he ne rau3te to þe stiere (steere, stere) þe wynde wolde þe bote ouerthrowe, Piers P. 8, 35. *Du.* stuur; *n. helm, rudder: O. Frs.* stiure: M. *H. Ger.* stiure; *n.: Ger.* steuer; *n.: Icel.* stýri; *n.*] v. steór-, steóres-mann.

steóra, stiéra, styra, an; *m. One who directs the course of a ship,* (a) lit.:—Steóra *gubernio,* Wrt. Voc. i. 48, 7: *gubernator,* 56, 17: *proreta,* ii. 69, 5: 75, 10. Swelce se stióra slêpe on midre sæ and forlure đæt stiórrôđur . . . Se biþ swîđe onlîc đæm stióran đe his stiórrôđor forliést on sæ *quasi dormiens in medio mari et quasi sopitus gubernator amisso*

clavo . . . Quasi clavum gubernator amittit, Past. 56, 3; Swt. 431, 30–36. Gelíc ðam scipe búton ǽlcum steóran, Basil admn. 6; Norm. 46, 21. (b) fig. :—God is steóra and steórróþer, forðæm hé reht and rǽt eallum gesceaftum, swá swá gód steóra (stióra, Cott. MS.) ánum scipe, Bt. 35, 3; Fox 158, 25. [Ilc ðhusent adde a meister wold, and under ðis ʒen steres ben, Gen. and Ex. 3413. *O. H. Ger.* stiuro *gubernator, nauclerus : Icel.* stjóri *a ruler* (poet.).] v. fore-, scip-steóra.

steóran, stióran, (and with umlaut) stiéran, steran, stíran, stýran; *p.* de. I. *to steer, guide a vessel* :—Sum [on] fealone wǽg stefnan steóreþ, Exon. Th. 296, 20; Crä. 54. Ic ǽfre ne geseah ǽnigne mann ðē gelícne steóran ofer stæfnan, Andr. Kmbl. 989; An. 495. Swíðe eáðe mæg on smyltre sǽ ungelǽred scipstiéra genóh ryhte stiéran, Past. 9; Swt. 59, 2. I a. fig. *to steer, guide, rule, direct* :—Se stiórþ ðam hrædwǽne eallra gesceafta *volucrem currum regit*, Bt. 36, 2; Fox 174, 20. Swá déþ ðæt mód, ðonne hit wacorlíce stiéreþ ðære sáwle *cum mens vigilanter animam regit*, Past. 56, 3; Swt. 433, 4. Stýrþ *regit*, Wülck. Gl. 254, 29. Steórdes *gubernasti*, Ps. Surt. ii. p. 188, 5. Se stýrde Dære mǽgþe *qui Deirorum provinciam gubernaret*, Bd. 4, 12; S. 581, 19 : 5, 23; S. 645, 38. Steóran and reccan ðone anweald ðe mē befæst wæs, Bt. 17; Fox 58, 27. Ða geornfulnesse ðe hē mid stióran scolde ðære sáwle and ðæm líchoman, Past. 56, 3; Swt. 431, 34. Þurh ðē ic ðys eówde stýran and rihtan [mihte], Blickl. Homl. 191, 28. Hē ða cyricean wæs reccende and stýrende *ecclesiam regens*, Bd. 5, 19; S. 639, 13. II. *to correct, restrain* a person (*dat.*) from wrong, (*gen.* or *prep.*) *give a right direction* to what is wrong :—Ic bēte sume leáse bóc oððe ic stýre (steóre, MS. H.) sumum stuntum menn *corrigo*, Ælfc. Gr. 28, 5; Zup. 173, 10. Se micla cræftiga hiertende tóscýfþ and egesiende stiérþ ofermētta mid ðære tǽlinge his hiéremonnum ðæt hé hié gebringe on lífe *magnus regendi artifex favoribus impellit, terroribus retrahit, ut auditores suos et descripto irreprehensibilitatis culmine restringat a superbia, et officium laudando, quod quaeritur componat ad vitam*, Past. 8, 1; Swt. 53, 16. Gif hē ðǽm rēceleásum stýrþ, ðonne sceal his steór beón mid lufe gemetegod. . . Wel déþ se ðe ungewittigum stýrþ mid swinglum, gif hē mid wordum ne mæg. Hit is áwriten: 'Ne biþ se stunta mid wordum gerihtlǽced,' Homl. Th. ii. 532, 11–15. Gif hē him sylfum stýrþ fram eallum stuntnyssum, Homl. Skt. i, 17, 22. Ðæt stýrþ (*checks*) ðam þurste, Lchdm. ii. 192, 11. Hē missenlíce monna cynne gielpes stýreþ, Exon. Th. 299, 20; Crä. 105. Swá biþ geóguðe ðeáw, ðǽr ðæs ealdres egsa ne stýreþ, 127, 25; Gú. 391. Gif bisceopas forgýmaþ, þæt hí synna ne stýraþ ne unriht forbeódaþ *if bishops neglect to restrain from sins and to forbid wrong*, Wulfst. 176, 29. Gif hē hit herede and on tyhte eft hé stiérde ðære gewilnunge *qui tamen laudans desiderium in pavorem vertit protinus*, Past. 8, 1; Swt. 53, 9. Iacobus his stírde *Jacobus prohibet*, 3, 1; Swt. 33, 10. Ðæt hē fram synnan gecyrre and óðrum mannum unrihtes stýre, L. Eth. vi. 42; Th. i. 326, 9: Wulfst. 308, 19. Mánfulra dǽda on ǽghwilcan ende stýre man swýðe, 309, 27. Gif seó wyrd swá hweorfan mót and ðú heore nelt stíran (steóran, Met. 4, 49), Bt. 4; Fox 8, 19. Stiéran sceal mon strongum móde, Exon. Th. 312, 13; Seef. 109. Stýran, 336, 18; Gn. Ex. 51. Ðæm sacerde náht ne fremaþ ðæt hē rihtwís beó gif hē ðám unrihtwísan nele hyra unrihtes stýran (cf. preósta nán ne wandige, ðæt hig ne bodigan ǽlcum men, hwæt him sig tó dónne and hwæt tó forgánne, Th. ii. 202, 11–13) *sacerdoti nihil prodest, quod ipse justus sit, si injustos pro injustitia eorum corrigere nolit*, L. Ecg. P. iii. 15 tit.; Th. ii. 196, 10. Se ðe wylle eard clǽnsian, ðonne mót hē georne ðyllíces stýran (steóran, MS. B.) *restrain such crimes*, L. C. S. 7; Th. i. 380, 9. Hē wolde ús mid liðnysse stýran, Homl. Th. i. 320, 10: Blickl. Homl. 63, 22. Him stýran cwom stefn *a voice came restraining Abraham from sacrificing Isaac*, Cd. Th. 204, 8; Exod. 416. Stýran his módes styrunge mid singalre gemetfæstnesse, Homl. Th. i. 360, 15. II a. *to keep back from what is good* :—Ic dysge dwelle and óðrum stýre *nyttre fóre I (night) lead the foolish astray and keep back others from a useful course*, Exon. Th. 393, 3; Rä. 12, 4. III. *to reprove, chide, rebuke* :—Se ðe steórþ þeóda *qui corripit gentes*, Ps. Lamb. 93, 10. Stiórde ł stiórend wæs him *comminatus est eis*, Mk. Skt. Lind. Rush. 8, 30. Seó menigo stýrde ðǽm blindan ðæt hē cleopode *the multitude rebuked him for calling out*, Blickl. Homl. 19, 5: 191, 12. Se hálga wer wordum stýrde unryhte ǽ (cf. Herod being reproved by John for Herodias his brother Philip's wife, Lk. 3, 19), Exon. Th. 260, 13; Jul. 296. Steórdon *increpabant*, Mt. Kmbl. Rush. 19, 13. Stiórdun *comminabantur*, Mk. Skt. Rush. 10, 13, 48. Swá hié him swýðor stýrdon, swá hé hlúdor cleopode, Blickl. Homl. 15, 21. Ná on ðínum yrre stýr ðú mē *neque in ira tua corripias me*, Ps. Lamb. 6, 2: Mt. Kmbl. 18, 15. God wolde stýran ðære nytennysse Cúðberhtes, and ásende án cild, ðæt hit his dyslícan plegan wíslíce ðreáde, Homl. Th. ii. 134, 5. IV. *to punish* :—Ðonne hý ágyltaþ him man stýre oððe mid swiðlícum fæstenum oððe mid teartum swingellum hý wylde *dum delinquunt, aut nimiis jejuniis affligantur, aut acribus verberibus coherceantur*, R. Ben. 54, 3. Ðonne beóde ic mínum gerēfan ðæt hí stýran ǽlcum ðara ðe ðis ne gelǽste . . . and on ðǽre steóre ne sý nán forgifnes, L. Edg. S. 1; Th. i. 272, 6. Swá hwilc ðissa (*various punishments*) swá man gerǽde; swá man mæg stýran,

and eác ðǽre sáwle gebeorgan, L. C. S. 30; Th. i. 394, 16. Hí sceoldan ðǽm unrihtdóndum mid grimnesse stēran; þeófum and mánswarum . . . sceolan ða dēman grimlíce stýran, Blickl. Homl. 63, 12–15. [Iesu Crist shall ben hæfedd to steorenn hemm, Orm. 1559. In yherde irened salt þou stere (*reges*) þa, Ps. 2, 9. Þu steorest te sea stream þ hit fleden ne mot fir þan þu markedest, Marh. 9, 34. *Goth.* stiurjan *to establish: O. Frs.* stiora, stiura *to steer; to hinder: O. H. Ger.* stiuren *gubernare, fulcire : Icel.* stýra, *to steer; to direct, govern.*] v. ge-, on-steóran (-stíran, -stýran); steór, steórend.

steór-bord, es; *n. Star-board, the right side of a ship looking forward* :—Hē lēt him ealne weg ðæt wēste land on ðæt steórbord and ða wídsǽ on ðæt bæcbord, Ors. 1, 1; Swt. 17, 10, 25. [Cf. *Icel.* stjórnborði: *Da. Swed.* styr-bord: *Du.* stuur-boord.]

steóre, an; *f. A regulation* :—Gif eówer hwilc forgýmeleásaþ and mē hýran nelle and emban ða steóran (steóra ?) swá beón nelle swá ic beboden hæbbe and on úrum gewritum stent, L. Ath. v. 11; Th. i. 240, 17. v. steór.

steórend, stýrend, es; *m.* I. *a ruler, governor* :—God, staðulfæst steórend, Andr. Kmbl. 2673; An. 1338. Stýrend, 241; An. 121. Drihten, ealra sceafta reccend and stýrend, Wulfst. 255, 18. II. *one who corrects, one who reproves* :—Stýrend *corrector, increpator*, Wrt. Voc. ii. 135, 82. v. steóran.

steórere, es; *m. A steerer* :—Hit wǽre swelce se stióra slēpe on midre sǽ . . . Ðæm stiórere biþ gelícost se mon ðe ongemong ðisses middangeardes costungum hine ágímeleásaþ, Past. 56, 3; Swt. 431, 31. [*O. H. Ger.* stiurari *gubernator, recuperator*.] v. steóra.

steóres-mann, es; *m. A steersman, one who guides a vessel, the captain of a vessel* :—Be ðon ðe mon on scipe bereáfod sý. Gif man beó æt his ǽhtan bereáfod, and hē wite of hwilcum scipe, ágyfe steóresman ða ǽhta, L. Eth. ii. 4; Th. i. 286, 17. [Steres-men *rulers of ten men*, Gen. and Ex. 3417. Twelue scipen weoren forloren, þa oðere weoren al todriuen, nes þer na steoresmon þat æuere aht cuðe þer on, Laym. 11985. Þe steoressmann aʒʒ lokeþþ till an steorne, Orm. 2135. *Swed.* styresman *a chief, ruler.*] v. steór-mann.

steórfa, an; *m.* I. *mortality, pestilence* :—Sceal áspringan wíde and síde stric and steorfa and fela ungelimpa, Wulfst. 86, 12: 159, 10. Gif hit geworde ðæt folce mislimpe þurh here oððon hungor, þurh stric oððe steorfan, L. I. P. 18; Th. ii. 324, 29. II. *flesh of animals that have died a natural death* :—Se ðe steorfan ete *qui morticinam ederit*, L. Ecg. P. iv. 27; Th. i. 212, 3. III. *a place where death has taken place* (?) :—Andlang mōres tó síferþingcsteorfan, Cod. Dip. B. i. 296, 34. [Stala and steorfa swiðe eow scal hene, O. E. Homl. i. 13, 29. *O. Sax.* man-sterbo : *O. H. Ger.* sterbo *pestis, cladis, pestilentia*.] v. fǽr-steorfa.

steorfan; *p.* stearf, *pl.* sturfon; *pp.* storfen *To die* :—Se ðe gelíð raðe hē styrþ oððe gēnunge hē áríseþ *he that takes to his bed (on the tenth day of the moon), soon will he die or he will be up again directly*, Lchdm. iii. 188, 21. Gif hrýðera steorfan, 54, 30. Annanias and Saphira mid fǽrlícum deáðe ætforan ðám apostolum steorfende áfeóllon, Homl. Th. i. 398, 34. [Se man þe nán gód ne heafde stærf of hungor, Chr. 1124; Erl. 253, 22. Wrecce men sturuen of hungær, 1137; Erl. 262, 27. Hi sturfe hungre, O. E. Homl. i. 233, 5. Caim starf (*died*), Gen. and Ex. 481. Summe storuen, 2975. Ilc was storuen, 3162. Steruyñ, *idem quod* deyyñ, Prompt. Parv. 474, col. 2. *O. Sax.* sterban: *O. L. Ger.* steruan: *O. Frs.* sterva: *O. H. Ger.* sterban.] v. á-steorfan.

steor-gleáw; *adj. Skilled in a knowledge of the stars* :—Steorgleáwra, tuncgelwítegana *mathematicorum*, Hpt. Gl. 467, 75.

steór-leás; *adj.* I. *without restraint, ungovernable, fierce* :—Sió rēþe oððe sió steórleáse *efferra*, Wrt. Voc. ii. 31, 16. II. *without regulation, profligate* :—Ðú cýþdest ðæt ðú nestest hwelces endes ǽlc angin wilnode ða ðú wēndest ðæt steórleáse men and rēceleáse wǽron gesǽlige and wealdendas ðisse worulde *quis sit rerum finis, ignoras, nequam homines atque nefarios, potentes felicesque arbitraris*, Bt. 5, 3; Fox 12, 35. III. *without instruction, foolish, ignorant* :—Þeáh hió (*the earth*) unwísum widgel þince, on stede stronglíc steórleásum men, Met. 10, 11. IV. *without rule, not living under a rule* :—Gif bescoren man steórleás (*not living under the rule of any religious house*) gange him on gestliðnesse, L. Wih. 7; Th. i. 38, 12. [Gif þu unel were, iwend þe from uuele, þi les þe ðu steorles losie on ende, O. E. Homl. i. 117, 35. Cf. *Goth.* libands usstiuriba ζῶν ἀσώτως, Lk. 15, 13. *Icel.* stjórn-lauss *unruly*.]

steór-mann, es; *m. A steersman, pilot, captain* :—Steórman *gubernio*, Wrt. Voc. i. 56, 18: *gubernator vel nauclerus*, 73, 79. Hera ðone steórman ac ná ǽrðan ðe hē becume gesundful tó ðære hýðe, Homl. Th. ii. 560, 22. [Stereman *proreta*, Wrt. Voc. i. 274, col. 2. He nom alle þa scipen and þa steormen alle, Laym. 28436. *Du.* stuur-man: *Icel.* stýri-maðr *a skipper, captain : Dan.* styr-mand *a mate: Swed.* styr-man.] v. steóres-mann, steór-rēþra.

steorn (?), e; *f. The forehead.* [*O. H. Ger.* stirna *frons.*] v. steornede.

steór-nægl (?), es; *m. The handle of a helm* :—Steórsceofol oððe [steór-?]nægl *clavus*, Wrt. Voc. i. 74, 3. [*O. H. Ger.* stiur-nagal *clavus.*]

steornede; adj. Having a big forehead; fig. bold, active :—Steornede (the word occurs in a list of adjectives denoting the possession of physical characteristics) frontalis vel calidus, Wrt. Voc. i. 45, 36. Steorrede (steornede?) frontalis, ii. 38, 55 : 151, 25.

Steórnes-healh. v. Streónes-healh.

steór-, stiér-, stýr-ness, e; f. Correction, discipline :—Hine sylfne dreágian mid stýrnysse ðære gástlícan steóre, Homl. Th. i. 360, 17. Hwîlon hê gewîtnaþ ðæs mannes gewitleáste mid stýrnysse ôðrum tô steóre, Homl. Ass. 62, 259. Stiérnesse disciplinam, Ps. Spl. T. 2, 12. [Cf. O. H. Ger. stiurida gubernatio.]

steór-oxa, an; m. A steer :—Steóroxa anniculus vel trio, Wrt. Voc. i. 23, 41. [Ger. stier-ochs a bull.]

steorra, an; m. A star :—Steorra stella, tungel sidus, Wrt. Voc. i. 41, 53. Swâna steorra hesperius, ii. 43, 39. Se hâra (hâta?) steorra canis vel canicula, stella quae Sirius vocatur, 128, 25. Se steorra ðe wê hâtaþ Ursa ne cymþ næfre on ðam westdæle, þeáh ealle ôþre steorran faren æfter ðære sunnan, Bt. 39, 13 ; Fox 232, 29–32. Se steorra (stearra, Lind.) ðe hî on eástdæle gesâwon, Mt. Kmbl. 2, 9. Steorra, se is cweden commeta, Bd. 4, 12 ; S. 581, 13. Beorhtnes scînendes steorran, 5, 12 ; S. 629, 5. Stiorran, Met. 28, 44. Ðone beorhtan steorran ðe wê hátaþ morgensteorra ; ðone ilcan wê hátaþ þone naman æfensteorra, Bt. 4 ; Fox 8, 2–4. Tâcna on steorrum, Lk. Skt. 21, 25. [O. Frs. stera : O. Sax. O. H. Ger. sterro : Goth. stairnô ; f. : O. H. Ger. sterno : Icel. stjarna ; f.] v. æfen-, dæg-, heofon-, morgen-, sæ-, scip-, swân-steorra.

steór-réðra, an; m. A steersman, skipper, captain :—Crist wæs· on ðæm scipe swâ se steórréþra ... Andreas âstâg on ðæt scip and gesæt be ðæm steórréþran, Blickl. Homl. 233, 4, 24 : 235, 23. v. steór-mann.

steór-róðor (-er, -ur), es; n. A rudder, lit. and fig. :—Steórróþer remus (an oar used for steering), Wrt. Voc. i. 48, 11. Steórróðer palmula, ii. 67, 68. Steórróðor, 116, 52. Steórróþur gubernaculum, i. 63, 52. God is steórróþer and helma clavus atque gubernaculum, Bt. 35, 3 ; Fox 158, 25. God æghwæs wealt mid ðæm helman and mid ðæm stiórróþre his gôdnesse Deus omnia bonitatis clavo gubernare credatur, 35, 4 ; Fox 160, 15. Steórróðre (stiór-, Cott. MS.), 35, 5 ; Fox 164, 28. Swelce se stióra slêpe and forlure ðæt stiórróður (clavum) ... Se dêþ swâ se stióra ðe ðæt stiórróðor forliésþ, Past. 56, 3 ; Swt. 431, 30–33. [O. H. Ger. stiur-ruodar gubernaculum, clavus, artemo.]

steor-sceáwere, es; m. I. an observer of the stars, an astronomer, astrologer :—Up on ðæm rodore ðara steorsceáwera Epicurii, Wrt. Voc. ii. 32, 4. [Cf. O. H. Ger. himil-scouwari mathematicus ; sterro-wartal magus.] II. a constellation (?) :—Steorrscêwere (sceorr-, Wrt.) constellationem, Wrt. Voc. ii. 79, 66. v. steor-wigle.

steór-scofl, e ; f. A rudder :—Steórsceofl gubernaculum, Wrt. Voc. i. 56, 46. Steórsceofol clavus, 74, 3.

steór-setl, es ; n. The steering-seat, the stern :—Steórsetl puppis, Wrt. Voc. i. 48, 10 : 56, 55 : 64, 5 : Ælfc. Gr. 9, 78 ; Zup. 75, 12. Scip oððe steórsetl puppis, 9, 28 ; Zup. 56, 10. Se Hælend weard on slæpe on ðam steórsetie erat in puppi dormiens (Mk. 4, 37), Homl. Th. ii. 378, 17.

steór-stefn, es ; m. The stern, poop :—Steórstefn puppis, Wrt. Voc. ii. 73, 28.

steort, es ; m. I. a tail, start (as in red-start, one of the names for ruticilla phoenicurus, also called fire-tail. Start, plough-start = plough-tail, v. Halliwell's Dict. Stark-naked is a corruption of start-naked) :—Steort cauda, Wrt. Voc. ii. 103, 20 : 129, 75. Se hâlga stert sacra spina, i. 283, 50. Ðære helle hund ongan fægenian mid his steorte, Bt. 35, 6 ; Fox 168, 17. Nym hyre (the adder's) steort (caudam), Ex. 4, 4. Sume wyrmas wǽren and sume fiscas ðe hæfden ân heáfod and monigne steort. Ða steortas, hê sǽde, ðæt hulpan ealle ðæs heáfdes, Shrn. 162, 14–16. II. a promontory, tongue of land (cf. Start Point in Devon, Start Island in the Orkneys) :—Andlang weges ðæt hit sticaþ on norðeweardum cynges steorte, Cod. Dip. Kmbl. iii. 48, 9. Of ðæm weall tô steorte, 464, 25. Be gemǽre ðæt on ðone steort ; of ðam steort on ða strǽt, 438, 22. Oð ðone steort ; fram ðam steorte andlang ðæs fûlan brôces, ii. 250, 22. Cf. Penwiht-steort the Land's End, Chr. 997 ; Erl. 135, 10. [Ðe leun drageð dust wið his stert ðer he steppeð, Misc. 1, 9. Stert of an appull, of a handle of a vessel, of a plow, Prompt. Parv. 474, col. 2. See also Cath. Angl. 363, nn. 2, 3. O. Frs. stert tail : Du. staart : O. H. Ger. sterz stiva : Ger. sterz tail ; plough-tail : Icel. stertr tail : Dan. stjert : Swed. stjert tail ; plough-tail.] v. rysc-steort.

steor-wigle, -wigl (?), es ; n. Prognostication by the stars, astrology :—Stiorwigle ꝉ mearcunge constellationem (cf. constellatio leáses spelles talu, Wrt. Voc. ii. 20, 68 ; and Span. constelacion prognostication of the stars), stiorwiglu constellationes, Hpt. Gl. 467, 78. Stiorwigl (-wiglunge?) astrologiam, 528, 64. v. steor-wiglung, wigle.

steor-wiglung, e ; f. Astrology :—Æfter steorwiglunge juxta constellationem, Anglia xiii. 33, 141. v. steor-wigle.

steór-wîrðe ; adj. Deserving reprobation :—Ðonne wê hwæthwugu stiórwierðes ongietaþ on ða ðe ûs underðiédde bióþ cum ea quae in subditis arguenda cognoscunt, Past. 28, 4 ; Swt. 194, 3.

stépan ; p. te To cause to take a step, to initiate :—Gistoepid initiatum, Wrt. Voc. ii. 112, 2. Gestéped, gehálgodne initiatum, 45, 70. Cf. stæppan ; p. stôp. v. (?) on-stépan.

stépan to bereave, stépan to exalt, stepe, -stéped, stépel, stépness, steppan, steppe-scôh, stêr. v. stîpan to bereave, stîpan to exalt, stæpe, stépan, stîpel, stîpness, stæppan, stæppe-scôh, stǽr.

stéran ; p. de. I. to cense, burn incense as a sacrifice :—Aaron stêrde mid thimiama, Num. 16, 47. Ozias wolde offrian and stêrde æt ðam weofode (Uzziah went into the temple to burn incense upon the altar of incense, 2 Chron. 26, 16), Homl. Ass. 58, 185. Nim ǽlc his stôrcillan and stêre ætforan Gode, Num. 16, 7. Stþérde (=stêrde) adoleret, sacrificaret, Hpt. Gl. 509, 59. Stêrden thurificarent, 513, 69. Tô stýrenne ad thurificandum, ad sacrificandum, 477, 66. II. to perfume a person as with incense :—Stêr (stýr, MS. B.) hyne mid ðære wyrte, Lchdm. i. 98, 19 : 206, 2. [þer ne schulen heo helle stenches stinken, þer me schal ham steoren mid guldene chelle, O. E. Homl. i. 193, 45.] v. stôr, stêring.

sterced-ferhþ. v. stærced-ferhþ.

stêring, e ; f. Incense :—Stêmendre stêrincge fragrantis incensi, Hpt. Gl. 441, 73. v. stêran.

ster-melda, an; m. The word occurs in the following apparently corrupt passage :—Gif frigman mannan forstele gif hê eft cuma stermelda secge an andweardne gecænne hine gif hê mǽge if a freeman steal a man ; if he (the man who has been stolen) come back to give information of the theft, let him make his charge against the thief when the latter is present ; let him (the thief) clear himself if he can, L. H. E. 5 ; Th. i. 28, 10. In the note on this passage stermelda is taken as steórmelda = delator fiscalis ; Schmid, on the other hand, gives the meaning ' delator qui rem, factum (stær) prodit.' Perhaps for stermelda might be written stelmelda, a sense which has been given in the translation above.

stern, stert, stete, stéþa. v. stearn, steort, stîle, stéda.

stic[c] (?); adj. Sticky, viscous :—Wið ômena geberste.... Sleah feówer scearpan ymb ða poccas and lǽt yrnan ðæt sticce (the sticky matter) ðe hit wille, Lchdm. ii. 100, 4.

stic-ádl, e ; f. Stitch, pain in the side :—Sticwærc, sticádl telum, i. dolor lateris, Wrt. Voc. i. 19, 23. v. stice.

sticca, an; m. I. a stick, peg :—Sticca gergenna (gergenna lignum teres, quo per duas ansas transmisso operculum firmatur ne excidat, Migne), Wrt. Voc. i. 287, 38 : ii. 41, 32. Se sticca (the tent-peg) him eode ût þurh ðæt heáfod in tô ðære eorþan, Jud. 4, 21. Styre mid sticcan, Lchdm. ii. 76, 26. Genim twegen sticcan feðerecgede and wrît on ǽgðerne sticcan be hwælcere ecge, i. 386, 4–6. Nim ǽnne sticcan and gníd tô sumum þinge, iii. 274, 3. II. a. the pointer of a dial :—Se sticca on ðæm dægmæle, Anglia viii. 317, 20. II. a spoon (cf. spôn) :—Lǽt yrnan ðæt blôd on grênne sticcan hæslenne, weorp ðonne ofer weg âweg, Lchdm. ii. 142, 20 : 144, 7 : 104, 7. Genim fíf sticcan fulle ecedes, i. 110, 21 : iii. 4, 18. Wring ðæt wôs of, ǽnne sticcan fulne, and huniges þrý sticcan fulle, 102, 14. Nim wîfes meolce þrý sticcæs fulla and cylepena ǽnne sticce fulne, 96, 27. [O. H. Ger. steccho palus, paxillus, fustis, clavus : Icel. stika ; f. a stick.] v. candel-, clader-, geoc-, plant-, regol-, seám-, stôr-, tôþ-sticca.

sticce sticky matter. v. stic[c]. Sticce a piece. v. stycce.

stice, es; m. I. a prick, puncture, stab, thrust with a pointed implement :—Se ðe ûs gehǽleþ from ðæm stice ûrra synna hê geðafode ðæt him mon sette ðyrnenne beáh on ðæt heáfud a peccatorum nos punctionibus salvans spinis caput supponere non recusavit, Past. 36, 9 ; Swt. 261, 13. Gif man þeóh þurhstingþ, stice gehwilce .vi. scillingas, L. Ethb. 67 ; Th. i. 18, 16. II. a pricking sensation, a stitch :—Gif stice bûtan innoðe sié, Lchdm. ii. 274, 28. Wið miltewærce and stice, 174, 4. Se hwîta stân mæg wiþ stice, 290, 10. Wið eágena hǽtan and stice, i. 352, 5. [Wið gestice, 393, 20.] [In his soule he hefde þe stiche of sore pine. ... Þeos stiche was þreuold, þet, ase þreo speres smiten him tô þer heorte, A. R. 110, 12–14. Stiches iþi lonke, H. M. 35, 26. Styche, peyne on þe syde telum, Prompt. Parv. 475, col. 1. Goth. stiks a point of time : O. Frs. steke a prick, stab : O. H. Ger. stih[h] ictus : Ger. stich a prick, stitch, puncture : Dan. stik a stab : Swed. stick a prick, stitch, stab.] v. fǽr-, in-stice ; stic-ádl, and next word.

sticel, es ; m. That with which a prick may be given, (stickle in stickleback ; cf. stickly prickly, Halliwell's Dict.) a sting, goad :—Ôðerne hê dráf mid sticele, ôðrum hê wiðteáh mid bridle illum stimulo impellere nititur, hunc freno moderatur, Past. 40, 3 ; Swt. 293, 1. Hê sǽwþ ðone sticel ðæs andan seminantur stimuli, 38, 7 ; Swt. 279, 9. þê mid stîðum âstyrest sticelum gǽlsan luxuriae stimulis te agitabis acutis, Dôm. L. 179. Ða gnættas mid swîþe lytelum sticelum hine deriaþ, Bt. 16, 2 ; Fox 52, 11. Sticelas ramnos, Blickl. Gl. [O. H. Ger. stihhil aculeus : Icel. stikill the pointed end of a horn.] v. sticels.

sticel ; adj. v. sticol.

sticels, es ; m. A goad, stimulus, thorn (lit. and fig.) :—Sticels aculeus, Wrt. Voc. i. 75, 2. Sticels (not sticel) vel gâðiscn, 15, 15. Mê is geseald sticels mînes lîchaman. ... Ic bæd mînne Drihten ðæt hê âfvrrode ðæs sceoccan sticels fram mê (there was given to me a thorn in the

flesh. . . . *I besought the Lord, that it might depart from me,* 2 Cor. 12, 7–8), Homl. Th. i. 474, 12–15. Sticelse *stimulo, monitione,* Hpt. Gl. 420, 45. Se yfela gást hine drehte mid deófollícum sticelsum, Homl. Skt. i. 18, 10. Sticelsas *rhamnos,* Ps. Spl. 57, 9. v. sticel.

stic-fód[d]er *a case for pegs* (? v. sticca, I), *a case for spoons* (? v. sticca, II), *a case made of twigs* (? cf. stic-tǽnel) :—Man sceal habban . . . sealtfæt, sticfódder, piperhorn, Anglia ix. 264, 19.

stician; *p.* ode. **I.** *trans. To stick, stab, pierce, prick* :—Oxa spæc and cwæð : 'Tó hwon sticast ðú mé,' Shrn. 30, 12. Mé on fæðme sticaþ hygegálan hond, Exon. Th. 394, 1 ; Rä. 13, 11. Hé (*the wounded elephant*) ða óþre elpendas sticade, Ors. 4, 1 ; Swt. 156, 13. Gé hyne (*Christ*) myd spere sticodon, Nicod. 13 ; Thw. 6, 35. Sticedon, Cd. Th. 297, 1 ; Sat. 510. Stycodon, Shrn. 147, 36. Hí ne mihte þorn stician, 66, 17. Stycigende *stimulosa,* Scint. 104, 6. **I a.** *to kill* (*to stick* is still used of killing pigs. Cf. sticung, II) :—Wé oþþe sticode beóþ oþþe on sǽ ádruncene *aut jugulamur aut mergimur,* Bd. 1, 13 ; S. 482, 1. Monige fanggene wǽron and heápmǽlum sticode *nonnulli comprehensi acervatim jugulabantur,* 1, 15 ; S. 484, 5. **I b.** *to thrust out* (cf. stingan) :—Sticode him mon ða eágan út *effossis oculis,* Ors. 4, 5 ; Swt. 168, 4. **I c.** *intrans.* :—Ðæt mé ongeán sticaþ, Exon. Th. 497, 20 ; Rä. 87, 3. **II.** *intrans.* (1) *To stick, remain fixed* :—Ðæs spácan sticaþ óþer ende on ðære felge, óþer on ðære nafe, Bt. 39, 7 ; Fox 222, 7. Lǽt ða ságlas stician ðǽron . . Ða ságlas sticiaþ eallne weg inn on ðám hringum . . . Simle ða ofergyldan ságlas sceolden stician on ðǽm gyldnum hringum, Past. 22 ; Swt. 171, 1–22. Mé on hreðre heáfod sticade, Exon. Th. 479, 10 ; Rä. 62, 5. On ðære róde sticodon mænige arewan, Chr. 1083 ; Erl. 217, 21. Sting ðín seax on ða wyrte, lǽt stician ðǽron, Lchdm. ii. 346, 12, 20 : Jud. 3, 22. (2) fig. *to be involved, be prevented from free action, lie encumbered* :—On hú ðióstrum hora seáþe ðara unþeáwa ða yfelwillendan sticiaþ *quanto in coeno probra volvantur,* 37, 2 ; Fox 188, 2. Sticiaþ gehýdde beorhte cræftas *latet obscuris condita virtus clara tenebris,* 4 ; Fox 8, 15. Ðæt ða synfullan sáwla sticien helle tómiddes, Salm. Kmbl. 344 ; Sal. 171. (3) *to be inherent* :—Seó godcundnys ðe on ðam men sticode, Homl. Th. ii. 386, 19. (4) *to be in possession of* (of demoniacal possession), *to lurk* :—Deófol ðé sticaþ on *daemonium habes,* Jn. Skt. 7, 20. 'Ðonne gesihst ðú hwæt ðǽron sticaþ' . . . Ðǽr gewende út of ðam fæte án næddre, Homl. Th. ii. 170, 19. Wé bebeódaþ ðam deóflum ðe on ðisum anlícnyssum sticiaþ, ðæt hí út faron, 496, 8. Se apostol cwæð tó ðam áwyrgedan gáste ðe hire on sticode, i. 464, 20. Ða deóflu ðe on ðam anlícnyssum sticodon, ii. 482, 8. **III.** of direction, *to run, lie* (cf. sceótan) :—Út æt ðæs croftes heáfod ðæt sticaþ on ðære lace, Cod. Dip. Kmbl. iii. 37, 24. Andlang weges ðæt hit sticaþ on norðeweardum cynges steorte . . . andlang weges ðæt hit sticaþ æt wícham, 48, 8–11. Ðonne swá ford ðæt hit sticaþ on miclancumb ; and of miclancumbe ðæt hit sticaþ on litlancumb, 405, 30. Ðonne tó ðam wuduwege ðæt hit sticaþ innan Nodre ; ðonne andlang Noddre ðæt hit sticaþ on Eatstánes landscare ; ðæt hit sticaþ up tó herpoðe, 446, 8–11. Wið súðan ða méde ðæt it sticaþ tóemnes ðam wiðigðyfelum, v. 194, 32. [*M. H. Ger. Ger.* stecken *to remain fixed.* Cf. *O. Sax.* stekan ; *p.* stak *to pierce, stab : O. Frs.* steka : *O. H. Ger.* stehhan ; *p.* stah *pungere.*] v. of-, tó-, þurh-stician.

sticol; *adj.* **I.** *lofty, reaching to a great height,* of a mountain :— 'Ic wille standan on ðisum steápum munte' . . . Moyses ðá ástáh tó ðam sticolan munte, Homl. Skt. i. 13, 9–12. Wæs án myrige dún . . . ful smeðe . . se streám arn of ðære sticolan dúne, 19, 108–115. Hét hí ástígan tó ánre sticolre dúne, 3, 235. **II.** *lofty, placed high, situated at a great height* :—Wé biddaþ ðæt ðú ástíge tó ðam sticelan scylfe, Homl. Th. ii. 300, 1. Martinus ástáh on ðam sticelan hrófe, 510, 7. Eraclius ástáh tó ðære sticolan upflóra, Homl. Skt. ii. 27, 67. Ástáh heofonan sticole *conscendit caelos arduos,* Hymn. Surt. 89, 8. **III.** *rough, rugged, difficult, steep* (Halliwell gives *stickle* as a Devonshire word = steep) :—Sticol *asper,* Wülck. Gl. 256, 32. Se weig is swíðe nearu and sticol, se ðe lǽt tó heofonan ríce . . . Ðonne máge wé ðurh Godes fultum ástígan ðone sticolan weg, ðe ús gelæt tó ðam écan life, Homl. Th. i. 162, 23–35. Se weg is rúm and fordheald, ðe tó déaðe lǽt ; se is neara and sticol, ðe tó lífe lǽt, R. Ben. 5, 21 : Shrn. 12, 19. On wyrmes líc sticoles (*rough, scaly*), Salm. Kmbl. 307 ; Sal. 153. Be westan róde óð sticelan stíg, Cod. Dip. Kmbl. iii. 406, 29. Sticule scylpas *scabri murices,* Germ. 399, 446. **III a.** *difficult, arduous* :—Sticol *arduam (rem),* R. Ben. Interl. 16, 1. [*O. L. Ger.* stecul *confragosus, fragosus, preruptus : O. H. Ger.* stechal *arduus, asper, fragosus, praeceps, praeruptus, abruptus.*]

stic-tǽnel *a wicker basket* :—Sticténel *fiscillus,* Wrt. Voc. ii. 108, 55. Sticténel *fiscilus,* 35, 37.

sticung, e ; *f.* **I.** *a pricking, piercing* :—Hié (*the elephants*) fóran wédende ǽgðer ge for ðæs flexes bryne ge for ðara nægla sticunge, Ors. 4, 1 ; Swt. 158, 8. **II.** *sticking* (pigs), *killing* ; cf. stician, I a :—On manegum stent ðæt se gafolswán sylle ǽlce geáre .xv. swýn tó sticunge . . . Gýme eác swán ðæt hé æfter sticunge his slyhtswýn wel behweorfe, L. R. S. 6 ; Th. i. 436, 12–16.

stic-wærc. v. stic-ádl.

stic-wyrt, e ; *f. Stitch-wort* ; stellaria holostea ; but the word glosses *agrimonia,* Wrt. Voc. i. 32, 2.

stiell, stiém, stiép, stiéra, stiéran, stiérness, stiernlíce. v. still, steám, stíþ, steóra, steóran, steóran, steórness, stiernlíce.

stif ; *adj. Stiff, unbending, rigid* :—Stífne *rigentem,* Germ. 394, 272. [He ches stiue here to shurte, O. E. Homl. ii. 139, 16. He (*the dead man*) biþ sone stif, Fragm. Phlps. 5, 45. Stif he wes on þonke, Laym. 2110. Sa strang and stijf in fight, C. M. 18140. Þat plaid (*plea*) was stif and starc and strong, O. and N. 5. *Du.* stijf : *M. H. Ger.* stíf : *Ger.* steif : *Dan.* stiv : *Swed.* styf.] v. stífian.

stí-ferh. v. stig-fearh.

stífian; *p.* ode *To be or to become stiff* :—Ic stífige *rigeo,* Ælfc. Gr. 26, 2 ; Zup. 154, 15. Ic stífie *obrigesco,* Wrt. Voc. i. 22, 32. Stífodan *rigebant, durescebant,* Hpt. Gl. 483, 68. v. á-stífian.

stíflcian. v. stýfician.

stíg, e ; *f. A path* (lit. and fig.), *footpath,* (*narrow*) *way* :—Orweg stíg (given already as a compound, orweg-stíg, *but* orweg *should be taken as adjective*) *devia callis,* Wrt. Voc. ii. 139, 57. Horweg stíg, 25, 25. Horuæge stíig, Txts. 56, 340. Strǽt wæs stánfáh, stíg wísode gumum ætgædere, Beo. Th. 646 ; B. 320 : 4433 ; B. 2213 : Andr. Kmbl. 1970 ; An. 987. Eástewearde andlang weges on hemléclǽge ; eástewearde andlang stíge on Ulfan treówe, Cod. Dip. Kmbl. iii. 437, 4. Of Heortwyllan on ða ealdan stíge ; ðæt andlang stíge, 438, 34. Leóht stíge mínre *lumen semitis meis,* Ps. Th. 118, 105. Stíge *calce* (*calle?*), Wrt. Voc. ii. 15, 66 : 95, 74. Gebígdre stíge *flexo tramite,* 149, 46 : Hpt. Gl. 493, 18. Fram stíge *tramite, via,* 486, 68. Tó rihtre stíge geteón *ad rectum tramitem revocare,* Bd. 5, 9 ; S. 623, 13 : 1, 12 ; S. 481, 8. Ðú ná forfleó [weg] hǽle se ðe nis búton mid stíge tó onginnenne *non refugias viam salutis que non est nisi angusto initio incipienda,* R. Ben. Interl. 6, 8. Be westan róde óð sticelan stíg ; ðonne be ðære stíge óð ða ealdan díc, Cod. Dip. Kmbl. iii. 406, 29. Of ðam stáne tó ðære grénan stíge, 38, 23. Ðýlæs ða gongen on suá frécne stíge ða ne mágon uncwaciende gestondan on emnum felda *ne, qui in planis stantes titubant, in praecipiti pedem ponant,* Past. 4, 2 ; Swt. 41, 7. Geseoh nú seolfes swæðe, swá ðín swát ágeát, blódige stíge, Andr. Kmbl. 2883 ; An. 1444. Stíga ðíne *semitae tuae,* Ps. Spl. 76, 19. Stíge (*semitas*) ðíne lǽr mé, Ps. Surt. 24, 4. Gif se níðsceapa nearwe stíge mé on swaþe séceþ *if the foe seek narrow paths in my track,* Exon. Th. 397, 24 ; Rä. 16, 24. Steáp stánhlíðo, stíge nearwe, enge ánpaðas, Beo. Th. 2823 ; B. 1409. [We sculde makien hi stiзes, O. E. Homl. i. 7, 1. He sende bi stiзen (weies, 2nd MS.) and by straten, Laym. 16366. Þiss Lamb iss þatt rihhte stih, Orm. 12916. Rihhteþþ Drihhtiness narrwe stiзhess, 9202. Sty, by pathe *semita, callis,* Prompt. Parv. 475. v. in Halliwell's Dict. *stie,* and cf. *Stye-head,* the pass from Borrowdale to Wastdale. *O. H. Ger.* stíga *semita, trames, callis : Icel.* stígr ; *m. a path, footway.* Cf. *Goth.* staiga *a path.*] v. medu-, mylen-stíg ; stíga.

stig (?), e ; *n. A wooden enclosure, a sty ;* but also part of a house, a hall (?) cf. stig-weard :—Gif cniht binnan stig sitte *if a servant sit within the hall* (?), Chart. Th. 612, 32. Stigo *vistrina* (*suestrina?* the word occurs at the head of a list 'de suibus'), Wrt. Voc. i. 286, 41. Stigu *auriola* (*oriola?* oriolum *porticus, atrium,* Migne), Txts. 38, 45. Cf. (?) forestige *vestibulum, introitum,* Hpt. Gl. 514, 59. Ondlong herpoðes on burghardes ánstigo ; ðonne forð tó bǽres ánstigon, Cod. Dip. Kmbl. ii. 172, 18. [Ase swin ipund ine sti, A. R. 128, 1. Stye *ara,* Wrt. Voc. i. 178, 14. Sty, swynys howus *ara, porcarium,* Prompt. Parv. 475. Þenk on helle stynkyng stye, H. R. 215, 3. Cf. *O. H. Ger.* stíga ; *f. cancelli, ara, ovile : Ger.* steige ; *f. hen-coop : Icel.* stía ; *f. a kennel ;* svína-stí *pig-sty : Dan.* sti *enclosure for swine, sheep, hens, etc. : Swed.* stia ; *f. sty for pigs, geese, etc.*] v. stigian.

stíga (?), an ; *m. :* stíge (?), an ; *f. A path* :—Faestin *vel* ánstígan, festin (-s, MS.) *vel* ánstíga *termofilas,* Txts. 104, 1042. v. stíg.

stígan; *p.* stáh, *pl.* stigon ; *pp.* stigen. **I.** *intrans. To go* (1) without implying ascent or descent :—Seó sunne stígþ on ða dæglan wegas wið hire uprynæs, *Phoebus secreto tramite currum solitos vertit ad ortus,* Bt. 25 ; Fox 88, 26. Of stíges *discedite,* Mt. Kmbl. Lind. 25, 41. Alle stígende (*discedentes*) from rehtwísnissum, Ps. Surt. 118, 118. (2) implying ascent, *to go from a lower to a higher level, to ascend, mount* :—Sió sunne ofer moncyn stíhþ á upweardes, Met. 13, 69. Bryne stígeþ heáh tó heofonum, Exon. Th. 233, 6 ; Ph. 520. Hálge gǽstas stígaþ tó wuldre, 234, 19 ; Ph. 542. Récas stígaþ ofer hrófum, 381, 5 ; Rä. 2, 6 : Ps. Th. 73, 22. Stigon ða þornas *ascenderunt spinae,* Mk. Skt. 4, 7. Sǽs up stigon ofer stæðweallas, Cd. Th. 83, 6 ; Gen. 1375. Ic wilnige ðæt ðeós sprǽc stigge on ðæt ingeðonc ðæs leorneres suǽ suǽ on sume hlǽdre, Past. proem. ; Swt. 23, 16. Ǽrðon up stige áncenned sunu, Exon. Th. 29, 17 ; Cri. 464. Sweart racu stígan onginneþ, Cd. Th. 82, 1 ; Gen. 1355. Geségon hí on heáhþu hláford stígan, Exon. Th. 31, 20 ; Cri. 498 : Shrn. 50, 15. Ðæt scip wile hwílum stígan ongeán ðone streám (*contra ictum fluminis conscendere*), Past. 58, 7 ; Swt. 445, 10. Ðá geseoþ Godes englas up stígende (*ascendentes*), Jn. Skt. 1, 51. (2 a) of getting into a vessel, etc., climbing a tree, etc. :—Hé stáh up on án treów *ascendit in arborem,* Lk. Skt. 19, 4.

Đā stāh hē on scip *ascendit navem*, Bd. 5, 9; S. 623, 27. Beornas on stefn stigon, Beo. Th. 429; B. 212. In ceól stigon, Andr. Kmbl. 697; An. 349. Đā gē on holm stigon, 858; An. 429. Leóde on wang stigon *they landed*, Beo. Th. 456; B. 225. Ǽr hē on bed stige, 1357; B. 676. Stīgan on wægn, Exon. Th. 404, 16; Rä. 23, 8. Hēt hē ǽnne mon stīgan on đone mǽst (*adscendere in arborem navis*), Ors. 4, 10; Swt. 202, 2. (3) Where the movement is downwards, *to descend*:—Ne stíhþ hē nyđer *ne descendat*, Lk. Skt. 17, 31. Đā stígaþ on helle *in infernum descenderent*, Past. 55, 2; Swt. 429, 26. Đā stāh and com smylte reng, Bd. 4, 13; S. 582, 34. Hié on sund stigon *they went down into the bed of the Red Sea*, Cd. Th. 198, 8; Exod. 319. Stíh ādún *descend*, Homl. Th. i. 580, 33. Ne stíge hē on his hús *non descendat in domum*, Mk. Skt. 13, 15. Đæt engel ufan of roderum stígan cwóme, Cd. Th. 248, 8; Dan. 510. Niþer stígende, of dūne stígende *descendentem*, Mt. Kmbl. 3, 16: Jn. Skt. 1, 51. II. *trans.* To ascend, mount :—Heáhlond stigon sibgemágas, Cd. Th. 202, 9; Exod. 385. Stealc hliþo stígan, Exon. Th. 498, 18; Rä. 88, 3. [The verb remained long in English and is used by Spenser: 'Ambition, rash desire to *sty*,' F. Q. ii. 7, 46. *Goth.* steigan: *O. Sax. O. L. Ger. O. H. Ger.* stīgan: *Du.* stijgen: *Ger.* steigen: *O. Frs.* stîga: *Icel.* stíga: *Dan.* stige: *Swed.* stiga.] v. ā-, fore-, ge-, ofer-stígan.

stige, es; *m. A going up* or *down* :—Drihtnes stige on heofonas up, Menol. Fox 129; Men. 64. v. niþer-, up-stige.

stigel, e; *f. A stile, set of steps for getting over a fence* :—Fram đam wōn stocce tō cinta stiogole, đanne fram cinta stiogole tō earnes beáme, Cod. Dip. Kmbl. ii. 73, 24. Stigole, iii. 227, 19. Stigele, 236, 25: v. 40, 6, 7, 10: 148, 1. Tō đære stigelæ tō đæs bisceopæs mearcæ, 84, 13, 16. Of đam seáđe in đa ealdan stihle; of đære stihle, iii. 386, 17–18. The word occurs also in compounds :—Đanon on đone bôchagan wiđ đere bôcstigele, v. 70, 27. [Ryght as they wolde han troden ouer a style, Chauc. Pard. T. 712. Style, where men gon over *scansillum, scansile*, Prompt. Parv. 475, col. 2. *O. H. Ger.* stiglia *a postern; posticium*.]

stigel-hamm, es; *m. An enclosure reached by a stile* (?) :—On stigel-hammas; of stigealhammum on wígferđes leáge, Cod. Dip. Kmbl. v. 289, 2.

stígend, es; *m. A sty, a small tumour on the edge of the eyelid* :—Stígend *ordeolus* (= *hordeolus*), Wrt. Voc. i. 20, 11. [Cf. *Norweg.* stig, sti, stigje.]

stígend, stígendlíc, stigenness. v. ā-, on-stígend, ofer-stígendlíc, ofer-, upā-stigenness.

stig-fearh *a young pig to keep in a sty* :—Ǽhteswāne gebyreþ stífearh, L. R. S. 7; Th. i. 436, 22.

stigian *to shut up in a sty* or *pen* :—Ođđe ic stigie, nyttes bicge, Salm. Kmbl. 402; Sal. 202. Swýn stigian, Anglia ix. 262, 2. [*Icel.* stía *to pen sheep*.] v. stig.

stigness, e; *f. A going down, a descent* :—Tō stígnisso *ad descensum*, Lk. Skt. Lind. 19, 37.

stigo. v. stig.

stig-ráp, es; *m. A stirrup* :—Stigráp *scansile*, Wrt. Voc. i. 84, 1. Stíráp, 23, 17. (In each case the word occurs in a list of words connected with riding.) Stírápas *scansilia*, 41, 34. [*O. H. Ger.* stega-reif: *Ger.* steg-reif: *Icel.* stig-reip.]

stigu. v. stig.

stigul, Wrt. Voc. i. 26, 45, *read* sāgul.

stig-weard, es; *m.* I. *a steward* (v. stig), *one who has the superintendence of household affairs ; especially matters connected with the table.* [The word, which is found generally with the form *stí-ward* and in late documents, occurs in Eadred's will, and in a connection which seems to shew the relative importance of the officer denoted by it. The king leaves to the archbishop 240 mancuses, to bishops and aldermen 120, to every *discđegn, hrægldegn*, and *biriele* 80, to every *stigweard* 30: Đænne an ic ǽlcan gesettan stigweard þritig mancusa goldes, Cod. Dip. B. iii. 75, 34.] :—Stíward *economus*, Wrt. Voc. i. 28, 13. Stíweard *discoforus, discifer*, ii. 140, 74. Đat lond đat Godríc míne stíward haueþ . . . Ǽlfwý mín stíward . . . Ǽlfnôđ mín stíward, Cod. Dip. Kmbl. iv. 268, 28–31. Se wæs đæs eorles stíward, Chr. 1093; Erl. 229, 6: 1096; Erl. 233, 6. Se đe mā manne in lǽde đonne hē sceole búton đæs stíwerdes leáfe and đæra feormera, Cod. Dip. Kmbl. iv. 278, 20. Mína cnihtas đa mína stíwardas witan, 59, 1. II. *fig. a steward, guardian* :—Mē þincþ betere đæt ic forlǽte đa gyfe and folgyge đam gyfan đe mē ǽgđer ys stíward ge đas welan ge eác hys freónscypes, Shrn. 176, 20. [Numbert, kinges stiward (*he is called* aldermon, l. 1420), Laym. 1451. Luue is heouene stiward, uor hire muchele ureoschipe, uor heo ne ethalt no þing, auh heo giueđ al þet heo haueđ, A. R. 386, 26. He (*the king*) called Aþelbrus, þat was stiward of his hus, Havel. 666. Putir far đe kinges stiward, Gen. and Ex. 1991. *Icel.* stí-varðr (*from English*).] v. next word.

stig-wita, an; *m. An officer of a household* (v. stig) :—Đa đe Sodoma and Gomorra golde berôfan bestrudon stigwitum *those who robbed Sodom and Gomorrah of gold, despoiled their houses of officers*, Cd. Th. 125, 14; Gen. 2079. Weallas beofiaþ ofer stíwitum *the walls tremble above the household*, Exon. Th. 383, 13; Rä. 4, 10. v. preceding word.

---

stihtan; *p.* te. I. *to dispose, arrange, regulate, direct, rule* :—Ic stihte (*disposui*) gekýþnysse mínum gecorenum, Ps. Lamb. 88, 4. Stapas on his heortan hē stihte *ascensiones in corde suo disposuit*, 83, 6. On đam ān and twentigan geáre đæs đe Willelm weólde and stihte Engleland, Chr. 1086; Erl. 219, 27. II. *to instigate, incite* :—Stihte hí Byrhtnôđ, bæd đæt hyssa gehwylc hogode tô wíge, Byrht. Th. 135, 34; By. 127. Ic heó tô þeófendum and tô gefiitum stihte, Wulfst. 255, 12. [*Du.* stichten : *O. H. Ger.* stiften *componere, concinnare* : *Icel.* stétta *to found, establish*.] v. ā-, fore-, ge-stihtan; stihtian.

stihtend, es; *m. A disposer, ruler* :—Þýstra stihtend (*the devil*), Exon. Th. 267, 23; Jul. 419. v. next word.

stihtere, es; *m. A disposer, director* :—Đæt hié geornlíce geđencen mid hū micelre giefe ofer him wacaþ se Scippend and se stihtere ealra gesceafta đonne hē hí nyle lǽtan tô hiera āgnum wilnungum *ut sollicita consideratione perpendant, Creator dispositorque cunctorum quanta super eos gratia vigilat, quos in sua desideria non relaxat*, Past. 50, 4; Swt. 391, 22.

stihtian; *p.* ode *To dispose, arrange, order, ordain, rule* :—Stihtaþ word his in dôme *disponet sermones suos in judicio*, Ps. Surt. 111, 5. Suíđe ryhte stihtaþ đone anwald se đe geornlíce conn ongietan đæt hē of him gadrige đæt him stǽlwierđe sié *potentiam bene regit, qui tenere illam noverit*, Past. 17, 5; Swt. 115, 2. Hē ealle gesceafta þurh his godcunde meht and þurh his écean snyttro æfter his willan receþ and stihtaþ, Blickl. Homl. 121, 16. Settaþ đa tô dômerum, đæt hié stihtien ymb đa eorđlican đing (*ut dispensationibus terrenis inserviant*), Past. 18, 2; Swt. 131, 8. Đý upplícan dôme stihtigende *superno dispensante judicio*, Bd. 4, 3; S: 567, 7. v. fore-, ge-stihtian; stihtan.

stihtung, e; *f. A disposition, arrangement, dispensation* :—Wæs đæt wunderlíco stihtung đære godcundan foreseónnesse *mira divinae dispensatio provisionis erat*, Bd. 5, 22; S. 644, 36. Hit wæs sweotole gesiéne, đæt hit wæs Godes stihtung, Ors. 6, 1; Swt. 252, 29. Eal seó stihtung wæs gefremed on đære sôþan onflǽscnesse for gefyllnesse đæs heofonlícan éþles, Blickl. Homl. 81, 28. Wæs đæs deóplíc eall word and wísdóm and đæs weres stihtung, Exon. Th. 169; 34; Gū. 1104. Mid wunderlícre stihtung (*dispensatione*) đære godcundan ārfæstnesse, Bd. 5, 22; S. 644, 11: 4, 29; S. 607, 42: Guthl. 2; Gdwin. 10, 20. Þurh godcunde stihtunge đære écan eádignysse him wǽre seó gifu forestihtod, 1; Gdwin. 10, 11: Bd. 5, 13; S. 633, 26. v. ā-, fore-, ge-stihtung.

stílan; *p.* de; *pp.* ed *To steel, temper, harden* :—Sum mæg stýled sweord, wǽpen gewyrcan, Exon. Th. 42, 28; Cri. 679. [Þat istelet (istelede, Bodl. MS.) irn tolimede hire, Jul. 58, 8. *Icel.* stæla *to steel, temper* ; sverð stælt mêđ eitri *a sword tempered with poison* ; cf. eitri herðr : *Germ.* stählen.]

stíle, es; *n. Steel* :—Stéli, steeli, stêl *accearium*, Txts. 37, 55. Staeli *ocearium*, 81, 1431. Stete *acerra* (? stéle *acearium*), Wrt. Voc. ii. 95, 56. Stýle *accearium*, 4, 29: 63, 34. Þeáh mec heard bíte stíđecg stýle, Exon. Th. 499, 11; Rä. 88, 14. Flinte ic eom heardra, đe đis fýr drífeþ of đissum strongan stýle heardan, 426, 26; Rä. 41, 79. Stýle gelícost, Beo. Th. 1975; B. 985. Heó oferbídeþ stánas, heó oferstígeþ stýle, Salm. Kmbl. 600; Sal. 299. [*Laym. A. R.* stel : *O. H. Ger.* stahal : *Icel.* stál.]

stíl-ecg; *adj. Steel-edged* :—Stíđ and stýlecg (*a sword*), Beo. Th. 3070; B. 1533.

stílen; *adj. Of steel, hard as steel* :—Đære stýlenan helle, Salm. Kmbl. 978; Sal. 490. Ne mihte ic of đære heortan heardne āđringan stýlenne stān, 1009; Sal. 506. [Wæs þe stelene brond swiđe brad and swiđe long, Laym. 7634. The stilen swerde, Parten. 256. *O. Frs.* stélen : *O. H. Ger.* stélin *ex calibe*.] .

still, stiell, es; *m. A leap, spring* :—Cyning engla munt gestylleþ, gehleápeþ hyllas . . . woruld ālýseþ þurh þone æþelan styll. Wæs se forma hlýp . . . wæs se ôđer stiell . . . se þridda hlýp . . . se feórđa stiell, Exon. Th. 45, 7–33; Cri. 715–728. v. stellan *to leap*.

stillan *to leap*. v. stellan.

stillan *to stall* [:—Hrýđer anstyllan, swín stigian, Anglia ix. 262, 1].

stillan; *p.* de. I. *to become still* or *calm* :—Đā stylde se storm sôna, and seó sǽ wearđ eft smylte, Shrn. 147, 9. Se æđeling hēt streámfare stillan, stormas restan, Andr. Kmbl. 3150; An. 1578: Salm. Kmbl. 796; Sal. 397. II. *to make still* or *calm, to still, pacify, appease, assuage* (with *dat.* or *acc.*) :—Đæt stilþ đam sáre, Lchdm. ii. 60, 5. Đæt sweord word gemaingfealdaþ mannes freóndscipe and stilleþ mannes feónd, Salm. Kmbl. p. 206, 2 : Salm. Kmbl. 268; Sal. 133. Cyning (*Christ*) ýdum stilde, wæteres wælmum, Andr. Kmbl. 902; An. 451. Đæt se đám ômum stille, Lchdm. iii. 182, 6. Beóþ đa elcran tô stillanne, 178, 14. [*O. Sax.* stillôn *to become quiet* ; stillian *to make quiet* : *O. H. Ger.* stillén *stupere, silere* : stillen *compescere, mitigare, mederi* : *Icel.* stilla *to still, calm, soothe, moderate*.] v. æt-, ge-, un-stillan ; stillian.

stille; *adj. Still, quiet.* I. in a physical sense, (1) of motion, (a) *without motion, at rest, not moving from a place, not disturbed* :—Seó sunne stôd stille ānes dæges lencge, Lchdm. iii. 262, 8. Swā hē stille stande, đǽr hine storm ne mæg āwecgan, Andr. Kmbl. 1003; An. 502. Stille on wícum siteþ, Exon. Th. 390, 26; Rä. 9, 7. Stille þynceþ lyft, 383, 14; Rä. 4, 10: 387, 5; Rä. 4, 74. Se monlíca (*the*

pillar of salt) stille wunode, Cd. Th. 155, 3; Gen. 2567. Wundum stille motionless from wounds, Beo. Th. 5653; B. 2830. Stánas sint stilre gecynde and heardre, Bt. 34, 11; Fox 150, 24. Seó sǽ ne mót heore mearce gebrǽdan ofer ða stillan eorþan, 21; Fox 74, 28. Twegen steorran standaþ stille, Lchdm. iii. 270, 17. Wit ðe ðisse strǽte stille þencaþ bídan, Cd. Th. 147, 9; Gen. 2436. Hí nýdde se tówarda winter ðæt hí stille wunodon swá hwǽr swá hí mihton *coegerat eos imminens hiems ut ubicumque potuissent quieti manerent*, Bd. 4, 1; S. 564, 39. Ðý lǽs fyrhtu stille (*quietos*) áwecce, Ps. Surt. ii. p. 202, 19. His wyrtruman wesan stille on staðole, Cd. Th. 251, 9; Dan. 561. *And fig.*: —Gif hé ne wolde lǽtan wrǽce stille, Exon. Th. 114, 10; Gú. 170. (b) *moving little* or *gently:*—Se man sceal swíþe stille beón *the patient must move about as little as possible*, Lchdm. ii. 148, 25. Oft stille wæter staðo brecaþ (cf. *still waters run deep*), Prov. Kmbl. 63. (c) *not easily moved* (?), *that will not run freely* (?):—Wǽte þicce and stille, Lchdm. ii. 138, 13. (2) of sound, (a) *silent:*—Ðeáh ðú stille sý and unrót *though thou be silent and sad*, Ap. Th. 15, 17. Se fæder hit gemǽnde stille *pater rem tacitus considerabat*, Gen. 37, 11. Hé hét ða Saducéiscan stylle beón *silentium inposuisset Sadducaeis*, Mt. Kmbl. 22, 34. *And fig.:*—Mid heortan stilre *corde tacito*, Hymn. Surt. 132, 30. Wén is ðæt eówer sum cweðe tó him sylfum on stillum geðohtum ..., Homl. Th. i. 580, 5. (b) *not loud:*—Mid stylre stemne, Homl. Th. ii. 410, 20. **II.** *quiet, unchanging, undisturbed, stable:*—Ðú ðe unstilla ágna gesceafta tó ðínum willan wíslíce ástyrest and ðé self wunast swíðe stille unáwendendlíc á forð simle *stabilis manens das cuncta moveri*, Met. 20, 16. **III.** *quiet, not vehement, gentle:*—Heó wæs on eallum þingum eáðmód and stille, Lchdm. iii. 430, 3. Ne ástyrige gé ðone stillan Drihten tó ǽnigre yrsunge, Homl. Th. i. 592, 3. Tó hwæm lócige ic búton tó ðǽm eáðmódum and tó ðǽm stillum *ad quem respiciam, nisi ad humilem et quietum?* Past. 41, 1; Swt. 299, 20. **IV.** *abstaining from, quit of.* v. stillness, IV:—Sió hí stille his þegnungæ óð biscopes dóm, L. Wih. 6; Th. i. 38, 11. [*O. Frs.* stille: *O. Sax.* stilli: *O. H. Ger.* stilli *quietus, tranquillus, serenus, immobilis, mitis, placidus.*] v. un-stille.

stillian. v. un-stillian; stillan.

stillness, e; *f. Stillness, quiet;* quies, Ælfc. Gr. 9, 27; Zup. 53, 9. **I.** in a physical sense, *absence of noise* or *disturbance:*—On ðisse tíde nihtlícre stillnesse *tempore isto nocturno quietis*, Bd. 4, 25; S. 601, 1. Windum stilnesse bebeódan, Blickl. Homl. 177, 17. Ðonne (*in church*) lǽrþ ús Godes engel stilnesse and gemetlíce sprǽce ... lǽrþ ús se deófol unstilnesse and ungemetlíce hleahtras and unnytte sprǽce, Wulfst. 233, 13–18. **II.** *quiet, silence:*—Stilnysse *taciturnitatis*, Hpt. Gl. 455, 54. Swígan ł stilnysse *taciturnitatem*, 503, 63. Hé mid stilnesse (*cum silentio*) his líf geendode, Bd. 4, 24; S. 599, 7. **III.** *absence of disturbance* or *molestation, tranquillity, peace, security:*—Stilnys *securitas, requies* ł *quietudo*, Hpt. Gl. 451, 43. Ðá he gewunelícan stilnesse Drihtne lifde *solito in silentio vacare Domino coepit*, Bd. 5, 9; S. 623, 31. Ðá hæfde Hannibal and Rómane án geár stilnesse (*quies a tumultu bellorum*) him betweónum ... On ðære stilnesse Scipia geeode ealle Ispanie, Ors. 4, 10; Swt. 198, 34. Ðú eart ofl of ðínre stilnesse áhworfen, Bt. 7, 1; Fox 16, 24. Gif wé ða stilnesse habbaþ, Past. pref.; Swt. 7, 9. Habbaþ eów stilnysse and sibbe, Homl. Th. i. 592, 6. Ða stylnysse middaneardlícære sibbe wé áwendaþ tó ýdelre orsorhnysse, ii. 540, 7. **IV.** *abstinence from, exemption from.* v. stille, IV:—Ðá ðá he lǽrde ðæt ðære ciricean ðegnas sceoldon stilnesse ðæra ðenunga habban (*be exempt from secular services*, cf. 129, 10), Past. 18; Swt. 130, 4. **V.** *that which appeases* (? cf. *O. Frs.* stilnese *nursing: Ger.* still-amme *wet-nurse: Swed.* stilla *to give fodder to cattle; to suckle a child*):—Stilnesse, gefylnesse *supplemento* (*supplementum viaticum, subsidium ad vitae necessaria*, Migne), Wrt. Voc. ii. 77, 9. [*O. H. Ger.* stilnissi *tranquillitas, silentium.*] v. un-stillnesse.

stíman, stéman, stýman; *p.* de *To emit a scent* or *vapour, exhale:*—Ic stéme oleo, Ælfc. Gr. 26, 1; Zup. 153, 2. Stémþ *exalet, i. redolet, spiret, fetet*, Wrt. Voc. ii. 144, 42: *fragrat, odorat, odorem dat*, 150, 34. Willsele stýmeþ swétum swæccum, Exon. Th. 212, 21; Ph. 213. Stémde *redolet*, Hpt. Gl. 516, 41. Unásecgendlíc brǽd stémde of hire gyrlum, Homl. Th. i. 444, 11: Homl. Skt. ii. 27, 110. Ne mihte nán wyrtbrǽd swá wynsumlíce stéman, 27, 113. Ðú stémenda *redolens*, Hymn. Surt. 47, 22. Stémendre *fragrantis, odorantis*, Hpt. Gl. 441, 72. Stémendes swæcces *nardi pistici*, 516, 38. Stémende *fragrantia*, 419, 52. Stémendum *fumigabundis*, 516, 30. Stémende *olentes, odorantes*, Wrt. Voc. ii. 150, 35. v. be-stéman, -stýman.

stíme (?) *a name given to a plant in* Lchdm. iii. 32, 19:—Stíme hǽtte ðeós wyrt, heó on stáne'geweóx. Cockayne says *water-cress*, in the note to the passage, but *nettle* in his glossary. Perhaps the alternative reading *stune* is the better, as it is said of the plant: stunaþ heó wærce ... wiðstunaþ heó áttre.

stíming, e; *f. Fragrance:*—Stémincge *fragrantia*, Hpt. Gl. 516, 40. Stémingce *fragrantiam, odorem*, 488, 28.

stinan (?); *p.* stan, *pl.* stánon; *pp.* stunen *To make a loud noise* [:—Gránode *vel* ásten (ástende? v. sténan) *rugiebam* (Ps. 37, 9), Blickl. Gl.]. v. stunian.

stincan; *p.* stanc, *pl.* stuncon; *pp.* stuncen *To emit a smell* or *vapour, exhale*, (1) where the kind of smell is not marked:—Stincþ *fragrat, i. odorat, i. odorem dat*, Wrt. Voc. ii. 150, 34. Stanc *exalcuit*, 29, 62. Stonc, 107, 54. Swá hý swýþost stincen *give out the strongest smell*, Lchdm. i. 206, 8. Ðæs stincendan *fumigabundi[s]*, Wrt. Voc. ii. 37, 20: 86, 40. Ðære stincendan *spirantis*, 75, 51. Stincende *fragrans*, 35, 73: 74, 65. Stincendi, 108, 76. (2) where the smell is a pleasant one:— Ic stince swóte oleo, Ælfc. Gr. 37; Zup. 220, 13. Swecca swétast, swylce on sumeres tíd stincaþ wyrta geblówene, Exon. Th. 178, 22; Gú. 1248. Stanc *redolet*, Hpt. Gl. 516, 41. Se líchoma stanc swá swóte, Shrn. 143, 28: 140, 13: Homl. Skt. i. 4, 347. (3) where the smell is an unpleasant one:—Hé stingð (stincð, MSS. B. C.) *faetet*, Jn. Skt. 11, 39. Ðæt oreð stincþ and áfúlaþ ðe ǽr wæs swéte on stence, Wulfst. 148, 7. Se líchoma stincþ fúle, Lchdm. ii. 236, 14: 220, 6. Stincþ, Exon. Th. 424, 1; Rä. 41, 32. Ongan se cealc mid ungemete stincan; ðá weard hé mid ðæm brǽþe ofsmorod, Ors. 6, 32; Swt. 288, 1. Him stód stincende steám of ðam múðe, Homl. Th. i. 86, 13, 10. Stingendum *putenti*, Hpt. Gl. 487, 64. [*O. H. Ger.* stinchan *odorem dare, odorare, fragrare, putere.*] v. ge-, tó-stincan; fúl-, swíð-, wel-stincende; swót-stencende; cf. stíman.

stincan; *p.* stanc, *pl.* stuncon *To spring, leap, move rapidly:*—Dust stonc tó heofonum, deáw feól on eorþan, Exon. Th. 412, 10; Rä. 30, 12. Se wyrm stonc æfter stáne, Beo. Th. 4565; B. 2288. [*Goth.* stigkwan withra *to proceed against: Icel.* stökkva *to spring, leap, take to flight.*] v. stencan.

sting, es; *m.* **I.** *a sting, stab, thrust made with a pointed instrument; the wound made by a stab* or *sting:*—Beslóh se þorn on ðone fót and swá strang wæs se sting ðæs þornes ðæt hé eode þurh ðone fót *the prick of the thorn was so hard, that the thorn went through the foot*, Guthl. 16; Gdwin. 68, 3. Lilla sette his líchoman beforan ðam stynge (*ante ictum pungentis*), Bd. 2, 9; S. 511, 24. Wið scorpiones stincg, Lchdm. i. 168, 3: 248, 21. Wið scorpiones stincg, genim ðás ylcan wyrte ... lege tó ðam stinge (cf. lege tó ðære wunde, 168, 7), 272, 22–24. **II.** v. in-, on-sting; stingan, I a.

stingan; *p.* stang, *pl.* stungon; *pp.* stungen. **I.** *to thrust something into:*—Sting ðín seax on ða wyrte, Lchdm. ii. 346, 12. Stingaþ stranglíc sár on his eágan, Wulfst. 141, 4. Nim án feðere, and stynge on hys múðe, Lchdm. iii. 130, 17. Wæs on slǽpe ætýwed ðæt hyre man stunge áne sýle on ðone bósum, Shrn. 149, 1. Crist hét stingan sweord in scæðe, Charter quoted by Lye. **I a.** *fig. to thrust one's self into the affairs of another, to exercise authority.* v. in-, on-sting:—Ná stinge nán mann on ðæt land, búton se hýred æt Xþes cyrcean, Chart. Th. 578, 6. Ic habbe ðæt geleornod, ðæt nán lǽwede man náh mid rihte tó stingan hine on ánre cirican, ná an án ðara ðinga ðe tó cyrcan belimpþ. And for ðí wé forbeódaþ eallan lǽwedan mannum ǽure ǽnne hláuordscipe ouer cyrcan, Cod. Dip. B. i. 137, 24. (Cf. *Icel.* Þú hefir mjök stungizk til þessa máls *thou hast meddled much with this case.*) **II.** *to prick with something, to sting, stab, pierce:*—Swá swá seó beó sceal losian, ðonne heó hwæt yrringa stingþ, Bt. 31, 2; Fox 112, 26. Stingeþ, Met. 18, 7. [Wyrm] stingeþ niéten, Salm. Kmbl. 308; Sal. 153. Hé mid gáre stang wlancne wícing, Byrht. Th. 135, 55; By. 138. Stincge *transfigat*, Anglia xiii. 37, 276. Gif þorn stinge man on fót, Lchdm. ii. 336, 20. Gif hine beón stingan, iii. 168, 13. Se lǽce his seax hwæt, ǽrðonðe hé stingan wille, Past. 26; Swt. 187, 6. Se cásere hine hét stingan mid írenum gyrdum, Shrn. 115, 24. Stingaþ hyne mid sáre on his eágan, L. E. I. prm.; Th. ii. 398, 19. [*Goth.* us-stiggan *to thrust out: Icel.* stinga *to sting, stick, stab.*] v. á-, be-, ge-, of-, on-, tó-, þurh-, under-stingan.

stintan, stióp, stiór, stiorc. v. styntan, steóp, steór, stirc.

stíp, stiép, es; *m. Deprivation* (?), *overthrow* (?):—Hé his torn gewræc on gesacum swíðe strengum stiépe, Cd. Th. 4, 27; Gen. 60. The passage refers to the expulsion of the angels from heaven. Cf. steóp-, á-stépness *orbitatio*, á-stýpan in Wulfst. 252, 11 : Wé wǽron ástýpede (bedǽled, MS. D.: ástýpte, Blickl. Homl. 107, 4) ðæs heofenlícan ríces. Grein suggests *overthrow* (cf. Milton's 'the dire event, That with sad overthrow and foul defeat Hath lost us Heaven'), *fall* as the meaning, and compares with *Icel.* steypa *to cast down, overthrow;* steyping *an overthrow.* Cf. also *Norweg.* stup *a precipice,* and see stúpian.

stípan *to deprive.* [*O. H. Ger.* stiufen *orbare.*] v. á-stípan; steóp-.

stípan; v. te. **I.** *to raise, build high, erect:*—Tó heofonum up hlǽdræ rǽrdon, strengum stépton stǽnenne weall ofer monna gemet, Cd. Th. 101, 2; Gen. 1676. **II.** *fig. to exalt, elevate, dignify, ennoble:*— Ic ðé on tída gehwone duguðum stépe, Cd. Th. 139, 7; Gen. 2306. Hé him frémum stépeþ, Exon. Th. 434, 10; Rä. 51, 8. Ðeáh ðe hine mihtig God mægenes wynnum stépte ofer ealle men, Beo. Th. 3438; B. 1717. Se feónd (*Nero*) his diórlingas duguþum stépte (cf. hé weorþode his deorlingas mid miclum welum, Bt. 28; Fox 100, 29) *dabat improbus verendis patribus indecores curules*, Met. 15, 8. Sinces brytta (*the king of Egypt*) hëht Abrahame duguðum stépan, Cd. Th. 111, 21; Gen. 1859: 142, 21; Gen. 2365. v. ge-, on-stépan; stípere, steóp.

stípel, es; *m. A tower:*—Stýpel *turris*, Wrt. Voc. i. 36, 39: 83, 32: Lk. Skt. 13, 4. Ðú ðencst tó gewyrcenne wundorlícne stýpel and swíðe heálícne; hoga ymbe ða gástlícan gestreón tó ðæs stýpeles getimbrunge

. . . Ne biþ ðes stýpol getimbrod mid ænigum weorcstáne, Basil admn. 2; Norm. 38, 6–14. Stêpel stræncðe *turris fortitudinis*, Ps. Lamb. 60, 4. Stêpeles *turris*, Hpt. Gl. 499, 60. Hine man byrigde æt ðam westende ðam stýfe (stýpele, MS. D.) ful gehende *he was buried at the west end (of the minster at Ely) quite close to the tower*, Chr. 1036; Erl. 165, 38. Ðæt hé gesáwe ða burh and ðone stípel (*the tower of Babel*), Gen. 11, 5. Stýpel, Homl. Th. i. 22, 19 : ii. 472, 25. Timbrian ánne stýpel *turrem aedificare*, Lk. Skt. 14, 28. Hé worhte of seolfre ænne heáhne stýpel and mid scínendum gymmum besette eall ðæt hús, and on ðære upfióra his cynestól geworhte, Homl. Skt. ii. 27, 29. On stýpelum *in turribus*, Ps. Spl. 47, 11 : 121, 7. [Hí clumben upp tô þe stêpel, Chr. 1070; Erl. 209, 9. Þá com se fír on ufenweard þone stêpel, and forbearnde ealle þe minstre, 1122; Erl. 249, 6.]

stípere, es; *m. A support, prop, pillar :*—Stípere *destina* vel *postis* vel *fulcimen*, Wrt. Voc. i. 26, 38. [Þe stipre þat is vnder þe vyne set May not bringe forþ þe grape, H. R. 135, 135. Cf. Heo wuned under þe chirche: ase uorte understipren hire, ȝif heo wolde uallen, A. R. 142, 16. Cf. *O. Frs.* stípe *a post*.] v. stípan.

stípness, stíran, stí-ráp. v. á-stépness, steóran, stig-ráp.

stirc, stiorc, styric, es ; *n. A stirk, calf, a young bullock* or *a heifer :*—Stirc *bucula, juvenca, vitula*, Wrt. Voc. ii. 126, 63. Styrc *juvencus*, i. 78, 44. Ðæt þridde stód ánum styrce (cealfe, MS. C. : cf. ðæs celfes gelícnyss belimpþ tô Lucan, 192) gelíc, Homl. Skt. i. 15, 183. Tô fēttum stiorce *ad vitulum saginatum*, Kent. Gl. 525. Stirc *buculam*, Wrt. Voc. ii. 12, 11 : 93, 12. Bringaþ án fætt styric *adducite vitulum saginatum*, Lk. Skt. 15, 23. [Styrk *boviculus*, Wrt. Voc. i. 204, 5. Styrk, neet, or heifer *juvenca*, Prompt. Parv. 476. *Ger.* stärke, sterke *a young cow that has not calved : M. H. Ger.* stirke, sterke.]

stirfan *to kill.* [*O. H. Ger.* ir-sterben *interficere, necare*.] v. á-styrfan.

stirfig; *adj. Pertaining to an animal that has died :*—Gif hwá ete styrfig flǽsc *si quis carnem morticinam ederit*, L. Ecg. P. iv. 27 ; Th. ii. 212, 17. [*O. H. Ger.* stirbig *mortalis, morticinus, moribundus.*]

stiria, stírian, stirigend-líc. v. styria, styrian, styrigend-líc.

stirnan (?); *p.* de *To be severe :*—Gistmægen (*the two angels with Lot*) styrnde (stýrde ? v. steóran) werode mid wíte, Cd. Th. 150, 22 ; Gen. 2495.

stirne; *adj. Stern, hard, austere, rigorous, severe :*—Ic wát ðæt ðú eart swíðe styrne mann *scio quia homo durus es* (Mt. 25, 24), Homl. Th. ii. 552, 31. Cyning sceal beón milde ðám gódum and styrne ðám yfelum, L. I. P. 2 ; Th. ii. 306, 1 : Wulfst. 267, 3. [God] hæfde styrne mód, gegremed grymme, Cd. Th. 4, 28 ; Gen. 60. [Se cyng heafde gifen þ abbotríce án Frenciscse abbot . . . hé wæs swíðe styrne man, Chr. 1070; Erl. 207, 32. Laym. *A. R.* sturne *. Orm.* stirne.]

stirninga; *adv. Sternly, inexorably :*—Ðæt wundor ðæt geond ðás woruld fareþ, styrnenga gǽþ, staðolas beáteþ, Salm. Kmbl. 565; Sal. 282.

stirn-líc; *adj.* I. *hard, harsh :*—Warna ðæt ðú nán þing styrn-líces ne sprece ongén Iacob *cave, ne loquaris contra Jacob quidquam durius*, Gen. 31, 29. II. *hard, unpleasant, severe* (of weather) :—Hwíltídum ðeós woruld is gesundful and myrige on tô wunigenne, hwílon heó is eác swíðe styrnlíc and mid mislícum þingum gemenged, swá ðæt heó biþ swíðe unwynsum on tô eardigenne, Homl. Th. i. 182, 35. Sceal áspringan here and hunger, bryne and blódgyte and styrnlíce styrunga, Wulfst. 86, 11. Seó heofone ús winþ wið, ðonne heó ús sendeþ styrnlíce stormas, 92, 17.

stirnlíce; *adv.* I. *sternly, hardly, harshly :*—Hé him ondwyrde and him swíðe stiernlíce stiérde *fregit eos responsionibus*, Past. 28, 6; Swt. 197, 19. Welig spycþ styrnlíce *diues affabitur rigide*, Scint. 78, 18. II. *inflexibly, rigorously :*—Cyning sceal eallum Godes feóndum styrnlíce wiðstandan, L. I. P. 2 ; Th. ii. 304, 20.

stirn-mód; *adj. Stern of mind :*—Stôpon styrnmôde (*the Hebrews proceeding against the Assyrians*), stercedferhðe, Judth. Thw. 24, 37; Jud. 227.

-stirre, -stirred, stirung. v. seofon-stirre, á-stirred, styrung.

stíþ; *adj. Stiff, hard.* I. in the following glosses :—Stíþ, rêþe *durus, crudelis, asper*, Wrt. Voc. ii. 142, 19. Stíð *inmitis*, Germ. 392, 33 : *rigens*, 393, 172. Stíðes *ardui, stricti*, Hpt. Gl. 416, 18 : *violentis, validis* ł *turbidis*, 440, 34. Stíðre *torridae*, 515, 46. On stíðum *in arto, duro, constricto*, 444, 15. II. of material, *stiff, firm,* (1) *strong, not bending easily, unyielding :*—Hit (*the sword*) on eorðan læg stíð and stýlecg, Beo. Th. 3070; B. 1533. Æsc byþ stíð staðule, ðeáh him feohtan on firas monige, Runic pm. Kmbl. 344, 25 ; Rún. 26. Stranga tor stíð wið feóndum *turris fortitudinis a facie inimici*, Ps. Th. 60, 2. Mec stíþne (*an anchor*), Exon. Th. 398, 17; Rä. 17, 9. Stíðe and rúge breóstroccas *renones*, Wrt. Voc. i. 40, 24. Hine mid stíðum ságlum beátaþ, Homl. Th. i. 432, 11 : 468, 32. Mid stíðum sticelum *stimulis acutis*, Dóm. L. 179. Se gestaþelade stíþe grundas *he fixed the firm foundations*, Exon. Th. 312, 4; Seef. 104. Ðeós wyrt hafaþ lange leáf and stíþe, Lchdm. i. 288, 15. Heó hafaþ máran leáf and stíðeran, 274, 7. (2) *of a thick consistency :*—Gif tô stíð sié *if the mixture be too stiff*, Lchdm. ii. 108, 17. Ðæt hit sý swá stíð ðæt hit wille wel clyfian, iii. 40, 13. II a. fig. (1) in a good sense :—Mé wæs strengðu strang stíþ on Dryhtne *fortitudo mea Dominus*, Ps. Th. 117, 14. Stan-

dan stíðe móde *to stand with unshaken soul*, 147, 6. Ic ðínes earmes ásecge stíþe strencðe, 70, 17. Ðone stíðan swioran fortredan *rigida colla victorum calcare*, Past. 33; Swt. 228, 8. (2) in a bad sense, *stiff* (as in *stiff*-necked) :—Gé wiðstandaþ ðam Hálgan Gáste mid stíðum swuran, Homl. Th. i. 46, 23. III. of persons, *hard, stern, inexorable, severe, austere :*—Ðú eart stíð man *homo austerus es*, Lk. Skt. 19, 21, 22. Heard ł stíð *durus*, Mt. Kmbl. Lind. 25, 24 : Past. proem.; Swt. 23, 24. Hé wæs swá stíð, ðæt hé ne rôhte heora eallra níð, ac hí môston ðes cynges wille folgian, gif hí woldon libban, Chr. 1086; Erl. 222, 31. Se man ðe tô ðon stíð biþ ðæt hé áðas sylþ ðæt hé tô náure sybbe fôn nelle *homo qui adeo durus sit ut juramenta praestet, se nullam pacem admittere velle*, L. Ecg. P. ii. 29 ; Th. ii. 194, 9. IV. of things that cause discomfort or require effort, e. g. weather, conflict, illness, punishment, *hard, severe, unrelenting, stubborn :*—Ðær wæs stíð gemót, Byrht. Th. 140, 40; By. 301. Gif seó untrumnes swá stíð beó, Lchdm. i. 260, 22. Sié ðær eác lufu, næs ðeáh tô hnesce; sié ðær eác rêðnes, næs ðeáh tô stíð, Past. 17; Swt. 127, 3. Hér wæs se stíþa winter, Chr. 1048; Erl. 171, 33. Beóþ ymbgyrde stranglíce tô ðysum stíðan gewinne, Homl. Skt. ii. 25, 341. Se dêmþ stíðne dóm ðám réceleásum *he will pass severe sentence on the careless*, Homl. Th. i. 320, 18. Gelácnian myd líðum lǽcedómum ðe myd stíðum *to cure with gentle remedies or severe*, Shrn. 189, 24. Wiþ ða stíþustan feferas, Lchdm. i. 114, 16. V. where conformity to a standard or rule is imposed, of discipline, mode of life, etc., *strict, rigid, severe, austere, hard :*—Se[ó] ealde ǽ næs swá stíð on ðám þingum swá swá Cristes godspel is, Boutr. Scrd. 22, 24. Ða on wéstenum wunigende woruldlíce éstas and gælsan mid stíðum lífe fortrǽdon, Homl. Th. i. 544, 28. Ðæt gáte hǽr getácnode ða stíþan dǽdbóte ðæra manna ðe heora sinna behreówsiaþ, Ælfc. Thw. 3, 36. Ðá ðá hí áxodon hú hé mihte swá stearce forhæfednysse healdan, hé andwyrde : 'Stíðran and wyrsan ic geseah,' Homl. Th. ii. 354, 24. VI. of speech whose subject-matter is unpleasing, *hard :*—Stíð is ðis word, hwá mæg hine gihêra, Jn. Skt. Rush. Lind. 6, 60. Cyning cunnode hwilc ðæs ǽðelinges ellen wǽre stíðum wordum : 'Ðú scealt mê onsecgan sunu ðínne,' Cd. Th. 172, 22; Gen. 2848. VII. *harsh* to the taste :—Ðeós wyrt biþ ðam góman stíð and wiðerrǽde for mete geþiged, Lchdm. i. 300, 10. Gemencged mid stíþum ecede, 156, 15. [*O. Frs.* stíth (*opposite of* teddre): *Icel.* stinnr *stiff, unbending, strong*.]

stíþe; *adv.* I. *strongly, very much, effectively :*—Cumaþ ealle engla þreátas stíðe ástyrode (*commoti* ; v. stíþ-mægen), Wulfst. 137, 14. Ðæt ðú míne stefne stíðe gehýre *exaudiet vocem meam*, Ps. Th. 54, 17. [Hou thai mai stithe stand igain the fend, Met. Homl. 4, 11.] II. *hardly, harshly, sternly, severely :*—Hú stíðe (*dure*) se landhláford sprǽc wið hig, Gen. 42, 30. Him ðæt stíðe geald fædera Lothes, Cd. Th. 125, 15 ; Gen. 2079. III. *austerely, strictly*, Homl. Th. ii. 146, 7.

stíþe, an (?); *f. A name given to lamb's cress, or to nettle* (cf. the lists of plants given in sections 45, 46, Lchdm. iii. pp. 30–36) :—Stíðe ðeós wyrt hátte, Lchdm. iii. 32, 23. v. stíþ.

stíþ-ecg; *adj. Of stiff* or *strong edge :*—Stíðecg stýle, Exon. Th. 499, 11 ; Rä. 88, 14.

stíþ-ferhþ, -frihþ; *adj.* I. *of firm, strong mind :*—Hálig Drihten, stíðferhð cyning, Cd. Th. 16, 10 ; Gen. 241. Stíðfrihþ, 7, 16; Gen. 107. Standaþ stíðferhðe (*Cherubim and Seraphim*), Andr. Kmbl. 1443; An. 722. Stíðferhþe hæleð higegleáwe, Chr. 975; Erl. 126, 24. II. *of stern mind :*—Stíðferhð cyning (*the Deity at the time of the deluge*), Cd. Th. 84, 32 ; Gen. 1406. Stíðferhð cyning steóre gefremede, ðá hé rêðemód reorde gesette eorðbúendum ungelíce, 101, 16 ; Gen. 1683.

stíþ-hugende; *adj. Of purpose stern :*—Stíðhugende hysas æt hilde, Byrht. Th. 135, 23 ; By. 122.

stíþ-hycgende; *adj.* I. in a good sense, *of firm, inflexible purpose, resolute :*—Stôpon tô ðære stówe stíðhycgende, Elen. Kmbl. 1429; El. 716. II. in a bad sense, *obstinate, stubborn :*—Hire Iudas oncwæð stíðhycgende : 'Ic ða stôwe ne can,' 1362; El. 683. Stíðhycgendum (*the multitude of unbelievers*), Andr. Kmbl. 1481; An. 742. III. *having hard, unpleasant thoughts :*—'Is mé feorhgedál leófre micle ðonne ðeós lífcearo.' Him ðá stefn oncwæð stíðhycgendum, 2858; An. 1431.

stíþ-hygd; *adj. Resolute, constant :*—Gé tô ðam lifgendan stáne stíðhygde staþol fæstniaþ, Exon. Th. 281, 30; Jul. 654.

stíþ-hygdig, -hýdig; *adj. Of stern purpose :*—Gestáh stíðhýdig (*Abraham when about to offer Isaac*) steápe dúne, Cd. Th. 175, 16 ; Gen. 2896. Stôpon stíðhýdige . . . þrungon þræchearde, Elen. Kmbl. 241; El. 121.

stíþian. v. á-, ge-, on-stíþian.

stíþ-líc; *adj.* I. *firm, strong :*—Stíðlíc stántorr (*the tower of Babel*), Cd. Th. 102, 14; Gen. 1700. II. of immaterial things, weather, conflict, discipline, penance, *hard, severe :*—Stíðlíc hreóhnys a *severe storm*, Homl. Th. ii. 18, 5. Wæs ðæra deófla gefeoht swíðe stíðlíc ongeán ða sáwle, 340, 30. Mót tô bóte stíðlíc dǽdbót, L. Pen. 3 ; Th. ii. 278, 8. Hí begunnon tô sleánne ǽlc heora ôðerne mid stíðlícum gefeohte, Jud. 7, 22. Swá swá hé strengest beón mihte ongeán ða stíðlícan scúras, Boutr. Scrd. 21, 6. III. of speech, *hard,*

harsh, severe :—Ne sceal nán mon geþrístlǽcan ðæt hé áht stíþlíces spræce ongeán his abbod, R. Ben. 16, 2. Sió æcs wient of dæm hielfe ðonne of dære ðreátunga gáþ tó stíðlíco word *ferrum de manubrio prosilit, cum de correptione sermo durior excedit*, Past. 21, 7; Swt. 167, 10. Sege ús for hwí ðú ús ðus stíþlíce word tó sprece, H. R. 7, 35. **IV.** *of persons, stern, hard, fierce :*—Ðá Ælfréd ðæt ofáxode, ðæt se here swá stíðlíc wæs, Shrn. 16, 8.

stíþlíce; *adv. Hardly, severely;* violenter, Hpt. Gl. 435, 60 : 514, 22 : rigide, Kent. Gl. 660. Stíðlícor *restrictius*, R. Ben. Interl. 6, 5. Stíþlícor *districtior*, i. *rigidior*, Wrt. Voc. ii. 141, 49. **I.** *firmly, without giving way :*—Ðás geweorc stondaþ stíðlíce, Exon. Th. 351, 28; Sch. 87. **II.** *strongly, effectually :*—Mé com stíðlíce tó móde *it was strongly impressed on my mind*, Anglia viii. 313, 3. Ðú stíðlíce eallum miltsadest, Ps. Th. 101, 12. **III.** *sternly, hardly, severely :*—Hwílon láreów mín áwecþ mé stíþlíce (*duriter*) mid gyrde, Coll. Monast. Th. 35, 31. Stíðlíce clypode wícinga ár, Byrht. Th. 132, 33; By. 25. Hé stíðlíce þrowode for úre ealra neóde, Wulfst. 126, 10. Hý fuhton stíðlíce ymbe ða hálgan sáwle, 236, 23. Hé hit sceal swíðe stíðlíce gebétan, L. E. I. 14; Th. ii. 412, 2. Hé wæs gescrýd wǽclíce and stíðlíce, Homl. Th. i. 330, 2. Hé swíðe stíðlíce leofode, ii. 38, 6. Stíðlíce drohtnigende, 354, 16. Hé stíðlícor mid untrumnyssum ofsett wæs, 120, 7. **IV.** *strictly :*—Ðæt líf stíðlíce healdan *to observe a course of life strictly*, R. Ben. 76, 4. [Hú hé stíðlucest hér on lífe libben mihte, Shrn. 12, 18.] [*Icel.* stinn-liga *strongly*.]

stíþ-mægen, es; *n. A strong force :*—Ðonne cumaþ upplíce eored-heápas stíþmægen ástyred *tum superum subito veniet commota potestas*, Dóm. L. 114. [Cf. Stið-imainede eorl, Laym. 25820.]

stíþ-mód; *adj.* **I.** *of constant mind, resolute :*—Strang and stíðmód gestáh hé on gealgan, Rood Kmbl. 79; Kr. 40. **II.** *of stern mind, stern :*—Stíðmód gestód wið steápne rond bealdor (*Beowulf*), Beo. Th. 5125; B. 2566. Him (*the people of Sodom*) tó sende stíðmód cyning (*God*) áras síne, Cd. Th. 146, 16; Gen. 2423. Se þeóden wæs strang and stíðmód, 279, 34; Sat. 248. Cyning stíðmód sý wið yfele, L. I. P. 3; Th. ii. 306, 26. Se stíðmóda cyning, Drihten ælmihtig, áwearp of ðam setle ðone módigan feónd, Wulfst. 145, 27. **III.** *of violent or fierce mind :*—Se stíþmóda (*Holofernes*) styrmde and gylede, módig and medugál, Judth. Thw. 21, 19; Jud. 25. **IV.** *of stubborn mind, stubborn, obstinate :*—Ðonne wurð seó heardnes stíðmódre heortan swíðe gehnexad þurh grimlíce steóra, Wulfst. 133, 17. [Cf. Arður stiðimoded kempe . . . Æuere wes Arður ærhðe bideled, Laym. 26022.]

stíþness, e; *f. Hardness, severity, force;* violentia, Hpt. Gl. 435, 76 : 516, 23 : duritia, 482, 66. **I.** *hardness, stiffness* in a physical sense :—Gif hwylc stíðnes on líchoman becume, genim ðás wyrte . . . lege tó ðam sáre, Lchdm. i. 132, 16. Wiþ æghwylce gegaderunga þe on ðam líchoman ácenned beóþ, genim ðás wyrte . . . lege tó ðam sáre, hit tófereþ ealle ða stíðnyssa, 140, 14 : 150, 10. **I a.** *fig. hardness* of heart :—Stíðnise heartes *duritiam cordis*, Mt. Kmbl. Lind. 19, 8. **II.** *firmness, constancy :*—Ða hnescan *vel* wácmód, ðæt synd ða ðe náne stíðnysse nabbaþ ongeán leahtras, Homl. Skt. i. 17, 40. **III.** *severity, strictness, hardness, rigour :*—Mid micelre carfulnysse stíðnyss seó sý gemetegud *magna sollicitudine districtio ipsa moderetur*, Scint. 123, 9. Beó him gesǽd eall seó stíðnys and earfoðnys ðe tó Gode lǽt *predicentur ei omnia dura et aspera per que itur ad Deum*, R. Ben. 96, 19. Ne hý mid weorces stíðnesse ofsette sýn *ut . . . ne violentia laboris opprimantur*, 75, 9. Ðæt wé mid sumere stíðnysse tó ðam gástlícum gefeohte ús gegearcian, Homl. Th. ii. 86, 12, 26 : 374, 15. Gif hwá ða stíðnysse áberan ne mæg ðe his scríft him tǽcþ *si quis austeritatem perferre nequeat, quam confessarius ejus ei praescripserit*, L. Ecg. P. iv. 60; Th. ii. 220, 25.

stíþ-weg, es; *m. A hard, rough way :*—Strong on stíðweg, Exon. Th. 384, 29; Rä. 4, 35.

stí-weard, -wita. v. stig-weard, -wita.

stóc (stoc ?). A word occurring mostly in local names, either alone or in compounds. The meaning seems, like that of *stów*, to be *place* (in the first instance perhaps a place fenced in, cf. (?) staca), and both words remain now only as names of places, *Stoke, Stowe*, or as parts of such names, Basing*stoke*, Tavi*stock*, Waltham*stow*. As may be seen from the Index to the Charters, Stóc occurs frequently, some of the references are here given :—Ðis is ðara þreora hída and .xxx. æcera bóc æt Stóce, Cod. Dip. Kmbl. iii. 190, 9 : 34, 12. Tó Stóce, 203, 21. Intó Stóce, 123, 8. In loco, qui celebri a soliculis nuncupatur æt Stóce uocabulo, 19, 32 : 33, 27. (With these two passages may be compared the following :—Apud locum ubi uulgari dicitur nomine æt Stówe, 323, 32.) In Stóce . . . in Súthstóce, 75, 25, 33. As an instance of a compound in which the word occurs may be given the following :—Sihtríc abbud on Tæfingstóce, vi. 196, 1. Hí Ordulfes mynster æt Tæfingstóc (Tefingstóce, MS. E.) forbærndon, Chr. 997; Erl. 134, 14. [Crist inn oþre stokess nemmneþþ þa þosstless hise breþre, Orm. 15694.] v. stóc-líf, -weard, -wíc.

stocc, es; *m.* **I.** *a stock, trunk, log :*—Stoc *truncus*, Wrt. Voc. i. 32, 42 : 80, 32 : axima, 287, 32. On ðone lytlan beorg ðǽr se stoc

stód . . . on gerihte tó ðam stocce on eásteweardan ðam leá, of ðam stocce súðrihte on ðære strǽt, Cod. Dip. Kmbl. ii. 250, 9-17. Tó ðam wón stocce, ðanne fram ðam wón stocce, 73, 22. Tó paðe stocce *to the sign-post* (?), v. 401, 37. Hé gehæfte hí on ánum micclum stocce and mid ísenum pílum heora ílas gefæstnode . . . Hí stódon stille on ðam stocce gefæstnode, Homl. Skt. i. 5, 386-402. Ic hæbbe of ðam stocce ðe his (*Oswald's*) heáfod on stód, ii. 26, 260. Oþ ðone calewan stoc, Cod. Dip. Kmbl. ii. 216, 1. Hé gefeól on ðone stocc be ðære stǽnenan strǽte ðe is geháten sacra uia, and tóbærst on feówer dǽlas. Ðá genáman men eft ðone stoc on weg and feówer syllíce stánas on ðære ilcan stówe álegdon, Blickl. Homl. 189, 12-15. Gé þeówiaþ fremdum godum, stoccum and stánum (*ligno et lapidi*), Deut. 28, 36. Stoccon *lignis*, 64. Tó stoccum, Cod. Dip. Kmbl. iii. 429, 7. **II.** *a wooden trumpet* (?) :—Béma ł stocc *tuba*, Mt. Kmbl. Lind. 6, 2. [*O. Frs.* stokk *a stock; stocks : O. L. Ger.* stokk *stipes : O. H. Ger.* stocch *truncus, stipes, lignum, cippus : Icel.* stokkr.] v. hand-, heáfod-, píl-stocc.

stoccen; *adj. Made of logs :*—Andlang Teóburnan tó ðære wíde herestrǽt; æfter ðære herestrǽt tó ðære ealde stoccene sancte Andreas cyricean tó ðe old wooden St. Andrew's church, Cod. Dip. Kmbl. iii. 73, 20. Cf. Stokenchurch in Oxfordshire, Stokenham in Devonshire.

stóc-líf, es; *n. A town, habitation :*—Stócclíf *oppidum, civitas*, Hpt. Gl. 500, 18. Se.mæg gedón ðæt ic sóftor eardian ǽgðer ge on ðisum lǽnan stóclífe (cf. Here have we no continuing city, Heb. 13, 14) ða while ðe ic on ðisse weorulde beó ge eác on ðam hécan háme ðe hé ús geháten hefþ *he can make me dwell more at ease both in this transitory habitation, while I am in this world, and also in that eternal home that he hath promised us*, Shrn. 163, 20. Se ðe ǽgðer wilt ge ðissa lǽnena stóclífe ge ðara écena háma, 164, 9. Cf. cot-, mynster-líf *for words in which* líf *is similarly used; and see* stóc.

stóc-weard, es; *m. A townsman :*—Stócweardum *oppidanis*, Hpt. Gl. 525, 49. v. stóc.

stóc-wíc, es; *n. A habitation, residence :*—On Casino ðam stócwíc *in the monastery at Monte Casino*, Earle, A. S. Lit. 200, 34. v. stóc.

stod *a post :*—Stod *propolim vel pertica*, Wrt. Voc. i. 16, 28. [A stake or a stode *palus*, Wülck. Gl. 600, 4. Stothe or post *posticulus*, Prompt. Parv. 478, col. 2.] v. duru-stod; studu.

stód, es; *n. A stud, a herd of horses :*—Stood *equartium*, Wrt. Voc. i. 23, 10. Ic geann mínon heáhdeórhunton ðæs stódes ðe is on Colingahrycge, Cod. Dip. Kmbl. iii. 363, 25. Ic gean mínum wífe healfes ðæs stódes æt Trostingtúne and mínum geféran healfes ðe mé mid rídaþ, and fó mín wíf tó healfum ðe on wealde is, and mín dohter tó healfum, iv. 300, 28. Ðat stód ðe ic ðér habbe, Chart. Th. 574, 20. [Asse . . . thou come of lither stode, P. S. 201, 2. Þe sulve stottes in þe stode, O. and N. 495. The hors of thilke stood Devoureden the mannes blood, Gow. 3, 204, 19. *O. H. Ger.* stuot *equaritia, grex equarum : Icel.* stóð ; *n. : Dan.* stod.]

stód-fald, es; *m. An enclosure for a stud of horses :*—Tó ðam aldan stódfalde; and ðonne fram ðam stódfalde, Cod. Dip. Kmbl. iii. 393, 21. Of ðam wylle on ðone stódfald; of ðam stódfalde, vi. 213, 21. Be norðan stódfaldan, iv. 66, 8. [*Dan.* stod-fold *an enclosure for horses*.]

stód-hors, es; *n. A stud-horse :*—Gyf mon mǽte ðæt hé feola stódhorsa habbe, Lchdm. iii. 176, 5. [*Icel.* stóð-hross.]

stodl *a post.* v. dur-stodl [*O. H. Ger.* turi-studil, -stuodil, -stodal *limen, postis : Icel.* stuðill *a prop, stay*]. v. stod, studu, *and next word.*

stodle (-a; *m.?*), an; *f. A slay, part of a loom :*—Hé sceal fela tówtóla habban . . . stodlan, Anglia ix. 263, 11. [Stodul or stedulle of wevynge *telarium* (cf. *Span.* telar *a loom*), Prompt. Parv. 476. Stodyll a toole for a wever, lame (cf. *lama* sleybrede, Wülck. Gl. 591, 28) *de tisserant*, Palsgrave (Halliwell's Dict.). Cf. *M. H. Ger.* stodel *pidonius* (*textoris*); in a gloss the word is further explained by *warfsteche.* v. Grff. vi. 654.] v. *preceding word.*

stód-mere, an; *f. A brood-mare, mare with a foal :*—Gif mon cú oþþe stódmyran forstele, and folan oþþe cealf of ádrífe, L. Alf. pol. 16; Th. i. 70, 24. [Ich am a ful stodmere, a stinckinde hore, A. R. 316, 15. Stodemere, Perceval 367 (Halliwell's Dict.). *Icel.* stóð-merr.]

stód-þeóf, es; *m. One who steals from a stud, a horse-stealer,* L. Alf. pol. 9; Th. i. 68, 5.

stofa, an; *m. A room for a warm bath :*—Stofa *balneum*, Wrt. Voc. ii. 101, 60. [*O. H. Ger.* stuba *a chamber that may be warmed : Icel.* stofa, stufa *a bathing-room that has a fire; a room.* The Romance languages borrowed from Teutonic, hence *Fr.* étuve : *Ital.* stufa : *Span.* estufa *a hot-house, bath-room*.] v. stuf-bæþ.

stofn, e; *f.* **I.** *a stem :*—Stoc *truncus*, stofn *stipes*, Wrt. Voc. i. 32, 43. [Þai thre stod on a stouen (stalke, stocke, other MSS.), C. M. 8036. Stovin a stump or stake; the part of a hawthorn left in a hedge after 'splashing' it, E. D. S. Pub. Leicestershire. *Icel.* stofn *a stem, stump of a tree.*] **II.** *a shoot of a tree :*—Stofna ł telgena *surculorum, virgultorum*, Hpt. Gl. 419, 65. Stofnes (stofne?), ówæstmas *surculos, ramusculos*, 409, 1. **II a.** *fig. offspring, progeny :*—Mid gestrénendlícere stofne *progenie propaganda*, 445, 64. [Stoven a sapling shoot from the stump of a fallen tree, E. D. S. Pub. B. 22, and Whitby

Gloss.] **III.** *a foundation :*—Swā gē āwurpon wāh of stofne *tamquam parieti inclinato*, Ps. Th. 61, 3. [*Icel.* stofn *a foundation;* stofna *to establish, lay the foundation of.*]

stól, es; *m.* **I.** *a stool, seat :*—Stool *tripes*, Wrt. Voc. ii. 122, 75. Gewyrc stól of prím treówum ... geót under ðone stól, Lchdm. ii. 76, 21–24. Man sceal habban ... sceamelas, stólas, Anglia ix. 264, 21. **II.** *the seat* (lit. and fig.) of one in authority, *the throne* of a king, *see* of a bishop :—Sóna se stól (*the throne*) scylfþ, L. I. P. 4; Th. ii. 308, 1. Stóles *cathedrae* (*pontificalis*), Hpt. Gl. 454, 33. Se sit swelce hē sitte on ðæm stóle ðæs forhwierfdan gemótes ... Se biþ beforan ðe on ðæm stóle sitt ðæm óðrum ðe ðær ymb stondaþ, Past. 56; Swt. 435, 24–28. Heofnes Wealdend ðe siteþ on ðam hālgan stóle, Cd. Th. 17, 16; Gen. 260: 19, 33; Gen. 300. Hū hē him strenglícran stól geworhte, 18, 15; Gen. 273. Geseón selfes stól herran ðínes, 36, 4; Gen. 566. Ofer stól *super cathedram* (*Mosi*), Mt. Kmbl. Lind. 23, 2. [*Goth.* stóls *a seat, throne : O. Sax. O. Frs.* stól : *O. H. Ger.* stól, stuol *sedes, sella, thronus : Icel.* stóll *a seat, throne, see.*] v. arce- (erce-), arcebiscop-, biscop-, brego-, cyne-, ealdor-, Eoforwíc-, éðel-, fealde-, friþ-, frum-, gang-, gebed-, gif-, gleow-, gum-, heáfod-, heofon-, hleów-, rodor-, þeóden-, yrfe-stól.

stole, an; *f.*: stol, es; *n.* (in Northern specimens) *A stole, long outer garment :*—Stole *stola*, Wrt. Voc. i. 81, 43. Stol wuldres gigeríde hine *stola glorie induit eum*, Rtl. 45, 29. Ðæt stol ǽriste *stolam primam*, Lk. Skt. Lind. 15, 22. Geonga in stolum (stollum, Rush.) Ɫ on ofer-slopum *ambulare in stolis*, 20, 46 : Mk. Skt. Lind. Rush. 12, 38. [*Icel.* stola ; *f. a stole.* From Latin.]

stom[m], stomer, stommettan, stomrian, stondan, stood. v. stam, stamer, stammettan, stamerian, standan, stód.

stópel, es; *m. A foot-step, mark left by the foot :*—Man dæghwamlíce ða moldan nimeþ on ðǽm lǽstum ... and nǽfre man ðǽre moldan tó ðæs feale ne nimeþ, ðæt mon ǽfre þurh ðæt mǽge á ðý máran dǽl on ðǽm stóplum gewercean (*make the footprints larger*) ... Forlǽt úre Drihten his ða hālgan fét ðǽr on ða eorþan besincan ... and swā núget on ðǽre eorþan ða stóplas onáþrycte syndon, Blickl. Homl. 127, 14–26. [Cf. *O. Sax.* stópo *foot-print.*]

stoppa, an; *m. A stop, a bucket, pail.* Halliwell gives *stop* a small well-bucket, and also *stoppe* a bucket, or milking-pail, as Norfolk words; the latter being still in use. 'The holy-water *stoppe* was a vessel containing holy-water placed near the entrance of a church, and was sometimes made of lead':—Stoppa *situla*, Wrt. Voc. i. 25, 10 : *bona* (?), 288, 2 : *botholicula*, ii. 126, 55 : *bothonicla*, 11, 20 : *bothonicula*, Txts. 42, 122. [*Prompt. Parv.* stoppe, boket *situla, haustrum*, stoppe, vessel for mylkynge *multra, multrale, multrum.*] v. buter-stoppa.

stoppian *to stop, close* an aperture. v. for-stoppian, Lchdm. ii. 42, 12. [From Latin (?).]

stór, es; *m. Frankincense, storax :*—Ðes stór *hoc thus*, Ælfc. Gr. 9, 33; Zup. 59, 14 : Wrt. Voc. i. 81, 25. Stór ðe biþ of gewringe *stacten*, 20, 28. Hí him geoffrodon gold and récels and myrran ... se stór getácnode ðæt hē is sóð God, Homl. Th. i. 116, 9. Áne hand fulle stóres, Lev. 2, 2. Nymeþ stór *sumite modicum storacis*, Gen. 43, 11. Ða ðe offrodon ðone stór *qui offerebant incensum*, Num. 16, 35. Brimne stór and hwítne rýcels, Lchdm. iii. 14, 21. [Encens, stor *olibanus*, Wrt. Voc. i. 140, 24 (13th cent.) 'Mj bene bi ydiȝt beuore þe ase þet stor.' þet stor huanne hit is ope þe uere smelþ zuete, Ayenb. 211, 17.] v. stéran.

stór; *adj. Great, strong, violent :*—Swā stór þunring wes, Chr. 1085; Erl. 219, 22. [Of þan fehte þe was feondliche stor, Laym. 85. Onkumen was Cadalamor ... wið ferding stor, Gen. and Ex. 842. Wunder wel starc and stor, O. and N. 1473. Stoor (store) or hard or boystows *austerus, rigidus*, Prompt. Parv. 477. See also *store* in Halliwell's Dict. *O. Frs.* stór : *O. L. Ger.* stóri *inclytus : Icel.* stórr : *Dan. Swed.* stor. Borrowed (?) from Scandinavian.]

storc, es; *m. A stork :*—Storc *ciconia*, Wrt. Voc. ii. 103, 81 : 14, 33 : i. 29, 19 : 77, 18 : 280, 24 : Ælfc. Gr. 7; Zup. 25, 6. Storc and swalewe heóldon ðone tíman heora tócymes, Homl. Th. i. 404, 25. [*O. H. Ger.* storah, storc *ciconia, ophimachus, ibis : Icel.* storkr.]

stór-cyll, e : -cylle, an; *f. A censer :*—Stórcyl *turibulum*, Wrt. Voc. i. 81, 27. .i. silfren stórcylle, Chart. Th. 429, 35. Se ðe bær ða stórcyllan tó ðære offrunge, Homl. Th. ii. 294, 20 : Homl. Ass. 58, 185. Nime eówer ǽlc his stórcillan, Num. 16, 6. Ða stórcyllan *haec turibula*, Ælfc. Gr. 14; Zup. 90, 41. Stórcillan, Lev. 10, 1.

storm, es; *m.* **I.** *a storm, tempest :*—Storm *nymbus*, Wrt. Voc. ii. 114, 70 : *procella*, i. 52, 62 : 76, 45 : *grando*, Blickl. Gl. Se swearta storm norðan and eástan, Met. 4, 22. Se stearca storm, 6, 11. Seó réþnes ðæs stormes *saevitia tempestatis*, Bd. 5, 1; S. 614, 9. Hē ofslóh on storme (*grandine*) wíngeardas heora, Ps. Spl. 77, 52. Mid ðý storme ðæs wintres *hiemis tempestate*, Bd. 2, 13; S. 516, 19. Ðæs stánhleoþu stormas cnyssaþ, Exon. Th. 292, 19; Wand. 101 : 307, 13; Seef. 23. Storma *nimborum*, Hpt. Gl. 439, 71. Stormum *nimbis*, Wrt. Voc. ii. 61, 36. On ðære hreón sǽ and on ðǽm miclan stormum, Past. 9; Swt. 59, 3. Gescyrped mid ðære rinde wið ða stearcan stormas, Bt. 34, 10; Fox 150, 8. Seó lyft áþyrþ ealle wolcna and stormas, Lchdm.

iii. 274, 10. **I a.** fig. *a storm* of arrows :—Strǽla storm scóc ofer scyldweall, Beo. Th. 6225; B. 3118. **I b.** *storm, disturbance, disquiet :*—Hwæt is ðonne ðæt ríce and se ealdordoom bútan ðæs módes storm, se biþ simle cnyssende ðæt scip ðære heortan, Past. 9; Swt. 59, 4. Swelce eác tóættécte ðisse gedréfnisse storm Sǽberhtes deáþ, Bd. 2, 5; S. 507, 6. Ða strongan stormas weoruldbisgunga, Met. 3, 3. **II.** *uproar, tumult :*—Storm up árás æfter ceasterhofum, cirm unlytel hǽdnes heriges, Andr. Kmbl. 2474; An. 1238. Storm up gewát heáh tó heofonum, herewópa mǽst, Cd. Th. 206, 30; Exod. 459. **III.** *violent attack,* cf. to *storm* a place :—Ðis is stronglíc, nú ðes storm becom, þegen mid þreáte (*of the harrying of hell*), Cd. Th. 288, 26; Sat. 387. Forstond ðú mec and gestýr him (*the devils*), ðonne storm cyme mínum gǽste ongegn, Exon. Th. 455, 32; Hy. 4, 58. [*O. Sax.* storm : *O. H. Ger.* sturm *procella, tempestas ; strepitus, agitatio, motus, seditio, tumultus : Icel.* stormr *a tempest ; tumult, uproar.*] v. styrman.

stór-sǽp, es; *n. Resin :*—Stórsæpes *resinae*, Hpt. Gl. 501, 1.

stór-sticca, an; *m. An incense-stick, rod for stirring the incense in the censer* (?) :—.i. silfren stórcylle mid silfrenum stórsticcan, Chart. Th. 429, 35.

stów, e; *f. A place.* The word remains either alone or in composition in place-names, e. g. *Stow* in Huntingdonshire, *Stowe* in Northamptonshire, Chep*stow* old *ceáp-stów* q. v. :—Stów *locus*, Wrt. Voc. i. 85, 31. **I.** *a place, spot, locality, site :*—Ðeó stów (*Calvary*) wæs gehende ðære ceastre, ðǽr se Hǽlend wæs áhangen, Jn. Skt. 19, 20 : Elen. Kmbl. 1347; El. 675. Nis ðæt heóru stów, Beo. Th. 2749; B. 1372. Wæs seó londes stów bimiþen fore monnum, óþþæt Meotod onwráh beorg on bearwe, Exon. Th. 110, 32; Gú. 117. Ða hwearf hē eft tó ðære leófan stówe *tunc reversus ad dilectae locum peregrinationis*, Bd. 5, 9; S. 623, 30. Teón wit of ðisse stówe, Cd. Th. 114, 30; Gen. 1912. Stópon tó ðære stówe ðe Dryhten ǽr áhangen wæs, Elen. Kmbl. 1428; El. 716. Tó ðam stówe (-um?) *ad loca*, Ex. 3, 8. Geseóþ ða stówe ðe se Hǽlynd wæs on áled, Mt. Kmbl. 28, 6. Ðæt hē ðær forgeáfe stówe mynster on tó timbrianne, Bd. 3, 24; S. 557, 26. On wéstum stówum *in desertis locis*, Mk. Skt. 1, 45. Hē gǽþ þurh unwæterie stówa, Lk. Skt. 11, 24. Muntas and móras and eác monige wéste stówa, Salm. Kmbl. 683; Sal. 341. **II.** *a place* on the body :—Gif ðú wille lim áceorfan ... gesceáwa ðú hwilc sió stów sié and ðære stówe mægen, for-ðon ðe ðara stówa sum raþe rotaþ, gif hire mon gímeleáslíce tilaþ, Lchdm. ii. 84, 22–25. Wið wífa earfoðnyssum ðe on heora inwerdlícum stówum earfeþu þrowiaþ ... wyrc tó sealfe, dó on wífa stówe, i. 338, 19–22. Lácnian ða sáran stówa, ii. 22, 3 : 70, 8 : 150, 16. **III.** *a place* which is built, *a house* or *collection of houses, a habitation, dwelling :*—Seó stów (*Ely*) wæs gehálgod ðam hālgan Petre, Chart. Th. 241, 2. On ðære stówe dura *in introitu tabernaculi*, Num. 12, 5. Ne onscunige ic nó ðæs neoþeran and ðæs unclǽnan stówe (*the prison of Boethius*), Bt. 5, 1; Fox 10, 15. Gange seó sócn intó ðære stówe (*the monastery at Ely*), Chart. Th. 243, 1. On ðære stówe (*the town of Zoar*) wē gesunde mágon bídan, Cd. Th. 152, 19; Gen. 2522. Ða sealde se cyning him wununesse and stówe on Cantwarabyrig *dedit eis mansionem in civitate Doruvernensi*, Bd. 1, 25; S. 487, 18. Hē ána gesæt dýgle stówe (*a hermitage*), Exon. Th. 111, 21; Gú. 130. Folc of eallum túnum and stówum, Bd. 2, 14; S. 518, 10. **IV.** *a place, position, station :*—Sió wyrd dǽlþ eallum gesceaftum stówa and tída, Bt. 39, 5; Fox 218, 33. Ða nú ryne healdaþ, stówe gesefnde, Cd. Th. 10, 21; Gen. 160. **V.** *a place* in a series :—Onfengon hí ða teóþan stówe on ehtnysse Godes cyrcena æfter Nerone, Bd. 1, 6; S. 476, 22. **VI.** *place, room, stead :*—Se ðe lifigende wǽre ðæs hádes hæfde mihte óþerne biscop his stówe tó hálgianne ðær se óðer forþférde *is qui superest consors ejusdem gradus, habeat potestatem alterum ordinandi, in loco ejus qui transierat, sacerdotem*, 2, 18; S. 520, 35. **VII.** *a place, passage* in a book :—Ic ðē sende ðæt spell ðē sylfum tó rǽdanne and on emtan tó smeágeanne and eác on má stówa tó wrítanne and tó lǽranne, Bd. pref.; S. 471, 11. [*O. Frs.* stó *a place : Icel.* eld-stó *a fire-place.*] v. ancor-, byrgen-, ceáp-, cot-, cwealm-, dóm-, eardung-, éðel-, folc-, freóls-, friþ-, fulwiht-, gemót-, gewin-, heáfod-, heg-, leger-, mold-, munuc-, mynster-, neáh-, onbíd-, oret-, pleg-, seonoþ-, spell-, sundor-, tintreg-, wæl-, wáfung-, weall-, wíc-, win-, wítnung-, wítung-, wræc-stów.

stówian; *p.* ode *To hold back, restrain :*—Stóuuigan *retentare*, Wrt. Voc. ii. 118, 72. [He sette stronge lasuen to steowien (stewe, 2nd MS.) his folke, Laym. 6266. Stew þine unwittie wordes, Marh. 6, 2 (and see note, p. 109). Lǽte me steowe (cf. stewe = A. S. stów, 145, 5) mi flesc, Misc. 193, 34. Beo stiward in oure stude til ȝe be stouwet (stowed, C-text MS Ɫ.; ruled, B-text) betere, Piers P. A-text 5, 39. ȝiff any man stow me this nyth I xal hym ȝeve a dedly wownde, Cov. Myst. (Halliwell's Dict.). Stowyñ or with stond idem quod stoppyñ, stowynge, stowwynge *obsistencia, resistencia*, Prompt. Parv. 478, col. 1.]

stów-líc; *adj. Local, relating to place,* (1) *occupying a place :*—God is ǽghwǽr, þeáh ðe se engel stówlíc sý. Nis se ælmihtiga Wealdend stówlíc, forðan ðe hē is on ǽlcere stówe, and swā hwider swā se stówlíca engel flíhþ, hē biþ befangen mid his andwerdnysse, Homl. Th. i. 348, 12–15. Stówlícere moldan *situ*, Germ. 391, 195. (2) *expressing*

relations of place :—Sume naman syndon *localia* ðæt synd stówlíce; ða geswuteliaþ gehendnysse oððe ungehendnysse, Ælfc. Gr. 5; Zup. 14, 18. Sume (*adverbs*) synd stówlíce, forðan ðe hí getácniaþ stówa, 38; Zup. 224, 12.

stówlíce; *adv. Locally, in respect of place* :—Ða Iudéiscan ðe on Crist gelýfdon wæron him gehendor stówlíce and eác ðurh cýððe ðære ealdan ǽ: wě wæron swíðe fyrlyne ǽgðer ge stówlíce ge ðurh uncýððe, Homl. Th. i. 106, 19–21.

strácian; *p.* ode *To stroke* :—Se lǽce grápaþ and strácaþ, ǽrðonðe hě stingan wille, Past. 26; Swt. 187, 5. Wildu hors, ðonne wě hié ǽresð gefangnu habbaþ, wě hié ðacciaþ and stráciad mid brádre handa *equos indomitos blanda prius manu tangimus*, 41, 4; Swt. 303, 10. Myd swýþe drígeon handum stráca geornlíce ðane innoþ, Lchdm. iii. 134, 17. [*O. H. Ger.* streichôn *demulcere*.]

strácung, e; *f. Stroking, caressing* :—Strácung *vel* ólǽcung *delinimentum*, Wrt. Voc. i. 54, 69.

strǽc, strec; *adj.* **I.** *strict, severe, rigorous, stern, hard* :—Hú se reccere sceal bión wið ðara yfelena undéawas strǽc for ryhtwíslícum andan *ut sit rector contra delinquentium vitia per zelum justitiae erectus*, Past. 17; Swt. 107, 6. Strǽc (strec, Cott. MSS.), 12; Swt. 75, 12. Ðæt se streca Déma ús geárige, Homl. ii. 126, 13. Ætforan ðæs gesíhðe ðæs strecan Déman, 124, 15. Streccere *rigidae, durae*, Hpt. Gl. 416, 16. Ðære strǽcan *asperrima*, Wrt. Voc. ii. 2, 24. Wě scoldon mid strecum móde stíðlícor libban and winnan wið leahtras, L. Ælfc. P. 12; Th. ii. 368, 18. Réðe and strece for ryhtwísnesse *justitiae severitate districti*, Past. 5, 1; Swt. 41, 19. Déde strece *actus strenuos*, Ps. Surt. ii. p. 201, 11. Tó ðæm strǽcstum (strecestum, MS. T.: strǽncstum, MS. A.) mynstermonna cynne *ad cenobitarum fortissimum genus*, R. Ben. 10, 4. **II.** *rigid, unyielding, obstinate, persistent.* v. strǽcness :—On óðre wîsan sint tó manianne ða ânfealdan strǽcan on óðre ða unbealdan. Ðæm anfealdan strǽcum is tó cýðanne ðæt hié bet [ne] truwien him selfum ðonne hié ðyrfen ðonne hí nyllaþ geðafan beón óðerra monna geðeahtes *aliter admonendi sunt pertinaces, atque aliter inconstantes. Illis dicendum est, quod plus de se, quam sunt, sentiunt, et idcirco alienis consiliis non acquiescunt*, Past. 42, 1; Swt. 305, 12–15. **III.** *violent, using force, uncompromising, vehement* :—Manig strec (strǽc, MS. B.) man wyle werian his man swá hwæðer him þincþ ðæt hě hine eáð áwerian mǽge. Ac wě nellaþ geþafian ðæt unriht, L. C. S. 20; Th. i. 388, 1. Heofena ríce þolaþ neáð, and strece (*violenti*) nimaþ ðæt, Mt. Kmbl. 11, 12. Godes ríce ðolaþ neádunge, and ða strecan mód hit gegrípaþ . . . Eal cristen folc sceal mid neádunge and strecum móde ðæt heofonlíce ríce geearnian, Homl. Th. i. 358, 25–35. Ða hǽþenan féngon tó wurðienne entas and strece woruldmen ðe mihtige wurdan on woruldafelum and egesfulle wǽran ða hwýle ðe hý leofedon, Wulfst. 105, 34. [Cf. strek *straightway* : He sal noght wend strek til purgatory bot even til helle, Pr. C. 3378. *M. H. Ger.* strac.] v. ânstrǽc, *and next word.*

strǽc, es; *n.* (?) **I.** *strictness, rigour* :—Strǽc *districtio, rigor*, Wrt. Voc. ii. 141, 48. Sý nátóðæshwón regoles strǽc gehealden *nullatenus districtio regule teneatur*, R. Ben. 61, 15. **II.** *violence, force* :—Hú mæg beón bútan strece and neádunge ðæt gehwá mid clǽnnysse ðæt gále gecynd þurh Godes gife ʒewylde? Homl. Th. i. 360, 1, 10. Hér man ýtte út Ælfgár eorl, ac hé com sóna inn ongeán mid strece þurh Gryffines fultum, Chr. 1058; Erl. 192, 36.

strǽc-líc; *adj. Rigorous, strict, severe* :—Gif hié ne beóþ gebundne mid strǽclíce láreówdóme *si hanc districtionis severitas non coarctat*, Past. 17, 9; Swt. 123, 17. Streclícere hǽse *rigid oimperio*, Hpt. Gl. 437, 4.

strǽclíce; *adv.* **I.** *strictly, sternly, vehemently* :—Gif him God ryhtlíce and strǽclíce (streclíce, Cott. MSS.) déman wile *si districte judicentur*, Past. 5, 3; Swt. 45, 20. Hwílum líðelíce tó ðreátigenne hwílum suíðlíce and strǽclíce tó ðráfianne *aliquando leniter arguenda, aliquando autem vehementer increpanda*, 21, 1; Swt. 151, 12. **II.** *violently, forcibly* :—Swá swá deáð streclíce ásyndraþ sáwle fram líchaman ealswá lufu Godes streclíce ásyndraþ mann fram middanearderne lufe *sicut mors uiolenter separat animam a corpore, ita dilectio Dei uiolenter segregat hominem a mundano amore*, Scint. 16, 14–16.

strǽcness, e; *f. Persistence, perseverance, pertinacity* :—Mid unáteriendlíc[r]e strecnysse *indefessa instantia* (*perseverantia*), Hpt. Gl. 434, 24. Hí mid ânrédnesse and mid strecnesse geearnodon heofona ríce, L. Ælfc. P. 13; Th. ii. 368, 29.

strǽd[a, -e?] *a pace, stride* :—Míle straedena *mille passus*, Mt. Kmbl. Lind. 5, 41.

strǽgdness. v. stregdness.

strǽgl (*from Latin* stragula ?), strǽl, strél, e; *f. A covering for beds, a rug, a mattress, bed* :—Strégl (g over a), strél *aulea*, Txts. 43, 249. Strél *stragua*, 99, 1907. Strél *vel* bedding *mataxa vel conductum vel stramentum*, Wrt. Voc. i. 59, 29. Strǽle mínum (-e ?) ic wǽte *stratum meum rigabo*, Ps. Spl. 6, 6. Strǽla *stragularum*, Hpt. Gl. 430, 67. Ealle strǽla his ðú ácyrdest *universum stratum ejus versasti*, Ps. Spl. 40, 3. [*Prompt. Parv.* strayle, bed clothe *stragula*.]

strǽl, strél, streál, es; *m.* : e; *f.* : strǽle, an; *f. An arrow, shaft,* *dart* (lit. and fig.) :—Ða genam hě his bogan and hine gebende and ðá mid geǽttredum strǽle ongan sceótan . . . Ða sóna mid ðan ðe se strǽl on flyge wæs, ða com swíðe mycel windes blǽd, ðæt seó strǽl wearð eft gecyrred, and ðá ðone ilcan mon, ðe heó ǽr from sended wæs, hě sceát, Blickl. Homl. 199, 17–23. Hě cwæð tó ðam deófle: 'Ðú heardeste strǽl tó ǽghwilcre unrihtnesse, 241, 3: Andr. Kmbl. 2380; An. 1191. Hě his costunge strǽle on ðam móde gefæstnode ðæs cempan. Hě mid ðære geǽttredan strǽle gewundod wæs . . . Ða hæfde hine seó deófollíce strǽl mid ormódnysse gewundodne, Guthl. 4; Gdwin. 28, 2–14. Swá seó strǽle byþ strangum on handa *sicut sagittae in manu potentis*, Ps. Th. 126, 5. Leóhtes strǽle ł leóma *lucis spiculum*, Hymn. Surt. 30, 6. Se mon wæs ofscoten mid his âgene strǽle mid ðý ðe hě wolde ðone fearr sceótan, Shrn. 83, 6. Hě forð onsendeþ biterne strǽl, Exon. Th. 48, 2; Cri. 765. Strǽlas *sagittae*, Ps. Spl. 63, 8: Blickl. Homl. 203, 9. Strǽlas, Ps. Surt. 56, 5. Strǽle beóþ scearpe *sagittae acutae*, Ps. Th. 119, 4: 143, 7. Scearpum strélum, 63, 4. Hě sende his strǽlas, 17, 14: Judth. Thw. 24, 34; Jud. 223. Ðá hét ic feá strǽla (*paucas sagittas*) sendan in ða burh innan, Nar. 10, 22. Hě sendeþ his strǽlo, Bd. 4, 3; S. 569, 20. Lǽteþ strǽle fleógan, farende flân, Exon. Th. 386, 4; Rä. 4, 56. Hě ða strǽle ðara áwergdra gásta him fram ásceáf, Guthl. 6; Gdwin. 42, 24. Strǽle bitere sendan, Ps. Th. 77, 11: 76, 14. [Strales bitere, Laym. 5695. *O. L. Ger. O. H. Ger.* strála: *f. sagitta, jaculum: M. H. Ger.* strâl, strâle; *m. f.* : *Ger.* strahl; *m.*] v. here-, wǽl-, wǽpen-strǽl.

strǽl-bora, an; *m. An archer* :—Strélbora *arcister*, Wrt. Voc. ii. 101, 8. Strélbora and scytta, 7, 32. Strélbora, 55.

strǽlian; *p.* ode *To shoot* :—Hí strǽliaþ hine *sagittabunt eum*, Ps. Spl. 63, 4.

strǽl-wyrt, e; *f. Club-moss* (?). Somner gives the word as glossing *callitrichon* :—Gif dolh fúlige, ceów strǽlwyrt on and gearwan, Lchdm. ii. 96, 9.

strǽt, e (*but uninflected forms occur*); *f.* **I.** *a road* :—Læg ân dríe strǽt þurh ða sǽ. And ðæt wæter stód an twá healfa ðære strǽte, Ex. 14, 21–22. Him þurh streámrǽce strǽt wæs gerýmed, Andr. Kmbl. 3159; An. 1582. Tó ðære ealdan strǽt; ondlong ðære strǽt, Cod. Dip. Kmbl. iii. 79, 30. On ða sealtstrǽt; andlang strǽt, 82, 26. Foldweg, cûþe strǽte, Beo. Th. 3272; B. 1634. Ceastre and torras and strǽta and brycge geworhte wǽron *civitates, farus, pontes, et stratae factae*, Bd. 1, 11; S. 480, 16. **II.** *a road in a town, a street, a paved road* :—Strǽt wæs stânfáh ... hié tó sele gangan cwômon, Beo. Th. 645; B. 320. Ða stânas ðæs temples licggeaþ æt ælcre strǽte ende *in capite omnium platearum*, Past. 18, 3; Swt. 133, 12. Loth sæt on ðære strǽt (*in foribus civitatis*) ... Híg cwǽdon: 'Wě willaþ wunian on ðære strǽt (*in platea*), Gen. 19, 1–2. Be ðisse strǽte, Cd. Th. 147, 8; Gen. 2436. Eode se apostol be ðære strǽt, Homl. Th. i. 60, 21: ii. 120, 16. Hě eode in burh, stóp on strǽte, Andr. Kmbl. 1969; An. 987. Enta ǽrgeweorc innan burgum strǽte stânfáge, 2473; An. 1238. Fenn strǽta *lutum platearum*, Ps. Spl. 17, 44. On strǽta hyrnum, Mt. Kmbl. 6, 5. On strǽton *in plateis*, Mk. Skt. 6, 56. Hí synd stǽnene mid ðám ðe man strǽta wyrcþ, Homl. Skt. i. 7, 134. Ða arn se ceorl geond ealle ða strǽt, Homl. Th. ii. 302, 8. [*O. Sax. O. L. Ger.* strâta: *O. Frs.* strête: *O. H. Ger.* strâza. *From Latin* strata.] v. cyne-, faroþ-, fird-, heáh-, here-, lagu-, mere-, port-, ranc-, sealt-, stân-strǽt.

strǽt, e; *f. A couch, bed* :—On bedding strǽte mínre *in lectum strati mei*, Ps. Spl. C. 131, 3. Ofer strǽte *super lectum*, 62, 7: 6, 6. [*From Latin.*]

strand, es; *n. A strand, shore* :—Strand *litus*, Wrt. Voc. i. 54, 24: *sablo*, Hpt. Gl. 502, 77. Se Hǽlend stód on ðam strande ... Ðæt strand getácnode ða écan staðolfæstnysse ðæs tówerdan lífes, Homl. Th. ii. 288, 30. Wudes ne feldes, sandes ne strandes, Lchdm. iii. 288, 1. Hí sǽton be ðam strande *secus littus sedentes*, Mt. Kmbl. 13, 48: Jn. Skt. 21, 4. Ða eode he be strande, Ap. Th. 7, 19. Gáþ tó ðære sǽ strande, Homl. Th. i. 64, 3. Urk mín húskarl habbe his strand eall forne gén hys âgen land, Cod. Dip. Kmbl. iv. 221, 6. Stranda *litorum*, Hpt. Gl. 449, 28. Strandum *litoribus*, 465, 9. [*O. Du.* strande, n.: *M. H. Ger.* strant; m.: *Icel.* strönd; f.] v. sǽ-strand.

strang; *adj. Strong*; fortis, Wrt. Voc. i. 83, 56: acer, vehemens, 17, 28: strenuus, ii. 74, 60. **I.** *of living beings*, (1) *strong, powerful, mighty* :—Hě wæs strang foreþingere *he was a powerful intercessor*, Homl. Skt. i. 5, 6. Fugel meahtum strang, Exon. Th. 40, 31; Cri. 647. Ðú eart mægenes strang, Beo. Th. 3692; B. 1844. Hú mæg man ingân on stranges (*fortis*) hús and hys fata hyne bereáfian, búton hě gebinde ǽrest ðone strangan (*fortem*)? Mt. Kmbl. 12, 29. Paminunde ðæm strongan cyninge *apud Epaminondam, strenuissimum imperatorem*, Ors. 3, 7; Swt. 110, 21. Wæron hér strange cyningas (*fortissimi reges*), Bd. 4, 2; S. 565, 30. Hě ys strengra (strǽngra, Rush. : strongra, Lind.) ðonne ic *est fortior me*, Mt. Kmbl. 3, 11: Lk. Skt. 11, 22. Wě wénaþ ðæt mon beó ðý strǽngra (strencra, Cott. MS.) ðe hě biþ micel on his líchoman, Bt. 24, 3; Fox 84, 7. Ic eom se strengesta (*fortissimus*) God ðínes fæder, Gen. 46, 3. Se strangesta cyning Æþelriþ *rex fortissimus Ædilfrid*, Bd. 1, 34; S. 499, 18. Ætýwan ðíne mǽrðe and ðíne strengestan hand, Deut. 3, 24. Feówer ða strengestan him betweónum

gespræcon, Ors, 3, 10; Swt. 138, 3. Of ðrîm folcum ðám strangestan Germanie *de tribus Germaniae populis fortioribus*, Bd. 1, 15; S. 483, 20. Of mînum strengestum feóndum, Ps. Th. 17, 18. (2) *strong, firm, resolute, hardy :*—Beó strang and staðulfæst *confortare et esto robustus*, Deut. 31, 7, 23. Ic wênde ðæt ic wære swîðe strong on manegum cræftum ac ic ongeat siððan ðû mê forlête hû untrum ic wæs *fortem me inter virtutes credidi, sed, quantae infirmitatis sim, derelictus agnovi*, Past. 65, 5; Swt. 465, 21. Ðætte ûre mód ðý fæstre and ðý strengre beforan Gode sié on ðæm cræftum *ut cor robustius in virtute solidetur*, 65, 6; Swt. 467, 9. Ðær wæs heáfde beslagen se strengesta martyr Scś Albanus, Bd. 1, 7; S. 478, 33. (3) *hard, severe, fierce, stern :*—Strang wæs and rêðe se ðe wætrum weóld, Cd. Th. 83, 8; Gen. 1376. Se þeóden wæs strang and stîðmód, 279, 34; Sat. 248. Petrus gecýðde ðæt hê wæs strengesð wið scylda, Past. 17, 6; Swt. 115, 17. Seó strengeste þeód *gens ferocissima*, Ors. 4, 11; Swt. 206, 34. II. of things, (1) *strong, able to resist force, firm :*—Ðû wære mê stranga tor, Ps. Th. 60, 2. Gê nânuht mid eów nabbaþ fæstes ne stronges ðætte þurhwunigean mæge, Ors. 2, 4; Swt. 74, 28. Æt strangum stâne, Ps. Th. 140, 8. Of ðissum strongan stýle, Exon. Th. 426, 25; Rä. 41, 79. Ðeós wyrt biþ cenned on fæstum landum and on strangum, Lchdm. i. 134, 19. Ðeós wyrt on Illyrico swîðost and strengost wexeþ, 284, 17. (2) *firm, valid, assured :*—Mê ðynceþ wîslîc, gif ðû geseó ða þing beteran and strengran ðe ûs bodade syndon, ðæt wê ðám onfón, Bd. 2, 13; S. 516, 10. (3) *strong in operation, effective, producing a great effect, potent :*—Ongeán swelce mettrymnesse mon beðorfte stronges læcedómes ... Is ðæm læce tô giémanne ðæt hê strangne læcedóm selle ðæm seócan, Past. 61, 2; Swt. 455, 26–29. Ða leáf syndon stranges swæcces, Lchdm. i. 310, 7. Gif ðû ðás wyrte sylst þicgean on strangon wîne, 172, 12. Strangre stemne, Cd. Th. 33, 24; Gen. 525. Ða recceras ætiéwaþ strangne andan *fortem zelum rectores exhibent*, Past. 21, 6; Swt. 164, 11. Ðæt is for hwî se góda læce selle ðam hálum men sêftne drenc and swêtne, and ôðrum hálum biterne and strangne, Bt. 39, 9; Fox 226, 12. Gelácnian mid ðæm drencum strangra wyrta gemanges, Past. 37; Swt. 269, 24. Se gewuna is strengra on ælcum worde, ðonne his regol sý, Ælfc. Gr. 30; Zup. 193, 2. Ðæt hié hæfden ðý strengran scyte *ne sagittarum jactus impedirentur*, Ors. 1, 10; Swt. 46, 13. Strongum helpum *validioribus auxiliis*, Rtl. 61, 11. (4) *strong, earnest :*—Ðá sealdon hî him strange manunge, Bd. 1, 12; S. 481, 13. (5) of that which is hard to bear, *hard, severe :*—Godes bebod, þeh hit strong wære, Ors. 6, 1; Swt. 252, 2. Hû strang hit biþ an helle tô biónne, Wulfst. 225, 12. Is se drohtað strang ðám ðe lagoláde lange cunnaþ, Andr. Kmbl. 626; An. 313. Strang wîte, Cd. Th. 155, 4; Gen. 2567. Ðæt sár biþ tô ðon strang, and hê næfþ nânne slæp ðonne hit strangost biþ, Lchdm. ii. 198, 25. Strang fefer, 226, 16. Com se stranga winter mid forste and mid snâwe and mid eallon ungewederon, ðæt næs nân man ðá on lîue, ðæt mihte gemunan swá strangne winter swá se wæs, Chr. 1046; Erl. 170, 32: Chart. Th. 163, 1. Hê âstealde swîðe strang gyld, Chr. 1040; Erl. 166, 20. Ealle ða gesetnessa ðe tô strange wæron and tô hearde, Ors. 5, 12; Swt. 244, 15. Wæs ðis gefeoht wælgrimre and strengre eallum ðam ærgedónum, Bd. 1, 12; S. 481, 25. Manig broc byþ mycle strengre ðonne tôðæce, ðeáh ic næfre nân strengre ne geðolode, Shrn. 185, 15. Ðonne biþ Drihtnes word rêðe gehýred, ðám synfullum stefna strangast, Wulfst. 256, 16. (6) of violent motion or action, *fierce, violent :*—Strong wind, Met. 7, 25. Strang storm, Lchdm. i. 326, 19. His tógan biþ ðearle strang, 364, 17. Se stranga rên, Ps. Th. 71, 6. Ðá gemunde hê ða strangan dæda ðara unmanna and ðæra woruldfrumena, Guthl. 2; Gdwin. 12, 27. Gif strongra storm and genip swýðor ðreáde, Bd. 4, 3; S. 569, 12. [*O. Sax.* strang: *Icel.* strangr: *O. H. Ger.* strengi.] v. byrðen-, for-, hyge-, lang-, mægen-, med-, ryne-, swîþ-, un-strang; strenge.

strange ; *adv.* I. *severely :*—Rîcum mannum man sceal strangor (*severius*) dêman ðonne ðám heánum, L. Ecg. C. 1; Th. ii. 132, 30. Ðeáh ðe ðæt wîte heardor and strangor dón sý *cum districtius agitur*, Bd. 1, 27; S. 490, 12. II. *strongly, violently :*—Seó sæ strange geondstyred on staþu beáteþ, Met. 6, 15: Soul Kmbl. 89; Seel. 45.

strang-hende, -hynde ; *adj. Strong of hand :*—Dauid is gecweden *fortis manum*, ðæt ys stranghynde, Ælfc. T. Grn. 7, 14.

strangian ; *p.* ode. I. *to grow strong, be strong, prevail, flourish :*—Ic strangige oððe geðeó *uigeo*, Ælfc. Gr. 26, 2; Zup. 154, 14. Strongaþ *praevaluit*, Ps. Surt. 51, 9. Ic strongade wið him *praevalui adversus eum*, 12, 5. Word unrehtwîsra strangadun (*praevaluerunt*) ofer ûs, 64, 4. Strangadan, swîðodon *invalescebant*, Wrt. Voc. ii. 74, 6. Strangedon, 46, 49. [Þet eower heorte erzian and eower feond strongian, O. E. Homl. i. 13, 28. *O. H. Ger.* strangên *confortari*.] II. *to make strong, confirm, comfort :*—Staþelige man and strangie and trumme hî georne, L. I. P. 4; Th. ii. 308, 3. Hê ðær wunode strangende hira heortan on geleáfan, Blickl. Homl. 249, 17. [Heo strangede þe walles, Laym. 4461.] v. ge-strangian; strangung.

strang-lîc ; *adj.* I. of persons, *strong, robust :*—Cniht, stranglîc on wæstme and wênlîc on nebbe, Ælfc. T. Grn. 16, 41. Hwæðer ðæt landfolc sí tô gefeohte stranglîc oððe untrumlîc *populum, utrum fortis sit an infirmus*, Num. 13, 20. II. of things, (1) *strong, firm, solid,*

---

able to resist force :—Næs nân ðæs stronglîc stân gefæstnod, ðæt mihte ðam miclan mægne wiðhabban, Cd. Th. 297, 14; Sat. 517. Ðeáh ðeós eorðe þince on stede stronglîc, Met. 10, 11. Minne stronglîcan stól, Cd. Th. 23, 27; Gen. 366. Sume bióþ beforan monna eágum gesewen swelce hié fæstlîcu and stronglîcu weorc wyrce *quidam quaedam ante humanos oculos robusta exerceant*, Past. 34, 6; Swt. 234, 19. (2) *requiring strength, laborious, hard :*—Nânne mon ðæs ne tweóþ ðæt se seó strong on his mægene ðe mon gesihþ ðæt stronglîc weorc wyrcþ *nemo dubitat esse fortem cui fortitudinem inesse conspexerit*, Bt. 16, 3; Fox 54, 29. (3) *hard to bear, severe :*—Ðis is stronglîc, nû ðes storm becom, Cd. Th. 288, 25; Sat. 387. Stingaþ stranglîc sár on his eágan, Wulfst. 141, 5. [*O. H. Ger.* strang-lîh *robustus*.]

stranglîce ; *adj. Strongly :*—Stranglîce *roborabiliter*, Wrt. Voc. ii. 84, 63. Stranglîce *fortiter*, stranglîcor *fortius*, stranglîcost *fortissime*, Ælfc. Gr. 38; Zup. 230, 15. I. *with power, with energy, strenuously, vigorously :*—Hê stranglîce rîxode and bewerode ðæt folc wið ða hæðenan leóda, Ælfc. T. Grn. 7, 7. Hê galdorcræftum wiðstód stranglîce, Andr. Kmbl. 333; An. 167: Exon. Th. 156, 15; Gû. 875. Hwæt getâcnaþ ða bân bûton stronglîce geworht weorc *quid per ossa nisi fortia acta signantur?* Past. 34, 6; Swt. 235, 16. II. *with violence, fiercely, vehemently :*—Hê byrnende from gebede swîceþ, stepeþ stronglîce, Exon. Th. 264, 34; Jul. 374. Se here ða burh besæton and hire stranglîce wið feaht, Chr. 1016; Erl. 156, 15. Hire mætte ðæt heó hæfde sweord on handa and ðæt heó stranglîce fuhte mid ðý, Shrn. 60, 30. Ðæt se wind swá stronglîce hrure on ða circan, ðæt ðær ne mihte nænig mon gestandan oððe gesittan, 81, 22. III. *boldly, bravely, hardily :*—Hî heora land stronglîce geeodan and freódóm onfêngon, Bd. 3, 24; S. 557, 46. Him gesewen wæs ðæt hê heardlîce and stranglîce spræce, 5, 13; S. 632, 25. IV. *firmly, in a manner to resist force :*—Hê biþ stranglîce wið ða getrymed, Past. 21; Swt. 165, 7. Tô ðon ðæt hê swá micle stranglîcor ârise swá hê hefiglîcor âfeóll *tanto post solidius surgeret quanto prius cecidisset*, 58, 5; Swt. 443, 32. V. *severely, sternly :*—On ðám is stronglîce tô ehtanne ða ðe him ne ondrædaþ wîtende syngian *in his interitus insequenda, qui non metuunt sciendo peccare*, Bd. 1, 27; S. 491, 37. Se man wæs stranglîce gewîtnad, Shrn. 73, 12. Se ðe swá stronglîce ða Iudéas þreáde, Blickl. Homl. 169, 7.

strang-mód ; *adj. Of strong mind, confident, resolute :*—God ða unstrangan ðyses middangeardes geceás, ðæt ða strangmódan, ðe on âgenum mihtum truwiaþ, gescende wæron, R. Ben. 138, 28.

strangness, e ; *f.* I. *strength :*—Strangnysse mîne *fortitudinem meam*, Ps. Spl. 58, 10. II. *force, violence :*—Of him is bodud Godes rîce and ealle on ðæt strangnysse wyrcaþ *ex eo regnum Dei euangelizatur, et omnis in illud uim facit*, Lk. Skt. 16, 16.

strangung, e ; *f.* I. *strengthening, invigorating :*—Hê (*Christ*) ne behófode nânes wæstmes ne nâre strangunge on ðære godcundnysse, Homl. Th. i. 150, 5. Mettas ðe célunge and strangunge mægen hæbben, Lchdm. ii. 176, 16. Ðæt lýft hê gesceóp tô ûres lîfes strangunge, Hexam. 4; Norm. 8, 17. II. *vigor :*—Helias lyfaþ git on lichaman mid langsumre strangunge, Homl. Skt. i. 18, 275. v. ge-strangung; strangian.

strapul, es ; *m. A covering for the leg, kind of trouser :*—Strapulas *tubruces (tubrucus* lanea ocrea, ocreis aut calceis coriaceis superimponi solita, Migne) vel *brace*, Wrt. Voc. i. 25, 61. [A strapylle *tibiale*, Wrt. Voc. i. 259, col. 2 (15th cent.). Straple of a breche, strappyl *femorale, feminale*, Prompt. Parv. 478. Þe strapils of breke *tribraca, femoralia*, Cath. Ang. 367. Sum wummon wered þe brech of heare and þe strapeles adun to hire uet ilaced ful ueste, A. R. 420, 5. Seide þat þey were liche to mares wiþ white legges up to þe þises, for þat tyme þe Longobardes usede strapeles wiþ brode laces doun to þe sparlyuer *asserens eos fore similes equabus, quarum cruretenus pedes sunt albi, eo quod Longobardi tunc temporis usque ad suras candidis fasceolis uterentur*, Trev. v. 355, 4.]

strê, streá-berige, streac, streál. v. streáw, streáw-berige, stearc, strǽl.

streám, es ; *m. A stream, current, flowing water ;* in the plural used of the sea in poetry :—Streám vel wǽto *irriguum*, Wrt. Voc. i. 28, 9. Streám *fluens*, ii. 149, 68: *alveus*, i. 54, 26. Streám, streúm *rema, reuma*, Txts. 92, 855. Streámum, streaumum, streámum *torrentibus*, 103, 2036. Hî on ðæs streámes brycge âbysgade wæron ... Scś Albanus eode tô ðære burnan ... ðá sóna âdrûgode se streám *fluminis ipsius occupabat pontem ... Sanctus Albanus accessit ad torrentem ... illico siccato alveo*, Bd. 1, 7; S. 478, 8–13. Hê wolde ða eá mid sunde oferfaran, ac hiene se streám fordráf, Ors. 2, 4; Swt. 72, 30. Ymbûtan ðone weall is se mæsta dîc, on ðæm is iernende se ungefóglecesta streám *fossa extrinsecus late patens, vice amnis circumfluit*, Swt. 74, 18. Ealle ða gewîtaþ swá swá wæteres streám, Blickl. Homl. 59, 20. Forðon seó stów on ôfre ðæs streámes (*super ripam fluminis*) wæs geseted, wæs his gewuna ðæt hê on ðone streám eode and on ðam streáme sencte, Bd. 5, 12; S. 631, 18–22. Humbre streámes *Humbrae fluminis*, 1, 25; S. 486, 17. On Trenton streáme *in fluvio Treenta*, 2, 16; S. 519, 31. Temese streáme *Tamense fluvio*, 2, 3; S. 504, 16: 2, 14; S. 518, 15. Gehlade âne cuppan fulle forð mid ðam streáme, Lchdm. iii. 74, 14. Hât gefæc-

cean ongean streáme healfne sester yrnendes wæteres, 12, 1. Sing ðis on yrnendum wætere, and wend ðæt heáfod ongeán streám, 70, 8. Ondlang ðæs streámes ... ondlang ðæs Doferdæles ongeán streám tó Wícforda, Cod. Dip. Kmbl. vi. 218, 29. Streámas stódon, Cd. Th. 206, 29; Exod. 459. Streámas wundon, Beo. Th. 430; B. 212. Wægas grundon, streámas styredon, Andr. Kmbl. 747; An. 374. Reáde streámas *the waters of the Red Sea,* Cd. Th. 196, 23; Ex. 296. Sealte streámas, Exon. Th. 206, 2; Ph. 120. Streámas, sealtýþa gelác, 308, 4; Seef. 34. [*O. Frs.* strám: *O. Sax.* stróm: *O. H. Ger.* stroum, strúm *alveus, amnis, torrens: Icel.* straumr.] v. brim-, eá-, ég-, égor-, fífel-, firgen-, fyrn-, lagu-, mere-, sǽ-, wǽg-, wæl-, wæter-, wille-streám.

streám-faru, e; *f. The going* or *flowing of a stream of water, a current :*—Se æðeling hét streámfare stillan *the prince bade the rush of waters cease,* Andr. Kmbl. 3150; An. 1578.

streám-gewinn, es; *n. The strife of waters :*—Bídaþ stille stealc stánhleoþu streámgewinnes, Exon. Th. 384, 12; Rä. 4, 26.

streám-líc; *adj. Of water :*—Ofer streámlícum ríðum *over rivers of waters,* Homl. Th. i. 444, 10.

streám-racu, e; *f. The bed* or *channel of a stream, a water-course :*—Streámracu *alveus,* Wrt. Voc. i. 54, 26. Streámrace *alveum,* ii. 4, 59. Him þurh streámræce strǽt wæs gerýmed, Andr. Kmbl. 3158; An. 1582. Fram streámracum óþ ðysse eorðan útgemǽru *a flumine usque ad terminos orbis terrae,* Ps. Th. 71, 8.

streám-rád, e; *f.* I. *the bed, course of a stream :*—Streám-raad, -rád, streúmrád *alveus,* Txts. 39, 129. II. *a watery road, the way across the sea :*—Sum streámráde con, weorudes wísa ofer wídne holm, Exon. Th. 296, 21; Crä. 54.

streám-ryne, es; *m. The running of a stream :*—Ðæt wæter swá genihtsumlíce út fleów ðæt hit arn streámrynes of ðam munte *the water flowed out so abundantly, that it ran streaming from the mountain,* Homl. Th. ii. 162, 8.

streám-stæþ, es; *n. A shore :*—Ofer streámstaðe stæppan *to land,* Cd. Th. 86, 21; Gen. 1434.

streám-weall, es; *m. A shore :*—Stáh ofer streámweall *he landed,* Cd. Th. 90, 12; Gen. 1494.

streám-wilm, es; *m. The boiling of the waters, surge; aestus :*—Streámwelm hwíleþ, Andr. Kmbl. 990; An. 495.

streáw, streów, strëu, strëw, es; *n. Straw, hay :*—Gærs oððe streów *foenum,* Ælfc. Gr. 4; Zup. 8, 3. Strëw, streów, streáw, 13; Zup. 83, 17. Strëwu, eglan *fistucam,* Wrt. Voc. ii. 36, 69: 72, 25. Ðæt streu (stré (*printed* sore, *but cf.* lytles strëes *festucae,* Mt. Kmbl. p. 15, 4), Lind.), Mt. Kmbl. Rush. 7, 4, 5. Sume hí cuwon heora gescý, sume streáw, Homl. Th. i. 404, 6. Bærne streúw, Lchdm. iii. 114, 7. [Þe cwene þet mid one strea brouhte o brune alle hire houses, A. R. 296, 12. *Havel.* stra: *Chauc.* stre, stree: *Piers P.* strawe: *O. Frs.* stré: *O. L. Ger.* stró; *gen.* strôs: *O. H. Ger.* strô, strao: *Icel.* strá.] v. sníd-, windel-streáw (-streów).

streáw-berige (streá-, streów-, strëu-), an; *f. A strawberry (plant* or *fruit)* :—Streáwberige *fraga,* Wrt. Voc. i. 67, 71. Streáberige, 31, 69. Streówberian wíse (streáwberge, MS. H.). Deós wyrt ðe man *fraga* and óðrum naman streáwbergean nemneþ, Lchdm. i. 138, 20. Streówberge *fraga,* Wrt. Voc. i. 286, 4: ii. 36, 59. Strëuberie *fascinium,* streúberian *fraga,* 38, 65, 66. Streáwberian wísan *fraga,* i. 79, 37. Streáwbergean leáf, Lchdm. ii. 350, 27. Streáwbergean wíse, 36, 11. Streáwberian wísan nioþowearde, 34, 24: 334, 11. Genim streáwberian nyþeweardan, iii. 2, 18. Streábergan *vel* eorþbergan *fragium* i. *pumorum,* Wrt. Voc. ii. 150, 30.

streáwberige-wíse, an; *f. A strawberry-plant* or *runner :*—Streábediewísan *framen,* Wrt. Voc. i. 31, 70. [*A* strebery-wyse *hec fragus,* a strebery *hoc fragum,* Wrt. Voc. i. 247, col. 1.]

streáwian, streówian; *p.* ode: strëwian; *p.* ede *To straw, strew :*—Ic strewige (streáwige, streówige) *sterno,* Ælfc. Gr. 28, 1; Zup. 165, 9. Wē streówiaþ (strewiaþ) axan uppan ûre heáfda, Homl. Skt. i. 12, 38. Streáwiaþ *evernenent* (*sternerent*?), Wrt. Voc. ii. 144, 30. Mid ðǽm hē strewede ðone weg, Past. 16; Swt. 103, 13. Sume of ðam treówum heówon and streówodon (streówedon, MS. A.: strewedon, MS. B.) on ðone weg, Mk. Skt. 11, 8. Strewodon (strëwedon, MS. A.), Mt. Kmbl. 21, 8. Streówodan, Blickl. Homl. 71, 8, 9. Ða hæþenan byrnende glêda streáwodon, Homl. Skt. i. 23, 35. Hē hēt streówian geond ða flór fela byrnende glêda, 8, 168. [*Orm.* strawwenn: *Chauc.* strawe: *Prompt. Parv.* strowiñ: *Goth.* straujan; *p.* strawida: *O. Frs.* strewa: *O. Sax.* strôedun, streidun, *p. pl.:* *O. L. Ger.* streidin *sterneret:* *O. H. Ger.* strewen, strouwen: *Icel.* strá.] v. be-, ge-streáwian, -streówian; strëgan.

streáwung, strec. v. strewung, stræc.

streccan; *p.* strehte, streahte; *pp.* streht, streaht, streced (v. streced-ness) *To stretch.* I. *to hold out, extend :*—Ðú strecst (*extendes*) ðíne handa, and óðer ðē gyrt, Jn. Skt. 21, 18. Strece ðǽrtó ðínne hiht, Homl. Th. i. 252, 7. II. *to spread out :*—Ðæt folc strehton (*straverunt*) hyra reáf on ðone weg, Mt. Kmbl. 21, 8: Mk. Skt. 11, 8: Lk. Skt. 19, 36. III. *to prostrate :*—Hē hine wæs on gebed streccende æt

líchoman ðæs Godes weres *prosternens se ad corpus viri Dei pia intentione,* Bd. 4, 31; S. 610, 29. [*O. H. Ger.* strecchen *extendere, porrigere, prosternere.*] v. ã-, ge-streccan.

strecednes, e; *f. A couch;* stratum :—Strecednes *stratum,* Ps. Lamb. 40, 4. Strecednysse mîne ic beþweá, 6, 7.

strec-líc, -líce, -ness. v. stræc-líc, -líce, -ness.

strêdan. v. stregdan.

strêgan *to strew :*—Græf golde strêgan (stregdan?), Exon. Th. 311, 25; Seef. 97. [*Goth.* straujan.] v. stregdan.

stregdan. [There are two verbs of this form, a strong and a weak. The conjugation is further complicated by the frequent loss of g, so that forms of the strong verb are found (?) belonging to two classes (cf. bregdan): while in the Northern Gospels strong and weak inflections are combined in the same word. The two verbs are here put together]; ic stregde, strigde, strêde, hē stregdeþ, strigdeþ, strêt; *p.* (*strong*) strægd, *pl.* strugdon and strêddon (v. strêdun, Mk. 11, 8: *but the form may be weak* = strægdon): (*weak*) stregde, strêdde, strugde (*North.*); *pp.* (*strong*) strogden: (*weak*) stregd, strêded, strêd *To strew, spread, scatter, sprinkle.* I. *to strew* something :—Se ðe ne somnigas streigdæs *qui non congregat, spargit,* Mt. Kmbl. 12, 30. Geswerc swê swê eácan strigdeþ (*spargit*), Ps. Surt. 147, 16. Monige ðæt wæter on ádlige men strêdaþ, Bd. 3, 2; S. 524, 32. Se wind se ðe ða bærnnisse in ða burg strægd *ventus qui urbi incendia sparserat,* 2, 7; S. 509, 28. Oðre ða telge strêdun (*sternebant*) on ðone woeg, Mk. Skt. Rush. 11, 8. Ðú somnas ðêr ðú ne strugdes (*sparsisti*) ... Ic somnigo ðǽr ic ne strugde (strægde, Rush.: strêdde, W. S. *sparsi*), Mt. Kmbl. Lind. 25, 24, 26. Ðæt áttor on eallum cyricum hē stregde (*aspersit*), Bd. 1, 8; S. 479, 35. Sió mǽngu strægdun hrægl heora on ðæm wege, sume telgran strægdun on ðæm wege, Mt. Kmbl. Rush. 21, 8. Nim ðæs hornes acxan and strêd, Lchdm. i. 334, 17. Strêd on hálig wæter *sprinkle holy water on,* iii. 56, 11, 18. On ðæs feóndes feax flána stregdan, Salm. Kmbl. 262; Sal. 130. Stregdende weter *aspergens aquas,* Ps. Surt. ii. p. 190, 9. Wæs heora lár sáwen and strogden betuh feówer sceátum middangeardes, Blickl. Homl. 133, 33. Wæs him morþorbed strêd, Beo. Th. 4864; B. 2436. II. *to sprinkle* a place with something :—Ðú strêdest (ástregdest, MS. T.) mē mid hysopon *asperges me hysopo,* Ps. Spl. 50, 8. Strêde man hit mid háligwætere *aspergatur aqua benedicta,* L. Ecg. P. iv. 38; Th. ii. 216, 1. III. *intrans. To scatter, disperse :*—Steorran strêdaþ of heofone, stormum ábeátne, Exon. Th. 58, 24; Cri. 940. Stregdaþ tóðas, Salm. Kmbl. 230; Sal. 114. Hí tó scipon strêddon *they dispersed to their ships,* Chr. 1010; Erl. 144, 3. IV. *to lay in order* (?) :—Streide *struere* (*struerem*?), strídae, streide *struere,* Txts. 99, 1910. v. ã-, be-, ge-, geond-, on-, under-stregdan, -strêdan.

stregdness, e; *f. Scattering, sprinkling :*—Mid strægdnesse (*aspersione*) ðæs wæteres, Bd. 5, 18; S. 635, 29.

strêl *a couch,* strêl *an arrow,* strême, strencan, strên. v. strǽgl, strǽl, strîme, stencan, streówen.

streng, es; *m.* I. *a string, cord, rope :*—Ráp *vel* strenc *funiculus, modicus funus,* Wrt. Voc. ii. 151, 67. Strengas *vel* bendas *lora,* 136, 77. Hē worhte swipan of strengon (*de funiculis*), Jn. Skt. 2, 15. (1) *a string of a musical instrument :*—Streng *fidis,* Wrt. Voc. i. 73, 54. On saltere týn strenga (*chordarum*), Ps. Spl. 32, 2. Strengum *fidibus,* Wrt. Voc. ii. 37, 22: 148, 71: Hpt. Gl. 520, 61. Mid týn strengum getogen hearpe, Ps. Th. 143, 10. (2) *a bow-string :*—Boga *arcus,* bogen (-an?) streng *anquina* (ar-?), Wrt. Voc. i. 35, 26. Strǽla storm strengum gebǽded, Beo. Th. 6226; B. 3117. (3) *in a ship, part of the rigging;* also *a cable.* v. ancer-streng *and cf. Icel.* strengr *in this sense :*—Ðæt scyp úte on ðære sǽ byþ gesund, gyf se streng (v. ancer-streng, l. 18) áþolaþ, for ðam hys byþ se óðer ende fast on ðære eorðan and se óðer on ðam scype ... Ðú scealt gefæstnian ðone streng on Gode, ðæt ðæt scyp healdan sceal ðínes módes, Shrn. 175, 21–31. Windas weóxon, strengas gurron, Andr. Kmbl. 748; An. 374. (4) *a ligament, string* (of the tongue) :—Wið ðam ðe se streng under ðare tunga tóswollen byþ, Lchdm. iii. 102, 2, 4, 5, 8. Strengce *nervo,* Hpt. Gl. 405, 73. Strenga *nervorum,* 475, 13. II. *fig. a line,. lineage* (cf. *Icel.* strengr, used of a narrow water-channel) :—Of ðam streng com Noe and his wíf, Ælfc. T. Grn. 3, 28. [*O. H. Ger.* strang *funus, funiculus :* *Icel.* strengr.] v. ancer-streng.

strengan; *p.* de *To make strong :*—[Þild birrþ ben wiþþ ihwillc mahht to beoldenn it and strengenn, Orm. 2614. Þe wepenn þ strengeð ham stalewurdlukest aзein me, Marh. 14, 19.] v. æt-strengan, á-strenged; strangian.

streng *strength.* v. strengu.

strenge; *adj. Severe, hard* (v. strang, II. 5) :—Hē his torn gewræc on gesacum swíðe strengum stiépe, Cd. Th. 4, 27; Gen. 60. v. strang.

-strenge, -strenged. v. tín-strenge, -strenged.

strengel, es; *m. One who strengthens* or *emboldens, a gallant leader :*—Nú sceal glêd fretan wígena strengel (*Beowulf*), Beo. Th. 6222; B. 3115.

strengest. v. strang.

streng-líc; *adj. Strong, firm :*—Hú hē him strenglícran stól geworhte, heáhran on heofonum, Cd. Th. 18, 14; Gen. 273. Cf. strang-lic.

strengra. v. strang.

strengðu (o); *indecl.*: strengð, e; *f. Strength:*—Strengð *acha*, i. *virtus*, Wrt. Voc. i. 17, 27. Seó strengð *vis*, Gl. Prud. 71. I. referring to living beings, (1) *strength, power to do, fortitude, power to bear, firmness, vigour* :—Strengþu heáfdes mínes *fortitudo capitis mei*, Ps. Th. 59, 6 : 117, 14. Mægnes strengðu, Exon. Th. 239, 23 ; Ph. 625. Módes strengð *fortitudo*, Wulfst. 51, 7. *Fortitudo*, ðæt is strængð oððe ánrédnyss, þurh ða sceal seó sáwul forbæran earfoðnysse mid ánrædum móde, Homl. Skt. i. 1, 165. Strængð *vigor*, Hymn. Surt. 10, 10. Strengcþ mín *fortitudo mea*, Ps. Spl. 17, 1. Strend *robur*, Kent. Gl. 795. Ic eów healde strengðu staþolfæstre, Exon. Th. 31, 3 ; Cri. 490. In ðære gæstes strengðu, 40, 14 ; Cri. 638. Beón wiðmeten ðínre strengðe *comparari fortitudini tuae*, Deut. 3, 24 : Ps. Spl. 38, 14. On strengðe horses, 146, 11. Mid strencgðe *cum potentia*, Ps. Th. 88, 11. Mid micelre strencðe áfylled hē worhte micele tācna, Homl. Th. i. 44, 23. Swā se fulfremeda wæstm biþ on fulre strencðe þeónde, ii. 76, 19. Se weard (*the angel at the gate of Eden*) hafaþ miht and strengðo, Cd. Th. 58, 22 ; Gen. 950. Ic ðíne strengþu (*virtutem*) singe, Ps. Th. 58, 16. Strengðe *fortitudinem*, Ps. Spl. 58, 18. Hī lǽrdon ðæt hī módes strengþo náman, Bd. 1, 12 ; S. 481, 5. (1 a) *the time when a man is strong, mature years:*—On mínum cildháde oððe on mínre geógoðe oððe on mínre strengðe oððe on mínre ylde, Anglia xi. 102, 2. (2) *violence, force:*—Hē ða ongeánwinnendan fæmnan mid micelre strengðe earfoðlíce ofercom, Ap. Th. 2, 5. Strenðe *violentiam*, Kent. Gl. 842. Hié ongunnon mid sweordum and mid strengþum þyder gān ; þohton ðæt hié woldan ofsleán ða apostolas, Blickl. Homl. 151, 1. II. referring to things, (1) *strength, efficacy, virtue, beneficial power* :—Hæfþ hit ða strængðe hyne tó gewyrmenne, Lchdm. i. 116, 1. Ðás sylfan strengþe heó hafaþ gewylled wið ðæs migþan earfoðlícnyssa, 284, 3. Hæfþ ðeós wyrt ealle heora strengða, 244, 1. (2) *of that which is hard to bear, strength, violence, severity, force:*—Ðī læs seó strengð ðære wyrte ða góman bærne, Lchdm. i. 316, 20. Wið áttres strenðe (strengðe, MS. B.), genim ðás wyrte . . . heó oferswíð ealle strenðe ðæs áttres, 114, 13–15. Ne mæg man ǽfre for his strengðe ðysne wyrttruman syllan þicgean on sundrum, 260, 18. Hē sceal upweard licgean, ðý læs hē ða strengþe ðyssæ lácnunge ongite, 300, 21. v. mægen-strengðu.

strengu (o); *indecl.*: streng, e ; *f. Strength*. I. referring to living beings, *strength, power, vigour, fortitude* :—Ðæs líchoman fæger and his strengo mæg bión áfyrred mid þreora daga fefre, Bt. 32, 2 ; Fox 116, 31 note. Dryhten strengo (*fortitudo*) folces his, Ps. Surt. 27, 8. Tor strengu, 60, 4. Ða medomnesse ðære stengio (-eo, Cott. MSS.) *dignitatem fortitudinis*, Past. 14, 5 ; Swt. 85, 23. Ðære gástlícan strenge hyht, Blickl. Homl. 135, 27, 34. Of ælre strengu (-o, Lind.), Mk. Skt. Rush. 12, 33. Strengo bistolen, mægene biuumen, Exon. Th. 410, 8 ; Rä. 28, 13. Strengo getrūwode ánes mannes, Beo. Th. 5074 ; B. 2540. Strenge, 3071 ; B. 1533. Full strenge *plenus fortitudine*, Rtl. 43, 34. Mid míne ágne mægene and strengo (-eo, Cott. MSS.), Past. 4 ; Swt. 39, 18 : Cd. Th. 98, 19 ; Gen. 1632. Strengeo, 150, 21 ; Gen. 2495. Hē gemunde mægenes strenge, Beo. Th. 2545 ; B. 1270. Strengum *vigorously*, Cd. Th. 101, 2 ; Gen. 1676. II. *of things*, (1) *strength, power* :—Mec wolcna strengu byreþ, Exon. Th. 390, 4 ; Rä. 8, 5. (2) *vigour, firmness* :—On strengo þeódscipes wlæc *in disciplinae vigore tepidus*, Bd. 1, 27 ; S. 492, 18. On færhæfdnesse strenge (strengeo, Cott. MSS.) *strange abstinentiae robore validi*, Past. 5, 1 ; Swt. 41, 14. (3) *virtue* :—Sint tó manianne ða mettruman tó ðæm ðæt hié gehealden ða strenge ðære geðylde *admonendi sunt aegri, quatenus patientiae virtutem servent*, 36, 9 ; Swt. 261, 2. [The word occurs often in a later MS., where *strengð(u)* is found in the earlier in the passages given under that word from Lchdm. i. Deades strenge warp him dun, Gen. and Ex. 714. Edmond uor ys strenge was ycluped Yrensyde, R. Glouc. 302, 7. *O. Sax. O. H. Ger.* strengî *robur, fortitudo*.] v. hilde-, mægen-, mere-, woruld-strengu(o).

streón, es; *n.* I. *gain, acquisition, treasure:*—Ðēr is strión ðín *ubi est thesaurus tuus*, Mt. Kmbl. Lind. 6, 21 : 12, 35 : Lk. Skt. Lind. 6, 45. Striónes *thesauri*, 12, 7, 5. Tilða ł stre (= streóna or streón) *quaestuum, lucrum*, Hpt. Gl. 452, 7. Ða ðe geléfeþ in striónum (on gistrión, Rush.) *confidentes in pecuniis*, Mk. Skt. Lind. 10, 24. Of striónum hiora *de facultatibus suis*, Lk. Skt. Lind. 8, 3. Strióna *thesauros*, Mt. Kmbl. Lind. 6, 20. [Gif þu hauest welþe . . . ahte nis non eldere stren (ayhte nys non ildre istreon, Jes. MS.), O. E. Misc. 113, 184.] II. *begetting (?), generating:*—Swā hwylc monn swā his wíf for intingan ánum brúceþ tó streónne (streónenne ? ; *other text has* bearna tó strýnenne) *si quis suam conjugem creandorum liberorum gratia utitur*, Bd. 1, 27 ; S. 495, 33 MS. T. [Crist he is sune, Noht after chesunge ac after strene ; for þan he him strende, alse þe sunne streneð liht, O. E. Homl. ii. 19, 24. The word is used also in the sense of *what is begotten, progeny, lineage, strain* :—Of hire owene streone (*race*), Laym. 2737. Streon (*offspring*) of a swuch strunde, Jul. 55, 16. Ne not ich none sunne þet ne mei beon lied to one of ham seouene oðer to hore streones, A. R. 208, 15. All follc wass þatt illke streon þatt Adam haffde strenedd, Orm. 27. Hiss stren shollde ben todrifenn, 16396. Þat holy

streon, O. E. Misc. 153, 217. Of God, nat of the streen of which they been engendered, Chauc. Cl. T. 157. Spenser uses the form *strene* in this sense.] III. *power (?)* :—Geþencaþ hwelc ðæs fǽsclícan gód sién and ða gesǽlþa ðe gē ungemetlíce wilniaþ ðonne mágon gē ongeotan ðæt ðæs líchoman fæger and his streón mágon (strengo mæg, Cott. MS.) beón áfeorrod mid þreora daga fefre *aestimate, quam vultis nimio corporis bona, dum sciatis hoc, quodcumque miramini, triduanae febris igniculo posse dissolvi*, Bt. 32, 2 ; Fox 116, 31. v. ge-streón ; streónan.

streón *a couch.* v. streówen.

streónan, (*but more often with umlaut*) striénan, strēnan, strīnan, strýnan; *p. de (with gen. acc.*). I. *to gain, acquire:*—On ðæm hiewe ðe hē sceolde his gielpes stiéran, on ðæm hē his striénþ. Mid ðý ðe hē sceolde his gestreón tóweorpan, mid ðý hē hié gadraþ, Past. 8 ; Swt. 55, 10. Strýneþ *foeneratur*, Wrt. Voc. ii. 38, 45. Se ðe him sylfum strýnþ *qui sibi thesaurizat*, Lk. Skt. 12, 21. Gif hē strióneþ allne middangeard *si lucretur universum mundum*, Lk. Skt. Lind. Rush. 9, 25. Nis eów forboden ǽhta habban, gif gē ða on riht strēnaþ, Blickl. Homl. 53, 28. Guman gylpe strýnaþ *men proudly lay up treasure*, Exon. Th. 445, 28 ; Dôm. 14. Hē hié gemyndgaþ ðara welegra ðe longe strīndon (striéndon, Hatt. MS.), and lytle hwíle brucon ; hú hrædlíce se fǽrlíca deáð hié on lytelre hwíle bereáfode ðæs ðe hié on longre hwíle mid unryhte striéndon (strīndon, Hatt. MS.), Past. 44 ; Swt. 332, 15–17. Striónas *thesaurizate*, Mt. Kmbl. Lind. 6, 20. Riht is ðæt gerēfan geornlíce tylian and symle heora hláfordan strýnan mid rihte, L. I. P. 12 ; Th. ii. 320, 13. Hē ūs fēran hēt gásta streónan, Andr. Kmbl. 662 ; An. 331. Se ðe his feore nyle hǽlo strýnan, Exon. Th. 96, 16 ; Cri. 1575. Tó striónanne *thesaurizandum*, Mt. Kmbl. p. 15, 1. Ðū ðe wǽre welena strýnende, L. E. I. prm. ; Th. ii. 398, 12. II. *to beget, generate, create* :—Gē strīnaþ suna and dohtra *filios generabis et filias*, Deut. 28, 41. Of ðysum þrim mannum, Noes sunum, eall ðes middangeard weard eft onwæcnod, þēh hyé Drihten on þreó streónde (*created them of three conditions*), Anglia xi. 3, 60. Seth strýnde suna and dohtra, Cd. Th. 69, 20 ; Gen. 1138 : 70, 13 ; Gen. 1152. Hē se wífe bearna strýnde, 70, 5 ; Gen. 1148 : 73, 8 ; Gen. 1201. Hié tósomne fērdon and bearna striéndon, Ors. 1, 10 ; Swt. 46, 10. Ðæt his bróðor nyme hys wíf and strýne him bearn, Mt. Kmbl. 22, 24. Hié sculon bearna striénan, Past. 51 ; Swt. 397, 10 : Ors. 4, 1 ; Swt. 154, 17 : Cd. Th. 59, 19 ; Gen. 966. Hē ongan óðres striénan bearnes be brýde, 68, 17 ; Gen. 1118. Strýnan, 71, 15 ; Gen. 1171. For intingan bearna tó strýnenne *creandorum liberorum gratia*, Bd. 1, 27 ; S. 495, 33. Ic wille ðæt hit gange on ða nýhstan hand mē, būtan hyra hwylc bearn hæbbe ; ðonne is mē leófast ðæt hit gange on ðæt strýned on ða wǽpned-healfe (*to the child born on the male side*), Cod. Dip. Kmbl. ii. 116, 15. [On hir he scal streonen (streni, 2nd MS.) þat scal wide sturien, he scal streonien (streoni, 2nd MS.) hire on ænne swiðe sellichne mon, Laym. 18844. Sikernesse streoneð зemelæste, A. R. 234, 3. All þatt streonedd wass þurrh Adam, Orm. 33. Behinden he (*elephants*) heom sampnen ðanne he sulen oðre strenen, O. E. Misc. 19, 609. Strenen *fornicantur*, Ps. 72, 27. See also Halliwell's Dict. *strain, strene*. *O. H. Ger.* striunen *lucrari*.] v. ge-streónan.

Streónes-halh *Whitby:*—On ðære stówe seó is gecweden Streóneshalh, Bd. 3, 24 ; S. 557, 2 : 4, 23 ; S. 592, 37. Hild abbodesse on Streónesheale, Chr. 680 ; Erl. 40, 13. Tymbrend ðæs mynstres ðe ys nemned Steórneshealh, Shrn. 148, 40. For the forms streanæs, streunaes, strenes, found in Bede's History, v. Txts. 489. In Bd. 3, 25 the word is explained by *sinus fari*.

streón-ful, streów. v. gestreón-ful, streáw.

streówen, streón, strēn, e ; *f.* I. *a couch, bed:*—In bed strēne mínre *in lectum stratus mei*, Ps. Surt. 131, 3. Strēne nīne *stratum meum*, 6, 7 : 40, 4 : 62, 7. Ðā hēht hē him streówne gegearwian (bedd gewyrcian, MS. B.) *jussit sibi stratum parari*, Bd. 2, 6 ; S. 508, 8. II. *a place where anything rests:*—Hord sceal in streónum bídan . . . hwonne hine guman gedǽlen *treasure shall remain in its places of rest . . . until men distribute it*, Exon. Th. 337, 22 ; Gn. Ex. 68.

streówian, streówung. v. streáwian, strewung.

streówness, e ; *f. Bedding, what is spread to lie on:*—Ðā bǽdon hine his discipulos ðæt hié mōstan hūru sume streównesse him under gedôn for his untrumnesse ; ða cwæð hē : 'Bearn, ne bidde gē ðæs ; ne gedafenaþ cristenan men ðæt hē elles dô, būtan swā hē efne on axan and on duste licge,' Blickl. Homl. 227, 12.

strēt, streððan, strēu, strewian. v. strǽt, be-streððan, streáw, streáwian.

strewung, e ; *f. What is spread to lie on, a couch* :—On bedde mínre strewunge *in lectum strati mei*, Ps. Lamb. 131, 3. [*O. L. Ger.* strewunga *stramentum*.]

stric, es ; *m.* (?) *Plague (?)* :—Eác sceal áspringan wíde and síde . . . stric and steorfa and fela ungelimpa, Wulfst. 86, 12. Stric and steorfa, orfcwealm and uncoðu, 159, 10. Gif hit geweorðe ðæt folce mislimpe þurh stric oððe steorfan, þurh unwæstm oððe unweder, L. I. P. 18 ; Th. ii. 324, 29. v. ge-stric.

strica, an ; *m.* I. *a stroke* of a pen, *a tittle, a mark, line:*—Án

strica oððe ân stæf ðære ealdan æ ne biþ forgæged *iota unum aut unus apex non praeteribit* (Mt. 5, 18), Homl. Th. ii. 200, 1 : Jud. 15. Strican ł mærcunge *characteres*, Hpt. Gl. 473, 13. Stricena *apicum*, stricum *characteribus, notis*, 512, 23, 52. Stricum *apicibus literarum*, 501, 56. **II.** *a streak, tract* :—Hit getîmaþ ðonne se môna beyrnþ on ðæm ylcan strican ðe seó sunne yrnþ, ðæt his trendel underscýt ðære sunnan tô ðam swîðe ðæt heó eall aþeóstraþ, Lchdm. iii. 242, 19. [Longe, croked strykes, Chauc. Astrolabe. Strek or ·poynt betwyx ij clausys yn a boke *liminiscus*, Prompt. Parv. 479. *Goth.* striks κεραία : *O. H. Ger.* strich *linea, nota, zona*. Cf. *Icel.* stryk *a stroke, dash*.]

strîcan ; *p.* strâc, *pl.* stricon ; *pp.* stricen. **I.** *to stroke, smooth, rub, wipe* :—Ne delfe hý nân man mid îsene and mid wætere ne þweá, ac strîce hý mid clâðe clǽne, Lchdm. iii. 30, 24. [Baldulf lette striken to þan bare lichen his bærd and his chinne *had his beard shaved off quite smoothly*, Laym. 20303. To make murrour bryʒt. Stryke theron blak sope, Rel. Ant. i. 108, 23 (15th cent.). Strekyñ or make pleyne *complano*, strekyñ or make playne be mesure *hostio*, strekyñ, as menn do cattys *palmito*, Prompt. Parv. 479, col. 2. To stryke a buschelle *hostiare*, Cath. Ang. 369. This pecke to conteyne stryken with a strykell as mutche as our standerd pecke holdeth upheaped, ib. note 1. To stryke a bed = to make it smooth, is quoted by Halliwell, who gives *strike* as a Devonshire word for to rub gently. *O. H. Ger.* strîhhan *linere, fovere*. Cf. *Icel.* strjúka *to stroke, rub, wipe* : *Dan.* stryge.] v. ymb-strîcan. **II.** *to make a stroke*. v. be-strîcan ; strica. **III.** *to go, move, run* :—Bûton ðæm rodere ðe ðâs rûman gesceaft æghwylce dæge ûtan ymbhwyrfeþ, strîceþ ymbûtan, Met. 20, 140. [Strikeð a stream ut of þ stanene þruh, Kath. 122, 2479. Comen alle strikinde of eauer euch strete *fit ex omni civitate concursus*, 35, 732. Hamun him to strac (wende to, 2nd MS.), Laym. 9318. Faraon strac inn aftterr Godes follc, Orm. 14810. þ blod strac adun of hire bodî, Marh. 5, 34 : 11, 7. Striken men þiderward, 17, 31. þe strunden þe striken (*ran*) adun of þine fet, O. E. Homl. i. 187, 28. A mous . . . stroke forth sternly and stode biforn hem alle, Piers P. prol. 183. See also Halliwell, *streke, strike* (2). The word is still used of motion as in *to strike* across a country. *O. H. Ger.* strîhhan *ire, meare* : *Ger.* streichen *to move, rush, rove*. Cf. *Icel.* strjúka *to go, rush* : *Dan.* stryge *to go*, stryge Landet om *to stroll about the country*.]

stricel, es ; *m.* **I.** *a strickle, an implement for smoothing corn in a measure*, v. strîcan, I :—Stricilum *trocleis, rotis modicis*, Txts. 100, 994. [*Hic modius* a buschylle, *hic corus* a mesur, *hoc* os[t]orium a strikylle, Wrt. Voc. i. 233, col. 2 (15th cent.). Strykylle *hostorium*, Cath. Ang. 369. In note 1 on this page are given the following : '*Rouleau* the round pin, stritchell, or strickle used in the measuring of cofn, etc. *Lorgaulté* the strickle used in the measuring of corne.' Cotgrave. 'When wee goe to take up corne for the mill, the first thinge wee doe is to looke out poakes, then the bushell and strickle.' Farming Books of H. Best, 1641. **II.** *that from which liquid flows* (? v. strîcan, II), *a breast that gives milk, a fount* :—Of stricele *ubere*, Germ. 390, 67. Of feówer stricelum *bis binis de fontibus*, Wrt. Voc. ii. 12, 39. v. tit-stricel.

strician *to knit, net*. [*O. H. Ger.* stricchen *nectere* : *Ger.* stricken.] v. ge-strician.

strîdan ; *p.* strâd, *pl.* stridon. **I.** *to stride* :—Strîdit *varicat*, Txts. 105, 2078. **II.** *to get by force* (?), *pillage, rob* :—Strâd (streád ? *from* strûdan) *conpilat*, Wrt. Voc. ii. 20, 14 : 96, 74. [Cf. *O. Sax.* strîdian *to dispute, contend* ; strîd *contest, strife* : *O. Frs.* strîda (wk.) *to contend* ; strîd *strife* : *O. H. Ger.* strîtan ; *p.* streit *pugnare, contendere, obtinere* ; strît *pugna, certamen*.] v. be-strîdan, *and next word*.

stride, es ; *m. A stride, pace* :—Faeðm *vel* tuegen stridi *passus*, Txts. 85, 1510. [Stryde *clunicatus*, strydyñ or steppyñ ovyr a thynge *clunico*, Prompt. Parv. 480.]

striénan. v. streónan.

strîman *to resist, oppose* :—Strîmendi *innixus*, Txts. 71, 1132 : *obnixus*, 81, 1404. [In some dialects, e.g. Northants, *to strime* = to stride. Could the verb have existed with the same double meaning as *strîdan*, q.v.?]

strîme, strême ; *adj. Having a current*. [*Icel.* streymr *having a current, running*.] v. swîþ-strîme.

strînan. v. streónan.

strînd, strýnd, e ; *f. A generation, stock, race, kin, tribe* :—Hé ne wæs of ðearfendum folce ac wæs æþelre strýnde *non erat de paupere vulgo, sed de nobilibus*, Bd. 4, 22 ; S. 591, 34. Wæs hê of æþelre strýnde Angelðeóde *de nobilibus Anglorum*, 5, 19 ; S. 637, 40. Of ðære cynelîcan strýnde *de stirpe regia*, 5, 7 ; S. 621, 8. Of Wôdenes strýnde (*stirpe*) monigra mǽgþa cyningcynn fruman lǽdde, 1, 15 ; S. 483, 30. Hié wǽron of Dauides cynnes strýnde, Blickl. Homl. 23, 28. His cynnes lâtwuá from ðon ðæt fore biþ his strýnde *tribunus, ab eo quod praessit tribui*, Rtl. 193, 15. In strýnd twoelfa *in tribus duodecim*, 78, 26. Doemende twoelf strýnda, Mt. Kmbl. Lind. 19, 28. Strýndum, Lk. Skt. Lind. 22, 30. [Of heore strund (owene streone, 2nd MS.), Laym. 2736. Strend toward *generatio futura*, Ps. 21, 32. Streon of a swuch strunde,

Jul. 55, 17. þet tu wite me wið ham (*deadly sins*) and alle heore strunden, A. R. 28, 7.] v. eormen-strýnd ; streónan, streón.

strîpan ; *p.* te *To strip*. [Erest he (*the devil*) strepte of him (*Job*) his shep, O. E. Homl. ii. 195, 28. Heo hau
ed istruped mine figer sterc naked, A. R. 148, 24. þu struptest and herhedest helle, Jul. 63, 16. Het strupen hire steortnaket, Kath. 1537. *O. H. Ger.* stroufen *stringere*.] v. be-strîpan (-strýpan).

strîð, es ; *m.* **I.** *struggle, fight, contest* :—Strange geneátas ða ne willaþ mê æt ðam strîðe geswîcan, Cd. Th. 19, 1 ; Gen. 284. **II.** *contention, dispute, strife of words* :—Hwæt scal ðê swâ lâðlîc strîð wið dînes hearran bodan ? 41, 28 ; Gen. 663. Ðone lâðan strîð, yfel and-wyrde, 36, 16 ; Gen. 572. [The word seems to occur only in that part of the Genesis which is supposed to be derived from an Old Saxon original, and to be a form borrowed from Old Saxon *strîd*. In the Liber Scintillarum *strîþlîce* glosses *districte*, 132, 9, and *strîðnysse* glosses *districtionis*, 123, 18 ; but these may be explained as errors for *stîplîce, stîðnysse* : the nominative of the latter glossing *districtio* occurs 123, 9.]

strîþ-lîce, -ness. v. preceding word.

strôd (strod), es ; *n.*? :—Andlang dîces on ðæt strôd ; eást andlang strôdes ; of ðam strôde on scagan, Cod. Dip. Kmbl. v. 230, 4. Ûtt þurh Wynnawudu on strôd norðweard (*the reference is to the same place in both charters*), 334, 32. On secglâges strôd ; of secglâhes strôde, iii. 79, 17. *The word occurs in local names*, Strôdwîc *Strudwick* (Northants), ii. 318, 30. Ðæt land æt Strôðistûne, iv. 288, 18. Perhaps it is left in *Strood* (Kent). [*O. H. Ger.* struot *silva*, Grff. vi. 751, Grmm. R. A. 635.]

-strôd. v. ge-strôd.

strôdgdness, e ; *f. Scattering* ; aspersio, Rtl. 122, 3. v. ge-strogdness.

strong. v. strang.

strop[p] *a strap, strop* :—Strop *vel* ârwiððe *struppus*, Wrt. Voc. i. 56, 37. [From Latin.]

-strowenness. v. ā-strowenness.

strûdan ; *p.* streád, *pl.* struden ; *pp.* stroden *To spoil, ravage, plunder, pillage, defraud* :—Hwæt is ðis manna ðe mînne folgaþ wyrdeþ, ǽhta strûðe, Elen. Kmbl. 1807 ; El. 905. Ðonne wê ûs for nôwiht dôþ ðæt wê earme menn reáfiaþ and strûðaþ in heora ǽhtum and heora gôdum *cum infirmiores spoliare et eis fraudem facere pro nihilo ducimus*, Bd. 3, 19 ; S. 548, 19. Fýnd gold strudon, Cd. Th. 121, 7 ; Gen. 2006 : Exon. Th. 436, 7 ; Rä. 54, 10. Hié tempel strudon, Cd. Th. 260, 18 ; Dan. 711. Hwâ ðæt hord strude, Beo. Th. 6244 ; B. 3126. Se ðone wong strude (MS. strade), 6139 ; B. 3073. Iudas hæfde onlícnesse ðara manna ðe willaþ Godes cyricean yfelian and strûdan, Blickl. Homl. 75, 24. Strûðende fýr, Cd. Th. 154. 15 ; Gen. 2556. [Cf. *O. H. Ger.* strutit *fraudat*, zi-strudida *destruxit*.] v. be-, ge-strûdan ; strýdan, *and following words*.

strude, Wrt. Voc. ii. 148, 26. v. next word.

strûdend, es ; *m. A spoiler, robber, usurer* :—Strûdend oððe grîpend *raptor*, Wrt. Voc. ii. 88, 69. Lǽnend *vel* strûde[nd] *fenerator*, 148, 26.

strûdere, es ; *m. A spoiler, robber* :—Strûdere *vel* reáfere *agressor*, Wrt. Voc. i. 19, 7. Strûderes *grassatoris*, Hpt. Gl. 513, 54. Strûderum *praedonibus, raptoribus*, 469, 74. [*M. H. Ger.* strudære.] v. woruld-strûdere ; strýdere.

strûdung, e ; *f. Spoliation, robbery, pillage* :—Deôflîce dǽda on stalan and on strûdungan, L. Eth. v. 25 ; Th. i. 310, 16 : vi. 28 ; Th. i. 322, 16. Utan forfleón stala and strûdunga (strûtunga, MS. C.), Wulfst. 115, 9 : 164, 1 : 129, 18.

strûta. v. strýta.

strûtian ; *p.* ode *To stand out stiffly or projectingly* :—Se hâlga wer hié (*the robbers who were trying to break into the church*) wundorlîce geband, ǽlcne, swâ hê stôd, strûtiendne mid tôle, ðæt hiera nân ne mihte ðæt morþ gefremman . . . Menn ðæs wundrodon, hû ða weargas hangodon, sum on hlǽddre, sum leát tô gedelfe, and ælc on his weorce wæs fæste gebunden, Swt. A. S. Prim. 87, 177. [Ne be þi winpil nevere so ʒelu ne so stroutende, Rel. Ant. ii. 15, 8 (13th cent.). His here strouted as a fanne, Chauc. C. T. 3315. Strowtyñ or ·bocyñ owte *turgeo*, Prompt. Parv. 480. *M. H. Ger.* strûzen. Cf. a-strout. 'A-strout. This word is still used in Somersetshire, explained by Mr. Norris, MS. Glossary, "in a stiff, projecting posture, as when the fingers are kept out stiff." The word occurs in Wright's Political Songs : The knif stant astrout, 336, 3. Further instances are : Hys yen stode owte astrote, Le Bone Florence of Rome, 2029. Bothe his eghne stode one strowte, Sir Isumbras.' Halliwell's Dict. The word *strut* is also used in the sense of *strife* : þair strut (*other MSS.* strife) it was unstern stith, C. M. 3461. *M. H. Ger.* strûz : *Ger.* strauss *strife, struggle*.]

strûtung, strycel. v. strûdung, stricel.

strýdan *to spoil, waste* :—Ðæs strýdendan (stryndedan, Wrt.) *prodiga* (cf. *O. H. Ger.* strutenti *prodigus*), Wrt. Voc. ii. 86, 51. v. ge-strýdan ; strýdere.

strýdere, es ; *m. A waster, prodigal* :—Strýdere *prodigus*, Wrt. Voc. ii. 68, 49. Stryndere (strýdere ?), 118, 28. v. preceding word.

strýnan, strýnd, stryndere, strýpan. v. streónan, strînd, strýdan, strýdere, strîpan.

3 O

strȳta, strūta, an; *m. An ostrich:*—Strȳta *strutio*, Wrt. Voc. ii. 121, 38. Strūta, i. 280, 4. [*O. H. Ger.* strūz *struthio*. From Latin.]

strȳððan, stubb. v. be-streddan, stybb.

studu, studu; *gen.* stude, studu; *dat.* stude(-u), stude, styde; *acc.* studu, studu(-o); *pl.* stude, stude(-a); *gen.* studa; *f. A post, pillar, prop, stud* (v. Halliwell's Dict. '*Stud* the upright in a lath and plaster wall, *Oxon.*'):—Áhēng hē ðone sceát on áne studu ðæs wæges (*in una posta parietis*). . . . Ðæt hūs forbarn nemþe seó studu ān (būtan ðære ānre stÿðe, MS. B.), Bd. 3, 10; S. 534, 28-35. Se lēg ðære studu (ða ilcan studu, col. 2) gehrīnan ne mihte. . . . Ðæt fýr eode andlang ðara nægla ðe seó studu (*destina*) mid gefæstned wæs and ðære stude nó ne onhrán (ða stuþo sceþþan ne meahte, col. 2). . . . Hī ða ða studu on ða cyrican setton. . . . Monige men of ðære ylcan styde (styþe, stuðe, MS. B., col. 2) sprytlan ācurfon, 3, 17; S. 544, 28-43. Hē hine onhylde tó ānre ðære studa ðe ūtan tó ðære cyrican geseted wæs ðære cyrican tó wraþe and ðær his gást āgæf (hē genom ða studu ðe seó cirice mid āwreþed wæs and on ðære styde stondende forðfērde) *adclinis destinae quae extrinsecus ecclesiae pro munimine erat adposita, spiritum vitae exhalaret ultimum,* S. 543, 37-41. Cypressus styde hié ūtan wreþedon and gyldne styþa hié ūton wreþedon, Nar. 5, 7, 8. Begēmþ stuðe (*or* stoðe) mīnre dure *observat postes ostii mei*, Kent. Gl. 281. [*Icel.* stoð; *f., pl.* stöðr, steðr, *later* stoðir, stuðir.] v. feor-, wræd-studu (-stuðu); stod, stuðan-sceaft.

stuf-bæþ, e; *n. A hot-air bath, vapour bath:*—Sile him drincan on stufbaþe, Lchdm. iii. 132, 13. Man machiaþ stufbæþ and baþeþe hine ðáron, 92, 21. v. stofa.

stulor; *adj. Furtive:*—Stulur *furtiva, clandestina, secreta*, Wülck. Gl. 245, 42; *furtiva*, Wrt. Voc. ii. 38, 30. **I.** *acting with stealth, stealthy:*—Seó hreóhnys is open costung, and seó smyltnys is stulor and dígele swica, Homl. Th. ii. 392, 24. **II.** *stolen:*—Wæteru stulre swēttran synd *aque furtiuae dulciores sunt*, Scint. 110, 11. [*Cf. O. Sax.* stulina *theft: O. H. Ger.* stulingun *clam: Icel.* stuldr *theft.*] v. next word.

stulorlíce; *adv. Furtively, stealthily;* furtim, Ælfc. Gr. 38; Zup. 238, 4.

-stun. v. ge-stun.

stund, e; *f.* **I.** *a stound* (used by Spenser and Fairfax, v. Nares, and still later in dialects, v. Halliwell), *a while, time, hour:*—Nis seó stund latu ðæt (*the hour will not be long in coming when*) ðē wælreówe wítum belecgaþ, Andr. Kmbl. 2422; An. 1212; Exon. Th. 156, 16; Gú. 875. Nó ic ða stunde bemearn, ne for wunde weóp *that* (*hard*) *time I bewailed not, nor wept for the wound*, Exon. Th. 499, 12; Rä. 88, 14. Æt stunda gehwam, 436, 30; Rä. 55, 9. **II.** *the hour appointed for a particular act, the signal which marks the hour:*—Geendedum gebedum sī swēged oþer tācn ł stund *finitis orationibus sonetur secundum signum,* Anglia xiii. 380, 215. On ðam fæce ðe stunda beón gehringede *in interuallo quo signa pulsantur,* 406, 952. Gecnyllendum ōþrum stundum *pulsatis reliquis signis,* 380, 219. Cf. tíd, **I c.** ¶ *adverbial use of cases or adverbial phrases,* cf. hwíl:—Hē word stunde āhóf *he spoke at once* (cf. *Ger.* zur Stunde), Andr. Kmbl. 832; An. 416: 2993; An 1499: Elen. Kmbl. 1445; El. 724; Ps. Th. 55, 11. Hē winnan nyle ǽnige stunde, Met. 25, 68. Ðū þoladest mægenearfeþu micle stunde, Exon. Th. 86, 21; Cri. 1411. Hwīlon hē on bord sceát, hwīlon beorn tǽsde, ǽfre embe stunde (*every now and again, from time to time*) hē sealde sume wunde, Byrht. Th. 139, 48; By. 271. Stundum (1) *at times, from time to time* [*Icel.* stundum: *Dan. Swed.* stundom *sometimes, now and then*]:—Stundum *punctis,* Germ. 398, 227. Ic ðīne strengþu stundum singe and ðīn milde mód morgena gehwylce, Ps. Th. 58, 16. Horn stundum song fúslíc leóð, Beo. Th. 2851; B. 1423. Ða ic sylf stundum gerád, stundum gereów(cf. *Icel.* stundum . . . stundum *sometimes . . . sometimes, now . . . now*), Cod. Dip. Kmbl. v. 331, 1. (2) *with exertions* or *pains* (v. ā-stundian, *and* cf. *Icel.* stund *in the sense of* care, pains, exertion; stundar *very, exceedingly;* stunda *to strive, take pains;* stundan *pains-taking;* stundliga *eagerly*):—Hē orod stundum teáh *he* (*the king Guthlac*) *drew his breath laboriously,* Exon. Th. 178, 17; Gú. 1245. (2 a) *with effort, earnestly, eagerly, fiercely:*—Stundum wrǽcon mægen æfter óðrum, Elen. Kmbl. 464; El. 232: 242; El. 121. Strong, stund-um rēþe *exceedingly fierce,* Exon. Th. 380, 41; Rä. 2, 3. Streámas staþu beátaþ, stundum weorpaþ on stealc hleoþa stáne and sonde, 382, 5; Rä. 3, 6. Mē strange stundum ongunnon *irruerunt in me fortes,* Ps. Th. 58, 3: 93, 6. Ic stefne tó ðē stundum (*earnestly*) cleopige, 85, 5: 97, 8. [*O. Sax. O. L. Ger.* stunda: *O. Frs.* stunde: *O. H. Ger.* stunta: *Icel.* stund.] v. orleg-, winter-, woruld-stund; stund-mǽlum.

stundian. v. ā-stundian.

stund-mǽlum; *adv.* **I.** *at intervals, gradually, little by little:*—Stundmǽlum *sensim,* Ælfc. Gr. 38; Zup. 228, 6: Zup. 236, 13: *sensim, paulatim,* Hpt. Gl. 451, 6: 469, 72: 482, 51. **II.** *at different times, alternately, now at one time now at another:*—Stundmǽlum *alternatim, singulatim, separatim,* 438, 53: *vicissim,* Ælfc. Gr. 38; Zup. 238, 4. Stuntmælum, R. Ben. Interl. 38, 10. [See *stoundmele* in Halliwell.]

stune, Lchdm. iii. 32, 19. v. stíme, *and next word.*

stunian; *p.* ode. **I.** *to crash, make a loud sound:*—Sum biþ

wíges heard, beadocræftig man ðǽr bord stunaþ *where the shield resounds,* Exon. Th. 295, 29; Crä. 40. Stunaþ eal geador winsum sanc *a pleasant song sounds all together* (*from the union of many voices*), Met. 13, 49. **II.** *to strike with a loud sound, crash, dash:*—Stíme (stune?) hætte ðeós wyrt . . . stond heó wið ǽttre stunaþ heó wærce stíðe heó hætte wiðstunaþ heó ǽttre *it resists poison, dashes on pain, stiff is it called, dashes against poison.* Lchdm. iii. 32, 22. Ðā wearð stearc storma gelác; stunede sió brūne ÿð wið óðre *one dark wave dashed against the other,* Met. 26, 29. [Later the word means *to confound, astonish, stupefy:*—If he hem stowned vpon fyrst, stiller were þanne alle þe heredmen, Gaw. 301. Stonyñ *stupefacio, percello,* Prompt. Parv. 476. Stonyd *attonitus,* Cath. Ang 365. Stoned ne basshed of no thyng be ye, Parten. 2940. Halliwell gives *stound* as a Northern word=to beat a drum. Cf. *Icel.* stynja *to groan;* stynr *a groan.*] v. stinan, ge-stun.

stunt; *adj. Foolish, stupid:*—Stunt *stultus,* Wrt. Voc. i. 47, 53. Stunt folc and unwís *popule stulte et insipiens,* Deut. 32, 6. Ic wæs stunt, and ic eom nū wís, Homl. Th. i. 432, 6. Ðū sprǽce swā swā ān stunt wíf, ii. 452, 31. Ðū stunta *fatue,* Mt. Kmbl. 5, 22. For eówer stuntan lage *per traditionem vestram,* Mk. Skt. 7, 13. Swā stunte nÿtenu *sicut bruta animalia,* Coll. Monast. Th. 32, 19. Cweþaþ ða ðe syndan stunte, ðæt mycel forhæfedness lytel behealde, Wulfst. 55, 23. [Mannkinn þatt wass stunnt and dill and skilllæs swa summ asse, Orm. 3714. *M. H. Ger.* stunz *dull: Icel.* stunt *short, scant, stunted.*] v. styntan.

stunt-líc; *adj. Foolish:*—Stuntlíc ys ǽnig þing swÿþor lufian ðænne God *stultum est aliquid plus amare quam Deum,* Scint. 17, 16. Hē nán þing stuntlíces ongeán God sprǽc *Job charged not God foolishly* (A. V.), Homl. Th. i. 472, 33. [Hwet is eure swa dusi and swa stuntlic swa is þet þe olde mon nule his mod to Gode awendan mid gode huhte, O. E. Homl. i. 109, 12.]

stuntlíce; *adv. Foolishly, stupidly:*—Stuntlíce fæst se ðe hine sylfne mid gálnysse befÿlþ, Homl. Th. ii. 100, 16. Hí nellaþ understandan hú stuntlíce hí dóþ, Homl. Skt. i. 17, 132. Hwæt is stuntlícor *quid est stultius?* Ælfc. Gr. 48; Zup. 279, 11.

stuntness, e; *f. Foolishness, folly, stupidity:*—Stultitia, ðæt is stuntnys, Wulfst. 52, 17. Ðysses middaneardes wÿsdóm is stuntnis ætforan Gode, Homl. Skt. i. 1, 228. Nelle ðū beón eádmód on wísdóm dínum ne geeádmét on stuntnesse (*stultitia*), Scint. 19, 13. Ðā āwende Crist ūre stuntnysse tó geráde, Homl. Th. i. 208, 19. Nū ðingþ ðam dysegan menn . . . ac hē ne understent nā his āgene stuntnysse, Hexam. 20; Norm. 28, 20. Gif hē him sylfum stÿrþ fram eallum stuntnyssum, Homl. Skt. i. 17, 23. [Fela stuntnesse beoð þer nan steore ne bið, O. E. Homl. i. 117, 22.]

stuntscipe, es; *m. Foolishness;* stultitia, Mk. Skt. 7, 22.

stunt-sprǽc, e; *f. Foolish speech:*—Þurh stuntspǽce *per stultiloquium,* Confess. Pecc.

stunt-sprǽce; *adj. Talking foolishly, foolish in speech:*—Stuntspǽcne *stultiloquum,* Scint. 97, 10.

stunt-wyrde; *adj. Using foolish words, foolish in speech:*—Se ðe wǽre stuntwyrde, weorðe se wíswyrde, Wulfst. 72, 17.

stúpian; *p.* ode *To stoop, bend the back:*—Gyf seó sunne hine (*the moon*) onǽlþ ufan þonne stúpaþ hē (*it has the light part curving downwards*) . . . for ðan ðe hē went ǽfre ðone hricg tó ðære sunnan weard, Lchdm. iii. 266, 20. Ðæt hē swā oft sceolde stúpian swā se cyning tó his horse wolde and ðonne se cyning hæfde his hrycg him tó hliépan *ut ipse acclinis humi regem superadscensurum in equum dorso adtolleret,* Ors. 6, 24; Swt. 274, 24. [Ha schulde stupin and strecche forð þat swire, Jul. 73, 11. Marie adun stupede, Misc. 53, 559: Fl. a. Bl. 697. He nimþ heide þet his tour ne hongi ne stoupi, Ayenb. 151, 6. To stoupe *nutare,* Rel. Ant. i. 6, col. 1 (14th cent.). Over þe table he gon stoupe, Alis. 1103. Layamon uses the verb transitively: Mon mæi mid strenðe stupen (stoupe, 2nd MS.) hine to grunde, 25950. [*O. Du.* stuypen *to bow.* Cf. *Icel.* stúpa (*st.*); steypa *to cause to stoop: Dan.* stupe *to fall: Swed.* stupa *to fall, tilt, lean forward;* stupande *sloping.*] v. stíp.

sturtan (? *vowel as in* murnan?); steart *To start, jump up:*—Sturtende (styrtende (*wk.*)? v. examples from *Middle English*) se halta gistód *exiliens claudus stetit,* Rtl. 57, 27. [Arður up sturte (storte, 2nd MS.), Laym. 23951. Pharaon stirte up, Gen. and Ex. 2931. Stirte forth, Havel. 873. Þe Romeyns sturte to anon her prince up to rere, R. Glouc. 212, 1.]

stút *a gnat, midge;* culex, Wrt. Voc. i. 23, 76: 77, 55. [His hors eren were so ful of gnattes and stoutes and of great flyes *aures equorum culicibus et ciniphibus ita sunt repletae,* Trev. v. 159, 9. Halliwell gives *stout* as a West Country word with an instance of its use. Perhaps some local names keep traces of the word, v. Cod. Dip. Kmbl. vi. 336, col. 2.]

stútere, es, *m.*?:—On stúteres hylle, Cod. Dip. Kmbl. v. 48, 10: 182, 10: 328, 10.

stuðan-sceaft, es; *m. A prop, stay:*—Ic gaderode stuþansceaftas, Shrn. 163, 5. Tó ðam ilcan wuda ðǽr ic ðǽs stuðansceaftas cearf, 14. [Cf. *Icel.* stoði (*wk.*) *a post;* styðja *a post.*] v. studu.

stuþu. v. studu.

stybb, stubb, stebb, es; *m. A stub, stump of a tree:*—Stybb *stirps,* Ælfc. Gr. 3; Zup. 7, 10. Ðes stybb *hic stirps,* 9, 58; Zup. 68, 8. Styb, Wrt. Voc. i. 33, 57; 80, 33. Treówwes steb *stipes,* 17, 7. Mid stybbe mid ealle *stirpitus,* Ælfc. Gr. 38; Zup. 239, 8. Æt ðæne ellenstyb; of ðam stybbe, Cod. Dip. Kmbl. iii. 24, 4. Andlang díces on ðone stubb, 10, 21. [*Icel.* stubbi, stubbr *a stump.*] v. ellen-, þorn-stybb (-stubb).

stycce, es; *n.* I. *a piece, bit:*—Stycce *frustrum,* Wrt. Voc. i. 82, 72. Sticce *offa,* 290, 47: *offa vel frustum,* 27, 18. Cnuca án sticce ðære wyrt, Lchdm. iii. 4, 21. Swé swé stycce hláfes *sic ut frusta panis,* Ps. Surt. 147, 17. Sticcum *frustris, partibus,* Wrt. Voc. ii. 151, 39. On lytlum sticcum leóðworda dæl reccan, Andr. Kmbl. 2974; An. 1490. Hit (*the veil of the temple*) on eorþan læg on twám styccum, Exon. Th. 70, 15; Cri. 1139. Hig curfon ðone ram eal tó sticceon (*in frusta*), Lev. 8, 20. Tó sticcon, 1, 6: Ex. 29, 17. Tó sticcum, Jud. 14, 6. Ðæt mon ðone disc tóbræce tó styccum, Bd. 3, 6; S. 528, 21. Hé feallende tóbærst on feówer sticca. Ða feówer sticca clifodon tó feówer stánum, Homl. Th. i. 380, 24. Hí tócurfon ðone líchaman on manugu sticceo. ... Ða gesomnodon hí ða sticceo, Shrn. 125, 10, 12. Þurh sticceo *per cola,* Wrt. Voc. ii. 69, 8. In sticco *frusta,* in sticce *frustatim,* 34, 32, 33. In sticce *frustatim,* 86, 78. On sticca *in frusta, in partes,* Hpt. Gl. 495, 30. Hé genam ða sticcu, Homl. Th. ii. 154, 19. II. *a small piece of money:*—Twá stycgce (stycas, Lind.) *duo minuta,* Mk. Skt. Rush. 12, 42. III. *a short space of time:*—Ðú á embe sicce (*after a bit*) fêhst eft on ða ilcan spræce ðe ðú ær spæce, Bt. 35, 5; Fox 164, 14. [Stucchen (sticches, 2nd MS.), Laym. 16703. To stucchen, Kath. 99, 1992. Smalliche be little stechches, Ayenb. 111, 14. *O. L. Ger.* stukki: *O. H. Ger.* stucchi *frustum, pars; obolum; spatium, tempus: Icel.* stykki *a piece.*] v. fell-, land-, molegn-, seolfor-stycce.

stycce-mǽlum (sticce-, stic-); *adv. In pieces, bit by bit, piecemeal:*—Styccimélum *particulatim,* Wrt. Voc. ii. 115, 81. Styccemælum *minutatim,* 54, 55. Sticcemælum, 77, 70. Sticmælum *frustratim, particulatim, minutatim,* 151, 37: *membratim, per singula membra,* Hpt. Gl. 407, 19. I. *to pieces, to bits:*—Þrié wulfas ánes deádes monnes líchoman styccemælum tóbrudon, Ors. 4, 2; Swt. 160, 21. Stánas sticmælum tóburston, Homl. Th. i. 108, 19. Hé sticmælum tóbræc ða anlícnysse, 464, 26. Ðæt húsel biþ sticmælum tódǽled, ii. 270, 33. II. *here and there, in different places:*—Styccimélum *passim,* Wrt. Voc. ii. 116, 60. On feáwum stówum styccemælum wíciaþ Finnas, Ors. 1, 1; Swt. 17, 5. Se cnoll is styccemælum mid wuda oferwexen, Blickl. Homl. 207, 27. Ðæs muntes cnoll is sticmælum mid wuda oferwexen, and eft sticmælum nid grénum felda oferbræded, Homl. Th. i. 508, 23. III. *little by little, by degrees, gradually:*—Ða ðýstru styccemælum swá ðicce wæron *tenebrae in tantum paulisper condensatae sunt,* Bd. 5, 12; S. 628, 12. Men dydon styccemælum ðæt hí ða moldan nómon *paulatim ablata terra,* 3, 9; S. 533, 22. Oþþæt ðú hí styccemælum áfedde mid ðý Godes worde *donec paulatim enutriti verbo Dei,* 3, 5; S. 527, 34. Sticcemælum, 1, 7; S. 477, 3: 1, 16; S. 484, 15: 5, 10; S. 624, 37. Ðone song hé gehýrde sticcemælum tó him neáléacan, 4, 3; S. 567, 43. Ða bleówan wit ða hylla and ástigon ðæron and scufon hig út on ða eá, and wit reówan sticcmælum mid uncrum fótum óð ðæt hig unc ásetton on óðre healfe ðære eá *then we inflated the bags, and mounted on them, and pushed them out into the river, and little by little we rowed with our feet, until they landed us on the other side of the river,* Homl. Ass. 205, 346.

stýfician; *p.* ode *To root up:*—Móna se ðridda weorca onginnan ná gedafanaþ bútan ðæt biþ geedcenned stífician *the third day of the moon is not good to attempt works, except to root up what has grown up again,* Lchdm. iii. 184, 18. [Cf. (?) *Icel.* stýfa *to chop off, curtail;* stúfr *a stump.*] v. á-stýfician, *and next word; and see* swetecian.

stýficung, e; *f. A clearing* (?):—Of ðære stýfycunge, Chart. Earle 248, 11. In ðone nordran stýfecing, Cod. Dip. Kmbl. iii. 399, 35. Stýfecinc, 18, 33.

stýle, stýl-ecg, stýled, stýlen, styll, styllan *to take a place,* styllan *to leap,* styllan *to stall.* v. stíle, stíl-ecg, stílan, stílen, still, stellan *to place,* stellan *to leap,* stillan.

styltan; *prs. subj.* (wið-)stylte; *p.* stylte, stylde, (for-)styldte; *pp.* stylted *To be amazed, confounded, be at a loss, be doubtful:*—Stylton *stupebant,* Mk. Skt. Lind. 6, 51. Styldon (stylton, Rush.), 1, 22. Hiá stylton *haesitantes,* Jn. Skt. Lind. Rush. 13, 22. [Cf. *O. H. Ger.* stullen *p.* stulta:—*Jumenta in partem alterum haeserunt* (stultun) *pavefacta,* v. Graff. vi. 676.] v. á-, for-, ge-, wið-styltan.

stýman. v. stíman.

styntan; *p.* te *To make* or *to become dull; hence to stint:*—Styntid *hebetat,* Wrt. Voc. ii. 110, 36. [In later English the verb is found transitive and intransitive:—þe qual gon to stunte, Laym. 31891. Menn sholldenn stinntenn to þewwtenn, Orm. 12844. Þe ueorðe hweolp is Idelnesse, þet is, hwo se stunt mid alle (*is utterly inactive*), A. R. 202, 10. Ystunt (*dulled*) is al my syht; This day me thuncheth nyht ... Stunt is all my plawe, Rel. Ant. i. 123, 18, 39 (14th cent.). God gan stable and stynte, Piers P. 1, 120. Of this cry they nolde neuere stenten, Chauc. Kn. T. 45. The preyere stynte, 1563. Styntyñ of werkynge or mevynge *pauso, desisto;* styntyñ or make a thynge to secyñ of hys werke or mevynge *obsto,* Prompt. Parv. 475–6. *Icel.* stytta *to shorten.*] v. á-, for-styntan; stunt.

stýpel, styr *a stir,* stýr, stýran, styrc, stýrend. v. stípel, ge-styr, steór, steóran, stirc, steórend.

styreness, e; *f.* I. *motion, movement:*—Mid his óðra lima styrenessa *aliorum motu membrorum,* Bd. 4, 11; S. 579, 27. Ic ealle níne styrenesse forleás *motum omnem perdidi,* 5, 6; S. 619, 19. Ðæt hors blon fram ðám unhálum styrenessum ðara [h]leoma *equus cessabat ab insanis membrorum motibus,* 3, 9; S. 533, 39. II. *a commotion, agitation, disturbance, perturbation,* (1) *in a physical sense:*—Styrnise michelo (*motus magnus*) geworden wæs in sae, Mt. Kmbl. Lind. 8, 24. Æfter styrenisse wætres *post motum aquae,* Jn. Skt. Rush. 5, 4. (2) *figuratively:*—Styrenise *tumultus,* Mk. Skt. Lind. 14, 2. Swá monigum and swá myclum styrenesse (-um?) wiþerweardra ðinga *tot ac tantis rerum adversantium motibus,* Bd. 5, 23; S. 646, 4. Styrenissum *perturbationibus,* Rtl. 59, 5. v. eorþ-, ge-, on-styreness.

styrfan, styrfig. v. stirfan, stirfig.

styria, styrige, styrga, styra, an; *m. A sturgeon;* but the word is used as the equivalent of several Latin names of fishes:—Styria *cragacus,* Wrt. Voc. ii. 105, 50: 15, 48. Styrga, styria, styra *porcopiscis,* Txts. 87, 1614. Styria, Wrt. Voc. ii. 68, 29. Styriga, i. 281, 59. Stiria, 65, 63. Styria *rombus,* 55, 61. Ælc seldsýnde fisc ðe weordlíc byþ, styria and mereswýn, Cod. Dip. Kmbl. iii. 450, 27. Andlang stræte út on styrian pól, vi. 9, 6. Mereswýn and stirian *delphinos et sturias,* Coll. Monast. Th. 24, 9. [*O. H. Ger.* sturo, sturjo *sturio, rombus, purro: Ger.* stör: *Du.* steur: *Icel.* styrja: *Dan.* stør: *Norweg.* størje. The Teutonic word was adopted in Romance speeches, and the French form is seen in English *sturgeon.*]

styrian; *p.* ede, ode *To stir, move:*—Ic styrige *moveo,* Ælfc. Gr. 26, 5; Zup. 156, 9. I. *intrans. To be in motion:*—Hé sig ofer ða deór and ofer ealle ða creópende ðe stiraþ on eorþan *praesit bestiis omnique reptili, quod movetur in terra,* Gen. 1, 26. Ealle ða þing ðe on eorðan stiriaþ ... Eall ðæt ðe styraþ and leofaþ, 9, 2, 3. Eall flæsc ðe ofer eorðan styrode, 7, 21. Streámas styredon, Andr. Kmbl. 747; An. 374. Ne stira ðú, sunne, of ðam stede, Jos. 10, 12. Hí ne móton swíþor styrian, Bt. 21; Fox 74, 8. Ða styriendan nétenu, 41, 5; Fox 252, 24. Hý wǽron styriende *commoti sunt,* Ps. Th. 47, 5. Styrendum *mobilibus* Mt. Kmbl. p. 8, 7. II. *trans. To put in motion:*—Styrede *agitabat* Wrt. Voc. ii. 10, 53: *exagitabat,* Txts. 180, 2. (1) *of physical movement:*—Hé styreþ ðone rodor and ða tunglu *coelum ac sidera movet,* Bt. 39, 8; Fox 224, 6: Exon. Th. 422, 29; Rä. 41, 13. Hí heora ágene stefne styriaþ, Met. 13, 49. Hé dyde ðæt án æren nædre hý styrede Wulfst. 98, 22. Ða stánas hí styredon for ðam swége, Bt. 35, 6; Fox 168, 1. Hé sceal gán and hyne styrian *he must walk and move about* Lchdm. i. 316, 17. (1 a) *to move* the strings of an instrument:—Ealle strengas se hearpere grēt mid ánre honda, ðeáh hé hié ungelíce styrige Past. 23; Swt. 175, 10. Ic míne hearpan genam and níne strenga styrian ongan, Wulfst. 255, 9. Hearpan stirgan, Exon. Th. 42, 8; Cri. 669. (1 b) *to put in violent motion, to stir up, disturb, agitate:*—I (*the storm*) streámas styrge, Exon. Th. 386, 31; Rä. 4, 70: 382, 11 Rä. 3, 9. Ðonne wind styreþ láð gewidru, Beo. Th. 2753; B. 1374 Hé hringsele hondum styrede, 5673; B. 2840. Styre mid sticcan, Lchdm. ii. 76, 25. [Streámas] styrgan, Exon. Th. 383, 29; Rä. 4, 18. Sel him styrgendne drenc, Lchdm. ii. 106, 25. Duruþegnum weard hildbed styred (*disturbed;* referring to the only course that seemed left to th cannibals, when the prison was found without their intended victims, viz to feed on the bodies of the dead prison-guards), Andr. Kmbl. 2186; An 1094. (2) *figuratively, to stir up, to excite, incite, rouse, move:*—Ó sædnysse stirgit *ad congeriem* (*satietatem*) *coartet,* Germ. 391, 30. Ná ðæra wǽtena ðe druncennysse styriaþ, Homl. Th. ii. 298, 19. Saca an wraca hé styrede gelóme, Wulfst. 106, 26. Gárulf Gúðere styrode, Fin Th. 37; Fin. 18. Swá sceal æghwelc láreów tó ánre lufan mid mislícur manungum his hiéremonna mód styrigean, Past. 23; Swt. 175, 12 (2 a) *to handle, treat, deal with:*—Secg ongan síð Beówulfes snyttrur styrian, Beo. Th. 1749; B. 872. (2 b) *to move, disturb, trouble, agi tate:*—Mid ðám bisgum ðe on breóstum styreþ mon on móðe, Met. 22 64. Ðara synfullena handa mē ná ne styrien, Ps. Th. 35, 11. Ða ð mē mid unryhte ǽnige styrian *qui insurgunt in me,* 108, 27. Swá bió módsefan of hiora stede styrede, Met. 7, 25. [*Laym. A. R.* sturien Orm. stirenn: *Ayenb.* sterie. Cf. *Icel.* styrr *stir, tumult, disturbance.* v. á-, be-, ge-, geond-, on-, ymb-styrian.

styric, styrigend. v. stirc, á-styrigend.

styrigend-líc; *adj. Moving:*—Hé styrigendlíces nán þincg findan n mihte, Homl. Skt. ii. 23 b, 735. Of styrigendlícum *mobilibus,* Germ 391, 26. God gesceóp eall libbende fisccinn and stirigendlíce *omne animam viventem atque motabilem,* Gen. 1, 21.

styring. v. styrung.

styrman; *p.* de. I. *of weather, to storm, rage:*—Hit ríne an sníwe and styrme úte *furentibus foris turbinibus hiemalium pluviaru vel nivium,* Bd. 2, 13; S. 516, 17. Styrmendum wederum, Bt. 7, 3

Fox 22, 5. **II.** of persons, *to storm, make a great noise, cry aloud, shout* :—Ic (*the wood pigeon*) búgendre stefne styrme (cf. ic hlúde cirme, l. 18), Exon. Th. 390, 25; Rä. 9, 7. Gehýr mín gebed nú ic stefne tó ðé styrme hlúde *exaudi vocem orationis meae*, Ps. Th. 139, 6. Mín stefn tó ðé styrmeþ Drihten *voce mea ad Dominum clamavi*, 141, 1. Stearcheort styrmde, stefn in becom heaðotorht hlynnan under hárne stán, Beo. Th. 5097; B. 2552. Holofernus hlóh and hlýdde, hlynede and dynede, ðæt mihten fira bearn feorran gehýran, hú se stíðmóda styrmde and gylede, Judth. Thw. 21, 19; Jud. 25. Styrmdon hlúde grame gúðfrecan, 24, 35; Jud. 223. Ic mid stefne ongann styrman tó Drihtne *voce mea ad Dominum clamavi*, Ps. Th. 76, 1. [Þe trouble wynde þat hy3t auster stormynge and walwyng þe see, Chauc. Boet. 29, 712. O.H. Ger. sturmen *tumultuari, perstrepere*: Ger. stürmen *to roar, rage*; *to take by storm*: Icel. styrma *to be stormy* (of weather); *to make a great noise, make much ado.* Layamon uses the verb in the sense *to attack violently* :—þat hæðene uolc mid muchelere strengðe sturmden (sweinde, 2nd MS.) þa Bruttes and driuen heom to ane munte, 18327. Þa Freinsce weoren isturmede & noðelas heo stal makeden, 1670.] v. be-styrman.

styrnan, styrne, styrnenga, stýr-ness, styrn-líc, styrnlíce, styrn-mód, -styrred, styrtan. v. stirnan, stirne, stirninga, steór-ness, stirn-líc, stirn-líce, -stirred, sturtan.

styrung, e; *f.* **I.** *motion* :—Sterung *gestus, motus corporis*, Hpt. Gl. 455, 44. Ðara unstillena gesceafta styring ne mæg nó weorþan gestilled, Bt. 21; Fox 74, 4. Monige beóþ blíðe and eác unblíðe . . . for ðæs blódes styringe and for líchoman medtrymnesse, Past. 27; Swt. 187, 24. Ðonne hí (*prepositions*) getácniaþ styrunge, ðonne beóþ hí gefeódde *accusativo*, Ælfc. Gr. 47; Zup. 274, 7. **I** a. *exercise, practice* :—Sió wiþerweardnes biþ wæru áscerred mid ðære styringe hire ágenre frécennesse *adversam fortunam videas ipsius adversitatis exercitatione prudentem*, Bt. 20; Fox 72, 6. **II.** of violent movement, (1) *literal, disturbance, agitation, commotion* :—Wearð mycel styrung (*motus*) geworden on ðære sæ, Mt. Kmbl. 8, 24. Árás micel styrung and hreóhnys on ðære sæ, Homl. Th. ii. 378, 14. Seó burh Naim is gereht ýðung oððe styrung, i. 492, 1. Æfter ðæs wæteres styrunge *after the troubling of the water* (A.V.), Jn. Skt. 5, 4. (2) fig. (a) *a disturbance, tumult* :—Ðe læs tó mycel styrung (*tumultus*) wurde on ðam folce, Mt. Kmbl. 26, 5. Blon sié styring *cessavit quassatio*, Ps. Surt. 105, 30. Ðæt wíf ðurh ða færlícan styrunge ne gýmde hire cildes, Homl. Th. i. 566, 8. Sceal áspringan bryne and blódgyte and styrnlíce styrunga, Wulfst. 88, 11. (b) *trouble* :—Wé sceolan on ælcne tíman and on ælcere styrunge mid ródetácne ða reðan áflían, Homl. Skt. i. 17, 143. (c) of the mind, *perturbaion, agitation, emotion* :—Stýrau his módes styrunge mid singalre gemetfæstnysse, Homl. Th. i. 360, 16. *Interjectio* geopenaþ ðæs módes styrunge mid behýddre stefne, Ælfc. Gr. 48; Zup. 278, 3. Gif wé ða unsceáðwíslícan styrunga on stæððignysse áwendaþ, Homl. Th. ii. 210, 30. v. á-, eorþ-styrung.

styðe. v. studu.

sú. v. sugu.

su-. For words beginning with *su-* followed by a vowel see *sw-*.

sub-diácon, es; *m. A sub-deacon* :—Hit is beboden subdiáconum and munecum, Blickl. Homl. 109, 25. v. under-diácon.

súcan; *p.* seác, *pl.* sucon; *pp.* socen *To suck* :—Ic súce sugo, Ælfc. Gr. 28, 5; Zup. 175, 4. Heó (*the air*) sýcþ ælcne wætan up tó hire, Lchdm. iii. 278, 7. Of ðæra cilda múðe ðe meolc súcaþ, Ps. Th. 8, 2. Ða breóst ðe ðú suce (*suxisti*), Lk. Skt. 11, 27 : Homl. Skt. i. 8, 125. Sucun (*suxerunt*) hunig of stáne, Ps. Surt. ii. p. 192, 43. Ðæt hig sucon, Deut. 32, 13. Ongunnon ealle ða næddran heora blód súcan, Homl. Th. ii. 488, 35. Súcende mid ealdum men *lactentem cum homine sene*, Deut. 32, 25. Ægðer ge men ge ða súcendan cild, Homl. Th. i. 246, 21. Of múðe súkendra (*lactantium*), Mt. Kmbl. Rush. 21, 16. [He moste suken, Laym. 13194. Vther þa 3æt sæc (soc, 2nd MS.) his moder, 12981. Þa tittes þ þu suke, 5026. Bi þeo tittes þet he sec, A.R. 330, 6.] v. á-, for-súcan, meolc-súcend; súgan, sícan.

súce. v. hunig-súce.

súcengra for súcendra, Ps. Spl. 8, 3.

sucga, an; *m. The name of a bird.* [In later times the word seems to apply to the *whitethroat,* which is called *hazeck* (Worcest.) and *hay sucker* (Devon), and to the *hedge-sparrow, isaac* or *hazock* (Worcest.), *segge* (Devon), E. D. S. Pub., Bird Names, pp. 23, 29. Chaucer uses *heysugge* (-*sogge, -soke*) of the sparrow : Thou (*the cuckoo*) mordrer of the hey-sugge, Parl. of F. 612. *Heges-sugge* (q. v.) is used to gloss the same word, *vicetula,* as *sucga* does.] :—Sucga, sugga, suca *ficetula*, Txts. 62, 422. Sucga, Wrt. Voc. ii. 35, 53. Sugga, i. 62, 43. Tó sucgan gráf, Cod. Dip. Kmbl. iii. 437, 27. [Sugge, br3d *curuca, linosa,* Prompt. Parv. 483, col. 2. Halliwell quotes *sugge* from Palsgrave.]

suchtyrga, suctyria. v. suhteriga.

sufel, es; *n. Anything, whether flesh, fish, or vegetable, eaten with bread, sowl* [' Anything used to flavour bread, such as butter, cheese, etc., is called *sowl* in Pembrokeshire,' Halliwell] ; *pulmentarium* :—Sile him fórmete on hláfe and on sufle and on wíne *dabis viaticum de gregibus et de area et torculari tuo* (the sufle corresponds to the gregibus, v. winter-sufel), Deut.

15, 14. Hæbbe gé sufol (*numquid pulmentarium habetis ?*) . . . Hé cwæð tó him : Lǽtaþ ðæt nett on ða swíðran healfe ðæs réwettes and gé gemétaþ, Jn. Skt. 21, 5–6. Wé gelýsaþ, ðæt genóh sý tó dæghwamlícum gereorde twá gesodene sufel (*cocta duo pulmentaria*). . . Gif mon æppla hæbbe oððe hwylces óþres cynnes eorðwæstmas, sý ðæt tó þriddum sufle. Sý ánes pundes gewihte hláf tó eallum dæge, R. Ben. 63, 10–15. Ðæt hiae simle ymb xii mónaþ gegearwien tén hund hláfa and swæ feola sufla, and ðæt mon gedéle tó ælmessan for míne sáwle, Chart. Th. 461, 11. [Ne þerf þet meiden s-chen nouðer bread ne suuel, A.R. 192, 18. Kam he neuere hom handbare, þat he ne broucte bred and sowel In his shirte, or in his couel, Havel. 767. I ne haue neyþer bred ne sowel, 1143. Þes two fishes ben souel to þes loves, Wicklif, Select Wks. i. 63. Sowvel, þat is mete to make potage and to medle among potage, ii. 137. Sowil, as thow knowe me to wiln (savoury meat, such as I love, A.V.), Gen. 27, 4. Alle that greden at thy gate . . . after fode, Parte with hem of thy payn of potage other of souel, Piers P. C. 9, 286. Forto haue my fylle of that frute I wolde forsake al other saulee (glossed by *edulium*), B. 16, 11. *Hoc potagium* a° potage, *hoc edulium* a° sowle, Wrt. Voc. i. 199, col. 2 (15th cent.). Sowylle, 266, col. 1 (15th cent.). *Edulia* sowell, Wülck. Gl. 579, 41 (15th cent.). Sowle *edulium, pulmentarium*, Cath. Ang. 349, col. 2. See the note there (from which the Wicklif passages have been taken), where from Andrew Boorde's Introd. to Knowledge is quoted, 'A gryce is gewd sole ;' and from Turner's Herbal, 'The most part vse Basil for a sowle or kitchen ;' and 'The fyrste grene leaues of elm tre are sodden for kichin or sowell as other eatable herbes be.' *Icel.* sufl *whatever is eaten with bread* : *Swed.* sofwel : *Dan.* sul *meat.* Cf. *O. H. Ger.* pi-sufili *pulmentum, polentum.*] v. lencten-, winter-sufel; ge-sufel ; *adj.,* syflig.

súgan; *p.* seáh, *pl.* sugon ; *pp.* sogen. **I.** *to suck* :—Ðú suge *suxisti*, Wrt. Voc. ii. 74, 49. Ðæt sió réðnes ðæs wínes ða forrotedan wunde súge and clænsige, Past. 17, 10 ; Swt. 125, 12. [In Txts. 64, 455 the entry *fellitat suggit* is perhaps all Latin, as the same form occurs again in a later glossary, where the termination of the verb is never -*it, fellitat,* i. *decepit, suggit,* beswícþ, Wrt. Voc. ii. 148, 29] **II.** *to fall in as the cheeks do when sucking* (?) :—Ðonne him on ðam magan súgeþ *when it is in his stomach as if it were sucked in,* Lchdm. ii. 192, 13 : 160, 1. [*O. H. Ger.* súgan : *Icel.* súga, sjúga.] v. á-, for-súgan ; súcan, sígan (sýgan) *to soak,* Lchdm. i. 134, 14.

súge, sugga, sugian. v. hunig-súge, sucga, swigian.

sugu, e; sú, e; *f. A sow* :—Sugu *scroffa,* Ælfc. Gr. 7; Zup. 25, 7 : *scrofa,* Wrt. Voc. i. 22, 73 : 286, 46 : ii. 120, 7. Sió sugu hí wille sylian on hire sole, Past. 54, 1; Swt. 419. 27. Suge *scrofe,* Wrt. Voc. ii. 92, 14. Suge sweard *vistula,* 124, 1. Mé (a badger) on bæce standaþ her swylce sweon lcorum (= hér swilce súe on hleórum, Grein) hlífiaþ tú éaran ofer eágum, Exon. Th. 396, 13; Rä. 16, 4. [A.R. suwe: *Ayenb.* zo3e: *Chauc. Piers P. Wick.* sowe : *Du.* zog: *Swed.* sugga: *O. H. Ger.* sú : *Ger.* sau: *Icel.* sýr; *acc.* sú: *Dan.* so.] v. gefearh-sugu.

suht, e ; *f. Sickness* :—Him yldo ne derede ne suht swáre, Cd. Th. 30, 24; Gen. 472. [This, the only instance of the use of the word, may be due to Old Saxon influence ; see the Héliand where the word occurs many times, in two of them with the same adjective as in the passage. The word is however widely spread: *Goth.* sauhts: *O. L. Ger.* suht *morbus*: *O. H. Ger.* suht *morbus, tabes*: *Ger.* sucht: *Icel.* sótt *sickness*: *Swed.* sut *affliction*: *Dan. Swed.* sot. It is found in the Cursor Mundi : Þai troud þat he moght þair broþer (*Lazarus*) hale of all his soght (mi3te make him hool to be, Trin. MS.), 14157; and Halliwell quotes a passage in which jaundice is called 3alow sou3t, Dict. 950.]

suhter-fæderan, -gefæderan ; *pl. m. Uncle and nephew* :—Hróþwulf and Hróðgár suhterfædran, Exon. Th. 321, 15 ; Víd. 46. Ða gódan twegen (Hróþgar and Hróþulf) sæton suhtergefæderan, Beo. Th. 2332; B. 1164. [Cf. the double meaning in *M. H. Ger.* veter, father's brother, brother's son.] v. next word.

suhter[i]ga, suhtriga, suhtria, an ; *m. A brother's son, a nephew* ; or, expressing the relation of those whose fathers were brothers, *a cousin* :—Suhterga *fratuelis,* Wrt. Voc. ii. 109, 16. Suchtyrga *fratuelis* i. *filius fratris,* 36, 4. Suctyrian *fratres patrueles, sic dictus est ad patres eorum, si fratres inter se fuerunt,* 39, 49. Ic (Abraham) com fædera ðín sibgebyrdum, ðú (*Lot*) mín suhterga, Cd. Th. 114, 9 ; Gen. 1901. His (*Abraham's*) suhtriga *Lot,* 122, 20 ; Gen. 2029. His suhtrian wíf, 106, 23 ; Gen. 1775. v. sweór.

sulh, suluh, sul[l] ; *gen.* sule, *but also* sules ; *dat.* sylg, sylh, syl ; *acc.* sulh, sul ; *n. pl.* sylh, syll ; *gen.* sula ; *dat.* sulum : *a weak genitive seems also to occur in* sylan scear ; *generally feminine, but see the genitive.* **I.** *a plough* :—Sulh *aratrum,* Wrt. Voc. i. 5, 2 : 289, 76. Sul, ii. 6, 19 : Ælfc. Gr. 17; Zup. 109, 18. Swá seó sulh ðone teóðan æcer gegá, L. Eth. ix. 7; Th. i. 342, 11 : L. Edg. i. 1 ; Th. i. 262, 9 : L. C. E. 8; Th. i. 366, 7. A be ðan wuda swá sulh and síðe hit gegán mæge, Cod. Dip. Kmbl. iii. 458, 20. Sule reóst *vomes,* Wrt. Voc. ii. 138, 72. Sules reóst, 25, 28 : 106, 20. Ðæs sules bodig, Lchdm. i. 402, 2. Sylan scear *vomer,* Coll. Monast. Th. 30, 29. Ðæt nán mon ne scyle dón his hond tó ðære sylg, Past. 51; Swt. 403, 2. Ælc man hæbbe

æt ðære sylh (syhl, MS.) .ii. wel gehorsede men, L. Ath. i. 16; Th. i. 208, 12. Tô syl ... mid ðære syl *ad aratrum ... aratro*, Coll. Monast. Th. 19, 15, 21. Man ða sulh forð drîfe, Lchdm. i. 404, 1. Mann ðe hys hand âsett on his sulh (suluh, Rush.), Lk. Skt. 9, 62. Hê hys sulh on handa hæfde, Ors. 2, 6; Swt. 88, 8. Hê sulh heóld, Shrn. 61, 18. Mid sul tô erianne, Salm. Kmbl. p. 186, 28. Heora sylh unrihte gangaþ *aratra eorum non recte incedunt*, Bd. 5, 9; S. 623. 12. Ðîne syll eodon, Homl. Th. ii. 450, 6. Þeáh hê erige his land mid ðûsend sula, Bt. 26, 3; Fox 94, 14. Sulum *aratris*, Wülck. Gl. 254, 6. II. In the following passage perhaps the word is used to denote *the quantity of land which could be cultivated with one plough*; v. sulincel and cf. plôg. *Caruca*, which occurs in the passage quoted below from the Laws, seems to have been used in this sense; e.g. in Florence of Worcester's description of the compilation of Domesday Book *quot carrucas* seems to represent *hû mycel landes* in the Chronicle; and later *sulh* is certainly so used, e.g. Ich þe ʒiue þritte solh of londe, Laym. 18779. Seouen sulʒene lond, 18789. Twenti sulhene lond, 13176. But the unit of assessment may have been *the plough with its team of oxen*. v. Seebohm, Vill. Comm., pp. 112-3. Sceóte man ælmessan ... swâ æt heáfde peninc, swâ æt sylh (*one MS. has* æt sulhgange. v. sulh-gang) peninc (cf. detur de omni caruca denarius vel denarium valens, et omnis, qui familiam habet, efficiat, ut omnis hirmannus suus det unum denarium, L. Eth. viii. 1; Th. i. 336, 24; *and see* sulh-ælmesse), Wulfst. 170, 20. [Gif þe suluh (ploh, MS. T.) ne erede, A. R. 384, 18. Þer cheorl draf his sulʒe ioxned swiðe fæire, Laym. 31811. Þe ilke þet zet þe hand aþe zuolʒ, Ayenb. 242, 31. The word is still used in Somerset, *zool*, v. E. D. S. Pub., W. Somerset Glossary.]

sulh-æcer, es; *m. A strip of land for ploughing.* v. Seebohm, Vill. Comm. s. v. æcer:—Eallum æhtemannum gebyreþ ... sulhæcer, L. R. S. 9; Th. i. 438, 1.

sulh-ælmesse, an; *f. Plough-alms, a contribution of one penny to be paid for every* sulh, v. sulh, II. It is first mentioned in the laws of Edward and Guthrum, and its payment is enjoined in those of succeeding kings. It was to be paid within fifteen days after Easter, or a penalty was incurred:—Sulhælmesse hûru fiftêne niht ofer Eástran, L. C. E. 8; Th. i. 366, 3. Gif hwâ sulhælmyssan ne sylle, gylde lahslit mid Deuum, wîte mid Englum, L. E. G. 6; Th. i. 170, 5: L. Ath. i. prm.; Th. i. 196, 10. Wê bebeódaþ ... sulhælmessan, and gif hit hwâ dôn nelle, sý hê âmânsumod, L. Edm. E. 2; Th. i. 244, 17. Gelæste man sulhælmessan ðonne .xv. niht beón onufan Eástran, L. Edg. i. 2; Th. i. 262, 17: L. Eth. v. 11; Th. i. 306, 31: vi. 16; Th. i. 318, 30. Sulhælmessan gebyreþ ðæt man gelæste be wîte æghwylce geáre ðonne .xv. niht beóþ âgân ofer Eástertîd, ix. 12; Th. i. 342, 31. Suluhælmessan, Shrn. 208, 29.

sulh-beám, es; *m. The curved hinder part of a plough, plough-tail:—* Sulhbeám *burris, curvamentum aratri*, Wrt. Voc. ii. 126, 79: *buris*, 12, 54: i. 15, 4. [Solowbeme *buris*, Wrt. Voc. i. 180, 29 (14th cent.?). Cf. plughbeme *buris*, 232, col. 2.]

sulh-gang, es; *m. A plough-gang* (pleuch-, *plough-gang* as much land as can be properly tilled by one plough, Jamieson's Dict. See too *pleuch-gate*, ib. Cf. for a similar use of *gang* in measurements *Icel.* sôlar-gangr = *a day*)? Æt heáfde peninc, æt sulhgange peninc, Wulfst. 170, 37. v. sulh II, sulung.

sulh-geside, es; *n. An appurtenance of a plough:—*Man sceal habban wængewædu, sulhgesîdu, Anglia ix. 264, 5. Cf. next word.

sulh-geteóh; *gen.* -teóges; *n. An implement belonging to a plough:* —Gegaderie hê ealle his sulhgeteógo tôgædere *let him collect together all the apparatus of his plough*, Lchdm. i. 400, 19.

sulh-geweorc, es; *n. Plough-work, making of ploughs:—*Tubal Cain smiðcræftega wæs and monna ærest sulhgeweorces fruma wæs ofer foldan (*Tubal Cain an instructor of every artificer in brass and iron*, A. V.), Cd. Th. 66, 19; Gen. 1086.

sulh-hæbbere, es; *m. One who holds a plough* (cf. hê his sulh on handa hæfde, Ors. 2, 6; Swt. 88, 8), *a ploughman:—*Sulhhæbbere *stibarius*, Wrt. Voc. ii. 79, 24. v. next word.

sulh-handla, an; *m. One who holds the handle of a plough, a ploughman*:—Sulhandla *stivarius, arator*, Hpt. Gl. 461, 71.

sulh-handle (-a; *m.?*), an; *f. A plough-handle, plough-tail:—*Sulhhandla (-e? v. handle *stiba*, Wrt. Voc. ii. 121, 10) *stiba*, Wrt. Voc. i. 15, 8. Sulhandlan *stivam*, Hpt. Gl. 470, 33.

sulian (?); *p.* ode *To sully* [:—Besutod (-sulod?) *obsoletum, sordidum*, Germ. 403, 26.] v. sylian.

sulincel, es; *n. A small portion of arable land:—*Sulincela *aratiuncula*, Wrt. Voc. ii. 6, 18. v. sulh, II.

sulung, e; *f. A Kentish word for a certain quantity of land, derived, like carrucata, from a name of the plough; from its origin it might mean so much land as could be cultivated by one plough.* From the first two passages given below it would seem that the *sulung* was equivalent to two hides (*manentes*), and later a *solanda*, which is probably the same word, is said 'per se habere duas hidas.' v. Seebohm, Vill. Comm., p. 54. But perhaps it is to be inferred that both hide and *sulung* were considered as on the same footing as regards the plough. Thus to the *gebûr* with his *gyrd landes*, i.e. one quarter of a hide, are to be given two oxen, L. R.

S. 4; Th. i. 434, 23, while a gift of half a *sulung* is accompanied by the further gift of four oxen, Chart. Th. 470, 9-14. v. Seebohm, pp. 138-9, and generally. In the Domesday Survey of Kent the assessment was given by *solins*, and the word remained in use. v. Pegge's Kenticisms, s.v. *sulling:* —Aliquam terrae partiunculam, hoc est duarum manentium ... ritu Cantiae *ân sulung* dictum, Cod. Dip. Kmbl. i. 249, 19. Terrae particula duarum manentium, id est, *ân sulung*, 250, 8. Yc gean intô Cristes cyrican on Cantwarabyrig ðæs landes æt Holungaburnan ... bûton ðære ânre sulunge ðe ic Sîferðe geunnen hæbbe, Chart. Th. 558, 27. Him man sælle ân half swulung ... and mon selle him tô ðem londe .iiii. oxan, and .ii. cý, and l. scêpa, 470, 8-14. Ðisses londes aran thrîe sulong æt hægethe thorne, Cod. Dip. Kmbl. i. 235, 7. Siendan feówer swulung ðæs londes ðe gebyreþ inntô Raculfe on Tænett ...; ðonne is ealles ðæs londes .xxv. swulunga and ân swulung on Ceólulfingtûne, iii. 429, 14-18. Ðæt lond æt Stânhâmstede (*Stanstead, in Kent*) .xx. swuluncga, i. ·292, 23. Se cyning (*Ethelbert of Kent*) sealde Wullâfe fîf sulung landes et Wassingwellan (*Washingwell, in Kent*) wið ðæm fîf sulungum et Mersahâm (*Mersham, in Kent*), ii. 66, 17-19. Twâ sulung æt Denetûne (*Denton, in Kent*), 380, 32.

sum; *indef. prn. Some.* I. one of many, *part of a whole*, used substantively and (1) governing in the genitive (a) a noun or pronoun, cf. the Gothic use of *sums*:—Wæs ic ðara monna sum *I was one of the men*, Chart. Th. 170, 7. Mê tô aldorbanan weorðeþ wrâðra sum, Cd. Th. 63, 18; Gen. 1034. Ðê wile beorna sum him geágnian, 109, 26; Gen. 1828. Ðæt is wundra sum ðara ðe geworhte wuldres aldor, 155, 14; Gen. 2572: 199, 28; Exod. 345: 200, 15; Exod. 357. Wæs Seón sum ðara kynincga, Ps. Th. 134, 11. Swâ swâ ûre sum *quasi unus ex nobis*, Gen. 3, 22. Wæs hira Matheus sum, Andr. Kmbl. 22; An. 11. Hê cýþde on sumre his bôca, Bt. 18, 2; Fox 64, 9. Hî woldon cuman tô sumere ðara stôwa, 34, 7; Fox 144, 9. Anlîc ðara his þegna sumum, 37, 1; Fox 186, 12. Fýr cymþ sume ðissa hærfesta (cf. the phrase *some or one of these days*), Wulfst. 205, 6. Manna cynnes sumne besyrwan, Beo. Th. 1430; B. 713. (b) a cardinal numeral, (a) *one of a company containing the number*:—Iacob fêrde hundseofontigra sum *omnes animae domus Jacob fuere septuaginta*, Gen. 46, 27. Hê âcîgde syfone ... eode eahta sum, Beo. Th. 6237; B. 3123. Hê twelfa sum hire âð sealde (*secum acceptis undecim comparibus suis*, p. 205), Chart. Th. 203, 2: L. Ath. i. 11; Th. i. 206, 3 note. (β) *one with a company containing the number*:—Hannibal ô, ̣fleáh feówera sum *Annibal cum quatuor equitibus confugit*, Ors. 4, 10; Swt. 202, 16. Gange hê feówra sum tô and beó him fîfta, L. Eth. ii. 4; Th. i. 286, 18. Hê com twelfa sum *cum duodecim lectis militibus venientem*, Bd. 3, 1; S. 523, 31. Wæs Agustinus feówertigra sum *socii ejus viri ut ferunt ferme quadraginta*, 1, 25; S. 486, 23. Com seofona sum (cf. ðæt deófol genam mid him ôþre seofon deóflo, St. And. 18, 7), Andr. Kmbl. 2623; An. 1313. Gewât xii-a sum ... se wæs on ðam ðreáte þreotteóða secg, Beo. Th. 4793; B. 2401. Fîftêna sum (cf. 3287; B. 1641, where Beowulf's companions, after one has been slain, are said to be fourteen), Beo. Th. 420; B. 207. (γ) *uncertain*:—Ðæt hê syxa sum ofslôge syxtig, Ors. 1, 1; Swt. 18, 7. (2) followed by *of*:—Sumne of ðám wîtegum *unum de prophetis*, Mk. Skt. 8, 28. Ðâ geneálæhton sume of Saducêum, Lk. Skt. 20, 27. (3) where the whole, of which the object denoted by *sum* is part, is to be inferred from the context:—Sigon ðâ tô slæpe: sum (*one of the sleepers*) sâre angeald æfenreste, Beo. Th. 2507; B. 1251. Habbaþ wê micel ærende ne sceal ðær dyrne sum (*any of the errands*) wesan, 548; B. 271. Sumne (*one of the creatures on the mere*) Geáta leód feores getwæfde, 2869; B. 1432. Sume (*some of the thanes*) ðær bidon, 806; B. 400. (4) where the word is quite indefinite, *some one*:—Sum tô lyt hafaþ, Salm. Kmbl. 688; Sal. 343. Ic sceal swelgan of sumes bôsme, Exon. Th. 395, 30; Rä. 15, 15. (5) where two members or two classes of the same group, or two parts of the same whole, are contrasted, *one ... another, some ... some*:—Ðonne lufaþ sum ðæt sum elles hwæt *one loves that, another something else*, Bt. 33, 2; Fox 122, 24. Hî gaderodon sum mâre sum læsse *alius plus, alius minus*, Ex. 16, 17. Eorle monigum Dryhten âre gesceáwaþ, sumum weána dæl, Exon. Th. 379, 17; Deór. 34. Sum heó hire on handum bær, sum hire æt heortan læg, Cd. Th. 40, 8-9; Gen. 636. Ânra gehwylc hæfþ syndrige gyfe fram Gode sume furðor ðonne sume *alius sic, alius vero sic*, R. Ben. 64, 10. Sume hî beóton sume hî ofslôgon *quosdam caedentes, alios uero occidentes*, Mk. Skt. 12, 5. Sió ungelîcnes hira gearnunga hié tiéhþ sume behindan sume, Past. 17; Swt. 107, 20. (6) where a series of individuals or of groups or of parts is enumerated:—Sum feóll wið ðone weg ... sum feóll ofer stânscyligean ... sum feóll on þornas ... sum feóll on gôd land; ân brohte þrîtigfealdne, sum syxtigfealdne, sum hundfealdne, Mk. Skt. 4, 4-8: Exon. Th. 42, 6-30; Cri. 668-680. Is se finta ... sum brûn, sum basu, sum splottum beseted, 218, 17; Ph. 296. Ânum hê sealde fîf pund, sumum twâ, sumum ân, Mt. Kmbl. 25, 15. Ðâ sende hê his þeów ... hê sende ôðerne ... eft hê sumne sende, Mk. Skt. 12, 2-5. Sume hî sædon ðæt hió sceolde forsceoppan tô león ... sume sceoldan bión eforas ... sume wurdon tô wulfan ... sume wurdon tô ðam deórcynne ðe mon hâtte tigris, Bt. 38, 1; Fox 194, 32 sqq.: 34, 7;

Fox 144, 7–9: Mt. Kmbl. 16, 14.   II. as an adjective (1) with a noun with or without a qualifying adjective, *a certain, some*, see also (5):—Sum man (*homo quidam*) hæfde twegen suna, Lk. Skt. 15, 11. Sum ægleáw man *quidam legis peritus*, 10, 25. Sum wítega of ðám ealdum, 9, 19: Bd. 3, 2; S. 524, 39. Ðeáh sum broc and sumu wiðerweardues hiera forwiernþ, Past. 50; Swt. 391, 35. Wæs him gegearwod sum heard harmscearu, Cd. Th. 28, 7; Gen. 432. Sum wæs æhtwelig æþeles cynnes ríce geréfa, Exon. Th. 243, 29; Jul. 18. On his heortan cwæð unhýdig sum *dixit insipiens in corde suo*, Ps. Th. 52, 1. Sumes hundredmannes þeówa, Lk. Skt. 7, 2. Sumes þinges wana, Bt. 34, 9; Fox 146, 18. Weorð forhwerfed ælc tó sumum dióre, 38, 1; Fox 196, 3. Hé com tó sumre stówe, Gen. 28, 11. For sumere twýrædnesse on cwertern ásend, Lk. Skt. 23, 19. (1 a) where two members of the same group are contrasted (*some ... other*):—Sume tunglu habbaþ scyrtran hwyrft, ðonne sume habban, Bt. 39, 3; Fox 214, 17. Sume láreówas sindon beteran ðonne sume, Homl. Th. ii. 48, 16. (2) with a pronoun where later English would use *some of*:—Hé gebád mid sumum ðæm fultume, Ors 3, 10; Swt. 140, 20. Læfdon hig hit sume *quidam ex eis*, Ex. 16, 20. Sume hí gelýsdon on deáde entas, Homl. Th. i. 366, 21. Sume gé (*quidam ex vobis*) ne gelýfaþ, Jn. Skt. 6, 64. Sume ða bóceras *quidam de scribis*, Mt. Kmbl. 12, 38. Ðá téð hié brohton sume, Ors. 1, 1; Swt. 18, 1. Ða sume wé nú gýmdon, Bd. 4, 7; S. 574, 27. (3) with *oðer*:—Sum óðer wítega, Homl. Th. i. 364, 18. Hé nales tó ídelnysse swá sume óþre ac tó gewinne on ðæt mynster eode, Bd. 4, 3; S. 567, 27. (4) with words denoting measure, *some* as still used with numerals, *one*; the use of *án*, and in later English of the indefinite article with numerals, may be compared with this use of *sum*:—'Ásend him twá scrúd and sum pund.' Se ðegen him andwyrde: 'Genim feówer scrúd and twá pund, Homl. Th. i. 400, 19. Genim ðysse wyrte sumne (*one*) gripan, Lchdm. i. 184, 18. Ða gegaderedon hí sum hund scipa, and fóron súð ymbútan and sum feówertig scipa norþ ymbútan, Chr. 894; Erl. 91, 4–6. Hié besǽton ðæt weorc útan sume twegen dagas, Erl. 93, 9. Ða wǽron hí sume tén geár on ðam gewinne, Bt. 38, 1; Fox 194, 7. Hý gán .xii. sume (twelfa sum, MS. B.), L. Ath. i. 11; Th. i. 206, 3. (4 a) where the number is indefinite, *some*:—Ða se Aulixes tó ðam gefiohte fór, ðá hæfde hé sume hundred scipa, Bt. 38, 1; Fox 194, 7. (5) adverbially or in adverbial phrases:—Se biscop is þeáh geset sumes (*in some degree*) tó máran bletsunge ðonne se mæssepreóst sý, L. Ælfc. P. 36; Th. ii. 378, 20. Sumes onlíce swá *velut*, Exon. Th. 214, 21; Ph. 242: Met. 8, 47. Swíðe gelíce, sumes hwæðre þeáh ungelíce (cf. *the corresponding prose* on sumum þingum ungelíce, Bt. 33, 4; Fox 128, 26), 20, 54. Sió eorðe hit helt and be sumum dæle swilgþ, Bt. 33, 4; Fox 130, 5: Met. 20, 96. Seó hæfþ sume dæle (cf. *som del* in Chaucer) læssau leáf, Lchdm. i. 144, 13. Æt sumum cyrre *once, on one occasion*, Ors. 1, 1; Swt. 17, 7: Cd. Th. 298, 25; Sat. 528. Sume síþe, Exon. Th. 20, 16; Cri. 318. Sumera ðinga eáðelícor *in some respects easier*, Homl. Th. i. 236, 11. [Goth. sums: O. Sax. O. Frs. O. H. Ger. sum: Icel. sumr.]

-sum *an adjective suffix* as in glad-*some*, win-*some*. [Goth. lustu-sams: O. Sax. O. H. Ger. lang-sam: Icel. frið-samr: O. Frs. hár-sum.] v. ang-, frem-, gehýr-, genyht-, lang-, lof-, luf-, sib-, wyn-sum *as examples*.

sumer (-or, -ur), es; *dat.* e; *m. Summer*:—Feówer tída synd getealde on ánum geáre ... *Aestas* is sumor, Lchdm. iii. 250, 10. On ðone nygeþan dæg ðæs mónðes (*May*) biþ sumeres fruma. Se sumor hafaþ hundnvgontig daga, Shrn. 83, 33. Sumor biþ sunwlitegost, Menol. Fox 473; Gn. C. 7. Beorht sumor, Cd. Th. 239, 23; Dan. 374. Sumer and winter; on sumera hit biþ wearm and on wintra ceald, Bt. 21; Fox 74, 23. Swá hǽttra sumor, swá mára ðunor and líget on geáre, Lchdm. iii. 280, 9. Gé witun ðæt sumor (-er, MSS. A. B. Lind. Rush.) ys gehende, Mt. Kmbl. 24, 32. Ær sumor on tún gá *before summer come*, Lchdm. iii. 6, 1. Yldum bringþ sigelbeorhte dagas sumor tó túne, Menol. Fox 176; Men. 89. Sumur, Exon. Th. 354, 58; Reim. 67. Ðonne on sumeres tíd sunne hátost scíneþ, 212, 12; Ph. 209. Ðú ðe ðam winterdagum selest scorte tída, and ðæs sumeres dahum langran, Bt. 4; Fox 8, 5. Swá hé in swoloþan middes sumeres wǽre *quasi in mediae aestatis caumate*, Bd. 3, 19; S. 549, 30. Wintres and sumeres *in winter and in summer*, Exon. Th. 200, 7; Ph. 37. Ic (*the fowler*) nelle fédan hig (*the hawks*) on sumera, forðamðe hig þearle etaþ, Coll. Monast. Th. 26, 9. Wiþ ðære sunnan hǽto on sumere, Bt. 34, 10; Fox 150, 9. Ðý sumera fór Ælfréd cyning út on sǽ, Chr. 875; Erl. 78, 5. Ðæs on sumera, 896; Erl. 94, 1. Ðý ilcan sumera, 897; Erl. 96, 14. Sumere, 885; Erl. 82, 25. Ofer ðone midne sumor (midne-sumor? cf. midne-dæg), 1006; Erl. 140, 5. Heó sý geworht ofer midne sumor, Lchdm. iii. 74, 11: Menol Fox 235; Men. 119. [O. Sax. O. H. Ger. sumar: O. Frs. sumur: Icel. sumar; *n.* (but earlier *m.*).] v. mid-, midde-, middan-sumer. See Grmm. D. M. c. 24.

sumer-hǽte, an; -hætu (o); *indecl. or gen. e*; *f. Summer heat*:—Gif ðære stówe neód oþþe gedeorf oðþe sumerhǽte hwylces eácan behófige *si loci necessitas uel labor aut ardor aestatis amplius poposcerit*, R. Ben. 64, 17. For ðære sumorhǽte, Ors. 3, 9; Swt. 132, 31. [Cf. Icel. sumar-hiti.]

sumer-lǽcan; *p.* -lǽhte *To draw near to summer*:—Wite gé ðæt hit sumorlǽhþ, Homl. Th. i. 614, 5.

sumer-lang; *adj. Long as in summer*, epithet of a day (cf. *live-long*):—Ic ásecgan ne mæg, þeáh ic gesitte sumerlongne dæg, eal þa earfeþu, Exon. Th. 272, 7; Jul. 495. Sumorlange dæg, 443, 29; Kl. 37. Ðú wercest sumurlange dagas swíðe háte, Met. 4, 19. [O. Sax. thiu niguða tíd sumarlanges dages, Hel. 3422. M. H. Ger. sumer-lanc.]

sumer-líc; *adj. Summer*:—Sumorlíc dæg *aestivus dies*, Wrt. Voc. i. 53, 28. Se sumerlíca sunnstede, Lchdm. iii. 250, 21. Mid ðære sumerlícan hǽtan, 252, 10. On sumerlícum tíman, Anglia xiii. 431, 939. [Eauer iliche sumerlich, Kath. 1663. O. H. Ger. sumar-líh *aestivus*: Icel. sumar-ligr.]

sumer-lida, an; *m.* [Lida, like the equivalent Icel. liði in *sumar-liði*, elsewhere refers to a single object, man or ship (v. lida, sǽ-, ýð-lida), but in the passage given below from the Chronicle the word seems to mean a fleet. Later in the same work liþ (q. v.), which seems taken from the Scandinavians, is used in this sense, e. g. ðæt lið ðæt on Sandwíc læg, 1052; Erl. 183, 40, can *sumer-lida* be intended to represent Norse *sumar-lið*? In one other place *sumer-lida* occurs, in company with words relating to the sea, and it there glosses *malleolus*; but here perhaps *sumer-loda* should be read, and *malleolus* be taken in the sense *shoot, twig* (see *spæc*); cf. O. H. Ger. sumar-lota, -lata *virgultum, palmes*. v. Anglia xiii. 330.] *A summer fleet*, one that sets forth in summer and returns in autumn :—Æfter ðissum gefeohte cuom micel sumorlida (tó Reádingum, MS. E.), Chr. 871; Erl. 74, 35. [Steenstrup takes the word to mean a force moving from its quarters in England, and leaving women, children, and goods behind there; but if Asser may be trusted, the reinforcement was from abroad. He says: 'quo praelio peracto, de *ultramarinis* partibus alius paganorum exercitus societati se adjunxit.'] Sumerlida *malleolus*, hýdscip *mioparo*, mæstcyst *modius*, Wrt. Voc. ii. 59, 25–27.

sumer-rǽdingbóc; *f. A lectionary for the summer*:—.ii. sumerrǽdingbéc, Chart. Th. 430. 16. [Cf. *Icel.* sumar-bók.] v. rǽding-bóc.

Sumer-sǽte, -sǽtan; *pl. The people or district of Somerset* :—Sumursǽtna se dǽl se ðær nióhst wæs ... Sumorsǽte alle and Wilsǽtan, Chr. 878; Erl. 80, 6–10. Mid Sumursǽtum, 845; Erl. 66, 21. On Dorsǽtum and on Sumærsǽton (Sumersǽtum, MS. C.), 1015; Erl. 152, 12. Ofer Sumersǽton and ofer Wealas, 1048; Erl. 180, 27. [He nom Sumersete, Laym. 21013. Dorsete and Wiltshire and Somersete also, R. Glouc. 3, 23.]

Sumersǽtisc; *adj. Of Somerset*:—Defenisces folces and Sumorsǽtisces, Chr. 1001; Erl. 137, 11.

sumer-selde, an; *f. A summer-house*:—Selde *proaula*, i. *domus coram aula*, sumerselde *zetas aestivales*, Wrt. Voc. i. 57, 47. [Cf. *Icel.* sumar-setr *a summer abode*.]

sumness, e; *f. Delay*:—Æfter monige ł longsum ł monigful sumnise (æfter micclum fæce, Rush.: fyrste, W. S.) *post multum*, Mt. Kmbl. Lind. 25, 19. [Cf. O. H. Ger. súmig *negligens*: súmheit *tardatio, negligentia*: Ger. säumniss *delay, stay*; säumen *to stay*; säumig *tardy*.]

sumor. v. sumer.

sumsende *humming, sounding* (of falling rain):—Hí (*the storm-clouds*) feallan lǽtaþ sweart sumsendu (suinsendu? v. swinsian) seáw of bósme, wǽtan of wombe, Exon. Th. 385, 19; Rä. 4, 47. [*Ger.* summen, sumsen *to hum, buzz*.]

sumur, sun-, suna. v. sumer, sunn-, sunu.

sund *sound.* v. an-, on-, ge-sund. [Sund, Ps. Th. 67, 20, *is an error for* ge-sund.]

sund, es; *n.* I. *power of swimming*:—Hé sealde ðám fixum sund and ðám fugelum fliht, Homl. Th. i. 16, 7: Hexam. 8; Norm. 14, 10. Dol biþ se ðe gǽþ on deóp wæter, se ðe sund nafaþ, ne gesegled scip, Salm. Kmbl. 449; Sal. 225. [Heore (*fishes*) sund is awemmed, Laym. 21326.] II. *the act of swimming*:—Hé on holme wæs sundes ðe sǽnra, Beo. Th. 2876; B. 1436. Hé ðé æt sunde oferflát *he beat you at swimming*, 1039; B. 517. Hé mid sunde (cf. *Icel.* með sundi) ða eá oferfaran wolde, Ors. 2, 4; Swt. 72, 29. Apollonius becom mid sunde tó Pentapolim, Ap. Th. 11, 6. Hié on sunde (cf. *Icel.* á sundi) tó ðære byrig fóron, Nar. 10, 28: Beo. Th. 3240; B. 1618. Ðú ðe wið Brecan wunne on sídne sǽ ymb sund flite *thou that didst strive with Brecan on the wide sea, didst contend in the matter of swimming*, 1019; B. 507. Flód on sund (cf. *Icel.* á sund) áhóf earce from eorðan, Cd. Th. 83, 32; Gen. 1388.   III. *sea, water*:—Streámas wundon, sund wið sande, Beo. Th. 431; B. 213. Ða wæs sund liden *then was the sea passed*, 452; B. 223. Se stán tógán, streám út áweóll ... sund grunde onféng, Andr. Kmbl. 3055; An. 1530. Sund unstille, Exon. Th. 338, 14; Gn. Ex. 78. Swelaþ sǽfiscas sundes getwǽfde (*the ocean having been dried up by the heat*), 61, 20; Cri. 987. Wǽglíþende setlaþ sǽmearas sundes æt ende *by the shore* (or *at the end of their swimming* (?)), 361, 6; Wal. 15. Ic on sunde áwóx ufan ýþum þeaht, 392, 6; Rä. 11, 3. Sǽmearas sunde getenge, Elen. Kmbl. 456; El. 228. Of nihtes sunde, Salm. Kmbl. 675; Sal. 337. Hié on sund (*the Red Sea*) stigon, Cd. Th. 198, 8; Exod. 319: Beo. Th. 1029; B. 512. Ðone ðe grund and sund, eorðan and hreó wǽgas, salte sǽstreámas ámearcode, Andr. Kmbl. 1494; An. 748. Hwá ðam sǽflotan sund wísode *who*

acted as pilot for the vessel, 762; An. 381: 976; An. 488. [Fiss on sund (watir, Trin. MS.), C. M. 621. *Icel.* sund *swimming; a sound: Dan. Swed.* sund *a sound, strait.*] v. syndig.

sund-búend, es; *m. A sea-dweller*, but the word, which occurs only in the plural, is used for *men, mankind;* cf. fold-búend :—Saturnus ðone sundbúende héton, hæleþa bearn, Met. 26, 48. Ðone Saturnus sundbúene hátaþ (cf. stiorran ðe wé hátaþ Saturnes steorra, Bt. 36, 2; Fox 174, 12), 24, 21. Hí (*acc.*) ne gesáwon sundbúende (cf. Hí (*the people of the golden age*) hió (*acc.*) nánwuht ne gesáwon, Bt. 15; Fox 48, 5), 8, 13. Ðæs ðe æfre sundbúend (*men*) secgan hýrdon, Exon. Th. 5, 22; Cri. 73. Ðæt ásecgan sundbúendum, 14, 19; Cri. 221.

sund-corn, es; *n. Saxifrage;* saxifraga granulata :—Sundcorn *saxifraga*, Wrt. Voc. i. 30, 55: 79, 25. Sundcorn. Deós wyrt ðe man *saxifragam* and óþrum naman sundcorn nemneþ. . . . Wið ðæt stánas on blǽdran wexen, genim ðás wyrte, Lchdm. i. 212, 7–11 (see the plate at the beginning of the volume). Sundcornes leáf, ii. 342, 9. Gif men weaxan stánas on ðære blǽdran, wyl sundcorn on ealaþ, 320, 6. Genim neogon piporcorna, fífténe sundcorn (*saxifragia*), iii. 18, 13.

sund-deáw (?), a plant name, *rosemary* :— Sundeáw (= sund-deáw? v. sund, II) *rosmarinus*, Wrt. Voc. i. 68, 77. *Sundew* remains as a name for *drosera rotundifolia*, v. E. D. S. Pub., Plant Names.

sund-flíte, Beo. Th. 1019; B. 507. v. sund, II.

sund-gebland, es; *n. The water's mingling*, used of the mere into which Beowulf plunged :—Se ðe meregrundas mengan scolde, sécan sundgebland, Beo. Th. 2904; B. 1450. Cf. ýð-gebland.

sund-gird, e; *f. A rod to measure the depth of water, a sounding-pole* :—Sundgyrd *bolis* (βολίς *sounding-lead*), Wrt. Voc. i. 63, 67: *bolidis*, 57, 7. Sundgerd in scipe *vel* metráþ *bolides*, ii. 102, 14. Sundgyrd on scipe *vel* metráþ *bolidis*, 126, 46: 11, 17. Cf. sund-líne, -ráp.

sund-helm, es; *m. A water-covering, the sea which covers* :—Mec sundhelm þeahte and mec ýþa wrugon, Exon. Th. 488, 4; Rä. 76, 1. Ic sundhelme ne mæg losian, 382, 13; Rä. 3, 10.

sund-hengest, es; *m. A sea-horse, a ship* :—Ceólum líðan, sundhengestum, Exon. Th. 53, 20; Cri. 853. Sǽlan sundhengestas, ealde ýðmearas, 54, 4; Cri. 863.

sund-hwæt; *adj. Active in swimming* :—Sǽfisca cynn swimmaþ sundhwate, ðǽr se swéta stenc út gewítaþ (-eþ?), Exon. Th. 363, 21; Wal. 57.

sund-lida (Th.), -liden (Grn.), Beo. Th. 452; B. 223. v. sund, III.

sund-líne, an; *f. A sounding-line* :—Sundlíne *cataprorates* (*cataprorates* linea cum massa plumbea qua maris altitudo tentatur, Migne), Wrt. Voc. i. 53, 8: 63, 66. Cf. sund-gird, -ráp.

sund-mere, es; *m. A place for swimming* :—On sundmere *in natatario*, Wrt. Voc. ii. 46, 50.

sundness. v. on- (an-) sundness.

sund-nytt, e; *f. The employment of swimming* :—Beówulf sundnytte dreáh *Beówulf swam*, Beo. Th. 4710; B. 2360.

sundor (-er, -ur); *adv.* **I.** *apart, aloof, by one's self, separately* :—Ne scealt ðú sunder beón from ðínum geférum on Ongelcyrican *tua fraternitas seorsum fieri non debet a clericis suis in ecclesia Anglorum*, Bd. 1, 27; S. 489, 11. Geseah se cyning heora sacerdas sundor stondon (*seorsum consistere*), 2, 2; S. 503, 38. Hé gesæt him sundor æt rúne, Exon. Th. 293, 3; Wand. 111: Andr. Kmbl. 2324; An. 1163. Gebærne wulfes ceácan and ða téþ sundor *burn the teeth by themselves*, Lchdm. ii, 102, 13. Se Hǽlend genam his twelf þegnas sundor of ðæm weorode, Blickl. Homl. 15, 7. Sundor ácígan *to call aside*, Elen. Kmbl. 1203; El. 603. **II.** *severally, each by himself* :—Sundor ánra gehwilc herige ðec *let each one severally praise thee*, Cd. Th. 239, 15; Dan. 370. Féran sceal sundor ánra gehwæs sáwl of líce, Exon. Th. 191, 24; Az. 93. Swá monig beóþ men ofer eorþan, swá beóþ módgeþoncas; ǽlc him hafaþ sundor sefan (sundor-sefan?), 344, 5; Gn. Ex. 169. Heó wile gesécan sundor ǽghwylcne feorhberendra, 420, 18; Rä. 40, 5: Salm. Kmbl. 130; Sal. 64. **III.** *in a manner different from others* :—Ilco ðoht óðer suindir áuráit *eundem sensum alius aliter expressit*, Mt. Kmbl. p. 3, 5. **IV.** *in a way that separates, asunder* :—Sundur gedǽlan líf wið líce *to part asunder life from body*, Beo. Th. 4836; B. 2422. Seó cwén bebeád cræftum getýde sundor ásécean ða sélestan (*to pick out the best workmen*), Elen. Kmbl. 2023; El. 1019: 813; El. 407. [*Goth.* sundrô: *O. Sax.* sundar (-or): *O. H. Ger.* suntar: *Icel.* sundr.] v. on-sundrum.

sundor-anweald, es; *m. Single authority, monarchy* :—Sunderanweald *monarchia*, Engl. Stud. xi. 66, 54. [*O. H. Ger.* suntar-walt *monarchia*.]

sundor-cræft, es; *m. A special power or art*, one possessed or exercised by an individual or a class :—Ða rícan on ðam woruldwelan nabbaþ nǽnne sundorcræft, Bt. 27; 2; Fox 98, 7. Seó wiht sundorcræft hafaþ, Exon. Th. 420, 14; Rä. 40, 3. Ðæt hý sundorcræfta sumne eác cunne *that each have some craft of his own that he knows*, L. I. P. 9; Th. ii. 314, 29. Sió gesceádwísnes is se sélesta sundorcræfta *reason is the best of distinguishing faculties* (as being the faculty peculiar to man; cf. hió is synderlíc cræft ðære sáwle, Bt. 33, 4; Fox 132, 10), Met. 20, 203.

sundor-cýþþ[u]; *f. Special, private knowledge* or *acquaintance, in-*

timacy :—Riht is ðæt mynecena ǽnige sundorcýþþe tó woruldmannum nabban, L. I. P. 15; Th. ii. 322, 34.

sundor-feoh; *n. Private property, private estate* :—Mín sundorfeoh on ðam neoþeran Hysseburnan, Chart. Th. 488, 10.

sundor-freódóm, es; *m. A special immunity, a privilege* :—Mid andweardum apostolícum sunderfreódómum *cum praesentibus apostolicis privilegiis* (153, 10), Cod. Dip. B. i. 155, 17: 154, 22. v. next word.

sundor-freóls, es; *m. A special immunity, a privilege* :—On ðissum sunderfreólse *priuilegio*, Cod. Dip. Kmbl. iii. 349, 26. Ðysne nínne sunderfreóls *hoc nostrum priuilegium*, 350, 12, 16, 32. v. preceding word.

sundor-gecynd *a peculiar nature* :—Hé hafaþ sundorgecynd, Exon. Th. 357, 18; Pa. 30.

sundor-genga, an; *m. One who goes by himself* :—Sum fearhrýþer ðæs óþræs ceápes geférscipe oferhogode, and him gewunode ðæt hé wæs geond ðæt wésten sundorgenga, Blickl. Homl. 199, 5. Cf. án-genga.

sundor-geréfland, es; *n. Land reserved to the jurisdiction of a geréfa* (?) :—On ðæm sundorgeréflande *in tribulano* (in the same glossary *in tribulanam* is rendered *in þa burh*) *territorio*, Wrt. Voc. ii. 45, 4. Cf. :—Ǽylmer habbe þat lond at Stonham þe ic hym er to hande let to reflande. And ic an Godric mine reue at Waldingfeld þa þritti acre ðe ic hym er to hande let, Chart. Th. 570, 34. *See al-o* geréf-mǽd.

sundor-gifu, e; *f. A special gift* or *grace, prerogative, privilege* :—For ðære sundorgife ðe him God sealde gumena ríce, Cd. Th. 254, 4; Dan. 606. Wé swylc ne gefrugnan ǽfre gelimpan, ðæt ðú in sundurgiefe swylce befénge, Exon. Th. 6, 6; Cri. 80. God monnum syleþ sundorgiefe *God gives to each man a special gift*, 293, 22; Crä. 5. Sindergife *privilegium*, Hpt. Gl. 466, 76. Ǽlc cræft hæfþ his sundorgife and ða gife and ðone weorþscipe ðe hé hæfþ hé forgifþ ǽlcum ðara ðe hine lufaþ *inest dignitas proſria virtuti, quam in eos, quibus fuerit adjuncta, transfundit*, Bt. 27, 2; Fox 96, 30. Sundorgife *prerogativa*, Wrt. Voc. ii. 66, 37. Syndergyfa, Hpt. Gl. 468, 53. [*O. H. Ger.* suntar-gepa.]

sundor-hálga, an; *m. A Pharisee*, (but in one passage it seems to mean) *a scribe* :—Twegen men . . . án wæs sunderhálga, and óðer wæs openlíce synful, Homl. Th. ii. 428, 3: 420, 34: 422, 3. Bóceras and sunderhálgan, Scint. 203, 3. Ða Fariséiscan and sundorhálgan (*scribes*) hine tó deáðe fordémdon, H. R. 9, 28. Manega ðæra sunderhálgena (*Pharisaeorum*), Mt. Kmbl. 3, 7. Ðæra wrítera and sundorhálgena, 5, 20. Sunderhálgena, Homl. Th. ii. 216, 26. Ða wǽron of sundorhálgon, Jn. Skt. 1, 24. [Þa sunderhalȝe and þa bocere, O. E. Homl. i. 245, 3. Cf. *O. H. Ger.* sundir-lebin *phrisaei*.]

sundor-irfe, es; *n. A private inheritance* :—Eal ðæt se rinca baldor sinces áhte oðde sundoryrfes, Judth. Thw. 26, 22; Jud. 340. Wilsumne regn wolcen bringeþ and donne áscádeþ God sundoryrfe *pluviam voluntariam segregabis, Deus, haereditati tuae*, Ps. Th. 67, 10.

sundor-land, es; *n. Separate land, an estate belonging to particular persons* (?) :— Tó hira sundorlande *ad prediolum*, Wrt. Voc. ii. 3, 51. Sundorland *predia*, 66, 75. The word occurs in an enumeration of boundaries, and Kemble explains it there as 'land set apart for special purposes' :—Ǽfter ðære stráte be ðære wǽllan on Sunderlond, Cod. Dip. Kmbl. iii. 118, 20.

sundor-líc; *adj. Special, peculiar* :—Ðám is sundorlíc sang tó singanne *singulariter canticum cantare*, Past. 52, 7; Swt. 409, 10. [*O. H. Ger.* suntar-líh *singularis*.] v. synder-líc.

sundorlíce; *adv. Apart, separately* :—Sundurlíce *seorsum*, Mk. Skt. Lind. 7, 33. v. synderlíce.

sundor-líf, es; *n. A private life* :—Hé sundorlíf (*vitam privatam*) and munuclíf wæs foreberende eallum ðám weolum ðæs eorþlícan ríces, Bd. 4, 11; S. 579, 7.

sundor-lípe. v. synder-lípe, *and next word.*

sundor-lípes; *adv. Separately, severally, specially* :—Sunderlípes *sequestratim, diverse, alternatim*, Hpt. Gl. 411, 18. Sunderlípas *separatim, singulariter*, 438, 40. [Weren þas þreo laȝe gewriten inne þa odre tablebreode sunderlipes *written separately on the one table* (*of stone*), O. E. Homl. i. 11, 32. Þu hauest iseid of euch a setnesse sunderlepes *of each order separately*, 261, 33. He cumed to elch man sunderlupes, ii. 5, 15. Ich habbe sunderliche (sunderlepes, MS. C.) ispeken of þeos þreo limes, A. R. 90, 5. Cf. Ðe almisse þe mon ded sunderlipe (*specially*) for to quemen ure drihten, O. E. Homl. i. 137, 18. O. Frs. sunder-lépis *specially*.] v. synder-lípes.

sundor-mǽd; *f. A separate, private meadow* :—Seó mǽd ðe ðártó gebyreþ wið Hummingtún seó his sundermǽd, Cod. Dip. Kmbl. v. 354, 30.

sundor-mǽlum; *adv. Singly, separately;* singillatim, Anglia xiii. 380, 217.

sundor-notu, e; *f. A special office* :—Gif ceorl geþeáh ðæt hé hæfde sundornote on cynges healle, L. R. 2; Th. i. 190, 17. v. next word.

sundor-nytt, e; *f. A special office, employment*, or *use* :—Ǽlc hæfþ sundornytte (sunder-, Hatt. MS.) *per officium diversa sunt*, Past. 34, 3; Swt. 232, 4. Hæfde Hróðgár seleweard áseted; sundernytte beheóld ymb aldor Dena, eóten weard ábeád, Beo. Th. 1339; B. 667. v. preceding word.

sundor-riht, es; *n. A special right, right peculiar to a class :*—Rôm-wara sundorriht *jus Quiritum*, Wrt. Voc. ii. 49, 11. Weala sunderriht, i. 20, 64.

sundor-seld, es; *n. A special seat, a seat that stands apart, a throne :*—Ðæt hê sundurseld wuldres nimeþ *ut solium gloriae teneat*, Ps. Surt. ii. p. 186, 27.

sundor-setl, es; *n. A residence apart, a hermitage :*—Hê ongan wilnian wêstenes and sundersetle[s?]... Hê leornode be ðâm anceran ðe on wêstene and on sundorsettlum heora lîf leofodon, Guthl. 2; Gdwin. 18, 20-24. Hê his fultum tô ðam sundorsetle sôhte, 3; Gdwin. 24, 2.

sundor-spræc, e; *f.* I. where a single person speaks privately with one or more, *private speech, a private conversation :*—Nero cwæð: 'Sege mê, Petrus, on sundorspræce, hwæt ðû ðence,' Homl. Th. i. 376, 27. Swâ swâ him (*Moses*) God silf dihte on heora sundorspræce, Ælfc. T. Grn. 3, 14. Cornelius Asina gefôr tô Hannibale tô sundorspræce *ad colloquium*, Ors. 4, 6; Swt. 172, 7. Ðâ nam Eugenia hî on sundorspræce, Homl. Skt. i. 2, 48. Ðâ clypode Herodes ða drý tungelwîtegan on sundorspræce, Homl. Th. i. 78, 17. Ðætte hê hæbbe his sundorspræce mid ðæm bilwitum *cum simplicibus sermocinatio ejus*, Past. 35, 4; Swt. 243, 16. Hý (*Hannibal and Scipio*) hiera sundorspræce (*colloquium*) tô unsibbe brohton, Ors. 4, 10; Swt. 202, 12. II. where many speak in private, *a private conference, council :*—Hî cômon ealle tôsomne tô heora sundorspræce, Homl. Th. ii. 250, 9.

sundor-stôw, e; *f. A separate place, a place set apart for a particular object :*—Ælcum ðara ðû gesettest his âgene sunderstôwe, Bt. 33, 4; Fox 128, 30.

sundor-weorþung, e; *f. Special honour, prerogative, privilege :*—Heó (*St. Michael's church*) nalles on goldes wlite and on seolfres ne scîneþ, ac on sundorweorþunge þurh godcundra mægen heó gewuldrad stondeþ, Blickl. Homl. 197, 9. Sundorweorðunge *prerogativam*, Wrt. Voc. ii. 65, 75. v. synder-weorþmynt.

sundor-wine, es; *m. A special friend, an intimate friend :*—Ne âswîc sundorwine, ac â symle geheald rihtum gerisnum, Exon. Th. 301, 34; Fä. 29.

sundor-wîs; *adj. Specially, singularly wise :*—Ænne giddum gearusnottorne... ðone hié ðære cwêne âgêfon, sægdon hine sundorwîsne, Elen. Kmbl. 1172; El. 588.

sundor-wundor, es; *n. A special wonder, that which especially excites wonder :*—Mê frôd wita sægde sundorwundra fela, Exon. Th. 313, 19; Môd. 2.

sund-plega, an; *m. Play in the water :*—Se tîreádga (*the Phenix*) twelf sîþum hine bibaþaþ... siþþan hine sylfne æfter sundplegan hefeþ on heánne beám, Exon. Th. 205, 12; Ph. 111. Se hærnflota (*the ship*) æfter sundplegan (*its journey across the sea*) sondlond gespearn, 182, 10; Gû. 1308.

sund-ráp, es; *m. A sounding line :*—Sundgyrd in scipe oððe [sund-] ráp, i. metráp *bolidis*, Wrt. Voc. ii. 11, 17. v. sund-gird.

sund-reced, es; *n. A sea-house*, a term for the ark :—Ðû (*Noah*) seofone genim on ðæt sundreced tûdra gehwylces, Cd. Th. 80, 28; Gen. 1335.

sundrian; *p.* ode *To sunder, separate.* [Scheaden þe eilen urom þe clene cornes, þet is, sundren god from vuele, A. R. 270, 28. Marie and Marthe weren sustren, auh hore lif sundrede, 414, 12. Nan ne mei sundrin from oðere, Kath. 1776. To sundren and mengen, Gen. and Ex. 468. O. H. Ger. suntarôn : Icel. sundra.] v. â-, ge-, on-, tô-sundrian; syndrian.

sund-wudu, a; *m. A ship :*—Sum mæg fromlîce ofer sealtne sæ sundwudu drîfan, Exon. Th. 42, 24; Cri. 677 : Beo. Th. 421; B. 208 : 3817; B. 1906.

suner *a herd.* v. sunor.

sunna, an; *m. The sun :*—Sôna eode sunna up, Gen. 32, 31 : Ps. Th. 148, 3. Sunne (-a, MS. J.), Ælfc. Gl. Zup. 297, 7. Sunna and mône, Nar. 28, 20. Ðæs sunnan âsprungnis oðþe ðære mônan, 28, 10. [The word is usually feminine in the Teutonic dialects, but masculine forms are found in *Goth.* sunna : *O. Sax.* sunno : *O. H. Ger.* sunno.] v. sunne.

Sunnan-æfen, es; *m. The evening before Sunday :*—On Sunnan-æfen *dominica uespera*, Anglia xiii. 396, 447. Gif esne ofer dryhtnes hæse wyrce an Sunnan-æfen efter hire setlgange ôð Mônan-æfenes setlgang, L. Wih. 9; Th. i. 38, 19. Hî læddon hine tô hiora hûstinge on Sunnan-æfen, Chr. 1012; Erl. 146, 34. [Giester sunneue, Chart. Th. 437, 18.] [*O. H. Ger.* sunnûn âband *vesper sabbati : Ger.* Sonnabend.]

sunnan-corn *gromel ;* lithospermon officinale, Lchdm. i. 314, 18 ; see the remark in Lchdm. ii. 407, col. 1.

Sunnan-dæg, es; *m. Sunday :*—Iúdagum Romani and eác Angli gehâlgedon on ðisra tungla gemynde heora dagas, and ðæne forman dæg hig hêton Sunnandæg, forðan heó ys ealra tungla wlitegost, and se dæg wæs ealra daga fyrmest on heora dagum, and nû ys on ûrum tîman, Gode lof ealles, Anglia viii. 321, 4-7. On ânum ðara restedaga se nû Sunnandæg is nemned *una Sabbati quae nunc Dominica dies dicitur*, Bd. 3, 17; S. 545, 30. Dômes dæg... se hâlgesta Sunnandæg... ðý dæge blissiaþ ða ðe Sunnandæges freóls heóldan, Wulfst. 244, 14-19. Æghwelce Sæternes dæg and

Sunnan, Shrn. 88, 33 : Lchdm. iii. 228, 4. Crist ârâs of deaðe on ðone Eásterlîcan Sunnandæg, Homl. Th. i. 216, 33. Men ne môton baðian Sunnandagum, L. Ecg. C. 35; Th. ii. 160, 27. Gif wê ða six Sunnandagas of âdôþ, Wulfst. 284, 4. ¶ The observance of the Sunday was enjoined by the laws. The time that had to be so observed was according to Wihtræd's Laws from sunset on Saturday to sunset on Sunday :—Gif esne wyrce an Sunnanæfen efter hire setlgange ôð Mônanæfenes setlgang, 9; Th. i. 38, 19; but later the time seems to have been extended, and to be from .3. on Saturday until dawn on Monday :—Healde man ælces Sunnandæges freólsunga fram nôntîde ðæs Sæternes-dæges ôþ ðæs Mônandæges lîhtinge, L. Edg. i. 5; Th. i. 264, 18 : L. Ælfc. C. 36; Th. ii. 362, 1 : Wulfst. 231. 9. During this time servile and free were forbidden to work under various penalties, the latter being liable even to a loss of freedom, L. In. 3; Th. i. 104, 6; L. E. G. 7; Th. i. 170, 15; the servile to a fine or to corporal punishment, *ib.*; and see L. Wih. 9-11; Th. i. 38, 18 : L. C. S. 45; Th. i. 402, 13; in general terms it is said :—[Ealra] Woroldlîcra weorca on ðam hâlgan dæge geswîce man georne, L. Eth. vi. 22; Th. i. 320, 12 : L. C. E. 15; Th. i. 368, 18. The only exception is the preparation of food :—Nân weoruldweorc, bûton mon his mete gearwige, L. E. I. 24; Th. ii. 420, 22. In case of necessity, however, and under certain conditions, travelling was allowed :—Gif hwam gebyrige ðæt hê nýde faran scyle, ðonne môt hê swâ rîdan, swâ rôwan, swâ swilce færelde faran swylce tô his wege gebyrige, on ða gerâd ðæt hê his mæssan gehýre and his gebedu ne forlæte, *ib.* More specifically there are prohibitions of Sunday trading :—Sunnandæges cýpinge gif hwâ âgynne, þolie ðæs ceápes and twelf ôrena mid Denum and .xxx. scitt. mid Englum, L. E. G. 7; Th. i. 170, 15 : L. Ath. i. 24; Th. i. 212, 15 : L. Eth. v. 13; Th. i. 308, 11 : of assemblies, except in case of extreme need :—Wê forbeóðaþ ælc folcgemôt, bûton hit for mycelre neódþearfe sî, L. C. E. 15; Th. i. 368, 17 : L. Eth. v. 13; Th. i. 308, 10 : vi. 44; Th. i. 326, 21 : of hunting :—Huntaðfara geswîce man georne, L. Eth. vi. 22; Th. i. 320, 12 : L. C. E. 15; Th. i. 368, 18; and compare the answer of the hunter in Ælfric's Colloquy :—Ic næs tôdæg on huntoðe, forðam Sunnandæg ys, Coll. Monast. Th. 22, 1 : of legal proceedings, L. E. Ğ. 9; Th. i. 172, 10-15. Theft on Sunday incurred a double fine, L. Ælf. pol. 5; Th. i. 64, 22-25. As to the religious observances connected with the day it is said :—Hit gedafenaþ ðæt gehwylce cristene men, ða þurhteón mâgon, on Sæternesdæg cume tô cyrcean, and him leóht mid bringe, and ðær æfensang gehýran and on ûhtan ðone ûhtsang, and on morgene mid heora offrungum cuman tô ðære mæssan symbelnysse. And ðonne hig ðyder cumen, ne sý ðær nân fâcn, ne nænig geflytu, ne nænig ungeþwærnes gehýred, ac smylte môde, æt ðære hâlgan þenunge, ægðer ge for hig sylfe ge for eal Godes folc þingien, ægðer ge mid heora gebedum ge mid heora ælmessan; and æfter ðære hâlgan þenunge him gehwâ hâm hwyrfe, and mid his freóndum and his nýhstum and mid ældeódigum hine gâstlîce gereordige, and hine wið oferæt and druncennysse beorge, L. E. I. 24; Th. ii. 420, 32 sqq. [*O. L. Ger.* Sunnun-dag : *Du.* Zon-dag : *O. H. Ger.* Sunnûn-dag : *Ger.* Sonn-tag : *Icel.* Sunnu-dagr.] v. Eáster-, Palm-Sunnandæg.

Sunnan-niht, e; *f. The night between Saturday and Sunday :*—Ælcum gesinhîwum gebyreþ ðæt hig hyra clænnysse healdon æfre Sunnannihte (*nocte diei Dominici*), L. Ecg. P. ii. 21; Th. ii. 190, 18. His lîc læg on byrgene ða Sæterniht and Sunnanniht... and hê ârâs of deaðe on ðone Eásterlîcan Sunnandæg, Homl. Th. i. 216, 27-33. Se ðe stalaþ on Sunnanniht... oððe on ðone Hâlgan Ðunresdæg, L. Alf. pol. 5; Th. i. 64, 22. Hû on Sunnannihtum nihtlîc wæcce tô healdenne sý. On Sunnandæge mon sceal hraðor ârîsan tô ûhtsange, R. Ben. 35, 2 : 42, 15 : Wulfst. 305, 23.

Sunnan-ûhta, an; *m. The time before day-break on Sunday ;* as an ecclesiastical term *the hour of matins on Sunday, or the service then held :*—'On Sunnandæg ðû cymst tô mê'... Se apostol on ðam Sunnanûhtan ærwacol tô ðære cyrcan com, Homl. Th. i. 74, 20. Gê sculon singan Sunnanûhtan, L. Ælfc. P. 44; Th. ii. 384, 4.

sunn-beám, es; *m. A sun-beam :*—Ealle ða niht stôd swylce beorht sunnbeám *tota ea nocte columna lucis stabat*, Bd. 3, 11; S. 535, 24 : Homl. Skt. ii. 26, 184. Him gæþ of se leóma swylce oðer sunnbeám, Lchdm. iii. 272, 5. Hwæt fremaþ ðam blindan seó beorhta sunbeám? Homl. Skt. i. 4, 275. Se rênboga cymþ of ðam sunbeáme and of wætum wolcne, Boutr. Scrd. 21, 26. v. sunne-beám.

sunn-bearu (-o), wes; *m. A sunny grove :*—Sunbearo lîxeþ, wuduholt wynlîc, Exon. Th. 199, 30; Ph. 33.

sunn-beorht; *adj. Bright with the sunshine :*—Hê his cýþþu eft, sunbeorht gesetu sêceþ *contendit solis ad ortus*, Exon. Th. 217, 10; Ph. 436.

sunn-bryne, es; *m. Sun-burn :*—Wiþ sunbryne, Lchdm. ii. 324, 16 : 300, 30.

sunn-deáw (?). v. sund-deáw.

sunne, an (sunnu, Cd. Th. 286, 14 ; Sat. 352, *and acc.* sunne, 147, 11 ; Gen. 2437 : *O. Sax. O. L. Ger.* have acc. *sunna*) ; *f.* I. *the sun :*—On ðam feórðan dæge gesceóp God twâ miccle leóht, ðæt is sunne and môna, and betæhte ðæt mâre leóht, ðæt is seó sunne, tô ðam dæge,

Lchdm. iii. 234, 6-8. Seó sunne is micle ufor ðonne se móna sý, 242, 10. Seó sunne is swíðe mycel; eall swá brád heó is, ðæs ðe béc secgaþ, swá eall eorðan ymbhwyrft, 236, 6. Ða (*at the creation*) wæs seó sunne seofon síðum beorhtre ðonne heó nú is, Shrn. 64, 19. Seó sunne (sunna, Lind.) byþ forsworcen, Mt. Kmbl. 24, 29. On sumera sunne scíneþ, Cd. Th. 233, 16; Dan. 276. Dæge sunnan *die sabbati*, Lk. Skt. Lind. 4, 16. **I a.** epithets or metaphors applied to the sun:—Háte scíneþ, blíeþ ðeós beorhte sunne, Cd. Th. 50, 19; Gen. 811. Swegles gim, sunne, Exon. Th. 212, 13; Ph. 209. Goldtorht sunne, 351, 11; Sch. 78. Heofones gim, wyncondel wera, sweglbeorht sunne, 174, 33; Gú. 1187. Sunne swegeltorht, Andr. Kmbl. 2497; An. 1250. Æðele sunne, Ps. Th. 103, 21. Sunne, mære tungol, sió æþele gesceaft, Chr. 937; Erl. 13, 16. *See also* candel, tapor. **I b.** forms used of the sun's course:—Seó sunne gǽþ be Godes dihte betweox heofenan and eorðan, on dæg bufon eorðan and on niht under ðysse eorðan, eall swá feorr ádúne on nihtlícre tíde under ðære eorþan swá heó on dæg bufon up ástíhþ, Lchdm. iii. 234, 18-22. Ðonne sunne on setle sié, ii. 346, 10. Ǽr sunne tó setle eode, Ex. 17, 12. Sunne setlgonges fús, Exon. Th. 174, 33; Gú. 1187. Sóna swá seó sunne sealte streámas heá oferhlífaþ, 206, 1; Ph. 120. Sunne gewát tó sete glídan, Andr. Kmbl. 2609; An. 1306. Sunne up on morgentíd glád ofer grundas... sió æþele gesceaft sáh tó setle, Chr. 937; Erl. 112, 13-17. Wé hátaþ ǽnne dæg fram sunnan upgang óð ǽfen; ac swá þeáh is on bócum geteald tó ánum dæge fram ðære sunnan upgange óð ðæt heó eft becume ðær heó ǽr upstáh, Lchdm. iii. 236, 1-5. Æfter sunnan setlgange, Gen. 28, 11: Ex. 22, 26. Æfter sunnan setlráde, Cd. Th. 184, 19; Exod. 109. **II.** used in phrases expressing exposure to the sun's heat or light, e.g. *in* or *out of the sun* :—Gelícge upweard wið hátre sunnan *let him lie on his back with his face turned towards a hot sun*, Lchdm. iii. 2, 10. Dríge on hátre sunnan, ii. 30, 19. Ryslas eáfisca on sunnan gemylte, 30, 1. Hé sæt út on sunnan, Shrn. 61, 24. Ásete on háte sunnan, Lchdm. ii. 252, 9: Exon. Th. 407, 34; Rä. 27, 4. Þeáh hine (*the sick man*) mon on sunnan lǽde, 340, 17; Gn. Ex. 112. **II a.** in the phrase *under sunnan* = in this world, cf. *sublunary* :—Hié ǽfre geseón under sunnan, Andr. Kmbl. 2025; An. 1015. Ðæt hit wurde, ðæt on eorðan geond ðás wídan weoruld wǽren swelce under sunnan, Met. 8, 42. **III.** used metaphorically :—Seó sóþfæste sunne, Exon. Th. 237, 9; Ph. 587. Mín se swétesta sunnan scíma, Juliana, 252, 21; Jul. 166. [*Goth.* sunnó: *O. Sax. O. L. Ger. O. H. Ger. Icel.* sunna. In the Scandinavian languages the ordinary word is *sól*, sunna is poetical: Sól heitir með mönnum, en sunna með goðum.] v. sunna; swegel, II.

sunne-beám, es; *m. A sun-beam* :—Hér æteówede cometa se steorra, and scán iii móuðas swilce sunnebeám, Chr. 678; Erl. 41, 5. v. sunn-beám.

sunn-feld, es *or* a; *m. Elysium* :—Sunfeld *Eliseum*, Wrt. Voc. ii. 32, 8. Hwǽr wuniaþ Enoc und Helias? Ic ðe secge, Malifica and Intimphonis (in tempis?), ðæt is, on sunfelda and on sceánfelda, Salm. Kmbl. p. 202, 1. (v. scín-feld.) [*O. H. Ger.* sunna-velt *Elysium*.]

sunn-folgend a plant-name (rendering the Latin *solisequia*), *heliotrope*, Wrt. Voc. ii. 120, 71. v. sólsece.

sunn-gang, es; *m. The course of the sun.* v. next word.

sunn-ganges; *adv. In the direction of the sun's movement, with the sun* :—Wende ðé ðonne iii sunganges, Lchdm. i. 400, 10. Bebeóde hé hine Gode geornlíce and hine gesénige, cyrre hine sungonges ymb, ii. 116, 9. To move with the sun was considered lucky, to move in the reverse direction unlucky; the latter method is consequently taken by witches in their ceremonies. So Spenser, 'She turned her contrary to the sunne... for she the right did shunne.' Cf. *Icel.* sólar-sinnis *with the sun* :—Þeir höfðu gengit sólarsinnis um goðahús, Droplaugarsona Saga II, 4. At sólu *prosperously;* and-sælis *against the course of the sun;* mostly used of witches or uncanny appearances :—Sá sauðamaðr Gró at hon gékk út, ok gékk andsælis um hus sín ok mælti erfitt mun verða at standa í mót giptu Ingimundarsona, Vatnsdæla Saga 59, 4. Cf. also Scotch *withershins*, see the examples in Jamieson's Dictionary.

sunn-gihte, es; *n.* (?) *A solstice* :—On ðone ylcan dæg (*June 24*) byþ *solstitia*, ðæt ys on úre geþeóde, sungihte, forðon ðe seó sunne standeþ on mydre lyfte... Ðonne gelympeþ ðæt wundorlíce on ðæs sumeres sungihte on mydne dæg, ðonne seó sunne byþ on ðæs heofones mydle, ðonne nafaþ seó sýl (*at Jerusalem*) nǽnige sceade; ðonne ðæs sungihtes beóþ þrý dagas forð áurnen, ðonne hafaþ seó sýl ǽrest lytle sceade, Shrn. 95, 29-96, 3. Cf. gebed-giht; *and see* sunn-stede.

sunn-líc; *adj. Solar* :—Ðæt sunlíce leóhtfæt *lampas Titanea*, Wrt. Voc. ii. 53, 24. On swá hwilcum sunlícum mónðe swá se móna geendaþ, Lchdm. iii. 250, 3. [*O. H. Ger.* sunna-líh.]

sunn-sceadu, we or e; *f. A sun-shade, veil, covering to keep off the svn* :—Sunsceadu *flammeolum* (*flameolum* curchyfe, Wrt. Voc. i. 238, col. 2), Wrt. Voc. ii. 149, 6.

sunn-scín *sun-shine* (? the word glosses *speculum*, Wrt. Voc. ii. 90, 14). v. scín.

sunn-scíne; *adj. Beautiful* or *splendid as the sun* :—Seó sunscíéne fǽmne, Exon. Th. 256, 9; Jul. 229.

sunn-set, es; *n. The place where the sun sets, the west* :—From sunsete (sunnsett, Lind.) *ab occasu*, Lk. Skt. Rush. 12, 54. Sunset *occidentem*, Mt. Kmbl. Lind. 24, 27. [Cf. *Icel.* sólar-seta, -setr *sunset;* sól-setr; *n. pl. sunrise and sunset.*]

sunn-stede, es; *m. A solstice* :—Sumor hæfþ sunnstede... winter hæfþ óþerne sunnstede... Gǽþ seó sunne norðweard óð ðæt heó becymþ tó ðam tácne ðe is gehǽten *Cancer*, ðǽr is se sumerlíca sunnstede... seó sunne cymþ eft súð tó ðam winterlícan sunnstede, Lchdm. iii. 250, 10-24. Ða Gréciscan onginnaþ hyra geár æt ðam sunnstede, 246, 19. God sette twegen sunnstedas, ðæne ǽnne on .xii. kl. Ian. and ðone óðerne on .xii. kl. Iulii, Anglia viii. 299, 16. v. sunn-gihte.

sunn-treów (?). In Wrt. Voc. i. 291, 3 *origia* is glossed by *suntreów*. Cockayne suggests, *oryza* sum treów, Lchdm. iii. 346, col. 1.

sunnu. v. sunne.

sunn-wlitig; *adj. Beautiful with the sun* :—Winter biþ cealdost, lencten hrímigost, sumor sunwlitigost, Menol. Fox 473; Gn. C. 7.

sunor (-er), e; *f. A herd of swine, a sounder* ('That men calleth a trip of a tame swyn is called of wylde swyn a *soundre;* that is to say, ȝif ther be passyd v. or vi. togedres.'—Halliwell's Dict.) :—Wæs unfeor suner *swína* (suner berga, Lind. *grex porcorum*) etende. Ða deóful bédun hinae: 'send úsic in ðás sunrae (suner, Lind. *gregem*) swína.'... Eode all siu suner niþerweardes in sae, Mt. Kmbl. Rush. 8, 30-32. Sunor... ðæt sunor, Lk. Skt. Lind. 8, 32, 33. [The word seems to be found in the Lombard sonar-pair, sonor-þahir *verres qui omnes alios verres in grege batuit et vincit;* see Grmm. Gesch. D. S. 483; Graff. 3, 202: and in the Frankish sonesti = *duodecim equas cum admissario, aut sex scrovas cum verre, vel duodecim vaccas cum tauro*, Grmm. Gesch. D. S. 383.]

sun-sunu; *m. A grandson* :—Gif his sunu and ðæs sunsunu, L. Wg. 11; Th. i. 188, 23.

sunu; *gen.* a, u; *dat.* a, u; *n. pl.* a, u, o: there are also weak forms *sing.* suna; *n. pl.* sunan; *gen.* sunena; *m.* **I.** *a son* :—Mín se gecorena sunu (sune, Rush.), Mt. Kmbl. 3, 17. Sum man hæfde twegen suna (suno, Lind. Rush.)... ealle his þing gegaderude se gingra sunu (suno, Rush.), Lk. Skt. 15, 11, 13. Sunu Healfdenes, Beo. Th. 1294; B. 645. Féng tó Beornica ríce Æþelfriþes suna, Bd. 3, 1; S. 523, 13. Swíþhelm, Seaxbaldes suna, 3, 22; S. 553, 42: 3, 24; S. 556, 26. Hwæðer hit sig ðínes suna, Gen. 37, 32. Word hiere suna, Elen. Kmbl. 443; El. 222: Exon. Th. 6, 34; Cri. 94. Heó ne gehýrde ná hyre leófan sunu stemne (but suna, 10, 24), Wulfst. 152, 16. Gif his sunu and his sunu sunu geþeóþ, L. Wg. 11; Th. i. 188, 10. Cyning ðe macode hys suna (sune, Lind.: sunu, Rush. *filio*) gyfta, Mt. Kmbl. 22, 2. Án mann hæfde twegen suna (suna, Lind.: sunes, Rush.); ðá cwæð hé tó ðam yldran suna, 21. 28: Beo. Th. 4055; B. 2025. Ic fare tó mínum sunu, Gen. 37, 35: Exon. Th. 40, 8; Cri. 635. Wille ic ásecgan sunu Healfdenes, mǽrum þeódne, mín ǽrende, Beo. Th. 694; B. 344. Ic ðé forgife sunu, Gen. 17, 16. Heó sunu (suno, Rush.) cende, Lk. Skt. 1, 57. Sege ðæt ðás míne twegen suna (suno, Lind.: sunæ, Rush.) sittan..., Mt. Kmbl. 20, 21: Cd. Th. 93, 24; Gen. 1551. Suno, 97, 19; Gen. 1615. Sunu, 199, 1; Exod. 332: 199, 19; Exod. 341. Hé worn gestrýnde suna and dohtra, Cd. Th. 74, 13; Gen. 1221. Hwæt suna hæfde Adam? .xxx. sunena and .xxx. dohtra, Salm. Kmbl. p. 184, 31-32. Hwí sceal ic beón bedǽled ǽgðer mínra sunena on ánum dæge? Gen. 27, 45: Lev. 7, 32. Zebedéis suna (suna, MS. A., Lind.: sunena, Rush.) módor *mater filiorum Zebedaei*, Mt. Kmbl. 27, 56. Sunana, p. 18, 14. Beóð Aarone and his sunum, Lev. 6, 20. Mid sunum ðínum, Cd. Th. 78, 28; Gen. 1300. Heora bearn blótan feóndum, sceuccum onsæcgean suna and dohtor, Ps. Th. 105, 27. Hire selfre sunu sweoloðe befæstan, bánfatu bærnan, Beo. Th. 2234; B. 1115. ¶ In expressions denoting degrees of descent :—Suna sunu *nepos, neptis*, þridda sunu *pronepus, proneptis*, Wrt. Voc. ii. 62, 35. Feórþa sunu *abnepos*, 4, 73: 8, 22. Fífta sunu *adnepos*, 8, 23. Suna sune *vel* bróðer sune *nepos*, feówerþe sune *abnepos*, fífte sune *adnepos*, sixte sune *trinepos*, i. 51, 71-77. Fæderan sunan *patrueles*, móddrian sunan *matrueles*, fæderon sunan *fratres patrueles*, 52, 1-4. **II.** used of animals :—Ðære myran sunu *equae filius*, Bd. 3, 14; S. 540, 30. Ðæs gores sunu *the beetle*, Exon. Th. 426, 11; Rä. 41, 72. [*Goth.* sunus: *O. Sax.* sunu (-o), *pl.* suni: *O. L. Ger.* sunu (-o), sun: *O. Frs.* sunu, sun, son; *pl. acc.* suna, sunar, sonen: *O. H. Ger.* sunu, sun; *pl.* suni: *Icel.* sonr; *pl.* sønir, synir, *acc.* sonu.] v. bisceop-, gást-, god-, hornung-, steóp-, sun-sunu.

sunu-cennicge (?) *one who bears a son, a mother* :—Sunucenn *genetrix*, sunuccennices *genetricis*, sunucennic *genetricis*, Rtl. 66, 23, 17, 11.

súpan; *p.* seáp, *pl.* supon; *pp.* sopen. **I.** *to sup, to take [fluid] into the mouth* :—Gif hé ðæt broð sýpþ, Lchdm. ii. 336, 16. Hé sáp (seáp, MSS. O, V.) of ðam calice blód, Homl. Skt. i. 3, 162. Súp ðæt wós, Lchdm. i. 86, 17. Hrefnim fót wel on wíne, súp swá ðu hátost mǽge, ii. 50, 25: 56, 2: iii. 48, 2. Seóð on wíne, súpe hit swá wearm and healde on his múðe, ii. 94, 20. Wyl on gáte meolce and súpe, ii. 100, 24. Þeáh ðú mid cuclere ðæt súpe, ðæt hylpþ, 184, 25. Genim fifleáfan seáw... syle him súpan, i. 86, 25, 28: 82, 23. Dó on swýþe gód beór, syle hyt him ðonne wlacu súpan, 196, 19. Hé gelǽhte ǽnne

calic and sealde his gingrum of tó súpenne, Homl. Th. ii. 244, 14. Hē scóf on hálig wæter of đam hálgan treówe, sealde đam ádligan of tó súpenne, Homl. Skt. ii. 26, 264. **II.** used figuratively:—Đeáh ic hine súpe, ic hine wille eft út áspíwan of mínum múđe, Past. 58; Swt. 447, 1. Đa đe ne suppas deáđ *qui non gustabunt mortem*, Mt. Kmbl. Lind. 16, 28. Ne mē se seáđ súpe mid múđe *neque urgeat super me puteus os suum*, Ps. Th. 68, 15. [To frete ar ful tyme were and þanne to sitten and soupen, Piers P. 2, 96. Soupe the lene broth, P. S. 324, 239. Soop up *absorbuit*, Wick. Apoc. 12, 16; sopen, *pp*., Ps. 123, 4. Me þoȝte Kaym tok Abelles blod and sop it op, Anglia i. 314, 473. Sowpone or sowpe *sorbeo, absorbeo; sowpynge sorbicio*, Prompt. Parv. 466, col. 2. [*Du.* zuipen *to drink, quaff: O. H. Ger.* súfan *sorbere: Ger.* saufen: *Icel.* súpa.] v. be-, ge-súpan; sopa, *and next word.*

súpe, an; *f.* (?) *A sup, draught:*—Súpe nigon súpan, Lchdm. ii. 102, 16. v. sopa.

supe *in* ic supe *sarcio*, Wrt. Voc. i. 288, 50, *read* (?) súwe, v. seowian.

sur glosses *lurco*, Wrt. Voc. ii. 70, 41, *read* (?) siir. v. sígere, sýr.

súr; *adj. Sour:*—Súr meolc *oxygala, acidum lac*, Wrt. Voc. i. 28, 2. Áwyl on súrum ealaþ, Lchdm. ii. 34, 15: 134, 10. Genim súrne æppel, 132, 15. Dó on súre flétan, 130, 12. Forʒá súr and sealtes gehwæt, 56, 23. Genim súre cruman berenes hláfes, 134, 8. Wínberian súre geseón, iii. 212, 24. [*O. H. Ger.* súr: *Icel.* súrr.]

súre, an; *f. Sorrel; rumex acetosa* (v. E. D. S. Pub., Plant Names, for terms in which *sour* is used to denote this plant):—Súrae *salsa*, Txts. 98, 974. Súre, Wrt. Voc. i. 68, 54: *saliunca*, ii. 119, 64. Wiþ cancerádle, súre, sealt . . ., Lchdm. ii. 108, 9: 266, 16. Wensealf, cersan, súran, 128, 14. Genim monnes súran, 124, 19. *See also* geáces súre *under* geác. [*Icel.* súra: *Dan.* syre. Cf. *Ger.* sauer-ampfer.] v. wudu-súre.

súr-eágede, -égede; *adj. Blear-eyed:*—Súreágede *lippus*, Wrt. Voc. i. 45, 57. Súrēgede, 75, 43. Súreágede (-egede, MS. H.), Ælfc. Gr. 30; Zup. 192, 10. v. súr-íge.

Surfe, Surpe; *pl. A Slavonic race inhabiting northern Germany; Latin forms are Sorabi, Soravi, Sorbi:*—Be norþaneástan Maroara (*Moravia*) sindon Dalamentsan . . . and be norþan Dalamentsan sindon Surpe, Ors. 1, 1; Swt. 16, 20. Surfe, Swt. 16, 33.

súrian *to sour.* [*O. H. Ger.* súrēn.] v. á-súrian.

súr-íge, -ége; *adj. Blear-eyed:*—Gif mon súrēge sié, Lchdm. ii. 2, 9: 36, 21. Đa súrígan eágan *lippos oculos*, Wrt. Voc. ii. 52, 55: 92, 22. [*O. H. Ger.* súr-ouger *lippus*, Grff. i. 123: *Icel.* súr-eygr; súrnar í augum *the eyes smart* from smoke; súr (applied to the eyes) *bleared*.] v. súr-eágede.

súr-milisc, -melsc; *adj. Having a mixture of sour and sweet in taste:*—Apulder *malus*, súrmilsc apulder *malus matranus*, swēte apulder *malomellus*, Wrt. Voc. i. 32, 48. Đa mettas đe strangunge mæʒen hæbben swá swá æppla nales tó swēte ac súrmelsce, Lchdm. ii. 176, 18.

súrness, e; *f. Sourness:*—Súrnesse *acredinis*, Wrt. Voc. ii. 6, 1.

súsl, es; n.: e; f. Torment, (1) where the word is certainly neuter:—Se seáđ đæs sing[alan] súsles, Nar. 50, 23. Súsles þegnum, Exon. Th. 275, 30: Jul. 558: 304, 18; Fä. 72. Hié đæt súsl þrowiende wǣron, Ors. 1, 12; Swt. 54, 26. In đæt swearte súsl (*hell*), Exon. Th. 142, 4; Gú. 639. Đa ungeendodan súslo đú byst þrowigende, Nicod. 29; Thw. 17, 12. Helle súslu *inferni supplicia*, Scint. 27, 8. (2) where the word is feminine:—Đeós hellíce súsl *hic tartarus*, Ælfc. Gr. 13; Zup. 86, 4. Fram đam ēcan hungre helle súsle, Ælfc. Gen. Thw. 3, 26. Geférlæhte on ánre súsle, Homl. Th. i. 132, 20. Faraþ hig on ēce súsle, and đa rihtwísan on đæt ēce líf *ibunt hi in supplicium aeternum, justi autem in vitam aeternam*, Mt. Kmbl. 25, 46. Hú hē synfullum súsle gefremme, Wulfst. 138, 9: Dóm. L. 153. (3) where the gender is uncertain:—Đē is súsl weotod, Cd. Th. 308, 14; Sat. 692: 257, 8; Dan. 654. Satan on súsle (*dat.* or *acc.*) gefeól, 309, 20; Sat. 712. Súsle geinnod, 3, 28; Gen. 42. Swingan, súsle þreágan, Exon. Th. 251, 9; Jul. 142. Súsl þrowian, Cd. Th. 5, 22; Gen. 75: 255, 9; Dan. 621. Súsel, 267, 21; Sat. 41. Hafastú máre súsel, 268, 33; Sat. 64. In súsla grund, Elen. Kmbl. 1885; El. 944: Exon. Th. 98, 8; Cri. 1604. On hwilcum súslum hē móste ēcelíce cwylmian, Homl. Th. i. 86, 2. Súslum beþrungen, Elen. Kmbl. 1896; El. 950: Exon. Th. 10, 8; Cri. 149. [*Grein compares the word with Icelandic forms, sýsl, sýsla business, sýsl painstaking, sýsla to do business, sýsliga busily.*] v. cwic-súsl; seóslig.

súsl-bana, an; *m. A torturing destroyer, one who tortures while he destroys:*—Swarte súslbonan (*devils*), Cd. Th. 305, 1; Sat. 640.

súsl-cwalu, e; *f. A destruction or death accompanied by torment:*—Đa átleásan geseóþ heora wíte and heora súselcwale hym tóweard, Wulfst. 238, 22. Đú scealt habban súselcwale á on ēcnysse, 241, 13.

súslen. v. cwic-súslen.

**súsl-hof, es; n. A place of torment, hell:*—Of helle, of đam súslhofe, Hy. 10, 31.

suster *a sister*, sustras, L. R. S. 5; Th. i. 436, 2, sutel. v. sweostor, sester, sweotol.

sútere, es; m. A shoemaker, souter (Scotch):—Sútere *sutor*, Wrt. Voc.

---

i. 74, 11. Sum sútere siwode đæs hálgan weres sceós . . . Anianus wæs geháten se ylca sútere, Homl. Skt. i. 15, 23, 27. Eówer sútere hē is *uester sutor est*, eówer súteres tól *uestri sutoris instrumenta*, Ælfc. Gr. 15; Zup. 105, 14. Gif hē smeáwyrhtan hæfþ, đam hē sceal tó tólan fylstan; sútere and óđran wyrhtan ælc weorc sylf wísaþ hwæt him tó gebyreþ, Anglia ix. 263, 18. Sútera hús *sutrina domus*, Wrt. Voc. i. 59, 3. [A sutare þet haueđ forloren his el, he secheđ hit anonriht, A. R. 324, 17. Euerych soutere þ^t wonyeþ in þe citee [of Wynchestre] þ^t halt shoppe, E. G. 358, 22. Euerych sowtere þ^t makeþ shon of newe roþes leþer, 359, 14 (14th cent.). More borynde þanne zouteres eles, Ayenb. 66, 12. Sowtare or cordewaner *sutor*, Prompt. Parv. 466, col. 2. *O. H. Ger.* sútári: *M. H. Ger.* sútære; schuoch-sútære (*from which Ger.* schuster): *Icel.* sútari *a tanner*. *From Latin* sutor.] Cf. scóh-wyrhta.

**[súþ]; cpve. súþra; spve. súþmest; adj. South, southern:*—Andlang đæs súđeran weges, Cod. Dip. Kmbl. iii. 408, 32. On đone sýđeran steđ . . . on đone norđere steđ, v. 148, 20. Đone súđran sunnstede, Lchdm. iii. 252, 15. Đone súđran steorran, 270, 18. On đæm súđmestan onwalde, Ors. 6, 1; Swt. 252, 15. Đa súđmestan Æthiopian hæfdon bryne for đære hǣte, 1, 7; Swt. 40, 5. ¶ Súþan *in combination with prepositions:*—Be-súđan sǣ *south of the sea*, Shrn. 145, 17. Him be-súđan, Cd. Th. 182, 1; Exod. 69. Nǽđer ne be-norđan mearce ne be-súđan, L. Ath. v. 5; Th. i. 232, 19. Be-súþan đæm múþan, Ors. 1, 1; Swt. 10, 8. Wiđ-súđan đone Sciringes-heal, Swt. 19, 18. Be đam wig-bede súþan *juxta altare ad austrum*, Bd. 5, 19; S. 641, 19. [*O. H. Ger.* sund- *and* Engl. súđ. sunn- *point to the* n *that has been lost from the English word*.] See the compounds which follow, and Cod. Dip. Kmbl. vi. 337, 338, for names of places in which *súþ* forms the first part.

súþ; *adv. In a southerly direction or position:*—Twelf míla brád súđ and norđ *ab austro in boream duodecim milia passuum*, Bd. 1, 3; S. 475, 19. Him is đæt heáfod súđ gewend and đa fēt norđ, Shrn. 66, 23. Syndon óđere eálond súđ fram Brixonte, Nar. 36, 7. Seó eá súþ đonan ligeþ, Ors. 1, 1; Swt. 8, 21: Salm. Kmbl. 382; Sal. 190. Fóron đá súþ ofer Temese, Chr. 851; Erl. 68, 1. Súđ ofer sǣ fóron, 897; Erl. 94, 28. Fóron súđ ymbútan, 894; Erl. 91, 5. Seó sunne cymþ eft súđ tó đam winterlícan sunnstede, Lchdm. iii. 250, 24: 260, 10: Cd. Th. 118, 16; Gen. 1966. Súđ ne norđ ofer eormengrund óþer nænig sēlra nære, Beo. Th. 1720; B. 858: Met. 10, 24. Súđ eást and west, 9, 42: 10, 5. Súđ west and eást, 14, 7. Swá heó (*the sun*) súđor biþ, swá hit swíþor winterlǣcþ, Lchdm. iii. 252, 2.

súþan; *adv.* **I.** *from the south:*—On đysum geare com micel sciphere hider súþan of Lidwiccum, Chr. 918; Erl. 102, 22. Gefaren tósomne súđan and norđan, Cd. Th. 120, 2; Gen. 1988. Gif hēr wind cymþ westan ođđe eástan, súđan ođđe norđan, 50, 11; Gen. 807. Súþan, Exon. Th. 55, 18; Cri. 885: 220, 23; Ph. 324. **II.** marking position, *to* or *in the south:*—Asia is befangen mid đǣm gársecge súþan and norþan and eástan, Ors. 1, 1; Swt. 8, 7. Ne dohte nǽđer đisse leóde ne súđan ne norđan, Chr. 1013; Erl. 149, 27. Healdaþ hine norđan and súđan on twá healfa twá hund wearda, Salm. Kmbl. 520; Sal. 259. [*O. L. Ger.* súthon *ab austro: O. H. Ger.* sundan: *Icel.* sunnan.]

súþan-eástan. **I.** *adv. From the south-east:*—Súþaneástan sunnan leóma cymeþ, Exon. Th. 56, 15; Cri. 901. **II.** in phrases marking position, *to the south-east:*—Be-súþaneástan (*ad Eurum*) đæm porte, Ors. 1, 1; Swt. 10, 9. On-suđaneástan đissum lande, Chr. 449; Erl. 13, 5. [*O. H. Ger.* sundan-óstan.]

súþaneástan-wind, es; m. A south-east wind; *euroafricus*, Wrt. Voc. i. 36, 14.

súþan-eásterne; *adj. South-eastern:*—Hē ferade súþaneásterne wind of heofenan *transtulit austrum de coelo*, Ps. Lamb. 77, 26. v. súþ-eásterne.

Súþan-hymbre, -humbre; *pl. The Southumbrians, the Mercians:*—Hēr Súþanhymbre (-humbre, Laud. MS.) ofslógon Æþelrēdes cwēne (cf. Æþelrēd Myrcna cyning, Bd. 4, 21; S. 590, 14), Chr. 697; Th. 67, cols. 1, 3. Hēr Cēnrēd fēng tó Súþanhymbre ríce (cf. Cēnrēd Myrcna ríce fore wæs. Bd. 5, 19; S. 636, 24), 702; Th. 67, col. 1. Úre cynecynn and Súđanhymbra eác, 449; Erl. 13, 21. v. Súþ-hymbre.

súþan-westan; *adv. From the south-west:*—Súđanwestan *ab affrico*, Wrt. Voc. ii. 98, 35: 4, 15: *a fafonio*, 99, 50.

súþanwestan-wind, es; m. A south-west wind; *africus*, Wrt. Voc. i. 36, 15.

**súþan-wind, es; m. A south wind:*—Súþanwind *auster vel nothus*, Wrt. Voc. i. 36, 9. Se đe hit mid súđanwinde onginne, đonne hæfþ hē sige, Lchdm. iii. 182, 3. Súþanwind (southenwind, Ps.) *austrum*, Ps. Surt. 77, 26. [A suđenwind blew đat day, Gen. and Ex. 3084. *Icel.* sunnan-vindr.]

**súþ-dǽl, es; m. A south part, the south:*—Súþdǽl *auster*, Ælfc. Gr. 8; Zup. 27, 7. Súđdǽles cwēn *regina austri*, Mt. Kmbl. 12, 42. Hig cōmon tó súđdǽle *ad australem plagam*, Gen. 13, 1. Tó súđdǽle *ad meridiem*, 14. Of súđdǽle Asiam, Ors. 1, 10; Swt. 44, 5. Hí on đam súþdǽle inn eodon, Homl. Th. i. 508, 9. Fram súþdǽle *a meridie*, Bd. 1, 1; S. 473, 12. Đære Asian súþdǽl *meridianam partem Asiae*, Ors.

1, 1; Swt. 14, 5. Ða súþdǽlas middangeardes, Bd. 1, 1; S. 473, 33. Ða súþdǽlas ðysses eálondes *australes partes Britanniae*, S. 474, 8. [Suþdale off þiss werelld is Mysinmbrion ȝehatenn, Orm. 16418. Cf. *O. H. Ger.* sunder-teil *dextera pars (templi)*.]

súþ-duru, a; *f. A south door :*—Wæs seó súðduru hwæðwega healfe máre, Blickl. Homl. 201, 15. [Cf. *Icel.* súðr-dyrr ; *pl. south doors.*]

súþ-eást ; *adv. South-east :*—Donua múða ðære eá scýt súðeást út, Ors. 1, 1 ; Swt. 22, 5 : Cd. Th. 42, 1 ; Gen. 667.

súþ-eástende, es ; *m. The south-east end :*—Ðæt (*India*) is se súþeástende ðisses middangeardes, Bt. 29, 3 ; Fox 106, 22.

súþ-eásterne ; *adj. South-eastern :*—Súðeásterne wind *eurus*, Ælfc. Gr. 4 ; Zup. 8, 2.

Súþ-Engle ; *pl. The people of the south of England :*—On Súð-Engla lage griðlagu ðus stent, L. Eth. vii. 9 ; Th. i. 330, 22.

superige. A plant name glossing *satirion*, Wrt. Voc. i. 32, 18. Cockayne takes the word to be the same as sæperige (q. v.), and the gloss to be a mistake, Lchdm. ii. 403, col. 1 ; but cf. *satirion* sanycle, Wülck. Gl. 613, 33, *saniculum* sanicle i. wudemerch, 554, 8.

súþerne ; *adj. Southern, coming from the south :*—Se óðer heáfodwind is súðerne, *auster* gehaten, Lchdm. iii. 274, 16 : Met. 5, 7. Se súðerna wind, Lchdm. iii. 276, 7 : Bt. 6 ; Fox 14, 23. Cwoen súðerne (súðernæs t súðdǽles, Lind.) *regina austri*, Lk. Skt. Rush. 11, 31 : Exon. Th. 480, 10 : Rä. 63, 9. Fram deófle súðernum *ab demonio meridiano*, Ps. Spl. 90, 6. Súþerne wind *austrum*, 77, 30 : *austrum, affricum*, Blickl. Gl. Sende se sǽrinc súþerne gār, Byrht. Th. 135, 47 ; By. 134. Hire (*the queen of Sheba*) olfendas bǽron súðerne wyrta, Homl. Th. ii. 584, 10. ¶ The word is often used in reference to things coming to England from the south of Europe, plants or medicine :—Genim súþerne cymen, Lchdm. ii. 184, 15. Ða súþernan finuglan, 142, 2. Súþerne popig, 212, 8. Súþerne rind *cinamomum*, iii. 301, col. 2 : *cinnamomum*, Wrt. Voc. ii. 131, 9. Dô ðone súþernan wermôd, ðæt is *prutene*, Lchdm. ii. 236, 19. Súðerne wudu *aprotanum*, Wrt. Voc. i. 79, 6. Súþerne wuda. Ðeós wyrt ðe man *abrotanum* and óðrum naman súðerne wuda nemneþ, Lchdm. i. 250, ˙16-18: iii. 12, 15 : 40, 5. Næglæs (cunæglæsse) hätte wyrt súðerne, ii. 106, 9. Oþer swilc *ameos* hätte súþerne wyrt, 192, 7. *Oxumellis* . . . drenc súþerne, 212, 6 : 254, 16. On ðam súðrenan *oxumelle*, 152, 1. Ðæt is súþerne lǽcedóm, 224, 14. On ðam súþernan lǽcedóme ðe hätte *oxumelle*, 248, 10. [*O. Frs.* suthern : *O. H. Ger.* sundirin *australis* : *Icel.* súðrænn.]

súþe-weard ; *adj. Southward, south :*—Tôemnes ðæm lande súðeweardum, Ors. 1, 1 ; Swt. 19, 1. From súþeweardum of norþeweardne, Bt. 16, 4 ; Fox 58, 12 : 18, 1 ; Fox 62, 1. On splott súðeweardne, Cod. Dip. B. iii. 336, 23. Ða gesǽtan súðewearde Bryttene, Chr. Erl. 3, 5. v. súþ-weard.

súþ-folc, es ; *n. A southern people, a people living south in relation to some other :*—Rômáne and eall súþfolc (ealle súþfolc, 146, 15), Lchdm. ii. 16, 1. Humbre stream tôsceádeþ súþfolc Angelþeóde and norþfolc, Bd. 1, 25 ; S. 486, 17. Eorldóm on Norðfolc and Súðfolc (*Suffolk*), Chr. 1075 ; Erl. 213, 5. Norðmen wǽron súðfolcum swice, Cd. Th. 120, 17 ; Gen. 1996. [Cf. *O. Sax.* súðar-liudi.]

súþ-gársecg, es ; *m. A southern ocean ; meridianus oceanus*, Ors. 1, 1 ; Swt. 8, 30.

súþ-gemǽre, es ; *n. A southern boundary :*—Hiera súþgemǽro licgeaþ tô ðam Reádan Sǽ, Ors. 1, 1 ; Swt. 10, 34.

Súþ-geweorc, es ; *n. Southwark :*—Ða cômon hý tô Súþgeweorce, Chr. 1052 ; Erl. 181, 3. [*Icel.* Súðr-virki.]

Súþ-Gyrwas (-e, -an) ; *pl. The southern division of the Gyrwas :*—Súþ-Gyrwa syx hund hýda, Cod. Dip. B. i. 414, 18. Súþ-Gyrwa ealdormon *princeps Australium Gyruiorum*, Bd. 4, 19 ; S. 587, 21. Súð-Gerwa, Shrn. 94, 20. Súð-Gyrwena, Lchdm. iii. 430, 14.

Súþ-hámtún *Southampton :*—Æt Súðhámtúne, Cod. Dip. Kmbl. vi. 49, 20. v. Hám-tún.

Súþhámtún-scír *Hampshire*, Cod. Dip. Kmbl. iv. 204, 16. [þe nywe forest þat ys in Souþhamtescyre, R. Glouc. 375, 9.]

súþ-heald ; *adj. Sloping* or *tending to the south :*—Rodor súðheald swifeþ swift, Met. 28, 17. Swá súðhealde swîþe hlimman *sicut torrens in austro*, Ps. Th. 125, 4. [*Icel.* súðr-hallr (*applied to the sun*).]

súþ-healf, e ; *f. The south side,* mostly, if not exclusively, in the phrase *on (ða) súþhealfe :*—On súðhealfe *ad meridianam plagam*, Num. 3, 29 : *contra meridiem*, Deut. 1, 7. On súþhealfe *a meridie*, Ors. 1, 1 ; Swt. 10, 26 : 14, 2. On súðhealfe ðære eás, Chr. 921 ; Erl. 108, 18 : 913 ; Erl. 102, 10. On súðhalfe Humbre streámes *ad meridianam Humbrae fluminis ripam*, Bd. 2, 16 ; S. 519, 19. Hî wendon ábútan Penwihtsteort on ða súðhealfe, Chr. 997 ; Erl. 130, 20. Hî wendon tô Lundene and dulfon áne mycele dîc on ða súðhealfe (on súðhealfe, MS. D.), 1016 ; Erl. 155, 9. On ða súðhealfe fram Babilonia *in dextera parte ab Babilonia*, Nar. 34, 17. On ða súðhealfe (*dexteriore parte*) landes Egiptua, 34. On ða súðhealfe gársecges *oceano dexteriore parte*, 36, 15. (Cf. *O. H. Ger.* sunder-teil *under* súþ-dǽl.) [þe an is a norðhalf, þe oðer a suðhalf, Laym. 15937. *O. H. Ger.* sund-, sundar-halpa *auster, meridies : Icel.* súðr-hálfa *the southern region.*]

Súþ-hymbre ; *pl. The Mercians :*—Súðhymbra (-humbra, Laud. MS.) rîce, Chr. 702 ; Th. 67, col. 3. Hér wæs Ósuuald ofslagen fram Pendan (and) Súþhymbrum (cf. fram ðam ylcan hǽþenan cyninge and ðære hǽþenan ðeóde Myrcna, Bd. 3, 9 ; S. 533, 11), 641 ; Erl. 27, 8. v. Súþan-hymbre.

súþ-land, es ; *n. A land lying to the south :*—Hé eardode on ðam súðlandum *in terra australi*, Gen. 24, 62. [He hæfde to dæle þat súðlond þat Locres wes icleped, Laym. 2111. *Icel.* súðr-land (*hence* Suther-land).]

súþ-mǽgþ, e ; *f. A southern tribe* or *province :*—Óðrum folcum ðara súþmǽgþa *caeteris australium provinciarum populis*, Bd. 3, 24 ; S. 557, 31. Hé eallum súþmǽgþum weóld and rîce hæfde óþ Humbre stream, 2, 5 ; S. 506, 10.

súþ-mann, es ; *m. A man living in the south :*—Súðmonna sinc (*those who carry off the treasure are said sécan súð*, 118, 16 ; Gen. 1995), Cd. Th. 121, 28 ; Gen. 2017 : 126, 4 ; Gen. 2096. [*Icel.* súðr-maðr.]

súþmest. v. súþ ; *adj.*

Súþ-Mirce ; *pl. The South Mercians :*—Súþ-Myrcna rîce, Bd. 3, 24 ; S. 557, 36.

Súþ-Peohtas, -Pihtas ; *pl. The South Picts :*—Súð-Pihtas (-Pyhtas, MS. E.), Chr. 565 ; Erl. 18, 4.

súþ-portic, es ; *m. A south porch :*—On ðam súðportice, Chr. 1036 ; Erl. 165, 39.

súþ-rador, -rodor, es ; *m. The south of the heavens :*—Súþrador *australis*, Blickl. Gl. Oþþæt seó sunne on súðrodor sǽged weorþeþ *postquam Phoebus equos in aperta refudit Olympi*, Exon. Th. 207, 14 ; Ph. 141.

Súþr-íg *the people* or *the district of Surrey :*—Cantwara him tô cyrdon and Súðríg and Súð-Seaxe, Chr. 823 ; Erl. 63, 20. Hî heafdon ofergán ealle Centingas and Súð-Seaxe and Súðríg and Bearrucscîre, 1011 ; Erl. 144, 28. v. next word.

Súþr-íge ; *gen.* [e]a, ena ; *pl. The people* or *district of Surrey :*—Cantware him tô cirdon and Súþríge and Súþ-Seaxe, Chr. 823 ; Erl. 62, 22. Cantwara rîce and Súþrígea and Súþ-Seaxna, 836 ; Erl. 66, 3. Súþrígea, 855 ; Erl. 70, 19. On Súþrígena lande be Temese streáme *in regione Sudergeona juxta fluvium Tamensem*, Bd. 4, 6 ; S. 574, 14. Ealhere mid Cantwarum and Huda mid Súþrígium (Súþrígum, MS. E.) gefuhton wiþ herige, Chr. 853 ; Erl. 68, 17. Of Cent ge of Súþrígum, 921 ; Erl. 107, 7. Fēngon tô West-Seaxna rîce and tô Súðrígean, 855 ; Erl. 71, 2. Tô Súðrígan, 836 ; Erl. 67, 3. Tô Godes ciricum in Súðrégum and in Cent, Cod. Dip. Kmbl. ii. 121, 8. Hé gewát on Súþríge (Súðrége, MS. E.) and on Súþ-Seaxe, Chr. 722 ; Erl. 44, 28. Fôron súþ ofer Temese on Súþríge (Súðríge, MS. E.), 851 ; Erl. 68, 2. *The word occurs in a Latin charter* . . . *In loco que appellatur Cyningestún in regione Súðrégie*, Cod. Dip. Kmbl. i. 318, 5. [Souþsex and Súþerei, Kent and Estsex, R. Glouc. 3, 21. Soþerey, 5, 23.]

súþ-rihte ; *adv. Due south :*—Seó eá irnþ ðonan súðryhte, Ors. 1, 1 ; Swt. 8, 17 : 17, 18, 19. Súðrihte, Cod. Dip. Kmbl. ii. 250, 17.

súþ-rima, an ; *m. A south coast :*—Ðý ilcan sumera forweard nô læs ðonne xx scipa mid monnum mid ealle be ðam súðriman, Chr. 897 ; Erl. 96, 15 : 1009 ; Erl. 141, 32. v. súþ-stæþ.

súþ-rodor, -sceáta. v. súþ-rador, sceáta, I.

Súþ-Seaxe, -Seaxan ; *pl. The people* or *district of Sussex :*—Him tô cirdon Súþ-Seaxe, Chr. 823 ; Erl. 62, 22. Of Eald-Seaxon cômon Súð-Sexa, 449 ; Erl. 12, 10. Súþ-Seaxan *meridiani Saxones*, Bd. 1, 15 ; S. 483, 24. Súþ-Seaxan ágen[n]e biscopas onfêngon, 5, 18 ; S. 635, 14. Súþ-Sexena landes is syufan þúsend hýda, Cod. Dip. B. i. 415, 1. Ælle Súþ-Seaxna cyning, Chr. 827 ; Erl. 62, 35. Súþ-Seaxna (Súð-, MS. E.) rîce, 836 ; Erl. 66, 3. Súð-Seaxna (Súð-Seaxena, MS. E.) cyning, 661 ; Erl. 34, 15. Hé gewát on Súþ-Seaxe and Ine gefeaht wiþ Súþ-Seaxum, 722 ; Erl. 44, 29. Eádulf cynges þegn on Súð-Seaxum, 897 ; Erl. 95, 3. Se here on Súð-Seaxum and on Bearrucscîre hergodon, 1009 ; Erl. 142, 22 : 998 ; Erl. 135, 21. Hér Ceólwulf gefeaht wið Súð-Seaxe, 607 ; Erl. 20, 27. Hî heafdon ofergán Súð-Seaxe and Súþríg and Bearrucscîre, 1011 ; Erl. 144, 27. [Folc læi inne Súð-sæxe, Laym. 15368. Souþsex (*a shire*), R. Glouc. 3, 21.]

Súþ-Seaxisc ; *adj. South-Saxon, of Sussex :*—Wulnôð cild ðone Súð-Seixiscan (-Seaxscian, col. 1 : -Seaxcisan, 260, col. 2), Chr. 1009 ; Th. 261, col. 2.

súþ-stæþ, es ; *n. A south shore, coast,* or *bank :*—West-Seaxna lond be ðæm súþstæðe, Chr. 897 ; Erl. 95, 9. v. súþ-rima.

súþ-wág, es ; *m. A south wall :*—Wið middan ðæs súðwáges, Homl. Th. i. 508, 15. Wið ðone súðwág tômiddes ðæs wáges, Blickl. Homl. 207, 15. [*Icel.* súðr-veggr.]

súþ-weard ; *adv. Southward, in a southerly direction, towards the south :*—Wilþ seó eá súþweard Eufrates *fluvius Euphrates tendens in meridiem*, Ors. 1, 1 ; Swt. 14, 10. Heó (*the sun*) cyrþ eft súðweard, Lchdm. iii. 250, 22 : 258, 13 : 252, 1.

súþ-weardes ; *adv. Southwards, in the south*, Met. 1, 4.

súþ-weg, es ; *m. A road lying to the south ;* in *pl. southern countries,*

the south :—Hié gesáwon of súðwegum fyrd Faraonis, Cd. Th. 187, 23 ; Exod. 155. [*Icel.* súðr-vegr ; in pl. *southern countries.*]

súþ-west ; *adv. South-west :*—Án ðæra gárena líþ súðwest (*in africum*), Ors. 1, 1 ; Swt. 24, 3.

súþ-westerne ; *adj. South-western :*—Se súðwesterna wind him on-geán stód, Apol. Th. 11, 3.

súþ-wind, es ; *m. A south wind,* Cd. Th. 196, 10 ; Exod. 289.

suto, -sutod, sutol, sutung, suwian. v. slítan, sulian, sweotol, slítung, swigian.

swá, swæ, swé (swé is the form in Ps. Surt. ; see also Txts. 600, col. 1. The form also occurs in Blickl. Homl. 23, 7). **I.** *rel. pron. As, that :*—Forgylde ðæt ángylde and ðæt wíte swá tó ðam ángylde belimpan wille, L. Alf. pol. 6 ; Th. i. 66, 3. Ðon gelíc swá lǽcas cunnon *such as doctors know,* Lchdm. ii. 192, 23. Brúcan swylcra yrmþa swá ðú unc ǽr scrife, Exon. Th. 373, 2; Seel. 102 : Homl. Th. ii. 162, 18. Yrfan hí swá hí wyrðe witan *let such inherit as they know to be entitled,* Chart. Th. 578, 9. Ne wíte hé ús swá neóde and hǽse gehýrsumodon, Guthl. prol. ; Gdwin. 4, 5. Ealne ðisne ymbhwyrft ðises middangeardes swá swá Oceanus útan ymbligeþ, Ors. 1, 1 ; Swt. 8, 2. **I a.** in combination with the *hw-* pronominal forms, *so,* as in whosoever, etc. :—Tó syllenne swá hwæt swá (suǽ huæt *quodcumque,* Lind.) heó hyne bǽde, Mt. Kmbl. 14, 7. Swá hwylc swá (suá huá, Lind. : swá hwá swá, Rush.) sylþ ánne drinc, 10, 41. Fram swá hwylcere untrymnesse swá hé on wæs, Jn. Skt. 5, 4. Swá hwylc man swá mildheortnesse nafaþ, Blickl. Homl. 13, 22. Swá hweðer swá hé wylle, L. Eth. i. 1 ; Th. i. 280, 16. Ðæt git ne lǽstan welhwilc ǽrende swá hé sendeþ, Cd. Th. 35, 15 ; Gen. 555. Folcrihta gehwylc swá his fæder áhte, Beo. Th. 5210 ; B. 2608 : Elen. Kmbl. 1287 ; El. 645. Swá hú swá hé mǽge *howsoever he can,* L. P. M. 2 ; Th. ii. 286, 25. Swá hwæder (hwyder, MS. A.) swá (suá huider, Lind. : hwider swá, Rush.) ðú færst *quocunque ieris,* Mt. Kmbl. 8, 19 : Lk. 9, 57: Blickl. Homl. 233, 33. *See other instances under the pronominal forms.* **II.** *demonst. pron. :*—Æt men fíftene peningas, and æt horse healf swá, L. Ff. ; Th. i. 224, 26. **III.** representing an adjective, generally one used with a verb of incomplete predication, *so, the same, such*—Hé gemétte ænne blindne mann, se wæs geboren swá, Homl. Skt. ii. 29, 52 : Cd. Th. 44, 33 ; Gen. 7, 8. Bebycggen ðone oxan and hæbben him ðæt weorð gemǽne, and eác ðæt flǽsc swá (i. e. *in common*), L. Alf. 23 ; Th. i. 50, 11. Ðæt hé wǽre heora munuc æt fruman and hí woldon hine habban swá deáðne *that he had been their monk at first, and they would have him so (their monk) when dead,* Homl. Th. ii. 518, 23. Cild ðiónde on eallum cræftum on cnihtháde and swá forþ eallne giógoþhád (*going on thriving all its youth*), Bt. 38, 5 ; Fox 206, 24. Gé wiþerwearde wǽron úrum gewunan and ealre Godes cyrícean swá (i. e. wiþerwearde), Bd. 2, 2 ; S. 503, 19. **III a.** swá swá *such as :*—Onlegena strengran swá swá is árom *stronger applications such as is copperas,* Lchdm. ii. 192, 22. **IV.** *adv.* (1) defined by that which precedes (a) of manner or condition (a) *so, in this or that way, thus :*—'Beón gegaderode ða wæteru' . . . Hit wæs ðá swá gedón, Gen. 1, 9, 15. Nis hit ná swá *it is not so* (*as you have said*), 18, 15. Hit ne mæg ná swá beón, Ex. 10, 11. Ðeáh hí his nǽfre ne geléfan, ðeáh it is swá, Bt. 36, 6 ; Fox 182, 17. Hé árás áblendum eágum, and his geféran hine swá (*in the manner mentioned*) blindne tó ðære byrig gelǽddon, Homl. Th. i. 386, 14 : 432, 11 : L. In. 21 ; Th. i. 116, 3. Hé hine dyde óðrum monnum suá (swǽ, Cott. MSS.) ungelícne, Past. 17 ; Swt. 113, 14. (β) *so, in the same way, in like manner :*—And swá forð (cf. *Germ.* und so weiter) *and so on, et caetera,* Ælfc. Gr. 18 ; Zup. 114, 5, and often. Se ealdor dyde hand swá gelíce *similiter fecit,* Th. An. 74, 4. *See* eal-swá. (b) of degree or extent, (a) where a high degree is implied, *so* (*exceedingly*) :—Ne gemétte ic swá mycelne geleáfan, Mt. Kmbl. 8, 10. Nán fullere ne mæg swá hwíte gedón, Mk. Skt. 9, 3. For hwon sǽdest ðú swá gemeleáslíce and swá wlætlíce ða ðing, Bd. 5, 9 ; S. 623, 9. Ðá ðú swá lustlíce gehérdest míne láre, Bt. 22, 1 ; Fox 76, 23 : 35, 3 ; Fox 158, 7. Ðonne hí heora gód on swá manige dǽlas tódǽlaþ, 33, 2 ; Fox 122, 25. (β) where the degree is definitely marked :—Se consul fór mid þrim hunde scipa . . . him cómon ongeán Punice mid swá fela scipa (*cum pari classe*), Ors. 4, 6 ; Swt. 176, 11. Se twelf síþum hine bibaþaþ . . . and swá oft of wyllgespryngum beorgeþ, Exon. Th. 205, 6 ; Ph. 108. Syx swá micel *to the same extent much six times, six times as much,* L. M. L. ; Th. i. 190, 3. (c) of cause (v. **V.** 6), *so, therefore, on that account :*—Hé him ðet land forbeád . . . and hé hit swá álét *he forbade him the land . . . and so he gave it up,* Chart. Th. 202, 12. (2) defined by that which follows, (a) of manner, *so, in such a manner* that :—Far mid him swá ðæt ðú ðó ðæt ic ðé bebeóde *vade cum eis, ita duntaxat, ut, quod tibi praecepero, facias,* Num. 22, 20. Ælc wíf sceolde gebídan swá ðæt heó ne cóme intó Godes temple, Homl. Th. i. 134, 16. Crist is Godes Sunu swá ðæt (*in such sort that*) se Fæder hine gestrýnde of him sylfum, 258, 26. Swá beclýsed ðæt nǽnig óþer hý onlúceþ, Exon. Th. 20, 26 ; Cri. 323. Wearþ ðæt geat belocen swá ðæt ða stánas feóllon tógædere, H. R. 103, 7. Gif eów swá líce þuhte utan gangan on ðissum carcerne, Blickl. Homl. 247, 1. Swá ðon gelícost ðe tóbrocen fæt, Lchdm. ii. 230, 25. Se mon biþ, ðæs ðe swá tó cweþanne

si, æghwæðer ge gehæfted ge freó, Bd. 1, 27 ; S. 497, 40. (b) of degree :—Swá ealde swá hié ðá wǽron hié gefuhton *as old as they then were, they fought,* Ors. 3, 11 ; Swt. 152, 16. Nys hyt swá stearc winter, ðæt ic durre lutian æt hám, Coll. Monast. Th. 19, 17. Swǽ opene scylde swá ðæt hé his bróðor ofslóge, Past. 34 ; Swt. 234, 2. Ðín mægen is swá mǽre, swá ðæt ǽnig ne wát ða deópnesse Drihtnes mihta, Hy. 3, 31 : Ors. 4, 10 ; Swt. 198, 15. Swá fullíce ðiónde . . . óþ ðe hé wyrþ ælces cræftes medeme, Bt. 38, 5 ; Fox 206, 22. Ða habbaþ beardas swá síde óð heora breóst, Nar. 38, 1. Súþ swá ðú hátost mæge, Lchdm. ii. 50, 25. (3) used indefinitely, *so* and *so :*—Ðeáh ðú nyte for hwí hé swá and swá dó *though thou know not why he act in this or that manner,* Bt. 39, 2 ; Fox 214, 13. (4) used emphatically, *so, exceedingly, as much as possible* :—Ongan hé hine baðian swá swátigne (*when perspiring profusely*), Ors. 3, 9 ; Swt. 124, 30 : Jud. Thw. 22, 19 ; Jud. 67. Ðú meaht swá wíde geseón, Cd. Th. 36, 1 ; Gen. 565 : 25, 30 ; Gen. 425. Sió onlícnes sendde mycel wæter swá sealt (*exceedingly salt*), Blickl. Homl. 245, 25. Genim ðás wyrte swá mearwe (*as tender as possible*), Lchdm. i. 192, 8 : 194, 2. Wel on swá hátum, ii. 50, 15. (5) with comparatives, *the,* (1) singly :—Oft wé mágon beón suá (swǽ, Cott. MSS.) nyttran æt him gif wé hié myndgiaþ hira gódna weorca *plerumque utilius apud illos proficimus, si eorum bene gesta memoramus,* Past. 32, 2 ; Swt. 211, 20. Beþe ða eágan, betere swá oftor *the oftener the better,* Lchdm. ii. 34, 16. Leng swá swíðor, Cd. Th. 60, 30 ; Gen. 985 : Beo. Th. 3712 ; B. 1854. (2) correlatives *the . . . the :*—Swá norðor swá smælre *the further north one goes, the narrower the land becomes,* Ors. 1, 1 ; Swt. 18, 29. Swá betere swá fǽtran and ferscran, Lchdm. ii. 196, 22. Swá háttra sumor, swá mára ðunor and liget, iii. 280, 9. Swá swá hé ús mærlícor gifeþ, swá wé him mærlícor þancian scylon ; swá þrymlícre ár, swá máre eádmódnes, Wulfst. 261, 19–21. Swá swá leng swá bet, Bt. 35, 3 ; Fox 160, 8. Swá swá hé lengra biþ, swá hí bióþ ungesǽligran, 38, 4 ; Fox 204, 15. Swá mycele swá ðú hér on worulde swýþor swincst. swá ðú eft bist on écnysse fæstlícor getrymed ; and swá myccle swá ðú ón ðisum andweardan lífe má earfoða drígast, swá myccle ðú eft on tóweardnesse geféhst, Guthl. 5; Gdwin. 32, 10–14. (2 a) with a comparative and a positive :—Ðæt hé suá micle wærlícor hine healde wið scylda swá hé gere witan mæg ðæt hé nó ána forwierð, Past. 28 ; Swt. 191, 10. **V.** *adverbial conjunction,* (1) with indic. (a) with a clause of comparison, *as :*—Ne biþ hé eall swá hé ǽr wæs, Bt. 34, 9 ; Fox 148, 8. Beóþ mildheorte, swá eówer fæder is mildheort, Lk. Skt. 6, 36. Hé gedreósan sceal, swá ðeós eorðe eall, Exon. Th. 124, 27 : Elen. Kmbl. 1761 ; El. 882. Hí mé ymbsealdon samod anlíce swá beón, Ps. Th. 117, 12. Héht onlíce, swá hé ðæt beácen geseah, tácen gewyrcan, Elen. Kmbl. 200 ; El. 100. (1 a) swá swá :—Eall ðæt ðe leofaþ beóþ eów tó mete, swá swá grówende wyrta ic betǽhte ealle eów, Gen. 9, 3. Gewurðe ðín willa on eorðan, swá swá on heofenum, Mt. Kmbl. 6, 10. Dón swá swá hý git dóþ, Bt. 16, 1 ; Fox 50, 2. (2) with indic. or subjunct. expressing an actual or possible result, *so that :*—Se consul fór tó Tarentan, swá Hannibal nyste, and ða burg ábrǽc, swá ða nyston ðe ðærinne wǽron, Ors. 4, 10 ; Swt. 198, 7–9 : 4, 11 ; Swt. 206, 3. Gif hwá stalie, swá his wíf nyte, L. In. 7 ; Th. i. 106, 15. Wesan swá him yldo ne derede, Cd. Th. 30, 22 ; Gen. 471 : 256, 12 ; Dan. 639. Bær hine seó brimwylf, swá hé ne mihte wǽpna gewealdan, Beo. Th. 3020; B. 1508. Se má eallum Angelcyningum Brytta ðeóde fornom, swá efne ðæs ðe hé mihte wiþmeten beón Saule, Bd. 1, 34 ; S. 499, 20. (3) with subjunctive, *as* (*if*) :—Iosue fleáh, swá hé áfyrht wǽre, Jos. 8, 15. Ðú hí betweónum wætera weallas lǽddest, swá hí wǽron on drígum, Ps. Th. 105, 9. Cweðan swá hé áuum sprece, Exon. Th. 84, 23 ; Cri. 1378. Nú is ðon gelícost swá wé ceólum líðan, 53, 16 ; Cri. 851. (4) with optative, *so :*—Swá ðýos dǽd for monnum mǽre gewurþe, Lchdm. iii. 60, 14. Ic ðæt geswerige þurh sóþ godu, swá ic áre æt him ǽfre finde, Exon. Th. 247, 19 ; Jul. 81 : Beo. Th. 875 ; B. 435. (5) with a conditional force, *provided that, if, so be that, so* (as in Shakspere : *So* it be new, there's no respect how vile ; v. Abbott, Shak. Gram. § 133) :—Nim, swá hit ðé ne mislícyge, Ap. Th. 20, 12. Ðú him ðet land forbeád, swá hé ǽniges brúcan wolde, Chart. Th. 202, 10. (6) marking a consequence, *so, therefore, on that account :*—Ic mæg rǽdan on his ríce ; swá mé ðæt riht ne þinceþ . . . , Cd. Th. 19, 11 ; Gen. 289 : 24, 22 ; Gen. 381 : Andr. Kmbl. 2657 ; An. 1330. (7) local, *where :*—On eallum Norþan-hymbrum ge eác on Pehtum swá Oswíes ríce wæs ðæs cyninges *quousque rex Osuin imperium protendere poterat,* Bd. 4, 3 ; S. 566, 30. Geseh hé bearwas standan, swá ǽr his blód ágeát, Andr. Kmbl. 2897 ; An. 1451 : 3163 ; An. 1584. (8) temporal, *as, when :*—Swá heó sǽ geseah, hé hió snióme fleáh, Ps. Th. 113, 3. Ic wát God ábolgen wyrð, swá ic him ðisne bodscipe secge, Cd. Th. 35, 10 ; Gen. 552. (9) marking the grounds of action, *as, since :*—Wé ðé lofiaþ, swá ðú hǽlend eart, Hy. 7, 116. (10) *although, yet :*—Swá hé þurh feóndscipe tó cwale monige démde, swá þeáh him Dryhten eft miltse gefremede, Elen. Kmbl. 994 ; El. 498: Cd. Th. 25, 10 ; Gen. 391. (11) in contracted clauses, *as, as* (*being*) :—Hwone hé lǽran scyle suá earmne, and hwane suá eádigne, Past. 26 ; Swt. 183, 9. Heora hláford weorðodon swá swá wuldres cyning (cf. hiora cyningas hí weorþodon for Godas, Bt. 38, 1 ; Fox 194, 16), Met. 26, 45. **VI.** swá . . . swá, (1)—where *swá*

occurs once with a demonstrative, once with a relative force, *so . . . as, so . . . that, as . . . as* :—Swâ forð swâ uncre wordgecwydu fyrmest wǽron *as far as ever our agreements went*, L. O. 11 ; Th. i. 182, 11. Swâ gelíc swâ ðû æt swǽsendum sitte, Bd. 2, 13 ; S. 516, 15. Suâ suíðe suâ hê of ðǽre ǽwe ne cerre *so as he turn not from the law*, Past. 23 ; Swt. 175, 4. Bûton hê suâ monige gecierre suâ hê mǽsð mǽge, 28 ; Swt. 191, 9. Hafa on mûþe swâ hât swâ ðû hâtost mǽge, Lchdm. ii. 50, 15. Swâ forð swâ ða óðre, Ælfc. Gr. 18 ; Zup. 114, 3. Ða unrôtnessa swâ ilce ofergáþ, swâ ðû cwist ðæt ða blissa ǽr dydon, Bt. 8 ; Fox 24, 33. Swâ wîde swâ wegas tólǽgon, Andr. Kmbl. 2469 ; An. 1236. Hê hine wolde swâ welige gedôn swâ hê his sunu wǽre, Shrn. 84, 14. Sôna swâ seó sunne sealte streámas oferhlifaþ, swâ se fugel gewîteþ, Exon. Th. 206, 1–6 ; Ph. 120. (1 a) swâ swâ :—Dô rysle tô swâ swâ sýn twâ pund *add lard so as there may be two pounds*, Lchdm. ii. 74, 1 : 250, 26. (2) correlative, (a) *either . . . or, as well . . . as* :—Onfôn swâ ēcum lífe swâ ēcum deáðe swâ ðû ǽr geworhtest swâ ēcum lífe swâ ungeondodon wîte *accipere sive vitam aeternam, sive mortem aeternam, prout antea fecisti ; sive vitam aeternam, sive infinitum supplicium*, L. Ecg. P. iv. 65 ; Th. ii. 226, 13. Ðæt heó gecure óðer ðæra, swâ heó forførðe, swâ heó ðám godum geoffrode, Homl. Skt. i. 8, 63 : 11, 33. Nim swâ wuda swâ wyrt swâ hweðer swâ ðû wille, Bt. 34, 10 ; Fox 148, 25 ; Wulfst. 108, 10. Smire mid ðǽre sealfe swâ niht swâ twâ swâ þearf sié *smear with the salve one night or two, as need be*, Lchdm. ii. 128, 1. Sié ðæt on cyninges dôme swâ deáð swâ líf swâ hê him forgifan wille *be it in the judgement of the king, as well death as life, as he will grant him*, L. Alf. pol. 7 ; Th. i. 66, 10. Hit biþ gewrecen swâ ǽr, swâ lator, Homl. Ass. 62, 253. Gilde swâ wer, swâ wîte, swâ lahslite, aa be ðam ðe seó dǽd sý, L. Eth. v. 31 ; Th. i. 312, 10. Ðonne môt hê swâ rîdan, swâ rôwan, swâ swilce fǽrelde faran swylce tô his wege gebyrige, L. E. I. 24 ; Th. ii. 420, 24. (b) *whether . . . or* :—Saga him, swâ hê wille swâ hê nelle, hê sceal cuman, Bd. 5, 9 ; S. 623, 11. Wê be him nâþor nyton, swâ hî libban, swâ hî deáde licgon, Homl. Skt. i. 23, 306. God lēt hî habban âgenne cyre, swâ hî heora Scyppend lufedon, swâ hî hine forlēton, Homl. Th. i. 10, 19 : 18, 30. Syle etân ǽr ðǽre tíde his tôcymes, swâ on dæge swâ on nihte, swæþer hyt sý, Lchdm. i. 364, 16. On swelce healfe swelce hié winnende beón woldan, swâ sûþ, swâ norþ, swâ eást, swâ west, Ors. 3, 5 ; Swt. 106, 13. (c) swâ hwæðer swâ . . . swâ *whether . . . or* :—Sete man ofer ðæne þriddan dæg, swâ hwæðer swâ heó beó fûl swâ clǽne, L. Aih. iv. 7 ; Th. i. 226, 31. (2 a) with the first *swâ* omitted, *or* :—Dēm ðû hî tô deáþe, swâ tô lífe lǽt, Exon. Th. 247, 33 ; Jul. 88. **VII.** *in combination with the particles* git, same, þeáh, þeána, *see those words.* [Goth. swē, swa : O. Frs. sâ : O. Sax. O. H. Ger. sô : Icel. svâ (*later* svô) : Dan. saa : Swed. så.] v. eal-swâ.

swǽ, swǽc[c]. v. swâ, swecc.

swǽfan (?) :—Sió gítsung ðe nǽnne grund hafaþ swearte swǽfeþ (swǽleþ? v. swǽlan) sumes onlíce efne ðám munte ðe nû monna bearn Etne hâtaþ se swefle byrneþ, Met. 8, 46–50. The Latin original has : Saevior ignibus Aetnae fervens amor ardet habendi, which is rendered in the prose version : Manna gítsung is swâ byrnende swâ ðæt fýr on ðǽre helle seó is on ðam munte ðe Ætne hātte, Bt. 15 ; Fox 48, 29. From comparison of these three passages, it seems that *swǽfeþ* should mean *burns*, while the form of the word suggests comparison with O L. Ger. suēvôn *in* berg suēvôt *mons coagulatus*, with O. H. Ger. sweibôn *volvere, ferri*, and later English *swayue* in :—He (*the whale that swallowed Jonah*) swengeþ and swayues to þe se boþem, Allit. Pm. 99, 253. All these verbs denote movement, a meaning which does not seem to suit *swǽfan* in the passage where it occurs.

Swǽfas, Swǽfe ; *pl. A Germanic people, the Suevi or Alamanni* ('um diese zeit (4th cent.) pflegt an die stelle des alten Suevennamens die benennung Alamannen einzutreten,' Grmm. D. S. 348), *the Swabians* :—Swǽfas forhergodon ealle Galliam *Alamanni Gallias pervagantes*, Ors. 6, 24 ; Swt. 276, 3. Wið norþan Donua ǽwielme and be eástan Rîne sindon Eást-Francan ; and be sûþan him sindon Swǽfas, on ôþre healfe ðǽre ié Donua ; and be sûþan him and be eástan sindon Bægware, se dǽl ðe mon Regnesburg hætt . . . Tô ðǽm beorgan ðe mon Alpis hætt licgaþ Begwara landgemǽro and Swǽfa, 1, 1 ; Swt. 16, 1–14. Engle and Swǽfe, Exon. Th. 321, 10 ; Wîd. 44. Mid Englum ic wæs and mid Swǽfum. 322, 10 ; Wîd. 61. Witta weóld Swǽfum, 319, 34 ; Wîd. 22. [O. H. Ger. Suâb Alamannus, Suâba, Suâpa *Suevi*.]

swǽlan ; *p. de To burn* (trans.) :—Onǽl ł swǽl ł bærn lændenu *ure renes*, Ps. Lamb. 25, 2. Hê sende of heofonum swǽlende lēg, Wulfst. 213, 6. [Heo heom letten swalen inne swærte fure (þe mahunes mid fure hii forswelde, 2nd MS.), Laym. 10188. Berned heore halles & swaleð heore bures, 6147. A bernene drake borwes swelde, 25594. Halliwell gives *sweal, swale* to burn.] v. be-, for-, ge-swǽlan (*read* -swǽlan) ; sâm-, unfor-swǽled, swelan.

swǽm, es ; *m. A trifler, vain, foolish person* :—Swǽm *nugator, inutilis, vanus*, Germ. 389, 32. Ic wylle ðæt Latona môder Apollinis and Diane fram mē gewîten, ðe Delo ákende, ðæs ðe ealde swǽmas gecýddon (*as the foolish triflers of old declared*), Anglia viii. 325, 29. Nû mæg hēr manna gehwilc gehýran hwet ðâs swǽmas wǽron ðe ûre yldra[n] him tô gebǽdon *now may every one hear in this account (of the gods) what these vain creatures were, that our forefathers prayed to*, H. Z. xii. 408, 15.

swǽman ; *p. de To trouble, afflict, grieve.* The verb occurs in this sense in later English :—Ofte hit timeð þat tat leoueste bearn sorheð and sweameð meast his ealdren, H. M. 35, 5. Þe engles beoð isweamed, þat seoð hare suster swa sorhfulliche afallet, 17, 20. Ure Louerd ne mei uor reouðe wernen hire, ne sweamen hire heorte mid wernunge, A. R. 330, 11. Þe swemande sorȝe soȝt to his hert, Allit. Pms. 54, 563. Cf. also : His hert began to melt For veray sweme of this swemeful tale, Lydgate (cited ib. ƒ. 199). Swemyn *molestor, mereo* ; sweem, swemynge *or* mornynge *tristicia, molestia, meror*, Prompt. Parv. 482, col. 1. In A. S. only the compound *â-swǽman* (q. v.) is found, apparently with the meaning *to become troubled or grieved.* To the instance given under *â-swǽman* may be added the following :—Swâ Sanctus Paulus cwæþ ðætte God hēte ealle ða âswǽman æt heofona ríces dura, ða ðe heora cyrican forlǽtaþ *God would bid all those grieve . . .*, Blickl. Homl. 41, 34. Sceolde se mín þearfa âswǽman (*have cause to grieve*) æt ðinre handa, Wulfst. 258, 2. Se sceocca sceall âswǽman æt ûs, gif wē ânrǽde beóþ on ûrum geleáfan, Homl. Skt. i. 17, 203. v. swâmian.

swǽpa, swêpa (-e, -o) ; *pl. Sweepings*, in compounds (not inserted in proper place) :—Æswǽpe (beánscalu) *quisquiliarum, surculi minuti*, Hpt. Gl. 420, 59. Âswǽpa *peripsema*, 504, 3. Geswǽpa *peripsema*, Wrt. Voc. ii. 65, 68. Geswêpa, geswǽpa (gen.-, MS.), 95, 18. Geswêpo, 76, 17. Bió hē gehealden for æscegeswáp *pro purgamento favillae deputetur*, Chart. Th. 318, 33. [O. H. Ger. â-sueipha *purgamenta, quisquilias*.] -swǽpe, -swâpe. v. hâd-, heorþ-, ymb-swǽpe.

swǽpels (*m.?*) ; swǽpelse, an ; *f. A wrap, garment* :—Swǽpels *amictus*, Ps. Surt. 106, 3. Ða swǽpelsan *amicula*, Wrt. Voc. ii. 3, 49. [Cf. Icel. sveipa *to wrap, swaddle* ; sveipa *a kerchief, hood* : Dan. suðbelse-barn *child in swaddling-clothes*.] v. swâpan.

swǽpig ; *adj. Fraudulent, deceitful* :—Swǽpige ł swicfulle *fraudulentas*, Hpt. Gl. 474, 17. v. ge-swip, swipor.

swǽr, swǽre, *and* swâr ; *adj.* [Halliwell gives *swēer* unwilling as a Northumbrian word, and *swere* dull, heavy, as a Durham one. In Jamieson's Dictionary the forms *sweir, swere, sweer, swear* are given with meanings lazy, indolent ; unwilling ; unwilling to give.] **I.** *heavy as a burden, of great weight* (lit. *or* fig.), *oppressive* :—Swǽr is seó byrðen ðe Godes bydel beran sceall, gif hē nele georne unriht forbeódan, L. I. P. 5 ; Th. ii. 308, 35 : Wulfst. 178, 8. Hē bið deófles tempel, and byrð swîðe swǽre byrðene on his bæce, Homl. Th. i. 212, 4. Ðæt swǽre gioc underlûtan, Met. 10, 20. His wæpnu syndon swǽre tô berenne, ac Cristes geoc is wynsum, Basil admn. 2 ; Norm. 36, 14. Sorh biþ swǽrost byrðen, Salm. Kmbl. 623 ; Sal. 311. Gif mîne synna wǽron âwegene on ânre wǽgan, ðonne wǽron hî swǽrran gesewene ðonne sandcorn on sǽ, Homl. Th. ii. 454, 24. **II.** *heavy, grievous, painful, unpleasant* :—Him yldo ne derede, ne suht swâre, Cd. Th. 30, 24 ; Gen. 472. Swâr leger, Exon. Th. 101, 21 ; Cri. 1662 : 201, 15 ; Ph. 56. Gebrec swâr and swîðlíc *a crash grievous and great*, 59, 19 ; Cri. 955. Ðæt hē swǽre ǽhweorfe hæftned hefige, Ps. Th. 125, 1. Dû þolades swâr gewin, Exon. Th. 86, 22 ; Cri. 1412. Geswencean mid swârum wîtum, Homl. Skt. i. 4, 181. Ða swâran (swǽran, *other MSS.*) wîta onfôn, 19, 46. Is swǽrra ðínra synna rôd, ðonne seó óþer wæs, ðe ic ǽr âstâg, Exon. Th. 91, 10 ; Cri. 1490. Nis ðys eall geswinc? and gyt mycele swǽrran ealle ða ungelimp ðe on ðysum lífe becumaþ, Hexam. 20 ; Norm. 28, 26. **III.** *heavy, sad, feeling or expressing grief* :—Ðæt swǽre *triste*, Wrt. Voc. ii. 88, 49. Mē is swǽre stefn, hefig, gnorniende *vox gemitus mei*, Ps. Th. 101, 4. **IV.** *of sin or evil, grave, grievous* :—Be hefigtýmum gyltum. Se bróðor se ðe mid swǽrra gylta hæfene bið gedered *de grauioribus culpis. Frater qui grauioris culpe noxa tenetur*, R. Ben. 49, 13. On scyldum swǽrum *in delictis*, Ps. Th. 67, 21. Gebundene swârum (*var.* swǽrum) gyltum, Anglia xi. 113, 38. Ða swǽran gyltas ðe hí âdrugon, Homl. Th. i. 340, 27. Dû micele swǽrran synna gefremodest, 54, 33. **V.** *of physical or mental inactivity, heavy, slow, dull, sluggish, slothful, indolent* :—Suær desis, Wrt. Voc. ii. 105, 79. Swǽr *deses*, 25, 12. Dû yfle esne and swǽr (swēr, Lind.) *serve male et piger*, Mt. Kmbl. Rush. 25, 26. Sum welig man wæs swangor and swǽr, and him wæs lâð þearfendum mannum mete tô syllenne, Wulfst. 257, 12. Nis hē swâr ne swongor *non est tarda*, Exon. Th. 220, 4 ; Ph. 315. On swârran ðisum lîchoman *in gravi isto corpore*, Hymn. Surt. 13, 15. **V a.** *inactive from weakness, enfeebled, weak* :—Mē is mín gást swǽr geworden *defecit spiritus meus*, Ps. Th. 142, 7. **V b.** *of sleep, heavy* :—Swâ fram slǽpe hwylc swǽrum árîse, Ps. Th. 72, 15. Gehefegod mid ðam swǽran slǽpe, Basil admn. 1 ; Norm. 34, 3. [Forr hefig & forr sware unngriþþ, Orm. 16280. Goth. swērs *grave, honoured* : O. Sax. swâri *grievous* (*sin, sickness*) : O. Frs. swēre : O. H. Ger. swâr, swâri *gravis, onerosus* : Ger. schwer : Icel. svârr (*a poetic word*) *heavy, grave.*] v. ge-swǽre.

swǽran ; *p. de To make heavy, to oppress* :—Eall se lîchama geswǽred byþ and gehefegud, Lchdm. iii. 120, 22.] [O. H. Ger. swâren *gravare, praegravare, opprimere* ; gi-swâren *gravare.*]

swǽre, swâre ; *adv. Grievously, oppressively* :—Eam ic swǽre geseald ðǽr ic ût swîcan ne mæg *traditus sum, et non egrediebar*, Ps. Th. 87, 8. Se hláford hefig gioc slêpte swâre on ða swyran sînra þegena, Met. 9, 56.

[Ne set me neuer naþing swa swere (sare, Bodl. MS.), Jul. 46, 10. *O. Sax.* ♀ *O. H. Ger.* swâro *graviter.*]

swǽr-líc; *adj. Grievous :*—Benedictus mid swǽrlícum heófungum bemǽnde, ðæt his leorningcild ðæs óðres deáðes fægnian sceolde, Homl. Th. ii. 164, 9. [*O. H. Ger.* swâr-líh *gravis.*]

swǽrlíce (swär-); *adv.*   **I.** of doing or bearing what is painful, *heavily, grievously :*—Nán man ne sceal his wífe geneálǽcan, siððan heó mid bearne swǽrlíce gebunden gǽþ, Homl. Th. ii. 324, 21. Hé sceolde hit mid fǽstne swárlíce gebêtan, Homl. Skt. i. 21, 261.   **II.** of sleeping, *heavily.* v. swǽr, **V** b :—Wé feóllon on slǽpe swárlíce, swylce wé on deáðe lágon, Homl. Skt. i. 11, 239. [*O. H. Ger.* swârlíhho *graviter.*]

swǽr-mód (swär-); *adj. Of an indolent, sluggish disposition :*—Sum welig man wæs prútswongor and swǽrmód, and him wæs láð þearfendum mannum mete tó syllenne, Wulfst. 257, 12 MS. D. v. swǽr, **V**, *and next word.*

swǽrmódness (swär-), e; *f. Sluggishness of disposition, slowness, dullness :*—Oft mon biþ suíðe wandigende æt ǽlcum weorce and suíðe lǽtrǽde, and wênaþ menn ðæt hit sié for suármódnesse and for unarodscípe, and biþ ðeáh for wisdóme and for wǽrscipe (*but the Latin is :* Saepe agendi tarditas gravitatis consilium putatur), Past. 20 ; Swt. 149, 15.

swǽrness (swär-), e ; *f.*   **I.** *heaviness* of a burden (lit. or fig.), *weight.* v. swǽr, I :—Hwí settest ðú ðises folces swárnysse (*pondus*) uppan mé ? Num. 11, 11. Ne mæg ic âna eówre swárnissa (*pondus*) and eówre saca ácuman, Deut. 1, 12.   **II.** *heaviness, want of readiness in moving, sluggishness,* v. swǽr, **V** :—Nán hǽfignes ðæs líchoman ne nán unþeáw ne mæg eallunga átíon of his móde ða rihtwísnesse . . . ðeáh sió swǽrnes ðæs líchoman and ða unþeáwas oft âbisegien ðæt mód mid ofergiotolnesse *non omne mente depulit lumen obliviosam corpus invehens molem*, Bt. 35, 1 ; Fox 154, 31.

swærnung, swarnung, v. swornian.

swǽs; *adj.*   **I.** (one's) *own; proprius.* v. swǽslíce, I :—Ðæt selegescot ðæt ic mé swǽs on ðé gehálgode *the tabernacle that I hallowed me as my own in thee*, Exon. Th. 90, 29 ; Cri. 1481.   **II.** the word, which occurs rarely in prose (see, however, the first passage cited), is used mostly in reference to the connection that belongs to relationship by blood or by marriage, or to dear companionship, and so often has the force of (one's) *own dear,* (one's) *dear :*—Ǽlþeódige mæn . . . swǽse mæn *foreigners . . . men of one's own race, natives,* L. Wih. 4 ; Th. i. 38, 2. Biþ him self sunu and swǽs fæder and eác yrfeweard *ipsa sibi proles, suus est pater et suus haeres,* Exon. Th. 224, 13 ; Ph. 375. Ic and mín swǽs fæder, Elen. Kmbl. 1032 ; El. 517. Mín ðæt swǽse bearn ! (cf. mín ðæt leófe bearn ! 166, 28 ; Gú. 1049), Exon. Th. 167, 1 ; Gú. 1053. Swǽs eft ongon (cf. fæder eft ongon etc., 7) his bearn lǽran, 302, 29 ; Fä. 43. Cwæð brýd tó beorne : ‘Mín swǽs freá,’ Cd. Th. 168, 15 ; Gen. 2783. Heó Adame hyre swǽsan were scencte, Exon. Th. 161, 11 ; Gú. 975. Wið fæder swǽsne, 39, 4 ; Cri. 617. Gif ðú sunu áge, oððe swǽsne mǽg, oððe freónd ǽnigne, Cd. Th. 150, 28 ; Gen. 2498 : 203, 11 ; Exod. 402. Heora swǽs cynn, Ps. Th. 105, 21. Geseh swǽsne geféran *he saw his own dear comrade,* Andr. Kmbl. 2018 ; An. 1011. Æfter swǽsne (*one's own dear lord*), Exon. Th. 289, 18 ; Wand. 50. Swǽse gesíþas *his own familiar comrades,* Beo. Th. 57 ; B. 29. Nǽnig swǽsra gesíða, 3872 ; B. 1934. Freónda má swǽsra and gesibbra *more of friends dear and near,* Exon. Th. 408, 34 ; Rä. 27, 22. Freóndum swǽsum and gesibbum, Cd. Th. 97, 13 ; Gen. 1612. Hé hêt hine (*Beowulf*) leóde swǽse sécean, Beo. Th. 3741 ; B. 1868. Mǽgburge swǽse and gesibbe *my kindred, dear and near ones* (or *dear and near kindred*), Exon. Th. 397, 19 ; Rä. 16, 22. Twá dohtor, swǽse gesweostor, 431, 29 ; Rä. 47, 3.   **III.** with a development of meaning similar to that in *kind* or *gentle ; gracious, kind, agreeable, pleasant* (used of persons or things). v. swǽs-líc :—Swǽs vel wynsum *eucharis,* Wrt. Voc. i. 61, 17. Líþe, swǽs *blanda,* ii. 127, 2. Tunge swǽse tóbrycþ heardnysse *lingua mollis confringit duritiam,* Scint. 8, 17. Drihten is niðum swǽs *suavis est Dominus,* Ps. Th. 99, 4. Ðú swǽs tó mé ðín eáre onhyld, 101, 2. Þeáh ðe ic on hylд gegange, ðænne swǽs wese *when it may be agreeable,* 131, 3. On sóðfæstra swǽsum múðe *in the gracious mouth of the just,* 117, 15. Weredum beóbreáde *vel* swǽsum *dulci favo,* Wrt. Voc. ii. 142, 9. Fram swǽsere tungan *a blanda lingua,* Kent. Gl. 159. Steorran forléton hyra swǽsne wlite *the stars resigned their sweet beauty,* Exon. Th. 71, 1 ; Cri. 1149. Sete swǽse geheald múðe mínum *set pleasant guard for my mouth,* Ps. Th. 140, 4. Beseoh on ðíne scealcas swǽsum eágum (*with gracious eyes*), 89, 18. Swǽsum wordum *dulcibus verbis,* Coll. Monast. Th. 32, 31 : *blandimentis,* Gl. Prud. 43 a. Swáse swegldreámas, Exon. Th. 82, 35 ; Cri. 1349. [*Goth.* swês ïðios ; swês ; *subst. property : O. Sax.* swâs (man) : *O. Frs.* swês *near, related : O. H. Ger.* swâs *familiaris, domesticus : Icel.* sváss *beloved, dear ; pleasant.*] v. ge-, un-swǽs ; swǽs-líc.

swǽse; *adv. Agreeably, pleasantly* [:—Geswǽse *blandide,* Wrt. Voc. ii. 127, 5].

swǽsend-dagas (swǽsing-); *pl. The ides ;* the Latin term seems to be so rendered from supposing it to be connected with the verb *edere ;* v. next word :—Swǽsingdagas *idus, ab edendo dicuntur,* Wrt. Voc. i. 53, 37. **Swǽsenddagas** *idus, ab edendo,* ii. 62, 27 : 48, 55.

swǽsende, es ; *but occurring almost always in pl.* swǽsendu (-a, -o) ; *n.*   **I.** *food, victuals, refection :*—Swǽsende *fercula,* Wrt. Voc. ii. 35, 19. Swǽsendo *fercula, cibaria,* 147, 83. Hé þanc gesægde ðá hé gereordod wæs : ‘Ðé ðissa swǽsenda Meotud leán forgilde,’ Andr. Kmbl. 771 ; An. 386. Ðæt hí on his hús ne eodon ne of his swǽsendum mete ðygedon *ne domum ejus intrarent neque de cibis illius acciperent,* Bd. 3, 22 ; S. 553, 28. Mid hígna suêsendum (*the articles of food are then given*). Ond ðás forecuedenan suêsenda all âgefe mon ðem reogolwarde, Txts. 444, 14–26. Ða ilcan wísan ou swǽsendum tó mínre tíde léstan (cf. hígon geformian tó mínre tíde, 449, 9), 450, 1. Suoesendo *agapem,* 39, 108. Swǽsendo, Wrt. Voc. ii. 2, 22. Ǽlmessum swǽsendo, 5, 35. Ða six Sunnandagas ðe wé swǽsendo on habbaþ *the six Sundays in Lent when we may take meat* (cf. nán dæg (*in Lent*) ne sý bútan Sunnandagum ânum, ðæt ǽnig mon ǽniges metes brúce ǽr ðære teóðan tíde oððe ðære twelfte, L. E. I. 37 ; Th. ii. 436, 6–8), Wulfst. 284, 5. Ðone mete and ða swǽsendo *dapes,* Bd. 3, 6 ; S. 528, 20.   ¶ In phrases :—Gán tó swǽsendum *to go to dinner ;* ire ad reficiendum, 5, 4 ; S. 617, 18. Sittan æt *or* tó swǽsendum *to sit at meat, take a meal :*—Hí æt beóde and æt swǽsendum sǽton *sederunt ad mensam,* 5, 5 ; S. 617, 10 : Cd. Th. 168, 7 ; Gen. 2779. Sittan tó his swǽsendum *residens ad epulas,* Bd. 3, 14 ; S. 540, 42 : 5, 5 ; S. 618, 17. Woldon wé tó úrum swǽsendum sittan *ceperamus uelle epulari,* Nar. 21, 12. Ðá hêt ic eallne ðone here ðæt hé tó swǽsendum sǽte and mete þigde *cenare militem jussi,* 23, 8. Swǽsende, swǽsenda þicgan *to take food :*—Sæt hé and swǽsende ðeah and dranc (sæt hé on swǽsendum and æt and dranc, MS. B.) *residebat, vescebatur, bibebat,* Bd. 5, 5 ; S. 618, 18. Hé on his hús eode and his swǽsendo ðeah *intravit epulaturus domum ejus,* Bd. 3, 22 ; S. 553, 30. Swá ðæt hé nǽfre mete onfêng ne swǽsendo ðeah *ita ut nihil unquam cibi vel potus perciperet,* 4, 25 ; S. 599, 29. Swǽsendo þicgean *jejunium solvere, prandere,* 5, 4 ; S. 617, 13, 16. Swǽsenda (up) girwan *to prepare a feast,* Judth. Thw. 21, 7 ; Jud. 9. Symbel ł swoese (swoesende ?) mín ic gearuade *prandium meum paravi,* Mt. Kmbl. Lind. 22, 4.   **II.** *flatteries, blandishments, fair speech.* v. swǽs, **III**, ge-swǽsness, swǽslǽcan :—Swêsendum *blanditiis,* Kent. Gl. 212. v. dæg-, undern-swǽsendu (-o).

swǽslǽcan; *p.* -lǽhte *To flatter, cajole, speak fair :*—Hió swêslêcþ *blanditur,* Kent. Gl. 194. v. ge-swǽslǽcan.

swǽs-líc; *adj. Kindly, pleasant, agreeable :*—Sárge gé ne sóhton, ne him swǽslíc word frófre gé sprǽcon, Exon. Th. 92, 20 ; Cri. 1511. Hé (*Antecrist*) winþ ongeán Godes gecorenan mid swǽslícum gifum. Hé sylþ ðam, ðe on hine gelýfaþ, goldes and seolfres genyhda, Wulfst. 196, 21. [*O. Frs.* swês-lik *familiaris : O. H. Ger.* swâs-líh *privatus, civilis, familiaris.*] v. un-swǽslíc, *and next word.*

swǽslíce; *adv.*   **I.** *properly.* v. swǽs, I :—Wé andettaþ swǽslíce and sóþlíce Fæder and Sunu and Hálígne Gást *confitemur proprie et veraciter Patrem et Filium et Spiritum Sanctum,* Bd. 4, 17 ; S. 585, 36.   **II.** of persons, *kindly, in a gracious, friendly manner, blandly ;* of things, *agreeably, pleasantly.* v. swǽs, **III** :—Ða nán lust yfel swǽslíce gewemþ *eos nulla voluptas mala blande corrumpit,* Scint. 3, 10. Him (*the good*) swǽslíce (cf. on ðæt frǽte folc (*the evil*) hé firene stǽleþ láþum wordum, 84, 17 ; Cri. 1375) sibbe geháteþ heáhcyning, Exon. Th. 82, 15 ; Cri. 1339. Nealles swǽslíce mé wæs síð álýfed *the way was not made easy for me,* Beo. Th. 6169 ; B. 3089. Cóman him tó and hine swǽslíce grêtton, Homl. Skt. i. 5, 210. Busiris wolde ǽlcne cuman swíþe árlíce underfón and swíþe swǽslíce wiþ gebǽran (*behave in a very friendly manner to him*), ac eft hé (*the guest*) sceolde beón ofslegen, Bt. 16, 2 ; Fox 52, 32. Tó fela manna is ðe þurh hiwunge eal óðer specaþ, óþer hý þencaþ . . . and swá gerâde mánswican on ða wísan swǽslíce swiciaþ (*deceive under an appearance of friendliness.* v. swǽslǽcan, swǽsness), Wulfst. 55, 6. Swǽslíce swicole *deceiving with fair words,* 79, 4 : 82, 2. [*O. Sax.* swâslíko *friendlily : O. H. Ger.* swâslíhho *familiariter.*]

swǽsness, e ; *f. Blandishment, fair speech :*—Swǽsnyssum *blandimentis, lenociniis,* Hpt. Gl. 481, 10. Gé Godes cempan, gé âwurpaþ eówerne cynehelm for ðám earmlícan swǽsnyssum (*the appeals made to your feelings*) ðissera heófiendra. Ne âwurpe gé eówerne sige for wífa swǽsnyssum, Homl. Skt. i. 5, 54–58. Ðæt sé ðissere worulde swǽsnyssa (*blanditias*) warnige, Scint. 216, 12. v. ge-swǽsness.

swǽsung, e ; *f. A making pleasant, an alleviation, a mitigation :*—Swǽsunga *fomenta,* Wrt. Voc. ii. 150, 7.

swǽs-wyrde; *adj. Of pleasant speech, pleasant in speech ;* facetus, Wrt. Voc. i. 61, 18.

swǽtan; *p.* te *To sweat.*   **I.** of the natural moisture of the skin :—Ðætte hé swa swíþe swǽtte swá hé iu swoloþan middes sumeres wǽre *quia ita, quasi in media aestatis caumate, sudaverit,* Bd. 3, 19 ; S. 549, 29. Sitte hé on bæþe óð ðæt hé swǽte . . . oþ hé wel swǽte, Lchdm. ii. 290, 1–6. Ðæt se mon swǽte swíþe, 332, 2 : iii. 8, 11. Hé ongan blácian and ungefôhlíce swǽtan, Homl. Th. i. 414, 12 : Wulfst. 141, 3.   **I a.** *to sweat* with hard labour, so *to toil :*—Ðæm ðe nú on gódum weorcum ne swǽt and suíðe ne suínceþ *qui nunc in bonis operibus non exsudat,* Past. 39, 2 ; Swt. 285, 13. Sume sceufon, sume tugon and swýðe swǽtton, óð ðæt hig geteorode wǽron, Shrn. 154, 27. Winnende *vel* swǽtende *desudans, i. laborans,* Wrt. Voc. ii. 139, 37.   **II.** *to*

sweat, send forth like sweat, to exude (of persons or things) :—Hí fleóþ and blóde hí swǽtaþ, Nar. 35, 33. Fýre swǽtaþ blácan líge *they sweat fire and flame*, Exon. Th. 385, 12; Rä. 4, 43. Mon geseah twegen sceldas blóde swǽtan (*sanguine sudare*), Ors. 4, 8; Swt. 188, 25. Hí gemétton ðone clúd swǽtende, Homl. Th. ii. 162, 6. **II a.** *to send forth blood, to bleed.* v. swát, II. 2 :—Hit ǽrest ongan swǽtan on ða swíðran healfe, Rood Kmbl. 39; Kr. 20. [*Icel.* sveita *to sweat.*] v. á-, be-, ge-swǽtan; swítan.

swǽp, es; *n.* **I.** *a track, the mark left by a moving body, a single footprint or a series of footprints* (lit. or fig.) :—Mé (*the plough*) biþ gongendre mín swǽd sweotol, Exon. Th. 403, 19; Rä. 22, 10. Ðonne fylge wé Drihtnes swæþe, Blickl. Homl. 75, 14: Rtl. 26, 5. Ðonne stæpþ se sacerd on ðone weg, ðonne hé on ðæt swæd ðara háligra winnaþ tó spyrigaune, Past. 13; Swt. 77, 20: pref.; Swt. 5, 16. Deáþ ne forlét nán swæþ ǽr hé geféhþ ðæt ðæt hé æfter spyreþ, Bt. 39, 1; Fox 212, 1: Met. 27, 14. Weard sáweþ on swæd mín (*the plough's*), Exon. Th. 403, 11; Rä. 22, 6. Swearte wǽran lástas, swaþu swíþe blacu, 434, 19; Rä. 52, 3. Ða swaðo wǽron útwearde ongunnen ðe on ðæm marmanstáne geméted wǽron, Blickl. Homl. 207, 11. Swylce mannes swaðu, ðon gelícost ðe ðǽr sum mon gestóde; and ða fótlástas wǽron swutole, 203, 35. Alle suæðo *omnes semite*, Rtl. 81, 20. Forlét úre Drihten his fét on ða eorþan besincan . . . leóhtfæt biþ á byrnende for ðara swaþa weorþunga, Blickl. Homl. 127, 31. Suoeðum, suæðum *semitis*, Rtl. 167, 1, 13. **II.** *a vestige, trace* :—Hwæt is elles ðis gewítendlíce sibb búton swelce hit sié sum swæd ðære écean sibbe *quod est enim pax transitoria, nisi quoddam vestigium pacis aeternae?* Past. 46, 5; Swt. 351, 25. v. bil-, dolh-, fót-swæþ; swaþu.

swǽp (?), swaþu (?) *a bandage, swathe* :—In swaþum *instita* (v. Jn. 11, 44 to which the gloss refers), Wrt. Voc. ii. 74, 17: 46, 51. v. sweþel, sweþian.

swǽpel. v. sweþel.

swǽðer, swaðer (= swá hwæðer, cf. *O. H. Ger.* sueder). **I.** *pronoun, Whichever of two* :—Swaðer uncer leng wǽre, Cod. Dip. Kmbl. ii. 113, 20, 25. Hwæðres ðara yfela is betere ǽr tó tilianne búton swæðres swæðer frécenlícre is *quae pestis ardentius insequenda est, nisi quae periculosius premit?* Past. 62; Swt. 457, 22. Dó swæþer ðú wille *do whichever you like*, Bt. 39, 4; Fox 218, 10. **II.** *in combination with* swá . . swá . . . *either* . . *or* . . . *whichever, whether* . . *or* :—Hé móste swá geceósan swá áweorpan swaþer (swæðer, *other MSS.*) hé wolde *licuit ei excusare aut suscipere*, R. Ben. 99, 15. Beón swæðer hig beón, swá (þe, *other MSS.*) sacerdhádes swá clerichádes, 110, 7. Gewylde man hine swaðor man mæge, swá cucenne swá deáþne, L. Edg. ii. 7; Th. i. 268, 17. Hí gefeallaþ on ða heortan suá nytt suá unnytt suæðer hié beóþ (*whether they be profitable or unprofitable*), Past. 15; Swt. 97, 2: 14; Swt. 85, 15. Biþ ǽlc gód weorc gód, sié swá open swá dégle, swæðer hit sié, 59; Swt. 451, 14. Wyl wermód swá drígne swá grénne swaþer hé hæbbe *boil wormwood, either dry or green, whichever you have*, Lchdm. ii. 296, 14. Deáh wé spirian swá mid lǽs worda swá mid má swæþer wé hit gereccan mágon, *though we use more or less words in our enquiry, according as we can explain the matter*, Bt. 35, 5; Fox 166, 12: 36, 7; Fox 184, 16. Hí móston dón swá gód swá yfel, swæþor swá hí woldon, 41, 2; Fox 246, 2.

swǽd-hlýpe, swǽþian, swǽðorian, swǽðrung, swagoþ, swalewe, swaloð. v. stæþ-hlípe, ge-swǽþian, swaðrian, ge-swǽðrung, swégan, swealwe, sweoloþ.

swámian; *p.* ode *To become dark* :—Rodor swámode ofer nidða bearn *heaven grew dark above the children of men*, i. e. *night came*, Exon. Th. 167, 33; Gú. 1069. v. á-swámian; swǽman.

swamm, es; *m. A fungus, mushroom*; also *a sponge* :—Suom, suamm *fungus*, Txts. 65, 938. Swamm oððe feldswam *fungus*, Wrt. Voc. ii. 36, 22. Swom *fungus, spongus, dicta ab uligine*, 152, 21. Ðes swam *hoc tuber* (cf. *tubera* taddechese (= toadstool), Wülck. Gl. 618, 4), Ælfc. Gr. 9, 18; Zup. 44, 1. Nym hláf and sealt and swamm, and cnuca hit eal tógadere, Lchdm. iii. 94, 21. Syle etan gebrǽdne swam, 142, 11. Sinwealte swammas *volvi*, Wrt. Voc. i. 30, 28. For mete heó sceal sume hwíle swamma brúcan; wundorlíce heó geeácnaþ, Lchdm. i. 346, 8. [*Goth.* swamms *a sponge*: *O. H. Ger.* swamm, swamp *fungus, tuber*: *Ger.* schwamm *sponge, fungus, excrescence*: *Du.* zwam: *Icel.* svöppr *a sponge*: *Dan.* svamp *sponge, fungus*: *Swed.* swamp.] .v. feld-, mete-swamm.

swan, swon, es; *m. A swan* :—Suan *holor*, Wrt. Voc. ii. 110, 42. Swan, 43, 7. Suon *olor*, 115, 45. Swon, ilfetu, 63, 40: *alvor*, 6, 55. Swann *olor*, i. 62, 12. Swan *diomedia*, 63, 14. Swanes feðre, Exon. Th. 207, 6; Ph. 137. *For instances of the word in local names, see* swonleáh, swonweg, Cod. Dip. Kmbl. iii. 48, 78. [*O. L. Ger.* swan: *O. H. Ger.* swan; *m.*, swana: *f.* cygnus: *Icel.* svanr.]

swán, es; *m.* **I.** *a herd*, particularly *a swineherd*; the herds of swine formed a very important item in the live-stock of the Anglo-Saxons. v. swín. For some account of the duties and rights of different kinds of swánas, see L. R. S. 6, 7; Th. i. 436 :—Suán *subulcus*, Wrt. Voc. ii. 121, 59: *flabanus*, 108, 72. Swán, 35, 66: *bubullus* (*-cus?*), in a list *de suibus*, i. 286, 58: ii. 11, 59. Hiene án swán (*subulcus*, Flor. Wig.) ofstang, Cht. 755; Erl. 48, 23. Hé (*Alfred*) on sumes swánes (the

swán is called *vaccarius* in the Latin Vita S. Neoti, but in other forms of the story, e. g. Matthew of Westminster's, he is *subulcus* and drives 'porcos ad solita pascua') húse his hleów gernde . . . Hit gelamp ðæt ðæs swánes wíf hǽtte hire ofen . . . and cwæþ tó ðan kinge: 'Wǽnd ðú ða hláfes ðæt heó ne forbeornen, for ðam ic geseó dæighwamlíce ðæt ðú micelæte eart, Shrn. 16, 13–20. Swána steorra (cf. swán-steorra) *hesperius*, Wrt. Voc. ii. 43, 39. Oxena hierdas *bobulcos*, swánas *subulcos*, 80, 18. Cúhyrdas *bubulcos*, swánas *subulcos*, Hpt. Gl. 464, 23. **II.** *a man, warrior* (? cf. *Icel.* sveinn) :—Ne gefrægn ic nǽfre wurdlícor æt wera hilde sixtig sigebeorna sél gebǽran, ne nǽfre swánas swétne medu (swa noc hwitne, Hickes) sél forgyldan, Fins. Th. 78; Fin. 39. [The form which in later English should be taken by the word is *swon*, and this is found in Palladius on Husbandry: Thy *swon* may se thaire (the pigs') nombr and up save The oppressed pigge, 3, 1086. It has not, however, come into modern English; the corresponding Scandinavian form, *Icel.* sveinn = *boy, lad, man, servant*, on the other hand, remains in *swain*. Early instances of its occurrence are: His sweyn (*also* swain) Leir forþ sende þat was hiredman hende, Laym. 3512. Þreo cnihtes and heore sweines, 18128. Erl ne barun, knict ne sweyn, Havel. 273. Cf. too *Dan.* svend *boy, lad, journeyman*: *Swed.* swen. *O. H. Ger.* swên, *like* swán; = *subulcus*.] v. æhte-, gafol-, in-swán.

swancor; *adj. Bending easily.* **I.** *of a horse* (cf. *Icel.* svangr used in the same connection), *slender, slim, active and graceful in movement* :—Þrió wicg swancor and sadolbeorhte, Beo. Th. 4356; B. 2175. [Jamieson gives *swank* slender; limber, agile: *swanking* supple, active: *swanky* tall and lank: *swanky* a strapping young countryman.] **II.** *pliant, supple* :—Hine Niðhád on néde legde swoncre seonobende *supple sinew-bands* (? *see* seonu-bend), Exon. Th. 377, 19; Deór. 6. [Cf. *M. H. Ger.* swankel: *Ger.* schwank *flexible, slim*: *Swed.* swank *a bend*; swank; *adj.* pliable, flexible; swank-rem *girth-leather*.] **III.** *without firmness, feeble, weak* :—Mín sául geweard swancur on móde ðæt ic on ðínre hǽlu hogode *defecit in salutari tuo anima mea*, Ps. Th. 118, 81.

swane-wyrt (?), Lchdm. ii. 74, 20.

swán-geréfa, an; *m. An officer whose duties were connected with the management of forests in respect to the pasturing of swine in them and to the use of wood. He seems to have been under the direct control of the alderman* :—Ða (at a gemót in 825) wæs tíolo micel sprec ymb wuduléswe tó súðtúne ongægum west on scýrhylte waldon ða swángeréfan ða lǽswe forður gedrífan ond ðone wudu geþiogan (-cgan, Thorpe) ðon hit aldgeryhto wéron ðon cuæd se biscop and ðara hína wiotan ðæt hió hin néren máran ondeta ðon hit ǽræded wæs on Aeðelbaldes dæge ðrim hunde swína mæst ond se biscop (and) ða hígen (tugen, Kemble) áhten twǽde ðæs wuda ond ðæs mæstes . . . In ða tiid wæs hama suángeréfa tó súðtúne and hé rád ðæt hé wæs et ceastre and ðone and gesceáwade suá hine his aldormon héht Eádwulf *there wanted this man a very great case about pasture in the wood at Sutton* (in Worcestershire). *The swain-reeves wanted to push the pasture and take the wood beyond the old rightful limits. The bishop and the counsellors of the brethren said, that they would never make further admission to them than was contained in the terms settled in Ethelbald's time* :—mast for three hundred swine, and the bishop and brethren should have two-thirds of the wood and of the mast . . . *At that time Hama was* swainreeve *at Sutton, and he rode to Worcester and watched the oath* (taken by the bishop in support of his case), *as his alderman Eadwulf* (Eadwulf dux is a witness to the charter) *bade him*, Cod. Dip. Kmbl. i. 278–279. See Kemble's Saxons in England, ii. 177; 81: and cf. the later swain-mote, which is a court touching matters of the forest.

swangor; *adj. Heavy in movement of the body or mind, slow, slothful, sluggish, indolent*, (a) physically :—Nis hé (*the Phenix*) swár ne swongor swá sume fuglas ða ðe late fugloþ lýft lácaþ fiþrum ac hé is snel and swift *non est tarda, ut volucres quae corpore magno incessus pigros per grave pondus habent, sed levis et velox*, Exon. Th. 220, 4; Ph. 315. (b) metaphorically :—Hé wæs swangor (prútswangor, MS. D.) and swǽr, and him wæs láð þearfendum mannum mete tó syllenne, Wulfst. 257, 12. Nalæs eallum monnum swongrium (swengum, MS. B.: suongrum, Bd. M.) and heora lífes ungemyndum *non omnibus desidiosis ac vitae suae incuriosis*, Bd. 5, 12; S. 630, 38. [*O. H. Ger.* swangar *gravidus, praegnans*: *Du.* zwanger: *Dan.* swanger.]

swangorness, e; *f. Heaviness, torpor, sloth, indolence, sluggishness* :—Ic wát ðæt swongorness hí ofsit and hí mid slǽwþe ofercymþ, Bt. 36, 6; Fox 180, 33. Ðæt is ðæt hé ða Godes gifa becnytte on ðæm sceáte his slǽwðe and hé for his swongornesse hié gehýde *pecuniam quippe in sudario ligare est percepta dona sub otio lenti torporis abscondere*, Past. 9; Swt. 59, 16. Ðæt is ðonne ðæt mon his eáge lǽte slápian ðæt mon for his unwísdóme and for his suongornesse ne mǽge ongietan ða undeáwas ðara ðe him underðiédde beóþ. Ne slǽpþ se nó fæste ac hnappaþ se ðe gecnáwan mæg hwæt tælwierðe biþ and suáðeáh for his módes swongornesse oððe recelíeste forwandaþ *cum se hieremenn somnum quippe oculi dare est intentione cessante subditorum curam negligere . . . Non autem dormire, sed dormitare, est quae quidem reprehenda sunt cognoscere, sed tamen propter mentis taedium dignis ea increpationibus non emendare*, 28; Swt. 195, 1–10.

swán-riht, es; *n. Law concerning the* swán (q. v.):—On manegum landum gebyreþ deópre swánriht, L. R. S. 6; Th. i. 436, 15.

swán-steorra, au; *m. The herd's star, the evening star* :—Suánsteorra *vesper*, Wrt. Voc. ii. 123, 42. Cf. swána steorra *under* swán.

swápan; *p.* sweóp; *pp.* swápen *To sweep.*   **I.** *trans.* (a) *To sweep with a brush* (lit. or metaph.) :—Ic swápe *uerro*, Ælfc. Gr. 28, 4; Zup. 169, 14. Ic sweóp gást mínne *scopebam spiritum meum*, Ps. Spl. 76, 6. (b) *to sweep, move* (*something*) *with the action of one sweeping* :—Swápeþ sigemêce mid ðære swíðran hond ðæt deófol gefeallaþ in sweartne lêg *he shall sweep the victorious blade with the right hand, so that devils shall fall into dark flame*, Exon. Th. 93, 24; Cri. 1531. [Mid beseme clene swopen *scopis mundatam*, O. E. Homl. ii. 87, 10. Me wule swopen þin hus, Misc. 176, 151 : Fragm. Phlps. 7, 6. Heó swopeð þe duste awei, A. R. 314, 6. Clensi and zuope þe herte, Ayenb. 109, 5. Chaucer has swope, swoope.]   **II.** *intrans. To sweep, have a sweeping motion, drive*; the form and much of the sense belong to *swoop* :—Húse on munte on swift wind swápeþ (cf. hûs on munte full ungemetlíc wind gestent, Bt. 12; Fox 36, 16) *montis cacumen protervus auster totis viribus urget*, Met. 7, 20. Cf. answeóp, -suaep *afflarat* (at-, ad-), Txts. 38, 32. Onsweóp, **43.** 235. Brim wíde wæðde, wælfæðmum sweóp, Cd. 208, 9; Exod. 480. Hê geseah swápendum (or under I. b) windum ðone lêg ðæs fýres ofer ðære burge wallas áhefenne (se wind ðæt fýr ofer ða wallas dráf, MS. B.) *cum ventis ferentibus globos ignis supra muros urbis exaltari conspiceret*, Bd. 3, 16; S. 542, 37. Cf. onsuápen *instincta*, Wrt. Voc. ii. 111, 79. Inswápen, 44, 35. [Swyíte swaynes ful swyþe swepen þertylle, Allit. Pms. 83, 1509.]   **III.** *to wrap.* v. be-, ymb-swápan. [*O. H. Ger.* sweifan *to swing.* Cf. *Icel.* sveipa (*wk.*) *to sweep, stroke; make a sweeping stroke* with a weapon ; *wrap, swaddle.*] v. á-, for-, tô-, ymb-swápan.

swár, swár-. v. swǽr, swǽr-.

swara *in* áþ-swara :—Ðes áðswara *hoc jus jurandum*, Ælfc. Gr. 14; Zup. 88, 6.

swarcan, swarcian, swáre. v. swearcan, swearcian, swǽre.

swarian. v. and-swarian. [*Icel.* svara *to answer* : *Dan.* svare. He wass wis to swarenn and to fraȝȝnenn, Orm. 8938. He called to his chamberlayn, þat cofly hym swared, Gaw. 2011.] v. áþ-swaring, swornian, swornung, sweart.

swaring (-ung), swarnian, swarnung, swart. v. áþ-swaring, swornian, swornung, sweart.

swaru (1) swer *in* an-*swer*. v. and-swaru. [Cf. *Icel.* svar; *n. answer* : *Dan.* svar. Forrhwi ȝho ȝaff swillc sware onnȝæn, Orm. 2422.] (2) *swearing, oath.* v. áþ-, mán-, mánáþ- (be mánáþsware *de perjurio*, L. Ecg. C. tit. 34 ; Th. ii. 130, 24) swaru. [Mid false sware, O. E. Homl. ii. 259, 35. Of alle sunnen . . . of sum uals word, of sware, A. R. 344, 3. He sahtnesse mid sware (treoðe, 1st MS.) hadde ifastned, Laym. 10893.]

swǽse. v. swǽs.

swát, es; *n.* [The passages in which the gender is marked are doubtful. Ðæt swót, Lchdm. iii. 98, 17, occurs in a late MS. ; ísen swát, ii. 296, 18, may be a compound ; ða swát, iii. 72, 28, may be a mistake for spátl, v. ii. 56, 15. Dutch has a neuter, German and Scandinavian have masculines.]   **I.** *sweat, perspiration* :—Seofoðe (*the seventh of the constituents from which Adam was made*) wæs deáwes pund, ðanon him (*Adam*) becom swát, Salm. Kmbl. 180. 15. Suát, Rtl. 192, 17. His swát (*sudor*) wæs swylce blôdes dropan, Lk. Skt. 22, 44. Of ealdum cláðum ðe beóþ eal on swáte, Homl. Ass. 35, 280. Swá ða swát (*but* ii. 56, 15 *has* spátl), beóþ missenlícu, Lchdm. iii. 72, 28.   **I a.** *that which exudes like sweat* :—Ðanne þeó brǽde geswáte nim ðæt swót *when the roast meat sweats, take that which exudes*, Lchdm. iii. 98, 17.   **I b.** *that which lies on anything as sweat lies on the skin* (?) :—Wiþ gongelwæfran bite, smít on ísen swát (isen-swát ?), ii. 296, 18.   **II.** *used of other moisture that comes from the body,* (1) *foam* :—Mið swáte *cum spuma*, Lk. Skt. Rush. 9, 39. (2) *blood* :—Saga mê hwæt ðæs lifgendan mannes gleng sý. Ic ðê secge ðæs deádan swát, Salm. Kmbl. 200, 10. Geseoh seolfes swæðe, swá ðín swát ágeát, blôdige stíge, Andr. Kmbl. 2881 ; An. 1443 = 2552 ; An. 1277 : Beo. Th. 5380 ; B. 2693. Him for swenge swát ǽdrum sprong forð under fexe, 5925 ; B. 2966. Beswyled mid swátes gange, Rood Kmbl. 45 ; Kr. 23. On róde ðú ðín blôd águte for heó and [hý] mid ðínum ðær æþelan swáte gebohtest, Wulfst. 255, 23. Cwealmdreóre, monnes swáte, Cd. Th. 60, 24 ; Gen. 986. Be sídan ðǽr Hǽlend his swát forlêt, 299, 6 ; Sat. 545 : Andr. Kmbl. 1935 ; An. 970 : Exon. Th. 88, 33 ; Cri. 1449.   **III.** *sweat that comes from labour,* hence *labour, toil* :—Ðǽr wæs suíðe suíðlíc gesuinc and ðǽr wæs micel swát ágoten and deáh ne meahte monn him of ániman ðone miclan rust *multo labore sudatum est, et non exivit de ea nimia rubigo ejus*, Past. 37, 3 ; Swt. 269, 12. On swáte ðínes andwlitan ðú brícst ðínes hláfes, Gen. 3, 19. Se man on gewinne and on swáte hê leofaþ, Blickl. Homl. 59, 30 : Cd. Th. 33, 8 ; Gen. 482. [*O. Sax. O. Frs.* swêt : *Du.* zweet ; *n.* : *O. H. Ger.* sweiz : *M. H. Ger.* sweiz *sweat* ; *blood* : *Ger.* schweiss ; *m.* : *Icel.* sweiti *wk. m.* : *Dan.* sved ; *m.* : *Swed.* swett ; *m.*] v. heaðu-, hilde-swát ; swǽtan, *and next word.*

-swát; *adj. in* ge-swát *sweaty, sweating* :—Ðara breósta biþ deáwig wǽtung, swá swá sié geswát, Lchdm. ii. 258, 18. Gebeþe ða hamma mid ðam stánbaðe ; ðonne hié sién geswáte, ðonne recce hê ða bán, 68, 6.

swátan (swatan ? v. Engl. Stud. viii. 479) ; *pl. Beer* :—Swátan *cervisia*, Wrt. Voc. i. 290, 62 : ii. 17, 25. Áwyl on súrum swátum oþþe on súrum ealað, Lchdm. ii. 34, 15. [Jamieson gives *swaits* new ale or wort ; but also *swats* new ale ; the thin part of flummery.]

swát-cláþ, es ; *m. A handkerchief, towel, napkin ;* sudarium :—Se apostol him ásende his swátcláð . . . Hê weard álýsed swá hraðe swí se swátcláð hine hrepode, Homl. Th. ii. 414, 21–25. [Cf. *O. H. Ger.* sweiz-tûh *sudarium, orarium* : *Ger.* schweiss-tuch *handkerchief* : *Icel.* sveita-dúkr *a napkin* : *Dan.* svede-dug *a handkerchief.*] v. swát-lín.

swát-fáh; *adj. Blood-stained* :—Oft æt hilde gedreás swátfág and sweordwund sec[g] æfter óðrum, Vald. 1, 5. Swátfáh syrce, Beo. Th. 2226 ; B. 1111.

Swá-ðeód, swaðor. v. Sweó-þeód, swæðer.

swaðrian, swæðorian ; *p* ode *To retreat, withdraw, subside* :—Geofon swaðrode . . . geótende gegrind grund eall forswealg, Andr. Kmbl. 3169 ; An. 1587. Hærn eft onwand . . . wǽdu swæðorodon, 1066 ; An. 533. Brimu swaþredon, ðæt ic sǽnæssas geseón mihte, Beo. Th. 1145 ; B. 570. v. swedrian.

swaþu, e ; *f. A track, trace, footstep, vestige ;* left in *swathe* a row of mown grass :—On Oliuetes dúne syndon nú gyt ða swæþe Drihtnes fótlásta . . ne mihte scó his swaðu beón ðǽm óðrum flórum geonlícod . . . ða his swaða syndon monnum tô ecre láre . . . men mihton sceáwian Drihtnes fôta swaðe, Shrn. 80, 35–81, 15. Næs bútan seó swaðu (*the trace of a wound, scar*) on, 05, 3. Wê sóðfæstes swaðe folgodon, Andr. Kmbl. 1346 ; An. 673. Him on swaðe fylgeþ *follows in his track, pursues him*, Salm. Kmbl. 186 ; Sal. 92 : Exon. Th. 397, 25 ; Rä. 16, 25 : 487, 23 ; Rä. 74. Hæleð lágon, on swaðe sæton *sat in the track, were left dead in the track of the retreating force*, Cd. Th. 125, 10 ; Gen. 2077 : 127, 21 ; Gen. 2114 : Andr. Kmbl. 2844 ; An. 1424. Hié (*the defeated Assyrians*) on swade reston, Judth. Thw. 26, 11 ; Jud. 322. On swaðe feóllon æðelinga bearn, Cd. Th. 120, 28 ; Gen. 2001. Hig unc ásetton on óðre healfe ðære eá, ðæt ða ne mihton uncre swaðe findon, Shrn. 42, 3. Nǽnige swaðe his *nullum ejus vestigium*, Bd. 4, 23 ; S. 595, 3. Þeáh ælda bearn lástas míne sécaþ, ic swaþe míne bemíþe, Exon. Th. 500, 26 ; Rä. 89, 12. Swæðe, Andr. Kmbl. 2880 ; An. 1443. Ða swaþe áwuniað reogollíces lífes *regularis vitae vestigia permanent*, Bd. 4, 3 ; S. 566, 43. Sweðe míne *vestigia mea*, Ps. Surt. 16, 5 : 17, 37. v. dolh-, fót-, swát-, weald-, wund-swaþu ; swæp.

swaþu ? :—Swína swaþu *suesta*, Wrt. Voc. i. 286, 56. Swína suadu (sceadu, Corp. Gloss.) *suesta, sivesta*, Txts. 98, 972.

swaþul, es ; *m. That which swathes or wraps* (? v. sweþel) :—Ðæs ne wêndon witan Scyldinga ðæt hit (*the hall*) manna ǽnig tôbrecan meahte nymþe líges fæðm swulge on swaþule *unless the flame's embrace swallowed up the house in its swathing fire*, i. e. *unless the house were completely wrapt in flames* (Thorpe would read *swaloðe* = heat, v. sweoloþ : Grein translates the word by *rauchqualm*; compare *Ger.* schwaden *vapour* : *M. H. Ger.* swadem : *O. H. Ger.* swedan *cremare*), Beo. Th. 1568 ; B. 782.

swátig; *adj.*   **I.** *sweaty* :—Ðá ongan hê hine baðian swá swátigne *cum sudans in amnem descendisset*, Ors. 3, 9 ; Swt. 124, 31. Godes engel mid handcláðe wípaþ ðíne swátigan limu, Homl. Th. i. 426, 31.   **II.** *bloody* :—Sweord wæs swátig, Beo. Th. 3143 ; B. 1569. Sweord and swátigne helm, Judth. Thw. 26, 20 ; Jud. 338. Ðú meaht geseón on mínre sídan swátge wunde, Exon. Th. 89, 19 ; Cri. 1459. [*M. H. Ger.* sweizec : *Ger.* schweissig : *Icel.* sweitugr.]

swátig-hleór ; *adj. Having a sweaty face* :—Ðú scealt swátighleór ðínne hláf etan (*in the sweat of thy brow shalt thou eat bread*, Gen. 3, 19), Cd. Th. 57, 27 ; Gen. 934.

swát-lín, es ; *n. A napkin, handkerchief* :—Swátlín *sudorium*, Wrt. Voc. ii. 73, 68. On ðæm swátlíne (*in sudarium*) ðe Xrist ymbe sprǽc on his godspelle, Past. 9 ; Swt. 59, 13. Ðín pund ðe ic hæfde on swátlín (*in sudario*) áléd, Lk. Skt. 19, 20. v. swát-cláþ.

swát-swaþu, e ; *f. A bloody track* :—Wæs sió swátswaþu Sweóna and Geáta, wælt ǽs wera, wíde gesýne, Beo. Th. 5884 ; B. 2946.

swát-þyrel, es ; *n. A pore* :—Swátþyrlu *pori* i. *spiramenta unde sudor emanat*, Wrt. Voc. i. 44, 25. [Cf. Swete-holle *porus*, Wrt. Voc. i. 209, 9. Swet-hole, Cath. Ang. 373, col. 2. *O. H. Ger.* sweiz-loh : *Ger.* schweissloch *a pore* : *Icel.* sweita-bora : *Dan.* swede-hul.]

swealwe, swealewe, an ; *f. A swallow* :—Suualuae, suualuuae, suualuue *progna*, Txts. 90, 828. Sualuuae, sualuae, sualuue *hirundo*, 68, 498. Swealwe, Wrt. Voc. ii. 43, 5. Swalowe, swaluwe, swalewe, Ælfc. Gr. 9, 3 ; Zup. 37, 7. Storc and swalewe, Homl. Th. i. 404, 25. Genim swealwan nest, Lchdm. ii. 100, 18. Swolwan, iii. 44, 13. Genim swealwan, gebærn tô ahsan, ii. 156, 8. Hú ða swalawan on him sæton ða sculdra ðæs hálgan weres Gúðláces, Guthl. 10 ; Gdwin. 52, 3–10. For instances of the word in local names, see Cod. Dip. Kmbl. vi. 338. [*O. H. Ger.* swalawa : *Icel.* swala.] v. hae-, heoru-, stæþ-swealwe.

swearc (?) ; *adj. Weak, feeble, faint.* v. next word. v. swearcan.

swearc-módness, e ; *f. Faintheartedness, pusillanimity* :—Fram swearcmódnesse gástes *a pusillanimitate spiritus*, Ps. Lamb. 54, 9. v. next word.

swearcan (?) *to grow dark* (?); metaph. *to grow faint, languish.* v. ā-swarċan, *the preceding and following words, and* sweorcan.

swearcian; *p.* ode. I. *to make* or *to become dark* :—Seó swearc-igende sunne and ða gescæafta samod ealne middaneard āðeóstrodon mid sweartre nihte for heora Scyppendes ðrowunge, Homl. Th. ii. 258, 15. II. *to make* or *to become troubled, to dismay.* v. ā-swarcian, *and preceding words.*

sweard, es; *m.* (?) *Sward* (= rind of bacon; cf. too green-*sward* the turf-covering of the earth), *skin, hide* :—Sweard *cutis,* fel *pellis,* Wrt. Voc. i. 283, 32–3. Sweard *cutis,* rib *costa,* heorte *cor,* ii. 16, 54–6. Swearth *cater,* 103, 22. Suge sweard *vistula,* 124, 1. Sweard *catrum,* 13, 52. [Swarde *or* sworde of flesche *coriana,* Prompt. Parv. 482. Turfe, sward of þe erþe *cespes,* 506. *O. Frs.* swarde *skin (of the head) : Du.* zwoord; *n. skin: M. H. Ger.* swarte, swart; *f. skin with hair on : Ger.* schwarte; *f. skin, rind: Icel.* svörðr; *m. the skin* (especially of the head); *hide* of walrus; gras-, jarðar-svörðr *green-sward.*]

swearm, es; *m.* A *swarm, crowd* :—Sue[a]rm *examen,* Wrt. Voc. ii. 107, 82. Swearm, 32, 17: 144, 43 (*examen* has been omitted here by Wright, see Wülck. Gl. 230, 6): Ælfc. Gr. 9, 12; Zup. 40, 14: *examen, multitudo,* Hpt. Gl. 437, 37: 496, 14. [*O. H. Ger.* swaram, swarm; *m. examen: Icel.* svarmr; *m. tumult: Swed.* swärm; *m. a swarm: Dan.* svärm *a swarm; rioting: Du.* swerm; *m. a swarm, crowd.*]

sweart; *adj.* I. *of colour, swart, swarthy, black, dark* :—Sweart *ater, teter; ceruleus,* Wrt. Voc. i. 46, 32, 53 (in a list of colours): *furvus,* ii. 34, 39, 40: *fuscus,* 38, 27: *luridus,* 53, 15: *pullus,* 57, 10: *niger,* Ælfc. Gr. 8; Zup. 27, 9: *caeruleus,* Hpt. Gl. 516, 14. Wudurēc sweart, Beo. Th. 6281; B. 3145. Hræfn sweart and sealobrūn, Fins. Th. 70; Fin. 35. On ðæm clife hangodan manige swearte sāula . . . and ðæt wæter wæs sweart under ðæm clife neoðan, Blickl. Homl. 209, 34–211, 1. Ðonne sweartan wolcnu (*nubes atrae*) him beforan gāþ, Bt. 6; Fox 14, 22. Engla and deófla, hwītra and sweartra, Exon. Th. 56, 9; Cri. 898. Mænigeo sweartra gāsta *spirituum deformium multitudo,* Bd. 5, 12; S. 628, 4. On sweartum stafum and atollīcun āwritene *tetricis descripta litteris,* 5, 13; S. 633, 8. Sweartran *furviores,* Wrt. Voc. ii. 37, 51. Hī āsettan ofer hyre ða sweartestan fyðra, L. E. I. prm.; Th. ii. 398, 27. II. *of absence of light or brightness, dark, black, gloomy* :—Óðer (beám) wæs swā wynlīc, wlitig and scēne . . . wæs se óðer eallenga sweart, dim and þȳstre, Cd. Th. 30, 35; Gen. 477. Eów is hām sceapen sweart sinnehte, Exon. Th. 142, 26; Gū. 650. Ða þeóstre ðære sweartan nihte, Bt. 4; Fox 6, 34. Ðære sweartan helle grund, Cd. Th. 22, 24; Gen. 345. Se ðe on þȳstre færeþ, on sweartre niht, Exon. Th. 54, 23; Cri. 873. Deorc gesweorc sinnihte sweart, Cd. Th. 7, 21; Gen. 109. Tó helle on ðone sweartan sīð, 45, 27; Gen. 733. On dīglum ł on sweartum dymnyssum *latibulis,* Hpt. Gl. 480, 28. Landa sweartost *hell,* Cd. Th. 31, 19; Gen. 487. III. *of absence of good, black* (crime), *dark, dismal* :—Gāstas twegen, óðer biþ golde glædra, óðer biþ grundum sweartra, Salm. Kmbl. 976; Sal. 488. Sweartes hæðendōmes *tetrae gentilitatis,* Hpt. Gl. 523, 41. Micel yfelnyss wæs on Iudēiscum mannum, ðā ðā hī syrwdon mid sweartum geþance (*with dark design*), hū hī Crist ācwealdon, Homl. Skt. i. 11, 318. Swā lange wæs hē hylt ðone sweartan nīð on his heortan, Homl. Th. i. 54, 13. Mānfulra heáp sweartne *the devils,* Salm. Kmbl. 299; Sal. 149. Ic fela gefremede sweartra synna, Exon. Th. 261, 10; Jul. 313: 270, 20; Jul. 468. Gē hellfirena sweartra geswīcaþ, 366, 4; Reb. 7. In ða sweartestan and ða wyrrestan wītebrōgan, Elen. Kmbl. 1859; El. 931. [*Goth.* swarts: *O. Sax.* O. *Frs.* swart: *O. H. Ger.* swarz: *Icel.* svartr.] v. fȳr-, swefel-sweart.

swearte; *adv. Darkly, dismally, evilly* :—Sió gītsung swearte swæfeþ onlīce ðam munte ðe monna bearn Etne hātaþ, Met. 8, 47 (v. swæfan). Satanus swearte geþohte (cf. Milton's 'dark designs,' and v. sweart, III), ðæt hē wolde on heofonum hēhseld wyrcan, Cd. Th. 287, 22; Sat. 371. Satanus swearte (*miserably*) þingaþ and ða atolan mid him wītum wērige, 292, 28; Sat. 447. Him ðæt swearte forgeald (*made grievous compensation*) Iudas innon helle, 301, 8; Sat. 578.

sweart-hēwen; *adj. Dark purple, violet-coloured* :—Ða sweart-hēwenan *cerula,* Wrt. Voc. ii. 20, 67.

sweartian; *p.* ode *To make* or *to become black* :—Ðanne sweartigaþ (sweratiged, MS.) hȳ (*the teeth*) and feallaþ (-eð, MS.), Lchdm. iii. 104, 17. Hē sweartade (*but see* spearcian), Cd. Th. 269, 24; Sat. 78. Ðā āræs se wind, and ða wolcnu sweartodon, Homl. Skt. i. 18, 151. Gesweartode *denigratos,* Hpt. Gl. 514, 32. [þ te hude snawhwit swartete as hit snarchte, Marh. 18, 14. *O. H. Ger.* swarzen *to become black: Icel.* svarta *to dye black.*] v. ā-sweartian.

sweart-lāst; *adj. Leaving a black track* :—Fugles wyn (*a pen*) stōp eft on mec (*a book*), sīþade sweartlāst, Exon. Th. 408, 12; Rä. 27, 11.

sweartness, e; *f.* I. *blackness* :—Sweartnysse *nigredine,* Hpt. Gl. 514, 50. II. *a black material* :—Sweartnesse *atramentum,* Wrt. Voc. ii. 84, 72: 5, 31.

swebban; *p.* swefde, swefede; *pp.* swefed. I. *to send to sleep, lull* :—Suebbo *sopio,* Wrt. Voc. ii. 120, 72. Ne hȳ lyft swefeþ, Exon. Th. 115, 19; Gū. 192. Swefed *sopitus,* Kent. Gl. 917. Wæs hē sæmninga mid leóhte slǣpe swefed, Guthl. 6; Gdwin. 42, 13. II.

of the sleep of death, *to put to death, kill* :—Hē swefeþ ond scendeþ, Beo. Th. 1204; B. 600. Ic hine sweorde swebban nelle, aldre beneótan, 1363; B. 679. Ne mōton wyt wrecan torn Godes, swebban synnig cynn, Cd. Th. 152, 35; Gen. 2531. [God sweueð hus mid þiestre nicht, O. E. Homl. i. 233, 33. He swefede þe mid þen sweiȝe, swote þu sleptest, Fragm. Phlps. 7, 42. O. *Sax.* an-swebian *to send to sleep, to cause to die:* O. H. *Ger.* int-swebben *sopire : Icel.* svefja *to lull, assuage.*] v. ā-, on-swebban; swefian, swefan.

swecc, swæcc, es; *m.* I. *a taste, flavour, savour* :—Ðæs (*the manna's*) swæc (*gustus*) wæs swilce smedema mid hunige, Ex. 16, 31: Bt. 34, 11; Fox 152, 1. Swæcces *nectaris, saporis,* Hpt. Gl. 488, 26. Būton swæcce (*sapore*) sealtes, Coll. Monast. Th. 28, 15. On swæce swylce grēne cystel, Lchdm. i. 108, 2. On swæcce swēttran ðonne beóna hunig, Homl. Th. ii. 136, 30: 144, 4. I a. *the sense of taste* :—Mid ūrum fíf andgitum . . . swæc and stenc, Homl. Th. i. 138, 27. Swæcc, ii. 550, 11: Wrt. Voc. i. 42, 60. II. *an odour, a scent, smell* :—Wundorlīces brǣdes swæc, Homl. Th. ii. 352, 15. Seó wundriende swētnes ðæs swæcces (*odoris*), Bd. 5, 12; S. 629, 20. Stēmendes swæcces *nardi pistici,* Hpt. Gl. 516, 38. Ðæt hūs wæs gefylled of ðære sealfe swæcce (*odore*), Jn. Skt. 12, 3. Gif ðū hyre blōsðman brȳtest, hē hæfþ swæc swylce ellen, Lchdm. i. 104, 20. Swecca swētast swylce stincaþ wyrta geblōwene, Exon. Th. 178, 20; Gū. 1247. Swæcca, 358, 11; Pa. 46. Swētum swæccum (*odoribus*), 212, 22; Ph. 214. Sweccum, Kent. Gl. 1016. II a. *the sense of smell* :—Swæc *odoratus* (in a list 'de homine et de partibus ejus'), Wrt. Voc. i. 282, 31: 64, 19. Stenc, swæc *olfactum,* swæc *odoratus,* ii. 62, 45, 46. [O. *Sax.* swek *an odour:* O. H. *Ger.* swehhi *odor.*]

sweccan *to smell.* [O. H. *Ger.* swehhen *olere, adolere, fragrare.*] v. ge-sweccan.

swediende. v. spediende.

swefan; *p.* swæf, *pl.* swǣfon; *pp.* swefen *To sleep.* I. *of natural sleep* :—Se ne slǣpeþ ne swefeþ (*or* III a) swȳðe *non dormitavit neque obdormiet,* Ps. Th. 120, 4. Hē swifeþ slǣpe gebiesgad, Exon. Th. 358, 1; Pa. 39. Hē sōfte swæf, Cd. Th. 12, 2; Gen. 179: 94, 19; Gen. 1564. Sceótend swǣfon, ða ðæt hornreced healdan sceoldon, Beo. Th. 1411; B. 703: 2564; B. 1280. Ðū-mōst sorhleás swefan, 3348; B. 1672: 238; B. 119. Geseah hē in recede swefan sibbegedriht samod ætgædere, 1462; B. 729: Exon. Th. 344, 25; Gn. Ex. 179. Swefan under swegles hleó, Andr. Kmbl. 1663; An. 834. Swefan on slǣpe, 1695; An. 851. II. *of the sleep of death* :—Se fǣge þegn æfter billes bite swefeþ, Beo. Th. 4127; B. 2060. Se wyrm ligeþ, swefeþ sāre wund, 5485; B. 2746. Swefaþ ða ðe beadogrīman bȳwan sceoldon, 4505; B. 2256. Hȳ deáðdrepe drihte swefeþ, synfullra sweót sāwlum lunnon, Cd. Th. 209, 7; Exod. 495. Hǣðene swēfon, deáðwang ridon, Andr. Kmbl. 2004; An. 1004. Hlāfurd sēcan oðþe hēr swefan, Vald. 1, 31. III. *metaphorically,* (a) *to denote lack of watchfulness* :—Ðonne se weard swefeþ, sāwele hyrde, Beo. Th. 3487; B. 1741. (b) *to denote cessation of activity* :—Swǣfon seledreámas, Cd. Th. 179, 29; Exod. 36. [Cf. þa sæ sweuede, Laym. 25548. *Icel.* sofa; *p.* svaf *to sleep: Dan.* sove: *Swed.* sofwa.] v. swebban.

swefecian. v. ā-swefecian, Wrt. Voc. ii. 31, 5: 77, 32. Cf. stȳfecian.

swefel, swefl, es; *m. Sulphur, brimstone* :—Swefl, swefel, swæfl *sulfur,* Ælfc. Gr. 9, 22; Zup. 49, 3. Swefel, Wrt. Voc. i. 37, 27. Ðæt sceal wrecan swefyl and sweart līg, Cd. Th. 145, 33; Gen. 2415. Se byrnenda swefl ðone munt (*Etna*) bærnþ, Bt. 16, 1; Fox 50, 4. Swefles *sulphuris,* Hpt. Gl. 489, 1. Níwes swefles fíf cuclermǣl, Lchdm. ii. 252, 21. Eallbyrnende rēnscūr mid swefle gemencged, Gen. 19, 24: Met. 8, 50. Swæfle, Boutr. Scrd. 22, 29, 32. Hit rīnde fȳr and swefl, Lk. Skt. 17, 29: Cd. Th. 153, 19; Gen. 2541. Hwylce þinc gelǣdest ðū (*the merchant*) ūs? . . . mæstlingc, ǽr and tin, swefel and glæs, Coll. Monast. Th. 27, 11: Lchdm. i. 200, 2. Swefl, ii. 56, 10. [*Goth.* swibls: *Du.* zwavel: O. H. *Ger.* swebal(-el, -il, -ul) *sweval sulphur: Ger.* schwefel: *Dan.* svovl: *Swed.* swafwel.]

swefel-rēc, es; *m. Sulphur-smoke, the smoke from burning sulphur* :—Rīneþ ofer ða synfullan swefelrēc *pluet super peccatores sulphur,* Ps. Surt. 10, 7. Cf. swefel-þrosm.

swefel-sweart (?); *adj. Dark with the smoke of sulphur* (?) :—Swefl-sweart *sulforia,* Wrt. Voc. ii. 121, 61.

swefel-þrosm, es; *m. The vapour or smoke of sulphur* :—Hē rȳnde ofer synfullan swefðrosm *pluit super peccatores sulphur,* Ps. Spl. 10, 7. Cf. swefel-rēc.

swefen, swefn, es; *n.* I. *sleep* :—Hit wæs deáðes swefn . . . menniscra morð, Cd. Th. 45, 1; Gen. 720. Hī slēpon swæfnum *dormierunt somnum,* Ps. Spl. 75, 5. Gif ic selle swefnu ł slǣp eágum mīnum *si dedero somnum oculis meis,* Ps. Lamb. 131, 4. II. *a dream* :—Hē rehte him his swefen (*somnium*) and bæd, ðæt hig him sǣdon, hwæt ðæt swefen beheóld, Gen. 41, 8. Him weard on slǣpe swefen ætȳwed, Cd. Th. 247, 13; Dan. 496. Swefn, 257, 7; Dan. 654. Hē ne wisse word swefnes sīnes, 223, 27; Dan. 126. Com on sefan hwurfan swefen wōma, 222, 25; Dan. 110: Elen. Kmbl. 142; El. 71. Óðer swefen hine mǣtte and hē rehte ðæt his brōðrum : 'Ic geseah on swefne (*per somnium*),' Gen. 37, 9. For ðære gesihðe ðe hē on ðæm swefne geseah, Past. 16;

Swt. 101, 18. Đá stód him sum mon æt đurh swefen (*per somnium*) . . . Đá hēt heó secgan đæt swefen, Bd. 4, 24; S. 597, 11–31. Swefn, Cd. Th. 159, 16; Gen. 2635 Tô āsecganne swefen, 224, 1; Dan. 129. Swefnu gefremminge habbaþ *dreams will have accomplishment*, Lchdm. iii. 186, 12. Swefenu, 196, 11. Swefna ýdele sint, 188, 21. Swefue (swæfna) gewisse synt, 186, 19, 27. Feor áweg gewítan swefna and nihta gedwymeru *procul recedant somnia et noctium fantasmata*, Hymn. Surt. 11, 29. Ic swefna cyst secgan wylle, Rood Kmbl. 1; Kr. 1. Hí āfēngon andsware on swefnum, Mt. Kmbl. 2, 12. On swefnum (soefnum, Lind.) gemynegod, 22: Homl. Th. i. 88, 15. Heó ādrǽfe swefnu *pellat sompnia*, Hymn. Surt. 37, 6. [Now God my swevene rede aright, Chauc. Nonne Pr. T. 76. Thanne gan I to meten a merueilouse sweuene, Piers P. prol. 11. Swevene or dreme *sompnium*, sweuene or slepe *sompnus*, Prompt. Parv. 483. O. Sax. sweban *a dream*: Icel. svefn, ȝöfn *sleep*: *a dream*: Dan. søvn *sleep*: Swed. sömn *sleep*.] v. un-swefen.

swefen-racu, e; *f. The interpretation of a dream*:—Galdorcræftas and swefenraca *incantationes et somniorum interpretationes*, L. Ecg. C. 29; Th. ii. 154, 29. v. next word.

swefen-reccere, es; *m. An interpreter of dreams, a diviner, sooth-sayer*:—Swefnreccere *conjectorem*, Wrt. Voc. ii. 15, 40. [Cf. *O. H. Ger.* troum-rechare *conjector*.] v. preceding word.

swefet, swefian *to lull*. v. swefot, ge-swefian.

swefian (?) *to move*. v. *passage given under* forþ-swebban (-swefian?) [cf. *O. H. Ger.* swebēn: *Ger.* schweben]. Cf. swífan.

sweflen; *adj. Sulphurous, of brimstone*:—Him stód swæflen líg of đam múđe, Homl. Th. i. 466, 26. Eđua đæt sweflene fýr, Ors. 2, 6; Swt. 88, 30. Hé eal đæt land mid sweflenum fýre forbærnde *Deus pluit super hanc terram ignem et sulphur, totamque regionem exustam aeterna perditione damnavit*, 1, 3; Swt. 32, 10: Ælfc. T. Grn. 4, 17. Swæflenum, Boutr. Scrd. 22, 32. Sweflenum þicnyssum *sulphureis flammarum globis*, Hpt. Gl. 499, 49.

swefn. v. swefen.

swefnian; *p.* ode.   *I. with acc. of person*, cf. mǽtan, *To appear in a dream to a person*:—Swá hwæt swá hine swefnaþ *whatever he dreams*, Lchdm. iii. 184, 9. Swá hwæt swá đé geswefnaþ, 154, 24. Āhicgan on sefan đínne hū đé swefnede, Cd. Th. 224, 5; Dan. 131. Đē heortan deópnyssa swefnian *te cordis alta somnient*, Hymn. Surt. 3, 19.   *II. with nom. of person, To dream*:—Gif đú swefnast đé twege[n] mōnan geseón *if you dream that you see two moons*, Lchdm. iii. 212, 25. [As sweveneth the hungrende and eteth (Isaiah 29, 8), Wick. Cf. *Dan.* søvne *to fall asleep*.]

swefnigend, es; *m. A dreamer*:—Hér gǽþ se swefnigend *ecce somniator venit*, Gen. 37, 19.

swēg, es; *m.*   *I. unregulated, confused sound, noise, din, crash*:—Suoeg, cirm *fragor*, Wrt. Voc. ii. 109, 27. Swǽg *clangor, sonitus*, Hpt. Gl. 451, 44. Ne wind ne wætres swēg, Blickl. Homl. 65, 19. Swēg on windes onlícnesse, 133, 15. Swēg innan đan heáfedan, Lchdm. iii. 92, 25. Wæs đeód on sǽlum sigefolca swēg, Beo. Th. 1292; B. 644: Cd. Th. 289, 26; Sat. 403. For gedréfednesse sǽs swēges (*sonitus*), Lk. Skt. 21, 25. Swoeges, Ps. Surt. 76, 18. Gebrece, swoege *fragore*, Wrt. Voc. ii. 33, 79. Mid micle swēge *cum maximo fragore*, Ors. 5, 10; Swt. 234, 3. Wiđ eárena swēge *for singing in the ears*, Lchdm. i. 350, 1. Nán monn ne gehiérde ne æxe hlem ne bíetles suēg, Past. 36; Swt. 253, 17. Hig fleóþ leáfes swēg (*sonitus folii volantis*), Lev. 26, 36. Micelne swēg unmǽtes wōpes, Bd. 5, 12; S. 628, 29. Swēgas (-es, MS.) *tonitrua*, Hpt. Gl. 452, 60. Swoegum *bombis*, Wrt. Voc. ii. 12, 8.   *II. regulated, modulated* or *articulate sound*, (a) *sound* made by living creatures, *voice, cry* or *note of a bird, song*:—Dyptongus is twýfeald swēg, Ælfc. Gr. 4; Zup. 7, 13. *Accentus*, đæt is on hwylcum stæfgefēge ǽlc word swēgan sceal, 50, 13; Zup. 290, 16. Swēg *tenor*, Hpt. Gl. 528, 21. Heofoncyninges stefn, wordhleóđres swēg, Andr. Kmbl. 186; An. 93. Swēg (*the voice of Moses*) swiđrode, Cd. Th. 197, 18; Exod. 309. Engla þreátas sigeleóđ sungon, swēg wæs on lyfte gehýred, Exon. Th. 181, 7; Gú. 1289. Biþ đæs hleóđres swēg (*the song of the phenix*) eallum songcræftum swētra, 206, 24; Ph. 131. Sume synd geworhte æfter gelícnysse ágenes swēges. titinnabulum belle, Ælfc. Gr. 5; Zup. 14, 2. Ganetes hleóþor and húilpan swēg, Exon. Th. 307, 9; Seef. 21. Swēga mǽste, 239, 9; Ph. 618. Tyrnende swēgas *rotatiles trocheos*, Germ. 403, 8. (b) *sound* made by means of an instrument. v. swēg-cræft. *voice*; *also the instrument*:—Đære býman swēg, Ex. 19, 19. Hearpan swēg, Beo. Th. 179; B. 89. Sume syndon geworhte æfter gelícnysse ágenes swēges. titinnabulum belle, Ælfc. Gr. 5; Zup. 14, 2. Swēg *classica*, Wrt. Voc. ii. 19, 67. On swēge (swōge, Ps. Surt.) býman *in sono tubae*, Ps. Spl. 150, 3. On swēge *in tympano*, 149, 3, MS. T. For đam swēge (*of the harp*), Bt. 35, 6; Fox 168, 1. Hearpan swinsigende swēg, Cd. Th. 66, 8; Gen. 1081. Heó gehýrde bellan swēg, Shrn. 149, 9: Homl. Th. ii. 156, 6. Swēgas *classica*, Wrt. Voc. ii. 131, 62. Dreámas (swēgas, MS. T.) *organa*, Ps. Spl. 136, 2.   ¶ In Wrt. Voc. ii. 110, 43: 43, 7, *swēg* glosses *hora*, because of the striking of a bell at the hours?   *III. a person*:—Be onfangenysse swēgea *de acceptione personarum*, Scint. 183, 17 (*swēg is used several times in the*

section under this heading to gloss persona). v. benc-, hearp-, here-, hilde-, morgen-swēg.

swēgan; *p.* de *To sound*.   *I. to make a noise*, (a) with the idea of movement, *to move violently with noise, to roar, rush, crash*:—Đonne sweiþ *cum insonuerit*, Kent. Gl. 12. Heora fyđera swēgaþ swá swá wæteres dyne, Wulfst. 200, 15. Æt đam forman gedelle swēgde út ormǽte wyllspring, Homl. Th. i. 562, 14. Swēgde swiđlíc wind of đam wēstene, ii. 450, 18. Đa wæterburnan swēgdon and urnon, Dóm. L. 3: Ps. Spl. 45, 3. Ǽrđan đe đæt scearpe swurd swēge tô his hneccan *descend with a crash upon his neck*, Homl. Skt. i. 19, 185. Swēgende *tumultuans*, Hpt. Gl. 528, 43. Com seó sǽ fǽrlíce swēgende, Homl. Th. i. 566, 7. Hē sette hine sylfne ongeán đam swēgendan líge, ii. 510, 8. Hē āsende swēgende fýr of heofonum, Homl. Skt. i. 2, 260. (b) without the idea of movement:—Swēgþ *tinnit*, Ælfc. Gr. 22; Zup. 128, 16. Swagoþ (swēgaþ? v. swēg, I) đa eáran, Lchdm. iii. 88, 5. Se heáf swēgde geond ealle đa ceastre, Ap. Th. 6, 10. Đæt ne sace óga on swēge *ne litis horror insonet*, Hymn. Surt. 9, 12. (b I) with a personal subject:—Drihten swēgþ *Dominus tonabit*, Cant. An. 10. God swēgde *Deus intonuit*, Ps. Spl. 28, 3: Ps. Lamb. 17, 14.   *II. of regulated, modulated sound, of speech, tone, music*:—Swēgþ eádmódnys on his stemne, Homl. Th. ii. 374, 11. *Consonantes*, đæt is samodswēgende, forđan đe hí swēgaþ mid đam fíf clypiendlícum, Ælfc. Gr. 2; Zup. 5, 17. Fæder stemn swēgde đus cweđende, Homl. Th. i. 104, 24; ii. 242, 8. Seó stefn đinre grētinge swēgde on mínum eárum, 202, 17. Heora bodunge swēg swēgde geond eall, Homl. Ass. 56, 144. Swēgde *increpuerit* (*musica*), Hpt. Gl. 445, 17. Ōþ đæt đæt forme tácn undernes swēge, Anglia xiii. 432, 953. Hwylc bóc is đæt đæt ne clypige and swēge, R. Ben. 133, 6. Đē úre stefn ǽrest swēge (*sonet*), Hymn. Surt. 7, 25. *Accentus*, đæt is swēg, on hwilcum stæfgefēge ǽlc word swēgan sceal, Ælfc. Gr. 50, 13; Zup. 290, 17. Sí swēged óþer tácn, Anglia xiii. 380, 215.   *III. to signify*:—Gregorius is Grēcisc nama, se swēigþ on Lēdenum gereorde Uigilantius, Homl. Th. ii. 118, 12. Biscop sceal beón ealle ofersceáwigende, swá swá his nama swēgþ, ii. 320, 7, 12. Swēgeþ, L. Ælfc. P. 37; Th. ii. 378, 28. [*Goth.* swōgjan *to groan*.] v. swōgan.

swēg-cræft, es; *m. The art of playing on a musical instrument*. v. swēg, II b:—Đá ongunnon ealle đa men hí herian on hyre swēgcræft . . . Apollonius cwæđ. 'Ic ongite cyst đín dohtor gefeól on swēgcræft, ac heó næþ hine ná wel geleornod,' Ap. Th. 16, 17–24.

swēg-dyne, -dynn, es; *m. A resounding din, crash*:—Heard gebrec, hlúd, unmǽte, swēgdynna mǽst (*the crack of doom*), Exon. Th. 59, 20; 955.

swēge; *adj. Sounding*:—Ungeswēge sang *diaphonia*, sum swēge (samswēge?) sang *canticum*, Wrt. Voc. i. 28, 35. v. án-, ge-swēge [: Geswēge *consona*, Wrt. Voc. ii. 134, 23. Of geswēgum *consona*, geswēgre *canora*, Anglia xiii. 132, 135, 137], swēt-, swíþ-swēge; hlúd-swēge; *adv.*

swegel, swegl, es; *n.*   *I. in a physical sense, heaven, sky*:—Đætte súđ ne norđ, be sǽm tweónum, ofer eormengrund, óþer nǽnig, under swegles begong, sēlra nǽre, Beo. Th. 1724; B. 860: 3550: B. 1773. Under swegles gang, Andr. Kmbl. 415; An. 208: 910; An. 455. Swefan under swegles hleó, 1664; An. 834: Elen. Kmbl. 1011; El. 507: Exon. Th. 38, 13; Cri. 606: 224, 11; Ph. 374. Swegles gim, heofontungol (cf. seó sunne, Bt. 35, 1; Fox 154, 29), Met. 22, 23. Swegles gim, sunne, Exon. Th. 212, 11; Ph. 208. Swegles leóht, gimma gladost, 218, 2; Ph. 288. Swegles leóma, 204, 26; Ph. 103. Swegles tapur, 205, 18; Ph. 114. On swegle *in the sky*, 34, 30; Cri. 550. Fareþ seþrum snell swegle tógeáhes, 206, 10: Ph. 124. Under swegle *under heaven*, 31, 27; Cri. 502: 210, 15; Ph. 186: Cd. Th. 85, 13; Gen. 1414: 105, 36; Gen. 1764: Beo. Th. 2160; B. 1078. Weorđeþ his (*the phenix*) hús onhǽted þurh hádor swegel (*cloudless sky*, cf. hádrum heofone, Met. 28, 48; or *bright sun*, v. III, and cf. hádor sægl, Andr. Kmbl. 2911; An. 1458), Exon. Th. 212, 19; Ph. 212.   *II. heaven*, (a) as the abode of the Deity:—Swegles ágend, Exon. Th. 34, 17; Cri. 543. Swegles aldor, Cd. Th. 53, 17; Gen. 862: 153, 18; Gen. 2540: 170, 4; Gen. 2807: Judth. Thw. 22, 31; Jud. 88. Swegles brytan, wuldres waldend, Cd. Th. 266, 17; Sat. 23. Swægles brytta, wihta wealdend, 272, 24; Sat. 124. Swegles gǽst *the Holy Ghost*, Exon. Th. 13, 16; Cri. 203. Swegles weard, Judth. Thw. 22, 27; Jud. 80. (b) as the abode of the blessed:—Nó đæs gilpan þearf synfull sáwel đæt hyre sié swegl ongeán, Exon. Th. 449, 11; Dóm. 69. Gástas sōhton swegles dreámas, engla ēđel, Andr. Kmbl. 1282; An. 641. Ic mæg swegles (or *under IV?*) gamen gehýran on heofonum, Cd. Th. 42, 18; Gen. 675. Swegles leóman, Cd. Th. 286, 13; Sat. 351. Swegles leóht, 266, 27; Sat. 28. Englas feredon sōþfæste sáwle innan swegles leóht, Chr. 1065; Erl. 198, 9. In swegles wuldre, Judth. Thw. 26, 26; Jud. 345. Gesǽlgum on swegle, Exon. Th. 101, 17; Cri. 1660: 137, 10; Gú. 557. Swegle benumene, 139, 23; Gú. 597. Sigorleán in swegle, Elen. Kmbl. 1242; El. 623. Đa đe swegl búan, Cd. Th. 6, 2; Gen. 82. On swegl faran, Exon. Th. 32, 15; Cri. 513.   *III. the sun* (but can *swegel* here = *segel*, *sigel* (q. v.)? cf. swegl = segl *a sail*, Cd. Th. 184, 10; Exod. 105: 182, 26; Exod. 81):—Heofontorht swegl gescynde under foldan fæþm, farende tungol, Exon. Th. 351, 1; Sch. 73. Swegl háte scán blác ofer burgsalo,

182, 3; Gû. 1304. Swegel byþ hâtost (on sumera), Menol. Fox 474; Gn. C. 7. **IV.** *music* (?). v. swegel-horn :—Đær (*in heaven*) wæs singal sang and swegles gong . . . Englas heredon hâlgan stefne Dryhten, dreám wæs on hyhte, Andr. Kmbl. 1738; An. 871. Eádige đær sittaþ mid swegle, Cd. Th. 305, 17; Sat. 648. v. Grmm. D. M. 708.

swegel-befealden; *adj. Heaven-surrounded, with heaven around :—* Hâfaþ wuldres bearn his seolfes seld sweglbefalden (-healden, Th.), lađaþ ûs đider tô leóhte, Cd. Th. 301, 28; Sat. 588.

swegel-beorht; *adj. Heaven-bright :—*Sweglbeorht sunne, Exon. Th. 174, 33; Gû. 1187. Cf. swegel-torht, heofon-beorht.

swegel-bôsm, es; *m. The interior of heaven, heaven :—* Hê biþ â rîce ofer heofenstôlas . . . sweglbôsmas heóld; đa wæron gesette wuldres bearnum, Cd. Th. 1, 18; Gen. 9.

swegel-candel[1], e; *f. The candle of the sky, the sun:—*Ær đæs beácnes cyme, sweglcondelle, Exon. Th. 205, 5; Ph. 108. Cf. heofon-candel.

swegel-cyning, es; *m. The king of heaven :—*Đæt ic wuldres God sêce, sweglecyning, Exon. Th. 167, 4; Gû. 1055. Sweglcyning, Cd. Th. 160, 30; Gen. 2689. Cf. heofon-cyning.

swegel-dreám, es; *m. Heavenly joy :—*Ufancundes engles of swegldreámum word, Exon. Th. 169, 21; Gû. 1098. Cheruphim and Seraphim on swegeldreámum, Andr. Kmbl. 1439; An. 720. Swâse swegldreámas gê (*the good at the day of judgment*) geseón môsten, Exon. Th. 82, 35; Cri. 1349. Cf. heofon-dreám.

swegel-horn; *Some kind of musical instrument :—*Sueglhorn *sambucus,* Wrt. Voc. ii. 119, 56. Swegelhorna *sambucorum, simphoniarum* (cf. simfonia, lignum concavum, Wrt. Voc. ii. 73, 60) i. *cithararum,* Hpt. Gl. 445, 19. [Cf. *Goth.* swigljôn *to pipe, play* the flute; swiglja *a piper, flute-player:* O. H. Ger. swegala *fistula, tibia, barbita, chelys, sistrum, calamus;* swegalari *tibicen, fidicen;* swegil-bein *cornus tibia (a wind instrument,* Grff. 3, 129).] v. swegel **IV,** *and next word.*

swegel-râd, e; *f. Music* (?) :—Scyl wæs hearpe, hlûde hlynede, hleóþor dynede, sweglrâd swinsade, Exon. Th. 353, 47; Reim. 29. [Cf. *O. H. Ger.* swegal-sang *music of the flute.*] v. preceding word.

swegel-torht; *adj. Heaven-bright :—*Swegeltorht sunne, Met. 29, 24. Beorht gewât sunne swegeltorht tô sete glîdan, Andr. Kmbl. 2497; An. 1250. Tunglu swegltorht, Exon. Th. 335, 31; Gn. Ex. 41. Wuldorfæstan wîc, sîd and swegltorht, Cd. Th. 2, 32; Gen. 28. Swegeltorhtan seld, 6, 27; Gen. 95. Cf. heofon-torht.

swegel-wered; *adj. Clothed with heavenly brightness :—*Siđđan morgenleôht, sunne swegelwered sûþan scîneþ, Beo. Th. 1216; B. 606. Cf. scîr-wered.

swegel-wuldor, es; *n. The glory of heaven :—*Đæt wit unc in đam êcan gefeán on sweglwuldre geseón môstun, Exon. Th. 173, 13; Gû. 1160. Cf. heofon-wuldor.

swegel-wundor, es; *n. A heavenly wonder, or a wondrous sound* (?). v. swegel, **IV** :—Se burgstede wæs gefylled swêtum stencum and sweglwundrum, eádges yrfestôl engla hleóđres *the dwelling-place was filled with sweet odours and with wondrous music* (?), *the blessed one's home with the voice of angels,* Exon. Th. 181, 13; Gû. 1292.

swêgend-lîc; *adj. Vocal, vowel :—I* and *u* beóþ âwende tô consonantes, gif hî beóþ tôgædere gesette ođđe mid ôđrum swêgendlîcum, Ælfc. Gr. 2; Zup. 6, 15.

sweger, swegr, e; *f. A mother-in-law :—*Sueger *socrus,* Wrt. Voc. ii. 120, 68. Sweger, i. 52, 8. Sweger, swegr, Ælfc. Gr. 11; Zup. 79, 18. Swegr (suegir, Lind.) on hyre snore, and snoru on hyre swegere (swegre, MS. A., Rush.: suoegir, Lind.), Lk. Skt. 12, 53. Sêa Maria is Godfæder snoru and Godes suna môdur and hâligra sâuwla sweger, Shrn. 118, 7. Sweger *socrum,* Wrt. Voc. ii. 72, 51. Đâ geseah hê Petres swegre (swægre, Rush.: suêr ł his wîfes môdor, Lind.) licgende, Mt. Kmbl. 8, 14. Snore ongên hyre swegre (swegran, MS. A.: swêr, Lind.), 10, 35. Swegere, Deut. 27, 23. [O. H. Ger. swigar: Ger. schwieger-mutter. Cf. *Goth.* swaihrô.]

swêg-hleóþor, es; *m. Sound, voice :—*Swêghleóþor (*rugitus magnus,* v. Anglia vi. 243) cymeþ, wôþa wynsumast, þurh đæs wildres mûđ; æfter đære stefne stenc ût cymeþ of đam wongstede, Exon. Th. 358, 8; Pa. 42. Swêg[h]leóþres geswin *the melody of vocal music,* 207, 5; Ph. 137.

swêging, e; *f. A sounding, sound, noise, roaring* (of the sea, etc.), *clanging* (of implements, etc.) :—Sûegungnisso (swêgung ł swêgnisso?) sæs *sonitus maris,* Lk. Skt. 21, 25. Swêgincga beátendra slecgea *sonitus tundentium malleorum,* Coll. Monast. Th. 31, 7. v. swêgan.

swegl. v. swegel.

swegle; *adj. Bright as the sun, splendid, brilliant,* (1) in a physical sense :—Đæt ic sceáwige swegle searogimmas, Beo. Th. 5491; B. 2749. (2) metaphorical, *celestial* :—Hê lîfes weg gesôhte swegle dreámas (cf. swegel-dreám), beorhtne boldwelan, Apostls. Kmbl. 64; Ap. 32. [O. Sax. swigli (sunnun lioht).]

swegle; *adv. Brightly, brilliantly, splendidly,* (1) in a physical sense :—Đonne sió reáde rôd ofer ealle swegle scîneþ on đære sunnan gyld, Exon. Th. 68, 13; Cri. 1103. Scîneþ sunna swegle hât, sôna gecerreþ ísmere ænlîc on his âgen gecynd (cf. đæt îs for đære sunna[n] scîman tô his âgnum gecynde weorþe, Bt. 39, 3; Fox 216, 1), Met. 28, 61. Sumor

swegle hât, Exon. Th. 338, 13; Gn. Ex. 78. (2) metaphorical :—Hŷ môtan his (*Christ's*) ætwiste brûcan, swegle gehyrste weorđian Waldend (cf. đonne scînaþ đa rihtwîsan swâ swâ sunne on hyra Fæder rîce, Mt. 13, 43), Exon, 24, 32; Cri. 393.

swegles æppel. Cockayne suggests *beetle nut,* Lchdm. ii. 32, 2 : 36, 5 : 56, 10 : 66, 8 : 308, 9, 22 ; and see glossaries to vols. ii, iii.

swêg-lic; *adj. Sonorous :—*Mid swêglicre stefne *sonora voce,* Anglia xiii. 412, 675.

swegl-siđe, Cd. Th. 184, 10; Exod. 105. v. segl, **I a.**

swegne *a net.* v. segne.

swegran *a ge-swegran cousins;* consobrimi i. ex sorore et fratre, vel ex duabus sororibus, Wrt. Voc. ii. 134, 18.

-swêgsumlîce. v. ge-swêgsumlîce.

swelan; *p. swæl, pl. swælon.* **I.** *to burn* (intrans.), *perish with heat :—*On fŷrbađe swelaþ sæfiscas sundes getwæfde, wægdeóra gehwylc wêrig swelteþ, Exon. Th. 61, 19; Cri. 987. **II.** *to burn* (of a hot sensation) :—Sió wund ongon, đe him se eorđdraca ær geworhte, swelan and swellan, Beo. Th. 5419; B. 2713. [Cf. *O. H. Ger.* suilizo *calor;* suilizôn *calere, arere.*] v. for-swelan; swælan.

swelc. v. swilc.

swelca, an; *m. A pustule, blister :—*Swelca *pustula,* Wrt. Voc. i. 19, 19. Cf. swellan.

swelgan; *p. swealh, pl. swulgon; pp. swolgen (with acc. or inst. (dat.))* *To swallow.* **I.** in a physical sense, (a) of taking food, etc., by living creatures :—Se draca hig swealh, and hig eft âspâw, L. E. I. prm.; Th. ii. 398, 40. Hê gefêng slæpendne rinc, bât bânlocan, synsnædum swealh, Beo. Th. 1490; B. 743. Hê (*a book-moth*) đâm wordum swealg, Exon. Th. 432, 15; Rä. 48, 6. Laures ceówe and đæt seáw swelge. Lchdm. ii. 230, 4. Syle đam cilde swelgan, i. 350, 14. Swylgende (-fende, Wrt.) drenc *a potion to be gulped down;* catapodia (=καταπότιον), Wrt. Voc. i. 20, 22. (b) of absorption or reception by inanimate things, *to swallow, take in, drink, absorb :—*Swâ sond rên swylgþ, Bt. 12; Fox 36, 13. Seó eorþe đæt wæter swilgþ, 33, 4; Fox 130, 6. Swelgeþ, Exon. Th. 439, 27; Rä. 59, 10. Eorđe wældreóre swealh hâlge of handum đînum, Cd. Th. 62, 19; Gen. 1016: 60, 22; Gen. 985. Eorđe swealh Sethes lîce *the earth closed over Seth's body,* 69, 32; Gen. 1144. Heofon rêce swealg (sealg, MS.) *the smoke mounted into the air,* Beo. Th. 6292; B. 3156. Fugles wyn (*the pen*) beámtelge (*ink*) swealg, Exon. Th. 408, 9; Rä. 27, 9. Ic (*a horn*) winde sceal swelgan of sumes bôsme, 395, 29; Rä. 15, 15. Hwîlum ic (*a fortress*) swelgan onginne beadowæpnum, 399, 7; Rä. 18, 7. (b 1) figuratively :—Đonne lîf and deáđ sâwlum swelgaþ (cf. đonne heofon and hel fira feorum fylde weorþeþ, 97, 17–20; Cri. 1592), 98, 7; Cri. 1604. **II.** figuratively, *to take in the mind, accept, imbibe* (wisdom) :—Swelhþ *affluit* (the passage to which the gloss belongs is Prov. 3, 13, where the Vulgate has: Beatus homo . . . qui affluit prudentia), Kent. Gl. 41. Đâ đâm wordum swealg brego *when the prince had heard those words,* Exon. Th. 196, 25; Az. 179. Hâliges lâre synnige ne swulgon, đeáh hê sôđra swâ feala tâcna gecŷđde. Andr. Kmbl. 1419; An. 710. Wile se Waldend, đæt wê wîsdôm â snyttrum swelgen, Exon. Th. 147, 32; Gû. 736. **III.** with the idea of violence or destruction, *to devour* (lit. or fig.), *to consume, engulf :—*Ic swelge wuda and wætre, Exon. Th. 499, 20; Rä. 88, 18. Lîg eal þigeþ eorþan æhtgestreón, græedig swelgeþ londes frætwe, 232, 16; Ph. 507. Swâ swylgþ seó gîtsung đa dreósendan welan đisses middangeardes, Bt. 12; Fox 36, 13. Đa đe swelgaþ folc mîn *qui devorant plebem meam,* Ps. Spl. 52, 5. Wælstreámas werodum swelgaþ, Cd. Th. 78, 31; Gen. 1301. Grundas swelgaþ Godes andsacan, Exon. Th. 97, 21; Cri. 1594. Nymþe liges fæđm swulge, Beo. Th. 1568; B. 782. [O. L. Ger. far-swelgan *absorbere:* O. H. Ger. swelgan *glutire: Icel.* svelgja *to swallow.*] v. for-, ge-, of-swelgan.

-swelge in ge-swelge[:—Geswelge *barathrum,* Hpt. Gl. 421, 30. Geswelgum *charybdibus, voraginibus,* 513, 29. Cf. swelwhe of a water or of a grownde *vorago,* Prompt. Parv. 482. *Icel.* svelgr; *m. a whirlpool.*] -swelge. v. grund-swelge.

swelgend, es; *m. A voracious person, a glutton, debauchee :—*Đes man is swelgend *ecce homo devorator,* Lk. Skt. 7, 34. Se swelgend, Alexander, Ors. 3, 7; Swt. 120, 16. v. swelge.

swelgend, es; *f., but also es; m. A place which swallows up* (lit. or fig.), *a very deep place, an abyss, a gulf, whirlpool :—*Đylæs hî forswelge sió swelgend đære upâhæfenesse *ipso elationis suae barathro devorantur,* Past. 57, 3; Swt. 439, 3. Seó grundleáse swelgend (gîtsunge) *vorans rapacitas,* Bt. 7, 4; Fox 22, 32. Swelgend *vorago,* Wrt. Voc. i. 54, 37: Kent. Gl. 449: Scint. 117, 9. Sweliend *barathrum,* Hpt. Gl. 529, 26. Swyliendes *voraginis,* 424, 31. Swelgendes, Anglia xiii. 28, 23. Swelgendi *voragine,*Wrt.Voc. ii. 124, 14. West tô đære swelgende; đonne fram đære swelgende, Cod. Dip. Kmbl. v. 281, 29. Tô swelgenc ; đanne fram swelgende, ii. 73, 27. Andlang brôces on đæt swelgend, iii. 460, 5. Andlang streámes on đone sweliend; of đam sweliende, 464, 27. v. ge-swelgend.

swelgendness, e; *f. A gulf, whirlpool :—*Swelgendnessum *carybdibus,* Wrt. Voc. ii. 18, 69. v. swelgness.

swelgere, es; *m. A glutton :—*Ic ne eom swâ micel swelgere đæt ic

ealle cynn metta on ânre gereordinge etan mǽge *non sum tam vorax, ut omnia genera ciborum in una refectione edere possim,* Coll. Monast. Th. 34, 35. [*O. H. Ger.* swelgari *glutto : Ger.* schwelger.] v. swelgend.

swelgness, e; *f. A whirlpool, gulf :*—Swelgnessum *carybdibus,* Wrt. Voc. ii. 86, 11. v. swelgendness.

sweliend, swell. v. swelgend, ge-swel.

swellan ; *p.* sweall, *pl.* swullon; *p.* swollen *To swell :*—Wið wunda ðe swellaþ, Lchdm. iii. 86, 16. Gif sino gescrince and æfter ðon swelle, ii. 68, 1. Gif fót oððe scancan swellan, iii. 38, 21. Sió wund ongon swelan and swellan, Beo. Th. 5419; B. 2713. Swellende blæddran *vesicae turgentes,* Ex. 9, 9,·10. Wiþ ǽlcre yfelre swellendre wǽtan, Lchdm. ii. 6, 26. Swellende yfele swilas, 264, 12. [*O. L. Ger. O. H. Ger.* swellan *tumere, turgere, obturgescere : Icel.* svella. Cf. *Goth.* uf-swalleins *inflatio.*] v. ā-, ge- (Lchdm. ii. 46, 9: 200, 22: 202, 5), tô-swellan.

swelling ; e; *f. A swelling,* used of a sail swelled out by the wind :— Gesión brecan ofer bædweg brimwudu, snyrgan under swellingum (cf. snel under segle, Andr. Kmbl. 1009; An. 505), Elen. Kmbl. 489; El. 245.

sweltan, swyltan, swiltan; *p.* swealt, *pl.* swulton; *pp.* swolten *To die a natural or a violent death :*—Swelte ic (*morior*) hér on lande, Deut. 4, 22. Wǽgdeóra gehwylc swelteþ, Exon. Th. 61, 22; Cri. 988. Swylteþ, 385, 33; Rä. 4, 54: 419, 13; Rä. 38, 5. Ne swylteþ *non obierit,* Wrt. Voc. ii. 88, 35. Swylt *moritur,* Jn. Skt. 21, 23. Hé swelt, Blickl. Homl. 245, 11. Gê sweltaþ, 8, 21. Ealle men sweltaþ, Bt. 18, 4; Fox 68, 13. Hí ne swyltaþ, Blickl. Homl. 47, 1. Hé swealt, Cd. Th. 70, 15; Gen. 1153. Swealt (sweolt, Thw.), Num. 20, 1. Hí swulton, Homl. Th. i. 84, 6: Cd. Th. 207, 10; Exod. 464. Ðæt ân man swelte for folce, Jn. Skt. 11, 50. Ðæt hyt wǽre beteré, ðæt ân man swulte, 18, 14. Ðæt hé swungen wǽre oþþæt he swylte, Blickl. Homl. 193, 4. Hwí lǽddest ðú ús ðæt wē swulton on ðisum wéstene, Num. 21; 5. Ic mæg sweltan bliðelíce *laetus moriar,* Gen. 46, 30 : Mt. Kmbl. 26, 35 : Ex. 10, 28. Sceal fǽge sweltan, Exon. Th. 335, 2; Gn. Ex. 27. Swyltan, Blickl. Homl. 59, 30. Se man scyle deádlíce swyltan (swiltan, MS. C.), Wulfst. 5, 9. Sweltende *obeuntem,* Wrt. Voc. ii. 64, 54. Beón swyltende, Blickl. Homl. 75, 33. ¶ *to die* by or of something, where the cause of death is expressed by a case or by a preposition with a noun :—Ne swelte ic mid sáre, Ps. Th. 117, 17. Ðú þurh deóra gripe deáþe sweltest, Exon. Th. 250, 11; Jul. 125. Gê sweltaþ deáðe, Cd. Th. 224, 28; Dan. 143. Draca morðre swealt, Beo. Th. 1789; B. 892: 5558; B. 2782. Hé forneáh hungre swealt, Ors. 4, 6; Swt. 170, 30. Hié hungre swultan, Blickl. Homl. 79, 15. Monige for hiora wundum swultan, Nar. 16, 9. Heora mænige máne swultan, Ps. Th. 77, 30. Tô ðam ðe hé deáðe swelte, L. Alf. 13; Th. i. 48, 2. Ðú sceald deáðe sweltan *morte morieris,* Gen. 2, 17 : L. Alf. 14; Th. i. 48, 4. Ic sceal æt ðé sweltan deáðe, Homl. Th. ii. 308, 27. Wundum sweltan, Byrht. Th. 140, 25; By. 293. Hí ondrǽdaþ him ðæt hí sceolan swyltan for ðam húsle, L. Ælf. E.; Th. ii. 392, 3. ¶ *to die* to anything, *become dead to,* have no further concern with :—Ðú scealt sweltan synna and Criste lybban, Homl. Skt. i. 3, 592. [*Laym. O. E. Homl.* swelten : *Orm.* swelltenn : *Chauc. Piers P.* swelte : *Goth.* swiltan : *O. Sax.* sweltan : *Icel.* svelta *to die* ; svelta hungri *to starve : Dan.* sulte *to starve* ; sulten *hungry.*] v. ā-, for-, ge-sweltan.

sweltend-líc ; *adj. Ready to die, about to die :*—Se wæs sweltendlíc *erat moriturus* (ready to die, A.V.), Lk. Skt. 7, 2. Wambe sweltendlíces flǽsces *uentrem moriture carnis,* Scint. 53, 2.

swemman ; *p.* de *To cause to swim, to bathe, wash.* [*Ger.* schwemmen *to water, wash, float : Dan.* svømme (heste) *to take* (horses) *into the water.*] v. be-swemman ; swimman.

swenc, es; *m. Trial, tribulation, affliction :*—In niðrung ł in suoenc deáðes *in damnationem mortis,* Lk. Skt. Lind. 24, 20. On swencum (suoenccum, Lind.: geswincum, W. S.) ł costungum mínum *in temtationibus meis,* Rush. 22, 28. In suoencum *in tribulationibus,* Rtl. 184, 4. v. ge-swenc.

swencan ; *p.* te ; *pp.* swenced, swenct (cf. *swinkt* = wearied, Comus v. 293) *To cause a person to labour, to cause trouble to* a person (a) where no good is implied, *to harass, vex, afflict, distress :*—Ic swencu hió *adfligam illos,* Ps. Surt. 17, 39. Hwí swencst ðú ðis folc . . . Pharaon swencþ ðín folc *cur afflixisti populum istum ? . . . Pharao afflixit populum tuum,* Ex. 5, 22-23. Ælc deáþlíc man swencþ hine selfne mid manigfealdum ymbhogum *omnis mortalium cura, quam multiplicium studiorum labor exercet,* Bt. 24, 1; Fox 80, 6. Eów nǽnig wiht ne deraþ ne ne swenceþ, Blickl. Homl. 239, 12. Suenceth *defatiget,* Wrt. Voc. ii. 106, 3. *Defatiget, lassat,* swenceþ, *flagellat,* 138, 16. Ða ðe mé swencaþ *qui tribulant me,* Ps. Th. 12, 5. Hwí swenctest ðú (*afflixisti*) ðínne þeów? Num. 11, 11. Man swencte ðæt earme folc ðe on ðam scipon lágon, Chr. 999; Erl. 135, 32. Hine wundra fela swe[n]cte on sunde, Beo. Th. 3024; B. 1510. Ða werigan gástas ðe mé swenctan and ðrycton *qui me premebant spiritus maligni,* Bd. 3, 11; S. 536, 37. On ðínre hátheortnesse ne swenc mé *ne in furore tuo corripias me,* Ps. Th. 6, 1. Beorge hé ðæt hé áwóh ne befó, ðý læs ðe hine mon swence swā hé óðerne man þohte, L. Eth. ii. 9; Th. i. 290, 8. Þeáh hine se wind woruldearfoþa swíðe swence, Met. 7, 50. Ðý læs ðe mon unmihtigne

man tô feor and tô lange for his āgenan swencte *lest a man of small means should be made to toil too far and too long for his own,* L. Eth. ii. 9; Th. i. 290, 4. Ne sceal nān mon siócne monnan gesárgodne swencan, ac hine mon sceolde lǽdan tô ðam lǽce, Bt. 38, 7; Fox 210, 20. Hé (*William I*) lét castelas wyrcean and earme men swíðe swencean, Chr. 1086; Erl. 222, 21. Ðú ðec sylfne ne þearft swíþor swencan *you need not trouble yourself any more,* Exon. Th. 245, 19; Jul. 47. Wítebendum swencan, Andr. Kmbl. 218; An. 109. Perseus wæs ealne ðone geár Rómāne swíþe swencende, Ors. 4, 11; Swt. 208, 13. Forhwon sindun gē swencende (*molesti*) ðam wífe? Mt. Kmbl. Rush. 26, 10. Fram unclǽnum gáste swenced beón *ab immundo spiritu vexari,* Bd. 3, 11; S. 536, 11. Mid ða ādle swenced *affectu incommodo,* 4, 31; S. 610, 21. Swā gewinnfullícum fyrdum swencte beón *tam laboriosis expeditionibus fatigari,* 1, 12; S. 481, 4: 2, 18; S. 520, 36. (b) where a good result is intended, *to mortify, chasten :*—Ða sylfan, ðe hí mid ðam wítum ðreágeaþ and swenceaþ (*adfligunt*), lufiaþ eác, Bd. 1, 27; S. 490, 18. Hí firenlustas forberaþ . . . swenceaþ hí sylfe, sāwle frætwaþ, Exon. Th. 150, 13; Gū. 778. Ða lāreówas sceolan heora āgenne líchoman swencean on forhæfdnesse, Blickl. Homl. 81, 6. [*O. E. Homl. A. R. Laym.* swenchen : *Orm.* swennkenn, swennchenn : *O. H. Ger.* swenchen *verberare*.] v. ge-swencan ; swincan.

swencedness. v. ge-swencedness.

sweng, es; *m. A blow, stroke :*—Sweng *ictus,* Ælfc. Gr. 11; Zup. 79, 6. Sweng oððe cnyssung *ictus,* 43; Zup. 255, 3. Eádweard ānne slóg swíþe mid his swurde, swenges ne wyrnde, Byrht. 135, 15; By. 118. Hé hond swenge ne ofteáh, Beo. Th. 3045; B. 1520. Him for swenge swāt ǽdrum sprong, 5924; B. 2966 : 5365; B. 2686. Ic mē gūðbordes sweng gebearh, Cd. Th. 163, 5; Gen. 2693. Weras him ondrēdon for ðære dǽde Drihtnes handa, sweng (*the stroke,* i. e. *the punishment threatened if Sara were not returned to Abraham*), 161, 26; Gen. 2671. Iacob swilt þrowode ðurg stenges sweng, Apostls. Kmbl. 143; Ap. 72. Hé feorhwunde líleát sweordes swengum, Beo. Th. 4761; B. 2386. Bord oft onfēng ýða swengas *oft the vessel's side received the billows' blows,* Elen. Kmbl. 478; El. 239. [In later English the word is used in a metaphorical sense similar to that of *stroke* in modern English, and may be compared with *M. H. Ger.* swanc, swang *a trick : Ger.* schwank : cf. also *Ger.* streich = *trick.* To wrastlen aȝein þes deofles swenges, A. R. 80, 8. Ȝef ha etstonden wulleð mine unwreste wrenches ant mine swikele swenges, wrestlin ha moten wið ham seoluen, Marh. 14, 12. Ȝif tweie men goþ to wrastlinge . . . and þe on can swenges swiþe fele . . . and þe oþer ne can sweng bute ane, O. and N. 795. Cf. *O. H. Ger.* swanch *swinging, stroke : M. H. Ger.* swanc, swang : *Ger.* schwang.] v. feorh-, headu-, heoru-, hete-, wæl-sweng.

swengan ; *p.* de *To cause to swing, to cause rapid movement, to swing, fling, dash, strike :*—Ða āhleóp ān leó of ðæs eorðscræfes þýstrum and hió swengde on hine . . . Ða eode uncer hláford sylf in ðæt scræf ða swengde sió lió sóna forð and forswealh hine *then a lion ran out from the darkness of the cavern and dashed on to him . . . Then our lord himself went into the cave ; then the lion dashed out at once and swallowed him up,* Shrn. 43, 9-18. Swengende discutiens, Wrt. Voc. ii. 141, 43. [He smat hine sare . . . æft he him to (to him, 2nd MS.) sweinde . . . dunt he him ȝef þane þridde, Laym. 8183. His sweord he sweinde bi his side, 21138. Swenged of þa hafden, 22839. He sweinde ham adun into helle grunde, A. R. 280, 13. Breid up þene rode stef and sweng him aȝean (*strike at him*), 290, 18. Þe drake rahte ut his tunge and sweinde hire in (*swung her into his mouth*) ant forswalh into his wide wombe, Marh. 10, 19. Swengyñ or schakyn as menne done clothys *excucio,* Prompt. Parv. 482. *Goth.* af-swaggwjan *to cause to waver, to shake one's confidence, make desperate.*] v. ā-, fram-, tô-swengan ; swingan.

swenge (?); *adj. Heavy, slothful.* v. swangor (b).

sweocol. v. swicol.

sweofot, es; *n. Sleep :*—Hé Hróðgāres heorðgeneátas slóh on sweofote, slǽpende fræt, Beo. Th. 3166; B. 1581: 4579; B. 2295. Hé (*the panther*) þreó nihta fæc swefeþ on sweofote, slǽpe gebiesgad, Exon. Th. 358, 1; Pa. 39. Ðonne hē selþ gecorenum his swefetu (cf. *the use of* swefen *in pl.*) ł slǽp *cum dederit dilectis suis somnum,* Ps. Lamb. 126, 3. [Þe king læi on sweuete, Laym. 17773. On sweouete, 17802. Ne þuhte hit þ ha weren deade, ah þ ha slepten a sweouete *dormientes potius quam extinctos putares,* Kath. 1427.] v. swefan.

sweogian. v. swigian.

sweogode glosses *praevaluit,* Ps. Spl. 51, 7, a mistake (?) *for* strongode.

Sweó-land, es; *n. The land of the Swedes, Sweden,* Ors. 1, 1; Swt. 19, 2. v. Sweó-ríce, -þeód.

sweoloþ, swoloþ (swóloþ?), es; *m. Heat, burning :*—Swoloð *aestus vel cauma,* Wrt. Voc. i. 53, 41. Swoloð (swaloð, MS. J.: sweoli, MS. W.) *cauma,* Ælfc. Gl. Zup. 306, 15. Ðes swolaþ (swoli, MS. W.) *hoc cauma,* Ælfc. Gr. 9, 1; Zup. 33, 12 note. Swoleðe *caumate,* Hpt. Gl. 482, 48: 495, 22. Hét Hildeburh æt Hnæfes āde hire selfre suna sweoloðe befæstan, bānfatu bærnan and on bǽl dón, Beo. Th. 2235; B. 1115. v. swelan, *and next word.*

sweoloþa, an; *m. Heat, burning :*—Hé swā swíþe swǽtte swā hē in

swoloþan middes sumeres wǽre *quasi in media aestatis caumate sudaverit,* Bd. 3, 19; S. 549, 30. Mid hǽtan and mid swoluðan *ardore et aestu,* Deut. 28, 22. v. preceding word.

sweolung (?), e; *f. Burning, inflammation:*—Biþ micel āþundenes and fefer mid sweolunga (sweopunga, MS. v. note on passage) ōmena *with inflammation from corrupt humours,* Lchdm. ii. 204, 25.

Sweón; *pl. The Swedes:*—Burgendan habbaþ Sweón be norþan him ... Sweón habbaþ be sūþan him ðone sǽs earm, Ors. 1, 1; Swt. 16, 31–34. Ðā Sweón heafdon weallstōwe geweald, Chr. 1025; Erl. 163, 11. Sacu Sweóna and Geáta, Beo. Th. 4936; B. 2472: (Swona, MS.), 5885; B. 2946: 5908; B. 2958. Ic wæs mid Sweóm, Exon. Th. 322, 4; Vīd. 58: 320, 19; Vīd. 31. Ðæs land hȳraþ tō Sweón, Ors. 1, 1; Swt. 20, 4. [*Icel.* Svíar. *The Latin form is* Suiones *in Tacitus, later* Sueones.] v. Sweó-land, -rīce, -þeód.

sweopung. v. sweolung.

sweór, swehor, es; *m.* **I.** *a father-in-law:*—Sueór *vetellus,* Txts. 106, 1099. Su[eó]r *socer,* 97, 1878. Sweór, Wrt. Voc. i. 52, 7: 72, 51: Ælfc. Gr. 8; Zup. 27, 13. Se wæs Caiphas sweór (sueór, Lind.), Jn. Skt. 18, 13: Gen. 38, 13. Sweór, swiór, Bt. 10; Fox 28, 13. Hǽðne wǽron begen, sweór and āþum, Exon. Th. 246, 22; Jul. 65. Ðā sende heó tō hire sweóre (*ad socerum suum*), Gen. 38, 25: 30, 25. Obab his sweór (*cognatum*), Past. 41, 5; Swt. 304, 9. Suehoras, sueóras *vitelli,* Txts. 104, 1062. Wæs Rōmeburg on fruman gehālgod mid brōðor blōde and mid sweóra (*the fathers of the Sabine women who were taken as wives by the Romans*), Ors. 2, 2; Swt. 66, 5. **II.** the word is also used to translate *consobrinus; a cousin:*—Sueór *consobrinus,* Wrt. Voc. ii. 104, 83. Gesweóras *consobrini,* sweór *consobrinus, filius patruelis,* 134, 17–20. Sw[e]ór *consobrinus,* 15, 2. [*Goth.* swaihra *father-in-law: O. H. Ger.* sweher, swēr *socer, levir: Ger.* schwäher.] v. sweger, suhtriga.

sweor, swer, swyr, es; *m. f. A column, pillar* (lit. or fig.), *that which is shaped like a pillar:*—Swer *columna,* Wrt. Voc. i. 26, 32: 81, 15. Ufeweard swer *epistilia,* ii. 30, 29. Ðū eart leóhtes swer, Blickl. Homl. 141, 1. Drihten swutelode him ðone weg on dæg þurh swert tācn on sweres gelīcnysse, and on niht swilce ān byrnende swer him fōr beforan, Ex. 13, 21: Homl. Th. ii. 196, 8. Mid ðȳ fȳrenan sweore on nieht and on dæg mid ðȳ sweore ðæs wolcnes, Past. 41, 5; Swt. 304, 7. On swere (swiorum, MS. T.) *in columna* (*nubis*), Ps. Spl. 98, 7. Þurh wolcnes swyr, Ps. Th. 98, 7. Hē geseah swer standan, and ofer ðone swer ǽrne onlīcnesse, Blickl. Homl. 239, 21. Greáte swā stǽnene sweras *uastitudine columnarum,* Nar. 36, 13. Hī hēton hine standan betwux twām stǽnenum swerum: on ðām twām swerum stōd ðæt hūs geworht. And Samson ... gelǽhte ðā sweras, Jud. 16, 25–29. Ðæt gēr is underwryðed mid þrīm swerum, ða synd ðus gecīged, id.' and non. and kl.', Anglia viii. 301, 37. Swyras (swioras, MS. T.: sweras, MS. C.) *columnas,* Ps. Spl. 74, 3. Sweoras gata *seras portarum,* Ps. Spl. T. 147, 2. Hire swyre *columnas ejus,* Ps. Th. 74, 3. Sweras unlytle, stapulas, Andr. Kmbl. 2985; An. 1495. [Sweor *columna,* Wrt. Voc. i. 92, 55.] [Grimm, R. A. 370, gives from a Swiss source ' an ein *schwiren* binden.']

sweora, swira, swyra, swura, an; *m.* **I.** *a neck:*—Sweora *collum,* Wrt. Voc. ii. 16, 51: *cervix,* 52. Foreweard sweora *capitium,* 45. Sweora *vel* swura *collum,* i. 43, 36. Swira *collum,* 283, 2: *cervex,* 3. Swyra *collum,* 64, 65: Soul Kmbl. 218: Seel. 111. Swiora smæl, Exon. Th. 486, 15; Rä. 72, 15. Ðā heó ðrycced wæs mid sāre hire sweoran ðæt heó oft cwǽde: 'Ic wāt ðæt ic be gewyrhtum on mīnum sweoran bere ða byrþenne ðysse ādle' *quia cum praefato dolore maxillae sive colli premeretur solita sit dicere:* 'scio, quia merito in collo pondus languoris porto,* Bd. 4, 19; S. 589, 22–26. Swile on hire sweoran *tumorem sub maxilla,* S. 588, 43. Tō his suiran getīged, Past. 2; Swt. 31, 18. Tō hys swyran (sweoran, MS. A.: suire, Lind.: swira, Rush.) geċnytt, Mt. Kmbl. 18, 6. Sweoran (sweoran, MS. A.: suiro, Lind.: swira, Rush.), Mk. Skt. 9, 42. Swioran *ceruice,* Lchdm. i. lxx, 9. Swiran *ceutro* (cf. *cervellum,* i. *ceutrum* brægen, Wrt. Voc. ii. 130, 31. *Ceutrum* þrotbolla, 131, 1), lxxi, 1. Underlūtan mid eówrum swiran ðæt deáþlīcne geoc, Bt. 19; Fox 68, 26: Met. 10, 19. Ðeáh hē him ðone stīðan swioran (swiran, Hatt. MS.) fortrǽde, Past. 33; Swt. 228, 9. Hē wȳscte ðæt ealle Rōmāne hæfden ǽnne sweoran (*unam cervicem*), Ors. 6, 3; Swt. 256, 27: Judth. Thw. 23, 5; Jud. 106. Hié sendon rāp on his sweoran (swyran, 20), Blickl. Homl. 241, 24. Is ymb ðone sweoran (*the neck of the phenix*) beága beorhtast, Exon. Th. 219, 10; Ph. 305. Oþ mannes swuran, Blickl. Homl. 245, 33: Gen. 41, 42: Deut. 28, 48. Swiran (swioran) *cladam,* Lchdm. i. lxx, 1 (see note). On ða swyran sīnra þegena, Met. 9, 56. **II.** of land, *a hause* (as in Esk *Hause*), *a col;* cf. ge-sweoru:—Dūna swioran *juga,* Wrt. Voc. ii. 48, 18. [*Sware, swire,* the neck, the declination of a mountain near the summit; the most level spot between two hills, Jamieson. Cf. *Icel.* Swíri, the local name of a neck-shaped ridge in western Iceland.] **III.** of water, *the part where the distance between opposite shores is least:*—Ofer swira sǽs (cf. ofer ðære sǽs mūðan, W. S.) *trans fretum maris,* Mk. Skt. Lind. Rush. 5, 1. On þūles sweran, Cod. Dip. Kmbl. iii. 97, 5. [Swiere (rimes with (wilde) diere), O. E. Homl. ii. 224, 146. Swore (rimes with (wilde) dore),

i. 169, 144. Sweore, 49, 28: A. R. 394, 19. Swire, 58, 7: Marh. 9, 8. Swure (swere, 2nd MS.), Laym. 4012. Sweore (swere, 2nd MS.), 26565. Sweore, swore (rimes with deore, dore beast), O. and N. 1125. Sweore, swore, suere, 73. Suere (rimes with ouerdere), R. Glouc. 389, 22. Swire (rimes with sire), Havel. 311. Swere (rimes with there), Gow. ii. 30, 17. *Icel.* svíri.] v. belced-sweora.

sweor-bān, es; *n. The neck-bone, the neck:*—Mīn Drihten, ðū ðīn hālige sweorbān geeádmēddest, Anglia xii. 505, 22. Oð swirbān *usque ad cervices,* Ps. Surt. ii. p. 190, 27: Ps. Spl. C. 128, 4. Onheldon eówerra feónda swyrbān, Shrn. 86, 22. [The swyrebane he swappes in sondyre, Morte Arthure (Halliwell).]

sweor-beáh; *gen.* -beáges; *m. A collar, band* or *chain for the neck, necklace:*—Myne *vel* sweorbēh *monile vel serpentinum,* Wrt. Voc. i. 40, 50. Swurbeáh *monile,* 74, 58. Swurbēh *murenula vel torques,* 16, 57. Ic ann ðære hlǽfdigan ānes swyrbeáges on hundtwelftigum mancussum and ānæs beáges on þrītegum mancussum, Chart. Th. 554, 1. Ǽnne sweorbeáh (on XL mancysan, on LXXX mancys), 501, 20, 31. Ic ðē forgife gyldenne swurbeáh *thou shalt have a chain of gold about thy neck* (A. V.), Homl. Th. ii. 436, 4, 16. Swurbeágas *crepundia* (cf. *crepundium,* i. *monile gutturis* myne, *crepundia* frætwunga, Wrt. Voc. ii. 136, 68–70), Ælfc. Gr. 13; Zup. 85, 9. Suirbēg[as] *monilia,* Rtl. 4, 3. Sweorbeágum ɫ halsmenum *monilibus, lunulis,* Hpt. Gl. 434, 63. Ic frætwode mīnne swuran mid mænigfealdum swurbeágum, Homl. Skt. i. 20, 57.

-sweorc. v. ge-sweorc.

sweorcan; *p.* swearc, *pl.* swurcon; *pp.* sworcen. **I.** in a physical sense, *to become dark, be obscured:*—Wedercandel swearc windas weóxon *the sun was darkened, the winds rose,* Andr. Kmbl. 744; An. 372. Swearc norðrodor won under wolcnum, woruld miste oferteáh, Exon. Th. 178, 33; Gū. 1253. **II.** figuratively of mental gloom, (a) of that which feels sadness, *to become troubled, gloomy, sad:*—Siteþ sorgcearig, on sefan sweorceþ, sylfum þinceþ, ðæt sȳ endeleás earfoða dǽl, Exon. Th. 379, 6; Deór. 29. Hē mōdsorge wæg, hreþer innan swearc, 165, 8; Gū. 1025. On hū grundleásum seaðe swinceþ ðæt sweorcende mōd *quam praecipiti mersa profundo mens hebet,* Met. 3, 2. (b) of that which causes sadness, *to become grievous, troublesome, saddening:*—Ne hine wiht dereþ, ādl ne yldo, ne him inwitsorh on sefan sweorceþ *nor in his mind springs gloomy care,* Beo. Th. 3478; B. 1737. [Swelled þe mære and swærkeð þa uðen, Laym. 22030. Swurken (þirkede (dirkede ?), 2nd MS.) under sunnen sweorte weolcnen, 11973. *O. Sax.* swerkan: Ni lāt thū thīnan sebon swerkan *do not be sad,* Hēl. 4042. *O. H. Ger.* swercan.] v. ā-, for-, ge-, tō-sweorcan.

sweorcend-ferhþ; *adj.* With the mind growing gloomy:*—Beornas (*the Assyrians after Holofernes' death*) stōdon ymbe hyra þeódnes trǽf sweorcendferhþe ... Ðā wæs hyra tīres æt ende, Judth. Thw. 25, 19; Jud. 269.

sweor-clāþ, es; *m. A cloth for the neck, a collar:*—Sweorclāþ *collarium,* Wrt. Voc. ii. 134, 48.

sweorcness. v. ge-sweorcness.

sweor-cops, es; *m. A neck-bond, pillory:*—Iuc oððe swurcops (sweor-) *bogia,* (*bogia* torques damnatorum quasi jugum bovis, Migne), Ælfc. Gl. Zup. 321, 2. Sweorcopsas *vel* handcopsas *boias, catenas,* Wrt. Voc. ii. 126, 43.

sweor-coþu, e; *f. A disease of the neck* or *throat, quinsy:*—Sweorcoþu *arterias,* Wrt. Voc. i. 19, 33. Wið sweorcoþe, Lchdm. ii. 2, 20: 44, 9. Various methods of treatment are given, 48, 4–28.

sweord, swurd, swyrd, es; *n. A sword:*—Sweord *framea,* Wrt. Voc. ii. 36, 11. Sweorde *mucrone,* sweordum *mucronibus,* sweord *macheram,* 54, 33–36. Sweord *gladius vel machera vel spata vel framea vel pugio,* i. 35, 7. Litel sweord *sica,* 13. Hiltleás sweord *ensis,* 33. Swurdes ord *mucro,* 15. Sweordes sceáð *classendis,* 34, 29. Swyrdes gyrdel *baltheus,* 40, 58. Ðæt ūs cwealm on be becume ne swurdes ecg *ne occidat nos pestis aut gladius,* Ex. 5, 4. Blōtan mid sweordes ecge, Cd. Th. 173, 6; Gen. 2857. Ðurh sweordes bite gedǽlan feorh wið flǽsce, Apstls. Kmbl. 68; Ap. 34. Hig feallaþ on swurdes (sweordes, MS. A.: suordes, Lind.: swordana, Rush.) ecge *cadent in ore gladii,* Lk. Skt. 21, 24. Standan mid ātogenum swurde, Jos. 5, 13. Hēr synt twā swurd (sweord, MS. A.: suordas, Lind.: sworde, Rush.) *ecce gladii duo,* Lk. Skt. 22, 38. Sweorda gelāc *the play of swords, battle,* Beo. Th. 2084; B. 1040. Sweorda lāfe *those whom the sword had spared,* 5865; B. 2936. ¶ The high esteem in which good swords were held in old times is marked in many ways. Their forging is in many legends said to be the work of other than human hands; so the sword which Beowulf seizes in Grendel's home is 'eald sweord eotenisc (cf. eald sweord eácen, 3330; B. 1663), ecgum dyhtig, ... giganta geweorc,' Beo. Th. 3120–9; B. 1558–62; and twice besides occurs the phrase 'eald sweord eotonisc,' 5225; B. 2616: 5950; B. 2979; see also 'enta ærgeweorc' applied to the workmanship of a sword, 3362; B. 1679. Cf. too the forging of Sigurd's sword in the Völsunga Saga. They are precious heirlooms, handed down through many years (v. epithet *eald* above); so Beowulf speaks of his sword as 'eald lāf,' Beo. Th. 2981; B. 1488, and the same phrase is used of the

sword wielded by one of his followers in the chief's defence, 1595; B. 795. In reference to the sword given by Beowulf to the Dane who had guarded his ship, it is said that the recipient 'syððan wæs on meodobence mádme ðý weorðra, yrfeláfe,' 3810; B. 1903; another sword is called 'Hreðles láf,' and of it is said 'næs mid Geátum sincmáðþum sélra on sweordes hád,' 4389–93; B. 2191–3; and later on mention is made of 'gomel swyrd, Eánmundes láf,' 5216; B. 2611; Hrunting, the sword which is lent to Beowulf, is 'án ealdgestreóna,' 2921; B. 1458. So, too, Byrhtnoth tells the Danes who demand tribute of him, that the tribute will take the form of 'ealde swurd,' used with unpleasant effect upon the invaders. The same point may be illustrated from other than poetical sources. Thus in Alfred's will it is said that he leaves 'Æþerède ealdormenn án sweord on hundteóntigum mancusum,' Chart. Th. 489, 32; in another will is the passage 'Freoðomunde fóe tó mínum sweorde, and ágefe ðæræt feówer ðúsenda,' 471, 23; another testator bequeathes his sword 'mid ðam sylfrenan hylte and ðone gyldenan fetels,' 558, 10; and another mentions the sword 'ðat Eádmund king mé selde on hund-tuelftian mancusas goldes and fóur pund silueres on ðan fetelse,' 505, 28. Indeed the sword is often mentioned in wills. The importance of the sword is further marked by its receiving a name. The sword with which Beowulf is armed for his attack on Grendel's mother is named Hrunting, and to the praise of this weapon the poet devotes several lines, Beo. Th. 2914–33; B. 1455–64; at a later period it is with 'Nægling ... gomol and grægmæl' that he fights, 5354; B. 2680. See, too, Wald. 4; Vald. 1, 3. And elsewhere the same point may be noted, e.g. in the Nibelungenlied. 'daz Nibelunges swert ... Palmunc was genant;' and this weapon plays a part in the drama to the last scene. In Scandinavian story there is Hákon's sword 'kvernbítr,' which king Athelstan gave him, and Egill has his sword that he called 'Naðr.' See, too, the story of the Cid and the two swords, Colada and Tizona, which he gave to his sons-in-law, the Infantes of Carrion, and which he claimed from them after their unworthy treatment of their wives, Chronica del Cid, c. cclii. Of the value of the sword and of the decoration bestowed upon it, of the shape or colouring, of the make, many epithets and phrases speak. In the Gnomic verses it is said, 'Gold geríseþ on guman sweorde,' Exon. Th. 341, 15; Gn. Ex. 126; and 'máðm in heaile, goldhilted sweord' is mentioned, 437, 27; Rä. 56, 14. See, too, the passages quoted under seolfor-hilt, -hilted. In the dragon's hoard are 'dýre swyrd,' Beo. Th. 6089; B. 3048: the sword which Beowulf seized in Grendel's retreat was golden-hilted, 3358; B. 1677, and on ðæm scennum scíran goldes þurh rúnstafas gesæd, hwam ðæt sweord geworht, írena cyst, ærest wære, wreoþenhilt and wyrmfáh,' 3390–3400; B. 1694–8. Beowulf lays aside his 'hyrsted sweord, írena cyst,' Beo. Th. 1349; B. 672: he gives a sword 'bunden golde,' 3805; B. 1901: his own sword is 'fáh and fæted,' 5395; B. 2700. Byrhtnoth's sword is 'fealohilte,' Byrht. Th. 136, 45; By. 166; and 'gerénod,' 35; By. 161. Beowulf's Nægling is 'grægmæl,' Beo. Th. 5357; B. 2681: the swords of the Hebrews are 'scírmæled,' Judth. Thw. 24, 38; Jud. 230: other swords are 'hring-mæled,' Cd. Th. 120, 10; Gen. 1992: Abraham girds himself 'grægan sweorde,' 173, 22; Gen. 2865: the Hebrews fight 'fágum sweordum,' Judth. Thw. 24, 18; Jud. 194: 25, 17; Jud. 264. The sword is 'bråd,' 26, 9; Jud. 318: Byrht. Th. 132, 12; By. 15: bråd and brúnecg, 136, 38; By. 163: it is 'gód,' 138, 58; By. 237; 'heard,' Beo. Th. 5966; B. 2987; 5269; B. 2638: Exon. Th. 325, 32; Víd. 120: 'heardecg,' Beo. Th. 2581; B. 1288: 'ecgum dyhtig,' 2578; B. 1287: Cd. Th. 120, 11; Gen. 1993: 'ecgum gecost,' Judth. Thw. 24, 39; Jud. 231: stýled, Exon. Th. 42, 28; Cri. 679. For some account of old swords, see Wright's The Celt, The Roman, and the Saxon, pp. 404–6, and Worsaae's Antiquities: see also Grmm. Gesch. D. S. p. 12. [O. Sax. O. Frs. swerd: O. H. Ger. swert: Icel. sverð.] v. gúð-, mál-, máðum-, stæf-, wǽg-sweord.

sweord (or sweorð) swearing. [O. H. Ger. swert, swart juramentum.] v. áþ-sweord.

sweord-bealu (-o), wes; n. Bale or hurt caused by the sword, Beo. Th. 2298; B. 1147.

sweord-berende; adj. (ptcp.) Sword-bearing:—Æðelingas sweord-berende, Cd. Th. 65, 2; Gen. 1060.

sweord-bite, es; m. The bite of a sword, wounding with a sword:—Áswebban þurh sweordbite to kill with the sword, Exon. Th. 278, 26; Jul. 603.

sweord-bora, an; m.   I. one who bears a sword for his own use, a swordsman:—Sweord spata vel pugio, swyrdbora spatarius, Wrt. Voc. i. 35, 8. Swurdbora, 84, 13. Swurdboran (gladiatorem) he gewordene gesihþ if (in a dream) he sees himself become a gladiator, Lchdm. iii. 204, 25. Sweordboran pugiles, Wrt. Voc. ii. 76, 46.   II. one who bears his lord's sword, a swordbearer:—Swá swá Eádmundes sweordbora hit reahte Æþelstáne cyninge, Swt. A. S. Prim. 83, 7. Totila ásende his swurdboran, Riggo geháten, gescrýdne mid his cynelícum gyrelum, Homl. Th. ii. 168, 12. [Cf. Icel. sverð-berari (translating lictor).]

sweord-fetels, es; m. A sword-belt:—Se cásere heora ælces sweordfætelsas hét forceorfan the emperor ordered the sword-belts of each of them to be cut, Homl. Skt. i. 23, 178. Cf. Ðat swerd on hundtwelf-

tian mancusas and fóur pund silueres on þan fetelse, Chart. Th. 505, 32. Ðæs swurdes mid ðam sylfrenan hylte ðe Wulfríc worhte and ðone gyldenan fetels, 558, 12. [Cf. O. H. Ger. swert-fezzil faidilus, vagidilus: Icel. sverð-fetill a sword-belt.] v. fetel.

sweord-freca, an; m. A warrior who uses a sword:—Hé ðæs wæpnes (the sword Hrunting) onláh sélran sweordfrecan, Beo. Th. 2940; B. 1468.

sweord-genídla, an; m. A foe armed with a sword:—Ðonne fyrdhwate on twá healfe tohtan sécaþ sweordgenídlan, Elen. Kmbl. 2359; El. 1181.

sweord-geswing, es; n. Striking with swords, an attack with swords:—Swyrdgeswing swíþlíc eówan to make a fierce attack, Judth. Thw. 25, 3; Jud. 240.

sweord-gifu, e; f. Gift of a sword:—Sceal sincþego and sweordgifu eówrum cynne álicgean taking of treasure and gift of sword shall fail for your race, Beo. Th. 5761; B. 2884.

sweord-gripe, es; m. Sword-grasp, seizing of swords:—Ðæt hí in winsele þurh sweordgripe sáwle forlétan so that in the banquet hall through seizing their swords they lost their lives, Exon. Th. 271, 26; Jul. 488.

sweord-hwíta, an; m. One who polishes a sword:—Gif sweordhwíta óðres mannes wǽpn tó feormunge onfó (cf. Si quelibet arma politori vel emundatori commissa sunt, L. H. I. 87, 3; Th. i. 593, 15), L. Alf. pol. 19; Th. i. 74, 8. Ic geann mínon swurdhwítan ðæs sceardan málswurdes, Chart. Th. 561, 22.

sweord-leóma, an; m. The glitter of swords:—Swurdleóma stód swylce eal Finnsburuh fýrenu wǽre there was flashing of swords, as if all Finnsburg were on fire, Fins. Th. 71; Fin. 35.

-sweordod. v. ge-swurdod.

Sweordoras (?); pl. m. A people of Mercia occupying a district of three hundred hides:—Sweordora þryú hund hýda (the name occurs in a list of districts in the land of the Mercians), Cod. Dip. B. i. 414, 21. [Mr. Birch suggests a connection with Swerford in Oxfordshire, and with the river Swere. Could the word contain as its second part the Çeltic dwr = water, seen in many river names, v. Taylor's Names and Places, p. 133, and mean the dwellers by the river Swere?]

sweord-plega; an; m. Sword-play, battle:—Æt ðam sweordplegan wíg forbúgan, Wald. 22; Vald. 1, 13.

sweord-rǽs, es; m. A sword-rush, an attack with swords:—Sweord-rǽs fornam, ðǽr se hálga gecrang wund for weorudum, Apstls. Kmbl. 118; Ap. 59.

sweord-slege, es; m. A sword-stroke, stroke with a sword:—Hyre sáwl weard álǽded of líce þurh sweordslege, Exon. Th. 282, 30; Jul. 671.

sweord-wegende part. Sword-bearing:—Swurdwege[n]de anbidian gehende saca mǽste getácnaþ (in a dream) to await men carrying swords betokens strifes at hand and very great ones, Lchdm. iii. 204, 28.

sweord-weras; pl. The name of a people (cf. the Suardones of Tacitus. v. Grmm. Gesch. D. S. 329):—Mid Seaxum ic wæs and mid Sweordwerum, Exon. Th. 322, 13; Víd. 62.

sweord-wigend, -wígende one who fights with a sword:—Sweord-wígendra síde hergas, Cd. Th. 194, 13; Exod. 260.

sweord-wund; adj. Wounded with the sword:—Oft æt hilde gedreás swátfág and sweordwund sec[g] æfter óðrum, Wald. 7; Vald. 1, 5.

sweord-wyrhta, an; m. A sword-wright, maker of swords, armourer:—Móna se án and twentigoða unnytlíce tó wyrcenne bútan swurdwyrhtan (but the word glosses gladiatoribus), Lchdm. iii. 194, 10.

-sweorf in ge-sweorf rasura ferri, ferrugo, Wrt. Voc. ii. 147, 65: 35, 32. [Cf. Icel. svarf filings.] v. ge-sweorf.

sweorfan; p. swearf, pl. swurfon; pp. sworfen To rub, scour, file:—Swyrfþ limat, Germ. 394, 274. Corfen sworfen cut and scoured (of the preparation of a wine-vat), Exon. Th. 410, 24; Rä. 29, 4. Mín heáfod is homere geþuren sworfen feóle, 497, 18; Rä. 87, 2. Cpds. with for, omitted in their place, are added here:—Forsweorfeþ elimat, i. mundat, Wrt. Voc. ii. 143, 1. Biþ forsworfen vel forgniden demolitur, extermi-natur, 138, 63. [In later English the verb has the sense of swerve = to turn (aside):—Swerve to no side, Gow. 3, 92. Þe dint swarf, Arth. and Merl. 9369. Heo swarf to Criste migravit ad Christum, Kath. 2181. Cf. Du. zwerven to wander, rove: O. Frs. swerva to move, go. For the old English verb, cf. Goth. af-swairban to wipe out; delere; bi-swairban to wipe: O. Sax. swerban to wipe: O. H. Ger. swerban tergere, extergere, siccare: Icel. sverfa to file.] v. á-, ge-sweorfan.

sweor-hnitu, e; f. A neck-nit, a nit that breeds at the back of the neck:—Sweorhnitu ursie, Wrt. Voc. i. 287, 48. Suernit (= sweorhnitu?) usia (cf. swínes lús usia, 122, 26), Wülck. Gl. 54, 34.

Sweó-ríce, es; n. Sweden:—Ðone sélestan sæcyninga ðara ðe in Swió-ríce sinc brytnade, Beo. Th. 4755; B. 2383: 4983; B. 2495. [Icel. Svía-ríki: Swed. Sverige.]

sweor-racenteáh; g. -teáge; f. A chain for the neck:—Swurracenteáh catelle, Wrt. Voc. i. 16, 64.

sweor-ród, e; f. A cross suspended from the neck:—Hé becwæð Wulfstáne ærcebiscope áne sweorróde (the Latin version has philacterium; cf. the use of this word for chains and medals worn by gladiators round

their necks as tokens of victory), Chart. Th. 551, 5. Óðrum líðlum sil-frenum swurródum, 429, 15.

sweor-sál *a collar.* v. sál, **V.**

sweor-sceacel, es; *m. A neck-shackle, pillory:*—Fótcopsa[s] *vel* sweorscacul *nerui*, boia, Wrt. Voc. i. 21, 15. v. sweor-cops.

sweor-teáh, -téh; *g.* -teáge, -tége; *f. A collar:*—Sweortéh *millus vel collarium*, Wrt. Voc. i. 23, 34. Sweorcláþ *vel* [sweor]tég *collarium*, ii. 134, 48. Swiortégum *collaribus*, vinculis, Hpt. Gl. 501, 38.

-sweoru. v. ge-sweoru, sweora, **II.**

sweor-wærc, es; *m. A pain in the neck:*—Lege on ðone sweorwærc, Lchdm. ii. 44, 22. Cf. sweor-coþu.

sweostor, swistor, swystor, swustor (-er, -ur); *indecl. in sing.; pl.* sweostor, sweostra, sweostru (u, y); *f. A sister.* **I.** of blood relationship:—Saga ðæt ðú sié sweostor mín, líces mæge, Cd. Th. 110, 3, Gen. 1832. Ðære swustur (suoester, Lind.: swester, Rush.) wæs Maria *huic erat soror nomine Maria*, Lk. Skt. 10, 39. Soester, Lind. 10, 40. Swuster, Gen. 12, 13. Seó yldre swyster, 19, 33. Sweostor bearna *nepotum*, Wrt. Voc. ii. 59, 70. Se wæs his sweostor sunu, Bd. 4, 16 ; S. 584, 16. Sweoster sunu, 2, 3 ; S. 504, 20. Swuster sunu, Byrht. Th. 135, 8 ; By. 115. Ðæt ðú gesecge sweostor mínre, Exon. Th. 172, 32 ; Gú. 1152. Oþer him sylfum, oþer his sweoster, Bd. 4, 6 ; S. 574, 13 : Homl. Th. ii. 546, 35. Gif hé geméteþ óðerne æt his swister, L. Alf. pol. 42 ; Th. i. 90, 28. Hé betæhte hý his swyster, Chr. 1048 ; Erl. 180, 23. Tó hyre gingran swuster, Gen. 19, 31. Forlét hé Pendan sweoster, Bd. 3, 7 ; S. 529, 29. Swustor (suoester, Lind. : swester, Rush.) *sororem*, Jn. Skt. 11, 5. Swuster, Gen. 25, 20. Hiera swostur (sweostor, swystor (-er), swustra) wærun Cuénburg and Cúþburh, Chr. 718 ; Th. pp. 70, 71. Neogone wǽran Nodþæs sweoster, Lchdm. iii. 62, 18. Ealle his swustra (suoester, Lind. : swæster, Rush.), Mt. Kmbl. 13, 56. Swustra (suoestro, Lind.: swester, Rush.), Mk. Skt. 6, 3. Swestro, Jn. Skt. Rush. 11, 3. Ic seah vi. gebróþor and hyra sweostor mid, Exon. Th. 394, 13 ; Rä. 14, 2. Ðe ne onfó swustru (swustra, MS. A.: suoestro, Lind.: swester, Rush.). Mk. Skt. 10, 30. **II.** of membership in a religious house:—Ætýwde sumre gódre swuster wundorlíc gesyhþ . . . Ðeós sweoster . . . , Bd. 4, 9 ; S. 576, 18–30. Seó gesomnung bróþra and sweostra, 4, 19 ; S. 589, 9. Ðá ongan heó on gesomnunge ðare sweostra sécan . . . Heó nǽnige andsware findan mihte, ðeáh ðe heó georne sóhte æt ðám swustrum, 4, 7 ; S. 574, 35, 40. Ðá geseah heó óþre sweoster (*sorores*) ymb hí restende . . . ðá áwæhte heó ealle ðá sweostera, 4, 23 ; S. 596, 5–14. [*Goth.* swistar : *O. Sax.* swestar : *O. Frs.* swester, suster : *O. H. Ger.* swestar : *Icel.* systir.] v. ge-sweostor ; ge-sweosternu.

sweót, es ; *n. A troop, band, squadron :*—Him on láste fór sweót Ebréa sigore geweorþod, Judth. Thw. 25, 38 ; Jud. 299: Ðý deáðdrepe drihte swǽfon, synfullra sweót sáwlum lunnon, Cd. Th. 209, 8 ; Exod. 496. Segn ofer sweóton, 185, 23 ; Exod. 127. Segen for sweótum, Elen. Kmbl. 247 ; El. 124. Sweótum in crowds, in shoals, Beo. Th. 1138 ; B. 567. Sunu Simeonis sweótum cómon (*came in bands*), Cd. Th. 199, 20 ; Exod. 341. Fífe fóran folc cyningas sweótum (*marched with their squadrons*), 119, 5 ; Gen. 1975. Moyses bebeád cígean sweót (*summon the bands*), 119, 25 ; Exod. 220.

sweóta (?). an ; *m. The scrotum :*—Sweótan marsem (= marsupium, v. Cockayne's remark, Lchdm. iii. 371, col. 1), Lchdm. i. lxxiv, 27.

Sweó-þeód, e ; *f. The Swedish people :*—Ne ic tó Sweóðeóde sibbe oððe treówe wihte ne wéne, Beo. Th. 5836 ; B. 2922. Swíðe mycel here ægðer ge landhere ge sciphere of Swaðeóde (Sweóðode, MS. F.), Chr. 1025 ; Erl. 163, 9. [*Icel.* Sví-þjóð.]

sweoþol. v. sweþel.

sweotol, swutol, switol, swytol, sutol (-ul, -al, -el); *adj.* Plain, manifest, evident, clear, patent :—Sweotul, gewis *evidens*, i. *manifestus*, patens, perspicuus, certum, Wrt. Voc. ii. 144, 35. Sweotol *evidens*, 29, 51. Seotol, 107, 42. **I.** of what may be clearly perceived by the senses, (a) by sight :—Biþ mín swæd sweotol, sweart on óþre healfe, Exon. Th. 403, 19 ; Rä. 22, 10. Wiht sweotol and gesýne, 420, 13 ; Rä. 40, 3. Him on eaxle weard syndolh sweotol, Beo. Th. 1638 ; B. 817. Ða fótlǽstas wǽron swutole and gesýne, Blickl. Homl. 203, 36. Fell hongedon sweotol and gesýne, Exon. Th. 394, 16 ; Rä. 14. 4. (b) by hearing :—Ðǽr wæs hearpan swég, swutol sang, Beo. Th. 180; B. 90. (c) by taste :—Ne sié on bergnesse tó sweotol ðæs ecedes scearpnes, Lchdm. ii. 224, 22. **II.** manifest to observation, that may be noticed by all, public, open, patent :—His nama wæs swutol geworden, Mk. Skt. 6, 14. Hit is on ús eallum swutol and gesýne, ðæt wé oftor brǽcan, ðonne wé béttan, Wulfst. 159, 5. Sweotol and geséne, Cd. Th. 170, 1 ; Gen. 2806. Hé wundra fela weorodum gecýðde sweotulra and gesýnra, Andr. Kmbl. 1129 ; An. 565. Swutelra, Menol. Fox 255 ; Men. 129. Sutelum *publicis*, Hpt. Gl. 525, 20. **III.** clear to the understanding, free from obscurity, plain, of proof, argument, indication, etc. :—Swutol is constat, Ælfc. Gr. 33; Zup. 206, 7 : liquet, Zup. 207, 6. Ðæt is swíþe sweotol tó ongitanne be sumum ǽdelinge, Bt. 16, 2; Fox 52, 18. Genóh sweotol is, ðætte gód word biþ betera ðonne ǽnig wela, 13 ; Fox 38, 22 : 36, 3; Fox 176, 27 : 36, 7; Fox 184, 5. Is on mé sweotul ðæt . . . it is plain from my case that . . . , Exon. Th. 275, 17; Jul.

55}. Biþ hit sweotol (swutul, Hatt. MS.), Past. 14 ; Swt. 83, 20. Swutol, 21 ; Swt. 153, 4. Ðæt wæs tácen sweotol it was a token that was an evident proof, Beo. Th. 1671 ; B. 833. Ðæt is swíþe swital (sweotol, Cott. MS.) on ðære týdrunge, Bt. 34, 12 ; Fox 152, 25. Wæs swytol, ðæt hé ǽr mihte wið deáð gebeorgan, Wulfst. 23, 15. Ðis eástorlíce gerýno ús ǽteóweþ ðæs écean lífes sweotole bysene, Blickl. Homl. 83, 8 : 99, 14. Tácen swutol, Cd. Th. 270, 12 ; Sat. 89. Orðancum swutulum *argumentis evidentibus* (apertis, manifestis), Hpt. Gl. 486, 21. Ðæt him biþ ungewitnode hiora yfel on ðisse worulde, ðæt is ðæt sweotoloste tácn (*the clearest indication*) ðæs mǽstan yfeles on ðisse worulde, Bt. 38, 3 ; Fox 200, 29. [Sutel (sotel, 2nd MS.) word *a clear message*, Laym. 1519. Bi Moisen is sutel and edcene, A. R. 154, 22. Wass full sutell and full sene, þatt . . . , Orm. 18862.] v. un-sweotol.

sweotole; *adv.* **I.** of a physical action, *clearly, without obstruction :*—Steorran geséon swá sutole swá on niht, Blickl. Homl. 93, 20. Gé sweotule geseóþ Dryhten faran, Exon. Th. 32, 13 ; Cri. 512. Sweotole on ðæs heþenes heáfod starian, Judth. Thw. 24, 8 ; Jud. 177. Ðonne sió sunne sweotolost scíneþ, Met. 6, 3. **II.** *in a manner open to general observation, evidently, openly, plainly, publicly :*—Wǽron heardingas sweotole gesamnod, Elen. Kmbl. 51 ; El. 26. Sweotule ða forweorðaþ (*their destruction will be seen by all*), Ps. Th. 101, 23. Sunne hire setlgang sweotule healdeþ, 103, 18. **III.** *openly, without reserve or concealment, plainly :*—Nis nú nán ðe ic him módsefan mínne durre sweotule ásecgan, Exon. Th. 287, 8 ; Wand. 11. **IV.** of thinking, knowing, stating, explaining, etc., *clearly :*—Sweotole ongitan, Bt. 33, 2 ; Fox 124, 34: Met. 26, 107. Sueotole, sweotule, Past. 7 ; Swt. 49, 2. Sweotule cunnan, Ps. Th. 118, 12. Sweotele gecnáwan, Bt. 3, 1 ; Fox 4, 29. Sweotole oncnáwan, Met. 12, 29. Swotole, Bd. 2, 12 ; S. 515, 20 : 3, 14 ; S. 540, 15. Swutele, swutole tócnáwan, Bt. 20 ; Fox 72, 15, 20. Be ðære sunnan sweotole geþencean, Met. 5, 1. Sweotole secgan, Met. 20, 182: Elen. Kmbl. 335 ; El. 168. Sweotole gecýðan, 1718 ; El. 861. Sweotole gereccan, Bt. 35, 3 ; Fox 160, 5. Sweotule, Met. 8, 2. Sweotule geséþan, Exon. Th. 15, 28 ; Cri. 243. Ða siex stafas sweotule bécnaþ, 407, 5 : Rä. 25, 10. Sweotolor, Bt. 34, 6 ; Fox 142, 3 : 11, 1 ; Fox 30, 29 : Met. 12, 23 : Shrn. 188, 31. Hwæðer ðú hit á sweotolor (*any more clearly*) ongiton mǽge, Bt. 34, 4 ; Fox 138, 16. Swá hé hit sweotolost and andgitfullícost gereccan mihte, Bt. procem. ; Fox viii, 4.

sweotolian, swutelian, swytelian ; *p.* ode. **I.** *to make clear or manifest, to shew, declare :*—Ælc gesceaft ðæt sweotolaþ, ðæt God éce is *Deum aeternum esse cunctorum degentium commune judicium est*, Bt. 42 ; Fox 256, 7. Hér swutelaþ on ðison cwyde hú Ædelréd geúðe ðæt Æðeríces cwyde standan móste, Chart. Th. 539, 20: 320, 24: 312, 8. Swytelaþ, 586, 25. Swetelaþ *expremit*, Kent. Gl. 1120. Ðæt ðæt man beháteþ, ðonne man fulluhtes gyrnþ, swytelaþ, ðæt man wile on ǽnne God gelýfan, L. I. P. 24 ; Th. ii. 338, 12. Hé ongan swutelian (*ostendere*) his leorningcnihtum, ðæt hé wolde faran, Mt. Kmbl. 16, 21. [He schawde and sutelede þ he wes soð godd, Kath. 1037. He schawde him and sutelede him seolf to hire, 1834. Þet hit sutelie in us hwuch was his lif, A. R. 382, 3.] **II.** *to become manifest :*—Ðin mycele miht manegum swutelaþ, Hy. 9, 32. [Hit schal sutelin (*become manifest*) sone, Jul. 18, 4. Þurh þis suteleð soð al þ ich segge, Kath. 1089.] v. ge-sweotulian.

sweotol-líc; *adj. Clear, plain :*—Gehýraþ hwæt God sylfa sǽde swytellícre (swutel-, MS. C.) segene, Wulfst. 45, 1.

sweotollíce; *adv. Clearly :*—Swutollíce *manifeste* and *manifesto*, Ælfc. Gr. 38 ; Zup. 235, 12. **I.** of a physical action, *clearly, plainly, distinctly :*—Hié sweotullíce geséon mihten ðære byrig weallas blícan, Judth. Thw. 23, 23 ; Jud. 136. Hí swutolíce (*manifeste*) engla sang gehýrdon, Bd. 3, 8 ; S. 532, 5. Swutollíce hé sprecþ *expresse loquitur*, Ælfc. Gr. 38 ; Zup. 228, 11. **II.** *openly, publicly :*—Ðæt heó swutollíce (*palam*) eallum cýdde, Bd. 4, 19 ; S. 588, 17. **III.** of perceiving, knowing, shewing, stating, etc., *clearly, plainly :*—Sweotollíce ongitan, Blickl. Homl. 97, 22 : 219, 36 : Bd. 5, 1 ; S. 614, 13. Sweotollíce, 4, 28 ; S. 607, 3. Swutollíce oncnáwan, Hy. 7, 90. Sweotollíce gecýðan, Elen. Kmbl. 1376 ; El. 690 : Blickl. Homl. 27, 26. Swutullíce, 181, 27 : Homl. Th. i. 76, 28. Him wæs gesǽd swutelíce, Gen. 15, 13. Sweotolícor gecnáwan, Exon. Th. 263, 26 ; Jul. 355. Swǽtolocor getǽcan, Shrn. 175, 34. Omarus sweotelícost sægde *Homerus luculentissimo carmine palam fecit*, Ors. 1, 11 ; Swt. 49, 15.

sweotolung, e ; *f.* **I.** *a manifestation :*—Ðes freólsdæg (*Epiphany*) is Godes swutelung gecweden, Homl. Th. i. 104, 29. **II.** *an explanation, definition :*—Ásmeáde swutelunge elucubratam definitionem (*manifestationem*), Hpt. Gl. 522, 47. **III.** *a declaration, setting forth, exposition, shewing :*—Hér onginþ seó bóc peri didaxeon (περὶ διδαξέων), ðæt ys seó swytelung hú fela géra wæs behúded se lǽcecræft, Lchdm. iii. 82, 1. **IV.** *evidence, testimony, declaration ; when written, a testament, title-deed, certificate, prescript :*—Hér is seó swutelung (*the will, testament*) hú Ælfhelm his áre and his æhta gefadod hæfþ, Chart. Th. 596, 5. Ðeós swutelung (*the evidence or testimony which has been recited in the previous part of the charter*) wæs ðǽrrihte gewriten and beforan ðam cincge gerǽdd, 540, 35. Wé habbaþ gedón swá swá ús swutelung (*evidence of your wish, mandate*) from eów com æt ðam ð. (*in*

respect to consecrating the bishop), 314, 1. Hí ða bóc tó swutelunge sealdan *they gave the charter as evidence (of a grant)*, 588, 14. Tó swutulunge ðæt man wite ðæt man clǽne bæc hæbbe (tó swutelunge ðæt man mid rihte fare, 9), L. A. G. 5; Th. i. 156, 5. Ic wille, ðæt ðú underfó ðás seofon lamb æt mé, ðæt hig tó swutelunge (*in testimonium*) beón, ðæt ic dealf ðisne pytt, Gen. 21, 30. Gyf ǽnig man sý, ðæt wylle ǽnig ðæra sócna him tó handa drægan, ic wylle ðæt hé cume beforan mé mid his sweotelunge (*with the evidence that substantiates his claim*), Cod. Dip. Kmbl. iv. 222, 32. Bringe hé swutelunge (switelunge, MS. D.), ðæt hé swá micel betǽht hæbbe, L. Edg. i. 4; Th. i. 264, 10. Ða gemǽtæ hé on ðam mynstre ða ylcan swutelunga (*evidences, title-deeds*) ðe his foregenga hæfde . . . Syððon se bisceop his swutelunge geeówod hæfde, Chart. Th. 302, 8–33. On ðissan þrim cyrografum ðe on ðissun ðrým mynstrum tó swytelungum gesette syndon, 233, 2. Swutelung[um] *adstipulationibus* (cf. *adstipulationibus* trymnessum, cýðnessum, Wrt. Voc. ii. 3, 63), Hpt. Gl. 525, 36. v. ge-swutelung.

Sweotolung-dæg, es; *m. Epiphany:*—Ðes dæg (viii. Idus Ian.) ge háten on bócum Sweotelungdæg, forðan ðe on ðisum dæge weard Crist mancynne geswutelod, Homl. Th. ii. 36, 20. Epiphania Domini *is translated by* Godes geswutelungdæg, i. 104, 18.

swer *a pillar*, swér *a mother-in-law*, swér *heavy*. v. sweor, sweger, swǽr.

swerian; *p.* swór (*but a weak* swerede *occurs*; cf. *Icel.* svarði *as well as* sór), *pl.* swóron; *pp.* sworen *To swear, make oath*.   I. absolute:—Se ðe sweraþ (swereþ, Ps. Th. Surt.) néhstan his *qui jurat proximo suo*, Ps. Spl. 14, 6. Ðæt land ðe ic fore swór heora fæderum *terram pro qua juravi patribus eorum*, Num. 14, 23. Ðæt land ðe ðú hira fæderum fore swóre, 11, 12. Hí wið mé sweórun *adversum me jurabant*, Ps. Surt. 101, 9. Ic secge eów, ðæt gé eallunga ne swerion, Mt. Kmbl. 5, 34. Hí mé hraþe æftter swerigean ongunnon, Ps. Th. 101, 6. Hé mót swerian for syxtig hída, L. In. 19; Th. i. 114, 11.   I a. *to swear* by *or* on:—Swá hwylc swá swereþ on temple . . . swá hwá swá swereþ on ðæs temples golde, Mt. Kmbl. 23, 16, 18, 20, 21. Swá swá ðú swóre on sóðfæstnysse ðíne, Ps. Spl. 88, 48. Ic swerige ðurh God *juro per Deum*, Ælfc. Gr. 38; Zup. 227, 4. Ne swerie gé þurh útencymena goda naman, Ex. 23, 13: Mt. Kmbl. 5, 34, 35. Ne swerigen gé nǽfre under hǽðene godas, L. Alf. 48; Th. i. 54, 23.   I b. *to swear* to anything:—Ðæt hí hit gegaderian and eft ágifan swá hí durran tó swerian, L. N. P. L. 57; Th. ii. 300, 2.   II. with an object, (1) a noun (pronoun):—Ða swóron hí swíðe, ðæt hit swá wǽre. Ða cwæð hé tó him: ‘Ac tó hwon sweriaþ git mán?’ Guthl. 14; Gdwin. 64, 6. Ic ne swór fela aþa on unriht, Beo. Th. 5470; B. 2738. Hé mé áðas swór, 949; B. 472. Hé him ðone áð swór, Gen. 24, 9. Ðone swergendan áð ðone hé swór *jusjurandum quod juravit*, Ps. Surt. ii. p. 199, 20. Wyrgdan, áð sweredan (áðsweredan?) *devotabant*, Wrt. Voc. ii. 26, 48. Se ðe mánáð swerige, L. Alh. i. 25; Th. i. 212, 18. Ðæs deádan mǽgas swerian uncéases áð, L. In. 35; Th. i. 124, 7. (1 a) *to swear* an oath by something:—Ða ðe áðas sweriaþ on hine, Ps. Th. 62, 9. Ic ǽne swór áð on hálgum, 88, 31. Gange ǽlc man ðæs tó gewitnesse ðe he durre on ðam háligdóme swerian, L. Eth. iii. 2; Th. i. 292, 14. Ic swór mǽne áðas miûra hláforda lífe, L. de Cf. 9; Th. i. 264, 11. (2) where the object is a clause that contains a statement of that which is confirmed by oath:—Ða ætsóc hé and swerede ðæt hé nǽfre ðone man ne cúðe *tunc coepit detestari et jurare quia non novisset hominem*, Mt. Kmbl. 26, 74. Hig swóron him betweónan, ðæt hig sibbe heóldon, Gen. 21, 31. Ða swóran hié swíðe, ðæt hié sóð sægdon, Nar. 25, 27. Swerige hé, ðæt hé him nán fácn on wiste, L. In. 56; Th. i. 138, 12: L. Alh. v. 12, 2; Th. i. 242, 4. Begite hé ðara .v. .i. ðæt him mid swerige, ðæt . . . i. 9; Th. i. 204, 11. Swerian (cf. gif hí ðone áð syllan ne durren, 304. 3) hí, ðæt him nǽfre áð ne burste, L. C. S. 30; Th. i. 392, 27. (2 a) *to swear* by, on . . . that . . . :—Swerian hí on ðam háligdóme, ðæt hig nellan nǽnne sacleásan man forsecgean, L. Eth. iii. 3; Th. i. 294, 4. Ic swerige þurh mé sylfne . . . ic ðé bletsige, Gen. 22, 16. Sweriaþ þurh Drihten, ðæt gé dón wið mé mildheortnisse, Jos. 2, 12. Ða ásweartode eall se king and swór under God ælmihtine and under ealle hálgan ðártó, ðæt hit næs ná his rǽd, Chart. Th. 340, 1. (3) where noun and clause both occur:—Swerige hé ðone áð, ðæt hé sý unscyldig, L. Ath. i. 23; Th. i. 210, 31. Ðæt Drihten swóre áð swíðe, ðæt God wolde sendan hungor, Wulfst. 209, 26. (3 a) with adjuration:—Áð swereþ engla þeóden þurh his sylfes líf, ðæt ðínes cynnes rím ne cunnon yldo, Cd. Th. 205, 5; Exod. 431. [*Goth.* swaran: *O. Sax.* swerian: *O. Frs.* sweria, swera, swara: *O. H. Ger.* swerien, sweren: *Icel.* sverja.] v. ā-, æt-, for-, ge-, óþ-swerian; swerigend-líc.

swerian; *p.* ede *To speak, talk:*—Oft ic fróde men gehýrde secgan and swerian ymb sume wísan hwæðer wǽre twegra strengra wyrd ðe warnung *I have often heard wise men speak and talk* (or? *swear, support what they said with oath*) *about a certain thing, whether of the twain were stronger, fate or caution*, Salm. Kmbl. 851; Sal. 425. v. and-swerian.

swerigend-líc; *adj. Pertaining to swearing:*—Sume (*adverbia*) synd *jurativa*, ðæt synd swerigendlíce, *per* ðurh . . . Má syndon swergendlíce *adverbia*, ac hwæt sceolon hí gesæde, nú wé swerian ne móton? Ælfc. Gr. 38; Zup. 227, 3–11.

swertling, es; *m. A tit-lark:*—Swertling *ficedula* (in later glossaries

 ficedula is translated *rooke*, Wülck. Gl. 583, 12: *nuthage = nuthatch*, 702, 32. See also *succa*), Wrt. Voc. i. 29, 10. v. sweart.

swerum, Wrt. Voc. ii. 56, 44, swerung, swés, swésende. v. swéte, áþswerung, swæs, swæsende.

swétan; *p.* te; *pp.* swéted, swét *To sweeten, make sweet*.   I. in a physical sense:—Nim hunig and swét ðone drænc, Lchdm. iii. 58, 30: ii. 120, 11. Swéte swíðe mid hunige, 216, 4. Swétedne, 111, 8, 15.   II. *to make pleasant:*—Hé (*the devil*) mec féran hét, ðæt ic ðé sceolde synne swétan, Exon. Th. 273, 32; Jul. 525. [Saullt þatt ure mete sweteþþ, Orm. 1649. Swetyñ or make a thynge swete to mannys taste *dulcoro*, Prompt. Parv. 483. *O. H. Ger.* suozen: *Icel.* sœta.] v. ge-swétan; swétian.

swéte; *adj. Sweet*.   I. in reference to the senses (lit. or fig.) (1) of taste:—Ðis ofet is swá swéte, Cd. Th. 41, 12; Gen. 655. Ðæt is for hwí se góda lǽce selle ðam hálum men séftne drenc and swétne, and óðrum hálum biterne and strangne, Bt. 39, 9; Fox 226, 11, 13. Swéte ofer hunig *dulcia super mel*, Ps. Spl. 118, 103. Gif hwá biteres hwes onberede, ðæt him þúhte beóbreád ðí swétre, Bt. 23 tit.; Fox xiv, 10. Sweótran ofer hunig, Ps. Surt. 18, 11. ¶ used substantively:—Wá eów ðe taliaþ ungód tó góde, biter ðing tó swéte and swéte belǽþaþ, Wulfst. 47, 7. (1 a) of food, *sweet* in *sweet*-meat, *delicate:*—Swéte mete ðápis, Wrt. Voc. ii. 28, 29. Se swéta mete ðe hié héton monna, Past. 17; Swt. 125, 19. Wyt æton swétne mete (*dulces cibos*), Ps. Th. 54, 13. Fram swétrum mettum *a cibis luculentioribus*, Wrt. Voc. ii. 6, 25. ¶ used substantively:—Hé forlét eall ðæt ðær lídes wæs and swétes *astu instructa vino epulisque deseruit*, Ors. 2, 4; Swt. 76, 14. Ys sáwl mín swétes gefylled *adipe et pinguedine repleatur anima mea*, Ps. Th. 62, 5. Ne mæg se flæschoma swéte forswelgan, Exon. Th. 311, 20; Seef. 95. (2) of smell, *sweet, fragrant:*—Ðǽr wæs swíþe swéte stenc, Blickl. Homl. 145, 29. Wyrta wearmiaþ, willsele stýmeþ swétum swæccum, Exon. Th. 212, 22; Ph. 214. Swétum wyrtum *with sweet-smelling herbs*, 241, 6; Ph. 652. Wynsumra steám, swétra and swíþra, 358, 15; Pa. 46. Of mûðe cwom swecca swétast, 178, 20; Gú. 1247. Ðara swétestena wyrta, Bd. 3, 8; S. 532, 20. (3) of freedom from unpleasant taste or smell, *sweet, pure, untainted:*—Mere in ðam wǽre fersc wæter and swéte genóg (*stagnum dulcissime aque*), Nar. 11, 26. Ða wæs ic gefeónde ðæs swétan wætres and ðæs ferscan *dulci aqua potata gaudio*, 12, 10. *Merum* hlúttor wín oðde swerum, *mero* wíne (l. (?) *mero* swétum wíne), Wrt. Voc. ii. 56, 44. Drince on swétum wætre, Lchdm. ii. 134, 23. Bæþ of swétum ferscum wæterum, 194, 10. (4) of sound, *sweet, harmonious:*—Swég ðæs swétan sanges, Bd. 5, 12; S. 630, 23. Swég eallum songcræftum swétra, Exon. Th. 206, 26; Ph. 132. Ða gehýrde hé ða swétestan stæfne, Bd. 4, 3; S. 567, 39.   II. in reference to the feelings, *sweet, agreeable, pleasant:*—Mé swéte and wynsum wæs ðæt ic oððe leornode oððe lǽrde *aut discere aut docere dulce habui*, Bd. 5, 24; S. 647, 27. Cristes onsýn on sefan swéte sínum folce, biter bealofullum, Exon. Th. 56, 29; Cri. 908. Hwæt déþ ðæt swéte word? Hit gemanigfealdaþ mannes freóndscipe and stilleþ mannes feónd (cf. *a soft answer turneth away wrath*), Salm. Kmbl. 204, 45. Geocc mín suoet ī éðe (wynsum, Rush. W. S.) is *jugum meum suave est*, Mt. Kmbl. Lind. 11, 30. Swoete and reht Dryhten *dulcis et rectus Dominus*, Ps. Surt. 24, 8. Ðú ðín swéte good sealdest þearfum, Ps. Th. 67, 11. Ða geógoðlustas ðe him swéte wǽron tó áræfnenne, Blickl. Homl. 59, 10. Hí mihton eáþe secgan sóþspell, gif him ða leásunga næron swétran, Bt. 35, 4; Fox 162, 16. Se swétesta láreów and se wynsumesta *doctor suavissimus*, Bd. 5, 22; S. 644, 3. Hwæt ðé sý hér on worlde swétast and leófast gesewen ðínra æhta, Blickl. Homl. 195, 20. Mín se swétesta sunnan scíma, Iuliana, Exon. Th. 252, 20; Jul. 166. Dóhtor mín seó dýreste and seó swéteste, 248, 11; Jul. 94. [*O. Sax.* swóti: *O. Frs.* swéte: *O. H. Ger.* suozi: *Icel.* sœtr.] v. hunig-, un-swéte; swét, swóte.

sweþel, sweoþol, es; *m. A swathe, wrap, band, bandage*; cf. *swaddling* band, clothes:—Sweþil *fascia*, Wrt. Voc. ii. 34, 74. Sneðelas, suedilas *instites*, Txts. 69, 1060. Sweþelas, Wrt. Voc. ii. 45, 48. Sweoþolas *fascia* [*e?*], 93, 69. Suuoeðles *institis*, Jn. Skt. Lind. 11, 44. Suaeðila *fasciarum*, Wrt. Voc. ii. 108, 18. Sweþila, 34, 76. Sweþela, 82, 36. Sweþelum *fasciarum*, 34, 21. Suithelon *institis*, Txts. 113, 72. ¶ of a funeral pile in whose fire the body is wrapped (?):—Wudurêc ástáh sweart ofer swioðole (swicðole, MS.) *the smoke rose black above the pile where Beowulf's body lay enwrapped*, Beo. Th. 6281; B. 3146, cf. swaþul. [Cf. Bondon wit a sueþelband (suadiling band, swaþeling bonde, other MSS.), C. M. 1343. A child in swetheleloutes, Met. Homl. 91, 14. *O. H. Ger.* swedili *malagma*.] v. sweðian.

sweðerian. v. sweðrian.

sweðian; *p.* swethede *To swathe, wrap*. [She swaþeþ (swetheled, suedeld, other MSS.) him wiþ cloþes, C. M. 11236. Swathyn chyldyr *fascio*, Prompt. Parv. 482. Swethed togeder, Pall. 149, 19.] v. besweðian (where add these passages, Lchdm. ii. 46, 32: 182, 19: 250, 18), bi-sweðian.

sweðrian, swiðrian, sweoðerian; *p.* ode (*some instances of the cpd.* ge-sweðrian, *omitted under that word, are given here*) *To retire, withdraw, abate, subside, decrease, fail, come to an end:*—Sweðraþ *facessit, discedit*,

Wrt. Voc. ii. 33, 30. Gesweðeriaþ *fatescunt* (fatiscere *dissolvi*, Migne), 96, 18. Mylt, sweþrede, áswand, áteorade *dissolvitur, desinit, discedit*, 147, 25. Gesuedrade, gesuidradae, gisuderadae *constipuisse*, Txts. 53, 525. Geswiðrade, Wrt. Voc. ii. 14, 71. Gesweþrade *constipuit*, i. *defecit*, 133, 63. Sweþeredan *fatescunt*, 37, 29 : *facescunt*, 91, 61. Gesueðradun, -suedradum *exoleuerunt*, Txts. 61, 786. *Exoliuerunt*, i. *tabuerunt, eruperunt, arripuerunt, uel* gesweþredon, Wrt. Voc. ii. 145, 82. Sweþriendum *facessante*, 33, 29. Sweðriende, 75, 20. **I.** in reference to concrete things :—Se bryne sweþraþ *the burning ceases*, Exon. Th. 213, 24 ; Ph. 229. Swég swiðrode *the sound ceased*, Cd. Th. 197, 18 ; Exod. 309. Cyre (cyrr?) swiðrode sǽs æt ende (*the sea no longer ebbed* (?), *it rolled back upon the Egyptians*), 207, 12 ; Exod. 465. Mere sweoðerade (*the sea subsided*), ýða ongin eft oncyrde, hreóh holmþracu, Andr. Kmbl. 930 ; An. 465. Dryhten forlét dægcandelle scíre scínan, sceadu sweðerodon, 1672 ; An. 838. Sweþredon, Exon. Th. 179, 16 ; Gú. 1262. Swiðredon, Cd. Th. 184, 27 ; Exod. 113. Ðonne ðú ongite ðæt ðæt geswel hnescige and swiþrige, Lchdm. ii. 208, 16. Ðæt fýr ongon sweðrian, Beo. Th. 5397 ; B. 2702. Swiðrian, Cd. Th. 8, 34 ; Gen. 134. **II.** in reference to abstract things :—Se longa gefeá ǽfre ne sweþraþ *the long joy never comes to an end*, Exon. Th. 238, 23 ; Ph. 608. Hwæþere him ðæs wonges wyn sweðrade *whether the delight in the plain was abating with him*, 123, 16 ; Gú. 323. Hild sweðrode, earfoð and ellen, Beo. Th. 1807 ; B. 901. Gif mægen swiðrade, Cd. Th. 193, 7 ; Exod. 242. Nó swiðrode ríce, 256, 12 ; Dan. 639. Him sweðradon synna lustas *sinful joys subsided in him*, Exon. Th. 109, 2 ; Gú. 84. Metod lét Babilone blǽd swiðrian, d. Th. 258, 30 ; Dan. 683. v. ge-sweðerian ; swaðrian, *and next word*.

sweðrung, e ; *f. Diminution, failure* [:—Ðæt tácnaþ wæstma gesweþrunge *that betokens a failure of crops*, Lchdm. iii. 180, 13.]

sweðung, swoðung, e ; *f. A poultice* :—Sweþing wiþ swile . . . gecnuwa ða wyrte, gemeng wið ǽges ðæt hwíte, beclǽm ðæt lim mid ðe se swile on sié, Lchdm. ii. 74, 24. Sealfae and sweþinge wið swylum, 6, 30. Gif hé sweðunga (swoðunga, R. Ben. Interl. 59, 11) gegearwode *si exibuit fomenta*, R. Ben. 52, 11. [O. H. Ger. swedunga *fomentum*.]

swétian; *p.* ede *To be sweet* or *pleasant* :—Ðætte ús biterige sió hreówsung, swá swá ús ǽr swétedon ða synna *that repentance may prove bitter to us, as before sins were sweet to us*, Past. 54, 5 ; Swt. 425, 14. v. swétan.

swétlécan. v. ge-swétléht.

swétlíce ; *adv. Sweetly, pleasantly* :—Swétlíce drincan ða word ðínes wísdómes *verba tuae scientiae dulciter haurire*, Bd. 5, 24 ; S. 649, 1.

swét-mete, es ; *m. A sweet-meat, delicacy* :—Of ðám swétmettum and of mistlícum dryncum ðæs líþes onwæcnaþ sió wóde þrág ðære wrænnesse, Bt. 37, 1 ; Fox 186, 16 : Met. 25, 40. v. swót-mete.

swétness, e ; *f. Sweetness* :—Swétnys *dulcedo*, Ælfc. Gr. 9, 3 ; Zup. 37, 6. Swétnesse *dulcedinis*, Wrt. Voc. ii. 28, 34. **I.** in reference to the sense (a) of smell, *fragrance* :—Mycel swétnys wundorlíces stences *fragrantia mirandi odoris*, Bd. 4, 10 ; S. 578, 13. Swétnes, 5, 12 ; S. 629, 20. Swétnysse stencg, 3, 8 ; S. 532, 18. In gistenc suoetnises *in odore suauitatis*, Rtl. 41, 30. Ic nardes stenc oferswíþe mid mínre swétnesse, Exon. Th. 423, 30 ; Rä. 41, 30. (b) of taste :—Suoetnis *ambrosea*, Wrt. Voc. ii. 100, 14. Ðæs monnan swétnes, Past. 17 ; Swt. 125, 23. Of bitternise in suoetnisse, Rtl. 114, 36. **II.** *sweetness, pleasantness, agreeableness* :—Seó swétnes ðæs hǽmeðþinges ðe hé ǽr lufode, Blickl. Homl. 59, 16. Hú micel is seó mycelnes ðínre swétnesse (*dulcedinis tuae*), Ps. Th. 30, 21. Mid ðære swétnesse ðínra bletsunga, 20, 3. Úre heortan gefyllan mid ðære swétnesse godcundra beboda, Blickl. Homl. 37, 8. Be swétnesse ðæs heofonlícan ríces, Bd. 4, 24 ; S. 598, 16. Ða worulðsælþa mid swíþe manigre swétnesse óleccaþ ðæm módum, Bt. 7, 1 ; Fox 16, 10. Beswícan þurh ða swétnesse ðara worda . . . þurh ða swétnesse ðara synna, Blickl. Homl. 55, 22, 24. Mid ða mǽstan swétnesse *maxima suauitate*, Bd. 4, 24 ; S. 596, 34.

swétole. v. sweotole.

swét-swége ; *adj. Of sweet sound, harmonicus, melodious* :—Mid swét-swégum leóþum *suavisonis carminibus*, Hymn. Surt. 58, 16.

swét-wyrde ; *adj. Agreeable of speech, bland* :—*Blandis sermonibus, lenis verbis* líþum *vel* swétwyrdum, Wrt. Voc. ii. 127, 4. *Balbus, qui vult loqui et non potest* wlips *vel* swétwyrda (*blandus seems to have been read?*), 125, 11.

swic (swíce ? *q. v.*), es ; *n. Deception, illusion* :—For swicum deóflicum *propter illusiones diabolicas*, Anglia xiii. 396, 441. [O. H. Ger. á-, bi-swih ; *pl.* -swicha ; *m.* : Icel. svik ; *n.* : Dan. svig *fraud, deceit*.] v. ǽ-, be-, ge-, lár-swic ; swíce.

swica, an ; *m.* **I.** *a deceiver* :—Swica *planus vel seductor*, Wrt. Voc. i. 47, 51. Se swica (*se ductor ille*) sǽde : 'Æfter þrým dagon ic áríse,' Mt. Kmbl. 27, 63. Seó smyltnys is stulor and dígele swica, Homl. Th. ii. 392, 25. **II.** *one who fails in fidelity or fealty, a traitor* :—Him man wearp on, ðæt hé wæs ðes cynges swica and ealra landleóda *that he was a traitor to his king and country*, Chr. 1055 ; Erl. 189, 4. Swá wurdon Willelmes swican geniðrade, 1075 ; Erl. 214, 17. [The suikes undergæton ð he (*Stephen*) milde man was, Chr. 1137 ; Erl. 261,

30. Ueond þet þuncheð freond is swike ouer alle swike, A. R. 98, **6.** Sweoke (*the false fiend*), H. M. 45, 34. Þus speken þeos swiken, . . . swa long heo hine lærde, þat he heom ileuede, Laym. 3816. Godard was þe moste swike . . . withuten on, þe wike Iudas, Havel. 423. Icel. dróttinsviki.] v. ǽ-, be-, fæder-, hláford-, mann-swica.

swícan; *p.* swác, *pl.* swicon ; *pp.* swicen. **I.** *to move about, wander* :—Oðer lifaþ lytle hwíle, swíceþ on ðisse sídan gesceafte, and ðonne eft mid sorgum gewíteþ, Salm. Kmbl. 737 ; Sal. 638. [O. H. Ger. swíhante *vagus*.] **II.** *to move away, depart, escape* :—Wiþ ðæt beón æt ne fleón, genim ueneriam and gehóh hý tó ðære hýfe ; ðonne beóþ hý wunigende and nǽfre ne swícaþ, Lchdm. i. 98, 2. Hé for mundgripe mínum scolde licgean lífbysig, bútan his líc swíce *unless his body had escaped (from my grasp)*, Beo. Th. 1937 ; B. 966. Eam ic geseald ðær ic út swícan ne mæg *traditus sum et non egrediebar*, Ps. Th. 87, 8. Hé biþ on ðæt wynstre weorud wyrs gesceáden, ðonne hé on ða swíþran hond swícan móte, Exon. Th. 449, 25 ; Dóm. 76. Sceal ánra gehwylc óðrum swícan, forðam Dryhten wile ðæt earme flǽsc eorðan betǽcan *each one must depart from other, for the Lord will commit frail flesh to earth*, Runic pm. Kmbl. 343, 14 ; Rún. 20. **II. a.** swícan from *to turn from, to withdraw favour* or *allegiance from, to rebel* :—Ða leóde him from swicon *the people renounced their allegiance to the king of the Elamites* (cf. recesserunt ab eo, Gen. 14, 4), Cd. Th. 119, 18 ; Gen. 1981. Nôhwæðere ælmihtig ealra wolde Adam and Euan árna ofteón ðeáh ðe hé him from swice *although he had withdrawn his favour from them* (perhaps hé = hié and swice is plural *though they had turned from him*, 58, 31 ; Gen. 954. **III.** *to desist from* (*dat.* or *prep.*), *cease from* :—Gif hé ðære hnappunge ne swícþ, ðonne hnappaþ hé óð hé wierð on fæstum slǽpe, Past. 28 ; Swt. 195, 11. Hé from gebede swíceþ, Exon. Th. 264, 33 ; Jul. 373. Á byþ on færylde, nǽfre swíceþ, Runic pm. Kmbl. 342, 26 ; Rún. 17. **IV.** *to deceive* :—Se ðe sweraþ nêhstan his and ná swícþ (*decipit*), Ps. Spl. 14, 6. Se swíceþ ða mengo *seducit turbas*, Jn. Skt. Rush. 7, 12. Ne nim ðú náne sibbe wið ðæs landes menn, ðe læs ðe hira ænig ðé swíce, Ex. 34, 15. **V.** *to fail in one's duty* to another, *be a traitor* to, *desert* :—Hwider hweorfaþ wé (*St. Andrew's followers*) hláfordleáse . . . gif wé swícaþ ðé *if we desert thee*, Andr. Kmbl. 814 ; An. 407. Nǽfre hit (*the sword*) æt hilde ne swác manna ænigum *it never failed any man in fight*, Beo. Th. 2925 ; B. 1460. Ðæt ðú Gode swíce *that thou prove traitor to God*, Andr. Kmbl. 1916 ; An. 960. Hé nele Gode swícan, Exon. Th. 265, 27 ; Jul. 387. Ða ríceste Frencisce men wolden swícan heora hláforde ðam cynge, Chr. 1087 ; Erl. 224, 3. Drihten mé swícan ne wile *the Lord will not desert me*, Ps. Th. 53, 4. [His men him suyken (*deserted*) and flugæn, Chr. 1140 ; Erl. 264, 14. Heo sworen swiken (*deceive*) þat heo nolden, Laym. 4101. Ðe hunte him (*the elephant*) wille swiken (*deceive*), O. E. Misc. 20, 637. Þas ilke nefre ne swiken (*ceased*) to brekene þa licome, O. E. Homl. i. 43, 9. Bute ʒef þu swike ham (*cease from such words*), Marh. 5, 4. Hwanne ich swike (*cease*), O. and N. 1459. Hy ne zuykeþ (*cease*) neure niʒt ne day, Ayenb. 157, 21. O. Sax. swícan : O. Frs. swíka : O. H. Ger. swíchan : Icel. svíkja : Dan. svige *to deceive, leave in the lurch* : Swed. swika.] v. á-, be-, ge-swícan ; swician.

swíce. v. swíce.

swic-cræft, es ; *m. Deception, treachery, fraud* :—Se þurh swiccræft (*by treachery*; but the Latin has *in seditione*) manslyht geworhte, Mk. Skt. 15, 7. Deóflíce dǽda on swiccræftan, L. Eth. v. 25 ; Th. i. 310, 18 : vi. 28 ; Th. i. 322, 18.

swic-dóm, es ; *m.* **I.** *deceit, fraud* :—Wæs swicdóm swíðra ðonne wísdom, and þúhte hwílum wísost se ðe wæs swicolast, and se ðe litelícost cúðe leáslíce hiwian unsóð tó sóðe, Wulfst. 128, 7 : 243, 13 : 52, 31. Swicdóm woruldwelena *deceptio divitiarum*, Mk. Skt. 4, 19. Mid syrewungum and swicdóme hé becom tó ðære cynelícan geðincðe, Homl. Th. i. 80, 34. Hí (*the Romans*) mid swicdóme hié (*the Sabine women*) begeáton, Ors. 2, 2 ; Swt. 64, 27 : Ælfc. T. Grn. 13, 20. Annanias and Saphira wurdon ofslegene for heora swicdóme, Homl. Ass. 59, 194. Hé (*Christ*) synne ne worhte ne nænne swicdóm on lífe, 47, 565. Hé hire sǽde þurh hire swicdóm begeaht, on hwam his strengð wæs, Jud. 16, 5. Se cyning swíðor micle wénende wæs ðæt hié ðonon fleónde wǽren ðonne hié ǽnigne swicdóm cýþan dorsten *the king thought it was far more probable that they were fleeing thence, than that they would venture to practise any ruse*, Ors. 4, 5 ; Swt. 76, 16. Swicdóma *deceptionum*, Hpt. Gl. 502, 18. **II.** *treachery, failure in loyalty, treason* :—Ða tugon hiene ðære burge witan ðæt hé heora swicdómes wið Alexander fremmende wǽre *the chief men of the town accused him of treasonable practices against them in his relations with Alexander* ; quasi urbem regi venditasset, Ors. 4, 5 ; Swt. 168, 17. Be hláfordsearwe (be cynincges swicdóme, MS. B.) *of treason*, L. Alf. pol. 4 ; Th. i. 62, 14. Hí sǽdon ðæt hí woldan cuman ðider for ðes cynges swicdóme *for the purpose of acting treacherously towards the king*, Chr. 1048 ; Erl. 178, 27. Wæs ðis land swíðe ástirad and mid mycele swicdóme áfylled *the land was much disturbed and filled with treason*, 1087 ; Erl. 224, 2. Wið ðam ðe hí ealle ánrædlíce bútan swicdóme (*without failure of their loyalty*) tó him (*Ethelred*) gecyrdon,

1014; Erl. 150, 13. **III.** *an offence*; scandalum :—Wá ðysum middangearde þurh swicdómas (*a scandalis*): neód ys ðæt swycdómas (*scandala*) cumon; þeáhhwæðere wá ðam menn ðe swycdóm (*scandalum,* þurh hyne cymþ, Mt. Kmbl. 18, 7. [Misdon þurh Beelzebubes swikedom, O. E. Homl. i. 55, 10. Þis nis nan swikedom, for þat weord ich hit halde, Laym. 8310. All þatt follȝheþþ swikedom, Orm. 3997. Þu me misraddest . . . Schild þi swikedom from þe lihte, O. and N. 163. Icel. svik-dómr *treason*.]

swice, es; *m.* **I.** *departure, escape.* v. swícan, **II** :—Helle hlinduru nágon hwyrft ne swice, útsíþ ǽfre *the gates of hell allow of no return or escape, of egress ever,* Exon. Th. 364, 30; Wal. 78. **I a.** *escape* from that which threatens to befall, *evasion* :—Ne biþ ðæs lengra swice sáwelgedáles ðonne seofon niht fyrstgemearces *there will not be a longer escape from death than a period of seven days,* Exon. Th. 164, 6; Gú. 1007. **I b.** *outcome, event, issue* :—Hé þenceþ ðæt his wíse þince untorcúþ biþ ðæs óþer swice ðonne hé ðæs fácnes fintan sceáwaþ *he thinks that his ways appear respectable; their event will be different when he observes the result of the fraud,* Exon. Th. 315, 15; Mód. 31. **II.** *deceit, fraud, treachery.* v. swícan, **IV, V** :—Hé ealle ða cyningas mid biswice (mid his swice, Cott. MS.) ofslóg *captos per dolum reges interfecit,* Ors. 3, 7; Swt. 114, 8. Hí on ðínum folce fácen geswipere syredan and tó swice hogedon *in plebem tuam astute cogitaverunt consilium.* Ps. Th. 82, 3: Exon. Th. 317, 6; Mód. 61. **III.** *offence, stumbling-block, snare*; scandalum :- Ðanun mæg áspringan seó mǽste sacu and se mǽsta swice ealra ungeþwǽrnessa *exinde grauissima occasio scandalorum oriri potest,* R. Ben. 129, 8. Hí settan mé swyce (swyþe, MS.) ðǽr ic síþade *juxta iter scandalum posuerunt mihi,* Ps. Th. 139, 5. [O. H. Ger. -swih; *pl.* -swihhi.] v. be- (*acc.* bigswicae, Lchdm. iii. 208, 12), hláford-, un-swice; swic.

swice, an; *f. A trap*, v. Swican *decipulam*, Hpt. Gl. 520, 30: Anglia xiii. 36, 263. [Þenne þe mon wule tilden his musestoch he bindeð uppon þa swike chese, O. E. Homl. i. 53, 21. A swyke *discipula,* Wrt. Voc. i. 221, col. 2 (15th cent.).]

swice; *adj.* **I** *deceitful, fraudulent* :—Hí widstandaþ ðam swican (*or subst.? v.* swica) Antecriste, Wulfst. 198, 14. [He minne fader biswak þurh swike his craftes (mid his luþer craftes, 2nd MS.), Laym. 14865.] **II.** *proving false to what is expected* :—Norðmen wǽron súðfolcum swice (i. e. *the southern people were deceived in their estimate of the northmen's power* (?); swice, as applied to the northmen, cannot mean *rebellious, renouncing allegiance,* for it was the southern peoples who had rebelled against the northern, v. 119, 8–18; Gen. 1976–1981). Cd. Th. 120, 17; Gen. 1996. **III.** *treacherous, failing in loyalty,* v. swícan, **V.** [Feren swike ðe sulden witterlike, Gen. and Ex. 2845.]

swice *and* (?) swicc, es; *m. A scent,* smell :—Suice, suicae osma (Gk. ὀσμή; cf. Span. husmo *smell, scent*; andar a la husma *to be on the scent*; husmear *to find out by smelling*), Txts. 83, 1468. Swice, Wrt. Voc. ii. 63, 57. Ðæt wæs swéte stenc . . . tó ðæm swicce men þrungon, Exon. Th. 359, 21; Pa. 66. v. swecc.

swícend, es; *m. A deceiver, betrayer* :—Se sáula swícend *the devil,* Homl. Ass. 196, 39: 197, 87. v. be-swícend.

-swicenness v. be-, ge-swicenness.

swic-full; *adj. Deceitful, fraudulent, crafty* :—Swicfulles *strophosae, callidae,* Hpt. Gl. 423, 61. Swicfullum *fraudulento,* 517, 45. Swicfulle *frivola, fraudulenta, falsa,* 444, 26. Swicfullum *fraudulentis,* 521, 31.

swician; *p.* ode. **I.** *to wander* :—Ðǽr hí swiciaþ on swíman, firenweorc beraþ, Exon. Th. 79, 33; Cri. 1300. Suicade, suicudae *spatiaretur,* Txts. 99, 1893. Hí ðurh cúþe stówe swicedon and fóron *per nota loca dispersi vagarentur,* Bd. 4, 4; S. 571, 4. Hí swycedan geond wésten *erraverunt in solitudine,* Ps. Th. 106, 3. Swicedan, 39. Swiciende *pervagatus,* Wrt. Voc. ii. 68, 79. **II.** *to depart, turn* :— Ná ic fram ðínum dómum dǽdum swicade *a judiciis tuis non declinavi,* Ps. Th. 118, 102. **III.** *to deceive* :—Mǽst ǽlc swicode and óðrum derede wordes and dǽde, Wulfst. 160, 3. Ne ǽnig ne syrwe ne óðrum ne swicie, 73, 12: 70, 5. Lytelíce swician, 55, 16. Ða men ðe ne dorstan for Godes ege swician . . . ða ðe cúðan swician and befician and mid leásbregdum earmum mannum derian, L. I. P. 12; Th. ii. 320, 21–26. Swiciende licceteras árísaþ and forlǽraþ tó manege, Wulfst. 89, 17. **III a.** *with prep.* on, ymb, *to practise deceit* in relation to a matter; cf. O. Sax. swíkan umbi :—Se ðe on mynstres ǽhtum mid fácne swicaþ *he who fraudulently deceives in the matter of a monastery's possessions,* Homl. Th. i. 398, 26. Annanias and Saphira swicedon on heora ágenum ǽhtum, 33. Se syrwienda deoful á swicaþ, embe mancyn *is ever practising deceit in respect to man,* Wulfst. 107, 23. Se sceaða georne swicode ymb ða sáwle, Cd. Th. 38, 15; Gen. 607. **IV.** *to offend;* also *to be offended;* scandalizare, scandalizari :—Gif ðín hand ðé swicaþ (*scandalizat*), Mt. Kmbl. 18, 8: 9: Mk. Skt. 9, 43, 45. Þeáh ðe ealle swicion ne swicige ic ðe ná *etsi omnes scandalizati fuerint sed non ego,* 14, 29. **IV a.** *to give offence by words, speak injuriously* :— Ná murcna ðú ná swica ðú *non murmures, non blasphemes,* Scint. 164, 16. [O. H. Ger. swichôn *vagari.*] v. á-, ǽ-, be-swician; swícan.

swicon, e; *f. Clearance from a criminal charge* :—Se ðe hereteáma betygen sié, hé hine be his wergilde álíese, oþþe be his were geswicne.

Se áð sceal bión healf be húslgengum. Ðeóf, siþþan hé biþ on cyninges bende, náh hé ða swicne *is not allowed the alternative of clearing himself by oath,* L. In. 15; Th. i. 112, 5. [Goth. swikns *innocent, clear of wrong-doing*; swiknei, swikniþa *purity*; swikneins *purification*: Icel. sykn *free from guilt, cleared from a criminal charge*; sykn, sykna *clearance from a criminal charge.*] v. ge-swicn; ge-swicnan.

-swicnan, -swicneful. v. ge-swicnan, ge-swicneful.

swicol, sweocol; *adj.* **I.** *deceitful, false, treacherous, crafty* :— Swicol *fallax vel mendax,* Wrt. Voc. i. 47, 50. (1) *of persons* :—Næs heó swicol nánum ðæra ðe hyre tó ðohte *she never deceived any one who trusted her,* Lchdm. iii. 428, 34. Se swicola Herod . . . cýdde syððan his fácenfullan syrewunge, Homl. Th. i. 82, 15. Ðæt swicole wíf (*Delilah*), Jud. 16, 8. Ða gescotu ðæs sweocolan feóndes *insidiantis hostis jacula,* Past. 56; Swt. 431, 5. Áfandod þurh ðone swicolan deófol, Ælfc. T. Grn. 10, 45. Ða swicolan *virum dolosum,* Ps. Th. 5, 6. Se ðe wæs swicolast and se ðe litelícost cúðe leáslíce hiwian unsóð tó sóðe, Wulfst. 128, 9. Swicolost, 268, 17. (2) *of things* :—Ðis líf is swá swicol, ðæt hit symble bepǽcþ, Homl. Skt. i. 5, 65. Ne sceole wé ná besettan úrne hiht on ðissum swicelum life, Homl. Th. i. 162, 18. Geseoh gif ic on swiculne weg oððe on unrihte eode *vide, si via iniquitatis in me est,* Ps. Th. 138, 21. **II.** *occasioning offence* (?). v. swice, **III,** swician, **IV,** swicol-líc, **II** :—Sóð biþ swicolost (switolost?), Menol. Fox 479; Gn. C. 10. [O. E. Homl. Laym. A. R. Havel. swikel: O. H. Ger. pi-swichal *subdolus:* Icel. svikall *treacherous.*] v. be-(bi-), un-swicol.

swicol-líc; *adj.* **I.** *deceitful, fraudulent* :—Swicollíce dǽda and ládlíce unlaga áscunige man swýðe; þæt is, false gewihta and wóge gemeta and leáse gewitnessa, L. Eth. v. 24; Th. i. 310, 12: vi. 28; Th. i. 322, 12. **II.** *occasioning offence.* v. swice, **III** ;—Ænig þing ungeþwǽrlíces and swicollíces (the Latin has *scandalorum spinas*), R. Ben. 38, 18.

swicollíce; *adv. With deceit, with guile, deceitfully, fraudulently, craftily* :—Hé cwæð ðæt hí wǽre wurdan ðæt hý ǽnig man tó swicollíce ne bepéhte mid leáslícre láre '*uidete, ne quis uos seducat,*' Wulfst. 88, 26: 55, 3. Ðæt wyrse is, ðæt hé swicollíce hiwige, swylce hé árfæstes módes sý, 53, 26. Aman smeáde swicollíce embe ðæt hú hé eall Iudéisc cynn fordyde *Haman plotted how to destroy all the Jewish race,* Homl. Ass. 96, 145.

swicolness, e; *f. Deceit, fraud, treachery* :—Míne synna ðe ic gefremede on mǽnan áðe and swicolnyssæ, Anglia xi. 102, 85. Antecrist lǽrþ unsóðfæstnysse and swicolnesse, Wulfst. 55, 12.

swicðole, Beo. Th. 6281; B. 3146. v. sweþel.

swícung, e; *f.* **I.** *deceiving, deluding, deceit, fraud, delusion* :— Mid swícunge deóflícre *inlusione diabolica,* Anglia xi. 117, 29. Swícunge ceápes *fraud in trade,* Lchdm. iii. 198, 31: 202, 13. For swícuncgum *propter illusiones,* R. Ben. Interl. 88, 5. **II.** *offence, occasion of stumbling;* scandalum :—Se ðe lufaþ bróðer his, swícung (*scandalum;* v. 1 Jn. 2, 10) on him nys, Scint. 14, 12. Neód hit ys ðæt cuman swícunga (*scandala*), swá þeáh wá ðam menn þurh ðæne swícung (*scandalum*) cymþ, 134, 2–3. [He (*false men*) ðe swiken, ðin agte wið swiking, ði soule wið lesing, O. E. Misc. 19, 602.] v. á-, ǽ- be-, hláford-swícung.

-swidung in ge-swidung, Lchdm. iii. 168, 2. v. swedring.

swífan; *p.* swáf, *pl.* swifon; *pp.* swifen. **I.** *to move in a course, wend, sweep* :—Hond hwyrfeþ geneahhe swífeþ mé geond sweartne *the hand passes over me* (a skin), Exon. Th. 394, 4; Rä. 13, 13. On ðære ilcan eaxe hwerfeþ rodor, recene scríþeþ, súðheald swífeþ swift (*sweeps swift*), Mét. 28, 17. Mannum þyncþ ðæt sió sunne on mere gange, under sǽ swífe, ðonne hió on setl glídeþ, 39. Sceal on ánum fét searoceáp (*a ship*) swífan, swíþe féran, faran ofer feldas, Exon. Th. 415, 6; Rä. 33, 7. [*Here are added examples of* á-swífan *omitted in their place* :—Ásuáb *exorbitans,* Wrt. Voc. ii. 107, 74. Áswífende *exorbitans, exorbitantes,* 31, 19, 31: 83, 7: 86, 10: exorbitantes, i. *circuientes, declinantes,* 145, 80.] **II.** *of a course of action, to come to take part in a matter* :—Ðá swáf Eánulf on wæs geréfa ðá genom eal ðæt yrfe him on ðæt hé áhte tó Tyssebyrig *then* (after the commission of a crime) *Eanulf, who was reeve, struck in or intervened, and took all the property from him* (the criminal) *that he owned at Tisbury,* Chart. Th. 172, 31. [O. Frs. swíva *to be uncertain:* Icel. svífa *rove, turn, sweep.* Cf. O. H. Ger. sweibôn *ferri, volvere, incitari.* Gothic has the verb sweiban; *p.* swaif (Lk. 7, 45) *with the meaning to cease, leave off.*] v. á-, on-, tó-swífan.

swift; *adj. Swift, fleet, that does or can move quickly* :—Suift *alacer,* Wrt. Voc. ii. 99, 76. Swift, 6, 51: *expeditus,* 145, 36: *celer,* Ælfc. Gr. 9, 18; Zup. 44, 9. Swyft *pernix,* 9, 64; Zup. 71, 2. Swift scip *archiromachus,* Wrt. Voc. i. 63, 30. Hé (*the phenix*) is snel and swift *velox est,* Exon. Th. 220, 8; Ph. 317. Ne se swifta mearh burhstede beáteþ, Beo. Th. 4521; B. 2264. Him on swift wind (cf. ungemetlíc wind, Bt. 12; Fox 36, 15) swápeþ, Met. 7, 20. Rodor swífeþ swift, 28, 17. Bufan ðam swiftan rodore, Bt. 36, 2; Fox 174, 15. Micel swég gǽþ of heora (*the stars*) swiftan ryne, Boutr. Scrd. 18, 43. Hors swiftne, Exon. Th. 400, 3; Rä. 20, 3: 487, 22; Rä. 74, 1. **Swifte**

ærendracan *veltes* (= *velites*), Wrt. Voc. i. 18, 23. Ic hæbbe swíþe swifte feþera, Bt. 36, 2; Fox 174, 4. Se móna is be sumum dæle swiftre ðonne seó sunne, Lchdm. iii. 248, 3. Ða (*Alfred's ships*) wæron ægðer ge swiftran ge unwealtran ge eác hiéran ðonne ða óðru, Chr. 897; Erl. 95, 13. Wind byþ on lyfte swiftust, Menol. Fox 464; Gn. C. 3. Gecunnian hwylc heora swiftost hors hæfde, Bd. 5, 6; S. 619, 1. Ealle ða menn ðe swyftoste hors habbaþ . . . Ðær beóþ ða swiftan hors ungefóge dýre, Ors. 1, 1; Swt. 20, 34–21, 6. v. ryne-swift.

swiftlere, es; *m.* *A slipper, shoe:*—Swiftlere *suptularis (suptalaris),* swiftlæras *suptalares,* Ælfc. Gl. Zup. 314, 15. Swyftleras *subtalares,* Coll. Monast. Th. 27, 31. Swifteleares, Wrt. Voc. i. 26, 19. [Cf. *O. H. Ger.* suftelara *talaria,* which Graff derives from Latin *subtalaris.* The English and German words seem to have the same origin.]

swiftlíce; *adv. Swiftly:*—Hredlíce † swiftlíce *velociter,* Ps. Lamb. 6, 11. Gálful líf swiftlíce (*celeriter*) gelæt tó ylde, Scint. 88, 19. Ðá férde his gást swyftlíce, Homl. Th. i. 452, 30. Zacheus swyftlíce of ðam treówe álíhte, 580, 34. Hí fleóþ swiftlíce, Wulfst. 200, 17.

swiftness, e; *f. Swiftness, fleetness, celerity:*—Hwá unlæredra ne wundraþ ðæs roderes færeldes and his swiftnesse, Bt. 39, 3; Fox 214, 16. Dysig se ðe getrúwaþ on his horses swiftnesse, Ps. Th. 32, 15. Hé swang ðone top mid swá micelre swiftnesse, Ap. Th. 13, 13. Da óðre deór ðe mihton hire ætfleón þurh heora fóta swiftnysse, Homl. Ass. 63, 280. Þurh ða swiftnysse (*the rapidity of the moon's motion*), Lchdm. iii. 248, 4. Uton behealdan ða wundorlícan swyftnysse ðære sáwle; heó hæfþ swá mycele swyftnysse, ðæt heó on ánre tíde besceáwaþ heofonan and ofer sæ flýhþ, Homl. Skt. i. 1, 123.

swift-ryne (?), es; *m. A swift course, rapid running of water:*—Singalrenes † swift[renes] *decursus,* Hpt. Gl. 418, 51.

swiftu (-o); *indecl. f. Swiftness:*—Hwá unlærdra ne wundrige rodres swifto? Met. 28, 3. v. swiftness.

swígan; *p.* de. I. *to be silent:*—God ná swígeþ *Deus non silebit,* Ps. Spl. 49, 3. Stiórdon him menigo ðætte hé suígde (*ut taceret*), Mk. Skt. Lind. 10, 48. Ðú bist suígende (swígende, Rush.), Lk. Skt. Lind. 1, 20. Geót swígende ðæt blód on yrnende wæter, Lchdm. ii. 76, 14 : 140, 26 : 290, 26 : 292, 25. Ðæt eall swígende gedó, 104, 10. Swígende (suígende, Hatt. MS.) hé cwæð on his móde . . . Ða swígendan (suígendan, Hatt. MS.) stefne se dígla Déma gehírde, Past. 4; Swt. 38, 16–20 : Blickl. Homl. 7, 16. Þú ána hí swígende tælst *thou alone by thy silence dost blame her,* Ap. Th. 16, 21. Hé oft ána sæt swígende múðe *saepe solus residens ore tacito,* Bd. 2, 9; S. 512, 13. Ðæt ánra manna gehwylc sceáwige hine sylfne swígende móde, Blickl. Homl. 57, 34. II. *to become silent from astonishment; stupere.* v. swígung, III, swige, III:—Swígdon † styldon *stupebant,* Mk. Skt. Lind. 1, 22. Stylton † suígdon, 6, 51. Suígdon (swígdon, Rush.), 10, 32. [*O. H. Ger.* swígén *silere, reticere:* for- (Ðeáh hé hit silf forswíge, his gegirla hine geswuteláþ, Ap. Th. 14, 3), ge- (*see* ge-swígde, -on, *given under* geswígian), óþ-swígan; swígian.

swíg-dæg, es; *m. A day on which silence was to be observed:*—Circlíce þeáwas forbeódaþ tó secgenne ænig spel on ðam þrým swígdagum, Homl. Th. i. 218, 31 : ii. 262, 16. [The three days referred to are the last three days of Passion Week. 'Besides the general injunction of silence in the ordinary business of life, and in various ritual matters, even the bells were to remain silent from the Thursday evening, which commemorated our Lord's betrayal, to the following Sunday morning. Nothing more, probably, was at first meant by this, than to impress a character of unusual solemnity upon the season, but it was eventually said that men were thus to be reminded of the time when the preaching of the Gospel wholly ceased; Jesus Himself being actually dead during most of it, and His disciples all along being dispersed panic-stricken.' Durand, quoted in Soames' Anglo-Saxon Church, p. 263. Cf. the injunction in the Ancren Riwle: Holdeð silence al þe swiðwike (swihende wike, MS. T.: swiwike, MS. C.) uort non of Ester euen, 70, 5–8. In German Good Friday is *der stille Freitag.*]

swíge (*but* swígea *occurs,* Scint. 82, 1), an; *f.* I. *silence, absence of speech:*—Hú se láreów sceal bión gesceádwís on his swígean (swigean, Cott. MSS.) and nytwyrðe on his wordum . . . Sió ungemetgode suíge (swigge, Cott. MSS.) ðæs láreówes on gedwolan gebringþ ða ðe hé læran meahte, Past. 15; Swt. 89, 3–10. Essaias cwæð, ðætte sió suýge (swigge, Cott. MSS.) wære ðære ryhtwísnesse fultum, 38; Swt. 279, 24. Sý heálíc swíge æt ðam gereorde, ðæt nánes mannes stefn gehýred ne sý bútan ðæs rædferes ánes, R. Ben. 62, 13. Ðá weard stilnes and swige geworden innon ðare healle, Ap. Th. 17, 6. Mé náwðer deág secge ne swíge, Exon. Th. 12, 23; Cri. 190. Náht framaþ, gif on eardungstówe swígea sý, Scint. 82, 1 : 213, 14. Be swígan . . . Hé for swígan mægene clypunge geswác . . . Leornerum for swígan hefígnesse seldhwænne leáf geseald sié tó sprecenne ymbe hálige spræca, R. Ben. 21, 8–17. Hí clumiaþ mid ceaflum, ðær hí sceoldan clypian; wá heom ðære swígan, L. I. P. 5; Th. ii. 308, 21 : Wulfst. 177, 1. Óðer ondréd ðæt hé forlure sprecende ða gestrión ðe hé on ðære swígean (swiggean, Cott. MSS.) geðencan meahte; óðer ondréd ðæt hé ongeáte on his swýgean (swiggean, Cott. MSS.) ðæt hé sumne hearm geswigode, Past. 7; Swt. 49, 19–22. Mid

suígean, 35; Swt. 237, 12. Mid swígan forberan *to bear in silence,* Homl. Th. ii. 164, 20. Heó swígan lufode, 546, 28. Wé cweðaþ ðæt sí best æfter Gode, ðæt man gemetigian cunne ge his spréce ge his swígan, Prov. Kmbl. 2. II. *silence, quiet, absence of noise;* also *a time of silence.* v. swíg-tíma :—Ne ærfæstness ne sib ne hopa ne swíge gegladaþ *nec pax nec pietas immo spes nulla quietis,* Dóm. L. 220. In swígean midre nihte *intempestive,* Wrt. Voc. ii. 46, 74. Swígan *conticinio* (cf. *conticinium,* ðonne ealle þing sweowiaþ on hyra reste, Lchdm. iii. 244, 2), 20, 30. III. *silence from astonishment, amazement; stupor.* v. fær-swíge, swígan, II, swígung, III. IV. *delay* (?). v. swígung, IV:—Suígo dyde ðe brýdgum *moram faciente sponso,* Mt. Kmbl. Lind. 25, 5. [*Or is this a different word?* cf. (?) *Icel.* svíg *a curve, circuit; sveigja to bend, sway.*] [*O. H. Ger.* swíga *taciturnitas, silentium.*]

swige; *adj.* I. *silent, not speaking:*—On óðre wísan mon sceal manigean ða swíðe swígean, on óðre wísan ða felaídspræcean, Past. 23; Swt. 174, 24. Ða ðe tó swíðe swíge (swigge, Cott. MSS.) . . . ða suíðe swígean (swiggean, Cott. MSS.) *taciturni . . . nimis taciti,* 38; Swt. 271, 6–10. Ðá wæs swígra secg (*Hunferth*) on gylpspræce (cf. Ðú worn fela, wine mín Húnferð, beóre druncen ymb Brecan spréce, 1064; B. 530), Beo. Th. 1964; B. 980. II. *silent, not making a noise, still :*—Wind wédende færeþ, and eft semninga swíge gewyrðeþ, Elen. Kmbl. 2548; El. 1275. Stille þynceþ lyft ofer londe, and lagu swíge, Exon. Th. 383, 16; Rä. 4, 11. Nis mín sele swíge, ne ic sylfa hlúd, 494, 1; Rä. 82, 1. v. swíþ-swíge.

swígen[n], e; *f. Silence, refraining from speech :*—Ðam láreówe sylfum deraþ hwílon his swígen, ac heó deraþ symle his underðeóddum, gif him biþ seó heofenlíce lár oftogen, Homl. Th. ii. 532, 4.

swigene ?:—Ðæs mannes bileofa is tó besceáwianne: ærest him is tó sellanne ðæt ðone innoð stille and smeþe, ne sié scearp ne tó afor ne slítende ne swigene, Lchdm. ii. 210, 21.

swigian, sweogian, sweowian, swugian, swuwian, sugian, suwian; *p.* ode. I. *to be silent,* (a) *of that which has voice:*—Ic suwige (swugige, swuwie) *taceo,* Ælfc. Gr. 26, 2; Zup. 26, 13. Swigaþ *silet (vipera),* Rtl. 125, 27. God ne swugaþ (swigaþ, Surt.) *Deus non silebit,* Ps. Th. 49, 3. Ðonne swíaþ (*silet*) hé (*the phenix*), Exon. Th. 207, 16; Ph. 142. Swigiaþ *conticiscent,* Wrt. Voc. ii. 14, 53. Ða ðe má swigiaþ (swugiaþ, Hatt. MS.) ðonne hié ðyrfen, Past. 38; Swt. 272, 24. Ða ðe swigiaþ (swugiaþ, l. 3), ðæt hié hié ne bodiaþ, 48; Swt. 365, 7. *Conticinium,* ðonne ealle þing sweowiaþ (suwiaþ, MSS. R. P.) on hyra reste, Lchdm. iii. 244, 2. Ic swigode (swygode, Spl.: sugode, Th.) *tacui,* Ps. Surt. 31, 3 : Exon. Th. 485, 16 : Rä. 71, 14. Ic swugode, swá swá se dumba, Ps. Th. 37, 13 : 49, 22. Deáh ðe seó tunge swigode, ðæt his líf wæs sprecende, Bd. 5, 12; S. 627, 30 : Ap. Th. 16, 19 : Cd. Th. 250, 15 ; Dan. 547. Hé suwode (swygode, MS. A.: swugode, MSS. B.C.: swigade, Rush.) *tacebat,* Mk. Skt. 14, 61 : Mt. Kmbl. 26, 63. Ða swigodon hí ealle and stille wæron *conticuere omnes,* Bd. 3, 11; S. 536, 31. Hí suwodon (swigedon MS. A.: swigadun, Rush.), Mk. Skt. 3, 4. Ne swiga (swuga, Th.: suwa, Lamb.) *dú ne sileas,* Ps. Spl. Surt. 38, 17. Ne swiga (swyga, Spl.) ðú *ne taceas,* Ps. Th. Surt. 82, 1. Ne swiga (swyga, Spl.: swuga, Th.) . . . ne suga *ne sileas . . . ne taceas,* Ps. Lamb. 27, 1. Ne swuga, Ps. Spl. 34, 25. Ðe læs ðú swuige *ne taceas,* 27, 1. Ic swigiende ealle ða niht áwunode, Bd. 5, 6; S. 619, 29. Ðú byst suwiende (swygende, MS. A.: suwigende, MSS. B. C.), Lk. Skt. 1, 20. (b) *of that which has not voice, not to make a noise :*—Hrægl mín swigaþ, Exon. Th. 389, 21; Rä. 8, 1. Ða ýða swygiaþ (swigadon, Surt.: swigedon, Spl.) *siluerunt fluctus ejus,* Ps. Th. 106, 28. II. *to be silent from astonishment, be amazed :*—Swigadun † stylton ofer lære his *stupebant super doctrina ejus,* Mk. Skt. Rush. 1, 22. III. *with an object* (gen. or acc.) *to be silent about something, to refrain from the mention of something :*—Gif ðú suwast hit and nylt folce his þearfe gecýðan, Wulfst. 283, 3. Hié nyllaþ geopenian ðæm syngiendum hiera unryht ac suigiaþ (swigiaþ, Cott. MSS.) ðara ðreáunga *iniquitatem peccantium nequaquam aperiunt, quia ab increpationis voce conticescunt,* Past. 15; Swt. 91, 11. Lyt swigode níwra spella se ðe næs gerád, Beo. Th. 5787; B. 2897. Hé ne suigige ðæs ðe nyttwyrðe sié tó sprecanne, ne ðæt ne sprece ðæt hé suigigean (swigian, Cott. MSS.) scyle *ne aut tacenda proferat, aut proferenda reticescat,* Past. 15; Swt. 89, 6–7. Hié mon sceal læran ðæt hí hwílum suigien (swugien, Cott. MSS.) ðæs sóðes *admonendi sunt, ut noverint nonnunquam vera reticere,* 35; Swt. 237, 9. [*O. Sax.* swigón: *O. Frs.* swigia.] v. for-, ge-swigian; swígan.

swigiendlíce, *adv. Silently, in silence :*—Sæt ic ána in ðam wéstenne . . . Ðá ongann ic swigiendlíce þencan be manegra munuca lífe, Homl. Ass. 204, 311.

swigness, e; *f. Silence; a time of silence :*—Cwyldtíd, swígnes *conticinium,* Wrt. Voc. ii. 135, 14. v. swíge, II, *and next word.*

swíg-tíma, an; *m. A time of silence :*—Seó niht hafaþ seofon tódæledny ssa . . . þridde ys *conticinium,* ðæt ys swítíma, Anglia viii. 319, 29. v. swíge, II, *and the preceding and following words.*

swígung, e; *f.* I. *silence, absence of speech :*—Hé (*John the Baptist*) ðam fæder (*Zacharias*) ða stefne ágeaf, ðá se heáhengel mid ðære swígunge fæstnunga geband ðone fæder, Blickl. Homl. 167, 11.

Hwanne besmát hine seó scyld ðære fealasprecolnesse? ... oþþe hú sceþede him seó synn ðære swígunga? 169, 7. Mid suígunga *cum silentio*, Rtl. 20, 15. Swígunge, Shrn. 41, 26. **II.** *silence, absence of noise* :— Martha ceigde Mariam suiugunga (swiunga, Rush.) and cwoeð *Martha vocavit Mariam silentio, dicens*, Jn. Skt. Lind. 11, 28. **II a.** *a time of silence.* v. swíge, II, and two preceding words :—Ðære swígunge *conticinio*, Wrt. Voc. ii. 24, 31 : 20, 29. Ih swígunge *in conticinio*, 47, 46. **III.** *silence* from astonishment, *amazement.* v. swígan, II, swíge, III :—Forstylton swígunge micelre *obstupuerunt stupore maximo*, Mk. Skt. Rush. 5, 42. **IV.** *delay.* v. swíge, IV :—Suígiunc dóes hláferd mín *moram facit dominus meus*, Mt. Kmbl. Lind. 24, 48. [*O. H. Ger.* swígunga *silentium.*] v. ge-swígung.

swilc, swelc ; *pron.* (the word can take the weak declension). **I.** where the word points to what has been already described, *such*, (1) used substantively, *that which has been already described, the like, the same* :— Ne biþ swylc (*the practice already described*) cwénlíc þeáw, Beo. Th. 3885 ; B. 1940. Ne biþ swylc earges síð, 5076 ; B. 2541. Ne sceolde ðé nán man swelces tó geléfan *no one would believe such a thing of you*, Bt. 5, 1 ; Fox 10, 2 : 19 ; Fox 68, 32. Hé ǽfre swylces geswíce, L. Ath. i. 6 ; Th. i. 202, 17. Heó áwiht swylces ne hýrdon, Elen. Kmbl. 1139 ; El. 571. Gif wífmen hwæt swylces derige, L. Alf. i. 236, 3 : Beo. Th. 1764 ; B. 880. Hæringcas and leaxas . . . and fela swylces (*et similia*), Coll. Monast. Th. 24, 13. Hæleða fela swelces and swelces wundraþ, Met. 28, 49. Be swilcum and swilcum ðú miht ongitan, Bt. 38, 1 ; Fox 196, 11 : Met. 26, 107. Wundorsióna fela secga gehwylcum ðara ðe on swylc staraþ, Beo. Th. 1997 ; B. 996 : 5589 ; B. 2798 : Met. 30, 18. Gif him (*a lunatic*) gelimpe ðæt hé man ofsleá . . . his mágas hine wið óðær swylc gescyldan *propinqui ejus eum contra simile quid servent*, L. Ecg. P. addit. 29 ; Th. ii. 236, 31. Swylcra síþfæt (*the journey of those just mentioned*), Exon. Th. 400, 12 ; Rä. 20, 9. Hú hé swylce ácwealde, Ps. Th. 108, 16. Oft ða swelcan (swylcan, Cott. MSS.) monn sceal forsión, Past. 37, 2 ; Swt. 265, 17. (2) used adjectively, *like that already described*, (a) agreeing with a noun :—Hine swelces gamenes gilpan lyste, Met. 9, 19. Swylces morðres, 32. Hig worhton óðer swilc þing *fecerunt quaedam similiter*, Ex. 7, 11. Hé ǽr ne síð óðre swylce láre gehýrde, Exon. Th. 169, 10 ; Gú. 1092 : Blick. Homl. 189, 22. Geþyld and ryhtwísnes and wísdóm and manege swelce cræftas, Bt. 34, 6 ; Fox 142, 1. Se is tó lytel swelcra láriówa, Met. 10, 55. Manegum swylcum (*talibus*) bigspellum hé spræc tó him, Mk. Skt. 4, 33. Manna sáulum hé gyfþ swilca gyfa. Ða swilcan gifa hí ne þurfon forlætan, Shrn. 192, 3. (b) predicatively :— Hió nǽfre siþþan swelc wæs *it (Rome) was never the same afterwards*, Ors. 6, 1 ; Swt. 252, 24. Gif hé suelc (swelc, Cott. MSS.) wǽre, Past. 16 ; Swt. 101, 10. Swelc wæs þeáw hira, Andr. Kmbl. 50 ; An. 25. Swylc, Beo. Th. 359 ; B. 178. Ðæt úre tída ne mihtan weorðan swilce, Bt. 15 ; Fox 48, 18. Swelce, Met. 8, 42. **II.** as an antecedent :— Swǽlc monn se ðe tó mínum ærfe fóe gedéle hé ælcum messepreóste binnan Cent mancus goldes, Cod. Dip. Kmbl. i. 351, 4. Ða com leóht swilc swá hí ǽr ne gesáwon, Homl. Skt. ii. 29, 263. Eal swylce seó mettrumnes biþ ðæs seócan mannes . . . swylc is ðæt líf ðysses middan-geardes, Blickl. Homl. 59, 31. Wǽre se man on swelcum lande swelce hé wǽre, Bt. 27, 3 ; Fox 98, 27. Ðæt hé ðone hláf on swilcere stówe áwurpe ðæt hine nán man findan ne mihte, Homl. Th. ii. 162, 25. Wé swylc ne gefrugnan gelimpan, ðæt ðú befénge, Exon. Th. 6, 7 ; Cri. 78. Ymb swelc tó sprecanne hwelc hit ða wæs, Ors. 1, 10 ; Swt. 48, 4. Swelce burg gewyrcan swelce sió wæs, 2, 4 ; Swt. 74, 8. Gif ic hæfde swilcne anweald, swylce God hæfþ, Bt. 38, 2 ; Fox 196, 19. Se wolde habban swilcne hlísan swá Benedictus, Homl. Th. ii. 162, 18 : Soul Kmbl. 278 ; Seel. 143. Hí ne þurhwuniaþ swelca, swelce hí ǽr tó cóman, Bt. 11, 1 ; Fox 30, 28. Swylcra yrmða swá ðú unc scrife, Soul Kmbl. 201 ; Seel. 102. Búton hé hæbbe swylce þéningmen ðe þeáwfæstnysse him gebeódon, Homl. Skt. i. pref., 62. **III.** in correlative clauses, swilc . . . swilc *such* . . . *as* :—Swylc biþ wedera cyst, swylc wæs on ðam fýre, Cd. Th. 238, 6 ; Dan. 350. Swylc scolde eorl wesan, swylc Æschere wæs, Beo. Th. 2661 ; B. 1328. Mid swelce hrægle hé in eode, mid swelce gange hé út, L. Alf. 11 ; Th. i. 46, 3. Swylce mǽla swylce hira mandryhtne þearf gesǽlde, Beo. Th. 2502 ; B. 1249. Eahtige hé hine selfne suelcne suelcne hé ondrǽtt ðæt hé sié, Past. 17 ; Swt. 119, 8. Séce swylcne hláford, swylcne hé wille, L. Ath. iv. 1 ; Th. i. 220, 24. Beóþ swylce (sǽlce, Lind.) gedréfednessa swylce (suelco, Lind.) ne gewurdon (*tales quales non fuerunt*), Mk. Skt. 13, 19 : Beo. Th. 6309 ; B. 3165. **IV.** containing both antecedent and relative, *such as* :—Ðonne ic wæs mid Iudéum, ic wæs swelc hié, Past. 16 ; Swt. 101, 6. Gestreón swilc ðǽr funden wæs, Cd. Th. 220, 5 ; Dan. 66. Ná hýrde wé ðæt ǽnig wurde hús árǽred swylic ðæt mǽre wæs, Anglia xi. 9, 30. Gódfremmendra swylcum giseðe biþ *to such as it shall be granted*, Beo. Th. 604 ; B. 299 : Met. 26, 87. Swilce wé ðé daga cígen *on such day as we call to thee*, Ps. Bèn. 19, 9. Eahtige hé hiene selfne swelcne hé ondrǽt ðæt hé sié, Past. 17 ; Swt. 118, 8. Hæfde his ende gebidenne swylcne hé ǽr æfter worhte, Judth. Thw. 22, 17 ; Jud. 65. Eall gedǽlan swylc him God sealde, Beo. Th. 145 ; B. 72. Ealle swylce hí habban scoldon, 3599 ; B. 1797. Cyningas swylce iú wǽron, Exon. Th. 510, 32 ; Seef. 83. Beaduþreáta mǽst

swylc cyning ymbsittendra meahte ábannan tó beadwe, Elen. Kmbl. 64 ; El. 32. **V.** in expressions relating to quantity or number, *so (as) much, so (as) many* :—Hwítes sealtes swilc swá mǽge mid feówer fingrum geniman *as much white salt as may be taken with four fingers*, Lchdm. ii. 130, 2. Swelc swá biþ þreó beána, 228, 5. Selle him twá swylc swylce man ǽt him nime, i. 400, 18. Mealwan seáwes þrý lytle bollan gemengde wiþ swilce tú wæteres (*twice as much water*), 214, 15. Genim wínes and eles swilc healf *take some wine and of oil half as much*, 180, 11. Medmicel pipores and óþer swilc cymenes *a moderate amount of pepper and an equal quantity of cummin*, 256, 5 : 134, 26. Feówertig daga nihta óðer swilc *forty days and as many nights*, Cd. Th. 83, 21 ; Gen. 1383 : Beo. Th. 3170 ; B. 1583 : Menol. Fox 279 ; Men. 141. [*Laym.* swilc, swulc, swulch ; soch, 2nd MS. : *Orm.* swillc : *A. R. Marh. O. and N.* swuch : *R. Glouc.* such : *Goth.* swa-leiks : *O. Sax.* sulík : *O. Frs.* se-lík, selk, sulk, sulch, suck : *O. H. Ger.* so-líh, su-líh, solh : *Icel.* slíkr.]

swilce, swelce ; *adv. conj.* **I.** *in like manner, also, as well, too* :— Se com swylce tó-dæg tó mé *ad me quoque hodie venire dignatus est*, Bd. 4, 3 ; S. 568, 17. Swylce hé brohte mycel feoh *attulit autem et summam pecuniae non parvam*, 4, 11 ; S. 599, 20. Hé wæs sóþ man, ðý hine dorste deófol costian ; swylce hé wæs sóþ God, ðý him englas þegnedon, Blickl. Homl. 33, 34. Swilce gelamp eft óðer wundor ðysum onlíc, 221, 18. Swilce óþre dæge ðæt ilce hié dydon, 241, 30 : Cd. Th. 81, 2 ; Gen. 1339 : 247, 24 ; Dan. 502. Swilce is seó feórðe *there is also the fourth*, 15, 14 ; Gen. 233. Wǽglíðende swilce wíf. heora *the seafarers, their wives too*, 86, 18 ; Gen. 1432. Swylce, Beo. Th. 226 ; B. 113. End suelce (suilcae, suilce) *atqueve*, Txts. 37, 75. Ic God herige and on God swylce gelýfe, Ps. Th. 55, 4. Ge swylce, Beo. Th. 4508 ; B. 2258. Hié hæfdon manige glengas ; eác swylce hié hæfdon wíf, Blickl. Homl. 99, 20. On ðære hálgan Ðrynnesse naman beó ðú hál, mid mínes láreówes geearnungum eác swylce gefultumod, Homl. Skt. i. 6, 40. Ná ðæt ǽnne ac eác swilce manige *non solum unum, sed etiam plures*, Coll. Monast. Th. 26, 19. Næs nó on gesundum þingum ánum, ac eác swylce on wiðer-weardum þingum, Blickl. Homl. 13, 8. Eác ic swylce on God gewéne, Ps. Th. 55, 4. Engla cynn and manna cynn and eác swylce werígra gásta, Blickl. Homl. 83, 12. Swylce eác feówer tída syndan, 35, 15. Hé helpeþ þearfan swylce eác wædlan *parcet pauperi et inopi*, Ps. Th. 71, 13 : Blickl. Homl. 75, 19 : Judth. Thw. 21, 14 ; Jud. 18 : 26, 20 ; Jud. 344. Swylce hé ús álésde, Blickl. Homl. 103, 13. Fífe cyningas, swilce seofene eác eorlas, Chr. 937 ; Erl. 112, 30. And ic ðé on hleóðre hearpan swylce eác gecwéme, Ps. Th. 107, 2. **II.** *so, in such manner, in a manner already described* :—Ðín mildheortnes is mycel wið heofenas, is ðín sóðfæstnes swylce wið wolcnum, Ps. Th. 56, 12. Lifge Ismael lárum swilce ðínum, Cd. Th. 141, 18 : Gen. 2346. Ne wé swylc ne gefrugnan ǽfre gelimpan, ðæt ðú in sundurgiefe swylce (*in such manner*) befénge, Exon. Th. 6, 7 ; Cri. 80. **III.** *as, like* :—Ðonne ic wæs mid Iudéum ic wæs swelce hié, Past. 16 ; Swt. 100, 7. Ne beó gé swylce líceteras *non eritis sicut hypocritae*, Mt. Kmbl. 6, 5. Genóh byþ ðam leorningcnihte ðæt hé sý swylce (*sicut*) hys láreów, and þeów swylce hys hláfurd, 10, 25. Se áwyrgda gást is heáfod ealra unrihtwísra dǽda, swylce unrihtwíse syndon deófles leomo, Blickl. Homl. 33, 8. Hyre twigu beóþ swylce swínen byrst, Lchdm. i. 156, 2. Weard gesewen swilce ánes mannes hand wrítende on ðære healle wáge, Homl. Th. ii. 434, 33. Steám up árás swylce réc, Elen. Kmbl. 1604 ; El. 804 : Andr. Kmbl. 178 ; An. 89. Hwylc biþ hé (*the body after death*) ðonne búton swylce stán, Blickl. Homl. 21, 26 : Homl. Th. i. 406, 14. Mé geweorð-ode wuldres ealdor swylce swá hé his módor eác geweorðode, Rood Kmbl. 181 ; Kr. 92. *See also passages under* swilc, II. **IV.** *as if* :—Se weard wið hine forwrégd swylce (suoelce, Lind.) hé his gód forspilde *quasi dissipasset bona ipsius*, Lk. Skt. 16, 1. Swelce hié cwǽden *as if they had said*, Past. pref. ; Swt. 5, 13. Men gehýraþ myccle stefne on heofenum, swylce ðǽr man fyrde trymme and samnige, Blickl. Homl. 91, 31 : Ps. Th. 101, 3. Ðæs temples segl sylf slát on tú, swylce hit seaxes ecg þurhwóde, Exon. Th. 70, 20 ; Cri. 1141. Hié on swíman lágon, swylce hié wǽron deáðe geslegene, Judth. Thw. 21, 23 ; Jud. 31. **V.** with words denoting measure, *about* :—Maria wunude mid hyre swylce (suǽlce, Lind. : swelce, Rush.) þrý mónþas *quasi mensibus tribus*, Lk. Skt. 1, 56. Se Hǽlend wæs on ylde swylce þrítigwintre *quasi annorum triginta*, 3, 23. Betuh ðæm clife on (ond ?) ðæm wǽtre wǽron swylce twelf míla, Blickl. Homl. 211, 3. [Sulch (ase, 2nd MS.) hit an liun were, Laym. 4085. Sulc (alse, 2nd MS.) he walde awede, 6486.]

swilcness, e ; *f. Quality* :—Sý gebróðrum reáf geseald be swilcnesse and staþele ðære stówe ðe hý on wuniaþ *secundum locorum qualitatem ubi habitant*, R. Ben. 89, 4. Ðysne wyrttruman syllan þicgean mid sumum óðrum mete gemencgedne be ðære swylcnysse ðe seó untrumnys ðonne byþ, Lchdm. i. 260, 20.

swile. v. swyle.

swilian *and* swillan *to swill.* **I.** *to wash* :—Ic þweá oððe ic swilige mín bed mid mínum teárum *lavabo lectum meum lacrimis meis*, Ps. Lamb. 6, 7. **II.** *to swill the mouth or throat, to gargle* :— Iagul swyleþ *gargarizat*, Wrt. Voc. ii. 40, 54. Seóh þurh cláð *and* swile mid ðæt geagl ; after ðam lǽcedóme gelóme mid ele swille ða

hracan, Lchdm. ii. 24, 25-27. Swille ðone geagal ... swille ða ceolan, 48, 19, 21. Gagul suille *gargarizet*, Wrt. Voc. ii. 109, 46. Sceal mon ðone geagl swillan, Lchdm. ii. 48, 15. Ðæt geagl tó swillanne, 24, 12, 28. [Kan ich dishes swillen, Havel. 919.] v. (?) ā-spýlian (-swylian?), be-swylian = *to wash* (not *to soil*), *and see next word*.

swiling *and* swilling, e; *f. A swilling, washing, gargling, gargle*:—Clǽsnunga and swiling wið hrúm and gillistrum, Lchdm. ii. 2, 3. Wyrc ðus swilinge tó heáfdes clǽnsunge ... habbe on múþe lange, ðonne yrnþ ðæt gillister út. Eft óþru swiling ... súpe wlæc and ðæt geagl swile and þweá his múð, 24, 14-23. Swille ða ceolan ... sýn ða swillinga hwílum háte, 48, 22. v. preceding word.

swillan, swilling, swilt. v. swilian, swiling, swylt.

swíma, an; *m.* I. *swimming in the head, dizziness, giddiness, vertigo*:—Hí āscamode swiciaþ on swíman *ashamed they wander dizzily*, Exon. Th. 79, 33; Cri. 1300. Wið ðone swíman, nim ... and cnuca ... wyrta ... ofgeót mid wætere ... nim ðone wǽtan and lafa ðín heáfod, Lchdm. iii. 48, 3. II. *a state of unconsciousness, a swoon*:—Licgan on swíman *to lie unconscious*, Judth. Thw. 21, 22; Jud. 30: 23, 5; Jud. 106. [For to wacken him (*Lazarus in the grave*) of his suime (swyme), C. M. 14201. Halliwell gives three instances of the word, in the following phrases, *to fall in swyme, to lie in swyme, to come as in swyme*. (In these four passages *swyme* rimes with *tyme*.) He also gives *swimy* = giddy in the head, as a Sussex word (v. also E. D. S. Pub. C. 4, where *swimy* or *swimy-headed* = giddy, is given as a Surrey word); and *swimer* a hard blow as used in Devonshire. O. Frs. swíma *giddiness, swoon*: Du. zwijm *swoon*: Icel. svími; liggja í svíma *to lie in a swoon*, slá í svíma *to stun*: Dan. svíme *a swoon*; svíme-slag *a stunning blow*.] v. heáfod-swíma.

swimman; *p.* swamm, *pl.* swummon; *pp.* swummen *To swim*:—Swimþ, swam *nat*, swimmende *nantes*, Wrt. Voc. ii. 61, 11, 13. Swam *nat*, 95, 80. I. *of living creatures moving in or on water*:—Swá swá fixas swimmaþ on wætere, Lchdm. iii. 272, 19: Exon. Th. 363, 21; Wal. 57. Ic on flóde swom, deáf under ýþe, 487, 17; Rä. 73, 4. Hié swumman ofer tó ðæm églande. Ðá hié ðá hæfdon feórðan dǽl ðære eá geswummen, Nar. 10, 29. Com tó lande lidmanna helm swymman, Beo. Th. 3252; B. 1624. Swimman hine geseón hearm getácnaþ, Lchdm. iii. 212, 18. Ðá geseah hé swymman scealfran on flóde, Homl. Th. ii. 516, 6. Teón ða wæteru forð swimmende cynn, Gen. 1, 20. II. *of a vessel moving on water*:—Secga geseldan swimmaþ on weg, Exon. Th. 289, 25; Wand. 53. Hine (*a vehicle*) oxa ne teáh, ne [hé] on flóde swom, 404, 28; Rä. 23, 14. Se swymmenda arc (*Noah's ark*), Homl. Th. ii. 60, 2. III. *of lying on the surface of water*:—Nim ompran neoþowearde ða ðe swimme, Lchdm. ii. 52, 19: 76, 5. Genim doccan ða ðe swimman wille, 88, 13. [O. H. Ger. swimman: Icel. svimma.] v. æt-, ge-, ofer-, óþ-swimman; -swemman.

swimmend-líc; *adj. Able to swim*:—Swymmendlíc *natatilis*, Ælfc. Gr. 9, 28; Zup. 55, 3.

swín, es; *n.* I. *a swine*. [As may be seen from the charters and the laws, swine were an important item in the livestock of the English. They were owned in large numbers (contrast the number held by the Norwegian Ohthere, *v. infra*), as appears from the passages given below, in which gifts of swine are recorded; references to their pasturage often occur, v. mæst, mǽstan, mǽsten; to the herd who had charge of them is assigned the second place in the list of those whose employments are defined in the Rectitudines Singularum Personarum, v. Th. i. 436; while the frequent occurrence of the word *swín* in local names, v. Cod. Dip. Kmbl. vi. 339, may be taken as further evidence. The value of swine, as compared with other domestic animals, is determined by the passages (*v. infra*) in the laws where the various animals are mentioned together.]:—Swín *porcus* vel *sus*, Wrt. Voc. i. 78, 36. Swín *sus*, 286, 43. Suove-taurili æt ðǽm geldum ðǽr wæs swín and sceáp and fear, ii. 31, 33: 86, 33. Mára ic eom and fǽttra ðonne āmǽsted swín, Exon. Th. 428, 9; Rä. 41, 105. Binnan cirictúne ǽnig hund ne cume, ne swín ðe má, L. Edg. C. 26; Th. ii. 250, 8. Emban úrne ceápgild: hors tó healfan pund ... And oxan tó mancuse, and cú tó .xx., and swýn tó .x. (*pence*), and sceáp tó scill., L. Ath. v. 6, 2; Th. i. 234, 1. Be ǽlces nýtenes weorðe gif hí losiaþ. Hors mon sceal gyldan mid .xxx. scill., myran mid .xx. scill., oxan mid .xxx. p̄., cú mid .xxiiii. p̄., swýn mid .viii. p̄., man mid punde, sceáp mid scill., gát mid .ii. p̄., L. O. D. 7; Th. i. 356, 5. Swínes smere *arvina* vel *adeps*, Wrt. Voc. i. 44, 20. Ðǽr wæs án swýna heord (suner berga, Lind.: suner swín, Rush. *grex porcorum*) ... Ða deófla hyne bǽdon ... 'Ásende ús on ðás swína heorde' ... And hig férdon on ða swín, Mt. Kmbl. 8, 30-32. Hé (*Ohthere*) hæfde tamra deóra syx hund .. Hé wæs mid ðám fyrstum mannum on ðǽm lande (*Norway*); næfde hé þeáh má ðonne twentig swýna, Ors. 1, 1; Swt. 18, 14. Ða ýtemestan leomo swína beóþ eáðmelte, Lchdm. ii. 196, 23. Mon selle tó Folcanstáne .x. oxan and .x. cý and .c. swína, Cod. Dip. Kmbl. i. 310, 27. Ic sello ðás lond ... and twá þúsendu swína ic sello mid ðém londe ii. 120, 15. Ic sello Berhtsige án híde bóclondes and ðǽrtó .c. swína, and geselle hió .c. swína tó Cristes cirican for mé and for míne sáwle and .c. tó Ceortesēge, 121, 3-6. Ðá hét ic gēniman swína micelne wrǽd

---

(*sues*) ... forðon ic wiste ðæt swín wǽron ðǽm elpendu ̄ lǽðe, Nar. 21, 23-26. Gif mon on his mæstene unálíefed swín geméte ... Gif mon nime æfesne on swýnum; æt þrýfingrum (*three fingers thick in fat*), ðæt þridde; æt twýfingrum, ðæt feórðe; æt þymelum, ðæt fífte, L. In. 49; Th. i. 132, 12-19. Gafolswán sylle ǽlce geáre .xv. swýn tó sticunge, L. R. S. 6; Th. i. 436, 13. II. *the image of a boar* as the crest of a helmet. Cf. swín-líca, eofor-cumbol, -líc:—Swín ofer helme, Beo. Th. 2577; B. 1286. Æt ðæm áde wæs éþgesýne swátfáh syrce, swýn eal-gylden, eofor írenheard, 2227; B. 1111. [Goth. swein: O. Sax. O. Frs. O. H. Ger. swín: Icel. svín.] v. gærs-, mere-, sliht-swín.

swinc, es; *n. Swink* (this form is used in the 16th century. v. Nares' Glossary), *labour, trouble, affliction*:—Erian se ðe hine gesihþ swincu mǽste him ongeán cumaþ *he that in a dream sees himself ploughing, very great troubles are coming upon him*, Lchdm. iii. 198, 28. Swinca *verberum*, Rtl. 40, 19. v. ge-swinc, swinc-full, -leás.

swincan; *p.* swanc, *pl.* swuncon; *pp.* swuncen. I. *to toil, labour, work with effort*:—Hwæt dést ðú on ðís folce? hwí swingst ðú ána? Ex. 18, 14. Hé nǽre ná ælmihtig, gyf him ǽnig gefadung earfoðe wǽre. His nama is *omnipotens*, ðæt ys, ælmihtig, for ðan ðe hé nræg eall ðæt hé wile, and his miht náhwár ne swincþ *his power nowhere works with effort*, Lchdm. iii. 278, 17. Unnytlíce wé swincaþ, ðonne wé ús gebiddaþ, gif ..., Bt. 41, 2; Fox 246, 21. Cumaþ tó mé ealle ðe swincaþ (wyrcaþ ꝥ winnes, Lind.: winnaþ, Rush. *laboratis*), Mt. Kmbl. 11, 28: Met. 4, 56. Búton Drihten timbriende hús on ýdel swingaþ (*laboraverunt*) ða ðe timbriaþ, Ps. Spl. 126, 1. Git (*Beowulf and Breca in their match*) seofon niht swuncon, Beo. Th. 1038; B. 517. Óðre swuncon (*laboraverunt*), and gē eodun on hyra geswinc, Jn. Skt. 4, 38. Swince *laboret*, Wülck. Gl. 250, 31. Swunce mǽre se ðe unriht gestreón on his handa stóde and læsse se ðe áriht on sprǽce *he in whose hand was unjust gain should take the greater trouble, he who made claim rightfully the less*, L. Eth. ii. 9; Th. i. 290, 4. II. a. *with prep. marking the end of the labour, to labour at, after, etc.*, anything:—Ne swincþ hé náuht æfter ðam hú hé foremǽrost seó; ne nán mon ne begit ðæt hé æfter ne swincþ, Bt. 33, 2; Fox 122, 33-35. Hé swanc for heofonan ríce mid singalum geswince, Homl. Skt. ii. 26, 111. Ðe læs ðe unmihtig man feorr for his ágenon swince, L. Ff.; Th. i. 226, 1. Ic wundrige hwí swá manige wíse men swá swíþe swuncen mid ðære sprǽce, Bt. 41, 4; Fox 250, 20. Ðú swíþor swincst on ðam spore, ðonne hí dón, 38, 5; Fox 206, 13. Suá hwá suá suincþ (swinceþ, Cott. MSS.) on ðæn ðæt hé leornige unþeáwas, Past. 36; Swt. 251, 4. Æfter ðam unrihte ðe hí an swincaþ, Ps. Th. 27, 5. Hé geseah hí on réwette swincende, Mk. Skt. 6, 48. Hí swincaþ wið synnum, Exon. Th. 150, 21; Gú. 782. Ða ðe meahton Godes friénd beón bútan gesuince hié 'suuncon (swuncon, Cott. MSS.) ymb ðæt hú hié meahton gesyngian *qui amici veritatis sine labore poterant, ut peccent laborant*, Past. 35; Swt. 239, 21. Ða race sóhton and ymb swuncon, Bt. 39, 4; Fox 216, 16. Hwý gē ymb ðæt unnet swincen, Met. 10, 21. Ne þearfe ic swíþe ymbe ðæt swincan, Bt. 35, 3; Fox 158, 8. II. *to be troubled, travail, be in difficulty or distress*:—Ic swince on mínre gránunge *laboravi in gemitu meo*, Ps. Th. 6, 5. On hú grimmum seáðe swinceþ ðæt sweorcende mód, Met. 3, 2. Ic swanc (*laboravi*) on mínre geómrunge, Ps. Lamb, 6, 7. Ðám wífum ðe æfter beorþre on sumum stówum swincan, Lchdm. i. 344, 2. [Cf. Ðonne se ufera dǽl ðæs líchoman on ǽnigum sáre oððe on earfeþum geswince, 332, 9.] II a. *of inanimate things*:—Gif se midwinter byþ on Seternesdeag, ðonne byþ wíudig lengten and westmas swincaþ and scép cwellaþ *the fruits of the earth will not thrive, and sheep will die*, Lchdm. iii. 164, 11. [The verb is common in Middle English and is used as late as Spenser's time.] v. be-swincan (*for* ge-swincan, *see under* II above); swencan.

swinc-full; *adj. Full of trouble* or *distress, disastrous*:—Ðæs ilcan geáres wæs swíðe hefelíc geár and swíðe swincfull and sorhfull geár binnan Englelande on orfcwealme, and corn and wæstmas swícon ætstandene, Chr. 1085; Erl. 219, 19. [Ðeos world is swincful, O. E. Homl. i. 7, 20. Þho (*the Virgin Mary*) wass swinnefull (*hard-working*) inn alle gode dedes, Orm. 2621.] v. geswinc-full, geswincfulnys.

swincgel. v. swingel.

swinc-leás; *adj. Without labour* or *toil*:—On ð̄ m ēcan life wé bútan geswince God heriaþ. Wé sceolon on andwerdum life hine herian, ðæt wé móton becuman tó ðære swincleásan herunge, Homl. Th. ii. 364, 9.

swinc-líc; *adj. Laborious, toilsome*[... ðe healdan ðone Sunnan-dæg fram ǽlcum geswinclícum worce, Wulfst. 294, 18.]

swincness. v. geswincness, Guthl. 12; Gdwin. 28, 23.

swind, Wrt. Voc. i. 44, 20. v. spind.

-swind. v. ǽ-swind.

swindan; *p.* swand, *pl.* swundon; *pp.* swunden *To waste away, languish, grow languid, be consumed*:—Se synfulla swindeþ *peccator tabescet*, Ps. Spl. 111, 9. Sáwel heora on yfelum swand *anima eorum in malis tabescebat*, 106, 26. Ealle oþþe hefige slǽpe swundon oþþe tó synne wacedon *omnes aut somno torpent inerti, aut ad peccata vigilant*, Bd. 4, 25; S. 601, 11. (v. ǽ-swind.) Swindan (*tabescere*) ðú dydest sáwle his, Ps. Spl. 38, 15: 118, 139. on ðam frumwylme heora gecyrrednesse hý Hí

sylfe fulfremede taliaþ, ac hý swíþe recene áwlaciaþ and swindende ácóliaþ, R. Ben. 135, 6. [Nede in swot and in swynk swynde mot the pore. Nede he mot swynde . . . that nath nout en hod his hed for te hude, P. S. 150, 2–4. O. H. Ger. swintan *tabescere, tabefieri, deficere, conticescere* : Ger. schwinden *to dwindle, decay, die away*.] v. á-swindan.

swínen; *adj. Of swine* :—Suínin *suellium*, Wrt. Voc. ii. 121, 72. Mid swínenum gore, Lchdm. i. 100, 11. Genim swínen (swýnes, MS. H.) smero, 114, 24. Ðæt hí eton swýnen flǽsc (ða swínnan, Ps. Surt. *porcina*), Ps. Th. 16, 14 : Shrn. 111, 7. Hyre twigu beóþ swylce swínen byrst, Lchdm. i. 156, 2. [*O. H. Ger.* swínin *porcinus, suillus.*]

swing. v. ge-swing ; swinge.

swingan; *p.* swang, *pl.* swungon ; *pp.* swungeþ. **I.** *to swinge, flog, beat, scourge,* (a) literal :—Ðás cild ic swinge *hos pueros flagello,* Ælfc. Gr. 7 ; Zup. 23, 21. Ic swinge *verbero,* ic eom beswungen *verberor,* 5 ; Zup. 9, 4. Gif hwylc wíf hire wífman swingþ (*flagellis verberavit*), L. Ecg. P. ii. 4 ; Th. ii. 184, 1. Hig swingaþ eów *flagellabunt vos,* Mt. Kmbl. 10, 17 : Mk. Skt. 10, 34. Ǽrest hiene mon swong *primo virgis caesus,* Ors. 4, 5 ; Swt. 168, 4 : Bd. 2, 6 ; S. 508, 13. Ða nam Pilatus ðone Hǽlend and swang (*flagellavit*) hyne, Jn. Skt. 19, 1. Hié hine swungon, Blickl. Homl. 23, 31. Mē weras slógon and swungon, Andr. Kmbl. 1927 ; An. 966. Ða deófol hine (*St. Anthony*) swungan, ðæt hē ne mihte hine ástyrigean, Shrn. 52, 27. Wiþ ðon ðe mon sié mónaþseóc ; nim mereswínes fel, wyrc tó swipan, swing mid ðone man, sóna biþ sēl. Amen, Lchdm. ii. 334, 2. Gyf hit cild sý oððe cniht, swinge hine man (*vapulet*), L. Ecg. P. iv. 52 ; Th. ii. 218, 31. Swingon *vapulare,* Lchdm. iii. 212, 2. Hē ða fǽmnan hēt nacode mid sweopum swingan, Exon. Th. 253, 30 ; Jul. 188 : 251, 8 ; Jul. 142. Hē byþ geseald ðeódum tó swingenne (tó swinganne, Rush. *ad flagellandum*), Mt. Kmbl. 20, 19 : Exon. Th. 99, 11 ; Cri. 1623. Hine mid swipum swingende geangsumiaþ, Homl. Th. i. 426, 22. Ðæt hē swá lange swungen wǽre ôþþæt hē swylte, Blickl. Homl. 193, 4. (b) metaphorical, *to chastise, afflict, plague* :—Ic ðreáge and swinge (swinge, Cott. MSS.) ða ðe ic lufige . . . God swingeþ (swingeþ, Cott. MSS.) ǽlc bearn ðe hē underfón wile, Past. 36 ; Swt. 253, 1–4. Ðone heó ǽr mid wítum swong, Exon. Th. 279, 22 ; Jul. 617. Mid monnum ne biþ swungne *cum hominibus non flagellabuntur* ; they are not plagued as other men, A. V., Ps. Surt. 72, 5. **II.** *to give a blow with the hand* :—Ðæt deófol cwæd : Swingaþ hine (*St. Andrew*) on his múd (cf. Sleáþ swingne (*St. Andrew*) ofer seolfes múd, Andr. Kmbl. 2601 ; An. 1302), Blickl. Homl. 243, 2. [Wæs] suungen *exalaparetur* (cf. wæs fýstslægenu *exalaparetur,* 32, 2), Wrt. Voc. ii. 107, 75. **III.** *without the idea of hurting, to whip a top, cream, etc., beat up* :—Mid geǽredre handa hē swang ðone top, Ap. Th. 13, 13. Genim mǽrcsápan and hinde meolc, mæng tósomme and swinge, Lchdm. iii. 4, 2. Swyng, 14, 32. Nime man sealt and þreora ǽgra geolcan, swinge hit swíðe tógædere, 40, 22. **IV.** *to strike, dash* :—Hē swang ðæt fýr on twá *he drove back the fire on either hand* (cf. that giswerk ward teswungan, bigan sunnun lioht hēdrôn an himile, Hēl. 5634), Cd. Th. 29, 12 ; Gen. 449. **V.** *to beat the wings* (?) :—Se fugel licgeþ lonnum fæst swíðe swingeþ *beats its wings violently* (?), Salm. Kmbl. 533 ; Sal. 266. Nis hearpan wyn, ne gód hafoc geond sæl swingeþ (*flaps its wings* (?) as it sits on the perch ; cf. the opening lines of the Poema del Cid, where one mark of the desolation of the Cid's home is that the perches are 'sin falcones e sin adtores :' or swingeþ = *flies* (?), soars, v. swengan, and cf. for the idea of movement : Bigan ūst up stīgan, swang geswerk an gemang, Hēl. 2243, and *Ger.* schwingen *to wing, soar,* schwinge *a wing, pinion* : *Dan.* svinge (*of a bird*) *to soar*) ne se swifta mearh burhstede beáteþ, Beo. Th 4520 ; B. 2264. [*O. Sax. O. H. Ger.* swingan : *O. Frs.* swinga.] v. be-, ge-, of-swingan ; swengan.

swinge, swynge (*both forms occur in the Pastoral*), an ; *f. A stripe, stroke.* **I.** literal, *a stroke with a scourge* or rod :—Scs. Petrus hine mid grimmum swingum swong and þreáde (*flagellis artioribus afficiens*) . . . Cwæþ him eác tó : 'Ic bende and swingan (*vincula, verbera*) ðrowade' . . . Ða wæs Laurentius mid ðæs Apostoles swingum (*flagellis*) swíþe gebylded ; cwom and eáwde mid hū miclum swingum (*verberibus*) hē ðreád wæs, Bd. 2, 6 ; S. 508, 12–24. Bedrífe hine (*n* wítepeów, v. Grmm. R. A. 703) tó swingum, L. In. 48 ; Th. i. 132, 10 : 54 ; Th. i. 138, 4. Ða hálgan men geðafedon on ðisse worlde monige swyngean and monige bendas and carcernu *sancti verbera experti, insuper et vincula et carceres,* Past. 30 ; Swt. 205, 12. **II.** metaphorical, *chastisement, afflicting stroke* :—Gefēged tó ðæm gefógstanum on ðære Godes ceastre būtan ðæm hiéwete ǽlcre suingean (swingan, Cott. MSS.) *sine disciplinae percussione,* Past. 36 ; Swt. 253, 20. Sunu mín ne ágiémeleása ðū Godes suingan (swingan, Cott. MSS.) *fili mi, noli negligere disciplinam Domini,* Swt. 253, 2. Ic neósiu in swingum (*verberibus*) synne heara, Ps. Surt. 88, 33. [With a swinge of his sworde [he] swappit hym in the face, Destr. Tr. 1271. *O. H. Ger.* swinga *flagellum : Ger.* schwinge *a winnow, fan.*] v. sweng, *and next word.*

swingel[l], e ; *and* swingel[l,e, an ; *f.* **I.** literal, (a) *a stripe, stroke* :—Hine man þreáge mid teartran steóre, ðæt is, him sīge on swingella wracu (*verberum vindicta*). Gif hē þurh ða swingella ne biþ geriht . . ., R. Ben. 52, 6–8. Mid teartum swingellum *acribus verberibus,* 54,

4. Geswencte on bendum and on swingelum (swinglum, MSS. C. V.) for ðam sóþan geleáfan, Homl. Skt. i. 5, 27. Swinglum, L. In. 48 ; Th. i. 132, 9, MSS. B. H. Wē witun ðē nellan on belǽdan swincgla ūs *inferre plagas nobis,* Coll. Monast. Th. 18, 24. (b) *a scourging, whipping, flogging* :—Gif hwā his hýde forwyrce and cirican geierne sié him sió swingelle (swingle, MS. B.) forgifen *if any one incur the punishment of flogging, and run to a church, let the flogging be remitted to him,* L. In. 5 ; Th. i. 104, 16. Hyne Drihten þreáde mid þearlwýslícere swingle. Ða eode hē tó ðam bysceope . . . and hym eówde ða lǽla ðæra(-e ?) swingellan ðe hē from Dryhtne onfēng, Shrn. 98, 14–18. Hē wēnde ðæt hē mid swinglan (*verberibus*) sceolde ða ánrēdnesse his heortan ánescian . . . Hē hine mid tintregum and mid swinglan oferswíþan ne mihte, Bd. 1, 7 ; S. 477, 43–478, 2. Hié hine swingaþ . . . and æfter ðære swinglan hié hine ofsleáþ, Blickl. Homl. 15, 11. Hē líchamlíce wrace mid swingelle þolige *vindicte corporali subdatur,* R. Ben. 48, 12. Ðonne áh se teónd áne swingellan (swingelan, MSS. B. H.) æt him (*the* wíteþeów), L. In. 48 ; Th. i. 132, 9. Gif hwylc wíf hire wífman swingþ and heó þurh ða swingle wyrd deád *si mulier aliqua ancíllam suam flagellis verberaverit, et ex illa verberatione moriatur,* L. Ecg. P. ii. 4 ; Th. ii. 184, 1. Hē hire swingele behēt, Homl. Skt. i. 9, 69. (c) *a scourge, rod, whip* :—Swinela *palmarum,* Hpt. Gl. 510, 40. (d) *a swingle, a stick to beat flax* [cf. a swinglestok *pessel,* the swingle *le pesselin,* to swingle the flax *estonger vostre lyn,* Wrt. Voc. i. 152, 39–44. A swyngelstok *excussorium, excudia,* Wülck. Gl. 581, 30 : *studia,* 614, 1. A swyndylstoc *exculidium,* a swyndilland *excudium,* 696, 7, 8. I bete and swyngylle flax, Rel. Ant. ii. 197, 34. See also Cath. Angl. 374-5 and the notes there. Cf. *Du.* zwingelen *to beat flax.* Halliwell gives *swingel* as a name in several dialects for the part of the flail that strikes the corn, and *batillus* is translated by a belle clapere *vel* swyngell, Wülck. 567, 39] :—Ic ða swingle (*but* spinle, MS. O. ; and the Latin text is *proiiciens quam gestabam colum*) mē fram áwearp, ðe ic seldon gewunode on handa tó hæbbene, Homl. Skt. ii. 23 b, 367. **II.** figurative, *chastisement, affliction* :—Wē scylen beón on ðisse ældeódignesse ūtane beheáwene mid suingellan . . . ðætte suā hwæt suā nū on ūsse ðætte ðæt áceorfe sió suingelle from ūs *nunc foris per flagella tundimur . . . quatenus quidquid in nobis est superfluum, modo percussio resecet,* Past. 36 ; Swt. 253, 18–22. Ðæt sár ðære suingellan (swingellan, Cott. MSS.) ðissa woruldbroca, Swt. 259, 2. Balthasar næs gemyndig his fæder swingle, Homl. Th. ii. 434, 27. Ða ðe him ondrēdaþ Godes swingellan . . . ða ðe suā áheardode beóþ ðæt hié mon mid nánre swingellan gebētan ne mæg, Past. 37 ; Swt. 263, 1–9. Ic eom nū tó swingellan gearu *ego in flagella paratus sum,* Ps. Th. 37, 17. Manifealde synt synfulra manna swingelan, 31, 12. Swingellan, 34, 15. Swyngla, Ps. Spl. 72, 5. Swinla *flagra,* Hpt. Gl. 527, 24. On swingelum *in verberibus,* Ps. Spl. 88, 32 : Homl. Th. i. 578, 25. Swinglum, Ps. Th. 88, 29. God ðurh mislíce swingla his folces synna gehǽlþ, Homl. Th. i. 472, 12. v. wind-swingla, *and preceding word.*

swingere, es ; *m. One who scourges* :—Nū ic (*mead*) eom bindere and swingere, sóna weorpere, Exon. Th. 409, 26 ; Rä. 28, 7.

swinglung, e ; *f. Giddiness, dizziness, vertigo* [cf. swingan, though the verb does not seem much used in the sense of modern *swing.* For the idea of *turning round,* seen in *vertigo,* cf. the following : He dude fóre of his cnihtes forte turnen þat hweol . . . ant het swingen hit swiftliche abuten' ant tidliche turnen, Jul. 58, 5. *See also* swengan, geswing] :—Swinglung *scottomia,* Wrt. Voc. i. 19, 20. Ðám mannum ðe swinclunge (swinglunge, MS. B.) þrowiaþ, Lchdm. i. 344, 6. [Cf. *Icel.* svingla *to rove : Dan.* svingle *to reel* ; svingel *giddy.* Cf. too *O. H. Ger.* swintilunga *vertigo.*]

swín-haga, an ; *m. An enclosure for swine* :—In ðone swínhagan ; of swínhagan, Cod. Dip. Kmbl. iii. 18, 33 : 399, 35.

swín-líca, an ; *m. The figure of a swine* or *boar* :—Wǽpna smið (ðone helm) besette swínlícum, Beo. Th. 2910 ; B. 1453. v. swín, II.

swin[n], es ; *m. Sound, melody* :—Swin, sang *melodia* (*Wright gives* swinsang *melodio ; perhaps* swinsung *should be read, but see the following gloss*), Wrt. Voc. ii. 57, 28. Swinne ł sangge *melodia,* Hpt. Gl. 467, 41. Swinn, dreám *melodiam,* 515, 42. [*From the same root as Latin sonus*?] v. ge-swin, *and following words* ; *and* cf. hlyn[n], hlynsian *for similar formation.*

swinsian; *p.* ode *To make a* (*pleasing*) *sound, make melody* or *music* :—Se fugel swinsaþ and singeþ swegle tógeánes *incipit illa sacri modulamina fundere cantus, et mira lucem voce ciere novam,* Exon. Th. 206, 9 ; Ph. 124 : 207, 11 ; Ph. 140. Swinsaþ sibgedryht swēga mǽste, 239, 8 ; Ph. 618. On psalterio ðe him swynsaþ oft *on the psaltery that oft makes music to him,* Ps. Th. 143, 10. Frætwe míne (*the swan's*) swinsiaþ, torhte singaþ, Exon. Th. 390, 8 ; Rä. 8, 7 : 55, 17 ; Cri. 885. Wit song áhófan hlúde bi hearpan, hleóþor swinsade, 325, 2 ; Víd. 105 : 353, 47 ; Reim. 29. Ðær wæs hæleþa hleahtor, hlyn swynsode (*a cheerful sound arose*), word wǽron wynsume, Beo. Th. 1227 ; B. 611. Sǽ swinsade *the sea made its music* (*but see* swinsung, II), Elen. Kmbl. 479 ; El. 240. Hearpan hlyn, swinsigende swēg, Cd. Th. 66, 8 ; Gen. 1081.

swinsung, e ; *f.* **I.** *melody, harmony* :—Suinsung *armonia,* Wrt.

Voc. ii. 100, 62: *melodium*, 113, 79. Dreám, swinsunge (-c?) *armonia*, 3, 29: 90, 61. Swinsung, Hpt. Gl. 498, 63. Gedrémere swinsunge *consona melodia*, 519, 6: *consona vocis harmonia* (*mo lulatione*), 467, 9. Wensumne swinsunge ł dreám *melodiam*, 438, 8. Bebudon him gif hé mihte ðæt hé in swinsunge leóþsanges ðæt gehwyrfde *praecipientes ei, si posset, hunc in modulationem carminis transferre*, Bd. 4, 24; S. 597, 35. Swinsunga *melos*, Wrt. Voc. ii. 57, 27. **II.** *sound* that is not harmonious :—Swinsunge sǽs *sonitus maris*, Lk. Skt. Rush. 21, 25. Wið eárena swinsunge and ungehýrnesse *for singing in the ears and hardness of hearing*, Lchdm. iii. 70, 23.

swinsung-cræft, es; *m. Music* :—Swinsungcræft *musicam*, Wrt. Voc. ii. 55, 29.

swio-, swió-. v. sweo-, sweó-.

swipa, swipe. v. swipu.

swipian, sweopian; *p.* ode *To scourge, strike, beat, lash* :—Hafaþ hé gyrde lange and ðone feónd sweopaþ, Salm. Kmbl. 185, MS. A.; Sal. 92. Rodor swipode meredeáða mǽst *the destroying sea lashed the skies*, Cd. Th. 207, 8; Exod. 463. [Icel. svipa *to whip; to move swiftly*] v. swippan.

swipor; *adj. Astute, cunning* :—Reáfaþ se snáw swiðor mycle ðonne se swipra (swiðra, Kmbl., but see Anglia i. 151) nið, Salm. Kmbl. 616; Sal. 307. [Swypyr or delyvyr *agilis*, swypyr and slydyr *labilis*, Prompt. Parv. 484. Cf. Icel. svipall *shifty*.] v. ge-swipor (*misprinted* ge-swip), -swiporness.

swippan; *p.* te *To scourge, beat, strike* :—Hafaþ hé gyrde lange and ðone feónd swipeþ, Salm. Kmbl. 185; Sal. 92. [The verb seems to be not uncommon in later English in the sense *to strike*, and also in that of *to move quickly* (Layamon also uses the noun *swipe* a stroke) :—He his sweord up ahof and adun sloh (swipte, 2nd. MS.), Laym. 23962. He braid ut his sweord and him to sweinde (swipte to þan kinge, 2nd MS.), 27627. He hine adun swipte, 16518. He his sweord swipte mid maine, 23978. He swipte þat hæfued of, 21425. Ich wulle his heued of swippen, 878. He lette his sweord adun swippen (hit adun swipte, 2nd MS.), 16510. Ine swifte wateres þe þet is isundred he is soue iswipt ford, A. R. 252, 20. He swipte hire of þ heaued *decollavit eam*, Kath. 2452. Heo swipten of mid sweord hire heaued *gladio percussa*, 2179. When þe saul fra þe body swippes, Pr. C. 2196. See also Halliwell's Dict. *swippe*, and cf. Dan. svipe *to smack, crack* a whip; Ger. schwippen *to whip*. Cf. also *swingan* and words related to it for connection of the ideas of striking and moving.] v. swipian.

swipu, e; swipu(-e), an; *f.* : swipa (?), an; *m.* **I.** literal, *a scourge, whip, rod* :—Suibæ *mastigia*, Txts. 78, 641. Swipe, Wrt. Voc. ii. 71, 22. Swipa (-u?) *anguilla vel scutica*, i. 21, 16. Sweopan *fla*[*g*]*ri*, ii. 37, 64. Áwundenre suipan, suiopan *verbere torto*, Txts. 104, 1051. Nim mereswínes fel, wyrc tó swipan, swing mid ðone man, Lchdm. ii. 334, 2. Ðám gelíc ðe Crist ádráfde mid swipe of ðam temple, L. Ælfc. C. 27; Th. ii. 352, 21. Suiopan, suipan *mastigium*, Txts. 77, 1276. Swipan, Wrt. Voc. ii. 55, 26. Hé worhte swipan (suuopa, Lind.: swiopa, Rush.) of strengon *flagellum de funiculis*, Jn. Skt. 2, 15. Sweopan, Salm. Kmbl. 219; Sal. 109. Hé worhte áne swipe of rápum, Homl. Th. i. 406, 7. Leádene swipa and óðre gepýlede swipa wurdon forð áborene, 424, 20. Swipena *flagrorum*, i. *flagellorum*, Wrt. Voc. ii. 149, 30: Hpt. Gl. 487, 58. Swipum *mastigiis, flagris*, 487, 49: *flagris*, Wrt. Voc. ii. 35, 70. Suiopum, 108, 74. Mid sweopum sleán, Exon. Th. 88, 18; Cri. 1442. Mid sweopum swingan, 253, 30; Jul. 188. Sweopum seolfrenum, Salm. Kmbl. 287; Sal. 143. Hí hine swuṅgon mid ísenum swipum, Guthl. 5; Gdwin. 36, 23. Mið swiopum (suuippum, Lind.) giðorscenne *flagellis caesum*, Mk. Skt. Rush. 15, 15. **I a.** *that with which a stroke is struck*, a sword (?), a javelin (?) :—Swypu *romphea*, Germ. 398, 189. Frome folctogan faraþ him tógegnes, habbaþ leóht speru lange sceaftas, swiðmóde sweopan, swenga ne wyrnaþ, deórra dynta, Salm. Kmbl. 243; Sal. 121. **II.** figurative, *affliction, chastisement* :—Swipu ne geneálǽcþ ðínum getealde *flagellum non appropinquabit tabernaculo tuo*, Ps. Lamb. 90, 10. Ne mæg heard sweopu weorðan húse ðínum on neáweste, Ps. Th. 90, 10. Ðǽre uplecan ðreá sweopon *supernae flagella districtionis*, Bd. 2, 5; S. 507, 2. Ic wylle swingan eów mid ðám smeartestum swipum, ðæt is, ic wítnige eów mid ðám wyrstan wíte, Wulfst. 295, 11. Synna suippum, Rtl. 42, 21. Suyppa ðínes uraððo, 8, 35. Syuipa, 41, 35. Syppo, 15, 25. Swipa *mastigias*, Hpt. Gl. 527, 27. [Gief he fend were, me sceolden eter gat ʒemete mid gode repples and stiarne swepen, O. E. Homl. i. 231, 21. Crist wrohhte an swepe, Orm. 15562. Icel. svipa *a whip*: Ger. schwippe *a lash, switch*. Cf. Prompt. Parv. sweype for a top, or scoorge *flagellum*.] v. preceding word.

swira, -swiria. v. sweora, ge-swiria *consobrinus*, Wrt. Voc. ii. 14, 73.

swirman; *p.* de *To swarm* (of bees) :—Ðonne hí (bees) swirman, Lchdm. i. 384, 23.

swital (-el). v. sweotol.

swítan (?); *p.* swát *in* for-swítan *to exhaust, impair, impoverish* land (?) :—Ðe lond æt Moran ic mid míne wífe bigat, and ic it siðen náwer ne forswát (-swác?) ne forspilde, Chart. Th. 584, 5. v. swǽtan.

swíþ; *adj.* **I.** *strong*, (1) of persons or personifications :—Metod mihtum swið, Cd. Th. 233, 32; Dan. 284; Andr. Kmbl. 2415; An.

1209: Exon. Th. 45, 8; Cri. 716. Ðý lǽs hé for wlence, mon móde swíð, of gemete hweorfe, 294, 34; Crä. 25. Hwæt wæs ðé, sǽ swíþa? forhwan fluge ðú swá? Ps. Th. 113, 5. Wyrd seó swíþe, 477, 16; Ruin. 25: Salm. Kmbl. 886; Sal. 442. Hé tóswengde þurh swíðes meaht líges leóman, Exon. Th. 189, 14; Az. 59. Ǽnne hæfde hé swá swíðne geworhtne, swá mihtigne on his módgeþohte, Cd. Th. 16, 33; Gen. 252. Hí swíðra oferstág weard, Exon. Th. 116, 3; Gú. 201. Biþ seó módor frommast and swíþost, 493, 1; Rä. 81, 23. (2) of things, (*a*) in reference to material things, (a) *producing a powerful effect* :—Swíð drenc wiþ áswollenum milte, Lchdm. ii. 256, 14. Ofgeót ðás wyrte mid swíþe beóre . . . wyl on swíþum beóre, 358, 14, 18. Stenc swíþra swæcca gehwylcum, Exon. Th. 358, 15; Pa. 46. Gif ðú wolde ðæt sió sealf swíðre sié, Lchdm. ii. 84, 8. Wylle swíþre medo . . . Wyrc swíðran (*the draught*), gif hé wille, 270, 7, 16. (β) *strong, violent* (of wind, stream, etc.) :—Swíþe hlimman *torrens*, Ps. Th. 125, 4. Gif swíþra wind árás *si flatus venti major adsurgeret*, Bd. 4, 3; S. 569, 10. (γ) *strong, not easily broken* :—Swíðne bogan, Ps. Th. 63, 3. (*b*) of immaterial things :—Ealdfeónda nið, searocræftum swíð, Exon. Th. 110, 25; Gú. 113. Wæs ðæt gewin tó swýð, tó láð and longsum, Beo. Th. 385; B. 191: 6163; B. 3085. Mid ðæm swíðan welme háðheortnesse, Met. 25, 46. Intó ðý swíðan slǽpe, Blickl. Homl. 205, 4. Þurh ða swíðan miht, Cd. Th. 237, 24; Dan. 342. Se willa biþ ðonne strengra ðonne ðæt gecynd. Hwílum biþ se willa swíþra ðonne ðæt gecynd, hwílum ðæt gecynd ofercymþ ðone willan, Bt. 34, 11; Fox 152, 11. Ðæt swýðre mægen wæteres, Ps. Th. 123, 4. ¶ *Swíþ* occurs often as part of proper names, either as the first or second element, v. Txts. 625, col. 1. **II.** The comparative is used where later English uses *right* (hand, side, etc.): —Swíðra *dexter*, Ælfc. Gr. 5; Zup. 13, 1. (1) With a noun :—Ðín swýðre eáge, ðín swíðre hand, Mt. Kmbl. 5, 29, 30. Ðú smítst ofer Aarones swýðre eáre . . . and ðæs swýðran fótes micclan tán, Ex. 29, 20. Hé sette his ða swíþ[r]an hand him on ðæt heáfod, Bd. 2, 12; S. 515, 19. Hé sette Ephraim on his swíðran hand . . . and Mannases on his winstran hand, ðæt wæs on Israhéles swíðran healfe . . . Hé hefde ðá his swíðran hand ofer Ephraimes heáfod, Gen. 48, 13, 14. Drihten mé ys on ða swýþran healfe, Guthl. 5; Gdwin. 36, 20. Ðú nymst ðone swýðran bóh, Ex. 29, 21. Gif hwá ðé sleá on ðín swýðre wenge, Mt. Kmbl. 5, 39. (2) Used without a noun, *the right hand, the right* :—Godes swýðra(-e?) forbeád Abrahame ðæt hé his suuu ne ofslóge, ac funde him ánne ram, Prud. 1 b. Ðǽne ðín seó swíðre sette *quam plantavit dextera tua*, Ps. Th. 79, 14. Tó swýðran *a dextris*, Ps. Spl. 15, 8. Hí ásetton hreód on hys swíðran, Mt. Kmbl. 27, 29. Ic sceáwade on ða swýðran *considerabam ad dexteram*, Ps. Th. 141, 4. Æt swýþrum þearfan *a dexteris pauperis*, Ps. Spl. 108, 30. Fram swýðrum ðínum *a dextris tuis*, 90, 7. [*Goth.* swinþs: O. Sax. swíði: O. Frs. swíth: M. H. Ger. swinde, swint *strong, quick*: Ger. ge-schwind: Icel. svinnr, sviðr *quick, wise*] v. earm-, for-, mód-, ofer-, un-swíþ.

swíðan; *p.* de; *but a strong form* swáð *also occurs.* **I.** *to make strong, give strength to, strengthen, support* :—Leng ne woldon Elamitarna aldor swíðan folcgestreónum, Cd. Th. 119, 16; Gen. 1980. Ongan Abimæleh Abraham swíðan woruldgestreónum, 164, 18; Gen. 2716. Swá reordeda manna mildost mihtum swíðed, 213, 9; Exod. 549. **II.** *to be strong, exercise strength, prevail* (?) :—Ic oforswíðrode ágen ł ongén ł swáð (= oferswáð? v. ofer-swíðan) hine *praevalui adversus eum*, Ps. Lamb. 12, 5. v. for-, ge-, ofer-, þurh-swíðan; swíðian.

swíðe; *adv. Very, much, exceedingly* :—Tó ðam swíðe *in tantum*, Hpt. Gl. 509, 34. Tó ðan swýðe *adeo*, Ælfc. Gr. 30; Zup. 193, 5. **I.** with adjectives, (1) of quantity :—Mid swíþe manigre swétnesse, Bt. 7, 1; Fox 16, 11: 11, 1; Fox 32, 34. Swíþe feáwa manna ongit, 19; Fox 70, 12. Swíðe lytle fiorme, Past. pref.; Swt. 5, 11. (2) of quality :—Hé biþ ðæra suíðe gemyndig, Past. 4; Swt. 37, 20. Ða swíðe swígean *nimis taciti*, 23; Swt. 174, 24. Swíþe heá ðúne, Blickl. Homl. 27, 16. Ús is swíþe uncúþ, 51, 35. Hé wæs swíðe welig *dives erat valde*, Lk. Skt. 18, 23. **II.** with adverbs or adverbial phrases :—Suíðe oft, Past. 3; Swt. 35, 9. Ðæt his láreów hine suíðe lythwón gemyndgige, 31; Swt. 207, 14. Ðá wundrade ic swíðe swíðe, pref.; Swt. 5, 19. Swíðe ðearle *vehementer nimis*, Gen. 17, 2. Drinc swýþe þearle, Lchdm. i. 78, 10. Swíþe eáþe . . . swíþe raþe, Blickl. Homl. 21, 17, 21. Swíþe lytelíce, Bt. 7, 1; Fox 16, 11. **II a.** in the superlative, *chiefly, especially, mostly* :—Seó bóc (*St. John's gospel*) hrepaþ swýðost ymbe Cristes godcundnysse, Homl. Th. i. 70, 1. Hwíþer wilt ðú mé swíþost lǽdan *whither especially wilt thou lead me?* Bt. 22, 2; Fox 78, 5. Þurh ofermétto ealra swíðost *most of all through pride*, Cd. Th. 22, 8; Gen. 337. Swíþost hé fór ðider for ðæm horschwælum *it was chiefly on account of the walruses that he went thither*, Ors. 1, 1; Swt. 17, 35. Swíðost hys spéda hý forspendaþ mid ðam langan legere, 21, 8. Ðæs hé wæs ealles swíþost tó hergenne, ðæt . . . *he was to be praised most of all for this, that* . . . , Blickl. Homl. 223, 27. Smire hine mid hrýþeres oþþe swíðost mid oxan geallan, Lchdm. ii. 44, 11. **III.** with verbs, intensifying their force :—Ne ðæt swíþe tó wundrianne is *it is not much to be wondered at*, Bd. 3, 9; S. 533, 24. Ðá arn ðæt wíf swíðe *then the woman ran fast*, Homl. Skt. i. 3, 650. Ælmyssan sylle hé

swýðe *eleemosynas reddat largiter*, L. Ecg. C. 3; Th. ii. 136, 34. Þicge hit swýðe, Lchdm. i. 80, 19. Seóð swýþe and ete swýþe *cook thoroughly and eat largely*, 82, 1. Ðæt Drihten swóre áð swíðe *solemnly swore an oath*, Wulfst. 209, 27. Ðæt wē his tó suíðe ne gítseden, Past. 3; Swt. 33, 18. Drihten is þearle swíþe tó herienne, Lchdm. iii. 436, 18. Hé þearle swíþe wearþ gegladod, 438, 27. Swá swýþe swá hé ðam cyninge wæs líciende, swá swýþe hē him sylfum mislícade, Bd. 5, 13; S. 632, 8. Mē swá swýþe ne lyst, swá . . ., Bt. 5, 1; Fox 10, 18. Hí swíþor clypodon *illi magis clamabant*, Mt. Kmbl. 27, 23. Nis ðé náuht swíþor *nothing affects you more*, Bt. 7, 1; Fox 16, 8: 7, tit.; Fox x, 13. Wē nellaþ be ðám ná swíðor áwrítan *we will not write further about them*, Homl. Th. ii. 466, 20. Wē willaþ furðor ymbe ðás emnihte swíðor sprecan . . . Embe ðis wē sprecaþ eft swíðor *we will say more about it later on*, Lchdm. iii. 240, 1, 7. Ða bróþra óþra weorca swýðor gýmdon *paid more attention to other works*, Bd. 3, 8; S. 532, 30. Swá hē him swíþor bebeád swá hí swíðor bodedon *quanto eis praecipiebat, tanto magis plus praedicabant*, Mk. Skt. 7, 36. Wæs hē swá micle swíðor on his móde gedréfed, swá his mód ǽr swíðor tó ðam woruld-sǽlþum gewunod wæs, Bt. 1; Fox 2, 27. Biþ ðý heardra ðe hit sǽ-streámas swíðor beátaþ, Cd. Th. 80, 10; Gen. 1326. Ðǽm módum ðe hí willaþ swíþost beswícan *the minds that they will most completely deceive*, Bt. 7, 1; Fox 16, 12. Ða hē hí swíðost forslagen hæfde *when he had inflicted a most severe defeat upon them*, 16, 2; Fox 54, 2. Ða hí swíðost worhton *when they were working hardest*, Homl. Th. i. 22, 22. Ðonne heó blēwþ swíðust *when it is in fullest blossom*, Lchdm. i. 160, 14. Forlǽtan unnytte ymbhogan swá hē swíþost mihte *as much as ever he could*, Bt. 35, tit.; Fox xvi, 27. Hiora scamiaþ swíþust ealles ða tó Sione hete swíðost hæfdon, Ps. Th. 128, 3. Swýþust ealra, 108, 28. Næfde se here Angelcynn ealles for swíðe gebrocod; ac hié wǽron micle swíþor gebrocede ond ceápes cwilde and monna; ealles swíþost mid ðæm ðæt manige ðara sēlestena cynges þēna forðférdon, Chr. 897; Erl. 94, 29-32. [The word is common in Middle English. O. *Sax.* swîðo: O. *Frs.* swîthe.] v. efen-, for-, ofer-, un-swíðe.

swíðestre. v. ofer-swíðestre.

swíþfæstness, e; *f. Violence, force* :—Þurh swíðfæstnesse his geþohtes *prae violentia cogitationis suae*, L. Ecg. C. 5; Th. ii. 138, 27.

swíþ-feorm; *adj.* I. *abounding in substance* :—Him ðá Abraham gewát ǽhte lǽdan golde and seolfre swíðfeorm and gesǽlig (cf. gewiton him ǽhta lǽdan, feoh and feorme, 99, 22; Gen. 1650), Cd. Th. 106, 12; Gen. 1770. II. *producing abundant sustenance, very fruitful* :—Beóþ góde wíngeardas and swíþfeorme mannum, Lchdm. iii. 162, 31. III. *violent.* v. next word :—Ic (*a storm*) wíde fēre swift and swíþfeorm, Exon. Th. 386, 35; Rä. 4, 72. Cf. swíþ-from.

swíþ-feormende *growing violent* :—Ða swíþfeormende *crudescentes*, Wrt. Voc. ii. 92, 23: 19, 42.

swíþ-ferhþ; *adj.* I. *of strong mind* or *soul* :—Snotor and swýðferhð (*Beowulf*), Beo. Th. 1656; B. 826. Swíðferhþe (*Beowulf's companions*), 990; B. 493. Hwæt swíðferhðum (*the Danes*) sēlest wǽre tó gefremmanne, 348; B. 173. II. *of violent mind, violent, impetuous* :—Geswearc ða swíðferð (*Juliana's father*), Exon. Th. 247, 13; Jul. 78. Oft bemearn swíðferhðes (*Sigemund*) síð snotor ceorl monig, Beo. Th. 1820; B. 908.

swíþ-ferom. v. next word.

swíþ-from; *adj. Exceedingly strong, of great energy* :—Hē (*the Deity*) biþ á ríce ofer heofenstólas heágum þrymmum sóðfæst and swíðfrom (-ferom, MS.; *but see also* swíþ-feorm) sweglbósmas heóld, Cd. Th. 1, 17; Gen. 9. Cf. Mín geswíþfroma (*addressing the Deity*), Anglia xii. 508, 1. v. next word.

swíþfromlíce; *adv. Strenuously, with great energy* :—Suíðfromlíce *naviter*, Wrt. Voc. ii. 114, 58.

swíþ-hwæt; *adj. Very strong*, Runic pm. Kmbl. 340, 13; Rún. 5. v. rǽd; *f.*

Swíþ-hún, es; *m. St. Swithin, bishop of Winchester, in which see he succeeded Helmstan, who died* 852. *In one MS. of the A. S. Chronicle, under the year* 861, *is the entry* :—Hér forðférde S. Swíðún biscop, Erl. 71, 20; *but in a charter of* 863, Swíðhún episcopus *is given as one of the witnesses*, v. Cod. Dip. Kmbl. v. 117, 22. *The name occurs often in the same connection in previous years* [For an account of him see Earle's Gloucester Fragments, and for the complete homily of which a fragment is given in that work, see Homl. Skt. vol. i. No. 21] :—Ðes Swýðún wæs bisceop on Winceastre, Homl. Skt. i. 21, 14. Se árwurða Swýðún (Swíðhún, Gloucester Frg.), 23. Æt Swýðúnes (Swíðhúnes, G. F.) byrgene, 98. Se smið andwyrde ðam árwurðan Swýðúne (Swíðhúne, G. F), 29. ¶ *For the name where there is no reference to the saint*, cf. ðæt suíðhúnincglond, Cod. Dip. Kmbl. i. 243, 10. Ab aquilone habens terminum suuealuue fluminis, a plaga oriente suíðhúninglond, a plaga occidentali ealhfleót, ab austro sighearding mēduue ond eac suíthúninglond, 250, 9-12.

swíþ-hycgende; *adj.* (*ptcpl.*) *Of strong purpose* :—Scealc monig swíðhicgende, Beo. Th. 1842; B. 919. Mágas ðara swíðhicgendra, 2036; B. 1016.

swíðian; *p.* ode. I. *to be* or *become strong, to prevail* :—Strangadan, swíðodon *invalescebant*, Wrt. Voc. ii. 74, 6. Strangedon, swíþedon, 46, 49. Ne wæs ðæt tó wundrianne ðeáh ðe ðæs cyninges bēne ðá hē mid Drihtne rícsade mid hine swíþode and genge wǽre *nec mirandum preces regis illius iam cum Domino regnantis, multum valere apud eum*, Bd. 3, 12; S. 537, 19. II. *to make firm, to fix* :—Suíðigaþ fígite, Wrt. Voc. ii. 108, 68. Swíþiaþ, 35, 60. v. for-swíðan (*under which* for-swíðede *is wrongly put*); swíðan.

swíþ-líc; *adj.* I. *very great, exceedingly great* :—Swíðlíc *grande, magnum*, Hpt. Gl. 434, 41. Samson gelǽhte ða sweras mid swíðlícre mihte and slóh hí tógædere *Samson apprehendens ambas 'columnas concussit fortiter columnas*, Jud. 16, 29. Hig cumaþ mid swíðlícum ǽhtum (*cum magna substantia*), Gen. 15, 14. II. *with the idea of violent disturbance, violent, strong* (*of storm, wind, etc.*) :—Reóhnys swýdlíc *tempestas valida*, Ps. Lamb. 49, 3. Swēgde swíðlíc wind of ðam wēstene, Homl. Th. ii. 450, 18. Heard gebrec, swár and swíðlíc, swēgdynna mǽst, Exon. Th. 59, 19; Cri. 955. For swíþlícum rēne, Bt. 12; Fox 36, 17. Wið swíðlícne flēwsan ðæs sǽdes, Lchdm. i. 220, 3. On wæterum swýðlícum *in aquis vehementibus*, Cant. Moys. 10. Hí sáwon swíðlíce rēnas, Boutr. Scrd. 21, 22. III. *of energetic, violent action, vehement, violent* :—Wið swíðlícne hracan, Lchdm. i. 270, 2. Him swyrdgeswing swíþlíc eówdon weras, Judth. Thw. 25, 3; Jud. 240. IV. *of that which affects the senses or the feelings, strong, intense, severe* :—Nǽfre ðú ðæs swíðlíc sár gegearwast, ðæt ðú mec onwende worda ðissa, Exon. Th. 246, 1; Jul. 55. Ða tēð cwaciaþ on swíðlícum cyle, Homl. Th. i. 132, 27. Ðonne hē on sumura for swíðlícre hǽtan geteorud byþ, Lchdm. i. 226, 22. Ðeós wyrt is hǽtre gecynde and swýðlícre, 236, 11. Strang tó swíðlícum drencum, Homl. Th. ii. 322, 15. God him sende swíðlíce ógan tý̂n cinna wíta, Ælfc. T. Grn. 5, 18. V. *of feeling, or emotion, intense, vehement* :—Hē mid swíðlícum luste his lífes gewilnode, Homl. Th. i. 86, 19. On swíðlícre blisse *in jubilo*, Ps. Lamb. 46, 6. VI. *of discipline or conduct, stern, severe, strict* :—Cildru behófiaþ swíðlícere steóre, Homl. Th. ii. 324, 33. Hē munucregol gesette mid swýðlícre drohtnunge, Basil prm.; Norm. 32, 6.

swíþlíce; *adv.* I. *very greatly, exceedingly* :—Se dēma wundrode swíðlíce (*vehementer*), Mt. Kmbl. 27, 14. Swá sárige on hiora móde and swá swíðlíce gedréfed *permotae*, Ors. 1, 10; Swt. 44, 30. Ic wát ðæt ðú woldest swíþe swíðlíce beón onǽled *quanto ardore flagrares*, Bt. 22, 2; Fox 78, 3. Ða wunode se, hálga wer on ancerlífe swíðlíce stíðe, Homl. Th. ii. 146, 7. II. *powerfully, energetically, strongly* :—Mē þincþ ðæt ðín gecynd and ðín gewuna flíte swíþe swíþlíce wiþ ðæm dysige, Bt. 26, 4; Fox 178, 28. III. *sternly, strictly, severely* :—Hwílum liðelíce tó ðreátianne, hwílum suíðlíce and strǽclíce tó ðráfianne, Past. 21; Swt. 151, 12. [Þe king him answerede swíðeliche fæire, Laym. 4421. O. *Sax.* swíðlíko (ēð giswerian).]

swíþlicness, e; *f. Excess*; nimietas, R. Ben. Interl. 73, 7.

swíþ-mihtig; *adj. Exceedingly mighty, of great might* :—Gesamnincga swíðmihtigra *synagoga potentium*, Ps. Th. 85, 13.

swíþ-mód; *adj.* I. *in a good sense*, (*a*) *great-souled, magnanimous, stout-hearted* :—Com ðá tó lande lidmanna helm (*Beowulf*) swíðmód swymman, Beo. Th. 3252; B. 1624. Swíðmód cyning, Cd. Th. 222, 5; Dan. 100: 225, 29; Dan. 161: 244, 18; Dan. 450. (*b*) *stern-minded* :—Ðone feónd swíðmód swipeþ, Salm. Kmbl. 185; Sal. 92. Folctogan faraþ him tógeánes, habbaþ swíðmóde sweopan, swenga ne wyrnaþ, 243; Sal. 121. II. *in a bad sense, of violent mind, arrogant, haughty, high-minded* :—Dryhtguman sine drencte mid wíne swíðmód (cf. stíþmoda, l. 19) since brytta (*Holofernes*), Judth. Thw. 21, 21; Jud. 30: 26, 22; Jud. 340. Swíðmód cyning (*Nebuchadnezzar after putting the three children in the furnace*), Cd. Th. 233, 1; Dan. 269: (*the king at the time of the dream*; cf. hē wæs wið God scyldig, 250, 20; Dan. 549), 242, 17; Dan. 529. Weard hé swíðmód in sefan for ðære sundorgife ðe him God sealde, 254, 3; Dan. 606. v. next word.

swíþmódness, e; *f. Greatness of soul, magnanimity* :—Be sigerfestnisse and swíþmódnisse úses Drihtnes mid ðæm hē ða hǽþnan ofercom, Anglia xi. 173, 12. Ne mágon hý̂ ðære tuugan gerecnisse ne hire mægnes swíðmódnisse áspyrian, Salm. Kmbl. 150, 4.

swíþness, e; *f. Strength, violence* :—Cyles swíþness *frigoris nimietas*, Anglia xiii. 397, 458. v. ofer-swíþness.

swíðor, swíðra, swíðrian. v. swíðe, swíþ, II, sweðrian.

swíðrian; *p.* ode. I. *to become* or *be stronger, to prevail* :—Ðæt wæter swíðrode swíðe ofer ða eorðan *aquae praevaluerunt nimis super terram*, Gen. 7, 19. Se hunger þearle swíðrode *praevaluerat fames in terra*, 12, 10. Hé swýðrode on ídelnysse his *praevaluit in vanitate ejus*, Ps. Lamb. 51, 9. Saulus micclum swýðrode *Saul increased the more in strength* (A. V. Acts 9), Homl. Th. i. 388, 3. Hyra stefna swíðredon *invallescebant voces eorum*, Lk. Skt. 23, 23. II. *to avail* :—Seó hálwende onsægedness[e] tó ēcre álýsnesse swíþrade and fromade *sacrificium salutare ad redemptionem valeret*, Bd. 4, 22; S. 592, 28. Swíþrian *valere*, swíþrigende *valens*, Hymn. Surt. 70, 3, 5. v. ofer-swíðrian.

swíþ-snel; *adj. Very quick :*—Sum biþ swíðsnel, hafaþ searolíc gomen gleódæda, leóht and leoþuwāc, Exon. Th. 298, 8 ; Crä. 82.

swíþ-sprecol; *adj. Proud in speech, speaking proud things :*—Ða swýðsprecelan tungan *linguam magniloquam*, Ps. Lamb. 11, 4.

swíþ-stincende; *adj. (ptcpl.) Emitting a strong scent :*—Swíþstincendre *flagrantior*, Wrt. Voc. ii. 38, 29.

swíþ-strang; *adj. Of great strength or force.* v. next word.

swíþ-stríme; *adj. Having a strong stream :*—Ðá com hē tō swíþ-strēmre (swíðstrangre, MS. B.) eá *pervenit ad flumen meatu rapidissimo*, Bd. 1, 7 ; S. 478, 4.

swíþ-swēge; *adj. High-sounding, heroic* (verse) :—Swíðswēgum metrum *heroico hexametro*, Hpt. Gl. 440, 12. Mid swíðswíum (= -swēg-um ?) sangum dreámes *dulcisonis (jucundis) melodiae*, 416, 1.

swíþ-swíge; *adj. Taciturn, too silent :*—Ða suíðsuígean (swíðe swígean, Cott. MSS.) ða felaídelsprǣcæn *nimis taciti, multiloquio vacantes*, Past. 23; Swt. 175, 24.

swí-tíma, switol. v. swíg-tíma, sweotol.

swíung, e ; *f. A spasm :*—Hramma *vel* swíung *spasmos*, Wrt. Voc. i. 19, 21.

swodrian; *p. ode To get drowsy, fall asleep :*—Ic hnæppode and ic swodrode *ego dormivi et soporatus sum*, Ps. Spl. 3, 5. [In his chaire he sat longe . . . a lutel he bigan to swoudri as a slep him nome. Þo þoȝte him in his swoudringe þat a whit coluere com, L. S. 439, 268. Cf. A day as he wery was, and a suoddrynge him nome . . . Seyn Cutbert to him com, R. Glouc. 264, 22. Halliwell gives *zwodder* = drowsy, dull, as a West-country word.] v. swaðrian, sweðrian.

swóg. v. swēg.

swógan; *p.* sweóg; *pp.* swōgen. **I.** *to make a sound, move with noise, rush, roar* (of wind, water, flame) :—Swógaþ windas, blāwaþ brecende bearhtma mǣste, Exon. Th. 59, 10; Cri. 950. Frǣtwe míne (a swan) swógaþ hlúde, 390, 7 ; Rä. 8, 7. Drihten lēt willeburnan on woruld þringan, ēgorstreámas swógan, Cd. Th. 83, 5 ; Gen. 1375. Fýr swōgende, 154, 17 ; Gen. 2557. Swōgende lēg, Beo. Th. 6282; B. 3145. Swōgende *strepente*, Wrt. Voc. ii. 74, 72. Ðǣm swōgendum, hleóðregendum *argutis*, 5, 36 : 86, 74. **II.** fig. *to move with violence, enter with force, invade.* v. in-swōgenness :—Ðæt nǣnig bisceop óþres bisceopscíre on swōge *ut nullus episcoporum parochiam alterius invadat*, Bd. 4, 5 ; S. 572, 32. [Þe soun of our souerayn þen swey in his ere, Allit. Pms. 104, 429. Cf. the noun in Mid. E. *swoughe*, *swoghe* = noise, e. g. of the see he herde a swoghe (Halliwell's Dict. q. v.), modern *sough* of the wind. But both verb and noun are used in the sense of swoon; for the verb v. geswōgen, and as later instances swowinde, A. R. 288, 25 ; he feol iswowen (-swoȝe, 2nd MS.), Laym. 3074 : for the noun see Stratmann and Halliwell. *O. Sax.* swōgan :—Swōgan quam engil, faran an feðerhamon, Hél. 5798.] v. ā-, ofer-, þurh-swōgan ; swēgan.

swógenness, swógung. v. in-swógenness, ge-swógung.

swól, es ; *m. (?), n. (?) Heat, burning :*—Suól *chaumos*, Wrt. Voc. ii. 103, 75. Swól *camos*, 17, 8 : i. 288, 41. Suóle *caumati*, ii. 103, 31. Swóle *caumate*, 22, 21. **I.** *of the heat of fire :*—Hē (the phenix) somnaþ swóles lāfe, gegædraþ bān gebrosnad æfter bǣlþræce, Exon. Th. 216, 16 ; Ph. 269. On swóle byrneþ þurh fýres feng fugel mid neste, 212, 23 ; Ph. 214. **II.** *of the sun's heat :*—Hē swā swíþe swǣtte swā hē in swóle (*caumate*) middes sumeres wǣre, Bd. 3, 19 ; S. 549, 30 MS. T. **III.** *of feverish heat :*—Sió ungemetlíce hǣto ðæs miltes cymþ of feferes swólle, Lchdm. ii. 244, 6. Hú se hāta maga swól þrowaþ, 160, 5 : 194, 12. [Cf. *Du.* zwoel *sultry*.] v. swólig.

swolgettan; *p.* te *To swallow, take into the throat :*—Ðonne sceal mon ðone geagl swillan gelóme on ðære ádle (quinsy), and swolgettan eced wiþ sealt gemenged, Lchdm. ii. 48, 16.

swólig (cf. dysig *for the form*), es ; *n. Burning, heat :*—Swólig *caumatio*, Wrt. Voc. ii. 130, 8. Hāt lyft and swólga (*sultriness ?*) bringaþ pūle on ðam milte, ðonne se mon wyrð tō swíþe forhæt, Lchdm. ii. 244, 7. [In a late MS. of Ælfric's Grammar and Vocabulary, *swoli, sweoli* translate *cauma*, Zup. 33, 12 note, 306, 15 note.]

swólig (?) ; *adj. Sultry.* v. preceding word.

swoloþ, swon-. v. sweoloþ, swan-.

swoncen-ferþ; *adj. ? :*—Hē (a man who has been hung) sígeþ swon-cenferð (swoncerferð *life having failed*, (?) v. swancor, I ; *or* sworcen-ferð *with darkened soul*, i. e. *dead* (?)), sāwle bireáfod, fealleþ on foldan, Exon. Th. 328, 29; Vy. 25.

swór *consobrinus*, -swora, -sworc, -sworcenness, -sworcenlíc, -soreness. v. sweór, māu-swara, ge-sworc, for-sworcenness, for-sworcenlíc, for-soreness.

sworettan; *p.* te *To draw a deep breath, to sigh, pant :*—Sworette *oscitavit*, Wrt. Voc. ii. 63, 64. Hē of inneweardre heortan swíþe sworete *ille intimo ex corde longa trahens suspiria*, Bd. 2, 1 ; S. 501, 14. Hē sume hwíle sæt and sworette *modicum suspirans*, 5, 19 ; S. 640, 29. Ða ūs nū bysmriaþ, ða ðe ǣr on ūrum bendum sworettan, Blickl. Homl. 85, 25. Ða ongan hē sworettan, swā swā eallunga gewǣced, on ðam oreðe belocen, Homl. Skt. ii. 23 b, 234. v. ā-sworettan.

sworettend-líc; *adj. Panting :*—Sworetendleca *anhela*, Wrt. Voc. ii. 9, 47.

sworettung, e ; *f. A deep drawing of the breath.* **I.** *as a sign of trouble, a sigh :*—From sworetunge mínum *a singultu meo*, Rtl. 20, 27. Heó nid wōpe and mid teárum wæs geondgoten and longe sworet-unge wæs teónde (*suspiria longa trahens*), Bd. 4, 23 ; S. 596, 10. Hē gemænigfealdode ða sworetunga ðām siccetungum, Homl. Skt. ii. 23 b, 201. **II.** *breathing hard* from illness or labour, *gasping, panting :*—Wið nearwre sworetunge, Lchdm. i. 340, 11. Hē mid langre sworetunge ðæt orð of ðām breóstum teáh, Guthl. 20 ; Gdwin. 80, 13. Hē wæs swíðe gewǣced on ðam langan geswince, and hē mid sworettungum wæs genyrwed, Homl. Skt. ii. 23 b, 770. Betwih ða[m] untruman sworettunga *inter aegra suspiria*, Bd. 3, 13 ; S. 538, 23.

swornian, swarnian ; *p.* ode *To coalesce :*—Suornodun, suornadur suarnadun *coaluissent*, Txts. 48, 198. Swornodon, Wrt. Voc. ii. 14,54. v. ā-swarnian.

swót; *adj. Sweet :*—Ðæt hūs gefylled wæs of suót stenc ðæs smirinese *domus impleta est ex odore ungenti*, Jn. Skt. Lind. 12, 3. Mid swótum wyrtum, Nar. 49, 8. [Þe swote bred of spices, A. R. 80, 2. His swote sauur, Marh. 4, 33. Þe swote Ihū, swottre þen euer ani þing, 11, 14. Se swiðe swote smeal, Kath. 1588. Swete Iesu, alre smelle swotest, 617. Aprille with his showres swoote, Chauc. C. T. prol. 1.] v. swēte, swótness.

swóte; *adv. Sweetly :*—Ic stince swóte *oleo*, Ælfc. Gr. 37: Zup. 220, 14. Se líchoma stanc swóte, Shrn. 143, 29. [Þu sleptest swóte, A. R. 238, 5. *O. H. Ger.* sōzo *suaviter*.]

swóðung. v. sweðung.

swót-líc; *adj. Sweet, savoury :*—Hū sió womb weorðe mid swótlec-ustum mettum gefylled *ut venter delectabiliter cibis impleatur*, Past. 43 ; Swt. 311, 8. [*O. H. Ger.* sōz-líh.] Cf. swētlíce, *and next word.*

swót-mete, es ; *m. A sweet-meat, delicacy :*—Nǣron ðá welige hámas ne mistlíce swótmettas, Bt. 15 ; Fox 48, 4. v. swēt-mete *and preceding word.*

swótness, e ; *f. Sweetness :*—Mycel swótnysse stænc, Shrn. 16, 1. In stencg suótnisses *in odore suavitatis*, Rtl. 88, 32. Suótnise stences, 65, 41. v. swētness.

swotoļe. v. sweotole.

swót-stence; *adj. Sweet-scented, odoriferous :*—Ambrosia elesealfe, *divino odore* ðære swótstencan, Wrt. Voc. ii. 2, 35.

swót-stencende (-stincende ?) *emitting a sweet odour :*—Suǣ ðæt rēcilc suótstencende stenc ic gisalde *sicut balsamum aromatizans odorem dedi*, Rtl. 65, 39.

swugian, swulung, -swundenness, swur-, swuster, swutol, swyft, swyft-lere, swylc. v. swigian, sulung, ā-swundenness, sweor-, sweostor, sweotol, swift, swiftlere, swilc.

swyld (?),e ;*f. A pang :*—Sár (þar, MS.) mē ymbsealde swylde (Grein sug-gests swylce) deáðes *trouble encompassed me, the pangs of death; circumde-derunt me dolores mortis*, Ps. Th. 114, 3. v. swelan, *and* cf. cwyld, cwelan.

swyle, es ; *m. A tumour, swelling, abscess :*—Swyle *apostema*, Wrt.Voc. i. 19, 35 : ii. 7, 68. Unwlitig swile . . . ðone ungeþwǣran swyle *tumor deformis . . . tumorem illum infestum*, Bd. 4, 32 ; S. 611, 17, 41. Se earm wæs on mycelne swyle gecyrred . . . ðeáh ðe se swyle ðæs earmes gesýne sí *brachium versum est in tumorem . . . tametsi tumor brachii manere videretur*, 5, 3 ; S. 616, 6, 38. Ðá āsweóll him se líchama . . . Ðá sóna eall se swyle gewāt fram him, Guthl. 16 : Gdwin. 68, 24. Wiþ innan-gewyrsmedum geswelle . . . lege on gelóme óþ ðætte open sié se swile, Lchdm. ii. 72, 24. Wiþ ceácena swyle and wiþ geagles swyle, 2, 19, 20. Wiþ ælcum heardum swile oððe geswelle, 70, 20. Wiþ deádum swile, 74, 12, 15. Wiþ springe . . . lege on ðone swile, 80, 17. Wið swylas, gāte tord ; smyre mid ða swylas ; hyt hý tódrífþ, and gedeþ ðæt hý eft ne ārísaþ, i. 354, 27. v. fǣr-, fót-, geagl-, hand-swyle.

swylfende, Wrt. Voc. i. 20, 22, swylian. v. swelgan, swilian.

swylt, es ; *m. Death, destruction.* **I.** *of the death of the body :*—Swylt hāligra *mors sanctorum*, Ps. Th. 115, 5. Ende becwom, swylt æfter synnum, Beo. Th. 2514 ; B. 1255. On galgan rídan, seomian æt swylte, Exon. Th. 329, 14 ; Vy. 34. Deáðberende gyfl (*the forbidden fruit*) ða sinhíwan tō swylte geteáh, 153, 10 ; Gū. 823. Swylt settan ðínum esnum *to put thy servants to death*, Ps. Th. 78, 2. Swylt ætfæstan, Andr. Kmbl. 2695 ; An. 1350. Swilt þrowian, Apstls. Kmbl. 142 ; Ap. 71. ¶ *The word often occurs with somewhat of a personal sense as the subject* of niman, forniman :—Ǣr ðec swylt nime, deáð for duguðe, Exon. Th. 257, 31 ; Jul. 255 : Elen. Kmbl. 892 ; El. 447. Ðǣr Seón cyning swylt dreórig fornam, Ps. Th. 135, 20 : Beo. Th. 2877 ; B. 1436. Ealle swylt fornam, druron dómleáse, deáðrǣs forféng, Andr. Kmbl. 1988 ; An. 996 : Exon. Th. 283, 5 ; Jul. 675 : 477, 19 ; Ruin. 27. **II.** *of the second death, thĕ perdition of the soul :*—Hí leahtrum fā, lēge gebundne, swylt þrowiaþ . . . ðæt is ēce cwealm, Exon. Th. 94, 14 ; Cri. 1540. [Cf. *Goth.* swulta-wairþja *lying at the point of death* : Icel. sultr *hunger, famine*.]

swylt (? = swylht, cf. swelgan ?), es ; *m. A whirlpool :*—Swyttes (swyltes ?) *gurgitis*, Hpt. Gl. 468, 72.

swylt-cwalu, e ; *f. Death-pang, death,* (1) *of the death of the body :*—

Þæs gástes síð æfter swyltcwale, Andr. Knıbl. 311 ; An. 156. (2) of the death of the soul :—Gif seó sáwl sceal mid deóflum drohtnoð habban . . . on swyltcwale and in sárum sorgum, Wulfst. 188, 4. Cf. deáþ-cwalu.

swylt-dæg, es ; *m. Death-day, day of death :*—Ær his swyltdæge, Cd. Th. 74. 12 ; Gen. 1221 : Beo. Th. 5588 ; B. 2798.

swylt-deáþ, es ; *m. Death :*—Ðú míne sáwle of swyltdeáðes láþum wiðlæddest *eripuisti animam meam de morte*, Ps. Th. 55, 11.

swymman, swýn, swynge, swyr, swyra, swyrd, swyrige (= *partiat*, R. Ben. Interl. 54, 4), swyster, -swystrenu, swytel, swýþ, sý, syb[b], sýcan. v. swimman, swín, swinge, sweor, sweora, sweord, scirian, sweoster, gesweosternu (-swistrenu), sweotol, swíþ, sí, sib[b], sícan.

-syd *in* ge-syd *a miry place*. [Halliwell gives *suddie* = miry, boggy. Cf. also *sod*. Cf. *O. H. Ger.* salz-suti *salsugo : Ger.* sudel *a puddle*.] Cf. seáþ.

syde, es ; *m. A decoction, the water in which anything has been seethed or boiled :*—Ðysse sylfan wyrte syde ðære tóþa sár gelíðigaþ, gyf hyne man swá wearmne on ðam múþe gehealdeþ, Lchdm. i. 280, 3. [*M. H. Ger.* sut : *Ger.* sud *seething ;* ab-sud *a decoction :* cf. *Icel.* soð *the broth or water in which meat has been sodden.*] v. seóðan.

sydung (*better* sidung, *under which form the word should be entered*), e ; *f. A regulation, rule :*—Sydung *regula*, Germ. 398, 217. Cf. Gesidode *determinabit*, 399, 431 : *conserit*, 469. Gesydod *concinna, conveniens, benecomposita*, 396, 321. Goth. sidôn *meditari*. *O. Sax.* gi-sidôn sorga *to cause sorrow* to a person : *O. H. Ger.* sitôn *machinari, disponere ;* gisitôn *instituere, destinare, conglutinare*. (*See* sídung, *where perhaps* sidung *should be read.*) v. sidu.

syfe. v. sife.

sýfer-ǽte ; *adj. Moderate in eating, sober, temperate :*—Sig se abbod clǽne and sýferǽte (sýfre, Wells Frgt.) *oportet eum esse castum, sobrium*, R. Ben. 119, 25. v. sýferness.

sýfer-líc (?) ; *adj. Sober, moderate :*—Séferlíce *sobriam* (but the termination of the Latin word is doubtful, v. note), Hymn. Surt. 16, 21.

sýferlíce ; *adv.* I. *with cleanliness, without impurity :*—Ðæt gé witen ðæt hit (*the preparation of the wafers for the mass*) clǽnlíce and sýferlíce gedón sý, L. E. I. 5 ; Th. ii. 404, 36. II. *soberly, purely, without excess or grossness :*—Sidefulle on ðeáwum and sýferlíce lybbende, Homl. Ass. 37, 327. III. *soberly, prudently, circumspectly :*—Ða cild ðe beóþ sýferlíce áfédde (cf. *the contrast in* l. 9, cild réceleáslíce áfédd) and wið unðeáwum gestýrede, Homl. Th. ii. 326, 17. Biddende séfcrlíce *precantes sobrie*, Hymn. Surt. 19, 11. [*O. H. Ger.* súbarlícho *ad sobrietatem.*]

sýferness, e ; *f. Sobriety, moderation, temperance, abstinence, purity :*—Sýfernys *abstinentia*, Wrt. Voc. i. 51, 7. Seó sýfernes þreáde ðæt werod cweðende ðæt hit ne fylígde ðære gálnesse *sobrietas increpat acies dicens ne sequantur luxuriam*, Prud. 46 a : 47 a–49 a. Seó sýfernes and óðre mægnu, 54. Rúmheortnys and sýfernys (*opposed to* gítsung and gífernes, 68, 15), Wulfst. 69, 1. Sýfernysse þearf *sinceritatis azima*, Hymn. Surt. 82, 31 : Scint. 42, 16. Ðære sýfernysse (*opposed to* drunkenness, v. l. 54) gód þodian, Homl. Ass. 146, 60. Mid micelre sýfernysse and gemetfæstnysse, and ná mid nánre oferfylle ne mid oferdrince, 144, 15. Sýfernysse (*opposed to* druncenscipe, l. 18), 145, 20 : Homl. Th. i. 360, 5. Ic brúce ðisum mettum mid sýfernysse (*cum sobrietate*), Coll. Monast. Th. 35, 5. Began ðá his geþanc tó sýfernysse (*opposed to* lust, v. 197, 75) gehwyrfan, Homl. Ass. 198, 96. [*O. H. Ger.* súbarnessi *purificatio, purgatio*.] v. un-sýferness.

sýfeþa, sýfian. v. sifeþa, seófian.

syfian ; *p.* de ; *pp.* ed *To provide with* sufel, q.v. [:—Gesyfiedne hláf, Wulfst. 170, 20. Brádne hláf well gesyfied, Chart. Th. 606, 3. *Icel.* syfiðr brauðhleifr.] v. ge-syfian ; syfiing.

syflige, an ; *f. A dish to be eaten with bread :*—Genihtsumian wé gelýfaþ twá gesodene syfian (oðde ?) sanda . . . twá sanda genihtsumiaþ *sufficere credimus cocta duo pulmentaria . . . duo pulmentaria cocta sufficiant*, R. Ben. Interl. 70, 11–15. v. next word.

syfling, e ; *f. Food to be eaten with bread :*—Syflyncge *pulmentario* (*pulmentarium quilibet cibus extra panem*, Migne), Hpt. Gl. 494, 57. Ðǽr feóll áðúne wearm hláf mid his syflinge, Homl. Th. i. 136, 18. Sind ða twá gesetnyssa, ðæt is sealmsang and wítegung, swylce hí syflinge wǽron tó ðám fíf berenum hláfum, ðæt is tó ðám fíf ǽlícum bócum, i. 188, 19. v. sufel, *and two preceding words*.

sýfre ; *adj. Sober, not giving way to appetite or passion, pure, temperate, circumspect :*—Sýfre (sýter, Wrt., *but see* Anglia viii. 451) *abstinens*, Wrt. Voc. i. 51, 8. Gif ðú drincst wín gemetlíce, sýfre (*sobrius*) ðú byst, Scint. 105, 17. Se mynstres hordere sí wís sýfre and ná oferettol *cellerarius monasterii sit sapiens, sobrius, non multum edax*, R. Ben. 54, 8. Sig se abbod clǽne and sýfre and mildheort *oportet eum esse castum, sobrium, misericordem*, 118, 26. Sidefull man . . . gesceádwís and sýfre, Homl. Th. i. 596, 32. Fæste ðæt mód sýfre *jejunet ut mens sobria*, Hymn. Surt. 63, 3. Séfre, 2, 32 : 27, 17. Mid sýfrum andgyte, Homl. Skt. ii. 23 b, 78. Swá swá Petrus cwæð : ' Beóþ sýfre and wacole' *be sober, be vigilant* (1 Pet. 5, 8), Homl. Th. ii. 448, 8. Clǽne and rihte and séfre *castique recti ac sobrii*, Hymn. Surt. 19, 5. Ða clǽnheortan . . . ða ðe

heora líchaman geclǽnsiaþ mid sýfrum þeáwum, Homl. Skt. ii. 23 b, 43. Clǽnust and sýfrust (*sincera*) gebedes átihtincg, Scint. 35, 14. [*O. Sax.* súbri : *O. H. Ger.* súbar, súbiri *mundus :* Ger. sauber : *Du.* zuiver *clean, neat.*] v. un-sýfre.

syge (*better* (?) sige), es ; *m. Sight, aim* (?) :—Scyppend hafa ðé tó hyhte and á sóð tó syge ðonne ðú secge hwæt *have God as your hope, and ever truth as your aim, when you say anything*, Exon. Th. 304, 2 ; Fä. 64. [Cf. (?) *Icel.* sigta *to aim at.*]

sýl, e ; *f. A pillar, column :*—Sós Arculfus sagaþ ðæt hé gesáwe on Hierusalem áne sýle . . . ðonne seó sunne byþ on ðæs heofones mydle ðonne nafaþ seó sýl nǽnige sceade . . . and swá ða dagas forð on sceortiaþ, swá byþ ðære sýle sceade lengra. Ðeós sýl cýþeþ ðæt Hierusalem ys geseted on myddre eorðan, Shrn. 95, 30–96, 5, 8. In sýle wolcnes *in columna nubis*, Ps. Surt. 98, 7. Ðære méder wæs on slǽpe ætýwed . . . ðæt hyre man stunge áne sýle on ðone bósum, 149, 2. Ercoles sýla *Herculis columnae*, Ors. 1, 1 ; Swt. 8, 26. Ðæt feoh ðe hié wiþ ðám sýlum sellan woldon, 4, 12 ; Swt. 210, 4. Ic getrymede sýle his *confirmavi columnas ejus*, Ps. Surt. 74, 4. [*O. Frs.* séle : *O. L. Ger. O. H. Ger.* súl *columna : Icel.* súla *a pillar.* Cf. *Goth.* sauls *a pillar.*] Cf. syll.

sýl = sylh v. sýll.   v. sulh, syll.

syla (= sylha), an ; *m. A ploughman :*—Syla *arator*, Hpt. Gl. 461, 72.

sylan. v. sulh.

syle. v. sylu.

sylen *a gift*, sylf, sylfor, sylfren. v. selen, self, seolfor, seolfren.

Syles eá *Selsey ; insula vituli marini*, Bd. 4, 13 ; S. 583, 8. v. seolh.

sylfring (*should be given under* seolfring), es ; *m. A silver coin :*—Þreó hund sylfringa *trecentos argenteos*, Gen. 45, 22.

sylh, Sýl-hearwa. v. sulh, Sigel-hearwa.

sylian ; *p.* ede *To sully, soil, pollute, defile :*—Hé on unscyldgum eorla blóde his sweord selede (cf. besyled, Bt. 16, 4 ; Fox 58. 18), Met. 9, 60. Sió sugu hí wille sylian on hire sole æfter ðæm ðe hió áðwægen biþ, Past. 54 ; Swt. 419, 27. [Þis sunne suleð þi sawle, H. M. 35, 15. Blind mon To þare diche his dweole fulieþ (*follows*) And falleþ and þar one sulieþ, O. and N. 1240. Mi sawle mit sunne isulet, Marh. 3, 14. Isuled, A. R. 396, 1. *O. Sax.* sulian : *O. H. Ger.* bi-sullen. Cf. *O. Frs.* sulenge *soiling :* Goth. bi-sauljan *to defile*.] v. be-sylian ; solian, sulian, sol, syle.

syll, e ; sylle, an ; *f.* I. *a beam that serves as a foundation or support, a sill, a basis, support :*—Grundstánas *cementum*, syll *basis*, fótstán *fultura*, Wrt. Voc. i. 61, 47–49. Syl *basis*, post *postis*, 86, 28, 29 : ii. 10, 74 : 101, 54. Syl *taber*, i. 289, 48. Copsus syl, *securis* [æx ?], ii. 133, 9. Cobsus syl, ætx [æcx [securis ?], 22, 48. Getimbrung *aedificium*, post *basis*, sylle *postis* vel *fulcimentum*, i. 47, 19–21. Ðá wolde hé hús timbrian mid his gebróðra fultume. Ðá bæd hé hí ánre sylle, ðæt hé mihte ðæt hús on ða sǽhealfe mid ðære underlecgan. Ðá gebróðra him behéton, ðæt hí woldon ðæt treów him gebringan. Ðá cómon hí and wurdon ðæs treówes ungemyndige ; ac God him ða sylle ásende mid ðam sǽlícum flóde, Homl. Th. ii. 144, 31–146, 4. Ðær fram sylle (*from the plank to which it was fixed*) ábeág medubenc monig, Beo. Th. 1555 ; B. 775. Ǽrest man ásnıeáþ ðæs húses stede, and eác man ðæt timber beheáwþ, and ða syllan man fægere geséþ, and ða beámas gelegþ, and ða ræftras tó ðære fyrste gefæstnaþ, Anglia viii. 324, 8. II. *figurative, a support, foundation :*—Ðonne hí ne beóþ mid nánre sylle underscotene ðæs godcundlícan mægenes *nullis fulti virtutibus*, Past. 1 ; Swt. 27, 17. [Sulle *bassis*, Wrt. Voc. i. 95, 38. Sylle of an howse *silla, soliva*, Prompt. Parv. 456. Til he came to the selle, upon the flore, Chauc. C. T. 3820. *Icel.* syll *and* sylla *a sill : Dan.* syld : *Swed.* syll. Cf. *Goth.* ga-suljan *to lay a foundation : O. H. Ger.* swelli ; *n. basis : Ger.* schwelle. *Also* (?) *Lat.* solea.]

syll *ploughs*, sylla *a giver*. v. sulh, sella.

sylla (= *sella? borrowed from Latin?*), an ; *m. A saddle :*—Sylla *sella*, sadolfelt *pella*, sadolboga *carpella*, Wrt. Voc. i. 291, 14–16.

sýlla, syllan, syllend, syl-líc, sylofren, syltan. v. sél, sellan, sellend, seld-líc, seolfren, siltan.

sylu, e, an ; *f. A miry place :*—Syle, sylen *volutabra*, Hpt. Gl. 486, 51. Syle, 506, 54. Ðis sint ða denstówa, bróchyrst and beaddan syla, Cod. Dip. Kmbl. ii. 318, 30. v. sol, syl-weg.

syl-weg, es ; *m. A miry way* (?) :—On sylweg ; andlang weges on ða hǽdihtan leáge, and swá on ðæt fúle slóh, Cod. Dip. Kmbl. iii. 262, 22. v. syle.

sýma, sýman, symbel *continual*. v. séma, síman, simbel.

symbel, symel, es ; *n.* I. *a feast, banquet, entertainment :*—Him (*Adam and Eve*) . . . and hyra eaferum swá weard sárlíc symbel, Exon. Th. 226, 15 ; Ph. 406. Him (*the blessed*) is symbel and dreám, 352, 12 ; Sch. 96. Se becom tó Prisce, ðær hé deófolgeldum geald. Ðá gelaþode he hine tó his symble. Ðá sǽde Marcellus him ðæt hé wǽre cristen, and him nǽre álýfed ðæt hé birgde ðara hǽþenra symbles, Shrn. 125, 28–31. Swefan æfter symble, Beo. Th. 238 ; B. 119. Symle, 2020 ; B. 1008. Ðonne árás hé fram ðam symle *surgebat a media coena*, Bd. 4, 24 ; S. 597, 7. Ðæt hámweorud tó symble gesomnod wæs and hé sæt mid him æt ðam symble *vicani coenantes epulabantur, resedit et ipse cum eis ad convivium*, 3, 10 ; S. 534, 26–28. Sittan æt symble,

Exon. Th. 413, 27; Rä. 32, 12: 314, 16; Mód. 15. Sittan tó symble, Cd. Th. 259, 33; Dan. 701: Beo. Th. 4214; B. 2104. Symle, 982; B. 489. Tó dam symle, Judth. Thw. 21, 12; Jud. 15. Ðǽr is Dryhtnes folc geseted tó symle, Rood Kmbl. 279; Kr. 141. Symbel (prandium) mín ic gearuade, Mt. Kmbl. Lind. 22, 4. Herodes symbel (cenam) worhte, Mk. Skt. Lind. 6, 21. Hé hét beran on dæt hús manegra cynna symbel, Shrn. 152, 25. Hé geaf mé sinc and symbel, Beo. Th. 4853; B. 2431. Symbel (ge)þicgan, 1242; B. 619: 2025; B. 1010. Symbel ymbsittan, 1132; B. 564. Symbel habban epulari, Ps. Th. 67, 2. Symbel ne álégon feasts failed not, Exon. Th. 352, 34; Reim. 5. Hwǽr cwom symbla gesetu? hwǽr sindon seledreámas, 292, 2; Wand. 93. Ðá wæs symbla mǽst geworden, 34, 31; Cri. 550: Beo. Th. 2469; B. 1232. Ðonne gecerres from symblum quando reuertatur a nuptis, Lk. Skt. Lind. Rush. 12, 36. II. a feast, religious festival:—Ðerh ðone dæg symbles (symbel, Lind.) per diem festum, Mk. Skt. Rush. 15, 6. Ðý ylcan dæge ealra wé healdaþ sancta symbel, Menol. Fox 397; Men. 200. [O. Sax. sumbal a feast, banquet: Icel. sumbl a banquet.]

symbel; adj. Of a feast or festival:—Simbel onsáh dæg sollempnis urgebat dies, Hymn. Surt. 96, 1. Gesettaþ dæg symbelne constituite diem sollennem, Ps. Lamb. 117, 27. Dæg symbelne hý dóþ ðé diem festum agent tibi, Ps. Spl. 75, 10. Ealle dagas simle omnes dies festos, 73, 9. v. symbelness.

symbel-calic, es; m. A chalice for use at festivals or at the solemnity of the Mass. v. symbelness, II:—Ic an Ðeódréd mín wíte massehakele ðe ic on Pauie bouhte and simbelcalice, Chart. Th. 515, 18.

symbel-cenness, e; f. The festival of a person's birth:—Of his symbelcenn' de ejus natalicio, Rtl. 80, 17. Symbelcen' dæt ué ðerh brúca natalicio perfrui, 78, 21. Symbelcenn' natalitiis, 93, 25. Ðaes symbelcennise wé bigóaþ cujus natalitia colimus, 65, 8: 79, 18. Symbelcen', 56, 13: 67, 8. [The meaning seems to require that the two parts of the compound should be separated, but the absence of inflexion in symbel where datives occur in the Latin seems to require the compound.]

symbel-dæg, es; m. I. a feast-day, a day of a banquet:—Æfter symbeldæge, Andr. Kmbl. 3052; An. 1529. Hé his líchoman wynna forwyrnde, symbeldaga, Exon. Th. 111, 34; Gú. 136. II. a festival, day of a religious feast:—Symbeldæg dies festus, Bd. 1, 27; S. 497, 1. Com ðyder mycel menigo for ðon symbeldæge, Blickl. Homl. 99, 29: Homl. Th. ii. 242, 21: Ps. Th. 117, 25. Se biscop sæt sume symbeldæge on ðære cierecan, Shrn. 78, 26. Ðone mǽron symbeldæg Drihtnes upstiges, Blickl. Homl. 131, 10. On symmeldæge (symbel-, MS. A.) per diem festum, Mk. Skt. 15, 6. His symbeldæg (natalitia) wé mérsiaþ, Rtl. 44, 28. Ic ðé symbledæg (diem festum) sette, Ps. Th. 75, 7. Symbeldagas dies festos, 73, 8.

symbel-gál; adj. Wanton with feasting:—Se ðe him wínes glæd wilna brúceþ, siteþ him symbelgál, Exon. Th. 449, 30; Dóm. 79.

symbel-geféra. v. simbel-geféra.

symbel-gereorde, n. A feast, banquet:—Biþ seó án snǽd sélre mycle tó þicganne ðonne him sýn seofon daga symbelgereordu, Salm. Kmbl. 816; Sal. 407.

symbel-gifa, an; m. A feast-giver:—Sáwla symbelgifa (the Deity), Andr. Kmbl. 2833; An. 1419.

symbel-hús, es; n. A banqueting-hall, dining-room:—Hé æteóweþ iów symbelhús (cenaculum) micel, Lk. Skt. Rush. Lind. 22, 12.

symbel-líc; adj. Of a feast or festival, solemn:—Dæge symellícum die sollempni, Anglia xiii. 390, 354. Gebedu symellíce orationes sollempnes, 417, 750. Daegas symbellíce dies festos, Ps. Surt. 73, 8.

symbellíce; adv. Solemnly:—Symbellíce solempniter, Rtl. 9, 7: 48, 40: Anglia xiii. 402, 539. Simbollíce, R. Ben. Interl. 98, 10.

symbelmónaþ-líc; adj. Pertaining to a month in which a solemnity was celebrated (?); the word translates comitiales in the gloss:—Ða symbelmónaðlícan ádla comitiales, Wrt. Voc. ii. 20, 39.

symbelness, e; f. Festivity, solemnity:—Symmelnysse festivitate, solemnitate, Hpt. Gl. 496, 17. Semelnyssa solemnia, festivitates, 500, 7. I. festivity, feasting:—Ðǽr ðurhwunaþ seó éce bliss; ne byþ ðǽr hungor ne þurst . . ., ac háligra symbelnys ðǽr þurhwunaþ á bútan ende, Wulfst. 143, 2. Symbelnes, Blickl. Homl. 65, 21. Hwǽr beóþ ðonne ða symbelnessa and ða ídelnessa and ða ungemetlícan hleahtras? 59, 17. II. a religious festival or solemnity:—Æftersanga symibolnys matutinorum sollempnias, R. Ben. Interl. 43, 2. In dege mérum symbelnisse (sollemnitatis) eówerre, Ps. Surt. 80, 3. Symelnysse, Ps. Spl. 80, 3. On dære Eástorlícan tíde symbelnysse in ipso tempore festi Paschalis, Bd. 3, 24; S. 557, 40. On ludéa symbelnysse (festivitate) wǽron geworden Drihtnes æfengereordu, Homl. Ass. 153, 40. Gérlíco symbelnise annua solemnitate, Rtl. 49, 25. Cuman tó ðære mæssan symbelnysse, L. E. I. 24; Th. ii. 420, 36: Bd. 1, 27; S. 496, 43: 2, 4; S. 505, 22. Ða symbelnessa mæssena sollemnia missarum, 4, 22; S. 592, 20. II a. festive nature:—Ðonne ealle dagas áteoriaþ, ðonne þurhwunaþ hé (Sunday) aa on his symbelnysse (it continues ever in its character of festival), Anglia viii. 310, 28.

symbel-tíd, e; f. A religious festival or solemnity:—Árwyrðe symbeltíd veneranda solemnitas, Rtl. 65, 1, 8. Eádges apostoles symbeltíde (festivitate), 47, 9. Symbbeltíd solempnitatem, 2, 27. Heald ða symbeltíde ðæs mónðes frumsceatta ðínes weorces, Ex. 23, 16. Árwyrðo symbeltído, Rtl. 49, 4. Symbeltídum sollennitatibus, 80, 31. Symeltídum, Anglia xiii. 397, 452. Symbeltído solemnia, Rtl. 49, 13: 50, 15: natalicia, 49, 25: 53. 1. Symbeltíde festa, 54, 11.

symbel-wérig; adj. Weary with feasting:—Wer (Noah) wíne druncen swæf symbelwérig, Cd. Th. 94, 19; Gen. 1564. Him symbelwérig (Abimelech) synna brytta þurh slǽp oncwæð, 159, 26; Gen. 2640.

symbel-wlanc; adj. Elate with feasting:—Siteþ symbelwlonc, lǽteþ wíne gewǽged word út faran, Exon. Th. 315, 32; Mód. 40.

symbel-wynn, e; f. The pleasure of feasting, the delight of the feast:—Gá nú tó setle, symbelwynne dreóh, Beo. Th. 3569; B. 1782.

symblan; p. ede; and symblian; p. ode To feast:—Hú mǽre ðín folc is, ǽlce dæge hit symblaþ, Ps. Th. 22, 7. Hió ofer hire suna symblaþ and blissaþ, 112, 8. Se weliga se ðe on dæm godspelle gesǽd is dætte ǽlce dæge symblede . . . Ða ðe ǽlce dæg symblaþ dives ille, qui epulatus quotidie dicitur splendide . . . epulando quotidie, Past. 43; Swt. 309, 3–9. Rihtwíse symbliaþ justi epulentur, Ps. Spl. C. 67, 3. Se weliga ǽlce dæge symblede (simblede, Cott. MSS.) dives epulabatur quotidie splendide, Past. 45; Swt. 337, 24. Mid ðý hí lange symbledon cum diutius epulis vacarent, Bd. 3, 10; S. 534, 30. Utan simblian epulemur, Wrt. Voc. ii. 143, 62. Symblendra swég sonus epulantis, Ps. Th. 41, 4.

symble, symblian, symel. v. simble, symblan, symbel.

symering-wyrt, e; f. The name of some plant:—Simæringcwyrt (symeringc-, Wrt. Voc. i. 79, 12) malua crispa, Ælfc. Gl. Zup. 310, 12. Simeringwyrt viola, Wrt. Voc. i. 68, 67.

symle, symlian, symlinga, symmel-dæg, sýn be, sýn sight, syn- ever-. v. simble, simblian, simblunga, symmel-dæg, sí, sín, sin-.

syn[n], e; f. I. with reference to human law or obligation, misdeed, fault, crime, wrong:—Se cyning his feóndum swíþe árede . . . Ðyslíc wæs seó syn (culpa) ðe se cyning fore ofslegen wæs, Bd. 3, 22; S. 553, 21. Hié georne smeádon hwæt sió syn wǽre ðe hié gefremed hæfdon wið ðam cásere, Elen. Kmbl. 828; El. 414. Ne synn ne sacu ne sár wracu nec scelus infandum, . . . aut Mars, aut ardens caedis amore furor, Ex. Th. 201, 10; Ph. 54. Ðá wæs synn and sacu Sweóna and Geáta then was their wrongdoing and strife between Swedes and Geats, Beo. Th. 4935; B. 2472. Senne facinus, Hpt. Gl. 519, 22. Synne stǽlan to charge with crime, Menol. Fox 569; Gn. C. 54. II. with reference to divine law, sin:—Heora synn (peccatum) ys swíðe gehefegod, Gen. 18, 20. Hé onfunde Godes ierre . . . ðeáh hé wénde dæt hit nán syn nǽre, Past. 4; Swt. 39, 6. Seó geofu wæs broht for ðære synne ðæs ǽrestan wífes . . . and seó synn wæs ádílegod, Blickl. Homl. 5, 4–6. Syn, 3, 7. Mænige líf bútan leahtre (crimine) habban mágon, bútan synne (peccato) hí ná mágon, Scint. 230, 12. Ælc ðe synne (peccatum) wyrcþ is ðære synne (peccati) þeów, Jn. Skt. 8, 34. Se ðe déþ áweg middaneardes synne (peccatum; synna, MS. A.: synna, MS. B. Lind. Rush.), 1, 29. Se hæfþ máran synne (synn, Lind.), 19, 11. Synne ne áspringaþ sins cease not, Exon. Th. 94, 11; Cri. 1538. Beóþ ðæs mannes synna gecwémran ðonne eal eorþlic goldhord, Blickl. Homl. 43, 21. Wé fela sinna didon, Hy. 7, 106. On synnum geboren, Jn. Skt. 9, 34. Of synnum mínum clǽnsa mé, Ps. Spl. 50, 3. Sennum, Ps. C. 38. Andettan synna, Mt. Kmbl. 3. 6. Senna, Blickl. Homl. 43, 14. [O. Sax. sundia: O. Frs. sende: O. H. Ger. sunta peccatum, culpa, noxa, nefas: Icel. synd.] v. fyrn-, heáh-, níd-syn[n].

syn-bót, e; f. Amends for sin, penance:—Bisceopum gebyreþ dæt hí ne beón tó feohgeorne æt synbóte, ne on ǽnige wísan on unriht ne strýnan, L. I. P. 10; Th. ii. 316, 32.

syn-byrðen[n], e; f. The burden of sin:—Hí synbyrþenne, firenweorc beraþ, Exon. Th. 79, 34; Cri. 1300. Ne þearf ðæs nán man wénan dæt his líchama móte ða synbyrþenna on eorþscrafe gebétan, Blickl. Homl. 109, 31.

syn-bysig; adj. Troubled in consequence of sin:—Hé heteswengeas fleáh ond ðǽrinne fealh secg synbysig, Beo. Zup. 2227.

syn-cræft, es; m. A sinful art:—Ne syncræftas (scyn-, other MS.) wé ne onhyrgen, Wulfst. 253, 10.

syn-dǽd, e; f. A sinful deed, sin, wicked act:—For syndǽda ðara eardendra ðe hire on lifdan a malitia inhabitantium in ea, Ps. Th. 106, 33. Se deófol ða syndǽda stǽleþ on ða gdas, Wulfst. 256, 7.

synder-ǽ; f. A separate, private law, law for an individual:—Syndurae privilegium, Rtl. 190, 19, col. 2. [O. H. Ger. suntar-éwa privilegium.]

synder-gifu. v. sundor-gifu.

synder-líc; adj. Separate, special, private:—And ðære synderlíc[an] ac privata, Wrt. Voc. ii. 9, 10. Ða synderlícan privatam, 75, 56. I. that is apart, separate, remote:—On senderlícum hulce in remoto (separato) tugurio, Hpt. Gl. 465, 43. II. private, that is done apart, not public:—Ða heáfodmenn on synderlícum geþeahte ðone sceat him sealdon, and bǽdon, dæt hí sǽdon, dæt ðæs Hǽlendes líc him wurde forstolen . . . Hí námon ðone sceatt, and swá þeáh on synderlícum rúnungum ðæt riht

eall ræddon, Homl. Ass. 79, 156–161.   **II a.** *private, without distinction, ordinary :*—On synderlícum dagum (cf. on weorcdagum *in contrast to* freólstídum, R. Ben. 37, 5 ; 36, 9) *diebus privatis,* R. Ben. Interl. 43, 2.   **III.** *that belongs to an individual* or *that is adapted to a particular purpose, not in common, special, peculiar, proper :*—Seó gesceádwisnes is synderlíc cræft dære sáwle, Bt. 33, 4 ; Fox 132, 10. Synderlíc gifu *praerogativa,* Hpt. Gl. 466, 42. Næfde se Fæder nán ðing synderlíces búton his Suna *the Father had nothing not in common with his Son,* Homl. Th. ii. 366, 12. Heora nán næfde siddan nán þincg synderlíces, ac didon him eal gemǽne, L. Ælfc. P. 20 ; Th. ii. 370, 36. For synderlícum wurðmente *privilegium, singularem honorem,* Hpt. Gl. 411, 30. Ðes miccla wurðmynt nis ná ealra manna, ac on synderlícum wurðmynte ðam gesǽligum mǽdenum and ðam clǽnum cnapum, Homl. Ass. 41, 431. Ánra gehwylc ðara apostola biþ geseted to his synderlícre stówe, Blickl. Homl. 143, 23. Hé ða syx dagas ǽr his þrowunga synderlíc weorc ǽlce dæge cýþde, 71, 30. God sealde heora ǽlcum synderlíce sprǽce, Ælfc. T. Grn. 4, 11. Ðonne wé for synderlecum synnum synderleca hreówsunga dóþ, Past. 53 ; Swt. 413, 28. Sume naman syndon *specialia,* ðæt synd synderlíce, ða ðe beóþ tódǽlede fram ðam gemǽnelícum, Ælfc. Gr. 5 ; Zup. 14, 6.   **IV.** *separated by superiority, singular, excellent, specially good :*—Ðys is synderlíc hǽcedóm wið eágena dymnysse, Lchdm. i. 178, 8. Synderlícere *singulari, speciali,* Hpt. Gl. 431, 23. v. sundor-líc.

synderlíce ; *adv.*   **I.** *apart, away from all others, in private :*—Synderlíce (*separatim*) hine Petrus and Iacobus and Iohannes and Andreas áhsodon, Mk. Skt. 13, 3.   **II.** *where many things are to be distinguished from each other, separately, severally, apart :*—Se án monn ongitt ðæt ðæt hé on óþrum ongit synderlíce (*in several ways*) ; hé hine ongit þurh ða eágan synderlíce, þurh ða eáran synderlíce, ðurh his rǽdelsan synderlíce, ðurh gesceádwísnesse synderlíce, Bt. 41, 5 ; Fox 252, 16–19. Synderlíce ánne gehwylcne hád God and hláford andettan wé synt geneádede *singulatim unamquamque personam Deum et dominum confiteri compellimur,* Ath. Crd. 19. Hine synderlíce ǽlc man beheóld, Homl. Skt. i. 23, 625. Ðara is ánra gehwylc synderlíce xxxtigum ðúsendum dǽla lengra ðonne eal middangeard, Salm. Kmbl. p. 150, 13. Heora ǽghwylc be heom sylfum synderlíce ðus cwæd, Homl. Ass. 162, 243.   **III.** *where one thing is to be distinguished from others of the same kind, specially, in particular* (as opposed to generally) :—Wé nemnaþ ealle ðing ǽgðer ge synderlíce ge gemǽnelíce ; synderlíce be ágenum naman, *Eadgarus,* gemǽnelíce, *rex* cyning, Ælfc. Gr. 5 ; Zup. 8, 9–11. *Animal* is ǽlc ðing ðe orðaþ, ðonne is synderlíce *homo* man, *equus* hors . . . ; gemǽnelíce *arbor* treów ; synderlíce *uitis* wíntreów, Zup. 14, 8–10. Þeáh heó synderlíce Iohannes gýmene betǽht wǽre, hwæðere heó drohtnode gemǽnelíce mid ðam apostolícum werode, Homl. Th. i. 438, 31 : ii. 112, 18–22. Hwí ne cwæd ðæt hálige gewrit be ðam men synderlíce, ðæt hé gód wǽre, swá swá hit cwæd mænigfealdlíce be ðam óþrum gesceaftum, ðæt hí góde wǽron ? Boutr. Scrd. 19, 18.   **IV.** *where the reference is to a single person or circumstance, only, exclusively, solely, to* or *by one's self :*—Ðæt word belimpþ synderlíce tó Gode ánum *that phrase belongs exclusively to God alone,* Homl. Th. ii. 236, 12. Hé him synderlíce (*to himself;* or (?) synderlíce, *adj.,* wíc *being used in plural*) wíc getimbrede *ipse sibi monasterium construxit,* Bd. 3, 19 ; S. 547, 30. Sume men ðæs wóses synderlíce (*by itself*) brúcaþ, Lchdm. i. 178, 11. Hú mæg ðǽr synderlíce ánes ríces monnes nama cuman *non fama hominum singulorum pervenire queat,* Bt. 18, 2 ; Fox 64, 1. Ðæt hors ic ðé synderlíce (*specialiter*) tó ǽhte geceás, Bd. 3, 14 ; S. 540, 28. Mæssige man áne mæssan sinderlíce for ðare neóde, ðe ús nú on handa stent, Wulfst. 181, 24. Hí hæfdon ǽlce dæge heora wítena gemót, and wǽron gesette synderlíce tó ðam ða senatores, Jud. p. 161, 32. Ná synderlíce for ðære ðeóde *non tantum pro gente,* Jn. Skt. 11, 52. Synderlíce on hyhte ðú gesettest mé *singulariter in spe constituisti me,* Ps. Spl. 4, 10. Ðonne hié synderlíce ðenceaþ hú hié selfe scylen fullfremodeste weorðan . . . mid ðý hí bereáfiaþ hié selfe ðara góda ðe hié wilniaþ synderlíce habban *cum sua lucra cogitant, ipsis se, quae privata habere appetant, bonis privant,* Past. 5 ; Swt. 41, 22–43, 1 : Swt. 45, 14. Nówuht him selfum synderlíce wilnian *nihil proprium quaerere,* 13 ; Swt. 77, 26. Senderlíce (*a Domino*) *proprie* (*uxor prudens,* Prov. 19, 14), Kent. Gl. 692.   **V.** *where degree is marked, specially, exceedingly, to a greater extent than in any other case, singularly :*—Syndrilíce *excellenter,* Rtl. 47, 1. Nalles ðæt án ðæt hé gód doo gemang óðrum monnum ac eác synderlíce suá suǽ hé on ðyncðum biþ furður ðonne óðre ðæt hé eác sié on his weorcum suá micle furður *ut non solum sit ejus operatio utilis, sed etiam singularis . . . sicut honore ordinis superat, ita etiam morum virtute transcendat,* Past. 14 ; Swt. 81, 22. Sum bróþor synderlíce mid godcunde gyfe gemǽrsod (*specialiter insignis*), Bd. 4, 24 ; S. 596, 30. Hé him synderlíce wilnade ðære wuldor, 5, 7 ; S. 620, 32. Hé hine lufode synderlíce, Homl. Th. i. 58, 6. Is synderlíce eallum Godes folce beboden ðæt hí heora gebeda lufian and ælmessan dǽlan, Homl. Ass. 164, 5. Se ðe synderlíce Cristes dýrling wæs, 151, 11. Ieremias ys úre wítega synderlíce, Ælfc. T. Grn. 9, 35. [Sunderliche, O. E. Homl. i. 11, 21 : 13, 1 :

261, 3 : A. R. 90, 5. *O. H. Ger.* sunderlícho *signanter, singulariter.*] v. sundorlíce.

synderlícness, e ; *f.* *Singularity, peculiarity :*—Forlǽtenre synderlícnysse *omissa specialitate* (*singularitate, peculiaritate*), Hpt. Gl. 413, 62.

synder-lípe ; *adj. Special, singular, separate :*—Senderlípes *speciali,* Hpt. Gl. 522, 63. Senderlípum *speciali, singulari,* 450, 66. Synderlýpum *peculiaribus,* Anglia xiii. 369, 62. Cf. án-lípe, *and see next word.*

synder-lípes ; *adv. Separately, singly :*—Sindorlípes *singillatim,* R. Ben. Interl. 47, 5. Senderlípes, Hpt. Gl. 484, 7. v. sundor-lípes.

synder-weorðmynt *a special honour, prerogative :*—Synderwurðmynt *praerogativa,* Wrt. Voc. i. 54, 61. Cf. sundor-weorþung.

syndig ; *adj. Skilled in swimming* (?) :—Sum byþ rynig ; . . . sum on londe snel, féðe spédig ; sum fealone wǽg stefnan steóreþ . . . ; sum biþ syndig, Exon. Th. 296, 28 ; Crä. 58. v. sund, I, II, *and* cf. *Icel.* syndr *able to swim.*

-synd-líc, syn-dolh, syndon. v. gesynd-líc, sin-dolh, sind.

syndrian ; *p. ode To sunder, separate :*—Eorþena langnyss ná syndraþ (*separat*), ða ðe sóð lufu geheód, Scint. 5, 13. Se ðe syndraþ fram leahtre, R. Ben. Interl. 117, 3. Ðæt God gegeadrade, monn ne suindria (*separet*), Mt. Kmbl. Lind. 19, 6. v. á-, ge-, tó-syndrian ; sundrian.

syndrig ; *adj.*   **I.** *separate, alone, not joined with others :*—Ic mé syndrig eom *singulariter sum ego,* Ps. Th. 140, 12. Wiþ fefre hylpþ syndrig marubie tó drincanne *to drink marrubium alone,* Lchdm. ii. 134, 27. Heáfdehtes porres [croppan] syndrigne sele þicgan, 230, 11. Nim syndrig sealt oððe wið weaxhláfsealfe gemeng, 246, 9. Áwyl ða wyrte and syndrigea betonican, neftan, etc., 76, 18.   **I a.** *standing apart, not accessible* (?) ; cf. synder-líc, I :—Hé (*Hannibal*) com tó Alpis ðǽm muntum . . . and ðone weg geworhte ofer munt Iof (munti fór MS. C.) Swá ðonne he tó ðæm syndrigum stáne com ðonne hét he hiene mid fýre onhǽtan and siþþan mid mattucan heáwan *ad Alpes pervenit . . . atque invias rupes igni ferroque rescindit,* Ors. 4, 8 ; Swt. 186, 18.   **II.** *special, set apart for a particular purpose :*—Sáwlsceat vel syndrig Godes lác *dano* (*dona* ?), Wrt. Voc. i. 28, 44. Ða Senatores dæghwamlíce smeádon on ánum sindrian húse, Jud. p. 161, 33.   **III.** *special, singular, extraordinary, remarkable for an unusual quality* or *for the unusual degree in which some quality exists :*—Ðæt is syndrig cynn, symle biþ ðý heardra ðe hit sǽstreámas swiðor beátaþ, Cd. Th. 80, 6 ; Gen. 1324. Him ða wæs syndrig ege ðǽr him ǽr wæs seó mǽste wyn, Ors. 2, 8 ; Swt. 92, 32. Míne þríe ða getreówestan frýnd ða wǽron míne syndrige treówgeþoftan (*my special confidants*), Nar. 29, 28.   **IV.** *of that which concerns a single person, private, own ; proprius, privatus :*—God, ðam syndrig (*proprium*) is dǽtte hé gimilsage, Rtl. 40, 19. Syndriges *propriae,* 33, 30. Be ðam ðæt munecas syndrige ǽhte næbben . . . Nǽnig nán ðing syndries ne áge *si debeant monachi proprium habere . . . Ne quis presumat aliquid habere proprium,* R. Ben. 56, 15–19 : L. I. P. 15 ; Th. ii. 322, 10. Fíf hída syndries landes . . . fíf hída gemǽnes landes, Cod. Dip. B. iii. 395, 28. Æfter syndrig mægn *secundum propriam virtutem,* Mt. Kmbl. Lind. 25, 15 : Ps. Th. 97, 2. Syndrige wyrðmenta *privilegia,* Hpt. Gl. 517, 1. Suindrig *propria,* Mt. Kmbl. p. 3, 9. From syndrigum *ex propriis,* Jn. Skt. Lind. 8, 44. Standan on syndrigum gebedum *to be engaged in private devotions,* Homl. Skt. ii. 26, 115. In syndrige *in propria,* Jn. Skt. Lind. 16, 32.   **V.** *separate, several, sundry, each separately :*—Moyses gebletsode ða twelf mǽgða ǽlce mid sindrige bletsunge, Deut. 33, 5. Hé syndrigne ácsode hwylces geleáfan hí wǽron *cujus essent fidei singuli, inquirebat,* Bd. 4. 17 ; S. 585, 13. Hwylcne ende syndrigo ðing (*singula*) hæbbende synd, 5, 23 ; S. 646, 6. Hig eodon and syndrie (*singuli*) férdon on hyra ceastre, Lk. Skt. 2, 3. Ongunnon suindrige (*or adv.* ?) ǽghwelc (*singuli*) cwoeða, Mt. Kmbl. Lind. 26, 22. Ic syndrigra (*singulorum*) hús and bedd geseah, Bd. 4, 25 ; S. 601, 9. Hé syndrigum geárum (*annis singulis*) hine neósode, 4, 29 ; S. 607, 12. Hig gesamnodon hig be sindrigum mǽgðum, Jos. 7, 16. Hé syndrigum (*singulis*) hys hand on settende hig gehǽlde, Lk. Skt. 4, 40. Scip ceigeþ syndrigum nomum *oues uocat nominatim,* Jn. Skt. Lind. 10, 3. Suindrigum his suá hwælc ðú eftsettes *singulis sua quaeque restitues,* Mt. Kmbl. p. 3, 11 : p. 4, 7. Þurh syndrige ðíne andsware ic ongeat, Bd. 4, 22 ; S. 591, 39.   **V a.** *in a distributive sense, one a-piece, one each :*—Ðá onfengon hig syndrige penegas (cf. ǽlc his pening, v. 9, the Latin in each case being *singulos denarios*), Mt. Kmbl. 20, 10. On septem epistolas canonicas ic sette syndrie béc (*libros singulos*), Bd. 5, 24 ; S. 648, 13. Dile, mintan and merce, syndrige sceafas geseóð, Lchdm. ii. 188, 24 : 228, 26. [*O. H. Ger.* sunderig *separatus, singularis, privatus, peculiaris.*]

syndrige ; *adv.*   **I.** *apart, separately, by one's self :*—Hé gefoerde in stówe unbýed syndrige (*seorsum*), Mt. Kmbl. Lind. 14, 13 : Mk. Skt. Lind. 4, 34. Syndrige áuunden *separatim involutum,* Jn. Skt. Lind. 20, 7.   **II.** *singly, one at a time :*—Ða ongunnon cuoeða him swyndria (*singillatim*), Mk. Skt. Lind. 14, 19. Ongunnon suindrige (*or adj.* ?) ǽghwelc (*singuli*) cwoeða, Mt. Kmbl. Lind. 26, 22. [*O. H. Ger.* sunderigo *separatim, seorsum, specialiter.*]

syndrigend-líc ; *adj. Separating :*—Adverbia discretiva synd syndrigendlíce, Ælfc. Gr. 38 ; Zup. 229, 7.

syndrig-líc; *adj. Special, singular, peculiar :*—Twegen cynelíce cnihtas mid syndriglícre (*speciali*) Godes gyfe wǽron gesigefæste, Bd. 4, 16; S. 584, 20. v. next word.

**syndriglíce; adv. I. specially, particularly :*—Ðæt hálige gewrit cýþeþ and syndriglíce (*specialiter*) Paules epistola, Bd. 1, 27; S. 489, 2. **II. singly, severally, one by one, of each one :*—Hé syndriglíce (*singillatim*) wæs fram him eallum frignende, Bd. 2, 13; S. 515, 40.

-syndrung. v. á-syndrung *divortium*, Wrt. Voc. ii. 28, 26.

syndur-ae, -sýne, syne-wealt. v. synder-æ, -síne, seonu-wealt.

syn-fáh; *adj. Stained with sin :*—Synfá men, Exon. Th. 67, 3; Cri. 1083.

syn-full; *adj. Sinful; used substantively, a sinner :*—Synful *peccator*, Wrt. Voc. i. 86, 63. Ðæt synfull gesyhþ *peccator videbit*, Ps. Th. 111, 9. Ic eom synful (synn-, Lind.) mann *homo peccator sum*, Lk. Skt. 5, 8. Synful, Jn. Skt. 9, 16. Þeáh ðe se mæssere synfull sý, L. Ecg. C. 7; Th. ii. 140, 1. Ðonne se synfulla his líf geendaþ, Blickl. Homl. 61, 2. Beó ðú milde mé synfullum, Lk. Skt. 18, 13. Ðæt gé gebiddan for mé ðam unwyrðestan synfullan, Anglia xi. 103, 95. On ðisse synfulran (*peccatrice*) cneórisse, Mt. Skt. 8, 38. Ða synfullan (synn-, Cott. MSS.) bytledon uppe on mínum hrygge, Past. 21; Swt. 153, 9: Blickl. Homl. 71, 35. Geseald on synfulra hand, Mt. Kmbl. 26, 45. Synnfullum mannum tǽcan, Blickl. Homl. 43, 15. Þeófas and synfulle men, 75, 28. Gesete him synfulle tó ealdrum *constitue super eum peccatorem*, Ps. Th. 108, 5. [*Icel.* synd-fullr.]

syngian; *p. ode To sin :*—Ic syngige *committo, admitto*, Ælfc. Gr. 37; Zup. 221, 8. Ic eom se lyðra man, se syngige swíðe genehhe, Hy. 3, 42. Ic singie nitende *peccavi nesciens*, Num. 22, 34. Gyf ðín bróðor syngaþ wið ðé *si peccaverit in te frater tuus*, Mt. Kmbl. 18, 15, 21. Ic ánum ðe syngode *tibi soli peccavi*, Ps. C. 47. Ðá sǽde him Plenius ðæt hé wóh bude, and miclum on ðǽm syngade, Ors. 6, 10; Swt. 264, 28. Wé singodon on úrum bréðer *peccavimus in fratrem nostrum*, Gen. 42, 21. Ne synga ðú *non moechaberis*, Ex. 20, 14. Ðe læs gé syngien (nelle gé syngian, Ps. Lamb.), Ps. Th. 4, 5. Se unrihtwísa cwyð ðæt hé wylle syngian (*ut delinquat*), 35, 1: Past. 17; Swt. 109, 17. Singian, Homl. Skt. i. 1, 88. Wið God singian *in Deum peccare*, Gen. 39, 9. Geopenian ðǽm syngiendum hiera unryht, Past. 15; Swt. 91, 11. Ðæt hié óþre syngiende rihtaþ, Blickl. Homl. 63, 24. [Hwenne þe muð suneȝeð on muchele ete, O. E. Homl. i. 153, 31. Þu sunegest . . . we suneȝieð, 17, 20, 36. Heo sunegede . . . heo makede him sunegen, A. R. 56, 1, 4. Þatt mann ne sinnȝheþþ nohht, Orm. 3970. Ine þri maneris me may zeneȝi, Ayenb. 20, 4. Ho so syngeþ (synegeþ, synneþ), Piers P. C-text, 11, 26. *O. Sax.* sundión : *O. H. Ger.* sunteón : *Ger.* sündigen : *Icel.* syndga.] v. for-, ge-syngian.

syngig (?); *adj. Sinful :*—Hwí flíhst ðú mé forealdodne syngigan (synnigan?), Homl. Skt. ii. 23 b, 192.

syn-grin *the toil or snare which a sin constitutes :*—Ðæt ús deófol of rihtan wege þurh deriende þýstra belǽdan ne mǽge, ne mid syngrinum tó swíðe gehremman *not hamper us too much with the snares of sin*, Btwk. 196, 19. Ðonne mæg se biscop ðæs mannes syngrina (*the toils of sin in which he is involved*) þurh Godes þafunge ðe swýðor gelíðian, Wulfst. 155, 26.

syngung, e; *f. Sinning :*—Ús is swíðe þearle tó éfstanne ðæt wé bewépan ðæt wé ǽr tó yfele gedydon, and ofer ðis ðǽre syngunge geswícan, Homl. Ass. 149, 137.

syn-leahter, es; *m. A sinful fault, a sin :*—Forbúgan ða synleahtras ðe ús forbodene synd, ðæt is unrihthǽmed and ǽrǽtas and oferdruncennessa, Wulfst. 134, 24.

syn-leás; *adj. Sinless, without sin :*—Hwylc eówer sí synleás (*sine peccato*), Jn. Skt. 8, 7. Crist þrowade for ús synleás, Wulfst. 121, 14: 151, 5. Ne biþ nǽfre nán man leahterleás ne synleás ealra þinga, 233, 24. Biþ oft synleás yfel gedoht ðǽm gódum *plerumque boni innoxie tentantur ad culpam*, Past. 54; Swt. 423, 3. Úre Drihten gescóp Adam háligne and clǽnne and synleásne, Wulfst. 153, 13. [*O. Sax.* sundi-lós.]

syn-léw, -leáw, e; *f. A sinful injury :*—Hér syndan þurh synleáwa sáre gelǽwede tó manege on earde, Wulfst. 165, 25. v. léw, lim-læw.

syn-líc; *adj. Sinful :*—Hé sceal scyldan cristenum mannum wið ǽlc ðara þinga ðe synlíc biþ, L. I. P. 7; Th. ii. 312, 24. Anbúgan tó nánum fúllícum and synlícum luste, Past. 14; Swt. 83, 15. Fyrenlusta and synlícra dǽda á má and má, Wulfst. 56, 7. Wé geáxiaþ nænig gód áwunigende and ealle worldlícu þing swíþe synlícu, Blickl. Homl. 109, 3. [Wǽron swíðe hefige and sinlíce gewinn betwux ðam Cásere of Sexlande and his sunu, Chr. 1106; Erl. 241, 23.] [*O. H. Ger.* sunt-líh *facinorosus, peccatorius : Icel.* synd-ligr.]

synlíce; *adv. Sinfully, wickedly :*—Hí sóhton synlíce sáwle míne, Ps. Th. 62, 8. Ða hǽðnan synlíce heora ða leásan godas mid mislícum deófolgeldum him laþodan on fultum, Blickl. Homl. 201, 30. Ðæs lífes ðe ðú mid leahtrum hafast ofslegen synlíce, Exon. Th. 90, 26; Cri. 1480. Se cyng and ða heáfodmenn lufedon swíðe and oferswíðe gítsunge on golde and on seolfre and ne róhtan hú synlíce hit wǽre begytan, Chr. 1086; Erl. 220, 6, 12. [*O. H. Ger.* suntlícho *impie*.]

syn-lust, es; *m. Sinful pleasure or desire, lust :*—Ic wæs swíðe onǽled mid ðǽre háþeortnysse ðæs synlustes, ðæt ic gewilnode bútan ceápe

ðæt hí mé tó geurnon, Homl. Skt. ii. 23 b, 337. Crist lǽrde, ðæt gehwá synnluste fæste wiðstóde ; Antecrist lǽrþ, ðæt gehwá his luste georne fulgange, Wulfst. 55, 11. Ða hlíwðe ðe hé ǽr þurh synlust gefremode, L. Edg. C. 16; Th. ii. 284, 5: Exon. Th. 17, 12; Cri. 269. Mancyn ðe nú is in ídelum gylpe and on synnlustum beswicen, Wulfst. 182, 13. Synlustum, Blickl. Homl. 57, 23. Synlustas fremman, Dóm. L. p. 30, 53.

synn. v. syn[n].

synnicge (-ecge), an; *f. A sinner, a sinful woman; peccatrix :*—Seó (*Mary Magdalen*) wæs ǽrest synnecge, Shrn. 107, 10.

synnig; *adj.* **I. in a religious sense, sinful, wicked :*—Ánra gehwylc, sóðfæst ge synnig, Exon. Th. 233, 11; Ph. 523. Se feónd and se freónd . . . synnig and gesǽlig, Elen. Kmbl. 1908; El. 956. Synnig wið sáwla nergend, Andr. Kmbl. 1841; An. 923. Hwí swigast ðú, synnigu tunge, Dóm. L. 67. Ðæs synnigan mód *peccantis mentem*, Past. 46; Swt. 357, 10. Sleáþ synnigne ofer seolfes múð, Andr. Kmbl. 2601; An. 1302. Synnig cynn (*the people of Sodom*), Cd. Th. 152, 35; Gen. 2531. Háliges láre synnige ne swulgon, Andr. Kmbl. 1419; An. 710. Beóþ ða syngan flǽsc scandum þurhwaden, Exon. Th. 78, 31; Cri. 1282. Fyrensulra ðreát, heáp synnigra *peccatores*, Ps. Th. 91, 6: Cd. Th. 145, 17; Gen. 2407. Hé biþ ðam yflum egeslíc tó geseónne, synnegum monnum, Exon. Th. 57, 18; Cri. 920. Syngum hondum, 70, 3; Cri. 1133: 84, 21; Cri. 1377. Ðú ðe ús synnige ádrífe fram dóme, Ps. Th. 107, 10. Hí hyra synnigan breóst beátaþ, Wulfst. 138, 12. Monige æfter ðæs líchoman scylde hí swá micle mæsðlícor gestaðoliaþ on gódum weorcum swá hí hí selfe synnigran ongietaþ, Past. 52; Swt. 411, 3. **II. in a legal sense, guilty, culpable.** v. scyldig :—Scyldig ł synnig *reus*, Mk. Skt. Lind. 14, 64. Synnig *culpabilis*, Rtl. 102, 7. Gif ceorl ceáp forstelþ . . . biþ se his dǽl synnig (scyldig, MS. H.) bútan ðam wífe ánum, L. In. 57; Th. i. 138, 17. Se ðe þeóf ofslihþ, se mót gecýðan mid áðe ðæt hé hine synnigne (scyldigne, MS. B.) ofslóge, 16; Th. i. 112, 8. Mon synnigne gefón æt openre scylde, 37; Th. i. 124, 22. [*O. Sax.* sundig : *O. H. Ger.* suntig *peccator, damnosus, noxius : Icel.* syndigr.] v. bær-, fela-, firen-, lyge-, un-synnig.

synnigness, e; *f. Sinfulness, guilt :*—Deáðsynnignise *reatum*, Rtl. 42, 33.

synoþ. v. seonoþ.

syn-rǽs, es; *m. A sinful impulse :*—Þence hé swíðe georne hwæt tó bóte mæge ongeán ǽlcne synrǽs, ðe þurh deófles sǽd ǽr weard áweaxen, L. Pen. 16; Th. ii. 284, 9.

syn-rust, es; *m. The foulness of sin :*—Synrust þweán and ðæt wom ǽrran wunde hǽlan *to wash away the foulness of sin and to heal the scar of the former wound*, Exon. Th. 81, 9; Cri. 1321. [Cf. the line in the Cathemerinon of Prudentius, 'quod limat aegram pectoris rubiginem.' v. Mod. Lang. Notes, May, 1889. Cf. also synne rust *peccati rubigo*, Scint. 4, 14.] Cf. syn-wund.

syn-sceaþa, an; *m. One who wickedly does harm, a malefactor, criminal, miscreant :*—Se synscaþa sceaþena þreáte éhstreám sóhte, Exon. Th. 282, 31; Jul. 70. Hié ne móste se synscaþa (*Grendel*) under sceadu bregdan, Beo. Th. 1418; B. 707. Ðone synscaðan gúðbilla nán grétan nolde, 1607; B. 801. Ða synsceaðan (*the heathens*) Godes tempel brǽcan and bǽrndon, Exon. Th. 44, 21; Cri. 706. Metod beslóh synsceaþan (*the apostate angels*) sigore and gewealde, Cd. Th. 4, 17; Gen. 55. Cf. mán-sceaþa.

syn-scyldig; *adj. Guilty of sin, wicked :*—Heortan wyrmas synscyldigra ceorfaþ and slítaþ *vermes sceler um mordebunt intima cordis*, Dóm. L. 168.

synt, -synto. v. sind, ge-synto.

syn-wracu, e; *f. The punishment of sin :*—Biþ him (*those in hell*) synwracu andweard, ðæt is éce cwealm, Exon. Th. 94, 15; Cri. 1540. Ðǽre synwræce siþþan sceoldon mægd and mægcgas morþres ongyldon, 153, 27; Gú. 832. Ic ne heóld teala, ðæt mé Hǽlend mín bibeád ; ic ðæs sceal geseón synwræce, 50, 2; Cri. 794.

syn-wrænness. v. sin-wrænness.

syn-wund, e; *f. A wound inflicted by sin :*—Ne syndon náne swá yfele wunda swá syndon synwunda, forðam þurh ða forwyrd se man écan deáðe, L. Pen. 4; Th. ii. 278, 17. Wé á sculon ídle lustas, synwunde, forseón, Exon. Th. 47, 18; Cri. 757.

syn-wyrcende *working sin, sinning, working iniquity :*—Synwyrcende (*the devil*), Elen. Kmbl. 1884; El. 944. Synwyrcende (*operantes iniquitatem*), ða ðe unrihtes æghwǽr þenceaþ, Ps. Th. 140, 11. Ansýna synnwyrcendra *facies peccatorum*, 81, 2.

**sype; m. Suction :*—Seó eorþe ðæt wæter helt and be sumum dǽle swilgþ, and for ðam sype heó biþ geleht, Bt. 33, 4; Fox 130, 6: Met. 20, 97. Cf. súpan, *and next word*.

sypian *to take in moisture :*—Glǽdenan rinde lytelra gedó þreó pund on glæsfæt; gedó ðonne ðæs scearpestan wínes tó .v. sestras, ásete ðonne on háte sunnan . . . ðæt hit sipige and socige .iiii. dagas, Lchdm. ii. 252, 11. Cf. súpan, *and preceding word*.

sypian (?), sipian (?); *p. ode To delay, be slow :*—Hé (*a sick person*) sipaþ and árísþ *tricabit et surget*, Lchdm. iii. 151, 2, 19, 28. (The reference is to an illness which begins on the 5th, 17th, or 27th day of the month.) Sypigende *senescens, frigescens*, Germ. 397, 345.

sẏr, *in the gloss* grundswylige, sẏr *senecio,* Wrt. Voc. i. 68, 42, *seems to have a meaning similar to that of* swylige. v. sur.

Syras, syrc, syrede(-on). v. Syre, serc, sirwan.

Syre, Syrie (?); *pl. The Syrians :*—Antiochus Sira cyning, Ors. 4, 11; Swt. 204, 24. Sennacherib Syria cyning, Homl. Th. i. 568, 2, 28. [*Goth.* Saur: *O.H.Ger.* Syr *Syrus.*] v. Syr-wàre.

syretum *latibulum* (= (?) syrwetum *latibulis; and see* siru-tûn), Wrt. Voc. ii. 54, 27.

syre-wrenc. v. siru-wrenc.

syrfe, an; *f. A* service-tree; sorbus :—Of caweldene tô ðære syrfan; ðonne of ðære syrfan tô healwícum, Cod. Dip. Kmbl. v. 262, 13. Ðonon tô ðan wôn stocce; and ðær tô wuda; ðonon on ða syrfan, vi. 234, 26. v. next word.

syrf-treów, es; *n. A* service-tree; sorbus :—In ðæt syrftreów; of ðam syrftreów in ðæt rûge mapel-treów, Cod. Dip. Kmbl. iii. 379, 22.

Syria (?) *Syria :*—Godes engel ofslôh ðæs Syrian cyninges here, Homl. Th. i. 570, 2. [*Goth.* Syria, Saura.]

sẏring, e; *f. Butter-milk :*—Hwæg serum, sẏring *raptura,* rynning *coagulum,* Wrt. Voc. i. 27, 68–70. Sẏring *baptua,* ii. 12, 64. Sceáphyrdes riht is ðæt hé hæbbe blæde fulle hweges oððe sẏringe ealne sumor, L. R. S. 14; Th. i. 438, 25. Cẏswyrhtan gebyreþ ðæt heó of wringhwæge buteran macige tô hláfordes beóde, and hæbbe ða sẏringe ealle bûton ðæs hyrdes dæle, 16; Th. i. 438, 33. [Cf. *Icel.* sẏra *sour whey* used as a drink instead of small beer.]

-syringas in Exsyringas, Exon. Th. 323, 22; Vid. 82.

Syrisc; *adj. Syrian :*—Naaman se Sirisca, Lk. Skt. 4, 27. Hi bædon Godes gescyldnysse wið ðone Syriscan here, Homl. Ass. 107, 170. [*O. H. Ger.* Sirisc *Arabicus.*]

Syro-fēnisc; *adj. Syro-phoenician :*—Wíf Sirofēnisces cynnes, Mk. Skt. 7, 26. [*Goth.* Saurini-fynikisks.]

syrwa, syrwan. v. siru, sirwan.

Syr-ware; *pl. The people of Syria, Syrians :*—Syrwara lond *Syria* Exon. Th. 209, 6; Ph. 166.

syððan, syx. v. siððan, six.

T

For the Runic T, see Tír.

tâ, (*contracted from*) tâhe, an; *f. A* toe :—Tâhae *allox,* Wrt. Voc. ii. 100, 8. Tâ, i. 71, 64. Sió micle tâ . . . sió æfterre tâ . . . sió midleste tâ . . . sió feórðe tâ . . . sió lytle tâ, L. Alf. pol. 64; Th. i. 96, 19–24. Swá mycle tâ . . . ðare mycclan tâan nægl, L. Ethb. 70, 72; Th. i. 20, 2, 5. Hê æthrân his swíðran þuman and ðæs wynstran fôtes miclan tân *tetigit pollicem manus ejus dextrae, similiter et pedis,* Lev. 8, 23. Tân and fingras *decies senos,* Wrt. Voc. ii. 27, 73. Ða miclan tân *alloces,* 5, 18. Ða tân scrincaþ (-eþ, MS.) up (*in gout*) *the toes shrink up,* Lchdm. iii. 48, 28. On ðan seofoþan mônþe ða tân and ða fingras beóþ weaxende, 146, 17. Gif heó mid ðám tân stæpeþ, 144, 15. Æt ðám ôðrum táum ealswá æt ðám fingrum, L. Ethb. 71; Th. i. 20, 3. Mid ðǽm táum *cum men'agris,* Lchdm. i. lxxiv, 21 (cf. lxxi, 13). Ofer hira handa þuman and ðæs swýðran fôtes micclan tân *super pollices manus eorum ac pedis dextri,* Ex. 29, 20. [*O. H. Ger.* zêha : *Icel.* tâ.] v. tân a toe.

tâ; *gen.* tân; *f.* **I.** *a* twig, shoot :—Tân ł twiga *vimina, virgulae,* Hpt. Gl. 428, 34. **II.** *a lot :*—Ðæt him dême seó tâ, gif hí hwæt dǽlan willaþ, Homl. Skt. i. 17, 86. Ða dǽldon ða cwelleras Cristes næl on feówer, heora ǽlcum his dǽl, swá him dêmde seó tâ, Homl. Th. ii. 254, 31. Hí wurpon ða tân betweox him, and bǽdon ðæt God sceolde geswutelian hwanon him ðæt ungelimp becôme. Ða com ðæs wítegan tâ upp, i. 246, 3–5. v. tân, *and cf. for a similar pair of forms* flá *and* flân.

tabule (-ele), an; *f.: also* tabula; *m.* **I.** *a table :*—Hæfdon hí mid him gehálgode fato and gehálgode tabulan on wigbedes wrixle *habentes secum vascula sacra et tabulam altaris vice dedicatam,* Bd. 5, 10; S. 624, 34. **II.** *a tablet, table on which to inscribe :*—Ðæra eúra getæl hæfþ seó tabule ðe wê mearkian willaþ, Anglia viii. 327, 41. On ánum leádenum tabulan (*but* áne leádene tabulan (*acc.*), 766), Homl. Skt. i. 23, 342. Ðás ðreó word stôdon on ánre tabulan. On ðære oðer tabelan wæs ðæt forme bebod: 'Ne hǽm ðú unrihtlíce,' Homl. Th. ii. 198, 5. Tabelan, 196, 34. Pilatus áwrát ðæs wítes intingan on ánre tabelan, 254, 24. Týn beboda áwrát se Ælmihtiga on ðám twám tabelum . . . Ða twá tabelan getácnodon ða twá bebodu, 204, 17–20. Twá stǽnene tabulan, Ex. 32, 15 : 34, 1. **III.** *a board which is struck to give a signal :*—Tabule æfter capitule byþ gecnucod *tabula post capitulum pulsatur,* Anglia xiii. 402, 536. Gecnucedre tabulan *pulsata tabula,* 390, 359 : 393, 397. [*O. H. Ger.* tavala, tabella *tabula, pugillaris.* From Latin.]

tacan; *p.* tóc *To take :*—Ða menn ealle hê tóc, and dyde of heom ðæt hê wolde (cf. ðamen hê áteáh swá swá hê wolde, MS. E.), Erl. 211, 20. Hê tóc swilce gerihta swá hê him gelagade (cf. hê nam swilce gerihta swá se cyng him geúðe, MS. E.), 1075; Erl. 212, 38. [*From Icel.* taka; *p.* tók.]

taccian (?); *p.* ode *To tame* [:—Getaccodon (-þaccodon? v. þaccian) *edomitis,* Germ. 402, 63].

tâcn, tâcen, es; *n. A token, sign :*—Tácne *dicimenta,* Wrt. Voc. ii. 106, 53 : 25, 57. Tâcn *indicia,* 44, 68. **I.** *a sign, significant form :*—Heofoncyninges tâcen *the cross,* Elen. Kmbl. 341; El. 171. Torht tâcen Godes *the sun,* Exon. Th. 204, 11; Ph. 96. Bûtan Godes tâcne (*the cross*), 271, 32; Jul. 491. Þurh tâcen ðære hálgan róde, Homl. Th. i. 62, 12. Tâcna torhtost, Elen. Kmbl. 327; El. 164. **Ia.** *an ensign* (lit. or fig.); cf. tácn-berend, -bora :—Tâcon *vexillum,* Rtl. 94, 7. Ic slôh grêne tâcne (*Moses' rod ;* Grein suggests *táne*) gársecges deóp, Cd. Th. 195, 23; Exod. 281. Swá swá sigefæst tâcon *veluti victricia signa,* Bd. 1, 8; S. 479, 24. Eal werod gehwyrfedum tâcnum (*versis signis*) fôron, Gl. Prud. 45 a. Hí ásetton tâcna heora *posuerunt signa sua,* Ps. Spl. 73, 6. **Ib.** *a token, a credential :*—Ne hê onfongen sí bûtan biscopes tâcne oþþe gewrite *ne absque commendatitiis litteris sui praesulis suscipiatur,* Bd. 4, 5; S. 572, 43. Ne ðú mê ôðíewest ǽnig tâcen ðe hê him tô sende, Cd. Th. 34, 20; Gen. 540. **Ic.** *a sign, monument :*—Hê hêt brycge gewyrcan his sige tô tâcne ðe hê on ðám síþe þurhteón þohte, Ors. 2, 5; Swt. 84, 4. **Id.** *a sign* of the Zodiac :—Ðonne ðære sunnan ryne beó on ðam tâcne ðe man *virgo* nemneþ, Lchdm. i. 164, 12. Ða twelf tunglena tâcna, iii. 242, 4. **II.** *a sign, distinguishing mark* (lit. or fig.) :—Tâcon *titulus,* Mt. Kmbl. p. 4, 3. Swylc wæs ðæs folces tâcen (*a practice which distinguished them, a distinct feature of their manners*), Andr. Kmbl. 58; An. 29. Hê onfêng torhtum tâcne (*circumcision*), Cd. Th. 143, 6; Gen. 2375. God him sealde tâcen (*posuit Dominus Cain signum*), ðæt nán ðæra ðe hine gemêtte hine ne ofslóge, Gen. 4, 15. **III.** *a sign to attract attention, a signal :*—Ðonne ætȳwþ mannes suna tâcn on heofonan, Mt. Kmbl. 24, 30. Cômon þrý men tô ðære hýde and ðær tâcn slógon (*gave a signal by striking*), Guthl. 11; Gdwin. 54, 24. Tâcen, 12; Gdwin. 58, 23. **IIIa.** *a sign of anything future, a prognostic :*—Ealle ða tâcno and ða forebeácnu ðe úre Drihten ǽr tôweard sægde, ðæt ǽr dômes dæge geweorþan sceoldan, Blickl. Homl. 117, 30. **IIIb.** *a sign, an action that conveys a meaning :*—Ðis sindon ða tâcna ðe mon on mynstre healdan sceal, ðǽr mon swígan haldan wile . . . Ðæs abbudes tâcen is ðæt mon his twêgen fingras tô his heáfde ásette and his feax mid genime, Techm. ii. 118, 1–5, *and often.* Treófugla tuddor tâcnum cýððon eádges eftcyme, Exon. Th. 146, 10; Gû. 707. **IV.** *a sign, indication, mark which shews condition or state :*—Nán tâcen ðerran tôcwýsednesse næs gesewen, Homl. Th. i. 62, 16. Nǽfre wommes tâcn eáwed weorþeþ, Exon. Th. 4, 18; Cri. 54. Ongietan be sumum tâcnum on his hiéremonna môde eal ðæt ðær gehýddes lutige, Past. 21; Swt. 153, 14. Witan ðæra tída tâcnu, Mt. Kmbl. 16, 3. **IVa.** as a medical term, *a symptom :*—Tâcnu ðære ádle, Lchdm. ii. 20, 26. Be tâcnum on roppe, 230, 16. Gif sié ða ceácan áswollen and sió þrotu and ðú ða tácn geseó, 46, 22. **V.** *a sign, symbol, emblem :*—Hwæt wille wê cweþan be ðam andweardan welan, ðe oft cymþ tô ðǽm gôdum, hwæt hê elles sié bûtan tâcn ðæs tôweardan welan, Bt. 39, 11; Fox 230, 12. Healdaþ mínne restedæg, hê ys tácn betwux mê and eów, Ex. 31, 13. Fugles tácen *the symbolical character of the phenix,* Exon. Th. 232, 22; Ph. 510. Ðæt wê ðý geornor ongietan meahten tírfæst tâcen, ðæt se fugel þurh bryne beácnaþ, 236, 14; Ph. 574. **VI.** *a sign which shews the truth or reality of anything, proof, demonstration, evidence :*—Ðæt biþ tâcn wísdômes, ðæt hine mon wilnige gehéran and ongitan, Bt. 38, 2; Fox 198, 22. Ðæt is swíþe sweotol tâcn ðam wîsan, ðæt hê ne sceal lufian tô ungemetlíce ðás woruldgesǽlþa, forðæm hí oft cumaþ tô ðǽm wyrstum monnum, 39, 11; Fox 230, 8. Him ðæt (*the writing on the wall*) tácen weard, ðæt hê ligeword gecwæð, Cd. Th. 260, 31; Dan. 718. Ðæt wæs tâcen sweotol, Beo. Th. 1671; B. 833. Hwæt dêst ðú tô tâcne, ðæt wê gelýfon, Jn. Skt. 6, 30. On ða ylcan tíid ðe hê (*David*) genam his (*Saul's*) spere on his getelde on niht, tô tâcne ðæt hê inne mid him slǽpendum wæs, Ps. Th. 35, arg. : Bd. 1, 1; S. 474, 36 : 2, 6; S. 508, 42 : 4, 28; S. 606, 41 : Blickl. Homl. 7, 15. Wê ðé ðás sǽlác brohton tíres tô tâcne, Beo. Th. 3312; B. 1654. Ic ðæs tâcen wege sweotol on mê selfum, Cd. Th. 54, 31; Gen. 885. Sancte Iohannes mycelnesse se Hǽlend tâcn sægde, *the Saviour shewed by his words the greatness of St. John,* Blickl. Homl. 167, 18. Ðǽr biþ on eádgum ðgesýne þreó tâcen somod, ðæs ðe hí hyra þeódnes wel willan heóldon, Exon. Th. 76, 7; Cri. 1236. Ic wêne ðæt ic ðé hæfde ǽr gereht be manegum tâcnum, ðætte monna sáwla sint undeáþlíce *tu idem es, cui persuasum atque insitum permultis demonstrationibus scio, menteis hominum nullo modo esse mortalius,* Bt. 11, 2; Fox 34, 33 : Elen. Kmbl. 1704; El. 854. **VII.** *a supernatural sign, miracle, prodigy :*—Ðis (*the turning of water into wine*) is ðæt forme tâcn ðe hê on his menniscnysse openlíce geworhte, Homl. Th. i. 58, 14. Ðisse fǽmnan menige weorc gástlícra mægna and monig tâcon heofonlícra wundra gewuniaþ gesǽde beón *hujus virginis multa solent opera virtutum et signa miraculorum narrari,* Bd. 3, 8; S. 531, 28. Hê (*Christ*) sôðra swá feala tâcna gecýðde, ðǽr hié tô sêgon, Andr. Kmbl. 1421; An. 711. Ic (*St. Michael*) gecýþe on eallum ðǽm tâcnum ðe ðǽr gelimpeþ, ðæt ic eom ðære stôwe hyrde, Blickl. Homl. 201, 8. On eallum tâcnum and forebeácnum ðe

God sende þurh hine, Deut. 34, 11. Gif ænig wítega secge tácnu and forebeácnu, 13, 1. Tácna, Homl. Th. i. 44, 24. Noldan hí ða torhtan tácen oncnáwan ðe him beforan fremede freóbearn Godes, Exon. Th. 40, 22; Cri. 642. Gesiáþ werc Dryhtnes ða set[t]e tácen ofer eorðan *videte opera Domini quae posuit prodigia super terram*, Ps. Surt. 45, 9. **VII a.** *a signal event, remarkable circumstance :*—Andsware cýðan tácna gehwylces ðe ic him tó séce *to give me an answer in reference to every remarkable circumstance about which I enquire of them* (cf. mé þinga gehwylc gecýðan, ðe ic him tó séce, 817; El. 409), Elen. Kmbl. 637; El. 319. Wé on gemynd witon ælra tácna gehwylc swá Tróiána þurh gefeoht fremedon, 1286; El. 645. [*Goth.* taikns; *f.*: *O. Sax.* tékan; *n.*: *O. Frs.* téken: *O. H. Ger.* zeihhan *signum, signaculum, nota, titulus, miraculum : Icel.* teikn, tákn *a token, sign, wonder.*] v. andgit-, bell-, fácen-, fore-, friðo-, luf-, sige-, sigor-, sóþ-, weá-, weder-, wer-, wundor-tácn.

tácn-berend, es; *m. A standard-bearer :*—Tácnberend *signifer*, Ælfc. Gr. 8; Zup. 27, 15.

tácn-bora, an; *m.* **I.** *a standard-bearer :*—Tácnbora *signifer*, *vexillifer*, Wrt. Voc. i. 35, 10: *signifer*, 84, 16. Tácenbora, Hymn. Surt. 113, 3. Tácnboran *draconarii* vel *vexillarii* vel *signiferi*, Wrt. Voc. i. 21, 66. **II.** *a leader, guide, director :*—Ðis is mín tácenbora ðe mé getǽhte ðæt ic tó ðé becom (*the word is used of the old fisherman who had directed Apollonius to the town,* v. p. 12), Ap. Th. 27, 22.

tácn-circul, es; *m. A circle or cycle which marks the date.* **I.** *the indiction, a cycle of fifteen years.* v. ge-ban :—Ðæm gǽre ðe wæs ágán fram Cristes ácennednesse eahta hund wintra and feówer and sixtig, and in ðam tácencircole ðæt twelfte geár (*the year of the indiction is the remainder after dividing* 864 + 3 *by* 15; *this remainder is* 12, *which agrees with the passage*), Chart. Th. 126, 3. **II.** *the lunar cycle of nineteen years; the place which any year occupies in the cycle is marked by the golden number of the year :*—Ðis wæs gewriten on ðam geáre ðe wæs ágán fram Cristes ácennednysse án þusend geára and án and sixtig geára, and on ðam tácncircule ðæt seofanteóðe geár (*the golden number of the year* 1061 *is the remainder after dividing* 1061 + 1 *by* 19; *this remainder is* 17, *which agrees with the number given in the passage*), Chart. Th. 390, 19.

tácnian; *p.* ode. **I.** *to make a mark upon something, to mark :*—Seó líget ðæt deófol bærneþ and tácnaþ, Salm. Kmbl. p. 148, 4. **II.** *to be a token or mark of something, to indicate, mark :*—Se steorra ðe wé hátaþ æfensteorra, ðonne hé biþ west gesewen, ðonne tácnnaþ hé æfen, Bt. 39, 13; Fox 232, 34. Ðysne dæg hié nemdon siges dæg; se nama tácnaþ ðone sige ðe Drihten wiþstód deófle, Blickl. Homl. 67, 14. Tácnendi *index*, Wrt. Voc. ii. 111, 40. **III.** *to indicate, point out :*—Hé þurh his láre éces lífes wegas sægde and tácnode, Blickl. Homl. 129, 18. **IV.** *to signify,* (a) *to express a meaning by means of figure or symbol, to express figuratively or symbolically :*—Hálige gewreotu ús tácniaþ ðás world þurh ðone mónan, Blickl. Homl. 17, 21. Hé bær him æcse and adesan on handa, tácnode (*signabat*) on ðam, ðæt hé tó gewinne on ðæt mynster eode, Bd. 4, 3; S. 567, 27. Tácnade Leoniða, hwelc moncwealm on Créca londe wæs, mid ðæm ðe hé sprecende wæs tó his geférum: 'Uton brúcan ðisses undernmetes swá ða sculon ðe hiora æfengíð on helle gefeccean sculon,' Ors. 2, 5; Swt. 84, 31. Ðæt hé sǽde and tácnode hwylcum deáðe hé wolde sweltan *hoc dicebat significans qua morte esset moriturus*, Jn. Skt. 12, 33: 21, 19. (b) *to be the figurative expression of, be a figure of something, to symbolize :* —Hwæt tácnaþ ðæt gold búton ða heánesse ðæs háligdómes *quid auro nisi excellentia sanctitatis exprimitur?* Past. 18; Swt. 133, 12. Hwæt tácnaþ Ezechhiel búton ða láreówas *cujus Ezechiel nisi magistrorum speciem tenet?* 21; Swt. 161, 8: Blickl. Homl. 79, 29: 17, 14. Cwæþ se godspellere ðæt leóht cyrde tó ðon blindan. Ðæt tácnaþ ðæt seó godcundnes onféng úre týdran gecynde, 17, 27. Hé cwæþ ðæt his þegnas dydon swá he him bebeád. Ðæt tácnaþ ðæt ðás láreówas ne sceolan Godes dómas náwþer ne ná wanian ne ne écan, 81, 3. Ðæt sweflene fýr tácnade hwelc gewinn ðe ðæm ðe nú sindon, Ors. 2, 6; Swt. 88, 30. **V.** *to indicate* what is future, *to portend :* —Hí (*two stars*) wítegan wǽron grimmes wæles . . . ðæt hí micel yfel mannum tóweard tácnedon (*signarent*), Bd. 5, 23; S. 645, 28. Bécnunge, tácniende *portendentes*, Wrt. Voc. ii. 66, 11. [þes fuȝel tacnede faie síð þes kinges, Laym. 2832. Tacnenn *to express symbolically*, Orm. 1639. Ðe blo tokeneð ðe waters wo, Gen. and Ex. 638. Toknyn or make tokene *signo*, Prompt. Parv. 495. *Goth.* taiknjan δεικνύναι: *O. H. Ger.* zeihhanón, zeihhanen *significare, indicare, monstrare: Icel.* tákna, teikna *to betoken,* mark, *denote.*] v. fore-, ge-tácnian; tácnan, tácnian.

tácnung, e; *f. Signification :*—Tácnunga *significationem*, Ps. Spl. 59, 4. **I.** *an indication, sign, characteristic mark, symptom :*—Lǽcedómas and tácnung on ðam roppe (cf. be tácnum on ðam roppe, 230, 16), Lchdm. ii. 164, 5. Be lyfte tácnungum *de aeris indiciis*, Nar. 3, 14. Hit nú is búton swylcum tácnungum ðæs yfeles ðe hit ǽr dyde *Aetna nunc tantum innoxia specie ad praeteritorum fidem fumat*, Ors. 2, 6; Swt. 90, 3. **II.** *an indication, evidence, proof :*—Wæs ðæs godcundan wundres sweotol tacnung (*indicium*), ðæt ðære fǽmnan líchoma bebyriged brosnian ne mihte, Bd. 4, 19; S. 587, 35. Ða hé mé in

tácnunge his lufan bebeád *quos mihi in indicium suae dilectionis commendaverat*, 2, 6; S. 508, 18. Gewuniaþ tó tácnuncge his mægenes gelómlíce wundor hǽlo geworden beón *ad indicium virtutis illius solent crebra sanitatum miracula operari*, 4, 3; S. 570, 9. **III.** *an indication of what is future, a presage, prognostic :*—Is seó stów nemned Heofenfeld wæs heó geára swá nemned for tácnunge ðæra tóweardra wundra vocatur *locus ille Heofenfelth, quod certo utique praesagio futurorum antiquitus nomen accepit*, Bd. 3, 2; S. 524, 34. Tó hwæm cumaþ hí elles búton tó tácnunge sorges and ánfealdes sáres *quid est aliud, quam futurae quoddam calamitatis indicium?* Bt. 7, 2; Fox 18, 21. **IV.** *a figurative representation, an emblem :*—Hwæt syndon ða woruldsǽlþa óþres búton deáþes tácnung? for ðam se deáþ ne cymþ tó nánum óþrum þingum búton ðæt hé ðæt líf áfyrre; swá eálc ða worulds`ǽlþa cumaþ tó ðam móde tó ðam ðæt hí hit beniman ðæs ðe him leófast biþ ðisse worulde, Bt. 8; Fox 26, 6. **V.** *direction, ordering :*—Ðás feówer heáfodríca wundon on feówer endum ðyses middangeardes mid unásecgendlícre Godes tácnunge *eadem ineffabili ordinatione per quatuor mundi cardines quatuor regnorum principatus fuerunt*, Ors. 2, 1; Swt. 60, 1. [Ða wes he awundred, what weore þis tacninge (*portent*), Laym. 15974. He tolde heom þa tacni[n]ge (*prophetic notice given in a dream*), 32126. Sette he up ðat ston for muniging And get on olige for tokning (*sign*; cf. Iacob lapidem erexit in titulum, fundens oleum desuper, Gen. 28, 18), Gen. and Ex. 1624. Ich wat al of þe tacninge (*signification*), O. and N. 1213. *O. H. Ger.* zeihhanunga *significatio, descriptio.*] v. ge-tácnung; tácning.

tácor (-ur), es; *m. A husband's brother, brother-in-law :*—Tacor (-ur) *levir*, Txts. 74, 598. Tácor, Ælfc. Gr. 8; Zup. 27, 20: *levir*, i. *frater mariti*, Wrt. Voc. i. 52, 31. Tácor, ðæt is brýdguma[n] bróðor *levirum*, ii. 84, 16. Tácor, 50, 30: Hpt. Gl. 498, 75. [*O. H. Ger.* zeihhor (-ir, -ur) *levir, frater mariti.*]

tádige, tádie, an; *f. A toad :*—Tádige *buffo*, Wrt. Voc. i. 24, 21. Tádie *rubeta*, 78, 57; Ælfc. Gr. 9, 3; Zup. 35, 3. [Tadde [*ru*]beta, Wrt. Voc. i. 91, 17. Liggeþ alse þe tadde deð in þere eorðe, O. E. Homl. i. 53, 14.]

tæbere (?) *some implement used in weaving :*—Tæbere *claus* (the word occurs in a list *de arte textoria* ; but in an almost identical list, p. 282, the form is *teltre*. v. teld-treów), Wrt. Voc. i. 66, 27. [Cf. (?) syl *taber*, Wrt. Voc. i. 289, 48, and *claus, lignum textorii* vel telde, ii. 131, 56.]

tǽcan; *p.* tǽhte *To shew.* **I.** *to offer to view, present :*—Tǽhte hé ðá ðam pápan sumne munuc ðæs nama wæs Andreas *cum monachum quemdam, nomine Andream, pontifici offerret*, Bd. 4, 1; S. 564, 4. Se ðe hæfþ .xx. hída, se sceal tǽcan .xii. hída gesettes londes, ðonne he faran wille. Se ðe hæfþ .x. hída, se sceal tǽcan, .vi. hída . . . Se ðe hæbbe þreó hída tǽce óðres healfes, L. In. 64–66; Th. i. 144, 5–11 MS. B. **II.** *to shew* an object to a person so that the object may be attained by the person, *to shew* a way, a place, etc. (1) lit. :—Ic tǽce sumum men his weg *dirigo*, Ælfc. Gr. 28, 5; Zup. 173, 8. Tǽceþ ús se torhta trumlícne hám, Cd. Th. 282, 29; Sat. 294. Him mon setl tǽhte and hé sæt æt ðam symble *he was shewn a seat, and sat at the feast*, Bd. 3, 10; S. 534, 28: 5, 19; S. 639, 35. Him freá tǽhte wegas ofer wésten, Cd. Th. 174, 5; Gen. 2873. Gewát him tó ðæs gemearces ðe him Metod tǽhte, 174, 29; Gen. 2885. Ðæs embe twá niht ðætte tǽhte God Elenan eádigre æþelust beáma, Menol. Fox 164; Men. 84: Elen. Kmbl. 1259; El. 631. (1 a) *without an object, to shew the way, direct :*—On niht hé tǽhte eów þurh fýr *nocte ostendens vobis iter per ignem*, Deut. 1, 33. (2) fig. :—Híg bugon eáde of ðam wege ðe ðú him tǽhtest *recesserunt cito de via, quam ostendisti eis*, Ex. 32, 8. Ða men ðe bearu habban him tǽcan hié lífes weg and rihtne gang tó heofenum, Blickl. Homl. 109, 17. (2 a) *without an object, to direct :*—Hwá tǽcþ ús teala and hwá sylþ ús ða gód ðe ús man gehǽt *quis ostendit nobis bona?* Ps. Th. 4, 7. **III.** *to shew* a person (*dat.* or *acc.*) the direction that must be taken, *to direct, to cause a certain direction to be taken*, the direction being marked by a preposition. (1) lit. :—On ðære stówe ðe him se stranga tó wordum tǽhte *on the place to which the Lord had directed him to go* (cf. 172, 24–; Gen. 2849–), Cd. Th. 175, 24; Gen. 2900. Nán man ne tǽce his getihtledan man fram him *let no one send his accused man away*, L. Ath. i. 22; Th. i. 210, 23: L. C. S. 28; Th. i. 392, 11. Tǽce him mon siððan tó nigcumenra manna húse, R. Ben. 97, 11. (2) fig. :—Niman hí ðone teóðan dǽl tó ðam mynstre and tǽcan him tó ðam nigoðan dǽle and tódǽle man ða eahta dǽlas on twá *let them take the tithe for the minster, let the next tenth fall to his share* (*let him be directed to take the next tenth*), *and let the remaining eight tenths be divided in two*, L. Edg. 3; Th. i. 264, 2. Ðú, fæder Augustinus, híg hæfst on ðínum bócum gesǽd, and ic gehwam wille ðærtó tǽcan ðe hiene his lyst má tó witanne *I will refer every one to the books, who desires to know more*, Ors. 3, 3; Swt. 102, 25. (2 a) where the dat. is omitted :—Seó ealde ǽ næs swá stíð on ðam þingum swá swá Cristes godspel is and tǽcþ tó ánum wífe *points to, directs a man to take, one wife*, Scrd. 22, 25. **IV.** *to shew* the course that must be followed, what should be observed, *to direct, appoint, prescribe, enjoin.* v. tǽcend :—Ðú tǽcst folce gemǽro ábútan ðone munt (*constitues terminos populo in circuitum*) and cwist:

'Warniaþ ðæt gē ne cumon tō nēh ðison munte,' Ex. 19, 12. Symle ðū tæhtest mildheortnesse, and ðæt man ōðrum miltsode, Homl. Th. i. 68, 23. Crist tæhte: 'Syllaþ ōðrum būtan ceápe,' Homl. Th. i. 412, 12. Eft hē him tæhte tō fultome ðæt hē him genáme áne íserne hearste-pannan *ei ad munitionem suam protinus subinfertus:* '*Et tu sume tibi sartaginem ferream,*' Past. 21; Swt. 161, 6. Hig didon hine on cweart-ern, ōð hig wiste, hwæt Drihten be him tæhte (*quid juberet Dominus*), Lev. 24, 12. Hē hine ælces þinges geclænsode, swā se pápa him tæhte *in the manner prescribed by the pope,* Chr. 1022; Erl. 161, 38. Ðā tæhte man hyre ðæt hió sciolde bringan his fæder gold *the court directed that she was to bring his father's gold,* Chart. Th. 289, 34. Ðæt hē him dǽdbōte tǽce *ut sibi poenitentiam praescribat,* L. Ecg. C. proem.; Th. ii. 130, 35. Ne sig nán ðing forlǽten ðæs ðe se regol tǽce on his fandunge, R. Ben. 104, 17. Bēte hē swā micel swā dēman tǽcan *quantum arbitri judicaverint,* Ex. 21, 22. Ðæt hȳ bētan swā swā bēc tǽcan, Wulfst. 165, 9. **V.** *to shew* to the mind by way of in-struction or of proof, *to teach.* (1) of persons:—Se Hálga Gást ðe tǽhþ rihtwísnysse, Homl. Th. i. 322, 5. Ǽfre se ðe áwent oþþe se þe tǽcþ of Lēdene on Englisc ǽfre hē sceal gefadian hit swā ðæt ðæt Englisc hæbbe his ágene wīsan *he that makes a translation from Latin into English, or he that in teaching turns Latin into English must use idiomatic English,* Ælfc. Gen. Thw. 4, 9. Ic ðē bebéode ðæt ðū ne forgite ðæt ðæt ic ǽr tǽhte . . . Ic ðē tǽhte ðætte ðæt wǽre ðæt hēhste gōd *maneant quae paullo ante conclusa sunt . . . nonne mon-stravimus ea vera bona non esse,* Bt. 34, 9; Fox 146, 13-19. Tǽc mē ðínne willan tō wyrcenne, 42; Fox 260, 11. Ic ðē mæg tǽcan ōþer ðing, 38, 3; Fox 198, 29. Ða mæssepreóstas sceolan heora scriftbēc mid rihte tǽcan and lǽran. Ða láreówas sceolan synnfullum mannum eádmōdlíce tǽcan and lǽran, ðæt hié heora synna cunnon onrihtlíce geandettan, Blickl. Homl. 43, 7-16. .xii. lahmenn scylon riht tǽcean Wealan and Ænglan . . . Ðolien ealles ðæs hý ágon, gif hí wōh tǽcen, L. O. D. 5; Th. i. 354, 9-11. Gif hwylc gódra wile his lytlingas hiom tō láre befæstan, hig sceolon him ēstlíce tǽcan, L. E. I. 20; Th. ii. 414, 10. Hē wile mōdum tǽcan, Cd. Th. 211, 17; Exod. 527. Hē wæs tǽcende dæghwomlíce binnan ðam temple, Homl. Th. i. 412, 29. (2) of things:—Seó emniht is swā swā wē ǽr cwǽdon on .xxiᵐᵃ. kl. April., swā swā ða geleáfullan rǽderas hit gesetton, and eác gewisse dægmǽl ūs swā tǽcaþ, Lchdm. iii. 256, 22. **VI.** *to shew, indicate, signify:*—Tǽhte *significat,* Jn. Skt. p. 8, 12: 21, 19: *indicaret,* Lk. Skt. p. 2, 14. Gif ðū hwæt be capitelhūse tǽcan wylle, Techm. ii. 122, 4: 118, 8, 17: 129, 3. v. be-, ge-, mis-tǽcan.

tǽcend, es; *m. One who prescribes or orders.* v. tǽcan, **IV:**—Gif hwylcum breþer hwæt hefelíces beboden sý underfō hē ða geboda his tǽcendes *si cui fratri aliqua gravia injunguntur, suscipiat jubentis imperium,* R. Ben. 128, 11.

tǽcing, e; *f.* **I.** *the pointing out of a course to be followed, direction, teaching.* v. tǽcan, **IV,** and previous word:—Hēr is seó ǽ, ðe ðū under hire tǽcinge winnan wylt, R. Ben. 96, 23. Sý him þreál geboden be regoles tǽcinge, 126, 4. Hē nolde nán ðing dōn be ðæs deófles tǽcunge, Homl. Th. i. 168, 26. Gif hē be bōca tǽcinge his líf gefadige, L. Eth. ix. 28; Th. i. 346, 17. Gif hwá nelle bētan æfter mínra biscopa tǽcinge, Chart. Erl. 230, 22. Gode þeówian æfter Sanctus Benedictus tǽcinge *according to the rule of St. Benedict,* Chart. Th. 549, 8: 227, 24: Lchdm. iii. 438, 20. Underfō hē ælcne regoles þeáw and tǽcinge; sig hē æfter Cristes bōce tǽcinge ðus geáxod, R. Ben. 104, 19. Þurh háligra bōca tǽcunge ūres Drihtnes willan mid gōdum dǽdum gefyllan, Homl. Ass. 144, 2. **II.** *teaching, doctrine:*—Swā ðæt wē þurhwunigen on Cristes láre and tǽcinge. R. Ben. 6, 1. *x* ána ongynþ of ðam stæfe í æfter ūðwitena tǽcinge, Ælfc. Gr. 2; Zup. 6, 5. Ðæra sind feówer æfter Priscianes tǽcinge, 24; Zup. 129, 16.

tǽcnan; *p.* [e]de. **I.** *to shew, present:*—Se ðe hæfþ .xx. hída se sceal tǽcnan (tǽcan, MS. B.) .xii. hída gesettes landes ðonne hē faran wille. Se ðe hæfþ .x. hída se sceal tǽcnan (tǽcan, MS. B.) .vi. hída gesettes landes. Se ðe hæbbe þreó hída tǽcne (tǽce, MS. B.) ōðres healfes, L. In. 64-66; Th. i. 144, 5-11. **II.** *to shew the road, point out* an object, *make known:*—Se him wægas tǽcneþ, Exon. Th. 434, 26; Rä. 52, 7. Tǽcne *indicet,* Jn. Skt. Lind. 11, 57. Taecnaendi (-endi) *index,* Txts. 70, 544. **III.** *to appoint, prescribe:*—Se mec wrǽde on legde, ðæt ic onbūgan ne mōt of ðæs gewealde, ðe mē wegas tǽcneþ, Exon. Th. 383, 26; Rä. 4, 16. v. tácnian, tǽcnian, tǽcan, tǽcnend.

-tǽcne. v. earfoþ-tǽcne.

tǽcnend, es; *m. One that shews* or *points out:*—Tǽcne[n]d *index,* Wrt. Voc. ii. 47, 74.

tǽcnian; *p.* ode *To shew, prove:*—Forðam ūs segþ ælc gesceádwísnes and ealle men ðæt ilce andettaþ ðæt God sié ðæt hēhste gōd forðam ðe hí tǽcniaþ ðæt eall gōd on him sý *ita vero bonum esse Deum ratio demonstrat, ut perfectum quoque in eo bonum esse convincat,* Bt. 34, 2; Fox 136, 6. v. tácnian, *and next word.*

tǽcning, e; *f. Shewing, proof:*—Ða cwæþ hē: 'Ic hit ðē ðonne wille getǽcan; ac ðæt án ic ðē bebeóde ðæt ðū þeáh for ðære tǽcninge ne forgite ðæt ic ǽr tǽhte' *atqui hoc verissima, inquit, ratione pate-*

faciam, maneant modo quae paullo ante conclusa sunt, Bt. 34, 9; Fox 146, 14. v. tácnung.

tæfl, e; *f.:* es; *n.* (?): tæfle, an (?); *f.* Properly *a board for the play-ing of a game.* But the word seems also used of *a game played on such a board:* cf. the use of the word *tables* at a later time:—Wyþ pleyynge at tables oþer atte chekere, R. Glouc. 192, 3. Kueade genienes of des and of tables huer me playþ uor pans, Ayenb. 45, 16. Tabulles *tabella* (15th cent.), Wrt. Voc. i. 202, col. 2. See also Strutt's Sports, Bk. iv, c. 2. The word seems to denote also *a die used in playing a game.* What was the precise nature of the games, to which this word and related forms are applied, does not appear; some of the references below would imply that games of chance are meant, and this would be in keep-ing with the love of gaming which Tacitus, Germ. c. 24, noticed among the Germans. But games of skill like chess may sometimes be meant. In Icelandic *tafl* is used of chess or draughts, as well as of dicing, and the Danes in England seem to have played chess (see Thrupp's Anglo-Saxon Home, c. xvi, sec. 7); and in O. H. Ger. *scah-zabel* = scacarium. Among the Welsh, too, was a game something like draughts, called *tawl-bwrdd* (Thrupp, p. 388):—Tæfil, tebl, teblae *alea,* Txts. 36, 6. Tæfl, Wrt. Voc. ii. 8, 7. *Incipit de alea.* Tæfl *alea,* ic tæfle tæflum *cotizo tesseris,* i. 284, 28, 31. Tæfel, 66, 47. Tæfel *alea,* cynningstán on tæfle *pirgus* (cf. O. H. Ger. zabel-bret *pirgus*), feðerscíte tæfel *tessere vel lepus-culae,* 39, 45-49. Tæslum *tesellum* (= tæflum *tessellis?* v. Wülck. Gl. 526, 5), ii. 93, 44. Dryhten dǽleþ sumum tæfle cræft, bleóbordes gebregd, Exon. Th. 331, 19; Vy. 70. Sum biþ hræd tæfle, sum biþ gewittig æt wínþege, 297, 25; Crä. 73. Hȳ twegen sceolon tæfle ymbsittan . . . habban him gomen on borde, 345, 2; Gn. Ex. 182. [Sum men pleoden on tæuelbrede (mid tauel, 2nd MS.), Laym. 8133. O. H. Ger. zabel; *n. alea,* wurf-zabel *alea, tessera:* Icel. tafl; *n. a game;* tafla *a piece used in a game.*] See the following words.

tæflan, tæflian; *p.* [e]de, ode *To gamble, game:*—Ic tæfle tæflum *cotizo tesseris,* Wrt. Voc. i. 289, 31. Ic tæfle *cotizo,* Wrt. Voc. ii. 16, 63. Tebliþ, tebleþ *cotizat,* Txts. 46, 178. Tæflaþ, Wrt. Voc. ii. 135, 36. [þe manne þat taueleþ and forleost þat game, O. and N. 1666. *Else-where the word means* to talk, argue:—Ich leote ham talkin and tauelin of godlec, Marh. 13, 31. Nefde hare nan tunge to tauelin (teuelin, MS. C.) a tint wið, Kath. 1247. Teuele he wið me, 820. *Icel.* tefla *to play at draughts, dice,* etc.]

tæfle (?); *adj. Given to play:*—Hond tæfles monnes *the hand of the gamester,* Exon. Th. 345, 8; Gn. Ex. 185.

tæflere, es; *m. A gamester, dicer, gambler:*—Teblere, teblheri *aleator, aleo,* Txts. 36, 7. Tæflere *aleator,* Wrt. Voc. ii. 8, 8: i. 66, 49: 284, 30. Wē lǽraþ, ðæt preóst ne beó hunta, ne hafecere, ne tæflere, ac plege on his bócum, swā his háde gebiraþ, L. Edg. C. 64; Th. ii. 258, 8. [*M. H. Ger.* zabelære *aleo.*]

tæfl-stán, es; *m. A die* or *a piece in a game* (tæfl):—Teblstán (tebel-) *calculus,* Txts. 47, 349. Tæflstán *calculus* (in a list 'de alea'), Wrt. Voc. i. 284, 29. Tæfelstán, 66, 48. Tæfelstánas *aleae,* 39, 46.

-tǽfran, tǽg. v. á-tǽfran, teáh.

tægl, es; *m. A tail:*—Oxan tægl biþ scitt. weorð, L. In. 59; Th. i. 140, 3. Foxes tægles se ýtemesta dǽl, Lchdm. i. 340, 22. Se ðrowend slihþ mid ðam tægle tō deáðe, Homl. Th. i. 252, 5, 10, 12. Ða beón beraþ ætterne tægel, Frag. Kmbl. 37; Leás. 20. Hí habbaþ tæglas ðam wyrmum gelíce ðe men hátaþ þrowend, Wulfst. 200, 14. [*Goth.* tagl; *n. hair:* O. H. Ger. zagel; *m. a tail:* Icel. tagl; *n. a (horse's) tail:* Norweg. tagl *horse-hair:* Swed. tagel *hair of mane* or *tail.*] v. cū-tægl.

tægl dye. v. telg.

tægl-hǽr, es; *n. A hair of an animal's tail:*—Gif ðū hafast mid ðē wulfes hrycghǽr and tæglhǽr ða ýtemæstan on síðfæte, būtan fyrhtu ðū ðone síð gefremest, ac se wulf sorgaþ ymbe his síð, Lchdm. i. 360, 21.

tæher, tæherende. v. teár, teárian.

tǽl, tel, es; *n. A tale, number, series:*—Heora tel biþ swā menigfeald, ðæt hit oferstíhþ sandceosles gerím, Homl. Th. i. 536, 33. Ðæra etendra tal *manducantium numerus,* Mt. Kmbl. Lind. 14, 21. Of tale *numero,* Jn. Skt. Rush. 6, 10. Tele *laterculo, numero,* Hpt. Gl. 442, 51. In tēnum talum *in decem numeros,* Mt. Kmbl. p. 3, 1. Cf. Forerim ł (fore-)tal *prologus,* p. 1, 1. [Hundred is ful tel, A. R. 372, 9. *O. Sax.* gēr-tal: *Icel.* tal; *n. a number, series.*] v. ge-, ofer-tæl; tæl-cræft, -mearc, -met; talu.

tǽl, e; *f.* (?) *Evil speaking, calumny, detraction:*—Tǽl *blasphemia, vituperatio,* Wrt. Voc. ii. 127, 9: *detractatio, vituperatio,* 139, 44. 'Ǽlc tǽl sié ánumen fram eów.' . . . Hit biþ unnyt ðæt mon tǽl ūtane forlǽte gif se yfela willa ðone onwald hæfþ ðæs ingeðonces '*omnis blasphemia tollatur a vobis.*' . . . *Frustra blasphemia ab exterioribus tollitur, si in interioribus malitia dominatur,* Past. 33; Swt. 222, 8-14. Ne frúne ic ðē for tǽle ne þurh teóncwide *I do not question you that I may detract or abuse,* An. 633. Hē þolaþ sárcwide secga . . . Ic bí mē secge ðis sárspell . . . Ic for tǽle ne mæg ǽnigne moncynnes ge-lufian, Exon. Th. 458, 1-26; Hy. 4, 93-106. Ðæt heó mec tǽle gerahte (-rǽhte? cf. ðæt hē ða hálgan weras hospe gerahte (-rǽhte?) *he calum-*

niated, 260, 21 ; Jul. 300) hēt mē fremdne god ofer ða ōþre ðe wē ǽr cūþon weorþian *that she attacked me with blasphemy, bade me honour a strange god above the others that we knew before*, 247, 4 ; Jul. 73. v. tāl.

-tǽl. v. leóf-tǽl.

tǽlan ; *p.* de. **I.** *to blame, rebuke, reprove, reproach, censure, accuse.* (1) *to blame* a person for what is wrong :—Ne ðreáþ ūs nān monn ne furðum ǽne worde ne tǽlþ *ne verbi quidem ab aliquo invectione laceramur*, Past. 17 ; Swt. 117, 22. Tǽlaþ ðegnas *accusant (pharisaei) discipulos*, Mk. Skt. p. 3, 14. Ðū mē tǽldesð and ðū mē cīddesð *me reprehendis*, Past. proem. ; Swt. 23, 10. Ða scamleásan Galatas suíðe openlíce Paulus tǽlde (*increpat*), 31 ; Swt. 207, 14. Hē lǽrde and tǽlde ealle men ðe worulde welan gaderiaþ mid unrihte, Ps. Th. 38, arg. Hí tǽldon hí *vituperaverunt*, Mk. Skt. 7, 2. Ðætte hiǽ tǽldun (*accusarent*) hine, Mk. Skt. Rush. 3, 2. Ðæt hié ongieten ðæt hié mon tǽle *that they may know that they are censured*, Past. 21 ; Swt. 151, 14. Se ðe ōðerne tǽlan wille, ðonne gange hē ǽrest on dígle stōwe and besceáwige hine sylfne, Wulfst. 233, 20. (2) *to blame* what is wrong in a person :—Ne tǽle ic nā micel weorc ne ryhtne onwald ac ic tǽle ðæt hine mon forðý up āhebbe on his mōde *non potestatem reprehendimus*, Past. 4 ; Swt. 41, 2-3. Ðonne gē eów selfum ondrǽdaþ ðæt ðæt gē on ōðrum tǽlaþ *dum sibi, quod increpat, timet*, 21 ; Swt. 159, 16. Hē tǽlde (*exprobravit*) hyra ungeleáffulnesse, Mk. Skt. 16, 14. Ða bóceras ðæt tǽldon, Homl. Th. i. 338, 20. Gif hē gesceádelíce hwilcu þing tǽle *si qua rationabiliter reprehenderit*, R. Ben. 109, 9. Leahtras tǽlan, 135, 18. Ðæt ðæt him mon on tǽlan wille *quod in eis reprehenditur*, Past. 31 ; Swt. 206, 6. Unþeáwas tǽlan and góde herian, Bt. 38, 3 ; Fox 200, 7 : Met. 19, 39. Tō tǽlenne, Bt. 27, 4 ; Fox 100, 19. **II.** *to speak evil of, blaspheme, revile, slander, calumniate, backbite :*—Eorl ðeórne mid teónwordum tǽleþ behindan, spreceþ fǽgere beforan, Fragm. Kmbl. 7 ; Leás. 4. Tǽleþ *blasveniat*, Wrt. Voc. ii. 73, 21. Ðis weorc heora ðe tǽlaþ (tēlaþ, Ps. Surt.) mē *þe werke of þa þat bacbite me* (Ps. 108, 20), Ps. Spl. 108, 19. For ðara stemne ðe mē hyspaþ and tǽlaþ *a voce exprobrantis et obloquentis*, Ps. Th. 43, 18. Of ðæm cristendóme ðe hié nū swíþost tǽlaþ, Ors. 2, 1 ; Swt. 64, 19. Ðū sǽte ongeán ðinne brōþor and tǽldest (tēldes, Ps. Surt., *detrahebas*) hine, Ps. Th. 49, 21. Hē his godu tǽlde, Exon. Th. 278, 16 ; Jul. 598. Hí tǽldon (tēldon, Ps. Surt. *detrahebant*) mē *me bakbate þai* (Ps. 108, 4), Ps. Spl. 108, 3. Hí mē tǽldon *exprobaverunt animam meam*, Ps. Th. 34, 8. Hig tǽldon ðæt land mid heora teónwordum *they brought up an evil report of the land* (A. V.), Num. 13, 33. Forðan ðe hig ðæt land tǽldon *by bringing up a slander upon the land* (A. V.), 14, 36. Ne hine ne tǽl, ne ne ter mid wordum, Basil admn. 5 ; Norm. 46, 11. Ne tǽl ðū ðinne Dryhten *thou shalt not revile the God*, L. Alf. 37 ; Th. i. 52, 29 : Ex. 22, 28. Þeora cynna syndon morþras ; ðæt is ðæt ǽrest, ðæt man tō ōþrum lǽþþe hæbbe, and hine hatige, and tǽle behindan him sylfum ; forðon seó synn biþ swíþe mycel, ðæt man ōþerne hatige and tǽle, Blickl. Homl. 65, 1-2. Tǽlan *carpere*, Wrt. Voc. ii. 19, 23 : 90, 11. Underlōh mē nū behreówsiendne, ðone ðe ðū ōð ðis audigendne and tǽlendne forbǽre, Homl. Th. ii. 418, 10. Tǽlendne wið ðǽm nēstan his dégullíce *dernlike his neghburgh bakbitand* (Ps.), Ps. Surt. 100, 5. Gebiddaþ for eówre ehteras and tǽlendum eów (*calumniantibus vos*), Mt. Kmbl. 5, 44. **III.** *to treat with contempt, to scorn, despise, insult, mock, deride, jeer at :*—Se stunta tǽlþ (*inridet*) lāre, Scint. 113, 18. Ðæt fǽsten tǽlþ God, Homl. Th. i. 180, 10. Se ðe tēleþ (*spernit*) mec, Jn. Skt. Rush. 12, 48. Tēld *deridet*, Kent. Gl. 718. Ða unrihtwísan tǽlaþ (cf. habbaþ on hospe, Met. 4. 44) ða rihtwísan, Bt. 4 ; Fox 8, 15. Tǽlde hē Rōmáne and hié swíþe bismrade mid his wordum *Romam infami satis notavit elogio*, Ors. 5, 7 ; Swt. 228, 19. Tēlde (*sprevit*) hine Herōdes, Lk. Skt. Lind. Rush. 23, 11. Tǽldon *sugillent*, Wrt. Voc. ii. 92, 18. Ðá tǽldon hí hine *inridebant eum*, Mk. Skt. 5, 40. Tǽldon *deridebant*, Lk. Skt. 8, 53. Hié hine on ðǽm tǽldon and bismrodan, ðæt hé his swā ánfealdne gegyrelan tōsnīðan sceolde, Blickl. Homl. 215, 9. Ealle ágynnaþ hine tǽlan (*inludere ei*), Lk. Skt. 14, 29. Sellas hine hǽdnum tō tǽlenne (*ad deludendum*), Mt. Kmbl. Lind. 20, 19. Hēhsacerdas tēlende (*ludentes*) cuoedon, Mk. Skt. Lind. Rush. 15, 31. Tǽlende *cavillantes*, Wrt. Voc. ii. 3, 60. Tēled is *calcatur*, 18, 48 : 83, 46 : *detractatur*, Kent. Gl. 924. Hē biþ tǽled fram swylcum mannum swylce ðǽre wyrte mihta cunnun *he is laughed at by such men as know the virtues of the plant*, Lchdm. i. 164, 6. [He is cnihtscipe tǽlden *they blamed his want of manhood*, Laym. 3801. Tælen *to reproach*, 3334. Giff mann wollde tælenn þatt (*reprove the sin*), Orm. 2033. Swuch he may telen of golnesse, O. and N. 1415. Icel. tæla *to delude, mock*.] v. be-, ge-tǽlan ; tǽlende, un-tǽled.

tǽl-cræft, es ; *m. Arithmetic :*—Mǽg geseón ǽlc man ðe telcræftas ǽnig gesceád can (*that knows anything of arithmetic*), ðæt hit māre is ðonne þreó hund geára syððan ðyllic feoh wæs farende on eorðan, Homl. Skt. i. 23, 699. v. getel-cræft ; rím-cræft.

-tǽle. v. leóf-, un-tǽle.

tǽlend, es ; *m.* **I.** *a reprover :*—Ðǽm tēlendum *reprehensoribus*, Mk. Skt. p. 2, 17. **II.** *a slanderer, backbiter, detractor :*—Swíþe seldon ǽnig man wile beón andetta, ðæt hē ǽfestig sý oððe tǽlend, Blickl. Homl. 65, 4. Ðone tǽlend *detrahentem*, Ps. Lamb. 100, 5. Mid

tēlendum *cum detractoribus*, Kent. Gl. 938. **III.** *a scorner, mocker, derider :*—Sēcþ tǽlend (*derisor*) wísdōm . . . gearwe synd tǽlendum (*derisoribus*) dōmas, Scint. 171, 13-14. Nelle ðū þreágean tǽlend (*derisorem*), 113, 12. Tēlend, Kent. Gl. 289.

tǽlende ; *adj.* (*ptcpl.*) **I.** *prone to blame, censorious :*—Ne beó hē tō tǽlende, L. E. l. 21 ; Th. ii. 416, 17 : Exon. Th. 305, 18 ; Fä. 90. Cf. Uton beorgan ūs wið tǽlnysse and wið twysprǽcnysse *caveamus nobis a vituperatione et a biloquio*, L. Ecg. P. iv. 66 ; Th. ii. 226, 31. **II.** *slanderous, backbiting :*—Ða ǽfstigan men and ða tǽlendan, Blickl. Homl. 65, 10.

tǽlere, es ; *m. A scorner, scoffer, mocker :*—Tēlerum *derisoribus*, Kent. Gl. 721.

tǽlg. v. telg.

tǽl-hleahtor, es ; *m. Scornful laughter, derision :*—Tǽlhlehter *derisio*, Wrt. Voc. i. 51, 4.

tǽling, e ; *f.* **I.** *reproof, rebuke :*—Hē egesiende stiérþ ofermētta mid ðære tǽlinge, Past. 8 ; Swt. 53, 16. Petrus anfēng Paules tǽlinge (*increpationem*), 19 ; Swt. 145, 18. Hié forberaþ ǽghwelce unryhte tǽlinge . . . hié forberaþ ðæt hié mid ðǽm sweorde hiera tungna tǽlinge ne sleáþ hira hlāiurdes ðeáwas *piae subditorum mentes ab omni se peste obtrectationis abstinentes praepositorum vitam nullo linguae gladio percutiunt*, 28 ; Swt. 199, 4. Hiera gefērena tǽlinge *reprehensionem proximorum*, 38 ; Swt. 273, 8. **II.** *evil-speaking, slander, calumny :*—Gif ðū gesihst fela penega tǽlincga oððe wærginga getācnaþ *if you see many pennies, it betokens calumnies or curses*, Lchdm. iii. 214, 16.

tǽlla (=telga? q.v.) :—Tǽllan *tyrso, vitibus*, Germ. 394, 280.

tǽl-leás ; *adj. Blameless :*—Biscepe gedafnaþ ðæt hē sié tǽlleás *oportet episcopum irreprehensibilem esse*, Past. 8 ; Swt. 53, 10.

tǽlleáslíce ; *adv. Blamelessly :*—Ðonne stæpþ se sacerd suíðe tǽlleáslíce on ðone weg *tunc sacerdos irreprehensibiliter graditur*, Past. 13 ; Swt. 77, 19.

tǽl-líc ; *adj. Blasphemous :*—Tǽllíce word *blasphemiae*, Mt. Kmbl. 15, 19, MS. A. v. tāl-líc.

tǽllíce ; *adv. Blasphemously, calumniously :*—Hē Criste wiðsóc and be ðam sōðan Gode tǽllíce sprecþ, Homl. Skt. i. 3, 249. v. un-tǽllíce ; tāllíce.

tǽl-mearc, e ; *f. A date :*—Sume ǽr sume sīð sume in ūrra ǽfter tælmearce tída gemyndum *some early, some late, some by the date in the memory of our times*, Exon. Th. 154, 27 ; Gū. 849.

tǽl-met, es ; *n. A measure expressed by number :*—Is tō ðǽre tíde tælmet hwíle seofon and twentig nihtgerímes *there is to that season a space of time expressed by the number twenty-seven if the reckoning be by days*, Andr. Kmbl. 226 ; An. 113.

tǽlness, e ; *f. Reproach, slander, calumny, detraction :*—Tēlnesse *sugillationis*, Wrt. Voc. ii. 87, 76. Sceomaes t̃ tēlnisses *confusionis*, Mt. Kmbl. p. 3, 10. Ða ðe mē tǽlnysse teónan ætfæstan *qui detrahunt mihi*, Ps. Th. 108, 28. Tēlnysse, 108, 3. Uton beorgan ūs wið tǽlnysse (*vituperatione*), L. Ecg. P. iv. 66 ; Th. ii. 226, 31 : Wulfst. 233, 19. Tō niomanne tēlnisse (*opprobrium*) mīne, Lk. Skt. Rush. 1, 25. Tēlnise t̃ sceoma calumniam, Lind. 3, 14. Tǽlnysse *detractio*, L. Ecg. C. proem. ; Th. ii. 132, 7. Tēlnisses weorlde *aerumnae saeculi*, Mk. Skt. Rush. 4, 19. Ða ðe tǽlnessa teónan wið heora ðam nēhstan nīð āhófan *detrahentem adversus proximum suum*, Ps. Th. 100, 4. Ðū tǽlnissum wiþ ða sēlestan sacan ongunne, Exon. Th. 254, 31 ; Jul. 205. Tǽlnyssa (tēlnisse, Ps. Surt.) *vituperationem*, Ps. Spl. 30, 16.

tǽlsum ; *adj. Numerous, harmonious, rhythmic :*—On tælsumum leúðe *carmine rythmico* (*numerali*), Hpt. Gl. 415, 55.

tǽlweorðlícness, e ; *f. Blameworthiness :*—Gē sweotolran gedóþ eówre tǽlweorðlícnesse (-wierð-, Cott. MSS.) *foedior vestra reprehensibilitas appareat*, Past. 8 ; Swt. 53, 15.

tǽl-wirðe, adj. ; *adv.* -wyrðe ; *adj. Blameworthy, reprehensible :*—Gecnāwan hwæt tǽlwierðe biþ *quae reprehendenda sunt cognoscere*, Past. 28 ; Swt. 195, 8. Tǽlwyrðes (-wierðes, Cott. MSS.), 195, 24. v. un-tǽlwirðe.

tǽlwirð-líc ; *adj. Blameable, reprehensible :*—Ðæt on ōðrum lande betst lícaþ ðæt biþ hwílum on ðam ōþrum tǽlwyrþlícost and eác miceles wítes wyrþe *quod apud alios laude, apud alios supplicio dignum judicetur*, Bt. 18, 2 ; Fox 64, 24. v. un-tǽlwirð-líc.

tǽlwirðlíce ; *adv. In a way that deserves censure, reprehensibly :*—Tǽlwyrðlíce *notabiliter*, Wrt. Voc. ii. 61, 16.

tǽlwirðlíceness. v. tǽlweorðlícness.

tǽman, tǽmes-píle. v. tēman, temes-píle.

tǽnel, es ; *m. A wicker basket :*—Taenil, tēnil *fiscilla* (-*cella*), Txts. 62, 403. Tǽnel, Wrt. Voc. ii. 35, 36 : *canistrum, vas vinetum*, 128, 18 : *cistella, capsilla, cartellum*, 131, 20 : *corbis vel qualus*, i. 24, 57. Litel tǽnel *quasillus*, 25, 6. Hē him on ðæt genam ǽnne lytelne tǽnel mid caricum gefylledne, Homl. Skt. ii. 23 b, 661, 714. Tǽnelas *fiscellos*, tǽnel *fiscellus*, Hpt. Gl. 497, 42, 43. Tǽnelum *fiscellis*, 468, 25 : Wrt. Voc. ii. 34, 4. [Tenel or crele *cartallus*, Prompt. Parv. 489. Cf. *Goth*. tainjō *κόφινος* : *O. H. Ger*. zeinna *canistrum, calathus, cartallus, fiscella* ; zeinnili *cartallus* : *Icel*. teinur ; *pl. f. a basket, creel*.] v. stic-tǽnel ; tān ; *and* cf. wilige, windel.

tǽnen; *adj. Of twigs :—*Tǽnene *sceptrinae* (sceptrum = *virga* in Aldhelm. v. Migne). Hpt. Gl. 483, 62. v. tán.

tæppa, an; *m. A tap :—*Ðonne ðú wín habban wille, ðonne dó ðú mid ðínum twám fingrum swilce ðú tæppan of tunnan onteón wille, Techm. ii. 120, 10. Tæppan teón, 12. [Hit behoueþ þet zuich wyn yerne by þe teppe ase þer is inne þe tonne, Ayenb. 27, 31. *Chauc.* tappe: *O. H. Ger.* zapfo; *m. duciculum, duciolus : Ger.* zapfen : Icel. tappi.] v. tæppian.

tæppa or **tæppe**, an; *m.* or *f. A band, ribbon, tape :—*Tæppan *tenia*, Wrt. Voc. i. 16, 63. [The tapes of hire white volupere, Chauc. C. T. 3241. Tappe *tenea*, Wrt. Voc. i. 196, col. 2 (15th cent.). Cf. *O. H. Ger.* teppi *sagum, tapetia.*]

tæpped, tæppet, es; *n. A covering* for a floor, wall, etc., *a carpet, hanging, coverlet ;* for a person, *a tippet :—*Án healf-hrúh tæppet *sipla* (*sipha ?* cf. in a list *de lectis et ornamentis eorum :—*Hec amphicapa, et tapeta ex utraque parte villosa. Hec sipha, idem est, 243, col. 1), Wrt. Voc. i. 40, 35. vii. oferbrǽdelsas and .ii. tæppedu, Chart. Th. 429, 26. Gemétum tepedum (*lectulum meum stravi*) *tapetibus pictis* (Prov. 7, 16), Kent. Gl. 200. [Cf. typet, tepet, Chauc. C. T. 233. Typitte *leripium*, Wrt. Voc. i. 238, col. 2. Typett, Prompt. Parv. 494. *O. H. Ger.* teppid(-th, -t), tepid(-t) *tapetium, saga cilicina.* From Latin.]

tæppel-bred, es; *n. A board covered with a carpet, a foot-stool :—*Fótscamel ł tæppelbred his fóta *scabellum pedum ejus*, Mt. Kmbl. Rush. 5, 35. Tæppilbred, 22, 44. [Cf. *O. H. Ger.* tepul *tapetum.*] v. preceding word.

tæppere, es; *m. One who sells wine, a tavern-keeper :—*Tæppere *caupus*, i. *tabernarius, qui vinum vendit*, Wrt. Voc. ii. 130, 3. Tæppere, wínbrytta *caupo, tabernarius*, i. 28, 10. Tæppere *caupo*, 74, 17 : Ælfc. Gr. 9, 3; Zup. 36, 13 : Scint. 226, 10. [*O. Frs.* tapper. Cf. *Icel.* tappr *a tapster.*] v. wín-tæppere; tæppian.

tæppestre, an; *f. A woman who sells wine, a hostess :—*Tæppestre *caupona*, Ælfc. Gr. 9, 3; Zup. 36, 13. [He knew the tavernes . . . and everych hostiler and tappestere, Chauc. C. T. 241.]

tæppet, tæppil-bred. v. tæpped, tæppel-bred.

tæppian; *p.* ode *To tap, put a tap into a cask :—*Gyf ðé gedrýptes wínes lyste, ðonne dó ðú mid ðínum swýþran scytefingre on ðíne wynstran hand, swylce ðú tæppian wille, and wænd ðínne scytefinger ádúne and twænge hine mid ðínum twám fingrum, swylce ðú of sumne dropan strícan wylle, Techm. ii. 125, 18. [*Icel.* tappa : *Ger.* zapfen.]

tær (?); *adj. Gaping, cleft* (?) :—Ða giniendan oððe tara *hiulcas*, Wrt. Voc. ii. 42, 49. Cf. (?) teran.

tǽsan; *p.* de *To tear to pieces, pull to pieces, tease* wool, *tear* a person's flesh with a weapon, *wound :—*Ic tótere oððe pluccige oððe tǽse (wulle added in MS. W.) *carpo*, Ælfc. Gr. 28, 4; Zup. 170, 13. *Carpsit, discerpsit, trahit, evellit, vel* tǽst, Wülck. Gl. 200, 5. (In Wrt. Voc. ii. 128, 76 a line is omitted.) Hwílon hé on bord sceát, hwílon beorn tǽsde; ǽfre embe stunde hé sealde sume wunde, ða hwíle ðe hé wǽpna wealdan móste, Byrht. Th. 139, 47; By. 270. Nim wulle, and tǽs hý, Lchdm. iii. 112, 8. [Þay (the does) were tened at þe hyȝe, and taysed to þe wattreȝ, Gaw. 1169. *But later forms seem also to point to a form* tásian :—Sheep, that is fulle of wulle upon his backe, they toose and pulle, Gow. i. 17, 8. Tosyn or tose wul *carpo*, Prompt. Parv. 497, and see note. I toose owlle and card het, Rel. Ant. ii. 197, 36 (15th cent.). Cf. *O. H. Ger.* zeisan; *p.* zias *carpere : O. Du.* teesen *to tease* wool : Dan. tæse.] v. á-, ge-tǽsan; tǽsl.

tǽse (?); *adj. Convenient, for general use* (?) :—Andlang herpoðes tó tǽsan mǽde and se hǽðfeld eal gemǽne, Cod. Dip. Kmbl. v. 78, 32. Tó tǽsan mǽde and se hǽðfeld eal gemǽne, 138, 19. v. (?) ge-tǽse, tǽs-líc.

tǽsl, tǽsel, e; *f. Teasel, teazle :—*Ðeós wyrt ðe man *camelleon alba*, and óþrum naman wulfes tǽsl (tǽsel, MS. B.) (cf. *wolf's-thistle*, E. D. S. Pub. Plant Names) nemneþ, hafaþ leáf wiþerrǽde and þyrnyhte, and heó hafaþ on middan sumne sinewealtne crop and þyrnyhtne, Lchdm. i. 282, 15. [Wilde tesel *virga pastoris*, Wrt. Voc. i. 141, 13 (13th cent.). Tasylle *carduus*, 191, col. 2 (15th cent.). Tasyl *carduus vel cardo fullonis*, Prompt. Parv. 487. Cloth . . . with taseles cracched, Piers P. 15,446. *O. H. Ger.* zeisala *carduus;* wolf(es)-zeisala *arnica.*]

tǽs-líc; *adj. Advantageous, good, convenient :—*Gewelgad ł tǽslícror (-or?) *potius*, Mt. Kmbl. Lind. 25, 9. v. next word.

tǽslíce; *adv. Conveniently :—*Sóhte huu hine teáslícor gesealla mæhte *querebat quomodo illum opportune traderet*, Mk. Skt. Lind. 14, 11. v. ge-tǽslíce.

tǽslum, Wrt. Voc. ii. 93, 44. v. teosol and tæfl.

tǽsness, tæso. v. ge-tǽsness, teosu.

tǽtan (?) *to gladden, make cheerful :—*Ful oft ðæt gegongeþ, ðætte wer and wíf in woruld cennaþ bearn, and mid bleóm gyrwaþ, tennaþ and tǽtaþ (*the father and mother try to make the child joyous, to amuse it;* Thorpe suggests *temiaþ* and *tǽcaþ*), Exon. Th. 327, 15; Vy. 4. [*Icel.* teita *to gladden, cheer;* teiti *gladsomeness, joy;* teitr *glad.*]

tættec (-a, -e ?) *a rag, tatter :—Dormitatio vestietur pannis seó sláp-*olnys byþ gescrýdd mid wácum tætticum, Homl. Ass. 9, 238. Nis se loddere mid his tættecon mín gelíca, Homl. Th. i. 256, 9. Cf. the

following passages from charters relating to the same land :—On tættucan stán (*in a later charter it is called* mægenstán, 291, 7), Cod. Dip. Kmbl. v. 112, 35. Tættucæn stán, 340, 3. Tættaces stán, 325, 30. Tædduces stán, 253, 4. Could the word mean *beggar ?* In the first mentioned charters *loddere sæccing* (*sæxcing*) occurs.

tágum, táhae. v. teáh, tá.

táh-spora, -spura, an; *m. The point of the toe* (?) :—Táhspura *calcis finis*, Wrt. Voc. ii. 127, 47. v. hand-spora, hél-spure, sporu.

tal *a number.* v. tæl.

tál, e; *f.: es; n.* (?) I. *evil-speaking, calumny, slander, vituperation, detraction :—*Tál *denotatio, detractio*, Scint. 83, 6. Tále *suggillationis* (*vituperationis*), Hpt. Gl. 527, 3), Anglia xiii. 37, 298. Tále *vituperationem*, Ps. Spl. 30, 16. Þurh tále *per detractionem*, Confess. Peccat. Ne tále ne dóþ *neque calumniam faciatis*, Lk. Skt. 3, 14. Ðurh ðis beóþ áwecte saca and tála *hinc suscitantur rixæ, detractiones*, R. Ben. 124, 18. Módignys ácenþ yfelsacunge, ceorunge, and gelómlíce tála, Homl. Th. ii. 222, 8. I a. *evil-speaking* in reference to the Deity, *blasphemy :—*Ælc synn and tál biþ forgifen mannum, ac ðæs Hálgan Gástes tál ne bið nǽfre forgifen *omne peccatum et blasphemia remittetur hominibus, Spiritus autem blasphemia non remittetur* (Mt. 12, 31), Homl. Th. i. 498, 22. Se cwyð tál ongeán ðone Hálgan Gást, se ðe mid unbehreówsigendre heortan þurhwunaþ on mándǽdum, 500, 15. Nán man ne beó swá dyrstig, ðæt hé ǽnig word oððe ǽnig (ǽnige ?) tál cweðe ongeán eówerum Gode, ii. 20, 28. II. *scorn, mock, derision, reproach :—*Tál and gebismerung *subsannatio et illusio*, Ps. Lamb. 78, 4. Þe læs ðe heó dó ðé on tále cuman feóndum ðínum *ne faciat te in obprobrium uenire inimicis tuis*, Scint. 177, 4. Ðæt man God tó tále habbe *that God be mocked*, Wulfst. 299, 14. Ðǽs word ðe Sennacherib ásende tó hospe and tó tále ðé and ðínum folce (*verba Sennacherib, qui misit ut exprobraret nobis Deum viventem*, 2 Kings 19, 16), Homl. Th. i. 568, 19. Tále *gannituræ, cachinnatione*, Hpt. Gl. 441, 2. Tále *subsannationem*, Ps. Lamb. 43, 14. III. *blame, censure, reproof :—*Ða bóceras ðæt tældon; ac heora tál næs ná of rihtwísnysse, Homl. Th. i. 338, 20. *Adjectiva* getácniaþ oððe herunge oððe tál (tále, MS. V. : tǽl, MS. T.), Ælfc. Gr. 5; Zup. 12, 11. [Cf. *O. H. Ger.* zála *periculum : Icel.* tál *allurement, device.*] v. tǽl, tǽlan.

talente, ax; *f. A talent :—*Hé ǽlce geáre gesealde twá hund talentana siolfres : on ǽlcere ánre talentan wæs .lxxx. punda, Ors. 4, 6; Swt. 170, 27. III M talentana, Swt. 180, 14. Swá fela talentena, 4, 10; Swt. 202, 22. [*O. H. Ger.* talenta ; *f. strong.*]

talian; *p.* ode. I. *to suppose* a thing (to be) such and such, *consider, reckon, account,* (a) where the object is a noun or pronoun :—Nó ic mé hnágran talige, ðonne Grendel hine, Beo. Th. 1359; B. 677. Ðæs ðe ic sóð talige, Andr. Kmbl. 3125; An. 1565. Talge, Exon. Th. 50, 3; Cri. 794. Hé hit swíðe unáberendlíc talaþ, Past. 33; Swt. 226, 18. Hé on ofslægenne talaþ, Bd. 4, 22; S. 591, 29. Hé talaþ hine sylfne wísne, Wulfst. 52, 29. Ða ðe hí sylfe wáce taliaþ, Homl. Th. ii. 374, 29. Ðæt hié taliaþ hálig, R. Ben. 9, 19. Talige hé hine sylfne wið God forworhtne, Wulfst. 155, 11. Hwæðer ðæt sié tó talianne wáclíc, Bt. 24, 4; Fox 86, 16. Gé beóþ mé talade and rímde on bearna stæl, Exon. Th. 366, 11; Reb. 10. (b) where the object is expressed by a clause :—Sóð ic talige, ðæt ic merestrengo máran áhte, Beo. Th. 1069; B. 532. Wén ic talige . . . ðæt ða Sǽ-Geátas sélran næbben tó gecesenne cyning ǽnigne, 3695; B. 1845. Wé fremful taliaþ, ðæt eal mynstres fadung on ðæs abbodes dóme stande, R. Ben. 125, 5. (c) where the supposition is expressed by a clause :—Ðú talas (*putas*), ðæt ic ne mǽge gebidda fader mín, Mt. Kmbl. Lind. 26, 53. Se man talaþ, ðæt hé ðonne hál sié, Lchdm. ii. 208, 6. Hwylc talge wé, ðæt se ende ðæs heora lífes wǽre, Blickl. Homl. 163, 5. (d) where the supposition is not expressed :—Nis ðis seó hell swá ðú talost and wénest, Bd. 5, 12; S. 628, 7. Gif ðín hige wǽre swá searogrim swá ðú talast, Beo. Th. 1193; B. 594. ¶ *with* swylce, tó, *to consider as :—*Ða áteorigendlícan ðing ðe heó nú tó sibbe and blisse talaþ, Homl. Th. i. 408, 26. Wá eów ðe taliaþ eów sylfe tó ðeódwitan *ve, qui sapientes estis coram oculis vestris*, Wulfst. 46, 26. Ne talode se ofermóda Phariseus tó suá micle mægene ða forhæfdnesse suá hé dyde, Past. 43; Swt. 313, 4. Heora líf is rihtor tó talianne tó ðcan deáðe, Wulfst. 25, 6. Tala ðé ðínne bróðor, swylce hé beó ðín lim, Basil admn. 5; Norm. 46, 11. Tó for náht taliende *parvi pendenda, neglegenda, ad nihilum judicanda*, Hpt. Gl. 418, 36. II. *to impute, ascribe, lay to the account of :—*Gif ðú talast tó ðínum geswince ðæt, ðæt ðú hæfst, Homl. Th. ii. 102, 29. Ne talige ic ðé ðæt tó nánre scylde *I do not impute it to you as any fault*, Shrn. 184, 21. Eádig se wer ðam ðe ne talode (*imputavit*) Drihten synne, Ps. Lamb. 31, 2. Ne tala ðú mé, ðæt ic ne cunne ðone intingan ðínre unrótnesse, Bd. 2, 12; S. 513, 40. Ne talige nán man his yfelan dǽda tó Gode, ac talige ǽrest tó ðam deófle, Homl. Th. i. 114, 18. III. *to reckon, enumerate :—*Tó talanna longsum is *enumerare longissimum est*, Mt. Kmbl. p. 7, 7. [*O. Sax.* talón : *O. Frs.* talia : *O. H. Ger.* zalón *considerare, reputare : Icel.* tala *to talk.*] v. ge-talian; tellan.

tál-líc; *adj.* I. *that conveys reproach, calumny*, etc., *calumnious, blasphemous :—*Þeáh hwá cweðe tállíc word ongeán mé, him biþ forgifen,

Homl. Th. i. 498, 24. Of ðære heortan cumaþ ... tállíce word (*blasphemiae*), Mt. Kmbl. 15, 19. Hí cwædon ðæt hé tállíce word spræce be Moyse and be Gode (*this man ceaseth not to speak blasphemous words against this holy place, and the law,* Acts 6, 13), Homl. Th. i. 44, 29: 46, 1. Se ðe ídele spellunge oððe tállíce word (*calumnies, backbiting*) lustlíce gehýrþ, 492, 19. **II.** *that deserves reproof, blameable, reprehensible :*—Gif ænig biþ mét teállíc *si quisque repertus fuerit reprehensibilis,* R. Ben. Interl. 54, 7. Nis ðæt cláne herigendlíc, ne ðæt gále tállíc, gif him steorran forgéfon, ðæt hí swá lyfedon, Homl. Skt. i. 5, 281. v. tǽl-líc, *and next word.*

tallíce; *adv. In' a way that deserves blame, reprehensibly :*—Tállíce *reprehensibiliter,* Wrt. Voc. i. 54, 46. Ne forseó gé Godes ðearfan, ðeáh ðe hí tállíce hwæt gefremman, Homl. Th. i. 332, 13. v. un-tállíce; tǽllíce.

talu, e; *and indecl.; f.* **I.** *a tale, talk, story, account :*—Leáses spelles talu *constelacion prognostication of the stars,* Wrt. Voc. ii. 20, 68. Ðá spræcon hí betwux him, and seó módor sæt hlystende hire tale ... Ðá se gingra bróðor ðis eall gehýrde fram ðam yldran bróðor hé sæde: ' Ic eom ðín bróðor be ðí[n]re tale,' Homl. Skt. ii. 30, 319-337. Ðæt se Ælmihtiga God gehýre ða talu ðe Syria cyning ásende tô hospe and tô edwíte his micclan mægenðrymme (*si forte audiat Dominus universa verba Rabsacis, quem misit rex Assyriorum, ut exprobraret Deum viventem,* 2 Kings 19, 4), Homl. Th. i. 568, 27. Mé ða treahteras tala wísedon, Salm. Kmbl. 10; Sal. 5. **II.** *talk, discussion, dispute :*—Tale(-u?) *disputatio, contentio, litigatio,* Hpt. Gl. 481, 60. Tale *disputationis, dissensionis,* 439, 57: *disputationis, certationis,* 459, 60. **III.** *a charge, claim :*—Ða heáhsacerdas sôhton tale ágén ðone Hælend *summi sacerdotes quaerebant aduersum Iesum testimonium,* Mk. Skt. 14, 55. Se ðe nánum ne derede, him man dyde talu, and hé wæs beswungen unscyldig for ûs, Basil admn. 4; Norm. 42, 27. Ðæt ælcere neóde beládung sý ádilegod ðæt hý þurh neóde náne tale tô syndrigre æhte næbben *that the excuse of necessity may be removed, so that they may not have any claim to private property on the ground of necessity,* R. Ben. 92, 5. Hé begeat swíðe mycelne sceatt of his mannan ðær hé mihte ænige teale tô habban oððe mid rihte oððe elles *where, rightly or otherwise, he could advance any claim to what he exacted,* Chr. 1085; Erl. 219, 11. **IV.** *an excuse, a defence :*—Míne gebróðra, hwilcere tale máge wé brúcan on his dôme, nû wé nellaþ búgan fram woruldlufe? Homl. Th. i. 580, 2: Lchdm. iii. 442, 3. Ðæt hý náne tale næbben, ðæt hý þurh nytennesse miśfón þurfen, 442, 10. Nabbe wé náne tale ongén ðe *we have no excuse to offer you; quid juste poterimus obtendere?* Gen. 44, 16. Hé ne mihte náne tale findan *he could not devise any defence,* Homl. Skt. i. 23, 624. Gif hé his yfelan dǽda mid leásum talum bewarian wile *si defendere uoluerit opera sua,* R. Ben. 52, 10. **V.** *as a law term, a case* (as regards either plaintiff or defendant), *an action,* cf. spræc :—Ongan tô specenne on ðat land ... Ðam cynge seó talu cûð wæs, Chart. Th. 302, 16. Édwine spæc on his ágene módor æfter sumon dǽle landes ... Ðá ácsode þe bisceop, hwá sceolde andswerian for his módor. Ðá sǽde Ðurcil Hwíta, ðæt hé sceolde, gif hé ða talu cûðe. Ðá hé ða talu ná ne cûðe, ðá sceáwode man þreó þegnas ðǽr ðær heó wæs ... Ðá ácsodon heó, hwylce talu heó hæfde ymbe ða land ... Ðá sǽde heó, ðæt heó nán land hæfde, ðe him áht tô gebyrede, 337, 2-24. Tale wyrðe *entitled to bring an action,* 266, 11. **VI.** *a tale, list, series :*—Talu *laterculus,* Wrt. Voc. ii. 53, 23. Ða talo *canones,* Mt. Kmbl. p. 2, 18. [*O. Sax.* gér-tala: *O. Frs.* tale *a* (*legal*) *case: O. H. Ger.* zala *numerus, series, catalogus, sententia, calculatio, supputatio: Icel.* tala *talk; tale, number.*] v. bóc-, folc-, hrægl-, of-, on-, rím-, tô-, wider-tala.

tam; *adj. Tame, the opposite of wild :*—Tam *subjugalis,* Wrt. Voc. ii. 73, 6. Wilde bár *aper,* tam bár *verres,* i. 22, 70-71. Seó leó, ðeáh hió wel tam sé, Bt. 25; Fox 88, 9. Tiles and tomes meares, Exon. Th. 342, 13; Gn. Ex. 142. Hé rít uppan tamre assene and hyre folan (*sittende on eosule and on folan sunu ðære teoma,* Rush.) *sedens super asinam et pullum filium subjugalem,* Mt. Kmbl. 21, 5. Wildu diór woldon stondan swilce hí tamu wǽron, Bt. 35, 6; Fox 168, 2. On ðære feórþan fléringa wæs ðæra tamra nýtena steall, Boutr. Scrd. 21, 9. Hé hæfde tamra deóra (*reindeer*) syx hund, Ors. 1, 1; Swt. 18, 10. Tame (*wudufuglas*), Bt. 25; Fox 88, 18: Met. 13, 44. [*O. H. Ger.* zam *subjugalis, domitus, mansuetus, mitis: Icel.* tamr *tame; ready for, used to.*]

tama, an; *m. Tameness :*—Ne þearf beorna nán wénan ðære wyrde, ðæt hió (*the lioness*) wel hire taman healde; ac ic tiohhie, ðæt hió ðæs níwan taman náuht ne gehicgge, ac ðone wildan gewunan wille geþencan hire eldrena, Met. 13, 23-28. Gif heó blódes onbirigþ, heó forgit sóna hire níwan taman, and gemonþ ðæs wildan gewunan hire eldrana, Bt. 25; Fox 88, 12.

tán, e; *m.* **I.** *a twig, sprout, shoot, branch :*—Tánas *arbusta,* Ps. Th. 79, 10: *vimina,* Germ. 390, 44: *antes,* Hpt. Gl. 496, 73. Ic on neorxna wonge ásette treów, ðæt ða tánas æpla bǽron, Cd. Th. 295, 7; Sat. 482. Tánum, fingerapplum *dactylis,* Hpt. Gl. 496, 64. Hé (*the phenix*) getimbreþ tánum and wyrtum nest on bearwe, Exon. Th. 227, 29; Ph. 430. Wudubearwas tánum týdraþ, 191, 6; Az. 84: 435, 17;

Rä. 54, 2: 458, 23; Hy. 4, 105. God gibloedsia gimeodomia ðás tánas missenlícra treóna *Deus benedicere dignare has frondes diversarum arborum,* Rtl. 95, 21. Beorc bereþ tánas bûtan tuddre, Runic pm. Kmbl. 342, 29; Rûn. 18. **I a.** *a stake* (? cf. *Icel.* teinn *a stake* to hang things on) :—Ðis syndan ða landgemǽre. Of ðam ealdan hornforda ... ádûn on ealda tán; swá anlang streámes on ealda hornford, Cod. Dip. Kmbl. iv. 45, 25. **II.** *a twig used in casting lots* [' Augury and divination by lot no people practise more diligently. The use of the lots is simple. A little bough is lopped off a fruit-bearing tree, and cut into small pieces; these are distinguished by certain marks, and thrown carelessly and at random over a white garment,' Tacitus' Germania, c. 10], *a lot;* also *a share that is determined by lot :*—Ða Eald-Seaxan næfdon ágenne cyning, ac monige ealdormen wǽron heora ðeóde foresette; and ðonne seó tíd gewinnes com, ðonne hluton hí mid tánum tô ðam ealdormannum, and swá hwylce heora swá him se tán ætýwde, ðonne gecuron hí ðone him tô heretogan, and ealle ðam fyligdon *non habent regem antiqui Saxones, sed satrapas plurimos suae genti praepositos, qui ingruente belli articulo mittunt aequaliter sortes, et quemcumque sors ostenderit, hunc tempore belli ducem omnes sequuntur,* Bd. 5, 10; S. 624, 22-26. Ðá wæs eall geador tô ðam þingstede þeód gesamnod; léton him ðá betweónum tán wísian hwylcne hira ǽrest óðrum sceolde tô foddorþege feores ongildan, hluton hellcræftum ... Ðá se tán gehwearf ofer ǽnne ealdgesíða, Andr. Kmbl. 2196-2210; An. 1099-1106. Hé sealde him wǽste land ðæt hí mid tháne getugan rihte *sorte divisit eis terram in funiculo di-tributionis,* Ps. Th. 77, 55. Nǽfre forlǽteþ Drihten firenfulra tán furðor gangan ðonne hé sóðfæstra settan wylle *never will the Lord let the lot of sinners go further than he will appoint the lot of the just;* non derelinquet Dominus virgam peccatorum super sortem justorum, 124, 3. Tán sendende *sortem mittentes,* Mt. Kmbl. Lind. Rush. 27, 35 : Jn. Skt. Lind. 19, 24. Hié ðysne middangeard on twelf tánum tôhluton and æghwylc ánra heora in ðæm dǽle [wunode?] ðe hé mid tán geeode *the apostles divided the world into twelve parts that were to be assigned by lot, and each one of them [remained?] in that part which he got by lot,* Blickl. Homl. 121, 7-9. Sendon tánas *miserunt sortes,* Lk. Skt. Lind. 23, 34. [*Goth.* tains *a twig, branch: O. H. Ger.* zein, zain *sarmentum, calamus, regula: Du.* teen *twig, osier: Icel.* teinn *a twig, sprout; a spit: Dan.* ten *a spindle: Norweg.* ten *a slender rod: Swed.* ten *spindle, rod.*] v. ác-, átor-, ellen-, hearm-, mistel-, wuldor-tán; tán; *adj.,* tá *a lot,* tǽnel *a basket.*

tán, e; *f. A toe :*—Tán *mentagra,* (seó) micele tán *allox,* Wrt. Voc. i. 45, 24, 25. Mid tánum *cum mentagris,* Lchdm. i. lxxi, 13 (cf. lxxiv, 21). [*O. Frs.* tâne; *f.: Du.* teen.] v. tá *a toe,* tánede; *and cf. the double forms* tán, tá *a lot.*

tán; *adj. Having branches, spreading,* used metaphorically of the offspring of a parent; cf. the use of *branch* in speaking of the members of a family :—Ic Ismael wille bletsian, swá ðû béna eart, ðæt feorhdaga on woruldríce worn gebíde tánum tûdre (*with a family that has many branches.* The passage in Genesis is : And as for Ishmael, I have heard thee : Behold, I have blessed him, and will make him fruitful, and will multiply him exceedingly; twelve princes shall he beget, and I will make him a great nation, 17, 20), Cd. Th. 142, 11 ; Gen. 2360. v. tán *a twig.*

tánages. v. tánian.

tánede; *adj. Having the toes diseased :*—Tánede *mentagricus* (the word occurs in a list of adjectives denoting diseases of the leg), Wrt. Voc. i. 45, 43 : ii. 58, 9. v. tán *a toe.*

tang, e; tange; *f. A pair of tongs :*—Tang *forceps,* Ælfc. Gr. 9, 55; Zup. 67, 3 : Wrt. Voc. i. 286, 78: ii. 33, 36: *delebra,* 138, 62. Tong *forceps,* 109, 6. Tange *forceps,* i. 86, 19. Tange *forcipis,* ii. 33, 35. Tangan, tange, Hpt. Gl. 417, 74. Ic hopige ðæt cherubin mid his gyldenan tange spearcan tô mínre tungan gebringan, Anglia viii. 325, 31. Tangan *forcipes,* Wülck. Gl. 241, 35 (omitted by Wright). Hí woldon mé gelǽccan mid heora byrnendum tangum, Homl. Th. ii. 352, 1, 5. Hí fýrene tangan him on handa hæfdon, Bd. 5, 12 ; S. 628, 42. [*O. L. Ger.* tanga *forceps: Du.* tang : *O.H. Ger.* zanga : *Icel.* töng.] v. fýrtang, Anglia ix. 263, 9, mǽl-tange, ísen-tanga (*read* -tange). v. Ælfc. Gr. Zup. 314, 9).

-tang *touching.* v. gader-, ge-táng; -tenge.

tán-hlyta, an; *m. One who divines by casting lots :*—Tánhlyta *sortilegus,* Wrt. Voc. i. 60, 13. v. tán, II.

tán-hlytere, es; *m. One who divines by casting lots :*—Tánhlytere *sortilegus,* Wrt. Voc. i. 57, 41. v. *preceding word.*

tánian(?) *to decide by lot :*—Tánages *decimatis,* Mt. Kmbl. Lind. 23, 23.

tannere, es; *m. A tanner* (?) :—Be eástan eá and tannera hole, Cod. Dip. Kmbl. ii. 411, 22.

tapor (-er, -ur); *m. A taper;* also *the wick of a lamp :*—Leóhtfæt *lampas,* candel *candela,* taper *papyrus* (cf. leóhtfæt *lucernarium,* weoce *papirus,* 26, 56), Wrt. Voc. i. 284, 35. Tapor *cereus,* 81, 32 : *cerastus,* ii. 130, 23. Swegles tapur *the sun,* Exon. Th. 205, 18; Ph. 114. Onfangenum tapere *accepto cereo,* Anglia xiii. 403, 548. Hé hiene onǽlþ mid ðǽm tapore (-ure, Hatt. MS.) ðæs godcundan liegges, Past. 36 ; Swt. 258, 13. *Acolitus* is gecweden se ðe candele oððe tapor byrþ, ðonne

mann godspell ræt, L. Ælfc. C. 14 ; Th. ii. 348, 4. Se sacerd gehálgodne tapor in ðæt wæter dēþ, Wulfst. 36, 5. Taperas *cerei*, Anglia xiii. 402, 529 : 403, 541. Ðrítig teapera, Chart. Th. 473, 32. Ðá com ðæs laudes menigu mid leóhtfatum and mid taperum, Homl. Th. ii. 474, 24. Taporas *cereos*, Germ. 395, 72. Taperas, Lchdm. iii. 202, 4.

tapor-æx, e ; *f. A small axe :*—Swá feorr swá mæg án taperæx beón geworpen út of ðam scipe up on ðæt land *quam longius de nave potest securis parvula, quam Angli vocant* tapereax *super terram projici*, Chart. Th. 317, 30. Habbe hē áne taperæx on his [handa], Chr. 1031 ; Erl. 162, 8. [*Icel.* tapar-öx (*borrowed from English*).]

tapor-berend, es ; *m. An acolyte* (v. tapor) :—Taporberend *accolitus*, Anglia xiii. 418, 759. Taporber[n]endum *accolitis*, 424, 840.

tappa, teappa ? :—Of rúwan beorge on teappan treów ; of tappan treów on westleás hagan, Cod. Dip. Kmbl. v. 277, 21. Teppan hýse, i. 194, 36. On teppen cnolle, iii. 415, 19. Ad Tapan halan, ii. 344, 6.

tara *tar*. v. teoru.

targe, an ; *f.* : targa, an ; *m. A targe, small shield* [apparently with the same development of meaning as *rand*, q.v. Cf. *O. H. Ger.* zarga *costa* (*aheni*) with the English word] :—Ic geann Ælmēre mīnen discðēne mīnes taregan, Cod. Dip. Kmbl. iii. 363, 12. Targa[n] *parma, scuto*, Hpt. Gl. 423, 50. Twá targan and twegen francan, Cod. Dip. Kmbl. iii. 304, 30. Targena *peltarum*, Hpt. Gl. 475, 64. [*Icel.* targa *a small round shield*. The word seems to have been taken into the Romance languages from Teutonic.] v. ge-targed.

-targed, tasol. v. ge-targed, teosol.

tawa (?) *an implement, a tool, an article for use in an employment.* [That towe (*part of a cart*) is toothed thicke, Pall. 159, 36. Tew of tyschynge *piscalia*, in plurali *reciaria*, Prompt. Parv. 490. Halliwell gives *tow*=tools, apparatus, as a word of the East of England. *O. Du.* touwe *the instrument of a weaver.*] v. ge-, web-tawa ; tawian.

tawian ; *p.* ode.   I. *to taw, dress or prepare* material :—Ðá bæd se Godes man ðæt him man íserngelóman mid hwæte ðyder brohte ðæt land mid tó tawienne. Ðá ðæt land ða getawod wæs and hē on gerisne tíd mid hwæte hit seów *ferramenta sibi ruralia cum frumento adferri rogavit, quod dum praeparata terra tempore congruo seminaret*, Bd. 4, 28 ; M. 366, 24. [Birrþ læredd mann þurrh spell mekenn þin berrte, and turrnenn itt and tawwenn itt and nesshenn itt, Orm. 15908. The sotter that tawith ȝure lethir, Rel. Ant. ii. 175, 24 (about 1308). Tewyā lethyr *frunio, corrodio*, Prompt. Parv. 490. *O. Du.* touwen *to curry leather* : *O. H. Ger.* zauwen, zouwen *exercere* (*ferrum*). Cf. *also* tew or tewynge of lethyr *frunicio*, Prompt. Parv. 489 : *O. H. Ger.* zawa *tinctura* : *Goth.* taui *work*. Teware *corridiator*, Prompt. Parv. 490 : *O. H. Ger.* zauwari *tinctorius*]. v. tewestre.   II. but the word seems to occur in the older time in reference to the ill-treatment of persons or things, *to intreat* shamefully *or evilly, treat* badly, *abuse, insult*. Cf. to *tew*=to trouble, vex, E. D. S. Pub. (Linc.), and see Halliwell :—Oft týne oðða twelfe (flotmen) ælc æfter óðrum scendaþ and tawiaþ tó bysmore ðæs þegnes cwenan and hwílum his dohtor oðða nýdmágan, ðæt hē on lócaþ ðe læt hine sylfne rancne and rícne, ær ðæt gewurde, Wulfst. 162, 20. Se deófol eów tawode þurh his drýmen swá swá hē wolde *the devil hath treated you as he pleased* (the persons addressed had been deprived in turn of the power of speech, motion, and sight) *by his wizards*, Homl. Th. ii. 486, 31. Hē heora burga forbærnde and hí tó bysmore tawode (tucode, MSS. C. V.) *he burnt up their cities and evilly intreated them*, Homl. Skt. ii. 25, 388. Hē Godes templ tawode tó bysmore *he had shamefully abused God's temple* (cf. l. 538), 25, 542. Ðæt folc hine hæfde swá yfele swilce hē sumes þinges scyldig wære ; and ealle men hine fram stówe tó stówe brudon, and tó wundre tawedon *treated him wondrous ill*, i. 23, 654. Ða ðe gefongne wæron hié tawedan mid ðære mǽstan unieðnesse ; sume ofslógon, sume ofswungon, sume him wið feó gesealdon. Ðá Rómáne ðæt geácsedan, ðá sendan hié ǽrendracan tó him . . . Ðá tawedan hié eft ða ǽrendracan mid ðæm mǽstan bismere, swá hié ða ôþre ǽr dydon, Ors. 4, 1 ; Swt. 154, 7-13. Ðæt hié hié mósten tawian mid ðære mǽstan bismrunge, 3, 3 ; Swt. 102, 21. v. ge-tawian ; teágan.

táxe (tádie ? *q.v.*), an ; *f. A toad :*—Táxan *rubetae, quae et ranae dicuntur*, Hpt. Gl. 450, 19.

te ; *prep. To :*—Ða mægenu weordaþ te færwyrðe (cf. tó færwyrde, 8), Past. 65 ; Swt. 463, 6. Heom te cwæþ *illis dixit*, Mt. Kmbl. Rush. 26, 21. Álēfed te habbanne, 14, 4. Te fullfremmanne, Past. 58 ; Swt. 445, 30 : 50 ; Swt. 391, 29. [*O. Sax.* te : *O. Frs.* te, ti : *O. L. Ger.* te, ti : *O. H. Ger.* za, ze, zi.] Cf. tó.

te-. v. te-flówan, -tredan, -weorpan *given under* tó-fiówan, -tredan, -weorpan. [*O. Frs.* te-, ti- : *O. H. Ger.* za-, ze-, zi-.] Cf. tó-.

te=þe *in* þætte.

teá *ten*. v. tín.

teáfor, es ; *n.*   I. *a pigment, material used for colouring, tiver* (red ochre for marking sheep (Suffolk), v. E. D. S. Pub. Old Farming Words, no. vi) :—Mētinge *pictura*, reád teáfor *minium*, Wrt. Voc. i. 46, 74. Teáfor *minium*, 75, 20. Tfaírf (=teáfre) *minio*, Germ. 400. 130. Meng swá ðu dēst teáfor, Lchdm. ii. 56, 6.   II. *a material used*

---

in making a salve :—Nim ladsar (*benzoin*) ðæt teáfur (*gum*), and galpani ôþres healfes panige whit, and gníd hyt tógadere mid wlacan ecede ; and nim ðanne ða sealfe and geót on ðæs seócys mannes eáre, iii. 88, 20. [In other dialects the word occurs with a meaning not easily connected with that of the English form. A somewhat similar connection, perhaps, is seen in the case of the different meanings of lybb, q.v.   *O. H. Ger.* zoubar ; *n. facinum, fascinatio, divinatio : Icel.* taufr ; *n. sorcery.* Cf. *O. L. Ger.* toufere *veneficus.* v. Grmm. D. M. 984.] v. tífran.

teáfor?, Exon. Th. 477, 27 ; Ruin. 31.

teág. v. teáh.

teágan ; *p.* teáde ; *pp.* teád *To dress, prepare :*—Íserngelóman ðæt land mid tó teágenne. Ðá ðæt land ða geteád wæs, Bd. 4, 28 ; S. 605, 33. Wel geteád alwe, Lchdm. ii. 226, 14. v. ge-teágan ; tawian.

teagor, es ; *n. The water from the eyes, tears :*—Teagor ýdum weól, háte hleórdropan, Exon. Th. 182, 23 ; Gú. 1314. [*Goth.* tagr *a tear.*] v. teár.

teáh, tǽh, tēh, tíh(-g) ; *gen.* teáge ; *f.*   I. *a tie, band :*—Teág, taeg *sceda*, Txts. 98, 964. Teáh, Wrt. Voc. i. 289, 36. Lege ðē his teáge an sweoran, Lchdm. iii. 42, 13. Hē cyning gebond fýrnum teágum, Exon. Th. 46, 7 ; Cri. 733. Liðewácum tagum (teágum ?, tānum ?, or tógum ? *as an alternative gloss to* lentis. v. tóh) (*alii*) *lentis viminibus* (*caedentes*), Hpt. Gl. 514, 70. [Teien togadere mid guldene teȝen, Laym. 20998. A teiȝ-doggue þat is in strongue teiȝe (*rimes with* eiȝe (*eye*)), L. S. 308, 301. He huld an hache harde wiþ teis, Jos. 504. *Icel.* taug ; *f. a rope, string.*] v. lád-, racent-, sweor-, web-teáh.   II. *a case, coffer, casket, box :*—Cest *vel* earc *cistella*, tǽg *mozytia vel* arcula, Wrt. Voc. i. 16, 38. Taeg *mantega* (=*mantica ?*), Txts. 35, 19 : 77. 1300. Tig, Wrt. Voc. ii. 55, 57. Hí ðás hálgan martyrrace on ánum leádenum tabulan mid stafon ágrófon, and ðæt gewrit mid twám inseglum on ánre teáge geinsegledon, Homl. Skt. i. 23, 344. Gemētton hí áne teáge, seó wæs geinsæglod mid twám insæglum . . . Man bær út ða teáge . . . Ðá fēng se portgeréfa tó ðære tége and hí sóna unhlidode, 23, 755-765. Búton hit (*the stolen property*) under ðæs wífes cǽglocan gebroht wǽre . . . ðæt is hire hordern and hire cyste and hire tége, L. C. S. 77 ; Th. i. 418, 22. Tégum, fóðrum *tepis* (=*thecis*), Txts. 101, 2010. [At hom is hire pater noster biloken in hire teye (*rimes with* eye (*eye*)), Misc. 191, 2. A riche tie Made all of gold and of perrie Out of the which she nam a ring, Gow. ii. 246, 19. Teye of a cofyr *teca*, Prompt. Parv. 487.] v. beorm-teáh.   III. *an enclosure, a close* (cf. *Icel.* teigr (teygr ?) *a close, paddock*) :—Hujus telluris termini . . . et aquilone meara-teág (=*horses' close*; cf. horsa croft, iii. 464, 3), Cod. Dip. Kmbl. i. 248, 12. Mansionem et clausulam, quam Angli dicunt *teáge*, que pertinet ad predictam mansionem, Chart. Th. 467, 19. Circumcincta est . . . in mer   brómteágh, ii. 49, 20.

teala, tealgor, teál-líc. v. tela, telgor, tál-líc.

tealt ; *adj.*   I. in a physical sense, *unsteady :*—Gif hí sculun nēðan on nacan tealtum, and se brimhengest brídles ne gýmeþ (cf. The floating vessel . . . Rode tilting o'er the waves, Milton, P. L. xi. 747), Runic pm. Kmbl. 343, 22 ; Kún. 21.   II. in a figurative sense, *unstable, not to be relied on, untrustworthy, precarious :*—Hú lǽne ðis líf is, hú tealt, Wulfst. 273, 7. Tealte syndon eorðan welan, 149, 8. Tealte beóþ eorðan dreámas, 264, 3. Tealte getrýwða sindon mid mannum, 82, 12 : 129, 6 : 159, 14. v. next word.

tealtian ; *p.* ode *To be unsteady, to shake, not to stand firm :*—Mid tealtendum grundwealle *nutabundo* (*titubando*) *fundamento*, Hpt. Gl. 497, 49. [Cf. þenne schal Niniue tylte to grounde. Allit. Pms. 102, 361. Feole temples tulten to þe eorþe, Jos. 100. *O. H. Ger.* zeltend rosz, zeltari *equus trutinans : Ger.* zelt amble ; zelter *palfrey : Icel.* tölta *to amble* ; tölt *an ambling pace.*] v. next word.

tealtrian ; *p.* ode *To shake, totter, stagger, be unsteady, to be in an uncertain* or *a precarious condition :*—Wē tealtrigaþ týdran móde hwearfiaþ heánlíce *we move with uncertain step and feeble mind, wander abjectly*, Exon. Th. 23, 19 ; Cri. 371. Ðý lǽs ðe ðæt eásterlíce gesceád tealtrige *lest the calculation of Easter be untrustworthy*, Anglia viii. 308, 4. Tealtrian mid fótum to *stagger*, Dial. 1, 4 (Lye). Ðý lǽs se steall cyricean tealtrian (taltrigan, Bd. M.) ongunne *ne status ecclesiae vacillare inciperet*, Bd. 2, 4 ; S. 505, 11. Tealtrian *vacillare, titubare*, Hpt. Gl. 529, 73. Tealtriendum ł gliddriendum *nutabundis*, 503, 3. Fela ôþera gesynto ða ðe him tealtriende (taltriendum, Bd. M.) gelumpon *alia quae periclitanti ei contigissent prospera*, Bd. 4, 22 ; S. 592, 21. Tealniende (tealtriende ?, tealtiende ?) *nutantes*, Ps. Lamb. 108, 10. [v. Skeat's Dict. s.v. *totter.*] v. preceding word.

teám ; es ; *m. A line* ; but the word which is used in the related dialects (v. *infra*) with a physical meaning is used in English figuratively.   I. *a line* of descendants, *offspring, progeny, family, children :*—Nán wer ne wífaþ, ne wíf ne ceorlaþ, ne teám ne biþ getýmed *children are not brought forth*, Homl. Th. i. 238, 1. Seó gelapung is úre ealra módor . . . hire teám nis ná líchamlíc ac gástlíc, 492, 8 ; Homl. Skt. i. 20, 9. Wuenumon and hire teám, Moruiw and hire teám and Wurgustel and his teám wuárun gefreód . . . Marh gefreóde Leðelt and ealle hire teám, Chart. Th. 626, 22-37. Ðæs teámes wæs tuddor gefylled unlytel dǽl eorðan gesceafta, Cd. Th. 97, 15 ; Gen. 1613. Berende in teáme *fecunda in sobole*,

Rtl. 110, 7. Hē Noe bearh and his wīfe and his teáme, Gen. 5, 31 note: Homl. Skt. i. 8, 18. Caines ofspring forweard ádrenced on đam deópan flóde . . . and of đam yfelan teáme ne com nán þing siđđan, Ælfc. T. Grn. 3, 27. Sēd ł teám *semen*, Mk. Skt. Lind. 12, 21, 22. Đæt folc týmde micelne teám on đam wēstene, Homl. Th. ii. 212, 17. Teám gestrýnan, 324, 11. Đreó wīteþeówe men mē salde bisceop and hire teám, Chart. Th. 152, 22. Fyllaþ eówre fromcynne foldan sceátas, teámum and tūdre, Cd. Th. 92, 27; Gen. 1535. ¶ of animals:— Beón týmaþ heora teám mid clǽnnysse, Homl. Th. ii. 10, 17. [Weóx swa Adames teám her, ne mahte hit na mon tellen, Jul. 61, 7. Drauh togedere al þene team under þe moder, A. R. 336, 15. Wurrþenn wiþþ childe, and tæmenn hire tæm, Orm. 2415. Ys foure sones . . . þys was a stalwarde tem, R. Glouc. 261, 4.] **I a.** *bringing forth children, child-bearing* :—Đonne wīf byþ teámes ætealdod, Homl. Ass. 20, 159. His wīf weard mid Esau and Iacob, and heó geswác đá teámes, 38, 339. [Weren bođe (*John's parents*) teames ateald, O. E. Homl. ii. 133, 32.] **II.** *a line* of animals harnessed together, *a team* :—Oxa on đam forman teáme (cf. oxa on frumteáme *imus*, ii. 48, 36) *imus*, on đam æfteran teáme *binus* (*bimus*), Wrt. Voc. i. 23, 47, 48. On đæm æftran teáme *bimus*, ii. 12, 70. v. feoþer-tíme, iuc-tíma, ge-týme. The old pictures represent the plough as drawn by two pairs of oxen one behind the other. Cf. My plowman . . . a teme (teome, MS. C.) shal he haue. Grace gaue Piers a teme, foure gret oxen, Piers P. B. 19, 256. **III.** as a legal term, (1) *vouching to warranty.* The word denotes one step in the proceedings of a suit for the recovery of property, which was found in one man's possession and claimed by another, who alleged that it had been stolen or had strayed from him. The peculiar character of the process to which it refers was determined by the formalities insisted upon by the law when property changed hands. At such a transaction the presence of witnesses was necessary (L. Ed. 1; Th. i. 158, 11: L. Edg. H. 4; Th. i. 258, 22: L. Edm. C. 5; Th. i. 253, 8: L. C. S. 23; Th. i. 388, 21: 24; Th. i. 390, 4), and one responsible person (*geteáma*), who according to Ine's laws must not be a *þeów man* (L. In. 47; Th. i. 132, 5), was to be fixed upon as representing the party that made the sale or transfer, and to him, if a question subsequently arose as to ownership, the new owner might refer (*tíman*) in support of his right; this referring the property to the party who had sold it was *teám.* In cases of undivided ownership the *geteáma* would be the person making the sale; in cases of joint ownership one of the parties would be taken. The proceedings in a suit in which *teám* was resorted to seem to have been somewhat as follows. The plaintiff, who made claim to property on the plea that it had been stolen from him, had to give security that he would carry on his case : Warige hine, se đe his ágen befóþ, đæt hē tō ǽlcan teáme hæbbe getrýwne borh, L. Eth. ii. 9; Th. i. 290, 6: Wil. I. 21; Th. i. 477, 11; the defendant had to declare how the property came into his hands, and to give security that he would produce his *geteáma* in court : Gif hwā befó đæt him losod wæs, cenne se đe hē hit æt befó hwanon hit him cōme, sylle on hand and sette borh (*pledge himself and find security*) đæt hē bringe his geteáman in đær hit besprecen biþ, L. Eth. ii. 8; Th. i. 288, 15. On the case being brought into court (which was to be held in *cynges sele*, L. H. E. 7; Th. i. 30, 18: 16; Th. i. 34, 7, or *kyninges burh :* Ælc teám beó on đæs kyninges byrig, L. Eth. iii. 6; Th. i. 296, 4), the plaintiff made oath, that he prosecuted his suit lawfully and fairly, L. O. 2; Th. i. 178, 10, and without malice, 4; Th. i. 180, 8; the defendant on his side made oath that he had had no part in the alleged robbery, but had acquired the property in a lawful manner, 3; Th. i. 178, 16, and was guiltless, 5; Th. i. 180, 14. He was now bound to produce witnesses of the transaction which resulted in his acquiring the property in dispute, or *teám* was denied him : Būton hē đara óđer (*certain witness*) hæbbe, nele him mon nǣnne teám geþafian, L. Edg. H. 4; Th. i. 260, 2. Ne beó ǽnig man ǽniges teámes wyrđe būton hē getrýwe gewitnysse hæbbe, L. C. S. 23; Th. i. 388, 20. Ne beó đǣr nán teám, 24; Th. i. 390, 6. If the witness was forthcoming, the *geteáma* had to be produced, and witness or oath again was called for to prove that the defendant's proceedings were correct : Wē cwǣdon, se đe týman scolde, đæt hē hæfde ungeligene gewitnesse đæs đæt hē hit on riht týmde, oþþe đone ađ funde đe se gelýfan mihte đe on sprece, L. Ed. 1; Th. i. 158, 16. If the *geteáma*, though living, were not brought, according to one regulation the defendant lost his case, and had to resign the property, L. H. E. 7; Th. i. 30, 9; according to another, if he could bring witness to prove the sale, he received the price of the property he had to give up, 16; Th. i. 34, 8. If the *geteáma* were dead other formalities were prescribed, L. In. 53; Th. i. 134, 17: L. Eth. ii. 9; Th. i. 290, 9. If all the requirements had been satisfied the property in question was handed over to the *geteáma :* Se đe yrfe bycge on gewitnesse, and hit eft týman scyle, đonne onfó se his đe hē hit ǽr æt bohte, L. Ath. i. 24; Th. i. 212, 12. Swā ic hit týme swā hit mē se sealde đe ic hit nū on hand sette, L. O. 3; Th. i. 180, 3: L. Eth. ii. 8; Th. i. 288, 20; and the defendant thereupon appealed to the *geteáma* to corroborate his statement of the case, 21. If the latter accepted the property, the former was cleared, and the *geteáma* himself was now in a similar position to that in which the defendant had stood, 22; but if he declined to receive it, and declared that it was not the property he had sold, then the defendant had to prove that it was : Gif se mon (*the geteáma*) onfón ne wille, and sægþ đæt hē him nǣfre đæt (*the property*) ne sealde, ac sealde óđer, đonne mót se gecýđan, se đe hit tíemþ, đæt hē him nán óđer ne sealde būton đæt ilce, L. In. 75; Th. i. 150, 7: cf. 35; Th. i. 124, 10. If however the case were not stopped, the process, in earlier times, was repeated until either there was a failure to produce a *geteáma* (v. teám-byrst), or the property was traced to some person whose right to its possession was undoubted : Gange se teám forđ óþþæt man wite hwǣr hē óđstande, L. Ed. 1; Th. i. 158, 15: L. Eth. ii. 9; Th. i. 290, 3. Betweox teáme gif hwā tō fēhþ, and ná furđor teám ne cenþ, ac ágnian wile, ne mæg mon đæs wyrnan, gif getrýwe gewitnes him tō ágenunge rýmþ, 290, 18. Later *teám* was necessary only three times : Týme hit man þrywa, æt đam feórđan cyrre ágnige hit, ođđe ágyfe đam đe hit áge, L. C. S. 24; Th. i. 390, 9. At one time also a change was made in the place where *teám* should be made : Be teámum. Hwilon stód đæt man sceolde þrywa týman đǣr hit ǽrest befangen wǣre, and syþþan fylgean teáme swá hwǣr swá man tō cende. Đá gerǣddan witan, đæt hit betere wǣre, đæt man ǽure týmde đǣr hit ǽrest befangen wǣre . . . đý lǽs đe mon unmihtigne man tō feor and tō lange for his ágenan swencte, L. Eth. ii. 9; Th. i. 288, 28. A case in which a defendant is cleared by his *geteáma*, who, however, cannot get himself cleared, is given Chart. Th. 206, 19 sqq. A woman had been stolen, and was found in the possession of one Wulfstan. Đá týmde Wulfstán hine (*the woman*) tō Æđelstáne; đá cende hē tēm and lēt đone forberstan. v. teám-byrst. Another case is mentioned where a bishop was not allowed *teám* : Ne móste se bisceop beón đara þreora nānes wyrđe đe eallum leódscipe geseald wæs on wedde, tale, ne teámes, ne áhnunga, 266, 11. (2) The word also occurs often in charters along with sac, sóc, toll, etc., where according to one definition it refers to the right to the forfeitures which were made in the suits where *teám* was resorted to : Theam, quod si aliquis aliquid interciebatur super aliquem, et ipse non poterat warrantum suum habere, erit forisfactura, et justicia similiter de calumpniatore, si deficiebat, sua erit, L. Ed. C. 22; Th. i. 452, 1. Donavi abbati . . . consuetudinem que dicitur teames, Chart. Th. 405, 1. v. teám-byrst. A different meaning is given elsewhere to the word. In Cod. Dip. Kmbl. iv. 202, 7 *teám* occurs, and in the Latin form of the charter is rendered by 'privilegium habendi totam suorum seruorum propaginem,' 203, 6. [O. Frs. tâm *a bridle; a line of descendants, progeny, family :* O. L. Ger. tôm *frenum : Du.* toom : O. H. Ger. zoum *funis, habena : Icel.* taumr *bridle, rein, cord.*] v. bearn-, frum- (v. II above), here-, leger-teám; tíman.

teáman. v. tíman.

teám-byrst, es; *m. The failure to produce a geteáma in a suit.* v. teám, III. I :—Đá týmde Wulfstán hine (*the stolen slave about whom the case had arisen*) tō Æđelstáne; đa cende hē tēm and lēt đone forberstan (*he admitted having sold the slave to Wulfstan, but would not declare from whom he had obtained it*) . . . Đá bæd Byrhferhđ ealdormann Æđelstán his wer for đam tēmbyrste, Chart. Th. 207, 4.

teám-full; *adj. Prolific, productive :*—Tudderfulle, teámfulle *fetose*, Wrt. Voc. ii. 148, 33. Sceáp heora teámfulle ł berende *oues eorum foetosae*, Ps. Lamb. 143, 13 : Ps. Spl. 143, 17.

teám-pól, es; *m. A breeding-pool :*—Up on Exan on đone neáran teámpól; đanon up on Exan; đonne of Exa[n] on đa smala[n] lace; of đære lace eft on Exa[n]; đanon up and lang Exa[n] on đone uferan teámpól, Cod. Dip. Kmbl. ii. 205, 8–11: iii. 441, 5–8.

tēan (?), tēgan (?); *p.* tēde *To grow tough or pliant :*—Tēdan lentescunt, Wrt. Voc. ii. 52, 57: 92, 77. v. tóan, tóh; *see also* (?)ge-teúgan.

teán-, teaper, teappa. v. teón-, tapor, tappa.

teár (= teahor), teór, tæher, teher, es; *m. A tear.* **I.** *a drop of water from the eye,* (1) caused by emotion, generally by grief :—Teár *flemen, flentium humor,* Wülck. Gl. 240, 13 : *lacryma,* Wrt. Voc. i. 43, 7. Teáras *lacrime,* 282, 55. Sealtes pund, đanon him (*Adam*) wǣron đa teáras sealte, Salm. Kmbl. 180, 16. Hruron him teáras, Beo. Th. 3749; B. 1872. Nalles for torne teáras feóllon, Elen. Kmbl. 2266; El. 1134. Pund saltes, of đon sindon salto tehero, Rtl. 192, 15. Mid teára ágotennysse, Lchdm. iii. 428, 10. Mid teára gytum, Blickl. Homl. 61, 20. Eágan gefyllede mid teárum, 189, 1. Wēpende mid teárum, 151, 20: Bd. 3, 14; S. 541, 3. Teárum mǣnan, Exon. Th. 285, 10; Jul. 285. Teárum geótan, 95, 34; Cri. 1567. Heó ongan mid hyre teárum (tæherum ł teárum, Lind.) hys fēt þweán, Lk. Skt. 7, 38. Teárum ł tehrum, Lind. 7, 44. Mid teherum (teórum, Rush.), Mk. Skt. Lind. 9, 24. Wēpende wēregum teárum, Andr. Kmbl. 118; An. 59. Wráđum teárum, Ps. Th. 59, 11. Tornlícum teárum, 125, 5. Sárige teáras, 55, 7. Teáras geótan *to shed tears,* Bd. 4, 28; S. 606, 14: Exon. Th. 11, 18; Cri. 173. (1 a) in plural, used for the feeling of which the tears are a sign, *grief, affliction :*—On deópnysse wōpes and teúra *profunditate fletus et lacrimarum,* Scint. 47, 4. Đū fēdest ús teára hláfe, and ús drincan gifest deorcum teárum, Ps. Th. 79, 5. Heó is fulneáh deád for teárum and for unrótnesse, Bt. 19; Fox 69, 30. Eua bær teáras on hire innoþe, Maria brohte đone ēcean gefeán eallum middangearde,

Blickl. Homl. 3, 12. Tehhero, Rtl. 40, 35. (2) caused by weakness. v. tíran:—Ðeós eáhsealf mæg wiþ ælces cynnes broc on eágon . . . wiþ tér, Lchdm. iii. 292, 2. Læcedómas wiðð eallum tiédernessum eágena . . . wið eágna teárum, ii. 2, 8. Wið eágena teára (-e, -as?), iii. 44, 29. **II.** *a tearlike drop* :—Ðá wearð beám monig blódigum teárum birunnen . . . sæp wearð tô swâte, Exon. Th. 72, 20 ; Cri. 1175. **II a.** *that which drops* or *exudes*, e. g. honey from a comb :—Balsames teár *opobalsamum*, Wrt. Voc. i. 33, 51. Swâ þicce swâ huniges teár *of the consistency of honey that has dropped from the comb*, Lchdm. ii. 74, 4. Genim balsami and huniges teáres emmicel, 28, 10, 4 : 108, 17. Gegadriende swâ swâ beón hunigcamb teáres *colligentes uti apes favu[m] nectaris*, Anglia xiii. 368, 46. Þynceþ þegna gehwylcum huniges bíbreád healfe ðý swétre, gif hê hwêne ær huniges teáre bitres onbyrgeþ, Met. 12, 10. Meng wið huniges teáre, Lchdm. iii. 46, 7. Nim huniges teár and merces sæd . . . mæng wið ðone teár, 4, 16. [*O. Frs.* tár : *O. H. Ger.* zaher : *Icel.* tár ; *n.*] v. bryne-, hunig-teár ; teagor.

-teáren. v. hunig-teáren.

tearflian ; *p.* ode *To wallow, roll over* :—On eorðan forgnyden fæmende hê tearflode (terflede, teorflede, later MSS.) *elisus in terram uolutabatur spumans*, Mk. Skt. 9, 20. [Cf. þe riȝt schul ryse to ryche reynynge, Truyt and treget to helle schal terve, L. H. R. 207, 311. *O. H. Ger.* zerben (sih) *to turn.*]

teár-geótende ; *adj.* (*ptcpl.*) *Tear-shedding, weeping* :—Adam myd teárg[e]ótendre hálsunge and myd mycelre stefne ðus cwæþ, Nicod. 30 ; Thw. 17, 27.

teárian ; *p.* ode *To shed tears* :—Tæherende (teherende, Rush.) wæs se Hælend *lacrimatus est Jesus*, Jn. Skt. Lind. ii. 35. [*Icel.* tárask *to shed tears.*]

teárig ; *adj.* **I.** *tearful, weeping* :—Teárigum sícetungum *lacrimosis singultibus*, Hpt. Gl. 421, 3. v. teár, **I. 1.** **II.** *watery, watering* (*of the eyes*) :—Gif mon biþ on wæterælfadle, ðonne beóþ him ða eágan teárige, Lchdm. ii. 350, 22. v. teár, **I. 2,** tíran.

teárig-hleór ; *adj. Having the cheeks wet with tears* :—Ic (*Hagar*) sceal teárighleór on wéstenne witodes bídan, Cd. Th. 137, 16 ; Gen. 2274. [Cf. *Icel.* tárug-hlýra *with tearful cheeks.*]

teár-líc, tearo. v. hunig-teárlíc, teoru.

teart ; *adj. Tart, sharp* (of pain, punishment, etc.), *severe* ; acer, asper :—Sticol oðße teart *asper*, Wülck. Gl. 256, 32. Ús ðincþ swiðe teart wîte ðæt án úre fingra on fýr becume, Homl. Th. ii. 590, 32. Ðæt hê ne ðurfe becuman tô ðam teartum bryne, 592, 17. Hê álýsþ mê fram teartum worde (*a uerbo aspero*), Ps. Lamb. 90, 3. Beó him gesæd ða teartan wítu, Homl. Th. ii. 344, 32 : Homl. Skt. i. 11, 82. Mid teartum wîtum getintregod, 8, 156. Mid teartum swingellum *acribus uerberibus*, R. Ben. 54, 4. Mislimp tearte *casus asperos*, Hymn. Surt. 16, 5. Teartere þræung *acrior correptio*, R. Ben. Interl. 59, 6. Hine man þreáge mid teartran steóre, R. Ben. 52, 6. Hê stíðran and teartran steóre underfó *majori uindicte subjaceat*, 71, 8. [Chaucer uses *tart* = sharp to the taste :—Poudre-marchaunt tart, Prol. 381.]

teart-líc ; *adj. Sharp, severe* :—Þeáh hwæt teartlíces on ðisum regule geset sý, R. Ben. 5, 11.

teartlíce ; *adv. Sharply, severely* :—Teartlíce *acriter*, Hpt. Gl. 477, 13 : 507, 53. Hê ðê tintregaþ teartlíce on wítum, Homl. Skt. ii. 25, 154. Hê beó teartlíce geswungen, Wulfst. 248, 13. Teartlícor *acrius*, Hpt. Gl. 515, 47. Teartlícur, Scint. 210, 12. Wê beóþ forswælede teartlícor *crememur acrius*, Hymn. Surt. 5, 15 : Homl. Th. i. 330, 34. Sý hê ealra teartlícost geþreád *acrius coherceatur*, R. Ben. 129, 10.

teartness, e ; *f. Sharpness, severity, asperity* :—Drihten herede Iohannes for ðære teartnysse his reáfes, forðan ðe hê wæs mid ofsendes hærum gescrýd wáclíce and stíðlíce, Homl. Th. i. 330, 1. For ðæs wyntres teartnysse, Homl. Skt. i. 11, 152. Teartnesse *acerbitatem, crudelitatem*, Hpt. Gl. 480, 56. Mid menigfealdum ðeówracena teartnyssum gebrégede, Homl. Th. i. 578, 27.

teart-numol ; *adj. Efficacious* :—Ðeós wyrt is swýþe scearpnumul (teart-, MS. B.) wið ðæt áttor, Lchdm. i. 152, 3. v. scearp-numol.

teáslíce, -teáw, tebl, teblere, tebl-stán, téder-, tédre, te-flówan, téfrung, tégan, tége, teging *tinctura*, tegðian, tegðung, teher, teherian, teigða, teigðian, teissum, tel. v. tæslíce, æl-, eal-teáw, tæfl, tæflere, tæfl-stán, tíder-, tídre, tô-flówan, tífrung, teán, teáh, telgung, teóþian, teóþung, teár, tíran, teóþa, teóþian, teosu, tæl.

tela, teala, teola, telo, tiolo ; *adv. Well.* **I.** *well, rightly, aright, correctly* :—Hê hine sceal níde tela læran. Ðý him is micel ðearf ðonne hê tela lærþ ðæt hê eác tela doo *dum commissis sibi cogitur bona dicere, ipsum prius necesse est, quae dixerit, custodire*, Past. 28, 3 ; Swt. 193, 12. Teala, Blickl. Homl. 75, 14. Ða slâwan sint tô manianne ðæt hié ne forielden ðone tíman for hiera slǽwðe ðe hié tela (tiola, Hatt. MS.) on dón mægen *pigri suadendi sunt, ne agenda bona, dum differant, amittant*, Past. 39, 1 ; Swt. 280, 20. Gif hí ðone frýdóm tela gehealdon . . . gif hí ðone frýdóm forheólden, Bt. 41, 3 ; Fox 208, 10. Hê ríce geheóld tela, Beo. Th. 4423 ; B. 2208 : 5468 ; B. 2737. Teala, Cd. Th. 74, 35 ; Gen. 1232. Læst eall tela, Beo. Th. 5320 ; B. 2663. Nú ic wát tela and ic onféng gewit mínes môdes *modo sanum sapio, recepi enim sensum*

animi mei, Bd. 3, 11 ; S. 536, 33. 'Geseoh ðæt ðú teala wite.' Cwæþ hê : 'Ne wêde ic' '*vide ut sanum sapias*.' '*Non,' inquit, 'insanio*,' 5, 13 ; S. 632, 32. Ðæt ic teala cunne ðín weorc healdan, Ps. Th. 118, 68 : Exon. Th. 336, 10 ; Gn. Ex. 46. Is wuldres leóht ontýned ðam ðe teala þenceþ, Cd. Th. 299, 29 ; Sat. 557 : Exon. Th. 347, 30 ; Sch. 20. Gif gê teala hycgaþ, Andr. Kembl. 3223 ; An. 1614. Beó nú on yfele, noldæs ær teala, Cd. Th. 310, 26 ; Sat. 733. Teala foresecgan, Ps. Th. 118, 172. Tela, Exon. Th. 432, 19 ; Râ. 49, 2. **II.** *well, perfectly, completely, thoroughly, certainly* :—Heald forð tela sibbe *continue without interruption to maintain peace*, Beo. Th. 1901 ; B. 948. Wudufuglas tela ætemede, Met. 13, 36. Ic ðé teala forgulde ealle ða gehât, Ps. Th. 65, 13. Ðær ðú mê teala hæle, 70, 2. Se ðe teala cúþe, Exon. Th. 349, 9 ; Sch. 43. Ic his bídan ne dear . . . nele ðæt ræd teale *I dare not await him . . . good counsel certainly will not require that*, 397, 8 ; Râ. 16, 16. **III.** *well, prosperously, happily* :—Geþeóh tela, Beo. Th. 2441 ; B. 1218. Ðú hulpe mín ðæt ic teala mihte, Ps. Th. 70, 20. Hê hêt ðæt teala wunian éce, 77, 68. Æfter ðæm Cartainenses wunnon on Sicilie ðær him seldon teola gespeów *cum adsidua nec umquam satis prospera adversus Siculos bella gererent*, Ors. 4, 5 ; Swt. 168, 20. **IV.** *well, in a beneficial* or *pleasant manner* :—Wê wæron hêr tela willum bewenede, Beo. Th. 3645 ; B. 1820. Ontýn ðinne múð, and ic hine teala fylle, Ps. Th. 80, 11 : 105, 5. Gif wê willaþ óþrum geleáffullum teala dón and helpan ðæs earman, Blickl. Homl. 75, 18 : 69, 17. Tala, Lk. Skt. 6, 27. **V.** marking degree, *very, to a great extent* :—Ic þigde tela micelne mete, Nar. 30, 25. Drincan tela micel, Lchdm. iii. 290, 12. Tela micel steáp, 294, 19. Teala, i. 374, 9. Teala líciendlíc, Ps. Th. 68, 13. Teala wynsume, 125, 2. Ða wæs tiolo micel sprēc, Chart. Th. 70, 17. Ic ðé an tela sincgestreóna *I give thee treasures in abundance*, Beo. Th. 2455 ; B. 1225. **VI.** as an exclamation, *well, good* :—Ða andswaredon hí : 'Nis hit lang tô ðon.' Cwæþ hê : 'Tela, utan wê ðære tíde bídan,' Bd. 4, 24 ; S. 599, 5. Cwæþ ic : 'Hwí ne sceolde mê swâ ðincan?' Ðá cwæþ hê : 'Telo', ðonne ðæt ðé swâ þincþ, ðonne ongit ðæt . . ., Bt. 38, 3 ; Fox 200, 22. v. un-tela ; til ; *and cf.* wel *for similar uses.*

télan. v. tǽlan.

teld, es ; *n. A tent, pavilion* ; left still in *tilt* of a cart :—On ðam telde (*tabernaculo*) heó ys, Gen. 18, 9. Eardungstôwa ł teld his *tabernaculum ejus*, Ps. Spl. 17, 13. Mon teld (geteld, MS. B.) ðærofer ábrædde (*tentorio majore extenso*), Bd. 3, 11 ; S. 535, 22. [And Ælfríc biscop I biquede míne teld and mín bedreáf þat ic best hauede út on mí fare mid mê, Chart. Th. 566, 32.] [Per Oswald sette his teld, Laym. 31384. Hengest bilæfde al his teld (hii lete stonde hire teldes, 2nd MS.), 16462. In here teld (on heora geteldum, Num. 16, 27) he (*Dathan and Abiram*) stonden, Gen. and Ex. 3769. Telte or tente *tentorium*, Prompt. Parv. 488. *O. H. Ger.* zelt ; *n. papillio* : *Ger.* zelt ; *n.* : *Icel.* tjald ; *n. a tent* : *Dan.* telt ; *n.*] v. ge-teld.

teldan ; *p.* teald, *pl.* tuldon ; *pp.* tolden *To spread a covering*. v. beofer-teldan ; teldian.

telde *a tent-peg* :—Claus (=*clavus*) *lignum tentorii vel* telde, Wrt. Voc. ii. 131, 56. v. teld-sticca, tildian.

teldian ; *p.* ode, ede *To spread* (a tent, an awning, a net, a snare, etc.) :—Teldat *conectit*, Wrt. Voc. ii. 105, 35 : 15, 36. Hí teldedon gryne and ða gehýddon *absconderunt mihi interitum laquei sui*, Ps. Th. 34, 8. [Þenne mon wule tilden his musestoch, O. E. Homl. i. 53, 20. At pleȝe he (*the devil*) teldeð þe grune of idelnesse . . . on þe grune þe þe werse haueð itelded . . . Drinch, þere teldeð þe werse þe grune of unrihte, ii. 211, 13–27. Tristre is þer me sit, oðer tildeð þe nettes, A. R. 334, 1. Weoren teldes itælded, Laym. 17489. Fantummes of fendes (*idols*) telded on lofte, Allit. Pms. 78, 1342. Sone watȝ telded up a tapit on tresteȝ ful fayre, Gaw. 884. Þei tildeden Absalon a tabernacle (*they spread Absalom a tent*, 2 Sam. 16, 22), Wick. A green an other hath for hem ytilde, Pall. 110, 164. *Icel.* tjalda *to spread a tent, to cover with an awning, stretch a covering over*.] v. teldan.

teld-sele (?) *a tent* :—Ganggeteld *papilio*, tyldsyle *tenda*, Wrt. Voc. i. 59, 12–13.

teld-sticca, an ; *m. A tent-peg* :—Geláhte seó wífman án ðæra teldsticcena geslóh ðá . . . ðæt se sticca him eode út þurh ðæt heáfod . . . Hê geseah hwär Sisara læg and se teldsticca sticode þurh his heáfod *tulit Iahel clavum tabernaculi . . . et clavum defixit in cerebrum . . . vidit Sisaram jacentem et clavum infixum in tempore ejus*, Jud. 4, 21, 22. [*O. H. Ger.* zelt-steccho *paxillus*. Cf. *Icel.* tjalds-nagli *a tent-peg*.]

teld-treów (?), es ; *n. A tent-peg* (?). *some implement in weaving* :—Teltreó *clus*, Wrt. Voc. ii. 104, 19. Teltré *claus*, 16, 34 : i. 282, 10. In the last instance the word occurs in a list *de textrinalibus*. v. telde, tæbere.

teld-wyrhta, an ; *m. A tent-maker* :—Paulus se ðe wæs on woruldcræfte teldwyrhta, Homl. Th. i. 392, 21.

télend, télere. v. tǽlend, tǽlere.

telg, tælg, es ; *m. A dye* :—Taelg *faex, fucus*, Wrt. Voc. ii. 109, 36 : 39, 3 (the entry is given, *fuscus* tægl oðße feax). Telg, deág *fucus*, telga *fucorum*, 36, 66, 67 : 70, 19 : 151, 52. Se weolocreáda tælhg

(tægl, MS. C.) *tinctura coccinei coloris*, Bd. 1, 1 ; S. 473, 20. Se reáda telg, Exon. Th. 408, 21 ; Rä. 27, 15. Telges *conquilii*, Wrt. Voc. ii. 19, 15. Telge *murice*, 57, 50 : *ostro*, 64, 37 : 87, 10. Telga *fucorum*, 88, 43. Ðætte Iosephes tunece wǽre telga gehwylces bleóm bregdende, Exon. Th. 357, 2 ; Pa. 22. v. æt-, beám-, weoloc-telg ; telgan.

telga, an ; *m.* A *branch*, *bough*, (a) literal :—Telge *ramus*, Mt. Kmbl. Lind. 24, 32 : Mk. Skt. Lind. 13, 28. Telgan *fronde*, Wrt. Voc. ii. 33, 60. Telgan *virgultum*, i. 39. 17. Unberende telgan *spadones*, 38, 8. Telgan gehladene, Exon. Th. 202, 28 ; Ph. 76. Telgu *rami*, Mk. Skt. Rush. 13, 28. Telgena *palmitum*, Wrt. Voc. ii. 66, 34. Telgum gescafenum *corticibus*, Hpt. Gl. 412, 41. Balzamum of ðæra treówa telgan (*ramis*) weól, Nar. 26, 21. Blǽda on treówes telgum, Cd. Th. 55, 10 ; Gen. 892 : 88, 24 ; Gen. 1470 : Exon. Th. 210, 19 ; Ph. 188. Beorc byþ on telgum wlitig, Runic pm. Kmbl. 342, 30 ; Rún. 18 : Ps. Th. 57, 8 : 103, 16. Telgo *frondes*, Mk. Skt. Lind. 11, 8 : *ramos*, 4, 32. Genim ðysse wyrte (*yarrow*) telgan, Lchdm. i. 198, 12 note. ¶ In the following passage Kemble and Leo take the word as meaning a strip of land (fallow), but as such a strip of land if fallow one year would not be so the next, its designation as the fallow strip would hardly serve the purpose of marking a boundary. *Telga* might rather refer to a branch distinguishable from the loss of its bark :—Andlang strǽte on ðone calewan telgan, Cod. Dip. Kmbl. i. 258, 7. See iii. xxxix, and Leo, Place Names, p. 66. (b) figurative :—Hé bær ða wétan ðære uncyste in ðæm telgan, ðone hé geteáh ǽr of ðan wyrtruman, Bd. 1, 27 ; M. 82, 14. Wróhtes telgan, Cd. Th. 61, 3 ; Gen. 991. Ealle ða telgan ðú gebræddest *extendisti palmites ejus*, Ps. Th. 79, 11. Telgo míno *ramos meos*, Rtl. 68, 32. v. wudu-telga ; telgor, telgra.

telgan *to dye* [:—Getelged oððe gedeágod *colerata*, Wrt. Voc. ii. 19, 14. Getelgode *fucate*, getelgod *fucatum*, 33, 58, 59. Getælged *colerata*, *fucata*, 134, 35.] v. twi-telged ; telgung.

telg-berend *that which produces a dye* :—Tæl(g)berend *ostriger*, Wrt. Voc. ii. 64, 72. v. telg.

telge (?) :—On xiiii nihte mónan is gód ǽlc telge tó anginnanne, Lchdm. iii. 178, 31. Cockayne refers the word to *telg* and translates *dyeing* ; but the passage at 190, 21, in which the same date is said to be 'eallum gód þingum gód' suggests a different meaning. The forms of the whole piece are corrupt.

telgian ; *p.* ode *To put forth shoots, to flourish* :—Treów telgade tír welgade *good faith flourished, glory abounded*, Exon. Th. 353, 57 ; Reim. 34.

telgor, tealgor, es ; *m.* : e ; *f.* A *plant, shoot, twig* :—On ðam dæge ðe God geworhte ǽlcne telgor on eorðan (*omne virgultum agri*), Gen. 2, 5. Telgre *vimen*, Engl. Stud. xi. 67, 95. Gif hwá mid him ðysse wyrte (*verbascum*) áne tealgre byrþ, ne biþ hé bréged mid ǽnigum ógan, Lchdm. ii. 176, 3. Tealgras *propagines*, Blickl. Gl. Ðeós wyrt (*wild gourd*) wið ða eorðan hyre telgra tóbrǽdeþ, Lchdm. i. 324, 3 note. [*Icel.* tjálgr ; *n.* a *prong*.] v. next word.

telgra, an ; *m.* A *shoot, branch, twig ; sucker of a root* :—Telgra *virgultum*, Wrt. Voc. i. 80, 4. Telgra *ramus (fici)*, Mt. Kmbl. Rush. 24, 32. Dó on ánne telgran (*morbeámes*), Lchdm. i. 332, 22. Of ánum stelan manega telgran weaxaþ, 276, 22. Ða telgran (ðæs wyrttruman), 318, 10. Telegran *antes, virgultus*, Hpt. Gl. 496, 71. Telgrum *viminibus, virgulis*, 483, 58 : *ramis*, Mt. Kmbl. Rush. 13, 32. Telgran *ramos*, 21, 8 : *surculos, virgulta*, Hpt. Gl. 433, 47. Genim ðysse wyrte (*yarrow*) telgran, Lchdm. i. 198, 12. Ðeós wyrt (*polium*) of ánum wyrttruman manega telgran ásendeþ, 276, 8. Ðeós wyrt (*wild gourd*) wið ða eorðan hyre telgran tóbrǽdeþ, 324, 3.

telgung, e ; *f.* *Dyeing*, or a *dye* :—Te[l]ging *tinctura* (cf. deáh *tinctura*, 40, 39), Wrt. Voc. i. 32, 8. Telgung *tinctorium*, 289, 13. Telgunge *tinctura*, ii. 89, 28.

tellan ; *p.* tealde ; *pp.* teald : also forms as from telian occur : ic telge, hí teliaþ ; *p.* telede ; *p.* teled. I. *to tell, narrate, recount, state* a case :—Þeáh ic hit lengre telle *though I make my story longer*, Chr. 1085 ; Erl. 218, 31. Dô ðæs leán tó ðám foresprecenan gódum ðe ic ðé ǽr tealde on ðriddan béc, Bt. 37, 2 ; Fox 190, 21. Se sunderhálga tealde his gódan dǽda, swilce God hí nyste, Homl. Th. ii. 428, 18. Swegen tealde ðæt his sciperes woldon wændon fram him *Swegen told (Beorn) that his (Swegen's) men would desert him (Swegen)*, Chr. 1046 ; Erl. 174, 13. Dauid tealde his ungelimp, and hú hé hine gebæd tó Gode, Ps. Th. 34, arg. Ða ungewiderunge ðe cómon swá wé beforan tealdon, Chr. 1086 ; Erl. 219, 33. Hí tealdon him (*Constantine*) ðæt he ðrowunga ðe úre Hǽlend ðrowede, H. R. 5, 21. Telle (*narres*) dínum suna hú oft ic hæbbe fordón ða Egiptiscan, Ex. 10, 2. Ute nú tellan (*let us state the case*) beforan swilcum déman swilce ðú wille *quovis judice contende*, Bt. 7, 3 ; Fox 20, 6. Ús sceamaþ hit nú máre tó tellanne *we are ashamed to tell any more of the matter*, Chr. 1050 ; Erl. 175, 39 : 1085 ; Erl. 218, 35. II. *to tell, count, reckon, compute, calculate* :—Hé teleþ (*computat*) ða andfengas ðe hine behéfe synt, Lk. Skt. 14, 28. Hé ne telþ hú miccle spéda wé áspendon, Homl. Th. i. 580, 17. Se lǽreów Béda telþ mid micclum gesceáde ðæt se dæg is xii. kl. Aprilis, 100, 13. 'Telle (*numera*) ǽlcne wépnedman' . . . Moises tealde (*numer-*

avit), Num. 3, 15, 16. Eallum ðe ðara cyninga tiide teledon *cunctis regum tempora computantibus*, Bd. 3 ; 1 ; M. 154, 10. Hí hluton, teledon *they cast lots and coun'ed*, Andr. Kmbl. 2207 ; An. 1105. Tele nú ða lenge ðære hwíle, Bt. 18, 3 ; Fox 66, 6. Tele nú ða gesǽlþa wiþ ðám sorgum *strike a balance between the happiness and cares*, 8, tit. ; Fox x, 22. Tele ðú ðæs mónan elde kl. Ian. óð ðæt ðú cume tó þrittiga ; fóh eft on ðone níwan, tele óð týne *starting from Jan. 1 with the number that marks the age of the moon on that day, count up to thirty ; begin then with the new moon, and count up to ten (the next Sunday after the date so reached will be Septuagesima Sunday)*, Lchdm. iii. 226, 30–228, 2. Telle ðæs steorran *numera stellas*, Gen. 15, 5 : Num. 1, 2, 3. III. *to reckon, account, consider*, (a) with an object having a noun, adjective, or phrase in apposition, *to consider* a thing such and such :—Hwam telle ic (*aestimabo*) ðás cneórysse gelíce ? Mt. Kmbl. 11, 16 : Lk. Skt. 7, 31. Ic Heaþobeardna hyldo ne telge Denum unfǽcne, Beo. Th. 4141 ; B. 2067. Ægleáwra mann ðonne ic mé tælige, Andr. Kmbl. 2967 ; An. 1486. Cyn ðara ðe hý ánsetlan teliaþ, R. Ben. 135, 4. Ic ðæt wénde and witod tealde, ðæt . . . , Exon. Th. 264, 1 ; Jul. 357. Ðone ic on firenum fæstne talde, Elen. Kmbl. 1815 ; El. 909. Ic nié ǽnigne . . . gesacan ne tealde, Beo. Th. 3551 ; B. 1773. Suá suá Saul ǽresð fleáh ðæt ríce and tealde hine selfne his suíðe unwierðne *sic Saul, qui indignum se prius considerans fugerat*, Past. 3 ; Swt. 35, 14 : Bd. 3, 14 ; S. 539, 42 : Beo. Th. 1592 ; B. 794 : 3625 : B. 1810. Gif se sacerd hine hreófligne tealde, Homl. Th. i. 124, 9. Hí hine oferhýdigne tealdon *eum notantes superbiae*, Bd. 2, 2 ; S. 503, 16. Hine Geáta bearn gódne ne tealdon, Beo. Th. 4375 ; B. 2184. Forcúþre is ðæt hé telle hine wísne, Wulfst. 59, 5. Ne mæg heó ús leáse tellan *mendacii arguere nos non potest*, Gen. 38, 23. Hine sylf ofer ealle men tellan, Chr. 1086 ; Erl. 222, 37. (b) with an object and prepositional phrase, *to consider* as (*tó, for, on*) :—Ne telle ic eów tó ðeówan *non dico vos servos*, Jn. Skt. 15, 15. Wé ðæt sylfe sár and wíte hyro on synne tellaþ *ipsam ei poenam suam in culpam deputamus*, Bd. 1, 27 ; S. 493, 25. Hig tellaþ mín wedd for náht *irritum facient pactum meum*, Deut. 31, 20. Ic ða geþeóde tó micclan gesceáde telede, Lchdm. iii. 442, 5. For náhte hé tealde ǽnig ðing tó biddenne búton gesihðe, Homl. Th. i. 158, 21. On bócum ðe ungelǽrede men þurh heora bilewitnysse tó micclum wisdóme tealdon *in books which unlearned men in their simplicity have considered as great wisdom*, 2, 22. Ðonne on úrum móde biþ ácenned sum ðing gódes, and wé ðæt tó weorce áwendaþ, ðonne sceole wé ðæt tellan tó Godes gyfe, and ðæt Gode betǽcan *consider it as God's grace, and attribute it to God*, 138, 23. Nis nú anweald tó tellanne tó sumum ðara héhstena góda ? . . . hwæðer nú gód hlísa sié for náuht tó tellenne ? Nis hit nán cyn, ðæt mon ðæt for náuht telle, Bt. 24, 4 ; Fox 86, 14–19. Se untweofealda biþ tó tellenne for fullfremod weorc, 36, 7 ; Fox 184, 24. (c) with a clause :—Hé tealde and wénde ðæt hé sceolde ða byldo his heortan ánescian *autumans se cordis ejus emollere constantiam*, Bd. 1, 7 ; S. 477, 43. Mid ðý hé tealde and hé wénde ðæt hé sweltan sceolde *cum se aestimasset esse moriturum*, 3, 27 ; S. 558, 41 : Cd. Th. 87, 3 ; Gen. 1443. Hú ne tealdan wit ðætte genyht wǽre gesǽlþa *nonne in beatitudine sufficientiam numeravimus* ? Bt. 35, 3 ; Fox 158, 12. Swá ðætte monige tealdon (*putarent*), ðæt heó gehǽled beón mihte, Bd. 4, 19 ; S. 589, 3 : Blickl. Homl. 117, 16. IV. *to impute* to (*dat.* or *prep.*), *ascribe, assign, put* a thing to a person's account :—Telle ic ða weorþmynd ðǽm wyrhtan næs ná ðé *ingenium mirabor artificis*, Bt. 14, 1 ; Fox 42, 18. Crist tealde ealne his wurðmynt tó his Fæder, Homl. Th. ii. 366, 16. Se wer ðam ðe ne tealde (*imputavit*) Drihten synne, Ps. Lamb. 31, 2. Ðæt ilce gér tó ðæs æfterfylgendan cyninges ríce teledon *idem annus sequentis regis regno adsignaretur*, Bd. 3, 1 ; M. 154, 12. Hí ealne ðone bryce uppon ðone cyng tealdon (cf. O. Sax. tellian an *to charge* ; *Icel.* telja á : see also on-talu) *they put all the breach of faith upon the king*, Chr. 1094 ; Erl. 230, 4. Ne tele ðú him ðis synn *ne statuas illis hoc peccatum*, Rtl. 44, 15. Telle hé ðæt Gode, næs him sylfum, L. E. I. 21 ; Th. ii. 416, 18. His niéhstena gód hé sceal tellan him selfum *he is to reckon as an item in the account of his own prosperity that of his neighbour* ; *sua commoda propinquorum bona deputare debet*, Past. 13 ; Swt. 79, 1. Se fulla anweald is tó tellanne tó ðám héhstum gódum *complete power is to be assigned to the class of highest goods*, Bt. 36, 7 ; Fox 184, 9. [O. Sax. tellian : O. Frs. tella : O. H. Ger. zellen ; *p.* zalta, zelita *numerare, computare, reputare, dicere, re'erre, narrare, notare, tribuere* : *Icel.* telja.] v. á-, be-, ge-, tó-tellan ; talian.

télnis, telo, teltré, tém, -téma, téman, -téme, témen. v. tælness, tela, teld-treów, teám, -tíma, tíman, -tíme, tímen.

Temes, Temese *the Thames*. In the declension both weak and strong forms are found. [In Latin, *nom.* Temis, Cod. Dip. Kmbl. i. 30, 12, Temes, ii. 23, 12 : *gen.* Tamisae, i. 98, 1 : *dat.* Taemise, 216, 25 : *acc.* Tamesim Bd. 1, 2 ; S. 42, 34 may be cited] :—Neáh ðære ié ðe mon hǽt Temes (Temese, MS. C.) *ad flumen Tamesim*, Ors. 5, 12 ; Swt. 238, 22. Sý eá hátte Temese, Chr. Erl. 5, 11. Ymbe heora landgemǽra : andlang Temese (on Temese, 8), L. A. G. 1 ; Th. i. 152, 18. Út on Temese ; ðonne ondlong Temese, Cod. Dip. Kmbl. iii. 438, 3–4. Fóron be Temese . . . be norþan Temese, Chr. 894 ; Erl. 92, 14, 20. Hí tugon hira scipu

up on Temese, 895 ; Erl. 93, 31. Hī nāmon him wintersetl on Temesan and lifedon of Eást-Seaxum, 1009 ; Erl. 143, 4.

temes(-is), es ; *m.* (? cf. lynis *for form and gender*) *A sieve.* [Temse *taratantarum*, Wrt. Voc. i. 200, col. 2 (15th cent.). Temze, temeze, temse, sive *setarium*, Prompt. Parv. 488. See also Halliwell, who quotes : ' Marcolphus toke a lytyll cyve or temse.' He gives, besides, '*temzer* a range or coarse searche' as an early Wiltshire word. Wright, in the note to the word in his Vocabulary, says that *temse* is still in use in the North of England. *O. Du.* tems. (The word seems to have been borrowed from a Teutonic source by French, which has *tamis* a sieve, *tamiser* to sift.) Cf. *O. H. Ger.* zemisa *furfures.*] v. next two words.

temesian, temsian *to sift* :—Hlāfo foregegearwad ł temised *panes propositionis* (cf. Tusser's Husbandry, 39, 10 : ' Some mixeth the rie with the wheat *Temmes* lofe on his table to haue for to eate.' In such a loaf the coarse bran only is removed. v. Glossary. *Temse-bread* is given in Ray's South and East-Country Words, E. D. S. Pub. B. 16), Mk. Skt. Lind. 2, 26. [Temzyn wythe a tymze, temsyn with a tenze *attamino, setario.* To tempse or syfte *taratantariso*, Prompt. Parv. 488. Cf. *temsing-chamber*, the sifting-room, Halliwell. *O. Du.* temsen *to sift.*] v. ge-temesed, *and preceding word.*

temes-pīle, an ; *f. A stake to support a sieve* [A ' temsynge staff ' = *cervida*, lignum quod portat cribrum, Prompt. Parv. 488, note 3] :—Man sceal habban syfa, hriddel, hērsyfe, tæmespīlan, fanna, Anglia ix. 264, 14. v. preceding words.

temian ; *p.* ede, ode *To tame* :—Ic temige *domo*, Ælfc. Gr. 24 ; Zup. 138, 2. Ic gewylde oððe temige, 36 ; Zup. 213, 14. Ic genyme mē briddas on hærīæste and temige hig, Coll. Monast. Th. 26, 5. Mon temeþ his unāliéfde lustas mid ðæm wordum ðære hālgan lāre, Past. 56 ; Swt. 433, 12. Gewylt, temaþ *domat, superat*, Wrt. Voc. ii. 141, 73. Hē ðone ealdan līchoman swencte and temede (*domabat*), Bd. 5, 12 ; S. 631, 36. Heora lāreówas ðe hī (wudufuglas) temedon, Met. 13, 39. Canst ðū temian (*domitare*) hig (*hawks*)? Coll. Monast. Th. 25, 21, 25. Wilde deór temian, Lchdm. iii. 200, 1 : 186, 21. Nȳtenu tymian, 184, 18. Temma *domare*, Mk. Skt. Lind. Rush. 5, 4. [*Goth.* ga-tamjan : *O. Frs.* tema : *O. H. Ger.* zemmen : *Icel.* temja.] v. ā-, ge-temian.

temised. v. temesian.

templ, tempel, es ; *n. A temple* :—Se wītga spræc suelce ðæt templ wære eal tōworpen ; hē cuæð . . . 'Tōworpne sint ða stānas ðæs temples,' Past. 18 ; Swt. 133, 10. ' Ðis templ wæs getimbrod on six and feówertigon wintron' . . . Hē hyt cwæð be hys līchaman temple, Jn. Skt. 2, 20, 21. Ðæt templ ealre clǣnnesse (*the Virgin's womb*), Blickl. Homl. 5, 19. Ofer ðæs temples heáhnesse, Mt. Kmbl. 4, 5 : 24, 1. On hālierne ł hergan, temple *sacello*, Hpt. Gl. 482, 37. Se Hǣlend com tō ðam temple, Jn. Skt. 8, 2. Wē wunedon wið Phogores templ *mansimus contra fanum Phogor*, Deut. 3, 29. Ðes tōwyrpþ Godes templ, Mt. Kmbl. 27, 40. On ðæt hālige Salemannes templ, Blickl. Homl. 17, 18. Ic lǣre ðæt ðæt tempel wē forleósan, Bd. 2, 13 ; S. 516, 33. Ōðre þeóda fela templa ātærdon, Homl. Th. ii. 574, 27. In Godes templum, Exon. Th. 131, 26 ; Gū. 461. Hī Godes tempel brǣcon and bærndon, 44, 24 ; Cri. 707. Templu ūre wē gehealdan, Scint. 16, 9. [*O. H. Ger.* tempal. For native words used before the Latin form was borrowed, v. hearh, ealh ; and cf. *Goth.* alhs : *O. Sax.* alah : *Icel.* hof, for similar terms in other dialects.]

templ-geat, es ; *n. The gate or door of a temple* :—Hē æt sumum sǣle stōd æt ðam tempelgeate, Wulfst. 49, 25.

templ-geweorc, es ; *n. A temple-building, temple* :—His þegnas āgunnon specan wið hine ymbe ðæt mǣre tempelgeweorc ðe ðǣr geworht wæs Gode tō wyrðmynte, Wulfst. 88, 17. Salomon wes se forma man ðe Gode tō lofe ǣrest on eorðan templgeweorc ārǣrde, 277, 25.

templ-hálgung, e ; *f. A consecration-festival* :—Ðā wǣron templhālgunga (*encenia*), Jn. Skt. 10, 22 : *schenofegias*, Engl. Stud. xiii. 27, 14.

templ-lic ; *adj. Pertaining to a temple*; the word translates *fanaticus* :—Hearhlīcre, ðæs hǣþenan, *vel* templīcre *fanatice*, i. *profani*, Wrt. Voc. ii. 147, 38. Templīcre ł dióflīcre *fanatica*, Hpt. Gl. 482, 25 : Anglia xiii. 34, 176.

temprian ; *p.* ode, ede. **I.** *to mix in due proportion, to mingle* :—Ic temprede (*potum meum cum fletu*) *temperabam*, Blickl. Gl. **II.** *to temper, regulate, moderate* :—Seó sunne gǣþ geond stōwa and tempraþ ða eorðlican wǣstmas ǣgðer ge on wæstme ge on rīpunge, Lchdm. iii. 250, 17. Hī nā tempreuon gyfernysse hǣtan *non temperauerunt gulae ardorem*, Scint. 107, 12. Bryne līchamena mid cealdrum ēstum tō temprigenne (*temperandus est*), 52, 2. [*O. H. Ger.* temp[a]rōn *obtemperare, temperare, medicare* : *Icel.* tempra. From Latin.] v. ge-temprian.

temprung, e ; *f. Tempering, moderation* :—Swā hwæt on temprunge byþ hālwende hit ys *quicquid temperamento fit salutare est*, Scint. 55, 1. Hafa ðū temprunge (*temperamentum*, i. *mediocritatem*), 172, 13. [*O. H. Ger.* temp[a]runga *temperantia, compositio.*]

temsian, tēn. v. temesian, tīn.

[**tendan** ; *p.* de *To kindle.*] [A gnast wule al þe brond tenden, O. E. Homl. i. 81, 7. Cwench hit er þen hit waxe and tende þe, A. R. 296, 21. It bigynneþ forto tiende, L. S. 314, 523. Itend of wreððe, Kath. 154. Teenden *incendere*, Wick. *Goth.* tandjan ; *Da.* tænde : *Swed.* tända.] v. ā-, on-tendan ; tennan.

-tendend, -tending, -tendness. v. ā-tendend, ā-tending, on-tendness.

tender *fuel* :—Tender *fomes*, Ælfc. Gr. 9, 26 ; Zup. 52, 11. Ðæt ne gehigeleás[t] mēte tender *ut non scurilitas inveniat fomitem*, R. Ben. Interl. 75, 17. v. tynder.

Tenet, Tǣnet[t] ; *also* Tenet-land *the isle of Thanet* :—Augustinus wæs cumende on Bretone ǣrest on Tenet ðam eálonde (Tenet-land, MS. B.) (*in insula Tanato*) . . . Is on eásteweardre Cent mycel eáland Tenet (*Tanatos insula*), ðæt is syx hund hīda micel . . . Ðæt eálond tōsceáðeþ Wantsumo stream fram ðam tōgeþeóddon lande, Bd. 1, 25 ; S. 486, 10-20. Hēr hǣðene men on Tenet ofer winter sǣton, Chr. 851 ; Erl. 67, 20 : 865 ; Erl. 70, 31. On ðyssum geáre Eádgār cyng hēt oferhergian eall Tenetland, 969 ; Erl. 125, 5. Tǣnet, Cod. Dip. Kmbl. iv. 232, 22. Inntō Raculfe on Tǣnett, iii. 429, 16. The following forms occur in Latin charters :—Tenid, i. 21, 1. Tenaet, 129, 18. Tanet, 118, 1. Tanat, vi. 189, 31. Tanatos insulam, iv. 237, 20. Insula Tanatorum, iii. 347, 15. Thanet, i. 13, 30 : 18, 15. Ðanet, v. 21, 19. Insula Thaeneti, i. 42, 16. Insula Thaenet, 116, 27.

tengan ; *p.* de *To press, hasten, hurry, proceed with haste or violence* :—Ðā tengde se Pharao æfter mid mycelre fyrde *then Pharaoh hastened after with a great army*, Homl. Th. i. 312, 3 : ii. 194, 16. Hē ðā þearle āblicged āweg tengde, 182, 2. Hē ontende ða burh and tencgde him ford syððan, Homl. Skt. ii. 25, 416. Se cǣsere tengde tō ðam botle, Homl. Th. i. 430, 23. Se fugol tō wuda tengde, ii. 162, 27. Æt suman cyrre tengde hē tō fyrde ongeán Persiscne leódscipe *on one occasion he was hastening to march against Persia*, i. 448, 32. Tengdon ða hæþenan mid wǣpnum tō ðam ǣwfæstum heápe, and slōgon ða cristenan, Homl. Skt. ii. 28, 66. Teng recene tō ðam fæstenne (*haste thee, escape thither*, Gen. 19, 22), Cd. Th. 152, 29 ; Gen. 2527. Hié hæfdon gecweden ðæt hié ealle emlīce on Latine tengden *they had agreed that they all in unbroken order would proceed to the attack of the Latins*, Ors. 3, 6 ; Swt. 108, 9. v. ge-tengan ; ge-tenge.

tennan (?) *to incite, encourage to effort* :—Ful oft ðæt gegongeþ, ðætte wer and wīf in woruld cennaþ beorn, and mid bleóm gyrwaþ, tennaþ and tǣtaþ, ōþþæt seó tīd cymeþ, ðæt ða geongan leomu, liffæstan leoþu, geloden weorþaþ (*the parents try to awaken the child's activity of body and mind, while it is still an infant*), Exon. Th. 327, 15 ; Vy. 4. [Thorpe would read *temiaþ*. Grein suggests comparison with *O. M. H. Ger.* Cf. Ih zeno sie *provocabo eos*, Grff. v. 685. Could *tendaþ* be read ? *Ontendan* and connected words are used figuratively ; see also *tendan.*]

tenys, Hpt. Gl. 513, 65. v. týnness.

teofonian ; *p.* ode *To associate, join* :—Ealswā teofanade se ðe teala cūþe ǣghwylc wiþ ōþrum ; sceoldon eal beran stīþe stefnbyrd, swā him se steóra bibeád, missenlīce gemetu (cf. þeáh ānra hwilc (*each of the elements*) wið ōþer sié miclum gemenged . . . fæste gebunden . . . mid bebode ðīne, Met. 20, 65-69). Exon. Th. 349, 8 ; Sch. 43. Swā teofenede se ðe teala cūþe dæg wiþ nihte . . . fisc wið ȳþum, 351, 18 ; Sch. 82.

teofrian ; *p.* ode *To allot* (?), *appoint* :—Ðone sylfan stān ðe hine wyrhtan āwurpan nū se geworden is hwommona heágost hālig Drihten tō wealles wraðe wīs teofrade (*he has appointed it to be the wall's support*) *lapidem quem reprobaverunt aedificantes, hic factus est in caput anguli : a Domino factum est illud*, Ps. Th. 117, 21. v. tiber (tifer).

teogopa (-eþa), teogoþian. v. teóþa, teóþian.

teoh [h], e ; *f.*; but also *m.* or *n. An association, a company, band* :—Besæt hē ðā sinherge sweorda lāfe weán oft gehēt earmre teohhe *with a migh'y host he besieged then those whom the sword had spared, to the wretched band woe he oft promised*, Beo. Th. 5868 ; B. 2938. Oððæt ic ðīnes earmes āsecge strencðe ðisse cneórisse eallum ðam teohhe ðe nū tōweard ys *donec annuntiem brachium tuum generationi omni, quae ventura est*, Ps. Th. 70, 17. Ðā hié gemitton teoche æt torre (*the people who were building the tower of Babel*), Cd. Th. 101, 26 ; Gen. 1688. Hēt tuddorteóndra teohha gehwylcre wæstmas fēdan *he bade each productive race bring forth fruits*, 59, 6 ; Gen. 959. [*M. H. Ger.* zeche ; *f. succession, association, company* : *Ger.* zeche.] v. next word.

teohhian, teohchian, teohgian, tihhian, teohian, teochian, tihian ; *p.* ode. **I.** *to suppose, consider, think*, (a) *with a clause* :—Ic tiohhie, ðæt hió ðæs taman nāuht ne gehicgge, Met. 13, 25. Gif hwā teochaþ (tiohhaþ, Cott. MSS.) ðæt hē æfæst sié *si quis putat se religiosum esse*, Past. 38 ; Swt. 281, 2. Swā hwæt swā hē swīþost lufaþ ðæt hē teohhaþ (tiohhaþ, Cott. MS.) ðæt him sié betst . . . ðonne ðe ðæt begiten hæfþ ðonne tihhaþ hē ðæt hē mǣge beón swīðe gesǣlig *quod quisque prae ceteris petit, id summum esse judicat bonum . . . beatum esse judicat statum, quem prae ceteris quisque desiderat*, Bt. 24, 3 ; Fox 84, 11-14. Tehhaþ, Fox 84, 16. Sume wēnaþ, ðæt . . . Sume teohhiaþ, ðæt . . . Manege tellaþ, ðæt . . ., 24, 2 ; Fox 82, 7-12 : 26, 2 ; Fox 92, 26 : Ps. Th. 11, 4. Hié tiohchiaþ ðæt ðæt (*silence*) scyle bión for eáðmēttum *tacere se aestimant ex humilitate*, Past. 41 ; Swt. 302, 3. Nān ðara gōda ðīn nis ðe ðū teohhodest (tiohhodes, Cott. MS.) ðæt hī ðīne beón sceoldan *nihil horum, quae in tuis computas bonis, tuum esse bonum monstratur*, Bt. 14, 2 ; Fox 42, 29. Se leása wēna ðara dysigena monna tiohhie, ðæt . . . *hominum fallax adnectit opinio*, 27, 3 ; Fox 98, 32. (b) *with tō, to consider as* :—Of gromra gripe, ðe ðū tō godum tiohhast *from the clutch*

of cruel ones, whom thou countest as gods, Exon. Th. 255, 17; Jul. 215.
Ælc mon tiohhaþ him ðæt tó sélestum goode ðæt ðæt hé swíþost lufaþ *every man considers that as his best good, which he most loves,* Bt. 33, 2; Fox 122, 23. Hí teohhiaþ ús him tó snædincgsceápum *aestimati sumus ut oves occisionis,* Ps. Th. 43, 23. Ðam wísan men com tó lofe and tó wyrðscipe ðæt se unrihtwísa cyning him teohhode tó wíte *cruciatus, quos putabat tyrannus materiam crudelitatis, vir sapiens fecit esse virtutis,* Bt. 16, 2; Fox 52, 27. Gif hé hit ne tiohchode eall tó anum *si utraque unum esse non decerneret,* Past. 49; Swt. 385, 34. (c) in other ways:—Teohgaþ *decreverit, cogitaverit,* Hpt. Gl. 412, 48. Ne biþ hé swá brád swá hé teohgaþ (tihhaþ, Cott. MS.), Bt. 30, 1; Fox 108, 12. **II.** *to purpose, determine, intend, appoint,* (a) with an accusative:—Man ús tyhhaþ twegen eardas *two dwellings are intended for us,* Hy. 7, 97. Oft ic léan teohhode hnáhran rince, Beo. Th. 1907; B. 951. (b) with an accusative and (implied) infinitive:—Swilce hé ná ða spræce ne mænde and tiohhode hit þeáh þiderweardes (*and yet he intended it to go in that direction*), Bt. 39, 5; Fox 218, 12. (c) with a clause:—Tó ðæm sóþum gesælþum ic tiohhie (tiohige, Cott. MS.) ðæt ic ðé læde, Bt. 22, 2; Fox 78, 7. Swá swá hé tiohhaþ, ðæt hit sié, 39, 6; Fox 220, 7. Nis nán gesceaft ðe he tiohhige (tiohhie, Cott. MS.) ðæt hió scyle winnan wiþ hire Scippendes willan . . . Hwæt wénst ðú gif ǽnegu gesceaft tiohhode ðæt hió wiþ his willan sceolde winnan hwæt hió mihte wiþ swá mihtiue swá wé hine gerehtne habbaþ *nihil est quod Deo contraire conetur . . . quid si conetur, num tandem proficiet quidquam adversus eum, quem potentissimum esse concessimus,* 35, 4; Fox 160, 21–27. Ðæt hé forðý reáfíge ðý hé tiohchie (teohhige, Cott. MSS.) ðæt hé eft scyle mid ðý reáflíace ælmessan gewyrcean *pro misericordia facienda peccare,* Past. 45; Swt. 341, 22. (d) with *tó:*—Swá hwæt swá ðú mé tó gyfe tihhie bring ðæt Gode tó onsægednysse *whatever you may intend as a gift to me, bring that as a sacrifice to God,* Homl. Ass. 123, 209. (e) with gerundial infinitive:—Ðær ðú ongeáte hwidre ic ðé teohhie (tiohige, Cott. MS.) tó lædenne *si, quonam te ducere aggredimur, agnosceres,* Bt. 22, 2; Fox 78, 1. Cildum ðe wé tiochiaþ úrne eard tó tó forlǽtanne, and hié tiochiaþ ús tó ierfeweardum tó habbanne, Past. 50; Swt. 391, 28. Hý teohhiaþ mé tó áfyrranne, Ps. Th. 39, 16. Hé tiohchode him mid tó fultemanne . . . hé teohchode hine tó lǽdanne on lífes weg, Past. 41; Swt. 305, 4, 5. His (*Ulysses*') þegnas for hiora eardes lufan tihodon hine tó forlǽtanne, Bt. 38, 1; Fox 194, 29. (f) undetermined:—Teohhaþ *distinat,* i. *disponit, contendit,* Wrt. Voc. ii. 141, 35. [Cf. *O. H. Ger.* gi-zehôn *instaurare, resarcire.*] v. ge-teohhian; teón (*wk.*).

-teohhung, teolian. v. fore-teohhung, tilian.

teol-þyrel, es; *n. A window :*—Teolþerla *fenestrarum,* Hpt. Gl. 409, 31. Cf. eág-þyrel.

teolung, teoma. v. tilung, tam.

teón (*from* teóhan); *p.* teáh, *pl.* tugon; *pp.* togen, tigen (v. of-teón) *To draw, pull :*—Ic teó *traho,* ic teó swýðe *pertraho,* Ælfc. Gr. 28, 5; Zup. 176, 5, 6. Teáþ *trahunt,* Wülck. Gl. 253, 32. **I.** (1) with the idea of horizontal movement, *to draw* along, *pull, drag :*—Ðú mé gebundenne mid fýrenum racenteágum týhst in éce fýr, Shrn. 117, 18. Heó teáh hyne (*Holofernes*) folmum wiþ hyre weard, Judth. Thw. 23, 1; Jud. 99. Ðá geseah ic monige ðara wérigra gásta fíf monna sáwla teón (*trahere*) on midde ða ðýstro . . . Tugon hí ða werígan gástas, Bd. 5, 12; S. 628, 32–36. Valerianus hét teón Ypolitum geond ðornas and brémelas, Homl. Th. i. 432, 34; Blickl. Homl. 241, 21. Se eádiga Andreas wæs togen, 241, 26. (2) where the movement is from within or from without, *to draw* a sword, blood, etc., *to haul* a net, *draw* in *or* out :—Ðú scealt, ðonne ðú on ðám sculdrum týhst blód, teón swíðe on ðære sídan, Lchdm. ii. 262, 26. Se iii tíhþ his fét suá hé inmest mæg . . . Hé tiéhþ his heáfod in tó him, Past. 35; Swt. 241, 11–21. Ða synfullan teóþ heora sweord *gladium evaginaverunt peccatores,* Ps. Th. 36, 13. Simon Petrus téh his nett on land, Jn. Skt. 21, 11. Teóh mid glæse oþþe mid horne, Lchdm. ii. 200, 13: 262, 5. Tæppan teón, Techm. ii. 120, 12. Teóþ út lange, Lchdm. iii. 16, 13. Onlegena út teónde ðone heardan swile, ii. 182, 16. Wæs on næs togen wundorlíce wægbora, Beo. Th. 2883; B. 1439. (3) where the movement is up or down, *to draw up or down, to draw* breath, *heave* a sigh, &c., *to hoist* a sail, *pull* a bell :—Mé tó grunde teáh feóndscaða, Beo. Th. 1111; B. 553. Hé oroð stundum teáh (cf. oroð up hlæden, v. 30), Exon. Th. 178, 17; Gú. 1245: Guthl. 20; Gdwin. 86, 16. Godwine eorl teáh up his segl, Chr. 1052; Erl. 183, 12. Hí tugon up heora segel, 1046; Erl. 174, 19. Ða apostolas tugon hié up and hié gesetton on ðæm fægran neorxna wange, Blickl. Homl. 143, 24. Tugon hié heora hrægl bufan cneów, Ors. 3, 5; Swt. 106, 16. Dó mid his handa, swylce hé wille áne hangigende bellan teón, Techm. ii. 118, 16. Heó longe swóretunge wæs teónde, Bd. 4, 23; S. 596, 10. (4) *to draw* to, *to attract :*—Ðære lyfte gecynd is ðæt heó téhþ tó ða rénas of ðæm sealtan sǽ, Shrn. 63, 27. (5) *to pull* the string of a bow, *strike* the strings of an instrument :—Ðære hearpan strengas se hearpere suíðe ungelíce tíehþ and styreþ, Past. 23; Swt. 175, 7. Ða teóþ heora suíðne bogan *intenderunt arcum,* Ps. Th. 63, 3. Togenum strengum, Ps. Th. 67, 24. (6) *to pull* a boat, *to row :*—On ða eá hí tugon up hiora scipu oþ ðone weald, Chr. 893; Erl. 88, 31: 895; Erl.

93, 31. Ðæt scip wile hwílum stígan ongeán ðone streám, ac hit ne mæg, búton ða rówend hit teón, ac hit sceal fleótan mid ðý streáme; ne mæg hit nó stille gestondan, búton hit ancor hæbbe, oððe mon mid róðrum ongeán tió, Past. 58; Swt. 445, 10–13. Hé ástígende on án scyp bæd hyne ðæt hé hit lythwón fram lande tuge . . . Hé cwæþ tó Simone: 'Teóh hit on dýpan,' Lk. Skt. 5, 3, 4. (7) *to draw, be of weight :*—Ðonne man sett ða synne and ða sáwle on ða wæge, and hý man wegeþ, swá man déþ gold wið penegas. And gif ða penegas teóþ swíðor ðonne ðæt gold, ðonne miswyrd ðam men hraðe. Swá biþ ðære sáwle and ðære synne; gif seó synn tíhþ swýðor ðonne seó sáwel, ðonne faraþ hý on forwyrd, Wulfst. 240, 1–6. (8) where there is no movement, *to pull, tug :*—Sume sceufon, sume tugon . . . and seó Godes fæmne hwædre stód. Ða brudon hig rápas on hyre handa and on hyre fét, and hig tugon myd ðam, and hig ne myhton hig ða git ánne fótlást furður áteon, Shrn. 154, 26–30. Se deófol wolde geniman ðone cnapan of Basilius handum, hetolíce teónde, Homl. Skt. i. 3, 443. **II.** *to bring, lead, put :*—Ða teáh hine Penda fyrde and here on, Bd. 3, 7; S. 529, 30: 1, 34; S. 499, 29. Penda teáh here wiþ Eást-Engle, 3, 18; S. 546, 14. 'Teóh eft ðíne hand on ðínne bósum.' Ða teáh hé hig ongeán, Ex. 4, 7. Héht eorla hleó eahta mearas on flet teón, Beo. Th. 2077; B. 1036. **II a.** with an idea of violence or compulsion :—Ða cwæð Iosue: 'Teóþ ða cynegas út of ðam scræfe,' Jos. 10, 22. Gif fáh mon cirican geierne, hine seofan nihtum mon út ne teó, L. Alf. pol. 5; Th. i. 64, 10. Belǽwende eów on gesamnungum and teónde tó cynegum, Homl. Th. ii. 540, 17. **III.** in various figurative senses, many of which may be rendered by words containing the root of *trahere* or of *ducere.* (1) *to teach, educate, bring up :*—Ic tý (teó, MSS. J. W.) oððe lǽre *imbuo,* ic teáh *imbui,* Ælfc. Gr. 28, 3; Zup. 166, 14. Hú lange týhst ðú ús and fédest teára hláfe *cibabis nos pane lacrymarum,* Ps. Th. 79, 5. Hwá teáh ðé ? . . . Se Hælend mé lǽrde mid onwrigenysse, Homl. Th. i. 378, 9. Hé iunge men teáh georne mid láre, swá ðæt ealle his geféran sceoldon sealmas leornian, Homl. Skt. ii. 26, 76. Wé lǽraþ ðæt preóstas geóguðe geornlíce lǽran and tó cræftan teón (*bring them up to crafts*), L. Edg. C. 51; Th. ii. 254, 26: L. Pen. 14; Th. ii. 282, 6. (2) *to draw* to *or* from, *attract, induce, seduce :*—Sió leáse gesælþ tíhþ ða ðe hiere tó geþeódaþ from ðæm sóþum gesælþum mid hiere ólecunge, Bt. 20; Fox 72, 7. Sió gecynd eów tíhþ tó ðam angite, ac eów tíhþ (teóhþ, MS. Bod.) gedwola of ðam angite, 26, 1; Fox 90, 7. Ðes middangeard wæs tó ðon fæger, ðæt he teáh men tó him þurh his fægernesse fram Gode, Blickl. Homl. 115, 11. Ðone mon sciele ealle mægene tó biscepháde teón ðe gástlíce liofaþ *ille modis omnibus debet ad exemplum vivendi pertrahi, qui spiritaliter vivit,* Past. 10; Swt. 60, 7. (3) *to draw* to one's self, *to take :*—Ic teó (nimo, Lind. Rush.) ealle þing tó mé sylfon, Jn. Skt. 12, 32. Sume hí teóþ *nominativum casum,* Ælfc. Gr. 33; Zup. 2068. Ne teáh Crist him ná tó on ðisum lífe land ne welan, Homl. Th. i. 160, 32: Ors. 5, 11; Swt. 236, 27. Hé æfter ðysum geþance teáh him elnunge tó ðæle *after this thought he in some measure took courage,* Homl. Skt. i. 23, 524. On ðæt geråd ðæt hié him Siciliam tó ne tugen ne Sardiniam *conditiones erant, ut Sicilia Sardiniaque decederent,* Ors. 4, 6; Swt. 180, 13. Ðæt hé hit on folcryht him tó teó, L. Ath. i. 9; Th. i. 204, 12. Ne teó se hláford ná máre on his æhte búton his rihtan heregeate, L. C. S. 71; Th. i. 412, 29. Ne teón hié nánwuht ðæs lofes tó him, Past. 44; Swt. 323, 1. (4) *to take* on one's self, *to assume :*—Hié him on teóþ, ðæt hié síen heortan lǽcas, Past. 1; Swt. 27, 1. Ðæt hé tió on hine selfne óðerra monna scylda, 16; Swt. 99, 1. Sanctus Paulus ðeóde lǽrde, ðæt hé him anwald on tuge, 40; Swt. 291, 20. Se him wæs on teónde ealdordóm ofer ða óþere, Ors. 2, 6; Swt. 88, 20. (5) *to bring, bring forth, produce, display :*—Meaht forð tíhþ heofoncondelle and holmas níd, Exon. Th. 349, 29; Sch. 53. Ða ðe plegaþ æt deádra manna líce and ælce fúlnysse ðær forð teóþ mid plegan, Homl. Skt. i. 21, 309. Ðú wið Criste wunne and gewin tuge, 267, 27; Jul. 421. Ða sceolde se ealdorman Ælfríc lǽdan ða fyrde, ac hé teáh forð ðá his ealdan wrenceas *he brought out his old tricks,* Chr. 1003; Erl. 139, 7. Hygewælmas (-os, MS.) teáh beorne on breóstum níð *envy produced fierce passions in the breast of the man,* Cd. Th. 60, 12; Gen. 980. Teón nú ða wæteru forð swimmende cynn . . . eall fisccynn ðe ða wæteru tugon forð (*produxerunt*), Gen. 1, 20, 21. Tó teónne forð ðone wísdóm ðære ealdan ǽ, Homl. Th. i. 190, 8. (6) *to bring, place :*—Sió ungelícnes hira geearnunga hié tíehþ sume behindan sume and hira scylda hé ðær gehabbaþ *variante meritorum ordine alios aliis culpa postponit,* Past. 17; Swt. 107, 20. Bisceop sceal scyldan cristenum mannum wið ælc ðæra þinga ðe synlíc biþ, and ðý hé sceal on æghwæt hine ðe swýðor teón (*he must the rather bring himself to everything, apply himself*), ðæt hé geornor wíte hú seó heord fare, L. I. P. 7; Th. ii. 312, 24. **IV.** *to draw* (ar in *to draw* nigh), *to go, proceed,* (1) intrans. :—Seó tó hám týhþ, Exon. Th. 416, 26; Rä. 35, 4. Hé ne mihte ongemong óþrum mannum bión, ac teáh tó wuda, Bt. 35, 6; Fox 168, 7. Hí tugon forð *they went on their way,* Homl. Th. i. 46, 11: ii. 490, 1. Fela hám tugon, Chr. 1096; Erl. 233, 23. Hira tungan tugon ofer eorðan *lingua eorum transivit super terram,* Ps. Th. 72, 7. Gif tósomne teó *if (hair-lip) draw together,* Lchdm. ii. 56, 9. (2) *with*

acc. *to go* a journey :—Æghwylcum đara đe mid Beówulfe brimláde
leáh, Beo. Th. 2107; B. 1051: 2669; B. 1332. Yldran ússe tugon
tongne síđ, Exon. Th. 228, 19; Ph. 440: 110, 28; Gú. 115. (3) figuratively :—Nú fandiaþ swelce wræccan and teóþ tó, woldon underión
đone weordscipe *such wretches press forward in their wish to receive the
honour*, Past. 7; Swt. 51, 22. [Laym. teon *to go, march*: Kath. teon
to pull: Gen. and Ex. ten *to go*; *to bring up*. Goth. tiuhan : O. Sax.
tiohan: O. Frs. tiá: O. H. Ger. ziohan *trahere, ducere, nutrire*.] v. á-,
be-, ge-, of-, ofer-, on-, óþ-, þurh-, wiđ-teón; for-, íđ-togen; teónd.

teón (*from* tíhan; *but the verb seems to have almost entirely given up
the conjugation to which this form would belong and to take that of* teón
from teóhan); *p.* teáh, *pl.* tugon; *pp.* togen, tygen *To accuse* a person
of something (acc. of person and gen. of charge, or charge expressed by
a clause) :—Đú mê stale týhst *furti me arguis*, Gen. 31, 32. Hwí tíhþ
úre hláford ús swá micles falses? 44, 7. Gif gê scyld on eów witen đæs
đe eów man tíhþ, Txts. 176, 10; Rtl. 114, 23: Exon. Th. 345, 13;
Gn. Ex. 187. Týhþ, Cd. Th. 36, 33; Gen. 581. Ic eom unscyldig æt
đære tihtlan đe N. mê tíhþ (týhþ, MS. B.), L. O. 5; Th. i. 180, 16.
Hý teóþ đê đæs đe hý sylfe habbaþ, Prov. Kmbl. 12. Hê teáh hiene
đæt hê his ungerisno spræce wiđ đa senatos *he* (Philip) *charged him*
(Demetrius, his son) *that he had spoken disparagingly of him to the
senate*, Ors. 4, 11; Swt. 206, 28. Đá tugon hié hiene, đæt hê heora
swicdómes wiđ Alexander fremmende wære, and hiene for đære tihtlan
ofslógon, 4, 5; Swt. 168, 16. Gif hine hwá hwelces teó, L. Alf. pol.
17; Th. i. 72, 6: 11; Th. i. 68, 19: L. In. 30; Th. i. 120, 18. Gif
hine man æniges þinges teó, L. C. S. 31; Th. i. 394, 28. Gif hine mon
tió gewealdes on đære dæde, L. Alf. pol. 36; Th. i. 84, 15: 31; Th. i.
80, 16. Gif man đone hláford teó, đæt hê be his ræde út hleópe, L. C. S.
30; Th. i. 394, 19. Gyf hine þreó men ætgædere teón, Th. i. 392, 23. Se
man đe man tuge *the man who shall have been accused*, L. Ath. iv. 6; Th. i.
224, 15. Gif hwá óđerne tión wille, đæt hê hwelcne ne gelæste đara đe hê
him gesealde, L. Alf. pol. 33; Th. i. 82, 5. [Goth. teihan to *shew*: O.
Sax. af-tíhan *to refuse*: O. H. Ger. zíhan *arguere*: Ger. *zeihen* to *accuse*:
Icel. tjá (*wk.*) to *shew*; cf. tiginn *distinguished*.] v. be-teón; teónd; tiht.

teón; *p.* teóde. I. *to make, frame, create, ordain, arrange,
contrive, bring about, construct*, (1) referring to material objects :—
Đysne wig đe đú đê tó wundrum teódest, Cd. Th. 228, 25; Dan. 208.
Thâ middungeard moncynnæs uard æfter tiáde (teóde, Bd. 4, 24; S. 597,
23) *dehinc terram custos humani generis creavit*, Txts. 149, 8. Helm
worhte wæpna smiđ, wundrum teóde, besette swínlícum, đæt hine bead-
omêcas bítan ne meahton, Beo. Th. 2909; B. 1452. Tó đam golde đe
hê him tó gode teóde *the gold that he had shaped for a god to himself*,
Cd. Th. 229, 13; Dan. 216. Se đás woruld teóde, Exon. Th. 335, 16;
Gn. Ex. 34: Andr. Kmbl. 1594; An. 798. (1 a) in a figurative expression :—
Đa heora tungan teóþ (*but the word may be from* teón to *draw*. v.
teón, I. 2), *as it seems also to govern* bogan *in the following clause*)
teónan gehwylce sweorde efenscearpe *exacuerunt ut gladium linguas
suas*, Ps. Th. 63, 3. (2) referring to immaterial objects :—Đæs đê þanc
sié đæt đú ús đás wrace teódest *for this be thanks to thee that thou didst
order this exile for us*, Cd. Th. 235, 21; Dan. 309. Him heáhcyning
fultum tióde *for him the high king contrived help*, 11, 11; Gen. 173.
Se đe ús đis líf tióde *he that framed for us this life*, Met. 20, 131.
Waldend him đæt wíte teóde, Exon. Th. 336, 4; Gn. Ex. 43. II. *to
furnish* with; instruere :—Mid beorhtnyssa ærnemergen þú tihst and mid
fýrum middæg *splendore mane instruis et ignibus meridiem*, Hymn. Surt.
10, 25. Nalæs hî hine læssan lácum teódan đonne đa dydon đe hine æt
frumsceafte forđ onsendon, Beo. Th. 86; B. 43. [M. H. Ger. zechen;
p. zechte *to arrange, contrive, bring about*.] v. fore-, ge-teón; teohhian.

teón. I. *hurt, damage, vexation* :—Đone on teón wigeþ feónd
his feónde *him* (the dog) *foe brings for the annoyance of his foe*, Exon.
Th. 433, 28; Rä. 51, 3. II. *insult, abuse, reproach, calumny* :—
Đá hine teóne wyrde (teónode and wyrgde? *see note*) Chus, Ps. Th. 7,
arg. Teóna *calumniarum*, Hpt. Gl. 506, 22. [Icel. tjón; *f. n. damage,
loss*.] v. níđ-geteón, and next word.

teóna, an; *m.* I. *damage, harm, hurt, mischief, annoyance,
trouble, vexation, detriment, loss* :—Mid đý hunige smire . . . ne biþ
sóna nán teóna *smear with the honey* . . . *there will be no hurt* (*from the
disease*) *directly*, Lchdm. ii. 104, 23: 156, 30. Đis weorc biþ deóflum
se mæsta teóna *this work will prove the greatest vexation to devils*,
Blickl. Homl. 47, 6. Hit him wyrþ tó teónan *it will turn to his hurt*,
51, 9. Ne him wiht gescód đæs đe hý him tó teónan þurhtogen hæfdon,
Exon. Th. 127, 36; Gú. 397: 269, 30; Jul. 458. Đæt behýded wæs tó
teónan cristenum folce *the cross had been hidden to the detriment of
Christians*, Elen. Kmbl. 1973; El. 988. Þohton đæt hié sceoldon
gewrecan hira teónan *they thought they would avenge the harm that
had been done them*, Chr. 921; Erl. 107, 17. Ymb đone teónan (*mischievous doctrine*) wæs gegaderad III hund biscepa and eahtatiéne hiene
tô âmânsumianne *conventus cccxviii episcoporum factus est, per quos
Arianum dogma exitiabile reprobatum est*, Ors. 6, 30; Swt. 282. 34.
Tiónan *infestationes*, Wrt. Voc. ii. 111, 61. Teónan, 45, 27. Se đe
hine fram swá monigum yrmđum and teónum (*tot ac tantis calamita-*

tibus) generede, Bd. 2, 12; S. 514, 19. Mid miclum teónum and
wítum, Ors. 5, 15; Swt. 250, 28 : Cd. Th. 36, 34; Gen. 581. Đæt tó
teónum weorþeþ, þeódum tó þreá, Exon. Th. 67, 20; Cri. 1091 : 75,
1; Cri. 1215. Synfull tôþum torn þolaþ teónum grimetaþ (*grievous-ly
groans*), Ps. Th. 111, 9. Ne mæg hê nó ryhtlíce geđyld læran búton hê
self geđyldelíce óđerra monna tiónan geđolige *neque potest veraciter bona
docendo impendere, si vivendo nescit aequanimiter alivia mala tolerare*,
Past. 33; Swt. 217, 4. On his tíman hæfdon men mycel geswinc and
swíđe manige teónan, Chr. 1086; Erl. 222, 20. II. *hurt that
comes from wrongful action, wrong, injury, wrongful action, iniquity,
offence, abuse, ill-usage, violence* :—Wolde hê đæt gyld ábrecan. Đa
hæþenan men hine mid teónan (*violence*) âweg ádrifon . . . Hê hit for
manna teónan gebrecan ne môste, Blickl. Homl. 221, 20-27. Ne dó ic
đê nænne teónan (teúne, Rush.) *non facio tibi injuriam*, Mt. Kmbl. 20, 13.
Se unrihtwísa wer wyle niman on teónan his néxtan dæde đeáh đe hê him
teónan ne geđó, Basil admn. 4; Norm. 44, 19. Đæt hê geþence đone
teónan (*injuriam*), đe wê him dydon, Gen. 50, 15 : Ps. Th. 102, 6. Se
đe úre ealra teónan wræce *he that should avenge the wrong done to us
all*, L. Ath. v. 7; Th. i. 234, 20 : 8, 3; Th. i. 236, 18 : Blickl. Homl.
33. 24 : Ors. 1, 11; Swt. 50, 12. Gê ne ongítaþ hú micelne teónan gê
dóþ Gode eówrum sceppende *nec intelligitis quantam conditori vestro
faciatis injuriam*, Bt. 14, 2; Fox 44, 31. Ic (*the devil*) đæs wealles geat
ontýne þurh teónan (*by means of the iniquity which I introduce into the
man's mind*), Exon. Th. 266, 22; Jul. 402. Ic fleáh hlæfdigan hete,
tregan and teónan, Cd. Th. 137, 15; Gen. 2274 : 226, 5; Sat. 497. Se
cyning ne gemunde đæra monigra teónena đe hiora ægđer óþrum gedyde
Astyages oblitus sceleris sui, Ors. 1, 12; Swt. 52, 22. Hê đa gefremedon
teónan (*factas injurias*) him eall forlêt, Bd. 3, 22; S. 553, 19. Teónan
and unriht *iniquitates nostras*, Ps. Th. 102, 12. III. *reproach,
insult, shame, calumny, abuse, contumely* :—Teóna *calumnia*, Hpt. Gl.
514, 64 : contumelia, Scint. 19, 4. Tióna, Kent. Gl. 345. Ic ehte mid
teónan *calumnior*, Ælfc. Gr. 25; Zup. 145, 1. Genimeþ his sæhta
Drihten mid mycclum teónan on him *the Lord will take from him his
possession with great shame to him*, Blickl. Homl. 53, 4. For teónan *for
shame*, 179, 12. Đa blæda đe ic đê on teónan geþah *the fruit which I
insulted you by taking*, Cd. Th. 54, 30; Gen. 885. Teónan đú wyrcst
ús mid đisse sage *haec dicens nobis contumiliam facis*, Lk. Skt. 11, 45.
Đa đe tælnessa teónan wiđ heora đam nêhstan âhófan *detrahentem
adversus proximum suum*, Ps. Th. 100, 4. Hí (*two well-born nuns*)
wæron æfter æþelborennysse oferhýdige and hearmcwydole, and đone
wer oft gedrehton. Đá cýdde se wer Benedicte, hú micelne teónan hê
forđyldegode mid đám mynecenum, Homl. Th. ii. 174, 10. Teónan
calumniae, Wrt. Voc. ii. 24, 49. Mid teónum gewæcende *afficientes
contumelia*, Lk. Skt. 20, 11. IV. *strife, discord* :—Eall đæra
Iudéiscra teóna árás þurh đæt hwí Drihten Crist se đe æfter flæsce sóđlíce
is mannes sunu eác swilce wære gecweden Godes sunu *all the strife of the
Jews arose from the question, why the Lord Christ, who according to
the flesh is truly son of man, should be called also son of God*, Homl. Th.
i. 48, 15. Oft wæron teónan wærfæstra wera weredum gemæne heardum
hearmplega. Đá ongan Abraham sprecan . . . 'Ne sceolon unc betweónan
teónan weaxan wroht wridian' (*facta est rixa inter pastores gregum
Abram et Lot . . . Dixit ergo Abram ad Lot* : 'Ne quaeso sit jurgium
inter me et te,' Gen. 13, 7, 8), Cd. Th. 113, 33-114, 12; Gen.
1896-1903. Symle teónan sêcþ yfel *semper jurgia quaerit malus*, Scint.
134, 12. Tiónan, Kent. Gl. 145. [The word remains in use in later
English, but gradually restricts the meaning to *pain, vexation*. Laym.
A.R. teone. Onn himm wrekenn him teone, Orm. 19866. Ne do he þe
neure swa muchelne teone ne wite, O. E. Homl. i. 15, 30. Wiđute
teone and treie, 193, 61. Hi hedden teone and seorewe, Misc. 89, 14.
Þu seist me boþe teone and schame, O. and N. 50. Teone ne tintreohe,
Kath. 402. Berninde of grome and of teone *furiis agitatus*, 1354. Mi
tene and min anger, Will. 552. Anger and tene, sorge and wo, Gen. and
Ex. 2992. Tyene *strife*, Ayenb. 66, 1. Nô word of jelousye or any
other teene, Chauc. Kn. T. 2248. In pure tene *in sheer vexation*, Piers
P. 6, 119. With trauaille and with tene, 135. Tene or angyr or
dyshese *angustia, tribulacio*, Prompt. Parv. 488. O. Sax. tiono *wrong,
evil*.] v. hyge-teóna; teóne, *and preceding word*.

teón-cwide, es; *m. Reproachful, abusive, insulting speech, blasphemy,
contumely, calumny, slander* :—Ne fríne ic đê for tæle ne þurh teóncwide,
Andr. Kmbl. 1266; An. 633. Þurh teóncwide *by their blasphemous
language* (saying that a miracle was wrought by magic), 1541; An.
772. Godscyld wrecan, teóncwide, Exon. Th. 254, 30; Jul. 205. Tión-
cwida *conviciorum*, Wrt. Voc. ii. 20, 44. Mid teáncuidum *contumelia*,
Lk. Skt. Lind. 20, 11. Hí ermþu gehêton tornum teóncwidum, Exon.
Th. 129, 10; Gú. 419. Cf. hearm-cwide.

teón-cwidian; *p.* ode, ede *To reproach, abuse, revile, calumniate* :—
Teóncwidedon *conviciebant*, Wrt. Voc. ii. 17, 58. Teóncwid[ed]on, 74,
33. Fore teáncuidendum ús *pro calumpniantibus nobis*, Rtl. 176, 33.
Cf. hearm-cwidian.

teónd, es; *m. One who draws* :—Heó behealdende wæs hwylcum
teónde hê upp âhafen wæs, Bd. 4, 9; S. 576, 34.

teónd, es; *m. An accuser:*—Gif wîteþeów mon betŷhþ . . . ðonne âh se teónd âne swingellan æt him, L. In. 48; Th. i. 132, 9. Eode se man sylf tô ðe man tuge, and hæbbe se teónd (se ðe tŷhþ, MS. B.) cyre, swâ wæterordâl swâ ŷsenordâl, L. Ath. iv. 6; Th. i. 224, 15. Tiónd, L. Eth. iii. 6; Th. i. 296, 3. Gylde man ðam teónde his ceápgyld, L. Edg. ii. 7; Th. i. 268, 19: L. Eth. i. 1; Th. i. 280, 20: 282, 3.

teóne, an; *f. Calumny, reproach:*—Teóne *calumnia*, Wrt. Voc. i. 21, 29. Wǣron hyra tungan getale teónan gehwylcre and tô yfele gehwam ungemet scearpe, Ps. Th. 56, 6. v. teóna.

teónere, es; *m. A calumniator:*—Hê geeádmê. ðane teónere *humiliabit calumniatorem*, Ps. Lamb. 71, 4.

teón-full; *adj.* **I.** *grievous, vexatious, troublous, woeful:*—Se teónfulla dæg (*the last day*), Wulfst. 187, 3. Hû geswincful and hû teónful ðis lif is *how full of travail and trouble this life is*, 273, 6. Ða teónfullan *infesta*, Wrt. Voc. ii. 88, 15. **II.** *of persons,* (1) *causing hurt* or *injury:*—Teónfullum on teso *so as to hurt the harmful* (those who were attending to the fiery furnace), Cd. Th. 232, 4; Dan. 255. (2) *causing vexation* or *annoyance, exasperating.* v. teónian, **I:**—Mægþ teónful *generatio exasperans*, Ps. Spl. 77, 10. **III.** *insolent, abusive, contumelious, contemptuous, calumnious:*—Teónful *injuriosus*, geflitful *contentiosus*, Wrt. Voc. i. 49, 32: 74, 32. Se mynstres hordere sî . . . nâ drêfend ne teónful (*non turbulentus, non injuriosus*), R. Ben. 54, 9. Ðû ne scealt nânne man wyrigan, ne nǣnne man tǣlan, ne teónful beón, Homl. Skt. i. 21, 359. Ys steór leás on mûþe teónfulles (*contumeliosi*), Scint. 114, 9. Teónfulle wê synd *contumeliosi sumus*, 155, 14. Wǣron hî æfter æþelborennesse oferhŷdige and hearmcwydole . . . Hî ðurhwunedon on heora teónfullum wordum *they persisted in their insolent language*, Homl. Th. ii. 174, 14. [In þa teonfulle (*destructive*) sǣ, Laym. 4585.]

teón-hete, es; *m. Harmful* or *wrongful hate, dire hostility:*—Wið ðam teónhete (*the hostility of the Egyptians in pursuit of the Israelites*), Cd. Th. 191, 34; Exod. 224. Wið teónhete, Ps. Th. 147, 2.

teónian; *p. ode.* **I.** *to vex, irritate.* v. teón-full, **II.** : —Hŷ teónedon † hig gremedon *irritaverunt* (*Moysen*), Ps. Lamb. 105, 16. **II.** *to reproach, revile, abuse, calumniate:*—Se ðe teónaþ þearlan tǣlþ Scyppende his *qui calumniatur pauperem, exprobrat factori ejus*, Scint. 156, 14: 178, 18. Ða hine (*David*) teóne wyrde (teónode and wyrgde? see note) Chus, Ps. Th. 7, arg. Ne teónian mê ða môdigan *non calumnientur me superbi*, Ps. Lamb. 118, 122. Teóniendum mê *calumniantibus me*, 121. [Hwon his briddes teoneð him *when its young ones vex it* (*the pelican*), A. R. 118, 10. Me teoneð mare þ . . . *quod altius me urit*, Kath. 550. I tene (*trouble*) hem no more, Allit. Pms. 60, 759. Þe naked to tene, Gaw. 2002. Alle wordes him tyeneþ and greueþ, bote yef hi ne by to god, Ayenb. 142, 28. Tyrauntz þat teneþ trewe men, Piers P. 15, 412. Tenyñ or urethyñ *irrito*, Prompt. Parv. 489. O. Frs. tiona, tiuna *to injure:* O. Sax. gi-tiunean *to harm.*] v. tînan.

teónlîce; *adv.* **I.** *in a manner that causes harm* or *trouble, grievously, miserably:*—Hî gedrêfde deópe weorðaþ . . . swylce teónlîce geteoriaþ, Ps. Th. 103, 27: Exon. Th. 226, 17; Ph. 407. **II.** *in a way that brings shame* or *affront, with insult* or *ignominy:*—Man sceal ða geóguðe lǣdan gehæft heánlîce and swâ bysmorlîce bringan of heora êðle and betǣcan eów teónlîce on hǣðenra hand, Wulfst. 295, 19. Sende on heora eorþan toscean teónlîce *he brought shame on them by sending frogs into their land*, Ps. Th. 104, 26. Ðencan hû hig hyne teónlŷcost âteón myhton *to devise how they might treat him with most ignominy*, Nicod. 14; Thw. 7, 7.

teón-lig, es; *m. Hurtful, destructive flame*, of the conflagration at the last day:—Eall þreó nimeþ fŷres wælm . . . teónlêg somod bærneþ þreó (*earth, sea, and sky*) eal on ân, Exon. Th. 60, 14; Cri. 969. Tiónlêg, Elen. Kmbl. 2556; El. 1279.

teón-rǣden[n], e; *f. Wrong, injury:*—Ðæt hig wrecan mihton heora teónrǣdenne mid tintergum on him (*ut reddamus ei* (Samson), *quae in nos operatus est*) . . . Hig woldon hine tintregian for heora teónrǣdene (*the wrong he had done to them*), Homl. Skt. ii. 25, 640 : Ælfc. T. Grn. 5, 18. Gif hê on gehwylcum teónrǣdennum (*injuriis*) geþyld lufige. . . Gê eác earfeþa and teónrǣdena (*injurias*) forberaþ, R. Ben. 27, 1, 21.

teón-smiþ, es; *m. A worker of hurt* or *wrong, an evil-doer:*—Wǣron teónsmiðas (*the evil spirits that persecuted Guthlac*) tornes fulle, . . . earme andsacan, Exon. Th. 114, 21; Gû. 176.

teóntig. v. hund-teóntig.

teón-word; *n. A word that conveys reproach, insult, abuse, calumny; a word that does wrong:*—Hig tǣldon ðæt land mid heora teónwordum *they slandered the land with their calumnies*, Num. 13, 33. Eorl ôðerne mid teónwordum tǣleþ behindan, spreceþ fægere beforan, Frag. Kmbl. 6; Leás. 4. Næs heó swâ nû æðelborene men synt mid ofermēttum âfylled . . . ne mid teónwordum *she was not, as nobly born men now are, filled with haughtiness . . . or with insolent words*, Lchdm. iii. 428, 33.

teorian; *p. ode.* **I.** *to tire* (intrans.), *faint, fail, cease:*—Treówgeþofta teoraþ hwîlum wâciaþ wordbeót *faithful comrade fails at times,* *feeble prove promises*, Exon. Th. 469, 21; Hy. 11, 5. Tiorade *desisse*, Txts. 57, 668. Teorode, Wrt. Voc. ii. 25, 37 : Exon. Th. 436, 29; Rä. 55, 8. Eágan mê teoredon *defecerunt oculi mei*, Ps. Th. 118, 82. Gif mon on langum wege teorige *if a man tire on a long journey*, Lchdm. ii. 16, 16. Lǣcedôm wiþ miclum gange ofer land ðŷ lǣs hê teorige, 16, 26. Be ðone ðe lâd teorie (*fail*). Ðeáh æt stæltyhtlan lâd teorie Ængliscan, L. O. D. 4; Th. i. 354, 13-14. Gif ðeós lâd teorie, 6; Th. i. 354, 31. **II.** *to tire* (trans.), *to cause to fail* or *faint:*—Gif mîne grame þenceaþ gâst teorian *if foes think to make my spirit faint*, Ps. Th. 141, 3. [Him trukeþ his iwit, him teoreþ (*fails*) his miht, Fragm. Phlps. 5, 38. O. Sax. far-terian *to destroy.*] v. â-, ge-teorian; teran.

teorig, teorigend-lîc, teorodness, teorung. v. un-teorig, â-teorigendlîc, ge-teorodness, â-, ge-teorung.

teors, es; *m. A tarse* (v. Halliwell's Dict.); *membrum virile:*—Teors *calamus*, herþan *testiculi*, Wrt. Voc. i. 65, 30. Teors *veretrum*, teors, ðæt wǣpen *vel* lim *calamus*, 283, 55, 56. Wið hærþena sâre and teorses, Lchdm. i. 358, 4. Smyre ðone teors and ða hærþan, ðonne hafaþ hê mycelne lust, 358, 19 : 350, 9. [O. H. Ger. zers *veretrum.*]

teoru(-o), teru(-o), tearo, taru; *gen.* teorwes, *also* tearos; *n.*: teora, tara, an; *m. Tar, resin, gum; also the wax of the ear:*—Teoru *gluten*, Txts. 67, 985. Teoru, teru *cummi*, 55, 616 : resina, 93, 1716. Blaec teoru (teru) *napta*, 79, 1360. Teru *bapis*, Wrt. Voc. ii. 125, 17 : *cummi*, 137, 44. Blæc teru *napta*, 60, 5. Tero *gluten*, 40, 25 : *napta*, 71, 35. Taru, Lchdm. ii. 312, 20. Wiþ teorwe, 132, 5. Meng wiþ sôte sealt, teoro, hunig, 76, 8 : 134, 11. Dô of ðînum eáran ðæt teoro, 112, 3. Meng wiþ pipor and wiþ teoran, 76, 7. [To maken a tur of tigel and ter, Gen. and Ex. 662. The tarre that to thyne sheep bylongeth, Piers P. C-text, x. 262. Terre *butumen*, Wrt. Voc. i. 227, col. 2 (15th cent.). Tere, 279, col. 2. Terre or pyk, Prompt. Parv. 489. Icel. tjara.] v. ifig-, scip-, treów-teoru (-tearo, -teora); tirwa.

teorung, e; *f. Fainting, failing, exhaustion:*—Sum gemyndleás wîf fêrde wôrigende geond wudas and feldas and ðǣr gelæg ðǣr hî seó teorung gelette *a certain witless woman went wandering about the woods and fields, and lay down where exhaustion prevented her going further*, Homl. Th. ii. 188, 15. v. â-, ge-teorung.

teosol(ul, -el), es; *m. A small squared piece of stone, a die:*—Tasul(-ol) *tessera*, Txts. 101, 2000. Tǣslum *tesellum* (*tessellis* in text, v. tǣfl), Wrt. Voc. ii. 93, 44. Tǣfles monnes, ðonne teoselum. weorpeþ, Exon. Th. 345, 9; Gn. Ex. 185. Tesulas *tesseras*, Txts. 114, 84. [From Latin.]

teosu, tesu, tǣsu(-o), wes; *m*(?). *n*(?). **I.** *hurt, injury:*—Ålet gewearf teónfullum on teso *the fire turned to the hurt of the harmful*, Cd. Th. 232, 4; Dan. 255. Lêcnade monigo of teissum † cualmum *curavit multos a plagis*, Lk. Skt. Lind. 7, 21. **II.** *wrong, fraud:*—Âlŷs mîne sâwle from ðǣre tungan ðe teosu wylle *libera animam meam a lingua dolosa*, Ps. Th. 119, 2. Biþ deófla wîse ðæt hî duguðe beswîcaþ and on teosu tyhtaþ *the devils' way is to seduce from virtue and to incite to wrong*, Exon. Th. 362, 9; Wal. 34. Ôðer hine lǣreþ ðæt hê healde Metodes miltse, ôðer hine tyhteþ and on tæso lǣreþ, Salm. Kmbl. 984; Sal. 493. v. next two words.

teosu-sprǣc, e; *f. Hurtful, deceitful speech:*—Se getynga wer on teosusprǣce *vir linguosus*, Ps. Th. 139, 11.

teoswian, teswian; *p. ode To hurt, injure, annoy:*—A hine ofslyhþ, T hine teswaþ, and hine on ða tungan sticaþ, Salm. Kmbl. 189; Sal. 94.

teóða, teogeða; *ord. num. Tenth*, (1) *marking order:*—Seó teóðe (teigða, Lind.) tîd *hora decima*, Jn. Skt. 1, 39. Ða wæteru wanedon ôð ðane teóðan mônð, and on ðam teóðan mônðe æteówdon ðæra munta cnollas, Gen. 8, 5. Wite cristenra manna gehwilc, ðæt hê his Drihtene his teóðunge, â swâ seó sulh ðone teóðan æcer gegâ, rihtlîce gelǣste, L. Eth. ix. 7; Th. i. 342, 11. See Seebohm's Village Community, p. 114. Ðŷ teogeþan dæge mônþes, Bd. 5, 23 ; S. 646, 15. In regula ða teiða *in canone decimo*, Mt. Kmbl. p. 3, 17. On ðone teogeþan dæg ðæs mônðes, Shrn. 102, 22. Teogþan, 84, 1. (2) *marking division:*—Syle ðone teóðan dǣl ealra ðînra wæsma, Deut. 14, 22. Ðŷ ilcan geáre gebôcude Æþelwulf cyning teóþan dǣl his londes ofer al his rîce Gode tô lofe and him selfum tô êcere hǣlo, Chr. 855; Erl. 68, 25 : Ex. 29, 40. Ðæs hereteámes ealles teóðan sceat Abraham sealde Godes bisceope, Cd. Th. 128, 5; Gen. 2122. Ðone têþan (*tenth*) dǣl, Bd. M.) dǣl, Bd. 4, 29; S. 608, 18. Ðîne teóðan sceattas âgyf ðû Gode, L. Alf. 38; Th. i. 52, 31. (2 a) *used substantively, a tithe:*—' Ic ðê wille gesyllan mîne teóðan (*decimas*)' . . . Gif wê cýre teóðan gesyllan nyllaþ, ûs ða nygon dǣlas biþ ætbrǣdene, and se teóða ân ûs biþ tô lâf[e], L. Ath. i. prm.; Th. i. 196, 20-26, cf. L. Edg. i. 3; Th. i. 264, 1-5.

teóðian, teogoðian; *p. ode.* **I.** *to take out a tenth part of anything:*—On eallum geáre sind getealde ðreó hund daga and fîf and sixtig daga; ðonne gif wê teóðiaþ ðâs geárlîcan dagas (*if we take a tenth of the days of the year*), ðonne beóþ ðǣr six and ðrîtig teóðingdagas, Homl. Th. i. 178, 21. **II.** *to take a tenth part and give it, to pay tithe of anything:*—Ic teóðie ealle mîne ǣhta, Homl. Th. i. 428, 25. Gê ðe teóðiaþ (teóðigaþ, MS. B.: tægþigaþ, Rush.) mintan and dile, Mt. Kmbl. 23, 23. Gê ðe teóþiaþ (teigðas, Lind.: iegdigas, Rush.)

ǽlce wyrte, Lk. Skt. 11, 42. Gē teogoðiaþ eowrne kymen, Past. 57; Swt. 439, 28. Teóðige hē eal ðæt hē áge, L. Pen. 15; Th. ii. 282, 22. Ús is wyrse ðæt wē úrne ceáp teóþian, gif wē willaþ syllan úre ðæt wyrste Gode, Blickl. Homl. 41, 7. Heáfodmen teóðian, Wulfst. 181, 18. Gif gē nellaþ teóðian ǽlc ðæra þinga ðe eów God lǽnþ. 297, 2: Homl. Th. i. 178, 30: ii. 608, 21. II a. *to grant* a·tenth:—Ðá ðá hē teóðode gynd eall his cyneríce ðone teóðan dǽl ealra his lande *quando decimam partem terrarum per omne regnum meum dare decreui*, Cod. Dip. Kmbl. v. 106, 21. v. ge-teóðian; un-teóðod.

teóðung(-ing), e; *f.*   **I.** *tithe, a tenth part,* (a) in passages not relating to the Christian church :—Hē sealde him ða teóðunge (*decimam*) of eallum ðám þingum, Gen. 14, 20. Of eallum þingum, ðe ðu mē sylst, ic bringe ðē teóðunga (*decimas*), 28, 22. Ic sylle teóþunga (tegðunge, Rush.: teigðuncgas, Lind. *decimas*) ealles ðæs ðe ic hæbbe, Lk. Skt. 18, 12. Abraham geaf ðam kincge Melchisedech ða teóðunga (*decimas*) of ðám ðingon ðe hē gewunnen hæfde, Prud. 56. • (b) with special reference to the English church. 'In A.D. 787 tithe was made imperative by the legatine councils held in England, which being attended and confirmed by the kings and ealdormen had the authority of witenagemots,' Stubbs' Const. Hist. i. 228. See also Kemble's Saxons in England, vol. ii, c. x. Accordingly laws of a later date and ecclesiastical writings contain injunctions for the payment of tithe :—Ic Æðelstan cyninge . . . eów bidde . . . ðæt gē of mínum ágenum gōde ágifan ða teóðunga, ǽgðer ge on cwicum ceápe ge on ðæs geáres eorðwæstmum; . . . and ða biscopas ðæt ilce dōn on heora ágenum gōde, and míne ealdormen and míne geréfan ðæt silfe. And ic wille ðæt bisceop and ða geréfan hit beódan eallum ðám ðe him híran sculon, ðæt hit tō ðam rihtan ándagan gelǽst sý . . . Gif wē ða teóðunga Gode gelǽstan nellaþ, hē ús benimeþ ðara nigon dǽla ðonne wē lǽst wēnaþ, L. Ath. i. prm.; Th. i. 194, 1–196, 7: L. Edm. S. 2; Th. i. 244, 15. Ðæt neádgafol úres Drihtnes, ðæt sýn úre teóðunga and cyriscsceattas . . . Ægðer ge earm ge eádig, ðe ǽnige teolunga habbe, gelǽste Gode his teóðunga mid ealre blisse, L. Edg. S. 1; Th. i. 270, 25–272, 2. Wile cristenra manna gehwilc, ðæt hē his Drihtene his teóðunge, á swá seó sulh ðone teóðan æcer gegá, rihtlíce gelǽste, L. Eth. ix. 7; Th. i. 342, 11. Godes ǽ ús bebýt, ðæt wē sceolon ealle ða ðing ðe ús gesceótaþ of úres geáres teolunge Gode ða teóðunge syllan, Homl. Th. i. 178, 28: Wulfst. 102, 20. Further, the time of payment and the penalties for neglect to pay were fixed:—Gif hwá teóðunge forhealde, gylde lahslit mid Denum, wíte mid Englum, L. E. G. 6; Th. i. 170, 1. Gif hwá teóðinge forhealde, and hē sí cyninges þegn, gilde .x. healfmearc, landágende .v. healfmearc, ceorl .xii. ōr, L. N. P. L. 60; Th. ii. 300, 9. Be teóðungum. Sý ǽlcere geóguðe teóðung gelǽst be Pentecosten, and ðara eorðwæstma be emnihte . . . and gif hwá ðonne ða teóðunge gelǽstan nelle, swá wē gecwedon habbaþ, fare ðæs cynges geréfa tō and ðæs bisceopes and ðæs mynstres mæssepreóst and niman unþances ðone teóðan dǽl tō ðam mynstre ðe hit tō gebyrige and tǽcan him tō ðam nigoðan dǽle; and tōdǽle man ða eahta dǽlas on twá, and fō se landhláford tō healfum, tō healfum se bisceop, L. Edg. i. 3; Th. i. 262, 19–264, 4: L. Eth. v. 11; Th. i. 308, 1: ix. 8; Th. i. 342, 14–23. Some information as to the destination of tithe is contained in the following:—Man ágife ǽlce teóðunge tō ðam ealdan mynstre ðe seó hýrnes tō hýrþ, L. Edg. i. 1; Th. i. 262, 6. Gif hwá þegena sý ðe on his bóclande cyricean hæbbe ðe legerstōw on sý, gesylle hē ðone þriddan dǽl his ágenre teóðunge intó his cyricean, i. 2; Th. i. 262, 13: L. C. E. 11; Th. i. 366, 25. Be teóðunge se cyng and his witan habbaþ gecoren and gecweden, ðæt þridda dǽl ðare teóðunge þe tō circan gebyrige gā tō ciricbóte, and ōðer dǽl ðam Godes þeówum, þridde Godes þearfum and earman (v. teóðung-sceatt) þeówetlingan, L. Eth. ix. 6; Th. i. 342, 6–9. Gange ǽgðer ge cyriscsceat ge teóðunge intó ðam hálgan mynstre, Chart. Erl. 236, 2. In a charter, which speaks of Edward as dead, a tithe of eight pennies from each hide is mentioned as due to Taunton :—Hér swutulaþ on ðisum gewrite hwylce gerihta langon intó Tántúne. Ðæt is . . . teóðung of ælcere híde eahta penegas, Cod. Dip. Kmbl. iv. 233, 8. v. æcer-, corn-teóðung.   **II.** *a tithing, an association of ten men* (ten such associations formed a *hynden*, q. v.). The word remains as the name of a local division in many of the southern counties, v. Stubbs' Const. Hist. i. 86, n. 2, but in the earlier time it seems to be personal. v. teóðung-ealdor, -mann :—Ðæt man funde ǽnne man ðǽr mǽre folc sig swá of áinre teóðunge ðǽr læsse folc sý *that one man should be provided alike where the population was large, as where it was so small that there was only one tithing to draw upon*, L. Ath. v. 4; Th. i. 232, 14. Ðæt wē ús gegaderian á emban ǽnne mōnað ða hyndenmenn and ða ðe ða teóðunge bewitan, v. 8, 1; Th. i. 236, 3. Ðæt ǽlc mon beó on teóðunge. Wē wyllaþ, ðæt ǽlc freó man beó on hundrede and on teóðunge gebroht, ðe láde wyrðe beón wylle oððe weres wyrðe, L. C. S. 20; Th. i. 386, 18–22. See Stubbs' Const. Hist. i. 85; Kemble's Saxons in England, vol. i, c. ix.

teóðung-ceáp, es; *m. Tithe-stock, stock paid as tithe*:—Geheraþ hwæt se æþela láreów sægde be manna teóþungceápe. Hē cwæþ: Nú neálǽceþ ðæt wē sceolan úre ǽhta and úre wæstmas gesamnian. Dōn wē ðonne Drihtne þancas ðe ús ða wæstmas sealde, and sýn wē gemyn-

dige ðæs ðe ús Crist sylfa bebeád. Hē cwæþ, ðæt wē symble emb twelf mōnaþ ágeáfon ðone teóþan dǽl ðæs ðe wē on ceápe habban . . . Úre Drihten bebeád, ðæt wē symle emb twelf mōnaþ gedǽlan ðone teóþan dǽl on úrum wæstmum and on cwicum ceápe, Blickl. Homl. 39, 10–20.

teóðung-dagas; *pl. Tithing-days, days amounting to a tithe of the year*, a term applied to the thirty-six week days in the six weeks of Lent from the first Sunday in Lent until Easter-day :—Gif wē teóðiaþ ðæs geárlícan dagas, ðonne beóþ ðǽr six and ðrítig teóðingdagas; and fram ðisum dæge (*the first Sunday in Lent*) ōð ðone hálgan Eásterdæg sind twá and feówertig daga; dō ðonne ða six Sunnandagas of ðam getele, ðonne beóþ ða six and ðrítig ðæs geáres teóðingdagas ús tō forhæfednysse getealde . . . Wē sceolon on ðisum teóðingdagum úrne líchaman mid forhæfednysse teóðian, Homl. Th. i. 178, 21–30: ii. 608, 20: L. E. I. 37; Th. ii. 436, 10. Ús gebyreþ, ðæt wē ǽlces þinges úre teóðunge rihtlíce Gode betǽcan; ðonne syndan ðás dagas (*fast days of Lent*) getealde for teóðingdagas innan geáres fæce, and wē sculan eác ða teóðunge wyrðlíce Gode gelǽstan, Wulfst. 102, 21.

teóðung-ealdor, es; *m. A chief of ten monks, a dean* :—Hwylce mynstres teóðingealdras (*decani*) beón sceolon. Gif seó gesefǽrden tō ðam micel sý, sýn gecorene of ðám sylfum gebrōðrum ða ðe gōdes gewittes sýn, and syn gesette tō teóðingealdrum (*constituantur decani*), R. Ben. 46, 6–10: 137, 17–20. Cf. teóðung, II.

teóðung-georn; *adj. Sedulous in paying tithes*:—Ælmysgeorn and cyricgeorn and teóþunggeorn tō Godes cyricean and earmum mannum *eleemosynas libenter erogans, et ad ecclesiam libenter frequens, et sedulo decimas erogans ecclesiae Dei ac pauperibus*, L. Ecg. C. prm.; Th. ii. 132, 15: Anglia xii. 518, 26.

teóðung-land, es; *n. Land that was subject to the payment of tithe* (?):—Ic fēng tō mínan londe and sealde hit ðon biscope ða fíf hída wið ðon londe æt Lidgeard wið fíf hídan and biscop and eal híwan forgeáfan mē ða feówer and án wæs teóðinglond *I resumed my land and sold it, the five hides to wit, to the bishop (of Winchester) for the land at Liddiard, for five hides, and the bishop and brethren granted me the four (free of tithe?) and one was subject to tithe*, Cod. Dip. Kmbl. ii. 135, 2–6. As may be seen from another charter, the land at Liddiard was in the hands of the bishop of Winchester, v. 144; and several names besides will be found common to the two charters. For the *teóðung* of a hide, see the last passage given under teóðung, I b.

teóðung-mann, es; *m.*   **I.** *one set over ten persons, a ruler of ten* :—Ic sette hig tō teóðingmannum *constitui eos decanos*, Deut. 1, 15. Geceós wíse men and sōðfæste . . . and gesete of him . . . teóðingmen (*decanos*), Ex. 18, 21.   **II.** *as a technical English term, the head of a tithing*, v. teóðung, II :—Wē cwǽdon be uncúðum yrfe, ðæt nán man nǽfde búton hē hæfde ðæs hundredes manna gewitnyssa oððe ðæs teóðingmannes, L. Edg. ii. 4; Th. i. 260, 1. Gyf neód on handa stande, cýðe hit man ðam hundredes men, and hē syððan ðam teóðingmannum, 2; Th. i. 258, 8.

teóðung-sceatt, es; *m. A tax of a tenth, a tithe*:—Teóþingsceat *decimatis*, Wrt. Voc. ii. 26, 36: 73, 44. Swá feala earmra manna swá on ðæs rícan neáweste sweltaþ, and hē him nele syllan his teóþungsceatta dǽl, ðonne biþ hē ealra ðara manna deáþes sceldig, Blickl. Homl. 53, 6. Mid ðam oftige ðæs neádgafoles ðe cristene men Gode gelǽstan sceoldon on heora teóðingsceattum, L. Edg. S. 1; Th. i. 270, 14.

teped, ter. v. tǽpped, ge-ter.

teran; *p.* tær, *pl.* tǽron; *pp.* toren *To tear, rend, bite, lacerate,* (1) literal :—Fealleþ on sídan ðæt ic (*a plough*) tōþum tere, Exon. Th. 403, 27; Rä. 22, 14. Hit tyrþ (*mordebit*) eal swá snaca, Scint. 105, 8. Teraþ *carpunt*, Germ. 395, 403. Gif swín deáde men teraþ (*lacerauerint*), L. Ecg. C. 40; Th. ii. 164, 38. Ðá tær hē his cláðas *scissis uestibus*, Gen. 37, 29, 34. Wyrmas gelíce mid ðǽm scillum gelíce mid ðē núþe ða eorðan slíton and tǽron *oribus scamisque suis humum atterentes*, Nar. 14, 12. Hæfdon híe tēð and híe mid ðǽm ða men wundodon and tǽron *habentes dentes quibus artus militum uiolabant*, 15, 9. Ða fuglas mid hiora clēum ða fixas tǽron, 16, 21. Hē ongon his hrægl teran, Exon. Th. 278, 10; Jul. 595. Feax teran *to tear the hair*, Judth. Thw. 25, 28. Ne sceal hē teran ne bítan swá swá wulf, Homl. Th. ii. 532, 9. Tō teorenne *lacerandum*, Txts. 172, 2. Terende weleras *mordens labia,* Scint. 78, 14. Teorende hine *discerpens eum*, Mk. Skt. Rush. 9, 26. Mid stíðstdum ł terendum tōðreómum *uálidis (uoracibus) gingiuis*, Hpt. Gl. 423, 43. (1 a) *to bite*, of pungent food, etc. :—Hē is swíðe biter on múþe and hē ðē tirþ on ða þrotan ðonne ðú his ǽrest fandast *talia sunt, ut degustata mordeant*, Bt. 22, 1; Fox 76, 29. (2') figurative :—Ne ðú hine ne tǽl ne ter mid wordum *do not back-bite*, Basil admn. 5; Norm. 46, 11. [Goth. dis-, ga-tairan : O. H. Ger. zeran.] v. á-, ge-, tō-teran.

Ter-finnas; *pl. Finns occupying country west of the White Sea* :—Ða Beormas hæfdon swíðe wel gebúd hira land . . . Ac ðara Terfinna land wæs eal wēste . . . Finnas, him þúhte, and ða Beormas sprǽcon neáh án geþeóde, Ors. 1, 1; Swt. 17, 29.

tergan. v. tirgan.

termen, es; *m. A term, fixed date :*—Gif ðú wille witan ðæt gemǽre

terminum septuagesimalis, ðonne tele ðū . . . ðonne on ðam teóðan stent se termen, ðæt gemǽre, Lchdm. iii. 228, 3. On noñ Aprilis byð se forma termen on ðam circule ðe ys *decennovenalis*, oððe *pascalis* gehāten, Anglia viii. 310, 42 : 323, 3. Ðæt gemǽre ðæs termenes pasche, 322, 34. On ðam termine ðære eásterlīcan tīde, 315, 19. Ymbe ðæne termen, 324, 29. [*Icel.* termin. From Latin.]

tero(-u), teso, tesulas, teswian. v. teoru, teosu, teosol, teoswian.

teter, tetr, es ; *m. Tetter, a cutaneous disease :*—Teter *balsis*, Txts. 43, 262 : Wrt. Voc. ii. 10, 61 : 125, 13 : *briensis*, i. 288, 5. Teter, tetr *inpetigo*, Txts. 69, 1047 : *petigo*, 85, 1550. Teter, Wrt. Voc. ii. 68, 3. Spryng *vel* tetr *papula vel pustula*, Txts. 88, 791. Se hæfþ teter (*inpetiginem*) on his līchoman, se hæfþ on his mōde gītsunga . . . Būtan tweón se teter būtan sāre hē ofergǽþ ðone līchoman, and suā ðeáh ðæt lim geunwlitegaþ, Past. 11 ; Swt. 71, 15–17 : Scint. 99, 10. On tetere *inpetigine*, Wrt. Voc. ii. 46, 27. Wið sceb and wið teter, Lchdm. i. 150, 5 : 234, 10. Wið teter, of andwlitan tō dōnne, 336, 3. *The form* tetra, *perhaps influenced by* lepra *which precedes it, also occurs :*—Ðonne becymþ of ðam yflum wǽtum oððe sió hwīte riéfþo þe mon on sūþerne *lepra* hǽt, oþðe tetra, oþþe heáfodhriéfðo, oððe ōman, Lchdm. ii. 228, 13. [A tetere *serpedo*, Wrt. Voc. i. 267, col. 2 (15th cent.). Cf. *O. H. Ger.* zitaroh *impetigo, scabies : Ger.* zitteroch ; zittermal *tetter, ring-worm.*]

tēþa, tēðed, te-treþ, te-tridit, te-weorpan, tewestre. v. teóða, ge-tēðed, tō-tredan, tō-weorpan, wull-tewestre.

tiber, tifer, es ; *n. A sacrifice, offering, victim :*—Wit fȳr and sweord habbaþ, hwǽr is ðæt tiber ðæt ðū torht Gode tō ðam bryngielde bringan þencest (cf. ic āxige hwǽr seó offrung sig; hēr ys wudu and fȳr *ecce ignis et ligna ; ubi est victima ?* Gen. 22, 7), Cd. Th. 175, 4 ; Gen. 2890. Ðū scealt mē onsecgan sunu ðīnne tō tibre *offeres filium tuum in holocaustum* (Gen. 22, 2), 172, 31 ; Gen. 2852. Se ðe on tifre gesald Drihten Hǽlend, 301, 1 ; Sat. 575. Hié Drihtne lāc begen brohton ; brego engla beseah on Abeles gield, cyning eallwihta, Caines ne wolde tiber· sceáwian (*ad munera illius* (*Cain*) *non respexit Dominus*, Gen. 4, 5), 60, 9 ; Gen. 979. Noe tiber onsægde (*obtulit holocausta*, Gen. 8, 20), 90, 29 ; Gen. 1502 : 108, 17 ; Gen. 1807. Hālig tiber (*Isaac*), 204, 6 ; Exod. 415. Ic on ðīn hūs gange and ðǽr tīdum ðē tifer onsecge . . . Ðǽs ic mid mūðe āspræc . . . ðæt ic ðē on tifrum forgulde ealle ða gehāt ðe ic mid mīnum welerum tōðǽlde *introibo in domum tuam in holocaustis . . . Haec locutum est os meum . . : Holocausta offeram tibi*, Ps. Th. 65, 12–13. Tiber, Cd. Th. 9, 2 ; Gen. 135. v. timber. [*O. H. Ger.* zepar, zebar *hostia, sacrificium, holocaustum : Ger.* ziefer *in* ungeziefer. Cf. *Icel.* tafn *a sacrifice, victim.* See Grmm. D. M. p. 36.] v. fyrd-(?), sige-, sigor-, wīn-tiber (-tifer).

tiberness, e ; *f. Sacrifice, destruction, immolation :*—Rǽde on his bōcum hwelce tibernessa ǽgðer ge on monslihtum ge on hungre ge on scipgebroce *let him read in his books what sacrifices of life there were by slaughter, famine, and shipwreck* (the Latin, which is not closely followed, has *qui caedem didicerunt*), Ors. 1, 11 ; Swt. 50, 18.

tican, Lchdm. ii. 60, 18, *read* tilian.

ticcen, es ; *n. A kid :*—Ticcen *hedus*, Wrt. Voc. i. 23, 1 : 78, 34 : *edum*, 288, 18 : ii. 30, 56. Ticcenes geallan, Lchdm. ii. 28, 21. Ðā nāmon hig ān ticcen and ofsnidon hit, Gen. 37, 31. Ic sende ðē ān ticcen (*hoedum*) of mīnre heorde, 38, 17, 20. Buccan wē offriaþ oððe ticcen, Homl. Th. ii. 210, 32. Ne sealdest ðū mē nǽfre ān ticcen (ticgen, Lind. : tycchen, *later MS.*), Lk. Skt. 15, 29. Ticcenu beóþ eáðmelte, Lchdm. ii. 196, 24. Bring mē twā ða betstan tyccenu (*hoedos*) . . . Heó befeóld his handa mid ðæra tyccena fellum, Gen. 27, 9, 16. Swā swā se hyrde āsyndraþ ða scēp fram tyccenum (ticgenum, Lind. : ticnum, Rush. : ticchenan, *later MS.*), Mt. 25, 32. The word occurs in local names, e. g. Ticcenes-, Ticnes-feld. v. Cod. Dip. Kmbl. vi. 342. [*O. H. Ger.* zicchīn, zicchī *hoedus : Ger.* zicke a kid.]

ticgende. v. tycgan.

ticia, an ; *n. A tick* (an insect infesting animals) :—Ticia *ricinus*, Txts. 109, 1130. [A teke *ascarida*, Wrt. Voc. i. 255, col. 1 (15th cent.). A tyke, Wülck. Gl. 566, 18. Tyke, wyrm, Prompt. Parv. 493. To fles ınt to fleye, to tyke ant to tadde, P. S. 238, 4. *O. Du.* teke : *M. H. Ger.* zeche, zecke : *Ger.* zecke. Cf. *the borrowed Romance forms, Fr.* tique : *Ital.* zecca.]

ticlum, Exon. Th. 420, 12 ; Rä. 40, 2. v. til.

tictator, es ; *m. The Anglicized form of Latin* dictator :—Hié him gesetton hīr[r]an lādteów ðonne hiera consul wǽre, ðone ðe hié tictatores hēton, and hié mid ðǽm tictatore micelne sige hæfdon, Ors. 2, 4 ; Swt. 70, 3.

tīd, e ; *f. Tide* (as in Shrove-*tide*, etc.), *time, hour ; tempus*, Wrt. Voc. i. 52, 39 : hora, 53, 17. **I. marking time when,** *time at which anything happens, time* or *date* of an event, *time, hour :*—Be ðam dæge and ðære tīde nān mann nāt . . . ge nyton hwænne seó tīd sig, Mk. Skt. 13, 32, 33. Ðā com his tīd ðæt hē sceolde of middangearde tō Drihtne fēran, Bd. 4, 3 ; S. 567, 13 : 4. 9 ; S. 577, 16. Tō morgen on ðisse ylcan tīde ic sende micelne hagol, Ex. 9, 18. Ðæt sylf his wæstmas tō rihtre tīde, Ps. Th. 1, 4. Hē on gerisene tīd mid hwǽte seów, Bd. 4, 28 ; S. 605, 34. On eallum tīdum secggan wē him þanc, Blickl. Homl.

103, 25. **I a.** *a proper time, time at which a thing can* or *ought to be done, time* (as in to be in *time*), *season, opportunity :*—Ðæt tīd wǽre stānas tō sendanne and tīd tō somnienne, Bd. 4, 3 ; S. 567, 9. Tīd is ðæt ðū fēre, Exon. Th. 179, 30 ; Gū. 1269. Hwīlum sié sprǽce tiid, Past. 38 ; Swt. 275, 17. Hē bīt ðære tīde, hwonne hē ðæs wierðe sié, ðæt hē hine besuīcan mōte, 33 ; Swt. 227, 11. On tīde hē sende hys þeów *at the season he sent a servant* (A. V.), Lk. Skt. 20, 10. Ðæt hē him on tīde mete sylle *to give them meat in due season* (A. V.), Mt. Kmbl. 24, 45. Tō tīde, Past. 63 ; Swt. 459, 12. Se ðe his ǽr tīde ne tiolaþ, ðonne biþ his on tīd untilad, Bt. 29, 2 ; Fox 106, 3. Ic ondette gīfernesse metes ǽr tīdum, and in tīde, ge eác ofer rihttīde, Anglia xi. 98, 24. Bi ðon herǽfter in heora tiid is tō secgenne *de quibus in sequentibus suo tempore dicendum est*, Bd. 3, 18 ; S. 546, 40. Ofer ða tīd ðæs sǽwetes, 4, 28 ; S. 605, 8. **I b.** marking a definite time in the day, *an hour :*—Hit wæs ðā seó teóðe tīd *hora erat quasi decima*, Jn. Skt. 1, 39. Ðā wæs neán seó syxte tīd, and þȳstro wǽron ofer ealle eorþan óð ða nigoþan tīde, Lk. Skt. 23, 44. Fram ðære sixtan tīde óð ða nigoðan tīd, Mt. Kmbl. 27, 45. Hē ūt eode embe ða sixtan and nigoðan tīde . . . embe ða endlyftan tīde, 20, 5–6. Ymbe ða nygoðan tīd clypode se Hǽlend, 27, 46. Ymb ða teóðan tīd dæges, Bd. 3, 27 ; S. 558, 12. Sele drincan on þreó tīda, on undern, on middæg, on nōn, Lchdm. ii. 140, 1. **I c.** as an ecclesiastical term, *a canonical hour, hour for a service, the service at such an hour :*—Ic sincge ǽlce dæg seofon tīda *psallo omni die septem synaxes*, Coll. Monast. Th. 18, 30. Wē lǽraþ ðæt man on rihtne tīman tīda ringe, L. Edg. C. 45 ; Th. ii. 254, 5. Gif preóst on gesetne tīman tīda ne singe, oþþe tīda ne singe, L. N. P. L. 36 ; Th. ii. 296, 3–4. **I d.** *a time at which a commemoration takes place, a tide, festival, anniversary :*—On ðone þriddan dæge ðæs mōnðes biþ ðæs hālgan pāpan tīd ðe is nemned Scē Antheri, Shrn. 47, 31 : 48, 5, *and often.* Tȳd, 150, 11 : 151, 17. Ðæs heáhengles (*St. Michael*) tīd, Blickl. Homl. 197, 4. Seó tīd (*the anniversary of a victory*), 205, 28. Cristes tīd *Christmas*, Lchdm. ii. 294, 27. Tō Scē Michaeles tīde *at Michaelmas*, Chr. 759 ; Erl. 54, 14. Se cyng nam ðǽr his feorme in ðære middewintres tīde, 1006 ; Erl. 140, 30. Ic bebeóde ðæt mon hiora tīd boega geuueorðie tō ānes dæges tō Ōsuulfes tīde *I enjoin that the anniversary of them both be kept on one day, on Oswulf's anniversary*, Chart. Th. 460, 1–7. Is ðeós tīd (*Easter*) ealra tīda hēhst and hālgost, Blickl. Homl. 83, 19. Beó ðām hālgum tīdan eallum cristenum mannum sib and sōm gemǽne, L. Eth. v. 19 ; Th. i. 308, 28. **II.** marking duration, (1) where the length of time is indefinite, *time, a period of time,* in pl. *times* (as in *our times*, etc.) :—Uncūþ biþ ǽghwylcum ānum men his līfes tīd, Blickl. Homl. 125, 7. Wē sceolan on ðisse sceortan tīde geearnian ēce ræste, 83, 2. Hē langre tīde ealle heora mǽgþe wæs geondfarende, Bd. 2, 20 ; S. 521, 26. On sibbe tīde *in time of peace*, 2, 16 ; S. 520, 10. Ðū ne oncneówe ða tīde ðīnre geneósunge, Lk. Skt. 19, 44. On ða tiid suā huelc suā biscephād underfēng, hē underfēng martyrdōm. On ða tiid wæs tō herigeanne ðæt mon wilnode biscephādes, Past. 8 ; Swt. 53, 18. Ic sume tīd fram ðē gewāt, Bd. 5, 12 ; S. 630, 29. Twelf wintra tīd *for the space of twelve years*, Beo. Th. 296 ; B. 147. Eálā ðæt wolde God ðæt ūssa tīda wǽren swelce, Met. 8, 40. Hē wæs him feor manegum tīdum (*for a long time*, A. V.), Lk. Skt. 20, 9. Ǽr eallum tīdum ācenned, Blickl. Homl. 31, 24. Ða ðe on mē gelýfaþ eallum tīdum on ēcnesse, 231, 4. Ðæt wæs geworden on Wulfheres tīdum, Bd. 3, 21 ; S. 551, 42. On ðām tīdum ārās Pelaies gedwild, Chr. 380 ; Erl. 11, 6. Sió wyrd dǽlþ eallum gesceaftum stōwa and· tīda, Bt. 39, 5 ; Fox 218, 33. (1 a) *time, condition of things :*—On ðam endenyhstan dagum ðissere worulde beóþ frēcenlīce tīda, Wulfst. 81, 12. (2) where the period is a definite one :—Ðā (*after the first act of creation*) eodon þrý dagas forð būton tīda gemetum (*without measurement of hours and days*) ; for ðan ðe tunglan nǽron gesceapene, Homl. Th. i. 100, 7. (2 a) *an hour* of the day :—Æfter lytlum fæce swylce ānre tīde, Lk. Skt. 22, 59. Healfre tīde fæc, Bd. 4, 3 ; S. 568, 1. On ānre tīde dæges *in the course of one hour*, Blickl. Homl. 31, 2. Steorran hié ætiéwdon ful neáh healfe tīd ofer undern, Chr. 540 ; Erl. 16, 4. Āne tīd dæges, 879 ; Erl. 80, 30. Hū ne synt twelf tīda ðæs dæges ? Jn. Skt. 11, 9. Feówer and twentig tīda . . . ðæt is ān dæg and ān niht, Lchdm. iii. 254, 13 : 260, 13–15. Æfter þrim tīdum gelǽd hyne tō bæþe, Lchdm. i. 302, 17. Ān wæcce hæfþ þreó tīda, Homl. Th. ii. 388, 14. (2 b) *one of the four seasons of the year :*—Hærfestlīcre tīde *autumnali* (*tempore*), Hpt. Gl. 496, 48. Oþ sumeres tīd, Bd. 4, 28 ; S. 605, 35. Feówer tīda syndan on ðæm geáre, Blickl. Homl. 35, 15. On lenctenlīcere emnihte wurdon geárlīce tīda gesette, Homl. Th. i. 100 3. Þurh ðæt gewrixle ðara feówer tȳda, ðæt ys lencten and sumer and herfest and winter, Shrn. 168, 12. Nihte and dæg ðū ðe gewissast and tīdena ðū selst tīda *noctem diemque qui regis et temporum das tempora*, Hymn. Surt. 6, 6. On wintregum tīdum, Ors. 1, 1 ; Swt. 12, 34. (2 b 1) *a season of the year :*—Se cyng gewende tō ðam middan wintra tō Wihtlande and wæs ðǽr ða tīd, and æfter ðære tīde gewende ofer sǽ, Chr. 1013 ; Erl. 149, 11–13. Gehealdaþ ðās tīd (*Lent*), Homl. Th. i. 180, 2. (2 c) *an age :*—Þreó tīda sind on ðysre worulde ; ān is seó ðe wæs būtan ǽ, óðer is seó ðe wæs under ǽ, seó ðridde is nū æfter Cristes tōcyme, Homl. Th. i. 312,

29.   **III.** as a grammatical term, *tense:*—*Verbum* ys word mid tíde and háde bútan case . . . Him gelimpþ . . . *tempus* tíd, Ælfc. Gr. 19; Zup. 119, 8–14. Tíd gelimpþ worde for getácnunge mislicra dǽda. Æfter gecynde synd þreó tída . . . andwerd tíd . . . forðgewiten tíd . . . tówerd tíd, 20; Zup. 123, 12–17. [*O. Sax. O. Frs.* tíd: *O. H. Ger. zít tempus, hora, aevum, saeculum: Icel.* tíð.] v. æfen-, án-, bed-, behreówsung-, bén-, blódlǽs-, cwyld-, cyric-, Eáster-, fæsten-, freóls-, fulwiht-, gebed-, gebyrd-, gefylling-, hærfest-, hancréd-, heáh-, heáhfreóls-, heófung-, hláfmæsse-, lencten-, merigen-, mete-, middæg-, morgen-, neáh-, nón-, riht-, symbel-, þrowung-, úht-, undern-, winter-tíd; hwíltídum; tíma.

tídan; *p.* de *To betide, befall, happen:*—Bisceopum gebyreþ ðæt symle mid heom wunian wel geþungene witan, . . . ðæt heora gewitan beón on ǽghwylcne tíman, weald hwæt heom tíde, L. I. P. 10; Th. ii. 316, 25. Gif ðan biscop[e] hwæt tíde, Cod. Dip. B. iii. 75, 6, 10, 13. [Þa tidde hit on an Wodnesdei, þet se king rad in his derfald, Chr. 1123; Erl. 249, 30. Ne tyt þe no part wiþ me, Marg. 308. What shulde us tyden? Chauc. M. of L. 337. Som tymes hym tit (bitit, B-text) to folwen hus kynde, Piers P. 14, 213, C-text. A meruellouse meteles me tydde to dreme, 11, 5, B-text. Tydyñ idem quod happyñ, Prompt. Parv. 493.] v. ge-, mis-tídan; tídung.

tíd-dæg, es; *m. The period of a person's life* (cf. the use of *dæg* = time, e. g. Gif ðú wistest on ðysum dínum dæge, Lk. Skt. 19, 42):—Enoses suuu ealra nigon hund wintra hæfde, ðá hé woruld ofgeaf, and týne eác, ðá his tíddæge rím wæs gefylled *when for his lifetime the number of years was completed,* Cd. Th. 71, 4; Gen. 1165.

tidder-. v. tíder-.

tíd-ege (?), es; *m. Fear of a time, fear of the time of death.* v. tíd, I:—Simle þreora sum þinga gehwylce ǽr his tídege (tide ge, MS.) tó tweón weorþeþ ádl oþþe yldo oþþe ecghete fǽgum fromweardum feorh óðþringeþ *ever in every case, before the fear of his end becomes doubtful (before his fear of death has lost any of its certainty?), one of three things, disease or age or violence, crushes the life out of the fey man, outward bound from this world,* Exon. Th. 310, 3; Seef. 69.

tíder-líc; *adj. Weak, frail:*—Se ðe gehielt his unsceadfulnesse and his gódan willan ðeáh hé hwæt tíederlíces oðóe yfelra weorca útan doo hé mæg ðæt æt sumum cierre bétan *si mentis innocentia custoditur, etiam si qua foris infirma sunt, quandoque roborantur,* Past. 34; Swt. 235, 23. In giscæf[te] tóderlícum *in sexu fragili,* Rtl. 51, 7. Tydderlícne líchoman hád *fragilem corporis sexum,* Hymn. Surt. 139, 13. Ic eom þurh míne tydderlíce gecynd líchamlíc man, Homl. Ass. 156, 123. Ðætte suæ fealo tóderlícro wé sindon suæ suíðe strongrum helpum wé sié áholpen *ut quanto fragiliores sumus, tanto validioribus auxiliis foveamur,* Rtl. 61, 9. v. tídre.

tíderness, e; *f.*   **I.** *weakness, frailty,* (a) *weakness* in a general sense, physical, mental, or moral:—Ne mæg úre tyddernes ðyder (*to heaven*) ástígan, Homl. Th. i. 138, 12: ii. 6, 29: 88, 18. Deós mennisce tyddernes biþ swá slídende swá glæs, ðonne hit scínþ and ðonne tóbersteþ, Shrn. 119, 22. Sió niht getácnaþ ða ðistro ðære blindnesse úrre tídernesse *per noctem caecitas nostrae infirmitatis exprimitur,* Past. 56; Swt. 433, 13. Tíddernysse *fragilitatis (humanae),* Hpt. Gl. 437, 31. Téderníse, Rtl. 45, 16 : 46, 32. For líchoman tídernesse (tíeder-, Hatt. MS.) *per imbecillitatem corporis,* Past. 10; Swt. 60, 10. Ðære tídernesse úres flǽsces wé beóþ underðiédde, 21; Swt. 159, 5. For ðæs módes týdernysse, Bt. 3, 2; Fox 6, 7: Blickl. Homl. 31, 30. Swá hwæt swá ic for unwísnesse and for tyddernesse (*fragilitate*) ágylte, Bd. 4, 29; S. 607, 29; Boutr. Scrd. 21, 17. Ðú wást, Drihten, ða menniscan tyddernysse, Blickl. Homl. 243, 30. (b) *the weakness of ill-health, infirmity:*—Gif hwylc mæssepreóst untruman men sprǽce forwyrne, and hé ðonne on ðære tyddernesse (*infirmitate*) swelte, L. Ecg. P. i. 2; Th. ii. 172, 28. Wiþ ælces dæges mannes tyddernysse inneweardes, Lchdm. i. 86, 16: ii. 196, 9. Lǽcedómas wið eallum tíedernessum eágena, 2, 6. Mid sáre geswenced, mid mislícum ecum and tyddernessum, Blickl. Homl. 59, 8. (c) *spiritual infirmity, sinfulness:*—Ægylt, mislimp *vel* tyddernes *excessus,* i. *culpa, delicta,* Wrt. Voc. ii. 145, 68. Ic eom andetta ealra synna ðara ðe ic æfre tó tíedernesse gefremede wið mínre sáwle þearfe, Anglia xi. 99, 89. Swá neár ende ðyssere worulde swá biþ unstrenge mennisc ðurh máran tyddernysse, Homl. Th. ii. 370, 17. v. innan-, innoþ-tíderness.

tíd-fara, an; *m. A traveller the time of whose journey is come* (?), or *one who journeys for a (short) time* (?):—Nú ðú (*the blessed soul immediately after death*) móst féran ðider ðú fundadest . . . eart nú tídfara tó ðam hálgan hám, Exon. Th. 102, 18; Cri. 1674.

tíd-genge; *adj. Current* or *lasting for a time:*—Tídgenge *menstruam,* Germ. 392, 10.

tíding. v. tídung.

tíd-líc; *adj.*   **I.** *lasting for a time, temporary, not eternal, of this world:*—Tyddre ys tídlíc miht *fragilis est temporalis potentia,* Scint. 215, 8. For tídlícre geswencednysse *pro temporali afflictione,* 149, 1. Þing tídlíc *rem temporalem,* 17, 9 : Rtl. 31, 28. Fram tídlícra þinga geþance, Scint. 34, 8. Tídlícum *temporalibus,* Rtl. 8, 9 : 18, 23 :

Anglia xiii. 381, 230.   **II.** *seasonable, opportune:*—Seó tídlíce oportunus, Wrt. Voc. ii. 64, 27: 80, 41. Ðú him mete sylest mǽla gehwylce and ðæs tídlíce tíd gemearcast *tu das escam illis in tempore opportuno,* Ps. Th. 144, 16.   **III.** *expressing relations of time, of time:*—Hwílon hé (*the word ut*) getácnaþ tíde . . . on ðissere stówe hé is *temporale adverbium,*ðæt is tídlíc, Ælfc. Gr. 44; Zup. 265, 19. Sume naman syndon *temporalia,* ðæt synd tídlíce, ða æteówiaþ tíman, 5; Zup. 14, 16. [*O. H. Ger.* zít-líh *temporalis, momentaneus: Icel.* tíð-ligr *temporal.*] v. un-tídlíc.

tídlíce; *adv.*   **I.** *for a time, temporarily:*—Yrsunge tídelíce (*but* tíde ne, MSS. O.T.) sceal mon gehealdan *iracundie tempus non reseruare,* R. Ben. 17, 6.   **I a.** *for time, in this world:*—Se ðe on ðisse worulde wel tídlíce (*temporaliter*) wealdt, bútan ende on écnysse ríxaþ, Scint. 182, 1.   **II.** *conveniently, at a suitable time:*—Hé sóhte ðætte tídlíce ðætte mæhte sellan hine (cf. hé sóhte hú hé eáðelícust hine gesealde, W. S.) *quaerebat oportunitatem ut traderet illum,* Lk. Skt. Rush. 22, 6.   **II a.** *seasonably, in a manner appropriate to a season:*—Seó dún wæs tídlíce gréne *the hill, as was natural to the season* (the date was June 22), *was green;* mons opportune laetus, Bd. 1, 7; S. 478, 21.   **III.** *in time, in good time, betimes, early, soon, quickly:*—Ic tídlíce tó mínre reste eode, for ðon ic wolde beón gearo æt sunnan upgonge, Nar. 30, 27. Ðæt gefremede Diulius hiora consul ðæt ðæt angin weard tídlíce þurhtogen *quod Duilius consul celeriter implevit,* Ors. 4, 6; Swt. 172, 3: 3, 1; Swt. 98, 14. Gif hió mon tídlíce tó bringþ *if it be brought in time,* 5, 13; Swt. 246, 34. Him spédlíce spearuwa hús begyteþ, and tídlíce turtle nistlaþ, Ps. Th. 83, 3: 105, 5. Ædre cymþ, tídlíce, ús Iulius mónað, Menol. Fox 260; Men. 131. Tídlícor, hrædlícor *maturius,* Wrt. Voc. ii. 55, 24. [Tídlike (*soon*) hem gan ðat water laken, Gen. and Ex. 1231. Let turnen hit tidliche (*swiftliche,* MS. C.), Kath. 1932 : Jul. 58, 6. *O. H. Ger.* zítlíhho *temporaliter, in tempore, mature.*] Cf. tímlíce.

tídlícness, e; *f. Opportunity:*—Tídlícnisse *opportunitatem,* Lk. Skt. Lind. 22, 6.

tídran. v. týdran.

tídre, tíedre, tédre, týdre, tiddre, tyddre, *and* tíder (? v. tidder, Hpt. Gl. 436, 59); *adj.*   **I.** *weak, fragile, easily broken:*—Tédre swá swá gangewifran nett, Ps. Th. 38, 12. Se wyrttruma byþ breáþ and tídre, ðonne hé gedríged byþ, Lchdm. i. 260, 7.   **II.** *weak, frail, of physical, mental, or moral weakness in persons:*—Ðæt hiw úre tyddran gecynde, Blickl. Homl. 29, 4. Seó godcundnes onféng úre týdran gecynde, 17, 27. Wé tealtrigaþ týdran móde, Exon. Th. 23, 20; Cri. 371. For úre eágena tyddernysse, Lchdm. iii. 232, 16. Ðæt týdre gewitt, Exon. Th. 2, 34; Cri. 29. Ða týdran mód, 147, 19; Gú. 729. Ða hildlatan holt ofgeáfon, týdre treówlogan, Beo. Th. 5686; B. 2847. Hwæt sind ða ðe ús biddaþ? Earme men, and tiddre, and deádlíce, Homl. Th. i. 256, 2. Tyddre, Boutr. Scrd. 22, 37. Nánre wuhte líchoma ne beoþ téderra ðonne ðæs monnes, Bt. 16, 2; Fox 52, 9. Ða hwítan líchoman beóþ mearuwran and tédran ðonne ða blacan and ða reádan, Lchdm. ii. 84, 21.   **II a.** *weak, having bad health, infirm:*—Gif wíf on ðon tédre sié *if a woman have that infirmity,* Lchdm. ii. 8, 25. Is ðæm lǽce tó giémanne ðæt hé swá líðne lǽcedóm selle ðæm seócan swá se týdra líchoma (*corpus debile*) mǽge ástandan, Past. 61; Swt. 455, 30. Gewǽht, tidder *fessa, fatigata,* Hpt. Gl. 436, 59.   **III.** *of immaterial things, frail, not lasting, fleeting:*—Hú lytel hé (*fame*) biþ, hú lǽne, hú tédre and hú bedǽled ælces gódes *quam sit exilis et totius vacua ponderis,* Bt. 18, 1; Fox 60, 29. Se wlite ðæs líchoman is swíþe flíónde and swíþe tédre and swíþe anlíc eorþan blóstmum *formae nitor ut rapidus est, ut velox, et vernalium florum mutabilitate fugacior,* 32, 2; Fox 116, 17. Ðis líf is lǽnlíc and tyddre and feallende and earm, L. E. I. prm.; Th. ii. 400, 16. Ðissere worulde wuldor gewítendlíc ys tyddre tídlíc miht *hujus saeculi gloria caduca est, fragilis temporalis potentia,* Scint. 215, 8. Týdrum *lubrico,* Germ. 401, 45. Sint swíþe tédre and swíþe hreósende ðás gesǽlþa *caduca felicitas,* Bt. 11, 2; Fox 34, 22. Tíedre (tédra, Cott. MS.), 20; Fox 72, 3. Tyddre weorþmyntas *fragiles honores,* tyddrum gefeohte *fragili bello vel inbecilla,* Wrt. Voc. ii. 150, 38–40. [*O. Frs.* teddre: *Du.* teeder.] v. un-tídre.

tíd-regn, es; *m. A seasonable rain:*—Drihten geopenaþ heofunan his sélustan goldhord and sent tídrénas on ðín land (*to give the rain unto thy land in his season; ut tribuat pluviam terrae tuae in tempore suo*), Deut. 28, 12.

tídrian; *p.* ode.   **I.** *of persons, to get weak* or *infirm* from illness or weariness:—Týdraþ ðis bánfæt ðis body grows weak, Exon. Th. 178, 5; Gú. 1239. Gif mannes fét on sýþe týdrien *if a man get footsore while travelling,* Lchdm. i. 84, 23.   **II.** *of things, to get or be frail, perishable:*—Ðæt sind ða getimbru ðe nó týdriaþ *those are the buildings that decay not,* Exon. Th. 103, 5; Cri. 1683. v. ge-tídrian.

tíd-sang, es; *m. A song used at a particular time, the service held at one of the canonical hours:*—Seofon tídsangas hí gesetton ús tó singenne dæghwamlíce . . . Se forma tídsang is úhtsang mid ðam æftersange ðe ðártó gebiraþ, prímsang, undernsang, middægsang, nónsang, æfensang, nihtsang. Ðás seofon tídsangas gé sculon singan, L. Ælfc. P. 31; Th.

ii. 376, 1–8 : L. Ælfc. C. 19 ; Th. ii. 350, 3–7. Wē syngaþ on ðone Ðunresdæg ūre tídsangas tōgædere . . . On ðone Frigedæg wē singaþ ealle ða tídsangas on sundor būton ðam ūhtsange ánum, 36 ; Th. ii. 358, 30–33. Wē lǽraþ ðæt man on rihtne tíman tída ringe, and preósta gehwilc ðonne his tídsang on circan gesēce, L. Edg. C. 45 ; Th. ii. 254, 6 : R. Ben. 67, 18 : Homl. Th. ii. 160, 19–24. Æt ǽlcan tídsange eal hīrēd āþenedum limum ætforan Godes weófode singe ðone sealm : *Domine, quid multiplicati sunt*, and *preces*, and *collecta*, Wulfst. 181, 26 : 171, 14. Ðonne bid hic hīwan tō tídsongum mīn gemund dōn, Chart. Th. 159, 9, 19. Tídsangas *canonica*, Wrt. Voc. ii. 128, 26. Se tídsang *matins*, R. Ben. 33, 1 : *complines*, 67, 10. v. tíd-þegnung.

tíd-sceáwere, es ; *m. An observer of times and seasons, an astrologer* :— Tídsceáwere *horoscopus* (horoscopus *astrologus, qui horas, maxime natales, inquirit vel considerat*, Migne), Wrt. Voc. i. 53, 18. v. tíd-ymbwlátend.

tíd-scriptor *a chronographer ;* chronographus, Wrt. Voc. ii. 131, 8. v. tíd-wrítere.

tíd-þegnung, e ; *f. Service performed at one of the seven canonical hours* :—Nū ic hæbbe be suman dǽle āhrepod ðe ðam dæghwamlícan tídþēnungum (*the services at the several hours are described in what precedes this remark*), Btwk. 220, 40. v. tíd-sang.

tídung, e ; *f. Tidings* :—Hí cýddan ðam cinge eall. Ðā wearð se cing swýþe blíðe [ðis]sere tídunge, Chr. 995 ; Th. 244, 38. [Ich þonkie mine drihte þissere tidinge, Laym. 24907. Gabriel brohte hire þe tidinge of Godes akenesse, H. M. 45, 7. Swilc tiding ðhugte Adam god, Gen. and Ex. 407. Ich mai bringe tidinge (tiþinge, Cot. MS.), O. and N. 1035. Tydyng, R. Glouc. 172, 1. Typing, 79, 11. No tale ne tidinge of þe worlde, A. R. 70, 19. *M. H. Ger.* zītunge : *Du.* tijding. Cf. *the forms* in -ende, -inde :—Þa come þe tidende (tidinge, 2nd MS.) þat Aganippus was dead, Laym. 3734. Tiðinde (tidinge, 2nd MS.), 5153. Neowe tidinde (tidinge, 2nd MS.) *fresh events*, 2052. Goddspell on Englissh nemmnedd iss . . . god tiþennde, Orm. D. 158. *Icel.* tíðindi *tidings ; an event : Dan.* tidende. The use of the word, even if its form be not borrowed from Scandinavian, seems to shew Scandinavian influence.] v. tídan.

tíd-weorþung, e ; *f. Worship at a particular time, service at one of the canonical hours* :—Hit nis nā tō gelýfanne, ðæt hý fæstende synd rihtlíce, būtan hý æfter hyra mæssan ðæs æfenes tídwurðunga gebíden, Homl. Ass. 140, 67.

tíd-wrítere, es ; *m. A chronicler, annalist* :—Tídwrítera *cronographorum*, Wrt. Voc. ii. 17, 68 : 75, 39. Týdwrítera *chronographorum, temporum scriptorum*, Hpt. Gl. 410, 58. v. tíd-scriptor.

tíd-ymbwlátend, es ; *m. An astrologer* :—Tídembwlátent *oroscopus*, Lchdm. i. lxi, 2. v. tíd-sceáwere.

tiéder-, tiédran, tiédre, -tiéfran, tiegle, tién. v. tíder-, týdran-, tídre, ā-tiéfran, tigele, tín.

tiér *distillation* (? cf. teár) ; *ornament, splendour* (? cf. *O. H. Ger.* ziarî, zierî, ceerî *ornamentum, venustas, decus*) ; *treasure* (? cf. *Icel.* taurar ; *pl. treasures*) ; *glory* (? v. tír) :—Nis nān wundor ðæt sió lyft sié wearm and ceald wǽt wolcnes tiér winde geblonden (cf. sió lyft is ǽgðer ge ceald ge wǽt ge wearm ; nis hit nān wunder, Bt. 33, 4 ; Fox 128, 35), Met. 20, 81.

tife, an ; *f. A bitch* :—Gif ðū wille ðæt wíf cild hæbbe oþþe tife hwelp, Lchdm. ii. 172, 21. [*Icel.* tefja *a bitch ;* tefja *to call a person a bitch : Dan.* tæve *a bitch : Swed.* tüfwa.]

tifer, -tífran. v. tiber, ā-tiéfran.

tífrung, e ; *f. Painting* :—Ðū leornodest onn ānum þóðere . . . ātéfred, ðað ðū meahtest beo ðære téfrunge ongytan ðises roðores ymbehwirft, Shrn. 174, 18.

tíg (?), es ; *m. An open place* (?) ; *a form occurring in composition with* fore, forþ. *For the former see* fore-tíge (*read* -tíg) ; *the instances of the latter are as follows* :—Forðtíges *vestubuli, atrii*, Hpt. Gl. 496, 28. On ðam forðtēge *in ipsis foribus*, Kent. Gl. 228. *Graff gives* zieh *forum, and Grimm, R. A.* 748, *cites* tie *a meeting-place, as a term of lower Saxony*.

Tíg, tíg *a case*. v. Tíw, teáh.

-tig *-ty, a numeral suffix in words denoting the decades ;* up to 60 such words are formed with a suffix only, from 60 to 120 hund is prefixed and tig suffixed, *hund-seofon-tig, hund-twelf-tig*. Other dialects make a distinction in the numerals at the same point. Gothic uses *tigus* (*pl. tigjus*) in the earlier, *-tēhund* in the later, O. Saxon -*tig* in the earlier, while 70 is given by *ant-sibunta ;* in *O. H. Ger.* the two forms are -*zug* and -*zô*. In *O. Frs.* and *Icel.* the same forms are used throughout. *Tig* is another form of the root seen in *ten* (*tehan*, g *for* h *according to Verner's Law*).

tígan ; *p.* de *To tie*, (a) literal :—Valerianus hēt beheáfdian on Ypolitus gesihðe ealle his hīwan, and hine sylfne hēt tígan be ðam fótum tō ungetemedra horsa swuran, Homl. Th. i. 432, 33. (b) figurative :—Nū ðū miht gehýran, hū ðes dǽl (*the conjunction*) tígþ ða word tōgædere, Ælfc. Gr. 44 ; Zup. 258, 10. [Heo wolden þa ban alle teien (tiȝe, 2nd MS.) togadere, Laym. 20997. Iteied (-tiȝed, 2nd MS.) tosomne, 25972. He teide ane clot to hire, A. R. 140, 7. Is þe latere dole euer iteied (-teiȝet) to ðe vorme, 14, 2. Tached oþer tyȝed, Allit. Pms. 14, 464. Kynges shulde taken transgressores and tyen hem faste, Piers P. 1, 96.] v. ge-, on-, un-tígan.

tige, tigel *a tile*, tigel *a trace*. v. tyge, hróf-tigel, tygel.

tigel-ærne(-a?), an ; *f.* (*m.*?) *A building made of brick* (?), *a building for making bricks* (?), *brick-kiln* (?) :—Forð on ða mearce in on ða tigelærnan, Cod. Dip. Kmbl. iii. 130, 29.

tigele, tigle, tiegle, an ; *f. A tile, brick* :—Tigule *tegula*, Txts. 101, 1992. Tigele *figulum*, Wrt. Voc. ii. 148, 79. Tigle *testula*, Germ. 391, 17 : *testa*, Ps. Spl. 21, 16. Mid weorcum clámes and tigelan *operibus luti et lateris*, Ex. 1, 14. Se weall is geworht of tigelan and eorðtyrewan *murus coctili latere atque interfuso bitumine compactus*, Ors. 2, 4 ; Swt. 74, 17. Genim swealwan, gebærn under tigelan tō ahsan, Lchdm. ii. 156, 9. Ða reádan tigelan gecnuwa tō duste, 114, 24. Nim sume tigelan (tiglan, Cott. MSS.) and wrít on hiere ða burg Hierusalem *sume tibi laterem, et describes in eo civitatem Jerusalem*, Past. 21 ; Swt. 161, 3, 9, 11. Tieglan (tiglan, Cott. MSS.), Swt. 161, 12, 20. Se ðe lǽrþ stuntne swylce se ðe belíme tigelan (*testam*) *whoso teacheth a fool is as one that glueth a potsherd together* (Eccl. 22, 7), Scint. 96, 19. Tigelan *lateres*, Wrt. Voc. ii. 51, 41. Tigelena gemet *a tale of bricks*, Ex. 5, 14. Tiglena *testularum*, Hpt. Gl. 499. 28. Tighelana *tegularum*, 459, 40. Tigelum, Exon. Th. 477, 28 ; Ruin. 31. Hig hæfdon tygelan (*lateres*) for stān, Gen. 11, 3. [*O. H. Ger.* ziagel, ziagalo *later, testa, imbrex : Icel.* tigl ; *n. a tile, brick.* From Latin.] v. þæc-tigele ; hróf-tigel (-tigele ? ; *perhaps for pl.* -tigla, -tiglan *should be read*).

tigelen ; *adj. Of pot* :—Fæt tigelen (-an ? ; lāmys, MS. C.) *vas figuli*, Ps. Spl. 2, 9. [*O. H. Ger.* ziagalin *laterinus, latericius*.]

tigel-fáh ; *adj. Many-coloured with tiles or bricks* :—Tigelfāgan trafu, Andr. Kmbl. 1683 ; An. 844.

tigel-getæl, es ; *n. A tale of bricks ;* laterum numerus :—Gē sceolon āgifan ðæt ilce tigolgetel, Ex. 5, 18.

tigel-geweorc, es ; *n.* I. *brickmaking* :—Ne sylle gē nān cef tō tigelgeweorce (*ad conficiendos lateres*), Ex. 5, 7. II. *work at making bricks* :—Āsettaþ him ðæt ilce tigelgeweorc ðe hig ǽr werhton *mensuram laterum, quam prius faciebant, imponetis super eos*, Ex. 5, 8. Tigulgeweorc, 16.

tigel-leáh ; *f. A brick-field* :—On tigelleáge, Cod. Dip. Kmbl. v. 267, 21.

tigel-stán ; es ; *m. A tile, pan-tile* :—Tigelstān *imbrex*, Engl. Stud. xi. 66, 50. [Cover hit wele with a teghellstane, Rel. Ant. i. 54, 30. Tielstoon, Wick. (Is. 16, 11). Tilston *tegula*, Wrt. Voc. i. 256, col. 1.]

tigel-wyrhta, an ; *m. A brickmaker, a potter* :—Fæt tygelwirhtan *vas figuli*, Ps. Lamb. 2, 9. Æcyr tigylwyrhtena *agrum figuli*, Mt. Kmbl. 27, 7. Tigelwyrhtena, 10.

tiger (?) *a tiger ; pl.* tigras, Nar. 38, 4 ; tigris, 12, 13 ; 15, 3. Deór ðe sind tigres gehátene . . . Ðās rēðan tigres, Homl. Th. ii. 492, 10–21. v. tigrisc.

tígere (?). v. bufan-tígere.

tíging, e ; *f. Tying, connection* :—Sume naman syndon *absolutivae*, ðæt synd ungebundene, ða ne behófiaþ nānre tíginge óðres naman, Ælfc. Gr. 5 ; Zup. 14, 14.

tigl *a trace*, tigle *a lamprey*, tigole. v. tygel, tygele, tigele.

tigrisc ; *adj. Of a tiger* :—Mid tigriscum fellum *tygridum pellibus*, Nar. 26, 14.

tigþian, tih(h)ian. v. tíþian, teohhian.

tiht, es ; *m. A charge, an offence with which one is charged ;* crimen :—Legerteám oððe tiht *flagitium*, Wrt. Voc. ii. 39, 34. Gif hwā cyninges borg ābrece, gebēte ðone tyht (tihtlan, MS. H.) swā him ryht wísie, L. Alf. pol. 3 ; Th. i. 62, 8. [*O. Frs.* tichta *accusation : O. H. Ger.* bi-ziht *nota ;* in-ziht *crimen*.] v. teón *to accuse*, and next word.

tihtan ; *p.* te *To charge* a person (*acc.*) with an offence :—Tyhte intentabat, Hpt. Gl. 519, 76. Hē hæþ gelǽd fulle láde æt ðan unrihtwífe ðe Leófgār bisceop hyne tihte *he has completely cleared himself of the offence with which the bishop charged him*, Chart. Th. 373, 33. Gif man óðerne sace tihte, L. H. E. 8 ; Th. i. 30, 11 : 10 ; Th. i. 30, 17 : L. Wih. 22 ; Th. i. 42, 3 : 23 ; Th. i. 42, 6 : 24 ; Th. i. 42, 10, 11. [Cf. *O. H. Ger.* in-zihtōn *criminari :* Ger. be-zichten, -zichtigen *to accuse*.] v. preceding word.

tihtan *to exhort*. v. tyhtan.

tiht-bisig ; *adj. Labouring under frequent accusations, often accused, and so of bad repute* :—infamatus et accusationibus ingravatus, L. Edm. C. 7 ; Th. i. 253, 23 : accusacionibus infamatus, L. H. I. ; Th. i. 567, 18. Cf., too, *the phrase* oft betygen, L. In. 18 ; Th. i. 114, 6 : 37 ; Th. i. 124, 21. One to whom the epithet applied was in an unfavourable position when brought into court, for he was forced to go to the threefold ordeal, and if he failed to clear himself was subject to a heavier penalty than others :—Gif hē tyhtbysig sý, gange tō ðæm þryfealand ordále . . . Gif hē fúl wurðe, æt ðam forman cyrre bēte ðam teónde twygylde . . . And æt ðam óðran cyrre ne sý ðǽr nān óðer bót būtan ðæt heáfod, L. Eth. i. 1 ; Th. i. 280, 9–282, 2. Niman ða tihtbysian

men ... and ǽlc tihtbysig man gange tó þryfealdan ordále, oððe gilde feówergilde, iii. 3; Th. i. 294, 6–11. Gif hwylc man sȳ swá tihtbysig and hine ðonne þreó men ætgædere teón, ðonne ne beó ðǽr nán óðer búton ðæt hé gange tó ðam þryfealdan ordále, L. C. S. 30; Th. i. 392, 22 (and see the whole section for the penalties). Be tihtbysigum. Se ðe tihtbysig sȳ, L. Edg. ii. 7; Th. i. 268, 13: L. C. S. 25; Th. i. 390, 17. Sȳ ǽlc man ðe tihtbysig nǽre ... ánfealdre láde wyrðe, 22; Th. i. 388, 9. tihte, tihten, tihtend, tihtend-líc, tihtere, tihting, tihtness. v. hól-tihte, tyhten, tyhtend, tyhtend-líc, tyhtere, tyhting, tyhtness.

tihtle, an; f. A charge, accusation:—Gif hit ánfeald tyh[t]le sȳ, dúfe seó hand æfter ðam stáne óð ða wriste, and gif hit þryfeald sȳ, óð ðæne elbogan, L. Ath. iv. 7; Th. i. 226, 16. Gif hit tihtle (tihtla, MS. B.) sī and lád forberste if a charge be brought, and the attempt to refute the charge fail, L. C. S. 54; Th. i. 406, 10: 57; Th. i. 406, 26. Swerige hé ðane áð (cf. next passage), ðæt hé sȳ unscyldig ðǽre tihtlan (tyhtelan) ... And ofgá ǽlc man his tihtlan mid foreáðe (cf. L. O. 2; Th. i. 178, 10: L. O. D. 6; Th. i. 354, 30), L. Ath. i. 23; Th. i. 212, 1–5. Ic eom unscyldig æt ðǽre tihtlan ðe N. mē tíhþ, L. O. 5; Th. i. 180, 16. Gif man folciscne mæssepreóst mid tihtlan belecge lādige hine swá swá diácon ðe regollíf libbe if a charge be brought against a secular priest, let him clear himself as a regular deacon would, L. Eth. ix. 21; Th. i. 344, 19: 22; Th. i. 344, 22. Tyhtlan (tihlan, MS. A.), L. C. E. 5; Th. i. 362, 7. Ne stent nán óðer lád æt tihtlan búte ordál betweox Wealan and Englan, L. O. D. 2; Th. i. 354, 1. Ðá tugon hié hiene ðæt hé heora swicdomes wið Alexander fremmende wǽre and hine for ðǽre tihtlan ofslógon they accused him of betraying them to Alexander, and on that charge slew him; hunc, quasi urbem Alexandro venditasset, necaverunt, Ors. 4, 5; Swt. 168, 18. Se ðe ða tihtlan áge the plaintiff, prosecutor, L. H. E. 10; Th. i. 30, 19. v. frum-, stæl-, wiðer-tihtle; tiht, and next word.

tihtlian; p. ode To charge with an offence, to accuse:—Gif man mæssepreóst tihtlige ánfealdre sprǽce, L. Eth. ix. 19; Th. i. 344, 11: 20; Th. i. 344, 15. Tihtlige (tihlige, MS. A.), L. C. E. 5; Th. i. 362, 12. v. be-tihtlian, ge-tihtlod.

Tiig. v. Tiw.

til; adj. I. good at anything, apt, capable, competent:—Hé wæs selfa til, heóld á ríce ēðeldreámas, Cd. Th. 97, 2; Gen. 1606: Beo. Th. 122; B. 61. Til sceal on ēðle dómes wyrcean, Menol. Fox 500; Gn. C. 20. Sum biþ beórhyrde gód, sum biþ bylda til hám tó habbanne, Exon. Th. 297, 29; Crä. 75. Till, Beo. Th. 5436; B. 2721. Hié wǽron an wíg gearwe ... efne swylce mǽla swylce hira mandryhtne þearf gesǽlde; wæs seó þeód tilu, 2505; B. 1250. Wǽron men tile, Cd. Th. 99, 11; Gen. 1644. Dióre gecēpte drihten Crēca Tróia burh tilum gesīðum, Met. 26, 20. [Cf. Goth. manna gatils (εὔθετος, aptus) in thiudangardja Guths a man fit for the kingdom of God, Lk. 9, 62.] II. good for anything, that serves a purpose, beneficial, serviceable, convenient, opportune:—His mildheortnyss is til mancynne, Ps. Th. 116, 2. Ys mín (a town's) innað til, wombhord wlitig, Exon. Th. 399, 11; Rä. 18, 9. Ne wæs ðæt gewrixle til, ðæt hié on bá healfa bicgan scoldon freónda feorum, Beo. Th. 2613; B. 1304. Áhte ic folgað tilne (a service that benefited me), Exon. Th. 379, 25; Deór. 38. Ðú mē þeódscipe lǽr ðínne tilne bonitatem et disciplinam doce me, Ps. Th. 118, 66. Gebiddaþ ealle hálige tó ðē on tilne tíman (in tempore opportuno), 31, 7. [Cf. Goth. dags gatils (εὐκαίρος, opportunus) a convenient day, Mk. 6, 21. Ei bigēteina til du wróhjan ina, Lk. 6, 7.] III. good, kind, gentle [cf. till = tame in Pegge's Kenticisms, E. D. S. Pub. Reprinted Gloss. C. 3]:—Til mon tiles and tomes meares a kind man is mindful of a gentle and tame horse, Exon. Th. 342, 12; Gn. Ex. 142. Him ðæs leán āgeaf Metend gumcystum til (liberally kind), Cd. Th. 108, 23; Gen. 1810. IV. good, excellent, (a) of moral good:—Til biþ se ðe his treówe gehealdeþ, Exon. Th. 293, 6; Wand. 112. Til sceal mid tilum the good shall be associated with the good, 334, 28; Gn. Ex. 23. Ðæt hió ðǽre cwēne oncweðan meahton swá tiles swá trāges, swá hió him tó sóhte, Elen. Kmbl. 649; El. 325. Tile and yfle the good and the evil (at the day of judgment), Cd. Th. 303, 10; Sat. 610. Hí (devils) duguðe beswícaþ and on teosu tyhtaþ tilra dǽda, Exon. Th. 362, 10; Wal. 34. Habbaþ freónda ðȳ má sóþra and gódra, tilra and getreówra, 409, 2; Rä. 27, 23. (b) of physical excellence:—Toscean teolum húsum on, cyninga cofum, eardedan, Ps. Th. 104, 26. V. til is found in proper names, see for examples Txts. 497. [Cf. O. H. Ger. zil: Ger. ziel aim, purpose.] v. tela, and next word.

til, es; n. I. use, service, convenience. v. til, II:—Gewritu secgaþ ðæt seó wiht (day) sȳ mid moncynne miclum ticlum (tielum? tilum?) sweotol and gesȳne, sundorcræft hafaþ, Exon. Th. 420, 12; Rä. 40, 2. II. goodness, kindness. v. til, III:—Mē on ðínum tile gelǽr ðæt ic teala cunne ðín sóðfæst weorc healdan in bonitate tua doce me justificationes tuas, Ps. Th. 118, 68. v. til-fremende.

til; prep. (used only in the North) To:—Fúsæ fearran kwómu æþþilæ til ánum (cf. fúse feorran cwómon tó ðam æðelinge, Rood Kmbl. 115; Kr. 58). Txts. 126, 13. Hé scóp ælda barnum heben til hrófe (cf. tó hrófe, Bd. 4, 24; M. 344, 11), 149, 6. Ðá cueð til (tó, Rush.)

him ðe Hǽlend tunc dicit illis Jesus, Mt. Kmbl. Lind. 26, 31. Huēr wiltū ðæt wē gearuiga ðē til eottanne (tó etanne, Rush.) Eástro ubi vis paremus tibi comedere Pascha? 26, 17. [The word retains its meaning in the Northern dialects, but otherwise it is used in reference only to time. O. Frs. til: Icel. til.]

tila well, Tile Thule. v. tela, Tyle.

til-fremmende doing good:—Tillfremmendra, Exon. Th. 440, 23; Rä. 60, 7. Cf. gód-fremmende.

tilia, tiliga, an; m. A husbandman, cultivator of land:—Tilia colonus, Wrt. Voc. i. 74, 66. Bigenga, tilia, inbúend colonus, i. incola, cultor, inquilinus, il. 134, 25. Tilia colonus, habitator, Hpt. Gl. 422, 60. Se merigenlíça tilia the labourer who came in the morning, Homl. Th. ii. 74, 30. Ðá sende hé tó ðam tiligum (tilium, MS. A. ad agricolas) his þeów ... Ðá cwǽdon ða tilian (coloni) ... Ðæs wíngeardes hláford fordēþ ða tiligean (tylian, MS. A. colonos), Mk. Skt. 12, 2, 7, 9. [Þe wise teolie prudens sator, O. E. Homl. i. 133, 10.] v. eorþ-, irþ- (yrþ-) tilia.

tilian, tiligan, tilgan, teolian, tiolian, tielian; p. ode To strive after or for some object. I. where the construction is not determined:—Hé higode oððe tilode nititur, Wrt. Voc. ii. 59, 69. Tioludun perstant, 117, 15. Tilege nitatur, 61, 56. Teolige decrevit, Hpt. Gl. 469, 50. Tilgende nisus, Wrt. Voc. ii. 60, 28. Tilgendum adnitentibus, 99, 32. Tillgendum, 6, 23. II. where the object of effort is not expressed, to strive to obtain, to labour, toil, procure with effort, provide, acquire, (1) where the person for whom the action takes place is not expressed:—Ic bebeóde eallum mínan gerēfan ðæt hí on mínan āgenan rihtlíce tilian and mē mid ðam feormian I command all my reeves, that they obtain revenue rightfully from my own property and maintain me therewith, L. C. S. 70; Th. i. 412, 21. Se ðe wæste scaðiende, weorðe se tiligende on rihtlícre tilðe, Wulfst. 72, 13. (2) with dat. of person for whom the effort is made:—Oxa teolaþ his hláforde, Homl. Th. i. 412, 3. Se ðe him sylfum teolaþ, 14. Se ðe him sylfum teolaþ, nā Gode, ne com se nā gyt binnon Godes wíngearde. Ða tyliaþ Gode, ða ðe ne sēcaþ heora āgen gestreón ðurh gȳtsunge, ii. 76, 32–34. Ðæt hē ða eorðan worhte and him ðēron tilode (he should provide for himself from it), Gen. 3, 23. Hit mǽre is ðonne ccc geára and lxxii wintra syððan ðyllíc feoh wæs farende on eorðan and ealle men hoem mid tilodon (procured for themselves what they wanted with that money; cf. Amang ðam feó ðe wē úre neóde mide bicgaþ, 706), Homl. Skt. i. 23, 703. Hé is wyrðe ðæt ðú him tilige he deserves that you exert yourself for him; dignus est ut hoc illi praestes, Lk. Skt. 7, 4. Preósta gehwilc tilige him rihtlíce and ne beó ǽnig mangere mid unrihte let every priest provide for himself honestly, and let none be a trader dishonestly, L. Edg. C. 14; Th. ii. 246, 23. Swá hwá swá ǽnige cýpinge on ðam dæge begæþ ... oððe ǽnig cræftig man him on his cræfte tylige (gets gain for himself by working at his craft), Wulfst. 296, 8. III. with gen. (1) of an object to be obtained by effort, (a) without reference to person for whom, to seek after, get after seeking, procure, make provision of:—Ðú wyfst and wǽda tylast you weave and make provision of garments, Homl. Th. i. 488, 26. Tilaþ ánra gehwilc āgnes willan (cf. winþ heora ǽlc on óþer æfter his āgenum willan, Bt. 21; Fox 74, 34), Met. 11, 83. Ælc man ðæs tiolaþ, hū hé on ēcnesse swincan mǽge, Ps. Th. 48, 7. Ða ðe on ðam beóþ ābisgode ðæt hié sibbe tiligaþ (tiliaþ, Cott. MSS.) qui faciendae pacis studiis occupantur, Past. 47; Swt. 363, 9. Ðæt hí unrihtes tiligeaþ, Ps. Th. 143, 9. Tilgaþ, Exon. Th. 230, 14; Ph. 472. Sume tiliaþ wífa for ðam ðæt hí þurh ðæt mǽge mǽst bearna begitan and eác wynsumlíce libban uxor ac liberi, qui jucunditatis gratia petantur, Bt. 24, 3; Fox 82, 25. Man tilode tó his hergeatwæn ðæs ðe man habban sceolde what was necessary for his heriots should be provided, Cod. Dip. Kmbl. iii. 352, 16. Mid his handcræfte hē teolode his and his gefērena forðdǽda, Homl. Th. i. 392, 16. Hí wunnon æfter wyrþscipe and tiledon (tiolodon, Cott. MS.) gódes hlísan mid gódum weorcum, Bt. 40, 4: Fox 240, 5. Ðæt hē suā tilige ðǽre orsorgnesse mid ðǽre ānfealdnisse ðætte hē ðone ymbeðonc ðæs wǽrscipes ne foflǽte ut sic securitatem de simplicitate possideant, ut circum-pectionem prudentiae non amittant, Past. 35; Swt. 237, 16. Ic an ðæs landes Ǽffan, and heó tilige uncer begea sáwla þearfe ðēron I grant the land to Ǽffe, and let her provide what is necessary for both our souls therefrom, Chart. Th. 495, 34: 497, 18. Laboratores syndon weorcmen ðe tilian sculon ðæs ðe eall þeódscype big sceall libban laboratores are workmen, that have to obtain by their efforts that by which all the nation has to live, L. I. P. 4; Th. ii. 306, 35: Beo. Th. 3651; B. 1823. Hē sceal fela tóla tilian he must procure many tools, Anglia ix. 262, 27: 261, 10. Seó lufu tuddres tó tilianne amor ortandi sobolis, Bd. 1, 27; S. 495, 38. (b) with dat. of person:—Paulus him sylfan nānes lofes ne tilade Paul took no praise to himself; nec Paulus sibi aliquid imputavit, R. Ben. 4, 5. Se here tilode him ðæs ðe hí behófdan the Danes provided themselves with what they needed, Chr. 1006; Erl. 140, 16. Hí heom metes tilodon, 1016; Erl. 157, 3: Hexam. 17; Norm. 26, 9. Ic lǽre ðæt ðú [ne?] fægenige óþerra manna gódes and heora ǽþelo tó ðon swíþe ðæt ðú ne tilige ðē selfum āgnes I advise you [not] to rejoice so much in other men's goodness and nobility, that you do not provide yourself with your own, Bt. 30, 1; Fox

108, 31. Ðæt man him ðurh fixnoðe bigleofan tilige, Homl. Th. ii. 208, 19. Tiliaþ eów freónda *get friends for yourselves*, i. 334, 27. Ðu scealt mid earfoðnyssum ðé metes tilian, 18, 15 : Homl. Skt. i. 23, 219. Noe ongan tó eorðan him ætes tilian *Noe began to provide himself with food from the earth*, Cd. Th. 94, 6 ; Gen. 1557. Him tilian fylle on fægum, Judth. Thw. 24, 26 ; Jud. 208. Him metes tó tylienne, Chr. 1052; Erl. 183, 20. (2) *of an object to which care, attention, is directed*, (a) in a general sense, *to care for, attend to, work for, provide for :*—Ðonne ðu tilast ðín on eorðan ne sylþ heó ðé náne wæ-tmas *when you try to get subsistence for yourself from the ground, it will give you no fruit*, Gen. 4, 12. Ðonne se sacerd his on ða ilcan wísan tielaþ (tiolaþ, Cott. MSS.) ðe ðæt folc dóþ *when the priest provides for himself in the same way that the people do*, Past. 18 ; Swt. 133, 8. Se ðe ne gýmþ ðæra sceápa ac tylaþ his sylfes *he that heeds not the sheep, but takes care of himself*, Homl. Th. i. 242, 1. Se ðe his ær tíde ne tiolaþ ðonne biþ his on tíd untilad *he that makes no provision for himself beforehand will be without provision when the time comes*, Bt. 29, 2 ; Fox 106, 3. Hé wæs fiscere and mid ðam cræfte his teolode, Homl. Th. i. 394, 2. Hé þearfendra þinga teolode *he attended to the concerns of the needy*, Ps. Th. 108, 30. Huntigan and fuglian and fiscian and his on gehwilce wísan tó ðære lænan tilian, Shrn. 164, 6. Lifes tiligan *to care for life*, Exon. Th. 81, 6 ; Cri. 1319 : Salm. Kmbl. 322; Sal. 160. Hié Norðanhymbra lond ergende wæron and hiera tilgende (*providing for themselves*), Chr. 876; Erl. 78, 15. (b) *in a special sense of medical care, to cure, treat, tend, attend to :*—Sceal ðæs módes læce ær tilian ðæs ðe hé wénþ ðæt ðone mon ær mæge gebrengan on færwyrde. Hwílum, ðeáh, ðær ðær mon óðres tiolaþ, ðær weaxð se óðer. Fordæm sceal se læce . . . tilian ðæs máran . . . Hwæðres ðara yfela is betere ær tó tilianne? Past. 62 ; Swt. 457, 10–22. Ðara stówa sum raþe rotaþ, gif hire mon gímeleáslíce tilaþ, Lchdm. ii. 84, 25. Tiloden (*curabant*) his lǽcas, Bd. 4, 32 ; S. 611, 19. Bútan his man tilige hé biþ ymb þreó niht gefaren *unless the patient be attended to, he will be dead in three days*, Lchdm. ii. 46, 18. Hú mon scyle gebrocenes heáfdes tiligean, 2, 4. Tilian, 56, 14. Hira man sceal tilian mid wyrtdrencum, 82, 16. Hwonan ic ðín tilian scyle *qui modo sit tuae curationis*, Bt. 5, 3 ; Fox 10, 35. IV. *with a dative, to cure, treat :*—Wísman gif heó tilaþ (*curet*) híre cilde mid ænigum wiccecræfte, L. Ecg. P. iv. 20; Th. ii. 210, 17. V. *with an accusative*, (1) *to gain, obtain :*—Se ásolcena ðeówa ðe nolde tilian nán ðing his hláforde mid ðam befæstum punde, Homl. Th. ii. 552, 29. (2) *to attend to, bestow care on, care for*, (a) in a general sense :—Se ðe ymbe ða eorðlícan spéda singallíce hogaþ, and ða écan gestreón ne teolaþ *he that is continually anxious about earthly wealth, and cares not for the eternal treasures*, Homl. Th. ii. 372, 23. (b) *of medical attention, to treat, attend to :*—His lǽcas hine mid sealfum lange teolodon, Guthl. 22 ; Gdwin. 96, 15. (c) *to till :*—Ðæt land tó tilianne, Chr. 1091 ; Erl. 228, 20. (c 1) *without object :*—Ðá man oððe tilian sceolde oððe eft tilða gegaderian, 1097; Erl. 234, 24. VI. *where the object for the sake of which an effort is made is pointed out by a preposition :*—Tó ðisum swicolum lífe wé swincaþ and tiliaþ and tó ðam tówerdan lífe wé tiliaþ hwónlíce *we labour and toil for this deceitful life, and for the future life we toil little*, Homl. Skt. ii. 28, 168. VII. *where the object of effort is expressed by an infinitive (simple or gerund), or a clause, to strive, attempt, endeavour, intend*, (1) with infin. :—Ðæt ðe wé bécnan tiliaþ, Met. 11, 79. Ic næfre ne teolade sittan on ánum willan mid ðam árleásum *cum impiis non sedebo*, Ps. Th. 25, 5. Ðá tilode hé ða stówe gecltensian *studens locum purgare*, Bd. 3, 23 ; S. 554, 26. Hé hine monnum gécyþan teolode, Blickl. Homl. 165, 31. (2) with gerund :—Ðu tilast (tiolast, Cott. MS.) wædle tó fliónne, Bt. 14, 2 ; Fox 44, 7 : 10; Fox 30, 1. Manege tiligaþ (tiliaþ, Cott. MS.) Gode tó cwémanne, 39, 10 ; Fox 228, 13. Ic tilode ðé tó lícianne, Ps. Th. 25, 3. Tylode, Bd. 5, 24 ; S. 649, 11. Hé tiolode (tilode, Cott. MSS.) hié betwux him tó tóscádanne, Past. 47 ; Swt. 363, 1. Hé teolode tó árísenne, Blickl. Homl. 219, 18. Hié ða londlióde tiolode má ússa feónda willan tó gefremmanne ðonne úrne *illi maiorem hosti quam mihi fauorem accomodantes efficere pergebant*, Nar. 6, 19. Swá hwylc man swá ðás scriftbóc tilige tó ábrecanne *quicunque confessionale hoc violare conatus fuerit*, L. Ecg. P. Addit. ; Th. ii. 238, 8. Ðæt hié tilgen (tiligen, Cott. MSS.) tó kýðanne, Past. 47 ; Swt. 363, 10. Hé sceal tilian suá tó libbanne *sic studet vivere*, 10 ; Swt. 61, 18. (3) *with a clause :*—Ða bilewitan sint tó herigenne fordæmðe hié simle suincaþ on ðæm ðæt hí tieligeaþ (tiliaþ, Cott. MSS.) ðæt hié ne sæculn leásunga secgan *laudandi sunt simplices, quod studeant numquam falsa dicere*, Past. 35 ; Swt. 237, 8. Ðín esne teolode ðæt hé ðíne sóðe word beeode *servus tuus exercebatur in tuis justificationibus*, Ps. Th. 118, 23. Ðæt wé teolian, ðæt wé sýn gearwe, Blickl. Homl. 125, 11. Uton teolian ðæt ús ðás tída ídle ne gewítan, 129, 36: 111, 18. Hé sceal tilian ðæt hé lícige *debet studere se diligi*, Past. 19 ; Swt. 147, 14 : L. E. I. 28 ; Th. ii. 424, 26 : Bt. 29, 3 ; Fox 106, 18 : Met. 16, 1. Tiligean, Ps. Th. 138, 17. Hé ne onginþ tó tilianne, ðæt hé ðæt weorð ágife, 48, 7. [Sculdest thu neure finden land tiled . . . War sæ me tilede, þe erthe ne bar nan corn, Chr. 1137; Erl. 262, 25, 39. To teoliende efter istreone, O. E. Homl. i. 133, 13.

Tulien after strene, ii. 155, 4. Heo tileden on eorðen, Laym. 1940. þat lond heo lette tilien, 2618. Ure Louerd tiled efter hore luue. A. R. 404, 14. Silence tilcð hire, and heo itiled bringeð forð uode, 78, 15. Ase lomen uorte tilien mide þe heorte, 384, 17. In swinc ðu salt tilen ði mete, Gen. and Ex. 363. Lond to tilie, R. Glouc. 21, 9. Heo swonke and tilede here lyflode, 41, 22. To taken his teme and tulyen (tilien, tilie) þe erthe, Piers P. 7, 2. Many wyntres men lyveden and no mete ne tulyeden (tylied, tiliden, tilieden, teleden), 14, 67. Ichave tyled him for that sore, Beves of Hamtoun (Halliwell's Dict.). Goth. ga-tilón *to obtain*: O. Sax. tilian (*with gen.*) *to obtain*: O. L. Ger. tilón *festinare, exercitari*: O. Frs. tilia *to till, to beget*: O. H. Ger. zilén *studere, conari, niti, contendere, moliri, adniti*; zilón (*with gen.*).]

tiliga. v. tilia.

till *a fixed point, station :*—Swá stent eal weoruld stille on tille, Met. 20, 172. On ðam gim ástíhþ on heofenas up hýhst on geáre and of tille ágrynt *in it (June) the sun mounts up into the skies highest in the year and declines from that point*, Menol. Fox 220 ; Men. 111. [Cf. O. H. Ger. zil *destinatum*: Ger. ziel.]

tillan; *p.* tilde *To touch, reach. In compounds* á-, ge-tillan; *instances omitted under those words are given here :*—Ðeáh ðe hé stæpe fulfremednysse átilþ (*adtingit*), Scint. 100, 15. Getilþ *contingat*, getilde *contigit*, Wrt. Voc. ii. 135, 9–13. Gif wé ðone hróf ðære heálícan eáðmódnesse getillan willaþ (*adtingere*), R. Ben. 23, 2. [þe niþer end tilde to his chinne, Brand. 24. He hadde a long berd þat tilled (tylde) to his wombe *habuit barbam prolixam usque ad ventrem*, Trev. v. 193. 8. Alle þat he miȝt tille, Fer. 59. O. H. Ger. [zillen]; *p.* zilta *tangit*.]

til-líc; *adj. Good, capable, able.* v. til, I :—þegn . . . tillíc esne . . . strong, Exon. Th. 436, 28; Rä. 55, 8: 480, 20; Rä. 64, 5.

tillíce; *adv. Kindly, graciously.* v. til, III, Exon. Th. 352, 28 ; Reim. 2.

til-módig; *adj. Noble-minded :*—Se eádga (*Abraham*) Drihtnes noman weorðade, tilmódig eorl tiber onsægde, Cd. Th. 113, 14; Gen. 1887. Ic ðé (*Abraham*) bidde ðæt ðu tilmódig treówa selle, ðæt ðu wille mé wesan freónd fremena tó leáne ðara ðe ic ðé gedón hæbbe, 170, 22 ; Gen. 2817. Heofona heáhcyning trymede tilmódigne (*Abraham*) : 'Ne lǽt ðu ðé ðín mód äsealcan,' 130, 27 ; Gen. 2166. Ða æðelingas . . . .xii. tilmódige (*the twelve apostles*), Apstls. Kmbl. 171 ; Ap. 86.

tilþ, e; *also* tilþe, an ; *f.* I. *labour which brings gain, by which acquisition is made, an employment*, (1) in a general sense :—Se ðe wǽre scaðiende weorðe se tiligende on rihtlícre tilðe *he that has been accustomed to steal, let him support himself by an honest employment*, Wulfst. 72, 13. (2) *with special reference to agriculture, tillage, cultivation, work on land :*—Se scádwís geréfa sceal witan ælcre tilðan tíman ðe tó túne belimpþ ; for ðam on manegum landum tilð biþ redre ðonne on óðrum ge yrðe tíma hrædra, ge mǽda rædran . . . ge gehwilc óðer tilð, Anglia ix. 259, 3–12. II. *gain from labour, produce of labour, acquisition*, (1) in a general sense :—Tilða ł stre[óna] *quaestuum*, Hpt. Gl. 452, 7. (2) *with reference to agriculture, crop, produce, fruit :*—Þurh mycele rénas, ðe ealles geáres ne ablunnon, forneáh ælc tilð on mersclande forférde, Chr. 1098; Erl. 235, 12. Ðæt land mid ðære tilðe ðe ðár ðænne on sý, Chart. Th. 329, 12. Ic geann ðæs landes mid mete and mid mannum and mid ealre tylðe swá ðǽrtó getilod biþ, 529, 18, *and often in the same will.* Fela tilða hám gæderian, Anglia ix. 261, 16. Ðá man oððe tilian sceolde oððe eft tilða gegaderian, Chr. 1097; Erl. 234, 25. Ealle eówre wæstmas and eorþlíce tilþa, Wulfst. 132, 14. [þe tilðe of rihtwisnesse, þæt is silence *cultus justiciae silencium*, A. R. 78, 15: Wick. Is. 32, 17. God sent þe sonne to saue a cursed mannes tilthe, Piers P. 19, 430. To sowe cockel with the corn So that the tilthe is nigh forlorn, Gow. ii. 190, 12. O. Frs. tilath *cultivation*.] v. ge-tilþ.

tilþe, an ; *f.* v. preceding word.

tilung, teolung, tiolung, tielung, e ; *f.* I. *striving, endeavour, effort, labour :*—On swelcum lǽnum weorþscipum ælces mennisces módes ingeþanc biþ geswenced mid ðære geornfulnesse and mid ðære tiolunga (tiluncga, Cott. MS.) *with the desire and striving for them*, Bt. 24, 3 ; Fox 82, 22. Hí swuncon on wingeardes biggencge mid gecneordlícere teolunge, Homl. Th. ii. 74, 33. Hí forgýmeleásodon ðæs écan lífes teolunge *they neglected striving after the life eternal*, 76, 2. Æfter níðum teolunge heara *secundum nequitias studiorum ipsorum*, Ps. Surt. 27, 4. II. *a pursuit, occupation, employment, business :*—Gestreón of ðære teolunge ðe hé him befæste *gain from the occupation he committed to them*, Homl. Th. ii. 552, 1. Sume teolunga sind ðe man begán mæg búton synnum . . . Petrus hæfde unpeolíce teolunge ær his gecyrrednysse, and hé for ði eft búton pleó tó his fixnoðe gecyrde, 288, 20–26. Se ríca man geswícþ his gebeórscipes, gif ða ðeówan geswícaþ ðæra teolunga, i. 274, 1. Gif se biscep self drohtaþ on ðam eorðlícum tielongum (tielengum, Cott. MSS.) *si presul ipse in terrenis negotiis versatur*, Past. 18; Swt. 133, 4. Getígede tó eorðlícum tielengum (tiolengum, Cott. MSS.) *deditae terrenis negotiis*, Swt. 135, 15. Gecorene tó Godes teolungum, Homl. Th. ii. 96, 1. Sécan ða gástlícan tylunga, 552, 10. Hé begæþ his hláfordes teolunga, i. 412, 4. Wé willaþ sprecan ymbe manna tilunga *ad hominum studia revertor*, Bt. 24,

4; Fox 84, 27. **III.** *care, attention, treatment, cure.* v. tilian, III. 2 b, IV:—Ðonne man tô wiccan tilunge sêce æt ænigre neóde, Wulfst. 171, 11. Hê his hǽlde sêcan wyle æt unâlýfedum tilungum, Homl. Th. i. 474, 21. Hê lǽrde ðurh ða tielunga (tiolunga, Cott. MSS.) ðæs Samaritaniscan (*per Samaritani studium*) ymb ðone gewundedan, Past. 17; Swt. 125, 7. **IV.** *gain that comes from labour, acquisition, fruit got by tilling the earth* :—Tilunge *quaestu, lucro,* Hpt. Gl. 419, 63. Swâ hwæt swâ hý gespariaþ on heora forhæfednessæ, and swâ hwæt swâ tôforan neádbehêfum belifen byþ on heora mægenes tilunge *whatever they save by their abstinence, and whatever over and above necessaries remains of acquisition by their ability,* R. Ben. 138, 17. Se gýtsere gýmþ grǽdelîce his teolunge, Homl. Th. i. 66, 10. Ða ðe ne sêcaþ heora âgen gestreón ac smeágaþ ymbe Godes teolunge (*gain to be made for God*), ii. 76, 35 : 558, 16. Ðû stunta, tô niht ðû scealt ðîn lîf âlǽtan; hwæs beóþ ðonne ðîne teolunga *whose shall thy gains be then?* Wulfst. 286, 24. Hî sceolon heora geáres teolunga Gode ðone teóðan dǽl syllan, Homl. Th. ii. 608, 22. Lâc of eorðan tilingum *de fructibus terræ munera,* Gen. 4, 3. Ete ælþeódig folc ðîne tilinga *fructus terræ et omnes labores tuos comedat populus quem ignoras,* Deut. 28, 33. Ǽgðer ge earm ge eádig, ðe ænige teolunga (tylunge, MS. F.) hæbbe, gelǽste Gode his teóðunga, L. Edg. S. 1; Th. i. 272, 1. [False teolunges, A. R. 208, 17. Þe wingeardes þet mot muche tilunge to uorte beren winberies, 296, 1. Fourty wynter folke lyued withouten tulyinge (tiliyng, tilynge), Piers P. 14, 63.]

tíma, an; *m. Time, hour;* tempus, Wrt. Voc. i. 76, 66 : hora, ii. 132, 67. **I.** *time when, time at which an event takes place* :—Hit wæs ðâ se tíma ðæt wînberian rípodon *erat tempus, quando jam praecoquae uvae vesci possunt,* Num. 13, 21. Swâ mon eorðan wæstmas hâm gelǽdeþ on rýpes tíman, Exon. Th. 214, 28; Ph. 246. Ðâ geweard hit on ðisum ilcan tíman oððe litle ǽr, ðæt . . . , Chr. 1009; Erl. 141, 28 : 1015; Erl. 152, 9. Thomas tô ðam tíman âgeán fêrde bûton bletsunga, 1070; Erl. 208, 9. Týman on âsettum týman, Homl. Th. i. 18, 26. On unâlýfedum tíman, ii. 94, 3. Gebiddaþ ealle hâlige tô ðê on tilne tíman (*in tempore opportuno*), Ps. Th. 31, 7. Ðonne hê nytwyrðne tíman ongiet tô sprecenne *cum opportunum considerat,* Past. 38; Swt. 275, 14. Ymbe ðone tíman ðe ðiss wæs *at the time when this was happening,* Ors. 4, 5; Swt. 168, 36. Tiéman, 4, 8; Swt. 186, 34. Ðæt hié ðoligen earfeðu ðǽm tímum ðe hié ðyrfen, Past. 36; Swt. 253, 10. Ne ðincþ mê nǽfre nânwuht swâ sóþlîc swâ mê þincþ ðîn spell ðên tímum (tîdum, Cott. MS.) ðe ic ða gehêre *cum tuas rationes considero, nihil dici verius puto,* Bt. 38, 5; Fox 204, 23. **I a.** *a time when a thing can or ought to be done, a proper time, opportunity* :—Ðonne ðæs ðinges tíma ne biþ ðæt hit mon sidelîce gebêtan mæge . . . ac ðonne se lâreów ieldende sêcþ ðone tíman ðe hê his hiéremenn sidelîce on ðreátigean mæge *cum rerum minime opportunitas congruit, ut aperte corrigantur . . . Sed cum tempus subditis ad correptionem quaeritur,* Past. 21; Swt. 153, 1–6. Ûs is tíma ðæt wê onwæcnen of slǽpe *hora est nos de somno surgere,* 63; Swt. 459, 33. Hwænne wylle gê singan? Þonne hyt tíma byþ, Coll. Monast. Th. 34, 5. Se wîsa hit ieldcaþ and bîtt tíman, Past. 33; Swt. 220, 10. Nis hit nân wundur, ðeáh se wîsa bîde his tíman, 38; Swt. 275, 13. Hê ðencþ ðæs tíman hwonne hê hit wyrs geleánian mæge *deteriora rependere, si occasio praebeatur, quaerat,* 33; Swt. 227, 23. **I b.** *time as in the phrases, in time, in good time, be-times; proper time* because soon enough :—Ealle ðâs ungesǽlða ûs gelumpon þurh unrǽdas, ðæt mann nolde him tô tíman (â tíman, MS. C.) gafol bêdan; ðonne hî mǽst tô yfele gedôn hæfdon, ðonne nam man grið and frið wið hî, Chr. 1011; Erl. 145, 2. Þeófas tô tíman (*forthwith*) forwurðan, bûton hig geswîcan, L. C. S. 4; Th. i. 378, 13. **I c.** *an appointed time* :—Mîn tíma ys gehende, Mt. Kmbl. 26, 18. Drihtenes engel com tô his tíman on ðone mere, and ðæt wæter wæs âstyred, Jn. Skt. 5, 4. **II.** *a period of time* :—His tíma ne biþ nâ langsum, Homl. Th. i. 4, 18. Hire tíma wæs gefylled, ðæt heó cennan sceolde, i. 30, 11. Ǽlces mannes tíma *the time that each man lives,* Anglia viii. 336, 27. **II a.** marking *date or limit, time during which certain events are happening, during which a particular person is living, etc.* :—On ðet gerâd ðet hê hæbbe ðone bryce ðes landes swâ lange swâ his týma sý *so long as he live,* Cod. Dip. B. iii. 106, 39. Hit wæs gewunelíc on ðam tíman, Homl. Th. i. 60, 26. On mínum tíman swâ on mînes fæder, L. Edg. S. 2; Th. i. 272, 28. On ûrum tíman, Chart. Th. 240, 11. On ðara heáhfædera tíman . . . on Moyses and on ðara wîtegena tíman, L. Ælfc. P. 6; Th. ii. 366, 7–8. Eall ðâs geeodon in ûssera tîda tíman, Exon. Th. 147, 12; Gû. 726. **II b.** *a season of the year* :—Feówer tíman beóþ . . . Uer ys lengtentíma, and hê gǽþ tô tûne on .vii. id. Febr. . . . Se óðer tíma hâtte aestas . . . Se þridda tíma ys autumnus . . . Se feórða tíma ys genemned hiemps, Anglia viii. 312, 14–31. On wintres tíman, ðæt is fram ðan anginne ðæs mônðes, ðe is Nouember gehâten, ôþ Eástran, R. Ben. 32, 10. On ælcne tíman, ge on wintra ge on sumera, 33, 20. **II c.** *an age of the world* :—Þrý tíman sind on ðyssere worulde; Ante legem, Sub lege, Sub gratia, Homl. Th. ii. 190, 1. Drý tíman synd getealde on ðissere worulde. Ân tíma wæs ǽr Godes ǽ . . . Óðer under Godes lage . . . Ðridde under Cristes âgenre gife, L. Ælfc. P.

6; Th. ii. 366, 6. **III.** as a grammatical term, *time of pronouncing a syllable, quantity* :—Ðæt rihtmetervers sceal habban feówer and twêntig tíman . . . Dactilus stent on ânum langum tíman and twâm sceortum, and spondeus stent of feówrum langum, Anglia viii. 314, 10–15 : 335, 14. **IV.** *time, condition of things* :—Æfter ðisum fæce gewurðan sceall swâ egeslíc tíma, swâ ǽfre ǽr ne wæs, Wulfst. 19, 3. Wâ ðâm wîfum ðe on ðam earmlícan tíman heora cild fêdaþ, 81, 7. [*Icel.* tími.] v. æfen-, ende-, gebyrd-, hærfest-, lencten-, mærsung-, nôn-, riht-, rîp-, sǽd-, swig-, þrowung-, un-tíma; tíd.

-tíma. v. here-tíma, ge-týma.

tíman; *p.* de. **I.** *to teem, be productive.* v. teám, **I.** (1) referring to a female, *to be with child, bear, bring forth* young :—Wâ ðâm wífum, ðe ðonne týmaþ and heora cild fêdaþ (*vae praegnantibus et nutrientibus,* Mt. 24, 19), Wulfst. 81, 6. Sindon sume gesceafta ðe týmaþ bûton hæmede, and biþ ægðer ge seó môder mæden ge seó dohtor; ðæt sind beón : hí týmaþ heora teám mid clǽnnysse, Homl. Th. ii. 10, 14–17. Lia underget ðæt heó leng ne týmde (*quod parere desiisset*), Gen. 30, 9. Ðonne heó (*the wife*) leng týman ne mæg, geswîcan hî hǽmedes, Homl. Th. ii. 94, 5. Heó týmende nâ leng beón ne mæg, Wulfst. 305, 29. (2) referring to a male, *to beget, have intercourse* with (wið) a woman :—Godes bearn týmdon wið manna dohtra and hig cendon *ingressi sunt filii Dei ad filias hominum illaeque genuerunt,* Gen. 6, 4. Ðâ bæd heó hire wer ðæt hê wið hire wylne týman sceolde (*ingredere ad ancillam meam,* Gen. 16, 2), Boutr. Scrd. 22, 23. Môste se bisceop niman him ân clǽne mæden and wið hý týman on âsettum tíman, L. Ælfc. C. 7; Th. ii. 346, 2 : Homl. Th. i. 18, 26. (3) where neither male nor female is specified, *to have offspring, bring forth* :—Fugelas ne týmaþ swâ swâ ôðre nýtenu, Homl. Th. i. 250, 22. Ðæt folc týmde micelne teám on ðam wêstene, Homl. Th. ii. 212, 17. Þeóda týmdon, Cd. Th. 75, 19; Gen. 1242. Têmaþ and wexaþ, 13, 1; Gen. 196. Týmaþ and tiédraþ, 91, 14; Gen. 1512. Feoh sceal on eorðan týdran and týman, Menol. Fox 557; Gn. C. 48. [Þe two tentaciuns þet temeð alle þe oðre, A. R. 220, 15. Elysabæþ ne mihhte tæmenn, Orm. 130. ʒif ha ne mei nawt teamen . . . ha cleopeð ham weolefulle þat teamen hare teames, H. M. 33, 22–25. Ghe sulde sunen and timen, and clepen it Smael, Gen. and Ex. 982. Aʒen þat þu (*the nightingale*) wilt teme, O. and N. 499.] **II.** as a technical term. v. teám, **III,** *to vouch to warranty* (acc. of that which is to be warranted and person vouched governed by tô), *to refer property* (acc.) to (tô) the person from whom it was obtained in support of the right of possession :—Gif sió hond tiémþ, sió ðone ceáp mon æt befêhþ, tô ôðrum men, L. In. 75; Th. i. 150, 6. Swâ hê hit âgnode, swâ hê hit týmde, L. Ed. 1; Th. i. 160, 8. Ðâ týmde Wulfstân ðone mann tô Ædelstâne, Chart. Th. 206, 25. Tǽme hê tô ðam mæn ðe him sealde, L. H. E. 16; Th. i. 34, 6. Ne môt forstolenne ceáp mon tiéman tô þeówum men, L. In. 47; Th. i. 132, 5. Se ðe yrfe bycge on gewitnesse and hit eft týman (mon teáman, *var. lect.*) scyle, L. Ath. i. 24; Th. i. 212, 13 : L. Ed. 1; Th. i. 158, 16. **II a.** in a general sense, *to refer* an opinion to the source from which it is derived in its support :—Benedictus ûs bôc âwrât leóhtre be dǽle ðonne Basilius, ac hê týmde swâ ðeáh tô Basilies tǽcinge for his trumnysse *for confirmation he referred to the teaching of Basil as the source from which he had drawn,* Basil prm. ; Norm. 32, 9. Benedictus týmde tô ðam regole ðe Basilius gesette, Homl. Skt. i. 3, 152. [*In later English* temen to = (1) *to betake one's self to* a place, *go to* :—To Albion þu scalt teman (wende, 2nd MS.), Laym. 1245 : 7174. (2) *to resort to, appeal to* in reverence or for help :—To hire he wolde teman (hire wolde he louie, 2nd MS.), Laym. 1265. Al hit trukeð us an hond þat we to temden, 16800. Gif þu temest (*appealest*) to þan rihten, and þu wult of Rome þolien æi dome, 24816. He temed him to þe king, Trist. 431. To witnesse temen, P. L. S. viii. 54. I hope to trede on þy temple & teme to þy seluen, Allit. Pms. 101, 316. (3) *to lead to* (?) :—Ic wolde iwiten to whan þis tocne wule ten, to wulche þinge temen, Laym. 9135.] v. ge-têman; un-tímende.

timber, es; *n.* **I.** *material* for constructing a house, ship, etc., *timber* :—Æfter siextegum daga ðæs ðæt timber (*arbores*) âcorfen wæs, ðǽr wǽron xxx and C scipa gearora, Ors. 4, 6; Swt. 172, 4. Ne sceal cyrcean timber (*ligna ecclesiae*) tô ænigum ôðrum weorce, L. Ecg. P. Addit. 16; Th. ii. 234, 16. Ðætte ne meahten godo beón ða ðe monna hondum geworhte wǽron of eorðlícum timbre oððe of treóm oððe of stânum *deos esse non posse, qui hominum manibus facti essent; dei creandi materiam lignum vel lapidem esse non posse,* Bd. 3, 22; M. 224, 15. Ǽrest man âsmeáþ ðæs hûses stede, and eác man ðæt timber beheáwþ, Anglia viii. 324, 8 : Lchdm. iii. 180, 8. **I a.** *material* of which anything is formed :—Sió lifer is blôdes timber and blodes his and fôstor, Lchdm. ii. 198, 2 : 160, 13. **II.** *a structure, building, edifice* :—Heó mid ðæm tô ðæm timbre (*aedificio*) gefæstnad wæs, Bd. 3, 17; S. 544, 31. Tô ðam heofonlícum timbre, 4, 3; S. 567, 12. In timbre *in aedificio,* Ps. Surt. 101, 8. Seó tîd gewât ofer timber (? tiber, MS.) sceacan middangeardes, Cd. Th. 9, 2; Gen. 135. Huulig timber *quales structurae,* Mk. Skt. Lind. Rush. 13, 1. Timbra *aedificiorum,* Ps. Surt. 128, 6. Ða burh manige menn mid heán tin:brum frættewodon

(*augustioribus aedificiis adornarunt*), Bd. 3, 19; S. 547, 24. **III.** *the building* of a house, ship, etc.:—Hé (*the sixth day of the moon*) is gód circan on tó timbrane, and eác scipes timber on tó anginnanne, Lchdm. iii. 178, 9. [*O. L. Ger.* timbar : *O. Frs.* timber: *O. H. Ger.* zimbar *materia, fabrica, structura, aedificium*: *Ger.* zimmer a *chamber*, *timber* : *Icel.* timbr. Cf. *Goth.* timreins *a building*, ga-timrjó *a building*.] v. an-, and-, boh-, bolt-, fugol-, fyrd- (?), heáh-, heofon-, hróf-, magu-timber ; ge-timbru.

timber-geweorc, es ; *n. Timber-work, preparation* or *cutting of timber for building* (?):—In bócholte timbergeweorc and widigunge *in beechholt the right to get timber for building and to cut wood for fuel*, Cod. Dip. B. i. 344, 12. v. timbran, **III.**

timber-hrycg, es ; *m. A wooded ridge* (?); as a local name *Timber-ridge* :—On timberhricges snád, Cod. Dip. Kmbl. v. 71, 1. Ofer fild-burnan on timberhrycg, iii. 463, 31. Timberrycg, 393, 27.

timberness, tim-bor. v. ge-, on-timberness, tym-bor.

timbran, timbrian ; *p.* ede, ode. **I.** *to build* (lit. or fig.), *construct* :—Ic timbrige *struo, construo*, Ælfc. Gr. 28, 5 ; Zup. 175, 11. Tóweorp hié, ne dú timbres (*aedificabis*) hié, Ps. Surt. 27, 5. Timbreþ Dryhten Sion, 101, 17 : Ps. Th. 146, 2 : Exon. Th. 450, 25 ; Dóm. 93. Gé timbriaþ (timbraþ, Rush.) wítegena byrgene, Mt. Kmbl. 23, 29 : Lk. Skt. 11, 47, 48. Ic timbrode setl ðín, Ps. Spl. 88, 5. Ða gódan weorc ðe hé ǽr timbrede, Past. 33 ; Swt. 215, 18. Hé burh timbrede, Cd. Th. 172, 6 ; Gen. 2840: Chr. 722 ; Erl. 44, 28. Timbrade, Ps. Th. 101, 14. Hié ceastra timbredon, Ors. 1, 10 ; Swt. 48, 10. Drehton ða hergas mid ðǽm æscum ðe hié ǽr timbredon. Ðá hét Alfréd cyng timbran langscipu ongén ða æscas, Chr. 897 ; Erl. 95, 7–11. Æfter ðæm hryre ðære upáhæfennesse hé ongan timbran eáðmódnesse, Past. 58 ; Swt. 443, 30. Wé ceorfaþ treówu on holte, ðæt wé hí eft up árǽren on ðæm botle, ðǽr ðǽr wé timbran willen, Swt. 445, 1 : Cd. Th. 64, 29 ; Gen. 1057. Weall stǽnenne timbran, 101, 34 ; Gen. 1692. On ðám telgum timbran nest, Exon. Th. 210, 20 ; Ph. 188. Ne mæg fira nán wísdóm timbran (timbrian, Bt. 12 ; Fox 36, 11, 8, 10), ðǽr ðǽr woruldgítsung beorg oferbrǽdeþ, Met. 7, 12. Uton timbrian ús ceastre *faciamus nobis civitatem*, Gen. 11, 4 : Ps. Th. 128, 2. Ecgbryht salde Reculf mynster on tó tymbranne (-ianne, MS. E.), Chr. 669 ; Erl. 34, 26. Timbriende *aedificans*, Ps. Surt. 146, 2. Timbrende *aedificantes*, 117, 22. Ðǽr wæs timbred templ, Nar. 37, 22 : Beo. Th. 620 ; B. 307. Bióþ timbrede cestre, Ps. Surt. 68, 36. **II.** *to instruct, edify* :—Hé nówiht elles dyde ðonne ðæt folc mid godcundre láre timbrede *nil aliud ageret quam plebem Christi verbo salutis instruere*, Bd. 2, 14 ; S. 518, 10. **III.** *to cut timber* (?). v. timber-geweorc, *and* cf. wudian :—Me mæig on sumera . . . bytlian . . . tymbrian, wudian, Anglia ix. 261, 11. [Letten þa kinges timbrien þa hallen, Laym. 5940. To timbren me mine crune, A. R. 124, 8. To timmbrenn himm an hus, Orm. 13368. Who tauȝte hem (*peacocks*) on trees to tymbre so heighe, Piers P. 11, 352. *Goth.* timrjan : *O. Sax.* ge-timbrón (-ian) : *O. L. Ger.* ge-timbran : *O. Frs.* timbra, timmera : *Du.* timmeren : *O. H. Ger.* zimbaren, zimbarón *aedificare, struere, instruere* : *Ger.* zimmern : *Icel.* timbra : *Dan.* tømre.] v. á-, be-, for-, ge-, in-, on-timbran (-ian), *and next word.*

timbrend, es ; *m. f. A builder, constructor* :—Se wæs timbrend (*constructor*) ðæs mynstres ðe gecweden is Médeshámstyde, Bd. 4, 6 ; S. 573, 40. Heó wæs seó ǽryste tymbrend ðæs mynstres ðe ys nemned Steórneshealh, Shrn. 148, 39.

timbrian. v. timbran.

timbrung, e ; *f. Building, a building* :—Ealdere timbrunga bóte *instructio*, níwe timbrung *constructio*, Wrt. Voc. i. 39, 58, 59. Timbrunga *domum exstructam*, Kent. Gl. 472. [Bileafden heo (*the builders of the tower of Babel*) heore timbrunge, O. E. Homl. i. 93, 23. Timbringe, 227, 4. Al is to his behefe and timbrunge toward his blisse, A. R. 124, 1.] v. ge-timbrung.

-**tíme** (v. teám, **I**, tíman, **I**, *and* cf. -bǽre). v. luf-, þweorh-tíme ; wróht-getíme.

-**tíme** (v. teám, **II**). v. feoþer-, ge-tíme (-týme).

tíme (v. tíma). v. un-tíme.

tímen (?) ; *adj. Belonging to a team.* v. teám, **II** :—Témen *bibina* (= *bis bina*?), Wrt. Voc. ii. 126, 8.

tímian. v. ge-, mis-tímian.

tímlíce ; *adv. In good time, soon* :—Ðú bǽde mé foroft Engliscra gewritena and ic ðé ne getíðode ealles swá tímlíce ǽr ðam ðe ðú mid geweorcum ðæs gewilnodest æt mé *you very often asked me for English writings, but I did not grant your request so very soon, not before you desired it from me with works*, Ælfc. T. Grn. 1, 16. [Ic mei longe libben and alle mine sunne timliche ibeten *repent of all my sins time enough*, O. E. Homl. i. 25, 13. Ase timliche as he hefde iherd þis (sone so he iherde þis, *other MS.*), Jul. 9, 5. He wolde timliche him speken wið, Laym. 31369. Bute ȝef þu þe timluker (*nisi maturius*) ure godes grete, Kath. 2086. *Icel.* tímaliga *timely, early.*] Cf. tídlíce.

tímness. v. un-tímness.

timpana, an ; *m. A tabret, timbrel* :—Hergaþ hine in timpanan *laudate eum in tympano*, Ps. Surt. 150, 4. Sellaþ timpanan, 80, 3. Plægiendra

timpanan *tympanistriarum*, 67, 26. Ic filigde ðé mid timpanum and mid hearpum, Gen. 31, 27. [*O. H. Ger.* timpana : *Icel.* timpan. From Latin.]

timpestere, es ; *m. A player on the timbrel* :—Timpestera (timpanestera ?) *tympanistriarum*, Ps. Lamb. 67, 26. [Cf. *O. L. Ger.* timparinna : *O. H. Ger.* tympinara ; *pl.*]

timple, an ; *f. Some implement used in weaving* :—Hé sceal habban fela tówtóla . . . flexlínan, spinle . . . presse, pihten, timplean, wifte, Anglia ix. 263, 12.

tín, tíen, tén, týn teá (*North.*) ten. **I.** as an adjective with a noun uninflected, except in the Northern specimens :—Tín dagas, Bd. 1, 23 ; S. 485, 24. Ðis is ðara týn hída bóc, Cod. Dip. Kmbl. v. 316, 33. Mid tíen bebodum, Past. 17 ; Swt. 125, 18. Tíen ceastro *Decapoleas*, Wrt. Voc. ii. 26, 14. Sume tén geár, Bt. 38, 1 ; Fox 194, 7. Týn þúsend (téno t teá ðúsendo, Lind. : tén þúsende, Rush.) punda, Mt. Kmbl. 18, 24. Gelíc ðám týn fǽmnum (téwm hehstaldum, Lind. : tén fémnan, Rush.), 25, 1. Mid týn (téum t ténum, Lind. : tén, Rush.) þúsendum, Lk. Skt. 14, 31. Týn (teá, Lind. Rush.) hreófe weras, 17, 12. Teá síðum, Lind. 15, 8. Fram wintrum ténum, p. 8, 4. Teá t téno hreáfo, p. 9, 8. Of téum hehstaldum, Mt. Kmbl. p. 19, 16. Teá monna lætwu *decanus*, Rtl. 193, 19. **II.** used as a substantive and declined, *nom.* -e, *gen.* -a, *dat.* -um. (1) alone :—Ðá gebulgon ða týne (téno, Lind. : ténu, Rush.) hí, Mk. Skt. 10, 41. Ða hildlatan, týne ætsomne, Beo. Th. 5687 ; B. 2847. Týna ealdor *decanus*, Wrt. Voc. ii. 138, 4. Næs tó ánum dæge, ne tó fífon, ne tó týnum, ne tó twéntigum, Num. 11, 19. Aldormonn ofer téno *decanus*, Rtl. 193, 21, 19. (2) governing a genitive :—Gif ðǽr beóþ týn rihtwísra, Gen. 18, 32. Hæfde se ealwalda engelcynna týne getrymede, Cd. Th. 16, 24 ; Gen. 248. Nigon hund wintra hæfde and týne, 71, 3 ; Gen. 1165. **II a.** *a set of ten* :—Týnum and twéntigum on ánum inne ætgædere restan *let them sleep by tens and twenties in one house*, R. Ben. 47, 7. **II b.** *the number ten* :—Ðis tal under him hæfis óðer tal ðe tó ténum wið forecyme (*a number that goes up to ten*), Mt. Kmbl. p. 3, 20. Tele dú óð ðæt ðú cume tó þrittiga, eft . . . tele óð týne (*count up to ten*), Lchdm. iii. 228, 2. [*Goth.* taihun : *O. Sax.* tehan : *O. Frs.* tian, tien : *O. L. Ger.* tén, tein, tian : *O. H. Ger.* zehan : *Icel.* tíu.]

tin, es ; *n. Tin* :—Tin *stagnum*, Wrt. Voc. i. 85, 10 : 286, 71 : Ælfc. Gr. 5 ; Zup. 15, 11. Ðæt tin, ðonne hit mon mid sumum cræfte gemengþ and tó tine gewyrcþ, ðonne biþ hit swíðe leáslíce on siolufres hiewe. Suá hwá ðonne suá lícet on ðære swingellan, hé biþ ðǽm tine gelíc inne on ðæm ofne, Past. 37 ; Swt. 269, 2–5. Tinnes *stagni*, Hpt. Gl. 431, 69. Ðiss folc is geworden mé tó áre and tó tine and tó leáde, Past. 37 ; Swt. 267, 17. Tin *stannum*, Coll. Monast. Th. 27, 11. [*O. H. Ger.* zin : *Icel.* tin.]

tin a beam. v. tinn.

Tína(-e ?), an *the river Tyne* :—Be Tínan ðære eá *juxta amnem Tinam*, Bd. 5, 21 ; S. 642, 36: Chr. 875 ; Erl. 76, 35.

tín-ámbre ; *adj. Containing ten 'ámbras'* :—Genim týnámberne cetel, Lchdm. ii. 86, 12.

tínan ; *p.* de *To vex, annoy, irritate, provoke* :—Se wellwillenda man wyle forberan gif hine man áhwǽr týnþ, oððe him tale gecwyð, Basil admn. 4 ; Norm. 44, 18. Ðá ðá se án (sunu) ðé týnde (cf. tirigde, l. 9), Homl. Th. ii. 30, 12. Hí yrsodon ̃t týndon Moyses *irritaverunt Moysen*, Ps. Spl. 105, 16, 8 : Blickl. Gl. : Cd. Th. 153, 24 ; Gen. 2543. Ne týn ðú ðíne neáhgebúras *non memor eris injuriae civium tuorum*, Lev. 19, 18. Ne ǽnig man óðerne ne tyrie ne ne týne ealles tó swýðe, Wulfst. 70, 9. Ne áblinnan wé, ðæt wé Gode cwémon and deófol týnan, Blickl. Homl. 47, 11. Ðæt hí ælpeódige men ne tyrian ne ne týnan, L. Eth. vi. 48 ; Th. i. 326, 28 : Wulfst. 309, 5. Gebiddaþ for eówerum ehterum and eów týnendum *orate pro persequentibus et calumniantibus vos* (Mt. 5, 44), Homl. Th. ii. 216, 17. v. teónian.

tinclian ; *p.* ode *To tickle* :—Náht swá onǽlþ and tinclaþ gecyndlima ðænne gemylt mete *nihil sic inflammat et titillat membra genitalia quam indigestus cibus*, Scint. 52, 5. Hé wiðstynt weorce se ðe tincligendre ná geþwǽrlǽcþ lustfullunge *resistit operi qui titillanti non accomodat delectationi*, 88, 9. [In Wycklif *tynclen* translates *tinnire*, 1 Sam. 3, 11 : 1 Cor. 13, 1.]

tind, es ; *m. A tine, prong, tooth* of an implement :—Tindas *rostri*, tindum *rostris*, Wrt. Voc. ii. 119, 30, 28. Deáh ánra gehwylc horn hæbbe .xii. tindas írene, and ánra gehwylc tind hæbbe synderlíce .xii. ordas, Salm. Kmbl. p. 150, 25. [Tindes *the rungs of a ladder*, A. R. 354, 20. Tynde *branch of a tree*, Allit. Pms. 3, 78. Tindes of harowis, Alex. 3908. A tynde *cremale* (a hook) ; a tynde of a beste, Cath. Angl. 389 (*where see several instances*). Tyynde, prekyl, tynde, pryke *carnica* ; tyndyt with tyndys *carnicatus*, Prompt. Parv. 494. Cf. *tine* stocks, the short crooked handles on the pole of a scythe, Halliw. Dict. *M. H. Ger.* zint *a spike, tooth* : *Icel.* tindr *a spike* ; also, *a peak.*] v. following words.

tindect. v. tindiht.

tindig ; *adj. Having spikes* or *prongs* :—Óstig gyrd *vel* tindig *scorpio* (scorpio *genus flagelli, ex virgis nodosis confecti, vel scutica in modum scorpionis aculeata*, Migne), Wrt. Voc. i. 20, 17. v. next word.

tindiht ; *adj. Having spikes* or *teeth, beaked* :—Tindicti (-ecte) *ros-*

tratum, Txts. 92, 868. Se câsere hine (St. Romanus) hêt stingan mid îrenum gyrdum tyndehtum, Shrn. 115, 25. v. preceding words.

tindting (tending?, tihting?) :—Tindtingce *suasionis, exhortationis*, Hpt. Gl. 485. 66.

Tine. v. Tîna.

tinen ; *adj. Of tin* :—Tinen *stagneus*, Ælfc. Gr. 5 ; Zup. 15, 11. Ælc calic gegoten beó, gylden oððe seolfren (oððe) tinen, ðe man húsl on hálgige, L. Edg. C. 41 ; Th. ii. 252, 21 note. Tynen, L. Ælfc. C. 36 ; Th. ii. 360, note 2. On tinum (= tinenum) fæte, Lchdm. ii. 236, 5. [With tynnen tonges, Pall. 152, 99. *O. H. Ger.* zinîn *stanneus*.]

tinend, es ; *m. One who vexes, annoys*, etc. v. tînan :—Gebiddaþ for eówerum êhterum and týnendum, Homl. Th. ii. 36, 16.

tin-feald ; *adj. Tenfold* :—Týnfealde deni, Ælfc. Gr. 5 ; Zup. 13, 15. Týnfealdum oððe twêntifealdum *deni aut viceni*, R. Ben. Interl. 54, 15. Þreowa on teónfealdum *ter denis*, Hymn. Surt. 104, 23. Feówer sîðo teáfald tal *quater denario numero*, Mt. Kmbl. p. 12, 12.

tinga, tingan, tingce. v. in-tinga, ge-tingan, tynge.

tinn, e ; *f.* (?) *A beam, rafter* :—Tiu *tignum*, Txts. 101, 2023. [Cf. (?) *O. H. Ger.* zinna *pinna*.]

tinnan *to stretch, extend* :—Blæd his blinniþ . . . lustum ne tinneþ *does not joyously extend*(?), Exon. Th. 354, 32 ; Reim. 54. Tinde bogan *tetendit arcum*, Blickl. Gl.

tîn-nihte ; *adj. Ten days old* :—On .x. nihtne mônan bidde swā hwas swa ðū wylle, hyt ðe byoþ gere. Se .x. nihta môna hê ys gôd tô standanne mid ædelum monnum, Lchdm. iii. 178, 19–21. Se ðe biþ âcenned on .x. nihtne ealdne mônan, se biþ ðrowere, 160, 28.

tin-strenge ; *adj. Having ten strings* :—On týnstrengum saltere *in decacordo psalterio*, Ps. Spl. 91, 3 : 143, 11. v. next word.

tîn-strenged ; *adj. Provided with ten strings* :—On týnstrengedum saltere, Blickl. Gl. Týnstrængedum, Ps. Lamb. 91, 4. Týnstrængdom, 143, 9. v. preceding word.

tinterg. v. tin-treg.

tin-treg, -terg, es ; *n.* : tin-trega, an ; *m. Torment* :—Ðær (*in heaven*) ne biþ nān besærgung ðæra mânfulra yrmðe, ac heora tintrega becymþ ðam gecorenum tô mâran blisse, Homl. Th. i. 334, 11. Nis ðǽr ne caru ne hreóh tintrega (cf. hreóge tintrega, Wulfst. 139, 30), Dôm. L. 261. Ðæt wæs helle tintreges mûþ *ipsum est os gehennae*, Bd. 5, 12 ; S. 630, 13. On ðâm grundum helle tintreges *in profundis tartari*, 5, 14 ; S. 634, 25. Ic on eorþan gebâd tintregan fela, Cd. Th. 296, 4 ; Sat. 497. Mê genihtsumiaþ ðâs tintrega, Blickl. Homl. 243, 26. Ðê sýn helle tintrega ontýned, Shrn. 79, 11. On ðissa tintrega stôwe *in locum hunc tormentorum*, Lk. Skt. 16, 28. For ðara tintregena mænigfyldnesse, Wulfst. 199, 6. Tintegrena *tormentorum*, Hpt. Gl. 415, 72. On tintregum gegripene *tormentis comprehensos*, Mt. Kmbl. 4, 24. Tintregum (tintergum, Lind.), Lk. Skt. 16, 23. Búton tintregum þeáh on helle-wîte, Homl. Th. i. 94, 6. Wrecan heora teónrædenne mid tintergum on him, Jud. 15, 10 : Exon. Th. 114, 33 ; Gú. 182. Ðonne hê ðara manna tintrego oferhiérde, Ors. 1, 12 ; Swt. 54, 27 : Bd. 5, 14 ; S. 635, 1 : Blickl. Homl. 243, 20. On ða wyrstan tintregu, 239, 10. In ða êcan tintregu, Wulfst. 185, 11. Tintergu, Exon. Th. 141, 3 ; Gú. 621. In tintergo *in gehennam*, Mt. Kmbl. Lind. 10, 28. Hié ealle worldlîce tintrega and ealle lîchomlîcu sâr oforhogodan, Blickl. Homl. 119, 19. Ðý læs ðe ðū þurh tintrega forwurðe, Homl. Th. i. 432, 9. Ic geseó, ðæt ðū ðâs tintregan gebýsmerast, 426, 5. Ân deófol árehte ânum ancran ðara synfulra sâwla tintregan and sûsla, Wulfst. 146, 19. [Eorðlíche tintreohen, O. E. Homl. i. 261, 16. Ne schal þe na teone ne tintreohen trukien, Kath. 403. Þu biþenche teonen and tintreohen, 1888. Cf. *Goth.* us trigôm ἐκ λύπης, 2 Cor. 9, 7. *Icel.* tregi *grief, woe*.]

tintregend, es ; *m. A torturer* :—Fram ðæm tintergendum (or *ptcpl.*?) *a tortoribus*, Wrt. Voc. ii. 2, 49.

tin-tregian, -tergian ; *p.* ode *To torment, torture, afflict* :—Ða ðe hê ne mæg fram rihtan geleáfan tô him gebîgan, ðonne tintregaþ hê ða on mænigfealde wîsan, Wulfst. 197, 7 : Blickl. Homl. 59, 31. Philippus hî miclum tintrade (tintergade, MS. C.) and bismrade, Ors. 3, 7 ; Swt. 118, 25. Se kâsere hine tintregode mid unâsecgendlîcum wîtum, Shrn. 116, 1. Ða ðe tintregedon ðone hâlgan wer, 73, 1. Hî tintregodon hine and forlêton hine sâmcucene *plagis impositis abierunt semivivo relicto*, Lk. Skt. 10, 30. Ða wîfmen hié swâ tintredon, Ors. 1, 10 ; Swt. 48, 13. Ðeáh ðe ðæt fýr tintregige ða unrihtwîsan, Homl. Th. ii. 590, 3. Hêt swingan and tintregian ðone Godes andettere *caedi Dei confessorem a tortoribus praecepit*, Bd. 1, 7 ; S. 477, 42. Tintergian, Shrn. 76, 33. Tinterga *torquere*, Mt. Kmbl. Rush. 8, 29. Decius gewende tô tintregienne ða cristenan, Homl. Th. ii. 424, 19. Tô tintreinne *torquendus, cruciandus*, Hpt. Gl. 482, 35. Ðær hê tintregd wearð : ærest hiene mon swong, ða sticode him mon ða eágan út, and siþþan him mon slôg ða handa of, ða ðæt heáfod, Ors. 4, 5 ; Swt. 168, 3. [Heo eow tin-traȝed and heow iswenchet, O. E. Homl. i. 13, 30. Cf. *O. Sax.* tregan *to trouble* : *Icel.* trega.] v. ge-tintregian ; tregian.

tintreg-lîc ; *adj. Tormenting, torturing, of hell* :—Be fyrhto ðæs tin-treglîcan (tintreganlîces, MS. B.) wîtes *de horrore poenae gehennalis*, Bd. 4, 24 ; S. 598, 16.

tintreg-stôw, e ; *f. A place of torment* :—Hî (*the devils*) ðone hâlgan wer gelæddon tô ðam sweartum tintrehstôwum, Guthl. 5 ; Gdwin. 38, 4.

tintreg-þegn, es ; *m. An officer who torments, an executioner* :—Tinter[g]ðegnum *lictoribus*, Wrt. Voc. ii. 3, 47. His dryhten hine salde tintergaþægnum (*tortoribus*), Mt. Kmbl. Rush. 18, 34.

tintregung, e ; *f. Torment, punishment* :—Tintregung *vel* wîte *tormentum*, Wülck. Gl. 178, 20. Hí ne mihton fram Gode þurh nâne tintregunga beón gebîgede, Homl. Th. i. 544, 2.

tîn-wintre ; *adj. Ten years old* :—.x. wintre cniht mæg bión þiéfðe gewita *a ten year old boy can be accessory to a theft*, L. In. 7 ; Th. i. 106, 18. Ðâ ðâ hê týnwintre on ylde wæs, Homl. Th. ii. 498, 28.

tîr, es ; *m. Glory, honour* :—Eów ys wuldorblæd torhtlîc tôweard and tîr gifeþe, Judth. Thw. 23, 35 ; Jud. 157. Tîr æt getohte, Byrht. Th. 134, 54 ; By. 104. Nis hêr (*in hell*) eádiges tîr ne worulde dreám, Cd. Th. 270, 20 ; Sat. 93. Ne biþ hira (*two twins*) tîr gelîc, Salm. Kmbl. 730 ; Sal. 364 : Exon. Th. 448, 11 ; Dôm. 52. Biþ týr scæcen, eorþan blædas, 447, 27 ; Dôm. 45. Tîres Wealdend (cf. wuldres Waldend, Cd. Th. 216, 27 ; Dan. 13) *the Deity*, Ps. Th. 79, 14. Tîres brytta, Judth. Thw. 22, 36 ; Jud. 93. Ðæt hý môstun tîres blæd êcne âgan, Exon. Th. 74, 27 ; Cri. 1212 : Andr. Kmbl. 210 ; An. 105. Tîres eádige *abounding in glory*, reges, Ps. Th. 71, 10 : Cd. Th. 91, 15 ; Gen. 1512 : Judth. Thw. 25, 22 ; Jud. 272. Tîres tô tâcne *in token of glory gained*, Beo. Th. 3312 ; B. 1654. Hê benam his feónd torhte tîre, Cd. Th. 4, 23 ; Gen. 58. Is ðæs wuldres ful heofun and eorðe, and eall heáhmægen tîre getâcnod, Elen. Kmbl. 1504 ; El. 754. Hwonne ûs lîffreá ðæt týdre gewitt tîre bewinde, Exon. Th. 3, 1 ; Cri. 29. Dryhten dǽleþ sumum gúþe blæd, sumum wyrp oððe scyte, torhtlîcne tiir, 331, 18 ; Vy. 70. Ðê tîr cyning and miht forgef, Andr. Kmbl. 970 ; An. 485. Hêr Æþelstân cyning and his brôþor ealdorlangne tîr (týr, *one MS.*) geslôgon æt sæcce, Chr. 937 ; Erl. 112, 3. Gê dôm âgon, tîr æt tohtan, Judth. Thw. 24, 19 ; Jud. 197. Æsca tîr æt gûðe, Cd. Th. 127, 10 ; Gen. 2108. Hê mê tîr forgeaf, wîgspêd wið wrâðum, Elen. Kmbl. 328 ; El. 164. Ða (*friends*) hyra týr and eád ýcaþ, Exon. Th. 409, 3 ; Rä. 27, 23. Ðû tîrum fæst nîða Nergend *thou Saviour of men, gloriously firm*, Cd. Th. 235, 27 ; Dan. 2 : Exon. Th. 354, 7 ; Reim. 42. [Þa kingges weoren deædde, heore duȝeðe todealde, here tir wes atfallen, Laym. 4237. *O. Sax.* tîr ; *see too* tîrlîce *gloriously* : *Icel.* týrr. Cf. (?) *O. H. Ger.* ziarî *decus*.] v. æsc-tîr, *and words in which* tîr *is the first component*.

Tîr, es ; *n. One form of the name of the Runic T ; it is also the name of the god corresponding to the Latin Mars, and apparently used also of the planet bearing his name* ; *as* Grimm *notices, the Runic symbol* ⋏ *resembles that used for the planet* ♂ :—Tîr byþ tâcna sum, healdaþ trýwa wel wið ædelingas, â byþ on færylde ofer nihta genipu, næfre swîceþ, Runic pm. Kmbl. 342, 21–26 ; Rûn. 17. *The other name of the rune is* Tî, v. Tîw, *the two forms* Tîr, Tîw *may be compared with Icelandic* Týrr ; *gen.* Týs (cf. *Dan.* Tirs-dag, Týr ; *gen.* Týs.

tîran ; *p. de To run with tears, to water* (*of the eyes*) :—Mê týraþ mîne eágan *lippio*, Ælfc. Gr. 30, 5 ; Zup. 192, 9. Ðær biþ wôp and tôða gebitt, for ðan ðe ða eágan týraþ on ðam micclum bryne, and ða têð cwaciaþ on swîðlîcum cyle, Homl. Th. i. 132, 26. Wiþ ðon ðe eágan týren (cf. wið eallum tiédernessum eágena, 2, 6), Lchdm. ii. 32, 28. Gif eágan týren, 34, 1 : 308, 19 : iii. 4, 23. Wið týrendum eágan, 4, 6 : i. 374, 3. Wið týrende eágan, i. 72, 14. v. teár, I. 2, tearig, II.

tîr-eádig ; *adj. Glorious* :—Tîreádig cyning (*the Deity*), Hy. 3, 2, 55 : (*Constantine*), Elen. Kmbl. 207 ; El. 104. Elene, tîreádig cwên, 1206 ; El. 605. Tîreádig and trâg (*Judas and the devil*), 1906 ; El. 955. Týreádig cyning (*the Deity*), Hy. 7, 56, 82. Se tîreádig (*the Phenix*), Exon. Th. 205, 1 ; Ph. 106. Tîreádigum men (*Hygelac*), Beo. Th. 4384 ; B. 2189. Torhte and tîreádige (*the twelve eapostles*), Apstls. Kmbl. 7 ; Ap. 4 : Andr. Kmbl. 4 ; An. 2 : 1329 ; An. 665. Tîreádige, hæleþ heaðurôfe on Brytene, Menol. Fox 26 ; Men. 13. Tîreádge, Exon. Th. 366, 10 ; Reb. 10. Þeóden hæfde him âlesen leóda dugeðe, tîreádigra twâ þúsendo ; ðæt wæron cyningas, Cd. Th. 189, 13 ; Exod. 184. Gâr-berendra, gúðfremmendra, tîreádigra, 192, 16 ; Exod. 232. [Cf. *Icel.* tír-göfugr-, -sæl (*poetical epithets of a hero*).]

tîr-fæst ; *adj. Of assured glory, glorious* :—From treówe becwom tîrfæst rîce Drihten ûre *Dominus regnavit a ligno*, Ps. Th. 95, 9. Cyning tîrfæst cystum gecýþed, Beo. Th. 1848 ; B. 922. Tîrfæst Metod, Cd. Th. 64, 2 ; Gen. 1044. Tîrfæst hæleð, bisceop se gôda . . . ðam wæs Cyneweard nama, Chr. 975 ; Erl. 126, 7. Tîrfæstne hæleð (*Moses*), Cd. Th. 181, 19 ; Exod. 63. Hwær ic tîrfæste treówe funde *ambulans in via immaculata*, Ps. Th. 100, 6 : Exon. Th. 473, 7 ; Bo. 11. Ðæt tîr-fæste tond, 202, 14 ; Ph. 69. Ongietan tîrfæst tâcen ðæt se torhta fugel þurh bryne beácnaþ, 236, 14 ; Ph. 574. Fyrd, tîrfæstra getrum, Menol. Fox 523 ; Gn. C. 32. Cf. blæd-, þrym-, wuldor-fæst.

tîr-fruma, an ; *m. The source of glory* or *the prince of glory, the Deity*, Exon. Th. 13, 21 ; Cri. 206.

tirgan, tirwian, tirigan, tirian ; *p.* tirgde, tirwede, tirigde *To vex, irritate, provoke, exasperate* :—Ic tyrige *lacesso*, Ælfc. Gr. 28, 1 ; Zup. 165, 12. Tirhþ *inridet*, Kent. Gl. 508. Tyrweþ *improperabit*, Ps. Lamb. 73, 10. Ða tredað ðec and tergaþ, and hyra torn wrecaþ, Exon.

Th. 119, 23; Gú. 259. Đa đe tyrwiaþ *qui exasperant*, Ps. Lamb. 65, 7: 67, 7. Hé tyride *exacerbavit*, i. *provocavit, adflixit*, Wrt. Voc. ii. 144, 56. Tyrgide *exacerbavit*, Hpt. Gl. 527, 51. Đæt wíf cwæđ, đæt heó wolde đone sunu đe hí tirigde awyrian, Homl. Th. ii. 30, 9. Hý tyrgdon (tyrigdon, Ps. Spl. 104, 26) *exacerbaverunt*, Blickl. Gl. Mē weras wordum tyrgdon, Andr. Kmbl. 1926; An. 965. Hí tyrgdon God mid gramlícum weorcum, Homl. Skt. i. 18, 52. Tyrwedon, tyrwadon ł gremedon, tyrwodan *exacerbaverunt*, Ps. Lamb. 77, 40, 41, 56. Hig mē tirigdon *ipsi me provocaverunt*, Deut. 32, 21. Hí hine mid heora wordum tirigdon, Homl. Th. ii. 454, 17. Earme ne tyrewiaþ *vex not the poor*, Wulfst. 50, 2. Ǽnig man óđerne ne tyrie ne ne týne, 70, 8. Đæt hí eldeódige menn ne tyrian ne ne týnan, 309, 4. Hé đás leóde mid here and mid ungylde tyrwigende wæs, Chr. 1100; Erl. 236, 2, Mægþ tyrwiende *generatio exasperans*, Ps. Lamb. 77, 8. [Tirgen *to get weary*, Misc. 12, 362. Tarien *to fatigue*, Chauc. Terren to wraþþe *provocare ad iram*, Wick. Deut. 4, 25. Terwyñ or make wery *lasso, fatigo*, terwyd *lassatus, fatigatus*, Prompt. Parv. 489. *O. Du*. tergen *to vex*: *Dan*. tærge *to exasperate, irritate*: *Ger*. zergen.] v. ge-tirgan.

tirging, tirwing, tiring, e; *f. Vexation, provocation, harassing emotion*:—Tyrging, tyring *zelus*, Blickl. Gl. [Terwynge *lassitudo, fatigacio*, Prompt. Parv. 489. *Du*. terging *provocation*.]

tiriaca, an; *m. A medicine*, properly *an antidote for poison*, cf. *tiriaca* drenc wyđ ǽttre, Wrt. Voc. i. 20, 20:—Tyriaca is gód drenc wiþ eallum innođtýdernessum, and se man se đe hine swá begæþ swá hit hér on segþ, đonne mæg hé him miclum gehelpan . . . Nime ǽne lytle snǽd đes tyriacan, Lchdm. ii. 288, 23–290, 3. [Low Latin *tiriaca* from Latin *theriaca*. In Mid. E. *triacle* = a sovereign remedy is common, see Skeat's note on Piers P. C. ii. 147.]

tirian, tirigan. v. tirgan.

tír-leás; *adj. Inglorious*:—Đara đe tírleáses (*Grendel's*) trode sceáwode, Beo. Th. 1690; B. 843. [Cf. *Icel*. tírar-lauss *inglorious*.]

tír-meahtig; *adj. Gloriously mighty*:—Tírmeahtig cyning (*the Deity*), Exon. Th. 72, 1; Cri. 1166: 209, 24; Ph. 175. [Cf. *Icel*. tírar-sterkr.]

tirwa, tirwe, an; *m. f. Tar, resin, gum*:—Tyrwa *bitumen*, tyrwan *bituminis*, Hpt. Gl. 488, 78, 77. On swǽce swylce tyrwe *smelling of resin*, Lchdm. i. 278, 2. Tirwan *resinae*, Hpt. Gl. 501, 4. Sumne dǽl tyrwan *modicum resinae*, Gen. 43, 11. Đú clǽmst mid tyrwan *bitumine linies*, 6, 14: Homl. Th. i. 20, 33: Ex. 2, 3. Croppas mid tyrwan gesodene, Lchdm. i. 224, 10. Hig hæfdon tyrwan (*bitumen*) for weallim, Gen. 11, 3. Teorwena, tyrwena *naptarum*, Hpt. Gl. 445, 29. Dó đonne đa tyrwan on *put the gums in*, Lchdm. iii. 14, 24. v. eorþ-tyrewa; teoru.

tirwan. v. ge-tirwan.

tirwen (?); *adj. Of resin*:—Tyrwene, stórsæpes, hryseles *resinae*, Hpt. Gl. 501, 1.

tirwian. v. tirgan.

tír-wine, es; *m. A glorious friend*, an epithet of the follower of a successful chief:—Se hláford biþ tó upáhæfen inne on móde for đæm anwalde đe him ánra gehwilc his tírwina tó fultemaþ, Met. 25, 21.

tiscge = disce:—In tiscge *in cateno*, Wrt. Voc. ii. 46, 53. Cf. in disce *in cateno*, 74, 29.

tit[t], es; *m. A teat, pap, breast*:—Tit *mamilla*, Wrt. Voc. i. 44, 13. Titt *uber*, Ælfc. Gr. 9, 18; Zup. 44, 2. Lege ofer đone wynstran tit, Lchdm. i. 192, 17. Tittas *mamillas*, lxxiv, 24 : Wrt. Voc i. 65, 7 : 283, 29 : ii. 56, 28. Wiđ tittia sár wífa, Lchdm. i. 112, 16. Titto (tito, Rush.) ł breósto *ubera*, Lk. Skt. 11, 27 : Rtl. 4, 17. [Þa tittes đæt þu suke, Laym. 5025. Bi þan titten (tyttes, 2nd MS.) anhon, 11936. Bi þeo tittes þet he sec, A. R. 330, 5. Teon þe tittes awei of þine breosten, Kath. 2098. A fostre wimman on was tette he sone aueđ lagt, Gen. and Ex. 2621. Tete *rimes with* swete (*I sweat*), Chauc. C. T. 3704; *with* lete, *pp*. of leten, Gow. i. 268, 3. Tete *uber*, Prompt. Parv. 489. *O. Du*. titte : *M. H. Ger*. zitze : *Ger*. zitze. The Teutonic form seems to have been borrowed by Romance languages, *Ital*. tetta, zizza : *Fr*. tette : *Span*. teta.]

tite-gár read (?) ategár:—Titegárum *phalarica, lanceis magnis* (cf. ategára *falarica, hasta*, 521, 6), Hpt. Gl. 425, 14. v. æt-gár.

-titelian. v. ge-titelian, *and next word*.

titelung, e; *f. A giving of the titles* or *headings*:—Titelung *recapitulatio*, Hpt. Gl. 433, 72.

tíþ, e; *f. Grant, cession, concession*:—Týþ *cessio*, Wrt. Voc. ii. 131, 6: *concessio*, 136, 11. Hý wǽron đē biddende mínra góda and đú hin symble tíđe forwyrndest *they were asking thee for my goods and thou didst ever refuse them the grant thereof*, Wulfst. 259, 11. Ne hæfde wit monig óđer hors đæt wē mihton đearfum tó týþe syllan *numquid non habuimus equos plurimos quae ad pauperum dona sufficerent?* Bd. 3, 14; S. 540, 27. Mid týþe and mid geþafunge Eádgáres cynenges, Cod. Dip. Kmbl. ii. 400, 23. Fela wundra gelumpon æt đæra apostola byrgenum đurh đæs Hǽlendes tíđe, Homl. Th. i. 384, 19. Hyre đæs Fæder on roderum tíđe gefremede, Judth. Thw. 21, 5; Jud. 6. v. next two words.

tíþe, tíþa (-e, -a; *masc*.: -u, -a, -e; *fem*.: -a; *pl*.) *in the phrases* tíþe(-a) beón, weorþan *to obtain* one's request, *to have granted the request* for something (*gen*.):—Sóna wæs gelǽred đætte hē wæs from Drihtne týþe

đǽre bēne đe hē bæd *statim edoctus impetrasse se quod petebat a Domino*, Bd. 4, 29; S. 607, 32. Myceles đú (*masc*.) bǽde, ac đú bist tíđa, Homl. Skt. i. 18, 284. Týđa, 3, 513. Đú (*Abraham*) đæs tíđa beó, Cd. Th. 142, 12; Gen. 2360. Hē ongann tó Gode wísdómes wylnian, and hē eác đæs tíđa weard, Wulfst. 277, 19. Đú (*Lot*) scealt đære bēne tíđa weorđan, Cd. Th. 152, 28; Gen. 2527. Þeáh đú (*Esther*) biddan wille healfne đone anweald . . . đú scealt beón tíþu đæs, Anglia ix. 33, 185. Heó ábæd æt Gode Godes willan tó đam đæt heó sunu hæfde, and heó sóna wæs tíđu (*other MSS*. tíđa), Homl. Ass. 38, 357. For swá hwæne swá heó bit, heó biþ tíđa simle, Homl. Skt. ii. 29, 274. Đæt ic (*a widow*) beó đæs tíđe đe ic bidde, Homl. Th. i. 566, 15. Ealles đæs đe gē biddaþ gē beóþ tíđa *omnia quaecunque petieritis in oratione accipietis*, Mt. Kmbl. 21, 22. v. bén-tíđe; tíþ, tíþian.

tíþian, tíghian; *p*. ode *To grant, concede*, (a) with gen. of that which is granted:—Bed Beorn đæt hē sceolde faran mid him tó đam cynge . . . and hē đæs tíđode, Chr. 1046; Erl. 174, 10. Treówe and hyldo tíđiaþ mē, Cd. Th. 152, 7; Gen. 2516. Đæt preósta gehwilc fulluhtes tíđige, sóna swá man his girne, L. Edg. C. 15; Th. ii. 246, 25. Nolde gē mē wǽda tíþian, Wulfst. 288, 33. Hit is swíđe geleáflíc, đæt hē hyre myceles đinges tíđian wylle, Homl. Th. i. 454, 2 : Gen. 18, 3. Hē náteshwón hire đæs tíđian nolde *qui nequaquam acquiescens operi nefario*, 39, 8. Ne hine mon on óđre wísan his bēne týþigean (tygþian, M. 220, 26) wolde *neque aliter quod petebat impetrare potuit*, Bd. 3, 21; S. 550, 43. (b) with acc. (?) the case is probably determined by the Latin:—Se him fultum tíþaþ *qui eis adjutorium prestitit*, Anglia xiii. 391, 366. Wísdóm lǽnende ł tíđiende litlingum *sapientiam praestans parvulis*, Ps. Lamb. 18, 8. (c) with a clause:—Nolde se cyning him tíđian đæt Israël férde forð ofer his gemǽru *qui concedere noluit, ut transiret Israel per fines suos*, Num. 21, 23. (d) used absolutely:—Đonne đú tíđast, Hy. 7, 56. Drihten mē gehírde and tíđode mē *exaudivit me Dominus*, Deut. 9, 19. Đá oferhogode hē đæt hē him æfter dyde oþþe wiernde oþþe tigþade, Ors. 6, 34; Swt. 290, 22. Ic gelýfe đæt hē wille đē tíđian, Homl. Skt. i. 21, 218: Homl. Th. i. 250, 2. Tó tíþienne *exaudienda est*, Wülck. Gl. 251, 6. [Leafdi, tuđe me mine bone, O. E. Homl. i. 207, 31. God haueđ herd þine bede and tiđed te bene *exaudita est oratio tua*, ii. 135, 7. Drihtin has þe tid (tidd, MS. G.) þi bon, C. M. 10966. All þaét ned uss iss Godess Gast uss tiþeþþ, Orm. 5365. O þing ich wolde bidde þe, þat þou me woldest tyþe (*rimes with* bliþe), R. Glouc. 114, 18.] v. ge-tíþian; tíþ, tíþe.

tit-stricel, es; *m. A nipple of the breast*:—Tit *mamilla*, meolce breóst *ubera*, tittstrycel *papilla*, Wrt. Voc. i. 44, 13–15. v. stricel, II.

titt. v. tit[t].

titul *a title, superscription*:—Titul ł merca *titulus*, Mk. Skt. Lind. 15, 26. [*O. H. Ger*. also borrows titul *in the same connection*:—Screib titul Pilatus síneru sahhu.]

Tíw, Tíg, Tí, es; *m.* **I.** *the god Tiw*, a Teutonic deity to whom amongst the Latin gods Mars most nearly corresponded:—Tíig *Mars, Martis*, Txts. 77, 1293. Tíg, Wrt. Voc. ii. 55, 56. Tuu (Tíw?), 58, 40. Đone Syxtum nēdde Decius se cásere Tíges (*Martis*) deófolgylde, Shrn. 114. 9. ¶ The word occurs oftenest in the connection in which it remains—in the name of one of the days:—On Tíwes-dæg *tertia feria*, R. Ben. 38, 6; R. Ben. Interl. 49, 14: Wulfst. 180, 25. On Tíwes-niht, Lchdm. iii. 146, 23. **II.** *one form of the name of the Runic T*; Tí is given as the name of the symbol ↑ in some alphabets, see Kemble on Anglo-Saxon Runes in Archæologia, vol. 28, pp. 338, 339. The word is probably to be recognized in the form *tyz*, which is given as the name of the Gothic T in the Vienna MS. containing a Gothic alphabet, and from it a Gothic Tius may be inferred. *O. H. Ger*. Ziu(-o) *the name of a god* (preserved in *M. H. Ger*. Zies-tag), *the name of a letter : Icel*. Týr *the name of a god* (*kept in* Týs-dagr), *name of a rune*. See Grmm. D. M. c. ix.] v. Tír.

tó; *prep. adv.* **I.** *with dat.* (1) with words expressing motion. (a) with verbs of coming, going, falling, etc., marking the end reached by that which moves, *to, at*:—Cómon twēgen englas tó đære birig, Gen. 19, 1. God him com tó, 20, 3: Mk. Skt. 5, 21. Hē férde tó đam munte, Gen. 19, 30. Féran tó đissum dimman hám, Cd. Th. 271, 27; Sat. 111. Bryne stígeþ tó heofonum, Exon. Th. 233, 7; Ph. 521. Conon gelende tó đære byrig, Ors. 3, 1; Swt. 98, 23. Néđan tó hilde, Cd. Th. 124, 11; Gen. 2061. Đá feóll hē tó đæs Hælendes fótum, Lk. Skt. 8, 41: 5, 8. Hē feóll tó foldan, Andr. Kmbl. 1835; An. 920. Búgan tó eorđan, Rood Kmbl. 84, Kr. 43. Nú sceal he faran tó incre andsware, Cd. Th. 35, 19; Gen. 557. (b) with verbs of bringing, bearing, drawing, sending, taking, etc., marking the end reached by that which is moved:—Mēce đone đín fæder tó gefeohte bær, Beo. Th. 4103; B. 2048. Hí him tó nimaþ mægeđ tó gemæccum, Cd. Th. 76, 17; Gen. 1258. Him fetigean tó sprecan síne, 161, 17; Gen. 2666. Hē hine lǽdde tó đam hálgan hám, 300, 19; Sat. 567. Hē him tó sende áras síne, 146, 15; Gen. 2422. Hē his gingran sent tó đinre sprǽce, 33, 6; Gen. 516. Sende se Fæder his sunu tó cwǽle, Homl. Th. ii. 6, 17. Hē tó áwylte stán tó hlide đære byrgene, Mt. Kmbl. 27, 60. (c) where the motion is directed to, but does not reach the object:—Hí tó đam hæđengilde bugon, Num. 25, 2. Ealle ábúgaþ tó đē, Hy. 7, 10. Hié onhnigon tó

ðam herige, Cd. Th. 227, 3; Dan. 181. Áhyld mē ðín eáre tô *inclina ad me aurem tuam*, Ps. Th. 70, 2. (1 a) with words implying motion :— Hig woldon tô Basan *ascenderunt per viam Basan*, Num. 21, 33 : Chr. 1036; Erl. 164, 26. Hê hēht him Abraham tô *he summoned Abraham to him*, Cd. Th. 112, 3; Gen. 1865 : 249, 18; Dan. 532 : Elen. Kmbl. 307; El. 154. (2) where the motion is figurative, (a) with words denoting change of condition, marking that to which a thing is changed, what a thing becomes, to what a thing is brought :—Hê heora wæter wende tô blôde, Ps. Th. 104, 25 : Cd. Th. 17, 13; Gen. 259. Heó alle forsceóp Drihten tô deóflum, 20, 14; Gen. 309 : Bt. 38, 1; Fox 194, 33. His gebed hweorfe tô fyrenum, Ps. Th. 108, 6. Forhwerfde tô sumum diôre, Met. 26, 87. Ðá weard hê tô deófle, Homl. Th. i. 12, 22 : Cd. Th. 20, 9; Gen. 305. Weorðan tô duste, Ps. Th. 89, 6. Ðú scealt tô frôfre weorþan leódum ðínum, Beo. Th. 3419; B. 1707. Weorðan tô wræce, Elen. Kmbl. 33; El. 17. Hî weorþaþ tô náuhte, Bt. 21; Fox 74, 36. Tô hwon sculon wit weorðan? Cd. Th. 50, 28; Gen. 815. Ic tô náwihte eom gebîged *ad nihilum redactus sum*, Ps. Th. 72, 17, 16. Paulus hine áwende of wôge tô rihte, Homl. Skt. ii. 29, 8. Swá is lâr and ár tô spôwendre spræce gelǣded, Exon. Th. 139, 14; Gú. 593. Ðam yfelan men ne becymþ tô nánum gôde, gif hê ðæs hálgan húsles unwurðe onbyrigþ, Homl. Th. ii. 278, 4. (b) with words denoting attainment, reaching to an object :—Fôn tô ríce *to come to the throne*, Chr. 871; Erl. 76, 3, and often. Ðǣ tô heortan grîpeþ ádl, Cd. Th. 57, 30; Gen. 936. (c) with verbs of attracting, alluring, drawing, forcing, etc. :—On ðæm weorce ðe hine nán willa tô ne spôn, Past. 33; Swt. 215, 10. Ðone fultum ðe hê him tô áspanan mehte, Ors. 3, 9; Swt. 126, 10. Þurh lâre spanan tô gefeán, Andr. Kmbl. 1195; An. 598. Ðæt ða sinhîwan tô swylte geteáh, Exon. Th. 153, 10; Gú. 823. Tô ðam gebede gebǣdon, Cd. Th. 228, 15; Dan. 202. v. ge-nýdan. (3) marking the end of extent, (a) marking the object reached :—Hî woldon witon hú heáh hit wǣre tô ðǣm hefone, Bt. 35, 4; Fox 162, 22. Ðanon wǣre tô helle duru hund þúsenda míla, Cd. Th. 310, 8; Sat. 723. Sió stôw ðe se weg tô ligþ, Bt. 33, 4; Fox 132, 37. Weg tô wuldre, Elen. Kmbl. 2297; El. 1150. Strǣte tô englum, Cd. Th. 282, 17; Sat. 228. (b) marking degree:—Gê etaþ tô fylle, Lev. 26, 5. Seóð tô feórðan dǣle, Lchdm. i. 188, 22. Seó sunne áþýstrode tô sweartre nihte, Homl. Skt. ii. 29, 11. Hê weard tô feore áfyrht *he was mortally afraid*, Homl. Th. i. 384, 7 : Homl. Skt. i. 7, 242. Fæsten tô berenan hláfe *a fast when nothing better than barley bread should be eaten*, Wulfst. 173, 10. Tô ánum mǣle fæstende *fasting to the point of taking but one meal in the day*, Homl. Skt. i. 20, 42. Gif man ðæt fýr sceal tô áhte ácwæncan, Wulfst. 157, 9. Tô náhte *not at all*, 190, 18 : 191, 3. Wǣron hié tô ðǣm gesárgode, ðæt hié ne mehton Súð-Seaxna lond útan berôwan, Chr. 897; Erl. 96, 8. Wæter-seócnyss hine ofereode tô ðan swîðe, ðæt . . . , Homl. Th. i. 86, 10. Wela ne mæg his hláford gehealdan tô ðon ðæt hê ne þurfe mâran fultumes, Bt. 29, 1; Fox 102, 16. (c) marking result attained, effect produced, *so as to produce* or *become, to* (the satisfaction, etc.). (1) where the object is concrete :—Tôbrecan tô styccum, Bd. 3, 6; S. 528, 21. Ceorfan tô sticcon, Lev. 1, 6. (2) where the object is abstract :—Ða ðe ealle gewîtendlîce ðing tô ðǣra apostola efenlǣcunge (*and so imitate the apostles*) forseóð for intingan ðæs ēcan lîfes, Homl. Th. i. 398, 23. Hannibal æt ðære ié gewîcade eallum Rômánum tô ðæm mǣstan ege (*which was the cause of very great terror to all the Romans*), Ors. 4, 9; Swt. 194, 8. Geweóx hê him tô wælfylle *he grew up to be a cause of destruction to them*, Beo. Th. 3427; B. 1711 : Salm. Kmbl. 747; Sal. 373. Gif hê hwæt tô gôde gefremede, Homl. Th. i. 332, 5 : 8, 9 : Exon. Th. 297, 1; Crä. 61. Dryhtne tô willan *to please the Lord*, Andr. Kmbl. 3280; An. 1643. Ðæs ðe gé him tô dare gedôn môtan, Exon. Th. 144, 2; Gú. 672 : 127, 36; Gú. 397. Tô wundre *so as to produce wonder, wondrously*, Homl. Skt. i. 23, 654. Tô þance, Andr. Kmbl. 2225; An. 1114 : Cd. Th. 32, 20; Gen. 506 : Beo. Th. 762; B. 379. Eal ða earfeþu ðe ic gefremede tô fácne, Exon. Th. 272, 10; Jul. 497. (4) marking the end towards which an action or object is directed, (a) with verbs of looking, listening (lit. and fig.) :—Beseoh tô mē *respice me*, Ps. Th. 12, 3. Tô heofenum beseoh, Elen. Kmbl. 166; El. 83. Ðá lôcode Petrus tô Paule, Blickl. Homl. 187, 34 : Beo. Th. 3313; B. 1654. Hî ðē tô hēraþ, Met. 4, 5. v. lôcian, hýran. (b) with verbs of pointing, directing :—Se Dryhtnes dôm wîsade tô nýdgedâle, Exon. Th. 129, 4; Gú. 415. Tǣcan tô, Cd. Th. 175, 22; Gen. 2899. (c) with verbs of urging, prompting, inciting, etc. :—Onbryrde tô godcundre lâre, Blickl. Homl. 33, 23 : Andr. Kmbl. 2237; An. 1120. Úsic lust hwæteþ tô ðærre mǣran byrig, 574; An. 287. (d) with words denoting destination, intention, etc. :—Hê monige dēmde tô deáðe, Elen. Kmbl. 997; El. 500 : Exon. Th. 247, 31; Jul. 87. Mec gesette Crist tô compe, 389, 3; Rä. 7, 2. His ríce ðǣr wē tô gesceapene wǣron, Homl. Th. ii. 6, 27 : Bt. 25; Fox 88, 7. (e) with words denoting address :—Ðá cwæð se Hǣlend tô him, Mt. Kmbl. 8, 4. Ðæt hié tô ðam beácne gehedu rǣrde, Cd. Th. 227, 23; Dan. 191. Ic clypige tô ðē, Ps. Th. 21, 2. Wîtebrôgan ðe ðú tô mē beótast, Exon. Th. 250, 35; Jul. 137 : Bd. 1, 27; S. 493, 30 : 5, 12; S. 628, 43. Habbaþ wē tô ðǣm mǣran ǣrende, Beo. Th. 545; B. 270. (f) with words denoting hostility :—Ðæt folc

---

mǣnde tô him Arone (*contra se et Aaron*), Past. 28, 6; Swt. 201, 4 : Ors. 3, 7; Swt. 120, 5 : Beo. Th. 5994; B. 3001: Ps. Th. 70, 22. Monige ðe tô mē feohtaþ *multi qui bellant me*, Ps. Th. 55, 3. Mē feóndas tô feohtaþ, 68, 17 : 58, 1. (g) with words denoting preparation, aptness, readiness, or the reverse :—Fýsan tô râde, Elen. Kmbl. 1960; El. 982 : Cd. Th. 173, 12; Gen. 2860. Hê ða leóde wenede tô wuldre, Andr. Kmbl. 3360; An. 1684. Gif hié tô ðam sîðe gyrwan, 1590; An. 796. Late tô ðam orlege, 94; An. 47. Tô gefeohte gearu, Num. 21, 33 : Elen. Kmbl. 45; El. 23. Ealdordôm tô hwônlîc tô swá micelre bodunge, Homl. Th. i. 38, 2. Gleáwast tô wîge and tô gewinne, Ors. 4, 1 ; Swt. 154, 33. (h) marking the object of a feeling or operation of the mind :—Se ðe næfþ lufe tô Godes sceápum, Homl. Th. i. 240, 18 : 334, 7. Ic hæbbe geleáfan tô Gode, Cd. Th. 34, 27; Gen. 544. Næs him tô ēðle wynn, Andr. Kmbl. 2326; An. 1164. Ne biþ him tô hearpan hyge, ne tô wîfe wyn, ne tô worulde hyht, Exon. Th. 308, 23–26; Seef. 44, 45. Abraham tô Gode cýðde hæfde, Homl. Th. ii. 190, 12 : 558, 1 : i. 10, 3. Cynengas ðe tô Gode lytelne ege hæfdon, Lchdm. iii. 442, 24. Ða ðe tô ðē egsan áhtan *qui timent te*, Ps. Th. 118, 79. Nân neát nyste nænne andan, ne nænne ege tô ôþrum, Bt. 35, 6; Fox 168, 10. Ðæt hê hæbbe clǣne heortan tô mannum, Wulfst. 239, 18. Hié hæfdon ungeþwǣrnesse tô eallum folcum, Ors. 6, 3; Swt. 258, 1 : Homl. Th. i. 38, 14. Swá hwæt swá gē habbaþ on eówrum môde tô ǣnigum men, 266, 30. Sió heánes ðe hié tô hopiaþ, Past. 41; Swt. 299, 5 : Met. 7, 44. Ðonne gelýfe ic tô Gode, ðæt hit ðam men gehelpe, Lchdm. ii. 290, 9; Chr. 1036; Erl. 165, 16. Hî hogedon tô nîðe, Ps. Th. 77, 20. Tô ðam beteran hycgan and hyhtan, Fragm. Kmbl. 82; Leas. 43. Tô swice þencan, Exon. Th. 317, 16; Mod. 61 : Beo. Th. 2281; B. 1138. Tô reáflâce rǣd áþencean *to devise counsel that has robbery for its object*, Ps. Th. 61, 10. Se cyning beþôhte swîðost tô Arpelles his ealdormenn, Ors. 1, 12; Swt. 52, 20. (i) marking a purpose to be effected, an end to be served, *to some end, for* some purpose :—Hê ásende ðone sunu tô úre álýsednesse, Homl. Th. ii. 6, 9. Ðæt folc geðafode ðæt sume leofodon tô wudunge and tô wæterunge, 222, 29. Álesen tô lâre, Elen. Kmbl. 571; El. 286. Ofn onhǣtan tô cwale cnihta feorum, Cd. Th. 229, 32; Dan. 226. Hê up áhôf bord tô gebeorge, Byrht. Th. 135, 40; By. 131. Hié tô gebede feóllon *they fell down to pray*, Cd. Th. 48, 18; Gen. 777; Andr. Kmbl. 2054; An. 1029. Hê genam on eallum dǣl ǣhtum sînum tô ðam gielde, Cd. Th. 90; Gen. 1501 : 175, 6; Gen. 2891. Hié werod læsse hæfdon tô hilde *a smaller band had they for battle*, Elen. Kmbl. 97; El. 49. Tô ðam ic eom ásend *therefore am I sent*, Lk. Skt. 4, 43. Tô hwan ys ðiss forspilled *to what purpose is this waste?* Mt. Kmbl. 26, 8. Tô hwan becôm ðú *wherefore art thou come?* 50 : Soul Kmbl. 34; Seel. 17. Tô ðam (ðon) ðæt *in order that, to the end that*, Ors. 1, 10; Swt. 48, 23 : Lchdm. iii. 438, 19 : Chart. Th. 436, 26. (j) marking an object for the benefit or service of which anything is intended, *for* :—Hê onfēng lîchoman gegyrelan tô his godcundnesse, Blickl. Homl. 9, 27. Hê hæfde xx elpenda tô ðæm gefeohte, Ors. 4, 1; Swt. 154, 30. Wênen hî him mâran mēde tô . . . Gif hî him mâran mēde tô ne wênaþ, Past. 59; Swt. 449, 12–13. Hê ðē worhte tô mē, Cd. Th. 50, 32; Gen. 817. Hê gewyrceþ tô wera hilde helm oþþe hupseax, Exon. Th. 297, 5; Crä. 63. Hié wǣpna náman tô ðon ðæt hié heora weras wrecan þôhton *they took arms for this reason, that they intended to avenge their husbands* (cf. Goth. du þē ei *pro eo quod*), Ors. 1, 10; Swt. 44, 32. (5) where position (lit. or fig.) is marked, (a) marking juxtaposition, *next to, at, by, alongside* :—Hî man bebyrigde tô hyre were *she was buried by her husband*, Homl. Th. i. 318, 1 : ii. 188, 5. Hê gesette ða hálgan rôde tô his heáhsetle swilce him tô geféran, H. R. 101, 10. Hié setton him tô heáfdum hilderandas, Beo. Th. 2488; B. 1242. Mid olfendes hǣrum tô lîce (*next the body*) gescrýdde, Homl. Th. ii. 506, 23 : Homl. Skt. i. 12, 36. Wyrm tô fýre *warm at the fire*, Lchdm. i. 374, 10 : Exon. Th. 393, 36; Rä. 13, 11. Tô hire freán sittan *to sit by her lord*, Beo. Th. 1287; B. 641. Symle hî sǣton ætsomne tô gereorde, Homl. Th. ii. 506, 22. Gesittan tô symble, Cd. Th. 259, 33; Dan. 701 : Judth. Thw. 21, 12; Jud. 15. Hiera súþgemǣro licgeaþ tô ðæm Reádan Sǣ, Ors. 1, 1; Swt. 10, 34 : 16, 13. Seó forme India liþ tô ðæra Sîlheorwena rîce, seó ôðer liþ tô Mēdas, seó ðridde tô ðam micclum gârsecge, Homl. Th. i. 454, 12–13. Þeáh ðe se Hálga Gást ne beó swutollîce genemned tô ðam Fæder and tô ðam Suna *along with the Father and the Son*, ii. 56, 29. (b) marking the place where an object is, *in, on* :—Ic cýðe ðam gerēfan tô gehwylcere byrig (þurh ealle mîne rîce, *other MS.*), L. Ath. i. prm.; Th. i. 194, 3. Hê gesette Iudas tô bisceope tô Godes temple, Elen. Kmbl. 2114; El. 1058. Tô horse *on horseback*, Exon. Th. 298, 7; Crä. 81. (c) fig., marking position or condition in which an object is placed :—Tô gewealde *in the power of, at the disposal of*, Cd. Th. 112, 7; Gen. 1867 : 132, 32; Gen. 2202 : 290, 15; Sat. 415. (d) with verbs of joining, adding to, cleaving, etc. :—Gesamnian sáwle tô lîce, Met. 17, 12. Hê sǣlde tô sande scip, Beo. Th. 3838; B. 1917. Geðeódde sum wer him tô, Homl. Th. ii. 504, 22. v. clifian, geþeódan, îcan. (e) marking order, *next to, after* :—Tô mînre mēder and geswystrum ðú mē eart se leófesta freónd *secundum matrem meam sororesque meas, acceptissime*, Nar. 1, 12 : Shrn. 108, 20. Scs Iohannes wæs ealra

manna se mǽsta and se hálgosta tô Criste seluum, 123, 6 : Homl. Skt. i. 6, 51 : Cd. Th. 17, 3 ; Gen. 254 : Ors. 2, 2 ; Swt. 66, 32. Hé wæs bufan eallum dǽm ðe on ðam ríce wǽron tô dǽm cyninge, 3, 11 ; Swt. 148, 5. Sió is mǽst tô Babilonia byrig, Nar. 33, 17. Ðú bist se ðridda man tô mé on mínum ríce, Homl. Th. ii. 436, 5, 17. Hé is geendebyrd tô Petre, 522, 2. (f) marking the position occupied, the purpose fulfilled by an object, *to, as, for* :—Wé habbaþ ús tô fæder Abraham *we have Abraham to our father,* Lk. Skt. 3, 8 : Mt. Kmbl. 14, 4 : Exon. Th. 245, 34 ; Jul. 54. Hig hæfdon heom tô gewunan, ðæt . . . , Mt. Kmbl. 27, 15. Ic hæbbe tô gewitnisse heofen and eorðan *testes invoco coelum et terram,* Deut. 4, 26. Hé hæfde Thesalium him tô fultume, Ors. 4, 1 ; Swt. 154, 30. Hié him ðæt gold tô gode noldon, Cd. Th. 228, 5 ; Dan. 197. Hé is tô freónde gód *he is good as a friend,* Exon. Th. 248, 28 ; Jul. 102. Ic genam hig tô wífe, Gen. 20, 12 : Bt. 8 ; Fox 24, 24. Hí him tô gewunon náman, ðæt . . . , Bd. 3, 5 ; S. 527, 7. Hé Agustinum him tô gespelian funde, Lchdm. iii. 434, 7. Ic clipie mé tô gewitnysse heofonan and eorðan, Deut. 30, 19. Him brego engla líg tô wræce sende, Cd. Th. 156, 6 ; Gen. 2584 : 21, 2 : Gen. 318. Hé sealde him tô bôte, ðæs ðe hé his brýd genam, gangende feoh, 164, 21 ; Gen. 2718 : 90, 24 ; Gen. 1500 : 124, 29 ; Gen. 2070. Eal folc fæste tô gemǽnelícre dǽdbôte, Wulfst. 180, 23. Hé is tô Cristes anlícnesse áset *divina positus vice dispensat,* Past. 13 ; Swt. 79, 10. Hé gearwaþ ðínne innoð his suna tô brýdbûre, Blickl. Homl. 9, 10. Tô lǽne *as a loan, on loan,* Deut. 15, 8 : Past. pref. ; Swt. 9, 7. Tô láfe *as a remnant, remaining.* v. láf, I. See also (j) below. (f 1) with verbs of making, appointing, being, accounting, naming, and the like, where often the preposition now has no representative, though *to, as, for* are sometimes used :—Mé feóndas geworhton him tô wæfersýne *they made me a spectacle for themselves,* Rood Kmbl. 61 ; Kr. 31. God ne gesceóp hine ná tô deófle . . . ac hé weard tô deófle *God did not create him a devil . . . but he became a devil,* Homl. Th. i. 12, 20. Hé him dyde bearn tô weorcþeówum *he made them slaves,* Cd. Th. 220, 21 ; Dan. 74 : 45, 6 ; Gen. 722 : Andr. Kmbl. 53 ; An. 27. Hig ne fundon hwæt hí him tô gylte dydon *they could not find what they could make a charge against him,* Lk. Skt. 19, 48. Ðam golde ðe hé him tô gode teóde, Cd. Th. 229, 13 ; Dan. 216 : Exon. 255, 18 ; Jul. 215. Hé sette hine on his húse tô hláfwearde *constituit eum dominum domus suae,* Ps. Th. 104, 17, 16 : 108, 5 ; Elen. Kmbl. 2111 ; El. 1057 : Blickl. Homl. 9, 5. God hine gesette manegum ðeódum tô fæder (*a father of many nations have I made thee,* Gen. 17, 5), Homl. Th. i. 92, 16. Hine gecés tô fæder and tô hláforde Scotta cyning, Chr. 924 ; Erl. 110, 14 : Cd. Th. 19, 3 ; Gen. 285 : Exon. Th. 3, 15 ; Cri. 36 : Andr. Kmbl. 647 ; An. 324. (v. *also* ge-hálgian, hálgian.) Beón tô tácnum, tô mete, Gen. 1, 14, 29. Næs him se swég tô sorge, Cd. Th. 322, 22 ; Dan. 264. Ða þeódlogan ðe taliaþ ðæt tô wærscype, ðæt . . . , Wulfst. 55, 15. Ne sete ðú him ðás dǽda tô synne, Homl. Th. ii. 34, 21. Heó hié sylfe tô ðeówene genemde, Blickl. Homl. 9, 23. Hine tô sylfcwale secgas nemnaþ, Exon. Th. 330, 24 ; Vy. 56. Deáh mon anweald and genyht tô twǽm þingum nemne, ðeáh hit is án, Bt. 33, 1 ; Fox 120, 20. (g) marking the place at which anything is sought, obtained, etc., *at, in* :—Sécean hilde tô Heorote, Beo. Th. 3984 ; B. 1990. Tô dûnscræfum drohtod sécan, Andr. Kmbl. 3077 ; An. 1541. (h) marking the source from which anything is sought, desired, expected, deserved, obtained, etc., *of, from* :—Ǽlcum ðe mycel gesealð is him man mycel tô sécþ *cui multum datum est, multum quaeretur ab eo,* Lk. Skt. 12, 48 : Elen. Kmbl. 638 ; El. 319. Wé sécaþ fultum tô ðé (*a Domino*), Ps. Th. 7, 11. Hí tô Rôme him fultumes bǽdon, Bd. 1, 12 ; S. 480, 22. Ðú iówan scolde ðæt him mon tô áscaþ, Past. 22 ; Swt. 173, 2. Ðú wilnodest tô ús ðæs gôdes ðe ðú tô him sceoldest, Bt. 7, 5 ; Fox 24, 3 : Past. 58 ; Swt. 447, 15 : Ors. 4, 6 ; Swt. 174, 24 : L. Ath. v. 8, 3 ; Th. i. 236, 15 : Wulfst. 277, 18. Girne hé tô Godes þeówum, ðæt . . . , 180, 11. Swá ic ðé wéne tô as *I expect of you,* Beo. Th. 2797 ; B. 1396 : 5836 ; B. 2922. Ne þurfon wé ná tô úrum mǽgum ne nán man tô his wífe ðencean tô ðam swýðe, ðæt him man æfter his fordsýþe tô ðam micel fore gedǽle, ðæt hí hine fram wítan álýsan *it is too much to expect of kinsmen or wife, that so much will be distributed for a man after his death as to release him from purgatory,* Wulfst. 306, 3. Ðonne môte wé ðæs tô Gode earnian bet *we must better deserve it of God,* 157, 2 : Ps. Th. 7, 3 : Ors. 5, 4 ; Swt. 224, 33. Hé geceápode tô ðæm senatum, ðæt hié ealle wǽron ymb hiene twywyrdige, 5, 7 ; Swt. 228, 17. Tô eorðan ǽtes tilian, Cd. Th. 94, 5 ; Gen. 1557 : 59, 31 ; Gen. 972. (i) marking the object on which an action takes effect, *to* (in to do something *to* anything) :—Hire man wôh tô ne dó, L. Edm. B. 7 ; Th. i. 256, 3 : Cd. Th. 136, 28 ; Gen. 2265. Gúðrǽsa fela ðara ðe hé geworhte tô West-Denum, Beo. Th. 3161 ; B. 1578. (j) marking agreement, likeness, *according to, at, after* :—Hié ús lǽrdon tô ðæm ðe hira willa wæs *secundum voluntatem suam erudiebant nos,* Past. 36 ; Swt. 255, 10 : Bt. 8 ; Fox 24, 24 : Homl. Th. i. 264, 23. Se ðe tô Godes bisene gesceapen is (cf. gesceapene æfter ðære biesene úres Scippendes, 17), Past. 36 ; Swt. 249, 22 : Cd. Th. 92, 14 ; Gen. 1528 : Gen. 1, 27. Uton wircean him sumne fultum tô his gelícnisse *faciamus adjutorium simile sibi,* 2, 18. Ðú wást ðæt ic symle tilode tô lífigenne tô ðínes múþes bebode *nosti quia ad tui oris imperium*

semper vivere studui, Bd. 4, 29 ; S. 607, 28. Hí folgodon Cristes láre tô ðære níwan ǽ (*according to the new law*), Ælfc. Gen. Thw. 2, 23. Ðú ða unstillan gesceafta tô ðínum willan ástyrast, Bt. 33, 4 ; Fox 128, 9. Tô hwylcum gemete *after what manner,* Blickl. Homl. 5, 7. ¶ in adverbial phrases, equivalent to adverbs in *-lice* ; but see also (f) :—Ic secge eów tô sôðum *ego autem dico vobis* (in v. 34 the same words are translated : Ic secge eów sôðlíce, Mt. Kmbl. 5, 32. Tô sôðum ic secge eów *amen dico vobis* (cf. sôðlíce *amen,* 10, 15), 8, 11. Tô sôðan, Ælfc. T. Grn. 1, 6. Hwæt eart ðú tô sôðe ? St. And. 28, 8. Tô wissan *praesertim,* tô sôðan í tô cúðan *pro certo, veraciter,* Hpt. Gl. 416, 40-43. Ic nát tô gewissan hwǽr hé wunaþ nú *I don't know for certain where he lives now,* Homl. Skt. i. 21, 31. (k) marking comparison, *compared to, in comparison with, beside* :—Ðes is úre God, and nis nán ôðer geteald tô him, Homl. Th. ii. 12, 30. (l) *in addition to, besides* :—Ðá sende hé æfter máran fultum, tô ðæm ðe ða burg ymbseten hæfdon (*in addition to the troops that had besieged the town*), Ors. 3, 7 ; Swt. 116, 23. Ðæt is his andweorc ðæt hé habban sceal tô ðam tólum, ðam þrím geférscipum biwiste *that is his material, that he must have in addition to the tools, provision for the three classes,* Bt. 17 ; Fox 60, 3. Candidus and Uitalis and fela ôþre tô him (*many others besides them*), Homl. Skt. ii. 28, 19. Tô ðam ðe ic on lífe geúðe *besides what I granted in my lifetime,* Chart. Th. 563, 22. (m) marking price or equivalence, *for, at* :—Hú ne becýpaþ hig twégen spearwan tô peninge *nonne duo passeres asse veniunt,* Mt. Kmbl. 10, 29. Ðis mihte beón geseald tô myclum wurðe (*multo pretio*), 26, 9. Geseald tô þrím hund penegum, Mk. Skt. 14, 5. Ic sealle eów hit, tô ðam wurðe ðe ic hit gebohte, Ap. Th. 10, 2. Heofonan ríce wæs álǽten Zachéo tô healfum dǽle his ǽhta, and sumere wudewan tô ánum feorðlinge, and sumum men tô ánum wæteres drence, Homl. Th. i. 580, 22-26. Hié hié selfe tô nôhte bemǽtan *they valued themselves at nothing,* Ors. 3, 7 ; Swt. 114, 37 : 3, 9 ; Swt. 128, 4. Ðises cwides hé geunn ðam híréde tô ðam forwyrdan (*as the price of, in return for, the agreement*), ðæt hí hine wel healdan, Chart. Th. 329, 29. Wit ðus barn ne magon wesan tô wuhte (*at any price, on any account*), Cd. Th. 52, 5 ; Gen. 839. (6) with the inflected infinitive, forming with the verb a phrase that is used (a) with a noun or its equivalent, (1) as a predicate expressing what shall or must be done to the object marked by the noun :—Mannes Sunu ys tô syllenne on manna handa *Filius hominis tradendus est in manus hominum,* Mt. Kmbl. 17, 22. Se anweald ne se weorþscipe ne beóþ tô wénanne, ðæt hit seó sôþe gesǽlþ sié. Swá hit is nú hrædost tô secganne be eallum ðǽm woruldgesǽlþum, ðæt ðǽr nánwuht on nis ðæs tô wilnianne seó, Bt. 16, 3 ; Fox 56, 27-31. (2) as attribute, (a) the verb having an active force :—Hé hæfþ anweald synna tô forgyfanne (*potestatem dimittendi peccata*), Mk. Skt. 2, 10. Ic hæbbe mihte ðe tô forlǽtenne (-nde, MS. C.) *habeo potestatem demittere te,* Jn. Skt. 19, 10 : Cd. Th. 18, 30 ; Gen. 280. Swá ús neód is tô dônne, L. Eth. vi. 42 ; Th. i. 326, 7. Tíd tô mildsiende *his tempus miserendi ejus,* Ps. Surt. 101, 14. (β) the verb having a passive force, the noun being the object of the action expressed by the verb :—Ic hæbbe ðone mete tô etanne ðe gé nyton *ego cibum habeo manducare, quem uos non scitis,* Jn. Skt. 4, 32. Ic hæbbe ðé tô secgenne (-anne, MS. A.) sum ðing *habeo tibi aliquid dicere,* Lk. Skt. 7, 40. Gif Drihten sylþ mé hláf tô etenne and reáf tô werigenne *si dederit Deus mihi panem ad vescendum et vestimentum ad induendum,* Gen. 28, 20. Ðæt hé genôh hæbbe tô etanne *quantum sufficit ad vescendum,* Ex. 16, 12. Nim ðæt ic ðé tô sillenne habbe, Ap. Th. 12, 2. Hé ðæt feoh tô sellanne næfde *he had not the money to give,* Ors. 3, 7 ; Swt. 116, 15. Tô for náht taliende *parvi pendenda, ad nihilum iudicanda,* Hpt. Gl. 418, 35. Suá suá sió leásung simle dereþ ðǽm secggendum, suá dereþ eác hwílum sumum monnum ðæt sôð tô gehiérenne *it harms some men that the truth should be heard ; audita vera nocuerunt,* Past. 35 ; Swt. 237, 11. Ðæm lǽreówe is tô wietanne, ðæt . . . , 63 ; Swt. 459, 6. (b) as object of a verb :—Hé ondréd ðyder tô farende (faranne, MS. A. : færenne, Lind. : færan, Rush.) *timuit illuc ire,* Mt. Kmbl. 2, 22. Álýfe mé tô farenne and bebyrigean mínne fæder, 8, 21. Ys álýfed on restedagum wel tô dônne (dôanne, Rush.) *licet sabbatis bene facere,* 12, 12. God geðafaþ Antecriste tô wyrcenne tácna, Homl. Th. i. 4, 30. Ne bud ðú mé ná ælmessan tô syllanne, Ps. Th. 39, 7. Ús gelustfullaþ tô sprecenne be ðan hálgan were, Homl. Th. i. 360, 29. Hig begunnon ðis tô wircanne, Gen. 11, 6. (c) adverbially, (1) with adjectives, (a) where the verb has an active force :—Ðæs gescý neom ic wyrðe tô berenne *cujus non sum dignus calceamenta portare,* Mt. 3, 11. Heora fét beóð swíðe hraðe blôd tô ágeótanne *veloces pedes eorum ad effundendum sanguinem,* Ps. Th. 13, 6. Fúse tô farenne, Beo. Th. 3614 ; B. 1805. (β) where the verb has a passive force, governing the noun qualified by the adjective :—Hwæðer is éðre tô secgenne ? Mk. Skt. 2, 9. Ðæt is nú hradost tô secgenne, Bt. 17 ; Fox 60, 14 : 16, 3 ; Fox 56, 29. Ðeáh heó gladu wǽre on tô lócienne, 6 ; Fox 14, 27 : Exon. Th. 57, 15 ; Cri. 920. Langsumlíc biþ ús tô gereccenne and eów tô gehýrenne ealle ða deópnyssa [*there seems here a mixture of two constructions,* 'these things are tedious to hear (tô gehýrenne),' *and* ' to hear (gehýran) *these things is tedious*], Homl. Th. i. 362, 32. Þeáh hé wyrðe ne sié tô álǽtanne *though he deserve not to be pardoned,* Cd. Th.

39, 9; Gen. 622. (2) with verbs, where the verb in the phrase expresses an action that the subject of the main verb intends (a) to be done :—Ût eode se sǽdere his sǽd tô sáwenne (ad seminandum), Mk. Skt. 4, 3. Gesceafta ðe hê gesceóp mannum tô ðeówianne, Ps. Th. 18, arg. Ne com ic rihtwíse tô gecígeanne, Mt. Kmbl. 9, 13. Mellitum hê sende tô bodianne (bodiende, 20, 19) fulluht, Chr. 604; Erl. 21, 19. Tô dónne rehtwísnisse *ad faciendas justificationes*, Ps. Surt. 118, 112. Tô ondetende *ad confitendum*, 141, 8. Gesend englas tô ontýnenne míne sefan and tô andswariende ðyssum árleásum, Nar. 40, 30. (β) to be suffered :— Cyning tô gefulliane com tô Rôme *the king came to Rome to be baptized*, Bd. 5, 7; S. 620, 26. (7) marking time, (a) marking a point of time at which anything takes place, *at* :—Tô midre nihte *at midnight*, Lk. Skt. 11, 5: Mt. Kmbl. 25, 6. Tô ðam ǽrdæge *at dawn*, Cd. Th. 190, 12; Exod. 198. Ðá áxode hê tô hwylcon tíman him bet wǽre. And hî sǽdon him, Gyrstandæg tô ðære seofoþan tíde se fefor hine forlêt, Jn. Skt. 4, 52. Ðæt hê him tô tíde gemetlíce gedǽle ðone hwǽte, Past. 63; Swt. 459, 12. Scyld gewát tô gesceaphwíle, Beo. Th. 52; B. 26. (1 a) where the time is determined by that which takes place :—Áswearc úre môd tô eówrum infǽrelde, Jos. 2, 11. Tô ðýssere dǽde weard ðæs cynges heorte áblicged, Homl. Th. ii. 474, 19. (b) marking a space of time in the course of which something takes place, *in the course of, in, on* :—Gê etaþ nǽs tô ánum dæge, ne tô twám, ne tô fífon, ne tô týnum, ne tô twéntigum, ac fullne mônoð, Num. 11, 20. Swá micel swá hê tô ðam dæge geðicgan mihte *as much as he could eat in the day*, Homl. Th. ii. 194, 34 : Lchdm. ii. 288, 26 : Homl. Th. ii. 288, 7. Wê wǽron tô ðæge ealle on ánnesse gemedemode, Blickl. Homl. 139, 26. Tô sunnedæge *in sabbato*, Jn. Skt. Lind. 7, 23. Tô heora symbeldæge (*at that feast*, A. V.), Mt. Kmbl. 27, 15. Tô ðisse næhte *in ista nocte*, Rush. 26, 31. Tô niht (cf. on ðisse nihte, Lk. 12, 20) ðú scealt ðín líf álǽtan, Wulfst. 286, 23. Hê biþ tô geáre deád *he will die in the course of the year*, Shrn. 83, 21. Nú tô geáre synd feówertýne epactas *in the present year there are fourteen epacts*, Anglia viii. 327, 10 : 329, 36. Tô dæge *to-day, at the present time*, Bd. 3, 16; S. 542, 35. (c) marking a space of time during which something continues, *for, during* :—Ðæt wæs tô swíðe scortre hwíle *that was for a very short time*, Past. 36; Swt. 255, 10 : Cd. Th. 31, 22; Gen. 489. Tô langum fyrste *for a long while*, Homl. Th. i. 388, 18. Tô wyrcenne tácna tô feórþan healfan geáre *to work miracles for three years and a half*, 4, 31. Hê worhte his weorc tô seofon nihtum, ii. 356, 5. Syððan tô twelf mônðum ne cymþ ðǽr nán ôðer scúr, Lchdm. iii. 254, 1. *See also* ealdor, feorh. (d) marking end of extent, *to* :—Hê frægn hú neáh ðære tíde wǽre . . . Ðá andswaredon hí : 'Nis hit lang tô ðon,' Bd. 4, 24; S. 599, 5 : Beo. Th. 5176; B. 2591 : 5683; B. 2845. Is tô ðære tíde tælmet hwíle seofon and twéntig nihtgerímes, Andr. Kmbl. 225; An. 113. Ðæt hit wǽre þrittig þúsend wintra tô ðínum deáðdæge, Soul Kmbl. 73; Seel. 37. **II.** with gen. (1) marking the object to or towards which motion takes place, *to, for* :—Gewát him se æðeling tô ðæs gemearces ðe him Metod tǽhte *the prince departed for the appointed place, which the Lord had shewed him*, Cd. Th. 174, 28; Gen. 2885. Gewát him Andreas gangan tô ðæs ðe hê gramra gemót gefrægen hæfde ôððæt hê gemétte be mearcpaðe standan stapul ǽrenne *Andrew went on his way towards the spot, where he had learned was the cruel ones' meeting, until he found standing by the path a brazen pillar*, Andr. Kmbl. 2120; An. 1061. Wôd hê tô ðæs ðe hê wínreced wisse *thither he made his way, where he knew the hall was*, Beo. Th. 1433; B. 714 : 3939; B. 1967 : 4811; B. 2410. Tô ðæs gingran þider ealle urnon ðǽr se éca wæs *thither ran all the disciples, to the place where the Eternal was*, Cd. Th. 298, 11; Sat. 531. Tô ðæs fôron Caldéa cyn tô ceastre forð ðǽr Israéla ǽhta wǽron *thither marched the Chaldeans, on to the city, where were the possessions of the Israelites*, 218, 19; Dan. 41. Cômon hildfrecan tô ðæs ða hæftas ǽr hearm þrowedon *they came where the captives had suffered*, Andr. Kmbl. 2142; An. 1072. Tô hwæs hí gearwe bǽron *whither they should bear their arms*, Cd. Th. 190, 1; Exod. 192. (2) marking position, *in, at* :—Hê wæs tô middes wætres *he was in mid stream*, Homl. Skt. ii. 30, 176. Hê him ðæs leán forgeald tô ðæs ðe hê on reste geseah Grendel licgan *he paid him the reward for it, where he saw Grendel lying on the couch*, Beo. Th. 3175; B. 1585. *See* tômiddes, **II.** (3) marking purpose, see also (5) :—Hié tô ðæs here samnodon, Andr. Kmbl. 2248; An. 1125. (4) marking extent or degree, *to the extent, to such a degree* :—Ðæt hê ðás hálgan tíde gehealde mid clǽnum fæstene tô ánes mǽles *that he keep this holy time with a pure fast to the extent of eating only once*, Homl. Th. 285, 1. Ná tô ðæs hwôn *nequaquam*, Deut. 13, 11. *See se*, **V** (b 1). (5) forming with nouns adverbial or prepositional phrases :—Tô gyfes *gratis*, Hymn. Surt. 37, 20. Ic ðē tô leánes ðinne noman mǽrsige *in recompense I will magnify thy name for thee*, Lchdm. iii. 436, 27. Womma tô leánes *in requital of sins*, Wulfst. 138, 12 : 139, 2. God him sylþ tô mēdes ðæt ēce líf, Homl. Skt. i. 12, 139 : St. Aud. 28, 20. Tô geflites *certatim, strenue*, Hpt. Gl. 408, 54: Ap. Th. 10, 5. Ðú dwollíce leofast swylce ðē tô gamenes *thou livest foolishly as if it were sport for you*, Homl. Ass. 6, 141. (6) marking time, (a) marking a point of time at which something takes place :—Etan tô middes dæges (*meridie*), Gen. 43, 16 : Ps. Th. 36, 6 :

Btwk. 216, 14. Tô middes mergenes, Lchdm. ii. 116, 7. Tô undernes, 194, 5. Tô nônes, 290, 7. Tô hwilces tíman, Homl. Th. i. 78, 18. Gif preóst tô rihtes tíman crisman ne feccé, L. N. P. L. 9; Th. ii. 290, 3. Tô ðises *now*, Jn. Skt. Lind. 2, 10. (b) marking a limit, *to, up to, until, till* :—Wæs hit ðá án tíd tô ǽfenes *it then was an hour to evening*, Nar. 13, 6. Tô ǽfenes *usque ad vesperam*, L. Ecg. C. 4; Th. ii. 138, 1 : Bd. 3, 23; S. 554, 32. Ðæt hí fæston tô nônes (*ad nonam usque horam*), 3, 5; S. 527, 9. (c) marking a space of time in the course of which something takes place, *at, in, on* :—Hí ǽton ǽne on dæg, and ðæt wæs tô ǽfennes, Bt. 15; Fox 48, 8. Ðæt mon hiora tíd boega geuueorðie tô ánes dæges tô Osuulfes tíde *that the anniversary of them both be celebrated on the same day, on Oswulf's anniversary*, Chart. Th. 460, 6. **III.** with acc. (1) marking direction or motion (lit. and fig.) :—Hê leát tô ðæs cáseres eáre, Homl. Th. i. 376, 28. Tô ða ríðe, ðon andlang ríðe, eft on sǽ, Cod. Dip. Kmbl. iii. 12, 21. Ða ðe hweorfan sceoldan tô ðis enge lond, Exon. Th. 3, 6; Cri. 32. Nô hý hine tô deáð déman môston, 135, 8; Gú. 521. (2) with the infinitive (cf. Gothic infinitive with *du*) with the same force as with the inflected infinitive :—Micel is tô secgan eall æfter orde, ðæt hê ádreág, Exon. Th. 134, 4; Gú. 502. Mǽl is mē tô féran, Beo. Th. 637; B. 316. Áfýsed biþ ágenne eard tô sécan, Exon. Th. 217, 5; Ph. 275. Hád tô hebban (hát tô hebbanne, Cd. Th. 236, 14; Dan. 321), 187, 27; Az. 37. Him sēlle þynceþ leahtras tô fremman, 266, 34; Jul. 408. Ne bisorgaþ hê synne tô fremman, 95, 13; Cri. 1556. Ðá ongan hê tô cweðan *coepit dicere*, Mk. Skt. 13, 5. Hê onsende worn ðæs werudes west tô féran, Cd. Th. 220, 25; Dan. 76. Hê tiolaþ ungelíc tô bión (biónne, Cott. MS.) ðam ôþrum, Bt. 39, 12; Fox 232, 7. Gié soecas mec tô cwella (cwellanne, Rush.) *quaeritis me interficere*, Jn. Skt. Lind. 8, 40. Hê sende ðegnas his tô geceiga (cēgan, Rush.) hiá sié gehlaðad *misit servos suos vocare invitatos*, Mt. Kmbl. Lind. 22. 3. (3) marking time :— Tô dæg *hodie*, Ps. Th. 2, 7 : Hy. 7, 76. Tô ǽfen *vespere*, tô morgen *mane*, Ex. 16, 12 : Cd. Th. 147, 12; Gen. 2438. **IV.** with instrumental, (1) marking end or purpose :—Hê com tô ðí ðæt hê wolde synna forgifan, Homl. ii. 226, 9. *See se*, **V.** Tô hwí *why*, Mt. Kmbl. 8, 26: 9, 4: 26, 65: Homl. ii. 134, 9. (2) marking end of extent (time) :—Næs lang tô ðý ðæt his broþor ðyses lǽnan lífes tíman geendode, Lchdm. iii. 434, 25. **V.** used adverbially, where a noun governed by the preposition might be supplied from the context, (1) where motion is expressed or implied :—Of ðære sôþan gesǽlþe cumaþ eall ða ôþre gôd, and eft tô, Bt. 34, 6; Fox 140, 17 : 25; Fox 88, 29 : 37, 2; Fox 188, 12. Gif twégen men fundiaþ tô ánre stôwe and habbaþ emnmicelne willan tô tô cumenne, 36, 4; Fox 178, 10. Lá leóf, hê is deád; gang tô and árær hine, Homl. Th. ii. 182, 10 : Beo. Th. 5290; B. 2648. Ðá férdon hí tô, Homl. Skt. ii. 30, 149. Seó eá on emtwá tôeode . . ., and seó eá eft tô arn, Homl. Th. ii. 212, 24. Hê tô forð gestôp dracan heáfde neáh, Beo. Th. 4568; B. 2289 : Byrht. Th. 136, 13; By. 150. Hê sende hys here tô *missis exercitibus suis*, Mt. Kmbl. 22, 7. Hê tô somnaþ ða ðe út gewitan, Ps. Th. 146, 2. Tô ná genealǽc *ne accesseris*, Scint. 65, 15. Tô lǽtan *to admit*, Past. 45; Swt. 337, 16. Wê tilien, ðæt wē tô môten, Exon. Th. 313, 5; Seef. 119. Tô sculon clǽne tô that place shall the pure go, 450, 26; Dôm. 93. Hine se cyning tô gelaþode, Bd. 5, 19; S. 640, 8. (2) with verbs of placing (lit. or fig.), adding, etc. :—Ða ilcan studu tô gesette tô trymnesse, Bd. 3, 17; S. 544, 22. Sume ic tô ýcte, pref. ; S. 472, 30. Tô ætýcean *superaddere*, 4, 30; S. 609, 33 : 1, 27; S. 490, 22. Be ðám wítan ðe witan tô lēdan, L. E. G. 5; Th. i. 168, 27 : Chart. Th. 370, 15. Swá hwæt swá ðú mǽre tô gedêst, Lk. Skt. 10, 35. (3) where position is marked :—Hú hié mid hiera wætrum tô licgeaþ *how they with their waters lie to one another*, Ors. 1, 1; Swt. 10, 5. (4) where direction is marked :—Ðǽr hý tô sēgun, Exon. Th. 31, 14; Cri. 495: Cd. Th. 232, 5; Dan. 255. Ðú úre unriht ásettest ðǽr ðú sylfa tô eágum lôcadest *posuisti iniquitates nostras in conspectu tuo*, Ps. Th. 89, 8. Wê beótiaþ tô, Blickl. Homl. 33, 27. (5) *in addition, besides, too* :—Ða styriendan nētenu habbaþ eall ðæt ða unstyriendan habbaþ, and eác mǽre tô, Bt. 41, 5; Fox 252, 26. Manegu ôþru gôd tô eác ðam *many other goods too in addition to those*, 34, 6; Fox 140, 32. Hæfde hê nigon hund wintra and hundseofontig tô, Cd. Th. 74, 18; Gen. 1224. Ne bæd hê nôt ðæt hê hiene mid ealle fortýnde mid gehále wáge, ac hê bæd dura tô (*he asked for a door to the wall*), Past. 38; Swt. 274, 23. **VI.** adverb, with adjectives or adverbs, *too* :—Hí sellaþ wið tô lytlum weorðe *they sell for too small a price*, Past. 59; Swt. 449, 14. Of tô micelre fylle, Lchdm. ii. 60, 19. Tô manega of ðam folce, Num. 25, 1. Wæs ðæt wíte tô strang, Cd. Th. 109, 8; Gen. 1819. Ðone ðe tô micelne andan hæfþ, ðú scealt hátan leó . . . ; and ðone sǽnan ðe biþ tô sláw, ðú scealt hátan assa, Bt. 37, 4; Fox 192, 18–20. Ðý læs hí hí tô up áhæbben, Bt. 39, 11; Fox 228, 23: Past. 13; Swt. 79, 17: 65; Swt. 461, 28. Ða untruman môd mon ne scyle tô heálíce lǽran, 63; Swt. 459, 4. Ðú hæfst ðara wǽpna tô hraþe forgiten, Bt. 3, 1; Fox 4, 21. Ne fare gê tô feorr, Ex. 8, 28: 19, 12. Ðæt man môte tô forð æfter luste libban and gýman ne ðurfe ná oferlíce swýðe ðæs ðe béc beódaþ *that living as a man pleases may be carried too far, and over much heed need not be taken of what books*

bid, Wulfst. 55, 17. [O. Frs. O. Sax. tō: Du. toe: O. H. Ger. zuo: Ger. zu.] v. hēr-, in-, þǣr-tō.

tō-, a prefix denoting separation, division, like Latin dis-, di-. [It occurs as late as the Authorized Version, in Jud. 9, 53, to brake. Cf. Goth. twis-: O. Frs. tō-, te-, ti-: O. Sax. te-, ti-: O. H. Ger. za-, zi-; zar-, zir-: Ger. zer-.]

tō-ǣtican to increase:—Swelce eác tōǣtēcte disse gedrēfnisse storm Sǣberhtes deáþ auxit autem procellam hujusce perturbationis etiam mors Sabercti, Bd. 2, 5; S. 507, 6. v. next word.

tō-ǣticness, e; f. An increase, augmentation:—Tōǣtýcnys augmentum, Bd. 3, 22; S. 553, 14. v. tō-īcness.

tōan (?), tōian (?). tōgian (?) to grow tough:—Tōadan lentescunt, Wrt. Voc. ii. 52, 57: 92, 77. v. tōh, tēan.

tō-bǣd (?-blǣd. v. tō-blǣdan) elevated, exalted:—Tōbǣdne i geuferodne elevatum, Ps. Lamb. 36, 35. Heó wyrd glǣdlíce on hyre heortan tobǣd, Anglia viii. 324, 16.

tō-beátan; p. -beót To beat to pieces, destroy by beating:—Hig gebundon đone bysceop be đǣm fótum on sumne fearr and đone gegremedon, đæt hē hleóp on unsmēđe eorđan and đam bysceope đæt heáfod tōbeót, Shrn. 152, 2. Com him swilc wind ongeán, swilce nán mann ær ne gemunde, and đa scipo ealle tōbeót, Chr. 1009; Erl. 142, 5. Scipia hēt ǣlcne hiéwestān tōbeátan omni murali lapide in pulverem comminuto, Ors. 4, 13; Swt. 212, 10. [Ure men hī tobetet they knock our men about, Laym. 3308. We tobeot his cheoken, A. R. 106, 24. Euer eosh man me tobeteþ, and hwanne heo habbeþ me ofslaʒe, O. and N. 1610.]

tō-beótiende. v. beótian.

tō-beran; p. -bær, pl. -bǣron; pp. -boren. I. trans. To carry off in different directions, carry off:—Hī tredaþ đec and tergaþ, tōberaþ đec blódgum lāstum (thy body will be torn to pieces), Exon. Th. 119. 25; Gū. 260. Đæt sǣd đe feóll be đam wege . . . wegfērende hit fortrǣdon, and fugelas tōbǣron (birds carried it off in all directions), Homl. Th. ii. 90, 15. Lǣtan hī his líchaman licgan bútan đǣre ceastre and woldon đæt hine fuglas tōbǣron, Shrn. 32, 6. Ealle đa líchoman đe wildeór ābiton, oþþe fuglas tōbǣron, oþþe fixas tōslitan, Blickl. Homl. 95, 16. Sýn his bearn tōboren wíde may his children be scattered far and wide; commoti amoveantur filii ejus, Ps. Th. 108, 10. [As he me in his fete tobere, Chauc. H. of F. ii. 60.] II. intrans. To move in different directions. separate:—Sió wund wile tōberan gif hió ne biþ gewriđen the edges of the wound will get further apart, if the wound is not bound up, Past. 17; Swt. 123, 15. v. next word.

tō-berenness, e; f. Difference; differentia, Wrt. Voc. ii. 28, 42.

tō-berstan; p. -bærst, pl. -burston; pp. -borsten. I. to burst asunder, to break (intrans). in two, or in pieces, be rent asunder:—Ic tōberste crepo, Ælfc. Gr. 24; Zup. 138, 5. Se heofon tōbyrst from đǣm eastdǣle óþ đone westdǣl, Blickl. Homl. 93, 22. Tōbirsteþ, Exon. Th. 420, 7; Rä. 39, 7. Se sceaft tōbǣrst the shaft was shivered, Byrht. Th. 135, 51; By. 136. Seó byrne tōbǣrst the corslet was rent, 135, 66; By. 144. Sum man feóll on ise đæt his earm tōbǣrst his arm was broken, Homl. Skt. ii. 26, 34. Seó eorþe tōbǣrst and đonan up wæs biernende fýr wið đæs hefones hiatu terrae flamma prorupit, Ors. 5, 10; Swt. 234, 7. Hē eode tō đǣre burge wealle, and fleáh út ofer, đæt hē eall tōbǣrst, 5, 12; Swt. 244, 3. Hē gefeól on đone stocc and tōbǣrst on feówer dǣlas, Blickl. Homl. 189, 13. Ān hridder tōbǣrst on emtwā, Homl. Th. ii. 154, 16. Stānas tōburston petrae scissae sunt, Mt. Kmbl. 27, 51. Đa scittelsas tōburston, Homl. Skt. i. 3, 348. Tōborstenum bendum ruptis vinculis, Lk. Skt. 8, 29. I a. to break out in sores. v. tō-borstenness:—Wið springas and wið tōborsten líc for carbuncles and for a body with breakings out, Lchdm. i. 272, 18. His líchama barn wiđútan mid langsumere hǣtan, and hē eal innan samod forswæled wæs and tōborsten, Homl. Th. i. 86, 5. II. to break out:—Tōberstaþ erumpunt, Wrt. Voc. ii. 144, 8. [His brest tobrosten, Chauc. Kn. T. 1833. O. Sax. tebrestan: O. H. Ger. zar-brestan crepare. discrepare: Ger. zer-bersten.]

tō-berstung, e; f. Bursting:—Đæs geswelles tōberstung, Lchdm. ii. 198, 10.

tō-bígende decrepit:—Tōbígende decrepito, Wrt. Voc. ii. 26, 26: 70, 4.

tō-blǣdan; p. de To inflate, puff up:—Sóđ lufu nā byþ tōblǣdd caritas non inflatur, Scint. 82, 10. v. next word.

tō-blāwan; p. -bleów; pp. -blāwen. I. to blow in different directions, scatter by blowing, blow away:—Hī beóþ duste gelícran đonne þǣt wind tōblǣwþ tamquam pulvis, quem projecit ventus a facie terrae, Ps. Th. 1, 5. Tōdrifen mid winde, swā weorþaþ axe giond eorþan eall tōblāwen, Met. 20, 106. On đam (helle) fýre gē beóþ tōblāwene, Homl. Skt. i. 7, 139. II. to inflate, puff up, distend with wind, swell, (a) lit. v. next word:—Gif se maga biþ tōblāwen, Lchdm. iii. 58, 13. [Himm wǣrenn fet and þeos tobollenn and toblawenn, Orm. 8080.] (b) fig. to cause the breast to swell with emotion:—Tōblāwen (superbie tumore) inflatus, Anglia xiii. 441, 1084: Hpt. Gl. 423, 23. Murcnungum tōblāwene quèstibus inflati, 421, 11. Tōblāwene mid módignysse, Scint. 84, 19: R. Ben. 124, 6. [Mid a lutel wind of a word toblowen and tobollen, A. R. 122, 16.]

tō-blāwenness, e; f. Inflation, distension:—Ungelýfendlíc tōblāwennys his innoð geswencte, Homl. Th. i. 86, 13.

tō-borstenness, e; f. A breaking out, abscess:—Hý đǣra innođa tōđuundennysse and tōborstennysse (ῥήγματα) gehǣleþ, Lchdm. i. 322, 22. v. tō-berstan, I a.

tō-brǣdan; p. de. I. to make broad, enlarge, extend, make great in size or number, (a) of material objects:—Hig tōbrǣdaþ hyra healsbēc dilatant philacteria sua, Mt. Kmbl. 23, 5. (b) of non-material objects, to make great, magnify, multiply, increase, improve the condition of a person:—Đeáh heora sý mycle mā đonne úre, þeáh đū ūs tōbrǣdest ongeán hý, and wið hī gefridast, Ps. Th. 11, 9. Đū tōbrǣdest heorte míne dilatasti cor meum, Ps. Spl. 118, 32. Tōbrēt dilatat, Kent. Gl. 648. Đū nā tōbrǣddest fýnd míne ofer mē, Ps. Spl. 29, 1: 4, 1. Đū đin sóđfæst weorc tōbrǣddest multiplicasti justitiam tuam, Ps. Th. 70, 20. Đa earfoðu mínre heortan synd swýđe tōbrǣd (dilatatae), 24, 15. II. to expand, extend, spread out, open wide, distend:—Gif đū đines scipes segl ongeán đone wind tōbrǣdest, Bt. 18, 32. Mid hū miclum gódum willan Dryhten tōbrēt (expandit) đone greádan his mildheortnesse ongén đa đe tō him gecierraþ, Past. 52; Swt. 405, 9. Hē tōbrǣdde (expandit) his feđeru, Deut. 32, 11. Tōbrǣd đíne handa swilce (đū) sceat āstrecce, Techm. ii. 122, 24. Tōbrǣd mūđ đín open thy mouth wide (A. V.), Ps. Spl. 80, 9. Āpened, tōbrǣd distenta, i. extenta, tōbrǣde destentat, Wrt. Voc. ii. 141, 22, 23. Tōbrǣddum apertis, 5, 15. Wē sǣton bócum tōbrǣddon, Salm. Kmbl. 863; Sal. 431. III. to extend, spread abroad, diffuse:—Đeós wyrt wið đa eorđan hyre telgran tōbrǣdeþ, Lchdm. i. 324, 3. Tō hwon wilnige gē, đæt gē eówerne naman tōbrǣdan ofer đone teóþan dǣl? Bt. 18, 1; Fox 62, 25. Đonne mæg hine scamian đǣre brǣdinge his hlísan for đam hē hine ne mæg furþum tōbrǣdan (tōbrēdan, Met. 10, 15) ofer đa nearwan eorþan āne brevem replere non valentis ambitum pudebit aucti nominis, 19; Fox 68, 25. His naman tōbrǣdan geond ealle eorþan, 30, 1; Fox 108, 12. God hafaþ his gemynd on heofonum and on eorđan tōbrǣd, Chr. 979; Erl. 129, 18. Binnan đǣm feówer hyrnum đises middangeardes is tōbrǣdd Godes folc sancta ecclesia per quatuor mundi partes dilatata tenditur, Past. 22; Swt. 171, 4. Tōbrǣdde diffusa, i. sparsa, dispersa, Wrt. Voc. ii. 140, 16. III a. intrans.:—Of đyson eahta deófles crǣftan ealle unþeáwas up āspringaþ and syđđan tōbrǣdaþ ealles tō wíde, Wulfst. 68, 17. [O. H. Ger. ze-breiten.]

tō-brǣdedness, e; f. Extent, an extensive place:—On tōbrǣdednesse i on brādnesse in latitudine, Ps. Lamb. 117, 5. On tōbrǣdednesse in latitudinem, 17, 20: Ps. Spl. 17, 22.

tō-brǣdness, e; f. Extent, breadth:—On tōbrǣdnysse in latitudine, Ps. Spl. 117, 5.

tō-brecan; p. -bræc, pl. -brǣcon; pp. -brocen To break, break in pieces:—Ic tōbrece frango, Ælfc. Gr. 28, 6; Zup. 176, 8: rumpo, 177, 4. Tōbrocen contrita, Hpt. Gl. 482, 67. I. in reference to material objects, to break in two, to break to pieces, break up, to separate into parts by striking or pulling:—Hē (the patch of new cloth) tōbrycþ hys stede on đam reáfe, and se slite byþ đe wyrsa, Mt. Kmbl. 9, 16. Đū mē tōbrǣce (disrupisti) bendas grimme, Ps. Th. 115, 7. Hē đone hlāf tōbræc on twā, Blickl. Homl. 181, 16. Đa ǣrenan scyttelas hē ealle tōbrǣc, 85, 7. Hē tōbrǣc hire (the lion's) ceaflas mid his barum handum, Ælfc. T. Grn. 7, 16. Hī đa gymstānas tōbrǣcon, Homl. Th. i. 60, 28. Hié đa scipu eall oþþe tōbrǣcon oþþe forbǣrndon, Chr. 894; Erl. 91, 25. Tōbrec hira scîncnyssa confringes statuas eorum, Ex. 23, 24: Lchdm. i. 370, 22. Tōbrec đinne hlāf and syle đone óđerne dǣl hungrium men break thy loaf in two and give one part to a hungry man, Homl. Th. i. 180, 4. Man sceolde tōbrecan his stef, Chr. 1047; Erl. 177, 7. Đa wildan hors scealden iornan and him đa time eall hī tōbrecan, Shrn. 72, 2. Tō gehwylcum bryce, hundes brægen āléd on wulle and đæt tōbrocene tō gewriþen, Lchdm. i. 370, 19. Wiþ ealdre wunde tōbrocenre, ii. 92, 1. Tōbrocen wērun sconco hiora frangeruntur eorum crura, Jn. Skt. Rush. 19, 31. Đa bytta beóþ tōbrocene rumpuntur utres, Mt. Kmbl. 9, 17. Heora scipu sume þurh oferweder wurdon tōbrocene, Chr. 794; Erl. 59, 22. II. to overthrow, break down, ruin, destroy, put into confusion, rout, (a) of material objects:—Ceaster heora đa tōbrǣc (destruxisti), Ps. Spl. 9, 6. Se đe tōbrǣc (destruebat) đone tempel Godes, Mt. Kmbl. Lind. 27, 40. Hyra setlu hē tōbrǣc (evertit), Mt. Kmbl. 21, 12: Mk. 11, 15. Wutun tiligean đæt wē heora burh tōbrecan móton accipient in vanitate civitates tuas, Ps. Th. 138, 17. Đæs ne wēndon witan Scyldinga, đæt hit (the hall) manna ǣnig tōbrecan meahte, Beo. Th. 1565; B. 780. Wæs đæt beorhte bold tōbrocen swíđe, 1999; B. 997. Weard folc tōtwǣmed, scyldburh tōbrocen, Byrht. Th. 138, 57; By. 242. Āne tōbrocene byrgenne sepulchrum dirutum, Ors. 4, 10; Swt. 202, 4. Eal đín carcern hē hafaþ tōbrocen, Blickl. Homl. 85, 22. Hreósaþ tōbrocene burgweallas, Exon. Th. 61, 1; Cri. 978. (b) of persons, to destroy, crush:—Ic tōbrǣce hí confringam eos, Ps. Spl. 17, 40. Đū hie tōbrǣce attrivisti eos, Past. 37; Swt. 267, 3. (c) of non-material objects:—Hit eallum đǣm senatum ofþyncendum đæt hē heora ealdan gesetnessa tōbrecan wolde (would overthrow their old laws), Ors. 5, 12; Swt. 244, 17. Đonne biþ se glencg āgoten and se þrym tōbrocen, Wulfst. 263, 8. Hit ongeat đæs wísdómes

lāre swíþe tótorenne and swíþe tóbrocenne, Bt. 3. 1; Fox 4, 31. **III.** *to take by assault :*—Tirus hē besæt and siþþan tóbræc and mid ealle tówearp *Tyrum oppressit et cepit*, Ors. 3, 9; Swt. 126, 17. Ða gigantas woldon tóbrecan ðone heofon *lacessentes coelum gigantes*, Bt. 35, 4; Fox 162, 12. Hēr wæs tóbrocen Rómána burh fram Gotum, Chr. 409; Erl. 11, 10. On ðissum geáre wæs Bæbbanburh tóbrocon, 993; Erl. 133, 1. **IV.** *to break* a promise, pledge, etc., *to infringe, violate :*—Swá hwá swá halt ðis write . . . hwá swá hit tóbreceþ, Chr. 675; Erl. 38, 27. Man his riht tóbræc, 975; Erl. 126, 17. Twēgen gebróðra tóbræcon ðone regol, Homl. Th. ii. 166, 34. Gif hē his bebod tóbrǽce, Homl. Ass. 60, 217. Wed synd tóbrocene oft and gelóme, Wulfst. 161, 12. **V.** *to break, interrupt :*—Wē tóbrecaþ úrne slǽp and gebiddaþ for eów, Homl. Ass. 51, 39. [The verb remains in the Authorized Version 'all to-brake his scull,' Jud. 9, 53. *O. Frs.* tó-breka (te-) : *O. L. Ger.* te-brekan : *O. H. Ger.* ze-brechen *disrumpere, confringere : Ger.* zer-brechen.] v. tó-brocen, un-tóbrocen.

tó-brédan, Met. 10, 15. v. tó-brǽdan, III.

tó-bregdan, -brédan ; *p.* -brægd, -brǽd, *pl.* -brugdon, -brúdon (-brudon?) ; *pp.* -brogden, -bróden (-broden? *in O. and N.* tobrode *rimes with* unsode). **I.** *to separate* (trans.) *by a quick movement.* (a) *to pull to pieces* (lit. and fig.) :—Hú ǽnig mæg gangan in húse stronges and fæst his tóbregdan (*diripere*), nymþe ǽr gebindaþ se stronge and ðonne hús his tóbrægdeþ (*diripiat*), Mt. Kmbl. Rush. 12, 29. Hē tóbrǽd (*dilaceravit*) áne león tó sticcum, Jud. 14, 6. Metod tóbrǽd monna sprǽce *the Lord destroyed the unity of human speech*, Cd. Th. 102, 5; Gen. 1695. Hié tóbrugdon blódigum ceaflum fira flǽschoman, Andr. Kmbl. 317; An. 159. Þrié wulfas ánes deádes monnes líchoman styccemǽlum tóbrúdon (*cadaver sparsum membratim reliquerunt*), Ors. 4, 2; Swt. 160, 21. Ða nicoras tóbrúdon hié, Nar. 11, 11. Hit ongeat his lāre swíþe tótorene and swíþe tóbrogdene, Bt. 3, 1; Fox 4, 31 note. Biþ se glenct ágoten and se þrym tóbróden, Vald. 263, 8 note. Ðæt hē wǽre from ðám hundum tóbróden, Shrn. 145, 4. (b) *to pull apart :*—Heora lima man ealle tóbrǽd ǽlc fram óðrum *their limbs were torn from one another*, Homl. Skt. i. 23, 72. Ða tóbrǽd Samson bēgen his earmas *Samson wrenched his arms apart*, Jud. 15, 14. **II.** *to separate* (intrans.) *by a quick movement, to break off, start* from sleep, cf. *Icel.* bregða svefni *to awake :*—Slǽpe tóbrægd folces weard, Cd. Th. 161, 15; Gen. 2665. Mid ðý heó ðá ðý slǽpe tóbrǽd *somno excussa*, Bd. 4, 23; S. 596, 5. Slǽpe tóbrugdon searuhæbbende, Andr. Kmbl. 3053; An. 1529. Ic gefrægn hǽleð slǽpe tóbrédan (-on, MS.), Judth. Thw. 25, 7; Jud. 247. **III.** *to separate by making a quick movement* with something (?) :—Oft hý wordum tóweorpaþ ær hý bacum tóbréden (*before they part and turn their backs on one another*, (?) cf. *Icel.* bregða hendi, fótum, etc.), Exon. Th. 345, 20; Gn. Ex. 192. [Hi eteþ flesch unsode swich wulves hadde hit tobrode, O. and N. 1008. The fend him tobrayd *illum daemonium dissipavit*, Wick. Lk. 9, 42. He tobraide his clothes, Gow. ii. 53, 11.]

tó-brítan ; *p.* te. **I.** *to break in pieces, crush, bruise* (lit. and fig.) :—Ic tóbrýte *tero*, Ælfc. Gr. 28, 1 ; Zup. 165, 14 : *confringo*, 28, 6 ; Zup. 176, 9. Ic tóbrýte hí *confringam eos*, Ps. Lamb. 17, 39. Ðú tóbrýtst hig *confringes eos*, 2, 9. Tóbrýt (*confringet*) Drihten cederbeám, Ps. Lamb. 28. 5 : 57, 7. Heó tóbrýt (*conteret*) ðín heáfod, Gen. 3, 15. Boga[n] tóbrýteþ, Ps. Spl. 45, 9. Teþ sinfulra ðú tóbríttest (*contrivisti*), 3, 7. Folc ðú tóbríttest (*confringes*), 55, 7. Gewít of ðære leásan anlícnysse, and tóbrýt hí eall and hire cræt samod, Homl. Th. ii. 496, 14. Tóbrýt (*contere*) earm ðæs synfullan, Ps. Lamb. 9 second, 15. Flǽsces tóbrýte (*terat*) módignesse, Hymn. Surt. 9, 22. Ne ús gedweld tóbrýte (*atterat*), 17, 24. Ðæt God úre helpe and tóbrýte ðisne here, Homl. Skt. ii. 25, 350. Ic bebeóde mínum þeówum þæt hí hí (*the idols*) ealle tóbrýton, i. 5, 236. Tóbrýtendes *confringentis*, Ps. Lamb. 28, 5. Tóbrýtendne (*conterentem*) deóful, Hymn. Surt. 115, 15. Boga heora biþ tóbrýt *arcus eorum confringatur*, Ps. Spl. 36, 16. Tóbrýt † tóbrocen *contrita, constricta*, Hpt. Gl. 482, 67. Tóbrýt *contritus*, 515, 5. Tóbrýttes *attritae, violatae*, 474, 75. Tóbrýtte † ofrorene *obruti, contriti*, 506, 6. Tóbréttum *quassatis, confractis*, 421, 39. **II.** *to crush with feelings of sorrow, to make contrite :*—Heortan ða tóbrýttan *cor contritum*, Ps. Lamb. 50, 19. Ða tóbrýtan on heortan *contritos corde*, 146, 3. [Corineus heom tobrutte ban and heora ribbes, Laym. 1602.]

tó-brítedness, e ; *f.* **I.** *a bruise, breach :*—Hē gewríð tóbrýtednyssa heora *alligat contritiones eorum*, Ps. Lamb. 146, 3 : 59, 4. **II.** *trouble, sorrow :*—Tóbrýtednys and ungesǽlignys *contritio et infelicitas*, 13, 3.

tóbrítend-líc ; *adj. Breakable :*—Ða tóbrýtendlícan *fragenda*, Wrt. Voc. ii. 150, 37.

tó-bríting, e ; *f. Crushing,* fig. *destruction :*—Tóbrýtincge forestæpþ ofermódignyss *contritionem precedit superbia*, Scint. 82, 12.

tó-brocen ; *adj.* (*ptcpl.*) *Suffering from eruptions :*—Wiþ innan tóbrocenum múðe, Lchdm. ii. 310, 19.

tóbrocen-líc ; *adj. Frail, perishable :*—Ðysse worulde wela is hwýlwendlíc and feallendlíc and tóbrocenlíc *the wealth of this world is transitory and decaying and frail*, Wulfst. 263, 13.

tó-brýsan *and* -brýsian ; *p.* de *To crush, break in pieces :*—Ic tóbrýse *tero*, Ælfc. Gr. 28, 1 ; Zup. 165, 14 MS. T. Ealle ðín bán ic tóbrýsige, Nar. 41, 20. Se ðe fylþ uppan ðysne stán hē byþ tóbrýsed (*confringetur*) ; and hē tóbrýsþ (*conteret*) ðone ðe hē onuppan fylþ, Mt. Kmbl. 21, 44. Ðú ealle míne bán tóbrísdest, Nar. 45, 5. Ealle his bán heó tóbrýsde, 44, 15. Tóbrýsiende *confringens*, Ps. Lamb. 28, 5. Gif hwá tóbrýsed sý *if any one be crushed* (*convulsus*), Lchdm. i. 122, 1. Tóbrýsede tigelan, Homl. Skt. i. 8, 169. [Ʒiff he wollde læpenn dun he nunnde tobrisenn all himm sellfenn, Orm. 12032. Al tobrised bac and þe, Havel. 1950. Tobrusede *brake in pieces*, Wick. (2 Kings 18, 4).]

tó-brýtan. v. tó-brítan.

tó-ceorfan ; *p.* -cearf, *pl.* -curfon ; *pp.* -corfen. **I.** *to cut to pieces, cut in two, cut up :*—Hē tócearf his basing on emtwá mid sexe, Homl. Th. ii. 500, 26. Hí tócurfon ðone líchaman on manugu sticceo, Shrn. 125, 10. Ða langnysse tóceorfan on pysena gelícnysse, Lchdm. i. 260, 15. Rammes lungen smæl tócorfen, 356, 21. Tócorfen *lacerata*, Wrt. Voc. ii. 53, 32. Þeáh ðe se beám beó tócoruen, H. R. 105, 15. **II.** *to cut off :*—Tócearf him ða éarelipprica *amputavit illi auricula*, Mk. Skt. Lind. 14, 47. [Til he wyste who couþe uche kyndam tokerve, Allit. Pms. 88, 1700. *O. Frs.* tó-kerva.]

tó-ceówan ; *p.* -ceáw, ol, -cuwon ; *pp.* -cowen To *chew to pieces, break up by chewing, masticate :*—Ðæt húsel biþ betwux tóðum tócowen, Homl. Th. ii. 270, 33. [Deoflen torendeð ham ant tocheoweð ham euch greot, O. E. Homl. i. 251, 12. Hit tocheoweð ant touret Godes milce, A. R. 202, 16.]

tó-cínan ; *p.* -cán, *pl.* -cinon ; *pp.* -cinen To *break* (intrans.) *into chinks, split, crack :*—Tócínit, tecínid *dehiscat*, Txts. 57, 653. Tócíneþ, Wrt. Voc. 25, 27 : *dehiscit*, 27, 15. Gif hit (*an egg*) ne tócíne, tósleah hwón *if it will not crack of itself, crack it slightly with a tap*, Lchdm. iii. 18, 2. Tócinan (-en ?) *rimosa*, Hpt. Gl. 529, 10. Gemette hē be wege sumne lícðrowere licgende eal tócínen (*the skin all cracked with the disease*), Homl. Th. i. 336, 9. [Hie drinkeð þat hie tochineð, O. E. Homl. ii. 199, 32. Þe stan tochan, i. 141, 17. Þæ heorte tochan (-chon, 2nd MS.), Laym. 21235. Þe roche tochon, Misc. 92, 77.]

tócir-hús ; *n. An inn ; diversorium* (*di-vertere* = tó-cirran *q. v.*) :—Tócirhús *diversorium*, Wrt. Voc. i. 38, 10.

tó-cirran ; *p.* de *To turn in different directions, to part :*—Æfter ðon ðe wit nú tócyrraþ and tógáne beóþ *postquam ab invicem digressi fuerimus*, Bd. 4, 29 ; S. 607, 20 MS. B. Hí mid mycelon unsehte tócyrdon *they parted on very bad terms*, Chr. 1094 ; Erl. 230, 6. Cf. tó-gán, -hweorfan.

tó-cleófan ; *p.* -cleáf, *pl.* -clufon ; *pp.* -clofen To *cleave asunder :*—Ic tócleófe (-clefe, MS. J.) *findo*, Ælfc. Gr. 28, 6 ; Zup. 178, 5. Ic tócláfe, Engl. Stud. xi, 65, 38. Gif ðú ǽnne stán tóclífst, ne wyrþ hē nǽfre gegaderod swá hē ǽr wæs, Bt. 34, 11 ; Fox 150, 26. Tóclýfþ *findit, i. rupit*, Wrt. Voc. ii. 148, 63. Ðonne God ðysne middangeard tócleófeþ, Blickl. Homl. 109, 35. Ða nýtenu synd clǽne ðe tócleófaþ heora cláwa, Homl. Skt. ii. 25, 55. Tócleáf *findit*, Wrt. Voc. ii. 37, 32. Se rēða kyning hine tócleáf on twá, Ælfc. T. Grn. 9, 21. Tócleófende *sulcans*, Wülck. Gl. 254, 21. Monnes cinbán gif hit biþ tóclofen, gesette mon .xii. scitt. tó bóte, L. Alf. pol. 50; Th. i. 94, 16. Ða sticcu ðæs tóclofenan hriddores, Homl. Th. ii. 154, 19. Oð ðone tóbrocenan beorg ðe ðær is tóclofen, Cod. Dip. Kmbl. ii. 251, 6. Æt ðam litlan tóclofenan beorge, iii. 421, 9. Tócleofenan, ii. 249, 26. [*In later English the word is used transitively and intransitively.* His ban tocluuen, Laym. 1920. Drihhtin toclæf þe sæ, Orm. 14798. He smot and toclef þat heued, R. Glouc. 186, 3. Mine herte shal tocleve, Chauc. T. and C. v. 613. Þe holi goste heuene shal tocleue, Piers P. 12, 141. Þe shell tooclef, Alis. (Skt.) 1009.] v. un-tóclofen.

tó-clifrian ; *p.* ode *To scratch* or *tear to pieces :*—Wæs tóclifrod *laniatur*, Germ. 398, 174. Hē unscrýdde hine ealne, and wylode hine sylfne on ðam þiccum brēmlum and þornum swá lange, ðæt hē eall tóclifrod árás, Homl. Th. ii. 156, 30.

tó-clipigend-líc ; *adj. Of address* or *appeal :*—O is tóclypigendlíc *abverbium*.

tó-clipung, e ; *f. Invocation, appeal :*—Ǽlc man biþ gefullod on naman ðære Hálgan Ðrynnysse and hē ne mót ná beón eft gefullod, ðæt ne sý forsewen ðære Hálgan Ðrynnysse tóclypung, Homl. Th. ii. 602, 3 : Homl. Skt. i. 12, 143 : Homl. Th. ii. 48, 15.

tó-cnáwan ; *p.* -cneów ; *pp.* -cnáwen To *discern, distinguish, know the difference between, understand :*—Tócnáweþ *discernit*, Blickl. Gl. Tócnáwen [beón] *dinosci, intellegi*, Wrt. Voc. ii. 140, 30. (1) with acc. :—Wē geseóþ þurh úre eágan and ealle ðing tócnáwaþ *by means of our eyes we see and distinguish all things*, Homl. Th. ii. 372, 27. Ðurh ða gesceádwísnesse wē tócnáwaþ good and yfel and gecósaþ ðæt gód and áweorpaþ ðæt yfel *per discretionem virtutes eligimus, delicta reprobamus*, Past. 11 ; Swt. 65, 22. Ða scearpþanclan witan ðe ðone twydǽledan wísdóm hlútorlíce tócnáwaþ, Lchdm. iii. 440, 29. Him is neód ðæt hē his ágene wódnesse tócnáwe *it is necessary for him to discern his own madness*, Homl. Th. ii. 110, 29. Cunne gē tócnáwan heofones hiw *faciem coeli dijudicare nostis*, Mt. Kmbl. 16, 3. Man mihte his líf tócnáwan *potuit ejus vita dinosci*, R. Ben. 108, 15 : Homl. Th. ii. 154, 25. Irre

oft ámirreþ monnes môd, ðæt hê ne mæg ðæt riht tôcnáwan, Prov. Kmbl. 28 : Homl. Th. i. 108, 23. Geseón and tôcnáwan ǽgðer ge gôd ge yfel *to see good and evil and know the difference between them*, 18, 4. Heora nân ne cúðe ôðres sprǽce tôcnáwan *not one of them could understand another's speech*, 318, 20. Heó ða môd ðe geopenaþ ðínra freónda and eác ðínra feónda, ðæt ðú hié miht swutele tôcnáwan . . . Mid hû micelan feó woldest ðú habban geboht, ðæt ðú swutole mihtest tôcnáwan ðíne frínd and ðíne fýnd, Bt. 20 ; Fox 72, 13–21. Hí cræftas and unþeáwas ne cunnon tôcnáwan *they cannot distinguish virtues and vices*, 36, 6 ; Fox 180, 30. Lǽcas cunnon ǽlces medtrumnesse ongitan and tôcnáwan *medicus aegritudinis modum dignoscit*, 39, 9 ; Fox 226, 17. Priscianus segþ ðæt man sceal tôcnáwan ǽlces dǽles mihte and getácnunge and swá undergytan hwæt hé sý ná be ðære declinunge *Priscian says, that we must distinguish the force and signification of each part of speech, and in this way, not by the declension, understand what it is*, Ælfc. Gr. 18 ; Zup. 111, 14. Nis nán ðing tôcnáwen on sôðre eáwfæstnesse, ðæt his láreówdôm ne gestaðelode, Homl. Th. i. 392, 18. (2) with acc. and appositive adjective :—Wé tôcnáwaþ his ríce and úre ríce ðǽr áwritene, ðær wé ær swilce be ôðrum mannum gereccednesse rǽddon *we discern his kingdom and our kingdom there described, where before we read the account as if about other men*, Homl. Th. ii. 64, 29. Ða tungelwítegan tôcneówon Crist sôðne mann *the astrologers discerned that Christ was really man*, i. 106, 33. (3) with a clause :—Gif wé gleáwlíce tôcnáwaþ, ðæt symmenda arc getácnode Godes geláðunge, Homl. Th. ii. 60, 2. On ðam múðe wé habbaþ swæcc, and tôcnáwaþ hwæðer hit biþ ðe wered ðe biter ðæt wé ðicgaþ, 372, 29. Ðæt ðeós menigu tôcnáwe, ðæt ðis hǽðengyld deófles biggeng is, i. 72, 3. Hú mihte Adam tôcnáwan hwæt hé wǽre, 14, 4. Tôcnáwan. ðæt ús is twyfeald neód, ii. 284, 23 : Lchdm. iii. 236, 10 : Homl. Ass. 107, 150. Ðus ðú miht tôcnáwan, hwænne nama cymþ of worde, hwænne word of naman, Ælfc. Gr. 36 ; Zup. 216, 5.

tô-cnáwenness, e ; *f. Knowledge, discernment, understanding, knowledge which appreciates the difference between things* :—Ne sind hí ðrý Godas . . . ac seó Ðrynnys is án sôð God . . . Ðeós tôcnáwennys is éce líf, Homl. Th. ii. 362, 32.

tô-cnyssan ; *p.* te ; *pp.* ed *To crush to pieces, smash, shatter* :—Ne forbrýte hê ná ðæt tôcnysede hreód (*arundinem quassatam*), R. Ben. 121, 6. [*O. H. Ger.* ze-cnussen *elidere*.]

tô-cumende ; *adj.* (*ptcpl.*) *Coming to a strange place, strange, foreign:* —Hé for Godes lufon eode tô reordum mid ðám tôcumendum mannum *for the love of God he took his meals with the strangers who came*, Shrn. 129, 27.

tô-cwæstednesse, e ; *f. Destruction* :—Geswác tôcwæstednys (-cwestedness, Ps. Lamb.) *cessavit quassatio*, Ps. Spl. 105, 29. [Cf. *Goth.* kwistjan *to destroy* ; kwisteins *destruction : Dan.* kvæste *to hurt*.]

tô-cweþan ; *p.* -cwæþ, *pl.* -cwǽdon ; *pp.* -cweden *To forbid, prohibit:* —Wé nellaþ secgan . . . for ðan ðe hyt tôcwǽdon ða wísan láreówas, and . . . ða hálgan bôceras forbudon tô secgenne, Homl. Ass. 24, 7. Tôcwedene *interdicta, prohibita*, Hpt. Gl. 421, 77. Wé forbeódaþ ordál and áðas (ordál and áðas ǽfre syndan tôcwedene, MS. B.) freólsdagum, Wulfst. 117, 14. Ordál and áðas and wífunga ǽfre sindan tôcwedene heáhfreólsdagum, L. Eth. vi. 25 ; Th. i. 320, 24 : v. 18 ; Th. i. 308, 24 : L. E. G. 9 ; Th. i. 172, 10. Cf. L. C. E. 17 ; Th. i. 370, 20.

tô-cwilman ; *p. de To afflict grievously, torment* :—Ða druncengeornan ná ðæt án ðæt hí on ðam tôweardan lífe mid écum tintregum tôcwylmede synt, ac eác hý synt on ðisum andweardan lífe mid mænigfealdum untrumnyssum gewǽhte, Homl. Ass. 146, 56.

tô-cwísan ; *p. de To shatter, break to pieces, crush, bruise* :—Ic tôcwýse *quasso*, Ælfc. Gr. 24 ; Zup. 137, 10. Ic tôcwýse *quatio*, tôcwýsde *quassi*, tôcwýsed *quassum*, 28, 4 ; Zup. 169, 6. Ofer ðare ðe hé fylþ hé tôcwýst (*comminuet*), Lk. Skt. 20, 18. Tôcwíesd, Ps. Lamb. 28, 6. Hê tôcwýseþ heáfdu *conquassabit capita*, 109, 6. Gimstánas tôcwýsan, Homl. Th. i. 60, 24. Ða wolde hê án eald hús tôcwýsan *he wanted to demolish an old house*, ii. 510, 12. Ætslád se hálga wer . . . swá ðæt hê forneán eal weard tôcwýsed, 512, 12. Sum cild bearn under ánum yrnendum hweóle and weard tô deáðe tôcwýsed, 26, 25 : 166, 20. Tôcwýsed hreód *arundinem quassatam*, Mt. Kmbl. 12, 20. Ðás gimstánas synd tôcwýsede for ýdelum gylpe, Homl. Th. i. 62, 6. ¶ The word seems used with a passive force in the following passage :—Feól se wáh uppan ðæs stuntan rǽdboran, þæt hê æll tôcwýsde and sum ôþer cniht samod, Homl. Skt. i. 8, 173.

tô-cwísedness, e ; *f. Crushed condition* :—Iohannes gegaderode ðæra gymstána bricas . . . Ða fǽrlíce wurdon ða gymstánas swá ansunde, ðæt furðon nán tácen ðære ǽrran tôcwýsednesse næs gesewen *that not even a trace of their having been crushed was visible*, Homl. Th. i. 62, 16.

tô-cyme, es ; *m. A coming to a place, coming, approach, arrival, advent* :—Uncer efenþeówa uncet sceolde út álǽdan, and uncer hláford ábád uncres tôcymes, Homl. Ass. 206, 385. Ǽr ðære tíde his (*an attack of convulsions*) tôcymes, Lchdm. i. 364, 16. Ǽr Antecristes tôcyme, Wulfst. 156, 7. Foran tô ðon tôcyme dômes dæges, Blickl. Homl. 35, 8. For Drihtnes cynedômes tôcyme, 87, 5. Deáþ mid his dígelan tôcyme, Homl. Ass. 54, 98. Gif se hírédes ealdor wiste ðæs ðeófes tôcyme, 54, 100. Ðone tôcyme ðæs Hálgan Gástes, Blickl. Homl. 131, 12. Folc sceal

gefeón on ðone his (*John the Baptist*) tôcyme, 167, 14. Se mæssepreóst, ðe se bisceop tô fundode . . . wyste his tôcyme, Homl. Skt. i. 3, 471. Hû hwósta missenlíce on níon becume . . . Se hwósta hæfþ manigfealdne tôcyme, Lchdm. ii. 56, 15. ¶ (1) *the coming of Christ to the world, the first or second Advent* :—Drihtnes tôcyme is his menniscnys. Hê com tô ús ðá ðá hé genam úre gecynd tô his Ælmihtigan Godcundnysse, Homl. Th. i. 600, 4. Swá byþ mannes Suna tôcyme, Mt. Kmbl. 24, 27, 37, 39. Hwilc tácn sí ðínes tôcymes, 24, 3. Ðes middangeard ðe hê mid his tôcyme fram synnum gehǽlde, Homl. Ass. 47, 561. Ǽr Cristes tôcyme, Blickl. Homl. 81, 27. Ða ðe Cristes tôcyme wiston, 81, 10. Ða hálgan wítegan wítegodon ǽgðer ge ðone ǽrran tôcyme on ðære ácennednysse, and eác ðone æftran æt ðam micclum dôme, Homl. Th. i. 600, 23. (2) *the anniversary of Christ's coming, Advent* :—Ðeós tíd ôð midne winter is gecweden *Adventus Domini*, ðæt is Drihtnes tôcyme, Homl. Th. i. 600, 4. Ðú scealt healdan ðone tôkyme mid ealre árwurðnesse, Lchdm. iii. 226, 7. [Efter Cristes tocyme, O. E. Homl. i. 89, 11.] v. hider-tôcyme.

tô-dǽl. v. tô-dál.

tô-dǽlan ; *p. de To divide, separate, distribute.* I. in the following glosses :—Ic tôdǽle *infindo*, Engl. Stud. xi. 66, 49 : *disclúdo*, Wrt. Voc. i. 39, 31. Ic tôdǽle *dispono*, ii. 141, 45. Tôdǽlan *findere*, 37, 33. Tôdǽlende *discrepantes*, 25, 60 : *dirimentes*, 28, 52 : *diremtas*, 28, 32 : 27, 48. Tôdǽled is *dispertitus est*, 26, 35 : 73, 26. Sient tôdǽlede *dirimuntur*, 28, 53. Tôdǽlede *discretas*, 28, 33. Tôdǽldum *dilotis*, 25, 49. Tôdaeldum, 106, 36. Tôscirid ł tôdǽled *summotum*, Hpt. Gl. 528, 12. II. *to divide* a whole into parts, (1) *trans.* :—Hê tôdǽleþ hyne *he shall cut him asunder* (A. V.) : dividet eum, Mt. Kmbl. 24, 51. Ðone tôdǽlaþ hí his feoh on fíf oððe syx, Ors. 1, 1 ; Swt. 20, 27. Ðone ánne noman (woruld) ðú tôdǽldest on feówer gesceafta, Bt. 33, 4 ; Fox 128, 28. Hê Reádne Sǽ tôdǽlde *qui divisit Mare Rubrum in divisiones*, Ps. Th. 135, 13. Hié heora here on tú tôdǽldon *agmine diviso in duas partes*, Ors. 1, 10 ; Swt. 46, 16. Stánas bióþ earfoþe tô tôdǽlenne, Bt. 34, 11 ; Fox 150, 24. On twá tedǽled ys intinga tô syngienne *bipertita est causa peccandi*, Scint. 140, 13. Tôdǽldu wæteru *divisas aquas*, Past. 53 ; Swt. 413. 27. (2) *intrans.* :—Hér tôdǽlde se foresprecena here on tú, Chr. 885 ; Erl. 82, 19. ¶ figuratively, *to destroy unity, make dissension in.* v. tô-dǽl, VIII :—Ǽlc ríce on hyt sylf tôdǽled byþ tôworpen. Gyf Satanas is tôdǽled on hine sylfne, hú stent his ríce ? Lk. Skt. 11, 17, 18. II a. *to divide* a whole by assigning the limits of the different parts :—Iosue ðone eard gewann and ealne tôdǽlde, Ælfc. T. Grn. 6, 8. Philippus and Herodes tôdǽldon Lysiam, and Iudéam feówrícum tôdǽldun, Chr. 12 ; Erl. 6, 4. Ða wearþ ðæt ríce tôdǽled on .v., 887 ; Erl. 86, 1 : 709 ; Erl. 42, 29. II b. *to divide* one number by another:—Tôdǽl ða twelf þurh fíf, Anglia viii. 328, 21 : 304, 40. III. *to divide* one thing from another, *part, separate,* (a) *trans.* :—Ðonne se líchama and seó sáwul hí tôdǽlaþ, Guthl. 20 ; Gdwin. 84, 13. Ðonne se earma líchama and seó wérige sáwul hí tôtwǽmaþ and tôdǽlaþ, Wulfst. 151, 11. Ongunnon ðæt monnes mágas hycgan, ðæt hý tôdǽldon unc, Exon. Th. 442, 14 ; Kl. 12. Hí ne mágon beón tôgædere genemnede, ac hí ne beóþ nǽfre tôdǽlede, Homl. Th. ii. 204, 28. (b) *intrans.* :—Swá tôdǽleþ se líchoma and seó sáwul, Wulfst. 149, 8. Nǽfre leófe ne tôdǽlað se láðe ne gemétaþ, 190, 2. IV. *to scatter, disperse* :—Drihten hig tôdǽlde of ðære stôwe geond ealle eorðan *the Lord scattered them abroad from thence upon the face of all the earth* (A. V.), Gen. 11, 8. Hê tôdǽlde ofermôdan *dispersit superbos*, Lk. Skt. 1, 51. Tôdael hié *dispertire eos*, Ps. Surt. 16, 14. Tôdǽlan heora geðeóde geond ðás woruld wíde, Ps. Th. 54, 8. Ealle his geféran ðurh ôþre stôwe tôdǽlede wǽron *omnes socii per alia essent loca dispersi*, Bd. 3, 27 ; S. 558, 37 : Gen. 10, 32. Wǽron tôdǽlede *dispargerentur*, Hpt. Gl. 518, 2. V. *to destroy* :—Ealle ðú his weallas wíde tôdǽldest *destruxisti omnes macerias ejus*, Ps. Th. 88, 33. Ne tôdǽldun (hí) ðeóde *non disperdiderunt gentes*, Ps. Surt. 105, 34. VI. *to distribute, give away* parts of a whole. v. tô-dál, VI :—Ic tôdǽle (*do*, Lk. 19, 8) healfne dǽl mínra gôda ðearfum, Homl. Th. i. 582, 2. Hê tôdǽlþ his herereáf *spolia ejus distribuit*, Lk. Skt. 11, 22. Sume ealle hyra þearfum Godes tôdǽlaþ *quidam omnia sua pauperibus Dei distribuunt*, Scint. 58, 12. Tôdǽlan werum tô wiste fǽges flǽschoman, Andr. Kmbl. 303 ; An. 152. Tôdǽlendes *distribuentis* (*dona*), Kent. Gl. 673. Hí wǽron tôdǽlende heora weoruldgôd syndrigum mannum, Bd. 1, 27 ; S. 489, 19. VII. *to divide* into shares, *to share* :—Sió sunne and se môna habbaþ tôdǽled butwuht him ðone dæg and ða niht swíþe emne, Bt. 39, 13 ; Fox 234, 5. VIII. *to divide, distinguish, separate, make a difference between* :—Hú wundorlíce Drihten tôdǽlde ðæt Egiptisce folc and ðæt Israhélisce folc, Ex. 11, 7. Beó nú leóht on ðære heofenan fæstnysse and tôdǽlon dæg and nihte, Gen. 1, 14. Hit hafaþ hát baþo ǽlcere yldo and háde ðurh tôdǽlede stôwe gescrǽpe (*per distincta loca accommodos*), Bd. 1, 1 ; S. 473, 22. IX. *to be different, be distinguished* from. v. tô-dál, V :—Sacerd náht tôdǽlþ fram folce *sacerdos nihil distat a populo*, Scint. 123, 19. Swá micelum swá tôdǽlan gewunaþ líf hyrdes fram hyrde *quantum distare solet uita pastoris a grege*, 120, 17. X. *to separate with the mind, discern, discriminate, distinguish.* v. tô-dál, IX :—Gif geþanc yfel fram gôdum angytes mid gesceáde tôdǽlþ *si mens mala a*

3 S 2

bonis intellectus ratione discernit, Scint. ·141, 7. In gómán ðær mon ðone smæc tódæleþ *in palato*, Wrt. Voc. ii. 48, 4. Nú tódælde Petrus swutelíce ðone sóðan geleáfan ðá ðá hé cwæð: 'Ðú eart ðæs lifigendan Godes sunu,' Homl. Th. i. 366, 31. Ðæt hig cunnon fægere tódælan hwæt byþ betwux *ab animali ad animale* and *ab inanimale ad inanimale*, Anglia viii. 313, 35. Tódæled *discretus*, Scint. 123, 1. **XI.** *to give forth, utter* (?) :—Ealle ða gehát ðe ic æfre hér mid mínum welerum tódælde (cf. mín gehát ðæt míne weleras ær gedældan, v. 12, where Ps. Spl. and Ps. Surt. have tódældon and the Latin is *vota quae distinxerunt labia mea*), Ps. Th. 65, 13. [*O. Sax.* te-dêlian : *O. Frs.* tó-dêla : *O. H. Ger.* ze-teilen *dividere, distribuere, dispertire, separare, spargere, scindere, distare* : *Ger.* zer-theilen : cf. *Goth.* dis-dailjan.]

tó-dæledlíce ; *adv. Separately,.not in connection* :—Tódæledlíce *sigillatim*, Ps. Spl. 32, 15 : *divise*, Ælfc. Gr. 38 ; Zup. 229, 9. Seó fífte declinatio gebígþ hire genitivum on *e* and *i* tódæledlíce (*the e and i are pronounced separately*), 7 ; Zup. 21, 14. Tódæledlícor *differentius*, i. *eminentius*, Wrt. Voc. ii. 140, 14. v. tó-dælendlíce.

tó-dæledness, e ; *f. Division, distinction, separation :* —Tódælednesse *discrimine*, Wrt. Voc. ii. 27, 63. **I.** *a division*, (a) *one of the different kinds of parts into which a whole may be divided* :—Feówertýne tódælednyssa synd on ðam dæge . . . Oðer tódælednysse hætte *momentum*, þridde minutum . . . feówerteóða *mundus*, Anglia viii. 318, 35-42. (b) *one of the parts into which a whole is divided* :—Wé wyllaþ tódælan ða abecedaria on twá tódælednyssa, 333, 5. **II.** *division, separation, break of connection or of continuity*, (a) local :—Fæder and Suna and se Háliga Gást búton ælcere tódælednesse (-ennesse, MS.), Shrn. 167, 34. (b) *temporal, intermission, interruption* :—Fram Eástron óð Pentecosten sý alleluia bútan tódælednesse (*sine intermissione*) gecweden, R. Ben. 39, 14. **III.** *a division, dividing-point, break, pause* :—Idus tódælednyssa ðæs mónðes, Ælfc. Gr. 13 ; Zup. 85, 6. *Cesuras*, ðæt synd ða tódælednyssa on ðám versum . . . Ða tódælednyssa on ðám versum synd feówer, Anglia viii. 313, 38.

tó-dælendlíc ; *adj. Divisible, separable :*—Swá tódælendlíc is líchama and sáwle, Wulfst. 264, 26.

tó-dælendlíce ; *adv. Separably, distinctly :*—Ealle ða tódælendlíce singende *omnia distincte psallendo*, Anglia xiii. 371, 78. Seó fífte declinatio gebígþ hire genitivum on *e* and *i* tódælendlíce, Ælfc. Gr. 7 ; Zup. 21, 14 note. v. tó-dæledlíce.

tó-dælness, e ; *f. A division, distinct part* :—Tódælnessa ðara wætera *divisiones aquarum*, Past. 53 ; Swt. 413, 26. In tódælnesse *in divisiones*, Ps. Surt. 135, 13.

tó-dál, -dæl, es ; *n. Division.* **I.** *a dividing* into parts, *partition* :—Mid þrýnum tódále *trina partitione*, Anglia xiii. 380, 217. **II.** *separation* :—Tódál *distractio*, Hpt. Gl. 500, 35. Sume naman synd *dividua*, ða getácniaþ tódál, Ælfc. Gr. 5 ; Zup. 13, 12. **III.** *a part of a whole, separate portion, section* :—Stæfcræft hæfþ þrítig tódál (cf. sum ðæra dæla is gehäten *nota*, 291, 9). Ðæt forme tódál is *vox* stemn . . . Sume tódál sindon *pedes*, Ælfc. Gr. 50 ; Zup. 289, 15-290, 13. Tódál *divisiones*, Kent. Gl. 766. Capitulas, ðæt is tódála angin, R. Ben. 42, 1. Ðæra ægðer on þrím tódálum wunaþ, Lchdm. iii. 440, 31. **IV.** *a mark which divides, dividing-point* :—Tódál *comma*, Engl. Stud. xi. 65, 9. *Distinctiones*, ðæt sind tódál, hú man tódælþ ða fers on rædinge. Se forma prica on ðam ferse is gehäten *media distinctio*, ðæt is on middan tódál . . . *Distinctio* is tódál, Ælfc. Gr. 50 ; Zup. 291, 2-7. Tódálæ *commate, incisione*, Hpt. Gl. 473, 22. Þurh fíftan fótes tódál *per penthemimerim*, 411, 12. Tódál *commata, incisiones, divisiones*, 411, 10. Tódála *incisiones*, Engl. Stud. xi. 66, 48. **V.** *distinction, difference.* v. tó-dælan, **VIII, IX** :—Tódál *differentia, divisio, distantia*, Hpt. Gl. 434, 48. Ná byþ tódál mæþa *nulla erit distantia personarum*, Scint. 184, 1. *Differentia*, ðæt is tódál betwux twám þingum, Ælfc. Gr. 50, 20 ; Zup. 293, 18. Micel tódál is betwux ðám gecyrredum mannum, Homl. Th. i. 398, 20 : 48, 35. Tódáles *differentiae, distantiae*, Hpt. Gl. 439, 1. Tódál *distantiam*, 438, 28. **VI.** *distribution.* v. tó-dælan, **VI** :—On rápincle tódáles (-dæles, Ps. Spl.) *in funiculo distributionis*, Blickl. Gl. Dihtung upplíces tódáles, Scint. 227, 8. On tódále gyfa mislíce onfóþ mislíce gyfa *in divisione donorum diversi percipiunt diversa munera*, 133, 8. **VII.** *scattering, dispersing.* v. tó-dælan, **IV** :—On tódále *effusione*, Wrt. Voc. ii. 142, 67. **VIII.** *dissension, want of union or peace.* v. tó-dælan, **II.** ¶ :—For ðam ðe ic com sybbe on eorþan sendan ; ne secge ic eów, ac tódál (*separationem*), Lk. Skt. 12, 51. **IX.** *discretion.* v. tó-dælan, **X** :—Nédbehéfes gerædes tódál *necessarie rationis discretio*, Anglia xiii. 375, 132. Fremfullum gesceádes tódále, 369, 52. Mid tódæle *cum discretione*, Scint. 81, 2. v. under-tódál.

tó-dállíc. v. un-tódállíc.

tó-déman ; *p. de To judge between, distinguish* ; *dijudicare* :—Tóscæt ł tódémeþ *dijudicat*, Ps. Lamb. 81, 1. Mihtig Freá eall manna cynn tódæleþ and tódémeþ *the mighty Lord will divide and will distinguish in his judgement between all mankind*, Dóm. L. 20.

tó-dihtnian ; *p. ode To dispose* :—Tódihtnodon *disposuerunt*, Blickl. Gl.

tó-dón ; *p.* -dyde. **I.** *to put asunder, divide, separate* :—Ðæt wæter and seó eorðe wæron gemengede óð ðone ðriddan dæg ; ðá tódyde

hí God, Hexam. 4 ; Norm. 8, 15. Gif hwylc wíf twégen gebróðra nimþ hire tó gemæccan, óþerne æfter óþrum, tódó man hig (*separentur*), L. Ecg. P. ii. 11 ; Th. ii. 186, 10. Tódó man hig on twá *separentur illi*, 19 ; Th. ii. 188, 27. **II.** *to undo, open* :—Tódyde *solvit, disligat*, Germ. 402, 39. Hí tódydon heora múð ongeán mé *aperuerunt in me os suum*, Ps. Th. 21, 11. [Þat deor todede (undude, 2nd MS.) his chæfles, Laym. 6507. Ic uulle mine riche todon allen minen dohtren, 2945.]

tó-dráfan ; *p. de To drive asunder, drive in different directions, drive away, expel, dispel, scatter, disperse* :—God ða hæðenan tódráfþ (*disperdet*), Jos. 3, 10. Seó sunne tódræfþ ða nihtlícan þeústru, Lchdm. iii. 234, 30. Háligra-manna ðe tódræfaþ ða leahtras and deófla heom fram, Homl. Skt. ii. 25, 703. Hé is sóð leóht ðe tódræfde ða þeóstra ðises lifes, Homl. Th. i. 144, 7. Hí mynstra tóstæncton, and munecas tódræfdon, Chr. 975 ; Erl. 127, 22. Ðæt hé tódræfe costnunga fram úre heortan, Homl. Th. i. 156, 23. Fela weard tódræfed Godes ðeówa, Chr. 975 ; Erl. 126, 12. Heora heriges wæs mycel ofslægen and eall tódræfed *cunctus eorum caesus sive dispersus exercitus*, Bd. 3, 18 ; S. 546, 36. Byþ seó heord tódræfed *dispargentur oves gregis*, Mt. Kmbl. 26, 31. Beón ða scép tódræfede, Mk. Skt. 14, 27. [A lutel windes puf mei al todreven hit, A. R. 254, 1. Of þan folck þe wes todrefed, Laym. 330.] v. tó-drífan.

tó-dræfedness, e ; *f. Dispersion, expulsion* :—Hé worhte áne swipe of rápum and hí ealle út áscynde. Ðeós tódræfednys getácnode ða tóweardan tóworpennysse, Homl. Th. i. 406, 8. On ðeóda tódræfednysse *in dispersionem gentium*, Jn. Skt. 7, 35. Tódræfednesse *dispersiones*, Ps. Lamb. 146, 2.

tó-dræfness, e ; *f. Division, difference* :—Tódroefnise wæs him bituién ymb ðæt *schisma erat in eis*, Jn. Skt. Lind. 9, 16.

tó-dreósan ; *p.* -dreás ; *pl.* -druron ; *pp.* -droren *To fall to pieces, fall away, decay* :—Ðæt goldgeweorc eall tódreás, swá swá weax gemylt æt fýre, Shrn. 156, 15. [He schal todreosen so lef on bouh, Misc. 94, 48. Alle þe bones beoþ todrore, 152, 182. Cf. *Goth.* dis-driusan.] Cf. tó-feallan.

tó-drífan ; *p.* -dráf, *pl.* -drifon ; *pp.* -drifen *To drive in different directions, drive away.* **I.** *to drive asunder, separate* :—Wit ætsomne on sæ wæron fíf nihta fyrst, óþþæt unc flód tódráf, Beo. Th. 1095 ; B. 545. **II.** *to scatter, disperse* :—Se wulf tódrífþ (tódrífeð, Lind. Rush.) ða scéap *lupus dispergit oves*, Jn. Skt. 10, 12. Wulfas tódrífaþ ðíne heorde, Blickl. Homl. 225, 18. Ðú hí wíde tódríf *disperge illos*, Ps. Th. 58, 11. Hé hí wolde on ðam wéstenne wíde tódrífan *ut prosterneret eos in deserto*, 105, 21. Wurde seó eorþe tódrifen mid ðam winde swá swá dust, St. 33, 4 ; Fox 130, 8 : Met. 20, 104. Licgaþ æfter lande loccas tódrifene, Andr. Kmbl. 2852 ; An. 1428. **III.** *to scatter, destroy* :—Hé hí on heora fácne fæste tódrífeþ *in malitiis eorum disperdet illos Dominus*, Ps. Th. 93, 22. Ðú míne feóndas tódrífe *disperdes inimicos meos*, 142, 12. **IV.** *to drive away, send elsewhere* :—Ðú ús tódrífe *repulisti nos*, Ps. Th. 59, 1. Fram áswengde *vel* tódráf *excussit*, i. *dejecit*, Wrt. Voc. ii. 146, 18. Dryhten áwearp hine ðá of ðam wuldre and wíde tódráf, Salm. Kmbl. 928 ; Sal. 463. Se ðe æfter rihte wille æfter spyrian swá deóplíce, ðæt hit tódrífan ne mæg monna ænig ne ámerran ænig eorðlíc þincg *quisquis vestigat verum, cupitque nullis ille deviis falli*, Met. 22, 3. Ða tódrífenan *actos*, Wrt. Voc. ii. 9, 58. **V.** *to drive away, dispel, put an end to* :—Gáte tord ða swylas tódrífþ, Lchdm. i. 356, 1. Ða springas hyt tódrífeþ, 7. Se hálga deófulgild tódráf and gedwolan fylde, Andr. Kmbl. 3372 ; An. 1690. Tódríf ðone mist ðe nú hangaþ beforan úres módes eágum *disjice nebulas*, Bt. 33, 4 ; Fox 132, 32 : Met. 20, 264. Weard se háta líg tódrifen and tódwæsced, Cd. Th. 238, 11 ; Dan. 353. [Al he todrof þes kinges here, Laym. 549. Hiss stren all shollde ben todrifenn and toskeʒʒredd, Orm. 16397. *O. Frs.* tó-dríva : *O. H. Ger.* ze-tríban *dispellere, dispergere, diverberare.*] v. tó-dræfan.

tó-dwæscan ; *p.* te *To extinguish* :—Weard se háta líg tódrifen and tódwæsced, Cd. Th. 238, 11 ; Dan. 353 : Exon. Th. 190, 2 ; Az. 67.

tó-dwínan ; *p.* -dwán *To vanish away, to burst and vanish* :—Seó eádiga fæmne hál fram him gewænte and eall sticmælum tódwán (-dwán ? *but both* þwíneþ *and* dwíneþ *occur*, Lchdm. i. 84, 25 : 82, 2) se draca út of ðan carcerne *the dragon burst all in pieces and vanished from the prison* (the Latin has : Crux crevit in ore draconis et in duas partes eum divisit. Cf. the later English version : His (*the dragon's*) bodi tobarst omidheppes, Marh. 10, 22), Homl. Ass. 175, 200. v. dwínan, for-dwínan.

tó-eácan ; *adv., prep.* **I.** *adv. In addition, besides* :—Hé beád his þegnum, ðæt hig lédon hira ælces feoh on his sacc and fórmete tóeácan (*datis supra cibariis in viam*), Gen. 42, 25. Ðú hæfst tóeácan eall ðæt ic ðé ær tealde, Bt. 10 ; Fox 28, 37. Ôþre fífe ic tóéke gestríónde *alia quinque superlucratus sum*, Mt. Kmbl. Rush. 25, 20. Micel gít hér tóeácan, Wulfst. 165, 21. **II.** *prep. with dat. In addition to, besides* :—Tóeácan ðæs landes sceáwunge, Ors. 1, 1 ; Swt. 17, 35. Tóeácan hiere hwætscipe and hiere monigfealdum duguþum, 1, 10 ; Swt. 46, 24. Ða breósð tóeácan ðæm boge *pectusculum cum armo*, Past. 14 ; Swt. 81, 25. Tóeácan ðám dómum, L. Ath. v. proem. ; Th. i. 228, 9. Tóeácan ðon ðe hine God sylf innan manode, Blickl. Homl. 217, 5. Tóeácan

dæm ðe hé hiénende wæs his folc, hé wæs sinþyrstende monnes blódes, Ors. 3, 9; Swt. 130, 30. v. þær-tóeácan; eáca.

tó-écness. v. tó-ícness.

tó-efnes, -emnes; *prep. with dat. On a level with, abreast of, alongside, beside*:—Andlang weges óð tóemnes ðære micelan díc *the boundary runs along the road until it comes on a level with the great dike,* Cod. Dip. Kmbl. ii. 251, 3. Ondlang bróces óð hyt cymþ tóemnes ðæm ealdan læghrycge, iii. 437, 17. *See other instances under* emn.

tó-endebyrdness, e; *f. Order, series, succession*:—Hé eallum mannum megena weorc mid wordum bodode. And tóendebyrdnesse his gesihþa ðám mannum ánum hit cýþan wolde, ðam ðe hine ácsodon for ðám luste inbryrdnesse *omnium opus virtutum praedicabat sermonibus. Ordinem autem visionum suarum, illis solummodo qui propter desiderium compunctionis interrogabant, exponere volebat,* Bd. 3, 19; S. 549, 20.

tó-fær, es; *n. A going away, departure, decease*:—Tófær his *excessum ejus,* Lk. Skt. Lind. 9, 31. v. next word, III.

tó-faran; *p.* -fór; *pp.* -faren. **I.** *to go in different directions, go off separately, part*:—On sumera tófór se here, sum on Eást-Engle, sum on Norðhymbre, Chr. 897; Erl. 94, 25. Ðá hié tógædere woldon, ðá com swá ungemetlíc rén, ðæt heora nán ne mehte nánes wæpnes gewealdan, and for ðæm tófóran, Ors. 4, 10; Swt. 194, 19. Tófóran on feówer wegas æðelinga bearn *they went off in four different directions,* Cd. Th. 102, 8; Gen. 1697. Ær ðam ðe his Apostolas tófarene wæron geond ealle eorðan tó læranne, L. Alf. 49; Th. i. 56, 4. **II.** *to disperse* (intrans.), *scatter*:—Swelce se bitresta smíc upp ástíge and ðonne wíde tófáre, Ors. 3, 11; Swt. 142, 21. Ær seó mengeo eft tófaran sceolde, Cd. Th. 100, 15; Gen. 1664. Oþ his fird tófaren wæs *until his army was dispersed,* Ors. 3, 11; Swt. 152, 21. Ðonne hié gind ðæt lond tófarene wæron, hié ðonne hié floccmælum slógan, 2, 5; Swt. 78, 12: 3, 7; Swt. 116, 29. **III.** *to go away, pass off, depart, become extinct.* v. tó-fær:—Syle drincan . . . ðæt yfel tófærþ, Lchdm. i. 118, 6. Syle drincan on wíne, eal ðæt áttor tófærþ, 122, 18. [The folk . . . shall *tofare* on every clyve, Anglia iii. 546, 146. *O. Sax.* te-faran *to disperse; to pass away: O. L. Ger.* te-faran *deficere: O. H. Ger.* ze-faran *dissolvi, praeterire, transire, perire, defluere.*] v. tó-féran, -gán, -gangan.

tó-feallan; *p.* -feóll; *pp.* -feallen *To fall to pieces, fall away, collapse, fall down*:—Ðá hié æt hiora theatrum wæron, ðá hit eall tófeóll (*collapsa est*), Ors. 6, 2; Swt. 256, 11. Ðá byfode seó eorðe, and stánas burstan, and stánweallas tófeóllan, Shrn. 67, 19: Homl. Th. ii. 216, 4. Him ða lima ealle tófeóllan *all his limbs fell off,* Shrn. 62, 3. [Scullen stanwalles biuoren him tofallen, Laym. 18867. Alls þatt temmple oferr hemm all tofelle, Orm. 16185. Per no guod ðed ne ys þet uolk toualþ (*populus corruet,* Prov. 11, 14), Ayenb. 184, 11. *O. Sax.* te-fallan *to fall down* (of a house): *O. H. Ger.* ze-, zer-fallan *cadere, concidere, diruere: Ger.* zer-fallen.] v. tó-dreósan; tó-fillan.

Tófe-ceaster *Towcester*:—Mon worhte ða burg æt Tófeceastre, Chr. 921; Erl. 107, 26. [Cf. Tófi, Tófa, *Scandinavian proper names.*]

tó-feng (?: *but cf. the expression* fón tó), es; *m. Taking, seizure*:—Se ðe ne sealde ús on gehæfte ł tófæncge (tó fæncge?) tóðum heora *qui non dedit nos in captionem dentibus eorum,* Ps. Lamb. 123, 6.

tó-féran; *p.* de. **I.** *to go in different directions, go off separately*:—Ær úres Drihtnes leorningcnihtas tóférdan, ealswá heom beboden wæs (cf. Mk. 16, 15), Wulfst. 21, 5: Homl. Th. i. 318, 3. **II.** *to disperse* (intrans.):—Ðá ðæt gafol gelæst wæs, ðá tóférde se here wíde swá hé ær gegaderod wæs, Chr. 1012; Erl. 147, 27. Hí geswicon ðære getimbrunge, and tóférdon geond ealne middangeard, Homl. Th. i. 22, 25: 318, 21. Hí tóférdon tó fyrlenum lande on swá manegum gereordum swá ðæra manna wæs, Ælfc. T. Grn. 4, 12. [Ða apostoli er þon þet heo toferden, O. E. Homl. i. 93, 8. *O. H. Ger.* ze-fuoren.] v. tó-faran, -gán.

tó-ferian; *p.* ede *To carry in different directions;* differte. **I.** *to remove, get rid of*:—Hit ðæt sár tófereþ, Lchdm. i. 114, 3: 108, 8: 130, 19: 190, 8. **II.** *to put off*:—Swá oft gebiddende ná raþe beóþ gehýrede úre ús dæda on eágan wé tóforan settan ðæt ðæt sylfe ðæt wé synd tóferede ná godcundre byþ geteald rihtwísnysse ac gyltes úres *quotiens orantes non cito exaudimur, nostra nobis facta in oculis proponamus, ut hoc ipsum quod differimur non divinae reputetur justitiae sed culpae nostrae,* Scint. 35, 10. **III.** *to digest;* digerere:—Ðæt seó dæges þigen tófered sý . . . and se maga gelýht, ðæt hé ðe eáð his wæccean healdan mæge *ut digesti surgant,* R. Ben. 32, 14.

tó-fesian; *p.* ede *To drive in different directions, disperse, scatter, rout*:—Gé eów tó gamene feónda áfillaþ oððe tófesiaþ swá fela swá gé reccaþ, Wulfst. 132, 21. Gé tófesede swíðe áfírhte oft litel werod earhlíce forbúgaþ, 133, 2.

tó-fillan; *p.* de *To cause to fall in different directions, to demolish, destroy, break to pieces*:—God heáfdas feónda gescæneþ and hé tófylleþ feaxes scádan ðe hér on scyldum swærum eodon *Deus conquassabit capita inimicorum suorum; verticem capilli perambulantium in delictis suis,* Ps. Th. 67, 21. [*Ger.* zer-fällen.] v. tó-feallan.

tó-fleógan; *p.* -fleág, *pl.* -flugon; *pp.* -flogen. **I.** *to fly asunder, fly to pieces*:—Hé slóh ða næddran, ðæt heó on viiii tófleáh *he struck the adder so that it flew into nine pieces,* Lchdm. iii. 34, 26. **II.** *to*

fly apart, to crack, have breakings out (of a diseased body):—Wið hreófe and wið tóflogen líc *for leprosy and for a body that has breakings out on it,* Lchdm. i. 352, 18. [*O. H. Ger.* ze-fliogan *dissipari.*]

tó-fleón; *p.* -fleáh, *pl.* -flugon; *pp.* -flogen *To flee in different directions, be dispersed in flight, flee away*:—Gif wæter on éaran swíðe gesigen sý, genim ðysse ylcan wyrte seáw, drýpe on ðæt eáre; sóna hyt tóflýð (-flíhð, MSS. H. B.) *the water will run away directly,* Lchdm. i. 188, 8. [Ða cnihtes alle weoren wide tofloȝen ut of þan wiðeruehte, Laym. 28668.]

tó-fleótan; *p.* -fleát; *pp.* -floten *To float in different directions, be dispersed by water, be carried away by water*:—Da brycge ðe forneáh eall tóflotan wæs *the bridge that was almost quite carried away,* Chr. 1097; Erl. 24, 299. [Mid te fleotinde word tofleoteð þe heorte, so þ longe þer efter ne mei heo beon ariht igedered togederes, A. R. 74, 29. Forstopped ouwer þouhtes, ase ȝe wulleð þ heo nout ne touleoten ȝeond te world, 72, 22. *O. H. Ger.* ze-fliozan *defluere, liqui, fatiscere: Ger.* zer-fliessen.]

tó-flówan; *p.* -fleów; *pp.* -flówen *To flow different ways, disperse in flowing, flow away*:—Ic tóflówe *defluo,* Ælfc. Gr. 28, 5; Zup. 175, 14. Tófleówan ł út urnan *defluxerant,* Hpt. Gl. 473, 37. **I.** *referring to material objects,* (a) *to flow in different directions, be dispersed*:—Iudas tóbærst on emtwá and his innoð tófleów, náteshwón gelógod on nánre byrgene, Homl. Th. ii. 250, 26. Him (*the stream*) on innan felþ muntes mægenstán . . . hé on tú siððan tósceáden wyrð . . . bróc biþ onwended of his rihtryne rýðum tóflówen, Met. 5, 20. (b) *to melt away, be destroyed*:—Swá hwæt swá ðeós gesyhþ oþþe hrepeþ, hyt tófléwþ swá ðæt ðær nánwiht belífeþ búton ðæt bán, Lchdm. i. 242, 26. Muntas swé swé wex tófleówun (*fluxerunt*), Ps. Surt. 96, 5. **II.** *metaphorically,* (a) *of want of concentration in the mind, to wander, be drawn hither and thither, be distracted*:—Nán wuht nis on ús ungestæððigre ðonne ðæt mód, for ðæm hit gewítt suá oft fram ús suá ús unnytte geðohtas tó cumaþ, and æfter ælcum ðara tófléwþ *nil in nobis est corde fugacius, quod a nobis toties recedit, quoties per pravas cogitationes defluit,* Past. 38; Swt. 273, 13. Hié nellaþ hié gehæftan and gepyndan hiora mód, ac hé læt his mód tóflówan on ðæt ofdele giémelíeste, 39; Swt. 283, 14. Gebyreþ oft ðæt hié beóþ suá micle ungestæððelicor tóflówene on hiera móde suá hié wénaþ ðæt hié orsorgtran beón mægen *quae tanto latius diffluunt, quanto se esse securius aestimant,* 38; Swt. 271, 18. (b) *to be separated, take different directions*:—Hú ungelíc spræc eode of ðissa tuéga monna múðe. . . . ðeáh heó an tú tefleówe, ðeáh wæs sió æspryng sió sóðe lufu, Past. 7; Swt. 49, 11. (c) *to spread*:—Suá willaþ ða synna weaxænde tóflówan gif hié ne beóþ gebundne mid láreówdóme, Past. 17; Swt. 123, 16. (d) *to pass away, be dissipated, scattered, rendered useless, brought to nothing*:—Ðæt wé gemundan ðæt úre dæde and úre geþohtas nalæs on ðisne wind on ídelnesse tóflówan (tóflówenne, Bd. M. 440, 24) ac tó dóme ðæs heán déman ealle gehealdene beón *ut meminerimus facta et cogitationes nostras non in ventum diffluere, sed ad examen summi judicis cuncta servari,* Bd. 5, 13; S. 633, 27. Of ðære tíde ongan se hyht and mægen Angelcynnes ríces tóflówan and gewanod beón *ex quo tempore spes coepit et virtus regni Anglorum fluere, ac retro sublapsa referri,* 4, 26; S. 602, 28. (e) *to separate in confusion, become disconnected*:—Ic ongite ðæt ealle gesceafta tófleówon swá swá wæter, and náne sibbe ne náne endebyrdnesse ne heóldon, gif hí næfdon ænne God ðe him eallum stiórde, Bt. 34, 12; Fox 154, 2.

tó-flówedness, e; *f. A flowing, flux;* fluxus, Ælfc. Gr. 11; Zup. 79, 1.

tó-flówende; *adj.* (*ptcpl.*) *Affluent, confluent*:—Ðæs tóflówendan welan *affluentibus prosperitatibus,* Past. 50; Swt. 391, 11. Tóflówendum confluentibus, Wrt. Voc. ii. 23, 62.

tó-foran; *prep. with dat., gen. Before.* **I.** *of place, in front of, in presence of,* (a) *preceding the case*:—Ealle þeóda beóþ tóforan (*ante*) him gegaderude, Mt. Kmbl. 25, 32. Hé ða hláfas bræc and sealde his leorningcnihtum, ðæt hí tóforan him ásetton, Mk. Skt. 6, 41: 8, 6. Hé ðæt ylce gefæstnode tóforan ðam pápan, Chr. 1070; Erl. 208, 19. (b) *following the case*:—Etaþ ðæt eów tóforan áset ys, Lk. Skt. 10, 8. Ic næbbe hwæt ic him tóforan lecge *non habeo quod ponam ante illum,* 11, 6. **II.** *of time, previous to,* (a) *with dat.*:—Tóforan eallum ðissum hig nimaþ eów *ante haec omnia inicient uobis manus suas,* Lk. Skt. 21, 12. Tóforan ðam Eástron, Chr. 1012; Erl. 146, 8. Tóforan ðam mónðe Auguste, 1013; Erl. 147, 15. (b) *with gen.*:—Hit wæs tóforan dæges, Nar. 16, 11. **III.** *marking degree, above, in a greater degree than*:—Synfulle tóforan eallum *prae omnibus peccatores,* Lk. Skt. 13, 2. Gé beóþ geblessod tóforan eallum óðrum mannum, Deut. 7, 14: Homl. Th. i. 444, 30. Assa is stunt nýten and tóforan óðrum nýtenum ungesceádwís, 208, 12. **IV.** *marking position or status, superior to*:—Ðæt hé sý tóforan óðrum mannum þurh his glencge geteald, Homl. Th. i. 528, 29. **V.** *marking preference*:—Habbaþ eów tóforan eallum ðingum ða sóðan lufe, Homl. Th. i. 606, 16: R. Ben. 55, 6. Tóforan eallum þingum wé myngiaþ, ðæt . . . , 58, 7: Wulfst. 239, 17. Tóforon, Bt. 42; Fox 260, 12. **VI.** *marking excess, over and above,*

beside, beyond :—Tóforan ðám *praeterea,* Ælfc. Gr. 38; Zup. 234, 9. Tóforon ðám oððe bútan ðám *praeter illa,* 47; Zup. 270, 9. Eall hit byþ oferflówendnyss and ídel tóforan ðisum (ðæt tóforan ðysum is) *quod supra fuerit, superfluum est,* R. Ben. 90, 5. Swá hwæt swá tóforan ðám neádbehéfum belifen byþ, 138, 16. Salomon forgeaf ðære cwéne swá hwæs swá heó gyrnde æt him, tóforan (*over and above*) ðære cynelícan láce ðe hé hire geaf, Homl. Th. ii. 584, 31. For fela gewissungum ðe seó án bóc hæfþ tóforan ðám óðrum *for many directions which that one book has, and the others have not,* Ælfc. T. Grn. 6, 40. [*Piers P.* to-fore : *Ayenb.* to-vore : *O. Sax.* te-foran : *O. Frs.* tó-fora.] v. foran.

tó-forlǽten; *ptcpl. Dismissed :*—Tóforlǽten [is] *dimittitur,* Hpt. Gl. 420, 52. v. *next word, and* tó-lǽtan.

tó-forlǽtenness, e; f. Intermission :—Bútan tóforlǽtennesse *sine intermissione,* R. Ben. Interl. 45, 11 : Homl. Th. i. 596, 15 : ii. 382, 1.

toft. A word apparently of Scandinavian origin, *Icel.* topt, tuft *a piece of ground, messuage, homestead; a place marked out for a house or building; in the special later Icelandic sense a square piece of ground with walls but without roof: Dan.* toft *an enclosed home-field.* It does not occur often in the earliest English, but it is found as the second part of many place-names in districts which were affected by the Danes. v. Taylor's Names and Places. In the Prompt. Parv. *toft* renders *campus;* in Piers Plowman it means an elevated piece of ground : I seigh a toure on a *toft,* Prol. 14; while later, according to Kennett, it is 'a field where a house or building once stood.' In the following passages it may mean *the enclosed ground in which the house stood :*—Healf ðæt land æt Súðhám, innur and úttur, on tofte and on crofte, Cod. Dip. Kmbl. iii. 317, 7. Nǽfre myntan ne plot ne plóh, ne turf ne toft, L. O. 13; Th. i. 184, 7; Lchdm. iii. 286, 23. [Ic an] intó ðe túnkirke on Mardingford .v. acres and áne toft and .ii. acres mædwe . . . And míne landseðlen here toftes tó ówen aihte, Cod. Dip. Kmbl. iv. 282, 26–29. Alle míne men fré, and ilk habbe his toft and his metecú and his metecorn. And ic an þe préstes toft intó þe kirke fré . . . And ic an Lefquéne fífténe acres and án toft . . . And Alfwold habbe, mid tón þe hé hér hauede, .xvi. acres mid tofte mid alle, Chart. Th. 580, 6–27. v. Grmm. R. A. 539.

tog, es; n. Strife, contention :—Ða fríðgeorne, ða ðe heá búta ęghwoelcum flíta and toge behaldan, Mt. Kmbl. Lind. 5, 9 note. [Cf. (?) *O. Frs.* toga *to treat with violence, pull about.*]

-tog. v. lang-tog (-toh), sceaft-tog.

toga *a leader* (only in compounds). [*O. Sax.* togo : *O. Frs.* toga : *O. H. Ger.* zogo : *Icel.* togi.] v. breóst-, folc-, here-toga.

tó-gædere, -gædre, -gadore; adv. Together.   I. marking union, association, joining, mingling, etc. :—Ealle ðú nemdest tógædere and héte woruld, Bt. 33, 4; Fox 128, 27: Met. 20, 56, 62. Gif ðú wið fýre foldan and lagustreám ne mengdest tógædere, 20, 112. Ðá com Godwine eorl and Swegen eorl and Harold eorl tógædere, Chr. 1048; Erl. 178, 19: Ps. Th. 94, 1: Homl. Skt. ii. 30, 430. Ða stánas feóllon tógædere, and wearþ geworht tó ánum wealle swá, 27, 88. Ða ýslan eft onginnaþ lúcan tógædere, geclungne tó cleowenne, Exon. Th. 213, 17; Ph. 225. Hlemmeþ tógædre grimme góman, 363, 30; Wal. 61. In Danai ðære ié Asia and Europe hiera landgemircu tógædre licgaþ, Ors. 1, 1; Swt. 8, 11. Heofon and eorðe hreósaþ tógadore, Andr. Kmbl. 2875; An. 1440.   II. marking hostile meeting :—Ða hí tógædere gán sceoldon ðá onstealdan ða heretogan ǽrest ðone fleám *when the battle should have been joined, the leaders were the first to fly,* Chr. 993; Erl. 132, 15: 998; Erl. 134, 18: Beo. Th. 5253; B. 2630. Ða hí tógædere cómon, ðá wolde se ealdorman beswícon ðone æþeling, and hí tóhwurfon bútan gefeohte, Chr. 1015; Erl. 152, 14: Ors. 4, 10; Swt. 202, 14. Hí ðær fæste tógædere fęngon *they attacked one another fiercely,* Chr. 999; Erl. 134, 25: 1001; Erl. 137, 12. Hí fęngon tógædere fæstlíce mid wǽpnum, Homl. Skt. ii. 25, 631. Hwænne hí tógædere gáras béron *when they should cross weapons,* Byrht. Th. 133, 48; By. 67. Ðá hí ǽrost tógedore geræsdon ðá man ofslóh ðes Cáseres geréfan *at the first encounter Caesar's lieutenant was slain,* Chr. pref.; Erl. 5, 7.   III. marking continuity :—Feówertig daga and feówertig nihta tógædere, Gen. 7, 4 : Homl. Th. i. 22, 3. Fæste .ii. dagas tógædere, gif him mægen gelǽste, Lchdm. ii. 218, 2 : 232, 19. [*O. Frs.* tó-gadera (-e).]

tógædere-weard; adv. In directions that will bring (people) together, will lead to meeting :—Ða hwíle ðe hié tógædereweard fundedon *while they were proceeding to meet one another*; Ptolemaeus occurrere bello Perdiccae parat, Ors. 3, 11; Swt. 146, 5. Ðá hié tógædereweard fóron ðá flugon Péne swá hié eft selfe sǽdon . . . ǽr hié tógædere geneálǽcten *when the armies were marching to meet one another, the Carthaginians fled, as they afterwards themselves said, before they were near meeting ;* Ap. Claudius tam celeriter Poenos superavit, ut ipse rex ante se victum quam congressum fuisse prodiderit, 4, 6; Swt. 170, 22 : 6, 36; Swt. 294, 21. Hé (= hié) hiera sundorsprǽce ðe hié betux ðǽm folcum tógædereweard gesprǽcan tó unsibbe brohton and hié tó gefeohte geredon *their conference, which they (Scipio and Hannibal) held after going to meet one another between the armies, they brought to a hostile conclusion and prepared themselves for battle,* 4, 10; Swt. 202, 12.

tó-gægnes. v. tó-geagnes.

tó-gǽlan; p. de To profane, violate :—Míne rihtwísnessa gif hig besmítaþ ł tógǽlaþ *si justitias meas profanauerint,* Ps. Lamb. 88, 32.

tó-gǽnan; p. de To utter, pronounce :—Hig spelliaþ ł hig tógǽnaþ and spræcaþ unrihtwísnesse *effabuntur et loquentur iniquitatem,* Ps. Lamb. 93, 4. Cf. gánian.

tó-gán; p. -eode; pp. -gán.   I. of living things, *to go in two different directions, to part, separate :*—Gif wíf and wer ǽne tógáþ, Homl. Th. ii. 324, 2. Apollonius and Hellanicus tóeodon mid ðisum wordum, Ap. Th. 8, 23. Mycel wæl feóll on ǽgðre healfe, and ða heras him sylfe tóeodan, Chr. 1016; Erl. 156, 20. Æfter ðon ðe wit nú tócyrraþ and tógáne beóþ *postquam ab invicem digressi fuerimus,* Bd. 4, 29; S. 607, 20 MS. B.   II. of material things, *to be sundered, to part :* —Ic tógá *dehisco,* Engl. Stud. xi. 65, 23. Hé slóh mid ánre gyrde on ða sǽ, and heó tóeode on twá, Wulfst. 293, 15 : Homl. Th. ii. 194, 19. Seó eá on emtwá tóeode, 2f2, 22. Ðá tóeodon ða stánas, and geopenode ðæt get, H. R. 103. 22.   III. *to go in many different directions, to disperse, go away :*—Ða wæteru tóeodon and wanedon *aquae ibant et decrescebant,* Gen. 8, 5. [Þe wlcne togað, O. E. Homl. i. 239, 25. þe rede see toeode, 141, 6. He smat Frolle uppen þæne hælm þat he atwa helden (toȝeode, 2nd MS.), Laym. 23980. *O. H. Ger.* ze-gán : *Ger.* zergehen.] v. tó-gangan, -gengan, -faran.

tó-gang; es; m. Access, approach :—His tógang (-gan, MS.) biþ ðearle strang, Lchdm. i. 364, 10. Sý getýþod gebróþrum tógang fýres *conceditur fratribus accessus ignis,* Anglia xiii. 307, 457. Nánne hæfþ tógang heortan onbryrdnyss *nullum habebit accessum cordis compunctio,* Scint. 173, 5.

tó-gangan; p. -gēng; pp. -gangen.   I. *to go in different directions, to part :*—Æfter ðon ðe wit nú betweoh unc tógongenne (tógangne, Bd. M. 372, 3) beóþ *postquam ab invicem digressi fuerimus,* Bd. 4, 29 ; S. 607, 20. Ðá hié betwih him tógangen (-gangende, Bd. M. 372, 20) wǽron *digredientes ab invicem,* S. 607, 36.   II. *to go away, pass away :*—Ne tógongeþ gumena hwylcum eáþe ðæt ic ðǽr ymb sprice *what I speak of does not easily pass away from any man* (it is a bow that speaks, and the reference is to a wound from a poisoned arrow), Exon. Th. 405, 30; Rä. 24, 10. v. tó-gán, -gengan, -faran.

tó-geagn; prep. adv. Towards, in the direction of an object :—Tógeaegn iornaþ iúh monn *occurrit uobis homo,* Mk. Skt. Lind. 14, 13. [Wes heom toȝæn (aȝein, 2nd MS.) þe kaissere, Laym. 9792. *Du.* tegen: *Ger.* zu-gegen.] v. *next word.*

tó-geagnes, -geagnæs, -geánes, -gǽnes.   I. prep. (1) with dat. before or after it. (a) where there is motion towards the object governed by the word; (α) without idea of hostility, *towards, so as to meet :*—Sittas (the translator has read sed ite as sedite, *and taken it as* sedete) cuoæðæþ ðegnum his ðætte tógeaegnes (-gægnes, Rush.) færes iúh *remain and tell his disciples that he will come to meet you,* Mk. Skt. Lind. 16, 7. Foerdon tógægnes him *processerunt obviam ei,* Jn. Skt. Lind. Rush. 12, 13. Mann cumende heom tógénes (tógeagnas him, Lind.) *hominem venientem obviam sibi,* Mt. Kmbl. 27, 32. Eode seó ceasterwaru tógeánes (-gægnas, Lind.) ðam Hǽlende, 8, 34. 'Ne cóme gé nó tógénes (-geánes, Cott. MSS.) mínum folce ðæt gé meahton standan on mínum gefeohte for Israhéla folce.' . . . Ðæt is ðonne ðæt hé fare tógeánes Israhéla folce him mid tó gefeohtanne, Past. 15; Swt. 89, 17, 21. Ðá eode se cining him tógeánes *egressus est rex in occursum ejus,* Gen. 14, 17. Symeon eode tógeánes ðam cilde . . . Symeon eode him tógeánes, Homl. Th. i. 136, 14, 34. Faraþ him tógénys (-geánes, MS. A.) *exite obviam ei,* Mt. Kmbl. 25, 6. Ðǽr him tógénes manige cómon, Andr. Kmbl. 1313; An. 657. Bær man him tógeánes ánre wydewan líc, Homl. Th. i. 60, 12 : Blickl. Homl. 67, 7, 10. (β) with idea of hostility, *against, to meet :*—Hí férdon tógeánes ðám hæðenum *they marched to meet the heathens,* Homl. Th. i. 504, 27. Ðá fyrdode hé him tógeánes, and wið him feaht, Chr. 835; Erl. 65, 24. Him ðǽr com tógeánes Byrhtnoð ealdorman mid his fyrde, and him wið gefeaht, 993; Erl. 132, 5. Ða scipu fóran tógénes him, 911; Erl. 100, 21. (b) where there is motion of the object governed by the word; (α) without idea of opposition, *in the way of, to meet the approach of, in readiness for, against the coming of :*—Biþ hit eft him tógeánes gehealden *it shall be preserved against his coming,* Blickl. Homl. 53, 14. Ðæt folc, ðæt ðǽr beforan férde, streówodan heora hrægl him tógeánes, 71, 8. Geseóþ ðæt hé ǽrest tó ðære sinoþstówe cymeþ and gesiteþ, and gif hé áriseþ tógeánes eów ðonne gé cumen (*si vobis adpropinquantibus adsurrexerit*), Bd. 2, 2; S. 503, 10 : Homl. Th. ii. 52, 13. Gástum tógeánes, Cd. Th. 146, 30; Gen. 2430. Gewít ðú áwyrgda in ðæt wítescræf, ðé is susl weotod gearo tógenes, 308, 15; Sat. 693. Gearwian ús tógénes grěne strǽte, 282, 15; Sat. 287. Tógeánes, Exon. Th. 450, 21; Dóm. 91. Ðǽr biþ oft open eádgum tógeánes heofonríces duru, 198, 17; Ph. 11. (β) with the idea of opposition, *against, for the purpose of resisting :*—Hér com Óláf cyng intó Norwegum, and ðet folc gegaderode him tógeánes and him wið gefuhton, Chr. 1030; Erl. 163, 17. Hé forlét his gingran tógeánes ðære ceáste *he left his lieutenant to oppose the tumult,* Homl. Skt. i. 7, 212. Hæfde hé Grendle tógeánes seleweard áseted, Beo. Th. 1336; B. 666. (c) marking the object towards or against which an action is directed :—

Ðá clypode se eádiga Godes ðeów him tógeánes *the blessed servant of God cried out addressing him*, Homl. Th. ii. 168, 17. Hí biton heora téð him tógeánes *they gnashed on him with their teeth* (Acts 7, 54), i. 46, 28. Huigon mid heáfdum heofoncyninge tógeánes, Cd. Th. 16, 3; Gen. 238. Ðæt hé gewyrce deórum dǽdum deófle tógeánes, Exon. Th. 310, 18; Seef. 76. Ne underféhþ hé ná gerýnu for him sylfum ac gecýðnysse tógeánes him sylfum, Homl. Th. ii. 276, 35. Hí cwǽdon gefeoht tógeánes ðære burhware *they declared war against the citizens*, i. 504, 13. (d) marking time, *on the approach of, towards*:—Tógeánes Eástron com ðæs pápan sande *the pope's legate came towards Easter*, Chr. 1095; Erl. 232, 27. (e) marking comparison or contrast:—Hú mæg manna eádmódnys beón mycel geþúht tógeánes his eádmódnysse ðe ælmihtig God is, Homl. Skt. i. 12, 288. (2) with acc.; instances of this government are rare, and the two following are doubtful:—Tógeánes his fýnd (feónd, MSS. C. U. W. : feónde, MS. D.) hé gǽþ *adversum inimicum pergit*, Ælfc. Gr. 47; Zup. 269, 6. Ðá geopenode seó sǽ tógeánes Moysen (*the declension of the word in the translation of Exodous is dat.* Moise; *acc.* Moise, Ex. 8, 8. Moises, 8, 25 : 4, 27 : 16, 2, etc.), Ælfc. T. Grn. 5, 26. II. *adv.* (1) *again, in return*:—Ic him óðerne (gár) eft wille sændan, fleógende fláne forane tógeánes, Lchdm. iii. 52, 25. Ðá hét se wiðersaca onfón ðæra hláfa, and ágifan ðam biscope tógeánes gærs . . . Basilius underféng ðæt gærs ðus cweðende : 'Wé budon ðé ðæs ðe wé sylfe brúcaþ, and ðú ús sealdest tó edleáne (cf. ðú sealdest ús tógeánes, Homl. Skt. i. 3, 220) ungesceádwísra nýtena andlyfene' *then bade the apostate to take the loaves, and to give the bishop grass in return . . . Basilius took the grass saying : 'We offered thee what we ourselves use, and thou hast given us as requital (thou hast given us in return) the sustenance of irrational beasts*,' Homl. Skt. i. 450, 2–8 : Homl. Skt. i. 3, 215. Cúðberhtus him tógeánes cwæð *Cuthbert said to them in reply*, Homl. Th. ii. 138, 34. Hió him andsware ǽnige ne meahton ágifan tógénes, Elen. Kmbl. 333; El. 167. Him tógénes ðá gleáwestan mǽldon *in reply the wisest said to him*, 1069; El. 536. (2) marking position or direction:—Heó gráp tógeánes *she made a clutch at him*, Beo. Th. 3006; B. 1501. Hé árás tógénes, Andr. Kmbl. 2021; An. 1013. [*Laym.* tó-ȝeines, -ȝænes : *Orm.* to-ȝænes : *Ayenb.* to-yens : *O. Sax.* te-gegnes.] v. þær-tógeánes.

tó-gecorenness, e; *f.* *Adoption*:—Tógicorenisse gást *adoptionis spiritum*, Rtl. 29, 28.

tó-gegnes. v. tó-geagnes.

tó-gehlytto *fellowship*:—Tógihlytto *consortio*, Rtl. 109, 31.

tó-geícendlíc; *adj.* *Adjective*:—Ða óðre naman synd *adjectiva*, ðæt synd tógeícendlíce, Ælfc. Gr. 8; Zup. 29, 4 : 9, 18; Zup. 44, 4.

tó-geíht; *adj.* (*ptcpl.*) *Added*:—*Interpolares* vel *additi*, ðæt synd ða tógeíhte dagas, Anglia viii. 306, 44. v. next word.

tó-geíhtness, e; *f.* *An addition, increase*:—*Adjectiones*, ðæt synt tógeíhtnyssa, Anglia viii. 302, 32.

tó-génes, togenness. v. tó-geagnes, for-togenness.

tó-gengan; *p.* de *To go different ways, to separate*:—Hié tógengdon on ðone grénan weald, sǽton on sundran, Cd. Th. 52, 9; Gen. 841. v. tó-gán, -gangan.

tó-geótan; *p.* -geát, *pl.* -guton; *pp.* -goten. I. *to diffuse, spread*:—Tógiút *diffundet* (*venena*), Kent. Gl. 914. Ðonne wé swíðe wíde út tógeótaþ ða láre *quando exterius late praedicationem fundimus*, Past. 48; Swt. 375, 10. Ǽr ðon sió yfele wǽte, se ðe on wintra gesomnad biþ, hié tógeóte geond óþera lima, Lchdm. ii. 228, 9. Tógoten is geofu in weolerum ðínum *diffusa est gratia in labiis tuis*, Ps. Surt. 44, 3. Mid ða Cristes cyricean, seó geond ealne middangeard tógoten is, Bd. 2, 4; S. 505, 26. II. *to pour away, to exhaust*:—Ádrugod and tógoten *dried up and exhausted* (said of an ointment), Lchdm. ii. 28, 7. v. tó-gotenness.

tó-gesceádan *to separate things from one another*; metaph. *to expound, interpret*:—Wæs ingunnen from Moyse and allum wítgum tógisceóde him in allum gewriotum ða ðe of him wérun *incipiens a Mose et omnibus prophetis interpraetabatur illis in omnibus scripturis quae de ipso erant*, Lk. Skt. Rush. 24, 27. v. tó-sceádan.

tó-geþeód[d]; *adj.* (*ptcpl.*) *Adjacent, contiguous, connected, adjoined*:—Ðæt eálond tósceádeþ Wantsumo streám fram ðam tógeþeóddan lande *insula, quam a continenti terra secernit fluvius Vantsumu*, Bd. 1, 25 ; S. 486, 20.

tó-geþeódende *adhering*:—Tógeþeóden(d)ne *adhaerentem*, Hpt. Gl. 485, 29.

togettan; *p.* te (used impersonally):—Togetteþ betweox sculdrum *there are spasms between the shoulders*, Lchdm. ii. 216, 22.

tó-gewegen; *adj.* (*ptcpl.*) *Applied*:—Ðæt se biscop ðæt tógeweghene fýr ðære cynelícan burghe onweg gewende *ut episcopus admotum ab hostibus urbi regiae ignem amoverit*, Bd. 3, 16; S. 542, 12.

togian; *p.* ode *To tug, drag, pull*:—Ða Godes wiðerwinnan ða fǽmnan genámon, út of ðære byrig ungerǽdlíce hí togoden, Homl. Ass. 178, 308. [Cf. *toggen to toy*, A. R. 424, 27 : Marh. 14, 6. Toggyñ *idem quod* strogelyn, toggyñ or drawyñ *tractulo*, toggynge, drawynge *attractulus*, Prompt. Parv. 495. *O. Frs.* toga *to pull about, treat roughly* :

O. H. Ger. zocchón *rapere*; zucchen *rapere : Icel.* toga *to draw, pull*.] v. for-togian; togung.

tógian. v. tóan.

tó-gínan; *p.* -gán; *pp.* -ginen *To yawn, gape, open* as the mouth does:—Eorðe tógaan and eall forswealh Dathanes weorod *aperta est terra, et deglutivit Dathan*, Ps. Th. 105, 15. Se stán tógán, streám út áweóll, Andr. Kmbl. 3044 ; An. 1525. Biþ ðæt heáfod tóhliden, handa tóliðode, geaglas tóginene, Soul Kmbl. 215; Seel. 110.

tó-glídan; *p.* -glád; *pp.* -gliden *To glide in different directions, glide away*. I. of a fluid :—Synt geárdagas forð gewitene, lífwynne gegliden, swá lagu tóglídaþ, Elen. Kmbl. 2536 ; El. 1269. II. of smoke, cloud, or the like, *to be dissipated, dispersed, dispelled, to disappear, vanish, pass away* :—Ðá gewinon his drýcræftas, swá swá réc ðonne hé tóglídeþ, oðde weax ðonne hit for fýre gemelteþ, Shrn. 135, 3. Hit biþ gelíc réna scúrum, ðonne hí of heofonum swýðost dreósaþ and eft raðe eall tóglídaþ, Wulfst. 149, 7 : 264, 2. Ðæt wolcn tóglád, Homl. Th. ii. 242, 11 : Chr. 979 ; Erl. 128, 7. Nihthelm tóglád, lungre leórde, Andr. Kmbl. 246 ; An. 123 : Elen. Kmbl. 156 ; El. 78. II a. metaph. of pain, care, or the like :—Sele drincan, sóna ðæt sár tóglít, Lchdm. ii. 356, 21. Ðenden him hyra torn tóglíde, Exon. Th. 345, 3; Gn. Ex. 182. Hyge weard mongum blissad, sáwlum sorge tóglidene, 71, 31 ; Cri. 1164. III. *to fall to pieces, collapse* :—Grundweal gearone, se tóglídan ne þearf, ðeáh hit wecge wind, Met. 7, 34. IV. *to slip away* :—Ðeáh ðe ðás cáseras him háton gewyrcean heora byrgene of marmanstáne and útan emfrætewian mid reádum golde, ðeáhhwæðere se deáð hit eal tódǽlþ ; ðonne biþ ðæt gold tósceacen, and ða gymmas tóglidene (*the gems have slipped from their settings*), Wulfst. 148, 18–24 : 263, 8. Gúðhelm tóglád, gomela Scylfing hreás *the war-helm slipped off, the aged Scylfing fell*, Beo. Th. 4967; B. 2487. [Þeo luue þat ne may her abyde . . . hit schal toglide, Misc. 94, 43. *O. Sax.* te-glídan *to pass away, come to nought*.]

tó-gotenness, e; *f.* *Diffusion, spreading, effusion*:—Wyþ ǽwyrdlan ðæs lichoman ðe cymeþ of tógotennysse ðæs geallan, Lchdm. i. 262, 11 : 270, 5. v. tó-geótan.

togung, e; *f.* *Spasm* :—Wið sína togunge, Lchdm. i. 136, 9, 19. v. togian.

tóh; *adj.* *Tough, tenacious, holding fast together*; *lentus* :—Tóh, tóch, thóch *lenta*, Txts. 73, 1198. Tóh, Wrt. Voc. ii. 49, 64. Ðæm tóu *lento*, 62. Ðæm tón *ab lentis*, 3, 48. I. *tough, pliant* :—Tóh (tóch, thóh) gerd *lentum vimen*, Txts. 75, 1207 : Wrt. Voc. ii. 50, 74. II. *tough, sticky, glutinous, clammy* :—Nim hwetstán brádne and gníd ða buteran on ðæm hwetstáne mid copore ðæt heó beó wel tóh, Lchdm. iii. 16, 22. [*Make*þ ȝoure speres toȝe and strang, Laym. 5865. Mid toȝen his mæine, 9319. Thei hadden towȝ cley for syment, Wick. Gen. 11, 3. Towhhe, not tender *tenax*, Prompt. Parv. 498. *Du.* taai *pliant, tough, clammy, sticky* : *O. H. Ger.* záhi *tenax* : *Ger.* záhe.] v. tóan, téan, tóhlíce.

tó-haccian; *p.* ode *To hack to pieces, cut to pieces* :—Sume hig wǽron on feówer dǽias tóhaccode, eall swá hig ðæs Hǽlendes tunecan on feówer tódǽldon, Homl. Ass. 186, 166. [To smale peces ich hym wolde tohakke, R. Glouc. 141, 14. *O. Frs.* tó-hakkia : *M. H. Ger. Ger.* zerhacken.]

tó-heald; *adj.* *Inclined*:—Tóhald *adclinis* vel *incumbens*, Txts. 37, 74. Tóheald *adclinis*, Wrt. Voc. i. 287, 74 : ii. 4, 41. Þeáh wuhta gehwilc wrigaþ tóheald, swíðe onhelded, wið ðæs gecyndes ðe him cyning engla æt frymðe getióde, Met. 13, 10. [Cf. *O. H. Ger.* zuo-hald *futurus, venturus*.]

tó-heáwan; *p.* -heów; *pp.* -heáwen *To hew to pieces, cut to pieces* :—Se cásere cwæð ðæt Basilla sceolde gebúgan tó ðam cnihte, oþþe hí man tóheówe mid swurde on twá, Homl. Skt. i. 2, 360. Man sceolde ða scipu tóheáwan, Chr. 1004; Erl. 139, 26. Wé synd ealle belǽwde tó úre lífleáste, ðæt wé beón tóheáwene mid heardum swurdum, Homl. Ass. 99, 255. [Turnus feol mid mechen toheawen, Laym. 178. The helmes thei tohewen and toschrede, Chauc. Kn. T. 1751. To zenne ne ssel he wende ayen, þaȝ me ssolde hine al toheawe, Ayenb. 178, 7. *O. Frs.* tóhawa : *M. H. Ger.* zer-houwen : *Ger.* zer-hauen.]

tó-higung, e; *f.* The word glosses *affectus*, Rtl. 18, 32 : 31, 40 : 7, 27 : *effectus*, 35, 37 : 63, 20. v. higian.

tohl. v. tól.

tó-hladan; *p.* -hlód; *pp.* -hladen *To disband, disperse* :—Ne meahte hié (*the builders of the tower of Babel*) gewurðan weall forð timbran, ac hié earmlíce heápum tóhlódon hleóðrunt gedǽlde *they could not combine to carry on the building of the wall, but, divided in speech, they miserably dispersed in troops*, Cd. Th. 101, 36; Gen. 1693. Cf. 235, 6; Dan. 302.

tó-hlecan (P); *p.* -hlæc; *pl.* -hlǽcon; *pp.* -hlocen *To disjoin, pull to pieces* :—Tóhlocene (tólocene? v. tó-lúcan; *but see also* hlec, hlecan. *In the glosses among which the word occurs intial* h *before a consonant does not seem to be inserted elsewhere, though it is twice omitted, in* lecum *rimosae*, 400, 69, wisligendre *sibilantis*, 394, 278) *diuulsa*, Germ. 398, 112.

tó-hleótan; *p.* -hleát, *pl.* -hluton; *pp.* -hloten *To divide into lots, to divide into parts for which lots are to be cast* :—Hý gedǽldan him mín

hrægl and ðæt tôhlutan *diviserunt sibi vestimenta mea et super vestem meam miserunt sortem*, Ps. Th. 21, 16. Hié (*the apostles*) ðysne middangeard on twelf tânum tôhluton, and æghwylc· ânra heora in dæm dæle ðe hê mid tân geeode manige þeóde ûrum Drihtne gestreónde, Blickl. Homl. 121, 8.

tôh-líc; *adj. Tough, tenacious.* v. next word.

tôhlíce; *adv. Toughly, tenaciously :*—Tôhlíce, thôlícae, tôchtlícae *uscide, viscide* (viscide *fortiter*, Migne), Txts. 107, 2170. Tôlíce *huscide*, 69, 1033.

tô-hlídan; *p.* -hlád, *pl.* -hlidon; *pp.* -hliden *To yawn, gape, open, crack* (intrans.), *split* (intrans.) *asunder :*—Tôhlád seó eorþe *terra dissiluit*, Ors. 3, 3; Swt. 102, 26. Tôhlád seó eorþe and wæs byrnende fýr up of ðære eorþan *flamma scisso terrae hiatu eructata*, 4, 2; Swt. 160, 24: Lchdm. iii. 428, 3. Se beorg tôhlád eorðscræf egeslíc *the hill yawned, an awful cave it grew*, Andr. Kmbl. 3173; An. 1589. Heofonas tôhlidon, Blickl. Homl. 105, 13. Tôhlídan *dehiscere*, Germ. 400, 482. Biþ ðæt heáfod tôhliden *the head shall be cloven*, Soul Kmbl. 213; Seel. 109. Hié gesáwon swelce se hefon wǽre tôhliden *coelum scindi velut magno hiatu visum*, Ors. 4, 8; Swt. 188, 26. Wæs ðæt beorhte bold tôbrocen . . . heorras tôhlidene *the hinges shewed gaping cracks*, Beo. Th. 2002; B. 999. Gimmas tôhlidene, Wulfst. 263, 8 note.

toh-líne, an; *f. A tow-line :*—Tohlíne *remulcum*, Wrt. Voc. i. 63, 64; *remulcus*, 57, 5. [Cf. Icel. tog *a rope, line*: Scott. tow *a rope of any kind*.]

tô-hlocen. v. tô-hlecan.

tô-hlystend, es; *m. A listener :*—On ðara tôhlystendra heortan . . . Hê gɔdeþ ða sprǽce unnytte ðæm tôhlystendum, Past. 15; Swt. 96, 8, 18.

tô-hnescian; *p.* ode *To soften away :*—Ðonne findest ðû ðæt hearde tôhnesced, Lchdm. ii. 250, 21.

tô-hopa, an; *m. Hope, expectation :*—Eádig byþ se wer ðe his tôhopa byþ tô swylcum Drihtne *beatus vir cujus est nomen Domini spes ejus*, Ps. Th. 39, 4. Hwæt is mín tôhopa *quae est expectatio mea?* 38, 9. On ðê ys eall ûre hæl and ûre tôhopa *Domini est salus*, 3, 7. Wâ eów welegum ðe eówer lufu eall and eówer tôhopa is on eówrum worldwelum, Past. 26; Swt. 180, 24. Sió lufu and se geleáfa and se tôhopa *fides spes et caritas*, 21; Swt. 167, 19, 25: Shrn. 179, 1: Bt. 10; Fox 30, 8. Se tôhopa ðære wræce, Bt. 37, 1; Fox 186, 23: Met. 25, 50. Ðû mé gesettest on tôhopan (*in spe*), Ps. Th. 4, 9: 15, 9. Ymbe ðone tôhopan ðe gê habbaþ on eów *de ea, quae in vobis est, spe*, Past. 22; Swt. 173, 9. Ealne his tôhopan sette hê on God, L. E. I. 21; Th. ii. 416, 17. [Nimeþ tohope to helme *sumentes galeam spei*, O. E. Homl. i. 155, 8. *O. L. Ger.* tô-hopa.] v. tô-hyht.

tô-hopian. v. hopian (tô).

tô-hopung, e; *f. Hope, expectation :*—Wæs eall heora myne fæst on tôhopunge ðæs êcean Drihtnes, Homl. Skt. i. 23, 155. v. tô-hopa.

tô-hreósan; *p.* -hreás; *pl.* -hruron; *pp.* -hroren *To fall to pieces.* **I.** of buildings, *to go to ruin :*—Monige ôþre ceastre tôhrorene wǽron *multis civitatibus conlapsis*, Bd. 1, 13; S. 482, 8. Mynstru tôrorene *coenobia diruta*, Anglia xiii. 366, 12. **II.** of flesh, *to decay, rot away :*—Beóþ ða líchaman tôhrorene (cf. gebrosnode, 148, 24) and tô duste gewordene, Wulfst. 263, 9. Beóþ fingras tôhrorene, Soul Kmbl. 219; Seel. 112. [Portchestre, al heo gunnen toreosen (todrese, 2nd MS.), mid fure and mid sehte foruaren, Laym. 9245. Þus Portchestre toræs (toreos, 2nd MS.), & nauere seodden heo ne aras, 9426.]

tô-hréran; *p.* de *To shake to pieces, to destroy :*—Tôhrêrde *diruit*, Hpt. Gl. 487, 75. Se grundweall ðara munta wæs tôhrêred *fundamenta montium conturbata sunt*, Ps. Th. 17, 7. Tôhrêrede *diruta, destructa*, Hpt. Gl. 459, 50.

tô-hrician; *p.* ode *To divide, separate, cut up :*—Tôhricod *secta*, Germ. 398, 183: *dissipatum*, 399, 303. Tôhricedum *resectis*, 398, 100. Cf. Hrycigende *resulcans*, 398, 144.

tohte, an; *f. A military expedition, war, battle :*—Nǽron ða twêgen tohtan sǽne, lindgelaces, land Persêa sôhton sîdfrome Simon and Thaddeus, Apstls. Kmbl. 150; Ap. 76. Gê dôm ágon, tír æt tohtan, Judth. Thw. 24, 19; Jud. 197. Ðæt wíf ðîn heáfod tredeþ mid fôtum sînum ðû scealt fiersna sǽtan tohtan *the woman shall tread thy head with her feet, thou shalt lie in wait to attack her heels*, Cd. Th. 56, 18; Gen. 914. Æt sæcce oferswîðan feónda gehwylcne, ðonne fyrdhwate on twá hæfde tohtan sêcaþ, Elen. Kmbl. 2358; El. 1180. [Cf. *O. Frs.* tocht-man *a leader*: *Ger.* zug *a march*: *Dan.* tog *expedition, march*; togt *a cruise, expedition*.] v. ge-toht, tyht, **II**, toga, teón, **IV**; *and* cf. fird.

tô-hweorfan; *p.* -hwearf, *pl.* -hwurfon; *pp.* -hworfen *To go in different directions, to part, separate.* **I.** of two persons or parties :—Ða cyningas cômon tôgædere and heora freóndscipe gefæstnodon . . . And hí tôhwurfon ðá mid ðisum sehte, Chr. 1016; Erl. 159, 5: 1091; Erl. 228, 8: 1093; Erl. 228, 39. **II.** of many persons, *to disperse :*—Eal seó fyrding tôhwearf, Chr. 1094; Erl. 230, 24. Ecgbryht lǽdde fierd wiþ Norþanhymbre, and hié him eáþmêdo budon, and hié on ðam tôhwurfon, 827; Erl. 64, 9. Siendon wê tôwrecene geond wîdne grund, heápum tôhworfene (-hworfne, Exon. Th. 186, 19; Az. 22) *we are scattered in exile through the wide world, dispersed in bands*, Cd. Th. 235, 6; Dan. 302. Cf. tô-cirran.

tô-hyht, es; *m. Hope, confidence, trust, glad expectation :*—Witena frôfur and eorla gehwam eádnys and tôhyht, Runic pm. Kmbl. 340, 10; Rûn. 4. Dæg byþ myrgð and tôhiht eádgum and earmum, 344, 12; Rûn. 24. Cf. tô-hopa.

tôian. v. tôan.

tô-ícness, e; *f. Increase :*—Mid ðý ðá seó gesetenes ðæs heofonlícan lifes dæghwamlíce tôêcnesse nom *cum vitae coelestis instituto quotidianum sumeret augmentum*, Bd. 3, 22; M. 226, 31. v. tô-ǽtícness.

tô-irnan; *p.* -arn, *pl.* -urnon; *pp.* -urnen *To run in different directions, run about :*—Þýstru ðû gesettest on þearle niht on ðære ealle wildeór wîde tôeornaþ *posuisti tenebras, et facta est nox; in ipsa pertransibunt omnes bestiae sylvarum*, Ps. Th. 103, 19. v. tô-rinnan.

tô-irnende; *adj. (ptcpl.) Running together :*—Ðá se Hǽlend geseah ða tôyrneudan menegu *cum videntet Jesus concurrentem turbam*, Mk. Skt. 9, 25.

tôl, es; *n.* **I.** *that by which one makes things* (cf. *Goth.* taujan *to make, do*), *a tool, implement, instrument*, (a) literal :—Tôl *ferramentum*, Wrt. Voc. i. 84, 60. Tool *instrumentum*, 21, 37. Tohl, ii. 49, 22. Mid· tôle *instrumento, materia*, Hpt. Gl. 443, 47. Ðý læs hié mid ðý tôle (*a surgeon's knife*) ðæt hâle líc gewierden, ðe hié sceoldon mid ðæt unhâle ǽweg âceorfan, Past. 48; Swt. 365, 11. Gif ðû ðîn tôl (*cultrum*) âhefst ofer hyt *if thou lift up thy tool upon it* (A. V.), Ex. 20, 25. Hwîlon befeóll ân sîde of ðam snæde intô ânum deópan seáðe. Benedictus wolde gefrêfrian ðone wyrhtan ðe ðæt tôl âmyrde, Homl. Th. ii. 162, 12. Wîglíce tôl *instrumenta bellica*, Hpt. Gl. 424, 28. Eówer sûteres· tôl *uestri sutoris instrumenta*, Ælfc. Gr. 15; Zup. 105, 15. Mid lâcniendlícum tôlum *instrumentis medicinalibus*, Hpt. Gl. 478, 2. Ðá cômon ða cempan (*the soldiers at the crucifixion*) mid cwylmbǽrum tôlum, and ðæra sceaðena sceancan tôbrǽcon, Homl. Th. ii. 260, 7. Be mynstres tôlum *de ferramentis monasterii*, R. Ben. 56, 2, 3. Sylle man ðam gebûre· tôl tô his weorce and andlâman tô his hûse, L. R. S. 4; Th. i. 434, 26. Ða nýdþearfe . . . ðæt is mete and drync and clâþas and tôl tô swelcum cræfte swelce ðú cunne ðæt ðê is gecynde, Bt. 14, 1; Fox 42, 6. Gif ðú nele ânum ôlæcan, forlæt eal ðæt ðú âge bûton wiste and wǽda and tô swylcum weorcum tôl swylce ðú cunne, Prov. Kmbl. 80. (b) metaph. :—Hwæt is hit elles bûtan getimbrunga and tôl hâligra manna (*instrumenta virtutum*), R. Ben. 133, 9. Ðis synt ða lâra and ða tôl gástlíces cræftes, L. E. I. 21; Th. ii. 418, 17. Ðû wâst ðæt nân mon ne mæg nǽnne cræft cýðan ne nǽnne anweald reccan bûtan tôlum and andweorce . . . Ðæt biþ cyninges andweorc and his tôl mid tô rícsianne, ðæt hê hæbbe his land fullmannod; hê sceal hæbban gebedmenn and fyrdmen and weorcmen. Hwæt ðû wâst ðætte bûtan ðisum tôlum nân cyning his cræft ne mæg cýðan . . . Ne mæg hê bûtan ðisum (*provisions of various kinds*) ðâs tôl gehealdan, ne bûtan ðisum tôlum nân ðara þinga wyrcan ðe him beboden is tô wyrcenne, Bt. 17; Fox 58, 28–60, 7. **II.** in a collective sense, *tools, machinery, apparatus :*—Decius cwæd : 'Æteówiaþ his gesihðum eal ðæt wîta tôl' (cf. eal ðæt þînungtôl, 424, 18). Ða wurdon hrædlíce forð âborene îsene clûtas, and îsene clâwa, and îsen bedd, and leádene swipa, Homl. Th. i. 424, 18. [*Icel.* tôl; *n. pl.* tools, cf. *Goth.* taujan *to do.*] v. pînung-, tow-, wîte-tôl.

tô-lǽtan; *p.* -lêt; *pp.* -lǽten *To let go in different directions, to cause to go different ways, to disperse, release, relax :*—Tôlǽt·e[þ] *relaxat*, Hpt. Gl. 405, 67. Gif mon sýþ gârleác on henne broþe and selþ drincan, ðonne tôlǽt hió ðæt sǽr (*costiveness*), Lchdm. ii. 276, 16. Hê forgiet hine selfne ðonne hê tôlǽtt and fægnaþ ongeagn ðara ôðerra word *oblitus sui in voces se spargit alienas*, Past. 17; Swt. 111, 10. Tôlǽtenum æddrum *laxis fibris*, Hynm. Surt. 102, 22. [*O. Sax.* te-lâtan *to scatter, disperse* (intrans.) : *O. H. Ger.* ze-lâzzan *desinere, deserere, liquefacere*: *Ger.* zer-lassen *to dissolve*.]

tô-lǽtenness, e; *f. Abandonment, a giving up :*—Ðeós wyrt ealle ealde and unlâcnigendlíce âdlu tôfereþ, swâ ðæt hê byþ gelácnud þeáh hê ǽr his hǽle on tôlǽtennesse wǽre *the patient will be cured, though before he had been in despair of his health*, Lchdm. i. 262, 3.

tolcendlíce; *adv. Wantonly :*—Tolcendlícor *petulantius*, Germ. 401, 41. v. following words.

tolcettan; *p.* te *To be wanton :*—Tolcetende ɫ fleardiende *infruticans, luxurians*, Hpt. Gl. 435, 36. v. next word.

tolcettung, tolgettung, e; *f. An incentive, incitement :*—Tolgetunge, ontyndnesse *titillationis*, Hpt. Gl. 520, 32. Tolgetunge *titillationum, accensionum*, 457, 73. v. preceding words.

tô-leoðian, -lêsan. v. tô-liðian, -lîsan.

tô-licgan; *p.* -læg, *pl.* -lægon; *pp.* -legen. **I.** *intrans.* of roads, rivers, etc., *to lie or run in different directions :*—Heó (*the Nile*) tôliþ on twâ ymb an îgland ðe mon hæt Meroen *the stream runs in two channels round the island of Meroen* ; faciens insulam nomine Meroën in medio sui, Ors. 1, 1; Swt. 12, 32. Ic wille ðara þreora landrîca gemǽre gereccan hû hié mid hiera wætrum tôlicgeaþ *I will describe the boundaries, in what different directions they run* ; ipsarum partium (*the three divisions of the world*) regiones significare curabo, Swt. 10, 5. Nû hæbbe wê gesǽd ymbe ealle Europe landgemǽro hû hí tôlicgaþ. Nû wille wê ymbe Affrica secgan hû ða landgemǽro tôlicgaþ *we have now told in respect to all the boundaries of the countries in Europe the several directions they take.*

Now we will tell of Africa how the different boundaries of the countries run, Swt. 24, 21–23. Ðær ða wegas tólicgaþ *where the roads run in different directions*, Cod. Dip. Kmbl. iii. 411, 21. II. *trans. To lie between, to lie and part, to divide, separate* :—Seó eá tóliþ Witland and Weonodland, Ors. 1, 1 ; Swt. 20, 6. Ælc ðæra spræca is tódæled on manega ðeóda, and ða sint tólegena and tódælda mid sǽ and mid wudum and mid muntum, Bt. 18, 2 ; Fox 62, 34.

tó-lísan ; *p. de To unloose, undo, dissolve* ; solvere, dissolvere, exsolvere, resolvere. I. *to undo that which is bound, release from a bond*, (a) literal :—Ðæt wíf tólýsde hire feax, Homl. Th. ii. 30, 16. (b) figurative, (1) *to release from captivity, difficulty, etc.* :—Drihten tólýseþ gecypsede *Dominus solvit compeditos*, Ps. Spl. 145, 6. Hé wæs gehyhtende ðæt hé sóna ðæs ðe hir mon gefullade his líchoman tólýsed wǽre *sperans quia mox baptizatus carne solutus esset*, Bd. 5, 7 ; S. 620, 36. Tólésed wǽran *extricaba[n]tur*, Wrt. Voc. ii. 83, 25. (2) *to do away with tension, relax, relieve* :—Hyt tólýseþ ða blǽdran and ða stánas forð gelǽdeþ, Lchdm. i. 270, 9. II. *to put an end to the connection between, to separate* :—Tólýsan líc and sáwle, Andr. Kmbl. 301 ; An. 151. Ðá tósceáden wearð líg tólýsed *then was the flame scattered, separated*, Exon. Th. 277, 23 ; Jul. 585. III. *to dissolve, put an end to, dissipate*, (a) *of concrete objects* :—Ðysse wyrte leáf tólýsaþ gehwylce yfele springas and heardnyssa, Lchdm. i. 262, 9. Scadu sweþredon tólýsed under lyfte, Exon. Th. 179, 17 ; Gú. 1263. (b) *of abstract objects* :—Ðære miltan sár hyt tólýseþ, Lchdm. i. 270, 11. Tólýseþ leóna mægen Drihten *molas leonum confringet Dominus*, Ps. Th. 57, 5. IV. *to dissolve, relax, destroy the force of, weaken* :—Ymhídignyssa ofðriccaþ ðæt mód, and unlustas tólýsaþ, Homl. Th. ii. 92, 15. Mid ðý ðe hié ðone drenc druncon, hraþe heora heorta wæs tólésed and heora mód onwended, Blickl. Homl. 229, 13, 18. Seó sáwul on flǽsclícum lustum biþ tólýsed, Homl. Th. i. 408, 16. Wǽrun míne ǽdra ealle tólýsde *renes mei resoluti sunt*, Ps. Th. 72, 17. V. *to desolate, destroy*. v. tó-lísedness, -lísend, -lísendlíc :—Nú syndon hí gewordene tólýsde *quomodo facti sunt in desolatione*, Ps. Th. 72, 15. VI. *to undo a bond*, (a) literal :—Ðá hét se apostol tólýsan ða rápas, Homl. Th. i. 464, 21. Ðonne tóslupan ða bendas and tólýsede wǽron *sunt vincula soluta*, Bd. 4, 22 ; S. 591, 13. (b) figurative :—Deáþes bend tóléseþ líffruma, Exon. Th. 64, 25 ; Cri. 1043. VII. *to discharge an obligation, to pay* :—Ic tólýsde ł ágeald *exolvebam*, Ps. Spl. 68, 6. VIII. *to break a connection* :—Seó geþeódnes ðæs heáfdes tóbrocen and tólýsed wæs *ut capitis junctura solveretur*, Bd. 5, 6 ; S. 619, 25. Ða tólýsdan geþeódnesse *dissolutam juncturam*, S. 620, 13. [O. H. Ger. ze-lósen *dissolvere, resolvere, dividere, dirumpere*.] v. un-tólísende.

tó-lísedness, e ; *f. Dissolution, desolation, dispersion* :—Tólésednes *dissolutio, dispersio*, Wrt. Voc. ii. 141, 40. Monige ðara bigengana ðonan gewitan for ðære burhge tólýsednesse (*ob desolationem*), Bd. 4, 25 ; S. 601, 35. On tólýsydnysse *in desolationem*, Ps. Spl. C. 72, 19.

tó-lísend, es ; *m. A destroyer, desolater* :—Wéstend, tólýsend *desolator, vastator*, Wrt. Voc? ii. 139, 34.

tó-lísendlíc ; *adj. Destructive, desolating* :—Mid glédum tólýsendlícum *cum carbonibus desolatoriis*, Ps. Lamb. 119, 4.

tó-lísing, e ; *f.* I. *dissolution, destruction* :—Geleáfan tólýsinge, Lchdm. iii. 206, 20. II. *release, redemption*. v. tó-lísan, I b :—Ðætte hé salde sáwel his lésnise ł tólésinc fore monigum *ut daret animam suam redemptionem pro multis*, Mk. Skt. Lind. 10, 45.

tó-lísness, e ; *f.* I. *dissolution, destruction* :—Sibbe tólésness, Blickl. Homl. 115, 16. II. *dissolution, death* :—Seó tíd mínre tólýsnesse and mínre forþfóre is swýþe neáh, Bd. 4, 29 ; S. 607, 21 : 4, 9 ; S. 577, 16.

tó-liðian ; *p. ode To dismember, disjoint* :—Ðá tóliðode se engel ðæt cild on ðam disce, Homl. Th. ii. 272, 18. Biþ ðæt heáfod tóhliden, handa tóliðode (-leoþode, Exon. Th. 373, 16), Soul Kmbl. 214 ; Seel. 109.

toll, es ; *n. m.* (?) *Toll, tax, custom, duty, due*. v. also IV :—Cynelíc toll *fiscale tributum*, Hpt. Gl. 440, 43. Nim ðone wecg, and syle tó tolle for mé and for ðé, Homl. Th. i. 512, 5. Æt hwám nimaþ cyningas gafol oððe toll *reges terrae a quibus accipiunt tributum vel censum?* Mt. Kmbl. 17, 25. Ðæs cáseres tolleras áxodon Petrus, ðá ðá hí geond ealne middangeard ðam cásere toll gegaderodon, ‘Wyle eówer láreów ǽnig toll syllan?’ Homl. Th. i. 510, 26–29. Se cyng ne róhte ná hú swíðe synlíce ða geréfan hit begeátan of earme mannon . . . Hý árérdon unrihte tollas, Chr. 1086; Erl. 220, 15. II. *that which is paid to individuals* :—Sume men syllaþ cyrcan tó hýre swá swá wáclíce mylna . . . ac hit ne gedafnaþ ðæt man dó Godes hús ánre mylne gelíc for lýðrum tolle, Homl. Skt. i. 19, 248–253. (Cf. *molta pensitatio quam a vasallis exigit dominus pro frumenti molitura in molendinis suis*, Migne.) Ðá hí nán þincg næfdon tó syllanne, ðá gyrnde hé ðæs wífes for ðam tolle (*passage money, fare*), il. 30, 168. III. *taking toll* :—Homl. Th. ii. 468, 10. Hé hine geseah sittan æt tolle, 18. Óðer is ðæt man him ðurh fixnoðe bigleofan tilige, and óðer ðæt man ðurh toll feoh gadrige *it is one thing for a man to get his living by fishing, and another to get money together by toll-taking*, 288, 20. IV. *as a technical*

term in England. In this connection *toll* is used to denote not only an amount payable to the king, but also freedom from the payment of such amounts. The word occurs not unfrequently in charters along with *sac, sócn, teám*, and other terms (v. Cod. Dip. Kmbl. i. xlv), and in the Latin version of an English charter is explained as ‘in uendendis et emendis mercibus a tolneto immunitas,’ Cod. Dip. Kmbl. iv. 203, 4–5. In like manner in the Laws of Edward the Confessor it is said : ‘Tol, quod nos vocamus theloneum, scilicet libertatem emendi et vendendi in terra sua,’ Th. i. 451, 30. Toll could be claimed by the king (1) on sales :—Si in strata publica seu in ripa emptorali quislibet mercauerit, thelon ad manum regis subeat ; quod si intus in curte praedicta (*the bishop of Worcester's*) quislibet emerit vel uendiderit, thelon debitum ad manum episcopi reddatur, Cod. Dip. Kmbl. ii. 119, 7–12. Cf. the grant by Edward in 904 of ‘villae mercimonium, quod Anglice ðæs túnes cýping appellatur,’ v. 158, 37 ; and that by ealdorman Æðelréd and Æðelfléd of a half of ‘ǽlc gerihta ðe tó heora hláforddóme gebyraþ on ceápstówe,’ 142, 33. The following passages give instances of the payment of toll :—Hér kýð on ðissere béc ðæt Leówine and his wíf gebohton Ælfilde tó feówer and sixtige penegon and Ælfríc Hals nam ðæt toll for ðæs kynges hand, Chart. Th. 635, 24 : 631, 28 : 639, 15 : 636, 2. Alword portgeréfa and Alwine fángon tó ðam tolle for ðæs cynges hand, 636, 30. Æilsig bohte ánne wífmann and hire sunu mid healfe punde, and sealde Æilsige portgeréfa and Maccosse hundredesmann .iiii. penegas tó tolle, 627, 14. Teolling gebohte Ælword and Édwine tó .vii. mancson tó cépe and tó tolle, and Ælword portgeréfa nam ðæt toll, 633, 2–7 : 639, 20–24. Æilgyuu álýsde Hig and Dunna and heora ofspring tó .xiii. mancson, and Æigulf portgeréfa and Godsuc námon ðæt toll, 638, 12–17. (2) *from ships coming into port*. For a list of such tolls see L. Eth. iv. 2 ; Th. i. 300 ; and for instances of tolls being remitted see Cod. Dip. Kmbl. i. 94, where the toll (*vectigal*) on one ship entering the port of London is remitted to bishop Aldwulf : i. 101, where the king remits ‘ nauis onustae transuectionis censum qui a theloneariis nostris tributaria exactione impetitur ; ut ubique in regno nostro libera de omni regali fiscu et tributo maneat.’ See also pp. 114, 116. In a charter of Cnut the tolls of Sandwich are the subject of grant : ‘nullus homo habet aliquam consuetudinem in eodem portu exceptis monachis aecclesiae Christi. Eorum autem est nauicula et transfretatio portus et theloneum omnium nauium cujuscumque sit et undecumque veniat,’ iv. 21. (3) *on transport by land or water*. See the last passage : ‘ Eorum est transfretatio portus.’ In another charter a grant of land carries with it ‘ theloneum aquarum,’ Cod. Dip. Kmbl. iii. 369, 25. In the charter inserted in the Chronicle under the year 963, *se toll* of certain streams is the subject of grant, Erl. 123, 2. See Kemble's Saxons in England, ii. 73–78. [O. Sax. tol[l] : O. H. Ger. zol[l] : Ger. zoll ; *m.* : *Icel.* tollr ; *m.* : *Dan.* told ; *m.*] v. scip-toll ; toln, *and following words*.

tollere, es ; *m. A toll-taker, tax-gatherer* :—Tollere *telonearius*, Wrt. Voc. i. 50, 56 : *theolonarius*, 74, 45. Matheus wæs tollere, Homl. Th. i. 324, 3 : ii. 288, 17. God hine áwende of tollere tó apostole, 468, 15. Ðone se Hælend geceás of woruldlícum tollere tó gástlícum godspellere, Homl. Skt. i. 15, 129. Ðæs cáseres tolleras áxodon Petrus . . . ‘Wyle eówer láreów ǽnig toll syllan?’ Homl. Th. i. 510, 27. [Ryche Pers þe tollere, H. S. 5816. I seiʒ tolleres in marketes, Piers P. prol. 220. Tollare or takare of tol *telonearius*, Prompt. Parv. 496.] v. tolnere.

toll-freó ; *adj. Free from toll, exempt from payment of toll* :—Tolfreó ofer ealle Engleland, wiðinne burhe and wiðútan, æt gárescépinge and on ǽfrice styde be wætere and be lande *per totam Angliam infra ciuitatem et extra, in omni foro et annuis nundinis et in omnibus omnino locis per aquam et terram, ab omni telonii exactione liberi sint*, Cod. Dip. Kmbl. iv. 209, 19.

toll-sceamol, es ; *m. A seat where a receiver of toll sits, a place for receiving contributions* :—Hé geseah ǽnne man sittende æt tollsceamule (*in teloneo*), Mt. Kmbl. 9, 9. Ðæt folc hyra feoh torfude on ðone tollsceamul (*in gazophilacium*), Mk. Skt. 12, 41, 43. v. toll-setl.

toll-scír, e ; *f. The office of taking toll, business of gathering taxes* :—Matheus árás and forlét his tollscíre *Matthew arose and gave up his occupation as tolltaker*, Homl. Th. ii. 468, 25.

toll-setl, es ; *n. A toll-booth, custom-house* :—Tolsetl *teloneum*, Wrt. Voc. i. 60, 36. Ðá geseah hé sittan sumne mannan æt tollsetle (*in teloneo*) ; in a tollbuðe, Wick. Mt. 9, 9), Homl. Th. ii. 468, 9. Matheus næfre æfter his gecyrrednysse æt tollsetle ne sæt, 288, 18. v. toll-sceamul.

toln, e ; *f. Toll* :—Hé begeat mid his sméhwrencan and mid his golde and seolfre eall dyrnunga æt Steorran, ðe ðá wæs ðæs kinges rǽdesman, ðæt him geweard se þridda pænig of ðære tolne on Sandwíc, Chart. Th. 339, 13 : 340, 35. [Heore is ðæt scip . . . and se tolne of ealle scipen *eorum est navicula . . . et theloneum omnium navium*, 318, 1.] [O. Frs. tolen, tolne ; *f.* ; tolna *to impose toll*: O. Sax. tolna *toll*: M. H. Ger. zoln.] v. *next word, and* toll.

tolnere, es ; *m. A toll-taker, tax-gatherer* :—Tolnere *telonearius*, Wrt. Voc. i. 50, 56 : *exactor*, Germ. 395, 48. [O. Frs. tolner : O. H. Ger. zolnare, zollanari *telonarius, publicanus*: Ger. zöllner.] v. *preceding word, and* tollere.

tó-lúcan; *p.* -leác, *pl.* -lucon; *pp.* -locen *To tear to pieces, wrench asunder, dislocate.* I. literal :—Ðæs ne wêndon witan Scyldinga, ðæt hit (*the hall*) manna ǽnig tóbrecan meahte, listum tólúcan, Beo. Th. 1566; B. 781. Forðon ðe míne innoþas on ðam fylle tólocene wǽron *eo quod interanea essent ruendo convulsa*, Bd. 5, 6; S. 619, 31. Sint mê leoð tólocen, líc sáre gebrocen, Andr. Kmbl. 2807; An. 1406. v. tóhlecan. II. figurative, *to root out, destroy* :—Ic hæbbe ðe gesette ofer ríce and ofer ðióda ðæt ðû hí tólúce and tóweorpe and forspilde and tóstence *constitui te super gentes et super regna, ut evellas et destruas et disperdas et dissipes*, Past. 58; Swt. 441, 31. [Hwil þ Marherete spec þus, me toleac hire, swa þ te reue fir þe stronge rune of þ blodi stream ne mahte for muchele grure lokin þiderweardes, Marh. 7, 11. Wâ is us þ we iseoð þi softe lich toluken swa ladliche, 6, 7. ʒef mi lich is toloken, 6, 12. Heo toluken þene king, and his leomen todrowen, Laym. 2602. Wilde deor limmel toluken ham, & tolimeden eauer euch lið from þe lire, Jul. 79, 5. Ich schal leoten toluken þi flesch þe fuheles of þe lufte *carnes volatilibus dilacerandas reiciam*, Kath. 2092. O. H. Ger. zilochan, -lohhan *devulsus, revulsus*.]

tó-lýsan, tom. v. tó-lísan; tam.

tóm; *adj. Empty*; figuratively, *free from.* Cf. leás :—Ðæt hý môstun mânweorca tóme lifgan and tíres blǽd êcne ágan (cf. the man farid inu an giwald Godes tionono tómig, Hêl. 2490), Exon. Th. 74, 26; Cri. 1212. [Tome saule (*animam inanem*) he filled with fode, Ps. 106, 9. Yee sal find þair tumbs tome (tume), C. M. 17798. Toom or voyde *vacuus*, Prompt. Parv. 496; temyñ or maken empty *vacuo, evacuo*, 488. Scott. toom, tume : *Icel.* tômr : *Dan.* tom.]

tó-mearcian; *p.* ode *To distinguish, describe* :—Tômearcode *distinxit*, Ps. Spl. 105, 32. Ðæt tall ymbehwyrft wǽre tômearcod *ut describeretur uniuersus orbis*, Lk. Skt. 2, 1. v. next word.

tó-mearcodness, e; *f. A description* :—Deós tómearcodnes (*descriptio*) wæs ǽryst geworden fram ðam dêman Cirino, Lk. Skt. 2, 2. v. *preceding word*, and tó-writenness.

tó-meldan *to destroy* peace *by talebearing, by spreading reports* :—Ðær is helle grund ðam ðe sibbe ful oft tómældeþ mid his mûþe (cf. Dante's Inferno, Canto 28, which describes the punishment of the sowers of scandal and schism), Exon. Th. 446, 22 ; Dôm. 26.

tó-middes; *prep.* (*adv.*) I. with dat. (1) marking rest, *in the mids. of, amidst,* (a) preceding the governed word :—Gewurde fæstnis tómiddes ðám wæterum *fiat firmamentum in medio aquarum*, Gen. 1, 6. Iosue hêt áhebban ôðre twelf stánas tómiddes ðam streáme (*in medio Jordanis alveo*), Jos. 4, 9. Tômiddes eów stód ðe gê ne cunnon *medius uestrum stetit quem uos non scitis*, Jn. Skt. i. 26. Hê stód ðær âna tómiddes eallum ðam folce, Homl. Skt. i. 23, 639. (b) following the governed word :—Hê stód him tómiddes, Homl. Skt. i. 23, 617. Ðær ic sylf beó him tómiddes, Homl. Th. ii. 284, 19. Sticaþ him tómiddes, Salm. Kmbl. 1010; Sal. 506. Holte tómiddes, Met. 13, 37 : Cd. Th. 21, 15; Gen. 324. (2) marking motion, *into the midst of* :—Hwænne ðû miht tó ðam folce becuman mid ealre ðínre fare tómiddes Hierusalem, Homl. Ass. 110, 259. Hine ðanon ealle átugan tómiddes ðære cýpinge, Homl. Skt. i. 23, 609. II. with gen. (here, perhaps, *middes* should rather be taken as noun governing the following word in the genitive). (1) marking rest, *in the midst of, in the middle of* :—Ðá fundon hié hiene tómiddes ðara wietena . . . ðá wæs hê gemêt sittende tómiddes ðara láreówa *invenerunt illum sedentem in medio doctorum . . . in medio doctorum sedens invenitur*, Past. 49; Swt. 385, 21–25. Ic sette míne hálgan stówe tómiddes eówre (*in medio vestri*), Lev. 26, 11. Tômiddes hyra *in medio*, Jn. Skt. 8, 3. Ðær ic beó tómiddes heora, L. E. I. 7; Th. ii. 406, 27. (2) marking movement, *into the midst of* :—Ðá hê hiene tómiddes ðæs wêstennes hæfde gelǽdd *in deserta perductus*, Ors. 6, 31; Swt. 286, 17. III. as adverb :—Hê áhte geweald ealles ðæs splottes ðár ðæt scræf wæs tómiddes, Homl. Skt. i. 23, 416. Sete on feówer healfe ðæs ceápes, and án tómiddes, Lchdm. iii. 56, 9. Álegdon ðá tómiddes mærne þeóden, Beo. Th. 6273; B. 3141.

tó-nama, an; *m. A surname, cognomen* :—His tónama wæs Cambises gecweden, Homl. Ass. 103, 25. 'Huættð ðê tónoma (*or* tó noma (*dat.*)? ; *Rush. has* noma) is?' And cuoeð tó him : 'Here tónoma mê is' *quod tibi nomen est? Et dicit ei* : '*Legio nomen mihi est,*' Mk. Skt. Lind. 5, 9. [Ðes wimman hadde on toname Magdalene, O. E. Homl. ii. 143, 13. Nu þu iherest of wuche gomen aras þer þe tonome . . . tonome ariseð ofte of lutle þing þe long ilasted, Laym. 9383. God gyueth the riche bowl towname (v. Lk. 12, 20–21), Piers P. C-text 13, 211. Ger. zu-name. Cf. Dan. til-navn.] v. next word.

tó-namian; *p.* ode *To surname* [:—Simon ðone getónomade (getornomade, MS.) stán *Simonem quem cognominauit Petrum*, Lk. Skt. Lind. 6, 14]. v. *preceding word*.

tó-nemnan; *p.* -nemde *To name separately, distinguish by name into parts* :—Hié ða þrié dǽlas on þreó tónemdon, Asiam, Europem, and Africam *they distinguished the three parts by the three names, Asia, Europe, and Africa*, Ors. 1, 1; Swt. 8, 4. Norþ ôþ ðone gársecg is eall Sciþþia lond binnan, þêh mon tónemne on twá and on þrittig þeóda *north up to the ocean is all Scythia, though it is divided into thirty-two nations,*

each having *its own name*, Swt. 14, 22. Swâ þeáh is tó geþencenne ðæt ða fíf þing þeáh hí tônemde sién mid wordum ðæt hit is eall án þing ðonne hí gegaderode beóþ *atqui necessarium est confiteri nomina quidem esse diversa, nullo vero modo discrepare substantiam*, Bt. 33, 1; Fox 122, 11.

tonian; *p.* ode *To thunder* :—Ic tonige *tono*, Ælfc. Gr. 24; Zup. 138, 3. [From Latin.]

tó-niman; *p.* -nam, *pl.* -námon; *pp.* -numen. I. *to take to pieces, divide* :—Hæfde se cyning his fierd on tû tónumen, Chr. 894; Erl. 90, 17. II. *to take away*, cf. æt-beran :—*Tollite portas, principes . . .* Ðæt byþ on Englisc : Gê ealdras, tónymaþ ða gatu, Nicod. 27; Thw. 15, 8.

tonwinto? The word occurs as a gloss to *adlido*, Txts. 39, 79.

topp, es; *m.* I. *a top, summit* :—Helmes top *apex, summitas galeae*, Wrt. Voc. i. 36, 1. II. *a lock of hair, tuft*; and fig. *a collection of rays of light* (?), as in the tail of a comet :—Se bróðor geseah eall ðæt hûs mid heofonlícre bryhto geondgoten, and hê ðær geseah fýrenne topp (*a stream of light* (?); cf. *Cometa* . . . men cweþaþ on Englisc, ðæt hit sié feaxede steorra, for ðæm ðær stent lang leóma of, Chr. 891; Erl. 88, 19. But, perhaps, *torr* should be- read, as the Latin has *turrim*; and the metrical version of the passage uses that word :—Heofonlíc leóma from foldan up swylce fýren tor ryht árǽred, Exon. Th. 180, 26; Gû. 1285) up of ðære eorþan tó heofones heánnysse, Guthl. 20; Gdwin. 88, 11. III. *a top* to play with (?) :—Mid gelǽredre handa hê swang ðone top mid micelre swiftnysse (the passage is obscure, and perhaps the Latin original has been mistranslated. Thorpe, p. 41, note, cites two Latin versions, one of which has 'accepto ceromate, cum docta manu circumlavit ei cum subtilitate'; the other 'accepto cyramoco, docta manu circulavit eum' : in each case the rubbing after the bath seems to be meant. But *swingan* (q. v.) elsewhere seems always used with the sense of *striking*, and hardly fits in with the meaning of the Latin), Ap. Th. 13, 13. [In later English the word seems mostly used of the hair at the top of the head, or of that which has a similarity with it, e. g. the leafy top of a tree :—Bi þone toppe (coppe, 2nd MS.) he hine nom, Laym. 684. Hongin bi þe toppe (teon bi þe top up, Bodl. MS.), Jul. 28, 6 : Piers P. 3, 139. Top ouer tail, Will. 2776. En vostre chef vus avet toup (*a top of heer*), Wrt. Voc. i. 144, 21 (13th cent.). Ne rohte he þeʒ flockes were lmeind bi toppes and bi here, O. and N. 428. His heer was by his eres ful round ishorn, His top was docked lyk a priest biforn, Chauc. Prol. 590. Top or fortop, top of the hed *aqualium*, Prompt. Parv. 496. Up to þe toppe from þe more, O. and N. 1422 : 1328. A top of flax *du lyn le toup*, Wrt. Voc. i. 144, 27. The word is used also of other things :—Teon seiles to toppa, Laym. 1339. Top or cop of an hey thynge *cacumen*, top of a maste *carchesia*, Prompt. Parv. 496. It is found, too, as the name of a plaything :—En la rue vus juvetz a toup (*a top of tre*), Wrt. Voc. i. 144, 25. Top of chylderys pley *trochus*, Prompt. Parv. 496. Sweype for a top *flagellum*, 482. O. Frs. top *a lock, tuft of hair* : Du. top *top, summit* : O. H. Ger. zopfe; *pl. cicinni, anciae* : Ger. zopf: *Icel.* toppr *a tuft* or *lock of hair*; *a top* of a mast : *Dan.* top *a top, summit*; *a tuft, crest*; *a top* to play with : *Swed.* topp *a top, summit*. The word was taken from the Teutonic into the Romance languages.]

tor *a tower; a rock.* v. torr.

tór; *adj. Difficult, hard.* v. tôr-begete, -cirre [& tat iss harrd & strang & tor and hefiʒ lif to ledenn, Orm. 6350. Erueð (tor, MS. T.) for te paien, A. R. 108, 9. An honful ʒerden beoð erueð for te breken (arn tor to breken, MS. T.), 254, 2. Tor for to telle, Will. 1428. Toor, 5066. O. H. Ger. zuor-, zuir-, zuur-, zûr-: *Icel.* tor-].

toran-eáge. v. toren-íge.

tór-begete; *adj. Hard to get* :—Gif hê beget and yt rinde, sió ðe cymþ of neorxnawonge, ne dereþ him nán ätter. Ðonne cwæþ se ðe ðás bóc wrât ðæt hió wǽre tórbegete, Lchdm. ii. 114, 3–6. Cf. êð-begete, *and see* tór.

tór-cirre; *adj. Hard to turn, hard to convert* :—Ða ðe wǽron ǽr swýðe heardes môdes and swýðe tórcyrres tó Crystes geleáfan, Shrn. 99, 1. Cf. earfoþ-cirre.

torcul glosses torcular, Mt. Kmbl. Rush. 21, 33. [O. H. Ger. torcul; *n.*; torcula; *f. torcular.*]

tord, es; *n. A turd, dung* :—Swínes tord, Lchdm. ii. 62, 22. Gâte tord, 122, 5. Genim níwe horses tord, 330, 27 : 148, 13. Genim culfran tord, 322, 9. v. weorf-tord; tyrdlu, *and next word.*

tord-wifel, es; *m. A dung-beetle* :—Ðær ðû geseó tordwifel on eorþan up weorpan, ymbfô hine mid twám handum mid his geweorpe, Lchdm. ii. 318, 15. [*Icel.* tord-yfill.] Cf. scearn-wifel.

tó-rendan; *p.* -rende *To rend in two, tear in pieces* :—Se hêh ða sacerd tóslát í tórende woedo his *summus autem sacerdos scindens vestimenta sua*, Mk. Skt. Lind. Rush. 14, 63. Wághrægl temples tóre[n]ded (tórended, Rush.) wæs in tuu *velum templi scissum est in duo*, 15, 38. Grin biþ tóránded *laqueus contritus est*, Ps. Th. 123, 7. [Wurmes wullen todelen þine þermes, lifre and lihte torenden, Fragm. Phlps. 6, 59. He is of þe tetore uolke, þet totereð his olde kurtel, and torendeð þe olde pilche, A. R. 362, 29. Haue ruþe of þi faire bodi, þt me ne lete hit noʒt þus torende, Marg. 28, 132. O. Frs. tó-renda.]

toren-íge; *adj. Blear-eyed* :—Gif hē wǽre torenīge (-igge, Cott. MSS.) oððe fleáh hæfde on eágan *si lippus fuerit, si albuginem habens in oculo*, Past. 11; Swt. 65, 5. Wiþ eágena sār, ðæt is ðonne ðæt hwā tornīge (toraneáge, MS. B.) sȳ *ad lippitudinem oculorum*, Lchdm. i. 108, 23. Wið eágena sāre, ðæt ys ðæt wē cweðaþ tornīge (-ēge, MS. H.) *ad epiphoras oculorum*, 156, 18.

torfian; *p.* ode. In the first instance *to throw with turf at* a person (cf. stǽnan), and then with stones or the like; so Icel. has tyrfa með grjóti ok með torfi, and Swed. tyrva med stenom. Afterwards in a more general sense *to throw*. I. *to throw* at an object, *strike* with a missile, *to stone* a person :—Seó clǽnnes ða fūlnesse mid flinte torfaþ *pudicitia libidinem cum saxo percutit*, Gl. Prud. 12 a. Ða deóflu mē swīðe geegsiaþ and eác swylce torfiaþ, Homl. Skt. i. 3, 424. Hī nāmon stānas, ðæt hī hine torfodon, Homl. Th. ii. 236, 21. Hī mid stānum torfodon ðone sóðfæstan Iacob, 300, 18. Hig nāmon stānas tó ðam ðæt hig woldon hyne torfian *tulerunt lapides, ut iacerent in eum*, Jn. Skt. 8, 59 : *ut lapidarent eum*, 10, 31. Ða leásan gewitan hine ongunnon ǽrest tó torfienne, Homl. Th. i. 50, 15. II. *to throw, cast*, (a) with acc. of thing thrown :—Hē geseah hū ðæt folc hyra feoh torfude on ðone tollsceamul, and manega welige torfodon fela *aspiciebat quomodo turba iactaret aes in gazophilacium, et multi diuites iactabant multa*, Mk. Skt. 12, 41. (b) without an object :—Ic torfige oððe sceóte *iacio*, Ælfc. Gr. 28, 6; Zup. 178, 16. Ða Frencisce men torfedon tówærd ðam weofode, Chr. 1083; Erl. 217, 17. [Samuel þe sticches toruede (tarueden, 1st MS.) oueral þan strede, Laym. 16703. Icel. tyrfa *to pelt* a person with something.] v. of-, tó-torfian; turf, *and next word*.

torfung, e; *f.* I. *a throwing of stones, stoning* :—Ðæt hine (*a slave who had absconded*) man lǽdde tó ðære torfunge, L. Ath. v. 6, 3; Th. i. 234, 8. Cf. Si fur servus homo sit, eant sexaginta et viginti servi et lapident eum, iii. 6; Th. i. 219, 13. v. Grmm. R. A. 693. II. *a throwing, casting, hurling* :—Hié his wǽran swīðe ehtende ge mid scotum ge mid stāna torfungum, Ors. 3, 9; Swt. 134, 16.

torht; *adj.* [The word with its derivatives is almost confined to poetry. It is, however, found not unfrequently as one of the components in proper names. v. Txts. 576: cf. *beorht* in the same class of words. See, also, *torhtness*.] *Bright, splendid*. I. of the brightness of light, literal or figurative, (a) referring to things in this world :—Æþelast tungla, torht tācen Godes ðe sun, Exon. Th. 204, 11; Ph. 96. Leóma leóhtade leóda mǽgþum torht, 15, 12; Cri. 235. Upheofon torhtne mid his tunglum *the firmament splendid with its stars*, 60, 13; Cri. 969. Heofon torhtne tungolgimmum, 71, 6; Cri. 1151. Heofonas torhte *the bright skies*, 58, 11; Cri. 934. Tungla torhtast *the sun*, Menol. Fox 219; Men. 111. (b) of heavenly brightness :—Wæs mē swegles leóht torht ontȳned, Exon. Th. 131, 19; Gú. 457. Wuldres leóht torht, 102, 17; Cri. 1674: Andr. Kmbl. 3222; An. 1614: Cd. Th. 299, 28; Sat. 557. II. of splendid appearance, *bright, beautiful, splendid*, (a) of living creatures :—Se torhta fugel (*the phenix*), Exon. Th. 236, 15; Ph. 574. Ða torhtan mægþ (*Judith* : cf. ides ælfscínu, 21, 11; Jud. 14), Judth. Thw. 22, 1; Jud. 43. Englas ælbeorhte, trume and torhte, Exon. Th. 55, 15; Cri. 884. (b) of inanimate objects :—Ðē is neorxna wang, boldwela fægrost . . . torht ontȳned, Andr. Kmbl. 209; An. 105. Ðæt torhte lond, Exon. Th. 199, 19; Ph. 28. Se torhta æsc, 429, 24; Rä. 43, 9. In ðære torhtan byrig, 34, 14; Cri. 542. Of ðam torhtan temple Dryhtnes, 12, 15; Cri. 186. Beám tānum torhtne, 435, 17; Rä. 54, 2. Him hildeðeór hof torht getæhte, Beo. Th. 631; B. 313. Torhtæ *vitreos, claros* (*gurgites*), Hpt. Gl. 406, 48. Tācna torhtost (*the cross seen by Constantine*; cf. ðæt wlitige treów, 330; El. 165), Elen. Kmbl. 327; El. 164. III. *splendid, glorious, noble, illustrious, having splendid qualities* or *properties*, (a) of persons :—Se torhta (*the Deity*), Cd. Th. 282, 29; Sat. 294. Árás se wuldormago, spræc tó his onbehtþegne, torht tó his gesīþe, Exon. Th. 179, 29; Gú. 1269. Bearn Godes, torhtes tírfruma[n], 13, 21; Cri. 206. Torhtne Drihten Hǽlend, Cd. Th. 301, 22; Sat. 575. Torhte and tíreádige twelfe *the twelve apostles*, Apstls. Kmbl. 7; Ap. 4 : Exon. Th. 366, 10; Reb. 10. (b) of things :—Wuldres blǽd torht, Cd. Th. 302, 5; Sat. 594. Seolf onfēng torhtum tācne (*circumcision*), 143, 6; Gen. 2375. Hé benam his feónd torhte tíre, 4, 23; Gen. 58. Ða hálgan duru heofona ríces torhte ontȳnan, Salm. Kmbl. 75; Sal. 38. Abraham wordum God torhtum cígde, Cd. Th. 108, 16; Gen. 1807. Noldan hī ða torhtan tācen (*Christ's miracles*) oncnāwan, Exon. Th. 40, 21; Cri. 642. Torhte frætwe, 211, 20; Ph. 200. In ðone torhtestan þrȳnesse þrym, 140, 29; Gú. 617. IV. of sight or voice, *bright, clear* :—Blind sceal his eágna þolian, oftigen biþ him torhtre gesihþe, Exon. Th. 335, 29; Gn. Ex. 40. Ðūhte him ðæt engel stígan cwōme and stefne ábeád, torhtan reorde, Cd. Th. 248, 10; Dan. 511. [O. Sax. torht : O. H. Ger. zoraht *clear, evident*.] v. freá-, geár-, gold-, heaðo-, heofon-, hilde-, hleór-, mǽre-, mere-, morgen-, randor-, sige-, sigel-, swegel-, wlite-, wuldor-torht.

torhte; *adv.* I. *clearly* :—Frætwe míne (*the swan's feathers*) swōgaþ hlúde, torhte singaþ, Exon. Th. 390, 9; Rä. 8, 8. Him torhte in gemynd his Dryhtnes naman dumba brohte, 440, 24; Rä. 60, 7. II. *beauti-*

fully, splendidly :—Hē anlícnesse geseh torhte gefrætwed, wlitige geworhte, Andr. Kmbl. 1430; An. 715. [O. H. Ger. zorahto *evidenter*.]

torhtian; *p.* ode *To make clear, shew* :—Tācnendi, torctendi *index*, Txts. 71, 1105. [Cf. O. H. Ger. gi-ougozorhtón *manifestare*.]

torht-líc; *adj. Splendid* :—Eów ys wuldorblǽd torhtlíc tóweard, Judth. Thw. 23, 35; Jud. 157. Torhtlícne tíir, Exon. Th. 331, 18; Vy. 70. [O. Sax. torht-lík.]

torhtlíce; *adv. Splendidly* :—Ðæt is sigedryhten ðe ðone sele frætweþ, timbreþ torhtlíce, Exon. Th. 450, 25; Dóm. 93. His mildheortnyss is ofer ús torhtlíce getrymed, Ps. Th. 116, 2 : Andr. Kmbl. 3358; An. 1683. [O. Sax. torhtlíko.]

torht-mód; *adj. Glorious, illustrious*; an epithet of the Deity, Judth. Thw. 21, 4; Jud. 6 : 21, 35; Jud. 93 : of Noah, Cd. Th. 90, 28; Gen. 1502.

torhtness, e; *f. Glory* :—Torhtnis, torchtnis *luculentum*, Txts. 75, 1243. Torhtnes, Wrt. Voc. ii. 51, 16. Gyf him þince (*seem in a dream*) ðæt his hús byrnþ, micel blǽd and torhtnes him byþ tóweard, Lchdm. iii. 170, 10.

tó-rinnan; *p.* -rann *To run in different directions, disperse* (intrans.) :—Suelce hit eall lytlum ríðum tórinne, Past. 38; Swt. 277, 13. [O. H. Ger. ze-rinnan : Ger. zer-rinnen.]

tó-rípan; *p.* te *To pluck in two, tear to pieces* :—Ðá hē fleáh ða tórȳpte hine ān brēmber ofer ðæt nebb. Ðá hē ætsacan wolde ða sǽde him mon ðæt tó tācne *when he fled, a bramble scratched him all over the face. When he wanted to deny* (*the charge brought against him*), *they told him this as a token*, Chart. Th. 172, 27. [v. Goth. raupjan *to pluck* : O. H. Ger. roufen *vellicare, runcare* : Ger. raufen.] v. rípan.

torn, es; *n.* [The word with its derivatives is almost confined to poetry; see, however, *torn-wyrdan*.] *Violent emotion of anger* or *grief* (cf. *teran*, and Goth. ga-taura *a rent*; ga-taurnan *to be torn*). I. of anger, (a) where there is just cause, *anger, indignation, wrath* :—Gewāt torne gebolgen dryhten Geáta (*Beowulf when the dragon ravaged the country*), Beo. Th. 4794; B. 2401. Ne mōton wyt on wǽrlogum wrecan Godes torn, Cd. Th. 152, 34; Gen. 2530 : 4, 24; Gen. 58 : 151, 13; Gen. 2508. Mē ðæt cynn hafaþ sáre ábolgen; nū mē Sethes bearn torn níwiaþ, 76, 16; Gen. 1258. Lífes leóhtfruma leng ne wolde torn prowigean *would not restrain his wrath*, 146, 14; Gen. 2422. (b) *un-righteous anger, rage* :—Wǽron teónsmiðas (*the evil spirits*) tō ees fulle, cwǽdon ðæt him Gúðlác earfeþa mǽst āna gefremede, Exon. Th. 114, 22; Gú. 176. Beóþ ða gebolgne . . . and heora torn wrecaþ *will wreak their rage*, 119, 24; Gú. 259. Synfull yrsaþ tóþum torn þolaþ teónum grimetaþ *peccator irascetur, dentibus suis fremet*, Ps. Th. 111, 9. II. of grief, *grief, affliction, trouble, distress* :—Cyning eallwihta Caines' ne wolde tiber sceáwian; ðæt wæs torn were hefig æt heortan, Cd. Th. 60, 10; Gen. 979. Hȳ twēgen sceolon tæfle ymbsittan, ðenden him hyra torn tóglíde, forgietan ðara geócran gesceafte, habban him gomen on borde, Exon. Th. 345, 3; Gn. Ex. 182. Ðǽr wæs wōpes hring torne bitolden, 34, 6; Cri. 538. Ðá wæs wōpes hring, hát heáfodwylm, ofer hleór goten; nalles for torne teáras feóllon, Elen. Kmbl. 2265; El. 1134. Hē lēt, torn þoliende, teáras geótan, Exon. Th. 165, 15; Gú. 1029. Inwidsorge ðe hié ǽr drugon and þolian scoldon, torn unlytel, Beo. Th. 1670; B. 833. Torn geþolode wine Scyldinga, weána gehwylcne, sídra sorga, 297; B. 147. Torn dreógan, Exon. Th. 131, 20; Gú. 458. Gristbitian mid tóðon torn þoligende *gnashing their teeth in despair*, Judth. 25, 21; Jud. 272. Abraham bæd him fa.umes . . . cwæð ðæt him wǽre weorce on mōde, sorga sárost. . . . Hié Abrahame treówa sealdon, ðæt hié his torn mid him gewrǽcon on wrāðum, Cd. Th. 122, 36; Gen. 2037. Ne sceal nǽfre his torn tó rycene beorn of his breóstum ácȳþan, nemþe hē ǽr ða bóte cunne mid elne gefremman, Exon. Th. 293, 7; Wand. 112. Torna gehwylces, Beo. Th. 4385; B. 2189. [O. Sax. torn *grief, affliction* : Du. toorn *anger* : O. H. Ger. zorn *commotio, zelus, fervor, ira, indignatio, dolor, molestia* : Ger. zorn.] v. gār-, lyge-torn, *and next word*.

torn; *adj. Causing violent emotions of grief* or *anger, grievous, distressing, bitter* :—Hī him ermþu gehēton tornum teóncwidum *they threatened him with misery in grievous words of insult*, Exon. Th. 129, 10; Gú. 419. Ic sceal godscyld wrecan, torne teóncwide (*grievous blasphemies*), ðe ðū tælnissum wiþ ða sélestan sacan ongunne, 254, 30; Jul. 205. Hī mē dǽdun (-m, MS.) torne télnysse, teónan mænige *detrahebant mihi*, Ps. Th. 108, 3. Ðæt wæs Hróðgāre hreówa tornost *it was to Hrothgar the bitterest grief*, Beo. Th. 4265; B. 2129. [O. Sax. torn *bitter* (*tear*).] v. torne, torn-líc.

torn-cwide, es; *m. A speech that causes grief, bitter, grievous, distressing words* :—Heora tungan torncwidum serwaþ swā oft nǽdran dóþ *acuerunt linguas suas sicut serpentes*, Ps. Th. 139, 3. Ongunnon gromheorte (*the evil spirits*) Godes orettan in sefan swencan, swiþe gehēton, ðæt hé in ðone grimman gryre gongan sceolde . . . ; woldun hȳ geteón mid torncwidum in orwénnysse Meotudes cempan, Exon. Th. 36, 25; Gú. 546.

torne; *adv. In a way that causes grief* or *distress, grievously, distressingly* :—Hē wíse dómas dēþ (ðām) ðe hér deorce ǽr teonan manige

torne geþoledan *facit judicium injuriam patientibus*, Ps. Th. 145, 6. Mē ys torne on mōde (cf. ys mē nū hige geómor, 22, 31 ; Jud. 87) *I am distressed in mind*, Judth. Thw. 22, 36 ; Jud. 93. Him ðæs wópes hring torne gemonade, Exon. Th. 182, 22 ; Gū. 1314. Heó mec torne tǽle gerahte (-rǣhte ?), 247, 3 ; Jul. 73.

torn-gemót, es ; *n. A meeting intended to cause trouble or molestation, an attack upon an enemy :*—Gif hē torngemót þurhteón mihte *if he could bring about a meeting with his foe*, Beo. Th. 2284 ; B. 1140.

torn-geníðla, an ; *m. A malignant, grievous, fierce enemy :*—Hēton hine ofer landsceare teón torngeníðlan, swā hié hit frēcnost findan meahton, Andr. Kmbl. 2462 ; An. 1232. Heó wǣron stearce, stāne heardran, noldon hire andsware ǣnige secgan torngeníðlan (*the Jews whom Elene asked about the cross*), Elen. Kmbl. 1132 ; El. 568. Hié (*the wicked after doomsday*) worpene beóþ in helle grund torngeníðlan, 2609 ; El. 1306.

torn-íge. v. toren-íge.

torn-líc ; *adj. Grievous, bitter :*—Ða hēr on tornlícum teárum (cf. wrēdan werk wōpu kūmian, tornon trahnon, Hēl. 5525) sāwaþ, Ps. Th. 125, 5. [*O. H. Ger.* zorn-līh *turbidus, iratus.*]

torn-mód ; *adj. Having the mind excited to anger, having rage in the heart :*—Gē (*the evil spirits*) mec nǣfre mótan tornmóde teón in tintergu, Exon. Th. 141, 2 ; Gū. 621. [Cf. *O. H. Ger.* zorn-muot *turbor ;* zorn-muotig *iracundus.*]

torn-sorh ; *gen.* -sorge ; *f. Anxious care :*—Tornsorgna ful eald ongon eaforan lǣran, Exon. Th. 304, 27 ; Fä. 76.

torn-word, es ; *n. A word that causes distress or grief, a contemptuous, scornful word :*—Hí mē hosp sprecaþ, tornworda fela, Exon. Th. 11, 17 ; Cri. 172. v. torn-wyrdan.

torn-wracu, e ; *f. Grievous revenge :*—Gē hēr āteóþ in ða tornwrǣce (*the destruction with which the evil spirits threatened Guthlac if he remained in his hermitage*) sigeleásne síð, Exon. Th. 120, 16 ; Gū. 272.

torn-wyrdan ; *p.* de *To address abusive words to, to vituperate :*—Hiera wíf him ongeán iernende wǣron, and hié wsýþe tornwyrdon, and ācsedon, gif hié feohtan ne dorsten, hwider hié fleón woldon ; ðæt hié óðer gener nǣfden būton hié on heora wíf hrif gewiton (*the Latin, however, is :* Uxores eorum obviam occurrunt, orant, in praelium revertantur : cunctantibus obscoena corporis ostendunt, quaerentes, num in uteros uxorum vellent refugere), Ors. 1, 12 ; Swt. 54, 2. v. torn-word.

toroc *a bung, stopper of a cask* (?) :—Toroc *dolua*, Wrt. ii. 141, 67. [*From Latin* (?) turachium *epistomium, dolii obturamentum*, Migne. Cf. (?), too, *French* douve *stave of a cask*. Another attempt at a meaning, however, may be suggested. Du Cange, who does not give *dolua*, gives *toroc* as a gloss for *gurgulio ;* if this were the same word as that in the A. S. gloss, perhaps the latter is *tó-roc ;* cf. ed-roc.]

torr, es ; *m.*   I. from Latin *turris, a tower ;* the native word is *stípel ;* q. v. :—Ðín nosu is suelc se torr (*turris*) on Liuano ðǣm munte, Past. 11 ; Swt. 65, 24 : Exon. Th. 266, 23 ; Jul. 402. Tor, Ps. Th. 60, 2 : Exon. Th. 180, 26 ; Gū. 1285. Ðā hēt hire fæder hí bewyrcean on ānum torre mid twelf ðeówennum, Shrn. 105, 33. Æt torre *at the tower* (*of Babel*), Cd. Th. 101, 26 ; Gen. 1688. Tó beácne torr, 100, 19 ; Gen. 1666 : Bt. 25, 4 ; Fox 162, 25. Monn getimberde torr (*turrem*), Mk. Skt. Lind. Rush. 12, 1. Torr (tor, Rush.), Lk. Skt. Lind. 14, 28. Ástāg Simon on ðone torr, Blickl. Homl. 187, 27. Hāt ðū mē ānne heáhne tor of mycclum beámum getimbrian, 183, 3. Hrófas sínd gehrorene, hreórge torras, Exon. Th. 476, 6 ; Ruin. 3 : Andr. Kmbl. 1684 ; An. 844. Ceastre and torras (*farus ;* v. fýr-torr) and strēta and brycge geworhte wǣron, Bd. 1, 11 ; S. 480, 16. Mid ceastrum ða ðe wǣron mid weallum and torrum (*turribus*) and geatum getimbrade, 1, 1 ; S. 473, 28 : Ps. Th. 47, 11 : 121, 7. On ðæs sǣs waroþe tó sūþdǽle ðanon ðe hí sciphere on becom [hí] torras (*turres*) timbredon tó gebeorhge ðæs sǣs, Bd. 1, 12 ; S. 481, 11. Ða torras and ða scylfas on him wǣron elpendas, Nar. 4, 16. *O. Frs.* thoer : *O. H. Ger.* turri, turra *turris.*] Cf. túr.   II. from Celtic, *a projecting rock, a tor :*—Torr *scopulus*, Wrt. Voc. i. 38, 20. Óð him (*the brook*) oninnan felþ muntes mægenstān ātrendlod of ðǣm torre (cf. micel stān wealwiende of ðam heáhan munte, Bt. 6 ; Fox 14, 29) *resistit rupe soluti objice saxi*, Met. 5, 17. Ǽrest on mercecumb (*in Dorset*), ðonne on grēnan pytt, ðonne on ðone torr æt mercecumbes ǽwielme, Cod. Dip. Kmbl. ii. 28, 32. On gyran torr (*in Devon*), iii. 412, 9. An horsa tor . . . on lytlan tor (*in Devon*), Cod. Dip. B. iii. 133, 10, 11. Stānrocca, torra *scopulorum*, Hpt. Gl. 449, 15. Torra *scopulorum*, 499, 68. Cf. Heáhtorra *alpium, montium*, 454, 42. v. fýr-, geat-, heáh-, mere-, seoh-, stān-torr.

torrebrande, Wrt. Voc. ii. 61, 44, read *torre* brande ; cf. *torribus* brandum, 94, 56.

tó-rýpan. v. tó-rípan.

tó-sǣlan ; *p.* de ; *impers. vb. To happen amiss* to a person (*dat.*) in respect to something (*gen.*), *to be lack* of something for a person :—Ne tósǽleþ him gūþgemótes siþþan ic þurh hylles hróf gerǣce *he* (*the dog*) *will not want for fighting, when I* (*the badger*) *reach through the hill's roof*, Exon. Th. 397, 26 ; Rä. 16, 25. Ic beom strong ðæs gewinnes gif ic stille weorþe gif mē ðæs tósǽleþ hí beóþ swíþran ðonne ic *I* (*the anchor*)

am strong for the struggle if I keep still ; if I fail in that they will be stronger than I, 398, 9 ; Rä. 17, 5. Tósǽle, Prov. Kmbl. 65.

tó-samne, -somne ; *adv. Together.*   I. with verbs of motion, where meeting takes place, (1) without hostility :—Ðā cóman ðǽr tósamne unārímedlíco mengeo, Blickl. Homl. 191, 9. Ǽr hí tósomne becómun *antequam convenirent*, Mt. Kmbl. 1, 18. Hēht tósomne ða heó sēleste wiste tó ðære hālgan byrig cuman, Elen. Kmbl. 2401 ; El. 1202. (2) with hostility :—Raðe ðæs ðe hié tósomne cómon *commisso praelio*, Ors. 4, 11 ; Swt. 208, 11. Fóron tósomne wrāðe wælherigas, Cd. Th. 119, 19 ; Gen. 1982.   II. with verbs implying collecting, assembling :—Beóþ ealle sǽfixas gegaderod tósomne *omnes pisces maris in unum congregabuntur*, Num. 11, 22. Hí tósomne eall werod clypedon *convocant totam cohortem*, Mk. Skt. 15, 16. Leóde tósomne bannan, Andr. Kmbl. 2188 ; An. 1095. Hēt ða tósomne síne leóde, Cd. Th. 245, 26 ; Dan. 469.   III. with verbs denoting joining, touching, mixing :—Tósomne gerǣt *congelaverat*, Wrt. Voc. ii. 133, 37. Tósomne cnyllaþ *conliserint*, 134, 66. Se wyrm gebeáh snúde tósomne, Beo. Th. 5129 ; B. 2568. Ða stānas bióþ earfoþe tó tódǽlenne and eác uneaþe tósomne cumaþ, Bt. 34, 11 ; Fox 150, 25. Hié him geblendon tósomne drync unheórne, Andr. Kmbl. 66 ; An. 33 : Exon. Th. 88, 11 ; Cri. 1438.   IV. of action, *in concert, at the same time :*—Ðā burston ða seofon weallas ealle tósomne, Homl. Th. ii. 212, 31. Eall þreó nimeþ fýres wælm tósomne, Exon. Th. 60, 8 ; Cri. 966. Englas hlýdaþ tósomne, 55, 14 ; Cri. 883 : Hy. 3, 16. Ðǽr gelǽde leng ne mihton geseón tósomne *the foes could not longer see one another*, Cd. Th. 190, 30 ; Exod. 207.   V. of uninterrupted time :—Moyses fæste feówertig daga and feówertig nihta tósomne, Homl. Th. ii. 100, 3. Tósomne, 198, 13. Hē fæste hwílum twēgen dagas, hwílum þrý tósomne, Shrn. 52, 20. Hit āgan rínan .xl. daga and .xl. nihta tósomne, Wulfst. 216, 33. [Heo ferden tosomne, Laym. 1393. Tosumne (togadere, 2nd MS.), 61. *O. Frs.* tó-samene : *O. Sax.* te-samne : *O. H. Ger.* zi-samane : *Ger.* zu-sammen.]

tó-samnian ; *p.* ode *To assemble, collect :*—Ðā bæd hē hine ðæt hē sumne dǽl landes æt him onfēnge, ðæt hē mihte mynster on getimbrian and Godes ðeówas tósomnian *he prayed him to receive from him a parcel of land, that he might thereon build a monastery and collect together servants of God*, Bd. 3, 23 ; S. 554, 11.

tó-sāwan ; *p.* -seów *To sow broadcast, scatter seed ;* fig. *to spread abroad, scatter, disperse,* (a) of concrete objects :—Sume hí cwǣdon, ðæt se líchoma ðe ǽne biþ tó duste gewend and wíde tósāwon, ðæt hē nǣfre eft tógædere ne cóme, Homl. Skt. i. 23, 376. Of Noes sunum ys tósāwen (*disseminatum*) eall mancynn ofer eorðan, Gen. 9, 19. Is micel dǽl ðæs mancynnes gehwǣr wíde tósāwen, Homl. Ass. 69, 94. (b) of abstract objects, *to disseminate* opinions, *distribute* favours, *sow* dissension :—Se manu ðe tósāwþ ungeþwǣrnysse betwux cristenum mannum, Homl. Th. i. 492, 14. Swā weordlíce wíde tósāweþ Dryhten his duguþe, Exon. Th. 299, 31 ; Crä. 110. Tósāwaþ (*labia sapientium*) *disseminabunt* (*scientiam*, Prov. 15, 7), Kent. Gl. 511. Ða fyrmestan bydelas ðe Godes lāre geond ðās land tósēowon, Homl. Ass. 56, 143. Seó leáse gesetnys ðe þurh gedwolmen wíde tósāwen is, Homl. Th. i. 438, 1.

tosca (-e) ; *f.* (?) ; *in the Ritual feminines sometimes end in* a), an ; *m. A frog :*—Sceomiende (*the glosser has taken* rubeta *as connected with* rubeo) ða ðió is ācueoden tosca *rubeta illa quae dicitur rana*, Rtl. 125, 27. Sette him heard wíte hundes fleógan and hí ǽtan eác yfle tostan (toscan ?) hæfdan hí eallunga ūt āworpen *immisit in eos muscam caninam, et comedit eos ; ranam, et exterminavit eos*, Ps. Th. 77, 45. Sende on heora eorþan toscean teónlíce *misit in terram eorum ranas*, 104, 26. [Cf. (?) *O. H. Ger.* zuscen *to burn* (so tosca *might refer to the venomous character of the animal*), cf. (?), *also, Swed.* tossa *a toad : Dan.* tudse.]

tó-scádan, -scægde. v. tó-sceádan.

tó-scǽnan ; *p.* de *To break to pieces :*—Bān ne tóscaenas (-scǽnas, Rush.) ꝉ ni gebraecgaþ *ꝉ of ním* (*os non comminuetis ex eo*, Jn. Skt. Lind. 19, 36. Ða feoturo forbræc ꝉ tóscǽnde (-sceǽnde, Lind.) *compedes comminuisset*, Mk. Skt. Rush. 5, 4. Ne furðon ān bāu nǣfde hē mid óþrum ac tóscǽnede ofer eall lāgon and tóworpene geond ða wídan eorþan *he had not even one bone along with another, but broken to pieces they lay in all directions and flung here and there throughout the wide world*, Homl. Skt. i. 23, 496. [Hí þe totorveþ . . . and þine fule bon toscheneþ, O. and N. 1120. In Layamon the word is intransitive :—Þu scalt toscæne mid mire eaxe . . . Corineus smat in enne stane . . . þe stan al tosceande (þat þe ston al tobrac, 2nd MS.), 2309-15.]

tó-sceacan, -scacan ; *p.* -sceóc, -scóc ; *pp.* -sceacen, scacen.   I. *to shake to pieces, shake violently, to disturb :*—Tóscæcþ *concutit, i. turbat, terreat*, Wrt. Voc. ii. 136, 47. Stefn Drihtnes tósceacende wēsten, Ps. Spl. 28, 7.   II. *to shake off, drive away, disperse :*—Ic tósceace *discutio*, Ælfc. Gr. 47 ; Zup. 277, 3. Hit ðæt āttor tósceaceþ, Lchdm. i. 352, 14 note. Hundes sceanca tósceaceþ ðone fefor, 362, 27. Hē tósceóc ðone líg of ðam ofne, swā ðæt ðæt fýr ne mihte him derigan, Homl. Th. i. 570, 14. Hē tóscóc ða dwollícan nytennysse, 602, 35. Módes slǽp tósceac *mentis somnum discute*, Hymn. Surt. 7, 23 : *dissice*, 19, 17. Biþ ðæt gold tósceacen, Wulfst. 148, 23 : 263, 9. [Gromes . . . þe totwicheþ and toschakeþ, O. and N. 1647. A wilde bor . . . man and houndes

. . . wiþ his taskes he al toshok, Beves 742. With shaking shal be toshaken pees, Wick. Is. 24, 20. Wynde may the plantes bigge toshake, Pall. 52, 240. The word is used also intransitively :—All þe worlde shall toshake, Anglia iii. 546, 156.]

tó-sceácerian; *p.* ode *To waste, devastate, scatter* :—Nú is eall mín heord tósceácerod *nunc omnis grex meus vastatus est;* they were scattered (Ezek. 34, 5), L. Ecg. P. iii. 16; Th. ii. 202, 28. Ðá wurdon hí ealle ðearle áfyrhte, and heora gesomnunga ealle wurdon sóna tósceácerode *then (at the coming of the emperor Decius) they (the Christians) were all very frightened, and their congregations were at once scattered*, Homl. Skt. i. 23, 23. v. sceácere.

tó-sceád, es; *n.* I. *a separating, distinguishing, distinction* :—Ne sié fram abbode háda tósceád on mynstre gehealden *non ab abbate persona in monasterio discernatur*, R. Ben. 12, 7. Mid ðæs nicelum dómes tósceáde *cum magna examinis discussione*, Anglia xiii. 375, 141. II. *the faculty of distinguishing* objects presented to the mind, *discrimination, discerning* :—Se Hálga Gást sylþ his gife ðám ðe hé wile. Sumum men hé forgifþ wítegunge, sumum tósceád gódra gásta and yfelra (*to one is given by the Spirit prophecy; to another discerning of spirits* (discretio spirituum), 1 Cor. 12, 10), Homl. Th. i. 322, 27. III. *difference, diversity* :—Hú micel scyle bión ðæt tósceád & hú mislíce mon scyle menn læran mid ðæm cræfte ðæs láreówdómes *quanta debet esse diversitas in arte praedicationis*, Past. 23; Swt. 173, 12. Biþ tósceád, swá swá se apostol sæde: '*Stella ab stella differt in claritate*,' Homl. Ass. 43, 486. Betwuh ðám þrím is swíþe micel tósceád, Bt. 42; Fox 256, 21. Dó sum tósceád betwuh mé and unrihtwísum folce *discerne causam meam de gente non sancta*, Ps. Th. 42, 1.

tó-sceádan, -scádan; *p.* -scéd, -sceád (in the Northern Gospels weak forms are found, and -sceáde occurs in Bede); *pp.* -sceáden. I. *to divide in two, separate* one thing from another, (1) literally, of local relations :—Swá swá sweord ða wunde tósceát on tú, Past. 60; Swt. 453, 17. Se stréam tósceádeþ súþfolc Angelðeóde and norþfolc *flumine meridiani et septentrionales Anglorum populi dirimuntur*, Bd. 1, 25; S. 486, 17. Neáh ðam sǽ ðe Engla land and Pehta land tósceádeþ *in vicinia freti quod Anglorum terras Pictorumque disterminat*, 4, 26; S. 602, 36. Hé tósceádes hiá betuih suá hiorde tósceádas scípo from ticgenum *separabit eos ab invicem, sicut pastor segregat oves ab haedis*, Mt. Kmbl. Lind. 25, 32. Ðætte God efne-gigedraþ monno ne tósceádeþ (tósceáda, Lind.) hé (*separet*), Mk. Skt. Rush. 10, 9. On ðæm dæge God tóscéd on twá eorðan and ðám, Shrn. 63, 24: 62, 35. Ðæt Severus onféng micelne dǽl Breotone and ðone mid díce tósceádde fram óþrum þeódum *ut Severus receptam Brittaniae partem vallo a caetera distinxerit*, Bd. 1, 5; S. 476, 3. Ðæt God gegeadrade monn ne tóslíte í tósceáda (*separet*), Mt. Kmbl. Lind. 19, 6. Tósceádende *segregans*, Ps. Surt. 67, 10. Hé (the stream) on tú tósceáden wyrð, Met. 5, 18. Tósceáden, Wulfst. 26, 2. Ða syndon Temese stréame tósceádene fram Centlande, Bd. 2, 3; S. 504, 16. Tósceádende mid Trentan stréame wiþ Norþ-Myrcum *discreti fluvio Treanta ab Aquilonalibus Mercis*, 3, 24; S. 557, 37. .iiii. fýr nówiht miclum fæce betwyh him tósceáden *quatuor ignes non multo ab invicem spatio distantes*, 3, 19; S. 548, 10. (2) figuratively, (a) *to divide* into parties, *cause division or dissension among* :—Hé tiolode hié betwux him tó tósceádanne . . . swá hé tóscéd ðara éhtera ánmódnesse and Paulus com gesund ðonon *inter semetipsos dividere studuit, quos contra se unitos vidit . . . facta in persecutorum unanimitate dissensio est, et divisa turba illaesus Paulus exivit*, Past. 47; Swt. 363, 1–8. (b) *to separate* contending parties *or* claims, *judge, decide between* :—God stód godum on gemange, and hé hí on midle tósceádeþ (*discernit*; he judgeth among the gods, A.V.), Ps. Th. 81, 1. Ic ne séce mín wuldor, is swá ðeáh se ðe sécþ and tósceát (*judicat*, Jn. 8, 50), Homl. Th. ii. 232, 8. Tósceád (*sors inter potentes*) dijudicat (Prov. 18, 18), Kent. Gl. 656. Tósceád *decerne (quod justum est*, Prov. 31, 9), 1134. Tósceád *discerne* i. *dijudica*, Wrt. Voc. ii. 140, 62. (c) *to make a distinction* between things, *to distinguish, treat or regard differently* :—Sóð lufu ne tósceát nænne be mæglícere sibbe *true love makes no distinction with respect to anybody on account of relationship*, Homl. Th. i. 128, 2. Se heofenlíca Fæder wuldraþ his bearn and tósceát his wuldor fram óðra manna wuldre ðearle unwiðmetenlíce *he distinguishes his glory beyond comparison from the glory of other men*, ii. 232, 9. Hwæt mǽnde Sanctus Paulus, ðá hé his láre suá cræftelíce tósceád (-scéd, Cott. MSS.) (*gave such different counsel in the two cases*), and ðone óðerne lǽrde, ðæt hé him anwald on tuge, óðerne hé lǽrde geðyld? Past. 40; Swt. 291, 20. Ðá ðá hé ðás eorðlícan sibbe tósceád (-scéd, Cott. MSS.) and ðá hefonlícan *cum terrenam pacem a superna distingueret*, 46; Swt. 351, 10. Ongiet georne hwæt sý gód oþþe yfel and tósceád simle *understand thoroughly what is good or evil, and always distinguish between them*, Exon. Th. 302, 34; Fä. 46. Tósceád íntingan mínne of ðeóde unhálgre *discerne causam meam de gente non sancta*, Ps. Spl. 42, 1. (d) *to separate* one thing from another with the mind, *to discern, distinguish, discriminate* :—Seó sáwul is on bócum manegum naman gecýged . . . Heó is *ratio*, ðæt is gesceád, ðonne heó tósceát, Homl. Skt. i. 1, 187. God gesyhþ ælces monnes geþanc, and his word and his dǽda tósceát

(*cernit*), Bt. 40, 7; Fox 244, 1. Mid ðære nose wé tósceádaþ (*discernimus*) stencas, Past. 11; Swt. 65, 20. Is micel niédðearf ðæt se reccere ða ðeáwas and ða unðeáwas cunne wel tósceádan *necesse est, ut rector animarum virtutes ac vitia vigilanti cura discernat*, 20; Swt. 149, 17. Mid hú micelan feó woldest ðú habban geboht, ðæt ðú swutole mihtest tócnáwan ðíne frínd and ðíne fýnd? Ic wát ðæt ðú hit woldest habban mid niiclan feó geboht, ðæt ðú hí cúþest wel tóscádan, Bt. 20; Fox 72, 22. Se ðe gesceádwísnesse hæfþ, se mæg tósceádan hwæt hé wilnian sceal and hwæt hé onscunian sceal, 40, 7; Fox 242, 18: Shrn. 167, 4. Tósceádan *discriminare*, Wrt. Voc. ii. 141, 2. Mǽden ácenned (*born on the first day of the moon*) biþ rihtlíce tósceádende (-ne, MS.), Lchdm. iii. 184, 6. (e) *to separate* things from one another, *to order, dispose, appoint* :—Ic tósceádo (-sceódo, Rush.) iuh suǽ tósceádde (-sceódo, Rush.) mé fæder mín ðæt ríc *ego dispono uobis sicut disposuit mihi pater meus regnum*, Lk. Skt. Lind. 22, 29. Tósceáda *disponere*, Mk. Skt. p. 2, 3. (f) *to separate* the parts of a confused whole, *to expound, interpret, render intelligible* :—Ðegnum his tósceáade (*disserebat*) alle, Mk. Skt. Lind. 4, 34. Tósceáade *interpraetabatur*, Lk. Skt. Lind. 24, 27. Tósceád (*dissere*) ús bisen, Mt. Kmbl. Lind. 13, 36: *edissere*, 15, 15. (g) *to discuss* :—Tósceádeþ *disputat*, Mt. Kmbl. p. 9, 18. II. *to separate in different directions, to scatter, disperse.* v. sceádan, I. 3 :—Manna bán Drihten tósceádeþ *Deus dissipat ossa hominum*, Ps. Th. 52, 6. Tóscádeþ, 67, 14. Meolc wið wíne gemencged ðæt áttor tósceádeþ, Lchdm. i. 352, 14. Stefn Drihtnes tósceádendis (*intercidentis*) lég fýres, Ps. Spl. 28, 7. Ðá tósceáda weard líg, tólýsed, Exon. Th. 277, 22; Jul. 584. III. *intrans. To be separated, to differ* :—Tósceádaþ *discrepent, distant*, Wrt. Voc. ii. 141, 15. His líf tóscéd (e *has been made out of* æ, *and* g *erased before it*: MS. B. *has* tosced : v. tó-sceocgan) fram ússa tída áswundenesse *vita illius a nostri temporis segnitia distabat*, Bd. 3, 5; M. 160, 25. [His lockes he toscædde, Laym. 30262. He wollde hire & te king todælenn and toshædenn, Orm. 19862. Englysche men usede þat tyme þe here of here ouerlyppes tosched & noȝt yschore, Trev. 3, 241. *O. H. Ger.* zi-sceidan *dividere, separare, segregare, discernere, distinguere. Cf. Goth.* dis-skaidan *differe, discernere*.] v. next two words.

tó-sceáden; *adj.* (*ptcpl.*) *Separate, distinct* :—Ælc þing ðe tósceáden biþ from óþrum biþ óþer, óþer ðæt þing, ðeáh hí ætgædere sién. Gif ðonne hwelc þing tósceáden biþ from ðam héhstan góde, ðonne ne biþ ðæt nó ðæt héhste gód *quod a qualibet e diversum est, id non est illud, a quo intelligitur esse diversum. Quare quod a summo bono diversum est sui natura, id summum bonum non est*, Bt. 34, 3; Fox 138, 2–5. v. tó-sceadenness.

tó-sceádend, es; *m. One who divides or separates* :—Tósceádend *discretor, divisor*, Wrt. Voc. ii. 141, 14.

tó-sceádenness, e; *f. Separation, distinction* :—Gé syndon clǽne, cwæð hé tó his þegnum, næs ná hwæðere ealle. Hér on ðysum cwide wæs ðæra apostola tósceádennys *here we have in these words a distinction made among the apostles*, Homl. Ass. 158, 162.

tó-sceocgan (?); *p.* -scægde *To stand out distinctly, be separated* from surrounding objects :—His líf tósceægde fram ússa tída áswundennysse *vita illius a nostri temporis segnitia distabat*, Bd. 3, 5; S. 526, 35. v. sceocgan; tó-sceádan, III.

tó-sceótan; *p.* -sceát, *pl.* -scuton *To rush in different directions, to disperse* (intrans.) *hurriedly, scatter* :—Tóscutan *dissiliunt*, Wrt. Voc. ii. 141, 9. Ðá tóscuton ða deóflu (cf. ða wǽron tóstencte ða wiðerweardan gástas *dispersi sunt spiritus infesti*, Bd. 5, 12; S. 629, 7; it is this passage in Bede which Ælfric is quoting), Homl. Th. ii. 352, 4. Ðá wǽron ða munecas swíðe áférede, nyston hwet heom tó dónne wǽre, ac tóscuton; sume urnon intó cyrcean, Chr. 1083; Erl. 217, 12. [Þe shell toshett (*burst asunder*) on þe schire ground. Whan it cofli tooclef þer crep oute an addre, Alis. (Skt.) 1008.]

tó-sceótan, Met. 27, 19, *is rather to be taken under* sceótan. *The passage is* :—Ungesǽlge men deáþ ǽr willaþ foran tó sciótan = tóforan sceótan *anticipate, rush in front of*; cf. ða ungesǽligan menn forsceótaþ deáþ foran, Bt. 39, 1; Fox 212, 3; *and see passages under* foran, foran-tó.

tó-sciftan; *p.* te *To divide for the purpose of distribution, to divide and distribute* :—Se cyng intó Wealan férde and his fyrde tóscyfte (*divided the force that the parts of it might take different routes*), and ðæt land eall þurhfór, swá ðæt seó fyrd eall tógædere com tó Snáwdúne, Chr. 1095; Erl. 232, 8. Se cyng lét tóscyfton ðone here geond eall ðis land tó his mannon *the king had the troops divided and quartered all over the country on his men*, 1085; Erl. 218, 8.

tó-scirian; *p.* ede *To separate, part* :—Tóscereþ *separat*, Kent. Gl. 575: 603: 727. Bíoþ tóscerede *separantur*, 669. Tóscirid í tódǽled *summotum*, Hpt. Gl. 528, 12. Tóscyrede *abjunctus*, Germ. 397, 441.

tó-scríðan; *p.* -scráð *To flow in different directions, be dispersed* :—Ðæt wæter unstille ǽghwider wolde wíde tóscríþan, wác and hnesce, ne meahte hit on him selfum ǽfre gestandan, Met. 20, 93. [*O. Sax.* ti-skrídan :—Thie neðal tiskréd, Hél. 5633.]

tó-scúfan; *p.* -sceáf *To thrust in different directions, thrust aside, scatter, disperse*, (1) literal :—Se ðone líg tósceáf hátan fýres, Cd. Th.

237, 20; Dan. 340: Exon. Th. 189, 6; Az. 55. Engel ðæt fýr tósceáf, 276, 11; Jul. 564. (2) figurative, *to do away, remove* :—Hé mid ælmessan ealle tóscúfeþ synna wunde, Exon. Th. 467, 28; Alm. 8. Tósceáf (-sceóf, Rush.) ða mæhtigo of sedle *deposuit potentes de sede*, Lk. Skt. Lind. 1, 52.

tó-sencende, Gen. 9, 11. v. tó-stencan.

tó-sendan; *p.* de.   I. *to send in different directions, send away, disperse, scatter* :—Áttru hit tósend *venena diffundet*, Scint. 105, 9. Hé tósende his geféran swilce for huntoðes intingan, Homl. Skt. ii. 30, 104. Hé tósende hí geond ealne middangeard, Homl. Th. i. 232, 5: 462, 15. Ðæra cnapena hundnigontig þúsenda hí tósendon tó gehwylcum leódscipum tó ðeówte *ninety thousand boys they sent away to all nations to slavery*, 404, 15. Ehtatýne sýþum hundteóntig þúsenda hí tósendon and wið feó sealdon wíde intó leódscipas, Blickl. Homl. 79, 23.   II. *to destroy* (?) :—Nabochodonosor com tó Hierusalem and ðæt manncyn ofslóh and ða burh tósende and ðæt tempel tówearp *destroyed* (the narrative in 2 Kings 25 or 2 Chron. 36 does not speak of the dispersion of the inhabitants of Jerusalem, but of the destruction of the city and the captivity of the inhabitants, so that *burh* seems to mean the city, not the citizens, and *tósende* = destroyed: v. 2 Kings 25, 9, 10; 2 Chron. 36, 17-20) *the city and demolished the temple*, Ælfc. T. Grn. 8, 17.

tó-seóðan; *p.* -seáð; *pp.* -soden *To boil to pieces* :—Seóð on cetele and wylle óþ ðæt hió sié eal tósoden, Lchdm. ii. 230, 8.

tó-séðan; *p.* de *To prove* :—Drihten, ðú ús sealdest gesceádwísnesse ðæt wé mágon tóséðan and tósceádan good and yfel, Shrn. 167, 3.

tó-settan; *p.* te *To set things apart from one another, to dispose; disponere* :—Se ðe tósetteþ † gestiht spǽca his on dóme *qui disponet sermones suos in judicio*, Ps. Lamb. 111, 5: Blickl. Gl. : Ps. Spl. 111, 5. Tósette *disposuit*, 83, 6: 104, 8.

tó-sígan; *pp.* -sigen *To fall to pieces, to decay, get worn out* :—Næren tósygene † forgnidene *non extricabantur*; ic tósíge † forgníde *extricor*, Hpt. Gl. 494, 36-39. Næs his reáf horig ne tósigen, Homl. Th. i. 456, 20. Binnon feówertig geára fæce næs nán man gelegerod on eallum ðam folce, ne heora reáf næs tósigen (cf. vestimentum tuum nequaquam vetustate defecit, et pes tuus non est subtritus, en quadrigesimus annus est. Deut. 8, 4), ii. 196, 14. [Þe bodi schal tosie (*printed* -fye), Spec. 101.]

tó-sittan; *pp.* -seten *To sit at a distance from one another, to be placed apart* :—Ðæs landes is .xliii. þeóda wíde tósetene for unwæstmbǽrnesse ðæs londes *gentes sunt quadraginta duae, propter terrarum infoecundam diffusionem late oberrantes*, Ors. 1, 1; Swt. 14, 18.

tó-slacian; *p.* ode *To relax, to make* or *to become remiss* :—Tóslacad (*qui mollis et*) *dissolutus* (*est in opere suo*, Prov. 18, 9), Kent. Gl. 638.

tó-sleán; *p.* -slóh, *pl.* -slógon; *pp.* -slegen *To strike to pieces, knock to bits* :—Tóslóg, tislóg *concidit*, Txts. 51, 516. Tóslóh, forheów *concidit*, Wrt. Voc. ii. 136, 16. (1) of material objects, (a) *to demolish, knock down* a building :—Þunor tóslóg heora hiéhstan godes hús *aedes salutis ictu fulminis dissoluta est*, Ors. 4, 2; Swt. 160, 18: 6, 14; Swt. 268, 29. Swíðlíc wind tóslóh ðæt hús æt ðam feówer hwemmum *a strong wind broke down the house at the four corners*, Homl. Th. ii. 450, 18. Ða hæþenan weras tóslógon his glæsenne calic ; ða gesomnode se bisceop ða brocu, Shrn. 114, 25. (b) *to divide in two by a blow* or *stroke* :—Hé tóslóh sǽ *interrupit mare*, Ps. Lamb. 77, 13. Gif hit (*an egg*) ne tócíne, tósleah hwón *if it will not crack, break it slightly with a blow*, Lchdm. iii. 18, 2. (2) of abstract objects, *to drive away* thoughts :—Ða yflan geþohtas ðe him on mód becumaþ hé sceal sóna on Criste tósleán . . . Ðonne hé hié tóslyhþ on Criste ðonne hé geðenceþ Cristes þrowunge and his wundra and mid ðǽm geþohtum áflýmeþ ða yfelan geþohtas *cogitationes malas cordi suo advenientes mox ad Christum allidere*, R. Ben. 18, 2-6. [*O. Frs.* tó-slá: *O. Sax.* te-slahan: *O. H. Ger.* zi-slahan: *Ger.* zer-schlagen.] v. un-tóslegen.

tó-slífan; *p.* -sláf *To split in two, cleave, cut to pieces* :—Tóslǽf, tócleáf *findit*, Wrt. Voc. ii. 37, 32: 93, 8. [Thai laiden on with swerdes clere, Helm and schield that stronge were Thai gonne hem al toschlíve. Gy of Warwike (in Halliwell's Dict.). *See* slífan, *where the later form of that verb is cited from* Prompt. Parv. 459, *but the reference is omitted*.]

tó-slítan; *p.* -slát, *pl.* -sliton; *pp.* -sliten *To tear in two, tear to pieces, tear asunder* :—Ic tóbrece oððe tóslíte *rumpo*, Ælfc. Gr. 28, 6; Zup. 177, 4. Ic tóslíte *scindo*, Zup. 178, 6 : *lacero*, 36; Zup. 214, 10: *lanio*, Zup. 216, 15.   I. *to tear in two, in pieces, rend* material, e. g. a garment, a bond :—Ðæt níua tóslítaþ *the new maketh a rent*, Lk. Skt. Lind. 5, 36. Se hēhsacerd tóslát † tórende (*scindens*) woedo his, Mk. Skt. Lind. 14, 63 : Past. 3; Swt. 35, 20. Hé tóslát (*disrupisset*) ða raceteága, Mk. Skt. 5, 4. Ne tóslíte ué ðæt cyrtel *non scindamus tunicam*, Jn. Skt. Lind. Rush. 19, 24. Ða hét ic eald hrægl tóslítan and habban wið ðǽm fýre *jussi scissas uestes opponere ignibus*, Nar. 23, 30. Ðæs temples wáhryft wearð tósliten on twēgen dǽlas fram ufeweardon óð nyþeweard *the vail of the temple was rent in twain from the top to the bottom*, Mt. Kmbl. 27, 51. Tóslitten wæs ðæt nett *rumpebatur retia*, Lk. Skt. Lind. 5, 6.   I a. *to give a torn appearance to* anything, *to serrate* (of leaves) :—Ðeós wyrt is gehwǽdon leáfum and tóslitenon, Lchdm. i. 290, 9.   I b. figuratively :—Hé ðone cræft bricþ

and ða orðancas ealle tóslíteþ, Salm. Kmbl. 147; Sal. 72. Gif ðé hwæt yfeles biþ, hraþe hyt byþ tósliten, swá wæs Abdias gyrdels ðæs wítegan, Lchdm. i. 328, 2.   II. *to rend, cleave, break asunder* that which is hard or bulky :—Ðú tóslíte wyllas and burnan *tu dirupisti fontes et torrentes*; thou didst cleave the fountain and the flood (A. V.), Ps. Spl. 73, 16. Hé tóslát stán *dirupit petram*; he opened the rock (A. V.), 104, 39. Hé tóslát sǽ *interrupit mare*, Ps. Lamb. 77, 13. Stánas tósliten † tóbrocen wéron *petrae scissae sunt*, Mt. Kmbl. Lind. 27, 51.   II a. figuratively :—Ða ic ðære heortan heardnesse mid geornfullícre fandunge tóslát *cum cordis duritia studiosis percunctationibus scinditur*, Past. 21; Swt. 155, 5.   III. *to tear the flesh, rend, bite, wound, lacerate*, generally of wounds made by animals, literally and figuratively :—Wurmas tóslítaþ heora líchaman mid fýrenum tóðum, Homl. Th. i. 132, 17. Ða líchoman ðe wildeór ábiton, oþþe fixas tóslitan, Blickl. Homl. 95, 16. Gif hund mon tóslíte oððe ábíte, L. Alf. pol. 23; Th. i. 78, 2. Ðæt se werewulf tó swíðe ne tóslíte, ne tó fela ne ábíte of godcundre heorde, L. I. P. 6; Th. ii. 310, 31. Ðe læs hig (*porci*) eów tóslýton (-slítas, Lind.) *ne dirumpant vos*, Mt. Kmbl. 7, 6. Tóslítan (-en, MS.) *discerpere, dilaniare*, Hpt. Gl. 423, 54. Ðam ðe tósliten (*bitten by a dog*) sý, Lchdm. i. 362, 25 : 370, 16. Se ðe tósliten beó *he that is bitten by a snake*, Num. 21, 8 : Homl. Th. ii. 240, 18. Swá swá sceáp from wildeórum beóþ fornumene, swá ða earman ceasterwaran tóslítene wǽron fram heora feóndum (*discerpuntur ab hostibus*), Bd. 1, 12; S. 481, 26. Scípo diówlíca ne forlǽt ðú onerninge ðætte wé sié tósliteno *oves diabolica non sinas incursione lacerari*, Rtl. 36, 1. Góman beóþ tóslitene, Soul Kmbl. 216; Seel. 110. Ða tóslitenan wunda heó forþrycceþ, Lchdm. i. 356, 14.   IV. *to tear asunder, part, separate* what has been joined, *sever* :—Mon eáþe tóslíteþ, ðætte nǽfre gesomnad wæs, Exon. Th. 380, 33; Rä. 1, 18. Sibbe tóslítaþ sinhíwan tú, 284, 16; Jul. 698. Ðæt God gegeadrade monn ne tóslíte *quod Deus conjunxit, homo non separet*, Mt. Kmbl. Lind. 19, 6.   V. *to pull to pieces, destroy the existence of* an object, abstract or concrete, *to destroy, dissipate* :—Ic undóe † tóslíto tempel ðis *ego dissolvam templum hoc*, Mk. Skt. Lind. Rush. 14, 58. Hý sǽlda tóslítaþ, Salm. Kmbl. 697; Sal. 348. Tóslát *destruit*, Mt. Kmbl. p. 16, 16. Tóslítende (eft gié tóslítas, Lind. Rush.) *Godes bebod rescindentes uerbum Dei*, Mk. Skt. 7, 13. Ríc tósliten biþ *regnum desolabitur*, Mt. Kmbl. Lind. 12, 25. Wæs semninga heofones smyltnes tósliten *subito interrupta est serenitas*, Bd. 5, 1; S. 613, 24. Ðurh ðæt wierð tóslieten (-sliten, Cott. MSS.) sió stilnes hiera hiéremonna módes and biþ gedréfed sió smyltnes hiera lífes *subditorum vitam dissipata quietis tranquillitate confundunt*, Past. 40; Swt. 289, 7.   VI. *to distract* the mind :—Hú oft sió bisgung ðæs ríces tóslít ðæt mód ðæs recceres *quod plerumque occupatio regiminis soliditatem dissipet mentis*, Past. 4; Swt. 37, 11.   VII. *intrans.* To be different :—Tóslíttaþ *discordat*, Mt. Kmbl. p. 2, 8. [*O. H. Ger.* ze-slízan *scindere, secidere, discerpere, lacessere, perdere, dissipare*.]

tó-slite, es ; *m. A rent, tear, laceration, wound made by scratching, cutting,* or *biting*. v. slítan, slite :—Gif hwá tóbrýsed sý, genim ðás wyrte . . . Eác swylce tóslíte heó gehǽleþ, Lchdm. i. 122, 3. [*O. H. Ger.* zur-, zi-sliz *discidium, repudium*.] v. tó-slítness, tó-slítan, III.

tó-slitere (?), es ; *m. One who tears in pieces* ; metaph. *one who causes dissension, a heretic* :—Tóslíterum (tóslitenum ? v. sliten) *hereticis*, Lk. Skt. p. 2, 11. v. next word.

tó-slítness, e ; *f.*   I. *a tearing in pieces, rending in pieces* :—Ungehéredre leoma tóslítnysse wundade *inaudita membrorum discerptione lacerati*, Bd. 1, 7; S. 479, 14.   II. fig. *dissension* :—Tóslítnisse (-slítnisse, Lind.) † *unsibbe dissensio*, Jn. Skt. Rush. 7, 43.

tó-slúpan; *p.* -sleáp, *pl.* -slupon; *pp.* -slopen *To slip apart* or *away, be relaxed, dissolved* :—Heó wæs tólésed † tóslopen *dissolvebatur, collabebatur*, Hpt. Gl. 502, 7. Tóslopen *remissus*, Germ. 393, 137: *dissipatum*, Wrt. Voc. ii. 139, 31. Ábogene, tóslopene *dimissa, i. humilia*, 140, 31.   I. of that which is bound, *to have the connection between several objects* or *that between the parts of the same object relaxed* :—Gif hé hí ne bunde mid his unábindendlícum racentum, ðonne tóslupan hí ealle *conjuncta naturarum ipsa diversitas dissociaret atque divelleret, nisi unus esset, qui quod nexuit contineret*, Bt. 35, 2; Fox 158, 1 : 34, 12; Fox 154, 3. Hæfþ God geheaþorode ealle his gesceafta, ðæt heora ǽlc wræðeþ óþer, ðæt hié ne móton tóslúpan, 21 ; Fox 74, 11. Se godcunda foreþonc heaþeraþ ealle gesceafta ðæt hí ne móton tóslúpan of heora endebyrdnesse *providentia suis quaeque nectit ordinibus*, 39, 5; Fox 218, 31. Mid wriþan gewriþen grundweall ná byþ tóslopen *loramento conligatum fundamentum non dissoluitur*, Scint. 200, 9. Gif se án gestæððega cyning ne staþelode ealla gesceafta, ðonne wurdon hí ealle tóslopene and tóstencte, and tó náuhte wurdon ealle gesceafta *quae nunc stabilis continet ordo, dissepta suo fonte fatiscant*, Bt. 39, 13; Fox 234, 27.   I a. *to be dissipated, destroyed* :—Smyre ða sár, hý tóslúpaþ, Lchdm. i. 268, 3. Mótan sǽs tóslúpan, iii. 36, 27.   II. of that which binds, *to be loosed, undone* :—Ðonne tóslupan ða bendas and tólýsede wǽron *sunt vincula soluta*, Bd. 4, 22; S. 591, 13, 22; 592, 7. Ða wénde heó ðæt seó snód tóslupe, ac heó áfunde ða snóde mid eallum cnottum fæste gewriðen, Homl. Th. ii. 28, 25. Gif hé ða (brídlas) lǽt tóslúpan *hic si frena remiserit*, Bt. 21; Fox 74, 33 : Met.

11, 80. Nú syndon Satanases bendas swýðe tóslopene, Wulfst. 83, 9. **II a.** of illness:—Seó fæstnys (costiveness) tóslýpeþ (-slípeþ, MS. B.), Lchdm. i. 164, 20. **III.** to get relaxed, (a) of material things:—Liþa tóslopene limbs relaxed in sleep, Hymn. Surt. 2, 10. (b) of non-material things, to be relaxed, get remiss:—Ðonne mon lǽt tóslúpan ðone ege nimia resolutione lenitatis, Past. 40; Swt. 289, 2. Ðæt un-geornfulle mód and ðæt tóslopene anima dissoluta, 39; Swt. 283, 12. Ðænne geþanc orsorh byð ágyfen on slǽwþe mód byð tóslopen cum mens secura redditur, in torporem animus laxatur, Scint. 92, 17. **IV.** to get paralysed, get powerless, (a) physically:—Ðær ða sina tóslúpaþ, Lchdm. ii. 280, 3. Ðá weard se líchama eal tóslopen, Homl. Th. i. 86, 25. Sum mǽden langlíce læg on legerbedde seóc, tóslopen on limum, sámcucu geðúht, ii. 510, 25. Se læg seofon geár tóslúpan on limum, Homl. Skt. i. 6, 255. (b) in reference to the mind:—Ðá weard heora heorte tóslopen and heora gást ne beláf on him dissolutum est cor eorum et non remansit in eis spiritus, Jos. 5, 1.

tó-slúping, e; f. Dissolution:—Tóslúpincg lífes dissolutio uitae, Scint. 68, 8.

tó-smeágan, -smeán; p. -smeáde To examine in detail, enquire into the several parts of a subject:—Betwuh ðám þrím is swíþe micel tósceád. Gif wit ðæt ealle sculon ásmeágan (tósmeágan, Cott. MS.), ðonne cume wit late tó ende ðisse béc, oððe nǽfre, Bt. 42; Fox 256, 21.

tó-snídan; p. -snád, pl. -snidon; pp. -sniden. **I.** to cut in two, cut in pieces, cut up:—Hé geteáh his seax and genam his sciccels ðe hé him on hæfde, tósnád ða hine on twá, and healfne sealde ðæm þearfan ... Ðá wǽron ðær manige men ðe ... hine bismrodan, ðæt hé his gegyrelan tósnídan sceolde, Blickl. Homl. 215, 5-10. Tósnidenre hreáþemúse blód, Lchdm. ii. 236, 17. Uppan ðám sticceon ðe ðær tósnidene beóþ membra quae sunt caesa, Lev. 1, 8. **II.** to cut away, cut off:—Sum mon tósnád (amputauit) him ðone ǽarlíprica, Mk. Skt. Rush. 14, 47. [O. Frs. te-snítha: O. H. Ger. za-snídan descindere, dirimere.]

tó-sócness and tó-sócnung gloss adquisitio:—In tósócnisse in adquisitionem, Rtl. 28, 35. Tósócnung adquisitio, 81, 14.

tó-somne, -somnian. v. tó-samne, -samnian.

tó-sprǽc, e; f. Speech addressed to a person, conversation:—Hine God hiéwcúðlicor on eallum ðingum innan lǽrde ðonne óðre menn mid his gelómlícre tósprǽce quem (Moses) de cunctis interius per conversationem cum Deo sedulam locutio familiaris instruebat, Past. 41; Swt. 304, 20. [O. H. Ger. za-sprácha eloquium.]

tó-sprǽdan; p. de To spread out, extend, expand, spread in different directions:—Seó henn tósprǽt hyre fyðera and ða briddas gewyrmþ, Anglia viii. 309, 26. Heó tósprǽt hire bósm ðær ðær ða rédan wuniaþ ..., and heó is genyrwed on ðone ende ðe ða gesceádwísan wuniaþ she expands her bosom where the fierce dwell ..., and is straitened in the quarter where the discreet dwell, Homl. Th. i. 536, 18. Tósprǽd ðíne fingras, Techm. ii. 122, 25. Ðæs mannes sáwl biþ on Gode tósprǽd, swá ðæt heó oferstíhþ middaneard, and eác hí sylfe, Homl. Th. ii. 186, 8. Stríc mid tósprǽddum handum niðer ofer ðíne breóst, Techm. ii. 119, 25. Wíf tósprǽddum loccum a woman with dishevelled locks, Lchdm. iii. 208, 10. [His holie lichame was tospred on þe holie rode, O. E. Homl. ii. 21, 29. Tosprad, 205, 33. He tospret touward ou his ermes, A. R. 402, 9. Þe Brutones þat were tosprad here and þere, R. Glouc. 134, 15. With open hede ... her hair tosprad, Gow. ii. 260, 4. O. L. Ger. te-spreidan dispergere: O. H. Ger. zar-, za-spreitan spargere, expandere, dispergere.]

tó-springan; p. -sprang, pl. -sprungon; pp. -sprungen To spring asunder, fly to pieces, to crack, burst open:—Tó ðám handum ðæt flǽsc tóspringaþ for chapped hands, Lchdm. iii. 114, 4. Se deófol wearp ǽnne stán tó ðære bellan, ðæt heó eall tósprang the bell flew all to pieces, Homl. Th. ii. 156, 10. Hí becómon tó ðam ísenan geate and ðæt tósprang ðærrihte him tógeánes they came to the iron gate, and it burst open straightway at their approach, 382, 12. Tósprang dissilit, Germ. 399, 272. Tóspringe crepet, 398, 112. [Er him þe herte tospringe, C. L. 593. O. H. Ger. zi-springan dissilire: Ger. zer-springen.]

tó-sprytting, e; f. Instigation. v. spryttan, II b.

tostan, Ps. Th. 77, 45. v. tosca.

tó-standan; p. -stód; pp. -standen. **I.** to stand apart, be distant; fig. to differ, be different:—Swǽ micel tóstondeþ eástdael from westdaele quantum distat oriens ab occasu, Ps. Surt. 102, 12. Tóstent, Blickl. Gl. Hú micel tóstent seó godspellíce sóðfæstnyss fram sceade ðære ealdan ǽ, Homl. Th. ii. 70, 29. Tóstænt differt, Wrt. Voc. ii. 140, 13. Tóstent discrepat, 141, 25: dispartire. Tóstandaþ distent, i. separent, 24. Tóstandendum mægna distantes vires, i. discordes, 26. **II.** to stand aloof, not to be forthcoming:—Be ðon ðe mon wíf bycgge, and ðonne sió gift tóstande. Gif mon wíf gebycgge, and sió gyft forð ne cume, L. In. 31; Th. i. 122, 3-6.

tó-stencan; p. -stencte; pp. -stenced, -stenct. **I.** to scatter the parts of a whole, disperse a number of objects gathered together:—Ðú tóstencst hig dissipabis eos, Ps. Spl. 143, 8. Se wulf cymþ tó ðam sceápum, sume hé ábítt, sume hé tóstencþ, Homl. Th. i. 240, 24: 238, 16. Ðínne líchoman geond ðisse ceastre lanan hié tóstenceaþ, Blickl.

Homl. 237, 5. Ðú tóstenctest feónd ðíne dispersisti inimicos tuos, Ps. Spl. 88, 11: 43, 13. Gif wind tó cóme, ðonne tóstencte hé ða lác sacrificium superveniens aura dispergeret, Past. 33; Swt. 217, 22. Se godcunda anweald hí (the builders of Babel) tóstencte, Bt. 35, 4; Fox 162, 24: Homl. Th. i. 318, 18. Tósteng (dissipa) þeóda ðe gefeoht willaþ, Ps. Spl. 67, 34. Ða lác tóstencan (dispergere), Past. 33; Swt. 219, 5. Tóstencud biþ ðæt éde, Mk. Skt. Rush. 14, 27. Tóstenced, Exon. Th. 16, 21; Cri. 256. Tóstencte dispersae, Wrt. Voc. ii. 140, 71. Ðá wǽron tóstencte (dispersi sunt) ealle ða wiðerweardan gástas, Bd. 5, 12; S. 629, 7: 1, 16; S. 484, 14: Bt. 39, 13; Fox 234, 27: Homl. Th. ii. 244, 34. **II.** to destroy the integrity of a whole, dissipate, bring to nought, overthrow:—Tóstencþ (disperdat) Drihten ealle weleras fácnfulle, Ps. Spl. 11, 3. Drihten tóstenceþ (dissipat) geþeaht ðeóda, 32, 10. Tóstencþ goldhord dissipabit thesaurum (Prov. 21, 20), Kent. Gl. 793. Se (Edwy) þurh his cildhádes nytenesse his ríce tóstencte and his ánnesse tódǽlde, Lchdm. iii. 434, 36. Tóstencton (dissipaverunt) unrihtwíse ǽ ðíne, Ps. Spl. 118, 126. Hí munucregol myrdon and mynstra tóstæncton, Chr. 975; Erl. 127, 21. Ofermódignyss seó ðe englas cúþe beswícan, micele má menn tóstencean (dissipare), Scint. 83, 13. Gif ys of mannum geþeaht ðis oððe weorc, sí tostenct (dissoluetur, dissipabitur) ... gif hit of Gode ys, gé ne mágon tóstencean (dissoluere) (Acts 5, 38, 39), 199, 2-4. Tóstencendes dissipantis (sua opera, Prov. 18, 9), Kent. Gl. 639. Ne biþ flód tóstencende, (-sencende, MS.) ða eorðan neque erit diluuium dissipans terram, Gen. 9, 11. Se yfela willa biþ tóstenced, swá récels beforan ðam fýre, gif mon ðæt weorc þurhtión ne mæg (potuisse miserius est) sine quo voluntatis miserae langueret effectus, Bt. 38, 2; Fox 196, 31. Sint tóstente dissipantur (cogitationes, Prov. 15, 22), Kent. Gl. 530. **II a.** intrans. To perish:—Hí tóstencton on ende disperierunt in Endor, Ps. Spl. 82, 9. v. next three words.

tó-stencedness, e; f. **I.** dispersion:—Drihten tóstencednyssa (dispersiones) somnigende, Ps. Spl. 146, 2. **II.** dissipation, destruction:—Hit is micel mægena tóstencennes (-stencednes, MS. T.) plurima destructio est, R. Ben. 128, 6. Ungeþyld is ealra mægna tóstencednys, Homl. Th. ii. 544, 6. Cometa, ðonne hé ætýwþ, ðonne tácnaþ hé hungor oððe cwealm oððe tóstencednysse ðæs eardes, Anglia viii. 321, 22. v. tó-stencan.

tó-stencend, es; m. One who dissipates or squanders, a prodigal:—Tóstencend prodigus, Lchdm. i. lxi, 7.

tó-stencness, e; f. Dispersion:—Tóstencnisse dispersiones, Ps. Surt. 146, 2.

tó-stician; p. ode To stab to pieces, wound severely by stabs, destroy by thrusts:—Funde hé hiene ǽnne be wege licgan mid sperum tósticad healfcucne invenit in itinere solum relictum, confossum vulneribus et extrema vitae efflantem, Ors. 3, 9; Swt. 128, 14. [Cf. Ger. zer-stechen.] Cf. tó-stingan.

tó-stincan; p. -stanc, pl. -stuncon To distinguish by smell:—Ðurh ða nosu wé tóstincaþ, hwæt clǽne biþ, hwæt fúl, Homl. Th. ii. 372, 30.

tó-stingan; p. -stang, pl. -stungon To prick to pieces, break by pricking:—Genim wulfes swýþre eáge and hyt tósting, Lchdm. i. 362, 2. Ðonne ðú ðæt geswel tóstinge oþþe sníþe, ii. 208, 20. [Olde men neddren tostyngeþ (sting them all to pieces, wound severely with their stings), Misc. 152, 177.] Cf. tó-stician.

tó-stregdan, -strédan. [For conjugation see stregdan.] **I.** trans. To disperse, scatter, destroy. The verb occurs mostly in glosses and renders the Latin verbs spargere, aspergere, dispergere, disperdere, dissipare, dispertire, destruere:—Mildheortnisse míne ic ne tóstregdo (-stréde, Ps. Spl., -stregde, C.) misericordiam meam non dispergam, Ps. Surt. 88, 34. Ic tóstréde, Scint. 230, 7. Tóstraigdes í tódríteþ dispergit, Jn. Skt. Lind. 10, 12. Tóstret (-straegdæþ, Lind.: -stregdes, Rush.), Lk. Skt. 11, 23. Tóstrigeded disperdet, Ps. Surt. 77, 38. Tóstrédeþ spargit, Ps. Spl. 147, 5: aspergit, Blickl. Gl.: dispersit, Ps. Th. 111, 8. Fǽgere weras tóstrédaþ ðone líg ðæt he mæg ná scedþan ðisse fǽmnan fair men scatter the flame, so that it cannot harm this virgin, Shrn. 130, 31. Ðú tóstrugde úsic dispersisti nos, Ps. Surt. 43, 12. Tóstregdyst, Ps. Spl. C. 43, 13. Hé tóstregde dispersit, 111, 8. Tóstrægd, Lk. Skt. Lind. Rush. 1, 51: Rtl. 177, 15. Tóstregd hié disperde illos, Ps. Surt. 53, 7: disperge, 58, 12. Tóstrigden (-stregdun, Ps. Spl. C.) wé hié disperdamus eos, 82, 5. Hé ne tóstrugde hié ne disperderet eos, 105, 23. Tóstrogden biþ dispertiatur, Mk. Skt. Rush. Lind. 3, 25. Ne biþ forléten stán ofer stáne se ðe ne sié tóstrogden (destruatur), 13, 2: Lk. Skt. Lind. 21, 6. Tóstrogden biðon (dispargentur) ða scípo, Mk. Skt. Lind. 14, 27. Ða ðe uoeron tóstrogden qui erant dispersi, Jn. Skt. Lind. 11, 52. Hié bióð tóstrogdne dispergentur, Ps. Surt. 58, 16: dissipentur, 67, 2. Tóstródne, 91, 10. Tóstréde synd (dispersa sunt) ealle bán míne, Ps. Spl. 21, 12. Geþancu and geþeahtu ðíne tóstrédde and tó náht getealde beón getácnaþ the dream betokens that your thoughts and counsels will be dissipated and counted for nought, Lchdm. iii. 214, 24. Scípa tóstrogdenra ovium dissipatorum, Rtl. 9, 38. **II.** intrans. To be dispersed, not to keep within proper bounds:—Ðonne ðæt mód flíhþ ðæt ðæt hit sié gebunden mid ege and mid láre, ðonne tóstrét (-strétt, Hatt. MSS.) hit on yfelre and on unnytte wilnunga and hæfþ ðæs suíðe. micelne hunger ut, quo se per

disciplinam ligare dissimulat, eo se esuriens per voluptatum desideria spargat, Past. 39; Swt. 283, 19.

tó-sundrian; *p.* ode *To separate:*—Hwanne hē tósundrode bearn Adames *quando separabat filios Adam*, Cant. M. ad fil. 8. v. tósyndrian.

tó-swápan; *p.* -sweóp *To disperse by a sweeping movement, to sweep apart* or *away:*—Se ðone líg tósceáf, tósweóp hine and tóswende þurh ða swíðan miht *he thrust back the flame on every side, swept and dashed it away by his strong might*, Cd. Th. 237, 23; Dan. 342. Tósweóp and tóswengde, Exon. Th. 189, 13; Az. 59.

tó-swellan; *pp.* -swollen *To swell out, grow big:*—Ic tóswelle *turgeo*, Ælfc. Gr. 26, 3; Zup. 155, 12. Tóswyllaþ *grossescunt, intumescunt*, Hpt. Gl. 447, 29. Se earm wæs swá swíþe greát and tóswollen *brachio in tantum grossescente*, Bd. 5, 3; S. 616, 23. Wiþ tótece ... ðæt tóswollene lim fram dære uferan healfe beþe, Lchdm. ii. 68, 13. Wæs án cnapa swíðe tóswollen þurh wyrmes slege, Homl. Th. ii. 514, 7: Homl. Skt. i. 3, 481. Ða tóswolnan *turgida*, Wrt. Voc. ii. 93, 7. Of ðam tóswollenum fótum (*feet swollen with dropsy*), Homl. Th. i. 86, 11. [Al ic æm toswollen, Laym. 17815. Heorte tobollen & toswollen, A. R. 282, 8. Toswelle *intumescere*, Wick. Jerem. 5, 22. Toswal; *p.* Mand. F. *O. H. Ger.* zi-suollan *tumida*.]

tó-swengan; *p.* de *To dash asunder, dispel by a stroke, drive apart.* v. tó-swápan. [Cf. Mid sweorde toswungen (tohewe, 2nd MS.), Laym. 8026.]

tó-sweorcan *to make dark:*—Beóþ tósworcene ł áþēstrede *obscurantur*, Hpt. Gl. 447, 36.

tó-swífan *to move off in different directions:*—Æghwilc oþer útan ymbclyppeþ, ðý læs hí tóswífen *each from without embraces other, lest they take their separate courses*, Met. 11, 36. v. Bt. 21; Fox 74, 11 *in* tó-slúpan, I.

tó-syndrian; *p.* ode *To separate;* fig. *to distinguish:*—Mid him hē tósyndraþ gif beteran ôðrum wē beóþ gemētte *apud ipsum discernitur si meliores aliis inveniamur*, R. Ben. Interl. 14, 8. Ðū settest on foldan swíðe feala cynna and tósyndrodest hig siddan, Hy. 9, 21; Btwk. 198, 6: Hy. 7, 65; Dom. L. 44, 65. Gescádene ł tósendrede *discretas, segregatas*, Hpt. Gl. 411, 21. v. tó-sundrian.

tot *a projection* (?):—Tot *artura*, Wrt. Voc. ii. 100, 73. [Cf. Þe eorþe aroos in þe manere of a tote (*in modum cumuli*), Trev. v. 163, 11 *note*. Tot, tote *a tuft*, Halliwell's Dict. Tute *a jutting out, projection;* ƚute *to jut out*, Jamieson. *Icel.* tota *a protuberance;* túttıa *to be swelled up: Dan.* tude *a spout.*] v. ge-tot; tot-rida.

tó-talu, e; *f. Reputation:*—Fore ðassum tótales intinge *pro hac reputationis causa*, Rtl. 102, 5.

tó-tellan; *p.* -teled *To distinguish in counting, count separately:*—Ān íglond ligþ út on gársecg ðǽr nǽngu biþ niht on sumera ne wuhte ðon mǽ on wintra dæg tóteled *an island lies out in the ocean, where in summer no night can be distinguished in reckoning time, any more than in winter day*, Met. 16, 15.

tó-teón; *p.* -teáh, *pl.* -tugon; *pp.* -togen.   I. *to pull to pieces, tear to pieces* (lit. and fig.):—Se wyrm ða tungan tótyhþ, Soul Kmbl. 234; Seel. 121. His æfterfolgeras feówertiene geár ðisne middangeard tótugon and tótǽron (*dilaniaverunt*), Ors. 3, 11; Swt. 142, 23. Ðám ðe ús mid tóðum tóteón woldan, Ps. Th. 123, 5. Biþ seó tunge tótogen (beóþ hira tungan tótogenne, Soul Kmbl. 222) on týn healfe, Exon. Th. 373, 25; Seel. 115.   II. *to pull away:*—Tótoghene *detracta*, Hpt. Gl. 515, 14. [Me þe sculde nimen and al tóteón mid horse, O. E. Homl. i. 9, 21. *O. H. Ger.* zi-ziohan ðistranere, *detrahere.*]

tó-teran; *p.* -tær, *pl.* -tǽron; *pp.* -toren *To tear to pieces:*—Ic tótere *lanio*, Ælfc. Gr. 24; Zup. 137, 2. Tóterað *discerpere*, Hpt. Gl. 520, 75. Beón tótoren *lacerari*, 527, 55.   I lit. *to tear to pieces a material:*—Ðū tótǽre (*concidisti*) mín hwíte ꞃægl, Ps. Th. 29, 11. Hē ðæs beran ceaflas tótær, Ælfc. T. Grn. 7, 15. Hē ðone pistol tótær, Homl. Th. ii. 122, 30. Hē tótær his tunecan, 45ʋ, 21. Hí tótǽron heora reáf, 454, 11. Fýrene næddran ðæt folc tótǽron, Num. 21, 6. Swilce hē tótǽre sum eáðelíc ticcen *quasi hoedum in frusta discerpens*, Jud. 14, 6.   II. metaph. *of violent feeling or action, to tear to pieces, to harass, distract, destroy:*—Gýtsung ealle middaneardes rícu tótyrþ *auaritia universa mundi regna discerpserit*, Scint. 99, 8. Welan ða sáwla tóteraþ mid pricungum mislícra geðohta, Homl. Th. ii. 88, 22. His æfterfolgeras feówertiene geár ðisne middangeard tótugon and tótǽron (*dilaniaverunt*), Ors. 3, 11; Swt. 142, 24. Be góde óþres ná sáriga ðū, for nánes gesun[d]fulnysse ðū sí tótoren *de bono alterius non doleas, nullius prosperitate lacereris*, Scint. 77, 9. Hit ongeat his láre swíþe tótorene ... se wísdóm sǽde ðæt his gyngran hǽfdon híne swá tótorenne, Bt. 3, 1; Fox 4, 31–6, 2. [Wolde he teteren roted fleshs ... auh tetereð and tolimeð cwike fleschs, A. R. 84, 5–8. Anne curtel þe wes swíðe totoren, Laym. 4994. Our lordes body they totere, Chauc. C. T. Group C. 474. Cf. *Goth.* dis-tairan.]

tóþ; *gen.* tóþes; *dat.* tēð; *inst.* tóþe; *pl.* toeð, tēþ, *and* tóþas; *m. A tooth, tusk:*—Tóð *dens*, Wrt. Voc. i. 64, 54. Tóþ, 282, 70. Forrotad tóð *dens putridus*, Kent. Gl. 966. Æt ðám feówer tóðum fyrestum,

æt gehwylcum .vi. scillingas; se tóð se ðanne bí standeþ .iv. scill.; se ðe ðonne bí ðam standeþ .iii. scill.; and ðonne sıþþan gehwilc scilling *for knocking out the four front teeth, for each a fine of six shillings: the tooth that stands next must be paid for with four shillings; that which stands next to this with three shillings; and then each tooth afterwards with a shilling*, L. Ethb. 51; Th. i. 16, 2–4. Tóð wið tēð *dentem pro dente*, Ex. 21, 24: Lev. 24, 20. Tóð fore tēð, L. Alf. 19; Th. i. 48, 21. Sete on ðone sáran tóþ, and hwílum ceówe mid ðý sáran tóþe, Lchdm. ii. 310, 16: Exon. Th. 495, 9; Rä. 84, 5. Gif hē tóð of ásleá, Ex. 21, 27. Tóð for tóð, Mt. Kmbl. 5, 18. Tēð *dentes*, tóða flǽsc *gingivae*, ða eahta forworden tēð betwux tuxum *adversi dentes*, Wrt. Voc. i. 43, 29–34. Wið ðæt cildum bútan sáre tēð wexen *to make teething easy for children*, Lchdm. i. 346, 13. Gif ða tēþ synd hole, ii. 310, 17. Oft mann smeáþ hwæðer tēþ bǽnene beón, Lchdm. iii. 104, 4, *and see whole article.* Heora tóþas wǽron gelíce horses twuxan, Guthl. 5; Gdwin. 34, 24: Exon. Th. 226, 18; Ph. 407. Mannes tóða beóþ on eallum his lífe .ii. and .xxx., Salm. Kmbl. 192, 13. Tóða sár, Lchdm. i. 72, 24. Tóþa wagung, 334, 9. Tóþa grystlung (grisbittung tóðana, Lind.) *stridor dentium*, Lk. Skt. 13, 28. Tóða gebitt, Homl. Th. i. 126, 20. Tóða geheáw, Cd. Th. 285, 18; Sat. 339. Bútan tóðum *suaeder*, Txts. 101, 1967. Hié (*walruses; so Icel.* tönn *is used of walrus-tusk*) habbaþ swíþe æþele bán on hiora tóþum; ða tēð hié brohton sume ðæm cyninge, Ors. 1, 1; Swt. 18, 1. Hē tóðum gristbitaþ *stridet dentibus*, Mk. Skt. 9, 18. Synfull tóþum torn þolaþ *peccatos dentibus suis fremet*, Ps. Th. 111, 9: Judth. Thw. 25, 21; Jud. 272. Toeð (tēþ, Ps. Spl.) synfulra, Ps. Surt. 3, 8: ii. p. 194, 19. Tēð, Deut. 32, 24. Tóð (tēð, Ps. Spl.: tóðas, Ps. Th.), Ps. Surt. 57, 7. Hí biton heora tēð mín tógeánes, Homl. Th. i. 46, 27. Tóþas, Exon. Th. 374, 5; Seel. 121: Salm. Kmbl. 230; Sal. 114. [*Goth.* tunþus: *O. Frs.* tôth, tond: *O. Sax. O. L. Ger.* tand: *O. H. Ger.* zand: *Icel.* tönn.] v. cweorn-, flǽsc-, fore-, grinde(-ig)-, wang-tóþ.

-tóþ, -tóþe *-toothed.* [*Icel.* -tannr.] v. blódig-tóþ, twisel-tóþe.

tóþ-ece, es; *m. Tooth-ache:*—Tóðæcce mē forwyrnde ælcre leornunga ... Ic wát ðæt manig broc byð mycle strengre ðonne tóðæce, ðeáh ic nǽfre nán strengre ne geðolode, Shrn. 185, 9–16. Lǽcedómas wiþ ðam uferan tóðece ge wiþ ðam niþeran, Lchdm. ii. 50, 7: 52, 6, 7. v. tóþwærc.

tó-þegnung, e; *f. Administration:*—Tóþēnung *amministratio*, Anglia xiii. 441, 1085.

tó-þenedness, e; *f. Distension :*—Tóþenednyssum *distentionibus*, Hpt. Gl. 529, 1.

tó-þerscan; *p.* -þærsc, *pl.* þurscon *To knock to pieces:*—Ðá com him swilc wind ongeán, swilce nán mann ǽr ne gemunde, and ða scipo ealle tóbeót and tóþræsc, Chr. 1009; Erl. 142, 5.

tóþ-gár, es; *m. A tooth-pick:*—Dó medmicel on ða eágan mid tóþgáre, Lchdm. ii. 36, 9. v. tóþ-sticca.

tó-þindan; *p.* -þand, *pl.* -þundon; *pp.* -þunden *To swell, grow big:*—Ic tóðinde *tumeo*, ðū tóðindst (-þintst, MSS. F. R.: -þindest, MS. U.: -þinst, MS. W.) *tumes*, hē tóðint *tumet*, Ælfc. Gr. 16; Zup. 107, 8–9.   I. in a physical sense:—Rif tóþand mǽdenes *alvus tumescit Virginis*, Hymn. Surt. 44, 1. Tóþindende *turgescens, intumescens (in cumulum)*, Hpt. Gl. 465, 11. Tóþunden *gravis*, Germ. 390, 142.   II. in a metaphorical sense, *to swell with pride, be puffed up, be arrogant:*—Tóþint *intumuerit, superbierit*, Hpt. Gl. 423, 25. Gif heora hwylc tóðint and hine on módignesse onhefþ and hē on ðam leahtre biþ onfunden *si quisque ex eis inflatus superbia repertus fuerit reprehensibilis*, R. Ben. 46, 16. Gif hwylc bróðor ongyten biþ tóþunden (*contumax*), 48, 3. Tóðunden oððe módig *contumax*, Ælfc. Gr. 9, 60; Zup. 69, 4. Is tóþundon (*inflammatum*) mín heorte, Ps. Lamb. 72, 21. Tóþundenys gylpes *tumentis jactantiae*, Hpt. Gl. 527, 36. Gif ænig mid tóðundene módignesse *si aliquis tumido supercilio inflatus*, Chart. Th. 319, 13. Ofermódignysse tóþondere tóblǽwen, Anglia xiii. 441, 1084. Hē hine mid tóðundenum móde forseah, Homl. Th. i. 330, 20: 450, 33. Tóþundenne and ástrehtne hneccan *tumentem et erectam cervicem*, Scint. 83, 17.

tóþ-leás; *adj. Toothless:*—Tóþleásera *edentularum*, Germ. 394, 305. [*O. H. Ger.* zan(e)-lôs *edentulus, edentatus: Icel.* tann-lauss.]

tóþ-mægen, es; *n. Strength of teeth* or *tusks:*—Eofor tóþmægenes trum, Menol. Fox 499; Gn. C. 20.

tóþ-rima, -reoma, an; *m. A gum:*—Tóþrima *gingifa*, Wrt. Voc. ii. 41, 22. Tóðreoma *ingua (gingiua?)*, i. 64, 55. Tóþriman *gingifa*, 282, 72. Wið tóþa sáre and tóðreomena, Lchdm. i. 318, 1, 4. Wið tóþreomena geswelle, 370, 29. Gníd gelóme ða tóðreoman, 346, 14. Mid slítendum tóðreomum *rabidis gingivis*, Hpt. Gl. 423, 45.

tó-þringan; *p.* -þrang, *pl.* -þrungon; *pp.* -þrungen *To press asunder, scatter by pressure:*—Hwílum ic wíde tóþringe lagustreáma full hwílum lǽte eft slúpan tósomne *sometimes I (the storm) drive wide apart the cups of the floods (i. e. the clouds), sometimes let them again glide together*, Exon. Th. 384, 34; Rä. 4, 37.

tóþ-sealf, e; *f. A tooth-salve:*—Wyrc ðus tóþsealfe: ofersǽwisc rind and hunig and pipor, meng tósomne, lege on, Lchdm. ii. 52, 3. Tóþsealfa, 4, 5.

tóþ-sticca, an; m. *A tooth-pick:*—Tóþsticca *dentile*, Wrt. Voc. ii. 138, 68. v. tóþ-gár.

tó-þunden. v. tó-þindan.

tó-þundenness, e; f. I. physical, *swollenness:*—Wiþ ðæra innoþa tóðundennysse, Lchdm. i. 282, 8: 198, 23. II. meta-phorical, *pride, arrogance, contumacy:*—Mid ðam áwyrigdan gáste tóþundennesse tóbláwen *maligno spiritu superbie inflatus*, R. Ben. 124, 5. Gif hé on tóþundennesse þurhwunaþ *si contumax fuerit*, 131, 8. For geþances tóþundennysse *propter mentis tumorem*, Scint. 183, 13. Ða eádmódan ðe náne tóðundennysse nabbaþ, Homl. Th. i. 550, 1.

tó-þundenlíce; adv. *Proudly, arrogantly:*—Tóþundenlíce arro-ganter, superbe, Hpt. Gl. 422, 8. Gif hwylc cræftigra manna for ðæs cræftes þingon hine tóþundenlíce onhefþ, R. Ben. 95, 5.

tó-þuniende *astonishing, amazing:*—Ðæm tóðuniendan *adtonito*, Wrt. Voc. ii. 4, 24. v. þunian.

tóþ-wærc, -wræc, es; m. *Tooth-ache:*—Læcedómas wiþ tóðwærce, Lchdm. ii. 50, 6, 8, 10, 21, 24. Wið tóþwræce, i. 370, 26. v. tóþ-ece.

tó-þwínan. v. tó-dwínan.

tóþ-wyrm, es; m. *A worm in a tooth:*—Wið tóþwærce, gif wyrm ete ða téð ... Wið tóðwyrmum ... læt reócan on ðone múð, dó blæc hrægl under, ðonne feallaþ ða wyrmas on, Lchdm. ii. 50, 10-20.

tótian; p. ode *To peep out, look;* Halliwell gives *toot*—to pry inqui-sitively, as a Northern word:—Se ceác oferhelede ða oxan ealle búton ða heáfudu tótodon út *the basin covered the oxen entirely, except that the heads peeped out;* luterem boves portant, qui facie exterius eminent, sed ex posterioribus latent, Past. 16; Swt. 105, 5. [Ech man þe cumeð pleie to toten (*look at*) oðer to listen, O. E. Homl. ii. 211, 20. Is hit so ouer vuel uor te toten (lokin, MS. T.) utward? ... Toten vt wiðuten vuel ne mei nouðer of ou, & nim ȝeme hwat vuel beo icumen of totinge, A. R. 52, 2-11. Euer se recluses toteð more utwardes, 92, 7. Aȝein kunde hit is, þ te deade totie, 50, 25. He bad me toten on þe tree, Piers P. 16, 22. He maketh him tote and pry, Gow. ii. 143, 6. He stod and totede in, Havel. 2106. Þanne totede y into a tauerne, Pl. Cr. 339. His bon toteden out, 425. *See also note on* totehylle, Prompt. Parv. 497, *and* tootere *speculator*, Wick. Is. 21, 6.]

tó-torfian; p. ode. I. *to fling in different directions, to toss about:*—Wæs ðæt scyp of ðam ýþum tótorfod (*jactabatur*), Mt. Kmbl. 14, 24. Cf. tó-weorpan. II. *to stone to pieces, destroy by throw-ing stones.* [Me þe sculde al toteon mid horse, oðer þe al totoruion mid stane, O. E. Homl. i. 9, 21. Stones hi doþ in heore slitte and þe totorveþ, O. and N. 1119.]

tó-trægelian; p. ode *To pull to pieces, pull away, strip:*—Tótræglion *exuunt*, Germ. 396, 267. v. trægelian.

tó-tredan; p. -træd, pl. -trædon; pp. -treden *To tread to pieces, trample upon:*—Tetridtid *deficit*, Txts. 56, 344. Tetridit *desicit* (*deficit?*), 57, 654. Tetreþ *desicit*, Wrt. Voc. ii. 139, 21. [Sum of þe sede werd totreden, O. E. Homl. i. 133, 22. Heo hit totreden mid horsen, Laym. 26771. Sixti hundred weoren totredene mid horsen, 27473. Wordliche þinges totreden & forhowien, A. R. 166, 22. Totrad *conculcavit*, Wick. Ps. 55, 2. O. L. Ger. te-tredan *conculcare:* O. H. Ger. zi-tretan: Ger. zer-treten.]

tot-rida, an; m. *That which swings on a projection, a swing* (?) *or a swinging figure* (?):—Totrida *oscida*, Wrt. Voc. i. 288, 52. Totridan *oscille*, ii. 63, 56: *oscillae*, Txts. 83, 1466. [Cf. scocga *oscille*, Grff. vi. 416: rita-scopha *oscilla*, 458: ii. 540. See Schmeller's Dict. 3, 310 and Diefenbach's Appendix to Du Cange, p. 402.] v. tot, *and* rídan, III.

tó-twǽman; p. de *To divide, separate, disjoin:*—Ic tótwǽme *disjungo*, Ælfc. Gr. 47; Zup. 277, 4. I. *to divide, stand between objects, separate one object from another:*—Gewurde fæstnis tómiddes ðam wæterum and tótwǽme (*dividat*) ða wæteru fram ðam wæterum. And God geworhte ða fæstnisse and tótwǽmde (*divisit*) ða wæteru, ðe wǽron under ðære fæstnisse, fram ðam ðe wǽron bufan ðære fæstnisse, Gen. 1, 6, 7. II. *to divide, part, dissociate, break the connection between:*—Sume hé (*the devil*) þurh graman tótwǽmþ, Homl. Th. i. 240, 26. Ðonne se líchama and seó sáuul hí tótwǽmaþ *when body and soul part*, Wulfst. 151, 11. Wé nellaþ ðs nǽfre tótwǽman *we do not wish to be separated*, Homl. Skt. i. 2, 71. Hí siredon hú hí hié tótwǽman mehten *Romani dolo divisere hostes*, Ors. 3, 10; Swt. 138, 7. Hié eft tótwǽmde wǽron, 3, 7; Swt. 118, 20. Loth férde fram eástdæle, and hig wurdon tótwǽmede (*divisi sunt*) heora ǽgðer fram his bréðer, Gen. 13, 11. Hí ne beóþ mid ǽnigum fǽce fram him sylfum tótwǽmede; on eallum weorcum hi beóþ tógædere, Homl. Th. i. 500, 5. III. *to disperse, scatter:*—Seó sunne tótwǽmþ ðære nihte þýstru mid hyre beorhtnysse, Anglia viii. 317, 6. Weard hér on felda folc tótwǽmed, Byrht. Th. 138, 57; By. 241. III a. *where the object is abstract:*—Beó ðam hálgan tídan eallum mannum sibb and sóm gemǽne and ǽlc sacu tótwǽmed *let every cause of strife be removed*, L. Eth. vi. 25; Th. i. 320, 29: L. C. E. 17; Th. i. 370, 11. IV. *to divide* with the mind, *distinguish, discern:*—Se apostol tótwǽmed ðæs gástes naman and ðæs módes, Homl. Skt. i. 1, 189. Tótwǽm t tósceád intingan mínne *discerne*

causam meam, Ps. Lamb. 42, 1. Tótwǽmendum (-þwǽm-, MS.) *distinguente, dividente, ordinante*, Hpt. Gl. 438, 54. Ne gemengende hádas ne edwiste tótwǽmende *neque confundentes personas, neque substan-tiam separantes*, Ath. Crd. 4. [Þe eorðe totwemde *the earth yawned*, Marh. 17, 28. Ure louerd totweamede his soule urom his bodie, A. R. 396, 20.] v. un-tótwǽmed.

tó-twǽmedness, e; f. *Division, want of union:*—Áwyrgede gástas beóþ his lǽtteówas and his geséran bútan ǽlcere tótwǽmednesse *accursed spirits will be his guides and comrades in close fellowship*, Wulfst. 194, 22.

tó-tyhting, e; f. *Instigation, prompting, suggestion:*—Ðisses geáres ða Scottas heora cyng Duncan ofslógon, and heom syððan his fæderan Dufenal tó cynge genámon, þurh ðes láre and tótihtinge hé weard tó deáðe beswicen, Chr. 1094; Erl. 231, 2.

tó-ward. v. tó-weard.

tow-cræft, es; m. *Skill in weaving* or *spinning:*—Heó (*the Virgin Mary*) weóx and weard fulfremed on gódra mægna heányssum, and heó ðá sóna gódum towcræftum onféng, swýðor ðonne ǽnig ðara ðe heora bearn wǽron ... Heó wolde beón iram ðære þriddan tíde óð ða nigoþan tíd ymbe hyre webbgeweorc, Homl. Ass. 126, 339. Cf. 132, 545 sqq. According to the Protevangelion, when a new veil for the temple had to be made, it fell to Mary's lot to spin the true purple, c. ix. 4. v. tow-hús, -líc.

tó-weard; adj. I. used attributively, (a) in an indefinite sense, *future, that is to come:*—Praesens tempus ys andwerd tíd ... *futurum tempus* is tówerd tíd, Ælfc. Gr. 20; Zup. 123, 17. Big ðam ege ðæs tóweardan dómes *de terrore futuri judicii*, Bd. 4, 24; S. 598, 15: Bt. 39, 11; Fox 230, 12. Tówurdre *futurae*, Hpt. Gl. 426, 48. Tó fleónne iram ðan tóweardan yrre *a futura ira*, Mt. Kmbl. 3, 7. On tóweardre worulde *in saeculo futuro*, Mk. Skt. 10, 30: Blickl. Homl. 15, 4. Hé nolde ongytan ðone tówerdon deáþ (*death that sometime will come*), 195, 17. Ða misweaxendan bógas of áscreádian, ðæt ða tóweardan ðeónde beón, Homl. Th. ii. 74, 13. Áwrítan ðám tówerdum mannum *to write for future generations*, Homl. Skt. i. 21, 11. (b) of the near future, *about to come, coming, at hand, approaching:*—Se tówarda winter *imminens hiems*, Bd. 4, 1; S. 564, 39. On ðære tóweardan tíde ðe ðá neálǽhte nidða bearnum, Cd. Th. 77, 30; Gen. 1283. Hwylc tówerd yfel ðú ðe on neáhnysse forhtast *quae ventura tibi in proximo mala formidas*, Bd. 2, 12; S. 514, 1. II. used predicatively, (1) referring to future circumstances, *toward* as in Shakespere, e. g. What might be *toward*, that this sweaty haste Doth make the night joint labourer with the day, Hamlet i. 1. (a) (*that is*) *to happen* or *be* some time or other, (*that is*) *to come:*—Se ðe æfter mé tówerd ys *qui post me venturus est*, Mt. Kmbl. 3, 11. Gif hé wiste on hwylcere tíde se þeóf tówerd wǽre, 24, 43. Georne wiste se Scyppend, hwæt tóweard wæs, Homl. Th. i. 112, 25. Hé nát hwæt him tóweard biþ *he knows not what is to happen to him*, Bt. 11, 1; Fox 32, 13. Hé wiste ðæt wíte ðæt him tóweard wæs, Blickl. Homl. 77, 29. Hé ys tóweard on micelre mǽgðe *futurus sit in gentem magnam*, Gen. 18, 18. Se ðe wæs tóweard tó ðisum middangearde, Homl. Th. i. 182, 24. Hé is tóweard tó démenne ðás world, Blickl. Homl. 81, 35. Ða þing ðe eów tówearde synd and hú eówer ǽlcon gebyreð ǽr his ende *quae ventura sunt vobis in diebus novissimis*, Gen. 49, 1. Eallum mannum, ðám ðe nú sint and ðám ðe tówearde sint, Deut. 29, 15. (b) *about to happen,* (*that is*) *to come soon, imminent, impending:*—Mid ðý hé ongeat ðæt him deáþes dæg tóweard wæs *cum diem sibi mortis imminere sensisset*, Bd. 4, 11; S. 579, 24. Tóweard ys ðæt Herodes sécþ ðæt cild tó forspillenne, Mt. 2, 13. Ðonne wambádl tóweard sié *when the disease is coming on*, Lchdm. ii. 216, 19. Tácn hú sió ádl tóweard sié, 256, 21. Hí gesáwon ðæt ðǽr tóweard wæs *they saw what was about to happen*, Lk. Skt. 22, 49. Eów ys wuldorblǽd tóweard *glory is about to come to you*, Judth. Thw. 23, 35; Jud. 157. Noe sægde, ðæt wæs þreálíc þing þeódum tóweard, Cd. Th. 79, 29; Gen. 1318. (c) *where the time is fixed, to take place, come to pass:*—On ðære nihte ðe ðæt gefeoht on merigen tóweard wæs, Homl. Th. i. 504, 21. (2) marking motion, *coming towards a place, approaching, about to come:*—Se Hǽlend geseah ðæt ðǽr wæs mycel mennisc tóweard (cf. se Hǽlend geseah ðæt micel folc com tó him *venit ad eum*, Jn. Skt. 6, 5), Homl. Th. i. 182, 5. Ða ongeáton hié ðæt se eádiga Michael ðǽr wæs tóweard *they then perceived that the blessed Michael had come there* (or *had been present* (?) cf. hí undergeáton ðæt Micahel ðæt tácen his andwerdnysse geswutelian wolde, Homl. Th. i. 506, 14), Blickl. Homl. 205, 2. (2 a) without inflection (or not adjective? v. III. 1 a):—Lócian hwæþer hé ðæt land gecneówe ðæt hié tóweard wǽron *speculari quam regionem teneret*, Ors. 4, 10; Swt. 202, 3. (3) marking position, *with the face towards* a person, *facing:*—Geseoh ðæt hé sié tóweard ðonne ðú in gange, Lchdm. ii. 352, 19. III. used appositively, (1) referring to future events, (a) where the futurity is indefinite:—Ða hálgan ǽr Cristes cyme hyne tóweardne sægdon *said he was to come*, Blickl. Homl. 81, 31: Homl. Th. i. 354, 26, 32. Hé him ðæt ríce tówerd sǽde *he told him that the kingdom was in store for him*, Guthl. 21; Gdwin. 96, 8. Hé forestihte ða

3 T

gecorenan tō đam ēcan līfe, for đan đe hē wiste hī sw·lce tōwearde *he knew they were to become such*, Homl. Th. i. 112, 32, 34. D·ihten ealle gōd him symle fremfullīce tōwearde dyde *the Lord ever had in store for him all good things to his advantage*, Lchdm. iii. 436, 23. ¶ Sometimes the word occurs without the inflexion that seems required, v. also **II.** 2 a; but perhaps in these cases the word should not be considered adjective. v. next word:—Wītgan hine tōweard sǽdon, Blickl. Homl. 71, 29. Ealle đa tācno & đa forebeácno đa đe ūre Drihten ǽr tōweard sægde, 117, 31. Hī geseóþ heora wuldor and heora wlite and blisse hym tōweard, Wulfst. 238, 21. (b) of an immediate future:—Hēr is ūre sylfra forwyrd tōweard getācnod *here is our own destruction shewn to be imminent*, Judth. Thw. 25, 30; Jud. 286. Se engel him sige tōweardne gehēht *the angel promised them that victory should be theirs* (*on the morrow*), Blickl. Homl. 201, 33; 117, 14. Aidan đām scypfarendum đone storm tōwardne sægde (cf. sōna đæs đe gē on scyp āstīgaþ ofer eów cymeđ mycel storm, 32), Bd. 3, 15; S. 541, 16. Hē foreseah Godes mynstrum micle frēcnesse tōwearde *monasteriis periculum imminere praevidens*, Bd. 3, 19; S. 549, 46. Hē wiste heora forwyrd hrædlīce tōweard, Homl. Th. i. 402, 12. (2) marking motion:—Đa leóde flugon đā hié đone here tōweardne wiston on đa burh Gerusalem *the people fled when they knew that the Roman army was on the march to Jerusalem*, Blickl. Homl. 79, 13. Hī gewunodon on gehwilcere byrig, ōđ đæt hī geāxodon đa apostolas tōwearde *they stopped in every town until they learned that the apostles were on the way thither*, Homl. Th. ii. 494, 2. [*O. Sax.* tō-ward.]

tō-weard; *prep. Toward, in the direction of.* **I.** *with gen.*:—Đa đe gāþ on ryhtne weg tōweard đæs hefonrīces, Past. 9; Swt. 59, 19. Hē wæs hym syđđan tōweard hys scypes farende, Homl. Ass. 190, 258. **II.** *with dat. or uncertain*, (1) preceding the governed word:—Đonne ærnaþ hȳ ealle tōweard đæm feó, Ors. 1, 1; Swt. 20, 36. Hī torfedon tōwærd đam weofode . . . and scotedon tōweard đam hāligdōme, Chr. 1083; Erl. 217, 17, 19. Crist wæs tōweard đære rōde gelǽd, Btwk. 214, 27. (2) following the governed word:—Hē eów onet tōweard *mors propinquat*, Met. 27, 8. (3) where *tō* precedes and *weard* follows (cf. *to* God *ward*, *to* us *ward* in A. V.):—Hē hine bær tō mynstre weard, Homl. Th. i. 336, 12: Wulfst. 302, 26. Hē went ǽfre đone hricg tō đære sunnan weard, Lchdm. iii. 266, 24. Tō scipan weard, Chr. 1009; Erl. 143, 11. Hī wendon him tō đære burge weard, 1048; Erl. 178, 1. Hī wǽron heom tō Lundene weard, 1052; Erl. 185, 4. Hē hēt đæt hē biheólde tō his Drihtne werd, Homl. Skt. ii. 31, 78.

tō-weardes; *prep. with dat. Towards.* **I.** preceding the case:—Hī fērdon tōwardes Ou, Chr. 1094; Erl. 230, 31. **II.** following the case:—Mīne frȳnd standaþ ongeán mē and synt mē tōweardes *amici mei adversum me appropinquaverunt et steterunt*, Ps. Th. 37, 11. Deáđ eów tōweardes onet, Bt. 39, 1; Fox 210, 27. Eów neálǽcþ se deáđ tōweardes, Wulfst. 231, 34. [*O. Sax.* tō-wardes.]

tōweard-līc; *adj. Future*:—Mē þincþ ic stande and his āgene stefne gehȳre swā swā hit tōweardlīc is tō gehȳranne, Homl. Skt. i. 23, 831. Đæt tācnaþ tōweardlīce firhto and brōgan, Lchdm. iii. 156, 10. God forgefe alle synne đīno ondweardlīca and tōweardlīca (*futura*), Rtl. 170, 11.

tōweardlīce; *adv. In the future, in time to come*:—Hē forecwæþ đæt hē tōwardlīce biscop beón sceolde *antistitem eum futurum esse praedixerat*, Bd. 4, 28; S. 606, 21.

tōweardness, e; *f.* **I.** *the time to come, the future*:—Đæt hē on tōweardnesse (*in futuro saeculo*) ēcelīce mid Criste rīcsian mōste, Bd. 3, 29; S. 561, 22. Swā đū on đisum andweardum līfe mā earfođa drīgast, swā myccle đū eft on tōweardnysse gefēhst, Guthl. 5; Gdwin. 32, 13. **II.** *a future coming*:—Ūre Drihten đæt gefylde, đæt hē þurh his đa hālgan tōweardnesse gehēt, Blickl. Homl. 119, 28.

tō-weccan; *p.* -wehte *To wake* (trans.) *up, stir up, arouse*:—Hū đa folc mid him fǽhþe tōwehton *how they stirred up strife amongst themselves*, Beo. Th. 5889; B. 2948.

tō-wegan; *p.* -wæg, *pl.* -wǽgon; *pp.* -wegen *To disperse, dispel*:—Heofones gim scīneþ, beóþ wolcen tōwegen *neu concreta nubes summoveat radios solis*, Exon. Th. 210, 11; Ph. 184.

toweht *a basket for putting wool in* (?):—The word occurs among terms connected with spinning and glosses *calatum* (= *calathus*? Calatum *is explained in Du Cange by* lignum piscatorum seu piscaria e lignis confecta, *a meaning which seems not to belong to the word here*), Wrt. Voc. i. 282, 17: ii. 16, 35.

tō-wendan; *p.* de *To overthrow, upset, subvert, overturn*:—Hē tōwende *evertit*, Hpt. Gl. 459, 52. Tōwendum *erutis, subversis*, 433, 44. **I.** with reference to material objects, (a) where the object is not of great extent, *to overthrow, demolish*:—Đā tōwende se hālga wer đæt deófolgild grundlunge, Homl. Th. ii. 164, 16. Đā tōwende se biscop đæt weofod, 508, 5. Hæfde se deófol tōwend đone weall *the devil had thrown down the wall*, 166, 19. Heora deófolgild wearđ tōwend, Homl. Skt. i. 22, 158. (b) where the object is of great extent, *to overthrow, destroy*:—God ealne đone eard tōwende *Dominus subvertit omnem regionem*, Gen. 19, 25. Đā tōwænde se cy..ing heora winsuman burh, Homl. Ass. 102, 8. Hī tōwendon đæt tempel, 68, 83. Hig heora burga tōwendon *subversis urbibus*, Num. 21, 3. Đæt ic đa burh ne tōwende *ut non subvertam urbem*, Gen. 19, 21. **II.** with reference to non-materia. ..jects, *to destroy by changing, to repeal* a law, *abrogate, abolish, overthrow, destroy*:—Crist tōwyrpþ đās stōwe and tōwent đa gesetnysse đe ūs Moyses tǽhte, Homl. Th. i. 46, 3. Hī woldon tōwendon ealle đa gesetnessa đe Domicianus hæfde ǽr geset, Ors. 6, 10; Bos. 120, 32 note. **III.** in a figurative sense:—Hāwa đæt se inra wind đē ne tōwende, Homl. Th. ii. 392, 32. [A sutare þet haueđ forloren his el, he towent euerich strea uort he beo ifunden, A. R. 324, 18. Mid þusendfeld wrenches he þe herte towendeđ, O. E. Homl. ii. 191, 26.]

tō-weorpan, -werpan, -worpan, -wurpan, -wyrpan; *p.* -wearp, *pl.* -wurpon; *pp.* -worpen *To throw in different directions, throw away, throw down, to scatter, disperse, destroy, overthrow*:—Tōwearp *discutit*, Wrt. Voc. ii. 28, 70. Tōwuorpon *destituunt*, tōworpne *destitutae*, 105, 81, 82. Tōwurpon, 25, 13. Tōworpenum *eruta*, 33, 16. *Destitutae, desertae, i. derelictae, vel* tōworpne, 139, 11. **I.** *to scatter* (lit. or fig.), *disperse*:—Se đe ne gaderaþ mid mē, hē tōwyrpþ (*spargit*), Mt. 12, 30. Hē sendeþ his strǽlo and hī tōweorpeþ (*dissipavit*), Bd. 4, 3; S. 569, 20. Đū ūs tōdrife and ūs tōwurpe geond werþeóda, Ps. Th. 59, 1. Hē đæt fȳr tōsceáf and đone līg tōwearp, Exon. Th. 276, 15; Jul. 566. Tōweorp đū đa đeóda *dissipa gentes*, Ps. Th. 67, 28. Đæt hē heora oferhȳd tōweorpe *ut superbiam eorum dissipet*, Bd. 4, 3; S. 569, 24. Mid đȳ đe hē sceolde his gestreón tōweorpan, mid đȳ hē hié gadraþ, Past. 8; Swt. 55, 11. Ūre bān syndon tōworpene *dissipata sunt ossa nostra*, Ps. Th. 140, 9. Đætte suno Godes, đa đe uoeron tōuorpen (*dispersi*), gesomnade in ān, Jn. Skt. Lind. 11, 52. **I a.** *to break in pieces, scatter the parts of a connected whole*:—Hē heora bendas tōwearp *vincula eorum disrupit*, Ps. Th. 106, 13. **II.** in a literal sense, *to overthrow*, (a) *to overturn* what is standing:—Hē āgeát đara mynetera feoh, and tōwearp hyra mȳsan (*mensas subvertit*), Jn. Skt. 2, 15. (b) *to throw down* what is set up, *destroy* a building, *demolish*:—Gif eówer godes miht đa cyrcan tōwurpan ne mæg, ic tōwurpe eówer tempel, Homl. Th. i. 70, 30. Đes tōwyrpþ (-wærpađ, Lind.) Godes templ, and hyt eft getimbraþ *qui destruebat templum Dei, et illud reaedificabat*, Mt. Kmbl. 27, 40. Đū tōwurpe weallfæsten his *deposuisti maceriam ejus*, Ps. Th. 79, 12. Ceaster heora đū tōwurpe (*destruxisti*), Ps. Spl. 9, 6. God tōwearp (*subvertit*) đa burga, Gen. 19, 25: Ors. 3, 7; Swt. 114, 2. Hē tōwearp đæt templ *Titus templum diruit*, 6, 7; Swt. 262, 20. Se godcunda anweald tōwearp đone torr (*the tower of Babel*), Bt. 35, 4; Fox 162, 25. Æþelburg tōwearp Tantūn đe Ine ǽr timbrede, Chr. 722; Erl. 44, 27. Hī tōwurpon đa heargas *destructis fanis*, Bd. 3, 30; S. 562, 15. Englas đæt hūs tōwurpon þurh gāstlīcne cræft, Homl. Th. ii. 510, 15. Đȳ læs eówer hūs windas tōweorpan, Exon. Th. 281, 22; Jul. 650. Hī mid æxum duru curfan, and teoledan đæt hī mid adesan ealle tōwurpan (*dejecerunt*), Ps. Th. 73, 6. Tōwurpan (-worpan, MS. A.: -weorpan, Rush.: -worpa, Lind.) Godes templ *destruere templum Dei*, Mt. Kmbl. 26, 61: Homl. Th. ii. 510, 13. Tōworpon, Bd. 2, 13; S. 517, 14. Tōwyrpan hira geweorc, Bt. 35, 4; Fox 162, 13. Ne biđ hēr lǽfed stān uppan stāne đe ne beó tōworpen, Mt. Kmbl. 24, 2: Mk. Skt. 13, 2. Wearđ Tirus seó mǽre burg eall tōworpenu *Tyrus excisa est*, Ors. 3, 9; Swt. 128, 28. Æfter tōworpenum templan *post deruta sacella*, Hpt. Gl. 467, 56. **III.** in a figurative sense, *to overthrow*, (a) where the object is a person, *to destroy the power of* a person, *to destroy*:—Hī tōweorp *destrue eos*, Ps. Th. 58, 11. Đæt đū tōwurpe feónd *ut destruas inimicum*, Ps. Spl. 8, 3. Hī wolde tōweorpan wuldres Aldor . . . đæt hē hī ne tōwurpe geond werþeóda *dixit ut disperderet eos . . . ne disperderet eos*, Ps. Th. 105, 19. Swā sint tō teweorpanne đa đe nān gōd ne dydon đurh đreáunge *qui nulla agere bona coeperunt, correctionis manu evertendi sunt*, Past. 58; Swt. 443, 33. Ic wolde tōwerpan bearn Hēlendes, Cd. Th. 270, 4; Sat. 85. Noldan hī tōworpan þeóde *non disperdiderunt gentes*, Ps. Th. 105, 26. Wutan hī tōwyrpan *disperdamus eos*, 82, 4. (b) where the object is not a person, *to overthrow* an institution, a practice, regulation, law, etc., *to put down, put an end to, destroy, make void, break, dissolve*:—Se đe tōwyrpþ ān of đysum bebodum *qui solverit unum de mandatis istis*, Mt. Kmbl. 5, 19. Hē ūre ǽ tōwyrpþ. Pilatus hym cwæđ: 'Hwæt ys đæt hē dēþ đæt hē mǽge eówre ǽ tōwerpan?' Hī cwǽdon: 'On restedagum hē hǽlþ,' Nicod. 2; Thw. 1, 23–27. Hē com tō đȳ đæt hē wolde ǽlc yfel tōwurpan, and ǽlc good ārǽran. Nū tōwyrpþ hē on ūs leahtras . . . Hē tōwyrpþ mōdignysse . . . and ealle unđeáwas hē tōwyrpþ, Homl. Th. i. 144, 28–32. Se wind tōweorpþ đære rosan wlite, Bt. 9; Fox 26, 19. Tōweorpeþ (-worpeþ, MS. B.), Salm. Kmbl. 149; Sal. 74. Đa heargas āīdlian and tōweorpan *fana profanare*, Bd. 2, 13; S. 516, 40. Hē wile ūre wītu tōweorpan *he will put an end to the pains we inflict*, Cd. Th. 289, 5; Sat. 393. Mīne āre tōweorpan *honorem meum repellere*, Ps. Th. 61, 4. Wutun symbeldagas Drihtnes on eorđwege ealle tōwurpan *comprimamus omnes dies festos Domini a terra*, 73, 8. Nelle gē wēnan đæt ic cōme tōwurpan (*solvere*) đa ǽ; ne com ic nā tōwurpan (-wearpan, MS. A.), ac gefyllan, Mt. Kmbl. 5, 17. Uton tōwurpan đās geflitu *dissolvamus has contentiones*, Coll. Monast. Th. 31, 23. Ælfhere hēt tōwurpon swȳđe manig munuclīf, Chr. 975; Erl. 127, 5. Oft becymþ se anweald đisse worulde tō swīþe gōdum

monnum for ðæm se anweald ðara yflana weorþe tóworpen *fit saepe, uti bonis summa rerum gerenda deferatur, ut exuberans retundatur improbitas,* Bt. 39, 11; Fox 228, 20. Ðone tóworpenan stal ðæs ríces *destructum regni statum,* Bd. 4, 26; S. 603, 8. Ðý læs tóworpen sién fyrngewritu, Elen. Kmbl. 860; El. 430. **IV.** *to throw out.* v. tóworpness, II :—Ðonne hió hié selfe tóweorpeþ út of hiere selfre *cum se extra semetipsam ejicit,* Past. 38; Swt. 277, 24. [Ne bið naut his *(the wise man's)* lare fremful, ʒif he mid wercan towerpeð his bodunge, O. E. Homl. i. 109, 7. Þatt temmple wass all þurrh hæþenn follc toworrpenn, Orm. 16277. *O. Frs.* tó-, ti-werpa: *O. Sax.* te-werpan *to scatter, to destroy: O. H. Ger.* zer-, ze-werfan *dissipare, dijicere, dispergere, destruere, demoliri: Ger.* zer-werfen.]

tó-weorpendlíc, -wyrpendlíc; *adj. Destructible:*—Tówyrpendlícne *destructilem,* Germ. 394, 348.

tó-wesness, -wesenness, -wisness, e; *f.* **I.** *separation, dissolution, divorce:*—Tówesnes *vel* tólésednes *dissolutio, dispersio,* Wrt. Voc. ii. 141, 40. Tówesnisse *defortii,* Txts. 181, 41. **II.** *difference, disagreement, discord, dissension:*—Hé sǽwþ ðone sticel ðæs andan óððæt ðær of áweoxþ tówesnes, and of ðære tówesnesse biþ ðæt fýr onǽled ðære feóunga ... Se se ðæt wæter út forléte wǽre fruma ðære tówesnesse *seminantur stimuli, oriuntur rixae, accenduntur faces odiorum ... Qui dimittit aquam, caput est jurgiorum,* Past. 38; Swt. 279, 9–13. Hú ʼunáberendlíc gylt sió tówesnes *(discordia)* biþ, 46; Swt. 349, 15. Wæs tówesnes geworden *orta dissensione,* Bd. 4, 12; S. 581, 15. Ðá sóhte Colemannus ðysse tówisnesse (-wesennesse, MS. B.) and ðysse unsibbe lécedóm *quaesivit Colmanus huic dissensioni remedium,* 4, 4; S. 571, 6. Ðonne hé him ondrǽt ða tówesnesse útane *dum humana foras jurgia metuunt,* Past. 46; Swt. 351, 23. [Cf. ge-weorþan *to agree: Goth.* gawairthi *peace.*]

towettan; *p.* te *To associate with:*—Riht is ðæt mynecena ne towettan woruldmannum ne ǽnige sundorcýððe tó heom habban ealles tó swíðe *(the other reading is* næfre wíð worldmen ǽnige gemánan worldlícre cýððe habban tó swíðe), L. I. P. 15; Th. ii. 322, 33.

tow-hús, es; *n. A spinning-house:*—Towhús of wulle *genitium* (= gynaeceum *locus seu aedes ubi mulieres lanificio operam dabant.* The women who worked were called *geniciariae pensiles,* Migne), Wrt. Voc. i. 59, 7. v. tow-cræft, -líc, -tól.

tó-wiðere, -wiðre; *prep. Against.* **I.** with dat. *in reply to:*—Hú mæg ic andsware findan wráþum tówiþere, Exon. Th. 12, 13; Cri. 185. **II.** with acc. *in opposition to:*—Wíg tówiþre, 341, 20; Gn. Ex. 129. [*Ger.* zu-wider.]

tow-líc; *adj. Pertaining to weaving:*—Towlíc weorc *textrinum opus,* Wrt. Voc. i. 26, 13: 82, 11.

tow-mýdrece, an; *f. A work-box, box for keeping materials connected with spinning or weaving* (?):—Án hræglcysð and án lytulu towmýderce, Chart. Th. 538, 21.

tó-worpenness, -worpedness, e; *f. Desolation, destruction:*—Heora tóworpennys *the destruction of the Jews by the Romans,* Homl. Th. i. 108, 3. Ða onsceonunge ðære tóworpennysse *abominationem desolationis,* Mt. Kmbl. 24, 15. Tóworpednysse (-worpennysse, MS. A.), Mk. 43, 14. Deós tódrǽfednys *(the driving the money-changers out from the temple)* getácnode ða tóweardan tóworpennysse ðurh ðone Rómāniscan here, Homl. Th. i. 406, 9. Ðær se Hǽlend beweópe ðære ceastre tóworpennysse, ðe gelamp æfter his ðrowunge, 402, 7: Homl. Ass. 46, 548.

tó-worpness, -wyrpness, e; *f.* **I.** *dispersion.* v. tó-weorpan, I :—On tówyrpnisse hǽðna *in dispersionem gentium,* Jn. Skt. Rush. Lind. 7, 35. **II.** *a throwing out, ejection.* v. tó-weorpan, IV :—Salde him mæhte gémnisse tó untrymnissum and tóworpnisse (-wyrpnise, Lind.) diówla *dedit illis potestatem curandi infirmitates et eiciendi daemonia,* Mk. Skt. Rush. 3, 15. [*O. H. Ger.* zi-worfnessi *desolatio.*]

tó-wrecan; *p.* -wræc, *pl.* -wrǽcon; *pp.* -wrecen *To drive in different directions, scatter, disperse:*—Weorðaþ tówrecene wíde ealle ða ðe unrihtes ǽror worhtan *dispergentur omnes qui operantur iniquitatem,* Ps. Th. 91, 8: 58, 15. Siendon wé tówrecene geond widne grund, heápum tóhworfene, Cd. Th. 235, 4; Dan. 301: Exon. Th. 186, 17; Az. 21: 16, 24; Cri. 258: Elen. Kmbl. 261; El. 131.

tó-writenness, e; *f. A detailed writing, a description:*—Se cásere sette gebann, ðæt wǽre on gewritum ásett eall ymbhwyrft. Deós tówritennys *(descriptio).* v. tó-mearcodness) wearð árǽred fram ðam ealdormenn Cyrino, Homl. Th. i. 30, 2.

tó-wríðan; *p.* -wráð *To twist different ways, to distort:*—Ic tówríðe *distorqueo,* Ælfc. Gr. 26, 3; Zup. 155, 15.

tow-tól, es; *n. An implement for spinning:*—Hé sceal fela towtóla habban, flexlínan, spinle ..., Anglia ix. 263, 10.

tó-wunderlíc *glosses admirabilis,* Ps. Spl. 41, 4.

tó-wurpan. v. tó-weorpan.

tó-wyrd, e; *f. Occasion:*—Ðá wǽron Seaxan sécende intingan and tówyrde heora gedáles wiþ Brittas *quaerentes occasionem divortii,* Bd. 1, 15; S. 483, 37.

tó-wyrpan, -wyrpendlíc, -wyrpness. v. tó-weorpan, -weorpendlíc, -worpness.

Trāci, Trācia, Trāciana. v. Þráceas.

tract; trachtere, tractere. v. traht; trahtere.

trǽf, es; *n.* **I.** *a tent, pavilion:*—Lǽdan ða torhtan mægð tó trǽfe ðam heán (cf. wæs seó hálige meówle on his búrgetelde, 22, 10; Jud. 57), Judth. Thw. 22, 2; Jud. 43: 25, 12; Jud. 255. Beornas stódon ymbe hyra þeódnes trǽf, 25, 19; Jud. 268. **II.** *a building:*—Tigelfágan trafu, torras. windige weallas, Andr. Kmbl. 1683; An. 844. [Cf. (?) *Icel.* traf *a fringe, hem:* in mod. usage, *a kerchief.*] v. hearg-, hell-, wearg-trǽf.

trǽgelian, trǽglian; *p.* ode *To pluck:*—Trǽglian *carpere,* Germ. 398, 84. Tó trǽgelgenne *carpendum,* 399, 388. [Cf. (?) *Lat.* tragula.] v. tó-trǽgelian.

trǽppan, trǽppe. v. treppan, treppe.

trǽg; *adj. Evil, bad:*—Tó trǽg, Exon. Th. 354, 37; Reim. 57. Se feónd and se freónd, tíreádig and trǽg, synnig and gesǽlig, Elen. Kmbl. 1906; El. 955. Ðæt hió ðære cwéne oncweðan meahton swá tiles swá trǽges swá hió him tó sóhte, 649; El. 325. [*O. L. Ger. O. H. Ger.* trāgi *iners, piger, segnis: Ger.* trāge: *Du.* traag. Cf. earh *for the double sense of* slow *and* bad.]

trǽg, e; trǽgu; *indecl.; f. Ill, affliction:*—Hé wénde him trǽge hnágre *he expected humiliating affliction for himself,* Elen. Kmbl. 1333; El. 668. [*O. L. Ger. O. H. Ger.* trāgî *ignavia, torpor.*] v. preceding word.

tragan = dragan, Jn. Skt. Rush. 21, 8, 11.

trǽge; *adv. Evilly, cruelly:*—Ðis is weorc ðara ðe oft wráðe mé trǽge tǽldan *hoc opus eorum, qui detrahunt mihi,* Ps. Th. 108, 20. [*O. H. Ger.* trāgo *tarde, segniter.*] v. trǽg.

traht, tract, es; *m.*: e; *f.* **I.** *a text, passage*; textus, tractus *(tractus* ecclesiastici cantus species, Migne) :—Æfter fyliaþ traht *sequitur tractus: 'Eripe me, Domine,'* Anglia xiii. 417, 743. Traht *tractus: 'Laudate Dominum,'* 425, 855. Mid trahte godspelles *cum textu euuangelii,* 416, 723. Nú bidde ic eów ðæt gé beón geðyldige on eówerum geðance óððæt wé ðone traht oferrǽdan mágon *I pray you to be patient in your thoughts until we have read the passage* (the passage is then given), Homl. Th. i. 166, 7. Ðæt man rǽde twá rǽdinga mid twám tractum and mid twám collectum, L. Ælfc. C. 36; Th. ii. 358, 19. **II.** *a treating of* a subject, *an exposition, a commentary:*—Traht *expositio,* i. *tractatio,* Wrt. Voc. ii. 145, 84. Ðes traht is langsum eów tó gehýrenne, ac wé willaþ nú úre sprǽce hér geendian, Homl. Th. ii. 536, 22: 70, 13: i. 248, 21. Trahte *commentis,* Wrt. Voc. ii. 19, 58: 94, 31. Wé oferrǽddon ðis godspel ... ac wé ne hrepodon ðone traht ná swíðor ðonne tó ðæs dæges wurðmynte belamp *we read the gospel, but we did not further touch the exposition* (or *text,* under I?) *than pertained to the honour of the day,* Homl. Th. i. 104, 6. Trahtas *commentariola,* Wrt. Voc. ii. 24, 51. Rǽde man ðære godcundan láre béc, and eác swá ða háligan trahtas *(expositiones)* ðe fram namcúþum fæderum geworhte synt, R. Ben. 33, 20. Trahta *commenta, documenta,* Hpt. Gl. 512, 32. [*O. H. Ger.* trahta *tractatio.* From Latin.] v. godspell-, sealm-traht, *and following words.*

trahtaþ, es; *m. A commentary:*—Trahtaðum *commentis,* Wrt. Voc. ii. 19, 58.

traht-bóc; *f. A book of exposition, a treatise, commentary:*—Gregorius gedihte manega hálige trahtbéc, Homl. Th. ii. 132, 15: i. 436, 10. Twá and hundseofontig bóca ðære ealdan ǽ and ðære níwan hé áwende ... búton óðrum menigfealdum trahtbócum ðe hé deópðancollíce ásmeáde, 15. Trahtere, es; *m. One who treats a subject, an expositor, interpreter, commentator:*—Mé ða treahteras tala wísedon, Salm. Kmbl. 9; Sal. 5. Treahteras *commentarii,* Wrt. Voc. ii. 15, 41. Fram trehterum *a commentariis,* 7, 28. Of flítendum trachterum *a vitiosis interpretibus,* Mt. Kmbl. p. i. 14. [*O. H. Ger.* pi-trahtari.] v. stǽr-trahtere; trahtnere.

trahtian; *p.* ode. **I.** *to expound, explain:*—Ðegnum his hé trahtade alle *he expounded all things to his disciples,* Mk. Skt. Lind. 4, 34. Se áwergda gást ongan Godes béc trahtian, and ðá sóna leáh, Blickl. Homl. 29, 29. **II.** *to discuss:*—Ðá ongunnon hý treahtigean, hwæter má mǽrlícra dǽda gefremed hæfde, ðe Philippus, ðe Alexander, Ors. 3, 9; Bos. 67, 3. [*O. H. Ger.* trahtón *tractare, reputare.*] v. getrahtian; trahtnian, traht.

trahtnere, es; *m. An expositor, commentator:*—Gregorius se trahtnere, Homl. Th. ii. 72, 21. Se trahtnere cwið, ðæt ðæt gyftlíce hús wæs ðryflére, for ðan ðe on Godes gelaðunge sind þrý stæpas gecorenra manna, 70, 16: i. 338, 16. Hieronimus se wísa trahtnere, Homl. Ass. 36, 296. v. trahtere.

trahtnian; *p.* ode. **I.** *trans. To expound, explain:*—Hægmon trahtnaþ ðis gospell, Homl. Th. i. 510, 26. Gregorius trahtnode ðis godspel, ii. 550, 1. Ic wolde eów trahtnian ðis godspel, ðe mann nú beforan eów rǽdde, i. 166, 3. Ðes cwyde is swíðor tó ondrǽdenne ðonne tó trahtnigenne, 332, 4: ii. 90, 5. **I a** *to give as explanation of (be)* something :—Wé sprǽcon be ðam sǽde ðe betwux ðam ðornum sprang ... Drihten sylf trahtnode be ðisum ðæt ðis sind ðe Godes word gehýraþ ac hí sind gebysgode mid heora welum *we spoke of the seed that sprang up among the thorns ... The Lord himself gave as explanation of this, that they are those that hear God's word, but are occupied with their*

wealth, Homl. Th. ii. 92, 7.   II. with prep. *to treat of* (*be, ymbe*):— Manega trahtnedon ymbe ðis angin *de hoc principio multi tractaverunt*, Anglia viii. 307, 7. Mid were æwfæstum trahtna (*tracta*) be hâlignysse, Scint. 200, 14. Nû wille wê be ðyssere freólstîde trahtnian, Homl. Th. i. 104, 9. Wê woldon gefyrn trahtnian be ðam lambe, i. 278, 11. [Nimeþþ gom off þiss þatt her iss trahhtnedd, Orrm. 11680.] v. â-, ge-, ofer-trahtnian; trahtian.

trahtnung, e; *f. Exposition, explanation, comment :*—Uton nû fôn on ðæs godspelles trahtnunge ðǽr wê hit forlêton *let us resume the exposition of the gospel, where we left it*, Homl. Th. i. 114, 35: ii. 72, 22. Ûs gedafenaþ ðæt wê mid ârfæstum geleáfan underfôn Drihtnes trahtnunge, 90, 4. Mid smeáþancelre trahtnunge *tenaci memoriae textu*, Hpt. Gl. 410, 65. Trah(t)nunge *commenta*, 479, 77. Mid gástlícum trahtnungum *commentariis, explanationibus*, 410, 24 : Homl. Th. ii. 2, 8.

trahtung, e; *f. Exposition, comment :*—Trahtunga *commenta*, Wrt. Voc. ii. 24, 50. Þracþungum (*l.* tractungum *or* trahtungum) *commentis, doctrinis*, Hpt. Gl. 482, 16. [*O. H. Ger.* trahtunga *retractatio*.]

traisc, tráisc (?); *adj.* In the following passage this word is used to translate *tragicus*, which, however, seems to have been taken as an adjective formed from a proper name. In another passage the same word is rendered by *trôiesc, trôisc* (q.v.) Trojan, perhaps the same meaning is intended here :—Æfter ðon hé eall geâr onwealh Norþanhymbra mǽgþe âhte nalas swâ swâ sigefæst cyning ac swâ swâ leódhata ðæt hê grimsigende forleás and hí on gelícnysse ðæs traiscan wæles wundade *dein cum anno integro provincias Nordanhymbrorum non ut rex victor possideret sed quasi tyrannus saeviens disperderet, ac tragica caede dilaceraret*, Bd. 3, 1 ; S. 523, 30.

tramet, es ; *m. A page :*—Bôc *liber*, stæf *littera*, leáf *folium*, tramet *pagina*, Wrt. Voc. i. 80, 75–78 : Ælfc. Gr. 7 ; Zup. 25, 5. Lâ hwylc tramet (*pagina*) is, oððe hwylc sprǽc ðæs godcundan lareoudômes, aðer oðþe ðære ealdan cýðnesse oþþe ðære nîwan, ðæt ne sý seó rihteste bysen ûran menniscan lífes, R. Ben. 133, 2. Tramod, R. Ben. Interl. 118, 2. Swâ fela trameta *tot paginae*, swâ fela leáfa *tot folia*, Ælfc. Gr. 18; Zup. 117, 12. Trametas *paginas*, Germ. 398, 181.

trandan (?) *to roll, move hastily :*—Trondendi *praeceps*, Txts. 89, 1668. [Cf. (?) *Icel.* trandill (as a nickname).] v. trendan, trendel.

treaflíce; *adv. Grievously, painfully :*—Eallum ðe deópe and ful treaflíce teónan þolian *omnibus injuriam patientibus*, Ps. Th. 102, 6. [Cf. (?) *Welsh* traf *stir, strain* ; trafu *to stir, agitate*.]

treágian ; *p. ode To repair, sew together :*—Treágiende *sarcientes, consuentes, componentes*, Hpt. Gl. 445, 73. Getreágede (-ode) *consuta*, 412, 38.

treahtere, treahtigean. v. trahtere, trahtian.

tredan ; *p.* træd, *pl.* trǽdon ; *pp.* treden.   I. *to tread, tread down, trample upon* (lit. and fig.) :—Ðû trides (*conculcabis*) león and dracan, Ps. Surt. 90, 13. Hê trit mid ðæm fêt *terit pede*, Past. 47 ; Swt. 357, 20. Hwîlum mec (*an animal's skin*) brýd trideþ fôtum, Exon. Th. 393, 27; Rä. 13, 6. Mê man tredeþ *conculcavit me homo*, Ps. Th. 55, 1 : Cd. Th. 56, 15 ; Gen. 912. Mê tredaþ feóndas mîne, Ps. Th. 55, 2 : Exon. Th. 119, 23; Gû. 259. Ðâ hét ic ðone here ðæt hié mid fôtum ðone snâw trǽdon *calcare militem niuem jubeo*, Nar. 23, 18 : Jos. 10, 24. Ða ðe mê trǽdan *conculcantes me*, Ps. Th. 56, 3. Ðæt hig hine trǽdun, Lk. Skt. 12, 1. Trêdun *proterunt*, Wrt. Voc. ii. 118, 2. Fôtum tredene, Bd. 3, 22; S. 552, 15.   I a. *with prep.* :—Anweald tô tredenne ofer snacan *potestatem calcandi supra scorpiones*, Lk. Skt. 10, 19.   II. *to tread upon, step upon, walk upon* :—Ðonne ic hrûsan trede, Exon. Th. 389, 22 ; Rä. 8, 1. Hió grundbedd trideþ, 493, 3; Rä. 81, 24. Se ðe môrland trydeþ, Elen. Kmbl. 1221; El. 612. Se fôtum tredeþ fiðru winda *qui ambulat super pennas ventorum*, Ps. Th. 103, 4. Ða ðe land tredaþ *those that move upon the earth* (Gen. 1, 28), Cd. Th. 13, 16 ; Gen. 203. Trǽdaþ, Exon. Th. 439, 5 ; Rä. 58, 5. Ðû flettpaðas nîne trǽde, Cd. Th. 165, 12; Gen. 2730. Hê wrǽclástas træd, Beo. Th. 2709 ; B. 1352. Meodowongas træd, 3291 ; B. 1643. Mearh moldan træd, Elen. Kmbl. 109 ; El. 55. Forð gán, foldweg tredan, Andr. Kmbl. 1550; An. 776. Gewât him se hearda sǽwong tredan, Beo. Th. 3933 ; B. 1964. Tredan elþeódigra foldan, Exon. Th. 329, 4; Vy. 29. Ic seah turf tredan .vi. gebrôðor, 394, 10; Rä. 14, 1. Ðû (*the serpent*) scealt ðînum breóstum bearm tredan brâdre eorðan, faran fêðeleás, Cd. Th. 56, 4 ; Gen. 907.   III. *in figurative senses, glossing Latin words* :—Sâwl gefylled trytt (*calcabit* ; tret, Kent. Gl. 1015) beóbreád *the full soul treadeth an honeycomb* (Prov. 27, 7), Scint. 50, 8. Tredaþ *terimus* (*otia temporum*), Wrt. Voc. ii. 78, 12. [*O. Frs.* treda : *O. L. Ger.* tredan : *O. H. Ger.* tretan : *Icel.* troða. Cf. *Goth.* trudan.] v. â-, be-, for-, ge-, of-, ofer-, tô-tredan; sin-tredende, *and following words.*

tredd. v. wîn-tredd.

treddan ; *p. de.*   I. *to tread under foot, trample upon* :—Treddun *proterunt*, Txts. 84, 749.   II. *to investigate, examine.* v. â-treddan :—Weord mê heorte forht ðæt ic ðîn hálig word tredde *a verbis tuis formidavit cor meum*, Ps. Th. 118, 161. [*O. H. Ger.* fortratta *proterit* ; trettenti *terens* : *Icel.* tratta ; *p.* ; traddr ; *pp.*]

treddian ; *p.* ode *To tread, step, walk :*—Raþe æfter ðon on fiôr feónd treddode, Beo. Th. 1455 ; B. 725. Cyning of brýdbûre trydde de, 1848 ; B. 922. Hié of ðam grimman gryre treddedon, Cd. Th. 243, 21 ; Dan. 439. Streámas ðû miht ou treddian eorðan gelíce *flumina pertransivit pede*, Ps. Th. 65, 5. [*O. H. Ger.* tretton *calcare, conculcare.*]

trede; *adj. Firm to tread on, that may be walked on :*—Sǽ cýðde hwâ hine gesette, tîrmeahtig cyning, for ðon hê hine tredne him ongeán gyrede, ðonne God wolde ofer sîne ýðe gân *ready for his coming the sea made itself firm for his tread, when God would walk over its waves*, Exon. Th. 72, 2 ; Cri. 1166.

tredel, es ; *m. A step :*—Tredelas *vel* stæpas *bases*, Wrt. Voc. i. 21, 48. [Grece or tredyl or steyre *gradus*, Prompt. Parv. 209. Tredyl or grece *gradus, pedalis*, 501. A tredel *subpedium*, Wülck. Gl. 614, 14: *suspendium*, 615, 3: *liciatorium*, 592, 33.]

tredend, es ; *m. One who treads :*—Tredend *calcatrix*, Wrt. Voc. ii. 127, 42.

trêg (treg ?), trîg (cf. (?) hêg, hîg *hay*, for the form), es; *n. A tray, trough :*—Trêg *alueolum*, Wrt. Voc. i. 290, 70. Nim ðæt reáde ryden, dô on trîg ; hæt stânes swîþe hâte, lege on ðæt trîg innan, Lchdm. ii. 340, 5–6. [Bye us vessel . . . Dysschys, cuppys, and sawsers, Bolles, treyes, and platers, Rich. 1490.] Cf. troh.

trega, an ; *m. Pain, grief, vexation, hurt, ill :*—Trega † anda ðînes hûses *zelus domus tuae*, Ps. Lamb. 68, 10. Tregan *injuriam* (cf. teónan, R. Ben. 17, 11), R. Ben. Interl. 20, 10. Ic fleah hlǽfdigan hete, tregan and teónan, Cd. Th. 137, 15 ; Gen. 2274. Ða twêgen trega (cf. ðyssa yfla hwæðer, 41), Met. 5, 42. Weá wæs ârǽred, tregena tuddor, Cd. Th. 60, 27 ; Gen. 988. [Mid ham is muruhðe moniuold wiðute teone and treie, O. E. Homl. i. 193, 61. Alkyn sorow and trey and tene, Pr. C. 7327. Al that whilom was murthe is turned to treie and tene, P. S. 340, 380. *Goth.* trigô ; us trigôm ἐκ λύπης, 2 Cor. 9, 7 : *O. L. Ger.* trego *dolor* : *Icel.* tregi *difficulty* ; *grief, sorrow*.] v. hell-trega, tin-treg, -trega.

tregian ; *p.* ode *To vex, trouble, afflict, grieve :*—Gif gê on unriht ne tregiaþ ne earme ne tyrewiaþ (*if ye oppress not the stranger, the fatherless, and the widow*, Jer. 7, 6), Wulfst. 50, 2. Ða ðe tregiaþ mê *qui tribulant me*, Ps. Spl. T. 3, 1. [Eall þis wæs God mid tô gremienne and ðás arme leóde mid tô tregienne, Chr. 1104 ; Erl. 239, 40. Quad Balaam: ‘For ðu tregest me,’ Gen. and Ex. 3975. Þai traied þe exacerbaverunt te, Ps. 5, 11. *O. Sax.* tregan *to trouble* : *Icel.* trega *to grieve*.] v. tin-tregian.

trehing (*but* þrihing *in Lambarde.* v. Schmid, A. S. Gesetz. 508). The form given in L. Ed. C. to the Scandinavian word, which in Icelandic appears as *þriðjungr* = the third part of a shire :—De treingis. Erant potestates super wapentagiis quas trehingas vocabant, scilicet, terciam partem provincie, et qui super ipsam provinciam dominabantur, trehing-gref . . . Et quod illi vocabant tria hundreda, vel iiii, vel plura, isti (*those of Danish England*) vocabant trehing. Et quod trehinge non poterat diffiniri, in scira servabatur, L. Ed. C. 31 ; Th. i. 455, 17–25. In Magna Carta, § 25, trethingii (*pl.*) occurs. The Anglicized form of the word probably began with þ, and Halliwell gives Thirdings as the term used of the Ridings. The present form, Riding, seems to have arisen from a confusion of the initial dental with the final sound of East, West, North.

trehtere. v. trahtere.

trem, trym *a step* :—Ic ðæt gehâte ðæt ic heonon nelle fleón fôtes trym *I vow that I will not flee hence one footstep* (cf. ðæt hê nolde fleógan fôtmǽl landes, 139, 57 ; Byrht. Th. 138, 68 ; By. 247. Fôtes trem, Beo. Th. 5044 ; B. 2525. *The form is probably to be recognized in a gloss given* Anglia viii. 33, 163 note, ægne trem *rendering* pedetemtim, *for which perhaps* fægre, tremmǽlum *might be read.* Cf. Hpt. Gl. 477, 78, *where the gloss for the same passage is* fægre ; fægre oððe fôtmǽlum *gradatim*, Wrt. Voc. ii. 40, 47 ; fôtmêlum *pedetemptim*, Txts. 90, 834; stæpmǽlum *gradatim*, Hpt. Gl. 497, 54. Cf. *also :* Ðonne wiðtremð hê and onhupaþ *gressum post terga revocet*, Past. 58 ; Swt. 441, 27.

tremes, tremesa, tremese, tremesse. v. trimes.

tremian, tremman *to confirm*, tremman *to step.* v. trymman, wiðtremman.

trendan (?) *to turn, roll.* v. sin-trendende *teres, but perhaps* sin-tredende *should be read*, see tredan = terere. [Let hym rollen and trenden with inne hym self the lyht of his inward syhte *in se revolvat intimi lucem visus*, Chauc. Boet. 100, 2835. *Chaucer also uses* bi-trenden, *and* un-trenden *occurs elsewhere.* Cf. *O. Frs.* trind, trund *round : Dan.* trind.] v. trandan, trinda, *and next word.*

trendel, es ; *m.*   I. *a circle, ring :*—Ân wunderlíc trendel (*mirabilis corona*) weard ateówed âbûtan ðære sunnan, Chr. 806; Erl. 60, 25. Gelden trendel *circulus aureus* (*in naribus suis*, Prov. 11, 22), Kent. Gl. 373. Brevis virgula (*the mark for short quantity*, i.e. ◡) ys ânes trendles dǽl ðus licgende, Anglia viii. 333, 29. On trendle *in rota*, Hpt. Gl. 471, 2. Stríc ðû mid ðínum scytefingre, swilce ðû trǽndel wyrce, Techm. ii. 129, 9. Trendla *circulorum* (the rings on a peacock's tail), Hpt. Gl. 419, 8.   I a. *a circle used in calculation :*—Ðǽs þing wê

geopeniaþ bet on đissum trendle (cf. gým đisses hwióles; hyt đê ætýwþ eall đæs mónan ryne, 33: *and:* Đás circulas synt behéfe preóstum, 44), Anglia viii. 328, 38. **I b.** figurative :—Trendel (*benedicens*) *coronam* (*anni*), Blickl. Gl. : Ps. Spl. T. 64, 12. **II.** where a surface, plane or spherical, is denoted, *a disk, orb :*—Đæs mónan trendel is symle gehâl, þeáh đe eall endemes eallunga ne scíne, Lchdm. iii. 242, 4 : Hpt. Gl. 418, 16. Đære sunnan trendel, Homl. Th. ii. 606, 12. Trendles *sphaerae*, trendel *sphaera*, Hpt. Gl. 489, 22, 23. Scínendne trendel heofones, Hymn. Surt. 22, 17. Trendlum *orbibus*, Hpt. Gl. 490, 76. **II a.** *a round place, a circus :*—Trendles, hringcsetles *circi*, Hpt. Gl. 488, 69. ¶ The word and the connected forms *trend, trind, trin* seem to occur in local names. v. Cod. Dip. Kmbl. vi. 343, 344. [*Wick.* trendil *sphaera : Prompt. Parv.* trendyl *troclea.* Trendel *giraculum*, Wülck. Gl. 586, 29 : trendell *catantrum*, 571, 19. Halliwell gives *trindle* = wheel as a Derbyshire word.] v. sin-trendel; *adj., and following words.*

trendeled; *adj. Made round :*—Tryndyled reáf *circumtectum*,Wrt.Voc. i. 40, 29. [*Panter* is blac mid wite spottes sapen al, wit and trendled als a wel, *Misc.* 23, 737.]

trendlian; *p.* ode *To trundle, roll.* [*Lefdis* letten teares treondlin (trendlen, MS. C.: trondlin, MS. B.), Kath. 2329. Þeȝ appel trendli from þon treowe, O. and N. 135. Hit trendeled doun, Allit. Pms. 2, 41. Be trendlid *volvi*, Wick. Jud. 7, 13. Trendelyn a round thynge *trocleo, volvo,* Prompt. Parv. 502.] v. á-trendlian; trendel.

treó. v. **treów.**

Treónta, Trênta, an; *The Trent :*—Andlang Trêntan, Chr. 1013; Erl. 147, 18. Man ofslóh ℱ .wine be Trêntan, 679; Erl. 41, 10. On Trênton (Treóntan, ℱ. M. 144, 14) streáme *in fluvio Treenta*, Bd. 2, 16; S. 519, 31. Mid Trêntan (Treóntan, Bd. M. 240, 1) streáme *fluvio Treanta*, 3, 24; S. 557, 37. Be Trêntan (Treóntan, Bd. M. 324, 15) đære eá *juxta fluvium Treanta*, 4, 21; S. 590, 14. On Trêntan; of Trêntan, Cod. Dip. Kmbl. iii. 396, 20. Đa brycge ofer Treóntan, Chr. 924; Erl. 110, 10.

treów, es; *n.* **I.** *a tree :*—Treów *arbor*, Wrt. Voc. i. 32, 26. Iung treów *arbustum*, 41. Wudu *silva*, áhæáwan treów *lignum*, 33, 56 : Ælfc. Gr. 8; Zup. 31, 13. Đæt treów wæs gód tó etanne, Gen. 3, 6. Treów (trêu, Lind.) *arbor*, Mt. Kmbl. 3, 10. Treów (trýw, MS. B.: treó, Lind.), 7, 17. Hit weard mycel treów (on treó miclum, Lind.: on tree miclum, Rush.) *factum est in arborem magnam*, Lk. Skt. 13, 19. Sunnan trió ágefeþ ondsware æt đæm upgonge . . . and đæt mónan trióv gelíce on niht dyde, Nar. 27, 16–19. Heó genam of đæs treówes wæstme, Gen. 3, 6. Æppelbære treów westm wircende, 1, 11. Tree *arborem*, Lk. Skt. Lind. 13, 6 : 19, 4. Gif man óđres wudu heáweþ unáliéfedne, forgielde ælc greát treów mid .v. scitt., L. Alf. pol. 12; Th. i. 70, 5. Oftost beóþ đa treówa getealde *feminini generis*, Ælfc. Gr. 6; Zup. 20, 14 : Ps. Spl. 95, 12. Treówu, Scint. 56, 17 : Ps. Th. 57, 8. Đa hálgan trió sunnan and mónan . . . and óþre treów, Nar. 27, 26–29. Treów, 32, 13. Triów, 28, 11. Treó sceolon brædan, Exon. Th. 343, 20; Gn. Ex. 160. Treó westmbêru *ligna fructifera*, Ps. Surt. 148, 9. Of đæra treówa wæstme, Gen. 3, 2. Triówa heánnisse, Nar. 28, 1. Betwih đæm rindum đæra trió, 27, 25. Trêa *lignorum*, Ps. Surt. 73, 5. Trêwna *arborum*, Mt. Kmbl. p. 15, 9. Đæra treówa (trýwa, MS. B.: trêuna, Lind.), Mt. Kmbl. 3, 10. Treóna, Rtl. 95, 23. Of đam treówum, Lind.: trêum, Rush.), Mk. Skt. 11, 8. Sumu treówa hé watrode, Past. 40; Swt. 293, 4: Nar. 27, 21. Treówa, Gen. 1, 29. Behealdaþ ealle trýwu (treówa, MS. A.: treó, Lind. Rush.), Lk. Skt. 21, 29. *The word occurs as the second part of many compounds, e. g.* æppel-, ceder-, corn-, cwic-, cyrs-, ele-, fíc-, gyr-, hwíting-, magdala-, palm-, persoc-, pín-, plúm-, ulm-, wín-, windel-treów; see also lád-, wudu-treów. **II.** a material, *wood :*—Hí worhton him anlícnyssa, sume of golde, sume of seolfre, sume of stánum, sume of treówe, Homl. Th. i. 22, 30. Hé hêt getimbrian cyrican of treówe, Chr. 626; Erl. 23, 40. Hé of treówe (treó, Bd. M. 138, 21) cyricean getimbrede, Bd. 2, 14; S. 517, 26. Monige of đam treówe (treó, Bd. M. 156, 5) đæs hálgan Cristes mæles spónas nimaþ, 3, 2; S. 524, 30. **III.** in a collective sense, *trees, a wood :*—Đa behídde Adam hyne on niddan đam treówe neorxena wanges *Adam hid himself among the trees of the garden; in medio ligni paradisi*, Gen. 3, 8. Hé (*the Phenix*) sylf biereþ in đæt treów innan torhte frætwe; đær se wilda fugel ofer heánne beám hûs getimbreþ (cf. hé heánne beám on holtwuda wunaþ, 209, 15; Ph. 171), Exon. Th. 211, 19; Ph. 200. **IV.** tree as in roof-tree, saddle-tree, *a piece of wood, a beam, log, stake, staff, cudgel :*—Scort sinewealt stán *vel* treów *cilindrus*, Wrt. Voc. i. 41, 35. Đonne seó sáwl hié gedæleþ wiþ đone líchoman, hwylc biþ hé đonne bûton swylce stán odđe treów (*a stone or a log*), Blickl. Homl. 21, 27. Of treówe de *stipite*, Wrt. Voc. ii. 27, 65. Gif mon mid treówe geslegen sié, Lchdm. ii. 8, 32 : 94, 23. Gé fêrdon mid swurdum and treówum mê gefón, Mk. Skt. 14, 48. Gewyrcean tor of treówum and of mycclum beámum, Blickl. Homl. 187, 12. Swá hwá swá getimbraþ ofer đisum grundwealle gold, odđe seolfor, odđe treówa, Homl. Th. ii. 588, 25. Treówu, 590, 13. Hié námon treówu and slógon on óþerne ende ísene næglas, Ors. 4, 1; Swt. 158, 4. v. fugol-, teld-, wægn-treów. **IV a.** *tree* as in

gallows-*tree*, ʻ*tree* used of the cross :*—Hælendes treów, Rood Kmbl. 50; Kr. 25. Wuldres treów, 28; Kr. 14. Đû đê on róde treów áhófe, Anglia xii. 506, 4 : Elen. Kmbl. 411; El. 206. Đurh treów ûs com líf, đá đá Crist hangode on róde, Homl. Th. ii. 240, 22. v. gealg-, wulfheáfod-treów, *and* ród. [*Goth.* triu *a tree; staff:* O. Sax. trio *a beam; the cross:* O. Frs. trê; *Icel.* trê *a tree; a beam; wood.*]

treów, trýw, e; *f.* The word is sometimes used in the plural with the force of the singular. **I.** *truth* to a promise or engagement, *faith* (as in good or bad *faith*, to keep *faith* with a person), *troth :*—Treów, sió geond bilwitra breóst áríseþ, Exon. Th. 343, 21; Gn. Ex. 160. Hálegu treów seó đú wid rodora weard healdest, Cd. Th. 127, 30; Gen. 2118. Wære gehealdan, treówe tácen, Andr. Kmbl. 427; An. 214. In swá hwylce tiid swá gê mid treówe (*truly*) tô mê on hyge hweorfaþ, and gê hellfirena geswícaþ, Exon. Th. 366, 1; Reb. 5. On treówe gelæton (-en?) *fidei commissum*, Wrt. Voc. ii. 148, 76. His treówe (treówa, Bd. M. 130, 27) and his gehât wid đê gehealdon . . . his treówes for feógýtsunge forleósan, seó đe dýrwurþre wære eallum máþmum *tibi fidem pollicitam servare . . . fidem suam, quae omnibus ornamentis pretiosior est, amore pecuniae perdere*, Bd. 2, 12; S. 514, 34–41. Đý læs ic mín gehât and míne treówe forleóse *ne fidem mei promissi praevaricer*, 4, 22; S. 592, 2. Ác fêreþ gelóme ofer ganotes bæd gársecg faudaþ hwæđer ác hæbbe ædele treówe *oft fares the oaken vessel over the gannet's bath; ocean proves whether the oak keeps excellent faith, i. e. whether the promise of safety, which its strength seems to give, is kept*, Runic pm. Kmbl. 344, 22; Rún. 25. Ne Hildeburh herian þorfte Eótena treówe, Beo. Th. 2148; B. 1072. Til biþ se đe his treówe gehealdeþ, Exon. Th. 293, 6; Wand. 112. Đu đær tírfæste treówe findest, 473, 8; Bo. 11 : Ps. Th. 100, 6. Đæt æfre on his dagum sceolde gewurđan swá lytle treówa, 13, arg. Mánum treówum woldon hié đæt feorhleán, fácne gyldan, Cd. Th. 187, 11; Exod. 149. Đu hæfst ongyten đa wonclan trúwa (treówa, Cott. MS.) đæs blindan lustes, Bt. 7, 2; Fox 18, 3. Đa đe mid tungan treówa gehâtaþ, fácenlíce þencaþ, Fragm. Knibl. 47; Leás. 25. Tír healdeþ trýwa wel wid ædelingas, Runic pm. Kmbl. 342, 22; Rún. 17. **II.** *truth* to a person, *fidelity, fealty, loyalty.* Cf. hold :—Đæs getreówan freóndes, đone mon lufaþ for treówum, Bt. 24, 3; Fox 82, 35. Dauid forbær đæt hê Saul ne dorste ofsleán for đæm ealdum treówum, Past. 28; Swt. 199, 3: 3; Swt. 37, 7. Cham ne wolde cýđan hyldo and treówa, Cd. Th. 96, 9; Gen. 1592. **III.** *the truth* of the stronger to the weaker, *grace, favour, help.* Cf. hold :—Treów wæs gecýþed, đætte Gúđlace God leánode ellen mid árum, Exon. Th. 129, 11; Gú. 419. Treówe *latibulo* (protection, faithful care; the passage in which the word occurs refers to the entrusting of his mother by Christ to St. John's protection), Hpt. Gl. 415, 57. Git mê sibblufan and freóndscipe cýđaþ, treówe and hyldo tídiaþ mê, Cd. Th. 152, 6; Gen. 2516: 34, 21; Gen. 541. Heó treówe gehêt *she promised God's favour*, 44, 25; Gen. 714. Hê treówa gehêt, his holdne hyge, 41, 8; Gen. 653. **IV.** *an assurance of faith or truth, word* (in to give or pledge one's *word*), *a promise, an engagement, a covenant, league :*—Hû þearf mannes sunu máran treówe *what need has a son of man of a better assurance?* Cd. Th. 204, 26; Exod. 425. Đa eorlas đe him treówe sealdon, 123, 17; Gen. 2046. Hê bæd hié đæt hié gemunden đara ealdena treówa *ad antiquorum jura foederum adhortatione persuadens*, Ors. 2, 5; Swt. 82, 9. Se đe his nýhstan swereþ, and hine mid treówum ne beswícþ *he that swears to his neighbour, and does not deceive him with assurances of good faith*, Ps. Th. 14, 6. For đám treówum đe đû genumen hæfdest tó Abrahame . . Đû him đæt gehête, đæt . . ., Cd. Th. 235, 26; Dan. 312. Se wæs ofslagen ofer aþas and treówa (*or under I*) *contra fidem jurisjurandi peremptus est*, Bd. 2, 20; S. 521, 17. Norþhymbre and Eást-Engle hæfdon Ælfrêde cyninge aþas geseald, and Eást-Engle foregísla vi; and þêh ofer đa treówa . . . fôron hié, Chr. 894; Erl. 90, 5. Ic eów treówa đæs míne selle, Cd. Th. 92, 28; Gen. 1535: 122, 35; Gen. 2037. Đû treówa selle, wæra đína, đæt đú wille mê wesan freónd, 170, 23; Gen. 2817. **V.** *faith* in something, *belief, trust, confidence :*—Treów in đê (*the Virgin Mary*) weordlícu wunade, Exon. Th. 6, 11; Cri. 82. Nó him for egsan earmra gæsta treów getweóde, 122, 25; Gú. 311 : 134, 28; Gú. 515. Đína água treówa and seó godcunde lufu and se tóhopa đé ne lætaþ geortrêwan be đam æcan lífe, Bt. 10; Fox 30, 8. Đa bebodu đe giet máran sint . . . đæt is, ryht dóm, and mildheortnes and treówa (cf. Mt. 23, 23 *where* geleáfa *renders* fides), Past. 57; Swt. 439, 31. Đa beraþ Godes fatu đa đe óđerra monna sáula underfoþ tó lædanne on đa treówa hira ágenra gearnunga *Domini vasa ferunt, qui proximorum animas perducendas in suae conversionis fide suscipiunt*, 13; Swt. 77, 4. Đû gelýfst đínum hláforde bet đonne đê selfum, and đínum gefêrum æmnwel and đê selfum; đû dêst swíđe rihte, mid đý đæt đú swá gooda treówa wit hí hefst, Shrn. 196, 25. Hê (*Noah*) hæfde him on hređre hálige treówa, Cd. Th. 201, 3; Exod. 366. Hê his treówa sceal, and his módgeþonc, má up đonne niþer habban tó heofonum, Met. 31, 18. [*Goth.* triggwa *a covenant:* O. Sax. trewa (*often pl.*): O. L. Ger. treuwa *foedus:* O. Frs. triuwe, treuwe: O. H. Ger. triuwa *fides, foedus.*] v. heáh-, hyge-, un-, wine-treów, *and next word.*

treówa, trýwa, an; *m. An assurance of good faith, a covenant.* v. treów, IV :—Se ēca treówa *the perpetual covenant* (cf. Ex. 31, 16), Wulfst. 210, 22. Nāðor ne wē on ðone here faran, ne heora nān tō ūs, būton man trýwan and gýslas betwýnan sylle friðe tō wedde, L. A. G. 4; Th. i. 156, 8. Cf. trúwa.

treówan, triéwan, trīwan, trýwan; *p.* de. I. *to trust* :—Irnaþ ealle endemes, ða ðe hiora ærninge trēwaþ *those who have confidence in their powers of running*, Bt. 37, 2; Fox 188, 10. Gehwylc hiora his ferhðe·treówde, ðæt hē hæfde mōd micel, Beo. Th. 2337; B. 1166. II. *to prove* one's self *true, to clear* one's self *of a charge of untrue conduct.* Cf. *Icel.* tryggva *to make firm and trusty* :—Gif hē (*a person accused of plotting against his lord*) hine selfne triówan wille, dō ðæt be cyninges wergelde, L. Alf. pol. 4; Th. i. 64, 2. Treówan, 33; Th. i. 82, 8 note. Trýwan, 19; Th. i. 74, 7 note. v. ge-treówan, -triéwan, -trýwan, mis-trīwan, or-trýwan, *and* treówian.

treów-bytt (?), e; *f. A wooden vessel* :—Flasce trinnubyttæ (triuuu-? =treów-, cf. trýwen byt *flasco*, 149, 33), *eadem et flascones*, Wrt. Voc. ii. 39, 78.

treów-cynn, es; *n. A kind of tree* or *wood, a tree, a wood* :—*Abies* ðæt treówcyn, Nar. 8, 21. Treócynn, Exon. Th. 472, 20; Bo. 2. Nim ælces treówcynnes dǽl ðe on ðæm lande sý gewexen, Lchdm. i. 398, 7. Hē ásmeáde be ǽlcum treówcynne fram ðam heágan cederbeáme oð ðæt hē com tō ðære lytlan ysopan *he spake of trees, from the cedar tree that is in Lebanon even unto the hyssop that springeth out of the wall* (1 Kings 4, 33), Homl. Th. ii. 578, 4. Oftost on treówcynne beóþ ða treówa getealde *feminini generis* and se wæstm *neutri generis*, Ælfc. Gr. 6, 9; Zup. 20, 14. Man worhte Noes earce of ðam treówcynne ðe is genemned Sem, Salm. Kmbl. p. 184, 16: Nar. 10, 13. Ðá ætýwde Drihten Moise ān treówcyn and hēt dōn ðæt treów on ðæt wæter, Ex. 15, 25.

treówe, triéwe, trýwe; *adj. True, faithful, trustworthy* :—Wæs hiera sib ætgædere, ǽghwylc ōðrum trýwe, Beo. Th. 2334; B. 1165. Hē sprǽc tō his onbehtþegne, tō his treówum gesīþe, Exon. Th. 179, 29; Gū. 1269. Hié ne beóþ nānum men getreówe (ne treówe, Bod. MS.), Bt. 7, 1; Fox 16, 17. Būton hē habbe twēgra trýwra manna gewitnesse, L. Eth. iii. 9; Th. i. 296, 18. Man namige .ii. trýwe þegnas, L. N. P. L. 57; Th. ii. 298, 31. His freónd se treówesta (getreówesta, Bd. M. 126, 30) *fidissimus amicus illius*, Bd. 2, 12; S. 513, 17. [*Goth.* triggws *true, faithful*: *O. Sax.* triuwi: *O. Frs.* triuwe: *O. H. Ger.* gi-triuwi: *Icel.* tryggr.] v. ge-, or-treówe (-triéwe, -trýwe).

treówen, trīwen, trýwen; *adj.* I. *of a tree* :—Hire hyrdeman sume āc āstáh, and his orf lǽswode mid treówenum helme, Homl. Th. ii. 150, 31. II. *of wood, wooden* :—Treówen *ligneus*, Ælfc. Gr. 5; Zup. 15, 14. Trīwen sceó *coturnus*, Wrt. Voc. i. 26, 21. Trýwen byt *flasco*, ii. 149, 33. On treówum mortere, Lchdm. iii. 180, 4. Trýwenan, i. 220, 11 : 230, 10. On treówenre cyste, Homl. Skt. i. 20, 69. On treówenre rōde, Nicod. 34; Thw. 20, 6. Hyre goldfágan treówenan cuppan, Chart. Th. 536, 18. Wirce treówene earce *facies arcam ligneam*, Deut. 10, 1. Godu treówene and stǽnene, 4, 28. On treówenum fatum, Ex. 7, 19. Treówenu fatu mon weorþige, Bt. 36, 1; Fox 172, 19. [*Wick., C. M.* treen. *Goth.* triiweins *wooden*.] v. pín-treówen.

treów-fæst; *adj. Faithful* :—Tunge mīn triówfest, Ps. C. 114. Treówfæst (treúfæst, Lind.) *fidelis*, Lk. Skt. Rush. 19, 17. Trewufæst (treuw-?), Mt. Kmbl. Lind. 25, 21. Wē on bōcum rǽdaþ be sumum treówfæstum wífe, Homl. Skt. i. 12, 179. Treófæsto, treófest *fideles*, Lk. Skt. Lind. 16, 11, 12. Wǽron his bebodu ealle treówfæste *fidelia omnia mandata ejus*, Ps. Th. 110, 5. [Trowfeste men, *O. E. Homl.* i. 89, 29. *Icel.* trú-fastr : *Dan.* troe-fast.] v. un-treówfæst.

treów-fēging, e; *f. A joining together of planks*; *commissoria, tabularum conjunctio*, Wrt. Voc. ii. 132, 31.

treów-fugol, es; *m. A tree-haunting bird* :—Treófugla tuddor, Exon. Th. 146, 9; Gū. 707.

treów-geþofta, an; *m. A faithful comrade, trusty companion* :—Monig biþ uncūþ treówgeþofta, Exon. Th. 469, 20; Hy. 11, 5. Treówgeþoftan (*St. Matthew and St. Andrew*), Andr. Kmbl. 2101; An. 1052. Ic mid mec gelǽdde mīne þrié ða getreówestan frýnd, ða wǽron mīne syndrige treówgeþoftan *assumpsi mecum tres fidelissimos amicos*, Nar. 29, 28. [Cf. *Icel.* trygg-vinr *a trusty friend*.]

treów-geweorc, es; *n. A wooden structure* :—Treówgeweorc on gelícnysse medmiceles hūses geworht *tumba lignea in modum domunculi facta*, Bd. 4, 3; S. 570, 16.

treów-gewrid, es; *n. A thicket of trees* :—Ys on Bretone land sum fenn unmǽtre mycelnesse . . . Ðǽr synd . . . manige eáland and hreód and beorhgas and treówgewrido, Guthl. 3; Gdwin. 20, 7.

treówian, triéwian, trýwian; *p.* ode. I. *to trust, confide* :—Ǽghwylcum ðe him on treówaþ *omnes qui confidunt in eis*, Ps. Th. 113, 17. Ða ðe treówiaþ (*confidunt*) on Drihtne, Ps. Spl. 124, 1. On mannan tō treówianne *confidere in homine*, Ps. Th. 117, 8. On ealdormen tō treówianne *sperare in principibus*, 9. II. *to be true* to a person :—Dōþ swā ic hāte ic eów treówige gif gē ðæt tācen gegāþ ðóð geleátan *do as I bid; I will be true* (or *gracious*, v. treów, III) *to you, if you use that sign* (*circumcision*), *true sign of belief* (cf. sete tācn sōð gif ðū wille on mē habban holdne freónd, 139, 17-22; Gen. 2311-2313), Cd. Th. 140, 7; Gen. 2324. III. *to prove* one's self *true, clear* one's self *from a charge of untrue conduct* :—Gif hwā ōðerne tión wille . . . gif hē hine treówian wille, in .xii. ciricum dō hē ðæt, L. Alf. pol. 33; Th. i. 82, 8. Gif hē hine triéwian wille, ðæt hē tō ðære lǽne fácn ne wiste, ðæt hē mōt, 19; Th. i. 74, 7. [Þenne he þe treowede alre best, þenne beswikes tu heom, Laym. 3413. Him þ ha treowed on, Kath. 1327, note. Þeo luue . . . þu treowest hire, Misc. 94, 42. Putifar trewið his wiwes tale, Gen. and Ex. 2037.] v. ge-treówian (-trýwian); treówan, treówsian, trūwian.

treów-leás; *adj.* I. *faithless*; *perfidus* :—Wēnstū ðæt ic sceole sprecan tō ðissum treówleásan men (*the sorcerer, Simon*), Blickl. Homl. 183, 32. Simon cwæþ: 'Ðis is ðæt mennisc ðe ealle mīne dǽda mid heora wordum onwendan.' Ðā cwæþ Neron tō Petre: 'For hwon wǽron gyt swā treówleáse?' 175, 26. Treówleásra *perfidorum*, Wrt. Voc. ii. 66, 54 : Wulfst. 186, 3. Ðara treówleásra (*perfidorum*) cyninga beboda, Bd. 1, 7; S. 476, 35. Trióleásra, Rtl. 59, 23. Trīwleásra, 24, 21. II. *without belief, infidel*; *infidelis* :—Se ðe ne gímþ ðara ðe his beóþ hē wiðsæcþ Godes geleáfan and hē biþ treówleás *qui suorum curam non habet, fidem negavit, et est infideli deterior*, Past. 18; Swt. 139, 3. [*O. Sax.* treu-lôs *perfidus*: *Icel.* trygg-lauss.] v. ge-treówleás.

treówleásnéss, e; *f. Faithlessness*; *perfidia* :—Trēuleásnis (-lēsnis) *perfidia*, Txts. 85, 1533. Ðonne lǽrþ ūs Godes engel smeáunge ymbe Godes beboda . . . ðonne lǽrþ ūs se deófol treówleásnesse Godes beboda (*unfaithfulness to God's commands*), Wulfst. 233, 19. v. ge-treówleásnéss.

treów-líc; *adj. True, faithful.* [*Icel.* trygg-ligr *trustworthy*.] v. getreówlíc, *and next word.*

treów-líce; *adv. Faithfully, truly* :—Ic dō swýðe treówlíce ymb hý *fiducialiter agam in eo*, Ps. Th. 11, 6. [Ich leote ham treowliche luuien ham, Marh. 13, 32. Þe luue is treouliche iuestned touward him, A. R. 218, 13. Þou3 3e be trewe of 3owre tonge and trewliche wynne, Piers P. i. 177. *O. H. Ger.* triulîho *fideliter* : Icel. tryggliga.] v. ge-, untreówlíce.

treów-loga, an; *m. One who fails to keep faith, one who fails in loyalty to his leader* :—Ða hildlatan holt ofgeáfon týdre treówlogan ða ne dorston ǽr dareðum lācan on hyra mandryhtnes miclan þearfe *those laggards in fight relinquished the wood, pitiful false ones to plighted faith, who dared not with darts sport in their liege lord's great need*, Beo. Th. 5686; B. 2847. [The treulogo (*Judas*), Hél. 4622.]

treów-lufu, e, an; *f. Faithful love* :—Wæs seó treówlufu (*the love of the disciples to Christ after the ascension*) hāt æt heortan, Exon. Th. 34, 7; Cri. 538.

treówness, e; *f. Trust, confidence* :—God ðū eart mīn frōfer, mīn trēwnes, and mīn tóhopa, Bt. 42; Fox 260, 15. v. or-treówness (-trýwness).

treów-rǽden[n], e; *f. The state* or *condition of being faithful* or *true* :—Swā ic ðē lǽre lǽst uncre wel treówrǽdenne *as I teach you, maintain our state of mutual faithfulness*, Cd. Th. 139, 5; Gen. 2305. Cf. hold-rǽden.

treówsian, trýwsian; *p.* ode. I. *to engage, pledge one's self* :—Him cōmon ongeán .vi. cyningas and ealle wið trýwsodon (wið hine getreówsodon, col. 1), ðæt hī woldon efenwy[r]hton beón on sǽ and on lande *six kings came to meet him, and all solemnly engaged to co-operate on sea and on land*, Chr. 972; Th. i. 225, col. 2. Se munuc ðe mynster næbbe cume tō scíre biscope and trýwsie (-ige) hine sylfne wið God and wið men ðæt hē þreó þing healdan wille, L. Eth. v. 6; Th. i. 306, 7: vi. 3; Th. i. 314, 25. II. *to prove* one's self *to be true, to clear* one's self *from a charge of untrue conduct* :—Gif hē hine trýwsian wylle, ðæt hē tō ðære lǽne fácn ne wiste, ðæt hē mōt, L. Alf. pol. 19; Th. i. 74, 7 note. [Þas weord ich wulle þe treosien þurh mine god *I will prove to thee the good faith of these words by an oath*, Laym. 8489. Trousien, 8315. The word also means *to trust* :—Þe king him treousede on, 9308.] v. ge-treówsian; treówan, treówian.

treów-steall, es; *n. A place where trees are planted, a plantation* :—Hit cymþ tō Wulfūnes treówstealle, Cod. Dip. Kmbl. iii. 404, 11. Oð ðæt treówsteall; ðonnon of ðan treówstealle, v. 297, 24. Oð Æðelstānes treówsteal, 298, 12. Cf. wæter-steall, *and next word*.

treów-stede, es; *m. A place where trees are planted* :—Iung treów *vel* treówstede *arbusta*, Wrt. Voc. i. 39, 8.

treów-teoru *resin* :—Trēuteru *bapis*, Wrt. Voc. ii. 101, 58 : *bapys*, 10, 76. v. teoru.

treówþ, triéwþ, trýwþ, e; *f. The word is used sometimes in plural with force of singular.* I. *truth, good faith, honour* :—Ðǽr dydon þeáh Rōmāne lytla triéwþa ðæt him ða wǽron lāðe ðe hiera hlāford beswican *there, however, the Romans acted a little honourably* (in hoc solo Romanis circa eum fortiter agentibus), *in that those who had betrayed their lord were detestable to them*, Ors. 5, 2; Swt. 218, 17. II. *fidelity* :—Heora gemynd þurhwunaþ for heora trýwðe wið God, Ælfc. T. Grn. 1, 12. III. *a covenant, an assurance of good faith* :—Ðis ys ðære treówðe blōd ðe Drihten eów behēt *hic est sanguis foederis, quod pepigit Dominus vobiscum*, Ex. 24, 8. In treówþe geþeóded gāstlíces

freóndscipes *spiritalis amicitiae foedere copulatus*, Bd. 4, 29; S. 607, 9. Mid ǽnigere treówðe *quolibet pacto*, Hpt. Gl. 469, 34. Treówða *foedera, pacta*, 404, 9. Treófða *foedera*, 416, 47. Hié nánra treówþa him ne wéndon búton ðæt hié nid ealle forweorþan sceolde *they expected no terms for themselves, but that they must entirely perish;* the Latin which this seems intended to translate is:—Non secus ac si capta esset, turbata civitas fuit, Ors. 4, 5; Swt. 166, 13. Ic gemunde mínra treówða ðe ic Abrahame behét *recordatus sum pacti mei*, Ex. 6, 5. Gif gé míne treówða gehealdaþ *si custodieritis pactum meum*, 19, 5. [O. H. Ger. ga-triuwida *confidentia;* missa-triuwida *diffidentia, suspicio : Icel.* tryggð *faith, truce.*] v. ge-, un-treówþ.

treów-þrág, e; *f. A season of good faith* or *trust :*—Men leahtras oft geceósaþ treówþrág is tó trág *men often prefer vice to virtue, the time when good faith is kept is all too short* (?), Exon. Th. 354, 37; Reim. 57.

treów-wæstm *fruit of a tree :*—Treówwæstmas wurdon ðære nihte þurh forste swíðe fornumene, Chr. 1110; Erl. 243, 2. Þurh wæstma forweorþenesse, ǽgðer ge on corne and eác on eallon treówwæstman, 1103; Erl. 239, 3.

treów-weorþung, e; *f. Tree-worship :*—Wé lǽraþ ðæt preósta gehwilc forbeóde treówwurþunga and stánwurþunga, L. Edg. C. 16; Th. ii. 248, 30. Cf. Wé forbeódaþ ǽlcne hǽðenscipe . . . ðæt is, ðæt mán weorðige . . . stánas oððe ǽniges cynnes wudutreówa, L. C. S. 5; Th. i. 378, 17-21. v. Grmm. D. M. c. 21.

treów-wyrhta, an; *m. A wood-wright, worker in wood, carpenter, joiner :*—Treówwyrhta *lignarius*, Wrt. Voc. i. 19, 10: 73, 29. Se Treówyrhta segþ :—Hwilc eówer ne notaþ cræfte mínon, ðonne hús and mistlíce fata and scypa eów eallum ic wyrce? Se Smiþ andwyrt :—Eálá Trýwwyrhta, for hwí swá sprycst ðú?, Coll. Monast. Th. 31, 9-17. Ic hæbbe treówwyrhtan *habeo lignarium*, 30, 1.

treów-wyrm, es; *m. A caterpillar :*—Hé salde treówyrme westmas heara *dedit erugini fructus eorum*, Ps. Surt. 77, 46. [A treworme *terudo, trunos*, Wrt. Voc. i. 223, col. 1.] v. leáf-wyrm.

trépe (?), trype (?), trýpe (?), es; *m. A troop, band :*—Blódige trépas í werodu *sanguineas acies*, Hymn. Surt. 47, 18. [*From Low Latin* tropus *or* trupa (?); cf. *Fr.* troupe : *Span.* tropa : *Ital.* truppa.]

treppan; *p.* te. I. *to tread :*—Hé trepeþ *terit*, Kent. Gl. 144. [Cf. Halliwell's Dict. trap *to tramp : Du.* trappen *to tread, trample : O. Frs. Ger.* treppe *a step.*] II. *to trap :*—Hió [tr]e[p]te *inretivit*, Kent. Gl. 211. v. be-træppan, *and next word.*

treppe, træppe (v. (?) colte-træppe (=*colt-trap?*) *ramnus*, Wrt. Voc. i. 285, 47), an; *f. A trap :*—Ic beswíce fugelas mid treppan *decipio aves decipula*, Coll. Monast. Th. 25, 15. [O. H. Ger. trapo *tenda.* From this Low Latin *trappa*, hence French *trappe*, which perhaps helps to determine the form of the later English word :—To lacchenn þe þurrh trapp, Orm. 12301. A mous caught in a trappe, Chauc. Prol. 145. A trappe *brida*, Wrt. Voc. i. 264, 8. Trappe for myce *muscipula, decipula*, trappe to take wythe beestys *tenabulum*, trappyd or betrappyd *decipulatus, illaqueatus*, Prompt. Parv. 499.]

tréu, tréw *a tree*, tréwan, tréwness, tribulaþ. v. treów, treówan, treówness, trifulian.

-tricce *in* ge-tricce (q.v.) *tractable* (?). [Cf. (?) *Du.* trekken *to pull : Dan.* trække.]

tridwet ? *in* getridwet spere *hasta*, Wrt. Voc. i. 35, 40.

triéwan. v. treówan.

trifet, es; *n.* Tribute :—Trifetum *tributis*, Kent. Gl. 426. [O. H. Ger. tribuz. From Latin.]

trifulian; *p.* ode *To pound, grind, triturate :*—Se ðe pílaþ *vel* tribulaþ *pilurus vel pistor*, Wrt. Voc. i. 20, 26. Gebæron tó ahsan, dô eced tó, trifula swíðe, Lchdm. ii. 150, 3. Menge eall tógædere, and trifolige, 186, 10. [*From Latin* tribulare.] v. ge-trifulian, *and next word.*

trifulung, e; *f. Grinding, pounding, threshing :*—In trifelunge *in tritura*, Wrt. Voc. ii. 46, 21.

tríg (trig?). v. trég.

trimes(-is), es; trimessa, an; *m. :* trimes[s], e; trimes[s]e, an; *f. :* þrimes; *gender uncertain.* I. as a weight, *a drachm :*—Genime ánes trymeses gewǽge, Lchdm. i. 74, 21. Ánre tremese (trymese, MS. H.) gewihte, 110, 9. Ánre tremesse wǽge, 72, 11. Genim áne (ánne, MS. O.) trymesan gewǽge, 78, 13. Nime áne trymessan fulle, 76, 6. Twégra trymesa, 78, 24. Twéga trymessa, 70, 15 : 72, 26. .iiii. trymesan, 76, 22 : 78, 8. Feówer trymessan, 76, 1, 10, 16. *De ponderibus incipit.* Solidos tres trymisas, Txts. 113, 80. II. as a coin, *a coin in English :*—Trymes *staterem*, Mt. Kmbl. Lind. 17, 27. Lidrine trimsas (trymsas) *asses scorteas (corteas)*, Txts. 38, 31. Liþerene trymsas *asses corteas*, Wrt. Voc. ii. 7, 18. (b) in England, *a coin of the value of three pence.* The gen. pl., þrimsa, þrymsa, occurs several times in the section headed Norðleóda laga, Th. i. pp. 186, 188. [O. H. Ger. drimisa (-issa), trimisa *dragma. From Latin* tremissis, tremisia.]

trinda, an; *m. A round lump, a ball :*—Geóte tó trindan . . . wyrce tó trindan, Lchdm. iii. 14, 10, 13. [Cf. Onn heffness whel all ummbetrin (*round about*), Orm. 17563. O. Frs. trind, trund *round : Dan. Swed.* trind; *omtrent about : Dan.* trindes *to grow round.*] v. trendan, trendel.

trinnu-byttae. v. treów-byt.

triumpha, an; *m. A triumph, the entry into Rome of a victorious general.* The following explanation of the term was inserted by Alfred in his translation of Orosius :—Ðæt hié triumphan héton, ðonne hié hwelc folc mid gefeohte ofercumen hæfdon, ðonne wæs heora þeáw ðæt sceoldon ealle hiera senátus cuman ongeán heora consulas æfter ðæm gefeohte, siex míla from ðære byrig, mid crætwǽne, mid golde and mid gimstánum gefrætwedum, and hié sceoldon bringan feówerféttes twá hwít. Ðonne hié hámweard fóran, ðonne sceoldon hiera senátus rídan on crætwǽnum wiðæftan ðæm consulum, and ða menn beforan him drífan gebundene ðe ðær gefongene wǽron, ðæt heora mǽrþa sceoldon ðý þrymlícran beón. Ac ðonne hié hwelc folc búton gefeohte on heora geweald geníéddon, ðonne hié hámweard wǽron, ðonne sceolde him man bringan ongeán of ðære byrig crætwǽn, se wæs mid seolfre gegiered, and ǽlces cynnes feówerféttes feós án, hiora consulum tó mǽrþe, Ors. 2, 4; Swt. 70, 22-35. The explanation is called forth by the passage : Heora án consul forsóc ðone triumphan, ðe him mon ongeán brohte . . . and sǽde, ðæt hié hæfden bet gewyrht, ðæt him mon mid heáfe ongeán cóme ðonne mid triumphan, 17-21. Hió nolde ðæt hié mon drife beforan ðæm triumphan, 5, 13; Swt. 246, 29. Noldan hié dôn ðone triumphan beforan hiora consulum, 4, 7; Swt. 182, 1 : 4, 10; Swt. 202, 24.

tríwen. v. treówen.

trod, es; *n. :* trodu, e; *f. A track :*—Be trode gestolenes yrfes. Gif mon trode bedrífþ forstolenes ýrfes of stæde on óðer, ðonne befæste mon ðæt spor landes mannum . . . Gif mon secge ðæt man ðæt trod áwóh drífe, ðonne mót se ðe ðæt yrfe áh trodað (trod óð ?) tó stæde lǽdan, L. O. D. 1; Th. i. 352, 3-11. Gyf him hundred bedrífe trod on óðer hundred, L. Edg. 4, 5; Th. i. 260, 3. Secga ǽnigum ðara ðe tírleáses trode sceáwode, hú hé on weg ðanon feorhlástas bær, Beo. Th. 1691; B. 843. [Þe dunes underuoð þe treden (trodes, MS. T.) of him suluen, A. R. 380, 26. Cf. treoden, l. 18. Yf thou trowyde . . . That thi witt . . . Commys of thiselfe and noȝt of Gode, That es grett pryde and fals trode, R. Brunne. Of his trodus no sygne ther nasse, Chron.Vilodun. Halliwell, from whose Dictionary the last two passages are taken, gives *trod*=footpath : see also E. D. S. Pub. Lincoln. Icel. troð; *n. a treading.* Cf. þe þet trodded wel and ofseched wel ut his owune feblesce, A. R. 232, 17.] v. wíg-, wiðer-trod.

trog, es; *m.* I. *a trough, tub, basin, vessel* for containing liquids or other materials :—Trog *albeus, genus vasis*, Txts. 109, 1140 : *canthera*, 49, 425 : Wrt. Voc. ii. 14, 7. Lege on hátne stán on troge, geót hwón wæteres on, Lchdm. ii. 326, 5 : iii. 30, 9. Dô on troh háte stánas, ii. 68, 5. Hé sende ðæt wæter in trog (*peluem*), Jn. Skt. Lind. Rush. 13, 5. Man sceal habban trogas, Anglia xiii. 264, 14. v. wín-trog. II. *a trough-shaped thing, a cradle, a boat :*—Cilda trog *conabulum*, Txts. 51, 492. Cf. ciltrog *cune*, 115, 154. Hé wæs biddende ánes lytles troges, ðæt hé mehte his feorh generian *exiguo contentus latere navigio*, Ors. 2, 5; Swt. 84, 15. III. *a water-pipe, conduit.* v. mylentrog. IV. *a basin of water* (?) :—Of ðæm forda on ðone sǽtroh, of ðæm troge on ðone hǽþenan byrgels, Cod. Dip. Kmbl. iii. 456, 32. Tó trogan, 434, 15 : 435, 11. [O. H. Ger. trog *alveus, alveolus, collectaculum, canalis : Icel.* trog.]

trog-hrycg *a ridge where there is a trough of water* (?) :—On trohhrycg, Cod. Dip. Kmbl. iii. 79, 17.

Trógia. v. Tróia.

trog-scip, es; *n. Some kind of boat.* The Latin words which it translates are *littoraria* and *tonsilla;* the ordinary meaning of the latter is, *a sharp-pointed pole stuck in the ground to fasten vessels to the shore,* so perhaps trogscip means *a boat fastened to the shore, to which another was moored :*—Trohscip *littoraria vel tonsilla*, Wrt. Voc. i. 56, 29 : *littoraria*, 48, 2 : 64, 4.

troh, tróh. v. trog, þróh.

Tróia, Trógia *Troy :*—Tróia, Gréca burg, áwésted wæs, Ors. 2, 2; Swt. 64, 20. Trógia burg barn, Bt. 16, 4; Fox 58, 4. Tróia burg ofertogen hæfde léga leóhtost, Met. 9, 16 : 26, 20. [O. H. Ger. Tróia : Icel. Trója.]

Tróiána (-e ?); *pl. The Trojans :*—Alra tácna gehwylc swá Tróiána þurh gefeoht fremedon, Elen. Kmbl. 1287; El. 645. Ymb ealra ðara Tróiána gewin, Ors. 1, 8; Swt. 42, 13. Ðær mǽre gewinn Gréca and Tróiána, 1, 11; Swt. 50, 9, 7 : Bt. 38, 1; Fox 194, 3. (In the corresponding passage in the metres, Met. 26, 12, Fox has *Tróia gewin*, while Grein gives *Tróiána*.)

Tróiánisc; *adj. Trojan :*—On ðæm Tróiániscan gefeohte, Ors. 1, 10; Swt. 48, 2 : 1, 11; 50, 24. [O. H. Ger. Tróiánisc.]

Tróiesc, tróisc; *adj. Trojan :*—Hé gelíce ðý Tróiescan (Tróiscan, Bd. M. 306, 20) wæle ealle ða landbigengan wolde út ámǽran *tragica caede omnes indigenas exterminare contendit*, Bd. 4, 16; S. 584, 6. [Of þan Troyscen monnen, Laym. 410.] v. traisc.

trondendi. v. trandan.

tropere, es; *m. One of the service books of the Church, that which contained the tropes (tropus* cantus ecclesiastici genus*);* troparium. v. Maskell's Monumenta Ritualia Ecclesiae Anglicanae, I. p. xxxvii :—.i.

tropere, Chart. Th. 430, 10. Đonne đû tropere haban wille, đonne wege đû đîne swî[þ]ran hand, and tyrn mid đînum swîþran scytefingre ofer đîne breóst foreweard, swilce đû notian wille, Techm. ii. 119, 10-12. [A tropere *troparium*, Wülck. Gl. 617, 38 : 755, 3. A tropery, 719, 34. A tropure, 648, 33 (all 15th cent. glosses).]

trúa. v. **trúwa.**

trucian; *p.* ode.   **I.** *to fail* in doing something :—Ne trucaþ heora nân âna đurh unmihte ac đurh gecynde ânre Godcundnysse hî wyrcaþ ealle æfre ân weorc *no one of them alone fails through want of power, but through the nature of one divinity they all work always the same work*, Homl. Th. ii. 42, 27. Cneów truciaþ *the knees fail*, Lchdm. ii. 242, 14.   **II.** *to fail* a person (*dat.*), *be wanting* in duty to a person :—Hê undergeat đæt his gesworene men him trucedan, and âgêfon hera castelas him tô hearme, Chr. 1090; Erl. 226, 32.   **III.** *to fail, come to an end* :—Trucaþ *periclitatur*, ic trucige *periclitor* (the passage is : Propria manu perire non licet, absque eo ubi castitas *periclitatur*; but the glosser seems to have taken the word to mean more than *is endangered*, and to have taken it as meaning *is lost*), Hpt. Gl. 468, 78-469, 1. [Him trucode ealle his mycele cræftes, Chr. 1131; Erl. 260, 2. Him trukeþ his iwit, Fragm. Phlps. 5, 38. Heo is afered leste þeo eorđe hire trukie, O. E. Homl. i. 53, 15. Heo trukieđ treođen to halden, Laym. 16861. þa iseh Hængest þ his help trukede, 16416. Wærc þe nauere nulle trukien, 17171. 3if bileaue him trukede, A. R. 230, 19. Ne schal him neauer tintreohe trukien *incredulos supplicio dampnat eterno*, Kath. 1796: 403. þis bold ... neuer truke ne schal, Misc. 97, 122. Til domes dai ne sal it troken, Al middelerd đerinne is loken, Gen. and Ex. 105.] v. **ge-trucian.**

trúgian. v. **trúwian.**

truht (trûht ?) *a trout* :—Truht *tructa*, Wrt. Voc. i. 55, 74 : 77, 64. [From Latin.]

trull. v. **turl.**

trum; *adj. Firm, strong;* firmus, Ælfc. Gr. 38; Zup. 236, 8.   **I.** *of material things*, lit. or fig. :—Hê is mê trum weall, Homl. Skt. i. 7, 127. Seó burh Asor wæs swîđe trum gefyrn and manegra burga heáfod *Asor antiquitus inter omnia regna haec principatum tenebat*, Jos. 11, 10. Trumre underweþincge *firmo fulcimento*, Wrt. Voc. ii. 148, 69. On trume stôwe *in locum munitum*, Ps. Th. 70, 2. Eálâ wæran đa ancras swâ trume and swâ þurhwuniende, Bt. 10; Fox 30, 10. Trume and torhte tungol, Exon. Th. 58, 11; Cri. 934. Ofer ealla truma ceastra ... Hwæt getâcniaþ đa truman ceastra *super omnes civitates munitas ... Quid per civitates munitas exprimitur?* Past. 35; Swt. 245, 6. Weal đý trumra, Exon. Th. 281, 23; Jul. 650. Biþ Drihten ûre se trumesta staþol, Blickl. Homl. 13, 10. Mid weallum and geatum and đâm trum-estum locum getimbrade *muris, portis, ac seris instructa firmissimis*, Bd. 1, 1; S. 473, 27.   **II.** *of living things*, (a) *strong, sound, having physical health or strength* :—Trum *validus* vel *vegetus*, Wrt. Voc. i. 51, 21. Gedafenaþ sacerde, đonne hê mannum fæsten scrîfeþ, đæt hê *wite* hwylc se man sig, trum þe untrum (*validus an invalidus*), L. Ecg. C. 1; Th. ii. 132, 25. Đonne se mon his lîchoman hælo forsihþ, đonne đonne hê wel trum biþ tô wyrceanne đæt hê wile, Past. 36; Swt. 249, 5. Wæs eft swâ ær on his lîce trum, Andr. Kmbl. 2953; An. 1479. Heorot hornum trum, Beo. Th. 2742; B. 1369. Eofor tôþmægenes trum, Menol. Fox 499; Gn. C. 20. Đa truman (cf. hâlan, l. 3) ... đa untruman *incolumes ... aegri*, Past. 36; Swt. 247, 5. (b) *strong, able to resist, fortified* against :—Wiđ eallum nædrum hê biþ trum, Lchdm. i. 92, 4. Wiđ eall næddercyn hê biþ trum, 244, 3. Trume wiđ deófla nîþum, Blickl. Homl. 171, 30. Sécaþ gê Drihten and gê beóþ teónan gehwylce ful trume, Ps. Th. 104, 4. (c) *in reference to moral qualities, strong, steadfast, firm* :—Ne biþ nân man trum đurh God, bûton se đe hine undergyt untrumne þurh hine sylfne, Homl. Th. ii. 392, 5. Iacobus trum in breóstum, Menol. Fox 266; Men. 134. Læt mê on đînum wordum weorđan trumne *confirma me in verbis tuis*, Ps. Th. 118, 28. God êcne and trumne, Cd. Th. 297, 30; Sat. 525. Englas trume and torhte, Exon. Th. 55, 15; Cri. 884.   **III.** *of non-material things, firm, stable, strong* :—Đæt môd ægđer ge trum ge untrum *animus et infirmus et fidelis*, Past. 51; Swt. 395, 3. Ân strica đære ealdan æ ne biþ forgæged, ôđ đæt hî ealle gefyllede beón. þus trum is seó ealde æ, Homl. Th. ii. 200, 2. Trum *ratum*, Hpt. Gl. 528, 25. Gif đû mid trumre heortan (*firmo corde*) gelýfest, Bd. 3, 13; S. 538, 43. Heó âhte trumne geleáfan, Judth. Thw. 9; Jud. 6. Eówer geleáfa biþ þe trumra, gif gê gehýraþ be Godes hâlgum, Homl. Th. i. 556, 27. v. med-, mis-, un-trum.

-trum. v. **ge-trum.**

truma, an; *m.*   **I.** *a troop* of soldiers. v. **trymman**, I. 6, II. 2 :—Truma *acies, exercitus*, Hpt. Gl. 477, 13. Hê fêrde mid fyrdlicum truman and đa burh geeode, Jos. 11, 10 : Homl. Ass. 113, 356. Truman *aciem*, Hpt. Gl. 426, 69. Hê gesette đa menn on ænne truman, đe mon hiora mægas ær on đæm londe slôg, Ors. 2, 5; Swt. 80, 19. Hê hæfde eahta and eahtatig coortana, đæt wê nû truman hâtaþ, 5, 12; Swt. 240, 33. Đa îsnodan truman *ferratas acies*, Wrt. Voc. ii. 147, 52.   **II.** *order* of troops, *array* :—Hê đæt folc bûton truman lædde *he led the army without keeping any order*, Ors. 4, 8; Swt. 188, 14.   **III.**

a support. v. **wyrt-truma.** [Breken Modredes trume, Laym. 28352. þat eadi trume of meidenes, H. M. 21, 33. Đu (*Jacob*) and đin trume ben ... to me welcume, Gen. and Ex. 1829. Hauelok wæs a ful god gome, He was ful god in eueri trome, Havel. 8.] v. fyrd-, ge-, scild-truma.

trumian; *p.* ode *To become strong, recover from illness* :—Đâ cwæþ hê đæt gewunalîce word đara frêfrendra : Truma đê hraþe and wel *dixit solito consolantium sermone : Bene convalescas et cito*, Bd. 5, 5; S. 618, 9. Hine gestôd sumu untrymnis ... sôna swâ hê trumian (*convalescere*) ongan, 4, 1; S. 564, 46. v. **ge-trumian.**

truming, e; *f. Gaining strength, recovery* :—Cwydas đôn truminge getâcnaþ, Lchdm. iii. 210, 30.

trum-lîc; adj.   **I.** *firm, strong, stable*, (a) *of material things*, lit. or fig. :—Tæceþ ûs se torhta trumlîcne hâm, burhweallas beorhte scînaþ, Cd. Th. 282, 30; Sat. 294. Đâ geseah ic gyldenne wîngeard trumlîcne and fæstlîcne *vineam solidam auro miratus sum*, Nar. 4, 28. Columnan swîđe trumlîce and fæste *columnae solidae*, 4, 21. (b) *of non-material things* :—Seó ealde gesetness ys eall swâ trumlîc, swâ swâ se Hælend sæde on his hâlgan godspelle, Jud. 15; Thw. 159, 29. Kynewyrđe ræd and trumlîc, Anglia viii. 308, 33. Đæt ôđer lîf đætte fæstre beón scolde and trumlîcre (*stabilior*), Past. 52; Swt. 411, 1.   **II.** *hortatory, of exhortation* :—Hê đam cyninge sende trumlîc ærendgewrit, Bd. 2, 17; S. 520, 19 note. v. next word.

trumlîce; adv.   **I.** *firmly, strongly, steadfastly* :—Trumlîce *firmiter*, Ælfc. Gr. 38; Zup. 236, 8. Đa gôdan weorc, đeáh đe hió beforan monna eágum đyncen trumlîce gedôn *etiam quae humanis oculis fortia videntur*, Past. 34; Swt. 237, 2. Đæt leód and lagu trumlîce stande, Wulfst. 74, 8. Eahta sweras syndon đe rihtlîcne cynedôm trumlîce up wegaþ, L. I. P. 3; Th. ii. 306, 19. Trumlîcor *firmius*, Rtl. 34, 26. Freóndscype trumlîcust (*firmissime*) wunaþ, Scint. 197, 18.   **II.** *in a way that encourages* (?) :—Ungeleáffullnîse trumlîce (*strongly* (?) : but the Latin is *clementer*) gedreáđ biđ, Mk. Skt. p. 5, 13.

trumme. v. **trymman.**

trumnaþ, es; m. Strengthening, confirmation :—Swilc God wyrceþ gæsta lîfes tô trumnaþe, Exon. Th. 147, 18; Gû. 729.

trumness, e; f.   **I.** *firmness, strength, certainty* :—Trumnesse *firmitatem*, Kent. Gl. 840. Đînes geleáfan trumnesse wê witon, Guthl. 5; Gdwin. 30, 18. Wê witon đæt manega clericas nyton hwæt byþ *quadrans*, ac wê willaþ his mihta and his trumnysse hêr geswutelian, Anglia viii. 306, 28.   **II.** *health* :—Đa truman sint tô manianne đæt hié gewilnigen mid đæs lîcuman trumnesse đæt him ne losige sió hælo đæs môdes đý đe læs him đý wirs sié gif hié đa trumnesse đære Godes giefe him tô unnyte gehweorfaþ *admonendi sunt incolumes, ut salutem corporis exerceant ad salutem mentis; ne, si acceptae incolumitatis gratiam ad usum nequitiae inclinent, dono deteriores fiant*, Past. 36; Swt. 247, 6-8.   **III.** *confirmation, support* :—Drihten trumnes mîn *Dominus firmamentum meum*, Ps. Spl. 17, 1 : 24, 15 : 72, 4. Đæra apostola tweónung næs nâ swâ swîđe heora ungeleáffulnys, ac wæs ûre trumnys, Homl. Th. i. 300, 34. Hê týmde tô Basilies tæcinge for his trumnysse, Basil prm.; Norm. 32, 10. Ealle trumnysse hláfes hê forcnáđ *omne firmamentum panis contrivit*, Ps. Spl. 104, 15.   **IV.** *a firm place, the firmament* :—Biđ trumnys on lande on heáhnyssum dûna *erit firmamentum in terra in summis montium*, Ps. Spl. 71, 16. Weorc handa his bodaþ trumnyss[e] ł staþol (*firmamentum*), Ps. Spl. 18, 1. v. **trymness.**

trus, es; n. Fallen leaves and branches or twigs as material for fuel :—.vi. fôđra truses ælce geáre, Cod. Dip. Kmbl. iii. 169, 10. [Icel. tros; *n. leaves and twigs from a tree picked up and used for fuel.*]

trúđ, es; m. A player on a trumpet, an actor, buffoon :—Trûđ *liticen*, Ælfc. Gr. 9, 12; Zup. 40, 7 : Wrt. Voc. i. 73, 66 (the word occurs in a list of terms connected with amusements). Com sum trûđ tô đæs bisceopes hîrede, se ne gýmde nânes lenctenes fæstenes, ac eode him tô kicenan, Homl. Skt. i. 12, 59. Trûþas *histriones*, gligmon *mimus, jocista, scurra, pantomimus, tumbere saltator*, Wrt. Voc. i. 39, 41-44. As an illustration of the character of the trûđ see Strutt's Sports and Pastimes, Bk. iii. c. 3, §§ 4, 7, where one picture is given of dancers accompanied by trumpeters, and another of a dancing bear attended by a trumpeter. [Icel. trúðr *a juggler.*] v. next word.

trúđ-horn, es; m. The trumpet of a trúđ, q. v. :—Trûđhorn *lituus*, Wrt. Voc. i. 73, 67. Trûđhornes *salpistae* (the passage is : Horrorem belli et classicae salpistae metuentes), Hpt. Gl. 422, 77.

trúw, e; f. Faith :—Đû hæfst ongyten đa wonclan trúwa đæs blindan lustes, Bt. 7, 2; Fox 18, 3. [O. H. Ger. trúa, trúwa *fides* : Icel. trú.] Cf. treów.

trúwa, trúa, an; m.   **I.** (*good*) *faith* :—Heriaþ ûrne Drihten, se đe ne forlæt on hine gelýfende and đa đe hihtaþ on his micclum trúwan, Homl. Ass. 112, 321. Đam ânum ic healde mînne trúwan æfre, Homl. Skt. i. 7, 56.   **II.** *faith, belief, confidence, trust* :—Se trúwa (trúa, Cott. MSS.) micelre orsorgnesse *fiducia magnae securitatis*, Past. 35; Swt. 243, 12. Be geleáfan oþþe trúwan *de fide*, Scint. 126, 16. For đam micclan geleáfan and for đam sôđan trúwan đe heó symle hæfþ tô

Gode, Homl. Ass. 29, 125. Hé hine gefullode mid fullum trúwan ðæt hé geleáfful wǽre, Ælfc. T. Grn. 17, 9. Se ðe mid dyslícum trúwan and mid gylpe sum wundorlíc ðing on Godes naman dón wile, Homl. Th. i. 170, 28. For ðæs cræftes trúwan (trúan) *from confidence on account of that art*, R. Ben. 95, 6 : 46, 16. Habbaþ Godes trúwan *have faith in God*, Mk. Skt. 11, 22 : Scint. 127, 1. Gif hopan trúwan wé nabbaþ *si spei fiduciam non habemus*, 33, 9. Habbaþ eów trúwan *habete fiduciam* (Mt. 14, 27), Homl. Th. ii. 388, 25. Hira godas on ðám hig trúwan hæfdon *dii eorum, in quibus habebant fiduciam*, Deut. 32, 37. Hig nefdon nánne trúwan tó nánum folce *they could not trust any people*, Nicod. 6 ; Thw. 3, 24. Gif heó it swá gehylt, swá ic hiræ trúwan tó hæbbe *as I have confidence in her (that she will do)*, Chart. Th. 527, 34. **III.** *a solemn assurance of good faith, a covenant, word :—* Se Frysa lét hine faran on his trúwan, Homl. Th. ii. 358, 22. Ic sette mín wedd on écne trúwan (*in foedus sempiternum*), Gen. 17, 19. Ic behét mínne trúwan *pepigi foedus*, Ex. 6, 4. **IV.** *faithful care, protection :—*Ic hine nam on mínne trúwan *ego in meam hunc recepi fidem*, Gen. 44, 32. [*O. Frs.* trouwa : *Icel.* trúa.] v. ge-, ofer-trúwa.

trúwian ; *p.* ode To trust, confide :—Ic trúwige *fido . . .*, ic trúwige *confido*, ic trúwode *confisus sum*, Ælfc. Gr. 33 ; Zup. 204, 14–16. **I.** with dat., *to trust* to :—Ðonne ða fortrúwodan him selfum tó suíðe trúwiaþ *dum protervi valde de se praesumunt*, Past. 32 ; Swt. 209, 6. Ða ðe hyra weorcum trúwiaþ, Exon. Th. 52, 24 ; Cri. 838. Ðá ðá ic him betst trúwode, Bt. 2 ; Fox 4, 12 : Beo. Th. 3991 ; B. 1993. Secgaþ ðæm welegum, ðæt hí tó wel ne trúwigen ðissum ungewissum welum (*sperare in incerto divitiarum suarum*), Past. 26 ; Swt. 181, 15. Heó ongan his wordum trúwian, Cd. Th. 40, 35 ; Gen. 649. **I a.** *to trust* something to a person :—Se Hælend ne trúgude hine sealfne him, Jn. Skt. Lind. 2, 24. **I b.** *to trust* to a person for something (clause with ðæt) :—Hygd bearne ne trúwode, ðæt hé wið ælfylcum éþelstólas healdan cúðe, Beo. Th. 4370 ; B. 2370. **II.** with gen. *to trust* in :—Geáta leód trúwode módgan mægnes, Beo. Th. 1343 ; B. 669. Hé his wísna trúwade, drohtes on ðære ádle, Exon. Th. 171, 30 ; Gú. 1134. Hwý hié ðara geearnunga hiora dígelnesse (and diégelnesse, Hatt. MS.) and ánette bet trúwien ðonne ðære hú hié óðerra monna mæst gehelpen *qua mente utilitati ceterorum secretum praeponit suum*, Past. 5 ; Swt. 46, 2. **II a.** with gen. and clause :—Hé wiðres ne trúwode ðæt hé sæmannum onsacan mihte *he did not trust in resistance, that he should be able to repel the seamen*, Beo. Th. 5899; B. 2953. **III.** with prepositions (be, on, tó), *to be confident of, trust* in, on, or to :—Ða ðe trúwiaþ on him *qui confidunt in eis*, Ps. Spl. 134, 18. Ealle his wǽpnu ðe hé on trúwude *universa arma in quibus confidebat*, Lk. Skt. 11, 22. Ða burhware trúwodon tó ðam wealle, Homl. Skt. ii. 25, 446. Trúa on Crist, Homl. Th. ii. 392, 34. Ðæt úre nán be him sylfum tó dyrstelíce ne trúwige *that none of us be over-confident about himself*, 82, 26. Ne trúwige nán man be ælmesdǽdum oððe on gebedum, bútan ðære foresǽdan lufe, i. 54, 11. **IV.** with a clause, *to trust* that :—Ic trúwige, ðeáh, ðæt sum wurðe ábírd þurh God, L. Ælfc. P. 3 ; Th. ii. 364, 17. [Muþe we wel trowen al . . . he misfoð, O. E. Homl. i. 67, 209. Þu ne wolldest nohht trowwenn mine wordess, Orm. 214. Wan hii þe troueþ alre best, Laym. 3413 (2nd MS.). Mon þe he wel trowede on, 2351. Wile he trowe me, Havel. 1656. *Chauc. Piers P. Wick.* trowe. *Goth.* trauan : *O. Sax.* trûón (*with gen.* or *prep.*) : *O. H. Ger.* trûen, trûwên (*same govt. as English*) confidere : *Icel.* trûa *to trust, believe* (*dat.* or *prep.*).] v. for-, ge-, or-trúwian ; ofer-trúwod ; treówian.

trúwung, tryccan, tryddian, trym, trymend-líc, trymeness, trymes, trynian, trymig. v. ge-trúwung, -trûgung, ge-tryccan, treddian, trem, trymmend-líc, trymness, trimes, trymman, un-trymig.

trymman, trymian ; *p.* trymede. **I.** *to make firm* or *strong,* (1) of material objects, *to construct strongly.* v. trum, I :—Ðæt hé trymede getimbro, Cd. Th. 18, 20 ; Gen. 276. Gé ðone weall ne trymedon ymbe hira hús *non opposuistis murum pro domo Israel*, Past. 15 ; Swt. 89, 19. (1 a) of non-material objects :—Se ðe him hálig gǽst wísaþ and his weorc trymaþ, Exon. Th. 124, 2 ; Gú. 333. Dagas syndon trymede *dies firmabuntur*, Ps. Th. 138, 15. (2) of physical health or strength, *to give strength to, strengthen.* v. trum, II a :—Hláf trymeþ heortan mannes *panis cor hominis confirmat*, Ps. Th. 103, 15. Onlegen tó trymmanne ðone magan and tó bindanne æfter útsihtan, Lchdm. ii. 180, 24. (3) of mental or moral strength, *to confirm, establish, give strength to* mind or heart. v. trum, II c :—Sóð Metod ðín mód trymeþ, Cd. Th. 170, 9 ; Gen. 2809. Hé trymede heora heortan mid Godes geleáfan, Blickl. Homl. 145, 21. Gǽst, se his hyge trymede, Cd. Th. 249, 23 ; Dan. 534. Engel hine elne trymede, Exon. Th. 113, 21 ; Gú. 161. Ðæt man Godes cyricean fæste tremede, ge lǽwede men ge hádode, Blickl. Homl. 43, 6. Ðæt hé hiera geleáfan trymede, Chr. 430 ; Erl. 10, 19. Ægðer óðrum trymede heofonríces hyht, Andr. Kmbl. 2104 ; An. 1053. Strangie man and trymme (trumme, L. I. P. 4 ; Th. ii. 308, 4) hí mid wíslícre Godes lage, Wulfst. 267, 21. Hé ongon his sefan trymman, Exon. Th. 169, 4 ; Gú. 1089. On ðæm medwísan is tó trymmanne (trymmianne, Cott. MSS.) swá hwæt suá hié ongietan mægen

ðæs godcundan wísdómes *in istis aedificandum est, quidquid de superna sapientia cognoscitur*, Past. 30 ; Swt. 203, 10. (3 a) as an ecclesiastical term, *to confirm.* v. un-trymed. (4) of abstract objects, *to corroborate, confirm* an agreement, a grant, testimony, statement, etc. v. trymmend, II :—Ic, Berhtwulf, ðás míne gesaldnisse trymme and fæstna in Cristes róde tácne, Cod. Dip. Kmbl. ii. 5, 33 : 47, 20. Gif ic cýðnisse trymmo *si ego testimonium perhibeo*, Jn. Skt. Lind. 5, 31. Ðæt trymeþ sió hálige ǽ, ðær hió cuæð, Past. 43 ; Swt. 309, 12. Wé trymmaþ *adstipulabimur*, Wrt. Voc. ii. 3, 28. Ic ðíne gewitnesse wordum trymede *servavi testimonia tua*, Ps. Th. 118, 168. Ðæt trymede sanctus Paulus, ðá hé cuæð ðæt . . . , Past. 11 ; Swt. 73, 2. Trymme hé eal mid wedde ðæt ðæt hé beháte, L. Edm. B. 5 ; Th. i. 254, 17. Trymmendre (*confirmante*) sprǽce, Mk. Skt. 16, 20. (5) *to give as surety :—*Trymide *commendabat*, Wrt. Voc. ii. 105, 22. Trymede, 15, 25. Hí geræddon ðæt man tremede gíslas on ǽgðer healfe, Chr. 1052 ; Erl. 187, 6. (6) *to trim, to set in firm order, array* troops. v. truma :—Hié hié bútan ðæm geate angeán Hannibal trymedon, Ors. 4, 10 ; Swt. 194, 17. Ðæt hié on morgenne hié forð trymedan ongeán heora feóndum, Blickl. Homl. 201, 35. Hí trymedon hí fæstlíce ongeán, Chr. 1048 ; Erl. 178, 31. Swylce ðær man fyrde trymme and samnige, Blickl. Homl. 91, 31. (6 a) of abstract objects, *to settle, arrange :—*Hé ðær ðone winter wunode and swá his síþfæt trymede and tó Róme com *ibi hiemem exigens sic Romam veniendi iter repetiit*, Bd. 5, 19; S. 639, 27. (7) *to strengthen* with words, *exhort, encourage, comfort :—*Hí hí mid wrǽðum wordum trymmaþ, Ps. Th. 63, 4. Drihten is swíðe mildheort, se ús trymede and lǽreþ ; hé cwæþ : 'Nelle ic ðæs synfullon mannes deáð,' Blickl. Homl. 97, 32 : Bd. 1, 23 ; S. 485, 39 : Andr. Kmbl. 927 ; An. 463. Heáhcyning sprǽce trymede tilmódigne, Cd. Th. 130, 27; Gen. 2166. Gé hyra sefan trymedon on frófre, Exon. Th. 83, 23 ; Cri. 1360. Swá hý hine trymedon, 110, 7 ; Gú. 104. Bégen gebróþru beornas trymedon, wordon bædon, Byrht. Th. 140, 49 ; By. 305. Ðíne láreówas, ða ðec tó góde trymmen, Exon. Th. 301, 4 ; Fá. 14. Lǽran sceal mon geongne monnan, trymman and tyhtan, 336, 10 ; Gn. Ex. 46 : 280, 33 ; Jul. 638. Wordum trymman, Andr. Kmbl. 856 ; An. 428. Ðá ongunnon hí hine geornlíce trymman and lǽran *coeperunt diligenter exhortari*, Bd. 5, 14 ; S. 634, 30. Trymian, Byrht. Th. 132, 17 ; By. 17. Ðú trymmende eard mec *exhortatus es me*, Ps. Surt. 70, 21. Tremegende monens, R. Ben. 4, 15. **II.** intrans. (?) (1) *to become strong :—*Monig sceal siþþan wyrt onwæcnan ; eác ðon wudubearwas tánum týdraþ trymmaþ eorðwelan *the woods teem with branches, grow strong* (?) *with the wealth of earth*, Exon. Th. 191, 7 ; Az. 84. (2) *to be arrayed.* v. truma :—Gáras trymedon, blicon bordhreóðan, býman sungon, Cd. Th. 187, 28 ; Exod. 159. Fór fyrda mæst, féðan trymedan, Elen. Kmbl. 70 ; El. 35. v. getrymman.

trymmend, es ; *m.* **I.** *one who strengthens* or *supports :—*Ðú mé wǽre trymmend *firmamentum meum*, Ps. Th. 70, 3. **II.** *one who makes a formal agreement.* v. trymman, I. 4 :—Trymmend *stipulatorem*, Wrt. Voc. ii. 88, 2.

trymmend-líc ; *adj.* Hortatory :—Trymendlíc *exortatorium*, Wrt. Voc. ii. 33, 17. Hé mid trymme[n]dlíce ǽrendgewrite hí gestrangode *epistola illos exhortatoria confortaverit*, Bd. 1, 23 ; S. 485, 15. Eác swylce ðæm cyninge hé sende trymmendlíce (-líc, Bd. M. 146, 9) gewrit *misit regi literas exhortatorias*, 2, 17 ; S. 520, 19.

trymmian. v. trymman.

trymming, e ; *f.* **I.** *a strengthening, confirming, establishing, edification :—*Se cyning ðæt mǽre hús (*the temple*) Gode betǽhte him and his folce tó trymminge and tó gescyldnysse wið ælces yfeles onscyte, Homl. Th. ii. 578, 22. Nú wylle wé eów secgan sum ðing ðe eów mǽge tó trymminge *that may serve for your edification*, Homl. Ass. 26, 50. Tó geleáfan trymminge *for the confirmation of belief*, 5, 111. Trimminge, Ælfc. T. Grn. 14, 8. **II.** *that which strengthens* or *supports,* (a) material, *a foundation :—*Curs móder áwyrtwalaþ trymmincge *the curse of the mother rooteth out foundations* (*firmamentum*, Ecclus. 3, 9), Scint. 174, 7. (b) non-material, *that which edifies :—*Wé wyllaþ sume óðre trimminge (*edifying matter*) be ðære mǽran Godes méder gereccan tó eówre gebetrunge, Homl. Th. i. 448, 9. v. ge-, ymb-trymming.

trymness, trymenes, e ; *f.* **I.** *firmness.* v. trumness, I :—Heora wítes ne biþ trymnes (trymenis, Ps. Surt.) *non est firmamentum in plaga eorum*, Ps. Th. 72, 3. Hiora trymnisse liomana *suorum firmitate membrorum*, Rtl. 32, 15. **II.** *that which makes firm, a support, prop,* (a) literal :—Man ða ilcan studu útan tó gesette tó trymnesse (wrǽðe, col. 1) ðæs wáges (*in munimentum parietis*) . . . tó trymnesse (fultume, col. 1) ðæs húses *in fulcimentum domus*, Bd. 3, 17 ; S. 544, 21–36. (b) figurative :—Drihten, ðú eart mín trymenes (-nis, Ps. Surt.) *Dominus firmamentum meum*, Ps. Th. 17, 1. Ðú eart mín trymnes (trymenis, Ps. Surt.), 30, 4. (c) *a firm place, fastness.* v. trumness, IV :—Biþ trymenis (*firmamentum* ; rodor, Ps. Lamb.) in eorðan in heánissum munta, Ps. Surt. 71, 16. **III.** *a strengthening, a confirmation,* (a) of a statement, agreement, etc. :—Trymnes *confirmatio, assertio*, Wrt. Voc. ii. 133, 27. Tó trymnisse *testamento*, Rtl. 191, 33. Trymnessum *adstipulationibus*, Wrt. Voc. ii. 1, 7 : 3, 63. (b) of or in a purpose, belief, etc. :—Ðá wæs

gestrangod Agustinus mid trymnysse ðæs eádigan fæder *roboratus confir- matione beati patris Agustinus*, Bd. 1, 25 ; S. 486, 13. (c) *a strength- ening by words, an exhortation :*—Trymnes *exortatio,* i. *monitio, doctrina,* Wrt. Voc. ii. 145, 77. Trymnises *exortationis,* Mk. Skt. p. 2, 5. Mid stefne his háligre trymenesse (trymnisse, Bd. M. 106, 26) and láre *voce sanctae exhortationis,* Bd. 2, 4 ; S. 505, 18. Trymnyssum *exhortationibus,* 1, 7 ; S. 477, 3. Trymenessum, 5, 22 ; S. 644, 6. v. ge-, un-trymness.

trymsas. v. trimes.

trymþ, e ; *f. Strength, support :*—Ealle getrymednesse ł trymðe hláfes hē forgnád *omne firmamentum panis contrivit,* Ps. Lamb. 104, 16. v. un-trymþ.

tryndyled, trȳw, trȳwa, trȳwan, trȳwe, trȳwen, trȳwian, trȳwsian, trȳwþ. v. trendeled, treów, treówa, treówan, treówe, treówen, treówian, treówsian, treówþ.

tú (*two*), tú (*thou*), tuá, tuáes. v. twēgen, þú, tweó, tweógan.

tucian (or tûcian ? ; in Piers P. (v. infra) *touked* occurs, but the form of the noun is *tokkere* as well as *touker,* Prol. 100 A-text, and Halliwell gives *tucker* = fuller as a western word) ; *p.* ode *To treat ill, to afflict, harass, vex :*—Unrihtwíse cyningas ðe ðis wérige folc wyrst tuciaþ (*quos miseri torvos populi timent tyrannos,* ða unrihtwísan cyningas . . . ðe ðis earme folc heardost ondræt, Bt. 36, 2 ; Fox 174, 26–29), Met. 24, 60. Hē heora fela ofslóh and tó sceame tucode *percussit Philisthiim ingenti plaga,* Jud. 15, 8 : Homl. Skt. ii. 26, 11. Hí man swang and tó ealre yrmðe tucode *they were scourged and treated to (afflicted with) every misery,* i. 23, 106. Hí man tó wæfersȳne tucode mid gehwilcum wítum, ii. 28, 129. Swingan and tó ealre sorge tucigan, i. 23, 715. Noldon hí ná cweþan ðæt hit wǽre wíte . . . and noldan nǽnne þingere sēcan, ac lustlíce hí woldan lǽtan ða rícan hié tucian æfter hiora ágnum willan *nec hos cruciatus esse dicerent, defensorumque operam repudiarent, ac se totos accusatoribus judicibusque permitterent,* Bt. 38, 7 ; Fox 210, 14. [Ure Louerd was on fele wise rewliche tuked, O. E. Homl. ii. 21, 32. He was so scheomeliche ituked and so seoruhfuliche ipined, A. R. 366, 3. Leccherie tukeð hire al to wundre & þreat to don hire schome, H. M. 17, 10. Ha tukeð ure godes to balewe & to bismere, Kath. 551. þu tukest wroþe and uvele Hwar þu miht over smale fuȝele, O. and N. 63. Cloth with taseles cracched, Ytouked and ytented, Piers P. 15, 447. Tuck *to pinch severely,* Devonshire : *to smart with pain,* Wilts., Halli- well's Dict. O. H. Ger. zocchôn *rapere, diripere.*] v. ge-, mis-tucian.

tuddor. v. túdor.

tude, an (?) ; *f. A shield :*—Tude *parma,* Hpt. Gl. 521, 9. Tudenarda (tudena, randa (?), tuderanda (?)) *scutorum,* 424, 5.

túdor, tuddor, es ; *n. That which grows from another* (used of animals or of plants), *offspring, progeny, product, fruit :*—Túdor oððe cyn *pro- pago,* Wrt. Voc. ii. 67, 33. On ða tíd wæs ofor eorþan tuddres æþelnes, Blickl. Homl. 115, 10. Hē tȳdreþ ǽlc túdor, Bt. 39, 8 ; Fox 224, 10. **I.** of human beings, (a) *a child :*—Tudder *pignus,* Ælfc. Gr. 9, 32 ; Zup. 59, 9. Bearn *vel* tudder *soboles* vel *proles,* Wrt. Voc. i. 51, 64 : *foetus,* i. *fructus, partus, filius, soboles,* ii. 148, 35. ' Ðú cennest sunu.' Mid ðȳ ðe heó gehȳrde ðone fruman ðæs godcundan tuddres, Blickl. Homl. 7, 20. Túdre *foetu,* Wrt. Voc. ii. 36, 34. Gyf hwylc wíf hæbbe on hyre innoðe deáðboren tuddur, Lchdm. i. 166, 4. Hyt ðæt tudder of ðam cwiðan gelǽdeþ, 296, 2. Tuddra *pignora,* Hymn. Surt. 52, 7. (b) in a general sense, *offspring, race, breed, family, children :*— Tuddor *prosapia,* Wrt. Voc. ii. 65, 71. Tudder (*maternae generationis*) *propago,* Hpt. 522, 30. Wē oncneówan ðæt ðæt tuddur ne grówan mihte of swylcum gesinscype *didicimus ex tali conjugio sobolem non posse suc- crescere,* Bd. 1, 27 ; S. 491, 5. Moncynnes tucodor, Exon. Th. 86, 32 ; Cri. 1417. Fruma ælda túdres, 151, 16 ; Gú. 796. Gódes túdres gesǽlig *bona sobole felix,* Bd. 3, 7 ; S. 529, 34 : 3, 18 ; S. 546, 39. Wæstm- bærnysse tuddres *faecunditatem sobolis,* 1, 27 ; S. 493, 8. Sunu gódes tuddres *filium bone indolis,* Scint. 177, 6. Ára ðínum earmum eorþan túdre (cf. help ðínum earmum moncynne, Bt. 4 ; Fox 8, 11), Met. 4, 31. Túdre fyllaþ eorðan, incre cynne, sunum and dohtrum, Cd. Th. 13, 2 ; Gen. 196 : 92, 27 ; Gen. 1535 : 107, 12 ; Gen. 1788 : 169, 18 ; Gen. 2801. Tó teónan manna túdre *to the hurt of mankind,* Exon. Th. 270, 3 ; Jul. 459. Ðæs teámes wæs tuddor gefylled unlytel dǽl eorðan gesceafta, Cd. Th. 97, 16 ; Gen. 1613. Ðonne ðæt flǽsc náuht elles ne sēcþ búton túdor *nisi fructum propaginis non quaerere,* Past. 51 ; Swt. 399, 5. God weorðaþ eorþan tuddor, Exon. Th. 43, 13 ; Cri. 608. **II.** of animals :—Wócor eorðan túdres *every kind of animal,* Cd. Th. 79, 18 ; Gen. 1313 : 86, 34 ; Gen. 1440. Setl ælcum eorðan túdre, 79, 3 ; Gen. 1305. Deáþ spyreþ æfter æghwelcum eorþan túdre, diórum and fuglum, Met. 27, 10. Treófugla tuddor cýðdon eádges eftcyme, Exon. Th. 146, 9 ; Gú. 707. Ðú seofone genim túdra gehwylces, Cd. Th. 80, 29 ; Gen. 1336. **II a.** of human beings and animals :—Tuddor bið gemǽne incrum (*the woman and the serpent*) orlegníð, Cd. Th. 56, 19 ; Gen. 914. Se egorhere eorðan tuddor eall ácwealde, búton ðæt earce bord heóld heofona freá, 84, 24 ; Gen. 1402. **III.** of plants :—Beorc byþ blǽda leás, bereþ tánas bútan túdre, Runic pm. Kmbl. 342, 29 ; Rún. 18. Brengþ eorþe ǽlcne westm and ǽlc túdor ǽlce geáre, Bt. 39, 13 ; Fox 234, 14 : Met. 29, 58. **IV.** metaphorical :—Weá wæs árǽred,

---

tregena tuddor, Cd. Th. 60, 27 ; Gen. 988. Ðonne mæg hē cennan ðæt túder ryhtes gedohtes (*prolem rectae cogitationis*), Past. 15 ; Swt. 97, 8. Ópre tuddru synna *cetere soboles peccatorum,* Scint. 112, 4. [Deor and fishshes and fugeles and here tuder, O. E. Homl. ii. 177, 17.] v. eorþ-, magu-, sige-túdor, *and next word.*

túdor (?) ; *adj. Prolific :*—Tuddre *fetose,* Wrt. Voc. ii. 148, 35. v. túdor-full.

túdor-fæst ; *adj. Prolific, fruitful :*—Túdorfæstum *foetosis,* Wrt. Voc. ii. 34, 15.

túdor-fóster, es ; *m. Nourishment of offspring :*—Æfter ðon tuddor- fóstre *vel of ðam siþborenum de post fetantes,* Wrt. Voc. ii. 138, 81.

túdor-full ; *adj. Prolific, fertile, fruitful :*—Tudderfulle, teámfulle *vel* tuddre *fetose,* Wrt. Voc. ii. 148, 34. On tudderfullum *fetosis, copiosis, fecundis,* Hpt. Gl. 484, 5, 7.

túdor-spéd, e ; *f. Abundance of offspring :*—Him engla helm tuddor- spéd onleác . . . lēt weaxan eft heora rímgetel, Cd. Th. 166, 24 ; Gen. 2752.

túdor-teónde *producing offspring* or *fruit :*—Hēt sǽs and eorðan tuddorteóndra teohha gehwilcre wæstmas fēdan, Cd. Th. 59, 5 ; Gen. 959 : 201, 14 ; Exod. 372.

tulge ; *cpve.* tylg ; *spve.* tylgest ; *adv. Strongly, firmly ;* but the word undergoes a similar change to that which is seen in the case of *swíðe* q. v., and is used with much the same force as that word :—Him beóþ under tungan tulge swearte ǽdra *he has under his tongue very black veins,* Lchdm. ii. 106, 23. Tylg *propensior* (-or from *-us* in Erfurt Gloss.), Txts. 84, 743. Ic bí mē tylgust secge ðis sárspell *I make this lament mostly about myself,* Exon. Th. 458, 5 ; Hy. 4, 95. [Nes þ naht wunderlic þ he þone deaþes deg swa unforht abad, for þon þe hit nes deaþes deg ac hit (his MS.) wes tylig Drihtnes blisse deg *it was not won- derful that he awaited the day of death so fearless, for it was not the day of death, but it was rather the day of the joy of the Lord,* Anglia x. 145, 160. Se ealles tylgest romanisce þeawe song in Godes circan *he sang chiefly after the Roman manner in God's Church,* 142, 36. (These two passages are from a MS. of the first half of the 12th century.) O. Sax. tulgo *very.* Cf. Goth. tulgus *steadfast ;* tulgitha *safety, a stronghold ;* tulgjan *to confirm.*]

tumbere, es ; *m. A tumbler, dancer, player :*—Gligmon *mimus, jocista, scurra,* gligman *pantomimus,* tumbere *saltator,* Wrt. Voc. i. 39, 42–44. Tumbere oððe gligman *histrio,* Ælfc. Gr. 9, 3 ; Zup. 35, 6. [The femi- nine form *tumbestre* occurs in later English : Herodias douȝter, that was a tumbestere, and tumblede byfore him, Halliw. Dict. Than comen tombesteres Fetys and smale, Chauc. Pard. T. 477. See Strutt's Sports and Pastimes, Bk. iii. c. v. § 3. Cf. A tumbler *saltator* in a list headed *nomina jugulatorum*), Wrt. Voc. i. 218, col. 1 : *saltatrix,* 216, col. 2 ; *and see* tumbullere *saltatrix,* in the note. Tumlare, tumblar *volutator, volutatrix,* Prompt. Parv. 506. Tumbelyster *tornatrix,* Wülck. Gl. 616, 47.] v. next word.

tumbian ; *p.* ode *To tumble, dance :*—Ðá tumbude (*saltavit*) ðære Herodidiscean dohtur beforan him, Mt. Kmbl. 14, 6. Tumbode, Mk. Skt. 6, 22. [þe wenche þat tombede (*v. r.* tomblede), Trev. iv. 365. Cf. Tumblide, Wick. Mt. 14, 6. Tumlyñ *voluto, volvo,* Prompt. Parv. 506. Eroud swore to here that tumbled yn the flore, Halliw. Dict.] v. preceding word.

tún, es ; *m.* **I.** *an enclosed piece of ground, a yard, court :*— Tuun *cors* (= *cohors*), Txts. 52, 281. Tún *choors,* Wrt. Voc. ii. 17, 32 : i. 291, 12. Yna (hýna ?) túnes tácen is ðæt ðú sette ðíne swýþran hand brádlinga ofer ðínne innoð, Tech. ii. 126, 15 (cf. gang-tún). Harewyrt lytelu oftost weaxeþ on túne (*in a garden*), Lchdm. ii. 132, 8. v. æppel-, apulder-, ber-, cafer-, cyric-, deór-, gærs-, gang-, leáh-, líc-, wyrt- tún. **II.** as a technical English term, (1) in its simplest form, *the enclosed land surrounding a single dwelling :*—Gif man in mannes tún ǽrest geirneþ .vi. scillingum gebēte ; se ðe æfter irneþ .iii. scillingas ; siþþan gehwylc scilling, L. Ethb. 17 ; Th. i. 6, 16. (2) where there were many dwellings, *a manor, vill,* ' *an estate with a village community in villenage upon it under a lord's jurisdiction,*' v. Seebohm's English Village Community, c. v. See also Kemble's Saxons in England, ii. c. vii : Stubbs' Const. Hist. s.v. town : Green's Making of England, c. iv : Cod. Dip. Kmbl. iii. p. xxxix ; in the last its frequent occurrence in English local names is noted :—Ego, Plegrēd, aliquam terre unculam emi et Eðelmóde, hoc est án healf tún, que ante pertinebat tó wilburgewellan, ðet land healf and healfne tún his terminibus circumcincta . . . hanc casam supranominatam ic, Eðelmód, Plegrēde donabo, Cod. Dip. Kmbl. ii. 66, 27–67, 3. Ic wille ðæt man frígæ hæalue míne men on ǽlcum túne for míne sáwlæ, and ðæt man dēle æal healf ðæt yrue ðæt ic hæbbæ on ǽlcum túne, iii. 273, 4–6. Gif in cyninges túne man mannan ofsleá, .L. scill. gebēte, L. Ethb. 5 ; Th. i. 4, 4. On eorles túne, 13 ; Th. i. 6, 9. Æghwilc man æt ðam túne, ðe hē tó hýre, L. H. E. 5 ; Th. i. 30, 1. Beó hē on carcerne on cyninges túne, L. Alf. pol. 1 ; Th. 60, 9 : Chr. 787 ; Erl. 56, 14. Gif se gereáfa ðis oferheald, gebēte .xxx. sciłł., and sié ðæt feoh gedǽled ðǽm þearfum ðe on ða[m] tún[e] synd, ðe ðis ungefremed wunie, L. Ath. i. prm. ; Th. i. 198, 12. Hē wæs on ánum

ðæs cyninges tûne nôht feor fram ðære foresprecenan byrig forðon ðe hê ðær hæfde âne cyricean and ân resthûs ... Ðæt eác swylce his ðeáw wæs on ôþrum cyninges tûne tô dônne *erat in villa* (in 544, 14 *tún* translates *vicus*) *regia non longe ab urbe de qua praefati sumus. In hac enim habens ecclesiam et cubiculum ...; quod ipsum et in aliis villis regis facere solebat*, Bd. 3, 17; S. 543, 20–29. Ciólulf sealde Eánmunde his mêge ðisne tuun (cf. Ego Cialulf dabo Eanmunde cognito meo aliquam partem terre iuris mei, hoc est in Dorobernia ciuitate, id est in longitudo .vi. uirgis et in latitudo .iii., 87, 27–31), Cod. Dip. Kmbl. ii. 89, 10. Ðis sind ðara feówer tûna londgemæra, iii. 77, 32. Ðær hê râd betwuh his hâmum oþþe tûnum (*villas*), Bd. 2, 16; S. 520, 11. v. tûn-cyrice, -gebûr, -geréfa, -incel, -land, -mann, -scîr, -steall, -stede; tûnes-mann, -tûningas. **II a.** where the residential character of the *tún* is the prominent one, the buildings or inhabitants being referred to :—Ðâ unga se tûn bernan, ðâ forburnon ealla ðara monna hûs ðe on ðæm tûne wæron, Shrn. 90, 3–5. Ðes tûn (*villa*) wæs forlæten, and ôþer wæs getimbred, Bd. 2, 14; S. 518, 11. Hê eode tô ðære cyricean ðæs tûnes (*villulae*), 5, 12; S. 627, 20. Hê hæfde ðæt biscrîce L. winð æt Scîreburnan, and his lîc liþ ðær on tûne (*or tûne* = cyrictûne?), Chr. 867; Erl. 72, 20. Ðone tûn ðe hê oftust on eardode gyt mon his naman cneóðeþ *cujus nomine vicus in quo maxime habitare solebat usque hodie cognominatur*, 2, 20; S. 522, 23. Wæs in ða tîd ðeáu Ongelcynnes folcum, ðæt ðonne mæssepreóst in tûn (*villam*) com, hî ealle gesomnodon Godes word tô gehŷranne, 4, 27; S. 604, 16. Ðæt cumende folc of eallum tûnum (*viculis*), 2, 14; S. 518, 9; 4, 27; S. 604, 26. Hê com tô ðæm ymbgesettum tûnum (*circumpositas ad villas*), and ðâm dwoliendum bodade, 604, 13. Se ðe reáfaþ man leóhtan dæge, and hê hit kŷþe tô þrîm tûnan, L. Eth. iii. 15; Th. i. 298, 12. Hê âslât ða tûnas ealle ymb ða burh *discissa viculis in vicinia urbis*, Bd. 3, 16; S. 542, 21. **III.** referring to the towns of Roman Britain :—On Swalewan streáme se ligþ be Cetereht tûne (*vicum Cataractam*; the Roman station, Cataractonium), Bd. 2, 14; S. 518, 15. Hêr Cynewulf and Offa gefuhton ymb Benesingtûn, and Offa nam þone tuun, Chr. 777; Erl. 54, 2. Cûþwulf feaht wiþ Bretwalas and iiii tûnas genom, 571; Erl. 18, 13. (See Green's The Making of England, c. iii.) Ceáwlin monige tûnas genom, 584; Erl. 18, 24. **IV.** in a general sense, *a habitation of men* :—Lengtentîma gæð tô tûne on .vii. id. Feb. (cf. sumor gæð tô mannum on .vii. id. Mai, 25) *spring comes to our dwellings on the 23rd of February*, Anglia viii. 312, 19. Se mônþ gæð on Sunnandæge on tûne (cf. sumer se mônð tô mannum, 14: 8), 304, 12. Cymeþ on ðam ylcan dæge ûs tô tûne forma mônað, Menol. Fox 16; Men. 8: 69; Men. 34. Folcum bringð morgen tô mannum mônað tô tûne Decembris drihta bearnum, 436; Men. 219. Yldum bringð sigelbeorhte dagas sumor tô tûne, 176; Men. 89. Bringð tiida lange ærra Lîða ûs tô tûne, Iunius on geard, 214; Men. 108. Oft mon fêreþ feor bî tûne (cf. *Icel.* fara um tûn *to pass by a house*) ðær him wât freónd unwiotodne *often a man travels far, passing the dwellings of men, and knows that he has no friend for himself in them*, Exon. Th. 342, 21; Gn. Ex. 146. Ær sumor on tûn gâ, Lchdm. iii. 6, 1, 3. Hwylce dæge ða môndas gân on tûn, Anglia viii. 304, 5, 25. Cymeþ scrîðan on tûn Maius, Menol. Fox 153; Men. 78. Lencten on tûn geliden hæfde, 56; Men. 28. On folc fêreþ October on tûn, 363; Men. 183. [The phrase is found in later English, e. g. Elde cumid to tune, Misc. 133, 534.] **V.** where the word is used to translate Latin forms, or refers to places not in England, (1) *the residence or estate of a single person, an estate, farm* :—Ðîn tûn *tua villa*, Ælfc. Gr. 15; Zup. 103, 7: Wrt. Voc. i. 84, 48. Hâtan his tûn ðæs anscôdan tûn *ejus habitaculum domum discalceati vocare*, Past. 5; Swt. 43, 17. Ðâ sende hê hine tô his tûne (*in uillam suam*; toun, Wick.), ðæt hê heólde his swŷn, Lk. Skt. 15, 15: Mt. Kmbl. 22, 5. Tûne *ad prediolum suum*, Anglia xiii. 36, 258. Neáh ðam tûne (*juxta praedium*; manere, Wick.) ðe Iacob sealde his suna, Jn. Skt. 4, 5. Sceall beón se læsta ðæl nŷhst ðæm tûne ðe se deáda man on lîð, Ors. 1, 1; Swt. 20, 33, 31. Wespasianus gefôr on ânum tûne bûton Rôme *Vespasianus in villa propria circa Sabinos mortuus est*, 6, 7; Swt. 262, 29. Hê gefôr on ðam ilcan tûne (*in eadem villa*) ðe his fæder dyde, 6, 8; Swt. 264, 4: Blickl. Homl. 219, 8–9. On ðone tûn (*villam*; toun, Wick.) ðe is genemned Gezemani, Mt. Kmbl. 26, 36. Ic bohte ænne tûn (*villam*; lond, Lind. Rush.: toun, Wick.: ferme, Tindal), Lk. Skt. 14, 18: Homl. Th. ii. 372, 19–21. Iosep sealde his gebrôðrum tûn (*possessionem*), Gen. 47, 11. Fegerne tûn timbrian, Shrn. 163, 16. Tûnas *territorii*, Wrt. Voc. ii. 76, 68. Hê gemenigfylde his spêda ægðer ge on tûnum ge on landum (*tam in aedibus quam in agris*), Gen. 39, 5. Hî nemnaþ hiora land and hiora tûnas be heora naman *invocabunt nomina eorum in terris eorum*, Ps. Th. 48, 10. (2) *a collection of dwellings, a village, town* :—Tuun vel ðrop *conpetum*, Txts. 54, 307. Tûn, þrop, Wrt. Voc. ii. 15, 7 (cf. *compitum*, i. *villa* þrop, 132, 55). Tûn *pagus*, i. 54, 2. Betfage se tûn, Blickl. Homl. 77, 15. In Bethania ðæm tûne, Mt. Kmbl. Rush. 26, 6. Of ðæm tuune (tûne, Rush.) on Galilées mêgð *a Cana Galilaeae*, Jn. Skt. Lind. 21, 2. Of Abian tûne (lond, Lind. Rush.) *de uice* (*uico has been read?*) Abia, Lk. Skt. 1, 5. Of ðæm tûne ðe Scariot hætte, Blickl. Homl. 69, 6: 211, 17: 221, 19: Homl. Th. ii. 54, 3. Hê eode on ðone tûn

ðe hâtte Dadissus, and ðær wunode ... Ðâ bæd hê ðæs tûnes hlâford, ðæt hê môste healdan heora æceras ... His suna wæron âfêdde on ôþran tûne, Homl. Skt. ii. 30, 213–217. Se resteþ on *uico longe*, ðæt is on ðæm langan tûne, Shrn. 76, 2. Ðeáh dû on tûn (*uicum*; lond, Lind. Rush.) gâ, Mk. Skt. 8, 26. Hê hêt ðone tûn (*uicum*) forbærnan, Bd. 5, 10; S. 625, 2. Bedrifen on ânne tûn *in cujusdan villulae casam deportatus*, Ors. 6, 34; Swt. 292, 1. Tûnas *oppida*, Wrt. Voc. ii. 64, 70. Com micel fŷrbryne on Rômeburg, ðæt ðærbinnan forburnon xiv tûnas *quatuordecim vicos flamma consumsit*, Ors. 6, 1; Swt. 252, 21. Fare wê on gehende tûnas (*uicos*; lond, Lind. Rush.: townes, Wick.), Mk. Skt. 1, 38: *villas*, Lk. Skt. 9, 12. [Halliwell gives *town* = court, farmyard, as a Devonshire word ; and in Jamieson's Dictionary *toun, town* = a farmer's steading, or a small collection of houses ; a single dwelling-house. 'Waverley learned from this colloquy, that in Scotland a single house was called a *town*,' Waverley, c. ix. O. Frs. tûn *a fence* : O. L. Ger. tûn *maceria* : Du. tuin *a fence* ; *a garden* : O. H. Ger. zûn *sepis, maceria* : Ger. zaun *a hedge* : Icel. tûn *an enclosure* within which a house is built ; *a farm-house with its buildings, homestead* : Norweg. tun *court, farmyard*.] v. burg-, neáh-, wîc-tûn ; tŷnan.

tûn-cressa, an ; *m.* : -cærse, -cerse, an ; *f. Town-cress* (v. E. D. S. Pub. Plant Names), *garden-cress, nasturtium* ; lepidium sativum :—Tuuncressa nasturcium, Txts. 79, 1259. Tûncærse, Wrt. Voc. ii. 60, 4, 64: i. 67, 70. Tûnkerse, 31, 50. Nim tûncersan sæd, Lchdm. ii. 90, 18.

tûn-cyrice, an ; *f.* A church in a tûn (q.v.) :—Habbe hê þat lond frê his day and his wîues, and after here bothere day intô þe tûnkirke, and þô men frê ... Þat lond schal intô tûnkirke ... and þô men frê, Chart. Th. 572, 20–33. Intô ðe tûnkirke on Mardingford, 593, 2.

tunece, an ; *f.* A tunic, coat :—Tunece *tonica*, Wrt. Voc. i. 284, 62. Tunice, Scint. 144, 7. Tunicæ *tunica*, Wrt. Voc. i. 39, 71. Hit ys mînes suna tunece, Gen. 37, 33 : Exon. Th. 357, 1 ; Pa. 22. Hî nâmon his tunecan (*tunicam*; cyrtel, Lind. Rush.) ; seó tunece wæs unâsiwod, Jn. Skt. 19, 23. Ðâ dyde hê on his tunecan (cyrtil (-el), Lind. Rush.), 21, 7 : Lk. Skt. 6, 29. Ðam ðe wylle niman ðîne tunecan (cyrtel l hrægl, Lind.: ðinne tonica, Rush.), læt him tô ðinne wæfels, Mt. Kmbl. 5, 40. Ðâ sende him mon âne blace hacelan angeán him on bismer, and eft hié him sendon âne tunecan ongeán, ða ðe hié tô gehêton, ðæt hê ealles bûton ârunge tô Rôme ne com (*the Latin seems to have been misunderstood, it is*: Senatus sagum, hoc est, vestem moeroris deposuit, atque antiquum togae decorem recuperavit), Ors. 5, 10 ; Swt. 234, 21–24, 31. Ðæt hê ûs forgeáfe ða undeádlîcan tunecan ðe wê forluron on ðæs frumsceapenan mannes forgægednysse, Homl. Th. i. 34, 29. Hió becwið hyre betstan dunnan tunecan, Chart. Th. 537, 31. Hió an Ceóldrýþe hyre blacena tunecena, swâ ðær hyre leófre beó, 538, 6. Se ðe hæfþ twâ tunecan (cyrtlas, Lind. Rush.), Lk. Skt. 3, 11 : Blickl. Homl. 169, 13. [O. H. Ger. tunihha *tunica*. From Latin.] v. ge-tunecod.

tûnes-mann, es ; *m.* A man living on a manor (tûn, q.v.) :—Gif hwilc tûnesman ænigne pænig forhæbbe, gilde se landrîca ðone pænig and nime ænne oxan æt ðam men (cf. L. Edg. i. 4 ; Th. i. 264, 9 : L. Eth. ix. 10 ; Th. i. 342, 25 in which 30 pence is fixed as a fine for not paying the *heord-penig* and *Rômfeoh*, 30 pence being the value of an ox according to L. Ath. v. 3 ; Th. i. 232, 7 : v. 6, 2 ; Th. i. 234, 1 : v. 8, 5 ; Th. i. 236, 31), L. N. P. L. 59 ; Th. ii. 300, 5. Tûnes-men, L. Edg. S. 13 ; Th. i. 276, 23. Cf. 8 ; Th. i. 274, 7. v. tûn-mann.

tunge, an ; *f.* [? *in the passage*: Âlés sâwle mîne fram tunge fâcenfulre *a lingua dolosa* (but in the next verse *linguam* is glossed by *tungan*, so that perhaps *tunge* is meant for nominative : O. L. Ger. and O. H. Ger., however, have strong as well as weak forms), Ps. Lamb. 119, 2], e; *f.* **I.** a tongue :—Tunge *lingua*, Wrt. Voc. i. 64, 56. Gif monnes tunge biþ of heáfde ôðres monnes dædum dôn, ðæt biþ gelîc and eágan bôt, L. Alf. pol. 52 ; Th. i. 94, 20 : Exon. Th. 373, 25 ; Seel. Ex. 115. His tungan (tungæs, Lind.: tunga, Rush.) bend *vinculum linguae eius*, Mk. Skt. 7, 35. Hê his tungan (tunga, Lind. Rush.) onhrân, 7, 33. Rômâne ðæm pâpan his tungon forcurfon, Chr. 797 ; Erl. 58, 13. **II.** *tongue*, (1) as representing the person who speaks with the tongue :—Sió tunge biþ gescinded on ðam lâriówdôme, ðonne hió ôðer lærð ôðer hió liornode, Past. 1 ; Swt. 27, 11. Seó tunge ðe swâ monig hâlwende word on ðæs Scyppendes lof gesette, Bd. 4, 24 ; S. 599, 11. Mîn tunge mærde ðîn weorc, Ps. Th. 70, 22. Âlýs mîne sâwle from ðære tungan ðe teosu wylle. Hwæt bið ðe seald from ðære inwitfullan tungan ? 119, 2, 3. Heora tungan sprecaþ fâcn, 5, 10. Wæron hyra tungan tô yfele gehwam scearpe, 56, 5. (2) representing the words expressed by the tongue, *words, speech, language* :—Hî mid tungan heora fâcenfullîce dydon, Ps. Spl. 5, 10. Mê inwit næs on tungan, Ps. Th. 138, 2. Fram swêsere tungan ûtoncumenre, Kent. Gl. 159. Ðâ betæhte Ecgferð on hâlre tungan (*in plain language*) land and bôc Dûnstâne, Chart. Th. 208, 11 : 272, 5. (v. hâl.) Wið andan land and wið ða micelan mannes tungan, Lchdm. i. 384, 22. Mid ðæm sueorde hiera tungna tælinge, Past. 28 ; Swt. 199, 6. (2 a) *a language, speech* :—Hî sprecaþ nîwum tungum, Mk. Skt. 16, 17. (3) representing power of speaking :—Ic hæfde ðe lætran tungan, Ex. 4, 10. **III.** *a tongue-shaped thing* :—Heard is mîn tunge, Exon. Th. 489, 16 ; Râ. 78, 8. Hit hafaþ tungan lange, **439,**

23; Rä. 59, 8. [*Goth.* tuggô : *O. Sax. O. L. Ger.* tunga : *O. Frs.* tunge :
O. H. Ger. zunga : *Icel.* tunga.] v. under-tunge; ge-tynge.

tún-gebúr, es; *m. A tenant in villenage, villein :*—Túngebúr *inquilinus* (cf. geneaot *inquilinis*, Txts. 71, 1117); geneát, Wrt. Voc. ii. 45,
57; bigenga *tilia*, inbúend *colonus*, i. *incola, cultor, inquilinus*, 134, 24),
Wrt. Voc. ii. 49, 56 : i. 18, 50.

tún-gel. v. tungol.

tún-geréfa, an; *m.*　　　I. *a reeve, steward, bailiff.* v. tún, II :—
Túngeréfa *villicus*, Wrt. Voc. i. 84, 50 : *villicus vel actor vel procurator
vel rector*, 18, 48. Ðá eodon hí on sumes túngeréfan gestærn and hine
bædon ðæt hé hí onsende tó ðam ealdormen ðe ofer hine wæs . . . Ðá
onféng hí se túngeréfa *intraverunt hospitium cujusdan villici, petierunt-
que ab eo, ut transmitterentur ad satrapam qui super eum erat . . . Suscepit eos villicus*, Bd. 5, 10; S. 624, 19–28. Ðá com hé tó ðam
túngeréfan se ðe his ealdormon wæs *veniens ad villicum qui sibi praeerat*,
4, 24; S. 597, 27. Ðá herede se hláford ðære unrihtwísnesse túngeréfan
(*uilicum*), Lk. Skt. 16, 8.　　II. *a praetor.* v. tún, V. 2 :—Ypolitus
wæs túngeréfa on Róme, Shrn. 117, 9 : 116, 9 : Homl. Th. i. 422, 11.
Hé hét betæcan ðone diácon ðam túngeréfan Ypolite, 426, 35.

tunge-þrum *a ligament of the tongue :*—Tungeðrum (undertunge-
þrum, lxxiv, 9) *sublinguae*, Lchdm. i. lxx, 9.

tung-full; *adj. Loquacious, talkative :*—Tungfull mann *linguosus
homo*, Scint. 81, 9. [Cf. *O.H. Ger.* zungal *linguosus*.]

tungilsinwyrt *white hellebore* (Cockayne), Lchdm. ii. 120, 2. Cf.
tunsing-wyrt.

tungl, tungla. v. tungol.

tunglen; *adj. Of the stars, sidereal :*—Seó tunglene heofon, Anglia
vii. 12, 109, 115. Tunglenes éþeles wlite *sidereae patriae decus*, Hymn.
Surt. 58, 2.

tunglere, es; *m. An astrologer, astronomer :*—Tunglera ł wiglera
Chaldaeorum, Hpt. Gl. 483, 5. Tunglera *mathematicorum* (the passage
is : Gentilitas, quae vitam veritatis expertem fato fortunae et Genesi
gubernari juxta mathematicorum constellationem arbitratur), Wrt. Voc.
ii. 79, 64 : 56, 68.

tungol (-ul, -el), tungl, es; *generally neuter, but pl.* tunglas *occurs :*
tungla, an; *m.*　　I. *a heavenly body :*—Tungel *sidus*, Wrt. Voc. i.
41, 54. Mænig tungul máran ymbhwyrft hafaþ on heofonum, Met. 28,
20. Saturnes steorra wandraþ ofer óþrum steorrum ufor ðonnᵹ ænig
óþer tungol, Bt. 36, 2; Fox 174, 14. Swá heofenes tunglu *sicut astra
coeli*, Deut. 10, 22. Sume tunglu habbaþ scyrtran hwyrft ðonne sume
habban, swá swá tunglu habbaþ ðe wé hátaþ wænes ðisla, Bt. 39, 3;
Fox 214, 17–19, 22. Tungl, Met. 28, 6, 12. Men sǽdon ðæt heofones
tungul (*astra*) hiora yfel flugon, Ors. 1, 8; Swt. 42, 24. Tungol, Exon.
Th. 58, 12; Cri. 934 : 204, 12 : Ph. 96. Tunglan *lumina*, Hpt. Gl.
446, 23 : Boutr. Scrd. 18, 31. Þás tunglan *haec sidera*, Ælfc. Gr. 14;
Zup. 90, 5. Tunglan nǽron gesceapene ǽr on ðam feórðan dæge. On
ðam feórðan dæge gesette se Ælmihtiga ealle tungla, Homl. Th. i. 100,
7–9. Saturnus yfmest is eallra tungla, Met. 24, 20. Se móna is ealra
tungla nyþemest, Boutr. Scrd. 18, 38. *Astronomia*, ðæt ys tungla gang,
Shrn. 152, 14. Æþelast tungla (*the sun*), Exon. Th. 204, 6; Ph. 93.
Under ᵹunglum *on earth*, Andr. Kmbl. 3; An. 2. Beheald ða tunglu
ðæs heán heofnes, Bt. 39, 13; Fox 232, 25 : Met. 29, 4. Tungl, 28, 5.
Tungel, Cd. Th. 132, 8; Gen. 2190.　　II. *a heavenly body other
than sun or moon, a star :*—Seó sunne and se móna and ealle tunglen
(tungla, MS. R.), Lchdm. iii. 246, 23. Gewíteþ sunne and móna and
eal tungla leóht áspringeþ, Blickl. Homl. 91, 23. Sunnan . . . móna . . .
tunglena (*siderum*), Hymn. Surt. 22, 29. Féran mid ðære sunnan betwyx ðám tunglum, Bt. 36, 2; Fox 174, 11. Sunnan leóma torht ofer
tunglas, Exon. Th. 7, 26; Cri. 107.　　III. *a planet* (including the
sun and moon) :—Ða seofon dweligendan tunglan (cf. steorran, 26) . . .
Þone yfemestan héton ða hǽþenan Saturnus . . . Se feórða is seó sunne
. . . Se scofoþa is se móna, Boutr. Scrd. 18, 29–38, 41. Tungel (*Saturn*),
Met. 24, 23. Tungol (*the sun*), Exon. Th. 350, 25; Sch. 48; Fox
937; Erl. 112, 14. Æðele tungol (*Venus*), Met. 29, 32. Móna, gǽstlic
tungol, Exon. Th. 44, 7; Cri. 699. Habbaþ ædele tungol emne gedǽled
dæg and nihte, . . . sunne and móna . . . þa wlitegan tungl, Met. 29,
35–39. Ða mǽran tungl, 9.　　IV. *a fixed star :*—Seó tungola
heofon, Boutr. Scrd. 18, 24, 28.　　V. *a group of stars, a constellation, division of the zodiac :*—Arthon hätte án tungol on norðdǽle, se
hæfþ seofon steorran . . . ðone hátaþ lǽwede menn carles wǽn. Se ne
gǽð nǽfre ádúne under ðyssere eorðan, swá swá óðre tunglar (tungla,
MS. R.) dóð . . . óðer tungel is on súððǽle ðysum gelíc, Lchdm. iii. 270,
9–15. Ðé is nú cúð ðes mónan færeld, on hwilcum tungle hé nú is oððe
on hwilce hé ðanon gǽð, Shrn. 173, 1. Under ðam tungle (*the zodiac*)
yrnð seó sunne and se móna and ða twelf tunglena tácna, Lchdm. iii.
242, 3. Hys geár is ðæt hé underyrne ealle ða twelf tunglan, 248, 21, 5.
[*Goth.* tuggl (uf tugglam, Gal. 4, 3; cf. under tunglum, Andr. Kmbl. 3;
An. 2.) *O. Sax.* tungal : *O. H. Ger.* zungal : *Icel.* tungl and tungli (*wk.*)
the moon : *Swed.* tungel *the moon*.] v. ǽdel-, heofon-, rodor-tungol.

tungol-ǽ; *f. Astronomy :*—Tungelǽ *astronomiam, legem astrorum*,
Hpt. Gl. 528, 60 : Anglia xiii. 38, 307.

tungol-bǽre; *adj. Starry :*—Tungelbǽrum *astriferis*, Hpt. Gl. 490,
75 : 493, 12.

tungol-cræft, es; *m. Star-craft, astronomy, astrology :*—*Astralo(g)ia*,
ðæt ys tungolcræft, Shrn. 152, 14. Tungelcræft *astronomia*, Hpt. Gl.
479, 47. Hí hí on tungolcræfte (*astronomiae*) lǽrdan, Bd. 4, 2; S. 565,
26. Wé rǽdaþ on tungelcræfte, ðæt seó sunne biþ hwíltídum þurh ðæs
mónelícan trendles underscyte áðýstrod, Homl. Th. i. 608, 31.

tungol-cræfta, an; *m. An astrologer, astronomer :*—Tungelcræftum
Chaldeorum, Wrt. Voc. ii. 18, 33 : 82, 6. v. next word.

tungol-cræftiga, an; *m. An astrologer, astronomer :* — Tungel-
cræftig[um ? v. *preceding word*] *caldeorum*, Wrt. Voc. ii. 20, 28. Þreó
tungolcræftegan cóman fram eástdǽles mǽgdum tó Criste, Shrn. 48, 17.
Ðreá tungelcræftigo, Rtl. 2, 15. Ða tungolcræftega (-cræftgu, Rush.)
Magi, Mt. Kmbl. Lind. 2, 1. Tungulcræftgum *Magis*, Rush. 2, 7, 16.
From drýum ł tungulcræftgum, Lind. 2, 16.

tungolcræft-wíse, an; *f. Astronomy :*—Tungelcræftwísan *astronomia*,
Wrt. Voc. ii. 5, 4.

tungol-gesceád, es; *n. Astrology, astronomy :*—Tungelgesceád *as-
trologia*, Hpt. Gl. 479, 60 : Anglia xiii. 38, 308.

tungol-gimm, es; *m. A starry gem, a star :*—Heofon ongeat, hwá
hine torhtne getremede tungolgimmum, Exon. Th. 71, 6; Cri. 1151.

tungol-wítega, an; *m. One who prophesies by means of the stars, an
astrologer :*—Tungelwítega *astrologus vel magus vel mathematicus*, Wrt.
Voc. i. 17, 14 : *mathematicus*, 60, 12. Ðá cómon ða tungolwítegan
(*Magi*) fram eástdǽle, Mt. Kmbl. 2, 1. Tungelwítegan, 2, 10 : Homl.
Th. i. 78, 5 : Chr. 2; Erl. 4, 28. Tuncgelwítegana, steorgléawra *mathematicorum*, Hpt. Gl. 467, 74. Æfter ðære tíde ðe hé geáxode fram
ðám tungolwítegum (*Magis*; drýum, Lind.), Mt. Kmbl. 2, 16. Hé
clypode on sundersprǽce ða tungelwítegan, 2, 7 : Homl. Th. i. 78, 17.

tung-wód; *adj. Tongue-mad, violent in speech :*—Uppstige sandfull
on fótum forealdudes swá wíf tungwód menn stillum *ascensus arenosus
in pedibus ueterani, sic mulier linguata homini quieto*, Scint. 223, 13.

tunice. v. tunece.

túnincel, es; *n. A small tún, small farmstead* or *estate :*—Túnyncel
butiuncula, Wrt. Voc. ii. 126, 82. Tó his túningclum *ad praediolum
suum*, túnincle *ad villam*, Hpt. Gl. 515, 63, 64.

-túningas; *pl. m. People of a tún* (?) :—Oþ ealdingctúninga mearce
oþ níwantúninga mearce, and of níwantúninga mearce *to the mark of
the people of Aldington, then to the mark of the people of Newington,
and from the mark of the people of Newington*, Cod. Dip. B. ii. 526, 7–8.
Wudetunnincga gemǽro, Cod. Dip. Kmbl. iii. 193, 10.

tuning-wyrt. v. tunsing-wyrt.

tún-land, es; *n. Land of an estate* or *a farm :*—Ðis sindon ða lond-
gemǽra ðæra túnlonda ðe intó Perscóran belimpaþ *these are the boundaries
of the lands forming the estate of Pershore*, Cod. Dip. Kmbl. iii. 76, 28.

tún-líc; *adj. Of a village, rustic :*—Túnlíc spǽc *comedia* (as if from
κώμη = *vicus*), Wrt. Voc. i. 27, 13.

tún-mann, es; *m. A man belonging to a tún :*—Túnman *villanus*,
Wrt. Voc. i. 84, 49. Furseus oncneów ða sáwle ; se wæs his túnman ǽr
on lífe (*he had lived on the estate* (tún) *belonging to Fursey's monastery*),
Homl. Th. ii. 344, 18. v. túnes-mann.

tún-melde, an; *f. Orach*; *atriplex hortensis :*—Túnmelde *crysolachan*,
i. *aureum olus vel atriplex*, Wrt. Voc. ii. 137, 6.

tunne, an; *f. A barrel, cask :*—Tunne *cuba*, Wrt. Voc. ii. 105, 56 :
17, 29 : *cupa*, i. 24, 54 : 83, 26 : *cantarus, ubi aqua mittitur, vel ydria*,
ii. 128, 11. Twá tunnan fulle hlútres aloð, Cod. Dip. Kmbl. i. 203, 8 :
Chr. 852; Erl. 67, 38. Tunnena *cuparum, modiorum*, Hpt. Gl. 488,
73 : *cuparum*, Wrt. Voc. ii. 18, 35. *Caupo* wínbyrels oððe on tunnum,
21, 13. Nim fela tunnan, and dó hí ðǽr on innan . . . Hí wurdon ða
gebrohte ealle tó ðam tunnum, Homl. Skt. i. 4, 259–307. [*O. Frs.*
tunne : *Du.* ton : *O. H. Ger.* tunna *ydria, crater* : *Ger.* tonne : *Icel.*
tunna : *Swed.* tunna : *Dan.* tønde. There are both Celtic and Low
Latin forms, *tunna* ; from which the English is taken is uncertain.] v.
wín-tunne.

tunne-botm, es; *m. The bottom of a cask :*—Tunnebotm (cf. byden-
botm *fundum*, in the same list ' nomina vasorum ') *tympanum*, the bottom
of a cask used as a drum ?, Wrt. Voc. i. 24, 55. [*Dan.* tønde-bund *the
bottom* or *the head of a barrel*.]

tún-rǽd, es; *m. A town-council :*—Man beád ðam túnrǽde ðe his suna
on áfedde wǽron ðæt man sceolde twégen cempan gescyrpan *an order
was given to the council of the town in which his sons had been brought
up, that two soldiers should be equipped*, Homl. Skt. ii. 30, 297.

tún-scipe, es; *m. The inhabitants of a tún :*—Cýþe hé hit ðonne hé
hám cyme; and gif hit cuce orf biþ mid his túnscipes gewitnysse on
gemǽne lǽse gebringe. Gif hé swá ne déð ǽr fíf nihtum, cýþan hit ðæs
túnes men ðam hundredes ealdre, L. Edg. S. 8; Th. i. 274, 26. Hé hét
ðone túnscipe eallne ofsleán and ðone tún forbærnan *mittens occidit vicanos
illos omnes, vicumque incendio consumpsit*, Bd. 5, 10; S. 625, 1. [Gif
twa men oþer iii coman ridend to an tun, al þe tunscipe flugæn for heom,
Chr. 1137; Erl. 262, 35.]

tún-scír, e; *f. Stewardship :*—Ágyf ðíne scíre ne miht ðú lencg tún-

scíre bewitan . . . Đonne ic besciced beó fram tūnscíre *redde rationem uilicationis tuae, jam enim non poteris uilicare* ᴠ . . *Cum amotus fuero a uilicatione,* Lk. Skt. 16, 2–4.

tunsing-wyrt, e ; *f. White hellebore :*—Tunsingwyrt. Deós wyrt đe man *elleborum album,* and óđrum naman tunsincgwyrt nemneþ, Lchdm. i. 258, 21–23 : iii. 302, col. 1. Tun[s]ingwyrt, ii. 68, 25. Cf. tungilsinwyrt. [*Tunsing* occurs, Cod. Dip. Kmbl. vi. 236, 15.]

tūn-steall, es ; *m. A farm-stead, farm-yard* (?) :—Ober đane ealdan tūnsteall, Cod. Dip. B. ii. 202. 7. On đone tūnsteal eástweardne, Cod. Dip. Kmbl. iii. 193, 14. Cf. hām-, mylen-steall, and *town-place* = farm-yard, which Halliwell gives as used in Cornwall.

tūn-stede, es ; *m. A village :*—Tūnstede *pagi,* Wrt. Voc. i. 36, 30.

tūn-weg, es ; *m. A road on a* tūn, *a private road :*—Ealles hereweg *publica via,* tuunweg *privata via,* Wrt. Voc. i. 37, 39–40. Tó tūnwcges ende, Cod. Dip. Kmbl. v. 281, 21. Đær tūnwegas út sceótaþ . . . þurh đone tūn, vi. 235, 6.

túr, es ; *m. A tower :*—Intó đam tūre on Lundene, Chr. 1100 ; Erl. 236, 31. Đone weall đe hí worhton onbútan đone túr, 1097 ; Erl. 234, 27. Sié ginyhtsumnisse in tūrum đínum *fiat habundantia in turribus tuis,* Rtl. 176, 13. [Manega mynstras and tūras gefeóllon, Chr. 1117 ; Erl. 246, 21. The use of the word in the Chronicle would be due to the Norman French, but in the Ritual to Latin ?] v. torr.

turf ; *gen. dat.* tyrf ; *pl.* tyrf *and* turf ; *f.* **I.** *a turf, sod, piece of earth with grass on it :*—Tyrf *gleba,* Wrt. Voc. i. 37, 20. Deós wyrt of ānre tyrf manega bógas āsendeþ, Lchdm. i. 290, 7. Hí đa flaxan gehýddon under ānre tyrf, Guthl. 15 ; Gdwin. 64, 16. Under āne (ānre ?) tyrf, 23. Ne turf ne toft *not a sod nor a field* (i. e. neither little nor much ?), Lchdm. iii. 286, 23. Tyrb *cespites,* Wrt. Voc. ii. 103, 69. Tyrf, 23, 18 : *glebe,* 40, 37. Genim feówer tyrf on feówer healfa đæs landes . . . drýpe on đone stađol đara turfa . . . bere đa turf tó circean and mæssepreóst āsinge feówer mæssan ofer đan turfon, and wende man đæt grēne tó đan weofode, and siþþan gebringe man đa turf đær hí ǽr wǽron . . . Nim đonne đa turf and sete đǽr ufon on, Lchdm. i. 398, 4–24. Turfum *glebulis,* Wrt. Voc. ii. 40, 36 : 80, 32. Turvum *glebulis, cespitibus,* Hpt. Gl. 470, 35. Đá gewrohte hē weall mid turfum (cf. vallum . . . de cespitibus, Bd. 1, 5), Chr. 189 ; Erl. 9, 25. On tyrf *in cespites,* Wrt. Voc. ii. 48, 16. Đa wæstmbǽre tyrf *feraces glebas,* 147, 51. **II.** *turf, greensward, the grassy surface of the earth :*—Blówendre tyrf *florei cespitis,* Wrt. Voc. ii. 149, 50. Sum stān mid đynre bewrigen *lapis obtectus cespite tenui,* Bd. 5, 6 ; S. 619, 20. Wǽter wynsumu of đære moldan tyrf brecaþ, Exon. Th. 202, 8 ; Ph. 66. Of disse eorþan tyrf, 222, 15 ; Ph. 349 : 423, 21 ; Rä. 41, 25. Ic seah tyrf tredan .vi. gebróđor, 394, 10 ; Rä. 14, 1. [*O. Frs. O. L. Ger.* turf : *O. H. Ger.* zurba *cespes, terra avulsa : Icel.* torf ; *n.* ; torfa ; *f. a turf, turf.*] v. eđel-, wangturf ; torfian.

turf-haga, an ; *m. An enclosed space covered with turf, a grassy enclosure :*—Ongan hē eorđan delfan under turfhagan (cf. wangstede, 1584 ; El 794), Elen. Kmbl. 1656 ; El. 830.

turf-hleów, es ; *n. A shelter built of turf* (?) :—Æfter furan on rischrīđig ; of rischrīđie on turfhleó ; of turfhleó æfter heáfdan on Pydewyllan, Cod. Dip. Kmbl. iii. 15, 26.

turl, trull *a ladle, scoop, trowel :*—Turl, scofl *trulla,* Wrt. Voc. ii. 122, 67. [*A trulle trulla,* Wülck. Gl. 617, 46. From Latin.]

turnian ; *p.* ode. **I.** *to turn* (intrans.), *revolve round an axis or centre :*—Đa ārleásan turniaþ on ymbhwyrfte, Homl. Th. i. 514, 23. Seó firmamentum tyrnþ symle onbútan ūs . . . and ealle đa steorran, đe hyre on fæste synd, turniaþ onbútan mid hyre, Lchdm. iii. 254, 16. Hwylces gecyndes is seó hcofon ? Symle turniende (*volubilis*). Gif heó turniende (*volubile*) is, hūmeta ne feald heó ?, Anglia vii. 108–110. Turnicng turniendre liđeran *vertigo rotantis (volventis) fundibuli,* Hpt. Gl. 422, 66. **II.** *of giddiness, to turn :*—Ad tornionem capitis. Þis ys se lācecræft be þan manne þat hym þing[þ] þ̄ hyt turnge ābótan hys heáfod, Lchdm. iii. 90, 8.] v. tyrnan.

turnigend-líc ; *adj. Revolving :*—Gif seó heofon turnigendlíc (*volubile*) is, Anglia vii. 12, 109 note. v. preceding word.

turnung, e ; *f. Turning, rotation :*—Turnunge *rotatu,* Wülck. Gl. 253, 14.

turtle, an ; *f. : but* turtla, an ; *m. also is found. A turtle-dove :*—Turtle *turtur,* Wrt. Voc. i. 29, 33 : 77, 43 : Ælfc. Gr. 5 ; Zup. 14, 2. Deós turtle *hic turtur,* 9, 22 ; Zup. 48, 16. Gemētt turtla nest him *invenit turtur nidum sibi,* Ps. Lamb. 83, 4. Geoffra mē tó lāce sume turtlan and sume culfran *sume mihi turturem et columbam,* Gen. 15, 9. Bringan tó lāce āne culfran and āne turtlan, Homl. Th. i. 140, 2. Bringe hē twā turtlan, Lev. 5, 7, 11 : 14 : Lk. Skt. 2, 24 : Homl. Th. ii. 210, 34. [Cf. *O. H. Ger.* turtul-tūba *turtur.* From Latin.] v. next word.

turtur, es ; *m. : turture, an ; f. A turtle-dove :*—Speara gemoeted him hūs and turtur nest *passer invenit sibi domum et turtur nidum,* Ps. Surt. 83, 4. Twēgen culfran briddas and twēgen turturan gemæccan, Blickl. Homl. 23, 27. Tuoe (twoege, Rush.) turturas *par turturum,* Lk. Skt. Lind. Rush. 2, 24. [*O. H. Ger.* turtur (Notker, Ps. 83, 4).]

Tuu, tuu, tuwa. v. Tīw, twēgen, twiwa.

tusc, tux, es : *a wk. pl.* tuxan *occurs ; m. A canine tooth* or *a molar tooth, a tusk :*—Tusc *genuino* (-*um*), Txts. 67, 961. Tux caninus, Wrt. Voc. ii. 127, 81. Monnes tux bið .xv. scitt. weorđ *the compensation to be paid for knocking out a man's canine tooth is* xv *shillings,* L. Alf. pol. 49 ; Th. i. 94, 12. Cf. L. Ethb. 51 ; Th. i. 16. Hundes tux, Lchdm. i. 370, 29. Se flǽsctoþ wiþæftan đone tux *gigra,* Wrt. Voc. ii. 42, 9. Mannes tuxas *canini vel colomelli,* i. 43, 31. Tuxas *canini,* ii. 16, 50 : 128, 21 : Lchdm. iii. 202, 19. Wiđ tóþwrǽce, hundes tuxas, i. 370, 26. Tuscum *genuinis,* cweorntóđum *molaribus,* Wrt. Voc. ii. 76, 39. Tuxum, 40, 44. Mid tuxum *ingenuis* (= *in genuinis ?*), 48, 50. Grindetóþum, tuxum *molaribus* (but see 76, 39 *ante*), 54, 46. Tuxum *ginguinis* ꝉ *ginguinibus* (the passage is : Ursorum gingivis carperentur), Hpt. Gl. 492, 1. Tuxum *dentibus* (*porcorum*), 507, 52. Heora (*the evil spirits'*) tóþas wǽron gelíce horses tuxuan, Guthl. 5 ; Gdwin. 34, 24. Hý habbaþ eoferes tucxas *habentes aprorum dentes,* Nar. 34, 32. Tuxan đara leóna *molas leonum,* Ps. Lamb. 57, 7. [*O. Frs.* tusk.] v. hilde-tusc, *and next word.*

tuscel, tuxl, es ; *m. A canine tooth* or *a molar tooth, a tusk :*—Gefóh fox, āsleah of cucum đone tuxl, lǽt hleápan āweg *catch a fox, knock out while alive the canine tooth, let the fox run away,* Lchdm. ii. 104, 12. Hý heora bán gnagaþ brynigum tuxlum (cf. byrnendum tóđum, Wulfst. 139, 11) *lacerant ignitis dentibus ossa,* Dóm. L. 14, 211. Tuxlas (*molas*) leóna tóbrycþ Drihten, Ps. Spl. 57, 6. [Twey tuxlys out of hys mouth set as of a bore, Octov. 929.] v. preceding word.

twā. v. twēgen.

twādæg-líc ; *adj. Lasting two days :*—Twādæglíc (twydæglíc, Bd. M. 350, 32) fæsten is genóh tó healdenne *biduanum sat est observare jejunium,* Bd. 4, 25 ; S. 600, 8. [Cf. *Ger.* zwei-tägig : *Icel.* tví-dægra *the name of a mountain desert taking two* '*dægr*' *to cross.*]

-twæccea. v. angel-twecca.

twǽde ; *adj. Doubled* (?), *containing two of three parts* of a whole ; the word occurs mostly as substantive, *two thirds, two parts of three :*—Wylle óþ sié twǽde bewylled đæs wóses (cf. bewyl óþ þriddan dǽl, 120, 15) *boil till two thirds of the juice are boiled away,* Lchdm. ii. 38, 11. Wylle óþ đæt se wǽta sié twǽde on bewylled (cf. 266, 31) *boil till the liquor be boiled down to two thirds,* 332, 17. Dó twǽde đæs wínes and þriddan dǽl đæs huniges *put two parts of wine to one of honey,* 306, 26. Dó đæs meluwes twǽde and đæs sealtes þriddan dǽl, 314, 5. Dó đæs huniges twǽde and đǽre buteran þriddan dǽl, 316, 7. Dó đæs swefles swilcan đara wyrta twǽde *to the quantity of sulphur put twice as much of the plants,* 78, 8. Se biscop and đa hígen āhten twǽde đæs wuda and đæs mǽstes, Chart. Th. 70, 29. Se cyning āh twǽdne dǽl (twegen dǽlas, MSS. B. H.) wetes, þriddan dǽl sunu oþþe mǽgas, L. In. 23 ; Th. i. 116, 15. [*O. L. Ger.* tuēdi *half : O. Frs.* twēde *two thirds, also half ;* twēdnath *two thirds.*] Cf. twi-dǽl.

-twǽfan. v. ge-twǽfan.

twǽman ; *p.* de *To divide, separate, part,* (1) *to prevent the joining of objects :*—Dyple (diple *signum in libris praesertim ecclesiasticis ad distinctionem opposium,* Migne) . . . þýs tāken gesetton đa ealdan wríteras on ciriclícum bócum, đæt hig twǽmdon ođđe ætýwdon đa gewitnyssa hāligra gewrita, Anglia viii. 334, 11. (2) *to part what has been joined :*—Man wite, đæt hý þurh mǽgsibbe tó gelænge ne beón, đe læs đe man eft twǽme đæt man ǽr āwóh tósomne gedydon (cf. hí (*William and his wife*) wǽron ᴢidden tótweamde for sibreden, Chr. 1127 ; Erl. 255, 20), L. Edm. B. 9 ; Th. i. 256, 10. (3) *to divide, cause dissension among :*—Đæt we ne lǽtan ūs deófol twǽman, Wulfst. 272, 24. (4) *intrans.* :—Wē nellaþ, Drihten, nǽfre fram đē twǽman, Homl. Skt. i. 11, 169. [Ic uulle mine kineþeode twemen mine bearnen, Laym. 2948. His attente is uorte unuestnen (tweamen, MS. C.) heorten, A. R. 252, 2. Ne mei unc nowđer līf ne deađ tweamin atwa, Marh. 5, 17.] v. ge-, tó-twǽman ; tó-twǽmedness, *and next two words.*

twǽmendlíce ; *adv. Separately :*—Twǽmendlíce *singulatim, separatim,* Hpt. Gl. 438, 51.

twǽming, e ; *f.* **I.** *division, separation, severing the connection between objects :*—Nis seó godcundnys gemenged tó đǽre menniscnysse, ne đǽr nān twǽming nys . . . Hē (*Christ*) þurhwunaþ on ānum hāde untótwǽmed, Homl. Th. i. 40, 34–39. Đǽr (*at the last day*) biþ seó twǽming rihtwísra manna and ārleá ra, 616, 28. Twǽming (*separation of man and wife*) is ālýfed đām đe lufiaþ swíđor đa heálícan clǽnnysse đonne đa hohfullan gālnysse, ii. 324, 3. Biđ ūs sēlre đæt wē his flǽsclícan lufe fram ūs āceorfon, and mid twǽminge (*by separation from him*) āwurpon, i. 516, 11. Úre Drihten forbeád đa yfelan twǽmincge betwux twām ǽwum, ii. 322, 32. **II.** *separation, distinction :*—Hē cwæđ '*đæs lifigendan Godes*' for twǽminge đæra leásra goda *he said* '*the living God*' *to distinguish him from the false gods,* Homl. Th. i. 366, 19.

twaltiga *palma,* Wrt. Voc. i. 80, 14, *apparently an error for* palmtwig, *q. v.*

-twanc (?). v. ge-twanc.

twā-nihte ; *adj. Two days old :*—On twānihtne mōnan far tó and bige land đæt đíne yldran āhton *when the moon is two days old, go and buy land that thy forefathers owned,* Lchdm. iii. 176, note 2.

twégen (twegen ? *In the later MSS. of the Gospels* tweigen *and* twegen

are found, but ei *may represent earlier* e, *e. g.* weig, Lk. 1, 79, eige, 2, 9; *or* ē, *e. g.* wreigende, 23, 10, wreigeð, 23, 14: *Layamon has* tweiȝe, tweien: *in the Ormulum the form is* tweȝȝen); *m.*: twā, twuā; *f.*: tū, tuu, twā; *n.*: *gen.* twēga, twēgea, tweágea, twíga, twēgera, twēgra (*later Gospels have* tweigre, tweire); *dat.* twām, twǣm. Besides these West Saxon are the following forms, *nom. acc.* twǣgen, twœgen, tuoegi, tuoege, tuőge, tuoe, tuē; *m.*: *f.* tuoege: *gen.* tuoega, tuoe, twēgen, tuoegara, twoegra, tuoera. *Two.* I. used adjectivally:—Tuēgen stridi *ea per passus*, Txts. 85, 1510. Twēgen (twǣgen, MS. E.) aldormen, Chr. 822; Erl. 62, 12. Twēgen englas, Gen. 19, 1. Óþre twēgen sealmas, R. Ben. 37, 11. Twǣgen míne mēgas, Cod. Dip. Kmbl. i. 310, 23. Twoegen gibróþæra, Txts. 127, 1. Middý wēron onfence fíf hláfo and twē fiscas, Mk. Skt. Lind. 6, 41. Brýda twā, Cd. Th. 65, 33; Gen. 1075. Twā þeóda ... twā folc, Gen. 25, 23. Sinhíwan twā, Cd. Th. 49, 9; Gen. 789. Dæt tweágea (twēgea, Hatt. MS.) bleó godweb, Past. 14; Swt. 86, 14. Of ðissa twēgea (tuēga, Hatt. MS.) monna mūðe, 7; Swt. 48, 10. Twēgra gebróðra bearn oððe twēgea gesweostra sunu and dohtor, Bd. 1, 27; M. 70, 4–5. Ðissa twēga yfela áuþer, Bt. 6; Fox 16, 2. Ys ðeós wyrt twēgea (twēgra, MSS. B. O.) cynna, Lchdm. i. 204, 9. Twēgra (tuoegara, Lind.: twoegra, Rush.: tweire, later MS.) manna gewitnes, Jn. Skt. 8, 17. Twoega nētna *duorum animalium*, Ps. Surt. ii. p. 189, 6. Tuoera scyldigra, Lk. Skt. p. 5, 14. Hié wærun on twǣm (tuǣm, l. 30) gefylcum, Chr. 871; Erl. 74, 16. His wífum twǣm, Cd. Th. 66, 26; Gen. 1090: Beo. Th. 2387; B. 1191. His twām gebróðrum, Gen. 9, 22: 19, 30. Twām (tuǣm, Lind.: twǣm, Rush.) hláfordum þeówian, Mt. Kmbl. 6, 24. On ðysum twām bebodum, 22, 40. On twām styccum, Exon. Th. 70, 15; Cri. 1139. Ic hæbbe twēgen suna, Gen. 42, 37. Heó geseah twēgen (tuoege, Lind.: twoege, Rush.) englas sittan, Jn. Skt. 20, 12: Lk. Skt. 10, 35. Twēgen (tuōge, Lind.: twoege, Rush.) briddas, 2, 24. Ymb twēgen mónðas, Chr. 871; Erl. 75, 28. Ðæt wæter stód an twā healfa ðære strǣte, Ex. 14, 22. Twā turtlan (tuoe (twoege, Rush.) turturas, Lind.) *par turturum*, Lk. Skt. 2, 24. Hé gelærde twuā mǣgþa, Shrn. 131, 26. Wē habbaþ twā (tuā, Hatt. MS.) bebodu, Past. 7; Swt. 48, 13. Twā eágan (tuoe ēgo, Lind.) hǣbbende, Mt. Kmbl. Rush. 18, 9. Bring mē twā ða betstan tyccenu, Gen. 27, 9. Ofer tū folc, Bd. 3, 21; S. 551, 33. II. used substantively, (1) absolutely:—Twēgen of his leorningcnihtum, Jn. Skt. 1, 35. Twēgen of eów, Mt. Kmbl. 18, 19. Ðær twēgen (tuoe, Lind.: twēge, Rush.: tweigen, later MS.) oððe þrý synt gegaderode, 18, 20. Twā (tuoege ł tuu wíf *duae*, Lind.: twā, Rush.) beóð æt cwyrne grindende, 24, 41: Lk. Skt. 17, 35. Tuu in líchome ānum, Rtl. 106, 32. Twēga sang *bicinium*, Wrt. Voc. ii. 13, 4. On twēgra gewittnesse (in mūð tuoe witnesa, Lind.: in mūþe twēgen gewitnesse, Rush.: tweigre, later MS.) *in ore duorum testium*, Mt. Kmbl. 18, 16. Ðā sende hē twēgen (tuoege, Lind.: twǣgen, Rush.) hys leorningcnihta, 11, 2. Ðara scipa tū (twā, MS. E.) hē genam, Chr. 882; Erl. 82, 11. Ðæt wē twā (tuu, MS. T.) oþþe ðreó gehýron, Bd. 3, 9; S. 533, 28. (1 a) distributively:—Hē sende hig twǣm *misit illos binos*, Lk. Skt. 10, 1. Hé sende hí twām and twām, Homl. Th. ii. 528, 27: 530, 1. Ða wuniaþ twām and þrím ætgædere (*bini aut terni*), R. Ben. 9, 15. Steorran of heofenan feóllan, næht be ānan oððe twām, ac swā þiclíce ðæt hit nán mann āteallan ne mihte, Chr. 1095; Erl. 231, 21. (2) with qualifying or defining words:—Wit Adam twā *we two, Adam and I*, Cd. Th. 290, 6; Sat. 411. Wer and wíf, hí tū beóþ in ānum líchoman, Bd. 1, 27; S. 491, 13. Hwelce twā synd wiþerweardran ðonne gód and yfel? Bt. 16, 3; Fox 56, 6. Wurdon ðam æðelinge bearn ātēded, freólícu tū, Cd. Th. 102, 30; Gen. 1708. Uncer twēga, 110, 9; Gen. 1835: Beo. Th. 5057; B. 2532. Ðonne him mon ðissa twēgea (tuēga, Hatt. MS.) hwæðer ondrǣt, Past. 27; Swt. 188, 9. Hwæðer ðara twēgra (twēga, Cott. MS.) þincþ ðē mihtigra? Bt. 36, 4; Fox 178, 15. Ðyssa twíga mǣst, Lchdm. iii. 28, 15. Mon dyde him twǣm ðone triumphan, Ors. 6, 7; Swt. 262, 25. Wið him twǣm, 6, 36; Swt. 294, 16. Betwih him twām, Bd. 1, 13; S. 482, 1. Andreas wæs óþer of ðām twām (tuǣm, Lind.: twǣm, Rush.) *erat Andreas unus ex duobus*, Jn. Skt. 1, 40. Be ðām neáhstan twām is æfter tō cweþanne, Bd. 4, 23; S. 594, 21. Him twām *duobus ex eis*, Mk. Skt. 16, 12. Ðā gebletsode Metod monna cynnes ða forman twā, fæder and mōder, Cd. Th. 12, 31; Gen. 194. Hé drāf of wícum dreórigmód tū, idese and his āgen bearn (*Hagar and Ishmael*), 169, 24; Gen. 2804. (3) in particular phrases:—Oþer twēga, oððe . . . oððe *either . . . or*, Bt. 11, 1; Fox 30, 26: 11, 2; Fox 34, 23. Deáh heó an tū tefleówe, Past. 7; Swt. 49, 11: Exon. Th. 70, 19; Cri. 1141: Chr. 885; Erl. 82, 19. Tōsliten on twā (tuu, Lind. Rush.), Mk. Skt. 15, 38. On twā (tū, Cott. MS.), Bt. 34, 11; Fox 150, 32. On twā (twuā, Cott. MS.), 38, 4; Fox 202, 27. Hí on twā fērdon *they parted*, Homl. Th. i. 388, 20. Ðæt wæter wearð tō twā tōdæled *divisa est aqua*, Ex. 14, 21. III. used in combination with other numerals, (1) with cardinals, (a) multiplicative:—Tū hund and þreó swylce þrittig eác wintra, Elen. Kmbl. 3; El. 2. Twā hund, 1264; El. 634. Twā (tuu, Lind.: tū, Rush.), hund elna, Jn. Skt. 21, 8. On twēgera hundred penega wurþe, 6, 7. Mid twām hundred penegon, Mk. Skt. 6, 37. Twā þúsendo, Cd. Th. 189, 14; Exod. 184. (b) added to the decades:—Twā (tuoege, Lind.: tū, Rush.)

and hundseofantig *septuaginta duos*, Lk. Skt. 10, 1. Hundseofontig tuoegi, Rtl. 113, 22. Nānne ðara twā and twēntigra monna, Ors. 6, 2; Swt. 256, 1. (2) with ordinals:—Se twā and feówertigeða sealm, R. Ben. 37, 14. Ðane twā and syxtigeþan, 36, 16. On ðære twā and twēntugoðan wucan, Mt. Kmbl. 8, 14, rubric. Mōna se twā and twēntigoðe, Lchdm. iii. 194, 17. IV. with the force of an adverb:—Hē tōdǣlde hig twā *divisit ea per medium*, Gen. 15, 10. Ðǣr wearþ micel gefeoht tuā (tuwa *in three MSS.*) on geáre, Chr. 885; Erl. 84, 7. Tū swā lange swā ða óðru *twice as long as the others*, 897; Erl. 95, 12. Nymaþ twā swā micel feós *pecuniam duplicem ferte*, Gen. 43, 12. Selle man him twā swylc swylce man æ per him nime, Lchdm. i. 400, 17. Seó hell ys twā swā deóp, and heó ys ealswā wíd, Wulfst. 146, 10. Seóð ðū hit twā swā swíðe swā hit ǣr wæs, Lchdm. iii. 12, 21. [*Goth.* twai; *m.* twōs; *f.* twa; *n.*; *gen.* twaddjē; *dat.* twaim; *acc.* twans; *m.* twōs; *f.* twa; *n.*: *O. Sax.* twêne; *m.* twā, twô; *f.* twê; *n.*; *gen.* twêiô; *dat.* twêm: *O. Frs.* twêne; *m.* twā; *f.* twā; *n.*; *gen.* twêra, twîra; *dat.* twām: *O. H. Ger.* zwêne; *m.* zwā, zwô; *f.* zwei; *n.*; *gen.* zweio, zweiio, zweiero; *dat.* zweim: *Icel.* tveir; *m.* tvær; *f.* tvau; *n.*; *gen.* tveggja; *dat.* tveim; *acc.* tvā; *m.* tvær; *f.* tvau; *n.*]

twelf, *generally indeclinable if used adjectivally and preceding the noun, but generally in other cases declined; nom. acc.* twelfe; *gen.* twelfa; *dat.* twelfum. *Twelve.* I. adjectival:—Ða twelf ðíne þeówas sind gebróðru, Gen. 42, 13. Wē twelf gebróðru wǣron ānes esnes suna, 32. Twelf (tuoelf *altered from* tuoelfo, Lind.: twelf, Rush.) tída ðæs dæges, Jn. Skt. 11, 9. Twelf wintra tíð, Beo. Th. 296; B. 147. Be twelf sealmum, R. Ben. 35, 6. Se tíreadga twelf síþum hine bibaþaþ, Exon. Th. 205, 2; Ph. 106: 202, 13; Ph. 69: Cd. Th. 285, 17; Sat. 339. Mid hys twelf leorningcnihtum (ðǣm twelfum ðegnum, Lind.: ðǣm twælf leorneras, Rush.), Mt. Kmbl. 26, 20. Ymbe twelf mónaþ *post annum*, L. Ecg. P. iv. 65; Th. ii. 224, 32. Tuoel ðegnas hē sendeþ, Mk. Skt. p. 2, 19. *In the following instance the word is inflected*:—Ān ðæra twelfa Drihtnes ðegena, Homl. Th. ii. 242, 15. I a. where the numeral follows the noun:—Ðā ongan hē sendan hālige weoras and geornfulle twelfe *holy men and diligent, twelve in number*; *viros sanctos et industrios . . . erant numero duodecim*, Bd. 5, 10; S. 623, 42. Hié getealdon fēðan twelfe, Cd. Th. 192, 2; Exod. 225. Míne suna twelfe, Salm. Kmbl. 30; Sal. 15. II. substantival, (1) absolutely:—Twelfe wǣron dǣdum dōmfæste, Apstls. Kmbl. 8; Ap. 4. Hé twelfa sum āð sealde *cum undecim comparibus suis sacramentum fecit*, Chart. Th. 203, 1. Hé com twelfa sum (*cum duodecim militibus*), Bd. 3, 1; S. 523, 31. Gewāt xii.-a sum, Beo. Th. 4793; B. 2401. Lond twelfum hērra fæðmrímes *per bis sex ulnas eminet ille locus*, Exon. Th. 199, 20; Ph. 28. Wē gefrunon twelfe under tunglum *we have heard of twelve men beneath the stars*, Andr. Kmbl. 3; An. 2. (2) with qualifying or defining words:—Hí twelfe (tuoelfo, Lind.), Lk. Skt. 8, 1. Hig twelfe (ða tuoelfo, Lind.) sǣdon him, 9, 12. Hé dyde ðæt hí twelfe mid him wǣron (ðætte hiá wēre twelfo mið him, Lind.), Mk. Skt. 3, 14. Hine āxodon ða twelfe, 4, 10. Ealra twelfa, Beo. Th. 6322; B. 3171. Eom ic ðara twelfa sum ðe hē gelufade, Exon. Th. 144, 20; Gū. 681. Hé wæs ān ðara twelfa (ān of ðǣm twelfum, Lind.), Jn. Skt. 6, 71. Ān of eów twelfum (ðǣm twelfum, Lind.), Mk. Skt. 14, 20. Hé ætýwde him twelfum (ðǣm tuoelfum, Lind.), 16, 14. Ðū ús twelfe trymman ongunne, Andr. Kmbl. 2837; An. 1421. Wē gesēgon eówre standan twelfe getealde, 1765; An. 885. *In the following instance the word is not inflected*:—Ðás twelf (tuelfe, Lind.: twælfe, Rush.) se Hǣlynd sende, Mt. Kmbl. 10, 5. [*Goth.* twalif: *O. Sax.* twelif: *O. Frs.* twelef, twilif, tolef: *O. H. Ger.* zwelif: *Icel.* tólf. These forms are declinable as in English.]

twelf-feald; *adj. Twelve-fold,* (1) with a noun:—Hí gegaderodon twelf wilian fulle. Ðæt twelffealde getel getācnode ða twelf apostolas, Homl. Th. i. 190, 11: 542, 4. Twelffeald geþungennes *duodenus apex*, twelffealdum setle *duodeno solio*, Wrt. Voc. ii. 142, 14, 13. (2) used substantively:—Be .xii.-fealdum āgife hē ðone ciricsceat, L. In. 4; Th. i. 104, 11. Forgilde hē mid twelffealdan, L. Eth. ix. 11; Th. i. 342, 28: Wulfst. 311, 6.

twelf-gilde; *adj. To be restored twelve-fold*:—Godes feoh and ciricean .xii.-gilde *the property of the church, if stolen, is to be restored twelve-fold* (the word, however, might be a noun = a restoration of twelve times the amount stolen, cf. ān-gilde; or adverb (dat. ?), cf. ix-gylde forgylde, 4; Th. i. 4, 3), L. Ethb. 1; Th. i. 2, 4.

twelf-hynde; *adj. As applied to a person, of the rank for which the wergild was twelve hundred shillings; applied to the wergild, that must be paid for a person of such rank.* As will be seen from the passages given below, the *twelfhynde man* was a þegn, and his importance, as marked by the wergild and otherwise, was six times that of the *ceorl*:—Ǣnig mǣgð, xii.-hynde oþþe twyhynde, L. Ath. v. 8, 2; Th. i. 236, 10. Be xii-hyndum men. Gif hē sié twelfhynde, L. Alf. pol. 31; Th. i. 80, 14. Gif hió sié cirlisc mid .lx. scill. gebēte . . . Gif hió sié xii-hyndu .cxx. scill. gebēte, 11; Th. i. 72, 15. Be twelfhyndes monnes wífe forlegenum. Gif mon hǣme mid twelf hyndes monnes wífe, hundtwelftig scill. gebēte ðam were . . . Cierliscum men feówertig scill. gebēte,

10; Th. i. 68, 8-12. Twelfhyndes monnes burgbryce .xxx. sciłł . . . Ceorles edorbryce .v. sciłł., 40; Th. i. 88, 9-11. Twelfhyndes mannes wer is twelfhund scyllinga (cf. Ceorles wergild is on Myrcna lage .cc. sciłł. Đegnes wergild is syx swā micel, L. M. L.; Th. i. 190, 1. Twelfhindus est homo plene nobilis, i. thainus cujus wera est duodecies .c. soł., L. H. 76, 4; Th. i. 581, 17. Twelfhinde, i. thaini, 70, 1; Th. i. 572, 22. See also L. W. l. 8; Th. i. 470, 14), L. E. G. 12; Th. i. 174, 13. Twelfhyndes mannes āđ forstent .vi. ceorla āđ; for đam gif man đone twelfhyndan man wrecan sceolde, hē biþ fullurecan on syx ceorlan, and his wergyld biþ six ceorla wergyld, L. O. 13; Th. i. 182, 19-22. xiihyndum men twyfealdlíce be đæs syxhyndan bôte, L. Alf. pol. 39; Th. i. 88, 4. Æt twyhyndum were mon sceal sellan tô monbôte .xxx. sciłł. . . . æt twelfhyndum .cxx. (cf. ad manbotam de twélfhindo, i. thaino .cxx. soł., L. H. 69; Th. i. 572, 19), L. In. 70; Th. 146, 14. Æt twelfhyndum were gebyriaþ twelf men tô werborge, L. E. G. 12; Th. i. 174, 18, 24. Cnut cing grēt . . . ealle mîne þegnas twelfhynde and twihynde, Chart. Th. 308, 16: Chart. Erl. 229, 20. ¶ In the following passage where the word is used without a noun perhaps *wer* may be supplied :—Hū man sceal gyldan twelfhyndes man (= twelfhyndes weres man *a man with a wergild of twelve hundred shillings*), L. E. G. 12; Th. i. 174, 12. v. six-, twi-hynde.

twelf-nihte; *adj. Twelve days old :*—On xii-niht[n]e mônan byþ gôd tô fēranne ofer sǣ, Lchdm. iii. 178, 26.

twelfta; *ord. num. Twelfth :*—Se twelfta *duodecimus*, Ælfc. Gr. 49; Zup. 282, 19. Môna se twelfta, Lchdm. iii. 190, 4. Hē wæs twelfta sylf, Andr. Kmbl. 1330; An. 665. ¶ Passages having reference to Twelfth-night, the twelfth day after Christmas, Epiphany :—Đæs (*the first of January*) embe fîf niht fulwihtiid ēces Drihtnes tô ūs cymeþ, đæne twelfta dæg tîreádige hæleþ hātaþ on Brytene, Menol. Fox 25; Men. 13. Đý twelftan dæge ofer Geohol *Epiphaniae*, Bd. 4, 19; S. 588, 8. Đys sceal on twelftan dæg, Rubc. Mt. Kmbl. 2, 1. On twelftan ǣfen, 2, 19. On Wôdnes dæg ofer twelftan dæg, 3, 13. Eádweard kingc com tô Westmynstre tô đam middan wintre . . . And hē forđfērde on twelftan ǣfen, and hyne man bebyrigde on twelftan dæig on đam ylcan mynstre, Chr. 1065; Erl. 196, 14-19.

twelftig. v. hund-twelftig.

twelf-wintre; *adj. Twelve years old :*—Úre Hǣlend đā hē wæs twelfwintre, Past. 49; Swt. 385, 20: Lk. Skt. 2, 42. Tuoelfwintro *duodennis*, p. 4, 4. Heó wæs twelfwintre *erat annorum duodecim*, Mk. Skt. 5, 42. Se wæs xii-wintre cniht, Shrn. 118, 13. Hē hæfde āne dohtor neán twelfwintre *filia unica erat illi fere annorum duodecim*, Lk. Skt. 8, 42. Man ne sparige nānan þeófe ofer .xii. pæningas' and ofer .xii.-wintre mann *no thief shall be spared above .xii. pence and above a twelve-year old person*, L. Ath. v. 1, 1; Th. 228, 13. Mon ne sparige nǣnne þeóf ofer .xii. winter (twelfwinterne, MS. B. L.) and ofer eahta peningas, 1, 1; Th. i. 198, 17. Gyf hine hwā āfylle ofer twelfwintre (ofer đæt hē biþ twelfwintre, MS. G.), L. C. S. 20; Th. i. 386, 22. Ælc man ofer twelfwintre sylle đone āđ, đæt hē nelle þeóf beón, 21; Th. i. 388, 6. Ælc man de beó ofer twelfwintre, Wulfst. 136, 17. *Perhaps in the last five passages* ofer twelfwintre *should be taken as a compound.* [*Goth.* twalib-wintrus.]

twengan; *p. de To pinch, squeeze, twinge :*—Gyf đe gedrýptes wînes lyste, đonne dô đū mid đînum swýþran scytefingre on đîne wynstran hand, swylce đū tæppian wille, and wænd đînne scytefinger ādúne and twængc hine mid đînum twām fingrum, swylce đū of sumne dropan strîcan wylle, Techm. ii. 125, 19. Cyrsena tāc[n] is đæt đū sette đînne winstran þúman on đînes lytlan fingres liđ and twenge hine siđđan mid đara swiþran hande, 124, 23. Twenge đū mid đînre swiđran neoþewearde þîne wynstran, 125, 1. [Þu havest clivres swiþe stronge, þu twengest þar mid so doþ a tonge, O. and N. 156. An holȝ stoc hwar þu þe miht hude þat me ne twenge þine hude, 1114. He twengde and schok hire bi þe nose, P. L. S. ix. 81. *O. H. Ger.* zwengen *remordere, praestringere*.]

twēntig, twēgentig; *num. Twenty.* **I.** used adjectively, with the inflexions of the plural adjective in gen. and dat., but also with singular gen. (1) alone :—Đis synd đara twēntiges hîda landgemǣra, Cod. Dip. Kmbl. iii. 429, 25. Mid twēntigum (twoegentigum, Rush.: tuoentigum, Lind.) þúsendum, Lk. Skt. 14, 31. On twēutigum fôtmǣlum, Elen. Kmbl. 1657; El. 830. Nǣs tô ánum dæge, ne tô twām . . . ne tô twēntigum, Num. 11, 19. Intô đýs twēntigum hîdum, Cod. Dip. Kmbl. v. 331, 1. (2) with other numbers, the inflection may be omitted if the noun does not immediately follow *twēntig :*—Nǣnne đara twā and twēntigra monna, Ors. 6, 2; Swt. 256, 2. Đæt mǣsten is gemǣne tô đam ān and twēntigum hîdum, Cod. Dip. Kmbl. v. 319, 29. Ymb twēntig . . . and fîf nihtum, Menol. Fox 371; Men. 187. **II.** used substantively, (1) alone :—Gif đǣr beóþ twēntig rihtwîsra, Gen. 18, 31. Ân twēntig is đara bôca đe Adeluuold gesealde *of the books that Athelwold gave there is a score*, Chart. Th. 244, 21. Wæs ic mid đe twēntig wintra, Gen. 31, 38. Næfde hē mā đonne twēntig sceápa and twēntig swŷnas, Ors. 1, 1; Swt. 19, 14. Hē hæfde twǣm lǣs đe twēntig wintra, Blickl. Homl. 215, 34. Twēntig (fîfe and twoegentig, Rush.: tuēntig, Lind.) furlanga,

Jn. Skt. 6, 19. (2) with other numbers :—Hundteóntig geára and seofon and twēntig geára, Gen. 23, 1. Seó menigu wæs ān hund manna and twēntig, Homl. Th. i. 296, 18. Onbîd hēr seofon and twēntig nihta, Blickl. Homl. 231, 5. (3) distributively :—Týnum and twēntigum on ānum inne ætgædere restan, R. Ben. 47, 7. [*Goth.* twai-tigjus : *O. Sax.* twēntig : *O. Frs.* twintege : *O. H. Ger.* zweinzug : *Icel.* tuttugu.]

twēntig-feald; *adj. Twenty-fold :*—Twēntigfeald getel *vicenarius*, Ælfc. Gr. 49; Zup. 285, 3. Twēntigfealde *uiceni*, 5; Zup. 13, 15.

twēntigođa; *ord. num. Twentieth :*—Se twēnteogođa (-tigođa) *uicesimus*, se ān and twēnteogođa *uicesimus primus*, Ælfc. Gr. 49; Zup. 283, 6. Môna se twēntigoþa . . . môna se ān and twēntigođa, Lchdm. iii. 194, 5-9. Se fîf and twēntugoþa dæg þæs mônþes, Nic. 1; Thw. 1, 11. On đære twā and twēntugođan wucan, Rubc. Mt. Kmbl. 8, 14. Đý twēntigþan dæge, Bd. 4, 5; S. 572, 7. On đone tū and twēntegđan dæge, Shrn. 93, 1 (*and often*). On đone fîf and twēntigođan dæg, 96, 11.

twēntig-wintre; *adj. Twenty years old :*—Ôđ hē sý twēntigwintre odđe gyt yldra, Wulfst. 3, 1.

tweó, twŷ; *gen.* tweón, twŷn; *m.* **I.** *doubt, uncertainty :*—Đonne đǣr ān tweó of ādôn biþ, đonne biþ đǣr unrîm āstyred *una dubitatione succisa innumerabiles aliae succrescant*, Bt. 39, 4; Fox 216, 18. 'Sum tweó mē hæfþ swiþe gedrêfed.' Đā cwæþ hē : 'Hwæt is se?' '*difficiliori ambiguitate confundor.*' '*Quaenam,*' *inquit* '*ista est?*' 41, 2; Fox 244, 14. Đū mē hæfst ārētne on đæm tweón đe ic ǣr on wæs be đam freódôme, Fox 246, 12. Wē habbaþ litelme gearowitan būton tweón, 41, 6; Fox 254, 10. Đonne secge ic eów būton ǣlcum tweón, 16, 1; Fox 50, 27. Đæt hē đæt on gehđu gesprǣce and tweón, Elen. Kmbl. 1332; El. 668. Tô tweón weorđan *to become doubtful*, Exon. Th. 310, 4; Seef. 69. Būtan tweón *without doubt, undoubtedly, doubtless, certainly*; *sine dubio*, Bd. 1, 7; S. 478, 7: 1, 25; S. 486, 26. Hwæđer wǣre twēgra būtan tweón strengra, Salm. Kmbl. 854; Sal. 426. Būton ǣlcum tweón *beyond all question*, Bt. 22, 2; Fox 78, 11: 21; Fox 72, 28: Met. 11, 1. Būton twŷn, R. Ben. Interl. 17, 4: Homl. i. 190, 18. Būta tuā *utique*, Mt. Kmbl. Lind. 9, 18. Ic wât đæt hine wile tweógan . . . Ne mæg se cyning đæne tweón eáđe gebētan? Wulfst. 3, 12. Đǣr seó wîse on tweón cyme *ubi res perveniret in dubium*, Bd. 1, 1; S. 474, 21. **I a.** where the subject of doubt is in the genitive :—Nis đæs nān tweó, đæt . . . *of this there is no doubt, that . . .*, Past. 6; Swt. 47, 10: Bt. 16, 3; Fox 54, 20. Nis đæs nān tweó (tweó, Cott. MS.) đæs . . ., 40, 1; Fox 234, 36. Đām englum nis nān tweó nānes đæra đinga đe hî witon, 41, 5; Fox 254, 10. Đæt hit heofoncyninges tācen wǣre, and đæs nān tweó nǣre, Elen. Kmbl. 342; El. 171. Đæt nǣre nǣnig manna đæt mihte đæra twēgra tweón (*the doubt about the two*, cf. 854; Sal. 426, given above) āspyrian, Salm. Kmbl. 870; Sal. 434. **I b.** where the subject of doubt is expressed by a clause :—Nis nān tweó, đæt đæs andwearda wela āmerþ đa men, Bt. 32, 1; Fox 114, 2. Hit is nān tweó, đæt . . ., 36, 3; Fox 178, 4. Nis nān tweó đæt hē forgifnesse syllan nelle đam đe hié geearnian willaþ *there is no doubt about his not being ready to grant forgiveness to those that are ready to deserve it*, Blickl. Homl. 65, 8. Him tweó þúhte, đæt hē Gode wolde geongra weorđan, Cd. Th. 18, 21; Gen. 276. Đæt hālige gewrit, đæt mē nis tweó đæt đū geara canst *sacra scriptura, quam te bene nosse dubium non est*, Bd. 1, 27; S. 489, 2: 4, 7; S. 575, 13. Him wæs on môde mycel tweó, hwæt hié be đære dorstan dôn, Blickl. Homl. 205, 10. Ná twŷ ys, đæt . . . *non dubium est, quod . . .*, Scint. 48, 10. Mid đý sumum monnum com on tweón hwæđer hit swā wǣre *cum hoc an ita esset quibusdam venisset in dubium*, Bd. 4, 19; S. 587, 26. **II.** *hesitation, delay :*—Būta tuiá đū onfindes *sine mora reperies*, Mt. Kmbl. p. 4, 4. **III.** *a doubtful state of things, state of indecision :*—On đæm tweón đe hié swā ungeorne his willan fulleodon đā becom him Antigonus mid firde on *in this state of indecision, in which they carried out his will so reluctantly, Antigonus fell upon them with an army*; *qui fastidiose ducem in disponendo bello audientes ab Antigono victi sunt*, Ors. 3, 11; Swt. 146, 24. [*O. Sax.* tweo : *O. H. Ger.* zweho *dubium, ambiguitas*.] v. un-tweó; tweógan (tweón), tweón *doubt*.

tweo-. v. twi-.

tweógan, tweón; *p.* tweóde. **I.** with impersonal construction, *to inspire doubt into* a person (*acc.*), (a) with gen. of object of doubt :—Wē witon đæt nānne mon đæs ne tweóþ, đæt se seó strong on his mægene đe mon gesihþ đæt stronglíc weorc wyrcþ, Bt. 16, 3; Fox 54, 28. Ne tweóþ mē đæs nāuht, 36, 3; Fox 176, 16: Exon. Th. 117, 13; Gū. 223. Nānne mon đæs tweógan ne þearf, đæt ealle men geendiaþ on đam deáþe, Bt. 11, 2; Fox 34, 34: 33, 1; Fox 120, 24. Tweógean, Blickl. Homl. 43, 1. (b) with a preposition :—Ymb đæt đe hiene tweóde, orn hé intô đæm temple, and frægn đæs Dryhten . . . Hié sculon, đonne hié ymb hwæt tweóþ, cyrran tô hiera āgnum inngeđonce, Past. 16; Swt. 102, 4-8. (c) with a clause :—Nǣnne mon ne tweóþ, đæt God sý swā mihtig, Bt. 35, 5; Fox 164, 4: 36, 3; Fox 176, 15. Ne đē nāuht ǣr ne tweóde, đætte God weólde ealles middaneardes, 35, 2; Fox 156, 30. Đēh đe hié ǣr tweóde, hwæđer hiene mon gefiéman mehte, Ors. 4, 9; Swt. 192, 15. And đæt đý lǣs tweóge, hwæđer đis sôþ sý, ic cýþe hwanan mē đæs spell côman *ut occasionem dubitandi subtraham, quibus auctoribus didicerim*

intimare curabo, Bd. pref.; S. 471, 20. Hine wile tweógan, hwæðer heó him sóð secge, Wulfst. 3, 7. Nǣnigne tweógean ne þearf, ðæt seó wyrd geweorþan sceal, Blickl. Homl. 83, 9. Tweógan, Bt. 37, 3; Fox 190, 8. (d) absolute :—Ic wât ðætte wile woruldmen tweógan, Met. 4, 51. II. *to feel doubt, to doubt, hesitate*, (a) with gen. of object of doubt :—Ne tweóþ ðæs nán (nænne, Cott. MS.) mon, Bt. 35, 5; Fox 164, 5. Ne mæg ic ðæs nó tweógan (twiógean, Cott. MS.), 34, 9; Fox 146, 26: 35, 4; Fox 160, 18. Ðæs tweógan ne þearf ænig, Exon. Th. 147, 13; Gú. 726. (b) with preposition :—Ic nât ymbe hwæt ðú tweóst, Bt. 5, 3; Fox 12, 13. Gif gé tweógaþ be ðæm ælmessum, Blickl. Homl. 41, 20. Ne tweóge ðis folc (*or acc.?*) be hire untrumnesse, 143, 12. Be ðam nis tó tweógenne, ac is tó gelýfanne, Bd. 3, 23; S. 555, 33. (c) with an infinitive :—Hí ne tweódon fêrende beón tô ðam êcan life, Bd. 4, 16; S. 584, 38. (d) with a clause :—Ic nâuht ne tweóge ðæt ðú hit mæge gelêstan, Bt. 36, 3; Fox 174, 31. Ðú cwist ðæt ðú nâht ne tweóge ðætte God ðisse worulde rihtere sié, 5, 3; Fox 12, 13. (e) absolute :—Hé swýðor tweóþ ðonne se ǣrra, Wulfst. 3, 10. Se ðe cuoeðas and ne tuáes ꝉ ne getuîga (ne twiás ꝉ ne twióge, Rush.) *qui dixerit et non haesitaverit*, Mk. Skt. Lind. 11, 23. Gif gé ne twígaþ *si non haesitaveritis*, Mt. Kmbl. Rush. 21, 21. Tuiáde *haesitabat*, Lk. Skt. Lind. Rush. 9, 7. Sume tweódun *quidam dubitaverunt*, Mt. Kmbl. Rush. 28, 17. Twiódun (tuiáton, Lind.) *haesitabant*, Jn. Skt. Rush. 13, 22. Ne tweóge *non cunctante*, Wrt. Voc. ii. 92, 49. Hé hine hêt ðæt hé ne tweóde, ac ðæt hé wǣre ânrǣd, Guthl. 4; Gdwin. 30, 7. Ðá ðæt folc ongan tweógan on heora heortan, Blickl. Homl. 143, 8. Tô tweónne *nutabundum*, Hpt. Gl. 459, 5. Tuîgendi *anceps*, Wrt. Voc. ii. 100, 40. Tweógende, 7, 2. Tweógende môd, Andr. Kmbl. 1542; An. 772. Nǣnig tweógende secgend *non quilibet dubius relator*, Bd. 3, 15; S. 542, 7. Tweógende cyninges *reges dubii*, 4, 26; S. 603, 17. Tweógende *hesitantes*, Wrt. Voc. ii. 43, 22 : 74, 19. [*O. Sax.* twehôn : *O. H. Ger.* zwehôn *dubitare, hesitare, cunctari*.] v. ge-tweógan, un-tweógende, -tweónde.

tweógend-líc; *adj. Doubtful, uncertain,* (1) where doubt is felt :—Tweógendlícre (*sine*) *ancipiti (ambiguitatis scrupulo veraciter credendum est)*, Hpt. Gl. 422, 32. Hé on tweógendlícan onbide wæs (*quem cunctantem*), hwæðer hé winnan dorste, Ors. 4, 11; Swt. 204, 28. (2) where doubt is caused :—Is tweógendlíc ðysse worulde wela, Wulfst. 263, 11. Tweógendlícra gewrita *Apocryphorum*, Hpt. Gl. 522, 48. v. un-tweógendlíc.

tweógendlíce; *adv. Doubtingly, doubtfully* :—Sume hí twîendlíce be his lífe sprǣcon, and ðæt cwǣdon, ðæt hí nyston hwæðer hé on Godes mihte ða þing worhte ðe þurh deófles cræft, Guthl. 17; Gdwin. 70, 16. v. un-tweógendlíce.

tweógung, tweóung, e; *f. Doubt* :—Ðú mê hæfst gefrýlsod ðære tweóunge mînes módes be ðære âcsunga ðe ic ðé âcsode, Bt. 41, 3; Fox 248, 25.

tweohsn, tweoxn *occurs in the place name* Tweoxneám = *between streams* :—Ðone hâm æt Winburnan and æt Tweoxneám (*Christchurch, in Hampshire*), Chr. 901; Erl. 96, 27. v. be-tweohsn; tweoxn.

tweó-líc; *adj.* I. *doubtful, uncertain* :—Hit biþ twýlíc, hwæðer hit on lífe âðolige, Homl. Th. ii. 50, 24. *Dubii generis,* ðæt is twýlíces cynnes, Ælfc. Gr. 6, 6; Zup. 19, 17. Ðǣr beóþ kende *homodubii*, ðæt beóþ twîlíce, Nar. 38: 35, 3 note. II. *ambiguous* :—Ðá and-wyrde hire se hâlga mid twýlícere sprǣce, Homl. Th. ii. 146, 14. [*O. Frs.* twî-lik (twi-?) *doubtful.*] v. un-tweólíc, *and next word*.

tweólíce; *adv.* I. *doubtfully, uncertainly* :—Tweólíce and unfæsdlíce hé âtíefreþ ðæs ðinges onlícnesse on his móde, Past. 21; Swt. 157, 13. II. *ambiguously* :—Ondwyrdon hié him tweólíce *responso ambiguo*, Ors. 4, 1; Swt. 156, 3. v. un-tweólíce.

tweó-mann, es; *m. A creature about which it is doubtful whether it be human* :—*Homodubii* hý syndon hâtene, ðæt beóþ twîmen, Nar. 35, 3. v. tweó-líc.

tweón *doubt* :—Nis nán twýn, ðæt eów ne beó forgolden *there is no doubt, but that you will be requited*, Homl. Th. ii. 444, 10. Búton tweónne *without doubt*, Bt. 36, 6; Fox 182, 9. v. tweó.

tweón *to doubt*. v. tweógan.

tweóne. I. *two;* only in combination with the preposition *be*, either immediately following it (v. be-tweónum) or being separated from it by the governed noun, the two words together in either case having the force of *between* :—Be sǣm tweónum, ofer eormengrund, Beo. Th. 1721; B. 858 : Exon. Th. 118, 10; Gú. 237. Be werum tweónum *among men*, Andr. Kmbl. 1116; An. 558. Hé wealdeþ be sǣ tweónum *dominabitur a mari usque ad mare*, Ps. Th. 71, 8. Cf. *O. H. Ger.* in zwiskên, untar zwiskên, *in later times* inzwischen, zwischen, *for a similar growth of adverb and preposition*. II. *double, not simple* :—Tweóne leóht *vel* deorcung *twilight, a mixture of light and darkness, crepusculum*, Wrt. Voc. i. 53, 3. v. tweónol, *and cf. O. H. Ger.* zwiski *biceps, non simplex, binus; so iz under* zuiskên *liehten ist : M. H. Ger.* zwischenlicht. [*Goth.* tweihnai : *Icel.* tvennr.]

tweónian, twînian, twýnian; *p.* ode. I. *impersonal with dat. or acc. of person, to cause doubt* :—Mê twýnaþ (tweónaþ, MS. H.) *ambigo*, Ælfc. Gr. 28, 6; Zup. 176, 13. Gyt mê tweónaþ,

Homl. Th. i. 72, 30. Gif hié giet ðǣr tweónaþ, Past. 16; Swt. 103, 9. Ðá twýnude (tweónode, MS. A.) him *haesitabat*, Lk. Skt. 9, 7 : Homl. Th. ii. 392, 5. Hwí twýnode ðé?, 17. (b) with gen. of object of doubt :—Ðý lǣs ðe hwam twýnige ðyssere gereccednysse, Homl. Th. i. 598, 31. Hú mæg ðe nú twýnian ðæs êcan leóhtes?, 160, 19. (c) with a preposition :—Gif hwam twýnige be ðam gemænelícum æriste, Homl. i. 132, 27. (d) with a gerundial infinitive :—Hwæt twýnaþ ðê, oþþe hwæt ondrǣst ðú ðê, ðone Hǣlend tô onfônne?, Nicod. 26; Thw. 14, 13. (e) with a clause :—Ðé ne twýnaþ nán ðing, ðæt ðú sáwle hæbbe, Homl. Th. i. 160, 21. Him twýnode be hwam hé hit sǣde *haesitantes de quo dicerit*, Jn. Skt. 13, 22. Ús ne þearf nã twýnian, ðæt wé gebyrian ne sceolon oððe heofonwarena cyninge oððe hellewîtes deóflum, Wulfst. 151, 19. II. *with nom. of person, to feel doubt, to doubt*, (a) absolute :—Se ðe nã twýnaþ on heortan his ac gelýfþ *qui non hesitauerit in corde suo sed crediderit*, Scint. 127, 1. Swá hwylc swá cwyþ . . . and on his heortan ne twýnaþ (tweónaþ, MS. A.), ac gelýfþ, Mk. Skt. 11, 23. Se is lytles geleáfan, se ðe hwæðhwega gelýfþ and hwæðhwega twýnaþ; se ðe mid ealle twýnaþ, hé is geleáfleás, Homl. Th. ii. 392, 17-19. Gif gé habbaþ geleáfan and ne twýniaþ (tweóniaþ, MS. A.), Mt. Kmbl. 21, 21. Hwí twýnedest (tweónedest, MS. A.) ðú *quare dubitasti?*, 14, 31. Sume hig tweónedon *quidam dubitaverunt*, 28, 17. (b) with gen. of object of doubt :—Ða beóþ âwyrigde ðe ðises twýniaþ, Homl. Skt. i. 5 107. Hé behâtes twîniende heofonlíces *ille promissi dubius superni*, Hymn. Surt. 103, 7. (c) with a preposition :—Nã twýna ðú âbútan ende *non dubites circa finem*, Scint. 27, 11. Ús is âlýfed be ðisum tô twýnienne, Homl. Th. ii. 520, 16. (d) with a clause :—Hé ârâs, on his môde tweónigende hú heó nihte Iordanes wæteru oferfaran, Homl. Skt. ii. 23 b, 680. [Ʒunge monnan mei tweonian hweðer hí moten a libban, O. E. Homl. i. 109, 14. Þa wile þe heo tweoneden þus, Laym. 907.] v. ge-tweónian; tweógan.

tweónigend, es; *m. One who doubts* or *hesitates* :—Twýnigend *hic et haec anceps*, Ælfc. Gr. 9, 55; Zup. 67, 9.

tweónigend-líc; *adj. Expressing doubt* :—Sume syndon *dubitativa*, ðæt synd twýnigendlíce (tweóniend-, MS. H.), Ælfc. Gr. 38; Zup. 228, 16. Twýniendlíce, 44; Zup. 261, 2.

tweónol, twýnol; *adj. Doubtful* :—Tweónul leóht *maligna lux* vel dubia, Wrt. Voc. i. 53, 6. Swá swá his genyþerung ungewiss ys swá eác forgyfenyss twýnol *sicut ejus damnatio incerta est, sic et remissio dubia*, Scint. 46, 1. Ðæt deorc ys oþþe twýnol *quod obscurum est aut dubium*, 222, 3.

tweónum. v. tweóne.

tweónung, twînung, twýnung, e; *f. Doubt, uncertainty, hesitation* :—Ðam men biþ módes tweónung, Lchdm. ii. 194, 3. Ðæra apostola tweónung be Cristes ǣriste, Homl. Th. i. 300, 33. Ðam deófle wæs micel twýnung, hwæt Crist wǣre, 168, 10. Ðý lǣs ðe ǣnig twýning eów derian mǣge be ðam líflícan gereorde, ii. 262, 24. Tw(e)ónunge *ambiguitatis, dubietatis*, Hpt. Gl. 422, 32. Went nú moncyn on tweónunga *men will be in doubt*, Bt. 4; Fox 8, 18. Hí búton ǣlcere tweónunge sceolon on êcnesse forwurðan, Homl. Ass. 145, 37. Bútan twýnunge *absque ambiguitate*, Ælfc. Gr. 272, 13 : *sine dubitatione*, R. Ben. Interl. 52, 12 : *sine scrupulo*, Anglia xiii. 367, 24. Gyt mê tweónaþ; ac gif ðú ðás deádan sceaðan ârǣrst, ðonne biþ mín heorte geclǣnsod fram ǣlcere twýnunge, Homl. Th. i. 72, 32. Twúnunge, twînunge *scrupulum, dubitationem*, Hpt. Gl. 504, 77. Sume syndon *dubitativa*, . . . ðás getâcniaþ twýnunge, Ælfc. Gr. 38; Zup. 229, 2. Ic mid tweóningum óðrum monnum bigleofan gesette *cum aliqua scrupulositate a nobis mensura victus aliorum constituitur*, R. Ben. 64, 11. Hí âdrǣfdon ealle twýnunga fram úre heortan, Homl. Th. i. 302, 3.

tweóung, tweowa. v. tweógung, twiwa.

twî *a twig*. v. twig.

twi-, *in composition with force of* two. v. following words. [*O. Frs. O. L. Ger.* twi- : *O. H. Ger.* zwi- : *Icel.* tví-.]

twîa v. twiwa.

twi-bête; *adj. Needing double compensation;* a term applied to an offence when from special circumstances the *bót* was twice that to be paid in an ordinary case :—Gif hwá nunnan mid hǣmedþinge oþþe on hire hrægl oþþe on hire breóst bútan hire leáfe gefô, sié hit twybête, (twibôte, MS. B. : twybôte, MS. H.) swá wé ǣr be lǣwdum men fundon (*in the case of a nun the bót for the offences referred to was twice that in the case of a lay woman; the case of the latter is the subject of sect.* 11 ; Th. i. 68, 13-70, 2), L. Alf. pol. 18; Th. i. 72, 10. Gif hwá lengctenbryce gewyrce . . . þurh ǣnige heálíce misdǣda, sý ðæt twybête (twibôte, MS. B.), L. C. S. 48; Th. i. 404, 1. [*O. Frs.* twi-bête (*with the same use as the English word*).] v. twi-bôte.

twi-bill, es; *n.* : twi-bile, es; *m. A two-edged axe* :—Twibill *bipinnis*, Wrt. Voc. ii. 12, 52. Twybill *bipennis*, i. 36, 5. Twilafte æx vel twibile *bipennis securis*, ii. 126, 28. Twybile (-bil, MS. W.) *bipennis*, Ælfc. Gr. 9, 28; Zup. 56, 9. Twibille *bipinnae* (=*bipenne*), Ps. Surt. 73, 6. Hé nam sum twibil and mid ðan þrý men tô deáðe ofslóh, Guthl. 12; Gdwin. 56, 23. Æcsa, twibilles (-as?) *bipennes* ꝉ *secures*, Hpt. Gl. 459, 2. [Twybyle (*printed* twybyl, Wrt. Voc. i. 196, 10) *bipennis* (in

a list 'nomina armorum'), Wülck. Gl. 654, 2. Twybyl *bisacuta*, 568, 21 (both 15th cent. glossaries). Twybyl, wryhtys instrument *bisacuta, biceps*; twybyl or mattoke *marra, ligo*, Prompt. Parv. 505. A twybylle *biceps, bipennis, bisacuta*, Cath. Angl. 398, and see note. The word remains in some dialects, v. E. D. S. Pub. West Somerset Dialect, under *two-bill*, and Halliwell's Dict. *twibil*.] v. next word.

twi-bille; *adj. Double-edged :—Bipennis* twibille *vel* stānæx (the double gloss seems to render the double character of the Latin word as adjective and noun; a little later (see preceding word) in the same glossary *bipennis* as noun is rendered by *twybill*), Wrt. Voc. i. 34, 60.

twi-bleó; *adj. Double-dyed :—*Of twibleóum derodine *bis tincto cocco*, Wrt. Voc. ii. 126, 30. Tweobleóm (twiblīum, Cott. MSS.), Past. 14; Swt. 83, 23. Tōeácan ðæm twiblión (-bleón, Cott. MSS.) godwebbe, Swt. 87, 18.

twi-bót (?) *double 'bót.'* Perhaps in the passages given under twibóte; *adv.* the word might be taken as a case of this noun. Cf. twi-gilde. [Cf. the Scandinavian law phrases, liggi i tvebóte, tväbötis drap, v. Grmm. R. A. 653. *Swed.* twe-böte a *double fine*.]

twi-bóte; *adj. Needing double compensation.* v. twi-bēte :—Se ðe stalaþ on Gehhol oþþe on Eástron oþþe on ðone Hālgan Ðunresdæg . . . , ðara gehwelc (*the offence in each of these cases*) wē willaþ sié twybóte, swā on Lenctenfæsten, L. Alf. pol. 5; Th. i. 64, 25. Gif ðisses hwæt gelimpe þenden fyrd ūte sié, oþþe in Lenctenfæsten, hit sié twybóte, 40; Th. i. 88, 12. v. next word.

twibóte; *adv. With double 'bót' :—*Gif hē ōðswerian nylle, gebēte ðone mǣnan áð twibóte, L. In. 35; Th. i. 124, 13. ii-bóte gebēte, L. Ethb. 3; Th. i. 4, 2 : 2; Th. i. 2, 9. v. twi-bót.

twi-browen; *adj. (ptcpl.) Twice-brewed :—*On twybrownum ealað, Lchdm. ii. 120, 10.

twi-bytme (?); *adj. Double-bottomed :—*On ðæt twigbutme del; of ðam delle on beran del, Cod. Dip. Kmbl. v. 28, 19. v. bytm.

twiccere, es; *m. One who pulls to pieces :—*Twickere *offarius vel particularius* (particularius *minister in monasteriis, qui cibos per partes dissecat singulis monachis*, Migne), Wrt. Voc. i. 27, 20. v. next word.

twiccian; *p.* ode *To twitch, pluck :—*Twiccaþ *villicat*, Wrt. Voc. ii. 97, 14. Sume (*ants*) hió twiccedan ða grasu mid heora mūðe, Shrn. 41, 2. Teóh him ða loccas, wringe ða eáran and ðone wangbeard twiccige, Lchdm. ii. 196, 13. Twiccian *carpere, arripere*, Wrt. Voc. ii. 128, 69. [Twykkyn, twychyn *tractulo*, Prompt. Parv. 505. *In Mid. E. the past is* twighte. Cf. *O. H. Ger.* zwecchôn *carpere* : *M. H. Ger. Ger.* zwicken.]

twicen, e : twicene, an; *f. A place where two roads meet :—*Twicen *ambitus*, Wrt. Voc. i. 37, 46. On twycenan (-cinan, MS. B.) *in biuio*, Mk. Skt. 11, 4. Of ðære mere on ða twycene; of ðære twycenan, Cod. Dip. Kmbl. iii. 77, 4. On ða smalan twichenan; and swā andlang twichenan, 240, 20. Tō ðere twichenen; of ðere twichene, 201, 27.

twidæg-líc. v. twǽdæg-líc.

twi-dǽl *a double portion, two parts out of three :—*Dō gegrundenne pipor on, and cropleác, hwǽtenes melwes twidǽl swilce ðæs pipores *twice as much wheaten meal as pepper, two parts of meal to one of pepper*, Lchdm. ii. 52, 22. Genim heorotcrop and saluian, bewyl twydǽl on wætre *boil away two parts out of three*, 50, 12. Cf. twǽde.

twi-dǽlan; *p.* de. I. *to divide in two :—*Twidǽledre *bifori*, twidǽledu (v. Wülck. Gl. 194, 24) *bifida, bis divisa*, Wrt. Voc. ii. 126, 13, 16. Ðone twydǽledan wísdóm, ðæt is andweardra þinga and gástlicra wísdóm, Lchdm. iii. 440, 29. Ðás twidǽledan *hanc bipartitam, divisam in duas partes*, Hpt. Gl. 434, 32. II. *to differ :—*Twydǽlþ *discrepat*, Scint. 125, 6. Hí cristenre lāre twydǽlaþ *christianae doctrinae dissentiunt*, 129, 10.

Twide, Tweode, an; *or indecl.* (cf. Humbre *for declension*); *f. The Tweed :—*In ófre Tweode (Tuidon, Bd. M. 360, 29) streámes *in ripa Tuidi fluminis*, Bd. 4, 27; S. 603, 34.

twidig. v. lang-twidig.

twi-ecge; *adj. Two-edged :—*Twiicce *biceps* (*gladius*, Prov. 5, 4), Kent. Gl. 87. Mid twyecgum *bipenne*, Ps. Th. 73, 6. Hæfde hē twiecge handseax *habebat sicam bicipitem*, Bd. 2, 9; M. 122, 12. Genim ðæt micle greáte windelstreáw twyecge, Lchdm. ii. 44, 5. Sweord twiecge *gladii ancipites*, Ps. Surt. 149, 6. [*O. H. Ger.* zwi-ekki.]

twi-ecgede; *adj. Two-edged :—*Twyecgede *anceps, biceps*, Ælfc. Gr. 9, 55; Zup. 67, 9, 10. Sworde twyecgedes *gladii ancipites*, Ps. Spl. 149, 6. Hæfde hē twigecgede (twyecge, MS.B.) handseax *habebat sicam bicipitem*, Bd. 2, 9; S. 511, 15. [*Icel.* tvī-eggjaðr.]

twiendlíce. v. tweógendlíce.

twi-feald; *adj. Twofold, double :—*Twyfeald *duplex*, Ælfc. Gr. 9, 61; Zup. 70, 2 : *geminus*, Wrt. Voc. ii. 42, 1 : 44, 21 : 41, 58. I. *as a multiplicative, twice as much, of twice the amount :—*Gyt synd manega getel on mislícum getācnungum . . . *duplex* twyfeald, Ælfc. Gr. 49; Zup. 287, 2. Ic ádreáh mycel broc mid Petre; nū is mín yfel twyfeald, nū Paulus ðæt ilce lǽreþ, Blickl. Homl. 175, 13. Twifealdum gāste (*Helisaeus Heliae*) *duplo* (*ditatus*) *spiritu*, Hpt. Gl 440, 47. Twifealdre gife *bino munere*, Wrt. Voc. ii. 126, 26 : Blickl. Homl. 101, 23.

Be twyfealdum ic forgylde *duplum*, Ælfc. Gr. 49; Zup. 286, 17 : L. Alf. 25; Th. i. 50, 23 : Homl. Th. ii. 562, 1. Hē him sylþ twifealdne mete (*cibos duplices*), Ex. 16, 29. 'Nymaþ twā swā micel fecs swā gē ǽr hæfdon' . . . Ðā nāmon hig twigfeald feoh '*pecuniam duplicem ferte*'. . . *Tulerunt ergo pecuniam duplicem*, Gen. 43, 12–15. II. *consisting of two items :—*Twyfealdre heolra *bilance*, Wrt. Voc. ii. 12, 1. Næbbe gē mid eów twyfeald hrægl (næbbe gē twā tunecan, Mt. Kmbl. 10, 10), Blickl. Homl. 233, 18. II a. *consisting of two parts, containing two elements :—*Ðæt twiefalde (twyfealde, Cott. MSS.) *gesuinc* . . . ðæt is ðæt hié ondrǽdaþ ðæt hí mon tǽlan wille . . . ; óðer is ðara *gesuinca* ðæt hí sēceaþ endeléase lādunga, Past. 35; Swt. 239, 4–8. Twufald intinge *duplex causa*, Mt. Kmbl. p. 1, 10. II b. *that belongs to one or other of two kinds :—*Ege is twyfeald, and ðeówdóm is twyfeald. Ān ege is būtan lufe, óðer is mid lufe . . . Swā is eác óðer ðeówt neádunge būton lufe, óðer is sylfwilles mid lufe, Homl. Th. ii. 524, 3–6. Wē tweofealdne deáþ ðrowiaþ, oþþe sticode beóþ, oþþe on sǽ ádruncene *oriuntur duo genera funerum, aut jugulamur, aut mergimur*, Bd. 1, 13; S. 482, 1. III. *doubtful, irresolute :* v. twifealdness, II :—Hē ða yfelan and ða twyfealdan geþōhtas forlēt (cf. hē hine hider and þyder gelómlíce on his móde cyrde, 28, 8; hē ðām tweógendum geþōhtum wiðstód, 18), and hine Scs Bartholomeus frēfrode, and hine hēt ðæt hē ne tweóde, ac ðæt hē wǽre ānrǽd, Guthl. 4; Gdwin. 30, 3–7. IV. *double* (as in *double dealing*), *not straightforward, deceitful.* v. twi-fealdness, III :—Dōmes dæg ārāfaþ ðæt cliwen ðære twifaldan (twyfealdan, Cott. MSS.) heortan *corda duplicitatibus involuta dissolvit*, Past. 35; Swt. 245, 22. Se ðe mid twyfealdum geðance tó mynsterlícre drohtnunge gecyrþ, and sumne dǽl his ǽhta dǽlþ, sumne him sylfum gehylt, . . . hē underfēhþ ðone āwyrgedan cwyde mid Annanian and Saphiran, ðe swicedon on heora āgenum ǽhtum, Homl. Th. i. 398, 28–33 : ii. 410, 32. Ðæt is syndrig yfel twiefealdra (twy-, Cott. MSS.) monna *est speciale duplicium malum*, Past. 35; Swt. 243, 24. Unclǣnu and twiefeald mód *impura corda*, Swt. 245, 12. V. *double* (as in bent *double*), *placed together :—*Ælc wāg biþ gebiéged twiefeald on ðæm heale *duplex semper est in angulis paries*, Past. 35; Swt. 245, 13. Ðæt yfelwillende mód gefielt hit self twyfeald oninnan him selfum, and sió twyfealdnes ðæs yflan willan hiene selfne twyfealdne gefielt oninnan him selfum *malitiosae mentis duplicitas sese infra se colligit*, 242, 6–9. [*O. Frs.* twi-fald : *O. L. Ger.* twi-veld, -fold : *O. H. Ger.* zwi-falt : *Icel.* tvī-faldr.] v. un-twifeald.

twifealdan. v. twifildan.

twifeald-líc; *adj. Double :—*Twyfealdlíc onbryrdnes eges and lufe, Homl. Th. i. 140, 16. Tuifallíco glædniso *geminata laetitia*, Rtl. 57, 2. Tuufallíce gāst *utriusque spiritus*, Mt. Kmbl. p. 14, 5. [Ðysra deáð wæs heora freóndan twyfealdlíc sār; ān, ðæt hí swā feárlíce ðises lífes losedan; óðer, ꝥ feáwa heora líchaman syððan fundena wǽron, Chr. 1120; Erl. 248, 12.] [*O. H. Ger.* zwifalt-líh : *Icel.* tvīfald-ligr.]

twifeald-líce; *adv. Doubly,* (1) *to twice the amount :—*On ðam sixtan dæge hig gaderodon twyfealdlíce *in die sexta collegerunt cibos duplices*, Ex. 16, 22 : L. Alf. pol. 39; Th. i. 88, 4. Gē gedóþ hyne helle bearn twyfealdlícor ðonne eów (*duplo quam vos*), Mt. Kmbl. 23, 15. (2) *in two ways :—*Ðis godspel mæg beón twyfealdlíce getrahtnod, ǽrest be Iudēiscum folce . . . , eft siððan be ǽlcum menn, Homl. Th. ii. 428, 5. Se biþ twyfealdlíce deád, se ðe on gódnysse unwæstmbǽre biþ, and on yfelnysse wæstmbǽre, 406, 18.

twifealdness, e; *f.* I. *doubleness, doubling.* v. twi-feald, I :—Geedlǽcend twyfealdnys *iterata dupplicatio*, Anglia viii. 331, 23. II. *irresolution.* v. twi-feald, III :—Of ðære leóhtmódnesse cymþ sió twiefealdnes and sió unbieldo *inconstantia ex levitate generatur*, Past. 42; Swt. 307, 3. III. *duplicity, deceitfulness.* v. twi-feald, IV :—Sió twyfealdness ðæs yflan willan *malitiosae mentis duplicitas*, Past. 35; Swt. 242, 8. Ða ðe nān sceadu ne geðiéstraþ ðære twiefaldnesse *quos nulla umbra duplicitatis obscurat*, Swt. 243, 23. Se iil getācnaþ ða twiefealdnesse ðæs unclǣnan módes ðe hit symle lytiglíce lādaþ *ericii nomine impurae mentis seseque callide defendentis duplicitas designatur*, Swt. 241, 8.

twi-ferclede. v. twi-fyrclede.

twi-fēre; *adj. Having two ways, accessible by two ways :—*Twifērum *bilustris* (cf. fǽreltu *lustra*, 53, 21, geondfērende *lustraturus*, 53, 54; *and see* un-fēre *invius*), Wrt. Voc. ii. 126, 22.

twi-fērlǽcan; *p.* -lǣhte *To dissociate :—*Ða ðe hí sylfe fram sōðre lufe twyfērlǽcaþ (-eþ, MS.) *qui semetipsos a caritate dissociant*, Scint. 6, 8.

twi-fēte; *adj. Two-footed :—*Twyfēte *bipes*, Ælfc. Gr. 9, 26; Zup. 51, 11 : 49; Zup. 287, 20. Sume biþ twiofēte, Bt. 41, 6; Fox 254, 27. [*Icel.* tvī-fættr.]

twifildan; *p.* de *To double :—*Ic twyfylde (-fealde, MSS. J. O. T. : -felde, MS. D.) *duplico*, Ælfc. Gr. 24; Zup. 138, 12 : 49; Zup. 287, 4. Twyfeldende mæssehacelan *duplicans casulam*, Anglia xiii. 406, 587. [Cf. *O. H. Ger.* zwifaltôn *geminare* : *Icel.* tvīfalda,]

twi-fingre; *adj. Two fingers thick,* term applied to the fat on swine :—Æt twyfingrum (spic), L. In. 49; Th. i. 132, 19.

twi-fiðerede; *adj. Double-winged, shaped as if with two wings* (?),

3 U

forked :—Twyfyrede (twyſyþerede, MS. C.: twifeðerede, MS. V.) *bisulcus* Ælfc. Gr. 49; Zup. 288, 11.

twi-fyrclede; *adj. Having two prongs, forked :*—Twyferclede *bifidus,* Ælfc. Gr. 49; Zup. 288, 10 note. [Cf. Wæs gesæwen swilce se beám (*the tail of a comet*) ongeánweardes wið ðes steorran ward fyrcliende wǽre *as if the tail were dividing in two, getting forked* (?), Chr. 1106; Erl. 240, 34. *Lat.* furculus *a fork with two or three prongs.*]

Twi-fyrd, -ford *Twyford,* a place-name occurring more than once in England and meaning *double ford :*—On ðære stówe ðe is cweden Æt Twyfyrde *in loco qui dicitur Ad Twifyrde, quod significat, ad duplex vadum,* Bd. 4, 28; S. 606, 5. Æt Twyfyrde, Cod. Dip. Kmbl. i. 114, 33. Tó Twyfyrde, iii. 203, 22. Of Twufyrde . . . æft on Twyfyrde, v. 147, 28–148, 22. On Twyfyrd; of Twyfyrde, iii. 444, 7. ¶ In Latin charters :—His nuncupantur uocabulis, Twyfyrde . . . , 153, 24. Apud Twyfird, v. 130, 31. *The form* Twyford *also occurs :*—Of Twyforde andlang Auene ðære eá swá ðæt mynstre stondeþ ofer Alne stream, vi. 220, 5. Cf. Circum fluuium Alne in loco qui dicitur Aet Tuiford, i. 29, 6. In loco qui Tuiforda appellatur, 74, 31.

twi-fyrede; *adj. Two-furrowed;* the word renders the Latin *bisulcus,* Ælfc. Gr. 49; Zup. 288, 11. [Cf. *O. H. Ger.* zwi-furhi, -furhig *bisulcus.*] v. furh.

twig, twí, es; *n. A branch, twig :*—Twig *ramus,* Wrt. Voc. i. 285, 80 : *palmes,* Jn. Skt. 15, 6. Hys twig (twi *later MS.*) byþ hnesce, Mt. Kmbl. 24, 32. His twí (twig, MS. A.: twi *later MS.*) biþ mearu, Mk. Skt. 13, 28. Ic eom swá ðæt twig, ðæt biþ ácorfen of ðam treówe, Homl. Skt. ii. 30, 191. Hé déþ ǽlc twig áweg, Jn. Skt. 15, 2. Of ðam twige (*Abel's murder*) ludon réðe wæstme, Cd. Th. 60, 28; Gen. 988. Heó brohte án twig (*ramum*) of ánum elebeáme, Gen. 8, 11 : Cd. Th. 88, 30; Gen. 1473. Gě synt twigu (*palmites*), Jn. Skt. 15, 5. Him ða twigu þiucaþ merge, Met. 13, 44. Twigu *arbusta* (twigges, Ps.), Ps. Spl. 79, 11 : Blickl. Gl. Ðonne ða twigo forburston, ðonne gewítan ða sáula niðer, ða ðe on ðæm twigum hangodan, Blickl. Homl. 211, 3. Tán ł twiga *vimina, virgulas,* Hpt. Gl. 428, 34. Twiga *asserum,* Wrt. Voc. ii. 10, 10. Eft spryttendum ðæm twigum (*virgultis*) ðæs Pelagianiscan wóles, Bd. 1, 21; S. 485, 5. Swilce se wudubeám ofersǽdmde ealne middangeard twigum and telgum, Cd. Th. 247, 28; Dan. 504: 248, 18; Dan. 515. Sume twigu hé lehte mid wætere, Past. 40; Swt. 293, 7. Hí námon palmtrýwa twigu (*ramos palmarum*), Jn. Skt. 12, 13. Genim wiþowindan twigu, Lchdm. ii. 34, 17. Sume seóþaþ ðære reádan netlan twigu, 218, 6. Twigo settende *propagines pastinans,* Wrt. Voc. ii. 78, 63. [Heo nomen þa twigga, O. E. Homl. i. 5, 2. He suinged him mið smele twige, 149, 1. þe uerþe tuyg, Ayenb. 22, 5. Twygge *virgula, ramusculus,* Prompt. Parv. 505: *vimen,* Wülck. Gl. 619, 27. *O. H. Ger.* zwíg, zwí: *Ger.* zweig.] v. ele–, ifig–, palm–, wín-twig; twigu, -twige.

twig-, twiga. v. twi-, twiwa.

twi-gǽrede; *adj. Cloven :*—Twygǽrede *bifidus,* Ælfc. Gr. 49; Zup. 288, 10. Cf. Bufan ðam hlince æt ðæs gǽredan (*pointed, angular*) landes ende, Cod. Dip. B. iii. 251, 42. v. gár, gára.

twige, twigea. v. twiwa.

-twige. v. líne-, þistel-twige.

twi-gedeágod; *adj. Double-dyed :*—Twigedeágodre deáge *bis tincto cocco,* Hpt. Gl. 431, 29.

twi-gilde (?), es ; *n. A double payment :*—Hé ágife twygilde (*or adverb* (?) v. *passages under* twigilde ; *adv., where, however, the word might be taken as a case of the noun ;* cf. án-gilde *which is a noun*), L. Eth. iii. 4; Th. i. 294, 20. [*O. Frs.* twi-ielde, v. Richthofen s. v. ield, and Grmm. R. A. 653.] v. next word.

twi-gilde; *adj. To be paid double :*—Cyricfrið ii-gylde, m[ynster]frið ii-gylde, L. Ethb. 1; Th. i. 2, 6. ii-gelde seó mund sý, 76; Th. i. 20, 13. [*Icel.* tví-gildr *of double value.*] v. next word.

twigilde; *adv. With a double payment :*—Gif ðeós lád teorie, gylde twygylde (cf. gylde ángyldes, l. 15), L. O. D. 6; Th. i. 354, 31. Gif þeów steleþ, ii-gelde gebête, L. Ethb. 90; Th. i. 24, 17. Bête hé ðam teónde twygylde, and ðam hláforde his were, L. Eth. i. 1; Th. i. 280, 20. Bête hé ðam teónde twygylde, and ðam hláforde his were, L. C. S. 30; Th. i. 394, 6. Sió bót biþ twysceatte (twyggylde, MS. B.) mâre *the bót shall be twice as much,* L. Alf. pol. 66; Th. i. 96, 31. v. two preceding words.

twigu (?), an; *f.; the forms in the Northern specimens may also be taken as weak,* tuigge, *pl.* tuiggo *A branch, twig :*—Steola *cauliculus,* twigu *ramunculus,* Wrt. Voc. ii. 129, 84. Twigge ł telge (telgra, Rush.) *ramus,* Mt. Kmbl. Lind. 24, 32 : Mk. Skt. Lind. 13, 28. Ðe tuigga *palmes,* Jn. Skt. Lind. 15, 6. Ða tuiggo (twigan *late southern MS.*) *palmites,* 15, 5. Telgo míno and twiggo *ramos meos, et rami,* Rtl. 68, 32. Twigena ordum, Salm. Kmbl. 286; Sal. 142. In tyggum his *in ramis ejus,* Mt. Skt. Lind. 13, 32. Tuiggo *ramos,* 21, 8. Telgo ł twiggo, Mk. Skt. Lind. 4, 32 : 11, 8. v. twig.

-twih (-twíh ?). v. be-twih. [Cf. *Goth.* tweihnai.]

twi-heáfdode; *adj. Double-headed :*—Twyheáfdede *anceps, biceps,* Ælfc. Gr. 9, 55; Zup. 67, 9, 10. Twyheáfdede oððe se ðe hæfþ twégen

---

líchaman *bicorpor,* 9, 21 ; Zup. 47, 17. [*O. H. Ger.* zwi-haupito *biceps* : *Icel.* tví-höfðaðr.]

twi-heolor, e; *f. A balance :*—Tuiheolore *bilance,* Wrt. Voc. ii. 102, 3. Twiwǽge *vel* (twi)heolore, 126, 20.

twi-híwe; *adj.* I. *of two forms* or *shapes :*—Twihíówe, swá swá biþ healf mon and healf fear *biformis,* Wrt. Voc. ii. 124, 31. Twihíwe *biformia,* 126, 12. II. *of two colours :*—Twihíwe *bicolor,* Wrt. Voc. i. 46, 34. Twihíwe godweb *coccum bis tinctum,* ii. 135, 44. Twifěrum *vel* (twi)híwum *bilustris,* 126, 22. v. next word.

twi-híwede; *adj.* I. *double-shaped, having two forms :*—Twyhíwede *biformis,* Ælfc. Gr. 49; Zup. 287, 9. II. *double-coloured :*—Twyhíwedum wurman *bis tincto cocco,* Hpt. Gl. 431, 30.

twi-híwian; *p. ode To assume two shapes, to dissimulate :*—Oððe hé nát oððe hé twyhíwaþ *aut ignorat, aut dissimulat,* Scint. 44, 8.

twi-hlidede; *adj. Double-lidded, having two openings :*—Twyhlydede *bipatens,* Ælfc. Gr. 49; Zup. 288, 6.

-twihn (-twíhn ?) *in* bi-twichn, Txts. 70, 546, bi-tuihn, 77, 1310.

twi-hweóle; *adj. Two-wheeled :*—Twihweólne *birotum,* Lchdm. i. lxii, 2.

twi-hwirft, es ; *m. A double course, double period :*—Twyhwyrftum (*printed* -hwyrhtum) *bilustris,* Hpt. Gl. 465, 40.

twi-hycgan (?) *to think differently, dissent, disagree :*—Twy iccende (= twyhycgende ?) *dissentiendo,* Anglia xiii. 367, 34.

twi-hynde; *adj.* As applied to a person, *of a rank for which the wergild was two hundred shillings ;* applied to the wergild, *that must be paid for a person of such rank.* As will be seen from the passages given below, the *twihynde man* was a *ceorl :*—Twelfhyndes mannes wer is twelf hund scyllinga. Twyhyndes mannes wer is twá hund sciłł. (*the article then deals with the case of the former, and concludes :* Eal man sceal æt cyriscum were be ðære mǽðe dón, ðe him tó gebyreþ, swá wé be twelfhyndum tealdan. Cf. too : Ceorles wergild is . . . ii hund sciłł. be Myrcna lage, L. Wg. 6; Th. i. 186, 11), L. E. G. 12; Th. i. 174, 14. Ænig mǽgð . . . xii-hynde oððe twyhynde, L. Ath. v. 8, 2; Th. i. 236, 11. Be twyhyndum were. Æt twyhyndum were mon sceal sellan tó monbóte .xxx. sciłł., L. In. 70; Th. i. 146, 12. Be twyhyndum men . . . Gif mon twyhyndne mon . . . ofsleá, L. Alf. pol. 29; Th. i. 80, 5–7. Cnut cing grêt . . . ealle míne þegnas, twelfhynde and twihynde, Chart. Th. 308, 16. v. six-, twelf-hynde.

twi-icce. v. twi-ecge.

twi-læpped; *adj. Having two skirts* or *lappets :*—Twilæpped scrúd *cinctus gabinus,* Wrt. Voc. i. 41, 5.

twi-lafte; *adj. Two-edged :*—Twilafte æx *bipennis securis,* Wrt. Voc. ii. 126, 27. Cf. (?) læppa or læfer.

twi-líc, twi-lí (-lí ?); *adj. Double, woven of double thread.* Cf. *twill* coarse linen cloth :—Tuili *biplex, duplex,* Txts. 109, 1151. Aenli *simplex,* tíli *bilex,* 115, 156, 157. [*O. H. Ger.* zui-líh *bilex* (*tunica*), *bissina* (*tunica*), *biplex* (*pannus*) : *Ger.* zwillich *ticking.*] Cf. þri-líc, *and see* twilíc-brocen.

twi-líc, -líce, -mann. v. tweó-líc, -líce, -mann.

twilíc-brocen; *adj. Woven of double thread and parti-coloured* (?) *or embroidered* (?) :—Hió becwið hyre twilíbrocenan cyrtel, Chart. Th. 537, 23. [Cf. (?) *Swed.* brokig : *Dan.* broget *parti-coloured.* *Jamieson* gives brocked, broukit, broked *variegated, having a mixture of black and white.* Cf. (?) Celtic forms *Welsh* brech *brindled : Irish* breacan *a plaid, tartan;* breacaim *I chequer, embroider.*]

twin; *adj. Twin, double :*—Twinnum *sangum geminis concentibus,* Hpt. Gl. 467, 31. [An had off twinne (*double*) kinde, Orm. 1361. He spacc off hise twinne kindess (*two, twin, natures*), 17478. On ilc he brend twin der (*Balaam offered on every altar a bullock and a ram,* Num. 23, 2), Gen. and Ex. 4020. Iosep gaf ilc here twinne srud (*to all of them he gave each man change of raiment,* Gen. 45, 22), 2367. On twinne half, 3248. *O. Frs.* twiska tuine kindem : *Icel.* tvinnr.] v. getwin.

twín, es; *n. Linen :*—Tuum (tuuín ?), tuigin, tuín *byssum,* Txts. 44, 138. Twín, Wrt. Voc. ii. 11, 14. Twiðræwen twín (*torta byssus*) . . . ðæt geðráwene twín, Past. 14; Swt. 87, 18, 42 : Swt. 89, 2. Of twispunnenum twíne línenum *torta bysso,* Swt. 83, 23. Mid geeðþráwenum twíne *cum bysso retorto,* Hpt. Gl. 431, 38 : Wrt. Voc. ii. 11, 70: Kent. Gl. 1145. Gescrýdd mid twíne (mið linnenom, Lind. *bysso*), Lk. Skt. 16, 19. [*Later the word is used as in mod. English.* A twines (twined, 2nd MS.) þræd, Laym. 14220. Twyne, threede *filum torsum vel tortum,* Prompt. Parv. 505. *Du.* twiju *twine, twist.*] v. twínen.

twinclian; *p. ode To twinkle :*—Se spearca ðara gódra weorca ðe tuinclaþ beforan mannum *cuncta, quae coram hominibus rutilant,* Past. 14; Swt. 87, 6. Ic ðæt lytle leóht geseah twinclian, Bt. 35, 3; Fox 158, 32.

twi-nebbe; *adj. Having two faces :*—Twynebbe *bifrontem,* Germ. 397, 448.

twínen; *adj. Of linen, linen :*—Bam (*read* ham, v. *the corresponding gloss* ham *subucula,* Hpt. Gl. 526, 30 : *at the same place,* byssina *is rendered by* línen) twínen *subucula bissina,* Anglia xiii. 37, 285. v. twín.

twing (twyng?) *what is pressed together* (?), *a mass, lump* :—Twinga ♅ *massas*, Hpt. Gl. 496, 70. v. next word.

twingan (?); *p.* twang; *pp.* twungen *To press, force* :—Se hrýnð ⟨tringaþ (twingeþ?), MS. M.⟩ muntas *qui tangit montes*, Ps. Spl. 103, 33. [I am twinged (twungen, MS. H.) and meked *incurbatus sum et humiliatus sum*, Ps. 37, 9. Whil þat twinges (*affligit*) me þe fo, 41, 10. *Ger.* zwingen.] v. twengan.

twinian. v. tweónian.

twi-nihte; *adj. Two days old* :—Twynihte grút, Lchdm. ii. 74, 9. v. twá-nihte.

twinn, twínung. v. twin, tweónung.

twin-wyrm (twin-?), es; *m.* The word glosses *buprestis* (= βούπρηστις *a poisonous beetle, which when eaten by cattle in the grass caused them to swell up*), Wrt. Voc. i. 24, 35.

twio-féte, twiógan, twio-ræde. v. twi-féte, tweógan, twi-ræde.

twi-ræde; *adj.* I. *of two minds, uncertain, undecided, irresolute* :—Geþenc be ðé selfum hwæðer ðú ænig ðing swá fæste getiohhod hæbbe ðæt ðé þynce ðæt hit næfre ðínum willum onwended weorþe . . . Oððe hwæðer ðú eft on ængum geþeahte swá twioræde sié ðæt ðé helpe hwæðer hit gewyrþe þe hit nó ne gewyrþe *consider in your own case whether you have so firmly determined anything, that it appears to you, that it will never with your consent be changed . . . Or again, whether in any plan you are so uncertain, that it may help you, if it is carried out, or if it is not*, Bt. 41, 3; Fox 250, 5-9. II. *of divided counsel, without unanimity* :—Ælc ríce ðe byþ twyræde on him sylfum *omne regnum divisum contra se*, Mt. Kmbl. 12, 25. [Bruttes weoren alle twiræde, heore teone wes þa mare, Laym. 19416.] v. ân-ræde, *and next word.*

twirédness, e; *f. Discord, dissension, disagreement* :—Sacu and twirédnyss (*strife, seditions*, Gal. 5, 20), Homl. Skt. i. 17, 26. Ðæt swá hweþer swá hit wære swá sibb swá twyrédnys betweónan Saxan and Myrcenum, ðæt ðæt mynster beó ðfre on sibbe, Cod. Dip. B. i. 156, 16. Se ðe sibbe Drihtnes twyrédnysse mid hátheortnysse tóbrycþ *qui pacem Domini discordiae furore rumpit*, Scint. 10, 2. God ná ys twyrædnysse (*dissensionis*) God, 134, 6. Be twirédnysse *de discordia*, 133, 17. Se wæs for sumere twyrédnesse (*seditione*) on cwertern ásend, Lk. Skt. 23, 19. Ða ðe ceáste and twyrædnysse styredon, Homl. Th. ii. 338, 11. Ðonne gé geseóþ gefeoht and twyrædnessa (*seditiones*), Lk. Skt. 21, 9. Twyrédnysse *dissensiones*, R. Ben. Interl. 109, 17. Twirédnesse *discordias*, Kent. Gl. 1124.

-twis, -twisa. v. ge-twis, -twisa.

twi-sceatte; *adv. To the extent of a double payment* :—Sió bót biþ twysceatte máre *the 'bót' shall be twice as much*, L. Alf. pol. 66; Th. i. 96, 31. [*O. Frs.* twi-skette.] Cf. twi-gilde.

twi-scyldig; *adj. Liable to a double penalty* :—Gif se frigea sunnandæge wyrce . . . þolie his freótes oþþe sixtig scill., and preóst sí twyscildig, L. In. 3; Th. i. 104, 7. Cf. twi-gilde.

twi-seht; *adj. Discordant, at variance* :—Twysehte *discordes*, Scint. 192, 13.

twi-sehtan (?) *to disagree, be at variance* :—Úðwitan gesihþ twysehtan (? Cockayne prints twyselican) hénðe getácnaþ *if in a dream a man sees philosophers disagree, it betokens humiliation*, Lchdm. iii. 204, 24.

twisehtness, e; *f. Discord, dissension, variance* :—Fram twysehtnysse yfele *a dissensionis malo*, Scint. 6, 12.

twisel; *adj. Forked, double.* [Twisil tunge *double tongue* (Ecclus. 5, 14), Wick. *O. H. Ger.* der onocentaurus bizeichinôt die zuislen zungin der mennisken.] v. following words.

twisel-tôðe; *adj. Having the teeth forked or double* :—Twiseltôðe *scinodens*, Wrt. Voc. i. 17, 15.

twisla, an; *m. The fork* of a river, road, etc. :—Of ðam mere on ðan lace ðær ða brócas twisliaþ; ðanne of ðæm twislan, Cod. Dip. Kmbl. v. 198, 34. [Twissel, twistle *that part of a tree where the branches separate*, Halliwell's Dict. *O. H. Ger.* zwisila *furca*. Cf. *Icel.* kvísl *a fork; fork of a river*.]

twisled; *adj.* (*ptcpl.*) *Forked* :—On ðone twisledan beám; of ðam twisledan beáme on ceorla geat; andlang mearce on ða twysledan âc, Cod. Dip. Kmbl. iii. 14, 1-4. Twisld corn *scandula* (scandella *genus annonae apud Italos, q. alii dicunt hordeum distichum esse, alii vero hordeum cantherinum*, Migne), Wrt. Voc. i. 38, 45. [Cf. *Icel.* kvíslatré *a forked tree*; twí-kvísladr *two-pronged*.] v. next word.

twislian; *p.* ode *To fork, branch* :—Ðær ða wegas twisligaþ, Cod. Dip. Kmbl. iii. 409, 4: iv. 66, 15. Ðær ða brócas twisliaþ, v. 198, 34. [Tunge fele-twiselende *dispertite lingue*; cloven tongues (Acts 2, 3), *O. E. Homl.* ii. 117, 29. Cf. *Icel.* kvísla *to branch*, of a tree, stream, etc.]

twisliht; *adj. Forked, branched* :—In ða twislihtran biricean, Cod. Dip. Kmbl. iii. 391, 21. [*O. H. Ger.* zwisillochti *bifurcus*.]

twislung, e; *f. Forking, branching, partition* :—Se þurh his cildhádes nytenesse ðis ríce tóstencte and his ânnesse tódælde . . . Æfter his forðsíþe Eádgár ealne Angelcynnes anweald begeat, and ðæs ríces twislunge eft tó ânnesse bróhte, Lchdm. iii. 436, 3.

twi-snæcce, -snæce, -snece; *adj. Double-pointed, cloven* :—Twysnæcce *bisulcus*, Ælfc. Gl. 49; Zup. 288, 11. (Cf. Snek *pessulum*, Wrt. Voc. i. 237, col. 2. Snekke or latche *clitorium, pessulum*, Prompt. Parv. 461. Snekk *obex, obecula*, Cath. Ang. 346 and see note. Sneck *a latch; a piece of land jutting into an adjoining field*, Halliwell's Dict. See also Jamieson's Dictionary *sneck*.) v. next word.

twi-snǽse; *adj. Double-pointed, cloven* :—Twysnésum *bisulcis*, Germ. 393, 73. v. snás, *and preceding word.*

twi-sprǽc, e; *f. Double speech, unfair speech, detraction* :—Fácon and éswico and æfisto and allo tuispréco *dolum et simulationes et invidias et omnes detractiones*, Rtl. 25, 25. [Sowen we defles sed . . . ivele word, hoker and scorn, . . . and cheast, and twispeche, and curs, and leasinges, . . . and alle swikele speches, Rel. Ant. i. 129, 24.]

twi-sprǽce; *adj. Double-tongued;* bilinguis, Ælfc. Gr. 49; Zup. 288, 7. With a metaphorical meaning, *deceitful in speech, false in speech,* (with pleasant words) *flattering,* (with envy) *detracting* :—Se ðe wǽre leássagol (twispǽce, MS. E.), weorðe se sóðsagol (sóðspǽce), Wulfst. 72, 16. Ne sýn wé tó tælende ne tó twigsprǽce *let us not be too free with calumnies and detractions*, 253, 6. Ne beó ðú nó tó tælende ne tó tweosprǽce . . . ac beó leófwende, Exon. Th. 305, 19; Fä. 90. Twisprǽce *a flattering* (*mouth*, Prov. 26, 28); (os) *lubricum*, Kent. Gl. 1007. Word twispéces *the words of a talebearer* (Prov. 18, 8); verba bilinguis, 636. Twisprǽce múð *the froward mouth* (Prov. 8, 13); os bilingue, 243. Gehega ðíne eáran mid þornigum hege, ðæt ðú ne gehýre lustum móde ðæra twysprǽcena word, Wulfst. 246, 10.

twisprǽcness, e; *f. Falseness in speech, detraction* :—Bebeorh ðé wið twisprǽcnysse *cave tibi a biloquio*, L. Ecg. C. proem.; Th. ii. 132, 10. Uton beorgan ús wið tælnysse and wið twysprǽcnysse and wið leáse gewitnysse *caveamus nobis a vituperatione et a biloquio et a falso testimonio*, L. Ecg. P. iv. 66; Th. ii. 226, 32. Twyspécnessæ, Wulfst. 290, 30. Ic ondette æfste, twysprǽcnesse and leásunge, Anglia xi. 98, 26. Ic andette tælnessa and twisprǽcnessa, leásunga and unriht gilp, L. de Cf 7; Th. ii. 262, 27.

twi-sprecan *to murmur* :—Hwisprendo ł tuispreccendo *murmurantes*, Jn. Skt. p. 4, 20. [Cf. *O. H. Ger.* zwi-sprehho *bifarius*.]

twi-spunnen; *adj.* (*ptcpl.*) *Double-spun, twice spun* :—Of twispunnenum twíne línenum *torta bysso*, Past. 14; Swt. 83, 23. v. twiþráwen.

twist *a branch, fork* (?) [The faucon moste fallen fro the twiste, Chauc. Squieres Tale, 442. A twyste *frons, ramus*, Cath. Ang. 399, and see note. Twist *the fourchure; a twig*, Halliwell's Dict. Cf. *Icel.* kvistr *a branch.*] v. candel-, mæst-twist; twisel.

twi-strenge; *adj. Two-stringed* :—Twistrenge *bifidus* (as if from *fides*), Ælfc. Gr. 49; Zup. 288, 10.

twi-telged; *adj.* (*ptcpl.*) *Double-dyed* :—Of twitælgedum *bis tincto*, Wrt. Voc. ii. 126, 31. Twitælgade *depploide*, Ps. Surt. 108, 29.

twi-þráwen; *adj.* (*ptcpl.*) *Double-twisted* :—Is beboden ðæt scyle beón twiðráwen (-ðráwen, Cott. MSS.) twín (*torta byssus*) on ðæm mæssegierelan, Past. 14; Swt. 87, 18. Ðæt tweoðráwene (twyðráwene, Cott. MSS.) twín, Swt. 89, 2. v. twi-spunnen.

twiwa, tweowa, twuwa, tuwa, tuwwa, tua, twiga, twigea, twige, twía; *adv. Twice* :—Hé hine twiwa (tuwa, MS. L.) mid fyrde gesóhte, Ors. 5, 2; Bos. 102, 37. Ðæt heó on geáre twigea (twiwa, MS. H.: tuwa, MS. B.) blówe, Lchdm. i. 320, 13. Hé gefeaht II (tweowa, MS. C.) wið ðone cyning, Ors. 6, 30; Swt. 280, 9. Tweowa on dæg *bis in die*, Coll. Monast. Th. 20, 17. Twuwa, Scint. 80, 11. Hú ne mynegodest ðú mé nú tuwa? Bt. 35, 2; Fox 156, 14. Tuwa (twiga, Bd. M.), Bd. 4, 1; S. 564, 16. Tuwa (twigea, Bd. M.) on geáre, 4, 5; S. 573, 6. Tuwa (tuiga, Lind.), Mk. Skt. 14, 30. Tua (tuwa, MSS. A. B. C.: twiga, Lind., Rush.), 14, 72. Ic fæste tuwa (tuigo, Lind.: twige, Rush.) on ucan, Lk. Skt. 18, 12. Æne oððe tua (tuwa, MSS. T. F.), R. Ben. 74, 20: Homl. Skt. i. 16, 80. Oftor ðonne tuwa (tuwa, *other MSS.*), Chr. 894; Erl. 90, 20. Twiga þriga *bis terque*, Wrt. Voc. ii. 126, 35. ¶ With numerals :—Sió gestód tuwa seofon hund wintra . . .; ðæt is III c wintra and I M, Ors. 6, 1; Swt. 252, 6. Tuwa fífe *binas quinquies*, Wrt. Voc. ii. 126, 25. Tuwa fíftig *bis quingentenum*, 11, 76. Twige, Hpt. Gl. 486, 77. Twía seofon beóþ feówertýne, Anglia viii. 302, 45. Twía líf beóþ týn, 328, 22. [*A. R.* twie, twien, twies: Laym. twien (twie, 2nd MS.): *Gen. and Ex.* twie: *O. E. Homl.* twiȝen, twies: *Orm.* twiȝȝess: *O. Frs.* twía, tuiia.]

twi-wǽg, e; *f. A balance* :—Twiwǽge *bilance*, Wrt. Voc. ii. 126, 20. [*O. H. Ger.* zwi-wâga *bilibris*.]

twi-weg, es; *m. A place where two roads meet* :—Twiweg *bivia vel bivium*, Wrt. Voc. i. 53, 57.

twi-wintre, -winter; *adj. Of two years* :—Twiwintre *biennis*, Wrt. Voc. ii. 126, 11 : binus, i. biennius, 126, 23. Twywintre *biennis, bimus*, Ælfc. Gr. 49; Zup. 287, 13, 18. Twiwinter *bimus vel biennis vel bimulus*, Wrt. Voc. i. 21, 58. Fram twywintrum cilde *a bimatu*, Mt. Kmbl. 2, 16: Homl. Th. i. 80, 16: 82, 11. Fram twiwintre fæce *a bimatu*, Ælfc. Gr. 49; Zup. 287, 19 note.

twi-wyrdig; *adj. Making contradictory or discordant statements, at*

variance in what is said :—Hié swā twywyrdige sindon *they disagree in what they say* (ille promisit futura meliora, isti asserunt meliora praeterita), Ors. 2, 5 ; Swt. 86, 8. Hē com tō Rōme and diégellīce gecéapede đæt hié ealle wǣron ymb hiene twywyrdige *cum Romam ipse venisset, omnibus pecunia corruptis seditiones dissensionesque permiscuit*, 5, 7 ; Swt. 228, 18. [Cf. *Icel.* tvī-mæli *a dispute, a discordant report*, one saying this, another that.]

twuwa, -twux, twȳ, twy-, twycene, twy-iccende, twȳn, twȳnian, twȳnigend-līc, twȳnol, twȳnung. v. twiwa, be-twux, tweó, twi-, twicene, twi-hycgan, tweón, tweónian, tweónigend-līc, tweónol, tweónung.

tȳ ; *indic. imper. subj. of* tȳn *to instruct.*

tyccen. v. ticcen.

tȳogan (?) *p.* togde (?) *To move quickly, quiver, palpitate* :—Tolcetende, brottetende (v. brogdettan *palpitare, vibrare*), ticgende *infruticans*, Hpt. Gl. 435, 37. [Cf. *O. H. Ger.* zucchen ; *p.* zuhta *rapere, eruere* : *Ger.* zucken *to shrug, writhe, palpitate* : *Icel.* tyggja *to chew* : *Dan.* tugge.] Cf. togian, togung, togettan.

tȳd *time*, tydder-, tyddre, tyddrian, tyddrung. v. tīd, tīder-, tīdre, tȳdran, tȳdrung.

tȳdran, tȳdrian ; *p.* ede *To propagate* :—Ic tyddrige (teddrige, MS. D.) *propago*, Ælfc. Gr. 36 ; Zup. 216, 14. **I.** *trans.* (a) *To bring forth, produce* :—Se godcunda foreþonc geednīwaþ and tȳdreþ (tīdreþ, Cott. MS.) ǣlc tūdor and hit eft gehȳt *nascentia occidentiaque omnia per simileis foetuum seminumque renovat progressus*, Bt. 39, 8 ; Fox 224, 10. (b) *to propagate, nourish, foster* :—Đīn hand plantode and tȳdrede ūre foregengan *plantasti eos*, Ps. Th. 43, 3. Hēr seó gālnese tȳdrode (tȳtrode, MS.) hir[e] cyn on hire sylfre *multitudinem vitiorum avaritia nigro lacte nutrit*, Gl. Prud. 57 b. Ælces landes gecynd is, đæt hit him gelīce wyrta tȳdrige (tȳdre, Cott. MS.) ; and hit swā đeþ ; friþaþ and fyrþraþ swīþe georne, Bt. 34, 10 ; Fox 148, 29. Wyrd seó swīđe ... heó wile late āđreótan, đæt heó fǣhđo ne tȳdre *it will be long before she is weary of fostering hate*, Salm. Kmbl. 898 ; Sal. 448. Telgran tīdrian *surculos pastinare* (*plantare, nutrire*), Hpt. Gl. 433, 48. Tȳdriende *pastinantem, rigantem*, 454, 19. Tytdriendum *propaganda*, Anglia xiii. 30, 75. Fācn wiþinnan tyddriende *dolum intus alentes*, Coll. Monast. Th. 32, 33. **II.** *intrans. To be prolific*, (a) *absolute* :—Tȳmaþ and tiédraþ, Cd. Th. 91, 14 ; Gen. 1512. Feoh sceal on eorđan tȳdran and tȳman, Menol. Fox 557 ; Gn. C. 48. Melce and tȳdrende *foetas*, Wrt. Voc. ii. 36, 32. (b) *with dat.* (*inst.*) *of that in which anything is prolific* :—Wǣstmum tȳdreþ, Exon. Th. 493, 18 ; Rä. 81, 32. Wudubearwas tānum tȳdraþ, 191, 6 ; Az. 84. Wǣstme tȳdraþ cederbeámas, Ps. Th. 103, 16. Tyddraþ, 64, 11. [Þenne men michel tuderiđ ... and here tuder swiđe wexeđ, O. E. Homl. ii. 177, 16. Þeȝȝte time wass all gan to tiddrenn and to tæmenn, Orm. 18307. Of hem ben tudered manig on, Gen. and Ex. 630.] v. ā-, on-tȳdran, ge-tyddrian ; tȳdred, un-tȳdrende, tȳdriend, tūdor.

tȳdre *weak*, -tȳdre. v. tīdre, on-, un-tȳdre.

tȳdred ; *adj.* (*ptcpl.*) *Provided with offspring* :—Heora sceáp wǣron swylce tȳdred *oves eorum foetosae*, Ps. Th. 143, 17. v. tȳdran.

tȳdrian *to bring forth*, tȳdrian *to get weak*. v. tȳdran, tīdrian.

tȳdriend, es ; *m. One that brings forth* :—Tȳdriend (tȳdriende ?) *fecundus*, i. *copiosus, fructuosus*, vel *habundans*, Wrt. Voc. ii. 148, 47. [Cf. (?) þe fule tuderende of flesliche lustes, O. E. Homl. ii. 55, 9.]

tȳdrung, e ; *f.* **I.** *propagation* :—Uneácniendlīcre tēdrunge *infecunda sterilitate*, Hpt. Gl. 430, 61. Ic ongite đæt ǣlc gesceaft willnaþ simle tō biónne ; đæt is swīþe swital on đ ære tȳdrunge, Bt. 34, 12 ; Fox 152, 25. [Cf. (?) Þis woreld ebbeđ þenne hit þat tuderinde widteođ *withholds its productivity*, O. E. Homl. ii. 177, 23.] **II.** *a branch* :—Tyddrung (tȳdrung, MS. T. : tiddrung, MS. V.) *ođđe* bōh *propago*, Ælfc. Gr. 36 ; Zup. 216, 15.

tyge, tige (v. *double forms* togen, tigen, *pp. of* teón), es ; *m.* **I.** *a pull, tug* :—Gange him tō mīnre byrgene and āteó āne hringan up, and gif seó hringe him folgaþ æt đam forman tige, đonne wāt hē đæt ic đē scnde tō him. Gif seó hringe nele up þurh his ānes tige, đonne ne sceall hē đīnre sage gelȳfan, Homl. Skt. i. 21, 43–48. Ārena tīum *remorum tractibus*, Hpt. Gl. 406, 70. **II.** *a dragging* :—Valerianus hine hēt teón geond đornas, and hē mid đam tige his gāst āgeaf, Homl. Th. i. 432, 35. **III.** *leading, conducting* :—Đone weterscype đe hē intō Nīwan mynstre geteáh, and him se tige sume mylne ādilgade (*the diverting of the water had ruined his mill*), Chart. Th. 232, 7. Tiga *aquae ductuum*, Hpt. Gl. 418, 49. **IV.** *a draught* of drink :—Hālwende tige drincan, Anglia viii. 321, 32. **V.** *a drawing* of an inference, etc., *a deduction* :—Wē wyllaþ embe đone geleáfan swiđor sprecan, forđan đe đises godspelles traht hæfþ gōdne tige *much good may be drawn from an examination of this gospel*, Homl. Th. i. 248, 21. Đis godspel hæfþ langne tige on his trahtnunge *the exposition of this gospel might be drawn out to a great length*, ii. 72, 22. Petrus āwrāt twēgen pistolas, hig hebbaþ langne tige tō geleáfan trimminge *much matter for the confirmation of belief may be drawn from them*, Ælfc. T. Grn. 14, 8. [Ete nu enes o dai and drinke o tige atte mete, O. E. Homl. ii. 67, 11. *O. H. Ger.* zug, zugi (*in cpds.*) *ductus, motus*.] v. of-, on-, wæter-tyge.

---

tyge-hōc, es ; *m. A hook to pull with*, the word occurs in a list of implements :—Scafan, sage, cimbīren, tigehōc, Anglia ix. 263, 2.

tyge-horn, es ; *m. A cupping-glass* :—Mid tigehorne, Lchdm. ii. 120, 17.

tygel, es ; *m. A strap to draw with, a trace* :—Tigel *tractorium* (cf. *tractorium* a trays, Wülck. 617, 7), Ælfc. Gl. Zup. 314, 16. [Tiȝel *tractorium*, Wrt. Voc. i. 92, 74. Þe reines oþer þe tiels, Trev. 4, 77. *O. H. Ger.* zugil *habena, lorum* : *Icel.* tygill *a strap, thong*.]

tygele (?), an ; *f. A lamprey* :—Tigle *murenula* (the word occurs in a list of the names of fishes ; *murenula* is elsewhere glossed by ǣl, 66, 5 : 281, 66 ; sǣ-ǣl, q. v.), Wrt. Voc. i. 55, 66. Cf. (?) preceding word.

tygele *a tile*, tyhhian. v. tigele, teohhian.

tyht, es ; *m.* **I.** *way, manner of conducting one's self, usage, practice* :—Ic đē giungne underfēng untȳdne and ungelǣredne and mē tō bearne genom and tō mīnum tyhtum getȳde ... Đū mē wǣre leóf ǣr đon đe đū cūþest mīnne tyht and mīne þeáwas *I received thee young, uninstructed and untaught, and took as my child and brought thee up to my ways ... Thou wast dear to me before thou knewest my way and my customs*, Bt. 8 ; Fox 24, 23–27. [Þat (*moderation*) is þeaw ant tuht forte halden, O. E. Homl. i. 247, 32. Cf. For þere ilke tuhtle (þinge, 2nd MS.) cnihtes weoren ohte, Laym. 24675. Elche untuhtle heo talden unwurđe, 24655.] **II.** *motion, move, march.* v. teón, **IV**, tohte, *and see passages from Layamon under* tyhtan, **I** :—Werod wæs on tyhte *the army was on the march*, Elen. Kmbl. 106 ; El. 53. Līg scrīþeþ ... brond biþ on tyhte, Exon. Th. 51, 7 ; Cri. 812. Fȳr biþ on tihte, 233, 16 ; Ph. 525. **III.** *in* ofertyht (?) *a covering, what is drawn over.* v. ofer-teón ; *and cf. Ger.* über-zug :—Þrong niht ofertiht londes frætwa đe, *the covering drawn over the land's decorations, pressed on*, Exon. Th. 179, 3 ; Gū. 1256. [*Goth.* us-tauhts *a carrying out, completion* : *O. H. Ger.* zuht *disciplina, eruditio, nutrimentum*.]

tyhtan ; *p.* te. **I.** *to draw, stretch* [:—Oferbrǣdels onbūtan getint *velamen in gyro tensum*, Anglia xiii. 421, 806]. [Tuhten is used in Layamon with the meaning of teón, **IV** :—Ure drihten heo bilæucđ, and tō Mahune heo tuhteđ, Laym. 27321. Troynisce tuhten (toȝe, 2nd MS.) tō þon Gricken, 810.] **II.** *but mostly in a metaphorical sense, to draw the mind to something, to incite, exhort, provoke, solicit, prompt, urge, persuade*, (1) *where the construction is uncertain* :—Ic tyhte *ortor*, Ælfc. Gr. 25 ; Zup. 144, 18 : *suadeo*, 26 ; Zup. 155, 6. Tyhto *sollicito*, tyhteþ, tyhtit *sollicitat*, tyhtan *sollicitare*, Txts. 97, 1887–3–9. Hē tihte *persuadet, docet*, Hpt. Gl. 491, 43 : *incitavit*, 511, 28. Tyhton *irridabant*, Txts. 73, 1152 : Wrt. Voc. ii. 45, 73. Tyctende (-i) *adridente*, Txts. 37, 70. Tyctendi *inlex*, 69, 1063. Tyhtende *adridens*, Wrt. Voc. i. 287, 70 : ii. 4, 39. (2) *where the object to which a person* (*acc*) *is exhorted, etc*, *is* (a) *marked by prep. on or tō* :—Ne tyht nān mon his hiéremonna mōd ne ne bielt tō gǣstlīcum weorcum *nulla subditorum mentes exhortatio sublevat*, Past. 18 ; Swt. 129, 10. Deófol tiht ūs tō yfele, Homl. Th. i. 174, 31. Ođer hine tyhteþ and on tæso lǣreþ, Salm. Kmbl. 983 ; Sal. 493. Hī (*devils*) on teosu tyhtaþ, Exon. Th. 362, 9 ; Wal. 34. God selfa tyhte (*suadente Deo*) Moyses on đone folgođ, Past. 7 ; Swt. 51, 21. Heó hyre leófe bearn georne lǣrde and tō gōde tihte, Lchdm. iii. 428, 29. Hine his yldran tō woruldfolgađe tyhton and lǣrdan *his parents urged him to temporal service*, Blickl. Homl. 211, 28. Hine tihtan tō his sāwle þearfe *eum hortari ad animae suae necessitatem*, L. Ecg. C. prm. ; Th. ii. 130, 40. Đreátian and tihtan (tyhtan, Cott. MS.) tō gōdum đeáwum, Bt. 38, 3 ; Fox 200, 8. Tyhtan and gremian tō spīwanne *to provoke to vomit*, Lchdm. ii. 184, 1. (b) *expressed by a clause* :—Iohannes đæt folc tihte, đæt hī ufor eodon fram đam deófles temple *John urged the people to go further away from the heathen temple*, Homl. Th. i. 70, 35. Đā tihte (*or* **III**) heora sum, đæt man đæs cnapan līc smyrian sceolde, ii. 28, 3. (c) *not expressed* :—Đū on ūs sāwle gesetnest and hī styrest and tihtest, Met. 20, 178. Lǣran sceal mon geongne monnan, trymman and tyhtan, Exon. Th. 336, 10 ; Gn. Ex. 46. Đæt se lǣréow sceolde beón miehtig tō tyhtanne on hālwende lāre *ut potens sit exhortari in doctrina sana*, Past. 15 ; Swt. 91, 15. **III.** *to suggest, bring to the mind* :—Swā hwæt swā þurh unclǣnnysse on þeáwum hit tiht (*se suggerit*), Hymn. Surt. 28, 31. Gif mid rīcan mannan wē wyllaþ sum þinc tihtan (*suggerere*), R. Ben. Interl. 53, 6. **IV.** *to instruct, teach.* v. ge-tyhtan. [Þe deofel heom tuhte tō þan werke, O. E. Homl. i. 121, 33. A þet wit cume forđ ant tuhte ham þe betere, 247, 6. Tuhten and teachen, 267, 15. Þet tu ne schuldest nout tuhten ne chasten þi meiden uor hire gult, A. R. 268, 21. Tihhtenn and turrnenn folc tō lefenn uppo Criste, Orm. 7048. *O. H. Ger.* zuhten, zuhtôn *nutrire, erudire* : *Ger.* züchten, züchtigen *to chastise* : *Dan.* tugte *to chastise, discipline*.] v. ā-, for- (fær-), ge-, leás-, mis-, on-tyhtan, *and following words.*

tyhten[n], e ; *f. An incitement, inducement, allurement, incentive, enticement* :—Tyhten, tyctin, thyctin *lenocinium*, Txts. 73, 1199. Tyhtend (tyhtenn ?) *allectio*, Wrt. Voc. ii. 3, 68. Tyhtinne, tyctinnae *incitamenta*, Txts. 69, 1074. Tyhtenne *lenocinia*, Wrt. Voc. ii. 50, 14 : *incitamenta*, 48, 70. Tihtennum *inlecebris*, 48, 67. Tyctinnum, Txts. 68,

tyhtend, tyhtiend, es; *m. One who exhorts, incites, instigates :*—Tyc-taend, tychtend *inlex*, Txts. 68, 509. Tyhtend *incentor*, Wrt. Voc. ii. 111, 58 : 44, 62 : 83, 39 : 94, 19. Tihtend *incentor, accensor, instigator :* tihtiend *adjutor, fautor*, Hpt. Gl. 495, 67, 70. v. yfel-tyhtend, for-tihtigend.

tyhtend-líc; *adj. That serves for exhortation, encouragement*, etc. (v. tyhtan), *hortative :*—Wĕ wyllaþ sume tihtendlíce sprǽce wið eów habban, Homl. Th. ii. 574, 20. Sume *adverbia* syndon *ortativa*, ðæt synd tihtendlíce, Ælfc. Gr. 38; Zup. 227, 16. Hĕ mid tihtendlícum wordum heora gewæhtan mód getrymde and gefrĕfrode, Homl. i. 562, 1.

tyhtere, es; *m. An inciter, instigator :*—Tyhtere *incentor*, Wrt. Voc. ii. 48, 48. Tihtere *leno*, i. 50, 55.

tyhting, e; *f. Persuasion, exhortation, encouragement, incitement, instigation, allurement, suggestion :*—Tihting *suasio*, Ælfc. Gr. 9, 3; Zup. 35, 10. Deófles costnung biþ on tihtinge . . . Deófol tiht ús tó yfele, ac wĕ sceolon geniman náne lustfullunge tó ðære tihtinge . . . Seó yfele tihting is of deófle, Homl. Th. i. 174, 30–35 : ii. 226, 29. Crist mid ðyssere tihtinge Petrum gehyrte, 374, 17. God hira mód onlíeht mid his fandunga and eác his tiehtinge (tihtinge, Cott. MSS.), Past. 35; Swt. 243, 22. For láre and for tiehtinge his ágenes firenlustes *persuasione luxuriae*, 50; Swt. 393, 7. Mid godcundre tihtincge *divino instinctu*, Anglia xiii. 384, 266. Mid welwyllendre tihtincga myngiende *benevola intentione hortando*, 448, 1179: Scint. 34, 1. Se ðe his bróðor hataþ ðurh ðæs deófles tiht-inge, Basil admn. 4; Norm. 44, 17. Tihtinga *incitamenta*, Hpt. Gl. 520, 35. Tychtingum, Wrt. Voc. ii. 111, 3. Hĕ micclum mid his bĕnum and tihtingum fylste *he helped much with his prayers and exhortations*, Homl. Th. ii. 126, 29. Se ðe óðerne tó leahtrum forspenþ, hĕ is man-slaga, ðonne hĕ ðæs óðres sáwle forpǽrþ þurh his yfelum tihtingum, 226, 32. Geþafian ðæs deófles tihtinga, 546, 11. [Defles tuihting, O. E. Homl. ii. 29, 2. Tihting, i. 229, 19.] v. tó-tyhting.

tyhtle *a charge*. v. tihtle.

tyhtness, e; *f. Instigation :*—Tyhtnesse *instinctu*, Wrt. Voc. ii. 86, 20. Tihtnesse, 46, 63 : 80, 28.

tyld-syle. v. teld-sele.

Týle *Thule :*—Án íglond . . . ðæt is Tíle háten (þe isle þat hy3t tile, Chauc. Boet. 3, 5. This form is used also in Trevisa, i. 325) *ultima Thule*, Met. 16, 15. [Icel. Tíle.] v. þýle (*the usual form*).

tylg, tylian. v. tulge, tilian.

tyllan; *p.* tylde *To draw, attract.* Found only in the compound *for-tyllan*, but see the following passages from later English. [Mi liht onswere tulde him upon me, A. R. 320, 13. Ne tulle 3e to þe 3ete none unkuðe harloz, 414, 5. As muche place as myd a þong ich may aboute tille, R. Glouc. 115, 18. Of þe purse þat seluer heo tulleþ, Misc. 188, 40. Ille felawes hafd maistri To tille this yong man to foli, Met. Homl. 113, 8. Þe world tyl hym drawes And tilles . . . þam þat him knawes, Pr. C. 1183. To þe scole him for to till (tille), C. M. 12175. He hauede . . . Al þe folk tilled intil his hond, Havel. 438. *Also, like* teón, *with sense of* proceed, go :—Twei leomes stode þere; The gryttere tylde Est . . . þe oþer hadde branches . . . And westward theu drowe, R. Glouc. 151, 20 : 152, 19. To gile ne to fraude wild he neuer tille, R. Brunne 128, 20. *Cf. also* tollen *to draw, attract :*—Þis tolled him toward þe, A. R. 290, 5. Ha tolliþ togederes *they draw, come together*, Marh. 14, 6. (See instances quoted, p. 110.) Swa mai mon tolli him to Lutle briddes, O. and N. 1627. To drawen or tollen *allicere*, Chauc. Boet. 2, 7. Tollyñ or mevyñ *incito, provoco, excito*, Prompt. Parv. 496.]

tylþ, tylung, týma, týman. v. tilþ, tilung, tíma, tíman.

tym-bor (?) *a revolving borer, an auger :*—Timbor *rotum vel taratrum*, bor *desile, scafa olatrum*, Wrt. Voc. i. 287, 9–11. [Cf. (?) tumbian: or (?) O. H. Ger. tûmón *rotari*.]

tymbran, týme, tymian, týn *ten*. v. timbran, tíme, temian, tín.

týn; *p.* týde, tydde (týdde?); *pp.* týd *To instruct, educate, teach :*—Ic tý oðde lǽre *imbuo*, Ælfc. Gr. 28, 3; Zup. 166, 14. Hĕ lǽrþ and hĕ týd heorde his *docet et erudit gregem suam*, Scint. 146, 7. Se wísdóm ðe hit lange ǽr týde and lǽrde, Bt. 3, 1; Fox 4, 30. Hĕ hine geornlíce týde and lǽrde hú hĕ drohtian sceolde *eum erudire studuit qualiter conversari debuisset*, Bd. 1, 27; S. 489, 5. Hí mycelne ðreát discipula on metercrǽfte and on tungolcrǽfte týdon and lǽrdan, 4, 2; S. 565, 26. Láreówas ðe hí (wudufuglas) týdon and temedon, Met. 13, 39. Swá hwilce men swá willnadon ðæt hí on hálgum leornungum týde wǽron hí hǽfdon gearuwe magistras ða ðe hig lǽrdon and tyddon *quicumque lectionibus sacris cuperent erudiri, haberent in promptu magistros qui docerent*, Bd. 4, 2; S. 565, 34 : 4, 3; S. 569, 6. Láreów ðú þele þeawas tý *doctor egregie mores instrue*, Hymn. Surt. 106, 5. Se ðe ðone mǽran noman abbodes underfehþ hĕ sceal mid twyfealdre láre ða wyldan and týn ðe him underþeódde synt *cum aliquis suscipit nomen abbatis dupplici debet doctrina suis preesse discipulis*, R. Ben. 11, 12. Hĕ scole gesette in ðære cneohtas tydde and lǽrde wǽron *instituit scholam in qua pueri literis erudirentur*, Bd. 3, 18; S. 545, 45. v. ge-týan, -týdan (*in each case read* -týn), and tedn, **III. 1.**

týnan; *p. de To teen, tine* (v. Halliw. Dict.), *close.* **I. to fence, enclose :**—Me mæig on sumera týnan, Anglia ix. 261, 11. Gif ceolas

gærstún hæbben gemǽnne oþþe óðer gedálland tó týnanne, and hæbben sume getýned hiora dǽl, sume næbben, L. In. 42; Th. i. 128, 6. **II. to close**, *shut* a door, *book :*—Middý hígna sæder týneþ ðæt duro *cum paterfamilias cluserit ostium*, Lk. Skt. Lind. 13, 25. Ðonne týnde hĕ his bĕc *clauso codice*, Bd. 4, 3; S. 569, 10. Middý ða duro uĕrun týndo *cum fores essent clausae*, Jn. Skt. Lind. 20, 19. **III. to close** a place, *prevent entrance into* a place, *shut up :*—Gié týndon ríc heofna *clauditis regnum coelorum*, Mt. Kmbl. Lind. 23, 13. **III a. to prevent** a person *granting access to others* (?), *render* a person *inaccessible :*—Týne hine Dryhten ðam ðe sár sprece sáwle mínre *may the Lord shut His heart to him that speaks evil against my soul*, Ps. Th. 108, 20. **IV. to close, conclude, bring to an end :**—Se hálga Willfriþ æfter .xlv. wintra ðæs onfongenan biscophádes ðone ýtemestan dæg týnde (*diem clausit extremam*), Bd. 5, 19; S. 636, 43. [An ancre nule nout tunen hire eiðurles a3ein deað of helle, A. R. 62, 17. Þa 3æten heo tunden uaste, Laym. 15320. Tynyñ *sepio :* tynyd or hedgydde *septus*, Prompt. Parv. 494. O. Frs. be-têna: O. Du. tuinen: O. H. Ger. zûnen *sepire :* Ger. zäunen.] v. á-, ǽ-, an-, be-, bi-, for-, ge-, on-, un-týnan; fore-týn(e)d; tún.

týnan *to vex*. v. tínan.

tyncen *a barrel* (?), *a bladder* (?) :—Ða gebeótode án his ðegna ðæt hĕ mid sunde ða eá oferfaran wolde mid twám tyncenum, Ors. 2, 4; Swt. 72, 30.

tyndeht. v. tindiht.

tynder, e; tyndren (-in), e (?); tyndre, an; *f.* **I. tinder, fuel** (lit. and fig.) :—Tyndir (-er) *napta, genus fomenti*, Txts. 80, 685. Geswǽlud spoon *vel* tynder *fomes*, Wrt. Voc. i. 39, 21. Tynder *fomes*, *incendium, astula minuta*, ii. 150, 4. Tyndrin, tyndirm (-in ?) *isca* (= esca *fomes*, Migne; cf. Span. yesca *tinder*), Txts. 72, 562. Tyndre *isica*, Wrt. Voc. i. 284, 21 : *isca*, ii. 45, 74 : *fomentam* (-*um* ?), 40, 7. Tyndre gódes cynnes *fomentum bone indolis*, Scint. 206, 17. Tindre *sica* (l. *isica* or *isca*), Wrt. Voc. i. 66, 38. Wĕ habbaþ ðone mǽstan dǽl ðære tyndran ðínre hǽle . . . nú ðú ne þearft ðĕ náuht ondrǽdan forðam ðe of ðam lytlan spearcan ðe ðú mid ðære tyndran gefĕnge lífes leóht ðĕ onlíehte *habemus maximum tuae fomitem salutis . . . nihil igitur pertimescas; jam tibi ex hac minima scintillula vitalis calor illuxerit*, Bt. 5, 3; Fox 14, 9–14. Tyndri *isica*, Txts. 116, 179. Of gecyndelícre tyndran *de ingenito fomite*, Wrt. Voc. ii. 139, 65. Tyndre *neptam*, 114, 59. Tynder, 60, 9. Tyndrum *fomitibus*, 33, 61. Deóful ná gewilnunge tyndran onǽlþ *diabolus non concupiscentiae fomenta succendit*, Scint. 210, 3. **II. a burner, an implement which burns :**—Mearcísern *vel* tynder *cauterium*, Wrt. Voc. ii. 129, 76. Tynder *furnus*, 149, 84. Tyndre *cautere*, Txts. 114. 100. Tund[e]ri, 111, 19. [He tinder nom and lette i þan nute-scalen don and fur þer on brohte, Laym. 29267. Of ston mid stel in of tunder, Misc. 17, 535. Tondre, tunder, Piers P. 17, 245. Tundyr *fungus, napta*, Prompt. Parv. 506. Du. tonder: O. H. Ger. zuntra; wk. f. fomes, isca: M. H. Ger. zunder; m. n.: Ger. zunder: Icel. tundr; n.: Dan. tønder: Swed. tunder. Cf. Goth. tundnan *to be set on fire*.] v. tender, tendan.

tynder-cyn[n], es; *n. Combustibles :*—Tyndercyn *matteoli* (v. spæc), Wrt. Voc. ii. 56, 66 : 78, 9.

tyndre, tyndren, -týne. v. tynder, ge-týne.

tynge; *adj. Skilful with the tongue, rhetorical :*—Tingcum *rhetoricis, facundis*, Hpt. Gl. 460, 41. [Cf. O. H. Ger. zungal *linguosus*.] v. ge-tynge.

týning, e; *f. A closing, fencing.* [Tynyñ or make a tynynge *sepio*, Prompt. Parv. 494.] v. be-, gafol-týning (-tíning); týnan.

týnness, e; *f. An enclosed place, a prison :*—Tênys (= týnnysse?) þrexwealdum (heó) tó geþeódde *Anastasia lautomiae liminibus haerescit*, Hpt. Gl. 513, 65 (cf. l. 57 *lautomiae* cwearternes). v. on-týnness.

týr, týran. v. tír, tíran.

tyrdlu, tyrdelu; *pl. n. Treddles* ('the droppings of sheep are called sheep's *tredles* in Somerset, *trattles* in Suffolk,' Lchdm. iii, Gl. *Treddle* excrement of rabbits, E. D. S. Pub. Old Farming Words. Halliwell quotes 'tak the *triddils* of an hare.' *Tyrdyl schepys* donge, Prompt. Parv. 494. Take scheps tridels or swynes muk, Rel. Ant. i. 53, 16) :—Haran tyrdlu, Lchdm. ii. 214, 4. Genim gáte tyrdlu, 72, 16, 27. Tyrdelu, 282, 7. v. tord.

tyrf, tyrgan, tyrging, tyrian, tyriaca, tyring. v. turf, tirgan, tirging, tirgan, tiriaca, tirging.

tyrnan; *p. de.* **I. to turn** (intrans.), *revolve* on an axis, *round* a centre :—Seó heofon tyrnþ onbútan ús swiftre ðonne ænig mylenhweól, Lchdm. iii. 232, 18 : 254, 11 : Boutr. Scrd. 18, 28 : Homl. Th. ii. 214, 29. Se firmamentum went on ðam twám steorran swá swá hweogel tyrnþ on eaxe, Lchdm. iii. 270, 22. Se cwyrnstán ðe tyrnþ singallíce and nǽnne fǽreld ne ðurhtíhþ, Homl. Th. i. 514, 20. Ða steorran ðe on ðam rodere standaþ tyrnaþ ǽfre ábútan mid ðam brádan rodere, Hexam. 7; Norm. 12, 32. Hí tyrndon mid bodige and heora fótwylmas áwendan ne mihton, Homl. Th. ii. 508, 19. Tyrn mid ðínum swíþran scytefingre *make circles with your right forefinger*, Techm. ii. 119, 11 : 126, 1. Tyrnende *rotante*, Hpt. Gl. 517, 9. **I a. figurative :—**

Tyrnende swēgas *rotātiles trocheos*, Germ. 403, 8. **II.** *to turn* (trans.), *to cause to revolve:*—Đā tyrndon đa hǣđenan hetelíce đæt hweowl, Homl. Skt. i. 14, 93. [*O. H. Ger.* turnen. From Latin.] v. be-, ymb-tyrnan; turnian.

tyrn-geat, es; *n. A turn-stile:*—Tô tyrngeate, Cod. Dip. Kmbl. iii. 405, 4.

tyrning, e; *f.* **I.** *a turning round:*—Tyrnincg turniendre liđeran *vertigo rotantis fundibuli*, Hpt. Gl. 422, 65. **II.** *roundness:*—Sinewealtre trendla tyrnincge *tereti circulorum rotunditate*, 419, 9. v. turnung.

tyrwa (-e), tyrwan, tyrwen, tyrwian. v. tirwa (-e), tirwan, tirwen, tirgan.

tysca, an; *m. A buzzard:*—Glida *milvus*, tysca *butzus*, Wrt. Voc. i. 280, 22, 23. Tysca *bizus*, ii. 126, 39. Cf. (?) tusc.

tyslian; *p.* ode *To dress:*—Ic secge đē, brôđor Eádweard, đæt gē dôþ unrihtlíce đæt gē đa Engliscan þeáwas forlǣtaþ đe eówre fæderas heóldon and hǣđenra manna þeáwas lufiaþ đe eów đæs lífes ne unnon and mid đam geswuteliaþ đæt gē forseóþ eówer cynn and eówre yldran mid đām unþeáwum đonne gē him on teónan tysliaþ eów on Denisc ābleredum hneccan and āblendum eágum. Ne secge ic nā māre embe đa sceandlícan tyslunge būton đæt ús secgaþ bēc đæt se beó āmānsumod đe hǣđenra manna þeáwas hylt on his lífe and his āgen cynn unwurþaþ mid đam *I tell you, brother Edward, that you do wrong to forsake the English customs that your fathers held and to love the customs of heathen men, that did not give you life, and that thereby you show that you despise your race and your forefathers, when to their shame you dress in Danish wise with bared* (? cf. blere *blurus, calvus*, Wrt. Voc. ii. 127, 13) *neck and darkened* (by hair falling over the eyes?) *eyes. I will say no more about that shameful fashion of dress, but that books tell us, that he is accursed, who holds the customs of heathen men in his life and thereby dishonours his own race*, Wanley Cat. pp. 121–122; see also Engl. Stud. viii. 62. Gedônum tācne gān and hī mid dægþernum tyslian gescýum *facto signo eant et se diurnalibus induant calciamentis*, Anglia xiii. 383, 260.

tyslung, e; *f. Dressing.* v. preceding word.

tȳtan; *p.* te *To stand out, be conspicuous* (?):—Ne tȳtaþ hēr tungul ac biþ tȳr scæcen *stars shall not shine forth, but glory shall have departed*, Exon. Th. 447, 26; Dôm. 45. [Cf. (?) Icel. tūta *a teat-like prominence*; tūtna *to be blown up*: Dan. tude *a spout*: Swed. tut: Du. tuit *a pipe, pike*.] Cf. tot.

tȳpa (-e). v. tīþe.

þ

For the Runic þ, see þorn.

þā; *adv. conj. Then, when.* When the word stands at the beginning of a clause and may be translated by *then*, the verb generally precedes its subject; if it is to be translated by *when* the subject generally precedes the verb. **I.** *then, at that time:*—Ic ofstikode hyne. Swīþe þrýste đū wǣre þā (*tunc*), Coll. Monast. Th. 22, 19. Đā wæs đæt Agustinus gelaþode tô his sprǣce Brytta bisceopas *interea Augustinus convocavit ad suum colloquium episcopos Brittonum provinciae*, Bd. 2, 2; S. 502, 5. Godes đeówas đā nāne landāre hleótan ne môston, Homl. Th. ii. 224, 4. On anginne đissere worulde menn môston lybban be heora lustum đā . . . wē ne môton lybban be ūrum lustum nū, Homl. Skt. i. 16, 233. Hit mæg eów nū fremian swā micclum swā hit đā mihte, Homl. Th. ii. 378, 12. Se stān đe đæt wæter đā of fleów, ii. 274, 1. Hē on fulluhte underfangen næs, forđan đe Martinus đā on neáwiste næs, 504, 24: Homl. Skt. i. 6, 112. On đære tíde đe Ehfrid and Æđelrēd wunnon, đā æt sumon gefeohte weard ân đegen āfylled, Homl. Th. ii. 356, 24. **II.** marking sequence,' *then, after that, thereupon:*—Đā cwæđ hē: 'Gā gē on mīnne wíngeard.' And hig đā ferdon, Mt. Kmbl. 20, 4. Sum iungling com mid gyrde tô mē, and wearp hī đā tô mē, Homl. Th. ii. 312, 17. Se engel mē lǣdde đā furđor . . . Efne đā æteówdon lígas . . . Ic đā beheóld đone ormǣtan líg, 350, 15–21: 456, 24–26. Eft đā on đære þriddan nihte middan hē gewāt of đisum lífe. Þā cômon eft englas and hine gelǣddon, 336, 2–5. Hwæt đā com sum man, 286, 19. Hwæt đa hæþenan þā hine bestôdon, Homl. Skt. ii. 28, 104. Þā se bisceop dyde up đone sanct, i. 21, 139. **III.** as adverbial connective, (1) of time, *when:*—Þā hē ūt eode embe underntíde, hē geseah đōre īdele standan, Mt. Kmbl. 20, 3: 3, 7. Hwæt hē dyde, þā hine seó menego þreáde, Blickl. Homl. 19, 11, 31: 5, 25. (2) of cause or reason, *when, since, as:*—Hī hēton hine secgan hweþer hē cristen wǣre, þā hē wilnode þyllíces, Homl. Skt. ii. 28, 106. Hwā mæg āuht ôþres cweþan būtan đū wǣre se gesǣligesta, đā đū mē wǣre ǣr leóf þonne cūþ, Bt. 8; Fox 24, 26. **III a.** where the form is doubled, or combined with đe; v. also **IV.** (1) marking time, *when:*—Crist sylf gefæstnode his sprǣce, þā đā hē spræc tô ānum Samaritaniscan wífe, Homl. Th. i. 482, 24. Đæt Drihten cwǣde tô Nichodēme, ān đæra ealdra, đā đā hē his lāre sôhte, ii. 238, 4, 9. Đā đe (*cum*) hē in āre wes, Ps. Surt. 48, 21: 106, 6. (1 a) where the two forms are separated:—Þā đā (2) marking cause or reason, *when, since, as:*—Đā cwæđ his gefēra, đæt hē gefyrn smeáde hwǣr hī bigleofan biddan sceoldon, đā đā hī đa fare

fērdon būton wiste, Homl. Th. ii. 138, 33. (3) marking condition, *when, if:*—Đæt hit wǣre geđūht đæs đe māre gemynd đæs fæder, đā đā se sunu, his yrfenuma, wæs gecíged đæs fæder naman, Homl. Th. i. 478, 11. **IV.** in correlative combinations, *then . . . when, when . . . then:*—Đā se cyng đæt hiérde, þā wende hē hine west, Chr. 894; Erl. 91, 9: 90, 22–24. Þā (*then*) þū cýþdest đæt þū nestest hwelces endes ǣlc angin wilnode, þā đū wēndest đæt steórleáse men wǣron gesǣlige, Bt. 5, 3; Fox 12, 34. Þā þā (*cum*) hē fæste feówurtig daga, þā ongan hyne syđđan hingrian, Mt. Kmbl. 4, 2. Đā đā hē ealdode, þā clypode hē his yldestan cniht him tô, Homl. Th. ii. 234, 22: 286, 6: 390, 19–22. Þā hē đā ūt faran wolde, þā hēt hē beóđan, Chr. 905; Erl. 98, 21: 894; Erl. 90, 33. Đā se wísdóm đā þis spell āreht hæfde, đā ongan hē giddian, Bt. -19; Fox 68, 19. Þā þe . . . þā sôna, Blickl. Homl. 163, 15. Þā geseah Abraham Drihtnes dæg, þā þā hē đæs gerýnu tôcneów, Homl. Th. ii. 234, 22. Se Frysa đā, þā đā hē hine gehæftan ne mihte, lēt hine faran, 358, 22. **IV a.** in combination with other demonstrative forms:—Mid đý đe heó gehýrde đone fruman, þā cwæþ heó þus, Blickl. Homl. 7, 20. Þær đū cýþdest đæt đū nystest mid hwilcan gerece God wylt đisse worulde, þā đū sǣdest đæt . . ., Bt. 5, 3; Fox 14. 4. ¶ *See* git, gita, gēn, gēna, nū *for other instances of the word*. [*Chauc. Piers P.* tho, thoo: *O. Frs.* thā: *O. Sax.* thô, thuo: *O. H. Ger.* dô: *Icel.* þā.]

þaca, þeaca, an; *m. A covering, roof:*—Đone song hē gehýrde tô him neálǣcan, đæt hē beeom tô đeacan đære cyricean (*ad tectum oratorii*), Bd. 4, 3; S. 567, 43. Cf. Bordđeaca, brodthaca *testudo*, Txts. 101, 1999. Bordþacan *latrariis*, Wrt. Voc. ii. 50, 52. v. ge-þaca *and* þæc.

þaccian; *p.* ode. **I.** *to pat, clap, strike gently*, with the open hand or the like:—Wildu hors, đonne wē hié ǣresđ gefangnu habbaþ, wē hié đacciaþ and stráciaþ mid brādre handa *equos indomitos blanda prius manu tangimus*, Past. 41; Swt. 303, 10. Đaccige hē hine selfne mid đǣm fiđrum his geđôhta *cogitationum alis semetipsos feriant*, 64; Swt. 461, 17. Ælc đara manna đe ôđerne swíđe lufaþ, hine lyst bet þaccian and cyssan đone ôđerne on bær líc, đonne đēr đær clāđas beotweóna beóþ, Shrn. 185, 31. [This carter thakketh his hors uppon the croupe, Chauc. C. T. 7141. Nicholas had . . . thacked hire about the lendes wel, 3304.] **II.** *to clap, put* one thing to another:—Nim đa wyrta and wyrce tôgadere . . . þacc yt þanne gelômelíce betwex đan scaldrun *take the herbs and work together . . . clap the mixture often between the shoulders*, Lchdm. iii. 118, 14.

þacian; *p.* ode *To thatch:*—Me mæcg in Agusto and Septembri and Octobri đacian, Anglia ix. 261, 17. [Thakkyn howsys *sartatego*, Prompt. Parv. 490. *M. H. Ger. Ger.* dachen *to roof*.] v. þeccan.

þadder; *adv. Thither, whither:*—Đadder (đider, Rush.) đes fǣrende is *quo hic iturus est*, Jn. Skt. Lind. 7, 35. [Perhaps a form due to Scandinavian influence. Cf. *Icel.* þađra *there.* But see *þæder*.]

þæc, es; *n.* **I.** *a roof:*—Þā gesēgon hī đone hræfn đa glofe teran uppe on ānes hūses þæce . . . Wilfriđ mid gyrde of đæs hūses hrôfe đa glofe gerǣhte, Guthl. 11; Gdwin. 54, 16–22: Ps. Th. 128, 4. Gē þearfum forwyrndon, đæt hī under eówrum þæce môsten in gebūgan, Exon. Th. 92, 6; Cri. 1504. Se đe on þæce siæ *qui in tecto*, Mt. Kmbl. Rush. 24, 17. Hē mycelne aad gesomnode on beámum and on ræfstrum and on wāgum and on watelum and on đacum *advexit plurimam congeriem trabium, tignorum, parietum, virgeorum, et tecti fenei*, Bd. 3, 16; S. 542, 23. Bodigaþ on þacum *praedicate super tecta*, Mt. Kmbl. Rush. 10, 27. Nam ic wyrđe đæt đū gā under þacu mīnne (*sub tectum meum*), 8, 8. **II.** *the material of which a roof is composed, thatch:*—Đæs hūses hrôf wæs mid đæce beþæht *culmen domus erat foeno tectum*, Bd. 3, 10; S. 534, 32. Đa tær đæt hors đæt đæc of đære cytan hrôfe, Homl. Th. ii. 136, 16. [*Chauc.* thacke (in houses of thacke) *thatch:* Prompt. Parv. thak for howsys *sartatectum : tectura, tegimen*, Wrt. Voc. i. 237, col. 1. *O. H. Ger.* dah *tectum, opertorium :* Ger. dach: *Icel.* þak *roof*.] v. fen-þæc; þaca.

þæcele, an; *f. A torch, light:*—Þæcile *fax*, Wrt. Voc. i. 284, 20. Đecele *facula*, ii. 77, 5. Đa fýr feóllon on đa eorþan swelce byrnende þecille *vise nubes ardentes de celo tanquam faces decidere*, Nar. 23, 26: 14, 15. Stôd se leóma him of swylce fýren đecelle (þecele, Bd. M. 476, 15) ongeán norđdǣle *portabant facem ignis contra aquilonem*, Bd. 5, 23; S. 645, 29. Đæccilla (đæcela, Rush.) *lucerna*, Lk. Skt. Lind. 11, 34: Mk. Skt. Lind. Rush. 4, 21. Þæcille (đæcella, Rush.), Jn. Skt. Lind. 5, 35. Đæccillæ (đæcela, Rush.) *lucernae*, Lk. Skt. Lind. 12, 35. Dryhtnes đecelan, Salm. Kmbl. 838; Sal. 418. Đæccillas *lampades*, Mt. Kmbl. p. 9, 20. Mid brondum ł đæccillum *cum facibus*, Jn. Skt. Lind. 18, 3. [Cf. Ælc beorn hæfde on heonde ane þechene bærninde, Laym. 8084.] v. fæcele.

þæcen. v. þecen.

þæc-tigele, an; *f. A tile for a roof:*—Þaectigilum *imbricibus*, Wrt. Voc. ii. 110, 56. [v. Halliwell's Dict. *thack-tiles*, and cf. Jamieson's Dict. *thack-stone*.]

þæder; *adv. Thither, whither:*—Gā đū and lǣde đis folc þæder þe ic þē ǣr sǣde *tu vade et duc populum istum, quo locutus sum tibi*, Ex. 32, 34. On mergen com se biscop þæder, Shrn. 139, 35. Hē þæder in

eode, 156, 13. [Cf. *Icel.* þaðra *there : Goth.* þaþrô *thence.*] v. þadder, þider, *and next word ; and* cf. hwæder.

þædres ; *adv. Thither :*—Hidres dædres (didres, Cott. MSS.) *hither and thither,* Past. 22 ; Swt. 169, 13. v. preceding word.

þæge, þage ; *pron. pl. They, these :*—Þæge twégen dagas, Lk. Skt. 11, 5 margin. Sume dæge wǽron hǽdene *erant gentiles quidam ex his,* Jn. Skt. 12, 20. Hê wyrcþ mâran donne þæge (þa, MS. A.) synt *majora horum faciet,* 14, 12. Saga mê hwanon wæs Adames nama gesceapen ? Ic dê secge, fram iiii steorrum. Saga mê, hwæt hâtton dage ? Salm. Kmbl. p. 180, 1. [*Laym.* þaie, þaye *they, the* (pl.), *those.*]

þéh *though.* v. þeáh.

þénan ; *p.* de *To moisten :*—Gif tô stíd sié, þén (*printed* þæm ; *but see* geþæn mid hunige, 144, 1) mid ðý hunige, Lchdm. ii. 108, 17. Þénda smerwunga wyrce of ele *make moist smearings of oil,* 182, 16. [Halliwell gives *thean* moist, damp, as a Westmoreland word ; and Jamieson has *thain, thane* with the same meaning.] v. of-þénnan (*read* -þénan ; *the form* ofþǽne *is subjunctive, not imperative*), þínan, þwǽnan, þân, þénian.

þénian, þénnan, þénne. v. þánian, þennan, þanne.

þǽr, þâr, þâra ; *adv. There, where.* **I.** local, (a) with demonstrative force, (1) *there, in that place :*—Hig cômon tô dære stôwe, and hê gebæd hine þær (*ibi*) tô Gode, Gen. 13, 4 : 18, 24. Gif dû þær (dér, Lind. : ðær, Rush. *ibi*) geþencgst dæt dín bróðor hæfþ ǽnig þing âgén dê, lǽt þær (*ibi*) dîne lâc beforan dam altare, Mt. Kmbl. 5, 23-24. Hê wæs âna þær (dér, Lind. : dér Rush.) *solus erat ibi,* 14, 23. Hê his bigleofan þær feccan sceolde, Homl. Th. ii. 156, 6. Gif þâr man ân bân finded unforbærned, Ors. 1, 1 ; Swt. 21, 12. Hê férde tô Bethania and lǽrde hî þâr (þér, MS. A.), Mt. Kmbl. 21, 17. Gê gegearwiaþ ûs þâra (dér, Lind. Rush.) *illic parate nobis,* Mk. Skt. 14, 15. Ic næs þâra (*ibi*), Jn. Skt. 11, 15, 31. Swîþe earfoþhâwe ac hit dæah þâra *very difficult to see, but still it is there,* Bt. 33, 4 ; Fox 130, 31. (2) *thither, to that place :*—Wæs Hæsten þær cumen mid his herge, Chr. 894 ; Erl. 91, 16. Ne mæg þér inwitfull ǽnig geséran, Cd. Th. 58, 18 ; Gen. 948 : Elen. Kmbl. 1467 ; El. 735. Ic dér cwom tô dam hringsele Hróðgâr grêtan, Beo. Th. 4023 ; B. 2009. (b) with relative force, (1) *where, in which place :*—Nellen gê goldhordian eów goldhordas on eorþan, þær (þâr MS. A. : dér ł huér, Lind. : þér, Rush. *ubi*) ôm and moðde hit fornimþ, and þær (þâr, MS. A. : dér, Lind. : þér, Rush.) þeófas hit delfaþ and forstelaþ : goldhordiaþ eów goldhordas on heofenan, þér (þâr, MS. A.) naðor ôm ne moðde hit ne fornimþ, and þár þeófas hit ne delfaþ ne ne forstelaþ, Mt. Kmbl. 6, 19-20. On wésten þær ǽr Adam forwearþ, Blickl. Homl. 29, 18 : 39, 5. On dære byrig þér se cyning ofslægen læg, Chr. 755 ; Erl. 50, 13. Sum feóll ofer stânscyligean þâr hit næfde mycele eorðan, Mk. Skt. 4, 5. (2) *whither, to which place :*—Ic lǽrde on temple þâr (þær, MS. A. : diddir, Lind. : dider, Rush. *quo*) ealle Iudéas tôgædere cômon, Jn. Skt. 18, 20. In dam éðle dér he ǽr ne cwom, Exon. Th. 27, 26 ; Cri. 436. Tô dam lande þér de lust myneþ tô gesécanne, Andr. Kmbl. 588 ; An. 294. (c) in correlative combinations, (1) þær . . . þær *there (where, thither, whither) . . . where (there, thither, whither) :*—þér (þâr, MS. A.) dîn goldhord is dér (þâr, MS. A.) is dîn heorte *ubi est thesaurus tuus, ibi est cor tuum,* Mt. Kmbl. 6, 21 : 18, 20. Sceáwa þér dust and drýge bân, þér þér dû ǽr gesâwe fægre leomu, Blickl. Homl. 113, 21. (2) where the two forms are not separated, and may be translated by *where :*—God gefîsle mid flǽsce þér þér dæt ribb wæs, Gen. 2, 21. Mín þén biþ þér þér (þâr þâr, MS. A.) ic eom *ubi sum ego illic minister meus erit,* Jn. Skt. 12, 26. Man môt hine gebiddan, beó þér þér he beó, Homl. Skt. i. 13, 67. (d) with a demonstrative and relative force, as in modern *where, whither :*—Ðæt hí geworhten stǽnene weal dér se câsere hét eorþwall gewyrcan, Bd. 1, 12 ; S. 481, 8. Ðæt hí woldan andlyfne niman dér hí hí findan mihton, 1, 15 ; S. 483, 39. Se monlíca wunode þér hié strang begeat wîte, Cd. Th. 155, 4 ; Gen. 2567. Hê nǽnne ne mæg gebringan þér he him gehêt, Bt. 32, 1 ; Fox 114, 4. Ðâ becom hit þér se cyningc feóll, Homl. Skt. ii. 26, 208 : Beo. Th. 718 ; B. 356. Far þér dû freónda wêne, Exon. Th. 119, 29 ; Gú. 262. **II.** metaph. usages, (1) *there, in that case, then :*—þér dû cýþdest dæt dû nystest . . . þá dû sǽdest dæt . . . , Bt. 5, 3 ; Fox 14, 2. Hû ne is se anweald þonne þær nâuht is *not, then, power in that case nought ?* 16, 2 ; Fox 54, 7. Gedence hê dæt hê biþ self suíde gelíc dam ilcan monnum de hê dér dreátaþ, Past. 17 ; Swt. 117, 16 : 54 ; Swt. 425, 22 : 12 ; Swt. 75, 13 (but see note on the last passage). (2) þær *then when, when, þær when :*—Dý læs hié selfe âcwelen dér dér hié ða ôdre lâcniaþ *ne alios medendo ipsi moriantur,* Past. 48 ; Swt. 371, 11. Dér dér ûs God forbeád *cum nos Deus prohiberet,* 59 ; Swt. 451, 5. Sîn dîne suna and dîne dohtra geseald ôdrum folce þér dû on lôcie (*videntibus oculis tuis*), Deut. 28, 32. Hí clumiaþ mid ceaflum þér hí sceoldan clypian, L. I. P. 5 ; Th. ii. 308, 21. (3) *in case that, if :*—Ðér wê ûs selfum dêmden donne ne dêmde ûs nô God *si nosmet ipsos dijudicaremus, non judicaremur,* Past. 53 ; Swt. 415, 5. Dér mín âgen folc mê hýran cûdan *si plebs mea audisset me,* Ps. Th. 80, 13 : Bt. 32, 2 ; Fox 78, 1 : 37, 3 ; Fox 100, 4 : 36, 2 ; Fox 174, 5. Geornor wê woldon beón forsugiende, þér wê for eówerre âgenre gnornunge môste,

Ors. 3, 8 ; Swt. 122, 10 : Exon. Th. 375, 20 ; Seel. 141 : Cd. Th. 279, 7 ; Sat. 234. Ðér Moyses ne hulpe *si non Moyses stetisset,* Ps. Th. 105, 19 : Past. 46 ; Swt. 355, 4 : Cd. Th. 49, 24 ; Gen. 797 : Ors. 2, 4 ; Swt. 70, 5. Ðâr dû nû gemyndest ða word de ic dê sǽde on dære forman bêc, donne miht dû be dâm wordum genóg sweotole ongitan dæt dæt dû ǽr sǽdest dæt þú nystest *si superiora concessa respicias, ne illud quidem longius aberit, quin recorderis, quod te dudum nescire confessus es,* Bt. 35, 2 ; Fox 156, 21. **III.** preparing the way for the subject, *there :*—Ðâ com þær rên and þær (þâr, MS. A.) bleówun windas *et descendit pluvia et flaverunt venti,* Mt. Kmbl. 7, 25, 27. Þâ æt sumum cirre cômon þær sex scipu tô Wiht, Chr. 897 ; Erl. 95, 18. Eálâ hwæt þér wæs fæger eádmódnes gemêted on dære â clǽnan fǽmnan, Blickl. Homl. 9, 21. þér is mid Estum ân mǽgd, Ors. 1, 1 ; Swt. 21, 13. **IV.** in combination with suffixed prepositions the word has the force of a pronoun ; see the forms given as compounds (though the attachment is rather slight, see e. g. *þær-on*) which follow. [*Laym.* þar, þare, þar ; *Orm.* þær : *A. R.* þer : *Gen. and Ex.* ðor : *Hav.* þor, þore : *O. Sax. O. L. Ger.* thâr : *O. Frs.* thêr : *O. H. Ger.* dâr, dâra. Cf. *Goth. Icel.* þar.]

þǽr-âbútan ; *adv. Thereabout, about that place :*—On Antiochian byrig and dǽrâbútan gehwær, Homl. Skt. ii. 25, 595. Tô dâm de dârâbútan (-onbútan, MS. A.) stôdon *circumstantibus,* Mk. Skt. 14, 69.

þǽr-æfter ; *adv. Thereafter, after that :*—Gif se terninus gescýt on sumon dæge dære wucan, donne byþ se sunnandæg þérsæfter Eústerdæg, Lchdm. iii. 244, 18. Hié âhebbaþ hié ofer hiera hiéremenn, and ne ondrǽdaþ done dôm de dǽræfter fylgþ, Past. 19 ; Swt. 145, 9.

þǽr-æt ; *adv. Thereat :*—His horsbær wæs fram his discipulum gehealden, and monige unfrume dǽræt hǽlo onféngon, Bd. 4, 6 ; S. 574, 7.

þéran (? þærran) *to dry, wipe :*—Hê dâ hêt geótan wæter on mundleówe and ongan his þegna fêt þweán and þéran (*other MSS. have* þar an, þær ana ; *the Latin in* Jn. 13, 5 *is* extergere. The word intended seems to be one corresponding to Icelandic *þerra,* which, as well as the form *mundlaug,* the equivalent of the rather uncommon *mundleów,* the modern version in that language uses in this passage) mid ðý lîne, ðe hê wæs begyrded, Homl. Ass. 155, 103.

þǽr-big ; *adv. Thereby, by that* (person or thing) :—Gif hwâ gefare donne ne gestríene, gif hê bróðor lǽfe, fô se tô his wîfe. Gif hê donne bearn dǽrbig(-bié, Cott. MSS.), *by the wife*) gestríene, donne cenne hê dæt dam gefarenan brêder de hié ǽr ǽhte, Past. 5 ; Swt. 43, 14.

þǽr-binnan ; *adv. Therein :*—Philippus þérbinnan ne mehte, Ors. 3, 7 ; Swt. 112, 36. Se bisceop bebeád, dæt hî heora lâc geoffrodon binnon dam temple, and hêt hî dérbinnon andbidigan, Homl. Th. i. 450, 25.

þǽr-bufan ; *adv. Besides, over and above that :*—Hê cwæd : 'Biscepe gedafnaþ dæt hê sié tælleás.' Ðérbufan (v. 1 Tim. 3, 2 sqq. *for the additional remarks referred to*) is geteald hwelc hê beón sceal, gif hê untǽlwierde biþ, Past. 8 ; Swt. 53, 10.

þearf *need,* **þærf** *leavened,* **þærh.** v. þearf, þeorf, þerh.

þǽr-in ; *adv. Therein, wherein :*—Hê wæs on Simones hûse, þérin geát dæt wíf ða deórwyrþan smerenesse on his heáfod, Blickl. Homl. 73, 3.

þǽr-inne. v. þér-út, -úte.

þǽr-mid ; *adv.* **I.** *therewith, with that :*—Ðâ geseah hê treów licgende, and dæt lytel ; ongan ðâ þérmid delfan, Homl. Skt. ii. 23 b, 767. **II.** temporal, *straightway, at the same time :*—Ðâ forceáw hê his âgenan tungan and wearp hine dérmid on dæt neb foran, Bt. 16, 2 ; Fox 52, 25.

þǽr-néhst ; *adv. Next to that :*—Godes grid is ealra grida geornost tô healdanne, and þérnéhst þæs cynges, L. Eth. vii. 1 ; Th. i. 330, 3.

þǽr-of ; *adv. Thereof, of or from that :*—Genim ðás ylcan wyrte, wyrc clýþan þérof, Lchdm. i. 196, 23. Ðæt ic macige mete dînum fæder þérof *ut faciam ex eis escas patri tuo,* Gen. 27, 9.

þǽr-ofer ; *adv. Thereover, over that :*—Se fiicbeám ofersceadaþ dæt lond, dæt hit under him ne mæg gegrówan, ne hê self nǽnne wæsdm dérofer ne bireþ, Past. 45 ; Swt. 337, 12. Hig tôdǽldon hys reáf, and wurpon hlot þérofer, Mt. Kmbl. 27, 35.

þǽr-on ; *adv.* **I.** *thereon :*—Hê com tô dam treówe, sôhte wæstm déron, and nǽnne ne gemétte, Homl. Th. ii. 408, 1. Se dêma hêt wyrcan âne hencgene and hêt hôn done bisceop þéron, Homl. Skt. ii. 29, 253 : Blickl. Homl. 71, 7. **II.** *therein :*—Hér is ân lytele burg, dǽr ic mæg mîn feorh on generian. Þér is ân lytel, and deáh ic mæg déron libban, Past. 51 ; Swt. 399, 24. Âwyrtwala grǽdignysse of dînre heortan, and âplanta þéron ða sôþan lufe, Homl. Th. ii. 410, 2. Segeþ þéron (*in the book*), dæt sum ríce man wǽre, Blickl. Homl. 197, 27. **III.** *thereinto :*—Ðû ne cymst þéron *non ingredieris eam,* Deut. 32, 52. Ðâ hêt hê gefeccan ǽnne ǽrenne oxan, and ða hâlgan déron dón, Homl. Skt. ii. 30, 422. **IV.** *thereof :*—And hê ne cûðe nân þing þárou (cf. næs heora nân de þâr ǽnig þing on cûðe, 41, 24), Gen. 39, 23.

þǽr-onbútan. v. þér-abútan.

þǽr-ongeán ; *adv. There against, on the contrary ; per contra :*—Englas cýðaþ dîne dǽda beforan Godes gesihðe, and deófol âwrît þérongén ealle dîne misdǽda, Wulfst. 248, 21.

þær-oninnan; *adv. Therein, thereinto :*—Healreced gewyrcean, and þæroninnan eall gedǽlan, Beo. Th. 142 ; B. 71.

þǽr-onufenan. v. ufenan.

þǽr-onuppan; *adv. Thereupon, thereon :*—Đa hǽþenan byrnende glêda streáwodon, and ðæronuppan deófle offrodon, Homl. Skt. i. 23, 35 : 13, 25.

þǽr-riht; *adj. Straight :*—Þærrihtum *strictis* (*but the passage glossed is* strictis mucronibus ; *the glosser seems to have given two senses of the word, as he gives* evaginatis *besides the English word*), Hpt. Gl. 495, 50. v. next two words.

þǽr-rihte; *adv. Straightway, forthwith, at once, immediately :*—Ðærrihte *confestim, continuo, statim, protinus,* Ælfc. Gr. 38 ; Zup. 229, 16–230, 1 : *mox,* Zup. 241, 6 : *confestim,* Scint. 236, 1. Gê gemétaþ þærrihte (sôna *statim,* Mt. Kmbl. 21, 2) getîgedne assan, Homl. Th. i. 206, 10 : 494, 13 : Mt. Kmbl. 3, 16 : 27, 51. Đá cwæð hê : 'Geweorðe leóht.' And leóht wæs þærrihte geworden, Lchdm. iii. 232, 9. Ðǽrryhte æfter rehte sanctus Paulus *paulo post subdit,* Past. 51 ; Swt. 395, 26. Hê wæs hálig þærrihte, swá hraðe swá hê mann wæs, Homl. Th. i. 200, 8. Sôna t ðáriht *statim,* Mk. Skt. Rush. 1, 20. v. next word.

þǽr-rihtes; *adv. Straightway :*—Þærrihtes *protinus,* Hymn. Surt. 92, 37 : 113, 35.

þærscan, þærsc-wald, -wold, þærst. v. þerscan, þerscold, dærst.

þǽr-tô; *adv. Thereto.* (1) *marking position or order, next, then :*—Ðara is se forma Maximianus, ðærtô se ôþer Malchus, and se ðridda þærtô Martinianus, Homl. Skt. i. 23, 3–5. (2) *marking addition, besides :*—Ic gesett hæbbe wel feówertig lárspella and sumne eácan ðærtô, Ælfc. T. Grn. 14, 1. Hê nôwiht ágnes hæfde bútan his cyricean and ðærtô feówer æceras, Bd. 3, 17 ; S. 543, 32. (3) *marking association :*—Ðá stôd ðære sunnan cræt mid feówer horsum on áne healfe ; on ôðre healfe stôd ðæs mônan cræt and ða oxan ðærtô, Homl. Th. ii. 494, 24. Hê becwað his láford his beste scip, and ða segelgeráeda ðártô, Cod. Dip. Kmbl. iii. 351, 25. (4) *where movement, lit. or fig., is implied :*—Ðæt hê ûs gebringe tô his êcan gebeórscipe, se ðe ûs ðærtô geladode, Homl. Th. ii. 378, 6. Ðá dǽlde se cásere ðæt rîce on feówer, and sette ðærtô feówer gebrôðra, i. 478, 20. Ðone ôþerne ðe hine ðærtô neádode, Homl. Skt. ii. 25, 227. (5) *thereto, for that end :*—Ûres Hǽlendes gerîp mænigfeald is and feáwa wyrhtan þærtô, Homl. Skt. ii. 29, 129.

þǽr-tóeácan; *adv. Besides, moreover :*—Hê ûrum gyltum miltsaþ, and ðærtôeácan ðæt heofenlíce rîce behǽt, Homl. Th. ii. 84, 8. • On ðæt gerád ðæt se eorl him tô handan lêt Uescam, and þærtôeácan ðes cynges men sacleás beón môston, Chr. 1091 ; Erl. 227, 9.

þǽr-tógeánes; *adv.* I. *local, opposite :*—Ic ðærtôgeánes stand-ende *ego e contra stans,* Coll. Monast. Th. 22, 15. Seó heofen . . . and seó eorðe þærtôgeánes, Homl. Skt. i. 13, 166. II. *on the contrary :*—Se gôda man biþ ðæs Hálgan Gástes templ. Swá eác ðærtôgeánes se fordôna man biþ deófles templ, Homl. Th. i. 262, 17 : Wulfst. 59, 3. III. *as an equivalent, as a set off, in return :*—Wê habbaþ heom geunnen . . . and hî ûs þártôgênes gifeþ . . . , Chart. Th. 436, 11–20. Se cyng ðone castel æt Bures gewann . . . Ðærtôgeánes se eorl ge-wann ðone castel æt Argentæs, Chr. 1094 ; Erl. 230, 9. IV. *in reply, in refutation :*—Ðonne cweðaþ hî : 'Hû magon ðás bán beón geedcucode ?' Ac wê cweðaþ þærtôgeánes, ðæt God mæg eal ðæt hê wile, Homl. Th. i. 236, 8 : Homl. Skt. ii. 27, 162.

þǽr-under; *adv. Beneath :*—Ealle ða ðe ofer ôðre bióþ heáfda ðara ðe ðærunder bióþ, Past. 18 ; Swt. 131, 24.

þǽr-uppan; *adv. Thereupon :*—Him wæs his myxen forlǽten, ðæt hê þæruppan sittan mihte, Homl. Skt. ii. 30, 200.

þǽr-ût; *adv. Thereout, outside :*—Moyses oft eode inn and ût on ðæt templ, for ðæm hê wæs ðærinne getogen tô ðære godcundan sceáwunga, and ðærût (ðærûte, Cott. MSS.) hê wæs ábisgod ymb ðæs folces ðearfe, Past. 16 ; Swt. 101, 25.

þǽr-ûte; *adv. Without, outside :*—Náhton hié náþer ne þærinne mete ne þærûte freónd, Ors. 2, 8 ; Swt. 92, 34. v. preceding word.

þǽr-wiþ; *adv.* I. *therewith :*—Hî sint þærwiþ gemengde, Bt. 33, 4 ; Fox 130, 29. II. *in return, in exchange :*—Drîfaþ hider eówre orf and ic sylle eów þærwið mete *adducite pecora vestra, et dabo vobis pro eis cibos,* Gen. 47, 16.

þǽr-ymbe; *adv. About that :*—Hig tô lyt þærymbe þenceaþ, Wulfst. 273, 1.

þǽr-ymbûtan; *adv. Thereabouts :*—Hê (*the Roman name*) com tô Parþum . . . hê wæs ðærymbûtan manegum folce swîþe egefull, Bt. 2 ; Fox 64, 13. On gehwylce healfe ðærymbûtan *circumquaque,* Bd. 3, 17 ; S. 543, 26.

þæslǽcan; *p.* -lǽhte *To agree, accord, fit :*—Þæslǽcan *congruant,* Wrt. Voc. ii. 133, 40. Þæslǽcende *congruentes, convenientes,* Hpt. Gl. 508, 5. v. ge-þæslǽcan.

þæs-lîc; *adj.* I. *suitable, seemly, becoming, fit, meet, congruous :*—Gehŷþlic, þæslîc *vel* gescrǽpe *commodus,* i. *honestus, congruus, utilis, aptus,* Wrt. Voc. ii. 131, 81. Ðæslîc (*operae*) *pretium* (v. operae pretium, *congruum* neádþearflíc, Hpt. Gl. 477, 38), Anglia xiii. 33, 163. Gif lîf his on wyrþscype sî wel þæslíc *si vita honore sit condigna,* Scint. 125, 5.

Swîðe þæslíc anginn menniscre álŷsednysse wæs ðæt se engel wearð ásend fram Gode tô ðam mǽdene, Homl. Th. i. 194, 27. Hit nis ná gedafenlíc ne þæslíc, ðæt ic ðe swá grimlíce forworht eom, ðæt ic ðîne anlícnysse sceáwige, Homl. Skt. ii. 23 b, 434. Náht þæslíces deáðe *nihil dignum morte,* Lk. Skt. 23, 15. Forbǽrn mid fŷre þæslícum (*congruo*), Hymn. Surt. 29, 25 : Hpt. Gl. 443, 15. Ðæt hí Godes þênunge mid þæslícere endebyrdnysse gefyldon, Homl. Th. i. 508, 29. Þurh þæslíce deáðbôte *per dignam poenitentiam,* Scint. 40, 2. Swá þæslíc folc and· him swá gecwǽme, Lchdm. iii. 434, 8. Ða gelimplícan, þæslíc *congrua,* i. *convenientia,* Wrt. Voc. ii. 133, 39. On þæslícum tîman *competentibus horis,* 132, 66. Þæslícum *congruis, aptis, opportunis,* Hpt. Gl. 437, 63. Hwæt mæg beón þæslícre *what can be more fitting ?* L. E. I. 27 ; Th. ii. 424, 5. II. *comely, fair, elegant :*—Mid þæslíce·getingnysse *elegante* (*pulchra*) *sententia* t *peritia,* Hpt. Gl. 528, 3. Ðá geseah ic tŷn geonge men genôh þæslíce on líchaman, Homl. Skt. ii. 23 b, 370. III. *accordant, in agreement, in harmony :*—Ða cŷðnessa nǽron þæslíce . . . And hyra cŷðnys næs þæslíc *convenientia testimonia non erant . . . Et non erat conveniens testimonium illorum,* Mk. Skt. 14, 56–59. v. ge-, un-þæslíc, *and next word.*

þæslíce; *adv.* I. *in that way, so :*—Þæslíce *ita,* Hpt. Gl. 417, 8. Hê ne dyde þæslíce ǽlcere þeóde *non fecit taliter omni nationi,* Ps. Lamb. 47, 20 : Blickl. Gl. Hié ðæt gewinn ðá þæslícost angunnan ðe hí hit ǽr ne angunnen *they began the contest then just as if they had never begun it before* ; sic quasi ex integro nova bella nascuntur, Ors. 3, 11 ; Swt. 150, 31. II. *suitably, fitly, meetly :*—Se dǽdbôte þæslíce deþ se ðe gylt his bôte lahlíce beheófaþ *ille poenitentiam digne agit, qui reatum suum satisfactione legitima plangit.* Scint. 46, 1 : R. Ben. 70, 21. Hê hlôd ðu flôwendan láre ðe hé eft þæslíce bealcette, Homl. Th. ii. 118, 22. Þæslícor *dignius,* Germ. 390, 33. v. un-þæslíc.

þæslícness, e ; *f. Fitness, meetness, agreement with what is right :*—Ða hláfordas hé manode ðæt hí milde wǽron heora ðeówum mannum mid þæslícnysse (v. Col. 4, 1 : Domini, quod justum est et aequum, servis praestate), Homl. Th. ii. 326, 27. v. un-þæslícness.

þæsma, an ; *m. Leaven :*—Þæsma godcundre rihtwísnesse *fermentum divinae justitiae,* R. Ben. 10, 18. [*O. H. Ger.* deismo *fermentum : Du.* deesem *leaven.*]

þæt, *pron.* v. se.

þæt; *adv. After that, then :*—Ǽrest ymbe heora landgemǽra : and-lang Temese, ðæt (ðonne *in other MS.,* v. l. 8) up on Legean, L. A. G. i ; Th. i. 152, 18. Ǽrest on Ucingford . . . þæt tô brocenan beorge ; swá tô Wuduforda ; þæt tô Luttes beorge ; . . . þæt ðurh ðone môr, Chart. Th. 186, 3–12. This use is very common in charters.

þæt; *conj. That.* I. introducing substantive clauses, (1) where the clause is equivalent to a noun in the nominative, and (a) stands as the subject of the verb in the main clause :—Genôh byþ ðam leorningcnihte þæt (þætte, Lind. Rush.) he sŷ swylce his láreów, Mt. Kmbl. 10. 25. On ðæs engles wordum wæs gehŷred þæt þurh hire beorþor sceolde beón gehǽled eall wîfa cynn, Blickl. Homl. 5, 2 ; : Andr. Kmbl. 181 ; An. 91. Hû mænige geár synt ðu on ðysum wêstene eardodest, Homl. Skt. ii. 23 b, 513. (b) where þæt or hit stands as subject in the main clause :—Hû hit beón mæg, þæt se Hálga Gást cumeþ ufan on ðe, Blickl. Homl. 7, 35. Þæt geweorþeþ on dômes dæge, þæt hê cymeþ tô dêmenne cwicum and deádum, 11, 3 : Andr. Kmbl. 1147 ; An. 574. Ðæt is gedafenlíc, ðæt ðú Dryhtnes word healde, Elen. Kmbl. 2334 ; El. 1168. Mid ðæm cræfte ðe ðá scondlícost wæs ; þæt wæs, þæt hié from heora wíc-stôwum under ðære eorþan dulfon, Ors. 2, 8 ; Swt. 90, 29. Nis þæt feor heonon, þæt se mere standeþ, Beo. Th. 2729 ; B. 1362. Hit is for seofon and feówertigum wintrum, þæt ic of ðære hálgan byrig ût fôr, Homl. Skt. ii. 23 b, 516. (c) where it further explains a noun in the main clause :—Ðæs gástes wæstmas synd ða gôdan ðeáwas, þæt se man lufige God . . . and beó gesibsum, Homl. Skt. i. 17, 53. Wæs ðæt weatâcen wîde gefrêge, þæt hié ðæs cnihtes cwealm gesôhton, Andr. Kmbl. 2243 ; An. 1123. (2) where the clause is equivalent to a noun in the accusative, and (a) stands as object to the verb in the main clause :—God spræcð þæt hit gôd wæs, Gen. 1, 4. Ic wât ðæt ðú eart gecŷðed, Elen. Kmbl. 1627 ; El. 815. (b) where it is in apposition to þæt or hit standing as object in the main clause :—Gif his sunu and ðæs sunsunu þæt begyten, þæt hê swá micle landes habbaþ, L. Wg. 11 ; Th. i. 188, 23. Wê leornedon æt him ðæt wê flugen . . . and eác ðæt, ðæt wê his ege ûs ne ondrêden, Past. 3 ; Swt. 33, 23. Ic þæt wât, þæt ús gescildeþ weoruda Dryhten, Andr. Kmbl. 867 ; An. 434. (c) where it further explains the object of the verb in the main clause :—Helmstán ða undǽde gedyde, ðæt hê Æðerêdes belt forstæl, Chart. Th. 169, 19 : Elen. Kmbl. 989 ; El. 496. Ælc man sylle ðone áð, ðæt hê nelle þeóf beón, L. C. S. 21 ; Th. i. 388, 7. Ic bebeóde wundor geweorðan, ðæt ðeós onlícnes eorðan sêce, Andr. Kmbl. 1461 ; An. 731. (2 a) where the subject of the clause is omitted, and the clause taken with the accusative of the main clause is equivalent to the accusative and infinitive construction :—Þeóf ðone ðe wê geáxian, þæt ful sŷ, L. Ath. v. 1 ; Th. i. 228, 13. Woruld-gerihta ic wille þæt standan on ǽlcum leódscipe, L. Edg. S. 2 ; Th. i. 272, 23. Se ðe mon gesihþ ðæt stronglíc weorc wyrcþ, Bt. 16, 3 ; Fox 54,

29 : Homl. Th. i. 234, 3. Gif gē gesāwen hwelce mūs þæt wǣre hlāford ofer ōþre mȳs, Bt. 16, 2 ; Fox 52, 2. Ealle ða weód ðe hē geseó ðæt ðām æcerum derigen, 23 ; Fox 78, 23. Swā fela manna swā mán wite þæt ungelygne sȳn, L. Ath. iv. 1 ; Th. i. 222, 10. (3) where the clause is equivalent to a noun in the genitive or dative. (a) where in the main clause is a verb, verbal noun, or adjective taking after it such a case, and (a) where the substantive clause stands as object :—Ǣnig ne wēnde, ðæt hē lifgende land begēte, Andr. Kmbl. 755 ; An. 378. Saga þonc ðæt ic his mōdor geweard, Exon. Th. 13, 30 ; Cri. 210. Is nū þearf micel þæt wē wīsfæsta wordum hȳran, Andr. Kmbl. 2335 ; An. 1169. Ic āhebbe mīne hand . . . þæt ic ne underfō ānne þwang *I lift my hand (in testimony of this) that I will not take a thing*, Gen. 14, 23. Ðonne hió geornast biþ þæt heó āfǣre fleógan, Ps. Th. 89, 10 : Elen. Kmbl. 536 ; El. 268. (a 1) where the subject is omitted :—Ælc mynetere ðe man tīhþ þæt fals feoh slóge, L. Eth. iii. 8 ; Th. i. 296, 12. (b) where the main clause contains a case of a pronoun in apposition to the substantive clause :—Ne ðurfon wē ðæs wēnan, þæt ūs wuldorcyning wille eard ālēfan, Cd. Th. 272, 5 ; Sat. 115. Nǣnne mon ðæs ne tweóþ, ðæt þe seó strong, Bt. 16, 3 ; Fox 54, 28. Hié þæs ðone willan næfdon, þæt hié heora noman hié benǣmon, Ors. 2, 8 ; Swt. 94, 7. Gif þæs geweorðe gesīd-cundne mannan, þæt hē unrihthǣmed genime, L. Wih. 5 ; Th. i. 38, 4 : Andr. Kmbl. 615 ; An. 308. Ne magon wē þæs wrace gefremman, þæt hē ūs hafaþ ðæs leóhtes bescyrede, Cd. Th. 25, 16 ; Gen. 394. (c) where the clause explains the noun in the main clause :—Wolde ic ānes tō ðē cræftes neósan, þæt ðū mē getæhte hū . . ., Andr. Kmbl. 969 ; An. 485. Gē wiðsócon sóðe and rihte, ðæt in Bethleme bearn cenned wære, Elen. Kmbl. 781 ; El. 391. (β) where the clause is equivalent to a phrase, preposition and noun, with adjectival force, (a) defining the noun in the main clause :—Hit wæs ðā se tíma, þæt wīnberian rīpodon *now the time was the time of the first-ripe grapes* (A. V.), Num. 13, 21. Nis seó þrāh micel, þæt ðē wærlogan swencan mōton *the time of your affliction is not long*, Andr. Kmbl. 215 ; An. 108. Nis seó stund latu, þæt ðē wælreówe wītum belecgaþ, 2423 ; An. 1213. Nū is se dæg cumen, þæt ūre mandryhten mægenes behófaþ, Beo. Th. 5297 ; B. 2646 : Val. 1, 9. (b) in apposition to a pronoun in the main clause :—Wæs seó hwíl þæs lang, þæt ic Gode þegnode *the time of my serving God was long (or þæs = so ?),* Cd. Th. 37, 5 ; Gen. 585. **II.** introducing clauses expressing end or purpose, *that, in order that :*—Sete ðíne hand ofer hí þæt (þætte, Lind. Rush.) heó hāl sȳ *ut salva sit,* Mk. Skt. 5, 23. Ðǣr se bisceop oft wæs, þæt hē fullade ðæt folc, Bd. 2, 14 ; S. 518, 15. Se deófol genam ðæt wíf him tō gefylstan, þæt hē ðone hālgan wer ðurh hī geswice, Homl. Th. ii. 454, 1. Hí cōmon him tō, þæt hí hine geeósodon, 7. **II a.** with a negative, *that . . . not, lest :*—Ic ne underfō ānne þwang, þæt ðū ne secge eft (*ne dicas*) : 'Ic gewelegode Abram,' Gen. 14, 23 : Lk. Skt. 8, 12. Waciaþ and gebiddaþ þæt (þætte, Lind. Rush.) gē on costnunge ne gān (*ut non intretis*), Mk. Skt. 14, 38. Hig ne eodon intō ðam dōmerne, þæt (þætte, Lind. Rush.) hyg nǣron besmitene *ut non contaminarentur,* Jn. Skt. 18, 28. **III.** introducing clauses denoting result, manner, kind, degree, (1) where no demonstrative word in the main clause is antecedent to the subordinate clause, *that, so that, so as* (with infin.) :—Hū mihtest ðū sittan on middum gemǣnum rīce, þæt ðū ne sceoldest ðæt ilce geþolian ðæt ōðre men ? Bt. 7, 3 ; Fox 22, 17: Homl. Skt. ii. 23 b, 522. Asyrie hæfdon LX wintra and ān hund and ān þūsend, þæt hit nā būton gewinne næs, Ors. 1, 8 ; Swt. 42, 4. Nis nǣnigu gecynd, . . . ðæt hē ne sȳ fȳres cynnes, Salm. Kmbl. 847; Sal. 423. Hē rād þæt he wæs et Ceastre *he rode so that he was at Worcester*, Chart. Th. 71, 11 : Andr. Kmbl. 1576 ; An. 789 : 1474 ; An. 738. Man gecwǣman ne mæg twām hlāfordum ætsomne, þæt hē ne forseó þone ōðerne, Homl. Skt. i. 17, 220, 224. þær is ān mǣgd þæt hí magon cyle gewyrcan, Ors. 1, 1 ; Swt. 21, 13. Tō ðām handum ðæt ðæt fel of gǣþ, Lchdm. iii. 114, 3. Hwylc man is þonne ǣfre, þæt hē wēne . . . *whatever man is there (of such a kind), that he can suppose . . .*, Wulfst. 214, 14 : Cd. Th. 227, 20 ; Dan. 189. Hyge wæs oncyrred, þæt hié ne murndon, Andr. Kmbl. 73 ; An. 37. Gif mon sié dumb oþþe deáf geboren, þæt hē ne mæge his synna andettan, L. Alf. pol. 14; Th. i. 70, 14. Æfter ðære gebysnunge wurdon ārǣrede muneclíf mid ðære gehealdsumnysse, þæt hí drohtnian on clǣnnesse, Homl. Th. i. 318, 8. Gewunige his fæstende, þæt hē wite þæt seó mæsse sȳ gesungen *let him continue his fast so, that he may know the mass has been sung*, L. E. I. 39; Th. ii. 438, 3. Gif ceorlisc man geþeó, þæt hē hæbbe .v. hīda landes, L. Wg. 9; Th. i. 188, 5 : 10 ; Th. i. 188, 7. Ða Gotan lǣssan hwíle hergedan, þæt hié þurh Godes ege þæt hié nāþer ne þa burg ne bærndon ne þæs þone willan næfdon . . . , Ors. 2, 8 ; Swt. 94, 5. (1 a) where the subject of the clause is omitted :—Nemne him mon .v. men, and begite þara .v. .I., þæt him mid swerige, L. Ath. i. 9 ; Th. i. 204, 11. Hwylc is manna þæt feores neóte . . . oððe hwylc manna is þæt his āgene sāwle generige, Ps. Th. 88, 41 : Elen. Kmbl. 750 ; El. 375 : Exon. Th. 273, 20 ; Jul. 519. (2) where the clause stands as relative to a preceding demonstrative word :—Hē lǣrde hig swā þæt (*ita ut*) hig wundredon, Mt. Kmbl. 13, 54. Þæt wíte wæs tō þæs strang, ðæt æghwelc man sceolde mid sāre on þās world cuman, Blickl. Homl. 5, 28. Swelc wæs

þeáw hira þæt hié ǣghwylcne ellþeódigra dydon him tō mōse, Andr. Kmbl. 51 ; An. 26. Swā is þære menigo þeáw, þæt . . . , 355 ; An. 178. Dȳn mægen is swā mære, mihtig Drihten, swā þæt ǣnig ne wāt eorð-būende þa deópnesse Drihtnes mihta, ne þæt ǣnig ne wāt engla hādes þa heáhnisse heofena kyninges, Hy. 3, 31–35. Gif his sunu and his sunu þæt geþeóþ, þæt hí swā micel landes habban, L. Wg. 11 ; Th. i. 188, 10. (2 a) where the subject of the clause is omitted :—Nis nǣnig swā snotor, ne þæs swā gleáw, þæt āsecgan mæge, Cd. Th. 286, 12 ; Sat. 351. (*See also se*, V, *swā.*) **IV.** introducing clauses expressing cause, reason :—Hwæt þence gē betwux eów, þæt (fordon, Lind. : forþon þæt, Rush.) gē hlāfas nabbaþ *why reason ye among yourselves, because ye have brought no bread ?* Mt. Kmbl. 16, 8. Byþ ðe meorð, þæt ðū ūs on lāde līðe weorðe, Andr. Kmbl. 551 ; An. 276. **V.** where the main clause is not expressed, (1) in narrative :—Ǣrþon ðe seó heánnes ðæs walles gefylled wǣre, þæt se cyning ofslegen wæs and þæt ylce geweorc Ōswalde forlēt (some form equivalent to *it happened* appears necessary before þæt, which word there is nothing in the Latin to suggest : Priusquam altitudo parietis esset consummata, rex ipse occisus opus idem Osualdo reliquit), Bd. 2, 14 ; S. 517, 31. Ðā æt nȳhstan mid fultume his freónda þæt hē gelȳfde, 3, 22 ; S. 552, 26 : 3, 24 ; S. 556, 21 : 4, 27 ; S. 604, 32 (cf. 3, 9 ; S. 533, 16–19 : 4, 3 ; S. 569, 1–3). And þæs embe āne niht ðæt wē Marian mæssan healdaþ, Menol. Fox 39 ; Men. 20 and often. Nō þæt ðín aldor Godes goldfatu in gylp beran, Cd. Th. 262, 34 ; Dan. 754 : 288, 9 ; Sat. 378 : 304, 24; Sat. 634. (2) in the titles of chapters :—Caput II. Ðæt se ǣrra Rōmwara Cāsere Breotene gesōhte, Bd. 1, 2 ; S. 475, 2 and often. (3) in exclamations :—Wā þæt ðes tōwyrpþ Godes templ, Mt. Kmbl. 27, 40. Eálā þæt nān wuht nis fæste stondendes weorces, Bt. 9 ; Fox 26, 21 : Met. 9, 55 : 18, 1. Eálā, mín Drihten, þæt ðū eart ælmihtig, 20, 1. Eálā, þæt ic eam ealles leás ēcan dreámes, Cd. Th. 275, 7 ; Sat. 168. **VI.** where the construction is elliptical :—Þrý dagas tō lāfe syndon þæt hié þē willaþ acwellan *three days remain before the day comes on which they mean to kill you*, Blickl. Homl. 237, 26. [O. Sax. that: O. Frs. thet : O. H. Ger. daz : Icel. at.] v. þætte ; conj.

þætte (= þæt þe ; v. se, IV. 3); *pron.* **I.** as a relative, *that, which :*—Ðæt ðū hyra frumcyn ícan wolde, þte æfter him cenned wurde, Cd. Th. 236, 9 ; Dan. 318 : 245, 32 ; Dan. 472. Ðæt hī ne forleóse his dreámes blǣd and his dagena rím and his weorces wlite and wuldres leán, þætte heofones cyning syleþ tō sigorleánum, Exon. Th. 97, 11 ; Cri. 1589. Metod fēt eall þte grōweþ, Met. 29, 70. Ðætte tǣlwyrðes sié, ðæt hié ðæt tǣlen, Past. 28 ; Swt. 195, 24. Wíslíce gē dyde, þte mannum bedígled wæs on eorðan þæt gē þæt on heofenas sōhtan, Blickl. Homl. 201, 1. **II.** combining antecedent and relative, *that which, what :*—On hire wæs gefylled þte on Cantica Canticorum wæs gesungen, Blickl. Homl. 11, 15. Dō ā þætte duge, Exon. Th. 300, 10 ; Fä. 4. Wā ðæm ðe gemonigfealdaþ ðæte (ðætte, Cott. MSS.) his ne biþ, Past. 44 ; Swt. 329, 18.

þætte (= þæt þe ; cf. eác wæs ðæt ðe beforan ðæm temple stód ceác, Past. 16 ; Swt. 105, 1, *and :* Ða wæs þte scyttelas wurdan tōbrocene, Blickl. Homl. 87, 5. þætte *is used in the same way as* þæt, q. v.) ; *conj. That.* **I.** introducing substantive clauses. (1) where the clause is equivalent to a noun in the nominative, and, (a) where the subject of the verb in the main clause :—Cūþ is þte Drihten fæstte, Blickl. Homl. 27, 23 : 87, 5. Weard undyrne cūð, gyddum geómore, þætte Grendel wan wið Hróðgār, Beo. Th. 305 ; B. 151. (a 1) where the subject of the substantive clause is omitted :—Nis eów forboden, þte ǣhta habban, gif gē ða on riht strēnaþ, Blickl. Homl. 53, 27. (b) where þæt or hit stands as subject in the main clause :—Hit is āwriten ðætte Dauid, ðā hē ðone læppan forcorfedne hæfde, ðæt hē slōge on his heortan, Past. 28 ; Swt. 198, 16. Is þæt þeódnes gebod, þte . . ., Exon. Th. 202, 13 ; Ph. 69. Ðæt gelimpan sceal, þætte lagu flōweþ, 445, 2 ; Dom. 1. Ðæt gesȳne weard, þætte wrecend lifde, Beo. Th. 2517 ; B. 1256. (c) where it further explains the subject in the main clause :—Ne biþ swylc cwēnlíc þeáw . . . , þætte freoðuwebbe feores onsæce leófne mannan, Beo. Th. 3888 ; B. 1942. Treów wæs gecȳþed, þætte Gūðláce God leánode, Exon. Th. 129, 12 ; Gū. 420 : Cd. Th. 223, 3 ; Dan. 114. (2) where the clause is equivalent to a noun in the accusative, and (a) stands as object to the verb in the main clause :—Hēr sagaþ se godspellere, þte Hǣlend wǣre lǣded on wēsten, Blickl. Homl. 27, 3 : 41, 34. Hæbbe ic gefrugnen, þte is feor heonan æþelast londa, Exon. Th. 197, 19 ; Ph. 1. (b) where it is in apposition to þæt or hit standing as object in the main clause :—Hí þæt ne gelȳfdon, þte liffruma āhafen wurde, Exon. Th. 41, 16 ; Cri. 656. (c) where it further explains the object of the verb in the main clause :—Bodan sægdon sóðne gefeán, þætte sunu wǣre Meotudes ācenned, Exon. Th. 28, 24 ; Cri. 451. Men gesēgon þeódwundor micel, þte eorðe āgeaf ða hyre on lǣgun, 71, 15 ; Cri. 1156. (3) where the clause is equivalent to a noun in genitive or dative :—Gode ælmiehtigum sí ðonc, ðætte wē nū ǣnigne onstal habbaþ lāreówa, Past. pref. ; Swt. 4, 1. Se ðæs onsóce, þte sóð wǣre mǣre mihta waldend, Cd. Th. 244, 21 ; Dan. 451. **II.** introducing clauses expressing end or purpose, *that, in order that :*— Beforan ðæm temple stód ǣren ceác, ðætte menn meahten hira honda

ðweán, Past. 16; Past. 105, 1. Sprec tô ðinum discipulum, þte sý geblissad heora heorte, and hié sýn ofergytende ðisse sǽwe ege, Blickl. Homl. 233, 36. **III.** introducing clauses expressing result, manner, kind, degree. (1) where no demonstrative word is antecedent to the subordinate clause, *that, so that*:—Hí wénaþ þ hí mægen eall ðæs gôd gegaderian tôgædere, þætte náu bûton ðære gesomnunga ne sié, Bt. 24, 4; Fox 86, 3. Daniel sægde him wíslíce wereda gesceafte, þte sóna ongeat cyning ord and ende ðæs ðe him ýwed wæs, Cd. Th. 225, 28; Dan. 161. Woldon hié feorhleán fâcne gyldan þte hê þ dægweorc dreóre gebohte *so that he should pay for that deed with blood*, 187, 14; Exod. 151. (1 a) where the subject of the clause is omitted:—Nis ǽnig man þætte swá bereáfod sié, Met. 22, 49. Nǽnig manna is þætte âreccan mæg, Andr. Kmbl. 1091; An. 546: Cd. Th. 210, 2; Exod. 509. (2) where there is a demonstrative form as antecedent:—Hê beóþ swá geþwǽra, þætte nô þ ân þ hí magon geféran beón, ac ðý furþor þ heora nân bûton ôþrum beón ne mæg, Bt. 21; Fox 74, 17. Ðǽr wæs swíþe swéte stenc swá þte ealle ða slǽpan ðe ðǽr wǽron, Blickl. Homl. 145, 29. Ðinne lîchoman hié tôstenceaþ swá þte ðin blôd flêwþ ofer eorðan swá swá wæter, 237, 6. (2 a) where the subject of the clause is omitted:—þ nis nân man, þte sumes eácan ne þurfe, Bt. 24, 4; Fox 86, 6. **IV.** where the main clause is not expressed:—Ðonne hî niðer âstígaþ tô âðweánne hiera niéhstena scylda, hié beóþ onlícost suelce hî beren ðone ceák . . ., ðætte (*the case is such, that*) suâ hwelc suâ inweard higige tô gangenne on ða dura ðæs êcean lífes, hê ondette ælce costunge, Past. 16; Swt. 105, 14. Æfter ðæm ðe Rômeburg getimbred wæs twâ hunde wintra and IIIIX, þætte (*it came to pass, that*) Cambisis fêng tô Persa rîce, Ors. 2, 5; Swt. 78, 2 : 4, 1; Swt. 154, 2. And ðæs embe fíf niht ðætte fulwiht tiid êces Drihtnes tô ûs cymeþ, Menol. Fox 22; Men. 11, and often. Eálá þte ðis moncyn wǽre gesǽlig, gif heora môd wǽre riht, Bt. 21; Fox 74, 40. Ðætte oft ðæs láreówdômes ðénung biþ swíðe untælwyrðlíce gewilnad, Past. 7, arg.; Swt. 47, 20, and often.

-þafa. v. ge-þafa.

þafet[t]ere, es; *m.* One who agrees or consents, one who is remiss in allowing:—Ðý læs se ðafetere, se ðe wile forgiefan ðæt hê wrecan sceolde, tô êcum wîtum geteð his hiéremenn *ne rector remittendo quod ferire debuit ad aeterna supplicia subditos pertrahat*, Past. 20; Swt. 149, 21. Ðæt hê swá stiére ðǽm ungeðyldegum irsunga, swá hê ðone hnescan ðafettere on receelêste ne gebrenge *sic ab impatientibus extinguatur ira, ut tamen remissis ac lenibus non crescat negligentia*, 60; Swt. 453, 25.

þafian; *p.* ode. **I.** to consent to, agree with, approve of, assent to, allow, permit. (a) with accusative:—Ic Beágmund ðis ðeafie and wríte, Chart. Th. 472, 22, 24, 28, 19, and often. Swâ hwylc swâ morþorslege þafaþ and hine man ðonne fremmeþ *quicunque ad homicidium consenserit, et id postea factum fuerit*, L. Ecg. C. 22; Th. ii. 148, 14. Heó hine monede ðæt hê weoruldhâd forlǽte and munuchâde onfênge. Ond hê ðæt well ðafode *he readily consented to it*, Bd. 4, 24; S. 598, 3. Ðê sint tû gearu swâ líf swâ deáð, swá ðe leófre biþ tô geceósanne; cýð hwæt ðû ðæs tô þinge þafian wille *say which alternative you mean to accept*, Elen. Kmbl. 1213; El. 608. Nǽfre ic ðæs þeódnes þafian wille mǽgrǽdenne *I will never consent to marriage with the prince*, Exod. Th. 249, 8; Jul. 108. (b) with dative:—Gê þafiaþ eówer fædera weorcum *consentitis operibus patrum uestrorum*, Lk. Skt. 11, 48. Ðafande woeron feh him tô seallanne *pacti sunt pecuniam illi dare*, Lind. 22, 5. (c) with a clause:—Gif hê þafaþ ðæt hê ût gâ of minstre *si consenserit, ut egrediatur de monasterio*, R. Ben. Interl. 98, 17. Þafodest ðû ðæt mê þeówmennen drehte, Cd. Th. 135, 21; Gen. 2246. Þafa ðæt ic ût âðô ðæt mot of ðínum eágan *sine eiciam festucam de oculo tuo*, Mt. Kmbl. 7, 4. Ða eorlas þafigan ne woldon ðæt hié forlêton leófne láreów, Andr. Kmbl. 804; An. 402. **II.** to submit to, bear, suffer, endure :—Ðê þincþ se earmra se ðæt yfel ðeþ ðonne se ðe hit þafaþ *miserior tibi injuriae illator, quam acceptor esse videretur*, Bt. 38,6; Fox 208, 19. Sum gewealdenmôd þafaþ in geþylde ðæt hê sceal, Exon. Th. 297, 20; Crä. 77. Eal ðæt hê for ûs þafode and ðolode, Wulfst. 23, 22. Ða eádigan martyras mænigfealde earfoðnyssa ðafedon, Homl. Skt. i. 23, 12, 89. Se þeódcyning ðafian sceolde Eofores ânne dôm, Beo. Th. 5919; B. 2963. Þafigan, Cd. Th. 227, 22; Dan. 190. Ic sceal þinga gehwylc þolian and þafian on ðinne dôm *I must suffer and submit to everything, as you decide*, Exon. Th. 270, 16; Jul. 466. Hié derede ægðer ge þurst ge hǽte, and ealne ðone dæg wǽron ðæt þafiende, Ors. 5, 7; Swt. 230, 17. **III.** to bear with, tolerate :—Hê ilde and ðafode ða scylda and ðeáh hê him gecýðde *et dissimulavit culpas, et innotuit*, Past. 21; Swt. 151, 22. [*Non me demergat tempestas* louerd ne þaue þu þat storm me duue, O. E. Homl. ii. 43, 15. ȝef ha ne letted me nawt, ah þauied ant þolied, Marh. 15, 19. Ne mahe ȝe nawt do me, bute þet he wule þeauien and þolien ow to donne, Jul. 19, 9. Þatt Godd ne þole nohht ne þafe laþe gastess to winnenn oferrhannd off uss, Orm. 5457. Euerilc husfolc ðe mai it ðauen on ger sep oðer on kide hauen, Gen. and Ex. 3139. Was neuere non þat mouhte þaue Hise dintes, noyþer knith ne knaue, Havel. 2696.] v. geþafian.

þafung, e; *f.* Consent, permission :—Be bisceopes þafunge *cum consensu episcopi*, L. Ecg. C. 26; Th. ii. 152, 3. Be his þafunge *permissionem suam*, R. Ben. Interl. 77, 6. Ðû wêndest ðæt seó weord ðas

woruld wende bûton Godes geþeahte and his þafunge, Bt. 5, 1; Fox 8, 32. Ne mæg se deófol mannum derian bûton Godes ðafunge, Homl. Skt. i. 17, 196. Þet weas mid Earnulfes þafunge (geþafunge, MS. A.), Chr. 887; Erl. 87. 3. [Vlesches fondunge goð to uord upe me þurch min þafunge, A. R. 344, note.] v. ge-þafung.

þage, þáh *though.* v. þæge, þeáh.

þametaþ ? :—Flôdas hafettaþ (þametaþ, MS. M.) handum *flumina plaudent manu*, Ps. Spl. 97, 8.

þan, þon; *adv.* **I.** *then, from that time, after that* :—Wæs wyrd ungemete neáh . . . nô þon lange wæs feorh æþelinges flǽsce bewunden, Beo. Th. 4838; B. 2423. [*Goth.* þan: *O. Sax. O. Frs. O. L. Ger.* þan.] **II.** *so, as* :—Wiþ ðæs ic wât ðu wilt higian þon ǽr þe ðú hine ongitest *towards it I know thou wilt hasten as soon as thou perceivest it*, Bt. 11, 2; Fox 34, 8. [*Cf. O. Sax.* than lango þe hê môsta is juguði neotan, Hêl. 3498.] **III.** with comparatives, in negative sentences. (a) with adjectives, (α) followed by ðonne or ðe, *any* :—Gif hió bearn gestríéne, nǽbbe ðæt ðæs ierfes þon (þe, MS. H.) mâre þe sió môdor *if she have a child, it shall not have any more of the property than the mother*, L. Alf. pol. 8; Th. i. 66, 20. On ôðrum ærne ðæt næbbe þon mâ dura ðonne sió cirice, 5; Th. i. 64, 15. Ne eart ðû þon leófre, ðonne se swearta hrefn ðon mâ ðe dear, *than the black raven*, Exon. Th. 370, 4; Seel. 52. Nǽfre hlísan âh meotud þan mâran þonne hé wið monna bearn wyrceþ weldǽdum, 191, 10; Az. 86. Hê ne úþe ðæt ǽnig ôþer man ǽfre mǽrða þon mâ gehêdde, ðonne hê sylfa, Beo. Th. 1012; B. 504. (α 1) where þon is preceded by wihte, *any at ull* :—Ne môt hê ðara hyrsta lǽdan of ðisse worulde wuhte þon mâre ðonne hê hider bróhte (cf. ne lǽt hê his nânwuht of ðís middanearde mid him mâre ðonne hê bróhte hider, Bt. 26, 3; Fox 94, 15-17), Met. 14, 10. (β) where the comparative takes the dative after it :—Hê ðâm ðe on sceare mâran wǽron on ðâm mægnum eáþmôdnesse and hýrsumnesse nôhte ðon læssa wæs *in respect to the virtues of humility and obedience he was not any less than those who were greater in the matter of the tonsure*, Bd. 5, 19; S. 637, 18. (γ) where neither particle nor case follows the comparative :—Næs ðâ wordlatu wihte þon mâre þæt se stân tôgân *then was there not any more delay at all in obeying the command, so that the stone split open*, Andr. Kmbl. 3043; An. 1524. Nâhte ic ðînre nǽfre miltse þon mâran þearfe *never had I any greater need of thy mercy* (*than I now have*), Judth. Thw. 22, 35; Jud. 92. (b) with adverbs, (α) followed by ðe :—Hê nât hwæt him tôweard biþ, þon mâ þe ðû wistest *he knows not what will happen to him any more than thou knowest*, Bt. 11, 1; Fox 32, 14. Wê his ne gefrêdaþ, þon mâ ðe mon his feax mæg gefrêdan bûtan his felle, Past. 18; Swt. 139, 20. Him ðæt nô ne derede, ðon mâ ðe ceald wæter, Shrn. 83, 17: Exon. Th. 364, 33; Wal. 80. (β) without ðe :—Ic ða word gehýrde and nôht ðon ǽr ðære ærninge blon *ego audiens, nihilominus coeptis institi vetitis*, Bd. 5, 6; S. 619, 15. Hê georne wiðsôc Iôsepes hûse ne þon ǽr geceás Effremes cynn *he utterly refused the house of Joseph, nor any more readily did he choose the race of Ephraim*, Ps. Th. 77, 67. Ǽfre ic ne hýrde þon cymlícor ceól gehladenne *I have not ever heard of a vessel any more fairly laden*, Andr. Kmbl. 721; An. 361. Ðâ ne wolde se pâpa ðæt geþafigean ne ða burhware ðon mâ *etsi pontifex concedere voluit, non tamen cives Romani potuere permittere*, Bd. 2, 1; S. 501, 33. Ne bewerede Penda ðon mâ gif hwylce men woldan Godes word lǽron ðæt hî ne môstan *nec prohibuit Penda, quin etiam verbum, si qui vellent audire, praedicaretur*, 3, 21; S. 551, 23. Ne ðon mâ se ðe gehât gehǽt, ne wêne hê ðæt hê sié â ðý neár hefonríce, gif hê hine from went ðâm gehâtum *nor any more let him that vows a vow suppose that he be ever the nearer heaven, if he turns from those vows*, Past. 51; Swt. 403, 2. Ne biþ sond þon mâ wið micelne rên hûses hirde *nor any more is the sand a guard for a house against much rain*, Met. 7, 20: 8, 23: 11, 69. Bûtan ðû ûsic þon ôfostlícor hreddan wille *if you do not save us any quicker*, Exon. Th. 17, 18; Cri. 272. (β 1) where âwiht or wuhte precedes þon, *any at all* :—Ðǽr nǽnegu biþ niht on sumera, ne wuhte þon mâ on wintra dæg tôteled tîdum, Met. 16, 14: 20, 108. Âwiht þon mâ, Ps. Th. 63, 7. [*O. Sax.* ni . . . than mêr the *not . . . any more than*. Cf. *Goth.* ni . . . þana mais: *O. H. Ger.* dana mêr.]

þân; *adj.* Moist :—Þa þânan *madentia*, Wrt. Voc. ii. 93, 71: 57, 12. [Thone, thoney=damp, is found as a word in E. Norfolk and of Midland counties in Marshall's Rural Economy (1795-6), and in Ray's North-country words (1691); v. E. D. S. Pub. Reprinted Glossaries, B. 3, 5, 15.] v. þânian, þanan.

þanan, þonan (-on, -un, -en); *adv.* **I.** with demonstrative force, *thence* :—Þanan *illic(-inc?)*, Wrt. Voc. ii. 110, 55. Ðonan *illinc*, 44, 54. (1) marking the point from which motion takes place :—Hê þanon (þonan, Rush.: þona, Lind.) eode, Mt. Kmbl. 4, 21. Þanon hé com on Iudêisce endas *inde exsurgens uenit in fines Iudaeae*, Mk. Skt. 10, 1. Hê wand up þanon, Cd. Th. 29, 7; Gen. 446. Hê fôr þanun (þanon, MS. A.: ðonan, Rush.), Mt. Kmbl. 11, 1. Þanun (-en, MS. A.), 12, 9. Monige þonan gewitan, Bd. 4, 25; S. 601, 34. Þonan, Exon. Th. 235, 9; Ph. 554. Hê ða hâlgan sâuwla þonon âlêdde, Blickl. Homl. 67, 19. Hât mîn blôd þonon âdrýgan, 183, 27. (1 a) followed by a relative particle, the two words together having force of relative :—Þider cuman, þonan þe

hit ǽr com, Bt. 25; Fox 88, 31. On ða rîcu, þonon þe hê ǽr sended wæs, Blickl. Homl. 9, 25. Ðý læs hê áfealle ðonon ðe hê fæsdlícost tô hopian scolde, Past. 51; Swt. 395, 11. (2) marking the point from or in regard to which direction or position is estimated :—Ðanon ðe hê blǽwþ him byþ nama geset *from the quarter that the wind blows is a name made for it*, Lchdm. iii. 274, 11. Ðæt flôd ys þanon tôdǽled on feówer eán *from that point the stream runs in four separate channels*, Gen. 2, 10. Ðá hê on botme stôd, ðá him þúhte ðæt þanon wǽre tô helle duru hund þúsenda míla gemearcodes, Cd. Th. 310, 7; Sat. 722. Hê ðǽr rom geseah unfeor þanon standan, 177, 9; Gen. 2927: Beo. Th. 3615; B. 1805. God wæs mín on ða swiðran, ðanon ic ne wende ǽfre tô aldre onsión míne, Elen. Kmbl. 696; El. 348. Hê sǽde ðæt ðæt land sié swíþe lang norþ þonan *he said that from that point the country stretches very far to the north*, Ors. 1, 1; Swt. 17, 4. Seó burh is west þonon from ðære stôwe on ánre míle, Blickl. Homl. 129, 3. (3) marking the place from which an action or operation proceeds :—Nalles þanon (*from hell*) gehéran in heofonum hálige dreám, Cd. Th. 284, 26; Sat. 327. Gesæt him be healfe . . . , þanon básnode hwæt him gúðweorca gifeðe weard, Andr. Kmbl. 2131; An. 1067. Hié ealle on yppan wunedon, þonen bídende ðæs Hálgan Gástes, Blickl. Homl. 133, 26. God wunaþ on ðære ceastre his ánfealdnesse; ðonan hé ðélþ manega gemetgunga eallum his gesceaftum, and þonon (-an, Cott. MS.) hê welt ealra, Bt. 39, 5; Fox 218, 18–21. Þonan án cyning rícsaþ . . . , ealra gesceafta waldeþ (cf. þǽr rícsaþ án cyning, se hæfþ anweald eallra ôþra cyninga *heic regum sceptrum dominus tenet*, Bt. 36, 2; Fox 174, 17), Met. 24, 31. (4) marking source, origin :—Mænige gefôþ hwælas and micelne sceat þanon (*inde*) begytaþ, Coll. Monast. Th. 25, 3. Þanon wôc fela geósceaftgásta, Beo. Th. 2535; B. 1265. Þonon Eómer wôc, 3925; B. 1960. Ðære wrænnesse wôðþræg . . . gedræfþ sefan ingehygd; þonan mǽst cymeþ unnetta saca, Met. 25, 43. (5) marking cause, reason :—Ne gehýrdest ðú Drihten cweþende, for þon þe ic eów sende swá swá sceáp on middum wulfum? þanon wæs geworden . . . ic bæd úrne Drihten ðæt hê hine æteówde, and hraþe hê mé hine æteówde, and hê mé tô cwæð . . . 'Ic sende tô ðé Andreas,' Blickl. Homl. 237, 30. Þonne God gangeþ for his folc . . . þanon eorðe byþ onhréred, Ps. Th. 67, 8. (6) temporal, *from that time, after that* :—Ðæt hé unæþele á forð þanan wyrð, Met. 17, 28. Hê forlæt his æþelo, and ðonan wyrþ anæþelad ôþ ðæt hé wyrþ unæþele, Bt. 30, 2; Fox 110, 22. Þanon forþ *exhinc*, Anglia xiii. 393, 404: *de cetero*, 439, 1059. Ðanon forþ *exinde*, 444, 1130. Hê ða gefeán ðæs heofonlíces éðles þanon forð geseón ne mihte, Wulfst. 1, 6. Siððan ongon Cain ceastre timbran . . . þanon his eaforan ǽrest wôcan bearn from brýde on ðam burhstede. Se yldesta wæs Irad *afterwards did Cain build a city . . . Not till after that were children born to his son (Enoch) in that town. The eldest was Irad* (v. Gen. 4, 17, 18), Cd. Th. 65, 4; Gen. 1061: 210, 14; Exod. 515. II. with relative force, *whence*. (1) referring to the point from which motion takes place :—Ic gecyrre on mín hús þanon (*unde*) ic út eode, Mt. Kmbl. 12, 44. Cunnaþ fýr eft tô his éðle, ðanon hit ǽror cwom, Salm. Kmbl. 834; Sal. 416. Hê gewát on Hibernia, ðonan hé ǽr com, Bd. 4, 25; S. 600, 13. Þonan, Exon. Th. 17, 12; Cri. 269. Hê tô ðæm fæderlícan setle eode, þonon hê nǽfre onweg ne gewát, Blickl. Homl. 117, 1. (2) referring to the point from or in regard to which direction or position is estimated :—On heofonas, þonon hé nǽfre won wæs, Blickl. Homl. 131, 17: 91, 5. (3) referring to the place from which an action or operation proceeds :—Hê hine sylfne hefeþ on heánne beám, þonan ýþast mæg síð bihealdan, Exon. Th. 205, 15: Ph. 113. (4) referring to source or origin :—Ðæt sum gestreón mé ic begyte þanon ic mé áféde *ut aliquod lucrum mihi adquiram, unde me pascam*, Coll. Monast. Th. 27, 21. (5) referring to cause or reason :—Hê má gewunode on his smiþþan sittan, ðonne hê wolde on cyricean singan. Þonon him gelamp ðæt sume men gewuniaþ cweþan *magis in officina sua residere, quam ad psallendum in ecclesia concurrere consuerat. Unde accidit illi, quod solent dicere quidam*, Bd. 5, 14; S. 634, 18. III. in correlative combinations :—Ðæt mé þincþ wiþerweard þing . . . ðætte þonan ðe hí teohhiaþ ðæt hí scylan eádigran weorþan, ðæt hí weorþaþ ðonan earmran, Bt. 26, 2; Fox 92, 24–27. Ðonon ðe hí útan bióþ áhæfene, ðanon hié bióþ innan áfeallene, Past. 50; Swt. 391, 12. [*O. Sax.* thanan: *O. Frs.* thana: *O. H. Ger.* danân *inde, illinc*.] v. *next word*, and þe, II. 1.

þanane; adv. I. *thence*. (1) local :—Ne gǽst ðú þanone (-ene, MS. A.: þonan, Rush.) *non exies inde*, Mt. Kmbl. 5, 26: Lk. Skt. 12, 59. Ðá gewát ic þanone, Homl. Skt. ii. 23 b, 422. Ðanonne, Judth. Thw. 23, 21; Jud. 132. (2) temporal, *after that* :—Rursum, dein vel þonane, Wrt. Voc. ii. 139, 63. (3) causal :—Ðonne mon lǽt tôslúpan ðone ege . . . , ðonne wierþ gehnescad ðonone sió ðreáung ðæs anwaldes, Past. 40; Swt. 289, 3. II. *whence* :—Wígheard tô Rôme wæs onsended, ðonone hí hider onsendon gewritu, Bd. 3, 29; S. 561, 3. [*O. L. Ger.* thanana: *O. H. Ger.* danana.] v. *preceding word*.

þanan-forþ. v. þe I. 6.

þanan-weard; adj. *Moving thence* :—Bebeád hê him, ðæt hê geara wiste, ðæt hê hine nǽfre underbæc ne besáwe, siþþan hê þononweard wǽre *lex dona coërceat, ne dum Tartara liquerit, fas sit lumina flectere*,

Bt. 35, 6; Fox 170, 9. [Cf. þeone Godd warp hire (*pride*) sone se ha iboren wes; & as ha nuste hwuch wei ha come þeneward, ne con ha neauer mare ifinden na wei aȝainward, H. M. 43, 8.]

þanc, es; m. I. *thought* :—On ðeóstrum ne mæg þances gehygdum ǽnig wíslícu wundur oncnáwan, Ps. Th. 87, 11. Þances gleáw þegn, Andr. Kmbl. 1113; An. 557. Þonces gleáw, Exon. 207, 19; Ph. 144. Þurh gemynda spéd, môde and dǽdum, worde and gewitte, wíse þance, Cd. Th. 118, 1; Gen. 1958. Ge þanc ge þeáwas, word and weorc georne gerihtan, L. P. M. 3; Th. ii. 288, 16. Drihten, úre môd gebíg, þanc and þeáwas on ðín gewil, Hy. 7, 78. [Þu þi þanc (þoht, 2nd MS.) al forhele, Laym. 4360. He put a swuc þonc in hire heorte, A. R. 222, 25.] II. *kindly thought, favour, grace* :—Oft hé þearfendra béne þance (*graciously*) gehýrde, Ps. Th. 101, 15. Ðis is landa betst, ðæt wit þurh uncres hearran þanc habban môston (cf. hie thuru thes késures thank ríki habda, Hél. 66), Cd. Th. 49, 22; Gen. 796. III. *agreeableness, pleasure, satisfaction; in phrases*, (a) æfter þance *according to what is agreeable, agreeably, pleasantly* :—Hé his líchoman forwyrnde woruldblissa . . . Him wæs Godes egsa mára in gemyndum ðonne hê menniscum þrymme æfter þonce þegan wolde *he refused his body worldly delights . . . There was too much fear of God in his mind for him to partake of human glory, following the dictates of pleasure*, Exon. Th. 112, 7; Gú. 140. (b) on þanc, tô þances, tô þance *to the satisfaction of a person, so as to please*, cf. *O. Frs.* tô thanke: *Icel.* til þakka eins, i þökk við einn *to one's liking: Ger.* zu Danke :—Hié nánwuht gódes ne magon Gode bringan tô ðances *nullum boni operis Deo sacrificium immolant*, Past. 46; Swt. 349, 8. Ðú hæfst tô þance geþénod ðínum hearan, hæfst ðé wið Drihten dýrne geworhtne (cf. habda ira Drohtine gethionôd te thanka, Hél. 506), Cd. Th. 32, 20; Gen. 506: Beo. Th. 763; B. 379. Se bisceop ðæs getíðode on ealra ðæra witena þanc *the bishop granted it to the satisfaction of all the witan*, Chart. Th. 303, 2. Cúð dyde Nergend ðæt Noe ðæt gyld on þanc ágifen hæfde (*the sacrifice had been well pleasing*), Cd. Th. 91; Gen. 1506. Him wíf sunu on þanc gebær *to his delight his wife bore him a son*, 167, 31; Gen. 2774. Ic ðé on hleóðre hearpan gecwéme . . . Ic ðé on þanc môte sealmas singan, Ps. Th. 107, 1: Andr. Kmbl. 3242; An. 1624. On þonc, Exon. Th. 402, 7; Rä. 21. 26. Hê of stánclife burnan leódum lǽdde on leófne þanc, Ps. Th. 135, 17. (c) on þance *pleasing, agreeable, grateful*, cf. *this thionost is im an thanke*, Hél. 118: *O. H. Ger.* in thanke, danche *gratus* :—On ðonce mé syndon ðíne word and ðín lufu *gratias ago benevolentiae tuae*, Bd. 2, 12; S. 513, 23. Mé is ðín cyme on myclum ðonce *gratus mihi est multum adventus tuus*, 4, 9; S. 577, 21: Exon. Th. 387, 22; Rä. 5, 9. Ðonne wǽron ǽgher gôde, ge ða ærran ge ðás æfterran, and nǽron nǽðere an þance *quid aliud colligi datur, nisi semper bona esse, sed ingrata*? Ors. 2, 5; Swt. 86, 10. Nǽnegum þúhte dæg on þonce gif sió dimme niht ǽr egesan ne brôhte (cf. þancwyrþre biþ ðæs dæges leóht for ðære egeslícan þióstro ðære nihte, Bt. 23; Fox 78, 28), Met. 12, 16. IV. *thanks* :—Gode ælmiehtegum sí ðonc, ðætte . . . , Past. pref.; Swt. 2, 18. Him ðæs þanc séd, Cd. Th. 68, 13; Gen. 1116: Hy. 7, 58: Andr. Kmbl. 2900; An. 1453. Ðisse ansýne Alwealdan þanc gelimpe, Beo. Th. 1861; B. 928. Swǽ gelǽrede biscepas, swǽ suǽ nú Gode ðonce wel hwǽr siendon, Past. prep.; Swt. 9, 4; 1; Swt. 27, 3: Andr. Kmbl. 2302; An. 1152. Ða gesceafta nǽron nánes ðonces ne nánes weorþscipes weorþe, gif hí heora unwillum hláforde hérden, Bt. 35, 4; Fox 160, 20. Hié ða lác þégon tô þance (*thankfully, gratefully*), Andr. Kmbl. 2225; An. 1114. Hé him dǽda leán gieldeþ, ðám ðe his giefe willaþ þicgan tô þonce, Exon. Th. 109, 26; Gú. 96. Hié on þanc curon ædelinges ést *they accepted Lot's kindness with thanks*, Cd. Th. 147, 20; Gen. 2442. Þanc ágan, habban *to have thanks, be thanked* for something (*gen., prep., or clause*) :—Ðæs áge þrynesse þrym þonc, Exon. Th. 37, 27; Cri. 599. Hafa árna þanc, Cd. Th. 147, 6; Gen. 2435. Hæfþ se þeówa ǽnigne þanc, forþam ðe hê dyde ðæt him beboden wæs, Lk. Skt. 17, 9. Þanc hafa, Iofes, ðæt ic ða môste oferwinnan, Ors. 4, 1; Swt. 156, 27. Þanc cunnan, witan [cf. He cuðe him ðerof wel gret ðhanc, Gen. and Ex. 1659. Sche . . . can hem therfore as moche thank as me, Chauc. Kn. T. 950] *to feel grateful, be thankful* for something (*gen.*) :—Ðám ðe þonc Gode witon ne cúþun, ðæs ðe hé on ðone hálgan beám áhongen wæs *to those who felt no gratitude to God for his sufferings, for his being hung on the cross*, Exon. Th. 67, 22; Cri. 1092: 74, 29; Cri. 1213. Ðú Waldende ðínre álýsnesse þonc ne wisses, 90, 5; Cri. 1474: 85, 5; Cri. 1386. Ðú ðæs ealles ǽnigne þonc ðínum nergende nysses on môde, 91, 29; Cri. 1498. God nele, ðæt him man his gifena þanc nyte, Wulfst. 261, 17. Þancas, þanc dôn *to give thanks*; gratias agere :—þanca dǽd *gratiarum actio*, Scint. 50, 5 :—Hé Gode þancas dyde *gratias agens*, Mk. Skt. 14, 23: Lk. Skt. 22, 17. Ðé ic þancas dô, forðam ðe ic ne eom swylce ôðre men, 18, 11: Jn. Skt. 11, 41: Scint. 50, 2, 3. Þanc ic dô, ðú gôda hyrde, forðon ðás sceáp mé efenþrowiaþ, Blickl. Homl. 191, 24. Þanc gegyldan [cf. Me him ne yeldeþ þonkes of his guodes, Ayenb. 18, 6] *to pay thanks, give from a feeling of gratitude, to reward* a service :—Him God wolde æfter þrowinga þonc gegyldan, ðæt hê martyrhád gelufade, sealde him snyttra, Exon. Th. 130, 23; Gú. 442. Þanc, þancas secgan [cf. To zigge grat þank, Ayenb. 18, 17] to

express *thanks* for something (*gen.*, *prep. clause*), *give thanks* :—Hē Gode his gōda ðanc sægde (*gratias agebat*), Bd. 3, 12; S. 537, 26: Cd. Th. 16, 4; Gen. 238: Andr. Kmbl. 2937; An. 1471: Blickl. Homl. 103, 25: 217, 34. Ne sæcgaþ ūs nēnne þanc, Homl. Skt. i. 3, 332. Saga ēcne þonc, ðæt ic his mōdor gewearþ, Exon. Th. 13, 28; Cri. 209. Wē sculon simle secgan Gode ðoncas for eów, Past. 32; Swt. 213, 10. ¶ *Þances*, *genitive*, *used alone or in combination with noun or pronoun*, *and having adverbial force* (cf. O. L. Ger. thankis *gratis* : O. H. Ger. danches *sponte, ultro, gratis*). (1) *thanks to a person on whom a result depends*, *by* (*one's*) *grace*, *favour* :—Ðæt næs nā eówres þances ac þurh God *it was not thanks to you but by God's will*; non vestro consilio sed Dei voluntate, Gen. 45, 8. Sege mē hwæþer se ðīn wela ðīnes þances swā deóre seó ðe for his āgenre gecynde *tell me whether that wealth of thine is so precious thanks to thee or from its own nature*, Bt. 13; Fox 38, 6. Hié rīcsedon næs ðeál mīnes ðonces *ipsi regnaverunt, et non ex me*, Past. 1; Swt. 27, 14. Godes þonces *by God's grace*, Chr. 897; Erl. 94, 29 : 883; Erl. 83, 18. (2) *where there is voluntary or unforced action, of* (*one's own*) *accord, with* (*a person's*) *consent, willingly, voluntarily* :—Hē him hiera ðonces gestiéran ne meahte *he could not restrain himself from them* (his vices) *of his own accord*, Past. 3; Swt. 35, 18. Gewilde man hī tō rihte þances oþþe unþances *let them be compelled to right whether they will or no*, L. Eth. ix. 40; Th. i. 348, 28. Hē nam sume mid him, sume þances, sume unþances *he took some of them with him, some willingly, others against their will*, Chr. 1066; Erl. 198, 36. Ðā þancodon hý ðyses Gode and mē swýþe georne, and heom eall ðis swýþe wel līcode, and cwædon ðæt heora þances ðis on ēcnesse stande *they said that they approve of the arrangement remaining in perpetuity*, Chart. Th. 117, 7. Ágenes þances *sponte*, Germ. 395, 64 : L. C. S. 75; Th. i. 416, 22. Gif hwā þeóf gemēte and hine his þances āweg læte būton hreáme *if any one come upon a thief and of his own accord let him get away without hue and cry*, 29; Th. i. 392, 14. Hwæþer ðe ðū hý forseó and ðīnes āgenes þonces hī forlēte būton sāre ðe ðū gebīde hwonne hī ðē sorgiendne forlētan *whether thou despise them and of thine own accord abandon them without a pang, or wait till the time comes when they abandon thee sorrowing*, Bt. 8; Fox 26, 12 : 7, 2; Fox 18, 13. Ðonne sió sául hire unðonces gebædd wierð ðæt yfel tō forlætanne ðæt hió ǽr hire āgnes ðonces gedyde, Past. 36; Swt. 251, 14. [Hi wenden alle fra þe king, sume here þankes and sume here unþankes, Chr. 1140; Erl. 265, 12. Bludeliche he wule herkien þet þe preost him leid on; ah þenne þe preost hine hat aʒefen þa ehte þon monne þet hit er ahte, þet he nulle iheren his þonkes *he will not listen to that if he can help it*, O. E. Homl. i. 31, 8. þe sulve mose hire þonkes wolde þe totose, O. and N. 70. Lordschipe wol not his thonkes han no felaweschipe, Chauc. Kn. T. 768.] (3) *where there is uncontrolled or independent action, at* (*one's*) *pleasure or will* :—Ðū wēndest ðæt seó wyrd ðās woruld wende heore āgenes þonces būton Godes geþeahte and his þafunge *thou didst suppose that fate turned this world at her own pleasure without the counsel and consent of God*, Bt. 5, 1; Fox 8, 31. (4) *where there is independent condition, in or of itself* :—Gif se weorþscipe and se anweald āgnes ðonces gōd wǽre, Bt. 16, 3; Fox 54, 9. Se anweald his āgenes ðonces gōd næs, ðā se gōd næs ðe hē tō com, 16. 4; Fox 58, 19. (5) *for* (*one's*) *sake* :—Wē biddaþ ðē ðæt ðū hit ūs ðīnes fæder þances forgife *we pray thee to forgive us it for thy father's sake*, Gen. 50, 17. Gedǽle hē ðæt wurð Godes þances *pretium Dei gratia distribuat*, L. M. I. P. 43; Th. ii. 276, 23: L. Pen. 14; Th. ii. 282, 11: L. E. I. 25; Th. ii. 422, 8, 9 : L. Ath. v. 8, 1; Th. i. 236, 8 : Wulfst. 238, 28 : Homl. Skt. i. 200: Lchdm. i. 400, 9. Ic ann ðæs landes intō mynstre Scā Marian þances, Chart. Th. 558, 33. Ungeniédde mid eówrum āgenum willan gē sculon ðencean for eówre heorde Godes ðonces nals na for fracedlecum gestreónum *providentes non coacte, sed spontanee secundum Deum, neque turpis lucri gratia, sed voluntarie*, Past. 18; Swt. 137, 20. Hié ða miclan feorme þigedon Cristes þonces ðe hié ǽr þigedon deófla þonces, Ors. 6, 21; Swt. 272, 22–24. [*Goth* thank fairhaitan χάριν ἔχειν, Lk. 17, 9: O. Sax. thank *grace, pleasure, thanks* : O. Frs. thank, thonk : O. H. Ger. danc, thanc *gratia* : Icel. þökk *pleasure, thanks*.] v. bealu-, fore-, ge-, hete-, hyge-, inge-, inwit-, nearu-, or-, searu-, un-þanc; un-þancness.

þanc-ful[l]; *adj.* I. *thoughtful* :—Mǽden carful þancful nytwyrþe clǽne *a maiden born on the ninth day of the moon will be careful, thoughtful, useful, chaste*, Lchdm. iii. 188, 14. II. *spirited; animosus* :—Cild ācenned (*born on the thirteenth day of the moon*), þancfull (*animosus*), þrīste, reáful, ofermōd, him sylfum gelīcignde, Lchdm. iii. 190, 13. III. *pleasing, agreeable*, cf. Icel. þekki-ligr *handsome, pleasant* :—Ðoncful *gratiosus*, Wrt. Voc. i. 61, 31. Þancfulle *idoneam*, ii. 44, 26. Wē hālsiaþ, God, ðæt þeów ðīn cync ūre . . . tō ðē . . . þancfull mǽge becuman *quaesumus, Deus, ut famulus tuus rex noster . . . ad te . . . gratiosus ualeat pervenire*, Anglia xiii. 381, 228. Þancfullust hýrsumnysse wæstm *gratissimus obedientie fructus*, 371, 84. IV. *thankful, grateful* :—Ðæt folc weard swā fægen his cystignessa and swā þancful, ðæt hig worhton him āne anlīcnesse of āre, Ap. Th. 10, 10. Beóþ þancfulle *grati estote* (Col. 3, 15), Homl. Th.

i. 606, 18. Wesaþ þancfulle þon Hǽlende eóweres andleofan, Blickl. Homl. 169, 16. V. *content, satisfied* :—Ðancful *contentus*, Wrt. Voc. ii. 16, 6. Þancfull, 24, 66. Ðæt hē ðoncfull sī stýre him ðæs bebodenan folces *contentus sit gubernatione creditae sibi plebis*, Bd. 4, 5; S. 572, 33. Ælþeódige bisceopas sýn ðoncfulle (*contenti*) heora gæstlīþnesse and feorme, S. 573, 3. Scottas wǽron ðancfulle (*contenti*) heora gemǽrum, 5, 23; S. 646, 36. [O. H. Ger. un-dancfol *ingratus*.] v. un-þancful[l].

þancfullīce; *adv.* *Thankfully, gratefully* :—Ðā ongeat Eustachius ðæt seó foresǽde costnung him ðā æt wæs, and þancfullīce hī underfēng, Homl. Skt. ii. 30, 144. v. scearp-þancfullīce.

þanc-hycgende *thoughtful* :—Hē, gumena nāt hwylc, þanchycgende ðǽr gehýdde deóre māðmas, Beo. Th. 4462; B. 2235.

þancian; *p. ode.* I. *to thank, give thanks, express in words or have in mind feelings of gratitude*. (1) *absolute* :—Drihten ðancode, ǽrðan ðe hē ða hlāfas tōbrǽce, Homl. Th. ii. 400, 16. Hē genam ðone calic þanciende *accipiens calicem gratias egit*, Mt. Kmbl. 26, 27. (2) *with dat. of person to whom thanks are given* :—Ðē þanciaþ Cristes þegnas, Hy. 7, 52, 49. Hī tō ðē cleopiaþ and ðē lofe þanciaþ *clamabunt et hymnum dicent*, Ps. Th. 64, 14. Hē Gode þancode, Mk. Skt. 8, 6: Homl. Skt. i. 3, 454. Hē feóll tō his fótum and him þancode *cecidit ante pedes ejus gratias agens*, Lk. Skt. 17, 16. Þearfan ic lǽrde ðæt hié Gode þancodon, Blickl. Homl. 185, 18. Wē sceolon him ðancian, Homl. Th. ii. 400, 18. Hig ðone hlāf ǽton Drihtne þanciende, Jn. Skt. 6, 23. (3) *with gen. of that for which thanks are given* :—Wē þanciaþ ðīnes weorðlīcan wuldordreámes, Hy. 8, 9. Hié þanciaþ þrymmes þrīstum wordum, Cd. Th. 242, 26; Dan. 425. Ne sceal hē beón tō georn deádra manna feós, ne tō lyt þancian heora ælmessan, Blickl. Homl. 43, 13. (4) *with dat. of person to whom thanks are given, and* (a) *gen. of thing for which* :—Hī Gode þonciaþ blǽdes and blissa, Exon. Th. 77, 14; Cri. 1256. Hē ðæs þancode Gode, Homl. Skt. i. 4, 237. Hié Gode þancudon ðæs siges, Blickl. Homl. 203, 33. Ðanca Gode ðīnre gesundfulnysse, Homl. Th. i. 400, 13. Þeáh hī his ðē ne ðancien, Ps. Th. 4, 8. Sceolde hē his Drihtne þancian ðæs leánes, Cd. Th. 17, 10; Gen. 257. (b) *with gen. of a pronoun and clause stating cause of thanks* :—Se gomela Gode þancode ðæs se man gespræc, Beo. Th. 2799; B. 1397: Elen. Kmbl. 1921; El. 962. Heó Gode þancode ðæs ðe hió sōð gecneów, 2276; El. 1139: Beo. Th. 1255; B. 625: Andr. Kmbl. 2022; An. 1013. Þoncade, Exon. Th. 148, 25; Gū. 750. Hī Gode þancodon ðæs ðe hī hyne gesundne geseón mōston, Beo. Th. 3257; B. 1626. Þancedon, 460; B. 227. (c) *the cause of thanks given in a clause introduced by ðæt* :—Ic ðancige ðē, ðæt ic ne eom nā swilce ōðre mannum, Homl. Th. ii. 428, 19. Hī þanceden þeódne, ðæt hit þus gelomp, Cd. Th. 298, 16; Sat. 534. Ðanca Gode, ðæt hē ðē gefultumode, Bt. 5, 3; Fox 14, 8. (5) *combining the construction of* (2) *and* (3) :—Ic þancige Gode and eów eallum ðe mē wel fylston, and ðæs friðes ðe wē nū habbaþ, L. Edm. S. 5; Th. i. 250, 4. II. *to express thanks by action, shew gratitude* :—Wē ðē freóndlīce wīc getǽhton ðū ūs leánest nū unfreóndlīce fremena þancast *as friends we assigned thee a dwelling, thou dost now unkindly requite us and shew thy gratitude for benefits*, Cd. Th. 162, 31; Gen. 2689. Sceolde hē mid lāce his clǽnsunge Gode ðancian *he should shew his gratitude to God for his cleansing by a gift*, Homl. Th. i. 124, 10. III. *to feel gratified, to rejoice* :—Þancaþ ł blissaþ *gratatur*, Hpt. Gl. 522, 60. Ðām ðe þanciaþ yfelum mīnum *qui gratulantur malis meis*, Ps. Spl. 34, 29. [O. Sax. thankian : O. Frs. thonkia : O. H. Ger. danchôn *satisfacere, benedicere, remunerare* : Icel. þakka.] v. ge-þancian.

þanc-metegung, *deliberation*. v. next word.

þanc-metung, e; *f.* *Deliberation, consideration* :— Gif hē mid ðancmetuncge (-metegunge, MS. B.: þoncmeotunge, Bd. M. 88, 4) and ðreodunge geþafaþ *si ex deliberatione consentit*, Bd. 1, 27; S. 497, 23.

-þancness. v. nearu-þancness.

þancol; *adj. Addicted to thought, acute* :—Cild ācenned (*born on the sixteenth day of the moon*) þancul (*efficax*; cf. scearpþancfullīce *efficaciter*, Scint. 206, 14; *and see* scearpþanclīce), staþolfæst, Lchdm. iii. 192, 8. Saga, þoncol mon, hwā mec bregde of brimes fæþmum, Exon. Th. 382, 17; Rä. 3, 12. v. deóp-, fore-; ge-, gearo-, hete-, hyge-, scearp-, searu-þancol; þancol-mód.

-þancollīce. v. deóp-þancollīce.

þancol-mód; *adj. Having the mind addicted to thought, of acute mind, wise, intelligent* :—Þancolmōd wer, þeáwum hýdig, Cd. Th. 102, 24; Gen. 1705. Seó gleáwe hēt hyre þinenne þancolmōde heáfod onwrīðan, Jud. Thw. 24, 5; Jud. 172. Ealle witen eorðbūende þoncolmōde ðæt hī ðǽr ne sint, Met. 19, 14.

þanc-snot[t]or; *adj. Wise in thought, wise* :—Þoncsnottor guma breóstgehygdum his bearn lǽrde, Exon. Th. 301, 19; Fä. 21. Fore there neidfaerae naenig uuiurthit thoncsnottura than him tharf sié, Txts. 149, 17.

þancung, e; *f. Thanking, thanks, thanksgiving* :—Gode sī lof and wuldor and dǽda þoncung ealra ðæra gōda ðe hē ūs forgifen hafaþ, Chart. Th. 136, 32. Sāwla þancung *thanksgiving by souls*, Hy. 9, 45. Ic ete mid micelre þancunge *manduco cum gratiarum actione*, Coll. Monast.

Th. 34, 29. Mid ealre þoncunga, Blickl. Homl. 31, 21. Hē underfēng ða lāc mid ðancunge, Homl. Th. ii. 170, 16. [Be] ðæncunge ðæm ðe wið ðýfðe fylstaþ, L. Edm. S. 5; Th. i. 250, 3. Ongan se bisceop ðancunge dôn Drihtne *episcopus gratias coepit agere Domino*, Bd. 2, 9; S. 511, 31 : 4, 23; S. 595, 19. Ðoncunge, 5, 19; S. 641, 2. Ic ðæs þoncunge dô Grēca herige, Nar. 2, 30. Þæs þancunga þine scealcas ealle hæfdan *all thy servants gave thanks for this*, Ps. Th. 101, 12. Wyrþe ðū eart, ðæt ðū onfô wuldor and dǣda þancunga, Blickl. Homl. 75, 2. Ðē ic sylle þancunga *tibi reddo gratias*, Ælfc. Gr. 15; Zup. 95, 15. Ðæt is tô wundrianne, ðæt hī swā lytle þoncunge wiston Iôsepe ðæs ðe hē hī æt hungre āhredde *it is wonderful that they felt so little gratitude to Joseph for saving them from famine*; hunc Ioseph, quem constituit Deus Aegyptiis conservatae salutis auctorem, quis credat ita in brevi eorum excidisse memoriae, Ors. 1, 5; Swt. 34, 32. v. þanc, IV.

þanc-weorþ, -wurþ, -wirþe; adj. *Thankworthy, deserving thanks, acceptable* :—Þurh ðære þancweorþan Cristes gyfe *through the help of Christ's grace, which is deserving of all thanks*, Lchdm. iii. 432, 23. Þancwurðre *gratuita (Christi gratia fretus)*, Hpt. Gl. 420, 76. Ûrum godum geoffrian ðancwurðe onsægednysse *to offer to our gods an acceptable sacrifice*, Homl. Th. i. 592, 34. Bûtan ðū him þoncwyrþe lâc onsecge, Exon. Th. 254, 17; Jul. 198. Ic eów secgan mæg þoncwyrþe þing, ðæt gê ne ðyrfen leng murnan on môde, Judth. Thw. 23, 33; Jud. 153. Þancwurðe gifa *grata (accepta) libamina*, Hpt. Gl. 415, 7. Gecwême (†) þancwurðe gife *grata munuscula*, 510, 71. Þancwurðe gratos, acceptos, caros, 416, 51. Ða ðe ic ðâm bigengum ðancwyrþe gelýfde *quae incolis grata credideram*, Bd. pref.; S. 472, 38. Smylte weder biþ ðý þancwyrþre *(gratius)* gif hit hwêne ǣr biþ stearce stormas . . . And þancwyrþre biþ ðæs dæges leóht for ðære egeslîcan þiústro ðære nihte, Bt. 23; Fox 78, 26–29. Þancwurðra *gratuita*, Hpt. Gl. 442, 26. Hē gearcode him gebeórscipe on his hûse, ac hē gearcode him micele þancwurðran gereord on his heortan, Homl. Th. ii. 468, 30. Þancwur[ðe]ste *gratissimum, acceptissimum, amantissimum*, Hpt. Gl. 441, 66. v. un-þancweorþ.

þancweorþlîce; adv. *Gladly, willingly, in a way that shews acceptance* :—Hī ðancweorþlîce *(gratanter)* wǣron fram him onfangene, Bd. 5, 10; S. 624, 2. Hē ðære gife ðancwurþlîce *(gratanter)* onfêng, 4, 30; S. 609, 9. Gif hē ǣr ne geæfstgode ðætte his brôður lâc wǣron ðancweorþlîcor onfongne ðonne his *nisi Cain invidisset acceptam fratris hostiam*, Past. 34; Swt. 235, 3. *In* Jn. Skt. 6, 11 þancwurðlîce dôn *translates gratias agere.*

þanc-word, es; n. *A word of thanks* :—Swā scrîþende hweorfaþ gleómen, þearfe secgaþ, þoncword sprecaþ, simle sumne gemêtaþ geofum unhneáwne, Exon. Th. 326, 32; Víd. 137.

þanêcan þe *whenever, as soon as ever* :—Ðonêcan þe heó ûtan behwerfed sié (cf. þonne hió ǣrest sié ûtan behwerfed, Met. 13, 77), Bt. 25; Fox 88, 34. Þeáh hî nū eall hiora lîf āwriten hæfdon, hū ne forealldodon ða gewritu þeáh and losodon ðonêcan þe hit wǣre swā some swā ða wrîteras dydon and eác ða þe hî ymbe writon *though they indeed had written all their life, yet would not the writings have become antiquated and have perished, as soon as ever it was done, in the same way as the writers did, and those too about whom they wrote*; quamquam quid ipsa scripta proficiant, quae cum suis auctoribus premit longior atque obscura vetustas? 18, 3; Fox 64, 28. Ac þonêcan (þan-, Bod. MS.) þe hê ðone anweald forlǣt, oððe se anweald hine, ðonne ne biþ hê ðam dysegan weorþ, 27, 1; Fox 94, 20. v. (?) êce.

þânian; p. ode *To be or to become moist* :—Þǣnie *madeo*, Wrt. Voc. ii. 58, 44. Ðâniaþ *madescunt*, 57, 39. v. þân.

þanne, þænne, þonne; adv. conj. *Then, when.* Generally if the subject follows the verb the word is to be rendered by *then*, if the subject precedes the verb, by *when*. [Þanne and þá differ in force; the former is used where the time of an action is indefinite, and is found with the future, the indefinite present and the indefinite past; the latter is used where a definite action has taken place. Cf. Þonne faraþ hig on êce susle, Mt. Kmbl. 25, 46, with : Ðā fêrde se ðe ða fîf pund underfêng, 25, 16. Þonne ðū fæste, smyra ðîn heáfod, 6, 17, with : Þá þá hê fæste feówertig daga, 4, 2. Symle ic gehýrde, þonne heofones gim west onhylde, Exon. Th. 174, 30; Gû. 1185, with : Þá hî ðis gehýrdon, hî fahnodon, Mk. Skt. 14, 11.] A.—demonstrative, *then*. I. of time, *then, at that time* :—Fôron hié bī swā hwaþerre efes swā hit þonne *(at the time of their going, whenever it was)* fierdleás wæs, Chr. 894; Erl. 90, 13. Ðæt geweorþeþ on dômes dæge . . . þonne forhtiaþ ealle gesceafta, Blickl. Homl. 11, 3 : 95, 29 : Exon. Th. 372, 21; Seel. 96. Þonne hî clypiaþ tô mê, and ic hî ne gehýre, Homl. Th. ii. 378, 2. Se deófol ðe beswâc ðone þeóf nele nâht on his ende gedafian, ðæt hê þonne gecyrre tô ðam Hǣlende, Homl. Skt. i. 19, 191. II. marking order or sequence, *then, after that*, (1) of time :—Swā hwylc swā morþorslege þafaþ, and hine man þonne fremmeþ *quicunque ad homicidium consenserit, et id postea factum fuerit*, L. Ecg. C. 22; Th. ii. 148, 14. Gang ǣr and gesybsuma wið ðînne brôðer, and þonne cum ðū syððan and bring ðîne lâc, Mt. Kmbl. 5, 24. Bûton hê gebinde ǣrest ðone strangan, and þonne hys hûs bereáfige, 12, 29. Nū wê faraþ tô Geru-

salem, and þonne beóþ gefylde ealle ða hâlgan gewreotu, Blickl. Homl. 15, 8. Se ðe gôd onginneþ, and þonne āblinneþ, 21, 34. Ðam ðe for his synnum onsǣgd weorþeþ, and þonne ā tô ealdre orleg dreógeþ, Exon. Th. 446, 28; Dôm. 29. Ealle ða hwîle sceal beón gedrync, oð ðone dæg ðe hî hine forbærnaþ. Þonne ðý ylcan dæge ðe hî hine tô ðæm āde beran wyllaþ, þonne tôdǣlaþ hî his feoh . . . Ðonne sceolon beón gesamnode . . . menn . . . þonne ærnaþ hý ealle . . . ; ðonne cymeþ . . . se ðæt swiftoste hors hafaþ tô ðæm ǣrestan dǣle, Ors. 1, 1; Swt. 20, 25–36. Âlecgaþ hî ðone mǣstan dǣl, þonne ôðerne, ðonne ðæne þriddan, Swt. 20, 31. Gê cweðaþ : 'Drihten, âtýn ûs.' Þonne cwyð hê: 'Ne can ic eów.' Ðonne ongynne gê cweþan. . . . Þonne segþ hê . . , Lk. Skt. 13, 25–27. Gif gê þonne git *(after that still)* nellaþ eów wendan tô mê, Homl. Skt. i. 13, 169 : Lk. Skt. 14, 32. Monige men syndon ðe cweþaþ ðæt hié on God gelýfon, and þonne hweþere *(and yet after saying so)* nellaþ āblinnan from heora unrihtum gestreónum, Blickl. Homl. 25, 5 : 55, 21. (2) of place or position :—Ǣt ðam feówer tôðum fyrestum . . . se tôð se þanne bî standeþ . . . se ðe þonne bî ðam standeþ . . . and þonne siþþan gehwilc, L. Ethb. 51; Th. i. 16, 3–4. Is se ðridda Martinianus, þonne se feórða Dionisius . . . þonne ðæs sixtan Seraphun nama is, Homl. Skt. i. 23, 5–6. II a. marking addition, *yet, besides* :—Hwæt mâre dêst ðū? Gewyslîce þænne mâre ic dô *certe adhuc plus facio*, Coll. Monast. Th. 19, 35. III. marking the succession of subjects treated of in narrative, *then, again* :—Næs ðæt þonne mǣtost mægenfultuma, ðæt him lâh þyle Hrôðgâres *and then* (the helmet and byrnie having been already spoken of) *that was not meanest of aids that Hunferth lent him*, Beo. Th. 2914; B. 1455. Ðænne (cf. And, 21; Men. 11; 38; Men. 19), Menol. Fox 46; Men. 23. IV. in a clause that is a qualification or contrast to a preceding clause, *then, yet, but* :—Feówertig daga, gif hit hysecild wǣre; gif hit þonne mǣdencild wǣre, . . . hundeahtatig daga, Homl. Th. i. 134, 18. Ða ðe mihton ðurhteón sceoldon bringan lamb and culfran. Gif þonne hwylc wîf tô ðam unspêdig wǣre, ðæt heó ðâs ðing begytan ne mihte . . ., 140, 2, 13 : Homl. Skt. i. 13, 163. Lîfes ic ðê geann, gif ðū gelýfst . . . Gif ðū þonne elles ðêst, ðū scealt deáþe sweltan, ii. 27, 73. Syndon ealle hǣþene godu hildedeóful; heofenas þænne *(autem)* worhte Drihten, Ps. Th. 95, 5. Ðæt hâlige gewrit ðæt cýþeþ . . . Ðonne is ðeáw ðæs apostolîcan setles *sacra scriptura testatur . . . Mos autem sedis apostolicae est*, Bd. 1, 27; S. 489, 5. Eác is swîðe micel þearf ðæt gê cýðon hū ungefôhlîcu scyld ðæt *(perjury)* is . . . Þonne habbaþ wê geāhsod ðæt hit sume men dôþ tô lytelre scylde; þonne nis hit nâ swā, ac is ân ðæra mǣstena scylda, L. E. I. 26; Th. ii. 422, 19–24 : Blickl. Homl. 175, 34. Twêgen beámas stôdon . . . ôðer wæs swā wynlîc . . . þonne wæs se ôðer sweart, Cod. Th. 30, 34; Gen. 477. Þeáh wê þillîco wîto witan, þonne hwæðere ne sceolon wê nǣfre geortrýwan be Godes mildheortnesse, L. E. I. proem.; Th. ii. 398, 42. Wê leorniaþ ðæt seó tîd sié dêgol . . . wê witon þonne hweþre ðæt hit nis nô feor tô ðon, Blickl. Homl. 117, 29. IV a. in an interrogative clause :—Wæs Cristes tôcyme ǣgðer ge hryre ge ǣrist. Hū ðonne? Homl. Th. i. 144, 27 : Exon. Th. 446, 30; Dôm. 30. V. marking a conclusion, inference or result based on a previous statement, *then, therefore, consequently* :—Ðæt ðonne *(from the statements already made)* biþ ðæs recceres ryht, ðæt hê ðurh ða stemne his lâriówdômes ætiêwe ðæt wuldor ðæs uplîcan êðles, Past. 21; Swt. 159, 22 : Blickl. Homl. 39, 23. Drihten cwæþ: 'Bringaþ gê eówerne teóðan sceat.' . . . Þonne sægþ on ðissum bôcum, ðæt Drihten sylf cwǣde, ðæt ðis menuissce cyn ne sceolde āgimeleásian, ðæt hié sealdon heora wæstma fruman for Gode, 41, 3. On ðone dæg hê sende ðone Hâlgan Gâst. Þonne forþon *(it may be inferred that on that account)* is hit swýðe micel cyn, ðæt gehwylc cristen man ðone dæg weorðige, L. E. I. 24; Th. ii. 420, 30 : Blickl. Homl. 63, 7. Hê mâ cêgde . . . ðæt is þonne *(we may infer)* ðæt wê sceolan beón gelǣrede mid ðysse bysene . . ., 19, 13 : 23, 9. Gifeón wê þonne *(for reasons contained in the preceding statement)* on þone gemânan Godes and manna, 11, 4 : 13, 24. Hæbbe ic geāhsod, ðæt hê wǣpna ne recceþ; ic ðæt þonne *(consequently)* forhicge, ðæt ic sweord bere tô gûþe, Beo. Th. 874; B. 435 : 3346; B. 1671. Ðū ûs wel dohtest. Gif ic þonne mæg ðînre môdlufan mâran tilian, ic béo gearo sôna, 3648; B. 1822. Hwylc beren mǣnde hê þonne elles bûton heofona rîce *what other barn can it be inferred that he meant, but heaven?* Blickl. Homl. 39, 27, 29. VI. marking a consequence dependent upon a hypothesis, *then, in that case*, (a) where the hypothesis is expressed in a clause introduced by *gif* :—Gif man frigne man gefô, þanne wealde se cyning . . ., L. Wih. 26; Th. i. 42, 15. Gif wê willaþ on Drihten gelýfan, þonne beó wê sittende be ðæm wege, Blickl. Homl. 23, 8 : 13, 10 : Mt. Kmbl. 24, 50 : Coll. Monast. Th. 29, 25. Gif wê deóplîcor ymbe ðis sprecaþ, þonne wêne wê ðæt hit wile ðincan ðâm ungelǣredum tô meuigfeald, Homl. Th. ii. 582, 24. Gif hwā cwyð ðæt hê lufige God, and his beboda ne hylt, hê biþ leás ðonne, 314, 31. Gif ðū wilt ðæt ðis feoh becume tô ðîure sâwle ðearfe, tôdǣl hit ðonne ðearfum, 484, 32. Gyf þonne Frysna hwylc ðæs morþorhetes myndgiend wǣre, þonne hit sweordes ecg sweðrian scolde, Beo. Th. 2216; B. 1106. Ðonne wêne ic tô ðê wyrsan geþingea, gif ðū Grendles dearst bîdan, 1054; B. 525. (b) where

the hypothesis is otherwise expressed:—Se ðe wille anwald ágon (= *if any one desires to have power*), þonne sceal hé ǽrest tilian ðæt hé his selfes áge anwald, Met. 16, 1. Se ðe feohtan ne dear mid Godes gewǽpnunge ongeán ðone feónd, hé biþ þonne mid ðám deófellícum bendum gewyld, Homl. Th. ii. 402, 18. (c) where the hypothesis is implied:—Wé sceolon ðone geleáfan mid gódum dǽdum gefyllan, þonne (*if we do so, then*) beó wé úrum Hǽlende fylgende, Blickl. Homl. 23, 10. Ic ðé lǽre, ðæt ðú hospcwide ne fremme; ðonne ðú geearnast ðæt ðé biþ éce líf seald, Elen. Kmbl. 1049; El. 526. Weorþiaþ gé eówerne God . . .; þonne gefylleþ Drihten eówer beren, Blickl. Homl. 41, 10. Lufian wé hine . . .; þonne ne lǽteþ hé ús nó costian, 13, 8, 26. Hwæt mǽnde hé elles, búton ðæt wé gefyllon ðæs þearfan wambe? Þonne (*if we do fill*, etc., then) ne hingreþ ús nǽfre, 39, 30. 'Hwæt ðéstú gif ic tó mergen middeges gebíde?' Hé cwæð: 'Sylf ic swelte þonne,' Homl. Skt. i. 3, 591. Ðes man is sóþfæst, ac þonne hwæþere git sindon bigswicon *this man is true, but yet (if that be so) then ye are deceivers*, Blickl. Homl. 187, 30. **VI a.** in questions, and referring to a condition contained in another sentence, *then, in that case:*—Wilt ðú syllan þingc ðín hér ealswá ðú hí gebohtest þǽr? Ic nelle. Hwæt þænne mé fremode gedeorf mín? Coll. Monast. Th. 27, 17. Hig beóþ tódǽlede. Hú mæg þonne hys ríce standan? Mt. Kmbl. 12, 26: Salm. Kmbl. 715; Sal. 357. **B.**—relative, *when.* **I.** of time. (1) of the time of a single action in the future:—Hwylc tácen biþ, þænne ealle ðás ðing onginnaþ beón geendud, Mk. Skt. 13, 4. Ðænne mannes sunu cymþ, gemēt hé geleáfan? Lk. Skt. 18, 8 : 13, 28. Ðonne ic cume tó ðé tǽc mé *quando veniam ad te, doce me*, Ælfc. Gr. 38; Zup. 224, 7. Ðonne se híredes ealdor ingǽð, gé standaþ þǽr úte, Lk. Skt. 13, 25. Ðonne ðú for unc ondwyrdan scealt, Exon. Th. 372, 5; Seel. 88. Hwænne wylle gé singan ǽfen oþþe nihtsangc? Þonne hyt tíma byþ, Coll. Monast. Th. 34, 5. Geþence mé, þonne ðé ðín wíse lície, Gen. 40, 14. Ic náme þænne ic cóme *veniens ego recepissem*, Mt. Kmbl. 25, 27. (2) referring to the times of an action which may occur an indefinite number of times, *when, at such times as:*—Þænne se yrþlingc unscenþ ða oxan, ic lǽde hig tó lǽse, Coll. Monast. Th. 20, 25. Bútan ðænne bises geboden weorþe, Menol. Fox 64; Men. 32. Eádige synt gé, þonne hí wyriaþ eów, Mt. Kmbl. 5, 11. Þonne ðú ðíne ælmessan sylle, ne bláwe man býman beforan ðé, 6, 2, 3, 5, 6. Symle hé sceal singan, ðonne hé his sweord geteó, Salm. Kmbl. 334; Sal. 166: Beo. Th. 46; B. 23: Andr. Kmbl. 503: An. 252: Exon. Th. 42, 18; Cri. 674. Saga ðæt ðú sié sweostor mín, þonne ðé leódweras fricgen (*whenever you are asked*), Cd. Th. 110, 5; Gen. 1833. Ðæt wǽron men fyrdhwate, þonne rond and hand helm ealgodon, Andr. Kmbl. 18; An. 9. Symle ic gehýrde, þonne heofones gim west onhylde, Exon. Th. 174, 30; Gú. 1185: 122, 11; Gú. 304: Cd. Th. 33, 21; Gen. 523. Ic ðonne (*dum*) mé hefie wérun, ic gegerede mec mid hǽran, Ps. Surt. 34, 13. (3) where the order in time of two circumstances is to be marked, *when, after:*—Eallum geleáffullum mannum englas þegniaþ, þonne hi habbaþ deófol oferswíþed, Blickl. Homl. 35, 3. Ðín ágen bearn frætwa healdeþ, þonne ðín flǽsc ligeþ, Cd. Th. 132, 5; Gen. 2188. Hwæt dó wé, þonne hé unc hafaþ geedbyrded oþre síþe, Exon. Th. 372, 29; Seel. 100. **II.** denoting a cause, *when, since, seeing that:*—Sindon monige tó ðreágenne, ðonne hié nellaþ ongietan hiera scylda, Past. 21; Swt. 159, 17. Ealle clǽne þingc ic ete. Swíþe waxgeorn eart ðú, þonne (*cum*) ðú ealle þingc etst, Coll. Monast. Th. 34, 31. Hí beóþ slítende wulfas, þonne hié for feós lufan earmne fordémaþ búton scylde, Blickl. Homl. 63, 10: Homl. Th. ii. 226, 31. Wén is ðæt hé wille bewitan his mennge on lífe ge on deáðe, þonne se lytla fugel ne befylþ on grin bútan Godes willan, Homl. Skt. i. 17, 188, 197. **II a.** in questions denoting the cause or reason for that not being done about which the question asks:—Hú lange wilt ðú bewépan Saules síð, þonne ic hine áwearp, ðæt hé leng ne ríxige *how long wilt thou mourn for Saul, seeing I have rejected him from reigning?* (A. V. 1 Sam. 16, 1) Homl. Th. ii. 64, 5. Hwá sceal tó his ríce fón, þonne hé bróðer næfþ, ne hé bearn ne beláefþ? 146, 19: i. 48, 12, 25. Hú mæg ic yrnan mid eów, þonne ic ne árás of ðysum bedde nú for nigon geárum? Homl. Skt. i. 21, 344. Hwæt wille wé furðor secgan hú se cásere his fyrdinge geendode, þonne hé forferde on ende, ii. 28, 118. **III.** *although:*—Ðú gelýfdest on mé, þonne ðú mé ne gesáwe *credidisti in me, cum ipse me non uideris*, Homl. Skt. ii. 24, 114. **IV.** denoting condition, case, *when, the case in which:*—*Iactantia*, ðæt is ýdel gylp; ðæt is ðonne se man biþ lofgeorn and mid lícetunge fǽrþ, Homl. Skt. i. 16, 302. Óðer deófolgild is, . . . ðonne se man forsihþ his Scyppendes beboda, 17, 50. Míne eágan synt ealra gelícast þonne esne biþ þonne his hláforde hēreþ *my eyes are most like the case of the servant obeying his lord*, Ps. Th. 122, 2. Ealle wé syndon ungelíce, þonne þe wé in heofonum hæfdon ǽrror wlite *we are all unlike what we were when in heaven we formerly had beauty*, Cd. Th. 274, 8; Sat. 151. Ðonne se móna wexeþ (*in its crescent condition*), hé biþ gelíc ðæm gódum men, Blickl. Homl. 17, 22. **C.** correlative, þanne . . . þanne *then . . . when, when . . . then*—Ðonne ðú ealle gedǽlde hæfst, þonne bist ðú ðé self wædla, Bt. 13; Fox 38, 34. Ðonne eów mislíciaþ ða mettrumnessa ðe gé on óðrum monnum geseóþ,

ðonne geðence gē hwæt gē sién, Past. 21; Swt. 159, 13-14, 19-21: Blickl. Homl. 17, 2-3. Þonne se móna wanaþ, þonne tácnaþ hē úre deáplícnesse, 17, 24: 19, 14-15, 28-29. Ðonne Godes becumaþ tó deáðe, ðonne gemētaþ hī yrfwyrdnysse, Homl. Th. ii. 526, 29-30: Exon. Th. 83, 7-10; Cri. 1352. Ðætte ðonne, ðonne hié ða untruman lácnian willaþ, ðætte hié ǽr gesceáwen, Past. 48; Swt. 370, 9. **D.** after comparatives, *than.* **I.** where the comparison is between different objects, (1) where the objects are expressed by single words or phrases:—Hē wæs ǽr þonne ic, Jn. Skt. 1, 15. Gē synt sēlran þonne manega spearuan, Mt. Kmbl. 10, 31. Ðē wæs leófra his sibb and hyldo þonne ðín sylfes bearn, Cd. Th. 176, 34; Gen. 2921: Andr. Kmbl. 2856; An. 1430. Leófre ys ús beón beswungen for láre þænne hit ne cunnan, Coll. Monast. Th. 18, 20: 24, 23. (1 a) where there is a negative with the comparative:—Næf þ nān mann māran lufe þonne ðeós ys, Jn. Skt. 15, 13. (2) where one or each object is expressed by a clause:—Sēlre biþ ǽghwǽm, ðæt hē his freónd wrece, þonne hē fela murne, Beo. Th. 2775; B. 1385. Ðē wǽre sélle, ðǽr ðú wurde fugel, þonne ðú ǽfre mon gewurde, Exon. Th. 372, 1; Seel. 85. (2 a) where there is a negative with the comparative:—Nis nǽnig māre mǽgen, þonne hē ðone áwyrgdan gāst oferswíþe, Blickl. Homl. 31, 31. Nyston beteran rǽd þonne hié ða behlidenan him tó lífnere gefeormedon, Andr. Kmbl. 2179; An. 1091. (2 b) in questions:—On hwam mæg se iunga rǽdran rǽd gemittan, þonne hē ðíne wísan word gehealde? Ps. Th. 118, 9. Hwæs wǽre mē māre þearf, þonne ic mid cilde wǽre? Gen. 25, 22. **II.** where the comparison is between the same object under different conditions:—Ácumendlícre byþ Sodoma lande on dōmes dæg þonne þǽre ceastre, Mt. Kmbl. 10, 15. Ic wylle cýpan hēr luflícor þonne ic gebicge ðǽr (*the price is higher in one case than in the other*), Coll. Monast. Th. 27, 19. Sceolan wē beón geornran ðæt wē Godes bebodu healdan, þonne wē úrne teónan gewrecan *our zeal to keep God's commands must be greater than our zeal to avenge our wrong*, Blickl. Homl. 33, 24. Nǽfre hlísan áh Meotud þan māran, þonne hē wið monna bearn wyrceþ weldǽdum *the glory is never greater than when working benevolently*, Exon. Th. 191, 11; Az. 86. Hē biþ on ðæt wynstre weorud wyrs gesceáden, þonne hē on ða swíþran hond swícan móte, 449, 24; Dóm. 76. **III.** where the comparative with þanne may be rendered by the positive preceded by *too* and followed by *for* with an infinitive or by an infinitive:—Seó is brādre þonne ǽnig man ofer seón mǽge *it is too broad for anybody to be able to see across*, Ors. 1, 1; Swt. 19, 19. Ðæt his mód wite, ðæt migtigra wíte wealdeþ, þonne hē him wið mǽge (*one too mighty for him to prevail against*), Cd. Th. 249, 1; Dan. 523. Him wæs Godes egsa māra in gemyndum, þonne hē menniscum þrymme þegan wolde (*too much fear of God for him to wish for human glory*), Exon. Th. 112, 6; Gú. 139. Deóplícor mid ús ðú smeágst, þonne yld úre ansón mǽge (*too deeply for our age to be able to take it in*), Coll. Monast. 33, 11. Se wæs mid his dǽdum snelra þonne hē mægenes hæfde *he was too quick in his actions to have enough strength for them;* celeritate magis quam virtute fretus, Ors. 2, 5; Swt. 78, 27. **IV.** where the adjective is in the positive, and the comparative required by þanne must be inferred:—Gód ys on Dryhten tó þenceanne, þonne on mannan wese mód tó treówianne *bonum est confidere in Domino, quam confidere in homine*, Ps. Th. 117, 8, 9. [O. H. Ger. danne.] v. þan; þe.

þanon, þanone, þār, þāra, þarf, þāriht, þār-riht, -rihte, þās, þasser, þassum. v. þanan, þanane, þǽr, þearf, þǽr-rihte, -riht, -rihte, þes.

þawenian. v. ge-þawenian.

þāwian (þawian?); *p.* ode *To thaw* (trans.):—Se þridda heáfodwind hātte zephirus . . . se wind tówyrpþ and ðáwaþ ǽlcne winter, Lchdm. iii. 274, 22. [Thowes *degelat* (*Deus*), Wrt. Voc. i. 201, col. 2 (15th cent.). Thowyñ or meltyñ, as snowe *resolvo*, thowyñ, as yce *degelat, resolvit*, thowe, of snowe or yce *resolucio, liquefaccio*, Prompt. Parv. 492. Her names . . . were almost ofthowed so, that of the lettres oon or two were molte away, Chauc. H. of Fame, iii. 53. Cf. O. H. Ger. douwen, dewen, *digerere, consumere:* Icel. þeyja *to thaw* (intrans.).]

þe; *indecl. particle.* **I.** as relative pronoun of any number, gender, or case, (1) where the antecedent clause does not contain a demonstrative:—Ic hit eom, þe wið sprece, Jn. Skt. 4, 26. Ðæt ðú ne sý gesewen fram mannum fæstende, ac ðínum Fæder þe ys on díglum: and ðín Fæder þe gesyhþ on dýglum hit ágylt ðé, Mt. Kmbl. 6, 18: Beo. Th. 5264; B. 2635. Idesa scēnost þe on woruld cóme, Cd. Th. 39, 18; Gen. 627. Swýðe manega synt þe þurh ðone weg faraþ, . . . swýðe feáwa synt þe ðone weg findon, Mt. Kmbl. 7, 13, 14. Gē þe yfle synt cunnun góde sylena syllan, 7, 11. Wið gehwylce yfelu þe on ðam innoðe dereþ, Lchdm. i. 280, 18. (2) where the antecedent clause con tains a demonstrative:—Hē fór tó ðæm íglande þe monn ðæt folc Mandras hǽtt (*the people of which are called Mandras*), Ors. 3, 9; Swt. 134, 5. Habbe hē ðone ilcan dōm þe (*the same sentence as*) se þe ðæt fals worhte L. C. S. 8; Th. i. 380, 22. *For other instances* v. se. (3) used in combination with the personal pronouns:—Saga hwæt ic hātte, þe ic lond reáfige, Exon. Th. 394, 6; Rä. 13, 14. Wē ðás word sprecaþ . . . þe wē in carcerne sittaþ, 2, 27; Cri. 25. Wē, þe ús befæst is seó gýming

Godes folces . . . *we, to whom is committed the care of God's people* . . . , L. E. I. 1; Th. ii. 402, 9. Fæder úre ðú þe eart on heofenum *Pater noster, qui es in coelis*, Mt. Kmbl. 6, 9. Ðú þe reccest, ðú nú behealdqui regis, intende*, Ps. Th. 79, 1. Ðonne we scrift ongit ðæs costunga ðe hé him ondetteþ *when the confessor hears the temptations of the man who confesses to him*, Past. 16; Swt. 105, 20. Ðære fǽmnan tíd þe hira (*whose*) noma wæs Scá Anatolia, Shrn. 102, 34. Scéaweras þe hira naman hér sint áwritene *viros, quorum ista sunt nomina*, Num. 13, 5: Lev. 11, 3. Ða men þe mon hiora mǽgas ǽr slóg, Ors. 2, 5; Swt. 80, 19. *For other instances see* hé. (4) where relative and antecedent are included in one form:—Eart ðú ðe tó cumenne eart? Lk. Skt. 7, 20. Wén ne brúceþ ðe can weána lyt, Runic pm. Kmbl. 340, 30; Rún. 8. Tó middes eów stód þe gé ne cunnon, Jn. Skt. 1, 26. Hér syndon þe ðíne deórlingas beón sceoldon, Homl. Skt. i. 23, 147. II. as adverb. (1) a relative adverb:—Ðonon ðe hí útan bióþ áhæfene, ðanon hié bióþ innan áfeallene, Past. 50; Swt. 391, 12. Ðæt úre ende geendige on God, þanon þe ús þæt angin com, Homl. Skt. i. 16, 8. On ðæs sǽs waroþe, ðanon ðe hí sciphere on becom, Bd. 1, 12; S. 481, 11. Þider ðe Stephanus forestóp, ðider folgode Paulus, Homl. Th. i. 52, 5. (2) before comparatives, (a) *any*. v. þan:—Ne ðearft ðú nó be ðǽm gesceaftum tweógan þe (þon, Cott. MS.) má þe be ðǽm óþrum *you need not doubt about those creatures any more than about the others*, Bt. 34. 10; Fox 148, 18: 34, 1; Fox 134, 15: L. Pen. 7; Th. ii. 280, 5: Homl. Skt. i. 7, 20. Nys mé ðýnes weales hǽmed nǽfre þe leófre þe mé nǽdre tóslýte, Shrn. 154, 22. Nis þeós woruld ðe geliccre ðære écan worulde þe is sum cweartern leóhtum dæge *this world is no more like the eternal world than a prison is like bright day*, Homl. Th. i. 154, 18. Nǽre hit þe geliccre ðære écean myrhðe, þonne biþ ðam menn þe sitt on cwearterne wið ðam menn þe fǽrþ frig geond land, Homl. Skt. i. 12, 107. Gif hwylc gód man from góde gewíte, ðonne ne biþ hé þe (þon, Cott. MS.) má fullíce gód (cf. *Goth.* ni magt þana mais fauragaggja wisan, Lk. 16, 2), Bt. 37, 3; Fox 190, 29. (β) = þý, *the*:—Swá biþ micle þe winsumre sió sóþe gesǽlð tó habbenne æfter ðam eormþum ðisses lífes, Bt. 23; Fox 78, 30. Hé hæfde giet ðe má unþeáwa þonne his eám nǽfde *avunculi sui erga omnia vitia ac scelera sectator, immo transgressor*, Ors. 6, 5; Swt. 260, 28. Swá þincþ ánra gehwæm sió sóðe gesǽlð þe betere and þý wynsumre, þe hé wíta má hér ádreógeþ, Met. 12, 20. Symle biþ þý heardra, þe hit sǽstreámas swýðor beátaþ, Cd. Th. 80, 8; Gen. 1325. Þe lǽs *lest*, Ex. 19, 21, 24: Mk. Skt. 4, 12: 13, 36. III. as conjunction. (1) introducing noun or adverb clauses, *that*, cf. þæt. (a) noun clauses:—Eác wæs ðæt ðe beforan ðǽm temple stód ǽren ceác, Past. 16; Swt. 105, 1. Heó ða fǽhðe wræc, þe ðú Grendel cwealdest, Beo. Th. 2672; B. 1334. Áras sceoldon wilspella mǽst gesecgan, ðe ðæt sigorbeácen méted wǽre, Elen. Kmbl. 1967; El. 985. Ðæt dysig is anlícost þe sum cild sié full hál geboren . . . *such folly is most like, that (just as if) a child were born quite healthy* . . . , Bt. 38, 5; Fox 206, 21. Hit is ðǽm gelícost þe ic sitte on ánre heáre dúne, Ors. 3, 11; Swt. 142, 13. (β) adverb clauses:—Hwæt is se manna þe ðú him cýþan woldest *quid est homo, quod innotuisti ei?* Ps. Th. 143, 4. Hé wolde ðæt ða tácn hé cúðon ðý swíþor tó buge, þe hé hæfde hiera ealdhláfordes sunu on his gewealde, Ors. 3, 11; Swt. 148, 32. Hé wæs sundes þe sǽnra, þe hyne swylt fornam, Beo. Th. 2877; B. 1436: Exon. Th. 432, 15; Rä. 48, 6. Hié ðæt gewinn ðæslícost angunnan, þe hí hit ǽr ne angunnen, Ors. 3, 11; Swt. 150. 31. (γ) in combination with other particles, where the combination may be rendered by a conjunction:—Ðeáh þe . . . swá ðeáh, Homl. Skt. i. 12, 106. Oþ þe (*until*) hyt eall álǽd biþ, Ors. 1, 1; Swt. 20, 31: Bt. 38, 5; Fox 206, 24. *See* þeáh, óþ, *and* se, V. (2) *than:*—Hé hæfde twǽm þǽs twéntig wintra, Blickl. Homl. 215, 34: Chr. 901; Erl. 96, 24. Ne hí hié selfe ðý beteran ne taligen, ðe ða óðre, Past. 44; Swt. 319, 18. *See* II. 2 a *above, and* þan. (3) *or*, (a) *alone*:—Þ hyt sy álýfed, þe ná? Mt. Kmbl. 22, 17: Lk. Skt. 7, 20. God ána wát hú hú gecynde biþ, wífhádes þe weres, Exon. Th. 223, 9; Ph. 357. Hwæðer wǽre twégra strengra, wyrd ðe warnung, Salm. Kmbl. 855; Sal. 427. (β) þe . . . þe *whether . . . or*:—Hwyder hé geléded sý, þe tó wíte, þe tó wuldre, Blickl. Homl. 97, 22. Ðe nyton hwænne ðæs húses hláford cymþ; þe on ǽfen, þe on midre nihte, þe on hancréde, þe on mergen, Mk. Skt. 13, 35. (β 1) hwæðer (*pronoun*) . . . þe . . . þe:—Ðæt ic wite hwæðer hit sig, þe sóð þe leás, ðæt gé secgaþ, Gen. 42, 16. Hwæðer ðincþ ðe ðonne, ðæt ða ðing sién, þe ðara sóþena gesǽlþa limu, ðe sió gesǽlþ self? Bt. 34, 6; Fox 142, 10. Hwæðer is ðé leófre, þe ðú nú onfó ða costnunga, þe neár ðínum ende? Homl. Skt. ii. 30, 131. (γ) hwæðer, þeáh . . . þe *whether . . . or*:—Hwæþer hé wacode ðe slépte, Bd. 2, 12; S. 513, 39. Hwæðer ðæs landes folc cristen wǽre ðe hǽðen, Homl. Th. ii. 120, 23. Ongitan hwæþer hit hysecild þe mǽdencild beón wille, Lchdm. ii. 172, 17: Exon. Th. 80, 16; Cri. 1307: Blickl. Homl. 117, 19. Ic nát þeáh ðú mid lignum fare, þe ðú Drihtnes eart boda, Cd. Th. 34, 4; Gen. 532. [*O. Sax.* the.]

þe=se, *in Northern Gospels*:—Ðe í hé *ipse*, Mt. Kmbl. Lind. 15, 24. Ðe ilca *ipse*, 3, 4. Ðe ðe *qui*, 3, 2. Ðe Hǽlend, Jn. Skt. Rush. 4, 2, 6.

peá, þeaca, þeaclíce, þeád, þeáf. v. þeów, þaca, þearllíce, II, þeód, þeóf.

þeáh, þáh, þéh, þeh; *adv. conj.* I. *yet, still, however, nevertheless:*—Ðeáh (ðéh, MS. A.) ic secge inc *verumtamen dico vobis*, Mt. Kmbl. 11, 22. Hé ðafode ða scylda, and ðeáh hé him gecýðde, Past. 21; Swt. 151, 23: Blickl. Homl. 55, 26. Hié hæfdon áþas geseald, and þéh ofer ða treówa fóron hié, Chr. 194; Erl. 90, 4. Dydon swá hwæþer swá hý dydon, ne dohte him náwþer; ðeáh hí sceoldon ðæt feorh álǽtan *let them do which they would, neither did them any good; they had nevertheless to lose their lives*, Bt. 29, 2; Fox 106, 2. Wǽron manige eác him þéh ic ða geðungnestan nemde *there were many besides them; however, I have named the chief*, Chr. 897; Erl. 95, 6. Ágife hé ðone teóþan sceat Gode, and dǽle þeáh his ælmessan forþ of ðon nigeoþan dǽlon *let him pay the tithe to God, and still go on distributing alms from the other nine parts*, Blickl. Homl. 53, 11. Ne magon ðis þeáh ealle men dón *all men, however, cannot do this*, 37, 34: Cd. Th. 44, 12; Gen. 708. Hwæt is ðe deórast þince hwæþer þe gold, þe hwæt? Ic wát þeáh gold, Bt. 13; Fox 38, 11. Gif ðú þeáh mínum wilt wordum hýran, Cd. Th. 35, 24; Gen. 559. Hé ne wisse word ne angin swefnes sínes, hét him secgan þeáh, 223, 28; Dan. 126. I a. combined with other particles, hwæðere, swá, se . . . þeáh:—Ðú ealle gesceafta ǽrest gesceópe swíðe gelíce, sumes hwæþre þeáh ungelíce, nemdest swá þeáh mid áne noman ealle tógædere, Met. 20, 52-56. Hwæþer (hwæþre?) ic ðe secge þeáh, ðæt . . . , Bt. 13; Fox 38, 7. Ac swá ðeáh wíse láreówas tódǽldon ðone praeteritum tempus, Ælfc. Gr. 20; Zup. 124, 1 : 38; Zup. 226, 1. And ábád swá þeáh (*nihilominus*) seofon dagas, Gen. 8, 12. Ic déme swá þeáh ða þeóde *verumtamen gentem ego judicabo*, 15, 14. Ðæt ðæs Hálgan Gástes þénung wǽre on ðære gyfe ðæs fullwihtes swá þeáh (*nihilominus*), nalles ðæs mannes, L. Ecg. C. 7; Th. ii. 140; 3: H.R. 101, 8. Sceolde hwæðere swá þeáh æþeling unwrecen ealdres linnan, Beo. 4876; B. 2442. [*Goth.* swê þauh.] *See* þeáh-hwæðere, *and* se, weald. II. *though, although*, (1) in clauses which express no uncertainty:—Þeáh (ðæch, Lind.: ðéh ðe, Rush.) se Hǽlend ne fullode quamquam Jesus non baptizaret*, Jn. Skt. 4, 2. Wǽron Rómware sóna gegearwod, ðeáh hié werod lǽsse hæfdon tó hilde, Elen. Kmbl. 96; El. 48. Þeáh hié ǽr ðæs écan lífes orwéne wǽron, hié synt nú swíþe blíþe, Blickl. Homl. 85, 27. (1 a) combined with þe:—Hí wǽron ðæs Hǽlendes gewitan, ðeáh ðe hí hine ðágyt ne cúðon, Homl. Th. i. 84, 4 : 82, 33. Þeáh ðe hé geong sý, Beo. Th. 3667; B. 1831. Ne beóþ gé tó forhte, þéh þe synnigra cynn swylt þrowode, Andr. Kmbl. 3217; An. 1611. (2) in hypothetical clauses, *though, if, even if:*—Ic ðé sylle swá hwæt swá ðú me bitst, þeáh (*licet*) ðú wylle healf mín ríce, Mk. Skt. 6, 23 : Bt. 18, 3; Fox 66, 10. Hwæt fremaþ ǽnegum menn, þeáh (ðáh, Lind.: ðeáh þe, Rush.) hé ealne middaneard gestrýne *si mundum universum lucretur*, Mt. Kmbl. 16, 26. On hwan mæg se mann módigan, þeáh hé wille *on what can man pride himself, even if he wishes?* Hom. Skt. i. 16, 371. Hwæt hæfst ðú æt ðam gifum, ðeáh hí nú éce wǽron? Bt. 13; Fox 38, 5. Nát þeáh ðú mid lignum fare, Cd. Th. 34, 2; Gen. 531. þæh, 281, 2; Sat. 265. Ðéh ðú þersce *si contuderis*, Kent. Gl. 1034. Þeáh man ásette twégen fætels full ealað oððe wæteres, hý gedóþ ðæt gelþer biþ oferfroren, Ors. 1, 1; Swt. 21, 15. (2 a) in combination with þe:—Þeáh (ðéh, Lind.: þéh, Rush.) þe ic scyle sweltan mid ðé, ne wiðsace ic ðé *etiamsi opportuerit me mori tecum, non te negabo*, Mt. Kmbl. 26, 35. Ðeáh þe etsi, 26, 33. Þeáh ðe (ðéh, Lind.: ðéh, Rush.), Mk. Skt. 14, 29. Þeáh ðe ðé man bere mete tóforan, hwónlíce ðé fremaþ ðæt ðú hine geseó, búton ðú his onbyrige, swá eác ðé ne fremaþ, þeáh ðe ðú ða hálgan láre gehýre, búton ðú hí tó gódum weorcum áwende, Homl. Th. ii. 402, 2-5. III. in correlative clauses:—Þeáh (ðáh, Lind. *etsi*) ic God ne ondrǽde, þeáh (*tamen*) ic wrece hig, Lk. Skt. 18, 4-5 : Bt. 13; Fox 38, 11. Ac þeáh ðú nú fier sié ðonne ðú wǽre, ne eart ðú þeáh ealles of ðam earde ádrifen, 5, 1; Fox 8, 35 : 7, 4; Fox 22, 26. Ðeáh ðe hé wið ða scyldgiendan swugode, hé hit him ðeáh suígende gesǽde, Past. 21; Swt. 151, 23. Þeáh þe man wafige wundorlíce mid handa, ne biþ hit þeáh bletsung . . . , Homl. Skt. ii. 27, 151. Þeáh lǽwedum mannum wíf sí ályfed, swá ðeáh hí ágan micele þearf, ðæt . . . , Wulfst. 305, 17. Ðeáh hé nǽre fullíce gefulwad, hweðre hé ðæt gerýne ðære hálgan fulwihte mid gódum dǽdum heóld, Blickl. Homl. 213, 13. Hwæþre hé getrymede heora geleáfan, þeáh hié ðæt word ne ongeáton, 17, 8. Ðaeh ðe . . . hweðre *quanquam . . . tamen*, Ps. Surt. 38, 7. Ðeáh hwæðere, þeáh heó synderlíce Iôhannes gýmenne betǽht wǽre, hwæðere heó drohtnode geménelíce mid ðam apostolícum werode, Homl. Th. i. 438, 31. [*Goth.* þauh: *O. Sax.* thôh: *O. Frs.* thâch: *O. H. Ger.* doh: *Icel.* þô.]

þeáh-hwæðere; *adv. conj. Yet, but, nevertheless, however :*—Ðeáh-hwæðere (*verumtamen*) ic secge eów, Mt. Kmbl. 11, 24: Lk. Skt. 10, 20. Ðeáhhwæðere (*autem*) gang tó ðære sǽ, Mt. Kmbl. 17, 27. Þeáhhwæþere, Blickl. Homl. 97, 25. Monige sint ðe mon sceal wærlíce lícettan, and ðeáhhwæðre eft cýðan, Past. 21; Swt. 151, 13. And hwǽdre him mæg wíssefa wyrda gehwylce gemetigian, . . . ðeáhhwæðre godcundes gástes brúcan *and yet can the wise-minded man moderate every fate for himself*, . . . *yet can he enjoy the divine spirit*, Salm. Kmbl. 883; Sal. 441. And þeáhhwæþere *et tamen*, Coll. Monast. Th. 29, 27. Þeáhhwæðere, Blickl. Homl. 31, 18. Þeáhhweþre, 93, 17.

Ac þeáhhwæþere *sed tamen*, Coll. Monast. Th. 18, 32 : Chr. 1009 ; Erl. 142, 26. Gyt þeáhhwæþere *adhuc tamen*, Coll. Monast. Th. 33, 9. Nyste þeáhhweðre hwæt hé him dón sceolde, Blickl. Homl. 215, 2. Gif hé ne áríst fordam ðe hé his freónd ys, þeáhhwæþere for hys onhrópe hé áríst, Lk. Skt. 11, 8. ¶ combined with swá :—And swá þeáh-hwæþere óþ ðone deáþ hé hine tintregaþ, Blickl. Homl. 59, 30. Ac swá ðeáhhwæðere seó menniscnys wæs æfre forestiht, Homl. Th. ii. 364, 25. Nolde ic cwic æfre swá þeáhhwæðere ðíne gewitnesse forlætan, Ps. Th. 118, 157.

þeaht, e ; *f. Counsel :*—Sum bisceop tó him férde, efne swá swá hé wære mid heofonlícre þeahte geléered, ðæt hé tó ðære spræce férde ðæs Godes mannes, Guthl. 17 ; Gdwin. 70, 8. Hí ræddon ðæt man hine gebunde, and óð deáð swunge. Nero, ðá ðá hé ðæs folces ðeaht geácsode, weard tó feore ásyrht, Homl. Th. i. 384, 7. On módes þeaht, Elen. Kmbl. 2482 ; El. 1242. v. ge-þeaht.

-þeahta, -þeahtend, -þeahtendlíc, -þeahtendlíce. v. ge-þeahta, -þeaht-end, -þeahtendlíc, un-geþeahtendlíce.

þeahtere, es ; *m. A counsellor :*—On .v. nihte mónan gang tó dínum þeahtere, Lchdm. iii. 170, 3. Ðæs cyninges þeahteras *regis consiliarii*, Bd. 2, 13 ; S. 516, 25. Gé yfelan þehteras ! ic næfre mé ne gebidde on eówer god, Nar. 42, 6. v. ge-, ræd-þeahtere.

þeahtian ; *p.* ode *To take counsel, to consult :*—Hié smeágeaþ and ðeahtigaþ on hiera módes rinde monig gód weorc tó wyrcanne, Past. 9 ; Swt. 55, 22, Hé mid his ealdormannum ðeahtode and sóhte hwæt he ðyssum ðingum tó dónne wære *cum suis primatibus curavit conferre, quid de his agendum arbitrarentur*, Bd. 2, 9 ; S. 512, 11. Ðá þeahtode þeóden úre módgeþonce, hú hé ða mæran gesceaft eft gesette, Cd. Th. 6, 21 ; Gen. 92. Hý þeahtodon hú hí mihton geniman míne sáwle *ut acciperent animam meam consiliati sunt*, Ps. Th. 30, 16. Weras þeaht-edon, Elen. Kmbl. 1091 ; El. 547. Hí þeahtedon ongén hine, hú hí hine fordón mihton *consilium faciebant aduersus eum, quomodo eum perderent*, Mk. Skt. 3, 6. Hí ðeahtodon embe ðæra apostola forwyrd, Homl. Th. i. 572, 30. Hé ða monnðwærnesse ðe hé ær ðurhtogen hæfde eft ðeahtigende on yfel gewend *mansuetudinem, quam tolerantes habuerunt, retractantes in malitiam vertunt*, Past. 33 ; Swt. 225, 22. Ðá wæron ða hæþenan betwih him ðeahtiende and sprecende, Bd. 5, 10 ; S. 624, 35. v. ge-, ymbe-þeahtian ; ræd-þeahtende.

þeahtung, e ; *f. Counsel, consultation :*—Tó ðæhtunge *consilio*, Mt. Kmbl. Lind. 27, 7. Ðætung *consilium*, 28, 12. Ðæhtung, p. 16, 14 : Mk. Skt. Lind. 3, 6 : 15, 1 : Lk. Skt. Lind. 7, 30. Ðæhtunge, Mk. Skt. Rush. 3, 6. v. for-, ge-, ræd-þeahtung.

þeána (combined with swá, se) ; *adv. conj. Yet :*—Pápa on Róme swá þeána gesette *papa Romanus tamen statuit*, L. Ecg. C. 7 ; Th. ii. 138, 36. Lífe ne gielpeþ hláfordes gifum, hýreþ swá þeána þeódne sínum, Exon. Th. 440, 6 ; Rä. 59, 13 : 106, 32 ; Gû. 81. Nó God wolde ðæt seó sáwl sár þrowade, lýfde se þeána ðæt hý him mid hondum hrínan mósten, 127, 3 ; Gû. 380.

þearf, e ; *f.* I. *need :*—Wé sceolan beón gemyndige Godes beboda, and úre sáwle þearfe, Blickl. Homl. 25, 27. Gemyndige úre sáula þearfe, 101, 16. God, ðe ælces monnes ðearfe wát, Bt. 39, 10 ; Fox 226, 25. Hé ealle can úre þearfe, Ps. Th. 102, 13. Seleþegn ealle beweotede þegnes þearfe, Beo. Th. 3598 ; B. 1797. Gleómen þearfe secgaþ, Exon. Th. 326, 31 ; Víd. 137. Ðæt hé ne ágæle gæstes þearfe, 51, 17 ; Cri. 817 : 298, 17 ; Crä. 86. Miltsa ðú ús, and gemyne ðú úre þearfa, Blickl. Homl. 225, 21. II. *need for or of something, which is expressed* (1) *by a genitive (a) of a noun, or of a pronoun referring to a noun :*—Hwylc ðearf is ðé húsles *quid opus est eucharistia* ? Bd. 4, 24 ; S. 598, 37 : Cd. Th. 54, 19 ; Gen. 879. Him wæs manna þearf, Beo. Th. 405 ; B. 201. Nænges þinges máre þearf nære, Blickl. Homl. 175, 9. Nalas þý þe úre Drihten ðæs wolcnes fultomes þearfe hæfde, 121, 13. Náhte ic ðínre miltse þon máran þearfe, Judth. Thw. 22, 35 ; Jud. 92. Drihten ðæs (*the ass*) áh þearfe, Blickl. Homl. 71, 1. (*B*) *of a pronoun that represents a clause :*—' Ic bidde ðé, ðæt ðú nyme ðé ládmenn.' Ðá cwæð hé : ' Nys mé ðæs nán þearf ' (*non est necesse*), Gen. 33, 15. Hí bædon ðæt hí móston on óðerne weg faran, and sædon ðæt him ðæs neód wære and eác þearf, Guthl. 14 ; Gdwin. 62, 6. Hwæs wære mé máre þearf, þonne ic mid cilde wære *quid necesse fuit concipere* ? Gen. 25, 22. Ðæs ánes ic áh þearfe, ðæt ðú mín freónd sig and ic ðíne miltse hæbbe *hoc uno tantum indigeo, ut inveniam gratiam in conspectu tuo*, Gen. 33, 15. Wé ðæs máran þearfe nágon, ðæt wé him æfre fram ábúgan, Homl. Skt. i. 23, 454. (*γ*) *of a pronoun that refers to a gerundial infinitive ; cf.* (3) :—Forþon nis mé ðæs þearf, cwæð Orosius, tó secgenne, Ors. 1, 11 ; Swt. 50, 15. (2) *by a clause :*—Him næs nán þearf (ðarf, Lind. *opus*), ðæt ænig man sæde gewitnesse be men, Jn. Skt. 2, 25 : 16, 30. Ús is eallum þearf, ðæt úre æghwylc óþerne bylde, Byrht. Th. 138, 41 ; By. 233. Is ðam weorce þearf, ðæt . . . , Exon. Th. 1, 21 ; Cri. 11. Ús is mycel ðearf, ðæt wé teolian, Blickl. Homl. 125, 11. Him wæs þearf micel, ðæt . . . , Cd. Th. 123, 32 ; Gen. 2054. Ic wéne ðæt hit sié nú ærest þearf, ðæt ic ðé gerecce hwær ðæt héhste gód is *nunc demonstrandum reor, quonam haec felicitatis perfectio constituta sit*, Bt. 34, 1 ; Fox 134, 3. Gé habbaþ micle ðearfe, ðæt gé simle wel dón, 42 ; Fox 258, 26. Ic áh mæste þearfe,

ðæt ðú mínum gáste gódes geunne, Byrht. Th. 136, 61 ; By. 175. (3) *by the gerundial infinitive :*—Monige menn angiennaþ smeágean suiðor ðonne him ðearf sié tó begonganne *nonnulli se in quibusdam inquisitionibus plus quam necesse est exercentes*, Past. 11 ; Swt. 67, 4. Ús is mycel þearf tó witenne, Blickl. Homl. 63, 5. Hwæt is ðæt ðæm men sý máre þearf tó þencenne ? 97, 19. Nis mé wihtæ þearf hearran tó habbanne, Cd. Th. 18, 25 ; Gen. 278. (4) *where that for which there is need is not expressed :*—Gé ðone hlísan habban tiliaþ ofer þióda má, þonne eów þearf sié, Met. 10, 22. Hit is eów uncúðre, ðonne gé þearfe áhton *you have less knowledge on the point than you have need of*, Wulfst. 292, 8. Ic ádráde, ðæt gé willan heora læs gýman, ðonne gé þearfa áhton, 297, 20. III. *needful things, what is needful :*—Ðá hét hé him heora ðearfe forgyfan *eis necessaria ministrari jussit*, Bd. 1, 25 ; S. 486, 29. Wé willaþ eów andlyfne syllan and eówre þearfe forgifan *quae victui sunt vestro necessaria ministrare curamus*, S. 487, 15. Æghwylc mon wile ðæt him Drihten selle ealle his þearfe, Blickl. Homl. 51, 15. Mé Dryhten sendeþ þurh monnes hond míne þearfe, Exon. Th. 121, 24 ; Gû. 293. IV. *what is required of a person, duty :*—Gif munuc wiðer-saca wurðe mid ealle, hé sí ámánsumod æfre, búton hé gebúge tó his þearfe, L. Eth. ix. 41 ; Th. i. 348, 33. Áfæstnie man symle georne on heortan godcunde þearfe (*duty towards God*), Wulfst. 75, 5. God sceáwaþ sylf, mid hwylcum geþance man tó cyrican fare, and hwæt ðær man dreóge wordes oððe weorces. And se ðe ðær ðæt déþ, ðæt his þearfa beóþ, se gegladaþ God, 279,1. Men forgýmdon Godes laga swýðor, ðonne heora þearfa wæron, 292, 13. V. *use, service, behoof, good, advantage, profit* [v. þearf-líc, II, *and cf.* Icel. þarfr *useful :* Dan. tarv. *behoof, good, benefit :* O. H. Ger. bi-darbi *utilis*] :—Nyttung *vel* þearf *vel* gewuna *usus*, Wrt. Voc. i. 54, 68. Ús wylla is, ðæt wé aa æfter úre ágenre þearfe geornlíce winnan *his will is that we ever strive diligently after our own profit*, Wulfst. 109, 8. For eówre þearfe mé sende God *pro salute vestra misit me Deus*, Gen. 45, 5. On ða geráð ðæt ðú ða eorþan sécan wille for gódra manna þearfe, Bt. 7, 3 ; Fox 22, 8. Bisceopas áscádaþ út of cyrican for heora ágenan þearfe ða, ðe heálíce hý sylfe forgyltan, Wulfst. 104, 11. Hé earfeþu geþolade fore þearfe þeódbúendra, lǽdlícne deáð leódum tó þearfe, Exon. Th. 72, 15 ; Cri. 1173. Ðæs múðes tunge sceal faran on ðara eárena ðearfe *ad usum suum auribus oris lingua concurrat*, Past. 34 ; Swt. 233, 8 ; Andr. Kmbl. 3302 ; An. 1654 : Beo. Th. 2916 ; B. 1456. Þonne wé biddaþ ongeán úre ágenre þearfe þonne forwyrnþ God ús ðæs ðe wé ungesceádwíslíce biddaþ *when we ask for what is opposed to our own good, God refuses us that which we ask indiscreetly*, Homl. Th. ii. 528, 8. Ðæt hí sýn gewordene bysen tó forwyrde swýðor þonne tó þearfe *that they have become an example to perdition rather than to profit*, L. I. P. 23 ; Th. ii. 334, 14. Úre ælc scute .iiii. pæng. tó úre gemǽn[r]e þearfe *each of us should contribute four pence to our common use*, L. Ath. v. 2 ; Th. i. 230, 16. Tó ðæs heres þearfe *to the service of the Danes*, Chr. 874 ; Erl. 76, 32. Eallum þeódscipe tó þearfe *for the good of the whole nation*, 1006 ; Erl. 141, 7. Mann wísdóm sprecþ manegum tó þearfe and tó rihtinge, Ælfc. T. Grn. 21, 27 : Wulfst. 32, 9. Úre Drihten ðe eallum mancynne can tó ðearfe (helpe, MS. E.), 14, 17. Godes hús séce hé gelóme him sylfum tó þearfe, 73, 16. Hit weard mancynne tó mycelre þearfe, 23, 5 : 119, 16. Tó þearfe *usefully, profitably*, 49, 2 : Byrht. Th. 138, 38 ; By. 232 : Menol. Fox 426 ; Men. 214. Se ðe ðæt déþ, hé déþ him sylfum mycle ðearfe, Wulfst. 113, 13 : 119, 11 : 303, 7. Fremmaþ gé leóda þearfe, Beo. Th. 5594 ; B. 2801. Ic wát ðæt ðú sécst míne ðearfe *I know that you seek my good*, Shrn. 182, 32. Gif eall gefærræden ðone ræd missrædaþ, and þeáh feáwa witena on ðam gefærscipe beón, ðæt ða þearfe wíslícor tócnáwan cunnon þonne sume, stande ðæra ræd ðe ða ðearfe geceósaþ, R. Ben. 116, 20. V a. *a useful thing, profitable employment :*—Ðæs hádes men ðe hwylum wæron nyttoste and geswincfulleste on godcundan þeówdóme and on bóccræfte, ða syndon nú unnyttaste, and ne swincaþ á swíðe ymbe ænige þearfe for Gode ne for worulde, L. I. P. 14 ; Th. ii. 322, 22. VI. *need, distress, straits, difficulty :*—Gif him þyslícu þearf gelumpe, Beo. Th. 5268 ; B. 2637 : 2504 ; B. 1250. Mec þearf monaþ, micel módes sorg, Exon. Th. 285, 21 ; Jul. 717. Gefultumend æt ælcere ðearfe *adjutor in opportunitatibus, in tribulatione*, Ps. Th. 9, 10. Swá ðæt se man ábrýd æt ælcere þearfe (cf. ábreóðe on ælcere néode, 59, 12), Wulfst. 323, 13. Gif ic æt þearfe ðínre scolde aldre linnan, Beo. Th. 2958 ; B. 1477. On hyra mandryhtnes miclan þearfe, 5691 ; B. 2849. Seó ecg geswác þeódne æt þearfe, 3054 ; B. 1525. Swylc sceolde secg wesan æt þearfe, 5411 ; B. 2709 : Byrht. Th. 140, 52 ; By. 307. Þonne weorðe ic mid eów æfre æt ðearfe, and eów ne forlæte æfre æt néode, Wulfst. 50, 5 : Ps. Th. 62, 7 : 70, 6. Ða ðe hine seóslige sóhtun on ðearfe, Exon. Th. 157, 30 ; Gû. 899. Wást ðú hú is gewand ymbe Creosos þearfe, ðá ðá hine Cirus gefangen hæfde ? Bt. 7, 3 ; Fox 22, 10. Hwæt miht ðú'on ða tíd þearfe gewépan ? Dóm. L. 176. Ðec nú for þearfum ðín ágen geweorc bídeþ, Exon. Th. 8, 3 ; Cri. 112. Ðæt bið for þearfum and for þreánýdum ǽrena biddaþ, 186, 3 ; Az. 14. Ðonne hwá tó his scrifte cymeþ, on ða geráð ðæt hé wille his þearfa tó him sprecan and his synna andettan, L. E. I. 31 ; Th. ii. 428, 9. [Gif hwa is swa sunful þet nulle his scrift íhalden, þenne segge ic eou, þet

nis hit nan þerf (*it is no use*), þet me her on þisse liue for his saule bidde pater noster, O. E. Homl. i. 9, 31. Alle þatt haffdenn ned and þarrfe to þin hellpe, Orm. 12247. *Goth.* þarba *need, want:* O. *Frs.* therve: O. *L. Ger.* therva *opus:* O. *H. Ger.* darba *privatio:* *Icel.* þörf.] v. feorh-, firen-, heáh-, nearu-, níd-, ofer-, sáwel-, un-, weá-, weoruld-þearf; þurfan.

þearf. v. þurfan.

þearfa; *adj.* I. *destitute of, needing* (with gen.):—Hrægles þearfa ic wreó mē wǣda leásne, Cd. Th. 53, 25; Gen. 866. [*Goth.* þarba (*with gen.*).] II. the word is generally used substantively, *a needy, poor person:*—Ðearfa *pauper*, wǣdla *egenus*, Wrt. Voc. i. 74, 21. Ðá sæt ðǣr sum þearfa æt ðǣm burggeate, Blickl. Homl. 213, 32. Ðá sæt ðǣr sum blind þearfa, 15, 16. Nǣ ðæt án ðæt hē wolde mann beón, ac eác swylce hē wolde beón þearfa for ús, Homl. Th. i. 140, 10. Fela sind ðearfan þurh hafenleáste . . . Sind ōðre ðearfan on gáste . . . on ðás wísan wæs Abraham ðearfa, and Dauid, se ðe hine sylfne geswutelode þearfan on gáste, þus cweðende: ' Ic eom wǣdla and þearfa.' Ða mōdigan rícan ne beóþ þearfan ne þurh hafenleáste ne on gáste, 550, 2–11. Nafa ðú nánes þearfan wedd mid dē nihtlangne fyrst, Deut. 24, 12. Gefyllan ðæs þearfan wambe, Blickl. Homl. 39, 29. Mec mon biþeahte mid þearfan wǣdum (*with the garments of a pauper*), Exon. Th. 87, 10; Cri. 1423. Se biscop nǣre miltsiende nánum Godes þearfan, Blickl. Homl. 45, 2. Ða gástlícan þearfan (ðaerfe, Lind.) *pauperes spiritu*, Mt. Kmbl. 5, 3. Eádige sind gē þearfan on gáste, Lk. Skt. 6, 20. Gener ðearfena *refugium pauperi*, Ps. Spl. 9, 9. Ðearfena and earmra manna *inopum*, Bd. 3, 6; S. 528, 17. Ic sylle ðearfum (*pauperibus*) healfe míne ǣhta, Lk. Skt. 19, 8. Syllan þearfon (*egenis*), Jn. Skt. 12, 5. Him gebyrode tō ðám þearfon (ðorfum, Lind.: ðarfum, Rush.), 12, 6. Ðæt hē dǣlde þearfum and wǣdlum, Homl. Skt. ii. 26, 59. Þearfum *matriculariis* (matricularius *a poor person supported by a church*), Wrt. Voc. ii. 86, 50: 57, 1. Þearfum *pauperculis, miseris*, Hpt. Gl. 458, 13. Hē dēmeþ fyrhte þearfan swylce hē þearfena bearn hǣleþ *judicabit pauperes, et salvos faciet filios pauperum*, Ps. Th. 71, 4. Þearfan ic lǣrde, ðæt hié heora wǣdle gefeán hæfdon, Blickl. Homl. 185, 17. Ic lǣre ge ða welegan ge þa þearfan, 107, 12. ¶ Besides enjoining almsgiving the church directly assisted the poor by assigning a certain proportion of the tithes to those whom it called *Godes þearfan*. Thus in general terms it is said:—Wē willaþ myngian freónda gehwylcne, ðæt hí Godes þearfan frēfrian and fēdan, L. Eth. vi. 46; Th. i. 326, 24; and in reference to tithe:—Þridda dǣl ðare teóðunge, ðe tō circan gebyrige, gā Godes þearfum and earmum þeówetlingum, ix. 6; Th. i. 342, 9: in return the poor were exhorted to intercede for the people whose alms they received:—Wē lǣraþ, ðæt preóstas, þonne hí ða ælmessan dǣlan, ða þearfan georne biddan, ðæt hig for ðæt folc þingian, L. Edg. C. 56; Th. ii. 256, 11. From other sources the poor derived benefit; certain fines were devoted to their use:—Gebēte hē .xxx. scitt., and sié ðæt feoh gedǣled ðǣm þearfum, ðe on ða[m] tūn[e] synd, L. Ath. prm.; Th. i. 198, 12. Gif feohbōt árīseþ, ðæt gebyreþ rihtlíce . . . tō þearfena hýdðe, L. Eth. vi. 51; Th. i. 328, 6. [*Ego egenus et pauper sum*, þet is : Ic eam þarua and wrecche, O. E. Homl. i. 115, 8. *Goth.* þarba *a poor person.*] v. ofer-, weoruld-þearfa; þorfa.

þearfan; *p.* de; *pp.* ed *To need, suffer need:*—Nū ðú ðæt swā openlíce ongiten hæfst, ne þearfe ic nū náuht swíþe ymbe ðæt swincan, Bt. 35, 3; Fox 158, 8. Úre ceaster is þearfende . . . wē þoliaþ ðone heardestan hungor, Ap. Th. 9, 7. Ðú, þíne þearfende *thou, needing food* (? v. þigen, II), Cd. Th. 149, 25; Gen. 2480. Ðæt ðū miltsige mē þearfendum, Exon. Th. 269, 13; Jul. 449. Ic ðē biddan wille miltse ðínre mē þearfendre, Judth. Thw. 22, 29; Jud. 85. Wē ðearfende þearle syndon *pauperes facti sumus nimis*, Ps. Th. 78, 8. Ða ðe hira hláf sellaþ ðæm synfullum ðe ðearfende beóþ, nalles for ðæm ðe hié synfulle beóþ ac for ðæm ðe hié menn beóþ and ðearfende beóþ *qui indigenti etiam peccatori panem suum, non quia peccator, sed quia homo est, tribuit*, Past. 44; Swt. 327, 8. ¶ *The present participle, as adjective or as substantive, often occurs,* (1) as adjective, *indigent, needy, poor:*—Hē sǣde ðæt hē folclíc man wǣre and ðearfende *rusticum se et pauperem fuisse respondit*, Bd. 4, 22; S. 591, 6. Widua ðiós ðærfen[de] (ðorfende, Rush.) *uidua haec pauper*, Mk. Skt. Lind. 12, 43. On ðearfendum life and on earmlícum *in humili et paupere vita*, Bd. 4, 13; S. 582, 23: 1, 15; S. 484, 7. Of ðearfendum folce *de paupere vulgo*, 4, 22; S. 591, 34. Ða ðearfendan láfe Brytta *pauperculae Brittonum reliquiae*, 1, 13; S. 481, 41. Ǣnig gemynd þearfendra manna, Blickl. Homl. 69, 10. Þearfendum mannum, 37, 20: 75, 23: 109, 14. Ðæt hē sealde sum þing þearfendum mannum (*egenis*), Jn. Skt. 13, 29. (2) as substantive, (a) *a poor person*; mostly in plural, *the poor:*—Ðone þearfendan áreccan *erigens pauperem*, Ps. Th. 112, 6. Eádge biðon ða ðærfendo (ðorfendo, Lk. Skt. Lind. 6, 20) *beati pauperes*, Mt. Kmbl. Lind. 5, 3. Þearfendra bēne *orationes pauperum*, Ps. Th. 101, 15: 108, 30. Hleó ðarfendra *refugium pauperum*, Rtl. 40, 25. Hē þearfendra ēhte *persecutus est hominem pauperem*, Ps. Th. 108, 16. Ðe ðe his ǣhta þearfendum (*pauperibus*) gedǣleþ, 111, 8. Ðearfendum, Past. 44; Swt. 327, 20. Ðarfendum *egenis*, Lk. Skt. Rush. 12, 5. Þearfendum and ælþeódigum

þeregrinis *et egentibus*, Cod. Dip. B. i. 155, 5. Hē fēdde þearfende, Homl. Skt. ii. 31, 53. Ða þearfendan Drihten gehýreþ, Ps. Th. 68, 34. (b) *a miserable person:*—Biþ ðæt þridde þearfendum (*the wicked*) sorg, Exon. Th. 79, 4; Cri. 1285. [*Goth.* ga-þarban matē ἀπέχεσθαι βρωμάτων, 1 Tim. 4, 3: O. H. Ger. darbēn *carere*.] v. be-, mete-, weoruld-, wine-þearfende; þearfedness; þearfend-líc; þearfan.

þearfedness, e; *f. Poverty:*—On wilsumlícre ðearfednesse *voluntaria paupertate*, Bd. 4, 3; S. 569, 3: Anglia x. 145, 163. Mid ðearfednesse ge mid heora ungelǣrednesse *paupertate ac rusticitate sua*, Bd. 4, 27; S. 604, 28. In ðearfednisse *in paupertate*, Ps. Surt. 30, 11.

þearfende. v. þearfan.

þearfend-líc; *adj. Poor,* (a) of persons, *indigent, destitute:*—Monnes bearn (*Guthlac in his hermitage*) swā þearfendlíc, Exon. Th. 128, 11; Gū. 402. (b) of things, *scanty, insufficient:*—Ðæt ðú ne forgite míne þearfendlícan gegirlan (cf. se fiscere tōslāt his wǣfels on twā and sealde Apollonige ðone healfan dǣl, 11, 27), Ap. Th. 12, 8.

þearfian; *p.* ode *To be in need:*—Hē þearfigendra sāwla gehǣleþ *animas pauperum salvas faciet*, Ps. Th. 71, 13. [*Icel.* þarfa; *p.* þarfaði.] v. be-þearfaþ, ge-þearfian; þearfan.

þearf-leás; *adj. Without having need* or *reason to do something.* v. þearf, II, þurfan, II. 4:—Þearfǽs hē syrwde ymbe Crist *he plotted against Christ, but he had no need to do it*, Homl. Th. i. 82, 20. Ic ðearfleás (þearfleáse, Job. Thw. 166, 22) hine geswencte *without having cause I afflicted him*, ii. 452, 16. [*Icel.* þarf-lauss *needless.*]

þearf-leáse; *adv. Needlessly, without cause.* v. preceding word.

þearf-líc; *adj.* I. *necessary:*—Lā hū þearflíc hit is *quanto magis*, Hpt. Gl. 454, 6. Nēd ł ðarflíc is *necesse esse*, Mt. Kmbl. Lind. 18, 7. Ðarflíc ł nēd is *necessarium est*, p. 13, 1. II. *useful, profitable,* v. þearf, V:—Behōflíc ł ðarflíc *utile*, Mt. Kmbl. p. 13, 6. Hēr is hālwendlíc lār and ðearflíc lǣwedum mannum, Wulfst. 134, 9. Ðæt is þearflíc gewuna, 104, 17: 108, 19: L. Ath. v. 8, 9; Th. i. 238, 18. On gódum lǣce biþ gelang þearflíc broces bōt, L. Pen. 9; Th. ii. 280, 13. Ðæt him ðearflíc nǣre, ðæt hē ðæs hálgan hǣse forhule his hláforde *that it would not be well for him to conceal the saint's bidding from his lord*, Homl. Skt. i. 21, 80. On gódan þeáwan and on þearflícan dǣdan, Wulfst. 121, 2. Sēlre ús is and ðearflícre, ðæt wē úre gyltas andetton, 136, 1. Ðarflícro (ðaroflícra, Rush.) is *utilius est*, Lk. Skt. Lind. 17, 2. Swā swā him þincæ ðæt mǣ þearflíicustþ sí, Chart. Th. 554, 36. [*Icel.* þarf-ligr *useful.*] v. beþearf-líc, nídþearf-líc.

þearflíce; *adv. Usefully, profitably, with profit, to good purpose:*—Wē mihton ðás hálgan rǣdinge menigfealdlícor trahtnian, ac ús swíþnaþ hwæðer gē magon mǣran deópnysse ðæron þearflíce tōcnáwan *whether you can with profit know the profounder parts of the subject*, Homl. Th. i. 556, 15. Angan listum ymbe þencean þearflíce hū hē þider meahte Crēcas oncerran, Met. 1, 60. [*Icel.* þarfliga *usefully.*]

þearflícness, e; *f. Poverty, neediness:*—Þærflícnys *paupertas*, Hpt. Gl. 438, 60. Þerflícnes *mendicitas*, Kent. Gl. 950. On þearflícnysse *in paupertate*, Ps. Spl. 30, 13: Scint. 127, 18: 148, 2. Þearflícnysse hē ondrēt *paupertatem ueretur*, 179, 8. Þearflícnysse lufian *paupertatem diligere*, Cod. Dip. B. i. 155, 7.

þearl; *adj.* (1) of persons, *severe, strict:*—Se ðearla and se ryhtwísa Dēma *districtus judex*, Past. 21; Swt. 167, 22. (2) of things, (a) pain, punishment, effort, and the like, *severe:*—Hē ðy wyrs meahte þolian ða þráge, ðā hió swā þearl becom, Met. 1, 77. Þreánýd þearl, Elen. Kmbl. 1404; El. 704. Wæs seó ádl þearl, Exon. Th. 160, 30; Gū. 951. And suā ðý ðearlan dōme hē forleás his mennisce *ut districto justoque judicio homo esse perderet*, Past. 4; Swt. 39, 23. Heó þrowedon þearl æfterleán, Cd. Th. 5, 24; Gen. 76. Þreá wǣron þearle, Exon. Th. 135, 4; Gū. 519: Ps. Th. 104, 12: 149, 7. Wēndon hié wera cwealmes, þearlra geþinga, Andr. Kmbl. 3194; An. 1600. Ðirst and hungor and ðearle gewin, Salm. Kmbl. 946; Sal. 472. (b) *utter, excessive:*—Þýstru ðū gesettest on þearle niht (*night utterly dark*) posuisti tenebras, et facta est nox, Ps. Th. 103, 19.

þearle; *adv. Severely, sorely, strictly, hard.* This word, as does *swíþe* (q. v.), tends to become an adverb of degree rather than one of manner or quality; where it qualifies words denoting pain, effort, or the like, it may be considered as keeping much of its old force, but even there it is used to translate Latin words marking degree; while in the case of words which do not convey such an idea, it becomes equivalent to *very, very much, exceedingly,* and the like. I. where there is the idea of pain, trouble, etc. (a) where the idea of manner is more prominent:—Þearle ys mē nū ðā, heorte ys onhǣted *matters go hardly with me now, my heart burns within me*, Judth. Thw. 22, 30; Jud. 86. Se ðe his þeóden ǣr þearle gerǣhte (*severely wounded*), Byrht. Th. 136, 29; By. 158. Hí fuhton ðearle *they fought hard*, Judth. Thw. 25, 16; Jud. 262: Chr. 937; Erl. 112, 23. Hí hungre wǣron þearle geþreátod, Andr. Kmbl. 2231; An. 1117: Beo. Th. 1124; B. 560: Rood Kmbl. 103; Kr. 52. Ðis is ðeóstra hám ðearle gebunden fæstum fýrclommum, Cd. Th. 267, 15; Sat. 38. Þearle hē dēmde *tantopere taxaverat*, Hpt. Gl. 454, 2. Ús stalu and cwalu . . . derede swýðe þearle *injured us very severely*, Wulfst. 159, 11. Ðæt hē him ðonne ðearlur (*districtius*) dēman scyle,

Past. 53; Swt. 419, 5. (β) where the idea of degree is more prominent, *very, very much, exceedingly, excessively* :—Sáwl mín gedréfed is ðearle *anima mea turbata est valde*, Ps. Spl. 6, 3. Geeádmēt ic eom ðearle (*nimis*), 37, 8 : Ps. Th. 78, 8. Þearle ic deorfe *nimium laboro*, Coll. Monast. Th. 19, 13. Forþōht þearle (cf. swíþe unrōt, Bt. 1; Fox 4, 4), Met. 1, 82. Ðæt folc wearð þearle geswenct mid ðam síðfate *taedere coepit populum itineris ac laboris*, Num. 21, 4 : Ps. Th. 103, 8 : Homl. i. 80, 14. II. *where there is no idea of pain, trouble, etc., very, to a great degree, very much, to a great extent, exceedingly* :—Geðancas þearle deópe *nimis profundae cogitationes*, Ps. Th. 91, 4. Þearle mildheort *multum misericors*, 144, 8 : Judth. Thw. 22, 23 ; Jud. 74. Swíðe gelýfed mann and ðearle eáwfæst, Homl. Th. ii. 306, 4. Ðú eall geworhtest þing þearle gōd (cf. swíþe gōde, Bt. 33, 4 ; Fox 128, 22), Met. 20, 45. Behēfe þearle *utilis valde*, Coll. Monast. Th. 27, 27 : 29, 31. Þearle deóplíce *valde profunde*, 32, 9. Þearle swíþe tō herienne, Lchdm. iii. 436, 18 : 438, 27. Hē geíhte folc his ðearle (*vehementer*), Ps. Spl. 104, 22. Þearle fremaþ cræft mín eów *multum prodest ars mea vobis*, Coll. Monast. Th. 28, 7 : Judth. Thw. 26, 3 ; Jud. 307. Ic ðē gemenigfilde swíðe þearle (*vehementer nimis*), Gen. 17, 2. Drīg swýþe þearle *dry very thoroughly*, Lchdm. i. 70, 10. Ðis godspel belimpþ swíðe þearle tō ðære mæran freólstíde *this gospel belongs very specially to the great festival*, Homl. Th. ii. 360, 10. Hig þearle etaþ *nimium comedunt*, Coll. Monast. Th. 26, 11. Gehwylc mē drincan sealde þearle *each gave me abundance to drink*, Exon. Th. 485, 1 ; Rä. 71, 7. Drinc swýþe þearle *drink very largely*, Lchdm. i. 78, 10. Hit on wolcnum oft þearle þunraþ, Met. 28, 25. v. for-þearle.

þearl-líc ; adj. *Severe, hard to bear* :—Ðá ðæt Andrea earmlíc þúhte, þeódbealo þearlíc tō geþolianne, ðæt hē swá unscyldig ealdre sceolde lungre linnan, Andr. Kmbl. 2273 ; An. 1138. Sceal se dæg weorþan, ðæt wē forð beraþ firena gehwylce ; ðæt biþ þearlíc gemōt (*a meeting that will be a severe ordeal for all*), Exon. Th. 447, 9 ; Dōm. 36. Deáþes cwealm, þearlíc wíte, 240, 25 ; Ph. 644. Þurh þearlíc þreá, 283, 10 ; Jul. 678.

þearllíce ; adv. I. *severely* :—Ðonne sint eác ðæm ilcan monnum suíðe ðearlíce (*dearlíce*, Cott. MSS.) tō recceanne ða godcundan cwidas *districte itaque contra illos divinae sententiae proferendae sunt*, Past. 37 ; Swt. 265, 22. Forðon is nēd, ðætte sume mid woningum, sume þearlícor (*ðearslícor*, Bd. S. 490, 11), sume liðelecor, synd gerehte *unde necesse est ut quidam damnis, quidam districtius, quidam levius, corrigantur*, Bd. 1, 27 ; M. 68, 5. II. *strictly, exactly, thoroughly* :—Drīg swýþe þearle (þeaclíce (þearlíce?), MS. O.), Lchdm. i. 70, 10. III. *violently* :—Swá biþ þe ðám heáclifum and torrum, ðonne hī hlífiaþ feor up ofer ða óðre eorðan, hý ðonne feallan onginnaþ and full þearlíce hreósan tō eorðan (*come with a great crash to the ground*), Wulfst. 262, 12. v. for-þearlíce.

þearl-mōd ; adj. *Of severe mind*, (1) *in a bad sense, stern, cruel* :—Hæfde his ende gebidenne unswǣslíce, swylcne hē ǣr æfter worhte, þearlmōd þeóden gumena (*Holofernes*), Judth. Thw. 22, 18 ; Jud. 66. (2) *in a good sense, severe in dealing with evil.* v. þearl, I :—Þearlmōd þeóden gumena (*the Deity*), 22, 34 ; Jud. 99.

þearl-wís ; adj. *Severe, strict* :—Ierre ðæs ðearlwísan dēman *districti iram judicis*, Past. 10 ; Swt. 63, 15. Beforan ðæm ðearlwísan dēman *apud districtum judicem*, 16 ; Swt. 105, 10 : Bd. 4, 25 ; S. 599, 36. Þearlwísere gýmene *districto regimine*, Hpt. Gl. 486, 61. God sylfa ðonne ne gýmeþ nǣnges mannes hreówe, ac biþ ðonne rēþra and þearlwísra ðonne ǣnig wilde deór, Blickl. Homl. 95, 30.

þearlwís-líc ; adj. *Severe, hard* :—Drihten hyne þreáde myd þearlwýslícere swingle *the Lord punished him with a severe flogging*, Shrn. 98, 15.

þearlwíslíce ; adv. *Severely, strictly* :—Ðreáge hē hine selfne ðearlwíslíce on his geðōhte *se districta animadversione corrigant*, Past. 64 ; Swt. 461, 20. Hié ða scyldigan þearlwíslíce dēmaþ, Blickl. Homl. 63, 20. Ðý læs hié wyrðen ðearlwíslecor gedēmede *ne districtius puniantur*, Past. 28 ; Swt. 191, 15.

þearlwísness, e ; f. *Severity, strictness* :—Seó ðearlwísnes ðæs heardan lífes *districtio vitae arctioris*, Bd. 4, 25 ; S. 599, 31. Hē hine wæs frignende mid ða apostolícan ðearlwísnesse *sciscitabatur apostolica districtione*, 2, 6 ; S. 508, 14.

þearm, es ; m. *A gut, an intestine* [*Tharm* = guts washed for making hogs' puddings, is given as a Lincolnshire word in Bailey's Dictionary ; with the meaning, ' material of which fiddle-strings are made,' it is given in E. D. S. Pub. Cumberland Glossary ; and in Jamieson's Dictionary *therm, tharme* = the intestines ; a gut prepared, especially as a string for a musical instrument] :—Þearm, þearm *intestinum*, Txts. 69, 1058. Þearm *fibra*, 63, 870 : Wrt. Voc. ii. 148, 55 : *intestinum*, 44, 2. Þearm *fibra*, þearma *fibrarum*, þearmas *fibrae*, 35, 39–41. Blind þearm *cecum*, 16, 59. Lǣcedōmas wiþ þearmes útgange, and gif men bilyhte sié ymb ðone þearm, Lchdm. ii. 170, 27. Þearmas *fibrae*, Wrt. Voc. i. 45, 16 : *intestina*, ii. 49, 50 : exta, Ælfc. Gr. 13 ; Zup. 85, 10. Ðearmas, Wrt. Voc. i. 71, 14. Smæle þearmas *ilia*, 44, 46. Þearma *fibrarum*, Hpt. Gl. 520, 62. Darmana, Txts. 111, 27. Þearmas *fibras*, Wrt. Voc. ii.

38, 5 : Hpt. Gl. 453, 14. [þærmes (þarmes, 2nd MS.), Laym. 818. Þermes, 18451. Þine þarmes þralinge, H. M. 35, 26. Thaarme or gutte *sumen, viscus*, Prompt. Parv. 490. A tharme *trutum*, Wrt. Voc. i. 247, 5 (15th cent.). O. Frs. thermar ; *pl.* : O. L. Ger. thermi ; *pl. exta* : O. H. Ger. darm *fibra* ; *pl.* darma *intestina, ilia* : Ger. darm : Icel. þarmr ; *pl.* þarmar : Dan. Swed. tarm *gut.*] v. bæc-, smeoru-, snædel-þearm, smæl-þearm, *and next word.*

þearme(, es ; n.? v. smæl-þearme) *the entrails* :—Tharme *viscera*, Txts. 107, 2140.

þearm-gewind, -wind, es ; m. The words seem to mean ' *that which enwraps the intestines,*' cf. *plecta* wæfelsa, gewynde, Hpt. Gl. 462, 64, but they are used to gloss *jugulam* (*-um?*), so should mean *the collarbone,* or *the hollow part of the neck above the collar-bone,* or *the throat* :—Gescyld ðearmgewind (ðearmwind, lxxiv, 24), breóstbán, breóst *tege jugulam, pectusculum, mamillas*, Lchdm. i. lxxii, 1.

þearm-gyrd *a belly-band, girth* :—Þearmgyrd *subligar* (the word occurs in a list of terms connected with horses), Wrt. Voc. i. 23, 16. [Cf. O. H. Ger. darm-gurtil *cingula.*] Cf. forþ-gyrd.

þeáter (*with declension like* winter?) *a theatre* :—Æt heora þeátra, Ors. 4, 1 ; Swt. 154, 2.

þeáw, es ; m. I. *a custom, usage, general practice* of a community :—Swá Iudéa þ[e]áw (ðeáu, Lind. : ðeów, Rush.) ys tō bebyrgenne *sicut mos Iudaeis est sepelire*, Jn. Skt. 19, 40. Hit wæs Iudisc þeáw, Blickl. Homl. 67, 8. Feówertig daga hit wæs þeáw (*mos*) ðæt man sceolde wēpan ǣlcne deádne mann, Gen. 50, 3. Siþþan wæs hiera (*the Amazons*) þeáw, ðæt hié ǣlce geáre tōsomne fērdon, Ors. 1, 10 ; Swt. 46, 8 : Beo. Th. 2497 ; B. 1246 : Andr. Kmbl. 50 ; An. 25. Wæs in ða tíd ðeáu Ongelcynnes folcum, ðæt . . . , Bd. 4, 27 ; S. 604, 15. Gyf hit on lande ðeáw sý, L. R. S. 3 ; Th. i. 432, 24. Hwæt ðeóde ðeáw sý, 4 ; Th. i. 434, 34 : 21 ; Th. i. 440, 21. Be ðære ðeóde ðeáwe ðe wē ðænne on wuniaþ, 440, 23. Ðara ðeóda þeáwas sint swíþe ungelíca, Bt. 18, 2 ; Fox 64, 22. Efenfela þeóda and þeáwa, Exon. Th. 334, 18 ; Gn. Ex. 18. Ðeóda ungelíca ægþer ge on sprǣce ge on ðeáwum, Bt. 18, 2 ; Fox 62, 29. II. *mode of conduct, custom, manner, practice, way, usage of a class or kind,* (a) referring to human beings :—Ne biþ swylc cwēnlíc þeáw, Beo. Th. 3885 ; B. 1940. Swá biþ geóguðe þeáw, Exon. Th. 127, 23 ; Gū. 390. Hýrena ðeáwe gē fleóþ *ye flee after the manner of hirelings*, Past. 15 ; Swt. 89, 14. Hē for eaxlum gestód Deniga freán, cúþe hē duguðe þeáw, Beo. Th. 724 ; B. 360. (b) *referring to animals* :—Hiora ðeáwe *suatim* (cf. *suatim, suarum more,* 77, 43), Wrt. Voc. ii. 88, 14. Ðú (*Nebuchadnezzar*) ne gewittes wást bútan wildeóra þeáw, Cd. Th. 252, 2 ; Dan. 572. (c) *referring to inanimate things* :—Ðæt mennisce mōd hæfþ wætres ðeáw (*aquae more*), Past. 38 ; Swt. 277, 6. Ælces mannes mōd hæfþ scipes ðeáw (*more navis*), 58 ; Swt. 445, 10. Nú ðú wást hwelce þeáwas ða woruldsǣlþa habbaþ . . . Gif ðē heora þeáwas líciaþ, Bt. ʒ, 2 ; Fox 18, 5–7. (d) *referring to all created things* :—God gesette unáwendendlícne sido and þeáwas eallum his gesceaftum, Bt. 21 ; Fox 74, 1 : Met. 11, 12. III. *a practice of religion, method of belief, way of thinking, legal usage* :—Gecynde riht *jus naturale*, þeáw vel wíse *solempnitas*, Wrt. Voc. i. 20, 32. Swylc wæs þeáw hyra *such was their religion*, Beo. Th. 359 ; B. 178. Ánmōde þeáwes *those who think alike ; unius moris* (cf. un[i]animes, Ps. Th. Surt.), Ps. Spl. 67, 6. Se forlēt his fulluht; and leouode on hǣðenum þeáwe, Chr. 616 ; Erl. 20, 40. Hē nǣnigne nýdde tō Cristenum ðeáwe (*ad Christianismum*), Bd. 1, 26 ; S. 488, 15. Tō reogollícum ðeáwe rihtra Eástrena *ad ritum Paschae canonicum*, 5, 22 ; S. 643, 38. Þis folc æfter ðeáwe tō húsle gange, Blickl. Homl. 207, 5. Heó ðone ðeáw ðæs Cristenan geleáfan (*ritum fidei*) healdan mōste, Bd. 1, 25 ; S. 486, 34. Hǣþennysse ðeáw forlǣtan *gentilitatis ritum relinquere*, 1, 26 ; S. 488, 12. Ðone ealdan ðeáw . . . ðý apostolican ðeáwe *inveteratam illam traditionem . . . apostolico more*, 5, 22 ; S. 644, 6–8. Hē hæfde beteran ðeáw, leóhtran geleáfan, Cd. Th. 256, 18 ; Dan. 642. Ða ðe on hǣðnum þeáwum dwelgende wǣron, Blickl. Homl. 201, 20. Geset is on cyrclícum þeáwum, Homl. Th. i. 150, 26. Hē áwrát áne bōc be cyrclícum ðeáwum, ii. 84, 23. Onfēngon hí rihtgelýfede ðeáwas on tō lifianne *susceperunt ritus vivendi catholicos*, Bd. 5, 22 ; S. 644, 23 : Hy. 9, 28. Ðeanne ánne wē hæfdon ðætte wē Cristene wǣron and swíðe feáwe ða ðeáwas *we should have the name only of being Christians, and very few of the practices of Christianity*, Past. pref. ; Swt. 4, 8. Þeáwas (*Epicuri*) *sectas*, Wrt. Voc. ii. 84, 67 : Hpt. Gl. 503, 59. IV. *a custom, habit, manner, mode of conduct* of an individual ; the pl. often may be rendered by *conduct, behaviour* :—Ðeáw wæs ðam ylcan biscope, ðæt hē ðæt weorc mā ðurh his fōta gange fremede, ðonne on his horsa ráde *moris erat eidem antistiti, opus magis ambulando quam equitando perficere*, Bd. 4, 3 ; S. 566, 31. Ælces gōdes þeáwas wísdōm gefyllþ ðone, ðe hine lufaþ, Bt. 27, 2 ; Fox 98, 2. Wá him ðæs þeáwes, gif hí unrǣdes ne geswícaþ, Exon. Th. 393, 11 ; Rä. 12, 8. Hē wæs swíþe yfel monn ealra þeáwa, búton ðæt hē wæs cēne *his conduct was very bad in every respect, except that he was brave*, Ors. 6, 14 ; Swt. 268, 27. Weorðe hē worda and dǣda, þeáwa and geþonca, ðæt hē ne forleóse his dreámes blǣd, Exon. Th. 97, 1 ; Cri.

1584. Ic geseó on eówres fæder þeáwum ðæt hé nys swá wel wið mé geworht, swá hé wæs gyrstandæg *I see by your father's behaviour that he is not so well disposed to me as he was yesterday*, Gen. 31, 5. Hé ongiet þe sumum ðingum oððe ðeáwum útanne ætiéwdum eall ðæt hié innan ðenceaþ, Past. 21; Swt. 155, 10. Wer gecorene on his ðeáwum *virum probum moribus*, Bd. 3, 23; S. 554, 9. Wæs hé swíðe geþungen on his ðeáwum *he was most excellent in his conduct*, Blickl. Homl. 217, 7: Judth. Thw. 23, 19; Jud. 129: Exon. Th. 126, 10; Gú. 369: 297, 14; Crä. 68. Þeáwum geþancul *habitually thoughtful*, Andr. Kmbl. 923; An. 462: Cd. Th. 102, 25; Gen. 1705. Lifian rihtum þeáwum *to live righteously*, 160, 4; Gen. 2646. Þeáwum lifian *to live virtuously*, Exon. Th. 319, 13; Víd. 11: Beo. Th. 4295; B. 2144. Ælc ðara ðe healdan wile hálige þeáwas *every one who will maintain habits of holiness*, Cd. Th. 92, 20; Gen. 1531. [O. E. Homl. þeau, þeu: A. R. þeau: Laym. þeauwes, þewes, þæwes; *pl.*: Orm. þæw: O. and N. Chauc. þewes; *pl.*: Prompt. Parv. thewe, maner or condycyon *mos*: O. Sax. thau: O. H. Ger. dau.] v. freoðo-, fulwiht-, leód-, mann-, mynster-, regol-, un-þeáw; ge-þýwe, un-geþeáwe.

þeáw *a slave.* v. þeów.

þeáw-fæst; *adj.* I. *of good manners, of well-ordered life, moral, virtuous* :—Loth hine fægre heóld, þeáwfæst and geþyldig, on ðam þeódscipe, Cd. Th. 116, 26; Gen. 1942: (*Abraham*), 161, 8; Gen. 2662. Wunige hé mid þeáwfæstum mannum *maneat cum bene moratis hominibus*, L. Ecg. P. i. 10; Th. ii. 176, 23. II. *gentle* :—Sumum hé syleþ monna milde heortan, þeáwfæstne geþóht, Exon. Th. 299, 28; Crä. 109. v. un-þeáwfæst.

þeáwfæstness, e; *f. Adherence to the rules of right conduct or method, discipline, obedience to rule* :—Þeáwfæstnesse *disciplinae*, Hpt. Gl. 432, 34. Be sealmsanges ðeáwfæstnesse *de disciplina psallendi*, R. Ben. 45, 2. Ða cild mid steóre and þeáwfæstnysse (*cum disciplina*) heora endebyrdnysse healdon, 116, 9: R. Ben. Interl. 106, 12. On háligre þeáwfæstnesse, Homl. Ass. 40, 406, 404: Homl. Skt. ii. 28, 138. Þéningmen ðe þeáwfæstnysse him gebeódon (cf. þénas ðe his willan gefyllaþ, 65), Homl. Skt. i. pref., 62. Regoles gehýrsumnesse and þeáwfæstnesse *regule oboedientiam et disciplinam*, R. Ben. Interl. 103, 16. Ðú hatast ðeáwfæstnysse (*disciplinam*, Ps. 50, 17), Homl. Th. ii. 532, 2.

þeáw-full; *adj. Moral, virtuous* :—Oft hig (*devils*) beswícaþ þeáwfulle weras (ða ðeáwfullan, MS. A.), Wulfst. 250, 4. [Heo Godd thonkeden mid þeufulle (witfolle, 2nd MS.), worden, Laym. 1797. Mid þeaufule talen schurted ou, A. R. 422, 19. Ne beo þu nawt tu trusti ane to þi meidenhad wiðuten oðer god and þawfulle mihte, H. M. 45, 4.] v. un-þeáwfull.

þeáwian *to serve.* v. þeówian.

þeáwian *to make (well) mannered.* [Wel ðewed, Gen. and Ex. 1914. So boner and þewed, Allit. Pms. 59, 733.] v. ge-þeáwian.

þeáw-leás; *adj. Ill-mannered, ill-conditioned* :—Swýn ðe cyrþ tó meoxe æfter his ðweále, þeáwleás nýten, Homl. Th. ii. 380, 11. [For lust hath leve, the lond is theweles, P. S. 255, 19.] †

þeáw-líc; *adj.* I. *usual, customary* :—Sum wît mid sealfe his fét smyrode, swá swá hit þeáwlíc wæs on ðære þeóde, Homl. Ass. 41, 439. II. *moral, figurative* :—Þeáwlíc[r]e spæce *tropologiae, figurati sermonis*, Hpt. Gl. 432, 13. Þeáwlícre spæce *tropologiam misticum, moralem*, 410, 43. Wé willaþ secgan hú ðäs läc tó ús belimpaþ æfter ðeáwlícum andgite, Homl. Th. i. 116, 33: ii. 110, 26: 210, 27: Wulfst. 234, 10. [O. H. Ger. dau-líh *moralis*.]

þeáwlíce; *adv. In accordance with good manners, properly* :—Gáþ þeáwlíce . . . and standaþ þeáwlíce *incedite morigerate . . . et state disciplinabiliter*, Coll. Monast. Th. 36, 1–5. Ðá hé ðæt hæfde ðeáwlíce (*rite*) gesett, Bd. 3, 19; S. 549, 37.

þec. v. þú.

þeccan; *p.* þeahte, þehte; *pp.* þeaht *To cover*, (1) *to cover* an object with something :—Ic wreó mé, leáfum þecce, Cd. Th. 53, 29; Gen. 868. Ðú ðín sylf þecest líc mid leáfum, 54, 15; Gen. 877. Se ðe heofen þeceþ wolcnum *qui operit coelum nubibus*, Ps. Th. 146, 8. Mec (*a horn*) þeceþ mon golde and sylfore, Exon. Th. 395, 2; Rä. 15, 1. Hé þeahte bearn middangeardes wonnan wæge, Cd. Th. 83, 10; Gen. 1377. Git eágorstreám earmum þehton, Beo. Th. 1031; B. 513. Saga hwá mec þecce, Exon. Th. 381, 21; Rä. 2, 14. Mec ongon hold gewēdum þeccan, 391, 13; Rä. 10, 4. Hine mid hrægle wryón and sceome þeccan, Cd. Th. 95, 3; Gen. 1573: 58, 7; Gen. 942. Wæstmum þeaht, 115, 20; Gen. 1922. Hleówfeðrum þeaht, 165, 31; Gen. 2740. Yþum þeaht, Exon. Th. 392, 7; Rä. 11, 4. Helmum þeahte, Cd. Th. 120, 3; Gen. 1989. (2) *to serve as covering to* an object. Earn ðeceþ (*tegit*) nest his, Ps. Surt. ii. p. 192, 31. Mec hrím þeceþ, Exon. Th. 490, 11; Rä. 79, 9. Forst and snáw eorþan þeccaþ, 215, 7; Ph. 249. Ic gealgan þehte *I was stretched upon the cross*, Andr. Kmbl. 1932; An. 968: Apstls. Kmbl. 44; Ap. 22. Ealne middangeard mereflód þeahte, Exon. Th. 200, 18; Ph. 42: Cd. Th. 8, 1; Gen. 117. Líca gehwilc ðara ðe lifes gást þeahte *every body that had within it a living spirit*, 77, 28; Gen. 1282. On hwelcum hí (*Weland's bones*) hlæwa hrúsan þeccan, Met.

10, 43. Sió filmen biþ þeccende and wreónde ða wambe, Lchdm. ii. 242, 17. ¶ In the following passages Grein suggests that the form is quite a different word = *comburere*, and Cosijn (P. B. 8, 574) takes it to be connected with þícgan (but see þecgan); but, perhaps, the verb may be the same here as in the previous instances, and used with much the same force as *wrap* in such a phrase as *wrapt* in flames :—Byrneþ þurh fýres feng fugel (*the phenix*) mid neste . . . þonne brond þeceþ heoredreórges hús, Exon. Th. 212, 27; Ph. 216. Hine äd þeceþ, 223, 26; Ph. 365. Seó hyre bearn gesihþ brondas þeccan, 330, 7; Vy. 47. Beágas sceal brond fretan, æled þeccean, Beo. Th. 6022; B. 3015. [To dyche and to thecche, Piers P. 19, 232. O. L. Ger. thekkan: O. H. Ger. decchen *tegere, operire, velare* : Icel. þekja *to cover*.] v. be-, (bi-), ge-, ofer-, un-þeccan; þeccend; þacian.

þecc-bryce, es; *m. A tile* :—Þeccbrycum *imbricibus*, Hpt. Gl. 459, 42. Cf. þæc-tigele.

þeccend, es; *m. One who covers* or *protects, a protector* :—Ðú eart þeccend (*protector*) mín, Ps. Th. 70, 5, 2.

þecel[l]e. v. þæcele.

þecen, e; *f. A roof* :—Þecen *vel* róf *tectum*, Wrt. Voc. i. 26, 34. Þæcen, 81, 9. Of daliscre þecene *dedalei tecti*, ii. 139, 68: Exon. Th. 493, 21; Rä. 81, 34. Swä swä spearwa on ðecene (on efese ł on þecene, Ps. Lamb.) *sicut passer in tecto*, Ps. Spl. 101, 8. Under míne þecene, Mt. Kmbl. 8, 8: Lk. Skt. 7, 6: Homl. Th. i. 126, 30: Mk. Skt. 13, 15: Exon. Th. 431, 18; Rä. 46, 2. Híg þecena (getimbrena ł þæcena, Ps. Lamb.) *foenum tectorum*, Ps. Spl. 128, 5. [O. L. Ger. thecina.]

þecgan; *p.* þegde; *pp.* þegde *To take, consume* :—Hine þegeþ þurst *he is consumed by thirst*, Lchdm. ii. 60, 7: 74, 22. v. ä-, ge-, of-þecgan; þicgan.

þecge (?), an; *f. A receptacle* (?) :—On hærfeste man sceal ðacian, ðecgean and fald weoxian, scipena behweorfan, Anglia ix. 261, 17. Cf. þicgan.

þéde, þédum, þéfel, þéfan-, þéfe-þorn. v. þeówan, þýfel, þífe-þorn.

þeflan *to pant, to be agitated* :—Þefiendra *anhelantium*, Hpt. Gl. 406, 8. Þefian *aestuare in animo*, Dial. 1, 9 (Lye). [Cf. (?) Icel. þefja *to smell*.]

þeften. v. þyften.

þegan (this seems the regular strong form for the verb which usually has weak forms in the present, þícgan, q. v.) *to take, accept* :—Him wæs Godes egsa mära in gemyndum ðonne hé menniscum þrymme þegan, wolde *there was too great fear of God in his thoughts for him to wish to get human glory*, Exon. Th. 112, 8; Gú. 140.

þegen, þegn, þeng, þén, es; *m.* I. *a servant, one who does service for another* :—Þén *minister*, Wrt. Voc. i. 82, 24. Swä hwylc swä wyle betweox eów beón yldra, sý hé eówer þén (*minister*), Mt. Kmbl. 20, 26: 23, 11. Ðá bæd hé his ðeng (*ministrum*), ðæt hé him stówe gegearwode. Ðá wundrade se ðeng, Bd. 4, 24; S. 598, 20. Geleáffull ðegn (esne, Rush.) þeów, W. S.) *fidelis servus*, Mt. Kmbl. Lind. 24, 45, 46. Ðá þénas ðe ðæt wæter hlódon, Jn. Skt. 2, 9. II. *where the service is of a public or official character, an officer, minister* :—Þegn *lictor*, Wrt. Voc. ii. 93, 10: 52, 59. Ðe læs se déma ðé sylle ðäm þéne (ðegne, Lind.: dægne, Rush.), Mt. Kmbl. 5, 25. Ða weorcgeréfan and ða þénas (*praefecti operum et exactores*) cwædon tó ðam folce: 'Pharao být, ðæt eów mann ne sylle leng näu cef,' Ex. 5, 10. Ða þeówas and ða þegnas *servi et ministri*, Jn. Skt. 18, 18. Þegna *lictorum*, Wrt. Voc. ii. 84, 24: 52, 41. Ábeád þeódcyning þegnum sínum, ombihtscealcum, Cd. Th. 112, 12; Gen. 1869. Ða ealdras sendon hyra þénas, ðæt hig woldon hine gefón, Jn. Skt. 7, 32. Sangeras and mæssepreóstas and manigfealdlíce ciricean þegnas, Blickl. Homl. 207, 32. II a. figurative :—Deófolgieldum, ðäm wyrrestum wítes þegnum, Exon. Th. 251, 29; Jul. 152. III. *where the service is military, a soldier* :—Án Ueriatuses þegn *unus ex iis* (one of the victorious Lusitanians), Ors. 5, 2; Swt. 216, 21. Þa þénas (ðegnas, Lind. Rush.) *milites*, Jn. Skt. 19, 2. Ðá hét he his ðegnas (*milites*) hine sécan . . . 'Ðone forhycgend úra goda ðú mé helan woldest swýþor ðonne mínum ðegnum (*militibus*) secgean,' Bd. 1, 7; S. 477, 7–20. Hí sealdon ðäm þegenum (*militibus*) micyl feoh, Mt. Kmbl. 28, 12. Ic hæbbe þegnas (ðeignas, Lind.) under mé *habens sub me milites*, 8, 9. IV. *a follower of a great man, a retainer* :—Þegn, gesíþa *cliens*, i. *socius*, Wrt. Voc. ii. 131, 70. Thegn, degn, þegu *adsaeculam*, Txts. 42, 101. Þegn, Wrt. Voc. ii. 4, 43. Ðæs ðegenes lof is ðæs hláfordes wurðmynt, Homl. Th. ii. 562, 6. Ðä wæs ðærinne Dauid mid his monnum. Ðä cleopedon his ðegnas him tó and hine lærdon ðæt hé hine ofslóge *illic cum viris suis Dauid inerat, cum eum viri sui ad feriendum Saul accenderent*, Past. 28; Swt. 197, 17. Árás se ríca (*Beowulf*), ymb hine rinc manig, þegna heáp, Beo. Th. 805; B. 400. Þegna, cnihta, forspillendra þéna *parasitorum, incniht parasitus*, Hpt. Gl. 504, 18–21. Þénum *parasitis, incnihttum clientibus*, 514, 52–54. Ðegnum *pedisequis*, Wrt. Voc. ii. 84, 75. IV a. figurative :—Ðú wäst hwelce þeáwas ða woruldsælþa habbaþ . . . Gif ðú heora þegen beón wilt, Bt. 7, 2; Fox 18, 6. V. *a follower of a teacher, a disciple* :—Iôhannes, se deóra þegn, Blickl. Homl. 67, 22. His þeguas

lǽddon him tô ðone eosol, 71, 5 : 15, 13. On Sancte Petres naman, Cristes ðegnes, 205, 14. Ðeignas his (his discipuli ł his þegnas, Rush.) discipuli ejus, Mt. Kmbl. Lind, 5, 1. Cuêdon tô ðeignum his, 9, 11. Ðegnum, 10 : Jn. Skt. Lind. Rush. 18, 19. **V a.** in poetry, borrowing the terms of war :—Sint geþreáde þegnas mîne (St. Andrew's disciples), geonge gûðrincas, Andr. Kmbl. 782 ; An. 391. **VI.** one engaged in a king's or queen's service, whether in the household or in the country, a thane. The word in this case seems gradually to acquire a technical meaning, and to become a term denoting a class (v. þegen-riht, -wèr), containing, however, several degrees. To illustrate the wider sense in which the word could be·used, when the þegen is spoken of in relation to the king, the following passages may be cited. In the Chronicle an. 897 'manige ðara sêlestena cynges þêna' includes two bishops and three aldermen, as well as a cynges þegn and a cynges horsþegn. In a charter Cnut greets 'ealle mîne þegnas, twelfhynde and twihynde' (the twihynde man is a ceorl : cf. too, 'ealne his leódscype, twelfhynde and twyhynde,' Chart. Erl. 229, 19), Cod. Dip. Kmbl. iv. 9, 30. The word seems general, too, in the passage, 'Weorðscipes wyrþe ǽlc be his mǽðe, eorl and ceorl, þegen and þeóden,' L. R. 1 ; Th. i. 190, 13. The more limited sense seems to belong to the word in the following :—Gif þegen geþeáh ðæt hê weard tô eorle, L. R. 5 ; Th. i. 192, 7. Cf. §§ 2, 6 ; and see L. In. 45, L. C. S. 72, L. M. L. given below. In some cases, too, it will be seen that the term implies military service, as when de militia regis juvenis is translated sum geong ðæs cyninges ðegin (see also other passages below from Bd. 4, 3, and 5, 13) ; in others, the service is that of the household, v. Bd. 4, 3 ; S. 567, 21, and bûr-, disc-, hrægl-þegen ; in others it is official work in the country. For the development of the class of thanes in England, see Stubbs' Const. Hist. s. v. thegn ; Kemble's Saxons in England, I. c. 7, II. c. 3 ; Schmid, A. S. Gesetz. s. v. þegen. (1) where the word is used of other than Teutonic peoples :—Wæs his (St. Martin's) fæder ǽrest cyninges þegn, and geðeáh ðæt hê wæs cininges þegna aldorman, Blickl. Homl. 211, 21. (Cf. His (St. Martin's) fæder wæs æðelboren, ǽrest cempa, and siððan cempena ealdor, Homl. Th. ii. 498, 25. St. Martin's father was a military tribune.) Þegn satrapa, Wrt. Voc. i. 42, 17. Xerxis þegn wæs hâten Marðonius, Ors. 2, 5 ; Swt. 82, 28. Wê sǽdon, ðæt hê wære ryhtwîsra ðâ ðâ hê ðeng wæs ðonne hê wǽre siððan hê kyning wæs. Ðâ ðâ hê ðegn wæs hê his feóud ne dorste ofsleán David rectior fuit in servitio, quam cum pervenit ad regnum. Servus adversarium ferire timuit, Past. 50 ; Swt. 393, 2–6. Tarcuinius ôðerne ðegn ongeán sende, Ors. 2, 3 ; Swt. 68, 17. Pharaones þegnas servi Pharaonis, Ex. 10, 7. His (Ulysses') þegnas him ne mihton leng mid gewunian, Bt. 38, 1 ; Fox 194, 27. Mænegum cyninges (king of Egypt) þegnum, Cd. Th. 111, 5 ; Gen. 1851. (1 a) where the Deity is the king served :—Metodes ðegn, Abraham, Cd. Th. 176, 6 ; Gen. 2907. Dryhtnes þegn (Guthlac), Exon. Th. 143, 22 ; Gû. 665. Ðû cyninges eart þegen geþungen, Andr. Kmbl. 1055 ; An. 528. Wuldres þegn, engel Drihtnes, Cd. Th. 136, 31 ; Gen. 2266. Ðæt is micel wundor, ðæt wolde þeóden þolian, ðæt wurde þegn swâ monig forlǽdd, 37, 30 ; Gen. 597. (1 b) figurative :—Fuglas þringaþ ymbe æþelne (the phenix), ǽghwylc wille wesan þegn and þeów þeódne mǽrum, Exon. Th. 209, 3 ; Ph. 165. (2) where the word applies to Englishmen or to other Teutonic peoples :—Hêr Hengest and Æsc gefuhton uuiþ Walas and hiera þegn ân wearþ ofslægen, Chr. 465 ; Erl. 12, 23. Gest hine clǽnsie sylfes âðe, swylce cyninges þeng, L. Wih. 20 ; Th. i. 40, 20. Lilla se cyninges ðegn him se holdesta minister regi amicissimus, Bd. 2, 9 ; S. 511, 22. Sum geong ðæs cyninges ðegin de militia ejus juvenis, 4, 22 ; S. 590, 33. Ðâ ondrǽdde hê andettan ðæt hê cyninges ðegen wǽre, ac sæde, ðæt hê folclîc man wǽre, and ðæt hê forðon in ða fyrd côme, ðæt hê sceolde cyninges ðegnum heora mete lǽdan timuit se militem fuisse confiteri ; rusticum se fuisse respondit, et propter victum militibus adferendum in expeditionem se venisse testatus est, S. 591, 5–9. Hê sǽgde ðæt hê wǽre cyninges ðeng ministrum se regis fuisse manifestans, 591, 38. Hê ðære ylcan cwêne ðeng (minister) wæs, 592, 13. Sum wer wæs on lǽwedum hâde ðæs cyninges ðegn vir in laico habitu atque officio militari positus, 5, 13 ; S. 632, 8. Þegn se ðe on handa bær ealowǽge, Beo. Th. 993 ; B. 494. Þegn Hrôðgâres, êgweard, 475 ; B. 235. Eádwold cynges ðegen, Chr. 905 ; Erl. 98, 28. Gif mon cyninges þegn beteó, gif hê hine lâdian dyrre, dô hê ðæt mid .xii. cininges þegnum. Gif man ðone man betŷhþ ðe biþ lǽssa maga (mâga?) ðonne se cyninges þegn, lâdige hê hine mid .xi. his gelîcena and mid ânum cyninges þægne, L. A. G. 3 ; Th. i. 154, 5–9. Gif cyninges þegn ætsace . . . gilde .x. healfmearc (cf. the next two sections), L. N. P. L. 51 ; Th. ii. 298, 7 : 58 ; Th. ii. 300, 3 : 60 ; Th. ii. 300, 9. Burgbryce mon sceal bêtan ealdormonnes .lxxx. sciłł., cyninges þegnes .lx. sciłł., gesîdcundes monnes landhæbbendes .xxxv. sciłł., L. In. 45 ; Th. i. 130, 9. Eorles heregeata syndon . . . And syþþan cyninges þegenes (þegnas, MS. G.) ðe him nŷhste syndon . . . And medemra þegna . . . And cyninges þegnes heregeata inne mid Denum ðe his sôcne hæbbe feówer pund. And gif hê tô ðam cyninge furðor cŷðde hæbbe . . ., L. C. S. 72 ; Th. i. 414, 4–20. Ðegenes lagu is, ðæt hê sŷ his bôcrihtes wyrðe, and ðæt hê ðreó ðinc of his lande dô, fyrdfæreld and burhbôte

and brycgeweorc. Eác of manegum landum mâre landriht ârîst tô cyniges gebanne, L. R. S. 1 ; Th. i. 432, 4–7. Ðegnes wergild is syx swâ micel (swâ ceorles). Ðonne biþ cynges ânfeald wergild .vi. þegna wer be Myrcna lage, L. M. L. ; Th. i. 190, 2–5. Hié (the Danes) sealdon (Ceólwulfe) ânum unwîsan cyninges þegne Miercna rîce tô haldanne, Chr. 874 ; Erl. 76, 27. Nân man nâge nâne sôcne ofer cynges þegen bûton cyng sylf, L. Eth. iii. 11 ; Th. i. 296, 23. Þæs cyninges þegnas (cf. ða men ðe mid ðam cyninge wǽrun, Erl. 48, 31), Chr. 755 ; Erl. 50, 3, 9. Hê wæs hyre (the queen's) ðêna hire hûses and hire gefêrscipes oferealdormonn erat primus ministrorum et princeps domus ejus, Bd. 4, 3 ; S. 567, 21. Se cyning gestôd æt ðam fŷre mid his ðegnum (ministris), 3, 14 ; S. 540, 34. Wê willaþ ðæt man namige on ǽlcon wǽpengetæce .ii. trŷwe þegnas and ǽnne mæssepreóst, L. N. P. L. 57 ; Th. ii. 298, 31. In the two following passages, though translations, the ideas are probably English :—Cyniugas ne magan nǽnne weorþscipe forþ bringan bûton heora þegna (servientium) fultume. Hwæt wille wê secgan be ðâm ðegnum (familiaribus ; cf. folgerum, l. 10), Bt. 29, 1–2 ; Fox 104, 12–15. Mid miclon gefêrscipe hiora þegna, and ða bióþ mid fetlum and mid gyldenum hyltsweordum and mid manigfealdum heregeatwum gehyrste, 37, 1 ; Fox 184, 4. **VI a.** a thane who served a bishop :—Wulfhere bisceopes ðegn, Chr. 1001 ; Erl. 136, 8. Ic Leófinc bisceop gebôcige sumne dǽl landes mînan holdan and getreówan þegene, ðam is Ægelric nama, for his eádmôdre gehêrsumnysse, Chart. Erl. 242, 11. Ic (Cnut) cŷðe, ðæt ic hæbbe geunnen him (archbishop Æthelnoth), ðæt hê beó his saca and sôcne wyrðe ofer his âgene menn and ofer swâ feala þegna swâ ic him tôlætan hæbbe, 233, 6. **VI b.** one engaged in the service of a republic :—Scipia, se betsta Rômâna þegn (se besta and se sêlesta Rômâna witena and þegena, MS. C.), Ors. 5, 4 ; Swt. 224, 24. **VII.** a person of rank, one of a class higher than the ordinary freeman (ceorl). v. þegen-boren :—Þegn primas, Wrt. Voc. i. 42, 14. Ðegn optimas, Ælfc. Gr. 9, 25 ; Zup. 50, 3. Ðeáh þrǽla hwylc hlâforde æthleápe, and hit æfter ðam geweorðe, ðæt wǽpngewrixl weorðe gemǽne þegene and þrǽle, gyf þrǽl ðæne þegen âfylle, licge ǽgylde ; and gyf se þegen ðæne þrǽl, ðe hê ǽr âhte, âfylle, gylde þegengylde, Wulfst. 162, 5–10. Ælc dohtig man on Kænt and on Sûð-Sexan, on þegenan and on ceorlan, Cod. Dip. Kmbl. iv. 11, 7. **VIII.** a brave man, noble man, good warrior ; vir fortis. v. þegen-lîc, -lîce, -scipe, III :—Gif gê swelce þegnas synt, swelce gê wênaþ ðæt gê sién, ðonne sceoldon gê lustlîce eówre âgnu brocu ârǽfnan, Ors. 3, 7 ; Swt. 120, 7. Ðæt wæs swîðe sweotol, ðæt hié ðâ wǽron beteran þegnas ðonne hié nû sién, ðæt hié ðæs gewinnes geswîcan noldon, 4, 9 ; Swt. 192, 32. **VIII a.** in poetry the word is used, like eorl, as a complimentary term for man, warrior :—Swylc sceolde secg wesan, þegn æt þearfe, Beo. Th. 5411 ; B. 2709. Se þegn (St. John) wæs on wynne, Exon. Th. 462, 21 ; Hö. 55. Þances gleáw þegn (St. Andrew), Andr. Kmbl. 1114 ; An. 557. Ðam þegne (Adam) ongan his hige hweorfan, Cd. Th. 44, 7 ; Gen. 705. Scyle âscian deóphŷdig mon . . . ne sceal ðæs âþreótan þegn môdigne, Exon. Th. 348, 1 ; Sch. 21. Ðæt micle mord menn ne þorfton, þegnas þolian, Cd. Th. 40, 18 ; Gen. 641. Wlance þegenas, unearge men, Byrht. Th. 137, 53 ; By. 205. Ne sceolon mê on ðære þeóde þegenas ætwîtan (cf. stedefæste hæleð, 139, 5 ; By. 249), 138, 15 ; By. 220. ¶ The word is applied to Christ :—Þegen mid þreáte, þeóden engla, Cd. Th. 288, 27 ; Sat. 388. [O. Sax. þegan : O. H. Ger. degan masculus, herus, miles, defensor : Icel. þegn.] v. ærn-, ambeht-, bed-, bûr-, burh-, cyric-, disc-, duru-, ealdor-, forþ-, gum-, hand-, heáh-, heal-, helle-, hrægl-, mægen-, mæsse-, magu-, mete-, scîr-, scôh-, sele-, tintreg-, weofod-, weoruld-, wíc-, wíf-þegen (-þegn).

þegen-boren; adj. Of gentle birth. v. þegen, VII :—Sŷ hê þegenboren, sŷ hê ceorlboren, L. O. D. 5 ; Th. i. 354, 20.

þegen-gilde, es ; n. The wergild for a thane :—Gyf þrǽl þegen fullîce âfylle, licge ǽgylde ; and gyf se þegen þæne þrǽl, ðe hê ǽr âhte, fullîce âfylle, gylde þegengylde, Wulfst. 162, 10. [Icel. þegn-gildi the wergild for a þegn.]

þegen-hyse; pl. -hyssas ; m. A follower, attendant :—Ðegnhyssas clientes (the passage in Aldhelm is : Ejusdem nefandae militiae tam calones et clientes . . . quam satrapae et proceres, Wrt. Voc. ii. 76, 73 ; 17, 74.

þegen-lagu, e ; f. Thane-law, the legal rights and privileges which attached to the rank of thane :—Se (the priest) ðe ðæs (concubinage) geswîcan wille and clænnesse healdan, hæbbe hê Godes miltse, and tô woruldwurðscipe sî hê þegenlage wyrðe as regards worldly dignity let him rank as a thane, L. C. E. 6 ; Th. i. 364, 16 : Wulfst. 270, 32. v. þegen-riht.

þegen-lîc; adj. Manly, brave, manful. v. þegen, VIII :—Andreas is gereht ðegenlîc, Homl. Th. i. 586, 11. Mâran lufe nimþ se heretoga on gefeohte tô ðam cempan, ðe æfter fleáme his wiðerwinnan ðegenlîce oferwinþ, ðonne tô ðam ðe mid fleáme ne ætwand, ne ðeáh on nânum gecampe nâht ðegenlîces ne gefremede, 342, 5.

þegenlîce; adv. Bravely, manfully, like a brave man, gallantly :—Beó ðû gehyrt and hicg þegenlîce tu confortare et viriliter age, Jos. 1, 18 : Homl. Skt. ii. 25, 248. Heó tô ðâm þegnon cwæð : 'Dôð þegnlîce and wel ; âbeódaþ mîne ǽrende tô ðam gemôte,' Chart. Th. 337, 36.

Gif hé ðegenlíce earfoðnysse forberþ, Homl. Th. i. 586, 19 : Homl. Skt. i. 11, 3. Ealle þeóda spræcon hū ðegenlíce hí fuhton, ii. 25, 324. Hé læg ðegenlíce ðeóðne gehende *he lay like a warrior close to his lord,* Byrht. Th. 140, 26; By. 294. [*O. L. Ger.* thegenlícho *viriliter : O. H. Ger.* thegan-, degan-lícho.] v. preceding word.

þegen-ræden[n], e ; *f. The condition of being a þegen, service :*—Þegnrædenne oððe híwrædenne *clientele,* Wrt. Voc. ii. 24, 44. Híréd-lícre þēnrædene *familiaris clientelae,* Hpt. Gl. 504, 46. Manige men of cyninges þegenrædene tó Cristes þeówdóme gecyrdon, Blickl. Homl. 173, 17. Cf. þegen-scipe.

þegen-riht, es ; *n. Thane-right, the legal rights and privileges which attached to the rank of thane* (e. g. Mæssepreóstes āð and woruldþegenes is on Engla lage efendýre . . . Twelfhyndes mannes (*a thane's*) āð forstent .vi. ceorla āð, L. O. 12, 13; Th. i. 182, 14–19) :—Se mæssepreóst biþ þegenrihtes wyrðe, L. O. 12; Th. i. 182, 17 ; L. Eth. v. 21 : vi. 5; Th. i. 306, 21 : vi. 5; Th. i. 316, 14. Gif ceorl geþeáh ðæt hé hæfde fullíce fíf hída ágenes landes . . . and sundernote on cynges healle, ðonne wæs hé ðononforð þegenrihtes weorðe, L. R. 2; Th. i. 190, 18. Gif massere geþeáh ðæt hé férde þrige ofer wídsǽ, se wæs þegenrihtes weorðe, 6; Th. i. 192, 10. v. þegen-lagu, -wer.

þegen-scipe, es ; *m.* I. *thaneship, the status of thane :*—Se dēma ðe óðrum wóh dēme . . . þolige hé his þegenscipes, L. Edg. ii. 3 ; Th. 266, 18 : L. C. S 15; Th. i. 384, 13. II. *a body of thanes :*—Scē Adrianes wæs ðæs cáseres ðegnscipes ealdorman, ðe Maximianus wæs nemned, Shrn. 59, 24. III. *bravery, manfulness, gallantry.* v. þegen, VIII, þegenlíc :—Beóþ nū gehyrte, and healdaþ mid ðegenscipe ða hálgan Godes ǽ, Homl. Skt. ii. 25, 258. Hé him eft his ríce tó forlét for his þegnscipe (*ob testimonium virtutis*), Ors. 3, 9; Swt. 132, 24. Alexander his ǽrestan ðegnscipe on ðon gecýþde, ðá hé ealle Crēcas mid his snyttro on his geweald geniédde, ealle ða ðe wið hiene gewin up áhófon *Alexander primam experientiam animi et virtutis suae, compressis celeriter Graecorum motibus, dedit,* Swt. 122, 32. IV. in that part of the Genesis which is thought to show Old Saxon influence, the word occurs with the meaning of *service to a lord,* like the Old Saxon *thegan-skepi :*—Nis mē on worulde mód ǣniges þegnscipes, Cd. Th. 51, 33; Gen. 836. On þegnscipe þeówian, 46, 15; Gen. 744. Hié þegnscipe Godes forgýmdon, 21, 19 ; Gen. 326. [*Icel.* þegnskapr *honour* (as in on one's *honour*); *liberality; allegiance of a* þegn.]

þegen-scolu, e ; *f. A band of thanes, a following :*—Þegenscole *clientele,* Wrt. Voc. ii. 18, 1.

þegen-sorh, -sorge ; *f. Sorrow for the loss of thanes :*—Grendel on reste genam þrítig þegna . . . Mǽre þeóden þegnsorge dreáh, Beo. Th. 263; B. 131.

þegen-weorod, es ; *n. A host of thanes :*—Ðæt wé tó ðam hýhstan hróf gestígan, ðǽr is geþungen þegnweorud, Exon. Th. 47, 6 ; Cri. 751.

þegen-wer, es ; *m. The wer-gild of a thane :*—Hé sý þegenweres and þegenrihtes wyrðe, L. Eth. v. 9 ; Th. i. 306, 21 : vi. 5 ; Th. i. 316, 14. Gif weofodþén se bóca tǽcinge his ágen líf rihtlíce fadige, ðonne sí hé fulles þegnweres and weorðscipes wurðe, ix. 28 ; Th. i. 346, 18. Cf. Mæsseþegnes and woruldþegnes wergyld is .ii. þúsend þrymsa, L. Wg. 5 ; Th. i. 186, 10. v. þegen-riht.

þégh, þegin, þegn, þegnen. v. þeóh, þegen, þignen.

þegnest (? *related to* þeguian *as O. Sax.* thionost, *O. L. Ger.* thianust, thienest, *O. H. Ger.* dionôst, *Icel.* þjónusta *are to verbs* thionôn, dionôn, þjóna *respectively; or* (?) þenest, þeónest *the English form corresponding to the nouns given above. The passage in which the word occurs separately refers to Germany, so perhaps the German form has been borrowed* (?) : *where it occurs as the first part of a compound the form is* þeónest ; *in the same passage, which is late,* eó *is written where* ē *is the more regular form, as* heót *for* hēt ; *perhaps, however, the Scandinavian form has influenced the English) Service :*—Þæs ilcan gēres fór Aldrēd biscop tó Colne ofer sǽ, and weard ðǽr underfangen mid mycclan weorðscipe, and him geaf ǽgðer þeneste ðe se biscop on Colone and se cásere, Chr. 1054; Erl. 189, 25. (This passage occurs in only one MS.) v. next word, and þegnisc.

þegnest-mann (?), es ; *m. A thane :*—Þás (a number of bishops, aldermen and others, who are named) and feola óþre kyninges þeónestmen (*but cf. Icel.* þjónustu-maðr *a liegeman*), Chr. 656; Erl. 33, 9. v. preceding word.

þegnestre, þēnestre, an ; *f. A female servant :*—þēnestran *cultricem, ministram,* Hpt. Gl. 438, 33. v. next word.

þegnian; *p.* ode; *pp.* od. I. *to serve* a person (*dat.*), do a person (*dat.*) *service, minister to, attend* upon :—Hwæðer ys yldra ðe se ðe ðénaþ ðe se ðe sitt *quis major est, qui recumbit? an qui ministrat?* Lk. Skt. 22, 27. On heáhsetle siteþ self cyning (God), and ðiós síde gesceaft þénaþ and þíówaþ, Met. 29, 77. Gif him árlíce esne þénaþ, Exon. Th. 430, 9 ; Rä. 44, 5 : 403, 28 ; Rä. 22, 14. Þénaþ *prosequitur,* Hpt. Gl. 451, 57. Hé (*a king*) bioþ swíþe anlíc ðara his þegna sumum ðe him þéniaþ, Bt. 37, 1 ; Fox 186, 12. Gif hē nǽre sóþ God, nā him englas ne þegnodon. On ðisse bysene is gecýþed, ðæt eallum geleáffullum mannum englas þegniaþ, ðonne hí habbaþ deófol oferswíþed, Blickl. Homl. 35, 1–4. Ic geornlíce Gode þegnode þurh holdne hyge, Cd. Th. 37, 7 ; Gen. 585.

Hē mē holdlíce ðegnade *mihi ministrabat,* Ps. Th. 100, 6. Ne com ic tó ðon on eorðan ðæt mē mon ðénode, ac tó ðon ðæt ic wolde ðegnian *filius hominis non venit ministrari, sed ministrare* (Mt. 20, 28), Past. 41 ; Swt. 301, 2. Iósep hæfde mycele gife æt his hláforde and þénode (*ministrabat*) him, Gen. 39, 4 : 40, 4. Englas him þegnedan, Blickl. Homl. 27, 22. Þegnedon, 33, 34. Þéna me *ministra mihi,* Lk. Skt. 17, 8. Ǽlc wēnþ ðæt hé þenige Gode *omnis arbitretur obsequium se praestare Deo,* Jn. Skt. 16, 2. Se biscop and se mæssepreóst, gif hí mid rihte willaþ Gode þeówian, ðonne sceolan hí þegnian dæghwamlíce Godes folce *the bishop and the priest, if they desire to serve God aright, must minister daily to God's people,* Blickl. Homl. 45, 30. Ne þúhte hit mē náuht rihtlíc, gif him sceoldan þeówe men þenigan (*þénian,* Cott. MS.), Bt. 41, 2 ; Fox 244, 27. On ðam húse hyra ðeáw wæs, ðæt hí ða untruman in lǽdan sceoldan, and him ætsomne ðénigean, Bd. 4, 24 ; S. 538, 29. I a. *where the instrument with which service is performed is given :*—Mec lǽdgeteónan þreátedon ; ic him þénode deóran sweorde swā hit gedéfe wæs *I served my foes with my good sword, as was fitting,* Beo. Th. 1125 ; B. 560. Ðás bóc Leófríc gef Scō Petro and eallum his æftergengum intó Exancestre Gode mid tó þénienne, Chart. Erl. 253, 12. II. *to serve* food, *to supply* wants, *minister* to necessities, *provide.* v. þegnung, V :—Ðæt ylce wæter eallum ðyder cumendum his heofonlíce gife genihtsumnesse ðegnaþ *aqua sufficientem cunctis illo advenientibus gratiae suae coelestis copiam ministrat,* Bd. 4, 28 ; S. 605, 32. Seó mýse is seó bóclíce lár, seó ðe ús ðénaþ lífes hláf, Homl. Th. ii. 114, 26. Martha gearwode ðam Hǽlende ǽfengereordu . . . Martha wæs geornful ðæt heó ðon Hǽlende tó gecwēmnesse þegnode (þénode, Lk. Skt. 10, 40) : heó him tó cwæþ : 'Hwý nelt ðú gēman ðæt mín sweostor mē lǽt áne þegnian (þénian, Lk. Skt.)?, Blickl. Homl. 67, 25–31. Malchus, ða þénunga ðe hé ðider bróhte, heom geornlíce þénode, ðæt hí be dǽle hí gereordodon, Homl. Skt. i. 23, 240. Se him ða gerýno ðénode (*ministrare solebat*) ðæs hálgan geleáfan, Bd. 3, 23 ; S. 554, 17. Hwænne gesáwe wē ðe hingrigendne oððe þyrstendne . . . and wē ne þénedon ðē?, Mt. Kmbl. 25, 44. Eallum Godes ðearfum man sceall weldǽda þénian, Homl. Th. i. 514, 5. Hé him bigleofan ðénian wolde, ii. 128, 29. Hé hēt hire þénian of his ǣtmetum, Homl. Ass. 110, 268. III. *to serve* an office, *administer, perform the duties of* an office :—Ðá hé ðā monig geár biscophád ðegnade *qui cum annis multis episcopatum administraret,* Bd. 3, 23 ; S. 555, 7. Ðénade, 4, 3 ; S. 566, 28, 40. Ðegnode Willferþ ða bisceopðénunge fíf geár *Vilfrid annos quinque officium episcopatus exercebat,* 4, 13 ; S. 583, 14. Medomlíce ðénian ða ðénunga *officium ministrare digne,* Past. 1 ; Swt. 27, 10. [Þe king him gon to þeinen, þæ quene bar to drinken, Laym. 30786. A þusen cnihtes þeineden þan kinge . . . þas beorn þa sunde from kuchene to þan kinge, 24595.] v. ge-, under-þegnian.

þegnisc (? cf. *-ska* nouns in Icelandic (?) ; or see þegnest (?)) *service :*—Þat sáwulgesceot sceulon ða canonicas habban, and swilce þēnisce dón for hig swilce hig ágon tó dóne, Chart. Th. 609, 16.

-þegnsum. v. ge-þēnsum.

þegnung, e ; *f. Service, ministration.* I. *service, good office done by one to another :*—Englas beóð tó ðegnunge gǽstum fram Gode hider on world sended, Blickl. Homl. 209, 23. Ne mihte se mānfulla ēhtere mid nánre ðénunge ðám lytlingum swā micclum fremian, Homl. Th. i. 84, 10. Þénunge *patrocinium* (*ut puellulas ad patrocinium vitae impendant,* Ald. 69), Hpt. Gl. 519, 2. Ða Ebrēiscan wíf cunnon þénunga *obstetricandi habent scientiam,* Ex. 1, 19. v. þignen. Hé brǽd hine on feala bleóna þurh deófles þegnunga *he changed himself into many forms by the devil's good offices,* Blickl. Homl. 175, 5. I a. *service rendered by things, use :*—Hé wæs lama and eallra his lima ðénunge benumen *deficiente omni membrorum officio,* Bd. 5, 5 ; S. 617, 38. In ðegnunge *in use;* in procinctu, Wrt. Voc. ii. 111, 16. Hé him beád his recedes hleów and þegnunge *he offered them the shelter and use of his house,* Cd. Th. 147, 19; Gen. 2442. I b. *use made of things :*—Nǽnig hí (*the cups*) hrínan dorste, ne ne wolde, būtan tó his neódþearflícre ðénunge (*ad usum necessarium*), Bd. 2, 16 ; S. 520, 8. II. *service* to a lord or master :—Á tó his (*king Oswine*) folgoþe and tó his ðénustan men cóman *ad ejus ministerium viri nobilissimi concurrerent,* Bd. 3, 14 ; S. 540, 12. Wuldres áras ðū tó þegnunge ðínre gesettest, Elen. Kmbl. 1474 ; El. 739. Oþþæt ðū gefylle ðíne þegnunge, tó ðære ðe ðú sended eart, Blickl. Homl. 233, 28. Pharao geþencþ ðíne þénunga *recordabitur Pharao ministerii tui,* Gen. 40, 13. Ðénunga *obsequia,* Wrt. Voc. ii. 62, 44. Gif him mon oftíhþ ðara þénunga, Bt. 37, 1 ; Fox 186, 10, 14. Þegnunga, Met. 25, 24, 32. Ðara gumena ðe him mid þegnungum þringaþ ymbe útan, 25, 28. For ðý ða óþra gesceafta þeówe sint, hí healdaþ hiora þénunga, Bt. 41, 3 ; Fox 248, 18. Þegnunga, Met. 11, 46. II a. *service, obedience, suit and service :*—Ðæt selfe wæter þegnunge gearwode beforan his fótum *the very water showed itself to be at his command by retreating before his feet* (cf. him gearu sóna þurh streámræce strǽt wæs gerýmed, symble wæs drýg: folde fram flóde, swā his fót gestóp, Andr. Kmbl. 3157–; An. 1581–), Blickl. Homl. 247, 10. III. *service* of an official, *office, official employment, ministry :*—Ðæs láreówdómes ðénung *praedicationis officium,* Past.

7; Swt. 47, 20. Hé wilnode ðære ðegnunga ðæs láriówdómes, Swt. 49, 15. Ðære clǽnan ðegnenga ðæs sacerdháde, Swt. 51, 2. Gefyldum dagum his (*St. Augustine*) dénunge *completis diebus officii sui*, Bd. 2, 3; S. 505, 3. Gif ðú wást ðæt ic unrihtlíce bisceopháde onfénge, ic lustlíce fram ðære déninge (*officio*) gewíte, 4, 2; S. 566, 6. Ic wæs gesett tó mínre þénunge (*officio*; office of cupbearer), Gen. 41, 13. Nǽnig sý beládod fram ðære kycenan þénunge *nullus excusetur a coquine officio*, R. Ben. 58, 14. Hé ðæs godspelleres þegnunga gefylde *he filled the office of evangelist*, Blickl. Homl. 167, 9. Ða eorðlican hláfordas sint tó ðæm gesette ðæt hié ða endebyrdnesse and ða ðegnunga hiora hiéredum gebrytnige *terrenae domus dominus famulorum ordines ministeriaque dispertiens*, Past. 44; Swt. 319, 20. **IV.** *the act of serving* in an official capacity:—Ðæt Leuíes mǽgð stóde beforan him on þénunge, Deut. 10, 8. **IV a.** *a service, an official performance, a service* of religion, *an office* of the church, *a ceremonial* or *ritual service*:—þénung *officium*, onsægung *immolatio*, Wrt. Voc. i. 28, 48–49. Þénunge *sacrificio*, Hpt. Gl. 521, 70. Ðis godspel belimpþ tó ðysses dæges ðénunge, Homl. Th. i. 104, 4. Byrgincge þénuncge *sepulturae officio*, Anglia xiii. 444, 1124. Nǽnigum heora álýfed sí ǽnige sacerdlíce ðénunge dón *nulli eorum liceat ullum officium sacerdotale agere*, Bd. 4, 5; S. 573, 4. Ðæt gé ða ðénunge fulwihte (*ministerium baptizandi*) æfter ðeáwe ðære hálgan Rómániscan cyricean gefyllan, 2, 2; S. 503, 21. Ða clǽnan þénunga *lauta* (*supernarum*) *munia* (*rerum*, Ald. 144), Wrt. Voc. ii. 90, 9 : 52, 51. Ðegnunge *munia*, 91, 32. Þénunge, 57, 8. Ðá his þénunga dagas (*dies officii eius*) gefyllede wǽron, Lk. Skt. 1, 23. Þénungum *culturis, ministeriis*, Hpt. Gl. 495, 27. Mænigfealdum þénungum *exequiis pluribus*, Wrt. Voc. ii. 144, 79. **IV b.** *a service, the formulas used in a service*:—Swíðe feáwa wǽron behíonan Humbre ðe hiora déninga cúðen understondan, Past. pref.; Swt. 3, 14. **V.** *service* of food, *a meal, food served, food, provision.* v. þegnian, **II,** and cf. gémung *for similar specialization of a general term*:—Ðá Drihtnes þénung wæs gemacod *cena facta*, Jn. Skt. 13, 2. Hé árás fram his þénunge *surgit a cena*, 4. Gearwa úre þénunga ðæt hig magon etan mid mé *instrue convivium, quoniam mecum sunt comesturi*, Gen. 43, 16: Mt. Kmbl. 26, 17. Ic wolde ðíne ðénunge sylf nú gearcian, gif ic mé mid féðunge ferian mihte, Homl. Th. ii. 134, 31. Malchus hæfde mid him eáþelícan fódan, and com tó his geséran . . . and ða eáðelícan þénunga ðe hé ðider bróhte heom ðénode, ðæt hí be dǽle hí gereordodon, Homl. Skt. i. 23, 233–240. Ðonne man fæste, ðonne dǽle man ða þénunga ðe man brúcan sceolde ealle Godes þearfan *when a man fasts, then let all the food that would have been used be distributed to the poor*, L. P. M. 3; Th. ii. 286, 28. Hé hét hire þénian of his ágenum þénungum and his ǽstmetum, ac heó nolde his sanda brúcan, Homl. Ass. 110, 269. **VI.** in a personal sense, *a following, retinue, train*, cf. folgaþ:—Seó hell and se deáð and heora árleásan þénunga wǽron áforhtode, Nic. 28; Thw. 16, 10. [Fer (werpð) manifeald þeninge *fire does service of many kinds*, O. E. Homl. i. 233, 26.] v. æfen-, bisceop-, cyric-, fulluht-, geár-, heáh-, líc-, mynster-, tíd-, tó-, úht-, weofod-, wic-þegnung (-þénung), *and following words.*

þegnung-bóc, e; f. *A service-book, a book giving the religious services that were to be performed*:—Leuiticus on Grécisc and ministerialis on Lýden, ðæt ys þénungbóc on Englisc, for ðam ðara sacerda þénunga sind ðár áwritene, Lev. pref. Sume úre ðéningbéc onginnaþ on Aduentum Domini, Homl. Th. i. 98, 26.

þegnung-fæt, es; n. *A vessel used in the service of the kitchen*:—Ðære kycenan wicþénas . . . heora þéningfata clǽne and hále ðam hordere betǽcen; se hordere eft ðære tóweardan wucan wicþénum ða ylcan þéningfata betǽce, R. Ben. 59, 6–12.

þegnung-gást, es; m. *A ministering spirit*:—Englas beóþ tó déninggástum fram Gode hider on worulde ásende, ðæt hí beón on fultume his gecorenum (*nonne angeli sunt administratorii spiritus, in ministerium missi propter eos, qui haereditatem capient salutis?* Heb. 1, 14), Homl. Th. i. 510, 15.

þegnung-hús, es; n. *A house in which an employment is carried on, a workshop*:—Þéninghúsum *officinis*, Germ. 394, 267.

þegnung-mann (þéning-, þénig-), es; m. **I.** in a general sense, *a serving-man, attendant*:—Wæs amang ðam Malchus heora ðénigmann, and ða eáðelícan þénunga ðe hé ðider bróhte heom geornlíce þénode, Homl. Skt. i. 23, 239. Árás Malchus heora þeningmann, and dyde eall swá his gewuna wæs, nam mid him sumne dǽl feós . . . ðæt feoh bær tó porte, 472–486 : 447. Se Hǽlend hét ða ðénigmen áfyllan six stǽnene fatu, Homl. Th. i. 58, 12. **II.** *a servant of a lord* or *king, a thane, minister*, (a) in a general sense:—Sint tó manienne ða ðe mildheortlíce sellaþ ðæt hié habbaþ, ðæt hié angieten ðæt hié sint gesette ðæm hefencundan Gode tó ðéningmannum tó dǽlanne ðás lǽnan gód (*ut a coelesti Domino dispensatores se positos subditorum temporalium agnoscant*), Past. 44; Swt. 321, 7. Án woruldcynincg hæfþ fela þegna; hé ne mæg beón wurðful cyninc búton hé hæbbe swylce þéningmen ðe þeáwfæstnysse him gebeódon, Homl. Skt. i. pref., 62. (b) as a technical English term:—Míne (*Alfred's*) ealdormenn and míne þénigmenn, Chart. Th. 490, 22. On cinges þéningmanna gemóte, Cod. Dip. Kmbl. vi. 80, 20. See Kemble's Saxons in England, ii. 47; Stubbs' Const. Hist. i. 186.

þegnung-weorod, es; n. *A body of attendants* or *serving-men:*—Manege of ðæs ealdormannes þénungwerode, Shrn. 154, 26.

þegu, e; f. *A taking, accepting.* v. beáh-, beór-, fód(d)or-, hring-, sinc-, wil-, wín-þegu; þicgan.

þeh = þec, þéh. v. þú, þeáh.

þel (þell), es; n. *A thin piece of wood* or *metal, a plank, plate:*—Gylden þel áslægen *bratea*, Wrt. Voc. ii. 12, 42. Weel *planca* (þell? þele? *the line is:* Corpus virgineum natat ceu *plana carina*, Ald. 199), 95, 79. Þeáh man gesette án brád ísen þell ofer ðæs fýres hróf . . . and þeáh man mid ðám hameron beóte on ðæt ísene þell, Wulfst. 147, 2–7. Ða wágas wǽron gyldne mid gyldnum þelum ánæglede fingres þicce *aurato parietes laminarum digitalium grossitudine*, Nar. 4, 25. Wǽpenu mid gyldenum þelum bewyrcean *arma aureis includere laminis*, 7, 12. [Cf. Icel. þili; *n. a plank.*] v. benc-, ceól-, wǽg-þel; þel-brycg, -fæsten; þelu; þiling, þille.

þel-brycg, e; f. *A bridge of planks:*—On herepaþ óþ ðelbrycge, Cod. Dip. B. iii. 682, 18. Of ðam brócæ in þælbricge; of þælbricge in héhstræte, Cod. Dip. Kmbl. vi. 60, 21. Cf. stán-brycg.

þele. v. þyle.

þel-fæsten(n), es; n. *A fortress of planks* (Noah's ark):—Nolde seó culufre under salwed bord syððan ætýwan on þellfæstenne, Cd. Th. 89, 17; Gen. 1482.

þelma, an; m. *A trap:*—Þelman *tendiculum*, Hpt. Gl. 429, 17. Cf. (?) þel.

þelma (?), an; m. *Heat:*—Se þelma and sió hǽto, Lchdm. ii. 82, 10. [Cockayne compares the word with *for-pylman*; but perhaps *welma* or *welm* should be read. The form *welm*, referring to the inflammation which 'þelma' denotes, occurs three times in the section.]

þelu. v. buruh-þelu; þel.

þén. v. þegen.

þencan; p. þóhte (þohte?) *To think.* **I.** absolute, *to meditate, cogitate, consider*:—Sceal scearp scyldwíga gescád witan worda and worca, se ðe wel þenceþ, Beo. Th. 584; B. 289. Ða leásan men treówa gehátaþ fægerum wordum, fácenlíce þencaþ, Fragm. Kmbl. 49; Leás. 26. Ðara sacerda ealdras þóhton ðæt hig woldon Lazarum ofsleán *cogitaverunt principes sacerdotum, ut Lazarum interficerent*, Jn. Skt. 12, 10: Blickl. Homl. 69, 26 : 77, 8. Weras þeahtedon and þóhton, Elen. Kmbl. 1094; El. 549. Ðá gunnon þencan ða bócerns *coeperunt cogitare scribae*, Lk. Skt. 5, 21. Ne mæg se flǽschoma, ðone him ðæt feorg losaþ, mid hyge þencan, Exon. Th. 311, 23; Seef. 96. Hé eode út on ðæt land þencende *egressus fuerat ad meditandum in agro*, Gen. 24, 63. **II.** where the thought is the object of the verb, *to think, have in the mind*:—Secge hé hwæt ic þence *let him say what my thoughts are*, Blickl. Homl. 181, 7. Ðeáh hwá mæge ongitan hwæt óþer dó, hé ne mæg witan hwæt hé ðencþ, Bt. 39, 9; Fox 226, 7. Gedó ðæt hý nægen dón ðæt yfel ðæt hý þencaþ and spreacaþ *decidant a cogitationibus suis*, Ps. Th. 5, 11. Weras ðe ðæt on geþóhtum þenceaþ: Wutun . . . , 138, 17. Ealle ða geþóhtas ðe hí þóhtan *omnes cogitationes eorum*, 145, 3. **II a.** where the thought is expressed:—Ðæt mæg beón, ðæt sume men þencan, 'hú mæg ic sécan ðæt gástlíce leóht?' Blickl. Homl. 21, 18. **III.** *to think, suppose, regard* as *an opinion* or *belief*:—Hé þenceþ ðæt his wíse wel hwam þince eal unforcúþ, Exon. Th. 315, 12; Mód. 30. Nǽnig heora þóhte, ðæt hé ðanon scolde gesécean folc, Beo. Th. 1386; B. 691. **IV.** *to think of, consider, employ the mind on* a subject, (1) where the subject of thought is in the accusative:—Ic ðíne sóðfæstnysse þence *meditabor in justificationibus tuis*, Ps. Th. 118, 117. Ðá þóhton hig ðis word, Lk. Skt. 9, 45. Him ðás þing þencendum *haec eo cogitante*, Mt. Kmbl. 1, 20. (2) where the subject of thought is in genitive:—Hé ðencþ ðæs tíman hwonne hé hit wyrs geleánian mæge *deteriora, si occasio praebeatur, quaerat*, Past. 33; Swt. 227, 23. Ðenc ðara worda mínra gebeda *intende voci orationis meae*, Ps. Th. 5, 1. Gif ðú ðone mon lácnian wille, þænc his gebǽra *consider his gestures*, Lchdm. ii. 348, 13. Hié nyllaþ ðæs ðencean, hú hié mægen nyttweorðuste beón hiera niéhstum, Past. 5; Swt. 45, 18. Wé móton ðæs þencan ðe egeslíc on ðissum bócum is gewriten, L. Ath. i. prm.; Th. i. 196, 23. (3) where the subject of thought is governed by a preposition, *to think* about, of, on a subject:—Ic ymb síþ sprǽce and on lagu þence, Exon. Th. 458, 9; Hy. 4, 97. Myccle swíðor wé sceolan þencan be ðǽm gástlícum þingum ðonne bé ðǽm lichomlícum, Blickl. Homl. 57, 13. Ic ðe on morgendæge þenceaþ *de crastino cogitare*, 213, 23. Onginnaþ ymb ða fyrde þencan, Cd. Th. 26, 18; Gen. 408. Hwæt is ðæt ðæm men sý máre þearf tó þencenne ðonne embe his sáuwle þearfe, Blickl. Homl. 97, 19. Gód ys on Dryhten tó þenceanne *bonum est confidere in Domino*, Ps. Th. 117, 8. (4) where the subject of thought is given in a clause introduced by an indirect interrogative:—Hé þencþ hú hé hine éþelícost beswícan mæge, Blickl. Homl. 55, 21. Hé þóhte hú hé him stól geworhte, Cd. Th. 18, 13; Gen. 272. Maria swígende ðóhte hwæt seó hálettung wǽre, Blickl. Homl. 7, 16. Hié þóhton hú hié hine ácwellan meahton, 241, 18 : Ps. Th. 72, 6. Is wén ðæt feala manna þence hwylcum edleáne hé onfó æt Drihtne, Blickl. Homl. 41, 14. Smeágean wé and þencan hwæt ðæt tácnode, 19, 4. Smeágan and þencan hwylce ðæs gódan mannes weorc and his dǽda

wǽron, 55, 12. (5) where the construction is uncertain :—Ðenð *excog- itat* (*de domo impii*, Prov. 21, 12), Kent. Gl. 775. **V.** *to direct the thoughts* to an object, (a) *to look* to with attention, *turn the thoughts* to : —Þenc nú swíðe geornlíce tó ðam ðe ic ǽr sáde *turn your thoughts very carefully to what I said before*, Shrn. 177, 35. Ðǽm welwillendum is tó secganne, ðonne hié gesióð hiera geférena gód weorc, ðæt hié eác ðencen tó him selfum *dicendum est benevolis, ut, cum proximorum facta conspiciunt, ad suum cor redeant*, Past. 34; Swt. 231, 11. Riht is ðæt munecas dæges and nihtes inweardre heortan á tó Gode þencan and geornlíce clypian *it is right that monks day and night ever earnestly direct their thoughts to God and diligently cry to him*, L. I. P. 14; Th. ii. 322, 3. (b) *to look* to with trust, expectation, *expect* of. Cf. Ger. zu-denken :—Næs heó swicol nánum ðæra ðe hyre tó ðóhte, Lchdm. iii. 430, 1. Ðá seonde hé ðæt man sceolde ða scipu tóheáwan; ac hí ábruðon ða ðe hé tó þóhte, Chr. 1004; Erl. 139, 26. Ne þurfon wé ná tó úrum mǽgum ne nán man tó his wífe ðencean tó ðam swýþe, ðæt him man æfter his forðsýþe tó ðam micel fore gedǽle, ðæt hí hine fram wítan álýsan, gif hé hér hine sylfne forgýmde *we need not expect so much of our kinsmen, and no man need expect so much of his wife, as that enough will be given for him after his death to redeem him from torment, if he neglected himself before*, Wulfst. 306, 4. (c) where purpose or intention is implied, *to turn the thoughts* to action, *to be bent* upon something, *have an intention* to do something :—Hí beóð gewæpnode on ða wísan, ðe man hors gewǽpnaþ, ðonne man tó wíge þencþ (*intends to go to war*), Wulfst. 200, 11. Feówer þing synt ealra þinga behéfost ðam árwyrðan men, ðam ðe þencþ tó ðam écan life, 247, 12. Hé tó gyrnwræce swíðor þóhte ðonne tó sǽlade *his thoughts were turned rather to vengeance than to voyage*, Beo. Th. 2282; B. 1139. Hí tó swice þóhton, and þrymcyning þeódenstóles berýfan, Exon. Th. 317, 6; Mód. 61. Gif hwylc mǽdenman mid gehádodum wunaþ, and heó tó ðam ylcan háde þence *si puella aliqua cum ordinatis habitet, et se eidem ordini destinet*, L. Ecg. P. ii. 17; Th. ii. 188, 10. **VI.** *to think* of something, where it is implied that effect will be given to the thought, *to determine, devise, mean, purpose, intend*, (a) with gen. :—Ne þence wé nánes yfeles nec ullas molimur insidias, Gen. 42, 31. Geheald mé, ðæt mé ne beswíce synwyrcende, ða ðe unrihtes ǽghwǽr þenceaþ *custodi me a scandalis operantium iniquitatem*, Ps. Th. 140, 11. Ða ðe mé ðenceaþ yfeles *qui cogitant mihi mala*, 34, 5. Ealle míne fýnd þóhton mé yfeles, 40, 8. Ðæt ic mán fleó and mid rihtheortum rǽdes þence, 93, 14. Ne mæg ðín ríce leng stondan, búton ðú heora forwyrde ðe geornor þence, Blickl. Homl. 175, 15. Ne þendú *ne moliaris* (*amico tuo malum*, Prov. 3, 39), Kent. Gl. 55. (b) followed by an infin., *to think* of doing something, *intend* to do :—Ic his swiðran hand settan þence *ponam manum ejus*, Ps. Th. 88, 22 : 107, 8 : 118, 109. Ic mé be healfe mínum hláforde licgan þence *I mean to lie by the side of my lord*, Byrht. Th. 141, 9; By. 319. Nó ic eów sweord ongeán óðberan þence . . . ac ic mínum Criste cwéman þence, Exon. Th. 120, 18–26; Gú. 274–277. Gif ðú úre bídan þencest, 119, 26; Gú. 260. Se ðe wrecan þenceþ freán, Byrht. Th. 139, 23; By. 258: Beo. Th. 3075; B. 1535: Cd. Th. 287, 9; Sat. 364. Ðonne wé tó hehselde hnígan þenceaþ, 277, 22; Sat. 208. Mid ðý hí wrecan þenceaþ *ad faciendum vindictam*, Ps. Th. 149, 7, 8. Hí unscyldige scotian þenceaþ *ut sagittent immaculatum*, 63, 3. Ic hine wríþan þóhte . . . ic hine ne mihte ganges getwǽman *I meant to bind him . . . but I could not stop him*, Beo. Th. 1933; B. 964 : 1483; B. 739. Hé ðæt gewrecan þóhte *he determined to punish that*, Cd. Th. 77, 13; Gen. 1274. Hié wyrnan þóhton Moyses mágum leófes síðes, 180, 27; Exod. 51. Hié wǽron wið ðæs fýres weard tó ðon ðæt hié hit ácwencean þóhton *ad extinguendum ignem concurrerunt*, Ors. 4, 10; Swt. 200, 17 : 1, 10; Swt. 44, 32. Se ðe gód beginnan þence hé ðæt angin on him sylfum ástelle *he who intends to begin reformation, let him make a beginning with himself*, Lchdm. iii. 438, 32. (b 1) with the gerundial infin. :—Hí ðǽr swá longe ðóhton tó beónne, Ors. 1, 14; Swt. 56, 22. Se ðe ða áre þænce tó þeófigenne oððe on óððre wís on tó áwendenne *qui quid illinc abstulerit sive in alium usum converterit*, Chart. Th. 177, 13. (b 2) with infin. omitted :—Ða Iudéas sóhton Iósep and ða twelf cnyhtas and Nichodemus . . . Ealle hig hig selfe bedýglodon . . . búton Nichodemus sylfa . . . Com hé tó hym . . . Eall swá gelíce Iósep æfter ðam hyne ætýwde, and heom tó com . . . Hig cwǽdon tó hym : 'Oncnáw nú ðæt hyt ðé lyt sceal fremian ðæt ðú tó þóhtest' (*know that it shall benefit you little, that you have determined to come to us*), Nicod. 12–13; Thw. 6, 14–38. Ðara ǽlces ðe ðæs wordes wǽre ðæt from Rómebyrg þóhte *of each one that should give expression to an intention of leaving Rome*, Ors. 4, 9; Swt. 190, 25. (c) followed by a clause :— Ða ðe swá þenceaþ, ðæt heó gehýden hǽlun míne *ipsi calcaneum meum observabunt*, Ps. Th. 55, 6. Heó ǽr þóhte ðæt heó Godes brýd wurþan wolde *antea statuerat, quo Dei sponsa fieret*, L. Ecg. P. ii. 17; Th. ii. 188, 13. 'Uton árísan and ácwellan ða apostolas' . . . Ða Iudéas ða árison, and hié ongunnon mid sweordum ðyder gán; þóhton ðæt hié woldan ofsleán ða apostolas, Blickl. Homl. 151, 1. (d) with an accus. to which a clause stands in apposition :—His ðegna ðreát ðe ðæt þence nú, ðæt hí his willan wyrcean georne *ministri ejus, qui facitis voluntatem ejus*, Ps.

Th. 102, 20. **VII.** *to think* of doing something with hope or expectation, *to desire, seek* :—Ðurh ða róde sceal ríce gesécan ǽghwylc sáwl, seó ðe mid Wealdende wunian þenceþ, Rood Kmbl. 240; Kr. 121. Hwæþer ðú ðonne ongite ðæt ǽlc ðara wuhta ðe him beón þencþ ðæt hit þencþ ætgædere beón gehál undǽled *quod autem subsistere ac permanere appetit id unum esse desiderat*, Bt. 34, 12; Fox 152, 26. Ðara gesǽlða wilniaþ ealle deáþlíce men tó begitanne, ðeáh hé ðurh mistlíce wegas ðencan tó cumanne, 24, 2; Fox 80, 31. **VIII.** *to think, call to mind, originate* in the mind :—Hié ðonne forhtiaþ, and feá þenceaþ hwæt hié tó Criste cwæðan onginnen *then will they fear, and few will think what to say to Christ*, Rood Kmbl. 228; Kr. 115. [Goth. þag(g)kjan; p. þáhta *to think, consider, consult, debate* : O. Sax. thenkian; p. þáhta : O. L. Ger. thenkan; tháhta : O. Frs. thanka, thenkia; p. thôgte : O. H. Ger. denchen; p. dáhta : Icel. þekkja; p. þátti *to perceive, know*.] v. á-, be-, bi-, for-, fore-, ge-, geond-, of-, under-, ymbe-þencan.

þencan in the following passage seems an error :—Sum on bǽle sceal brondas þencan (*Thorpe would read* þeccan; *Grein suggests* sumne on bǽle sceal brond áswencan. Cf. ge-swencan), Exon. Th. 329, 33; Vy. 43.

þénda, Lchdm. ii. 182, 16. v. þǽnan.

þende; *conj. While* :—Ðendi hé ðæt þóhte engel Drihtnes æteáwde him *haec eo cogitante angelus Domini apparuit ei*, Mt. Kmbl. Rush. 1, 20. Þende hé ðá gespræc *adhuc eo loquente*, 17, 5. Ðende wæs hé sprecende i ða hwíle hé spræc *adhuc ipso loquente*, Lind. 26, 18. Ðende ðonne (mid þý þonne þende, Rush.) wæs ðe Hǽlend in Bethania *cum autem esset Jesus in Bethania*, 26, 6. Þende *regente* (perhaps here the word is the beginning of a rendering of the absolute construction, as in the previous passages; or it might be (?) a mistake for *þeódne*), Germ. 403, 35. [Goth. þandé, þandei *while, as long as*; *since*: O. H. Ger. danta *quia, ideo*.] v. next word.

þenden. I. *conj. While.* (1) where the periods of the actions marked by the verbs in the conjoined clauses are co-extensive, *as long as, (all the) while* (*that*) :—Ic Drihtne singe þenden ic wunige on woulddreámum *psallam Deo, quamdiu ero*, Ps. Th. 103, 31. Byþ his sóþfæstnys mǽre þenden þysse worulde wunaþ ǽnig dǽl *justitia ejus manet in seculum seculi*, 111, 3: 101, 10: Cd. Th. 93, 9; Gen. 1542: 56, 7; Gen. 908. Ne þearft ðú ðé wiht ondrǽdan, þenden ðú míne láre lǽstest, 130, 33; Gen. 2169: Beo. Th. 574; B. 284. Mon. mæg gelácnian, þenden of ðǽre lifre sió blódsceáwung geondgét ealne ðone líchoman, Lchdm. ii. 222, 9. Heó wǽron leóf Gode, ðenden heó his word healdan woldon, Cd. Th. 16, 18; Gen. 245. Þenden, 73, 5; Gen. 1200: 194, 3; Exod. 255: 216, 17; Dan. 8: Beo. Th. 59; B. 30: 1141; B. 57: Exon. Th. 157, 34; Gú. 901. Þenden, 37, 8; Cri. 590: 50, 14; Cri. 800. Þendon, Andr. Kmbl. 3422; An. 1715. Þynden, 2648; An. 1325. Þenden wé on eorðan eard weardigen, Exon. Th. 48, 15; Cri. 772 : Ps. Th. 105, 5. (2) where the verbs of the conjoined clauses denote contemporaneous actions. v. II :—Hý sceolon tæfle ymbsittan þenden him hyra torn tóglíde *they shall sit at their play, while their grief slips away*, Exon. Th. 345, 3; Gn. Ex. 182. (3) where the period of the action of the verb in the first clause is included within that of the verb in the subjoined clause, *while, at some time during the period when* :—Gif ic ǽnegum þegne þeódenmádmas forgeáfe, þenden wé on ðan gódan ríce sǽton, Cd. Th. 26, 22; Gen. 410. Hé frægn ða mænigeo hwæt hine gemǽtte, þenden reordberend reste wunode, 223, 21; Dan. 123. **II.** *adv. Meanwhile* :—Heorot innan wæs freóndum áfylled, nalles fácnstafas þeód-Scyldingas þenden fremedon, Beo. Th. 2043; B. 1019. Dǽdum mildheort, þenden geðyldig, Ps. Th. 85, 14: 91, 13. v. preceding word.

þenedness, þeneness, þénest, þenestre, þeng. v. tó-þenedness, á-þeneness (ðenenis *is given in* Ps. Surt. ii. p. 194, 15, *but* áðenenes *in* Txts. 411, 48), þegnest, þegnestre, þegen.

þengel, es; *m. A prince* :—Segncyning, manna þengel, Cd. Th. 188, 24; Exod. 173. Hringa þengel (*Beowulf*), Beo. Th. 3018; B. 1507. [Icel. þengill *a prince* (only in poetry).] Cf. fengel, strengel.

þenian, þénian, þénisc, þéning. v. þennan, þegnian, þegnisc, þegnung.

þennan, þenian; *p.* þenede. **I.** *to stretch, spread out, extend, bend* (*a bow*) :—Ic míne handa tó ðé hebbe and ðenige *expandi manus meas ad te*, Ps. Th. 87, 9. Bogan his ðeneþ *arcum suum tetendit*, Ps. Surt. 7, 13. Ic míne handa tó ðé þenede *expandi manus meas ad te*, Ps. Th. 142, 6. Ða synfullan ðenedon (*intenderunt*) bogan, Ps. Surt. 10, 3. Ðene (*praetende*) mildheortnisse ðíne weotendum ðec, 35, 11. Ðænne ðone swiðran earm swá hé swíþast mæge *let him stretch out the right arm as hard as he can*, Lchdm. iii. 22, 11. Swá hwider swá se cining his ríce þennan wolde *whithersoever the king could extend his power*, Anglia x. 142, 47. Hé ða fǽmnan hét nacode þennan and mid sweopum swingan *he bade stretch the maiden out naked and scourge her with whips*, Exon. Th. 253, 29; Jul. 187. Þenian *to stretch on the cross*, Rood Kmbl. 103; Kr. 52. Ðennende ðú áðenes bogan ðínne *tendens extendes arcum tuum*, Ps. Surt. ii. p. 190, 5. **II.** *to prostrate, overthrow* :—Ðæt hé þenede hig on wéstene *ut prosterneret eos in deserto*, Ps. Spl. 105, 25. **III.** *to strain, make an effort, exert one's self, press on* (v. Gothic) :—Ðæt geswinc his sýðfætes ne understande

mid hrædestan ryne þenigende arn (*he exerted himself in running*), for
ðam ðe hé gewilnode hine geðeódan ðam ðe ðǽr fleáh, Homl. Skt. ii.
23 b, 186. [*Goth.* sik ufþanjan *se extendere* (Phil. 3, 14): *O. Sax.*
sie netti thenidun, Hél. 1155: *O. L. Ger.* thenan *intendere, extendere*:
O. H. Ger. dennen *extendere, expandere, distendere*: *Icel.* þenja *to stretch,
extend.*] v. á-, be-, ge-þennan, -þenian.

þenning, e; f. *Stretching, extension*:—Be Cristes earm[a] þenninge
and his honda on róde, Anglia xi. 172, last line.

þénsum, þénung, þeó. v. ge-þénsum, þegnung, se.

þeód, e; f.　　I. *a nation, people*:—Ðeód winþ ongén þeóde *consurget gens in gentem*, Mt. Kmbl. 24, 7. Of ðám frumgárum folc
áwæcniaþ, þeód unmǽte, Cd. Th. 138, 15; Gen. 2292. Eást-Engla
cyning and seó þeód gesóhte Ecgbryht cyning, Chr. 823; Erl. 62, 24.
Eal seó þeód ðe on Eást-Englum beóþ, L. A. G. prm.; Th. i. 152, 3.
Myrcena ðeód onféng fulluht, Lchdm. iii. 430, 21. Ðeós þeód (*the
Jews*), Elen. Kmbl. 934; El. 468. Ðá wæs þeód (*the citizens of
Mermedonia*; cf. burhwaru, 2189; An. 1096) gesamnod, Andr. Kmbl.
2198; An. 1100. Cham ys fæder ðære Cananéiscre þeóde, Gen. 9, 18.
Ðǽr wæs micel unþuærnes ðære þeóde (*the Northumbrians*) betweox
him selfum, Chr. 867; Erl. 72, 8. Mid ðǽm ieldstan witum mínre
þeóde, L. In. prm.; Th. i. 102, 6. Þióde aldor, Dauid, Ps. C. 146. In
lond ðara ðeáde *in regionem Gerasenorum*, Mt. Kmbl. Lind. 8, 28.
Hér Édwine kyning wæs gefulwad mid his þeóde, Chr. 627; Erl. 24, 2.
Ic déme ða þeóde (*gentem*, the Egyptians), Gen. 15, 14. Clǽnsie man
ða þeóde, L. E. G. 11; Th. i. 174, 2. Ealla óðræ Cristne ðióda, Past.
pref.; Swt. 7, 5. Of ðám frumgárum twá þeóda (*the Moabites and the
Ammonites*) áwócon, Cd. Th. 158, 11; Gen. 2615. Þeóda gentes,
Ps. Th. 65, 7. Þeóde, 78, 1: 113, 10. Manegra þeóda fæder *pater
multarum gentium*, Gen. 17, 4. Ofer þeóda gehwylce, Beo. Th. 3414;
B. 1705. Drihten, ðeóda waldend, Cd. Th. 238, 27; Dan. 361.
Eardas rúme Meotud árǽrde for moncynne, efenfela þeóda and þeáwa
(i. e. *each people has its own customs*), Exon. Th. 334, 18; Gn. Ex. 18.
Ðiéda *gentium*, Ps. Surt. 17, 44: ii. p. 192, 17. On ðeódum *inter gentes*,
Ps. Th. 107, 3. Ofer ealle þeóde *super gentes*, 65, 6. Hí þreátiaþ ymbsit-
tenda óþra þeóda, Met. 25, 14. Lǽraþ ealle þeóda *docete omnes gentes*,
Mt. Kmbl. 28, 19.　　I a. where the general term is used, but only a
part of the people is actually concerned:—Sió þeód geseah in Hierusalem,
godwebba cyst ufan eall forbærst, Exon. Th. 70, 6; Cri. 1134. Inne on
healle wæs ðeód on sǽlum, Beo. Th. 1291; B. 643. Heó ðæs áð lǽdde
on ealre ðeóde gewitnesse tó Æglesforda, Chart. Th. 202, 3. Æþelréd
Norþanhymbra cyning wæs ofslægen from his ágenre þeóde, Chr. 794;
Erl. 58, 5.　　I b. in pl. *the gentiles*:—Se þeóda lǽreow Paulus,
Homl. Th. i. 96, 35; Shrn. 58, 33. Þara þeóda (ðeáda, Lind.) Galilea,
Mt. Kmbl. Rush. 4, 15.　　I c. *a race*:—Giganta cyn . . . ðæt wæs
fremde þeód écean Dryhtne, Beo. Th. 3387; B. 1691.　　I d. in a
general sense, particularly in pl. *people, men*:—Gif ðú eáðmódne eorl
geméte, þegn on þeóde (*among men*), Exon. Th. 318, 7; Mód. 79: 176, 4;
Gú. 1204. Ðæt wé siþþan forð ða sǽllan þing móten geþeón on þeóde,
23, 31; Cri. 377: 8, 33; Cri. 127: 208, 23; Ph. 160. Cristes þegnas
biddaþ God áre ealre þeóde; ðú him tíðast, swá ðú eádmód eart ealre
worlde, Hy. 7, 55. Grécas . . . Egiptisce þeóda . . . Romani and Englisce
þeóda, Anglia viii. 309, 19–21. Þeóda wlítaþ . . . hú seó wilgedryht
wildne weorþiaþ, Exon. Th. 221, 28; Ph. 341. Hé þeóda gehwam
(*to every one on earth*) hefonríce forgeaf, Cd. Th. 40, 19; Gen. 641:
Exon. Th. 429, 4; Rä. 42, 8. Geþola þeóda þreá *endure men's
oppression*, Andr. Kmbl. 213; An. 107. Se ðisne ár hider onsende
þeódum tó helpe (*to help people*), 3209; An. 1607. Is wíde cúð
ðeódum, ceorlum and eorlum, Menol. Fox 61; Men. 30. David wæs
swíðe geðancol tó ðingienne þiódum sínum wið ðane Sceppend, Ps.
C. 7.　　II. in a local sense, *the district occupied by a people, a
country*:—Án hearpere wæs on ðære þeóde ðe Thracia hátte, Bt. 35, 6;
Fox 166, 28. Se wæs on ðære ðeóde ðe hátte Babilonige, Cd. Th. 226,
16; Dan. 172. Ða beorgas onginnaþ in Narbonense ðære ðeóde, Ors.
I, I; Swt. 22, 20. In ðær ðeáde *in Galilaeam*, Jn. Skt. Lind. 4, 45.
Aulixis hæfde twá ðióda under ðam Kásere. Ða ðióda wǽron hátene
Iþacige and Rétie, Bt. 38, 1; Fox 194, 4.　　III. *a language*.
v. ge-þeóde:—Þeáh ðe seofon men sittan on middanearde, and heó mihton
sprecan on æghwylcere þeóde ðe betwux heofonum and eorðan wǽre,
ðara is twá and hundseofontig, Wulfst. 214, 29. [*Goth.* þiuda *a nation,
people*; pl. *the gentiles*: *O. Sax.* thiod, thioda *a people*; in pl. *men*:
O. L. Ger. thiad *gens, natio*: *O. Frs.* thiade *people, men*: *O. H. Ger.*
diot, diota *gens, populus, plebs, natio*: *Icel.* þjóð *a nation, people*; in a
local sense, *a land, country*.] v. el- (æl-), eást-, gum-, heáh-, neáh-,
norþ-, sige-, Sweó-, wer-þeód, irmen-þeóde; in-geþeóde.

þeód-. As the first part of several compounds (see below) *þeód* has
the force of *general, great*; a similar use is found in *O. Sax.* and *Icel.*
The form is also found in proper names, e. g. Ðeód-bald, Bd. 1, 34;
S. 499, 33. Ðeód-ríc, Bt. 1; Fox 2, 5. Þeód-Scyldingas, Beo. Th.
2042; B. 1019. Cf. regn-.

þeódan, þiédan, þídan, þýdan; p. de *To join* (trans. or intrans.),
attach:—Be ðám ðe wið ða dǽdbétendan ðeódaþ *de is qui junguntur*

excommunicatis, R. Ben. 50, 9. Ðonne hý sume mid geficum wið ðone
ánne þeódaþ and leásettaþ, sume wið ðone óþerne *dum adulantur
partibus*, 125, 2. Ða woruldgesǽlþa hú hié simle tó ðám gódum ne
ðeódaþ ne ða yfelan góde ne gedóð ðe hié hié oftost tó geðeódaþ *fortuna
nec se bonis semper adjungit, et bonos, quibus fuerit adjuncta, non efficit,*
Bt. 16, 3; Fox 56, 33. Ðá weóxon ða fýr swýþe and hí tógædere
þeóddon and samnedon óþ ðæt ðe hí wǽron on ǽnne unmǽtne lég
geánede *crescentes ignes usque ad invicem sese extenderunt, atque in
immensam adunati sunt flammam*, Bd. 3, 19; S. 548, 20. Nán bróðer
wið óþerne ne þeóde, ne mid his geþeódrǽdenne ne lette on unþæslícum
tíman *neque frater ad fratrem jungatur horis incompetentibus*, R. Ben.
74, 23. Ðæt hé hiene nánwuht ne áhebbe ofer his gelícan ne from
hiera geférrǽdenne ne ðiéde *quia per elationem se minime a proximorum
societate disjungit*, Past. 46; Swt. 349, 5. Þæt wé ús georne tó Gode
þýdon *that we diligently attach ourselves to God*, Blickl. Homl. 115, 21.
Mid cnottum (wǽre) þeód *nexibus nodaretur*, Hpt. Gl. 481, 31. Þiód
subjugatae, subjunctae, 519, 4. [Cf. *Icel.* þýða *to associate, attach*.]
v. á-, be-, ge-, óþ-, under-þeódan (-þiédan, -þídan, -þýdan).

þeód-bealu, wes; n. *Great ill, grievous ill*:—Þeódbealu on þreó healfa
(*referring to three elements in the misery of the lost*; cf. *O. Sax.*
thiod-arbêdi, *applied to the expulsion from Eden*), Exon. Th. 78, 2;
Cri. 1268. Andrea þúhte þeódbealo þearlíc tó geþolianne, ðæt hé swá
unscyldig ealdre sceolde linnan (cf. *O. Sax.* thiod-quâlu, *applied to the
crucifixion, and to the agony in the garden*), Andr. Kmbl. 2273; An.
1138. Cf. þeód-þreá.

þeód-búend[e]; pl. *Those living in nations, mankind, men*:—Hé
(*Christ*) earfeþu geþolade fore þearfe þeódbúendra, lǽðlícne deáð leódum
tó helpe, Exon. Th. 72, 16; Cri. 1173. Hé geðingade þeódbúendum
wið fæder swǽsne fǽhþa mǽste, 39, 3; Cri. 616: 84, 11; Cri. 1372.

þeód-cwén, e; f. *A great queen, an empress*:—Þeódcwén *the empress
(Elene)*, Elen. Kmbl. 2310; El. 1156. v. next word.

þeód-cyning, es; m.　　I. *the king of a whole nation, a monarch,
an independent sovereign*. [Ei má þá kalla þjóðkonunga er skattkonungar
eru, Edda. Ef hann (*Harold Fairhair*) vill leggja undir sik allan Noreg
ok ráða því ríki jafnfrjálsliga, sem Eiríkr konungr Svíaveldi, eða Gormr
konungr Danmörku, þá þykkir mér hann mega heita þjóðkonungr,
Haralds Saga, c. 3.]:—Þeódcyning (*the king of Egypt*; cf. folcfreá,
III, 7; Gen. 1852), Cd. Th. 112, 11; Gen. 1869. Ðeódcyning
(*Ongentheow*), Beo. Th. 5932; B. 2970. Se ðeódcyning (*Hrothgar*),
4294; B. 2144. Ðiódcyning (*Beowulf*), 5151; B. 2579. Æt þearfe
þeódcyninges, 5382; B. 2694. Ðæs þeódkyninges (-kynges, MS. D.)
(*Edward the Confessor*), Chr. 1066; Erl. 198, 15. Fore þrymme ðeódcyninges æniges on eorðan, Apstls. Kmbl. 36; Ap. 18. Gewiton hié
feówer þeódcyningas (cf. Thadal rex gentium, Gen. 14, 1) þrymme
micle, Cd. Th. 118, 14; Gen. 1965. Ðǽr beóþ þearfan and þeódcyningas
(*paupers and monarchs*; pauperque potensque), Dóm. L. 161. Wé Gár-
Dena in geárdagum þeódcyninga þrym gefrunon, Beo. Th. 3; B. 2. Ond
swá micel wundor and wæfersín wæs mínes weoredes on fægernisse ofer
ealle óþre þeódkyninges ðe in middangearde wǽron *fuitque inter
uarietates spectaculorum in conspiciendo talem exercitum, qui ornatu
pariter ac uiribus inter gentes eminebant*, Nar. 7, 19.　　II. *the king
of all nations, the monarch of the world, the Deity*:—Bútan ðe þeódcyning (cf. Exon. Th. 367, 25 which has here éce Dryhten), ælmihtig
God ende worulde wyrcan wille, weoruda Dryhten, Soul Kmbl. 24; Seel.
12. [*O. Sax.* thiod-kuning (*used of Christ and of Herod*): *Icel.* þjóðkonungr.] Cf. þeóden.

þeóddon *served.* v. þeówan.

þeód-egesa, an; m. *A terror that affects whole nations, a mighty,
general terror*:—Ðonne mægna cyning on gemót (*at the day of judgement*) cymeþ, þeódegesa biþ hlúd gehýred, Exon. Th. 52, 16; Cri. 834.

þeóden, es; m.　　I. *the chief of a þeód* [cf. dryhten, dryht *for
connexion of þeóden, þeód*], *a prince, king*; the word is used almost
exclusively in poetry, but occurs once in the Laws in an alliterative
phrase:—Ælc be his mǽðe, eorl and ceorl, þegen and þeóden, L. R. 1;
Th. i. 190, 14. Eádmund cyning, Engla þeóden, Chr. 942; Erl. 116, 7.
Cyning, þeóden Scyldinga, Beo. Th. 3746; B. 1871. Gúðcyning, Wedera
þeóden, 4661; B. 2336. Ðeóden gumena (*Holofernes*), Judth. Thw. 22,
18; Jud. 66. Hér Eádgár wæs, Engla waldend, tó cyninge gehálgod . . .
on ðam xxx wæs ðeóden gehálgod, Chr. 973; Erl. 124, 28. Se mondryhten, se eów máðmas geaf . . . hé oft gesealde helm and byrnan, þeóden
his þegnum, Beo. Th. 5730; B. 2869: Cd. Th. 158, 34; Gen. 2627.
Ríce þeóden, 161, 31; Gen. 2673: 222, 24; Dan. 109. Mǽre þeóden,
Beo. Th. 259; B. 129: 3434; B. 1715. Wealhþeów ðeóðnes dohtor,
4354; B. 2174: 3678; B. 1837: 2174; B. 1085. Þrý wǽron on ðæs
þeódnes byrig, ðæt hié noldon hyra þeóðnes dóm þafigan onginnan, Cd.
Th. 227, 18; Dan. 188. Þeóðnes (*Constantine*) willan, Elen. Kmbl.
534; El. 267. Ásecgan suna Healfdenes, mǽrum þeódne, mín ærende,
aldre dínum, Beo. Th. 695; B. 385: Cd. Th. 221, 25; Dan. 93. Þegnas
þeódne sægdon, 228, 20; Dan. 205. Þeóden mǽrne þegn, winedryhten
his, wætere gelafede, Beo. Th. 5435; B. 2721: 5570; B. 2788. Leófne
þeóden, ríces hyrde, 6151; B. 3079. Mǽrne þeóden, hláford leófne,

6274; B. 3141. **II.** *a great man, a lord, chief:*—Úre þeóden (*Byrhtnoth*) líð, eorl on eorðan, Byrht. Th. 138, 39; By. 232: 135, 18; By. 120. Þrymfæst þeóden (*Noah*), Cd. Th. 200, 27; Exod. 363. Þeóden leófesta, Andr. Kmbl. 575; An. 285: (*Guthlac*), Exon. Th. 163, 1; Gú. 987. Þurh ðæs þeódnes word, 174, 2; Gú. 1171. Eorl Beówulfes wolde freádrihtnes feorh ealgian, mæres þeódnes, Beo. Th. 1598; B. 797: 3259; B. 1627. Seó ecg geswác þeódne (*Beowulf, not yet a king*) æt þearfe, 3054; B. 1525. Hé læg ðegenlíce ðeódne (*Byrhtnoth*) gehende, Byrht. Th. 140, 27; By. 294. Mec ides freán sealde, holdum þeódne, swá hió hâten wæs, Exon. Th. 479, 7; Rä. 62, 4. Hæleð, þeódnas þrymfulle, þegnas wlitige, Andr. Kmbl. 725; An. 363. **II a.** *referring to other than men:*—Fuglas þringaþ ymbe æþelne, æghwylc wylle wesan þegn and þeów þeodne mærum (*the Phenix*), Exon. Th. 209, 4; Ph. 165. **III.** *referring to the Deity,* (1) *to God:*—Wæs freá eallum leóf, þeóden his þegnum, Cd. Th. 5, 31; Gen. 80: 37, 29; Gen. 597: 218, 4; Dan. 34. Þeóden, rodera waldend, 73, 10; Gen. 1202. Freá ælmihtig, mære þeóden, 52, 34; Gen. 853. Swegles aldor, ríce þeóden, 53, 21; Gen. 864. Engla þeóden, 205, 6; Exod. 431. Swegles ealdor, þearlmôd þeóden gumena, Judth. Thw. 22, 34; Jud. 91. Se þíoden, Met. 11, 80. Þegnas þrymfæste þeóden heredon, Cd. Th. 2, 7; Gen. 15. (2) *to Christ:*—Lífes ceápode þeóden moncynne, Exon. Th. 68, 1; Cri. 1097. Se brego mæra tô Bethania, þeóden þrymfæst, his þegna gedryht geladade, 29, 3; Cri. 457. Crist, cyninga wuldor, mære ðeóden, Menol. Fox 4; Men. 2. Se drihten, ðe deáð for ús geþrowode, þeóden engla, Cd. Th. 306, 19; Sat. 666: Elen. Kmbl. 971; El. 487. [*Goth.* þiudans βασιλεύς: *O. Sax.* thiodan (*used of God and Christ, as also of earthly rulers*): *Icel.* þjóðann (*poet.*) *a king, ruler; a great man.*]

þeódend-líc. v. under-þeódendlíc.

þeóden-gedál, es; *n.* *The separation from a lord* by his death:—Ellen biþ sélast ðam ðe sceal dreógan dryhtenbealu, behycgan þeódengedál . . . se wât his sincgiefan holdne biheledne, Exon. Th. 183, 8; Gú. 1324.

þeóden-hold; *adj.* *Faithful to a lord, loyal:*—Þegn þeódenhold, Andr. Kmbl. 767; An. 384. Petrus and Paulus ðeódenholde ðrowedon on Rôme, Menol. Fox 243; Men. 123. Wígend unforhte, þeódenholde, Cd. Th. 189, 10; Exod. 182. Hé wígena fand æscberendra .xviii. and .ccc. eác þeódenholdra (þeonden, MS.), 123, 10; Gen. 2042. Hé mid wuldre geweorðode þeódenholde, 183, 5; Exod. 87. Cf. dryhten-hold, Cd. Th. 137, 32; Gen. 2282.

þeóden-leás; *adj.* *Without a lord, deprived of one's prince:*—Hié hira beággyfan banan folgedon þeódenleáse, Beo. Th. 2210; B. 1103. Cf. hláford-leás.

þeóden-máðum, es; *m.* *A treasure given by a prince:*—Gif ic (*Satan*) ænegum þegne þeódenmádmas forgeáfe, Cd. Th. 26, 20; Gen. 409.

þeóden-stól, es; *m.* *The seat of a king, a throne:*—Þrymcyning þeódenstôles berýfan, Exon. Th. 317, 8; Môd. 62. Ymb þeódenstôl hý þringaþ, 25, 7; Cri. 397: 319, 16; Víd. 13.

þeód-eorþe, an; *f.* *The whole inhabited earth, the world:*—Hwæt sceoldon dé (*the guilty soul*), þeódeorðan fýlnes (cf. Exon. Th. 368, 7), úre ælmessan? Wulfst. 240, 15.

þeód-feónd, es; *m.* *The arch-enemy:*—Se þeódfeónd, Antecrist sylfa, Wulfst. 83, 16. Hé fordeþ ðæne þeódfeónd and on helle grund besenceþ, 86, 20: 85, 19: 54, 20.

þeód-fruma, an; *m.* *A prince of a people, a lord, ruler:*—Ðæt hí þíowien swilcum þíodfruman (hláforde, Bt. 39, 13; Fox 234, 29), Met. 29, 94. Cf. land-fruma.

þeód-gestreón, es; *n.* *A great treasure:*—Brúc ðisses beáges, and ðisses hrægles neót, þeódgestreóna, Beo. Th. 2440; B. 1218. Nalæs hí hine læssan lácum teódan, þeódgestreónum, 87; B. 44.

þeód-guma, an; *m.* *A chief man of a people, a great man:*—Ða þeódguman (cf. eorlas æscrôfe, 26, 20; Jud. 337), Judth. Thw. 26, 17; Jud. 332: 24, 26; Jud. 208. [*O. Sax.* thiod-gumo:—Thiodgumo, mâri mahtig Krist, Hél. 2576. The word is also used of John the Baptist.]

þeód-here; *gen.* -her(i)ges; *m.* *The army of a nation, the military force of a people:*—Þeódherga wæl *the slain of the nations who fought,* Cd. Th. 130, 15; Gen. 2160.

þeód-herpaþ, (-æþ, -oþ), es; *m.* *The highway, public road:*—On ðæne þeódherpað, Cod. Dip. Kmbl. iii. 24, 2. Þeódherpoð, v. 157, 14, 16. On ðone þeódherpað west on herpað, Chart. Erl. 330, 5. [Cf. *O. H. Ger.* diet-uuec *via publica,* Grff. i. 669: *Icel.* þjóð-braut, -gata, -leið, -vegr *a high road.*] Cf. þeód-weg.

-þeódig. v. el- (æl-) þeódig. [*O. Sax.* eli-thiodig: *O. H. Ger.* eli-diotic.]

þeódisc; *adj.* *Belonging to a people, gentile:*—Þeódisce *gentiles,* Anglia xiii. 37, 268. [Þa þeodisce men (þe Romanisse, 2nd MS.), Laym. 5838. *O. L. Ger.* thiudisca liudi *germania: O. H. Ger.* diutisk *teutonicus.* Cf. *Goth.* þiudiskô ἐθνικῶς, *gentiliter.*] v. el- (æl-) þeódisc, *and next word.*

þeódisc, es; *n.* *A language:*—Deáh hit gebyrige ðæt ða ûtemestan ðíoda eówerne naman up âhebban and on manig þeódisc eów herigen *licet remotos fama per populos means diffusa linguas explicet,* Bt. 19; Fox 68, 30. Þíodisc, Met. 10. 26.

þeód-land, es; *n.* **I.** *an inhabited district, a region, country:*—Fromcyme folde weorðeþ, þeódlond monig, ðíne gefylled, Cd. Th. 106, 4; Gen. 1766. Ðá becwom ic on Caspiam ðæt lond; ðá wæs ðær seó wæstmberendeste eorþe ðæs þeódlondes, and ic swíðe wundrade ða gesǽlignesse ðære eorðan *Caspias portas peruenimus, ubi cum fertilissimarum regionum admirarer felicitatem,* Nar. 5, 21. Tô wrítanne be ðæm þeódlonde Indie *scribendum de regionibus Indie,* I, 15. Wé neálêhtan ðæm þeódlonde (*regioni*), 26, 12. Hé forþférde on Middel-Englum on ðam ðeódlande (*regione*) ðe is nemned on Feppingum, Bd. 3, 21; S. 551, 35. On ðam ðeódlande (*regione*) ðe is gecýged Élíge, 4, 19; S. 588, 1. Gotan geþrungon þeódlond monig, Met. 1, 3. Sculon âgan eaforan ðíne þeódlanda gehwilc, Cd. Th. 133, 15; Gen. 2211. On Cantwara mægþe and eác on ðâm ðeódlandum ðe ðærtô geþeódde wæron (*in contiguis eidem regionibus*), Bd. pref.; S. 471, 26. Ðá férdon wé on ôþer þeódlond India *in alias Indie profecti regiones,* Nar. 22, 2. Wé fram dæge tô ôþrum geáxiaþ ungecyndelíco wítu geond þeódland (*throughout the world*) tô mannum cumene, Blickl. Homl. 107, 26. Hé wearð wíde geond þeódland geweorðad, Chr. 959; Erl. 119, 23: Exon. Th. 19, 26; Cri. 306. **II.** *the continent (?):*—Fýr cymþ and hit gefealþ ǽrest on Sceotta land . . . and hit ðonne færþ on Brytwealas . . . and ðonne hit færþ on Angelcyn . . . Ðonne hit færþ súð ofer sǽ geond ðæt þeódland (on ða þeódland, 215, 18), and hit ðær forbærnþ ðæt mancyn, swá hit hér ǽr dyde, Wulfst. 205, 13. [*Icel.* þjóð-land *a country.*]

þeód-líc; *adj.* *Of a people:*—Ðeódlíc nama *gentile nomen,* Ælfc. Gr. 9, 45; Zup. 65, 6. [*O. H. Ger.* diete-lîh.]

þeód-lícettere (*Antichrist*): *m.* *An arch-hypocrite:*—Se þeódlícetere (*Antichrist*) hit gehíwaþ swá ðæt læst manna wât, hú hé him wið ðone ðeódfeónd gescyldan sceal, Wulfst. 54, 18.

þeód-loga, an; *m.* *An arch-impostor, a great liar:*—Ða gôdan Godes þegnas sædan, ðæt hé (*Simon the sorcerer*) luge, and hý geswutelodon, ðæt hit eal leás wæs, ðæt se þeódloga sæde, Wulfst. 99, 23. Antecrist lǽrþ unsôðfæstnysse and swicolnesse . . . and swá dôð ða þeódlogan, ðe taliaþ ðæt tô wærscype, ðæt man cunne lytelíce swician and mid unsôðe sôð oferswíðan, 55, 15.

þeód-mægen, es; *n.* *A tribal force:*—Þridde þeódmægen (*the tribe of Simeon, which came third*), Cd. Th. 199, 21; Exod. 342. Cf. folcmægen, fôr æfter ôðrum, 199, 31; Exod. 347.

þeód-mearc, Cd. Th. 187, 33; Exod. 158, *read* þeód mearc.

þeódness, e; *f.* *A junction, joining:*—Gedafenlíc þeódnys *habilis conjunctio,* Wrt. Voc. i. 54, 60. Þeódnysse *copulam, conjunctionem,* Hpt. Gl. 481, 51. v. ge-, under-þeódness.

þeód-sceaþa, an; *m.* *A criminal against the community, a spoiler of the community, a great criminal or spoiler:*—Wâc biþ se hyrde, ðe nele ða heorde bewerian, gyf ðær hwylc þeódsceaða sceaðian onginneþ. Nis nân swá yfel sceaða swá is deófol sylf. Ðonne môton ða hyrdas beón swíðe wacore, ðe wið ðone þeódsceaðan folce sceolon scyldan, L. C. E. 26; Th. i. 374, 22–28: Wulfst. 191, 6–13. Þeódsceaða, fýrdraca, Beo. Th. 5369; B. 2688: 4545; B. 2278. Se þeódsceaða (*famine*), Andr. Kmbl. 2232; An. 1117. Gyf God ne gescyrte ðæs þeódsceaðan (*Antichrist*) lífdagas, Wulfst. 86, 17. God biddan, ðæt hé ús gescylde wið ðæne þeódsceaðan (*Antichrist*), 80, 6. Ðider (*to hell*) sculon þeófas and ðeódsceaðan, 26, 18: 165, 36: Exon. Th. 98, 20; Cri. 1610. Lácende lêg láðwende men þreáð, þeódsceaþan, 97, 25; Cri. 1596. [*O. Sax.* thiod-skaðo (*the devil*).] Cf. folc-, leód-sceaþa.

þeód-scipe, es; *m.* *A people:*—Him cierde tô eall se þeódscype on Myrcna lande *all the people of Mercia,* Chr. 922; Erl. 108, 25. Eal þeódscype hine hæfde for fulne cyng, 1013; Erl. 148, 36. Þes þeódscype *the English,* Wulfst. 163, 19. Se ðeódscype *the Jews,* 147. Cyning sceal geþeón and his þeódscype eác swá, 266, 21. *Oratores* syndon gebedmen, ðe sceolon for ðæne cyngc and for ealne þeódscipe þingian georne. *Laboratores* syndon weorcmen, ðe tilian sceolon ðæs, ðe eall þeódscipe big sceal lybban, 267, 10–15; L. I. P. 2; Th. ii. 304, 15: 4; Th. ii. 306, 33–36. Hí lêon ealles ðeódscipes geswincg forwurðan, Chr. 1009; Erl. 142, 12: 1048; Erl. 178, 23. An hé (*king Eadred*) his sâwla tô anlíesnesse, and his ðeódscype tô þearfe sixtýne hund punda, Cod. Dip. B. iii. 75, 1. On ðam þeódscipe (*the people of Sodom*), Cd. Th. 116, 27; Gen. 1942. Wið þeódscipe Assiriæ, 15, 11; Gen. 231. Hí nimaþ úre land and úrne þeódscipe (*gentem*), Jn. Skt. 11, 48: Guthl. 12; Gdwin. 58, 11. Hét se cyng âbannan ût ealne þeódscipe, Chr. 1009; Erl. 142, 25. Hú heó rihtlícost heora þeódscipe gehealdan mehton, Chart. Th. 139, 22. Þeódscypas winnaþ heom betweónan, Wulfst. 86, 7. Fela mǽrra manna of manegan þeódscipan, Chr. 1049; Erl. 172, 24. Æfter sumum þeódscipum byþ ðes saltus on .xv. kl. Decembris, Anglia viii. 309, 18. Cf. folc-, leód-scipe.

þeód-scipe, es; *m.* *Connexion, association, fellowship.* v. þeód-ness, þeódan:—Uton witan hwá hine ðæs wurðscipes cúðe ðe hé sceolde gestandan on ðam rímcræfte. Ic wât gere, ðæt hé ys þeódscipes wyrðe *it is entitled to be connected with arithmetic,* Anglia viii. 308, 23. Ðæt wé gésine ne sýn Godes þeódscipes, metodes miltsa *that we lack not fellowship with God, the Maker's mercies,* Cd. Th. 211, 19; Exod. 528.

Næfre ðū geþreátast ðīnum beótum, ðæt ic þeódscype ðīnne lufie, Exon. Th. 253, 10 ; Jul. 178.

þeód-scipe, es ; m. **I.** *teaching, instruction :*—Ðeódscipe ðīn hē mē lǣrde *disciplina tua ipsa me docebit*, Ps. Surt. 17, 36. **I a.** *instruction, being taught :*—Ðū fiódes ðeódscipe and ðū āwurpe word mīn efter ðē *odisti disciplinam et projecisti sermones meos post te*, Ps. Surt. 40, 17. **I b.** *testimony :*—Forebodan bið ðis godspell in ðeódscip ł cýðnise (*in testimonium*) allum cynnum, Mt. Kmbl. Lind. 24, 14. **II.** *what is taught* or *enjoined, a rule, regulation, law, injunction :*—Ðū him ǣrest ne sealdest, æfter ðam apostolícan ðeódscipe, meolc drincan, Bd. 3, 5 ; S. 527, 33. **II a.** *a collection of regulations, law, religion :*—Swā swā bī ðan ealdan ðeódscipe ða ūttran weorc wǣron behealden, swā on ðam nīwan ðeódscype . . . *sicut in Testamento* (v. **I b**) *veteri exteriora opera observantur, ita in Testamento novo* . . . , 1, 27 ; S. 494, 30. Ealle ða þing ðe hālige men writon on ealdum oþþe on neówum þeódscipe, Blickl. Homl. 133, 2. .vii. gebróðor geþrowedon deáþ for ðære ealdan ǣ bebode . . . Ðā cwæþ se cniht (*the seventh brother*) : 'Ic sylle mīnne līchoman for ūssa fædera ðeódscipe, swā mīne bróþor dydon,' Shrn. 111, 20. Ic geseah manige góde and on Godes þeódscipe wel heora líf lǣddon *alios fuisse narrabat verae religionis cultores*, Guthl. 17 ; Gdwin. 70, 24. Ðū hine þeódscipe ðīnne lǣrest *Je lege tua docueris eum*, Ps. Th. 93, 12. **III.** *discipline, a disciplinary regulation :*—On strengo þeódscipes and þreá tō wlæc *in disciplinae vigore tepidus*, Bd. 1, 27 ; S. 492, 18. Æfæstnia untrymnisse hire mægne ðeátscip[es] *muniat infirmitatem suam robore disciplinae*, Rtl. 110, 3. On reogollícne ðeódscipe *observatione disciplinae regularis*, Bd. 3, 3 ; S. 526, 9. Hē micele gýminge hæfde mynsterlícra ðeódscipa *curam non modicam monasticis exhibebat disciplinis*, 3, 19 ; S. 547, 28. Reogollícum ðeódscipum underþeódde *regularibus disciplinis subditus*, 4, 24 ; S. 598, 21 : 3, 19 ; S. 547, 20. In cyriclícum ðeódscipum and in mynsterlícum heálíce intimbred *ecclesiasticis ac monasterialibus disciplinis summe instructus*, 5, 8 ; S. 621, 34. Þætte ūs fæstern gidíi ðōhto ūsra heofonlícum gilǣr ðeádscipum *ut nobis jejunium proficiat, mentes nostras coelestibus instrue disciplinis*, Rtl. 14, 20. On mynstrum hē leornade gāstlíce ðeódscipas, Shrn. 50, 26. **IV.** *(regular) custom, (proper) mode of conduct :*—Bēte ðara ǣghwelc nid ryhte þeódscipe ge mid were ge mid wīte *let him make amends for each in the regular way both with* wer *and with* wīte, L. Alf. pol. 2 ; Th. i. 62, 4. Wæs Godes lof hafen þrymme micle ōþ ðisne dæg mid þeódscipe (*with proper observance?* or *among the people?* v. þeódscipe *a people*), Exon. Th. 284, 10 ; Jul. 695. Hē wolde habban ða ðenunga ðeáwas and ðeódscipe tō lǣranne, Past. 17 ; Swt. 121, 18. Ða men, ðe bearn habban, lǣran hié ðam rihtne þeódscipe, and him tǣcean lifes weg and rihtne gang tō heófonum, Blickl. Homl. 109, 17. Fæderas ic lǣrde, ðæt hié heora bearnum þone þeódscipe lǣrdon Drihtnes egsan (*fathers, bring up your children in the nurture and admonition of the Lord*, Eph. 6, 4), 185, 19. **V.** *learning, knowledge, understanding :*—Nis in him ðiódscipe *non est in eis disciplina* ; neither is there any understanding in them (Deut. 32, 28), Ps. Surt. ii. p. 194, 41. Nis nū fela folca ðætte fyrngewritu healdan wille, ac him hyge brosnaþ, ídlaþ þeódscype (*or under* **IV** ?), Exon. Th. 304, 13 ; Fä. 69. Hē wæs on godcundlícan þeódscipe getýd and gelǣred (*sacris litteris et monasticis disciplinis erudiebatur*) . . . Hē wæs twā geár on ðære leornunge, ða hæfde hē his sealmas geleornode, Guthl. 2 ; Gdwin. 18, 11. Ðū mē ðīnne lǣr ðīnne tilne and wísdōmes word *bonitatem et disciplinam et scientiam doce me*, Ps. Th. 118, 66. Hē forget hine selfne and ða lāre and ðone ðiódscipe ðe hē geliornode, Past. 50 ; Swt. 393, 17. Heó hēht gefetigean forðsnotterne, and his lāre geceás ðurh þeódscipe (*on account of his learning?* or *with a view to learning?* The Latin has : *convocans virum disciplinatum*), Elen. Kmbl. 2331 ; El. 1167.

þeód-stefn, es ; m. *A stock, people :*—Betere is tō gebídanne ǣnne dæg mid ðē, ðonne ōðera on þeódstefnum þūsend mǣla, Ps. Th. 83, 10. Cf. leód-stefn.

þeód-þreá *a great calamity :*—Hié wordum bǣdon, ðæt him gāstbona geoce gefremede wið þeódþreáum (*the injuries inflicted by Grendel*), Beo. Th. 358 ; B. 178. Cf. þeód-bealu.

þeód-weg, es ; m. *A highway :*—In þiódweg ; æftær þiódwege, Cod. Dip. Kmbl. v. 187, 30. On ðeódweg norð ofer ðone weg, 42, 30. [*Icel.* þjóð-vegr *a high road.*] Cf. þeód-herpaþ.

þeód-wíga, an ; m. *A mighty warrior :*—Se þeódwíga (*the panther*) . . . ellenróf, Exon. Th. 357, 33 ; Pa. 38.

þeód-wita, an ; m. **I.** *one of the wise men of a nation, one whose knowledge fits him for a place in the councils of the nation, a senator :*—Senatores, ðæt synd þeódwitan, Jud. p. 161, 32. Ðā wǣron þeódwitan (leód-, MS. H.) weorðscipes wyrðe, L. R. 1 ; Th. i. 190, 12. Be ðeódwitan. Cyningan and bisceopan, eorlan and heretogan, gerēfan and dēman, lārwitan and lahwitan gedafenaþ mid rihte ðæt hí þeódwitan weordan, L. I. P. 5 ; Th. ii. 308, 12. **II.** *a man of great wisdom* or *learning, a sage :*—Wā eów ðe taliaþ eów sylfe tō ðeódwitan *ve, qui sapientes estis coram oculis vestris*, Wulfst. 46, 26. **II a.** *used of a poet :*—Se þeódwita Virgilius, Anglia viii. 320, 30. Oft ða þeódwitan ðus heora meteruers gewurðiaþ, 332, 15. **II b.** *used of a historian or philosopher or man of science :*—An þeódwita wæs on Britta tídum, Gildas hātte, Wulfst. 166, 17.

Manega þing wē mihton of þeódwitena gesetnysse geícean, Anglia viii. 321, 24. [Cf. *Icel.* þjóð-skáld, -smiðr *a great poet, craftsman.*] Cf. leód-wita.

þeód-wrecan *to avenge thoroughly, take great vengeance on :*—Grendles módor gegān wolde sorhfulne sīð, sunu þeódwrecan (*Ettmüller would read* suna deáð wrecan ; *but perhaps the force of* þeód- *here and its composition with a verb may be illustrated by the case of* full-, *which is compounded with verbs, and has the force of* per- ; *see the verbs in the Dictionary. The parallel between* full- *and* þeód- *might be further illustrated from compound adjectives in Icelandic, e.g.* full-glaðr *and* þjóð-glaðr, full-góðr *and* þjóð-góðr), Beo. Th. 2561 ; B. 1278.

þeód-wundor, es ; n. *A great wonder, mighty miracle :*—Men gesēgon þeódwundor micel, ðætte eorðe āgeaf ða hyre on lægun, Exon. Th. 71, 14 ; Cri. 1155.

þeóf, es ; m. *A thief* [the secrecy implied by the word is marked in the following passage from the Laws dealing with injury done to a wood : Fýr biþ þeóf . . . sió æsc biþ melda, nalles þeóf, L. In. 43 ; Th. i. 128, 19–23. Cf. *Goth.* þiubjō ἐν κρυπτῷ] :—Þeóf *fur*, scaþa *latro*, Wrt. Voc. i. 74, 23. Gyf se hīredes ealdor wiste on hwylcere tíde se þeóf (ðeáf, Lind. *fur*) tōwerd wǣre, Mt. Kmbl. 24, 43. Ealle ða ðe cōmun wǣron þeófas (ðeáfas, Lind.) and sceaþan (*fures et latrones*) . . . Þeóf (ðeáf, Lind. *fur*) ne cymþ būton ðæt hē stele and sleá, Jn. Skt. 10, 8–10. Þeóf ðe on þýstre færeþ, on sweartre niht, Exon. Th. 54, 21 ; Cri. 872 : 432, 10 ; Rä. 48, 4. Ðeóf sceal gangan in ðýstrum wederum, Menol. Fox 543 ; Gn. C. 42. Ðǣr þeófas (ðeáfas, Lind. *fures*) hit delfaþ and forstelaþ, Mt. Kmbl. 6, 19. On helle beóþ þeófas and gītseras ðe on mannum heora ǣhta on wōh nimaþ, Blickl. Homl. 61, 21. Hēr syndan rýperas and reáferas and woruldstrúderas and ðeófas and þeódscaðan, Wulfst. 165, 36. Þeófum *grassatoribus*, Wrt. Voc. ii. 40, 35. Ealle niht ic (*the ox-herd*) stande ofer ða oxan waciende for þeófan (*propter fures*), Coll. Monast. Th. 20, 29. ¶ The passage last cited suggests a state of society in which property was not very secure, and the suggestion seems borne out by the many passages, dealing with thieves, that are to be found in the Laws. Thieving was so far common, that the law enacted : Gif feorrancumen man oþþe frǣmde būton wege gange, and hē ðonne nāwðer ne hrýme, ne hē horn ne blāwe, for þeóf hē biþ tō prófianne, L. Wih. 28 ; Th. i. 42, 23 : L. In. 20 ; Th. i. 114, 15 ; and on such a scale was it conducted that according to the numbers of the depredators acting together were different terms used of them : Ðeófas wē hātaþ óð .vii. men ; from .vii. hlóð óð .xxxv. ; siþþan biþ here, L. In. 13 ; Th. i. 110, 13. The frequency of this particular form of crime may also be inferred from the later enactment : Wē wyllaþ ðæt ǣlc man ofer twelfwintre sylle ðone āð, ðæt hē nelle þeóf beón ne þeófes gewita, L. C. S. 21 ; Th. i. 388, 6. But far stronger measures than the exacting of such an oath were in force. The law made provision for the pursuit of thieves, L. Edg. H. 2 ; Th. i. 258, 6, and imposed penalties on those who, being summoned, or hearing the hue and cry, neglected to take part in the pursuit, 3 ; Th. i. 258, 14 : L. C. S. 29 ; Th. i. 392, 17 : while a reward was given to him who seized a thief : Se ðe þeóf gefēhþ, hē āh .x. scitl., L. In. 28 ; Th. i. 120, 5. To let a thief go, when caught, was a crime, L. In. 36 ; Th. i. 124, 14 ; so, also, to allow him, when discovered, to escape without raising hue and cry, L. C. S. 29 ; Th. i. 392, 14 : to harbour a thief, except in those cases where the right of asylum might for three or nine days be extended to him, was to become liable to the fate of a thief, L. Ath. iii. 6 ; Th. i. 219, 6 : iv. 4 ; Th. i. 224, 4 : v. 1, 2 ; Th. i. 228, 21 ; to fight for him was equally penal, v. 1, 3 ; Th. i. 228, 23 : v. 8, 3 ; Th. i. 236, 18. And the laws which affected the thief himself were very severe. Any one above the age of twelve, who was caught stealing property above the value of eight pence, was liable to capital punishment, L. Ath. i. 1 ; Th. i. 198, 15 ; according to other regulations, for a theft which, on conviction, rendered the thief liable to be slain, the limit of age was made fifteen years, L. Ath. v. 12, 1 ; Th. i. 240, 28, and the limit of value was twelve pence, L. Ath. v. 1, 1 ; Th. i. 228, 12 : v. 12, 3 ; Th. i. 242, 8. The extreme penalty was not in all cases exacted ; but in case of repeated conviction there was to be no remission, L. Ath. v. 1, 4 ; Th. i. 230, 3. Cf. too the passages : Gesēce ǣbera þeóf ðæt ðæt hē gesēce, oððe se ðe on hláfordsearwe gemēt sý, ðæt hí nǣfre feorh ne gesēcen, būton se cyningc him feorhgeneres unne, L. Edg. ii. 7 ; Th. i. 268, 22 : L. C. S. 26 ; Th. i. 390, 27. Sý hē þeóf, and þolige heáfdes and ealles ðæs ðe hē āge, L. Edg. S. 11 ; Th. i. 276, 13. The kinds of death mentioned in L. Ath. iii. 6 ; Th. i. 219, are throwing from a rock or drowning in the case of a free woman ; in the case of a *servus homo*, stoning by slaves ; in that of a *serva ancilla*, burning. Further a thief who was taken in the act, or taken in flight, or who resisted, instead of being handed over to justice (on cyninges bende, L. In. 15 ; Th. i. 120, 4 : se cyning āh ðone þeóf, 28 ; Th. i. 120, 6), might be slain without the intervention of the law, and the death called for no ' wergild,' L. Wih. 25 ; Th. i. 42, 13 : L. In. 12 ; Th. i. 110, 7 : 16 ; Th. i. 112, 7 : 35 ; Th. i. 124, 6 : L. Ath. i. 1 ; Th. i. 198, 20 ; and in cases of flight or resistance the fact that the value of the stolen property was less than twelve pence was to be no bar to the slaying, L. Ath. v. 12, 3 ; Th. i. 242, 10. He who struck down a thief in public was rewarded : Se ðe þeóf fylle beforan ōðrum mannum, ðæt

hē wǣre of ūre ealra feó .xii. pænᵹ đe betera for đære dǣda and đon anginne, L. Ath. v. 7; Th. i. 234, 22. Short of death were the punishments of selling into slavery, of imprisonment, fine, and mutilation: Gif man frigne man æt hæbbendre handa gefó, đanne wealde se cyning þreora ānes: oþþe hine man cwelle, oþþe ofer sǣ selle, oþþe hine his wergelde ālése, L. Wih. 26; Th. i. 42, 15. Gif þeóf sié gefongen, swelte hē deáđe oþþe his líf be his were man āliése, L. In. 12; Th. i. 110, 8. Gif man þeóf on carcerne gebringe, đæt hē beó .xl. nihta on carcerne, and hine mon đonne ālýse ūt mid .cxx. scitt., L. Ath. i. 1; Th. i. 198, 21. Cutting off the hand or foot of a 'cirlisc þeóf' is mentioned, L. In. 18; Th. i. 114, 5: 37; Th. i. 124, 20. The same punishment is mentioned, L. C. S. 30; Th. i. 394, 10; and in aggravated cases the more severe sentence was passed, that the eyes were to be put out, and the nose, ears, and upper lip to be cut off, ib. An instance of punishment for theft, in which the eyes were put out and the ears cut off after (wrongful) conviction is given, Homl. Skt. i. 21, 265. If the thief managed to escape, he was declared an outlaw: Beó se þeóf ūtlah wiđ eall folc, L. C. S. 30; Th. i. 394, 24. v. Grmm. R. A. 635 sqq.; Schmid, A. S. Gesetz. s. v. Diebstahl. [Goth. þiubs: O. Sax. thiof: O. Frs. thiaf: O. H. Ger. diob: Icel. þjófr.] v. beó-, gold-, mann-, mūs-, regn-, sǣ-, stód-, wergild-þeóf; infangene-þeóf; þífþ.

þeóf, e; f. Theft:—Nā dón þeófæ non facere furtum, R. Ben. Interl. 19, 12. [O. H. Ger. diuba furtum.]

þeóf-denn, es; n. A thieves' cave:—Andlang weges tô đam þeófdenne, Cod. Dip. Kmbl. iii. 15, 28.

þeófend, þeófent, e; f. (the word seems to occur only in the plural) Theft:—Of hearte ūtgaas . . . điófunta de corde exeunt . . . furta, Mt. Kmbl. Lind. 15, 19. Điófunto (-ento, Rush.) furta, Mk. Skt. Lind. 7, 22. Wiđ þeófentum, Lchdm. iii. 58, 1. Ic heó tô þeófendum and tô geflitum stihte, Wulfst. 255, 11. Ne leásunga tô sæcganne, ne þeófenda tô begangenne, 253, 8. Ne dóe đū điófonto ł stalo non facies furtum, Mt. Kmbl. Lind. 19, 18: Lk. Skt. Lind. 18, 20. Đióf[]nto, Rtl. 103, 3.

þeófe-þorn. v. þífe-þorn.

þeóf-feng, es; m. Seizing of thieves; the Latin rendering of the term in Charters is comprehensio (or captio) furis (-um). I. The word seems to denote the obligation of one who holds land to arrest and bring to justice those who committed theft on that land, and occurs generally in connection with the burdens from which land, when granted, was relieved:—Ic forgyfe đisne freóls tô đære hālgan stówe æt Scíreburnan, đæt hit sý gefreód alra cynelícra and alra dómlícra þeówdóma, ge þeóffenges ge æghwelcre [un]iéđnesse ealles worldlíces broces, nymđe fyrde and bryceweorces, Chart. Th. 125, 11. Đæt hit (the monastery at Horton) sý gefreód ealra cynelícra and ealdordómlícra þeówdóma, ge þeóffengces ge æghwylcere uneáđnesse ealles woroldlíces broces būton fyrdsócne and burhgeweorce and bryggeweorce, 389, 28. Corresponding cases in Latin charters are the following:—Ego Ecgberhtus . . . hanc libertatem donabi aecclesiae . . . , ut omnes agros sint libera ab omni regali seruitio (then follows a list of exemptions), . . . et ab omnibus difficultatibus regalis uel saecularis seruitutis, cum furis comprehensione intus et foris, praeter pontis constructione et expeditione liberata permaneat, Cod. Dip. Kmbl. i. 288, 5. Terra predicta liber et securus omnium rerum permaneat, id est, regalium et principalium tributum, et ui exactorum operum siue poenalium causarum, furisque comprehensione, et omni saeculari grauidine, ii. 28, 22. Ui exactorum operum et penalium rerum, principali dominatione, furisque comprehensione, et cuncta seculari grauidine . . . secura et immunis, 65, 14. Omnium regalium debitorum et principalium rerum, caeterarumque causarum, furisque comprehensione, et ab omnium saecularium seruitutum molestia secura et inmunis, 95, 33. Furum comprehensione, iii. 277, 4. Captio furum, iv. 2, 26. II. In other passages, however, the word implies advantage, and seems to refer to the right to receive the fines which might be exacted in case of conviction for theft. For such emoluments cf. Gif frigman stelþ . . . cyning āge đæt wíte and ealle đa æhtan, L. Ethb. 9; Th. i. 6, 2. Ealle wītu (in cases of theft) sint gelíce, . . . cxx. scitt., L. Alf. pol. 9; Th. i. 68, 7: L. Ath. i. 1; Th. i. 198, 23. Gif þeuw stele . . . hine man ālése .lxx. scitt., L. Wih. 27; Th. i. 42, 20. Hine man his wergelde ālése, 26; Th. i. 42, 17: L. In. 12; Th. i. 110, 8. These emoluments of the crown are made the subject of grant:—Concedo consuetudines, ut ab omnibus apertius et plenius intelligantur Anglice scriptas, scilicet, mundbryce, feardwítæ, fihtwíte . . . þiésphang, hangwíte, gryđbryce . . . toll et teám, aliasque omnes consuetudines quae ad me pertinent, Chart. Th. 384, 24. Terram liueram ab omni seruitute, cum omnibus quae ad eam sunt rite pertinentibus, cum furis comprehensione, et cum omnibus rebus quae ad aecclesiam Sancti Andreae pertinent, cum campis, etc., Cod. Dip. Kmbl. ii. 109, 21.

þeóf-gild, es; n. Payment made in the way of fine or compensation by one convicted of stealing:—Swerian hí đæt him nǣfre āđ ne burste, ne hē þeófgyld ne gulde (i. e. that he had never been convicted of stealing), L. Eth. i. 1; Th. i. 280, 13: iii. 4; Th. i. 294, 13: L. C. S. 30; Th. i. 392, 28. [Cf. Icel. þýfi-gjöld fine for theft.]

þeófian (and þeófan? The Lindisfarne gloss has đæt đū ne forstele ł ne forđiófe, Mk. 10, 19; the Kentish Glossary, điófende furtivus; but this might imply the form điófian, cf. tācnendi and tācnian: cf. also, for both

force and form of the participle styrende agitatam, Mt. Kmbl. Lind. 11, 7) to thieve, steal:—Se đe đa āre þænce tô þeófigenne qui quid illinc abstulerit, Chart. Th. 177, 13. Điófende weteru stolen waters; aque furtive, Kent. Gl. 309. [O. H. Ger. thaz sie mit stalu nan nirzuken noh inan thar githiuben, Ot. iv. 36, 12.] v. ge-þeófian.

þeóf-mann, es; m. A robber, bandit, brigand:—Ān hirde wæs Ueriatus hāten, and wæs micel þeófmon Viriathus homo pastoralis et latro, Ors. 5, 2; Swt. 216, 7.

þeóf-scip, es; n. A pirate-vessel:—Đeófscip (thēb-) mimopora (= myoparo), Txts. 79, 1316: Wrt. Voc. ii. 55, 67.

þeóf-scolu, e; f. A gang or band of thieves:—Gif đū wǣre wegfêrende and hæfdest micel gold on đe, and đū đonne becóme on þeófsceole (þiófscole, Cott. MS.), đonne ne wêndest đū đe đínes feores, Bt. 14, 3; Fox 46, 26.

þeóf-scyldig; adj. Guilty of theft:—Stent đonne þeófscyldig se đe hit on handa hæfþ, L. Eth. ii. 9; Th. i. 290, 16.

þeóf-slege, es; m. Thief-slaying:—Be đeófslege. Se đe þeóf ofslihþ, L. In. 16; Th. i. 112, 6.

þeóf-sliht, es; m. Thief-slaying:—Be đeófslihte. Se đe þeóf slihþ, hē mót āđe gecýđan đæt hē hine fleóndne for þeóf slóge, L. In. 35; Th. i. 124, 4.

þeóf-stolen; adj. (ptcpl.) Stolen, taken by thieves:—Swā ic sprǣce drífe . . . swā mē þeófstolen (forstolen, MS. H.) wæs đæt orf, L. O. 2; Th. i. 178, 14: L. O. D. 8; Th. i. 356, 12. Æt ǣlcon đeófstolenan orfe, L. Ff.; Th. i. 226, 2. [Icel. þjóf-stolinn. Cf. M. H. Ger. diep-, diup-stâle: Ger. dieb-stahl.]

þeófþ, þeóft. v. þífþ.

þeóf-wracu, e; f. Punishment for theft:—Gif hē eft ofer đæt stalie . . . sleá man hine on đa þeófwrace, L. Ath. v. 1, 4; Th. i. 230, 4.

þeóging, e; f. Profiting, thriving, progress, advancement:—Þeógincg đín swutul sý eallum profectus tuus manifestus sit omnibus; that thy profiting may appear to all (A. V. 1 Tim. 4, 15), Scint. 203, 8. Þeóginc (profectus) mannes gyfu Godes ys, 132, 17. Swylcre þeógincge tanti profectus, Anglia xiii. 372, 94. Þeógincgum profectibus, Scint. 210, 1. [O. H. Ger. dîhunga provectus.] v. þeón.

þeóh; gen. þeós; dat. þeó; pl. þeóh; gen. þeóna; dat. þeón; n. A thigh:—Þeóh, thêgh coxa, Txts. 54, 295. Þeóh, Wrt. Voc. ii. 15, 6. Þeóh femur, femoris, ys swā đeáh eft gecweden femen, feminis, Ælfc. Gr. 9, 22; Zup. 49, 10. Inneward þeóh femen, þeóh coxa, ūtanweard þeóh femur, Wrt. Voc. i. 44, 60–62. Þeóh femur, þeóh coxa, hype clunis, 71, 46–49. Þeóh femor, innewerd þeóh femina, þeóhscanca coxa, 283, 63–65. Þeóh vel hype femur, ii. 148, 18. Đæt đeóh getácnode his cynn, Homl. Th. ii. 234, 33. Gif þeóh gebrocen weorđeþ, .xii. scillingum gebête, L. Eth. 65; Th. i. 18, 13. Gif man þeóh þurhstingđ, stice gehwilce .vi. scillingas, 67; Th. i. 18, 16. Gif monnes þeóh biþ þyrel, geselle him mon .xxx. scitt. tô bóte; gif hit forad sié, sió bót eác biþ .xxx. scitt., L. Alf. pol. 62; Th. i. 96, 13. Đā æthrān hē his sine on his þeó tetigit nervum femoris ejus, Gen. 32, 25. 'Hæbbe eówer ǣlc his sweord be his đeó.' Đonne mon hæfþ his sweord be his đió, đonne . . . , Past. 56; Swt. 433, 11: Exon. Th. 431, 2; Rä. 45, 1. Under mín þeóh subter femur meum, Gen. 24, 2: 47, 29: Ps. Th. 44, 4. Bind on đæt winstre þeóh up wiđ đæt cennende lim, Lchdm. ii. 328, 22. Þeóh bathma, i. femora, Wrt. Voc. ii. 125, 28. Đeeoh (đyóh, lxxiv, 3) bathma, femora, i. lxx, 2. Gif men his đeóh acen, 78, 23: ii. 66, 4. Hyre (the bee's) đa rūwan þeóh wurđaþ swýđe gehefegode, Anglia viii. 324, 13. Bāna, þeóna coxarum, Wrt. Voc. ii. 75, 27: Lchdm. i. 208, 3. Đæra đeóna sār, 80, 2. On þeóh in femoribus, Anglia xi. 117, 25. Smyre đa þeóh, sóna hý beóþ hāle, Lchdm. i. 354, 20: ii. 64, 26: Ors. 1, 7; Swt. 38, 3. [O. Frs. thiach: O. L. Ger. thio femur: O. H. Ger. dioh femur, femen, coxa: Icel. þjó.]

þeóh-ece, es; m. Thigh-ache:—Lǣcedóməs wiþ þeóhece, Lchdm. ii. 6, 6: 64, 26.

þeóh-gelǣte, es; n. A thigh-joint, the meeting of the thigh with the part of the body above it:—Ersendu mid đām đeóhgelǣtum (þeóhsconcum, lxxiv, 19) nates cum femoribus, Lchdm. i. lxx, 10. [Cf. O. H. Ger. lidi-gilâz artus, compago.] v. ge-lǣte.

þeóh-geweald; pl. n. Genitalia:—Đa þeóhgeweald mid đæm þeóhhweorfan genitalia cum genuclis, Lchdm. i. lxxiv, 20. v. ge-weald.

þeóh-hweorfa, an; m. A knee-joint; genuculum (cf. cneów-wyrste geniculi, Wrt. Voc. i. 44, 70). v. preceding word.

þeóh-sceanca, an; m. A thigh-shank, the upper part of the leg:—Earsendu nates, þeóh femur, þeóhscanca coxa, Wrt. Voc. i. 65, 36–38: 283, 61–65. Đa hypbān đa earsenda mid đæm þeóhsconcum catacrinas, nates cum femoribus, Lchdm. i. lxxiv, 19. [O. Frs. thiach-schonk. Cf. Icel. þjó-leggr the thigh-bone.]

þeóh-seax, es; n. A short sword that could be worn on the thigh:—Þeóhsaex semispatium (= -spathium), Wrt. Voc. ii. 120, 26. Sweord macheram, þeóhseax senspatium, 96, 29. Cf. hup-seax.

þeóh-wærc, -wræc, es; m. A. Pain in the thighs:—Wiđ þeóhwræce . . . smyre đa þeóh, sóna hý beóđ hāle, Lchdm. i. 354, 19.

þeón [from þíhan; and this from an earlier nasal stem, of which traces are preserved in the past forms, where g has replaced h by Verner's law] :—Đunge polleceret, Wrt. Voc. ii. 66, 40. Fród fæder freóbearn lǣrde

wordum wísfæstum, ðæt hé wel þunge, Exon. Th. 300, 9; Fä. 3. *See also the passages given under* ge-þingan; ofer-þeón; ge-, heáh-, wel-þungen; on-þungan, Exon. Th. 497, 3; Rä. 85, 23 (*omitted in its place*)]; *p.* þáh *and* þeáh, *pl.* þigon *and* þugon; *p. pr.* þiende *and* þeónde; *pp.* þigen *and* þogen *To thrive, grow, flourish, prosper :—*þíhþ *cluit, pollet, viget, nobilitat*, Wrt. Voc. ii. 131, 75. þáh *pubesceret*, 66, 22. þeó *vigeat*, Wülck. Gl. 257, 17.   I. of persons in respect to either physical or moral growth. (1) absolute :—Se his yldrum ðáh tó frófre *he grew up a comfort to his parents*, Cd. Th. 67, 28; Gen. 1107. Sunu weóx and ðáh, 138, 30; Gen. 2299. þág, 167, 25; Gen. 2771. Ðæt cynn þeáh, Wulfst. 13, 11. Heó ðurh mægðhád mærlíce þeáh, Homl. Skt. i. 2, 3. Cnæplingc weóx ł þéh *puer pollesceret*, Hpt. Gl. 466, 60. Hyhtful *vel* ðíendi *indolis*, Wrt. Voc. ii. 111, 54. þíonde, 45, 58. (2) *where that, in which the growth, etc., takes place, is stated :—*Se gǽst þíhð in þeáwum, Exon. Th. 126, 10; Gū. 369. Sume on ǽghwam þeóþ *quidam in utrisque pollent*, Scint. 221, 1. Hé weóx under wolcnum, weorþmyntum þáh, Beo. Th. 16; B. 8. Se Hǽlend þeáh on wísdóme and on ylde *Jesus proficiebat sapientia et aetate*, Lk. Skt. 2, 52 : Homl. Skt. i. 2, 23. þeáh hwá þeó on eallum welum and on eallum wlencum, Bt. 19; Fox 68, 31. þió, Met. 10, 28. Ðeóude on cræftum *virtutibus pollens*, Past. 9; Swt. 59, 11. Ðíonde, Bt. 38, 5; Fox 206, 22. Ðiiende on wæstum *proficiens incrementis*, Rtl. 38, 41. (3) *where that, in relation to which the growth, etc., takes place, is stated :—*Monge lifgaþ gyltum forgiefene, nales Gode þígaþ, Exon. Th. 130, 3; Gū. 432. Wǽron hálige sacerdas Gode ðeónde, Homl. Th. i. 544, 11.   II. of things abstract or concrete :—Andgyt þýhð *sensus uiget*, Scint. 52, 8. þýhð (*virginitatis gratia*) *adolescit*, Hpt. Gl. 436, 67. Ic þéh óþþæt ic wæs yldra, Exon. Th. 485, 2; Rä. 71, 7. Se síð ne þáh ðam ðe unrihte inne gehýdde wræte *the journey did not turn to the profit of him who unrighteously had hidden treasure within*, Beo. Th. 6109; B. 3058 : 5665; B. 2836. Hine (him?) se cwealm ne þeáh, Exon. Th. 278, 30; Jul. 605. þeáh, bleów (*gratia*) *floruerit*, Hpt. Gl. 441, 48. His wæstmas genihtsumlíce þugon (*uberes fructus ager attulit*, Lk. 12, 20), Homl. Th. ii. 104, 15. Se líchama þeóude on strangum breóste, on fullum limum and hálum, i. 614, 11. Ðás wanunge getácnaþ se wanigenda dæg his (*John's*) gebyrdtíde, and se ðeónda (*increasing, lengthening*) dæg ðæs Hǽlendes ácennednysse gebícnaþ his ðeóndan mihte, 358, 4. Betweox óðrum mægenum bið ðeónde (ðíonde, Hatt. MS.) sió earnung ðæs geswenctan flǽsces *inter virtutes ceteras afflictae carnis meritum proficit*, Past. 14; Swt. 86, 25 : Hpt. Gl. 420, 37. þeónde *florentis* (*pudicitiae*), 511, 50. [Here tuder swíðe wexeð and wel þieð, O. E. Homl. ii. 177, 18. He was þogen on wintre and on wastme, 177, 15. His welðe ðeg, Gen. and Ex. 2012. Wexen he (*they*) and ðogen wel, 2542. So wex here erue, and so gan ðen, 803. So mot I the, Chauc. N. P. T. 156. Theen or thryvyn *vigeo*, Prompt. Parv. 490. Goth. þeihan : O. L. Ger. O. Sax. thíhan : O. H. Ger. díhan *proficere, pollere, florere, crescere, excellere*.] v. for-, fore-, ge-, mis-, ofer-, on-þeón.

þeón; *p.* þeóde *To do, perform, effect :—*Wé ðæt ǽbylgð nyton, ðæt wé gefremedon, þeódon bealwa wið ðec ǽfre, Elen. Kmbl. 805; El. 403. v. ge-þeón.

þeón *to press.* v. þeówan.

þeónest-mann. v. þegnest-mann.

þeór, es (?), e (?), gender is uncertain : in the following passages, which might be decisive, the forms are doubtful :—Wiþ þeóre drenc, and eft wiþ þǽre (*if þǽre refers to* þeór *the word would be feminine, but perhaps* þeore *should be read ; cf. the text :* Wyrc gódne ðeórdrenc . . . Wiþ þeóre and sceótendum wenne, 324, 15–25) and sceótendum wenne and eft beþing wiþ þam (*the beþing is for* þeór, v. 326, 3, *so if* þam *refers to* þeór *the word is masc. or neuter*) gif þeór gewunige on ánre stówe, Lchdm. ii. 300, 30. Drenc wiþ þeóráðle . . . gif hé on þam innoþe biþ þonne ádríʃþ hine ðǽs drinc út (hé *and hine may, perhaps, be taken as referring to* þeór, *as ádl is fem.*), 118, 1–12. The meaning, too, is doubtful. It seems to denote an inflamed swelling or ulcer ; it is mentioned in connection with wens (Lchdm. ii. Bk. ii. §§ 30, 31); in reference to the eye it is said to be the same as 'fig' (ii. 38, 5), and is mentioned in close connection with the same disease (iii. 30, 3–16); the same prescription is good 'wið ðam micclan líce and wið ðprum giccendum blece and þeórgeride' (iii. 70, 28); þeórwyrt is used against hreófl (ii. 78, 13); purgative and emetic drinks are used for its cure (ii. 115, 23), and bleeding (118, 21 : 120, 12), cupping (120, 16), and fomentation (326, 3) are prescribed :—Gif þeór gewunige on ánre stówe, Lchdm. ii. 326, 1 : iii. 30, 6. Drenc gif þeór sié on men, ii. 354, 16 : iii. 28, 13. Sealf wiþ þeóre . . . Wiþ þeóre on fēt, ii. 118, 12, 28. Wið ðeóre, iii. 20, 15 : 28, 7, 19: 30, 3, 13. v. next word.

þeór-ádl, e ; *f. Some disease.* v. preceding word :—Wiþ þeóráðle on eágum ðe mon gefigo hǽt, Lchdm. ii. 38, 5. Lǽcedómas wið þeórádlum . . . Drenc wiþ þeóráðle, 116, 1, 13 : 118, 1, 18 : 172, 30. Drencas and sealfa wið þeórádlum, 12, 1, 3.

þeorcung, Anglia xiii. 398, 475 : 400, 508. v. deorcung.

þeór-drenc, es ; *m. A drink for* þeór-ádl, Lchdm. ii. 324, 18 : iii. 28, 25. v. þeór-ádl.

þeorf; *adj.*   I. *unleavened,* used substantively, *unleavened bread :—*Gehafen hláf *fermentacius panis*, ðeorf *azimus*, Wrt. Voc. i. 41, 15–16. þeorf *azimum*, ii. 6, 8. Sýfernysse þearf *sinceritatis azima*, Hymn. Surt. 82, 31. þeorfne hláf ðú scealt etan *vesceris azymis*, Ex. 34, 18 : Homl. Th. ii. 264, 16. Etaþ þeorf *azyma comedetis*, Ex. 12, 15, 18. Doege ðara ðorofra (ðefra, Rush.) mæta *die azymorum*, Mt. Kmbl. Lind. 26, 17. Lactuca hätte seó wyrt ðe hí etan sceoldon mid ðám þeorfum hláfum, Homl. Th. ii. 278, 26, 18. On ðærfum biluitnisse *in azymis sinceritatis*, Rtl. 25, 19. Healdaþ þeorfe mettas *observabitis azyma*, Ex. 12, 17. Hí worhton þeorfe heorðbacene hláfas *fecerunt subcinericios panes azymos*, 12, 39 : Lev. 8, 2 ; Homl. Th. ii. 210, 34 : 264, 2.   I a. in the Lindisfarne gloss *fermento* is glossed by ðærfe, Mt. 16, 6, 11.   II. of milk, *fresh* (as opposed to sour ? cf. Icel. þjarfr, of water, *fresh* as opposed to salt), *skim* (? so Cockayne) :—Dó on beór swá on wín swá on þeorfe meoluc, Lchdm. ii. 270, 29. [Bræd all þeorrf wiþþutenn berrme, Orm. 997. þerue kakez, Allit. Pms. 57, 635. þerf bred, Mand. 19, 1. Of þerf brede *de azymo pane*, Trev. v. 9, 6 : Wick. Gen. 19, 3. Therf, not sowyryd *azimus*, Prompt. Parv. 490. Aº tharf bred *panis siliginus, sigalinus*, Wrt. Voc. i. 198, 8, 9. O. H. Ger. derb brót *azymus :* Icel. þjarfr *unleavened ; fresh* (water).]

þeorf-dæg, es ; *m. A day on which unleavened bread was to be eaten :—*Ðæt geríst preóstum tó witanne hwæt beó betwyx Eástron and ðeorfdagum. Eásterdæg wæs se forma dæg on ðære ealdan ǽ, þonne se móna wæs .xiiii., and ða seofon dagas, ðe ðǽr æfter wǽron, wǽron gecíged *dies azimorum*, Anglia viii. 330, 19.

þeorf-hláf, es ; *m. A loaf of unleavened bread :—*Hí ǽton þeorfhláfas, Jos. 5, 11.

þeorfling, es ; *m. An unleavened loaf :—*Ðeorflingas *azimos*, Wrt. Voc. ii. 6, 32. [þerrfling bræd iss clene bræd, forr þatt itt iss unnberrmedd, Orm. 1590.]

þeorfness, e ; *f. Unleavenedness* ; metaphorically, *freedom from impurity, purity :—*Ðonne wé búton yfelnysse beorman on ðeorfnysse sýfernysse and sóðfæstnysse faraþ, Homl. Th. ii. 212, 1. On ðeorfnyssum sýfernysse and sóðfæstnysse, 278, 25.

þeorf-symbel, es ; *n. The feast of unleavened bread :—*Ðú ytst þeorfsymbel, Ex. 23, 14.

þeór-gerid, es ; *n. The inflammation accompanying* þeór (?) :—Gódne morgendrænc . . . wið ðam micclan líce and wið óþrum giccendum blece and þeórgeride and ǽghwylcum áttre, Lchdm. iii. 70, 28. v. þeór ; *and* cf. (?) O. H. Ger. rito *febris.*

þeorscwold. v. þerscold.

þeór-wærc, es ; *m. The pain caused by* þeór (q. v.) :—Wiþ þeórwærce, Lchdm. ii. 120, 7.

þeór-wenn, es ; *m. An inflamed wen, a carbuncle* (?) :—Wiþ þeórwenne, Lchdm. ii. 342, 16.

þeór-wyrm, es ; *m. A worm in a boil :—*Wiþ þeórwyrme on fēt, Lchdm. ii. 12, 2 : 118, 25.

þeór-wyrt, e ; *f. Ploughman's spikenard* ; *inula conyza :—*Wiþ hreófle . . . þeórwyrt, Lchdm. ii. 78, 13. Wyrc gódne ðeórdrenc . . . þeórwyrt, 324, 20. Ðýorwyrt, iii. 28, 27.

þeóster-cofa, an ; *m. A dark chamber,* used of the place where a person or thing is buried :—Under neólum niðer næsse gehýdde in þeóstorcofan, Elen. Kmbl. 1662; El. 833. Ðæt heó ðis bánfæt beorge bifæste, láme bilúce líc orsáwle in þeóstorcofan, Exon. Th. 173, 29; Gū. 1168. v. þeóster-loca.

þeóster-full (þístre-, þrýstre-) ; *adj. Full of darkness, dark, obscure :—*Þeósterfull wæter *tenebrosa aqua*, Ps. Spl. 17, 13. Ðín líchama byþ þýsterfull (*tenebrosum*), Mt. Kmbl. 6, 23. þéstreful *tenebrosa, obscura*, Hpt. Gl. 483, 53. Of þrýstreʃulre *de latebroso, tenebroso*, 458, 52. Se engel mé lædde tó ánre þeóstorfulre stówe . . . ealle ða ðeóstorfullan stówe, Homl. Th. ii. 350, 15–26. Hé geseah swilce án ðeóstorful dene, 338, 5. Ðeósterfulle wununga, i. 68, 4. þésterfulle dimhoua *latebrosa latibula*, Hpt. Gl. 446, 5.

þeóster-fullness, e ; *f. Darkness, obscurity :—*þéstrefulnysse *latebras, tenebras*, Hpt. Gl. 488, 33.

þeóster-líc; *adj. Dark :—*Ðæs muntes cnoll mid þeósterlícum genipum oferhangen wæs, Homl. Th. i. 504, 30. [Cf. þe clene of herte þet hier ssolle ysy him be byleaue, ac alneway þiesterliche, Ayenb. 244, 10.]

þeóster-loca, an ; *m. A dark enclosure, a tomb :—*In byrgenne bidende wæs under þeósterlocan, Elen. Kmbl. 967; El. 485.

þeósterness, e ; *f. Darkness :—*Wearð micel þeósternes ofer eallne middangeard *tetra nox obducta terris est*, Ors. 6, 2; Swt. 256, 16. Ðá com ðære nihte þýsternys, Homl. Ass. 203, 265. þǽsternes (cf. þióstro, Met. 21, 40), Bt. 34, 8; Fox 146, 4. On ðýsternesse *in obscuro*, Ps. Spl. 10, 2. [A. R. þeosternesse : Orm. þeossterrnesse : Gen. and Ex. ðisternesse : Piers P. þesternesse : O. L. Ger. thiusternussi.]

þeóstre *darkness.* v. þeóstru.

þeóstre, þeóster (-or, -ur) ; *and* þiestre, þístre, þýstre ; *adj. Dark.*   I. in a physical sense, *without light :—*Ðis (*hell*) is ðeóstræ hám, Cd. Th. 267, 14; Sat. 38. Ðá hangode swíðe þýstru wæter on ðám wolcnum *tenebrosa aqua in nubibus*, Ps. Th. 17, 11. Wæs se óðer beám eallenga

sweart, dim and þýstre, Cd. Th. 30, 36; Gen. 478. Þýstre genip, 9, 9; Gen. 139. Se þeóstra, Wulfst. 186, 4. Niht seó þýstre, Judth. Thw. 21, 25; Jud. 34. Hit wearð þýstre tenebrae factae erant, Jn. Skt. 6, 17. On óþre healfe ys þýstre land, Shrn. 120, 20. Ða fûlnessa ðæs ðýstran ofnes foetorem tenebrosae fornacis, Bd. 5, 12; S. 629, 21. Under ðam scûwan ðære ðýstran nihte sub nocte per umbras, S. 628, 15. Wæs heora sum ðýstran onsýne (tenebrosae facii), 5, 13; S. 633, 3. In ðære sweartan niht and in ðære þýstran, Nar. 15, 1. In ðam þýstran hám (hell), in ðam neólan scræfe, Exon. Th. 283, 21; Jul. 683. Þýstre land (hell), Cd. Th. 46, 1; Gen. 737. Sume ðara ðýstra gásta quidam spirituum obscurorum, Bd. 5, 12; S. 628, 40. Þeóstrum nihtum, Bt. 7, 4; Fox 22, 28. Deóf sceal gangan in ðýstrum wederum, Menol. Fox 544; Gn. C. 42. Niht biþ wedera þeóstrost, Salm. Kmbl. 621; Sal. 310. II. metaphorically, of absence of spiritual or mental light, or of cheerfulness:—Gif ðín eáge byþ deorc, eall ðín líchama byþ þýstre (ðióstor, Rush.), Lk. Skt. 11, 34. Biþ seáð ðam fyrenfullan deóp ádolfen, deorc and ðýstre, Ps. Th. 93, 12. Tódríf ðone þiccan mist, ðe wið ða eágan foran ûsses módes hangode, hefig and þýstre, Met. 20, 266. On hû ðióstrum horaseáþe ðara unþeáwa, Bt. 37, 2; Fox 188, 1. On ðás þeóstran weorulde, Exon. Th. 86, 18; Cri. 1410. Ðióstur (caecatum) habbas ge heorta iówre, Mk. Skt. Rush. 8, 17. Breóst innan weóll þeóstrum geþoncum, Beo. Th. 4653; B. 2332: Elen. Kmbl. 623; El. 312. [Laym. þe þestere (þustere) niht: Orm. þestterr: O. and N. bi þeostre nihte: Ayenb. þiestre: O. Sax. thiustri: O. Frs. thiustere.]

þeóstrian, þêstrian, þiéstrian, þîstrian, þýstrian; p. ode. I. to make dark or dim, to make the eye less capable of seeing, dim the sight:—Se dæg bleut and ðióstraþ hiora eágan, Bt. 38, 5; Fox 206, 5. II. to grow dark or dim:—His eágan þýstrodon caligaverunt oculi ejus, Gen. 27, 1: 48, 10. Ðýstrodan, Bd. 4, 10; S. 578, 19. Geseah ic onginnan ðýstrian ða stówe vidi obscurari incipere loca, 5, 12; S. 628, 10. [Þa þestrede þe dai, Chr. 1135; Erl. 260, 32. Steorren sculen þeostren, O. E. Homl. i. 143, 20. Heó þeóstred (make dark) ham suluen, A. R. 94, 20. Aras a ladlich weder, þeostrede (þustrede, 2nd MS.) þa wolcne, Laym. 4575.] v. á-, for-, fore-, ge-, of-þeóstrian (-þióstrian, -þêstrian, þiéstrian, -þîstrian, -þýstrian).

þeóstrig; adj. Dark:—All líchoma ðín ðióstrig (tenebrosum) biþ, Mt. Kmbl. Lind. 6, 23; Lk. Skt. Lind. 11, 34. Ðióstrig ł blind hearta caecatum cor, Mk. Skt. Lind. 8, 17. Ðurh ðrióstrie wegas per vias tenebrosas, Kent. Gl. 21.

þeóstru (sometimes written þr- instead of þ-) and þiéstru, þîstru, þýstru; f.: and þeóstre, þýstre; n. [cf. O. Sax. thiustri; n.] Darkness (lit. and metaph.); dimness of sight (lit. or metaph.); like the Latin tenebrae, which it translates, it is often used in the plural:—Ðær wæs deorc þeóstru, Ps. Th. 87, 6. Leóht and þeóstro, Cd. Th. 239, 27; Dan. 376. Þióstro, Met. 21, 41. Gif ðæt léht, ðætte in ðé is, þeóstru sint, ðæt þeóstre hû micel biþ, Mt. Kmbl. Rush. 6, 23. On ðæt gemǽre leóhtes and þeóstro, Bt. 35, 6; Fox 170, 13. Mid þýstro genipum, Blickl. Homl. 203, 8: 209, 33. On ðýstres onlícnisse . . . on leóhtes onlícnisse, Salm. Kmbl. p. 144, 30. For ðære egeslícan þióstro ðære nihte, Bt. 23; Fox 78, 29. Ne gǽþ hé on þeóstro, Blickl. Homl. 103, 31. In þýstro, Exon. Th. 432, 10; Rä. 48, 4. Mid þýstro, Cd. Th. 148, 1; Gen. 2450. Wið eágena þýstru and genipe, Lchdm. i. 366, 13. On þeóstre, Exon. Th. 87, 11; Cri. 1423: 94, 27; Cri. 1546. Þeóf ðe on þýstre færeþ, on sweartre niht, 54, 22; Cri. 872. Mid þýstre, 462, 20; Hö. 55: 470, 12; Hy. 11, 14. Þeóstru tenebre, Wrt. Voc. i. 76, 48. Þeóstru wǽron, Gen. 1, 2. Becômon ðicce ðeóstru, Homl. Th. ii. 194, 4. Swá ðáþ ðe ðeóstro ðínre gedréfednesse, Bt. 6; Fox 14, 30. Beóþ þeóstra gewordene, Blickl. Homl. 93, 18. Ðære nihte þióstro hí onlîhtaþ, Bt. 38, 5; Fox 206, 5. Ðióstro, Jn. Skt. Lind. Rush. 6, 17. Gif ðæt leóht, ðe on ðé is, synt þýstru (ðióstræ, Lind.: þeóstru, Rush.), hû mycle beóþ ða þýstru (ðióstro, Lind.), Mt. Kmbl. 6, 23: Lk. Skt. 11, 35. Ðæt þýstro eów ne befón ut non tenebrae uos compræhendant, Jn. Skt. 12, 35: 1, 5. Æfter ðæm clammum helle þeóstra, Blickl. Homl. 83, 22: Exon. Th. 143, 28; Gû. 668. Se beorhta dæg tódrǽfþ ða dimlícan þeóstru ðære sweartan nihte . . . Crist ûs fram deófles ðeóstrum álýsde . . . ' Uton áwurpan þeóstra weorc . . . ,' Homl. Th. i. 604, 1–5. Ðara þióstra ðisse worulde, Bt. 36, 2; Fox 174, 26. Þýstra (ðióstrana, Lind. Rush.) anweald, Lk. Skt. 22, 53: 11, 36. Ðiéstra dæg and mistes, Past. 35; Swt. 245, 5. On þýstra bealo, Exon. Th. 76, 32; Cri. 1248. Þrýstra wrǽce, 37, 15; Cri. 593. Hé gedǽlde ðæt leóht fram ðám þeóstrum, Gen. 1, 4; Cd. Th. 8, 21; Gen. 127: Blickl. Homl. 65, 17. Þióstrum, Bt. 39, 3; Fox 214, 30. Ðýóstrum, Mt. Kmbl. Lind. 8, 12. Mid ðæm ðiéstrum (ðístrum, Hatt. MS.) ðisses andweardan lífes . . . on ðæm ðístrum (ðiéstrum, Hatt. MS.), Past. 65; Swt. 64, 8, 12. Ðýstrum, Ps. Th. 106, 9. ¶ The acc. sing. and pl. are given together, as often the two cannot be distinguished:—God hét ða þeóstra niht, Gen. 1, 5: Blickl. Homl. 17, 36. Deós India hæfþ on ânre sídan þeóstru, Homl. Th. i. 454, 14. On ða ûttran þýstru (in þeóstra ðær ýtterre, Rush.), Mt. Kmbl. 25, 30. Ðara deófla þeóstro hé oforgeát mid his leóhte, Blickl. Homl. 85, 8, 21. Áweorpan ða ðióstro his módes, Bt. 35, 6; Fox 166, 26: Met. 24, 56: Rtl. 37, 9. On ða ûttran þýstro (in ðióstre ðæt

ýtemǽst ł ýterræ, Rush.), Mt. Kmbl. 22, 13: 8, 12. Þýstro (ðióstro, Lind.: ðióstru, Rush.) tenebras, Jn. Skt. 3, 19. Þistro, Bt. 3, 2; Fox 6, 10. Ðístro, Past. 56; Swt. 433, 13. Þýstru, Ps. Th. 103, 19. Þrióstre senna, Hy. 8, 28. [Of þeóstran de tenebris, O. E. Homl. i. 131, 12. Hit luveþ þuster and hateþ liht, O. and N. 230. O. Sax. thiustria; f.: thiustri; n.] v. carcern-, hinder-þeóstru.

þeóstrung, þýstrung, e; f. Darkness, gloom, obscurity:—Hí on ðære þýstrunge hine swencton, Guthl. 5; Gdwin. 36, 14. Næhtes [ðió]strung noctis caliginem, Rtl. 182, 35.

þeótan and þútan; p. þeát, pl. þuton. I. to howl like a wolf:—Wulf ðýtt lupus ululat, Ælfc. Gr. 22; Zup. 129, 1. Hwílum hí ðuton eall swá wulfas, Shrn. 52, 29; Bt. 38, 1; Fox 194, 36. Sume hí tó wulfum wurdon . . . hió þióton ongunnon, Met. 26, 80. Ðeótende swá swá wulf, Homl. Th. i. 374, 9. Hwílum swá swá þeótende wulf, hwílum swá beorcende fox, Shrn. 141, 12. II. of other sounds:—Þeótende murmurans, Germ. 399, 417. Hlówende, þútende bombosa, Wrt. Voc. ii. 126, 51. Ðære þútendan bombose, 11, 71. Him on gafol forlét ferðfriþende feówer wellan scíre sceótan on gesceap þeótan (or gesceap-þeótan. ? v. gesceap, III, and þeóte. The passage describes a calf sucking from its mother; if þeótan is an infinitive, it must refer to the sound made by the milk coming from the teat, but perhaps gesceap-þeúte may be a compound noun meaning the teat), Exon. Th. 420, 2; Rä. 39, 4. [Bigunnen to þeoten and to ȝellen alle þe untrume weren, Marh. 22, 29. Wummone wrod is wuluene . . . ne ðeð heo bute þeoteð, A. R. 120, 12. Ȝeinde ȝurinde & þeotinde wið reowfule reames queruloso gemitu deplorantes, Kath. 161. Giff mann wollde tælenn þatt, and hutenn hire & þutenn, Orm. 2034. O. H. Ger. diozan stridere, fremere, strepere, mussare: Icel. þjóta to whistle (of the wind, etc.), to howl (of a wolf), to rush: Dan. tude to howl. Cf. Goth. þut-haurn, -haurnjan.] v. á-þeótan; ge-þeót, -þot, and next word.

þeóte, an; f. A pipe or channel through which water rushes:—Þeóte canalis, Wrt. Voc. i. 38, 17: fistula, 39, 56. Of þeótan (þeóte, Wrt. Voc. ii. 76, 4, the passage glossed is the same) tubo, Hpt. Gl. 418, 61. Þeótan organa, Wrt. Voc. ii. 64, 51: 97, 24. Þeótan, wæterþrûh cataractae, 13, 15. Ealle heofones þeótan wǽron mid wætere gefylde, Wulfst. 206, 17. In stefne ðeótena (cataractarum) ðínra, Ps. Surt. 41, 8. Þeótum fistulis, Wrt. Voc. ii. 108, 67: 35, 59. Hé wundorlíce mid þeótum wæter ût âteáh, Homl. Skt. ii. 27, 32. Hé ûs ontýneþ heofones þeótan, Blickl. Homl. 39, 31: 51, 11. [O. H. Ger. watar-dioza cataractae. Cf. Icel. þjótandi the name of an artery.] v. líc-, wæter-þeúte, and previous word.

þeów, es; þeówa, an; m. A servant; often with the stronger sense of slave; servus, famulus, mancipium:—Ic Bêda Cristes ðeów and mæsse-preóst Baeda famulus Christi et presbyter, Bd. pref.; S. 471, 7. Se ðe wyle betweox eów beón fyrmest, sý hé eówer þeów (ðeá ł ðegn servus, Lind.), Mt. Kmbl. 20, 27: 18, 26: 10, 24. Se Godes þeów the priest, Blickl. Homl. 49, 3. Metodes þeów (Abraham), Cd. Th. 146, 29; Gen. 2429. Drihtnes þeów (Guthlac), Exon. Th. 121, 8; Gû. 285. Þegn and þeów þeódne mǽrum, 209, 3; Ph. 165. Þeów mancipium, Wrt. Voc. ii. 80, 31. Ðeówa servus, Ælfc. Gr. 5; Zup. 12, 18. Þeówa, Wrt. Voc. i. 50, 15. Cham biþ þeówena þeówa (servus servorum) his gebróðrum . . . beó Chanaan Semes þeówa (servus), Gen. 9, 25–26. Wé synd ealle ðíne þeówas . . . Sig se mín þeówa, ðe ðone læfyl forstæl, and fare gé frige, 44, 17–18. Ðú góda þeówa, Lk. Skt. 19, 17: Mt. Kmbl. 25, 23. Se yfela þeówa, 24, 48. Se hláford and se þeówa gelíce clypiaþ tó ðam heofonlícan Fæder, Homl. Th. ii. 326, 28. Gif ðeówe wyle Godes þeówa beón if one wishes to enter a monastery, L. Ecg. C. 25; Th. ii. 150, 28. Biþ hé deófles ðeówa, Homl. Th. i. 172, 20. Hé biþ ðæra ǽhta ðeówa, 66, 7. Fram Gode hé is send, and hé is Godes þeówa, Blickl. Homl. 247, 19. Ðæs Godes þeówes synna, 49, 6. Moises gelíca mínes þeówes, Num. 12, 7. Ðæs þeówan hláford, Lk. Skt. 12, 46. Ic cweðe tó mínum þeówe (ðeua, Lind.), Mt. Kmbl. 8, 9. Geseoh hû ðás men ðínum ðeówe dóþ, Blickl. Homl. 229, 23: Ps. Th. 118, 49. Gecum tó mínum ðeówan Saulum, Homl. Th. i. 386, 19: Exon. Th. 157, 19; Gû. 894. Ðissum ðeá (famulo) ðínum, Rtl. 103, 13. Ðiosne ðeá hunc famulum, 97, 4. Sêc ðínne þeów, Drihten, Blickl. Homl. 87, 31. Ðone unnyttan þeówan, Mt. Kmbl. 25, 30: Homl. Th. i. 64, 17: ii. 578, 26. Wit syndon Cristes þeówas, Blickl. Homl. 187, 21: Wulfst. 157, 19. Eálá gé míne ðeówan beóþ getreówe o mea mancipia, estote fideles, Ælfc. Gr. 15; Zup. 102, 3. Ðeás servi, Mt. Kmbl. p. 18, 7. Ða þeówan drincaþ medo, Ors. 1, 1; Swt. 20, 17. Micel menigu Godes deówa (ðiówa, Hatt. MS.), Past. pref.; Swt. 4, 11. Wítniendra þiówa lictorum, Wrt. Voc. ii. 52, 77. Seó myccle menigo heora þeówa, Blickl. Homl. 99, 34. Ðæra þeówa (ðeána, Lind.) hláfurd, Mt. Kmbl. 25, 19. Ân ðæs bisceopes þeówena, Jn. Skt. 18, 26: Wulfst. 199, 22. Ðæm earmestan Godes þeówum, ðe ða cyrican mid godcundum dreámum weorþiaþ, Blickl. Homl. 41, 26. Ðá clypode hé his tŷn þeówas, Lk. Skt. 19, 13. Ðeá ðíno famulos tuos, Rtl. 100, 22: 170, 31. Ðíuwas (ða ðiówe, Rush.) ancillas, Lk. Skt. Lind. 12, 45. ¶ Slavery, which is mentioned by Tacitus (Germania, cc. 24, 25) as existing among the Germans, is recognized by the earliest English laws, and early traces of

it are to be found in the English slaves whom Gregory saw at Rome. It was a condition that was due to many causes. The fortune of war might put life and liberty at the disposal of another, as in the case of the Northumbrian, Imma, who, falling into the hands of a hostile Mercian, was by him sold to a Frisian, Bd. 4, 22. Kidnapping, to judge by Theodore's Penitential, was not unknown: Si quis Christianus alterum Christianum vagantem reppererit, eumque furatus fuerit, ac vendiderit, Th. ii. 50, § 5; and cf. Earme men beswicene and hreówlíce besyrwde, and út of ðisan earde gesealde swýðe unforworhte fremdum tó gewealde, Wulfst. 158, 13. Freedom might be forfeited as the punishment of crime; e. g. Gif hwá stalie on gewitnesse ealles his híredes, gongen hié ealle on þeówot .x.-wintre cniht mæg bión þiéfðe gewita, L. In. 7; Th. i. 107, 16; and cf. Wulfst. 158, 14. Gif se frigea on Sunnandæg wyrce, þolie his freótes, L. In. 3; Th. i. 104, 6. See also L. Eth. vii. 16; Th. i. 332, 18. v. wíte-þeów. Again, the power which one relative had over another was at times exercised to enslave the latter. A child of less than seven years might, in case of need, be sold by its father: Se fæder his sunu, gif him mycel neód byþ, hé hine mót on þeówet gesyllan óð ðæt hé biþ .vii. winter; ofer ðæt, bútan ðæs suna willan, hé hine ne mót syllan, L. Ecg. C. 27; Th. ii. 152, 17: L. Th. P. 19, 28; Th. ii. 19, § 28. Cf. L. Alf. 12; Th. i. 46, 12. The sale of kindred is elsewhere, and not without occasion, denounced: Gif hwylc cristen man his ágen bearn, oððe his néhstan mǽg wið ǽnigum wurðe sylle, næbbe hé nánne gemánan mid cristenum mannum, ǽr hé hine álýsed hæbbe of ðam þeówdóme, L. Ecg. P. 26; Th. ii. 212, 8; cf. Wé witan ful georne, hwǽr seó yrmþ geweard, ðæt fæder gesealde bearn wið weorðe, and bearn his módor, and bróðor óþerne fremdum tó gewealde, Wulfst. 161, 6. Further, slavery was at times entered into voluntarily; such cases seem contemplated in Theodore's Penitential: Homo .xiii. annorum sese potest servum facere, Th. ii. 19, § 29; and that such cases did occur may be seen from the following passage: Geatfleda geaf freóls . . . ealle ða men ðe heó nam heora heáfod for hyra mete (cf. On .xii. mónðum ðú scealt sillan ðínum þeówan men .vii. hund hláfa and .xx. hláfa, bútan morge[n]metum and nónmetum, Salm. Kmbl. p. 192, 18) on ðam yflum dagum, Chart. Th. 621, 9. And besides the causes enumerated there was that which must have been the most efficient—birth; the child of slaves was itself a slave: cf. the phrase in the document last cited, in which freedom is given to certain persons and to 'eall heora ofsprinc, boren and unboren.' See also þeów-boren. The terms used in connection with the slave shew him to be the property of his master: Gif þeúw stele and hine man ácwelle, ðam ágende hine man healfne ágelde, L. Wih. 27; Th. i. 42, 20. Gif hwylc man his ǽht (servum) ofslyhþ, L. Ecg. P. ii. 3; Th. ii. 182, 29: L. M. I. P. 11; Th. ii. 268, 9. Wǽron ðǽr ðreó wíteþeówe men búrbærde and ðreó ðeówderbde; ða mé salde bisceop tó ryhtre ǽhta, and hire teám, Chart. Th. 152, 22. Bought and sold like an animal, his treatment in other respects was that of an animal. Tacitus (Germania, c. 25) had remarked that the Germans often killed their slaves on the impulse of passion, and that it was done with impunity. The same might be said of the English: Gif hwylc man his ǽht (servum suum) ofslyhþ, and hé náne gewitnysse næbbe ðæt hé forworht sig, bútan hé hine for his háþeortnesse and for gýmeleáste ofslihþ, L. Ecg. P. ii. 3; Th. ii. 182, 29: L. M. I. P. 11; Th. ii. 268, 9: L. Th. P. 21, 12; Th. ii. 23,'§ 12. Gif hwylc wíf for hwylcum lyþrum andan hire wífman swingþ, and heó þurh ða swingle wyrð deád, and heó unscyldig biþ, L. Ecg. P. ii. 4; Th. ii. 182, 32: L. M. I. P. 12; Th. ii. 268, 11: L. Th. P, 21, 13; Th. ii. 24, 1. The inferiority of the slave is marked in many ways by the law. The price of redemption in the case of the þeów who stole was seventy shillings, L. Wih. 27; Th. i. 42, 20; in the case of the free man it was 120 shillings, L. Ath. i. 1; Th. i. 198, 23. Ðeówæs wegreáf sé .iiii. scillingas, L. Ethb. 89; Th. i. 24, in the case of the ceorl it is six shillings, 19; Th. i. 8, 1. Gif þeów steleþ, .ii. gelde gebéte, 90; Th. i. 24, 17. Gif frigman fréum steld, .iii. gebéte, 9; Th. i. 6, 2. So, too, in the matter of punishments; where the freeman can pay a fine, the slave pays with his hide, i.e. is scourged; see L. In. 3; Th. i. 104, 2: L. E. G. 7; Th. i. 172, 1: 8; Th. i. 172, 6: L. C. S. 45; Th. i. 402, 15: L. In. 13: 15; Th. i. 40, 7, 11. Gif þeów man fúl wurðe . . . swinge hine man þriwa, L. Ath. i. 19; Th. i. 208, 22. Or mutilation was inflicted, where a freeman was fined, L. Alf. pol. 25; Th. i. 78, 14. The manner in which the punishment of death was executed was an ignominious one—stoning by slaves, L. Ath. iii. 6; Th. i. 219, 13: v. 6, 3; Th. i. 234, 8. The slave could not be vouched to warranty, L. In. 47; Th. i. 132, 5; and he was not allowed the holidays given to freemen, L. Alf. pol. 43; Th. i. 92, 3. Three days, however, in the year were granted, the Monday, Tuesday, and Wednesday before Michaelmas: Sit omnis servus liber ab opere illis tribus diebus, quo melius jejunare possit, et operetur sibimet quod vult, L. Eth. viii. 2; cf. Wulfst. 181, 18; and one of Alfred's laws speaks of fragments of time in which it was possible for the slave to earn something: Æghwæt ðæs ðe ðeówum monnum ǽnig mon for Godes noman geselle, oþþe hié on ǽnegum hiora hwílsticcum geearnian mǽgen, L. Alf. pol. 43; Th. i. 92, 12. It was thus possible for a slave to acquire property, and the church endeavoured

to render his possession secure: Ne biþ álýfed æt ðam þeówan his feoh tó nimanne, ðæt hé mid his swynce begiteþ, L. Ecg. P. Addit. 35; Th. ii. 238, 6: L. Th. P. 19, 30; Th. ii. 19, § 30. Throughout the influence of the church seems to have been exerted in favour of the slave. The sale of slaves into heathen lands was denounced: Gif hwá cristene man on hǽðendóm sylle, se ne biþ wurðe ǽnigre reste mid cristenum folce, bútan hé gebycge eft hám ongeán, ðæt hé út sealde, L. M. I. P. 43; Th. ii. 276, 20; see, too, L. Th. P. 42, 3, 4; Th. ii. 50, §§ 3, 4: L. Ecg. E. 150; Th. ii. 124, 2: and probably freedom was not unfrequently granted at the suggestion of the church. Cf. such expressions as: Geatfleda geaf freóls for Godes lufa and for heora sáwla þearfe, Chart. Th. 621, 3. Ðá freóde Folcerd Agelwine his man and his ofsprinc Criste tó lofe and Séa Maria, and his sáwle áliésednisse, 634, 20. Cf. too, L. Wih. 8; Th. i. 38, 15, and L. In. 3; Th. i. 104, 2. Gif þeów mon wyrce on Sunnandæg be his hláfordes hǽse, sié hé frioh. To the same effect is L. C. S. 45; Th. i. 402, 18: Gif hláford his þeówan freólsdæge nýde tó weorce, þolige ðæs þeówan, and beó hé syþþan folcfrig. See on the question of slavery Kemble's Saxons in England,' i. c. 8, Andrews' Old English Manor, c. 3, Grimm's R. A., pp. 300 sqq. [Goth. þius; pl. þiwós; m.: O. H. Ger. deo: Icel. þý; n.] v. esen-, níd-, under-, weorc-, wíte-þeów; lád-teów, and following words.

þeów, e; þeówe, an; f. A female servant or slave:—Seó forespecene Cristes þeówe praefata Christi famula, Bd. 4, 9; S. 577, 13: 4, 10; S. 578, 5: 4, 23; S. 592, 36. Seó Cristes þeówe, Guthl. 20; Gdwin. 92, 2. Án menen ł þeówæ (ðíua, Lind.) ancilla, Mt. Kmbl. Rush. 26, 69. Ðínre þeówan sunu filium ancillae tuae, Ps. Th. 85, 15: 115, 6. On Marian ðínre þeówan, Blickl. Homl. 157, 3. Ðió famulam, Rtl. 103, 40. Ðióe, 104, 18. Ðá wǽron ða Cristes ðeówe út gangende of cyricean egressae de oratorio famulae Christi, Bd. 4, 7; S. 575, 3. Ða hús ðara untrumra Cristes ðeówna casulas infirmarum Christi famularum, 3, 8; S. 531, 33. Ne gǽd hió út swá þeówena gewuna ys non egredietur, sicut ancillae exire consueverunt, Ex. 21, 7. Án from ðǽm ðíówum una ex ancillis, Mk. Skt. Lind. Rush. 14, 66. [Goth. þiwi: O. Sax. thiu; thiwa: O. H. Ger. diu; diwa: Icel. þýr.] v. þeówen.

þeów; adj. Servile, not free, bond:—Ðes ðeówa mann hic manceps, Ælfc. Gr. 9, 55; Zup. 67, 2. Gif þeów monn wyrce on Sunnandæg be his hláfordes hǽse, sié hé frioh, L. In. 3; Th. i. 104, 2. Ðeów swán and ðeów beócere, L. R. S. 6; Th. i. 436, 19. Ðeáh hwá bebycgge his dohtor on þeówenne, ne sié hió ealles swá þeówu swá óðru mennenu, L. Alf. 12; Th. i. 46, 13. Þeów mennen, Agar, Cd. Th. 135, 22; Gen. 2246. On þeówum dóme ł tó ðeówan wæs gesealed in servuum venundatus est, Ps. Lamb. 104, 17. Se ðe sleá his ágenne þeówne esue, L. Alf. 17; Th. i. 48, 12. Feówertýnewintre man hine sylfne mæg þeówne gedón (se servum facere), L. Ecg. C. 27; Th. ii. 152, 27. Hit þurh ǽnne þeówne mon geypped weard quadam ancilla indice, Ors. 3, 6; Swt. 108, 31. Gif hé þeów oþþe þeów mennen ofstinge, L. Alf. 21; Th. i. 50, 3: Cd. Th. 134, 32; Gen. 2233. Gif hwylc swíþe ríce cyning nǽfde nǽnne frýne mon on eallon his ríce, ac wǽron ealle þeówe . . . Gif him sceoldan þeówe men þénigan, Bt. 41, 2; Fox 244, 24–27. Ealla gesceafta hé hæfde getiohhod ðeówe (þeówu, Cott. MS.) búton englum and monnum, ða óðru gesceafta þeówe sint, 41, 3; Fox 248, 16–18. Gé giet tó dæge wǽron Somnitum ðeówe hodie Romani Samnio servirent, Ors. 3, 8; Swt. 122, 12. Þeówe men ða ðrig dagas beón weorces gefreóde, Wulfst. 181, 18. Seofæn þeówæ mæn, Chart. Th. 163, 10. Ne freó ne þeówe, Cd. Th. 166, 12; Gen. 2746. Freóra and þeówra, 166, 26; Gen. 2753. Míne wealas eriaþ mea mancipia arant, mínra þeówra manna æceras meorum mancipiorum segetes, mínum ðeówum mannum (mancipiis) ic dǽle penegas, míne þeówan men (mancipia) ic ðreáge, fram mínum þeówum mannum ic eom gefultumod, Ælfc. Gr. 15; Zup. 101, 19–102, 5. Þeówe men manode se apostol . . . Ða hláfordas hé manode ðæt hí milde wǽron heora ðeówum mannum, Homl. Th. ii. 326, 21–27. Nelt ðú nán ðing yfeles habban . . . ne yfele cild, ne yfele ðeówe men, 410, 16. [Cuð me 3ef þu art foster of freo monne oðer þeow wummon, Marh. 4, 2. Heo weren þeowe, Laym. 334. ʒonge and olde, þewe and freo, Al. 3. O. H. Ger. frie getuon nals teuue.] v. wíte-þeów, and preceding words.

þeówa. v. þeów; m.

þeówan, þéwan, þíwan, þýwan, þýgan, þeón, þían, þýn, and þeówian, þíwian, þýwian; pres. ic þý, hé þýþ; p. þeówde, þéwde, þíwde, þýwde, þýgde, þeóde, þýde; ppr. þýwende, þíende; pp. þéd, þýd. I. to press:—Hwílum mec (an animal's skin) wonfeax wale wegeþ and þýð, Exon. Th. 393, 31; Rä. 13, 8. [Hé mec (a cup) fin]grum þýð, 480, 24; Rä. 64, 6. Þýde compressit (the line in Aldhelm is: Dulcia sed Christi compressit labra labellis), Wrt. Voc. ii. 95, 33. Ðýde, 19, 67. Hé ðá hit eft sette on ðæt ylce þyrh and þýde mid his fét, Homl. Skt. i. 21, 72. Hé sum fæc ðone swyle mid ðýgde (or middýgde? mid þýde, Bd. M. 382, 29) aliquandiu tumorem horum adpositione comprimere curabat, Bd. 4, 32; S. 611, 41. Sceal mon ðam men mid drĩum handum ða handa and ða fét gnídan swíðe and þýn with dry hands must the hands and feet be rubbed hard and squeezed for the man, Lchdm. ii. 182, 9. Ðæs mannes fét and handa man sceal swíþe þýn, 182, 25. II. to

press on, urge on, drive :—Weard æt steorte wegeþ mec (*a plough*) and
þýd, Exon. Th. 403, 10; Rä. 22, 5. Se mec on þýd æftanweardne,
480, 2; Rä. 63, 5. Hwílum ic (*a storm*) sceal tó staþe þýwan (þyran,
MS.) flintgrægne flód, 383, 30; Rä. 4, 18. Þéwende (? þerende, MS.)
inruens, Wrt. Voc. ii. 111, 12. III. *to press* with a weapon, *to
stab, pierce :*—Ðá hét hé him his seax árǽcan, and hine sylfne hetelíce
ðýde, Homl. Th. i. 88, 10. Ðá ðýde se cwellere hine bæftan mid
átogenum swurde, ii. 478, 19. Se fear arn him tógeánes and hine ðýde,
ðæt hé his feorh forlét, Homl. Skt. i. 12, 73. Hí hine ufan mid ísenum
geaflum ðýdon ... Hí mid heora forcum hine ðýdon, Homl. Th. i. 430,
5–11. IV. *to press, threaten, rebuke :*—Seó módinys on horse ðýwð
ðæt folc *superbia in equo minatur turbis*, Gl. Prud. 31 b. Hé þýwþ
(*arguet*) ðysne middaneard be synne, Jn. Skt. 16, 8. Hé ondrǽde ða
þeówwrace ðe Drihten þurh his wítigan ðýwð ðus cweþende *metuat
prophete comminationem per quem dicit Deus*, R. Ben. 51, 14. Se ðe
brimu bindeþ, brúne ýða þýð and þreátaþ, Andr. Kmbl. 1039; An. 520.
Hé ne þíwaþ *non comminabitur*, Ps. Lamb. 102, 9. Hé ðýwaþ mé
increpabit me, 140, 5. Gif ic ðæt gefricge, ðæt ðec ymbsittend egesan
þýwaþ, Beo. Th. 3659; B. 1827. Hé hine þeówde tó ofsleánne, Homl.
Ass. 112, 342. Ðíwde, Homl. Th. ii. 174, 32. Ðýwde, 308, 16. Seó
ofermódnes þýwde (*minatur*) ðæt folc, Gl. Prud. 31 a. Ðú hine þíwe
commineris, 114, 10. Næs se folccyning ymbsittendra ðe mec
dorste egesan ðeón, Beo. Th. 5465; B. 2736. Þeówígende, þéwende
minax, Ælfc. Gr. 9, 60; Zup. 69, 7. Ic gá út þýwende (*minando*)
oxan ... Ic hæbbe sumne cnapan þýwende (*minantem*) oxan mid
gádísene, Coll. Monast. Th. 19, 15, 27. V. *to oppress, subjugate :*
—Þéde *mancipium*, Wrt. Voc. ii. 56, 70. Cf. geðédum *subjugatis*, 121,
69. Þédum *teste* (? the passage is :—Teste tyranni (*Holofernes*) capite),
Hpt. Gl. 525, 53. [Þat he miȝte þat liþere folc so þewe, P. L. S. 24, 57.
Goth. ga-þiwan *to pierce; to subject*; ana-þiwan *to subject* : *O. H. Ger*.
theuwe *humiliat*; gi-diota, -dieti *confracti* : *M. H. Ger*. diuwen *to
oppress* : *Icel*. þjá *to con͡strain; chastise, afflict*.] v. á-, for-, ge-, þurh-
þeówan (-þéwan, -þíwan, -þeón, -þían, -þýan, -þýn); *and see* þyddan.

þeówan ; *p*. þeówde, þeódde *To serve* (with dat.):—Ðæt sind ða
gecostan cempan, ða ðam cyninge þeówaþ, Exon. Th. 107, 22; Gú. 62.
Hé Dryhtne þeówde, 146, 20; Gú. 712. Israhéla folc on hæftnéde
Babiloniscum cyninge þeówde, Homl. Th. ii. 84, 27 : 66, 9. Hé Drihtne
ðeówde *Domino servierat*, Bd. 4, 24; S. 599, 9. Ic him geornlícor
ðeódde *illis impensius servire curavi*, 2, 13; S. 516, 9. Ic beðead
ðeówum mannum, ðæt hí getreówlíce heora hláfordum þeówdon, Homl.
Th. i. 378, 33 : Homl. Skt. i. 2, 85. Ða óðre beóþ frige, ðeáh ðe hí on
lífe lange ǽr ðeówdon, Homl. Th. ii. 326, 33. Ðæt hið þeówdon Godes
ciricum, Blickl. Homl. 185, 29 : Bd. 4, 11; S. 579, 15. Ðeówdun
servierunt, Ps. Surt. 80, 7. Hí hǽþenum bigangum ðeówdon *paganis
cultibus provincia serviebat*, Bd. 4, 13; S. 582, 5. Hí swá frige Drihtne
gefeónde ðeóddon (þeówodon, Bd. M. 240, 13) *sic liberi Christo servire
gaudebant*, 3, 24; S. 558, 1. v. þeówian.

þeów-beócere. v. þeów; *adj*.

þeów-boren ; *adj. Slave-born, born of parents in slavery :*—Ne sceal
hé (*the abbot*) ðone æþelborenan settan beforan ðane þeówborenan, gif se
þeówborena ǽr on ðæm mynstre wæs, bútan hé for hwylcum gesceáde hit
dó *non preponatur ingenuus ex servitio convertenti nisi forte aliqua
rationabilis causa existat*, R. Ben. 12, 13.

þeów-byrde, -berde ; *adj. Of servile birth :*—Wéron ðǽr ðreó wíteþeówe men búrbærde, and ðreó ðeówberde, Chart. Th. 152, 20.

þeów-cnapa, an ; *m. A servant-lad :*—His ðeówcnapena án weard
þearle áwéd; ðá sette Martinus his handa him onuppon, and se feónd
fleáh forht for ðam hálgan, and se ðeówa siððan gesundful leofode, Homl.
Th. ii. 510, 27.

þeów-dóm, es ; *m. Service;* in an unfavourable sense *servitude, slavery,
bondage, thraldom :*—Ðes þeówdóm *haec seruitus*, Ælfc. Gr. 9, 33 ; Zup.
60, 7. Þeówdóm *mancipatio*, Wrt. Voc. i. 59, 57 : *famulatus*, ii. 147,
35. Þeówdómum *famulatibus*, 34, 3. Ðeówdóm is twyfeald ... Is óðer
ðeówt neádunge bútan lufe, óðer is sylfwilles mid lufe, se gedafenaþ
Godes ðeówum, Homl. Th. ii. 524, 3. I. in the more favourable
sense :—For lufan ðæs godcundan ðeówdómes, Bd. 4, 23; S. 593, 31.
Godes þeówdómes, Blickl. Homl. 23, 18. Ne þearst ðú nó wénan, ðæt
ða wlitegan tungl ðæs þeówdómes áþroten weorðe, Met. 29, 40. Embichta
ðeáðdóme *obsequium servitutis*, Mt. Kmbl. p. 8, 3 : Rtl. 9, 13. Wyrta
ðeówdóme manna, Ps. Spl. 103, 15. Swá hwá swá ðæs wyrþe biþ, ðæt
hé on heora ðeówdóme beón mót, ðonne biþ hé on ðam hélustan freóðóme,
Bt. 5, 1; Fox 10, 13. On Godes ðone sóþan þeówdóm, Blickl. Homl.
45, 24; Elen. Kmbl. 402; El. 201. Hé hine sylfne on ðeówdum (-dóm, Bd.
M. 450, 29) gesealde ðara muneca *monachorum famulatui se contradens*,
Bd. 5, 19; S. 637, 12. I a. *service* of the church, *divine service :*—On
mynstre ðǽr lytel þeówdóm sý, L. C. E. 3; Th. i. 360, 22. Nú habbaþ
hig ðæt mynster gesett mid preóstan, and willaþ ðǽr habban þeówdóm
eall swá man hæfþ on Paules byrig on Lundene, Chart. Th. 370,
22. II. in the less favourable sense :—Gif se Godes þeów nelle
ðǽre cyrican on riht þeówian, ðæt hé ðonne mid lǽwedum mannum
onfó ðæs heardestan þeówdómes, Blickl. Homl. 49, 5. Se freódóm ðæs

unáræfnedlícan þeówdómes *freedom from the intolerable bondage*, 137,
13. From deófles þeówdóme, 65, 33 : 73, 8. Hí synd of miclum dǽle
heora sylfes anwealdes, hwæþere of miclum dǽle hí syndon Angelcynnes
ðeówdóme betǽhte (*Anglorum sunt servitio mancipati*), Bd. 5, 23 ; S.
647, 4. Nales ðæt án ðæt men hié mehten áliésan mid feó of þeówdóme,
ac eác þeóda him betweónum bútan þeówdóme gesibbsume wǽron, Ors.
1, 10; Swt. 48, 34. Ealle ða men ðe hié on ðeówdóme hæfdon, hié
gefreódon, 4, 9; Swt. 190, 31 : L. Alf. prm.; Th. i. 44, 5. In néðhérnisse
† in ðeádóme ic bégo *in servitutem redigo*, Rtl. 6, 9. Þeówdóm þolian,
Cd. Th. 135, 9; Gen. 2240 : 136, 24 ; Gen. 2263. Ðeáðdóm *captivitatem*,
Lk. Skt. p. 10, 16. [Heo woneð inne þeowedome (þeu-, 2nd MS.),
þrelwerkes doð, Laym. 454. I þeowwdom unnderr laferdd, Orm. 3611.
Leden ut of þeoudome (þeowedom, MS. C.), A. R. 218, 28. Leaden in
to þeowdom, H. M. 5, 5.] v. ǽfen-þeówdóm.

þeówdóm-hád, es ; *m. Service :*—Monige hí sylfe and heora bearn má
gyrnaþ on mynster and on Godes ðeówdómhád tó syllanne ðonne hí synd
bigongende woruldlícne camphád *plures se suosque liberos, depositis armis,
satagunt magis accepta tonsura monasterialibus adscribere votis, quam
bellicis exercere studiis*, Bd. 4, 23; S. 647, 8.

þeówe. v. þeów; *f*.

þeówen, þíwen, [n]e ; þeówene, an ; *f. A female servant* or *slave, a
handmaid :*—Ic eom Drihtnes þeówen (*ancilla*, Lk. 1, 38), Blickl. Homl.
9, 20. Ðá com tó hym án þeówyn (-en, MS. A.) *accessit ad eum una
ancilla*, Mt. Kmbl. 26, 69. Seó Godes ðeówen, Bd. 4, 9; S. 576, 14:
Homl. Skt. ii. 23 b, 192. Nergendes þeówen (*Judith*), Judth. Thw. 22,
23; Jud. 74. Ic Luba eáðmód Godes ðíwen, Chart. Th. 475, 21. Hé
sceáwode ða eáþmódnesse his þeówene, Blickl. Homl. 7, 4. Swá eágan gáð
earmre þeówenan (*ancillae*), ðonne heó on hire hláfdigean handa lócaþ, Ps.
Th. 122, 3. Heó hié sylfe tó ðeówene genemde, Blickl. Homl. 9, 24.
Þeówene, 89, 12. Gif hwá ásleá his þeówe oþþe his þeówenne (-ene,
MS. H.) ðæt eáge út, L. Alf. 20; Th. i. 48, 24. Þeówne (*or adj*.?)
bernam, Wrt. Voc. ii. 12, 24. Ðeáh hwá bebycgge his dohtor on þeówenne, L. Alf. 12; Th. i. 46, 12. Se fæder hire sealde áne þeówene (*servam*),
Gen. 29, 29. Ðíoenne *famulam*, Rtl. 104, 2 : 34, 10 (see Skeat's collation). Ðíoen, 25. Sió Affra and hire þreó ðeówena ... Sió Affra wæs
ǽrest forlegor wíf mid hire þeówenum, Shrn. 115, 3–5. Ðá hét hire fæder
hí bewyrcean on ánum torre mid twelf ðeówennum, 106, 1. Þeówenna
bernas, Wrt. Voc. ii. 95, 59 : 12, 17. Ic hæbbe þeówas and þeówena
(*servos et ancillas*), Gen. 32, 5. v. efen-þeówen.

þeówene. v. preceding word.

þeówet(-ot, -ut), þeówt, [t]es ; *m. Service;* in an unfavourable sense,
servitude, bondage, slavery :—Is óðer ðeówt neádunge bútan lufe, óðer
is sylfwilles mid lufe, se gedafenaþ Godes ðeówum, Homl. Th. ii. 524,
5. I. of voluntary service :—Githro sǽde ðæt Moyses on dyslícum
gesuincum wǽre mid ðæs folces eorðlícan ðeówote *quod terrenis populorum negotiis stulto labore deserviat*, Past. 18; Swt. 131, 14. II.
of forced service :—Ælc ðeówt biþ geendod on ðisum andweardan lífe,
bútan ðæra ánra ðe synnum ðeówiaþ; hí habbaþ écne ðeówt, Homl. Th.
ii. 326, 30. Sí þreora án for his feore, wergild, éce þeówet, hengenwíte-
nung, L. Eth. vii. 16; Th. i. 332, 18. Of þeówetes húse *de domo servitutis*, Deut. 6, 12. Þeówettes, 13, 10. Se synfulla ðeówaþ ðam
wyrstan ðeówte *the sinner is a slave to the worst slavery*, Homl. Th. ii.
228, 10. Of þeówete út álǽdan *de servitute eruere*, Ex. 6, 6. Þeówette,
Deut. 5, 6. Of þeówte gelǽdan, Ælfc. T. Grn. 5, 19. Of þeówte álísan,
2, 11. On þeówote gebringan *to enslave*, Ors. 3, 9; Swt. 128, 29.
Ðeówte, Homl. Th. ii. 190, 90. Tó ðeówte gelǽdan, 66, 34. Hé
wolde ðæt folc habban ongeán tó his lande tó his láðum þeówte, Ælfc.
T. Grn. 5, 26. Bige ús tó ðæs cynges þeówette *eme nos in servitutem
regiam*, Gen. 47, 19. Tó þeówte bebycggan *to sell into slavery*, L. Alf.
12; Th. i. 46, 12 MS. H. On þeówete standan *to remain in bondage*,
L. Ath. v. 12, 2 ; Th. i. 242, 5. Ðære wylne sunu wunaþ eal his líf on
ðeówte, Homl. Th. i. 110, 29. Drihtenes áre oððe deófles þeówet, Hy.
7, 98. Tódǽlan freót and þeówet *to distinguish between freedom and
slavery*, i. e. between the free and the bond, L. C. S. 69; Th. i. 412, 10.
On þeówot gangan, L. In. 7 ; Th. i. 106, 17. On þeówot sellan, Ors. 3,
7; Swt. 112, 30. Þeówet, L. Ecg. C. 27 ; Th. ii. 152, 17. [Cf. Swa
summ þu þeowwtesst tin eorþlike laferrd swa shall þin sune himm þeowwtenn, butt iff he wurrþe lesedd ut off hiss þeowwdomess bandess, Orm.
43–46.]

þeówet-dóm, es ; *m. A service :*—Georne ymb ealle ða ðeówutdómas
(ðíowot-, Hatt. MS.) ðe hié Gode dón sceoldon, Past. pref.; Swt. 2, 10.

þeówet-líc ; *adj. Servile :*—Þeówtlíc (þeówet-, þeówot-) *servilis*,
Ælfc. Gr. 9, 28 ; Zup. 55, 1. Þeówtlícum inhírede *vernacula clientela*,
Hpt. Gl. 483, 71. Mid þeówetlícum móde *seruili mente*, Scint. 63, 13.
Þeówtlícne líchoman *servile corpus*, Hymn. Surt. 50, 12. Gif wé ðeówt-
lícera weorca, ðæt sind synna, geswícaþ, Homl. Th. ii. 208, 6. Ðæt
Sunnandæg freóls beó fram þeówetlícum weorcum, Wulfst. 292, 7.

þeówetling, es ; *m. A (poor) slave ;* the diminutive form seems to be
depreciatory in this case :—Quintianus wæs grǽdig gítsere, deófles þeów-
etlincg *a miserable slave of the devil*, Homl. Skt. i. 8, 6. Þeáh ðe hé
brúce brádes ríces hé is earm ðeówtling ná ánes hláfordes *though he exer-*

cise extensive power, he is a poor miserable slave, and not of a single master, Homl. Th. ii. 228, 11. Ælc hysecild ægðer ge æþelboren ge þeówetling, i. 92, 1. Ælces mannes þeówetlingas ða ðrý dagas weorces beón gefreóde, Wulfst. 171, 19. Be teóðunge . . . þridda dǽl gá þearfum and earman þeówetlingan, L. Eth. ix. 6; Th. i. 342, 9. Þearfena helpan and þeówetlingan beorgan, L. I. P. 11; Th. ii. 318, 26. Þeówetlingum *servulis,* Hymn. Surt. 25, 18. Þeówtlingas *servulos,* 124, 13 : 125, 5. v. níð-þeówetling.

þeówetscipe, es; *m. Service* :—Hé ealle ðæs regoles bebodu and fulfremednysse ðæs munuclícan þeówtscypes geheóld, Homl. Skt. ii. 23 b, 26.

þeów-hád, es ; *m. The condition of a servant, service* :—Heó háligryfte onféng and Godes ðeówháde *she took the veil and accepted the condition of a servant of God; accepto velamine sanctimonialis habitus,* Bd. 4, 19; S. 587, 42. Hé Godes ðeówháde and sceare onféng *accepta tonsura,* 5, 12; S. 627, 26.

þeówian; *p.* ode. **I.** *to serve* (of animate or inanimate objects), *be a servant* or *slave.* (1) in the more favourable sense, (a) absolute :—Ðiós síde gesceaft þénaþ and þiówaþ, Met. 29, 77. Gehérsumendre stilnesse † þieówiende *quiete,* Hpt. Gl. 413, 20. (b) followed by dat. of the person or institution served :—Ic (*an animal's skin*) dryhtum þeówige, Exon. Th. 394, 9; Rä. 13, 15. Him ánum ðú þeówast (þěwige, Rush.) *illi soli servies,* Mt. Kmbl. 4, 10. Ðam (*God*) þeówiaþ ealle, ða ðe þeówiaþ . . . ge ða ðe hit witon, ðæt hié him þeówiaþ, ge ða ðe hit nyton, Bt. 21 ; Fox 72, 30-32 : Exon. Th. 106, 34; Gú. 40. Hú ne þeówode ic ðé for Rachele *nonne pro Rachel servivi tibi?* Gen. 29, 25 : Lk. Skt. 15, 29. Hí ne mihton elles beón, gif hé ne þiówedon hiora fruman, Bt. 39, 13; Fox 234, 30. Þiówoden, Met. 29, 99. Ða ðe fram cildháde Gode þeówodon, Homl. Th. ii. 78, 17. Him ánum ðú þeówa, Blickl. Homl. 27, 21. Eall ðeós eorðe Gode þeówie, Ps. Th. 99, 1. Sume secgaþ ðæt se milte ðám sinum þeówige, Lchdm. ii. 242, 22. Ðæt we ðíwgen him *ut serviamus illi,* Ps. Surt. ii. p. 199, 26. Ne mæg nán man þwám hláfordum þeówian (ðeówigan, Rush.), Mt. Kmbl. 6, 24 : Lk. Skt. 16, 13. Drihtne on dǽdum þeówian, Blickl. Homl. 31, 11. Gif hí mid rihte willaþ Gode þeówian, ðonne sceolan hí þegnian Godes folce, 45, 30. Ðǽre cyrican þeówian, 49, 4. Hé nolde Gode þeówian, Cd. Th. 17, 24; Gen. 264. ¶ In special reference to the services of religion :—Þeówian his Drihtne swá wel swá hé (*the monk*) betst mæge, L. Eth. v. 6; Th. i. 306, 9. Ic ðǽr Englisce scole gesette, ðe ǽfre for úre þeóde Gode þýwian scolde, Chart. Th. 116, 35. Þeówigende (ðió hérde Gode, Rush.) on fæstenum and on hálsungum, Lk. Skt. 2, 37. Wæs heó Drihtne ðeówiende on ðam mynstre, Bd. 3, 8; S. 531, 15. On ðam mynstre wǽron fíf bróþra oþþe syxe Drihtne ðeówiende, 4, 13 ; S. 582, 23. (c) with dat. of practice in which a person labours, *to be devoted* to, *attend* to, *bestow pains on, work* at :—Hé wæs manod ðæt hé his ðam gewunelícan wæccum and gebedum geornlíce ðeówode *admonitus est vigiliis consuetis et orationibus indefessus incumbere,* Bd. 3, 19; S. 547, 15. Sealmsangum hig þeáwian *psalmodiis inserviant,* Anglia xiii. 373, 117. (d) with acc. (?) of service done, *to perform* a service :—Ealle ða ðénunga ðe (*acc. or dat.?*) wé nú ðiówiaþ and wyrceaþ *quod in actione servemus,* Past. 34; Swt. 233, 10. (2) in the less favourable sense, (a) absolute :—Eálá gé ðeówan . . . ne ðeówige gé tó ansýne, Homl. Th. ii. 326, 24. Gif ðú þeów bigst, þeówie hé six gér and beó him freoh on ðam seofoðan, Ex. 21, 2. Gilde hé, þeówige hé *whether he pay or serve* (*as a slave*), L. Eth. vii. 17 ; Th. i. 332, 19. (b) with dat. of that which is served :—Hé biþ ðæra ǽhta ðeówa ðonne hé him eallunga þeówaþ (*he is entirely in bondage to them*), Homl. Th. i. 66, 7. Se synfulla ðeówaþ ðam wyrstan ðeówte *the sinner is a slave to the worse slavery,* ii. 228, 10. Pharao áh ægðer ge eów and eówer land . . . Hig cwǽdon : 'We þeówiaþ blíðelíce ðam cynge,' Gen. 47, 25. **II.** *to enslave, reduce to a state of slavery, deprive of freedom* :—Ðæt hé ús þeówige *ut violenter subjiciat servituti nos,* Gen. 43, 18. Sý ǽlc cirice on Godes griðe and on ðæs cynges and on ealles cristenes folces, and ǽnig man heonanford cirican ne þeówige, L. Eth. v. 10; Th. i. 306, 27 : vi. 15; Th. i. 318, 26. [Þa hwile þu þeowest þire sunne, O. E. Homl. i. 25, 1. Heo hine beden þat he nomen heom to þrallen & heo him wolden þiwien (hii him wolde be þeouwe, 2nd MS.), Laym. 10015.] v. ge-þeówian, þeówan.

þeówincel, es; *n. A young slave, a slave* :—Ðiówincelu *familici* (the word has been taken as if connected with *famulus*), Ps. Surt. ii. p. 186, 15.

þeów-líc, adj. *Servile* :—Þeóulíc *servilis,* Ælfc. Gr. 9, 28; Zup. 55, 1 MS. W. [Ressteda33 off alle þewwlike dede, Orm. 4177. O. H. Ger. deo-líh.]

þeówling, es ; *m. A slave* :—Þeówlincgas ða þrý dagas ǽlces weorces beón frige, Wulfst. 173, 23. Cf. þeówetling.

þeów-men[n]en. v. þeów; *adj.*

þeów-níd, e ; *f. Violence or force that enslaves or subdues, oppression, enslavement* :—His suhtriga (*Lot*) þeównýd þolode ; bæd hé (*Abraham*) ða rincas ðæs rǽd áhicgan, ðæt his hyldemǽg áhred wurde, Cd. Th. 122, 21; 2030. Wé nú hǽðenra þeównéd (þreanýd, Exon. Th. 187, 1; Az. 28), 235, 18; Dan. 308: Elen. Kmbl. 1536; El. 770. For þreáum and for þeónýdum (for þearfum and for þreánýdum, Exon. Th. 186, 4; Az. 14) *on account of afflictions and oppression,* Cd. Th. 234, 19; Dan. 294.

þeówot, þeówracian, þeówracu, þeówt, þeówut. v. þeówet, þeów-[w]racian, þeów-[w]racu, þeówet.

þeów-weorc, es ; *n. Servile work, work to be done by a slave* :—Gif hwá freót forwyrce . . . sý hé ðæs þeówweorces wyrðe ðe ðǽr tó gebyrige *if any one forfeit his freedom . . . let him have such servile work assigned him as pertains thereto,* L. Ed. 9; Th. i. 164, 12. Gif esne ofer dryhtnes hǽse þeówweorc wyrce an Sunnanǽfen (v. þeówet-líc, *last passage*), L. Wih. 9; Th. i. 38, 18.

þeów-[w]racian; *p.* ode *To threaten* :—Ne on écnysse hé þeówracaþ *neque in aeternum comminabitur,* Ps. Spl. 102, 9. v. next word.

þeów-, þíw-, þýw-[w]racu, e, an; *f. A threat, threatening* :—Martianus hét hí gebúgan tó his deófolgyldum, ðe lǽs ðe hí fordémede wurdon; ac Iulianus ne róhte ðæs réðan þýwrace (*cared not for the cruel one's threat*), Homl. Skt. i. 4, 114. Þreále oððe þeówraca[n?] *invectionis, inlationis,* Hpt. Gl. 448, 52. Heó nǽs áfyrht for his þeówracan, Homl. Skt. i. 7, 87. For ðeówracan sweartra deófla, Homl. Th. ii. 142, 32. Hé ondrǽde ða þeówwrace ðe Drihten þurh his wítigan ðýwþ *metuat prophete comminationem,* R. Ben. 51, 13. Basilius cýdde ðæs réðan cáseres ðeówrace, Homl. Th. i. 450, 17. *Uae* getácnaþ hwílon wánunge, hwílon ðeówracan (þeówrace, MS. D. : þíwrace, MS. C. : ðíwwrace, MS. U.), hwílon wyrigunge, Ælfc. Gr. 48; Zup. 278, 17. Ðíne ðeówracan synd hwílwendlíce, Homl. Skt. i. 14, 100. Mid menigfealdum ðeówwracena teartnyssum gebrégede, Homl. Th. i. 578, 27. Hé ne mihte mid nánum þeówracan ða cristenan geegsian, 564, 2. Mid þíwracum *minis,* Scint. 63, 8. Þeówwracan *minas,* Wülck. Gl. 252, 19. Hí him ne ondrédon hǽðenra cyninga þeówracan, Homl. Th. ii. 44, 12. Ic forseó ðíne þeówracan, Homl. Skt. i. 7, 124. Ic gehýre hyra egeslícan þíwracan, 3, 432. v. þeówan, **IV.**

þeox *a spear* :—Ísenum bǽrsperum † þioxum *ferratis venabulis,* Hpt. Gl. 423, 68. [Cf. (?) Thyxyl *ascia,* Prompt. Parv. 491, and see note there. Thyxylle, Wrt. Voc. i. 234, 18. O. H. Ger. dehsa *ascia;* dehsísen *confertorium;* dehsala *ascia, ferrum confertorium: Icel.* þexla *an adze.*]

þerende *inruens,* Wrt. Voc. ii. 111, 12, (se) þe rende (? *from* rennan), or (?) þéwende. v. þeówan, **II.**

þerh. v. þurh.

þero ? :—.vii. hríðru and six weðeras and .xl. cýsa and .vi. lang þero and þrítig ombra rúes cornes, Chart. Th. 40, 8.

þerscan; *p.* þærsc, *pl.* þurscon; *pp.* þorscen *To thrash.* **I.** *to strike, beat, flog, scourge* :—Ðú ðe rehtlíce ðersces synfullo *qui juste verberas peccatores,* Rtl. 43, 9. Ða wéregan neát ðe man drífeþ and þirsceþ, Elen. Kmbl. 716; El. 358. Se ðunor ðæt deófol ðrysceþ mid ðære týrenan æcxe, and hit drífeþ tó ðære írenan racenteáge ðe his fæder on eardaþ, Salm. Kmbl. p. 148, 6. Sume hiá ðurscun *quosdam caedentes,* Mk. Skt. Lind. 12, 5. Hí þurhsun (þurcsun, MS. A.) his nebb *percutiebant faciem ejus,* Lk. Skt. 22, 64. Ða hié hé hine mid stengum ðyrscan, Shrn. 55, 10. Ongunnun sume mid fýstum hine slá † ðarsca (*caedere*), Mk. Skt. Rush. 14, 65. Ðærscende hine stánum *concidens se lapidibus,* Lind. 5, 5. Swoelce lyft ðerscende (*verberans*), Rtl. 6, 7. **II.** *to thrash corn* :—Hé corn ðærsc and ðæt windwode, Shrn. 61, 19. Mænige inweorc wyrcean, ðerhsam, wudu cleófan, Anglia ix. 261, 25. Flór on tó þerscenne *area,* Wrt. Voc. i. 37, 59. **III.** *to pound, batter* :—Ðéh ðú þercce . . . swá berecorn ðercce[n]dum si contuderis (*stultum in pila*) *quasi ptisanas feriente,* Kent. Gl. 1034-6. Ðerscaþ ðone weall mid rammum, Past. 21; Swt. 161, 6. [Þín þrosshenn corn, Orm. 1530. Goth. þriskan *triturare :* O. H. Ger. drescan *triturare : Icel.* þryskva (*wk.*) *to thrash corn.*] v. be-, ge-, tó-þerscan.

þerscel, es ; *m. An implement for thrashing corn, a thrashle, threshel* (v. Halliwell's Dictionary), *a flail* :—Þerscel *tritorium,* Wrt. Voc. i. 16, 36 : 34. 49 : *bainus,* ii. 115, 2. Ðerscel, 12, 73. [O. H. Ger. driscil *tribula.*]

þerscel-flór, e ; *f. A threshing-floor* :—Hé áfeormaþ his þyrscelflóre (*aream suam*), Mt. Kmbl. 3, 12. v. þirsce-flór.

þerscold, þerxold, þrexold, þersc-wold, -wald, es ; *m. A threshold* :—Oferslege oððe þerexwold (þræx-, þreox-, þerx-wold, ðrexold) *limen,* Ælfc. Gr. 9, 12 ; Zup. 40, 15. Þrexwold, Wrt. Voc. i. 85, 65. Þerxwald, 290, 16. Þerscwold oððe duru, ii. 52, 5. Write on ðínum þerscolde (*limine*), Deut. 6, 9. On ðam þerxolde, Ex. 12, 22. Fram deáþes ðrecswalde (þirsc-, Bd. M. 398, 23) *ab ipso mortis limite,* Bd. 5, 6 ; S. 618, 34. Of ðæs portices dura ðærscwolde, Blickl. Homl. 207, 11. Ofer ðone ðerscold, Past. 13; Swt. 77, 22. Ðerscwold, Homl. Skt. ii. 23 b, 413. Þerscwold, Shrn. 141, 17. Þeorscwold, Bt. 21 ; Fox 74, 26. Þyrscwold, Met. 11, 68. Þerxwold, Lchdm. ii. 142, 12. Þrexwealdum *liminibus,* Hpt. Gl. 513, 66. Ðærscwaldas *limina,* Bd. 5, 7; S. 620, 27. [Icel. þresköldr.]

þes, þæs ; *m. :* þeós, þiós, þiús ; *f. :* þis, þiss, þys ; *n. demons. pron. This* :—*Iste,* þes (þæs, MS. F.), ys æteówiendlíc, and ðǽr biþ, ðǽr man swá bícnaþ be him ; *ille,* hé, ne biþ ðǽr ætforan andwerd, ðǽr men swá be him clypaþ . . . *ille* hé, *ipse* hé sylf, *iste* ðes, *hic* ðes, Ælfc. Gr. 93, 8-13. **I.** used adjectively. (1) alone with a noun :—Eal þes middangeard, and þás windas, and þás regnas, Blickl. Homl. 51, 19.

Ðes eorl, Beo. Th. 3409; B. 1702. Þes Paulinus, Chr. 627; Erl. 25, 5. Ymbhwyrft þes, Exon. Th. 424, 21; Rä. 41, 42. Þeós (ðiós, Lind. Rush.) sealf *unguentum istud*, Mk. Skt. 14, 5. Þeós (ðiús, Lind.: ðiós, Rush.) stefn *uox haec*, Jn. Skt. 12, 5. Þeós wundrung, Exon. Th. 6, 24; Cri. 89. Þiós eorðe, Met. 20, 118. Snytry ðiós *sapientia haec*, Mt. Kmbl. Lind. 13, 54. Þis word *verbum istud*, Mt. Kmbl. 28, 15. Þis (ðis, Lind.) godspel *hoc evangelium*, 24, 14. Þisses middangeardes, Blickl. Homl. 27, 17. Þisses lifes, Cd. Th. 68, 21; Gen. 1120. Þysses, Beo. Th. 397; B. 197: Blickl. Homl. 31, 3. Þyses, 115, 5. Heofones þisses, Met. 24, 3. Þisse woerulde *saeculi istius*, Mt. Kmbl. 13, 22: Blickl. Homl. 17, 17. Ðás, 117, 35. Ðiós (þás, Rush.) *istam*, Mt. Kmbl. Lind. 15, 15. Ðæt folc þis wundor geseah, Blickl. Homl. 15, 29. On þýs geáre, 119, 2. Mid þisse sealfe and mid þýs drence, Lchdm. ii. 118, 17. Mid ðýs beácne, Elen. Kmbl. 184; El. 92. On þis middanearde, Bt. 26, 3; Fox 94, 16. Fram þis wígplegan, Byrht. Th. 141, 2; By. 316. Ealle þás gód cumaþ, Blickl. Homl. 29, 10. Ðás men, 189, 28. Þissa leóda land, Andr. Kmbl. 535; An. 268. Þyssa, Met. 7, 54. Þeossa, Blickl. Homl. 15, 13. Ealra ðeassa portweorona gewitnisse, Cod. Dip. Kmbl. ii. 3, 11. Worda þissa, Exon. Th. 246, 6; Jul. 57. Of þissum lioðobendum, Cd. Th. 24, 23; Gen. 382. Þyssum, Andr. Kmbl. 175; An. 88: Blickl. Homl. 25, 9. Þisum, Met. 20, 255. Þysum, 29, 98: Blickl. Homl. 145, 5. Þeossum, 95, 11 : 135, 31. Þiossum, Met. Einl. 4. Þás folc sleán, Cd. Th. 151, 10; Gen. 2506. Ðás dæda, Blickl. Homl. 31, 20. Ðás word, 177, 33. Þ́es, 5, 30. (1 *a*) where objects are contrasted, *this* as opposed to *that*, one as contrasted with others :—Ðonne hí eów éhtaþ on þysse (ðissær, Lind.) byrig, fleóþ on óðre, Mt. Kmbl. 10, 23. Þis leóht wé habbaþ wið nýtenu gemǽne, ac ðæt leóht wé sceolan sécan, ðæt wé mótan habban mid englum gemǽne, Blickl. Homl. 21, 13. Ánra gehwylc hæfþ syndrige gife of Gode, sume þás gife, sume óðre gife, Homl. Ass. 34, 242. (2) with numerals or adjective forms used substantively :— Ðæt fæsten þyses feówertiges daga, Blickl. Homl. 35, 5. On þyssum feówertigum nihta, 35, 17. For ðisum (ðeosun, Hatt. MS.) ilcan is geséd, Past. 17; Swt. 120, 9. Ðioson, Swt. 125, 6. Be ðýs ilcan, 22; Swt. 168, 19. Þás þyllíce mé tugon tó helle, Homl. Skt. i. 4, 290. Mon ðissa twéga hwæðer ondrǽtt suíður ðonne óðer, Past. 27; Swt. 189, 9. Menn þisra seofona héddon, Homl. Skt. i. 23, 137. (3) where the noun is qualified by an adjective :—Þes ealda man, Blickl. Homl. 43, 33 : Cd. Th. 7, 11; Gen. 104. Ðæs andwearda wela, Bt. 32, 1; Fox 114, 2. Þeós swíðre hand, Cd. Th. 195, 22; Exod. 280 : Blickl. Homl. 5, 29. Þiós, Bt. 5, 3; Fox 14, 4. Ðiós unstille gesceaft and þeós (þiós, Cott. MS.) hwearfiende, 39, 6; Fox 220, 23. Þis mennisce cynn, Blickl. Homl. 17, 14. Ne þyses lǽnan welan, ne þyssa eorþlícra geofa, 21, 11. On ðissum andweardan dæge, 171, 3 : Cd. Th. 271, 27; Sat. 111. Ðeossum, 271, 20; Sat. 108. Þysum, Hy. 3, 53. Be þisse ondweardan tíde, Blickl. Homl. 15, 4. Þeosne andweardan dæg, Homl. Skt. ii. 23 b, 579. Þýs uferan Sunnandæge, Blickl. Homl. 119, 15. Hwelc þæs flǽsclícan gód sién, Bt. 32, 1; Fox 116, 28. (3 *a*) with a numeral used adjectively :—Intó ðýs twéntigum hídum, Cod. Dip. Kmbl. v. 331, 1. On þýs ylcum þrím dagum, Lchdm. iii. 76, 26. II. used substantivally, (1) pointing out a person or object :—Þes ys smiðes sunu, Mt. Kmbl. 13, 55 : Jn. Skt. i. 34. Nys þes Iósepes sunu? Lk. Skt. 4, 22. Hé wæs geháten Zosimus. Ðes on ánum mynstre drohtnode, Homl. Skt. ii. 23 b, 22. Æfter ðyses forðsíðe, 25, 142. Hwanon ys þysum (ðissum, Lind.) þes wísdom? Mt. Kmbl. 13, 54, 56. Tó hwon lǽddest ðú hider þeosne? Blickl. Homl. 85, 25 : 87, 1. Wé þissa wundra gewitan sindon; eall þás geeodon in ússera tída tíman, Exon. Th. 147, 11; Gú. 725. Þysum (*for these men*) is tó gearcienne ða reþestan wíta, Homl. Skt. 24, 21. Ðeossum ða *his qui*, Ps. Surt. 30, 24. Þassum, Mt. Kmbl. Rush. 8, 32. (1 *a*) *this, the present* :—Ǽr ðissum (ðysum, Cott. MSS.), Past. pref.; Swt. 7, 16. Of þisson forð áwa tó worulde *ex hoc nunc et usque in seculum*, Ps. Th. 120, 7. Ðyssum, 130, 5. Ðyssan, 113, 25. Oð ðiss (ðis, Cott. MSS.) *hitherto*, Past. 23; Swt. 173, 14. Þis, Homl. Skt. i. 22, 44. (1 *b*) where there is a contrast between two objects, *this* as opposed to *that* or *the other* :—Ðeós wyrt ys twégea cynna; þonne ys þeós reád . . .; þonne ys óðer byterre on byrgincge, Lchdm. i. 320, 15. Þeós . . . seó óþer, Exon. Th. 91, 9; Cri. 1489. Of þysum on þæt, Ps. Th. 74, 8. ¶ Þis, like þæt, is used with the substantive

verb in reference to a subject of any gender or number :—Þis is mín se leófa sunu, Blickl. Homl. 29, 28. Þis ys se dæg, Ps. Th. 117, 22 : Cd. Th. 195, 7; Exod. 273. Eart ðú þis, Drihten? 298, 22; Sat. 537. Þis is seó eorðe, 107, 10; Gen. 1787. Þis (ðiós, Lind. Rush.) is eówer tíd *haec est hora uestra*, Lk. Skt. 22, 53. Þis is landa betst, Cd. Th. 49, 21; Gen. 795. Þiss wǽron ealle Créca leóde, Ors. 3, 1; Swt. 100, 13. Þis sint ða ðe sceolon standan *hi stabunt*, Deut. 27, 12 : Jos. 12, 1. Þis synd ða bebodu and dómas and laga *haec sunt judicia atque praecepta et leges*, Lev. 26, 46: Num. 3, 2. Þe þis sint hira naman *quorum ista sunt nomina*, 1, 5. Ne synt ná þis wódes mannes word *haec uerba non sunt daemonium habentis*, Jn. Skt. 10, 21. Sint þis ða gód and ðæt edleán ðe ðú gehéte? Bt. 3, 4; Fox 6, 19. (2) where the pronoun refers to that which has just been stated :—' Ðú cennest sunu.' Ðá cwæþ heó : ' Hú mæg þis geweorþan?' Blickl. Homl. 7, 21. Hwá ne wafaþ ðæs ðonne se fulla móna wyrþ ofertogen mid þióstrum? oððe eft ðæt ða steorran scínaþ beforan ðam mónan and ne scínaþ beforan ðære sunnan? Ðisses hí wundriaþ, Bt. 39, 3; Fox 214, 31. Þisses, Exon. Th. 15, 18; Cri. 238. Hwanun wát ic þis? Lk. Skt. 1, 18. Swá lange swá ge ðis dydon, Blickl. Homl. 169, 21: Exon. Th. 39, 24; Cri. 627. For þis (þý, Cott. MS.) is se cwide sóþ *for this reason is the saying true*, Bt. 36, 7; Fox 184, 18. Þiss, Mt. Kmbl. 28, 14. Wé ðiss (ðis, Cott. MSS.) feáwum wordum sǽdon, Past. 3; Swt. 33, 6 : 22; Swt. 169, 3. Wé sculon ðissa ǽgðer underðencean, 7; Swt. 49, 23. ¶ Referring to a circumstance which serves to mark time :—Æfter ðrim móndum ðises (*the circumstance just mentioned*), Homl. Th. ii. 496, 29. Æfter þisson, Jn. Skt. 11, 7. Betwux ðisum, Homl. Th. i. 480, 27. Æfter þiossum, Blickl. Homl. 239, 32. (2 *a*) where the pronoun refers to a statement immediately following :—Þis næs gecweden be Criste, ðæt his fót æt stáne oþspurne, Blickl. Homl. 29, 30. Þis þinceþ riht, ðæt ðú ðé áferige of þisse folc-sceare, Cd. Th. 149, 17; Gen. 2476: 294, 2; Sat. 465. For ðeosum wæs geworden . . . for ðæm, Past. 15; Swt. 91, 26. [*O. Sax.* thius; *f.*; thit; *n.*: *O. Frs.* this, thisse; *m.*; thius, thisse; *f.*; thit, this; *n.*: *O. H. Ger.* dese; *m.*; desiu; *f.*; diz; *n.*: *Icel.* þessi; *m. f.*; þetta; *n.*]

þéwan; *v.* þeówan, þý, þeówan.

þicce; *adj. Thick.* Þicce *condensa*, i. *spissa*, Wrt. Voc. ii. 135, 64. Þicce and þynne, Exon. Th. 424, 8; Rä. 41, 36. I. of substances, (1) of liquids or moist materials, *thick, viscous* :—Gegníd on gewleced wæter, oþ ðæt hit sié swá þicce swá huniges teár, Lchdm. ii. 74, 4. Náne óþre wǽtan ðæt þicce and stille sié, 138, 13. Oþ hit sié swá þicce swá þynne bríw, 314, 3 : 316, 24. Swá þicce swá molcen, 332, 18. Wyl on swípum beóre ðæt hit sié þicce, 358, 19. Þat hé út hræcþ byþ swýþe þicce, iii. 126, 11. Ðicce, ii. 262, 21. Tósoden and þicge geuruen, 230, 8. Mid þiccere wǽtan, 280, 4. Ne drince hé þicce wín (cf. þynne wín, l. 18), 254, 26. Of þiccum *lento* (*defruto*), Hpt. Gl. 408, 38. 'Wá ðæm ðe gaderaþ an hine selfne ðæt hefige fenn (*densum lutum*) . . . Ðæt is ðonne ðæt mon gadrige ðæt ðicke (ðicce, Cott. MSS.) fenn (*densum lutum*) on hine, Past. 44; Swt. 329, 19. (2) of solid material, *dense* :—Sió eorþe is hefigre and þiccre þonne óþra gesceafta, Bt. 33, 4; Fox 130, 19. Þicre, Met. 20, 134. II. of air, cloud, darkness, etc., *thick, dense* :—Þicce genip (*nubes densissima*) oferwréh ðone munt, Ex. 19, 16. Ǽr se þicca mist þinra weorde, Met. 5, 6. Sió þicce ǽrlyft *gravis*, Wrt. Voc. ii. 41, 74. Þiccre *crassae, densae* (*noctis*), Hpt. Gl. 446, 25. Tódríf ðone þiccan mist, Met. 20, 264. Anlíce swá ðú bærne þornas þyre þicce fýre, Ps. Th. 117, 12. Ða ðýstru swá ðicce wǽron *tenebrae in tantum condensatae sunt*, Bd. 5, 12; S. 628, 12. Becómon ðicce ðeóstru, Homl. Th. ii. 194, 3. III. where objects are placed close together, *thick, dense* :—Gif hǽr tó þicce sié, Lchdm. ii. 156, 8. Of þiccum (*thickly planted?*) felde *de denso campo*, Wrt. Voc. ii. 138, 59. Ðú lǽtst mé on þicne wudu, Bt. 35, 5; Fox 164, 13. On þiccon bearwum, Lchdm. i. 322, 25. Intó ðam wudu ðǽr hé þiccost wæs, Homl. Skt. ii. 30, 31. In ðone þiccestan wudu, Shrn. 118, 16. III *a.* *growing thickly, abundant* :—Geborh Godes bringeþ tó genihte wæstme weordlíce and wel þicce (*or adv.?*) *montem Dei, montem uberem; mons coagulatus, mons pinguis*, Ps. Th. 67, 15. Seó eá (*Nile*) gedéþ mid ðæm flóde swíþe þicce eorþwæstmas, Ors. 1, 1; Swt. 12, 36. IV. marking dimension, (1) in a general sense, *thick, stout* :—Dó on ánne þicne (þynne, MS. H.) línenne cláð, Lchdm. i. 240, 21. Lege on þone þiccestan cláð oþðe on fel, ii. 200, 11. (2) of more exact measurement, *thick* :—Hí woldon witon hú heáh hit wǽre tó ðæm hefone, and hú ðicke (þicce, Cott. MS.) se hefon wǽre, oððe hwæt ðǽr ofer wǽre, Bt. 35, 4; Fox 162, 22. Se weall wæs .xx[x]. fóta ðicce *is locus murum triginta pedes latum habuit*, Ors. 4, 13; Swt. 210, 30. Seó eá oferfleów mid fótes þicce flóde, 1, 3; Swt. 32, 6. Ða wágas wǽron gyldne mid gyldnum þelum ánǽglede fingres þicce *auratos parietes laminarum digitalium grossitudine*, Nar. 4, 26. [*O. Sax.* thikki: *O. Frs.* thikke : *O. H. Ger.* dicchi *crassus, spissus, densus, torosus, grossus, frequens* : *Icel.* þykkr.] *v.* þicness.

þicce; *adv.* I. marking closeness in the texture or composition of a whole, *closely* :—Þicce gewefen hrægel *pavidensis*, Wrt. Voc. i. 40, 11. II. marking closeness of separate objects, *thickly, densely, closely* :—Ðá flugon ða lægetu swylce fýrene strǽlas tó ðæm þicce ðæt . . .,

3 Y

Blickl. Homl. 203, 10. Swā þicce is þeó heofon mid steorrum áfylled on dæg swā on niht, Lchdm. iii. 234, 31. Wæl þicce gefylled *the corpses lying thick on the ground*, Cd. Th. 130, 16 ; Gen. 2160. Swā þicce hié āweóllon swā æmettan *they swarmed as thick as ants*, Nar. 11, 12. **III.** marking action that occurs with frequency or with little intermission:—Feónda feorh feóllon đicce, Cd. Th. 124, 20 ; Gen. 2065. Hió spræc him þicce tō *she spoke to him again and again*, 43, 1 ; Gen. 684. **IV.** marking abundance, *thickly* :—Lege đæt dust swíþe þicce on clāđ, Lchdm. ii. 148, 15 : 340, 21. Wearđ beám monig blōdigum teárum birunnen reáde and þicce, Exon. Th. 72, 22 ; Cri. 1176. [*O. Sax.* thikko (mid thiodu gisetan) : *O. H. Ger.* diccho *dense, frequenter, saepe.*] v. þiclíce.

þiccet[t], es ; *n. A place where there is dense growth* (v. þicce, **III.**), *a thicket* :—On þyccetum *in condensis*, Ps. Lamb. 117, 27. Stefn Drihtnes āwríhþ þiccettu (þiccetu, Ps. Lamb.) *vox Dominis revelabit condensa*, Ps. Spl. 28, 8.

þiccian ; *p.* ode *To thicken* (trans. and intrans.), *to make* or *to become thick*, of persons, *to throng* :—Ic điccige *denso* and *denseo*, Ælfc. Gr. 37 ; Zup. 220, 8. Þiccaþ *densescit, spissat*, Wrt. Voc. ii. 138, 74. Đā þiccodan þider semninga þa Ismaheli, Shrn. 38, 4. [Hit bicometh to a thikke blod . . . neoȝe dayes hit thicketh so, Wrt. popl. science 139, 3. Thykkyñ or make thykke, as wodys *condenso*, thykkyñ or make thykke, as lycurys *spisso, inspisso*, Prompt. Parv. 491. *O. H. Ger.* dicchēn *glomerare, grossescere, crebriscere.*]

þicness. v. þicnes.

þiccol(-ul) ; *adj. Stout, corpulent* :—Þiccol *corpulentus*, Wrt. Voc. i. 83, 47. Điccul, 51, 13.

þic-feald ; *adj. Dense, close* :—Þicfealdum þreátum *spissis cohortibus*, Hpt. Gl. 413, 1. v. next word.

þicfildan. v. ge-þicfyldan (*l.* geþycfyldan *densere*, Germ. 401, 21). v. preceding word.

þicgan ; *p.* þah, þeah, *and* þigde, þigede, *pl.* þǣgon, þēgon, *and* þigdon, þigedon ; *pp.* þegen, *and* þiged. **I.** *to take, receive, accept* :—Hē him brād syleþ lond tō leáne, hē hit on lust þigeþ, Exon. Th. 331, 31 ; Vy. 76. Hié đa lác þēgon tō þance, Andr. Kmbl. 2225 ; An. 1114. Đeáh he māđmas þēge, Elen. Kmbl. 2516 ; El. 1259. Đæt hý beágas þēgon, Exon. Th. 283, 29 ; Jul. 687. Heó hafaþ gefreód đa men đe heó þigede æt Cwæspatrike, Chart. Th. 621, 18. Welan þicgan, Exon. Th. 331, 1 ; Vy. 61. Feoh þicgan, 332, 7 ; Vy. 81. His giefe þicgan tō þonce, 109, 26 ; Gū. 96. Lāfe þicgan, 498, 9 ; Rä. 87, 10. Ne gē đæt geþyldum þicgan woldan, 131, 12 ; Gū. 454. Se æđeling gehwilcan feoh and feorh beád, and heó nǣnig þicgan noldan, Chr. 755 ; Erl. 51, 5. **II.** *to take* food, poison, medicine, etc., *to eat* or *drink, consume* :—Hū đæt ne gemylt, đæt se maga þigeþ, Lchdm. ii. 158, 16. Þigđ, 186, 21. Fýr þigeþ lænne líchoman, Exon. Th. 213, 4 ; Ph. 219. Nō hē þigeþ mete, 215, 27 ; Ph. 259 : 357, 28 ; Pa. 35. Of đam mete đe wē þicgaþ, Bt. 34, 11 ; Fox 150, 35. Đonne hig mete þicgeaþ *cum panem manducant*, Mt. Kmbl. 15, 2. Hí hyra hlāf þicgaþ, Mk. Skt. 7, 5. Hē on his hūs eode and his swǣsendo đeah *intravit epulaturus domum ejus*, Bd. 3, 22 ; S. 553, 30. Đæt hē nǣfre oftor swǣsendo đeah (*reficeret*), 4, 25 ; S. 600, 16. Swā đæt hē nǣfre mete onfēng ne swǣsendo đeah *ita ut nihil unquam cibi vel potus perciperet*, S. 599, 30. Wē medu þēgon, Beo. Th. 5260 ; B. 2633 : Judth. Thw. 21, 15 ; Jud. 19. Hí wiste þēgon, Andr. Kmbl. 1186 ; An. 593. Hié fira flǣschoman þēgon, 49 ; An. 25. Hí þēgun æppel, Exon. Th. 226, 8 ; Ph. 402. Đæt hē mæte đygde *ad prandendum*, Bd. 5, 4 ; S. 617, 11. Israhēl đigde đæs lambes flǣsc, Homl. Th. ii. 278, 18. Þigde *consumeret, biberet*, Hpt. Gl. 450, 32. Þigede, Guthl. 4 ; Gdwin. 26, 18 : 5 ; Gdwin. 34, 7. Đygede, Bd. 5, 4 ; S. 617, 17. Wit eaples þigdon, Cd. 290, 7 ; Sat. 411. Hē sumum liéfde tō đicgganne đætte hē nolde đæt hí ealle đigden, Past. 59 ; Swt. 451, 29. Þigedan, Ors. 3, 6 ; Swt. 110, 1 : 6, 21 ; Swt. 272, 23. Đæt hí of his swǣsendum mete ne đygedon *ne de cibis illius acciperent*, Bd. 3, 22 ; S. 553, 28. Ceorf nygan penegas, and đige đa, Lchdm. iii. 8, 2. Nǣfre gē beódgereordu unārlíce eówre þicgeaþ, Cd. Th. 91, 29 ; Gen. 1519. Ne hē nāht fūles ne þicge (*comedat*), Jud. 13, 4. Wiþ þon þe mon þicge ātor, Lchdm. ii. 110, 24. Đæt hí mōston onfōn and đicgean đa foresetenysse hlāfas *ut panes propositionis acciperent*, Bd. 1, 27 ; S. 496, 14 : 5, 4 ; S. 617, 14. Nolde ic mid þæm men mīnne mete đicgean *cum hoc simul non edebam*, Ps. Th. 100, 5. Đicgan, Bd. 4, 19 ; S. 588, 12 : Homl. Th. ii. 244, 11 : 40, 13. Syle đone wyrttruman đam seócan þicgean . . Gyf đū đäs wyrte sylst þicgean on strangon wíne, Lchdm. i. 172, 10-13. On drince þicgean, 198, 25. Genim đäs ylcan wyrte, seóđ on hunige, syle þiggean (þiggcan, MS. H. : þicgan, MS. B.), 150, 9. Symbel þicgan, Beo. Th. 2025 ; B. 1010. Đonne āliéfþ hē đæm siócan eal đæt đæt hine lysđ tō đōnne and tō đycganne, Past. 50 ; Swt. 391, 25. Biđ seó ān snǣd sēlre tō þicganne, Salm. Kmbl. 813 ; Sal. 406. Se forbeád blōd tō þicgenne, Ælfc. T. Grn. 4, 43. Đicgendum (điccendum, Rush.) miđ him and etendum *discumbentibus cum eis et manducantibus*, Mk. Skt. Lind. 14, 18. [Þet mon to muchel ne þigge on ete and on wete, O. E. Homl. i. 105, 3. *But later the word means* to beg :—He haueth me do mi mete to thigge, Havel. 1373. Beggyn or thyggyn

mendico, Prompt. Parv. 28. Cf. thyggynge or beggynge *mendicacio*, 490. Thiggand *egenus*, Ps. 39, 18. See also Halliwell's Dict. and Jamieson's Dict. *O. Sax.* thiggean (*wk.*) *to receive, to ask* : *O. H. Ger.* diggen (*wk.*) *impetrare, petere, expetere* : *Icel.* þiggja ; *p.* þâ, *pl.* þâgu ; *pp.* þeginn *to receive, accept.*] v. ge-þicgan, þegan ; þegu.

þiclíce ; *adv. Thickly, in great numbers, in quick succession* :—Đā hié gesāwan đa deádan men swā þiclíce tō eorþan beran, Ors. 3, 10 ; Swt. 138, 25. Steorran of heofenan feóllan, nāht be ānan ođđe twām, ac swā þiclíce đæt hit nān mann āteallan ne mihte, Chr. 1095 ; Erl. 231, 21. v. ful-þiclíce ; þicce.

þicness, e ; *f.* **I.** referring to the consistency of matter, *thickness, viscosity.* v. þicce, I :—Cnuca mid wíne on huniges þicnysse, Lchdm. i. 126, 12. Gyf hwā mycelne hracan þolige, and hē đone him eaþelíce fram bringan ne mæge for đycnysse, 284, 24. Seóþ ōþ đæt đæt hæbbe huniges þicnesse, ii. 190, 5. **II.** referring to the lack of transparency, *thickness, obscurity, cloud, darkness.* v. þicce, II :—Genipu þicnysse *nubes et caligo*, Ps. Spl. 96, 2. Tegānre þicnysse *rupto tenebrarum situ*, Germ. 388, 43. Of nyþerhreósendre þicnysse *deciduo imbre*, 390, 79. Wē ne magon for đære fyrlynan heáhnysse and đæra wolcna đicnysse and for ure eágena tyddernysse hí (heofenan) nǣfre geseón, Lchdm. iii. 232, 16. Đa þicnyssa smíces stigon upp *the clouds of smoke rose up*, Homl. Skt. i. 23, 36. **III.** *a thicket.* v. þicce, III :—On đicnessum *in condensis*, Ps. Lamb. 117, 27. Đicnyssa *condensa*, 28, 9. **IV.** referring to dimension, *thickness, depth, a thick body.* v. þicce, IV :—Hreóflícre þicnesse *elephantina callositate*, Hpt. Gl. 519, 31. Hit nǣfde eorþan þicnesse *non habebat altitudinem terrae*, Mk. Skt. 4, 5. Sweflenum þicnyssum *sulphureis flammarum globis*, Hpt. Gl. 499, 41. [*O. H. Ger.* diknissa *densitas.*]

þídan, þiddan. v. þeódan, þyddan.

þider, þieder ; *adv. Thither, whither*, where motion is expressed or implied. **I.** as absolute demonstrative, *thither, to that place* :—Ne færst đū þider (*illuc*), Deut. 1, 37. Đa đe hine þider lǣddon, Gen. 39, 1. Þyder (đider, Lind.) faran *illuc ire*, Mt. Kmbl. 2, 22. Hē com þyder (đidir, Lind. : þidera, Rush.), Jn. Skt. 18, 3. Đyder (đidder, Lind. : đider, Rush.), 11, 8. Hí tō đon đider (*illo*) sende wǣron, đæt hí sceoldon đæt gyldene mynet mid him geniman đætte đider (*eo*) of Kent com, Bd. 3, 8 ; S. 530, 40. Đā fērde hē đyder, Blickl. Homl. 225, 7. Nū þyder ingongaþ and mē ætstondaþ, 207, 2. Uton mid him þyder geond gān, Homl. Skt. i. 23, 748, 321. Đæt gisefđe đe đone þyder ontyhte, Beo. Th. 6164 ; B. 3086. Hit witena nān þider (cf. þǣr, Bt. 32, 3 ; Fox 118, 9) ne sēceþ *no wise man goes thither to look for it*, Met. 19, 8. Þider wǣron fūse, Cd. Th. 190, 9 ; Exod. 196. Hē þyder folc samnode, 230, 5 ; Dan. 228 : Blickl. Homl. 67, 20. Se síþfæt is þyder tō lang, 231, 26. Ærende wē þyder habbaþ, 233, 11. **I a.** in an indefinite sense :—Đæt hió on ænige healfe ne heldeþ ; ne mæg hió hider ne þider sīgan, Met. 20, 164. On healfa gehwǣr, sume hyder, sume þyder, Elen. Kmbl. 1093 ; El. 548. **II.** as antecedent :—Đā ferede hine Godes hand þider, þǣr hine men siđđan āredon, Shrn. 57, 5. Uton ācerran þider, þǣr hé sylfa sit, Cd. Th. 278, 6 ; Sat. 217. Gingran þider ealle urnon, þǣr se ēca wæs, 298, 11 ; Sat. 531. Þider cuman, þonan þe hit ǣr com, Bt. 25 ; Fox 88, 30. **III.** in correlative clauses, *thither . . . whither* :—Đider becuman . . . đieder đe he wilnaþ, Past. 11 ; Swt. 65, 16. Đyder đe hé sylfa tōweard wæs æfter deáþe, đider hé his mōdes eágan sende ǣr his deáþe *ubi erat futurus post mortem, ibi oculos mentis ante mortem misit*, Bd. 5, 14 ; S. 634, 41. Þider đe Stephanus forestōp, đider folgode Paulus, Homl. Th. i. 52, 6. **IV.** where antecedent and relative are contained in the one form, *to the place to which, whither* :—Cuman þyder (đidder, Lind.) ic fare *quo ego vado venire*, Jn. Skt. 8, 21. Đū mōst fēran þider đū fundadest, Exon. Th. 102, 12 ; Cri. 1671 : Met. 26, 119 : 13, 3. Đæt heó mē gerihte þyder hire willa wære, Homl. Skt. ii. 23 b, 509. **V.** as a relative, *whither* :—Tō heofenum, þider hié witon đæt hē āstāg, Blickl. Homl. 125, 29. Hē tō heofenum lōcade, þyder his mōdgeþanc ā geseted wæs . . . tō Drihtne þyder hē fēran sceal, 227, 17–22. v. þæder.

þideres, þidres ; *adv. Thither* :—Đæt sió ūterre ābisgung đissa worldđinga đæs monnes mōd gedrēfđ and hine scofeđ hidres đidres, đæt hē āfilþ of his āgnum willan *quod cor externis occupationum tumultibus impulsum a semetipso corruat*, Past. 22 ; Swt. 168, 13. Hidres þidres, Bt. 40, 5 ; Fox 240, 21. Hē lange hyderes and þyderes sēcende fōr, Homl. Skt. ii. 23 b, 702. v. þædres.

þider-inn, -in ; *adv. Into that place*, (1) where motion is expressed or implied :—Þeáh hwā his āgen spere sette tō ōđres mannes hūses dura, and hē þiderin (-in, MS. B.) ærende hæbbe, L. C. S. 76 ; Th. i. 418, 5. Ic mē þyderinn eode, Homl. Skt. ii. 23 b, 500. Hié þider-inn wǣron gesamnode *they were got together into the place*, Blickl. Homl. 207, 36. (2) of other relations :—Eal seó sōcna đe đærto hēreþ and đæt land þiderinn *the land belonging to it*, Chart. Th. 547, 2. Ic wille đæt se cyng beó hlāford đæs mynstres and đære landāra đe ic þyderinn becweden hæbbe (*that I have bequeathed to the monastery*), 547, 32. His bēc ealle hē cwæđ þyderin, 550, 23. Ōsaníg gange þyderin, 550, 19.

þider-leódisc; adj. Of that people:—Hé geleórde on Burgenda mǽgðe, and hé wæs bebyrged mid micle wópe ge Angelcynnes monna ge þiderleódiscra, Shrn. 134, 24.

þider-weard; adv. Thitherward, in that direction, towards that place or point:—Iosue férde mid his fyrde þiderweard ascendit Iosue et omnis exercitus cum eo, Jos. 10, 7. Ðá hé þiderweard seglode as he sailed towards that port, Ors. 1, 1; Swt. 19, 24. Ealle þiderweard éfeston all hastened towards the spot, Guthl. 1; Gdwin. 8, 20. Hié wǽron floc-mǽlum þiderweard they were flocking to the place, Ors. 4, 10; Swt. 200, 19: 5, 13; Swt. 246, 21. Ðá hé ðyderweard wæs when he was on the way to it, Homl. Skt. ii. 30, 179: Chr. 1009; Erl. 142, 3. Beheóld Abraham þyderweard Abraham looked in that direction, Gen. 19, 27. Beseah hé þiderweard, Homl. Skt. i. 23, 499. Þinga gehwilc þiderweard fundaþ, Met. 13, 14. Wuhta gehwilc wilnaþ þiderweard, 20, 159.

þiderweardes; adv. Thitherwards.—Wæs se cyng þiderweardes on fǽre . . . þá hé þá wæs þiderweardes and sió óþeru fierd wæs hámweardes the king was on the march thither . . . When he was on the way thither and the other troops were on the road home, Chr. 894; Erl. 90, 32. Swá heó ǽr dyde þyderweardes as she did before when on the way to that place, Homl. Skt. ii. 23 b, 724. Ðá ongon hé sprecan swíþe feorran ymbúton, swilce hé ná þa sprǽce ne mǽnde, and tiohhode hit þeáh þiderweardes (towards that point), Bt. 39, 5; Fox 218, 12.

þidres, þiédan, þiéfan, þiéfe-feoh, þiéfþ, þiéstru. v. þideres, þeódan, ge-þiéfan (read -biéfan), þífe-feoh, þiéfþ, þeóstru.

þífe-feoh stolen goods:—Gif þiéfefeoh (forstolen feoh, MS. H.) mon æt ciépan befó, L. In. 25; Th. i. 118, 13. [Cf. Icel. þýfi; n. stolen goods.]

þífe-, þeófe-, þéfe-, þýfe-, þéfan-þorn, es; m. Buckthorn:—Ðeófeðorn, thébanthorn ramnus, Txts. 93, 1710. Þífeþorn, Wrt. Voc. i. 33, 43. Þéfeþorn, 68, 34. Þýfeþorn ramnus vel sentix ursina, 39, 23. Þéfan-ðorn, coltetræppe ramnus, 285, 47. Þéfauþorn, Lchdm. iii. 312, 15: 352, 12: 354, 24. Nim ðéfeþorn, iii. 56, 27. Þéfeðorn ramnum, Ps. Spl. T. 57, 9. [Wiclif uses thevethorn in the passage last cited, as also in Jud. ix. 14; see, too, Ps. 57, 10, and Prompt. Parv. thevethorn the ramnus. Thief is given as a word for bramble in E. D. S. Leicestershire Glossary. O. H. Ger. dêpan-dorn ramnus.] v. þúfe.

þiffe?:—Defruto ł felde ł þiffe (þíse? Cf. (?) theve, brusch, Prompt. Parv. 490; or þífele (?). The passage glossed is lento careni defruto, in which the first word is glossed by of þiccum, but in the margin by of þiccum þéfele. Cf. too Wrt. Voc. ii. 138, 59 de lento fruto of þiccum felde), Hpt. Gl. 408, 50.

þífþ, þiéfþ, þýfþ, þeófþ, þeóft, e; f. I. theft, act of thieving:—Be ánre nihtes (nihte, MS. B.) ðiéfðe (þýfte, MS. B.: þýfðe, MSS. G. H.). Gif hit bið nihteald þiéfð (þýfð, MS. H.) if a day has elapsed since the theft was committed, L. In. 73; Th. i. 148, 10. Móna se syofoða . . . þýfð gestrangaþ, Lchdm. iii. 186, 22. Gif hwa stalie on gewitnesse ealles his híredes, gongen hié ealle on þeówot .x.-wintre cniht mæg bión þiéfðe (þýfðe, MSS. B. H.) gewita (cf. wǽron cradolcild gehéowode þurh wæl-hreówe unlaga for lytelre þýfðe, Wulfst. 158, 15), L. In. 7; Th. i. 106, 18. Betygen þiéfðe (þífðe, MS. H.), 37; Th. i. 124, 22. Be ðýfðe betogenum. Gif hwa þífðe betogen sý, L. Ed. 6; Th. i. 162, 16. Onsacan ðære þiéfðe (þeófðe, MS. B.), L. In. 46; Th. i. 130, 14. Se ðe þýfðe forworht wǽre openlíce, L. Ath. v. 1, 4; Th. i. 228, 25. Gif man leúd ofsleá an þeófðe, licge bútan wyrgelde, L. Wih. 25; Th. i. 42, 13. Be ðeófes onfenge æt ðiéfðe, L. In. 28; Th. i. 120, 4: 37; Th. i. 124, 20. Þýfðe, L. Ath. i. 3; Th. i. 200, 20. Æt openre þýfðe, L. C. S. 26; Th. i. 392, 3. Ðá geácsode se biscop ðæt ða bécc forstolene wǽron, bæd ðara bóca geornlíce . . . man gerehte ðam biscope ða forstolenan bécc, and bóte æt ðære þýfðe, Chart. Th. 265, 10. For þeófte oþþe for manslihte, L. Wil. ii. 1; Th. i. 489, 6. Gif hé ða þiéfðe gedierne, L. In. 36; Th. i. 124, 17. Ðæt hý on heora mǽge náne þýfðe (þeófðe) nyston, L. Ath. i. 13; Th. i. 206, 2: iv. 4; Th. i. 224, 6. Man forgã þýfðe (-a), i. 20; Th. i. 210, 3. Ealles folces þing byþ ðe betere æt ðám þýfðum, v. 8, 9; Th. i. 238, 20. Ðæncunge ðæm ðe wið ðýfðe fylstaþ. Ic þancige Gode and eów eallum ðæs frides ðe wé nú habbaþ æt ðæm þýfðum, L. Edm. S. 5; Th. i. 250, 5. II. what is stolen, theft:—Tó ðý ðæt earm and eádig móte ãgan ðæt hý mid rihte gestrýnaþ, and þeóf nyte hwær hé þýfðe (þeófte, MS. C.) befæste, þeáh hé hwæt stele, L. Edg. S. 2; Th. i. 274, 3. [O. E. Homl. Laym. A. R. þeofðe: R. Glouc. þufþe: Gen. and Ex. ðefte: Ayenb. þiefþe: Chauc. thefte. O. Frs. thiufthe, thiufte: Icel. þýfð, þýft.]

þigaþ, Exon. Th. 130, 3; Gú. 432. v. þeón.

þigen, e; f. I. the taking of food, partaking, eating or drinking:—Ne sý him gemǽne þigen mid gebróðrum geþafod non permittatur ad mense communis participationem, R. Ben. 69, 13. Ðæs hálgan húsles ðygen partaking of the eucharist, Homl. Th. i. 266, 17. Se frumsceapena man weard ãdrǽfed of neorxenawanges myrhðe for ðæs forbod-enan bigleofan, 118, 25. Lactuca is biter on ðigene lettuce is bitter in the eating, ii. 278, 27. Mid unãlýfedre ðigene, 332, 1. Æt ðære ðigene (at the Passover), 280, 34. Sý hé ãscyred fram gemǽnre mýsan þigene from eating at the common table; a mensa, R. Ben. 49, 15; 70, 4. Sý ã on ðære þigene forhefednes let there ever be moderation in taking wine, 65, 3. Wið ǽttres ðigne, Lchdm. i. 150, 3. His gereordes þigene hé ána underfó refectionem cibi solus accipiat, R. Ben. 49, 6: Wulfst. 284, 25: Homl. Th. ii. 98, 30. Ðurh ðæs hálgan húsles þygene ús beóþ úre synna forgyfene, i. 266, 8. Ðurh ánes æpples ðigene through eating an apple, ii. 330, 33. Ða oferflówendlícan ðygene excessive eating and drinking, i. 360, 13. II. what is taken, food, meat or drink:—þigen edulium, Hpt. Gl. 513, 63. Ðæt seó dæges þigen tófered sý and seó hǽte ðære þigene oferslegen ut digesti surgant, R. Ben. 32, 14. Ne sý him nánre óðere þigene getíðod let him have no other food given him, 69, 21. Werede ðigene nectareum edulium, Hpt. Gl. 413, 38. Ða hálgan ðigene (the eucharist) onfón, Homl. Th. ii. 280, 29. Heora þigne ge-healdan to retain their food, Lchdm. i. 90, 12. Þygne, 8, 6. Ðú ðæs werðeóde wræccan láste freónda feásceaft gesóhtest þíne þearsende (þíne for þigne? needing food; or þíne pron. (v. þín, III) thy men being in need), Cd. Th. 149, 25; Gen. 2480. Fram eallum ðám þigenum ðe hracan oþþe innoþ tó miclum luste getýhþ, R. Ben. 138, 14. Áwendan úrne swæcc fram unãlýfedum, ðigenum, Homl. Th. ii. 374, 5. v. blód-þigen; þicgan.

þiging, e; f. The taking of anything to eat or drink, eating or drinking:—Of metta and of drincena þiginge, Lchdm. ii. 244, 12.

þignan to eat:—Hý ðýnde depastus est eam, Ps. Spl. C. 79, 14.

þigen[n], þínen[n], þinnen[n], e; f. I. a female servant, female attendant, handmaid:—Ðignen pedisequa, Wrt. Voc. ii. 116, 63. Þínen, i. 282, 15: ancilla, ii. 4, 12. Þínen, wyln abra, i. ancilla, i. 17, 26. Þýnen vernacula, servula, ancilla, Hpt. Gl. 498, 20. Sum þínen (ðig-nen, Lind.) a certain maid, Lk. Skt. 22, 56. Sió ðignen (ðegnen, Rush.) durehaldend ancilla ostiaria, Jn. Skt. Lind. 18, 17. Ic eom Godes ðínen behold the handmaid of the Lord (Lk. 1, 38), Homl. Th. i. 200, 10: Homl. Skt. ii. 23 b, 237. Heó cwæð tó him: 'Ic eom deófles ðínen, Shrn. 140, 18. Þínene ancillae, Gen. 35, 25, 26: Scint. 229, 6. Þí-nenne, Ps. Lamb. 115, 6: 85, 16. Þinnenre (-ne?), Ps. Spl. 85, 15. Þinnenne abrâ, Wrt. Voc. ii. 87, 49. Seó abbudisse eode mid ánre hire ðignenne (cum una sanctimonialium feminarum) . . . Ðá hét heó hire ðínenne (ministram) gãn, Bd. 3, 11; S. 536, 18–27. Þínenne, Judth. Thw. 24, 4; Jud. 172. Heó hæfde áne þínene (ancillam), Gen. 16, 1. Áne hire þínena unam e famulabus suis, Ex. 2, 5. Þínennum pedisequis, þínenna pedisequas, Wrt. Voc. ii. 67, 9, 10. Þínum vernaculis, Hpt. Gl. 523, 26. Þínenne vernaculas, 404, 56. Þínena ancillas, Lk. Skt. 12, 45: Gen. 33, 2: Homl. Th. ii. 478, 10. II. used with the meaning of byrþ-þignen, a mid-wife:—Se cyning cwæþ tó ðám þínenum ðe ðám Ebréiscean wífun þénodon (obstetricibus Hebraeorum) . . . Ða þínena (obstetrices) him ondrédon God, Ex. 1, 15, 17, 20, 21. v. beorþor-(written beoþor-), byrþ-, duru-, in-þignen (-þínen).

þiht; adj. Tight, firm, strong. This word seems to be the second part in each of the two compounds found in the following charm:—Ge-hwér férde ic mé ðone mǽran magaþíhtan mid ðysse mǽran meteþíhtan ðonne ic mé wille habban and hám gãn, Lchdm. iii. 68, 17. [Thyht, hool fro brekynge integer, solidus; thyhtyñ or make thyht integro, consolido, solido, Prompt. Parv. 491. Halliwell gives thiht close, compact, as an Eastern counties word. M. H. Ger. dîhte: Ger. dicht: Icel. þéttr.]

Þíla(-e). v. Þýle.

þílian, þillian, þillan to plank, lay planks as in making a bridge:—Ðá hét Maxentius oferbricgian ða eá eal mid scipum, and syððan ðylian swá swá ðóne bricge, Homl. Th. ii. 304, 22. Tó þilianne plancas ponere, Cod. Dip. B. iii. 659, 33. Tó þillianne, 5, 8, 10, 14. Tó þelliene, 26. Tó þillanne, 28. Tó þyllanne, 24. (The section is headed: Ðis is ðære bricge geweorc on Hróuecæstre.) [O. H. Ger. gi-dillôn internere (pontes): Ger. dielen to board, plank: Icel. þilja to cover with deals, to board, plank.] v. next two words.

þiling, e; f. A boarding, flooring, something composed of planks:—Breda þiling vel flór on tó þerscenne area, Wrt. Voc. i. 37, 59. Hig fæstniaþ ðone stepe þurh ða þilinge (deck; cf. Icel. þiljur; pl. the deck), Shrn. 35, 15. v. wǽh-þiling.

þille, an; f. A boarding, flooring, floor:—Ðille tabulata, tabulamen, Wrt. Voc. ii. 122, 8, 10. Pille tabulamen, i. 290, 73. [O. H. Ger. dilla; f.; dil, dillo; m. planca, ima pars navis, pluteus, tabula parietis: Ger. diele: Icel. þilja; f. a deal, plank, planking.) v. þel.

þillíc. v. þyllíc.

þín; pron. poss. I. attributive, thy, thine, (1) with noun alone:—Tó becume þín (ðín, Lind.) ríce. Gewurðe þín willa, Mt. Kmbl. 6, 10. Þínes fæder God, Gen. 31, 29. Far of þínum lande and of þínre mǽgðe and of þínes fæder húse, 12, 1. Þíne dura belocenre, Mt. Kmbl. 6, 6. (1 a) where the noun is to be inferred:—Ða ilcan ðe habbaþ nú heora ágnes þances forlétan, nales þínes, Bt. 7, 2; Fox 18, 13. (1 b) strength-ened by ágen:—þín ágen geleáfa þé hæþ gehǽledne, Homl. 15, 26. Þurh þíne ágene gémeleáste, Bt. 5, 1; Fox 10, 1. Gif ðæt þíne ágne (ágnan, Bod. MS.) welan wǽron, 7, 3; Fox 20, 17. (2) where the noun is qualified by an adjective:—For þínum ídlan gilpe, Blickl. Homl. 31, 14. For þínum gódan willan . . . ða leán eallra þínra gódena weorca, Bt. 7, 3; Fox 22, 14–16. Þurh þíne æðelan hand, Hy. 7, 5. Ða mód

ðínra getreówra freúnda . . . nimaþ hí heora men mid him and lǽtaþ þíne feáwan getreówan mid þé, 20; Fox 72, 14–17. (3) where a demonstrative pronoun is used with the noun :—Þes þín sunu, Lk. Skt. 15, 30. Þín se fægresta fæþm *that fairest bosom of thine*, Blickl. Homl. 7, 24. Sege mé hwæþer se þín wela swá deóre seó, Bt. 13; Fox 38, 6 : Met. 20, 29. Áles þíne þa líófan gesceft, Hy. 8, 33 : Ps. Th. 90, 7. Hí ðæt þín fægere hús forbærndan, 73, 7. Ðæt wé ðæt yrfe þín herige, 105, 5. (4) used in the genitive where the personal pronoun might be expected :—On þínes silfes hand, Hy. 7, 83. Þurh þínes sylfes geweald, Exon. Th. 466, 26; Hö. 127. Þínre sylfre sunu, 21, 23; Cri. 339. Mid þínes ánes geþeahte *with the counsel of thee alone*, Bt. 33, 4; Fox 128, 19 : Met. 20, 40. ¶ In poetry the pronoun may be separated from the words to which it belongs :—Blǽd is árǽred geond wídwegas, wine mín Beówulf, þín ofer þeóda gehwylce, Beo. Th. 3414; B. 1705. Ðæt ic mægburge móste þínre rím miclian, Cd. Th. 134, 6; Gen. 2220. Gewít þú þinne eft waldend sécan, 138, 16; Gen. 2292. II. used predicatively, *thine* :—Gilpan ðæt heora fægernes þín sié, Bt. 14, 1; Fox 40, 22. Nán ðara góda þín nis, 14, 2; Fox 42, 29. Ealle míne þing synt þíne (ðíno, Lind.), Lk. Skt. 15, 31. Hig wǽron þíne (ðíno, Lind.), Jn. Skt. 17, 6. III. used substantively, *thine* :—Nis sceat ðæs ic þínes áhredde *not a penny of what I saved of thine*, Cd. Th. 129, 16; Gen. 2144. Ðónne þú and þíne beóþ álýsde *when thou and thine are released*, Lchdm. i. 328, 25. Þú ðás werðeóde gesóhtest þíne þearfende (*thy men being in need* (?); v. þigen, II), Cd. Th. 149, 25; Gen. 2480. Ealle míne synt þíne, and þíne (ðíno, Lind.) synt míne, Jn. Skt. 17, 10. [*Goth.* þeins : O. L. Ger. O. Frs. thín : O. H. Ger. dín : Icel. þinn.]

þínan; *p.* þán; *pp.* þinen **To get moist** or **damp** :—Dó on næsc, hæbbe him on, ðý læs hit þíue, Lchdm. ii. 36, 8. v. of-þínen; þænan.

þincan, þínd. v. þyncan, ge-þind.

þindan; *p.* þand; *pp.* þunden. I. **to swell up** :—Þindeþ him se milt *his milt swells up*, Lchdm. ii. 232, 11. Þint sió lifer, 198, 23. Gif innoþ þinde, i. 354, 1. Þindan, Exon. Th. 431, 17; Rä. 46, 2. Se streám ongan tó þindenne ongeán swilce hit wǽre án heáh dún (*ad instar montis intumescentes*), Jos. 3, 16. Hit biþ þindende, Lchdm. ii. 210, 22. Gif ómihte blód and yfel wǽte on ðam milte sié þindende, þonne sceal him mon blód lǽtan, 252, 25 : 168, 11. II. figurative, *to swell with indignation, pride*, etc. :—Synful yrsaþ, tóþum torn þolaþ, þearle þindeþ (in this and the next passage the Latin verb is *tabescere*, but *tumescere* seems in each case to have been read), Ps. Th. 111, 9. Ic þand (*I was angry*) wið ðan ðe hí teala noldan þínre sprǽce spéd gehealdan, Ps. Th. 118, 158. Ðindende weleras *labia tumentia*, Kent. Gl. 1002. v. á-, tó-þindan; for-þunden; ge-þind.

þínen. v. þignen.

þing, es; *n.* I. **a thing**, (1) **a single object**, material or immaterial :—Hweðer ðú wéne ðæt ǽnig ðing on ðisse worulde swá gód sié, Bt. 34, 1; Fox 134, 6. Ðonne ða líf þing ealle gegadorade beóþ, ðonne beóþ hit eall án ðing, and ðæt án þing biþ God, 33, 2; Fox 122, 18. Ðæt ilce ðú miht geþencan þe ǽlcum ðinge, ðæt nán þing ne biþ swelce hit wæs, siþþan hit wanian onginþ, 34, 9; Fox 148, 9 : Met. 20, 37. Ǽghwilc þing ðe on ðís andweardan lífe lícaþ, 21, 28. Ǽlc þing ðe líf hæfde, Gen. 7, 22. Ic seah sellíc þing singan, Exon. Th. 413, 10; Rä. 32, 3. Hefon and eorþe and sǽ and ealle ða þing ðe on ðæm syndon, Blickl. Homl. 91, 21. Ðinga scæpend *rerum creator*, Rtl. 180, 9. Wé ágyltaþ þurh feówer þing, þurh geþóht and þurh word and þurh weorc and þurh willan, Blickl. Homl. 35, 14. (1 a) of particular classes of objects, (a) **a thing** of value, **property, a thing** for sale : generally in pl., *things, goods* :—Him eallum wæs gemǽne heora ðing, Homl. Th. i. 316, 9. Nán man nán þing ne bycge ofer feówer peninga weorð ne libbende ne licgende, L. C. S. 24; Th. i. 390, 2. Breng ðing *offer munus*, Mt. Kmbl. Lind. 8, 4. Him eallum wǽron heora ðing gemǽne, Homl. Th. ii. 506, 18. 'Búton ðú mé sylle sum ðinra þinga' . . . se apostol cwæð : 'Hafa mínne stæf,' 416, 34. Hú Wulfgyð gean hire þinga æfter hire fordsíðe, Chart. Th. 563, 3. Ðingum *muneribus*, Mt. Kmbl. p. 14, 2 : Lk. Skt. Lind. 21, 4. Ða teúþan sceattas ge on lande ge on óþrum þingum ge on óþrum gestreónum, Blickl. Homl. 51, 8. Hié mid miclum þingum hámweard fóran *ingentem praedam ad classem devexerunt*, Ors. 4, 6; Swt. 176, 27. Ðe ealle his þing bewiste *qui praeerat omnibus quae habebat*, Gen. 24, 2. Ðinga ł geafa *munera*, Mt. Kmbl. Lind. 2, 11. Ða felarícan bróhton micele ðing, Homl. Th. i. 582, 14. Hé hæfde ǽr his ðing þearfum gedǽlede, ii. 500, 24. Hé becwæð his ðincg, Homl. Skt. i. 19, 211 : 18, 414 : 9, 41. Hé on swilce weorc áspende his ðing, ii. 31, 68. Wilt ðú syllan þingc ðíne? Coll. Monast. Th. 27, 15. (β) **a thing** to eat :—Eal ða wǽtan þing, and ða smerewigan, and eal swéte þing, Lchdm. ii. 210, 27. Mid wyrmendum þingum swilc swá pipor, 2 : 82, 4, 15. (2) **a thing** that is done, **an action, a proceeding, way** of conduct :—Án þing ðé is wana (*one thing remains for you to do*); gesyle eall ðæt ðú áge, Mk. Skt. 10, 21. Plyhtlíc þingc hit ys gefón hwæl, Coll. Monast. Th. 24, 21. Hé on axan and on duste licge : gif ic eów óþres ðinges bysene onstelle ðonne ágylte ic, Blickl. Homl. 227, 15. Wes ðú gemyndig Marian þinga *be mindful of Mary's conduct*, 67, 33. Æfter þissum þingum hyra fæderas dydon ðam wítegum *in these ways their*

fathers treated the prophets, Lk. Skt. 6, 23. Tó morgen déð Drihten ðás þing, Ex. 9, 5 : Mk. Skt. 11, 33 : 7, 8 : Ps. Th. 28, 4. Ealle ða þing ðe wé ofor his bebod gedydon, Blickl. Homl. 91, 16 : 131, 33. (3) **a thing** that happens, **an event, what takes place** :—Nǽnges þinges mǽre þearf nǽre ðonne his unriht yppe wurde *nothing better could happen than that his wickedness should become manifest*, Blickl. Homl. 175, 9. Þás þing ealle geweorþan sceoldan, 109, 8 : Homl. Th. ii. 538, 3. Ðara þinga (ðingana, Lind., Rush.) race ðe on ús gefyllede synt, Lk. Skt. 1, 1. Æfter þeossum þingum, Blickl. Homl. 95, 11. (4) where the word has much the same force as a cognate accusative, or where the meaning of the indefinite **þing** is determined by a verb :—Hé hine ǽlces þinges geclǽnsode ðe him mann on sǽde *he cleared himself of every charge that was brought against him*, Chr. 1022; Erl. 161, 37. Hú manigfeald þing Drihten geþrowode *what manifold suffering the Lord endured*, Blickl. Homl. 91, 11. Ic sceal þinga gehwylc þolian, Exon. Th. 270, 15; Jul. 465. Hine betellan æt ǽlc ðæra þinga þe him man on léde, Chr. 1048; Erl. 180, 12. Ealle ða þing ðe hálige men writon, Blickl. Homl. 133, 1. (5) **a thing, circumstance** ; in combination with an adjective nearly the same as the neuter of the adjective used as substantive, or as an abstract noun formed from the adjective :—Is ðæt earmlíc þing, ðæt his gebídan ne magon burgsittende, Met. 27, 16 : 28, 53. Þreálíc þing, réðe wíte, Cd. Th. 79, 28; Gen. 1318. Æfter sóðum ðincge *according to the truth*, Homl. Th. ii. 230, 14. Swá hit ágǽð mid sóðum ðincge *as it actually happens*, Homl. Skt. i. 17, 109. Wundorlíc ðingc, Lchdm. i. 112, 13. Manegu ðíglu ðing sindon tó smeágeanne, Past. 21; Swt. 153, 13. On ðǽm sélran þingum and on ðǽm gesundrum *in secundis rebus*, Nar. 7, 26. Næs nó on gesundum þingum ánum ac eác swylce on wiðearweardum þingum *not only in prosperity but also in adversity*, Blickl. Homl. 13, 7 : 35, 33 : Exon. Th. 337, 1 ; Gn. Ex. 58. Tó écum ðingum *to eternity*, Homl. Th. i. 16, 18 : 616, 21 : 568, 25. Tó sóþan þingon *truly*, Homl. Skt. i. 23, 736. Mid unrihtum þingum *per fraudem*, L. Ecg. P. ii. c. 13; Th. ii. 180, 23. Be fullum ðingum *fully*, Wulfst. 51, 11 : 57, 8. (6) **state, condition** :—Ealles folces þing byþ ðe betere æt ðám þýfðum *the condition of the whole people will be the better in the matter of the thefts*, L. Ath. v. 8, 9; Th. i. 238, 20. Séna hine gelóme; his þing biþ sóna sélre, Lchdm. ii. 344, 19 : Exon. Th. 378, 1 ; Deór. 9. Tó hwan ðínre sáwle þing (sið, Exon. Th. 368, 11) siððan wurde, Soul Kmbl. 39 ; Seel. 20. Cýð hwæt ðú ðæs tó þinge þafian wille *declare to which* (*life or death*) *thou wilt assent as thy condition*, Elen. Kmbl. 1212; El. 608. (7) **a thing, matter, subject** of consideration or enquiry :—Ánes þinges ic ðé wolde ácsian, Bt. 34, 1 ; Fox 134, 5 : Blickl. Homl. 117, 20. Uton ðás þing geþencean, 97, 1. (8) **concern, affair** :—Hé þearfendra þinga teolode *he attended to the concerns of the needy*, Ps. Th. 108, 30. Sió geornfulnes eorðlícra ðinga *terrena studia*, Past. 18; Swt. 128, 15. Hé mínre geðylde þingum wealdeþ *ab ipso est patientia mea*, 61, 5. On menniscum ðingum *in human affairs*, Blickl. Homl. 213, 6. Hé wolde beón embe his þincg, Homl. Skt. i. 6, 120. (9) **a cause, sake, account, reason**; in the phrase *for . . . þingum* :—Þinge *causam*, Wrt. Voc. ii. 20, 12. For ǽnegum þinge *quacumque ex causa*, Mt. Kmbl. 19, 3. Árís tó mínum þinge (*in causam meam*), Ps. Th. 34, 22. For hira þinge *because of them*, Deut. 28, 34. For ðan miceles blódes þinge, Lchdm. iii. 140, 30. For feós þinge *pecuniae causa*, L. Ecg. P. addit. 20; Th. ii. 234. 30. Búton forlegennysse þingum *excepta fornicationis causa*, Mt. Kmbl. 5, 32. For þisum þingum *igitur*, Bt. 26, 2; Fox 92, 19. For ðám þingum *for those reasons*, Homl. Skt. ii. 23 b, 12. For þrím þingum Hǽlend eode on wésten, Blickl. Homl. 29, 19. For moniga monna ðingum, Past. 5; Swt. 41, 22 : Ps. Th. 50, arg. For mínes wífes þingon *propter uxorem meam*, Gen. 20, 11 : 43, 30. For lósefes þingon, 39, 5 : Homl. Skt. i. 23, 304. For his sceatta ðingon *for the sake of his money*, Basil admn. 9 ; Norm. 52, 29. For ðæs áðes þingum *propter juramentum*, L. Ecg. P. ii. 29; Th. ii. 194, 12. For mínon þingon, for eówrum þingou *propter me, propter vos*, Jn. Skt. 12, 30 : 11, 15. For mínum ðingum, Bt. 7, 3; Fox 20, 3. For ðínum þingum, 7, 2; Fox 18, 28. Þurh þon þingum ðú eart eádig on écnesse, Nar. 46, 23. Incan *vel* þing *causas, res*, Wrt. Voc. ii. 130, 12. (10) **an object, a purpose** :—Gode wé cyrican betǽcaþ tó ðám þingum, ðæt cristene men ðǽrtó faran magan and ðǽr heora neóda tó Gode mǽnan and synna forgifenesse biddan, Wulfst. 278, 19. Hé weard man geboren tó ðám þingum, ðæt hé mid his ágenum feore mancynn álýsde of deófles gewealde and of helle wíte, 16, 11. þingum *purposely*, Exon. Th. 472, 10; Rä. 61, 14. (11) **a relation, respect** :—Hé ðone welegan wǽdlum efnmǽrne geðafaþ ǽlces þinges *in every respect*, Met. 10, 32, 50. Unmǽle ǽlces þinges, Exon. 21, 12; Cri. 333. Ne wéne hé nánes ðinges hine selfne beteran, Past. 17; Swt. 107, 16. Sóð hit sǽdon sumera ðinga (*in some respects*), Homl. Th. i. 190, 16 : 236, 11. Ðæt ic ðé geþeó þinga gehwylce, Hy. 4, 12. God hine gebletsode on eallum þingum, Gen. 24, 1. Wæs heó on eallum þingum þe eáþmóddre, Blickl. Homl. 13, 3. On ænigum þingum cræftig, 49, 28. On eallum ðingum gehýrsum, Bd. 2, 2 ; S. 514, 17. (12) **a condition** :—Hié bǽdon friðes, ac hit Scipia nolde him álíefan wið nánum óþrum þinge bútan hié him ealle hiera wǽpeno ágeáfen, Ors. 4, 13; Swt. 210, 20. (13) **a way, means, wise**; mostly in phrases, every *way*, by no

means, in any *wise*, etc. :—Ðæt ðæt nænig ðing ne gedafenade *quia nulla ratione conveniat*, Bd. 2, 12 ; S. 514, 38. Ne mihte hine nán man þurh ǽnig þing (*by any means*) áteón, Homl. Skt. i. 4, 194. Gif gē mīne lima þurh ǽnig þing gehǽlan magon, 5, 198 : Wulfst. 49, 7. Þurh ǽlc þing *by all means*, L. I. P. 2 ; Th. ii. 304, 13. Oeghwelce ðinga *omni modo*, Wrt. Voc. ii. 115, 50. Ðā wolde hē ǽlce þinga ðæt gyld ābrecan, Blickl. Homl. 221, 21. Hwæðer wēn wǽre ðæt wē ǽnige ðinga furþon ðæt eálond gesēcean mihte *si forte insulam aliquo conamine repetere possemus*, Bd. 5, 1 ; S. 613, 29. Ne hī his bēnum ǽnige ðinga geþafigean woldan *nor would they in any wise assent to his prayers*, 2, 1 ; S. 502, 14. Ne magon gē ǽnige ðinga lífes hlāfe onfón *nullatenus valetis panem vitae percipere*, 2, 5 ; S. 507, 20 : Beo. Th. 4738 ; B. 2374 : Homl. Skt. ii. 23 b, 721. Ðæt hē nāne ðinga ðæt ryht tō suīðe ne bodige *ut ne recta quidem nimie proferantur*, Past. 15 ; Swt. 95, 17. Hē nāne þinga beór ne drince, Lchdm. ii. 88, 11. Nǽnig þinga, L. E. I. 21 ; Th. ii. 406, 21. Nǽnige ðinga *nequaquam*, Bd. 2, 5 ; S. 507, 23 : 5, 6 ; S. 619, 8 : *nullatenus*, 1, 27 ; S. 495, 20. Hié nǽnige þinga ongeán lōcian ne mihton, Blickl. Homl. 203, 10 : Homl. Skt. ii. 23 b, 12 : Met. 10, 16. Mid nānum ðingum *by no means*, Past. 21 ; Swt. 167, 24. Mid ðām þingum *by those means*, Lchdm. ii. 208, 26. Ōðero ðingo *alioquin*, L. Skt. Lind. 14, 32. (14) *thing*, as in some*thing*, any*thing*, etc. :—Nān þing gēnes *nihil virens*, Ex. 10, 15. Gē ne biddaþ mē nānes þinges *me non rogabitis quicquam*, Jn. Skt. 16, 23. Mid ǽnige þinge *in aliquo*, Chart. Th. 422, 28. Ge on mete, ge on hrægle, ge on ǽghwylcum þinge, Blickl. Homl. 219, 30. Nyste ic nān þing þises *I knew nothing of this*, Gen. 21, 26 : Lk. Skt. 9, 36 : Mt. Kmbl. 26, 72. Styrigendlīces nān þincg findan, Homl. Skt. ii. 23 b, 735. Sum ðing miccles gebīcnodon ða tungelwītegan Homl. Th. i. 118, 20 : ii. 24, 19. Hæfð se mann ealra gesceafta sum ðing, i. 302, 19. Canst ðū ǽnig þing *scis tu aliquid?* Coll. Monast. Th. 20, 37. Beó ðæt þinga ðæt hit beó *be it what it may*, Btwk. 222, 8. Hūru þinga *praesertim*, Ælfc. Gr. 38 ; Zup. 238, 6. Ærost þinga *first of all*, Wulfst. 32, 9 : L. I. P. 10 ; Th. ii. 316, 11. Raþost þinga *at the earliest*, L. C. S. 24 ; Th. i. 390, 14. II. *a meeting, court* :—An medle oþþe an þinge, L. H. E. 8 ; Th. i. 30, 12. Hē Freán gesihð faran tō þinge (*the meeting held at the day of judgement*), Exon. Th. 57, 32 ; Cri. 927. Þing gehēgan *to hold a meeting*, Andr. Kmbl. 314 ; An. 157 : 1859 ; An. 932. Þing sceal gehēgan frōd wiþ frōdne, biþ hyra ferð gelīc, Exon. Th. 334, 19 ; Gn. Ex. 18. Ic wið Grendel sceal āna gehēgan ðing, Beo. Th. 856 ; B. 426. [*O. Sax.* thing *res* ; thing-hûs *court-house* : *O. Frs.* thing *res* ; *meeting, court* : *O. H. Ger.* ding *res, substantia, negotium* ; *concio, conventus, concilium* : *Icel.* þing ; *pl.* things, *articles* ; þing *an assembly, meeting* ; *Norweg.* stor-thing *parliament.*] v. breóst-, brýd-, cīpe-(cýpe-), cyric-, ge-, woruld-þing.

þingan ; *p. de* To *invite, address* :—Hē him thinget *invitat se*, Wrt. Voc. ii. 49, 39. Þinge *interpella*, Hymn. Surt. 127, 14. [Cf. *O. H. Ger.* dingen ; *p.* dingta *conducere, convenire* ; gi-dingen *appellare.*] v. ge-þingan ; un-þinged.

þingere, es ; m. I. *an advocate, intercessor* :—Þingere *advocatus*, Wrt. Voc. ii. 99, 39. Ðingere, 4, 48 : *interventor*, Rtl. 79, 36. Beón ðingere for óðerra scylde *intercessor fieri pro culpis aliorum*, Past. 10 ; Swt. 63, 20. Ðæs wordes (*Paraclete*) andgit is swā mon cweþe þingere, Blickl. Homl. 135, 33. Ðæt heó ūs sȳ milde þingere wið ūrne Drihten, 159, 33. Ic beó eówer þyngere tō Gode, Shrn. 155, 2 : Homl. Ass. 137, 701. Mid ða gife his ðingeres *gratia suo intercessori*, Bd. 4, 29 ; S. 608, 3. Mid þingere *cum advocato*, Wrt. Voc. ii. 79, 56 : *advocato*, interpellatore, Hpt. Gl. 466, 72. Hī noldon nǽnne þingere sēcan *defensorum operam repudiarent*, Bt. 38, 7 ; Fox 210, 13. Ða þingeras (*oratores*) þingiaþ nū ðæm ðe læssan þearfe āhton, Fox 208, 25. Þingeras wið Drihten, Cod. Dip. Kmbl. i. 114, 18 : Wulfst. 240, 10. II. *a priest*, who in his office intercedes for the people. v. þingian, I a :—Preóst *vel* þingere *clericus*, Wrt. Voc. i. 42, 24. [*O. Frs.* thingere : *O. H. Ger.* dingari *advocatus.*] v. cyrc-, fore-, ge-þingere.

þingestre, an ; *f. A female advocate* :—Þæt heó ūs beó þingestre tō ðam heofenlīce mǽgenðrymme, Homl. Ass. 137, 698.

þing-gemearc, es ; *n. Measuring* (*time*) *by events* (? cf. other compounds of *gemearc*, e. g. fōt-, geár-gemearc, where the first part determines the character of the measurement, measurement by feet, by years ; in the case of almost all such compounds it is an (adverbial) genitive that is found) :—Ðā wæs āgangen tū hund and þreó geteled rīmes swylce þrittig eác þinggemearces wintra *measuring by the events that had happened two hundred and thirty-three years would be counted as past, things had been going on for two hundred and thirty-three years*, Elen. Kmbl. 6 ; El. 3. Ðā wæs first āgán þinggemearces būtan þrim nihtum *things had gone on till there remained only three days of the allotted time*, Andr. Kmbl. 295 ; An. 148.

þingian ; *p.* ode. I. *to intercede, ask favour, supplicate, plead, intervene*, (1) absolute :—Ðǽr Satanus þingaþ, Cd. Th. 292, 28 ; Sat. 447. (2) *to intercede* for a person (*dat.* or for *with dat. or acc.*) :—Ic for mīnes Godes hūse þingie, Ps. Th. 121, 9. Ða þingeras þingiaþ ðǽm ðe læssan þearfe āhton ; þingiaþ ðǽm ðe man yfiaþ, and ne þingiaþ ðǽm ðe ðæt yfel dóþ *oratores pro his, qui grave quid perpessi sunt, misera-*

---

tionem judicum excitare conantur, Bt. 38, 7 ; Fox 208, 26. Ic (*Christ*) eów þingade, ðā mē on beáme beornas sticedon, Cd. Th. 296, 29 ; Sat. 509. Him (*himself*) ðingode David, and tō Drihtne gebæd, Ps. C. 26 : Elen. Kmbl. 985 ; El. 494. Gif hē wyle him sylfum þingian *si pro seipso supplicare velit*, L. Ecg. P. iv. 62 ; Th. ii. 222, 25. Hwæðer his māgas him fore þingian willon *num amici ejus pro eo intervenire vellent*, L. Ecg. P. addit. 29 ; Th. ii. 236, 32. (3) *to ask* for (*for with acc.*) a person that some favour may be granted (*clause*) :—Ðæt hē sceolde for hī ðingian (*supplicatu obtineret*) ðæt hī ne ðorftan fēran, Bd. 1, 23 ; S. 485, 36. (4) *to intercede, plead* before a person :—þinga for ðeódne ǽr ðam seó þráh cyme, ðæt hē ðec āworpe of woruldrīce, Cd. Th. 252, 33 ; Dan. 588. (5) *to intercede* for a person (*dat.* or for *with dat. or acc.*) to or with another (*tō, mid, wið*) :—Ðā spæc ic him fore and þingade him tō Ælfrēde cinge. Ðā lȳfde hē ðæt hē mōste beón ryhtes wyrðe for mīnre forspǽce, Chart. Th. 169, 30. Wǽre þearf ðæt him mon þyngode tō ðam rīcum, Bt. 38, 7 ; Fox 208, 29. Se ðe biþ ðone monn ðæt him ðingie wið óðerne ðe hē biþ eác ierre *cum is, qui displicet, ad intercedendum mittitur*, Past. 10 ; Swt. 63, 12. Hū mæg ǽnig man tō his hlāforde óðrum þingian, gif hē his hlāforde sylf hæfð ābolgen, L. I. P. 21 ; Th. ii. 332, 5. Earmum ðingian tō ðam rīcan, Homl. Th. ii. 558, 2. Tō ðingienne þiódum sīnum wið ðane Sceppend, Ps. C. 7. (6) *to make intercession* to (*tō*) a person that something may be granted (*clause*) :—Þingode Dauid tō Dryhtne, ðæt . . . , Ps. C. 146. I a. referring to intercession to the Deity. v. þingere, II :—Ða ðe on heofenum syndon, hī þingiaþ for ða ðe ðyssum sange fylgeaþ, Blickl. Homl. 45, 36. Swā oft swā hig clypiaþ tō Criste, and for folces neóde þingiaþ, L. C. E. 4 ; Th. i. 362, 4. Ic for ðē þingode, Homl. Skt. i. 5, 416. Būton sum hālga mē þingie tō ðam Hǽlende, ii. 26, 255. Ðingige, Homl. Th. ii. 518, 34. Hē cleopie tō Godes hālgum, and bidde ðæt hig him tō Gode þingien, L. E. I. 23 ; Th. ii. 420, 10. Hū dearr hē ðingian óðrum monnum, and nāt hwæðer him selfum geðingod bið *quomodo aliis veniam postulat qui, utrum sibi sit placatus, ignorat?* Past. 10 ; Swt. 63, 9 : Homl. Th. ii. 388, 4 : 528, 15 : i. 174, 10 : L. I. P. 21 ; Th. ii. 332, 6. For heora campwered gebiddan and tō Gode ðingian *ad exorandum Deum pro milite*, Bd. 2, 2 ; S. 503, 40. For hine ðingian and for sibbe his ðeóde *ad supplicandum pro pace gentis ejus*, 3, 24 ; S. 556, 43. Ðæt hī for mīnum untrumnessum ðingian mid ða upplīcan ārfæstnesse *ut pro meis infirmitatibus apud supernam clementiam intervenire meminerint*, pref. ; S. 472, 35. II. *to make terms, settle*, (1) absolute :—Āge hē þreora nihta fierst him tō gebeorganne, būton hē þingian wille, L. Alf. pol. 2 ; Th. i. 62, 2. (2) *to settle* a dispute :—Siððan ic ða fǽhðe feó þingode, Beo. Th. 945 ; B. 470. Feorhbealo feó þingian, 315 ; B. 156. (3) *to settle* the terms of an agreement, *to agree* that . . . :—Būton hiora hwæðer ǽr þingode ðæt hī ðit āngylde healdan ne þorfte, L. Alf. pol. 19 ; Th. i. 74, 11. (4) *to settle* with (*wið*) a person, *to come to terms, be reconciled*. v. (6) :—Heom man raþe ðæs wið þingode soon *after people came to terms with them*, Chr. 1001 ; Erl. 136, 32. Hē sceal þingian wið ðone ðe hē ābylgþ *debet reconciliari ei quem offendebat*, L. Ecg. P. addit. 19 ; Th. ii. 234, 27. (5) *to make terms* for :—Be ðǽm ðe for ordāle ðingiaþ. Gif hwā þingie for ordāl, þingie on ðam ceápgilde, and nǽht on ðam wīte, L. Ath. i. 21 ; Th. i. 210, 15. Weorpe ðæt neát tō honda oþþe fore þingie *let the beast be handed over, or terms settled for it*, L. Alf. pol. 24 ; Th. i. 78, 10. (6) *to make terms* for a person (*dat.* or for) with (*wið*) another, *to reconcile*. v. (4) :—Gif gesīðcund mon þingige wið cyning for his inhīwan, oððe wið his hlāford for þeówe, L. In. 50 ; Th. i. 134, 2. Þinga þē wið God *concilia tibi Deum*, L. Ecg. P. iv. 66 ; Th. ii. 226, 17. Ðæt wē ūs beþencan and wið God sylfne þingian, Wulfst. 166, 35. (6 a) where the person with whom is not stated :—Ne sié him nō ðý þingodre *none the more shall the case be settled for him*, L. In. 22 ; Th. i. 116, 12. III. *to settle to do something, to determine*. v. ge-þingan, geþingian, II :—Hū hē him on ðás world þingian ongan *how he settled for himself to come into this world*, Blickl. Homl. 105, 8. IV. *to speak, discourse* :—Ðē ða wordcwydas wittig Drihten on sefan sende ; ne hȳrde ic snotorlīcor, on swā geongum feore, guman þingian, Beo. Th. 3691 ; B. 1843. IV a. *with geán, ongeán, wið, to address, accost*. v. þingian :—Him brego engla geán þingade, Cd. Th. 62, 5 ; Gen. 1009. Iudas hire ongeán þingode, Elen. Kmbl. 1214 ; El. 609 : 1330 ; El. 667 : Exon. Th. 116, 20 ; Gū. 210. Hyre se wræcmæcga wið þingade, 258, 5 ; Jul. 260 : 268, 9 ; Jul. 429 : Andr. Kmbl. 612 ; An. 306 : 1264 ; An. 632 : Elen. Kmbl. 154 ; El. 77. [Do we mid ure weldede þingen us wið ure helende, O. E. Homl. ii. 43, 30. To þingen uss wiþþ ure Godd, Orm. 8997. O. Sax. thingôn : Hie gêng im wið thena heritogon mahlian, thingôn wið thena thegan kēsures, Hēl. 5725. O. Frs. thingia *placitare* : O. H. Ger. dingôn *concionare, judicare, disceptare, pacisci* : Ger. dingen *to bargain for, agree on* : Icel. þinga *to hold a meeting ; to consult about, discuss.*] v. for-, fore-, ge-, óð-þingian.

þingiend. v. fore-þingiend.

þing-leás ; *adj. Exempt* :—Unscyldigo and ðingleáso from ðissum synne *innocentes et immunis ab hoc crimine*, Rtl. 114, 7. [O. Frs. thing-lôs.]

þing-ræden[n], e; *f. Intercession, advocacy, pleading, intervention, mediation*, (1) in a general sense:—Đa apostoli hí ástrehton æt đæs ealdormannes fótum, biddende đæt đa hǽđengildan nǽron for heora intingan ácwealde ... Đá cwæđ se ealdorman: 'Wundor mé đincþ eówer đingræden,' Homl. Th. ii. 484, 14. God heora synne đurh his (*Job's*) đingrǽdene forgeaf, 458, 4: 292, 1. Gif đú geþafian nelt þingrǽdenne (*pleading on behalf of a lover*), Exon. Th. 250, 13; Jul. 126. (2) of intercession to the Deity. v. þingian, I a:—Mid đínre (*Stephen's*) þingrǽdene *tuo interventu*, Hymn. Surt. 46, 24. Þissere for þingrǽdene *hujus obtentu*, 139, 27. Þurh his hálgena þingrǽdene, Chart. Erl. 231, 28. Þurh his móder đingrǽdene, Homl. Th. i. 450, 26. Þa þingrǽdene for đam folce, ii. 536, 11. Mid đínum (*St. Andrew's*) þingrǽdenum *tuis intercessionibus*, Hymn. Surt. 126, 8. Þurh heora menigfealdan þingrǽdena, Homl. Th. i. 556, 19. v. fore-þingræden.

þing-stede, es; *m. A place where a meeting (v. þing, II) is held*:—On đam þingstede (*in the place to which Christ had summoned his disciples to speak with them for the last time*). Cf. tó Bethauia þeóden his þegna gedryht gelađade; hý đæs láreówes word ne gehyrwdon, hyra sincgiefan, 29, 2-9; Cri. 456), Exon. Th. 31, 17; Cri. 497. Ic gefrægn leóde tósomne bannan ... Đá wæs tó đam þingstede þeód gesamnod, Andr. Kmbl. 2197; An. 1100. [An that hús innan, þǽr Pilatus was an thero thingstedi, Hél. 5307. *O. H. Ger.* ding-stat *forum, conciliabulum: Icel.* þing-staðr *place where a þing is held*.] v. next word.

þing-stów, es; *f. A place of meeting, a public place*:—*Compitum*, i. villa vel þingstów vel þrop, Wrt. Voc. ii. 132, 55. In sprēc[stów] t in đingstów *in foro*, Mt. Kmbl. Lind. 20, 3. From đingstówe (-stów, Lind.) *a foro*, Mk. Skt. Rush. 7, 4. v. geþing-stów, *and preceding word.*

þingung, e; *f. Intercession, intervention, mediation*:—Þingunge *interventu*, Wrt. Voc. ii. 111, 28 : 48, 76. Hí on friþe wunedon þurh đære cwéne þingunge, Homl. Ass. 101, 313. ¶ *Especially intercession to the Deity*:—Đæt ic mid eallum đone wæstm árfæstre đingunge geméte *ut apud omnes fructum piae intercessionis inveniam*, Bd. pref.; S. 472, 39. Hé đære eádigan Marian fultumes and đingunge bæd, Homl. Th. i. 448, 19. Gif wé for synfullum mannum gebiddaþ and hí đære đingunge unwurđe synd, ii. 528, 12. Biddaþ eów þingunge æt đysum martyrum, i. 88, 33: ii. 110, 30. Eádges Ióhannes đincgunge (*intercessione*), Rtl. 46, 30: 51, 16. Ús tó þingunge, Chart. Th. 240, 24. Đurh đa bróþorlícan đingunge *per intercessionem fraternam*, Bd. 4, 22; S. 592, 22: 5, 19; S. 640, 42: Homl. Th. i. 76, 22. Đǽr nǽnige þingunga ne beóþ, Blickl. Homl. 95, 30. Wé biddaþ þingunga æt hálgum mannum, Homl. Th. i. 174, 9. Mid hira đingengum, Past. 10; Swt. 63, 15. Þurh bisceopes þingunga *ex episcopi interventu*, L. Ecg. P. i. c. ix; Th. ii. 170, 17. v. fore-, ge-þingung.

þinne, þió *a slave*, *þig, pron.* (Jn. Skt. Lind. Rush. 4, 5), þio-, þió-, þióen. v. þynne, þeów, se, þeo-, þeó-, þeówen.

þír *a female servant*:—Đír t sió đíguen (đír t đegnen, Rush.) *ancilla*, Jn. Skt. Lind. 18, 17. [*The Scandinavian form* þýr?]

þirda, þirding, þirel, þirlian. v. þridda, þridding, þyrel, þyrlian.

þirran, þierran *to dry, wipe*. [*O. H. Ger.* derren *torrere, exsiccare*: ar-derren *arefacere: Icel.* þerra *to dry, wipe dry*. Cf. *Goth.* ga-þairsan *to wither*.] v. á-þierran.

þirsce-flór *a threshing-floor*:—Hig cómon tó đære þirsceflóre *venerunt ad aream*, Gen. 50, 10. v. þerscel-flór.

þirscwald, þis, þísl, þislíc. v. þerscold, þes, þíxl, þyslíc.

þistel, es; *m. A thistle*:—Þistel, thistil *cardu[u]s*, Txts. 47, 384. Đystel, Wrt. Voc. i. 79, 56. Þistel *carduus*, 31, 53. Se onscunienda þystel *carduus orrens*, ii. 22, 43. Se unbráda þistel *scolimbos*, i. 69, 12. (Se unbráde thistel, Lchdm. iii. 305, col. 1. Brád thistle *erithius*, 302, col. 1.) Þúfe þistel *sow thistle*, Lchdm. ii. 312, 20. Genim đæs scearpan þistles moran, 314, 11. Þistles blóstm *thistle-down*; *pappus*, Wrt. Voc. i. 32, 23. Þistlum *card[u]is*, ii. 128, 63. ¶ The word is found in compounds which are names of places. Þistel-beorh, Cod. Dip. B. iii. 396, 33: þistel-mere, Cod. Dip. Kmbl. iii. 82, 15: þistel-leáh, iv. 49, 2. [*O. H. Ger.* distil: *Icel.* þistill.] v. þú-, wudu-þistel.

þistel-geblǽd *a blister caused by the prick of a thistle*:—Wiđ þorngeblǽd, wiđ þys[tel]geblǽd, Lchdm. iii. 36, 22.

þistel-twige, an; *f. A goldfinch or some other bird that eats thistledown*. [In E. D. S. Pub. Bird Names, p. 58, *thistle-finch* is given as a name of the goldfinch, and Halliwell quotes: 'Carduelis a linnet, a thistlefinch' (1585). Cf. *O. H. Ger.* distil-finco *carduelis*: *Ger.* distel-fink *a linnet: O. H. Ger.* distil-ziu *carduelus*:]—Þisteltuige, distiltige *cardella*, Txts. 47, 381. Þisteltwige, Wrt. Voc. i. 281, 19.

þístra, þrístra *a trace (?), part of an animal's harness*:—Þrístra conjuncta, Wrt. Voc. ii. 136, 34. Þrístra, i. 16, 9. [For similar double form cf. þeóstru, for meaning cf. (?) *Goth.* þinsan: *O. H. Ger.* dinsan *trahere*: 'Bavarian dünsel *a twisted withy or other thin branch, used to bind rafts of wood to the shore*,' Cod. Dip. Kmbl. iii. xlii; *and see Du Cange coniuncta*.]

þístru, þiú (Mt. 24, 15). v. þeóstru, se.

þiustra, Wrt. Voc. ii. 100, 18; *according to form the word might*

belong to þeóstru, *but it glosses* ambulas, *the meaning of which, according to the dictionary*, is endive or chicory.

þíwan, þíwen, þíwracan. v. þeówan, þeówen, þeó[w]-wracu.

þíxl, þísl, e; þísle, an; *f.* I. *a beam* or *pole of a waggon*; temo: used, also, like *temo*, to denote a constellation, the Bear:—Wǽnes đísl (wægne þíxl (wægnes?), 100, 72) *archtoes*, Wrt. Voc. ii. 7, 23. Þísl *temo* vel *arctoes*, i. 16, 24: *themon* (in a list 'de plaustris'), 284, 46. Þístle *temo* (*Wülcker prints* þísle *themon*, Gl. 295, 14), 66, 53. Díxl *arquamentum*, Txts. 109, 1147. Tunglu đe wé hátaþ wǽnes đísla, Bt. 39, 3; Fox 214, 19. Án đara tungla woruldmen hátaþ wǽnes þísla, Met. 28, 10. Þíxlum, díxlum, díslum *temonibus*, Txts. 101, 2007. II. *a pole* (?):—Of đære ác in đa heortsole; of đære sole in đa þísle; of đære þísle eft in đa mýđan, Cod. Dip. Kmbl. iii. 380, 6. [*O. L. Ger.* thísla: *wk. temo: Du.* dissel *axle-tree*; dissel-boom *beam* or *pole of a carriage: O. H. Ger.* díhsel, díhsila, dísala; *f. temo: Ger.* deichsel; *Icel.* þísl; *Swed.* tistel-stång *coach-pole.*]

þó, þóae. v. þohe.

þocerian; *p.* ode *To run to and fro, run about*:—Þoceraþ *cursat, currit, cursitat*, Wrt. Voc. ii. 137, 53. Þocerodan (*vitae late praeconia*) *cursant*, 95, 19: 19, 65. Sitte him đín mód on mínum hrǽdwǽne, þocrige him on minne weg *mea semita, meis vehiculis revertaris*, Bt. 36, 1; Fox 174, 1. [Cf. *Icel.* þoka *to move*.]

þoddettan; *p.* te *To push, strike, batter*:—Đa deóflu þoddetton đa earme sáwle and héton hý út faran of đam líchaman, Wulfst. 235, 15. Þoddetton *pulsent*, Germ. 399, 264. v. þyddan.

þoden, es; *m. A violent wind, a whirlwind; also, a whirlpool; turbo*:—Þoden *alcanus* (l. *altanus*), Wrt. Voc. ii. 100, 3: *altanus*, i. 17, 34. Đoden *turbo*, Ælfc. Gr. 9, 3; Zup. 37, 10. Þoden đa nán ne tócwíse oþþe worigende tówurpon windas *turbo quam nullus quatit aut vagantes diruunt venti*, Hymn. Surt. 142, 26. Cumendum swá þoden tó tóstæncanne mé *venientibus ut turbo ad dispergendum me*, Cant. Ab. 14. Đonne sió geornfulnes eorđlícra đinga ábisgaþ đæt ondgit and áblent đæs módes eágan mid đære costunge đæm folce, suǽ suǽ dust déđ đæs líchoman eágan on sumera mid đodene (đodne, Hatt. MS.) *dum pastoris sensus terrena studia occupant, vento tentationis impulsus ecclesiae oculos pulvis caecat*, Past. 18; Swt. 128, 17. Hér wǽron rēđe forebēcna cumen ofer Norđhymbra land . . . đæt wǽron ormēte þodenas and ligrescas, Chr. 793; Th. 101, 5, col. 1. Þurh đæs windes blǽs, đe swýđlíce đa heánnyssa đæs roderes scecđ mid his þodenum, Anglia viii. 320, 34. Lageflódum þodenum *ceruleis turbinibus*, Wrt. Voc. ii. 130, 38. [Swa þode (þodde, 2nd MS.) on felde þenne he þat dust heȝe aȝiueð from þere eorđe, Laym. 27645.]

þóe. v. þohe.

þoft (?), e; þofte, an; *f. A rower's bench*:—Scipsetl *transtra*, þofta (þoftan?) *trastra* vel *juga*, Wrt. Voc. i. 48, 15. Þoftan *transtra*, 56, 41: 63, 43. (All three occur in lists 'de navibus.') [*Du.* doft; *f. a rower's bench: Icel.* þopta; *wk. f. a rowing bench.* Halliwell gives thoft-fellow a fellow-oarsman.]

-þofta, -þoftian. v. ge-þofta [*Icel.* þopti *a bench-fellow*], ge-þoftian.

þoft-ræden[n], e; *f. Fellowship*:—Đú hopast đæt đú hæbbe đoftrǽdene tó đam áwyrigedan deófle, đonne đú bǽde đæt hé đé ásende his englas tó mínre đære, Homl. Th. ii. 416, 14. v. ge-þoftræden.

þoftscipe, es; *m. Fellowship*:—Sum bróđor . . se him wæs on gástlícum (-re, MS.) þoftscipe geþeóded, Guthl. 10; Gdwin. 52, 5: 14; Gdwin. 62, 2. v. ge-þoftscipe.

þohe, þóe, þó; *gen.* þón; *f. Clay*:—Thóhae, thóae *argilla*, Txts. 36, 3. Đó, Wrt. Voc. ii. 6, 16. Þóe *creta*, 136, 78. [*Goth.* þáhó πηλός: *O. H. Ger.* dáha *argilla, testa: Ger.* thon: *Icel.* þá *muddy ground*.] v. next word.

þóhiht, þóiht; *adj. Clayey*:—Đóihte *argillosa*, Wrt. Voc. ii. 6, 15.

þóht, es; *m. Thought, mind*:—Pund gefe of đon is đóht monnes *pondus gratiae, inde est sensus hominis*, Rtl. 192, 23. Suǽ líchomes suǽ đóhtes (*anime*) hǽlo, 99, 13. Háles đóhtes *sane mentis*, Mk. Skt. Lind. 5, 15. Mid þóhtes wilnunga besmiten *desiderio cogitationis coinquinatus*, L. Ecg. C. 5; Th. ii. 138, 15. In alle đóht đínne *in tota mente tua*, Mt. Kmbl. Lind. 22, 37. In đon ilco đócht *in eundem sensum*, p. 9, 18. Unstađolfæstnis đóhta *instabilitas mentium*, Rtl. 192, 21. [*Icel.* þóttr, þótti.] v. ge-þóht.

þol. v. þoll.

þole-byrde; *adj. Bearing patiently, patient, long-suffering*:—Þolebyrde mann *patiens homo* . . . Wer þolebyrde *vir patiens*, Scint. 13, 11, 13. [He beđ þoleburde, O. E. Homl. ii. 79, 25.]

þolebyrdness, e; *f. Patience, long-suffering, endurance*:—Tó þolibyrdnysse þrowunga strange *ad tolerantiam passionum fortes*, Scint. 3, 8. [On giwer þoleburdnesse *in patientia vestra*, O. E. Homl. ii. 79, 9.]

þole-mód; *adj. Of a patient disposition, patient, long-suffering*:—Þolemód *longanimis*, Hpt. Gl. 437, 43. Heó wæs þolemód and gestǽđþig on hire gebǽran, and ne geseah hí nán man yrre, Homl. Ass. 127, 367. Þolemód on wiþerweardum þingum *patient in adversity*, R. Ben. 26, 18. His mon fandige hwæđer hé þolemód (þolo-, Wells Frag.) sý and geþyldig *probetur in omni patientia*, 99, 4. Þolomód *patiens*, Scint.

8, 12, 14. Ðæt se mann beó geðyldig and ðolomód (þol-, MSS. U. D.), Homl. Skt. i. 16, 335. Geðyldig and ðolmód, 17, 55. On ðære écan worulde, ðe gewelgaþ ða þolmódan, Homl. Th. ii. 456, 2. [Gordoille wes þolemod, Laym. 3141. Katerine wes þuldi & þolemod, Kath. 173. Þolemod is þe þet þuldeliche abereð wouh þet me deð him, A. R. 158, 4. Cf. Icel. þolin-móðr patient : Dan. taal-modig.]

þole-mód (?) f. patience :—Patientia ðæt is ðolmód, Homl. Skt. i. 16, 334, MS. D. [Habbe we edmodnesse and þolemod, O. E. Homl. i. 69, 266.] v. next word.

þolemódness, e; f. Patience, long-suffering, endurance :—Patientia ðæt is geðyld and þolmódnys gecwæden, Homl. Skt. i. 16, 334. Se geþyldiga man mid his þolmódnysse his sáwle gehylt, ii. 28, 146. Ða getreówfullan ealle lífes wiðerweardnesse forþyldigian scylun, be hiora þolemódnesse (þolo-, MS. T.) is þus áwriten, R. Ben. 27, 8, 13. Wurðigan ða gódan þeáwas . . . geþyld and þolemódnysse, Guthl. 2 ; Gdwin. 18, 16. [Þolemodnesse and edmodnesse, of mild and meek heart, A. R. 158, 2. Þolemodnesse, H. M. 41, 10 : Ayenb. 68, 4.]

þolian; p. ode To thole (still used in some dialects), suffer, endure. I. to suffer what is evil, punishment, reproach, illness, grief, etc. :—Hwílon forlidenesse ic þolie aliquando naufragium patior, Coll. Monast. Th. 27, 1. Þolige, Exon. Th. 499, 18 ; Rä. 88, 17. Hé þreánýd þolaþ, Beo. Th. 573 ; B. 284. Synfull tóþum torn þolaþ peccator dentibus suis fremet, Ps. Th. 111, 9. Þoliaþ wé þreá on helle, Cd. Th. 25, 5 ; Gen. 389. Hí hosp þoliaþ contumeliam toleravit, Hpt. Gl. 506, 25. Þú þoladest ferres, i. sustinebas, contuleras, Wrt. Voc. ii. 147, 55. Ðú þolades mægenearfeþu, Exon. Th. 86, 19 ; Cri. 1411. David his ehtnesse ðolade, Past. 28 ; Swt. 197, 17. Hé þeównýd þolode, Cd. Th. 122, 21 ; Gen. 2030. Hí bryne þolodon, Rood Kmbl. 296 ; Kr. 149. Ða ðe elþeódigra edwít þoledon, hæþenra hosp, Judth. Thw. 24, 30 ; Jud. 215. Þoledan and þrowedan luebant, Wrt. Voc. ii. 53, 19. Þeh ðú drype þolie, Andr. Kmbl. 1910 ; An. 957. Gif wíf ðone fléwsan ðæs wætan þoligen, Lchdm. i. 308, 2. Fela þinga þolian fram yldrum multa pati a senioribus, Mt. Kmbl. 16, 21. Þoligean, Lk. Skt. 24, 26. Ðæt micle morð þolian, Cd. Th. 40, 18 ; Gen. 641. Þeówdóm þolian, 135, 9 ; Gen. 2240. Hýndo þolian, 198, 18 ; Exod. 324. Torn þolian, Beo. Th. 1669 ; B. 832. Wítu þolian, Andr. Kmbl. 2828 ; An. 1416. Ðæt ðam weligan wæs weorc tó þolianne, Exon. Th. 276, 21 ; Jul. 569. Hé lét, torn þoliende, teáras geótan, 165, 15 ; Gú. 1029. Syle ðam þoligendan ðicgean . . . ðú hine gelácnast wundorlíce, Lchdm. i. 220, 17 : 17 : 188, 1. Mid tóðon torn þoligende, Judth. Thw. 25, 21 ; Jud. 272. I a. to suffer, undergo, submit to discipline, treatment :—Se ðe ðysne lǽcedóm þolaþ, Lchdm. i. 300, 20. Ðæt wíf ðe on blódryne wæs fram manegum lǽcum fela þinga þolode, Mk. 5, 26. Tó ðolienne ðínne willan, Ps. C. 90. I b. of things which are used to do hard work :—Seó ecg geswác þeódne æt þearfe ; þolode ǽr fela hondgemóta, Beo. Th. 3055 ; B. 1525. II. to suffer a person, bear with, tolerate a condition of things, let come to pass :—Swá lange swá ic mid eów beó, swá lange ic eów þolige (ðola, Lind. : ðolo, Rush.), Mk. Skt. 9, 19. Þolie (ðola, Lind. : ðolo, Rush.), Lk. Skt. 9, 41. Ðæt is micel wundor, ðæt hit God wolde þolian, ðæt wurde þegn swá monig forlǽdd, Cd. Th. 37, 29 ; Gen. 597. III. to suffer lack or loss of something (gen.), to lose what one has, to fail to get what one desires ; in many cases the loss or failure is the result of wrong either done or suffered by the subject of the verb, to forfeit, be (wrongfully) deprived of :—Ic ðolige sumes ðinges careo, Ælfc. Gr. 26, 2 ; Zup. 154, 16. Ic ðolige mínes feós careo mea pecunia, ðoligende his þinges carens sua re, 41 ; Zup. 250, 11. Ðonne þolie ic ðus miceles ðæs ðe míne foregengan hæfdon in that case I shall be (unfairly) deprived of thus much of what my predecessors had, Cod. Dip. Kmbl. iii. 327, 16. Gif ðú Drihten forgitst, ðú ðolast ðære écan méde, Homl. Th. i. 140, 32. Ðolaþ carebit (benedictione), Prov. 20, 21. Gé þoliaþ ðæs ðe eów God behét for eówre ungehírsumnisse, Deut. i. 40. Hý (evil spirits) háma þoliaþ, Exon. Th. 115, 22 ; Gú. 193. Ic þolade gódes ealles, 457, 16 ; Hy. 4, 84. Hé férde swá swá his forcúða fæder, and his lífes ðolode and his lǽnan ríces, Homl. Skt. i. 18, 231. Hé (Job) hæfde his wíf, þeáh hé his bearna þolode, ii. 30, 204. Þolade caruerit, Wrt. Voc. ii. 23, 83. Ne forgit ðú deáð, ðý lǽs ðú þolie ðæs écan lífes, Prov. Kmbl. 17. Þeáh God wille habban hys willan tó forlǽtan, and hé ðæs eft þolige, Ors. 1, 5 ; Swt. 34, 36. Þolige hé his wǽpna and his ierfes, L. Alf. pol. 1 ; Th. i. 60, 14 : L. Edg. i. 4 ; Th. i. 264, 15. Þolie se þeówa his hýde oþþe hýdgyldes . . . þolie se frigea his freótes, L. In. 3 ; Th. i. 104, 4, 6. Þolige se déma, ðe óðrum wóh déme, á his þegenscipes, L. Edg. ii. 3 ; Th. i. 266, 17 : Chart. Th. 606, 30 : Homl. Th. ii. 94, 33. Hefonríces þolian, Cd. Th. 40, 3 ; Gen. 633 : Exon. Th. 402, 8 ; Rä. 21, 26. Blind sceal his eágna þolian, Gn. Ex. 39. III a. with a preposition :—Þolige hé be healfre ðære bóte, L. Alf. pol. 11 ; Th. i. 68, 19. IV. intrans. To hold out, exercise endurance, endure, not to give in :—of things, to last, continue to be serviceable :—Ic tó aldre sceal sǽcce fremman, þenden ðis sweord þolaþ, Beo. Th. 4992 ; B. 2499. Gif mín (an anchor's) steort þolaþ if my tail can stand the strain on it, Exon. Th. 398, 16 ; Rä. 17, 8. G[esǽlig ?] biþ ðæt, ðonne mon him

sylf ne mæg wyrd onwendan, ðæt hé ðonne wel þolige, 459, 16 ; Hy. 4, 117. Æt ðearfe þolian, unwáclíce wǽpna neótan, Byrht. Th. 140, 53 ; By. 307 : 137, 45 ; By. 201. [Goth. þulan to tolerate, endure : O. Sax. tholian, tholón (trans. acc. and gen., and intrans.) to suffer, endure, lose, hold out : O. L. Ger. tholón pati, sustinere : O. Frs. tholia : O. H. Ger. dolên, dolôn pati, sustinere, tolerare, luere : Icel. þola ; p. þolði.] v. á-, for-, ge-, mid-þolian.

þoligend, þoligendlíc, þolibyrdness. v. mid-þoligend, un-þoligendlíc, þolebyrdness.

þoll, es ; m. A thole or thowl, a peg in the side of a boat to keep the oar in place :—Þoll scalmus, Wrt. Voc. i. 63, 79. Thol, ii. 120, 15. [Tholle, cartepynne cavilla, Prompt. Parv. 492. Du. dol a thole : Icel. þollr a wooden peg ; esp. the thole of a row-boat : Dan. tol a thole ; tolle-gang a row-lock.]

þolle, an ; f. A frying-pan :—Hwer I þollan sartaginem, Hpt. Gl. 503, 16. v. fýr-þolle.

þol-mód, þolo-byrde, þolo-mód, þon, þonan, þonc, þon-écan, þonne, þonon. v. þole-mód, þole-byrde, þole-mód, þan, þanan, þanc, þan-écan, þanne, þanan.

Thómas Thomas :—Thómas án of ðám twelfon, Jn. Skt. 20, 24. Þómas, 26. Ðómas, 28. Thómas genédde, Apstls. Kmbl. 99 ; Ap. 50. Scé Thómas týd ðæs apostoles, Shrn. 155, 28. Nergend Thómase forgeaf éce ríce, Menol. Fox 444 ; Men. 223. Cf. Hé sǽde þómé (thómase, later MS.), Jn. Skt. 20, 27. Hé nýdde ðysne Thómam, ðæt hé weorðode sunnan deófolgild, Shrn. 156, 9 : Mk. Skt. 3, 18.

þoot, Txts. 64, 444. v. wóþ.

Þór the Scandinavian form of a name which in English is Þunor (-er), one of the gods, Thor :—Nú secgaþ sume ða Denisce men on heora gedwylde, ðæt se Iouis wǽre, ðe hý Þór hátaþ, Mercuries sunu, ðe hí Óðon namiaþ ; ac hí nabbaþ ná riht : for ðan ðe wé rǽdaþ on bócum, ge on hǽþenum ge on cristenum, ðæt se hetula Iouis tó sóðan is Saturnes sunu, Wulfst. 107, 8–13. Þór and Ówðen, ðe hǽðene men heriaþ swíðe, 197, 19. Fled (fleó ?) þór (? þr̄, MS.) on fyrgen hǽfde (fyrgenheáfde ?), Lchdm. iii. 54, 17. v. Þunor.

porch. v. þurh.

þorfa ; adj. Destitute, poor ; used as a substantive, a needy person :—Of ðorfum de egenis, Jn. Skt. Lind. 12, 6. [Icel. þurfi or þurfa wanting, in need of. Cf. Goth. ga-þaurbs continens.] Cf. þearfa.

þorfan ; p. te To need :—Ne ðorfeþ (-æþ, Lind.) non indiget, Jn. Skt. Rush. 13, 10. Ne ðo[r]feþ ða ðe hálo sint tó lǽce non egent qui sani sunt medico, Lk. Skt. Lind. 5, 31. Cf. þearfan, and see next word.

þorfend, es ; m. : þorfende ; adj. (ptcpl.) used substantively. A needy person, a poor person :—Wæs sum ðærfe I ðo[r]fond (mendicus) . . . wæs deád se ðorfendo (mendicus), Lk. Skt. Lind. 16, 20, 22. Ofer armne and ðorfend super egenum et pauperem, Rtl. 175, 33. Ðorfendo pauperes, Mk. Skt. Lind. 14, 7. Eádgo ða ðorfendo beati pauperes, Lk. Skt. Lind. 6, 20. Ic sello ðorfendum do pauperibus, 19, 8 : Mk. Skt. Lind. Rush. 14, 5. Ðorfendum I nǽfigum (ðarfendum, Rush.) egenis, Jn. Skt. Lind. 12, 5. Cf. þearfende, and see preceding and following words.

þorfendness, e ; f. Poverty, destitution :—In ðorfendnisse in paupertate, Rtl. 105, 11.

þorf-fæst ; adj. Useful :—Ne on eorðo ne in feltûne I on mixenne ðor[f]fæst is neque in terram neque in sterculinium utile est, Lk. Skt. Rush. Lind. 14, 35. Ðor[f]fæst utilis, Rtl. 192, 7. Sié ðor[f]fæsta ûs prosint nobis, 91, 27. [Cf. All þatt hemm wass þurrfe, Orm. 9628. Icel. þurf-samr helping.] Cf. þearf, V, þearf-líc, II, and next word.

þorf-leás ; adj. Useless :—Ðe ðor[f]leása ðegn inutilis servus, Mt. Kmbl. Lind. 25, 30. Ðor[f]leáse I som ðíor gefeóllon néh strǽt quaedam: ceciderunt secus viam, 13, 4. Esnas ðor[f]leáse (-leóse, Rush.) wé sindon servi inutiles sumus, Lk. Skt. Lind. 17, 10. Cf. þearf-leás, and preceding word.

þorh. v. þurh.

þorian (?) to dare :—Thorie dosmui (domui ?), Wrt. Voc. ii. 141, 82. [Icel. þora to dare.]

þorn, es ; m. A thorn, the prickle of a plant or a plant on which such prickles grow :—Þorn spina, Wrt. Voc. i. 33, 44 : 80, 22 : tribulus, 33, 45 : dumus, ii. 25, 70. On ða þyrnan westeweardes, ðær se mycla þorn stód, Cod. Dip. Kmbl. iii. 404, 13. Tó hafucðornæ ; of ðam þornæ on ðone brádan stán . . . on hælnes þorn ; of ðam þorne on done bróc, v. 348, 21. On weocan þorn ; of ðam þorne, vi. 92, 3. Ðornas sentes, Wrt. Voc. ii. 120, 28. Þornas, i. 33, 41 : 80, 19. Hí wundon cynehelm of þornum plectentes coronam de spinis, Mt. Kmbl. 27, 29 : Exon. 88, 27 ; Cri. 1446. Of ðæm hylle ðæt swá be ðǽm .IIII. þornan ; of ðǽm þornan be ðǽm heáfdon, Cod. Dip. Kmbl. iii. 263, 31. Þornas and brémelas spinas et tribulos, Gen. 3, 18. Sume feóllon on þornas ; and ða þornas weóxon and forðrysmudon ða, Mt. Kmbl. 13, 7 : Mk. 4, 7. Swá ðú bærne þornas fýre sicut ignis in spinis, Ps. Th. 117, 12. Áðió hé of lande ða þornas and ða fyrsas and ðæt fearn and ealle ða weód ðe hé gesió ðæt ðám æcerum derigen liberat arva fruticibus, falce rubos filicemque resecat, Bt. 23 ; Fox 78, 22 : Met. 12, 3. [The word is found in many

local names. v. Cod. Dip. Kmbl. vi. 341.] ¶ *The name of the letter* þ *was* þorn:—Þ byþ þearle scearp, Runic pm. Kmbl. 339, 13; Rún. 3. [*Goth.* þaurnus : *O. Sax. O. Frs. O. L. Ger.* thorn *spina, dumus*: *O. H. Ger.* dorn : *Icel.* þorn *a thorn; the name of the letter* þ.] v. appel-, brēmel-, gemǣr-, hǣg-, haga-, lûs-, mǣr-, pól-, pric-, set-, slāh-, þîfe-þorn, *and following words*.

þorn-geblǣd *a blister caused by the prick of a thorn*, Lchdm. iii. 36, 21.

þorn-grǣfe, an; *f. A thorn-copse*:—Andlang ðære þorngrǣfan, Cod. Dip. Kmbl. v. 148, 4.

þornig; *adj. Thorny, full of thorns.* v. þorn:—Se yrðling lufaþ ðone æcer ðe æfter ðornum and brēmelum genihtsume wæstmas āgifþ swiðor ðonne hē lusige ðone ðe ðornig næs, Homl. Th. i. 342, 8. Gehega þîne eáran mid þornigum hege, Wulfst. 246, 9. [*O. H. Ger.* dornig: *Ger.* dornig.]

þorniht; *adj. Thorny, full of thorns* (v. þorn) *or briars*:—Þorniht *senticosus*, Wrt. Voc. i. 33, 41. Tô ðæm þornihtan heáfodlonde, Cod. Dip. Kmbl. iii. 263, 32. On ðam þornehtan dûne, 421, 24. On ða þornihtan leáge, v. 389, 14. Ðæm ðornihtum *senticosis* (velut rosa senticosis exorta surculis, Ald. 18, 14), Wrt. Voc. ii. 77, 47. [*O. H. Ger.* dornohti *spinosus*: *Ger.* dornicht.]

þorn-rǣw, e ; *f. A row of thorn-bushes*:—On ða þornrǣwe, Cod. Dip. Kmbl. iii. 77, 28. On ða ealdan þornrǣwe, 199, 33, 34.

þorn-rind, e ; *f. The bark of a thorn-tree*:—Huutbeámes rinde and þornrinde gecnûa tô duste, Lchdm. ii. 52, 1.

þorn-stybb, -stubb, es ; *m. The stump of a thorn-tree*:—Tô ðæm þornstybbe; of ðam þornstybbe, Cod. Dip. Kmbl. v. 252, 28. Tô ðan þornstybbe, vi. 8, 33, 37. On ðonæ þornstub, v. 291, 11. On ðone þornstyb; of ðam stybbe, Cod. Dip. B. iii. 169, 33.

þorof. v. þeorf.

þorp, þrop, es ; *m.* Perhaps the idea at first connected with the word is that of an assemblage, cf. the use in Icelandic : Maðr heitir einnhverr . . . þorp ef þrír ero, Skáldskaparmál ; þyrpast *to crowd, throng*; þyrping *a crowd*: later the word may have been used of the assemblage of workers on an estate, and also of the estate on which they worked; all three ideas seem to be implied in one or other of the following glosses :—Tuun, þrop, ðrop *conpetum*, Txts. 53, 557: Wrt. Voc. ii. 15, 7. *Compitum* i. *villa vel* þingstôw *vel* þrop, 132, 56. Þrop *fundus*, i. 37, 51. The idea of an estate belongs to the word in Gothic : Þaurp ni gastaistald ἀγρὸν οὐκ ἐκτησάμην, Neh. 5, 16. In the end the meaning came to be *hamlet, village*, in which sense it remained for some time in English, e.g. : Ic Ædgar gife freodom Scē Petres mynstre Medeshamstede of kyng and of biscop, and ealle þa þorpes þe ðærto lín : ðæt is, Æstfeld and Dodesthorp and Ege and Pastun, Chr. 963; Erl. 121, 40. He com to Bethfage, swo hatte þe þrop, O. E. Homl. ii. 89, 13. Ther stod a throp . . . in which that poure folk hadden her bestes and her herbergage, Chauc. Cl. T. 199. Thorp, litell towne or thoroughfare *oppidum*, Prompt. Parv. 492. The word is now obsolete, but it remains in a great many local names, either alone or in composition ; though, as such names are found mostly in those parts of England which were affected by the Danes, its occurrence in them may be due rather to Scandinavian than to English influence. v. Leo, Anglo-Saxon Names of Places, p. 43 sqq. ; Taylor's words and Places, s.v. [*Goth.* þaurp: *O. Frs.* thorp, therp: *O. L. Ger.* thorp, tharp: *Du.* dorp: *O. H. Ger.* dorf *villa, vicus, praedium, oppidum, municipium* : *Icel.* þorp *a hamlet, village*.]

þost, es ; *m. Dung, ordure;* with this meaning *thoste* (according to a MS. glossary cited by Halliwell) is used in Gloucestershire:—Wyrc drenc of hwîtes hundes þoste, Lchdm. i. 364, 5. Bærn hundes ðost and gnîd smale, 7. Nim drîgne hundes þost, 11 : ii. 48, 8. [þost, thoste *stercus*, Ps. 82, 11. An horse thoste Rä. S. 237, 14. As a thost in the weie totreden, Wick. Ecclus. 9, 10. Ass uryne and swynes thost, Pall. 116, 348. Thoste or toord *stercus*, Prompt. Parv. 492. *O. H. Ger.* dost *stercus, coenum.*]

-þot. v. ge-þot. [Cf. *Icel.* upp-þot *a great stir*.]

þoterian; *p.* ode *To howl, wail, cry out*:—Þotraþ *clamat*, Wrt. Voc. ii. 21, 12. Geómriende hell þoteraþ *gemens infernus ululat*, Hymn. Surt. 84, 34. Gē wēpaþ and þoteriaþ *plorabitis et flebitis*, Scint. 167, 3. Hî ðotorodon swilce ôðre wulfas, Homl. Th. ii. 488, 27. v. þeótan.

þoterung, e ; *f. Howling, wailing, crying*:—Stefn wæs gehýred wôp and mycel þoterung (þoterung, MS. A.) *vox audita est, ploratus et ululatus multus*, Mt. Kmbl. 2, 18: Homl. Th. i. 80, 19. Ne āblinð grânung and þoterung (on helle), 68, 7. Geómerung and singal þoterung, Wulfst. 114, 27. Hē weóp swîðe biterlîce and hē feóll tô Iôhannes fôtum mid geómerunge and þoterunge, Ælfc. T. Grn. 18, 32. Hē symle clypode mid swîðlîcere þoterunge : 'And wâ dissere burhware,' Homl. Th. ii. 302, 12.

þôþer (-or, -r), es ; *m. A ball, sphere*:—Thôthur, thôthor *pila*, Txts. 87, 1584. Ðôþor, Wrt. Voc. ii. 68, 17. Þôðer, i. 86, 6. Þôþor, 287, 15. Ðôþer *pila vel sfera*, 39, 51. Þôþer *ballum*, ii. 125, 14. Ðû leornodest ðone cræft ðe wē hâtaþ geometrica ; on ðam cræfte ðû leornodest onn ânum þôðere oððe on æpple ātēfred, ðæt ðû meahtest be

ðære têsfrunge ongytan ðises rodores ymbehwirft . . . Ðû leornodest be ânre lînan wæs āwriten anlang middes ðæs þôþeres . . . Ðû secgst ðæt ðû ymbe ða lînan wite ðe on ðam þôðere ātēfred wæs . . . Ic wolde witan hweðer ðû eác wite ymbe ðone þôðer ðe seó lýne on āwriten is, Shrn. 174, 16–175, 1. Ðā āgan se cyngc plegan wið his geféran mid þôðere, and Apollonius yrnende ðone ðôðor gelǣhte, Ap. Th. 13, 1–3.

Thrâceas, þrâcie (?); *pl. The Thracians*:—Ðrâcia cyning, Met. 26, 22, 59, 7. Dorus Thrâcea cyning, Ors. 3, 11 ; Swt. 152, 3. *In other passages Latin forms occur*, Traci, Thraci :—Be westan ðære byrig sindon Traci, 1, 1 ; Swt. 22, 8. Hê wæs farende on Thraci and hié tô him gebîgde Thracas *domuit*, 3, 9; Swt. 124, 9: 4, 11 ; Swt. 204, 16. *Another form is* Trâciane; *pl.*:—Trâciana *Traciarium (provincias*, Ald. 64, 10), Wrt. Voc. ii. 85, 74. *The name of the country is given as* Trâcia, Thrâcia :—On Trâcia (Thrâcia, MS. C.) ðæm londe, Ors. 3, 7 ; Swt. 114, 15. Lysimachus befêng Thrâciam *Thracia Lysimacho data*, 3, 11 ; Swt. 142, 33.

þracian. v. ā-, au-, on-þracian.

þracu; *gen.* þræce; *f.* I. not in a bad sense, *power, force*:—Þracu (-a, MS.) wæs on ôre, heard handplega, hægsteald môdige, wîgend unforhte, Cd. Th. 198, 22 ; Exod. 326. Sigores tâcn wið þeóda þræce *a token of victory against the power of nations*, Elen. Kmbl. 369 ; El. 185. Se câsere hêht bannan tô beadwe, beran ût þræce . . . wǣron Rômware sôna gegearwod *the emperor bade give the summons to war, bade put forth their power* (?) . . . *At once were the Romans prepared*, 90; El. 45. Geceósan swā þrymmes þræce swā þrýstra wræce *to choose either the power of glory or the misery of darkness*, Exon. Th. 37, 14; Cri. 593. Oft wē ofersēgon þeóda þeáwas, þræce môdigra *the power of the proud*, 118, 12; Gû. 238. II. in a bad sense, *violence*:—Oft hî þræce rǣrdon . . . feóndscipe rǣrdon . . . hâlge cwelmdon . . . bærndon gecorene, Exon. Th. 243, 18; Jul. 12 : 262, 16; Jul. 333. [*O. Sax.* môdthraka.] v. ādl-, æsc-, bǣl-, ecg-, flân-, gâr-, gûð-, hild-, holm-, lîg-, môd-, wǣpen-, wîg-þracu; þrece.

þræc. v. ge-þræc, *and preceding word*.

-þræc. v. on-þræc.

þræc-heard; *adj. Brave in battle*:—Þrungon þræchearde, Elen. Kmbl. 245 ; El. 123.

þræc-hwîl, e ; *f. A time of suffering, a hard time*:—Ongan ðâ hreówcearig sâr cwânian . . . 'Ðû mec þreádes þurh sârslege . . .' Hine seó fǣmne forlêt æfter þræchwîle, Exon. Th. 275, 22 ; Jul. 554. [Cf. *Icel.* þrekaðr *wearied, exhausted*.]

þræc-rôf; *adj. Valiant*, Cd. Th. 122, 22 ; Gen. 2030.

þræc-wig, es ; *m. Hard fighting*:—Þurstige þræcwîges, Cd. Th. 189, 9 ; Exod. 182.

þræc-wudu, a; *m. A spear*:—Helm, byrne, þræcwudu, Beo. Th. 2496; B. 1246.

þrǣd, es ; *m. A thread*:—Ðrêd *filum*, Wrt. Voc. ii. 108, 59 : i. 66, 28. Þrǣd, ii. 35, 44 : i. 81, 65 : *fila*, 282, 11. Se gyldna ðrêd *bratea fila*, ii. 89, 37 : 12, 3. Þrǣd mē (*a coat of mail*) ne hlimmeþ, ne æt mē hrisil scrîþeþ, Exon. Th. 417, 18; Rä. 36, 6. Cnyte mid ânum ðrǣde, Lchdm. i. 218, 20. Mid ânum reádum þrǣde, 100, 19. Mid wyllenan þrǣde, ii. 310, 22. Him ne hangaþ nacod sweord ofer ðam heáfde be smalan þrǣde, Bt. 29, 1; Fox 102, 28. Þrǣda *filorum*, Hpt. Gl. 494, 18. Āþrāwenum ðrǣdum *contortis*, Wrt. Voc. ii. 21, 18. Webb byþ gefylled mid þrǣdum *tela consummatur filis*, Scint. 216, 2. [*O. L. Ger.* thrâd *filum*: *O. Frs.* thrêd: *Du.* draad: *O. H. Ger.* drât: *Ger.* draht *Icel.* þrâðr: *Dan.* traad.] v. col-, gold-, hefeld-, rihtung-, weall-þrǣd þrâwan.

þrǣft *a quarrel, dispute, contention, chiding*:—Siteþ symbelwlonc searwum lǣteþ wîne gewǣged word ût faran þrǣfte þringan þrymme gebyrmed æfæstum onǣled oferhygda ful *flushed with the feast he sits, affected with wine, words he guilefully lets fare forth, crowd out with quarrel in their train, leavened as he is with pride, inflamed with ill-will, full of overweening*, Exon. Th. 316, 1 ; Môd. 42. [*Icel.* þrapt *quarrel* : þrefa *to wrangle*. Jamieson gives *thrafftly* in a chiding or surly manner.] v. (?) þrafian.

þrǣgan (cf. *Goth.* þragjan, *and for conjugation cf.* plegan); *p.* de *To run, proceed in a course*:—Sume tungul lǣsse gelîðaþ, ða ðe lâcaþ ymb eaxe ende, oððe micle mâre geférað, ða hire midore ymbe þearle þrægeþ (-aþ?) (cf. sume tunglu habbaþ lengran ymbhwyrft ðonne sume habban, and ða lengestne ðe ymb ða eaxe middewearde hwearfaþ, Bt. 39, 3; Fox 214, 24), Met. 28, 24. Ðǣr him eoh fore mîlpaðas mæt, môdig þrægde, Elen. Kmbl. 2524; El. 1263. Ic seah hors swîþe þrægan, Exon. Th. 400, 4; Rä. 20, 3.

þrǣge. v. wǣpen-þrǣge.

þrǣl, es ; *m. A thrall, slave, servant*:—Ðe yfle ðrael *malus servus*, Mt. Kmbl. Lind. 24, 48. Allra ðrǣl † esne *omnium servus*, Mt. Skt. Lind. Rush. 10, 44. Se ðe ðoeð synne ðrǣl is synnes, Jn. Skt. Rush. 8, 34. Ne cwedo ic iów ðrǣlas (ðrǣllas, Lind.), for ðon ðrǣl (ðrǣll, Lind.) nât hwæt wyrceð hlâford his, 15, 15. Wē witan ðæt þurh Godes gyfe þrǣl weard tô þegene, and ceorl tô eorle, L. Eth. vii. 21 ; Th. i. 334, 8. Ðeáh þrǣla hwylc hlâforde æthleápe and of cristendôme tô wîcinge weorðe, and hit

æfter ðam eft geweorðe, ðæt wǽpngewrixl weorðe gemǽne þegene and þrǽle, gyf þrǽl ðæne þegen fullíce áfylle, licge ǽgylde ealre his mǽgðe; and gyf se þegen ðæne þrǽl, ðe hé ǽr áhte, fullíce áfylle, gylde þegengylde, Wulfst. 162, 5–10. Oft þrǽl ðæne þegen, ðe ǽr wæs his hláford, cnyt swýðe fæste and wyrcþ him tó þrǽle, 163, 1. Gebéte þrǽl mid his híde, þegn mid .xxx. scillingan, 181, 9. Ðe hláferd ðrǽles ðæs *dominus servi illius*, Mt. Kmbl. Lind. 24, 50. Ðrǽles (ðrǽlles, Lind.), Lk. Skt. Rush. 12, 46. Ic cuoeðo ðrǽle mínum, Lind. 7, 8. Hé sende óðerne ðrael, Mk. Skt. Lind. 12, 4. Gif Englisc man Deniscne þrǽl ofsleá, gylde hine mid punde, and se Denisca Engliscne eal swá, gif hé hine ofsleá, L. Eth. ii. 5; Th. i. 286, 24. Þrǽles ne móton habban ðæt hí ágon on ágenan hwílan mid earfeðan gewunnen, Wulfst. 158, 38. Antecristes þrǽlas, 55, 9. Ðonne beó gé ealle þrǽlas *tunc eritis omnes servi*, Coll. Monast. Th. 29, 25. [From Icel. þrǽll.]

þrǽl-riht, es; *n. Thrall-right*; in pl. *the legal rights and privileges which belonged to the thrall* :—Freóriht wǽron fornumene and ðrǽlriht generwde . . . Frige men ne mótan wealdan heora sýlfra, ne faran ðǽr hí willaþ, ne áteon heora ágen, swá swá hí willaþ; ne þrǽlas ne móton habban ðæt hí ágon on ágenan hwílan mid earfeðan gewunnen, ne ðæt heom on Godes ést góde men geuðon and tó ælmesgife for Godes lufan sealdon, Wulfst. 158, 15.

þrǽs *a fringe, border* :—Ðrǽs, liste *limbus*, Txts. 75, 1228. Ðrǽs, þrǽs oresta, 85, 1455. Ðrǽs, Wrt. Voc. ii. 63, 51. Þrǽs *instita*, i. 26, 10. Ðrǽsi *lymbo*, Txts. 75, 1264. Liste oððe þrǽs *lembum*, listum oððe þrǽsum *limbus(-is?)*, Wrt. Voc. ii. 50, 68, 69.

þrǽsce, an; *f. A thrush* :—Ðrostle *trita*, ðraesce *truitius*, Wrt. Voc. ii. 122, 79. [Cf. Icel. þröstr; *gen.* þrastar *a thrush*.] v. þrysce.

þrǽscende. v. þrǽstan.

þrǽst. v. ðærst.

þrǽstan; *p.* te. I. *to twist, writhe, roll about* :—Ðæt hors on misenlíce dǽlas hit wond and ðrǽste *cum diversas in partes se torqueret*, Bd. 3, 9; S. 533, 36. Hé misenlícum styrenessum ongan his limu ðrǽstan *diversis motibus coepit membra torquere*, 3, 11; S. 536, 15. II. *to torture, torment, harass, plague, afflict* :—Ǽnne of ðám mannum ðe hí on ðam fýre bærndon and ðrǽston *unum de eis quos in ignibus torrebant*, 3, 19; S. 548, 48. Ðætte Bryttas hié sylfe ðrǽston (*contriverint*) on ingefeohtum, 1, 22; S. 485, 11. Mé þrǽscende (þrǽstende?), Homl. Skt. ii. 23 b, 554. Hé grimme sáre ongan ðrǽsted beón (*torqueri*), Bd. 5, 13; S. 632, 19. Ða unríman mænigeo ðrǽste wǽron *innumerabilis multitudo torqueretur*, 5, 12; S. 628, 4. Missenlícum cwealmnyssum ðrǽste *diversibus cruciatibus torti*, 1, 7; S. 479, 13. III. *to press, constrain* :—Tó nirwienne í tó þrǽstenne *artandum, constringendum*, Hpt. Gl. 480, 32. [Is] þrǽst *compellitur, coartatur*, 469, 20. [In later English the word seems mostly used intransitively, *to press in, on, out* :—Monie þurles, þer þet water þrest in, A. R. 314, 14. Þreaste smoke ut, Marh. 9, 6. He þraste to þan fihte, Laym. 27644. Moni þusenden þrasten ut of telden, 26318. Heo þresten in uppon me *irruerunt super me*, A. R. 220, 31. Mine cnihtes scullen þræsten (þreaste, 2nd MS.) biforen me, Laym. 23373. He thurgh the thikkeste of the throng gan threste, Chauc. Kn. T. 1754.] v. á-, for-, ge-þrǽstan.

þrǽstedness, þrǽstness. v. for-þrǽstedness, for-, ge-þrǽstness.

þrǽsting, e; *f. Torment, affliction* :—Swá hé sceal etan ðætte hiene sió gewilnung ðære gífernesse of his módes fæstrædnesse ne gebrenge, ne eft sió ðrǽsting (ðrǽsðing, Hatt. MS.) ðæs líchoman ðæt mód ne áscrence mid upáhæfennesse *ne aut illos appetitus gulae a mentis statu dejiciat, aut istos afflicta caro ex elatione supplantet*, Past. 43; Swt. 316, 7.

þrǽxwold. v. þerscold.

þrafian; *p.* ode. I. *to urge, press* :—Ic ðrafige *urgeo*, Ælfc. Gr. 26, 3; Zup. 155, 12. Gif ic míne heorde tó swíðe þrafige on gancge and swence hig ealle hig sweltaþ ánes dæges *si greges meos plus in ambulando fecero laborare, morientur cuncti una die*, R. Ben. 120, 20. Mec mín freá þrafaþ on þýstrum, hætst on enge, Exon. Th. 383, 1; Rä. 4, 4. Se biscop sceal þrafian ða mæssepreóstas mid lufe ge mid láþe, ðæt hié healdan Godes ǽwe on riht, Blickl. Homl. 45, 8. II. *to reprove, rebuke, correct*. v. þrafung :—Se Hǽlend on manegum wísum ðrafode and áfandode his gingran, and geedlǽhte ðæt ðæt hé ǽr tǽhte tó fulre láre, Homl. Th. ii. 296, 22. Drihten, ne þreá ðú me ne ne þrafa on ðínum yrre *Domine, ne in ira tua arguas me*, Ps. Th. 37, 1. Hwílum liðelíce tó ðreátiaénne, hwílum suíðlíce and strǽclíce tó ðrafianne *aliquando leniter arguenda, aliquando vehementer increpanda*, Past. 21; Swt. 151, 12. [Cf. (?) Goth. þrafstjan *to exhort, encourage, comfort*.] v. (?) þræft.

þrafung, e; *f. Reproof, rebuke, censure* :—Þrafunge *argumenti* (v. þrafian, II), Hpt. Gl. 487, 20. Ðæt is ðonne swelc mon mid forewearde orde stinge, ðæt mon openlíce and unforwandodlíce on óðerne rǽse mid tǽlinge and mid ðrafunga *ex mucrone quippe percutere, est impetu apertae increpationis obviare*, Past. 40; Swt. 297, 13. Se Hǽlend æteówde hine sylfne cucenne his gingrum æfter his ǽriste on manegum ðrafungum, Homl. Th. i. 294, 16. Hí (*Job's friends*) mid manegum ðrafungum hine (*Job*) geswencton, ii. 454, 21. v. níd-þrafung.

þrág, þráh, e; *f.* I. *a time, season* :—Ðonne seó þrág cymeþ wefen wyrdstafum, Exon. Th. 183, 9; Gú. 1324. Wergendra tó lyt

þrong ymbe þeóden, ðá hyne sió þrág becwom, Beo. Th. 5759; B. 2883. Ǽr ðam seó þráh cyme, ðæt hé ðec áworpe of woruldríce, Cd. Th. 252, 34; Dan. 588. Nis seó þráh micel, ðæt hí ðé swencan móton, Andr. Kmbl. 214; An. 107. Ða æfterwritenan lǽcedómas ne sculon on áne þráge tó lange beón tó gedóne, Lchdm. ii. 186, 12. Nis ðæt eówer ðæt gé witan ða þráge and ða tíde *non est vestrum nosse tempora vel momenta*, Blickl. Homl. 117, 24. II. *having reference to the condition of things at any time, time as in good, bad, hard, etc. times* :—Hú seó þrág (*the happy time just described*) gewát, swá heó ne wǽre, Exon. Th. 292, 7; Wand. 95. Is ðeós þrág ful strong, ic sceal þinga gehwylc þolian, 270, 13; Jul. 464. Onwæcnaþ sió wóde þrág ðære wrǽnnesse and gedréfþ hiora mód *libido versat avidis corda venenis*, Bt. 37, 1; Fox 186, 18 : Met. 25, 41. Ic mé þyslícre ǽr þráge ne gewénde *I did not expect such a time as I have had*, Exon. Th. 269, 21; Jul. 453. Wéndon hié þearlra gehinga, þráge hnágran, Andr. Kmbl. 3195; An. 1600. Óð ðæt rímgetæl reðre þráge daga forð gewát, Cd. Th. 85, 26; Gen. 1420. Hé ðý wyrs meahte þolian ða þráge, ðá hió swá þearl becom (cf. Ðá hit gelomp ðæt hé on swá micelre nearunesse becom, Bt. 1; Fox 2, 26), Met. 1, 77. III. *adverbial uses*. Cf. hwíl :—Þráge *interim*, Wrt. Voc. ii. 110, 76. Hé þráge mid ús wunode *he dwelt with us for a time*, Blickl. Homl. 131, 19 : Exon. Th. 208, 24; Ph. 160 : Ps. Th. 81, 5 : 111, 4 : Met. 20, 134. Tódríf ðone mist ðe þráge nú hangode hwýle, 20, 264. Hé þráge siððan wícum wunode, Cd. Th. 108, 25; Gen. 1811 : 74, 5; Gen. 1217. Hit þráge sceal in sondhofe siþþan wunian, Exon. Th. 173, 30; Gú. 1168. Swelge hé ða ðráge ðe (*while, as long as*) hé mǽge, Lchdm. ii. 284, 14. Geǽrndon hí sume ðráge *they raced for some time*, Bd. 5, 6; S. 619, 9. Ealle þráge *all the time*, Ps. Th. 101, 25 : Exon. Th. 324, 2; Víd. 88 : Judth. Thw. 25, 2; Jud. 237 : Apstls. Kmbl. 60; Ap. 30. Ðú sǽgdest ðæt ic sceolde lifigan lange ðráge, Ps. Th. 118, 116. Bád sunu Lameches sóðra geháta lange þráge, Cd. Th. 86, 5; Gen. 1426 : 153, 25; Gen. 2544 : 252, 4; Dan. 573 : Beo. Th. 108; B. 54 : Andr. Kmbl. 1580; An. 791. Wǽran hí ǽr on hæþenra hæfteclommum lange þrága, Chr. 942; Erl. 116, 17. Þéh mín líchama lytle ðráge on niðerdǽlum eorðan wunige, Ps. Th. 138, 13. Swá ic þrágum (*at times, sometimes*) winne, hwílum . . . hwílum, Exon. Th. 386, 26; Rä. 4, 67 : 381, 1; Rä. 2, 4 : 494, 6; Rä. 82, 4 : Cd. Th. 271, 29; Sat. 112 : Elen. Kmbl. 2475; El. 1239. Wæter wynsumu mónþa gehwam bearo geondfaraþ þrágum (*at appointed times*), is ðæt þeódnes gebod, ðætte twelf síþum ðæt tírfæste lond geondlǽce laguflóda wynn, Exon. Th. 202, 11; Ph. 68 : Ps. Th. 138, 11. [Habben an alpi þraȝe summe lisse, O. E. Homl. i. 35, 10. He tah hine aȝein ane þrowe, Laym. 640. God þraȝhe *a good while*, Orm. 3475. Lat me nu habbe mine proȝe (*rimes with* oȝe), O. and N. 260. Sume þroȝe *for a while*, 478. Or he reste hym ony thrawe, Rich. 5062. Liþe me a litel proȝe, Horn. 336. Þrawe, Havel. 276 : R. Brun. 180, 11 : Alis. 3836. Thi pynes lastes bot a thrawe, Met. Homl. 142, 2. Throwe, Ch. M. of L. T. 953. Many a throwe, Ch. Yem. T. 941. Any throwe, Monk's T. 3326. Throwe, a lytyl wyle *momentum*, Prompt. Parv. 493.] v. earfoþ-, ryne-, treów-þrág.

þrág-bisig; *adj. Occupied for a time* (?), *periodically employed* (?) :—Ic sceal þrágbysig þegne mínum hýran georne, Exon. Th. 387, 6; Rä. 5, 1. The subject of the riddle is a millstone, and the Latin riddles on which the English one is based seem to suggest that the epithet might refer to running; Aldhelm has : Par labor ambarum . . . altera currit; Symphosius : Non desinit ille moveri. v. Prehn's Rätsel des Exeterbuches. But the verse requires þrág, while the verb, þrægan, has a short vowel; and þrág seems always (?) used in the sense of *time*.

þrág-mǽlum; *adv. From time to time, at times, at intervals* :—Ic wæs nýde gebǽded, þrágmǽlum geþreád, ðæt ic ðé sóhte, Exon. Th. 263, 3; Jul. 344. Ne meahton hió word forðbringan, ac hió þrágmǽlum þíoton ongunnon, Met. 26, 80. Hit on wolcnum oft þearle þunraþ, þrágmǽlum eft ánforlǽteþ (cf. hit hwílum þunraþ, hwílum ná ne onginþ, Bt. 39, 3; Fox 214,) 28, 55. Ðrágmǽlum, Andr. Kmbl. 2461; An. 1232.

þrang (?) *a throng, crowd* :—Wæterberendra þran[gum] *lixarum coetibus*, Hpt. Gl. 427, 15. [Grete thrang of men, Pr. C. 4704. A þral in þe þrong, Allit. pms. 42, 135. Du. drang *a crowd*: M. H. Ger. dranc : Ger. drang : Icel. þröng.] v. ge-þrang; þringan.

þráwan; *p.* þreów; *pp.* þráwen *To throw* (v. throw, thraw *to turn wood, to twist*; throwster *one that throws or winds silk* or *thread*; throwing-clay *clay that will work on the wheel*, Halliw. Dict. See also, E. D. S. Pub. Holderness, Lincolnshire and Huddersfield dialects, *thraw, thrown* : Jamieson's Dict. *thraw*), *twist* :—Ic samod þráwe *contorqueo*, Ælfc. Gr. 26, 3; Zup. 155, 16. I. *trans. To twist, rack, torture* :—Hé hét hí on hencgene ástreccan and ðráwan swá swá widðan, Homl. Skt. i. 8, 113. Hé hét hine hón on hengene, and mid hengene ðráwan tó langere hwíle, Homl. Th. ii. 308, 31. II. *intrans. To twist, turn round*, (1) *to take a different direction* :—Se líg sóna ðreów ðwyres wið ðæs windes ðráwunge and ðone turned round in a contrary direction towards the wind, Homl. Th. ii. 510, 8. (2) *to turn round, revolve* :—Þráwende *rotante* (*fusa*, Ald. 175, 34), Wrt. Voc. ii. 93, 78. (3) *to curl* :—Þráwendum í cyrpsiendum loccum *crinibus crispantibus*, Hpt. Gl. 435, 9. [Haremarken þrauwen

mid winde, Laym. 27359. *But Layamon uses the word intransitively also of movement :*—Of his horse he þreou (cf. anan swa ich lihte of blonken, 793), 807. Þa cheorles up þreowen (þreuwen, 2nd MS.) *the churls started up,* 12321. Þrawen wyth a þwong, Gaw. 194. *The word, however, early gets the meaning of* throwing :—Horn þreu þe ring to grunde . . . 'Palmere trewe, þe ring þat þu þrewe,' Horn 1160-72. Ded he threow him to grounde, Alis. 2425. In fire saltou thrawe þam *in igne dejicies eos,* Ps. 139, 11. *O. L. Ger.* thrãan *rotare: Du.* draaijen *to turn, twist: O. H. Ger.* drãjan ; *wk. tornare, torquere : Gen.* drehen.] v. ã-, be-, ge-, ge-ed-, þurh-þrãwan ; twi-þrãwen.

þrãwing-spinel, e ; *f. A curling-iron, crisping-pin :*—Þrãwincspine í hǽrnǽdla *calamistro,* Hpt. Gl. 435, 7 : 513, 75 : 526, 46.

þreá, þrawu ; *gen.* þreá ; *pl.* þreá ; *f.:* þreá ; *gen.* þreán, *also* þreás (?) ; *m. ; also neuter.* I. rebuke, reproof, threat :—Þrauuo, thrauu, trafu *argutiae,* Txts. 41, 200. Se ðe ege healdeþ eallum þeódum and his þreá ne sí ðǽr for ãwiht *qui corripit gentes, non arguet ?* Ps. Th. 93, 10. For ðínre þreá *ab increpatione tua,* 75, 5. Hé mid heardre ðreá hí on sprǽc *aspera illos invectione corrigebat,* Bd. 3, 5 ; S. 527, 11. Ðreán *adversione,* Wrt. Voc. ii. 2, 29. Ðǽm scamleásan ne wyrd nó gestiéred bútan micelre tǽlinge and miclum ðreán *impudentes ab impudentiae vitio non nisi increpatio dura compescit,* Past. 31 ; Swt. 205, 23. For ðínum þreán and for ðínum yrre *ab increpatione tua, ab inspiratione spiritus irae tuae,* Ps. Th. 17, 16. Ða him þreá ðíne þearle ondrǽdaþ *ab increpatione tua fugient,* 103, 8. Hé for him þreá geaf kyningum *corripuit pro eis reges,* 104, 12. Gé hláfordas, dóð ge eówrum monnum ðæt ilce, and gemetgiaþ ðone ðreán *vos domini eadem facite illis, remittentes minas,* Past. 29 ; Swt. 203, 1. Ne hí Agustinus lárum ne his bénum ne his ðreám *(increpationibus)* gebáfigean woldan, Bd. 2, 2 ; S. 502, 14. Ðreá þeódum eáwan *ad faciendas increpationes in populis,* Ps. Th. 149, 7. II. *chastisement, correction, punishment, an infliction* that has been deserved, justifiable *severity :*—Se egsan þreá *the pain caused by the terror of the day of judgement,* Exon. Th. 65, 34 ; Cri. 1064. Seó lufu ðæt gemet ðǽre ðreá *(the punishment to be imposed for stealing)* dihtaþ, Bd. 1, 27 ; S. 490, 21. On strengo þeódscipes and þreá tó wlǽc *in disciplinae vigore tepidus,* S. 492, 18. Ðǽre uplecan ðreá sweopon *supernae flagella districtionis,* 2, 5 ; S. 507, 2. Æfter ðǽre ðreá *(flagello)* 4, 31 ; S. 611, 1. Ðæt weorþeþ þeódum tó þreá, ðám ðe þone Gode ne cúþun, Exon. Th. 67, 21 ; Cri. 1092. Ðoliaþ wé þreá on helle, Cd. Th. 25, 5 ; Gen. 389. Þurh egsan þreá, Exon. Th. 83, 32 ; Cri. 1365. Næs ǽnig ðæt mec þus bealdlíce bendum bilegde, þreám forþrycte, 273, 22 ; Jul. 520. III. *an infliction* (where no idea of correction is implied), *evil, ill, pang, plague, calamity, affliction :*—Tó ne geniólaecað tó ðé yfel and ðreá *(flagellum)* ne geneólaeceþ getelde þínum, Ps. Surt. 90, 10. Hí gesomnadon in mec ðreá *(flagella),* 34, 15. Heó fleón gewãt þreá *(ill treatment,* cf. Gen. 16, 6) and þeówdóm, Cd. Th. 136, 24 ; Gen. 2263. Geþola þeóda þreá *bear the ills inflicted on thee by the gentiles,* Andr. Kmbl. 213 ; An. 107. Swylt ealle fornom . . . þurh þearlíc þreá *death carried off all . . . by a terrible calamity (shipwreck),* Exon. Th. 283, 10 ; Jul. 678. Þreá wǽron þearle, þegnas grimme, 135, 4 ; Gú. 519. Monge ðreá *(flagella)* synfulra, Ps. Surt. 31, 10. Wé ðec for þreáum and for ðeonýdum *(for þearfum and for þreanýdum,* Exon. Th. 186, 3) ãrna biddaþ, Cd. Th. 234, 18 ; Dan. 294. Bonan mǽndon ðæt hý monnes bearn þreám oferþunge and him tó earfeðum ãna cwóme gif hý him ne meahte mãran sãrum gyldan gyrnwrǽce *the murderous spirits made moan, that a child of man would have surpassed them in afflictions* (i. e. *would have caused them greater miseries than they had done to him), and alone would have come to their distress, if they could not requite their misery on him with greater pains,* Exon. Th. 128, 10 ; Gú. 402. Þreám forþrycced þurh ðæs þeódnes word *grievously oppressed by the prince's words (which announced his death),* 174, 1 ; Gú. 1171. Hí beág ymb mín heáfod þreám *(painfully* or *with reproaches ?)* biþrycton, 88, 26 ; Cri. 1446. Hé Godes ðeówdóm miccle swíðor lufode þonne ða ídlan þreás ðisse worlde *he loved God's service much more than the vanities and vexations of this world,* Homl. 211, 27. III a. *in reference to* inanimate things :—Sunne weard þreám ãþrysmed *the sun was miserably darkened (at the crucifixion),* Exon. Th. 70, 5 ; Cri. 1134. Wind nearwe geheaðrod, þreám forþrycced *the wind, straitly confined, strictly repressed,* Elen. Kmbl. 2551 ; El. 1277. [Hie nimed swo bittere þrowes, þat hie ne mai hire muð holden, O. E. Homl. ii. 181, 2. A thrawe hire cam, Alis. 616. Wa geres us thol hard traues (thrawes, MS. C.), Met. Homl. 36, 16. In his harde þrowe, L. H. R. 150, 18. On his last þrowe, Ass. B. 533. Throwe, womannys pronge *erumpna,* Prompt. Parv. 493. *O. Sax.* thrã *(in* thrã-werk) *: O. H. Ger.* drauua, drouua, drôa *animadversio, comminatio, mina : Icel.* þrá *a throe, pang.*] v. bróh-, cwealm-, heáh-, mód-, þeód-þreá, *and next word.*

þreágan, þreán, *and* þreáwian (v. þreápian) ; *p.* þreáde [*in* Bt. 38, 1 ; Fox 196, 7 *a form occurs that might be a strong past of* þreán, *on the analogy of* þwean, sleán :—Ðæt gewit wæs swíþe sorgiende for ðám ermþum ðe hí drógan ; cf. *the rendering of the same passage in the metres :* Ðæt mód wæs swíðe sorgum gebunden for ðæm earfoþum ðe him on sǽton, Met. 26, 97. *But, perhaps,* drugon *should be read, as*

Latin is : Mens super monstra, quae *patitur,* gemit] ; *pp.* þreád. I. *to reprove, rebuke, reproach :*—Ic hine þreáge (ðreá, Lind. : ðriá, Rush.) and forlǽte *corripiam illum et dimittam,* Lk. Skt. 23, 22. Ne þreáge (drégu, Rush. : þreá, Spl. C. : þrǽwie, Spl. T.) ic eów *non arguam te,* Ps. Th. 49, 9. Þreáge (ðréu, Surt. : ðreáge, Spl.), 49, 23. Ðú ðreást (ðreádes, Surt. : þreádest, Spl.) ðeóda *increpasti gentes,* 9, 5. Gif ðú ðreást (dreast, MS.) *si corripueris,* Kent. Gl. 714. Ðreád *corripit,* 514 : *arguit,* 290. Ðú oferhýdige þreádest (ðreádes, Surt.) *increpasti superbos,* Ps. Th. 118, 21. Abraham þreáde Abimelech mid wordum *Abraham increpavit Abimelech,* Gen. 21, 25 : Andr. Kmbl. 3371 ; An. 1689. Hé ðreáde ðæne wind, Lk. Skt. 8, 24 : 23, 40. Ðreáde *corripit,* Past. 21 ; Swt. 151, 20. Þreádon *increpabant,* Mt. Kmbl. 19, 13. Ne þreá ðú mé *ne arguas me,* Ps. Th. 6, 1 : 37, 1. Þreá hine openlíce *publice argue eum,* Lev. 19, 17 : R. Ben. 13, 9. Mé sóðfæst gerecce (ðreád, Surt.) and þreáge (ðreáð, Surt.) *corripiet me justus et increpabit me,* Ps. Th. 140, 7. Ðreágan *redarguere,* Past. 2 ; Swt. 31, 12. Hé ongan hine þreágean (þreágan, MS. B.) *coepit increpare eum,* Mk. Skt. 8, 32. Ðreiga, Mt. Kmbl. Rush. 16, 22. Hé sceall stídlícor þreán *(arguere),* R. Ben. Interl. 15, 1. Monige sindon suíðe líðelíce tó ðreáceanne *nonnulla sunt leniter arguenda,* Past. 21 ; Swt. 157, 24. Sindon monige suíðe suíðe tó ðreágeanne ðæt hí gehiéran ðreágende of ðæs láriówes múðe hú micle byrdenne hié habbaþ on hiera scyldum *nonnulla sunt vehementer increpanda, ut quanti sit ponderis culpa ab increpantis ore sentiatur,* Swt. 159, 16-18. Ðá andwyrde se óðer ðreágende 'the other answering rebuked him (Lk 23, 40), Homl. Th. ii. 256, 12. Ðreágende wer *uir objurgans,* 530, 28. Wæs hé fram ðám bróþrum ðreád *corripiebatur a fratribus,* Bd. 5, 14 ; S. 634, 10. Wé beóþ þreád *corripiemur,* Ps. Spl. 89, 12. II. *to punish* one who deserves punishment, *to chastise* by way of discipline, with a view to amend, *to chasten, correct :*—Ða ðe ic lufige, ða ic ðreáge and beswinge, Homl. Th. i. 470, 26. God beswingð and þreáð ða ðe hé lufaþ, ii. 548, 18 : Exon. Th. 63, 23 ; Cri. 1024. Lég þreáð þeódsceaþan, 97, 25 ; Cri. 1596. Wé sculon men ðreágean swá swá ða gódan fæderas gewuniaþ heora bearn ða hí for heora synnum ðreágeaþ and swingaþ and hwæðere ða sylfan ðe hí mid ðám wítum ðreágeaþ and swenceaþ lufiaþ eác *sic nos fidelibus tenere disciplinam debemus, sicut boni patres filiis solent, quos et pro culpis verberibus feriunt, et tamen ipsos quos doloribus adfligunt amant,* Bd. 1, 27 ; S. 490, 15-18. Ðú mé þreádes þurh sãrslege, Exon. Th. 275, 7 ; Jul. 546. Drihten hyne þreáde myd þearlwýslícere swingle for his ungehýrsumnysse, Shrn. 98, 14. Hé hine sylfne þreáge swíðe þearle mid forhæfednesse ǽtes and drinces, L. Pen. 14 ; Th. ii. 282, 18. Synrust þweán, hine sylfne þreán, Exon. Th. 81, 10 ; Cri. 1321. Mid þýstrum þreáh, Ps. Th. 104, 24. Ðregende ðreáde mec Dryhten *castigans castigavit me Dominus,* Ps. Surt. 117, 18. Hé him eáwde mid hú miclum swingum hé ðreád and wítnod wæs, Bd. 2, 6 ; S. 508, 24. Hé (the man who will not give tithes) bið mid wítum þreád æfter his deáþe, Blickl. Homl. 49, 25. Synfulle *(those in purgatory)* beóþ þreád, Elen. Kmbl. 2590 ; El. 1296. III. *of undeserved punishment, to torture, torment, afflict, distress, vex, oppress :*—Seó wyrd þreáþ ða unscildigan and nimeð ne þreáþ ðám scildigum *fortuna premit insonteis debita sceleri noxia poena,* Bt. 4 ; Fox 8, 13. Se hine mid miclum wítum þreáde, ðæt hé Criste wiðsóce, Shrn. 93, 33. Ðreáde, 118, 19, 21. 'Þreá hig lóca hú ðú wylle.' Sarai hig ðá geswencte, Gen. 16, 6. Ic hálsige ðé ðæt ðú mé ne þreáge *(torquens),* Mk. Skt. 5, 7 : Lk. Skt. 8, 28. Ðrége *urgeat,* Ps. Surt. 68, 16. Ðá hét hé hí ãhón be hire loccum and hí þreágean mid missenlícum wítum, Shrn. 75, 21 : 104, 16. Swingan and þreágan, Exon. Th. 251, 9 ; Jul. 142. Cóme ðú ús tó þreágenne *(torquere),* Mt. Kmbl. 8, 29. Ðreágende *torquens,* Kent. Gl. 662. Mid sumre untrumnesse his líchaman ðreád *quadam infirmitate corporis arreptus,* Bd. 3, 19 ; S. 547, 12. (Wé) biað þreáde *aporiamur (aporiare ad angustiam reducere,* Migne), Wrt. Voc. ii. 100, 44 : 7, 6. (Wǽron) þreád *(cruciatibus) artabantur, stringebantur,* Hpt. Gl. 484, 10. III a. *where the subject of the verb is not a person :*—Seó langung hine þreáde, Blickl. Homl. 113, 14. Gif strongra storm and genip swýþor ðreáde *si procella fortior aut nimbus perurgeret,* Bd. 4, 3 ; S. 569, 12. [Praghand *castigans,* Ps. 117, 18. What if pretty þryuande be þrad *(punished),* Allit. Pms. 60, 751. *O. Sax.* gi-þrôôn *corripere* (Lk. 23, 22, v. *first passage in* I *above*) *: O. H. Ger.* drauwen, drouwen *arguere, redarguere, increpare, minari, minitari.*] v. ge-þreán.

þreágend, es ; *m. One who reproves* or *corrects :*—Þreágendes *(vox) correctoris (amici),* Hpt. Gl. 527, 48.

þreágung, þreáwung (v. þreápung), þreáung, þreáng, e ; *f.* I. *reproof, rebuke :*—Þreáunge *castigationis (censura),* Hpt. Gl. 476, 48. For ðære strenge ðínre þreáunga, Ps. Th. 38, 11. Of þreáunga (þregunge, MS. T. : ðreánge, Surt.) ðínre *ab increpatione tua,* Ps. Spl. 17, 18 : 79, 17 : 103, 8. Fram ðreáwunge (ðreáung, Surt.), 75, 6. Hú gescéadwís se reccere sceal bión on his ðreáunga *quae esse debet rectoris discretio correptionis,* Past. 21 ; Swt. 151, 5. Ðreáunge *correptionibus,* Swt. 155, 5. Ðreánge *increpationem,* Ps. Surt. 37, 15. Ðreángum *increpationibus,* 38, 12. Ðreánge *increpationes,* 149, 7. II. *a threat :*—Ælc gleáw mód hit gewarenaþ ægðer ge wiþ heora þreáunga ge wið ôlecunga *prudentia nec formidandas fortunae minas, nec exoptandas facit esse*

blanditias, Bt. 7, 2; Fox 18, 24. **III.** *chastisement, punishment :—*
Ic wæs beswungen ǽlce dæg and þreáung (*castigatio*) mín on dægrǽde,
Ps. Spl. 72, 14. Ðæt ic ídel heonone ne hwyrfe míne synna ou þreágunge
berende *that I may not go hence with nothing accomplished, bearing my
sins to punish me*, Homl. Skt. ii. 23 b, 672. **III a.** *correction :—*
Æfter deáþe nán þreágincge ys leáf *post mortem nulla correctionis est
licentia*, Scint. 48, 16. [O. H. Ger. drowunga, drôunga *animadversio,
comminatio.*]

þreahs. v. þreax.

þreál, e; *f. Correction :—*Þreál *correctio*, Wrt. Voc. ii. 135, 81. **I.**
*correction by words, reproof, rebuke :—*Ðreál *correptio*, Kent. Gl. 1061.
Þreále *invectionis*, Hpt. Gl. 448, 52. Hé (*John the Baptist*) ða heardheortan ðeóde mid stearcre ðreále and stíðre myngunge tó lífes wege
gebígde, Homl. Th. i. 362, 34. **II.** *correction by acts, chastisement,
punishment, discipline :—*Gif hé bétan nele underlicgge hé rihtlícre þreále
si non emendaverit, discipline regulari subjaceat, R. Ben. 56, 13. Ðú
(*Belshazzar*) noldest ðé warnian þurh ðínes fæder ðreále, Homl. Th. ii.
436, 8. Mistlíce þreála gebyriaþ for synnum, bendas oððe dyntas . . .,
L. Pen. 3; Th. ii. 278, 25. Hine man mid líchamlícum þreálum gewylde,
R. Ben. 57, 12 : 58, 10. Ðonne wurð seó heardnis stíðmôdre heortan-
swíðe gehnescad þurh grimlíce steóra and heardlíce ðreála, ðe ic on
mancyn sǽnde, Wulfst. 133, 19.

þreá-líc; *adj. Miserable, woeful, calamitous :—*Godes ágen bearn
héngon fæderas ûsse; ðæt wæs þreálíc geþôht, Elen. Kmbl. 851; El. 426.
Wæs þreálíc þing (*the deluge*) þeódum tôweard, rêðe wíte, Cd. Th. 79,
28; Gen. 1318. Ða apostolas þrowedon folcbealo ðreálíc, mǽrne
martyrdôm, Menol. Fox 248; Men. 125.

þreán. v. þreágan.

þreá-níd, es; *n.: e; f. Force or compulsion that punishes* or *causes
misery, affliction that comes from punishment :—*Ic hit leng ne mæg helan
for hungre; is ðes hæft tô ðan strang, þreányð ðæs þearl *this imprisonment is so hard, so severe the pain of my punishment*, Elen. Kmbl. 1404;
El. 704. Þrowigean þreáníed micel fýres wylm *to suffer much torturing
violence, the fervor of fire*, Cd. Th. 229, 7; Dan. 213. Þreányð þolian,
Beo. Th. 573; B. 284: Exon. Th. 187, 1; Az. 28. Þreánêd, 270, 12;
Jul. 464. Blíðheort wunode eorl in þreánêdum *cheerful the man remained in his misery*, Andr. Kmbl. 2530; An. 1266. Wê ðec for
þearfum and for þreánýdum ǽrena biddaþ *we pray thee for mercy on
account of our needs and afflictions*, 186, 4; At. 14: Beo. Th. 1668;
B. 832. Ðone feónd hê gefetrode fýrnum teágum, biþeahte þreánýdum
(*with penal restraints*), 359, 11; Pa. 61. Þreánêdum beþeaht, Elen.
Kmbl. 1764; El. 884.

**þreá-nídla, es; *m. Painful constraint, restraint of punishment,
oppression :—*Béc âmyrgaþ môdsefan of ðreánýdlan ðisses lífes *books
bring the mind to mirth from the painful pressure of this life*, Salm.
Kmbl. 481; Sal. 241. Ðonne wyrd and warnung winnaþ mid hira
ðreánýdlan hwæðerne âðreóteþ ǽr *when fate and prudence strive, each
with its own hard constraint, which of the two tires first ?* 857; Sal. 428.
Nealles sylfes willum ac for þreánêdlan, Beo. Th. 4450; B. 2224. Hê
þeóstra þegnas þreáníedlum bond *he bound the ministers of darkness with
penal restraints*, Exon. Th. 143, 29; Gú. 668.

þreáníd-líc; *adj. That entails painful violence, calamitous, afflictive :—*
Micel is ðæt ongin and þreáníedlíc ðínre gelícan ðæt ðú forhycge hláford
ûrne *great is the undertaking and calamitous for the like of thee to
despise our lord*, Exon. Th. 250, 16; Jul. 128.

þreáp (?) *a troop, band :—*Þreápum *commanipularibus, sociis* (*perhaps
heápum should be read*, cf. efenheápum *commanipularibus*, Wrt. Voc. ii.
20, 27; or þreátum; v. þreát: *but þreáp may have a double sense as
þreát has* (see, too, þreápian, þreátian); *in later English it remains with
the meaning* strife, contest, e. g.: Wituten þrep (ani enuy, alle chidyng)
or strijf, C. M. 13310. This þrepe (*the siege of Troy*) for to leue, Destr.
Tr. 9845: *perhaps, also, in sense of troop :—*An feondes trume . . .
þe saules . . . awarieþ al a-þrep (*in a troop?* or = *Ital.* a gara) al so
wulues doþ þe sceip, Misc. 149, 85. Halliwell gives *thrap* to crowd, as an
Essex word; Hpt. 477, 52: 487, 33.

**þreápian; *p. ode To rebuke, reprehend :—*Oft gelimpeþ, ðonne hê tó
suíðe and tó ðearllíce ðreápian (ðreáwian, ðreátian, Cott. MSS.) wile his
hiéremenn, ðæt his word beóþ gehwyrfedo tó unnyttre ofersprǽce *plerumque contingit, ut, dum culpa subditorum cum magna invectione corri-
pitur, magistri lingua usque ad excessus verba pertrahatur*, Past. 21; Swt.
165, 17. [Þrepe *arguere*, Ps. 93, 10. Himm birrþ þræpenn wiþþ skill
onnʒænness alle sinness *he must with discretion contend against all sins*,
Orm. 5744. Whan ʒe aʒens the prechur threpe *when ye blame the
preacher* (quotation in Halliwell's Dict.). Ha þreapeð aʒein þe, Kath. 1916.
Bihat al þ tu wult, þreap (*threaten*) þrefter inoh, 1499. In þraldom to
þrepe (*contend*) with þe werld, Destr. Tr. 12134. Þai þrappit with
stormys, 2003. They threpide wyth the throstille, D. Arth. 930. See
also Halliwell's Dict. *threap, thripe*; Jamieson's Dict. *threpe*. Cf. Al þet
fortune may þreapny (*threaten*) an do, Ayenb. 84, 20.] v. þrípel, *and
next word.*

þreápung, e; *f. Rebuke, reproof :—*Ðæt geðreátade môd bið suíðe

*raðe gehwierfed tô fióunga gif him mon tô ungemetlíce mid ðǽre
ðreápunga (ðreáwunga, Cott. MSS.) oferfylgð suíður ðonne mon ðyrfe
*correpti mens repente ad odium proruit, si hanc immoderata increpatio,
plus quam debuit, affligit*, Past. 21; Swt. 167, 14. [Þrepyng *strife*, Allit.
Pms. 43, 183. Cf. Cheaste. Þes boʒ him todelþ ine .vij. oþre boʒes . . .
þe zixte þreapninge (*threatening*) . . . Efterward comeþ þe þreapnynges
and beginneþ þe medles and þe werres, Ayenb. 65–66.] v. two preceding
words.

þreát, es; *m. **I.** *a troop, band, crowd, body of people, swarm,
press, throng*, (1) indefinite :—Þreát *turba*, Wrt. Voc. ii. 137, 29. Ðreát
(ðreátt, Rush.), Mk. Skt. Lind. 3, 32. Ðreót (ðæt folc ł ðreátas, Lind.)
turbae, Lk. Skt. Rush. 3, 10. Menigo ðreád (monige ðreátas, Rush.)
multa turba, Mk. Skt. Lind. 3, 7 : 5, 21 : Lk. Skt. Lind. 8, 40. Þreát
chorus, Wrt. Voc. ii. 17, 33 : i. 291, 13. Se ðreát (*caterva*) ðara Godes
ðeówa, Bd. 4, 7; S. 574, 34. His ðegna ðreát *ministri ejus*, Ps. Th.
102, 20. Heofonengla þreát, Exon. Th. 57, 34; Cri. 928. Wítgena
weorod, wífmonna þreát, 462, 7; Hö. 48. Þreátes *classis*, Wrt. Voc. ii.
24, 78. Þreáte *examine*, 33, 26. Gesomnadum ðreáte (*coetu*) bisceopa,
Bd. 4, 17; S. 585, 12 : Blickl. Homl. 95, 6. Se wæs on ðam ðreáte
þreotteóða secg, Beo. Th. 4803; B. 2406. Cyning þreáte fôr, herge tô
hilde, Elen. Kmbl. 102; El. 51 : Cd. Th. 288, 27; Sat. 388. Hió
þrungon on þreáte *they pressed in a crowd*, Elen. Kmbl. 657; El. 329.
In ðreáte *in choro*, Ps. Surt. 149, 3 : 150, 4. Hí gesomnodan mycelne
ðreát discipula *congregata discipulorum caterva*, Bd. 4, 2; S. 565, 25.
Wyrma þreát, Cd. Th. 285, 12; Sat. 336. Gif hê on þreát cymeþ, Exon.
Th. 380, 4; Rä. 1, 2. Ofer ðreótt, Mk. Skt. Rush. 8, 2. Menigo ł
ðreátas *turbae*, Mt. Kmbl. Lind. 13, 2. Engla þreátas, Blickl. Homl. 11,
12. Ealle ða mycclan þreátas ðe him mid fêrdon, 99, 35. Þurh þreáta
geþræcu, Exon. Th. 417, 17; Rä. 36, 6. Mid engla ðreátum *ducibus
angelis*, Bd. 4, 23; S. 596, 12. Þreátum *festis choreis*, Wrt. Voc. ii.
148, 14. Weras ðreátum and þrymmum þrungon and urnon, Judth.
Thw. 23, 39; Jud. 164. Meara þreátum, Exon. Th. 119, 19; Gú. 257.
Ðreáttum *turbis*, Rtl. 95, 6. (2) *in a more definite sense :—*Ðreát *turma
.i. xxxii equites*, Jn. Skt. Lind. 18, 12 margin. Ðes þreát *haec cohors*,
Ælfc. Gr. 9, 44; Zup. 64, 12. Cohors, *d. milites vel* þreát, Wrt. Voc. ii.
136, 1. Þreát *falanx, multitudo militum, cohors*, 147, 6. Of þreáte *ex
falange*, 29, 66. Ðreóte, 107, 59. Ðæs dêman cempan gegaderodon
ealne ðone þreát (ðreád, Lind. *cohortem*), Mt. Kmbl. 27, 27. Þicfealdum
þreátum eóroda *spissis legionum cohortibus*, Hpt. Gl. 413, 1. **II.**
violence, compulsion, force, oppression, punishment, ill-treatment. v.
þreátend :—Is ðeós þrág ful strong, þreát ormǽte; ic sceal þinga gehwylc
þolian, Exon. Th. 270, 14; Jul. 465. Hê wæs gebunden fýre and líge;
ðæt wæs fæstlíc þreát (*a punishment that pressed on him without remission*), Cd. Th. 284, 22; Sat. 325. Gotan eástan sceldas lǽddon þreáte
(*by force* or (?) *with their army*) geþrungon þeódlond monig, Met. 1, 3.
Stôdan him âbútan swearte gástas and mid micclum ðreáte (*with great
violence*) him onsigon, Homl. Th. i. 414, 9. Ða ðe hæfdon sum þing
lytles tô bigleofan, ðæt gelæhton reáferas and of ðam mûðe him âbrudon
unmǽðlíce mid þreáte, Homl. Ass. 68, 73. Mid swíðlícum þreáte, Ælfc.
T. Grn. 21, 12 : Homl. Skt. ii. 28, 105 : 29, 217. Godes ǽ forgægan
for his gramlícan ðreáte, 25, 220. Ne forhtige gê for ðæs fyrnfullan
þreátum (*cruelties*, or (?) *troops*), 25, 260. Hié ealle worlde weán and
ealle þreátas (*all the woes of the world and all miseries*) oferhogodan . . .
hié ealle worldlíce tintrega and ealle lichomlícu sár oforhogodan, Blickl.
Homl. 119, 16. [Riden ut þritti þusend, þe þræt (*throng*) wes þa mare,
Laym. 9791. Listed wich þreat (*punishment, trouble*, cf. God wile his
swerd dragen, 22), Dauid setted uppen us, O. E. Homl. ii. 61, 20. Þrat
moste I þole and unþonk, Allit. Pms. 93, 55. 'Herekempen scullen þi
lond wasten . . .' þis iherde þe king, þræt (*threat*) þas kaiseres, Laym.
22582. For scrið ne ðret *neither for entreaty nor threat*, Gen. and Ex.
2021. Ne recche ich noht of þine þrete, O. and N. 58. Grete wordis
and moche grym þrete, Destr. Tr. 2595. Hire sire and hire dame þreteþ
hire to bete, nule heo forgo Robin for al heore þrete, Misc. 190, 84.
M. H. Ger. drôz *annoyance, molestation*. Cf. Icel. þraut; *f. a struggle,
labour, hard task*.] v. â-, beadu-, beorn-, eóred-, ge-, gúð-, here-,
heofon-, íren-, mægen-, mearc-, sige-, wǽg-þreát.

**þreátend, es; *m. A violent person, one using violence* or *compulsion :
—*Ðæm ðreátende *violenti*, Mt. Kmbl. Lind. 11, 12. Ðæm nêdende ł
ðæm ðreátende *volenti* (l. *violenti*), 5, 42. Ðæm ðreáddende *angarianti*,
p. 14, 17. v. þreát, II.

þreátian; *p. ode. **I.** *to urge, press :—*Ðreátade *urguet*, Wrt.
Voc. ii. 124, 21. (1) *to oppress, afflict, vex, trouble, exercise, harass :—*
Ðú ðreást ða ðeóda ðe ûs ðreátigeaþ, Ps. Th. 9, 5. Mec lâðgeteónan
þreátedon þearle *my foes harassed me sorely*, Beo. Th. 1124; B. 560.
Wyrd . . . for ðý cymþ tô ðæm gôdan, ðæt hió ôþer twêga dô, oððe hine
þreátige tô ðon ðæt hê bet dô ðonne hê ǽr dyde, oððe him leánige ðæt hê
ǽr tela dyde *fortuna . . . remunerandi exercendive bonos causa deferatur*,
Bt. 40, 1; Fox 236, 3. Ðreátende *maceratus*, Wrt. Voc. ii. 113, 49:
55, 43. (2) *to urge a person to something, press for something, force
to do something :—*For ðí ic ðreátige ðê tó ûra goda offrunge, ðæt ðis folc,
ðe ðú beþôhtest, forléton ða ídelnysse ðínre láre, Homl. Th. i. 592, 31.

Seó wyrd ðe þreátaþ ða yflan tô wîtnianne *fortuna quae justo supplicio malos coercet*, Bt. 40, 2; Fox 236, 25 note. Hê þreátode hine tô hǽþenscipe, Shrn. 33, 10. Æghwylc hine þreátode æfter ðâm bôcum *every one tormented him for the books*, 123, 29. Ða cempan hine ðreátodon ðæt hê his lâc ofsrian sceolde *the soldiers urged him to offer his sacrifice*, Homl. Th. i. 416, 27. Men ðreátian and tihtan tô gôdum ðeáwum for ðam ege ðæs wítes *ad rectum supplicii terrore deducere*, Bt. 38, 3; Fox 200, 7. Sceolan ða bisceopas men georne þreátigean, and him bebeódan, ðæt hí Godes dômas on riht healdan, Blickl. Homl. 47, 35. Ongan se câsere hine ðreá:ian tô hǽðengylde, Shrn. 121, 12. Ða fǽmnan Simfronius ongan þreátian his suna tô wífe *that virgin (St. Agnes) Simfronius attempted to force to be wife to his son*, 56, 5. Geneáded ł þreátod *coacta*, Hpt. Gl. 508, 22. II. *to reprove, rebuke*:—On wuda ðû wildeór wordum þreátast *increpa feras silvarum*, Ps. Th. 67, 27. Geðence hê ðæt hê biþ self suíðe gelíc ðâm ilcan monnum ðe hê ðǽr ðreátaþ and hênð *aequales se ipsis fratribus, qui corriguntur, agnoscant*, Past. 17; Swt. 117, 16. Se ðe brúne ýða þreátaþ *he that rebukes the waves* (cf. geðreáðade tô sae *increpavit mari*, Mt. Kmbl. Lind. 8, 26), Andr. Kmbl. 1039; An. 520. God þreátode (*arguit*) ðê, Gen. 31, 42. Ða þreátode (*increpavit*) se fæder hine, 37, 10. Ðonne se lâreów sēcð ðone tíman ðe his hiéremen on ðreátigean (ðreágean, Cott. MSS.) mǽge *cum tempus subditis ad correptionem quaeritur*, Past. 21; Swt. 153, 6. Liðelíce tô ðreátianne (ðreátigeanne, Cott. MSS.) *leniter arguenda*, Swt. 151, 11. III. *to threaten*:—Hê þreátaþ ðone earman mid his eágum *oculi ejus in pauperem respiciunt*, Ps. Th. 9, 29. Hí þreátiaþ eall moncynn mid hiora þrymme *ore torvo comminantes*, Bt. 37, 1; Fox 186, 6: Met. 25, 13. [*In later English the forms from* þrētan, þrietan (e. g. p. þrette) *occur, though in the earlier time this form seems very rare.* v. þrítan. He gon þretien swíðe, þat al he wolde heom todrive, Laym. 17300. Mine þralles me þretiað (*threaten*), 493. Þe king þræted Brutun, þat . . ., 504. Summe prætteden heore ueond, 27131. Oluhnen oðer þreaten, A. R. 248, 8. He þrette us for to smiten, 366, 16. He bigon to þreatin hire *vehementius adversus eam in vocem erupit*, Kath. 2078. Þreatin *minari*, 626. To þrete *to complain* (cf. pleny, 548), Allit. Pms. 17, 560. Þat þretes (*reproves*) þe of þyn unþryfte, 89, 1728. Euereuch man me mid stone þreteþ (*ill-treats*), O. and N. 1609. Sho was adrad, for he so þrette (*threatened*), Havel. 1163: Gen. and Ex. 2023. An canticle ðæt ðreated (*rebuked*) ðo men, 4125. Ne threte (*arguis*) me, Ps. 6, 2. He watȝ þreted (*abused*) and þef called, Gaw. 1725. Of thralles y am thrat (*ill-used?*), P. S. 158, 17.] v. â-, ge-þreátian; þrítan; þreát-nian; þreátung; þreótan.

þreát-mǽlum; *adv. In troops, in crowds*:—Þreátmēlum *manipulatim*, Wrt. Voc. ii. 113, 38.

þreátnian; *p.* ode *To urge, force, compel*:—For hwilcum ðingum neádaþ se deófol eów ðæt gê cristene men tô his biggengum ðreátniaþ *for what reasons does the devil compel you to force Christian men to his worship?* Homl. Th. i. 424, 3. [Myd word he þretneþ muche, and lute deþ in dede, R. Glouc. 457, 14. Disciplis thretenyden (*comminabantur*) to men offrende, Wick. Mk. 10, 13.] v. þreátian.

þreátung, e; *f.* I. *compulsion, force, violence, oppression, ill-treatment.* v. þreátian, I:—Hí bestungon him on mûþ mid mycelre ðreátunge ðone fûlan mete, Homl. Skt. ii. 25, 34. Pilatus hê hæfde on þreátunge ðy hê hiene selfne ofstong *Pilatus tantis angoribus coarctatus est, ut sua se manu transverberaverit*, Ors. 6, 3; Swt. 258, 10. Hié heora land tô bismere oferhergodan, and him ðæs nǽnige bôte dydon bûton ofermôdlíce wíg and þreátunge *they harried their land, and for that they made them no amends, but in their arrogance made war on them and harassed them*, Blickl. Homl. 201, 24. II. *rebuke, reproof.* v. þreátian, II:—Mid ðreátunge *correptionibus*, Past. 21; Swt. 154, 5. Ðonne of ðære ðreátunga gâþ tô stíðlíce word *cum de correptione sermo durior excidit*, Swt. 167, 10. Ðurh ðæt îsern is getâcnod ðæt mægen ðara ðreátunga *per ferrum increpationis fortitudo signatur*, Swt. 163, 24. II a. *correction*:—Tô ðam yflum cymþ rêþu wyrd tô edleáne his yfla oððe tô þreátunge and tô lâre ðæt hé eft swâ ne dô *fortuna aspera puniendi corrigendive improbos causa deferatur*, Bt. 40, 1; Fox 236, 8. III. *threatening*:—On ðam geáre gegaderade Eádward cyng mycele scypferde on Sandwíc þurh Magnus þreátunge on Norwegon (v. Saga Magnûs gôða, cc. 37, 38: Magnús konungr gerði sendimenn til Englands . . . en þat stôð á brêfum . . . ' Vil ek, at þû gefir upp ríkit fyrir mêr; en at öðrum kosti mun ek sœkja til með styrk hers '), Chr. 1046; Erl. 171, 25. [Ihorde þe king of þisse herde þreting, Laym. 22582. Vre Louerd hefde ifuld him of his þreatunge *comminatione tua replesti me*, A. R. 156, 3. Þreting ne bene, Misc. 156, 17.]

þreáung. v. þreágung.

preá-weorc, es; *n. Pain inflicted as a punishment*, used of the misery of hell, as in O. Sax. the phrase *thrâ-werk tholôn*:—Wit hearmas, þreáweorc þoliaþ, and þŷstre land, Cd. Th. 45, 35; Gen. 737.

þreáwian, þreáwung. v. þreágan, þreágung.

þreax, þreahs *rottenness*:—Þreahs *caries*, Wrt. Voc. ii. 20, 56. Swâ swâ forrotod þreax, Basil admn. 7; Norm. 48, 20.

prece, es; *m. Force, oppression*; the result of oppression, *weariness*,

---

exhaustion:—Ðǽr synt tô sorge ætsomne gemenged se þrosma (þrosmiga. Wulfst. 138, 26) líg and se þrece gicela *there to their sorrow are mingled together the stifling flame and the violence of cold*; frigora mista simul ferventibus algida flammis, Dom. L. 191. Hneppade sâwle mîn for ðrece *dormitavit anima mea prae taedio*, Ps. Lamb. 118, 28. [O. Sax. wâpanthreki *force of arms*: Icel. þrekr; *m.*: þrek; *n. strength, fortitude*; þrekinn *enduring*; þrekaðr *wearied, exhausted*.] v. þracu.

þrecswald, þreiga, þremma, þreó, þreo-. v. þerscold, þreágan, þrymma, þrí, þri-.

preodian, þridian; *p.* ode. I. *to deliberate, take thought*:—Hê on his môde ðôhte and ðreodode ðæt hê wolde eall Angolcyn of Breotone gemǽrum áflýman *totum genus Anglorum Brittaniae finibus erasurum se esse deliberans*, Bd. 2, 20; S. 521, 28. Hê þreodode and smeáde on his môde, hwæt hê embe ðæt dyde, Homl. Ass. 124, 242. Ic frôd þrágum þreodude, Elen. Kmbl. 2475; El. 1239. Weras þeahtedon, þrydedon and þôhton, 1094; El. 549. II. *to deliberate, hesitate*:—Þrydaþ *hesitat*, Wrt. Voc. ii. 137, 35. Ne þreodode hê fore þrymme ðeódcyninges ǽniges on eorðan, ac him êce geceás líf *he did not hesitate before the glory of any king on earth, but (at once) chose life eternal*, Apstls. Kmbl. 35; Ap. 18. v. ymb-þreodiende, *and next word.*

preodung, þridung, e; *f.* I. *deliberation*:—Þridung *discrepatio*, Wrt. Voc. ii. 25, 63. Gif hê mid þancmetunge and ðreodunge (ðrydunge, MS. B.) geþafaþ *si ex deliberatione consentit*, Bd. 1, 27; S. 497, 23. II. *hesitation*:—Geseah hê sume earme sâwle ût fundigende of hyre líchaman, ac heó ne dorste ût gân, forð ðæt heó geseah ða áwyrgedan gâstas beforan hyre standan. Ða cwæð ân ðæra deófla tô hyre: 'Hwæt is ðín þriding? hwí nelt ðû ût gân?' Wulfst. 140, 13. Tweógendlícere tweónunge þrydunge *ancipiti ambiguitatis scrupulo*, Hpt. Gl. 422, 34. v. ymb-þreodung.

preohtig. v. þrohtig.

preóhund-wintre; *adj. Three hundred years old*:—Hê wæs on ðisum lífe þreóhundwintre and fíf-and-sixtigwintre, Gen. 5, 23.

preó-niht; *pl. Three days*:—Hê þreónihta (or? þreó nihta, þreó *being undeclined after the manner of* feówer, *etc.*) fæc swefeþ, Exon. Th. 357, 34; Pa. 38.

preosel-lîc. v. þrisel-lîc.

preótan; *p.* þreát; *pp.* þroten *To weary*:—Ic ðê bydde ðæt ðê ne ðreóte, ne ðú ða sprêce ðǽr ne forlête *I pray thee that it may not weary thee, and that thou do not leave the conversation there*, Shrn. 188, 20. [Goth. us-þriutan *to trouble, be troublesome to*: O. H. Ger. bi-, gi-, irdriozan *to weary, trouble*: Ger. ver-driessen: Icel. þrjóta *to lack, want.*] v. â-þreótan, un-âþreótende, â-þrotennes, -þrotsum; þreát, þreátian.

preó-teóða, þreotteóða *thirteenth*:—Se þreotteóða (priot-, þret-, preó-) *tertius decimus*, Ælfc. Gr. 49; Zup. 282, 20. Se wæs on ðam ðreáte þreotteóða secg, Beo. Th. 4804; B. 2406. Paulus is se ðreotteóða ðyses heápes, Homl. Th. ii. 520, 30. Seó ðreotteóðe mǽigð, i. 396, 4. On ðære þrytteóðan wucan, Mt. Kmbl. 11, 20 rubc. On ðæm þreotteóðon geáre, Ors. 4, 10; Swt. 200, 33. Þrytteóðan, Homl. Skt. i. 6, 272.

preó-tíne *thirteen*:—Ðreótêno, Salm. Kmbl. 581; Sal. 290. Ðreótýne, Menol. Fox 229; Men. 116. Þreottýne *tredecim*, Ælfc. Gr. 49; Zup. 281, 11: Bd. 1, 23; S. 485, 23. Ðǽr syndon betweónan ðâm twâm mynstrum ðreottýne míla ámetene, 4, 23; S. 596, 26. Þreótiénum *terdenis*, Wrt. Voc. ii. 82, 38.

preótíne-geáre; *adj. Thirteen years old*:—Seó fǽmne wæs .xiii. geáre, Shrn. 153, 32.

preoxwold, þrepel, þrescwald, þrexweald. v. þerscold, þrípel, þerscold.

prí, þrie; *m.*: þrie, þrió, þré; *f. n.* (ðreá, ðriá, ðreó, ðrió *in North.*); *gen.* þreóra, þrióra (*and* ðreána *in North.*); *dat.* þrim (þrím? ðrïim *in North.*, *but cf. Goth.* þrim), *later* þreom. *Three*:—Tres þrý gebyriaþ tô masculinum and femininum, *tria* þreó tô neutrum, Ælfc. Gr. 49; Zup. 281, 3. I. *used adjectively*:—Þrié Scottas cuômon, Chr. 891; Erl. 88, 5. Ða þreó clystru ðæt sind þrí dagas, Gen. 40, 12. Ða þrí windlas ðæt sind þrí dagas, 18. Ða brôðor þrý, Cd. Th. 122, 28; Gen. 2033. Þreó godas, Hy. 10, 44. Ða þrê (ðreánan, Blickl. Homl. 145, 31. Nú synt þreó (ðrió, Lind., Rush.) gêr, Lk. Skt. 13, 7. Tô ðara ðreóra burga ânre, Past. 21; Swt. 167, 17: Ors. 1, 1; Swt. 10, 4. Þrióra, Swt. 4, 10. On þrim (ðriim, Lind.: ðrim, Rush.) dagon, Mk. Skt. 15, 29: Jn. Skt. 2, 19, 20. Þrim, Lk. Skt. Lind. 4, 25: Mt. Kmbl. Lind. 27, 63. On ðæm þrim geárum on þrim folcgefeohtum, Ors. 3, 9; Swt. 128, 21. Of ðâm þrim sunum, Anglia xi. 2, 37. On ðýs ylcum þrim dagum, Lchdm. iii. 76, 26. Þreom nihton ǽr Candelmæssan, Chr. 1078; Erl. 215, 28. Hié ða þrié dǽlas on þreó tônemdon, Ors. 1, 1; Swt. 8, 3: Swt. 10, 3. Míne þrié ða getreówestan frŷnd, Nar. 29, 27. He gestrînde þrí suna, Gen. 6, 10: Ex. 2, 2. Lǽn me þrý (ðreó, Lind.: ðriá, Rush.) hlâfas, Lk. Skt. 11, 5. Þrý (þreó, Rush.) dagas and þreó (þreó, Rush.) niht, Mt. Kmbl. 12, 40. Ymbe þreó mônað, Ors. 5, 11; Swt. 238, 11. On ðrió wísan, Past. 53; Swt. 417, 20. Þrió míla, Ors. 4, 10; Swt. 194, 7. Ymb þreó niht, 3, 11; Swt. 152, 19: Elen. Kmbl. 1663; El. 833: Gen. 29, 2. Þreó eardungstôw (ðreá hûso, Lind.: ðreó selescotu, Rush.), Mt. Kmbl. 17, 4. Ðreá hûsa (ðriá hûs, Rush.), Mk. Skt. Lind. 9, 5. Þrió mydd hringa, Ors. 4, 9; Swt. 190, 12. Ðrió

gecynd, Bt. 33, 4; Fox 132, 3: 35, 6; Fox 168, 19. Gif hé ðás þreó þing ne déð, Ex. 21, 11. Þreó gér, Lk. Skt. 4, 25. II. used substantivly, (1) absolutely:—Ðær twégen oððe þrý (þreó, MS. A.: ðreó, Lind.: þreó, Rush.) synt gegaderode, Mt. Kmbl. 18, 20. Beóð fífe on ánum húse tódǽlede, þrý (ðrió, Lind.: ðriá, Rush.) on twégen and twégen on þrý (þreó, MS. A.: ðrió, Lind., Rush.), Lk. Skt. 12, 52. Ðæt ǽlc word stande on twégra oððe þreóra (ðreá, Lind.: þreó, Rush.) gewittnesse, Mt. Kmbl. 18, 16. Ðæt wé twá oððe ðreó gehýron, Bd. 3, 9; S. 533, 29. (1 a) distributively:—Ða wuniaþ twám and þrim ætgædere, R. Ben. 9, 15. (2) with qualifying or defining words:—Þa þrý cómon, Cd. Th. 221, 24; Dan. 93. Wé þrý, 242, 3; Dan. 413. Ða módhwatan þrý, 238, 21; Dan. 413. Cómon þrý gelǽrede weras . . . hí ða ealle þrý tógædere grétton ðone cyngc, Ap. Th. 19, 22: Homl. Th. ii. 384, 4. Ða þreó ðé ne lǽtaþ geortrēwan, Bt. 10; Fox 30, 8. Hwylc ðara þreóra (ðísra ðreána, Lind.), Lk. Skt. 10, 36: Homl. Th. i. 288, 27. Of ðisum þrim Noes sunum, Gen. 9, 19. Of him þrim, Anglia xi. 2, 45. Betwuh ðam þrim, Bt. 42; Fox 256, 20. Se ðe ðás ðreó hæfþ, 14, 2; Fox 44, 26. (3) in the phrase on þreó:—On þreó tónemnan, Ors. 1, 1; Swt. 8, 3. On ðreó tódǽlan, Bd. 5, 12; S. 627, 21. III. in combination with other numerals, (1) with cardinals, (a) multiplicative:—Þreó hund fǽdma, Gen. 6, 15. Þreó hund wera, Jud. 7, 6. On þisum þrim hundrydum, 7. Wiþ þrim hundred (ðriim hundum, Lind.) penegon, Jn. Skt. 12, 5. (b) added to the decades:—Þreó and twéntig, Ex. 32, 28. Þreo-and-hundeahtatig-wintre, 7, 7. (2) with ordinals:—Se þreó-and-syxtigeða, R. Ben. 37, 16. Móna se þrí-and-twéntigoða, Lchdm. iii. 194, 21. [Goth. þrija; n.; gen. þrijé; dat. þrim; acc. þrins; m. f.; þrija; n.: O. Sax. thrie, threa; dat. thrim: O. L. Ger. thrie, m.; thriu; n.; dat. thrim: O. Frs. thré, m.; thria; f.; thriu; n.; gen. thríra; dat. thrium, thrim, threm: O. H. Ger. drí; m.; drío; f.; driu; n.; gen. drío; dat. drim, drin: Icel. þrír; m.; þrjár; f.; þrjú; n.; gen. þriggja; dat. þrim(r), þrem(r); acc. þrjá; m.; þrjár; f.; þrjú; n.]

þría. v. þriwa.

þri-beddod; adj. Having three beds or couches:—Búr þrybeddod triclinium, Wrt. Voc. i. 58, 5.

þridæg-líc; adj. Lasting three days:—Þreodæglíc fæsten jejunium triduanum, Bd. 4, 25; S. 600, 8.

þri-dǽled; adj. Divided into three parts, tripartite:—Þreodǽledes tripertiti, Hpt. Gl. 511, 10. Þreodǽled tripartitam, 438, 27. Þreodǽlede tripertitas, in tribus partibus divisas, 451, 15. [Icel. þrí-deildr.]

þridda, þirda (in North.) third:—Se ðridda tertius, Ælfc. Gr. 49; Zup. 282, 16. I. as an ordinal:—Se forma . . . se óðer . . . se þrydda (ðirda, Lind.: þridde, Rush.), Mt. Kmbl. 22, 26. Se þridda (ðirdda, Lind.: ðirda, Rush.), Mk. Skt. 12, 21. Twégen men . . . mǽg wæs his ágen þridda, hé feórða sylf, Cd. Th. 173, 29; Gen. 2868: Elen. Kmbl. 1707; El. 855. Heofonwaru and eorðwaru, helwaru þridde, Hy. 7, 95. Ðǽre þriddan eá nama, Gen. 2, 14. On nánum heolstrum heofenan, oþþe eorþan, oþþe sǽ þriddan, Homl. Th. ii. 146, 32. Ðý þryddan dæge (ðe ðirda dæg, Lind.), Mt. Kmbl. 16, 21. On ðære þriddan (ða ðirdda, Lind.: ðirda, Rush.) wæccan, Lk. Skt. 12, 38. Æfter ðon ðridan dæge, Blickl. Homl. 181, 2. Nán þridde be him sylfum ne lét hé he admitted no third hypothesis about himself, Homl. Skt. i. 23, 633. I a. marking degrees of relationship:—Þridde fæder proavus, þridde móder proavia, Wrt. Voc. i. 51, 55, 56. Þridde fæder abavus, 72, 21. Mínes fæderan þridda fæder abpatruus meus, ii. 8, 24. Þridda sunu pronepus, 62, 36. II. fractional. v. twǽde:—Bewyl óþ þriddan dǽl, Lchdm. ii. 120, 15. Seóþe tó þriddan dǽle, i. 98, 7. [Goth. þridja: O. Sax. thriddio: O. Frs. thredda: O. H. Ger. dritto: Icel. þriþi (gen. þriþja).]

þridding (?), e; f. The doing of a thing for the third time (? Halliwell gives thirding with this meaning as a Suffolk word):—Ðirding (but the word has been altered to ðirde. v. Skeat's collation) scipdrincende tertio naufragantem, Rtl. 61, 31.

þridung, þrie-, þrielig, þriétan. v. þreodung, þri-, þrilig, þrítan.

þri-ex; n. ? :—On ðæt þri ex; of ðam þri exe, Cod. Dip. Kmbl. iii. 436, 28.

þri-feald; adj. Threefold, triple:—Ic cwæþ ðæt sió sáwul wǽre þriofeald, Bt. 33, 4; Fox 132, 2. Þriefald, Met. 20, 183. Ðrifald trinus, Rtl. 111, 8: Mt. Kmbl. p. 14, 6. Ðryfeald, Homl. Th. ii. 606, 24. Ðryfeald triplex, Ælfc. Gr. 9, 61; Zup. 70, 2: 49; Zup. 284, 17. From ðæm þriefealdan (triplici) brægene, Wrt. Voc. ii. 80, 57. On þreofealdum húse iu triclinio, 45, 80. Þreofealdum fæce terna intercapedine, Hpt. Gl. 462, 76. Mid þreofealdre lencge terna proceritate, 445, 7. Ðú þriefalde on ús sáwle gesettest, Met. 20, 176. Þryfealdne (þreo-, MS. B.) áð . . . þryfealde (þri-, MS. B.) láde mid þryfealdan foráðe, L. C. S. 22; Th. i. 388, 12–15. Ða þriefealdan sáwla, Bt. 33, 4; Fox 130, 39. Ðrifaldo ternos, Rtl. 193, 33. [O. Frs. thri-fald: O. H. Ger. dri-falt: Icel. þrí-faldr.]

þrifealdlíce; adv. Triply:—Ðriof[e]ealdlíce tripliciter, Kent. Gl. 839. Þriefealdlíce (þry-, MS. B.: þri-, MS. H.), L. Alf. pol. 39; Th. i. 88, 3.

þri-feoðor; adj. Triangular:—Ðrifeoðor, ðrifedor, trifoedur triquadrum, Txts. 103, 2052.

þri-fête; adj. Having three feet:—Þryfête tripes, Ælfc. Gr. 9, 26; Zup. 51, 12: 49; Zup. 287, 20. Þriefête rícelsfæt cythropodes, Wrt. Voc. ii. 15, 60. [Icel. þrí-fættr.]

þrifildan; p. de To triple:—Ic þryfylde triplico, Ælfc. Gr. 49; Zup. 287, 4. [Icel. þrísalda.]

þri-fingre; adj. Three fingers thick:—Æt þryfingrum (spic is added in MS. B.), L. In. 49; Th. i. 132, 18.

þri-fingre; adv. By a distance equal to the breadth of three fingers:—Gif se ord sié þreofingre ufor ðonne hindeweard sceaft, L. Alf. pol. 36; Th. i. 84, 17. Cf. Grmm. R. A. 101.

þri-flére; adj. Having three floors, three-storied:—Ðæt gyftlíce hús wæs ðryflére, for ðan ðe on Godes geladunge sind þrý stæpas gecorenra manna, Homl. Th. ii. 70, 17.

þri-fótede, -fótad; adj. Three-footed:—Þrifótede tripes, Ælfc. Gr. 49; Zup. 287, 20 note. Þryfótad fæt trisilis, Wrt. Voc. i. 25, 30.

þri-fyrede; adj. Three-furrowed; the word renders Latin trisulcus, Ælfc. Gr. 49; Zup. 288, 12 note.

þriga. v. þriwa.

þri-gǽrede; adj. Cloven into three parts, three-pronged:—Þrygǽrede (þreo-, þrio-) trifidus, Ælfc. Gr. 49; Zup. 288, 10. v. gár, gára.

þri-geáre; adj. Three years old:—Se onféng fulwihte ða hé wæs þrigeáre cniht, Shrn. 119, 19. [Icel. þrí-ærr.]

þri-geáre, es; n. A space of three years:—Þrigeáre (þreóra geára ferst) triennio, Hpt. Gl. 519, 15. [Icel. þrí-æri; n.]

þri-gilde; adj. To be paid threefold:—Cleroces feoh .iii. -gylde, L. Ethb. 1; Th. i. 2, 6. [Icel. þrí-gildr of threefold value; þrí-gilda to pay threefold.] v. twi-gilde; adj. and subst.

þri-gilde; adv. (or case of a noun þri-gilde. v. twi-gilde; subst. and adv.) With a treble payment:—Gif man inne feoh genimeþ, se man .iii.-gelde gebéte, L. Ethb. 28; Th. i. 10, 1. Gylde hé hit þrygylde, L. A. G. 3; Th. i. 154, 11. Gange hé tó ánfealdum ordále oþþe gilde .iii.-gylde, L. Eth. iii. 4; Th. i. 294, 15.

þri-heáfdede; adj. Three-headed:—Þryheáfdede triceps, Ælfc. Gr. 9, 55; Zup. 67, 11. [Icel. þrí-höfðaðr.]

þrihing for (?) þriþing. v. trehing.

þri-híwede; adj. Having three forms:—Ðryhíwede triformis, Ælfc. Gr. 49; Zup. 287, 10.

þri-hlidede; adj. Three-lidded, having three openings:—Ðryhlidede tripatens, Ælfc. Gr. 49; Zup. 288, 6.

þri-hyrne; adj. Three-cornered, triangular:—Ðæt sǽd byþ þreohyrne, Lchdm. i. 316, 10.

þri-hyrnede; adj. Triangular:—Þryhyrnede triangulus, Ælfc. Gr. 49; Zup. 289, 4. [Icel. þrí-hyrndr.]

þri-leáfe, -léfe, an; f. Trefoil (cf. later, three-leaved grass, triple grass. v. E. D. S. Pub. Plant Names):—Geáces súre vel þrileófe trifolium, Wrt. Voc. i. 30, 24.

þrilen; adj. Woven with three threads:—Þrylen hrægel trilicis vestis, Wrt. Voc. i. 40, 19.

þrili. This word has the form of an i-stem adjective in the glosses ðrili trilex, Txts. 35, 29; drili triplex, 115, 158; and that þril- is the main part of the word seems suggested by þrilen (q.v.), by þrielig in þrielig hrægil triligium, Wrt. Voc. i. 289, 53, and by later English þrile, e.g. An God, þrile in þreo hades, A. R. 26, note a; þrumnesse þreofald ant anfaldte, þrile i þreo hades, Marh. 11, 27. þrille-hod trinity, C. L. 1239. Cf. too, O. H. Ger. drilero triplici (catena). On the other hand it might seem that the form is þri-li from comparison with aen-li simplex, Txts. 115, 156; cf. too, O. H. Ger. dri-líh drilex (tunica), dri-líha trilicem (tunicam): Ger. drillich ticking. Perhaps the word has been influenced by the Latin which it translates. v. twi-líc, and next word.

þri-líc; adj. Threefold:—Án myhtylíce and þrylíc hádelíce unus potentialiter trinusque personaliter, Hymn. Surt. 29, 13: 55, 13: 105, 15. Eálá ðú ðrilíc godcundnyss, 133, 5. Ðé þrylícne and ǽnne, 146, 32. v. preceding word.

þrilig; adj. Woven with three threads:—Þrielig hrægil triligium, Wrt. Voc. i. 289, 53. v. þrili.

þri-líðe (?); adj. Having three months named Líða, a term applied to the year in which a fourth summer month was intercalated; the passage in which the Latinized form of the word occurs is as follows: Quotiescunque communis esset annus, ternos menses solares singulis anni temporibus dabant, cum vero embolismus, hoc est xiii mensium lunarium annus occurreret, superfluum mensem aestati apponebant, ita ut tunc tres menses simul Lida nomine vocarentur, et ob id annus thrilidus cognominabatur habens quatuor menses aestatis, ternos, ut semper, temporum caeterorum. Beda de temporum ratione, c. 13. v. Grmm. Gesch. D. S. c. vi.

þrimen a third:—Nim sealtes þrymen, Lchdm. ii. 124, 4. [O. Frs. thrimen (-in) amounting to a third; thrimenath a third part.]

þri-milce, es; m. (?) The early name for the month of May:—Se fífta mónað is nemned on úre geðeóde Ðrymylce, for ðon swylc genihtsumnes

wæs geó on Brytone and eác on Germania lande, of ðæm Ongla ðeód com on ðás Breotone, ðæt hí on ðæm mónðe þriwa on dæge mylcedon heora neát (Bede's Latin is: Thrimilci dicebatur, quod tribus vicibus in eo per diem pecora mulgebantur; talis enim erat quondam ubertas Britanniae vel Germaniae, e qua in Britanniam natio intravit Anglorum, De temp. rat. c. 13), Shrn. 77, 37. Ðonne Ðrymelces mónað bið geendod ðonne bið seó niht eahta tída lang, 87, 28. Ðrymlce mónað, Chr. Erl. Introd. xxxi, margin.

þrim-feald; adj. Threefold:—On ðam þrimfealdan (þry-, MSS. B. L.) ordále, L. Ath. i. 4; Th. i. 202, 4. On ðam þrimfealdum (þry-) ordále, 6; Th. i. 202, 13. Æt þrimfealdre (þryfealdre, 17) spræce, L. Eth. ix. 19; Th. i. 344, 13. Be ánfealdum simplum, be twyfealdum duplum, be þrimfealdum triplum, Ælfc. Gr. 49; Zup. 286, 18. [Cf. Ileafan on þa halʒa þreomnesse, O. E. Homl. i. 99, 34. Þe heuenliche þremnesse was mid him, ii. 137, 7. Þrumnesse, A. R. 160, 10. Þrimm-nesse, Orm. 11177. Cf. also O. Frs. thrim-dél (threm-) a third.]

primsa, þrindende, Exon. Th. 431, 23; Rä. 46, 5. v. trimes, þindan (?), þrintan (?).

þrinen; adj. Threefold; trinus:—God ánfeald and samod þrynen (trinus), Hymn. Surt. 105, 3. God þrynen and án, 115, 37: 137, 31. Þrynenum gebede trina oratione ... Mid þrynum tódále trina partitione, Anglia xiii. 380, 214, 217. God ðæne ðrynenne on ánnesse and ænne on ðrynnesse wé andettaþ Deus quem trinum in unitate et unum in trinitate confitemur, Wanl. Cat. 292, col. 1. [Cf. Icel. þrinnr.]

þring. I. a press, crowd. [Utforen al þan dringe (þringe, 2nd MS.), Laym. 14966. Amidden þan þrunge (þringe, 2nd MS.), 27524. Cf. Among þe þrenge of sipmen, 2229 (2nd MS.). Myd wel muchel þrynge, Misc. 86, 72. Cf. No þrung of folc, A. R. 162, 8.] v. eofor-, ge-þring. II. (or þryng ?) what presses or confines :—Þryng cannalis, Wrt. Voc. ii. 128, 5. [Cf. Icel. þröng a strait, a narrow place.]

þringan; p. þrang, pl. þrungon; pp. þrungen. I. trans. To press, crowd, throng:—Ðás menegeo ðe ðringaþ turbae te comprimunt, Lk. Skt. 8, 45. Ðæt folc hine þrang, Homl. Th. ii. 394, 17. Þrungun torquent, Wrt. Voc. ii. 122, 56. Hí þrungon (geðringdon, Lind.: on ðrungun, Rush.) comprimebant illum, Mk. Skt. 5, 24. Ðú gesyxst ðás menigu ðe ðringende (ðringende on ðec, Rush.) uides turbam comprimentem te, 31. II. to throng, press round, upon, crowd together :—Hý ymb þeódenstól þringaþ georne, Exon. Th. 25, 8; Cri. 397: 208, 30; Ph. 163. Fugla cynn on healfa gehwone heápum þringaþ contrahit in coetum sese genus omne volantum, 221, 18; Ph. 336. Gelíc sumum ðara gumena ðe him geornost mid þegnungum þringaþ ymbe útan, Met. 25, 28. Wer-gendra tó lyt þrong ymbe þeóden, Beo. Th. 5758; B. 2883. Ðá him ðæt folc swiðost an þrang ubi se obrui a circumfusa multitudine persensit, Ors. 3, 9; Swt. 134, 18. Duguð samnode, hildfrecan heápum þrungon, Andr. Kmbl. 252; An. 126. III. to press, move with violence, eagerness or hurry, press on, press forward, force a way :—On hú grund-leásum seáðe ðæt mód þringþ ... hit þringþ on ða fremdan þístro tendit in externas ire tenebras, Bt. 3, 2; Fox 6, 7-10: Met. 3, 7. Sum on oferhygdo þryme þringe (þrymme þringeþ? cf. below Homl. Skt. ii. 25, 781: Rä. 4, 61), Exon. Th. 314, 34; Mód. 24. Hé on ðæt weorod þrong for ðon ðe him wæs leófre ðæt hiene mon ofslóge ðonne hiene mon gebunde he pressed into the host (of the enemy), because he would rather be slain than made prisoner, Ors. 5, 12; Swt. 244, 12. Se ðe mid gebeóte and mid micclum þrymme þrang intó ðam temple, Homl. Skt. ii. 25, 781. Him arn on lást, þrang þýstre genip dark cloud made its resistless way, Cd. Th. 9, 9; Gen. 139: Exon. Th. 179; Gú. 1255. Wræccan þrungon (pressed forward), 461, 28; Hö. 42: Elen. Kmbl. 245; El. 123. Hí þrungon and urnon ongeán ða ðeódnes mægþ, Judth. Thw. 23, 40; Jud. 164. Tó weallgeatum wígend þrungon, Andr. Kmbl. 2408; An. 1205: Beo. Th. 5913; B. 2960. Tó ðam swicce men on healfa gehwone heápum þrungon, Exon. Th. 359, 24; Pä. 67. Hé lǽteþ word út faran, þræfte þringan, 316, 1; Mód. 42. Ic gewíte þringan þrymme micle, 386, 13; Rä. 4, 61. Ne þurfon gé nó hogian on ðæm anwealde, ne him æfter þringan, Bt. 16, 1; Fox 50, 30. Hé lét wille-burnan on woruld þringan, Cd. Th. 83, 2; Gen. 1373. Tó ðam wícum hí cwómon hlóþum þringan, Exon. Th. 156, 1; Gú. 868. Ðá ongan ic nýdwræclíce gemang ðam folce wið ðæs folces (temples ?) þringan, Homl. Skt. ii. 23 b, 405; Judth. Thw. 25, 8; Jud. 249. [(1) His sporis he gynneth in hors thryng, Alis. 2388. Cumpanyes thringen thee, Wick. Lk. 8, 45. Gif eiþer oþer faste þringe, O. and N. 756. To noght he thrange (redegit) Israele, Ps. 77, 59. Liknes of þa to noght thryng saltou (rediges), 72, 20. I am to noghte thrungen, 22. (2) Þe folc cumþ fastlice and elce deʒie þicce þringeð, O. E. Homl. i. 237, 29. A thousand of men thrungen togyderes, Piers P. 5, 517. (3) Iudas him com þrynge, Misc. 42, 177. Into þe deuelez þrote man þryngeʒ bylyue, Allit. Pms. 43, 180. Carrais him on þrong (Carais to þrong, 2nd MS.), Laym. 10652. Through her hert the sword throng, Gow. iii. 262, 7. Þrungen euchan biuoren oþer forte beo bihefdet, Jul. 67, 11. Binnen heo þronge (alle in þronge, 2nd MS.), Laym. 9421. O. Sax. thringan (trans- and intrans.): O. H. Ger. dringan urgere, stipare: Icel. þryngva. Cf. Goth.

þreihan.] v. á-, æt-, be-, for-, ge-, of-, on-, óþ-, tó-, ymb-þringan; út-áþrungen.

þri-nihte; adj. Three days old :—Gif hé biþ ácenned on .iii.-nihte mónan, Lchdm. iii. 160, 20: 176, 22, and note 2. [Icel. þrí-nættr.] v. twi-nihte.

þrinna. This seems a Scandinavian form [cf. Icel. þrennar tylftir three twelves ; e. g. þrennar tylftir eigu at dæma málit, Njála c. 144] :—Ládige hé hine mid þrinna .xii., L. Eth. iii. 13; Th. i. 296, 29.

þrinness, þriness, e; f. Trinity, mostly in the special sense the Trinity :—Ðæs mannes sáwl hæfð ðære hálgan þrynnysse anlícnysse; for ðan ðe heó hæfð on hire ðreó ðing ... Is hwæðere se man án man, and ná ðrynnys, God ... þurhwunaþ on ðrynnysse háda and on ánnysse ánre godcundnysse; nis ná se man on ðrynnysse wunigende, swá swá God, Homl. Th. i. 288, 17-35. Ðeós þrynnys is án God, 10, 7. Is seó hálige þrinnis on ðisum þrim mannum, Ælfc. T. Grn. 2, 8. For ðí is gecweden ' uton wyrcan,' ðæt wǽre geswutelod ðære hálgan þrynnysse weorc on ánnysse. Seó hálige þrynnys is undergiten on ðam worde ' uton wyrcan,' Boutr. Scrd. 19, 12. Ðrínes trinitas, Ps. Surt. ii. p. 202, 23. Þrynes, Exon. Th. 24, 4; Cri. 379. Þrynis, 286, 3; Jul. 726. Þrynysse þrym, 37, 26; Cri. 599. Of ðæm mægene ðære hálgan þrynesse, Blickl. Homl. 29, 12. On ðære hálgan þrynnysse, 249, 23. Mid þrym fingrum man sceall sénian for ðære hálgan þrynnysse (ðrymnysse, MS. U.; v. Middle English quoted under þrim-feald), Homl. Skt. ii. 27, 156. Clypung tó ðære hálgan Ðrynnisse invocatio ad sanctam trinitatem, Hymn. Surt. 1, 1. Wé andettaþ ... ðrynnesse in ánnesse and ánnesse on ðære ðrynnesse, Bd. 4, 17; S. 585, 36. For ða háligan ðrinesse, Rtl. 114, 17. Þrynesse, Blickl. Homl. 205, 30. [O. H. Ger. drinissa.]

þrintan; p. þrant, pl. þrunton; pp. þrunten To swell :—Þrinteþ, Exon. Th. 315, 1; Mód.. 24. v. á-þrintan.

prió, þrióstrig. v. þrí, þeóstrig.

þripel, es; m. An instrument of punishment, a kind of cross :—Eculeus vel þrýpel genus tormenti, Wrt. Voc. ii. 142, 25. Unhêh þrêpel eculeus (equuleus patibulum, furca cui decollatorum martyrum cadavera affigebant, Migne), i. 21, 18. v. þreápian, and next word.

þripel-úf (?) an instrument of punishment :—Wæarhród vel þrýpelúf eculeus vel catasta, Wrt. Voc. i. 55, 52. v. þripel.

þri-rêðre; adj. Having three banks of oars; used substantively trireme :—Ðá næfde hé má scipa ðonne án; ðæt wæs ðeáh þrerêþre, Bt. 38, 1; Fox 194, 10. Þrierêþre ceól, Met. 26, 27. Án .C. ðara miclena þrierêðrena centum triremes, Ors. 3, 1; Swt. 96, 27: 5, 13; Swt. 246, 6. [O. H. Ger. dri-ruodri.]

prisce. v. þrysce.

þri-scite; adj. Triangular, three-cornered :—Ispania land is þryscyte Hispania trigona est, Ors. 1, 1; Swt. 24, 1. Sicilia is ðryscýte Sicilia tria habet promontoria, Swt. 28, 2. On ðone þryscýtan crundel, Cod. Dip. Kmbl. v. 374, 26. [Cf. Icel. þrí-skeyta a triangle.]

þrisel; adj. Divided into three. v. twisel, and next word.

þrisel-líc; adj. Tripartite :—Ðǽr beóþ men ácende þreosellíces híwes nascuntur homines tripartito colore, Nar. 35, 29. v. twisel.

þri-slite, -slitte (?); adj. Three-forked, three-pointed :—Hæfdon ða wyrmas þriesli[ ]te (a letter has been erased before the t, see note, and Anglia i. 510, iv. 151) tungan cum trisulcis linguis, Nar. 14, 12. Cf. next word.

þri-snæcce, -snæce, -snece; adj. Three-pointed, cloven in three :—Þrysnece (-snæcce, -snæce) tungan hæfþ seó næddre trisulcam linguam habet serpens, Ælfc. Gr. 49; Zup. 288, 12. v. twi-snæcce.

þríste and þrist; adj. Bold, (1) in a good sense :—Móna se sixta ... se ðe bið ácenned, þríste, mǽre, Lchdm. iii. 186, 15. Ic ofstikode bár. Swíþe þrýste (audax) ðú wǽre ðá, Coll. Monast. Th. 22, 19. Þríste sceal mid cénum, Exon. Th. 337, 8; Gn. Ex. 61. Gewát hé (Andrew) þríste on geþance, Andr. Kmbl. 473; An. 237. Elene, þríste on geþance, Elen. Kmbl. 533; El. 267. Eorl unforcúð, elnes gemyndig, þríst and þrohtheard, Andr. Kmbl. 2529; An. 1266. Þríst, þonces gleáw, Exon. Th. 207, 19; Ph. 144. Geþinga ús þrístum wordum, 21, 30; Cri. 342: Cd. Th. 242, 27; Dan. 425. Ic ǽnig ne mêtte þrístran geþóhtes, mægþa cynnes, Exon. Th. 275, 14; Jul. 550. (2) in a bad sense, bold, presumptuous, audacious, shameless :—Ðý læs hé tó þríste sié for ðý underfenge his láreówdómes ne doctrinam praesuntio extollat, Past. proem.; Swt. 23, 23. Ðonne he wilnaþ on his móde ðæt hé sciele rícsian hé bið swíðe forht and swíðe behealden; ðonne hé hæfð ðæt hé habban wolde, hé bið swíðe þríste mens principari appetens fit ad hoc pavida, cum quaerit, audax cum pervenerit, 9; Swt. 57, 4. Mǽden ofermódig, þríste on líchaman mid manegum werum, Lchdm. iii. 190, 16. (Cf. mǽden módig. dyrstig, manega weras wilnigende, 25.) Ic (the devil) wênde þríste geþoncge, ðæt ic ðe meahte þlást earfeþum áhwyrfan from hálor, Exon. Th. 264, 2; Jul. 358. Forhwon beóð suǽ ðríste ða ungelǽredan ðæt hí underfón ða heorde ðæs láriówdómes ab imperitis pastorale magisterium qua temeritate suscipitur? Past. 1; Swt. 25, 16. Ða ðe tó ðam þríste sýn, ðæt hig God oferseóð, Wulfst. 270, 23. Hié wǽron womma ðríste, inwitfulle, Cd. Th. 77, 9; Gen. 1272. Tó frece, synna þríste, 155, 31; Gen. 2581. Wǽron Sodomisc cynn synnum þríste, 116, 13;

Gen. 1935. Wed gesyllan eallra unsnyttro, þrístra geþonca, Elen. Kmbl. 2569; El. 1286. Ic þrísta sum þeófes cræfte, Exon. Th. 486, 24; Rä. 72, 20. [Ʒif he were swa þriste, and he hit don durste, þ he heom wolde leaden, Laym. 356. Þer þe dusie mon bið þriste, O. E. Homl. i. 117, 23. Ne helpþ noht þat þu beo so þriste, ich wolde fihte bet mid liste þan þu mid alle þine strengþe, O. and N. 171. To uvele we beoþ al to þriste, P. L. S. 8, 10. O. Sax. thrísti: Ger. dreist.] v. ellen-, gár-, un-, wíg-þríste(-þríst).

þríste; adv. Boldly, (1) in a good sense, confidently, without apprehension, fear, hesitation, reserve:—Hé þríste geneðde on óðre dælas, Apstls. Kmbl. 100; Ap. 50. Hé þríste bebeád, ðæt hié his láre lǽston, Andr. Kmbl. 3303; An. 1654: Elen. Kmbl. 818; El. 409. Ne wæs ǽnig ðæt nē þus þríste hrínan dorste, Exon. Th. 273, 4; Jul. 511. Heó ne meahte þríste geþencan, hú ymb ðæt sceolde she could not think with confidence of the event, 378, 6; Deór. 12. Ic mundbyrd on ðé þríste hæfde in te confirmatus sum, Ps. Th. 70, 5. Ic ðín bebod þríste gelýfde, 118, 66. (2) in a bad sense, without sense of shame, presumptuously, audaciously:—Heó þríste ongan wið Sarran swíðe winnan, Cd. Th. 135, 10; Gen. 2240. Gē him þríste oftugon ye had no misgivings when ye refused them help, Exon. Th. 92, 18; Cri. 1510. (3) in the Psalms þríste seems used several times with an intensive force, much as swíðe is used:—Þa þearfendan þríste Drihten gehýreþ holdlíce hears attentively and graciously; exaudivit pauperes Dominus, Ps. Th. 68, 34. Hé þearfena bearn þríste hǽleþ completely saves; salvos faciet filios pauperum, 71, 4: 81, 4: 112, 6. Ealle hine þeóda þríste heriaþ greatly praise; magnificabunt, 18. Þríste ongunnon georne slépan ða ðe on horsum wǽron, 75, 5. Ðú míne geðóhtas þríste oncneówe thou didst thoroughly know my thoughts; intellexisti cogitationes meas, 138, 2. Hé þearfendra ðríste ēhte he persecuted the poor exceedingly; persecutus est hominem pauperem, 108, 16.

þríst-full; adj. Presumptuous:—Þrístfulle presumptuosi, Anglia xiii. 369, 55.

þríst-hycgende; adj. (ptcpl.) Thinking or intending boldly, firm of purpose, having bold resolve:—Ðú geþóhtest þrísthycgende, ðæt ðú ðinne mægðhád Meotude sealdes bútan synnum, Exon. Th. 18, 24; Cri. 288. On þeóde geþeón, ðæt hē wese þrísthycgende, 336, 17; Gn. Ex. 50.

þríst-hygdig, -hýdig; adj. Bold-minded, courageous:—Þióden þrísthýdig, Beo. Th. 5612; B. 2810. Nergend ðrísthýdigum Thómase forgeaf ēce ríce, bealdum beornwigan bletsunga his, Menol. Fox 443; Men. 223. Sum biþ æt þearfe þrísthýdigra þegn mid his þeódne, Exon. Th. 298, 1; Crä. 78. [Cf. O. Sax. thríst-mód thegan (Peter).]

-þrístian. v. ge-þrístian.

þríst-lǽcan; p. -lǽhte; pp. -lǽht To become bold, to dare, presume:—Wē þrístlǽcaþ biddan audemus rogare, Hymn. Surt. 111, 34. Hié sint tó manianne ðæt hié nó ðý swíður wið hié ne ðrístlǽcen (ðrísd-, Hatt. MS.) admonendi sunt, ne contra eos audaciores fiant, Past. 28; Swt. 196, 5. Be ðære árfæstan Godes cennestran mildheortnysse þrístlǽcende, ic mē of ðære stówe ástyrede, ðe ic ðis gebæd, Homl. Skt. ii. 23 b, 457. [Awah þet he efre wulle þristelechen oðer biþenchen mid his fule heorte þe heo wulle underfon swa heʒ þing swa is Cristes licome in his sunfulle buke, O. E. Homl. i. 25, 30.] v. ge-þrístlǽcan.

þrístlǽcness, e; f. Boldness, audacity, temerity, presumption:—Ic eom ondetta . . . ðrístlǽcnesse mínra synna, Anglia xi. 98, 22. Gyf man þurh þrýstlǽcnysse man fullaþ si quis ex temeritate aliquem baptizaverit, L. Ecg. P. addit. 30; Th. ii. 236, 34.

þrístleásness (?), e; f. Want of boldness:—Ic eom andetta . . . þrístleásnyssa mínra synna, Anglia xi. 101, 34. v. preceding word.

þrístlíce, þrístelíce; adv. Boldly, confidently, (1) in a good sense:—Hé sprӕc þrístlícor mid hine confidenter ait, Gen. 44, 18. (2) in a bad sense:—Ða underðióddan sint tó manianne ðæt hié ðara undeáwas ðe him ofergesette bióð tó swíðe and tó ðrístelíce (ðrísdlíce, Hatt. MS.) ne eahtigan admonendi sunt subditi, ne praepositorum suorum vitam temere judicent, Past. 28; Swt. 196, 1. Oft þeóf þrístlíce sorgleáse hǽleð forfēhð, Exon. Th. 54, 21; Cri. 872. Ðú (the devil) þrístlíce þeóde lǽrest, Andr. Kmbl. 2371; An. 1187.

þrístling (?), es; m. A bold person; found in the local name which occurs in the following passage:—On þrístlinga dene; of þrístlinga dene ufeweardre, Cod. Dip. Kmbl. iii. 82, 28. v. þríste.

þrístness, e; f. Boldness, presumption, temerity:—Ðrísnes praesumptio, Kent. Gl. 1169. Mid þrýstnesse dyrstigere praesumptione temeraria, Anglia xiii. 383, 262. On þrístnysse in temeritate, Scint. 139, 3. Hē yfel þurh þrístnysse gefremede malum per audaciam perpetravit, 40, 5.

þrí-strenge; adj. Three-stringed:—Þrystrenge (þreo-, þrío-) trifidus, Ælfc. Gr. 49; Zup. 288, 10. v. twi-strenge.

þrítan; p. te. I. to weary:—Ðæt folc weard þrít and þearle geswenct mid ðam síðfæte taedere coepit populum itineris ac laboris, Num. 21, 4. II. to urge, press, force:—Seó wyrd ðe þriétaþ (-eþ?) ða yflan tó wítnianne fortuna quae justo supplicio malos coercet, Bt. 40, 2; Fox 236, 25. [Icel. þreytask to be exhausted; þreytt tried, exhausted; Dan. træt wearied.] v. á-, ǽ-þríetan (-þrítan) þreátian.

þrítig, þrittig; num. Thirty:—Þrittig, þrítig triginta, Ælfc. Gr. 49; Zup. 281, 17. I. used substantively as a neuter. (1) governing

a noun in the genitive, when the inflections are gen. -es, dat. -um. (a) alone:—Ðam sceal .xxx. scitt. tó bóte (cf. in next line: Ðam sculon .v. scitt. tó bóte), L. Alf. pol. 56; Th. i. 94, 28. Hwæt gif ðær beóð þrítig? God cwæð: Ne dô ic him ná láð, gif ðær beóþ þrítig rihtwísra, Gen. 18, 30. Þrittig fæðma biþ se arc on heáhnisse, 6, 15: Jn. Skt. 6, 19. Ymb þrittig wintra, Bt. 39, 3; Fox 214, 25. Ymb þrítig wintergerímes, Met. 28, 25. Ymb þrítig geárgerímes, 29. Hé genam þrítig þegna, Beo. Th. 246; B. 123. Dô hí ealle tógædere, ðæt þrítig seolforsticca, Anglia xi. 8, 19. Wintra ðrittih (ðrítig, Rush.), Lk. Skt. Lind. 3, 23. Hé wæs ðrítiges geára eald, Past. 49; Swt. 385, 15. Þrítiges míla brád, Chr. 893; Erl. 88, 29. Ðrittiges heáh elngemeta, Cd. Th. 79, 8; Gen. 1308. xxx.-tiges manna mægencræft, Beo. Th. 764; B. 379. Ða hæþenan ðrittigum síþa mǽre weorud hæfdon, Bd. 3, 24; S. 556, 22. (β) in combination with other numerals:—Þreó and þrítig geára, Cd. Th. 296, 16; Sat. 503. Eahta and þrittig (ðrittih, Lind.: ðrítig, Rush.) wintra, Jn. Skt. 5, 5. Hé ríxode twá læs xxx geára, Chr. 641; Erl. 27, 16. Mid feówer hunde scipa and þrítigum, Ors. 4, 6; Swt. 172, 31. (2) as a plural with gen. -a:—Com se cyning þrítiga sum (þrittigum sum, MS. E.) ðara monna ðe in ðam here weorþuste wǽron, Chr. 878; Erl. 80, 20. II. used adjectively, (1) alone:—Þrítig þúsend wera, Jos. 8, 3. Þrítigon síðon tricies, Ælfc. Gr. 49; Zup. 286, 2. Cf. þrittig síðon seofon beóð twá hundred and týn, Anglia viii. 303, 7. Þrítig síðon twelf, 29. Þrittigum sýþum hundteóntig þúsenda, Blickl. Homl. 79, 25. Ða þryttig scyllingas, Mt. Kmbl. 27, 3. (2) in combination with other numerals:—Ðæt is ealles .xxx. and vi. peningas, Anglia xi. 8, 18. Mid þrym and ðrittigum mannum, Homl. Skt. i. 5, 128. Mid ðám aþelestum ceastrum ânes wana ðrittigum civitatibus viginti et octo nobilissimis, Bd. 1, 1; S. 473, 26. ¶ In the following passage the construction is unusual:—On þrýtiges wintres ylde, Anglia xi. 2, 26. II a. of age, thirty (years old):—Se Hǽlend wæs þrítig ðá hine maun fullude, Anglia xi. 3, 77. III. used in forming ordinals:—Se wæs fram Agusto ðridde eác ðrittigum tricesimus tertius ab Augusto, Bd. 1, 6; S. 476, 17. [Goth. þrins tiguns (acc.): O. Sax. thrítig: O. Frs. thrítich: O. H. Ger. drízug: Icel. þrír tigir.]

þrítig-feald; adj. Thirty-fold:—Mid þrittisfealdne hēhnysse tricena altitudine, Hpt. Gl. 445, 8. Sum berð þrittigfealdne wæstm, Homl. Ass. 21, 175: Mt. Kmbl. 13, 8. Þrítigfealdne, Mk. Skt. 4, 20. Ða habbaþ þrittigfealde mēde, Homl. Ass. 21, 179.

þrítigoða; num. adj. Thirtieth:—Se þrittigoða (ðrítogoða, þritteogoða, þreotteogaþa) tricesimus, Ælfc. Gr. 49; Zup. 283, 10. Ðý ðrittigoþan geáre, Bd. 5, 23; S. 647, 29.

þrítig-wintre, -wintra; adj. Thirty years old:—Iósep wæs þrítigwintre triginta annorum erat Joseph, Gen. 41, 46: Lk. Skt. 3, 23: Homl. Th. i. 26, 3. Ða ðá Crist wæs þrítigwintra (or þrítig wintra), Homl. Th. ii. 38, 25.

þriwa, þrywa, þreowa, þriowa, þriuwa, þriga, þrige, þría; adv. Thrice, three times:—Þriwa (þreowa) ter, Ælfc. Gr. 49; Zup. 285, 14: 38; Zup. 232, 7: Exon. Th. 207, 20; Ph. 144. Þriwa on gēre tribus vicibus per singulos annos, Ex. 23, 14: ter in anno, 17. Ne sint ðæt þreó godas þriwa genemned, ac is ân God, Hy. 10, 44. Þriwa (ðriga, Lind.: þriowa, Rush.) ter, Mt. Kmbl. 26, 34. Þriuwa, Rush. 75. Þriwa (ðria, Lind.: ðrige, Rush.) þriwa (ðriga, Lind.: ðrige, Rush.), Lk. Skt. 22, 61. Þriwa (þrywa, MS. A.), Jn. Skt. 13, 38: 21, 14. Þriga, Wrt. Voc. ii. 126, 35. Ðriga, Bd. 1, 13; S. 481, 42. ¶ With numerals:—Cwed þriwa nigon síþan, Lchdm. i. 202, 11. Þriwa seofon beóð ân and twēntig, Anglia viii. 302, 43. Þriwa feówer beóð twelf, 328, 21. [Þrie twenti sixty, H. M. 23, 29. Þrie he eode abuten, Laym. 17432. Þreie (þries, 2nd MS.), 26066. Þrien, 14352: þreoien, 14338. Þries, A. R. 106, 18. Þriʒess, Orm. 1149. Þriʒʒess, 5945. Thrie, Alis. 1263. Þrye, R. Glouc. 191, 14. Þries, Ayenb. 35, 11. O. Sax. thriwo, thriio: O. L. Ger. thrío: O. Frs. thría, thriia.]

þri-wintre, -wintra, -winter; adj. Of three years, three years old:—Thriuuintri (ðriuuintri, Corpus Gl.) steór prifeta, Txts. 86, 780. Þrywintre (-wintra) triennis, trimus, Ælfc. Gr. 49; Zup. 287, 13, 18. Þriwinter trimus, vel triennis, vel trimulus, Wrt. Voc. i. 21, 59. Þrywinter triennis, 23, 53. Wæs cnihtcild sume ne wæs yldre ðonne ðrywintre erat puer trium circiter, non amplius, annorum, Bd. 4, 8; S. 575, 27. Geoffra mē ân þriwintre hrýðer and ǽnne þriwintre ramm and ânne þriwintre gát sume mihi vaccam triennem et capram trimam et arietem annorum trium, Gen. 15, 9. Hé ásende him tó ân ðrywintre cild, Homl. Th. ii. 134, 7. [Icel. þrí-vetr three years old.]

þroc, es; n. I. a throck (v. E. D. S. Pub., Cheshire Gloss., where is quoted: 'The Throck is the piece of Timber on which the suck (share) is fixed.' Academy of Armory by Randle Holmes. Also spelt thruck):—Dentale, s. est aratri pars prima in qua uomer inducitur quasi dens sule reóst vel þroc, Wrt. Voc. ii. 138, 72. (v. Wülck. Gl. 219, 4.) II. a table:—Mynetera þrocu hē tóbræc mensas nummulariorum euertit, Mk. Skt. 11, 15. [Cf. O. H. Ger. druh; f. cippus, compes.]

þróh glosses rancor:—Rancor þróh (thróch, Erfurt Gl.) vel invidia, vel odium (tróh rancor, Corpus Gl.), Txts. 92, 874. v. next word.

þróh; adj. glosses rancidus :—Of ðrón æsðancan rancida invidia, Anglia xiii. 33, 156. Swá ðrógum tam rancidis (v. Hpt. Gl. 472, 61 : tam rancidis (fetidis, amaris, s. invisis, abominatis) þrón, biterum, mid swá biterum), 148. v. preceding word.

þroht, es; m. Oppression, affliction, hardship :—Ic hit leng ne mæg helan for hungre, is ðæs hæft tó ðan strang, þreánýd ðæs þearl, and ðes þroht tó ðæs heard, Elen. Kmbl. 1405; El. 704. [Cf. Icel. þróttr fortitude.]

þroht; adj. Oppressive, grievous :—Him sorgendum sár óðclífeþ, þroht þeódbealu, Exon. Th. 78, 2; Cri. 1268. Ellen biþ sélast ðam ðe oftost sceal dreógan dryhtenbealu, deópe behycgan þroht þeódengedál, 183, 8; Gú. 1324.

þroht-heard; adj. I. strong under afflictions, having fortitude or endurance in trouble :—Ne geald hé (Stephen) yfel yfele, ac his ealdfeóndum þingode þrohtheard he requited not evil with evil, but strong to bear his sufferings he interceded for his foes, Elen. Kmbl. 985; El. 494. Blíðheort wunode eorl unforcúð elnes gemyndig, þríst and þrohtheard in þreánédum, Andr. Kmbl. 2529; An. 1266. Héton lǽdan út þrohtheardne þegn, woldon ellenrófes mód gemiltan; hit ne mihte swá, 2781; An. 1393. Þegnas þrohthearde þafigan ne woldon, ðæt hié forléton leófne láreów, 803; An. 402. II. grievously hard :—Wæs se leódhete þrohtheard, Andr. Kmbl. 2279; An. 1141.

þrohtig (?); adj. Enduring, firm, persevering, laborious :—Ðrohtig (in the MS. e is written over o) pervicax, Txts. 87, 1556. Ic eom swifte ðonne hé, þrágum strengra, hé þreohtigra, Exon. Th. 494, 7; Rä. 82, 4. [Cf. Icel. þróttigr powerful.]

þrop, þrosle. v. þorp, þrostle.

þrosm, es; m. I. smoke, vapour :—Se þeóstra þrosm, Wulfst. 186, 4. On forsworcennesse sweartes þrosmes and ðæs weallendan pices, 139, 1 : Dóm. L. 199. Eft átogenum ðara fýra ðrosmum . . . eác fúlnes wæs mid ðæs fýres ðrosme retractis ignium vaporibus . . . et foetor cum eisdem vaporibus, Bd. 5, 12; S. 628, 24–26. Ða ðe þrosme beþeahte in þeóstrum sǽton, Exon. Th. 8, 11; Cri. 116 : Elen. Kmbl. 2593; El. 1298. Ða biteran récas, þrosm and þýstro, Cd. Th. 21, 18; Gen. 326. Se þrosma (but see þrosmig) líg, Dóm. L. 191. II. darkness, a dark space :—Sweart þrosm onáslít tetrum chaos inlabitur, Hymn. Surt. 13, 36. Betwux ús and eów is gefæstnod micel ðrosm (inter uos et nos chaos magnum firmatum est, Lk. 16, 26), Homl. Th. i. 332, 17. III. in Germ. 398, 230 þrosm glosses chautérem. v. swefel-þrosm; þrysman.

þrosmig; adj. Smoky, vaporous :—Ðǽr synd sorhlíce tósomne gemencged se þrosmiga líg and se þrece gycela (frigora mista simul ferventibus algida flammis, Dóm. L. 25, 95), Wulfst. 138, 26.

þrostle, þrosle, an; f. A throstle, singing-thrush :—Ðrostle trita, Txts. 103, 2062 : turdella, 2068. Þrostle, Wrt. Voc. i. 281, 16 : merula, 62, 45 : 77, 19. Ðrostle, ii. 55, 61. Þrosle merula vel plara, i. 29, 57. Án blác þrostle flicorode ymbe his neb, Homl. Th. ii. 156, 22. Of ðam leá on þrostlan wyl, Cod. Dip. Kmbl. v. 345, 3. [M. H. Ger. drostel.] Cf. þrysce.

-þrot. v. ǽ-þrot, á-þrotsum.

þrot-bolla, an; m. The gullet, windpipe :—Ðrotbolla gurgulio, Ælfc. Gr. 9, 3; Zup. 35, 7 : Wrt. Voc. ii. 110, 15. Þrotbolla, i. 43, 41 : 64, 62 : 282, 82 : ceutrum, ii. 131, 1. Eal þrotbolla chautrum, i. 43, 42 : ii. 22, 59. Gif monnes þrotbolla biþ þyrel, gebéte mid .xii. scill., L. Alf. pol. 51; Th. i. 94, 18. Ðrotbollan gurgilioni, Lchdm. i. lxx, 9. Þrotbollan gurguliones, Wrt. Voc. ii. 40, 45 ; Hpt. Gl. 490, 20. [Nu schal forrotien . . . þi þrotebolle þat þu mide sunge, Misc. 178, 173. And by the throtebolle he caught Aleyn, Chauc. Reeve's T. 353. The throtebolle epyglotum, Wülck. Gl. 580, 21 (15th cent.). Throte bolle frumen hominis est, rumen animalis est, ipoglotum, Cath. Angl. 386.]

þrotu, an; f. The throat :—Þrotu guttur, Wrt. Voc. i. 43, 39 : Ps. Lamb. 5, 11 : Scint. 97, 16 : Lchdm. ii. 46, 22. Þeós ðrotu hoc guttur, Ælfc. Gr. 9, 22; Zup. 49, 3. On ðære ðrotan, Lchdm. ii. 2, 18. On ða þrotan, Bt. 22, 1; Fox 76, 30. Hé (Judas) gewráð ða forwyrhtan ðrotan, seó ðe lytle ǽr belǽwde Drihten, Homl. Th. ii. 250, 16. Wouldcara forsmoriaþ ðæs módes ðrotan, 92, 11. Þrotan gurguliones, Wrt. Voc. ii. 82, 52. [O. H. Ger. droza, drozza gurgulio.] v. æsc-, eofor-þrotu(-e).

-þrowen in á-þrowen, read á-dropen.

þrowend, es; m. A scorpion :—Hí habbaþ tæglas ðam wyrmum gelíce ðe men hátaþ þrowend, Wulfst. 200, 15. Scorpius, ðæt is þrowend, Lchdm. iii. 246, 1. Se wyrm ðrowend slihþ mid ðam tægle tó deáde . . . Ondréd ðe ðone ðrowend . . . Bið hiht geæt̄trod mid ðæs ðrowendes tægle, Homl. Th. i. 252, 4–11. Se ðe gegrípð þrowend (scorpionem), Scint. 86, 11 : 225, 4. Þrowendra regulorum, serpentium, Hpt. Gl. 450, 17.

þrowende(-as P); pl. The Thronds (?), people in North Norway (Icel. Þrændir : Norw. Thrönder) :—Mid Þyringum ic wæs and mid Þrowendum and mid Burgendum, Exon. Th. 322, 17; Víd. 64.

þrowere, es; m. I. a sufferer :—Gif mann bið ácenned on .x. nihta ealdne mónan se bið þrowere, Lchdm. iii. 156, 27. v. líc-þrowere. II. a sufferer for religion, a martyr :—Ðe fruma ðrowere protomartyr, Rtl. 197, 9. Ðroweres ðínes martyris tui, 75, 41. Ðæs þroweres gemynd Scī Ypolyti, Shrn. 117, 8. Scē Ciricius tíd ðæs

þroweres, Chr. 916; Th. i. 190, col. 2. Ðrowres, Rtl. 50, 15. Ðrowre martyrem, 2. Monge Godes þrowera, Exon. Th. 113, 5; Gú. 153 : 111, 25; Gú. 132. Ðrowara ðínra martyrum tuorum, Rtl. 63, 16, 34. Ðrowerana, 44, 32. Wuldrigo ðrowras gloriosos martyres, 75, 34.

þrowian (þrówian ?), þreowian (þreówian ?); p. ode To suffer :—Ic ðrowige patior, Ælfc. Gr. 29; Zup. 186, 9. I. to suffer as opposed to to act :—Verbum ys word . . . getácniende oððe sum ðing tó dónne oððe sum ðing tó þrowigenne oððe náðor, Ælfc. Gr. 19; Zup. 119, 10. II. to suffer what is painful. (1) with acc. :—Mid gewyrhtum ic ðás þrowige, Blickl. Homl. 89, 7. Ðú ne þrowast nǽnige þrowunge, 157, 14. Wíf ácenþ bearn and þrowaþ micel earfoþu, Bt. 31, 1; Fox 112, 2. Hungor hí þrowiaþ famem patientur, Ps. Spl. 58, 7 : Andr. Kmbl. 562; An. 281 : Exon. 98, 30; Cri. 1615. Hé ðæs gewinnes weorc þrowade, Beo. Th. 3447; B. 1721. Hé drepe þrowade, 3183; B. 1589 : Exon. Th. 256, 10; Jul. 229. Hé for ælda lufan fela þrowade, 69, 10; Cri. 1118 : Blickl. Homl. 23, 35. Hí ermða þrowodan, 17, 17. Ðrowedon, Menol. Fox 244; Men. 123. Hí heora scylde wíte ðrowedon poenas sui reatus luerent, Bd. 4, 26; S 602, 14. Swá oft swá wé óht uneáþes þrowian æt yfflum monnum, Blickl. Homl. 33, 22. Þǽh þe ealle ǽswice þrowige on þé ic nǽfre þrowe si omnes scandalizati fuerint in te, ego numquam scandalizabor, Mt. Kmbl. Rush. 26, 33. Híra untrynnesse hé sceal ðrowian on his heortan ex affectu cordis alienae infirmitati compatitur, Past. 10; Swt. 61, 16. Éce wíte ðrowian, Homl. Th. i. 66, 14. Sceame þrowian, Soul Kmbl. 98; Seel. 49. Sár þrowian synna tó wíte, Exon. Th. 77, 1; Cri. 1250. Wrace þrowian, biterne bryne, Andr. Kmbl. 1230; An. 615. Wóp ðrowian, heáf under heofonum, Salm. Kmbl. 934; Sal. 466. Torn þrowigean, Cd. Th. 146, 14; Gen. 2422. Þrowigean þreánéd micel, fýres wylm, 229, 6; Dan. 213. Manega earfoðnesse fram Iudéum ic wæs ðrowiende, Blickl. Homl. 237, 10. (2) without acc., generally to suffer martyrdom :—Ic þrowode, Cd. Th. 296, 17; Sat. 503. Mín Drihten, áne tíd on róde ðú þrowodest, Blickl. Homl. 243, 28. Godes sunu on róde galgan þrowode, 27, 28 : Elen. Kmbl. 841; El. 421 : Rood Kmbl. 165; Kr. 84. Ðrowode, Menol. Fox 167; Men. 85. Þrowode martyrizavit, Wrt. Voc. ii. 55, 14. His mæssepreóst þreowude mid him, Shrn. 124, 1. Þrowedon agonizarunt, Wrt. Voc. i. 3, 6. Þreowedan, 81, 50. Hú Drihten wolde cuman tó ðære stówe ðe hé on þrowian wolde, Blickl. Homl. 15, 5. Hé wolde þrowian for ealra manna hæle, 63, 22; 77, 13. Hié lǽddun hine ðæt hé þrowigan salde duxerunt eum ut crucifigerent, Mt. Kmbl. Rush. 27, 31. Ys mannes sunu fram him tó þrowigenne (þrowende bið, Rush.) Filius hominis passurus est ab eis, Mt. Kmbl. 17, 12. Tó ðrowienne, Homl. Th. i. 82, 27. Ðrouande passurum, Lk. Skt. p. 6, 9. Þrowigende laturi, Wrt. Voc. ii. 86, 39 : 52, 46. ¶ The past participle is used as if the verb were a causative = to make to suffer, to crucify :—Æfter ðonne ðe hé þrowad wæs after he was crucified, Mt. Kmbl. Rush. 27, 44. Cf. Geðrowod under Pilate, Homl. Th. ii. 596, 15. Ðone geðrowodan Crist, 292, 13. (2 a) with gen. of instrument inflicting death :—Hé sceolde deófolgeldum geldan, oððe sweordes þrowian suffer death by the sword, Shrn. 129, 3. III. to suffer for something, pay for, atone for :—Ic ðrouuiu persolvio, Wrt. Voc. ii. 117, 16. Ðrowode expe[n]disset, throuadae expendisse[t], Txts. 61, 783. Þrowode expendisset, Wrt. Voc. ii. 29, 63. Hé þrowade ðæs þeówes sleacnysse he suffered for the slowness of the servant, Shrn, 43, 15. Þoledan and þrowedan luebant, Wrt. Voc. ii. 53, 19. Ðú scealt þrowian ðinra dǽda gedwild thou shalt expiate the error of thy deeds, Cd. Th. 57, 2; Gen. 921. Þrowgende luendi, Wrt. Voc. ii. 94, 51 : 52, 65. [Crist þrouwede deð, O. E. Homl. i. 17, 29. Þrowede, ii. 101, 9. Hwi walde he þrowin as he dude, Kath. 1135. He ðrowede and ðolede, Gen. and Ex. 1180. O. H. Ger. drôen, druoen pati.] v. á-, efen-, ge-þrowian.

þrowiend-líc, þrowigend-líc; adj. I. capable of suffering :—Ðá wearð hé (Christ) gesewenlíc on úrum gecynde and þrowigendlíc, Homl. Th. i. 120, 26 : ii. 6, 32. Ðis is ðín gecynd ðus ðrowigendlíc, ðe ic of ðé genam, 256, 28. Hélias wæs ús mannum gelíc, ðrowiendlíc, swá swá wé, 330, 16. II. as a grammatical term, passive :—Passiva verba, ðæt synd þrowiendlíce word, Ælfc. Gr. 19; Zup. 121, 1. Eal swá gáð ða óðre ðrowigendlícan word, 27; Zup. 161, 15. v. un-þrowi(g)endlíc.

þrowing, þreowing, e; f. I. suffering as opposed to doing :—Verbum is word, and word getácnaþ weorc oððe ðrowunge oððe geþafunge, Ælfc. Gr. 5; Zup. 9, 3. Him (the verb) gelimpþ significatio, ðæt ys getácnung, hwæt ðæt word getácnige, dǽde oððe þrowunge oððe náðor, 19; Zup. 119, 14. II. suffering which is painful :—Ic geteorode on ðære þrowunga, Ps. 38, 11. Ðú ne þrowast nǽnige þrowunge on ðínum líchoman, Blickl. Homl. 157, 15. Wæs monigu ðrowunga from swíðe monigum lǽcum fuerat multa perpesa a compluribus medicis, Mk. Skt. Rush. 5, 26. Hé gehýrde heora þrowunga he heard of their sufferings (they had been struck blind), Blickl. Homl. 153, 35. II a. as a medical term, a painful symptom :—Tácn ðæs ofercealdan magan, ðæt ða men ne þyrst, ne hí swól gefélaþ on magan, and ne biþ him ǽnig wearm þrowung getenge, Lchdm. ii. 194, 13. III. suffering that is undergone for the sake of religion, suffering of persecution, cross (in the phrase to take up one's cross) :—Him

God wolde æfter þrowinga þonc gegyldan, ðæt hé martyrhád móde gelufade, Exon. Th. 130, 22 ; Gú. 442. Lǽdæ ðróunc his and fylge meh *tollat crucem suam et sequatur me*, Mk. Skt. Lind. 8, 34. Ðrowung (ðrowunge, Rush.), Lk. Skt. Lind. 9, 23. Ðróung (ðrowunge, Rush.), 14, 27. Se ðe in þrowingum þeódnes willan dreógeþ, Exon. Th. 125, 18 ; Gú. 356 : 148, 26 ; Gú. 750. Gehýran heora þrowunga *to hear of the sufferings of St. Peter and St. Paul*, Blickl. Homl. 173, 2. **III a.** *suffering which ends in death, passion, martyrdom* :—Ðrouinges *martyrii*, Rtl. 64, 18. Ðróunges *passionis*, 50, 23. Ic, eówer emnðeówa and Cristes ðrowunge gewita, Past. 18 ; Swt. 137, 16. Se ðe biþ gemyndig Drihtnes þrowunge and his ǽriste, Blickl. Homl. 83, 14. Ða móddru on heora cildra martyrdóme þrowodon . . . neód is ðæt hí beón efenhlyttan ðæs écan edleánes, ðonne hí wǽron geféran ðære ðrowunge, Homl. Th. i. 84, 20. On hwæs tíman hé ðrowunge underhnige *in whose time he had submitted to martyrdom*, ii. 506, 31. Ðæt hé tó ðrowunge becóme *ad martyrium pervenire*, Bd. 1, 7 ; S. 478, 12. Be Cristes ðrowunge *de passione dominica*, 4, 24 ; S. 598, 13. Drihten ús mid his þrowunga álésde, ðá hé on róde galgan ástág, Blickl. Homl. 97, 10 : 35, 7 : 81, 31. Ðrowenge *passione*, Rtl. 50, 4. Hé ongan árweorþian ða ðrowunge háliga martyra *honorem referre incipiens caedi sanctorum*, Bd. 1, 7 ; S. 479, 1. Þurh his þrowinga, Exon. Th. 29, 29 ; Cri. 470 : 69, 33 ; Cri. 1130. **III b.** *the anniversary of a martyr's suffering* :—On ðone feówer and twéntigoðan dæg ðæs mónðes byð Sči. Crissoȝones týd and þrowung, Shrn. 151, 17, 31. Þreowung, 114, 21. [Vre clateres halie passiun, þet is his halie þrowunge, O. E. Homl. i. 119, 26. Inntill þrowwinnge and pine, Orm. 15205. Cheosen er licomes hurt þen soule þrowunge, A. R. 372, 6. Wiðuten ðrowing and figt, Gen. and Ex. 1317. O. H. Ger. druuunga *passio*.] v. following words.

þrowing-ræding, e ; f. *A martyrology* :—Sí rǽdd þrowungrǽding *legatur martyrologium*, Anglia xiii. 385, 286.

þrowing-tíd, e ; f. **I.** *the time at which a person suffered martyrdom* :—Fram ðissere worulde fruman óð Xpes þrowungtíd, ðæt is six þúsend geára and .c. geára and lviii geára, Anglia xi. 7, 18. Weorðian wé on ðissum andweardan dæge Sancte Petres þrowungtíde, Blickl. Homl. 171, 4. **II.** *the anniversary of the time when some one suffered* :—Ðeós tíd fram ðisum andwerdan dæge (*fifth Sunday in Lent*) óð ða hálgan Eástertíde is gecweden Cristes ðrowungtíd, Homl. Th. ii. 224, 19. On ælces geáres ymbryne ymbe his ðrowungtíde, i. 564, 24.

þrowing-tíma, an ; n. *A time of suffering* :—Ðonne mín ðrowungtíma cymþ, ðonne geswuteläþ seó menniscnys hire untrumnysse, Homl. Th. ii. 56, 2.

þrúh (also þrýh, Bd. S. 580, 14) ; gen. dat. þrýh, and dat. þrúh ; f. : dat. þrúge ; m. n. *Wood or stone hollowed out*. **I.** *a trough, pipe, conduit* :—Ðrúh, thruuch, thruch *tubo*, Txts. 103, 2067. Þrúh *vel* mylentroh *canalis*, Wrt.Voc. ii. 128, 16. Of þrýh t þeótan *tubo*, Hpt. Gl. 418, 61. Of ðam bróce in ðæt þrúh ; of ðam þrúge, Cod. Dip. Kmbl. iii. 380, 3. Ðá gesomnodon ða sticceo hí in ða þrúh, þurh ða ðe ðæt wæter flëow ; ðá ne meahte ðæt wæter flówan, Shrn. 125, 12. **II.** *a box, chest* :—Fiscella spyrte t þrúh, Germ. 400, 492. **III.** *a coffin, sarcophagus, tomb* :—Þrúh *sarcofagum*, Wrt. Voc. i. 49, 28. Ðúrh, 85, 78. Ðrúh oððe ofergeweorc *mausoleum*, 85, 76. Ðá gearwodan hí his líchoman tó bebyrigeanne on stǽnenre ðrúh (byrgenne stǽnenne ðrúh, MS. T.) . . . ðá wæs se líchoma sponne lengra ðære ðrýh (ðonne seó ðrýh, MS. B.) . . . Hí tóætýcton lengeo ðære ðrýh . . . Ðá ðóhton hí ðæt hí óþre ðrýh (ðúrh, MS. B.) sóhton . . . Ðá wæs seó ðrýh (þrúh, Bd. M. 296, 28) gemëted gerisenlicre lengo . . . seó ðrúh wæs ðam líchoman lengre *cujus corpori tumulando praeparaverant sarcofagum lapideum . . . invenerunt hoc mensura palmi longius esse sarcofago . . . addiderunt longitudini sarcofagi cogitabant aliud quaerere loculum . . . Inventum est sarcofagum illud congruae longitudinis*, Bd. 4, 11 ; S. 580, 3–14. Wæs him ðrúh (*loculus*) gegearwod, 5, 5 ; S. 617, 39. Ðá stód on ðære stówe sum stǽnen ðrúh . . . Ðá lédon ða þegenas ðone Hǽlend ðǽron . . . Hí ða ðrúh geinnsegelodon, Homl. Th. ii. 262, 1–11. Þréh *sarcophagi*, Hpt. Gl. 499, 58. Of þríh *de tumba*, 450, 73. Se engel áwylte ðæt hlid of ðære ðrýh . . . Crist mihte, belocenre ðríh, faran of middangearde, Homl. Th. i. 222, 8–13. Hí gemëtton níwe ðrúh of marmanstáne on cyrcan wíson gesceapene . . . Æt ðære hálgan þrýh sind getídode heofonlíce lácnunga, 564, 19–31. On eallhwítre ðrúh of marmstáne geworht, Cod. Dip. Kmbl. iii. 60, 21. Of ðære stǽnenan þrýh ðe stent wiðinnan, Homl. Skt. i. 21, 22. On treówene ðrúh *ligneo in locello*, Bd. 4, 19 ; S. 588, 21, 25, 31, 34. On ða stǽnenan ðrúh *in sarcophago*, S. 589, 40. Ða ðúrh (*loculum*) be him gesett, 5, 5 ; S. 618, 6. Hí his líc gedydon on þrúh, Blickl. Homl. 191, 33 ; Guthl. 20 ; Gdwin. 84, 7, 14. Hé worhte áne ðrúh on hwítum seolfre tó ðæra apostola líce, Homl. Th. ii. 498, 3. ¶ The word seems left in local names, *Thrubrook, Througham*, v. Cod. Dip. Kmbl. vi. 342. [Me leit þene licome in þare þruh, O. E. Homl. i. 51, 5. Strikeð a stream ut of þ stanene þruh (*de sepulchro*), Kath. 2480. Ine stonene þruh biclused . . . þeos þruh, A. R. 378, 12 : Misc. 51, 511. In throghes *in sepulcris*, Ps. 67, 7. Thurhwe stone, throwe or throwstone *sarcofagus*, Prompt. Parv. 493. A trughe mau-

seolum, Cath. Angl. 386, *and see note there.* Icel. þró ; f. *a trough* ; stein-þró *a stone coffin*.] v. wæter-þrúh.

þrum. v. tunge-þrum. [Thrumm of a clothe *filamen*, Prompt. Parv. 493. Throm *licium*, Wrt. Voc. 235, 5. O. H. Ger. drum, thrum *meta, finis*.]

þrust-fell, es ; n. *A cutaneous disease, leprosy* :—Blaec thrustfel *bitiligo*, Txts. 46, 398. [Goth. þruts-fill *leprosy* ; þruts-fills *leprous*. Cf. Icel. þroti *a tumour*.]

þrútian ; p. ode *To swell with pride or .anger* :—Hé ásende his swurdboran, Riggo geháten, gescrýdne mid his cynelícum gyrelum, swilce hé hit sylf wǽre. Ðá gesæt Benedictus forn ongeán ðam Riggon, ðe mid ðam leáslícum getote inn eode ðearle ðrútigende (*he entered in a very pompous manner*), Homl. Th. ii. 168, 16. Hé cwæþ hire þus tó mid þrútigendum móde (*angrily, passionately*), Homl. Skt. i. 10, 273. [Cf. Icel. þrútinn *swoln* ; reiði-þrútinn *swoln by anger* ; þrútna *to swell*.]

þrúting, e ; f. *A swelling* of the mind from anger, etc., *angry emotion* :—Hé befrán mid mycelre ðrútunge, hwæt se brýdguma wǽre, Homl. Skt. i. 7, 76.

þrý, þry-, þrýan. v. þrí, þri-, á-þrýan, ge-þrýde.

þryccan ; p. þrycte, þryhte ; pp. þrycced, þryht. **I.** trans. *To press, crush, oppress, repress, trample* :—Sittaþ mánfulle on heáhsetlum and hálige under heora fótum þryccaþ *perversi resident celso mores solio, sanctaque calcant colla*, Bt. 4 ; Fox 8, 14 : Met. 4, 38. Ðæt sió manung hié ne ðrycte *ne admonitio eos concuteret*, Past. 32 ; Swt. 213, 22. Ða gástas ðe me swenctan and ðrycton *qui me premebant spiritus*, Bd. 3, 11 ; S 536, 37. Ðrycce se magister ða belde *reprimatur praecipitatio*, Past. 61 ; Swt. 455, 21. Swá hié se stán and seó eorþe þrycce, Blickl. Homl. 75, 9. Hé mid wédenheortnesse módes ðrycced wæs *mentis vesania premebatur*, Bd. 2, 5 ; S. 507, 4. Untrumnesse ðrycced and hefigod *infirmitate pressus*, 4, 24 ; S. 598, 25. **II.** intrans. *To press, force a way* :—Wé ðás wíc magun fótum áfyllan ; folc in ðriceþ meara þreátum and monfarum, Exon. Th. 119, 18 ; Gú. 256. [He wænde mid his crucche us adun þrucche, Laym. 19483. Þre at þe fyrst þrast he þryȝt to þe erþe, Gaw. 1443. A þral þryȝt in þe þrong, Allit. Pms. 42, 135. *To thrutch* is still used in some dialects ; see E. D. S. Pub. Lancashire and Cheshire Glossaries, where see also *thrutchings* = whey squeezed out whilst the cheese is under pressure. O. H. Ger. drucchen *premere, comprimere*.] v. bi-, ge-, of-, on-þryccan.

þrycness, e ; f. *Oppression, affliction, tribulation* :—Biþ ðrycnisse micelu *erit tribulatio magna*, Mt. Kmbl. Rush. 24, 21. In ðrycnisse *in tribulationem*, 9. [O. H. Ger. thrucnessi *pressura* (Ju. 16, 33).] v. ge-, of-þrycness.

þrydian, þrýdge, þrydlíce, þrydung, þrýh. v. þreodian, þrýþig, þrýþlíce, þreodung, þrúh.

ðryhte, in Mt. Kmbl. Rush. 27, 31, seems an error for ðý ryfte which glosses clamyde in the same passage of the Lindisfarne Gloss.

þrylen, þryl-hús, þrym. v. þrilen, þyrl-hús, þrymm.

þrym-cyme, es ; m. *A glorious coming* :—Ic (*Guthlac*) on móde mäð monna gehwylcne þeódnes þrymcyme (*the coming of the angel* (wuldres wilboda) *each evening to Guthlac*), Exon. Th. 177, 20 ; Gú. 1230.

þrym-cyning, es ; m. *The king of glory, the Deity* :—Ðú, sigora waldend, þeóda þrymcyning, Met. 20, 205. Þrymcyning rícne, Exon. Th. 317, 7 ; Mód. 62 : Elen. Kmbl. 986 ; El. 494. Cf. wuldor-cyning.

þrym-dóm, es ; m. *Glory* :—Ðæt eorðlíce mægn ðe tó dóme (þrymdóme, MS. D.) cumen is, Wulfst. 254, 14. v. þrymness.

þrymen. v. þrimen.

þrym-fæst ; adj. *Glorious, majestic, illustrious, mighty*, (1) as epithet of the Deity :—Mægencyninga Meotod, þrymfæst þeóden, Exon. Th. 58, 31 ; Cri. 944. Se brego mæra, þeóden þrymfæst (*Christ*), 29, 3 ; Cri. 457 : Andr. Kmbl. 645 ; An. 323. (2) in other connections :—Eorl unforcúð . . . þeóden þrymfæst, Andr. Kmbl. 957 ; An. 479. Þrymfæst þeóden (*Noah*), Cd. Th. 200, 27 ; Exod. 263. Ic (*the cross*) þrymfæst hlifige under heofonum, Rood Kmbl. 166 ; Kr. 84. Se wyrm (*a bookworm*) forswealg þrymfæstne cwide, Exon. Th. 432, 11 ; Rä. 48, 4. Þeóda þrymfæste (*angels*), Cd. Th. 2, 6 ; Gen. 15. Þeóda þrymfæste, 114, 22 ; Gen. 1908 : 158, 10 ; Gen. 2615.

þrym-full ; adj. *Glorious, magnificent, illustrious, mighty* :—Nergendes þeówen ðrymfull (*Judith*), Judth. Thw. 22, 23 ; Jud. 74. Wǽre ðú (*the body*) ðe wiste wlonc . . . , þrymful, Exon. Th. 369, 12 ; Seel. 40. Ic (*a storm*) ástíge strong, stundum rëþe, þrymful þunie, 380, 42 ; Rä. 2, 4 : 386, 25 ; Rä. 4, 67. Ic bidde ðinne þrymfullan cynescype, Homl. Skt. i. 23, 793. Þeódnas þrymfulle, þegnas wlitige, Andr. Kmbl. 725 ; An. 363. Þegnas þrymfulle (*the disciples*), Exon. Th. 34. 12 ; Cri. 541.

þrym-líc ; adj. *Magnificent, splendid, glorious* :—Ða apostolas cwǽdon ðæt hit (*the temple*) wǽre þrymlíc geweorc and fæger, Homl. Th. 77, 32. Swíðe mycel cyrice and þrymlíc, 125, 20. Þrecwudu þrymlíc, Beo. Th. 2496 ; B. 1246. Lidweardas þrymlíce, Andr. Kmbl. 489 ; An. 245. Ðrymlíc swǽsendo, Judth. Thw. 21, 7 ; Jud. 8. Swá hé ús mærlícor giseþ, swá wé him mærlícor þancian scylon ; swá þrymlícre ár, swá märe eádmódnes, Wulfst. 261, 21. Sceoldon hié ða menn beforan him drífan gebundene ðe gefongene wǽron, ðæt heora mærþa sceoldon ðý þrymlícran beón, Ors. 2, 4 ; Swt. 70, 30.

3 Z

þrymlíce; *adv.* *Magnificently, splendidly, gloriously*:—Hú þrymlíce ðú (*God*) ðíne gife dǽlest, Andr. Kmbl. 1093; An. 547: Elen. Kmbl. 1558; El. 781: Exon. Th. 18, 23; Cri. 288. Cyning þrymlíce of his heáhsetle scíneþ, 232, 30; Ph. 514. Wæter wynsumu bearo geondfaraþ þrymlíce, 202, 11; Ph. 68: Menol. Fox 153; Men. 78. Án and þryttig geára he ríxode þrymlíce on Hierusalem, Homl. Skt. i. 18, 470.

þrymm, es; *m.* I. *a host, great body of people, a force, multitude*:—Eall heofonlíc þrym (cf. ðæt heofonlíce werod, l. 9) hire tócymes fægnian wolde. Eác we gelýfaþ ðæt Drihten sylf hire tógeánes cóme *all the heavenly host would rejoice at her advent. We believe, too, that the Lord himself would come to meet her*, Homl. Th. i. 442, 13. Ðé þanciaþ þúsenda fela, eal engla þrym ánre stefne, Hy. 7, 50: Cd. Th. 267, 11; Sat. 36. Ealle ábúgaþ tó ðé, ðúra engla þrym, Hy. 7, 11. Seó heá dugud and se engla þrym, Exon. Th. 65, 33; Cri. 1064. Glæd gumena weorud, . . . heofonduguða þrym, 101, 7; Cri. 1655. He wile cuman in wolcne and mid engla þrymme, Blickl. Homl. 121, 19. He ásende Rapsacen mid micclum drymme (*with a great army*, A.V. Is. 36, 2), Homl. Th. i. 568, 6: ii. 304, 6: Homl. Skt. ii. 25, 531. Se ðe mid micclum þrymme (cf. he com mid werode, 763) þrang intó ðam temple, 781. Se hundredes ealdor com mid mycclum þrymme, 841. Heora godas ne mihton hí gescyldan wið mínne drymm (*host or power?*), Homl. Th. i. 568, 10. Hý forheówan Heaðobeardna þrym, Exon. Th. 321, 21; Vid. 49: 461, 14; Hö. 35. Cyning (*God*) on gemót cymeþ þrymma mǽste, 52, 15; Cri.834. Ðú (*Christ*) ǽr wǽre eallum geworden worulde þrymmum, 14, 10; Cri. 217. Of ðǽm engelcun þrymmum *from the angelic hosts* (or *glories?*), Blickl. Homl. 5, 13: 21, 15. Weras and wíf, wornum and heápum, dreátum and þrymmum þrungon and urnon, Judth. Thw. 23, 40; Jud. 164. Se ðe herga þrymmas on geweald gebrǽc, Cd. Th. 127, 14; Gen. 2110. I a. *a great body* of water:—Flóda þrym (*the host of waters*) sealte sǽstreámas sǽlde habbaþ *commoveatur mare et plenitudo ejus*, Ps. Th. 95, 11. Ýþa drym *the host of waves*, Beo. Th. 3841; B. 1918. Swá wætres þrym ealne middangeard mereflód þeahte *cum diluvio mersisset fluctibus orbem*, Exon. Th. 200, 16; Ph. 41: Andr. Kmbl. 3070; An. 1538. We þuruh flóda þrym faraþ *transivimus per aquam*, Ps. Th. 65, 11. Com æfter niht lagustreámas (=es?) wreáh þrym mid þýstro *night covered the great mass of water with darkness*, Cd. Th. 148, 1; Gen. 2450. II. *force, power, might*:—Ðǽr wæs módigra mægen forbéged, wígendra þrym, Andr. Kmbl. 3142; An. 1574: 6; An. 3. Clang wæteres þrym *the water's might withered*, i. e. *the water was frozen*, 2522; An. 1262. On ðære fyrde wǽron feówertig þúsenda and seofon þúsenda swýðe gewæpnode, and cómon ðá mid þrymme tó Iudéiscum cynne, Homl. Skt. ii. 25, 334. Hié wið Drihtne dǽlan meahton wíc werodes þrymme *by the might of their band*, Cd. Th. 2, 31; Gen. 27. Eall ðæt ða þeódguman þrymme (*by force or gloriously?*) geeodon, Judth. Thw. 26, 17; Jud. 332. Se mec mæg écan meahtum, geþeón þrymme, Exon. Th. 427, 14; Rä. 41, 91. Bewyl þrimme (*strongly, thoroughly*) ðæt ealo on ðære wyrte, Lchdm. ii. 276, 14. He þrymmum (*mightily, with power*) cwehte mægenwudu mundum, Beo. Th. 476; B. 235. Seraphinnes cynn unáþreótendum þrymmum singaþ *the seraphim with unwearying powers sing*, Exon. Th. 24, 22; Cri. 388. III. *glory, majesty, magnificence, greatness, grandeur*:—Mín þrym is from eastewearde middangearde oþ ðæt westanweardne *majestas mea pervenit ab occidente usque in orientem*, Nar. 25, 24. Drihtenes þrym *the majesty of the Lord*, 274, 34; Sat. 164: Exon. Th. 37, 26; Cri. 599: Judth. Thw. 22, 30; Jud. 86. Wæs him (*the fallen angels*) forbíged þrym, wlite gewemmed, Cd. Th. 5, 12; Gen. 70: 306, 11; Sat. 662. Lof wíde sprang, miht and mǽrðo, þrym unlytel, Apstls. Kmbl. 16; Ap. 8. Þín heáhsetl is þrymmes áfylled, Wulfst. 254, 18. Wuldres déma, ðrymmes hyrde, Judth. Thw. 22, 15; Jud. 60: Blickl. Homl. 65, 32. On ðone gefeán ðæs heofonlícan þrymmes, 63, 27. Þremmes, 73, 34. Gif him (*a king*) geberede, ðæt him wurde oftogen þrymmes and wǽda and þegnunga, Met. 25, 32. Mið ðý cymeþ in drymme his *cum uenerit in majestate sua*, Lk. Skt. Rush. 9, 26, 31: 21, 27: Bd. 3, 22: S. 552, 16: Exon. Th. 106, 22; Gú. 45: Hy. 8, 40. Babilon ðe ic self átimbrede tó kynestóle and tó drymme *Babylon quam ego aedificavi in domum regni*, Past. 4; Swt. 39, 17: Homl. Th. ii. 432, 32. Mín werod fóran ymb me úton mid þrymme (*with magnificent array*), and herebeácen and segnas beforan me lǽddon, Nar. 7, 16. He fór mid drymme and mid prasse, Homl. Skt. i. 23, 26: Elen. Kmbl. 658; El. 329: Bt. 37, 1; Fox 186, 7: Met. 25, 13. Ne preodode he fore þrymme ðeódcyninges ǽniges, Apstls. Kmbl. 35; Ap. 18: Exon. Th. 112, 7; Gú. 140. Hiá geségon drymm (*drym*, Rush.) his *uiderunt majestatem ejus*, Lk. Skt. Lind. 9, 32: Exon. Th. 63, 23; Cri. 1024: 234, 17; Ph. 541. Ðínes mihtes þrym *potentiam tuam*, Ps. Th. 70, 18: Exon. Th. 349, 19; Sch. 48. Þone þrym and þa fægernesse ðæs temples *the magnificence and beauty of the temple*, Blickl. Homl. 77, 30. We Gár-Dena in geárdagum þeódcyninga þrym gefrunon, hú ða æþelingas ellen fremedon, Beo. Th. 4; B. 2. Þrymmas weóxon duguda dreámhæbbendra, Cd. Th. 5, 32; Gen. 80: Menol. Fox 468; Gn. C. 4. Eallra þrymma God, Elen. Kmbl. 1036; El. 519. Cyninga setl þrymmum (*magnificently*) gefrætewad, Wulfst. 253, 22. Heágum þrymmum *most*

gloriously, Cd. Th. 1, 16; Gen. 8. He hié álædde of helle grunde on ða heán þrymmas (*the high glories*) heofona ríces, Blickl. Homl. 67, 22. IV. denoting a glorious, magnificent person *or* object:—Ealra cyninga þrym (*the Deity*), Hy. 7, 45: Eleu. Kmbl. 1629; El. 816. Ealra þrymma þrym, Exon. Th. 45, 28; Cri. 726. Rodera þrim, heofona heáhfreá, 26, 28; Cri. 423. Wuldres þrym, 6, 13; Cri. 83. Ðú ðe sitst ofer engla drymm (*qui sedes super cherubim*, Is. 37, 16) (*or* (?) drymm=*host*), Homl. Th. i. 568, 15. Mec (*a hurricane*) þrymma sumne one *of glorious things*, Exon. Th. 383, 2; Rä. 4, 4. [Her throme fourti thousand men thai founde (quoted in Halliwell). A god man on þat throm, C. M. 7423. Cf. Heo folc funden feouwer þrumferden (fouruald ferde, 2nd MS.), Laym. 1356. *O. Sax.* heru-thrummi *in* mid heruthrummeon *violently*; *Icel.* þrymr *an alarm, noise* (*poet.* of battle); used, too, in cpds. denoting a warrior.] v. cyne-, éðel-, god-, heáh-, heofon-, here-, hilde-, hyge-, mægen-, ofer-, wuldor-þrymm.

þrymma, an; *m.* A *strong* or *great man, a warrior*:—Þrymman sceócan, módige maguþegnas, morðres on luste, Andr. Kmbl. 2280; An. 1141. [Cf. *Icel.* þrymr *glorious*; and the poet. cpds. in þrym-, denoting a warrior.] v. hilde-þremma.

þrymness. v. heáh-, mægen-þrymness; *and* cf. þrym-dóm.

þrym-ríce, es; *n.* A *glorious kingdom, heaven*:—Drihten wolde cuman of ðam cynestóle and of ðæm þrymríce hider on ðás world, Blickl. Homl. 105, 11.

þrymsa. v. trimes.

þrym-seld, es; *n.* A *throne*:—Ofer drymseld *super thronum*, Ps. Surt. 9, 5: 88, 30. Þrymseld *thronos*, Lchdm. i. lxxiii, 22.

þrym-setl, es; *n.* A *throne*:—Heofon ys Godes þrymsetl (*thronus*), Mt. Kmbl. 5, 34. On Godes þrymsetle, 23, 22. Se cásere seóll of his drymsetle, Shrn. 76, 31. Beforan þrymsetle Cristes *ante tribunal Christi*, Anglia xiii. 387, 311: Blickl. Homl. 101, 29. *Throni* sind þrymsetl, Homl. Th. i. 342, 34. Ge sitton ofer þrymsetl (*thronos*) démende twelf mǽgða Israhél, Lk. Skt. 22, 30: Blickl. Homl. 31, 8. [Þrimsetles *troni*, O. E. Homl. i. 219, 10.]

þrym-sittende; *adj.* (*ptcpl.*) *Dwelling in glory, inhabiting heaven*:—Seó þrynis þrymsittende, Exon. Th. 286, 3; Jul. 726. Þegn þrymsittendes wuldorcyninges, Andr. Kmbl. 834; An. 417: 1056; An. 528. Sié ðe þrymsittendum þanc, Elen. Kmbl. 1618; El. 811: Exon. Th. 239, 19; Ph. 623. Écne God þrymsittendne, 268, 20; Jul. 435.

þrym-wealdend; *adj. Glory-ruling, ruling heaven*:—Seó Hálige Ðrynnys ðe is þrymwealdend God, Homl. Th. ii. 316, 4. Þrimwealdend, Homl. Skt. ii. 27, 156. Se drimwealdenda Scyppend, Homl. Th. i. 112, 10. We sceolon biddan ða hálgan ðæt hí ús þingion tó ðam þrymwealdendum Gode, Homl. Skt. i. 21, 288.

þrynen, þryness, þryng, þryþel. v. þrinen, þrinness, þring, þrípel.

þryscan *to press.* v. ge-, of-þryscan.

þrysce, an; *f.* A *thrush*:—Þryssce *strutio*, Wrt. Voc. i. 63, 2. Þrisce *trutius*, 281, 23. [Þrusche and þrostle, O. and N. 1659. Thryshe *mauiscus*, Wülck. Gl. 595, 20. *O. H. Ger.* drosca.] v. þræsce, þrostle.

drysceð, Salm. Kmbl. p. 148, 6. v. þerscan.

þrysman(-ian); *p.* de, ode *To choke, stifle, suffocate*; fig. *to keep in subjection*:—Alexander .xii. geár ðisne middangeard under him þrysmde and egsade *Alexander per duodecim annos trementem sub se orbem ferro pressit*, Ors. 3, 11; Swt. 142, 22. [*O. Frs.* thresma, tresma *to choke, stifle, strangle*.] v. á-, for-, of-þrysman(-ian).

þrýste, þrystig, þrýstru, -þryt, -þrytness. v. þríste, þyrstig, þeóstru, ǽ-þryt, á-þrytness.

þrýþ, e; þrýþu (? *indecl.* v. mód-þrýþu); *f. Force, power, strength*; the word seems to occur only in the plural, *forces, troops, hosts*:—Of ðam stáne wæter cwóman swýþe wynlíce wætera þrýðe (*the waters' forces*); *eduxit aquam de petra, et eduxit tamquam flumina aquas*, Ps. Th. 77, 18. Heofon weardiaþ ufan wætra drýðe *the waters' forces guard heaven above*, 103, 3. Sóna wǽrun geworht wætera drýþe, 148, 5. Wætra þrýþe stille stondaþ, Exon. Th. 210, 12; Ph. 184. Eorlas fornóman asca þrýþe, wǽpen wælgífru *hosts of spears, weapons ravenous for slaughter, have swept off the men*, 292, 15; Wand. 99. Þrýþa dǽl *some forces* (?), 481, 15; Rä. 65, 4. Eóredciestum hí faraþ, folca þrýþum, 220, 27; Ph. 326. Beornþreát monig faraþ folca þrýþum, eóredcystum, 358, 26; Pa. 51. Æfter him folca þrýðum sweótum cómon, Cd. Th. 199, 18; Exod. 340. Wæteregsa stód þreáta þrýðum *the terrible waves stood in battalions*, Andr. Kmbl. 751; An. 376. Ecga þrýðum *with hosts* (or *force?*) *of swords*, 2298; An. 1150. ¶ Þrýþum *vehemently, mightily, fiercely, greatly*:—Teónleg þrýþum bærneþ þreó eal on án, grimme tógædre, Exon. Th. 60, 15; Cri. 970. Ic seah wiht (*a cask*), wombe hæfde micle þrýþum geþrungne, 495, 3; Rä. 84, 2. Ic wiht (*bellows*) geseah, womb wæs þríþum áþrunten, 419, 7; Rä. 38, 2. Þrýðum dealle, Beo. Th. 992; B. 494. ¶ Þrýþ *is used in the formation of many proper names.* v. Txts. 638. [*Icel.* Þrúðr *the name of a daughter of Thor and Sif*; it is used in the formation of proper names.] v. hilde-, hyge-, wæter-þrýþ; mód-þrýþu.

þrýþ-ærn, es; n. A splendid house, a palace:—Næfre ic ænegum men ǽr ǽlýfde ðrýþærn (cf. heáhsele, 1298; B. 647) Dena. Hafa nú húsa sélest, Beo. Th. 1318; B. 657.

þrýþ-bearn, es; n. A mighty youth:—Ic ǽfre ne geseah ǽnigne mann, þrýðbearn hæleð, ðé gelícne, steóran ofer stæfnan, Andr. Kmbl. 987; An. 494.

þrýþ-bord, es; n. A strong shield, E'en. Kmbl. 302; El. 151. [Cf. Icel. þrúð-hamarr the mighty hammer of Thor.]

þrýþ-cyning, es; m. A mighty king (the Deity), Andr. Kmbl. 872; An. 436. Cf. þrym-cyning.

þrýþ-full; adj. Mighty, strong, powerful:—Fóron æfter burgum þegnas þrýðfulle, oft hí þrǽce rǽrdon, Exon. Th. 243, 17; Jul. 12. Ic (the devil) bebeóde bearnum mínum, þegnum þrýðfullum, ðæt hié ðé hnǽgon, Andr. Kmbl. 2659; An. 1331.

þrýþ-gesteald, es; n. A splendid abode:—Þeódnes þrýðgesteald (heaven), Exon. Th. 22, 19; Cri. 354. Cf. wuldor-gesteald.

þrýþian. v. ge-þrýþian.

þrýþig (?); adj. Mighty, powerful, strong:—Hæleð onetton módum þrydge (þrýðge?), Cd. Th. 119, 28; Gen. 1986. [Icel. þrúðigr doughty; and cf. þrúð-módigr heroic of mood.] v. þrýþlíce for d instead of ð, and next word.

þrýþ-líc; adj. Mighty, powerful:—Rinc manig, þrýðlíc þegna heáp, Beo. Th. 805; B. 400. Ðrýðlíc, 3258; B. 1627.

þrýþlíce (?); adv. Mightily:—Bissextus ðe on gewunan hæfþ ðæt hé binnan ðam feórðan geáre ealle ðære wucan dagas þrydlíce (þrýþlíce?) æthríne, Anglia viii. 302, 14. Hé oft gesealde healsittendum helm and byrnan swylce hé þrydlícost (þrýþlícost?) ðhwǽr feor oððe neáh findan meahte he often gave to his followers helm and corslet such as for greatest strength anywhere far or near he could find (cf. for similar use of the adverb: Hé sóhte, hú hé sárlícast meahte feorhcwale findan, Exon. Th. 276, 25; Jul. 571), Beo. Th. 5731; B. 2869. [Or, perhaps, þrydlíce = deliberately, might be read. v. þreodian.]

þrýþ-swíþ; adj. Exceedingly powerful:—Mǽre þeóden unblíðe sæt, þolode ðrýðswýð, þegnsorge dreáh the great prince sat cheerless, he, mighty, suffered, grief for his thanes' loss he endured, Beo. Th. 262; B. 131. Þrýðswýð beheóld, mǽg Higeláces, hú se mánscaða gefaran wolde, 1477; B. 736. [Cf. the proper names Æþel-swíþ, Beorht-swíþ.]

þrýþ-weorc, es; n. A splendid, mighty work:—Þrýðweorc (a statue; v. the description: Wrætlíce wundorágrǽfen anlícues engla . . . torhte gefrætwed, wlitige geworht . . . anlícnes engelcynna ðæs brémestan, 1423-35; An. 712-8), Andr. Kmbl. 1546; An. 774.

þrýþ-word, es; n. A brave word, noble speech:—Ðá wæs eft swá ǽr (cf. word wǽron wynsume, 1228; B. 612) inne on healle þrýðword sprecen, ðeód on sǽlum, sigefolca swég, Beo. Th. 1290; 643.

þú; pers. pron. Thou. I. alone:—Ðis land ðe þú gesihst, Gen. 13, 15. Hwæt eart þú þe þýn ansýn is swylce ánes sceaþan, and hwæt ys ðæt tácen þe þú on uppan þínum exlum byrst? Nicod. 32; Thw. 18, 19. Gewít þú, Abraham, féran . . . þú scealt Isaac mé onsecgan, Cd. Th. 172, 24; Gen. 2849; Andr. Kmbl. 1899; An. 952. Ic áscige ðé, ðú Boetius, hwí þú swá manigfeald yfel hæfdest? Bt. 27, 2; Fox 96, 12. Eá lá þú mín Drihten God, hwæt gifst þú mé? Ger. 15, 2. Westú gearo, Bd. 5, 19; S. 640, 44. Scealtú ceól gestígan, Andr. Kmbl. 439; An. 220. Hié woldon þín onbídan, Blickl. Homl. 233, 27. Se ðe nold þé (ðec mið, Lind. Rush.) wæs qui erat tecum, Jn. Skt. 3, 26. Ne biþ þec mǽlmete, nymþe móres græs, ne rest witod, Cd. Th. 252, 7; Dan. 575. Þú gesyxst þás menigu þé (ðec, Lind.: on ðec, Rush.) ðringende, Mk. Skt. 5, 31. Se hálga gást on þé (ðeh, Lind.: ðec, Rush.) becymþ, Lk. Skt. 1, 35. Þec Sarre áh, Cd. Th. 137, 8; Gen. 2270. Ne forlǽte ic þé, 136, 10; Gen. 2256. I a. used reflexively:—Ne ondrǽd þú þé, Gen. 15, 1: Lk. Skt. 1, 30. Ðú hafast þé on fyrhðe eorles ondsware, Andr. Kmbl. 1013; An. 507. Him þú þé ofset on hand, Cd. Th. 33, 11; Gen. 518. Ásend þé (ðeh, Lind.: þec, Rush.) nyþer, Mt. Kmbl. 4, 6. Þonne þú þé gebidde, 6, 6. II. strengthened with self or ána:—Þú sylf ne gesyhst þæne beám on þínum ágenum eágan, Lk. Skt. 6, 42. Þú (Juliana) sylfa meaht gecnáwan, Exon. Th. 262, 32; Jul. 341: Cd. Th. 36, 12; Gen. 570. Ðú eart seolfa geong, Andr. Kmbl. 1010; An. 505. Þú meaht þé self geseón, Cd. Th. 38, 23; Gen. 611. Þæt þú þa beorhtan ús sunnan onsende, and þé sylf cyme, Exon. Th. 8, 8; Cri. 114. Hwí swingst þú ána? . . . Ne miht þú ána hit ácuman, Ex. 18, 14, 18. Þú ána canst ealra gehygda, Andr. Kmbl. 135; An. 68. Þé wæs leófra his hyldo, þonne þín sylfes bearn, Cd. Th. 176, 34; Gen. 2921. Lufa þínne néhstan swá þé sylfne (ðec seolfne, Lind.), Mt. Kmbl. 19, 19. III. combined with þe to express the relative:—Fæder úre þú þe eart on heofenum Pater noster, qui es in coelis, Mt. Kmbl. 6, 9. Drihten þú þe míne fæderas on þínre gesihþe eodon, God þú þe mé féddest Deus, in cujus conspectu ambulaverunt patres mei, Deus, qui pascit me, Gen. 48, 15: Elen. Kmbl. 1448; El. 726. Wé þé þanciaþ, þe þú hafest on gewealdum hiofen and eorþan, Hy. 8, 12. Eálá þú Hǽlend þurh þíne þrowunga þe þú getuge tó þé ealle ða sáwla, H. R. 15, 3. [Goth. þu; gen. þeina; dat. þus; acc. þuk: O. Sax. thu; gen. thín; dat. thí; acc. thik: O. Frs. thu; gen. thín; dat. acc. thi: O. H. Ger. dû; gen. dîn; dat. dir; acc. dih: Icel. þú; gen. þín; dat. þér; acc. þik.] v. gé, git.

þúf, es; m. A tuft. I. applied to foliage:—Þúfum crinibus (the passage is: Dum virgas steriles atque superfluas flammis de fidei palmite concremant, ut concreta vagis vinea crinibus silvosi inluviem poneret idoli), Germ. 402, 71. v. þúf-bǽre, and following words. II. the crest of a helmet (?). v. Lydus de magistrat.:—καλοῦσι δὲ αὐτὰς οἱ μὲν Ῥωμαῖοι ἰούβας, οἱ δὲ βάρβαροι τουφάς. v. next section. III. a kind of standard, made with tufts of feathers:—Illud genus vexilli, quod Romani Tufam (tufa genus vexilli ex confertis plumarum globis, v. Du Cange s. v.), Angli vero Tuuf (v. ll. thuuf, thuf, Txts. 137, 1), ante eum ferri solebat (the A.-S. version has only:—Him mon symle ðæt tácen beforan bær), Bd. 2, 16. Ðá wæs þúf hafen, segen for sweótum, Elen. Kmbl. 246; El. 123. Sunu Simeones sweótum cómon, þúfas wundon ofer gárfare, Cd. Th. 199, 22; Exod. 342. Hié gesáwon þúfas þunian, 187, 32; Exod. 158. v. sige-þúf.

þúf-bǽre; adj. Bearing foliage, leafy:—Bóh þúfbǽres pintreówes frondentis pini stipitem, Hpt. Gl. 458, 67.

þúfe; adj. Tufted, having leaves in tufts (?), bushy:—Þúfe þistel sow thistle, Lchdm. ii. 312, 20. v. ge-þúf, þífe-þorn, and preceding and following words.

þúfian; p. ode To become leafy or bushy:—Þúfaþ and wridaþ frutescit, Wrt. Voc. ii. 38, 13.

þúfig; adj. Full of leaves, with thick foliage:—Þúfigum frondosis, Wrt. Voc. ii. 38, 14.

þúft, es; m. A place full of bushes:—Gewrid oþþe þúftas frutecta, Wrt. Voc. ii. 38, 25. v. þýfel.

þuhsian, þux[s]ian; p. ode To make misty, dark:—Eall upheofon biþ sweart and gesworcen and swýðe geþuhsod (cf. Dóm. L. 8, 105, which has geþuxsað), deorc and dimhíw and dwolma sweart, Wulfst. 137, 9. [Cf. Icel. þoka fog, mist.]

þullic. v. þus-líc.

þúma, an; m. The thumb:—Ðúma, thúma, thúmo pollux, Txts. 89, 1617. Swá greáte swá ðín þúma, Lchdm. iii. 18, 25. Ic com mid handa on ðone stán drífan, and se ðúma gebrocen wæs, Bd. 5, 6; S. 619, 24. Gif se þúma biþ of áslægen, ðam sceal .xxx. scitt. tó bóte. Gif se nægl biþ of áslegen, ðam sculon .v. scitt. tó bóte, L. Alf. pol. 56; Th. i. 94, 28. Gif man þúman of áslæhþ, .xx. scitt. Gif þúman nægl of weorðeþ, .iii. scitt. gebéte, L. Ethb. 54; Th. i. 16, 9. Hé æthrán his swíðran þúman (pollicem manus ejus dextrae), Lev. 8, 23. Þúman pollices, Wrt. Voc. ii. 82, 48: Ex. 29, 20. [Mid te þume, A. R. 18, 14. Þe nayle of þe þoume, Ayenb. 43, 14. tó the thowme, Rel. Ant. i. 190, 22 (end of 14th cent.). Thombe, Chauc., Piers P. Thowmbe, Prompt. Parv. 492. In other glossaries of 15th cent. it is spelt thome, Wrt. Voc. i. 184 (where also thombe): 207, col. 2: thowme, 186, col. 1: thombe, 179: 247, col. 2 (in same glossary also) thumb, 246, col. 1. O. Frs. thúma: O. L. Ger. thúmo: O. H. Ger. dúmo: Dan. tomme: Swed. tumme.] v. þýmel.

þumle entrails:—Tharme, thumle viscera, Wrt. Voc. ii. 123, 72.

-þunca, -þuncan, þundende, þune-líc, þuner. v. æf-þunca, be-þuncan, þunian, þunor-líc, þunor.

þung, es; m. A poisonous plant, (vegetable) poison; the word is used to translate aconitum, eleborus, mandragina, as well as the more general term toxa (cf. letali toxa = mortali veneno, Hpt. Gl. 427, 54):—Þung, woedeberge eleborus, Wrt. Voc. ii. 107, 12: 29, 21. Þung mandragina, 59, 42: aconita, i. 31, 58: aconitum, 67, 16: toxa, 68, 26: coxa (r. toxa), 67, 15. Þung toxa or toxicum (printed toxi pang), 289, 52. Gif mon þung ete, áþege buteran and drince; þung gewít on ða buteran. Eft wiþ ðon, ásleá him mon fela scearpena on ðam scancan, ðonne gewít út ðæt áttor þurh ða scearpan, Lchdm. ii. 154, 1-4. Scalf wiþ ðam miclan líce . . . þung . . ., 78, 25. Ámber fulne holenrinda and æscrinda and þunges, 332, 16. Nim ðone miclan þung, 154, 14. Thungas, þungas aconita, Txts. 36, 23. Þungas, Wrt. Voc. ii. 4, 20. v. cluf-þung; f.

þunge, -þungen. v. þeón, ful-, ge-, heáh-þungen.

þungenness, e; f. Excellence, virtue:—Mid hú monigum médum mín fæder and mín móder mé [wǽron] biddende, ðæt ic forléte míne (ge-?)þungenesse (the speaker wished to become a monk), Shrn. 36, 26. v. ge-þungenness.

þunian; p. ode. I. to stand out, be prominent, be lifted up, stick up:—On ðam forman dæge on ðam middangeard þunaþ gesceapen primo dierum quo mundus extat conditus, Hymn. Surt. 4, 4. Þunie (þu me, Th.) him gewinnes wearn ofer wealles hróf may much strife be lifted up for it above the top of its wall; circumdabit eam super muros ejus iniquitas, Ps. Th. 54, 9. Hié gesáwon fyrd wegan . . . þúfas þunian they saw the host march . . . saw the standards lift their tops above the ranks, Cd. Th. 187, 32; Exod. 158. Þindan and þunian, þecene hebban, Exon. Th. 431, 17; Rä. 46, 2. I a. fig to be lifted up, be proud, cf. colloquial to be stuck up:—Wǽre ðú (the body) ðé wiste wlanc, þrymful ðunedest, Soul Kmbl. 79; Seel. 40. v. on-þuniau, and cf. þennan. II. to make a noise, to sound, resound, creak:—Ic

(*a storm*) ástîge strong, þrymful þunie, Exon. Th. 380, 42 ; Rä. 2, 4. Sundwudu þunede *the ship's timbers creaked*, Beo. Th. 3817 ; B. 1906. Þunode oððe hleóþrede *increpuerit*, Wrt. Voc. ii. 44, 14. Dynedan and þunedan *crepitabant*, 21, 17. Ðære thundendan (thuniendan ? *but cf.* (?) *Icel.* þundr *a name of Odin* ; þund *the name of a mythical river*) *bombosae* (*vocis mugitum*), Wrt. Voc. ii. 77, 59. v. tô-þuniende, þunung. ge-þun, þunor, *and cf.* *Lat.* tonare, tonitrus.

þunor (-ar, -er, -ur), es ; *m.* **I.** *thunder* (implying not only sound but also striking) ; tonitrus, fulmen :—Þunor *tonitruum vel tonitrus*, Wrt. Voc. i. 52. 45 : 76, 34 : Blickl. Homl. 91, 34. Ðuner (ðunor, Rush.), Jn. Skt. Lind. 12, 29. Ðunar byð hlûdast, Menol. Fox 467 ; Gn. C. 4. Ðunor cymð of hætan and of wætan . . . seó hæte and se wæta winnaþ him betweónan mid egeslîcum swêge, and ðæt fŷr âbyrst ûs ðurh ligett . . . Swâ hâttra sumor, swâ mâra ðunor and lîget on geáre. Ða þuneras (þunras, MS. R.) . . . on Apocalipsin . . . ne belimpaþ tô ðam ðunere (þunre, MS. R.) ðe on ðyssere lyfte oft egeslîce brastlaþ. Se byþ hlûd for ðære lyfte brâdnysse, and frecenfull for ðæs fŷres sceótungum, Lchdm. iii. 290, 2-15. Þunor tôslôg heora godes hûs *aedes Salutis ictu fulminis dissoluta est*, Ors. 4, 2 ; Swt. 160, 18. Án þunor tôslôg hiora Capitoliam *fulmine Capitolium ictum*, 6, 14 ; Swt. 268, 29. Hiene ofslôg ân þunor *fulmine ictus interiit*, 6, 29 ; Swt. 278, 17. Ðunres bearn *filii tonitrui*, Mk. Skt. 3, 17. Þunres slege *a clap of thunder*, Nicod. 23 ; Thw. 13, 3. Þunres slege *fulgura*, Ps. Spl. T. 96, 4. Stefne ðunures micles, Rtl. 47, 22. God âsende rên mid ðunore, and manega menn mid ðam ðunore swulton, Homl. Skt. i. 15, 93. Beóþ myccle þuneras on heofnum, Blickl. Homl. 93, 15. Ðâ sceolde hê sendan ðunras and lŷgetu, Bt. 35, 4 ; Fox 162, 13. **II.** *one of the Teutonic gods, to whom, among the Roman, Jupiter seems to have been considered most nearly to correspond* ; hence Jupiter is translated by Þunor :—Þunor oððe Ðûr *Joppiter*, Wrt. Voc. ii. 47, 33. Þunor, 93, 59. Þuner *Jovem*, 112, 5. **II a.** *it is mostly in connection with the fifth day of the week that the word occurs* :—On ðam fîftan dæge ðe gê Ðunres hâtaþ, Homl. Th. ii. 242, 23. Ðunres-dæges nama is of Ðunor, Anglia viii. 321, 16. On ðone hâlgan Ðunres-dæg, L. Alf. 5 ; Th. i. 64, 24. Tô ðam hâlgan Þurres-dæge, Homl. Skt. ii. 23 b, 621. Gang on þunres-æfen (*Wednesday evening*), Lchdm. ii. 346, 10. *It is found also in local names*, e. g. Ðunres-feld, Ðunres-leáh, Cod. Dip. Kmbl. vi. 342. [Þa Þunre heo ȝiuen þunres dæi (þoris dai, 2nd MS.), Laym. 13929.] **III.** *a thane of king Egbert of Kent* :—Ermenrêd gestrŷnde twêgen sunu ða syððan wurðan gemartirode of Ðunore, Chr. 640 ; Erl. 26, 4. See for more details of the event thus recorded, Lchdm. iii. 422 sqq., and the Latin charter, Cod. Dip. Kmbl. iv. 236. [*O. L. Ger.* Thuner :—Ec forsacho Thuner ende Uuôden : *O. Frs.* thuner, tonger ; Thunres-dei : *O. H. Ger.* donar ; Toniris tac : *Icel.* Þôrr. See Grmm. D. M. c. 8.] v. þôr, þúr.

þunor-bodu *a gilthead* (a kind of fish) :—Ðunorbodu *sparus*, Wrt. Voc. i. 55, 71 (in a list 'nomina piscium').

þunor-clæfre, -clæfre, an ; *f.* Bugle ; ajuga reptans (cf. þundre clovere *consolida media*, Wrt. Voc. i. 140, 68, and *consound* in E. D. S. Pub. Plant Names) :—Þis is seó æðeleste eáhsealf . . . Genim . . . ðunor-clæfran blôsman, Lchdm. ii. 4, 7. Ðunorclâfran, i. 374, 4.

þunor-lîc ; *adj. Thunderous, of thunder* :—Þune[r]lîcum cirme *tonitruali fragore*, Hpt. Gl. 451, 47.

þunor-râd, e ; *f. Thunder, a peal of thunder* :—Ne biþ þær lîget . . . ne þunerâd (þunor, Wulfst. 139, 31) *non fulmina*, . . . tonitru, Dôm. L. 16, 263. Ðâ com þunerâd and lêgetsleht and ofslôh ðone mæstan dæl, Shrn. 57, 35. Ðâ wæs geworden mycel þunorrâd, Blickl. Homl. 145. 29. Ðonne þunorrâd biþ, ne sceþeð ðam men ðe ðone stân (*agate*) mid him hæfð. Lchdm. ii. 296, 30 : iii. 374, col. 2. From stefne ðunur-râde (þunurâde, Spl.) *a voce tonitrui*, Ps. Surt. 103, 7 : 76, 19. Þunur-râda ðînre *tonitrus tui*, Ps. Spl. 76, 17. Hió âhôf ðæt heáfod of ðære mŷsan somod mid ðære þunorâde, Lchdm. iii. 374, col. 2. Biddaþ Drihten, ðæt his þunorrâda (*tonitrua*) geswîcon, Ex. 9, 28, 33, 34. Gif lígette and ðunorrâde (*tonitrua*) eorþan and lyfte brêgdon, Bd. 4, 3 ; S. 569, 12. Þunerâda, Hpt. Gl. 509, 22. Ðunorrâda hlynn, Wulfst. 186, 3. Mycel mægen lîegetslyhta and þunerâda, Lchdm. iii. 374, col. 2. Gôd wið lígetta and wið þunorrâda, ii. 290, 16. Hê worhte þunorrâda on heofonum *intonuit de caelum Dominus*, Ps. Th. 17, 13 : Ex. 9, 23. [Cf. *Icel.* reið *a clap of thunder*, from the notion of Thor *driving* through the air. See Grmm. D. M. c. 8.]

þunorrâd-lîc ; *adj. Thunderous, of thunder* :—Of þunerâdlîcan cerme *tonitruali fragore*, Hpt. Gl. 451, 46.

þunorrâd-stefn, e ; *f. A voice of thunder* : — Wæs þunurrâdstefn strang on hweóle *vox tonitrui tui in rota*, Ps. Th. 76, 14.

þunor-wyrt, e ; *f. Thunder-plant* (v. E. D. S. Pub. Plant Names), *house-leek* ; sempervivum tectorum :—Nime þunorwyrt, Lchdm. ii. 118, 2. [On plants that were a protection against thunder, see Grmm. D. M. pp. 167, 1147.]

þunres dæg. v. þunor, II a.

þunrian ; *p.* ode *To thunder* :—Hê is mægenþrymmes God. and hê þunraþ ofer manegum wæterum *Deus majestatis intonuit, Dominus super

multas aquas, Ps. Th. 28, 3. Hit ðunraþ *tonat*, Ælfc. Gr. 22 ; Zup. 128, 17. Hit hwîlum þunraþ, Bt. 39, 3 ; Fox 44, 34 : Met. 28, 55. Seó menio sædon ðæt hyt þunrode (*tonitruum factum esse*), Jn. Skt. 12, 29. Þunerode of heofonum Drihten *intonuit de coelo Dominus*, Ps. Spl. T. 17, 15. [*O. H. Ger.* donarôn.]

þunring, e ; *f. Thundering, thunder* :—Swâ stôr þunring and lægt wes, swâ ðæt hit âcwealde manige men, Chr. 1085 ; Erl. 219, 22.

þunung, e ; *f. A creaking, a rattle* :—Þununge *crepitum*, Wrt. Voc. ii. 21, 25. Þununga *crepundiorum*, 23, 64.

þun-wang, e : -wange, -wenge, an ; *f.* (and *n. ?*) Wange, wenge *are both found neuter, though also the plurals* wangas, wangan *occur*) A *temple* :—Þunwang *timpus*, Wrt. Voc. i. 42, 50. Þunwencge (-wenge, -wange) *timpus*, Ælfc. Gr. 9, 32 ; Zup. 59, 5 : 298, 2. Gif ic on þunwange gereste *si dedero requiem temporibus meis*, Ps. Th. 131, 4. Bufan his þunwengan *supra tempus capitis ejus*, Jud. 4, 21. Þunwonge *tympora*, Wrt. Voc. i. 282, 44. Þunwonga sâr *dolor timporum*, ii. 143, 34 : Lchdm. i. 156, 22. Þunwongena *timporum*, Wrt. Voc. ii. 87, 61. Þunwangena, Anglia xiii. 37, 291. Þunwængum (-wengum, Spl. C.) *timporibus*, Ps. Lamb. 131, 5. Ðunwoengum, Rtl. 181, 13. Þunwange *malas*, Wrt. Voc. ii. 57, 30. Smire ðone man mid on þa þunwonge, Lchdm. ii. 334, 15. Smyre ða ðunwonga, i. 216, 8. Gnîd on ða þunwunge, 380, 15. Smire ða þunwangan mid, ii. 20, 8. Þunwongan, 306, 2. [Þungana, Lchdm. iii. 292, 22. Lay on the forheyd and on the thunwanges, Rel. Ant. i. 54, 26, 43 (quoted in Halliwell's Dict.). Thunwonge of mannys heede *tempus*, Prompt. Parv. 493. Thunwange *tempus*, Cath. Angl. 387, and see note. Thonwangnes, Wrt. Voc. i. 185, col. 2 (15th cent.). *O. H. Ger.* dun-wengi : *Icel.* þunn-wangi : *m.* ; þunn-wengi ; *n.*] v. (?) þynne *and* wange.

þúr, es ; *m.* Thor, the god who most nearly corresponded to Jupiter ; hence Jupiter is translated by Þúr :—Þunor oððe Ðúr *Joppiter*, Wrt. Voc. ii. 47, 33. Þúres môdur *Latona*, 53, 4. On Galienus dagum ðæs kâseres hêt Necetius Rôme burge gerêfa hî lædan tô Þúres deófulgeldum, Shrn. 128, 9. Ðys godspel sceal on Þúres-dæg, Rubc. Jn. 7, 40. On Þúres-dæg, Rubc. Jn. 5, 30. The word is found also in local names, e. g. Ðúres-leáh, Ðúrgârtûn, Cod. Dip. Kmbl. vi. 342. [A. R. Þurs-dei.] v. þunor, þôr.

þuren. v. ge-þuren.

[þurfan] ; *prs.* ic, hê þearf, ðú þearft, *pl.* wê þurfon ; *p.* þorfte ; *subj. prs.* ic þurfe, þyrfe, *pl.* þurfen, þyrfen ; *prs. ptcpl.* þurfende, þyrfende *To need.* **I.** *to be in need*, *have need* of something, (1) absolute :—Gif ðú clâþa þe mâ on hæfst, þonne ðú þurfe, Bt. 14, 1 ; Fox 42, 15. Ðú gæderast mâre, þonne ðú þurfe (þyrfe, Cott. MS.), 14, 2 ; Fox 44, 8. Nis hit gôd, ðæt hié sién on ðam lâðe leng, þonne ðú þurfe, Cd. Th. 243, 3 ; Dan. 430. Sam hî þyrfon, sam hî ne þurfon, hî willaþ þeáh, Cd. Th. 26, 2 ; Fox 92, 30. Ða þurfende *pauperes*, Mt. Kmbl. Rush. 5, 3. Þyrfendra *egentum*, Wrt. Voc. ii. 142, 69. (2) with gen. of thing needed :—Beó ðe þe ðînum, and læt mê þe mînum ; ne gyrne ic ðînes, ne ðú mînes ne þearft (ðærft, Lchdm. iii. 288, 9), L. O. 13 ; Th. i. 184, 16. Ne ðearf hê nânes þinges bûton ðæs, ðe hê on him selfum hæfþ, Bt. 24, 4 ; Fox 86, 8 : Cd. Th. 204, 27 ; Exod. 425. Hwæt ðurfe (ðurfu, Lind.) wê leng gewitnisse *quid adhuc egemus testibus*, Mt. Kmbl. Rush. 26, 65. Ða þurfon swîþe lytles, ðe mâran ne willniaþ þonne genôges, Bt. 14, 2 ; Fox 44, 13. Ne ðorfte hê nâ mâran fultumes, 26, 2 ; Fox 92, 22. Hî his sume ðorfton, Past. pref. ; Swt. 9, 16. Ðæm ðe micles ðorfton . . . ðæm ðe lytles ðyrfe, 44 ; Swt. 325, 5-7. Swâ welig ðæt hê nânes þinges mâran ne þurfe, Bt. 24, 2 ; Fox 82, 4. (3) with acc. :—Mûþa gehwylc mete þearf, Exon. Th. 341, 12 ; Gn. Ex. 125. Mete bygeþ, gif hê mâran (*or gen. ?*) þearf, 340, 14 ; Gn. Ex. 111. **II.** *to need* to do something. (1) where a want has to be satisfied, a purpose to be accomplished, or the like :—Ðú meaht ðê self geseón, swâ ic hit ðê secgan ne þearf, Cd. Th. 38, 24 ; Gen. 611. Gif hit sié sumor, ðô wermôdes sædes dust tô ; gif hit sié winter, ne þearft ðú ðone wermôd tô dôn, Lchdm. ii. 180, 29. Hê ne ðearf nâ faran fram stôwe tô stôwe, Homl. Th. i. 158, 4. Hwæt ðurfon (þurfe, Bod. MS.) wê nú mâ sprecan ? Bt. 24, 4 ; Fox 86, 22. Hî witan, hwær hî eáfiscas sêcan þurfan (*where they must seek them, if they are to find them*), Met. 19, 25. Syle mê ðæt wæter, ðæt mê ne þyrste, ne ic ne ðurfe hér feccan, Jn. Skt. 4, 15. Ðŷ læs wê leng sprecen ymbe ðonne wê þyrfon (wê ne þyrfen, Cott. MS.), Bt. 34, 2 ; Fox 136, 14. (2) where the need is based on grounds of right, fitness, law, morality, etc., *to be bound* to do something because it is right, etc. :—Nô ðú mîne þearft hafelan hŷdan *the duty of burying me will not fall upon you*, Beo. Th. 895 ; B. 445. Gif hê gewitnesse hæbbe, ne þearf hê ðæt geldan (*he is not bound by law to pay*), L. Alf. 28 ; Th. i. 52, 3. Ne þearf hê him onfôn, L. In. 67 ; Th. i. 146, 4. Mê ðæt riht ne þinceþ, ðæt ic ôleccan þurfe Gode, Cd. Th. 19, 12 ; Gen. 290. Ðæt ðú ne wêne, ðæt ðú Iudéa leásungum gelŷfan þurfe *that you may not think, that you are bound to believe the Jews' false tales*, Blickl. Homl. 177, 35. Hê swîðor his môd gebint tô ðæm unnyttran weorcum, ðonne hê ðyrfe (*more than is fitting for him*), Past. 4 ; Swt. 37, 21. Ðonne mon mâ fæst, ðonne hê ðyrfe (*more than religion requires*), 43 ; Swt. 313, 2. Gif ða gyltas tô ðam hefelîce beón, ðæt hê tô bisceopes dôme tæcan þurfe (*he must do it because the church has prescribed such a course*), L. Ecg. P. i. 11 ; Th. ii. 176,

30. Gedón hí ðæt hira synna ne ðyrfen (*need not, because of the divine ordinance*) bión gesewene æt ðæm nearwan dóme, Past. 53; Swt. 413, 16. (3) with the idea of compulsion, or where the inevitability of a consequence is expressed; in some cases the word might be taken almost as an auxiliary, of much the same force as *shall; to be obliged, be compelled by destiny*:—Gé ne þurfon hér leng wunian *you shall not be obliged to stop here longer*, Ex. 9, 28. Nis ðæt þonne nænig man, ðæt þurfe ðone deópan grund ðæs hátan léges gesécean, Blickl. Homl. 103, 14. Næs him ænig þearf, ðæt hé sécean þurfe *there was no need to force him to seek*, Beo. Th. 4984; B. 2495. Feallaþ ofor ús, ðæt wé ne þurfon ðysne ege leng þrowian, Blickl. Homl. 93, 34. Þý læs gyt láð Gode weorðan þyrfen *lest the inevitable consequence, your becoming hateful to God, follow*, Cd. Th. 36, 26; Gen. 577. Þý læs ða týdran mód ða gewitnesse wendan þurfe, Exon. Th. 147, 21; Gú. 730. Náuht ðæs ðe hé ondréde, ðæt hé forleósan þorfte (*should be obliged to lose*), Bt. 26, 2; Fox 92, 22. Hwý him on hige þorfte (*should necessarily follow*) á þý sæl wesan, Met. 15, 9. Hú hé ðisse worulde wynna þorfte læsast brúcan *how he should be least under the necessity of enjoying the delights of this world*, Exon. Th. 122, 20; Gú. 308. Ðæt ðæt micle morð menn ne þorfton þolian *that men would not necessarily have to suffer that great perdition*, Cd. Th. 40, 17; Gen. 640. (4) *to have good cause or reason for doing something*:—Ðú sorge ne þearft beran on ðínum breóstum, Cd. Th. 45, 28; Gen. 733. Ic ðé scylde, ne þearft ðú forht wesan, 131, 5; Gen. 2171: Blickl. Homl. 191, 18: Beo. Th. 3353; B. 1674. Ðú ðec sylfne ne þearft swíþor swencan, gif ðú God lufast, Exon. Th. 245, 18; Jul. 46. Ðæt is genóg sweotol, ðætte nánne mon ðæs tweógean ne þearf, Bt. 11, 2; Fox 34, 34: Blickl. Homl. 41, 36: 83, 9. Ne þearf ðæs nán mon wénan, 101, 13: 109, 30. Ne ðarf mon ná ðone medwísan læran, ðæt hé ða lotwrencas forléte, forðonðe hé hié næsþ, Past. 30; Swt. 203, 15. Ne þearf hé gefeón *he will have no cause to rejoice*, Cd. Th. 92, 4; Gen. 1523: Exon. Th. 449, 9; Dóm. 68: Beo. Th. 4016; B. 2006. Mé wítan ne ðearf Waldend, 5475; B. 2741: Cd. Th. 165, 7; Gen. 2728. Ne ðurfe wé ceorian, Homl. Th. ii. 438, 27. Ne þurfan gé nóht besorgian, hwæt gé sprecan, Blickl. Homl. 171, 18. Ne þurfe gé beón unróte, 135, 24. Ne þurfun gé wénan, Exon. Th. 142, 16; Gú. 645. Ne þurfon mé hæled ætwítan, Byrht. Th. 139, 4; By. 249. Ic eów secgan mæg, ðæt gé ne ðyrfen leng murnan, Judth. Thw. 23, 33; Jud. 153. Sume him ondrædaþ earfoþu swíþor þonne hý þyrfen, Bt. 39, 11; Fox 228, 24. Gif hé náne æhta næfde, ne þorfte hé nánne feónd ondrædan . . . Gif ðú swelces nánuuht næfdest ne þorftest ðú ðé nánwuht ondrædan, 14, 3; Fox 46, 23–28. Nó hé ðære feohgyfte scamigan þorfte, Beo. Th. 2057; B. 1026: 2147; B. 1071. Hwær hé ðara nægla swíðast wénan þorfte *where he had most reason to expect that he should find the nails*, Elen. Kmbl. 2206; El. 1104. Nó wé þus swíðe swencan þorftan, þær ðú freónda lárum hýran wolde, Exon. Th. 129, 19; Gú. 423. Ne þorfton hí hlúde hlihhan, Cd. Th. 5, 17; Gen. 73. Hí gearowe wæron deáde sweltan, gif hí ðorfton (*if the occasion demanded it*), Homl. Th. ii. 130, 5. (5) where the need arises from an advantage to be gained, or purpose to be served, *to be use, to be good for a person to do something*:—Ne þearf ic yrfestól bytlian *it is too good or use for me to build an hereditary seat*, Cd. Th. 131, 14; Gen. 2176. Ne þearft ðú sæce ræran, Elen. Kmbl. 1876; El. 940. **III.** *to owe*, cf. sculan, **I**:—Ne þearf ic N. sceatt ne scilling, ne pænig ne pæniges weorð; ac eal ic him gelæste ðæt ðæt ic him scolde, L. O. 11; Th. i. 182, 9. [*Goth.* þarf, *pl.* þaurbum; *prs.*; þaurfta; *p.*; þaurbands; *prs. ptcpl.*: *O. Sax.* tharf, *pl.* thurðun; *prs.*; thorfta; *p.*: *O. Frs.* thurf, thorf, *pl.* thurvon; *prs.*: *O. H. Ger.* darf, tharf, *pl.* durfun, thurfun; *prs.*: dorfta; *p.*: *Icel.* þurfa; þarf, *pl.* þurfum; *prs.* þurfti; *p.*; þurfandi; *prs. ptcpl.*] v. be-þurfan; þearfan, þearfian, þorfan.

þurh, þurg, þuruh, þorh, þorch, þerh, þerih, þærh; *prep. Through.* **A.** with acc. v. also **C.** **I.** local, (1) marking motion into and out at the opposite side :—Þorh (dorh, ðorh) ludgaet *per seudoterum*, Txts. 84, 741. Ðurh ða duru wé gáð in *per hostium intramus*, Ælfc. Gr. 47; Zup. 269, 18: Cd. Th. 29, 8; Gen. 447. Gangaþ inn þurh (ðerh, Lind.) nearwe gaet, Mt. Kmbl. 7, 13: Lk. Skt. 18, 25. Syllan drincan þurh þyrel, Exon. Th. 485, 1; Rä. 71, 7. Ðá férde hé þurh (ðerh, Lind.) hyra mydlen, Lk. Skt. 4, 30. Wé þuruh fýr faraþ and þuruh flóda þrym *transivimus per ignem et aquam*, Ps. Th. 65, 11. Hé wæs on breóstum wund þurh ða hringlocan, Byrht. Th. 132, 29; By. 145. (1 a) where the preposition follows the governed word :—Duru, ðe ic wæs þurh hider onsended, Blickl. Homl. 9, 1. (2) marking motion over *or* in, cf. geond :—Hé férde þurh ða ceastre and ðæt castel bodiende, Lk. Skt. 8, 1. Hé ástyraþ ðis folc, lærende þurh ealle ludéam, 23, 5. Hé hleóprede þurh hátne líg, Exon. Th. 185, 4; Az. 2. Ic þurh ðín hús middan eode *perambulabam in medio domus meae*, Ps. Th. 100, 2. Ðæt fýr nimeþ þurh foldan gehwæt, Exon. Th. 62, 18; Cri. 1003. **II.** temporal, marking continuity, *through, for, during* :—Ðurh twégen dagas *per biduum*, Bd. 4, 19; S. 589, 2. Þurh ealne dæg *tota die*, Ps. Th. 73, 21. Þuruh, 87, 9. Ðorh syndrie neht *per singulas noctes*, Surt. 6, 7. Þurh scírne dæg, Exon. Th. 439, 15; Rä. 59, 4. Þurh lytel fæc, 115, 6; Gú. 185. Þurh ælda tíd, 152, 11; Gú. 807. Þurh

ealra worulda woruld, Ps. Th. 71, 5. **III.** other relations, (1) marking the agent, *through, by* :—Þorch (dorh, ðorh) byrgeras *per vispelliones*, Txts. 86, 760. Ðerih, 151, 6. Wá ðam menn þurh ðone ðe (ðe ðorh hine, Lind.) byð mannes Sunu beláewed, Mt. Kmbl. 26, 24: Chr. 1014; Erl. 151, 8. Hié hié wendon ealla ðurh wíse wealhstódas on hiora ágen geðióde, Past. pref.; Swt. 7, 4. Seó hergung wæs ðurh Alaricum geworden, Bd. 1, 11; S. 480, 11. Hé ða bisceopðéninge ðurh hine sylfne (*per se*) ðénian ne mihte, 4, 23; S. 594. 27. Wiste Cúðberhtus eal ðe ðam wíse, and wolde þurh hine sylfne hí geneósian (*would visit her in person, the visit should be made by himself*), Homl. Th. ii. 142, 11. Gif hwá ymb cyninges feorh sierwie þurh hine oþþe þurh wreccena feormunge (*by his own direct acts or by the harbouring of criminals*), L. Alf. pol. 4; Th. i. 62, 15. (1 a) preposition following case :—Wá ðam ðe hig þurh (ðerh, Lind. Rush.) cumaþ *uae illi per quem ueniunt*, Lk. Skt. 17, 1. (2) marking the means or instrument, *through, by, by means of, by use of* :—Swá hé spræc þurh hys hálegra wítegena múð (*per os sanctorum*), Lk. Skt. 1, 70. Hé ðurh ðæra wealhstóda múð ðam cyninge bodade, Homl. Th. ii. 128, 21: 148, 12. Ic þurh múþ sprece mongum reordum, Exon. Th. 390, 13; Rä. 9, 1. Þurh his sylfes múð, 464, 6; Hö. 83: Andr. Kmbl. 1301; An. 651. Tódæl þurh seofon *divide by seven*, Anglia viii. 304, 41. Cnuca hý þurh hý selfe *pound it by itself* (*per se*), Lchdm. i. 130, 4: 192, 17. Wé ðæt gehýrdon þurg hálige béc, Apstls. Kmbl. 126; Ap. 63. Þurg wítgena wordgeryno, Elen. Kmbl. 577; El. 289. Hié lufodon wísdóm and ðurh ðone hié begeáton welan, Past. pref.; Swt. 5, 14. Hé geférde þurh feóndes cræft, Cd. Th. 29, 21; Gen. 453: 1, 21; Gen. 11: Blickl. Homl. 17, 11. Þuruh, Ps. Th. 70, 1. Hé hié tó heofona ríce laþode þurh his wundorgeweorc and þurh ða godspellícan láre, Blickl. Homl. 7, 9: Andr. Kmbl. 1949; An. 977. Ðæt hí heora synna wítnade and bétte ðurh fæsten and ðurh wópas and ðurh gebedo, Bd. 4, 25; S. 599, 25. Wé witon unrím ðara monna ðe ða écan gesælða gesóhtan nallas ðurh ðæt án ðæt hí wilnodon ðæs líchomlícan deáðes ac eác manegra sárlícra wíta hié gewilnodon wið ðam écan lífe *multos scimus beatitudinis fructum non morte solum, verum etiam doloribus suppliciisque quaesisse*, Bt. 11, 2; Fox 36, 3. Gif hine mon geyflige mid slege oþþe mid bende oþþe þurh wunde, L. Alf. pol. 2; Th. i. 62, 4. (3) marking the efficient cause or reason, *through, in consequence of, as the result of, by reason of, on account of* :—Heofonríces duru belocen standeþ þurh ða ærestan men, Blickl. Homl. 9, 2. Wæs micel unfrið þurh sciphere, Chr. 1001; Erl. 136, 2. Gif seó hringe nele up þurh his ánes tige, Homl. Skt. i. 21, 47: Ps. Th. 64, 11. Ðurh Æþelrédes hæse (*jubente Ædilredo*) Wilfriþ hine tó biscope gehálgode, Bd. 4, 23; S. 594, 29: Andr. Kmbl. 3038; An. 1522. Ic þurh his willan ásend wæs *Dei voluntate missus sum*, Gen. 45, 8: Exon. Th. 194, 3; Az. 133. Þurh cléne gecynd *in consequence of a pure nature*, Hy. 9, 11: 7, 24. Ðá mihte heó wíde geseón þurh (*in consequence of*) ðæs láðan láre, Cd. Th. 38, 3; Gen. 601: 39, 25–27; Gen. 631–2. Þurg, Apstls. Kmbl. 25; Ap. 13. Hé ðurh his gylt on ðám inrum þeóstrum befeóll, Homl. Th. ii. 556, 20: Cd. Th. 21, 29, 30; Gen. 331–2. Þurh ða eáðmódnesse mid geleáfullum hé gefylde ðysne middangeard, Blickl. Homl. 11, 7. Hé mætte, and hé rehte ðæt his bróðrum; þurh ðæt hig hine hatedon ðe swíðor (*quae causa majoris odii seminarium fuit*), Gen. 37, 5. Hit wearð gelet þurh ðæt ðe Magnus hæfde micelne scypcræft, Chr. 1048; Erl. 173, 7: Wulfst. 161, 1. Wearð ðær ðerih ðuruh sum þing fleám ástiht, Chr. 998; Erl. 134, 19. Þurh hwæt ðú ðus hearde ús eorre wurde, Elen. Kmbl. 799; El. 400. Man þurh ælc þingc rihtwísnesse lufige, Wulfst. 266, 18. (4) marking motive or feeling that prompts action, *through, from* :—Ðurh (ðerh, Lind.: ðærh, Rush.) andan hine sealdon ða heáhsacerdas, Mk. Skt. 15, 10. Se forhátena spræc þurh feóndscipe, Cd. Th. 38, 21; Gen. 610. Ic Gode þegnode þurh holdne hyge, 37, 7; Gen. 586: Ps. Th. 77, 38. Hyre þurh yrre ágeaf andsware fæder feóndlíce, Exon. Th. 249, 25; Jul. 117. Gif wé þurh eáþmódnesse eall áræsnaþ, Blickl. Homl. 13, 91. Hí fricgaþ þurh fyrwet, Exon. Th. 6, 30; Cri. 92. (5) marking the circumstance which renders state or action possible or right, *through, in, in virtue of, by right of* :—Ðæt Martinus wære wyrðe ðæs hádes, and ðæt folc gesælig ðurh swelcne biscop *that Martin was worthy of the office, and the people happy in such a bishop*, Homl. Th. ii. 506, 9. His blód ágeát God on galgan þurh his gástes mægen *in virtue of his spirit's strength*, Cd. Th. 299, 16; Sat. 550. Hé fæste feówertig daga þurh his mildsa spéd, 306, 23; Sat. 668. Heó hit þurh monnes geþeaht ne sceáwode, 38, 12; Gen. 605. Him bearn Godes déman wille þurh his dæda spéd, 304, 2; Sat. 623: 301, 30; Sat. 589. (6) marking manner, state, *in, by, in the character of, by way of* :—Þorch (dorh, ðorh) óbst *per anticipationem*, Txts. 84, 757. Ðurh endebyrdnesse singan *per ordinem cantare*, Bd. 4, 24; S. 597, 6. Ðæt fýr ábyrst út ðurh ligett *in the shape of lightning*, Lchdm. iii. 280, 6: Elen. Kmbl. 2210; El. 1106. Ácenned in middangeard þurh mennisc heó *born into the world in human shape*, 12; El. 6. Onsýne þurh cnihtes hád *visible in the form of a youth*, Andr. Kmbl. 1824; An. 914. Hnígan mid heáfdum þurh geongordóm *to bow the head as vassals*, Cd. Th. 46, 12; Gen. 743. Ne can ðara idesa ówðer þurh gebedscipe beorna neáwest, 148, 35; Gen. 2467. Hé ðolode ðurh wíte (*as punishment*) ða ýttran

blindnysse . . . Hé ðolaþ þeóstra ðurh wrace, Homl. Th. ii. 556, 19–21. Wundorgiefe þurh goldsmiþe *wondrous gifts in the goldsmith's art*, Exon. Th. 331, 24; Vy. 73. Stód him sum man æt ðurh swefen (*per somnium*), Bd. 4, 24; S. 597, 11: Cd. Th. 159, 16; Gen. 2635: 160, 21; Gen. 2653. Him synna brytta þurh slǽp (*in sleep*) oncwæd, 159, 28; Gen. 2641. (7) marking accompanying circumstances of an action, *in, with*:—Ðú scealt þurh wóp and heáf on woruld cennan, þurh sár micel, sunu and dohtor, Cd. Th. 57, 4–7; Gen. 923–4. Líg þurh lust geslóh micle mǽre ðonne gemet wǽre, 231, 19; Dan. 249. Ðara ðe hyra lífes þurh lust brúcan, Exon. Th. 127, 19; Gú. 388. Ða wácran ðás woruld healdaþ, brúcaþ þurh bisgo, 311, 6; Seef. 88. (8) marking aim, *with a view to*:—Hé Drihten mid hondum genom þurh edwít (*with a view to disgrace him*), Cd. Th. 307, 17; Sat. 681. Heó his láre geceás ðurh þeódscipe (*with a view to instruction (?), in order to be instructed; or on account of his learning (?)*), Elen. Kmbl. 2331; El. 1167. (9) with verbs of swearing, adjuring, etc., *through, by, in*:—Sume synd *jurativa*, ðæt synd swergendlíce, *per* ðurh: *juro per Deum* ic swerige ðurh God, Ælfc. Gr. 38; Zup. 227, 3. Ic swerige þurh mé sylfne *per memetipsum juravi*, Gen. 22, 16. Ðæt gé ne swerion ne þurh heofon . . . ne þurh eorðan . . . ne ðú ne swere þurh ðín heáfod, Mt. Kmbl. 5, 34–36: Cd. Th. 205, 10; Exod. 433: Elen. Kmbl. 1369; El. 686. Ðú deópe áðe þurh ðínes sylfes sóð benemdest, Ps. Th. 88, 42. Ic ðec hálsige þurh gǽsta weard, Exon. Th. 174, 14; Gú. 1177. Hý þurh mínne noman bǽdan, 92, 12; Cri. 1507. Eallum ðǽm ðé mé gecégaþ þorh ðínne noman, Shrn. 105, 6. Ic ðé háte þurh ða hehstan miht, Cd. Th. 308, 18; Sat. 694. (10) marking extent:—Hwí is ðis fæsten þus geteald þurh feówertig daga *why is this fast reckoned at forty days?* Homl. Th. i. 178, 19. **B.** with dat. v. also **C.** **I.** local, (1) marking motion into and out at the opposite side:—Englas flugon swilce ðurh ánre dúna inð ðære heofenan, Homl. Th. ii. 349, 19. Gif ðǽr biþ án hwem open forlǽten, ðæt se here þurh ðam infær hæbbe, 432, 5. Ðerh middum, Lk. Skt. Lind. 4, 30. (2) marking motion over or in:—Ic wæs getogen þurh ðisse ceastre lanum, Blickl. Homl. 243, 29. **II.** in other relations, (1) marking means or instrument:—Geufered þurh lárewlícum basincge *exaltatus melote*, Hpt. Gl. 440, 71. Heó wolde þurh his mynegungum hire mód getrymman, Homl. Th. ii. 146, 10: 448, 27. Ðurh ðínum drýcræftum, 414, 4. Þurh ðam eárplættum, 248, 25. Ðurh twám gewritum, Wulfst. 230, 3. Ðerh múðe háligwara, Lk. Skt. Lind. 1, 70. (2) marking cause:—Hé næs ácweald ðurh ðam heálícan fylle, Homl. Th. ii. 300, 19. Seó geladung ys weaxende þurh ácennedum cildum and waniende þurh forðfarenum, Lchdm. iii. 238, 2. (3) marking manner, state:—Hé ðæt weorc ðæs godspelles má ðurh his fóta gange fremede ðonne on his horsa ráde (*more on foot than on horseback*), Bd. 4, 3; S. 566, 32. Ða com úre Drihten þurh wolcnum (*in clouds*), Blickl. Homl. 145, 35. **C.** in the following passages both acc. and dat. are used:—Ðá áxode se ealderman ðone hæftling, hwæðer hé ðurh drýcræft oððe ðurh rúnstafum his bendas tóbrǽce, Homl. Th. ii. 358, 10–11. Hí sume þurh freónda fultum and ælmesdǽdum, and swiðost þurh hálige mæssan beóð álýsede, 352, 25–27. Ðurh ða treówu and ðam streáwe and ðam ceafe sind getácnode leóhtlíce synna, 590, 12–14. **D.** with gen.:—Wé beóð geclǽnsode þurh ðæs hálgan húselganges, Homl. Th. ii. 266, 23. **E.** as adverb; see also the following compounds:—Hé sǽ tóslát and hí fóran þurh, Ps. Th. 77, 15. Ðǽr wæs fleóhnet ymbe ðæs folctogan bed áhongen, ðæt se bealofulla mihte wlítan ðurh, and on hyne nǽnig monna cynnes, Judth. Thw. 22, 5; Jud. 49. [O. E. Homl. þurh, þurch, þuregh: Laym. þurh, þorh: Orm. þurrh: A. R. þurh, þuruh: Gen. and Ex. ðurg: Havel. þoru: R. Glouc. þoru, þoru: Chauc. thurgh: Piers P. þorows, thorw: Goth. þairh: O. Sax. thurh, thuru: O. L. Ger. thurh, thuru(-o): O. Frs. thruch: O. H. Ger. durh, duruh (-ah, -eh).]

þurh-. With words expressing motion the prefix signifies *through, over*; in other cases it implies *thoroughness, completeness, continuity*; with adjectives of quality it has an intensive force. It is often a rendering of the Latin prefix *per-*; sometimes of *trans-*.

þurh-beorht; *adj. Very bright, splendid*, (1) lit.:—Heora nebwlíte ongann tó scínenne swilce seó þurhbeorhte sunne, Homl. Skt. i. 23, 820. (2) fig.:—Swá micele máran eádmódnysse ðú sý þurhbeorht (*perspicuus*), swá mícele swá máran wurþnysse foresett ðú eart, Scint. 22, 17. Yrfweardnes mín þurhscínendlíc ł þurhbeorht (*praeclara*) is, Ps. Lamb. 15, 6.

þurh-bitter; *adj. Very bitter, exasperating*:—Þweor mǽgþ and tyrwiende ł þurhbitter *generatio prava et exasperans*, Ps. Lamb. 77. 8.

þurh-bláwen; *adj. (ptcpl.) Inspired*:—Mid forewitigum þurhbláwen gáste *presago afflatus spiritu*, Anglia xiii. 370, 65.

þurh-borian *to bore through, perforate*:—Ða wolde ic witan hwæðer ða gelícnissa wǽron gegotene ealle swá hé sǽde; hét hié ðá þurhborian *simulacra quae an solida essent scire ego cupiens omnia perforavi*, Nar. 20, 1. [O. H. Ger. durh-porón *perforare, terebrare*.]

þurh-brecan *to break through*:—Wordes ord breósthord þurhbræc, Beo. Th. 5577; B. 2792. [O. H. Ger. durh-brehhan *dissecare*.]

þurh-brengan *to bring through*:—Hé tóslát sǽ and hé þurhbróhte (*perduxit*) hig, Ps. Lamb. 77, 13. [O. H. Ger. durh-bringan *perferre*.]

þurh-brogden; *adj. (ptcpl.) Transported*:—Ðorhbrogden *trajectus*, Wrt. Voc. ii. 122, 63.

þurh-brúcan *to enjoy thoroughly*:—Hwylc manna þurhbrýcþ (*perfruitur*) mettum búton swæcce sealtes, Coll. Monast. Th. 28, 15.

þurh-burnen; *adj. (ptcpl.) Thoroughly burnt, burnt through*:—Bærn swá ðæt hit sí þurhburnen, Lchdm. iii. 40, 11.

þurh-clǽnsian *to cleanse thoroughly*:—Þurhclǽnsaþ (ðerhclǽnsade, Lind. *permundavit*) he will throughly purge, Mt. Kmbl. Rush. 3, 12.

þurh-creópan *to creep through*:—Swá swá mon melo sift, ðæt melo ðurhcrýpþ (þurg-, Cott. MS.) ǽlc þyrel, Bt. 34, 11; Fox 152, 2.

þurh-delfan *to dig through, bore through, pierce*:—Ic ðurhdelfe *perfodio*, Ælfc. Gr. 28, 6; Zup. 179, 10. Þurhdelfeþ, Ps. Th. 79, 15. Hý þurhdulfon (*foderunt*) míne handa and míne fét, 21, 15. Þurhdol[fen] *confossa, transfixa*, Hpt. Gl. 501, 29.

þurh-dreógan *to carry through, perform, pass time*:—Árísende óþre þurhdreógan *surgentes cetera peragant*, Anglia xiii. 423, 825. Nihte þurhdreógan *noctem peragere*, 394, 420.

þurh-drífan. **I.** *to drive through, pierce, transfix*:—Him man ǽgðer þurhdráf mid ísenum næglum ge fét ge handa, Wulfst. 22, 21. Þurhdrifon hí mé mid næglum, Rood Kmbl. 91; Kr. 46. Hé lét hine sylfne bindan and him ǽgðer þurhdrífan mid næglum ge fét ge handa, Wulfst. 110, 15: Exon. Th. 68, 27; Cri. 1110. Dolgbennum þurhdrifen, Andr. Kmbl. 2793; An. 1399. Míne handa mid næglum þurhdrifene, Homl. Th. i. 220, 17. Þurhdryfene, L. E. I. 21; Th. ii. 416, 30. **I a.** *fig. to penetrate, permeate, imbue*:—Ðeáh ic ǽr mid dysige þurhdrifen wǽre, Elen. Kmbl. 1410; El. 707. **II.** *to drive violently; perpellere*:—Word spearcum fleáh, ðonne hé út þurhdráf (*when he sent out his words vehemently, exclaimed vehemently*), Cd. Th. 274, 33; Sat. 163. [He let þurhdriuen þe spaken mid gadien, Kath. 1920. Wes mon þurhdriuen upon þe rode *homo cruci affixsus est*, 1198.]

þurh-dúfan *to dive through*:—Hé wæter up þurhdeáf, Beo. Th. 3243; B. 1619.

þurh-etan *to eat through, eat out*:—Se wyrm ða eágan þurheteþ, Soul Kmbl. 236; Seel. 122. Áholad, þurhetan (-en ?) *exesum vel comessum*, Wrt. Voc. ii. 144, 76. Swyrd ómige, þurhetene, Beo. Th. 6090; B. 3049.

þurh-fær, es; *n. An inner, secret place*:—Ælc synful on his þurhfærum (*penetrabilibus*) byð bedíglod, Scint. 39, 2. v. þurh-fére, -farenness, -faran, **IV.**

þurh-fæstnian *to transfix*:—Ðorhfæstnadon *transfixerunt*, Jn. Skt. Lind. Rush. 19, 37.

þurh-faran. **I.** *to go through* or *over, to traverse; pertransire*, (1) *trans.*:—Burnan þurhfór (ł -fǽrþ) sáwla úre wénunga þurhfór sáwla úre wæter *torrentem pertransivit anima nostra, forsitan pertransisset anima nostra aquam*, Ps. Spl. 123, 4. Hé ðæt land eall þurhfór, Chr. 1095; Erl. 232, 8: 1097; Erl. 233, 38. (2) *intrans. To pass*:—Mid ðí ðe ðú þurhfǽrst (*pertransires*) on wéstene, Ps. Spl. 67, 8. On anlícnysse þurhfærþ man *in imagine pertransit homo*, 38, 9: 102, 15. Ðǽr scipu þurhfaraþ (ðorhfearaþ, Surt.) *illic naves pertransibunt*, 103, 26. **II.** of a weapon, *to pierce, pass through*:—His sword ðíne sáwle þurhfærþ, Lk. Skt. 2, 35. Ísen þurhfór sáwla his, Ps. Spl. 104, 17. **III.** *to pass beyond, transcend*:—Hefonas hé ðurhfór (*transcendit*). Past. 16; Swt. 99, 23. **IV.** *to penetrate*:—Sió stefn ðæs láriowes ðurhfærþ ða heortan ðæs gehírendes *illa vox auditorum cor penetrat*, Past. 14; Swt. 81, 9: 21; Swt. 155, 11: Bt. 13; Fox 38, 27. Ðeáh ðú ðæt héhste ðurhfare *cum summa penetras*, Past. 65; Swt. 467, 1. Þurhfare *penetret*, Anglia xiii. 378, 192. Ðæt word ðære láre ne mæg ðurhfaran ðæs wǽdlan heortan *egentis mentem doctrinae sermo non penetrat*, Past. 18; Swt. 137, 6. Þurhfarende *penetrans*, Hymn. Surt. 84, 9. [O. H. Ger. durh-faran *transire, permeare, penetrare*.] v. þurh-féran.

þurh-farenness, e; *f. An inner, secret place; penetrale*:—On þurhfarennyssum cyninga heora *in penetralibus regum ipsorum*, Ps. Spl. 104, 28. v. þurh-fær, -fére, -faran, **IV.**

þurh-féran. **I.** *to penetrate* or *over*:—Ðæt geðyld ðurhférde ðara leahtra truman *patientia medias acies transit*, Gl. Prud. 26 b. Hé þurhférde hǽðenre þeóde eard, Shrn. 155, 34. Hí þurhférdon ealle ða land . . . óððæt hí cómon ðǽr hé wunode, Homl. Skt. ii. 30, 231. Þurhférende (*humida cum siccis*) *pervadens* (*caerula plantis*), Wrt. Voc. ii. 96, 38. **II.** *to penetrate, get into*:—Hér Rodla ðurhférde (*penetravit*) Normandi mid his here, Chr. 876; Th. i. 145, col. 3. [He þe þurhferde dead, Kath. 1142.] v. þurh-faran.

þurh-fére; *adj. That may be passed through* or *over, passable, pervious*:—Geat þurhfére *porta pervia*, Hymn. Surt. 112, 9. *The neuter used substantivally translates* penetrale:—On þurhférun *in penetralibus*, Mt. Kmbl. 24, 26. v. þurh-fær.

þurh-fleón *to fly through*:—Cume án spearwa and hrædlíce ðæt hús þurhfleó *adveniens unus passerum domum citissime pervolaverit*, Bd. 2, 13; S. 516, 18.

þurh-fón *to get through, penetrate*:—Heó ðone fyrdhom þurhfón ne mihte láþan fingrum, Beo. Th. 3013; B. 1504.

þurh-gán. I. *to go over* or *through* :—Fixas þurhgáð (*perambulant*) paðas sǽs, Ps. Spl. 8, 8. Ic wille ðurhgán orsorh ðone here, Homl. Th. ii. 502, 11. II. *of a weapon, to pass through, pierce* :— Hé sette his swurdes ord tógeánes his innoðe, and feól he on uppon, ðæt him ðurheode (*or* him ðurh eode, *under* þurh, B. I (1)), Homl. Th. ii. 480, 15. His swurd sceal ðurhgán ðíne sáwle, i. 146, 8. III. *to penetrate, permeate, pervade* :—Seó eorðe byð mid ðam winterlícan cyle þurhgán, Lchdm. iii. 252, 7. [Heo þurheoden Francene þeode, Laym. 5217. Lǽten heom þurhgon al þa duȝeðe, 19645. Þeȝȝ sholldenn all þurrhgan þiss middellærd, Orm. 12860. *Goth.* þairh-iddja ; *p.* : *O. H. Ger.* durh-gán *pertransire, penetrare.*] v. next word.

þurh-gangan. I. *to go over* or *through, perambulate* :—þurhgangende *perambulante*, Ps. Spl. 90, 6. II. *to pierce* :—Ne forhtast ðú ðé on dæge flán on lyfte, ðæt ðé þuruhgangan gáras on ðeóstrum, Ps. Th. 90, 6. [*Goth.* þairh-gaggan : *O. H. Ger.* durh-gangan.]

þurh-gefeoht, es ; *n. War* :—Þorhgefeht, þorgifect *perduellium*, Txts. 85, 738.

þurh-geótan. I. *to pour over, cover by pouring* (lit. and fig.) :— Ðú þurhgute hine gedréfednysse *perfudisti eum confusione*, Ps. Spl. 88, 44. Ic mid ða liffæstan ýþe ðurhgoten wæs *vitali unda perfusus sum*, Bd. 5, 6 ; S. 620, 18. II. *to fill, saturate* :—Ðonne se sacerd gehálgodne tapor in ðæt wæter déð, ðonne wyrð ðæt wæter mid ðam hálgan gáste ðurhgoten, Wulfst. 36, 6. II a. *to fill, imbue, inspire* :—Gleáwnysse þurhgoten, Elen. Kmbl. 1920 ; El. 962. [*O. H. Ger.* durh-giozan *perfundere.*]

þurh-gléded ; *adj.* (*ptcpl.*) *Thoroughly furnished with burning coals* :—Wæs se ofen onhǽted, ísen eall ðurhgléded, Cd. Th. 231, 8 ; Dan. 244.

þurh-hǽlan *to heal thoroughly; persanare* :—Ealle ða þincg, ðe on ðæs mannes líchoman tó láðe ácennede beóþ, heó ðurhhǽleþ (þur-, MS. O.), Lchdm. i. 124, 22.

þurh-hǽlig ; *adj. Very holy* :—Þurhháligere gerde *sacrosancti viminis* (Moses' rod), Hpt. Gl. 409, 70. Þurhháliges blódes *sacrosancti cruoris*, 503, 46. Tó ðam þurhháligum haligdóme Drihtnes líchaman and blódes *ad sacrosanctum sacramentum corporis et sanguinis Domini*, Wanl. Cat. 79, 4.

þurh-heflg ; *adj. Very heavy; praegravis*, Dial. 2, 3 (Lye).

þurh-hwít ; *adj. Very white* :—Þurhhwít *candidus*, Wülck. Gl. 163, 6 (omitted in Wrt. Voc. i. 46, 30).

þurh-irnan *to run through* :—Þurharn *cucurrit*, Wrt. Voc. ii. 137, 60. Ðæt swurd ðe ðæra cildra lima þurharn, Homl. Th. i. 84, 18.

þurh-lǽran *to persuade* :—Nele God wrecan yfelnysse se andettan gyltas þurhlǽrþ *non uult Deus ulcisci malitiam, qui confiteri delicta persuadet*, Scint. 38, 12.

þurh-lǽred ; *adj. Very learned* :—Þurhlǽred *vel gleáw expertus*, i. *multum peritus*, Wrt. Voc. i. 22, 35.

þurh-láð ; *adj. Very hateful, odious* :—Þurhláð *odiosus*, Wrt. Voc. i. 28, 66.

þurh-leóran *to pass through* :—Ðorhleórdun *pertransierunt*, Ps. Surt. 76, 18.

þurh-lócung, e ; *f. A looking through* or *over, a preliminary examination* (?) *of a book; but the word glosses prohemium* :—Ðurhlócung *prohemium*, forespǽc *praefatio*, Wrt. Voc. i. 51, 38. [Cf. Illc an ferrs to þurrhlokenn offte, Orm. dedic. 68.]

þurh-rǽsan *to rush through* :—Hwílum ic þurhrǽse, Exon. Th. 384, 31 ; Rä. 4, 36.

þurh-sceótan *to shoot through, transfix, pierce* :—Þurhscét *transfigat*, Hpt. Gl. 526, 3. Hé his byrnsweord getýhþ, and ða líchoman þurhsceóteþ, Blickl. Homl. 109, 35. Ðǽr wearð Alexander þurhscoten mid ánre flán . . . hé þurh ðæt folc geþrang ðæt hé ðone ilcan ofslóg ðe hiene ǽr þurhsceát *in eo praelio sagitta trajectus eatenus pugnavit, donec eum, a quo vulneratus esset, occideret*, Ors. 3, 9 ; Swt. 134, 22–27. Besyrian ðone earman and þurhsceótan ða unscæðfullan heortan, Ps. Th. 36, 13. Ðá wurdon hí mid deófles flán þurhscotene, Homl. Th. i. 62, 28. [*O. Frs.* thruch-skiata : *O. H. Ger.* durh-sciozan.]

þurh-scíne ; *adj. Transparent* :—Þurhscýne stán *specularis*, Wrt. Voc. i. 38, 30. [Cf. *O. Frs.* thruch-skínich.]

þurh-scínendlíc ; *adj. Splendid;* praeclarus, Ps. Lamb. 15, 6. v. þurh-beorht.

þurh-scrípan. I. *of physical movement, to pass through, glide through* :—Synd twelf tácna on ðam forespreceuan circule ðe seó sunne þurhscríð, Anglia viii. 298, 18. II. *to go through a subject, examine, consider; perlustrare* :—Ǽlc ðæra ðe wyle ða eásterlícan blisse mǽrsian, ne sceal hé náðer ne ðæs lambes flǽsc hreáw etan, ne gesoden, ac gebrǽd ; ðæt ys, ðæt hé ne sceal þurh menniscnysse wísdom þurhscríðan ða hálgan flǽscennysse úres Drihtnes (*he shall not by the aid of human wisdom examine* (?) *our Lord's nature according to the flesh*), ne on him gelýfan swylce hé sý ánfeald man búton his godcundnysse, ac wé sceolon gelýfan ðæt hé ys sóð man and sóðlíce God, Anglia viii. 324, 1. [Al þa londes ic scal þurhscriðen (þorhride, 2nd MS.), Laym. 10887.]

þurh-scyldig ; *adj. Very guilty* :—Hí (*the Jews who plotted against Christ*) synd þurhscyldige for heora syrwunge, Homl. Skt. i. 11, 321.

þurh-sécan. I. *to make search for, seek out* :—Þurhsécende *conquirens*, Scint. 209, 3. [II. *to search through, examine* (?), *as in later English* :—He þurhseched al þe soule, O. E. Homl. ii. 191, 28. Twa Goddspelless uss birrþ þurrhsekenn, Orm. 242. He hefde al þ́ lond ouergan and þurhsoht *peragratis provincie finibus*, Kath. 519. Þe poyson þe veynes so þorwsouȝte, R. Glouc. 151, 11. *O. H. Ger.* durh-suohhan *to search through.*]

þurh-seón *to see through, see into, penetrate with the sight* (lit. or fig.) :—God geseóþ and þurhseóþ ealle his gesceafta, Bt. 41, 1 ; Fox 244, 11. Þurhsyhþ, Met. 30, 16. Gif hwá biþ swá scearpséne, ðæt hé mæge hine (*Alcibiades*) þurhseón, swá swá Aristoteles sǽde ðæt deór wǽre, ðæt mihte ǽlc wuht þurhseón . . . gif ðonne hwá wǽre swá scearpséne, ðæt hé mihte ðone cniht ðurhseón, ðonne ne þúhte hé him nó innon swá fæger, swá hé útan þúhte si, *ut Aristoteles ait, lynceis oculis homines uterentur, ut eorum visus obstantia penetraret, nonne introspectis visceribus, illud Alcibiadis superficie pulcerrimum corpus, turpissimum videretur*, Bt. 32, 2 ; Fox 116. 19–25. Wé sceolon gleáwlíce þurhseón ússe hreþercofan heortan eágum, Exon. Th. 81, 24 ; Cri. 1328. [He þurhsihð elches mannes þanc, O. E. Homl. ii. 222, 90 : i. 165, 90. Þe blake cloð is wurse to þurhseon, A. R. 50, 16. *O. H. Ger.* durh-sehan *visu penetrare.*]

þurh-seón *to strain through, penetrate* :—Ealle ða fúllnessa ðæs fúllan ofnes and ðæs þeóstran ðe mec ǽr ðurhseáh *omnem foetorem tenebrosae fornacis, qui me pervaserat*, Bd. 5, 12 ; M. 430, 6.

þurh-sleán. I. *to smite through, strike through* (lit. and fig.) :— Hé his byrnsweord getýhþ and ðás world ealle þurhslyhþ, Blickl. Homl. 109, 34. Hire swiora næs þurhslagen, Homl. Skt. i. 12, 235. Ðá weard heó mid micelre sárnysse ðurhslegen, Homl. Th. ii. 30, 21. II. *to smite* :—Ic ðerhslæ ł hríno ðone hiorde *percutiam pastorum*, Mk. Skt. Lind. Rush. 14, 27. [*O. Frs.* thruch-slá : *O. H. Ger.* durh-slahan *percutere, pulsare.*]

þurh-smeágan, -smeán *to search through, inquire into, examine into, investigate* :—Ða ðe mid carfulre gýmene gástlíce bebodu þurhsmeágeaþ *qui solerti cura spiritalia precepta perscrutantur*. Ðá hé ða seofon cræftas ealle hæfde þurhsmeáde, Shrn. 152, 18. Hé ríxade ofer Englæland, and hit mid his geápscipe swá þurhsmeáde (*made such a thorough inquisition*), ðæt næs án híd landes innan Englælande, ðæt he nyste hwá heó hæfde, oððe hwæs heó wurð wæs, Chr. 1086 ; Erl. 222, 10. Þurhsmeágean *perscrutari*, Scint. 32, 11.

þurh-smúgan. I. *of movement* (lit. or fig.), *to creep through, move slowly through* :—Se wyrm ða tungan tótýhþ, and ða téð þurhsmýhþ, Soul Kmbl. 235 ; Seel. 121. Ðæt gér, ðe man hǽt *solaris*, þurhsmíhþ *Zodiacum* ðone circul on þrim hund dagum and fíf and syxtigum, Anglia viii. 303, 22. II. *to go carefully through a subject, go over the details* :—Hé sceal snotorlíce smeágean and georne þurhsmúgan ealle ða ðing ðe hláforde magan tó rǽde *he must prudently consider and diligently go over in his mind all those matters which may be to his lord's advantage*, Anglia ix. 259, 18. Nú wille wé úre sprǽce áwendan tó ðam iungum munecum ðe heora cildhád habbaþ ábisgod on cræftigum bócum . . . Hig habbaþ áscrutnod Serium and Priscianum, and þurhsmogun Catus cwydas *they have gone carefully through Cato's Disticha*, Anglia viii. 321, 28.

þurh-spédig ; *adj. Very wealthy* :—Ðǽr eardode sum þurhspédig mann (cf. *of the same person* sum ríce man and for worlde ǽhtspédig, Blickl. Homl. 197, 27), Homl. Th. i. 502, 8.

þurh-stician *to stick through, pierce, transfix* :—Ðorhsticadun *transfixerunt*, Jn. Skt. Lind. 19, 37. [Cf. Heo þuruhstihten Isboset adun into þe schere (*percusserunt eum in inguine*, 2 Sam. 4, 6), A. R. 272, 12. *O. Frs.* thruch-steka : *O. H. Ger.* durh-stehhan *confodere, transfigere.*]

þurh-stingan *to stab through, pierce, thrust through* :—Gif man þeóh þurhstingð *if the thigh is thrust through*, L. Ethb. 67 ; Th. i. 18, 16. Þurhstind, 32 ; Th. i. 12, 1. Hé ðurhstong ðone cyninges ðeng and ðone cyning gewundade *tanta vi hostis ferrum infixit, ut per corpus militis occisi etiam regem vulneraret*, Bd. 2, 9 ; S. 511, 24. Þurhsting his eáre *perforabis aurem ejus*, Deut. 15, 17. Ðæt hé hine selfne ne ðurhstinge mid ðý sweorde unryhthǽmedes *ne luxuriae se mucrone transfigeret*, Past. 43 ; Swt. 313, 8. Hé hét hine mid sweorde þurstingan, Shrn. 131, 33. Þurhstungen *confosa*, Wrt. Voc. ii. 24, 42. [Weren his fet mid irnene neiles þurhstungen, O. E. Homl. i. 147, 32.]

þurh-swimman *to swim through* or *over, pass by swimming* :— Ðorhsuimmaþ *tranant*, Wrt. Voc. ii. 122, 74.

þurh-swíðan *to prove very strong* :—Hé þurhswídde on ídelnesse *praeualuit in vanitate*, Ps. Lamb. 51, 9.

þurh-swógan *to press through, penetrate, pervade* :—Ealle ða fúllnessa ðæs ðýstran ofnes ðe mé ǽr ðurhsweógh(-swég, Bd. M. 430, 6 note) *omnem foetorem tenebrosae fornacis, qui me pervaserat*, Bd. 5, 12 ; S. 629, 21.

þurh-teón. I. *to carry through, get a proposal accepted, a request granted* :—Ðá hé ðæt (*his proposal*) uneáþe þurhteáh þurd dum

aegre impetraret ab ea, Bd. 4, 11; S. 579, 17. Lucius bæd ðæt hē cristen gedōn wǣre, and hē þurhteáh ðæt hē bæd (*by a later hand this is turned into* him wærð tiþod ðæt hē bæd), Chr. 167; Erl. 8, 15. II. *to carry out* a plan, orders, etc., *give effect to* an intention:—Bið oft synleás yfel geðōht ðǣm gōdum, ðonne hī hit mid weorcum ne ðorhtióð, Past. 54; Swt. 423, 4. Hē nōhwæþer ðyssa (*neither of these plans*) gefremede ne ðurhteáh *ne aliquid horum perficeret*, Bd. 5, 9; S. 622, 23. Mennisclīc is ðæt mon on his mōde costunga ðrowige on ðæm luste yfles weorces, ac ðæt is deófullīc ðæt hē ðone willan ður[h]teó, Past. 11; Swt. 71, 15. Gif hē ðæt þurhtió, ðæt hē getihhod hæfþ, Bt. 34, 7; Fox 144, 4. Cweþan ðæt sió godcunde foretiohhung getiohhod hæfde ðæs ðe hió ne þurhtuge, 41, 3; Fox 248, 21. Hī nōhwæþere heora willnunge habban ne ðurhteón magan *in neutro cupitum passunt obtinere propositum*, Bd. 5, 23; S. 647, 3: Ors. 1, 2; Swt. 30, 22. Ðæt hē ðæt mihte mid ðȳ māran ealdorlīcnesse ðurhteón and gefremman, Bd. 5, 21; S. 642, 30. III. where continuous action is implied, *to carry through, carry on to a* (*successful*) *end, to accomplish, perform;* of evil actions, *to perpetrate*:—Se cwyrnstān ðe tyrnð singallīce and nǣnne færeld ne ðurhtīhþ, Homl. Th. i. 514, 20. Micel tōsceád is betwuh ðære synne, ðe mon longe ymbsireð, and ðære ðe mon fǣrlīce ðurhtiéhð, Past. 56; Swt. 435, 6. Hī ðæt yfel þurhtióþ (þurgtióð, Cott. MS.) *scelus perficiunt*, Bt. 38, 2; Fox 196, 34. Ðurhteáh *patraverat*, Wrt. Voc. ii. 67, 51. Ða scylde ðe se him self ǣr nyste se hié þurhteáh *culpam, quam nescit ipse etiam, qui perpetravit*, Past. 15; Swt. 91, 14. Silla wið Marius heardlīce gefeaht þurhteáh (*fought and won*) and hiene gefliémde *Sulla Marium gravissimo praelio tandem vicit*, Ors. 5, 11; Swt. 236, 21. Ðonne gē ymb ðæt ān gefeoht alneg ceoriað ðe eów Gotan gedydon, hwȳ nyllaþ gē geþencan ða monegan ǣrran ðe eów Gallie oftrædlīce bismerlīce þurhtugon *the many former fights that the Gauls often fought and won against you to your disgrace*, Ors. 3, 11; Swt. 142, 9. Hī lǣrdan hine ðæt hē ða fōre ðurhtuge *they persuaded him to perform* (perficere) *the journey*, Bd. 5, 19; S. 637, 27. Ðæt ðæt mōd ðurhtuge swelce synne, Past. 56; Swt. 435, 4. Ne mæg se ælmihtiga Wealdend þurhteón ðæt hē dō his ðeówan rīce *cannot the almighty Ruler accomplish the enriching of his servants?* Homl. Th. i. 64, 17. Wē ne magan for ūre tyddernysse þyllīc fæsten þurhteón *we cannot on account of our weakness accomplish such a fast*, Wulfst. 285, 27. Þurhtión (þurg-, Cott. MS.) ðæt yfel ðæt hī lyst *cupita perficere*, Bt. 38, 2; Fox 196, 28, 32. Þurhtión nāwuht goodes (cf. nān gōd dōn, Bt. 37, 1; Fox 186, 27), Met. 25, 59. Ðæt hié nāne mildheortnesse þurhteón ne mehtan, Ors. 2, 1; Swt. 64, 17. Ðā wearð eft Ianes duru andōn, þēh ðǣr nān gefeoht þurhtogen ne wurde *though no battle had actually been fought;* nulla bella sonuerunt, 6, 1; Swt. 254, 19. Ne biþ ðǣr sin ðurhtogen *peccatum perpetratum non est*, Bd. 1, 27; S. 497, 21: Exon. Th. 128, 1; Gū. 397: 270, 1; Jul. 458. Þurhtogen *conlatum* (v. þurhtogenness), Wrt. Voc. ii. 134, 41. Mid ðȳ ðurhtogenan weorce, Past. 48; Swt. 367, 12. Ða ðe ða ðurhtogenan (cf. geworhtan, 53; Swt. 413, 3) synna wēpaþ *qui peccata deplorant operum*, 23; Swt. 176, 22. III a. of continuous but uncompleted action, *to carry on, continue*:—Hē swā six and twēntig daga ðæt færeld þurhteáh swilce hē tō sumum menn mid gewisse fōre *so for six and twenty days he continued the journey, as if he were with certainty travelling to some one*, Homl. Skt. ii. 23 b, 159. IV. where a result is marked, *to bring to a successful issue, to achieve, bring about, bring to pass*:—Gif hē torngemōt þurhteón mihte *if he could bring about a meeting*, Beo. Th. 2284; B. 1140. His sige tō tācne ðe hē ðurhteón þōhte *as a monument of the victory that he thought to achieve*, Ors. 2, 5; Swt. 84, 5. On Criste ānum is ealles siges fylnes þurhtogen *in Christ alone does the fullness of all victory come to pass*, Blickl. Homl. 179, 7. V. *to afford*:—Hit wæs geset on ðære ealdan ǣ, ðæt ða ðe mihton ðurhteón sceoldon bringan ānes geáres lamb, and āne culfran . . . Gif hwylc wīf tō ðam unspēdig wǣre, ðæt heó ðās ðing begytan ne mihte . . ., Homl. Th. i. 138, 35. Āne feorme swā gōde swā hī bezte þurhteón magon *a refection as good as ever they can afford*, Chart. Th. 531, 15. Gif hwā ne mage ðurhteón ða spēda ðæt hē gesewenlīc lāc Gode offrige *if any man cannot afford such means, that he may offer a visible gift to God*, Homl. Th. i. 584, 2. Sȳ him gefultumad and frōfor þurhtogen *solacia accomodentur eis*, R. Ben. 85, 18. VI. *to go through, undergo*:—Swā swā wīf ācenþ bearn and þrowaþ micel earfoþu æfter ðam ðe heó ǣr micelne lust þurhteáh, Bt. 31, 1; Fox 112, 3. Hefige geswincu wē þurhteón, ðæt wē tō heofenan āstīgan magan, Scint. 101, 11. Gif hē ða beþinge þurhteón ne mæge, Lchdm. ii. 340, 10. VII. *to draw, drag*:—On wītu helle mann gālnys þurhtȳhð *in poenas tartari hominem libido pertrahat*, Scint. 89, 5. Ða ðe ōþre tō unrihtwīsnysse lǣrende þurhteóð (*pertrahunt*), 192, 4.

þurh-þeówan, -þian, -þygan, -þȳn *to thrust through, pierce through, transfix*:—Ic ðurhðȳ (-þȳge, MS. J.) *perfodio*, Ælfc. Gr. 28, 6; Zup. 179, 10. Seó clǣnnys ðurhðȳð (*transfigit*) ða gālnysse mid swurde, Gl. Prud. 13 b. Hē siwode scōs and ðurhþide his hand, Homl. Skt. i. 15. 24. Ðurhðȳde, Homl. Th. i. 452, 14. Hī þurhðȳdon (ðurhðȳgdon, Ps. Lamb. 21, 17) mīne handa *foderunt manus meas*, ii. 16, 23. Dauides þegnas hine (*Absalom*) þurhðȳdon, Homl. Skt. i. 19, 223. Ðā wolde hē þurhþȳn hī mid swurde, 12, 225. Þurhþiende *transverberans, transfigens*, Hpt. Gl. 411, 66. Þurhþēd *confossa, transfixa*, 501, 29. Mīn bān bið mid sārnysse þurhðȳd *os meum perforatur doloribus* (Job 30, 17), Homl. Th. ii. 456, 12. Ōþre wǣron mid stengum þurhðȳde, i. 542, 28. v. þeówan, *and* þurh-þyddan.

þurh-þrāwan *to twist through* [:—Se wǣte of hūse dropaþ on stān . . . and ðane stān þurhþurleþ and þurhþreáwþ, Lchdm. iii. 104, 11].

þurh-þyddan *to thrust through, pierce through*:—Ðā com sum cempa swīþe gewǣpnod, and hyne sōna þurhþydde, Homl. Skt. i. 3, 273. v. þyddan, *and* þurh-þeówan.

þurh-þȳn. v. þurh-þeówan.

þurh-þyrel; *adj. Pierced through, perforated*:—Gif hrif wund weorðeþ, .xii. scill. gebēte. Gif hē þurhþirel weorðeþ, .xx. scill. gebēte, L. Ethb. 61; Th. i. 18, 7. Gif sió lendenbrǣde biþ on bestungen, geselle .xv. scill. tō bōte; gif hió biþ þurhþyrel (-þyrl, MS. B.), ðonne sceal ðǣr .xxx. scill. tō bōte, L. Alf. pol. 67; Th. i. 98, 3. Cf. þurh-wund.

þurh-þyrelian, -þyrlian *to pierce through, make a hole through, perforate*:—[Se wǣte of hūse dropaþ on stāne . . . and ðane stān þurhþurleþ, Lchdm. iii. 104, 11.] 'Ðurhðyrela ðone wāg.' Ðā ic ðone wāh ðurhðyreludne hæfde '*fode parietem.*' *Cum fodissem parietem*, Past. 21; Swt. 153, 17. Ðæt mon ðurhðyrelige ðone weall . . . Hē cuæð: 'Ðā ic hæfde ðone weall ðurhðyrelod, Swt. 155, 1–3. Þurhþyrlige his hlāford his eáre *dominus perforabit aurem ejus* (Ex. 21, 6), L. Alf. 11; Th. i. 46, 10. Wǣron ða eáran him þurhþyrelode *perforatis auribus*, Nar. 26, 30.

þurhtogenness, e; *f. A religious reading in monasteries, especially after meals;* collatio:—Þurhtogenessa and gesetnessa heora līfes *conlationes patrum et instituta vite eorum*, R. Ben. Interl. 118, 7. Cf. þurh-togen *conlatum*, Wrt. Voc. ii. 134, 41.

þurh-trymman *to confirm thoroughly, corroborate*:—Werc cȳðnisse ðerhtrymmaþ of mē *opera testimonium perhibent de me*, Jn. Skt. Rush. Lind. 10, 25.

þurh-ūt; *prep. adv. Throughout, quite through*:—Ðæt spere him eode þurhūt, Homl. Skt. i. 12, 55. Hē fōr þurhūt Eoferwīc, Chr. 1066; Erl. 200, 33. [Mid helle sweordes al snesien ham þuruhut, A. R. 212, 23. He sahede hire þurhut, Marh. 22, 11. Ðis lond ðurgut he charen, Gen. and Ex. 3704. Ane stræte þurhut al þis kinelond, Laym. 4826. Sunne þurhut forleten, O. E. Homl. i. 23, 10. Þurhut gode and þurhut clene on mode, O. and N. 879. *Ger.* durch-aus.]

þurh-wacol; *adj. Very watchful, vigilant*:—Wacul *vigil* vel *vigilans*, ðurhwacul *pervigil*, Wrt. Voc. i. 46, 3. Þurhwacol, 75, 66. Þurhwacol *pernox*, Ælfc. Gr. 9, 65; Zup. 71, 15. (1) in reference to persons:—Þurhwacol emhīdignys *pervigil sollicitudo*, Hpt. Gl. 426, 51. Hē ābād on ðam legere . . . þurhwacol on gebedum, Homl. Th. ii. 516, 30. Wē hālsiaþ eów ðæt gē beón on gebedum þurhwacule *hortamur vos orationibus pervigiles existere*, Cod. Dip. B. i. 154, 36. Hī on heora gebedum wunodon þurhwacole ōð midde niht, Homl. Skt. i. 11, 44. Hī ealle ða niht mid hālgum sprǣcum ðæs gāstlīcan līfes ðurhwacole āspendon, Homl. Th. ii. 184, 14. Hī heom weardas setton, þurhwacole menn, Homl. Skt. i. 11, 147. (2) in reference to time, *very wakeful, quite sleepless*:—Hine gedrehte singal slǣpleást, swā ðæt hē þurhwacole niht būton slǣpe ādreáh, Homl. Th. i. 86, 17. [Cf. *Goth.* þairh-wakan *pernoctare, vigilare*: *O. H. Ger.* durh-wahhēn; durh-wacha *pervigilium.*]

þurh-wadan. I. *to pass through*:—Ða hyssas þrȳ wylm þurhwōdon, Cd. Th. 245, 16; Dan. 464. II. of a weapon (lit. or fig.), *to pierce* through, *penetrate*:—Ðæt swurd þurhwōd wyrm, ðæt hit on wealle ætstōd, Beo. Th. 1785; B. 890: 3139; B. 1567: Byrht. Th. 140, 31; By. 296. For ðām næglum ðe ðæs Nergendes fēt þurhwōdon, Elen. Kmbl. 2139; El. 1066. Swylce hit seaxes ecg þurhwōde, Exon. Th. 70, 21; Cri. 1142. Ða syngan flǣsc, scandum þurhwaden, 78, 32; Cri. 1283. [*O. H. Ger.* durh-watan *pertransire.*]

þurh-wæccendlīc; *adj. Very vigilant*:—Mid þurhwæccendlīcan mōde, Homl. Skt. ii. 23 b, 43.

þurh-werod (?); *adj. Very sweet*:—Hwylc manna þurhwerodum (*dulcibus; perhaps þurh* is an error of the scribe brought about by the *þurh* of the following word) þurhbrȳcþ mettum būton swæcce sealtes, Wrt. Voc. i. 9, 21.

þurh-wlītan *to look through, penetrate with the sight*:—Glæs ðæt mon mæg eall þurhwlītan, Exon. Th. 79, 2; Cri. 1284. Wē ne magun hygeþonces ferð eágum þurhwlītan, 82, 1; Cri. 1332.

þurh-wrecan *to thrust through*:—Sumne heó mid sweorde ofslōgen, sumne mid spiten betweón felle and flǣsce þurhwrǣcon, Homl. Ass. 171, 39. Ōð hielt þurhwrecen (*ense*) *capulo tenus* (*per utraque latera*) *adacto*, Wrt. Voc. ii. 86, 69.

þurh-wund; *adj. Wounded by a weapon which has passed quite through*:—Gif mon biþ on hrif wund, geselle him mon .xxx. scill. tō bōte; gif hē þurhwund biþ, æt gehweðerum mūðe .xx. scill., L. Alf. pol. 61; Th. i. 96, 11. [Cf. Sinness þatt stinngenn and þurrhwundenn all þatt bodig and tatt sawle. Orm. 17443.] Cf. þurh-þyrel.

þurh-wunian. I. *to continue, last, not to come to an end, not to*

pass away :—Godes ege þurhwunaþ â worlda world *timor Domini permanens in seculum seculi*, Ps. Th. 18, 8. His ríce þurhwunaþ on écnesse, Blickl. Homl. 65, 16. Hús rihtwísra þurhwunaþ (*permanebit*), Scint. 73, 2, 16. Heora gemynd þurhwunaþ â tô worulde, Ælfc. T. Grn. 1, 11. God, se ðe ǽfre þurhwunode bûton ǽlcum anginne, 2, 3. Þurhwunedan *duraverunt*, Wrt. Voc. ii. 28, 58 : Wülck. Gl. 256, 2. Úre nâ þurhwunedun fæderas *nostri non mansere parentes*, 6. Þa leornian on eorþan ðæra ûs cýþ þurhwunige on heofenum *illa discere in terris quorum nobis notitia perseueret in coelis*, Scint. 218, 13. Lang mid þingum úrum þurhwunian (*durare*) wē nâ magan, 183, 4. Âdl þurhwunigende *languor perseuerans*, 153, 17. **II.** *to continue* in a place, with a person, *to remain, not to leave* :—Ic þurhwunode (*perseveravi*) on ðam munte feówertig daga and feówertig nihta, Deut. 9, 9. Gē synt ðe mid mē þurhwunedon (ðerhwunadon(-un) *permansistis*, Lind., Rush.) on mínum geswincum, Lk. Skt. 22, 28. Þurhwunedan munecas on Xpes cyrican *monks have continued to live* (*permanserunt*) *in Christchurch*, Chr. 995 ; Th. i. 244, 29. Ðæt hē symle on ûs eardige, and wē on him þurhwunian (*permaneamus*), Scint. 16, 11. Ðæt gē þurhwunion lange on ðam lande, Deut. 4, 40. Eálá wǽran ða ancras swâ trume and swâ þurhwuniende, ðonne mihte wē ðý éþ geþolian swâ hwæt earfoþnessa swâ ûs on becôme *haereant ancorae, precor ; illis namque manentibus, utcumque se res habeant, enatabimus*, Bt. 10 ; Fox 30, 10. **III.** *to continue* in a condition, *not to change* ; where purpose or effort is implied, *to persevere, persist, hold out*, (1) absolute :—Se þurhwunaþ (ðerhwunes ꝼ ðerhwunia wælla, Lind.) ôð ende, se byþ hâl *qui perseveraverit usque in finem, hic salvus erit*, Mt. Kmbl. 10, 22 : Scint. 90, 2 : Blickl. Homl. 21, 36. Se ðe ôð ende þurhwunaþ (*sustenuerit*), Mk. Skt. 13, 13. Nâ ongynnendum mēd ys behâten ac þurhwunigendum (*perseverantibus*) ys geseald, Scint. 91, 3, 1. (2) where the condition is given by a complementary noun or adjective :—Ǽfre hē biþ ânes môdes, and glæd þurhwunaþ, Homl. Th. i. 456, 25. Heó þurhwunode mǽden, 24, 27. Heó onwealg on hiere onwalde æfter þurhwunade *manet adhuc et regnat incolumis*, Ors. 2, 1 ; Swt. 62, 24. Heó â clǽne þurhwunode, Blickl. Homl. 3, 18. Hē þurhwunode unspecende and mihteleás forð ôð ðone Ðunresdæg, Chr. 1053 ; Erl. 186, 23. Is rihtost ðæt hē ðananforð wydewa þurhwunige, L. I. P. 22 ; Th. ii. 332, 32. (3) where the condition is given in a phrase :—On gôde on ðam hē ongan ôþ ende hē þurhwunaþ *in bono quo coepit usque in finem perdurat*, Scint. 227, 15. Gyt git þurhwuniaþ on incre ânwilnesse *ye still persist in your obstinacy*, Blickl. Homl. 187, 33. His (*Adam's*) bendas wǽron onlýsde . . . Eua ðâgyt on bendum þurhwunode, 89, 6. Hié þurhwunian on rihtum geleáfan, 77, 19. Hē nolde þurhwunian on ðære sôðfæstnisse ðæs sôðfæstan Godes sunu, Ælfc. T. Grn. 2, 41. Se ðe on ðam gesǽlþum ðurhwunian ne môt, Bt. 2 ; Fox 4, 15. **IV.** *to continue* an action, *persevere* with or in, *not to desist from, not to leave off* :—Eua ðâgyt on wôpe þurhwunode, Blickl. Homl. 89, 6. Hié forþ on heora yfelum þurhwunedon, 79, 8. Ðâ hig þurhwunedon (*perseuerarent*) hine âxsiende, Jn. Skt. 8, 7. On ðam gewinne þurhwunian, Bt. 37, 1 ; Fox 186, 31 : Met. 25, 70. Þurhwunian on fulfremedlícum weorcum, Blickl. Homl. 77, 19. Þurhwunian his bēne *persistere petitioni sue*, R. Ben. Interl. 95, 16. Hē þurhwunigende mid gebedum wæs Drihtnes lôf singende, Blickl. Homl. 231, 9. Drihten eallum mannum þurhwunigendum on sôþre andetnesse cwæþ, 171, 15. [An lond þer he mihte þurhwunian (wonie, 2nd MS.), Laym. 1384. Þu wið Godd þurhwunest in alre worlde world, Kath. 663. Cf. *Goth.* þairh-wisan *manere, permanere*.]

þurhwunigendlíce ; *adv. Perseveringly, persistently, continuously* :—Þurhwunigendlíce begýman hit gedafenaþ môd úre *perseveranter intendere oportet animum nostrum*, Scint. 33, 18.

þurh-wunung, e ; *f.* **I.** *continued dwelling, residence* :—Ôþer cyn is muneca, ðæt is wēstensetlan, ðe feor fram mannum gewítaþ . . . geefenlǽcende Élian and Iôhannem, ða þurhwununge on wēstenes innoþe heóldon, R. Ben. 134, 14. **II.** *perseverance, persistence, constancy* :—Be þurhwununge . . . Mægen gódes weorces þurhwunung ys *de perseverantia . . . Virtus boni operis perseuerantia est*, Scint. 90, 1–14. Gif hē behǽt staðolfæste þurhwununge *si promiserit de stabilitate sua perseuerantiam*, R. Ben. 97, 20.

þurruc. **I.** *a small ship* :—Þurruc *cumba vel caupolus* (the word occurs in a list of names for different kinds of ships), Wrt. Voc. i. 56, 30. **II.** *the bottom part of a ship* (?) :—Se æften-stemn *puppis*, þurruc *cumba* (cf. scipes botm *cimba vel carina*, 56, 32), bytme *carina*, scipes flôr *tabulata navium*, Wrt. Voc. i. 63, 37–40. In this instance the word seems to mean rather part of a ship than the whole, and in this sense it is used later. It occurs in the Persones Tale : 'Smal dropes of water, that enteren thurgh a litel crevis in the *thurrok*, and in the botom of a ship.' Tyrwhitt in explanation quotes the following : 'Ye shall understande that there ys a place in the bottome of a shyppe, wherin ys gathered all the fylthe that cometh into the shyppe, and it is called in some contre of thys londe a *thorrocke* . . . Some calle yt the *bulcke* of the shyppe.' See also *thurrok* of a shyppe *sentina*, Prompt. Parv. 493.

þúrs-dæg. v. þúr.

þurst, es ; *m.* Thirst (lit. and fig.) :—Ne biþ ðǽr hungor ne þurst, Blickl. Homl. 65, 19 : Exon. 101, 20 ; Cri. 1661. Beóð ðe hungor and

þurst hearde gewinnan, 118, 27 ; Gú. 246. Hungor se hâta ne se hearde þurst, 238, 33 ; Ph. 613. Se hâta þurst, 430, 6 ; Rä. 44, 3. Ne biþ se ðurst gefylled heora gítsunga, Bt. 7, 4 ; Fox 22, 31. Ðú woldest ûs ofsleán mid þurste (*siti*), Ex. 17, 3. On ðurste mínum hí drencton nié mid ecede, Ps. Spl. 68, 26 : 103, 12. On hungre and on þurste hē biþ âféded, Blickl. Homl. 59, 35 : Homl. Th. i. 392, 7. Drihten âsent hungor on eów and þurst and næcede, Deut. 28, 48. Ðeós wyrt þyrstendon ðone þurst gelíþigaþ, Lchdm. i. 268, 12. [*Goth.* þaurstei ; *f.* : *O. Sax. O. L. Ger.* thurst : *O. H. Ger.* durst : *Icel.* þorsti.] v. ungemet-þurst, þyrst.

þurstig ; *adj. Thirsty* (lit. and fig.) :—Hē sylfa þursti wæs *ipse sitiens*, Nar. 8, 4. Þurstig wyll *bibulus fons*, Scint. 13, 12. Swâ swylgþ seó gítsung ða dreósendan welan, for ðam hió hiora simle biþ ðurstegu, Bt. 12 ; Fox 36, 14. Swâ hwâ swâ sylþ ceald wæter drincan ânum þurstigan menn, Homl. Th. i. 582, 24 : ii. 106, 15 : Wulfst. 287, 20. Hwænne gesâwe wē ðē þurstine ? 288, 21. Þurstige múðe, Ps. Th. 61, 4. Þurstige þræcwíges, Cd. Th. 189, 9 ; Exod. 182. Heolfres þurstge, Exon. Th. 373, 24 ; Seel. 114. [*O. H. Ger.* durstig.] v. þyrstig.

þuruh. v. þurh.

þus ; *adv. Thus, in this manner, degree, etc.* **I.** where the manner, etc., is determined by what precedes, (1) with verbs :—'Mín Drihten, gestranga míne heortan. Ðus gebiddende ðam hâlgan Andrea Drihtnes stefn wæs geworden, Blickl. Homl. 245, 3 : Exon. Th. 236, 6 ; Ph. 570 : 43, 9 ; Cri. 686. Ðus (*sic*) unc gedafnaþ ealle rihtwísness gefyllan, Mt. Kmbl. 3, 15 : Lk. Skt. 24, 46. Hú mæg ðis ðus geweorþan ? Blickl. Homl. 21, 20. Ðâs dǽda þus gedône from Drihtne, 31, 20. Cucler fulne þus geworhtes drincan, Lchdm. ii. 182, 23. Lǽtaþ þus sinite usque huc, Lk. Skt. 22, 51. (2) with adjectives :—Ðis wíf wæs âfundyn on unrihton hǽmede. Moyses ûs bebeád ðæt wē sceoldon þus geráde mid stánum oftorfian, Jn. Skt. 8, 5 : Deut. 4, 32. Þuss gerádum âdle, Anglia xiii. 434, 995. Þus manige men, Beo. Th. 679 ; B. 337. Ic nǽfre ðé gemǽtte þus mēðne, Exon. Th. 163, 3 ; Gú. 988 : 376, 19 ; Seel. 376 : 447, 4 ; Dôm. 34. (3) with adverbs :—Nô wē ðé þus swíðe swencan þorftan, Exon. Th. 129, 18 ; Gú. 423 : 268, 16 ; Jul. 433 : Judth. Thw. 22, 36 ; Jud. 93. **II.** where the manner, etc., is determined by what follows, (1) with verbs :—God spræc þus : 'Ic eom Drihten þín God,' Ex. 20, 1. Se engel þus cwæþ : 'Wes ðú hâl,' Blickl. Homl. 5, 3 : Andr. Kmbl. 124 ; An. 62 : Mt. Kmbl. 2, 5. Þus sindon hâten fæder and môdur, ðæs wē gefrægen habbaþ . . . Maria and Iôseph, 1371 ; An. 686. Sôðlíce þus wæs Cristes cnéores *now the birth of Jesus Christ was on this wise*, Mt. Kmbl. 1, 18. Ic wēne þus, ðæt . . . , Exon. Th. 468, 8 ; Phar. 4. Ðú ða sâwle þus gesceópe, ðæt hió hwearfode on hire selfre (cf. swâ ðú gesceópe ða saûle, ðæt hió sceolde hwearfian on hire selfre, Bt. 33, 4 ; Fox 132, 11), Met. 20, 205. Þuss, Gen. 2, 16. (2) with adjectives :—Ic wēne ðæt ðú nǽfre tô ðus mycles mægnes lǽcedômum becôme swylcum swâ ic gefregn ða ðe fram Æscolapio férdon, Lchdm. i. 326, 5. **III.** used in place of a definite expression :—Ðâ cwæð Petrus : 'Beceápode gē ðus micel landes ?' Heó andwyrde : 'Geá, leóf, swâ micel' *Peter answered unto her, Tell me whether ye sold the land for so much ? And she said, Yea, for so much* (Acts 5, 8), Homl. Th. i. 316, 32. [*O. Sax. O. Frs.* thus.]

þúsend. **I.** as a numeral noun, neuter and fem. (v. Ps. Th. 118, 72, and cf. cognates), *a thousand* ; *gen.* þúsendes, *pl.* þúsendu (-o, -a, -e) ; *also sometimes in the multiples, though, perhaps, in these cases the whole number is to be considered as singular*, e. g. Tele ða lenge ðære hwíle wið tēn þúsend wintra . . . Tele nú ðæt tēn þúsend geára wið ðæt éce líf, Bt. 18, 3 : Fox 66, 6–10. Wæs þreó þúsend ðæra leóda âlesen, Elen. Kmbl. 569 ; El. 285 : Blickl. Homl. 119, 3. v. under (2) other examples. (1) without other numerals, (a) governing a genitive :—Þúsend wintra biþ swâ geostran dæg *mille anni sicut dies hesterna*, Ps. Th. 89, 4 : Exon. Th. 230, 1 ; Ph. 364. Fealleþ ðé on ða wynstran wergra þúsend, Ps. Th. 90, 7. Mænigfeald þúsend môdblissiendra *millia laetantitium*, 67, 17. Ân þúsend manna, Ors. 1, 10 ; Swt. 46, 34. Ðeáh hē erige his land mid ðúsend sula, Bt. 26, 3 ; Fox 94, 14. Erigan æcera þúsend, Met. 14, 5. Ôð ðæt hē þúsende disses lífes wintra gebídeþ *postquam vitae jam mille peregerit annos*, Exon. Th. 208, 5 ; Ph. 151. Hē ofslôg fela þúsend monna, 6, 13 ; Swt. 268, 17. Hē heora monig ðúsend ofslôg, Ors. 3, 7 ; Swt. 110, 33. Ðurh ðæs bodunge gelýfdon fela ðúsend manna, Homl. Th. ii. 296, 22. Manega ðúsenda engla, 334, 16. Heora fela ðúsenda gefongen wæs, Ors. 3, 4 ; Swt. 104, 11. Ic ðé þúsenda þegna bringe, Beo. Th. 3662 ; B. 1829. (b) where the genitive of the objects numbered is not given :—Hwæþer ðis þúsend sceole beón scyrtre ðe lengre, Blickl. Homl. 119, 6. Ðæt forme þúsend, ðæt ys seó forme yld, Anglia viii. 335, 45. Þúsendes ealdor *ciliarcus*, Wrt. Voc. i. 18, 10. Æfter ðam þúsende biþ se deófol unbunden, Wulfst. 243, 23. On þúsende ðære cneórisse *in mille generationes*, Ps. Th. 104, 8. Hié ðone here geflémdon and his fela þúsenda ofslôgon, Chr. 911 ; Erl. 100, 28 : Ors. 3, 7 ; Swt. 118, 8 : Cd. Th. 289, 23 ; Sat. 402 : 290, 26 ; Sat. 421. Hē fór mid monegum þúsendum, Ors. 5, 4 ; Swt. 224, 19. Ic mē nâ ondrǽde þúsendu folces *non timebo millia populi*, Ps. Th. 3, 5. Betere ðonne mon mē geofe ðúsende goldes and seolfres *super millia auri et*

argenti, 118, 72. (2) with other numerals as multipliers, (a) alone:— Twá þúsend, Ælfc. Gr. 49; Zup. 282, 12: Ors. 2, 4; Swt. 76, 30. iiii þúsend monna, 2, 5; Swt. 80, 13. v þúsend wera, Chr. 508; Erl. 15, 18. Syx þúsend olfenda, Homl. Th. ii. 458, 18. Wæs Rómána eahta þúsend feohtendra, Ors. 4, 1; Swt. 158, 11: 4, 9; Swt. 192, 24. Týn þúsend punda, Mt. Kmbl. 18, 24. Tén ðúsend, Ps. Surt. 90, 7. Endlefan þúsend monna, Ors. 2, 5; Swt. 78, 24. Feówertýne þúsend sceápa, Homl. Th. ii. 458, 17. xvi þúsend punda, Chr. 994; Erl. 133, 27. Ðæt wǽre þrítig þúsend wintra, Exon. Th. 369, 5; Seel. 36: Salm. Kmbl. 544; Sel. 271. cxi þúsend, Chr. 71; Erl. 9, 2. Án hund þúsend manna and hundeahtatig ðúsend, Homl. Skt. i. 18, 403. Ðæt wǽron fíéftiéne þúsend monna, Ors. 3, 9; Swt. 128, 22. Twá þúsend, Mk. Skt. 5, 13: Cd. Th. 189, 14; Exod. 184. Twá ðúsendu swína, Chart. Th. 481, 5. Twá þúsenda, 471, 22: Jos. 7, 3. Ðá férdon þreó þúsenda feohtendra wera, 7, 4. iii þúsendo (-a, MS. E.) londes, Chr. 648; Erl. 26, 16. Ágefe hé feówer ðúsendo, Chart. Th. 471, 24. v þúsendu wera, Chr. 508; Erl. 14, 17. Fíf þúsendo. Andr. Kmbl. 1181; An. 591. Hé him gesealde seofon þúsendo, Beo. Th. 4397; B. 2195. Týn þúsendo, Ps. Th. 90, 7: 67, 17. Téno ðúsendo (þúsende, Rush.), Mt. Kmbl. Lind. 18, 24. Geselle et ðem londe .x. ðúsenda, Chart. Th. 465, 30. Cantwara him gesealdon xxx þúsenda, Chr. 694; Erl. 43, 21. Ðǽr wæs ofslagen eahtatig þúsenda, Ors. 5, 8; Swt. 232, 2. Hundeahtatig þúsenda, 2, 5; Swt. 78, 17. Án hund þúsenda gehorsedra, 3, 9; Swt. 124, 34: Cd. Th. 310, 9; Sat. 723. Hund þúsenda landes and locenra beága, Beo. Th. 5981; B. 2994. cxi þúsenda, Chr. 71; Erl. 8, 2. Ðone sang ðe nán mon elles singan ne mæg, búton ðæt hundteóntig and feówertig and feówer ðúsendo, Past. 52; Swt. 409, 10. Weard tú hund þúsenda ofslægen, Ors. 2, 5; Swt. 78, 28. Ðá com him ongeán twá hund þúsenda monna, 3, 9; Swt. 132, 30. His heres wæs seofon hund þúsenda, 2, 5; Swt. 80, 18. viii c þúsenda, Swt. 80, 4. Ðæt wæs nigon x hund þúsenda, Swt. 84, 29. þúsend ðúsenda ðénodon him, Homl. Th. i. 348, 2. Tó twǽm ðúsendum, Mk. Skt. Lind. Rush. 5, 13. Tén ðúsendum, Ps. Surt. 67, 18. Mid týn þúsendum cuman ágén ðone ðe him ágén cymþ mid twéntigum þúsendum, Lk. Skt. 14, 31. Fíf hund þúsendum *quinquagenis milibus*, Hpt. Gl. 426, 11. (b) in combination with hundreds, tens, units:—Twá þúsend wintra and twá hund and twá and feówertig geára gerímes, Anglia viii. 336, 1. Feówer þúsend wintra and feówer hund and twá and hundeahtatig, Ors. 1, 14; Swt. 58, 9. Wǽron ágán .v. þúsend wintra and .cc. wintra, Chr. 11; Erl. 7, 2. v. þúsend wintra and cc. and xxvi, 33; Erl. 7, 10. Gersenes hírêdes wǽron seofon þúsenda and fíf hundredu ... Gaathes hírêdes wǽron eahta þúsendo and six hundredu ... Merarîes hírêdes wǽron six þúsendo and twá hundrydo ... ðá wǽron hira twá and twéntig þúsenda, Num. 3, 21-39. Rómána wæs án c and án m ofslagen *Romanorum mille centum periere milites*, Ors. 4, 6; Swt. 176, 14. Ðæt wæs v hund monna and án m, 5, 12; Swt. 240, 34. II. as an adjective indecl.:—þúsend getýme oxena and þúsend assan, Homl. Th. ii. 458, 18. Mid þúsend gemetum *mille modis*, Wülck. Gl. 254, 44. On hund þúsend wintrum . . . on syx þúsend wintrum, Anglia viii. 335, 46-336, 20. On six þúsend wintrum, Wulfst. 244, 2. Tén ðúsend sîðan hundfealde ðúsenda, Homl. Th. i. 348, 3. III. the word is sometimes used of value without expressing the unit (cf. the Icelandic use of *hundrað*); see the passages (quoted above), Chr. 648; Erl. 26, 16: 694; Erl. 43, 21: Beo. Th. 4397; B. 2195: 5981; B. 2994: Chart. Th. 465, 30: 471, 24; Ps. Th. 118, 72. [*Goth.* þúsundi; *f. n.:* *O. Frs.* thûsend: *O. L. Ger.* thûsint: *O. Sax.* thûsundig: *O. H. Ger.* dûsunt, tûsunt; *f. n.: Icel.* þúsund; *f.* (later *f.* and *n.*)]

þúsend-ealdormann, es; *m. A captain of a thousand men:*—Þúsend-ealdermen *chiliarcho*, Hpt. Gl. 515, 76.

þúsend-feald; *adj. Thousand-fold, a thousand:*—Ðæt þúsendfeald getæl is fulfremed, Wulfst. 243, 26, 23. Þúsendfealdre gegaderunge *millena congerie*, Hpt. Gl. 416, 63. Ðæt wǽron þúsendfealde onsægednyssa, Homl. Th. ii. 576, 8. Ongǽn þúsendfealde deriende cræftas *contra mille nocendi artes*, Wrt. Voc. ii. 135, 30: Hpt. Gl. 424, 45. [Mid þusendfeld wrenches he þe herte towended *per mille meandros agitat quieta corda*, O. E. Homl. ii. 191, 26.]

þúsend-gerím, es; *n. Numeration by thousands, counting with the unit a thousand:*—Ðría ðreóténo ðúsendgerímes *thirty-nine thousand*, Salm. Kmbl. 582; Sal. 290.

þúsend-getæl, es; *n. The number a thousand:*—Þúsendgetel biþ fulfremed, and ne ástíhþ nán getel ofer ðæt, Homl. Th. i. 188, 34.

þúsend-híwe; *adj. Of a thousand shapes:*—Þúsendhíwe *milleformes*, Coll. Monast. Th. 32, 29.

þúsend-líc; *adj. Numbered by thousands:*—Ðúsendlícre *milleno*, Wrt. Voc. ii. 57, 40. Ðæt hé ús gescylde wiþ ða þúsendlícan cræftas deófles costunga, Blickl. Homl. 19, 16.

þúsend-mǽle (?); *adj. A thousand each, a thousand:*—Ðúsendmǽle *mellena*, Wrt. Voc. ii. 58, 19. Betere is tó gebídanne ánne dæg mid ðé ðonne óðera on þeódstefnum þúsendmǽla, Ps. Th. 83, 10.

þúsend-mǽlum; *adv. In thousands:*—Weras and wíf somod wornum and heápum þrungen and urnon þúsendmǽlum, Judth. Thw. 23, 40;

Jud. 165: Cd. Th. 190, 8; Exod. 196: 304, 18, Sat. 632. Him ymb flugon engla þreátas þúsendmǽlum, 300, 23; Sat. 569: 279, 11; Sat. 236: 296, 28; Sat. 509: Andr. Kmbl. 1744; An. 874.

þúsend-mann, es; *m. A captain of a thousand men:*—Gesete of him þúsendmen and hundrydmen *rulers of thousands and rulers of hundreds* (A. V.), Ex. 18, 21, 25.

þúsend-ríca, an; *m. A ruler of a thousand men:*—Þúsendríca *millenarius*, Wrt. Voc. i. 18, 9.

þus-líc, þul-líc; *adj. Such:*—Nǽfre adeáwde ðuslíc (swylc, W. S.), Mt. Kmbl. Lind. 9, 33. Ðuslíc *talem*, 18, 5. Mǽhto ðullíco *uirtutes tales*, Mk. Skt. Lind. 6, 2. Ðuslícra is ríce Godes *talium est regnum Dei*, 10, 14. Of ðuslícum cnæhtum *ex hujusmodi pueris*, 9, 37. Mid ðullucum (ðuslícum, Rush.) monigum bîspellum *talibus multis parabolis*, 4, 33. Ðuslícum fultumum, Rtl. 64, 33. Ðuslíco (-u, Rush.) monigo gíé ðóas *hujusmodi multa facitis*, Mt. Skt. Lind. 7, 13. Ðe fæder ðullíco (ðuslíco, Rush.) soecað *pater tales quaerit*, Jn. Skt. Lind. 4, 23. [Of þulliche wepnen, O. E. Homl. i. 255, 15. Þeos and swuche (þullich, MS. C.) oþre, A. R. 8, 7. Gon and iseon swuch (þullich, MS. C.), 10, 13. Of swuche (þullic, MS. C.), 82, 3. Swuche (þulliche, MS. C.), men, 84, 20. Þulli, Marh. 7, 27: H. M. 9, 25. Þullich, Kath. 847. Þellich, Ayenb. 6, 12.]

þútende. v. þeótan.

þú-þistel, es; *m. Sow-thistle:*—Þúðistel (-þistil) *lactuca*, Txts. 73, 1179: Wrt. Voc. ii. 50, 57. Cf. þúfe.

þuuf, þuxsian. v. þúf, þuhsian.

þwǽle (or -a?; *m.*), an; *f. A band, fillet:*—Ðuaelum *taenis*, Txts. 101, 1991. Thuélan *vittas*, 107, 2120. From its form the word, apparently, should mean *towel*, cf. *O. H. Ger.* dwahila, dwehila; *f. mantile, mappula, manutergium: M. H. Ger.* dwehele, dwéle: *Du.* dwaal *a towel*; *a shroud.*

þwǽnan; *p. de To soften by moisture, ointment, etc., to soften:*—Rysele oþþe gelyndo wiþ gárleác gemenged and on álêd ðone swile þwǽnþ, Lchdm. ii. 72, 5. Ðæt (*the ointment*) ða áheardodan swilas bêt and þwǽnþ, 246, 17. Ðá hé ðam feaxe onfêng ðæs hálgan heáfdes ðá wæs hé monad ðæt hé tó gesette and sum sæc ðone swyle mid ðýgde and ðwénde (ðwǽnde, MSS. B. T.) *admonitus, cum accepisset capillos sancti capitis, adposuit, et aliquandiu tumorem horum adpositione comprimere ac mollire curabat*, Bd. 4, 32; S. 611, 41. Sceal mon mid ûtyrnendum drencum áteón út ða horhehtan wǽtan. þwǽne mid ðý ǽrest, Lchdm. ii. 222, 26. Gif ðú wylle mannes wambe þwǽnan, i. 82, 11. v. á-, geþwǽnan, *and* cf. (?) þénan.

þwǽre, an; *f. An instrument for beating or stirring:*—Thuaere, thuêrae, thuêre *tudicla, tudica*, Txts. 103, 2072. v. þweran.

þwǽre; *adj. Gentle, agreeable:*—Scs Arculfus sǽde ðæt ðǽr hangade úþmǽte leóhtfæt and ðwǽre (*a lamp giving an agreeable light?*), Shrn. 81, 17. [Gif hé on Tíwesdæg biþ ácenned, se biþ ǽwerd on his life and biþ mán and ðwǽre (*effeminate?, but perhaps* manþwǽre *should be read, the text is late*), Lchdm. iii. 162, 11.] v. efen-, ge-, mann-, un-þwǽre.

þwǽrian. v. ge-, nid-þwǽrian; þwárian.

þwǽrlécan; *p.* -lǽhte *To consent:*—Þwǽrlǽhte *consentiret*, Hpt. Cl. 465, 63. v. ge-þwǽrlécan.

þwǽrness, e; *f. Agreement:*—Hí him ðǽr eádmêdo budon and þwǽrnessa (geþuǽrnesse, MS. A.), Chr. 827; Erl. 65, 7. v. geþwǽrness.

þwang, es; *m.: e; f. A thong, strip of leather:*—Ðwangc *corrigia*, Wrt. Voc. i. 84, 2. Grênre hýde, þwanges *recentis corii*, Hpt. Gl. 483, 31. Ic ne underfô ánne þwang (*corrigiam caligae*), Gen. 14, 23. Mid ðuongum (ðwongum, Rush.) *sandalis*, Mk. Skt. Lind. 6, 9. Ðæs ne eom ic wyrðe ðæt ic his sceóna þwanga (ðuongas, Lind.) þwongas, Rush.) búgende uncnytte *cujus non sum dignus procumbens soluere corrigiam calciamentorum ejus*, Mk. Skt. 1, 7. [*Orm. Laym.* þwang, þwong: *R. Glouc.* þong: *O. H. Ger.* dwang *frenum*.] v. brídel-, ôl-, scóhþwang; þweng.

þwárian; *p.* ode *To bring into agreement, make harmonious:*—Hé gemetgaþ ða feówer gesceafta, ða hé þwaraþ and gewlitegaþ (geþwǽraþ and wlitegaþ, Cott. MS.), hwílum eft unwlitegaþ and on óþrum híwe gebrengþ and eft geedníwaþ *elementa in se invicem temperat, et alterna commutatione transformat*, Bt. 39, 8; Fox 224, 9. [Cf. *O. H. Ger.* twârôn *misceri*.] v. þwǽrian.

þwarm. v. þwearm.

þwástrian (=? hwástrian, q. v.) *to murmur, speak low:*—Þis ic spece nú gyt mid swá miccle ege ðæt mé þinceþ ðæt mé sió tunge stomrige nis hit gyt forðun ðæt ic þwástrian durre *I dare not yet even speak low*, Shrn. 42, 35.

þweál, es; *n. m.* I. *washing:*—Ðhuehl, thuachl *delumentum*, Txts. 55, 641. Þweál, Wrt. Voc. i. 61, 20: ii. 25, 18: *delumentum*, i. *lavatio*, 138, 52: *lustramenium*, Hpt. Gl. 483, 20. Ðeáh swín ðwægen sié, gif hit eft filþ on ðæt sol, ðonne biþ hit fúlre ðonne hit ǽr wæs, and ne forstent ðæt ðweál náuht, Past. 54; Swt. 421, 3. Hwæt forstent him ðæt ǽrre ðweál (*lavatio*), 21. Ðæt wæter his bâna ðweáles *aqua lavacri*, Bd. 3, 11; S. 536, 6. Clǽnsunge ðweáles and bæþes *lavacri purificationem*, 1, 27; S. 495, 16. Be weres þweále *de viri lotione*, L.

Ecg. C. xxvi. tit. ; Th. ii. 130, 10. Æfter fóta ðweále *post pedum lavationem*, Anglia xiii. 392, 392: R. Ben. 83, 23. For ðæs reáfes þweále, 91, 4. Swýn ðe cyrþ tó meoxe æfter his ðweále, Homl. Th. ii. 380, 11. Se Hǽlend hí áþwóh mid þweále, 242, 29: Blickl. Homl. 147, 22. Mid þweále ðæs hálgan fulluhtes, Lchdm. iii. 434, 2. Eádig ðú eart ðe onfénge ðone þweál mínre gife, Homl. Skt. ii. 30, 111. Þweálu clǽnes wæles *lavacra puri gurgitis*, Hymn. Surt. 52, 13. Þweála (þwealu, MS. A.) calica *baptismata calicum*, Mk. Skt. 7, 8. [Forhabbe hé hyne wyð ǽlc þweald, Lchdm. iii. 134, 25.] II. *what is used in washing, ointment.* (Cf. Icel. þvál *a kind of soap*, þvæla *to wash with soap:* Swed. twål *hard soap.*) v. þweán, II :—Pund ðuahles *librum unguenti*, Jn. Skt. Lind. 12, 3. [*Goth.* þwahl; *n. lavacrum: O. H. Ger.* dwahal.] v. fót-, hand-, heáfod-þweál.

þweán; *p.* þwóh, *pl.* þwógon; *pp.* þwagen, þwægen, þwegen, þwogen. I. *to wash.* (1) with object of that which is to be cleansed :—Petrus cwæð tó him : ' Ne þwyhst (ðuðas, Lind.: ðwǽs, Rush.) ðú nǽfre míne fét.' Se Hǽlend cwæþ: ' Gif ic ðé ne þweá (áðóa, Lind.: ðwǽ, Rush.), nǽfst ðú nánne dǽl myd mé, Jn. Skt. 13, 8. Ðú ðwehst (ðwǽs, Surt.) mé *lavabis me*, Ps. Lamb. 50, 9. Hé his handa ðwehþ (ðwéð, Surt.), Ps. Th. 57, 9. Ne þweáð (ðwás, Lind. : thuáð, Rush.) hí hyra handa, Mt. Kmbl. 15, 2. Gif ic þwóh (geðuóg, Lind. : ðwóg, Rush.) eówre fét, Jn. Skt. 13, 14. Ic þwóh (ðwóg, Surt.), Ps. Th. 72, 11. Ic ín ða eá ástáh and of ðam wætere míne handa þwóh, Homl. Skt. ii. 23 b, 502. Hé híre fét mid his teárum þwóh, 744. Hé þwóh Aaron and his suna, Lev. 8, 6. Heó hí ðwóhg, Bd. 3, 9; S. 534, 13. Ðæt sylfe wæter ðæt hí ða bán mid ðwógan, 3, 11; S. 535, 33. Ðæt wæter wæs gedréfed, ðonne ðǽr micel folc hiera fét and honda on ðwógon, Past. 16; Swt. 105, 22. Þwógan, Blickl. Homl. 149, 6. Þwógon, Gen. 43, 24. Þweah (ðuah, Lind. : þwah, Rush.) ðíne ansýne, Mt. Kmbl. 6, 17. Ne þweh ðú ná míne fét áne, Jn. Skt. 13, 9. Þweáð eówre fét, Gen. 19, 2. Ne beðearf búton ðæt man his fét þweá (áðóa, Lind. : ðwǽ, Rush.), Jn. Skt. 13, 10. Búton hí hyra handa þweán, Mk. Skt. 7, 3. Scealt ðú ðweán (geðóas, Lind.: ðú mé ðwoege, Rush.) míne fét, Jn. Skt. 13, 6. Heó wolde hig þweán æt ðam wætere, Ex. 2, 5. Hé underséhð ðæt fenn ðara ðweándra, Past. 16; Swt. 105, 24. (2) with object of that which is to be cleansed away :—Mid hú micle elne ǽghwylc wille synrust þweán, Exon. Th. 81, 9; Cri. 1321. II. *to anoint.* v. þweál, II :—Ðuah heáfud ðín *unge caput tuum*, Mt. Kmbl. Lind. 6, 17. [*Goth.* þwahan: *O. Sax.* thwahan: *O. H. Ger.* dwahan: *Icel.* þvá.] v. á-, be-, ge-þweán.

þwearm, es; *m.* (?) *A cutting instrument* :—Thuearm, duæram, þuarm *scalprum*, Txts. 94, 891.

þwénan. v. þwǽnan.

þweng, e; *f.* (?) *A band* :—Ðuencgu (þwænge, Rush.) *philacteria*, Mt. Kmbl. Lind. 23, 5. [*Icel.* þvengr *a thong.*] v. þwang.

þweora, an; *m.* *Crossness, peevishness* :—Ælc ðweora and ǽlc ierre and unweorðscipe ... sié ánumen fram eów *omnis amaritudo, et ira, et indignatio ... tollatur a vobis* (Eph. 4, 31), Past. 33; Swt. 222, 8. His múð hé sceal from ǽlcum þweoran (*or adj.?* v. L. E. I. 21; Th. ii. 416, 33) and yflum wordum gehealdan *debet os suum a malo vel pravo eloquio custodire*, R. Ben. 18, 7. [Cf. *Goth.* þwairhei *indignatio* (in the verse just given): *Icel.* þver-leikr *crossness.*] v. þweorh, III.

þweores, þwires, þwyres. I. *across* as opposed to along, *athwart, transversely, crosswise* as opposed to lengthways :—Lege bred þweores ofer ða fét *lay a board across over the feet*, Lchdm. ii. 342, 6. Ðonon þweores ofer ðone beorh, Cod. Dip. Kmbl. v. 353, 16. Adam wæs on længe fíf and hundnigontiges fingra lenge ofer þweoras ða fingras (i. e. *taking the breadth, not the length, of the fingers*), Anglia xi. 2, 28. Andlang ðære þorngrǽfan þwyres ofer Hysseburnan on gósdæne; ðonne andlang ðæs weges ðe lið andlang gósdæne þwyres ofer in waldes weg, Cod. Dip. Kmbl. v. 148, 4-7. Þwyres ofer ðrý crundelas; ofer ða strét; þwyres ofer ða dúne, 13, 32 : vi. 226, 15. Hé hét ænne weall þwyres ofer eall ðæt lond ásettan from sǽ óþ sǽ, Ors. 6, 15; Swt. 270, 13. Binnan ðam díce ðe wé gemynegodon ðæt Severus hét ðwyres ofer ðæt eálond gedícian *intra vallum quod Severum trans insulam fecisse commemoravimus*, Bd. 1, 11; S. 480, 19. Ðá wolde hé þurhþýn hí þwyres mid ðam swurde *then he wanted to thrust her through, from one side to the other, with the sword*, Homl. Skt. i. 12, 225. Hí wurdon áworpene intó ðam byrnendum ofne gebundene ðwyres (*bound across, with their arms bound to their sides*), Homl. Th. ii. 312, 1. II. *on the flank* :—Ðá hét hé ðæt hiere (*the serpent*) mon mid ðǽm palistas þwyres on wurpe. Ðá weard hiere mid ánum wierpe án ribb forod, Ors. 4, 6; Swt. 174, 10. Hé gesette twá folc diégellíce on twá healfa ... and bebeád ðǽm twám folcum, ... ðæt hié on Reguluses fird on twá healfa þwyres on fóre (*that they should attack Regulus's army on both flanks*), 176, 3. Hannibal him com þwyres on, 4, 8; Swt. 188, 15. III. *perversely, wrongly* :—Godes wiþerwinnan ðe willaþ ǽfre þwires, Jud. Thw. 157, 30. Se ðe his neáxtan hataþ, se bið geháten ðæs áwyrgedan deófles bearn, ðe wyle ǽfre ðwyres, Basil admn. 4; Norm. 44, 14. [*O. H. Ger.* tweres *oblique*; cf. *Ger.* quer, quer über: *Icel.* þvers, þvers um *across:* *Dan.* tværs, tvært over *across.*] v. next word.

þweorh, þwerh, þwyrh; *adj.* I. *crooked, cross,* (1) for the literal sense see þweores. (2) fig. :—Ðuer wig *perversa via*, Kent. Gl. 772. On ðweorum wige *in via perversi*, 812. Þwuru (þweoru, MS. A.) beóþ on gerihte *erunt prava in directa*, Lk. Skt. 3, 5. II. *adverse, opposed* (cf. *Icel.* þver-úð *discord*). v. þweorian, þweor-líc, II :—Þwyr oððe wiðerrǽde *adversus*, Ælfc. Gr. 38; Zup. 240, 1. Gif gé beóð þwyre tó ðisum, Homl. Skt. i. 11, 94. Ungeþwǽre and þwyre him betwýnan *at variance among themselves*, 13, 236. III. *cross, angry, bitter.* v. þweora :—Ðǽm þweorum (þreorum, Wrt.) *rancidis*, Wrt. Voc. ii. 80, 59. v. þróh. IV. *perverse, wrong, evil, depraved, froward* :—Heorte ðuerh (þweor, Spl. : þweorr, Lamb.) *cor pravum*, Ps. Surt. 100, 4. Cyn ðuerh (þweor, Spl. Lamb.) *genus pravum*, 77, 8. Þwyr geþanc *prava mens*, Scint. 68, 3. Mann þwyr *homo peruersus*, 134, 11. Þwyr mód *proteruus animus*, 19. Hé wæs þwyr on dǽdum, Homl. Th. i. 534, 2. Seó híwrǽden is swíðe ðwyr *domus exasperans est*, ii. 530, 29. Decius se þweora heóld ríce, Homl. Skt. i. 23, 12. Eálá þwyre (þweóre, MS. A.) cneóres *O generatio perversa*, Mt. Kmbl. 17, 171. Þwure (þweore, MS. A.), Lk. Skt. 9, 41. Seó ðwyre sáwul, Homl. Th. i. 408, 13. Ðæt ðwyre mód, 410, 21. Ðwerre heortan *peruersi cordis*, Kent. Gl. 612. Mid þweorum (ðý ðweoran, Surt.) *cum perverso*, Ps. Spl. 17, 28. Hé eall ðurh his unrihtdǽde mid ðweorum life áþýstrade *universa prave agendo obnubilavit*, Bd. 5, 13; S. 633, 33. On þwerre sprǽce *in locutione perversa*, Confess. Peccat. Ðweran *perverso*, Kent. Gl. 142. On bogon þweorne (ðone ðweoran, Surt.) *in arcum pravum*, Ps. Spl. 77, 63. On ðóðre wísan siut tó manienne ða bilwitan on óðre ða ðweoran and ða lytegan *quomodo admonendi simplices et versipelles*, Past. 35; Swt. 237, 5. Ða ðweoran hí ofslógon, Homl. Th. i. 232, 7. Ða ðwyran beón geðreáde, ðæt hí tó Godes rihte gebúgan, ii. 96, 5. On þweorra (ðwyrra, Wells Frag.) manna (*pravorum*) gewit, R. Ben. 119, 13. Ðwyrra, Homl. Th. i. 552, 35. Ðweorum *pravis*, Wülck. Gl. 251, 13. Mid þweorum ðeáwum, Homl. Th. i. 302, 30. From þweorum and yfelum wordum, L. E. I. 21; Th. ii. 416, 33. Geðyldig wið ðwyrum mannum, Homl. Th. ii. 514, 11. Ða ðwyran *improbos*, R. Ben. Interl. 15, 10. ¶ On þweorh *wrongly, evilly* :—Ðæt hí ðý máre wíte hæbben ðe hí gere witon ðæt hí on ðweorh dóð, Past. 55; Swt. 429, 9. Hé ongeat ðæt hé hæfde on ðweorh gedón, ðæt seó mægþ wæs bútan biscope, Bd. 3, 7; S. 530, 22. Ðonne hig eów tela tǽcean, and him sylf on ðweorh dóð, L. E. I. 21; Th. ii. 418, 7. Hié on þweorh sprecaþ, Cd. Th. 145, 30; Gen. 2413. [*Goth.* þwairhs *angry:* *O. H. Ger.* dwerah *transversus*; in duerh *in transversum:* *Icel.* þverr *cross, transverse.*] v. ire-þweorh.

þweorh-furh (?) *a cross furrow, a rough place* :—Þuerhfyri *salebrae* (cf. *O. H. Ger.* furihi *salebras; sulcos*), Txts. 95, 1761. Ða unsméþan ðwerfuru *salebrosos* (*complanans*) *anfractus* (Ald.), Wrt. Voc. ii. 78, 26. Þwyrhfero *anfractus* (the passage in Aldhelm is: Errabundis anfractibus exorbitans), 83, 6. Ðweorhfyro, 2, 20. *See also* þwyres fura *salebroso* (the passage is : Genus explanat salebroso pagina versu), 90, 60.

þweorian, þwyrian; *p.* ode *To be opposed, adverse to* (wið), *to be at variance.* v. þweorh, II :—Ic ðwyrige oððe ic wiðerige *adversor*, Ælfc. Gr. 25; Zup. 145, 18. On sibbe is sulfremednyss ðǽr ðǽr nán ðing ne þwyraþ (*there are no conflicting elements*), Homl. Th. i. 552, 21. Ic eom sóðfæstnys, ac ðás ðweorigaþ wið mé *these men are opposed to me*, 380, 8. Ne mæg ðeós offrung beón on ðære heortan ðe mid gýtsunge oððe andan gebysgod bið, for ðan ðe hí ðwyriaþ wið ðone gódan willan *they are adverse to the good will*, 584, 20. Oððe hí his fét gesóhton, him and Gode gehýrsumigende, oððe gif heora hwylc ðwyrode (*if any one of them was adverse*), hé his andweardnysse forfleáh, 560, 10. Ða heáfodmen wiðcwǽdon and symle ðwyrodon *ever proved adverse*, ii. 260, 2. Wiðersaca[n]dan † þw[r]eredon *apostataverant*, Hpt. Gl. 510, 50. Ðá ongunnon Pharisei him betwýnan ðwyrian *the Pharisees began to be at variance among themselves*, Homl. Th. ii. 298, 28.

þweor-líc, þwyr-líc; *adj.* I. *reversed, contrary, opposite* :—Ða word, ðe synd *passiva*, beóð *activa* gif se r byþ áweg gedón; ... twá dǽdlíce word synd ðe habbaþ þwyrlíce getácnunge; ðæt ðe geendaþ on o getácnaþ þrowunge, and ðæt ðe geendaþ on or getácnaþ dǽde, Ælfc. Gr. 19; Zup. 122, 17. Bið swíðe þwyrlíc, ðæt ðearfa beó módig *it is quite reversing the proper order of things for a needy person to be proud*, Homl. Skt. i. 13, 123. II. *adverse.* v. þweorh, II :—Antecrist is gereht ðwyrlíc Crist, Homl. Th. i. 4, 22. III. *perverse, evil, depraved* :—Hé wæs ácenned búton synne, and næs nán ðing ðwyrlíces on him, Homl. Th. i. 176, 5. Fram þwyrlíce sprǽce *a pravo eloquio*, R. Ben. Interl. 21, 9. Gesamnodon gehwylce ðwyrlíce wiðercoran, Homl. Th. i. 468, 5. Ðonne ðwyrlícra manna heortan beóð geemnode, 362, 26. Þwyrlícra *pravorum*, R. Ben. Interl. 107, 11.

þweorlíce; *adv.* I. *awry, askew, in reversed order* :—Þwyrlíce færð æt ðam húse ðǽr seó wyln bið ðære hlǽfdian wissigend and seó hlǽfdige bið ðære wylne underðeódd, Homl. Skt. i. 17, 10. II. *in a way that offers opposition, obstinately, flatly* (of refusal) :—Hé wiðcwæð þwyrlíce (*flatly;* cf. *Icel.* synja, neita þverliga; *or angrily*, v. þweorh, III) and hí mid gedréfedre ǽbilignysse him fram ádráf, Homl. Th. ii. 24, 29. III. *perversely, evilly* :—And suá ðeáh hé mid wón

weorcum hit tô ðweorlíce ne fremeþ ðeáh hê hit on his môde forlǽtan ne mǽge *et quamvis prava non exerceat opere, ab his tamen non evellitur mente*, Past. 11 ; Swt. 73, 13. *Enervatius, i. debilius* sleaclíce, þweorlíce ; *enerviter*, wâclíce, *turpiter*, Wrt. Voc. ii. 143, 55. Þwyrlíce lybbende *praue uiuendo*, Scint. 45, 18.

þweorness, e ; *f.* I. *crookedness* (fig.) v. þweorh, I. 2 :— Ðwyrnyssa beóð gerihte, Homl. Th. i. 362, 22. II. *opposition.* v. þweorh, II. þweorlíce, II :—*Aduersus* ongeán mid þwyrnysse (.i. *discordia*, MS. W.), Ælfc. Gr. 38 ; Zup. 239, 14. Ðá sceorede ðágyt se yldesta hǽðengylda mid nivcelre þwyrnysse *the chief idolater still refused most flatly*, Homl. Th. i. 72, 10. III. *perversity, iniquity, evil, depravity* :—Mycel is seó þwyrnes (cf. *abundabit iniquitas*, 8), Wulfst. 82, 17. Fram þwyrnysse *a prauitate*, Scint. 32, 3. For heora lifes ðwyrnysse, Homl. Th. ii. 530, 24. Forbeóde hê ða þwyrnesse hyra ungeþeahtes *prohibeant pravorum prevalere consensus*, R. Ben. 119, 9. Hrædlíce bið se Dêma tô úrum bênum gebíged, gif wê fram úrum ðwyrnyssum beóð gerihtlǽhte, Homl. Th. ii. 124, 35 ; Lchdm. iii. 276, 18. Manna þwyrnyssa *hominum prauitates*, Scint. 44, 9.

þweorscipe, es ; *m. Perversity, iniquity, depravity* :—Se bið ðæm ísene gelíc inne on ðæm ofne, se ðe for ðære suingellan nyle his ðweorscipe forlǽtan, ac ofan his níhstan his líhte, Past. 37 ; Swt. 269, 6.

þweor-timbre (?) ; *adj.* Cross-grained (?), *stubborn* :—Ic wât ðæt ic ǽr ne síð ǽnig ne mêtte þrístran geþôhtes ne þweorhtimbran (*Grein suggests* -tîmran, v. *next word*) mægþa cynnes, Exon. Th. 275, 15 ; Jul. 550.

þweor-tíme ; *adj.* I. *given to opposition, contentious.* v. þweorh, II :—Ðone rêþan, ðe biþ þweortême, ðû scealt hâtan hund, nallas mann *ferox, atque inquies linguam litigiis exercet? cani comparabis*, Bt. 37, 4 ; Fox 192, 16. II. *given to evil, wicked, depraved* :— And ðeáh ðonne hê ongiete ða scylda ðara ðweortíemena, ðonne geðence hê ðone ealdordôm his onwe:ldes *cum pravorum culpa exigit, potestatem sui prioratus agnoscat*, Past 17 ; Swt. 107, 12. Ðǽm gôdum hê sceal mid wordum stýran and ðǽm þweortýmum mid swingellum *honestiores animos verbis corripiat, improbos autem verberum castigatione*, R. Ben. 13, 20.

þwer. v. þweorh.

þweran ; *p.* þwær, *pl.* þwǽron ; *pp.* þworen, þuren :—*To twirl, stir.* [*O. H. Ger.* dweran ; *p.* dwar ; *pp.* dworan *miscere*.] v. â-, ge-þweran ; þwirel.

þwinan ; *p.* þwân, *pl.* þwinon ; *pp.* þwinen *To get less, dwindle, be reduced* (of a swelling) :—Beþe ða fêt and smyre, ðonne þwînaþ (-eþ, MS.) hý sôna (*the swelling goes down*), Lchdm. i. 84, 25. Ðonne þwînaþ ða áswollena sina, ii. 282, 8. Tácn ðæt se swile þwînan ne mæg, ne út yrnan on ðære lifre, 162, 3 : 212, 9. Cf. dwînan.

þwirel ; es ; *m. A stick for whipping milk* :—Meolc *lac*, fliéte *verberatum*, molcen *lac coagolatum*, þwiril *verberaturium*, Wrt. Voc. i. 290, 26–30. [*O. H. Ger.* dwiril : *M. H. Ger.* twirel, twirl : *Ger.* querl, quirl *a twirling-stick* ; querlen *to beat·up* : *Icel.* þyrell *a whisk* to whip milk ; flauta-þyrell *a stick for whipping milk* ; þyrla *to whirl* ; cf. þwara *a stick used to stir up a cauldron*.] v. þweran.

þwires. v. þweores.

þwítan ; *p.* þwât, *pl.* þwiton, þweoton ; *pp.* þwiten *To thwite* (still in some dialects, e. g. Lancashire), *to cut, cut off* :—þwîteþ, Exon. Th. 354, 50 ; Reim. 63. Monige of ðam treówe ðæs hâlgan Cristes mǽles spônas and sceafþan nimaþ (spônas ðwîtaþ, MS. B.) *multi de ipso ligno sacrosanctae crucis astulas excidere solent*, Bd. 3, 2 ; S. 524, 31. Monige of ðære ilcan styþe spônas ðweoton and sceafþan nômon (ðæt geþwît náman, MS. B.) *astulis ex ipsa destina excisis*, 3, 17 ; S. 544, 44. col. 2 (sprytlan ácurfon, col. 1). Genim ðone wyrttruman, delf up, þwît nigon spônas on ða winstran hand, Lchdm. iii. 292, 2. [Telwyð or thwytyð *abseco, reseco*, Prompt. Parv. 488. To thwyte *dolare*, Cath. Angl. 388, and see note. Cf. *Icel.* þveita *a small axe*, þvita *a kind of axe*, þveit, þveiti *a cut-off piece, a p···el of land*.] v. â-, for-þwítan ; ge-þwit.

þwur, þwyr, þwyr-, þwyre, þý, þýan. v. þweor, þweor-, þweores, þe, þeówan.

þý-dæges ; *adv. On that day, then* :—Gif ðǽr byð ân ofer ða seofon, ðonne ðæcgað ðæt ðæt se mônð gǽd on Sunnandæg on tûne ; gif ðǽr beóð ofer ða seofon twá oððe þreó, feówer oððe fífe oððe syxe, wite ðú tô sôðe ðæt ðýdæges cymð se mônð tô mannum, Anglia viii. 304, 13 : 310, 39.

þyddan ; *p.* de *To strike, thrust, push* :—*Impingere* on besettan (*in margin* on þidden, ic on þydde), Hpt. Gl. 505, 46. Ðá dydde Æfner hine mid hindewerde sceafte on ðæt smældearme ðæt hê wæs deád *percussit eum Abner aversa hasta in inguine, et transfodit eum, et mortuus est*, Past. 40 ; Swt. 295, 17. Se assa þidde his hláfordes fôt þearle tô ðam hege *junxit asina se parieti et attrivit sedentis pedem*, Num. 22, 25. Ðá ábrǽd Aoth his swurd and hine hetelíce þidde swá ðæt ða hiltan eodon in tô ðam innoðe *Aoth tulit sicam, infixitque eam in ventre ejus, tam valide, ut capulus sequeretur ferrum in vulnere*, Jud. 3, 21. Ðæt is ðæt mon mid hindewearde sceafte ðone ðydde ðe him oferfylge *aversa hasta persequentem ferire est*, Past. 40 ; Swt. 297, 14. [Þenne þudde ich in

ham luueliche þohtes, Marh. 14, 7. Þa þudde ha uppon þe þurs feste wið hire fot, 12, 17. He þudde (þraste, 2nd MS.) frommard his breoste, Laym. 1898.] v. þurh-, wiþ-þyddan ; þoddettan ; *see also* þeówan.

þyder, þýfe. v. þider, þífe.

þýfel, es ; *m. A bush ; a thicket ; a leafy plant* :—*Frutectum, i. arborum densitas vel ramus* (*ramnus ?*) þýfel, *frutices, ramos* (*ramnos ?*) þyrne, *frutex, frutecta* þýfel, Wrt. Voc. ii. 151, 42–45. Þýfel *frutex*, i. 33, 42. Ðýfel, ℮o, 20. Þýfel *spartus*, 32, 40 : *spina, sentrix*, 33, 47. Of þiccum þêfele, Hpt. Gl. 408, 38. On ðone hundes þýfel ; of hundes þýfele forþ on ðone þorn, Cod. Dip. Kmbl. iii. 425, 29. Andlang paðes on ðone hyndes þýfel ; of ðæm þýfele andlang weges, vi. 36, 4. Þýfelas *frutecta*, Wrt. Voc. i. 39, 9. His þýfelas † twygu *arbusta ejus*, Ps. Lamb. 79, 11. Þýfela *vel* boxa *belsarum*, Wrt. Voc. ii. 125, 44. Genim ðysse wyrte, ðe wê león fôt nemdon, fíf ðýfelas bútan wyrttruman, Lchdm. i. 98, 16. [Smale fuȝele þat fleoþ bi grunde and bi þuvele, O. and N. 278.] v. brêmel-, rysc-, sceald-, widig-þýfel ; þúf.

þýfen (?) ; *adj. Bushy* :—Þýften (*printed* rypleu) *sparteus* (cf. þýfel *spartus*), Germ. 399, 467.

þyften, e ; *f. A female servant* :—Þyften *verna, famula, servus*, Hpt. Gl. 470, 9. Þeftau *vernacula, servula, ancilla*, 461, 56. [Þe oðer is ase lefdi ; þeos is ase þuften, A. R. 4, 11. Mi lauerd biseh his þuftenes mekelec *respexit humilitatem ancillae suae*, H. M. 45, 12.] v. ge-þofta.

þýgan. v. þeówan.

þyhtig ; *adj. Strong, firm* :—Sweord ecgum þyhtig, Beo. Th. 3121 ; B. 1558. v. hyge-, un-þyhtig.

þylc ; *pron. Such* :—Þes þylc fela spycð *iste talis multum loquitur*, Scint. 80, 19. Gif hê áwiht þylces dô *si tale quid fecerit*, L. Ecg. C. 15 ; Th. ii. 142, 27. Þylces fela his *similia*, Coll. Monast. Th. 27, 11. Ânne þilícne lytling *unum parvolum talem*, Mt. Kmbl. 18, 5. Hwæt is þes þe þam ic þilc gehýre *de quo audio ego talia*, Lk. Skt. 9, 9. Manega ôþre þylce (þyilíce, MS. A.) gê dóð *alia similia his facitis multa*, Mk. Skt. 7, 8. Feáwa synd ða þylce gebedu habban *pauci sunt qui tales orationes habeant*, Scint. 33, 3. [*Laym. Chauc.* þilk : *R. Glouc.* þulk : *Icel.* því-líkr.]

þyl-cræft, es ; *m. Rhetoric* :—Þelcræft *rethorica*, Hpt. Gl. 479, 55. v. þyle.

þyld *patience* :—On ðylde iówre settas gê sáwle iówre *in patientia uestra possidebitis animas uestras*, Lk. Skt. Rush. 21, 19. [Þild to þolenn unnseollþe, Orm. 2603. Þild *patientia*, Ps. 9, 19. *O. H. Ger.* dult.] v. ge-þyld.

þyldig ; *adj. Patient* :—Strong and ðyldig *fortis et patiens*, Rtl. 101, 8. [Þuldi and þolemod, Kath. 174. *O. H. Ger.* dultíg *patiens*.] v. ge-, un-þyldig.

þyldigian ; *p.* ode *To endure* :—Wel þyldigende hî beóð *bene patientes erunt*, Ps. Spl. 91, 14. v. â-, for-, ge-ðyldigian (-þylgiau).

þyle, es ; *m. An orator, spokesman* :—Gelǽred þyle fela spǽca mid feáwum wordum geopenaþ *doctus orator plures sermones paucis verbis aperit*, Scint. 119, 3. Þylas *oratores*, Wrt. Voc. ii. 63, 1. *As a proper name* þyle *is found in* Exon. Th. 320, 5 ; Víd. 24 :—Þyle weóld Rondingum. ¶ In Beowulf *the þyle* of the Danish king is mentioned : —Húnferþ þyle, Beo. Th. 2335 ; B. 1165. Þyle Hrôðgáres, 2917 ; B. 1456. In two passages it is noted that he sat at his lord's feet :—Húnferð maþelode ðe æt fôtum sæt freán Scyldinga, 1002 ; B. 499 : 2335 ; B. 1165. He is the only one of the courtiers who is actually stated to have addressed Beowulf, so that the duty of leading the conversation seems to have fallen to him. If a gloss in Wrt. Voc. ii. 25, 31—*scurris* hof-ðelum—may be read *de scurris* of ðelum (= ðylum) or hofðylum, perhaps his function was something like that of the later court jester, and the style of his attack on Beowulf hardly contradicts the supposition. [*Icel.* þulr ; cf. þylja *to say, chant*.] v. þyl-cræft.

þyle, Thíla *Thule*, some island in the north-west of Europe :—Be westan-norðan Ibernia is ðæt ýtemeste land ðæt man hǽt Thíla (*insula Thule*), and hit is feáwum mannum cúð for ðære oferfyrre, Ors. 1, 1 ; Swt. 24, 20. Ðæt íland ðe wê hátaþ Thýle, ðæt is on ðam norþwestende ðisses middan-geardes *ultima Thule*, Bt. 29, 3 ; Fox 106, 23. Thíle hätte ân ígland be norðan þysum íglande, syx daga fær ofor sǽ, Lchdm. iii. 260, 2. v. Týle.

þylian, þyilíc, þyl-líc, þylman. v. þilian, þyl-líc, þys-líc, for-þylman.

þýmel, es ; *m. A thumbstall, fingerstall, thimble* :—Wiþ scurfedum nægle . . . wyrc þýmel tô, and lege eald spic onufan ðone nægl, Lchdm. ii. 150, 6. [Themyl *digitale*, Wülck. Gl. 578, 29 (15th cent.). The-melle, thymbylle, thymle *digitale, parcipollex, pollicium*, Cath. Ang. 383, where see note. Thymbyl *theca*, Prompt. Parv. 491. *Icel.* þumall *the thumb of a glove*.]

þýmel ; *adj. A thumb thick*, applied to the fat of swine :—Æt þýmelum, L. In. 49 ; Th. i. 132, 19.

þýn *to press.* v. þeówan.

þyncan ; *p.* þúhte. I. *to seem, appear.* (1) where the subject of the verb is expressed :—Ðynceþ him swíðe leoht sió byrðen ðæs láreuwdômes *pondus magisterii levius aestimant*, Past. proem. ; Swt. 24, 9. Mê ðeós (rôd) heardra þynceþ, Exon. Th. 91, 9 ; Cri. 1489 : 383, 14 ; Rä. 4, 10 : Met. 12, 8. Þincð, Bt. 23 ; Fox 78, 25 : Met. 12, 18. Þis

þinceþ riht micel, ðæt . . ., Cd. Th. 149, 17; Gen. 2476. Mē ðæt riht ne þinceþ, ðæt . . ., 19, 11; Gen. 289. Þynceþ, Andr. Kmbl. 1218; An. 609. Hī ne wundriaþ mæniges þinges ðe monnum wunder þynceþ, Met. 28, 82. Ðæs ðe mē þynceþ, Andr. Kmbl. 944; An. 472: Ps. Th. 101, 3. Him ða twigu þincaþ merge, Met. 13, 44. Hý wyrðe þinceaþ, Beo. Th. 742; B. 368. Lytel þūhte ic leóda bearnum, Exon. Th. 87, 14; Cri. 1425. Ne þūhte hē him nó innon swā fæger swā hē ūtan þūhte. Þeáh ðū nū hwam fæger þince, ne biþ hit nó ðý raþor swā, Bt. 32, 2; Fox 116, 24. Hē ðūhte him selfum suīðe unlytel se parvulum non videbat, Past. 17; Swt. 113, 12. Hire þūhte hwītre heofon and eorðe, Cd. Th. 38, 7; Gen. 603: 111, 4; Gen. 1850: Beo. Th. 1688; B. 842: Met. 12, 15. Tō lang hit him þūhte, hwænne hī tōgædere gāras bēron, Byrht. 133, 47; By. 66. Him ðæt wræclīc þūhte, Cd. Th. 233, 4; Dan. 270. Ðæt wundra sum monnum þūhte, ðæt . . ., Exon. Th. 133, 13; Gū. 489: 169, 27; Gū. 1101. Hié ðam were geonge þūhton men, Cd. Th. 146, 27; Gen. 2428. Ealle brimu blōdige þūhton, 214, 20; Exod. 572: Andr. Kmbl. 880; An. 440: Beo. Th. 1737; B. 866. Ðý ðe hý him sylfum sēllan þūhten, Exon. Th. 455, 24; Hy. 4, 54. Hē þenceþ ðæt his wīse welhwam þince unforcūþ, 315, 13; Mōd. 30. Þeáh hit lang þince, Met. 10, 66. Hwæt eów sēlest þynce, Elen. Kmbl. 1062. Hié wilniaþ ðæt hié ðyncen ða betstan, Past. 18; Swt. 134, 18. Hwelc wīte sceal ūs tō hefig ðyncan? 36; Swt. 255, 3. Hī woldon mē swīþe bitere þincan, Bt. 22, 1; Fox 76, 19. Swā hit þincan mæg, L. I. P. 19; Th. ii. 326, 36. (2) where the subject is not expressed, as in methinks :—Swā mē ðincþ, Bt. 33, 1; Fox 120, 21: 36, 3; Fox 176, 30. Swā ðē ðyncþ, 38, 2; Fox 196, 22. Hū þincþ eów quid vobis videtur, Mt. Kmbl. 21, 28. Þincþ him genóg on ðam ðe hī binnan heora ægenre hȳde habbaþ, Bt. 14, 2; Fox 44, 22. Dēm ðū hī tō deáþe, gif ðē gedafen þince, Exon. Th. 247, 32; Jul. 87. Ðeáh monnum swā ne þince, Bt. 39, 8; Fox 224, 17. (2 a) where the verb is followed by a clause :—Mē þincþ ðæt hit hæbbe geboht sume leáslīce mǣrþe, Bt. 24, 3; Fox 82, 24. Ne þynceþ mē gerysne, ðæt wē rondas beren, Beo. Th. 5299; B. 2653. Hwæt þincþ ðē ðæt ðū sý?, Jn. Skt. 8, 53. Þyncþ him ðæt hē næbbe genóg, Bt. 33, 2; Fox 124, 4. Him selfum þincþ ðæt hē nænne næbbe, swā swā manegum men þincþ ðæt hē nænne næbbe, 29, 1; Fox 104, 8. Wrætlīc mē þinceþ, hū seó wiht mæge wordum lācan, Exon. Th. 414, 11; Rä. 32, 18. Þinceþ ðē miht ðū libban potes vivere? Bd. 5, 6; S. 619, 40. Ðūhte heom ðæt hit mihte swā, Cd. Th. 266, 14; Sat. 22. Ne þūhte gerysne rodora wearde, ðæt Adam leng āna wǣre, 11, 9; Gen. 169. Ðeáh ūs þince ðæt it on wóh fare, Bt. 39, 8; Fox 224, 20. Higesnotrum mæg þincan, ðæt . . ., Met. 10, 8. II. to seem fit :—Swā mycel swā ðē þince as much as to you seems good, Lchdm. ii. 74, 2. Dō swā ðē þynce, gif ðū frygnen sié, Elen. Kmbl. 1078; El. 541. [Goth. þug[g]kjan : O. Sax. thunkian : O. H. Ger. dunchan : Icel. þykkja.] v. ge-, mis-, of-, on-þyncan.

þyncþ[u]; f. Honour, dignity, rank :—Suā suǣ hē on ðyncðum (ge-ðyncðum, Cott. MSS.) bið furður ðonne óðre sicut honore ordinis superat, Past. 14; Swt. 81, 23. v. ge-þingþu.

þȳnde. v. þignan.

þyng, es; m. (?) Growth, progress, profit :—Mið ðynge proficiendo, Rtl. 83, 40. v. ge-þynge.

þyn-hlǣne; adj. Wasted, shrunk :—Ða gescruncenan and ða þyn-hlǣnan marcida, Wrt. Voc. ii. 57, 23.

þynne; adj. Thin :—Ðæt ic reccan mōste þicce and þynne, Exon. Th. 424, 8; Rä. 41, 36. Ic dō sum ðing ðinne tenuo, Ælfc. Gr. 24; Zup. 137, 9. I. of dimension, (1) thin, lean, the opposite of fat or stout :—Þynne monn galbus, Wrt. Voc. ii. 42, 11. Ne mæg him se līchoma batian, ac bið blāc and þynne, Lchdm. ii. 206, 11. (2) thin, the opposite of thick :—Mid ðynre tyrf bewrigen obtectus cespite tenui, Bd. 5, 6; S. 619, 20. Hē hæfde midmycle neosu ðynne vir naso pertenui, 2, 16; S. 519, 34. Seóh þurh þynne hrægl, Lchdm. ii. 290, 4. Seó wyrt hafaþ þynne leáf, Lchdm. i. 288, 16. (3) thin, the opposite of broad :—Ðæt seó ðynneste dolhswaþo ætȳwde ut tenuissima cicatricis vestigia parerent, Bd. 4, 19; S. 589, 19. II. of density, (1) where the parts of a whole are not close together, thin :—Oft of ðinnum rēnscūrum flēwð seó eorðe, Homl. Th. ii. 466, 7. In sceagan ðær hé þynnest is, Cod. Dip. Kmbl. iii. 391, 15. (2) applied to liquids, air, etc., thin :—Sum ūtgang biþ þynne, sum mid þiccum wǣtum geondgoten, Lchdm. ii. 276, 24. Hit sié þicce swā þynne brīw, 314, 4. Lyft is līchamlīc gesceaft, swȳðe þynne, iii. 272, 17. Þynne wīn, ii. 254, 18. Snāw cymð of ðam þynnum wǣtan, iii. 278, 23. Hē elles ne ðeah nemne medmicel hlāfes mid ðynre meolce lac novum in phiala ponere solebat, et post noctem ablata superficie crassiore, ipse residuum cum modico pane bibebat, Bd. 3, 27; S. 559, 35. Ǣr se þicca mist þynra weorðe, Met. 5, 6. III. fig. (1) thin, weak, feeble :—Hwilc sié sió gecynd ðæs līchoman, hwæðer hió sié strang ðe heard . . . ðe hió sié hnesce and mearwe and þynne, Lchdm. ii. 84, 14. Ðynre eþunge ānre ætȳwde ðæt hé lifes wæs halitu tantum pertenui quia viveret demonstrans, Bd. 5, 19; S. 640, 24. (2) delicate, fine :—Andgyt þēnunge gearwigende Gode þȳhð symle and þynne hit byð sensus officium exhibens Deo uiget semper et tenuis fit, Scint. 52, 9. [O. H. Ger. dunni tenuis : Icel. þunnr.]

þynness, e; f. I. thinness, slightness of density. v. þynne, II. 2 :—Metta meltung and þynnes, Lchdm. ii. 198, 3. II. weakness. v. þynne, III. 1 :—Hit gehǣld ða þynnysse ðære gesihðe, Lchdm. i. 134, 27 (see note).

þynnian; p. ode To make or to become thin :—Ðynnade obtenuerat, Txts. 182, 80. Ðæt þicce horh ðū scealt mid ðām lǣcedómum wyrman and þynnian, Lchdm. ii. 194, 22. [O. H. Ger. gi-dunnót attenuatus; dunnēn rarescere : Icel. þynna to make thin.] v. ā-, ge-þynnian.

þynnol(-ul); adj. Lean, meagre :—Ðynnul macilentus, Wrt. Voc. i. 51, 16.

þynnung, e; f. Thinning, making thin :—Lǣcedōmas ðe þynnunge mægen hæbben and smalunge, Lchdm. ii. 260, 23.

þyn-wefen; adj. Thin-woven :—Thynwefen hrægl levidensis, Wrt. Voc. ii. 54, 17.

þyrel (from þyrhel, v. þurh), þyrl, es; n. A hole made through anything, an aperture, orifice :—Ælces kynnes mūð vel ðyrl orificium, Wrt. Voc. i. 19, 57. Ic borige terebro. ðyrl foramen, 84, 65 : Ælfc. Gr. 9, 12; Zup. 40, 16. Is on ðam wāge ðyrl geworht est foramen in pariete, Bd. 4, 3; S. 570, 17. Gif ān þyrl (foramen) open byð forlǣten, ðanon fram feóndum beó inn āgan, Scint. 140, 6. Ðæt wǣre ðyrel on middum ðam hweóle, Shrn. 81, 13. Gerȳme ðæt ðæt þyrel (the aperture made by a lancet) tō nearo ne sié, Lchdm. ii. 208, 25. Stōl niþan ðyrele, 76, 22. Ðæt īsen (a scythe) becōm swymmende tō ðam snǣde and tō ðam þyrle ðe hit ǣr of āsceát, Homl. Th. ii. 162, 14. Swā swā mon melo sift; ðæt melo ðurhcrȳpþ ǣlc þyrel, Bt. 34, 11; Fox 152, 2. Þurh nǣdle þyrel (ðyril, Lind.) per foramen acus, Mk. Skt. 10, 25 : Wrt. Voc. ii. 73, 1. Ðyrl, Lk. Skt. Lind. 18, 25. Gif wyrm þyrel gewyrce . . . drype on ðæt þyrel, Lchdm. ii. 114, 14. Drincan syllan þurh þyrel, Exon. Th. 485, 1; Rä. 71, 7 : 397, 18; Rä. 16, 21. Ne furþon ān þyrl (foramen) būton cræfte mīnon (the smith's) ðū ne miht dōn, Coll. Monast. Th. 31, 17. Hē ðæt īsen sette on ðæt ylce þyrl, and hit fæste stōd, Homl. Skt. i. 21, 71. Þyrel foramina, Wrt. Voc. ii. 149, 73. Þyrlum finistris, 148, 60. Ðyrelum foraminibus, Bd. 3, 17; S. 544, 30, col. 2. Ðurh ða ðyrlo, 544, 32, col. 1. Ic hét hié þurhborian . . . hét ic eft ða ðyrelo mid golde forwyrcean, Nar. 20, 3. [A. R. þurl a window : Ayenb. þerle.] v. eág-, ears-, hūn-, næs-, nos-, teol-þyrel (-þyrl, -þerl), and next word.

þyrel; adj. Perforated, having a hole or holes, pierced through :—Gif eáre þirel weorðeþ, .iii. scitt. gebēte, L. Ethb. 41; Th. i. 14, 6 : 49; Th. i. 14, 15. Þyrel, 45 ; Th. i. 14, 10. Gif monnes þeóh biþ þyrel (þyrl, MS. B.), L. Alf. pol. 62; Th. i. 96, 13 : 63; Th. i. 96, 16. Gif se wāh bið ðyrel if the wall have a hole through it, Past. 21; Swt. 157, 17. From ðyrelan stāne, Cod. Dip. Kmbl. ii. 29, 2. On ðone þȳrlan stān, iii. 406, 11. Þyrlum, 436, 34. Se ðe mēdsceattas gaderaþ, hē legeþ hié on ðyrelne pohchan (in sacculum pertusum), Past. 45; Swt. 343, 20. Gif ðegna hwelc ðyrelne kylle brōhte tō ðȳs burnan, bēte hine georne, 65; Swt. 469, 10. Heáfodwunde tō bōte. Gif ða bān beóð būtū þyrel (þyrle, MSS. B. H.) . . . Gif ðæt ūterre bān bið þyrel . . ., L. Alf. pol. 44; Th. i. 92, 14. Gif būtū þyrele sién, L. Ethb. 47; Th. i. 14, 12. Hē eówaþ ūs his þyrlan handa, Wulfst. 90, 6. [O. H. Ger. durchil pertusus.] v. þurh-þyrel.

þyrelian, þyrlian; p. ode. I. to make a hole through, pierce through, perforate :—Þirlie his hláford his eáre mid ānum ǣle dominus perforabit aurem ejus subula, Ex. 21, 6. Þirlige, Lev. 25, 10. II. to make hollow; fig. to make vain :—Áidlie ł þyrlie obunco (? obunco is glossed by ymbclipe, Wrt. Voc. i. 22, 31), Engl. Stud. xi. 66, 66. Þyrliaþ cavantur, evacuantur, hol cava, Wrt. Voc. ii. 129, 62. [He lette þurlen his scheld, A. R. 392, 24. To þurlin godes side wið speres ord, Jul. 41, 14. With a spere was thirled his brestboon, Ch. Kn. T. 1852. To hem þat his herte þirled, Piers P. i. 172. Thyrlyñ (thryllyn) or peercyñ penetro, terebro, perforo, Prompt. Parv. 491.] v. þurh-þyrelian.

þyrelung, e; f. Perforation, piercing through :—' Ðurhðyrela ðone wāg'. . . Hwæt is sió ðyrelung ðæs wāges 'fode parietem.' . . . Quid est parietem fodere? Past. 21; Swt. 153, 25. [In his side þurlunge, O. E. Homl. i. 207, 13.]

þyrel-wamb; adj. Having the stomach pierced :—Þyrelwombne, Exon. Th. 490, 13; Rä. 79, 11.

þyrfende. v. þurfan.

þyringas; pl. The Thuringians :—Wōd weóld þyringum, Exon. Th. 320, 17; Vīd. 30: 322, 16. Mid Eást-þyringum, 323, 30; Vīd. 86. Maroara habbaþ bewestan him þyringas, Ors. 1, 1; Swt. 16, 11. v. Grm. Gesch. D. S. c. xxii.

þyrl, þyrlian. v. þyrel, þyrelian.

þyrl-hūs, es; n. A turner's shop :—Þyrlhūs tornatorium, Wrt. Voc. i. 58, 45.

þyrn-cin, es; n. (?) A small prickly plant (-cin diminutive suffix?), a thistle :—Cwyst ðū gaderaþ man wīnberian of þornum oððe fīcæppla of þyrncinum (tribolis), Mt. Kmbl. 7, 16.

þyrne, an; f. A thorn-bush :—Þyrne, thyrnae dumus, Txts. 57, 710. Þorn oþþe þyrne dumus, Wrt. Voc. ii. 25, 70. Þyrne frutices, ramos, 151, 44: dumus, i. spina, spineta, Wülck. Gl. 225, 23. Hē geseah ðæt seó

þyrne (*rubus*) . . . næs forburnen, Ex. 3, 2, 3. On ða þyrnan westewarde
ðǽr se mycla þorn stód, Cod. Dip. Kmbl. iii. 404, 12. On ða rúgan
þyrnan ; of ðǽre þyrnan on ða brémbelþyrnan, 419, 12. On gáte
þyrnan ; of ðǽre þyrnan on blace þyrnan, vi. 2, 5. On ða blacan þyrnan ;
of ðǽre þyrnan, 220, 20. On ða ealdan þyrnan, Cod. Dip. B. iii. 336,
25. þyrnan *dumos*, þyrnum *dumis*, Wrt. Voc. ii. 27, 43, 44. [Cf. *Icel.*
þyrnir a *thorn*.] v. brémel-, brér-, mǽr-þyrne.

þyrnen ; *adj. Of thorns* :—Hé hæfde fiþru swylce þyrnen besma, Shrn.
120, 28. Hí mid þyrnenum helme his heáfod befengon, Homl. Th. ii.
252, 26. Þyrnenne helm (ðyrnenne bég, Lind., Rush.) *spineam coronam*,
Mk. Skt. 5, 17 ; Jn. Skt. 19, 5. Þyrnenne cynehelm (sigbég of ðornum,
Lind., Rush.) *coronam de spinis*, 2. Ðyrnenne beág, Past. 36 ; Swt. 261,
14. Þyrnenne, Exon. Th. 69, 27 ; Cri. 1127. Ðone ðyrnenan helm,
Homl. Th. ii. 254, 10. Þyrnenan, Wulfst. 124, 5. [*O. H. Ger.* thornen :
O. H. Ger. durnín *spineus*.]

þyrnet[t], es ; *n. A place full of thorns, a thicket of thorn-bushes* :—
Þirnetum *spinetis*, of ácynnendlícum ł fexedum þyrnetum *de spinetis
nascentibus* (*gignentibus*), Hpt. Gl. 463, 32–36. Of þiccum þyrnetum
senticosis, spinosis, 436, 47.

þyrniht ; *adj. Thorny, prickly* :—Ðeós wyrt hafaþ leáf . . . þyrnyhte,
and heó hafaþ sumne sinewealtne crop and þyrnyhtne, Lchdm. i. 282,
14–17. Þynne leáf and ða hwónlíce þyrnihte, 288, 17. [*O. H. Ger.*
dornoht *spinosus* : *Ger.* dornicht.]

þyrran. v. þyrrian.

þyrre ; *adj.* I. *dry, lacking water* :—On þyrran mǽræ, Cod. Dip.
Kmbl. v. 117, 5. II. *lacking sap or moisture* :—Þornas þyre
(þyrre ? *but cf. O. L. Ger.* thiori holt), Ps. Th. 117, 12. III. as
a medical term, *dry* :—Hine drecep þyrre hwósta, Lchdm. ii. 264, 13.
Eallum þyrrum líchomum hǽmedþing ne dugon ac swíþost þyrrum and
cealdum ; ne dereþ hit hátum and wǽtum, 222, 28–30. [*Goth.* þaursus
dry, withered : *O. H. Ger.* durri *aridus, siccus, torridus* : *Icel.* þurr :
O. L. Ger. thurritha *dryness*.]

þyrrian(-an ?) *to make* or *to become dry* :—Corfen, sworfen, cyrred,
þyrred, Exon. Th. 410, 25 ; Rä. 29, 4. [Cf. *Goth.* thaurseith mik
I *thirst* : *O. Sax.* thorrón *to be withered up, consumed* : *O. H. Ger.* dorrên
arescere : dorren *arefacere*.] v. for-þyrrian.

þyrs, es ; *m. A giant, an enchanter, a demon* :—Ðyrs, heldióbul *Orcus*,
Wrt. Voc. ii. 115, 64. Ðyrs sceal on fenne gewunian ána innan lande,
Menol. Fox 545 ; Gn. C. 42. Þyrses *Caci*, Hpt. Gl. 445, 2. Þyr[ses]
colossi, Hpt. Gl. 445, 2. Gehégan ðing wiþ þyrse (*Grendel*), Beo. Th.
856 ; B. 426. Ealdum þyrse (þyrre, MS.), Exon. Th. 425, 29 ; Rä. 41,
63. Þyrsa oððe wyrmgalera *Marsorum*, þyrsas ł wyrmgaleras *Marsi*,
Hpt. Gl. 483, 13–15. Cf. Wrt. Voc. ii. 82, 9. Ánige þyrsas *Cyclopes*,
Wrt. Voc. ii. 22, 37. Ðyrsa *Cyclopum*, 21, 72. [Com þe þurs Maxence,
Kath. 1858. Ichabbe isehen þene þurs of helle, Marh. 11, 7. Thykke
theefe as a thursse, Halliwell's Dict. Ther shal lyn lamya (satyr, A. V.),
that is a thirs (thrisse), or a beste hauende the bodi lic a womman and
horse feet, Wick. Isaiah 34, 15. *O. H. Ger.* durs *Dis, daemonium* : *Icel.*
þurs a *giant*.] v. orc-þyrs.

þyrscel, þyrscwold. v. þerscel, þerscold.

þyrstan ; *p.* te *To thirst*. I. used impersonally, (1) with acc. of
person :—Mé þyrst (ic ðyrsto, Lind., Rush.) *sitio*, Jn. Skt. 19, 28. Ne
þyrst ðone nǽfre ðe on mé gelýfð (se ðe gileféð(-es, Lind.) on mec ne
ðyrsteð(-es, Lind.) ǽfre, Rush.) *qui credit in me non sitiet umquam*, 6, 35.
Mine, sáwle þyrst *sitivit anima mea*, Ps. Th. 41, 2. Ða men ne þyrst,
Lchdm. ii. 194, 12. Mé þyrste (mec þyrste, Rush.) *sitivi*, Mt. Kmbl.
25, 35. Mínne þegn þyrste and mínne here, Nar. 8, 11. Hine ðyrstte,
Past. 36 ; Swt. 261, 16. Ðæt mé ne þyrste (þ ic ne ðyrste(-o, Lind.),
Rush.) *ut non sitiam*, Jn. Skt. 4, 15. Cume tó mé se ðe hine þyrste (se
ðe ðyrsteð, Lind. ; gif hwelc þyrste, Rush.) *qui sitit veniat ad me*, 7, 37.
Ongan ðone oferhýdgan þyrstan on deáþ, Shrn. 130, 1. Drihten ealle
ða gefylde, ða ðe hié on eorþan léton hingrian and þyrstan for his noman,
Blickl. Homl. 159, 17. (1 a) with acc. of person and gen. of object of
thirst :—Ðeáh ðæt folc ðyrste ðǽre láre, Past. 2 ; Swt. 31, 7. (2) with
dat. of person :—Ðyrste sáwle mínre *sitivit anima mea*, Ps. Spl. 41, 2 :
62, 2. Swá hwam swá ðyrste, cume tó mé, Homl. Th. ii. 274, 3. II.
with nom. of person suffering thirst ; see also extracts from Northern
Gospels in I. (1) absolute :—Mín sáwl on ðé þyrsteþ *sitivit in te anima
mea*, Ps. Th. 62, 1. Ðyrsteþ sáwul mín, Ps. Surt. 42, 3 : 62, 2. Hwænne
gesáwe wé ðé þyrstendne, Mt. Kmbl. 25, 37, 44. Heó þyrstendon ðone
þurst gelíþigaþ, Lchdm. i. 268, 11. (2) with gen. (or acc. ?) of object of
thirst :—Flǽsc ðonne hit God þyrst *caro tunc Deum sitit*, Scint. 54, 6.
Eádige ða ðe þyrstaþ rihtwísnysse (*iustitiam*), 49, 17. Ða ðe rihtwísnesse
þyrstaþ (lǽt hig þyrstan, MS. A.), Mt. Kmbl. 5, 6. Ðú ðe þyrstende
wǽre monnes blódes, Ors. 2, 4 ; Swt. 76, 33. Hió ðyrstende wæs on
symbel mannes blódes, 1, 2 ; Swt. 30, 27. [*O. Sax.* thurstian (*impers.*) :
O. H. Ger. dursten (*pers. and impers.*) : *Icel.* þyrsta (*impers.*), cf. *Goth.*
þaursjan (*impers.*) *to thirst*.] v. ge-þyrst, of-þyrsted, sin-þyrstende :
þyrre.

þyrstig ; *adj. Thirsty* :—Ic wæs ðyrstig *sitivi*, Mt. Kmbl. Lind. 25, 35.
Ðyrstende ł ðrystig (þyrstigne, Rush.) *sitientem*, 37. v. þurstig.

þys, es ; *m. A storm* :—Seó orsorhnes gǽþ scýrmǽlum swá þæs windes
þys *prosperam fortunam videas ventosam*, Bt. 20 ; Fox 72, 5 note.
[*Icel.* þyss *uproar, tumult* ; þysja *to rush*.]

þys-líc, þyl-líc ; *pron. Such.* I. used adjectivally, (1) qualifying
a noun :—Gif him þyslícu þearf gelumpe, Beo. Th. 5267 ; B. 2637.
Wénst ðú ðæt ðé ánum þyllíc (þellucu, Cott. MS.) hwearfung and þillíc
(þillícu, Cott. MS.) unrótnes on becumen, and nánum óþrum móde swelc
ne on becóme, Bt. 8 ; Fox 24, 35. Ic mé þyslícre þráge ne gewénde,
Exon. Th. 269, 20 ; Jul. 453. Þyslícne þegn, 316, 7 ; Mód. 45 : Elen.
Kmbl. 1087 ; El. 546. Ða ðe ðyllícne gylt þurhteóð, Homl. Ass. 148,
122. Ða ongan hé forð sendan þyllíce stemne and þus cwæð . . . , Homl.
Skt. ii. 23 b, 190. Ðyllíce, 204. Þislíc ǽrende se pápa eft onsende and
ðás word cwæð . . . , Blickl. Homl. 205, 22. Ne geceás ic nó ðis fæsten,
ac ðyllíc fæsten ic geceás : brec ðæm hyngriendum ðínne hláf, Past. 43 ;
Swt. 315. 13. Ðæt mód þillíc sár cweþende wæs, Bt. 5, 1 ; Fox 8, 24.
Mid þyslíce þreáte hláford fergan, Exon. Th. 32, 23 ; Cri. 517. Þyllícu
þing syndon gereht, Homl. Ass. 199, 142. Ðyllecum unrihtum, Bt. 16,
4 ; Fox 58, 10. Ðyslíce gife and swá mycle *tanta taliaque dona*, Bd. 2,
12 ; S. 514, 13. Ðyllíca giefa, Past. 5 ; Swt. 41, 13. Ðyllíce gyltas,
Homl. Ass. 149, 132. Ðyllíce weorc, Homl. Skt. i. 17, 28. Manega
óþre þyllíce ðing *alia similia his multa*, Mk. Skt. A. 7, 8. (2) predicative :—Ðæt seó onwrihgnes ðyslíc wǽre *revelationem hujusmodi esse*, Bd.
3, 8 ; S. 531, 37. Ðyslíc mé is gesewen ðis líf *talis mihi videtur vita*, 2,
13 ; S. 516, 13. Ðonne ðás ðyslíc ðé tó cume, 2, 12 ; S. 514, 22.
Sió onsýn biþ þyslícu, Lchdm. ii. 348, 21. Þyllíc bið se ende ðæs
líchoman fǽgernesse, Blickl. Homl. 59, 21 : Homl. Th. i. 88, 10. Þyllíc
byð ðæt cyn *haec est generatio*, Ps. Th. 23, 6. Ic nǽfre ðé þyslícne
gemétte, þus meðue, Exon. Th. 163, 2 ; Gú. 987. Þás tácno þyslíco
syndon, Blickl. Homl. 109, 6. Ða ðe ðyllíce beóð, Past. 5 ; Swt. 41, 20 :
Homl. Ass. 146, 63. II. used substantively :—Þes þyllíca sý
gemyngod *hic ammoneatur*, R. Ben. 48, 6. Gif ówiht þislíces gelimpe
si hujus simile quid acciderit, L. Ecg. C. 15 ; Th. ii. 144, 2. Ðises hí
wundriaþ and manies þyllíces, Bt. 39, 3 ; Fox 214, 32 : Ps. Th. 9, 31 :
Homl. Th. ii. 158, 2 : Homl. Skt. i. 12, 275 : ii. 28, 106. Ymb ðyllíc
is tó geðencenne, Past. 3 ; Swt. 21, 20. Þyllíc, Homl. Skt. ii. 28, 119.
Sege hwænne ic ǽfre ǽr þillíc ðé gedyde *dic, quid simile unquam fecerim
tibi*, Num. 22, 30. Nǽfre wé ǽr þyllíc ne gesáwon *numquam sic vidimus*,
Mk. Skt. 2, 12 : Ors. 4, 4 ; Swt. 104, 3. Hwam beóð ðás ðyllecan gelícran
quibus isti sunt similes? Past. 33 ; Swt. 226, 23. Ðás ðyllíce bringaþ
gestreón, Homl. Th. ii. 550, 35 : Homl. Skt. i. 4, 290. For ðyllecum
næs hé geunrótsod, Bt. 16, 4 ; Fox 58, 8. Mid þyllícum and mid
manegum þyllícum, Ps. Th. 10, 7. Ða ðe heora lustum folgiaþ and
ðyllíce ádreógaþ, Homl. Ass. 196, 45. Wé gehýrdon þyllíce gereccan,
Homl. Skt. i. 6, 184, 189. [*O. E. Homl. Kath. A. R. H. M.* þullich :
Ayenb. þellich.] Cf. þus-líc.

þýster, þýstre, þýstrian, þýstru, þýstrung. v. þeóster-, þeóstre, þeóstrian, þeóstru, þeóstrung.

þýþel, Ps. L. 79, 11, -þýtiþ, þýwan, þýwen. v. þýfel, á-þýtiþ, þeówan,
þeówen.

U

For the Runic U, see úr.

u the letter u :—*Mortuus* on twám uum, Ælfc. Gr. 31 ; Zup. 197, 16.

úder an udder, a breast :—Of údrum *uberibus*, Kent. Gl. 203. [Iddyr
or uddyr of a beeste, pappe uber, Prompt. Parv. 258. *O. Du.* uder,
uyder : *Du.* uijer : *O. H. Ger.* útar, útiro *uber* : *M. H. Ger.* iuter, úter :
Ger. euter : *Icel.* júgr : *Dan.* yver : *Grk.* οὖθαρ.]

úf, es ; *m. An owl* ; the word also glosses *vultur* :—Uuf *bubo*, Wrt.
Voc. ii. 102, 28. Úf, i. 29, 45. Ðes úf *hic vultur*, Ælfc. Gr. 9, 22 ;
Zup. 48, 17 note. [*O. H. Ger.* úvo *bubo* : *Icel.* úfr *some kind of bird*.]
v. húf ; þripel-úf ?

úf, es ; *m. The uvula* :—Múðes hróf *palatum*, úf *sublingua*, Wrt. Voc.
i. 64, 59. Cf. Undertungan *sublinguae*, 282, 79. [*From Latin* uva ?]
v. húf.

ufan ; *adv.* I. *from above, down*, (1) where motion is expressed or
implied :—God him sende ufan greáte hagolstánas *Dominus misit super
eos lapides magnos de coelo*, Jos. 10, 11 : Blickl. Homl. 51, 12. Him
feóll ufan flǽsc *pluit super eos carnes*, Ps. Th. 77, 27. Seó lyft týhð
ðone wǽtan tó hyre neoðan and ða hǽtan ufon, Lchdm. iii. 280, 3. Ic
eom engel Godes ufan síþende, Exon. Th. 258, 7 ; Jul. 261. Hine ufan
neósade Meotud, 159, 24 ; Gú. 931 : Beo. Th. 3005 ; B. 1500. Se ðe
ufa cuom *qui desursum venit*, Jn. Skt. 3, 31. Se Hálga Gást cumeþ ufan
on ðé, Blickl. Homl. 7, 35. Ic on andwlitan sígan lǽte wǽllregn ufan
wídre eorðan, Cd. Th. 81, 24 ; Gen. 1350. Hé ða eágan þurheteþ ufon
on ðæt heáfod *it eats its way through the eyes down into the head*, Exon.
Th. 374, 7 ; Seel. 122. Ic férde tó foldan ufan *from edle*, Cd. Th. 296,
2 ; Sat. 496. Hí feóllon ufon of heofonum, 20, 11 ; Gen. 308. Ufan
cumende of heánisse *oriens ex alto*, Ps. Surt. ii. p. 199, 40 : Cd. Th. 248,
7 ; Dan. 509. Of roderum ufan onsended, 237, 14 ; Dan. 337 : Exon.
Th. 368, 20 ; Seel. 27. (2) where an action is directed from a higher

to a lower point :—Seó sunne lócaþ ufan on helle, Salm. Kmbl. p. 200, 2. Ufan engla sum cȳgde . . . ufan of roderum wuldergāst mælde, Cd. Th. 176, 7-16; Gen. 2908-2911. Ðæt eów ne bið ufan ālȳfed, Exon. Th. 138, 31; Gū. 584. Godwebba cyst ufan eall forbærst, Exon. Th. 70, 13; Cri. 1138. II. marking position, *above, at the top* :—Synd ðǽr þrȳ porticas . . . fægere ufan oferworhte and oferhrȳfde. Seó cyrice is ufan open . . . and þeáh ðe ðæt hús ufan open sȳ, hweþre hit biþ ā þurh Godes gife ufan wiþ ǽghwilc ungewidro gescylded, . . . and nǽfre nǽnig man ða lǽstas ufan oferwyrcean ne mihte, Blickl. Homl. 125, 24-35 : 19, 27 : Exon. Th. 219, 17; Ph. 308 : 446, 14; Dóm. 22. Æscholt utan grǽg, Beo. Th. 665; B. 330. Hæfdon hī Dryhtnes leóht ufan (*above, in heaven*) forlēten, Cd. Th. 269, 7; Sat. 69. Hér is fȳr micel ufan and neoðone, 24, 9; Gen. 375. On ǽlcere stówe he is hire emnneáh ge ufan ge neoþon, Bt. 33, 4; Fox 130, 23. ¶ Associated with prepositions :—Saturnus ȳfemest wandraþ ofer eallum ufan ōþrum steorrum, Met. 24, 24. On ðám ufan stōdon scyttan, Nar. 4, 15. On ðam seaðe ufan hē hūs getimbrode, Guthl. 4; Gdwin. 26, 9. Ðā gefeóll hē on his earm ufan, Bd. 3, 2; S. 525, 2. [*Icel.* ofan *from above ;* ofan ā *upon : O. Sax.* bi-oban.] v. be-, on-ufan ; ufane, ufenan.

ufan-cund ; *adj.* *Heavenly, celestial ;* supernus :—Segnbora ðæs ufancundan kyninges, Blickl. Homl. 163, 22. Ufancundes engles word, Exon. Th. 169, 19; Gū. 1097. Engel ufancundne, 176, 26; Gū. 1216. Ufancundne ege *fear from heaven,* 143, 8; Gū. 658. Ðonne hē ongiete ðone ufancundan willan *cum superna voluntas agnoscitur,* Past. 7; Swt. 51, 8. Āras ufancunde *celestial messengers,* Exon. Th. 31, 29; Cri. 503.

ufane(-en(n)e) ; *adv.* I. *from above, down,* (1) where motion is expressed or implied :—Hié sǽdon ðæt hit ufane of ðære lyfte cōme, Ors. 3, 6; Swt. 108, 30. Hē dēd ðæt fȳr cymð ufene, Wulfst. 97, 21. Steorran hreósaþ ufene of heofonum, 93, 8. (2) where an action is directed from a higher to a lower point :—Ufone sceal ðæt heáfod gíman ðæt ða fēt ne āslíden *caput debet ex alto providere, ne pedes torpeant,* Past. 18; Swt. 131, 25. Clypigende ufenne, Homl. Skt. i. 9, 25. I a. *from above, from heaven* :—Ic eom ufane *ego de supernis sum,* Jn. Skt. 8, 23. Ōð ge sȳn ufene (*ex alto*) gescrȳdde, Lk. Skt. 24, 49. II. marking position, *above* :—Ne hire on nánre ne mót neár ðonne on ōðre stōwe gestæppan, strĩceþ ymbūtan ufane and neoðane efenneáh gehwǽðer, Met. 20, 141. [*O. Sax.* obana *from above : O. H. Ger.* obana *desuper,* superne, super, supra.] v. ufan.

ufan-weard ; *adj.* The word may be translated by *top of* (the noun with which it agrees) :—Sōna wǽron wit on his heánesse on ðam wealle ufanweardum *statim fuimus in summitate ejus* (i. e. muri), Bd. 5, 12; S. 629, 18. On ðysse dūne ufanweardre *in hujus* (*montis*) *vertice,* 1, 7; S. 478, 25. Ðā gesáwon wē westan ðone leóman sunnan and se leóma gehrān ðǽm treówum ufonweardum *videmus ab occidente jubare fulgentibus Phebi radiisque percussa arborum cacumina,* Nar. 28, 25. [*Icel.* ofan-verðr. Cf. *O. Sax.* oðan-wardan ; *adv.*] v. ufe-weard.

ufemest. v. ufera *and* ufor.

ufenan. I. *adv. From above* :—Se ðe ufenan com *qui desursum uenit,* Jn. Skt. 3, 31. Hē dēd ðæt fȳr cymð ufenon, Wulfst. 97, 21 note. Seó landfyrd com ufenon and trymedon hig be ðam strande, Chr. 1052; Erl. 184, 24. II. *prep. with acc. Above, besides* :—Ufenan eall ðis *insuper,* Dom. L. 10, 144 : 18, 271. Ufenon eal ðis, 14, 212. Hē ða bóc hire tō lēt . . . , and ufenan ðæt hire āð sealde, Chart. Th. 203, 1. ¶ þǽr on ufenan *thereupon* :—Hē cwæð, ðæt ān culfre him fluge wið ðæt heáfod, swilce heó ðǽr on ufenan settan wolde, Homl. Ass. 198, 112. [þe munt þe Vther wes ufenan (þar Vther lay ouenan, 2nd MS.), Laym. 18337. Swa deð ǽlc witer mon þa neode cumeð uuenan, 28501. He smat hine uuenen (ouenan, 2nd MS.), þat hæued, 18090. þe eotend smat þer an ouenan (ouenan, 2nd MS.), 26051. He smat in enne stane þer Locrin stod vuenan, 2314. *O. H. Ger.* obenan.]

ufene. v. ufane.

ufer(r)a ; *cpve. :* ufemest ; *spve.* I. *local, upper, higher ; upmost, highest* :—Ufre scrūd *an upper garment ;* ependeton, Wrt. Voc. i. 59, 52. Ðæt uferre hrif, Lchdm. ii. 224, 8. On ðam uferan dǽle ðæs heáfdes *in superiore parte capitis,* Bd. 5, 2; S. 614, 45. Ðone wísdóm ðara uferrena gásta *supernorum spirituum scientiam,* Past. 3; Swt. 32, 13. Ðam uferum (uferrum, Ps. Surt.) *superioribus,* Ps. Spl. 103, 14. Ða uferan (uferran, Ps. Surt. 103, 3) *superiora,* Blickl. Gl. On ðam midne ðæge bið seó sunne on ðam ufemestum ryne stígende, Homl. Th. ii. 76, 18. II. *temporal, later, after.* v. uferian, II, ufeweard, II :—Oððe eft uferran dōgore oððe ðonne *either at a later day or at the time,* Past. 38; Swt. 281, 12. Hē ðe teóþan dæge him ðone Hálgan Gást onsende . . . on ðǽs hálgan tíde ðe nū ðȳs uferan Sunnandæge bið *he sent them the Holy Ghost on the tenth day . . . at the holy time which will be on the Sunday after next,* Blickl. Homl. 119, 12. Ðȳ lǽs hit monn uferan dōgore wrǽce *ne quis eum umquam ulcisci meditaretur,* Ors. 4, 5; Swt. 168, 6. Gif eówre bearn eów befrĩnaþ eft on uferum dagum *quando interrogaverint vos filii vestri cras,* Jos. 4, 6. On uferan

dagum, Wulfst. 88, 20 : Chart. Th. 356, 7. Uferan dōgrum, Beo. Th. 4407; B. 2200 : 4773; B. 2392. On uferum tídum, Lchdm. iii. 438, 15. On uferan tídan, Wulfst. 89, 1. [þe ufere (ouere, 2nd MS.) hond habben of þan kinge, Laym. 1520. An uuere daʒe (þar after, 2nd MS.), 27794. þe huuemeste bou, O. E. Homl. ii. 219, 15. *O. H. Ger.* obero, oberoro : *Icel.* efri *upper ;* of time, *later.*] v. yfemest.

uferian ; *p.* ode. I. *to elevate, make higher* :—Ufered (uffred, MS.) *sublimatus,* Hpt. Gl. 473, 42. II. *to make later, to delay.* v. ufera, II :—Mín hláfurd uferaþ hys cyme *moram facit dominus meus venire,* Mt. Kmbl. 24, 48 : Lk. Skt. 12, 45. [*O. H. Ger.* obarōn *differre.*] v. ge-uferian.

ufe-weard, uf-weard ; *adj.* I. *local, upper ;* generally may be translated by *upper part of* (the noun which it qualifies) ; used substantively, *upper part* :—Is se hals grēne nioþoweard and ufeweard *the lower and upper parts of the neck are green,* Exon. Th. 218, 23; Ph. 299. Ufeweard swer *epistilia,* Wrt. Voc. ii. 30, 29. Ufeweard eáre *pinnula,* i. 282, 62. Ufeweard eáre, 43, 15. Ufeweard lippa *labium,* niðera lippe *labrum,* 43, 24-5. Ufeweard exle ðæs æftran dǽles *ola,* 43, 46. Eal ufeweard nosu *columna,* 43, 18. Up tō ānre dūne tō ufeweardum ðam cnolle *ad verticem montis,* Jud. 16, 3. Ða eágan bióð on ðam líchoman foreweardum and ufeweardum *oculi sunt in ipsa honoris summi facie positi,* Past. 1; Swt. 29, 13. Ðreó stōdon æt ufeweardum ðǽm mūðan, Chr. 897; Erl. 95, 23. Fram his hnolle ufewerdan, Homl. Th. ii. 452, 26. Hē geseah Drihten on ufeweardre ðære hlǽdre *vidit Dominum innixum scalae,* Gen. 28, 13. Seó stōw is on Oliuetes dūne ufeweardre, Blickl. Homl. 125, 19. Fram ufeweardon ōð nyþeweard *a summo usque deorsum,* Mt. Kmbl. 27, 51 : Mk. Skt. 15, 38. On ufeweardum *at the top,* Gen. 6, 16. On ufeweardan, Homl. Th. i. 536, 9. Fram ufeweardan ōð neoþeweardan, ii. 496, 26. Būtan ðam heáfde ufweardum *except the upper part of the head,* Homl. Skt. i. 18, 353. Of ufeweardum bergum *de vertice,* Wrt. Voc. ii. 27, 74. Se mōna gehrān mid his scíman ðǽm triówum ufeweardum, Nar. 30, 8. II. *temporal, later, latter part* of a time. v. ufera, II :—On foreweardne sumor and eft on ufeweardne hærfest *in the early part of the summer and again in the latter part of autumn,* Chr. 913; Th. i. 186, col. 2.

ufon. v. ufan.

ufor ; *cpve. :* ufemest ; *spve. adv. Higher ; highest* :—Ufor *superius,* ufemest *supreme,* Ælfc. Gr. 38; Zup. 240, 10. I. *local,* (a) *at or to a greater height* :—Seó sunne stígþ ufor and ufor, Bt. 25; Fox 88, 27. Saturnus wandraþ ofer ōþrum steorrum ufor ðonne ǽnig ōþer tungol, 36, 2; Fox 174, 14. Seó sunne is micle ufor (furþor, MS. R.) ðonne se nōna, Lchdm. iii. 242, 11. Naþor ne ufor (ufror, MS. R.) ne nyðor, 254, 17 : 266, 18. Swā hī ufor fērdon, Homl. Th. ii. 548, 15. Gif se ord sié ufor ðonne hindeweard sceaft, L. Alf. pol. 36; Th. i. 84, 17. Se earn flȳhð ealra fugela ufemest, Homl. Skt. i. 15, 198. (b) where distance rather than height is marked (cf. to go *up* country), *farther from a coast, from a spot* :—Hēr fōr se here ufor on Froncland *in this year the Danes made their way further inland in France,* Chr. 881; Erl. 82, 4 : Beo. Th. 5895; B. 2951. Ðā hēt ic hī hwæthwega ufor gān *I bade them retire somewhat from the spot,* Homl. Th. ii. 32, 22 : i. 70, 35. (c) *of position, higher, at or to a more honourable place* :—Lā freónd site ufur *amice ascende superius,* Lk. Skt. 14, 10. II. *metaphorical, higher* :—Hē bið suā micle sēl gehíered suā hē ufor gestent on his lífes geearnungum, Past. 14; Swt. 81, 17. Nabbaþ hī nān gód ofer ðæt tō sēcanne, ne hī nānwuht ne magon ne ufor ne ūtor findan, Bt. 34, 12; Fox 154, 16 : Exon. Th. 427, 8; Rä. 41, 88. III. *temporal, later.* v. ufera, II :—Fíf nihtum ufor *five days later,* Menol. Fox 355; Men. 179 : 68; Men. 34. Ðonne ymb .iii. niht gesēcæn hiom sǽmend, būton ðam ufor leófre sió ðe ða tihtlan āge *then after three days let them seek themselves an arbitrator, unless the prosecutor would liever do it later,* L. H. E. 10; Th. i. 30, 19. [All þiss icc seʒʒde her uferr mar, Orm. 1715. *Icel.* ofarr *higher up ;* of time, *later.*]

uf-weard, uht. v. ufe-weard, wiht.

úht, es; *m. The time just before daybreak* :—Ðā hit wæs foran tō úhtes *antelucanum demum tempus,* Nar. 15, 31. Gang eft tō ðonne dæg and niht furþum scáde on ðam ilcan úhte, Lchdm. ii. 346, 14. [*O. Sax.* adro an úhta *primo mane* (Mt. 20, 1).] v. next word.

úhta, an ; *m.* I. *the last part of the night, the time just before daybreak* :—Þis wæs on úhtan eall geworden ǽr dægrēde, Cd. Th. 294, 2; Sat. 465. On úhtan mid ǽrdæge, Beo. Th. 252; B. 126 : Andr. Kmbl. 469; An. 235 : 2775; An. 1390 : Elen. Kmbl. 209; El. 105. Syle drincan ǽr úhton, Lchdm. iii. 20, 2. Cymð on úhtan eásterne wind, Cd. Th. 20, 26; Gen. 315 : 289, 31; Sat. 406 : Exon. Th. 443, 24; Kl. 35. On úhtan *very early in the morning ;* ualde mane (Mk. 16, 2), ualde diluculo (Lk. 24, 1), Exon. Th. 459, 17; Hö. 1 : 460, 14; Hö. 17. Úhtna gehwylce, 287, 3; Wand. 8 : 471, 24; Rä. 61, 6. II. *as an ecclesiastical term, the time at which the earliest of the seven canonical services was held, the time of nocturns* :—De nocturna celebratione. On úhtan wē sculon God herian, ealswā Dauid cwæð : 'Media nocte surgebam ad confitendum tibi,' Btwk. 220, 17. Hit gedafenaþ ðæt gehwylce cristene men on Sæternesdæg cume tō cyrcean, and ðǽr

æfensang gehýran, and on ûhtan ðone ûhtsang, L. E. G. 24 ; Th. ii. 420, 35. Eallum cristenum mannum is beboden ðæt hí ealne heora líchoman seofon síþum gebletsian mid Cristes róde tácne, ǽrest on ǽrnemorgen . . . seofoþan síþe on ûhtan, Blickl. Homl. 47, 19. [Godess enngell comm himm to onn uhhtenn þær he sleppte, Orm. 2484. Hi slo3en and fu3ten þe ni3t and þe u3ten, Horn 1376. Ruddon of þe dayrawe ros upon u3ten, when merk of þe mydny3t mo3t no more last, Allit. Pms. 64, 893. *Goth.* air ûhtwôn πρωὶ ἔννυχα λίαν, Mk. 1, 35. *O. Sax.* adro an ûhton *primo mane*, Mt. 20, 1. *O. H. Ger.* uohta *diluculum : Icel.* ótta *the last part of the night.*] v. mæsse-, sunnan-ûhta ; ûht, *and following words.*

ûhtan-tíd, e ; *f. The time of early morning :*—Ûhtan(-en, MS.)-tíd *matutinum*, Wrt. Voc. i. 53, 7. On ûhtan-tíde *matutinis horis*, Bd. 4, 12 ; S. 581, 14. On ûhtu-tíd (ûhte-, Lind.) *galli cantu*, Mk. Skt. Rush. 13. 35. [Crist ras onn uhhtenntid, Orm. 5832. *Icel.* óttu-tíðir *matins.*] v. ûht-tíd.

ûhtan-tíma, an ; *m. The time of nocturns :*—On ûhtan-tíman, Btwk. 194, 14.

ûht-cearu, e ; *f. Care that comes in the early morning*, Exon. Th. 442, 4 ; Kl. 7.

ûht-floga, an ; *m. A creature that flies in the early morning :*—Ðæs wyrmes denn, ealdes ûhtflogan, Beo. Th. 5513 ; B. 2760.

ûht-gebed, es ; *n. A prayer repeated in the early morning, matins :*—Ûhtgebed *vel* þénung *matutinum officium*, Wrt. Voc. i. 28, 29. Se eádiga wer his ûhtgebedum befeal, Guthl. 6 ; Gdwin. 42, 12.

ûht-hlem[m], es ; *m. A din made in the early morning* (the noisy conflict of Beowulf and Grendel ; cf. dryhtsele dynede . . . reced hlynsode . . . swêg up âstâg, 1540–1569 ; B. 767–782), Beo. Th. 4019 ; B. 2007.

ûht-líc ; *adj.* I. *of early morning ;* matutinus :—Tó ðæm ûhtlícum *ad matutinum*, Ps. Spl. T. 29, 6. On ûhtlícum *in matutino*, 100, 9. II. *of matins :*—Fram ðære tíde ðæs ûhtlícan lofsanges *a tempore matutinae laudis*, Bd. 4, 12 ; S. 537, 23 : 4, 7 ; S. 575, 2. Tó ûhtlícum lofsangum *ad matutinales laudes*, Anglia xiii. 382, 243.

ûht-sang, es ; *m. One of the services of the church, nocturns* or *matins :*—Hû fela sealma on nihtlícum tídum tó singenne synt. On wintres tíman is se ûhtsang þus tó beginnenne . . . *quanti psalmi dicendi sunt nocturnis horis. Hiemis tempore premisso in primis versu . . .*, R. Ben. 33, 7. Seofon tídsangas hí gesetton . . . Se ôðer is ûhtsang mid ðam æftersange ðe ðærtó gebiraþ, L. Ælfc. P. 31 ; Th. ii. 376, 5 : L. Ælfc. C. 19 ; Th. ii. 350, 6. Of ðære tíde ûhtsanges *ex tempore matutinae synaxeos*, Bd. 4, 19 ; S. 588, 13 : Shrn. 94, 32. Hwâ âwecþ ðé tó ûhtsancge (*ad nocturnos*)? Coll. Monast. Th. 35, 27. Tó ûhtsange, tó æftersange *ad nocturnam, ad matutinam*, Anglia xiii. 396, 449 : 401, 523. Ðæt ðære nihte tó láfe sié æfter ðam ûhtsange *quod restat post vigilias*, R. Ben. 32, 17. From Eástron oð ða kalendas Nouembris sý se ǽrest ðæs ûhtsanges swâ gemetegad, ðæt lýtel fæc gehealden sý betwyh ðæm ûhtsange and ðæm dægrédsange, and upâsprungenum dægrédsang sý begunnen *a Pasca usque ad kalendas Novembris sic temperetur hora vigiliarum agenda, ut parvissimo intervallo custodito mox matutini, qui incipiente luce agendi sunt, subsequantur*, 32, 19–33, 1 : 34, 7. Ða þénunga ðe wé habbaþ on Godes þeówdôme tó mæssan, and tó ûhtsange, and tó eallum tídsangum, L. Ælfc. P. 30 ; Th. ii. 374, 34. Ic sang ûhtsang *cantavi nocturnam*, Coll. Monast. Th. 33, 25 : Anglia xiii. 380, 220. Ûhtsang singan *nocturnas laudes dicere*, Bd. 4, 24 ; S. 599, 4. On ûhtan ðone ûhtsang gehýran, L. E. I. 24 ; Th. ii. 420, 35. Tíd ûhtsanga *hora vigiliarum*, R. Ben. Interl. 37, 12. Be nihtlícum ûhtsangum *de nocturnis vigiliis*, 46, 9. [Ure Leafdi uhtsong siggeð oþisse wise, A. R. 18, 19. Uhtsong bi nihte ine winter, ine sumer iþe dawunge, 20, 19. Da33sang and uhhtennsang, Orm. 6360. *O. H. Ger.* ûhti-sang *orgia : Icel.* óttu-söngr *matins.*] v. next word.

ûhtsang-líc ; *adj. Of nocturns :*—Ûhtsanglíc lof *nocturna laus*, Anglia xiii. 436, 1014.

ûht-sceaþa, an ; *m. One who robs in the night* or *early morning :*—Eald ûhtsceaða . . . nacod níððraca nihtes fleógeþ, Beo. Th. 4534 ; B. 2271.

ûht-þegnung, e ; *f.* v. ûht-gebed.

ûht-tíd, e ; *f. Early morning time, the time before daybreak :*—Ûhttíd *vel* beforan dæge *matutinum*, Wrt. Voc. ii. 58, 64. On ûhttíde *in matutino*, Ps. Spl. 48, 15. Moyses bebeád eorlas on ûhttíd folc somnigean, frecan árísan, Cd. Th. 191, 17 ; Exod. 216. v. ûhtan-tíd.

ûht-wæcce, an ; *f. A nightly vigil :*—Be ðám ûhtwæccum *de nocturnis vigiliis*, R. Ben. 40, 10.

ûle, an ; *f. An owl :*—Úlae *cavanni*, Txts. 47, 378 : *ulula*, 107, 2150. Úle *noctua, ulula*, 81, 1382 : *ulula*, Wrt. Voc. i. 281, 7 : ii. 62, 35 : *noctua*, i. 281, 6 : ii. 60, 35 : *noctua vel strinx*, i. 77, 41 : *strix vel cavanna* vel *noctua vel ulula*, 29, 11. Ne ete gê úlan (*noctuam*), Lev. 11, 16. Úlena (*cavannarum*, Wrt. Voc. ii. 87, 69 : 19, 20 : Hpt. Gl. 526, 62. [*O. H. Ger.* ûwila, ûla *noctua, ulula, bubo ; Icel.* ugla.]

ulm-treów, es ; *n. An elm-tree :*—Ulmtreów *ulmus*, Wrt. Voc. i. 32, 63. [*Wiclif uses* ulmtree, Is. 41, 19.]

uma, huma, an ; *m.* I. *a weaver's beam :*—Uma *scapus*, Wrt. Voc. i. 66, 25 : *scafus*, 282, 8 (in each case the word occurs in a list of terms connected with weaving). Huma *scafus*, Corpus Gl. ed. Hessels 106, 206 (the word is omitted by Wright in Voc. ii. 120, 26, and in Txts. 97, 1832). II. *the name of some plant :*—Genim uman, and medmicelne bollan fulne ealað ; bewyl ðæt ealo on ðære wyrte, Lchdm. ii. 276, 12.

umbor, es ; *n. A child :*—Hwæt wit tô willan umbor wesendum ǽr ârna gefremedon, Beo. Th. 2378 ; B. 1187. Ða ðe hine forð onsendon ænne ofer ýðe umbor wesende, 92 ; B. 46. Meotud âna wât hwǽr se cwealm cymeþ ðe heonan of cýþþe gewíteþ umbor ýceþ ðá ǽr âdl nimeþ dý weorþeþ on foldan swâ fela fíra cynnes *the Lord only knows what becomes of the pestilence that departs away from the land. He increases the children then, before disease carries them off* (a great many children are born before pestilence returns to a country), *so it happens that there are so many of mankind on earth*, Exon. Th. 335, 9 ; Gn. Ex. 31. Cf. cniht-wesende.

un-. *The prefix* (1) *expresses negation ;* (2) *gives a bad sense, as in* un-dǽd, un-dôm, un-lagu, un-lǽce ; (3) *reverses an action, as in* un-bindan, un-dôn ; (4) *is intensive* (?) v. un-hár.

un-âbeden ; *adj. Unbidden, unasked :*—God beád mancynne ðæt hí hine biddan sceoldon, and hê wile syllan unâbeden ðæt ðæt wê ûs ne wêndon þurh ûre bêne, Homl. Th. ii. 372, 16 : L. O. 8 ; Th. i. 180, 28.

un-âberendlíc ; *adj. Unbearable, intolerable :*—Unâberendlíc gyhða, Homl. Th. i. 86, 11. On ðam ne eardaþ nán eorðlic mann for ðam unâberendlícum (unâberiendan, MS. R.) bryne, Lchdm. iii. 260, 23 note. Unâberendlíce *intolerabilem*, Ps. Lamb. 123, 5. Unâberendlíce broc, Bt. 39, 10 ; Fox 228, 4. Dyslícu gehât and unâberendlícu *vota stulta et intoleranda*, L. Ecg. C. 19 ; Th. ii. 146, 33.

un-âberendlíce ; *adv. Unbearably, intolerably :*—Ðæt hê ðæt ryht tô suíðe and tô ungemetlíce and tô unâberendlíce ne bodige *ne recta nimie et inordinate proferantur*, Past. 15 ; Swt. 95, 18. Ðæt hí tô unâberendlíce ne beóden *ne plus justo jubeant*, 28 ; Swt. 189, 19. Wê wǽron unâberendlíce fornumene, Homl. Th. ii. 416, 12.

un-âberiende. v. un-âberendlíc.

un-âbígendlíc ; *adj. Inflexible :*—Unabêgendlícre *inflexibili*, Wrt. Voc. ii. 45, 23.

un-âbindendlíc ; *adj. That cannot be unbound, indissoluble :*—Hine gebindaþ ða wôn wilnunga mid heora unâbindendlícum racentum *quem vitiosae libidines insolubilibus adstrictum retinent catenis*, Bt. 16, 3 ; Fox 56, 18 : 33, 4 ; Fox 130, 31. v. un-onbindendlíc.

un-âblinn, es ; *n. A not ceasing :*—Ic geseó ða mânfullan smeáunge ðínre heortan ; manna kynnes costere hafaþ âcenned on ðé ða unablinnu (-blinnunge ?) ðæs yfelan geþôhtes *I see the wicked device of thy heart ; the tempter of the race of men hath begotten in thee those incessant recurrences of* (that never ceasing from) *that evil thought*, Guthl. 7 ; Gdwin. 46, 10.

un-âblinnende ; *adj.* (*ptcpl.*) *Unceasing :*—Mid unâblinnendre stemne *incessabili voce*, Hymn. ad Mat. 4.

un-âblinnendlíc ; *adj. Unceasing, incessant :*—Ðǽr wæs unâblinnendlíc staþolfæstnys Godes herunge æghwylcne dæg and eác nihtes, Homl. Skt. ii. 236, 86.

un-âblinnendlíce ; *adv. Unceasingly, incessantly, without ceasing, without intermission :*—Unâblinnendlíce *incessanter*, Rtl. 3, 17 : 23, 32 : *indesinenter*, Past. 13 ; Swt. 77, 20. Heó ðurh syx singal geár ðære ylcan hefignesse âdle unâblinnendlíce won *per sex continuos annos eadem molestia laborare non cessabat*, Bd. 4, 23 ; S. 595, 18 : Homl. Skt. i. 19, 9 : ii. 23 b, 328, 155. Hí gebiddaþ unâblinnendlíce, i. 5, 446 : Blickl. Homl. 123, 16. Gnættas ǽgþær ge ða men ge ða nýtenu unâblinnendlíce þíniende wǽron *ciniphes nusquam evitabiles*, Ors. 1, 7 ; Swt. 36, 31.

un-âbreeendlíc ; *adj. Inextricable :*—Ða unâbrecendlícan *inextricabilem*, Wrt. Voc. ii. 43, 69.

un-âcenned ; *adj. Unbegotten :*—Ðú eart unâcenned Fæder, hê is Sunu of ðé ǽfre âcenned, Homl. Th. i. 464, 34.

un-âcnycendlíc ; *adj. Not to be knocked up* (of bonds), *indissoluble :*—Unâcnycendlícre sibbes bende *insolubili pacis vinculo*, Rtl. 108, 21.

un-âcumendlíc ; *adj. Unbearable, intolerable, impossible to be borne, excessive :*—Be unâcumenlícra (-endlíca, 8, 24) ðinga gebode. Gif hwylcum brêþer hwæt hefelíces and unacumenlíces beboden sý *si fratri impossibilia jubentur. Si cui fratri aliqua gravia aut impossibilia* (unâcumendalíce, R. Ben. Interl. 114, 5) *injunguntur*, R. Ben. 128, 9–11. Ða wiðercoran unâcumendlíce hǽtu þrowiaþ, Homl. Th. i. 532, 1. Unâcumenlícum *inextricabili, infatigabili, inextinguibile*, Hpt. Gl. 497, 68. Unâcumendlícum hagelum *inexhaustis* (*inconsummatis, investigabilibus*) *imbribus*, 414, 63.

un-âcumendlícness, e ; *f. Unbearableness, impossibility to be borne :*—Unâcumenlícnesse *impossibilitatis*, R. Ben. Interl. 114, 9.

un-âcwencedlíc ; *adj. Unquenchable, inextinguishable :*—On helle unâcwencedlíces fýres *in gehennam ignis inextinguibilis*, Mk. Skt. 9, 45. On unâcwencedlícum fýre, Lk. Skt. 3, 17. On unâcwencedlíc fýr, Mk. Skt. 9, 43.

un-ádrugod; *adj. Undried* :—On nîwne weall unádrugodne and unástîdodne, Past. 49; Swt. 383, 32.

un-ádrysendlîc; *adj. Unquenchable, inextinguishable* :—Unádrysendlîc *inextinguibilis*, Mk. Skt. Lind. Rush. 9, 43: Rush. 9, 45. v. next word.

un-ádrysnende; *adj. Not to be quenched* or *extinguished* :—Mid dæccille his unádrysnendre *cum lampade sua inextinguibili*, Rtl. 106, 10.

un-ádrysnendlîc; *adj. Unquenchable, inextinguishable* :—Unádrysnendlîc *inextinguibilis*, Mk. Skt. Lind. 9, 45: Lk. Skt. Lind. 3, 17 (unádryssenlîc, Rush.). v. preceding word.

un-ádwæsced; *adj. Unquenched, unextinguished, never extinguished* :—Þár bið unádwæsced fýr *there shall be fire that is not quenched*, Homl. Ass. 168, 115, 129.

un-ádwæscedlic; *adj. That is not quenched* or *extinguished* :—On dæt unádwescedlîce fýr *into the fire that is not quenched*, L. E. I. prm. ; Th. ii. 394, 17. Of dæm unádwæscedlîcum lîgum, Th. ii. 396, 5.

un-ádwæscendlîc; *adj. Unquenchable, inextinguishable* :—On middel dæs unádwæscendlîcan lîges *in medium flammarum inextinguibilium*, Bd. 5, 12; S. 628, 2. On unádwæscendlîcum fýre, Mt. Kmbl. 3, 12: Homl. Th. i. 526, 22: Homl. Skt. i. 17, 32.

un-æmetta, -æmta, an; *m. Want of leisure for doing something, occupation, business* :—Unémetta *negotia* (unemotan *negotio*, Ep. Gl. 680), Txts. 81, 1371. Gif hit sié se dæt land hæbbe dæt hê dis forgýmeleásie bûton hit hæres unæmetta sié, donne . . . *if it happen that he who has the land neglect this arrangement, unless occupation in connection with the Danes be the cause of the neglect, then* . . . , Chart. Th. 159, 7. Gif man hwylc metrum cild tô mæsspreóste bringe, donne fullie hê hit sôna, and for nænigum unæmtan ne forlæte [dæt] hê hit ne fullie *if any sick child be brought to a priest, let him baptize it at once, and do not let him be prevented by any occupation from baptizing it*, L. E. I. 17; Th. ii. 412, 22. Gif hwâ mid hwylcum unæmtan genýd sý, dæt hê tô dære mæssan cuman ne mæge, 39; Th. ii. 438, 1. Misenlîce intingan and unæmtan oft gelimpaþ *diversae causae impediunt*, Bd. 4, 5; S. 573, 7.

un-æmtig(i)an *to prevent a person being at leisure, to deprive of leisure* :—Ne lyste dê wîfes de dê on nânum þincgum ne ábysige ne dê ne unæmtige tô dînum wyllan *do not you desire a wife that may worry you in nothing, and may not prevent you being at leisure to follow your own will*, Shrn. 183, 12.

un-æt, es; *m. Excessive eating, revelling* :—Hê begæd unætas and oferdrincas and gálscipe *comessationibus vacat et luxuriae atque conviviis*, Deut. 21, 20.

un-æþelboren; *adj. Not nobly born, not of noble birth* :—Gif se æþelborena done unæþelborenan oferþýhd, sý hê gemedemad furdur be his geearnungum donne se unæþelborena, R. Ben. 12, 15–17. Ealle cristene men, ægder ge rîce ge heáne, ge ædelborene ge unædelborene, and se blâford and se deówa, ealle hî sind gebrôdra, Homl. Th. i, 260, 20. [Æþelboren [nobi]lis (v. 85, 60), unæþelboren *ignobilis*, Wrt. Voc. i. 95, 16.]

un-æþele; *adj.* I. of persons, *not noble*, (1) as regards birth :—Gedence hê simle, sié suá ædele suá unædele, Past. 14; Swt. 85, 15. Hwî ofermódige gê ofer ôþre men for eówrum gebyrdum, nû gê nánne ne magon métan unæþele, ac ealle sint emnædele, gif gê willaþ done fruman sceaft geþencan, Bt. 30, 2; Fox 110, 16: Met. 17, 17. Sume beóþ swîde æþele and wîdcûþe on heora gebyrdum, ac hî beóþ mid wædle ofþrycte, dæt him wære leófre dæt hî wæran unæþele donne swâ earme, Bt. 11, 1; Fox 32, 1. Æþele and unæþele *nobiles, ignobiles*, Bd. 5, 7; S. 621, 14. Ge æþele ge unæþele *tam nobiles quam privati*, 5, 23; S. 647, 7. Frige and þeówe, ædele and unædele, Ap. Th. 12, 20. Unaedilra (-sa, MS., aedilra, Ep. Erf.) *gregariorum*, Txts. 67, 993. Unæþelra, Wrt. Voc. ii. 41, 7. Leófre mê is dæt hê (*the king*) mê tô deápe gesylle donne unæþelra (*ignobilior*) man, Bd. 2, 12; S. 513, 27. (2) as regards character :—Wæs se cyning æþelre gebyrde deáh de hê on dæde unæþele wære *erat rex natu nobilis quamlibet actu ignobilis*, Bd. 2, 15; S. 518, 37. Ælc man de allunga underþeóded biþ unþeáwum . . . wyrþ anæþelad ôþ dæt hê wyrþ unæþele (*degener*), Bt. 30, 2; Fox 110, 22: Met. 17, 28. II. of things, *ignoble, mean, infamous* :—Unæþelre ádle. *degeneri languore*, Wrt. Voc. ii. 138, 33. Mid dý unæþelan gidde *cum infami eulogio*, 87, 40: 19, 16: 137, 42.

un-æþellîce; *adv. Ignobly* :—Ic cûþe sumne brôþor . . . wæs hê geseted on æþelum mynstre, ac hê unæþelîce his lîf lifede *novi fratrem . . . positum in monasterio nobili, sed ipsum ignobiliter viventem*, Bd. 5, 14; S. 634, 9.

un-æþelness, e; *f. Ignobility*; *ignobilitas, infamia*, Dial. 2, 23.

un-ætspornen; *adj. That is not hindered* :—Unætspornenum fôtum *inoffensis pedibus*, Dial. 1, 9.

un-æwisc; *adj. Modest, bashful* :—Unêwisc *pudicus*, Wrt. Voc. i. 65, 1.

un-áfandod; *adj. Untried, unproved* :—Ic wolde witan hû dû dæt ongytan woldest, hweder de dû woldest unáfand(o)des geleáfan de áfandud witan, Shrn. 181, 2.

un-áfeohtendlîc; *adj. Not to be overcome* :—Unáf(e)ohtendlîc *ineluctabilis*, Wrt. Voc. ii. 48, 52. Unáfæhtendlîc *inexpugnabile*, Rtl. 92, 18.

un-áfunden; *adj.* I. *not found out, undiscovered* :—Hê hæfde ænne lícdrowere belocen on ânum clyfan, and hine dær áfedde unáfunden ôd dæt, Homl. Skt. i. 3, 482. II. *not tried* :—Unáfundenum *inexperto*, Wrt. Voc. ii. 44, 28.

un-áfýled; *adj. Undefiled* :—Unáfîlede wegas *inpolluta via*, Ps. Lamb. 17, 31.

un-áfylledlîc; *adj. Insatiate* :—Hî (*hell and avarice*) habbaþ unáfylledlîce grædignysse dæt hî fulle ne beód næfre, Homl. Skt. i. 16, 285.

un-áfyllendlîc; *adj. Insatiable* :—Ic hæfde unáfyllendlîce gewilnunga, Homl. Skt. ii. 23 b, 341. [Unafillendliche gredinesse, O. E. Homl. i. 103, 17.]

un-áfyllendlîce; *adv. Insatiably*, Homl. Skt. ii. 23 b, 329.

un-ága, an; *m. One without possessions, a poor person* :—Hê of cordan mæg done unágan weccan *suscitans a terra inopem*, Ps. Th. 112, 6.

un-ágæledlîce; *adv. Unremittedly* :—Hié sceoldan mancynne bodian; swâ dæt cûþ gewearþ dæt hié dæt scoþþan dydon unágæledlîce, Blickl. Homl. 121, 5.

un-ágán; *adj. Not lapsed, with the time of its lease not run out* :—Ego Ealdulf . . . quandam ruris particulam . . . cuidam militi nomine Leofenad . . . largitus sum . . . et post uitae suae terminum duobus tantum haeredibus immunem derelinquat; quibus defunctis, aecclesiae . . . restituatur. Ðis is seó gerædnes de Ealdulf hæfd gerád tô setnesse. da hwîle dis land unágán sê *as long as the lease of the land runs*, Cod. Dip. Kmbl. iii. 295, 22–33. Nû gewrîte ic Cyneswîde dæt dreóra hîda lond on dreóra monna daeg . . . and êc ic hire lête tô . . . dæt twêga hîda lond . . . da hwîle hit unágaen seó, ond Cyneswît hit tô nængum ôdrum men ne lête da hwîle hit unágaen sê, bûtun tô hire bearna sumum . . . Ond ic biddu dæt dis dreóra hîda lond and êc dæt twêga, donne hit ágáen seó (*when its lease has run out*), dæt hit sê ágefen intô Clife; and êc ic and all hîgen hâlsigaþ ûsse æfterfylgend, dæt heora nænig dæt gefe gewonige, ær hit swâ ágáen sî, swâ hit on dissum gewrite stondeþ, ii. 100, 12–29.

un-ágen; *adj. Not one's own, not in a person's possession* or *under his control* :—Gehiéren da eádmôdan hû êce dæt is dæt hié wilniaþ, and hû gewîtende and hû unágen dæt is dæt hié onscuniaþ *audiant humiles, quam sint aeterna, quae appetunt, quam transitoria, quae contemnunt*, Past. 41; Swt. 299, 9. Gif man widuwan unágne (*a widow of whom he is not the guardian*) genimeþ, L. Ethb. 76; Th. i. 20, 13, and see note. Hié sculon suâ micle êstelîcor dælan suâ hié ongietaþ dæt him lænre and unágenre bið dæt hié dær dælaþ *tanto humiliter praebeant, quanto aliena esse intelligunt, quae dispensant*, Past. 44; Swt. 321, 9.

un-ágifen; *adj. Not given up, not repaid* :—Nolde Sigelm tô wigge faran mid nânes mannes scette unágifnum *Sigelm would not go to battle with the money owing to any man unpaid*, Chart. Th. 201, 24.

un-águnnen; *adj. Not begun, without a beginning* :—On unágunnenre Godcundnysse and on ongunnenre menniscnysse of *Divinity without beginning and of humanity with beginning*, Homl. Th. iii. 292, 16.

un-álîefed; *adj. Unallowed, illicit, unlawful* :—Be unáliéfedes mæstennes onfenge. Gif mon on his mæstene unáliéfed swîn gemête, L. In. 39; Th. i. 132, 11. Unálýfedre willnunge *inlicitae concupiscentiae*, Bd. 1, 27; S. 495, 9. Ne sceal hê náht unáliéfedes dôn, ac dæt dætte ôdre menn unáliéfedes dôt hê sceal wêpan *qui nulla illicita perpetrat, sed perpetrata ab aliis deplorat*, Past. 10; Swt. 61, 14. Eall dæt hié unáliéfedes denceaþ, 21; Swt. 155, 12. Unálýfedne gesynscipe *inlicitum conjugium*, Bd. 3, 22; S. 553, 25. Ðone unálýfedan bryne mînra leahtra, Homl. Skt. ii. 23 b, 331. Wiþsacaþ dám unálýfdum gestreónum, Blickl. Homl. 53, 23. Unáliéfede geþôhtas *cogitationes illicitas*, Past. 13: Swt. 77, 21.

un-álîfedlîc; *adj. Not allowable, unlawful, illicit* :—Unálýfedlîc þing *hoc nefas*, Ælfc. Gr. 9, 25; Zup. 51, 2. Swýnen flæsc Iudéum unálýfedlîc ys tô etanne, Ps. Th. 16, 14: Homl. Th. ii. 456, 35. Hî ne môstan for him náht unálýfedlîces begangan, Shrn. 65, 11. On hûs gehwyrfed unálýfedlîcra scylda *in inlecebrarum cubilia conversae*, Bd. 4, 25; S. 601, 14.

un-álîfedlîce; *adv. Unlawfully* :—Ne sceal mon unálýfedlîce gelustfullian *non concupiscere*, R. Ben. 16, 19. Unálýfedlîce *illicite*, Dial. 2, 2.

un-álîfedness, e; *f. What is not allowed, licence, licentiousness* :—Lîchomlîcre unáléfednesse *corporalis inlecebre*, Wrt. Voc. ii. 135, 83. Úrum lîcumlîcum unálýfednessum deówigende *carnis inlecebris servientes*, Bd. 4, 25; S. 601, 37: 5, 6; S. 618, 39: 5, 14; S. 634, 5. Hî dis lîf forseód and ealles dysses lîfes unálýfednessa, R. Ben. 136, 31.

un-álîfendlîc; *adj. Unallowable, illicit* :—Unálýfendlîcum *illecebrosis, inlicitis*, Hpt. Gl. 505, 42. From unálîfendlîcum *ab inlecebrarum*, Wrt. Voc. ii. 2, 12: 44, 13.

un-álîfendlîce; *adv. Unallowably, unlawfully* :—Unálýfendlîce *inlicite*, Scint. 141, 4.

un-álîsendlîc; *adj. Not to be remitted* or *forgiven, without remission* :—Bið his scyld unálýsendlîc, Homl. Th. i. 500, 18.

un-ámánsumod; *adj. Unexcommunicated, relieved from sentence of excommunication* :—Ðá wæron da ámánsumedan mynecena bebyrigede

... Benedictus hēt mæssian for đām mynecenum ; cwæđ đæt hī siđđan unāmānsumode wæron, Homl. Th. ii. 174, 28.

un-āmelt; *adj. Unmelted :*—Unāmaelte (-melti, Erf.) smeoruue *pice, saevo,* Txts. 87, 1581.

un-āmeten; *adj. Unmeasured.* (1) *not having determined limits :*—Unāmeten is se Fæder, unāmeten is se Sunu, (unāmeten is se Hālga Gāst) ... Ne synt þrȳ unāmetene ... ac is ān unāmeten *immensus Pater, immensus Filius, immensus Spiritus Sanctus ... Non tres immensi ... sed unus immensus,* Ath. Crd. §§ 9, 12. (2) *very great, immense, boundless :*—God unāmetenre ārfæstnysse *Deus immense pietatis,* Anglia xi. 112, 2. Nis ūs nān gemet on đam ǽrran bebode, forđan đe wē sceolon ūrne Scyppend lufian mid unāmetenre lufe, Homl. Th. ii. 314, 12 : Homl. Skt. i. 16, 254.

un-anbundenlīc. v. un-onbundenlīc.

un-andcȳđigness, e ; *f. Ignorance :*—Scyld unondcȳđignesse mīnre *delicta ignorantiae meae,* Ps. Surt. 24, 7. [Cf. *O. H. Ger.* ant-kundig *expertus.*]

un-andergilde; *adj. Not to be paid back* (?), *that may be retained* (?) :—Geđenc nū hwæt đīnes āgnes seó ealra đissa woruldǽhta and welena, odđe hwæt đū đǽron āge unandergildes, gif đū him sceádwíslíce æfter spyrast. Hwæt hæfst đū æt đām gifum đe đū cwist đæt seó wyrd eów gife, and æt đām welum, đeáh hī nū éce wǽron ? *age enim, si jam caduca et momentaria fortunae dona non essent, quod in eis est, quod aut vestrum umquam fieri queat, aut non perspectum consideratumque vilescat,* Bt. 13 ; Fox 38, 1–5. v. andergilde.

un-andet; *adj. Unconfessed :*—Ǽnig man mid unandettan heáfodleahtrum hūsles ne ábyrige, ac andette and bēte, Wulfst. 71, 7.

un-andgitfull; *adj. Not intelligent, without understanding :*—Unondgetfulle *insensati,* Ps. Surt. ii. p. 195, 11. Đæt đa andgytfullan mid worda lāre tō Godes willan gemyngode sȳn and đa unandgytfullan mid gódum dǽdum getrymede *ut capacibus discipulis mandata Domini verbis proponat, et simplicioribus factis suis divina precepta demonstret,* R. Ben. 11, 16. Đǽm unandgytfullum (*infirmis intellectibus*) đæt gāstlíce angyt is earfoþe tō understandenne, 66, 19.

un-andhéfe; *adj. Insupportable :*—Byrþenne hæfige and unandhoife *onera gravia et inportabilia,* Mt. Kmbl. Rush. 23, 4. [Cf. *O. H. Ger.* ant-heffen *sustentare.*]

un-andweard; *adj. Not present :*—Drihten nolde līchamlíce sīđian tō đæs cyninges untruman bearne, ac unandweard mid his worde hine gehǽlde, Homl. Th. i. 128, 17.

un-andwendlíc; *adj. Immovable, unchangeable* ; *immobilis,* Bt. 39, 6 ; Fox 220, 16 note. v. un-āwendendlíc.

un-andwís; *adj. Inexperienced :*—Unandwís *inexpertus,* Wrt. Voc. ii. 84, 13 : 46, 76. Đȳ unandwísan *inexperto* (the passage, however, is *in experto terrore,* Ald. 34), 79, 39.

un-ānrǽdness, e ; *f. Inconstancy :*—Hē (the devil) nǽnige mehte wiđ ūs nafaþ, būton hwylc man þurh đa unānrǽdnesse his mōdes him wiđstandan nelle, Blickl. Homl. 31, 34.

un-anwendenlíc. v. un-onwendenlíc.

un-āpínedlíce; *adv. With impunity :*—Unāpínedlíc(e) *inpune,* Rtl. 113, 36.

un-ār, e ; *f. Dishonour :*—Tō unāre *to the dishonour (of God),* Anglia xi. 98, 45. Hē wēpende mǽnde đa unāre đe him mon būton gewyrhton dyde *deplorans injurias suas,* Ors. 5, 12 ; Swt. 240, 9 : Ps. Th. 68, 7. [*O. L. Ger. O. H. Ger.* un-ēra *dedecus, contumelia, damnum.*]

un-āræfnad; *adj. Intolerable, insupportable :*—Đa unāræfnodan wurþaþ tōbrocenne *quae non toleranda sunt, dirimantur,* L. Ecg. C. 19 ; Th. ii. 146, 4. Þurh đa wǽdlan stōwe wætres and þurh đa unāréfndon lond wildeóra and wyrma *per immania et egentia plerumque aquarum, per aliquot serpentium ferarumque loca,* Nar. 26, 8.

un-āræfnedlíc; *adj. Intolerable, impossible to bear :*—Būton hit unǽrǽfnedlíc sȳ tō ofercumenne đa þing đe ūs synd fram đē forestihtode *unless it be beyond our powers of endurance to overcome the things that are fore-ordained for us by thee,* Homl. Skt. ii. 30, 133. Seó unāræfnedlíce byrþen synna, Blickl. Homl. 75, 9. Se freódōm đæs unāræfnedlícan þeówdōme, 137, 13. Mid unāræfnedlíce þurste geswencte, Nar. 9, 17 : 8, 21. Wæter unāræfnedlíc *aquam intolerabilem,* Ps. Lamb. 123, 5. Þurh đa lond đe đa unāréfnedlícan cyn nædrena in wǽron *in execrabilia serpentum genera,* Nar. 6, 22.

un-āræfnendlíc; *adj. Intolerable :*—Unāræfnendlíc *intolerabile,* Bd. 5, 12 ; S. 627, 38. Unāræfnendlíce fūlnes *foetor incomparabilis,* S. 628, 25. Weter unāréfnendlíc *aquam intolerabilem,* Ps. Surt. 123, 5. Be đām tintregum unāræfnendlícum (*intolerabilibus*), Bd. 5, 12 ; S. 628, 5. Grimme hergunge and unāræfnedlíce *acerbas atque intolerabiles irruptiones,* 3, 24 ; S. 556, 5.

un-āreccendlíc; *adj. Not to be related, indescribable :*—Unāreccendlíc blis *inenarrabile gaudium,* Scint. 26, 15.

un-āreht *undiscussed, not expounded :*—Nū hæbbe wē gereht be welan and be anwealde, and đæt ilce wē magon reccan be đām þrīm đe wē unāreht (unreht, Cott. MS.) habbaþ *similiter ratiocinari de honoribus, gloria, voluptatibus licet,* Bt. 33, 2 ; Fox 124, 21.

---

un-ārian *to dishonour :*—Se unārade sčs Georgies anlícnysse, Shrn. 73, 13. v. ge-unārian.

un-ārímed; *adj. Unnumbered, numberless, countless :*—Unārímed mengeo, Blickl. Homl. 199, 1. Seó unārímede menigo, 87, 18. Mid đȳ unārímedan weorode, 25, 35. Mid hū miclan feó woldest đū habban geboht ...? Ic wolde mid unārímedum feó gebycgan *quanti aestimabis* ...? *Infiniti,* Bt. 34, 9 ; Fox 146, 11. Unārímede untrumnessa, Blickl. Homl. 209, 13. Unārímedum *numerosis,* Hpt. Gl. 408, 67 : Bt. 1 ; Fox 2, 11. Be đǽm unārímdum cynnum *de innumeris generibus,* Nar. 1, 17.

un-ārímedlíc; *adj.* I. *innumerable, countless :*—Se cásere gegaderode unārímedlíce fyrde ... seó fyrd wæs unārímedlíc đe hē gegaderad hæfde, Chr. 1050 ; Erl. 173, 21–24. Seó unārímedlíce menigo hāligra sāula, Blickl. Homl. 87, 6. Be đǽre unārímedlícan mengeo his weoredes, đæs wæs būton unārímedlícan feþum ... *de innumerabili exercitu, in quo fuerint permultae peditum copiae* ... Nar. 4, 10–12. Cōman tōsamne unārímedlíco mengeo, Blickl. Homl. 199, 9. Hī genāmon unārímedlíco herereáf, Chr. 473 ; Erl. 12, 26 : 584 ; Erl. 18, 25. II. *boundless, infinite, shewn in countless instances :*—Đæt mid Drihtne sig unārímedlícu mildheortnys *quod apud Dominum sit innumerabilis misericordia,* L. Ecg. P. i. 9 ; Th. ii. 176, 14. Hē bæd hié đæt hié gemunden đæs unārímedlícan freóndscipes đe hié hæfdon on ealddagum, Ors. 2, 5 ; Swt. 82, 19.

un-ārímedlíce; *adv. Innumerably :*—Đa geswinc đe hē fela wintra dreógende wæs unārímedlíce oft (*times without number*), Ors. 5, 4 ; Swt. 224, 29.

un-ārlíc; *adj.* I. *dishonourable, disgraceful, shameful :*—Đīn mōdor gewíteþ of weorulde þurh scondlícne deáđ and unārlícne and heó ligeþ unbebyrged in wege fuglum tō mete and wildeórum *mater tua miserando turpissimoque exitu sepultura carebit, iacebitque in uia praeda auium ferarumque,* Nar. 31, 29. Sægde Lameh unārlíc spel (*a tale of shame*) : 'Ic honda gewemde on Caines cwealme,' Cd. Th. 66, 31 ; Gen. 1092. II. *unkindly :*—Unārlíce yrfebec *a will in which nothing is left to nearest relatives :* inofficiosum testamentum, Wrt. Voc. ii. 49, 17.

un-ārlíce; *adv.* I. *disgracefully, shamefully :*—Mē þeówmennen drehte dǽdum and wordum unārlíce, Cd. Th. 135, 29 ; Gen. 2250. II. *mercilessly, cruelly :*—Nǽfre gē mid blōde beódgereordu unārlíce eówre þicgeaþ, Cd. Th. 91, 28 ; Gen. 1519.

un-ārodscipe, es ; *m. Inactivity, spiritlessness :*—Oft mon biđ suíđe wandigende æt ǽlcum weorce and suíđe lætrǽde, and wēnaþ men đæt hit sié for suārmōdnesse and for unarodscipe, and biđ đeáh for wísdome and for wærscipe *saepe agendi tardatas gravitatis consilium putatur,* Past. 20 ; Swt. 149, 15.

un-ārweorþian *to dishonour :*—Ic ārwurđige (-weorđige, Jn. Skt. 8, 49, MS. A.) mīnne Fæder and gē unārwurđiaþ (-weordđodon, Jn. Skt. MS. A.) mē *honorifico Patrem meum et uos inhonoratis me,* Homl. Th. i. 442, 21.

un-ārweorþness, -wirþness, e ; *f. Disrespect, irreverence :*—Unārwyrđnyss *irreverentia,* Scint. 224, 1.

un-āsaedde. v. un-āsedd.

un-āscended; *adj. Unharmed, not to be harmed :*—Unāscended fruma *incorruptibile principium,* Jn. Skt. p. 1, 12. Unāscendedo *interos,* Rtl. 114, 7. Unāscendado, 101, 36 ; 172, 3 : 179, 1.

un-āscirigendlíc; *adj. Inseparable :*—Đǽre Hālgan þrynnysse is ān godcundnyss, and ān gecynd, and ān willa, and ān weorc unāscyrigendlíce (*inseparable ;* or *inseparably* ?), Homl. Th. i. 326, 27.

un-āscirod; *adj. Not separated :*—Unāscyrod *inremota, s. inseparata,* Wülck. Gl. 253, 3.

un-āscruncen; *adj. Unwithered, unfading :*—Unāscryncan *immarcessibilem,* Rtl. 24, 32.

un-āsecgende; *adj. Not to be told, unspeakable, ineffable :*—Mid unāsecgendre swētnysse *cum ineffabili dulcedine,* Bd. 4, 3 ; S. 568, 3.

un-āsecgendlíc; *adj.* I. *implying greatness, beyond the powers of language to describe, unspeakable, indescribable, ineffable :*—Hwæt wundor is, gif se ælmihtiga God is unāsecgendlíc ? Homl. Th. i. 286, 26 : 322, 9 : ii. 232, 5 : Homl. Skt. i. 1, 33 : Elen. Kmbl. 929 ; El. 466. Unāsæcgendlíc clǽne girȳno *ineffabile sacramentum,* Rtl. 33, 8. Unāsægcgendlíc, 35, 15. Đæt unāsecgenlíce wræc, Blickl. Homl. 25, 24. Unāsecggenlíce, 65, 21. Đæt wæs unāsecgendlíc ǽnigum menn hū mycel đæs folces wæs *it was impossible for any man to say how much people there was,* Chr. 1011 ; Erl. 145, 14. God unāsecgendlícere mildheortnesse *Deus inestimabilis misericordie,* Anglia xi. 112, 1. Wundriende đǽre unāsecgendlícan gesǽlignesse đǽra manna, đe him God forgifþ ealle heora scylda, Ps. Th. 31, arg. Æfter his unāsecgendlícum foreþonce, Bt. 39, 5 ; Fox 220, 2. Unāsæcendlícum *inenarrabili,* Rtl. 38, 5. Unāsecgendlícum, Blickl. Homl. 87, 21. Mid unāsecgendlícre wurđmynte, Ap. Th. 10, 21. Hē hæfde fulneáh unāsecgendlícne sige, Bt. 16, 2 ; Fox 54, 1 : Homl. Th. i. 532, 1, 2. Hī nāmon unāsecgendlíce here hūđe, Chr. 1046 ; Erl. 171, 1. Đa unmǽtan tyntregu and đa unāsecgendlícan wīta, L. E. I. prm. ; Th. ii. 396, 35. Hē nam of hire eall đæt heó āhte on golde and on seolfre and on unāsecgendlícan þingum (*things innumerable*), Chr. 1042 ; Erl. 169, 21. Unāsæcendlícum costum *ineffabilibus modis,* Rtl. 108, 27. II. *not proper to tell, not to be told :*—Đa unāsecgendlícan *nefandas,* Wrt. Voc. ii. 61, 40.

un-ásecgendlíce; *adv. Unspeakably, in a way that cannot be told, ineffably* :—Háligne Gást forþleóꝛende of Fæder and of Suna unásecgendlíce (*inenarrabiliter*), Bd. 4, 17 ; S. 586, 14.

un-ásedd; *adj. Unsated* :—Unásaedde (-seddae) *inopimum*, Txts. 71, 1102. Unásedde, Wrt. Voc. ii. 48, 81.

un-áseolcendlíc; *adj. Eager, energetic, vehement* :—Of geornum *subnixis*, unáseolcendlícum (-seoclendlícum, MS.) menegungum *hortamentis* (the whole passage is: Puberem subnixis precibus et inauditis blandimentorum hortamentis flectere nitebantur, Ald. 46), Hpt. Gl. 485, 50. v. un-ásolcenlíce, á-seolcan.

un-áseowod; *adj. Unsewed, without seam* :—Seó tunece wæs unásiwod (-seowod, MS. A.) *erat tunica inconsutilis*, Jn. Skt. 19, 23.

un-áseðendlíc (-sédendlíc ? v. sédan); *adj. Insatiable* :—Unáseðendlíc *insaturabilis*, Kent. Gl. 471. Unáseðenlíc *insatiabilis*, 522 : 1031 : *insaturabilis*, 1087.

un-ásmeágendlíc; *adj. Unsearchable, past finding out, inscrutable* :—Seó godcundnys is unásmeágendlíc, Homl. Th. ii. 232, 4 : Homl. Skt. i. 1, 33. Ðín myldheortnys is swíðe mycel and unásmægendlíc, 3, 548. Ðære sáwle brógan, unásmeágendlícu yrmðu (*misery beyond the power of man to explore*), Wulfst. 249, 19. Hé mé gefrætwode mid unásmeágendlícra wurðfulnesse, Homl. Skt. i. 7, 31. Ða óðre heofenan, ðe bufan hyre synd and beneoðan, synd mannum unásmeágendlíce (*are beyond the reach of men's investigation*), Lchdm. iii. 232, 23. His lára and his drohtnunga sind ús unásmeágendlíce, Homl. Th. i. 392, 23.

un-ásolcenlíce; *adv. Not lukewarmly, heartily, energetically, with vigour* :—Gif ðæt gebodene bið gefremed unforhtlíce and unsleaclíce and unásolcenlíce *si quod jubetur non trepide, non tarde, non tepide efficiatur*, R. Ben. 20, 19.

un-áspringende; *adj. Unfailing* :—Geunne mé ðæt ðis wæter sý mé tó fulwihtes bæþ unáspringende (*fiat mihi haec aqua fons baptismi indeficiens*, Ass. 217, 326), Nar. 46, 9.

un-áspyrigendlíc; *adj. That cannot be investigated, that cannot be learnt by inquiry* :—Unásperiendlíc *in[inve]stigabilis*, Kent. Gl. 91. Unásporiendlíc, Dial. 2, 16.

un-ástíðod; *adj. Not made firm* :—Gif mon on níwne weall unádrugodne and unástíðodne micelne hróf and hefigne onsett, ðonne ne timbreþ hé nó healle ac hryre *quod structuris recentibus necdum solidatis, si tignorum pondus superponitur, non habitaculum sed ruina fabricatur*, Past. 49 ; Swt. 383, 32.

un-ástyrigendlíc; *adj. Motionless* :—Ic fór of dúne on ða eorðan, and forneáh eallunga unástyrigendlíc bútan gáste læg, Homl. Skt. ii. 23 b, 576. Beón hig unástyriendlíce (*immobiles*) swylce stán, Cant. M. 16 (Ex. 15, 16).

un-ástyrod; *adj. Unmoved*; *inmotus*, Hymn. Surt. 11, 4.

un-ásundrodlíc; *adj. Inseparable*; *inseparabilis*, Rtl. 122, 10 : 109, 13.

un-áswundenlíce; *adv. Not languidly, not slowly, promptly* :—Ða ðóhte hé ðæt hé sceolde weorulde wiþsacan, and ðæt unáswundenlíce swá gedyde (*non hoc segniter fecit*), Bd. 4, 3 ; S. 567, 23. Heó ðæt weorc unáswundenlíce gefylde *opus non segniter implevit*, 4, 23 ; S. 593, 36.

un-átalodlíc; *adj. Unnumbered, innumerable* :—Gyltingum unátaladlícum *delictis innumerabilibus*, Rtl. 124, 42.

un-áteald; *adj. Uncounted* :—Gif se dæg bið forlæten unáteald, ðǽrrihte áwent eal ðæs geáres ymbrene ðwyres, Boutr. Scrd. 28, 32. v. un-teald.

un-átellendlíc; *adj. Innumerable* :—Seó fyrd wæs unátellendlíc ðe hé gegaderod hæfde, Chr. 1049 ; Erl. 172, 23. Hí bereáfedan hí æt eallon ðan gærsaman ðe héo áhte ; ða wǽron unátellendlíce, 1043 ; Erl. 168, 34. Ðam se fæder becwæð gersuman unáteallendlíce, 1086 ; Erl. 221, 8. Míne unátellendlíce (*innumerabilia*) beón ic oncnáwe gyltas, Anglia xi. 118, 62.

un-átemed; *adj. Untamed, unsubdued* :—Severus micelne dǽl Breotone mid díce tósceádde fram óþrum unátemedum ðeódum, Bd. 1, 5 ; S. 476, 3.

un-átemedlíc; *adj. Untameable* :—Forðon ðe ða men wǽron unátemedlíce and heardes módes and ellreordes *eo quod essent homines indomabiles et durae ac barbarae mentis*, Bd. 3, 5 ; S. 527, 25.

un-áteoriende; *adj. Unwearying, indefatigable* :—Unáteoriendum þénungum *indefessis* (*infatigabilibus*) *famulatibus*, Hpt. Gl. 463, 8.

un-áteorigendlíc; *adj.* I. *indefatigable, unwearied* :—Mid unáteriendlíce strecnysse *indefessa instantia*, Hpt. Gl. 434, 22. II. *that shall not fail, unending, imperishable* :—Hí befæston Godes láre heora underþeoddum tó unáteorigendlícan gafele, Homl. Th. i. 544, 18. Wé habbaþ unáteorigendlíce sáule, 96, 18. Wé ðe sind éce on úrum sáwlum, and eác beóð on líchaman unáteorigendlíce æfter ðam gemǽnelícum ǽriste, ii. 462, 30.

un-áteorigendlíce; *adv.* I. *indefatigably* :—Unáteorien[d]líce *infatigabiliter*, Hpt. Gl. 424, 17. II. *unceasingly, without failing* :—Unáteoriendlíce *incessabiliter*, R. Ben. Interl. 22, 13. God ðe wé unátirendlíce (*incessanter*) ondrǽden, Chart. Th. 316, 33.

un-áteorod; *adj. Unwearied, unexhausted, unfailing* :—Unáteorodne *inexhaustam* (*indefessam, indeficientem*), Hpt. Gl. 463, 18.

un-áþreótende; *adj. Unwearying, inexhaustible* :—Hí unáþreótendum þrymmum singaþ, Exon. Th. 24, 21 ; Cri. 388.

un-áþroten; *adj. Unwearied, persevering* :—Sindon tó séceanne stronge and unáðrotene láreówas and ðurhwuniende *fortes perseverantesque doctores quaerendi sunt*, Past. 22 ; Swt. 171, 9.

un-áþrotenlíce; *adv. Unweariedly, unceasingly* :—Hí sint tó manienne ðæt hí unáðrotenlíce ða gedónan synna gelǽden beforan heora módes eágan *admonendi sunt, ut incessanter admissa ante oculos reducant*, Past. 53 ; Swt. 413, 14. Ealne ðisne andweardan welan hí swíþe unáþrotenlíce sécaþ, Bt. 32, 3 ; Fox 118, 21 : 39, 13 ; Fox 234, 7.

un-átweógendlíce. v. un-tweógendlíce.

un-áwǽscen; *adj. Unwashed* :—Unáwǽscen wull *lana succida* vel *sucilenta*, Wrt. Voc. i. 61, 8. Unáwaxen wul *lana sucida*, ii. 54, 6.

un-áwégendlíc; *adj. Immovable, unshaken* :—Seó gefæstnung staðelfæst and unáwǽgendlíc mid þurhwuniende rihte beó gefæstned *confirmatio stabilis et inconcussa perseverantissimo jure consolidetur*, Chart. Th. 319, 9.

un-áwemmed; *adj. Unstained, undefiled, immaculate* :—Unáwoemmed *immaculatus*, Rtl. 24, 42 : 29, 11, 15. Unáwemdo *eunuchi*, Mt. Kmbl. Lind. 19, 12.

un-áwemmedlíc; *adj. Incorruptible, immaculate* :—Eádignisse unáwoemmedlícum *beata immortalitate*, Rtl. 33, 12.

un-áwemmedness, e; *f. Incorruption* :—Crist gewát of deáðe tó life, and of brosnunga tó unáwemmednysse, and of wíte tó wuldre, Anglia viii. 330, 10.

un-áwend, -áwended; *adj. Unchanged, unaltered* :—Hláfordes rihtgifu stande ǽfre unáwend (-áwended, MS. B.), L. C. S. 82 ; Th. i. 422, 3 : Cod. Dip. Kmbl. iv. 231, 17. Ic wille ðæt se fréols stonde unáwent, 219, 20.

un-áwendedlíc, -áwendlíc; *adj. Unchangeable, fixed, invariable* :—Ðú ðe ealle ða unstillan gesceafta tó ðínum willan ástyrast and ðú self simle stille and unáwendedlíc ðurhwunast *qui stabilis manens das cuncta moveri*, Bt. 33, 4 ; Fox 128, 10. Unáwendedlíce, unáwendlíce *fixa*, Ælfc. Gr. 43 ; Zup. 254, 17 note. v. un-áwendendlíc.

un-áwendende; *adj. Unchanging* :—Se ús gesette sido unáwendendne, Met. 11, 13. v. next word.

un-áwendendlíc; *adj. Unchangeable, unalterable, invariable* :—Heora nán nǽfre of ðam háde ðe hé is ne áwent, forðan ðe God is unáwendendlíc, Homl. Th. ii. 606, 27 : Bt. 35, 2 ; Fox 158, 4. God is ealra ðinga reccend and hé ána unáwendendlíc wunaþ and eallra ðara áwendendlícra welt *rerum orbem mobilem rotat, dum se immobilem ipsa conservat*, 35, 5 ; Fox 166, 9. Sió godcunde foreteohhung is ánfeald and unáwendendlíc (*simplex immobilisque*), 39, 6 ; Fox 220, 16. God gesette unáwendendlícne sido, 21 ; Fox 74, 1. Eorþan ðú sealdest unáwendendlíce *terram dedisti immobilem*, Hymn. Surt. 19, 33. Gif hí beóð *participia*, ðonne beóð hí . . . *mobilia*; gif hí beóð naman, ðonne beóð hí . . . *fixa*, ðæt is unáwendendlíc, Ælfc. Gr. 43 ; Zup. 254, 17. Ða habbaþ ða écean reste and unáwendendlíce welan, Homl. Skt. i. 22, 219.

un-áwendendlíce; *adv. Unalterably, without possibility of change* :—Ic nát hwæþer hit eall gewyrþan sceal unáwendendlíce, ðæt hé wát and getiohhod hæfþ. Ða cwæþ hé : 'Ne þearf hit nó eall gewiorþan unáwendendlíce; ac sum hit sceal geweorþan unáwendendlíce, Bt. 41, 3 ; Fox 248, 30–250, 2.

un-áwidlod; *adj. Uncontaminated, undefiled* :—Lombes unáwidlades *agni incontaminati*, Rtl. 24, 40. Erfeueardnisse unáwidlad *hereditatem incontaminatam*, 32.

un-áwirded; *adj. Uninjured, uncorrupted* :—Unáwerded *incorruptibilis*, Jn. Skt. p. 1, 12. Unáwoerdedo *inlaesos*, Rtl. 102, 31.

un-áwriten; *adj. Unwritten* :—Thomes ðrowunge wé forlǽtaþ unáwritene, Homl. Th. ii. 520, 9. Unáwritten *cautionem*, Lk. Skt. Lind. 16, 6.

un-bældo. v. un-bildu.

un-banden; *adj. Released from bonds* :—Æfter ðam þúsende byð se deófol unbanden (-bunden ?), Anglia viii. 336, 15.

un-beald; *adj. Not bold, not confident, irresolute* :—Oft gebyreþ ðæm manðwæran, ðonne hé wierd ríce ofer óðre men, ðæt hé for his manðwǽrnesse ásláwaþ and wierð tó unbald (-beald, Hatt. MS.), forðæm sió unbieldo and sió manðwǽrnes bióð swíðe anlíce *nonnunquam mansueti, cum praesunt, vicinum et quasi juxta positum torporem desidiae patiuntur*, Past. 40 ; Swt. 288, 1. On óðre wísan sint tó manianne ða ánfealdan strǽcan, on óðre ða unbealdan . . . Ðæm unbealdum is tó cýðanne hú giémeleáse hié bióð ðonne hié hié selfe tó swíðe forsióð *aliter admonendi sunt pertinaces, atque aliter inconstantes . . . istis intimandum est, quod valde se despicientes negligunt*, 42 ; Swt. 305, 12–16. Wénde ic ðæt ðú ðý wærra weorþan sceolde, and ðý unbealdra, Exon. Th. 268, 4 ; Jul. 427. [Laym. un-bald.]

un-bealu; *gen. -beal(u)wes*; *n. Innocence* :—Mid unbealuwe ealre heortan *in innocentia cordis mei*, Ps. Th. 100, 2.

un-bealufull; *adj. Innocent, harmless* :—Of unbealafullum feoh geræcan, L. I. P. 12 ; Th. ii. 320, 26 : Wulfst. 83, 13.

un-beboht; *adj. Unsold:*—Hé hæfde tamra deóra unbebohtra syx hund, Ors. 1, 1 ; Swt. 18, 10.

un-bebyriged; *adj. Unburied:*—Ðín môdor ligeþ unbebyriged *mater tua sepultura carebit,* Nar. 31, 30 : Shrn. 40, 4. Heora líchaman licgaþ unbebyrgede (-byrigde, MS. F.), Wulfst. 199, 10. [Unbiburiet *inhumatus,* Kath. 2243.]

un-beceás ; *adj. Not giving occasion to litigation, indisputable, incontestable :*—Bidde hé ða hond ðe ðæt ierfe hafaþ, ðæt hé him gedó ðone ceáp unbeceásne (*that he shew the chattel to be his by incontestable right*), L. In. 53 ; Th. i. 136, 7.

un-becrafod ; *adj. Not subjected to claims :*—Ðær se bônda sæt unbecrafod *where the husband dwelt without having had any claims made upon him,* L. C. S. 73 ; Th. i. 414, 22. v. un-crafod.

un-becweden ; *adj. Unbequeathed, not left by will :*—On ælcum þingum ðe ðær unbecweden bið, on bôcum and an swilcum lytlum, Chart. Th. 538, 24. Ðæt land æt Sendan and æt Sunnanbyrg unbecwedene and unforbodene wið ælcne man, 208, 38.

un-beden ; *adj. Unbidden, unasked :*—Sume preóstas . . . unbedene gaderiaþ hí tô ðam líce, swá swá grædige rémmas ðár ðár hí hold-geseóð, L. Ælfc. P. 49 ; Th. ii. 386, 2. [Toc Crist unnbedenn and unnbonedd to mælenn, Orm. 17081.]

un-befangenlíc; *adj. Incomprehensible :*—God is uñásecgendlíc and unbefangenlíc, Homl. Th. i. 286, 27.

un-befliten ; *adj. Uncontested, indisputed :*—Ðá wæs hígen and hláforde lond unbefliten éghwæs and seoððan á óð his daga ende, Chart. Th. 48, 1 : 481, 14 : 483, 3.

un-befohten ; *adj. Unfought, unopposed :*—Ðá wénde se here . . . ðæt hié mehten faran unbefohtene ðér ðér hié wolden, Chr. 911 ; Erl. 100, 23 : Byrht. Th. 133, 28 ; By. 57.

un-beföndlíc (?). v. un-beseóndlíc.

un-begán ; *adj.* I. *uncultivated :*—Unbegánum *incultis* (*arvis,* Ald. 200), Wrt. Voc. ii. 96, 2 : 47, 38. Deós wyrt byþ cenned on dúnum and on unbegánum stôwum, Lchdm. i. 230, 4 : 238, 17. II. *unadorned :*—Unbegán *inculta, non ornata,* Hpt. Gl. 435, 26.

un-begrípendlíc ; *adj. Incomprehensible :*—Se myccla mægenþrym and se unbegrípendlíca, Blickl. Homl. 179, 9. Unbegrípendlíc and ungesýnelíc God, 185, 31.

un-begunnen ; *adj. Without beginning :*—Sum ic eom is edwistlíc word and gebyraþ tô Gode. ánum synderlíce, forðan ðe God is ǽfre unbegunnen and ungeendod on him sylfum and ðurh hine sylfne wunigende, Ælfc. Gr. 32 ; Zup. 201, 9 : Wrt. Voc. i. 70, 1 : Homl. Th. ii. 204, 12 : Homl. Skt. i. 1, 16 : Homl. Ass. 25, 25. Hé wæs ǽfre unbegunnen Scyppend, Hexam. 1 ; Norm. 4, 3.

un-beheáfdod ; *adj. Unbeheaded :*—Ic eów læte unbeheáfdod, Homl. Skt. i. 23, 185.

un-behéfe ; *adj.* (or *subst.* ?) *Unsuitable, inconvenient, unprofitable :*—Unbehéfe *incommodum,* Wrt. Voc. ii. 85, 35. [Al þat ure sowle and ure lichame beð unbiheue, O. E. Homl. ii. 7, 30. *Also subst.* Hie turnden fro him hem seluen to unbihefe, 121, 26.] Cf. un-bryce.

un-behelendlíce ; *adv. Without the possibility of concealment :*—Ðæt bið eallum open unbehelendlíce, ðæt man ǽr hæl, Wulfst. 138, 3.

un-behelod ; *adj. Uncovered, naked :*—Hé læg on his getelde unbehelod (*nudatus*), Gen. 9, 21, 22.

un-behreówsigende ; *adj. Unrepenting, impenitent :*—Se ðe mid unbehreówsigendre heortan þurhwunaþ on mándǽdum, Homl. Th. i. 500, 15.

un-belimp, es ; *n. Mischance, accident :*—Of unbelimpum (*fortunae*) *casibus* (*oppressos,* Ald. 42), Hpt. Gl. 478, 25.

un-beorhte ; *adv. Not brightly :*—Ealle steorran weorþaþ gebirhte of ðære sunnan, sume þeáh beorhtor, sume unbeorhtor (*less brightly*), Bt. 34, 5 ; Fox 140, 6. Sume beorhtor, sume unbyrhtor, 33, 4 ; Fox 132, 21.

un-bereáfigendlíc; *adj. Not to be taken away :*—Syle mé ðæt unbereáfigendlíc gebæd ðínre fulfremednysse, Homl. Skt. ii. 23 b, 242.

un-berende ; *adj.* I. *not bearing, barren, sterile :*—Elizabeth wæs unberende (-berend, Lind. *sterilis*), Lk. Skt. 1, 7. Unbeorendu (*sterilis*) cende monige, Ps. Surt. ii. p. 186, 17. Ne biþ mid eów nán þing unberendes ne on mannum ne on nýtenum, Deut. 7, 14. Ðæt unberende treó hé genimes *palmitem non ferentem fructum tollet,* Jn. Skt. Lind. 15, 2 margin. Unbeorende *sterilem,* Ps. Surt. 112, 9. Eádige syndon ða men ða ðe wæron unberende, Blickl. Homl. 93, 30. Unberende telgan *spadones,* Wrt. Voc. i. 38, 58. II. *unbearable :*—Byrðenna unbǽrende *onera importabilia,* Mt. Kmbl. Lind. 23, 4. [Goth. un-bairands *barren.*]

un-berendlíc ; *adj. Unbearable, intolerable :*—On ðam ne eardaþ nán eorðlíc mann for ðam unberendlícum bryne, Lchdm. iii. 260, 23.

un-berendness, e ; *f. Barrenness, sterility :*—Unberendnise *sterilitas,* Rtl. 118, 1. Unbeore[n]dnisse *sterilitatem,* Ps. Surt. 34, 12.

un-besacen ; *adj.* I. *of persons, unmolested by litigation :*—Ðær se bônda sæt uncwyd and unbecrafod, sitte ðæt wíf and ða cild on ðam ylcan unbesacen, L. C. S. 73 ; Th. i. 414, 23. II. *of things, not made the subject of litigation, uncontested :*—Ðæt ðæt land swá

unbesæccen gange intô ðære cyrican swá hit ða on dæg wes ða hit man him tô lǽt *that the land pass into the possession of the church as uncontested as it was on the day when it was let to him,* Chart. Th. 159, 24. Ðá sealde hé Æþelrige unbesacen land on hand, ðæt hé þanonforð syþþan ðǽron ne sprǽce *he gave the land up to Æþelrige uncontested, so that thenceforth he would not lay claim to it,* 289, 31. Hió ðæt land hæbben unbesacen wið ælce haud (*not liable to suits from any side*) ða hwíle ðe hió lifgean, and gif Ælfw leng sió, ðonne sý hit hyre unbesacen, Cod. Dip. Kmbl. ii. 150, 22-25 : L. C. S. 80 ; Th. i. 420, 21. Hé him gedó ðone ceáp unbesacene *let him make the chattel secure from being the subject of litigation,* L. In. 53 ; Th. i. 136, 7 note. v. un-forboden.

un-besceáwod; *adj. Inconsiderate, heedless :*—Se ðe unbesceáwud ys tô specenne hé ongytt yfele *qui inconsideratus est ad loquendum sentiet mala,* Scint. 78, 7. Unbesceawad, Kent. Gl. 433: Unbesceáwode *inprovida* vel *inconsiderata,* Wrt. Voc. i. 55, 13.

un-besceáwodlíce ; *adv. Inconsiderately, heedlessly :*—God swýþor tô yrsunge unbesceáwudlíce forþclypian ðænne foresceáwudlíce tô synna forgyfenyssa innlaþian *Deum potius ad iracundiam inconsiderate prouocare quam prouide ad peccaminum ueniam inuitare,* Anglia xiii. 370, 76.

un-bescoren ; *adj. Unshorn, without the tonsure :*—Sume sídfeaxe gáþ, ðæt seó bescorene hálignes ne sý weorþre ðenne seó unbescorene, R. Ben. 135, 29.

un-besenged ; *adj. Unsinged, unscorched, unburnt :*—Se bið swýðe clǽne ǽlcere synne, se ðe ðæne bryne ðurhfærð unbesencged (-sænged, MS. C.), Wulfst. 25, 19.

un-beseóndlíc ; *adj. Incomprehensible :*—God on ðrymme unbeseóndlíc (-föndlícne ?) *Deum majestate incomprehensibilem,* Bd. 3, 22 ; S. 552, 16.

un-besmiten ; *adj. Undefiled, unpolluted, unsullied, pure :*—Unbesmiten weg *impolluta via,* Ps. Spl. 17, 32. Gif heó unbesmiten (*impolluta*) tô him cyrre, L. Ecg. Addit. 12 ; Th. ii. 234, 6 : Nar. 41, 11. Ðæs unbesmitenan líchaman úres Drihtnes, Homl. Skt. ii. 23 b, 113 : Blickl. Homl. 155, 32 : 3, 15 : Homl. Skt. i. 4, 69 : 23 b, 503. Healdaþ eówre handa unbesmitene (*innoxias*), Gen. 37, 22.

un-besorh ; *adj. Not the object of care, that one does not care about :*—Ðá hét se cyning clypian him tô unbesorge men (*men that he didn't care about*), Homl. Th. ii. 486, 9. v. be-sorg.

un-bétcd ; *adj. For which amends has not been made :*—Nænig bihelan mæg on ðam heardan dæge wom unbéted, Exon. Th. 80, 25 ; Cri. 1312.

un-beþôht ; *adj. Unreflecting, inconsiderate :*—Micle hrædlícor hí wǽren áðwægene ðæra scylda mid ðære hreówsunga, gif hí fǽrlecor syngoden unbeðóhte *citius delicta poenitendo abluerent, si in his sola praecipitatione cecidissent,* Past. 56 ; Swt. 435, 2.

un-beþirfe. v. un-biþirfe.

un-beweddod ; *adj.* I. *unbetrothed :*—Gif hwá lið mid unbeweddudre fǽmnan *si quis dormierit cum virgine necdum desponsata,* Ex. 22, 16. Gif hwá fǽmnan beswíce unbeweddode and hire mid slǽpe, L. Alf. 29 ; Th. i. 52, 5. Unbeweddod mǽden *puellam virginem, quae non habet sponsum,* Deut. 22, 28. II. *unmarried :*—Unbeweddod *innuba,* Wrt. Voc. i. 52, 35. Mǽden seó ðe unbeweddud ys *uirgo quae innupta est* (1 Cor. 7, 34), Scint. 69, 3. Gif Maria unbeweddod wǽre and cild hæfde, ðonne wolde ðæt folc mid stánum hí oftorfian, Homl. Th. i. 196, 11.

un-bewilled ; *adj. Not boiled away :*—Seóþ on wætre óþ ðæt ðæs wætres sié þridda[n] dǽl unbewelled, Lchdm. ii. 248, 18.

un-biddende ; *adj. Not praying, without praying :*—Gif hé nele biddan ðæs ecan leóhtes, hé sitt ðonne blind be ðam wege unbiddende, Homl. Th. i. 156, 4.

un-bildu(-o) ; *indecl. f. Want of boldness, weakness, irresolution, inconstancy :*—Sió unbieldo and sió mandwǽrnes bioð swíðe anlíce *weakness and gentleness are very much alike,* Past. 40 ; Swt. 288, 1. Of ðære leohtmódnesse cymð sió twiefealdnes and sió unbieldo *inconstantia ex levitate generatur,* 42 ; Swt. 307, 3. Sió unfæsðrǽdnes and sió unbieldo ðara geðóhta *cogitationum inconstantia,* Swt. 308, 5. Ðonne hié of unwísdóme oððe of wácmódnesse and of unbieldo oððe of untrymnesse módes oððe líchoman gesyngaþ *cum sola ignorantia vel infirmitate delinquant,* 21 ; Swt. 159, 1. Ða lytelmódan and ða undrístan ðonne hié ongietaþ hiera unbældo and hiera unmiehte *pusillanimes dum nimis infirmitatis suae sunt conscii,* 32 ; Swt. 209, 7.

un-bindan ; *p.* -band, *pl.* -bundon ; *pp.* -bunden *To unbind, untie :*—Ne eom ic wyrðe ðæt ic unbinde (*soluam*) his sceóþwang, Jn. Skt. 1, 27. ' Æfter his beháte ic ðé unbinde ' . . . . Se engel hine ða unband, Homl. Th. i. 466, 31. Swá hwæt swá ðú unbindst (*solveres*) ofer eorðan, ðæt byð unbunden (*solutum*) on heofonum, Mt. Kmbl. 16, 19 : 18, 18. Ðæs fæder tungan his nama unband, Homl. Th. i. 352, 31. Álésde ł unband *soluit,* Ps. Lamb. 104, 20. Unband *dissoluit,* Cant. Abac. 6. Hiá onfundun fola gibundenne, and unbundun hine, Mk. Skt. Rush. 11, 4. Sceal se láreów hine unbindan fram ðam ecum wíte, swá swá ða apostoli líchamlíce Lazarum álýsdon, Homl. Th. i. 234, 14, 9. Ælr ðon God heó ðæs wræces unbindan wolde, Anglia xi. 2, 24. Æfter þúsend geárum bið Satanas unbunden *post mille annos soluetur Satanas,* Wulfst. 83, 6. Beón unbunden *dissolui* (*a peccato*), Scint. 38, 12. Hí wurdon

unbundene, Homl. Th. ii. 20, 8. From synna bendum unbundeno a peccatorum vinculis absolutus, Rtl. 7, 13. [Cf. Goth. and-bindan: O. Sax. ant-bindan: O. H. Ger. int-, in-bindan.] v. on-bindan.

un-birnende; adj. Without burning, without being on fire, Beo. Th. 5089; B. 2548.

un-bisc[e]opod; adj. Unconfirmed :—Wē lǽraþ . . . . þæt ǽlc cild sý gefullod binnon .xxxvii. nihtum, and þæt ǽnig man tó lange unbiscopod ne wurðe, L. Edg. C. 15; Th. ii. 246, 28. Unbiscpod (-biscopod, MSS. C. E.), Wulfst. 120, 15. Wē secgaþ eów, þæt ǽlc cild sceall beón binnon þryttigum nihtum gefullod . . . Ne nǽnne man man ne lǽte unbisceopod tó lange . . . And witan ða ðe cildes onsón æt fulluhte oððe æt bisceopes handum, þæt hí hit on rihtum geleáfan gebringan, 300, 16–30. [Longe beon unbishoped, A. R. 204, 29.]

un-biþirfe; adj. Useless, vain, unprofitable :—Ðú hafast unbiþyrfe ofer witena dóm wísan gefongen you have taken an unprofitable course contrary to the judgement of wise men, Exon. Th. 248, 18; Jul. 97. Ða (false gods) sind geásne góða gehwylces, ídle, orfeorme, unbiþyrfe, ne ðær freme mēteþ fira ǽnig, 255, 21; Jul. 217. [O. Sax. un-bitherbi: O. H. Ger. un-biderbi inutilis, vanus, inanis.]

un-blanden; adj. Unmixed :—Unblonden non mixtum, Rtl. 68, 30.

un-bleoh; adj. Not coloured, clear, bright, splendid :—Is mín land foremǽre and mē swýðe unbleó haereditas mea praeclara est mihi, Ps. Th. 15, 6. Hwæt mæg beón heardes hēr on lífe wið ðam ðú móte gemang ðam werode eardian unbleoh on ēcnesse (but there is no corresponding word in the Latin, which is : Quid durum saeclo consetur in isto, utque illas inter liceat habitare cohortes ?), Dóm. L. 302. Cf. ungebleoh.

un-bletsung, e; f. Cursing :—Fela is ðæra ðe . . . . embe bletsunga oððe unbletsunga leohtlíce lǽtaþ, and nā understandaþ . . . 'Quodcumque benedixeritis et cetera,' L. I. P. 6; Th. ii. 310, 36.

un-blinnendlíce; adv. Incessantly :—Unblinnendlíce dón wæs incessabiliter acta est, Bd. 1, 6; S. 476, 26.

un-bliss, e; f. Unhappiness, grief, sorrow, misery :—Mycel is mē unbliss mínra dýrlinga miss, Homl. Skt. i. 23, 271. Nú wē beóð blíðe, and eft on micelre unblisse, Homl. Th. i. 184, 3. Manege unblisse and micele sorga becómon ðám Iudéiscum æfter Cristes slege, Homl. Ass. 79, 179.

un-blíðe; adj. I. sad, sorrowful, grieved :—Unblíðe tristis, Mt. Kmbl. Rush. 19, 22 : Beo. Th. 261; B. 130. Giómormód, unblíðe, 4529; B. 2268. Wæs hē swýðe unblíðe . . . Ðá geseah Gúðlác ðone bróþor sárig, Guthl. 9; Gdwin. 50, 6. Beón in unblíðum móde moestus esse, Mt. Kmbl. Rush. 26, 37. Ðám unblíðum (tristibus) sint tó cýðanne ða gefeán ðe him gehátene sindon . . . Gehíeren ða unblíðan (tristes) ða leán ðæs gefeán ðe hié tó hopiaþ . . . Monige beóð ðeáh blíðe and eác unblíðe (laeti vel tristes) for ðæs blóðes styringe, Past. 27; Swt. 187, 16–24 : 61; Swt. 455, 10. Hú blinde hí (the envious) beóð, ðonne hí beóð unróte for óðerra monna gódan weorcum and for hira ryhtum gefeán beóð unblíðe quantae caecitatis sint qui alieno provectu deficiunt, aliena exultatione contabescunt, 34; Swt. 231, 17. Hý áswindaþ vel heó beóþ unblíþe contabescunt, i. exsiccant, Wrt. Voc. ii. 134, 74. Weorod eall árás, eodon unblíðe, weóllon teáras, Beo. Th. 6054; B. 3031 : Cd. Th. 223, 29; Dan. 127. Gemynd hê ða ungelimp ðe hê hæíde on his wrecsíðe and ne byð þeáh nā ðe unblíðre (not less glad), Shrn. 204, 11. Ðonne hwylcum men gelimpeþ ðe his leóf fæder gefarþ, ne mæg ðæt nā beón ðæt ða bearn ðe unblíðran ne sýn it cannot be that the children are not the sadder, Blickl. Homl. 131, 25. II. unkind, shewing ill-will or displeasure, stern, angry :—Gif ēgo ðín unblíðe sê si oculus tuus nequam fuerit, Mt. Kmbl. 6, 23. Ðá weard unblíðe Abrahames cwên hire worcþeówe, wrāð on móde, heard and hréðe, Cd. Th. 136, 16; Gen 2259. Him unblíðe andswarode wulfheort cyning, 224, 10; Dan. 134. III. unquiet, not peaceful :—Gidreáð ðe unblíðo corripite inquietos, Rtl. 11, 37. [O. H. Ger. un-blídi tristis.]

un-blíðemēðe; adj. Sadhearted, sorrowful :—Unblíðemoede moestus, Mt. Kmbl. Lind. 26, 37.

un-blódig; adj. Bloodless :—On unblódium gefeohte incruento prelio, Germ. 395, 16.

un-boht unbought, free :—Unboht ł unceáped gratis, Mt. Kmbl. Lind. 10, 8. Sacleás ł unsynnig ł unbocht gratis, Jn. Skt. Lind. 15, 25.

un-boren; adj. Unborn :—Se ðe unborenum cildum líf sylð, Homl. Skt. i. 23, 429. Ða unborenan bearn, Past. 48; Swt. 367, 20. [Goth. un-baurans.]

un-brád; adj. Not broad, narrow :—Eall swā brád seó sunne is swā eall eorðan ymbhwyrft, ac heó þingð ús swýðe unbrád, Lchdm. iii. 236, 8. Se unbráda þistel scolimbos, Wrt. Voc. i. 69, 12. On brǽde, ðǽr hit brádest is, fíf geurda, and ðǽr hit unbrádost is, ánne geurde, Chart. Th. 156, 29. Ðǽr ðæt land unbrádest is, ðǽr hit sceol beón eahtatýne fóta brád, 236, 8.

un-brēce; adj. Unbreakable, indestructible :—Flint unbrǽcne, Exon. Th. 1, 11; Cri. 6. Tír unbrǽcne, Apstls. Kmbl. 172; Ap. 86.

un-brice, un-briéce. v. un-bryce, un-brýce.

un-brocheard; adj. Tender, delicate :—Unbrocheard vel sēfta delicatus,

i. tenerus, Wrt. Voc. ii. 138, 39. Hwî ne miht ðú ongitan, ðætte ǽlc wuht cwíces biþ innanweard hnescost and unbrocheardost quid, quod mollissimum quodque, sicuti medulla est, interiore semper sede reconditur ? Bt. 34, 10; Fox 150, 6.

un-brosnigendlíc; adj. Incorruptible, imperishable :—His líchama wæs grápigendlíc, and ðeáhhwæðere unbrosnigendlíc; hê æteówde hine grápigendlícne and unbrosnigendlícne, Homl. Th. i. 230, 26 : 300, 10: Homl. Skt. ii. 27, 146. Ðú unscrýddest ðe ðone brosnigendlícan mann and ðē gescrýddest ðone unbrosnigendlícan mann, 30, 114: Homl. Ass. 45, 521. On ðam gemǽnelícum ǽriste beóð úre líchaman geedcennede tó unbrosnigendlícum líchaman, Homl. Th. i. 394, 33.

un-brosnung, e; f. Incorruption :—Beóð úre líchaman geedcynnede tó unbrosnunge, ðæt is tó ēcum dingum, Homl. Th. i. 394, 27. Áwende fram brosnunge tó unbrosnunge, ii. 206, 2.

un-bryce; adj. Unbroken, inviolate, uninjured :—Hwæþre his meahta spēd hálig wunade, dóm unbryce, þeáh hê deáþes cwealm ræfnan sceolde, Exon. Th. 240, 21; Ph. 642. Hyre wæs mægen unbrice, 256, 22; Jul. 235.

un-brýce; adj. (or subst. ?) Useless, unprofitable :—Unbrýce, unbrýce, unbrycci ineommodum, Txts. 69, 1050. Unbriéce incommodum, inutile, Wrt. Voc. ii. 44, 76. Nyle hê ða dǽrstan him dón unbrýce faex ejus non est exinanita, Ps. Th. 74, 8. [But calleth hym yn the gospel ryche, As unkynde and unbryche, Halliwell's Dict. Goth. unbrūkjai skalkōs servi inutiles, Lk. 17, 10.] Cf. un-behēfe.

un-brýde. v. next word.

un-brygd (?), es; m. A not unfair turn, fair dealing (?) :—Swā ic hit hæbbe, swā hit se sealde, ðe tó syllanne áhte, unbrýde and unforboden, and ic hit āgnian wille tó āgenre ǽhte so I have it, as he gave it, who had the right to give, without fraud and unforbidden, and I mean to possess it as my own property, L. O. 13; Th. i. 184, 4. Cf. brægd, brygd (bryd). Or, perhaps, unbrýde = un-brigde without liability to be reclaimed; cf. Icel. brigð a right to reclaim, chiefly of landed property.

un-bunden; adj. Not bound :—Nelle ic (a bow) unbunden ǽnigum hýran nymþe searosǽled, Exon. Th. 406, 10; Rä. 24, 15. Gif hê hine bescíre unbundene . . . Gif hê hine gebinde and ðonne bescíre, L. Alf. pol. 35; Th. i. 84, 7.

un-burh (?). v. un-býing.

un-býed; adj. Uninhabited, desert :—Unbýed is styd disertus est locus, Mk. Skt. Lind. Rush. 6, 35. Wēstig ł unbýed deserta, Mt. Kmbl. Lind. 23, 38. In unbýedum londæ in deserto, p. 9, 14.

un-býing (?) a solitude :—Unbyergo (-býengo? or -byrego, from -burh? Cf. un-lond) solitudines, Rtl. 1, 17.

un-byrged; adj. Unburied :—Se cásere bebeád ðæt hine man forlēte unbyrgedne, Shrn. 57, 1.

unc; dat. : unc, uncet (-it), acc. : uncer; gen. Us two, me and thee, me and him. (1) alone :—'Hwæt wylle gyt ðæt ic inc dó?' Ðá cwǽdon hí : 'Syle unc ðæt wit sitton, ān on ðíne swýðran healfe and ōþer on ðíne wynstran,' Mk. Skt. 10, 37. Hê sǽde me eall, Gen. 41, 13. Ðú mē behēte hál ðæt ðæt ðú mē sealdest, on ða gewitnesse ðe unc ðā mid wæs, L. O. 7 ; Th. i. 180, 24. Unc is his hyldo þearf, Cd. Th. 41, 30; Gen. 664. Æfter ðon ðe wit nū betweoh unc tógongenne beóþ, ne geseó wit unc ofer ðæt in ðysse weorulde, Bd. 4, 19; S. 607, 20. Beforan ungc, 5, 12; S. 628, 15. Mid ðý ic unc wēnde ingangende beón, S. 629, 39. Wit unc werian þóhton, Beo. Th. 1085; B. 540. Gif hê forhigeþ uncet fyrenfulle, Shrn. 42, 27. Ðá sende hê uncerne esenþeówan mid unc, ðæt hê uncet sceolde út álǽdan . . . ðā ne mihte hê unc gesión, 43, 1–5. Sege mínum brēðer ðæt hê dǽle uncer ǽhta wið mē, Lk. Skt. 12, 13. Wit ðe uncer ǽrdǽdum onfóð, 23, 41. Ðú hæfst yfele gemearcod uncer sylfra síð, Cd. Th. 49, 14; Gen. 792. Mid uncer āgene swurde, Shrn. 39, 25. Wit gerehton bi ealre uncer fóre, 43, 34. Uncer lāþette ǽgðer óðer, 39, 22. Ne nǽfre uncer āwþer his ellen cýðde, Exon. Th. 496, 29; Rä. 85, 22. (2) with numeral forms :—Ic wið ðe sceolde for unc ánum twām ǽrendsprǽce ābeódan, Exon. Th. 472, 12; Rä. 61, 15. Unc mǽran twáin, 496, 6; Rä. 85, 10. Bismærædu ugket men bā ætgadre, Txts. 126, 8. Hwæðer uncer twēga, Beo. Th. 5057; B. 2532 : Cd. Th. 110, 9; Gen. 1835. Ic rǽd sprece bēgra uncer, 115, 4; Gen. 1914. (3) with the name of the person who is associated with the speaker :—Sceolde unc Adame (for me and Adam) yfele gewurðan ymb ðæt heofonríce, Cd. Th. 25, 1; Gen. 387. Is ðæt land healf ðæs cinges, healf uncer Brentinges, Cod. Dip. Kmbl. iii. 422, 11. Uncer Grendles of me and Grendel, Beo. Th. 4009; B. 2002. [Laym. Marh. Gen. and Ex. unc: Orm. uunc baþe : Kath. bituhten unc tweien (us twa, v. r.): Laym. O. and N. hwæðer unker. Goth. ugkis; dat.: ugkis, ugk; acc.; ugkara; gen.: O. Sax. unk; dat. acc.; unkerð; gen.: O. H. Ger. unker (zweio): gen.: Icel. okkr; dat. acc.; okkar; gen.] v. wit, uncer.

un-cáfscipe, es; m. Inactivity, sluggishness; ignavia :—Ðá fēng Nero tó ríce; se æt nēxtan forlēt Brytene īgland for his uncáfscipe (cf. se næht freomlíces ongan on ðære cynewísan, ac . . . hê Breotona ríce forlēt nihil omnino in re militari ausus est . . . Brittaniam pene amisit, Bd. 1, 3; S. 475, 20), Chr. 47; Erl. 7, 26.

un-campróf; *adj. Unwarlike, not bold in battle :*—Uncamprófes *inbellis,* Germ. 399, 420.

un-capitulod; *adj. Not provided with titles to the several sections :*—Hyt is tó witanne hwî ðeós feórþe bóc sig uncapitulod nú þa ǽrran béc synt gecapitulode *sciendum est, quare liber hic quartus sit sine capitulis, cum priores libri capitulis instructi sint* (v. pp. 170, 180, 194, where the titles to the sections of bks. I, II, III are given), L. Ecg. P. iv; Th. ii. 204, 1.

un-ceáped. v. un-boht, *and cf.* un-cípe.

un-ceápunga; *adv. Without payment* or *recompense;* gratis :—Nó ic wið feohsceattum ofer folc bere Drihtnes dómas, ac ðe unceápunga orlæg secge, Cd. Th. 262, 18; Dan. 746.

un-ceás, -ceást, es (*but* ceás *and* ceást *are both fem.*) *Absence of quarrel, inhostility :*—Se ðe þeóf slihð hé mót áðe gecýðan ðæt hé hine fleónde for þeóf slóge, and ðæs deádan mǽgas him swerian unceáses (-ceástes, MS. H.) áð *the kinsmen of the dead man shall swear to the slayer an oath that they will have no quarrel with him,* L. M. 35; Th. i. 124, 8. Cf. the similar phrase in reference to the seizing of a thief: Ða mǽgas him (*the captor*) swerian áðas unsǽhða, 28; Th. i. 120, 6.

un-cenned; *adj. Not begotten :*—Wuldor Fæder ðam uncænnedan *gloria Patri ingenito,* Hymn. Surt. 120, 13.

uncer; *pron. poss. Of us two, our* (of two persons) :—Uncer hláford hióld hiora olfendu and ábád uncres tócymes . . . wit geségon ðæt uncer efenþeów wæs forworden . . . and se uncer hláford ábád uncres tócymes . . . sió lió forswealh uncerne hláford . . . Wit geseágon uncre feónd forwordene, Shrn. 43, 2–21. Uncres gewinnes, Exon. Th. 254, 1; Jul. 190. Of uncrum wege, Bt. 40, 5; Fox 240, 18. Of uncrum feó, Bd. 3, 14; S. 540, 8. Uncerne hwelp, Exon. Th. 380, 31; Rä. 1, 16. Crist wát uncre clǽnnysse, Shrn. 40, 20: 42, 3; Cd. Th. 139, 4; Gen. 2304. Uncre eágan, Mt. Kmbl. 20, 33: Homl. Skt. ii. 30, 374. Mid uncrum fótum, Shrn. 42, 1: Gen. 31, 16. For uncera sáule, Cod. Dip. Kmbl. iii. 304, 33. [*O. Sax.* unka: *Icel.* okkarr.]

uncet. v. unc.

un-cípe; *adj. Given without payment, gratuitous :*—Sió uncýpe *gratuita* (*Dei gratia,* Ald. 78), Wrt. Voc. ii. 88, 9. [Cf. *Icel.* ú-keypis *gratuitously.*] v. un-ceáped.

un-clǽmod; *adj. Rough-cast, unsmoothed :*—Unclǽmodum *impolitis,* Germ. 398, 258.

un-clǽne; *adj.*   I. in a physical sense, *unclean, foul, filthy.* v. un-clǽnness, clǽne.   I a. as applied to animals or things, *unclean, not fit for food :*—Seó æ monig ðing bewereþ tó etanne swá swá unclǽne (*inmunda*), Bd. 1, 27; S. 494. 33. Hwæt gif hit unclǽne (*immundi*) beóþ fixas? Ic wyrpe ða unclǽuan út, and genime mé clǽne tó mete, Coll. Monast. Th. 23, 15. Ða óðre synd unclǽne (*polluta*), Lev. 11, 12. Be swýnum and be óðrum unclǽnum nýtenum *de porcis et de aliis impuris animalibus,* L. Ecg. C. 40, tit.; Th. ii. 130, 31.   II. in a moral sense, *unclean, impure :*—Unclǽne *incestus vel impurus,* Wrt. Voc. i. 50, 13: *incestus,* 51, 35: 72, 13. Se unclǽna (*inmundus*) gást, Mk. 1, 26. Woruldmonna seó unclǽne gecynd, Exon. Th. 63, 9; Cri. 1017. Besmitene mid ðem unclǽnan firenluste, Blickl. Homl. 25, 8. Wæs sum man unclǽne (*inmundum*) deófol hæbbende, Lk. Skt. 4, 33. Unclǽne ingeþoncas, Exon. Th. 80, 33; Cri. 1316. Unclǽnra *inpudicarum,* Wrt. Voc. ii. 45, 14.

un-clǽnlíc; *adj. Uncleanly, impure :*—Cunnunga ða unclǽnlíco gifliǽ *contactus inlicitorum fugat,* Rtl. 110, 1.

un-clǽnlíce; *adv. Impurely :*—Swá hwilc man swá Godes weorc clǽnlíce wirceþ, hé bið écelíce gehealden. Se ðe hit unclǽnlíce wyrceþ, hé bið áwyrged intó helle, Homl. Ass. 168, 121. Wé wilniaþ mid úrum hláforde clǽnlíce sweltan, swíðor ðonne unclǽnlíce mid eów lybban, Homl. Th. i. 432, 26.

un-clǽnness, e; *f.*   I. in a physical sense, *uncleanness, impurity, foulness, squalor :*—Suǽ huæd in húsum ðás ýd eft ástrægde beuærle unclǽnnisse *quicquid in domibus haec unda resperserit careat inmunditia,* Rtl. 121, 36. Fúle unclǽnnessa *olidos* (*ergastulorum*) *squalores,* Hpt. Gl. 509, 75.   II. in a moral sense, *uncleanness, impurity, obscenity :*—Láð unclǽnnys *detestanda obscenitas,* Hpt. Gl. 506, 74. Se roccere sceal beón simle clǽne on his gedóhte, ðætte nán unclǽnnes (*immunditia*) hine ne besmíte, Past. 13; S. 75, 20: Rtl. 97, 29. Wrǽnre unclǽnnysse *lascivae obscenitatis,* Hpt. Gl. 505, 38. Hwá unclǽnnisse líf álífde, Exon. Th. 448, 31; Dóm. 62. Unclǽnnysse *spurcitia,* Hpt. Gl. 439, 8.

un-clǽnsian; *p.* ode *To defile, pollute :*—Unwyrtrumias ł unclǽnsias *eradicetis,* Mt. Kmbl. Lind. 13, 29. Unclǽnsia *inquinare,* p, 17, 12. v. ge-unclǽnsian, *and next word.*

un-clǽnsod; *adj. Not purified :*—Ðý læs ǽnig unclǽnsod dorste on swá micelne háligdóm fón ðære clǽnan ðegnenga ðæs sacerdhádes *ne non purgatus adire quisque sacra ministeria audeat,* Past. 7; Swt. 51, 1. v. un-geclǽnsod.

un-clǽnu(-o); *f. Uncleanness, impurity :*—Fulle sint unclǽno *pleni sunt inmunditia,* Mt. Kmbl. Lind. 23, 25. Fulla sint all ł ég̅huelc unclǽnæ *plena sunt omni spurcitia,* 27.

un-cnyttan; *p.* te *To unknot, untie :*—Ðæs ne eom ic wyrðe ðæt ic his sceóna þwanga búgende uncnytte *cujus non sum dignus procumbens soluere corrigiam calciamentorum eius,* Mk. Skt. 1, 7: Lk. 3, 16. (Wǽron) uncnytte (*vinculorum ligamina*) *enodarentur, solverentur,* Hpt. Gl. 482, 59.

un-coðu, e; *f.:* -coða, an; *m. Disease :*—Ús stalu and cwalu, stric and steorfa, orfcwealm and uncoða (*murrain and disease*) derede swýðe þearle, Wulfst. 159, 10. Gé gehwilce uncoðe gehǽldon, Homl. Th. i. 64, 23. Hé mid ísene ðone uncoðan (ða uncoðe, *v. rr.*) áceorfe, R. Ben. 52, 19. Orfcwealm oþðon mancwealm þurh fǽrlíce uncoða, Wulfst. 170, 2.

un-cræft, es; *m. An evil art, ill practice :*—Gif hé þurh gedrinc oððe þurh óðerne uncræft man ácwelle *si ex ebrietate vel alia prava arte hominem occiderit,* L. Ecg. P. iv. 68, 22; Th. ii. 230, 28. Gyf hit geweorðe ðæt man mid tyhtlan and mid uncræftum sacerd belecge, L. C. E. 5; Th. i. 362, 8. Utan sume getrýwða habban ús betweónan bútan uncræftan, Wulfst. 167, 5.

un-cræftig; *adj. Powerless :*—Se earma flýhð ·uncræftiga slǽp sleác mid sluman slincan on binder *somnus iners torporque gravis, desidia pigra cessabunt,* Dóm. L. 239.

un-crafod; *adj. With no claim made upon one :*—Se ðe sitte uncrafod on his áre on lífe, ðæt nán man on his yrfenuman ne sprece æfter his dæge *he that dwells on his property without any claims being made on him in his lifetime, that no man shall bring an action against his heir after his death,* L. Eth. iii. 14; Th. i. 298, 9. v. un-becrafod.

un-cristen; *adj. Not Christian :*—Ðeáh ðe hí ðágyta uncristene wǽron *thought they were not yet Christians,* Bd. 4, 16; S. 584, 9 note.

un-cumlíðe; *adj. Inhospitable :*—Se Hǽlend sprǽc tó sumum weligum men, ðe . . . him wæs láð þearfendum mannum mete tó syllenne, and hé wæs uncumlíðe, Wulfst. 257, 14.

un-cúþ; *adj. Unknown;* incognitus, Ælfc. Gr. 33; Zup. 205, 10.   I. *unknown, strange :*—Wæs Breotone eálond Rómánum uncúþ (*incognita*), Bd. 1, 2; S. 475, 3: Beo. Th. 4434; B. 2214. Gif men uncúð swyle on gesitte, Lchdm. i. 194, 27. Ðæt wǽre gelǽht án uncúð geong man, Homl. Skt. i. 23, 613: Ors. 6, 31; Swt. 286, 22. Heó on wéstenne gewunade eallum monnum uncúð, Shrn. 107, 24. Mon uncúþes andwlitan and uncúþes gegyrlan *hominem vultus habitusque incogniti,* Bd. 2, 12; S. 513, 35. Firum uncúþ, hwí..., Met. 4, 39. Word áres uncúþes, Exon. Th. 175, 5; Gú. 1190. Nis ðæs nán tweó. Ac ic wolde nú ðæt ðú mé sǽdest hwæthwegu uncúþes, Bt. 34, 6; Fox 142, 24: Beo. Th. 1757; B. 876. Ne fyligeaþ hig uncúþum (*alienum*), for ðam ðe hig ne gecneówun uncúðra (*alienorum*) stefne, Jn. Skt. 10, 5.   Uncúðum gode *deo ignoto,* Homl. Skt. ii. 29, 23. Be uncúðum yrfe (cf. ignotum pecus, L. Edm. C. 5; Th. i. 253, 7), L. Edg. H. 4; Th. i. 258, 21. Ðá álede ic mínne kyne-gyrlan and mé mid uncúþe hrægle gegerede (*I went incognito*), Nar. 18, 2. Hwá gifþ ðam uncúþan lífes fultum, Ap. Th. 11, 15. Gehýrde hé óðerne sang swilce uncúðne, Homl. Th. ii. 334, 16. Uncúðne weg, Met. 13, 58: Cd. Th. 181, 9; Exod. 58: Beo. Th. 2825; B. 1410. Drihten sent uncúðe þeóde ofer eów ða ðe gé ne cunnon *ducet te Dominus in gentem, quam ignoras,* Deut. 28, 36. Geopenigean uncúðe wyrd, hwǽr hé ðara nægla wénan þorfte, Elen. Kmbl. 2202; El. 1102. Nime man uncúþ sǽd æt ælmesmannum, Lchdm. i. 400, 17. Uncúð ádle *pestilentiae,* Mt. Kmbl. Lind. 24, 7. Cf. Se hwíta stán mæg wið eallum uncúþum (*unknown, and so caused by witchcraft?*) brocum, Lchdm. ii. 290, 11. Ðǽr him folcweras fremde wǽron, wine uncúðe, Cd. Th. 110, 32; Gen. 1847. Ðǽm folce seldsiéne and uncúðe wǽron wínes dryncas, Ors. 2, 4; Swt. 76, 12. Mínra firena ðe mé uncúðe wǽron *delicta ignorantiae meae,* Ps. Ben. 24, 6. Ðás ðe sint unncúðo *haec ignoras,* Jn. Skt. Lind. 3, 10. Hié uncúðra ǽngum ne willaþ feóres geunnan *they will grant no stranger life,* Andr. Kmbl. 355; An. 178: Cd. Th. 163, 14; Gen. 2698. Se útancuměna munuc ðe of uncúðum eardum cymð *si quis monachus peregrinus de longinquis provinciis supervenerit,* R. Ben. 109, 4. Gif wé scomiaþ ðæt wé tó uncúðum monnum (*men we do not know*) suelc sprecen, Past. 10; Swt. 63, 6. Oft ic nú miscyrre cúðe sprǽce, and þeáh uncúðre ǽrhwílum fond, Met. 2, 9.   II. *unknown, not understood :*—God sealde heora ǽlcum synderlíce sprǽce, ðæt heora ǽlcum wæs uncúð, hwæt óðer sǽde, Ælfc. T. Grn. 4, 11.   III. *unknown, uncertain :*—Ðære tíde ðe ús uncúþ is *ejus quod nobis incertum est temporis,* Bd. 2, 13; S. 516, 15. Heora sylfra forþfóre ðære tíd[e] is uncúþ *suum exitum, cujus hora incerta est,* 4, 3; S. 568, 21: Blickl. Homl. 125, 7. Us is swíþe uncúþ hwæt úre yrfeweardas getreówlíces dón willon, 51, 35: 119, 7. Clypiaþ gyt hlúdor uncúð þeáh ðe hé slǽpe (*cry aloud . . . peradventure he sleepeth,* 1 Kings 18, 27), Homl. Skt. i. 18, 119. Monig biþ uncúþ treówgeþofta teoraþ hwílum wáciaþ wordbeót *many a thing is uncertain, trusty comrade sometimes fails, weak prove words of promise,* Exon. Th. 469, 19; Hy. 11, 4. Fægere word ðis synd ðe gé bróhton, ac hí níwe syndon and uncúþe *pulchra sunt verba quae adfertis, sed nova sunt et incerta,* Bd. 1, 25; S. 487, 10. Ðonne cuman fǽrlíce on urcúðum tídum tó mynstre cumaþ *incertis horis supervenientes hospites,* R. Ben. 85, 9.   IV. *ungentle, unkind, hostile, harsh, unfriendly.* v. un-cúþlíce :—Bróga cwom egeslíc and uncúð, ealdfeónda níð, Exon. Th. 110, 23; Gú. 112. Móna

se ehtoða ... cild ácenned uncúð (*unfriendly ?*), strang, Lchdm. iii. 188, 3. Wē genēðdon eafoð uncúþes (*Grendel*); ūþe ic swíþor, ðæt ðú hine selfne geseón mōste, feónd fylwērigne, Beo. Th. 1924; B. 960. Sceaþa eáweþ uncúðne níð, 558; B. 276. Mec ongon hreówan ðæt mín hondgeweorc on feónda geweald fēran sceolde, sceolde uncúðne eard cunnian, sáre síþas, Exon. Th. 86, 34; Cri. 1418. [*Goth.* un-kunþs *ignotus*: *O. H. Ger.* un-kund *ignotus, incognitus, peregrinus, agrestis, incertus*: *Icel.* ú-kunnr *unknown*.]

un-cúþlíc; *adj. Unknown, strange, uncanny*:—Ða stānas sint ealle swíðe góde of tó drincanne wiþ ealle uncúþlícu þing, Lchdm. ii. 290, 14.

un-cúþlíce; *adv. Unkindly*:—Ðam elþeódigan and útancumenan ne lǽt ðú nó uncúðlíce wið hine ne mid nánum unrihtum ðú hine ne drecce (*peregrino molestus non eris*, Ex. 23, 9), L. Alf. 47; Th. i. 54, 21. [He spacc till hiss moder þuss unncuþliʒ (v. Jn. 2, 4), Orm. 14341. *Icel.* ú-kunnliga *like a stranger*.] v. un-cúþ, IV.

un-cwaciende; *adj. Without shaking* or *tottering*:—Ða ðe ne magon uncwaciende gestondan on emnum felda *qui in planis stantes titubant*, Past. 4; Swt. 41, 7.

un-cweden; *adj. Unsaid, revoked*:—Uncwedene yrfebēc *ruptum testamentum*, Wrt. Voc. i. 20, 42.

un-cweþende; *adj.* I. *not having speech*:—Ðeáh ðe gesomnod sý eal ðætte heofon oððe hel oððe eorðe ǽfre ácende, and ánra gehwylc ge ðæra cweðendra ge ðæra uncweðendra hæbbe gyldene býman on múðe, Salm. Kmbl. p. 152, 9. II. *not having a voice, inanimate*:—Hweþer ðú ongite ðæt ða uncweþendan gesceafta wilnodon tó biónne on ēcnesse swá swá swá men gif hí mihton *ea quae inanimata esse creduntur, nonne quod suum est quaeque simili ratione desiderant ?* Bt. 34, 11; Fox 150, 17. [Cf. Waldandes dóð unqueðandes só filo antkennian scolda ... erða ... bergós ... stēnós, Hél. 5663.]

un-cwid[d]; *adj. Undisturbed by charges, in undisputed possession*:—Se ðe sitte uncwýdd and uncrafod on his áre on lífe, L. Eth. iii. 15; Th. i. 298, 9. Ðǽr se bōnda sǽt uncwýd (-cwýdd, MS. G.) and unbecrafod (cf. ubi bunda manserit sine calumpnia, L. H. I. 14, 5; Th. i. 526, 3), L. C. S. 72; Th. i. 414, 22. [Cf. *Icel.* ú-kvíðinn *unconcerned*.] v. cwiðan.

un-cwisse; *adj. Speechless*:—Ðære tungan onstyrenesse beswicade (*linguae motu caruit*). Ðá wǽron ðrý dagas and ðreó niht fulle ðæt heó wæs uncwisse, Bd. 4, 9; S. 577, 18.

un-cyme; *adj. Mean, paltry, poor*:—On uncymre byrigenne geseted *ignobili traditus sepulturae*, Bd. 1, 33; S. 499, 7. Wæs his æþeleste rǽst on nacodre eorðan. Ða bǽdon hine his discipulos ðæt hié mōstan húru sume uncyme streównesse him under gedōn for his untrumnesse, Blickl. Homl. 227, 12. Ne hæfde wit monig óðer uncymran hors *nunquid non habuimus equos viliores plurimos ?* Bd. 3, 14; S. 540, 26.

un-cynde; *adj. Unnatural*:—Nim swá wuda swá wyrt of ðære stōwe ðe his eard and æþelo biþ on tó weaxanne and sette on uncynde stōwe him, ðonne ne gegrēwþ hit ðǽr náuht, Bt. 34, 10; Fox 148, 27. v. un-gecynde.

un-cynlíc; *adj. Unsuitable, improper*:—Ðæt wǽre uncynlícre, gif God næfde on eallum his ríce náne frige gesceaft, Bt. 41, 2; Fox 244, 28 note.

un-cyn[n]; *adj. Unsuitable, unfitting, improper*:—Ðæm ne is uncynn mæht bið sald *cui non inmerito potestas datur*, Lk. Skt. p. 3, 3.

un-cýpe. v. un-cípe.

un-cyst, e: -cyste, an; *f. A vice, defect, fault.* I. *of the body, a disorder*:—Wið wífa earfoðnyssum; ðás uncyste Grēcas hátaþ hystem cepnizam, Lchdm. i. 334, 18. Tó eallum uncystum ðe on gōmum beóð ácenned, 348, 12. II. *of diction, a fault, solecism*:—Ðære uncyste sylocismi, laudacismi, ða uncyste barbarismi (the passage is: Inter Scillam soloecismi et barbarismi baratrum ... scopulosas lautacismi collisiones, Ald. 80), Wrt. Voc. ii. 88, 27–33: 52, 49. III. *of morals, a vice, fault*:—Ðæt on ús ne sý gemēted nǽnigu stōw ǽmetig gástlícra mægena, ðæt ðǽr mæge yfelu uncyst on eardian, Blickl. Homl. 37, 10. Deós deáþberende uncyst (*envy*), 65, 13. Hē bær ða wǽtan ðære uncystan (-cyste, Bd. M. 82, 13) in ðam telgan *portat in ramo humorem vitii*, Bd. 1, 27; S. 495, 26. Ða uncyste ðære ánwielnesse *vitio obstinationis*, Past. 6; Swt. 47, 16. Gif ðú nán gód dōn nelt Gode tó wurðmynte, ðonne geswutelast ðú mid ðære uncyste ðíne yfelnysse, Homl. Th. i. 142, 2. Fýr ǽleþ uncyste, Exon. Th. 233, 17; Ph. 526: 81, 27; Cri. 1330. Gif hwylce uncysta on biscopum gemētte sýn *si qua sunt in episcopis vitia*, Bd. 1, 27; S. 492, 17. Ða unsýfernysse uncysta *rudera vitiorum*, 4, 3; S. 569, 32: 1, 27; S. 495, 32. Uncysta *passionum*, Wrt. Voc. ii. 77, 33. Ða men ðe ðyssum uncystum (*covetousness, envy, lust*) fylgaþ, Blickl. Homl. 25, 9. Hwá ongyt his uncysta *delicta quis intelligit ?* Ps. Th. 18, 11. Sume wealdaþ ealle uncysta and leahtras on him sylfum, Homl. Th. i. 344, 35. III a. *the vice of avarice, niggardliness, parsimony, want of liberality.* v. un-cystig:—Ðises mannes (*the rich man who gave nothing to Lazarus*) uncyst and upáhefednys hine besencte on cwicsūsle, Homl. Th. i. 328, 22. Spærnesse ł uncyste *frugalitatis*, Hpt. Gl. 425, 66. Ne hē uncysta ná begange *nec avaritie studeat*, R. Ben. 55, 3. [*O. H. Ger.* un-kust *vitium, scelus, dolus.* Cf. *Icel.* ú-kostr *a fault*.]

un-cystig; *adj. Niggardly, parsimonious, not liberal*:—Uncystig *frugus*, Wrt. Voc. ii. 109, 18: 36, 5: *frugi* vel *parcus*, i. 47, 37: *parcus*, Ælfc. Gr. 28, 7; Zup. 180, 13. Uncystig oððe spærhynde *frugi*, 9, 78; Zup. 74, 12. Fæsthafol oððe uncystig *tenax*, Wrt. Voc. i. 76, 5. Ne sǽde ðæt hálige godspel (Lk. c. 16) ðæt se ríca reáfere wǽre, ac wæs uncystig and mōdegode on his welum, Homl. Th. i. 328, 19. 'Gē noldon on mínum naman tíðian' ... Ðonne faraþ ða uncystigan intó ēcere cwicsūsle, ii. 108, 30: Wulfst. 289, 8. Ða uncystgan hē cysta lǽre, swá hē ða cystgan on merringe ne gebringe; ond swá eft ða rūmmōdan fæsthafolnesse lǽren, swá hí ða uncystegan on yfelre hneáwnesse ne gebrengen *sic tenacibus infundatur tribuendi largitas, ut tamen prodigis effusionis frena minime laxentur; sic prodigis prædicetur parcitas, ut tamen tenacibus periturarum rerum custodia non augeatur*, Past. 60; Swt. 453, 27–29. [*O. H. Ger.* un-kustig *rudis, impurus, dolosus, improbus*.]

un-cýðig; *adj. Ignorant, unacquainted*:—Wittende *sciens* ... uncýðig ł unwittende *ignorans*, Lk. Skt. p. 7, 18. Ðá wundrade heó ymb ðæs weres snyttro, hú hē swá geleáfful on swá lytlum fæce ond swá uncýðig ǽfre wurde gleáwnysse þurhgoten *she wondered at the man's wisdom, how in so little space and (previously) so ignorant he should ever become so full of belief, saturated with prudence*, Elen. Kmbl. 1918; El. 961. Elnes uncýðig *ignorant* (i. e. *devoid*) *of strength*, Exon. Th. 175, 23; Gú. 1199. [*Icel.* ú-kunnigr *unacquainted*: *Ger.* un-kundig.] v. on-cýðig; un-andcýðigness.

un-cýððu(-o); *indecl.*: -cýðð, e; *f.* I. *ignorance*:—Ne sprǽc hē (*Moses*) hit nó forðýðe his mód áuht geníerwed wǽre mid ðære uncýððe ðæs síðfætes *neque enim Moysi mentem ignorantia itineris angustabat*, Past. 41; S. 304, 17. Mín sceal of líce sáwul on síðfæt. nát ic sylfa hwider, eardes uncýðþu (*in ignorance of the land to which it is bound*), Exon. Th. 284, 22; Jul. 701. II. *a country not one's own, a strange land*:—Siþþan se eþel úðgenge weard Adame and Euan ... ðá hý on uncýððu scofene wurdon, on gewinworuld, Exon. Th. 153, 18; Gú. 827. [Þe soule is her in uncuðke ... and nout eðcene hwuch heo schal iwurðen in hire owune riche. þet fleshe is her et home, A. R. 140, 17–20.]

un-dǽd; e; *f. An ill deed, evil action, a crime, misdeed*:—On yfelan gedance and on undǽde, Wulfst. 165, 5. Ðá Helmstán ða undǽde gedyde ðæt hē Æðerēdes belt forstæl *when Helmstan committed the crime of stealing Æthered's belt*, Chart. Th. 169, 19, 28. Yflo uerco ł undǽdo *mala opera*, Jn. Skt. Lind. 3, 19. Scyldig and mánful mid undǽdum eall gesýmed *sceleratis impius actis*, Dóm. L. 58. Man deófol georne forbúge and his undǽda ealle oferhogie, Wulfst. 68, 12. [*O. H. Ger.* un-tát *delictum, macula, flagitiosum*: *Ger.* un-that.]

un-dǽftelice. v. un-gedǽftlíce.

un-dǽled; *adj. Undivided, not separated*:—Hit þencþ ætgædere beón gehál undǽled, forþam gif hit tódǽled biþ, ðonne ne biþ hit nó hál, Bt. 34, 12; Fox 152, 27. Ða hwíle ðe seó sáwul and se líchoma undǽlde beóþ, 34, 9; Fox 148, 5.

Undalan; *pl. The name which remains as Oundle, a town in Northamptonshire*:—Fērde hē forþ on his mynstre ðe hē hæfde on Undalana mǽgþe (*in provincia Undalum*), Bd. 5, 19; S. 641, 16. On ðære mǽgþe seó is gecýged In Undalum *in provincia quae vocatur In Undalum*, S. 636, 43. Wilferð biscop forðférde in (on v. r.) Undalum, Chr. 709; Erl. 45, 1. In Latin charters the form is Undale:—Uillam Undale ... de ipsa uilla Undale, Cod. Dip. Kmbl. iii. 93, 1, 8. Uillam de Undale, v. 6, 22. In later English it is Undela:—Ic gife ðone tún ðe man cleopeþ Undela, Chr. 963; Erl. 122, 4.

un-deáded; *adj. Not deadened*:—Wiþ springe ge ádeádedum ge undeádedum, Lchdm. ii. 8, 7.

un-deádlíc; *adj. Immortal, undying, imperishable, endless*:—God hálig and undeádlíc (*immortalis*), Rtl. 169, 17. Hē wunaþ undeádlíc, se ðe wæs deádlíc, Homl. Th. i. 150, 22. Se mann wǽre ǽfre undeádlíc, gif hē his Drihtne gehýrsumode, Hexam. 15; Norm. 22, 27. Undeádlíc, wyrm *the worm that never dies*, Homl. Skt. i. 4, 385. Tó onfónne ðæs undeádlícan gegyrlan on neorxna wange, Homl. Ass. 142, 105. Hí wǽron gehátene ealle *immortalis*, þæt sindon undeádlíce, Jud. Thw. p. 162, 31. þurh undeádlíce worulda *per immortalia secula*, Anglia xi. 119, 77. v. un-deáþlíc.

un-deádlícness, e; *f. Immortality*:—Úre ǽhta sind ēce on heofenum, ðǽr ðǽr undeádlícnys rícsaþ, Homl. Th. ii. 484, 28. Hyht hiora undeádlícnise (*immortalitate*) full is, Rtl. 86, 22: Homl. Th. i. 544, 3. Hæfde God ðæs mannes sáwle gegódod mid undeádlícnysse ... we ne forluron ná ða undeádlícnyssæ, 20, 1–4: Bd. 1, 27; S. 493, 4: 3, 21; S. 551, 3. v. un-deáþlícness.

un-dearninga(-unga), -deornunga; *adv. Without secrecy* or *concealment, openly*:—Elene for eorlum sprǽc undearninga, ides reordode hlúde for herigum, Elen. Kmbl. 809; El. 405: Fins. Th. 45; Fin. 22. Undearnunga, Elen. Kmbl. 1237; El. 620. Ðú ofer ealle undearnunga ðíne bearn sprecest and bealde cwyst *locutus es in aspectu filiis tuis et dixisti*, Ps. Th. 88, 16. Ic seah wyhte twá undearnunga plegan, Exon. Th. 429, 9; Rä. 43, 2. Gekýþe hē ðæt hē ðæt feoh undeornunga his cúðan ceápe in wíc gebohte, L. H. E. 16; Th. i. 34, 10.

un-deáþlíc; *adj. Immortal :*—Se líchoma bið ðonne undeáþlíc, þeáh hé ǽr deáþlíc wǽre, Blickl. Homl. 21, 31. Se ðe com deáðlíc tó ðissum middangearde ... hé árás undeáðlíc, Homl. Th. i. 222, 12, 18. Wé sprecaþ ymbe God, deáðlíce be undeáðlícum, 286, 8. Monna sáwla sint undeáþlíce (undeáðlíca, Cott. MS.) and éce, Bt. 11, 2 ; Fox 34, 33. v. un-deáðlíc.

un-deáþlíce; *adv. Immortally, to immortality :*—Úre Drihten on ðam ðriddan dæge undeáþlíce of deáðe árás, H. R. 5, 24.

un-deáþlícness, e ; *f. Immortality :*—Bið úre deáðlíca líchama áwend tó undeáðlícnysse, Homl. Th. ii. 70, 4. v. un-deáðlícness.

un-deáw; *adj. Without dew :*—Gewyrc ða wyrt on morgenne ðonne hió gedeáw sié, sume beóð undeáwe, Lchdm. ii. 92, 15.

un-declínigendlíc; *adj. Indeclinable :—Nihil* náht *indeclinabile,* ðæt is, undeclínigendlíc, Ælfc. Gr. 9, 8 ; Zup. 39, 6 : 38 ; Zup. 223, 1 : 44 ; Zup. 258, 1. *Indeclinabilia,* ðæt synd, undeclíniendlíce, 9, 78 ; Zup. 75, 3.

un-deógollíce. v. undígellíce.

un-deóp; *adj. Not deep, shallow* (lit. and fig.) :—Nis ðæt rǽdlíc ðing, gif swá hlútor wæter hlúd and undióp tóflóweþ æfter feldum óð hit tó fenne werð, Past. 65 ; Swt. 469, 6. Ðý lǽs mon má geóte on ðæt undíópe mód ðonne hit behabban mæge ðæt hit ðonne oferflówe *ne cum angusto cordi incapabile aliquid tribuitur, extra fundatur,* 63 ; Swt. 459, 14. [Sume hi diden in crucethus ð is in an cæste þat was scort and nareu and undeþ, Chr. 1137 ; Erl. 262, 9.]

un-deópþancol; *adj. Not given to think deeply, shallow :*—Nú smeáð sum undeópðancol man hú God mæge beón ǽghwǽr ætgædere, and náhwár tódǽled, Homl. Th. i. 286, 29.

un-deór(-deóre?); *adj. Not dear, cheap, common :*—Undeór hit is *vile valet,* Wrt. Voc. i. 28, 61. Ðæt hié mon ná undeórran weorðe móste lésan ðonne hié mon be ðam were geeahtige, L. Alf. pol. 32 ; Th. i. 82, 1. Hé nemde ða undíórestan wyrta ðe on wyrttúnum weaxe and ðeáh swíðe welstincenda *cum decimari minima diceret, extrema quidem de oleribus maluit sed tamen bene olentia memorare,* Past. 57 ; Swt. 439, 32. [Undeore he makeð God, þet for eni worldliche luue his luue trukie, A. R. 408, 14. *O. H. Ger.* un-tiuri *vilis : Icel.* ú-dýrr *cheap, of little value.*]

un-deóre; *adv. Cheaply, at a small cost :*—Undeóre hé bohte *vile vendidit,* Wrt. Voc. i. 28, 63. Gá seó wǽge wulle tó .cxx. p̄. and nán man hig ná undeóror ne sylle, L. Edg. ii. 8 ; Th. i. 270, 4. Ðæt sý undeóror geseald ðonne hit woroldmannum gewunelíc sý *vilius detur quam ab aliis secularibus,* R. Ben. 95, 17. Swylce mon undeórest bicgan mæge *quid vilius comparari potest,* 89, 17.

un-deornunga. v. un-dearnunga.

under; *prep. adv. Under.*   **I. with dat.** (1) local, without motion to bring one object under another, (a) where one object has another vertically above it :—Ða wæteru ðe wǽron under ðære fæstnisse, Gen. 1, 7. Under heofenum, 6, 17. Heó áléde ðone sunu under sumum treówe, 22, 15. Ic ge·seah ðé ða ðú wǽre under ðam fíctreówe, Jn. Skt. 1, 48. (a 1) where one object is supported by another :—Mearh under módegum, Elen. Kmbl. 2383 ; El. 1193. Ðæt scip wæs yrnende under segle, Ors. 1, 1 ; Swt. 19, 34 : Andr. Kmbl. 1009 ; An. 505. Wedera leód heard under helme, Beo. Th. 689 ; B. 342. Cwom Wealhþeów gán under gyldnum beáge, 2330 ; B. 1163. (b) where one object is at the lower part of another, *under, at the foot of :*—Wæs bát under beorge, Beo. Th. 427 ; B. 211. Ða com of móre under misthleoþum Grendel gongan, 1425 ; B. 711. Under weallum, Cd. Th. 146, 6 ; Gen. 2418. v. neoþan. (c) where an object is surrounded, covered, shut in, etc. by another, *under, within :*—Heora andwlitan inbewrigenum under loðum, Cd. Th. 95, 29 ; Gen. 1586. Under lindum, 192, 7 ; Exod. 228. Under gyrdelse, Exon. Th. 436, 34 ; Rä. 55, 11 : 431, 3 ; Rä. 45, 2. Under heolstorlocan bídan *to wait in prison,* Andr. Kmbl. 288 ; An. 144 : Beo. Th. 3860 ; B. 1928. Heó under breóstcofan bearn ácende, Hy. 10, 16. Hwæþer him yfel þe gód under wunige *whether evil or good dwell within the mind,* Exon. Th. 82, 4 ; Cri. 1333. (d) where an object is surrounded by others, *among :*—Ne mehton ða senátus nǽnne consul under him findan, Ors. 4, 10 ; Swt. 196, 10. Sang se wanna fugel under deoredsceaftum, Cd. Th. 119, 23 ; Gen. 1984. (2) local, where motion is implied :—Mec mín freá sendeþ under sǽlwonge, Exon. Th. 382, 27 ; Rä. 4, 2. (3) figurative, (a) marking subordination, subjection, rule, etc. :—Sete hig under Aarone, ðæt hig þenigeon him ... Beón hig þénas under Aarone and his sunum, Num. 3, 6, 9. Ða ðe under Alexandre fyrmest wǽron, Ors. 3, 11 ; Swt. 142, 21. Aulixes hæfde twá ðióda under ðam Kásere, Bt. 38, 1 ; Fox 194, 4 : Met. 26, 5. Under Rómwarum, Hy. 10, 26. Burga fífe wǽran under Norðmannum gebégde, Chr. 942 ; Erl. 116, 15. Ic eom man under anwealde gesett, and ic hæbbe þegnas under mé, Mt. Kmbl. 8, 9. Ealle ða rícu ðe him under beóð, Bt. 16, 1 ; Fox 50, 3. Bútan ðam dǽle ðe under Dena onwalde wæs, Chr. 901 ; Erl. 96, 23. Under hæþenra hyrda gewealdum, Exon. Th. 44, 19 ; Cri. 705. Eáþmódgiaþ eów sylfe under ðære mihte Godes handa, Blickl. Homl. 99, 3. Oðer tíd is seó ðe wæs under ǽ ; seó ðridde ... is gecweden under Godes gife, Homl. Th. i. 312, 31. Cild ic eom

under gyrde (*sub virga*) drohtniende, Coll. Monast. Th. 34, 21. (b) marking protection, shelter :—Under mundbyr[d]e *sub pretextu,* Wrt. Voc. ii. 79, 84 : 84, 15. Under wealla hleó, Cd. Th. 259, 13 ; Dan. 691. (c) marking pretence :—Under intingan *sub obtentu* (Mk. 12, 40), Wrt. Voc. ii. 73, 43. (d) marking exposure, suffering :—Hú se mánscaða under fǽrgripum gefaran wolde, Beo. Th. 1480 ; B. 738. Under stormum, Exon. Th. 476, 21 ; Ruin. 11. Fela ðæs ðe hé ádreág under nídgysta nearwum clommum, 134, 21 ; Gú. 511. Under Godes egsan, 146, 2 ; Gú. 703. (e) marking rank, degree :—Under hire selfre hió biþ ðonne, ðonne heó lufaþ ðás eorþlícan þing, Bt. 33, 4 ; Fox 132, 17. (f) marking circumstances or conditions under, among, or during which something takes place :—Be ðam men ðe bið húsl forboden and under þam (*interim*) forðfǽrð, L. Ecg. P. i. 13, tit. ; Th. ii. 170, 25 : *interea,* 13 ; Th. ii. 178, 15 : Chr. 876 ; Erl. 78, 12 : 1046 ; Erl. 173, 5. Hé him gehét ðæt hé his ríce wið hiene dǽlan wolde and hiene under ðæm ofslóg *Titum, mox ut in societatem regni adsumpsit, occidit,* Ors. 2, 2 ; Swt. 66, 12. Under ðæm ðe hé him onwinnende wæs *while he was warring upon them,* 1, 2 ; Swt. 30, 5. Under ðæm gewinne hié genámon friþ *in the course of the struggle they made peace ; pace armis quaesita,* 1, 10 ; Swt. 46, 7. Swá wæs ðæt hié under ðære sibbe tó ðære mǽstan sace become, 4, 7 ; Swt. 182, 28 : 4, 12 ; Swt. 210, 10 : Chr. 865 ; Erl. 70, 33. Wé sceolan under ðæm feówerte[g]oþan geríme syllan ðone teóþan dǽl úre worldspéda *we must during that forty days give the tithe of our worldly wealth,* Blickl. Homl. 35, 18. (g) marking manner :—Under earhfǽre bannan tó beadwe *to summon to war by sending round an arrow* (v. Grmm. R. A. 162), Elen. Kmbl. 87 ; El. 44.   **II. with acc.,** (1) local, where motion is expressed or implied, (a) where one object comes to have another vertically above it :—Ne eom ic wyrðe ðæt ðú gá under míne þecene, Lk. Skt. 7, 6. Sume steorran gewítaþ under ða sǽ, Bt. 39, 3 ; Fox 214, 26. Stefn in becom under hárne stán, Beo. Th. 5100 ; B. 2553. Hió ðæt líc ætbær under firgenstreám, 4263 ; B. 2128. Ða námon hig ánne stán and lédon under hine, Ex. 17, 12. Lǽd under earce bord eaforan ðíne, Cd. Th. 80, 23 ; Gen. 1333. Under helm drepen biteran strǽle, Beo. Th. 3495 ; B. 1745. ¶ Combined with on :—Lecgan uppan ðone stán and on under, Lchdm. iii. 38, 18. (b) where one object comes to the lower part of another :—Weorod eodon unblíðe under Earna nǽs wundur sceáwian ; fundon on sande sáwulleásne ðone ðe him hringas geaf, Beo. Th. 6055 ; B. 3031. (c) where one object comes to be surrounded, covered, shut in, etc. by another :—Hé gelædde brýd under burhlocan, Cd. Th. 153, 12 ; Gen. 2537 : Andr. Kmbl. 1879 ; An. 942. Under heolstorhofu hreósan, Elen. Kmbl. 1524 ; El. 764. In under eoderas, Beo. Th. 2068 ; B. 1037 : Cd. Th. 147, 25 ; Gen. 2445. Under sceát, 124, 17 ; Gen. 2064 : Exon. Th. 436, 21 ; Rä. 55, 4. (d) where extension under a surface is implied :—God under roderas feng wolde ðæt eorðe geseted wurde woruldsceafte, Cd. Th. 6, 33 ; Gen. 98 : 71, 5 ; Gen. 1166. Under heofenes hwealf, Beo. Th. 1156 ; B. 576 : 4033 ; B. 2015. Under swegles begong, 1724 ; B. 860 : 3550 ; B. 1773 : An. 415 ; An. 208. Siððan ǽfenleóht under heofenes hádor beholen weorþeþ *after the evening light has died out everywhere beneath the sky,* Beo. Th. 832 ; B. 414. (2) figurative, (a) marking subordination, subjection, rule, etc :—Under hand hǽðenum déman *in subjection to a heathen ruler,* Cd. Th. 220, 14 ; Dan. 71. Gewát him Abraham under Abimelech ǽhte lǽdan, 158, 22 ; Gen. 2621. Under ánes meaht ealle forlǽtan, Exon. Th. 294, 30 ; Crä. 23. Hí wǽran geseald under sweordes hand *tradentur in manus gladii,* Ps. Th. 62, 8. (b) marking exposure, suffering :—Se eów in hæft bedráf, under nearowe clom, Exon. Th. 138, 2 ; Gú. 570. (c) in various other senses :—Ne þurfon gé wénan ðæt gé ðæt orceápe sellon, ðæt gé under Drihtnes borh syllaþ, Blickl. Homl. 41, 13. Hí him his forwierndon and hit under ðæt lǽdedon for ðon ðe hé æt ðæm óþrum cirre sige næfde *they refused him the triumph, and sheltered themselves under the excuse, that he had not on the other occasion been victorious,* Ors. 5, 2 ; Swt. 216, 31. Under monnes híw *in human form,* Exon. Th. 144, 22 ; Gú. 682. Ne swerigen gé nǽfre under (cf. þurh, Ex. 23, 13) hǽðene godas, L. Alf. 48 ; Th. i. 54, 23. Se king swór under God ǽlmihtine and under ealle hálgan ðærtó, Chart. Th. 340, 2. Symle byð under dæg and niht feówer and xxᵗⁱᵍ tída, Lchdm. iii. 260, 12. Ðú áhst tó fyllene ðíne seofen tídsangas under dæg and niht, Wulfst. 290, 18.   **III.** adverbial ; see also the compound forms given below. (1) where one object has another vertically above it :—An treów ðæt mæge .xxx. swína under gestandan *a tree so big that thirty swine can stand beneath,* L. In. 44 ; Th. i. 130, 3. His hors wearð under ofscoten *his horse was killed under him,* Chr. 1079 ; Erl. 216, 25. Sume crupon under *some crept beneath,* 1083 ; Erl. 217, 22. Hé hét fýr under bétan, Homl. Th. i. 4, 393. Ðǽr ðæs Hǽlendes fótlástas syndon under, Shrn. 81, 29. (2) where relative height is marked, *below, beneath :*—Hé funde wynleásne wudu, wæter under stód, Beo. Th. 2837 ; B. 1416. (3) *down* as opposed to up :—Wið ðone ðe him mete under ne gewunige *if his food do not remain down,* Lchdm. ii. 190, 1 : 198, 23. [*Goth.* undar : *O. Sax.* undar : *O. Frs.* under : *O. H. Ger.* untar : *Icel.* undir.] v. þǽr-under.

under-ágenlíc; *adj. The word glosses subnixus in :* Underágenlícum beadum *subnixis precibus,* Rtl. 182, 33.

under-andfónd *glosses* susceptor *in:* Underondfóendo *susceptores,* Rtl. 193, 11.

under-bæc; adv. **I.** where there is motion of a person in the direction towards which the back is kept turned, *backwards,* (a) approaching an object :—Sem and Iafeth eodon underbec *Shem and Japhet went backward ;* incedentes retrorsum, Gen. 9, 23. (b) motion from :—Ðá eodon hig underbæc *they went backward ;* abierunt retrorsum, Jn. Skt. 18, 6. Hwílum ic underbæc bregde nebbe, Exon. Th. 498, 5 ; Rä. 87, 8. Feallan underbæc *to fall backwards,* Homl. Th. ii. 392, 8 : Homl. Skt. i. 12, 63. Sceófan underbæc, 14, 88 : 18, 345 : Homl. Th. ii. 300, 15. **II.** marking retreat, where there is motion of a person in the direction to which his back has been turned, *back :*—Diabolo non dicitur : ' Uade retro me,' sed : ' Uade retro '. . . Crist cwæð tó ðam deófle : ' Gá ðú underbæc.' Deófles nama is gereht, nyðerhreósende. Nyðer hé áhreás and underbæc hé eode ðá ðá hé wæs áscyred fram ðære heofonlícan blisse, Homl. Th. i. 172, 30–35 : Wrt. Voc. ii. 71, 70. Nú næfð Israêl nánne stede wið his fýnd ac flíhð underbæc *nec poterit Israel stare ante hostes suòs, eosque fugiet,* Jos. 7, 12. Underbæc cyrran *to turn back,* Exon. Th. 405, 2 ; Rä. 23, 17 : Ps. Lamb. 34, 4. (Under bæce, Ps. Spl. 34, 5.) Ðú gehwyrfdest míne fýnd underbæc *in convertendo inimicum meum retrorsum,* Ps. Th. 9, 3. **III.** where an action is directed towards a point behind the agent's back, *behind, back :*—Ðú forwurpe mín word underbæc fram ðé *projecisti sermones meos postea,* Ps. Th. 49, 18 : Homl. Th. ii. 532, 3. Ne beseoh ðú underbæc *noli respicere post tergum,* Gen. 19, 17 : Cd. Th. 154, 28 ; Gen. 2562 : Jos. 8, 20. Ðá beseah hé hine underbæc wið ðæs wífes, Bt. 35, 6 ; Fox 170, 14. Háwian underbæc *respicere retro* (Lk. 9, 62), Past. 51 ; Swt. 403, 2. **IV.** where the point from which something proceeds is behind the recipient :—Ðín eáran gehíraþ underbæc *thine ears shall hear a word behind thee* (Is. 30, 21) ; aures tuae audient verbum post tergum monentis, Past. 52 ; Swt. 405, 26 : 407, 12. v. next word.

under-bæcling ; adv. Back :—Ðonne gecerraþ míne fýnd underbæcling *tunc conuertentur inimici mei retrorsum,* Ps. Lamb. 55, 10. Underbæclinc, Blickl. Gl.

under-beginnan *to attempt :*—Nú þincþ mé ðæt ðæt weorc is swíþe pleólíc mé oþþe ænigum men tó underbeginnenne, Ælfc. Gen. Thw. 1, 14. v. under-ginnan.

under-beran *glosses* supportare, sustinere, subsistere :—Underbearaþ *subportantes,* Rtl. 13, 35. Underbær *sustinuit,* 27, 31. Underbeara *subsistere,* 7, 38. Underberende *supportantes,* Scint. 24, 1.

under-bígan *to subject :*—Underbéged *subjectus,* Rtl. 125, 25 : Mt. Kmbl. p. 3, 10. v. under-búgan.

under-brédan *glosses* substernere *in :* Underbrǽddon gegerelo *substernebant uestimenta,* Lk. Skt. Lind. 19, 36.

under-bregdan *to spread under :*—Úre eáre hí him on niht underbrédaþ and mid óðran hí wreóð *unam aurem sibi noctem substernunt, de alia se cooperiunt,* Nar. 37, 12.

under-búgan *to submit :*—Paulus underbeáh swurdes ecge, Homl. Th. i. 382, 6. Úre Hǽlend ródehengene underbeáh, 66, 7.

under-burh *a suburb :*—Of Gomorra underburgum *de suburbanis Gomorrhae,* Deut. 32, 32.

underburh-ware ; *pl. The inhabitants of a suburb :*—Of underburhwarum *de suburbanis,* Cant. M. ad fil. 32. v. preceding word.

under-cirran *glosses* subvertere *in :* Undercerrende *subvertentem,* Lk. Skt. Lind. Rush. 23, 2.

under-crammian *to stuff full below ;* suffercire :—Hí mid byrnendum glédum ðæt bed undercrammodon, Homl. Th. i. 430, 4.

under-creópan *to enter surreptitiously ;* surrepere :—Ðá wæs ðæs wítegan cnapa mid gítsunge undercropen *avarice crept into the heart of the prophet's servant,* Homl. Th. i. 400, 16. Ðæt ne feónd ús undercreópe (*but the Latin is* subripiat), Hymn. Surt. 12, 28.

under-cuman *glosses* subvenire, succedere *in :* Undercwom *succedente,* Mt. Kmbl. p. 8, 9. Undercyme *subveniat,* Rtl. 66, 35. Undercymende *succedente,* 37, 35.

under-cyning, es ; *m. A dependent, tributary king, one who rules under another.* Cf. þeód-cyning :—Cyning *rex,* lytel cyning oððe undercyning *regulus,* Ælfc. Gr. 5 ; Zup. 16, 19. Kyning *basileus,* undercyning *regillus,* Wrt. Voc. i. 17, 54. Sum undercyning wæs *erat quidam regulus,* Jn. Skt. 4, 46, 49 : Homl. Th. i. 128, 5. Ic Offa, Myrcena kining . . . Ic Aldréd, Wigraceastres undercining (*subregulus*), Cod. Dip. Kmbl. i. 186, 13. Griffin swór áðas ðæt hé wolde beón Eádwearde kinge hold underkinge, Chr. 1056 ; Erl. 190, 35. Tiberius hæfde anweald ofer eall Rómána ríce, and him wæs undercyning Heródes, Nicod. 1 ; Thw. 1, 8. Ðe Hǽlend stód befær ðone undercynige (*praesidem*), and gefrægn hine ðe undercynig (*praeses*), Mt. Kmbl. Lind. 27, 11, 21. Tó undercyningum *ad praesides,* 10, 18 : Mk. Skt. Lind. 13, 9. Ic ðǽr gemétte Týtum and Vespasianum ðýne (*Tiberius'*) getrýwestan undercyningas, Homl. Ass. 191, 285. [*Icel.* undir-konungr.]

under-delf *glosses* suffossum, Ps. Spl. 79, 17. v. under-holung.

under-delfan *to dig under, undermine, dig out* (lit. or fig.) :—Ic underdelfe *subfodio,* Ælfc. Gr. 28, 6 ; Zup. 179, 11. Æt dura hé under-

delfeþ (*suffodiet*) fótwylmas ðíne, Scint. 196, 8. Ðú beswice oððe underdulfe (*supplantasti*) onárísende on mé under mé, Ps. Lamb. 17, 40. Seáþ hé geopnode and hé underdealf ðæne *lacum aperuit et effodit eum,* 7, 16. Hé nolde geþafigan ðæt man hys hús underdulfe *non sineret perfodi domum suam,* Mt. Kmbl. 24, 43 : Lk. 12, 39. Hé nolde geþafian ðam þeófe náteshwón ðæt hé underdulfe ðigellíce his hús, Homl. Ass. 50, 13. Ne beóþ underdolfene t ne beóþ forscrænste stæpas his *non supplantabuntur gressus ejus,* Ps. Lamb. 36, 31. ¶ underdelfan *glosses* suffocare, Mt. Kmbl. Lind. 13, 7, 22 : Mk. Skt. Lind. Rush. 4, 7, 19.

under-diácon, es ; *m. An under-deacon, a sub-deacon :*—Underdiácon *subdiaconus,* Wrt. Voc. i. 42, 26 : Rtl. 194, 9. Subdiaconus is sóðlíce underdiácon, se ðe ða fatu byrð forð tó ðam diácone, and mid eádmódnysse þénaþ under ðam diácone æt ðam hálgan weofode mid ðam húselfatum, L. Ælfc. C. 15 ; Th. ii. 348, 9.

under-dón *to put under ;* subjicere :—Ðone wudu ðe man ðæt fýr sceal underdón *ligna quibus subjiciendus est ignis,* Lev. 1, 12.

under-drencan *glosses* suffocare *in :* Underdrencdo *suffocati,* Mk. Skt. Lind. 5, 13.

under-drifenness *glosses* subjectio *in :* Of underdrifenise diówla *de subjectione daemonum,* Lk. Skt. p. 6, 16.

under-etan *to eat away below, to sap ;* subedere :—Ðæt mennisce mód bið undereten and áweged of his stede ðonne hit se wind strongra geswinca ástyroþ, Bt. 12 ; Fox 36, 17.

under-fang *glosses* susceptor, Ps. Spl. 3, 3 : 17, 3 : 45, 7 : 143, 2.

under-fangelnes *glosses* susceptio *in :* Underfangelnes heáfdes mínes *susceptio capitis mei,* Ps. Lamb. 107, 9.

under-fangenness, e ; *f. Undertaking, assumption :*—Mid underfangennysse menniscnysse *assumptione humani,* Ath. Crd. 35.

under-feng, es ; *m. Undertaking, acceptance :*—Ðýlæs hé for ðý underfenge (*the undertaking the office of teacher*) his eádmódnesse forlǽte, oððe eft his líf sié ungelíc his ðenunga, oððe hé tó ðríste and tó stíð sié for ðý underfenge his láreówdómes *ne aut humilitas accessum* (*ad culmen regiminis*) *fugiat ; aut perventioni vita contradicat ; aut vitam doctrina destituat ; aut doctrinam praesumtio extollat,* Past. proem. ; Swt. 23, 22.

under-flówan *to flow under :*—Wæs ic neoþan wætre, flóde underflówen *beneath had I water, the flood flowing under me,* Exon. Th. 392, 3 ; Rä. 11, 2.

under-folgoþ, es ; *m. An office under a superior :*—Hé (*Julian*) sǽde ðæt nán cristen man ne móste habban nænne his underfolgoþa (sunderfolgeþa, Swt. 286, 5), Ors. 6, 31 ; Bos. 128, 2.

under-fón ; *p.* -íeng, *pl.* -féngon ; *pp.* -fangen. **I.** *to receive, to have given, to get :*—Ne underfó ic náne beorhtnesse æt monnum *claritatem ab hominibus non accipio,* Jn. Skt. 5, 41. On ðam lande ðe ðú underféhst *in the land which the Lord giveth thee* (A.V.) ; in terra quam acceperis, Deut. 28, 8. Eálá ðæt hit is gód ðæt mon micelne welan áge, nú se nǽfre ne wyrþ orsorg ðe hine underféhþ *O praeclara opum mortalium beatitudo, quam cum adeptus fueris, securus esse desisti,* Bt. 14, 3 ; Fox 46, 34. Hú micelne unweorþscipe se anwald brengþ ðam unmedeman, gif hé hine underfégþ, 27, 2 ; Fox 96, 10. Mid ðam casu (*ablative*) byð geswutelod, swá hwæt swá wé underfóð æt óðrum . . . *ab hoc homine pecuniam accepi* fram ðisum men ic underféng feoh, Ælfc. Gr. 7 ; Zup. 23, 7–11. Ða Godes þeówas ðe ða sceattas underfóð ðe wé Gode syllaþ, L. Edg. S. 1 ; Th. i. 272, 15. Ðá underféng Iudas ðæt folc æt ðam bisceopum, Jn. Skt. 18, 3. Ne cwæð hé nú, 'ðú héte mé,' ac 'forgeáfe mé' ; mid ðam worde is seó gifu geswutelod ðe hé on ðære menniscnysse underféng. Seó menniscnys wæs underfangen fram ðam godcundum worde, Homl. Th. ii. 364, 11–14. Hí heora méde underféngon, 80, 2 : i. 68, 30. Ðæt hé ne cnytte ðæt underfongne feoh on ðam swætline, Past. 9 ; Swt. 59, 13. **II.** *to receive, submit to* a rite, etc. :—Godes ðeów, se ðe hád underféhð *God's servant, who takes orders,* Homl. Th. ii. 48, 31. Ðá wé fulluht underféngan, Wulfst. 167, 1. **III.** *to receive* a person, (1) *to receive* for the purpose of entertaining, sheltering, harbouring, etc. :—Florus férde him tógeánes and ða ædelan Godes menn underféng tó him, and foresceáwode ða wununge, Homl. Skt. i. 6, 138. Ðæt nán man nænne man ne underfó ná leng ðonne þreó niht, L. C. S. 28 ; Th. i. 392, 9. Ðæt náðor ne hý ne wé ne underfón óðres wealh, ne óðres þeóf, ne óðres gefán, L. Eth. ii. 6 ; Th. i. 288, 4. (2) *to receive* for safe conduct, custody, etc. :—Ðá underféngon ðæs déman cempan ðone Hǽlend on ðam dómerne, Mt. Kmbl. 27, 27. Náh tó farenne Wylisc man on Ænglisc land bútan gesettan landmen, se hine sceal æt stæde underfón, and eft ðǽr bútan fácne gebringan, L. O. D. 6 ; Th. i. 354, 25. (3) *to receive* as a servant or dependent :—Be ðon ðe óðres mannes man underfó fram ðisum men ic underféng feoh. Ne underfó nán man óðres mannes man bútan ðæs leáfe ðe hé ǽr fyligde, L. Ed. 10 ; Th. i. 164, 14 : L. Ath. i. 22 ; Th. i. 210, 20 : iv. 1 ; Th. i. 220, 18. 'God underféng his cnapan Israhêl.' Mid ðam naman syndon getácnode ealle ða ðe Gode gehýrsumiaþ, ða hé underféhð tó his werode, Homl. Th. i. 204, 13. (4) *to receive, admit* into a society :—Hé sí underfangen on gegæðrunge *suscipiatur in congregatione,* R. Ben. Interl. 97, 4. (5) *to receive* as a master, *to submit to :*—Gif se anweald of his ágenre gecynde gód wǽre,

ne underfēnge hē næfre ða yfelan ac ða gódan, Bt. 16, 3; Fox 54, 23. (6) *to receive, admit the claims of*:—Se ðe eów underfēhð, hē underfēhð mē, and se ðe mē underfēhð, hē underfēhð ðone ðe mē sende, Mt. Kmbl. 10, 40. Ic com on mīnes Fæder naman and gē mē ne underfēngon. Gyf ōðer cymþ on his āgenum naman, hyne gē underfóð, Jn. Skt. 5, 43. (6 a) *to receive, admit the force of* a person's words, *accept* testimony:—Hē cýð ðæt hē geseah and gehýrde, and nān man ne underfēhþ his cýðnesse, Jn. Skt. 3, 32. Se ðe mīne word ne underfēhð, hē hæfþ hwā him dēme, 12, 48. Ne underfóð ealle menn ðis word, Mt. Kmbl. 19, 11. **IV.** *to receive* what is offered, *to accept*:—Drihten gebed mīn hē underfēng (-fang, MS.) *Dominus orationem meam suscepit*, Ps. Spl. 6, 9. Hē nolde nāne frēfrunge underfón *noluit consolationem accipere*, Gen. 37, 35. **V.** of things, *to receive, serve as a receptacle for*:—Underfó *receptet (the passage is :* Cadaver nequaquam sepulchri sarcophagus receptet, Ald. 52), Hpt. Gl. 496, 11. **VI.** *to receive* or *accept* an office, a duty, etc., *to take upon one's self, to undertake* a labour, task, etc., (a) where the object of the verb is a word implying action or effort:—Regulus underfēng Cartaina gewinn *Regulus, bellum Carthaginense sortitus*, Ors. 4, 6; Swt. 174, 1 : 2, 2; Swt. 66, 13. Ðæt ilce (ða ðegnunga ðæs lāreówdómes) ðæt hē untælwierðlíce ondrēd tó underfoonne, Past. 7; Swt. 48, 19. Sió giémen ðæs underfangenan lāreówdómes *suscepta cura regiminis*, 4; Swt. 37, 13. Ðā hié gewin hæfdon underfongen *bella suscepta*, Ors. 4, 9; Swt. 192, 29. Ǽr hē hæbbe godcunde bóte underfangen, L. Edm. S. 4; Th. i. 248, 25. Underfangenre andwealhnysse *adeptae integritatis*, Hpt. Gl. 465, 70. (b) where the object of the verb denotes that in respect to which action or effort is needed:—Gif wīfman hīwrǽdene underfēhð *si mulier familiam susceperit*, L. Edg. C. 25; Th. ii. 272, 7. Ðonne hié monna heortan underfóð tó lǽronne, Past. 21; Swt. 161, 12. Hig underfóð ðis folc mid ðē *ut sustentent tecum onus populi*, Num. 11, 17. On ða rīcu ðe hē underfangen hæfde, Bt. proem.; Fox viii, 8. **VII.** *to receive* what is burdensome, *undergo, bear*:—Gē underfóð eówere unrihtwísnissa (*ye shall bear your iniquities*, A.V.), ðæt gē witon mīne wrace *recipietis iniquitates vestras, et scietis ultionem meam*, Num. 14, 34. **VIII.** *to take surreptitiously, to steal*:—Gyf feoh sý underfangen (or ? under fangen), Lchdm. iii. 286, 4. [O. E. Homl. Laym. Kath. Gen. and Ex. under-fon : Orm. unnderr-fon; A. R. under-von : Piers P. under-feng; p.: Ayenb. onderving : O. H. Ger. untar-fāhan.] v. under-niman.

under-fōnd glosses susceptor, Ps. Lamb. 3, 4: 53, 6: 90, 2.

under-fóndlíc; adj. *To be received*:—Se underfónlíca *suscipiendus*, R. Ben. Interl. 97, 8.

under-fylgan glosses subsequi in : Underfylgdon (-fyligdon, Rush.) *subsecutae*, Lk. Skt. Lind. 23, 55.

under-gān *to undermine, ruin*:—Gif hwylcne man deóful tó ðam swýþe undergān hæbbe *si diabolus hominem aliquem adeo perdiderit*, L. Ecg. P. iv. 14; Th. ii. 208, 12. Ne sý nān eorðcund cyning mid gītsunge tó ðæm swīþe undergān, Lchdm. iii. 444, 3. [Ðis maidenes redden . . . hu he migten undergon (*deceive*) here fader, Gen. and Ex. 1147.]

under-gangan *to undergo* :—Ic undergange *subeo*, Ælfc. Gr. 37; Zup. 217, 17. [Me birrþ beon fullhtnedd att tin hannd þin blettsinng tunnderrganngenn, Orm. 10661.]

under-geoc; adj. *Accustomed to the yoke, tame*:—Ofer ðone fola suna undergeocas (cf. on folan sunu ðære teoma, Rush.) *super pullum filium subjugalem*, Mt. Kmbl. Lind. 21, 5.

under-gerēfa, an; m. *An under-officer*:—Gerēfa *consul*, undergerēfa *proconsul*, Wrt. Voc. i. 18, 5. Se undergerēfa *the pro-prefect*, Homl. Skt. i. 4, 332 : 7, 216.

under-geþeóded subject:—Nǽnig ealdormonna ne ūs undergeþeódedra (-endra, MSS. B. H.), L. In. proem.; Th. i. 102, 11. v. under-þeódan.

under-ginnan; p. -gann; pl. -gunnon; pp. -gunnen *To begin, attempt* :—Ic gedyrstlæhte ðæt ic ðās gesetnysse undergann *I ventured to attempt this work* (the translation of a Latin work), Homl. Th. i. 2, 27. Gregorius ūs trahtnode ðyses godspelles dígelnysse ðus undergynnende : 'Dryhten ūs gewilnaþ . . . etc.', 608, 9.

under-gitan; p. -geat, pl. -geáton; pp. -giten *To understand, perceive, know*:—Ic gefrēde oððe undergyte *sentio*, Ælfc. Gr. 30, 2; Zup. 190, 11. Ic undergyte *perpendo*, 28, 7; Zup. 181, 6. Gif folces man syngaþ þurh nytenyss[e] and his gylt undergit (*et cognoverit peccatum suum*), Lev. 4, 28. Fram hyra wæstmum gē hí undergytaþ (*cognoscetis*), Mt. Kmbl. 7, 16. Ðā Samson heora syrwunga undergeat and ārās, Jud. 16, 3; Homl. Th. i. 62, 30. Ðā undergeat Noe, ðæt ða wæteru wǽron ādrūwode ofer eorðan, Gen. 8, 11. Ðā Ulfcytel ðæt undergeat, Chr. 1004; Erl. 139, 25. Ðā Eádwine eorl and Morkere eorl ðæt undergeáton, 1066; Erl. 198, 39; Ors. 3, 7; Swt. 112, 26. Hig ne undergēton (-geáton, MS. A.) ðæt hē tealde him God tó fæder *non cognouerunt quia patrem eis dicebat*, Jn. Skt. 8, 27. Ne uudergiton (-geáton, MS. A.) (*cognouerunt*) hys leornigcnihtas ðās þing ǽrest, 12, 16. Gif hwam gelustfullaþ tó witanne hwæt sý *quadrans*, ðonne undergite hē ðæt *quadrans* byð se feórða dǽl, Anglia viii. 298, 10. Understandan

capere, intelligere, undergite *capiat*, Hpt. Gl. 437, 28. Undergitende heortan *cor intelligens*, Deut. 29, 4. Beón undergiten *colligi, cognosci, intelligi*, Hpt. Gl. 460, 13. Þe beóð undergitene *noscuntur, intelleguntur*, 430, 11.

under-hebban glosses accipere, sustollere, sublevare in :—Onfoeng ł underhóf *accepit*, Mt. Kmbl. Lind. 8, 17. Genom ł underhóf *sustulit*, Jn. Skt. Lind. Rush. 5, 9. Underhóf *subleuasset*, 6, 5. Underhebendum ēgum *subleuatis oculis*, Lind. 17, 1. Ðæt uē sié underhefen *sublevari*, Rtl. 80, 1.

under-hlystan renders subaudire in : Subaudio ic underhlyste, *subaudis* ðu underhlyst, *subaudit* hē underhlyst, Ælfc. Gr. 26; Zup. 151, 2-4.

under-hlystung renders subauditio in : Subauditionem et personam, ðæt ys, underhlystunge and hād, Ælfc. Gr. 26; Zup. 151, 1.

under-hnígan; p. -hnāh; pl. -hnigon; pp. -hnigen. **I.** *to descend beneath, go lower than* a place:—Grundum ic hrīne, helle underhnīge, heofonas oferstīge, Exon. Th. 482, 23; Rä. 67, 6. Hwīlum ýða ic sceal underhnígan, 386, 29; Rä. 4, 69. **II.** *to submit to* what is laborious or painful, *be subjected to* evil, *undergo* punishment, etc., (a) with acc. of that which is undergone:—Ðonne hí ða scandlícan lustas ðisses middangeardes mid hira módes willan underhnígaþ *cum turpi hujus mundi desiderio humanae mentis voluntas substernitur*, Past. 52 ; Swt. 405, 3. For intingan hērsumnesse ic hāten geþafode ðæt ic ðone hād underhnāh (-nágh, S. 566, 8) þeáh ðe ic unwyrðe wǽre *obedientiae causa jussus subire hoc quamvis indignus consensi*, Bd. 4, 2; M. 260, 8. Hí underhnígon ðone hwílendlícan deáþ *mortem subiere temporalem*, 4, 16; S. 584, 37. Hē underhnīge menniscne þeówdóm *se humano servitio subjiciat*, L. Ecg. P. Addit. 18; Th. ii. 234, 24. Ic eom nýded ðæt ic sceal hraþe deáþ underhnígan *ad articulum subeundae mortis compellor*, Bd. 3, 13; S. 538, 26. Beheáfdunge underhnīgan *capitalem sententiam subire*, Hpt. Gl. 477, 74. (b) with dat. :—Sixtus underhnáh swurdes ecge, and his twēgen diáconas samod, Homl. Th. i. 420, 17. Hē ðam deáðe underhnáh and ðone deófol oferswýðde, Homl. Skt. i. 16, 115. Regolícore stýre hē underhnīge *disciplinae regulari subjaceat*, R. Ben. Interl. 19, 1. Ðonne sceal hē underhnígan ðære steóre regollícre lāre, R. Ben. 16, 4. Hí sceoldon underhnígan nacodum swurde, Homl. Skt. i. 5, 28. Hēt se rēða cwellere hine underhnígan swurdes ecge, Homl. Th. i. 428, 8. (c) case uncertain :—Hē sāres wīte underhnīhð *doloris poenae succumbit*, Scint. 12, 9. Ðū galgan underhnige, Anglia xii. 506, 1. Nyste heora nān on hwæs tíman hē ðrowunge underhnige, Homl. Th. ii. 506, 31.

under-holung glosses suffossum in : Underholunga *suffossa*, Ps. Lamb. 79, 17.

under-hwítel, es; m. *An under-whittle*; ragana, Wrt. Voc. i. 59, 28. [Migne gives *racana* vêtement déchiré, de peu de valeur.]

under-hwrædel. v. under-wrǽdel.

under-ícan glosses subjungere in : Underýcende *subjungentes*, Anglia xiii. 385, 292.

un-derigende; adj. *Harmless, innocent*:—Nocens, derigende, is nama and *participium*, and innocens, underigende, of ðam gefēged ys ǽfre nama, Ælfc. Gr. 9, 38; Zup. 62, 16. Underigende handum *innocens manibus*, Ps. Spl. 23, 4. Hió mid wíflíce niðe wæs feohtende on ðæt underiende folc, Ors. 1, 2; Swt. 30, 19.

un-derigendlíc; adj. *Harmless, innoxious*:—Se Hālga Gāst hí ealle onǽlde mid undergendlícum fýre, Homl. Th. i. 298, 6.

under-irnan *to under-run, run beneath* :—Hē underyrnþ ealle ða twelf tācna, Lchdm. iii. 248, 1. Ðonne seó sunne hí hæfð ealle underurnen, 246, 10. ¶ As a gloss of succurrere :—Underiorn *succurre*, Rtl. 43, 17.

under-lādteów, es; m. *A subordinate ruler*, applied to the consuls in comparison with the kings:—Him ða Rómáne æfter ðæm lādteówas (underlātteówas, MS. C.) gesetton ðe hié consulas hēton ðæt heora ríce heólde ān geár ān monn *igitur regibus urbe propulsis, Romani consules creauerunt*, Ors. 2, 2; Swt. 68, 2.

under-lǽded glosses subductus, Lk. Skt. Lind. 5, 11 : sublatus, Mt. Kmbl. p. 3, 10.

under-lecgan. I. *to underlay, support* :—Ic underlecge *fulcio*, Ælfc. Gr. 30, 2; Zup. 190, 5. Ðā bæd hē hí ānre sylle, ðæt hē mihte ðæt hūs mid ðære underlecgan, Homl. Th. ii. 144, 33. Ðeáh hit mid nāne anwald ne sié underlēd *cum nulla potestate fulcitur*, Past. 17; Swt. 113, 25. Ðonne bið se elnboga underlēd mid pyle and hnecca mid bolstre *pulvillo cubitus vel cervicalibus caput jacentis excipitur*, 19; Swt. 143, 17. **II.** *rendering* supponere, substernere, etc. :—Ic underlecge *subpono*, Ælfc. Gr. 28, 3; Zup. 167, 17 : 47; Zup. 276, 9. Underlegdon *substernebant*, Lk. Skt. Lind. 19, 36. Of underlēdum brandum *suppositis torribus*, Hpt. Gl. 489, 6. [O. H. Ger. unter-leccen *fulcire*.]

under-licgan *to be subject, submit, yield*:—Ne mæg se preóst mannum ðingian, gif hē synnum underlíð, Homl. Th. ii. 320, 21 : Homl. Skt. i. 11, 155. Ne underlicge wē synnum *nec succumbamus vitiis*, Rtl. 82, 25. Underlicgge hē þreále *discipline subjaceat*, R. Ben. 56, 12. Ða underðíeddan mon sceal lǽran ðæt hí him eáðmódlíce underlicgen *subditi*

admonendi sunt ut humiliter subjaceant, Past. 28 ; Swt. 189, 20. [O. H. Ger. unter-ligan *subjacere*.]

under-lihtan *glosses* sublevare *in*: Uē underlihtad sié *sublevemur*, Rtl. 51, 23. Ðætte uē sié underlihtad *sublevari*, 72, 3.

underling, es ; *m.* (the word seems to occur only in late texts) *An underling, a subordinate, a subject*:—Heó (*the Jews*) sydðen æfre unwurðe wæron on heora lífdagen and get synden underlinges, Homl. Ass. 194, 50. Eádward kyng grēt Harald eurl and alle his undurlynges (*omnes meos ministros*) in Herefordeshíre, Cod. Dip. Kmbl. iv. 218, 14. [Leir king scal beon eouwer lauerd . . . & Aganippus ure king scal beon his underling, Laym. 3657. Inobedience, þet is . . . underling þet ne buhð nout his prelat, A. R. 198, 18.]

under-lútan ; *p.* -leát, -luton ; *pp.* -loten *To stoop beneath something in order to raise or support it, to support, bear, submit to*:—Ða ðe beóð mid hira ágnum byrðennum ofðrycte, ðæt hié ne magon gestondan, hié willaþ lustlíce underfón óðerra monna, ond unniéðige hié underlútaþ mid hira sculdrum óðerra byrðenna *qui ad casum valde urgetur ex propriis, humerum libenter opprimendus ponderibus submittit alienis*, Past. 7 ; Swt. 51, 25. Hē árás underléat ðæt bēr eode *ille surrexit sublato grauato abiit*, Mk. Skt. Lind. Rush. 2, 12. Eálá ofermódan! hwí gē wilnigen ðæt gē underlútan mid eówrum swiran ðæt deáþlícne geoc *quid o superbi colla mortali jugo frustra levare gestiunt?* Bt. 19 ; Fox 68, 26. Hwí eów á lyste mid eówrum swiran selfra willum ðæt swǣre gioc underlútan, Met. 10, 20.

undern, es ; *m. The third hour of the day, nine in the morning* ; in later English (v. infra) it is used of the *sixth hour*, a use it seems to have in *undern-rest*, q. v.:—Undern *tertia*, middæg *sexta*, Wrt. Voc. i. 53, 11. Undern is dæges þridde tíd, Btwk. 214, 33. Ðæs hírēdes ealdor gehýrde wyrhtan on ærnemerigen, eft on undern . . . Se ærmerigen wæs fram Adam óð Noe, se undern fram Noe óð Abraham, Homl. Th. ii. 74, 7–19. Tó undernes, Lchdm. ii. 194, 6. Byð seó sceadu tó underne and tó nóne seofon and twēntigoþan healfes fótes *the shadow at nine and at three is twenty-six and a half feet long*, iii. 218, 3. and often. Æt underne . . . ǣr underne, Blickl. Homl. 93, 22, 36. Ǣr undern . . . ofer undern *mane . . . vespere*, Lev. 6, 20. On ða þriddan tíd dæges, ðæt is on undern, Shrn. 79, 35. Wē ētaþ on ðam Sunnandagum on undern and on ǣfen, Homl. Skt. i. 12, 3. Sele drincan on þreó tída, on undern, on middæg, on nón, Lchdm. ii. 140, 1. Fram hancrēde óð undern, Homl. Th. i. 74, 21 : Chr. 538 ; Erl. 16, 2. Ofor undern, Blickl. Homl. 93, 15. Healfe tíd ofer undern, 540 ; Erl. 16, 4. Óð heáne undern *usque ad tertiam plenam*, R. Ben. 74, 11. [Abuten undern deies . . . abute swucke time alse me singeð messe (from prime oðet midmareȝen, hwenne preostes singeð heore messen, MS. C.), A. R. 24, 11. So ha dede at undren and and at midday also (Mt. 20, 3), Misc. 33, 22. At þon heye undarne (Acts 2, 15), 56, 657. It was the thridde our (that men clepen undrun), Wick. Mk. 15, 25. The time of undern of the same day, Ch. Cl. T. 260. *But the word sometimes denotes a later hour*:—Bi þis was undren (under, undrin) on þe dai (*the sixth hour*, Lk. 23, 44), C. M. 16741. Undorne, 19830. The our was as the sixte or undurn, Wick. Jn. 4, 6. An orendroñ, ornedrone *meredies*, Cath. Angl. 261, where see note. *See also the later English forms given under* undern-mǣl, -mete, -tíd. O. Sax. undorn *the third hour*: O. Frs. ond, unden (*and see Richthofen Wtbch.*) : O. H. Ger. untarn *midday* : Icel. undorn *nine o'clock* A. M. or *three o'clock* P.M. ; *a meal*. Cf. Goth. undaurni-mats ἄριστον. As in the case of *mǣl* = meal, the word seems to have come to denote the eating that takes place at the time, which at first the word denoted. v. Halliwell's Dict. *aandorn*, and see the forms in other dialects in Cl. & Vig. Icel. Dict. s.v. undorn.] v. *compounds with* undern-.

un-derne. v. un-dirne.

under-neoþan, -nyþan ; *prep. adv. Underneath*:—Þurhscoten mid ánre flán underneoðan ðær breóst *sagitta sub mamma trajectus*, Ors. 3, 9 ; Swt. 134, 23. Ðú nymst cealfes blód mid ðínum fingre on ðæs weofodes hyrnan and gítst ðæt óðer undernyðan (*reliquum sanguinem fundes juxta basim altaris*), Ex. 29, 12. [Þet fotspure þe wæs undernæðen his fote, Chr. 1070 ; Erl. 209, 8.] v. neoþan.

under-neoþemest *lowest*:—Eorðe stód ealra gesceafta underniþemæst (under niþemæst? Cf. sió eorðe is nioþor ðonne ǣnig óðru gesceaft, Bt. 33, 4 ; Fox 130, 20), Met. 20, 135.

undern-gereord, es ; *n. A morning meal, breakfast*:—Underngereord *prandium*, ǣfengereord *cena*, nónmete *merenda*, Wrt. Voc. i. 38, 12. Æt his underngereorde ǣr hē tó ðæm gefeohte fóre, Ors. 2, 5 ; Swt. 84, 34. Gif wē fæstaþ and ðæt underngereord tó ðam ǣfengifle healdaþ, ðonne ne bið ðæt nán fæsten, L. E. I. 38 ; Th. ii, 436, 28. Heora underngereordu and ǣfengereordu hié mengdou tógædere, Blickl. Homl. 99, 22.

undern-gifl, es ; *n. Food eaten in the morning, breakfast*:—Ðonne ðú hæbbe gegearwod underngifl (-giefl, Hatt. MS.) oðþe ǣfengifl *cum facis prandium aut coenam*, Past. 44 ; Swt. 322, 19.

under-niman ; *p.* -nam, *pl.* -námon ; *pp.* -numen. **I.** *to take surreptitiously, to steal*. v. under-fón, VIII.:—Gif feoh sý undernumen (under numen?), Lchdm. i. 392, 8. **II.** in figurative senses, (1) *to take into the mind, receive* what is said, taught, etc. :—Gehādede men hit sceolon him ásecgan, undernimð se ðe wile, Wulfst. 305, 20. Hē deóplíce undernam Drihtnes láre æt him, Homl. Skt. ii. 29, 76. Abraham undernam hefiglíce ðás word *dure accepit hoc*, Gen. 21, 11. Ðis sind ðæra apostola word, undernimaþ, hí mid carfullum móde, Homl. Th. i. 236, 4 : H. R. 7, 29. Ne underfóð ealle menn ðis word . . . Undernyme se ðe undernyman mæge (cf. ne underneomeð (uoð, MS. B.) nawt þis ilke word alle . . . Hwase hit me underneomen, underneome, H. M. 19, 27) *non omnes capiunt verbum istud . . . Qui potest capere, capiat*, Mt. Kmbl. 19, 11–12. Man mæg swíðe eáðe witan, se ðe hit underniman wile, ðæt hit riht nis . . ., Wulfst. 305, 1 : Homl. Ass. 26, 53. [Hire fader hefde iset hire to lare and heo undernom (-ueng, MS. R.) hit wel, Kath. 117.] (2) *to take upon one's self*:—Gif ðú leornian wille hú ðæt gewurðan mæge, ðonne undernim ðú leorningcnihtes híw, Homl. Th. i. 590, 20. [We þis feht habbeoð undernumen buten Arðures rede, Laym. 26734. To poure iheorted eni heih þing to undernimen ine hope of Godes helpe, A. R. 202, 6. Hardy to greate þinge ondernime, Ayenb. 83, 19.] (3) *to blame, resent* (?) :—Ðá undernam Godwine eorl swýðe ðæt on his eorldóme sceolde swilc geweorðan, Chr. 1052 ; Erl. 179, 16. [He cometh not to the light, that his workis be not undirnomun *ut non arguantur opera ejus*, Wick. Jn. 3, 20. Impacient is he that wil not ben iaught ne undernome of his vices, Chauc. Pers. T. Whoso undernymeth me hereof, Piers P. 5, 115. Underneme *reprehendo, deprehendo, arguo* ; undernemynge *deprehensio, reprehensio*, Prompt. Parv. 511.]

under-niþemæst. v. under-neoþemest.

undern-mǣl, es ; *n. Morning-time*:—On undernmǣl, Beo. Th. 2860 ; B. 1428. An undernmǣl, Homl. Skt. ii. 30, 319. [Ther walkith noon but the lymytour himself in undermēles and in morwenynges, Chauc. W. of B. T. 19. Undermele *postmeridies, postmessimbria, merarium*, Prompt. Parv. 511. Cf. In an undermele tyde *meridiano tempore*, Trev. v. 373, 9. See also Nares' Dict. *undermeal*.]

undern-mete ; *m. Food eaten in the morning, breakfast*:—Undermete *prandium*, ǣfenmete *cena*, Wrt. Voc. i. 290, 65. Uton brúcan ðisses undernmetes swá ða sculon ðe hiora ǣfengifl on helle gefeccean sculon *prandete tamquam apud inferos coenaturi*, Ors. 2, 5 ; Swt. 86, 1. Undernmete *prandium*, Mt. Kmbl. Rush. 22, 4. [An orendroñ mete, ordrone mete *merenda* ; to ete orendroñ mete *merendare*, Cath. Angl. 261, where see note. Goth. undaurni-mats ἄριστον, *prandium*.]

undern-rest, e ; *f. Rest in the morning*:—Ðá hē árás on dæge of undernræste (*postquam de meridiana quiete surrexerunt*, Bede's Vita Cudberct, c. 35), Shrn. 64, 7.

undern-sang, es ; *m. The service at the third hour of the day, tierce*:—Undernsang *tertia*, R. Ben. 39, 19 : 40, 6 : L. Ælfc. C. 19 ; Th. ii. 350, 6. Undernsanges gebed *tertie oratio*, R. Ben. Interl. 47, 10. Æt ǣfensonge and æt undernsonge, Chart. Th. 137, 34.

undern-swǣsendu ; *pl. Breakfast*:—Ðæt hē mid ðý biscope sǣte æt his undernswǣsendum (*ad prandium*), Bd. 3, 6 ; S. 528, 13.

undern-tíd, e ; *f.* **I.** *the third hour of the day, nine o'clock* A.M.:—Ðá wæs undertíd *erat hora tertia*, Mk. Skt. 15, 25 : Homl. Th. i. 314, 22 (see Acts 2, 15). Úres andgites merigen is úre cildhád, úre cnihthád swylce undertíd, on ðam ástíhð úre geógoð, swá swá seó sunne dēð ymbe ðære ðriddan tíde, ii. 76, 15. From undertíde (underne tíde, S. 592, 7), ðonne mon mæssan oftost singeþ *a tertia hora quando missae fieri solebant*, Bd. 4, 22 ; S. 328, 32. Embe underntíde *circa horam tertiam*, Mt. Kmbl. 20, 4. Ǣrest on ærnemorgen, óþre síþe on underntíd, Blickl. Homl. 47, 17: 133, 27. **II.** *the service at the third hour*:—Wē sungon underntíde and dydon mæssan *cantavimus tertiam et fecimus missam*, Coll. Monast. Th. 33, 31. [Hit is undertid (Acts 2, 15), O. E. Homl. i. 91, 2 : Mk. Skt. 15, 25, col. 2. Þe soðe sunne iðe undertid was istien on heih, A. R. 400, 15. Þan was it underntide (undrin-, undir-) o þe dai *about the sixth hour* (Acts 10, 9), C. M. 19830.]

undern-tíma, an ; *m. The third hour of the day, nine o'clock* A.M.:—On undern wē sculon God herian, forðam on underntíman Crist wearð tó deáþe fordēmed . . . And eft com se Hálga Gást on underntíman ofer ða apostolas, Btwk. 214, 26–30. [Godess Gast com i firess onnlicnesse an daȝȝ att unndernntime, Orm. 19458.]

under-plantian *glosses* supplantare *in*: Ðú underplantedest *supplantasti*, Ps. Spl. 17, 41. Dysig byð underplantud *stultus supplantatur*, Scint. 169, 2.

under-sceótan ; *p.* -sceát, *pl.* -scuton ; *pp.* -scoten. **I.** *to move to a place beneath, to intercept*:—His (*the moon's*) trendel underscýt ðære (ða, MS. R.) sunnan tó ðam swíðe ðæt heó eall áþeóstraþ, Lchdm. iii. 242, 20. **II.** *to under-prop, support*:—Hí ne beóð mid nánre sylle underscotene ðæs godcundlícan mægenes *nullis fulti virtutibus*, Past. I ; Swt. 27, 17. Cf. under-stingan.

under-scyte, es ; *m. Intercepting, intervention*:—Se móna mæg þurh his underscyte ða sunnan áþeóstrian, Lchdm. iii. 242, 25. Wē rædaþ on tungelcræfte ðæt seó sunne bið hwíltídum þurh ðæs mónelícan trendles underscyte áðýstrod, Homl. Th. i, 608, 32.

under-sécan ; *p.* -sóhte *To investigate* ; discutere (cf. *discutiens*, i.

judicans, querens, Wrt. Voc. ii. 141, 42) :—Ðæt is ðæt hié ðara ðing ðe him underdiódde bióð for ðam ege ánum ðæs innecundan déman under-séce *est subjectorum causas pro sola interni judicis intentione discutere*, Past. 13 ; Swt. 79, 8. Ðæm láreówe is swíðe smeálíce tó underséceanne be ðæm weorcum ðara ofertrūwedena *subtiliter ab arguente discutienda sunt opera protervorum*, 32 ; Swt. 209, 12. [Huo þet heþ þise yefþe, he onderzekþ þe redes þet me him yefþ, Ayenb. 184, 23. *Ger.* unter-suchen.]

under-serc, es ; *m. An under-garment* :—Undersyrc *colophium* (= *colobium*), Wrt. Voc. ii. 22, 45.

under-singan *renders* succinere *in :* Ic undersinge *suceino*, Ælfc. Gr. 28, 7 ; Zup. 181, 2.

under-sittan *renders* subsidere *in :*—Ic undersitte *subsideo*, Ælfc. Gr. 26, 5 ; Zup. 157, 5.

under-smeágan *glosses* subrepere, subripere *in :* Undersmǽge *subripiat*, R. Ben. Interl. 71, 6. Undersmége *subrepat*, 72, 9. v. next word.

under-smúgan ; *p.* -smeáh, *pl.* -smugon ; *pp.* -smogen *To creep under, come upon unawares, surprise :*—Ídelnysse underþeódde gálscype under-smýhð *otio deditos luxuria subripit*, Scint. 89, 8. Ne undersmúge gítsunge yfel *non subripiat* (but other MS. *subrepat*) *avaritie malum*, R. Ben. Interl. 95, 7. Ðæt ǽnig þinc ne undersmuge on wege gesyhð ne *quid forte subripuerit in via visus*, 113, 11. Swá hý nǽfre mid oferfylle undersmogene and beswicene ne weorðan *ne subrepat satietas aut ebrietas*, R. Ben. 64, 19.

under-standan ; *p.* -stód, *pl.* -stódon ; *pp.* -standen. I. *to understand, have insight into :*—Ðú genóh wel understentst ðæt ic ðé tó sprece, Bt. 13 ; Fox 38, 1. Se godcunda foreþonc hit understent eall swíþe ryhte . . . wé ne cunnon ðæt riht understandan, 39, 8 ; Fox 224, 19–21. Gecýðnessa ðíne ic ongeat ł understód *testimonia tua intellexi*, Ps. Lamb. 118, 95. Understand ðás gesihðe *intellige visionem* (Dan. 9, 23), Homl. Th. ii. 14, 9. Ðam men is gemǽne mid englum ðæt hé understande i. 302, 22. Eal ðæt syndon micle and egeslíce dǽda, understande se ðe wille, Wulfst. 161, 9. Understande se ðe cunne, 162, 12. Snotornys, þurh ða seó sáwel sceal hyre Scippend understandan, Homl. Skt. i. 1, 157. Ðæt wé magon understandan ða þing ðe ðú specst *ut possimus intelligere quae loqueris*, Coll. Monast. Th. 32, 15. Swá clǽne hió (*learning*) wæs óðfeallenu on Angelcynne ðæt swíðe feáwa wǽron behionan Humbre ðe hiora ðéninga cúðen understandan, Past. pref. ; Swt. 3, 14. Gehwá ðe his ágene þearfe wille understandan, L. Eth. vi. 27 ; Th. i. 322, 9. I a. *with prep.* ymbe :—Ne mæg nán gesceaft fulfremedlíce smeágan ne understandan ymbe God . . . Englas ne magon fulfremedlíce understandan ymbe God, Homl. Th. i. 10, 2–5. II. *to understand, perceive, know certainly :*—Understand be ðam hú se ælmihtiga God hí ealle gesceóp búton antimbre *know certainly concerning them, how that the almighty God created them all without matter*, Homl. Skt. ii. 25, 178. Understandaþ eác georne, ðæt deófol ðás þeóde dwelode, Wulfst. 156, 7. Understanden (beón) *deprehendi*, Hpt. Gl. 526, 18 : *conjici*, 469, 30. [I do gowe to understonden *ego notifico nobis*, Cod. Dip. Kmbl. iv. 218, 15.] III. *to understand* in such and such a sense, *to conceive of, consider :*—Ne understand ðú hit mé tó unrihtwísnesse *do not consider it as unrighteousness in me*, Ps. Th. 21, 2. Gif ic eáðmódlíce mé sylfne ne understóde, ac mín mód on módignesse anhófe *si non humiliter sentiebam, sed exaltavi animam meam*, R. Ben. 22, 18. III a. *with prep.* be :—Ǽwfæstlíce understandende be úre ealra ǽriste *piously conceiving of the resurrection of us all*, Homl. Skt. ii. 25, 472. III b. *to accept as correct :*—Ðis ylce understand be ðam óðrum dagum *take the same rule as applicable in the case of the other days*, Anglia viii. 304, 29. IV. *to observe, notice, consider :*—Understand (*or* I) rǽdere, hwæt seó rǽding cwyð, Anglia viii. 309, 1. Understand míne sprǽce *animadverte sermonem* (Dan. 9, 23), Homl. Th. ii. 14, 9. Ðæt tó understandenne ealle gedémdon *hoc adtendendum cuncti decreuerunt*, Anglia xiii. 371, 90. Ðæt geswinc his sýðfætes ne understandende hé mid hrædestan ryne arn, Homl. Skt. ii. 23 b, 186. On dehter ná understandendre *in filia non aduertente* se, Scint. 225, 7.

under-standenness *glosses* substantia *in :* Understondennisse *substantia*, Rtl. 31, 40.

under-standing, e ; *f. Intelligence :*—On andgyte inran understandincge *sensu interioris intelligentie*, Scint. 221, 13.

under-stapplian *glosses* supplantare *in :* Understappla ł forscrænc hine *supplanta eum*, Ps. Lamb. 16, 13.

under-staþolfæst. v. un-staþolfæst.

under-stingan *to under-prop, support :*—Understungen and áwreðed mid ðýs hwílendlícan onwalde *fultus temporali potentia*, Past. 17 ; Swt. 113, 11. Cf. under-sceótan.

under-stregdan *to under-strew :*—Hé wæs nacod and on carcern onsænded, and ðær wæs understregd mid sǽscellum and mid scearpum stánum *he was stripped and sent to prison, and there had sea-shells and sharp stones strewed under him*, Shrn. 51, 13.

under-þegnian *glosses* subministrare *in :* Underþénaþ *subministrat*, Scint. 5, 6.

under-þencan *to look into, consider :*—Wé sculon swíðe smeálíce ðissa ǽgðer underðencean *hoc in utrisque subtiliter intuendum*, Past. 7 ; Swt. 49, 23.

under-þeód ; *adj.* (*ptcpl.*) *used substantively. Subject, subordinate :*—On óðre wísan sint tó manianne ða underðióddan, on óðre ða ofer-gesettan. Ða underðiéddan (-ðióddan, Cott. MSS.) mon sceal lǽran ðæt hié elles ne sién genǽt *aliter admonendi sunt subditi, atque aliter praelati. Illos ne subjectio conterat*, Past. 28 ; Swt. 189, 14. Ðonne ðæt mód ðara underðiédra (-ðiéddra, Cott. MSS.) hwæthwugu ryhtlíces ongitan mæg *subditorum mens cum quaedam recte sentire potuerit*, 19 ; Swt. 147, 1. On his (*the abbot's*) underðeóddera módum *in discipulorum mentibus*, R. Ben. 10, 18. Se láreów sceal ǽrest on him sylfum ælcne leahter ádwǽscan, and sidðan on his underðeóddum, Homl. Th. i. 320, 30. Landfranc wæs geháded on his ágenum biscopsetle fram eahte biscopum his underðióddum, Chr. 1070 ; Erl. 206, 5. [Prost scal spenen among al his underþede, O. E. Homl. i. 85, 14.] v. next word.

under-þeódan, -þiédan, -þídan ; *p.* de. I. *to subject, subjugate, render subject*, (1) *with dat.* :—Se líchoma hine him (*the devil*) underðiód mid ðære lustfulnesse . . . Swá swá sió nædre lǽrde Euan on wóh and Eue hí hire underðiód[d]e mid lustfulnesse, swá swá líchoma *caro se delectatione subjicit . . . Unde et ille serpens prava suggessit, Eva autem quasi caro se delectationi subdidit*, Past. 53 ; Swt. 417, 24–27. Hí hí underþiódaþ þíneówum, Bt. 37, 1 ; Fox 186, 28. God ðú ðe mé sealdest ðæt ic meahte swylc wíte dón mínum feóndum, and mé swylc folc underþýdes (-þeodes, Ps. Surt.) *Deus qui das vindictas mihi, et subdidisti populos sub me*, Ps. Th. 17, 45. Hé ús underþeódde úre folc *subjecit populus nobis*, Ps. Th. Spl. Surt. 46, 3. Claudius Orcadus Rómana cynedóme underþeódde, Chr. 47 ; Erl. 6, 26. Ðú mé folc mænig underþeóddest *subjiciens populum meum sub me*, Ps. Th. 143, 3. Tó ðara hláforda dóme ðe hé hine ǽr underþeódde (-þiódde, Met. 25, 66), Bt. 37, 1 ; Fox 186, 29. Ic míne sáwle wylle Gode underþeódan *nonne Deo subdita erit anima mea ?* Ps. Th. 61, 1 : Met. 25, 63. Hí druncennesse and oferhýdo . . . wǽron heora swiran underðeóddende (*subdentes*), Bd. 1, 14 ; S. 482, 27. Reogollícum ðeódscipum underþeóded *regularibus disciplinis subditus*, 4, 24 ; S. 598, 21. His anwealde underþeóded, Bt. 26, 3 ; Fox 94, 15. Ælc mon ðe underþeóded (-þiéded, Met. 17, 23) bið unþeáwum, 30, 2 ; Fox 110, 20. Underþeód (-þýded, Met. 16, 4), 29, 3 ; Fox 106, 19. Næs him nó ðý læs underþeód eall ðes middangeard, 16, 4 ; Fox 58, 10. Hé wæs him underþeód (-ðióded, Lind., Rush.) *erat subditus illis*, Lk. Skt. 2, 51. Underþeód (-þýd, Th. : -ðióded, Surt.) beó ðú Drihtne *subditus esto Domino*, Ps. Spl. 36, 6 : *subjecta*, 61, 5. Him se mǽsta dæl weard underþiéded, Ors. 1, 10 ; Swt. 44, 5. Gif hé wiðcwǽde ðæt hé nǽre underðiódd (-ðídd, Hatt. MS.) his Scippende *si auctoris imperio obedire recusaret*, Past. 7 ; Swt. 50, 13. Him wesan underþýded, Exon. Th. 138, 13 ; Gú. 575. Syndan né fremde cynn underþeódde *mihi allophili subditi sunt*, Ps. Th. 107, 8. Deófolseócnessa ús synt underþeódde *daemonia subjiciuntur nobis*, Lk. Skt. 10, 17. Ealle ðás mǽgþe Æþelbalde on hýrsumnesse underþeódde syndon *hae omnes provinciae Ædilbaldo subjectae sunt*, Bd. 5, 23 ; S. 646, 27. Ðám ánum ðe Gode underþeódde syndon mid myclum hádum, Blickl. Homl. 109, 22. Swǽsum wordum underþeódde (*dediti*), Coll. Monast. Th. 32, 33. Ða ðe him underðiédde (-ðídde, Cott. MSS.) bióð *subjecti*, Past. 4 ; Swt. 39, 7. Wé ealle ðære hnescnesse úres flǽsces beóð underðiédde (-ðidde, Cott. MSS.) *cuncti corruptionis nostrae infirmitatibus subjacemus*, 21 ; Swt. 159, 6. Eall ða ðing ðe hire underþiéd sint, sint underþiéd ðam godcundan foreþonce, Bt. 39, 6 ; Fox 220, 20. Ða ealdormen beóð Gode underðýdde, Ps. Th. 46, 9. (2) *with a preposition :*—Ðú underþeódest folc mín under mé *subdis populum meum subter me*, Ps. Spl. 143, 3. Gé underþiódaþ eówre héhstan medemnesse under ða eallra nyþemestan gesceafta *vos dignitatem vestram infra infima quaeque detruditis*, Bt. 14, 2 ; Fox 44, 33. Ðú underþeóddest folc úre mé *subdis populos sub me*, Ps. Spl. 17, 49. Ne wæs ǽfre ǽnig cyning ðæt má heora landa him tó gewealde underþeódde, Bd. 1, 34 ; S. 499, 23. Underðeódende folc under mé *subjiciens populos sub me*, Ps. Surt. 143, 2. (3) where that to which there is subjection is not stated :—Ðonne hé underðiód *quando summiserit*, Kent. Gl. 1004. Ðæt hit ungedæfenlíc sig, ðæt se dǽdbéta hine ná on ða wísan ðissa woroldlícra þinga ne underþeóde *quod indecorum sit, poenitentem in re mundanorum horum negotiorum se non cohibere*, L. Ecg. P. i. 7 ; Th. ii. 174, 25. Ús is tó gelýfenne ðæt se Hǽlend þyder cóme, næs nó genéded, ne underþeóded, ac mid his wyllan, Blickl. Homl. 29, 15. Underþeód *dedito*, Hpt. Gl. 509, 62. II. *to subject, cause to endure, render liable :*—Hefigran scylde and hefigran wítum hé hine underðiét *poenae gravioris culpae se subjicit*, Past. 54 ; Swt. 421, 6. Ðá hét se cásere ðone diácon miclum wítum underþeódan, Shrn. 56, 34. III. *to subjoin, add :*—Hé underþeódde and him sédde ðæt se dæg swíþe neáh stóde his forþfóre *subjunxit diem sui obitus jam proxime instare*, Bd. 4, 3 ; S. 568, 15. IV. *to support :*—Underþióð *subnixa, suffulta*, Hpt. Gl. 467, 21. Underþeódne *subnixum*, 507, 57.

under-þeódendlíc *renders* subjunctivus *in :* Subjunctivum, ðæt ys underðeódendlíc, Ælfc. Gr. 15 ; Zup. 98, 23. *Subjunctivae*, ðæt sind underþeódendlíce, 46 ; Zup. 267, 7.

under-þeódness, e; *f. Subjection, submission*:—For yrmþo ðære underþeódnysse *ob aerumnam subjectionis*, Bd. 4, 16; S. 584, 41. Wite hé ðæt hé micle eáðmódra beón sceal on regoles underðeódnysse *sciens se multo magis discipline regulari subditum*, R. Ben. 112, 1. Micle swýþor is tó hálsienne Drihten mid ealre eáðmódnesse and mid ealre underþeódnysse (*cum omni humilitate*), 45, 18. Him gehýrsumiaþ óðra engla werod mid micelre underðeóduysse, Homl. Th. i. 342, 34: 346, 34: Ælfc. T. Grn. 1, 31. Hí him gehétan eáþmóde hýrnysse and singale underþeódnysse *subjectionem continuam promittebant*, Bd. 1, 12; S. 480, 27.

under-þeów, es; *m. One reduced to slavery, one who serves under or is subject to another, a slave, servant*:—Hé geniédde Arhalaus ðone lǽteów ðæt hé wæs his underþeów, Ors. 5, 11; Swt. 238, 2. Ða búrgware bǽdon ðæt hié mósten beón hiera underþeówas ðá hí hié bewerian ne mehton *petentes, ut quos belli clades reliquos fecit, saltem servire liceat*, 4, 13; Swt. 212, 5: 2, 8; Swt. 92, 23. Ealle wurdon Iuliuse underþeówas *Caesar omnes ad deditionem compulit*, 5, 12; Swt. 242, 27.

under-þídan, -þiédan, -þýdan. v. under-þeódan.

under-tódál *renders* subdistinctio *in* : *Subdistinctio*, ðæt is undertódál, Ælfc. Gr. 50, 14; Zup. 291, 5.

under-tunge *glosses* sublingua *in* : Undertungan *sublinguae*, Wrt. Voc. i. 282, 79.

under-tungeþrum. v. tunge-þrum.

under-weaxan *glosses* succrescere *in* : Underwexaþ *succrescunt*, Scint. 104, 8.

under-wed[d], es; *n. A pledge, security*:—'Gif ðú mé sylst underwedd (*arrhabonem*), óð ðæt ðú mé sende ðæt ðú mé behǽtst' . . . 'Hwæt wilt ðú tó underwedde (*pro arrhabone*) nyman?' . . . Iudas sende án tyccen wið his hirde, ðæt hé sette ðæt underwedd, Gen. 38, 17-20. Gylde hé ðæt yrfe oþþe underwed lecge, L. O. D. 1; Th. i. 352, 8: 8; Th. i. 356, 10.

under-wendan *glosses* subvertere · *in* : Hé underwende *subuertat*, Scint. 196, 6.

under-wrǽdel *glosses* subfibulum *vel* subligaculum, Wrt. Voc. i. 40, 61.

under-wreðian, -wreoðian, -wriðian *to support, sustain*:—Ic underwreðige (-wreoðige, MSS. F. O.) *fulcio*, Ælfc. Gr. 30, 2; Zup. 190, 5. Underwreðie, Engl. Stud. xi. 65, 34. Ðú underwreðdes *sustentas*, Rtl. 45, 9. Underwreoðaþ his untrumnesse *sustentat inbecillitatem suam*, Kent. Gl. 644. Ealle stówa Drihten ymbfēhþ and neoþan underwreþeþ, Blickl. Homl. 23, 21. Man ða ræftras tó ðære fyrste gefæstnaþ and mid cantlum underwriðaþ, Anglia viii. 324, 10. Hé mid criccum his fēðunge underwreðode, Homl. Th. ii. 134, 25. Hí underwriðedon his handa *sustentabant manus ejus*, Ex. 17, 12. Ðæt hí underwreþigen *ut leuent*, Germ. 390, 173. Ða ðe bet cunnon, sceolon gǽman óðra manna, and mid heora fultume underwryðian, Homl. Th. ii. 282, 2. Sceancan míne mē tó underwreðigenne on yfel strange wǽron *crura mea ad me sustinendum in malum fortes fuere*, Anglia xi. 117, 23. Mid gódum weorcum underwreðed *bonis actis fultus*, Past. 19; Swt. 141, 18: Homl. Skt. ii. 23 b, 228: Hpt. Gl. 430, 36; Rtl. 76, 3. Treów wyrtrumum underwreðyd, Runic pm. Kmbl. 341, 30; Rûn. 13. Underwreoþod ǽghwanone *circum fultus undique*, Hymn. Surt. 46, 12. Hyt ys underwryðed mid þrím swerum, Anglia viii. 301, 37. Mid týn rihtingum underwriðode, 304, 32: Homl. Th. i. 444, 35. Underwreðdedo *suffulti*, Rtl. 71, 17. ¶ *The word glosses* supponere *in* : Drihten underwriðaþ ł [under]set handa his *Dominus supponet manum suam*, Ps. Lamb. 36, 24.

under-wreðung, e; *f. Support, sustentation*: — Underwreþung líchaman *sustentatio corporis*, Scint. 56, 10. Trumre underwreþincge *firmo fulcimento*, Wrt. Voc. ii. 148, 69.

under-writan *to subscribe, sign*:—Wē ealle mid Cristes ródetácn fæstnedon and underwritan *nos omnes subscripsimus*, Bd. 4, 17; S. 586, 16.

under-wriðian. v. under-wreðian.

under-wyrtwalian *glosses* supplantare *in* : Ðú underwyrtwæledæst *supplantasti*, Ps. Spl. T. 17, 41.

un-dígollíce ; *adv. Not secretly, openly, clearly, plainly*:—Se dígla Dēma him swíðe undígellíce (-deógollíce, Hatt. MS.) (*aperte*) geondwyrde, Past. 4; Swt. 38, 19. Þus spræc God gefyrn, hit is swá ðeáh swá gedón swýðe neáh mid ús . . . and undígollíce, Homl. Skt. i. 13, 177.

un-dilegod ; *adj. Not blotted out, not effaced*:—Swá se wrítere, gif hé ne dilegaþ ðæt hé ǽr wrát, ðeáh hé nǽfre má náuht ne wríte, ðæt bið ðeáh undilegod, ðæt hé ǽr wrát *neque enim scriptor, si a scriptione cessaverit, quia alia non addidit, etiam illa, quae scripserat, delevit*, Past. 54; Swt. 423, 33.

un-dirne, -dierne, -dyrne ; *adj. Not hidden, discovered, revealed, manifest*:—Gif mon áfelle on wuda wel monega treówa and wyrð eft undierne (-dyrne, MS. B.), L. In. 43; Th. i. 128, 20: 44; Th. i. 130, 3. Ðæt weard underne eorðebúendum, ðæt Meotod hæfde miht, Cd. Th. 265, 1; Sat. 1. Wíde weard wyrd undyrne, Apstls. Kmbl. 84; Ap. 42. Ðá wæs Grendles gúðcræft gumum undyrne, Beo. Th. 255; B. 127:

4004; B. 2000. Bið him synwracu andweard undyrne *the punishment of sin shall stand revealed before them*, Exon. Th. 94, 16; Cri. 1541. Nú is undyrne werum, hú ða wihte hátne sindon, 429, 19; Rä. 43, 15. Ic wordum wemde wyrd undyrne, Andr. Kmbl. 2959; An. 1482. Ic gearwe wát ðæt ðē (*God*) siendan ealle wísan (*printed* wifan) undierne and cúðe ðínre ðære hálgan þrynesse, Anglia xi. 97, 7. *Nuncupatio est* undyrne yrfebēc, Wrt. Voc. ii. 62, 24.

un-dirne ; *adv. Openly, clearly, plainly*:—Weard ylda bearnum undyrne cúð ðætte Grendel wan wið Hróðgár, Beo. Th. 303; B. 150: 825; B. 410.

un-dóm, es ; *m. Unjust judgement*:—Wá ðam ðe rǽreþ unriht tó rihte and undóm dēmeþ earnum tó hýnðe *vae qui condunt leges iniquas; et scribentes, injustitiam scripserunt ; ut opprimerent in judicio pauperes* (Is. 10, 1-2), Wulfst. 47, 26: 128, 10: 268, 1. Se ðe unlage rǽre oþþe undóm gedēme for lǽdde oþþe for feohfange, L. C. S. 15 ; Th. i. 384, 9. Hí geútlageden ealle Frencisce men ðe ǽr unlage rǽrdon and undóm dēmdon and unrǽd rǽddon, Chr. 1052; Erl. 186, 2. Wē cýðaþ dēman and gerēfan, ðæt hig ágan þearfe, ðæt hí unrihtes geswícan and náhwár þurh undóm for feó ne for freóndscipe forgýman heora wísdóm, Wulfst. 267, 28. Weard ðes ðeódscipe swíðe forsingod þurh undómas, 130, 4.

un-dómlíce ; *adv. With bad judgement, indiscreetly*:—Hyrde oþþe unbindan undómlíce ondrǽde oþþe gewríþan *pastor vel absoluere indiscrete timeat uel ligare*, Scint. 202, 14.

un-dón ; *p.* -dyde ; *pp.* -dón *To undo*. I. *to undo* that which is closed, *to open*:—Ðá heó ðone windel undyde *aperiens fiscellam*, Ex. 2, 6. Ðá undyde hira án his sacc *aperto sacco*, Gen. 42, 27. Hé undyde his múð, Homl. Th. i. 548, 14. Ðá undydon wē úre saccas *aperuimus saccos nostros*, Gen. 43, 21. Ðæt hé undó his eágan, Anglia viii. 317, 5. II. *to undo* that which is bound, *to release*, (1) literal:—Beón þreó niht ǽr man ða hand undó, L. Ath. i. 23; Th. i. 212, 4. (2) figurative, *to release, absolve*:—From allum úsig synnum undó *ab omnibus nos peccatis absolve*, Rtl. 42, 3. Ða ðe synna racenteg gifæstnigaþ milsa ðínre árfæstnisse undóe *quos delictorum catena constringit miseratio tuae pietatis absoluat*, 40, 23. Ðæs on .ix. nihton ðæt wed undó hē mid rihtan gylde *nine days after let him release the pledge by lawful payment*, L. O. D. 1; Th. i. 352, 9. III. *to undo* that which closes, *to open* a door, etc.:—Undó mē duru sóðfæstra *aperite mihi portas justitiae*, Ps. Th. 117, 19: 23, 7. Undónde *reserando* (*valvam*), Hpt. Gl. 478, 11. Ðá wearþ eft Ianes duru undón (andón, Swt. 254, 17) *apertus est Janus*, Ors. 6, 1; Bos. 116, 25. Undónum remmingum *apertis obstaculis*, Hpt. Gl. 489, 73. IV. *to undo* that which binds or fastens, *to undo* a bolt, a knot, etc.:—Godes engel undyde ða locu ðæs cwearternes, Homl. Th. i. 572, 26. Ic ne am wyrðe ðætte ic undóe (*soluam*) his ðuong scóes, Jn. Skt. Lind. 1, 27. Undón (undóa, Lind.) ł loesan þwongas *soluere corrigiam*, Mk. Skt. Rush. 1, 7. V. *to undo* what has been done, *to abrogate, destroy*, (1) where the object is material:—Ic undóe tempel ðis *ego dissoluam templum hoc*, Mk. Skt. Lind. Rush. 14, 58. (2) where the object is not material:—Ðet hyra nán næ undó ðet ic tó ðám hálgum mynstrum gedón hebbe, Chart. Th. 232, 28. Ne mæg undóa ða gewriota *non potest solui scriptura*, Jn. Skt. Rush. 10, 35. Nællas gié woenæ forðon ic cuom tó undóenne (*solvere*) ae, Mt. Kmbl. Lind. 5, 17. Ðætte ne sē undóen ae *ut non soluatur lex*, Jn. Skt. Lind. Rush. 7, 23. v. on-dón.

un-drēfed ; *adj. Untroubled, not made turbid*:—Gē gedrēfdon hiora wæter mid iówrum fótum, ðeáh gē hit ǽr undrēfed druncen, Past. 2 ; Swt. 31, 3.

un-drifen ; *adj. Not driven*:—Ælc ceápscip frið hæbbe, ðe binnan múðan cuman[mæg ?], þēh hit unfriðscyp sý, gyf hit undrifen bið (*si non sit abacta tempestatibus* (Lat. vers.). For the fate of what was driven, cf. such a grant as the following:—Ic cíðe eów ðæt Urk habbe his strand . . . and eall ðæt tó his strande gedryuen hys, Cod. Dip. Kmbl. iv. 221, 5-8), L. Eth. ii. 2 ; Th. i. 284, 21.

un-druncen ; *adj. Not drunk, sober*:—Hé suá micle bet his ágen dysig oncnēw suá hé undruncenra wæs *he recognized his own folly so much better as he was more sober*, Past. 40; Swt. 295, 8. [*Icel.* ú-drukkinn.]

un-drysnende *inextinguishable*; inextinguibilis, Mt. Kmbl. Lind. 3, 12.

un-dyrne. v. un-dirne.

un-eácniendlíc ; *adj. Unproductive, sterile*; infecundus, Hpt. Gl. 430, 57.

un-earh ; *adj. Undaunted, intrepid, fearless*:—Unǽrh *impavidus, intrepidus*, Hpt. Gl. 502, 61. Sum cásere wæs on ðám dagum unearh on gefeohtum, Homl. Skt. ii. 27, 47. Gif mann bið ákenned on .xxii. nihta ealdne mónan se bið unearh fihtling, Lchdm. iii. 158, 11. Ðǽr mihton geseón Winceastre leódan rancne here and unearhne, ðæt hí be hyra gate tó sǽ eodon, and mǽte and mádmas ofer .L. míla hini fram sǽ fættan, Chr. 1006; Erl. 140, 26. Wendon forð wlance þegenas, unearge men, Byrht. Th. 137, 54; By. 206.

un-earhlíc ; *adj. Intrepid, dauntless*:—Hé cwæð tó ðam cásere unearhlícere stemne, Homl. Skt. i. 23, 164.

un-eáðe *and* un-iéðe (-éðe, -íðe, -ýðe); *adj.* I. of that which is not easy to do, *difficult, hard :*—Nis ðæt uneáðe ealwealdan Gode tô gefremmanne, Andr. Kmbl. 409; An. 205. Hê sæde ðæt se cræft uníeðe wære tô gehealdenne *praedicit quia difficile capitur*, Past. 52; Swt. 409, 20. Hit is uníeðe tô gesecgenne hû monege gewin wæron, Ors. 1, 12; Swt. 52, 8. II. of that which is not easy to bear, *troublesome, unpleasant, grievous :*—Se líchoma on ðone fúlostan stenc bið gecyrred ... and hê byð uneáðe ælcon men on neáweste tô hæbbenne, Blickl. Homl. 59, 15. Uneáðe mê is ðis *I am in a great strait* (2 Sam. 24, 14), Homl. Skt. i. 13, 247. Ðæt folc hine hæfde swâ yfele swâ hê sumes þinges scyldig wære ... and him wæs swâ uneáðe amang ðam, and him ða eágan floterodon, and bitere teáras âlêton, 23, 654. For hwý sceal ænigum menn ðyncan tô rêðe oððe tô uníeðe ðæt hê Godes suingellan geðafige *cur asperum creditur, ut a Deo homo toleret flagella?* Past. 36; Swt. 261, 20. Seó wîse wæs mîne (in mê, *v. l.*) on twâ healfa unêþe *quae res dupliciter me torsit*, Nar. 9, 23. Him bið unêþe þurst getenge *he will be oppressed by troublesome thirst*, Lchdm. ii. 174, 23. Wamb ungewealden and unýþe, 242, 5. Unýþe *molestus*, Wrt. Voc. ii. 56, 12. Swâ oft swâ wê ôht uneáþes þrowian æt yfflum monnum, Blickl. Homl. 33, 22. Hê Gode þancie ealles ðæs ðe hê him forgeaf, ægðer ge ýðran ge unýðran, L. E. I. 29; Th. ii. 426, 11. Hê was underfange[n] of ðam hâdesmannum ðe him ealra uneáþest was, ðæt was clerican *he was received by those of the clergy that it was most distasteful to him to be received by, that is by the secular clergy* (cf. Aþelwold dráf ût ða clerca of þe biscoprîce, 963; Th. i. 220, 19), Chr. 995; Th. i. 244, 6. III. of that which is not readily done, to which one is not easily moved, and so is little done :—Ðû gionga, bió ðê uníðe tô clipianne and tô lǽranne *do not let it be an easy matter to you to call and to teach;* adolescens loquere vix, Past. 49; Swt. 385, 10. [þeih hem be uneáðe ne sal nafre eft Crist þolien deað for lesen hem of deaðe, O. E. Homl. ii. 225, 183. Corineus was uneðe and wa on his mode, Laym. 2259.]

un-eáðe; *adv.* I. where a thing is not easily done, *with difficulty :*—Se weliga uneáþe (-eáða, Lind.) gæþ in heofuna rîce *dives difficile intrabit in regnum coelorum*, Mt. Kmbl. Rush. 19, 23. Swîðe uneáðe (-eáða, Lind.) i hefige, Mk. Skt. Rush. 10, 23; Lk. Skt. Lind. 18, 24. Hê uneáðe âwæig com, and him ðær micel forfêrde, Chr. 1052; Erl. 181, 18. Swîðe strang gyld, ðæt man hit uneáðe âcom, 1040; Erl. 166, 21. Ða lufe mon mæg swîðe uneáþe oððe nâ forbeódan, Bt. 35, 6; Fox 170, 11. Ongit hê swâ micle mâran sige on him selfum swâ hê uniéð wiðstôd *he will feel so much greater victory in himself as he had greater difficulty in withstanding*, Past. 52; Swt. 407, 26. Ðisse âðle fruman mon mæg ýþelíce gelácnian ... and æfter uneð, gif hió bið unwislíce tô lange forlǽten, Lchdm. ii. 232, 17. Cumaþ ælle tô ânum hláforde, sume æð sume uneð, Shrn. 187, 15. II. where a thing is not easily borne, *grievously, hardly :*—Sume uneáþe gedrycnede (gedrehte, MS. C.) âweg côman *turpi macie exinanitos adflictosque pestilentia dimiserit*, Ors. 3, 3; Swt. 102, 10. Ríc heofna uneáðe geðolas *regnum caelorum vim patitur*, Mt. Kmbl. Lind. 11, 12. III. where a thing is not readily done, *unwillingly, hardly :*—Ðâ geþafedon ðæt uneáþe ða his gesacan *quod cum adversarii inviti concederent*, Bd. 2, 2; S. 502, 24. Ðâ underfêng hê hig uneáðe *vix fratre compellente suscipiens*, Gen. 33, 11. IV. with a force only slightly removed from a negative, *hardly, scarcely, only just :*—Uneáþe cwic ætberstende *vix vivus evadens*, Coll. Monast. Th. 27, 3. Uneáðe Isaac geendode ðâs sprǽce ðâ com Esau *vix Isaac sermonem impleverat, venit Esau*, Gen. 27, 30. Hê uneáþe ðurh hine sylfne oþþe ârîsan oþþe gangan mihte *vix ipse per se exsurgere aut incedere valeret*, Bd. 4, 31; S. 610, 19. Uneáþe ic mæg forstandan ðine âcsunga and cwist þeáh ðæt ic ðê andwyrdan scyle *vix rogationis tuae sententiam nosco, ne dum ad inquisita respondere queam*, Bt. 5, 3; Fox 12, 15. Ic hit mæg uneáþe mid wordum gereccan *sententiam verbis explicare vix queo*, 20; Fox 70, 27. Uneáþe ænig com tô ende ðære sprǽce *ad rem ... cui vix exhausti quidquam satis sit*, 39, 4; Fox 216, 16. Uneáðe (*pretium scorti*) vix (*est unius panis*, Prov. 6, 26), Kent. Gl. 163. Ungeáþe (uneáþe, Cott. MS.), Bt. 35, 3; Fox 158, 28. [It wass tïll ennde brohht unnæþe and all wiþþ mage, Orm. 16289. þu me hauest sore igramed þat ic mai uneaþe speke, O. and N. 1605. Cf. He spac uneðes, so e gret, Gen. and Ex. 2341. *Chauc.* unnethe, unnethes, and v. Halliwell's Dict. *unnethe*.]

un-eáðelíc; *adj.* I. *difficult* to do, *impossible :*—Uneáðelíc ðæt ys mid mannum *apud homines hoc impossibile est*, Mt. Kmbl. 19, 26. Uneáþelíc, Mk. Skt. 10, 27. Líg fýres on ceafa yrnende æthabban ys uneáþelíc *flammam ignis in paleas currentem retinere est impossibile*, Scint. 57, 7. II. *difficult* to bear, *grievous, troublesome :*—Ne heó (*a sin*) næfre ne þince eów tô ðan hefig ne tô ðan uneáðelíc ne tô ðam fracodlíc, ðæt gê æfre lǽton ænig ðing ungeandett, Wulfst. 135, 12. Uneþelícne wæterbollan *a grievous dropsy*, Lchdm. ii. 204, 13. Ðás onfóað uneðelíc (-eðlíc, Lind.) dôm *hi accipient prolixius judicium*, Mk. Skt. Rush. 12, 40.

un-eáðelíce; *adv.* I. *with difficulty :*—Mid ðý wit ðæt uneþelíce ðurhtugan ðæt hê ðæs geþafa beón wolde *cum hoc difficulter impetraremus*, Bd. 5, 4; S. 617, 17. Hê geseah ðæt hê unýþelíce (*difficulter*)

mihte ða heánnesse ðæs cynelícan môdes tô eáðmôdnesse gecyrran, 2, 12; S. 512, 27. II. *with trouble* or *inconvenience, under difficulties :*—Hê uniéþelíce æfter wudum fôr and on môrfæstenum, Chr. 878; Erl. 78, 33. Ða scipu wurdon swîðe uneðelíce âseten *the ships were stranded in a most inconvenient manner*, 897; Erl. 95, 29.

un-eáðelícnes, e; *f. Difficulty :*—Ðâ wæs mycel uneþelícnes geworden be his byrignesse *facta difficultate tumulandi*, Bd. 4, 11; S. 580, 8.

un-eáðlácn[e?], -lǽcne; *adj. Not easily cured :*—Bið ðonne se milte uneáþlǽcne, ðonne ðæt blôd âheardaþ on ðæm ǽdrum, Lchdm. ii. 250, 5. Cyrnelu uneáðlácnu, 240, 21. Ða dolh beóþ uneáðlácnu, 242, 10. Uneáðlácnu, 242, 3. Seó wǽte wyrcþ uneáþlácna âdla, 226, 15.

un-eáðlǽce; *adj. Not easily cured :*—Gif hit bið of yfelre inwǽtan hit bið ðe uneáþlǽcra, Lchdm. ii. 258, 27.

un-eáðmilte; *adj. Not easily digested, indigestible :*—Sió melt mete wel, swîþost ða ðe hearde beóð and uneáðmylte, Lchdm. ii. 220, 23.

un-eáðnes, e; *f.* I. *uneasiness of mind, anxiety, trouble, grief, difficulty :*—Hê ealle ða word gehýrde, and ǽfre wæs his uneáðnys wexende, Homl. Skt. i. 23, 621. Ne bið ðær sâr ne gewinn, ne nǽnig uneþnes, ne sorg ne wôp, Blickl. Homl. 103, 35. Hê swýþe weóp and mid mycelre uneðnysse his eágospind mid teárum leohte. Ðá frêfrode hine Gúthlác and him cwæð tô: 'Ne beó ðú nâ geunrôtsod, forþon ne bið mê nǽnig uneðnysse ðæt ic tô Drihtne fare, Guthl. 20; Gdwin. 82, 2–8. Hí on wôpe wǽron and hí on uneáðnysse sprǽcon, Homl. Skt. i. 23, 247. Ðú manigfeald yfel hæfdest and micle uneþnesse on ðam rîce, Bt. 27, 2; Fox 96, 13. Ealle angnysse and uneáðnysse, Lchdm. iii. 156, 13. Gif hit geberige ðæt hê ða unǽtnessa âbidan scel, Chart. Th. 509, 33. Hwæt wylt ðú tô mêde gesyllan ðam ðe ðe fram ðissum uneðnyssum âlýseþ? Shrn. 16, 29. Ðú canst mîne yrmþa, ðú mê wǽre symble on fultume on mînum unýðnyssum, Guthl. 21; Gdwin. 94, 11. II. *severity, harshness :*—Ða ðe ðǽr gefongne wǽron hié tawedan mid ðære mǽstan uniéðnesse; sume ofslôgon, sume ofswungon, sume wið feó gesealdon, Ors. 4, 1; Swt. 154, 8.

Unecung? The word occurs in a list of territorial names :—Unecung(a?)ga (Ynetunga, p. 415; Unecung-ga, p. 416) twelf hund hýda, Cod. Dip. B. i. 414, 26.

un-efn, -efen, -emn, -emne (?); *adj. Unequal, unlike, dissimilar, diverse, irregular :*—Hû ðǽr wæs unefen racu unc gemǽne, ic onfêng ðîn sâr ðæt ðú môste gesǽlig mînes éþelríces neótan, Exon. Th. 89, 20; Cri. 1460. Dysigra monna môd bið suîðe unemn and suîðe ungelíc ... Ac ðara monna môd bið spîðe unemn, for ðæm hit gedeð hit self him selfum suîðe ungelíc for ðære gelômlícan wendinge, for ðæm hit næfre eft ne bið ðæt hit ǽr wæs *cor stultorum dissimile erit ... Cor vero stultorum dissimile est, quia, dum mutabilitate se varium exhibet, numquam id, quod fuerat, manet*, Past. 42; Swt. 306, 12–18. Sume word synd gehâtene *anomala* oþþe *inequalia*. *Anomalus* is unemne, *inequalis* ungelíc, Ælfc. Gr. 32; Zup. 199, 3. Ða unefne i ungelíco burna woegas *diversos rivulorum tramites*, Mt. Kmbl. p. 2, 9.

un-efne; *adv. Unequally, diversely :*—Swâ unefne is eorþe þicce *sicut crassitudo terrae*, Ps. Th. 140, 9.

un-efnlíc; *adj. Unequal, diverse :*—Unefenlícra *diversarum*, Mt. Kmbl. p. 7, 5.

un-endebyrdlíce; *adv. In a disorderly manner, without order, irregularly :*—Gif hê unendebyrdlíce onet mid ðære sprǽce *si inordinate ad loquendum rapitur*, Past. 15; Swt. 93, 18. Ðonne ðæt môd bið forlǽten and onstyred and tôdǽled ungedafenlíce and unendebyrdlíce on uneðeáwas *si inordinatis dimissa motibus mens vitiis dissipatur*, 43; Swt. 315, 7. Unendebyrdlíce *inordinate*, Scint. 101, 14; 191, 3.

un-ered; *adj. Unploughed :*—Unered land *rus*, Wrt. Voc. i. 37, 49.

un-éðe, un-éwisc. v. un-eáðe, un-ǽwisc.

un-fácne, -fǽcne; *adj. Without deceit, without fraud :*—Unfaecni, -fêcni *non subscivum*, Txts. 81, 1386. Unfǽcne (printed -sæcne), Wrt. Voc. ii. 60, 16. Gif man mǽgð gebigeþ ceápi, geceápod sý, gif hit unfácne is, L. Ethb. 77; Th. i. 22, 2. Gif man mannan ofsleá, unfácne feó gehwilce gelde (*there should be no fraud as regards anything given in payment of the* wergild), 30; Th. i. 10, 4. Ic Heaþobeardna hyldo ne talige Denum unfǽcne, freóndscipe fæstne, Beo. Th. 4143; B. 2068. Hæbbe hê him twêgen oþþe þreó unfácne ceorlas tô gewitnesse, L. H. G. 16; Th. i. 34, 4.

un-fæderlíce; *adv. In an unfatherly manner :*—Saturnus wæs swâ wælhreów, ðæt hê fordyde his âgene bearn ealle bútan ânum and unfæderlíce macode heora líf tô lyre, Wulfst. 106, 6.

un-fæge; *adj. Not fey, not appointed to die :*—Mæg unfǽge eáðe gedîgan weán and wræcsîð, se ðe Waldendes hyldo gehealdeþ *out of misery and exile may easily come one not appointed to die, who possesses God's favour*, Beo. Th. 4571; B. 2291. Wyrd oft nereþ unfǽgne eorl, ðonne his ellen deáh, 1150; B. 573. [*Icel.* ú-feigr *not fey.*] v. un-fǽglíc.

un-fæger; *adj. Not fair, not beautiful, foul, ugly, horrid :*—Sió gefrêdnes mæg gefrêdan ðæt hit líchoma biþ, ac hió ne mæg gefrêdan hwæþer hê bið ðe blac ðe hwît, ðe hê unfæger, Bt. 41, 4; Fox 252, 12. þincð his (*a dead man's*) neáwist lâþlíco and unfæger, Blickl. Homl. 111, 30. Him of eágum stôd líge gelícost leóht unfæger *from*

Grendel's eyes there shot a horrid light like flame, Beo. Th. 1459; B. 727. Se unfægere *larbata* (cf. hreófe *larbatos*, 86, 64: egisgríma *lar-bula*, 112, 21), Wrt. Voc. ii. 95, 68. [*Goth.* un-fagrs *ingratus* : Icel. ú-fagr *ugly*.]

un-fægere; *adv. Unpleasantly, ungently, terribly, cruelly* :—Hé ðæt unfægere wera cneórissum gewrecan þóhte, Cd. Th. 77, 11; Gen. 1273. Gripon unfægre under sceát werum scearpe gáras, 124, 16; Gen. 2063. Sampson hewis doun of þa hirdis, hurtis þam unfaire, Alex. (Sk.) 1224, [and see Glossary.]

un-fægerness, e; *f. Foulness, ugliness, abomination* :—Unfegernis slitnese *abominatio desolationis*, Mt. Kmbl. Lind. 24, 15.

un-fǽglíc; *adj. Not indicating impending death* :—Ðæt is tácn ðínre hæle; swá swá lǽca gewuna is, ðæt hé cweþaþ ðonne hió seócne mon gesióþ, gef hé hwelc unfǽglíc (ungefǽglíc, Cott. MS.) tácn (*a symptom which does not indicate that a disease is mortal*) him on geseóþ: mé þincþ nú ðæt ðín gecynd flíte swiþe swiþlíce wiþ ðæm dysige *id, uti medici sperare solent, indicium est erectae jam resistentisque naturae*, Bt. 36, 4; Fox 178, 27. v. un-fǽge.

un-fǽhð, e; *f. Absence of hostility*; the word refers to the abstention from the prosecuting of the feud, which under certain conditions it would be allowable for the kinsmen of a man to follow up :—Se ðe þeóf geféhð, hé áh .x. sciłł. ... and ða mǽgas him swerian áðas unfǽhða (cf. unceáses áð, 35; Th. i. 124, 8, the circumstances in the two cases being similar), L. In. 28; Th. i. 120, 6. v. un-fáh.

un-fǽle; *adj. Evil, ill, bad* :—Unfǽle (*printed* -sǽle), gemáh *improbus*, Wrt. Voc. ii. 45, 16. (1) applied to living objects :—Hí wéndon ðæt hit unfǽle gást (*phantasma*) wǽre, Mk. Skt. 6, 49. *Satiri* vel *fauni* vel *selini* vel *fauni ficarii* unfǽle men, wudewásan, unfǽle wihtu, Wrt. Voc. i. 17, 20. Unfǽle men *sa'tyri* vel *fauni*, wudewásan *ficarii* vel *invii*, 60, 23-4. [Gif þe unfele man his wille folgeð, and teð him to unwrenches, O. E. Homl. ii. 79, 27. Þe laþe gast cwelleþþ hemm þurrh his unnfæle þeowwess, Orm. 8034. Idisse wildernesse beoð monie vuele bestes (un-feale bestes monie, MS. T.), A. R. 198, 2. Зef heo is atbroide þenne heo is unfele and forbrode, O. and N. 1381.] (2) applied to inanimate objects :—Ofet unfǽle (*the forbidden fruit*), Cd. Th. 45, 7; Gen. 723. Unfǽle *dira* (the passage is : *dira vinculorum ligamina*, Ald. 44), Anglia xiii. 34, 178. [Þat water is unfǽle, Laym. 22018. Þat lond is grislich and unfele, þe men beoþ wilde and unisele, O. and N. 1003. Þe stude (*hell*) is swiþe unvele (*rimes with* hele = *heal*), Misc. 73, 45. Cf. A seolcuð mere ... mid uniuele þingen, Laym. 21744.]

un-fæst; *adj. Not firm, unstable, unsteady, weak* :—Hú ne is ðé nú genóh sweotole gesǽd ðæt seó wyrd ðé ne mæg náne gesǽlþa sellan, for ðam ðe ǽgher is unfæst ge seó wyrd ge seó gesǽlþ *manifestum est, quod ad beatitudinem percipiendam fortunae instabilitas aspirare non possit*, Bt. 11, 2; Fox 34, 21. Hwæt getácnaþ ðonne ðæt flǽsc búton unfǽst weorc and hnesce ... ? Oft ðeáh gebyreþ ðætte sume on monegum weorcum unfæste beóð ongietene *quid per carnes nisi infirma quaedam ac tenera acta signantur? Et plerumque contigit, ut quidam in nonnullis suis actibus infirmi videantur*, Past. 34; Swt. 235, 14-17. Ðonne ðæt mód bið on monig tódǽled, hit bið on ánes hwæm ðe unfæstre *impar quisque invenitur ad singula, dum confusa mente dividitur ad multa*, 4; Swt. 37, 15. [*O. H. Ger.* un-festi *infirmus*.]

un-fæstende; *adj. Not fasting* :—Ðæt ǽnig unfæstende man húsles ne ábirige, L. Edg. C. 36; Th. ii. 252, 1.

un-fæstlíce; *adv. Not firmly, uncertainly, vaguely* :—Ðonne mon smeáð on his móde ymb hwelc eorðlíc ðing, ðonne deð hé swelce hé hit átifre on his heortan, and swǽ tweólíce and unfæstlíce hé átifreþ ðæs ðinges onlícnesse on his móde ðe hé ðonne ymb smeáð, Past. 21; Swt. 156, 13.

un-fæstrǽd[e], -rǽd ; *adj. Infirm of purpose, inconstant, unstable, weak* :—Unfæstrǽd *inconstans*, Wrt. Voc. ii. 49, 4. Ðá ongon hé ǽresð herigean on him ðæt ðæt hé fæsðrǽdes wiste and sóna æfter ðon suíðe líðelíce hierd[d]e ða ðe hé unfæsðrǽde (unfæstrǽdes, Cott. MSS.) wisse *prius in eis, quae fortia prospicit, laudat, et caute monendo postmodum, quae infirma sunt roborat*, Past. 32; Swt. 213, 9. Ða ungestæððegan and unfæsðrǽdan *inconstantes*, 23; Swt. 177, 4: 42; Swt. 305, 11.

un-fæstrǽdness, e; *f. Instability, inconstancy, levity* :—Hié wéndon ðæt hé nyste hiera leohtmódnesse and hiera unfæstrǽdnesse *dum de ipsa levitate motionis praedicatori suo se incognitos crederent*, Past. 32; Swt. 214, 2. On heora wandlunga hié gecýþdon heora unfæstrǽdnesse, Bt. 7, 2; Fox 16, 32 note.

un-fáh; *adj. Not regarded as a foe*, used of the kinsmen of a criminal when not involved in the feud which their kinsman's guilt occasioned :—Gif hwá heonanforð ǽnigne man ofsleá, ðæt hé wege sylf ða fǽhðe ... Gif hine seó mǽgð forlǽte ... ðonne wille ic ðæt eall seó mǽgð sý unfáh, bútan ðam handdǽdan, L. Edm. S. 1; Th. i. 248, 2-7. [*O. Frs.* un-fách.] v. un-fǽhþ.

un-fealdan; *p.* -feóld *To unfold, unroll* :—Unfealdaþ *replicant*, i. *reuoluunt*, Scint. 140, 2. Hé ða bóc unfeóld *reuoluit librum*, Lk. Skt. 4, 17.

un-feferig; *adj. Not feverish* :—Syle drincan on wíne, gif hé unfeferig sý; gif hé on fefere sý, syle drincan him on wætere, Lchdm. i. 164, 19.

un-félende; *adj. Unfeeling, callous* :—Yfele swilas unfélende, Lchdm. ii. 264, 13.

un-feor[r]; *adv. Not far off.* I. marking position, *at no great distance off*, (1) where the point from which the distance is measured is given by an adverb :—Ðǽr wæs unfeorr (-feor, MS. A.) án swýna heord *erat non longe ab illis grex porcorum*, Mt. Kmbl. 8, 30. Ðá geseah hé deófol ðǽr unfeor standan, Blickl. Homl. 227, 24. (2) with dative :—Ðá hé wæs unfeor ðam húse *cum non longe esset a domo*, Lk. Skt. 7, 6. Ðæt is unfeor ðære byrig Neapoli *quod est non longe a Neapoli*, Bd. 4, 1; S. 563, 30 : Cd. Th. 125, 22; Gen. 2083. Unfeor herge *haud procul a delubro*, Hpt. Gl. 493, 36. (3) with dative and adverb :—Se rinc him ðǽr rom geseah unfeor þanon standan, Cd. Th. 177, 9; Gen. 2927. (4) with preposition :—Hí wǽron unfeor fram lande *non longe erant a terra*, Jn. Skt. 21, 8. Fram ðam mynstre unfeor wæs ðære abbudissan mynster *a quo (monasterio) non longe illa monasterium habebat*, Bd. 3, 11; S. 536, 1. (5) where the point from which distance is measured is implied :—Wutað ðætte unfeorr sié *scitote quod in proximo sit*, Mk. Skt. Lind. 13, 29. II. with verbs of motion, (*to*) *no great distance* :—Hig wendan unfeorr út on Wealas *they marched a short distance into Wales*, Chr. 1055; Erl. 190, 12.

un-feormigende *inexpiable* :—Ðá onhrán mín mód hǽlo andgit mid mé sylfre þencende ðæt mé ðone ingang belucen ða onfeormeganda mínra misdǽda *the inexpiable circumstances of my misdeeds had closed the entrance for me*, Homl. Skt. ii. 23 b, 426. v. feormian, III.

un-fére; *adj. Incapacitated, disabled, infirm, feeble* :—Se wæs Æþel-stánes biscopes gespelia syððan hé unfére wæs, Chr. 1055; Erl. 190, 21. [Þa iwærd þe king unfere. Swa þe king seoc læi ... ne mihte he þer of beon hæl, Laym. 6780. Al unfer he it (*Moses's leprous hand*) fond, Gen. and Ex. 2810. Þat licour þat to dele unto þe unfere, L. H. R. 115, 277. See also Halliwell's Dict. Icel. ú-fœrr *disabled*.]

un-flitme; *adv. Without dispute* :—Fin Hengeste elne unflitme áðum benemde ðæt hé ða weálafe árum heólde (*Fin confirmed with oaths the terms he made with Hengest, and there was no dispute about the terms which were settled*), Beo. Th. 2198; B. 1097.

un-forbærned; *adj. Unburnt, not burnt up, not consumed by fire* :—Ðǽr is ðeáw, ðonne ðǽr bið man deád, ðæt hé líð inne unforbærned mid his mágum and freóndum mónað, ... hwílum healf geár ðæt hí beóð unforbærned, Ors. 1, 1; Swt. 20, 19-24. Gyf man án bán findeþ unforbærned, hí hit sceolan miclum gebétan, Swt. 21, 12. Tiburtius eode ofer ða byrnendan gléda unforbærnedum fótum, Homl. Skt. i. 5, 380.

un-forboden; *adj. Unforbidden, not prohibited, free from any moral or legal hindrance* :—Ðæt hí móston him beran unforboden flǽsc, Homl. Skt. ii. 25, 91. Swá ic hit hæbbe, swá hit se sealde, ðe tó syllanne áhte, unforboden (*no one had a right to forbid the entering into possession of the property*), L. O. 13; Th. i. 184, 5. Unforboden and unbesacan, Cod. Dip. Kmbl. iv. 234, 20. Gebohte se arcebisceop æt Ælfhége ðæt land æt Sendan mid .xc. pundum, and æt Sunnanbyrg mid .cc. mancussan goldes, unbecwedene and unforbodene wið ǽlcne man tó ðære dægtíde; and hé him swá ða land geágnian derr, swá him se sealde ðe tó syllenne áhte, Chart. Th. 208, 38.

un-forbúgendlíc; *adj. Unavoidable, inevitable* :—Unforbúgendlíc *in-evitabile*, Hpt. Gl. 440, 40.

un-forbúgendlíce; *adv. Without turning aside, constantly, fixedly* :—Ic cwæð tó hire geornlíce and unforbúgendlíce behealdende and cweðende : Eálá ... , Homl. Skt. ii. 23 b, 431.

un-forburnen; *adj. Unburnt, not consumed by fire* :—Se wind ábær ðone líg tó ðæs cyninges botle, swá ðæt him ne belǽfde nán þing unforburnen, and hé sylf earfoðlíce ðam fýre ætbærst, Homl. Th. ii. 480, 7.

un-forcúþ; *adj. Not despicable, not ignoble, not wicked, honourable, noble, good* :—Ic eom heard and strong, forðsíðes from, freán unforcúð, Exon. Th. 479, 22 : Rä. 63, 2. Hér stynt unforcúð eorl ðe wile gealgian eþel ðysne, Byrht. Th. 133, 16; By. 51. Eorl unforcúð elnes gemyndig, Andr. Kmbl. 2527; An. 1265. Nǽfre ic sǽlidan sélran mette ... ic wille ðé, eorl unforcúð, biddan, 949; An. 475. Ðegn unforcúð, Menol. Fox 338; Men. 170. Hé þenceþ ðæt his wíse þince eal unforcúþ, Exon. Th. 315, 14; Mód. 31. Cweðan ealle ðæt unforcúðe ðe him on standeþ egsa Dryhtnes *dicant qui timent Dominum*, Ps. Th. 117, 4.

un-forcúþlíce; *adv. Nobly, excellently* :—Metode geþungon Abraham and Loth unforcúðlíce, swá him from yldrum ǽðelu wǽron, Cd. Th. 103, 9; Gen. 1715.

un-fordyt[t]; *adj. Unobstructed, unstopped* :—Ða unfordyttan gemágnesse *obstinatam importunitatem* (*garrulitatem*), Hpt. Gl. 491, 24.

un-forebyrdig; *adj. Impatient* ; *inpatiens*, Scint. 8, 13.

un-fored. v. un-forod.

un-foresceáwod; *adj. Unconsidered, hasty, without due consideration* :—Næs hit ná fǽrlíc geðóht oððe unforesceáwod rǽd, ðæt se ælmihtiga God ðysne middangeard gesceóp, ac wæs ǽfre æt fruman on his écum rǽde, Hexam. 14; Norm. 22, 5.

un-foresceáwodlíc; *adj. Hasty, inconsiderate, rash* :—On scyterǽs oþþe on fǽrfyll, unforesceáwodlíc *in preceps*, Wrt. Voc. ii. 47, 44. v. un-forsceáwodlíc.

un-forfeored (un-forfored (?). v. un-forodlíc; *also* ungebrocenre *extricabili*, 33, 7 : perhaps in each case *inextricabilis* should be read, cf. untósliten *inextricabilis*, 110, 60) ; *adj. Unbroken* ; extricabile, Wrt. Voc. ii. 145, 22. v. forod.

un-forgifen ; *adj.* I. *unforgiven :—*Ealle scylda ðe wið God beoð ungebétta beoð unforgifne on dómes dæge, Past. 33; Swt. 220, 17. II. *not given in marriage* (cf. Goth. fra-gifts *espousal* ; Icel. ú-gefinn *unmarried*) :—Unforgifenum *innupti*, Wrt. Voc. ii. 45, 19.

un-forgitende ; *adj. Not forgetful, mindful :—*Ðínra gewinna and earfoða ic eom unforgitende, Guthl. 19; Gdwin. 76, 22.

un-forgolden ; *adj. Unremunerated, not paid for :—*Nafa ðú áne niht unforgolden ðæs weorc ðe ðé wirce *do not leave unpaid for a night the work of him that works for thee*, Lev. 19, 13.

un-forhæfedness, e ; *f. Incontinence :—*Gýfernyss móder ys unforhæfednysse (*incontinentiae*), Scint. 89, 14. Unforhæfdnysse, Bd. 1, 27 ; S. 493, 36.

un-forhladen ; *adj. Unexhausted:—*Unforhladenum *inexaustis*, Wülck. Gl. 255, 39.

un-forht ; *adj. Not frightened, not afraid, fearless, intrepid :—*Ðæt gedyld stent unforht betweónan ðara leahtra truman *patientia inter acies vitiorum intrepida stat*, Gl. Prud. 17 b. Hwæt eart ðú, ðú ðe swá unforht ús tó eart cumen ? Nicod. 28; Thw. 16, 33 : Homl. Skt. i. 18, 262 : Cd. Th. 199, 7; Exod. 335 : Exon. Th. 278, 21 ; Jul. 601 : Rood Kmbl. 218; Kr 110. Se Hælend unforht áxode, Homl. Th. ii. 246, 13 : Exon. Th. 255, 5; Jul. 209. On wicge sæt ombeht unforht, Beo. Th. 579 ; B. 287. Him seó unforhte ágeaf andsware, Exon. Th. 251, 18; Jul. 147. Unforhte móde hé genealæhte ðære stówe, Blickl. Homl. 67, 1. Se man hýwaþ hine sylfne mihtine and unforhtne, Wulfst. 53, 15. Wígend unforhte, Cd. Th. 189, 6; Exod. 180 : Byrht. Th. 134, 5; By. 79. Hí unforhte and blíþe underhnigon deáþ *mortem laeti subiere*, Bd. 4, 16; S. 584, 37. Ðæt hí ðý baldran and ðý unforhtran wæron (ðæt heora compweorodes mód ðý unforhtre beón sceolde, col. 2) *sperantes minus animos militum trepidare*, 3, 18 ; S. 546, 24.

un-forhte ; *adv. Fearlessly :—*Hé wille leóde etan unforhte, Beo. Th. 892 ; B. 444.

un-forhtigende ; *adj. Not fearing, fearless :—*Hé wolde leódum bodian on fyrlenum lande unforhtigende, Homl. Th. ii. 140, 29.

un-forhtlíce ; *adv. Fearlessly, without fear :—*Unforhtlíce *non trepide*, R. Ben. 20, 18. Hé unforhtlíce ða stræle ðara áwerigdra gásta him fram ásceáf, Guthl. 6 ; Gdwin. 42, 24. Twá swalewan hí setton unforhtlíce on ða sculdra Gúðlaces, 10 ; Gdwin. 52, 9 : Homl. Th. i. 508, 1 : ii. 558, 30. Hé ðý unforhtlícor done deáþ áræsnode, Shrn. 129, 21.

un-forhtmód ; *adj. Fearless :—*Ic unforhtmód ðæs drences onfó, Homl. Th. i. 72, 17. Sixtus unforhtmód tó his preóstum clypode : 'Míne gebróðra, ne beó gé áfyrhte, and eówer nán him ne ondræde ða scortan tintregunga,' 416, 6.

un-forlæten ; *adj. Not left :—*Unforlætne *non relicto*, Mk. Skt. Lind. Rush. 12, 20.

un-formolsniendlíc ; *adj. Undecaying, incorruptible :—*Unforwurdenlícne ł [unfor]molsniendlícne (*or* [un]molsniendlícne ?) *incorruptam, immarcescibilem*, Hpt. Gl. 407, 37.

un-formolsnod ; *adj. Undecayed :—*His líchama lið unformolsnod, Th. An. 124, 4.

un-formolten ; *adj. Unconsumed, undigested :—*Se wítega wæs gehealden unformolten on ðæs hwæles innoðe, Homl. Th. i. 488, 7.

un-forod(-ed) ; *adj. Unbroken, inviolate :—*Werige hine se Fræncisca mid unforedan áþe, L. W. ii. 3 ; Th. i. 489, 25. Wé sceolon healdan ðone bróðerlícan bend unforedne, Homl. Th. i. 260, 29. v. next word.

un-forodlíc ; *adj. Indissoluble :—*Unforedlícre racent[e]ágæ *inextricabili collario*, Hpt. Gl. 455, 9. Unforedlícum bende *inextricabili* (*in-dissolvibili*) *repagulo*, 462, 73. Unforadlíce *inextricabile* (*vinculum*), 521, 75.

un-forrotigendlíc ; *adj. Not liable to decay, imperishable, incorruptible :—*Beó his calic of clænum antimbre geworht unforrotigendlíc, gylden oððe seolfern oððe tinen, L. Ælfc. C. 22 ; Th. ii. 350, 23. Unforrotiendlíc, 36 ; Th. ii. 360, 42.

un-forrotodlíc ; *adj. Not liable to decay, incorruptible, imperishable :—*Unforrotedlíces *immarcescibilis, imputribilis*, Hpt. Gl. 467, 45.

un-forsceáwodlíce ; *adv.* I. *unexpectedly :—*Óþ ðæt ðe hig (wildeór) cuman tó ðam nettan unforsceáwodlíce *usque quo perveniant ad retia improvise*, Coll. Monast. Th. 21, 17. II. *without fore-thought, without consideration :—*Ne getímode Thóme unforsceáwodlíce ðæt hé ungeleáfful wæs, ac hit getímode þurh Godes forsceáwunge, Homl. Th. i. 234, 19. Gif hé hit æne and unforsceáwodlíce gedyde *si semel et inconsiderate fecerit*, L. Ecg. C. 39 ; Th. ii. 164, 24. v. un-foresceáwodlíc.

un-forswæled ; *adj. Unburnt, unscorched :—*Ic geseó feówer weras gangende onmiddan ðam fýre ungewemmede and unforswælede *ego video quatuor viros . . . ambulantes in medio ignis, et nihil corruptionis in eis est* (Dan. 3, 25), Homl. Th. ii. 20, 15.

un-forswigod ; *adj. Not passed over in silence, not omitted :—*Án weorc hé hæfde unforswigod . . . ðæt wæs sealmsang *one work he never allowed to pass in silence . . . that was psalmsinging*, Homl. Skt. ii. 23 b, 35.

un-forswíped ; *adj. Unconquered :—*Ic ðæs þoncunge dó ðæm unforswýþdum úrum weorode *ago gratias inuicto exercitui nostro*, Nar. 2, 31.

un-fortredde *not destroyed by treading* ; a name given to a plant that can grow in trodden paths, *knot-grass* ; polygonum aviculare :—*Pilogonus et sanguinaria* ðæt is unfortredde, Wrt. Voc. i. 68, 66. Unfortrædde. Ðeós wyrt ðe man *proserpinacam* and óðrum naman unfortredde nemneþ, heó bið cenned gehwær on begánum stówum, Lchdm. i. 112, 4-7. [Cf. way-grass, E. D. S. Plant Names : *O. H. Ger.* wege-trat *centenodia* ; umbi-trat *serpinacia* ; ana-tret *proserpinaca*.] v. next word.

un-fortreden ; *adj. Not destroyed by treading :—*Unfortreden wyrt *appoligonius* (= *polygonum*), Lchdm. iii. 299, col. 2. v. preceding word.

un-forwandigendlíce ; *adv. Unhesitatingly, freely, without regard to fear* or *shame :—*Gif ðú wundrige ðæt swá scamfæst fæmne swá unforwandigendlíce ðás word áwrát, ðonne wite ðú ðæt ic hæbbe þurh weax áboden, ðe náne scame ne can, ðæt ic silf ðé for scame secgan ne mihte, Ap. Th. 21, 9.

un-forwandodlíc ; *adj. Undeterred by fear* or *shame, fearless, free :—*Ðæt hé wiðstande mid his spræce ðam unryhtwillendum ðe ðyses middangeardes waldaþ mid freóre and unforwandodlícre stefne *voce libera hujus mundi potestatibus contraire*, Past. 15 ; Swt. 89, 23. Ðonne wénaþ hié ðæt hié sprecen for unforwandodlícre and orsorglícre ryhtwísnesse *se credunt loqui per libertatem rectitudinis*, 41 ; Swt. 302, 5.

un-forwandodlíce ; *adv.* I. *without swerving, directly :—*Forðrihte, unforwandedlíce *indeclinabiliter, inevitabiliter* (ad destinatum indeclinabiliter dirigit locum, Ald. 2), Hpt. 406, 4. II. *unexpectedly, suddenly :—*Unforwandedlíce *ex improviso, extemplo, subito*, Hpt. Gl. 457, 35. III. *with a disregard of fear, unhesitatingly, freely, fear-lessly :—*Ne durron ryht freolíce læran and unforwandodlíce sprecan *loqui libere recta pertimescunt*, Past. 15 ; Swt. 89, 12 : 41 ; Swt. 302, 2. Ðæt mon openlíce and unforwandodlíce on óðerne ræse mid tælinge *impetu apertae increpationis obviare*, 40 ; Swt. 297, 12. Æghwylc cristen man dó swá him þearf is . . . unforwandodlíce his synna gecýþe, L. Eth. v. 22 ; Th. i. 310, 6 : Wulfst. 180, 6 : Homl. Ass. 141, 69. IV. *rashly, recklessly, inconsiderately, heedlessly :—*Unrædlíce, unforwandedlíce *in-consulte, inconsiderate*, Hpt. Gl. 474, 57 : 509, 64. Ðær bið dæghwomlíce wóp . . . and endeleás cwylming, tó ðam Egeas onet unforwandodlíce, Homl. Th. i. 592, 17.

un-forwealwod ; *adj. Unwithered, undecayed :—*Bringan Drihtne unforwealwod wæstm gódra weorca, Blickl. Homl. 73, 25.

un-forwordenlíc ; *adj. Undecayed, uncorrupt :—*Unforwurdenlícne *incorruptam*, Hpt. Gl. 407, 36.

un-forworht ; *adj. Not criminal, innocent :—*Wæron earme men beswicene and út of ðisan earde gesealde swýðe unforworhte fremdum tó gewealde, Wulfst. 158, 13. Úre hláfordes gerædnes is ðæt man cristene menn and unforworhte of earde ne sylle, L. Eth. v. 2 ; Th. i. 304, 15. Se ðe hit áwende æt unforworhtum þingum *he who sets aside the grant when there is no criminality on the part of the grantee* (cf. the phrase frequent in Oswald's charters : Gif hwá búton gewyrhtum hit ábrecan wille, iii. 21, 30, and often. See also, in another of Oswald's charters : Si quid praefatorum delicti praevaricantis causa defuerit jurum, praevari-cationis delictum secundum quod praesulis jus est emendet, aut illo quo antea potitus est dono et terra careat, vi. 125 ; and see Kemble's Saxons in England, i. 311), Cod. Dip. Kmbl. ii. 408, 5. [*O. Frs.* un-forwrocht *not forfeited.*] v. for-wyrcan.

un-forworht [*different from preceding word.* v. fór-wyrcan (*l.* for-), *and* cf. *O. H. Ger.* furi-wurchen *obstruere*] ; *adj. Unobstructed, without hindrance, free* ; the term is used of land that after several lives was to revert to the grantor, and seems to render the word *immunis* in the Latin charters :—On ða geråd, weorce hé ðæt hé weorce, ðæt ðæt land seó unforworht intó ðære hálgan stówe (the Latin previously in the same charter is : Ad usum primatis in Weogornaceastre redeat inmunis. See also the passage : Tellus episcopali restituatur cathedrae absque ullius controversiae obstaculo, iii. 232, 24), Cod. Dip. Kmbl. ii. 396, 33 : 397, 29 : 384, 22. (The formula is common in Oswald's charters. See Cod. Dip. Kmbl. i. xxxiii, and Kemble's Saxons in England, i. 312.)

un-fracodlíce ; *adv. Not dishonourably, honourably, virtuously :—*Ic wilnode andweorces tó ðam weorce ðe mé beboden wæs tó wyrcanne, ðæt wæs, ðæt ic unfracodlíce and gerísenlíce mihte steóran and reccan ðone anweald ðe mé befæst wæs *materiam gerendis rebus optavimus, quo ne virtus tacita consensceret*, Bt. 17 ; Fox 58, 27.

un-frætewod ; *adj. Unadorned, unpolished :—*Unfratewode *inculta*, Germ. 396, 180.

un-fremful ; *adj. Unprofitable, not advantageous :—*Unfremful bið ðæt folc beó bútan steóre oððe bútan æ him eallum tó hearme, Homl. Skt. i. 13, 126. Unfremful *imperfectum* (incomplete, not of use), Hpt. Gl. 524, 66.

un-fremu, e; *f. Hurt, loss, damage, detriment :*—Hū nyt biđ đæt, đeáh đū đē ealne middaneard and ealle eorđan wille gestrýnan, gif đū đínre sáwle unfreme and forlorenesse gewyrcst? Anglia xi. 8, 29. Đū blǽda náme on treówes telgum, and mē on teónan ǽte đa unfreme, Cd. Th. 55, 12; Gen. 893. [Đe man noteđ wel his ȝiepshipe, þe birgeđ him seluen wiđ his aȝene soule unfreme, and erneđ after his soule freme, O. E. Homl. ii. 195, 9.]

un-freóndlíce; *adv. In an unfriendly manner :*—Wē đē freóndlíce wíc getǽhton, đū ús leánest nū unfreóndlíce, Cd. Th. 162, 30; Gen. 2689.

un-fricgende *not questioning :*—Mē sægde đæt wíf hire wordum selfa unfricgendum *the woman of her own accord told me without my asking*, Cd. Th. 160, 12; Gen. 2649.

un-friþ, es; *n.* I. *absence of peace, hostilities :*—Hér wæs micel unfriđ on Angelcynnes londe þurh sciphere, and wel gehwǽr hergedon and bærndon *in this year there were constant hostilities in England through the Danes, and they harried and burned pretty well everywhere*, Chr. 1001; Erl. 136, 1. Hér áspón Æđelwald đone here tó unfriđe, đæt hié hergodon ofer Mercna land *in this year Ethelwold enticed the Danes to hostilities, so that they went across Mercia harrying*, 905; Erl. 98, 14. Hē behét đæt hē nǽfre eft tó Angelcynne mid unfriđe cumon nolde *he promised that he would never again come and disturb the peace of England*, 994; Erl. 133, 33. Se cyng bæd Godwine eorl faran intó Cent mid unfriđa, ac se eorl nolde ná geđwǽrian đǽre infare, forþan him wæs láđ tó ámyrrenne his ágenne folgađ, 1048; Erl. 178, 8. For unfriđe *on account of hostilities*, L. N. P. L. 56; Th. ii. 298, 26. Hié ne dorston forþ bí đǽre eá siglan for unfriþe; for đǽm đæt land wæs eall gebún on óþre healfe đǽre eás *they durst not sail on past the river for fear of being attacked; for the land was all cultivated on the other side of the river*, Ors. 1, 1; Swt. 17, 22. II. *referring to the king's peace, the state of being out of the king's peace :*—Fare se ealdorman tó; gif hē nelle, fare se cyning tó; gif hē nelle, licȝe se ealdordóm on unfriđe (*the old Latin version renders this :* adeat aldremannus; si nolit, rex; si nolit, sit pars illa praeter pacem), L. Eth. ii. 6; Th. i. 286, 34. [Membriz hefde inomen griđ, ah sone he makede unfriđ, Laym. 2557. O. Frs. on-frede, un-fretho : O. H. Ger. un-fridu : Ger. un-friede : Icel. ú-friđr.]

unfriþ-flota, an; *m. A hostile fleet :*—Se[o] unfriđflota wæs đæs sumeres gewend tó Rícardes ríce, Chr. 1000; Erl. 137, 5.

unfriþ-here, es; *m. A hostile army, an army that is carrying on hostilities :*—Com se ungemetlíca unfriđhere tó Sandwíc, Chr. 1009; Erl. 142, 16. On đissum geáre wæs đet gafol gelǽst đam unfriđehere, 1007; Erl. 141, 13.

unfriþ-land, es; *n. A hostile country, a country with which hostilities are being carried on :*—Gyf Æđelrédes cynges friđman cume on unfriđland (*terram hostilem*, Latin version), and se here đǽrtó cume, hæbbe friđ his scip and ealle his ǽhta, L. Eth. ii. 3; Th. i. 286, 7.

unfriþ-mann, es; *m. A man of a country not at peace with another, a man of a hostile country :*—Gif hē his ǽhta bere geman[g] đara unfriđmanna ǽhta intó húse, þolie his ǽhta *si pecuniam suam inter pecuniam unfriđmannorum, i. e. pacem non habentium, in domo mittat, perdat pecuniam suam* (Lat. vers.), L. Eth. ii. 3; Th. i. 286, 11. [O. Frs. unfreth-monn.] v. preceding word.

unfriþ-scip, es; *n.* I. *a ship which is carrying on hostilities :*—Đam cynge com word đæt unifriđscipa lǽgen be westan and hergodon, Chr. 1046; Erl. 173, 5. II. *a ship belonging to a hostile country :*—Ælc ceápscip friđ hæbbe đe binnan múđan cuman(-e?), þeh hit unfriđscyp sý, gyf hit undrifen biđ *omnis ceapscip, i. e. navis institoris, pacem habeat, quae in portum veniet, licet navis sit inimicorum, si non sit abacta tempestatibus* (Lat. vers.), L. Eth. ii. 2; Th. i. 286, 21.

un-fród; *adj.* I. *not old :*—Đá wæs gegongen guman(-ū, MS.) unfródum (cf. geongum, 5712; B. 2860) earfođlíce, đæt hē on eorđan geseah đone leófestan bleátne gebǽran, Beo. Th. 5635; B. 2821. II. *not wise, ignorant, rude.* [Goth. un-fróþs *foolish :* Icel. ú-fróđr *ignorant.*] v. next word.

un-fródness, e; *f. Ignorance, rudeness :*—Unfródnyssa (cf. edwítu, R. Ben. 97, 7) geþyldelíce beran *difficultatem patienter portare*, R. Ben. Interl. 95, 14.

un-from; *adj. Not strong, feeble, weak :*—Đæt hē sleac wǽre, æđeling unfrom, Beo. Th. 4382; B. 2188. Eágan đíne gesáwon đæt ic ealles wæs unfrom on ferhþe *imperfectum meum viderunt oculi tui*, Ps. Th. 138, 14.

un-fúl; *adj. Not foul, good;* but the word glosses *insulsum*, Mk. Skt. Lind. Rush. 9, 50.

un-fulfremed; *adj. Imperfect :*—*Praeteritum imperfectum*, đæt is unfulfremed forđgewiten, Ælfc. Gr. 20; Zup. 124, 3. Đæt hí didon unfulfremed (*inperfectum*) forlǽtende, R. Ben. Interl. 24, 1. Đa đing đe hē unfullfremed gemētte *ea quae minus perfecta reperit*, Bd. 4, 2; S. 566, 2.

un-fulfremedness, e; *f. Imperfection :*—Đæt hí murkien for hira unfullfremednesse *ut imperfectionis suae taedio tabescant*, Past. 65; Swt.

467, 13. Unfulfremednisse míne (*inperfectum meum*) gesēgun ēgan đín, Ps. Surt. 138, 16.

un-fulfremming, e; *f. Imperfection :*—Unfulfremmingce míne *imperfectum meum*, Ps. Lamb. 138, 16.

un-fúliende; *adj. Incorruptible :*—Unfúliendre clǽnnysse *imputribilis pudicitiae*, Hpt. Gl. 467, 46.

un-fúliendlíc; *adj. Incorruptible :*—Unfúliendlícere gecynde *imputribilis naturae*, Hpt. Gl. 419, 36.

un-fullod; *adj. Unbaptized :*—Swá hwylc mæssepreóst se đe wite đæt hē unfullod sý, fullige man hine *omnis presbyter, qui noverit quod non sit baptizatus, baptizetur*, L. Ecg. C. 7; Th. ii. 138, 23. Be unfullodon mæssepreóste, Th. ii. 128, 17.

un-fulworht; *adj. Unfinished, uncompleted, imperfect :*—Đa đe ... swá hwylce bysiga swá hý on handa hæfdan unfulworhte lǽtaþ *ex occupatis manibus quod agebant inperfectum relinquentes*, R. Ben. 20, 3.

un-fyrn; *adv.* I. *of past time, not long ago :*—Weorþodan wē nū unfyrn for tēn nihtum đone symbeldæg foran tó đyssum ondweardan dæge *not long ago now, ten days from to-day, we celebrated the festival*, Blickl. Homl. 131, 9. II. *of future time, before long :*—Secgas míne gearwe sindon; đa đe unfyrn faca feorh ætþringan, Andr. Kmbl. 2741; An. 1373. Nū ic fundige tó đē of đisse worulde; nū ic wát đæt ic sceal ful unfyr[n] faca, Exon. 454, 32; Hy. 4, 42. [Cf. Icel. ú-forn *not old.*] Cf. un-gefyrn, -geára.

un-gænge, ungc, un-geæhtendlíc. v. un-genge, unc, un-geeahtendlíc.

un-geéwed; *adj. Unmarried :*—Uniéwedan *innuptis*, Hpt. Gl. 525, 17.

un-geandet[t]; *adj. Unconfessed :*—Đæt gē nǽfre ne lǽton ǽnige synne ungeandet ... đæt gē lǽton ǽnig đing ungeandett ... đæt se deófol eów nǽge náht on tó bestelenne ungeandettes, Wulfst. 135, 9–32.

un-geára; *adv.* I. *of past time, not long ago, lately :*—Ic wæs ungeára on niht ábysgod on wæccum *nuper occupatus noctu vigiliis*, Bd. 4, 25; S. 600, 39. Đæt wæs ungeára, đæt ic ǽnigra mē weána ne wēnde đote gebídan, Beo. Th. 1868; B. 932. II. *of the future, before long, soon :*—Đone egesfullan dómes dæg, se cumeþ nū ungeára, Blickl. Homl. 101, 28. Ungeára nū, Cd. Th. 289, 9; Sat. 395: Beo. Th. 1209; B. 602. Đū ungeára deáþe sweltest, Exon. Th. 250, 8; Jul. 124. Cf. un-fyrn.

un-gearu; *adj.* I. *not ready, not prompt, indisposed* to act :*—Se sixta leahter is *accidia* geháten, đæt is slǽwđ on Englisc, đonne đam menn ne lyst nán gód dón and hē biđ ǽfre ungearu tó ǽlcere duguđe, Homl. Skt. i. 16, 299. [Ungearu to elchere duȝeþe, O. E. Homl. i. 103, 28.] II. *not ready, not in a fit state* for use :*—Đý læs sió earc sí ungearo tó beranne *ut ad portandum arcam nulla mora praepediat*, Past. 22; Swt. 173, 11. Đý læs hine ǽnig wuht gǽlde ungearowes (-ewes, Cott. MSS.), đonne mon đa earce beran scolde *ut, cum portari arcam opportunitas exigit, portandi tarditas nulla generetur*, Swt. 171, 23. II a. of land, *uncultivated :*—Gúđlác đæs wídgillan wéstenes đa ungearawan stówe đǽr gemētte, Guthl. 3; Gdwin. 20, 10. III. *not ready, not prepared* for attack :*—Wē đē beóđ holde, gif đū ús hýran wilt, oþþe đec ungearo (-geára ?) eft gesécaþ, Exon. Th. 119, 9; Gú. 252. Hē on ungearone đone Osríc mid his fyrde becom and hine mid ealle his weorude ádylgode *Osricum erumpens subito cum suis omnibus imparatum cum suo exercitu delevit*, Bd. 3, 1; S. 523, 26. Ælc here hæfd đý læssan cræft đonne hē cymđ, gif hine mon ǽr hit cume; for đæm hē gesihđ đa gearwe đe hē wēnde đæt hē sceolde ungearwe findan. Him wǽre đonne iéđre đæt hē hira ǽr gearra wēnde, đonne hē hira ungearra wēnde, and hí gearuwe mētte *dum contra ictum quisque paratior redditur, hostis, qui se inopinatum credidit, eo ipso, quo praevisus est, enervatur*, Past. 56; Swt. 433, 27–31. Þeóf forfēhđ slǽpe gebundene eorlas ungearwe, Exon. Th. 54, 27; Cri. 875. Andra besierede đæt folc đe hié ymbseten hæfde on ánre niht ungearwe *exercitum incautum Andro oppresserat*, Ors. 4, 5; Swt. 170, 2. Hié fóron út nihtes and cómon on ungearwe men, Chr. 921; Erl. 106, 13. Hē nihtes on ungearwe hí on bestæl *ex improviso adgredi et insperatas circumvenire maluerit*, Ors. 1, 10; Swt. 46, 34. Hié on Ahtēne ungearwe becóman and hié gefiémdon *Agesilaus improviso bello supervenit*, 3, 1; Swt. 98, 15. Đæt hē on đa burgware on ungearwe becóme *quibus repente incautam urbem opprimeret*, 4, 5; Swt. 166, 32: 4, 10; Swt. 196, 25. Hē on ungearwe on Ahtēne mid firde gefór, 3, 7; Swt. 118, 20. [O. H. Ger. un-garo *imparatus.*]

un-geárwyrd; *adj. Not honoured, not respected :*—Ungeárwyrd *intemerata*, Wrt. Voc. ii. 45, 25.

un-geáþe = un-eáþe, Bt. 35, 3; Fox 158, 28.

un-geáxod; *adj. Unasked :*—Hē ungeáxod clypode : 'Ic eom cristen,' Homl. Th. i. 428, 6.

un-gebeard[e ?], -bierde, -bird (-byrd); *adj. Beardless, young :*—Ungebyrd *investis*, Wrt. Voc. ii. 47, 28: 92, 54. Ungebarde hysse *effebo hircitallo* (cf. beardleás hysse, Hpt. 487, 78), ii. 82, 32. Đa ungebyrdan heápas *investes catervas*, 44, 41. [O. H. Ger. un-giparta *sine barba, impubis.*]

un-gebeorhlíce; *adv. Not safely, rashly* (?), *intemperately* (?) :—

Lufiaþ, gê weras, eówere wíf on æwe; ne beó gê bitere him ungebeorhlíce (*nolite amari esse ad illas*, Col. 3, 19), Homl. Th. ii. 322, 26.

un-gebét[t]; *adj.* 　　I. of things, *unamended, uncorrected;* in reference to sin, *not amended through the penance prescribed by the church :*—Scylde ðe an hiera ealra gewitnesse gedón wæs and ðágiett ungebétt (-bét, Hatt. MS.) *culpam quae apud eos et perpetrata fuerat, et incorrecta remanebat*, Past. 32 ; Swt. 210, 7. Gyf hê ænigne gylt ungebét hæfð, L. Ælfc. C. 32 ; Th. ii. 354, 29. Ealle scylda, ðe wið God beóð ungebétta, beóð unforgifne on dómes dæge, Past. 33 ; Swt. 220, 17. 　　II. of persons, *unatoned* because 'bót' has not been made :—Ðá wæs hê disse spæce, ægðer ge on lífe and æfter, ungeládod ge unwreht *he was*, both when alive and afterwards, *uncleared from this charge and unatoned* (i. e. neither was his innocence proved nor was the case settled by the payment of 'bót'), Chart. Th. 540, 4. 　[*Icel.* ú-bættr *unatoned*]

un-gebierde; v. un-gebearde.

un-gebíged; *adj. Unbent :*—Unibígedre *inflexi*, Hpt. Gl. 476, 23.

un-gebígendlíc; *adj. Inflexible;* in grammar, *indeclinable :*—Ungebígendlíc *inflexibile*, Hpt. Gl. 425, 34. Ðás naman synd *indeclinabilia*, ðæt synd ungebígendlíce . . . *nugas* is ungebígendlíc on declínunge, Ælfc. Gr. 9, 25 ; Zup. 51, 2-6.

un-gebleoh; *adj. Of different colours, unlike :*—Ungebleoh *discolor*, Ælfc. Gr. 9, 21 ; Zup. 47, 16 : *discolor*, i. *dissimilis*, Wrt. Voc. ii. 140, 79.

un-gebletsod; *adj. Unblessed :*—Sume ic funde bútan Godes tácne, gýmeleáse, ungebletsade, Exon. Th. 271, 34 ; Jul. 492.

un-geblýged; *adj. Undismayed :*—Him tæringa ádl in gewód ; hê on elne swá þeáh ungeblýged bád beorhtra geháta blíþe in burgum, Exon. Th. 158, 23 ; Gú. 913. [Cf. þa iwarð þat folc swíðe abluied (*stupebant omnes*, Acts 2, 7), O. E. Homl. i. 89, 31. *O. H. Ger.* plúcheit *diffidentia : M. H. Ger.* er-bliugen *to frighten : Icel.* bljúgr *shy : Dan.* bly : *Swed.* blyg. See also Diefenbach's Gothic Dict. i. 307, § c.] Cf. á-blícgan.

un-geboden; *adj. Unsummoned, unbidden :*—Þreó mótlæþu ungeboden on .xii. mónþum *the tenant attended three courts without summons in the year*, Chart. Th. 433, 22. Perhaps the word is to be found in the phrase de placito ungebendro (ungebendeo, MS. R. = ungebodene?), L. Eth. iv. 4 ; Th. i. 301, 21. Cf. Tribus principalibus mallis, qui vulgo *ungeboden ding* vocantur . . . tria plebiscita, quae dicuntur *ungeboten* . . . tria judicia per annum, quae dicuntur judicia *non indicta*, Grmm. R. A. 823.

un-geboht; *adj. Unbought :*—Ic hér on sóðre gewitnesse stande, unábeden and ungeboht, L. O. 8 ; Th. i. 180, 28.

un-geboren; *adj. Unborn :*—Ge for geborene ge for ungeborene, L. A. G. proem. ; Th. i. 152, 6.

un-gebrocen; *adj. Unbroken :*—Ungebrocenre *extricabili*, Wrt. Voc. ii. 33, 7. v. un-forfeorod.

un-gebrocod; *adj. Unafflicted, uninjured :*—Ðonne wê manna líchaman derigaþ, búton wê ðære sáwle derian magon, ða líchaman þurhwuniaþ on heora áwyrdnysse . . . Ðonne hí gelýfaþ þæt wê godas sind . . . wê forlætaþ ðone líchaman ungebrocodne, and cépaþ ðære sáwle, Homl. Th. i. 464, 6.

un-gebrosnendlíc; *adj. Incorruptible :*—Ða ungebrosnendlícan limo *incorrupta membra*, Bd. 4, 30 ; S. 609, 29.

un-gebrosnod; *adj. Uncorrupted, undecayed :*—Ungebrosnad *incorruptus*, Bd. 3, 6 ; S. 528, 29. Ðá wæs heó swá ungebrosnad gemêted swá heó ðý ilcan dæge wære forðféred, Shrn. 94, 36. His handa siondan ungebrosnode in ðære cynelícan ceastre, 114, 1 : Chr. 641 ; Erl. 27, 11 : Homl. Th. ii. 568, 24.

un-gebrosnung, e ; *f. Incorruption :*—Ungebrosnunge onféhð *incorruptionem recipit*, Scint. 71, 2.

un-gebunden; *adj. Unbound :*—Sume syndon *absolutiuae*, ðæt synd ungebundene, Ælfc. Gr. 5 ; Zup. 14, 13.

un-gebyde, un-gebyrd. v. un-gebyrde, un-gebearde.

un-gebyrde; *adj. Not natural, uncongenial :*—Ælc gesceaft flíhþ ðætte him wiþerweard biþ and ungebyrde (? -byde, Fox) and ungelíc, Bt. 34, 11 ; Fox 150, 23.

un-gebyredlíc; *adj. Unsuitable, incongruous :*—Ungebyredlíc *incongruum*, Rtl. 179, 34.

un-gecirred; *adj. Unconverted :*—Ðý læs ðe ænig ungecyrred woroldman mid ungewitte regules geboda ábræce, Lchdm. iii. 442, 1.

un-geclænsod; *adj. Uncleansed, unpurified :*—Swá hwá swá ungeclænsod byð, hê gefrêt ðæs fýres æðm, Homl. Th. i. 616, 23 : L. E. I. 44 ; Th. ii. 440, 21. v. un-clænsod.

un-gecnáwen; *adj. Unknown :*—Hê fela þinga forðteáh ðe ðam folce ungecnáwe[n] wæs and ungewunelíc, Ap. Th. 17, 13.

un-gecnirdness, e ; *f. Negligence, want of diligence :*—Menige sind ðe ðurh ungecnyrdnysse ðisum ðeówan (*the slothful servant in the parable*) geefenlæcaþ, Homl. Th. ii. 552, 35.

un-gecóplíc; *adj. Unfit, inconvenient, troublesome;* importunus :—Saca mid ungecóplícum *quarrels with rude fellows*, Lchdm. iii. 200, 18. v. un-gedáfenlíc.

un-gecóplíce; *adv. Unsuitably, unseasonably :*—Ongecóplíce *importune*, Scint. 80, 14.

un-gecoren; *adj.* 　　I. *unchosen, unselected;* used in reference to those who swore along with another, when they were not selected by the party making oath from a number of persons named to him, as was the case in the *cyre-áð*, q. v. :—Ðæt hê ðone áð funde, gif hê mæhte, ungecorenne, ðe se onspeca on gehealden wære. Gif hê ðone ne mehte, ðonne namede him man six men and begéte ðara syxa ænne æt ánum hrýðere, i. e. *if he could bring those to swear with him, that the claimant was satisfied with, there was no need to nominate persons from whom he was to choose; if he could not, then six men were to be nominated and from them he was to get one for every ox (or its equivalent) that was in dispute*, L. Ed. 1 ; Th. i. 158, 20. Ðonne mót hê syxa sum ungecorenra, ðe getrýwe sýn, ðone áð syllan, L. O. D. 1 ; Th. i. 352, 12. 　　II. *reprobate, evil :*—For ðissum lænan lífe ic sylle ðæt unlæne, for ðyssum ungecorenum ðæt gecorene, Wulfst. 264, 19. Ðá burhware (*of Jerusalem*) him (*Christ*) wæron for heora ungeleáfan and mándædum swíþe forhogde and ungecorene, Blickl. Homl. 77, 28. [*Goth.* un-gakusans ἀδόκιμος, *reprobus.* Cf. *Icel.* ú-kjörligr *wretched*.] v. next word.

un-gecost; *adj. Bad, evil, vicious :*—On ungecostum ðeáwum *moribus improbis*, Bd. 5, 23 ; S. 647, 1. v. un-cyst, *and preceding word.*

un-gecwême; *adj. Unpleasant, disagreeable;* ingratus, Scint. 38, 15.

un-gecýd[d], -gecýðed ; *adj. Undeclared :*—Gif céap ofer .v. niht ungecýd on gemǽne læse wunaþ, L. Edg. S. 9 ; Th. i. 276, 1.

un-gecynde; *adj. Unnatural :*—Nim swá wuda swá wyrt of ðære stówe ðe his eard and æþelo biþ on tó weaxanne, and sette on ungecynde stówe him, ðonne ne gegrêwþ hit ðær náuht, Bt. 34, 10 ; Fox 148, 27 note. Hié hæfdun hiera cyning áworpenne Ósbryht and ungecyndne (*not of the royal race ;* cf. 'non de regali prosapia progenitum,' Asser) cyning underféngon, Chr. 867 ; Erl. 72, 10. [We scullen of londe driuen unicunde (*foreigners*), Laym. 18429.] v. next word.

un-gecyndelíc; *adj.* 　　I. *unnatural, not in accordance with the nature of a thing :*—Ungecyndelíc is ælcre wuhte, ðæt hit wilnige deáþes, Bt. 34, 11 ; Fox 152, 7. 　　II. *not natural, supernatural :*—Ungecyndelíc fýr cymð færunga on eówre burga, Wulfst. 297, 13. 　　III. *unnatural, contrary to nature, monstrous :*—Hit is ungecyndelícu ofermódgung ðæt se monn wilnige ðæt hine his gelíca ondræde *contra naturam superbire est, ab aequali velle timeri*, Past. 17 ; Swt. 109, 11. Swíþe ungecyndelíc yfel, ðæt ða bearn sieredon ymbe ðone fæder, Bt. 31, 1 ; Fox 112, 12. Gecyndelícra synna oþþe ungecyndelícra, L. de Cf. 6 ; Th. ii. 262, 24 : Anglia xi. 98, 19. On ungecyndelícum þingum *in rebus naturae contrariis*, L. M. I. P. 40 ; Th. ii. 276, 7 : Anglia xi. 3, 78. Wê fram dæge tó óþrum geáxiaþ ungecyndelíco wítu and ungecyndelíce (-cynelíce, MS.) deáþas tó mannum cumene, Blickl. Homl. 107, 26. Hwæt wære ungecyndlíce, gif God næfde on eallum his ríce náne frige sceaft under his anwealde, Bt. 41, 2 ; Fox 244, 28.

un-gecyndelíce; *adv. Unnaturally :*—And sæ hê dêð beón ungemetlíce and ungecyndelíce swíþe ástyrode, Wulfst. 196, 3.

un-gedæftlíce, -gedæftelíce; *adv. Unseasonably, unsuitably :*—'Ðæt ðú lære ægðer ge gedæftlíce ge ungedæftlíce (-dæfte-, Cott. MSS.).' Ðeáh hê cuæde un[ge]dæftelíce, hê cuæð ðeáh ær gedæftelíce '*insta opportune, importune.*' *Dicturus importune praemisit opportune*, Past. 15 ; Swt. 97, 16.

un-gedæftness, e ; *f. Importunity, unseasonableness :*—Ðonne sió ungedæftnes hit ne cann eft gedæftan *si habere importunitas opportunitatem nescit*, Past. 15 ; Swt. 97, 19.

un-gedafenlíc; *adj. Unbecoming, unseemly, unmannerly :*—Ungedafenlíc *indecens*, Ælfc. Gr. 14 ; Zup. 87, 12. Ðæt hit ungedafenlíc sig *quod indecorum est*, L. Ecg. P. i. 7 ; Th. ii. 174, 22. Ðæt man intó circan ænig þingc ne lógige, ðæs ðe ðártó ungedafenlíc sí, L. Edg. C. 27 ; Th. ii. 250, 11. Mid ungedafenlícre and unwærlícre ofersprǽce *loquacitatis incauta importunitate*, Past. 15 ; Swt. 95, 19. Saca mid ungedafenlícum *quarrels with unmannerly fellows* (v. un-gecóplíc), Lchdm. iii. 204, 20. Wê oft ymb ungedafenlíce wísan smeágeaþ, Past. 18 ; Swt. 139, 22. Gif preóst on circan ungedafenlíce þingc gelógige, gebéte ðæt, L. N. P. L. 26 ; Th. ii. 294, 12.

un-gedafenlíce; *adv. Unbecomingly, unseasonably, inordinately, in an unseemly manner, indecently :*—Oðer ðara irsunga bið tó ungemetlíce and tó ungedafenlíce átyht on ðæt ðe hió mid ryhte irsian sceall *illa ira in hoc, quod debet, inordinate extenditur*, Past. 40 ; Swt. 293, 13. Ðæt hwelc man ungemetlíce and ungedafenlíce wilnige ðæt hê scile his blísan tóbrǽdan, Bt. 18, 2 ; Fox 64, 20. Boda Godes word ægðer ge gedafenlíce ge ungedafenlíce *preach God's word both in season and out of season*, Homl. Ass. 12, 306. Be ðam men ðe ungedafenlíce hǽmð *de homine qui turpiter fornicatur*, L. Ecg. P. ii. 6 tit. ; Th. ii. 180, 9.

un-gedafenlícness, e ; *f. Inconvenience :*—Ungedafenlícnyssum *inopportunitatibus*, Ps. Spl. C. second 9, 1. v. gedafenlícness.

un-gedafniendlíc; *adj. Unseemly, indecent :*—Unidafniendlíc fúlnes *indecens obscenitas*, Hpt. Gl. 492, 60.

un-gedéfe; *adj. Troublesome, disagreeable :*—Cild ácenned ungedéfe, ofermód, felasprecol, Lchdm. iii. 192, 22. Hér ys seó bót hú ðú meaht ðíne æceras bétan, gif hí nellaþ wel wexan oþþe ðær hwilc ungedéfe þing on gedón bið, i. 398, 2. Mannum ungedéfum *hominibus importunis*,

Scint. 38, 15. Ungeþeáwfæstan and ða ungedéfan þreán *indisciplinatos et inquietos arguere*, R. Ben. Interl. 15, 1.

un-gedéfelíce; *adv. Unfitly, in a way that ill suits the conditions of a case:*—Wæs ðam yldestan ungedéfelíce mæges dædum morþorbed stréd . . . Hædcyn his mæg ofscét, bróðor óðerne *for the eldest unfitly, by a kinsman's deeds, was the death-couch spread . . . Hæthcyn with his arrow slew his kinsman, brother slew brother*, Beo. Th. 4862; B. 2435.

un-gedered(-od); *adj. Unhurt, uninjured:*—Ic bidde ðé, u ca peruica, . . . ðæt ðú mé gegearwie, ðæt ic sý ungedered fram ættrum and fram yrsunge *te precor, uica peruica, ut ea mihi prestes, ut a uenenis et ab iracundia interus sim*, Lchdm. i. 314, 10. Hé æfter ðam drence ansund and ungederod ðurhwunode, Homl. Th. i. 574, 12. Ne sceal hé ungederod ðæs écan lífes brúcan, ii. 336, 20. Hé wunade betwux eallum deórcynne ungederod, i. 486, 35: Homl. Ass. 17, 169. Án man mihte faran ofer his ríce mid his bósum full goldes ungederad, Chr. 1086; Erl. 222, 5. Ðære ungederedan *inlibatae*, Wrt. Voc. ii. 44, 16. Seó leó heóld ðæt cild ungedered, Homl. Skt. ii. 30, 183.

un-gedrehtlíce; *adv. Unweariedly, indefatigably;* infatigabiliter, Wrt. Voc. ii. 48, 51.

un-gedyrstig; *adj. Faint-hearted, diffident:*—Ða unmódigan and ða ungedyrstigan wénaþ ðæt ðæt suíðe forsewenlíc sié ðætte hié dód and fordon weorðaþ oft ormóde *pusillanimes vehementer despecta putant esse, quae faciunt, et idcirco in desperatione franguntur*, Past. 32; Swt. 209, 10.

un-geeahtendlíc; *adj. Inestimable:*—Mid ða sylfan mycelnysse ðæs ungeæhtendlícan (ungeendedlícan and [un]geeahtendlícan, MS. B.) gerýnes *ipsa inaestimabilis mysterii magnitudine*, Bd. 1, 27; S. 496, 11.

un-geendigendlíc; *adj. Indefinite, infinitive:*—Gif ic cweðe: Nescio, quis hoc fecit, ðonne byð se quis infinitivum, ðæt is, ungeendigendlíc, Ælfc. Gr. 18; Zup. 113, 16: 116, 14. Ðæt fífte gemet is *infinitivus*, ðæt is ungeendigendlíc, forðan de ðær ne byð nán spræc geendod, 21; Zup. 126, 7.

un-geendod; *adj.* I. *endless, without end, not coming to an end:*—God is æfre unbegunnen and ungeendod, Ælfc. Gr. 32; Zup. 201, 10: Homl. Th. i. 8, 27: Homl. Skt. i. 1, 16. Ðær is ðæt éce blis and ðæt ungeendode ríce, Blickl. Homl. 25, 30, 24. Gif ðú getælest ða hwíle ðisses hwílendlícan wið ðæs ungeendodan lífes hwíla, Bt. 18, 3; Fox 66, 5. Swá écum lífe swá ungeendodon wíte *sive vitam aeternam, sive infinitum supplicium*, L. Ecg. P. iv. 65; Th. ii. 226, 14. Geond ungeendode worulde, Homl. Th. i. 76, 7. Ðæra gesælða ðe him ungeendode becuman sculon *felicitas, quae sine transitu attingitur*, Past. 52; Swt. 407, 30. II. *infinite, very great* in number, extent, etc.:—Se hine slóh on ðæt næsþyrl, ðæt ðær út fleów ungeendod blód, Shrn. 112, 31. Ungeendodre lengo *infinitae longitudinis*, Bd. 5, 12; S. 627, 36. Ungeendedum forbeácnum *infinitis prodigiis*, Hpt. Gl. 490, 67.

un-geendodlíc; *adj. Infinite:*—Nis nó tó metanne ðæt geendodlíce wiþ ðæt ungeendodlíce *infiniti atque finiti nulla poterit esse collatio*, Bt. 18, 3; Fox 66, 13. v. un-geeahtendlíc.

un-gefǽglíc, un-gefǽrum. v. un-fǽglíc, un-gefére.

un-gefandod; *ptcpl. Not tried, not experienced:*—Sint tó manigenne ða ðe ðonne giet ungefandod habbaþ flǽslícra scylda *admonendi sunt peccata carnis ignorantes*, Past. 52; Swt. 407, 19: 409, 16, 22.

un-gefaren; *adj. Untravelled, without a road:*—On ungefarenum and on wæterigum *in invio et in aquoso*, Blickl. Gl. (Ps. 62, 3: 106, 40). v. un-gefére, -geféred, -geférne.

un-gefeálíce; *adv. Joylessly, miserably:*—Beornræd féng tó ríce and lytle hwíle heóld and ungefeálíce, Chr. 755; Erl. 52, 3.

un-gefége; *adj. Unsuitable, absurd;* ineptus, Wrt. Voc. i. 61, 38. [Ferde he hauede inoh, muchel and unifeie (onimete, 2nd MS.), Laym. 5573. O. H. Ger. un-gifógi *importunus, enormis*.] v. un-gefóg.

un-geféled; *adj. Without feeling, without sensation, insensible:*—Ða lǽcedómas ðe wé lǽrdon ðæt mon dyde tó ðære ungefélan heardnesse ongunnenre on ðære lifre, Lchdm. ii. 212, 15.

un-geféled; *adj. Not possessed of feeling, insensible:*—Ðonne seó ungefélde áheardung ðære lifre tó langsum wyrð, Lchdm. ii. 210, 3. Gif ðæt líc tó ðon swíþe ádéadige, ðæt ðær nán gefélnes on ne sié, ðonne scealt ðú eal ðæt deáde and ðæt ungefélde of ásníþan, 82, 27.

un-gefére; *adj.* I. *lit. impassable:*—Ungefére wel wegleás pæð *invium*, Wrt. Voc. i. 53, 61. On ungeférum *in invio*, Ps. Spl. C. 106, 40. Mid wéstenum and ungefǽrum londum, Bt. 18, 2; Fox 62, 36. II. *fig. impervious, impenetrable:*—Ða mód ðe Dryhtne ungeféru sint *mentes Deo impenetratae*, Past. 35; Swt. 245, 23.

un-gefére; *adv. Impassably:*—Ungefére [im]pervie, Wrt. Voc. ii. 61: 69, 17.

un-geféred; *adj. Unapproached, inaccessible:*—Feldas and wudu and dúna, ða wǽron monnum ungeférde for wildeórum and wyrmum, Nar. 20, 11. In án nearo fæsten micel ungeféredra móra *in angustias inaccessorum montium*, Bd. 4, 26; S. 602, 20. v. un-gefaren.

un-geférendlíc (?); *adj. Inaccessible, difficult of access:*—Fóran wé þurh ða ungeférenlícan (-férend-?) eorþan, Nar. 17, 7.

un-geférlíc; *adj. That cannot be united or that separates;* applied to war in which those, who naturally should be comrades, are opposed, *civil,*

social:—Wearþ ofer ealle Italia ungeférlíc unsibb *sociale bellum tota commovit Italia*, Ors. 5, 10; Swt. 232, 31: 5, 10 tit.; Swt. 5, 31. Ungeférlíces *dissociabile*, Wrt. Voc. ii. 141, 39. v. next word.

un-geférlíce; *adv. In civil war:*—Hé .v. gefeoht ungeférlíce (wel cynelíce gefeaht and, MS. C.) þurhteáh *bella civilia quinque gessit*, Ors. 5, 13; Swt. 244, 25. v. preceding word.

un-geférne; *adj. Impassable:*—In ungefoernum *in invio*, Ps. Surt. 106, 40. In ðæm ungefoernan, 62, 3.

un-gefeþered; *adj. Unfeathered:*—Ungefeþeredne *inplumem*, Wrt. Voc. ii. 48, 21.

un-gefóg(-fóh); *adj.* I. *immense:*—Hí námon sceattas genóge sylfrene and gyldene ungefóge, Homl. Skt. i. 23, 199. II. *in a bad sense, intemperate, immoderate excessive:*—Seó þwyre sáwul on hwílwendlícum bricum biþ ungefóh, Homl. Th. i. 408, 15. Hefigtýme leahter is ungefóh fyrwitnys, ii. 374, 3. Hé wæs mid ungefóhre gýtsunge ontend, i. 414, 5. [He sloh þer uniuoȝe, moni and inoȝe, Laym. 21793. Noldest þu nefre ben inouh, buten þu hefdest unifouh, Fragm. Phlps. 7, 23. O. Frs. un-efóg.]

un-gefóge; *adv. Immensely, exorbitantly:*—Ðær beóð ða swiftan hors ungefóge dýre, Ors. 1, 1; Swt. 21, 6.

un-gefóglíc; *adj. Immense, enormous,* (1) *in a physical sense:*—Ungefóhlíc hreám *immensus clamor*, Greg. Dial. 1, 9. Ymbútan ðone weall is se mǽsta díc, on ðam is iernende se ungefóglecesta stréam *fossa extrinsecus late patens vice amnis circumfluit*, Ors. 2, 4; Swt. 74, 18. (2) *in a moral sense:*—Is swíðe micel þearf ðæt gé georne mǽnra ǽða stýran, and eówrum hýremonnum cýðon, hú ungefóhlícu scyld ðæt is, L. E. I. 26; Th. ii. 422, 20.

un-gefóglíce; *adv. Excessively, intemperately, immoderately:*—Hí ongann ungefóhlíce swǽtan, Homl. Th. i. 414, 12. Hí mid eorþlícum teolungum ungefóhlíce hí gebysgiaþ, 524, 14: Boutr. Scrd. 20, 11.

un-gefrǽge; *adj. Unheard of:*—Ungefrǽge *inauditum*, Wrt. Voc. ii. 46, 66: 80, 60. [Cf. Icel. ú-frægr *not famous*.]

un-gefrǽgelíc; *adj. Unheard of, unusual, extraordinary:*—Gyf hyra (*gallinarum*) hwylc man æthríneþ, ðonne forbærnaþ hí sóna eall his líc; ðæt syndon ungefrǽgelícu (unge frelicu, un ge fræ licu, MSS. v. Anglia i. 332) lyblác, Nar. 34, 3. Ða deór habbaþ eahta fét, and wælkyrian eágan, and twá heáfda . . .; ðæt syndon ungefrǽgelícu (-fregelicu, MS. T.) deór, 34, 8.

un-gefrǽglíce; *adv. In an unheard of manner, to an unheard of extent, unusually, extraordinarily:*—Catulus swá ungefrǽglíce forcwæd Nonium *Catullus Nonium strumam appellat*, Bt. 27, 1; Fox 94, 32. Swíþe ungefrǽglíce upáhafen on his móde, 37, 1; Fox 186, 8. Se hearpere wæs swíþe ungefrǽglíce gód, 35, 6; Fox 166, 29.

un-gefrætwod; *adj. Unadorned:*—Ungefrætwodu *incompta*, Wrt. Voc. ii. 44, 3.

un-gefrédelíce; *adv. With insensibility, callously:*—Hié beóð tó ðreágeanne and tó swinganne mid swá micle máran wíte suá hié ungefrédelícor beóð áheardode on hiera unðeáwum *tanto acriori invectione feriendi sunt, quanto majori insensibilitate duruerunt*, Past. 37; Swt. 265, 16.

un-gefremed; *adj. Not accomplished, not done:*—Ungeffremed *infectum*, Wrt. Voc. ii. 87, 36. Ungefremed, L. Ath. i. proem.; Th. i. 198, 13.

un-gefullod, -gefulwad; *adj. Unbaptized:*—Gif ungefullod cild fǽrlíce bið gebróht tó ðam mæssepreóste, ðæt hé hit mót fullian sóna, ðæt hit ne swelte hæðen, L. Ælfc. C. 26; Th. ii. 352, 15: Homl. Th. ii. 50, 20. Ðeáh ðe he ungefullod gyt farende sý, 500, 35. Hine swá fǽrlíce deáð fornam, ðæt hé ungefullad forðférde. Ða Sanctus Martinus ðæt geseah . . . him wæs ðæt swíþe myccle weorce ðæt hé swá ungefulwad forðféran sceolde, Blickl. Homl. 217, 18–23.

un-gefullod; *adj. Unfulfilled:*—Ðære béne ungafullodrc, Exon. Th. 441, 7; Rä. 60, 14.

un-gefylled; *adj. Unfilled, unsatisfied:*—Ic eom getogen tó fremdum þeáwum ðurh ða ungefyldan (-gefylledan, Cott. MS.) gítsunge woruldmonna *nos ad constantiam, nostris moribus alienam, inexpleta hominum cupiditas alligabit?* Bt. 7, 3; Fox 20, 26.

un-gefylledlíc; *adj. Insatiable:*—Ðam ungefylledlícan *insatiabili*, Ps. Lamb. 100, 5: Nar. 42, 12.

un-gefyllendlíc; *adj. Insatiable;* insatiabilis, Scint. 50, 8: 110, 16.

un-gefynde; *adj. Not to be found or provided as food* (?) (cf. (?) *the phrase* mete findan *to provide food*):—Se æcer ðe stent on clǽnum lande, and bið unsmǽðmbǽre oððe ungefynde corn bringð oððe deáf *terra, quae exculta sterilem segetem gignit*, Past. 52; Swt. 411, 19.

un-gefyrn; *adv. At no distant date, before long, soon:*—Ðú áfindst his mihte ungefyrn on ðé sylfum, Homl. Skt. i. 23, 153. Eallum folce ðæs swíðe ungefyrn (*very soon after that*) hé geswutelian wolde hwæs gehwá gelýfan sceolde, i. 23, 405. v. un-fyrn.

un-gegearwod, -gegered; *adj. Not dressed:*—Ungigearuad woede *non vestitum veste*, Rtl. 108, 1. Ungegeradne, Mt. Kmbl. Rush. 22, 11.

un-gegrét; *adj. Ungreeted:*—Hé wolde tó ðam mynstre faran and his gebróðra grétan, forþan hé ǽr fram heom ungegrét gewát, Guthl. 3; Gdwin. 22, 20.

un-gehádod; *adj. Not ordained, not in holy orders :*—Be unge-hádedan mædene. Gif hwylc mædenman mid gehádodum wunaþ, and heó tô ðam ylcan háde þence . . . ne biþ heó nâ wið God unscyldig, þeáh heó ungehádod wǽre *de puella non ordinata. Si puella aliqua cum ordinatis habitet, et se eidem ordini destinet . . . non erit insons coram Deo, etiamsi non sit ordinata,* L. Ecg. P. ii. 17 ; Th. ii. 180, 19–188, 9–12. Gewylces ungehádodes wîfes tâcen is . . . , Techm. ii. 129, 18. [Artu ihoded oþer þu cursest al unihoded, O. and N. 1178.] v. un-hádod.

un-gehǽledlíc; *adj. Incurable ;* insanabilis, Ps. Surt. ii. p. 195, 21.

un-gehǽlendlíc; *adj. Incurable ;* insanibilis, Ps. Surt. ii. p. 194, 17. v. un-hǽmed.

un-gehǽmed; *adj. Unmarried :*—Ungehǽmed *innupta,* Hpt. Gl. 434, 37. v. un-hǽmed.

un-gehǽplíc; *adj. Unsuitable, incongruous :*—Ungehǽplíc (-geþǽslíc, Wrt., *but see* Anglia viii. 452) incongruus, Wrt. Voc. i. 61, 39.

un-gehálgod; *adj. Unhallowed, unconsecrated :*—On ungehálgedum Cristes mǽle *in cruce non consecrata,* L. Ecg. C. 34 ; Th. ii. 158, 36. Mid wîne ungehálgudum, Anglia xiii. 422, 818. Ungehálgod fýr *ignem alienum,* Lev. 10, 1. v. un-hálgod.

un-gehâten; *adj. Not promised :*—Ðæt ungehâten is sceal beón ge-endod, Blickl. Homl. 189, 27.

un-geheáfdod; *adj. Not come to a head :*—Gif se slyte blind bið and mid ðam geswelle ungeheáfdud, ðonne lege ðû ða wyrte ðærtô, sóna hit sceal openian, Lchdm. i. 92, 26.

un-gehealdsum; *adj. Incontinent :*—Se óðer heáfodleahter is ge-cweden forliger oððe gálnyss, ðæt is ðæt se man ungehealdsum sý on hǽmede, and hnesce on môde tô flǽsclícum lustum, Homl. Th. ii. 220, 4. Ðæt gê (*maidens*) wislíce lybbon and wel geþeáwode beón . . . nâ tô unge-healtsume (-heald-, *in one MS.*), Homl. Ass. 47, 575.

un-gehealdsumlíce; *adv. Incontinently :*—Ðæt eald wíf sceole ceorles brúcan ungehealtsumlíce, Homl. Ass. 20, 159.

un-gehealdsumness, e ; *f. Incontinence :*—Hí (*a widow or widower marrying again*) sculon dǽdbôte dôn for heora ungehealdsumnesse, L. Ælfc. P. 43 ; Th. ii. 382, 34. Ðes þeódscype þurh ungehealdsumnesse âwyrd is, Cod. Dip. Kmbl. iii. 349, 7.

un-gehende; *adv. Not near, at a distance, far off :*—Se ðe tô ðam ungehænde sý, ðæt hé dæghwamlíce his circan gesêcan ne mæge, Homl. Ass. 144, 8.

un-gehendness, e ; *f. Remoteness, distance :*—Sume naman syndon *localia,* ðæt synd stôwlíce, ða geswuteliaþ gehendnysse oððe ungehend-nysse, Ælfc. Gr. 5 ; Zup. 14, 19.

un-geheort; *adj. Disheartened, without courage :*—Ðá ðá hí gesâwon swâ mænigfealde ôgan on mistlícum wîtum, ðá wurdon hí sóna unge-heorte (*they lost heart*), and deófle offredon, Homl. Skt. i. 23, 62. v. un-gehirt.

un-gehîred; *adj. Unheard of :*—Ungehêredre leoma tôslítnysse wund-ade *inaudita membrorum discerptione lacerati,* Bd. 1, 7 ; S. 479, 13.

un-gehírness, e ; *f. Hardness of hearing, deafness :*—Wiþ eágwærce and wiþ ungehýrnesse, Lchdm. ii. 316, 1. Wið eágena dimnessa, wið eárena swinsunge and ungehýrnesse, iii. 70, 23.

un-gehîrsum; *adj. Inattentive to what is said, unsubmissive, dis-obedient :*—Hû lange wylt ðû beón ungehîrsum *usque quo non vis subjici mihi ?* Ex. 10, 3. Oððe hé bið ânum gehýrsum, and ôðrum ungehýrsum *aut unum sustinebit, et alterum contemnet,* Mt. Kmbl. 6, 24. Ðis is uncer ungehîrsuma sunu, hé forhogaþ ðæt hé hîre uncre lâre *filius noster iste protervus et contumax est, monita nostra audire contemnit,* Deut. 21, 20. Ne forlǽte hé ða ungehîérsuman (-hîr-, Hatt. MS.), Past. 12 ; Swt. 74, 16. v. un-hîrsum.

un-gehîrsumness, e ; *f. Want of submission, disobedience :*—Môd-ignys âcenð forsewennysse and ungehýrsumnysse, Homl. Th. ii. 222, 7. [Pane stede þe se deofel of hafel þurh unihersamnesse, O. E. Homl. i. 221, 30. Cf. *O. H. Ger.* un-gihôrsamî *inobedientia.*] v. un-hîrness.

un-gehîrsumod; *adj. Not subject, disobedient :*—Ungehýrsumude in-oboedienti, R. Ben. Interl. 12, 8. v. ge-hýrsumian, II, un-hîrsumness.

un-gehirt; *adj. Disheartened, cowardly :*—Ða ungehyrtan of heora wege âflýman, Wulfst. 192, 24. v. un-geheort.

un-gehíwod; *adj.* I. *not formed, without form :*—On ðam ungehíwodum antimbre ðe hé ða gesceafta of gesceóp *in materia informi creavit omnia,* Btwk. Scrd. 18, 15. Tô gescippenne ðæt ungehíwode antimber *ad formandam informem materiam,* 19, 3. II. *not feigned, unfeigned :*—Unihíwidre (*gloriosa*) *non fictae* (*puritatis palma,* Ald. 24), Hpt. Gl. 447, 46.

un-gehleóþor; *adj. Dissonant, discordant :*—Ungeswêga *vel* [un]-gehleóþre *vel* ungerâde *dissona,* i. *discordantia, incongrua,* Wrt. Voc. ii. 141, 37. v. ge-hleóþ (*read* -hleóþor).

un-gehrepod; *adj. Untouched, intact :*—Ne þorfte Adam deáðes on-byrian, gif ðæt treów môste standan ungehrepod, Homl. Th. i. 18, 25. God wolde ðæt hî ungehrepode on ðam scræfe slêpon, Homl. Skt. i. 23, 317.

un-gehrinen; *adj. Untouched, intact :*—Seó studu ungehrinen (*intacta*) fram ðam fýre âwunede, Bd. 3, 10 ; S. 534, 36.

un-gehwǽde; *adj. Not slight, considerable, much :*—Gif mete sý âwyrd and ungehwǽde mylcen, Lchdm. ii. 142, 14.

un-gehwǽrness, -gehýrness, -gehýrsum, -gehyrt. v. un-geþwǽrness, -gehírness, -gehîrsum, -gehirt.

ungel, es ; *m.* (?) *Fat :*—Ungel *arvina,* Wrt. Voc. i. 71, 11. Mid ungle ł mid fætnysse lamba *cum adipe agnorum,* Cant. M. ad fil. 14. Beó mîn sâwul gefylled swâ swâ mid rysle and mid ungele *sicut adipe et pinguedine repleatur anima mea* (Ps. 63, 5), Homl. Th. i. 522, 35. [*Du.* ongel ; *m.* suet.]

un-gelácnod; *adj. Uncured :*—Næs nǽnig untrum ðæt hé ungelácnod fram him fêrde, Guthl. 15 ; Gdwin. 66, 16. v. un-lácnod.

un-gelâdod; *adj. Not acquitted, uncleared* of a charge :*—Ðâ wæs hé ðisse spǽce, ægðer ge on lîfe ge æfter, ungelâdod ge ungebêtt, Chart. Th. 540, 4.

un-geleccendlíc; *adj. Irreprehensible ;* inreprehensibilis, Scint. 119, 11.

un-gelǽred; *adj. Untaught, unlearned, ignorant, unskilled :*—Un-gelǽred *idiota,* Wrt. Voc. i. 55, 48. Swîðe eáðe mæg on smyltre sǽ ungelǽred scipstiéra genôh ryhte stiéran *quieto mari recte navem et im-peritus dirigit,* Past. 9 ; Swt. 59, 1. Ceahhetung swâ swâ ungelǽredes folces *cachinnum quasi vulgi indocti,* Bd. 5, 12 ; S. 628, 30. Dysine and ungelǽredne ic ðé underfêng, Bt. 7, 3 ; Fox 20, 9. Tô hwon ðû sceole for ôwiht ðysne man habban ungelǽredne fiscere (*St. Peter*), Blickl. Homl. 179, 14. Ungelǽrede wê syndon *idiotae sumus,* Coll. Monast. Th. 18, 8. Forhwon beóð ǽfre suǽ ðríste ða ungelǽredan ðæt hí underfôn ða heorde ðæs láriówdômes *ab imperitis pastorale magis-terium qua temeritate suscipitur ?* Past. 1 ; Swt. 25, 16. v. un-lǽred.

un-gelǽredlíce; *adv. Without instruction, ignorantly, in an un-disciplined manner :*—Swîþe unwíslíce and ungelǽredlíce (*indocte*) gê dydon, ðæt gê scooldan on feówernihte mônan blôd lǽtan, Bd. 5, 3 ; S. 616,·13. Ungelǽredlíce (*indisciplinate*) nâ geþwǽrlǽce mûþ ðín, Scint. 136, 2.

un-gelǽredness, e ; *f. Uninstructedness, ignorance, inexperience, rudeness :*—Monige sindon mê swîðe onlíce on ungelǽrednesse *sunt plerique mihi imperitia similes,* Past. proem. ; Swt. 25, 8. Hê hié ðreáde for hira ungelǽrednesse *pastorum imperitia increpatur,* 1 ; Swt. 27, 24. Mid ðearfednesse ge mid heora ungelǽrednesse *paupertate ac rusticitate sua,* Bd. 4, 27 ; S. 604, 28.

un-gelâdod; *adj. Uninvited :*—Drihten nolde gelâdod sîðian tô ðæs cyninges bearne . . . and hê wæs gearo ungelâdod tô sîðigenne mid ðam hundredes ealdre, Homl. Th. i. 128, 18.

un-geleáf; *adj. Unbelieving :*—Ne magon ðǽr eard niman ungeleáfe menn *qui non credunt inhabitare in eo,* Ps. Th. 67, 19. [Cf. þu art unlef mine worde *non credidisti uerbis meis,* O. E. Homl. ii. 125, 24.]

un-geleáfa, an ; *m. Unbelief :*—Hê wundrode for heora ungeleáfan (*incredulitatem*), Mk. Skt. 6, 6 : Blickl. Homl. 77, 27. Ungeleáuon, Chr. 616 ; Erl. 22, 21. For ungeleáfe heora, Mt. Kmbl. Rush. 13, 58 : 17, 20. [*O. Sax.* un-gilôðo : *O. H. Ger.* un-giloubo. Cf. *Goth.* un-galaubeins.]

un-geleáfful[1] ; *adj. Unbelieving, incredulous :*—Ne beó ðû un-geleáfful (-full, MS. C.) *noli esse incredulus,* Jn. Skt. 20, 27. Eálá ungeleáffulle cneórys, Mk. Skt. 9, 19. Eálá gê ungeleáffulle cneóres, Mt. Kmbl. 17, 17. Hê ungeleáfful wæs Cristes ǽristes, Homl. Th. i. 234, 20. Ne sý mê nân man tô ungeleáfful þe ðam þingum wrîtende ðe ic gehýrde, Homl. Skt. ii. 23 b, 16. Ða deófolgyldan ðe ðǽgyt ungeleáf-fulle wǽron, Homl. Th. i. 70, 24. Tô beswícenne ungeleáffulra manna heortan, Blickl. Homl. 189, 8. Hê æteówde ða wunda ðǽm ungeleáf-fullum mannum, 91, 2.

un-geleáffullíc; *adj.* I. *unbelieving, incredulous :*—Gif hwâ ðises ne gelýfð hê ys ungeleáfulíc, Jud. 15, last line. II. *unbelievable, incredible :*—Ic wât ðæt hit wile ðincan swýðe ungeleáffullíc ungelǽredum mannum, Lchdm. iii. 270, 7. v. un-geleáflíc.

un-geleáffullíce; *adv. Incredibly ;* incredibiliter, Scint. 54, 9 : Basil admn. 7 ; Norm. 48, 20.

un-geleáffulness; e ; *f. Unbelief, incredulity :*—Gefylst mînre unge-leáffulnysse (*incredulitatem*), Mk. Skt. 9, 24. Ne dô ðu æfter heora ungeleáffulnesse, Blickl. Homl. 237, 9. Ungeleáfulnesse, 241, 34. For hyra ungeleáffulnysse, Mt. Kmbl. 13, 58 : 17, 20. Hê tǽlde hyra unge-leáffulnesse, for ðam ðe hí ne gelýfdon ðâm ðe hine gesâwon of deáþe ârîsan, Mk. Skt. 16, 14.

un-geleáflíc; *adj. Incredible :*—Ðæt ân þing wǽre ungeleáflíc on ðǽre race geset, Homl. Th. ii. 520, 12. Ðæt wile þincan ungeleáflíc eallum ðǽm ðe ða stôwe on uferum tîdum geseóð, Lchdm. iii. 438, 14 : Chr. 1036 ; Erl. 165, 9.

un-geleáfsum; *adj. Unbelieving, infidel, not Christian :*—Se ðe ðam suna is ungeleáfsum ne gesyhþ hê lîf *qui incredulus est filio non uidebit vitam,* Jn. Skt. 3, 36. Ða ungeleáfsuman ðeóde gesêcan *incredulam gentem adire,* Bd. 1, 23 ; S. 485, 33. Gif hî ungeleáfsume (*infideles*) wǽron, hê hí laþede ðæt hí onfêngon ðam gerýne Cristes geleáfan, 3, 5 ; S. 526, 30. Manige hǽþne men ungeleáfsume, Blickl. Homl. 129, 24. Ðâm ungeleáfsumum (*infidelibus*) nôht biþ clǽne, Bd. 1, 27 ; S.

494, 40. Hē ungeleáfsume (*incredulos*) tō Cristes geleáfan getrymede, 3, 19; S. 547, 10.

un-geleáfsumness, e; *f. Unbelief, infidelity, heathenism* :—Monige on Angeldeóde, mid ðý hī ðāgyta on ungeleáfsumnysse (*infidelitate*) wǽron, Bd. 1, 27; S. 491, 22.

un-gelíc; *adj. Unlike, different, dissimilar, diverse* :—Ungelíc *dispar*, Ælfc. Gr. 9, 17; Zup. 43, 2: *dissimile*, Kent. Gl. 512. Ðē is ungelíc wlite siððan ðú lǽstest mīne lāre *you have a different beauty, since you followed my teaching*, Cd. Th. 38, 26; Gen. 612: 222, 29; Dan. 112. Ðonne is ungelíc be ðon ēcan lífe *now with the life eternal it is different*, Blickl. Homl. 97, 28. Ungelíc is ūs *our lots are different*, Exon. Th. 380, 5; Rä. 1, 3. Hē tiolaþ ungelíc tō biónne ðam ōþrum, Bt. 39, 12; Fox 232, 7: Cd. Th. 23, 9; Gen. 356. Hit is ungelíc ūrum gecynde, Met. 20, 33. Unlícum hāde *dispari sexu*, Hpt. Gl. 461, 5. Ungelíce, Wrt. Voc. ii. 28, 41. Hē reorde gesette eorðbūendum ungelíce, Cd. Th. 101, 21; Gen. 1685. Se ðe bīspell secgan wolde, ne sceolde fón on tō ungelíc bīspell ðære sprǽce ðe hē ðonne sprecan wolde *cognatos, de quibus loquimur, rebus oportere esse sermones*, Bt. 35, 5; Fox 166, 20. Syndon dryhtguman ungelíce, Exon. Th. 314, 32; Mōd. 23. Wē syndon ungelíce ðonne ðe wē in heofonum hæfdon wlite *we are different from what we were when we had beauty in heaven*, Cd. Th. 274, 7; Sat. 150. Hē cwæþ ðæt hē gesāwe ungelíce bēc him berende beón ðurh ða gōdan gāstas oþþe ðurh ða gālan *quod codices diversos per bonos sive malos spiritus sibi videt offerri*, Bd. 5, 13; S. 633, 24. Hī wilnaþ þurh ungelíce earnunga cuman tō ānre eádignesse, Bt. 24, 1; Fox 80, 9. Ealle gesceafta ðú gesceópe him gelíce, and eác on sumum þingum ungelíce, 33, 4; Fox 128, 26; Met. 20, 55.

un-gelíca, an; *m. One not like another* :—Ic hæbbe ōðerne lufiend, ðīnne ungelícan (*a very different person from you*), Homl. Skt. i. 7, 28. [Ever ich am þin unilike, O. and N. 806.] v. ge-líca.

un-gelíce; *adv. Not in like manner, differently, diversely* :—Hwæðer ðú mæge gemunan ðætte ǽlces monnes ingeþanc wilnaþ tō ðære sōþan gesǽlþe tō cumenne, ðeáh hē ungelíce hiora earnige *meministine intentionem omnem voluntatis humanae, quae diversis studiis agitur, ad beatitudinem festinare?* Bt. 36, 3; Fox 176, 21. Ða strengas se hearpere swíðe ungelíce styreþ, and mid ðý gedēð ðæt hī nāwuht ungelíce ðǽm sōne ne singaþ ðe hē wilnaþ. Ealle hē grēt mid ānre honda, ðeáh hē hié ungelíce styrige *chordas tangendi artifex, ut non sibimetipsi dissimile canticum faciat, dissimiliter pulsat. Chordae uno quidem plectro, sed non uno impulsu feriuntur*, Past. 23; Swt. 175, 7–10. Is hám sceapen ungelíce englum and deófium, Exon. Th. 56, 11; Cri. 899: 56, 34; Cri. 910: 83, 29; Cri. 1363: 283, 31; Jul. 688. Bið ðám ōþrum ungelíce willa geworden, 77, 28; Cri. 1263. Biþ ðam ōþrum ungelíce, se ðe on eorþan eáðmōd leofaþ, 317, 18; Mōd. 67: Elen. Kmbl. 2611; El. 1307: Exon. Th. 380, 14; Rä. 1, 8.

un-gelíclíc; *adj. Unseemly, improper* :—Ne hē cnihtlíce gālnysse næs begangende, ne ungelíclíce ōlæcunge, ne.leáslícetunge, Guthl. 2; Gdwin. 12, 17.

un-gelíclíce; *adv. Improperly* :—Hū ðone cealdan magan ungelíclíce lyste, Lchdm. ii. 160, 7.

un-gelícness, e; *f. Difference, dissimilarity, diversity* :—Ic cwæð ðæt ǽghwelc mon wǽre ōðrum gelíc ācenned, ac sió ungelícnes hiera earnunga hié tíhð sume behindan sume . . . Hwæt ðonne ða ungelícnesse ðe of hiera undeáwum forðcymeþ, se godcunda dōm gedencð *omnes homines natura aequales genuit, sed variante meritorum ordine alios aliis culpa postponit. Ipsa autem diversitas, quae accessit ex vitio, divino judicio dispensatur*, Past. 17; Swt. 106, 18, 22. For ðære ungelícnesse ðara hiéremonna sculun beón ungelíc ða word ðæs lāreówes, 23; Swt. 175, 2.

un-gelífed; *adj. Not possessed of belief, unbelieving, infidel* :—Se ungelýfeda Ualens genam Godes circean of ðām Godes þeówum, Homl. Skt. i. 3, 318. Hī ofslógon swíðe ða hǽðenan, ðæt ðǽr nān ne belāf ðæra ungelýfedra cucu, Homl. Th. ii. 212, 33. Paulus spræc swíðe egeslíce be ungelýfedum mannum : hē cwæð : 'Ða ðe Godes ǽ ne cunnon, and būton Godes ǽ syngiaþ, hī eác būton Godes ǽ losiaþ,' 52, 22 : i. 460, 26 : Blickl. Homl. 63, 22. Ungelýfdum, 55, 32.

un-gelífed; *adj. Unallowed, illicit* :—Ðurh ungelýfedne willan *per inlicitam voluptatem*, Bd. 3, 19; S. 548, 29.

un-gelífedlíc; *adj. Incredible, marvellous* :—Swā ðæt nān wundor [nis] ne eác ungelýfedlíc þincg, Homl. Skt. ii. 23 b, 39. Nis nān tō ðam ungelýfedlíc spel, gif hē hyt segð, ðæt ic him ne gelíffe, Shrn. 196, 18. Ðæs wealles micelness is ungeliéfedlíc tō secgenne *murorum ejus vix credibilis relatu magnitudo*, Ors. 2, 4; Swt. 74, 14. Hit is ungeliéfedlíc tō secganne *incredibile dictu est*, 3, 9; Swt. 134, 15. Ungeliéfedlíc is ǽnigum menn ðæt tō gesecgenne *pene incredibile apud mortales erat*, 2, 4; Swt. 74, 7. Hē hæfde āne swíðe wlitige dohter ungeliéfedlícre fægernysse, Ap. Th. 1, 9. For ðam ungelífedlícan wlite ðæs mǽdenes, 3, 12. Hē un-geliéfedlícne micelne weg on ðæm dæge gefōr, Ors. 3, 9; Swt. 124, 27. v. un-gelífedlíc.

un-gelífend, es; *m. An unbeliever* :—Se ðe ungeléfend (-en, Lind.) is *qui incredulus est*, Jn. Skt. Rush. 3, 36 : 20, 27.

un-gelífendlíc; *adj. Incredible, extraordinary* :—Ungelýfendlíc tō

blāwennys his innoð geswencte, Homl. Th. i. 86, 12. Ic ðē mæg tǽcan ōþer ðing ðe dysegum monnum wile ðincan get ungeléfendlícre (-léfed-, Cott. MS.) *hoc quod dicam, non minus mirum videatur*, Bt. 38, 3; Fox 198, 30.

un-gelífness (?), e; *f. Unbelief* :—Fore ungeleáffulnisse ł ungeléfenise hiora *propter incredulitatem illorum*, Mt. Kmbl. Lind. 13, 58.

un-geligen. v. un-gelygen.

un-gelimp, es; *n. m. Misfortune, mishap* :—Ǽlc ungelimp cymð of deófle *omne infortunium venit a diabolo*, L. Ecg. P. iv. 66; Th. ii. 226, 26. Ðonne mē hwylc ungelimp becymð, Ps. Th. 39, 18. Him cymð ege and ungelimp, 13, 9. Wæs swíðe hefelíc geár and swíðe sorhfull geár . . . and swā mycel ungelimp on wæderunge swā man nāht ǽdelíce geþencean ne mæg, Chr. 1085; Erl. 219, 21. Hwā is swā heardheort ðæt ne mæg wēpan swylces ungelimpes? 1086; Erl. 219, 40. Ðæt mīne fýnd ne gefeón mīnes ungelimpes, Ps. Th. 34, 23, 24. Hī blissedon on mīnum ungelimpe, 34, 15. Wē sceolon ǽgðer ge on gelimpe ge on ungelimpe cweðan : 'Ic herige mīnne Drihten on ǽlcne tíman, Homl. Th. i. 252, 13 : Homl. Skt. i. 16, 251. Ic andette mīne scylda and seófige mīn ungelimp, Ps. Th. 21, 2 : Homl. Th. i. 584, 5. Ðā geáxodon þrý cyningas eal his ungelimp, ii. 454, 6. Wēpendlíc tíd wæs ðæs geáres, ðe swā manig ungelimp wæs forðbringende, Chr. 1086; Erl. 220, 23. Gif ūs ungelimpas on ǽhtum getīmiaþ, Homl. Th. ii. 328, 27. Fela ungelimpa gelimpð ðysse þeóde, Wulfst. 162, 12. Him becōmon fela yrmða on eallum ungelimpum, Ælfc. T. Grn. 20, 43. Ðæt se man geunrōtsige ongeán God for ungelimpum ðises andweardan lífes, Homl. Th. ii. 220, 17. On ungelimpum . . . on gesǽlðum, Homl. Skt. i. 16, 348. Hē geþafaþ ðæt ða gōdan habban unsǽlþa and ungelimp on mænegum þingum *bonis dura tribuat*, Bt. 39, 2; Fox 214, 4. Hē (Job) ða ungelimp geáxod hæfde, Homl. Th. ii. 450, 30.

un-gelimplíc; *adj. Unseasonable, unhappy, unfortunate* :—Ungelimplíc slápolnys *lethargia*, Wrt. Voc. i. 75, 63. (Cf. 46, 1 where two entries seem confused, v. next word.) Ungelimplíce gewyderu, Wulfst. 172, 18. Ða ungelimplícan *inepta*, Wrt. Voc. ii. 48, 53. Wē oft ongytaþ ðæt ārīseþ þeód wið þeóde and ungelimplíco gefeoht (*unhappy wars*) on wōlícum dǽdum, Blickl. Homl. 107, 28.

un-gelimplíce; *adv. Unseasonably, unhappily* :—Ungelimplíce slápol *lethargus vel letargicus*, Wrt. Voc. i. 75, 62. Hér is ðære lyfte fágetung ðurh mislíce stormas ðe ungelimplíce becumaþ, Homl. Th. ii. 538, 33.

un-gelygen; *adj. Not lying, true* :—Būtan ðæs gerēfan gewitnesse . . . oþþe ōðres ungelygenes (-lig-, *v. l.*) mannes, L. Ath. i. 10; Th. i. 204, 19. Ungeligenes, i. 12; Th. i. 206, 10 note. Ðæt hē hæfde ungelígene gewitnesse . . . ðæt hē gelǽdde ungelígne gewitnesse, L. Ed. 1; Th. i. 158, 16, 19. Swā fela manna swā man wite ðæt ungelygne sýn. . . . And sién heora áðas ungelygenra manna þe ðæs feós wyrðe, L. Ath. iv. 1; Th. i. 222, 10, 13. Hæbbe hē ðæs portgerēfan gewitnesse oþþe ōðera ungelígenra manna ðe man gelýfan mæge, L. Ed. 1; Th. i. 158, 12. Ðú tēhtest mē swā ungelygena gewittnesse swā ic nān ōðer dōn ne mæg būte ic nǽde scall hym gelífan, Shrn. 201, 17. v. un-lygen.

un-gemaca, an; *m. Not a match, not an equal* :—Ungemaca *impar*, Ælfc. Gr. 9, 17; Zup. 43, 1 : Germ. 389, 76. [þæ drake elcches wurmes unimake *unlike all other serpents*, Laym. 17961.]

un-gemæc[e]; *adj. Unlike, unequal, dissimilar* :—Ungemæccre wurman *dispari murice*, Wrt. Voc. ii. 141, 19. [O. L. Ger. un-gimac *infestus* : O. H. Ger. un-gimah[h] *dispar*.]

un-gémean. v. un-gímen.

un-gemǽte; *adj. Immeasurable, immense* :—Weard þurh ðæt ungemǽte orfcwealm, Chr. 1115; Erl. 245, 17. [Unimete festen and to michel forhefednesse, O. E. Homl. i. 101, 29 : 253, 11. Þe ferde wes swa muchel, þat heo wes unimete, Laym. 4964. Her is chele and hete and hunger unymete, Misc. 73, 50. In his unimete blisse, A. R. 40, 13. O. H. Ger. un-gimǽzi *inaequalis*.] v. un-mǽte, *and next words*.

un-gemǽte; *adv. Immeasurably, immensely* :—Mid ungemǽte miclum ege geslægene *timore immenso perculsos*, Bd. 5, 12; S. 627, 14 note.

un-gemǽtlíc; *adj. Immense, excessive* :—Mid ungemǽtlíce gewilnunge anwaldes *dominationis libidine*, Ors. 1, 2; Swt. 28, 27. [Cf. Swa unimeteliche þu swanc, O. E. Homl. i. 281, 18. Unimeteliche and unendliche more, A. R. 398, 25.]

un-gemeaht (?); *adj. Weak* :—Hū micle unmihtegran (ungemihtran, Bod. MS.) hī wǽron, gif hī his nān gecynde næfdon, Bt. 36, 5; Fox 180, 4. v. meaht; *adj.*

un-geméde; *adj. Disagreeable, discordant, adverse*, Exon. Th. 315, 2; Mōd. 25. [Cf. O. H. Ger. un-gimōti *dispendium, damnum, contumelia, injuria*.] v. un-mēde, *and next word*.

un-gemédness, e; *f. Adversity, calamity* :—From ungimoednisum ðætte wē sié ālēsed *ab adversitatibus liberari*, Rtl. 63, 29. [O. H. Ger. un-gimōtnissi *humilitas*.]

un-gemeltness, e; *f. Indigestion* :—Gebeorh ðæt hié ungemeltnesse ne þrowian, Lchdm. ii. 184, 11.

un-gemenged; *adj. Unmixed, unmingled* :—Hit is gecynd ðære godcundnesse ðæt hió mæg beón ungemenged wið ōþre gesceafta būton ōþerra gesceafta fultume *ea est divinae forma substantiae, ut neque in*

externa dilabatur, nec in se externum aliquid ipsa suscipiat, Bt. 35, 5; Fox 166, 5. v. un-menged.

un-gemet, es; *n.*   **I.** *immensity, an immense number :*—Ealles his heres wæs swelc ungemet ðæt mon eáðe cweþan mehte ðæt hit wundor wære hwær hié wæteres hæfden ðæt hié mehten him þurst of ádrincan *ut exercitui immensaeque classi vix ad potum flumina suffecisse memoratum sit*, Ors. 2, 5; Swt. 80. 7. Hé heora ungemet ofslóg; þe ðæm mon mehte witan, ðá hé and ða consulas hié átellan ne mehton *quot millia hominum interfecta ipse consul ostendit; qui numerum explicare non potuit*, 3, 10; Swt. 140, 29.   **II.** *immoderation, excess :*—Of ungemete ælces þinges, wiste and wæda, wíngedrinces, Met. 25, 38. Ðæt hé ne wilnige wynsumran wyrde ðonne hit gemetlíc sié, ne eft tó rēþre; for ðæm hé ne mæg náþres ungemet ádrióhan, Bt. 40, 3; Fox 238, 22.   **II a.** adverbial uses of cases, *ungemetes, (mid) ungemete, ungemetum, to excess, without measure, excessively, immensely, very :*—Ungemetes wel, Beo. Th. 3589; B. 1792. Wese ðín esne on ðē ungemete blíðe *servus tuus laetabitur*, Ps. Th. 108, 27 : 115, 2 : 141, 7 : 143, 17 : Beo. Th. 5436; B. 2721. Ic bidde ðínre ansýne ungemete georne, mid ealre gehygde heortan mínre *deprecatus sum faciem tuam in toto corde meo*, Ps. Th. 118, 58 : 108, 3 : 115, 1. Ungemete neáh, Beo. Th. 4832; B. 2420 : 5450 : B. 2728. Hió wile weahsan mid ungemete *sine mensura dilatata*, Past. 11; Swt. 71, 17. Mid ungemete (cf. ungemetlíce, Bt. 38, 1; Fox 194, 25), Met. 26, 62. Se mid ungemete gernde anwalda ofer óþre *ardens cupiditate dominandi*, Ors. 3, 11; Swt. 148, 29 : 4, 5; Swt. 166, 25 : 6, 3; Swt. 256, 28. Ða folc bútú on feferáðle mid ungemete swulton *gravissima pestilentia uterque exercitus angebatur*, 4, 10; Swt. 198, 35. Ðá ongon se cealc mid ungemete stincan, 6, 32; Swt. 288, 1 : Homl. Skt. i. 23, 230. Ungemetum rēðe, Runic pm. Kmbl. 341, 2; Rún. 3 : 341, 15; Rún. 11. Ungemettan fæste mid cludum ymbweaxen *mirae asperitatis*, Ors. 3, 9; Swt. 132, 10. Ungemetum georne, Ps. Th. 118, 107 : 142, 1. Ungemetum swíðe, 118, 67. Eágan ungemetum wēpaþ, Dóm. L. 12, 193. [From *mesure* into unimete, A. R. 74, 28. *O. H. Ger.* un-gimez.]

un-gemet. *Where the word seems to be used with an adjective or with an adverbial force, it is given, as in the case of* ungemet-hleahtor, *as part of a compound :* cf. *O. Sax.* un-met (*with adjectives*) : *O. H. Ger.* un-mez, Grff. ii. 898–9. Cf. ungesceád-micel.

ungemet-ceald; *adj. Excessively cold :*—Winter bringeþ weder ungemetcald, Met. 11, 59.

un-gemete, un-gemetegod, un-gemetegung, ungemetelíce. v. un-gemet, II a, un-gemetgod, un-gemetgung, un-gemetlíce.

un-gemetfæst; *adj.*   **I.** in a moral sense, *immoderate, immodest, intemperate :*—Ðá forseah se Catulus hine, for ðam hé hine wiste swíþe ungesceádwísne and swíþe ungemetfæstne, Bt. 27, 1; Fox 96, 5. Ðám monnum ðe beóþ neátenum gelíce, ðæt beóð unrihtwíse and ungemetfæste, 14, 1; Fox 42, 4.   **II.** in reference to physical things, (1) *immoderate, excessive :*—Hwílum cymð of ungemetfæstre hǽto, hwílum of ungemetfæstum cyle, Lchdm. ii. 56, 16. Hwílum of ungemætfæstre hǽto, hwílum of ungemetfæstum cyle, hwílum of ungemetlícre wǽtan, hwílum of ungemætlícre drígnesse, iii. 72, 29. (2) applied to the stomach, *irretentive :*—Ðæs hátan magan ungemetfæstan tácn sindon, ðonne hé bið mid ómum geswenced, ðam men bið þurst getenge, Lchdm. ii. 192, 25 : 160, 4. [*O. H. Ger.* un-gimezfast *immoderatus, immodestus.*]

ungemet-fæst; *adj. Extremely firm :*—Ðǽr hé mæge findan eáðmētta stán ungemetfæstne, grundweal gearone (cf. on ðam fæstan stáne eáðmētta, Bt. 12; Fox 36, 22), Met. 7, 33.

un-gemetfæstlíc; *adj. Irretentive :*—Ðis sint tácn ðæs hátan magan ómihtan ungemetfæstlícan, Lchdm. ii. 192, 24. v. un-gemetfæst, II. 2.

un-gemetfæstness, e; *f. Intemperance :*—Ðú wilt cweþan ðæt wrænnes and ungemetfæstnes hí ofsitte *sed transversos eos libido praecipitat, sic quoque intemperantia fragiles*, Bt. 36, 6; Fox 182, 2.

ungemet-geneahhie; *adv. Extremely :*—Mē fyrenfulra rápas ungemetgeneahhie oft beclyptan, Ps. Th. 118, 61.

ungemet-gímen[n], e; *f. Excessive care :*—Se rēþa rēn, sumes ymbhogan ungemetgēmen (cf. se rēn ungemetlíces ymbhogan, Bt. 12; Fox 36, 19), Met. 7, 28.

un-gemetgod; *adj. Immoderate, excessive, intemperate, indiscreet :*—Ðonne sió ðreáung bið ungemetgad *cum increpatio immoderate accenditur*, Past. 21; Swt. 165, 18. Sió ungemetgode suíge *indiscretum silentium*, 15; Swt. 89, 9. Ne durre wé ðás bóc ná miccle swíðor gelengan, ðí læs ðe heó ungemetegod sý, Homl. Th. ii. 520, 4. Ungemetegod lufu, 220, 6 : Homl. Skt. i. 16, 276. Tunge ungemetegud(-ad) *lingua immoderata*, Scint. 78, 10 : Kent. Gl. 507. Mid ðære ungemetgodan smeáunge, Past. 11; Swt. 67, 8. Ða ungemetgodan sprǽce *immoderatam locutionem*, 38; Swt. 281, 1. His ungemetegodan lufe, Homl. Skt. i. 3, 363. Beweóp se ylca apostol ungemetegodra manna líf, ðus cweðende : 'Heora wamb is heora god,' Homl. Th. i. 604, 27.

un-gemetgung, e; *f. Want of moderation, excess, intemperance :*—'Coða becumaþ.' Efne hér is foresǽd manna líchamana ungemetegung and geswenceduys, Homl. Th. ii. 538, 30. Ðý læs ðæt innegeðonc sié gebunden ðære heortan for ðære ungemetgunge ðæs ymbehogan ðæra

---

úterra ðinga (*per moderatam cordis intentionem non impeditur*), Past. 18; Swt. 141, 8.

ungemet-hleahtor, es; *m. Immoderate laughter :*—Hú micele mā wēnestú ðæt hé mid yrre ða út áweorpe of his temple, ðe mid unnyttum gesprǽcum and mid ungemethleahtrum ða stówa, ðe tó Godes þeówdóme gehálgode wǽron, fýlaþ and besmýtaþ, L. E. I. 10; Th. ii. 408, 32.

ungemet-lange; *adv. Excessively long*, Cd. Th. 20, 23; Gen. 313.

un-gemetlíc; *adj.*   **I.** *immoderate, inordinate, excessive, too great :*—Ungemetlíco forgifnis . . . ungemetlícu irsung *inordinata remissio . . . effrenata ira*, Past. 20; Swt. 149, 9–11. Ungemetlícu sprǽc *immoderata loquacitas*, 43; Swt. 309, 2. Se rēn ungemetlíces ymbhogan, Bt. 12; Fox 36, 19 : 18, 1; Fox 60, 24. Ðú woldest brúcan ungemetlícre wrænnesse *voluptariam vitam degas*, 32, 1; Fox 114, 20. For ungemetlícum cyle, 33, 4; Fox 130, 34. Mid ungemetlícre gítsunge *intemperans cupido*, Txts. 180, 1 : Ors. 1, 2; Swt. 30, 28. Of ungemetlícre drígnesse, Lchdm. ii. 56, 17. Ungemætlícre, iii. 72, 30. Of ðam ungemetlícan gegerelan, Bt. 37, 1; Fox 186, 16. Hé onsent ofer hig ungemetlíce hǽto ðære sunnan, Ps. Th. 10, 7. Ðá hié angeátan ðæt hé ungemetlíc gafol wið ðæm friþe habban wolde *cum intolerabiles conditiones pacis audissent*, Ors. 4, 6; Swt. 174, 24. Ungemetlíca meteśócna and ungemetlíce unlustas, Lchdm. ii. 174. 27. Ða ungemetlícan hleahtras, Blickl. Homl. 59, 18.   **II.** *immense, very great :*—Ungemetlíc *inmane*, Wrt. Voc. ii. 48, 25. Gif hit full ungemetlíc wind gestent, Bt. 12; Fox 36, 15. Ungemetlíc moncwealm *incredibilium morborum pestis*, Ors. 6, 23; Swt. 274, 11. Wæs ungemetlíc wæl geslægen, Chr. 867; Erl. 72, 15. Ðá com se ungemetlíca unfriðhere, 1009; Erl. 142, 16.   **III.** *not of the same measure, diverse :*—Ungemetlícra *diversarum*, Mt. Kmbl. p. 7, 5. v. un-metlíc.

un-gemetlíce; *adv.*   **I.** *immoderately, beyond measure, excessively, too (much) :*—Se ðe wile ungemetlíce gesceádwís beón, Past. 11; Swt. 67, 6. Tantalus ðe ungemetlíce gífre wæs, Bt. 35, 6; Fox 168, 33. Gé wilniaþ eówerne hlísan ungemetlíce tó gebrǽdanne, 18, 1; Fox 62, 18. Ne nán preóst ne drince ungemetelíce, L. Ælfc. C. 29; Th. ii. 352, 28. Hú ungemetlíce gē bemurciaþ, Ors. 1, 10; Swt. 48, 17. Ongan hió hine lufian, and hiora ǽgþer óþerne swíþe ungemetlíce, Bt. 38, 1; Fox 194, 25. Ðæt hé ðæt ryht tó swíðe and tó ungemetlíce (*nimie et inordinate*) ne bodige, Past. 15; Swt. 95, 17. Ðæt hié tó ungemetlíce ne forweaxen *ne immoderatius excrescant*, 18; Swt. 141, 6 : 21; Swt. 167, 14.   **II.** *immensely, exceedingly, very greatly :*—Ðá weard Cain ungemetlíce yrre *iratus est Cain vehementer*, Gen. 4, 5. Seó wæs ungemetlíce micel *serpens mirae magnitudinis*, Ors. 4, 6; Swt. 174. 4. Sió eá hæfde ungemetlíce ceald wæter *praefrigidus amnis*, 3, 9; Swt. 124, 29. Ic com swíþe ungemetlíce ofwundrod *vehementer admiror*, Bt. 13; Fox 40, 4. Isaac wundrode ungemetlíce swíðe *Isaac ultra quam credi potest admirans*, Gen. 27, 33. [*O. H. Ger.* un-gimezlíhho *hyperbolice.*] v. un-metlíce.

ungemet-lytel; *adj. Exceedingly little :*—Ðæt ðeós eorðe sié eall for ðæt óþer ungemetlytel, Met. 10, 9.

un-gemetness, e; *f. Extravagance :*—Ungemetnisse *dementiam*, Txts. 180, 3.

ungemet-scearp; *adj. Excessively sharp :*—Wǽron hyra tungan ungemetscearpe, Ps. Th. 56, 5.

ungemet-þurst, es; *m. Excessive thirst :*—Se háta maga ungemetþurst þrowaþ, Lchdm. ii. 160, 4.

ungemet-wæcce, an; *f. Excessive wakefulness :*—Monige ádla . . . on unmóde and on ungemetwæccum, Lchdm. ii. 176, 2.

ungemet-wæl, es; *n. Very great carnage :*—Ðǽr wæs ungemetwæl geslægen, Chr. 867; Erl. 73, 14.

ungemet-wilnung, e; *f. Excessive desire :*—Ungemetwilnung ǽtes and slǽpes, Dóm. L. 30, 44.

un-gemidlod; *adj. Unbridled :*—Swá swá módig hors, ðe ungemidlod byð, Ælfc. T. Grn. 17, 22. Seó ofermódnes ungemidledum (-odon) horse fleáh *superbia effreni volitat equo*, Gl. Prud. 29 a. v. un-midlod.

un-gemilt; *adj. Undigested :*—Heald georne ðæt se mete sí gemylt . . . , for ðan ðe se ungemylta mete him wyrcð mycel yfel, Lchdm. ii. 284, 4.

un-gemód; *adj. Disagreeing, contentious, at variance; discors :*—Ðǽm ungesibsuman is tó cýðanne ðæt hié wieten ðætte swá lange swá hié beóð from ðære lufe áðíed hiera niéhstena and him ungemóde beóð . . . *admonendi sunt dissidentes, ut noverint, quod . . . quamdiu a proximorum caritate discordant*, Past. 46; Swt. 349, 7. Ða ungemódan . . . , ða gemódan *discordes . . . , pacati*, 23; Swt. 177, 9.

un-gemódigness, e; *f. Dissentiousness :*—Ðæt wæs ungerím, ðæt þurh deófles ungemódignesse intó helle behreás, Wulfst. 8, 15 note. v. next word.

un-gemódness, e; *f. Contentiousness, indisposition to agree :*—Ða ungesibsuman sint tó manianne ðæt hié witen ðæt hié nó on tó ðæs monegum gódum cræftum ne ðióð ðæt hié ǽfre mægen gástlíce bión gif hié ðurh ungemódnesse ágiémeleásiaþ ðæt hié ánmóde bión nyllaþ on ryhte and on góde *discordes admonendi sunt, ut sciant, quia, quantislibet*

virtutibus polleant, spiritales fieri nullatenus possunt, si uniri per concordiam proximis negligunt, Past. 46; Swt. 344, 9.

un-gemolsnod; *adj. Uncorrupted, undecayed :*—Hé healdeþ ða deádan líchoman ungemolsnode under eorðan, óþ ðæt hí eft cuce árísaþ, Shrn. 82, 21.

un-gemunecod; *adj. Not made a monk :*—Unhádod man and ungemunecod *homo non ordinatus nec monachus*, L. Ecg. C. 12; Th. ii. 142, 4.

un-gemynd *distraction* or *confusion of mind, dementedness :*—Wiþ ungemynde and wið dysgunge, Lchdm. ii. 142, 1, 4: 14, 16. Wiþ deófle and ungemynde, 352, 7. Wiþ heáfodece and wiþ ungemynde and wiþ ungehýrnesse, 314, 25. Cf. gemynd-leás.

un-gemyndig; *adj. Unmindful, forgetful :*—Ungemyndig *immemor*, Ælfc. Gr. 9, 21; Zup. 47, 14. Ne byð æfre God ungemyndig ðæt hé niltsige manna cynne *numquid obliviscetur misereri Deus ?* Ps. Th. 76, 8. Swá hwá swá ungemyndig (*immemor*) sié rihtwísnesse, Bt. 35, 1; Fox 156, 10: Met. 22, 55. Hé wæs ungemyndig ðæs hálgan gewrites, Homl. Th. i. 82, 13. Ungemyndig *oblitum*, Germ. 388, 36. Ungemyndigne, 388, 24. Hí wurdon ðæs treówes ungemyndige, ac God wæs his gemyndig, Homl. Th. ii. 146, 2.

un-gemyndum, Bd. 5, 12; S. 630, 38, un-geneð. v. un-gímende, un-geníð.

un-genge; *adj. Impracticable, useless, vain :*—Gé ungænge gedydon bebod Godes *irritum fecistis mandatum Dei*, Mt. Kmbl. Rush. 15, 6. [*Icel.* ū-gengr *not fit to walk on.* Cf. *Ger.* un-gangbar *not current, impracticable.*]

un-geníd[d]; *adj. Unforced, uncompelled :*—Ne mæg ic náne cwica wuht ongitan . . . ðe ungeneð lyste forweorþan *nihil invenio, quod, nullis extra cogentibus, abjicial manendi intentionem*, Bt. 34, 10; Fox 148, 14. Ðæt ealle gesceafta hiora ágnum willum ungenédde him wæron underþeódde, 35, 5; Fox 164, 29. Ungeniédde (-nîdde, Cott. MSS.) mid eówrum ágenum willan gē sculon ðencean for eówre heorde *providentes non coacte, sed spontanee*, Past. 18; Swt. 137, 19: Ors. 5, 15; Swt. 250, 14.

un-geocian *to unyoke :*—Ic ungeocige oððe tótwæme *disjungo*, Ælfc. Gr. 47; Zup. 277, 3.

un-georne; *adv.* I. *unwillingly, reluctantly :*—Ðá on ðæm tweón ðe hié swá ungeorne his willan fulleodon *qui fastidiose ducem in disponendo bello audientes*, Ors 3, 11; Swt. 146, 24. II. *without diligence, negligently :*—Ðæt hé tó ungeorne bewiste hwæt hé on þeóstrum dyde, Blickl. Homl. 183, 23. [*O. H. Ger.* un-gerno: *Ger.* un-gern: *Icel.* ū-gjarna *unwillingly.*]

un-geornful[l]; *adj. Not diligent, negligent, careless, remiss :*—Ðæt ungeornfulle mód hyngreð *anima dissoluta esuriet*, Past. 39; Swt. 283, 11. Ðý læs hine se wærscipe geléde on ealles tó micle hátheortnesse, oððe eft sió ánfealdnes hine tó ungeornfulne gedoo tó ongietanne, ðý læs hé weorðe besolcen *quatenus nec seducti per prudentiam calleant, nec ab intellectus studio ex simplicitate torpescant*, 35; Swt. 239, 2. Hyne nán man geseah ungeornfulne tó Cristes þeówdóme, Guthl. 20; Gdwin. 92, 19.

un-geréd; *adj.* I. *stupid, rude, unskilled, foolish, ignorant :*—Walah *sive* ungeréd *barbarus*, Wrt. Voc. ii. 75. Gif se sacerd bið ungeréd ðæs láreówdómes *sacerdos si praedicationis est nescius*, Past. 15; Swt. 91, 24. Sum ungeréd mann . . . nolde gán tó ðam axum on ðone Wódnes-dæg, Homl. Skt. i. 12, 41. Dysig bið se wegférenda man, se ðe ninið ðone smeðan weg, ðe hine mislæt . . . Swá eác wé beóð ungeráde, gif wé lufiaþ ða hwílwendlícan lustas, Homl. Th. i. 164, 10. Ða dwollícan béc rædaþ ungeráde menn, ii. 444, 25. Deáh ða dysegan and ða ungerádan his gelýfan nyllan, Wulfst. 305, 14. Se ðe ungerádum oððe ungeðyldigum stýrð, Homl. Th. i. 306, 5. II. *discordant, disagreeing, at variance :*—Ungeráde *dissona (sermonum procacitate*, Ald. 59), Wrt. Voc. ii. 85, 20: 26, 69. Næles ungeráde *non dissona (sententia*, Ald. 65), 86, 12: 60, 69. Dissona .i. *discordantia, incongrua*, ungeswēga *vel* ungeráde, 141, 37. Simle bióþ ða gódan and ða yflan ungeþwære betwyh him, ge eác hwílum ða yflan bióþ ungeráde betwuh him selfum *ut probis atque improbis nullum foedus est, ita ipsi inter se improbi nequeunt convenire*, Bt. 39, 12; Fox 230, 27: Ors. 2, 7; Swt. 90, 6. Ða lätteówas wæron Agustuse ungeráde, 6, 1; Swt. 254, 18. Ðonne se abbod and se práfost ungeráde beóð and him betwyx sacaþ *dum contraria sibi invicem abbas prepositusque sentiunt*, R. Ben. 124. Ic sceal nū mid ungerádum wordum gesettan, þeáh ic geóhwílum gecoplíce funde *carmina qui quondam studio florente peregi, moestos cogor inire modos*, Bt. 2; Fox 4, 7.

un-geréd, es; *n.* I. *stupidity, folly, unreason :*—Fela dyslíce dæda deriaþ mancynne oððe for ánwylnysse oððe for ungeráde; swá swá sume menn dóð, ðe dyslíce fæstaþ ofer heora mihte . . . Nū gesettan ða hálgan fæderas ðæt wé fæston mid geráde, Homl. Skt. i. 13, 92. II. *discord, disagreement, variance :*—Ðætte án sibb Godes lufe bútan ælcum ungeráde ūs suíðe fæste gebinde *tunc sola nos in aedificio concordia caritatis liget*, Past. 36; Swt. 253, 22.

un-gerádness, e; *f. Disagreement :*—Gyf hyne méte, ðæt hé

äwiht beran geseó, ðæt byð ungerádnes, Lchdm. iii. 170, 20. v. un-gerædness.

un-geræd (?); *adj. Stupid :*—Ungeræd[e?] *insipidus*, stunt *stultus*, Wrt. Voc. i. 47, 42. (*The MS. has* ungeræd, v. Wülck. 165, 16.)

un-gerǽdelíce; *adv.* Roughly, rudely :—Ða Godes wiðerwinnan ða fǽmnan genámon, ūt of ðære byrig ungerǽdelíce hí togoden, Homl. Ass. 178, 307. [Þe weregede gastes hine uniredlíce (*or see* un-gerýdelíce?) underfangeð mid stearne swupen, O. E. Homl. i. 239, 10.]

un-gerǽdness, e; *f. Discord, disagreement, variance :*—Betux Agathocle and his folce weard ungerǽdnes *in exercitu Agathoclis orta est seditio*, Ors. 4, 5; Swt. 170, 15. Sóna swá hié him betweónum ungerǽdnesse up áhófon swá forwurdon hié ealle *discordia exitio fuit*, 5, 3; Swt. 222, 19: 6, 6; Swt. 262, 14. For his feóndum gebidde hé, mid ðám ðe wið hyne ungerǽdnysse hæbben, L. E. I. 21; Th. ii. 418, 15.

un-gerec[c], es; *n. Disorder, tumult, violence :*—Ungerecc (-rec, Rush.) *tumultus*, Mt. Kmbl. Lind. 26, 5. Ungerece *impetu* (cf. *O. H. Ger.* Mit mihhilu ungirehhu *magno impetu*, Mk. 5, 13), Rush. 8, 32. Hé ôðerne cyninges þegn in ðæm ungerecce ácwealde *in ipso tumultu alium de militibus peremit*, Bd. 2, 9; M. 122, 24. [*O. H. Ger.* un-gireh *tumultus, seditio, impetus, inquietudo, passio.*]

un-gereccan *to repel a charge from, to clear :*—Gif hit man him on gerecce, and hé hine ungereccan ne mæge, L. Ath. iv. 1; Th. i. 222, 4.

un-gereclíc; *adj. Disorderly, tumultuous, ungovernable :*—Seó menego tácnode ða flǽsclícan willan and ða ungereclícan uncysta, Blickl. Homl. 19, 6.

un-gereclíce; *adv. Without order, tumultuously, without restraint :*—Se ðe ungereclíce liofaþ and his gecynd nyle healdan, ne biþ náuht *est enim quod ordinem retinet, servatque naturam; quod vero ab hac deficit, esse etiam derelinquit*, Bt. 46, 6; Fox 182, 21. Ic ongite ðæt ealle gesceafta tôfleówon swá swá wæter and náne sibbe ne náne endebyrdnesse ne heóldon, ac swíþe ungereclíce tôslupen and tô náuhte wurden, gif hí næfdon ǽnne God ðe him eallum stiórde and racode and rædde *vel ad nihilum cuncta referuntur, et uno veluti vertice destituta, sine rectore fluitabunt*, 34, 12; Fox 154, 3.

un-gerédelíce. v. un-gerýdelíce.

un-geregnod; *adj. Unornamented :*—Massehakele þæt is ungerénad, Chart. Th. 515, 26.

un-gereord; *adj. Not having an intelligible language :*—Ungereord *barbarus*, Wrt. Voc. ii. 125, 22.

un-gereordedlíc, -gereordlíc; *adj. Insatiable :*—Ungereordedlícne (-gereo[r]dlícre, Ps. Spl. C.) *insatiabili*, Ps. Surt. 100, 5.

un-gereordod; *adj. Unfed, not having had a meal :*—Se déma ungereordod sæt bútan ǽlcere ðénunge unþances fæstende, Homl. Skt. i. 19, 91.

un-gerian. v. un-girwan.

un-gerím, es; *n. A countless number, an immense number* or *quantity :*—Feala ôðra gôdra þegna and folces ungerím, Chr. 1010; Erl. 143, 23. Ðara wæs ungerím, Shrn. 48, 31. Ðara ys fornéan ungerím, Ælfc. Gr. 5; Zup. 18, 3. Ðæt hé gegaderige ungerím ðissa welena, Bt. 26, 3; Fox 94, 13. Hire olfendas bǽron ungerím goldes, Homl. Th. ii. 584, 11. Ungerím feós syllan, Homl. Skt. i. 12, 101. Cf. un-rím.

un-gerím; *adj. Countless, numberless, innumerable, incalculable, immense :*—Ðǽr is ungerím fæc betweox hyre and ðære eorðan, Lchdm. iii. 254, 12. Ðæt wæs ungerím (ungerímlíc, MS. E.), ðæt intô helle behreás, Wulfst. 8, 15. Cômon ða hǽðengildan mid ungerímum folce, Homl. Th. ii. 494, 16. S. Anastasius, scs Basilius and ungeríme ôðre, L. Ælfc. C. 6; Th. ii. 344, 30: Ælfc. Gr. 9, 21; Zup. 46, 14. Ôðre ungeríme, 9, 37; Zup. 62, 5: 9, 38; Zup. 63, 8. Tô gefremminge ungerímra tácna, Homl. Th. i. 310, 17. Ungerímum *innumeris*, Wülck. Gl. 255, 14. Se deófol . . . wyrcð ungeríma wundra, Homl. Th. i. 4, 16. Ungeríme hûðe *numerosas praedas*, Hpt. Gl. 522, 20. [Mikell follc and unigerím iss onn erþe, Orm. 18993.] Cf. un-rím; un-getel.

un-gerímed; *adj. Unnumbered, innumerable :*—Mid ðære ungerímedan mænigo *innumerabilis multitudo*, Bd. 5, 12; S. 628, 4 note.

un-gerímedlíc; *adj. Innumerable :*—Ðá geseah hé ða mycelan and ða ungerímedlícan ferde his feónda, H. R. 3, 15.

un-gerímlíc; *adj. Innumerable, incalculable*, Wulfst. 8, 15 note. v. un-gerím.

un-gerípod; *adj. Immature, premature :*—Unirípedes (deáþes) *immaturae (mortis*), Hpt. Gl. 507, 38. On ungerípedum freódôme and unstæððigum þeáwum, Ælfc. T. Grn. 17, 12. Hé forfleáh ungerípedan deáð, Homl. Th. i. 390, 31.

un-gerisende; *adj. Unbecoming, indecent :*—Ungerísendre æfesne *indecens obscenitas*, Hpt. Gl. 492, 59.

un-gerisene, -gerisne; *adj.* I. *unsuitable, inappropriate :*—Ungerisenu *indecens (est stulto gloria*, Prov. 26, 1), Kent. Gl. 977. Nis ungerisne ðæt wé án wundor of monegum ásecgan *nec ab est unum e pluribus miraculum enarrare*, Bd. 3, 2; S. 524, 38. II. *unseemly, indecent :*—Ðý læs hé ôwiht unwyrþes oððe ungerisenes dyde mid his mûþe *ne aliquid indignum suae personae vel ore proferret*, Bd. 4, 11; S. 579, 26. Ungerisnre bysene ðū hâtest hié wítnian *you order them to be punished*

in a way that inflicts indignity upon them, Blickl. Homl. 189, 31. Un-gerysenre æfsna *indecens obscenitas,* Hpt. Gl. 492, 62.

un-gerisene, es; *n.,* or un-gerisenu; *indecl. f.* **I.** *inconvenience, disagreeableness :*—Hit ðé biþ oððe ungetǽse oððe frécenlíc, eall ðæt ðú ofer gemet dést . . . seó ofering ðé wurþ oþþe tó ungerisenum oþþe tó plió, Bt. 14, 1; Fox 42, 16. Wið scurfendum næglum, gebærned hundes heáfod and seó acxe ðǽron gedón; ða ungerisnu hyt on weg áfyrreþ, Lchdm. i. 370, 10. **II.** *unseemliness, indignity, disgrace :*—Tó æwisclícum bismer .i. ungerisne *ad infame dedecus,* Hpt. Gl. 507, 8. Hé teáh hiene ðæt hé his ungerisno sprǽce *he accused him of speaking unbecomingly of him;* velut sui proditorem, Ors. 4, 11; Swt. 206, 29. Wege hé ða ungerisenu (*contumeliam,* Latin version), L. Ath. iv. 1; Th. i. 222, 7. Gif mín geréfa ungerysena gebýt áðer oþþe túnes-mannum oþþe heora hyrdon, L. Edg. S. 13; Th. i. 276, 27. Him is leófre ðæt hé leóge ðonne him mon ǽnigra ungerisna tó wéne *eligit falsa de se jactari, ne mala possit vel minima perpeti,* Past. 33; Swt. 217, 16. Bið ðæt sǽd ágoten tó unclænnesse and tó ungerisnum *ad immunditiam semen effundit,* 15; Swt. 97, 11.

un-gerisenlíc; *adj. Unseemly, dishonourable, base :*—Is ðæt ungerisenlíc wuldor ðisse worulde and swíþe leás *gloria quam fallax, quam turpis est,* Bt. 30, 1; Fox 106, 30. Ðeáh ðe ful monige mid gerisenlícum weorcum árísen from eorðan, mid ungerisenlícum gewilnungum ðissa worolddinga hié hié selfe ǽlecgeaþ on eorðan *etsi honesta actione nonnulli quasi a terra se erigunt, ambitione tamen inhonesta semetipsos ad terram deponunt,* Past. 21; Swt. 157, 8. Áðó of his móde ungerisenlíce ymbhogan, Bt. 29, 3; Fox 106, 19. Hwæt ungerisenlícre sié ðonne ðæt *quo quid turpius excogitari potest?* 30, 1; Fox 108, 6.

un-gerisenlíce; *adv. In an unsuitable, unseemly* or *unbecoming manner, with indignity, dishonourably, basely :*—Ungerisenlíce *inconvenienter,* Wrt. Voc. ii. 43, 64. Hé sceal tilian ðæt hé ne sié tó ungerisenlíce underþeód his unþeáwum, Bt. 29, 3; Fox 106, 19. For ðære gewilnunga woroldgielpes hé onlýt ungerisenlíce tó ðissum eorðlícum, suá ðæt neát for gífernesse onlýt tó ðære eorðan, Past. 21; Swt. 157, 2. Dauid, ðá hé ðone læppan forcorfenne hæfde, suíðe suíðlíce hreówsade ðæt hé him (*Saul*) ǽfre suá ungerisenlíce (-risen-, Cott. MSS.) geðénigan, sceolde, 28; Swt. 199, 18. Hé beþǽhte hí intó his búre, and hí man ðǽrinne ofslóh ungerisenlíce (*they were basely slain*), Chr. 1015; Erl. 152, 1.

un-gerisenness, -gerisness, e; *f. Unseemliness, shame :*—Unirisnysse *dedecus,* Hpt. Gl. 507, 35.

un-gerýde; *adj. Rough, violent :*—Se egeslíca swég ungerýdre sǽs, Wulfst. 137, 7. [Unirude duntes wið mealles istelet, O. E. Homl. i. 253, 12. Cf. An unrude raketehe, 249, 24. Unseollþe unnride inoh forr to dreȝhenn, Orm. 4784. Oferrcumenn wiþþ nan unnride strenncþe, 12527. Ðis fís (*the whale*) ðat is unride (*rimes with* wide), Misc. 16, 505: 20, 631: (*rimes with* side), 646. A kowel ful unride (*rimes with* shride = scrýdan), Havel. 964. Þe unrideste wunde þat men may see, 1985. Þen rewis þe king of unride (-rode) werkis, Alex. (Sk.) 871. Þou has ragid with unryd gestis, 460. See also Halliwell's Dict. *unride.*] v. following words.

un-gerýde, es; *n. A rough place :*—Ungerýdu beóð on sméðe wegas *aspera erunt in uias planas,* Lk. Skt. 3, 5.

un-gerýdelíce; *adv. Violently, with impetus :*—Cwæð se Hǽlend: 'Ic geseah ðone sceoccan swá swá scínende líget feallende adún dreórig of heofonum,' for ðam ðe hé áhreás ungerýdelíce, Hexam. 10; Norm. 18, 7. Ða felga bióþ fyrrest ðære eaxe, for ðæm hí faraþ ungerýdelícost (-réðe-, Cott. MS. v. (?) un-gerǽdelíce), Bt. 39, 7; Fox 222, 21. [Þe meiden reat him mitte raketehe unrudeliche, Jul. 54, 1. Ha þe dintede unrideli o rug, O. E. Homl. i. 281, 27. Þer as þe rogh rocher unrydely watȝ fallen, Gaw. 1432. If any of his feris raged with him unridly, Alex. (Skt.) 638. Þen rekils it unruydly, & raynes doune stanys, 566. And oferr warrp þær i þe flor unnriddliȝ þeȝȝre bordess, Orm. ii. p. 419.]

un-gerýdness, e; *f. Violence, tumult :*—Ungerýdnyss and gewinn *tumultus et conluctatio,* Scint. 82, 2.

un-gesadelod; *adj. Not saddled :*—Eahte hors, feówer gesadelode and feówer ungesadelode (unsadelode, MS. G.) . . . feówer hors, twá gesadelode and twá ungesadelode (unsadelode, MS. G.), L. C. S. 72; Th. i. 414, 5–10.

un-gesǽlhþ, e; *f. Unhappiness, misery :*—For ungesǽlhðe ðissere earman þeóde, Chr. 1057; Erl. 192, 26. v. next word.

un-gesǽlig; *adj. Unhappy, unfortunate :*—Ungesǽlig *infelix,* Wrt. Voc. i. 74, 30. Ðæs ungesǽligan *infausti,* ii. 47, 56. **I.** *of persons, unhappy,* (a) *suffering, misfortune, calamity, etc. :*—Ne meht ðú cweðan ðæt ðú earm sé and ungesǽlig (*te existimari miserum*), Bt. 8; Fox 24, 23. Ðæt is seó mǽste unsǽld ðæt mon ǽrest weorþe gesǽlig and æfter ðam ungesǽlig *in omni adversitate fortunae infelicissimum genus est infortunii, fuisse felicem,* 10; Fox 26, 31. Gif ðú gesihst hwylcne swíþe ungesǽligne mon and ongitst ðeáh hwæthwegu gódes on him, hwæþer hé sié swá ungesǽlig swá se ðe ealninga nǽfþ . . . ac hú þyncþ ðé be ðam ðe nánwuht gódes nǽfþ, gif hé hæfþ sumne eácan yfeles; se ðú wilt secgan sié ungesǽligra ðonne se óðer *si miseriae cujuspiam*

bonum aliquid addatur, nonne felicior est eo, cujus pura ac solitaria sine cujusquam boni admissione miseria est? . . . *Quid si eidem misero, qui cunctis careat bonis, praeter ea, quibus miser est, malum aliud fuerit annexum, nonne multo infelicior eo censendus est, cujus infortunium boni participatione relevatur?* 38, 3; Fox 200, 14–20. Sedechias se ungesǽliga kining, ðe man gelǽdde on bendum tó Babiloniam birig, Ælfc. T. Grn. 8, 11. Ða unþeáwas nǽfre ne bióþ unwítnode . . . ða yfelan bióþ simle ungesǽlige, Bt. 36, 1; Fox 172, 26: Met. 27, 18. Ungesǽlge, Exon. Th. 75, 4; Cri. 1216. Wá lá wá ðæt ða ungesǽligan menn ne magon gebídon hwonne hé (*death*) him tó cume, Bt. 39, 1; Fox 212, 1. Ára mé, ungesǽligost ealra wífa, Blickl. Homl. 89, 22. (b) *suffering want of moral good :*—Deófol sǽwð unwísdóm and gedéð þurh ðæt, ðæt ungesǽlig man wísdómes ne gýmeþ, Wulfst. 52, 27 note. Se ungesǽliga gýtsere wile máre habban ðonne him genihtsumaþ, ðonne hé furðon orsorh ne brícð his genihtsumnysse, Homl. Th. i. 64, 33. Hié beóð suíðe ungesǽlige ðonne hié yfeliaþ for ðæm ðe óðre menn gódigaþ . . . Hwá mæg beón ungesǽligra *quantae infelicitatis sint, qui melioratione proximi deteriores fiunt* . . . *Quid istis infelicius?* Past. 34; Swt. 231, 18–22. Hí synt earmran and dysigran and ungesǽligran ðonne ic hit árecan mæge, 32, 3; Fox 118, 28: Met. 19, 42. (c) *causing unhappiness :*—Ungesǽlig *infelix* (a son that bringeth reproach, Prov. 19, 26), Kent. Gl. 716. **II.** *of things,* (a) *unfortunate, calamitous :*—Ðis ungesǽlige geár *infaustus ille annus,* Bd. 3, 1; S. 523, 32. (b) *unprofitable, evil :*—Se ungesǽliga gewuna beláf of hǽðenra manna biggenge, Homl. As. 146, 47. Ðeó árleása scylda wé gehýrdon—ungesǽlig mǽrsunge his (*Herod's*) gebyrdtíde . . . Wé ne móton úre gebyrdtíde tó nánum freólsdæge mid ídelum mǽrsungum áwendan, Homl. Th. i. 480, 34. [*Laym. A. R.* un-iseli.] v. un-sǽlig.

un-gesǽliglíce; *adv.* **I.** *unhappily, miserably :*—Hé (*Judas*) hine sylfne áhéng and swá ungesǽliglíce tó écan deáðe wæs geníðerad, Homl. Ass. 158, 164. Hé hine seluan éce hellewíte ungesǽliglíce getilaþ, Chart. Th. 117, 23. **II.** *wickedly.* v. un-gesǽllíce :—Swá ungesǽliglíce Iudas ðam láreówe deáð sǽtade, swá him eall his líf tó ungesǽlðum weard, Homl. Ass. 161, 225.

un-gesǽligness, e; *f. Unhappiness, calamity, misery :*—Ungesǽlignys *infelicitas,* Ps. Spl. 13, 7. Seó ungesǽlignys becom on ðæt folc, ðæt hig ðone Hǽlend geféngon and on róde áhéngon, Nicod. 1; Thw. 1, 12. Wæs se dóm oncyrred Euan ungesǽlignesse, ðæt heó cende on sáre and on unrótnesse, Blickl. Homl. 3, 8. Hé ða ðeóde fram langre wónesse and ungesǽlignysse (*infelicitate*) álýsde, Bd. 2, 15; S. 519, 10.

un-gesǽllíce; *adv. Unhappily, miserably, wickedly;* improbe :—Hæfst se yfela gást ungifa . . . and ða hé déð ðam mannum ðe ungesǽlíce him gehýrsumiaþ, Wulfst. 52, 12. [His sune ðe uniseliche (onselliche, 2nd MS.) luuede, his deden weoren forcuðe, Laym. 7022.] v. un-sǽle.

un-gesǽlþ, e; *f.* **I.** *unhappiness, illfortune, calamity :*—Ealle ðæs ungesǽlða ús gelumpon þurh unrǽdes, Chr. 1011; Erl. 145, 1. Biþ simle ða eówre gesǽlþa on sumum þingum ungesǽlþa (unsǽlþa, Cott. MS.), Bt. 29, 1; Fox 102, 20. Sume secgaþ ðæt sió wyrd wealde ǽgðer ge gesǽlþa ge ungesǽlþa ǽlces monnes, 39, 8; Fox 224, 13. Ne meaht ðú nó mid sóþe getǽlan ðíne wyrd for ðam leásum ungesǽlþum (unsǽlþum, Cott. MS.) ðe ðú þrowast, 10; Fox 28, 2. Eall his líf tó ungesǽlðum and tó ermðum weard, Homl. Ass. 161, 226. Áfyr fram ðé ða unnettan ungesǽlþa and ðone yflan ege ðisse worulde, Bt. 6; Fox 14, 33. **II.** *unhappiness* which consists in absence of moral good :—Ða yfelan habbaþ ðrió ungesǽlþa (unsǽlþa, Cott. MS.); án is ðæt hí yfel willaþ, óþer ðæt ðæt hí magon, þridde ðæt hí hit þurhtióþ *triplici infortunio necesse est urgeantur, quos videas scelus velle, posse, perficere,* Bt. 38, 2; Fox 196, 33. Nán man ne dear for árwyrðnesse ðæs ánsetlan leahtras tǽlan; him synt eác ða ungesǽlþa leófran, ðæt hé hý nyte, ðænne hé hí lácnige, R. Ben. 135, 18. [Þurh him (*Adam*) deð com in þis middenerd and oðer uniselðe, O. E. Homl. i. 171, 197. For heora uniselðe (*wickedness*), Laym. 2545.] v. un-sǽlþ.

un-gesawen. v. un-gesewen.

un-gesceád, es; *n. Indiscretion, unreason :*—Ða ðe on ðam sylfum cildum mid ungesceáde geháðeortaþ *qui in ipsis infantibus sine discretione exarserit,* R. Ben. 130, 7. Hé on ánum dæge mid ungesceáde forspild þreóra daga andlifene, Homl. Ass. 145, 30 : Lchdm. iii. 442, 32.

un-gesceádig; *adj. Indiscreet, unreasonable, irrational :*—Hwá is manna tó ðam ungesceád and ungewittig, ðæt hé ðæm cyninge his áre ætrecce for ðí ðe his geréfa forwyrht biþ? Lchdm. iii. 444, 7.

un-gesceádlíc; *adj. Indiscreet, irrational :*—Ungesceádlíc swígea *indiscretum silentium,* Scint. 213, 13. Mid ungesceádlícum þinge *in re irrationali,* L. Ecg. P. ii. 6; Th. ii. 184, 10.

un-gesceádlíce; *adv. Unreasonably, excessively :*—Ungesceádlíce (-sceáde-) *irrationabiliter,* R. Ben. 54, 13 note. Tácn ðæs ungesceádlíce cealdan magan (cf. ðæs ofercealdan magan, 192, 25: 194, 11), Lchdm. ii. 160, 4. v. next word.

ungesceád-micel; *adj. Excessively great :*—Ǽled wæs ungesceádmicel, Cd. Th. 231, 6; Dan. 243. Cf. un-gemet-, *and see preceding word.*

un-gesceádwís; *adj.* **I.** *not acting according to reason, un-*

reasonable, irrational, unwise, foolish :—Hé hine wiste swíþe ungesceád-wísne and swíþe uugemetfæstne, Bt. 27, 1 ; Fox 96, 4. Ic wundrige hwí men sién swá ungesceádwíse dæt hié wénan dæt dis andwearde líf mæge done monnan dôn gesǽligne, 11, 2 ; Fox 34, 36 : 39, 9 ; Fox 226, 9. Deáh ungesceádwísum monnum swá ne þince, 39, 8 ; Fox 224, 16. **II.** *not possessed of reason, irrational :*—Ǽlc gesceaft, ǽgder ge gesceádwís ge ungesceádwís, Bt. 42 ; Fox 256, 7. Ic eom ofwundrod hwí eów þince dære ungesceádwísan gesceafte gôd betere donne eówer ágen god, 13 ; Fox 40, 5. On dara ungesceádwísra niétena gesibsum-nesse, Past. 46 ; Swt. 349, 24. Hé hine gehwyrfde tô ungesceádwísum neátum *in irrationale animal hunc vertit,* 4 ; Swt. 39, 22.

un-gesceádwíslíc ; adj. *Indiscreet, imprudent, unreasonable, extrava-gant :*—Dætte hé ne dôhte náwuht ungesceádwíslíces ne unnetlíces *nec indiscretum quid vel inutile cogitet,* Past. 13 ; Swt. 77, 12. Gif wé hwæt ongietaþ on him ungesceádwíslíces gedoon *si qua ab eis inordinate gesta sunt,* 32 ; Swt. 211, 22. v. un-sceádwíslíc.

un-gesceádwíslíce ; adv. *Indiscreetly, unreasonably, foolishly :*—Da de hiora ágen ungesceádwíslíce healdaþ *qui sua indiscrete tenuerunt,* Past. 44 ; Swt. 329, 11. Se gítsere and se de woruldwelan lufaþ ungesceádwís-líce, Swt. 331, 8. Donne wé biddaþ ongeán úre ágenre þearfe, donne forwyrnd God ús dæs de wé ungesceádwíslíce biddaþ, Homl. Th. ii. 528, 9. Ongesceádwíslíce *inrationabiliter,* R. Ben. Interl. 61, 11.

un-gesceádwísness , e ; f. *Unreasonableness, foolishness :*—Hwæt segst dú dæt sié forcúþre donne sió ungesceádwísnes ? hwî geþafiaþ hí dæt hí bióð dysige ? hwî nyllaþ hí spyrigan æfter cræftum and æfter wísdôme ? *quid enervatius ignorantiae caecitate ? an sectanda noverunt ?* Bt. 36, 6 ; Fox 180, 31, 35. Hé lǽt his môd tôflôwan on dæt ofdele giémeliéste and ungesceádwísnesse æfter eallum his willum *anima neglectam se inferius per desideria expandit,* Past. 39 ; Swt. 283, 15.

un-gesceapen ; adj. **I.** *unshapen, unformed :*—Interjectio is ân dǽl sprǽce getácniende dæs môdes gewilnunge mid ungesceapenre stemne (*voce incondita*), Ælfc. Gr. 48 ; Zup. 277, 17. He is sôd Scyppend, de da ungesceapenan eáhhringas (*the eyes of the man who was born blind*) geopenode, Homl. Th. i. 474, 8. **II.** *uncreated :*—Ungesceapen (*increatus*) is se Fæder, ungesceapen is se Sunu, and is ungesceapen se Hálga Gást, Ath. Crd. 8.

un-gescended ; adj. *Uninjured :*—Ungiscended *inlessa,* Rtl. 146, 23.

un-gesceþþed ; adj. *Uninjured, entire :*—Wæs his líchama gemêted ungesceþþed *corpus inventum est inlesum,* Bd. 3, 19 ; S. 550, 11.

un-gescrêpe, -gescræpe ; adj. *Inconvenient, unfit, useless :*—Unbriéce, ungescræpe *incommodum, inutile,* Wrt. Voc. ii. 44, 76. Ǽghwylcre menniscre eardunge ungescræpe *humanae habitationi minus accommodus,* Bd. 4, 28 ; S. 605, 20.

un-gescrêpness , e ; f. *Inconvenience :*—Seó ungescrêpnes dæs sáres fram heora eágan gewát *doloris incommodum ab oculis amoverent,* Bd. 4, 19 ; S. 589, 37.

un-gescrêpu -o) ; f. *or* un-gescrêpe ; n. *Inconvenience, an inconvenient thing :*—Mid dý dá se foresprecena brôþor langre tíde dyllíc ungescrêpo woon (*dyllíce ungescrêpo wonn ?*) *cum tempore non pauco frater praefatus tali incommodo laboraret,* Bd. 4, 32 ; S. 611, 22.

un-gesegnod, -gesênod ; adj. *Not marked with the sign of the cross :*—Gif dæt deófol mêteþ ungesênodes mannes mûd and líchoman, and hit donne on forgitenan mannes innelfe gewîteþ, Salm. Kmbl. p. 148, 10.

un-gesêne. v. un-gesíne.

un-geseónde ; adj. *Not seeing, blind :*—Gyf dú on foreweardon sumera þigest hwylcne hwelpan donne gyt ungeseóndne, ne ongitest dú ǽnig sár, Lchdm. i. 368, 26.

un-gesewen, -gesawen ; adj. *Unseen, invisible :*—Dá dá da tungel-wîtegan done cyning gecyrdon, dá weard se steorra him ungesewen, Homl. Th. i. 108, 29. Done ungesewenan (*invisibilem*) engel, Past. 36 ; Swt. 257, 8. Odre ungesawene þing mon môt mid âde gewyrdan, L. O. D. 7 ; Th. i. 356, 6.

un-gesewenlíc ; adj. *Invisible :*—Seó eorde wæs æt fruman eall ungesewenlíc, for dam de heó eall wæs mid ýdum oferdeht, Hexam. 5 ; Norm. 10, 17. Heora (*angels*) ungesewenlíce gecynd, Homl. Th. i. 538, 28. Se ungesawenlíca feónd, Wulfst. 52, 8. Unisæwenlícere mihte *invisibili potestate,* Hpt. Gl. 482, 69. God menniscum eágum ungesewenlícne, Bd. 3, 22 ; S. 552, 17. Da ôdre heofenan synd un-gesegenlíce, Lchdm. iii. 232, 23. Mid dære gewilnunge dara ungesewen-lícra dinga, Past. 16 ; Swt. 98, 3 : Bt. 21 ; Fox 72, 30 : Met. 11, 5. Wid mínum wiþerwinnam gesewenlícum and ungesewenlícum, Bt. 42 ; Fox 260, 11. Hé offrige da ungesewenlícan lác, Homl. Th. i. 584, 3.

un-gesewenlíce ; adv. *Invisibly, without being seen :*—God cymd ungesewenlíce tô geswǽsre heortan, Homl. Th. ii. 316, 4.

un-gesib[b] ; adj. **I.** *not related, strange :*—Hé bid fremede Freán ælmihtigum, englum ungelíc (*ungesibb,* MS. B.) âna hwearsaþ, Salm. Kmbl. 69 ; Sal. 35. Ic (*the cuckoo*) under sceáte ungesibbum weard eácen gǽste, Exon. Th. 391, 20 ; Rä. 10, 8. **II.** *not at peace, at variance :*—Da twá mǽgþa da de betwih him ungeþwǽre and ungesibbe wǽron *provinciae quae ab invicem discordabant,* Bd. 3, 6 ; S.

528, 32 : Blickl. Homl. 225, 6. [Cf. Betere weare sǽhte þene swulc unisibbe, Laym. 9845.]

un-gesibsum ; adj. *Prone to discord, quarrelsome :*—Lôca hwylc cristen man sý ungesibsum, man âh on dam dæge hine tô gesibsumianne, Wulfst. 295, 4. On ôdre wísan sint tô manigenne da gesibsuman, on ôdre da ungesibsuman *aliter admonendi sunt discordes, atque aliter pacati,* Past. 46 ; Swt. 345, 6. Dǽm ungesibsumum is tô cýdanne *admonendi sunt dissidentes,* Swt. 348, 5.

un-gesibsumness , e ; f. *Proneness to discord, quarrelsomeness, discord :*—Wé magon gecnáwan on dara ungesceádwísra niétena gesibsum-nesse hú micel yfel sió gesceádwíslíce gecynd durh da ungesibsumnesse gefremeþ *si solertes aspicimus, concordando sibi irrationalis natura indicat, quantum malum per discordiam rationalis natura committat,* Past. 46 ; Swt. 351, 1.

un-gesilt ; adj. *Unsalted :*—Swínen smeru ungesylt, Lchdm. i. 146, 20. v. un-silt.

un-gesíne ; adj. *Invisible :*—Ungesêne wearþ *disparuit,* Wrt. Voc. ii. 106, 43.

un-gesínelíc ; adj. *Invisible :*—Seó ungesýnelíce sáwl, Blickl. Homl. 21, 25. Ungesýnelícne God, 185, 31. Flǽsclíce men da ungesýnelícan ne magon angytan, Wulfst. 2, 4.

un-gesoden ; adj. *Unsodden :*—Nim dú da ylcan wyrte ungesodene, Lchdm. i. 92, 29.

un-gesôm ; adj. *At variance :*—Ǽfter sumum fyrste wurdon hí un-gesôme, Philippus and Arethe, Homl. Th. i. 478, 25. [Hit itit þat wif and were beoþ unisome, O. and N. 1522.]

un-gestæddig ; adj. *Inconstant, unstable :*—Dæt ungestæddige folc *mobile vulgus,* Bt. 39, 3 ; Fox 216, 2. Dam ungestæþþegan and dam gálan dú miht secgan dæt hí biþ gelícra unstillum fugelum donne gemet-fæstum monnum *levis atque inconstans studia permutat ? nihil ab avibus differt,* 37, 4 ; Fox 192, 22. Da gesceaftas de wé embe sprecaþ, dæt heó ús þince ungestæddíge, hý habbaþ sumne dǽl gestæþinesse, Shrn. 168, 30. Da ungestæddegan and unfæsdrǽdan *inconstantes,* Past. 23 ; Swt. 177, 3. Nánwuht nis on ús unstilre and ungestæddigre donne dæt môd *nil in nobis est corde fugacius,* 38 ; Swt. 273, 11.

un-gestæddiglíce ; adv. *Unsteadily, without stability :*—Donne mon da fæstrǽdnesse his môdes innan forlíst, donne bid hé hwílum swíde ungestæddíglíce ástyred útane on his limum *qui statum mentis perdidit, subsequenter foras in inconstantiam motionis fluit,* Past. 47 ; Swt. 359, 7. Hié beóð suá micle ungestæddelícor tôflôwene on hiera môde suá hié wénaþ dæt hié stilran and orsorgran beón mægen *quae tanto latius diffluunt, quanto se esse securius aestimant,* 38 ; Swt. 271, 17.

un-gestæddigness , e ; f. *Unsteadiness, levity, want of firmness :*—Se hæfd singalne sceabb se de nǽfre ne blind ungestæddignesse *jugem habet scabiem, cui carnis petulantia sine cessatione dominatur,* Past. 11 ; Swt. 70, 3. Gif hé eallunge forberan ne mæg for hira ágnum undeáwum and for hiera ungestæddignesse *qui pro infirmitate sese abstinere vix possunt,* 28 ; Swt. 199, 9.

un-gestreón , es ; n. *Ill-gotten treasure :*—Da wôhgeornan woruldrícan mid heora golde and seolfre and eallum ungestreónum, Wulfst. 183, 9.

un-gestroden ; adj. *Not subjected to forfeiture* or *confiscation of goods :*—Swǽse mæn ciriclícæs gemánan ungestrodyne þoligen *natives shall forfeit the communion of the church but without being subjected to forfeiture of goods* [cf. gestrod *proscriptionem* (the passage in Aldhelm is : Proscriptionem rerum et patrimonii jacturam), Wrt. Voc. ii. 81, 67], L. Wih. 4 ; Th. i. 38, 3.

un-geswêge ; adj. *Dissonant, discordant :*—Ungeswêge *dissona,* Hpt. Gl. 513, 51. Ungeswêge sang *diaphonia,* Wrt. Voc. i. 28, 34. Un-geswêgre *dissona,* Hpt. Gl. 505, 76. Ungeswêga *vel* ungehleóþre *dissona,* i. *discordantia,* Wrt. Voc. ii. 141, 36. Ungeswêgium *absonis,* Germ. 392, 12.

un-geswencedlíc ; adj. *Unwearied, indefatigable :*—Mid ungeswenced-líce luste heofonlícra gôda *infatigabili coelestium bonorum desiderio,* Bd. 5, 12 ; S. 631, 35.

un-geswícendlíce ; adv. *Unceasingly, incessantly :*—Ungeswícendlíce *indesinenter,* Scint. 28, 7 : *incessanter, jugiter,* 131, 8 : *incessabiliter,* R. Ben. 19, 4 note.

un-geswuncen ; adj. *Unlaboured :*—Ungeswuncenre *inelaborate,* Wrt. Voc. ii. 49, 59.

un-gesýnelíc. v. un-gesínelíc.

un-getǽse ; adj. *Inconvenient, disagreeable, troublesome, obnoxious :*—Ungetǽse *infestus,* Wrt. Voc. ii. 45, 46. Odde hit dé deraþ odde hit dé unwynsum biþ, odde ungetǽse (-getǽse, Cott. MS.) odde frêcenlíc *aut injucundum, aut noxium,* Bt. 14, 1 ; Fox 42, 13. Gilpes dú girnest ? ac dú hine ne miht habban orsorgne, for dam dú scealt habban simle hwæt-hweg wiþerweardes and ungetǽses (-getǽses, Cott. MS.) *gloriam petis ? sed per aspera quaeque distractus, securus esse desistis,* 32, 1 ; Fox 114, 20. Da cyningas de æfter Rômuluse rícsedon wǽron forcúdran donne hé wǽre, and dǽm folcum lâdran and ungetǽsran, Ors. 2, 2 ; Swt. 66, 26.

un-getǽse , es ; *n. An inconvenience, a trouble :*—Gif hé dǽm gehiér-

suman mannum næfde geteohchad his éðel tô sellanne, hwié wolde hé hié mid ænegum ungetæsum læran? *nisi correctis haereditatem dare disponerit, erudire eos per molestias non curaret*, Past. 36; Swt. 251, 24. Mid hú monigfaldum ungetæsum and mid hú heardum brocum ús swingaþ and ðreágaþ úre worldcunde fædras *quam dura carnales filios disciplinae flagella castigent*, Swt. 253, 24.

un-getæslíce; *adv. Inconveniently, incommodiously :*—Ðonne ðæt scyp ungetæslícost on ancre rít, Shrn. 179, 17.

un-getæsness, e; *f. Inconvenience :*—Ungetæsnesse *incommoditate*, Wrt. Voc. ii. 44, 30.

un-getel; *adj. Innumerable :*—On þisum and on manegum and on ungetelum (ge)óþrum intingan ic syngade *in istis et in multis atque innumeris aliis causis peccavi*, Confess. Peccat. Cf. un-gerím.

un-getemed; *adj. Untamed :*—Se wilda fola hæfde getácnunge ealles óðres folces, ðe wæs ðágyt hæðen and ungetemed, Homl. Th. i. 208, 23. Tígan tô ungetemedra horsa swuran, 432, 33.

un-getemprung, e; *f. Rough weather;* intemperies, Anglia xiii. 397, 461.

un-geteón, es; *n. Foul injury :*—Wið nétana ungetiónu . . . Engel se ðe ásetted is ofer nétno úsra gihalda ða ðætte ne mæg diól onræd ða (*ut non poterit diabolus inequitare illa*), Rtl. 119, 15.

un-geteoriendlíce; *adv. Indefatigably :*—Ðis synt ða lára and ða tól gástlíces cræftes, gif hig from ús dæges oððe nyhtes ungeteoriendlíce begongenne beóð . . ., L. E. I. 21; Th. ii. 418, 18.

un-geteorod; *adj. Unwearied, unfailing, unexhausted :*—Ungeteorudne goldhord on heofenum *thesaurum non deficientem in caelis*, Lk. Skt. 12, 33. Ungeteorodne, Homl. Skt. ii. 23 b, 92. Ungetyradne *inexhaustam*, Hpt. Gl. 463, 19.

un-getése. v. un-getése.

un-geþæslíc; *adj. Unfit, unsuitable,* Wrt. Voc. i. 61, 39. v. un-gehæplíc.

un-geþanc, es; *m. n. Evil thought :*—Bútan hé mid fulre dǽdbóte his ungeþanc gebéte, R. Ben. 21, 6. Áfyrsiaþ of mínre gesyhðe ða ungeðanc eówra heortena *auferte malum cogitationum vestrarum ab oculis meis*, Wulfst. 48, 20.

un-geþancfull; *adj. Unthankful, ungrateful :*—Ðú man, tô hwan eart ðú mé swá ungeþancfull mínra gifena? Wulfst. 259, 1: 241, 4.

un-geþeaht, es; *n. Evil counsel :*—Forbeóde hé and álecge ða ðwyrnysse heora ungeþeahtes *prohibeant pravorum prevalere consensum*, R. Ben. 118, 10.

un-geþeahtendlíce; *adv. Inconsiderately, unadvisedly :*—Nalæs hé sóna and ungeþeahtendlíce ðám gerýnum onfón wolde ðæs Cristenan geleáfan *non statim et inconsulte sacramenta fidei Christianae percipere voluit*, Bd. 2, 9; S. 512, 6.

un-geþeáwe; *adj. Not in accordance with one's habits :*—Se biscop bæd ðone hálgan wer ðæt hé scolde tô gereorde fón mid him; and hé swá dyde, þeáh hit his life ungeþeáwe wǽre, Guthl. 17; Gdwin. 72, 27.

un-geþeáwfæst; *adj. Illregulated of conduct :*—Ungeþeáwfæstan *indisciplinatos*, R. Ben. Interl. 14, 16. v. un-þeáwfæst.

un-geþeód; *adj. Separate, disjoined :*—Tóføran ðá (*after the confusion of tongues*) on feówer wegas ædelinga beóð ungeþeóde (cf. hý beóð geþeóde þeódscipum on gemang betwyx heáhfæderas and hálige wítegan *vatidicis junctos patriarchis atque prophetis*, Dóm. L. 18, 282), Cd. Th. 102, 11; Gen. 1698.

un-geþinged; *adj. Undetermined, unsettled :*—Se egeslíca dæg, se cymð ofer ealle eorðwaran ungeðinged (*the time is not fixed and known beforehand;* repentina dies illa), Past. 43; Swt. 317, 12. v. un-þinged.

un-geþungen; *adj. Vile, base, ignoble :*—Ðú ungeþungena hund, Nar. 42, 12.

un-geþwǽre; *adj.* **I.** *not in harmony, at variance, discordant, not in agreement :*—Ungeþwǽra *discordator, discors*, Wrt. Voc. ii. 140, 77. Simle bióþ ða gódan and ða yflan ungeþwǽre betwyh him, ge eác hwílum ða yflan bióþ ungeráde betwuh him selfum, ge furþum án yfel man bið hwílum ungeþwǽre him selfum *ut probis atque improbis nullum foedus est, ita ipsi inter se improbi nequeunt convenire*, Bt. 39, 12; Fox 230, 26–29. Gód and yfel bióþ simle ungeþwǽre betwux him and simle on twá willaþ *bonum malumque adversa fronte dissideant*, 37, 3; Fox 190, 13. Fýr and wæter and manega oþra gesceafta ðe beóþ á swá ungeþwǽra betwux him swá swá hí beóþ, 21; Fox 74, 16. Ða twá mǽgþa, ða ðe betwih him ungeþwǽre and ungesibbe wǽron *provinciae, quae ab invicem discordabant*, Bd. 3, 6; S. 528, 31: Blickl. Homl. 225, 6. Ðá wǽron ungeþwǽre preóstas on ánum his mynstra; ða hé wolde sibbian . . . Se biscop ða ungeðwǽran preósta ðreáde, Homl. Th. ii. 516, 4–15. Ðæt ic móste ofercuman ða þeóda ðe mé ungeðwǽre wǽron, Ps. Th. 15, 2. **II.** *given to discord, quarrelsome :*—Ðætte on óðre wísan sint tô manianne ða geðwǽran (cf. ða gesibsuman, 6) on óðre ða ungeðwǽran (cf. ða ungesibsuman, 6) *quomodo admonendi sunt discordes et pacati*, Past. 46; Swt. 344, 5. **III.** *disagreeable, troublesome, vexatious :*—Hé sum fæc ðone ungeþwǽran swyle ðwénde *aliquamdiu tumorem illum infestum mollire curabat*, Bd. 4, 32; S. 611, 40. Ðú hine ongeáte on eallum þingum unweorþne ðæs anwealdes, swiþe sceamleásne and ungeþwǽrne (ungewǽrne, Bod. MS.) búton ælcum gódum þeáwe (*the. Latin*

is : Cum in eo mentem nequissimi scurrae respiceres), Bt. 27, 2; Fox 96, 19. v. un-þwǽre.

un-geþwǽre, es; *n. A disturbance, dissension :*—Ðýles ungerec ł ungeþwǽre in ðæm folce gewyrde *ne forte tumultus fieret in populo*, Mt. Kmbl. Rush. 26, 5.

un-geþwǽrian; *p. ode. To disagree, be at variance, differ :*—Ic ungeðwǽrige *dissentio*, Ælfc. Gr. 30; Zup. 190, 13. Anda fram gódan willan ungeþwǽregaþ *invidia a bona voluntate discordat*, Scint. 143, 3. Hé ongeat ðæt hí on monegum ðingum Godes cyricean ungeþwǽredon *vitam ac professionem minus ecclesiasticam in multis esse cognovit*, Bd. 2, 4; S. 505, 22. Ungeþwǽrudon *discordarent*, Anglia xiii. 367, 34.

un-geþwǽrlíce; *adv. Ungently, crossly :*—Ðá andswarode heó hire ungeþwǽrlíce: 'Ðeáh ðe God ðínne wer æt ðé genáme, hwæt sceal ic ðæs dón?' Homl. Ass. 121, 153.

un-geþwǽrness, e; *f.* **I.** *discord, dissension, disagreement, division, quarrel :*—Ungeðwǽrnes *discordia*, Wülck. Gl. 255, 17. Seó ungeðwǽrnes wundode ða geðwǽrnesse *discordia vulnerat concordiam*, Gl. Prud. 77: 78. Ðonne weaxaþ ða ofermétta and ungeþwǽrnes (cf. þonan mǽst cymeþ . . . unnetta saca, Mét. 25, 44), Bt. 37, 1; Fox 186, 19: Homl. Th. ii. 220, 32. Ungeðuǽrnis *vecordia*, Rtl. 163, 1. Ungehwǽrnys (=-þwǽrnys) *simultas*, Hpt. Gl. 495, 59: 522, 16. Ungeþwǽrnes wæs geworden on ðære menigeo for him *dissensio facta est in turba propter eum*, Jn. Skt. 7, 43: 10, 19. On ðisum geáre árás seó ungehwǽrnes on Glæstingabyrig betwyx ðam abbode and his munecan, Chr. 1083; Erl. 217, 1. Ðýlæs ænegu ungeþwǽrnes on his ágnum ríce áhafen wurde *prius quam adversa fama novas res domi moliretur*, Ors. 2, 5; Swt. 82, 30. Ne mihte hé mid ðone cyning . . . sibbe habban; ac swá mycel ungeþwǽrnys and unsibb betwih him árás (*ingravescentibus causis dissensionum*), ðæt hí heora fyrd gesomnedon, Bd. 3, 14; S. 539, 35. Be ungeþwǽrnysse wið his néhstan *de discordia cum proximo suo*, L. Ecg. P. ii. 27 tit.; Th. ii. 182, 1. Sii his wunung on hellewíte mid ðám ðe symle on ælcre ungeðwǽrnesse blissiaþ, Cod. Dip. Kmbl. iii. 129, 27. Se swicola feónd sǽwð ungeðwǽrnysse betwux mancynne, Homl. Th. ii. 318, 19. Ða hwíle ðe hé ænige ungeþwǽrnysse hæbbe on his heortan wið his ðone néhstan *quamdiu simultatem ullam in corde suo cum proximo suo habet*, L. Ecg. P. ii. 27; Th. ii. 192, 28. Bilewite cild ne hylt langsume ungeþwǽrnysse tô ðam ðe him derode, Homl. Th. i. 512, 15. Hié ægþer hæfdon ungeþwǽrnesse ge betweónum him selfum ge tô eallum folcum *they were at variance both among themselves and with all nations*, Ors. 6, 3; Swt. 258, 1. 'Þeód áríst ongeán þeóde.' Mid ðisum wordum hé foresǽde manna ungeðwǽrnyssa, Homl. Th. ii. 538, 26, 17. **II.** *trouble, disquiet.* v. ungeþwǽre, **III** :—On hú grundleásum seáðe ðæt mód þringþ, ðonne hit bestyrmaþ ðisse worulde ungeþwǽrnessa (*terrenis flatibus aucta crescit in immensum noxia cura*), Bt. 3, 2; Fox 6, 9. v. un-þwǽrness.

un-geþyld, e; *f. : es; n.* [v. ge-þyld] *Impatience :*—Hú mycel Godes geþyld is, and hú mycel úre ungeþyld is, Blickl. Homl. 33, 26. Ungeðyld *impatientiae culpa*, Past. 43; Swt. 309, 2. Sió ungeðyld, Swt. 311, 21: 33; Swt. 220, 66. For ðæm unwrence ðære ungeðylde . . . for ðæm undeáwe ðære ungeðylde *per vitium impatientiae*, 33; Swt. 214, 20, 23: Swt. 224, 2. Mid ungeðylde (-geðylde, Hatt. MS.), 43; Swt. 310, 15. Gif hé wyrþ on ungeþylde *cum dederit, impatientiae manus*, Bt. 11, 1; Fox 32, 33. Ðætte ðæt mód ne berǽse on ungeðyld *ne ad impatientiam spiritus erumpant*, Past. 43; Swt. 313, 21. Þurh ungeþyld *per intolerantiam*, Scint. 150, 1. [O. H. Ger. un-gidult, -gidultî; f. impatientia.]

un-geþyldig; *adj. Impatient :*—Ðæt wæs ungeþyldig heretoga . . . hé wǽpn gegráp mid tô campienne, ǽr ðon ðe hé tô his líchoman leomum becóme, Blickl. Homl. 165, 33. Se ðe ðysne lǽcedóm þolaþ, hé sceal upweard licgean, ðý læs hé ungeþyldig (*if he is impatient*) ða strengðe ðyssæ lǽcnunge ongite, Lchdm. i. 300, 21. Se ðe biþ ungeþyldig, and ceoraþ ongeán God on his untrumnysse, Homl. Th. i. 472, 8. Se dysega ungeðyldega all his ingeðonc hé geypt, Past. 33; Swt. 220, 9. Swá ungeþyldige ðæt hí ne magon nán earfoþu geþyldelíce áberan, Bt. 39, 10; Fox 228, 2 note. On óðre wísan siut tô manianne ða ungeðyldegan (-geðyldgan, Hatt. MS.) and on óðre ða geðyldegan . . . Ðæm ungeðyldegum (*impatientibus*) is tô sæcganne, Past. 33; Swt. 214, 3–6. Se ðe ungerádum oððe ungeðyldigum stýrð, Homl. Th. i. 306, 5. Æghwelc monn bið onfunden swǽ micle læs geláered ðonne óðer swǽ hé bið ungeðyldegra *tanto quisque minus ostenditur doctus, quanto minus convincitur patiens*, Past. 33; Swt. 216, 3.

un-geþyldiglíce, -geþyldelíce; *adv. Impatiently :*—Ðá scylde hé ongeán swíþe geþyldelíce, Fox 68, 1.

un-geþyre; *adj. Dissentient* (?) :—Ungeðyre *discensor*, Txts. 57, 684. Cf. (?) ge-þuren, *pp.* of geþweran, *and* un-geþwǽre.

un-getímu; *f. or* **un-getíme, es;** *n. Mishap, misfortune :*—On ðæm dagum wǽron ða mǽstan ungetína (cf. (?) un-geteón: ungetíma *is the reading of the other MS. here and in the following passages*) on Rómánum, ǽgðer ge on hungre ge on moncwealme *duo vel maxima omnium malorum abominamenta, fames et pestilentia, fessum urbem corripuere*, Ors. 2, 4; Swt. 70, 7. Ðæt hellefýr wæs geswiðrad, swá ealle ungetína (-getíma, MS. C.) wǽron *Sicilia requiem malorum, nisi nunc, nescit*, 2, 6; Swt.

90, 2. Hwelce ungetíma (-getíma, MS. C.), ǽgðer ge on monslihtum ge on hungre ge on scipgebroce ge on mislícre forscapunge, 1, 11; Swt. 50, 18.

un-getogen; adj. Uneducated:—Fisceras and·ungetogene menn geceás Drihten him tô leorningcnihtum, and hí swá geteáh, ðæt heóra lár oferstáh ealne woruldwísdóm, Homl. Th. i. 576, 28. Hé geceás siððan woruldlíce úðwitan, ac hí môdegodon, gif hé ǽr ne gecure ða ungetogenan fisceras, 578, 14.

un-getreów, -getreówe, -getrîwe, -getrýwe; adj. Untrue, unfaithful, faithless:—Wǽrleás mon and ungetreów, Exon. Th. 343, 27; Gn. Ex. 163. Ungetreówe infidus, Wrt. Voc. i. 49, 31: ii. 91, 58: 47, 25: infidelis, i. 74, 28. Se ðe sý folce ungetrýwe (-getrîwe, MS. G.), L. Edg. ii. 7; Th. i. 268, 14: L. Eth. i. 4; Th. i. 282, 29, 30. Gyf hwylc man sý swá ungetrýwe ðam hundrede, L. C. S. 30; Th. i. 392, 21. Be ungetreówum mannum. Gyf hwylc man sý ðe eallum folce ungetrýwe sý, 33; Th. i. 396, 13. Ungetrýwan (-getreówan, MS. B.) men ceóse man ánfealdne áð, 22; Th. i. 388, 11. Hit is ungeleáful cynren and ungetreówe bearn generatio perversa est et infideles filii, Deut. 32, 20. Mé áblendan ðás ungetreówan woruldsǽlþa, Bt. 2; Fox 4, 9. Hé sett his dǽl mid ðám ungetreówum (infidelibus), Lk. Skt. 12, 46. [O. H. Ger. un-gitriuwi infidelis.]

un-getreówness, e; f. Unfaithfulness, infidelity:—Se ðe forlǽt ðone cele ungetreównesse quisquis amisso infidelitatis frigore vivit, Past. 58; Swt. 447, 6.

un-getreówþ, e; f. Bad faith, breach of good faith:—Hér sýn on lande ungetrýwða (-treówða, MS. B.: -trîwða, MS. C.) micle for Gode and for worulde, Wulfst. 160, 6.

uu-getrum; adj. Infirm:—Manige men bióþ ungetrume (untrume, Cott. MS.) ǽgþer ge on môde ge on líchoman, Bt. 39, 10; Fox 226, 36.

un-getýd; adj. Untaught, unskilled, rude:—Ðæt ungetýde folc (rudis ille populus) nolde geliéfan, Past. 50; Swt. 389, 33. Æt ðæm ungetýdum folce apud imperitum vulgus, 48; Swt. 365, 22. Ðæt hié ne scolden forhyggean ðone geférscipe ðara synfulra and ðara ungetýdra, 16; Swt. 105, 15. v. ge-týn (-týan).

un-getýdd; adj. Untaught, unskilled:—Seó bóc wæs yfele gehwyrfed and gyt wyrs fram sumum ungetýddum (a quodam imperito) gerihted, Bd. 5, 24; S. 648, 24. v. ge-týdan.

un-getynge; adj. Unapt of speech, not eloquent:—Ic eom ungetinge on sprǽce incircumcisus sum labiis, Ex. 6, 12. v. next word.

un-getyngfull; adj. Ineloquent:—Ungetingfullum infantissimo, ineloquentissimo, Germ. 392, 3.

un-gewǽpnod; adj. Unarmed:—Ungewǽpnad inermis, Wrt. Voc. ii. 48, 13. Ðá hét se cyning healdan Martinum, ðæt hé wurde áworpen ungewǽpnod ðam here, Homl. Th. ii. 502, 14. Ðá geseah Æþelfrið heora sacerdas sundor stondon ungewǽpnade, Bd. 2, 2; S. 503, 39.

un-gewær (?); adj. Incautious, inconsiderate (v. un-geþwǽre, III). v. un-wær.

un-geweald impotence, inability to control. The word occurs only in the genitive, with the force of an adverb. I. where an action is done without the actor's intending it, unintentionally, not wilfully, involuntarily:—Hé wrǽc his ungewealdes on ðære byrig hiora misdǽda, Ors. 6, 5; Swt. 262, 2. Gif hwá his cild ofslihð tô deáðe ungewealdes (praeter voluntatem), L. Ecg. P. ii. 1; Th. ii. 182, 21: L. M. I. P. 8; Th. ii. 268, 1: L. Alf. pol. 13; Th. i. 70, 9. Ungewealdes, Past. 21; Swt. 167, 1. Suelce hé hit ungewisses oððe ungewealdes doo agit velut nesciens, 33; Swt. 215, 11. Ðonne hé of yfelum willan ne gesyngaþ, ac of unwísdóme and ungewealdes oððe ungewealdes cum non malitia, sed sola ignorantia, delinquitur, 21; Swt. 157, 25. Se ðe hine nédes ofslóge oþþe unwillum oþþe ungewealdes, L. Alf. 13; Th. i. 46, 23. Gif man unwilles oþþe ungewealdes ǽnig þing misdéð, L. Eth. vi. 52; Th. i. 328, 21. Eówre synna ðe gé geworhton gewealdes oþþe ungewealdes, Wulfst. 135, 30. II. where something happens that is not controlled or brought about by a person:—Úre gást biþ swíþe wíde farende úrum unwillum and úres ungewealdes . . . ðæt biþ þonne wé slápaþ our spirit wanders far independently of our wishes or control . . . that is when we sleep, Bt. 34, 11; Fox 152, 4. Is dǽm tô cýðanne ðæt hí hié hié warenigen ǽgðer ge wið ða ungemetlícan blisse ge wið ða ungemetlícan unrótnesse, for ðæm hira ǽgðer ástyreþ sumne unðeáw, ðeáh hié ungewealdes cuman of ðæs líchoman medtrymnesse, Past. 27; Swt. 189, 3. Gif him gewealdes gebyrige oððe ungewealdes ðæt hé on ðæs hwæt befoo ðe wið his willan sié, 28; Swt. 199, 22. [þurh uniweald per impotenciam, O. E. Homl. ii. 63, 6. O. Frs. un-ewald.]

un-gewealden; adj. Not under control, disordered (?):—Ðonne for miclum cele wamb sié ungewealden (cf. Lǽcedómas tô wambe gemetlícunge, 164, 3), Lchdm. ii. 228, 23. Wamb ungewealden and unýþe . . . tunge ungewealden and unsmeþe, 242, 5–9. Ðonne se man mete þigð, ðonne áwyrpð hé eft and hæfð ungewealdene wambe and ða micgean, 204, 10.

un-geweaxen; adj. Ungrown, not grown up:—Ungeweaxenra deáþ acerva mors (i.e. the death of the young), Wrt. Voc. ii. 8, 31.

un-geweder, es; n. Bad weather, storm, tempest:—Se stranga winter

mid forste and mid snáwe and mid eallon ungewederon, Chr. 1046; Erl. 170, 33. Hefigtýme geár on ungewederan, ðá man oððe tilian sceoldon oððe eft tilða gegaderian, 1097; Erl. 234, 24. v. un-weder.

un-gewemmed; adj. I. physical, unspotted, immaculate, uncorrupted, uninjured:—Se líchoma wæs geméted ungebrosnod and ungewemmed corpus incorruptum inventum est, Bd. 4, 19; S. 588, 38. Hé ungewemmed of ðam hátum bæðe eode, Homl. Th. i. 58, 28. Seó hálge stód ungewemde wlite, Exon. Th. 277, 33; Jul. 590. Eall ða hrægel ungewemmed (intemerata) wǽron, Bd. 4, 30; S. 608, 40. v. un-forswǽled. II. moral, undefiled, unstained, inviolate, immaculate:—Ungewǽmmed ic beó immaculatus ero, Ps. Spl. 18, 14. Ungewǽmmed inviolata, Hymn. Surt. 54, 25: incorruptibilis, Jn. Skt. p. 1, 6. Uniwemmedes inlaesae (virginitatis), Hpt. Gl. 435, 53: inlibatae, 511, 47. Uniwemmedre immunem, 507, 49. On ungewemmedum mægðháde, Homl. Th. i. 58, 8. Hé his wíf him betǽhte ungewemmed, Gen. 20, 14. III. uncertain:—Ungeuuemmid infractus, Wrt. Voc. ii. 111, 36. [O. H. Ger. un-giwemmit immaculatus, inlibatus.] v. un-wemmed.

un-gewemmedlíc; adj. Incorruptible:—Uniwemmedlícre clǽnnysse immarcescibilis (imputribilis) pudicitiae, Hpt. Gl. 467, 47.

un-gewemmedlíce; adv. Uncorruptly, purely, inviolably:—Ðæt ða dómas and ða gesetnysse, ða ðe fram hálgum fæderum árǽdde and gesette wǽron, ðæt ða fram eallum ús ungewemmedlíce (incorrupte) healdene wǽron, Bd. 4, 5; S. 572, 19.

un-gewemmedness, e; f. Purity:—Rihtlíc is mé swá besmitenre fram ðínre clǽnan ungewemmednesse beón áscirod, Homl. Skt. ii. 23 b, 438.

un-gewemness, e; f. Freedom from pollution:—Uniwemnysse immunitatis, Hpt. Gl. 434, 28.

un-gewendendlíc; adj. Unchangeable, invariable, chronic (of disease):—þeáh man sý on hwylcre ungewendendlícre (-dedlícre, MS. H.) ádle, Lchdm. i. 328, 20.

un-gewéned; adj. Unexpected:—Se here wæs cumende ungewénedre tíde on herfeste legio inopinata tempore autumni adveniens, Bd. 1, 12; S. 480, 41. Of ungewénedum ex improviso, Lchdm. iii. 200, 23. Of ungewéndum, 204, 19. [O. H. Ger. un-giwánit inopinatus.] v. un-wéned.

un-gewérigod; adj. Unwearied:—Ungewérigadre geornfullnysse indefessa instantia, Bd. 4, 3; S. 568, 14.

un-gewid(e)re, es; n. Bad weather, storm, tempest:—Hit biþ wiþ ǽghwylc ungewidro gescylded, ðæt ðǽr nǽfre nænig dǽl regnes ne ungewidres in cuman ne mæg, Blickl. Homl. 125, 31–3. Gif ðé þince ðæt ðú óþerne máran lǽcedóm dón ne durre for ungewiderum, Lchdm. ii. 254, 2. Is ðeós woruld on stormum and on ádlum and on ungewyderum, Wulfst. 273, 9. Mycel orfes wæs ðæs geáres forfaren ǽgðer ge þurh mistlíce coða ge þurh ungewyderu, 1041; Erl. 169, 9: Lchdm. iii. 210, 26. [Bið his erd ihened eiðer ȝe on herȝunge ȝe on hungre ȝe on cwalme ȝe on uniwidere, O. E. Homl. i. 115, 36. O. Sax. un-giwideri (wið ungiwidereon allun standan, Hél. 1813): O. H. Ger. un-giwitiri tempestas, procella, hiems, ventus.] v. un-widere.

un-gewiderung, e; f. Bad weather:—Syððan com, þurh ða mycclan ungewiderunge ðe cômon, swýðe mycel hungor ofer Engeland, ðæt manig hundred manna earmlíce deáðe swulton þurh ðone hungor, Chr. 1086; Erl. 219, 23.

un-gewild; adj. Unsubdued, unsubjected:—Ungewyld indomita, ineffrenata, Hpt. Gl. 461, 49. Ungewyldre ǽwnunge effrenatae jugalitatis, 434, 25. Ungewylde indomitos, 457, 76. v. ge-wildan (-wyldan).

un-gewilde; adj. Not in subjection:—Heora gecynd wæs him ungewylde their nature was not in subjection to them, Hexam. 17; Norm. 26, 2. Gif ǽnig leódscype wæs ungewylde ðam Cásere, ðonne send hé him tô swá fela eóroða ðe mihton gebígan ðæt mennisc him tô, Jud. Thw. p. 161, 35. Hé underþiédde him selfum monege þeóda ðe ǽr wǽron Rômánum ungewilde, Ors. 6, 30; Swt. 284, 6. v. ge-wilde (-wylde).

un-gewildelíc; adj. Not to be subdued, unyielding:—Hæbbe se mann heardheortnysse and ungewyldelíc mód . . . ðonne forsearaþ swíðe hraðe ðæt hálige sǽd on his heortan, Homl. Th. ii. 92, 2.

un-gewill[1]; adj. Displeasing, not with the good will of a person:—Se arcebiscop leáfe æt ðam cynge nam, ðeáh hit ðam cynge ungewill wǽre, ðæs ðe men léton, Chr. 1097; Erl. 234, 16. [þeáh hit ðam arcebiscope swyðe ungewille wǽre, 1120; Erl. 248, 21. Cf. Halde we us from unwil, O. E. Homl. i. 69, 264.]

un-gewintred; adj. Not adult:—Be ungewintredes wífmannes nédhǽmde. Gif mon ungewintrædne wífmon tô niédhǽmde geþreátige, sié ðæt swá ðæs gewintredan monnes bót, L. Alf. pol. 26; Th. i. 78, 16.

un-gewirded; adj. Uninjured:—Ne mæg him bitres wiht gedëran, ac gescylded á wunaþ ungewyrded þenden woruld stondeþ, Exon. Th. 210, 5; Ph. 181.

un-gewís; adj. Uncertain, unknown:—For ðam ðe him cúþ forþfór tôweard wǽre and ungewiis (ungewiss? q. v.) seó tíd ðære ylcan forðfôre eo quod certus sibi exitus, sed incerta ejusdem exitus esset hora futura, Bd. 3, 19; S. 547, 16.

un-gewislíc; *adj. Uncertain, unknown, uncommon* :—Wênst ðú ðæt hit hwæt níwes sié oððe hwæthwugu ungewislíces ðæt ðe on becumen is *novum credo aliquid inusitatumque vidisti*, Bt. 7, 2 ; Fox 16, 27 note.

un-gewisness, e ; *f. Uncertainty, ignorance* :—Ða ðe ðurh ungewisnysse (*per ignorantiam*) synne fremmaþ . . . ða ðe him ne ondrædaþ witende (*sciendo*) syngian, Bd. 1, 27 ; S. 491, 36. Swá hwæt swá on hyre unclænnysse ðurh ungewisnesse (*per ignorantiam*) gelumpe, 4, 9 ; S. 576, 28.

un-gewiss, es ; . *n.*   I. *uncertainty, ignorance, unconsciousness* :— Se ðe his sylfes blôd on spátl mid ungewisse forswelge *qui sanguinem proprium inscius cum saliva sorbuerit*, L. Ecg. C. 40 ; Th. ii. 166, 5. Gif eall folc syngaþ þurh ungewiss (*through ignorance* (A. V.) ; *per imperitiam*), Lev. 4, 13.   Swá hwæt swá wé þurh ungewiss oððe þurh hwylce dysignesse gedôn habban, Homl. Ass. 143, 136.   ¶ *Ungewisses in ignorance, unintentionally, unconsciously, unwittingly* :—Ða scylda ðe ic ungewisses geworhte ne gemun ðú ; ðæt synt ða ðe ic wênde ðæt nán scyld nære *delicta ignorantiae meae ne memineris*, Ps. Th. 24, 6. Swelce hé hit ungewisses oððe ungewealdes doo *agit velut nesciens*, Past. 33 ; Swt. 215, 10. Ðonne hé of unwísdôme and ungewisses oððe ungewe4ldes gesyngaþ *cum sola ignorantia delinquitur*, 21 ; Swt. 157, 25.   II. *what is uncertain* or *unknown* ; *incertum* :—Ða gebende án scytta his bogan and áscêt ána flán swylce on ungewis (*cf. O. H. Ger.* in ung(i)uis *a casu*) *vir quidam tetendit arcum, in incertum sagittam dirigens* (1 Kings 22, 34), Homl. Skt. i. 18, 220. Ðæt hié on swá micle nêþinge and on swá micel ungewiss ægðer ge on sæs fyrhto ge on wêstennum wildeóra ge on þeóda gereordum, ðæt hié hiene æfter friþe sôhton *that they ventured upon so much that was hazardous and so much that was unknown in respect to both the terrors of the sea and the deserts with their wild beasts and the languages of nations to seek him and get peace*, Ors. 3, 9 ; Swt. 136, 24.   Ongewiss and díglu wísdômes ðínes ðú swutelodest mê *incerta et occulta sapientiae tuae manifestasti mihi*, Ps. Spl. 50, 7.   v. next word, II a.   III. the word also glosses *ignominia*. v. next word, III :—Gefyl ansýne heora of ungewisse *imple facies eorum ignominia*, Ps Spl. 82, 15 : Blickl. Gl.

un-gewiss; *adj. Uncertain* :— Ungewis *incerta*, Wrt. Voc. ii. 48, 17.   I. of persons, *not having knowledge, ignorant* :—Gif hê hit nát, hwelce gesælþa hæfþ hé æt ðam welan, gif hé biþ swá dysig and swá ungewiss ðæt hê ðæt witan ne mæg *si nescit, quaenam beata sors esse potest ignorantiae caecitate?* Bt. 11, 2 ; Fox 34, 26.   Ungewiss com se deófol tô Criste, and ungewiss hê eode áweig ; for ðan ðe se Hælend ne geswutelode ná him his mihte, Homl. Th. i. 176, 9–11.   Ðá ðá ic ðær lange stôd ungewis mínes færeldes, ii. 350, 26.   II. of things, (a) *not known, of which there is not certain knowledge* :—On ðínes lífes ryne ðe ðé is ungewiss, Basil admn. 8 ; Norm. 52, 8.   Ungewisse þingc and dýgelnyssa wísdomes ðínes *incerta et occulta sapientiae tuae*, Ps. Lamb. 50, 8.   (b) *not conveying certain knowledge* :—Ðú stunta, on hwilce wísan sceole wê ðé gelýfan and ðínum ungewissum wordum ? Homl. Skt. i. 23, 697.   III. *ignominious.* v. preceding word, III :— Ungewis *ignominiosus*, Kent. Gl. 715.   [*O. H. Ger.* un-giwiss *incertus, inexpertus, fortuitus.*]

un-gewítendlíce; *adv. Without passing away, permanently* :—Gehíéren hí ðæt ðás andweardan gôd bióð from ælcre lustfulnesse swíðe hrædlíce gewítende and swá ðeáh sió scyld ðe hí ðurh ða lustfullnesse ðurhtióð ungewítendlíce bið ðurhwuniende mid wræce *audiant quod bona praesentia et a delectatione citius transitura sunt, et tamen eorum causa ad ultionem sine transitu permansura*, Past. 58 ; Swt. 441, 21.

un-gewítfæstness, e ; *f. Madness* :—Him cymð brægenes ádl and ungewítfæstnes him bið *he will be out of his senses*, Lchdm. ii. 222, 3.

un-gewítfull; *adj.*   I. *foolish, insensate* :—Eálá gé ungewítfullan Galata, hwá gehefgode eów *O insensati Galatae, quis vos fascinavit?* Past. 31 ; Swt. 207, 14.   Welan and weorþscipes hí willniaþ, and ðonne hí hine habbaþ, ðonne wênaþ hí swá ungewítfulle ðæt hí habban ða sôþan gesælþa, Bt. 32, 3 ; Fox 118, 30.   II. *mad, insane, not in one's senses* :—Ða wôdðraga ðæs ungewítfullan monnes se læce gehæld *furor insanorum ad salutem medico reducitur*, Past. 26 ; Swt. 183, 21. Sume ða untruman wæron dumbe, sume ungewítfulle, Homl. Ass. 180, 364.

un-gewítfulness, e ; *f. Madness, insanity* :—Saules ungewítfulnes (-full-, Hatt. MS.), Past. 26 ; Swt. 185, 1.

un-gewítlíc; *adj. Senseless, foolish* :—Ungewítlíce word, Lchdm. ii. 176, 2.

un-gewítnigendlíce; *adv. With impunity*; impune, Ælfc. Gr. 38 ; Zup. 233, 6 note.

un-gewítnod; *adj. Unpunished*, (1) *of the person to whom punishment might be given* :—Ne beó gé on nánre leásre gewitnysse, for ðon ðe se leása gewita ne bið hé næfre ungewítnod, L. E. I. 27 ; Th. ii. 424, 1 : Homl. Ass. 148, 10.   Ne wên ðú ná ðe ðæt ðú ungewítnod beó, Homl. Skt. ii. 25, 159.   Wé beóð mid Gode swá micle suíðor gebunde suá wé for monnum orsorglícor ungewítnode syngiaþ bútan ælcre wrace (*quanto apud homines inulte peccamus*), Past. 17 ; Swt. 117, 23.   (2) *of the fault for which punishment might be given* :—Ðæt him biþ unge-

wítnode hiora yfel on ðisse worulde, Bt. 38, 3 ; Fox 200, 28. v. un-wítnod.

un-gewit[t], es ; *n.*   I. *madness, insanity* :—Nán unhál cild, ne deáf, ne blind, ne ungewittes, Homl. Ass. 179, 322.   Hié sindon suá micle wærlícor tô oferbúganne suá mon ongiet ðæt hié on máran ungewitte beóð *tanto caute declinandi sunt, quanto et insane rapiuntur*, Past. 40 ; Swt. 295, 22.   Gif hwylc man of his gewitte feólle . . . Gif man hine ofsleá on ðam ungewitte, L. Ecg. P. addit. 29 ; Th. ii. 236, 31. Se yfela gást on ungewitte his (*Saul's*) môd áwende, Homl. Skt. i. 18, 11.   Sende ðé Drihten on ungewitt and blindnysse *percutiat te Dominus amentia et coecitate*, Deut. 28, 28.   II. *folly, stupidity* :—Ðý læs ðe ænig ungecyrred woroldman mid nytnesse and ungewitte regules geboda ábrêce, and ðære tale brúce, ðæt hê misfênge ðý hé hit sêlre nyste, Lchdm. iii. 442, 2.   Hí mid heora gedwolspræce eall folc ámyrdon. And Theodosius, ðá hé swilce ungewitt ælce dæge gehýrde, hê weard sárig, Homl. Skt. i. 23, 370.   [*O. H. Ger.* un-giwizzi *insipientia.*]

un-gewittig; *adj.*   I. *mad, insane* :—Ic wát ðæt gê wênaþ ðæt ic ungewittige môde (*insana mente*) sprece, Bd. 4, 8 ; S. 576, 1.   II. *foolish, senseless* :—Gif cinges gerêfena hwylc gyltig biþ, hwá is manna tô ðam ungesceád and ungewittig, ðæt hê ðæm cyninge his áre ætrecce for ðí ðe his gerêfa forwyrht biþ ? Lchdm: iii. 444, 8.   III. *not having reason, irrational* :—Hí (*the innocents*) wæron gehwæde and ungewittige ácwealde, Homl. Th. i. 84, 21.   Beóð ða ungewittigan cild gehealdene on ðam fulluhte þurh geleáfan ðæs fæder and ðære mêder, ii. 50, 35.   Ða yfelan men ne magon cuman þider, ðider ða ungewittigan gesceafta wilniaþ tô tô cumenne, Bt. 36, 5 ; Fox 180, 3.   [*O. Sax.* un-giwittig *unwise.*]   v. un-wittig.

un-gewittiglíce (-witte-) ; *adv. Unwisely, foolishly*; stolide, Gr. Dial. 2, 3.

un-gewittigness, e ; *f. Foolishness*; stultitia, Gr. Dial. 2, 31.

un-gewlitig; *adj. Not bright, not brilliant* :—Ealle ða ðing ðe beorhte beóð ðonne seó sunne hym on scínaþ, hí lýhtaþ ongeán ; ac ða ðe ungewliige (-wlitige?) beóð, ða ne lýhtaþ náwiht ongeán ða sunnan, þeáh heó hym on scíne, Shrn. 180, 15.   v. un-wlitig.

un-gewlitigian *to disfigure, deform, deprive of beauty* :—Hê gewliitegaþ and gegeraþ ælle gesceafta and æft ungewliteaþ, and ungeraþ, Shrn. 198, 12.   v. un-wlitigian.

un-geworht; *adj.*   I. *not made* :—Gif Hê geworht wære, ne wurde Hê næfre ælmihtig God . . . Hê wæs æfre ungeworht, Homl. Skt. i. 1, 69.   II. *not finished* :—Ungeworht *infectum*, Wrt. Voc. ii. 45, 17.

un-gewriten; *adj. Unwritten* :—Ungewriten yrfe *intestata hereditas*, Wrt. Voc. i. 20, 41 : ii. 49, 20.   v. un-writen.

un-gewuna, an ; *m. A bad custom, evil practice* :—Se eorðlíca gefêrscipe hiene tiéhð on ða lufe his ealdan ungewunan *ad vetustatem vitae per societatem secularium ducitur*, Past. 22 ; Swt. 169, 9.   Ða ðe ðone ungewunan hæfdon, ðæt hí heora wíf glengdan swá hí weofoda sceoldan, geswícan ðæs ungewunan, L. I. P. 23 ; Th. ii. 336, 20.   Se bið siwenîge ðonne his môd and his andgit ðæt gecynd áscirpð and hê hit ðonne self gescint mid his ungewunan and wôm wilnungum *lippus est, cujus sensum natura exacuit, sed conversationis pravitas confundit*, Past. 11 ; Swt. 69, 9.   Æt ðam unþeáwe ðe dysige men on ungewunan healdaþ (*which foolish men observe as a custom, and a bad one it is*), Wulfst. 305, 9.

un-gewuna; *adj. Unaccustomed, unused* :—Micel gedál is on ðam mægene ðæs ðe sié gewin (-wun?) þrowungum and ðæs ðe sié ungewuna swelcum þingum, Lchdm. ii. 84, 19.   [Cf. *O. H. Ger.* un-giwon *inusitatus, insolens, novus.*]

un-gewunelíc; *adj.*   I. *unusual, unwonted* :—Ðæs wundredon men, for ðí ðæt hit wæs ungewunelíc, Homl. Th. i. 184, 30. Fela fægera þinga ðe ðam folce ungecnáwe[n] wæs and ungewunelíc, Ap. Th. 17, 14. Hwílum gebyrede swíþe ungewunelíc and ungecyndelíc yfel, ðæt ða bearn sieredon ymbe ðone fæder, Bt. 31, 1 ; Fox 112, 12.   Hwæthwegu ungewunelíces *novum aliquid inusitatumque*, 7, 2 ; Fox 16, 27.   Ðá geseah hé ealle ða cytan mid heofonlíce leóhte gefylde.   Hê ðá wæs forhtlíce geworden for ðære ungewunelícan gesihþe, Guthl. 21 ; Gdwin. 94, 23.   II. *unfrequented, uninhabitable* :—Seó stôw wæs swá wêsten and swá dígle, ðæt næs ná ðæt án ðæt heó wæs ungewunelíc, ac eác swilce uncúð ðam landleódum him sylfum, Homl. Skt. ii. 23 b, 106. Án ðæra ðæla is ungewunelíc for ðære sunnan neáweste ; on ðam ne eardaþ nán eorðlíc mann for ðam unberendlícum bryne, Lchdm. iii. 260, 21 note.   v. un-gewynelíc.

un-gewunelíce; *adv. Unusually, in an unwonted manner* :—Cometae synd gehátene ða steorran ðe færlíce and ungewunelíce æteówiaþ, Lchdm. iii. 272, 4.   Ðære sæ gemengednyssa and ðæra ýða swêg ungewunelíce gyt ne ásprungon (*it has not been unusual for them to occur already*), Homl. Th. i. 610, 12.   [*O. H. Ger.* un-giwonalíhho *insolite*.]

un-gewuni[g]endlíc; *adj. Uninhabitable* :—Án ðæra ðæla is ungewuniendlíc (-wunig-, MS. P. : unwuniendlíc, MSS. L. R.) for ðære sunnan neáweste ; on ðam ne eardaþ nán eorðlíc mann, Lchdm. iii. 260, 21. v. un-wuni[g]endlíc.

un-gewyld, -gewylde. v. un-gewild, -gewilde.

un-gewynelíc; adj. Unusual, unwonted :—Mid ege ungewynelícum timore insolito, Anglia xiii. 411, 651. v. un-gewunelíc.

un-gewyrht in the phrases be ungewyrhtum undeservedly, not according to one's deserts; gratis, Ps. Surt. 34, 7, 19: 68, 5: Ps. Spl. C. 108, 2: 118, 161.

un-gifeðe; adj. Not granted :—Ús wæs á syððan milts ungyfeðe, Beo. Th. 5835; B. 2921.

un-gifre; adj. Harmful, unfortunate :—Ǽr gð sceonde fremmen, ungifre yfel ylda bearnum, Cd. Th. 149, 5; Gen. 2470. v. gifre.

un-gifu, e; f. An evil gift :—Hǽfð se yfela gást hērongeán seofonfealde ungifa . . . ða yfelan ungifa ðæs árleásan deófles syndan ðus genamode insipientia, stultitia . . ., Wulfst. 52, 7-20: 58, 14.

un-gild, -gilde, es; n. An improper or excessive tax :—Hē ǽfre ðás leóde mid here and mid ungylde tyrwigende wæs, for ðan ðe on his dagan ǽlc riht áfeóll, and ǽlc unriht up árás, Chr. 1100; Erl. 236, 2. Ðis wæs swiðe geswincfull geár þurh manigfeald ungyld, 1098; Erl. 235, 11. Ús ungylda swýðe gedrehton, Wulfst. 159, 12. Ðis wæs swiðe hefigtýme geár . . . on ungyldan ða nǽfre ne áblunnon, Chr. 1097; Erl. 234, 25.

un-gilda, an; m. One who is not a full member of a guild :—Æt ǽlcon rihtgegyldan áne byrðene wudes and twá æt ðam ungyldan (cf. (?) for a difference between those in the same gild, Hæbbe ǽlc gegilda .ii. sesteras mealtes, and ǽlc cniht ánne, 613, 32), Chart. Th. 606, 16.

un-gilde; adj. Not entitled to wergild :—Gif se friðman fleó oþþon feohte, and nelle hine cýþan, gif hine man ofsleá, licge ungylde, L. Eth. ii. 3; Th. i. 286, 14. Homo qui aliquem innocentem affliget in via regia, si jaceat, jaceat in ungildan ǽkere [as the technical name of the crime here referred to was forsteal (cf. si in via regia fiat assaltus super aliquem, forestel est, L. H. I. 80, § 2; Th. i. 586, 2), the passage seems to be a Latin equivalent for the following : Gif hwá forsteal gewyrce, . . . gif hē sylf gewyrce ðæt hine man áfylle, licge ǽgilde, L. Eth. vi. 38; Th. i. 324, 21-24; so that the phrase licgan in ungildan œcere seems to be equivalent to licgan ungilde], L. Eth. iv. 4; Th. i. 301, 23. [Icel. ú-gildr for whom no wer-gild is to be paid.] v. ǽ-, or-gilde.

un-gímen[n], e; f. Carelessness :—Þurh ungémænne synne (ðurh gýmeleáste, col. 1) per culpam incuriae, Bd. 3, 17; S. 544, 24, col. 2: 2, 7; S. 509, 19. Ungýmenne, 4, 25; S. 599, 20. Ðurh ungýmenne per incuriam, 4, 9; S. 576, 28.

un-gímende; adj. (ptcpl.) Careless, negligent :—Ða ðing ðe se Dryhtnes wer geseah nales eallum monnum suongrum and heora liifes ungémendum [ungemyndum for (?) ungýmendum, Bd. S. 630, 38] sæcgan wolde haec quae viderat idem vir Domini non omnibus desidiosis ac vitae suae incuriosis referre volebat, Bd. 5, 12; M. 434, 5. v. preceding word.

un-gin[n]; adj. Not ample, contracted :—Á sceal ðæs heánan hyge hord onginnost, Exon. Th. 346, 18; Gn. Ex. 206.

un-girwan, -girian; p. -girwde, -girede To strip, divest : — Hē gewlitegaþ and gegeraþ ælle gesceafta and æft ungewliteaþ and ungeraþ, Shrn. 198, 13. Gúðlác hine sylfne ungyrede, Guthl. 16; Gdwin. 88, 16. Hē hine ungyrede ðæs godcundan mægenþrymmes, Blickl. Homl. 103, 2. Hiǽ ungeredun (exuerunt) hine and gegearwadun (induerunt) hine his ágene wēde, Mt. Kmbl. Rush. 27, 31. Ðeodoricus wæs ungyred and unscód, Shrn. 85, 32. v. on-girwan.

un-glæd; adj. Dull, cheerless :—Swá eác se súþerna wind hwílum miclum storme gedrēfeþ ða sǽ ðe ǽr wæs smylte wedere glæshlútra on tó seónne; ðonne heó swá gemenged wyrð mid ðan ýdum, ðonne wyrþ heó swíþe hraðe ungladu, þeáh heó ǽr gladu wǽre on tó lócienne si mare volvens turbidus Auster misceat aestum, vitrea dudum, parque serenis unda diebus, mox, resoluto sordida coeno visibus obstat, Bt. 6; Fox 14, 26. [Goddes glam to hym (Jonah) glod, þat hym unglad made, Allit. Pms. 94, 63. Icel. ú-glaðr.]

un-glædlíc; adj. Stern, implacable :—Stíð, grimm, unglædlíc inmitis, atrox, implacabilis, Germ. 392, 33. v. glæd, III.

un-glædnes glosses imperitia, Wrt. Voc. ii. 46, 25, but un-glǽwnes perhaps should be read. v. un-gleáwness.

un-gleáw; adj. I. of persons, without understanding, without skill, not sagacious, ignorant, blind (fig.) :—Ungleáw imperitus, Wrt. Voc. i. 55, 49. Ungleu caecus, Mt. Kmbl. Lind. 15, 14. Synt gē þus ungleáwe (inprudentes)? Ne ongyte gē ðæt . . . ? are ye so without understanding? Do ye not perceive that . . . ? Mk. Skt. 7, 18. Ungleáwe inertes, Wrt. Voc. ii. 45, 32. Hit ne biþ seó ylce ádl, þeáh ðe ungleáwe lǽcas wēnan ðæt ðæt seó ylce healfdeáde ádl sí, Lchdm. ii. 284, 24. I a. where that in which there is want of skill is expressed :—Ðá wæs ic ungleáw ðæs geþeódes ðara Indiscra worda . . . ðá rehte hit mē se bisceop, Nar. 29, 14. Wē ðæs londes ungleáwe and unwíse (imprudentes; but the Latin is not literally translated) wǽron, 10, 6. II. of things :—dull, not apt for service :—Sweord gebrǽd gód gúðcyning, gomele láfe, ecgum ungl[e]áw (dull of edge; cf. sió ecg gewác, bát unswíðor ðonne his ðiódcyning þearfe hæfde, 5148-; B. 2577-), Beo. Th. 5121; B. 2564. [Icel. ú-glöggr not clever.]

un-gleáwlíce; adv. Without understanding, without sagacity, un-

wisely, imprudently :—On his heortan cwæð unhýdig sum ungleáwlíce ðæt God nǽre dixit insipiens in corde suo, Non est Deus, Ps. Th. 52, 1. Gif hē ðære styringe ne wiðstent, ðonne gescient hē ða gódan weorc ðe hē oft ǽr on stillum móde ðurhteáh, and suá ungleáulíce for ðæm scyfe ðære styringe suíðe hrædlíce tówierpð ða gódan weorc ðe hē longe ǽr foretonclíce timbrede qui, dum perturbationi suae minime obsistunt, etiam si qua a se tranquilla mente fuerant bene gesta, confundunt, et improviso impulsu destruunt, quidquid forsitan diu labore provido construxerunt, Past. 33; Swt. 215, 17. [O. H. Ger. un-glaulíhho insolenter.]

un-gleáwness, e; f. Want of understanding, unskilfulness, foolishness, blindness (fig.) :—Unglǽdnes (-glǽwnes?) imperitia, Wrt. Voc. ii. 46, 25. Sió ungleáwnes biþ on ðē selfum, ðæt ðú hit ne canst on riht gecnáwan, Bt. 39, 10; Fox 226, 33. Ongleáwnis imperitia, Scint. 5, 5. [Un]gleáwnysse rusticitatis, Hpt. Gl. 529, 16. Hē nǽfre for his unglaunesse (ungleáwnesse, MS. T.) and for his unscearpnesse ða ðēnunge on riht geleornian mihte nullatenus propter ingenii tarditatem potuit ministerium discere, Bd. 5, 6; S. 620, 7. Of ungleáunesse imperitia (os stultorum pascitur imperitia, Prov. 15, 14), Kent. Gl. 520. Ungleównise heartæs caecitatem cordis, Mk. Skt. Lind. Rush. 3, 5.

un-gleáwscipe, es; m. Want of understanding, foolishness; imperitia, Scint. 83, 16.

un-glenged; adj. Unadorned :—Unglenied inculta, non ornata, Hpt. Gl. 435, 25.

un-gníðe; adj. Not scanty, liberal, abundant :—Monigfealde sind geond middangeard gód ungníðe (-gnyde, MS.) ðe ús dǽleþ tó feorhnere Fæder ælmihtig manifold and abundant are the goods which for our life's support the Father almighty gives on earth, Exon. Th. 359, 31; Pa. 70. v. gneáð.

un-gód; adj. Not good, evil, bad :—Seldan hē bið eald; ungódan deáðe hē swylt, Lchdm. iii. 184, 23. On ylde ungódum deáðe heó swylt, 188, 28. [Dede unngod and unnclene, Orm. 16739. O. H. Ger. un-guot : Icel. ú-góðr.]

un-gód, es; n. Evil, ill :—Wá eów ðe taliaþ ungód tó góde and gód þing tó yfele vae qui dicitis malum bonum, et bonum malum, Wulfst. 47, 6. Heó firenaþ mec wordum, ungód gæleþ, Exon. Th. 402, 25; Rä. 21, 35. [Nis þing so god þat ne mai do sum ungod, O. and N. 1364.]

un-grápigende; adj. Not handling, that does not handle :—Hí habbaþ ungrápigende handa manus habent, et non palpabunt (Ps. 115, 7), Homl. Th. i. 366, 27.

un-grēne; adj. Not green :—Folde wæs ðágyt græs ungrēne, Cd. Th. 7, 36; Gen. 117.

un-grið violation of peace, hostility (Lye). [Forr sware unngriþþ þatt heþenn follc þær wrohhte, Orm. 16280.]

un-grund; adj. Bottomless, boundless, immense :—Ðæs heriges hám eft ne com ealles ungrundes ǽnig of that host, all boundless as it was, not one came home again, Cd. Th. 209, 32; Exod. 508. [Cf. Icel. ú-grunnr not shallow; ú-grynni; n. boundlessness, in phrases like ú-grynni hers, liðs, manna.]

un-grynde; adj. Bottomless, deep, Exon. Th. 354, 21; Reim. 49. v. preceding word.

un-gyld, -gylde. v. un-gild, -gilde.

un-gyltig; adj. Innocent :—Hit God wræc on him, swá oft swá hié mid monnum ofredan, ðæt hié mid hiera cucum onguldon, ðæt hié ungyltige cwealdon, Ors. 4, 7; Swt. 184, 9. [Un-gilti innoxius, Wick. Gen. 37, 22. Ongylty immunis, innocens, Prompt. Parv. 365.]

un-gýmen[n]. v. un-gímen[n].

un-gyrdan; p. de To ungird :—Se cyning ungyrde hine ðá his sweorde rex discinxit se gladio suo, Bd. 3, 14; S. 540, 35. Se cásere hēt hine ungyrdan and bewǽpnian, Homl. Skt. ii. 30, 409. Gif him þince ðæt hē sý ungyrd, broc ðæt biþ, Lchdm. iii. 172, 12.

un-gyrian. v. un-girwan.

un-hádian; p. ode To deprive of orders :—Sý hē unhádod ordine suo privetur, L. Ecg. C. 6; Th. ii. 138, 22. Sume wyllaþ ðæt hē sig eft unhádod nonnulli volunt ut denuo ordine caveat, 3; Th. ii. 136, 36. [He him plihte he wolden unhadien Costanz . . . þe abbed unhadede his broðer, Laym. 13169.] v. on-hádian.

un-hádod; adj. Not ordained :—Unhádod man homo non ordinatus, L. Ecg. C. 12; Th. ii. 142, 3. v. un-gehádod.

un-hádung renders exordinatio in :—Unhádunge exordinationes, R. Ben. Interl. 110, 8.

un-hǽl, e; f. v. un-hǽlan. v. un-hǽlu, ge-unhǽlan.

un-hǽlp, e; f. Bad health, sickness, weakness, infirmity :—Líchaman unhǽlð ormǽte mægenu sáwle tóbrycð corporis debilitas nimia uires anime frangit, Scint. 54, 17: 107, 11. Ne beþurfon lǽces ða ðe hále synd, ac ða ðe unhǽlþe habbaþ (qui male habent), Lk. Skt. 5, 31. [Ne elde ne unhelðe, O. E. Homl. ii. 35, 6: Misc. 108, 113.]

un-hǽlu; indecl. : un-hǽl, e; f. I. bad health, disease, sickness, infirmity, (a) of persons :—Se oferdrenc fordēð ðæs mannes gesundfulnysse and unhǽl becymð of ðam drence, Ælfc. T. Grn. 21, 38. Oferfyll bið ðæs líchaman unhǽl, Wulfst. 242, 4. Búton ðē unhǽl oððe yld derige, 247, 34. Ða diófla gelǽrdon hié, ðæt ða ðe on unhǽle wǽran, ðæt hié

hále for hié cwealdon, Ors. 4, 4; Swt. 164, 17. Gif ðú wile hál beón, drinc ðé gedeftlíce; ælc oferfyl fét unhǽlo, Prov. Kmbl. 61. Unhǽlo *languorem*, Mt. Kmbl. Lind. 4, 23: *crucem*, 16, 24. Ðám ðe under hý migaþ . . . ða unhǽle heó gehǽlþ, Lchdm. i. 360, 9. Ðás unhǽle (*blotch*), ii. 76, 16. Untrymmnise t unhǽlo *infirmitates*, Mt. Kmbl. Lind. 8, 17. (b) of animals, *unsoundness* :—Gif mon hwelcne ceáp gebygeþ and hé ðonne onfinde him hwelce unhǽlo on binnan .xxx. nihta, L. In. 56; Th. i. 138, 11. [Licome unhele, O. E. Homl. i. 7, 23. Uu-hæle and ælde, Laym. 11546. Unnhal þurrh unnride unnhæle, Orm. 4779. O. H. Ger. un-hailî *insania*. Cf. *Goth.* un-haili; *n. ill-health.*] II. *misfortune, mishap* :—Sorge ne cúðon, wonsceaft wera, wiht unhǽlo, Beo. Th. 241; B. 120. [Envye that sory is of other mennes wele and glæd is of his sorwe and unhele, Chauc. Doct. T. 116. *Icel.* ú-heill *mishap.*]

un-hǽmed; *adj. Unmarried* :—Unhǽmedo *innuba* (*voluit dotales linquere pompas*, Ald. 195), Wrt. Voc. ii. 95, 50. v. un-gehǽmed.

un-hál; *adj. In bad health, sick, weak, infirm, unhealthy, unsound,* (a) in reference to persons :—Ðér ðæt heáfod bið unhál *languente capite*, Past. 18; Swt. 129, 7. Ic eom unhál *infirmus sum*, Ps. Th. 6, 2. Unhál *debilis*, Mk. Skt. Lind. Rush. 9, 43. Ðe unhála *languidus*, Jn. Skt. Lind. Rush. 5, 7: Lchdm. i. 360, 18. Ne mæg se unhála ðam hálan gelíce byrðene áhebban, L. C. S. 69; Th. i. 412, 8. Ðý læs hié mid ðý tóle ðæt hále líc gewierden ðe hié sceoldon mid ðæt unhále áweg áceorfan *dum per hoc in se sana perimunt, per quod salubriter abscindere sauciata debuerunt*, Past. 48; Swt. 365, 12. Eágan mé syndon unhále *oculi mei infirmati sunt*, Ps. Th. 87, 9: 108, 24. Fét míne unhále (*inbecilles*), Anglia xi. 116, 22. Sume habbaþ bearn genôge, ac ða beóþ hwílum unhále oþþe yfele and unweorþ, Bt. 11, 1; Fox 32, 8. Ða hálan . . . ða unhálan *incolumes . . . aegri*, Past. 36; Swt. 247, 4. Ðara unhálra t ádligra *languentium*, Jn. Skt. Lind. Rush. 5, 3. For hwí se gôda lǽce selle ðæm unhálum, sumum líþne drenc, sumum strangne *cur aegri quidam lenibus, quidam vero acribus adjuvantur*, Bt. 39, 9; Fox 226, 12. Alle unhále *omnes male habentes*, Mt. Kmbl. Lind. 8, 16: Homl. Skt. i. 21, 155. Ealle ða unhálan, Mk. Skt. i, 32. (b) of animals :—Gif man áfindeþ his ǽhte, syððan hé hit gebohte ha[f]eþ, unhál, L. O. 7; Th. i. 180, 21. Ðæt hors blon fram ðám unhálum (*insanis*) styrenessum ðara leoma . . . and sóna árás hál and gesund, Bd. 3, 9; S. 533, 38. [Adam bicom unmihti and unhol, O. E. Homl. ii. 35, 9. *Laym. Orm.* un-hal: *Goth.* un-hails: *O. H. Ger.* unheil *insanus.*]

un-hálgod; *adj. Unhallowed, unconsecrated* :—Fela dropena unhálgodes eles, Lchdm. i. 380, 5. Gif preóst on unhálgodon húse mæssige, L. N. P. L. 13; Th. ii. 292, 16. v. un-gehálgod.

un-hálig; *adj. Unholy* :—Of unháligre þeóde *de gente non sancta*, Ps. Lamb. 42, 1.

un-hálwendlíc; *adj. Incurable* :—Fyll unhálwendlíc *casus insanabilis*, Scint. 80, 8. Þeáh man sý on hwylcre unhálwendlícre ádle, Lchdm. i. 328, 21. Hira wín is dracena gealla and næddrena ǽttor unhálwendlíce *fel draconum vinum eorum et venenum aspidum insanabile*, Deut. 32, 33.

un-handworht; *adj. Not made with hands* :—Ic unhandworht tempel getimbrie *ego templum non manu factum aedificabo*, Mk. Skt. 14, 58. [*Goth.* un-handuwaurhts.]

un-hár; *adj. Very grey* (un- *seems to have here the unusual force of an intensive*) :—Hróðgár, eald and unhár (cf. *the epithets elsewhere applied to him,* gamolfeax, 1220; B. 608: blondenfeax, 3586; B. 1791), Beo. Th. 719; B. 357.

un-heáh; *adj. Not high, low* :—Unhéh (*printed* unhela, *but see* Anglia viii. 450) þrepel *eculeus*, Wrt. Voc. i. 21, 18. Unhége sceós *talares*, i. 26, 23. Faraþ tô feldlandum and dúnlandum and tô unhéheran landum *venite ad campestria atque montana et humiliora loca*, Deut. 1, 7. Hwǽr se weall unhéhst sý, Homl. Th. i. 484, 10.

un-heánlíce; *adv. Not in an abject manner, gallantly* :—Hé un-heánlíce hine werede *he defended himself gallantly*, Chr. 755; Erl. 48, 33.

un-hearmgeorn; *adj. Inoffensive* :—Se Hálga Gást com ofer Criste on culfran híwe for ðí ðæt hé wolde getácnian mid ðam ðæt Crist wæs on ðære menniscuysse swíðe líðe and unhearmgeorn, Homl. Th. ii. 44, 20.

un-hége, un-hela. v. un-heáh.

un-helian; *p.* ede *To uncover, reveal* :—Nis nán þing oferheled ðe ne beó unheled *nihil opertum est quod non reueletur*, Lk. Skt. 12, 2. [God dede, þet wule adeaden, forworpeð hire rinde, þet is, unheleð hire . . . Þe figer . . . schal adruwien rindeleas, þuruh þet hit is unheled, A. R. 150, 8–20. Hire hede unhelid was, Alex. (Skt.) 3450. If his hous be unheled (-hiled), Piers P. 18, 319.]

un-heóre, -heórlíce, -hére. v. un-híre, -hírlíc, -híre.

un-hered; *adj. Unpraised* :—Þeáh hé seó ánum gehered, ðonne biþ hé óþrum unhered, Bt. 30, 1; Fox 108, 15.

un-herigendlíc; *adj. Not praiseworthy* :—Se bið unherigendlíc ðe unnyt leofaþ, Homl. Th. ii. 406, 17.

un-hérsum, un-híere, -híorde, -híore. v. un-hírsum, un-híre.

un-híre, -heóre, -hére, -híere, -híore, -hýre; *adj. Fierce, savage, cruel, deadly, dire, dreadful, frightful* :—Unhíere *carolios*, Wrt. Voc. ii. 19,

59. Unhére (*printed* unkere), 94, 36: *carolios, atrox, inobediens,* 129, 17. Unhére, sceþðende *caustica, nocens,* 130, 12. Wælgrim, unhére *funestus, crudelis, perniciosus,* 151, 64. Unhýri, unhiórde *trux,* Txts. 100, 983. Unhýre *funesta: i. scelesta, criminosa,* Wülck. Gl. 245, 1. Ðæs unhiéran *cruentae,* Wrt. Voc. ii. 24, 34. Ða unhióran *infestos,* 47, 59. (1) *of living creatures* :—Weard unhióre (*the fire-drake*), Beo. Th. 4818; B. 2413. Grendeles môdor . . . wíf unhýre, 4247; B. 2120. Se (*Ishmael*) bið unhýre, orlæggífre, wiðerbreca wera cneórissum *hic erit ferus homo, manus ejus contra omnes* (Gen. 16, 12), Cd. Th. 138, 5; Gen. 2287. Ne gémde hé ná swá swýðe hú hé ǽræfnede ðæs unhýran cwelres hand, Shrn. 129, 9. Ða unhiéran *torvam* (*gypsam*), Wrt. Voc. ii. 96, 12. Hió hyne scyldeþ wið unhýrum nihtgengum and wið egeslícum gesihðum, Lchdm. i. 70, 5. (2) *of things* :—Egl unheóru *a cruel talon,* Beo. Th. 1978; B. 987. Weder unhióre *hard weather,* Met. 29, 65. Him geblendon drýas drync unheórne, se onwende wera ingeþanc, Andr. Kmbl. 68; An. 34. [*O. Sax.* un-hiuri·: *O. H. Ger.* un-hiuri, -hiur *dirus: Icel.* ú-hýrr *unfriendly-looking, frowning.*]

un-híre, -híore; *adv. Fiercely* :—Hé lócaþ unhióre, swíðes wingeþ, gilleþ geómorlíce, Salm. Kmbl. 532; Sal. 265.

un-hírédwist *renders* infamiliaritas, Scint. 203, 13.

un-hírlíc; *adj.* I. *fierce, savage,* (1) *of living creatures* :—Mera mengeo on onsióne máran and un[hý]rlícran ðonne ða elpendas, Nar. 11, 1. (2) *of things* :—Ðá cwom ðær swíðe micel wind and tô ðæs unheórlíc se wind geweóx ðæt hé ðara úra getelda monige áfylde *tum euri uenti tanta uis flantis exorta est, ut omnia tabernacula nostra euerterit,* Nar. 22, 28. Unhiérlíc storm of ðæm munte ástág, Blickl. Homl. 203, 7. II. *dismal, doleful* :—Ic forht and unrôt ðás unhýrlícan fers onhefde mid sange *carmina prae tristi cecini haec lugubria mente,* Dôm. L. 11. [Leper mas bodi ugli and lathe and unherly, Metr. Homl. 129, 26. *O. H. Ger.* un-hiurlícha *eumenides: Icel.* ú-hýrligr *frowning.*]

un-hírsum; *adj. Disobedient, inattentive* :—Wæs hé náwiht unhýrsum his yldrum, Guthl. 2; Gdwin. 12, 14. v. un-gehírsum.

un-hírsumlíce; *adv. Disobediently* :—Ic ne dyde árleáslíce ne un-hýrsumlíce wið mínne Drihten *nec impie gessi a Deo meo,* Ps. Th. 17, 21.

un-hírsumness, e; *f. Disobedience* :—Ðurh gewyrht sumre unhýrsum-nesse *per meritum cujusdam inobedientiae,* Bd. 4, 6; S. 573, 38: 5, 6; S. 619, 22. Ǽt ðæs ǽrestan mannes unhýrsumnesse, Blickl. Homl. 85, 31. For heora unhýrsumnesse Godes beboda, 95, 8. Gefriða mé wið ðises folces unhýrsumnesse *eripies me de contradictionibus populi,* Ps. Th. 17, 41. Hé déð unhiérsumnesse Gode, Past. 54; Swt. 421, 32. v. un-gehírsumness.

un-hírwan *to speak very ill of, calumniate* :—Ne ǽnig man ôþerne bæftan ne tæle ne hyrwe (unhyrwe, MS. C.), Wulfst. 70, 15.

un-hiþy. v. un-hýþig.

un-híwe; *adj. Formless* :—Unhíwe *informia,* Germ. 399, 259.

un-híwed *glosses* discolor, Mt. Kmbl. p. 3, 19.

un-hleówe; *adj. Chill* :—Unhleówan wǽg *the chill wave,* Cd. Th. 209, 4; Exod. 494.

un-hlidian; *p.* ode *To uncover, to remove the lid* or *covering from something* :—Féng se portgeréfa tô ðære tége and hé hí unhlidode, Homl. Skt. i. 23, 765. Seó byrgen wæs open geworden and unhlidod, Wulfst. 214, 19.

un-hlís[e?]; *adj. Of evil repute, disreputable* :—On unhlísum wige *infami via,* Kent. Gl. 475.

un-hlísa, e; *m. Ill-fame, evil report, discredit, infamy* :—Unhlísa *infamia,* Wrt. Voc. i. 76, 3. Gode swá gecwéme þurh hálige drohtnunge, ðæt him nán unhlísa ne fylge þurh ǽnigne fracodscipe, R. Ben. 141, 4: L. Ælf. P. 31; Th. ii. 376, 24. Ðǽm hádum ðe mon nánes unhlýsan ǽt wénan ne þorfte, L. E. I. 12; Th. ii. 410, 9. Ðá cwæð Eugenia, ðæt heó eáþe mihte ðæs forligeres unhlísan hí beládian, Homl. Skt. i. 2, 205. Tô unhlísan *infamiam,* Wrt. Voc. ii. 44, 48. Sume swíðe yfele férdon . . . and yfele geendodon on heora unhlísan, Ælfc. T. Grn. 8, 10. Ða wíf ðe heora ǽwe healdaþ wið unhlísan, Homl. Ass. 39, 376: 108, 208.

un-hlísbǽre; *adj. Disreputable* :—Tô ðǽm unhlísbǽrum *ad infame,* Wrt. Voc. ii. 9, 19.

un-hlíseádig; *adj. Disreputable* :—Ðæs unhlíseádgan *infamis,* Wrt. Voc. ii. 44, 49.

un-hlísful; *adj. Disreputable* :—Unhlísful *infamis,* Wrt. Voc. i. 76, 2. Unhlísfullum gydde *infami elogio,* Hpt. Gl. 524, 76. Unhlísfullest *infamis,* 505, 12.

un-hlísig; *adj. Disreputable* :—Unhlísie *infames,* Kent. Gl. 24.

un-hlytm *an ill-sharing* (?) :—Hengest wunode mid Finne . . . un-hlitme (finnel unhlitme, MS.) *Hengest dwelt with Fin and his lot was not a happy one* (? v. hlytm, *and cf. Icel.* ú-hlutr, -hluti *harm, hurt*), Beo. Th. 2262; B. 1129.

un-hneáw; *adj. Not niggardly, liberal, bounteous,* (1) *giving liberally* :—Sumne hý gemétaþ geofum unhneáwne, Exon. Th. 326, 36; Víd. 139. Ælfwine hæfde heortan unhneáweste hringa gedáles, 323, 3; Víd. 73. (2) *given liberally, abundant* :—Geofum uuhneáwum, 43, 10; Cri. 686. [*Icel.* ú-hnöggr.]

un-hoga; *adj. Unwise, foolish :*—Unhogo (-hogu, Rush.) *inprudentes*, Mk. Skt. Lind. 7, 18. Cf. wan-hoga.

un-hold; *adj.* I. *unfriendly, hostile :*—Ðæt dyde unhold mann *inimicus homo hoc fecit*, Mt. Kmbl. 13, 28. Hé ástealde swíðe strang gyld . . . and him wæs ða unhold eall ðæt his ǽr gyrnde, Chr. 1040; Erl. 166, 22. Weard ðeóden unhold þeóden ðám ðe ǽhte geaf, Cd. Th. 218, 4; Dan. 34. Hé him ða sætnunge gewearnode ðæs unholdan cyninges *regis sibi infesti insidias vitavit*, Bd. 2, 12; S. 515, 12. Ealle his ǽhta unholde fýnd, ríce réðe mann gedǽle, Ps. Th. 108, 11. Hé mé álýsde of huntum unholdum, 90, 3. II. *unfaithful, disloyal :*—Se unholda ðeówa, Homl. Th. ii. 556, 18. [Monies monnes sare iswinc habbeð oft unholde, O. E. Homl. i. 161, 36. Unholde uorureten þe strencðe of his soule *alieni comederunt robur ejus*, A. R. 222, 10. *O. Sax.* un-hold *unfriendly, hostile : O. H. Ger.* un-hold *inimicus*, unholde; *pl. eumenides.*] v. next word.

un-holda, an; *m. A fiend :*—Hé his áras hider onsendeþ, ðí læs unholdan wunde gewyrcen, Exon. Th. 47, 29; Cri. 762. [*Goth.* un-hulþa; *m.;* uu-hulþó; *f. an evil spirit : O. L. Ger.* un-holdo *a devil: O. H. Ger.* un-holda *diabolus : Ger.* un-hold *a fiend, devil.*]

un-hrædsprǽce; *adj. Not ready of speech :*—Ic eom unhrædsprǽce *incircumcisus sum labiis*, Ex. 6, 30.

un-hreóflig; *adj. Not leprous :*—Ne gedyde se sacerd ðone man hreófligne oððe unhreófligne, Homl. Th. i. 124, 24.

un-hrór; *adj. Not stirring :*—Unhrórum neátum *immobilibus animantibus*, Bt. 41, 5; Fox 254, 14. [Cf. *O. H. Ger.* un-gi[h]ruorig *immobilis.*]

un-húfed; *adj. (ptcpl.) With the head made bare, with the head shaved :*—Unhúfed *decalvata* (the passage is : Quamvis flava caesaries raderetur et per publicum decalvata traheretur, Ald. 62), Hpt. Gl. 510, 13. v. húfian.

un-hwearfiende; *adj. (ptcpl.) Unchanging, immutable :*—Nánwuht woruldíces fæstes and unhwearfiendes beón ne mæg, Bt. 8; Fox 26, 11.

un-hwílen; *adj. Not temporary, eternal :*—Him is symbel and dreám éce, unhwýlen, Exon. Th. 352, 13; Sch. 97. Unhwílen, Elen. Kmbl. 2461; El. 1232. Ðǽr bið symle gearu freónd unhwílen, Andr. Kmbl. 2309; An. 1156. Ic mé sylfum wát æfter líces hryre leán unhwílen, Exon. Th. 167, 27; Gú. 1066. Hé him éce geceás langsumre líf, leóht unhwílen, Apstls. Kmbl. 40; Ap. 20.

un-hygdig; *adj. Foolish :*—On his heortan cwæð unhýdig sum ungleáwlíce *dixit insipiens in corde suo*, Ps. Th. 52, 1. Hwí nǽre ðu genóg earm and genóg unhýdig (-hýdig ?), þeáh ðé þúhte ðæt ðú welig wǽre, ðonne ðú óþer twéga oððe hæfdest ðæt ðú noldest, oððe næfdest ðæt ðú woldest ? Bt. 26, 1; Fox 90, 30 note. Cf. wan-hygdig.

un-hyldu(-o); *f. Disfavour, unfriendliness :*—Ða habbaþ his unhyldo ðe hit him bryttian sceoldon *iram merentur, qui dispensatores sunt*, Past. 44; Swt. 321, 4. Him is unhyldo Waldendes witod, Cd. Th. 45, 20; Gen. 729. þurh hine wurdon manege geypte ðe mid heora rǽde on ðes cynges unheldan (=hyldum ?) wǽron (*who were hostile to the king*), Chr. 1095; Erl. 232, 20. [*O. Sax.* un-huldi : *O. H. Ger.* un-huldî.]

un-hýre, un-hýrsum. v. un-híre, un-hírsum.

un-hýðig; *adj. Without that which is advantageous or beneficial, unhappy :*—Hwí nǽre ðú ðonne genóg earm and genóg unhýþi (*printed* -hiþy ; -hydig, Cott. MS.), þeáh ðé þúhte ðæt ðú welig wǽre, ðonne ðú óþer twéga oððe hæfdest ðæt ðú noldest, oððe næfdest ðæt ðú woldest ? Bt. 26, 1; Fox 90, 30. Gewát beorn unhýðig (*Guthlac's disciple when he had just lost his master*), Exon. Th. 181, 32; Gú. 1302. Hié unhýðige gecyrdon luste belorene láðspell beran, Andr. Kmbl. 2157; An. 1080.

un-inseglian; *p. ode To unseal :*—Hí (*a casket*) nán man ne uninsæglode ǽr hí ealle ðyder cómon . . . Se portgeréfa hí uninsæglode, Homl. Skt. i. 23, 762. Hig uninseglodon ðæt loc and cǽgan, Nicod. 14; Thw. 7, 11.

un-íðe, un-kere. v. un-eáðe, un-híre.

un-ípian; *p. ode To disquiet, molest :*—Ðonne mé unýþgiende wǽron *cum me molesti essent*, Ps. Spl. T. 34, 15.

un-lácnigendlíc; *adj. Incurable :*—Unlácnigendlíce ádlu, Lchdm. i. 262, 1.

un-lácnod; *adj. Uncured :*—Hé hæfð on his nebbe opene wunde unlácnode, Past. 9; Swt. 61, 4. v. un-gelácnod.

un-lǽce, es; *m. An unskilful physician :*—Hú unlǽcas (cf. unwíse lǽcas, 232, 8) wénaþ ðæt ðæt sié lendenádl, Lchdm. ii. 164, 8.

un-lǽd, -lǽde; *adj.* I. *poor, miserable, unhappy, unfortunate :*—Unlǽde bið and ormód se ðe á wile geómrian, Salm. Kmbl. 699; Sal. 349. Óðer bið unlǽde on eorðan, óðer bið eádig, 731; Sal. 365. Him mæg eádig eorl eáðe geceósan mildne hláford ; ne mæg dón unlǽde swá, 784; Sal. 391. Lócaþ fram ðam unlǽdan hláford *his lord turns his looks from the unhappy man*, 765; Sal. 382. II. *in a moral sense, poor, miserable, wretched :*—Hé þincð swíðe dysig man and swíðe unlǽde ðe nele hys andgyt ǽcan ða hwíle ðe hé on ðisse weorulde byð, Shrn. 204, 24. Unlǽde bið on eorðan, unnyt lífes se þurh ðone cantic ne can Crist geherian, Salm. Kmbl. 41; Sal. 21. Rǽd biþ nyttost, yfel

unnyttost, ðæt unlǽd nimeþ, Exon. Th. 341, 3; Gn. Ex. 120. Helle gǽst, earm and unlǽd, 279, 19; Jul. 616. Se unlǽda (-e, MS.) Iudas, se ðe hine tó deáþe belǽwed hæfde, Wanl. Catal. 134, col. 1. Swá heó ðæs unlǽdan (*Holofernes*) eáþost mihte wel gewealdan, Judth. Thw. 23, 3; Jud. 102. Ða þrowunga ðe hé ádreág æt ðæm unlǽdan folce Iudéa, Blickl. Homl. 97, 16. Ðæt wíte, ðæt ðon unlǽdon geteohhod biþ; him wǽre betere ðæt hé nǽfre geboren nǽre, 25, 24. Ða unlǽdan (*the chief priests who wished to kill Lazarus*, Jn. 12, 10), 77, 9. Gesǽlige beóð ða ðe ðam fyliaþ, and unlǽde beóð ða ðe ðam widsacaþ, Wulfst. 264, 21. Gé sind unlǽde earm[r]a geþóhta, Andr. Kmbl. 1487; An. 745. Unlǽdra (*the cannibals*) eafoð, 59; An. 30: (*evil spirits*), 283; An. 142. Hé æt ðǽm unlǽdum Iudéum manig bysmor geþrowade, Blickl. Homl. 23, 30: 85, 1. [Þo unlede (þese wikkede, MS. V.) fode, Al. (T.) 333. Hu he is unlede (*miserable*) þat foleweþ quene rede, Misc. 122, 337: O. and N. 1644. *Goth.* un-léds *poor.*] v. un-lǽdlíce.

un-lǽde; *adj. Stray* (?) :—Ðá forstæl hé ða unlǽdan oxan, Chart. Th. 172, 21.

un-lǽdlíce; *adv. Miserably, wretchedly :*—Hé (*Judas*) hine sylfne swíðe earme and unlǽdlíce of ðære gemánan ealra Godes gecorenra ádilgode, Homl. Ass. 153, 48. Hé hine sylfne swíðe unlǽdlíce áhéng and swá ungesǽliglíce tó écan deáðe wæs geníðerad, 158, 163.

un-lǽgne. v. un-lígne.

un-lǽne; *adj. Not transitory, permanent :*—For ðissum lǽnan lífe ic sylle ðæt unlǽne, Wulfst. 264, 18.

un-lǽred; *adj. Untaught, unlearned, ignorant :*—Gelaered oððe unlaered *doctus vel indoctus*, Mt. Kmbl. p. 1, 6. Hú mæg unlǽred déma óðerne lǽran? L. I. P. 19; Th. ii. 326, 32. Ðætte unlǽrede (-lǽrde, Hatt. MS.) ne dyrren underfón láreówdóm *ne venire imperiti ad magisterium audeant*, Past. 1; Swt. 24, 14. Tó ðon ðæt unlǽrede sýn gelǽrede *ut indocti doceantur*, Bd. 1, 27; S. 492, 24. Ðú ongytest ðæt hié syndon unlǽrede men, Blickl. Homl. 183, 7. Hwá unlǽredra ne wundraþ . . .? Bt. 39, 3; Fox 214, 15: Met. 28, 1. v. un-gelǽred.

un-lǽt; *adj. Not slow, quick, ready, active :*—Unlǽt *non pigra*, Wrt. Voc. ii. 61, 61. Wíga, unlǽt láces (*death*), Exon. Th. 164, 5; Gú. 1007. Hræd and unlǽt, 436, 9; Rä. 54, 11.

un-láf, e; *f. A child not left by a father at his death, a child born after the father's death, a posthumous child :*—Unláb *posthumus*, Wrt. Voc. ii. 117, 67. Unláf, 69, 2: 93, 70.

un-lagu, e; f. I. *violation of law, illegality, injustice :*—Mid unlage *contra justitiam*, Cod. Dip. Kmbl. iv. 198, 15: 224, 11. Ðeáh ðe Harold ðæt land mid unlage út nam, 274, 29. Ic nelle gedafian ðæt man hym ǽnige unlage beóde *nec impune feram quod aliquis ei injuriam inferat aut molestiam*, 196, 27: vi. 187, 22. Se ðe unlage rǽre oþþe undóm gedéme, L. C. S. 15; Th. i. 384, 9: Chr. 1052; Erl. 186, 2. Ðæt man rihte lage up árǽre and ǽlce unlage áfylle, L. Eth. v. 1; Th. i. 304, 11. Fela unrihta and yfelra unlaga árysan, Chr. 975; Erl. 127, 30. Æt unlagum *unlawfully* (cf. Icel. at úlögum *in a lawless manner*), L. C. S. 61; Th. i. 408, 18. Swicollíce dǽda and lǽðlíce unlaga áscunige man, ðæt is, false gewihta . . . and leáse gewitnesa, L. Eth. v. 24; Th. i. 310, 12: vi. 28; Th. i. 322, 13. Ðæt man rihte laga up árǽre, and ǽghwylce unlaga georne áfylle, vi. 8; Th. i. 316, 26: L. C. S. 1; Th. i. 376, 8: 11; Th. i. 382, 7: Wulfst. 156, 13. Hé ne róhte ná hú manige unlaga hí dydon, Chr. 1086; Erl. 220, 13. II. *a bad law :*—Man beháteþ, ðonne man fulluhtes gyrneþ, ðæt man aa wile deófol áscunian, and his unlaga forbúgan and ealle his unlaga áweorpan, Wulfst. 144, 10. Cradolcild wǽron geþeówode þurh wælhreówe unlaga for lytelre þýfðe, 158, 14. [*Icel.* ú-lög; *pl. lawlessness, injustice.*]

un-land, es; *n. What is not land :*—On ðam fíftan dǽle healfum londes and unlondes (*sea, marsh, etc.*), Bt. 18, 1; Fox 62, 23. Tó ðam unlonde (*the whale, by whose side seamen,* '*deeming him some island,*' *moor their bark*), Exon. Th. 361, 3; Wal. 14.

un-landágende; *adj. Not owning land :*—Gesíðcund mon unlandágende, L. In. 51; Th. i. 134, 9.

**un-lár, e; f. Evil teaching, incitement to evil :*—Ða ðe nú deófle fyligaþ and his unlárum, Wulfst. 19, 17: 37, 12. Se ðe gehealt Godes beboda, and forbúhð deófles unlárum, L. I. P. 21; Th. ii. 330, 28. Ðæt hé forsace and forbúge deófles unlára, Wulfst. 32, 15: 144, 9.

un-leahtorwirþe; *adj. Unblameable :*—Godes ǽ is swíðe unleahtorwyrðe *lex Domini inreprehensibilis*, Ps. Th. 18, 7.

un-leánod; *adj. Not repaid :*—Ic wille, gif ic ǽnigum menn ǽnig feoh unleánod hæbbe, ðæt míne mágas ðæt geleánian, Chart. Th. 491, 5.

un-leás; *adj. Not false, true :*—Unleás *non frivola, non falsa*, Hpt. Gl. 432, 21. Se bið unleás forscrencend ðe is leahtras forscrencð, Homl. Th. i. 586, 23. Wé willaþ eów gereccan óðres mannes gesihðe ðe unleás is, ii. 332, 26. Gelýf hys hálgum, for ðam hí wéron swíðe unleáse gewitan, Shrn. 199, 14. Unleásra manna sægena, 195, 29. Hweðer ðe ðince Honorius wísra oððe unleásera ðonne Crist, 196, 27. Wísran and unleáseran ðegnas, 197, 8.

un-leáslíce; *adv. Not falsely, truly :*—Heó geseah ðæt his bodung unleáslíce gefylled wæs, Homl. Th. i. 42, 29. Gif wé ðæt geðyld on úrum móde unleáslíce healdaþ, ii. 546, 1. Hé cwæð ðis unleáslíce, 386,

10. Mid ânfealdnysse sprece hê ǽfre unleáslíce, L. Ælfc. C. 30 ; Th. ii. 354, 5.

un-leóf; *adj. Not dear, not beloved, odious, hateful :*—Sodomware, Gode unleófe, Cd. Th. 148, 6 ; Gen. 2452. Gigantmæcgas, Gode unleófe, Metode láðe, 77, 1 ; Gen. 1268. Wíglâf seah on unleófe (*the followers who had failed Beowulf in his need*), Beo. Th. 5719 ; B. 2863. [Al þat is on unlef and unqueme, hit is þat oðer iqueme, O. E. Homl. ii. 189, 25. *Goth.* un-liubs *not beloved : O. H. Ger.* un-liup *insuavis, non optatus.*]

un-leoþuwác; *adj. Inflexible, intractable, implacable :*—Unlioþuwác (-lidouuác, -liuduuác) *intractabilis*, Txts. 69, 1079. Unlipewác (in-, MS.), Wrt. Voc. ii. 48, 72. Ðâ wæs mín hláford in micle hâtheortnysse, and hê wæs swíðe unlioðewác geworden wið mê, and hê gebrægd his swurd and wolde mê ofsleán, Shrn. 39, 14. [*O. H. Ger.* un-lidoweih *invulsus, inplicabilis, implacabilis.*]

un-leoþuwácness, e; *f. Inflexibility, implacability :*—Unlioþuwácnis *infestatio*, Wrt. Voc. ii. 110, 65. Unleoþwácnes, 45, 42.

un-libbende; *adj. Not living, dead :*—Se dyde monig wundor ge lybbende ge unlybbende, Shrn. 127, 22. v. un-lifigende.

un-líchamlíc; *adj. Incorporeal :*—Seó sâwul is unlíchomlíc, Homl. Skt. i. 1, 176. Unlíchamlíce *incorporalia*, Ælfc. Gr. 5 ; Zup. 11, 19 : 9, 21 ; Zup. 47, 2.

un-lícwirþe; *adj. Unpleasing, disagreeable :*—Nimaþ mê mid eów ; ne beó ic nâ eów unlícwyrðe, Homl. Skt. ii. 23 b, 374.

un-lífed; *adj. Unallowed, illicit :*—Unliéfedo *illicitum*, Past. 51 ; Swt. 397, 30. Sió unliéfde byrðen *pondus illicitum*, Swt. 401, 5. Fleón ðone unliéfedan bryne úres líchoman *illicita carnis incendia declinare*, Swt. 397, 36. Ða ðe ðæt unliéfde herigaþ *qui illicita laudant*, 55 ; Swt. 427, 11.

un-lífes; *adv. Not alive :*—Hê is nû unlífes, Homl. Skt. i. 18, 203.

un-lifigende; *adj. Not living, dead, defunct :*—Unlífigendes fêt and folma, Beo. Th. 1492 ; B. 744. Heáfod Holofernus unlyfigendes, Judth. Thw. 24, 9 ; Jud. 180. Unlifigendes, Elen. Kmbl. 1754 ; El. 879. Siteþ eorl ofer óðrum unlifigendum (*Beowulf*), Beo. Th. 5809 ; B. 2908. Ðæt biþ drihtguman unlifigendum æfter sêlest, 2782 ; B. 1389. Hê aldorþegn unlyfigendne, deádne wisse, 2621 ; B. 1308. Fore gileáffullum unlifigendum *pro fidelibus defunctis*, Rtl. 173, 37. Fore deádum ł un-lifiendum, Jn. Skt. p. 4, 20. On heora ealdfeóndum unlyfigendum, Judth. Thw. 26, 8 ; Jud. 316. v. un-libbende.

un-lígne; *adj. Not to be denied* or *rejected, incontrovertible :*—Biscopes word and cyninges sié unlǽgne bûton âðe *a bishop's word and a king's is to be accepted without an oath*, L. Wih. 16 ; Th. i. 40, 12. Þissa ealra âð sié unlǽgnæ, 21 ; Th. i. 42, 1. v. lígnian, and cf. *O. H. Ger.* lougeníg *negativus.*

un-lísan; *p.* de.   **I.** *to unloose, undo :*—Ic bidde ðê ðæt ðû mê unlýse ða insæglunge, Homl. Skt. i. 3, 537.   **II.** *to release :*—Hê beád ðæt man sceolde unlêsan ealle ða menn ðe on hæftnunge wǽron, Chr. 1086 ; Erl. 223, 38.

un-líþe; *adj. Ungentle, harsh, severe, cruel :*—Ðê tô heortan hearde gripeþ ádl unlíðe, Cd. Th. 57, 32 ; Gen. 937. Geliðewáca ðisne unlíðan cyle, Homl. Skt. i. 11, 192. Scearpnyssa beóð âwende tô smêðum wegum, ðonne ða yrsigendan mód and unlíðe gecyrraþ tô manðwǽrnysse, Homl. Th. i. 362, 30.

un-lofod; *adj. Unpraised :*—Ne lǽt ðû unlofod ðæt ðû swutele ongite ðæt lícwyrðe sý ; ðǽr ðê âuht tweóge, lofa ðæt gemetlíce, Prov. Kmbl. 62.

un-lûcan; *p.* -leúc *To unclose, open :*—Godes engel unleác ðæt cweartern, Homl. Skt. ii. 25, 839. *Hostiarius* is ðære cyrcean durewerd, se sceal ða cyrcan unlûcan geleáffullum mannum, and ðâm ungeleáffullum belûcan wiðûtan, L. Ælfc. C. 11 ; Th. ii. 346, 29. Nân man ne dorste ða duru unlûcan, Homl. Ass. 113, 360.

un-lust, es; *m.*   **I.** *absence of desire, disgust, disinclination.* (a) *want of appetite :*—Lǽcedôm gif men unlust (cf. *Ger.* Unlust *zum Essen*) sié getenge, Lchdm. ii. 16, 15 : 150, 17. Wiþ metes unluste, 184, 15 : 28, 5. Wiþ unluste and wlætan þe of magan cymð, 158, 12. Wiþ sáre and unluste ðæs magan, se ðe ne mæg ne mid mete ne mid drincan beón gelácnod, 158, 17. Hié þrowiaþ ormǽtne þurst and metes unlust, 230, 19. Ungemetlíca metesôcna and ungemetlíce unlustas and císnessa, 174, 28. [Þi mahe wið unlust warpeð hit (*food*) eft ut, H. M. 35, 31. *Icel.* û-lyst *a bad appetite.*] (b) *disinclination to action, listlessness :*—Hié þrowiaþ unlustes (*the slothful servant*) ðolaþ neádunge þeóstra ðurh wrace, se ðe ǽr lustlíce forbær his unlustes (or *under* **III** *?* cf. 552, 12) þeóstra, Homl. Th. ii. 556, 22. [ȝæn unnlusst and forrswundennleȝȝc, Orm. 4562. He doth alle thing with slaknes and excusacioun, and with ydelnes and unlust, Chauc. Pers. T. (de accidia). *Goth.* wairþan in unlustau ἀθυμεῖν.]   **II.** *want of pleasure, joylessness, weariness :*—For unluste *prae taedio*, Ps. Spl. T. 118, 28. Hê wylleþ hine on ðam wíte, wunaþ unlustum (cf. lustum, on lustum), Salm. Kmbl. 538; Sal. 268. [*O. H. Ger.* un-lust *taedium, fastidium.*]   **III.** *an evil pleasure, lust :*—Ne unlust on hire mód ne becom, Homl. Th. ii. 10, 10. Ða gestód hine micel líchamlíc costung . . . hê âwende ðone unlust tô sárnysse, 156, 32.   Unlusta

voluptatum, Scint. 106, 10.   Ealle hyra unlustas hí sceolon gebêtan sylf-wylles on ðyssum lífe, oððe unþances æfter ðyssum lífe, Homl. Th. i. 148, 27. Þurh unlustas *libidinibus*, L. Ecg. C. 5, tit. ; Th. ii. 128, 15. Winnan wið leahtras and unlustas forseón, L. Ælfc. P. 12 ; Th. ii. 368, 19.

un-lustian. v. ge-unlustian.

un-lybba (and un-lybbe, an ; *f.,* or un-lybb ; *dat.* -lybbe ; *n. ?*), an ; *m.*   **I.** *poison :*—Sumum men wæs unlybba geseald, ac hit ne mihte hine âdýdan, Homl. Th. ii. 178, 11. Unþeáwas weaxaþ on yfel, swa swa âtres unlibba on men ; swá hê leng ðæs áttres þigene bedíhlaþ, swá wyrð his untrumnes mâre, R. Ben. 135, 16. Unlybbe *delatera* (*deletera ?*), Wrt. Voc. ii. 138, 55. Hí rǽddon ðæt hí mid áttre hine âcwealdon ; gemengdon ðâ unlybban tô his drence, Homl. Th. ii. 158, 15. Hê ðygde unlybban on his mete, 504, 14. Hê ðone unlybban hâlsode, and hine ealne gedranc, i. 72, 24, 19 : L. Pen. 7 ; Th. ii. 280, 6 : 6 ; Th. ii. 280, 2 : Wulfst. 150, 5. Wiþ unlybbum, Lchdm. ii. 292, 30.   **II.** *poison* used for purposes of witchcraft, *witchcraft, sorcery :*—Gif wíf drý-cræft and galdor and unlibban wyrce . . . Gif heó mid hire unlybban man âcwelleþ *si mulier artem magicam, et incantationes, et maleficia exerceat . . . si maleficiis suis aliquem occiderit*, L. Ecg. C. 29 ; Th. ii. 154, 8–11. Ðæs flǽsces weorc . . . hǽðengild oððe unlybban (*veneficia*, Gal. 5, 19), Homl. Skt. i. 17, 25. Ðínre môdor fela unlybban *matris tuae veneficia multa* (2 Kings 9, 22), 18, 333. [Cf. *Icel.* û-lyfjan *poison.*] v. ge-unlybba, lybb, *and next word.*

unlyb-wyrhta, an ; *m. A poison-maker, one who prepares poisons for purposes of witchcraft, a sorcerer :*—Unlybwyrhta *veneficus*, Wrt. Voc. i. 74, 39. Wyccan and wælcyrian and unlybwyrhtan, Wulfst. 298, 19. Antecrist hæfð mid him drýmen and unlybwyrhtan and ðâ ðe cunnan galder âgalan, 194, 18. Unlibwyrht[en]a wiccecræft *maleficorum* (*veneficorum*) *necromantia*, Hpt. Gl. 501, 62.

un-lyft, e; *f. Bad air, malaria :*—Rômane and eall sûðfolc worhton him eorþhûs wið ðære unlyfte, Lchdm. iii. 16, 2.

un-lygen; *adj. Unlying, truthful :*—Ceápige man on ðæs portgerêfan gewitnesse oþþe óðres unlygenes mannes, L. Ath. i. 12 ; Th. i. 206, 10. [*Icel.* û-lyginn. Cf. *O. H. Ger.* lugin *mendax.*] v. un-gelygen.

un-lyt; *n. No little, much :*—Gê mânes unlyt wyrceaþ, Ps. Th. 61, 9.

un-lytel; *adj.*   **I.** *of size, extent, not little, great :*—Wê magon tôcnâwan be hyre leóman ðæt seó sunne unlytel is, Lchdm. iii. 236, 11. Wolcen unlytel, Ps. Th. 77, 16. Unlytel dǽl foldan, Cd. Th. 154, 4 ; Gen. 2550. Unlytel dǽl eorþan gesceafta, 97, 17 ; Gen. 1614. Unlytel leádes clympre, Exon. Th. 426, 17 ; Rä. 41, 75. Hê geseah sweras unlytle, Andr. Kmbl. 2985 ; An. 1495.   **II.** *of quantity, amount, number, not little, not few in number, much :*—Tô miclum bryne sceal wæter unlytel, Wulfst. 157, 9. Menigo, folc unlytel, Elen. Kmbl. 1740 ; El. 872 : 565 ; El. 283 : Andr. Kmbl. 2542 ; An. 1272 : Beo. Th. 1000 ; B. 498. Se eorl com mid unlytlan weorode, Chr. 1068 ; Erl. 206, 10. Mycel feoh and unlytel *summam pecuniae non parvam*, Bd. 4, 11 ; S. 579, 20 : 4, 5 ; S. 571, 35. Micle lâc and unlytle ælmessan, Wulfst. 278, 5. Ic him gestrýnde unlytel folc, Homl. Th. i. 592, 31. Hyre wer lǽfde unlytle ǽhta on lande and on feó, Homl. Skt. i. 2, 155.   **III.** *of quality, degree, not little, great,* (a) *of persons :*—Hê ðûhte him selfum suíðe unlytel and suíðe medeme *se parvulum non videbat*, Past. 17 ; Swt. 113, 12.   (b) *of things :*—Storm, cirm unlytel, Andr. Kmbl. 2476 ; An. 1239. Dôm unlytel, Beo. Th. 1775 ; B. 885 : Apstls. Kmbl. 16 ; Ap. 8. Wundor unlytel, Cd. Th. 250, 26 ; Dan. 522. Unlytel spell *a tale of serious import*, 145, 14 ; Gen. 2405. Torn unlytel, Beo. Th. 1670 ; B. 833. Wundur unlytel *mirabilia*, Ps. Th. 104, 5. [Cf. *Icel.* û-lítill.]

un-lytel, es ; *n. No small amount, much :*—Eft weard folces unlytel, Wulfst. 10, 17. Hê lǽfde þære wudewan unlytel on feó and on óðrum ǽhtum, Homl. Ass. 108, 200 : Chr. 921 ; Erl. 106, 14.

un-mǽg, es ; *m. One who is not a kinsman, an alien ;* or *a bad kinsman :*—Unmǽgas, Wald. 102 ; Vald. 2, 23.

un-mǽge; *adj. Not of kin, not related, alien :*—Ic wæs unmǽge gyst môdor cildum *factus sum hospes filiis matris meae*, Ps. Th. 68, 8.

un-mǽgness, e ; *f. Inability, weariness :*—Ǽþrot, unmǽgnes, ǽmelnes *fastidium*, Wrt. Voc. ii. 146, 46. v. un-maga.

un-mǽgþlíc. v. un-mǽþlíc.

un-mǽle; *adj. Spotless,* (1) in a physical sense, *without marks* or *spots :*—Æt ânes heówes cý, ðæt heó sý eall reád oððe hwít and unmǽle, Lchdm. iii. 24, 14. (2) in a moral sense, *immaculate, virgin :*—Unmaelo *virgo*, Wrt. Voc. ii. 123, 75. Ðurh ðingunge his ðære eádigan nieder ðære unmǽlan fǽmnan Scâ Marian *per intercessionem beatae suae genetricis semperque virginis Mariae*, Bd. 5, 19 ; S. 640, 42. Hât unmǽlne mon gefeccean swígende ongeán streáme healfne sester yrnendes wæteres, Lchdm. iii. 10, 31. Þê (*the Virgin Mary*) unmǽle ǽlces þinges, Exon. Th. 21, 11 ; Cri. 333. Hê on fǽmnan âstâg, mæged unmǽle, 45, 18 ; Cri. 721.

un-mǽne; *adj.*   **I.** *free from evil, pure,* (1) of persons :—Þeáh þe þû welig beó þû nâ byst unmǽne fram gylte (*inmunis a delicto*), Scint. 179, 10. Sancta Maria, ides unmǽne, Hy. 10, 14. Fram ealre synne unmǽne *ab omni peccato inmunes*, Scint. 67, 20. (2) of an oath, *without*

perjury :—Se áþ is clǽne and unmǽne, L. O. 6 ; Th. i. 180, 18. **II.** *free from, exempt from, not sharing in :*—Ályséde fram bende ǽlces mennissces dómes, fram ǽlcere gærsuman woruldlícra brúcunga clǽne and unmǽne *nexu humanae conditionis exuti, ab omni munere secularium functionum immunes,* Cod. Dip. B. i. 154, 16. [*O. Frs.* un-mên *unperjured : O. H. Ger.* dhiu unmeina magad Maria : *Icel.* ú-meinn *harmless.*]

un-mǽre ; *adj. Not illustrious, inglorious :*—Þeáh hé on ðam lande seó mǽre ðonne biþ hé on óþrum unmǽre *fit, ut quem tu aestimas gloriosum, pro maxima parte terrarum videatur inglorius,* Bt. 30, 1 ; Fox 108, 16.

un-mǽrlíc ; *adj. Ignoble :*—Hwæþer ðé þynce unweorþ and unmǽrlíc seó gegaderung *obscurumne hoc, atque ignobile censes esse ?* Bt. 33, 1 ; Fox 120, 29.

un-mǽte ; *adj. Immense, enormous, excessive :*—Þeód unmǽte, Cd. Th. 138, 15 ; Gen. 2292. Síde herigeas, folc unmǽte, Andr. Kmbl. 1305 ; An. 653 : Menol. Fox 11 ; Men. 6. Gebrec unmǽte, Exon. Th. 59, 18 ; Cri. 954. Ðæt unmǽte gestreón goldes and seolfres, Blickl. Homl. 99, 28. Ðæt unmǽte sár weóx *augescente dolore nimio,* Bd. 3, 9 ; S. 533, 32 : 4, 25 ; S. 599, 43. Micelne swég unmǽte wópes *sonitum immanissimi fletus,* 5, 12 ; S. 628, 29. Ðæt mægen ðære unmǽtan (*immensi*) hǽto . . . on middel ðæs unmǽtan (*infesti*) cyles, S. 627, 41, 42 : Homl. Skt. ii. 23 b, 573. Bóc unmǽtre (*enormis*) micelnesse, Bd. 5, 13 ; S. 633, 5 : Guthl. 3 ; Gdwin. 20, 1 . Mid unmǽtan here, Chr. 1068 ; Erl. 206, 21. Mid unmǽte ege geslægene *timore immenso perculsus,* Bd. 5, 12 ; S. 627, 14. On ǽnne unmǽtne lég geǽnede *in inmensam adunati flammam,* 3, 19 ; S. 548, 21. Giefe unmǽte, Exon. Th. 273, 16 ; Jul. 517. Ðær synd unmǽte móras, Guthl. 3 ; Gdwin. 20, 4. Ðǽr synt ða unmǽtan tyntregu, L. E. I. prm. ; Th. ii. 396, 34. Ic wæs on unmǽtum costnungum winnende, Homl. Skt. ii. 23 b, 578. [Mid unmete drunche, O. E. Homl. i. 103, 9. *O. H. Ger.* un-mázi *immensus, ingens.*]

un-mǽþ, e ; *f. Transgression, wrong.* v. mǽþ, IV :—Ða discipulas wǽron on heora módgeþance swíðlíce áfyrhte and gedréfde, swá hit nǽnig fyren wæs (unmǽþ næs, MS. F. i. e. *it was perfectly right that they should be troubled*), Homl. Ass. 162, 234. [Min is þe guld and þe unmeþ, Fl. a. Bl. 675. Evrich þing mai leosen his godhede mid unmeþe and mid overdede, O. and N. 352.]

un-mǽþlíc ; *adj. Not in due measure, immoderate, excessive :*—Of gítsunge beóð ácennede . . . leás gewitnyss and unmǽðlíc neádung, Homl. Th. ii. 220, 11. [So hit unmedluker is, wunnen aʒean þe uestluker, A. R. 238, 17.]

un-mǽþlíce ; *adv.* **I.** *immoderately, out of measure :*—Seó wydewe mænigfealde sceattas hyre unmǽðlíce beád (*was immoderate in her offers*), Homl. Skt. i. 2, 147. Hé hét ðone bisceop unscrýdan and unmǽðlíce (*or* II) swingan, ii. 29, 231. **II.** *inhumanely.* v. mǽþlíce :—Ða ðe hæfdon sum þing lytles tó bigleofan, ðæt gelǽhton reáferas and of ðam múðe him ábrudon unmǽðlíce mid þreáte, Homl. Ass. 68, 73. Ǽlc lǽhte of óðrum ðone mete of ðam múðe swíðe unmǽgðlíce, Ælfc. T. Grn. 21, 11. [For lutle ich mei makien to muchelin unmeaðeliche, ʒef me hut hit, Marh. 15, 9. ʒeieð luddre and unmedluker, A. R. 266, 1.]

un-mǽtlíc ; *adj. Immense :*—Laforas unmǽtlícre micelnisse *capri ingentis forme,* Nar. 15, 2. [*O. H. Ger.* un-mázlíh *incomparabilis.* Cf. *Icel.* ú-mátaliga *immoderately.*]

un-mǽtness, e ; *f. Immenseness, excess :*—Fore unmǽtnysse ðæs gewinnes *ob nimietatem laboris,* Bd. 3, 8 ; S. 532, 31. Mid unmǽtnesse miceles stormes *tempestatis impetu,* 5, 12 ; S. 627, 40. Ða cwom micel snáw . . . ða ic ða unmǽtnisse and micelnisse ðæs snáwes geseah, ða ðúhte mé ðæt ic wiste ðæt hé wolde ealle ða wícstówe forfeallan *cadere mox . . . immense ceperunt nives quarum aggregationem metuens ne castra cumularentur,* Nar. 23, 14.

un-maga, an ; m. : *a person without means, a needy person :*—Se maga and se unmaga ne beóð ná gelíce, ne ne magon ná gelíce byrðene áhebban . . . and ðý man sceal gescádlíce tóscádan . . . welan and wǽdle, L. Eth. vi. 52 ; Th. i. 328, 16 : L. C. S. 69 ; Th. i. 412, 6. Ðis is mihtiges mannes and freóndspédiges dǽdbótlihtingc, ac án unmaga ne mæg swilc gefordian, L. P. M. 4 ; Th. ii. 288, 20. Ðú ne scealt nǽfre gelíce déman . . . ðam strangan and ðam unmagan, L. de Cf. 3 ; Th. ii. 260, 25. Ðearfan and unmagon *pauperem et inopem,* Ps. Spl. 36, 15. **II.** *a person who cannot maintain himself, one who is dependent upon others :*—Mardocheus hæfde Hester for dohtor, for ðan hire deád wæs ge fæder ge módor, ða ða heó unmagu (-maga, *v. l.*) wæs, Homl. Ass. 94, 86. Gif hwá óðrum his unmagan óðfǽste, L. Alf. pol. 17 ; Th. i. 72, 4. [On-mawe, Fer. i. 2658. *O. H. Ger.* un-mag *segnis, dissolutus ; parvulus : Icel.* ú-magi *one who cannot maintain himself,* e. g. *a child.*]

un-manig ; *adj. Not many, few :*—Æfter unmonegum geárum *post aliquot annos,* Bd. 3, 17 ; S. 544, 9 : 5, 18 ; S. 636, 18. Unmonigum dagum *non multis diebus,* Jn. Skt. Lind. 2, 12. Ymbe unmanige dagas, Guthl. 2 ; Gdwin. 18, 26 : 3 ; Gdwin. 22, 17 : 5 ; Gdwin. 34, 13.

Unmonige *paucos,* Mt. Kmbl. p. 15, 7. Unmonige fiscas *paucos pisciculos,* Mt. Kmbl. Rush. 15, 34. [*O. H. Ger.* un-manig.]

un-mann, es ; m. **I.** *a bad man, an inhuman person :*—Swá fela ðúsend engla mihton eáðe bewerian Crist wið ðam unmannum (*those who came to seize Jesus*), gif hé ðrowian nolde sylfwilles for ús, Homl. Th. ii. 246, 30. [Cf. *O. H. Ger.* un-mennisco : *Ger.* un-mensch : *Icel.* ú-mannan *a person fit for nothing.*] **II.** *one who is not a mere man, a hero :*—Ða gemunde hé ða strangan dǽda ðara unmanna (*perhaps* iumanna *should be read ; the Latin is :* Valida priscorum heroum facta reminiscens. v. geó-, iú-mann) and ðæra woruldfrumena, Guthl. 2 ; Gdwin. 12, 27.

un-meagol ; *adj. Feeble, insipid :*—Unmeagol *emellus,* Wrt. Voc. ii. 143, 30. *Insipidum, quod saporem non habet, hoc est* unmeagle *sive* ǽmelle, 49, 37.

un-meaht, -meht, -mieht, -miht, e ; *f. Weakness, lack of power :*—Heora unmiht and heora untrymð is swíðe gemanifealdod *multiplicatae sunt infirmitates eorum,* Ps. Th. 15, 3. Biþ geond fingras cele and cneówa unmeht, Lchdm. ii. 258, 14. Gif ðé þince ðæt ðú maran lǽcedóm dón ne durre for unmihte ðæs mannes, 254, 1 : Homl. Skt. i. 13, 21. Ða ongan ic þencan ðæt mé ðæt gelumpe for ðære wíflícan unmihte, ii. 23 b, 411. Ðurh unmihte, Homl. Th. ii. 42, 27. Hié ongietaþ hiera unbældo and hiera unmiehte (-mihte, Cott. MSS.) *infirmitatis suae sunt conscii,* Past. 32 ; Swt. 209, 8. Hú magan ða cyningas forhelan hiora unmihte, ðonne hí ne magan nǽnne weorþscipe forþbringan búton heora þegna fultume ? Bt. 29, 1 ; Fox 104, 13. Oð unmihte, Ps. Th. 106, 17. Ne bióð ðæt náne mihta ðæt mon mæge yfel dón, ac beóþ unmihta, Bt. 36, 7 ; Fox 182, 28. Hió his unmehta hine gemyndgaþ *infirmitatis memoriam ad mentem revocat,* Past. 65 ; Swt. 465, 32. Wurþaþ hig þurh ðæs niettruman unmihta beswicene, Wulfst. 285, 11. [Monnes unmihte *hominis infirmitatem,* Kath. 1022. *Goth.* un-mahts : *O. H. Ger.* un-maht *inbecillitas, inpotentia, infirmitas : Icel.* ú-máttr.] v. módunmeaht.

un-meaht ; *adj. Impossible :*—Nǽniht unmæht bið iúh *nihil inpossibile erit vobis,* Mt. Kmbl. Lind. 17, 20. [Un-maht *impotent,* A. D. 297.]

un-meahtelíc (-mihte-, -miht-) ; *adj. Impossible :*—Eów ne byð ǽnig þing unmihtelíc *nihil inpossibile erit vobis,* Mt. Kmbl. 17, 20 : Lk. Skt. 1, 37. Unmihtlíc is *inpossibile est,* 17, 1. Ða þing ðe mannum synt unmihtelíce, 18, 27. [*O. H. Ger.* un-mahtlíh *impossibilis : Icel.* ú-máttuligr *impossible.*] v. un-meahtiglíc.

un-meahtig, -mehtig, -mihtig ; *adj.* **I.** *not mighty, weak, impotent, of little power or means :*—Unmihtig *inpos,* Ælfc. Gr. 9, 31 ; Zup. 58, 2 : *inpotens,* 14 ; Zup. 87, 13. Ys Drihtnes hand unmihtig (*invalida*)? Num. 11, 23. Gif ic beó bescoren, ðonne beó ic unmihtig (*recedet a me fortitudo mea et deficiam*), Jud. 16, 17, 19. Mannes fultum is unmihtig and ídel *vana salus hominis,* Ælfc. T. Grn. 11, 41. Ðé læs ðe unmihtig man feorr for his ágenon swince, L. Ff. ; Th. i. 224, 27. Tó ánum mǽdene unmihtigum tó wífe, Homl. Skt. i. 10, 257. Ðý læs mon unmihtigne man tó feor for his ágenan swencte, L. Eth. ii. 9 ; Th. i. 290, 3. Ongit hú unmihtige ða yfelan men beóþ *vide quanta vitiosorum hominum pateat infirmitas,* Bt. 36, 5 ; Fox 180, 2 : 29, 1 ; Fox 104. 12 : 36, 2 ; Fox 174, 27. Unmehtige, Met. 24, 62. Ne beþǽce man hine sylfne, ðæt hé secge, ðæt ǽnig hád sý unmihtigra ðonne óðer, Homl. Th. i. 284, 17. Swá hwæt swá unmihtigre bið, ðæt ne bið ná God, 228, 27 : Hexam. 3 ; Norm. 6, 7, 5. Hú micle unmihtegran hí wǽron, Bt. 36, 5 ; Fox 180, 4. **II.** *impossible :*—Ðis unmæhtig is *hoc impossibile est,* Mt. Kmbl. Lind. 19, 26 : Mk. Skt. Lind. Rush. 10, 27. Unmǽhtigo (-mǽhtge, Rush.) *inpossibilia,* Lk. Skt. Lind. 18, 27. [Adam bicom unmihti, O. E. Homl. ii. 35, 8. Þilke unmyʒty tyraunt, Chauc. Boet. 13, 242. *Goth.* un-mahteigs *weak ; impossible : O. H. Ger.* un-mahtíg *invalidus, infirmus, imbecillis : Icel.* ú-máttigr *weak, infirm.*]

un-meahtiglíc ; *adj.* **I.** *weak :*—Hé biþ unmehtiglíc, Lchdm. ii. 60, 8. Unmihtiglíc, iii. 74. 23. **II.** *impossible :*—Unmæhtiglíc (-iclíc, Lind.) *inpossibile,* Lk. Skt. Rush. 1, 37 : 17, 1. Suíðe unmæghtiglíc, Lind. 18, 24. v. un-meahtelíc.

un-meahtigness, e ; *f. Weakness, impotence :*—Sume men secgeaþ, ðæt heó him unmihtignesse and untrumnysse on gebríncge, Lchdm. i. 248, 23. Heó fremaþ wið ða unmihticnysse ðæs migdan and wið ðæra innoða ástyrunga, 272, 16.

un-medume (-ome, -eme) ; *adj. Unmeet, unfit, unworthy :*—Se ðe him ondrǽdan sceal ðæt hé unmedome (-eme, Cott. MSS.) sié *hoc indignus pertimescat,* Past. 11 ; Swt. 73, 21. Him ðúhte selfum ðæt hé wǽre swíðe unmedeme *parvulum se in suis oculis viderat,* 17 ; Swt. 112, 11. Ic swíðe unmedodum nédde tó Dryhtnes líchoman, Anglia xi. 99, 70. Hwylc ðæt unmedeme gód wæs *quae sit imperfecti boni forma,* Bt. 34, 1 ; Fox 134, 5. Drihten, ðú ðe eall medemu geworhtest and náht unmedemes, Shrn. 165, 31. Hát mé unmedemre ða duru beón untýnede, Homl. Skt. ii. 23 b, 447. Ongitan hú micelne unweorþscipe se anwald brengþ ðam unmedeman, Bt. 27, 2 ; Fox 96, 10. Ða ðe unmedome bióð tó ðære láre for unwisdóme *quos a praedicatione imperfectio prohibet,* Past. 49 ; Swt. 375, 18. Gestíran ðære willunge ðǽm unmedemum, ðæt hiera nán ne durre grípan on ðæt ríce *ne imperfecti culmen arripere*

regiminis audeant, 4; Swt. 41, 5. Ða ðe hé unmedume gemétte ðes Godes geleáfan, Anglia x. 141, 18. [Cf. O. H. Ger. un-metamî *intemperies*.]

un-medumlíce; adv. *Unmeetly, unworthily*:—Is swíðe frécendlíc ðæt ðæm húsle hwá ungeclǽnsod and unmedomlíce onfoo, L. E. G. 44; Th. ii. 440, 22. Minne hád ic hæbbe unmedumlíce gehealden, L. de Cf. 9; Th. ii. 264, 10. Unmeodomlíce, Anglia xi. 99, 60.

un-meltung, e; f. *Indigestion*:—For unmeltunge, Lchdm. ii. 254, 1.

un-menged; adj. *Unmixed*:—Se unmengeda *non mixta*, Wrt. Voc. ii. 59, 78. Hé (*Adam*) of ðære eorðan selfre unmængedre gesceapen wæs, Anglia xi. 1, 9. v. un-gemenged.

un-mennisclíc; adj. *Inhuman*:—Wé hérdon on ealdum spellum, ðæt sum sunu ofslóge his fæder, ic nát húmeta, búton wé witon ðæt hit unmennislíc (-líce, Cott. MS.) dǽd wæs *nimis e natura dictum est, nescio quem filios invenisse tortores*, Bt. 31, 1; Fox 112, 16.

un-met[t], es; n. *Excess*:—Of ðam unmetta and ðam ungemetlícan gegerelan, of ðam swétmettum and of mistlícum dryncum (cf. of ungemete ælces þinges, wiste and wǽda, wíngedrinces, and of swétmetann (-mettum?), Met. 25, 38), Bt. 37, 1; Fox 186, 16. [Cf. O. H. Ger. un-mez:—In *guotis unmezze in luxuria*.] v. next words.

un-met[t], -mete; adj. *Without measure, immense, excessive*:—Wæs seó éhtnysse unmetre eallum ðam ǽrgedónum *quae persecutio omnibus anteactis immanior fuit*, Bd. 1, 6; S. 476, 23. [O. H. Ger. un-mez, -mezzi *immensus, ingens*.]

un-metlíc; adj. *Immoderate, excessive*:—Hí swá unmetlícre (-um?) ege fóron, Guthl. 5; Gdwin. 36, 3.

un-metlíce; adv. *Immensely, exceedingly*:—Ða columnan wǽron unmetlíce greáte heáhnisse upp *columnae ingenti grossitudine atque altitudine*, Nar. 4, 22. v. un-gemetlíce.

un-micel; adj. *Not great, little*, Greg. Dial. 2, 15.

un-midlod; adj. *Unbridled, unrestrained*:—Unmidled *effrenus*, Wrt. Voc. ii. 142, 60. Ða upáhǽfenan weorðaþ unmidlode and áðundene geniédde mid hiera upáhæfenesse *elatos effrenatio impellit tumoris*, Past. 41; Swt. 302, 10. Swá ða ofermódan ne weorðen unmidlode *ut superbis non crescat effrenatio*, 60; Swt. 453, 21. v. un-gemidlod.

un-miht, un-mihtan. v. un-meaht, ge-unmihtan.

un-milde; adj. *Ungentle, harsh, rude*:—Gif hé is unmilde and oferhýdig *si inmitis et superbus est*, Bd. 2, 2; S. 503, 7. [Hæþenndom iss unnmeoc and all unnmilde, Orm. 9880. Þu (*the owl*) art unmilde, O. and N. 61. *Goth.* un-milds ἄστοργος: O. H. Ger. un-milti *inmitis*: Icel. ú-mildr.]

un-mildheort; adj. *Hard-hearted, merciless, pitiless*:—Se déma betǽcþ ða unrihtwísan ðam unmildheortan wítnere, Homl. Ass. 8, 203. Ðe unmiltheortne welige *inmisericordem divitem*, Lk. Skt. p. 9, 2. Mé cóman tó Sílhearwan . . . hí wǽron unmildheorta, and mé tugon tó ðære sweartan helle, Homl. Skt. i. 4, 289.

un-milts, e; f. *Sternness, wrath*:—Hæbbe hé Godes unmiltse *may the wrath of God abide on him*, Cod. Dip. Kmbl. ii. 4, 2: Cod. Dip. B. ii. 315, 21.

un-miltsigendlíc; adj. *Unpardonable*:—Hwî wæs ðæs heáhengles syn unmiltsigendlíc, and ðæs mannes miltsigendlíc? Btwk. Scrd. 17, 21.

un-miltsung, e; f. *Want of consideration, impiety* towards God, *pitilessness* towards men:—Gif hié genunan willaþ hiora ieldrena unmiltsunge ðe hié tó Gode hæfdon, ge eác him selfum betweónum *if they will remember their forefathers' impiety to God, and pitilessness among themselves*; recolant majorum suorum tempora sceleribus exsecrabilia, dissensionibus foeda, Ors. 2, 1; Swt. 64, 16.

un-mirigþ. v. un-myrhþ.

un-mód, es; n. *Despondency, dejection*:—Of ðæs magan ádle cumaþ monige ádla . . . on unmóde and on ungemetwæccum, Lchdm. ii. 176, 1. [O. H. Ger. un-muot *perturbatio*: Ger. un-muth.]

un-módig; adj.   I. in a depreciatory sense, *without courage, fainthearted, pusillanimous*:—On óðre wísan sint tó manianne ða módgan, on óðre ða unmódgan and ða undrístan (*pusillanimes*) . . . ða unmódgan and ða ungedyrstigan wénað ðæt ðæt suíðe forsewenlíc sié ðætte hié dóð, and fordón weorðaþ oft ormóde, Past. 32; Swt. 200, 1–12.   II. in a good sense, *not proud, diffident, humble*:—Geclǽnsa mé ða hwíle ðe ic on ðisse worulde sí, and gedó mé unmódigne, Shrn. 171, 1.

un-módigness, e; f. *Pride, arrogance* (un- giving a bad sense):—Ic ondette ofermétto and unmódennesse (nesse *is written above* móden), Anglia xi. 98, 33.

un-molsniendlíc. v. un-formolsniendlíc.

un-murn; adj. *Untroubled*:—Hí slǽp hiora [swǽfon?] sylfum unmurne *dormierunt somnum suum*, Ps. Th. 75, 4. [Cf. Aylmar aȝen gan turne wel modi and wel murne, Horn. 704. O. H. Ger. morna *moeror*. French morne. See also murcen.] v. next word.

un-murnlíce; adv. *Carelessly, without compunction, without anxiety*:—Brond ǽleþ ealdgestreón unmurnlíce, Exon. Th. 51, 9; Cri. 813. Blódig wæl eteþ áugenga unmurnlíce, Beo. Th. 903; B. 449. Se ðe unmurnlíce mádmas dǽleþ, egesan ne gýmeþ, 3516; B. 1756.

un-myndlinga; adv.   I. where an act is not intended or expected by the doer of it, *undesignedly, without meaning* to do something:—Nis hit nán wundor ðeáh hwá wéne ðæt swylces hwæt unmyndlinga gebyrige, þonne hé ne can ongitan for hwí God swylc geþafaþ *nec mirum, si quid ordinis ignorata ratione, temerarium confusumque credatur*, Bt. 39, 2; Fox 214, 9. Gif hé unmyndlunge (*without having previously intended to do it*) ceáp áredige út on hwylcere fare, L. Edg. S. 8; Th. i. 274, 23. Maurus arn uppon ðam streáme unmyndlunge (*unaware of what he was doing*), swilce hé on fæstre eorðan urne . . . undergeat æt néxtan ðæt hé uppon ðæm wætere arn, and ðæs micclum wundrode, Homl. Th. ii. 160, 9. Hí unmyndlinga (*unintentionally*) swíðe fǽsthealdne weorcstán upp áhwylfdon, Homl. Skt. i. 23, 423.   II. where an act is unexpected by the object of it, *unexpectedly*:—Hé hiene spón ðæt hé on Umenis unmyndlænga (*de insperato*) mid hære become, Ors. 3, 11; Swt. 146, 8. Bútan hit swá limpe ðæt hwylc cuma unmyndluncga cume, R. Ben. 67, 12. Gif him ǽfre unmendlinga geberede ðæt . . . , Met. 25, 30. [Hire wone is to cumen bi stale, ferliche and unmundlunge hwen me least weneð, O. E. Homl. i. 249, 20. Þe ȝeape wrastlare mid þen ilke turn mei his fere unmunlunge aworpen, A. R. 280, 10.]

un-mynegod; adj. *Undemanded*:—Gif preóst geárgerihta unmynegode lǽte *if a priest leave the yearly dues without payment asked*, L. N. P. L. 43; Th. ii. 296, 15. v. mynegian, II c.

un-myrhþ, e; f. *Sadness, misery*:—On unmyrhðe his líf geendian, Wulfst. 148, 9.

un-myrige; adj. *Unpleasent, unfair*:—Unmyrge plega *collidium* (= colludium; cf. colludium, turpis ludus, Corp. Gl. ed. Hessels 35, 643), Wrt. Voc. ii. 134, 72. [Ne beo þe song never so murie þat he ne schal þinche unmurie ȝef he ilesteþ over unwille, O. and N. 346.]

unna, an; m.: unne, an; unn, e; f.   I. *grant, allowance, permission*:—Ic cýðe eów ðæt hit is mín fulla unna, ðæt heó becweðe hire land *I declare to you that she has my full permission to bequeathe her land*, Cod. Dip. Kmbl. iv. 200, 27 : 223, 24. Hit is mín unna and mín fulle leáfe ðæt hé dihte privilegium, vi. 203, 23. Mid unnan Godes and his hálgena . . . mit unnan híredes, ii. 58, 23, 25. Se wæs tó Eoferwícceastre be cinges unnan and ealra his witena tó ærcebisceope gehálgod, Chr. 971; Erl. 125, 36. Habban hí ðone feórðan pening be mínre unnan *omnem quartum nummum fratribus reddendum censeo*, Cod. Dip. Kmbl. iii. 61, 16. Be mínre unne and gife habban hí and wealdan *meo concessu et dono habeant et possideant*, iv. 200, 7. Hé mid his unne tó Scotlande fór, Chr. 1093; Erl. 229, 20. Bútan hé ðæs abbodes unnan begite *nisi ea abba jubeat*, R. Ben. 94, 9. Hé eów sige forgeaf þurh unnan ðæs Ælmihtigan, Homl. Th. i. 506, 27 : Homl. Skt. i. 3, 556.   II. *willingness* to give, *pleasure* in doing something:—Gelǽste hé Gode his teóðunga mid ealre blisse and mid eallum unnan . . . Gif hé hit mid unnan and fulre blisse dón wolde, L. Edg. S. 1; Th. i. 272, 2, 13.   III. *a grant, what is given*:—Se ðe ðás gyfu and ðisne unnan wille Gode and sancte Petre ætbrédan, Cod. Dip. Kmbl. iv. 276, 31.

unnan; prs. ic, hé an[n], pl. wé unnon; p. úðe.   I. *to grant a person* (*dat.*) *something* (*gen.*), *to give, allow*:—Gé gehírað hwæs ic Gode ann, L. Ath. i. prm. ; Th. i. 194, 14. Ic an Eádwearde ðæs landes, Chart. Th. 487, 18, 32. Ic ðe an tela sincgestreóna, Beo. Th. 2455; B. 1225. Ðæs steápes onféhð ðe hé an *he receives the cup to whom God gives it*, Ps. Th. 74, 7. Gif mé Waldend an lengran lífes, Cd. Th. 110, 18; Gen. 1840. Hæfde gefohten foremǽrne blǽd, swá hyre God úþe, Judth. Thw. 23, 16; Jud. 123. Gif ic mínum eágum unne slǽpes *si dedero somnum oculis meis*, Ps. Th. 131, 4. Ðæt mé unne God écan dreámes, Exon. Th. 454, 13; Hy. 4, 32. Eal folc geceás Eádward tó cynge; healde ða hwíle ðe him God unne, Chr. 1041; Erl. 169, 5. On ða gerád ðe ge mé unnan mínes, L. Ath. i. prm. ; Th. i. 198, 1 note. Gif hié him ðæs ríces úþon, Chr. 755; Erl. 50, 17. Ic feores ðe unnan wille, Exon. Th. 254, 4; Jul. 192 : Andr. Kmbl. 292; An. 146. Hé ða bóc unnendre handa hire tó lét *librum bona voluntate dimisit*, Chart. Th. 202, 36. Bóc and land ðelácan unnendere heortan, 376, 5. Unnende móde, 126, 22.   I a. with dat. of person and clause:—Him God úðe, ðæt hé hyne sylfne gewræc, Beo. Th. 5741; B. 2874. Þenden lifes weard unnan wolde, ðæt hé blǽdes hér brúcan móste, Exon. Th. 158, 2 ; Gú. 902.   II. *to wish something* (*gen.*) *to a person* (*dat.*):—Ða ðe mé yfeles unnon *them that wish me evil* (A. V.); qui cogitant mihi mala, Ps. Th. 39, 17. Ne dyde ic, ðæs ic ðe weán úðe (*because I wished you woe*), Cd. Th. 163, 3; Gen. 2692. Geweard ðætte Perse gebudan frið eallum Créca folce, næs ná for ðæm ðe hié him ǽnigra góda úþen (*non quod misericorditer fessis consuleret*), Ors. 3, 1; Swt. 98, 31. [Gledieð alle wid me, ðæt me god unnen, Marh. 21, 22. Þine feond þe þe ufel unnen, Laym. 28117. Þe mire dohter wel on, 11928. Þu hit (*sorrow*) myht segge swyhc mon þat [hit] þe ful wel on, wyþute echere ore he on þe muchele more (he wolde þad þu heuedest mor, 2nd Text), Misc. 116, 238].   III. *to wish something* (*gen.*) *for a person* (*dat.*), *to like a person to have something*:—Se arcebiscop wénde ðæt ðæt biscoprice sum óðer mann ábiddan wolde, ðe hé his wyrs truwude and úðe (*somebody else, that he would have been worse pleased should have it*), Chr. 1043; Erl. 169, 28. Oft hit gesǽleþ ðæt his ǽhta weorþaþ on ðæs

onwealde ðe hē ǽr on his lífe wyrrest úþe *it often happens that his property gets into the power of the man that when alive he would have been least pleased should have it*, Blickl. Homl. 195, 4. **III a**. with dat. of person and a clause:—Ðæt is, ðæt hwā fare mid his mōde æfter his niéhstan, and him unne ðæt hē tō ryhte gecierre *that is, that a man go in spirit after his neighbour, and be glad that he turn to right*, Past. 46; Swt. 349, 14. **IV**. *to* like a condition of things, *to be pleased* :— Hē ne úþe, ðæt ǽnig ōþer man mǽrða mā gehédde ðonne hē sylfa, Beo. Th. 1010; B. 503. Úþe ic swíþor ðæt ðū hine selfne geseón mōste *I should have been much better pleased, that you could have seen the creature himself*, 1925; B. 960. Ne meahte hē, ðeáh hē úðe wel, on ðam frumgāre feorh gehealdan *he could not keep life in the prince, though he would have been well pleased to do it*, 5703; B. 2855. [Hwer ich habbe iwiket, ich on wel þ ȝe witen, Kath. 1744. O. H. Ger. unnan: Icel. unna.] v. ge-, of-unnan.

unne. v. unna.

un-neáh; *adj. Distant, far* :—In lond unnēh *in regionem longinquam*, Lk. Skt. Lind. 15, 13 : 19, 12.

un-neáh *not near*. **I.** as adv. *far* :—Syndan ealle hí fram ǽ ðínre unneáh gewiten *a lege tua longe facti sunt*, Ps. Th. 118, 150. **II.** as prep. *far from* :—Óþlæ unnēg, Txts. 127, 1.

un-nédige, un-nēh, un-net. v. un-nídige, un-neáh, un-nyt.

unnend, es; *m. One who grants* :—Unnend ꝉ forgefend *prestabilis*, Rtl. 5, 10.

un-nídige; *adv. Without compulsion, willingly* :—Ða ðe beóð mid hira ágnum byrðennum ofðrycte ðæt hié ne magon gestondan, hié willaþ lustlíce underfón ōðerra monna, ond unniédige hié underlútaþ mid hira sculdrum ōðerra byrðenna tōeácan hiera ágnum *qui ad casum valde urgetur ex propriis, humerum libenter opprimendus ponderibus submittit alienis*, Past. 7; Swt. 52, 25. Oft hit gebyreþ ðætte manige men bióþ swā ungetrume ǽgþer ge on mōde ge on líchoman ðæt hí ne magon ne nān god dón, ne nān yfel nyllaþ unnēdige, Bt. 39, 10; Fox 228, 1.

un-nídunga; *adv. Without necessity* or *compulsion* :—Hē ðurh his ágene geornfulnesse gesyngaþ unniédenga *desiderio peccatur*, Past. 37; Swt. 265, 12.

un-níþing, es; *m. Not a rascal, an honest man* :—He beád ðæt ǽlc man ðe wǽre unníðing sceolde cuman tō him, Chr. 1087; Erl. 226, 2.

un-nyt[t]; *adj. Useless, vain, idle, unprofitable* :—Unnyt sprǽc *fabula*, Wrt. Voc. ii. 146, 64. Nān brōðor ymbe ídelnesse and unnette sprǽce (unnytte sprǽca, *v. l.*) beó . . . ne biþ hē nā him ānum unnyt *ne frater vacet otioso et fabulis . . . non solum sibi inutilis est*, R. Ben. 74. 15–18. Unlǽde bið on eorðan, unnyt lífes, se þurh ðone cantic ne can Crist geherian, Salm. Kmbl. 42; Sal. 21. Nis him nān wuht unnyt ðæs ðe hē gesceóp, Bt. 39, 5; Fox 218, 17. Ðes wída grund stōd ídel and unnyt, Cd. Th. 7, 14; Gen. 106: Beo. Th. 830; B. 413. Unnet gelp. Met. 10, 17. Adó of his mōde fela ðara ymbhogona ðe him unnet sié (cf. ungerísenlíce ymbhogan, Bt. 29, 3; Fox 106, 19), 16, 6: 22, 10. Hit wæs unnet gebod, Bt. 41, 3; Fox 246, 32. Se unnytta and forhogoda *inrita*, Wrt. Voc. ii. 48, 65. His word beóð gehwyrfedo tō unnyttre ofersprǽce, Past. 21; Swt. 165, 17. Ðý unnyttan *nugaci*, Wrt. Voc. ii. 94, 69. Gehæft mid ðære unnyttan lufe ðisse middangeardes, Bt. 34, 8; Fox 144, 25. Unnytne gefeán, Met. 5, 27. Unnytne andan, Bt. 39, 3; Fox 214, 33. Unnetne, Met. 28, 52. Hwæþer ðæt sié tō talianne wáclíc and unnyt *num imbecillum, ac sine viribus aestimandum est?* Bt. 24, 4; Fox 86, 16. Ðone ídelan hlísan and ðone unnyttan gilp, 19; Fox 68, 21. Ðone unnyttan þeówan *inutilem servum*, Mt. Kmbl. 25, 30. Hí lufiaþ ðæt hí sýn ídle and unnytte *inutiles facti sunt*, Ps. Th. 13, 4. Unnetta saca *vain disputes*, Met. 25, 44. Heora hæþenan gild wǽron ídelu and unnyt, Blickl. Homl. 223, 2. Áfyr fram ðé ða yfelan sǽlþa and ða unnettan, and eác ða unnettan ungeselþa, Bt. 6; Fox 14, 32. Ðonne ðæt mōd bið on monig tōdǽled, hit bið on ánes hwæm ðý unnyttre . . . Oft ðonne mon forlēt ða fæstrǽdnesse . . . hine spænd his mōd tō swíðe monegum unnyttum weorce . . . Hē swíður his mōd gebint tō ðam unnyttan (-nyttran, Hatt. MS.) weorcum ðonne hē ðyrfe, Past. 4; Swt. 36, 14–21. Rǽd biþ nyttost, yfel unnyttost, Exon. Th. 341, 2; Gn. Ex. 120. Ðæs hādes men ðe hwýlum wǽron nyttoste . . . syndon nū unnyttaste, L. I. P. 14; Th. ii. 322, 21. [*O. E. Homl.* un-nit, -net, -nut : A. R. un-nut, -net : Orm. un nitt : Goth. un-nutis *inutilis* : O. H. Ger.* un-nuz[z] *inutilis, cassus, otiosus, ignavus* : Icel. ú-nýtr.] v. next word.

un-nyt[t], es; *n.* **I.** *a vain thing, vanity, frivolity* :—Ne geríseþ ǽnig unnytt mid bisceopum, ne doll ne dysig, L. I. P. 9; Th. ii. 314, 30. Ðū hātodest ða ðe beeodon ídelnesse and ða ðe unnyt worhton *odisti observantes vanitates supervacue*, Ps. Th. 30, 6. Wē lǽraþ ðæt man æt ciricwæccan ǽnig unnit ne dreóge, L. Edg. C. 28; Th. ii. 250, 13 : 26; Th. ii. 250, 5 : 65; Th. ii. 258, 12. Ðonne mæg hē ongitan ðæt yfel and ðæt unnet, ðæt hē ǽr on his mōde hæfde, Bt. 35, 1; Fox 154, 26. Hwý gē ymb ðæt unnet swincen? Met. 10, 21. Hwí smeágaþ hí unnytt *quare meditati sunt inania*, Ps. Th. 2, 1. Unnyttu ꝉ ídelnyssa sprǽcon ánra gehwilc *vana locuti sunt unusquisque*, Ps. Lamb. ii. 3. **II.** *an evil thing, iniquity* :—Hí unnyt sæcgeaþ *loquentur iniquitatem*, Ps.

Th. 93, 4. [On unnet *in vain*, O. E. Homl. i. 107, 3. He isihð and ihercð oðerhwule unnut, and spekeð umbe hwule, A. R. 352, 28.]

un-nyt[t], e; *f. Ill use, disadvantage, hurt* :—Gif hié ða trumnesse ðære Godes giefe him tō unnytte (-nyte, Hatt. MS.) gehweorfaþ *si incolumitatis gra'iam ad usum nequitiae inclinent*, Past. 36; Swt. 246, 8. Ic andette eal ðæt ic ǽfre mid eágum geseáh tō gítsunge oððe tō tǽlnesse, oþþe mid eárum tō unnytte gehýrde, oþþe mid mínum múðe tō unnytte gecwæd, L. de Cf. 8; Th. ii. 264, 1–2. Lā hwæt fremaþ cyrichatan cristendóm on unnyt *see what Christianity can do to the disadvantage of the church's foes*, Wulfst. 67, 19. v. nytt.

un-nytlíc; *adj. Useless, unprofitable* :—Ðysse wyrte wyrttruma is unnytlíc (-net-, *v. l.*), Lchdm. i. 258, 4. Ðætte hē ne ðōhte nāht ungesceádwíslíces ne unnytlíces (-net-, Hatt. MS.) *nec indiscretum quid vel inutile cogitet*, Past. 13; Swt. 76, 12. [O. H. Ger. un-nuzlîh *inusitatus*.]

un-nytlíce; *adv.* **I.** *uselessly, vainly, to no purpose* :—Unnytlíce *inaniter*, Wrt. Voc. ii. 48, 43 : *nugaciter*, 80, 33 : 60, 62. Ðý læs hié unnytlíce forweorpen ðæt ðæt hié sellen for hira hrædhýdignesse *ne praecipitatione hoc, quod tribuunt, inutiliter spargant*, Past. 44; Swt. 321, 17 : 15; Swt. 95, 24. Unnytlíce wē swincaþ, gif wē his nabbaþ ðý mǽran danc, Bt. 41, 2; Fox 246, 21. **II.** *to ill purpose* :—Suā hié egeleáslícor and unnytlícor brúcaþ Godes giefe *quo bonis Dei male uti non metuunt*, Past. 36; Swt. 247, 10.

un-nytlícness, e; *f. Uselessness, unserviceableness* :—Wið ðæra eárena unnytlícnysse, and wið ðæt man wel gehýran ne mæge, Lchdm. i. 212, 3 : 214, 20.

un-nytness, e; *f. Uselessness, frivolity, vanity, triviality* :—Unnytnis *nugacitas*, Wrt. Voc. ii. 115, 5. Unnytnes, 60, 24. Ærendwrecan unnytnesse *nugigerelus*, 60, 21. Hí on unnytenesse gewordene synt *inutiles facti sunt*, Ps. Spl. T. 13, 4. Ðæt on ðam hālgan Sunnandæge nān man hine tō unnytnesse tō swíðe ne geþeódde, Wulfst. 227, 6. Wē forbeódaþ ðæt ge geflítu, ge plegan, ge unnytte word, ge gehwelce unnytnesse in ðām hālgan stōwum tō dónne, L. E. I. 10; Th. ii. 408, 23.

un-nytwirðe; *adj. Not fit for use, useless, unprofitable, unserviceable* :—Hē nis ðæt ān unnytwurðe *non solum sibi inutilis est*, R. Ben. Interl. 83, 9. Wē beóð gehātene yfele þeówan and unnytwyrðe, Homl. Ass. 57, 150. Unnytwyrþe gewordene hig synt *inutiles facti sunt*, Ps. Lamb. 52, 4.

un-nytwirðlíce; *adv. Uselessly, vainly, to no purpose, unprofitably* :—Ðæra wiðercorenra wíte tiht oft heora mōd unnytwurðlíce tō lufe, Homl. Th. i. 332, 29. Ágyldan gesceád ealra ðæra ýdelnyssa ðe hí unnytwurðlíce nū begāð, ii. 220, 31. Se forlýst ða gife ðe hē unnytwurðlíce underfēng, 556, 16.

un-ofercumen; *adj. Unsubdued* :—Unobercumenre (-ofaercumenrae, -ofercumenrae) *indigestae*, Txts. 71, 1097. Unofercumene *indigeste*, Wrt. Voc. ii. 45, 52.

un-oferfēre; *adj. Not to be crossed, impassable* :—Unoferfoere *intransmeabili*, Txts. 73, 1144. Unoferfēre, Wrt. Voc. ii. 45, 68.

un-oferhrēfed; *adj. Not roofed over* :—Seó cirice is ufan open and unoferhrēfed, Blickl. Homl. 125, 26, 31.

un-oferswíðed; *adj. Unconquered* :—Unoferswíþed hiht *invicta spes*, Hymn. Surt. 123, 34. Ðín geþyld wē cunnon unoferswýþed, Guthl. 5; Gdwin. 30, 19. Ðú unoferswýðda Alexander in gefeohtum *invicte belli Alexander*, Nar. 29, 9.

un-oferswíðedlíc; *adj. Unconquerable* :—Sume men wǽron unoferswíþedlíce, swā ðæt hí nān ne mihte mid nānum wíte oferswípan *quidam suppliciis inexpugnabiles*, Bt. 39, 11; Fox 230, 1. v. un-oferswíðendlíc.

un-oferswíðende; *adj. Unconquerable, invincible* :—Geoffra ðíne lāc ðam unoferswíðendum Apolline, Homl. Skt. i. 14, 35.

un-oferswíðendlíc; *adj. Invincible* :—Unoferswíþendlíc weorud *invincibilis exercitus*, Bd. 1, 15; S. 483, 16.

un-oferwinnende (?); *adj. Not to be overcome* :—Ða unoferwinnene (-winnende (?), -wunnene (?). v. un-oferwunnen) *ineluctuabilis*, Wrt. Voc. ii. 84, 54 : 46, 78.

un-oferwinnendlíc; *adj. Invincible, unconquerable* :—Unoferwinnendlíce (-wunnendlíce, *v.l.*) here *invictissimum exercitum*, Ors. 3, 7; Swt. 112, 7. Unoferwinnendlíce halsbearga *loricam inextricabilem (inexpugnabilem*), Hpt. Gl. 424, 34.

un-oferwrigen; *adj. Not covered over* :—Ða sceame mínes líchaman hæbbende unoferwrigene, Homl. skt. ii. 23 b, 208.

un-oferwunnen; *adj. Unconquered* :—Ic ða mōste oferwinnan ðe ǽr wǽron unoferwunnen *qui ante hac invicti fuere viri, hos ego in pugna vici*, Ors. 4, 1; Swt. 156, 28.

un-ofslegen; *adj. Unslain* :—His brōðor geendode his líf on sibbe unofslegen, Homl. Th. ii. 544, 31.

un-onbindendlíc; *adj. Not to be unbound, indissoluble* :—Mid un-anbindendlícum racentum *irresoluto nexu*, Bt. 25; Fox 88, 6. v. un-ābindendlíc.

un-onwendedlíc; *adj. Unchangeable* :—Ne wyrð seó burh nǽfre onwend, ða hwíle ðe God byð unonwendedlíc on hire midle, Ps. Th. 45, 4.

un-onwendendlíc; *adj. Unchangeable, immutable :*—God āna unanwendendlíc wuniaþ *se immobilem conservat*, Bt. 35, 5; Fox 166, 9 note: Met. 20, 17 : 24, 43.

un-onwendendlíce; *adv. Unchangeably, immutably, without variableness :*—Ðæt hí gedencen hū hrædlíce se eorðlíca hlísa ofergǽd, and hū unanwendendlíce se godcunda ðurhwunaþ *ut pensent, humana judicia quanta velocitate evolant, divina autem quanta immobilitate perdurant*, Past. 59; Swt. 447, 30. Ic nāt hwæþer hit eall gewyrþan sceal unanwendendlíce, ðæt hē getiohhod hæfþ, Bt. 41, 3; Fox 248, 30 note: 250, 1, 2 note. Se wísa mon eall his líf lǽt on gefeán unonwendendlíce, 12; Fox 36, 24.

un-orne; *adj. Simple, plain, poor, mean, humble :*—Dunnere, unorne ceorl, Byrht. Th. 139, 18; By. 256. [Crist warrþ unnorne and wrecche and usell child forr þatt he wollde uss alle maken riche, Orm. 3368. He warrþ an unnorne and wrecche mann, 4884. Crist wass unncuþ ȝet, and unnwurrþ, and unnorne, 16163. Heo beo ful unorne, oðer of feir elde, A. R. 424, 5. ȝef hire laverd is forwurþe and unorne at bedde and at borde, O. and N. 1492. Horn nis noȝt so unorn; Horn is fairer þane beo he, Horn. 330. ·He brohte hine uppen unorne mare, þet bitacneð ure unorne fleis, O. E. Homl. i. 85, 3. Hiss (*John the Baptist*) fode wass unnorne, forr nass nan esstemete þær, Orm. 828 : 11548. Unnorne mete and wæde, 6337. Þet heo ne grucchie uor none mete, ne uor none drunche, ne beo hit neuer so unorne, A. R. 108, 2. Ower cloðes beon unorne, and warme, and wel iwrouhte, 418, 17. Mi stefne is bold and noht unorne, heo is ilich one grete horne, O. and N. 317. Swa (*as Nichodemus did*) to lefenn upponn Crist wass rihht unnorne læfe, Orm. 16809.] v. next word.

un-ornlíc; *adj. Poor, plain, mean :*—Hí nāmon him ealde gescý and unornlíc scrúd *they took old shoes and mean apparel ; tulerunt calceamenta perantiqua, induti veteribus vestimentis*, Jos. 9, 5·. [Arrchelauss flæh inntill oþerr land, and tære he wass unnorneliȝ (*in mean estate*), Orm. 8251. Cf. Lætenn swiþe unorneliȝ and litell off þe sellfenn, Orm. 3750: 7525 : 4886. Me wore leuere i wore lame, þanne men . . . him onne handes leyde unornelike, or same seyde, Havel. 1941.] v. preceding word.

un-pleólíc; *adj. Not dangerous, without risk*, (1) *as regards physical hurt :*—Unpleólíce hit bið on lytlum scipe and on lytlum wætere, ðonne on miclum scipe and on miclum wætere, Prov. Kmbl. 29. (2) *as regards moral hurt :*—Augustinus cwæð, ðæt unpleólíc sý þeáh hwá lǽcewyrte ðicge; ac ðæt hē tǽþ tó unālýfedlícere wíglunge, gif hwá ða wyrta on him becnitte, būton hē hí tó ðam dolge gelecge, Homl. Th. i. 476, 3. Sume teolunga sind ðe man earfoðlíce mæg būton synnum begān. Petrus hæfde unpleólíce teolunge, and hē būton pleó tó his fixnoðe gecyrde, ii. 288, 25. Ðás tācna sind dígle and unpleólíce, i. 306, 30.

un-pleólíce; *adv. Without danger, without risk, safely :*—Ða óðre apostoli to Godes hǽse leofodon be heora lāre unpleólíce; ac ðeáh hwæðere Paulus āna nolde ða ālýfdan bigleofan onfón, ac mid āgenre teolunge his neóde foresceáwode, Homl. Th. i. 392, 20.

un-rǽd, es; *m.* I. *evil counsel, ill-advised course, bad plan, folly :*—Scipia sǽde, ðæt hit (*the building of a theatre*) wǽre se mǽsta unrǽd and se mǽsta gedwola *dicens, inimicissimum hoc fore bellatori populo ad nutriendam desidiam, lasciviaeque commentum*, Ors. 4, 12; Swt. 210, 1. Eádríc gewende ðone cyning ongeán. Næs nān māra unrǽd gerǽd ðonne se wæs *Eadricus per dolum fecit exercitum Anglorum redire. Non fuit pejus concilium factum in Anglia de tali re*, Chr. 1016; Erl. 157, 22. Ðæs unrǽdes (*the building of the tower of Babel*) stíðferhð cyning steóre gefremede, Cd. Th. 101, 15; Gen. 1682. Gif ðú unrǽdes ne geswícest, Exon. Th. 249, 31; Jul. 120: 393, 14; Rä. 12, 10: 410, 6; Rä. 28, 12. Hí geeácnodon heora yfel and God mid weorcum gegremedon . . . swā ðæt hig Eglone þeówodon for heora unrǽde, Jud. 3, 14; Thw. p. 162, 29. Hý férdon on unriht and unrǽde fyligdon *ambulaverunt post vanitatem* (Jeremiah 2, 5), Wulfst. 49, 9. Absalon férde forð mid his unrǽde, and wolde his āgenum fæder feores benǽman, Homl. Skt. i. 19, 215. Ðæra hālgena líc woldon hí besencan on flóde, ac se ælmihtiga Scyppend wiðslóh ðam unrǽde. Sum wíf wæs ðe wiste heora unrǽd, ii. 29, 324. Tó his (*Lucifer's*) unrǽde gefæstnod, Ælfc. T. Grn. 2, 44: 4, 10: Cd. Th. 43, 33; Gen. 700. Hé intó Englelande mid māran unrǽde férde ðone him behófode, Chr. 1093; Erl. 229, 3. Wæs gesǽd ðæt hē wǽre on ðam unrǽde, ðæt man sceolde on Eást-Sexon Swegen underfón, Chart. Th. 539, 27. Agathocles gehydde untreówlíce wið hiene . . . On ðære hwíle ðe hē ðone unrǽd þurhteáh, Ors. 4, 5 ; Swt. 170, 13. Ne wend ðú ðé on ðæs folces unrǽd and unriht gewil *non sequeris turbam ad faciendum malum* (Ex. 23, 2), L. Alf. 41 ; Th. i. 54, 6. Ðæs engles mód ðe ðone unrǽd (*rebellion against God*) ongan ǽrest fremman, Cd. Th. 3, 3; Gen. 30. Hyra freá ǽrest unrǽd (*the setting up of the image*) efnde, 227, 13; Dan. 186. Hí þégun æppel unrǽdum (*ill-advisedly*), Exon. Th. 226, 9; Ph. 403. Ealle ðás ungesǽlþa ús gelumpon þuruh unrǽdas, Chr. 1011 ; Erl. 145, 22. II. *disadvantage, prejudice, hurt :*—Ic andette . . . ǽlcne glænȝc ðe tó mínes líchaman unrǽde ǽfre belimpe, L. de Cf. 7; Th. ii. 262, 28: Anglia xi. 98, 28. Hé helle outȳneþ ðam ðe líces wynne fremedon on unrǽd, Exon. Th. 364, 14;

Wal 70. Hí drugon heora sylfra ēcne unrǽd, Cd. Th. 116, 16 ; Gen. 1937. [*Laym.* un-ræd *ill-counsel.* Þat child his unred to rede wend, O. and N. 1464. Iacobes sunes deden unred, Gen. and Ex. 1906. *O. H. Ger.* un-rāt : *Icel.* ú-rāð; *n. bad counsel, an ill-advised step.*]

un-rǽden[n], e; *f. An ill-advised action :*—Hé ða unrǽden folmum gefremede, ofslóh bróðor sínne, Cd. Th. 60, 16 ; Gen. 982. [Cf. *Icel.* ú-rāðan *an ill-advised step.*]

un-rǽdfæstlíce; *adv. Unadvisedly, without heeding good counsel :*—Twā geár hē ríxode unrǽdfæstlíce, Homl. Skt. i. 18, 456.

un-rǽdlíc; *adj. Ill-advised, foolish, vain :*—Hié hæfdon plegan and oforgedrync, and dyslíce and unrǽdlíce hālsunga. Blickl. Homl. 99, 21. [*O. H. Ger.* un-rātlíh *inconsultus : Icel.* ú-rāðligr *inadvisable, inexpedient.*]

un-rǽdlíce; *adv. Unadvisedly, inconsiderately :*—Unrǽdlíce *inconsulte*, Wrt. Voc. ii. 80, 78 : 44, 79 : *inconsulte, inconsiderate*, Hpt. Gl. 474, 57 : 509, 64. Hé begann tó lufienne leahtras tó swíðe mid his cnihtum, ðe unrǽdlíce férdon on heora ídelum lustum, Ælfc. T. Grn. 17, 14. On ðám ænglum ðe unrǽdlíce mōdegodon, Homl. Skt. i. 13, 183. Ús gedafenaþ ðæt wē hit wēnon swíðor ðonne wē hit unrǽdlíce gesēþan, Homl. Th. i. 440, 31. [*O. H. Ger.* un-rātlícho *inconsulte : Icel.* ú-rāðliga.]

unrǽd-síþ, es ; *m. A foolish, unprofitable way :*—Ic unrǽdsíþas ōþrum stýre nyttre fōre, Exon. Th. 393, 2 ; Rä. 12, 4.

un-reht *not treated*, un-reht *wrong.* v. un-āreht, un-riht.

un-reordian (?) *to speak ill of, to abuse :*—Swā firenfulle fācnum wordum heora aldorǽgn unreordadon (on reordadon? v. on, B. III. 5), Cd. Th. 269, 1 ; Sat. 66.

un-rētan. v. ge-unrētan.

un-rēðe; *adj. Not fierce, gentle :*—Wē rǽdaþ be ðære culfran gecynde, ðæt heó is unrēðe on hire clawum, Homl. Th. ii. 44, 25.

un-rētu (-o); *f. Anxiety, disquiet :*—Ðā bǽdon mec míne geféran ðæt ic on swā micelre mōdes unrēto and nearonisse mec selfne mid fæstenne ne swencte *rogantibus amicis ne me anxietate et jejunio condeficerem*, Nar. 30, 23. v. un-rōt.

un-ríce; *adj. Not rich or not powerful, poor, humble :*—Ða ðe unríce synd and hafenleáse þearfan *pauperiores*, R. Ben. 104, 7. Mín geréfa oþþe ǽnig óðer man ríccre oþþe unríccre, L. Edg. S. 13 ; Th. i. 276, 26. Ic wāt ðæt ðú hefst ðone hláford ðe ðú treówast bet ðonne ðē silfum, and swā hefð eác manig ðæra ðe unrícran hláford hefð ðonne ðú hefst, Shrn. 196, 11. [*Icel.* ú-ríkr.]

un-riht; *adj. Wrong, evil, bad, unjust, unlawful, depraved, perverse :*—Ic eom geþafa ðæt hit nāuht unriht wǽre ðæt mon ða yfelwillendan men hēte nētenu *fateor nec injuria dici video vitiosos in belluas mutari*, Bt. 38, 2 ; Fox 196, 17. Ic geléfe ðætte ǽlc unriht wítnung sié ðæs yfel ðe hit dēþ, ðæs ðæs ðe hit þafaþ *apparet, illatam cuilibet injuriam non accipientis, sed inferentis esse miseriam*, 38, 6 ; Fox 208, 20. Unreht, 37, 2 ; Fox 188, 7. Se yfla, unrihta willa wōhhǽmetes *voluptas*, Met. 18, 1. Of unrihtum wege *de via iniquitatis*, Ps. Th. 106, 16. Gif hwá genìéd sié oþþe tó hláfordsearwe oþþe tó ǽngum unryhtum fultume, L. Alf. pol. i. 60, 5. Unryhtre ǽ *unlawful marriage*, Exon. Th. 260, 14 ; Jul. 297. Hē gedwolan fylde, unrihte ǽ *idolatry*, Elen. Kmbl. 2081 ; El. 1042. Ic wundrige for hwí swá rihtwís dēma gecige unrihte gife wille forgifan, Bt. 38, 3 ; Fox 202, 16. Tuoege wōhfullo ł unrehto *duo nequam*, Lk. Skt. Lind. 23, 32. Unrǣhto (-rehte, Rush.) ł wōh *prava*, 3, 5. Ða unrihtan men, Blickl. Homl. 231, 10. Āblinnan fram heora unrihtum gestreónum, 25, 5. Cynewulf benam Sigebryht his ríces for unryhtum dǽdum, Chr. 755 ; Erl. 48, 19. Mid hiera unryhtum bisenum *per exemplum pravi operis*, Past. 9 ; Swt. 59, 18. Fram sumum unrihtum lāreówum *a quibusdam perversis doctoribus*, Bd. 2, 15 ; S. 518, 30. Unrihte wegas ealle *omnem viam iniquitatis*, Ps. Th. 118, 104. Unrihte gemeta and wōge gewihta āweorpe man, Wulfst. 70, 3. Ða unrehtan *iniqua*, Ps. Surt. 9, 24. [*O. Frs.* un-riucht : *O. Sax.* un-reht : *O. H. Ger.* un-reht *improbus, injustus, iniquus, vitiosus : Icel.* ú-rēttr.] v. following words.

un-riht, es; *n.* I. *wrong, evil, iniquity, injustice :*—Unriht *injuria*, Wrt. Voc. ii. 49, 26. Hira unriht (*iniquitas*) wearð untýned, Ps. Th. 72, 5: Blickl. Homl. 175, 9. Hwylc unryht mæg bión māre *quae potest iniquior esse confusio?* Bt. 39, 9 ; Fox 224, 28. Unrihtes feala *iniquitates*, Ps. Th. 54, 3. Gē ðe unrihtes wyrceaþ *qui operantur iniquitatem*, 52, 5 : 58, 2 : 70, 3. Ða ðe unrihtes þenceaþ, 140, 11 : Blickl. Homl. 111, 11. Ðǽm mannum ðe heora synna and unrihtes geswícaþ . . . and nǽfre tó unrihtum ne gewendaþ, 193, 22 : Elen. Kmbl. 1029 ; El. 516. Ic him ðæs unrihtes (*seeking to destroy Jesus*) andsæc fremede, 941 ; El. 472. Ðæt hý bealodǽde, ǽlces unryhtes gescomeden, Exon. Th. 80, 5 ; Cri. 1303. Ða oferhýdegan, ðe mé unrihte (*or adv.*) grētan *injuste iniquitatem fecerunt in me*, Ps. Th. 118, 78 : Cd. Th. 78, 12 ; Gen. 1292. Full mið unrehte *plenum iniquitate*, Lk. Skt. Rush. 11, 39. Mid unryhte *wrongfully*, Chr. 823 ; Erl. 62, 23. Mid unrihte, Ps. Th. 55, 1 : 68, 28: Andr. Kmbl. 3116 ; An. 1561. Ne dēm nān unriht *non injuste judicabis*, Lev. 19, 15. Ðonne gē unriht wirceaþ *quando feceritis malum*, Deut. 31, 29. Eallum ðe unriht wyrceaþ *qui operantur iniquitatem*, Ps.

Th. 58, 5. Hí wyrceaþ unriht (*injustitiam*), 93, 4. Unriht dón, Bt. 38, 3; Fox 202, 12: Cd. Th. 217, 16; Dan. 23. Grendel unriht æfnde, Beo. Th. 2512; B. 1254. Unryht fremian, Exon. Th. 79, 16; Cri. 1291: Ps. Th. 118, 51. Unriht (*iniquitatem*) sprecan, 72, 6: 62, 9: Homl. Th. ii. 452, 6. Hé mid listum speón idese on ðæt unriht (*taking the forbidden fruit*), Cd. Th. 37, 13; Gen. 589. Hit wæs mid unriht him of genumen, Chr. 1072; Erl. 211, 8. Heó on unriht þóhtan *injusta cogitatio eorum est*, Ps. Th. 118, 118: Beo. Th. 5471; B. 2739: Elen. Kmbl. 1161; El. 582: Wulfst. 158, 10. Ðý læs hié on unryht hǽmen *propter fornicationem*, Past. 16; Swt. 99, 14: 51; Swt. 397, 19. Ne gemune ðú ealdra unrihta *ne memineris iniquitates antiquas*, Ps. Th. 78, 8: Wulfst. 156, 10. Babylonie mid monigfealdum unryhtum and firenlustum libbende wǽron, Ors. 2, 1; Swt. 64, 7. Unrihtum, Bt. 16, 4; Fox 58, 10: Blickl. Homl. 109, 20: Cd. Th. 259, 2; Dan. 685. Him gyldeþ God ealle ða unriht (*iniquitates*) ðe hí geearnedan, Ps. Th. 93, 22. Ǽr man áweódige ða unriht and ða mánweorc, Wulfst. 243, 19. **II.** *a defect :*—Ða ðe mid unrihte heora gecyndes beóþ geuntrumade *quae naturae suae vitio infirmantur*, Bd. 1, 27; S. 494, 21. [*O. Frs.* un-riucht : *O. Sax.* un-reht : *O. H. Ger.* un-reht *iniquitas, injustitia, injuria, nefas.*]

unriht-cyst, e; *f. Vice, excess :*—Ic andette ... unrihtgilp and ídel word and unrihtcysta and ælcne glængc (cf. ic ondette ... unnyttes gilpes bigong and ídle glengas, uncyste ..., Anglia xi. 98, 27) ðe tó mínes líchaman unrǽde ǽfre belimpe, L. de Cf. 7; Th. ii. 262, 28.

unriht-dǽd, e; *f. Evil-doing :*—Hé eall ðurh his unrihtdǽde áþýstrade *universa prave agendo obnubilavit*, Bd. 5, 13; S. 633, 33. God wyle ðæt Sunnandæg freóls beó fram eallum unrihtdǽdum and þeówetlícum weorcum, Wulfst. 292, 7. Cf. yfel-dǽd.

unriht-dǽde; *adj. :* unriht-dǽda, an ; *m. Evil-doing ; an evil-doer :*—Se synfulla and se unrihtdǽda *peccator et iniquus*, Ps. Lamb. 9 second, 3. Cf. yfel-dǽde, -dǽda.

unriht-déma, an ; *m. An unjust judge :*—Ða unrihtdéman, ðe démaþ ǽfre be ðám sceattum and swá wendaþ wrang tó rihte, Wulfst. 203, 25 : 298, 19.

unriht-dóm, es; *m. Wrong, iniquity :*—Hié for ðám cumble on cneówum sǽton, efndon unrihtdóm, swá hyra aldor dyde ... hyra freá unrǽd efnde, Cd. Th. 227, 7; Dan. 183.

unriht-dónde; *adj. Evil-doing ; substantive, an evil-doer :*—Him wæs beboden ðæt hí sceoldan ðǽm unrihtdóndum stéran, Blickl. Homl. 63, 12. Cf. yfel-dónde.

un-rihte; *adv. Not rightly, unjustly :*—Unrihte wé dydon *injuste egimus*, Ps. Spl. 105, 6 : 118, 78. Heora sylh unrihte gangaþ *aratra eorum non recte incedunt*, Bd. 5, 9; S. 623, 12. [*O. L. Ger. O. H. Ger.* un-rehto *injuste, improbe.*]

unriht-feóung, e; *f. Evil hatred, unjustifiable hate :*—Hwý gé ǽfre scylen unrihtfeóungum eówer mód drǽfan? Met. 27, 1. v. fióung.

unriht-gestreón, es; *n. Unrighteous gain :*—Ða ðe heora sylfra sáula forhycgaþ for feós lufan, and unrihtgestreón lufiaþ, Blickl. Homl. 63, 8.

unriht-gewil (? or unriht gewil). v. *first passage under* gewil.

unriht-gewilnung, e; *f. Evil desire :*—Ða ðe swíðe hrædlíce beóð oferswíðde mid sumre unryhtgewilnunge (cf. l. 33: ða ðe mid fǽrlíce luste bióð oferswíðde *qui repentina concupiscentia superantur*) *qui subito motu peccant*, Past. 56; Swt. 429, 30.

unriht-gilp, es; *m. n. Vainglory :*—Ic andette unrihtgilp and ídel word, L. de Cf. 7; Th. ii. 262, 27. Cf. ídel-, leás-gilp.

unriht-gítsung, e; *f. Greed, covetousness :*—Ða welan, and ðæt mycele gylp, and seó unrihtgítsung, and ðæt man ðæm earman forwyrne, ðæt is eal swíþe mycel synn beforan Gode, Blickl. Homl. 53, 21; Wulfst. 290, 26. Hé woruldsǽlþa ðe onlǽnde æfter his bebodum tó brúcanne, nallas ðinre unrihtgítsunga gewill tó fulfremmanne, Bt. 7, 5 ; Fox 24, 10. Ðæt wé gescildan ús wiþ ða eahta heáhsynna ... ðæt is morþor and stala, máne áþas and unrihtgítsunge ..., Engl. Stud. viii. 479, 96.

unriht-hǽman; *p. ðe To cohabit unlawfully, to commit adultery or fornication :*—Hé déð ðæt heó unrihthǽmð ; and se unrihthǽmð ðe forlǽtene genimð *facit eam moechari; et qui dimissam duxerit, adulterat*, Mt. Kmbl. 5, 32. Se unrihthǽmð *moechatur*, Lk. Skt. 16, 18. Heó unrihthǽmð, Mk. Skt. 10, 12. Ne unrihthǽm ðú *ne adulteris*, 10, 19: Homl. Th. ii. 208, 15. Ne unrihthǽme ðú *non moechaberis*, Deut. 5, 18: Mt. Kmbl. 5, 27. Fram unrihthǽmendum mæssepreóste *a fornicante presbytero*, L. Ecg. C. 17, tit. ; Th. ii. 128, 29.

unriht-hǽmdere, es ; *m. An adulterer :*—Mid unrehthǽmderum *cum adulteris*, Ps. Surt. 49, 18. v. unriht-hǽmend, -hǽmere.

unriht-hǽmed, es ; *n. Unlawful cohabitation, illicit intercourse, adultery, fornication :*—Cwéna geligr *vel* unrihthǽmed *adulterium*, Wrt. Voc. i. 21, 33. Forligr flǽsces unrihthǽmed is *fornicatio carnis, adulterium est*, Scint. 87, 14. Unrihthǽmed wífes *fornicatio mulieris*, 86, 15 : Met. 9, 6. Mid ðý sweorde unrihthǽmedes (-hǽmdes, Cott. MSS.) *luxuriae mucrone*, Past. 43; Swt. 313, 9. On ðæt hnesce bedd ðæs gesinscipes, næs on ða heardan eorðan ðæs unrighthǽmdes, 51 ; Swt. 397, 23. Ne ða unfæsðrádan ðe ne magon hira unryhthǽmdes geswícan *neque*

adulteri, neque molles, neque masculorum concubitores, Swt. 401, 28. Se yfela willa unrihthǽmedes (*voluptas*) gedrǽfð fulneáh ælces monnes mód ... sceal ælce sáwl forweorðan æfter ðam unrihthǽmede, Bt. 31, 2 ; Fox 112, 24: Met. 18, 10. On unrihthǽmede *in adulterio*, Jn. Skt. 8, 3. Be mónaðádles hǽmede and be óþrum unrihthǽmede (*de alio pravo coitu*), L. Ecg. C. 16, tit. ; Th. ii. 128, 26 : Shrn. 99, 4. Monige hié gehealdaþ wið unryhthǽmed *multi scelera carnis deserunt*, Past. 51 ; Swt. 399, 7. Hé onscunede unrihthǽmed *recusabat stuprum*, Gen. 39, 10. Gif wer unrihthǽmed fremeþ wiþ óþer wíf, Blickl. Homl. 185, 25. Unrihthǽmedu *adulteria*, Mt. Kmbl. 15, 19 : Bt. 16, 4 ; Fox 58, 1.

unriht-hǽmed; *adj. Adulterous :*—Unrihthǽmede mæn tó rihtum lífe mid synna hreówe tó són, L. Wih. 3 ; Th. i. 36, 18. Mid unrihthǽmedum *cum adulteris*, Ps. Spl. 49, 19.

unriht-hǽmend, es ; *m. An adulterer :*—Þeófum and mánswarum and unrihthǽmendum, Blickl. Homl. 63, 13.

unriht-hǽmere, es ; *m. An adulterer, a fornicator :*—Unrihthǽmere *adulter*, Kent. Gl. 169. Se ðe wísaþ on ðam forlǽtenum wífe bið unrihthǽmere geháten fram Gode, Homl. Th. ii. 322, 35. Unrihthǽmeras *adulteri*, Lk. Skt. 18, 11: Wulfst. 298, 16 : Homl. Th. ii. 324, 7 : Homl. Ass. 147, 94: *fornicarii*, Homl. Skt. i. 17, 38.

un-rihtlíc; *adj. Unrighteous, unjust, wicked, wrongful :*—Unryhtlícu iersung, ðæt is ðæt mon iersige on óðerne for his góde (*on account of his prosperity*), Past. 27 ; Swt. 189, 8. Hié him andwyrdon ðæt hit gemálíc wǽre and unryhtlíc ðæt swá oferwlenced cyning sceolde winnan on swá earm folc *responderunt, stolide opulentissimum regem adversus inopes sumsisse bellum*, Ors. 1, 10 ; Swt. 44, 11. Gif ælces mannes líf ǽfre sceole swá gán ðæt hé mæge forbúgan bysmorlíce dǽda, ðonne bið unrihtlíc ðæt ða unrihtwísan onfón wítnunge for heora wóhnysse, Homl. Skt. i. 17, 231: Homl. Th. i. 292, 5. Ne lufa ðú gítsunga ne unrihtlíce welan ... Beó ðé swíðe ælfremed ælc unrihtlíc gestreón, Basil admn. 9; Norm. 52, 18–21. For ðínum gódan willan ðú wéndest ðæt ðé nánwuht unrihtlíces on becuman ne mihte, Bt. 7, 3 ; Fox 22, 15. Þing unrihtlíc *rem injustam*, Ps. Spl. 100, 3. Hé ne mæg ðurhteón ðæt unryhtlíce weorc *nequaquam usque ad opus nefarium rapitur*, Past. 11; Swt. 73, 6. Mé egleþ swýðe ða unrihtlícan gefeoht ðe betwux ús sylfum syndan, L. Edm. S. prm. ; Th. i. 246, 24. Be unrihtlícum hǽmedum *de pravis coitibus*, L. Ecg. C. 21, tit. ; Th. ii. 130, 4. [*O. H. Ger.* un-rehtlíh : *Icel.* ú-réttligr.]

un-rihtlíce; *adv. Unrighteously, unjustly, wickedly, wrongfully :*—Unrihtlíce *injuste*, Ps. Spl. 68, 6 : 118, 78. Nóht unryhtlíce *non injuste*, Past. 39 ; Swt. 285, 3. Gé unrihtlíce libbaþ *inique agetis*, Deut. 31, 29. Wé syngodon, wé dydon unrihtlíce, Homl. Th. ii. 420, 26: Wulfst. 160, 4: Blickl. Homl. 59, 19. Unrihtlíce hí mé hatiaþ *odio iniquo oderunt me*, Ps. Th. 24, 17. Se wæs unrihtlíce ofslagen ofer áþas and treówa, Bd. 2, 20 ; S 521, 17. Se wæs unrihtlíce ádrǽfed, Chr. 1022; Erl. 161, 36. Hé nǽnigne man unrihtlíce fordémde, Blickl. Homl. 223, 32. Se ðe unwǽrlíce and unryhtlíce gewilnige *qui incaute expetiit*, Past. proem. ; Swt. 23, 15. For ðam sceatte ðe hé lufode unrihtlíce, Basil admn. 9 ; Norm. 54, 15. Ælc ðe hǽmð búton rihtre ǽwe, hé hǽmð unrihtlíce, Homl. Th. ii. 208, 16. [*O. H. Ger.* un-rihtlíhho : *Icel.* ú-réttliga.]

unriht-lust, es ; *m. Improper desire :*—Ðú eart scyldigra ðonne wé for ðínum ágnum unrihtlustum, Bt. 7, 5 ; Fox 24, 7.

unriht-lyblác, es ; *n. m. Sorcery :*—Ne galdorsangas ne unrihtlyblác wé onginnen, Wulfst. 253, 11. v. lyb-lác.

un-rihtness, e ; *f. Iniquity, injustice, wrong :*—Unrehtnise *iniquitatis*, Rtl. 42, 27. Unrehtnises, 174, 10. Ðú heardeste strǽl tó ǽghwilcre unrihtnesse, Blickl. Homl. 241, 4. Ða ðe wyrcaþ unrihtnesse (*injustitiam*), Ps. Lamb. 93, 4. Unrehtnisse *iniquitatem*, Lk. Skt. Rush. 13, 27.

unriht-weorc, es ; *n. Improper work :*—Be Sunnandæges unrihtweorcum *de profanis operibus die Dominico*, L. Ecg. C. 35, tit. ; Th. ii. 130, 25.

unriht-wíf, es ; *n. A woman of bad character :*—Godwine hæfð gelǽd fulle láde æt ðan unrihtwífe ðe Leófgár bisceop hine tihte, Chart. Th. 373, 32.

unriht-wifung, e ; *f. Unlawful matrimony :*—Hé forlét ða unrihtwífunge *abdicato connubio non legitimo*, Bd. 2, 6 ; S. 508, 30.

unriht-willend, es ; *m. :* or -willende *adj.* (*ptcpl.*) *An ill-disposed person, an unrighteous person :*—Ðæt hé wiðstande mid his sprǽce ðám unryhtwillendum ðe ðyses middangeardes waldaþ *hujus mundi potestatibus contraire*, Past. 15 ; Swt. 89, 22.

unriht-wilnung, e ; *f. Improper desire, cupidity, concupiscence :*—Ánra gehwylces unrihtwillnung (*cupiditas*) on ðyssum fýre birneþ, Bd. 3, 19; S. 548, 27. Ða ðe mid sumere unryhtwilnunga beóð fǽringa oferswíðede *qui repentina concupiscentia superantur*, Past. 23 ; Swt. 179, 2.

un-rihtwís; *adj. Unrighteous, unjust, evil :*—Unrihtwís *injustus*, Wrt. Voc. i. 75, 70. Unrihtwís dóm ðæt se hálga wer swá ðrowode, Homl. Th. i. 596, 24. Se ðe ys on lytlum unrihtwís- (*iniquus*), Lk. Skt. 16, 10. Se unrihtwísa *injustus*, Ps. Spl. 35, 1. Se unrihtwísa déma *judex iniquitatis*, Lk. Skt. 18, 6. Se unrihtwísa cásere Neron, Bt. 16, 4; Fox 58, 2 : 28 ; Fox 100,

25 : Met. 15, 1. Of handa unrihtwíses (*iniqui*), Ps. Spl. 70, 5. Cýderas unrihtwíse *testes iniqui*, 26, 18. Unrihtwíse *injusti*, Lk. Skt. 18, 11. Ða unryhtwísan *impii*, Past. 11 ; Swt. 65, 12. Ða unrihtwísan tǽlaþ ða rihtwísan *justus tulit crimen iniqui*, Bt. 4 ; Fox 8, 15. Ða unrihtwísan cyngas *tyranni*, 37, 1 ; Fox 186, 26 : 36, 2 ; Fox 174, 26. Ic hatode ða gesamnunge unrihtwísra (*malignorum*), Ps. Th. 25, 5. Se áwyrgda gást is heáfod ealra unrihtwísra dǽda, swylce unrihtwíse syndon deófles leomo, Blickl. Homl. 33, 7. Hé wæs nid unrihtwísum (-rehtuísum, Lind. : -rehtwísum, Rush. *iniquis*) geteald, Mk. Skt. 15, 28. Be ðǽm ofermódum and ðám unrihtwísum cyningum, Bt. 37, 1 ; Fox 186, 1 : Met. 25, 2. [*Icel.* ú-rēttwíss.]

un-rihtwís[u] (?) : -rihtwíse (?), an ; *f.* *Unrighteousness, iniquity :*—Árfest eallum unrihtwísum (*iniquitatibus*) dínum, Ps. Spl. 102, 3. [*Icel.* ú-rēttvísi ; *f.* *unrighteousness.*] v. riht-wís (?) ; *f.*

un-rihtwíslíce ; *adv.* *Unrighteously :*—Ic cwæð tó ðǽm unrihtwísum : 'Ne dó gé unryhtwíslíce' *dixi iniquis* : '*Nolite inique agere*,' Past. 54 ; Swt. 425, 21. [*Icel.* ú-rēttvísliga.]

un-rihtwísness, e ; *f.* *Unrighteousness, iniquity, injustice :*—Nis nán unrihtwísnys (*injustitia*) on him, Jn. Skt. 7, 18. Unryhtwísnys (*iniquitas*) ríxaþ, Mt. Kmbl. 24, 12. Unrihtwísnys, Ps. Spl. 7, 3 : 35, 3. Ðonne hwæm hwæt cymþ máre ðonne ðē þincþ ðæt hé wyrþe sié, ne biþ sió unryhtwísnes nó on Gode, ac sió ungleáwnes biþ on ðē selfum, Bt. 39, 10 ; Fox 226, 32. Sió duru ðære únryhtwísnesse *janua iniquitatis*, Past. 21 ; Swt. 157, 22. Unrihtwísnesse (-rehtwísnesses, Lind.) túngeréfa *vilicus iniquitatis*, Lk. Skt. 16, 8, 9. Ðú ágiltst fædera unrihtwísnysse (*iniquitatem*) hira bearnum, Ex. 34, 7 : 20, 5. Ða ðe unrihtwísnesse wyrceaþ *qui faciunt iniquitatem*, Mt. Kmbl. 13, 41 : Blickl. Homl. 89, 16. Árfæst eallum mínum unrihtwísnessum, 89, 3. Unrehtuísnísum *iniquitatibus*, Rtl. 169, 29. God hæfð árásod úre unrihtwísnissa, Gen. 44, 16 : Blickl. Homl. 87, 29.

unriht-wrigels, es ; *n. A veil of error :*—Hié wǽron stǽnenre heortan and blindre, ðæt hié ðæt ongeotan ne cúðan, ðæt hié ðǽr gehýrdon, ne ðæt oncnáwan ne mihton, ðæt hié ðǽr gesáwon ; ac God áfyrde him ðæt unrihtwrigels (cf. ðone unrihtan wrigels, Wulfst. 252, 4) of heora heortan (cf. Their minds were blinded ; for until this day remaineth the same vail untaken away in the reading of the old testament ; which vail is done away in Christ . . . The vail is upon their heart. Nevertheless when it shall turn to the Lord, the vail shall be taken away, 2 Cor. 2, 14–16), Blickl. Homl. 105, 30.

unriht-wyrcend, es ; *m.* : or -wyrcende ; *adj.* (*ptcpl.*) *An evil-doer,* or *evil-doing :*—Ic ne ineode on ðæt geþeaht unrihtwyrcendra *cum iniqua gerentibus non introibo*, Ps. Th. 25, 4. Mid ðám unrihtwyrcendum *cum operantibus iniquitatem*, 27, 3. Belocen ðǽm synnfullum mannum and ðǽm unrihtwyrcendum, Blickl. Homl. 61, 11.

unriht-wyrhta, an ; *m. An evil-doer, a worker of iniquity :*—Gewítaþ fram mé ealle unrihtwyrhtan (*operarii iniquitatis*), Lk. Skt. 13, 27. Unryhtwyrhtan, Past. 1 ; Swt. 27, 23. Unrihtwyrhtan *iniqui*, Ps. Th. 118, 86.

un-rím, es ; *n. A countless number, an incalculable number* or *amount,* (1) *without a following genitive :*—Ðonne án tweó of áðon biþ, ðonne biþ unrím ástyred *ut una dubitatione succisa innumerabiles aliae succrescant*, Bt. 39, 4 ; Fox 216, 19. (2) *with a genitive plural :*—Ðæt is unrím on ealra cwycra *illic reptilia, quorum non est numerus*, Ps. Th. 103, 24 : Shrn. 65, 24. Hié in ðære eá áweóllon swá ǽmettan, swilc unrím heora wæs, Nar. 11, 14. Him com unrím wildeóra ðǽrtó, Shrn. 118, 16 : Met. 20, 190. Him gelýfde leóda unrím, 26, 40. Reced weardode unrím eorla, Beo. Th. 2480 ; B. 1238. Ðæt is herga mǽst, eádigra unrím, Exon. Th. 352, 3 ; Sch. 92. Mid unríme þegna and eorla, Met. 25, 7. Þeáh hé áge ǽhta unrím, 14, 4. Betwuh ðperra unrím ǽwyrdleana *inter alia detrimenta innumera*, Bd. 1, 3 ; S. 475, 21 : Andr. Kmbl. 1408 ; An. 704. Hé gehét unrím máþma *promisit se ei innumera ornamenta largiturum*, Bd. 3, 24 ; S. 556, 8 : Exon. Th. 245, 12 ; Jul. 43. Wé witon unrím monna *multos scimus*, Bt. 11, 2 ; Fox 36, 2. Hé ofslóh unrím Walana, Chr. 605 ; Erl. 21, 26 : Cd. Th. 194, 15 ; Exod. 261 : 220, 13 ; Dan. 70 : Exon. Th. 270, 23 ; Jul. 469. Wíta unrím, Cd. Th. 22, 4 ; Gen. 335 : 48, 15 ; Gen. 776. (2 a) *with the verb in the plural :*—Beóð ðé áhylded fram wíta unrím, grimra gyrna ðe ðé gegearwad sind, Exon. Th. 252, 33 ; Jul. 172. (2 b) *in the following the construction is peculiar, the word seeming indeclinable ; v. next word, ¶ :*—Nalæs mid ánes mannes geþeahte, ac mid gesægene unrím geleáffulra witena *non uno quolibet auctore, sed fideli innumerorum testium adsertione*, Bd. pref. ; S. 472, 25. Bútan óþrum læssan unrím ceastra *praeter castella innumera*, 1, 1 ; S. 473, 28. (3) *with a sing. gen. of word implying multitude :*—Unrím heriges, Chr. 937 ; Erl. 112, 31. Cnósles unrím, Exon. Th. 430, 15 ; Rä. 44, 9. Ðæt his (*of that race*) unrím á in wintra worn wurðan sceolde, Cd. Th. 236, 21 ; Dan. 324. Eác ðám wæs unrím óðres mánes (cf. ðæt wæs tó-eácan ðþrum unárímedum yslum, Bt. 1 ; Fox 2, 11), Met. 1, 44. Ðǽr wæs wunden gold on wǽn hladen, ǽghwæs unrím, Beo. Th. 6261 ; B. 3135. Sió hálige cirice unrím folces befēhð mid ánfealde geleáfan *innumeros sanctae ecclesiae populos unitas fidei contegit*, Past. 15 ; Swt. 95, 7 : Exon. Th.

36, 1 ; Cri. 569. Hé geaf him gúðgewǽda ǽghwæs unrím, Beo. Th. 5241 ; B. 2624. [*O. Sax.* un-rím (engilô).] v. un-gerím, *and following words.*

un-rím *and* un-ríme ; *adj.* *Innumerable, incalculable, not to be numbered :*—Unrím getæl *ingens numerus*, Nar. 9, 13. Folc unrím (*or pl. ?*) þrymfæste twá þeóda áwócon, Cd. Th. 158, 9 ; Gen. 2614. Werod, mægen unríme, Elen. Kmbl. 121 ; El. 61. Hyra fromcynn swá unríme weorðan sceolde, Exon. Th. 188, 4 ; Az. 40 : 187, 26 ; Az. 36. Ðǽr is máðma hord, gold unríme, Beo. Th. 6016 ; B. 3012. Mid ða unríman mænigeo *innumerabili multitudo*, Bd. 5, 12 ; S. 628, 4. Wǽron on ðyssum felda unríme gesomminge *erant in hoc campo innumera conventicula*, S. 629, 24. Monige sindon geond middangeard, unrímu cynn, Exon. Th. 355, 38 ; Pa. 2 : 389, 5 ; Rä. 7, 3. ¶ In the following passage the word seems indeclinable, unless *unrím-gód* = an immense, incalculable good, may be taken as a compound ; cf. unrím-folc, and see preceding word (2 b) :—Se symle leofaþ gehwǽr on unrím gódum *qui innumeris semper vivit ubique bonis*, Bd. 2, 1 ; S. 500, 23. v. un-gerím ; *adj.*

unrím-folc, es ; *n. An innumerable people :*—Gif hé underfēnge ðone ealdordóm swelces unrímfolces búton ege *si ducatum plebis innumerae sine trepidatione susciperet*, Past. 7 ; Swt. 51, 12. Cf. síd-, wíd-folc.

un-rípe ; *adj. Unripe, immature :*—Unrípe deáð *immatura mors*, Wrt. Voc. i. 39, 20. Ða unrípan *immatura*, Hpt. Gl. 518, 22. [*O. H. Ger.* un-rifi *immaturus.*]

un-rót ; *adj.* I. *sad, sorrowful, troubled, gloomy :*—Unrót *tristis*, Wrt. Voc. i. 51, 1 : 83, 37. Hé ongann beón unrót (*moestus*). Ða sǽde se Hǽlynd : 'Unrót (*tristis*) is mín sáwl,' Mt. Kmbl. 26, 37, 38. Unrót *contristatus*, Ps. Th. 37, 6 : Exon. Th. 73, 2 ; Cri. 1183 : 166, 3 ; Gú. 1037. Geómormód, earg and unrót, eallum bidǽled dugeþum and dreámum, 86, 14 ; Cri. 1408. Hwæðer ðú ǽfre áuht unrót wǽre ðá ðá ðú gesǽlgost wǽre *inter illas abundantissimas opes numquam ne animum tuum concepta ex qualibet injuria confudit anxietas ?* Bt. 26, 1 ; Fox 90, 21. Ða andsworode ðæt unróte mód, 3, 4 ; Fox 6, 18. Se Hǽlend hine unróte geseah *uidens illum Jesus tristem factum*, Lk. Skt. 18, 24 : Exon. Th. 177, 28 ; Gú. 1234. Ða unrótan *mestam*, Wrt. Voc. ii. 54, 42. Wǽron hig swíðe unróte (*tristes*), Gen. 40, 6 : Judth. Thw. 25, 29 ; Jud. 284. Middaneard geblissaþ, and gé beóð unróte *mundus gaudebit, vos autem contristabimini*, Jn. Skt. 16, 20 : Blickl. Homl. 135, 15, 25. Ne beóð gé unróte, ac gefeóþ mid mē, 191, 22 : 225, 14. Higum unróte módceare mǽndon, Beo. Th. 6288 ; B. 3148. Ða men (*men with pain in the spleen*) beóð mægre and unróte, Lchdm. ii. 242, 3. On óðre wísan sint tó manianne ða gladan (blíðan, l. 14), on óðre ða unrótan (*tristes*), Past. 27 ; Swt. 186, 13. Hé geseah ða men ealle unróte (*moestos*) ðe him æt wǽron, Bd. 5, 5 ; S. 618, 6. II. *displeased, harsh, angry :*—For hwig syndon gé swá unróte ongeán mē ? Is hyt for ðam ðe ic ábǽd ðæs Hǽlendes líchaman æt Pilate ? Nicod. 13 ; Thw. 6, 29. v. þurh-unrót.

un-rótian. v. ge-unrótian.

un-rótlíc ; *adj. Gloomy :*—Unrótlíc heofon *triste coelum*, Mt. Kmbl. Lind. 16, 3. v. next word.

un-rótlíce ; *adv. Gloomily, sadly :*—Reádaþ unrótlíce ðe heofun *rutilat triste coelum*, Mt. Kmbl. Rush. 16, 3. Unrótlíce dóþ *exterminant* (but perhaps the word is adjective, as the passage to which the gloss belongs is Mt. 6, 16 : Nolite fieri sicut hypocritae tristes : exterminant facies suas), Wrt. Voc. ii. 72, 21 : 30, 64.

unrót-mód ; *adj. Sad at heart :*—Hé for ðære geómrunga ðæs óþres deáþes leng on ðam lande gewunian ne mihte ; ac hé unrótmód of his cýþþe gewát, Blickl. Homl. 113, 12.

un-rótness, e ; *f. Sadness, sorrow, trouble, gloominess :*—Unrótnys *tristitia*, Wrt. Voc. i. 83, 42. Ðætte sió unrótnes, ðe hé for ðæm yflan weorcum hæbbe, gemetgige ðone gefeán ðe hé for ðǽm gódan weorcum hæfde, Past. proem. ; Swt. 24, 3. Unrótnyss (*tristitia*) gefylde eówre heortan, Jn. Skt. 16, 6. Eówer unrótnys (-ródtnis, Lind.) byð gewend tó gefeán, 16, 20. Nis ðǽr ǽnig sár gemēted, ne ádl, ne ece, ne nǽnig unrótnes, Blickl. Homl. 25, 31. Ic hit wiste be sumum dǽle, ac mē hæfde ðiós unrótnes ámerredne, ðæt ic hit hæfde mid ealle forgiten ; and ðæt is eác mínre unrótnysse se mǽsta dǽl, ðæt . . . *eaque mihi etsi ob injuriae dolorem nuper oblita, non tamen ante hac prorsus ignorata dixisti ; sed ea ipsa est vel maxima nostri caussa moeroris ; quod . . . ,* Bt. 36, 1 ; Fox 172, 2–4. Se fífta leahtor is *tristitia*, ðæt is ðissere worulde unrótnyss ; ðæt is ðonne se man geunrótsaþ ealles tó swýðe for his ǽhta lyre . . . Twá unrótnyssa synd ; án is ðeós yfele, and óðer is hálwende, ðæt is ðæt se man for his synnum geunrótsige, Homl. Skt. i. 16, 289 : Homl. Th. ii. 220, 16 : Wulfst. 68, 15. Ðé is frófre máre ðearf ðonne unrótnesse *medicinae tempus est, non querelae*, Bt. 3, 3 ; Fox 6, 15. Hí weorþaþ gerǽfte mid ðære unrótnesse and swá gehæfte *moeror captos fatigat*, Bt. 37, 1 ; Fox 186, 21 : Met. 25, 48. Gefeá búton unrótnesse, Blickl. Homl. 65, 18 : 85, 33. On wópe and on unrótnesse and on sáre his líchoma sceal hér wunian, 61, 1 : 3, 9. Hé hig funde slǽpende for unrótnesse (*prae tristitiam*), Lk. Skt. 22, 45. Se heora unrótnesse ealle gewríðeþ *qui alligat contritiones eorum*, Ps. Th.

146, 3. Hé hiene on unrótnesse oððe on ormódnesse gebringð, Past. 21; Swt. 166, 12. Of ðæs magan ádle cumaþ . . . micla murnunga and unrótnessa bútan þearfe, Lchdm. ii. 174, 26. Mid manegum unrótnessum Dauid wæs ofðrycced under Sawle, Ps. Th. 38, arg. Ðæm oferblíðum is tó cýðanne ða unrótnessa (*tristia*) ðe ðæræfter cumaþ, and ðám unblíðum sint tó cýðanne ða gefeán (*laeta*) ðe him gehátene sindon, Past. 27; Swt. 187, 15.

un-rótsian; *p.* ode. **I.** *to be sad, to be sorrowful :*—Hé unrótsade *contristatus*, Mk. Skt. Lind. Rush. 3, 5. Ðá unrótsodon helware, Homl. Skt. i. 4. 292. Ðæt gehwá for his synnum unrótsige mid sóðre dædbóte, Homl. Th. ii. 220, 20. Ðú lærdest ðæt wé ne unrótsodon, þeáh úre spéda wanodon, Shrn. 167, 12. Hé ongann unrótsian *coepit contristari*, Mt. Kmbl. 26, 37. Hiá ongunnon unrótsia (-rótsiga, Rush.) *illi coeperunt contristari*, Mk. Skt. Lind. 14, 19. Unrótsande wæs *contristatus est*, Mt. Kmbl. Lind. 14, 9. **II.** *to make sad* or *sorrowful :*—Alle gidroefede † unrótsade (unróðsad † gestyred, Lind.) wérun *omnes conturbati sunt*, Mk. Skt. Rush. 6, 50. v. ge-unrótsian.

un-rúh; *adj. Not rough, smooth :*—Cyrtil unrúh † smoeðe *tunica inconsutilis*, Jn. Skt. Lind. Rush. 19, 23.

un-ryne, es; *m. An ill-running, diarrhœa :*—Gif ðú ðás wyrte sylst þicgean on strangon wíne, heó ðæs innoðes unryne gewríð, Lchdm. i. 172, 13.

un-sac (-sæc?); *adj. Free from any charge :*—Unsac hé wæs on lífe *no charge was brought against him while alive*, Lchdm. iii. 288, 6. v. sac, on-sæc, *and cf. Icel.* ú-sekr *not guilty*.

un-sadelod. v. un-gesadelod.

un-sæd; *adj. Unsatisfied, insatiable :*—Unsædre heortan *insatiabili corde*, Ps. Th. 100, 5.

un-sæd, es; *n. Bad seed :*—Ealle unþeáwas áweallaþ of deófle, and hé ðæt unsæd sáweþ tó wíde, Wulfst. 40, 23.

un-sægd, -sæd; *adj. Unsaid :*—Wé hit lætaþ unsæd, Wanl. Cat. 6, 13.

un-sæl, es; *m. Unhappiness :*—Ða deóflu wæron on miclum unsælum (v. sæl, IV. ¶), and ða englas wæron on swíðe micelre blisse, Wulfst. 236, 26. [Unsel him wes on mode, Laym. 30541. Sum unsel heom is ihende, O. and N. 1263. Þer heo þolyeþ al unsel, Misc. 146, 90. A draȝte of unsele *an unfortunate draught*, i. e. poison, Alex. (Skt.) 1106. On unsele oðer an untime *at an improper season* or *time*, Rel. Ant. i. 131, 43. Cf. *Icel.* ú-sæla *unhappiness*.]

un-sælan; *p.* de *To untie, unbind, loose :*—Git moeteþ æsul gesælde and folan mid hire, unsæleþ (*solvite*), Mt. Kmbl. Rush. 21, 2. Onlésed, unsæled *desolutus*, i. *liberatus*, Wrt. Voc. ii. 139, 29. Unmidled *vel* unsæled *effrenus*, 142, 60.

un-sæle; *adj. Evil, wicked :*—Unsæle, gemáh *improbus*, Wrt. Voc. ii. 45, 16. [Crist warrþ unnorne and wrecche and usell child, Orm. 3668. Holde ich no mon for unsele (*miserable*) otherwhyle that he fele sum-thyng that him smerte, Rel. Ant. i. 113, 13. *Goth.* un-séls πονηρός: *Icel.* ú-sæll *unhappy*.]

un-sælig; *adj.* **I.** *of persons, unhappy, unblest, miserable* as being evil :—Deófol sæwð unwísdóm, ðæt unsælig man wísdómes ne gýmeþ, Wulfst. 52, 27. Ðú miltsige mé (*a devil*), ðæt unsælig (ic) ne forweorþe, Exon. Th. 269, 14; Jul. 450. Hí (*the good*) fore góddædum blissiaþ, ða hý (*the wicked*) unsælge ær forhogdun tó dónne, 79, 9; Cri. 1288. Hæleð unsælige (*the unbelieving Jews*), Andr. Kmbl. 1122; An. 561. **II.** *of things, unhappy, bringing misery :*—Æppel ðæt unsælga (cf. Milton : the fruit whose mortal taste brought death into the world, and all our woe), Cd. Th. 40, 10; Gen. 637. Þe unseli Senei, A. R. 174, 1. Hwa se is swa unseli, þat he þis soð schunie, Kath. 1793. Unnseliȝ mann amm ic wurrþenn, Orm. 4812. Ðat folc unseli (*the people of Sodom*), Gen. and Ex. 1073. Unsely wrecche, Chauc. second N. P. T. 468. *O. H. Ger.* un-sálig *infelix*.] v. un-gesælig.

un-sælþ, e; *f. Unhappiness, misfortune, misery :*—Ðæt is seó mæste unsæld on ðís andweardan life, ðæt mon ærest weorþe gesælig and æfter ðam ungesælig *in omni adversitate fortunae infelicissimum genus est infortunii, fuisse felicem*, Bt. 10; Fox 26, 30. Him wære ealra mæst unsælþ ðæt, ðæt se fyrst wære óþ dómes dæg *licentiam infelicissimam, si esset eterna*, 38, 4; Fox 204, 16 Hwelc mæg him mare unsæld becuman *quid eorum mente infelicius?* Past. 45; Swt. 340, 4. Hié wilniaþ óþera manna unsælþa and him cymð sylfum ðæt ylce *infelicitas in viis eorum*, Ps. Th. 12, 7. Hé hwílum selþ ða gesælþa ðæm gódum and ðæm yflum unsælþa . . . hwílum hé eft geþafaþ ðæt ða gódan habbaþ unsælþa and ungelimp and ða yfelan habbaþ gesælþa *qui saepe bonis jucunda, malis aspera, contraque bonis dura tribuat, malis optata concedat*, Bt. 39, 2; Fox 214, 1-5: 10; Bt. 28, 8. [He fleh mid muchele unsælðe, Laym. 4748. Al for hire onselþe (*wickedness*), 2nd MS. 2545. To þolenn illc unnsellþe, Orm. 1561. Unnsellþe, 4811. Sum unselþe heom is ihende, O. and N. 1263. Ðo wex unselðe on hem . . . dolc, sor, and blein, Gen. and Ex. 3026. *O. H. Ger.* un-sálida *infelicitas, dementia*.] v. un-gesælþ.

un-sæpig; *adj. Not sappy, sapless :*—Treówa gif hí beóð on fullum mónan geheáwene, hí beóð heardran tó getimbrunge, and swíðost gif hí beóð unsæpige geworhte, Homl. Th. i. 102, 24.

un-samwræde; *adj. Not united, opposed, contrary :*—Gif ða gódan ðonne simle habbaþ anweald, ðonne nabbaþ ða yfelan næfre nænne, for ðam ðæt gód and ðæt yfel sint swíþe unsamwræde *nam cum bonum malumque contraria sint, si bonum potens esse constiterit, liquet imbecillitas mali*, Bt. 36, 3; Fox 176, 2. v. sam-wrædness.

un-sár; *adj. Not sore, without soreness* or *pain :*—Se teter bútan sáre hé ofergæð ðone líchoman . . . ; se giecða bið suíðe unsár, Past. 11; Swt. 71, 19. Gníd mid ða tóðreoman, hí beóð clæne and unsáre, Lchdm. i. 346, 15. Ðæt geswel wyrð unsárre *the swelling becomes more free from pain*, ii. 208, 4.

un-sáwen; *adj. Not sown :*— ii. æceras, óðerne gesáwene, and óðerne unsáwene, L. R. S. 10; Th. i. 438, 5.

un-scæpfull, -scæþþig, -scæþþende. v. un-sceapfull, -sceaþþig, -sceaþþende.

un-sceád[e]líce; *adv. Unreasonably :*—Gif hwylc bróðor unsceádelíce hwæs bidde *si quis frater aliqua inrationabiliter postulat*, R. Ben. 54, 13. v. un-gesceádlíce.

un-sceádwíslic; *adj. Unreasonable, irrational :*—Gif wé ða unsceádwíslícan styrunga on stæððignysse áwendaþ, Homl. Th. ii. 210, 30. v. un-gesceádwíslic.

un-sceamfæst; *adj. Shameless, impudent :*—Unsceamfæst *impudens*, Wrt. Voc. i. 86, 57. Unscamfæst *impudens, inverecundus, sine pudore*, Hpt. Gl. 472, 37. [Onschamefæst *inpudens, inverecundus, effrons*, Prompt. Parv. 367.]

un-sceamfulness, e; *f. Shamelessness, immodesty, lasciviousness :*—Unsceomfulnise (-scomfulnisse, Rush.) *inpudicitia*, Mk. Skt. Lind. 7, 21.

un-sceamiende *not being ashamed :*—Ðæt hé mæge fore eágum eorðbúendra unscomiende ædles brúcan bysmerleás, Exon. Th. 81, 17; Cri. 1325.

un-sceamig; *adj. Not to be confounded, unabashed :*—Is on mé sweotul ðæt ðú unscamge æghwæs wurde on ferþe fród *in me is it plain, that thou, O woman not to be confounded! hast become in everything sagacious in mind*, Exon. Th. 275, 18; Jul. 552. [*O. H. Ger.* scamig *confusus, erubescens*; unscameg ze uuerdenne *not to be put to shame*.]

un-sceamlíc; *adj. Immodest, shameless :*—Ic hí ástyrede mid fúllícum gespræcum. Hí míne unsceamlícan gebæra geseónde mé on heora scip námon tó him, Homl. Skt. ii. 23 b, 377. [*O. H. Ger.* un-scamalíh *impudens*.]

un-sceamlíce; *adv. Shamelessly :*—Ðæt [hié] mid ðám hæleðum hæman wolden unscomlíce, Cd. Th. 148, 19; Gen. 2459. [*O. H. Ger.* un-scamalícho *impudenter*.]

un-sceandlíce (?); *adv. Shamelessly :*—Ic mé unsceandlíce [*the un-* has been erased (*properly.* v. sceandlíce) *in one MS.*], swá swá ic gewuna wæs, tómiddes heora gemengde, Homl. Skt. ii. 23 b, 372.

un-scearp; *adj. Not sharp :*—Unscearp wín, Lchdm. ii. 212, 4. v. scearp, II.

un-scearpness, e; *f. Want of sharpness, dullness :*—For his ungleáwnesse and for his unscearpnesse *propter ingenii tarditatem*, Bd. 5, 6; S. 620, 7.

un-scearpsíne; *adj. Not sharpsighted :*—Ealdes mannes eágan beóþ unscearpsýno . . . þus mon sceal unscearpsýnum sealfe wyrcean tó eágum, Lchdm. ii. 30, 27-32, 1.

un-sceaþfull (-scæþ-, -sceþ-); *adj. Innocent :*—Se ðe æfter ðæm higaþ ðæt hé eádig sié on ðisse worulde, ne biþ hé unsceaþful (-full, Cott. MSS.) *qui festinat ditari, non erit innocens*, Past. 44; Swt. 331, 15. Hé ðe unscæðfull býð mid his handum *innocens manibus*, Ps. Th. 23, 4. Unsceðfull wið ða unsceðfullan, 17, 25 : 24, 19. Ða wegas ðæra unsceðfulra *vias immaculatorum*, 36, 17. Ða unscæðfullan heortan *rectos corde*, 36, 13. [Shep iss all unskaþefull, Orm. 1176.]

un-sceaþfullíce; *adv. Innocently :*—Ða unsceaðfullíce (-sceð-, Cott. MSS.) libbaþ *qui innocenter vivunt*, Past. 37; Swt. 263, 7.

un-sceaþfulness, e; *f. Innocence :*—Se ðe gehielt his unsceaðfulnesse and his gódan willan *ut mentis innocentia custoditur*, Past. 34; Swt. 234, 22. Ðý læs hí forlætan hiora unsceaþfulnesse (-sceð-, Cott. MS.) *desinet colere forsitan innocentiam*, Bt. 39, 10; Fox 228, 5. Unsceaðfulnisse *innocentiam*, Ps. Surt. 17, 25. Unsceðfulnisse, 7, 9 : 40, 13 : 100, 2.

un-scelleht. v. un-scilliht.

un-scende, -scynde; *adj. Without disgrace, honourable, noble :*—Ælfheres láf (*a coat of mail*) golde geweorðod, ealles unscende, æðelinges reáf, Wald. 96; Vald. 2. 20. Gife unscynde *a noble gift* (the nails from the cross), Elen. Kmbl. 2400; El. 1201 : 2492; El. 1247. Eów Dryhten geaf dóm unscyndne, 730; El. 365. Se him dóm forgeaf, unscyndne blæd, Cd. Th. 263, 16; Dan. 763. [*O. H. Ger.* un-scant *non ignominiosus*.]

un-scended; *adj. Unharmed, uncorrupted :*—Erfeweardnisse unscende *hereditatem incorruptam* (v. 1 Pet. 1, 4), Rtl. 24, 32. Unscendede hond *manum inlesam*, 102, 37.

un-scendende; *adj. Innocent :*—Unscendende ic am *innocens ego sum*, Mt. Kmbl. Lind. 27, 24.

un-sceód. v. un-scóg(i)an.

un-sceótan *to open :*—Unsceót *vel* geopena *exentera,* Wrt. Voc. i. 61, 13. v. an-sceótan.

un-sceþþende ; *adj. Innocent, harmless :* — Unsceþþende ic eam *innocens ego sum,* Mt. Kmbl. Rush. 27, 25. Onfóh mîne sáwle, for ic wæs unsceðpende and clǽnheort, Shrn. 139, 22. Ða bilehwitnysse ðæs unsceþþendan (*innocentis*) lîfes, Bd. 1, 26 ; S. 487, 40. Ðæt hê ðære unsceþþendan (*innocuae*) ylde cilda ne árede, 2, 20 ; S. 521, 25. Ðone mildheortan and ðone unsceþþendan Crist, Blickl. Homl. 3, 11. Hî ða unscæþþendan (*innoxiam*) ðeóde forhergodon, Bd. 4, 26 ; S. 602, 6. Unsceaþþiendra fordémednesse *proscriptionibus innocentium,* 1, 6 ; S. 476, 25. Ðæt hê mæge fordón ða unsceðþendan *ut interficiat innocentem,* Ps. Th. 9, 28.

un-sceþþig, -scæþþig ; *adj. Innocent, harmless :*—*Innocens* unsceððig (-scæððig) is æfre nama, Ælfc. Gr. 43 ; Zup. 253, 16. Beó se cristena man unsceaðþig and bilewite, Homl. Th. i. 142, 10. Hér com Ælfréd se unsceððiga æþeling, Chr. 1036 ; Erl. 164, 25. Hys ðæt synnige blód wæs ágoten on ða wrace hyre ðæs unsceððian blódes, Shrn. 155, 8. Hê sǽde ðæt án gehwǽde wolcn upp ástige mid ðære unscæðþigan (*not threatening storm*) lyfte, Homl. Skt. i. 18, 150. Cain his ágenne bróðor rihtwîsne and unscæððigne ácwealde, Boutr. Scrd. 20, 41. Búton hî wǽron swá eádmóde and swá unscæððigne swá ðæt cild wæs, Homl. Th. i. 512, 12. Culfran sind swîðe unscæððige fuglas, i. 142, 8. Hî ða deór swá getemedon, ðæt hî mid him unscæððige (*harmless*) wunodon, ii. 492, 14. Betwuh ðám unscæððigum *inter innocentes,* Ps. Th. 25, 6 ; Homl. Th. i. 88, 33. [He ne wollde nohht unshaþiȝ wimmann wreȝhenn, Orm. 2889.]

un-sceþþigness, e ; *f. Innocence, harmlessness :*—Æfter gerisenre áre heora unscæþþignysse *juxta honorem innocentibus congruum,* Bd. 2, 20 ; S. 522, 7. On unscæððignysse heortan *in innocentia cordis,* Ps. Spl. 100, 2. Gyt hê hyit his unscæððignysse *adhuc retinens innocentiam,* Homl. Th. ii. 452, 15 : 210, 29. Habban ða unscæððignysse on heora móde ðe cild hæfð, i. 512, 18. Ne funde hé on him náne synne ac unscæððignysse, Boutr. Scrd. 20, 25. Unsceaðþignysse, Ps. Spl. 36, 39. [Menn þatt cwemmdenn Godd þurrh unnshaþiȝnesse, Orm. 58.]

un-scilliht ; *adj. Not shell* (of fish) :—Fixas unscellehte, Lchdm. ii. 88, 9.

un-scirped ; *adj. Not dressed :*—Monno unscirped *hominem non vestitum,* Mt. Kmbl. Lind. 22, 11.

un-scód. v. next word.

un-scóg[i]an *to unshoe, take off the shoes :*—Unsceógien hî gebróþor *discalcient se fratres,* Anglia xiii. 413, 683. Ðonne bið ús suíðe frácodlîce óðer fót unscód *quasi unius pedis calceamentum cum dedecore amittit,* Past. 5 ; Swt. 45, 14. Hê wæs gelǽded and ungyred and unscód, Shrn. 85, 32. Nyme ðæt wîf his gescý of his fótum, and nemne hine ǽlc man unsceóda (*discalceatus*), Deut. 25, 10. Unsceóde *discalciati,* Anglia xiii. 416, 735. Unscódum fótum, Wulfst. 170, 16.

un-scoren ; *adj. Unshorn, unshaven :*—Locc unscoren *coma vel cirrus,* Wrt. Voc. i. 42, 45. Hî beón unscorene *sint inrasi,* Anglia xiii. 408, 609. Hî lange tîd eodon ealle unscorene and sîdfeaxe, Ap. Th. 6, 12. [*Icel.* ú-skorinn.]

un-scortende ; *adj. Not failing, not running short :*—Strión unscortende *thesaurum non deficientem,* Lk. Skt. Lind. Rush. 12, 33.

un-scrýdan ; *p. de To undress, strip, divest :*—Unscrýdde *exfibulat,* i. *exsolvit,* Wrt. Voc. ii. 145, 24. Byþ unscrýdd *exuitur,* Scint. 226, 9. Unscrídde *exutos, nudatos,* Hpt. Gl. 423, 52. Unscrýdde, Homl. Skt. i. 11, 146. (1) with acc. of person :—Baðiendra manna hûs, ðær hî hí unscrêdaþ *in adopteriлm,* Wrt. Voc. i. 37, 6. Hine man sóna unscrýde and ða reáf nime ðe hé ǽr notode *mox exuatur rebus propriis quibus vestitus est,* R. Ben. 101, 22. Hê hét hine unscrýdan, Homl. Th. i. 432, 3 : 424, 12. Gif hwá his lîc forstǽle, nolde hê hine unscrýdan, 220, 8. (2) with acc. of person and dat. of garment :—Hê hine unscrídde ðam healfan scicelse, Ap. Th. 12, 22. Hî unscrýddon hyne hys ágenum reáfe, Mt. Kmbl. 27, 28, 31 : Mt. Skt. 15, 20 : Homl. Th. ii. 252, 24, 29.

un-scyld, e ; *f. Innocence :*—For unscylde *propter innocentiam,* Ps. Spl. 40, 13. [*O. Frs.* un-skelde : *O. H. Ger.* un-sculd *innocentia.*]

un-scyld, e ; *f. A grievous fault :*—Gif mîne fýnd ne rícsiaþ ofer mê, ðonne beó ic unwemme, and beó geclǽnsod fram ðǽm mǽstum scyldum ; ac gif hî mê ábyggiaþ, ðonne ne mæg ic smeágan mîne unscylda, Ps. Th. 18, 12.

un-scyldig ; *adj.* **I.** *innocent, guiltless :*—Unscyldig *insons,* Ælfc. Gr. 9, 39 ; Zup. 63, 16. Mid werum unsyeldigum unscyldig (*innocens*) ðú bist, Ps. Spl. 17, 27 ; Andr. Kmbl. 2275 ; An. 1139. Hér wearð Ecgbriht abbud unscyldig ofslegen, Chr. 916 ; Th. i. 190, col. 2. Ne cweþe ic ná ðæt ðæt yfel sié ðæt mon helpe ðæs unscyldigan (-scyldgan, Cott. MS.), Bt. 38, 7 ; Fox 210, 4. Sweord besyled on unscyldigum (-scyldgum, Met. 9, 59) blóde, 16, 4 ; Fox 58, 18. Se ðe unscildigne man belǽwe *qui percutiat animam sanguinis innocentis,* Deut. 27, 25. Unscildigne and rihtwîsne ne ofsleh ðú *insontem et justum non occides,* Ex. 23, 7. Ðú woldest ðone besmitan ðe ðú nánwiht yfles on nystest. Tó hwon lǽddest ðú ðeosne freóne and unscyldigne hider ? Blickl. Homl. 87, 1 : Exon. Th. 143, 11 ; Gú. 659. Unscyldigne, synna leásne, Elen. Kmbl. 990 ; El. 496. Hú ne is se yfelwillende and yfelwyrcende ðe ðone

unscyldgan wîtnoþ ? Bt. 38, 6 ; Fox 208, 11, 15. Ne syle ðú unscyldigra sáwla deórum ðe ðé andettaþ *ne tradas bestiis animas confitentes tibi,* Ps. Th. 73, 18. Ðǽm wǽre máre þearf ðe ða óþre unscyldige yfelaþ, ðæt mon bǽde ðæt him mon dyde swá micel wîte swá hî ðám óþrum unscyldegum dydon, Bt. 38, 7 ; Fox 208, 30 : Met. 4, 36. Hî unscyldige scotian þenceaþ *ut sagittent immaculatum,* Ps. Th. 63, 3. Seó wyrd þreáþ ða unscildigan (*insontes*), Bt. 4 ; Fox 8, 13. **I a.** *innocent* of a crime, charge, (1) with gen. :—Ðet hê wæs unscyldig ðæs ðe him geléd wæs, Chr. 1052 ; Erl. 187, 20. Unscyldigne eofota gehwylces, Elen. Kmbl. 845 ; El. 423. (2) with preposition :—Ic eom unscyldig, ǽgðer ge dǽde ge dihtes, æt ðære tihtlan ðe N. mê tîhð, L. O. 5 ; Th. i. 180, 15. Sind mænige rihtwîse unscyldige wið heáfodleahtras, Homl. Th. i. 342, 9. **I b.** *guiltless* in relation to (*wið*) a person. v. un-scyldigness :—Ic eom unscyldig wið ðás mîne fýnd *ego in innocentia mea ingressus sum,* Ps. Th. 25, 1. Ðú ne bist unscyldig wið mê gif ðú on îdelnesse cígst mînne noman *nec habebit insontem Dominus eum, qui assumpserit nomen Domini frustra* (Ex. 20, 7), L. Alf. 2 ; Th. i. 44, 8. **II.** *innocent, not accountable* for an ill result, *not responsible :*—Gif oxa ofhnîte wer oþþe wíf, ðæt hié deáde sién . . . Se hláford bið unscyldig (*the owner of the ox shall be quit,* A. V. Ex. 21, 28), gif se oxa hnitol wǽre . . . and se hláford hit nyste, L. Alf. 21 ; Th. i. 48, 29. His hláford bið unscildig *dominus bovis innocens erit,* Ex. 21, 28. Unscyldig ic eom fram ðyses rihtwîsan blóde *innocens ego sum a sanguine justi hujus,* Mt. Kmbl. 27, 24. Ne ofsleh ðú unscildine mannan . . . ðis ic dyde mid bilewitnysse *num gentem ignorantem et justam interficies? . . . in simplicitate cordis mei feci hoc,* Gen. 20, 4. [*O. Sax.* un-skuldig : *O. Frs.* un-skeldech : *O. H. Ger.* un-sculdîg *innocens, indebitus, nil meritus : Icel.* ú-skyldigr *not due.*]

un-scyldiglíc ; *adj. Innocent, not obnoxious, unobjectionable :*—Ungerisnre bysene ðú hátest hié wítnian, ah mê þynceþ unscyldiglícre ðæt him man heáfod of áceorfe búton óðrum wítum, Blickl. Homl. 189, 32.

un-scyldigness, e ; *f. Innocence :*—Dauid sang ðisne sealm be his unscyldnesse wið (v. un-scyldig, **I b**) his sunu, Ps. Th. 25, arg. Æfter unscyldignisse mînre *secundum innocentiam meam,* Ps. Spl. 7, 9 : Rtl. 48, 40.

un-scynde. v. un-scende.

un-seald ; *adj. Ungiven :*—Seó séleste gyrd is gyt unseald, Homl. Ass. 131, 495.

un-sealt ; *adj. Without salt, insipid :*—Gif ðæt sealt unsealt (*insulsum*) biþ, Mk. Skt. 9, 50. Unsaltera *insulsior,* Wrt. Voc. ii. 48, 6.

un-sefuntig, -seofuntig (= hund-seofontig) *seventy :* — Unsefuntig *septuaginta,* æfter unseofuntigum *post septuaginta,* Mt. Kmbl. p. 2, 3, 11 : Lk. Skt. p. 6, 15. Unseofontigum, p. 6, 14. [Cf. *O. Sax.* ant-sibunta.]

un-seht ; *m. f. n. Disagreement :*—Hî macodon mǽst ðet unseht betweónan Godwine eorle and ðam cynge, Chr. 1052 ; Erl. 187, 27. [Mǽst þis unsehte wæs forþan þe se cyng fylste his nefan, Chr. 1116 ; Erl. 245, 29. He mid unsehte fram þam cynge for . . . Hi mid unsehte tohwurfon, 1106 ; Erl. 240, 20, 25. For þære unsehte þe he hæfde wið France, 1112 ; Erl. 243, 32. For þes cynges unsehte of France, 1117 ; Erl. 246, 6. Mid unsæhte, 1123 ; Erl. 250, 26. *Icel.* ú-sátt, -sætt *disagreement.*] v. seht, *and next word.*

un-seht ; *adj. Not in agreement, in hostility, at variance :*—Eádríc cild and ða Bryttas wurdon unsehte and wunnon heom wið ða castelmenn on Hereforda *Eadric and the Welsh broke out into hostility* (against William. v. Florence of Worcester, who says that Edric summoned two Welsh kings to help him and laid waste Hereford. The same writer, under the year 1070, notes that Edric was reconciled with William) *and fought with the garrison at Hereford,* Chr. 1067 ; Erl. 203, 40. Sóna ðæræfter wurdon unsehte se cyng and se eorl *directly after the king and the earl fell out,* 1102 ; Erl. 238, 6. [Heo weron unsæhte and heo weren unsome, Laym. 3930. Þou and his sone woxen unsauȝt (*fell out*), and þou sloug him þere, Jos. 433. Folk that were unsaught toward her king (*at variance with their king*) for his pillage, Gower iii. 153, 26. *Icel.* ú-sáttr *disagreeing, unreconciled.*] v. seht ; *adj.*

un-seldan ; *adv. Not seldom, frequently :*—Ðone sang wê sungon unseldan mid heom, Homl. Skt. i. 21, 264. Oft and unseldan, L. E. G. proem. ; Th. i. 166, 9 : Btwk. 222, 2. Oft and unseldon, L. Pen. 2 ; Th. ii. 278, 5. [*Icel.* ú-sjaldan.]

un-seþe, Wrt. Voc. ii. 150, 80. v. un-sóþ.

un-settan *to displace, put down :*—Tó unsettanne (-setanne, Rush.) *ad deponendum,* Mk. Skt. Lind. 15, 36.

un-sewenlíc (?) ; *adj. Invisible :*—Hwæt wênst ðú be ðære unsewenlícran wyrde (Cott. MS. *has* unwénlícran, *and the Latin is :* Quid reliqua, quae, cum sit *aspera*), Bt. 40, 2 ; Fox 236, 24. v. un-gesewenlíc.

un-sib[b], e ; *f.* **I.** *unfriendliness, unkindliness, enmity :*—Unsib *simultas,* Wrt. Voc. ii. 120, 62. Swá mycel ungeþwærnys and unsibb árás ingravescentibus causis dissensionum, Bd. 3, 14 ; S. 539, 35. Ðæt ðridde is unsibbe fýr, ðonne wé ne forhtigaþ ðæt wê ða mód ábylgean úra ðæra nýhstena *tertium dissensionis, cum animos proximorum offendere non formidamus,* 3, 19 ; S. 548, 17 : Anglia xi. 101, 37. Unsibbe *simultate,* Wrt. Voc. ii. 87, 11. Unsibbe *simultatem,* 83, 38. God ús

lǽrð sibbe and wynsumnesse, and deófol ús lǽrð unsibbe and wróhte, Homl. Ass. 168, 112: Cd. Th. 281, 13; Sat. 271. Ic andette mínes módes mordor and unsibbe and ofermódignesse, L. de Cf. 8; Th. ii. 262, 32. Oferfyll ne murneþ ne for fæder ne for mēder, ne for nánum gesibban men. Ealle unsibba hit wyrcð, Wulfst. 242, 8. **II.** *strife, hostilities, war :*—Gif hié gemunan willaþ hiora ieldrena wǒlgewinna and hiora monigfealdan unsibbe *recolant majorum suorum tempora, bellis inquietissima*, Ors. 2, 1; Swt. 64, 15. Hié ðæt heóldun mid micelre unsibbe, and tú folcgefeoht gefuhton, and ðæt lond oft forhergodon, and ǽghwæþer ðperne oftrædlíce út dræfde, Chr. 887; Erl. 86, 10. Ðæt hí wæron unsibbe and gefeoht fram heora feóndum onfónde *quia bellum ab hostibus forent acceptiri*, Bd. 2, 2; S. 503, 30. His ii suna ymb ðæt ríce wunnon, and ða unsibbe mid gefeohte dreógende wǽron, Ors. 2, 7; Swt. 90, 17. Æfter hū monegum wintrum sió sibb gewurde ðæs ðe hié ǽ[re]st unsibbe wið monegum folcum hæfdon, Ors. 4, 7; Swt. 182, 18. **III.** *division, variance, disagreement, disunion :*—Unsib (*dissensio*) áuorden wæs in ðær menigo fore hine, Jn. Skt. Lind. 7, 43. Unsib *seditio*, Wrt. Voc. ii. 120, 30. Forlǽtaþ ða úterran sibbe, and habbaþ ða innerran fæste, ðætte eówer unsibb geeádmēde ðæs synnigan mód (*ut peccantis mentem vestra discordia feriat*), Past. 46 ; Swt. 357, 9. Ðá sóhte Colemannus ðysse unsibbe (*dissensioni*) læcedom, Bd. 4, 4; S. 571, 6. Hé hiera sundorsprǽce tó unsibbe bróhton *their colloquy led to no agreement ;* infecto pacis negotio, Ors. 4, 10; Swt. 202, 13. [Betere his sahte þaue onsibbe, Laym. 9845, 2nd MS. *Goth.* un-sibja *iniquitas :* O. H. Ger. un-sippe *seditio.*]

un-sibbian ; *p.* ode *To disagree :*—Unsibbaþ *desidet*, i. *discordat*, Wrt. Voc. ii. 139, 20. Unsibbade *desidebat*, 25, 16.

un-sibsumness, e; *f. Want of tranquillity, anxiety :*—Mið unsibsumnise gedroefede *anxietate turbati*, Jn. Skt. p. 6, 1.

un-sidefull ; *adj. Immodest :*—Unsideful *impudicus*, Wrt. Voc. i. 51, 33 : O. E. Homl. i. 300, 29, 30.

un-sidefulness, e ; *f. Immodesty, immorality :*—Se fífta unþeáw is ðæt wíf beó unsydefull. Unsydefulnys bið sceamu for worulde, and ðæt unsydefulle wíf bið unwurð on life, O. E. Homl. i. 300, 30.

un-sidu, a ; *m. A bad habit, vicious custom, mal-practice :*—Sóð is ðæt ic secge, ǽtære man unlaga on lande oððe unsida lufige tó swíðe, ðæt cymð ðære þeóde tó unþearfe, L. I. P. 4; Th. ii. 308, 8. Nis eác nán wundor, þeáh ús mislimpe, forðam wē witan ful georne, ðæt . . . weard þes þeódscype swýðe forsyngod . . . þurh hæþene unsida, Wulfst. 164, 2. Áne misdǽda hē dyde þeáh tó swíðe, ðæt hé ælþeódige unsida lufode, Chr. 959; Erl. 121, 1. [*Icel.* ú-siðr.]

un-sigefæst ; *adj. Not victorious, unsuccessful :*—Ða ðe God wurdodon sigefæste wǽron symle on gefeohte ; ða ðe fram Gode bugon tó bysmorfullum hǽðenscype wurdon gescynde and á unsigefæste, Homl. Skt. i. 18, 44.

un-silt ; *adj. Unsalted :*—Unsilt (-slit, MS.) smeoro *saevo*, Wrt. Voc. ii. 119, 45. Unsylt smeoru, Lchdm. iii. 18, 5. v. un-gesilt.

un-síþ, es ; *m.* **I.** *an evil, ill-advised expedition :*—Weard ofslegen Ecgfridus on his unsíðe, ðá ðá hē on Peohtum begann tó feohtanne tó dyrstelíce ofer Drihtnes willan (cf. Ecgfrid, cum temere exercitum ad vastandam Pictorum provinciam duxisset, multum prohibentibus amicis, extinctus est, Bd. 4, 26), Homl. Th. ii. 148, 16. **II.** *a mishap, misfortune :*—Heó ðurh wódnysse micclum wæs gedreht . . . heó ǽr ðon eáwfæst leofode, ðeáh ðe se unsíð hire swá gelumpe, Homl. Th. ii. 142, 12. [Þu (*the owl*) ne singst never þat hit nis for sume unsiþe (*mishap*), O. and N. 1164.]

un-sleac, -sleac ; *adj. Not slack, not lazy, strenuous, active, diligent :*—Unsleac *inpiger*, Wrt. Voc. ii. 47, 30 : 93, 15 : Kent. Gl. 140. [*Icel.* ú-slakr.]

un-slæclíce ; *adv. Not slackly, not languidly, strenuously :*—Gif ðæt gebodene bið gefremed unsleaclíce (*non tarde*), R. Ben. 20, 18.

un-slǽpig ; *adj. Sleepless :*—Unslǽpige *insomnes*, Wrt. Voc. ii. 48, 20.

un-slǽwð, Past. 45 ; Swt. 341, 4, *seems an error for* un-sǽlð, *the reading of the Cott. MSS.*

un-slǽw, -slǽw, -sleáw ; *adj. Not slow, not sluggish, active, ready, quick :*—Unslǽw *impiger* vel *praepes*, Wrt. Voc. i. 49, 34. Unsleáw *inpiger*, 74, 34. Hé hine sylfne getengde in Godes þeówdóm æscróf, unslǽw, Elen. Kmbl. 403 ; El. 202. Se ðe wǽre full slǽw, weorðe se unslǽw tó cyrican, Wulfst. 72, 15. Ðone ðe him on weorcum gecwēmde elne unslǽwe, Exon. Th. 159, 7 ; Gú. 923. Wígan unslǽwne (*St. Andrew*), Andr. Kmbl. 3419; An. 1713. Hí slógon tógædere unslǽwe mid wǽpnum, Homl. Skt. ii. 25, 375.

un-slǽwlíce ; *adv. Not slowly, not sluggishly, actively :*—Hié sculon gehiéran ðætte is geháten ðæm monnum ðe lustlíce and unslǽwlíce lǽraþ ðæt ðæt hié ðonne cunnon (*qui in hoc, quod jam obtinuit, corporis vitio non tenetur*), Past. 49; Swt. 381, 1.

un-sleac, -sleáw, -slit. v. un-slǽc, -slǽw, -silt.

un-slúped ; *adj.* (*ptcpl.*) *Unloosed :*—His tungan bend weard unslýped *solutum est uinculum linguae ejus*, Mk. Skt. 7, 35.

un-sliten ; *adj. Unrent :*—Ðæt cyrtel wæs unslitten, . . . Cuoedon : Ne tóslíte (*scindamus*) uē hiá *the coat was unrent . . . They said : Let us not rend it*, Jn. Skt. Lind. 19, 23. [*Icel.* ú-slitinn.]

un-slopen ; *adj.* (*ptcpl.*) *Unloosed :*—Æfter þúsend geárum bið Satanas unbunden . . . and nú syndon Satanases bendas swýðe tóslopene (unslopene, MS. H.), Wulfst. 83, 9.

un-smeoruwig ; *adj. Not fatty* or *greasy :*—Genim unsmerigne healfne cýse, Lchdm. ii. 292, 23.

un-smēþe ; *adj. Not smooth, rough, uneven :*—Unsmēðe *scabra*, Wrt. Voc. ii. 97, 13. Unsmēðe hrægel *birrus*, i. 40, 25. Eoh bið útan unsmēðe treów, Runic pm. Kmbl. 341, 27 ; Rún. 13. Tunge unsmēþe, Lchdm. ii. 242, 10. Wē habbaþ hrepunge, ðæt wē magon gefrēdan hwæt bið smēðe, hwæt unsmēðe, Homl. Th. ii. 372, 33. His unsmēðan (*leprous*) líces, 512, 6. Ne unsmēðes wiht, Exon. Th. 199, 15 ; Ph. 26. Unsmoeði *scabro*, Wrt. Voc. ii. 120, 24. Ðære unsmēþan *elefantinosa*, 142, 82 : 31, 8. Hē hleóp on unsmēðe eorðan, Shrn. 152, 1. Ða unsmēþan tungan smirewan, Lchdm. ii. 238, 25. Ðeós wyrt bið cenned on unsmēþum stówum, i. 160, 18. Ða unsmēþan *salebrosos*, Wrt. Voc. ii. 78, 25. [Unnsmeþe þurrh bannkes and þurrh græfess, Orm. 9209.]

un-smēþness, e ; *f. Roughness :*—Unsmēðnes *callosita*, Wrt. Voc. ii. 18, 36 : 82, 54. Unsmēþnes, 127, 55.

un-smóþe (-smóþ? *but* see sófte ; *adj.*) ; *adj. Rough :*—Unsmóþi *aspera*, Wrt. Voc. ii. 101, 15. Unsmóðe, 7, 33. v. smóþ, un-smēþe.

un-snotor, -snottor ; *adj. Unwise, foolish :*—Unsnotor *insipiens*, Ælfc. Gr. 9, 38 ; Zup. 62, 15 : Ps. Lamb. 91, 7. Unsnoter *inprudens*, Wrt. Voc. i. 76, 13. Se unsnotera ł se unwita *insipiens*, Ps. Lamb. 13, 1. Se unsnotera ł se dysega, 48, 11. Ðæt bið swíþe dysig man and unsnottor on his life, se þe lufaþ ðás eorþlícan welan and ne lufaþ God ðe hit him eal sealde, Blickl. Homl. 195, 24. Fú dysega man and ðú unsnottra, 49, 35. Unsnotterra *insipientum*, Rtl. 86, 14. Unsnoterum *insipientibus*, Ps. Lamb. 48, 21 : L. Ælfc. C. 23 ; Th. ii. 352, 2.

un-snotorlíce ; *adv. Unwisely, imprudently :*—inprudenter, Ælfc. Gr. 38 ; Zup. 223, 15 : 228, 10.

un-snotornyss, e ; *f. Folly :*—Tó unsnotornysse *ad insipientiam*, Ps. Lamb. 21, 3. Þurh unsnotornesse, Wulfst. 166, 25.

un-snyterness, e ; *f. Folly :*—Tó unsnyternesse *ad insipientiam*, Ps. Spl. T. 21, 2. v. snytre.

un-snytro (-u) ; *f. Folly :*—Gefylled mið unsnytro *repleti insipientia*, Lk. Skt. Lind. 6, 11. Hosp unwísum ł unsnytro ðú sealdest mē *opprobrium insipienti dedisti me*, Ps. Spl. T. 38, 12. Worda eallra unsnyttro ǽr gesprecenra, Elen. Kmbl. 2567 ; El. 1285. Hé his selfa ne mæg for his unsnyttrum ende geþencean, Beo. Th. 3472 ; B. 1734 : Met. 9, 11. Hé unsnytrum (*foolishly, unwisely*) Andreas hēt áhón, Exon. Th. 260, 35 ; Jul. 300. Unsnyttrum, 251, 14 ; Jul. 145 : 153, 25 ; Gú. 831 : Elen. Kmbl. 1900; El. 947.

un-soden ; *adj. Unsodden, unboiled :*—On unsodenan hunige, Lchdm. iii. 40, 6. Sceápes hóhscancan unsodenne, ii. 38, 8. Genim ða ylcan wyrte unsodene, i. 198, 15 : Exon. Th. 488, 18 ; Rä. 76, 8.

un-sófte ; *adv.* **I.** *not at ease, in discomfort.* v. sófte, **II :**—Gif men fērlíce wyrde unsófte, Rtl. 114, 24. **II.** *not gently, hardly, severely :*—Hwǽr mon unsófte getilaþ on forewearde ða ádle *in case severe treatment is used in the early stages of the disease*, Lchdm. ii. 260, 15. Ða ðe hine unsófte ádle gebundne gesóhtun, Exon. Th. 155, 10 ; Gú. 858 : 83, 16 ; Cri. 1357. Hí wrehton unsófte ealdgeníþlan, Judth. Thw. 24, 37 ; Jud. 228 : Blickl. Homl. 203, 18 **III.** *hardly, with difficulty, with trouble :*—Wē hit unsófte mid longsceaftum sperum ofscotadon *vix ipsis defixa est venabulis*, Nar. 15, 28. Ic ðæt unsófte ealdre gedígde, Beo. Th. 3314 ; B. 1655 : 4287; B. 2140 : Elen. Kmbl. 263 ; El. 132 : Exon. Th. 168, 20 ; Gú. 1080. [Þer is þe sunfulle unsófte to beon, Misc. 91, 25. O. H. Ger. un-samfto *difficulter, aegre.*]

un-sóftlíce ; *adv. Ungently, hardly :*—Ualerianus áwēd hrýmde : ‘ Eálá ðú, Laurentius, unsóftlíce tíhst ðú mē gebundenne mid byrnendum racenteágum,’ Homl. Th. i. 434, 7.

un-sóm, e ; *f. Disagreement :*—Gyf hyra ǽnig wið ǽnigne mon ǽnige unsóme hæbbe, ðæt hē wið ðone geþingie . . . Man sceal ǽlce unsóme and ealle geflytu gestyllan, L. E. I. 36 ; Th. ii. 434, 2–7.

un-sorh ; *adj. Without care, without anxiety, secure :*—Ðá ongeat hé ðæt ðær wæs godcundlíc mægen ondweard, and hé ðære mildheortnesse unsorh ábád, Blickl. Homl. 217, 29.

un-sóþ, es ; *n. Untruth, falsehood :*—Mid unsóðe sóð oferswíðan, Wulfst. 55, 16. Leáslíce híwian unsóð tó sóðe, 128, 9. Ic nelle secgan unsóð on mē sylfne, Homl. Skt. i. 12, 195.

un-sóþ ; *adj. Untrue, false :*—Unsóþe (Wright gives *unseþe*, but Wülcker Voc. 243, 15 *unsoþe*) sage (-a, Wülcker) *falsa dicta*, Wrt. Voc. ii. 150, 80. [*Icel.* ú-sannr *untrue*.]

un-sóþfæst ; *adj.* **I.** *untruthful, unveracious :*—Ic silf geseah ðæt ðæt mē unsóðfæstran men sædon, ðonne ða wǽron ðe ðæt sédon ðæt wit ymb sint, Shrn. 204, 12. **II.** *unjust, unrighteous :*—Unsóðfæstne wer *virum injustum*, Ps. Th. 139, 11. Wē unsóðfæste wǽron *injuste egimus*, 105, 6. Sóðfæst fore unsóðfæstum *justus pro injustis*, Rtl. 21, 32.

un-sóþfæstness, e ; *f. Unrighteousness, injustice : —* Unsóðfæstnys *injustitia*, Ps. Th. 54, 9. Ðes sóðcuoed is and unsóðfæstnise in ðæm ne is *hic verax est, et injustitia in illo' non est*, Jn. Skt. Lind. 7, 18.

un-sóþian. v. ge-unsóþian.

un-sóþsagol; *adj. Speaking falsely, lying, mendacious :*—Unsóðsagul *falsidicus* vel *falsiloquus*, Wrt. Voc. i. 47, 47. Unsóðsagol *falsidicus*, 76, 20. Se smið ne dorste secgan ðæs gesihðe ænigum menn, nolde beón gesewen unsóðsagul (-ol) boda, Homl. Skt. i. 21, 58.

un-spannan; *p.* -speónn *To unclasp, unfasten :*—Unspeón *exfibulat*, i. *exsolvit*, Wrt. Voc. ii. 145, 24. [Cf. Seint Iohan ine his iborenesse unspennede his feder tunge into prophecie, A. R. 158, 14.] v. on-spannan.

un-spéd, e; *f. Want, indigence, penury :*—Unspéde *inopiae*, Ps. Spl. 43, 27. For unspéda *prae inopia*, 87, 9. Ðíos of unspoed (unspoedum, Rush.) hire alle ða ðe hæfde sende *haec de paenuria sua omnia quae habuit misit*, Mk. Skt. Lind. 12, 44. [It fel to mikel unspede *it turned out very unfortunate*, C. M. 15420. O. Sax. un-spôd *disadvantage:* O. H. Ger. un-spuot.]

un-spédig; *adj.* I. *without means, poor, indigent :*—Ðes and ðeós and ðis unspédige *hic et haec et hoc inops*, Ælfc. Gr. 9, 56; Zup. 68, 1. Gif hwylc wíf tó ðam unspédig wære ðæt heó ðás ðing begytan ne mihte, Homl. Th. i. 140, 3. Generigende unspédigne *eripiens inopem*, Ps. Spl. 34, 12: Blickl. Gl. Hé geendebyrde ðone unspédigan fiscere ætforan ðam rícan cásere, Homl. Th. i. 578, 9. Eádge biðon ða ðaerfe, ðæt is unspoedge menn *beati pauperes*, Mt. Kmbl. Lind. 5, 3 note. Ða rícostan men drincaþ myran meolc, and ða unspédigan drincaþ medo, Ors. 1, 1; Swt. 20, 17. For yrmðe unspédig[ra] *propter miseriam inopum*, Ps. Spl. 11, 5. Ðonne ðú geseó geongran man ðonne ðú sý, and unwísran, and unspédigran, Prov. Kmbl. 31. Þeáh hý sýn on ðyson woruldsælþon ða unspédgestan, Ors. 1, 2; Swt. 30, 4. II. *barren, poor, unproductive :*—Hié gesæton sorgfulne land, eard and éðyl unspédigran fremena gehwylcre, Cd. Th. 59, 12; Gen. 962. [O. H. Ger. un-spuotig.]

un-spiwol; *adj. Not emetic :*—Unspiwol drenc, Lchdm. ii. 274, 11, 20. Unspiule dreuceas, 170, 11.

un-sprecende; *adj. Not speaking, unable to speak, without speech, speechless :*—Unsprecende cild *infans* vel *alogos*, Wrt. Voc. i. 50, 41: 73, 11: Ælfc. Gr. 9, 37; Zup. 61, 7. Hé (*Christ in Simeon's arms*) ðá gyt on ðære menniscnysse unsprecende wære, Homl. Th. i. 142, 26. Færinga sáh hé niðer spræce benumen, and þurhwunode swá unsprecende, Chr. 1053; Erl. 186, 23. Unsprecende forneán *almost speechless*, Homl. Skt. i. 3, 481. Ða unsprecendan cild, Homl. Th. ii. 116, 14: 50, 15.

un-stæfwís; *adj. Illiterate :*—Unstæfwís *inlitterata*, Germ. 393, 82.

un-stæðdig; *adj.* I. *not steady, remiss, irregular :*—Sum munuc wæs unstæðdig on Godes lofsangum, Homl. Th. ii. 160, 19. II. *unstable, inconstant, fickle :*—Wé sceolon fyligan úrum Heáfde fram ðissere unstæðdigan worulde tó his staðelfæstan ríce, Homl. Th. ii. 282, 21. Ðises lífes gewilnung gelæc ða unstæðdian tó manegum leahtrum, Homl. Skt. i. 5, 67. III. *unsteady, unstaid, not sober, light, wanton :*—Weard hé gegripen mid ðære gálnysse his unstæðdigan heortan, Homl. Ass. 110, 247. Hé his sylfes geweóld on ungerípedum freódóme and unstæðdigum þeáwum, Ælfc. T. Grn. 17, 13. Ða unstæðdigan hleápunge ðæs mædenes (*the daughter of Herodias*), Homl. Th. i. 480, 35.

un-stæðdigness, e; *f.* I. *unsteadiness, instability, inconstancy :*—Sun munuc mid gemáglícum bénum gewilnode ðæt hé móste of ðam munuclífe . . . Ðá weard se hálga wer geháthyrt ðurh his unstæðdignysse, and hét hine áweg faran, Homl. Th. ii. 176, 18. II. *want of sedateness, levity, wantonness :*—Of gálnysse cumaþ ungemetegod lufu and eágena unstæðdignys, Homl. Th. ii. 220, 7. Se grimlíca deófol lærd dyrstignysse and gebringð réceleáse men on unstæðdignesse wordes and weorces, Wulfst. 54, 3. Mid módes unstæðdignysse, 60, 1.

un-staþolfæst; *adj.* I. *unsettled, not remaining in one place, not stationary :*—Ða twám and þrím dagum geond missenlícra monna húsum wuniaþ, æfre unstaþolfæste (*numquam stabiles*) and woriende, R. Ben. 9, 23. II. *unsettled, desirous of change :*—Sum munuc weard unstaðolfæst on his mynstre, and gewilnode ðæt hé móste of ðam munuclífe, Homl. Th. ii. 176, 14. III. *not steadfast, unstable, not enduring, easily moved :*—Unstaðolfest weorc *opus instabile*, Kent. Gl. 369. Hí nabbaþ wyrtruman on him ac beóð unstaðolfæste *non habent radicem in se sed temporales sunt*, Mk. Skt. 4, 17. Understaþolfæste (unstaþolfæste? cf. ðæt ungestæðdige folc, Bt. 39, 3; Fox 216, 2) *mobile vulgus*, Met. 28, 69. IV. *unsettled in mind, wavering :*—Ðone unstaðolfæstan bróðor and ðone tweónigendan *fratrem fluctuantem*, R. Ben. 51, 3. [þe twafalde mon is unstaþelfest (*inconstans*) on alle his weies, O. E. Homl. i. 151, 29. Unstaðeluest bileaue, A. R. 208, 16.]

un-staþolfæstness, e; *f. Instability, inconstancy :*—Wolcnes pund, ðanon him wæs his módes unstaðelfæstnes geseald, Salm. Kmbl. p. 180, 11. Unstaðolfæstnis ðóhta *instabilitas mentium*, Rtl. 192, 21. Se hálga wer swíðe mid wordum ðreáde his unstaðolfæstnysse, Homl. Th. ii. 176, 17. v. preceding word.

un-stedefull; *adj. Unstable, apostate :*—Mið englum unstydfullum *cum angelis apostaticis*, Rtl. 121, 21.

un-stedefulness, e; *f. Instability ;* but the word glosses *infestatio :*—Aelc unstydfulnis ðæs uncláenes gástes *omnis infestatio inmundi spiritus*, Rtl. 122, 24. v. on-stedefullness.

un-stenc, es; *m. A bad smell, stench :*—Hý mid nosan ne magon náht geswæccan bútan unstences ormætnesse *foetor ingenti complet putredine nares*, Dóm. L. 207. Náht elles gestincan búton unstenca ormætnessa, Wulfst. 139, 8.

un-stillan. v. ge-unstillan.

un-stille; *adj. Not still, unquiet :*—Unstille *inquies*, Ælfc. Gr. 9, 26; Zup. 52, 4. I. *of motion, not at rest, moving :*—Ðæt wæter unstille æghwider wolde tóscríþan, ne meahte hit on him selfum æfre gestandan, Met. 20, 92. Sund unstille, Exon. Th. 338, 14; Gn. Ex. 78. Swift wæs on fóre, dreág unstille winnende wéga, 434, 24; Rä. 52, 5. Eall ðiós unstille gesceaft, Bt. 39, 6; Fox 220, 23. Ðæt unstille hweól (*velox rota*) ðe Ixion wæs tó gebunden, 35, 6; Fox 168, 31. Ða unstillan woruldgesceafta, Met. 11, 19. Ðara unstillena gesceafta styring ne mæg nó weorþan gestilled, ne eác onwend of ðam ryne ðe him geset is, Bt. 21; Fox 74, 3. Ðú ðe ealle ða unstillan gesceafta tó ðínum willan ástyrast *qui das cuncta moveri*, 33, 4; Fox 128, 9 : Met. 20, 14. II. *liking movement* (lit. or fig.), *unquiet, restless ;* in a bad sense, *unruly :*—Hé cwæð ðæt sió tunge wære unstille yfel *lingua, inquietum malum*, Past. 38; Swt. 281, 7. Eh byð unstyllum æfre frófur, Runic pm. Kmbl. 343, 9; Rún. 19. Ða unstillan (*inquietos*) hé sceal þreágean, R. Ben. 13, 12. Nánwuht nis on ús unstilre and ungestæðdigre ðonne ðæt mód *nil in nobis est corde fugacius*, Past. 38; Swt. 273, 11. III. *unquiet, disturbed :*—Ðæt mæden hæfde unstille niht, Ap. Th. 18, 27. IV. *not at peace, troubled :*—Hé wæs fram ðam áwyrgedan gáste unstille ; and swá swýþe hé hine drehte, ðæt hé his sylfes nænig gemynd ne hæfde, Guthl. 13; Gdwin. 60, 12. [O. H. Ger. un-stilli *inquiens, vacillans, inquietus.*]

un-stillian. v. ge-unstillan.

un-stillness, e; *f.* I. *absence of rest, motion :*—Unstilnis *agitatio*, Wrt. Voc. ii. 99, 55. II. *disturbance, noisiness, clamour :*—Gif hé ðurhwunaþ cnucigende, ðonne áríst se híredes ealdor for ðæs óðres onhrópe, and him getiðaþ ðæs ðe hé bitt, ná for freóndrædene, ac for his unstilnysse, Homl. Th. i. 248, 33. Lærd ús se deófol unstilnesse and ungemetlíce hleahtras, and unnytte spræce, Wulfst. 233, 18. III. *tumult, bustle, commotion :*—Hé ne mihte ða unstillnesse ðara onfeallendra menigeo áberan *tumultus inruentium turbarum non ferret*, Bd. 3, 19 ; S. 549, 32. IV. *disturbance, breach of peace :*—On ðæs wífes gebærum onfundon ðæs cyninges þegnas ða unstilnesse (*the king had been attacked and killed*), Chr. 755; Erl. 50, 3. Læcedemonie hæfdon man-unstillnessa ðonne hié mægenes hæfdon *Lacedaemonii, inquieti magis quam strenui*, Ors. 3, 1; Swt. 98, 34. V. *restlessness, unruliness :*—Hé geseah ðæt hig niorwedom mid fæstenum and mid gebedum hiora líchaman unstillnesse, Shrn. 37, 3. VI. *disquietude, disturbance of mind, trouble :*—Se ðe his bróðor hataþ, hé hæfd unstilnesse and swýðe drófi mód, Basil admn. 4; Norm. 44, 16.

un-strang; *adj. Not strong, weak, feeble :*—Unstrang *invalidus*, Wrt. Voc. i. 51, 22 : 83, 57. Heó (*Judith*) wæs lytel and unstrang, Homl. Ass. 114, 411. Hwæt is se intinga ðæt án þúsend manna ðe ne magon ástyrian, swá unstrang swá ðú eart? Homl. Skt. i. 9, 110. Á sceal man ðam unstrangan men líðelícor déman ðonne ðam strangan, L. C. S. 69; Th. i. 412, 4. Hyra handa wæron unstrange hine tó ácwellanne, Shrn. 117, 31. Hé (*Peter*) mid his gange getácnode ægðer ge ða strangan ge ða unstrangan on Godes folce. Cristes geladung ne mæg beón búton strangum, ne búton unstrangum. Ðá ðá him twýnode, ðá getácnode hé ða unstrangan. Hwæt sind ða unstrangan? Ða sind unstrange ðe sláwe beóð tó gódum weorcum, Homl. Th. ii. 390, 15-25. Ðæt ða unstrangan (*infirmi*) ofersýwede heora þeówdóm ne forfleón, R. Ben. 121, 23. Sý fultum geseald ðam wácmódum and ðam unstrangum *inbecillibus procurentur solacia*, 58, 18. Hæla ða unstronga *sanare infirmos*, Lk. Skt. Lind. Rush. 9, 2. Deáh hwá anweald hæbbe, gif óþer hæfþ máran, beþearf se unstrengra ðæs strengran fultumes *si quid est, quod in ulla re imbecillioris valentiae sit, in hac praesidio necesse est egeat alieno*, Bt. 33, 1; Fox 120, 18. Hwæt is unstrengre ðonne se mon ðe bið tó ungemetlíce oferswíþed mid ðam tédran flæsce, 36, 6; Fox 182, 3 : Homl. Th. ii. 370, 16. Óþ ðæt hié (*inflammations*) unstrangran weorþan, Lchdm. ii. 178, 14. Hú ne miht ðú geseón ælce dæge ðæt ða strengran nimaþ ða welan of ðam unstrengrum, Bt. 26, 2; Fox 92, 15. v. strang, and next word.

un-strenge; *adj. Weak :*—Paulus cwæð : 'Wé strange sceolon beran ðæra unstrengra byrdene' *debemus nos firmiores imbecillitates infirmorum sustinere* (Rom. 15, 1), Homl. Th. ii. 390, 26. v. strenge.

un-stydfull. v. un-stedefull.

un-styri[g]ende; *adj. Not moving, unmoving, stationary :*—Monige sint cwucera gesceafta unstyriende, swá scylfiscas sint . . . Ða styriendan nétenu habbaþ eall ðæt ða unstyriendan habbaþ, and eác máre tó . . . For ðæm sint ðás sceafta ðus gesceapene ðæt ða unstyriendan hí ne áhebben ofer ða styriendan, Bt. 41, 5; Fox 252, 20-31.

un-styrigendlíc; *adj. Not to be stirred, not to be carried :*—Byrðenna unstyrendelíco *onera inportabilia*, Mt. Kmbl. Lind. 23, 4.

un-súr; adj. Not sour :—Eala forgá and meoloc þicge unsúre, Lchdm. ii. 292, 30. [Icel. ú-súrr.]

un-swǽs; adj. Unpleasant, disagreeable :—Settan mē ðǽr mē unswǽsost (or adv. ?) wæs posuerunt me in abominationem sibi, Ps. Th. 87, 8. [Icel. ú-svást veðr bad weather.]

un-swǽse; adv. Unpleasantly. v. preceding word.

un-swǽslíc; adj. Unpleasant, ungentle :—Hē hæfde his ende gebidenne unswǽslícne, Judth. Thw. 22, 17; Jud. 65.

un-swefen, es; n. A bad dream :—Sing ðis ylce gebed on niht ǽr ðū tō ðínum reste gā, ðonne gescylt ðē God wið unswefnum ðe nihternessum on menn becumaþ, Lchdm. iii. 288, 22.

un-sweotol; adj. Not evident, not to be seen, not discernible :—Nān ðara gesceafta ne mæg bión būton ōþerre, ðeáh hió unsweotol sié on ðære ōþerre, Bt. 33, 4; Fox 130, 26. Ne mæg hirá ǽnig būton ōþrum bión, ðeáh hí unsweotole (or adv. ?) somod eardien, Met. 20, 146.

un-swēte; adj. Unsweet, (1) of taste, bitter, sour :—Hí mē geblendon bittre tōsomne unswētne drync ecedes and geallan, Exon. Th. 88, 12; Cri. 1439. (2) of smell, offensive, fetid :—Ðonne ne biþ se þost tō unswēte tō gestincanne, Lchdm. ii. 48, 14. [O. Sax. un-swōti : O. L. Ger. un-suōti molestus : O. H. Ger. un-suozi.]

un-swice, es; m. (or -swic, es; n. ?) Good faith, absence of deceit or treachery :—Ðá gyrnde hē grides and gísla, ðæt hē mōste unswican intó gemōte cuman and út of gemōte he required safeconduct and hostages, that he might come to the meeting and go from it without treachery, Chr. 1048; Erl. 180, 7. [Cf. Icel. verða fyrir svikum to be exposed to treachery.]

un-swicende; adj. Not failing in duty to others, faithful, loyal :—Ic (Cnut) cýðe eów ðæt ic wylle beón hold hláford and unswícende tó Godes gerihtum and tó rihtre woroldlage, Chart. Erl. 229, 22. Hig áþas swóron and gíslas saldan ðæm cynge and ðæm eorle, ðæt heó him on allum þingum unswícende beón woldon, Chr. 1063; Erl. 195, 17. v. next word.

un-swici[g]ende; adj. Unfailing, that does not deceive, loyal :—Griffin swór áðas ðæt hē wolde beón Eádwearde kinge hold underkincg and unswicigende, Chr. 1056; Erl. 190, 35. Ðæt ðú wið Waldend heólde fæste treówe; seó ðē freoðo sceal weorðan áwa tó aldre unswiciendo, Cd. Th. 204, 25; Exod. 424.

un-swicol; adj. Not false, not treacherous, honest :—Unswicel non falsa, Hpt. Gl. 432, 24. Uton beón eádmóde and sóðfæste and unswicole and rihtwíse, Wulfst. 109, 13. [Icel. ú-svikull guileless.]

un-swíþ; adj. Not strong, weak :—Gif drenc sié tó unswíþ, Lchdm. ii. 270, 15: iii. 18, 22. v. swíþ, I. 2 a. a.

un-swíðe; adv. Not strongly, weakly :—Sió ecg bát unswíðor ðonne his ðiódcyning þearfe hæfde, Beo. Th. 5150; B. 2578.

un-sýferlíc; adj. Impure, uncleanly :—Sume synna beóþ swíþe unsýferlíce, ðæt se man wandaþ ðæt hē hí æfre ásecgge, Blickl. Homl. 43, 17. [O. H. Ger. un-súbarlíh squalidus.]

un-sýferness, e; f. Impurity, uncleanness (physical or moral) :—Se ðe forgýmeleásige gehálgod húsl, ðæt him sig unsýfernys (sordes) on, L. Ecg. P. iv. 44; Th. ii. 216, 18. Ðǽr unsýfernes on ne sý ne unclǽnnes, L. E. I. 5; Th. ii. 406, 1. Fulle bâna deádra and æghwilcre unsýfernissæ (omni spurcitia), Mt. Kmbl. Rush. 23, 27. Monige wilþsócan ðære unsýfernysse deófolgylda abrenunciata sorde idolatriae, Bd. 3, 21; S. 551, 21. Fram unsýfernyssum ðara ǽrrena mâna, 3, 23; S. 554, 27. On unsýfernyssum betwih deófolgyldum in sordibus inter idola, 3, 30; S. 562, 18: 3, 1; S. 523, 23: 5, 19; S. 639, 23. Ða unséfernessa ðe ðǽr beóþ sió lifer áwyrþ út and ðæt clǽne blód gesomnaþ, Lchdm. ii. 198, 5.

un-sýfre; adj. Impure, unclean, foul (physically or morally), (1) physical :—Gif hió swíþor unsýfre weorpe (weorþe?), clǽnsa mid hunige, Lchdm. ii. 210, 2. Wíc unsýfre (a prison), Andr. Kmbl. 2622; An. 1312. Unsýfra olidarum, Wrt. Voc. ii. 65, 15. (2) moral :—Forhwan ðú ðæt selegescot, ðæt ic mē on ðē gehálgode, þurh firenlustas fúle synne unsýfre (or adv. ? Cf. O. H. Ger. un-súbro sordide) besmite, Exon. Th. 90, 34; Cri. 1484. Synfulra weorud, swâ fúle swâ gǽt, unsýfre folc, Exon. Th. 75, 35; Cri. 1232. Ær se unsýfra (Holofernes) womfull onwóce, Judth. Thw. 22, 24; Jud. 76. Be ðam sacerde ðe hine sylfne besmít þurh unsýfre sprǽce (impuro sermone), L. Ecg. C. 5, tit.; Th. ii. 128, 14. Þurh unsýfre sprǽce per turpiloquium, 5; Th. ii. 138, 4. [O. H. Ger. un-súbar, -súbiri fedus, immundus, sordidus.]

un-syn[n], e; f. Not guilt, not crime :—Ne hûru Hildeburh herian þorfte Eótena treówe; unsynnum (with no faults on her part, undeservedly; gratis. Cf. un-synnig, II) wearð beloren leófum æt ðam lindplegan, bearnum and bróðrum, Beo. Th. 2149; B. 1072.

un-syngian; p. ode To exculpate, prove innocent; purgare :—Hine móton his mǽgas unsyngian, L. In. 21; Th. i. 116, 8.

un-synnig; adj. I. innocent, guiltless, without sin :—Sacleás l unsynnig, Jn. Skt. Lind. 15, 25. Crist symle unsynnig wunode, Homl. Th. ii. 524, 35. Ne ðúhte him tó huxlíc, ðæt hē mid gesceáde hine betealde unsynnine, 226, 12. Dauid miclum his ágenes herges pleáh, ðǽr hē ymb his getreówne ðegn unsynnigne sierede, Past. 3; Swt. 37, 8.

Gif esne óðerne ofsleá unsynnigne, L. Ethb. 86; Th. i. 24, 11 : L. Alf. pol. 29; Th. i. 80, 6: 35; Th. i. 84, 2: Beo. Th. 4185; B. 2089. Unsynnige insontem, Wrt. Voc. ii. 46, 22. Ús men secgaþ, ðæt hí unsynnige beón, ðeáh ðe hí mettas him on mūð bestingon on swilcum fæstendagum, Homl. Th. ii. 330, 30: Mt. Kmbl. Lind. 5, 3 note. Ða unsinnigo innocentes, 12, 7. II. undeserved :—Seó his unsynnige cwalu wæs gewrecen, Shrn. 93, 13. [O. Sax. un-sundig : O. H. Ger. un-suntíg insons, innocens, inculpabilis : Icel. ú-syndig.]

un-tǽle; adj. Blameless, without reproach :—Ðeáh ðe nǽfre ne wurde syððan mancynne gemiltsod, ðeáh wǽre Godes rihtwísnys eallunga untǽle, Homl. Th. i. 112, 19. Sió wiþerweardnes biþ simle untǽlu and wracu áscirred mid ðære styringe hire ágenre frēcennesse videas adversam fortunam sobriam, succintamque, et ipsius adversitatis exercitatione prudentem, Bt. 20; Fox 72, 5. Ic ða leóde wát ǽghwæs untǽle, Beo. Th. 3734; B. 1865. Sýn hý swâ gecorene, ðæt hý untǽle sýn and sacerdhâdes þurh ealle góde cysta wyrþe, R. Ben. 140, 6. Heó hæfd twâ ðing untǽle for Gode, sinscipe and eádmódnysse, Homl. Ass. 40, 399.

un-tǽled; adj. Unblamed :—Ðý læs hié forlǽten untǽlde óðerra monna yfele unðeáwas ne pravos hominum mores nequaquam redarguant, Past. 46; Swt. 351, 20.

un-tǽllíce; adv. Blamelessly, without reproach :—Ðonne birð se sacerd suíðe untǽllíce áwriten ðara fædra naman on his breóstum, ðonne hē singallíce geðencð hiera lífes bisene, Past. 13; Swt. 77, 17 : 5; Swt. 45, 12. Hí wǽron rihtwíse and heóldon Godes beboda untǽllíce erant justi incedentes in omnibus mandatis Domini sine querela (Lk. 1, 6), Homl. Th. i. 200, 35. v. un-tállíce.

un-tǽlwirðe; adj. Not blameable, irreprehensible, praiseworthy :—Is geteald hwelc hē beón sceal, gif hē untǽlwierðe bið quae sit irreprehensibilitas ipsa, manifestat, Past. 8; Swt. 53, 11. Ǽw Dryhtnes untǽlwyrðe (inre braehensibilis), Ps. Surt. 18, 8. Ðæt hē gecnáwe óðerra monna weorc untǽlwierðe (-wyrðe, Cott. MSS.) ut laudabilia aliorum facta cognoscant, Past. 30; Swt. 205, 5. Ðæt hié ðæs ðe untǽlwyrðran wǽren, 32; Swt. 215, 1.

un-tǽlwirðlíce; adv. Blamelessly, laudably :—Ðætte oft ðæs láreówdómes ðénung bið swíðe untǽlwyrðlíce (-wierð-, Cott. MSS. laudabiliter) gewilnad, and eác swíðe untǽlwierðlíce (laudabiliter) monige beóð tó geniédde, Past. 7; Swt. 47, 20. Ðæt ilce ðæt hē untǽlwyrðlíce (-wierð-, Cott. MSS. laudabiliter) ondréd tó underfónne, ðæt ilce se óðer swíðe hergeondlíce (laudabiliter) gewilnode, Swt. 49, 18.

un-tala. v. un-tela.

un-tállíce; adv. Blamelessly, without reproach :—Hē ealle ðæs regoles bebodu untállíce geheóld, Homl. Skt. ii. 23 b, 26. v. untǽllíce.

un-tamed, Wrt. Voc. ii. 142, 40. v. un-temed.

un-tamlíc (?); adj. Untameable :—Untamlíc (printed untamcul) indomabilis, Germ. 397, 11.

un-teala. v. un-tela.

un-teald; adj. Uncounted :—Gyf se dæg byð forlǽten unteald, ðǽrrihte áwent eall ðæs geáres ymbryn þwyres, Lchdm. iii. 264, 12.

un-tealt; adj. Steady :—Ðá hēt Ælfréd cyning timbrian lange scipu ongeán ðás ǽsceas; . . . ða wǽron ǽgðer ge swiftran, ge unteal[t]ran, ge eác heárran, ðonne ða óðru, Chr. 897; Th. 177, 1, col. 2. v. tealt.

un-tela; adv. (but in some cases it seems a noun ?) Not well, ill, badly :—Ða scamleásan nyton ðæt hié untela dóð, búton hit mon him sæcge impudentes se delinquere nesciunt, nisi a pluribus increpentur, Past. 3; Swt. 206, 1. Swâ micle hí onfóð ðǽr mâre wíte, swâ hí hēr gearor witon ðæt hí untela dóð, and [hit] ðeáh nyllað forlǽtan tanto illic graviora tormenta percipiat, quanto hic malum non deserit, etiam quod ipsa condemnat, 55; Swt. 429, 19. Ðeáh hine hwâ áhsode, for hwí hē swâ dyde, ðonne ne mihte hē hit nâ gereccan, ne geþafa beón nolde, ðæt hē untela dyde requiretur delictum ejus, nec invenietur, Ps. Th. 9, 35. Hē wát ðæt hē untela dǽd faciant quae non fuisse gerenda decernant, Bt. 39, 12; Fox 230, 29. Getímige ús tela on líchaman, getímige ús untala, symle wē sceolon ðæs Gode ðancian, Homl. Th. i. 252, 15. Wiþ ðon ðe men mete untela melte and gecirre on yfele wǽtan, Lchdm. ii. 226, 5. Ne forsuwa ðú nâ ðæt untela gedón sý, Prov. Kmbl. 44. Hwætst tó untala dyde hē (hwæt dyde untale, Rush.) quid mali fecit? Mt. Kmbl. Lind. 27, 23. v. tela.

un-temed; adj. Untamed, wild :—Untemed (Wright prints untamed, but see Wülcker 226, 14), wilde edomitus, Wrt. Voc. ii. 142, 40. Untemed hors, Ps. Spl. C. 32, 17. [Wick. un-temid.] v. temian.

un-teorig; adj. Untiring, unceasing :—Rodor recene scríþeþ, súðheald swifeþ swift, untiorig, Met. 28, 17.

un-teóðod, -tiogoðad; adj. Untithed :—Gē tiogoðiaþ eówre mintan and eówerne dile, and lǽtaþ untiogoðad ðætte diórwyrðre is eówra óðra ǽhta, Past. 57; Swt. 439, 29.

un-þǽslíc; adj. Unsuitable, unseemly, unbecoming, unfit :—Unðǽslíc indecens, Ælfc. Gr. 14; Zup. 87, 12. Ne gedafenaþ biscope ðæt hē beó on dǽdum folces mannum gelíc. Geswíc swâ unðǽslíces plegan, Homl. Th. ii. 134, 13. Ús ne gedafenaþ ðæt wē úrne líchaman, ðe Gode is gehálgod on fulluhte, mid unþǽslícum plegan gescyndan, i. 482, 9. On

unþæslícum tíman *horis incompetentibus*, R. Ben. 74, 23. Tó ðan un-þæslícum *ad ineptas*, Hpt. Gl. 510, 35. Onþæslíce míslára *inportunas suggestiones*, Scint. 33, 19.

un-þæslíce; *adv. Unsuitably, in an unseemly manner :*—Ðæt nán þing unþæslíce ne gelympe on nánes limes þénunge, Homl. Skt. i. 1, 204.

un-þæslícness, e; *f. Unseemliness, impropriety :*—Menn dæftaþ heora hús, gif hí sumne freónd onfón willaþ tó him, ðæt nán unðæslícnys him ne ðurfe derian, Homl. Th. ii. 316, 8.

un-þæslícu; *indecl. f. Incongruity, absurdity :*—Eáþe is tó understandenne of hwylcum antimbre ðeós unþæslícu ásprincð ðisse miclan tóþundennesse *quod quam est absurdum facile advertitur, quia materia ei datur superbiendi*, R. Ben. 124, 13.

un-þanc, es; m. *disfavour, displeasure, anger, ill-will :*—Oft ða unwaran láreówas for ege ne durron cleopian, ondrædaþ him sumra monna unðonc *saepe rectores improvidi humanam amittere gratiam formidantes loqui pertimescunt*, Past. 15 ; Swt. 89, 12. Hine on unðanc R eorringa gesêceþ, Salm. Kmbl. 197 ; Sal. 98.   **II.** *an unpleasing act, a displeasure, an offence, annoyance :*—Cweðe gê ðæt ic eów dide æfre ǽnigne unþanc ? Ap. Th. 26, 3. Nú ic wolde ðê ðone unþanc mid yfele leánian, Gen. 31, 29. Hê bræc ðæne palant æt Neomagan and eác fela óðra unþanca hê him dyde, Chr. 1049 ; Erl. 172, 22. Hê wolde geofan him ðone castel, ðæt hê mihte syððan dæghwamlíce his unwinan unþancas dôn, 1075 ; Erl. 212, 16.   **III.** *not thanks, displeasure expressed in words :*—Ðá ágeaf hê ðæt feoh tó unðances (*he gave back the money without getting any thanks*), and his eác hæfde micelne dem *talentum cum sententia damnationis amisit*, Past. 49 ; Swt. 379, 9. Ðá wæs Hannibale æfter hiera hǽðeniscum gewunan ðæt andwyrde swíþe láð and him unþanc sǽde ðæs andwyrdes *abominatus dictum Annibal*, Ors. 4, 10 ; Swt. 202, 7.   ¶ The word occurs most frequently in the genitive, with adverbial force, where something is done without a person's consent or good-will, (1) *absolute, unwillingly, without consent, on compulsion ; ingratis :*—Niman hí unþances (*without the person's consent*) ðone teóðan dǽl, L. Edg. i. 3 ; Th. i. 264, 1. Hê ðone deófol ádrǽfde of ðam preóste . . . Se deófol, ðe hine ǽr unðances forlêt, hine sôna gelǽhte, Homl. Th. ii. 170, 11. Ðú miht forleósan unðances ða ðing ðe áteorian magon, ac gif ðú sylf for Gode gôd byst, ðæt ðú ne forlýst nǽfre unðances, 410, 26–28. Far ðê frig ; nis ná úre gewuna ðæt ǽnig man unðances tó Gode gecyrre, 416, 32. Se cyning sende æfter Amane, and hê unþances com, Homl. Ass. 99, 247 : Hexam. 20 ; Norm. 28, 23. Unþances fæstende, Homl. Skt. i. 19, 92. Ealle hyra unlustas hí sceolon gebêtan sylfwylles on ðyssum lífe, oððe unþances æfter ðyssum lífe, Homl. Th. i. 148, 28 : Homl. Skt. i. 17, 31. Se mægðhád sceal beón geoffrod be his ágenum cyre, ðæt seó lác beó leófre ðonne hê wǽre, gif hê unðances wǽre, Homl. Ass. 33, 237. Gewilde man hí tó rihte þances oððe unþances, L. Eth. ix. 40 ; Th. i. 348, 38. Hê nam sume mid him, sume þances, sume unþances, Chr. 1066 ; Erl. 198, 37. (2) *with noun or pronoun, without (a person's) consent, not of (one's own) accord, against (one's) will :*—Ðá gerád Æþelwold ðone hám æt Winburnan ðæs cynges unþances (*bútan ðæs cyninges leáfe*, MS. A.), Chr. 901 ; Erl. 97, 12. On ðám castelan ðe hí ǽr ðes eorles unþances begiten hæfdon, 1091 ; Erl. 227, 10. Ðá weard hê gecristnod his mága unþances, Homl. Skt. ii. 31, 24. Scealt ðú ðínes unþances ðone hord ámeldian, Homl. Skt. i. 23, 716. Án his manna wolde wícian æt ânes búndan húse his unðances, Chr. 1048 ; Erl. 177, 36. Þeáh him ðæt word ofscute his unnþances (*licet verbum illud improviso exprimerit*, 1055 ; Erl. 189, 6. Ðonne sió sául hire unðonces gebǽdd wierð ðæt yfel tó forlǽtenne, Past. 36 ; Swt. 251, 12. Ðis folc ðe úre unþances faran wyllaþ, Ex. 14, 5. Hí heora unðances hié begeáton, Ors. 2, 2 ; Swt. 64, 27. Unþonces, 5, 13 ; Swt. 244, 21. Unþances, Jud. 11, 33. [þat him wes mucheles unðonc (mid mochel onþong, 2nd MS.], Laym. 22370. Hit is þe an unðonke, 11769. Seoruwe uor luve of eie worldliche þinge, oðer uor eni unðonc, A. R. 202, 12. For þeft and for þrepyng unþonk may mon haue, Allit. Pms. 43, 183 : P. S. 327, 90: Chauc. T. and C. 5, 699. Unthank come on his heed, Reeve's T. 162. Hy wyteþ and zyggeþ onþank, Ayenb. 69, 15. A king of Britaine hauede heo bewedded al hire unðonkes, Laym. 4032. Bettre iss to þe mann to don all hiss unnþannkess god þan ifell hise þannkess, Orm. 7194. *O. H. Ger.* un-dankes *ex necessitate, invite.* Cf. *Icel.* ú-þökk; *f.* reproach, censure.]

un-þancful[1]; *adj. Unthankful, ungrateful :*—Ic wæs micles tó unðoncful Gode mínes gewittes and mínre hǽle and ealra ðara gôda ðe ic on lifde, Anglia xi. 99, 67. Hê is gôd ofer unþancfulle (unðoncfullum, Lind.) *benignus est super ingratos*, Lk. Skt. 6, 35. v. þanc-ful[1].

un-þancweorþ, -wirþe; *adj. Ungrateful; ingratus.* (1) *not agreeable, unacceptable :*—Gemǽgnys is ðam sôðan Dêman gecwême, þeáh ðe heó mannum unðancwurðe sý, Homl. Th. ii. 126, 3.   (2) *thankless :*—Wê wǽron unðancwurðe and wendon ús fram Criste, ac hê ús gesôhte, Basil admn. 4; Norm. 42, 5. God, se ðe dæghwomlíce getíðaþ weldǽda unðancwurðum (cf. *ipse benignus est super ingratos*, Lk. 6, 35), Homl. Th. ii. 418, 23.

un-þearf, e; *f. Disadvantage, hurt, harm, detriment :*—Gif ðú heora untreówa onscunige, oferhoga hí and ádríf hí fram ðê, for ðam hí spanaþ ðe tó ðínre unþearfe *si perfidam perhorrescis, sperne atque abjice perniciosa ludentem*, Bt. 7, 2 ; Fox 18, 10. Ðæt wyrð ðære þeóde eall tó unþearfe, L. I. P. 4 ; Th. ii. 308, 3, 9 : Wulfst. 267, 30. Deófol má and má manna forlǽrde and getíhte tó heora ágenre unþearfe, 10, 4. Ðú lutodest on ðam láðum cristendôme ðám godum tó teónan and mê tó unþearfe, Homl. Skt. i. 5, 414. Hê gegaderode his folc tó ðæs cynges unþearfe, ac hê wæs gelet (hê gaderode his folc þan cyngce tó unþearfe hê þôhte, ac hit weard heom seolfan tó mycclan hearme, MS. D.), Chr. 1075 ; Erl. 213, 17. Hê fêrde for his bróðær unþearfe intó Normandige he (*William*) *went to Normandy on account of the injury his brother had done him* (see Henry of Huntingdon), 1091 ; Erl. 227, 5. Nú hæfð se yfela gást seofontealde ungifa, ðæt sýn unþearfa managra manna, Wulfst. 52, 9. [*Icel.* ú-þörf *harm.*]

un-þearf; *adj. Needless, useless :*—Unðærfe ðing *nequaquam*, Mt. Kmbl. Lind. 2, 6.   [*Icel.* ú-þarfr *useless, bad; ú-þarfi needless.*]

un-þearfes; *adv. Needlessly, without cause :*—Heora fêt beóð swíðe hraðe blôd tó ágeótanne unþearfes for yflum willan, Ps. Th. 13, 6. v. þearf-leás.

**un-þeáw, es ; m. A bad habit, an evil practice, a vice, fault :*—Médsceattas áwendaþ wôiíce tó oft ða rihtan dômas, and seó yfelnyss becymð ofer eallum folce ðǽr ðǽr se unþeáw orsorhlíce ríxaþ, Ælfc. T. Grn. 20, 34. Ðæt is ðara monna unþeáw ðæt hí nyton hwæt hí send *sese ignorare hominibus vitio venit*, Bt. 14, 3 ; Fox 46, 8. Nán hæfignes ðæs líchoman, ne nán unþeáw ne mæg eallunga áteon of his môde ða rihtwísnesse . . . ðeáh sió swernes ðæs líchoman and ða unþeáwas oft ábisegien ðæt môd mid ofergiotulnesse, 35, 1 ; Fox 154, 29. Má dereþ monna gehwylcum módes unþeáw ðonne mettrymnes lǽnes líchoman, Met. 26, 112. Suá ðú meaht ǽlcne unðeáw on ðǽm menn ǽresð be sumum tácnum ongietan . . . Siððan bið sió duru ðære unrihtwísnesse ontýned *uniuscujusque peccati prius signa forinsecus, deinde janua apertae iniquitatis ostenditur*, Past. 21 ; Swt. 157, 19. Mon sceal ðone unþeáw of mynstre áwyrtwalian, ðæt nǽnig ne gedyrstlǽce ǽnig ðing tó syllenne bútan ðæs abbodes hǽse *hoc vitium amputandum est de monasterio, ne quis presumat aliquid dare sine jussione abbatis*, R. Ben. 56, 16. Hwæt is sáwla hǽlo búte rihtwísnes ? oððe hwæt is hiora untrymnes búte unþeáwas ? *quid aliud animorum salus videtur esse, quam probitas ? quid aegritudo, quam vitia ?* Bt. 39, 9 ; Fox 226, 19. Ne sié hê tó ungerisenlíce underþeód his unþeáwum *nec victa libidine colla foedus summittat habenis*, 29, 3 ; Fox 106, 19 : Met. 16, 4. Hê wæs swíþe gefylled mid unþeáwum and firenlustum *homo flagitiosissimus*, Ors. 6, 3 ; Swt. 256, 23 : Ps. Th. 7, 13. Wê sceolon faran fram unðeáwum tó gôdum ðeáwum, gif wê willaþ faran tó ðam êcan lífe, Homl. Th. ii. 282, 23. On unðeáwum *in abusione*, Ps. Spl. 30, 22. Lufie mon ðone man, and hatige his unþeáwas, Bt. 39, 1 ; Fox 212, 8 : Met. 27, 32 : Bt. 29, 3 ; Fox 106, 27 : Met. 16, 24. Ýdel bið seó lár ðe ne gehǽlð ðære sáwle leahtras and unðeáwas, Homl. Th. i. 60, 35. Ðá wolde hê forbúgan ða unðeáwas ðe menn begáð, ii. 38, 4 : 154, 12 : Chr. 1067 ; Erl. 204, 31. Ðý lǽs hié forlǽten untǽlde óðerra monna yfele unðeáwas (yfle ðeáwas, Cott. MSS.) *ne pravos hominum mores nequaquam redarguant*, Past. 46 : Swt. 351, 21. Gê ða Engliscan þeáwas forlǽtaþ ðe eówre fæderas heóldon, and hǽðenra manna þeáwas lufiaþ, and mid ðam geswuteliaþ ðæt gê forseóð eówer cynn and eówere yldran mid ðám unþeáwum, ðonne gê him on teónan tysliaþ eów on Denisc ábleredum hneccan and áblendum eágum, Engl. Stud. viii. 62, 4. [Ðat unþeáw . . . þat ilke unhende flesches brune, H. M. 9, 27. Sparuwe cheatereð euer and chirmeð . . . Moni ancre haueð þet ilke unþeau, A. R. 152, 23. He þaht hit weren for undeawe, þ he hire weore swa unwourd, Laym. 3064. To hatenn all þatt Godd iss lef and lufenn alle unnþæwess, Orm. 17782. Him is loþ everich unþeu, O. and N. 194.]

un-þeáwfæst; *adj. Of bad habits, vicious, ill-mannered, ill-conditioned :*—Hit is bysmorlíc dǽd, ðæt ǽnig man ǽfre swá unþeáwfæst beón sceole, ðæt hê ðone múð ufan mid mettum áfylle, and on óðerne ende him gange ðæt meox út, Engl. Stud. viii. 62, 15. Onþeáwfæste *indisciplinatorum*, Hpt. Gl. 526, 75. Hwam becumaþ wunda oððe eágena blindnyss búton ðám unðeáwfæstum ðe wôdlíce drincaþ, and heora gewitt ámyrraþ? Homl. Ass. 6, 144. v. un-geþeáwfæst.

un-þeáwful[1]; *adj. Undisciplined, ill-conditioned :*—Unþeáwfulra *indisciplinarium*, Wrt. Voc. ii. 47, 10: 87, 75.

un-þeccan; *p. -þehte To uncover :*—Hiá unðehton ðæt hús *nudauerunt tectum*, Mk. Skt. Lind. 2, 4.

un-þeód; *adj.* —Ne lyste ðê fægeres wífes and wel gelêredes and seó ðínum willum and wel unþeód (underþeód? *subject*), Shrn. 183, 10.

un-þinged; *adj. Uninvited, sudden, unexpected :*—Ðý lǽs iów gemête se rêða and se egeslíca dæg, se cymð ofer ealle eorðwaran undinged, swǽ swǽ grin and superveniat in vos repentina dies illa. *Tamquam laqueus superveniet in omnes, qui sedent super faciem omnis terrae*, Past. 43 ; Swt. 316, 12. Hí ofer cume unþinged deáð, ástígon heó on helle lifigende *veniat mors super eos, et descendant in infernum viventes*, Ps. Th. 54, 14. Dol biþ se ðe him his Dryhten ne ondrǽdeþ ; cymeþ him se deáð un-

þinged, Exon. 312, 8; Seef. 106: 335, 18; Gn. Ex. 35. v. un-ge-þinged.

un-þingod; *adj. Unatoned, unsettled:*—Swā eác se ðe ōðrum bismer cwið, oððe dēð, ðeáh hē geswíce, and hit nǽfre eft ne dō, ðeáh hit bið gedōn, ðæt hē dyde, and unðingad, gif hē hit ne bēt *neque qui contumelias irrogat, si solummodo tacuerit, satisfecit*, Past. 54; Swt. 423, 35.

un-þolemódness, e; *f. Impatience:*—þurh unðolemódnesse *per inpatientiam*, Confess. Peccat.

un-þoligendlíc; *adj. Intolerable:*—Nāht unþoligendlíore *nihil intolerabilius*, Scint. 208, 14.

un-þorffæst; *adj. Useless, needless:*—Unðor[f]fæst bidda *ineptum rogare*, Rtl. 179, 34. v. þorf-fæst.

un-þríste; *adj. Timid, diffident, faint-hearted:*—Ða unmódgan and ða unðrístan (*pusillanimes*)... Ða lytelmódan and ða unðrístan, ðonne hié ongietaþ hiera unbældo and hiera unmiehte, hié weorðaþ oft ormóde, Past. 32; Swt. 209, 5-8. Ða unðríestan (-ðrístan, Cott. MSS.), Swt. 211, 15.

un-þrowi[g]endlíc; *adj. Incapable of suffering, impassible:*—Se ðe is unðrowigendlíc on his godcundnysse, Homl. Th. i. 116, 27: 120, 25. Seó Godcundnys ne mihte nān ðing þrowian, for ðan ðe heó is unðrowigendlíc, ii. 6, 30. Unðrowigendlíc, 270, 32.

un-þurhsceótendlíc; *adj. Impenetrable:*—Mid ðý unþurhsciótendlíore gescyldnesse gescyld mē *inpenetrabili tutela me defende*, Lchdm. i. lxix, 5.

un-þurhtogen; *adj. Not carried through, not performed:*—Hwæt wēne gē hwæt sió ðurhtogene unrihtwísnes geearnige, nū sió unðurtogene ārfæsðnes swā micel wíte geearnaþ *quid mereatur injustitia illata, si tanta percussione digna est pietas non impensa*, Past. 44; Swt. 329, 14.

un-þwǽre; *adj. At enmity, not in agreement:*—Gif ðū gemanst ðæt ðín bróðor sig unþwǽre wið ðē *si recordatus fueris quod frater tuus simultatem tecum habet*, L. Ecg. P. ii. 27; Th. ii. 194, 1. v. un-geþwære.

un-þwǽrian. v. ge-unþwǽrian.

un-þwǽrness, e; *f. Discord, dissension, disagreement:*—Ðǽr wæs micel unþuǽrnes (-ðwǽrnesse, MS. E.) ðære þeóde betweox him selfum, Chr. 867; Erl. 72, 8. Unðwērnesse *discordias*, Kent. Gl. 155.

un-þwagen, -þwægen, -þwegen, -þwogen; *adj. Unwashed:*—Ða hwíle ðe hig unþwogene beóð, L. Ælf. E.; Th. ii. 392, 14. Unþwogenum (-þwagenum, MS. A.: -ðwægnum, Rush.: -ðuegenum, Lind.) handum *manibus non lotis*, Mk. Skt. 7, 2. Unþwogenum (-ðuēnum, Lind.: -ðwegenum, Rush.), Mt. Kmbl. 15, 20. Unðweánum, p. 17, 11. [*Goth.* un-þwahans.]

un-þyhtig; *adj. Weak:*—Unðyhtge (-þyotgi, -dyctgi) ēgan *vitiato oculo*, Txts. 107, 2133.

un-þyldig; *adj. Impatient:*—Hí bióþ swā unþyldige ðæt hí ne magon nān earfoþa geþyldelíce āberan, Bt. 39, 10; Fox 228, 2. [*O. H. Ger.* un-dultig *inpatiens*.]

un-þyldlícness, e; *f. Difficulty:*—Ðā wæs mycel unþyldlícnes geworden be his byrignesse *facta difficultate tumulandi*, Bd. 4, 11; S. 580, 8 note.

untíd-fyll[u]; *f. Unseasonable repletion, excessive drinking at improper times:*—Be oferfylle. *Ve, qui consurgitis mane ad bibendum*, *etc.* Wā eów, hē cwæð, ðe lufiaþ untídfylla and ǽr on morgen oferdrenc dreógaþ, Wulfst. 46, 14.

untíd-gewidere, es; *n. Unseasonable weather:*—Ðises ylcan géares wǽron swíðe untídgewidera, and for ðí geond eall ðis land wurdon eorðwæstmas eall tó medemlíce gewende, Chr. 1095; Erl. 232, 35.

un-tídlic; *adj. Unseasonable:*—Untídlíc *intemporalis*, Germ. 394, 316. Nā, swā hit gewuna is, of untídlícan gewideran, ðæt is, of wǽtum sumerum and of drýgum wintrum and of rēðre lenctenhǽte, and mid ungemǣtre hærfestwǽtan *non, ut adsolet, temporum turbata temperies, hoc est, aut siccitas hiemis, aut repentinus calor veris, aut humor aestatis, vel autumni divitis indigesta illecebra*, Ors. 3, 3; Swt. 102, 5. [*O. H. Ger.* un-zítlih *inportunus*.]

un-tídlíce; *adv. Unseasonably, at a wrong time:*—Eall ðæt mon untídlíce onginþ, næsþ hit nō æltæwne ende, Bt. 5, 2; Fox 10, 27. [*O. H. Ger.* un-zítlicho *immature*.]

un-tídre; *adj. Not weak, firm, strong:*—Him wæs hyge untyddre, Andr. Kmbl. 2506; An. 1254.

untíd-sprǽc, e; *f. Unseasonable speech:*—Unnytte dǽde and untídspǽca forhogian, L. I. P. 14; Th. ii. 322, 10.

untíd-weorc, es; *n. Unseasonable work, work done at a wrong time:*—Geswícan untídweorces (*work done on Sunday*), Wulfst. 209, 27. Geswícan untídweorca, 221, 19.

un-tígan; *p. de To untie, unbind, loose:*—Ic unbinde oððe untíge *soluo*, Ælfc. Gr. 28; Zup. 177, 7. Ne untígð (*soluit*) eówer ǽlc on restedæge his oxan oððe assan fram ðære binne? Lk. Skt. 13, 15. Gyt gemētaþ as-an folan getíged... untígaþ hyne. Gif inc hwā āhsaþ hwí gyt hyne untígeaþ... Ðā hig hine untígdon, ðā cwǣdon ða hláfordas: 'Hwí untíge gē ðæne folan,' 19, 30-33: Mk. Skt. 11, 2-4. Petrus ðone ryððan untígde, Homl. Th. i. 374, 2. Ðā sende God his apostolas tó

gebundenum mancynne, and hēt hí untígan. Hú untígdon hí ðone assan? 208, 4-6. Untýgaþ hí, 206, 11. Hwæt dō gyt ðone folan untígende? Mk. Skt. 11, 5. Se ðe gesyhð assan clipiende oððe untíende (-tí[g]edne?) yrnan, Lchdm. iii. 198, 12. Ðonne wē sind geladode, ðonne sind wē untígede, Homl. Th. i. 210, 7, 9.

un-tilod; *adj. Without provision made:*—Se ðe his ǽr tíde ne tiolaþ ðonne biþ his on tíd untilad *who makes no provision for himself beforehand, for him will there be no provision made when the time comes*, Bt. 29, 2; Fox 106, 3.

un-tíma, an; *m.* I. *a wrong time, an improper time:*—Se lǽce ðonne hē on untíman lācnaþ wunde, hió wyrmseþ and rotaþ *secta immature vulnera deterius infervescunt*, Past. 21; Swt. 153, 2. Hē wilnaþ hǽlo tó late and on untíman, ðonne hē ǽr nolde hié gehealdan, ðā ðā hē hí hæfde *salus infructuose ad ultimum quaeritur, quae congruo concessa tempore utiliter non habetur*, 36; Swt. 249, 8. Ælc ðæra manna ðe yt oððe drincð on untíman on ðam hālgan lenctene oððe on rihtfæstendagum, Homl. Skt. i. 12, 76: Anglia xi. 113, 26. II. *a bad time, an unhappy condition of things, a mishap* (cf. *French* malheur):—Ic ásende ofer eówer land ǽlcne untíman, ðæt bið egeslíce greát hagol, se fordēð eówre wæstmas, and unásecgendlíce þunras..., Wulfst. 297, 7. [*Continentia*, þat is, þat man þe spuse haveð, his golliche deden widteo, swo hit be untíme, Rel. Ant. i. 132, 18. Vres misseide, oðer in untime, A. R. 344, 3. A man schulde not ete in untyme, Chauc. Pers. T. In vntyme ne shulde no bourde on bedde be, Piers P. 9, 186. *Icel.* ú-tími *a wrong time* (koma í útíma *to come too late*); *an evil time, mishap*.]

un-tíme; *adj. Unhappy, unfortunate, ill-timed:*—Se dysiga dranc būtan bletsunge and eode him ūt. Man slætte ðā ǽnne fearr, and se fear arn him tógeánes, and hine ðýde ðæt hē his feorh forlēt, and gebohte swā ðone untíman drenc, Homl. Skt. i. 12, 74. v. un-tíma, II, un-tímness.

un-tímende; *adj. Not productive, barren:*—Sarai wæs untýmende (*sterilis*); næfde heó nān bearn, Gen. 11, 30: Jud. 13, 2: Boutr. Scrd. 22, 22. Hit is swíðe ungedafenlíc ðæt forwerode menn and untýmende gifta wilnian, ðonne gifta ne sind gesette for nānum ðinge būton for bearnteáme, Homl. Th. ii. 94, 12. Eádige synd ða untýmendan *beatae steriles*, Lk. Skt. 23, 29. v. tíman, I.

un-tímness, e; *f. Misfortune, unhappiness:*—Ic sende on eówrum hūsum cwealm and hungor and untímnesse and fýr, ðæt forbærnð ealle eówre welan, Wulfst. 207, 18. v. un-tíma, II, un-tíme.

un-tiogoðað, un-tiorig, un-tíþe. v. un-teóþod, un-teorig, un-tygþa.

un-tóbrocen; *adj. Not broken in pieces:*—Gif gē ðone bend healdaþ sóðre bróðerrǽdene untóbrocenne, Homl. Th. ii. 318, 5. Hē sum þing hæfde untóbrocen, Homl. Skt. i. 5, 258.

un-tóclofen; *adj. Uncloven:*—Ða ðe synd gehófode on horses gelícnysse untóclofenum clawum, Homl. Skt. ii. 25, 45.

un-tódǽled; *adj. Undivided, unseparated:*—Is ðæt full gód ðæt eall ætgædere is untódǽled, Bt. 34, 9; Fox 146, 28. Hē biþ ānfeald untódǽled, 33, 2; Fox 122, 18, 21. Se God is simle on ānum untódǽled, 34, 6; Fox 142, 22: Wulfst. 21, 19. Willnade se cyning ðæt se wer him syndriglíce untódǽlede geférscipe (*individuo comitatu*) lāreów wǽre, Bd. 5, 19; S. 639, 3.

un-tódǽledlíc; *adj. Indivisible, inseparable:*—Ánfeald and untódǽledlíc *simplex indivisumque natura*, Bt. 33, 1; Fox 120, 10: 34, 7; Fox 144, 19. Hí ðrý ān God untódǽledlíc, Homl. Th. i. 150, 15: 248, 9: 500, 29. Ðære Hālgan Ðrynnysse weorc is ǽfre untódǽledlíc, 498, 35. *Littera* is se lǽsta dǣl on bócum and untódǽledlíc, beóð ða stafas untódǽledlíce, Ælfc. Gr. 2; Zup. 4, 19-5, 3. Ðæt gecynd gefēhþ ða friénd tógædre mid untódǽledlíore lufe, Bt. 24, 3; Fox 84, 2.

un-tódǽledlíce; *adv. Indivisibly, inseparably:*—Seó ðe hæfð ðæs ðreó ðing on hire tógædre wyrcende untódǽledlíce, Homl. Th. i. 288, 25: 500, 14: 368, 2.

un-tódǽledness, e; *f. Undividedness:*—Indivisio, ðæt is untódǽlednyss, Anglia viii. 318, 17.

un-tódǽllíc (-dǽl-); *adj. Indivisible, inseparable:*—Tódǽl ða twā, ðonne byð ān tó láfe; ðæt ys untódállíc, Anglia viii. 318, 30. Untódǽllíore[ro] *inseparabili, indivisibili*, Hpt. Gl. 430, 50.

un-tóledenlíce; *adv. Unremittingly:*—indesinenter, Gr. Dial. 2, 8.

un-tólísende; *adj. Inextricable:*—Ðý untólýsendum *inextricabili*, Wrt. Voc. ii. 78, 70. Untólésende *inextricabilem*, 76, 48.

un-tósceacen; *adj. Undisturbed, undestroyed:*—Swā lange swā God wolde ðæt Cristen geleáfa mid Engolcynne untósceacen weóxa, Chart. Th. 127, 11. Swā lange swā God wylle ðæt Cristen geleáfa mid Angelcynne untósceacan wurðe, 390, 35.

un-tóslegen; *adj. Not beaten to pieces:*—þeáh ðæt scyp sí ūte on ðære sǣ on ðām ýðum, hyt byð gesund untóslegen, gyf se streng aþolaþ, for ðam hys byð se óðer ende fæst on ðære eorðan and se óðer on ðam scype, Shrn. 175, 22.

un-tósliten; *adj. Not torn asunder:*—Untósliten *inextricabilis*, Wrt. Voc. ii. 110, 60: *inima* (?), 45, 37. Hí heóldon his tunecan untóslitene, Homl. Th. ii. 254, 32. Untóslitenum *indisruptis*, Wrt. Voc. ii. 88, 31: 47, 13.

un-tósprecendlíc; *adj. Ineffable:*—Rícsiendum ūrum Dryhtne, ðæm

hiéhstan and ðæm untósprecendlícan (cf. regnante Domino nostro, summo et ineffabili rerum Creatore omnium, 106, 19) ealra þinga and ealra tída Scippende, Chart. Th. 124, 8 : 388, 24.

un-tótwǽmed ; adj. Undivided, unseparated : — Nis Cristes godcundnys gerunnen tó ðære menniscnysse, ac hé þurhwunaþ þeáh á on écnysse on ánum háde untótwǽmed, Homl. Th. i. 40, 30. Ðæra weorc is symle untótwǽmed, ii. 366, 20.

un-trǽglíce ; adv. Well, honestly : —Ásecaþ ða ðe snyttro mid eów hæbben, ðæt mé þinga gehwylc þríste gecýðan untrǽglíce, ðe ic him tó séce, Elen. Kmbl. 819 ; El. 410.

un-treów, e ; f. Bad faith, faithlessness, perfidy, fraud : —Mé ðás woruldsælþa blindne on ðis dimme hol forlǽddon, and mé ðá berýpton rǽdes and frófre for heora untreówum, Met. 2, 13. Gif ðú heora untreówa onscunige, oferhoga hí, and ádríf hí fram ðé, for ðam ðú spanaþ ðé tó ðínre unþearefe si perfidam perhorrescis, sperne atque abjice perniciosa ludentem, Bt. 7, 2 ; Fox 18, 8. Heó ðæt leóht geseah ellor scrídan, ðæt hé hire þurh untreówa tácen iéwde, Cd. Th. 48, 10 ; Gen. 773. [O. Sax. un-trewa : O. H. Ger. un-triuwa fraus.]

un-treówe, -trýwe ; adj. Untrue, not faithful : —Gyf hwylc man sý untrýwe ðam hundrede, L. C. S. 30 ; Th. i. 392, 21 note. [Goth. untriggws iniquus : Icel. ú-tryggr faithless, untrustworthy.]

un-treówfæst ; adj. Unfaithful, untrustworthy : — Untrýwfæst flecti facilis, Germ. 401, 27. Hí cwǽdon tó ðam Hǽlende : ' Wé wyton ðæt ðú of forlygere wǽre ácenned ; and óðer ys, ðæt ðýn cynn ys on Bethleem swýþe untreówfæst ; and þrydde ðæt ðýn fæder and ðýn módor flugon of Egiptan lande for ðam ðe hig nefdon nánne trúwan tó nánum folce,' Nicod. 6 ; Thw. 3, 22.

un-treówlíce ; adv. With bad faith, perfidiously : —Agothocles gedyde untreówlíce wið hiene, ðæt hé hiene on his wárum beswác and ofslóg per Agathoclem insidiis circumventus, occisus est, Ors. 4, 5 ; Swt. 170, 9. Ðá bæd hé ðæt mon dyde beforan him ðone triumphan. Ac him Rómáne untreówlíce his forwierndon, and hit under ðæt ládedon for ðon ðe hé ǽr sige næfde, 5, 2 ; Swt. 216, 31.

un-treówsian ; p. ode. I. to defraud : —Ne untreówsige gê nó eów betweoxn nolite fraudare invicem, Past. 16 ; Swt. 99, 14. II. to offend : —Ðonne beóð manega untreówsede tunc scandalizabuntur multi, Mt. Kmbl. 24, 10 note. v. ge-untreówsian.

un-treówþ, e ; f. Bad faith, perfidy : —Ða Dænescan, ðe wæs ǽrur geteald eallra folca getreówast, wurdon áwende tó ðære mǽste untríwðe and tó ðam mǽsten swicdóme ðe ǽfre mihte gewurðan, Chr. 1086 ; Erl. 223, 7. Antigones forlét ðæt setl. Ac Umenis him wénde from Antigones hámfærelte micelra untreówða Antigonus ab obsidione discessit. Sed nec sic Eumeni spes firma aut salus certa, Ors. 3, 11 ; Swt. 146, 21. Adam týhð mé untryówða, Cd. Th. 36, 33 ; Gen. 581. Agothocles gedyde untreówlíce wið hiene . . . Gif hé ðá ða áne untreówþa ne gedyde, from ðæm dæge hé mehte bútan gebroce Cartaina onwald begietan, Ors. 4, 5 ; Swt. 170, 11. [Icel. ú-trygð falseness, faithlessness.]

un-trum ; adj. Weak, sick, ill, infirm : —Untrum infirmus, Wrt. Voc. i. 75, 45. Untrum ic eom infirmus sum, Ps. Spl. 6, 2 : Mt. Kmbl. 25, 36. Næs ðæra leóda ǽnig untrum non erat in tribubus eorum infirmus, Ps. Th. 104, 32. Þeówa untrum servus male habens, Lk. Skt. 7, 2. Se ðe ǽr untrum wæs qui languerat, 7, 10. Ðá weard hé untrum on feforádle, Blickl. Homl. 217, 15. Nǽnig næs tó ðæs untrum, ðæt hé sóna hǽlo ne onfénge, 223, 23. Ðæt flǽsc is untrum caro infirma est, Mt. Kmbl. 26, 41. Man hwylcne dǽl his hrægles tó untruman men bróhte, ðæt hé weard hál geworden, Blickl. Homl. 223, 25. Untrume ealle wǽran infirmati sunt, Ps. Th. 106, 11. Wæs ðǽr on neáweste untrumra manna hús, on ðam hyra ðeáw wæs ðæt hí ða untruman in lǽdan sceoldan erat in proximo casa, in qua infirmiores induci solebant, Bd. 4, 24 ; S. 598, 27. v. ge-untrum.

un-trumian ; p. ode. I. to make weak, weaken : —Ic untrumige infirmo, Ælfc. Gr. 47 ; Zup. 276, 7. II. to be or to become weak : —Ná ic untrumge non infirmabor, Ps. Spl. 25, 1. Ðæs bróðer untrumade cujus frater infirmabatur, Jn. Skt. Rush. 11, 2. [Þa was þe king swíðe untrumed, Laym. 15037.] v. ge-untrumian ; un-trymman.

un-trumlíc ; adj. Weak : —Hwæðer ðæt landfolc sí tó gefeohte stranglíc oððe untrumlíc populum, utrum fortis sit an infirmus, Num. 13, 20.

un-trumness, e ; f. Weakness, sickness, illness, infirmity : —Freneticus se ðe þurh sleápleáste áwêt, frenesis seó untrumnys, Wrt. Voc. i. 75, 61. Untrumnys ægritudo, Bd. 1, 27 ; S. 494, 18 : infirmitas, 3, 12 ; S. 537, 12. Ðá gestód his wíf untrumnes on hire eágan ingruente oculis caligine subita, 4, 10 ; S. 578, 18. Líchomlícre untrumnesse ðrycced corporea infirmitate pressus, 4, 24 ; S. 598, 25. Of untrumnysse (infirmitate) ðæs gecyndes, 1, 27 ; S. 494, 13. Mid his módes untrumnesse (infirmitate), Past. 54 ; Swt. 423, 21. Hér Eádsige forlét ðet biscopríce for his untrumnisse, Chr. 1043 ; Erl. 169, 23. Mid ðære untrumnesse (fever) swíðe geswenced, Blickl. Homl. 227, 8. Mihtig ǽlce untrumnesse tó hælenne, 223, 22. Underwreoðaþ his untrumnesse sustentat inbecillitatem suam, Kent. Gl. 644. Gemænigfylde synd untrumnyssa (infirmitates) heora, Ps. Spl. 15, 3 : 102, 3. On manegum gemetum geneósaþ God

manna sáwla . . . hwíltídum mid untrumnyssum, Homl. Th. i. 410, 28. On feforádle and on mislícum óþrum untrumnessum, Blickl. Homl. 209, 11. v. un-trymness.

un-trymed ; adj. Unconfirmed : —Se ðe him biþ unfullod oððe untrymed qui ipse non baptizatus vel non confirmatus sit, L. Ecg. C. 7 ; Th. ii. 140, 19.

un-trymig, -trymmig ; adj. Weak, sick, infirm : —Líchoma is untrymig (infirma), Mk. Skt. Lind. Rush. 14, 38. His sunu untrymig uæs filius infirmabatur, Jn. Skt. Lind. 4, 46 : 11, 2. Untrymig is infirmatur, 11, 3. Untrymmig infirmus, Mt. Kmbl. Lind. 25, 36, 43. Ða ðe wǽrun untrymige (-trymig, Lind.) qui infirmabantur, Jn. Skt. Rush. 6, 2. Ofer untrymigum super aegrotos, Mk. Skt. Lind. Rush. 16, 18. Fore untrymigum pro infirmis, Rtl. 177, 19.

un-trymig[i]an to become weak, sick, infirm : —Ða ðe untrymigdon qui infirmabantur, Jn. Skt. Lind. 6, 2.

un-trymigu(-o) ; f. Weakness, sickness, infirmity : —Ðæt heá gegême all unhǽl and all untrymmigo ut curarent omnem languorem et omnem infirmitatem, Mt. Kmbl. Lind. 10, 1.

un-trymman, -trymian ; p. ede To be or to become weak, sick, ill, infirm : —His sunu untrymede (unntrymade, Lind.) filius infirmabatur, Jn. Skt. Rush. 4, 46. Ðæs bróðer untrymade, Lind. 11, 2. Hé ongann untrymmia coepit egere, Lk. Skt. Lind. 15, 14.

untrymness, e ; f. Weakness, sickness, illness, infirmity : —Hwæt is sáwla hǽlo, búte rihtwísnes ? hwæt is hiora untrynnes búte unþeáwas ? quid vero aliud animorum salus videtur esse, quam probitas ? quid aegritudo, quam vitia ? Bt. 39, 9 ; Fox 226, 18. Ðé untrymnes ádle gongum bysgade, Exon. Th. 163, 7 ; Gú. 990. Of untrymnesse módes oððe líchoman infirmitate, Past. 21 ; Swt. 159, 1 : 56 ; Swt. 435, 15. In untrymnisse wæs ðú lécedôme in infirmitate sis medicina, Rtl. 105, 11. Hé hǽlde ǽghwilce ádle and ǽghwilce untrymnisse (infirmitatem), Mt. Kmbl. Rush. 10, 1. Ða untrymnesse hiera heortan ic wolde getrymman cordis infirmitatem munimus, Past. 4 ; Swt. 41, 4 : 10 ; Swt. 61, 16. Untrymnise, Rtl. 49, 30. Se ðe ne mægi gidrouia untrymnissum úsum qui non possit conpati infirmitatibus nostris, 91, 7. Lǽcedômas wið eallum untrymnessum heáfdes, Lchdm. ii. 2, 1. v. untrumness.

un-trymþ, e ; f. Weakness, sickness, infirmity : —Heora unmiht and heora untrymð is swíðe gemanifealdod multiplicatae sunt infirmitates eorum, Ps. Th. 15, 3. Gif hwylc wíf seteþ hire bearn ofer hróf oððe on ofen for hwylcere untrymðe hǽlo (alicujus morbi sanandi causa), L. Ecg. C. 33 ; Th. ii. 156, 36.

un-tweó ; gen. -tweón ; m. Not doubt, certainty : —Bið untweó (-treo, MS.) ðæt ðǽr Adames cyn cwíþeþ gesárgad there is no doubt that Adam's race will lament afflicted, Exon. Th. 59, 31 ; Cri. 961.

un-tweód ; adj. Not inspired with doubt, unwavering : —Hé hæfde him on innan ellen untweódne, Andr. Kmbl. 2485 ; An. 1244.

un-tweógende, -tweónde ; adj. Undoubting, unhesitating, unwavering, certain : —Gif wé hæfdon ǽnigne dǽl untwiógendes andgites swá swá englas habbaþ si divinae judicium mentis habere possemus, Bt. 41, 5 ; Fox 254, 7. Hí óþ heora lífes ende untweógende móde þurhwunodan, Blickl. Homl. 171, 13. Hé nǽfð gearone willan untweógendne tó ðæm weorce, Past. 54 ; Swt. 423, 26. Hyht untweóndne, Elen. Kmbl. 1592 ; El. 798. Ðæt wé ðý untweógendran be ús gelýfden ðæt wé be ðæm leorniaþ, Shrn. 67, 24.

un-tweógendlíc ; adj. Certain : —Untuéndlíc sind certi sumus, Lk. Skt. Lind. 20, 6.

un-tweógendlíce ; adv. I. without feeling doubt, certainly, unhesitatingly : —Ic hit untweógendlíce gelýfde in tó gesettanne eam indubitanter inserendam credidi, Bd. 4, 22 ; S. 592, 30. Eallum mannum þurhwuniggendum in tintregum untweógende, Blickl. Homl. 171, 16. Hié untweógendlíce wéndon ðæt heora hláford wǽre on heora feónda gewealde, Ors. 3, 9 ; Swt. 134, 27. Ægðer ðara folca wénde untweógeudlíce ðæt hié sceoldon on ða eorþan besincan, 4, 2 ; Swt. 160, 29 : 4, 5 ; Swt. 166, 13. Ðætte hié ðý fæsðlícor and ðý untweógendlícor gelífden ðara écena ðinga ut ad aeternorum fidem certius roboretur, Past. 50 ; Swt. 389, 35. Þéh hý gelýfdan be his segene, ðe hit ǽr geseah, untweógendlícor (-átweógeudlícor, MS. C.), ðonne ða heora segene ǽt gelýfdon, ðe æfter heom ácende wǽron, Wulfst. 2, 12. II. so as not to cause doubt, unequivocally, indubitably : —Hé his ǽrendracan ásende tó ðære ðeóde, and him untweógendlíce secgan (say in a way that should leave no room for doubt) hét, ðæt hié ðǽr sccolden, oþþe ðæt lond æt him álésan, oþþe hé hié wolde mid gefeohte fordôn missis legatis qui hostibus parendi leges dicerent, Ors. 1, 10 ; Swt. 44, 8. v. tweógend-líc.

un-tweólíc ; adj. Undoubted : — Untweólícere indubitata, Hpt. Gl. 411, 38.

un-tweólíce ; adv. I. undoubtedly, indubitably, certainly : —Smyre ðone seócan ; untweólíce ðú hyne álýsest, Lchdm. i. 302, 4. Hé grípð untweólíce ðæt behátene ríce, Homl. Th. i. 360, 25 : Homl. Ass. 97, 184. Heó getácnode untweólíce ða hálgan geládunge, 114, 412. Untwílíce, Ælfc. T. Grn. 3, 39. Se Sunu is gást and hálig untwýlíce, Homl. Th. i. 282, 30. Untwýlíce ðú líhst, 378, 6. Ða ungeleáffullan

untwýlíce forwurðaþ on ēcnesse, ii. 60, 15 : 110, 27 : Basil admn. 4; Norm. 44, 12 : 5 ; Norm. 46, 18. **II.** *without feeling doubt, with certainty :*—Ealle ða geleáffullan fæderas sǣdon untwýlíce and geþwǣrlehton on ðam ánum, ðæt God gescypð ǣlces mannes sáwle, Homl. Skt. i. 1, 85.

un-tweónigende ; *adj. Not to be doubted, indubitable :*—God untwýnigendre mildheortnesse *God of mercy which must not be doubted* (but the Latin is : Deus inestimabilis misericordie), Anglia xi. 115, 45.

un-twifeald ; *adj.* **I.** *not double* (v. twi-feald, **IV**), *simple, sincere, honest, pure :*—Nis nán scild trumra wið ðæt tuiefalde gesuinc ðonne mon sié untwiefeald (-twy-, Cott. MSS.) *nil est ad defendendum puritate tutius,* Past. 35 ; Swt. 239, 10. Se untweofealda willa bioþ tó tellenne for fullfremod weorc, Bt. 36, 7 ; Fox 184, 24. Ne magon wē nǣfre gereccan ðone yfelan mon clǣnne and untwifealdne *malos esse, pure atque simpliciter nego,* 36, 6 ; Fox 182, 18. Gif hí góde beóþ and hláfordholde and untwifealde *si probi sunt,* 14, 1 ; Fox 42, 24. Hí untweofealde treówa gehealdaþ, Met. 11, 95. **II.** *not double, united, without division :*—Ðeáh hē hwelcne wæstm forð brenge gódes weorces, gif hē ne bið of gódum willan and of untwiefaldre (-twy-, Cott. MSS.) lufan ongunnen, ne bið hē náwuht *qui etsi boni operis fructus in suis actionibus proferunt, profecto nulli sunt, quia non ex unitate caritatis oriuntur,* Past. 47 ; Swt. 359, 17. [*O. H. Ger.* un-zwifalt.]

un-twílíce, -twýlíce, un-twýnigende. v. un-tweólíce, un-tweónigende.

un-týd ; *adj. Ignorant, uninstructed, unskilled :*—Dysig bið se lǣce and untýd ðe wilnaþ ðæt hē óðerne mon gelǣcnige, and nát ðæt hē self bið gewundad *improbus et imperitus est medicus, qui alienum mederi appetit, et ipse vulnus, quod patitur, nescit,* Past. 48 ; Swt. 371, 6. Ic ðē giungne underfēng untýdne and ungelǣredne, Bt. 8 ; Fox 24, 24.

un-týdre, es ; *m. An evil growth, evil progeny, a monstrous birth :*—Ðanon untýdras ealle onwócon, eotenas and ylfe and orcneas, swylce gigantas (cf. Milton : Where nature breeds, Perverse, all monstrous, all prodigious things, Abominable, P. L. Bk. 2), Beo. Th. 222 ; B. 111.

un-týdrende ; *adj. Not propagating :*—Swínes blǣdran untýdrendes, ðæt is gylte, Lchdm. ii. 86, 23.

un-tygþa(-e), -týþa(-e) ; *adj. Unsuccessful in obtaining a request :*—For ðæm ne meahte Balaham geearnian ða Godes giefe ðe hē biddende wæs, ðá hē Israhēla folc wirgean wolde and for hine selfne gebiddan, for ðæm hē wearð untygða ðe hē hwierfde his stemne nales his mód *hujus correptionis donum idcirco Balaam non obtinuit, quia ad maledicendum pergens vocem, non mentem mutavit,* Past. 36 ; Swt. 257, 18. v. tíþe (*where read* tygþe(-a), týþe(-a) : cf. *O. Sax.* tugiðon).

un-týnan ; *p. de.* **I.** *to unclose, open :*—Euplis bær Cristes godspel in fóðre . . . Ðá untýnde Eplius ðæt Cristes godspel, Shrn. 116, 33. Án ðara cempa mið spere sidu his untýnde (*aperuit*), Jn. Skt. Lind. Rush. 19, 34. Hē untýnde ealle ða bernu, Gen. 41, 56. Hí untýndon heora goldhordas (*apertis thesauris*), Mt. Kmbl. 2, 11. Ðæt hǣ ádelle wæterpyt oþþe betýnedne untýne, and hine eft ne betýne *si quis aperuerit cisternam, et foderit, et non operuerit eam* (Ex. 21, 33), L. Alf. 22 ; Th. i. 50, 6, note. Untýne insigloe *aperire signaculum,* Rtl. 29, 17. Gié geseáð ðæt heofun untýned (*apertum*), Jn. Skt. Lind. i, 51. Hát ða duru beón untýnede, Homl. Skt. ii. 23 b, 448. Untúned bóc *aperto codice,* Mt. Kmbl. p. 4, 1. Byrgenna untuende (-túnede ?) ꝺ untýned wēron, Lind. 27, 52. **II.** *to disclose, lay open :*—Untēnd *aperiet* (*stultitiam*), Kent. Gl. 452. Hira unriht wearð eall untýned, Ps. Th. 72, 5. **III.** the word is used to gloss *solvere* and *inhiare* in the following :—Se ðe untýnes ꝺ tōslittes (*solverit*) ēnne of bebodum ðissum, Mt. Kmbl. Lind. 5, 19. Hē untýnde (*solvebat*) ðone Sunnadæg, Jn. Skt. Rush. Lind. 5, 18. Ðætte eardlíc lusto wiðsæcgende giliorniga wē untýna (*inhiare*) heofonlíco, Rtl. 34, 20. v. on-týnan.

un-týned ; *adj. Unfenced :*—Ceorles weorðig sceal beón wintres and sumeres betýned. Gif hē bið untýned . . . , L. In. 40 ; Th. i. 126, 14.

un-wáclíc ; *adj. Not mean, not poor, noble, splendid :*—Gegiredon ád unwáclícne, Beo. Th. 6268 ; B. 3138.

un-wáclíce ; *adv.* **I.** *not weakly, resolutely, without faltering :*—Ðæt hí æt þearfe þolian sceoldon, unwáclíce wǣpna neótan, Byrht. Th. 140, 54 ; By. 308. Ic beó gearo sóna unwáclíce willan ðínes, Exon. Th. 245, 25 ; Jul. 50. **II.** *not meanly, nobly, splendidly :*—His aferan eald brytteðon unwáclíce, Cd. Th. 258, 12 ; Dan. 674.

un-wǣded ; *adj. Not clothed :*—Monno unwēded mið wēde *hominem non vestitum veste,* Mt. Kmbl. Lind. 22, 11.

un-wǣr ; *adj.* **I.** *not on one's guard, unaware, unprepared :*—Gif ðē man scotaþ tó, ðú gescyltst ðē, gif ðú hit gesihst ; gif ðú unwǣr bist, ðú bist ðe swíðor geswenct, Homl. Th. ii. 538, 11. Hí cweþaþ ðæt tó worde, ðæt se biþ on geþance wǣrast and wísast, se ðe óðerne can raðost ásmeágan and oftost of unwǣran sum ðing gerǣcan, Wulfst. 55, 22. Perpena on ðone cyning ungearone (unwǣrne, MS. C.) becom *Perperna Aristonicum inproviso bello adortus,* Ors. 5, 4 ; Bos. 104, 26. Ðý lǣs ðe se smíc derige ðám unwarum, Homl. Th. ii. 418, 5. **II.** *unwary, heedless, incautious, inconsiderate :*—Módignys is endenēxt gesett, for ðan ðe se unwæra on ende oft módegaþ on gódum weorcum, Homl. Th. ii. 222, 4. Þencð se unwara eall swá deófol hine lǣrð, Wulfst. 298, 32. Unwaɼe weorude, Exon. Th. 363, 25 ; Wal. 59.

Deófol wile beswícan ðone unwaran, Homl. Th. i. 16, 22 : Blickl. Homl. 55, 23. Hig fordrencton ðone unwæran Loth, Gen. 19, 35. Ða unwaran *indocti et praecipites,* Past. proem.; Swt. 25, 12. Ða unwaran lǣreówas *rectores improvidi,* 15 ; Swt. 89, 10. His word beóð góde geðúhte unweran (-warum,' MS. C.) mannum, Wulfst. 54, 17. Deófol ðēð swýðe lytelíce, ðæt hē ongyt unwære (-ware, MS. C.) menn, 11, 16. Unware *inexpertos, incautos,* Hpt. Gl. 498, 67. Ða ðe galdorcræftas begangaþ and unwære men beswícaþ, Blickl. Homl. 61, 24. Unware, 185, 2. Unuuere *incautos,* Kent. Gl. 902. Tó fordónne ða unwaran, Basil admn. 2 ; Norm. 34, 30. Oft ðonne se hirde gǣd on frēcne wegas, sió hiord ðe unwærre bið gehríst *cum pastor per abrupta graditur, ad praecipitium grex sequitur,* Past. 2 ; Swt. 31, 1. **III.** adverbial uses :—Hí unwares (*unawares, unexpectedly*) cómon, and hē fyrst næfde, ðæt hē his fyrde gegadrian mihte, Chr. 1004 ; Erl. 139, 21. Unwæres, 1093 ; Erl. 229, 5. Ðá com Harold heom ongeán on unwaran (cf. *Icel.* at ūvörum *unexpectedly*), 1066 ; Erl. 200, 38. Ðá com Harold on unwær (cf. *Icel.* koma á úvart *to take by surprise*) on ða Normenn, Erl. 201, 26 : 202, 7 : 1043 ; Erl. 168, 32. Hí cómon unwær on heom, 1050 ; Erl. 175, 32 : 1067 ; Erl. 205, 25. [He wes to unwar, Laym. 7810. Sunnen sleað þeo unwarne soule, A. R. 274, 5. *Icel.* ú-varr.]

un-wæres. v. un-wær, **III.**

un-wærlíc ; *adj. Unwary, incautious, heedless :*—Suá suá unwærlícu and giémeleáslícu sprǣc menn dweleþ *sicut incauta locutio in errorem pertrahit,* Past. 15 ; Swt. 89, 8. Oft ðæt mægen ðære láre wierð forloren, ðonne mon mid ungedafenlícre and unwærlícre ofersprǣce ða heortan gedweleþ ðara ðe ðǣrtó hlystaþ *saepe dictorum virtus perditur, cum aꝺud corda audientium loquacitatis incauta importunitate laevigatur,* Swt. 95, 19. Ðeáh ðú fela unwærlícra worda gesprǣce, Exon. Th. 254, 6 ; Jul. 193. [*Icel.* ú-varligr *unwary.*]

un-wærlíce ; *adv. Unwarily, incautiously, without caution ; heedlessly :*—Ic lǣre ðæt hira nán ðara ne wilnie ðe hine unwærlíce begá ; and se ðe hí unwærlíce gewilnige, ondrǣde hē ðæt hē hí ǣfre underfēnge *ut haec, qui vacat, incaute non expetat ; et qui incaute expetiit, adeptum se esse pertimescat,* Past. proem.; Swt. 23, 14. Geðence se láriów ðæt hē unwærlíce (*incaute*) forð ne rǣse on ða sprǣce, 15 ; Swt. 95, 9. Gif sió wund bið unwærlíce gewriðen *cum fractura incaute colligatur,* 17 ; Swt. 123, 18. Se ðe ðone wuda unwærlíce (*incaute*) hiéwð, 21 ; Swt. 167, 15. Ða eode hē on íse unwærlíce *dum incautius in glacie incederet,* Bd. 3, 2 ; S. 525, 1. Ða ðe unwærlíce and gēmeleáslíce Gode hýraþ, Blickl. Homl. 63, 22 : 57, 9 : Exon. Th. 363, 34 : Wal. 63 : L. Ælfc. P. 7 ; Th. ii. 366, 13. Him com ongeán Hanno unwærlíce, and ðær ofslagen wearð, Ors. 4, 10 ; Swt. 200, 4 : Chr. 1068 ; Erl. 206, 9. Þænne gyltas unwærlíce [wē] forgyfaþ *dum culpas incaute remittimus,* Scint. 149, 8. [Ne ne wite hie a wiche halue ne a wiche wise he hem wile bisette, þanne he hem unwarliche (*unexpectedly*) his dintes giueð, O. E. Homl. ii. 191, 32. *Icel.* ú-varliga *unwarily.*]

un-wærness, e ; *f. Heedlessness, want of caution, imprudence :*—Þurh ðás unwærnysse hē gebringð hine on helle, Wulfst. 299, 7.

un-wærscipe, es ; *m. Heedlessness, inconsideration, imprudence :*—Ða gē forluron þurh unwærscipe, Homl. Th. i. 68, 4.

un-wǣscen ; *adj. Unwashen :*—Nim sigelhweorfan unwǣscene, Lchdm. ii. 108, 24.

un-wæstm, es, e ; *m. f. n.* **I.** *an evil growth, a bad plant, a tare, weed :*—Unwæstm (ða weód, Rush.) zizania, Mt. Kmbl. Lind. 13, 38. Huona hafes unwæstm (ðæt weód, Rush.) *unde habet zizania ?* 13, 27. Gié geadrias ðæt unwæstm, 13, 28. Ða unwæstma zizania, 13, 30, 40. Ðara wunwæstma zizaniorum, 13, 36. **II.** *bad growth, failure of crops :*—Eów unwæstm þurh unweder gelóme gelimpeþ, Wulfst. 133, 6. Gyf hit geweorðe ðæt on þeódscype becume heálíc ungelimp, unwæstm oððon unweder, orscwealm oððon mancwealm, 170, 1. Gif hwæt fǣrlíces on þeóde becymð, beón hit miswyderu oððon unwæstmas, 271, 3. Ús unwidera for oft weóldon unwæstmas, 129, 5 : 159, 13.

un-wæstmbǣre ; *adj. Unproductive, barren, sterile :*—Unwæstmbǣre elebeám *oleaster,* Wrt. Voc. i. 33, 19. On ðisum dæge ácende seó unwæstmbǣre móder ðone mǣran wítegan, Homl. Th. i. 356, 4. Se ðe on gódnysse unwæstmbǣre bið, ii. 406, 19. Se ðe eard seteþ unwæstmbǣrre *qui habitare facit sterilem,* Ps. Th. 112, 8. Unwestembǣre tēdrunge *infructuosa* (*infecunda*) *sterilitate,* Hpt. Gl. 430, 56. Hí woldon mē gedón unwæstmbǣrne, swá swá se ðe bútan ǣlcum yrfewearde byð, Ps. Th. 34, 12. Unwæstmbǣre wíf *sterilem,* Ps. Lamb. 112, 8. Ic wyrce on land unwæstmbǣre, Homl. Th. ii. 102, 34. Sume treówu hé cearf, ðý lǣs hié tó ðæm forweóxen ðæt hié forseareden, and ðý unwæstmbǣrran wǣren, Past. 40 ; Swt. 293, 7.

un-wæstmbǣrness, e ; *f. Unproductiveness, barrenness, sterility :*—Unwestmbǣrnys *steriitas* vel *infoecunditas,* Wrt. Voc. i. 53, 45. Unwæstmbǣrnys *sterilitas,* 76, 79. On hungre is geswutelod ðære eorðan unwæstmbǣrnys, Homl. Th. ii. 538, 31. For unwæstmbǣrnesse ðæs londes *propter terrarum infoecundam diffusionem,* Ors. 1, 1 ; Swt. 14, 18. On his ácennednysse hē ætbrǣd ðære mēder hire unwæstmbǣrnysse, Homl. Th. i. 352, 30.

un-wæstmberendlíc ; *adj. Barren, sterile :*—Seó stów is unwæstmber-

endlīcu for ðǽra næddrena mænigeo *loca illa sterilia sunt propter multitudinem serpentium*, Nar. 34, 28.

un-wæstmberendness, e; *f. Barrenness, sterility* :—Mē mīne fýnd āscufon fram ðǽre hālgan onsegdnysse for mīnre unwæstmberendnysse, Homl. Ass. 126, 329.

un-wæstmfæst; *adj. Barren, sterile* :—Unwæstmfæst ðara godcundra mægena, Blickl. Homl. 163, 6.

un-wæstmfæstness, e; *f. Barrenness, sterility* : — Sōna seó unwæstmfæstnes fram him fleáh, Blickl. Homl. 163, 17.

un-wæterig; *adj. Without water, dry* :—On unwæterium *inaquoso*, Ps. Lamb. 62, 3. On unwæterige stōwe, 77, 17. þurh unwæterige (-wæterie) stōwa *per loca inaquosa*, Lk. Skt. 11, 24.

un-wandiende; *adj. Unhesitating* :—Ða ðe unwandiende ðara scyldegena gyltas ofslōgen *qui delinquentium scelera incunctanter ferirent*, Past. 49; Swt. 381, 25.

un-warnod; *adj. Unwarned* :—Gif preóst ōðerne unwarnode lǽte ðæs ðe he wite ðæt him hearmian wille, L. N. P. L. 33; Th. ii. 294, 25.

un-wealt; *adj. Not given to roll, steady* :—Ða scipu wǽron ǽgðer ge swiftran, ge eác unwealtran, ge eác hiéran ðonne ða ōðru, Chr. 897; Erl. 95, 14 (v. note, p. 320). v. wiltan.

un-wearnum; *adv. Without hindrance* :—Hē slǽpendne rinc slāt unwearnum, Beo. Th. 1487; B. 741: Exon. Th. 309, 27; Seef. 63. v. wearn.

un-weaxen; *adj. Not grown up, young* :—Him be healfe stōd hyse unweaxen, cniht on gecampe, Byrht.'Th. 136, 17; By. 152. Cild unweaxen, Chr. 975; Erl. 126, 5. Ðus mē fæder mīn unweaxenne (*when a boy*) wordum lǽrde, Elen. Kmbl. 1055; El. 529. Se eorl wolde sleán eaferan sīnne unweaxenne (*Isaac*), Cd. Th. 204, 1; Exod. 412. Isaac bearn unweaxen, 173, 34; Gen. 2871. Hē hēt ealle ārīsan geonge ... Ðā upp āstōdon eaforan unweaxne, Andr. Kmbl. 3252; An. 1629.

un-weder, es; *n. Bad weather, tempest* :—Nū cweðaþ sume men ðæt se mōna hine wende be ðan ðe hit wuderian sceal on ðam mōnðe; ac hine ne went nǽfre nāðor ne weder ne unweder of ðam ðe him gecynde is, Lchdm. iii. 268, 4. Ðǽr ne cymð storm ne nān unweder ðæt ðam corne derie, Homl. Th. i. 526, 30. Heálīc ungelimp, unwæstm oððon unweder, Wulfst. 170, 1. Hī synd geneádode mid stormum ðæs unwederes (-wedres, MS. F.) *tempestatibus acti*, Ælfc. Gr. 44; Zup. 260, 12. Hē geðreáde ðæt wind and hroeðnise ł unwoeder ðæs wætres *increpavit ventum et tempestatem aquae*, Lk. Skt. Lind. 8, 24: p. 5, 18. Eów unwæstm þurh unweder gelōme gelimpeþ, Wulfst. 133, 7. Ūs unwedera for oft weóldan unwæstma, 159, 12. Eall ðæt geár wæs swīðe hefigtýme on unwæderum, Chr. 1041; Erl. 169, 9. [Unweder (*the plague of hail*), Gen. and Ex. 3058. *Icel.* ū-veðr *bad weather, storm*.] v. un-geweder, un-widere.

unweder-līce; *adv. In a way that indicates bad weather, threateningly* :—Tō dæg hit býð hreóh weder; ðeós lyft scīnð unwederlīce *hodie tempestas, rutilat enim triste coelum*, Mt. Kmbl. 16, 3.

un-wegen; *adj. Unweighed* : — Ðera ōðera wyrta ælces healues penincges gewihta, and vi pipercorn unwegen, Lchdm. i. 376, 7.

un-wemlíc; *adj. Spotless, pure, virgin* :—Mid ðam unwemlīcan cǽgan *virgineo clave*, Wrt. Voc. ii. 91, 75.

un-wemme; *adj.* I. of concrete objects, *spotless, without blemish, without defect, uninjured* :—Lamb unwemme *agnus absque macula*, Ex. 12, 5 : Lev. 9, 2, 3. Se æþela wong æghwæs onsund wið ýðfare gehealden stōd hreóra wǽga unwemme *ille locus, cum diluvium mersisset fluctibus orbem, exsuperavit aquas*, Exon. Th. 200, 25; Ph. 46. Hē eft mid his unwemmum līchoman hine gegyrede, Blickl. Homl. 89, 35. Hē gelǽdde ðæt folc ealle unwemme ofer ða Reádan sǽ *he led the people all of them uninjured over the Red Sea*, Btwk. 196, 2. II. of abstract objects, *uninjured, inviolate* :—Cyninges handgrið stande unwemme, L. E. G. 1; Th. i. 166, 21 (cp. L. Eth. vi. 14; Th. i. 318, 25). Godes cyrican wē sculan ǽfre lufian and nǽfre derian wordes ne weorces, ac gridian hý symle and healdan unwemme, Wulfst. 67, 17. Hī eodon of ðam fýre feorh unwemme, Exon. Th. 197, 7; Az. 186. III. in a moral sense, *undefiled, pure, immaculate, perfect* :—For ðī ic weorðe unwemme (*immaculatus*) beforan him, Ps. Th. 17, 23 : 18, 12. Sié heorte mīn unwemme, Ps. Surt. 118, 80. Se Hǽlend betwux synfullum unwemme fram ælcere synne ðurhwunode, Homl. Th. i. 356, 14. On unwemmum (ðæm unwemman, Surt.) wege *in via immaculata*, Ps. Th. 100, 1. Unwemme weg, Ps. Surt. 17, 33. Flǽsc unwemme, Exon. Th. 26, 18; Cri. 418. Hié scoteden ðone unwemman, Ps. Surt. 63, 5. Unwemme synt ðīne wegas *impolluta via ejus*, Ps. Th. 17, 29. Ða ðe unwemme (*immaculati*) on hiora Dryhtnes ǽ gangaþ, 118, 1. Weagas unwemra, Ps. Surt. 36, 18. Unwæmme, Ps. Th. 17, 31. Ðæt wē ūrne līchaman and ūre sāwle swā unwemme him āgeofan on dōmes dæg, swā hē hine ǽr gesceóp, Blickl. Homl. 103, 22. II a. of virginity, *pure, immaculate* :—Unwemme *immunis (carnali spurcitia*, Ald. 21), Hpt. Gl. 442, 5 : Homl. Skt. i. 7, 59. Fram unwemme fǽmnan ácenned, Blickl. Homl. 167, 21. Æt Sancta Maria ðǽre unwemman fǽmnan, 105, 20. Heó lufode ðone Hǽlend ðe hī heóld unwemme, Homl. Skt. i. 20, 18: Exon. Th. 19, 13; Cri. 300. Unwemme *immunes, incon-*

---

taminati, Hpt. Gl. 447, 43. Hēr syndon inne unwemme twā dohtor mīne, ne can ðara idesa ōwðer gieta þurh gebedscipe beorna neáwest, Cd. Th. 148, 30; Gen. 2864. [*Goth.* un-wamms *sine macula, immaculatus* : *O. Sax.* idis un-wamma; *acc.*] v. next word.

un-wemmed; *adj. Unspotted, undefiled, immaculate* :—Ðū wǽre symle fǽmne oncnāwen, and ðīne līchaman hæbbende clǽne and unwemmed (-wæmme, MS. G.), Homl. Skt. ii. 23 b, 437. On wege unwæmmedum *in via immaculata*, Ps. Spl. 100, 1, 7. Unuoemedo *immaculatam*, Rtl. 104, 18. [*A. R. Orm. Ps. Chauc.* un-wemmed.]

un-wemming, e; *f. Incorruption, immortality* :—Ðám unwemmincge sēcendum *his incorruptionem quaerentibus* (Romans 2, 7), Scint. 41, 10.

un-wemness, e; *f. Purity* :—Ðæt ða clǽnan and ða unwemman hira clǽnnysse and hiora unwemnysse forð gehīóldan, Homl. Ass. 207, 422. [*Cf. Goth.* un-wammei *sinceritas*.]

un-wendness. v. ge-unwendness.

un-wēne; *adj.* I. *hopeless, not having hope* or *expectation*. Cf. or-wēne :—Wæs ðǽr ān cnapa geǽttrod þurh næddran, swīðe tōswollen þurh ðæs wyrmes slege, unwēne his līfes, Homl. Th. ii. 514, 7. Hē gehǽlde ānre wydewan sunu ðe unwēne læg, Homl. Skt. i. 6, 103. Hē fond hlingende, fūsne on forðsīþ, freán unwēnne, Exon. Th. 171, 4; Gū. 1121. II. *not hoped for, unexpected* :—Gyf him þince ðæt hē æt forðgewitenum men āhtes onfō, of unwēnum (*ex improviso* : v. un-gewēned) hym cymeþ gestreón, Lchdm. iii. 170, 7. Forðon hiá unwoene (unwoen, Lind.) sint mæhte in him *propterea inopinantur uirtutes in illo*, Mk. Skt. Rush. 6, 14. [*Icel.* ū-vænn *hopeless, not to be expected*. Cf. *O. H. Ger.* un-wān *desperatio*.]

un-wēned; *adj. Unhoped for, unexpected* :—Ic him eft his rīce āgeaf, and ðā ðære unwēndan āre ðæs rīces (de hē him seolfa ne wēnde) ðæt hē ðā mē eall his goldhord æteówde *regna Poro restitui, qui, ut ei insperatus honor donatus est, mihi thesauros suos manifestavit*, Nar. 19, 23. On unwēnedum forþsīþe hē beóð gegripene *inprouiso exitu rapiuntur*, Scint. 181, 12. v. un-gewēned.

un-wēnlíc; *adj. Not giving grounds for hope, unpromising* :—Ðā ðū ðē selfum ðūhtest unwēnlīc *when you did not seem to yourself to have much chance of success* ; cum esses parvulus in oculis tuis, Past. 17; Swt. 113, 9. Hié oft gebidon on lytlum staþole and on unwēnlīcum (*a slight foundation and one that gave little hopes of success*), Ors. 4, 9; Swt. 192, 34. Hwæt wēnst ðū be ðære gōdan wyrde, ðe oft cymþ tō gōdum monnum on ðisse worulde, hweðer ðis folc mæge cweþan ðæt hit sié yfel wyrd? ... Hwæt wēnst ðū be ðære unwēnlīcan wyrde ðe oft þrietaþ ða yflan tō wītnianne, hwæðer ðis folc wēne ðæt ðæt gōd wyrd sié? *quid vero jucunda fortuna, quae in praemium tribuitur bonis, num vulgus malam esse decernit? ... quid reliqua, quae, cum sit aspera, justo supplicio malos coercet, num bonam populus putat?* Bt. 40, 2; Fox 236, 24 note. [*Magað* unwānlīk, Hēl. 4959. *Icel.* ū-vænligr *leaving little hope of success*.]

un-wēnunga; *adv. Unexpectedly* :—Men cwǽdon ðonne him hwæt unwēnunga gebyrede, ðæt ðæt wǽre weás gebyred, Bt. 40, 6; Fox 242, 4. [*Goth.* un-wēniggō *repentine*.]

un-weód, es; *n. A noxious weed* (lit. or metaph.) :—Seó eorðe ūs wind wið, ðonne heó forwyrneþ eorðlīces wæstmes and ūs unweóda tō fela āsendeþ, Wulfst. 92, 19. Man sceal ælc unriht mid rihte bētan and unweód āweódian and gōd sǽd ārǽran, 73, 2.

un-weorclíc; *adj. Unsuitable for work* :—Seó niht hafaþ seofon tōdǽlednyssa ... seó feórðe is *intempestivum*, ðæt ys mid niht oððe unworclīc tīma, Anglia viii. 319, 30.

un-weorþ, -wurþ, -wyrþ, -wierþe, -wyrþe; *adj.* I. *of no value* :—Mid deórwyrþum reáfum ne beóþ hý gescrýdde, ac mid unweorþum, R. Ben. 137, 9. Ðæt heora heortan mid wācum mettum and unweorþum ne sýn ofersýmede, 138, 11. II. *of no dignity, little esteemed* :—Gif munuc eaðhylde bið, þeáh hine man wācne and unweorðne talige *si omni vilitate contentus sit monachus*, R. Ben. 29, 3, 6. Gif munuc hine sylfne ýttran and unweorðran talaþ ðonne ǽnigne ōþerne *si omnibus se inferiorem et viliorem credat*, 29, 11. Æt ðǽm feórðan cirre hié sendon Hannan heora ðone unweorðestan þegn, and hē hit ābæd *novissime Annonis, minimi hominis inter legatos, oratione meruerunt*, Ors. 4, 7; Swt. 182, 13. Nime ge ðā ðe unweorþuste sién *them who are least esteemed* (1 Cor. 6, 4), Past. 18; Swt. 131, 7. III. *unworthy, not of sufficient merit* :—Ic ðone hād underhnāg, ðeáh ðe ic unwyrþe wǽre *quamvis indignus consensi*, Bd. 4, 2; S. 566, 8. Nemne God mē earmum and unwyrþum (*misero mihi et indigno*) gemiltsian wylle, 3, 13; S. 538, 35. Sum ungesceádwīs man hine sylfne āhēng ... Martinus hine unwurðne of deáðe ārǽrde, Homl. Th. ii. 506, 1. III a. with gen. of that of which one is unworthy :—Ðý læs ǽnig lāreówdōm underfōn durre ðara ðe his unwierðe sié *ne temerare sacra regimina, quisquis his impar, audeat*, Past. 3; Swt. 33, 8. Ic am unwyrðe micles hērnisse *ego sum indignus tanti officii*, Rtl. 98, 16. Mānsceaða, feores unwyrðe, Exon. Th. 97, 27; Cri. 1563. Saul ǽresð fleáh ðæt rīce and tealde hine selfne his suīðe unwierðne (*indignum se prius considerans*), Past. 3; Swt. 35, 18. Ðū hine ongeáte unweorþne ðæs anwealdes, Bt. 27, 2; Fox 96, 18. Hié woldon selfe fleón ða byrðenne suā micelre scylde, ða

ðe his unwierðe wǽron *indigni quique tanti reatus pondera fugerent*, Past. 2; Swt. 31, 15. Hié woldon habban gódne hlísan, þeáh hí his unwyrþe sién, Bt. 18, 1; Fox 60, 26. Gif anweald becymþ tó ðam men ðe his ealra unweorþost biþ, 16, 1; Fox 48, 34. **IV.** *worthless, bad, contemptible, despicable, ignoble*:—Hwæþer ðé þynce unweorþ and unmǽrlíc seó gegaderung? *obscurum hoc atque ignobile censes esse?* Bt. 33, 1; Fox 120, 29. Ðú wilt habban ealle fǽgere ðing and ácorene, and wilt ðé sylf beón wáclíc and unwurð, Homl. Th. ii. 410, 20. Unwurð scop *tragicus vel comicus*, Wrt. Voc. i. 60, 9. Unweorþe scopas *tragedi vel comedi*, 39, 39. Hí syndon ǽwisce on líchoman and unweorðe *sunt publicato corpore et inhonesto*, Nar. 38, 13. Sume habbaþ bearn genóge, ac ða beóþ yfele and unweorþ *alius prole laetatus, filii filiaeve delictis moestus illacrymat*, Bt. 11, 1; Fox 32, 9. Geþenc nú hwæþer ǽnig mon beó á ðý unweorþra ðe hine manige men forsióþ. Gif ðonne ǽnig mon á ðý unweorþra biþ, ðonne biþ ǽlc dysi man ðe unweorþra ðe he mǽre ríce hæfþ ǽlcum wísum men . . . Se anweald ne mæg his wealdend gedón nó ðý weorþran, ac he hine gedéþ ðý unweorþran (wyrsan, Bod. MS.) *si eo abjectior est, quo magis a pluribus quisque contemnitur, . . . despectiores potius improbos dignitas facit*, 27, 2; Fox 98, 8–14. Se eallra wyrresta and se eallra unweorþesta mon *pessimus*, 14, 3; Fox 46, 21. **IV a.** *with dat. of person by whom one is considered worthless*:—Xersis wearþ his ágenre þeóde swíþe unweorþ *Xerxes contemtibilis suis factus*, Ors. 2, 5; Swt. 84, 23. Ǽlcum witum láþ and unweorþ *omnibus invisus*, Bt. 28; Fox 100, 28: Met. 15, 6. Se ídela gylp ús beó ǽfre unwurð (-wyrð, MS. U.), Homl. Skt. i. 16, 367. Philippus him dyde heora wíg unweorð *Philip made light of their fighting power*, Ors. 3, 7; Swt. 118, 3. Bisceopum gebyreþ, ðæt hí ǽghwylc gefleard heom unwyrð lǽtan, L. I. P. 10; Th. ii. 316, 28. Romulus and ealle Rómware óþerum folcum unweorðe wǽron, Ors. 2, 2; Swt. 66, 16. **V.** *ignominious, dishonouring*:—Mid ealre ðare unwurð[r]este scame beó hé gescænt *ignominiosissima confusione subsannetur*, Chart. Th. 318, 34. [Unwurð *of no value*, A. R. 94, 4. Uvel strengþe is lutel wurþ, Ac wisdom ne wurþ never unwurþ, O. and N. 770. Crist wass unnwurrþ (*little esteemed*), Orm. 16163. He bið unworþ, þe mon þe litul ah, Laym. 3464. Þu maht to pi were iwurden þe unwurdere, H. M. 33, 12. Yhealde for uyl and onworþ, Ayenb. 132, 24. O. H. Ger. un-werd *ignobilis, contemtibilis, obscurus, dejectus*: Icel. ú-verðr *un-worthy*.]

un-weorþe; *adv.* *Unworthily*:—Unwyrðe *indigne*, Mk. Skt. Lind. 10, 14: 14, 4. Gif he ðæs hálgan húsles unwurðe onbyrigð, Homl. Th. ii. 278, 5. [Cf. Goth. un-wairþaba *unworthily*.]

un-weorþian; *p. ode.* **I.** *to dishonour, disgrace*:—Hú ne unweorþast ðú ðé selfne, ðæt ðú winsð wiþ ðam hláfordscipe ðe ðú self gecure? Bt. 7, 2; Fox 18, 29. Seó cwén, ðe ðin word forseah, ne unwurðode ðé ǽnne, ac ealle ðine ealdormenn *non solum regem laesit regina, sed et omnes principes* (Esther 1, 16), Homl. Ass. 93, 53. Gé unworðadun mec *uos inhonoratis me*, Jn. Skt. Rush. 8, 49. Ðæt he God ne unwurðige, Homl. Skt. i. 13, 86. Ðæt man unweorðige ða ðe godcunde láre wyrdan, Wulfst. 168, 7. Unweorðian *dehonestare*, Wrt. Voc. ii. 76, 15: 26, 40. Ða swelcan monn sceal unweorðian mid ǽlcre unweorðnesse *sine dedignatione dedignandi sunt*, Past. 37; Swt. 265, 18. Forsewen and geunwurþod, Homl. Th. i. 24, 4. **II.** *to become dishonoured*:—Unwurðiaþ *vilescunt*, Hpt. Gl. 462, 53. Unwurðie *vilescat*, 420, 13. [We unwurðeð ure Drihten, wurðeð þe deuel, O. E. Homl. ii. 181, 29. He shameþþ þe and unnwurrþeþþ, Orm. 18285. To onworþi, Ayenb. 22, 18. Icel. ú-virða *to slight*.] v. ge-unweorþian.

un-weorþlíc; *adj.* **I.** *of little value or importance, humble*:—Ða hláfordas and ða recceras scoldon ðencean ymb ðæt hélicuste and ða underðióddan scoldon dón ðæt unweorðlícre *a subditis inferiora gerenda sunt, a rectoribus summa cogitanda*, Past. 18; Swt. 131, 10. **II.** *that has little honour, not famous or splendid, poor*:—Hié lange wǽron ðæt dreógende ǽr heora áðer mehte on óþrum sige gerǽcan, ǽr Alexander late unweorðlícne sige (*a by no means famous victory*) gerǽhte *commissoque praelio diu anceps pugna tandem tristem pene victoriam Macedonibus dedit*, Ors. 3, 9; Swt. 134, 8. **III.** *ignoble, disgraceful, infamous*:—Mid ðý unæþelan gydde ym[b]f unweorðlícan *cum infami eulogio*, Wrt. Voc. ii. 137, 43. God ða mǽstan ofermétto genidrode mid ðære bismerlícestan wrace and ðære unweorðlícostan (*tormenta turpia*), Ors. 1, 7; Swt. 38, 5. [Ȝif þu art unwurðlich (*of little account*), H. M. 33, 11. Þe man þoleþ þet he by uoullíche ydraȝe, and ase persone onworþlych (v. unweorþ, II), Ayenb. 132, 35. Icel. ú-virðiligr *contemptible*.]

un-weorþlíce; *adv.* **I.** *unworthily, in an unsuitable manner*:—Him is micel ðearf ðæt hié geornlíce gedéncen ðæt hié tó unweorþlíce ne dǽlen ðæt him befæsð bið *necesse est, ut sollicite perpendant, ne commissa indigne distribuant*, Past. 44; Swt. 321, 14. Unwurðlíce, Cd. Th. 28, 23; Gen. 440. **II.** *with indignity, with contempt, ignominiously*:—Hé wearð self unweorðlíce ofslagen *Domitianus interfectus est; cujus cadaver ignominiosissime sepultum est*, Ors. 6, 9; Swt. 264, 15. Hé heora ǽrendracan swá unweorðlíce forseah, ðæt hé heora self onseón nolde *legatos ad se missos injuriosissime etiam a conspectu suo abstinuit*, 4, 8; Swt. 186, 7: 3, 10; Swt. 140, 3. Hí heóldon ðæt gold unwurðlíce

they held the gold in contempt, Homl. Th. i. 326, 24. **III.** *with indignation*:—Ða se Hǽlend hí geseah unwurðlíce (-weorð-, MS. A.) hé hit forbeád *quos cum uideret Iesus indigne tulit*, Mk. Skt. 10, 14. Sume hit unwurðlíce (-weorð-, MS. A.) forbǽron *erant quidem indigne ferentes*, 14, 4. [Unworthly þou wroght . . . when þou was bowne with a brande my body to shende, Alex. (Skt.) 869. O. H. Ger. un-werdlícho *indifferenter, indigne*: Icel. ú-virðiliga *scornfully*.]

un-weorþness, e; *f.* *Indignity, contempt, disgrace*:—Mid unweorðnesse *dedignatione*, Past. 37; Swt. 265, 18. Hé his ríce mid micelre unweorðnesse and mid micelre uniéðnesse gehæfde, Ors. 6, 24; Swt. 276, 1. Hé his onféng mid micelre unweorðnesse *a quo arrogantissime exceptus est*, 6, 30; Swt. 280, 12. Ða wón wyrd ǽgþer ge on ðara uurihtwísra anwealda heánesse, ge on mínre unwurþnesse and foreseuwenesse, Bt. 5, 1; Fox 10, 22. [Onworþnesse (despit) is wel grat zenne, Ayenb. 19, 35. O. H. Ger. un-werdnissa *contempt*.]

un-weorþscipe, es; *m.* **I.** *dishonour, disgrace*:—Hwæþer ðú nú mǽge ongitan hú micelne unweorþscipe se anweald brengþ ðam unmedeman? *videsne quantum malis dedecus adjiciant dignitates?* Bt. 27, 2; Fox 96, 9. **II.** *indignation*:—Ierre and unweorðscipe *ira et indignatio*, Past. 33; Swt. 222, 10. v. next word.

un-weorþung, e; *f.* **I.** *disgrace, shame*:—Sýn gescrýdde mid gescendnysse and unwurþunge ðe yfel sprecaþ ofer mé, Ps. Spl. 34, 30. **II.** *indignation*:—Hit bið unnyt ðæt mon unweorðunga forlǽte *frustra indignatio tollitur*, Past. 33; Swt. 222, 12. [Icel. ú-virðing *disgrace*.]

un-wered; *adj.* *Unprotected*:—Wit baru standaþ unwered wǽdo; nys unc wuht beforan tó scúrsceade, Cd. Th. 50, 21; Gen. 812.

un-wérig; *adj.* *Not weary, fresh*:—Æt níxtan wurdon hí ealle geteorode, and hé ána unwérig him æfter fyligde, Homl. Skt. ii. 30, 34. Gif mon on mycelre ráde weorþe geteorad, nime betonican . . . ; ðonne bið hé sóna unwérig, Lchdm. i. 76, 8. Hé hét ðæt mon ðæt fæsten brǽce and on fuhte dæges and nihtes, simle án legie æfter óþerre unwérig *cum alias aliis legiones dies noctesque succedere sine requie cogeret*, Ors. 5, 11; Swt. 238, 9.

un-werod; *adj.* *Not sweet*:—Wæter ðý unwerodre tó drincanne, Past. 58; Swt. 447, 19.

un-widere, es; *n.* *Bad weather, tempest*:—Ús unwidera weóldon unwæstma, Wulfst. 129, 4: 159, 12 note. [O. H. Ger. un-witari *tempestas.*] v. un-weder, -gewidere.

un-widlod; *adj.* *Unpolluted*:—Unwidlad *inpollutus*, Rtl. 90, 34.

un-wil[1]; *n.* *Absence of good will, dislike, despite, repugnance, reluctance*; the genitive, with adverbial force, *against one's will, not willingly, without one's consent, without intention, involuntarily*, is (almost) the only case used. (1) alone:—Gif hé hit dide unwilles *si praeter voluntatem id fecerit*, L. Ecg. P. ii. 1; Th. ii. 182, 13. Unwilles wé magon forleósan ða hwílwendlícan gód, ac wé ne forleósaþ nǽfre unwilles ða écan gód, Homl. Th. i. 576, 7–9: Hexam. 17; Norm. 26, 3. Gif ðú mé unwilles gewemman dést, Homl. Skt. i. 9, 90. Hé ðǽr wunode ða niht unwilles, se ðe sylfwilles nolde, Homl. Th. ii. 184, 13: L. Eth. vi. 52; Th. i. 328, 21. (2) *with pronouns*:—Þeáh ðú mé geoffrige mínes unwilles, ic beó þeáh unscyldig, Homl. Skt. i. 9, 87. Hire unwilles *invita ipsa*, L. Ecg. P. ii. 15, tit.; Th. ii. 180, 27. Heó wæs hire unwilles fram him *ab eo invita aberat*, 15; Th. ii. 186, 29. Heora unwilles, L. Edg. S. 2; Th. i. 274, 5. [Ich mot nede, ant neoðeles min unwil hit is to dou al þ ti wil is, Marh. 13, 3. Ha wes him ihondsald þah hit hire unwil were, Jul. 7, 12. Heo wes ihondsald al hire unwilles, 6, 5.] v. next word.

un-willa, an; *m.* *What displeases, displeasure, what is not desired*:—Nafa ðú tó yfel ellen, ðeáh ðé sum unwilla on become; oft brincð se woruld ðone willan ðe bið eft, Prov. Kmbl. 40. Hé drýhð deófles willan and Godes unwillan, Wulfst. 12, 13. ¶ *the word occurs mostly in dat.* (*sing. or pl.*) *with adverbial force*, unwillan, unwillum *against one's will, unwillingly, not voluntarily, without one's consent, in despite of one*. (1) alone:—Se ðe monnan nédes ofslóge, oððe unwillum, L. Alf. 13; Th. i. 46, 22. Hí sealdon unwillum áþas, Met. 1, 24. Ród ðe ic unwillum on beom gefæstnad, óðer ðe ic gestág willum mínum, Exon. Th. 91, 12; Cri. 1491: 360, 11; Wal. 4. Se ðe mid his willan bið besmiten . . . Se ðe onwillan (*invitus*) bið besmiten, L. Ecg. C. 5; Th. ii. 138, 7. (2) *with pronouns*:—Ðec ðín sáwl sceal mínum unwillan (-willum, Soul Kmbl. 125) oft gesécan, Exon. Th. 370, 22; Seel. 63. Ic áscige ðé, forhwí ðú ðæt ríce ðínum unwillan (-willum, Cott. MS.) forléte? Bt. 27, 2; Fox 96, 14. Sǽton ða Gotan on lande, sume be ðæs cáseres willan, sume his unwillan, Ors. 6, 38; Swt. 298, 5. Hé for ðam ege his unwillum ðonan wende, 4, 5; Swt. 166, 8. Nis nán syn þeáh man his unwillum blódes byrige of his tóðum, L. Ecg. C. 40; Th. ii. 166, 27. Úre gást biþ swiþe wíde farende úrum unwillum (*independently of our will*), Bt. 34, 11; Fox 152, 4. Godes anweald nǽre full eádiglíc, gif ða gesceafta hiora unwillum him hérden, 35, 4; Fox 160, 19: Ps. Th. 44, 16: Ors. 6, 13 tit.; Swt. 6, 3. Heora bégra unwyllum, Shrn. 204, 6. [He wuneð on wanrede and þoleð his unwille, O. E. Homl. ii. 123, 6. O. Sax. un-willeo:—An Godas unwilleon, Hél.

2460. *O. H. Ger.* un-willo *nausea*; sīnen unwillen *against his will:* *Icel.* ú-vili, at úvilja eins *against one's will.*] v. preceding word.

un-willende; *adj. Unwilling, not desiring or intending:*—Ic hit unwillende dô, Homl. Ass. 180, 353. Ðæt hí ne hlîpen unwillende on ðæt scorene clif unðeáwa *per multa, quae non appetunt, iniquitatum abrupta rapiuntur*, Past. 33; Swt. 215, 7. [*Icel.* ú-viljandi *unwilling, not intending.*]

un-wilsumlîce; *adv. Against one's will, not of one's own accord:*—Se sceal nýde on helle duru unwilsumlîce geniþerad gelæded beón *necesse habet in januam inferni non sponte damnatus introduci,* Bd. 5, 14; S. 634, 20.

un-windan; *p.* -wand, *pl.* -wundon; *pp.* -wunden *To unwind, unwrap* what is wrapped up:—Ðâ hêt hê unwindan ðæs cnihtes lîc, Homl. Th. i. 66, 24.

un-wine, es; *m. An unfriend, enemy:*—Ðæt hê mihte his unwinan unþancas dôn, Chr. 1075; Erl. 212, 16. Gif ic ongên ne cume, þæt þú it nêfre ne lêt weldon mîne unwinan æfter mê þe mid unrichte sitteð ðêron and nyttað it mê êuere tô unðanke, Chart. Th. 584, 10. [Wreken hine of his unwines, Laym. 1628. For to beon itempted of þe unwine of helle, A. R. 178, 27. Eð were ure lauerd to awarpen his unwine (*diabolus*), Kath. 1221. Herode wass unnwine wiþþ Filippe, Orm. 19 38. *Icel.* ú-vinr *an enemy.*]

un-wîs; *adj.* I. *unwise, foolish, stupid:*—Unwîs *insipiens*, Wrt. Voc. i. 76, 11: Ps. Spl. 91, 6: Ps. Th. 73, 17: Deut. 32, 6. Se unwîsa, Ps. Spl. 13, 1: 52, 1. Ðû wâst ðæt ic eom unwîs hyges *tu scis insipientiam meam*, Ps. Th. 68, 6. Unwîs *glebo*, Wrt. Voc. ii. 109, 81. Hié sealdon ânum unwîsum cyninges þegne Miercna rîce tô haldanne, Chr. 874; Erl. 76, 27. Unwîse on folce and dysige *insipientes in populo et stulti*, Ps. Spl. 93, 8: Ps. Th. 73, 21: Blickl. Homl. 59, 22: Homl. Skt. i. 17, 70. Monige men bioð ðe noldon ðone hlîsan habban ðæt hié unwîse (-wiése, Hatt. MS.) sién *sunt nonnulli, qui aestimari hebetes nolunt*, Past. 11; Swt. 66, 3. Earfoðtæcne unwîsra gehwæm (cf. dysgum monnum, Bt. 33, 4; Fox 130, 28), Met. 20, 148. Þeáh hió unwîsum wîdgel þince, 10, 10. Neátum ðæm unwîsum *jumentis insipientibus*, Ps. Surt. 48, 13: Blickl. Homl. 89, 9. Ic lære ge snottre ge unwîse, 107, 12. On ðæs unwîsestan lâre, L. Alf. 41; Th. i. 54, 8. Gê ðe on folce unwîseste ealra syndon *qui insipientes estis in populo*, Ps. Th. 93, 8. II. *ignorant:*—Gif hê hit nât, hwelce geselþa hæþþ hê æt ðam welan, gif hê biþ swâ dysig and swâ unwîs ðæt hê ðæt witan ne mæg? *si nescit, quaenam beata sors esse potest ignorantiae caecitate?* Bt. 11, 2; Fox 34, 26 note. Unwîse lâreówas cumaþ for ðæs folces synnum. Forðon oft for ðæs lâreówes unwîsdôme misfaraþ ða hiéremenn, and oft for ðæs lâreówes wîsdôme unwîsum hiéremonnum bið geborgen. Gif ðonne ægðer bið unwîs *pastorum saepe imperitia meritis congruit subjectorum: quia quamvis lumen scientiae sua culpa exigente non habeant; districto tamen judicio agitur, ut per eorum ignorantiam hi etiam, qui sequuntur, offendant*, Past. 1; Swt. 29, 3–6. Wênaþ unwîse læcas ðæt ðæt sié lendenâdl, ac hit ne bið swâ, Lchdm. ii. 232, 8. Unwîsum *ignaris*, Wrt. Voc. ii. 44, 23: *imperitis*, Mt. Kmbl. p. 2, 1. II a. with gen. *ignorant* of something:—Hê wæs ðære godcundan æfestnysse unwîs *divinae erat religionis ignarus*, Bd. 1, 34; S. 499, 22: 2, 20; S. 521, 22: 4, 13; S. 582, 18. Wê ðæs londes ungleáwe and unwîse wæron, Nar. 10, 7. [*Goth.* un-weis *idiota*; un-weis bi *ignorant of:* *O. Sax.* un-wîs *foolish:* *O. H. Ger.* un-wîs *insipiens, brutus, hebes, fatuus:* *Icel.* ú-vîss *foolish.*] v. on-unwîs.

un-wîsdôm, es; *m.* I. *folly, stupidity:*—Unwîsdôm *stultitia*, Mk. Skt. Lind. Rush. 7, 22. *Insipientia*, ðæt is unwîsdôm, Wulfst. 52, 17. From onsiéne unwîsdômes (*insipientiae*) mînes *fra face of mine unwisdome* (Ps.), Ps. Surt. 37, 6. Tô unwîsdôme *ad insipientiam*, Ps. Spl. 21, 2. Mid unwîsdôme gefyllede, Lk. Skt. 6, 11. Hit com of ðæs abbotes unwîsdôme, ðæt hê misbeád his munecan on fela þingan, Chr. 1083; Erl. 217, 3. Se ðe samnaþ ungemætlîce weolan, for his unwîsdôme (*stultitia*) sylle hê ðone þriddan dæle þearfum, L. Ecg. P. addit. 7; Th. ii. 232, 24. Se wîsdom is ðæs Hâlgan Gâstes gifu; deófol sæwð ðærtôgeánes unwîsdôm, Wulfst. 52, 26. II. *ignorance:*—Unwîsdômes blendnise *ignorantiae cecitate*, Rtl. 38, 9. Sôna swâ hí heora môd âwendaþ from Gode, swâ weorþaþ hî âblende mid unwîsdôme *ubi oculos a summae luce veritatis ad inferiora dejecerint, mox inscitiae nube caligant*, Bt. 40, 7; Fox 242, 31. For ðæs lâreówes unwîsdôme *pastorum imperitia*, Past. 1; Swt. 29, 4. Hê of yfelum willan ne gesyngaþ, ac of unwîsdôme (*non malitia, sed sola ignorantia*), 21; Swt. 157, 25. [We habbet idon unwisdom, Laym. 3383. Bihold i þine soule two þinges—sunne and ignorance; þet is, unwisdom and unwitenesse, A. R. 278, 7. *O. H. Ger.* un-wîstuom *ignavia, insapientia.*] v. on-unwîsdôm.

un-wîslic; *adj. Unwise, foolish:*—Ðæra unwîslicra geþanca, Homl. Skt. ii. 23 b, 526. [*Icel.* ú-vîsligr *foolish.*]

un-wîslîce; *adv. Unwisely, foolishly:*—Unwîslîce *insipienter*, Past. 15; Swt. 93, 21: Elen. Kmbl. 586; El. 293. Ðâ beseah Lothes wîf unwîslîce underbæc, Gen. 19, 26. Ðisse âdle fruman mon mæg yþelîce gelácnian, and æfter unêð, gif hió bið unwîslîce tô lange forlæten, Lchdm. ii. 232, 18. [*O. H. Ger.* un-wîslîcho *inmature.*]

un-wîsness, e; *f.* I. *ignorance:*—Swâ hwæt swâ ic for unwîsnesse âgylte *quicquid ignorantia deliqui*, Bd. 4, 29; S. 607, 29. Hî þurh unwîsnesse (*per ignorantiam*) gesyngodon, 1, 27; Bd. 491, 29. II. *wickedness:*—In ðærfum yfelgiornisse and unwîsnisse *in fermento malitiae et nequitiae*, Rtl. 25, 19.

un-wita, e; *m. A foolish, stupid, witless person, a fool:*—Se unwita *insipiens*, Ps. Lamb. 13, 1. Eorp unwita, Exon. Th. 433, 21; Rä. 50, 11. Gebîg fram unwitan (*insensato*), and ðû nâ wiþerast on stuntnysse his, Scint. 188, 11. Wê læraþ ðæt preósta gehwilc tô sinoðe gefædne man tô cnihte and nænigne unwitan ðe disig lufige, L. Edg. C. 4; Th. ii. 244, 14. Gif hit unwitan ænige hwîle healdaþ bûtan hæftum, hit ðurh hrôf wadeþ, bærneþ boldgetimbru, Salm. Kmbl. 821; Sal. 410. [Gif eni unweote acseð ou of hwat ordre ȝe beon, A. R. 8, 22. Oðre þurh wicchecreftes biȝuliað unweoten (-witen, MS. R.), Kath. 1054. Unweoten buten wit, Marh. 6, 11. *O. H. Ger.* un-wizzo *inscius, ignavus:* *Icel.* ú-viti *an idiot, a witless person:* ú-vita *senseless, insane.*]

un-witende; *adj. Unwitting, not knowing, not aware* of what is done, *unconscious:*—Hê moñig tâcen self gedyde, þêh hê hié unwitende dyde, Ors. 5, 14; Swt. 248, 14. Hê oft unwitende slôg mid his heáfde on ðone wâg, 5, 15; Swt. ²50, 12. Gelícost ðam ðe hê hiene selfne unwitende (witende? *the Latin is:* Ut voluntariam sibi conscivisse mortem putaretur) hæfde âwierged, 6 36; Swt. 294, 11. [Nyme oþre manne þinges onwytinde and wyþoute wylle of þe lhorde (*without the knowledge or consent of the owner*), Ayenb. 37, 5. Unwiting this preost of his false craft, Chauc. Ch. Y. T. ¹320. *O. L. Ger.* sô ik it uuitandi dâdi, sô uuuitandi: *Icel.* ú-vitaudi *not knowing, unconscious, not intending.*]

un-wiþerweard; *adj. Not adverse, not in opposition:*—His folgeras, swâ hié unwiðerweardran and gemôdran beóð, swâ hié fæstor tôsomne beóð gefægde tô gôdra monna hiénðe *sequaces illius, quo nulla inter se discordiae adversitate divisi sunt, eo in bonorum gravius nece glomerantur*, Past. 47; Swt. 361, 20.

un-wiþmetenlic; *adj. Incomparable, not to be compared:*—Is his eádmôdnys ¹s unwiðmetenlîc, Homl. Skt. i. 16, 119. Ðis hâlige mæden, Godes môdor, is unwiðmetenlîc eallum ôðrum mædenum, Homl. Th. i. 442, 29. Hire geðincðu ôðra hâlgena unwiðmetenlîce sind, 446, 5.

un-wiþmetenlîce; *adv. Incomparably, beyond compare, indescribably:*—Unwiðmetenlîce *incomparabiliter, inenarrabiliter*, Hpt. Gl. 414, 29: Homl. Th. i. 64, 18. Ðes symbeldæg oferstîhð unwiðmetenlîce ealra ôðra hâlgena mæssedagas, 442, 27: ii. 232, 10.

un-wîtnigendlîce; *adv. With impunity:*—Unwîtnigendlîce oððe bûʼan wîte *inpune*, Ælfc. Gr. 38; Zup. 233, 6.

un-wîtnod; *adj. Unpunished:*—Hî wênaþ ðæt ðæt sié sió mæste geselþ, ðæt men seó âlêfed yfel tô dônne, and sió dæd him môte beón unwîtnod *vel licentiam, vel impunitatem scelerum putant esse felicem*, Bt. 38, 5; Fox 206, 8. Ða ðe him biþ unwîtnode eall hiora yfel on ðisse worulde habbaþ sum yfel hefigre ðonne ænig wîte sié, ðæt is, ðæt him biþ unwîtnod hiora yfel *improbi cum supplicio carent, inest eis aliquid ulterius mali, ipsa impunitas*, 38, 3; Fox 200, 25–28. Ða unþeáwas næfre ne bióþ unwîtnode, 36, 1; Fox 172, 25. Hê geþafade ða scylde unwîtnode, Past. 17; Swt. 123, 6. v. un-gewîtnod.

un-wîtnung, e; *f. Impunity:*—Gyltes unwîtnung *sceleris impunitas*, Scint. 235, 5.

un-witod, -wiotod; *adj. Uncertain:*—Unwuted *incertus*, Rtl. 6, 5: 106, 15. Wel mon sceal wine healdon on wega gehwylcum; oft mon fêreþ feor bî tûne, ðær him wât freónd unwiotodne (*where he cannot look for a friend*), Exon. Th. 342, 23; Gn. Ex. 146.

un-wittig; *adj. Without wit* or *understanding*, (1) not in a bad sense:—Ge weras, ge wîf and ða unwittigan cild, Homl. Ass. 29, 122. (2) in a bad sense:—Wel déð se ðe unwittigum stýrð mid swinglum, gif hê mid wordum ne mæg. Hit is âwriten: 'Ne bið se stunta mid wordum gerihtlæced,' Homl. Th. ii. 532, 13. Þeówian unclænum deóflum and ðâm unwittigum heargum, Homl. Skt. ii. 30, 52. [Þat nan ne beo so wilde, nan swa unwitti, þat word talie ær he ihere minne horn, Laym. 786. Stew þine unwittie wordes, Marh. 6, 2. *O. H. Ger.* un-wizzig *insipiens, inprovidus, insanus.*] v. un-gewittig.

un-wittol; *adj. Ignorant:*—On manegum beó ðu swylce unwittol *in multis esto quasi inscius*, Scint. 80, 12.

unwit-weorc, es; *n. A work of folly, foolish work:*—Wê habbaþ nêdþearfe ðæt wê tô lange ne fylgeon unwitweorcum, Blickl. Homl. 111, 2.

un-wlite, es; *m. Disgrace*; dedecus, Wrt. Voc. ii. 27, 35: 26, 70: 85, 28.

un-wlitig; *adj. Not beautiful, ugly, foul:*—Unwlitig *deformes*, Wrt. Voc. ii. 28, 68. Unwlitig swile *tumor deformis*, Bd. 4, 32; S. 611, 17. Hû fægerne and hû wlitigne monnan ic hæbbe âtæfred, swâ unwlitig wrîtere swâ swâ ic eom *pulchrum depinxi hominem pictor foedus*, Past. 65; Swt. 467, 19. Ðes and ðeós unwlitige *hic et haec dedecor*, Ælfc. Gr. 9, 21; Zup. 47, 15. Simle ðæt unwlitige wlitigaþ ðæt wlitige *ever does the fair make fair the unfair*, Shrn. 165, 34. v. un-gewlitig.

un-wlitigian; *p.* ode. I. *to make ugly, deprive of beauty,*

disfigure :—Đa hē gewlitegaþ; hwīlum eft unwlitegaþ, Bt. 39, 8; Fox 224, 9. **II.** *to become ugly* :—Se wǽta āstīgđ tō đǽm lime, đonne āsuild hit and āhefegaþ and unwlitegaþ *humor ad virilia labitur, quae cum molestia dedecoris intumescunt*, Past. 11; Swt. 73, 10. v. ge-unwlitigian, un-gewlitigian.

un-wlitigness, e; f. *Ugliness, disfigurement* :—Semninga gehrān hē his eágan; đā gemētte hē hit swā hāl swā swā him nǽfre nǽnig swyle ođđe unwlitnes on ætȳwde (*ac si nil unquam in eo deformitatis ac tumoris apparuisset*), Bd. 4, 32; S. 612, 7.

un-wlitigung, e; f. *Disfiguring* :—Sió unwlitegung *deformatio*, Wrt. Voc. ii. 28, 62.

un-wrǽne; adj. *Not lustful* :—Lǽcedōmas gif man tō wrǽne sié oþþe tō unwrǽne, Lchdm. ii. 14, 25; 144, 20.

un-wrǽst, -wrǽste; adj. *Weak, poor, sorry, miserable, wretched* :—Forcūđlic ł unwrǽste *absurdum*, Hpt. Gl. 455, 50. Eálā hū leás and hū unwrēst is đysses middaneardes wela, Chr. 1086; Erl. 220, 40. Hē weard him on đam unwrǽstum (unwrǽste, Th. 321, 10) scipe (cf. uneáđe ætburstan, p. 320, col. 2) and fērde ofer sǽ, Chr. 1051; Th. 319, 3. Hī hī selfe lēton ægþer ge for heáne ge for unwrǽste, Ors. 3, 1; Swt. 98, 23. [Gif þǽr wǽre hure an unwreste wrenc (*a miserable trick*) þ he mihte get beswicen anes Crist, Chr. 1131; Erl. 260, 4. Ur lif wes unwreast . . . his lif was haliȝe, O. E. Homl. i. 237, 12. Đe unwreste herde *iners pastor*, ii. 39, 17. Þenne þat hæfd is unwrǽst þe hælp (hæp? heop, 2nd MS.) is þæ wurse, Laym. 16307. Þatt tu unn-orneliȝ off þe sellfenn læte, and halde þe forr unnwrǽste, Orm. 4889. Giff þu tellest all þin witt unnwrǽste, 4909. Ge muwen icnowen þet he is eruh and unwrest, A. R. 274, 16 (cf. Heo beođ to woke and tō unwreste iheorted, 268, 7). Hwet nu, unwreste men and wacre þen eni wake *quid vos ignavi et degeneres*, Kath. 1260. To binden faste upon an asse swiþe unwraste, Havel. 2820. Hit schal beo a thyng unwreste, heved of cok, breost of man, crop as best, Alis. 620.] v. next word.

un-wrǽstlíce; adv. *Weakly, absurdly* :—Đys hīw ealde ūđwitan gesettan āgēn đam þingum đe Zenodotus unwrǽstlíce gesette, Anglia viii. 334, 17. [Gif þu werest te erest wocliche (unwreastliche, MS. T.), A. R. 294, 5.]

un-wrecen; adj. **I.** *of a person, unavenged* :—Sceolde æþeling unwrecen ealdres linnan, Beo. Th. 4877; B. 2443. **II.** *of crime, unpunished* :—Gif hī đæs wilniaþ, þæt him hiora yfel unwrecen sié be đæs gyltes andefne, Bt. 38, 7; Fox 210, 7.

un-wrenc, es; m. **I.** *an evil trick, a malicious wile, a wicked artifice* :—Đisne unwrenc (*the device practised by Potiphar's wife*) heó geþōhte, Gen. 39, 16. Hē (*Antichrist*) bið eal unwrenca full, Wulfst. 97, 16. Mid đam unwrencan bið Antecrist eal āfylled, 54, 15. **II.** *an evil practice, a vice* :—For đæm unwrence đǽre ungeđylde *per impatientiae vitium*, Past. 33; Swt. 215, 19. [He teđ him to unwrenches (*evil practices*) to stele ođer refloc . . , O. E. Homl. ii. 79, 28. Þet is his unwrench (*artifice*) . . . he egged þe to a þing, þet þunched god, A. R. 268, 16. Ne spedestu noht mid þine unwrenche (*your vile tricks*), O. and N. 169.]

un-wreón; p. -wráh, -wreáh, pl. -wrigon, -wrugon; pp. -wrigen, -wrogen *To uncover* (lit. or fig.) *what is covered, to reveal* :—Hē un-wrīhþ þiccetu *revelabit condensa*, Ps. Lamb. 28, 9. Þæt mǽden unwreáh hire heáfod, Ap. Th. 26, 14. Hē unwreáh his rihtwīsnesse *revelavit justitiam suam*, Ps. Lamb. 97, 2. Hī unwreogon đæt hūs *nudaverunt tectum*, Mk. Skt. Rush. 2, 4. Đa lícmen his neb unwrugon, Homl. Th. ii. 334, 31. Unwreóh (*revela*) Drihtne weg đīnne, Ps. Spl. 36, 5. Nǽnig gedēgled đæt ne sē eft unwrigen (*nihil opertum quod non revelabitur*), Mt. Kmbl. Lind. 10, 26. Unwrigen *retectum, discoopertum*, Germ. 389, 11. Ūre misdǽda bioþ ealle opene and unwrigene beforan ūs, Wulfst. 225, 23. Unwrogene (*revelata*) synd stađolas ymbhwyrftes eorđan, Ps. Lamb. 17, 16. Đa deópan þing beóđ unwrogene, Anglia viii. 334, 7. v. on-wreón.

un-wrigedness (-wrigenness?), e; f. *A revelation* :—Of unwriged-nesse (*lectio*) *de apocalipsi*, R. Ben. Interl. 42, 16. v. on-wrigenness.

un-writen; adj. *Unwritten* :—Ne lēt ic đæt unwriten, Bd. pref.; S. 472, 26. Hī for heora slǽwþe forlēton unwriten đara monna þeáwas and heora dǽda đe on hiora dagum foremǽroste wǽron, Bt. 18, 3; Fox 64, 34. v. un-gewriten.

un-wrítere, es; m. *A bad, incorrect writer* :—Mycel yfel dēđ se unwrítere, gyf hē nele his wōh gerihtan, Ælfc. Gr. pref.; Zup. 3, 24.

un-wríþan; p. -wráþ; pp. -wriđen *To untwist, unbind* (lit. or fig.) :—Unwrīþan *distringere*, Scint. 232, 2. v. on-wríþan, *and next word*.

un-wriđen; adj. *Not bound* :—Đæt hē mid ungemetlícre grimsunge his hīeremonna wunda tō suíđe ne slīte, ne eft for ungemetlícre mild-heortnesse hié ne lǽte unwriđena *ut neque multa asperitate exulcerentur subditi, neque nimia benignitate solvantur*, Past. 17; Swt. 125, 16.

un-wunden; adj. *Not wound* :—Unwunden gearu *glomus*, Wrt. Voc. i. 59, 36.

un-wundod; adj. *Not wounded*, Cd. Th. 12, 10; Gen. 183.

un-wuni[g]endlíc; adj. *Uninhabitable* :—Beóđ twēgen dǽlas on twā healfa đam gemetegodum dǽle unwuniendlíce, for đan đe seó sunne ne cymđ him nǽfre tō, Lchdm. iii. 262, 2. v. un-gewuni[g]endlíc.

un-wurþ. v. un-weorþ.

un-wynsum; adj. *Unpleasant* :—Đeós woruld hwīltīdum is myrige on tō wunigenne, hwīlon heó is swíđe styrnlíc, and mid mislícum þingum gemenged, swā đæt heó bið swíđe unwynsum on tō eardigenne, Homl. Th. i. 184, 1. Hit đe unwynsum (*injucundum*) bið, Bt. 14, 1; Fox 42, 13. Ǽlc wyrd, sam hió sié wynsum, sam hió sié unwynsum *omnis fortuna vel jucunda vel aspera*, 40, 1; Fox 236, 2. Rēþu wyrd and unwynsumu, 40, 2; Fox 238, 2. Hē hine gegyrede mid hǽrenum hrægle swíþe heardum and unwinsumum, Blickl. Homl. 221, 24. [O. H. Ger. un-wunnisam *incultus, invenustus*.]

un-wynsumness, e; f. *Unpleasantness* :—Se stenc wearđ āwend tō wynsumum brǽde, and eall seó unwynsumnyss him wearđ tō blysse, Homl. Skt. i. 4, 215. Se mann gewyrđeþ tōswollen and tō stence āwended mid unwynsumnysse, Basil admn. 8; Norm. 50, 23.

un-wyrcan; p. -worhte *To undo, destroy* :—Ic þurh manslihtas mē scyldigne dyde wiđ đē, mín Hǽlend, đā đā ic đīn handgeweorc unwyrcan dorste, Anglia xi. 113, 34.

un-wyrd, e; f. *Bad fortune, misfortune* :—Gyf him þince đæt hē leád habbe, sum unwird him byđ tōweard, Lchdm. iii. 170, 5. Đæt mē nū þyncþ đætte ic đǽs unwyrd ārǽfnan mæg đe mē on becumen is *ut jam me imparem fortunae ictibus non arbitrer*, Bt. 22, 1; Fox 76, 13.

un-wyrþ. v. un-weorþ.

un-wyrttrumian; p. ode *To unroot, pluck up by the roots, root up* :—Đȳ lǽs gié unwyrtrumias (*eradicetis*) đone huǽte, Mt. Kmbl. Lind. 13, 29.

un-ymbwendedlíc; adj. *Unmoved, unalterable* :—Unymbwoendedlíc *inmotus*, Rtl. 164, 34.

un-ȳþe, un-ȳþian. v. un-eáþe, un-īþian.

up (ūp?), upp; adv. *Up.* **I.** where motion takes place, (a) from a lower to a higher point, (α) from the (earth's) surface to a point above it :—Hī eodon up tō đǽre dūne *ascenderunt verticem montis*, Num. 14, 40. Đa āstāh eft sunnan up *ascendit super occasum*, Ps. Th. 67, 4. Gewende se engel up, Homl. Skt. ii. 27, 100: i. 21, 56. Hē geseah windum đone rēc up ofer đǽre burge wallas āhefenne, Bd. 3, 16; S. 543, 2. Đā genam hine God mid sáwle and mid līchaman up in đone heofon, Salm. Kmbl. p. 182, 14. Āteó hē āne hringan up of đare þrȳh . . . Gif seó hringe nele up þurh his ānes tige, Homl. Skt. i. 21, 44, 47. (β) to the (earth's) surface from a point beneath it :—Seó burh, đǽr sunne up on morgen gǽđ, Salm. Kmbl. p. 186, 4. Ođ đæt seó sunne eft becume đǽr heó ǽr up stāh, Lchdm. iii. 236, 5. Nīwe steorra wæs upp yrnynde, Bd. 4, 12; S. 581, 14. Se mōna up eode, Nar. 30, 7. Hī delfaþ gold up of eorþan, Nar. 35, 8. Wolde đæt se hālga wer wurde up gedōn, Homl. Skt. i. 21, 136, 138, 140: Bd. 3, 7; S. 529, 24. Nime hē upp his mǽg *let him take his kinsman up from the grave*, L. Eth. iii. 7; Th. i. 296, 10. Đa ancras upp teón, Bd. 3, 15; S. 541, 40. Wiđ đon đe men blōd upp wealle þurh his mūþ, Lchdm. i. 74, 14. Beforan his fōtum wæs wyl upp yrnende, Bd. 1, 7; S. 478, 27. (a 1) where the motion is from sea to land :—Mid đȳ wē upp cōman tō lande, and ūre scyp eác swylce fram đam ȳþum upp ābǽron *cum evadentes ad terram, naviculam quoque nostram ab undis exportaremus*, Bd. 5, 1; S. 614, 10. On đissum eálonde com hē ǽrest upp on Westseaxum *primum Gevissorum gentem ingrediens*, 3, 7; 529, 9. Hē wæs ādrifen þā hē up on Frysena land *pulsus est Fresiam*, 5, 19; S. 639, 20. Be cīepemonna fōre up (upp, MS. H.) on londe, L. In. 25; Th. i. 118, 11 note. Hī cōmon up on Limene mūþan mid .ccl. scipa, Chr. 893; Erl. 88, 25. Hī up cōmon æt Leptan đǽm tūne *ad Leptim oppidum copias exposuit*, Ors. 4, 10; Swt. 202, 9. Đā wē up cōmon *when we landed*, Homl. Skt. ii. 30, 325: Chr. 860; Erl. 70, 25. Þēh đa menn up ætberstan intó đǽre byrig, L. Eth. ii. 2; Th. i. 286, 2. Se here hiene on niht up bestæl, Chr. 865; Erl. 70, 34. Cnut com tō Sandwíc, and lēt đǽr up đa gíslas, 1014; Erl. 151, 9. Hēt đā up beran æþelinga gestreón, Beo. Th. 3844; B. 1920. (a 2) marking arrival, or coming into notice (cf. colloquial to turn *up*). Cf. a. β :—Đæs ymb .iii. niht ridon .ii. eorlas up, Chr. 871; Erl. 74, 6. Up ābrecaþ *erumpunt*, Wrt. Voc. ii. 144, 7. (a 3) *up a river, against the stream* :—On đa eá hī tugon up hiora scipu ōþ đone weald, Chr. 893; Erl. 88, 31. (b) where a body remains in the same place but moves in an upward direction :—Đā ārās hē upp, Jn. Skt. 8, 7, 10. Hē up āsæt, Bd. 5, 12; S. 627, 14. For hwý đæt fȳr fundige up, Bt. 34, 11; Fox 150, 19. Gǽd seó eá up, and oferflētt đæt land, Lchdm. iii. 252, 24. Up hleápende *exoriens*, Wrt. Voc. ii. 144, 9. Hī (*the plants*) up sprungon, Mt. Kmbl. 13, 5. **II.** marking direction, (a) of physical action :—Abraham beseah upp, Gen. 18, 2. Đīues brōđor blōd clypaþ up tō mē of eorđan, 4, 10. Hē lōcade upp on heofon, Bd. 4, 9; S. 577, 20. Lōciaþ nū ealle up, Nar. 28, 26. (b) of mental action :—Langaþ đē āwuht up tō Gode, Cd. Th. 32, 1; Gen. 497. (c) marking measurement :—Habbe hē his strand upp of sǽ and ūt on sǽ, Cod. Dip. Kmbl. iv. 221, 7. **III.** marking position, (a) *up, on high* :—Hī (*beams*) man mæg up fēgean (*lignum ad summa levatur*, Past. 58; Swt. 445, 3.

Wǽron ða wealdleðer swá up getíged, swá swá hig urnon tó heofenum up, Shrn. 156, 12. Iosue hí up áhéng on fíf wácum bógum *Iosue suspendit eos super quinque stipites*, Jos. 10, 26. Heó stód upp on ánre upflóra, Homl. Skt. i. 18, 341. .vii. upp hangene bella, Chart. Th. 430, 4. Lyft up geswearc, Cd. Th. 207, 4; Exod. 461. Ða tánas up æpla bæron, 495, 7; Sat. 482. Up in heofonum, 284, 26; Sat. 327: Exon. Th. 281, 11; Jul. 644. In roderum up, 22, 17; Cri. 353. (b) *erectly :*—Áhó on up standende twig, Lchdm. i. 332, 15. Up standende herebeácn *pira*, Wrt. Voc. i. 41, 43. (c) *up, to a high point :*—Gif se móna urne swá up swá seó sunne déð, Lchdm. iii. 248, 6. Hió cymþ swá up swá hire ýsemest gecynde bið, Bt. 25; Fox 88, 27. **IV.** marking separation, as in to cut *up*, break *up :*—Up álióde *evulsum, abscisum*, Hpt. Gl. 474, 36. Hé ða eá upp forlét an feówer hund eá and on lx . . . and æfter ðæm Eufrate hé eác mid gedelfe on monige eá up forlét *fluvium per magnas concisum deductumque fossas in quadringentos sexaginta alveos comminuit . . . etiam Euphratem derivavit*, Ors. 2, 4; Swt. 74, 1-5. **V.** in figurative expressions :—Ðú áhefst upp mín heáfod *exaltans caput meum*, Ps. Th. 3, 2. Ðæt hý hý upp ne áhófen for heora welum, 48, arg. Áhafen up *elevatus in sublime*, Kent. Gl. 1118: Bt. 16, 1; Fox 48, 29. Ðe læs ðé God up bréde ðone godspellícan cwide *lest God bring up the words of the Gospel against thee*, Wulfst. 248, 9: 249, 3. Ðér bær Godwine eorl up his mál *Earl Godwine brought his case up* or *forward*, Chr. 1052; Erl. 187, 19. Syþþan up cymð deófles costnung *orta tribulationi*, Mk. Skt. 4, 17. Se wæs up cymen in Palestina mægðe *he was a native of Palestine*, Shrn. 141, 6. Ne hebbe gé tó up eówre hornas, Past. 54; Swt. 425, 22. Se mann ána gǽþ uprihte; ðæt tácnaþ ðæt hé sceal má þencan up ðonne nyþer, Bt. 41, 6; Fox 254, 30: Homl. Skt. i. 1, 58. Eahta sweras rihtlícne cynedóm up wegaþ, L. I. P. 3; Th. ii. 306, 20. [*O. Sax. O. L. Ger.* up: *O. Frs.* up, op: *Icel.* upp: *O. H. Ger.* úf, cf. *Goth.* iup.] v. uppe.

up, upp; *adj. That is above, that is on high :*—Neoman ús tó wynne weoroda Drihten, upne écne gefeán, Cd. Th. 277, 4; Sat. 199. Gé synd uppe godu (uppe-godu? v. up-godu), ealle upheá and ǽðele bearn *dii estis et filii excelsi omnes*, Ps. Th. 81, 6. v. up-ness.

up-áhafenness, e; *f. Uplifting, elevation.* **I.** literal :—Upáhafenes handa mínra *eleuatio manuum mearum*, Ps. Lamb. 140, 2. Seó symbelnyss ðære hálgan róde upáhefennysse, Homl. Skt. ii. 23 b, 399. **II.** metaphorical, (a) *exaltation :*—Drihten ys mín upáhafenys *Dominus exaltatio mea*, Ex. 17, 15. His mód bið áfédd mid ðære smeáunga ðære wilnunga óðerra monna hiérnesse and his selfes upáhæfenesse, Past. 8; Swt. 55, 6. (b) *exultation :*—Welerum upáhafennysse *labiis exultationis*, Ps. Spl. 62, 6. (c) *arrogance, pride :*—Hwæt is ðonne forcúðre ðonne sió upáhæfennes (-hæfenes, Hatt. MS.)? . . . Hwæt mæg hiérre bión ðonne sió sóðe eáðmódnes? *quid elatione dejectius . . . Quid humilitate sublimius?* Past. 41; Swt. 300, 18. Ðá wæs gehroren sió upáhæfenes Paulus, and sóna æfter ðæm hryre ðære upáhafennesse hé ongan timbran eáðmódnesse, 58; Swt. 443, 29. Hér is úres módes upáhafennes; ac ðǽr is ðære þýstro dymnes, L. E. I. proem.; Th. ii. 394, 12. Élc upáhafenes omnis *arrogans* (cf. 242), Kent. Gl. 547. Ðonne ðæt mód ðenceþ gegrípan him tó upáhafenesse (-hef-, Hatt. MS.) ða eáðmódnesse, Past. 8; Swt. 54, 12. For gilpe and for upáhafenesse *elationis intentione*, 9; Swt. 55, 2. For his (Haman) upáhafennesse, Homl. Ass. 96, 135. Ða upáhafenesse (*Nebuchadnezzar's*) God getælde, Past. 4; Swt. 39, 20. Upáhefenysse *insolentiam, superbiam*, Hpt. Gl. 526, 73. v. up-áhefedness, -hefness.

up-áhefedlíce; *adv. Proudly, arrogantly :*—Upáhefedlíce *arroganter, superbe*, Hpt. Gl. 422, 8.

up-áhefedness, e; *f. Elevation.* **I.** literal :—*Exaltatio sancte crucis*, ðæt is on Engliscre spræce upáhefednyss ðære hálgan róde, for ðan ðe heó wæs áhafen on ðam dæge, Homl. Skt. ii. 27, 140. **II.** metaphorical, (a) *exaltation :*—þurh ðás clypunge is gesweotolad, ðæt ǽlc upáhefednes ásprincð of módignesse cynrene *ostendit nobis omnem exaltationem genus esse superbie*, R. Ben. 22, 13. (b) *exaltation :*—Upáhæfdnes mín *exaltatio mea*, Ps. Lamb. 31, 7. (c) *arrogance, pride :*—Ðisses mannes (*Dives'*) uncyst and upáhefednys, Homl. Th. i. 328, 22: ii. 560, 20: Homl. Skt. i. 16, 163. Aman wearð gehýnd for his upáhefednysse, Homl. Ass. 101, 322. Hé (*Lucifer*) wæs fordón þurh ða miclan upáhefednysse, Homl. Th. i. 12, 21. Upáhefednesse forfleón *elationem vel jactantiam fugere*, R. Ben. 18, 23: L. E. I. 21; Th. ii. 418, 12. v. up-áhafenness.

up-áspringness, e; *f. Uprising :*—Upáspringnes i eástdǽl *ortus*, Ps. Lamb. 102, 12.

up-ásprungenness, e; *f. Uprising, origin :*—Ymbe ðises bissextus upásprungnysse wé wyllaþ rúmlícor iungum cnihtum geopenian, Anglia viii. 306, 14.

up-ástigenness and -ástígness, e; *f. An ascent*, (1) *a going up, an ascension :*—Be ðisse drihtenlícan upástigennesse, Blickl. Homl. 117, 6. Æt ðære upástígnesse, 121, 13: 171, 9: Shrn. 78, 10: 79, 29: Nar. 30, 14. Ða hálgan upástigenesse on heofonas, Blickl. Homl. 119, 36. Upástígnesse, 81, 11. (2) *a means of going up :*—Ic on ðam wealle

---

nǽnige duru ne eághþyrl ne uppástígnesse geseón mihte, Bd. 5, 12; S. 629, 15. Ðú ðe setst genipu upástígnesse ðinne *qui ponis nubem ascensum tuum*, Ps. Lamb. 103, 3. Stapas i upástigenesse *ascensiones*, 83, 6.

up-cund; *adj. Supernal, celestial :*—Onǽled mid ðý upcundan leóhte *illustratus superno lumine*, Past. 49; Swt. 379, 24. Cuman tó ðam upcundan ǽþelan ríce, Exon. Th. 17, 10; Cri. 268. Ðæt wé magon upcund ríce gestígan, 348, 27; Sch. 34. Upcundra ceastergewarena *supernorum civium*, Hpt. Gl. 423, 5. Upcundra eádegum setlum *sedibus superum beatis*, Dóm. L. 303. [*In* Bd. 4, 23; S. 595, 40 *for* upcundne swég *read* uppe cúðne swég; v. Bd. M. 340, 5.]

up-cyme, es; *m. Up-coming, rising, up-springing :*—Upcyme wylla *the springing up of the fountains*, Cd. Th. 240, 12; Dan. 385. From sunnan upcyme *a solis ortu*, Ps. Surt. 49, 1: 106, 3. Naehte upcyme *noctis exortum* (l. *exortu*, v. Hymn. Surt. 2, 21), ii. p. 202, 1. Ic monnum sceal ýcan upcyme eádignesse *for men I shall increase the upspringing of happiness*, Exon. Th. 413, 3; Rä. 31, 9.

up-eard, es; *m. A dwelling on high :*—Ic eom síþes fús upeard niman, Exon. Th. 166, 31; Gú. 1051.

up-ende, es; *m. The upper end, top end :*—Se steorra Ursa is swíþe neáh ðam upende ðære eaxe *summo vertice mundi flectit rapidos Ursa meatus*, Bt. 39, 13; Fox 232, 33: Met. 29, 18. At ðas akeres upende, Cod. Dip. Kmbl. iii. 434, 2. Oð ðære foryrðe upende, 419, 33.

up-engel, es; *m. An angel of heaven :*—Upengla fruma, Andr. Kmbl. 451; An. 226. Upengla weard, Menol. Fox 417; Men. 210.

up-færeld, es; *n. (m.?) A journey up, an ascension :*—Eal heofonwaru wundrode ðysre fǽmnan upfæreldes, Homl. Th. i. 444, 1.

up-feax; *adj. Having hair at the top, bald in front;* recalvus, Wrt. Voc. i. 288, 57.

up-fléring, e; *f.* **I.** *an upper floor :*—Seó upfléring tóbærst under his fótum, Homl. Th. ii. 164, 3. **II.** *an upper chamber :*—Hí ástigon upp on áne upfléringe (*coenaculum*, Acts 1, 13), Homl. Th. i. 296, 9: 314, 7.

up-flór, a, e: flóre, an; *f.:* es; *m. An upper floor* or *story, upper chamber :*—Upflór *solarium*, Wrt. Voc. i. 83, 31. Heó hæfde hig behíd on hire upflóra (*solario domus suae*), Jos. 2, 6. Gesamnodon hí on sumre upflóra (*de tecto et solario spectantes*) ealle ða heáfodmen, Jud. 16, 25. Wuniende on ánre upflóra (*in coenaculo*), Homl. Th. i. 314, 5: ii. 184, 26: 164, 2: Homl. Skt. i. 10, 58, 64, 81: 18, 341: ii. 27, 31, 67. Seó wádewe wunode on clǽnnysse æfter hire were on hyre upflóre, Homl. Ass. 108, 204. Ða yldestan Angelcynnes witan gefeóllan of ánre upflóran (*solario*), Chr. 978; Erl. 127, 10. Sume férde upp on ðone upflóre, 1083; Erl. 217, 18.

up-gang, es; *m.* **I.** *a going up, rising* of a heavenly body :—Sunnon upgong æt middan sumere *ortus solis solstitialis*, Bd. 5, 12; S. 627, 34. Æfter sunnan upgonge, L. Alf. 25; Th. i. 50, 20. Ǽr sunnan upgange, Lchdm. ii. 306, 17. Æt sunnan upgonge, Nar. 27, 17. Fram sunnan upgange óð hire setlgang, Ps. Th. 49, 2: 112, 3. Uppgange, 106, 3. Tóforan mónan upgonge, Nar. 13, 9. Hí (*the constellations of the zodiac*) gefyllaþ twá tída mid hyra upgange oððe nyðergange, Lchdm. iii. 246, 8. **II.** *a going up,* (a) to land from sea, *a landing.* v. up, I. a 1 :—Hí forwerndon heom ǽgðer ge upganges ge wæteres, Chr. 1046; Erl. 171, 5. (b) from the coast inland, *an incursion :*—Hí námon him wintersettl on Temesan . . . Ðá æfter middan wintra hí námon ǽnne upgang út þurh Ciltern and swá tó Oxneforda, 1009; Erl. 143, 9. **III.** *a way of going up :*—Hié gerýmdon ðone upgang and geworhtan, Blickl. Homl. 201, 17. [*O. H. Ger.* úf-gang *ortus: Ger.* auf-gang: *Icel.* upp-gangr: *Dan.* op-gang *ascent; stairs.*] v. next word.

up-gange, an; *f. Landing.* v. up-gang, II. a :—Hí bǽdon ðæt hí upgangan ágan mósten *they asked for leave to land*, Byrht. Th. 134, 20; By. 87. [*Icel.* upp-ganga; *f. a going ashore; a landing-place.*]

up-gemynd, es; *n. Mindfulness of what is above, thought directed heavenward :*—Hé hæfde hlúttre lufan, éce upgemynd engla blisse, Andr. Kmbl. 2129; An. 1066.

up-godu (-o); *pl. n. The gods above :*—Upgodo *superi*, Wrt. Voc. ii. 80, 19.

up-hafenness, e; *f. Elevation :*—Uphefenes honda mínra *elevatio manuum mearum*, Ps. Surt. 140, 2. v. up-áhafenness.

up-heáfod, es; *n. A top end (?) :*—Tó crofte tó ðan upheáfdan (*to the croft, to the top end of it?*); of ðan upheáfdan, Cod. Dip. Kmbl. vi. 79, 10. v. heáfod.

up-heáh; *adj.* **I.** *tall, lofty :*—Wæs hé .x. fóta upheáh *pedum non amplius decem statura altior*, Nar. 26, 28. Ða trió meahte beón hundteóntiges fóta upheáh *he pedum centum alte erant arbores*, 27, 28. Sindon dúna upheá, Exon. Th. 443, 15; Kl. 30. Wǽron hié swá greáte swá columnan, ge eác sume uphýrran (*proceriores*), Nar. 14, 5. **II.** fig. *lofty, noble :*—Ealle upheá and ǽðele bearn *filii excelsi omnes*, Ps. Th. 81, 6. **III.** *upright :*—Ðá genam Sanctus Martinus hine be his handa and upheáh árǽrde, Blickl. Homl. 219, 20. Ðám treówum ðe him gecynde biþ upheáh tó standanne, Bt. 25; Fox 88, 22. [*Icel.* upp-hár *high, tall.*] v. up-lang.

up-heald, es; *n.* *Upholding, support, maintenance:*—Ic eom ðæs mynstres mund and upheald, Cod. Dip. Kmbl. iv. 232, 7. [Crist, Hælennde and hellpe and god upphald, Orm. 9217. *Icel.* upp-hald; *n. support, maintenance: Dan.* op-hold.]

up-hebbe, an; *f.* *A coot* (so called because it lifts up its tail when moving over the water, Grein):—Uphebbean hûs *fulicae domus,* Ps. Th. 103, 17.

up-hebbing, e; *f.* *Uplifting, uprising:*—Uphebbing *ortum,* Lk. Skt. Lind. 8, 8.

up-hefness, e; *f.* *Exaltation:*—Gâstlîcre uphefnesse *extaseos,* Wrt. Voc. ii. 31, 70.

up-heofon, es; *m.* *The heavens, the sky:*—Eall upheofon bið sweart and gesworcen, Wulfst. 137, 8. Beofaþ ealbeorhte gesceaft ... dyneþ upheofon, Exon. Th. 448, 25; Dôm. 59. Eálá middaneard ... eálá upheofon, Cd. Th. 275, 6; Sat. 167. Ðû geworhtest eorþan frætwe and upheofen; ðæt is heáh geweorc handa ðînra *terram tu fundasti; et opera manuum tuarum sunt coeli,* Ps. Th. 101, 22. Eorðan ic bidde and upheofon, Lchdm. i. 400, 3: Exon. Th. 60, 12; Cri. 968: Andr. Kmbl. 1596; An. 799. Se gâst upheofon gesôhte *spiritus astra petit,* Bd. 2, 1; S. 500, 20. [Cf. *O. Sax.* up-himil: *O. H. Ger.* ûf-himil: *Icel.* upp-himinn.] Cf. up-rodor.

up-hûs, es; *n.* *An upper chamber:*—Uphûses *cenaculi,* Wrt. Voc. ii. 24, 67.

up-lang; *adj.* **I.** *tall, high:*—Wæron hié nigon fôta uplonge *pedum alti .ix.,* Nar. 22, 6. **II.** *upright:*—Ongeán sunnan upweard licge hê ... ðonne uplang âsitte, Lchdm. ii. 18, 16: iii. 2, 12. Sæweall uplang gestôd, Cd. Th. 197, 7; Exod. 303: Beo. Th. 1523; B. 759. Uplong, Exon. Th. 495, 16; Rä. 85, 4. v. up-heáh.

up-legen, e; *f.* *A hair-pin:*—Uplegen *discriminale,* Wrt. Voc. ii. 141, 1. Uplegene *vel* feaxpreónas *discriminalia,* i. 17, 2.

up-lendisc; *adj.* *Uplandish, country* (as opposed to town), *rural, rustic:*—Uplendisc *forensis* (*forensis qui foras est,* Migne), Germ. 389, 41. Eft begann sum uplendisc mann egeslîce hrŷman tô ðâm ârleásum burhwarum ... Ðâ arn se ceorl geond ealle ða strǣt hrŷmende, Homl. Th. ii. 302, 4–8. Wê wyllaþ ðisne circul âmearkian, ðæt se uplendiscea preóst (cf. Chaucer's: Poure persoun dwellyng uppon lond) wite his naman; mæg beón ðe glǣdre his heorte ðe hê sum þing hêrof undergyte, Anglia viii. 317, 38. Ic wêne, lâ, uplendisca preóst, ðæt ðû nyte hwæt beó atomos, 318, 14. Ðû byst uppan lande mid wîmmannum oftor ðonne ic beó ... Ic hit gehŷrde secgan, ðæt ðâs uplendiscan wîf wyllaþ oft drincan, Engl. Stud. viii. 62, 12. Wê witon ðæt ðâs þing þincaþ clericum and uplendiscum preóstum genôh mænigfealde, Anglia viii. 321, 25. [Oplondisch men wol lykne hamsylf to gentilmen ... The Saxon tonge ys abide scarslych wiþ feaw uplondysch men, Trev. i. c. 59. Uplondysche mann *villanus,* Prompt. Parv. 512, where see note.]

up-lic; *adj.* **I.** *on high,* (1) *referring to this world:*—Ðonne bið gefylled eall uplîc lyft ǣtrenum lîge, Wulfst. 138, 5. Ûre Drihten gesceóp ... ða upplîcan heofenan ... and ðæt upplîce lyft, Hexam. 4; Norm. 6, 20–24. (2) *referring to heaven:*—Ðæt wuldor ðæs uplîcan êðles *supernae patriae gloriam,* Past. 21; Swt. 159, 23. Tô gefeán ðǣre upplîcan ceastre, Bd. 1, 7; S. 479, 15. In ðam uplîcan êðle, Exon. Th. 225, 20; Ph. 392. Uplîcne hâm, Cd. Th. 287, 5; Sat. 362. Ðara uplîcra burhwara gefêrscipe, Blickl. Homl. 197, 16. **II.** *lofty, sublime:*—Upplîcan *anagogen,* upplîc andgyt *supernum intellectum,* Hpt. Gl. 506, 17–19. Uplîcum andgite *anagogen,* Anglia xiii. 28, 15. (Cf. *Anagogen, celsissimo intellectu,* Wrt. Voc. ii. 75, 37. Gâstlecum andgite *anagogen,* i, 10. Þara godcundan heáhstan *anagogen,* 9, 7.) God mid ðǣm uplîcum and mid ðǣm dîeglum ðingum hira môd onlíéht mid ðæm scîman his giefe, Past. 35; Swt. 243, 20. **III.** *celestial:*—Se uplîca Dêma, Blickl. Homl. 95, 33: Chr. 979; Erl. 1169, 17. Gif hine gecîst sió uplîce gifu *quem superna gratia elegit,* Past. 7; Swt. 51, 4. Cuæð sió uplîce stemn tô Moyse, 11; Swt. 63, 23. Seó upplîce ârfæstnys, Bd. 3, 13; S. 538, 31. Mid ðam uplîcan mihte geþreád, 1, 7; S. 478, 40. Sprecan ðe ðâm upplîcan dômum Godes, 5, 19; S. 640, 34. [*O. H. Ger.* ûf-lîh *supernus.*]

up-lyft (*for gender see* lyft) *the air above:*—Ðû wealdan miht eorðan mægen and uplyfte, wind and wolcna, Btwk. 196, 29; Hy. 9, 6. Ðû geworhtest heofonas and eorðan, eardas and uplyft, 198, 4; Hy. 9, 19.

up-ness, e; *f.* *Height:*—Ðû ðe oferwrîhst mid wæterum hire upnyssa *qui tegis aquis superiora eius,* Ps. Lamb. 103, 3. v. up; *adj.*

upon; *adv.* *From above:*—Swâ fæger dropa ðe on ðâs eorðan upon dreópaþ *sicut stillicidia stillantia super terram,* Ps. Th. 71, 6.

upp. v. up.

uppan (-on); *prep. dat. acc.* **I.** *dat.* (1) *where there is rest or motion on a surface, upon, on:*—Moises wæs lange uppan ðam munte, Ex. 32, 1. Gâ uppan Sinai dûne and stand uppan ðǣre dûne ufeweardre. Ne cume nân mann uppan ðǣre dûne, 34, 2–3. Hê bæd æt Gode ðæt hé him sealde wæter uppan ðǣre dûne, Homl. Skt. i. 19, 112. Geoffra hyne uppon ânre dûne *offeres eum in holocaustum super unum montium,* Gen. 22, 2. Hê ǣtstôd uppon ânum beáme, Chr. 978; Erl. 127, 11. Hê rît upp n tamre assene *sedens super asinam,* Mt. Kmbl.

21, 5. Ðâ wearð Eustatius uppon his horse and his gefeóran uppon heora *Eustace got on his horse and his companions on theirs,* Chr. 1048; Erl. 177, 38. Hê bær his tunecan, and âlêde uppon ðâm twâm deádum, Homl. Th. i. 74, 2. Ðû byst uppan lande (*up country, in rural districts;* cf. Chaucer's 'poure persoun dwellyng uppon londe'; *and see* uplendisc) oftor ðonne ic beó, Engl. Stud. viii. 62, 9. To ǣlcen cyrcean up lande, Chr. 1086; Erl. 223, 36. (2) *marking relative height, above:*—Him uppan wæs rôd ârǣred, Elen. Kmbl. 1768; El. 886. Ðonne bist ðû ofer uppan rodere (cf. ðonne bist ðû bufan ðam rodore, Bt. 36, 2; Fox 174, 15), Met. 24, 27. (3) *of time* (the case is doubtful in some instances), (a) *upon, at:*—Hê Ansealme uppon Pentecosten his pallium geaf, Chr. 1095; Erl. 232, 30. Uppon Scê Michaeles mæssan ætŷwde ân selcûð steorra, 1097; Erl. 234, 9, 19, 32. (b) *after:*—Ðys sceal on Sunnandæg feówertŷne nyht uppan Eástron, Jn. Skt. 10, 11 rubc. On ðisum geáre wǣron Eástron on viii kal. Apr., and ðâ uppon Eástron on Scê Ambrosius mæsseniht, ðæt is .ii. non. Apr. ..., Chr. 1095; Erl. 231, 17. (4) *marking object of attack:*—Ða hǣðenan men hergodan uppon ðâm Xþenan mannan, Chr. 1086; Erl. 223, 11. (5) *marking ground of trust, upon* trust, honour, etc.:—Se cyng him nâþer nolde ne gîslas syllan ne uppon trŷwðan geunnon ðæt hê mid griðe cuman môste and faran, Chr. 1095; Erl. 231, 24. **II.** *acc.* (1) *where there is motion on to a surface, upon, on to:*—Ðû gîtst ðæt blôd uppan ðæt weofod *fundes sanguinem super altare,* Ex. 29, 20: Mt. Kmbl. 26, 7. Se ðe fylð uppan ðysne stân, 21, 44. (2) *of time.* v. I. 3. (3) *marking object of attack, upon, against:*—Hê fôr uppon heora brôðer and uppon ðone eorl wann, Chr. 1095; Erl. 231, 8–10. Se cyng his fyrde beád, and uppon ðone eorl tô Norðhymbran fôr, Erl. 231, 26. Heora ǣgðer uppon ôðerne tûnas bærnde, 1094; Erl. 230, 12. Hí ealne ðone bryce uppon ðone cyng tealdon *they laid all blame for the breach upon the king,* Erl. 230, 4. (4) *marking addition:*—Ða bodan cŷðdon ðæt his brôðer grið and forewarde eall æfterwæd, bûtan se cyng gelǣstan nolde eall þet hî on forewarde hæfdon ǣr gewroht, and uppon ðæt (*in addition to that, on the top of that*) hine forsworenne clypode, bûton hê ða forewarde geheólde, Chr. 1094; Erl. 229, 31. [*O. E. Homl. Laym. A. R. O. and N.* uppen: *O. Sax.* uppan: *O. Frs.* uppa: *O. H. Ger.* ûfan.] v. on-, þær-uppan.

uppan = yppan *in uppende proferens,* Anglia xiii. 423, 836. v. ge-upped.

uppe; *adv.* **I.** *up, above, on high:*—Ðêh hê uppe seó, Cd. Th. 281, 2; Sat. 265. Salte sǣstreámas and swegl uppe, Andr. Kmbl. 1498; An. 750. Hræfn uppe gôl, Elen. Kmbl. 104; El. 52. Uppe ofer rodere (cf. bufan ðam rodore, Bt. 33, 4; Fox 130, 15), Met. 20, 124. Wearð ætŷwed uppe on roderum steorra on staðole, Chr. 978; Erl. 126, 23. Hê geseah âne hlǣdre standan æt him on eorðan. Oðer ende wæs uppe on hefenum, Past. 16; Swt. 101, 19. Saul hine wolde sêcean uppe on ðæm munte, 28; Swt. 197, 13. Wǣron ða lâc forbærndu uppe on ðæm altere, 33; Swt. 222, 24. Uppe on ðam eaxlegespanne, Rood Kmbl. 17; Kr. 9. Fugel uppe sceal lâcan on lyfte, Menol. Fox 536; Gn. C. 38. Mynster tô timbrianne on heánum môrum uppe (*in montibus arduis ac remotis*), Bd. 3, 23; S. 554, 20. **I a.** *referring to heavenly bodies, up:*—Oþ ðæt sunne uppe sié, Lchdm. ii. 346, 22. Næs se môna ðâgyt uppe *quum luna erat oritura,* Nar. 29, 22. On winterlîcre tîde hí (*the Pleiades*) beóð on niht uppe and on dæg âdûne, Lchdm. iii. 272, 2. **I b.** *where there is motion from the sea up to the land.* v. up, I. a 1:—Gif hê his scip uppe getogen hæbbe, L. Eth. ii. 3; Th. i. 286, 8. Se ciépemonna fore uppe on londe. Gif ciépemon uppe on folce ceápie, L. In. 25; Th. i. 118, 11. Wǣron ða men uppe on londe of âgâne, Chr. 897; Erl. 95, 24. **I c.** *marking arrival.* v. up, I. a 2:—Ðâ se cyng geáxode ðæt se here uppe wæs *when the king learned that the Danes had appeared upon the scene,* Chr. 1016; Erl. 157, 13. **I d.** *referring to heaven:*—Hí wiston Drihten êcne uppe, Cd. Th. 227, 31; Dan. 195. Eádige ðǣr uppe sittaþ, 305, 16; Sat. 647. Ys ûre se hâlga God on heofondreáme uppe mid englum *Deus noster in coelo sursum,* Ps. Th. 113, 11: Cd. Th. 273, 25; Sat. 142: Exon. Th. 24, 19; Cri. 387: 239, 30; Ph. 629: Fragm. Kmbl. 86; Leás. 45. Wê mid englum uppe wǣron, Cd. Th. 289, 2; Sat. 391. Ne uppe on heofone ne niðer on eorðan *neque in coelo sursum nec in terra deorsum,* Deut. 4, 39. Wê syngodon uppe on earde, Cd. Th. 279, 1; Sat. 231. Uppe on roderum mid englum, Exon. Th. 90, 4; Cri. 1468: Hy. 3, 30. **II.** *marking discovery.* v. yppan:—Hê hí gemartirode swâ hê dyrnlîcost mihte, and hê geðôht hæfde ðæt hî ðǣr nǣfre uppe ne wurdan, ac ðurh Godes mihte hí ðanon gecŷdde wurdon, Lchdm. iii. 424, 31. **III.** *marking effectual action.* (Cf. *Icel.* uppi *vera to take place.*):—Ðara ðe wile ânra hwylc uppe bringan (*bring it to pass*), ðæt ðû ðǣre gyldnan gesihst Hierusalem weallas blîcan, Salm. Kmbl. 466; Sal. 233. [Her uppe, Orm. 1169. Þer uppe, A. R. 94, 12. Uppe on, *O. E. Homl.* i. 5, 2: Laymı. 17495. *O. Sax.* uppa (-e), thâr uppa an: *Icel.* uppi, uppi â, uppi î.]

uppe-godu (?); *pl.* v. up; *adj.*

uppe-land, es; *n.* *Up-country, country* as opposed to town, *rural districts:*—Ðæt ǣlc man ðe wǣre unnîðing sceolde cuman tô him of

porte and of uppelande, Chr. 1087; Erl. 226, 3. v. up-lendisc, uppan, I. I. last two passages.

uppe-weardes; *adv. Upwards :*—Nim mid dínum twâm handum uppeweard[n]es, Lchdm. iii. 38, 10. v. up-weardes.

uppian; *p.* ode *To mount up, rise :*—Ðæt wæter, ðonne hit bið gepynd, hit miclaþ and uppaþ and fundaþ wið ðæs ðe hit ær from com *ad superiora colligitur*, Past. 38; Swt. 277, 7.

uppon. v. uppan.

up-riht; *adj.* I. *upright, erect :*—Ic uppriht âstôd, Beo. Th. 4191; B. 2092. Mannum hê gesealde uprihtne gang, Homl. Th. i. 276, 4. II. *lying with the face turned upwards.* Cf. up-weard :—Upriht âstreht *supinus*, Hpt. Gl. 457, 33. [*O. H. Ger.* úf-reht *erectus*: *Icel.* upp-réttr.]

up-rihte; *adv.* I. *uprightly, erectly :*—Mê þûhte ðæt mín sceáf ârise and stôde uprihte, Gen. 37, 7. Mann âna gǽþ uprihte, Bt. 41, 6; Fox 254, 30: Homl. Skt. i. 1, 57. Ðá árás se cnapa and uprihte eode, 6, 41. II. *right up, exactly overhead, in the zenith :*—Gǽð seó sunne uprihte (upp-, MS. P.) on ðam sumerlícan sunnstede on middæge, Lchdm. iii. 258, 15.

up-rodor (-er), -rador, es; *m.* I. *the firmament on high, the visible heavens, the sky :*—Wolde hê ðæt him eorðe and uproder and síd wæter geseted wurde woruldgesceafte, Cd. Th. 7, 1; Gen. 99. Ðás woruld, eorðan ymbhwyrft and uprodor, 179, 10; Exod. 26: 205, 2; Exod. 429. Eorðan sceátas and uprodor, Exon. Th. 312, 6; Seef. 105. Eorðan and uprodor, 69, 32; Cri. 1129: Cd. Th. 182, 15; Exod. 76. Ealne ymbhwyrft and uprador, Elen. Kmbl. 1459; El. 731. Hwílum cerreþ on uprodor ælbeorhta lêg, Met. 29, 51. II. *heaven :*—Wæs Gúðláces gǽst gelǽded in uprodor fore onsýne êces Dêman, Exon. Th. 148, 34; Gú. 754. Hê lǽdeþ eádige gástas on uprodor, Cd. Th. 212, 25; Exod. 544. In uprodor, 177, 33; Exod. 4. Cf. up-heofon.

up-ryne, es; *m. A coming up, rising* of a heavenly body, *coming* of day :—Wiþ hire (*the sun's*) uprynæs, Bt. 25; Fox 88, 27. Fram sunnan upryne *a solis ortu*, Ps. Spl. 106, 3. Uprine, 112, 3. Æfter sunnan setlgange ǽr mônan upryne, Lchdm. i. 330, 18. Ymb ðæs dæges uppyrne *circa exortum diei*, Bd. 4, 8; S. 576, 11.

up-spring, es; *m.* I. *an upspringing, rising* of a heavenly body, *coming* of day or night :—Upspryng *ortus* (*solis*), Ps. Spl. 103, 23. Fram ðære sunnan upspringes anginne *a solis ortus cardine*, Hymn. Surt. 50, 2. Fram ðære sunnan upspringe, Anglia viii. 317, 10. Upsprince, Ps. Spl. 49, 2. Eásterne wind, *subsolanus* gehâten, for ðan ðe hê blǽwð fram ðære sunnan upspringe, Lchdm. iii. 274, 15. Nihte of upspringe *noctis exortu*, Hymn. Surt. 2, 20. Ná manega dagas, ac ân, se nát nǽnne upspring ne nâne geendunge, Homl. Th. i. 490, 18. Þurh ðæs steorran upspring, 108, 5. II. *a rising* of water, *breaking forth :*—On upspri[n]c (*diluvii*) *inruptionem*, Anglia xiii. 32, 124. III. *birth :*—Hê gestrýnde Cainan. Æfter ðes upspringe (*post ejus ortum*) hê leofode eahtahundgeáre and fíftêne geár, Gen. 5, 10. IV. *what springs up :*—Lígloccode upspringas *flammicomos ortus*, Wrt. Voc. ii. 149, 10.

up-sprungenness, e; *f. Defect;* in reference to the sun, *eclipse :*—*Eclypsis solis*, ðæt is sunnan ásprungennysse (uppsprungennes, MS. B.), Bd. 3, 27; S. 558, 10.

up-stige, es; *m.* I. *ascension, mounting :*—Nis bútan tweón tô understandenne se upstige and se niþerstige (*the ascending and descending on Jacob's ladder*) on nâne ôþere wîsan, bútan ðæt heofona ríces upstige mid eádmôdnesse geearnod bið and mid ofermêttum forwyrht, R. Ben. 23, 6-9. Hê becom tô ðæm heáhsetle ðære rôde; on ðæm upstige (*by the ascent of the cross*) eall úre líf hê getremede, Blickl. Homl. 9, 36. Seó dún stent . . . twelf míla on upstige fram ânre byrig *there is an ascent of twelve miles from the town to the hill*, Homl. Th. i. 502, 6. ¶ especially *the ascension* of Christ *to heaven* :—Ðone mǽron symbeldæg Drihtnes upstiges, Blickl. Homl. 131, 11: Exon. Th. 41, 13; Cri. 655. Æfter upstige êcan Dryhtnes, 44, 31; Cri. 711: 38, 31; Cri. 615: Blickl. Homl. 137, 23: Homl. Th. i. 324, 31. Æfter Cristes upstige tô heofonum, 98, 24: ii. 380, 24: H. R. 3, 4. Uppstige on heofonas, Bd. 3, 17; S. 545, 23. II. *an ascent, a way of ascending :*—Uppstige sandfull *ascensus arenosus*, Scint. 223, 13. Se seteþ wolcan upstige his *qui ponit nubem ascensum suum*, Ps. Surt. Lamb. 103, 3. [Cf. *O. H. Ger.* úf-stîc *ascensus : Icel.* upp-stiga.]

up-stígend, es; *m. One who ascends :*—Ða gecoreno upstígendo *electos ascensores*, Rtl. 193, 33.

up-wæstm *growth upwards, stature :*—Se cyningc hêt bringan îsenne scamol; se wæs emnheáh ðæs mannes upwæstme; ðæt wæs twelf fæðma lang *jussit rex fieri scamnum ferreum secundum statum ejus. Artifices tulerunt mensuram ejus quae erat cubitorum duodecim*, Anglia xvii. 113, 9. [Cf. *Icel.* up-vöxtr *growth, tallness.*]

up-waras (-an, -e); *pl. The dwellers above, the celestials :*—Tô upwarum *ad superos*, Wrt. Voc. ii. 9, 67.

up-weard; *adj.* I. *turned upwards :*—Ongeán sunnan upweard licge hê *let him lie on his back with his face to the sun*, Lchdm. ii. 18, 13: iii. 2, 10. Licge hê upweard æfter ðon gôde hwíle, ii. 318, 14. Hê

sceal upweard licgean, i. 300, 20. Mon on bedde dæges upweard ne licge, ii. 26, 19. Álege ðone man upweard, 342, 5. Hê mid bǽm handum upweard (*with his face turned upwards?* or adverb? *he stretched his hands up.* v. upweardes) plegade, Elen. Kmbl. 1609; El. 806. Nis ðæt gedafenlíc ðæt se môdsefa monna ǽniges niþerheald wese, and ðæt neb upweard, Met. 31, 23. Hê âsette his sweord upweard and ðâ hyne sylfne ofstang *he placed his sword with the point up, and then stabbed himself*, Shrn. 132, 10. Nioþan upweardne on nearo fêgde, Exon. Th. 479, 11; Rä. 62. 6. For ðam gelômlícum ðeáwe his gebeda, swâ hwǽr swâ hê sæt, ðæt his gewuna wæs ðæt hê his handa upwearde hæfde ofer his cneówa *ob crebrum morem orandi, semper ubicumque sedens, supinas super genua sua manus habere solitus sit*, Bd. 3, 12; S. 537, 25. II. *moving upwards.* v. up, I. a. β :—Ðæt leóht ðe wê hâtaþ dægrêd cymð of ðære sunnan, ðonne heó upweard bið, Lchdm. iii. 234, 29. v. upheáh, -lang, *and next word.*

up-weard; *adv. Upwards, up.* (1) of motion, (a) *from a lower to a higher point :*—Ðá gewende eal se sang upweard tô heofenum, Homl. Th. ii. 548, 14: Elen. Kmbl. 1609; El. 806 (? v. preceding word). (b) *up* into a country. v. up, I. a 1 :—Swegen wende intó Humbran múðan, and swâ uppweard andlang Trêntan, Chr. 1013; Erl. 147, 18. (2) *of reckoning, in the calendar, upward, backward :*—Swâ fela daga tell ðú fram Martius mônðes ende upweard . . . Rím swâ fela daga upweard fram pridie Kl. Martii, and ic ðê secge tô gewissum, ðonne ðú cymð tellende tô .vii. id. Martii, ðonne gemêtst ðú ðǽr lunam primam, Anglia viii. 327, 9-13. Tellaþ þreó and twêntig daga fram æfteweardum Martium upweard, 329, 28. [Cnihtes eoden upward, cnihtes eoden adonward, Laym. 15244. Kasten upward (*sursum*) . . . dranen dunewardes, Kath. 1964. To climben upward, A. R. 72, 20. Ha biheold uppard, Jul. 74, 14. Reccnedd uppwarrd (*back*) and dunnwarrd, Orm. 2056.]

up-weardes; *adv. Upwards :*—Hê onginþ of ðam wyrttrumum, and swâ upweardes grêwþ ôþ ðone stemn, Bt. 34, 10; Fox 150, 2. Hê biþ upweardes (cf. swâ sprincþ hê up, Bt. 25; Fox 88, 24), Met. 13, 54. Hió stîhþ â upweardes, 13, 62. Hê hæfde his handa upweardes, Blickl. Homl. 227, 16. Hê his handa wæs uppweardes brǽdende wið ðæs heofones *manus ad coelum tendens*, Ors. 4, 5; Swt. 166, 19.

up-wearp. [*Icel.* upp-varp.] v. sǽ-upwearp.

up-weg, es; *m. The way to heaven :*—Wæs Gúðláces gǽst gelǽded on upweg, Exon. Th. 180, 15; Gú. 1280: 184, 6; Gú. 1340: Andr. Kmbl. 1659; An. 832. Hî dôm hlutan, eádigne upwæg, Menol. Fox 383; Men. 193. [*O. Sax.* up-weg.]

up-yrne. v. up-ryne.

ur (occurring only as it is represented by the U-rune); *adv. Formerly :*—⑁ (= ur) wæs geára (cf. iú (geó) . . . geára) geógoðhádes glǽm; nú synt geárdagas forð gewitene, lífwynne geliden, Elen. Kmbl. 2530; El. 1266. ⑁ wæs longe laguflôdum bilocen lífwynna dǽl, feoh on foldan, Exon. Th. 50, 25; Cri. 806. v. or.

úr, es; *m. A kind of ox, a bison;* urus: *also the name of the U-rune :*—Úr (⑁) byþ ânmôd and oferhyrned, feohteþ mid hornum mǽre môrstapa, Runic pm. Kmbl. 339, 7; Rún. 2. (The rune is written without representing a word, Exon. Th. 284, 32; Jul. 706.) [*Goth.* úraz *name of the U-rune : Icel.* úrr *a kind of ox; úr the name of the U-rune : O. H. Ger.* úr-ohso: *Ger.* auer-ochse.]

úre; *gen. pl. of personal pronoun of first person. Of us :*—Adam can yfel and gôd, swâ swâ úre sum (*quasi unus ex nobis*), Gen. 3, 22. Ús is eallum þearf, ðæt úre æghwylc ôþerne bylde, Byrht. Th. 138, 42; By. 234: Beo. Th. 2776; B. 1386. Úre ealra bliss eardhæbbendra *laetantium omnium nostrum habitatio*, Ps. Þh. 86, 6. Weorð ðú úre gemyndig *memor fuit nostri*, 113, 21. Gemiltsa úre *miserere nostri*, Ps. Spl. 122, 4. Gif ðú úre bídan þencest, Exon. Th. 119, 26; Gú. 260. ¶ used as a possessive :—Wê sceolan syllan ðone teóþan dǽl úre worldspêda, and wê sceolan úre daga ðone teóþan dǽl on forhæfdnesse lifgean, Blickl. Homl. 35, 19, 20. Geþencean úre sáula þearfe, 95, 24. Úre synna forgifnessa, 97, 14. Fram ðam heáhsetle úre Gescyppendes, 11, 29. v. ús.

úre; *adj. pronoun.* I. *our :*—Úre *noster*, Ælfc. Gr. 15; Zup. 93, 17. Úre Drihten, Blickl. Homl. 11, 21. Fæder úre (úrer, Lind.) *Pater noster*, Mt. Kmbl. 6, 9. Úre se trumesta staþol, Blickl. Homl. 13, 10. Tô úre Drihtnes mêder, 15, 2. On naman Godes úres, Ps. Spl. 19, 5. Beorhtnes blîdan Drihtnes úres, Ps. Th. 89, 19. Seó rihteste bysen úran (úres, MS. F.) menniscan lifes, R. Ben. 133, 4. Ða blindnesse úre ælþeódignesse, Blickl. Homl. 23, 2: 77, 14. Mid eallre úre heortan megolnesse, 65, 23. Úrum Hǽlende fylgende, 23, 11. Deóre Drihtne úrum, Cd. Th. 17, 17; Gen. 261. Mid úre ânre sáule, Blickl. Homl. 91, 16. Úrne dæghwamlícan hláf, Mt. Kmbl. 6, 11. Álese wê úre sáule, Blickl. Homl. 101, 10: 33, 13. Ge wê ge úre fædoras, Gen. 44, 34. Sió ân ræst eallra úrra (úra, Met. 21, 14) geswinca, Bt. 34, 8; Fox 144, 27. Be ðære hǽlo úrra sâwla, L. In. pref.; Th. i. 102, 8: Exon. Th. 154, 26; Gú. 848: Blickl. Homl. 131, 1. Úra synna forlætnesse, 35, 36. Úrum fæderum, Deut. 5, 3. Forgyf ús úre gyltas, swâ swâ wê forgyfaþ úrum gyltendum, Mt. Kmbl. 6, 12. II. *predicative, ours :*—Ðonne bið úre seó yrfeweardnes, Mk. Skt. 12, 7. Ðonne wê ðæm ðearfum

hiera niéddearfe sellaþ, hiera ǽgen wē him sellaþ, nalles ūre, Past. 45; Swt. 335, 18. III. where in place of an inflected form of the adjective the genitive *ūre* might be expected :—Nis ðæt mín miht ne nǽniges ūres, Blickl. Homl. 151, 29. Gē habbaþ gecýðed ðæt gē ūres nānes ne siendon *ye have shown that ye are of no one of us;* nullius vos esse monstratis, Past. 32; Swt. 211, 14. Gif hwelc forworht monn cymð and bitt ūrne hwelcne, 10; Swt. 63, 1. Ūrum sceal sweord and helm . . . bám gemǽne, Beo. Th. 5312; B. 2659. v. ūser.

úre-lendisc; *adj. Of our country :—*Ūrelendisc *nostras,* Ælfc. Gr. 15; Zup. 93, 17.

úrer. v. ūre; *adj.*

úrig-feþera; *adj. Wet-feathered, with dewy plumage :—*Earn ūrigfeþera, Judth. Thw. 24, 27; Jud. 210. Úrigfeðera earn, Elen. Kmbl. 57; El. 29. Ūrigfeðra, 221; El. 111. Úrigfeþra, Exon. Th. 307, 17; Seef. 25. [Cf. *Icel.* úr *drizzling rain;* ūrigr *wet;* úrig-toppi *dewy-mane* (epithet of a horse in a verse).] v. deáwig-feþere.

úrig-lást; *adj. Making a dewy track, walking the wet earth :—*Sum sceal on feþe on feorwegas nýde gongan, and his nest beran, tredan ūriglást elþeódigra frécne foldan, Exon. Th. 329, 4; Vy. 29.

ús; *dat.:* ús, úsic; *acc.:* úser, usser; *gen.; pron. pl. first person. To us, us, of us :—*Wel ūs wæs on Egipta lande *bene nobis erat in Aegypto,* Num. 11, 18. Wē habbaþ ūs tó fæder Abraham, Lk. Skt. 3, 8. Ūs ys betere *expedit nobis,* Jn. Skt. 11, 50. Ūs nis nā álýfed, 18, 31. Ūs neód is, L. Eth. vi. 42; Th. i. 326, 7. Hæle ūs (úsic, Lind. Rush.) *salva nos,* Mt. Kmbl. 8, 25. Ne geléd ðū ūs (úsih, Lind.) on costnunge, ac álýs ūs (úsich, Lind.) of yfele, 6, 13. Gif ðū ūs (úsig, Lind.: úsic, Rush.) ūt ádrífst, ásende ūs (úsig, Lind.: úsic, Rush.) on ðás swína heorde, 8, 31. Hē ūs álésde of deófles þeówdóme, Blickl. Homl. 73, 7: Cd. Th. 25, 8; Gen. 390: Andr. Kmbl. 530; An. 265. Úsic, Bd. 5, 1; S. 614, 10: Ps. Th. 64, 3; Cd. Th. 162, 4; Gen. 2676: Exon. Th. 3, 2; Cri. 30: Beo. Th. 5270; B. 2638. Þeáh ðe úser feá lifgen, 188, 8; Az. 42. Hē cwom úser neósan, Beo. Th. 4155; B. 2074. Geóca úser, Cd. Th. 234, 14; Dan. 292. Helpe usser, Ps. Th. 67, 20. ¶ gen. used as a possessive (v. ūre), *our :—*Úser yldran, Cd. Th. 234, 26; Dan. 298. [*Goth.* uns, unsis; *dat. acc.; gen.: O. Sax. O. Frs.* ūs; *dat. acc.:* úser; *gen.: O. H. Ger.* uns; *dat.* unsih; *acc.; gen.: Icel.* oss; *dat. acc.*] v. wē, unc, ūre, *and next word.*

Úse, Wúse, an; **Ús** (or **Úse;** *indecl.?*), e; *f.* The name of several rivers in England, Ouse :—Andlang Úsan óð hí cómon tó Bedanforda, Chr. 1010; Erl. 143, 33. Of Úsan up on Wilbaldes fleót . . . On Úsan; andlang Úsan (*the charter refers to Northamptonshire*). Cod. Dip. Kmbl. iii. 454, 14-25. Landgemǽre æt Ollanēge . . . In on Úse; andlang Úse, 170, 22-31. Tó Úse stæðe (cf. uulgare prisco usu nomen imposuerunt Use, l. 2), v. 226, 14. Up on Úsan óð Wætlinga-strǽt, L. A. G. i; Th. i. 152, 10. Eall hira land betwuh dícum and Wūsan, Chr. 905; Erl. 98, 20. Betwyx Úsan and Trēntan, 1069; Erl. 207, 16. See, too :—In Úsanmere *Ousemere* (in Warwickshire), Cod. Dip. Kmbl. iii. 375, 9. In provincia Usmerorum (*some part of Worcestershire*), i. 154, 20. In aliis multis locis; hoc est . . . aet Stūre in Usmērum, 173, 18, 34. Of Úsmere . . . on Úsmere, vi. 68, 14.

úser, usser; *adj. pron. Our :—*Nergend úser, Cd. Th. 34, 11; Gen. 536. Drihten úser, Ps. Th. 59, 1. Drihten usser, 54, 8: Cd. Th. 53, 3; Gen. 855. Usses Dryhtnes ród, Exon. Th. 67, 7; Cri. 1085. Endeláf usses cynnes, Beo. Th. 5619; B. 2813. Módes usses, Met. 21, 12. Ne meaht ðū in usse mǽgþe ne on ussum gemánan wunian, Bd. 2, 5; S. 507, 27. On eallum ussum cynne, Blickl. Homl. 151, 12. Ussum móde, Exon. Th. 2, 32; Cri. 28. Mid usse líchoman, 47, 14; Cri. 755. Hláf úserne (úsenne, Lind.) *panem nostrum,* Mt. Kmbl. Rush. 6, 11. Freán úserne, Beo. Th. 5997; B. 3002: Andr. Kmbl. 680; An. 340. Úsa ꝥ úserne (úse ꝥ úserra, Rush.) *nostram,* Jn. Skt. Lind. 3, 11. Úsra (úsern, Rush.) *nostram,* 10, 24. Hálne dō kyninge usserne, Ps. Lamb. 19, 10. Usserne God *Deum nostrum,* Ps. Th. 98, 5. Oþ usse tíde, Bd. 2, 16; S. 519, 37. Wē usse gesihþ upp áhófan, 5, 1; S. 613, 32: Exon. Th. 464, 23; Hö. 91. Úse yldran, 160, 20; Gú. 946. Ðæt ussa (úre, Bt. 15; Fox 48, 18) tída wǽren swylce, Met. 8, 40. Usse sáula, 21, 35. In ussera tída tíman, Exon. Th. 147, 12; Gú. 725. Mildsa sáulum ussa leóda, Bd. 3, 12; S. 537, 31. Goda ussa gield, Exon. Th. 251, 16; Jul. 146: 279, 26; Jul. 619. Godum ussum, 252, 26; Jul. 169. On ussum sáwlum, 80, 29; Cri. 1314. Tó ussum wǽpnum, Nar. 21, 19. Forgef ús scylda úsra, suǽ uoe forgefon scyldgum úsum, Mt. Kmbl. Lind. 6, 12. Heó beswác yldran usse, Exon. Th. 226, 31; Ph. 414. Ussa sáula, Met. 23, 11. [*Goth.* unsar: *O. Sax.* ūsa: *O. Frs.* unse, úse: *O. H. Ger.* unsar.] v. uncer.

úsic. v. ús.

út; *adv.* I. where there is motion, lit. or fig., *out, beyond the bounds within which a thing is enclosed,* (1) with verbs of going. (*a*) without words determining whence or whither motion proceeds :—Hē lǽteþ word ūt faran, Exon. Th. 315, 35; Mód. 41. Uton gán ūt *egrediamur foras,* Gen. 5, 8 : 27, 3 : Cd. Th. 148, 24; Gen. 2461. Ðá eodon hig ūt án æfter ánum *unum post unum exiebant,* Jn. Skt. 8, 9. Cume án spearwa ðurh óþre duru in, ðurh óðre ūt gewite, Bd. 2, 13; S. 516, 18.

(*a* 1) *out* on an expedition :—Wæs Eádmund cyng gewend ūt, and gerád ða West-Seaxan, Chr. 1016; Erl. 155, 13. (*a* 2) *out,* in the sense of leaving a place :—Be ðám ðe ūt faraþ, hwæðer hí mon eft underfón scyle, R. Ben. 53, 6. (*a* 3) *out* to the closet :—Gif mon ne mǽge ūt gegán, Lchdm. ii. 276, 12 : 230, 21, 23. v. úte, II. 1 a. (*a* 4) of the passage of time, *out,* with the idea of coming to an end :—Ūt gangendum ðam mónþe ðe wē Aprelis hátaþ, Lchdm. iii. 76, 14. (β) with words denoting whence motion proceeds :—Ūt áfaren of dínes fæder eþele, Bt. 5, 1; Fox 8, 29 : Cd. Th. 216, 14; Dan. 6. Lēt of breóstum word ūt faran, Beo. Th. 5096; B. 2551. In tó gemóte cuman, and ūt of gemóte, Chr. 1048; Erl. 180, 7. Fleógan of húse ūt, Cd. Th. 87, 2; Gen. 1442. Gangan ūt of earce, 89, 29; Gen. 1488. Ða ðe ūt gongaþ of múþe, Bd. 1, 27; S. 494, 34. Moyses oft eode inn and ūt on ðæt templ, Past. 16; Swt. 101, 24. Ic of ðé ūt síðode, Soul Kmbl. 110; Seel. 55. Ðǽr ic ūt swícan ne mæg *non egrediebar,* Ps. Th. 87, 8. Hionan ūt wítan, Met. 24, 52. (γ) with words denoting whither motion proceeds :—Ic wæs ūt ácymen on ælþeódig land *advena fui in terra aliena,* Ex. 2, 22. v. ūt-ácumen. Fleáh cásere ūt on Crēcas, Met. 1, 21. Hē eode ūt on ðæt land, Gen. 24, 63. Cnut wende him ūt þurh Buccingahámscíre intó Beadafordscíre, Chr. 1016; Erl. 154, 6. (1 a) with verbs that imply going :—Ic ne mæg ūt áredian, Bt. 35, 5; Fox 164, 14. Heó forlēt hyre hæftlingas ūt, Homl. Th. i. 228, 17. Word ðe hē ūt forlēt, Blickl. Homl. 59, 19. Gif mec se mánsceaða of eorðsele ūt gesēceþ, Beo. Th. 5024; B. 2515. Hí bediccodon ða burh ūton ðæt nán mann ne mihte ne inn ne ūt, Chr. 1016; Erl. 155, 11. Heó wolde ūt þanon feore beorgan, Beo. Th. 2589; B. 1292. (2) where motion (lit. or fig.) is caused, with verbs of bearing, casting, driving, releasing, etc. :—Geóte man ðone wǽtan ūt *liquor effundatur,* L. Ecg. C. 39 ; Th. ii. 164, 7. Hwæthugu of cyricean ðurh stale ūt ábregdan, Bd. 1, 27; S. 490, 5. Deófolseócnessa ūt tó ádrífanne, Mk. Skt. 3, 15. Ða landbigengan ūt ámǽran, Bd. 4, 16; S. 584, 7. Ūt tó anýdenne *expellendum,* Scint. 210, 13. Hē ūt áwearp ða sceamolas, Blickl. Homl. 71, 18. Ic mægenbyrðenne hider ūt ætbær, Beo. Th. 6176; B. 3092. Hié ne mehton ða scipu ūt brengan, Chr. 896; Erl. 94, 10. Ðone æþeling ðe hē ūt flēmde, Chr. 725; Erl. 45, 31. Álǽd mē ūt of ðyssum bendum, Blickl. Homl. 87, 34. Sum lytel cniht sweart teáh ðone bróðor of ðære cirican ūt, Shrn. 65, 18. Ne mæg nán man of mínre handa ūt álinnan, Deut. 32, 39. God bebeád ðæt hí sceoldon álýsan hysecild ūt mid fíf scyllingum, Homl. Th. i. 138, 16. Hé hine of earfoðum ūt álýsde, Ps. Th. 90, 15. Ðæt land eode eft intó ðære stówe ðe hit ūt álǽned wæs, Cod. Dip. Kmbl. iv. 267, 6. Ðeáh ðe Harold ðæt land mid unlage ūt nam, 274, 29. Nǽnig mon his geþóht openum wordum ūt ne cýðe *nemo palam pronunciet,* Nar. 28, 30. Hit nǽnig mon ūt cýþan ne móste *no man might spread the news of it abroad,* 32, 17. Ūt mǽran, 32, 22. (2 a) figurative, as in to carry *out, to an end,* marking completeness. v. ūt-cwealm. (3) *out, forth,* as in to break *out :—*Se wielm ðæs innoþes ūt ábiersð, Past. 11; Swt. 71, 9. Streám ūt áweóll, Andr. Kmbl. 3045; An. 1525. Wiþ ūt áblegnedum ómum, Lchdm. ii. 10, 5 : 98, 25. Ðǽr blód and wæter ūt bicwóman, Exon. Th. 69, 1; Cri. 1114. Geseah stream ūt ðonan brecan of beorge, Beo. Th. 5084; B. 2545. Ðætte seó wǽte ūt fleówe, Bd. 4, 19; S. 589, 1. Hí of mínre sídan swát ūt guton (gotun, MS.), Exon. Th. 88, 33; Cri. 1449. Cleopaþ se alda ūt of helle, Cd. Th. 267, 7; Sat. 34. (4) with the idea of removal from the place in which a thing is fixed, to knock *out,* pull *out,* etc. :—Ic ūt ádelfe *effodio,* Ælfc. Gr. 28, 6; Zup. 179, 11. Þafa ðæt ic ūt ádó (*ejiciam*) ðæt mot of ðínum eágan . . . Ádó ǽrest ūt ðone beám of ðínum ágenum eágan, Mt. Kmbl. 7, 4-5. Ūt ástingan, Chr. 797; Erl. 59, 43. Ðū ðe ūt átuge (*extraxisti*) mē of innoðe, Ps. Lamb. 21, 10. Áteon ūt ða wǽtan, Lchdm. ii. 222, 25. Gif hwá sleá his weales eáge ūt oððe his wylne, lǽte hig frige for ðam eágan ðe hē ūt ádyde, Ex. 21, 26 : L. Alf. 20; Th. i. 48, 25. Ðá sticode him mon ða eágan ūt *effossis oculis,* Ors. 4, 5; Swt. 168, 4. (5) with verbs of summoning :—Hē hine ácigde ūt, Bd. 2, 12; S. 513, 19. Ában ðū ða beornas ūt of ofne, Cd. Th. 242, 34; Dan. 429. Ūt *from this world,* Salm. Kmbl. 962; Sal. 480. (5 a) summoning to service :—Ðá hēt se cyng ábannan ūt ealne þeódscype of West-Seaxum, Chr. 1006; Erl. 140, 8. Hēt se cyning bannan ūt here . . . Ðá hí þider ūt cómon (cf. ðone here ðe ðam cynge mid wæs, Erl. 181, 8), 1048; Erl. 180, 1: Exon. Th. 120, 12; Gú. 270. (6) *out, away from home, abroad :—*Gif hē unmundlunge ceáp áredige ūt on hwylcere fare, būton hē hit ǽr cýdde ðá hē ūt rád, L. Edg. S. 8; Th. i. 274, 23. (7) *out, away from land :—*Hweðer gē eówer hundas and eówer net ūt on ðæ sǽ lǽdon? Bt. 32, 3; Fox 118, 14: Met. 19, 19. Hié ūt óðreówon, Chr. 897; Erl. 96, 7. Ūt feor on Wendelsǽ, Met. 26, 30. Wit on gársecg ūt aldrum neððon, Beo. Th. 1079; B. 537. Guman ūt scufon wudu, 436; B. 215; Chr. 897; Erl. 96, 7. Nacan ūt áþringan, Exon. Th. 474, 31; Bo. 39. II. where there is not motion, *out.* (1) *outside :—*Ne beóð hí ūt fram ðé átýnde *non excludantur,* Ps. Th. 67, 27. (1 a) *not within doors, not in the house, abroad :—*Niman hí him wíf and heora andlyfene ūt onfón *sortiri uxores debent, et stipendia sua exterius* (not in a monastery) *accipere,* Bd. 1, 27; S. 489, 18. v. ūt-wǽpnedmann. (2) *on the surface :—*Byrgennum ūt hwítum *monumentis dealbatis,* Mt. Kmbl. p. 19, 12. (3) *out, away from land :—*Orcadas ða

eálond, ða wæron út on gársecge bútan Breotone *Orcadas insulas ultra Brittaniam in oceano positas*, Bd. 1, 3; S. 475, 13 : Ps. Th. 96, 1 : Met. 16, 12. (4) figurative, *externally :*—Se ðe út wel lærð mid his wordum, hé onféhð innan ðæs inngeðonces fǽtnesse *qui exterius praedicando benedicit, interioris augmenti pinguedinem recipit*, Past. 49; Swt. 381, 4. [*Goth. O. Sax. Icel.* út : *O. H. Ger.* úz.] v. þær-, þurh-út.

út-ácumen, -cymen[e] ; *adj. Stranger, alien, foreign :*—Dēmaþ ǽlcon men riht, sí hit burga man, sí hit útácymene (*peregrinus*), Deut. 1, 16. Se útácymena (útancumena, *v. l.*) munuc ðe of uncúðum eardum cymð *si quis monachus peregrinus de longinquis provinciis supervenerit*, R. Ben. 108, 4. Gé wǽron útácymene (*advenae*) on Egipta lande, Lev. 19, 34; 25, 23. Eallum and mágum and útácymenum *omnibus et propinquis et extraneis*, Scint. 3, 14. Útácymene and ælþeódige *aduenas et peregrinos*, 137, 16. Útácymene *peregrinos*, Lev. 23, 22. Wræccan ł útácumenan *aduenas*, Ps. Lamb. 145, 9. v. útan-cumen, -cymene.

úta-cund (úta = útan or úte ; v. innan-, inne-cund) ; *adj. Foreign, alien, strange :*—Útacund cynn *alienigena*, Lk. Skt. p. 9, 8. Ðes útacunda, Lind. 17, 18. On útacund *in alieno*, 16, 12. Útacund *alienum* . . . ðara útacundra *alienorum*, Jn. Skt. Lind. 10, 5. From útacundum *ab alienis*, Mt. Kmbl. Lind. 17, 25 : Rtl. 168, 13.

utan *let us.* v. witon.

útan (-on) ; *adv. prep.* **A.** *adv.* **I.** *from without :*—Weard mē on hige leóhte útan and innan, Cd. Th. 42, 21 ; Gen. 677. Gif ðú wénst ðæt him áhwonan útan cômon ða gód ðe hé hæfþ, ðonne wǽre ðæt þing betere, ðe hit him fram cóme, ðonne hé, Bt. 34, 3 ; Fox 136, 26. Ælfréd com útan (úton, MS. E.) mid fierde, Chr. 885; Erl. 82, 23. Ǽghwylcne ellþeódigra ðara ðe ðæt eáland útan sóhte, Andr. Kmbl. 56 ; An. 28. Cumaþ of eálandum útan kyningcas, Ps. Th. 71, 10 : 79, 13. **II.** *without, on the outside.* (1) where action, stated or implied, may be thought of as operating on an object from without :—Hát wæs him útan wráðlíc wíte, Cd. Th. 23, 6 ; Gen. 354 : 285, 23 ; Sat. 342. (1 a) with *ymb, be,* as prepositions or prefixes of verbs :—Pontius hæfde ðone consul mid his folce útan befangen, Ors. 3, 10 ; Swt. 140, 22 : Met. 13, 7. Ǽghwilc óþer útan ymbclyppeþ, Met. 11, 35 ; Exon. Th. 423, 2 ; Rä. 41, 15. Hí hine útan ymbðringaþ, Salm. Kmbl. 256 ; Sal. 127. (2) where action takes place outside an object :—Hí bedícodon ða burh úton, Chr. 1016 ; Erl. 155, 11. Hé ðone búr útan beeode, 755 ; Erl. 48, 30. Land belicgan úton, Cd. Th. 15, 7 ; Gen. 229. Besittaþ hié útan, Past. 21 ; Swt. 161, 4 : Chr. 894 ; Erl. 93, 9 : 918 ; Erl. 104, 1. Hié hine ðær útan besǽton, 894 ; Erl. 92, 23. Ðæt nán neód sý útan tó farenne *ut non sit necessitas vagandi foris*, R. Ben. 127, 7. Se ðe sceal healdan folc útan wið feóndum, Ps. Th. 120, 4. Se fugel ymbseteþ útan líc hálgum stencum, Exon. Th. 212, 3 ; Ph. 204. Útan ymbestandne mid unríme þegna, Met. 25, 7. Úton, Bt. 37, 1 ; Fox 186, 3. Cyrican wyrcean ymb ða cyrican útan ðe hé ǽr worhte, Bd. 2, 14 ; S. 517, 30. (3) *on the outside, on the surface :*—Útan (*a foris*) wlitige, innan fulle deáðra bána, Mt. 23, 27 : Runic pm. Kmbl. 341, 26 ; Rún. 13 : Blickl. Homl. 197, 11. Ðæt treów biþ úton gescyrped mid ðære rinde, Bt. 34, 10 ; Fox 150, 7 : Beo. Th. 3011 ; B. 1503. Úton tó gesett tó trymnesse ðæs húses, Bd. 3, 17 ; S. 544, 35 : Exon. Th. 233, 26 ; Ph. 530. Úton hié wǽron elpendbánum geworhte, Nar. 5, 5 : Exon. Th. 474, 31 ; Rä. 41, 47. Beámas útan ofǽtes gehlædene, Cd. Th. 30, 3 ; Gen. 461. On ðysse eorðan úton *on the face of the earth*, Ps. Th. 64, 6. Innan and útan eorðan líme gefæstnod, Cd. Th. 80, 1 ; Gen. 1322 : Beo. Th. 1552 ; B. 774 : Exon. Th. 62, 21 ; Cri. 1005 : 219, 2 ; Ph. 301. (3 a) figuratively, *outwardly :*—Gé ætýwaþ mannum útan (*a foris*) rihtwíse, Mt. Kmbl. 23, 28. Deáh hé fæger word útan ætýwe, Fragm. Kmbl. 32 ; Leás. 18. (4) with *ymb* or *be* and verbs of motion or rest, *about, round :*—Ðæt hé hine ǽghwonon útan ymbsáwe (cf. behealde hé on feówer healfe, Bt. 19 ; Fox 68, 21), Met. 10, 4. Útan behwerfed, Bt. 25 ; Fox 88, 35 : Met. 13. 77, 78. Hé ǽlce dæg úton ymbhwyrfþ ealne ðisne middaneard, 39, 3 ; Fox 214, 16 : Met. 28, 4, 13. Hié ne mehton Súð-Seaxna lond útan berówan, Chr. 897 ; Erl. 96, 9. Ætýwdon twégen steorran ymb ða sunnan útan, Bd. 5, 23 ; S. 645, 23. (5) *away from land :*—Eálond útan, Beo. Th. 4657 ; B. 2334. **B.** *prep. with gen. Without, outside of.* v. útan-bordes, -landes. [*O. Sax.* útan: *O. H. Ger.* úzan *foras, a foris : Icel.* útan *from outside ; outside.*] v. be-, on- [Lchdm. ii. 292, 27), wiþ-, ymb-útan ; *útane, and compounds with* útan *as prefix.*

útan-bordes ; *adv. Abroad :*—Man útanbordes wísdóm and láre hieder on lond sóhte *people abroad came hither in search of learning*, Past. pref. ; Swt. 3, 11. [*Icel.* útan-borðs *overboard : Dan.* uden-bords. Cf. *Goth.* útana (*with gen.*) : *O. H. Ger.* úzan (*with gen.*) : *Icel.* útan (*with gen.*).] v. útan-landes.

útan-cumen, -cymen[e] ; *adj. Come from without.* **I.** *from another land, foreign, alien, strange :*—Útancuman *advena*, Wrt. Voc. i. 74, 64. Ðǽr nán útancymen (úiencumen, Cott. MS.) mon cuman ne dorste, Ors. 5, 2 ; Swt. 218, 1. Se útancumena munuc ðe of uncúðum eardum cymð, R. Ben. 109, 4. Dam útþeódigan and útancumenan (útcymenan, MSS. G. H.) ne lǽt ðú nó uncúðlíce wið hine, L. Alf. 47 ; Th. i. 54, 20. Ne hyrwe gé útancymenne man (*advenam*), Lev. 19, 33. Ælþeódige men and útancumene swýðe ús swencaþ, Wulfst. 91, 19. Gé

wǽron útancymene (*advenae*) on Egipta lande, Deut. 10, 19. Útancumenra *exterorum* i. *peregrinorum*, Wrt. Voc. ii. 145, 62. Þurh útancymen[r]a goda naman *per nomen externorum deorum*, Ex. 23, 13. Útancumene and elþeódige ne geswenc ðú, L. Alf. 33 ; Th. i. 52, 14. **II.** *belonging to another :*—Gif útancymene (*alienus*) oxa óðres oxan gewundaþ, Ex. 21, 35. v. út-ácumen.

útane (-one, -ene) ; *adv.* **I.** *where there is motion* (lit. or fig.) *to an object, from without :*—Útene *extrinsecus*, Wrt. Voc. ii. 145, 21. Him biþ se wela útane cumen, and hé ne mæg útane náuht ágnes habban, Bt. 27, 2 ; Fox 98, 7, 8. Ic nolde ðæt ðú wéndest ðæt Gode áhwonan útane cóme his gódnes, 34, 2 ; Fox 136, 23 : 34, 19, 34 ; Fox 144, 20. Ðý læs ðonne hié oferhyggaþ ðæt hié sién oferreahte útane mid óðerra manna lárum hié sién innan gehæfted mid ofermétum *ne dum aliorum suasionibus foris superari despiciunt, intus a superbia captivi teneantur*, Past. 42 ; Swt. 307, 6. Him mon útane of óðrum londum an wann, Ors. 3, 7 ; Swt. 110, 28. **II.** *where there is not movement to an object.* (1) *outside :*—Se here ða burh útone besǽton, Chr. 1016 ; Erl. 156, 14. Se rodor hine hæfþ útane (cf. se rodor ðás rúman gesceaft útan ymhwyrfeþ, Met. 20, 137), Bt. 33, 4 ; Fox 130, 22. Ðætte wé scylen beón on ðisse ælðeódignesse útane beheáwene mid suingellan, tó ðæm ðæt wé sién geféged tó ðæm gefógstánum on ðære Godes ceastre *quia nunc foris per flagella tundimur, ut intus in templum Dei postmodum disponamur*, Past. 36 ; Swt. 253, 18. (2) *on the outside, on the surface :*—Se wielm ðæs innoðes út ábiersð, and wierð tó sceabbe, and moniga wunda útane wyrcð, Past. 11 ; Swt. 71, 10. (3) *out, at sea :*—Ða gerǽdde se cyng ðæt man gegaderode scipu . . . and hí sceoldan cunnian gif hí muhton ðone here áhwǽr útene betræppen, Chr. 992 ; Erl. 131, 27. (4) *outwardly, externally :*—Ðonne hé ongit be sumum ðingum oððe ðeáwum útone (-anne, Hatt. MS.) ætiéwdum (*signis exterius apparentibus*) eall ðæt hié innan ðenceaþ, Past. 21 ; Swt. 155, 10. Útane, 28 ; Swt. 195, 22. Gif munuc inne on his heortan eáðmód bið, and ná ðæt án, ac eác swylce útone mid his líchoman eáðmódnesse gebýcnige, R. Ben. 31, 3. (5) *with ymbe, about :*—Ða ymbe ðæt útene forðférde Decius *about that time Decius died*, Homl. Skt. i. 23, 348. [*O. H. Ger.* úzana.] v. útan.

útan-landes ; *adv. Abroad, in distant countries :*—Þeóda ðe eard nymaþ útanlandes *gentes qui habitant fines terrae*, Ps. Th. 64, 8. [Cf. In outenland *in terra aliena*, Ps. 136, 4. Utenerdes *in foreign lands*, Gen. and Ex. 956. Laban ferde fro Caram into utenstede, 1741. *Icel.* útan-lands, -lendis *abroad ;* útanlands-maðr, -siðir *a foreigner, foreign customs.*] v. útan-bordes.

útan-weard ; *adj. Outside, exterior ;* may be translated, *the outside of* the noun with which it agrees :—Útanweard þeóh *femur*, Wrt. Voc. i. 44, 62. Fram ðæm múþan útanweardum, Chr. 893 ; Erl. 88, 32. Hlǽw ymbehwearf útanweardne, Beo. Th. 4583 ; B. 2297. Útaweard fingeras *extremum digiti*, Lk. Skt. Lind. 16, 24. ¶ adverb :—Útaword *deforis*, Mt. Kmbl. Lind. 23, 25, 26 : Lk. Skt. Rush. 11, 39. [*Icel.* útan-verðr.] v. úte-, út-weard.

út-cwealm, es ; *m. Utter destruction :*—Útcualm *internicium bellum dicitur, quo nullus remanet*, Wrt. Voc. ii. 111, 83.

út-cymen. v. út-cumen.

út-drǽf, e ; *f. Ejection, expulsion :*—Ða onscunode se Eádsige Aðelwold, and ealle ða munecas ðe on ðam mynstre wæron, for ðære útdrǽfe ðe he gedyde wið hí, Homl. Skt. i. 21, 85.

út-drǽfere, es ; *m. One who drives out :*—Útdrǽfere *exterminator*, út ádrifen *exterminatus*, Wrt. Voc. i. 51, 45.

úte ; *adv. Outside, without.* **I.** *where there is motion to the outside :*—Ne com se here ofter eall úte of ðæm setum ðonne tuwwa, Chr. 894 ; Erl. 90, 19. Ðæt hé up heonon úte mihte cuman, Cd. Th. 27, 10 ; Gen. 415. Móste ic úte weorþan, 23, 34 ; Gen. 369. **I a.** fig. with the idea of degradation, *out of one's position :*—Bútan ðam ánum ðe for heora leahtrum of hyra endebyrdenesse útor (uttor, Wells Fragm.) áscofene synd *exceptis his quos abbas degradaverit*, R. Ben. 115, 9. **I b.** *out, into another's possession :*—Weard ðæt land úte and hæfdon hit cynegas *ablatum est in manibus regum*, Chart. Th. 271, 27. **II.** *on the outside :*—Ic eom úte *ego foris sum*, Ælfc. Gr. 38 ; Zup. 242, 5. (1) *outside* a house, any enclosed place, etc. :—Petrus sæt úte (*foris*) on ðam cafertúne, Mt. Kmbl. 26, 69 : Lk. Skt. 1, 10. Tó ðam wiggendum ðe ðǽr unróte úte (*outside the tent*) wǽron, Judth. Thw. 25, 29 ; Jud. 284. Gé standaþ ðǽr úte (uuta, Lind. *foris*), Lk. Skt. 13, 25 : Jn. Skt. 18, 16 : 20, 11 : Bd. 2, 12 ; S. 513, 30 : Blickl. Homl. 201, 18 : 217, 35. His líchoma wæs úte bebyriged néh cyricean *positum corpus ejus foras juxta ecclesiam*, 3, 8 ; S. 504, 31. Mycel menigu ymb hine sæt, and tó him cwǽdon : 'Hér is ðín módor úte (*foris*),' Mk. Skt. 3, 32. (1 a) in a special sense. v. út, I. 1. a 3 :—Sum coþu is ðære wambe, ðæt ðone seócan monnan lysteþ útganges, and ne mæg ðonne hé úte betýned bið (*when he is at the closet*), Lchdm. ii. 236, 3. (1 b) *out, not residing* in a place :—Ðæt muneca gehwylc, ðe úte sý of mynstre . . . ; gebúge intó mynstre, L. Eth. 5, 5 ; Th. i. 306, 2. (1 c) in reference to persons :—Ðæt mód mæg findan on innan him selfum ealle ða gód ðe hit úte sécþ, Bt. 35, 1 ; Fox 154, 25. (1 d) where the locality is non-material :—Ðam ðe úte synt ealle þing on bigspellum gewurþaþ, Mk.

Skt. 4, 11. Nú sind wé úte belocene fram ðam heofenlícan leóhte, Homl. Th. i. 154, 13. Ðǽr wæs Evan wóp úte betýned, Blickl. Homl. 7, 14. (2) *outside, on the outer side* :—Gé ðæt úte is calices geclǽnsiaþ, Lk. Skt. 11, 39. (3) *out, out of doors, in the open air* :—Se cyng hét him úte setl gewyrcean *rex, residens sub divo*, Bd. 1, 25 ; S. 486, 38. Hí slépon úte on triówa sceadum, Bt. 15 ; Fox 48, 12 : Met. 8, 27. Gnættas cómon ofer eall ðæt land, ge inne ge úte, Ors. 1, 7 ; Swt. 36, 30. Ic seah wyhte twá úte plegan, Exon. Th. 429, 10 ; Rä. 43, 2. (4) *out, away, at a distance* :—Úttor *exterius*, Ælfc. Gr. 38 ; Zup. 240, 7 : Exon. Th. 426, 35 ; Rä. 41, 84. (4 a) *out, away from habitations, in open country* :—Hé ne mihte on ða ceastre gán, ac beón úte (*foris*) on wéstum stówum, Mk. Skt. 1, 45. On burgum beóþ blóstmum fægere, swá on eorðan hég úte on lande, Ps. Th. 71, 16. Hé genam hine æt eówde úte be sceápum, 77, 69. (4 b) *out, from home on service* :—Hié wǽron simle healfe æt hám, healfe úte, Chr. 894 ; Erl. 90, 18. Hí lágon úte ealne ðone herfest on fyrdinge, 1006 ; Erl. 140, 9. (4 c) *out, not in one's own country, abroad* :—Him leófre wæs ðæt hé úte wunne ðonne hé æt hám wǽre, Ors. 3, 7 ; Swt. 110, 30. (4 d) *out, away from land* :—Án ígland ðæt is úte on ðǽre sǽ, Chr. 895 ; Erl. 93, 24. Ðá sǽton hié úte on ðam íglande, 918 ; Erl. 104, 11. Gefeaht Scipia wið Hannibal úte on sǽ, Ors. 4, 11 ; Swt. 204, 36. (5) *marking degree or extent* :— Hí nánwuht ne magon ufor ne útor (*beyond*) findan, Bt. 34, 12 ; Fox 154, 16. Ðám ðe him ðás woruld úttor lǽtan, ðonne ðæt éce líf, Exon. Th. 109, 28 ; Gú. 97. [*O. Sax. O. Frs.* úta : *Icel.* úti.] v. þǽr-úte ; út, útan.

útera ; *cpve.* **útemest, útmest** ; *spve. adj. Outer, outmost.* I. *of position or order* :—Seó útre wamb *venter*, Wrt. Voc. i. 45, 21. Gif ðæt úterre (úttere, MS. B. : útre, MS. H.) bán bið þyrel, L. Alf. pol. 44 ; Th. i. 92, 15. Hié forgeátan ðara útera gefeohta *they forgot the foreign wars*, Ors. 2, 6 ; Swt. 88, 24. Wurpaþ hyne on ða úttran (útteran, MS. A. : ðǽm útmestum, Lind.) þýstro *mittite eum in tenebras exteriores*, Mt. Kmbl. 22, 13 : 25, 30 (wútmestum, Lind.). Oð tó útmeste *usque ad extremum*, Rtl. 55, 36. Ða útemestan ðíoda *the most distant nations*, Bt. 19 ; Fox 68, 29. In útmestum *in extremis*, Mk. Skt. Lind. 5, 23. II. *external, not of the inner man* :—Úre mann úttra *noster homo exterior*, Scint. 53, 20. Hú se láreów ne sceal ða inneran giémenne gewanian for ðǽre úterran ábisgunge (*exteriorum occupatione*), ne eft ða úterran ne forlǽte hé for ðǽre inneran . . . ðý lǽs hé sié gehæft mid ðam úterran ymbhogan, Past. 18 ; Swt. 127, 8–14. For ðǽre úttran geornfulnesse woruldlícra dǽda *pro industria exteriori*, Bd. 5, 13 ; S. 632, 8. Ðætte wé swá lufigen ðisne úterran and ðisne eorðlícan fultum, Past. 50 ; Swt. 389, 2. Ða úttran weorc wǽron behealden *exteriora opera observantur*, Bd. 1, 27 ; S. 494, 30 : Scint. 60, 4. Þeáh hé mé ðara úterrena gewinna gefreóde, þeáh winnaþ wið mé ða inran unrihtlustas, Ps. Th. 15, 7 : Past. 18 ; Swt. 139, 23. Ðara úterra weorca, Swt. 127, 12 : 141, 8. Þeáh ðe ic næbbe ða úttran lác, ic geméte on mé sylfum hwæt ic lecge on weófode ðínre herunge, Homl. Th. i. 584, 15. [*O. Frs.* útera : *O. H. Ger.* úzero.] v. innera.

úter-mere, es ; *m. Outer-sea, open sea* :—Hié forfóron him ðone múðan on útermere, Chr. 897 ; Erl. 95, 22. [*Cf. Icel.* út-sjár.]

úte-weard ; *adj. Outward, extreme* ; *may sometimes be translated* on the outside of, at the extremity of, *the noun to which it refers* ; *sometimes is used substantively, the outward part, extremity* :—Úteweard (*del*) *crepido*, Wrt. Voc. i. 34, 27. Se munt is mycel úteweard *the hill presents a large surface*, Blickl. Homl. 207, 26. iiii míla fram ðǽm múðan úteweardum *four miles from the outside of the mouth*, Chr. 893 ; Erl. 88, 32. Ða geféngon hié ðara þreóra tú æt ðǽm múðan úteweardum, 897 ; Erl. 95, 26. Hé sý onfangen on úteweardre endebyrdnesse *in ultimo gradu recipiatur*, R. Ben. 53, 11. Hé ðencð on ðam oferbrǽdelse his módes . . . Ac on úteweardum his móde hé liéhð him selfum, Past. 9 ; Swt. 55, 18–24. Heó hafaþ langne wyrtruman and ðone úteweardne sweartne *it has a long root, and that black on the outside*, Lchdm. i. 304, 2. Ðú smítst his blód ofer úteweard Aarones swýðre eáre *sanguinem ejus pones super extremum auriculae dextrae Aaron*, Ex. 29, 20. Úteweard nosterle *pinnulae*, Wrt. Voc. i. 43, 22. Smyra ða eágan útewarde, Lchdm. i. 374, 10. ¶ *with preps. forming prepositional or adverbial phrases* :—Ðes eard (*England*) nis swá mægenfæst hér on úteweardan ðǽre eorðan brádnysse, Homl. Sk. i. 13, 107. Ðeáh hine man wácne and unweorðne talige and an úteweardum forlǽte and tó úteweardum medemige *si omni vilitate vel extremitate contentus sit monachus*, R. Ben. 29, 4. [*O. Frs.* úta-werd.] v. útan-, út-weard.

út-fær, es ; *n. A going out, egress, exit* :—Útfær *egressio*, Ps. Lamb. 18, 7. On útfære *in exitu*, 73, 5. Ðæt wé symle ðone mǽran gylt forfleón þurh útfære ðæs lǽssan, Homl. Th. i. 484, 8. Ðeáh heó nán útfær ne gemét, 410, 10. On útfærum heora *in egressibus suis*, Ps. Lamb. 143, 13.

út-færeld, es ; *n. A going out* :—*Exodus* on Grécisc, *exitus* on Lýden, útfæreld on Englisc, Ex. Thw. tit. Útfæreld his fram Fæder *egressus ejus a Patre*, Hymn. Surt. 44, 17. Hí ǽr Moyse and hys folce ðæs útfæreldes wyrndon, Ors. 1, 7 ; Swt. 38, 19. Útfæreld *exitum*, Ælfc. Gr. 30 ; Zup. 193, 8. Ne fare hé út tó gefeohte ne him nán man útfæreld beóde (*he is not to be called upon to leave home*), Deut. 24, 5.

út-faru, e ; *f. A going out, going abroad* or *out of doors* :—Ðæt nán neód ne sý munecum útan tó farenne, for ðý ðe seó útfaru nán þing ne framaþ hira sáulum *ut non sit necessitas monachis vagandi foris quia omnino non expedit animabus eorum*, R. Ben. 127, 8. [*Icel.* út-fór.]

út-fór, e ; *f. A going out* from the body, *an evacuation* :—Be drencum and útfórum, Lchdm. ii. 14, 30.

út-fús ; *adj. Ready to sail* :—Þǽr æt hýðe stód hringedstefna útfús, Beo. Th. 65 ; B. 33.

út-gang, es ; *m. A going out, exit, egress* :—*Exitus, finis, effectus, terminus, egressus* útgong, endestæf, Wrt. Voc. ii. 144, 83. Útgang *egressio*, Ps. Spl. 18, 6 : *exitus*, 118, 136. (1) *a going out* of a place, *egress, exit* :—Ná ðæt hé Criste útganges rýmde, Homl. Th. i. 222, 9. Be útgonge (*egressu*) folces of Ægypta lande, Bd. 4, 24 ; S. 598, 11. Ðú mé ne dést tó útgonge íc ne mæg *you will not make me go out, and I cannot*, Shrn. 141, 21. Útgang ðínne and ingang Dryhten gehealde *Dominus custodiat introitum tuum et exitum tuum*, Ps. Th. 120, 7. Þurh earmlícne deáþ and þurh sárlícne útgang ðæs mánfullan lífes, Guthl. 2 ; Gdwin. 14, 21. Útgong heonan, Exon. Th. 282, 10 ; Jul. 661. (1 a) *the right of egress* :—Ingong and útgong, Chart. Th. 578, 26. (2) *a coming out* from a position within a body :—Lǽcedómas wið þearmes útgange, and wið bæcþearmes útgange, Lchdm. ii. 170, 27, 29. (2 a) *in a special sense, evacuation* of the body :—Sum coþu is ðǽre wambe ðæt ðone seócan monnan lysteþ útganges, Lchdm. ii. 236, 3. (3) *in reference to time, the going out* of a period, *the conclusion, end* :—Se ǽresta Mónandæg æfter útgange ðæs mónþes Decembris *the first Monday after December has gone out*, Lchdm. iii. 76, 18. (4) *a place by which anything comes out, an exit, passage* :—On útgange burnan *in exitus aquarium*, Ps. Th. 106, 34. Næfð útgang sió stów, Lchdm. ii. 218, 17. (4 a) *in a special sense, of part of the body* :—*Viscera* inilve, *meatis* útgang, *anus* bæcþearm, Wrt. Voc. i. 283, 59. Ða swylas ðe beóð on mannes handum oððe on óþrum limum oððe ymb ðone útgang, Lchdm. i. 356, 17 : 364, 20. (4 b) *a privy. Cf.* forþ-gang :—In útgeong ł in feltún (innun útgongum, Rush.) *in secessum*, Mk. Skt. Lind. 7, 19. (5) *what comes out* of a body, *an evacuation* :—Sceáwige mon hwylc se útgang sié þe micel þe lytel, Lchdm. ii. 218, 12 : 200, 1 : 220, 6. Gesceáwa ælce dæge ðæt ðín útgong and micge sié gesundlíc, 226, 20, 22. Be ðǽre coþe ðe se mon his útgang þurh ðone múð him fram weorpe, 236, 12. Næs þurh ða micgean þ̄ ne ac eác þurh óþerne útgang, 250, 11. [*O. Frs.* út-gong : *O. H. Ger.* úz-gang *exitus, egressus, eventus ; diarria, dysenteria : Icel.* út-gangr, -ganga *a going out ; a passage.*] v. út-geng.

út-gársecg, es ; *m. The ocean at the horizon, the ocean at a distance from land.* v. út, II. 2 :—Tungol (*the sun*) on ǽfenne útgársecges grundas pæþeþ *the sun at even holds its way beneath the depths of utmost ocean*, Exon. Th. 350, 29 ; Sch. 70. [*Cf. Icel.* út-haf.]

út-gefeoht, es ; *n. Foreign war* :—Ðætte Bryttas sume tíd gestildon fram útgefeohte *ut Brittones, quiescentibus ad tempus exteris bellis*, Bd. 1, 22 ; S. 485, 11.

út-gemǽre, es ; *n. An extreme boundary* :—Of eorðan útgemǽrum *a finibus terrae*, Ps. Th. 60, 1. Oþ ðysse eorðan útgemǽru *ad terminos orbis terrae*, 71, 8.

út-geng, es ; *m.* (or ? -genge, an ; *f.* v. genge) *An outlet, exit* :—Tó útgengum weogas *ad exitus viarum*, Mt. Kmbl. Rush. 22, 9. [*Cf. Icel.* út-ganga ; *wk. f.*] v. út-gang.

út-healf, e ; *f. The outside, exterior* :—Úthealf ðæs beddes *sponda* (v. sponda, est exterior pars lecti, 242, col. 2), Wrt. Voc. i. 41, 28. [*Cf. Icel.* út-hálfa *the outskirts.*]

út-here ; *gen.* -her(i)ges ; *m. A foreign army* :—Se here férde swá hé sylf wolde, and seó fyrding dyde ðǽre landleóde ælcne hearm, ðæt him náðor ne deáhte ne innhere ne úthere, Chr. 1006 ; Erl. 140, 13. Ða scipu sceoldan ðisne eard healdan wið ælcne úthere, 1009 ; Erl. 141, 25.

úþgendra. v. next word.

úþ-genge ; *adj. Fugitive, transitory, not to be retained, passing out of one's possession* :—Se eþel úðgenge wearð Adame and Euan, eardríca cyst beorht óðbróden *that country could no more be held by Adam and Eve, the choicest realm was taken away from them*, Exon. Th. 153, 12 ; Gú. 824. Ðær wæs Æschere feorh úðgenge *there life fled from Æschere*, Beo. Th. 4253 ; B. 2123. Ðæs éðel wǽre éce tó gelýfanne on heofonum, nalæs on eorþlícre frætwædnysse, on gewítendre and on úþgengre *cujus sedes aeterna non in vili et caduco metallo, sed in coelis esset credenda*, Bd. 3, 22 ; S. 552, 20. Ðæt hié ne ástigan on ofermédu, ne úþgendra (-gengra ?) welena tó wel ne truwodon, Blickl. Homl. 185, 14. [*Cf. Goth.* unþa- in unþa-þliuhan *to escape.*]

út-hleáp, es ; *n. The fine for allowing a culprit to escape* (cf. L. In. 36 ; Th. i. 124, 14) :—Úðleáp, Chart. Th. 411, 30 : 359, 3 (*printed* -leaw). The word occurs in a list of privileges granted by the king.

úþ-mǽte ; *adj. Immense, very great* :—Ðǽr hangade úþmǽte leóhtfæt, byrnende dæges and nihtes ofer ðara Drihtnes fóta swaða (cf. Hangaþ ðǽr eác bufan ðǽm lástum geregnod swíþe mycel leóhtfæt . . . and bið á dæges and nihtes byrnende, Blickl. Homl. 127, 29), Shrn. 81, 17.

úþ-wita, -weota, an ; *m. A person distinguished for wisdom or learning* in general or in a special branch, *a philosopher, scribe, geo-*

metrician, etc.:—Se gomola, eald úðwita (cf. fród fæder, módsnottor, 300, 4; Fä. 1), Exon. Th. 304, 6; Fä. 66. Úðweota *a councillor, senator*, Andr. Kmbl. 2211; An. 1107. Úðuuta *philosophus*, Wrt. Voc. ii. 117, 24. Cato wæs openlíce úþwita, Bt. 19; Fox 70, 8: Met. 10, 50. Epicurus se úþwita, Bt. 24, 3; Fox 84, 21. Úre úþwita Plato, 33, 3; Fox 126, 35: 35, 1; Fox 156, 9: Met. 22, 54. Úðwita *sophista*, Ælfc. Gr. 7; Zup. 24, 8. Gleáwum úðwitum and getincgum *gymosophistis et rhetoribus*, úðwita *gymnosophista*, Hpt. Gl. 479, 6-9. Án swíþe wís mon ongan fandigan ánes úþwitan and hine bismerode, for ðam hé hine swá orgellíce up áhóf and bodode ðæs ðe hé úðwita (*philosophus*) wǽre; ne cýðde hé hit mid nánum cræftum . . . Ðá wolde se wísa mon his fandigan, hwæðer hé swá wís wǽre swá hé self wénde ðæt hé wǽre, Bt. 18, 4; Fox 66, 27-33. Án úðuutta *unus scriba*, Mt. Kmbl. Lind. 8, 19. Úðwitan *sophistae*, Hpt. Gl. 449, 46. Ðá clypode se apostol ðone úðwitan Graton, Homl. Th. i. 60, 31. Ðæs ðe ús secgaþ béc, ealde úðwitan (*historians*), Chr. 937; Erl. 115, 18: *astronomers*, Menol. Fox 329; Men. 166. Úþwitan (*philosophers*) secgaþ ðæt sió sáwul hæbbe ðrió gecynd, Bt. 33, 4; Fox 132, 3: Met. 20, 184: Homl. Skt. i. 1, 96. Úðweotan (*the Jewish scribes and elders*), Elen. Kmbl. 943; El. 473. Úðwuta *scribae*, Mt. Kmbl. Lind. 15, 1. Úðuta (-wutu, Rush.), Mk. Skt. Lind. 1, 22. Úðwutto (-wuta, Rush.), Lk. Skt. Lind. 22, 66. Swá swá úþwitena gewuna is *ut geometrae solent*, Bt. 34, 4; Fox 138, 28. Sume of úðuutum (-wutum, Rush.) *quidam de scribis*, Mk. Skt. Lind. 7, 1. Wǽ iúh uudutum, Mt. Kmbl. Lind. 23, 29. Ic sende tó iúh wítgo and snotre menn and úðuto (*scribas*), 23, 34. Úðwiotan his *seniores suos*, Ps. Surt. 104, 22: 118, 100. [Magy wærenn uþwitess swíþe wise, Orm. 7083.]

úþ-witian; *p.* ode *To study philosophy*:—Ic úðwitige oððe ic smeáge embe wísdóm *philosophor*, Ælfc. Gr. 25; Zup. 146, 2.

úþ-witigung, e; *f. The study of philosophy, philosophy*:—Ðæt heó on woruldwýsdóme wǽre getogen æfter Gréciscre úðwytegunge . . . Heó þeáh on wísdóme and tó úðwytegunge, Homl. Skt. i. 2, 20-23. Befæst tó woruldlícre láre and tó úðwitegunge, 4, 185. Hé cwæð him tó: 'Nú ic hæbbe ðé oferðogen on úðwitegunge.' Se biscop him andwyrde: 'God forgeáfe ðæt ðú úðwitegunge beeodest,' Homl. Th. i. 448, 34: Homl. Skt. i. 3, 210.

úþ-witlíc; *adj. Philosophical*:—Ðære úðwitlícan *acathemice*, Wrt. Voc. ii. 79, 10. Ðære úðwiottelícan, 9, 13. Ða úþwitlícan *gymnica*, 91, 21: 41, 39. Úþwitlícum *gimnicis artibus*, 42, 34.

útian; *p.* ode *To put out.* (1) *to put a person out of a place, to expel, remove*:—Ðæt ænig man ciricþén ne útige búton biscopes geþehte, L. Eth. v. 10; Th. i. 306, 28. Gif man preóst of circan on unriht útige, L. N. P. L. 22; Th. ii. 294, 2. (2) *to put a thing out of one's possession, to alienate*:—Gif preóst ciricþingc útige, L. N. P. L. 27; Th. ii. 294, 14. Úttige, Cod. Dip. Kmbl. iv. 208, 10. [O. Frs. útia: O. H. Ger. úzon *to put out*.] v. ge-útian.

út-irnende; *adj. Running out of the body.* (1) *of medicine, purging, purgative*:—Wyrtdrenc ðe ne bið útyrnende, Lchdm. ii. 282, 9: 170, 25. Sele him wyrtdrenc útyrnende, 280, 17. Útyrneande, 336, 1. Mid swelcum útyrnendum drencum, 23: 82, 17. (2) *of a disease, diarrhoeic*:—Disse ádle fruman mon mæg gelácnian on ða ilcan wísan ðe ða útyrnendan, Lchdm. ii. 232, 17. (3) *of persons, suffering from diarrhoea or dysentery*:—Hú mon ða útyrnendan men scyle lácnian, Lchdm. ii. 278, 16. v. út-ryne, *and next word.*

út-irning, e; *f. A flux*:—In útiorningc (úttiornende, Rush.) blódes *in profluuio sanguinis*, Mk. Skt. Lind. 5, 25.

út-lád, e; *f. Carriage out of a place, the right to carry things out of a place*:—Mid inláde and mid útláde *cum inductione et eductione*, Cod. Dip. Kmbl. iv. 209, 5.

út-lǽs, we; *f. Out-pastures, pasture-land away from the house*:—Seó útlǽs, Cod. Dip. Kmbl. vi. 214, 14, 21.

út-laga, an; *m. An outlaw*:—Útlaga *exlex*, Ælfc. Gr. 9, 62; Zup. 70, 5: *exul*, 9, 10; Zup. 39, 14; Wrt. Voc. i. 50, 58: 74, 26. Hé scel beón útlaga wið mé, Wulfst. 296, 10. Útlagen (-an? -ne?) *extorrem*, Hpt. 412, 73. Se ðe Godes útlagan hæbbe on geweálde, L. Eth. ix. 42; Th. i. 350, 1. Wé beódaþ ðæt útlagan Godes and manna of earde gewítan, L. C. S. 4; Th. i. 378, 11. Riht is ðæt ða útlagan weorþan, ðe tó Godes rihte gebúgan nellan, Wulfst. 269, 5. Útlagan *exules*, Hymn. Surt. 5, 25. [*Icel.* út-lagi.] v. út-iah.

út-lagian; *p.* ode *To outlaw, banish, proscribe*:—Útlagode mann Ælfgár eorl, Chr. 1055; Erl. 189, 3: 1069; Erl. 207, 7. Norðhymbra útlagodon heora eorl Tostig, 1064; Erl. 194, 14. Wið ðam ðe hí ǽfre ǽlcne Denisc[n]e cyning útlagede of Englalande gecwǽdon, 1014; Erl. 150, 15. [*Icel.* út-lægja *to banish.*] v. ge-útlagian.

út-lagu (?), e; *f. Outlawry*:—Útlaga, L. C. S. 13 tit.; Th. i. 382, 17. Æt eallan utlaga (-an? v. út-lah, **III**) þingan *de omnibus utlarie rebus*, W. ii. 3; Th. i. 489, 20.

út-lah; *adj. Out-lawed*; substantively, *an outlaw.* **I.** *of a person in respect to his own country*:—Gif hé man tó deáðe gefylle, beó hé útlah, L. E. G. 6; Th. i. 170, 10: L. Edg. H. 3; Th. i. 258, 19: L. Eth. i. 1; Th. i. 282, 15: L. C. S. 49; Th. i. 404, 11: Chr. 1048; Erl. 180, 3. Sý hé útlah (-laga, MS. B.), L. C. S. 45; Th. i. 400, 18. Se ðe

útlages weorc gewyrce (cf. *Icel.* göra útlaga verk), 13; Th. i. 382, 18. Gif hwá ámánsodne oþþe útlahne (ámánsumodne oþþe útlagene, MS. B.) hæbbe and healde, 67; Th. i. 410, 18. Se cyng cwæð hine útlage and ealle his suna, Chr. 1052; Erl. 181, 10. **I a.** *where it is stated with respect to whom one is an outlaw*:—Beó hé útlah wið God and ámánsumod fram eallum Cristendóme, Chart. Erl. 231, 15: Wulfst. 271, 24. Sý hé útlah (-laga, MS. B.) wið God and wið men, L. C. S. 39; Th. i. 398. 25. Beó se þeóf útlah wið eall folc, L. Eth. i. 1; Th. i. 282, 9: L. C. S. 30; Th. i. 394, 24. **II.** *of a person in respect to a country not his own*:—Hí ǽfre ǽlcne Deniscne cyng útlah of Englalande gecwǽdon, Chr. 1014; Erl. 150, 33. Ælc ðara landa ðe ǽnigne friðige ðæra ðe Ænglaland hergie beó hit útlah wið ús and wið ealne here, L. Eth. ii. 1; Th. i. 284, 18. Gif heora menn sleán úre ǽhta, ðonne beód hý útlage ge wið hý ge wið ús, ii. 7; Th. i. 288, 10. **III.** *calling for outlawry*:—Gif se Englisca beclypaþ Frenciscne mid útlagan þingan *si Anglicus appellet Francigenam de utlagaria*, W. ii. 3: Th. i. 489, 22. [*Icel.* út-lagr, út-laga.]

út-land, es; *n.* **I.** *a foreign country*:—Hé ðíne gemǽru gemiclade, ðú on útlandum áhtest sibbe *qui posuit fines tuos pacem*, Ps. Th. 147, 3. **II.** *out-lying land.* v. in-land. [Outlandes *foreign lands*, Mand. F. 3212: *Icel.* út-lönd *foreign countries; the outlying fields.*]

út-lenda, an; *m. A foreigner, stranger, not a native.* v. in-lenda:—Útlenda *extorris, alienus*, Hpt. Gl. 415, 76: *exul*, i. *peregrinus, alienus*, Wrt. Voc. ii. 146, 27. Exterres, i. *exules, peregrini* útlendan, *extranei* wreccean, 146, 5. v. next word.

út-lende; *adj. Foreign, strange, not native*:—Útlende ic eom and ælðeódig *advena ego sum et peregrinus*, Ps. Spl. 38, 17. Iacob útlænde (*accola*) wæs on eorðan Cham, 104, 21. Hé mǽnde be his feóndum ægðer ge inlendum ge útlendum, Ps. Th. 2, arg. [O. H. Ger. úz-lenti *exul*: *Icel.* út-lendr *foreign.*]

út-lendisc; *adj. Outlandish, foreign*; substantively, *a stranger*:—Sí hé landes man, sí hé útlendisc (*peregrinus*), Lev. 24, 22. Ðæt útlendisc man inlendiscan derie, L. O. D. 6; Th. i. 354, 28. Útlendisc *exul*, Ælfc. Gr. 9, 10; Zup. 39, 15. Útlendiscum *extraneo*, Scint. 193, 16. Hig noldon ðæt útlendiscum þeódum wǽre ðes eard þurh ðæt ðe swíðor gerýmed ðe hí heom sylfe ǽlc óðerne forfóre, Chr. 1052; Erl. 184, 31. Hé útlændisce hider in tihte, 959; Erl. 121, 3. [*Icel.* út-lenzkr *foreign.*]

út-líc; *adj. External, foreign*:—For ermþo ðære útlecan underþeódnesse (*subjection to those without*), Bd. 4, 16; M. 308, 30. Hé his ðeóde fram útlícre hergunge (*ab externa invasione*) álýsde, 4, 26; S. 603, 20.

útmest, uton, úton. v. útere, witon, útan.

út-ryne, es; *m. A running out* (*excursus*) *tó helle*, Hymn. Surt. 44, 21. Ðæs blódes útryne, Lchdm. i. 294, 17. Is se útryne (*what runs out*) swilce blódig wæter, ii. 202, 1. Útryne *exitum*, Scint. 224, 6. Útrynas *exitus*, Blickl. Gl.: Ps. Spl. 106, 33. Útrinas, 106, 35. [O. Frs. út-rene.]

út-scyte, es; *n. An out-shoot, outlet, place where a stream or road runs into another*:—Be bróce óð Pippelriðiges útscyte, Cod. Dip. Kmbl. v. 330, 20. 'Faraþ tó wega útscytum' . . . Útscytas ðara wega sind áteorung woruldlícera weorca, Homl. Th. i. 526, 11-14.

út-scytling, es; *m. A stranger*:—Mid útscytlinge ne dó ðú rǽd *cum extraneo ne facias consilium*, Scint. 200, 4.

út-siht, e; -sihte, an; *f. Diarrhoea, dysentery*:—Útsiht *diarria*, blódig útsiht *dissenteria*, Wrt. Voc. i. 19, 52, 53: ii. 141, 3. Wið útsihte, Lchdm. i. 114, 6: iii. 18, 1: 46, 13. Wið útsihte; ðysne pistol se ængel bröhte tó Róme ðá hý wǽran mid útsihte micclum geswæncte, 66, 6. Tácn be útsihte, ii. 170, 18. Gyf hé on útsihte sý, i. 260, 24. Wespasianus gefór on útsihte *Vespasianus profluvio ventris mortuus est*, Ors. 6, 7; Swt. 262, 28. Æfter útsihtan, Lchdm. ii. 180, 25. For útsihtan, 254, 3: 276, 22. Þurh ða wambe útsihtan, 224, 5. Wið útsiht and wið ðæs innoðes ástyrunge, i. 254, 7: iii. 294, 7. Hé bið gód wið lengtenádle and wið útsiht (*contra dysenteriam et diarrhoeam*), L. Ecg. C. 38; Th. ii. 162, 23. v. mete-útsiht.

útsiht-ádl, e; *f. Diarrhoea, dysentery*:—Sió útsihtádl cymð manegum of tó miclum útgange, Lchdm. ii. 278, 7. Wið útsihtádle, 320, 11.

út-síþ, es; *m. A going out* (lit. or fig.); *excessus*, Ps. Lamb. 115, 2: *exitium*, Wrt. Voc. ii. 144, 84: Hpt. Gl. 503, 35. Gǽst útsíþes geon the *spirit eager for departure from this world*, Exon. Th. 178, 9; Gú. 1241. Nágon hwyrft ne swice, útsíþ ǽfre ða ðǽr in cumaþ *those who come in there never have return or escape, never egress*, 364, 31; Wal. 79.

út-wǽpnedmann, es; *n. A stranger, outsider*:—Hí útwǽpnedmonna freóndscipes ceápiaþ *externorum sibi virorum amicitias comparent*, Bd. 4, 25; S. 601, 18.

út-wærc, es; *m. Dysentery, painful evacuation*:—Se útwærc, Lchdm. ii. 278, 4. Wyrd ðæt tó útwærce, 278, 15. Wiþ útwærce, 174, 1: 234, 30: 276, 20.

út-waru, e; *f. Defence away from home*:—Gif ceorlisc man geþeó ðæt hé hæbbe .v. hída landes tó cynges útware, L. Wg. 9; Th. i. 188, 6: L. R. 3; Th. i. 190, 21.

út-weald, es; *m. An outlying wood*:—An útwalda, Cod. Dip. Kmbl. ii. 73, 36.

út-weard; *adj. Outward, tending to the outside :*—Eoten wæs útweard, Beo. Th. 1526; B. 761. Dynt mið honde uutearde *alapam,* Jn. Skt. Lind. 18, 22. Wæs gesýne ðæt ða swaðo wǽron ǽrest útwearde ongunnen, Blickl. Homl. 207, 12. [*O. Frs.* út-ward.] v. útan-, úteweard, *and next word.*

út-weardes; *adv. Outwards, towards the outside :*—Suá bið sió costung ǽresð on ðæm móde, and ðonne fēreþ útweardes tó ðære hýde, óð ðæt nió út ásciet on weorc, Past. 11; Swt. 71, 5.

út-wícing, es; *m. A foreign pirate :*—Hugo eorl weard ofslagen innan Anglesēge fram útwíkingan, Chr. 1098; Erl. 235, 6.

W

wá. I. *adv. Woe, ill :*—Ða mē grame wǽron and mē wá dydon (cf. *Goth.* wai-dēdja), Ps. Th. 118, 38. (1) with dat. of person :—Ðē byþ ǽfre wá *it shall be ever ill with thee,* Nicod. 26; Thw. 14, 12: Beo. Th. 369; B. 183: Exon. Th. 444, 25; Kl. 52: Blickl. Homl. 61, 2. Him biþ æt heortan wá, Salm. Kmbl. 210; Sal. 104. Him wæs ǽghwǽr wá, Cd. Th. 285, 24; Sat. 342. Bið ðam men full wá, 40, 5; Gen. 634. Hí ne mihton ásecgan, hú wá ðám sáwlum byð, Wulfst. 147, 17. Ðæt him nǽsfre ǽr nǽre swá wá swá swá him ðā wæs, 235, 19. Ne weorðe ðē nǽsfre tó ðæs wá, ðæt ðú ne wēne betran andergilde, Prov. Kmbl. 41. (2) with gen. of the source of ill :—Wæs gehwæþeres waa, Met. 1, 25. (3) with dat. of person, and (a) gen. of source :—Ðæm folce wæs ǽgþres waa, ge ðæt . . . , ge eác ðæt . . . , Ors. 3, 7; Swt. 114, 31. Him wæs gehwædres wá, ge . . . ge . . . , Elen. Kmbl. 1253; El. 628. (b) with a clause :—Him bið wá on his móde, ðæt gē swá ánræde beóð, Homl. Skt. i. 17, 167. Ðá wæs ðam deófle waa on his móde, ðæt se man sceolde ða myrhðe geearnian, Hexam. 17; Norm. 24, 22. II. *interject.* (1) *woe, alas ;* vae, (a) with dat. of person :—Wá (wǽ, Lind. Rush.) ðam menn *uae homini illi,* Mk. Skt. 14, 21. Wá eów ðe hlihaþ, Blickl. Homl. 25, 24. Wá mē forworhtum, Exon. Th. 280, 20; Jul. 632. Waa ieów welegum, Past. 26; Swt. 181, 23. (b) with dat. of person and (a) gen. of cause of ill :—Wá ðæs gestreónes ðam ðe his mǽst hafaþ, Wulfst. 45, 19. Wá heom ðæs wærscipes, 268, 19; Hy. 2, 6; Exon. Th. 393, 11; Rä. 12, 8. Wá mē (*heu mihi*) ðære wyrde, Ps. Th. 119, 5. (β) with preposition :—Wá mánfullan (*ve impio*) for his misdǽdan, Wulfst. 45, 15. Wá (wǽ, Lind.) ðysum middangearde þurh swicdómas *vae mundo a scandalis,* Mt. Kmbl. 18, 7. ¶ *combined with* lá, wá lá, wá lá wá. (1) *well-a-way, well-a-day* (*Laym.* wa la wa : *A. R. O. and N.* wo la wo : *Chauc.* wai la wai) :—Wá lá! áhte ic mínra handa geweald, Cd. Th. 23, 32; Gen. 368. Wá lá ðære yrmðe and wá lá ðære woruldscame, Wulfst. 163, 3. Wá ús lá, Blickl. Homl. 153, 26. Wá lá wá *eheu,* Wrt. Voc. ii. 32, 44. Wá lá wá hú ic greów . . . , wá lá on hú micelre genihtsumnysse ic hwílum wæs, Homl. Skt. ii. 30, 189–193. Wá lá wá ðæt is sárlic *heu, proh dolor!* Bd. 2, 1; S. 501, 14. Wá lá wá ðæt ða ungesǽligan menn ne magon gebídan hwonne hē him tó cóme, Bt. 39, 1; Fox 212, 1. (2) expressing anger or contempt, *ah ;* vah :—Wá lá wá *euge, euge,* Ps. Lamb. 39, 16. Wá ðæt ðes tówyrpð Godes tempel, Mt. Kmbl. 27, 40. Wá (wǽ, Lind. Rush.) se tówyrpð ðæt tempel *va qui destruit templum,* Mk. Skt. 15, 29. [*Goth.* wai *vae:* *O. Sax. O. H. Ger.* wē : *Icel.* vei.] v. wei; wáwa, weá.

waa, waac. waad, waar. v. wá, wác, wád, wár.

wác; *adj.* I. *yielding, not rigid, pliant, fluid :*—Waac *lentus,* Wrt. Voc. i. 61, 35. Wæter, wác and hnesce (cf. ðæt hnesce and flówende wæter, Bt. 33, 4; Fox 130, 3), Met. 20, 93. Wác hreód ðe ǽlc hwiða windes mæg áwecggan, Past. 42; Swt. 306, 6. Gerd wǽcc ł bifiende (hreád ðæt wagende, Rush.) *harundinem quassatam,* Mt. Kmbl. Lind. 12, 20. Byrhtnóð wand wácne æsc (*the pliant ash-shaft*), Byrht. Th. 132, 68; By. 43. Iosue hí up áhēng on fíf wácum bógum *Iosue eos suspendit super quinque stipites,* Jos. 10, 26. II. *weak, feeble, wanting mental or moral strength, wanting courage :*—Wác bið se hyrde funden tó heorde, ðe nele ða heorde ðe hē healdan sceal mid hwæinæ bewerian, L. C. E. 26; Th. i. 374, 22. Wác bið ðæt geðanc on cristenum men, gif hē ne cann understandan þurh rihtne geleáfan ðæne ðe hine gescóp, Wulfst. 20, 9: Cd. Th. 40, 34; Gen. 649. On gewitte tó wác, Andr. Kmbl. 423; An. 212. Ne tó wác wiga, ne tó wanhýdig, Exon. Th. 290, 18; Wand. 67. Ðæt wæs wíglic werod : wác ne grētton in ðæt rincgetæl rǽswan herges, Cd. Th. 192, 18; Exod. 233. Ic, Ælfríc, munuc and mæssepreóst, swá þeáh wáccre ðonne swilcum hádum gebyrige, Homl. Th. i. 2, 12. Hæfde hire wácran hige Metod gemearcod, Cd. Th. 37, 16; Gen. 590. Sume láreówas sindon beteran ðonne sume ; sume sind wáccran, swá swá wē beóð, Homl. Th. ii. 48, 17. III. *poor, mean, not of great value or in high esteem ;* vilis. v. wác-líc, -ness :—Mid wáces olfendes hǽrum gescrýdde, Homl. Th. ii. 506, 23. Ðone wácan assan hē geceás, i. 210, 15. ii forealdode rǽdingbēc swíðe wáke, and .i. wác mæssereáf, Chart. Th. 430, 31. Hē ðis wáce forlēt, líf ðis lǽne, Chr. 975; Erl. 124, 31: Exon. Th. 53, 25; Cri. 856. Swá tealte beóð eorðan dreámas, and swá wáce syndan ǽhta mid mannum, Wulfst. 264, 4. Ða wácan fugelas, Homl. Th. ii. 462, 25. Hwí forgifð God

ðam wácum wyrtum swá fægerne wlite, 464, 16. Hwí dēst ðú ðē sylfe ðurh wáce þeáwas swilce ðú wyln sý, Homl. Skt. i. 8, 44. Hit is on worulde á swá leng swá wácre ; men syndon swicole, and woruld is ðe wyrse, Wulfst. 83, 10. Seó stów (*Abingdon*) næs wáccere ðonne (*inferior to*) formænig ðara ðe his yldran ǽr gefyrþredon, Lchdm. iii. 438, 11. Ælc man sylð on forandæge his góde win, and ðæt wáccre ðonne ða gebeóras drunciaþ, Homl. Th. ii. 70, 26. Gedroren is ðeós duguð eal, wuniaþ ða wácran, Exon. Th. 311, 4; Seef. 87. Fyrmest manna *primas,* wácost manna *infimas,* Ælfc. Gr. 9, 25; Zup. 50, 3. Ne eart ðú wácost (*minima,* Mt. 2, 6) burga, Homl. Th. i. 78, 14. On reáfe wáccust *habitu vilissimus,* Scint. 21, 7. Hwí wēnst ðú, ðonne nú ða wácestan gesceafta eallunga ne gewítaþ, ðæt seó seóleste gescaft mid ealle gewíte? Shrn. 198, 19. [*O. Sax.* wēk : *O. H. Ger.* weih *lentus, mollis, liquens, imbecillis, debilis : Icel.* veikr.] v. leoþu-, wund-wác.

wác, es; *n. A weakness :*—Nyste ic on ðám þingum ðe ðú ymbe specst fúl ne fácn, ne wác ne wom tó ðære dǽigtíde ðe ic hit ðē sealde, ac hit ǽgðer wæs ge hál ge clǽne búton ǽlcon fácne, L. O. 9; Th. i. 182, 3.

wacan ; *p.* wóc ; *pp.* wacen *To wake ;* but occurring mostly in the sense *to come into being, be born, spring :*—Sió mǽgburg ðe ic æfter wóc *the family from which I sprang,* Exon. Th. 401, 34; Rä. 21, 21. Abrahame wóc bearn of brýde *to Abraham a child was born of his wife,* Cd. Th. 167, 10; Gen. 2763: Beo. Th. 3925; B. 1960. Of ðam eorle wóc unrím þeóda, Cd. Th. 99, 15; Gen. 1646: 98, 29; Gen. 1637: Beo. Th. 2535; B. 1265. Ðæm feówer bearn in worold wócun, 119; B. 60. Wócon, Cd. Th. 131, 31; Gen. 2184. Þanon his eaforan wócan, bearn from brýde, 65, 5; Gen. 1061. Ǽr him sunu wóce, 70, 25; Gen. 1158. [He awoc (woc, 2nd MS.) of slæpe, Laym. 25566. Ðe king woc, Gen. and Ex. 2111. Aboute þe middel of þe nith wok Ubbe, Havel. 2093.] v. á-, on-wacan.

wacan *a watch.* v. wacen.

wáce ; *adv. Weakly.* (1) *feebly, faintly, without boldness :*—Ic mínum gewyrhtum wáce trúwige *I have feeble trust in my own merits,* Anglia xii. 502, 9 : Exon. Th. 52, 24; Cri. 838. (2) *feebly, inefficiently, without energy, remissly :*—Nú syndon cyrcan wáce gegrídode *churches are very inefficiently protected,* L. I. P. 25; Th. ii. 340, 11. Wē tó wáce hýraþ úrum Drihtne *we are too remiss in obedience to our Lord,* Wulfst. 91, 13 : Exon. Th. 50, 13; Cri. 799. Wē rihte getrýwða healdaþ tó wáce *we are too remiss in keeping good faith,* Wulfst. 91, 17. Hí míne heorde wáce begímdon, 190, 21. Ic wáccor hýrde Dryhtne ðonne mín rǽd wǽre, Exon. Th. 453, 18; Hy. 4, 16. Gif hē wáccor hý behwyrfð ðonne ðæt hē tó ágenum teleþ, L. Edg. S. 1; Th. i. 272, 10. [*O. H. Ger.* weiho *enerviter.*]

wacen (-an, -on, -un), e; *f.* I. *wakefulness, sleeplessness :*—Ðone intingan ðínre unrōtnisse and ðínre wacone (wǽcene, Bd. M. 128, 23) *tuae moestitiae et insomniorum causam,* Bd. 2, 12; S. 513, 41. II. *a watch, vigil :*—'Wel ðú dēst ðæt ðú nalæs ðē slǽpe forgeáfe, ac má woldest wæccan (weacenum, Bd. M. 354, 7) and gebedum ætfeolan.' Cwæþ hē : 'Ic wát ðæt mē ðæs is micel ðearf, ðæt ic hálwendum weacenum ætfeole,' Bd. 4, 25; S. 601, 1–3. III. *a watch, a division of the night :*—Ðíu feórða waccen (feórþe ðære wacone, Rush.) *quarta vigilia,* Mt. Kmbl. Lind. 14, 25. Ymb ða feárða wacune (wacan, Lind.) *circa quartam uigiliam,* Mk. Skt. Rush. 6, 48. On ða æfterra wacone (waccane, Lind.) *in secunda uigilia,* Lk. Skt. Rush. 12, 38. IV. *a watch, guard :*—Haldende wacone (wacana, Lind.) næhtes *custodientes uigilias noctis,* 2, 8. V. *a rousing, an incitement :*—Wacana mægna incitamenta virtutum, Rtl. 63, 36. v. on-wacan; *f.* ; wæcen.

wacian ; *p.* ode *To watch, wake :*—Ic wacige *uigilo,* Ælfc. Gr. 41; Zup. 245, 10. (1) *to remain awake, not to sleep :*—Gif wē tó lange waciaþ, wē áteoriaþ, Homl. Th. i. 488, 34. Ic waecade *vigilavi,* Ps. Surt. 101, 8. Hwæðer hē wacode ðe slēpte, Bd. 2, 12; S. 513, 39. On middere nihte gewurdon on slǽpe Pictauienscisce bepǽhte, ðæt of ealre ðære menigu án man ne wacode, Homl. Th. ii. 518, 26. Ealle oþþe hefige slǽpe swundon, oþþe tó synne wacedon *omnes aut somno torpent inerti, aut ad peccata vigilant,* Bd. 4, 25; S. 601, 12. Sceal se man wacyan ealle ða niht, ðe ðone drenc drincan wille, Lchdm. iii. 6, 4. (1 a) *of the eye, to be freed from obstruction, to open :*—Gif eágan forsetene beóð, genim hræfnes geallan . . . drýp on ðæt eáge . . . ðonne wacaþ ðæt eáge (*the eye opens again*), Lchdm. iii. 2, 24. (1 b) *to be alert :*—Se sláwa ongit hwæt him ryht bið tó dónne, swelce hē ealneg wacige, and swá ðeáh hē áslǽwaþ, for ðæm ðe hē náwuht ne wyrcð *piger enim recte sentiendo quasi vigilat, quamvis nil operando torpescat,* Past. 39; Swt. 283, 7. Hē wecð hine selfne, ðæt hē wacie on ðære geornfulnesse gódra weorca (*ut studio bonae actionis evigilent*), 64; Swt. 461, 14. Wacige, 461, 16. Ðæt heó mihte beón ácenned, and wacian, and árísan, and faran of stówe tó oþerre, Blickl. Homl. 19, 22. (2) *to keep one's self awake* or *alert because there is special need of attention, to watch, be on the watch, be on guard :*—Ic ðē tó wacie (waecio, Ps. Surt.) *ad te vigilo,* Ps. Th. 62, 1. In ídelnisse węciaþ ða haldaþ hié *in vanum vigilant qui custodiunt eam,* Ps. Surt. 126, 1. Gif hē wiste hwænne se þeóf cuman wolde, witodlíce hē wacude (*uigilaret*), Lk. Skt. 12, 39. Hine twēgen ymb weardas wacedon, Exon. Th. 109, 6; Gú. 86. Wacodon menn,

swá swá hit gewunelíc is, ofer án deád líc, Homl. Skt. i. 21, 290 : Blickl. Homl. 149, 6. Geheald húsa sélest, . . . waca wið wráþum, Beo. Th. 1324; B. 660. Waciaþ (*vigilate*) and gebiddaþ eów, Mt. Kmbl. 26, 41. Wacigeaþ, 24, 42. Hé beóde ðam durewearde, ðæt hé wacige, Mk. Skt. 13, 34. Is micel ðearf ðæt se reccere geornlíce wacige (*solerter invigilet*), Past. 19; Swt. 141, 13. Ic bidde eów, ðæt gé wacian mid mé, Blickl. Homl. 139, 20. Ne mihtest ðú áne tíde wacian, Mk. Skt. 14, 37. Wacigean, Mt. Kmbl. 24, 43. Man sceal wacigean and warnian, Wulfst. 90, 2. Tó wacene *ad vigilandum*, Rtl. 85, 1. Ic stande ofer hig waciende (*vigilando*) for þeófan, Coll. Monast. Th. 20, 29. Hé wæs waciende on gebede *erat pernoctans in oratione*, Lk. Skt. 6, 12. Se þeów ðe hláford sint wacigenne (*uigilantem*), Scint. 116, 9. Hyrdas wæron waciende and nihtwæccan healdende ofer heora heorda, Lk. Skt. 2, 8. (2 a) in a bad sense, *to watch, be on the watch* to injure :—Wacaþ se ealda, Fragm. Kmbl. 61; Leás. 32. [Þe herdes þe wakeden ouer here oref . . . were herdes wakiende and wittende here oref, O. E. Homl. ii. 31, 22–27. Ðus agen alle gode herdes to wakegen gostliche, 41, 5. Festen, wakien, A. R. 6, 8. His cnihtes wakeden alle nihte, Laym. 9859. Þat haveth fele nihtes waked, Havel. 2999. His liche was waked, Gen. and Ex. 2516. Þet uolk þet late louieþ to soupi, and to waki be niȝte, Ayenb. 52, 18. O. Sax. O. L. Ger. wakôn : O. H. Ger. wahhôn. Cf. Goth. wakan : O. H. Ger. wahhên : Icel. vaka.] v. á-, be-, morgen-, ofer-, þurh- (v. Blickl. Homl. 227, 7) wacian.

wácian ; *p.* ode. **I.** of persons, *to be or become weak, want resolution* or *courage.* v. wác, **II** :—Ðonne se heretoga wácaþ, ðonne biþ eall se here swíðe gehindred, Chr. 1003; Erl. 139, 12. Be ðam mihte man oncnáwan, ðæt se cniht nolde wácian æt ðam wíge, Byrht. Th. 132, 2; By. 10. **II.** of things, *to be or become weak, not able to endure, to fail* :—Ne wáciaþ ðás geweorc, Exon. Th. 351, 26 ; Sch. 86. Teoriaþ hwílum, wáciaþ wordbeót, 469, 22 ; Hy. 11, 6. **III.** *to become poor or mean.* v. wác, **III** :—Wachiaþ *vilescunt*, Hpt. Gl. 462, 52. [Þa ældede þe king and wakede an aðelan (failede his mihte, 2nd MS.), Laym. 2938. Heo weoren swa drunken, þ wakeden heore sconken, 13466. Bruttes wokeden (*lost heart*) þa, 26996. His heorte gon to wakien, 19978. Þi strengþe wokeþ, Misc. 101, 15. Piers P. wakie, wokie *to soften* : O. H. Ger. weihhên, weihhôn *infirmari, emarcescere.*] v. á-, ge-wácian ; wæcan.

wác-líc ; *adj. Poor, mean, of little dignity* or *worth, paltry* : v. wác, **III** :—Wáclíc *vilis*, Wrt. Voc. i. 28, 64 : Hpt. Gl. 523, 74: *inutile, contemptum*, 470, 22. Ðú wilt habban ealle fægere ðing and ácorene, and wilt ðé sylf beón wáclíc and unwurð, Homl. Th. ii. 410, 20 : 372, 8. Hwæþer ðæt nú sié tó talianne wáclíc and unnyt ðætte nytwyrþost is eallra ðissa woruldþinga ? *num imbecillum, ac sine viribus aestimandum est, quod omnibus rebus constat esse praestantius ?* Bt. 24, 4 ; Fox 86, 16. Wé mihton eów secgan áne lytle bysne, gif hit tó wáclíc næere, Homl. Th. i. 40, 27. Wáclíc bið him swá lytel tó sendenne, 400, 20. Hí wædliende on ánum wáclícum wæfelse férdon, 62, 29. Him þúhte tó wáclícre dæde, ðæt hé fordyde hine ænne, Homl. Ass. 96, 142. Ðæt gecynd ðe hí ær wáclíc tealdon, Homl. Th. i. 38, 30. Manega Lazaras gé habbaþ . . . Ðeáh ðe hí sýn wáclíce geðúhte, 334, 30. Wudehunig and óðre wáclíce ðigena, 352, 8. Sume men syllaþ cyrcan tó hýre swá swá wáclíce mylna, Homl. Skt. i. 19, 249. On wáclícum ðingum wícnian *to perform menial offices*, ii. 170, 25. Wáclícum *foedis*, Germ. 395, 78. Hí unrædlíce férdon on heora ídelum lustum and wáclícum gebærum, Ælfc. T Grn. 17, 16. [*Icel.* veik-ligr *vilis.*] v. un-wáclíc.

wáclíce ; *adv.* **I.** *weakly, feebly* :—Wáclíce *enerviter*, Wrt. Voc. ii. 29, 32 : *enerviter, turpiter*, 143, 56. **II.** *poorly, meanly, cheaply* :—Eówer reáf ne beó tó ranclíce gemacod, ne eft tó wáclíce, ac weriȝe gehwá swá his háde tó gebyriȝe, L. Ælfc. C. 35 ; Th. ii. 358, 7. Gehwam sceamaþ, gif hé geladod bið tó woruldlícum gyftum, ðæt hé wáclíce gescrýd cume, Homl. Th. i. 528, 23. Wáclícor *vilius*, R. Ben. Interl. 92, 4. *Diminutiva* syndon wanigendlíce . . . *bene* wel, and of ðam is *belle* ná ealles swá wel, *bellissime* ealra wáclícost, Ælfc. Gr. 38 ; Zup. 231, 4. [Gif þu werest te wocliche, A. R. 294, 5. The poure þat beoð wacliche iȝeouen and biset nuele, H. M. 9, 18. O. H. Ger. weihlícho *enerviter.*] v. un-wáclíce.

wác-mód, *adj.* **I.** *of weak disposition, morally weak* :—Ða hnescan (*vel* wácmód, *written above the line*), ðæt synd ða ðe náne stíðnysse nabbaþ ongeán leahtras, Homl. Skt. i. 17, 40. **II.** *faint-hearted, pusillanimous* :—Gif yrmð getímaþ wácmód ná wuna ðú sí *calamitas contigerit, pusillanimis non existas*, Scint. 172, 6. Crist lærde ðæt man tó wácmód (cf. Mt. 24, 6 : Mk. 13, 7) ðonne ne wurde, Wulfst. 89, 6. On óðre wísan sint tó monianne ða ofermódan, on óðre wísan ða earmheortan and ða wácmódan (*pusillanimes*), Past. 32 ; Swt. 209, 3. Beó hit eal mid gemete ðe læs ðe ða wácmódan beón ormóde *omnia mensurate fiant propter pusillanimes* (for ðam wácmódum, R. Ben. Interl. 82, 7), R. Ben. 74, 1. Sý fultum geseald ðám wácmódum and ðám unstrangum, ðæt hí mid unrótnesse ða hýrsumnesse ne ðón *imbecillibus procurentur solacia, ut non cum tristitia hoc faciant*, 58, 17. Secgaþ ðám wácmódum, ðæt hí beón gehyrte, and nánðing ofdrædde *say to them that are of a fearful heart, Be strong, fear not* (Is. 35, 6), Homl. Th. ii. 16, 15. [O. Sax. wêk-mód.]

wácmódness, e ; *f.* **I.** *weakness of character, moral weakness* :—Ðý læs sió scyld, ðe hiene costaþ, for his luste and for his wácmódnesse hine ofersuíðe *ne vitium, quod tentat, mollitie delectationis subigat*, Past. 13 ; Swt. 79, 22. **II.** *faintheartedness, want of courage, pusillanimity, cowardice* :—Ignauia, ðæt is wácmódnys, Wulfst. 52, 18. Se fífta leahtor is unrótnys ðissere worulde. Of ðam bið ácenned wácmódnys, . . . and his sylfes orwénnys, Homl. Th. ii. 220, 19. Of wácmódnesse and of unbieldo oððe of untrymnesse módes oððe líchoman *infirmitate*, Past. 21 ; Swt. 159, 1. Gedréfde mid wácmódnesse *pusillanimitate turbatos*, 32 ; Swt. 213, 6. For wácmódnesse *from want of courage*, 40 ; Swt. 289, 3. Ongeán módstaðolnysse and módes strencðe se deófol sendeþ wácmódnesse and lýðerne earhscype, Wulfst. 53, 12. **III.** *weakness, feebleness* :—Sí forescéawod wácmódnyss (*inbecillitas*), nateshwón heom (*old men and children*) stíðnis regoles ná sí gehealdan on fódum, R. Ben. Interl. 68, 14. Untrumera wácmódnesse, 72, 3. [Cf. O. H. Ger. weih-môtî *pusillanimitas, teneritudo.*]

wácness, e ; *f. Meanness of condition, mean estate ; vilitas*, v. wác, **III** :—Horsþénes wácnys (*printed* wænys) *mulionis vilitas*, Hpt. Gl. 438, 70. Mid ealre wácnisse hylde *omni vilitate contentus*, R. Ben. Interl. 33, 14. Hwí forgifð God ðam wácum wyrtum swá fægerne wlite, . . . búton for ðan ðe wé sceolon mid wácnysse and sóðre eádmódnysse ða heofenlícan fægernysse geearnian, Homl. Th. ii. 464, 18. Hí bædon, ðæt ða gymstánas (*gems which had been pebbles before a miraculous change*) áwendon tó heora wácnysse, i. 68, 19. [Þat te strengðe of þe helpe mi muchele wacnesse, O. E. Homl. i. 273, 14. Þe ueond þurh hire (*Eve's*) word understond hire wocnesse, A. R. 68, 6.]

-wacnian. v. á-, on-wacnian ; wæcnan.

wacol (-ul, -el); *adj. Watchful, vigilant* :—Wacol *vigil*, Wrt. Voc. i. 75, 64. Wacul *vigil vel vigilans*, 46, 2. Ðes and ðeós wacole (-ele) *hic et haec uigil*, Ælfc. Gr. 9, 8 ; Zup. 39, 3. Ða ðe cariaþ mid wacelum móde hú hí óðra manna sáwla Gode gestrýnan, Homl. Th. ii. 378, 2. Gewinn wið ðone wacolan feónd, 560, 28. Wacele (-ole) beón on gódum weorcum, Homl. Ass. 53, 86. Wacule (-ole), R. Ben. 2, 7. Mótan ða hyrdas beón swíðe wacole, Wulfst. 191, 12. Uigilantius, ðæt is on Englisc wacolre, Homl. Th. ii. 118, 13. [O. H. Ger. wachal *uigil* : Icel. vökull.] v. ær-, þurh-wacol.

wacollíce ; *adv. Watchfully, vigilantly* :—Hé (*Gregory*) wæs swíðe wacol on Godes bebodum, and hé wacollíce ymbe manegra ðeóda þearfe hogode, Homl. Th. ii. 118, 15.

wacon. v. wacen.

wacor ; *adj. Watchful, vigilant* :—Se ðe wære slápol, weorðe se ful wacor, Wulfst. 72, 14. Beó ðú wacor *esto vigilans*, Past. 58 ; Swt. 445, 20. Sint tó manienne ða ðe hiera synna onfunden habbaþ, ðætte hié mid wacore móde (*vigilanti cura*) ongietan . . , 52 ; Swt. 405, 8. Ðonne móton ða hyrdas beón swíðe wacore, L. C. E. 26 ; Th. i. 374, 27: L. I. P. 6 ; Th. ii. 310, 27. [Uigilaui, ich was waker, seið Dauid, A. R. 142, 25. Wyþ þeoues þu most beo waker and snel, Misc. 97, 150. Wakyr *pervigil*, Prompt. Parv. 514. O. H. Ger. wachar *vigil, pervigil* : Icel. vakr *watchful, alert ; nimble.*] v. eád-wacer ; wæccer.

wacorlíce ; *adv. Watchfully, vigilantly, carefully* :—Sint tó læranne ða ofersprǽcean ðæt hié wacorlíce (*vigilanter*) ongieten . . . , Past. 38 ; Swt. 277, 4. Ðonne ðæt mód wacorlíce stiéreþ ðære sáwle *cum mens vigilanter animam regit*, 56 ; Swt. 433, 4. Is ús swíðe wocorlíce tó geðenceanne *vigilanti consideratione pensandum est*, 49 ; Swt. 385, 24.

wacsan. v. wæscan.

wác-scipe, es ; *m. Remissness* :—Ðæt hí stýran ælcum ðara ðe ðis ne gelǽste and mínra witena wed ábrecan mid ænigum wácscipe wille, L. Edg. S. 1 ; Th. i. 272, 7. Cf. wáce (2).

wacu *a waking, wake, watch.* [Heo hefde ileaned one wummone to one wake on of hore weaden, A. R. 314, 27. Heó haveþ daies care and nihtes wake, O. and N. 1590.] v. niht-wacu.

wád, es ; *n. Woad*, a plant much used for dyeing, which circumstance may account for the appearance of the word as a gloss to some of the following Latin words :—Ðis wád *hic sandyx*, Ælfc. Gr. 9, 69 ; Zup. 72, 14. Wyrt oððe wád *sandix* (the passage to which this gloss belongs is Vergil Eclogae, iv. 45, quoted by Aldhelm), Wrt. Voc. ii. 87, 33. Wád *sandix*, i. 32, 6 : 68, 70 : 79, 42. Waad *fucus*, 32, 7. Dolhsealf. Genim wádes croppan, Lchdm. ii. 94, 11. Of wáde ł hæwenre deáge *ex hyacintho* (cf. wáde *iacincto*, Anglia xiii. 29, 52. Cf. O. H. Ger. weitîn *iacinctus*), Hpt. Gl. 431, 26. Wið bryne, wád wyl on buteran, smire mid, Lchdm. ii. 132, 1, *and see* i. 174, 1–5. Man mæg on hærfeste wád spittan, Anglia ix. 261, 16. ¶ The growth of woad seems marked by the occurrence of the word in such forms as *wád-beorh, wád-denu, wád-lond* in charters :—Of ðære díc on wádbeorgas ; of wádbeorgan, Cod. Dip. Kmbl. iii. 77, 15. Æt wádbeorhe, 82, 29. On wádbeorh ; of wádbeorhge, 232, 36. On wáddene ; andlong wáddene, vi. 137, 12. Ðæt wáðlond, iii. 390, 17: 381, 5. [O. Frs. wede : O. H. Ger. weit *sandix*.]

wadan ; *p.* wód, *pl.* wódon ; *pp.* waden *To go, pass, proceed.* **I.** of actual movement, (a) absolute :—Wód wíges heard, . . . and wið ðæs beornes stóp, Byrht. Th. 135, 38 ; By. 130 : 139, 13 ; By. 253. Brimmen wódon, 140, 29 ; By. 295. Ðá com hæleða þreát wadan, Andr. Kmbl.

2543; An. 1273. Gesión wadan wǽgflotan, Elen. Kmbl. 491; El. 246. (b) with prepositions:—Hit ðurh hróf wadeþ, Salm. Kmbl. 824; Sal. 411. Ic wód ofer waþema gebind, Exon. Th. 287, 34; Wand. 24. Wægn ne be grunde wód, 404, 29; Rä. 23, 15. Hit ofer eall wód and eode, Nar. 15, 22. Ðæt feórðe cyn wód on wǽgstréam, Cd. Th. 197, 22; Exod. 311. Hé wód þurh ðone wælréc, Beo. Th. 5315; B. 2661. Ðe wód under wolcnum, 1432; B. 714. Wódon wælwulfas west ofer Pantan, ofer scír wæter, Byrht. Th. 134, 38; By. 96. Ðis leóhte beorht cymeþ ofer misthleoþu wadan ofer wǽgas, Exon. Th. 350, 9; Sch. 61. Gewát him se æðeling wadan ofer wealdas, Cd. Th. 174, 30; Gen. 2886. On sǽ wadan, 51, 22; Gen. 830. Hé lét his francan wadan þurh ðæs hysses hals, Byrht. Th. 135, 59; By. 140. (c) with acc. of the way traversed: —Gé wadaþ wídlástas, Andr. Kmbl. 1353; An. 677. Hé wód (woð, MS.) geócrostne síð, Cd. Th. 254, 23; Dan. 616. Wadan wræclástas, 272, 17; Sat. 121: Exon. Th. 286, 23; Wand. 5. II. fig.:—Ða ðe on eallum ðingum wadaþ on hiora ágenne willan, and æfter hiora líchoman luste irnaþ, Bt. 41, 2; Fox 246, 23. Ða men ðe on eallum þingum wadaþ on heora ágenum willan, and on heora lustum heora líf áspendaþ, Homl. Skt. i. 17, 239. Ðæt seó wyrd on ðínne willan wóde, Bt. 20; Fox 72, 19. [O. Frs. wada: O. H. Ger. watan: Icel. vaða.] v. an-, ge-, geond-, ofer-, on-, þurh-wadan.

wád-sǽd, es; n. Woad-seed:—Línsêd sáwan, wádsǽd eác swá, Anglia ix. 262, 11.

wád-spitel a woad-spade, Anglia ix. 263, 6. v. spitel.

wadung, e; f. Going, travelling:—Ús sceamaþ tó secgenne ealle ða sceandlícan wíglunga ðe gé dwǽsmenn drífaþ oððe on wífunge oððe on wadunge (see, for instance, Lchdm. i. 328, 330, where the virtues of various parts of a badger in case of journeying are stated, and 102, ii. 154 for similar passages in reference to mugwort. Cf. also: Sind manega mid swá miclum gedwylde befangene, ðæt hí cépaþ be ðam mónan heora fær, Homl. Th. i. 100, 23), Homl. Skt. i. 17, 102.

wé, wæbb, wæbbung. v. wá, web, webbung.

wǽcan; p. wǽhte; pp. wǽht, wǽced To weaken, afflict, oppress:— Se foresprecena hungur Bryttas swýþe wǽhcte Brittones fames praefata magis magisque adficiens, Bd. 1, 14; S. 482, 16. Ðý lǽs his yrre ús yrmþum swence and wǽce ne ejus ira nos damnis affligat, 4, 25; S. 601, 40. Scealt ðú ðínne líchaman þurh forhæfdnysse wǽccan, Guthl. 5; Gdwin. 32, 9. Ðá hé mid swinglum and tintregum wǽced wæs cum tormentis afficeretur, Bd. 1, 7; S. 477, 45. Mid ðý seó mǽgð wǽced wæs mid wæle provincia cum clade premeretur, 3, 30; S. 561, 37. Mid ða ádle wǽced and swenced quo affectus incommodo, 4, 31; S. 610, 20: Exon. Th. 410, 27; Rä. 29, 5. Ða men beóþ mid hriþingum swíþe strangum wǽcede, Lchdm. ii. 258, 3. [O. H. Ger. weihen; p. weihta mulcere, enervare.] v. á-, ge-, on-wǽcan; wácian.

wǽcca. v. hálig-wǽcca.

wǽccan; p. wǽhte To watch, wake; except in the Northern specimens the verb seems to occur only in the present participle, wacian (q.v.) being used elsewhere:—Wæccaþ (-as, Lind.) gé vigilate, Mt. Kmbl. Rush. 24, 42. Wæcceþ (wæcas, Lind.), 26, 41. Wæccas, Mk. Skt. Lind. Rush. 13, 37. Ðæt hé wæcce (gewæhte, Lind.) ut uigilet, Rush. 13, 34. Suá huoeðer wé woæca ⁊ wé slépa sive vigilemus sive dormiamus, Rtl. 28, 37. Wæcca hé walde (hé wæcende beón walde, Rush.) vigilaret, Mt. Kmbl. Lind. 24, 43. Walde wæcce (wæca, Lind.), Lk. Skt. Rush. 12, 39. For hwon hé wæccende sǽte quare pervigil sederet, Bd. 2, 12; S. 513, 38: Cd. Th. 191, 12; Exod. 213: Beo. Th. 1420; B. 708. Hé wæccende ða niht on hálgum gebedum áwunode, Guthl. 5; Gdwin. 34, 14. Of scondlícum geþóhte ðæs wæccendan (vigilantis) up cymeþ seó bysmrung slǽpendes . . . ðæt hé wæccende ðóhte, ðæt hé nó witende árǽfnode, Bd. 1, 27; S. 497, 5-9. Heó wæs wæccende dæges and nihtes, Blickl. Homl. 137, 22. Mid wæccende gýmen[ne], L. E. I. prm.; Th. ii. 400, 31. Se fand wæccendne wer, Beo. Th. 2540; B. 1268. Wæccende, 5674; B. 2841. Hé hêt mec wæccende wunian, Exon. Th. 422, 18; Rä. 41, 8. Ðæt gé wæccende wearde healden, 282, 13; Jul. 662. Ða þeówas ðe se hláford wæccende (-o, Lind.: wæcende, Rush. uigilantes) gemêt, Lk. Skt. 12, 37: Blickl. Homl. 145, 6. [Þ heo wecchinde ham werien, Marh. 15, 33.] v. ge-wæccan; þurh-wæccende, Lk. Skt. Lind. 6, 12.

wǽcce, an; f. I. wakefulness, sleeplessness:—Gif men sié micel wǽce getenge, popig gegníd, smire ðínne andwlitan mid, . . . raþe him biþ sió wæcce gemetgod, Lchdm. ii. 152, 12-14. Wæcæ, 16, 19. Dæges and nihtes ic swanc on hǽtan and on wæccan die noctuque aestu urebar, fugiebatque somnus ab oculis meis, Gen. 31, 40. Tó slǽpe. Gáte horn under heáfod geléd, weccan (wæccan, MS. B.) hé on slǽpe gecyrreþ, Lchdm. i. 350, 21. Hí singale wæccan þrowiaþ, ii. 258, 7. Hú micel sár, and hú micele wæccan, and hú micle unrótnesse hé hæfþ, Bt. 31, 1; Fox 110, 30. II. where the wakefulness is intentional, watching, watchfulness, a watch, vigil:—Wæcce vigilia, Wrt. Voc. i. 75, 65: excubia, Engl. Stud. xi. 65, 28. Gé sceolon witan, ðæt twá wæccan synd; án is ðæs líchaman, óðer ðæs módes. Ðæs líchaman wæcce is ðonne wé waciaþ on cyrcan æt úrum úhtsange, ðonne óðre men slápaþ . . . Ðæs módes wæcce is micele betere, ðæt se man hogie hú hé gehealden beó

wið ðone deófol, Homl. Ass. 51, 35-49: R. Ben. 35, 2. Man wacaþ tó oft on unnyt . . . ; and micle betere is ælcum cristenum men, ðæt hé náne wæccan æt cyrican næbbe, ðonne hé ðǽr wacyge mid ænigan gefleorde. Ac se ðe rihtlíce his wæccan healdan wylle, . . . wacie hé and gebidde hine georne, ðonne fremaþ him seó wæcce, Wulfst. 279, 11-17. Gif hwelc mon fæste oþþe nytte (Cockayne alters to nihte, but this is unnecessary; see beginning of preceding passage) wæccan dó, Shrn. 104, 29. Tó wæccum ad excubias, vigilias, Hpt. Gl. 488, 37. On hálgum wæccan vigiliis sanctis, Bd. 4, 25; S. 600, 15. Wæcceum, Ps. Th. 76, 4. Wæccan excubias, Wrt. Voc. ii. 92, 48. Weardsetl oððe wæccan, 30, 11. Gif hwá his wæccan (vigilias) æt ænigum wylle hæbbe, oððe æt ænigre óðre gesceafte, búton æt Godes cyricean, L. Ecg. P. iv. 19; Th. ii. 210, 11. III. a division of the night, a watch:—Drihten com tó his leorningcnihtum on ðære feórðan wæccan. Án wæcce hæfð þreó tída; feówer wæccan gefyllað twelf tída; swá fela tída hæfð seó niht, Homl. Th. ii. 388, 13. On ðære æfteran wæccan in secunda uigília, Lk. Skt. 12, 38. Embe ða feórðan wæccan, Mk. Skt. 6, 48. [Noðing ne maked wilde uleschs tommure þen deð muche wecche; vor wecche is ine holie write ipreised. . . . Ure Louerd teihte us wecche, A. R. 144, 1-9. Temien hire fleschs mid wecchen, 138, 6. Wiþþ fassting, and wiþþ wecche, Orm. 1451. O. H. Ger. wacha: Icel. vaka.] v. cyric-, niht-, úht-, ungemet-wæcce; wacen.

wæccend (?), es; m. A watcher, watchman:—Ne mæg hí cynlíce wæccend . . . weard gehealdan in vanum vigilant qui custodiunt eam, Ps. Th. 126, 2.

wæccendlíc. v. þurh-wæccendlíc.

wæccer, wæcer; adj. Vigilant, watchful:—Þurh niht wæcer (printed wæter) pernoctans (Lk. 6, 12), Wrt. Voc. ii. 74, 42. Mid wæccere (wæccre, Bd. M. 84, 2) móde is tó smeágeanne vigilanti mente pensandum est, Bd. 1, 27; S. 496, 2. v. wacor.

wæcen, e; f. A waking, watch:—Wecen vigilia, Wrt. Voc. i. 46, 4. Waecene vigilias, Ps. Surt. 76, 5. v. wacen.

wæcer, wæcian, Wǽclinga ceaster, Wǽclinga strǽt. v. wæccer, wacian, Wætlinga ceaster, Wætlinga strǽt.

wæcnan; p. ede To waken, arise, spring:—Ne wæs hit lenge, ðæt se ecghete (secg hete, MS.) æfter wælníðe wæcnan scolde, Beo. Th. 171; B. 85. Of idese biþ eafora wæcned, Cd. Th. 144, 20; Gen. 2392. [Þat ter walde wakenen of wif and weres somninge worldes weole, H. M. 31, 5. Þu art walle of waisdom, ant euch wunne wakeneð ant waxeð of þe, Marh. 11, 1. He began to wakne, Havel. 2164. Ther wakeneth in the world wondred ant newe, P. S. 152, 17. Also transitive:—Itt iss waccnedd off slæp þurh þatt te faderr stireþþ itt and waccneþþ, Orm. 5845. Thai wakned Crist, Met. Homl. 134, 9. Goth. ga-waknan to become awake: Icel. vakna.] v. á-, on-wæcnan, and next word.

wæcnian. v. á-, on-wæcnian, and preceding word.

wæd, es; n. A ford, shallow water, water that may be traversed (cf. wadan, and the forms wade, wath in place-names, e.g. Biggles-wade, Longwathby); poet. a body of water, sea:—Bí wædes ófre, Exon. Th. 360, 22; Wal. 9. Wyllelm king lǽdde scypferde and landfyrde tó Scotlande . . . him sylf mid his landfyrde férde inn ofer ðæt wæd (æt ðam gewæde, MS. E. Cf. wath a ford, Jamieson's Dict.), Chr. 1073; Erl. 211, 25. Wit on sǽ wǽron, óþ ðæt unc flód tódráf, wado weallende, Beo. Th. 1096; B. 546: 1166; B. 581. Sǽholm oncneów ðæt ðú gife hæfdes . . . wǽdu swæðorodon, Andr. Kmbl. 1066; An. 533. Wé on sǽbáte ofer waruðgewinn wada cunnedon faroðrídende, 878; An. 439: Beo. Th. 1021; B. 508. Ðonne ic (a swan) wado drêfe when I trouble the waters (i.e. swim), Exon. Th. 389, 24; Rä. 8, 2. [A wathe vadum, flustrum, Cath. Angl. 410, and note: O. H. Ger. wat, furt vadum: Icel. vað a ford.] v. ge- (geueada vada brevia, Wrt. Voc. ii. 123, 17), mearc-, seolh-wæd.

wǽd, e; f.: wǽde, es; n. I. referring to the dress of human beings. (1) a weed (as in palmer's, widow's weeds), an article of dress, a garment:—Martinus mê bewǽfde mid ðyssere wǽde, Homl. Th. ii. 500, 34. Ne cume hé búton his oferslipe, ne hé þénige búton ðære wǽde, L. Edg. C. 46; Th. ii. 254, 11. In wǽde (vestimentum) ald . . . from wǽde (vestimento), Mt. Kmbl. Lind. 9, 16. Gehrán woede (wédum, Rush.) his tetigit uestimentum ejus, Mk. Skt. Lind. 5, 27. Ungiearuad woede gímungalícum non vestitum veste nuptiali, Rtl. 108, 1. Woede háluende vestimentum salutare, 103, 22. Hé næfþ ða neódþearfe áne, ðæt is wist and wǽde, Bt. 33, 2; Fox 124, 17. Woedo uestimenta, Mk. Skt. Lind. 9, 3. Ic wæs nacod, nolde gé mê wǽda tíþian, Wulfst. 288, 33. Wǽda leásne, Cd. Th. 53, 27; Gen. 867: 256, 2; Dan. 634: Met. 25, 32. Ðú wǽda tylast, Homl. Th. i. 488, 26. Of ungemete wiste and wǽda, Met. 25, 39. Hé hine gescyrpte mid eallum ðám wlitegestum wǽdum, Bt. 25, 39. Fox 100, 26: Cd. Th. 58, 5; Gen. 941. Hí hine wǽdon bereáfodon, Homl. Th. i. 430, 2. Gif dynt sweart sié búton wǽdum if a blow cause a bruise in a part not covered by the clothes, L. Ethb. 59; Th. i. 18, 3. Binnan wǽdum in a part covered by the clothes, 60; Th. i. 18, 5. Ofer wǽda míne super vestem meam, Ps. Spl. 21, 17: Cd. Th. 52, 20; Gen. 846: Met. 8, 23. Forlǽt eal ðæt ðú áge búton wiste and wǽda, Prov. Kmbl. 80. Mið ðý gewearp woedo

(giwēdo, Rush.) his *proiecto uestimento suo*, Mk. Skt. 10, 50. Hē sette uoedo (giwēdo, Rush.) his *ponit uestimenta sua*, Jn. Skt. Lind. 13, 4 : Mk. Skt. Lind. 11, 8. Wit baru standaþ unwered wǣdo, Cd. Th. 50, 21 ; Gen. 812. Sylle mon him wist and wǣdo, Exon. Th. 336, 12 ; Gn. Ex. 336. (2) in a collective sense, *clothing, dress :*—Līchoma forđor is đon wēde *corpus plus est quam vestimentum*, Mt. Kmbl. Lind. 6, 25. Đæt gād ne wǣre wiste ne wǣde, Cd. Th. 222, 11 ; Dan. 103. Đæt gebyreþ tō wǣde and tō wiste đam đe Gode þēowian, L. Eth. vi. 51 ; Th. i. 328, 7. Heó wæsceþ his warig hrægl and him syleþ wǣde nīwe, Exon. Th. 339, 25 ; Gn. Ex. 99. II. of other covering, equipment, or dressing. v. ge-wǣdian :—Wǣde *mataxa* (cf. strǣl *vel* bedding *mataxa vel corductum vel stramentum*, i. 59, 29), Wrt. Voc. ii. 59, 28. Wǣde *antemne* (= sail ? *rigging* ? v. wǣde-rāp ; *and* cf. *Icel.* vāð *sail* (poet.)), 100, 29. Strengas gurron, wǣdo gewǣtte, Andr. Kmbl. 749 ; An. 375. Se wǣlisca (hafoc) wǣdum and dǣdum his ǣtgiefan eáđmōd weorþeþ, Exon. Th. 332, 25 ; Vy. 90. Wuldres trēów wǣdum geworđode, Rood Kmbl. 29 ; Kr. 15. [*O. Sax. O. L. Ger.* wādi ; *n. clothing* : *O. Frs.* wēde, wēd ; *n.* : *O. H. Ger.* wāt ; *f. amictus, vestimentum, vestis, vestitus* : *Icel.* vāð ; *f. a piece of stuff ; a garment.*] v. heađu-, here-, lim-, līn-wǣd ; ge-wǣde.

wǣd-brēc ; *pl. f. Breeches, a covering for the loins :*—Wǣdbrēc *perizomata vel campestria vel succinctoria*, Wrt. Voc. i. 25, 62 : *perizomata vel campestria*, 81, 64. Hig siwodon fícleáf and worhton him wǣdbrēc (*perizomata*), Gen. 3, 7.

-wǣde, -wǣded. v. ǣ-wǣde, un-wǣded.

wǣdelness, e ; *f. Poverty, want, indigence, penury :*—Wǣdlnes *inedia*, Wrt. Voc. ii. 44, 50. For weþelnysse (wǣdelnesse, Bd. M. 298, 25) woruldgóda *prae inopia rerum*, Bd. 4, 12 ; S. 581, 9. Þurh weþelnysse (wǣdelnesse, Bd. M. 68, 4) *ex inopia*, I, 27 ; S. 490, 9. Of wǣdlnysse (wēdelnisse, Ps. Surt.) *de inopia*, Ps. Spl. C. 106, 41 : 87, 10. On wǣdlnysse (wēdelnisse, Ps. Surt.) *in mendicitate*, 106, 10. Đonne đæs sellendan mōd ne cann đa wǣdelnesse (*inopiam*) geđolian, Past. 44 ; Swt. 325, 14. Wēdelnisse, Ps. Surt. 43, 24. v. wǣter-wǣdelness ; wǣdl.

wǣde-rāp, es ; *m. A stay, halyard ; pl. rigging :*—Segelgyrdas *antemnas*, wǣderrāp (wǣderrāp, Wrt.) *rudentum* (the passage is : Antemnas solvens de parte rudentium, Ald. 213), Wrt. Voc. ii. 97, 30. Untōslitenum wǣderāpum (the passage is : Quod nostrarum carbas antennarum *indisruptis rudentibus* feliciter transfretaverint, Ald. 80), 88, 32. [*O. H. Ger.* wāt-reif *rudens.*]

wǣdian *to clothe, dress.* [*O. Sax.* wādian *to clothe* : *O. H. Ger.* wāten *vestire, induere* : *Icel.* vǣða.] v. ge-wǣdian.

wǣdl (v. P. B. viii. 535), e : wǣdle, an ; *f. Poverty, want :*—Wēdl *penuria*, Wrt. Voc. ii. 117, 2. I. *poverty, indigence, want, penury :*—Þǣr þǣr word synd fela gelōme ys wǣdl (*egestas*), Scint. 78, 9 : Dōm. L. 265 ; Wulfst. 139, 31. Seó mennisce wǣdl, đe nǣfre gefylled ne biþ wilnaþ ǣlce dæg hwæthweg đises woruldwelan, Bt. 26, 2 ; Fox 94, 2. Wǣdel, Exon. Th. 238, 30 ; Ph. 212. Of wǣdle weán *de inopia*, Ps. Th. 106, 40 : Exon. Th. 201, 12 ; Ph. 55. Þearfan ic lǣrde đæt hié heora wǣdle gefeán hæfdon, Blickl. Homl. 185, 18. Hí wilniaþ đa heafene đysse gestreónfullan wǣdle, R. Ben. 136, 1. Hié for wǣdle weorđon on murcunga, đæt hié eft ongiennen giétsian for hiera wǣdle *ad murmurationem proruunt, sed cogente se inopia usque ad avaritiam devolvuntur*, Past. 45 ; Swt. 341, 2–4 : Ps. Th. 87, 9. Wǣre đū on wǣdle, sealdest mē wilna geniht, Soul Kmbl. 284 ; Seel. 146. Mid wǣdle and mid henþe ofþrycte *angustia rei familiaris inclusi*, Bt. 11, 1 ; Fox 30, 33. Đæt hē hláfes ne gȳme, gewende tō wǣdle and đa wiste widsæce (*choose want as his portion and refuse the food*), Elen. Kmbl. 1230 ; El. 617. Đonne hié gefylden and gebēten đa wǣdle hiera hiéremonna *dum subjectorum inopiam satiant*, Past. 18 ; Swt. 137, 12 : 44 ; Swt. 325, 11 : Bt. 13 ; Fox 38, 32. Đū tilast wǣdle (*indigentiam*) tō fliónne, Bt. 14, 2 ; Fox 44, 7. Đa hreósendan welan ne magon eówre wǣdle (*indigentiam*) eów fram áđōn, ac gē ēcaþ eówre ermđe (wǣdle, Cott. MS.) mid đam đe hí eów tō cumaþ, 26, 2 ; Fox 94, 8–10. Hē wilnaþ welan and flíhđ đa wǣdle (*penuriam*), 33, 2 ; Fox 122, 33. Đe læs đe þurh wǣdle and hæfenleáste đære ǣfestnesse welm áwlacige, Lchdm. iii. 442, 19. Wēdle *egestatem*, Kent. Gl. 316. Đǣr is wyrma slite and ealra wǣdla gripe, Wulfst. 114, 24. ¶ weak forms :—Gē þēowiaþ eówrum feóndum and Drihten āsent hungor on eów and þurst and næcede and ǣlce wǣdlan *servies inimico tuo, quem immittet tibi Dominus, in fame et siti et nuditate et omni penuria*, Deut. 28, 48. Man sceal gesceádlíce tōsceádon ylde and geóguđe, welan and wǣdlan, L. Edg. C. 4 ; Th. ii. 262, 5. Ia. with gen. of that which is wanting :—Wǣdl hláfes, Greg. Dial. 2, 21. Hit tácnaþ nýtena wǣdla, Lchdm. iii. 180, 21. II. *unproductiveness, barrenness :*—Cumaþ seofen swíđe wæstmbǣre geár and swíđe welige . . . and đæræfter cumaþ ōđre seofene mid swá micelre wǣdle (*tantae sterilitatis*) and hungre, đæt man forgitt đa ǣrran geár, Gen. 41, 30. Hē đæs landes wæstmbǣrnesse đara syfan geára sǣde, and đara ōþera syfan geára wǣdle (*agrorum sterilitatem*), Ors. 1, 5 ; Swt. 34, 10. [Al þat god of þisse londe we sculen leden mid us, and heo bilæuen wrecches, and wælde (= wædle) heom scal fulien, Laym. 1002. *O. H. Ger.* wātali *egestas.*]

wǣdla. I. as adjective, *poor, needy, indigent :*—Wǣdla *egenus*,

---

Wrt. Voc. i. 50, 54 : 74, 22. Oehtende wes mon đearfan and wēdlan *persecutus est hominem pauperem et mendicum*, Ps. Surt. 108, 17. Ia. with gen. of what is wanting, *wanting*, (1) of persons :—Ne geseah ic his sǣd, đæt wǣre hláfes wǣdla *non vidi semen ejus egens panem*, Ps. Th. 36, 24. Wurdon menn wǣdlan hláfes, 104, 14. (2) of things, *deficient in, poor in*:—Wæs seó stōw ge wæteres wǣdla ge eorþwæstma *erat locus et aquae et frugis inops*, Bd. 4, 28 ; S. 605, 18. Þurh đa weallendan sond and þurh đa wǣdlan stōwe wæteres and ǣlcere wǣtan *per ferventes arenas et egentia humoris loca*, Nar. 6, 9 : 26, 8. Ib. *begging :*—Hē sæt blind wiđ đone weg wǣdla (*mendicans*), Mk. Skt. 10, 46. II. as predicative adjective or substantive, *poor, needy ; a poor, needy person :*—Ic eom wǣdla (wēdla, Ps. Surt.) *egenus sum*, Ps. Th. 85, 1 : *egens*, 87, 15. Hē wearđ wǣdla *coepit egere*, Lk. Skt. 15, 14. Đā hē wǣdla (*mendicus*) wæs, Jn. Skt. 9, 8. Se welega đæt hē is wǣdla, Homl. Th. ii. 88, 27. Đonne se mon wǣdla biþ, hē wilnaþ welan, Bt. 33, 2 ; Fox 122, 32 : Exon. Th. 91, 22 ; Cri. 1496. Se se on his gǣste biđ wǣdla, Past. 44 ; Swt. 325, 14. Đa đe đæs welan gītsiaþ, hí biđ symle wǣdlan and earmingas on hyra mōde, Prov. Kmbl. 50. Gif eall þises middaneardes wela cōme tō ānum men, hū ne wǣron đonne ealle ōþre men wǣdlan ? . . . Đonne đū ealle geđælde hæfst, đonne bist đū đe self wǣdla, Bt. 13 ; Fox 38, 20–35. III. as substantive, *a poor, needy person, a beggar :*—Sum welig man wæs . . . and sum wǣdla (*mendicus*) wæs . . . Se wǣdla forđférde, Lk. Skt. 16, 19–22. Se reófla wǣdla, Homl. Th. i. 330, 10. Đearfa and wǣdla hergaþ noman đinne *pauper et inops laudabunt nomen tuum*, Ps. Surt. 73, 21. Geđeaht wǣdlan (wēdlan, Ps. Surt.) *consilium inopis*, Ps. Spl. 13, 10. Hē hine on wǣdlan hȳwe æteówde, Homl. Skt. i. 23, 221. Hié nānne mon geweligian ne magon, būton hié ōþerne gedōn tō wǣdlan (*sine ceterorum paupertate*), Bt. 13 ; Fox 40, 1. Ic gewirce eów tō wǣdlan *visitabo vos in egestate*, Lev. 26, 16. Đȳ læs hwā him self weorđe tō wǣdlan, Past. 44 ; Swt. 325, 7. Hē ālȳseþ đæne wǣdlan (*profitus inopem*, Ps. Th. 71, 12 : (wēdlan, Ps. Surt.) *egenum*, 34, 11. Sōna swā đū geseó nacodne wǣdlan, Blickl. Homl. 37, 21. For yrmđum đæra wǣdlena (wēdlena, Ps. Surt.) *propter miseriam inopum*, Ps. Th. 11, 5. Dēđ Drihten dōmas đe wǣdlum weorđaþ *faciet Dominus judicium inopum*, 139, 12. Hē đone welegan wǣdlum efnmǣrne gedēđ, Met. 10, 31. [Scullen þe wædlen alle iwurđen riche, Laym. 5872. Þa weoleʒen and đa weađlen, 427. Riche men and wedlen, 497. Wrecche and wæđle and usell mann, Orm. 5638 : 7732 : 7770 : 7889 : *O. H. Ger.* wātal, wādal *egens.*] v. nīd-wǣdla.

wǣdlian ; *p.* ode. I. *to be poor, indigent, needy, in want :*—Ic wǣdlige *egeo*, Ælfc. Gr. 26, 2 ; Zup. 154, 15. Hē wēdlaþ *egebit*, Kent. Gl. 835. Se đe wǣdlaþ *qui indiget*, 333. Đa welegan wǣdledon (wēdladon, Ps. Surt.) and eodon biddende *divites eguerunt*, Ps. Th. 33, 10. Beóđ welige hwīlwendlíce, đæt gē ēcelíce wǣdlion, Homl. Th. i. 64, 16. Đā wurdon hí dreórige on mōde, đæt hí wǣdligende on ānum wǣclícum wǣfelse férdon, 62, 28. Ia. *to be in want of something, to lack, not to have enough :*—Leádes đa men wǣdliaþ, and goldes genihtsumiaþ *plumbo egent, auro habundant*, Nar. 31, 4. Wēdliende hláf *egens panem*, Ps. Surt. 36, 25. II. *to beg :*—Se đe sæt and wǣdlode *qui sedebat et mendicabat*, Jn. Skt. 9, 8. Mē sceamaþ đæt ic wǣdlige *mendicare erubesco*, Lk. Skt. 16, 3. Hí wǣdlian (wēdlien, Ps. Surt.) *mendicent*, Ps. Spl. 108, 9. Sum blind man sæt wiđ đæne weg wǣdligende (*mendicans*), Lk. Skt. 18, 35. Wǣdliende, Blickl. Homl. 17, 31, 34. Hē wēdlat *mendicabit*, Kent. Gl. 731. [Þe king wæilien (wǣdlien ? *to go as a beggar*) agon wide ʒeon þas þeoden, Laym. 28880. *O. H. Ger.* wādalōn *evagari.*]

wǣdlig ; *adj. Poor, needy, destitute :*—Hē wacode ealle đa niht mid đam wǣdlian hreófian, Homl. Skt. i. 3, 486. Hē on mislícum yrmđum mannum geheólp, wǣdligum and wanscrýddum, Homl. Th. ii. 500, 17.

wǣdlness. v. wǣdelness.

wǣdlung, e ; *f.* I. *poverty, indigence, want :*—Đār is geómerung and wǣdluncg, Wulfst. 114, 27. Hine (*Lazarus*) geswencte seó wǣdlung, and áfeormode ; đone ōđerne (*Dives*) gewelgode his genihtsumnys, and bepǣhte, Homl. Th. i. 332, 9. Of wǣdlunga *de inopia*, Ps. Spl. 106, 41. On wǣdlunga *in mendicitate*, 106, 10. Þearfan hē lǣrde đæt hí on lífes wǣdlunge geđyldige beón, Homl. Th. ii. 328, 15. Ne đū ne wēn nā đæt ic āht underfenge for ǣnegum welan, ac symle on wǣdlunge lyfde, Homl. Skt. ii. 23 b, 341. II. *begging :*—Hē đa wanspēdigan cristenan ne geđafode đæt hí openre wǣdlunge underđeódde, ac hē gemanode đa rícan đæt hí đæra cristenra paupertate mid heora spēdum gefrēfrodon *he would not allow the destitute Christians to be subject to public begging, but admonished the rich to succour with their wealth the poverty of the Christians*, Homl. Th. i. 558, 26.

wǣfan ; *p. de To wrap up, clothe :*—Utan wǣfan nacode, Wulfst. 119, 6. [*Goth.* bi-waibjan *to clothe.* In later English the verb expresses motion :—Þe ivele gost weueđ wide and wandređ (*vadit*, v. Mt. 12, 43), O. E. Homl. ii. 85, 33. Ich smet of Modred is hafd þat hit wond (wefde, 2nd MS.) a þene wald, Laym. 28049. Þa cnihtes wefden up þa castles ʒæte, 19003. Cf. *O. H. Ger.* za-weibōn *dispergere* ; weibōn *fluere, fluitare, agitari* : *Icel.* veifa *to wave, vibrate.*] v. be-, ymbe-wǣfan, *and next word.*

wǽfels, es; *m.* *A covering, wrap, cloak, veil* :—Wǽfels *tegmen*, Ælfc. Gr. 9, 12; Zup. 41, 1. Wǽfelses t scýtan *sindonis*, Hpt. Gl. 494, 13. Wǽfel(se), basincge *chlamide*, 456, 46. Under wǽfelse *velamento, indumento*, 457, 24. Mid gewefenum wǽfelsa *consuta plectra*, 462, 63. Hí wǽdligende on ánum wáclicum wǽfelse férdon, Homl. Th. i. 62, 29. On wǽfelse (*tegmine*) fyþera đínra, Ps. Spl. 35, 8. Oferbrǽdels t wǽfels *opertorium*, Ps. Lamb. 101, 27. Đam đe wylle niman đíne tunecan, lǽt him tó đínne wǽfels (*pallium*), Mt. Kmbl. 5, 40 : Gen. 39, 12 : 24, 65 : Ap. Th. 11, 27. Ælmesgedál dǽle man gelóme, mete đám ofhingredum, wǽfels đám nacedum, Wulfst. 74, 4. Wǽfels *pallium*, Kent. Gl. 968.

wǽfer-gange, an; *f.* *A spider* :—Wǽfyrgange (gongeweafre, Ps. Surt.) *aranea*, Ps. Spl. 89, 9. v. gange-wifre.

wǽfer-geornness, e; *f.* *Eagerness to see sights* :—Mǽssepreóstas ne sceolon fremdra manna túnas, ne hús, for nánre wǽfereornnysse sécan, L. E. I. 13; Th. ii. 410, 19.

wǽfer-hús, es; *n.* *A theatre, amphitheatre* :—Hé lǽdde hí tó đam wǽferhúse, đǽr đa deór wunodon, beran and león, đe hí ábítan sceoldon, Homl. Skt. ii. 24, 49.

wǽfer-líc; *adj.* *Of a theatre* :—Wǽferlíce glencgu *theatrales pompas*, Hpt. Gl. 407, 42. v. wafor-líc.

wǽferness, e; *f.* *Public exhibition, display, show* :—On wǽfernysse t wǽfersǽne *per publicum* (the passage is : Quamvis flava caesaries raderetur, et per publicum decalvata traheretur, Ald. 62), Hpt. Gl. 510, 11.

wǽfer-sín, -sién, -sýn, -seón, e; *f.* *A sight, show, spectacle* :—Wǽfersýn *spectaculum*, Wrt. Voc. i. 55, 44. Đæt ic him wǽfersýn wǽre *factus sum illis in parabolam*, Ps. Th. 68, 11. Ond swá micel wundor and wǽfersién wæs mínes weoredes on fægernisse *fuitque inter uarietates spectaculorum in conspiciendo talem exercitum*, Nar. 7, 18. Wǽfersǽne *spectaculi*, Hpt. Gl. 508, 28. Wǽfersýne, 487, 47. Wǽfersǽne *spectaculo*, 412, 1. Mid wundurfulre wǽfersǽne *stupendo spectaculo*, 470, 76. Wǽfersýne, Bd. 3, 3; S. 525, 38 : 5, 12; S. 628, 8. Hé bebeád his folce đæt hí tó đyssere wǽfersýne (*a man trying to fly*) cómon, Homl. Th. i. 380, 15. Eall wered đe æt đisse wǽfersýnne wǽron, Lk. Skt. 23, 48. On wǽfersǽne (v. wǽferness) *per publicum*, Hpt. Gl. 510, 12. Hí woldon đa gymstánas tócwýsan on ealles đæs folces gesihđe tó wǽfersýne, Homl. Th. i. 60, 25 : 542, 32. Hí mé geworhton him tó wǽfersýne, Rood Kmbl. 61; Kr. 31. Wé for úrum synnum tó swylcere wǽfersýne synd, Homl. Skt. ii. 25, 158. Wǽfersǽne *spectaculum*, Hpt. Gl. 435, 49 : 501, 46. Se dæg mé ætýwde swíđe micele wǽfersýne, Shrn. 41, 15. Tó đissum wæferseónum, Blickl. Homl. 187, 15. [*O. H. Ger.* wabar-siuni *spectaculum.*]

wǽfer-stów, e; *f.* *A place for spectacles, an amphitheatre* :—Weaferstówa *amphitheatrum*, Lchdm. i. lxi, 9. v. wafung-stów.

wǽfre; *adj.* I. *flickering, wavering, quivering* :—Wylm đæs wǽfran líges (cf. Icel. vafr-logi), Cd. Th. 231, 2; Dan. 241. II. fig. *wavering, languishing* :—Him wæs geómor sefa, wǽfre and wælfús, Beo. Th. 4831; B. 2420. Hé ne meahte wǽfre mód forhabban in hreþre, 2305; B. 1150. III. *active, nimble* (? cf. the force of the old adjective *quiver*) :—Wearđ him tó handbanan wælgæst wǽfre, Beo. Th. 2666; B. 1331. [Cf. Uten uorsien þisne midelard and his wouernesse (*instability?*), Anglia i. 31, 18. *M. H. Ger.* waberēn *vacillare : Icel.* vafra *to hover about.*] v. wafian.

wǽfs. v. wæps.

wǽfþ, wæft, e; *f.* *A sight, show, spectacle* :—Wǽsđ *vel* wǽfersýn *spectaculum*, Wrt. Voc. i. 55, 44. Hwá mæg forbæran đæt hé swylcre wǽfte ne wundrige, đætte ǽfre swylc yfel gewyrþan sceolde under đæs ælmihtigan Godes anwealde *quae fieri in regno potentis omnia Dei nemo satis potest admirari*, Bt. 36, 1; Fox 172, 14. v. wafian.

wǽg *a way*, wǽg *a wall*. v. weg, wág.

wǽg, es; *m.* I. *movement, cf. Goth.* wēgs *motus* (*in mari*) :—Đú his ýþum miht ána gesteóran, đonne hí on wǽge wind onhréreþ *motum fluctuum ejus tu mitigas*, Ps. Th. 88, 8. II. *a wave, water, the wave, sea* :—Fámig winneþ wǽg wiđ wealle, Exon. Th. 383, 33; Rä. 4, 20. Wídfæđme wǽg, Andr. Kmbl. 1065; An. 533. Þurh wǽges wylm, Exon. Th. 283, 14; Jul. 680 : Elen. Kmbl. 459; El. 230. Wǽges weard, Andr. Kmbl. 1263; An. 632. Wǽges weard, 1201; An. 601. Ýđ wiđ lande winneþ, wind wiđ wǽge, Met. 28, 58. Staþelas wiđ wǽge, wætre windendum, Exon. Th. 61, 8; Cri. 981 : 351, 23; Sch. 84. Oft ic (*an anchor*) sceal wiđ wǽge winnan and wiþ winde feohtan, 398, 1; Rä. 17, 1. Mec upp áhóf wind of wǽge, 392, 19; Rä. 11, 10 : 405, 10; Rä. 23, 21. Wiht (*an ice-floe*) cwom æfter wǽge líþan, 415, 22; Rä. 34, 1. Feówertýne gewiton mid đý wǽge in forwyrd sceacan, Andr. Kmbl. 3186; An. 1596 : Cd. Th. 206, 25; Exod. 457. Wonnan wǽge *with the dark wave*, 83, 13; Gen. 1379. Wǽg *aquam*, Hpt. Gl. 418, 28. Hié scufon wyrm ofer weallclif, léton wǽg niman, flód fæđmian frætwa hyrde, Beo. Th. 6256; B. 3132. Sum fealone wǽg stefnan steóreþ, streámráde con, Exon. Th. 296, 19; Crä. 53. On sealtne wǽg, 361, 30; Wal. 27 : Cd. Th. 236, 19; Dan. 323. Gewát se fugel earce sécan ofer wonne wǽg, 88, 8; Gen. 1462. Windas

weóxon, wǽgas grundon, Andr. Kmbl. 746; An. 373 : 911; An. 456 : 3088; An. 1547. Hreó wǽgas, salte sǽstreámas, 1496; An. 749. Wonne wǽgas, Cd. Th. 8, 4; Gen. 119. Wiđ ýđfare gehealden hreóra wǽga, Exon. Th. 200, 24; Ph. 45. Wrǽclíce syndon wǽgea gangas, đonne sǽstreámas swíđust flówaþ *mirabiles elationes maris*, Ps. Th. 92, 5. Wága gurgites, Hpt. Gl. 464, 76. Féran ofer wǽga gewinn, Andr. Kmbl. 1863; An. 934. Ealle đa đe onhrêraþ hreó wǽgas on đam brádan brime, Exon. Th. 194, 19; Az. 141. Wadan ofer wǽgas, 350, 9; Sch. 61. Flód fealewe wǽgas, Andr. Kmbl. 3177; An. 1591. Fealwe wǽgas (wegas?), Exon. Th. 289, 11; Wand. 46. [*Goth.* wēgs *a wave : O. Sax.* wág : *O. Frs.* wēg : *O. H. Ger.* wág *liquor, gurges, vorago, pontus, aequor, lacus, fretum : Icel.* vágr *a wave, sea.*] v. fífel-, módig-, sǽ-wǽg.

wǽg (*see also* wǽge), e; *f.* I. *a weight,* (a) as a general term :—Byrden odđe wǽg *pondus*, Ælfc. Gr. 9, 32; Zup. 58, 17 note. Genim đære ylcan wyrte ánre tremesse wǽge, Lchdm. i. 72, 11. Genim twéga trymessa wǽge, 70, 15. Þreóra trymessa wǽge, 72, 26 : 74, 4. Habbaþ emne wǽga *aequa sint pondera*, Lev. 19, 36. (b) *as a definite weight, a wey* :—Án wég spices and cǽses, Cod. Dip. Kmbl. i. 312, 8. Selle mon uuége cǽsa, 293, 11. .i. wǽge cǽsa, .i. wǽge speces, 296, 35. .ii. wǽga spices and cǽses, 299, 18. .iii. wǽga, 311, 3. (c) fig. :—Đa gewunelícan wǽge (*pensum*) heora đeówdómes hig náteshwón forgímeleásion, R. Ben. 78, 11. II. *an implement for weighing, a balance* :—On wǽge beóđ áwegene *statera ponderabuntur*, Scint. 97, 7. Weh on wǽge, Lchdm. i. 374, 15. Gelícere wǽge *aequa bilance*, Hpt. Gl. 512, 76. Tó wǽge t tó disce *ad mensam*, Lk. Skt. Lind. Rush. 19, 23. Đonne man sett đa synne and đa sǽwle on đa wǽge, Wulfst. 240, 1. Wǽga *trutina* . . . lytle wǽga *momentana* vel *statam*, Wrt. Voc. i. 38, 38, 42. [Nicodemus brouhte an hundred weien of mirre and of aloes, A. R. 372, 7. Sevene waxpund makieþ onleve ponde one waye, twelf weyen on fothir, Rel. Ant. i. 70, 22. A weye of Essex chese, Piers P. 5, 93. Seint Austin deđ þeos two bođe in one weie, A. R. 60, 10. Me seal weʒe þet word er hit by yzed . . . Zoþnesse halt þise riʒtuolle waye . . . Þis waye ne ssel hongi of þis half, ne of yend half, Ayenb. 256, 6–10. *O. H. Ger.* wâgi (*dat.*) *pondere;* wâga *pondus, libra, statera, lanx, trutina : Icel.* vág *a weight;* vágir *pl. scales, a balance.*] v. pening-, pund-, twi-, wull-wǽg; wǽge-tunge.

wǽgan; *p.* de *To vex, harass, afflict* :—Hé het hí swingan, wítum wǽgan, Exon. Th. 251, 10; Jul. 143. Đæt gé mec tó wundre wǽgan mótun (cf. erlós skulun wégian mi te wundrun, dôt mi wîties filu, Hêl. 3088), 124, 22; Gú. 341. [*O. Sax.* wêgian : *O. H. Ger.* weigen *vexare, afficere, affligere, exagitare.*] v. ge-wǽgan.

wǽgan; *p.* de *To deceive, delude* :—Ne gewurđe hit đæt ic on đǽm hálgum gerecednyssum wǽge, Homl. Skt. ii. 23 b, 18. Bepǽhst *vel* wǽgest *deludis*, i. *decipis*, Wrt. Voc. ii. 138, 53. Uuêgiđ *fefellit*, 108, 46. Wǽgeþ *fefellit*, i. *eludit*, 35, 28. Wǽgđ *mentitur*, Kent. Gl. 414 : *fallit*, 933. Gif hwylc bróđor wǽgđ and misfêhđ on boduncge sealma ođþe rǽdincge *si quis dum pronuntiat psalmum fallitur lectionem*, R. Ben. 71, 5. Gesuícas t wǽges *mentientes*, Mt. Kmbl. Lind. 5, 11. Wǽgde *vel* bepǽhte *fefellit*, i. *delusit*, Wrt. Voc. ii. 148, 27. Ne hine nówiht his geleáfa wǽgde, Bd. 4, 32; S. 612, 3. Weleras wǽgendes *labia mentientis*, Scint. 95, 4. Wǽgendre gesǽlignesse *vel* bepǽcendre *fallentis fortunae*, Wrt. Voc. ii. 146, 73. Wǽgende welere *labium mentiens*, Kent. Gl. 596. Wǽged *delusus* (v. Mt. 2, 16), Wrt. Voc. ii. 71, 57 : 26, 29. Wǽged wæs *deluditur*, 95, 63 : 27, 26. Wǽged *ludificatus*, 86, 22. v. á-, be-, ge-wǽgan.

wǽg-bora, an; *m.* *A wave-bearer, a creature that lives beneath the waves* :—Wundorlíc wǽgbora, Beo. Th. 2884; B. 1440.

wǽg-bord, es; *n.* *A wave-board, a plank of a vessel* :—Đú of eorđan wæstmum wiste under wǽgbord (cf. lǽd under earce bord, 80, 23; Gen. 1333; be útan earce bordum, 81, 33; Gen. 1354) gelǽde, Cd. Th. 81, 4; Gen. 1340.

wǽg-deór, es; *n.* *A sea-beast* :—Wǽgdeóra gehwylc swelteþ, Exon. Th. 61, 21; Cri. 988.

wǽg-dropa, an; *m.* *A wave-drop, a salt tear* (?) :—Hé háte lét teáras geótan, weallan wǽgdropan, Exon. Th. 165, 17; Gú. 1030.

wǽge (*see also* wǽg), an; *f.* I. *a weight,* (a) as a general term :—Byrden odđe wǽge *pondus*, Ælfc. Gr. 9, 32; Zup. 58, 17. Hæbbe ǽlc man rihte wǽgan and rihte gemetu *pondus habebis justum et verum et modius aequalis et verus erit tibi*, Deut. 25, 15. (b) *as a definite weight, a wey* :—Gá seó wǽge (wǽg, MS. G.) wulle tó .cxx. þ., and nán man hig ná undeóror ne sylle, L. Edg. ii. 8; Th. i. 270, 3. II. *an implement for weighing, a balance, scale* :—Deós wǽge odđe scalu *lanx*, Ælfc. Gr. 9, 73; Zup. 73, 10. Wǽge *trutina*, 36; Zup. 215, 18 : *statera*, Scint. 81, 12 : 110, 12. Libra, đæt is pund odđe wǽge, Lchdm. iii. 246, 1. Gelícere wǽgan *in equilibrio*, 234, 5 : 238, 26. Ælc đæra đinga đe man wihđ on wǽgan, Ælfc. Gr. 13; Zup. 84, 3. Áwegene on ánre wǽgan, Homl. Th. ii. 454, 23 : 436, 12. On wǽgum (wégum, Ps. Surt. Spl.) *in stateris*, Ps. Lamb. 61, 10. v. efen-wǽge.

wǽge, wǽg[e], es; *n.* *A cup* :—Wégi *poculum*, Wrt. Voc. i. 290, 82. Sume ic geteáh, tó geflíte fremede . . . beóre druncne; ic him byrlade wróht of wége, đæt hí in wínsele þurh sweordgripe sáwle forlêtan of flǽschoman, Exon. Th. 271, 24; Jul. 487. Fǽted wǽge, dryncfæt deóre,

Beo. Th. 4499; B. 2253. Hě mandryhtne bær fǽted wǽge, 4553; B. 2282. [*O. Sax.* wâgi, wêgi *a vessel.* Cf. (?) *O. H. Ger.* bah-weiga; *f. ferculum, discus, lanx* : *Icel.* veig; *f. strong drink.*] v. bǽde-, deáþ-, ealo-, líþ-wǽge (-wége, -wég).

wægen. v. wægn.

wǽge-tunge, an; *f. The tongue of a balance :*—Wǽgetunge (or wǽge tunge, v. wǽg, II) *examen,* Wrt. Voc. i. 38, 41. [*Ger.* wage-zunge.]

wǽg-fær, es; *n. A sea-journey :*—Ic ðě ongitan ne meahte on wǽg-fære, Andr. Kmbl. 1845; An. 925.

wǽg-fæt, es; *n. A water-vessel, a cloud :*—Won wǽgfatu, lagustreáma full (*cups*), Exon. Th. 384, 33; Rä. 4, 37.

wǽg-faru, e; *f. A sea-passage, passage through the sea* (the passage through the Red Sea) :—Nū se ágend up árǽrde reáde streámas in rand-gebeorh, syndon ða foreweallas fægre gestépte, wrætlícu wǽgfaru, óð wolcna hróf, Cd. Th. 196, 27; Exon. 298.

wǽg-flota, an; *m. A wave-floater, a ship :*—Hū ðū wǽgflotan sund wísige, Andr. Kmbl. 973; An. 487. Gesión brecan ofer bæðweg brim-wudu myrgan, sǽmearh plegan, wadan wǽgflotan, Elen. Kmbl. 491; El. 246: Beo. Th. 3818; B. 1907.

wǽg-hengest, es; *m. A wave-steed, a ship :*—Hě bát gestág, wǽg-hengest wræc, Exon. Th. 181, 34; Gū. 1303. Hí gehlódon hildesercum wǽghengestas, Elen. Kmbl. 472; El. 236. [Cf. *Icel.* vág-marr *a ship.*]

wǽg-holm, es; *m. The billowy sea :*—Gewát ofer wǽgholm flota fámigheals, Beo. Th. 439; B. 217.

wǽg-líþend, es; *m.* : -líþende; *ptcpl. A sea-farer; sea-faring :*—Wénaþ wǽglíþende, ðæt hý on eálond sum eágum wlíten, Exon. Th. 360, 26; Wal. 11. Ne móston wǽglíðendum wætres brógan hrínon, ac hié God nerede, Cd. Th. 84, 9; Gen. 1395: Beo. Th. 6297; B. 3159. Hæleð langode, wǽglíþende, hwonne hié of nearwe stæppan mósten, Cd. Th. 86, 17; Gen. 1432. [*O. Sax.* wâg-líðand.]

wægn, wægen, es; *m. A waggon, wain, carriage, vehicle :*—Wægn *vehiculum,* Wrt. Voc. ii. 123, 40. Wǽn *plaustrum,* Wrt. Voc. i. 66, 51: 284, 43: *plaustrum vel carrum,* 16, 19: 85, 69. Mid ðý hě ða se wǽn (wægn, MS. T.) com ðe man ða bán on lǽdde *cum venisset carrum in quo ossa ducebantur,* Bd. 3, 11; S. 535, 17 note. Hě ofer wǽg gewát, wǽn æfter ran, Runic pm. Kmbl. 343, 32; Rún. 22. Wægnes hweól *rotam,* Ps. Th. 82, 10. Wǽnes weð (swæð? pæð?) *orbita,* Wrt. Voc. i. 37, 47. Ánes wǽnes gangweg *actus,* 37, 37. On wǽnes eaxe hwearfaþ ða hweól, and sió eax byrþ eallne ðone wǽn, Bt. 39, 7; Fox 220, 27: 39, 8; Fox 224, 6. Wǽne *carruca,* Hpt. Gl. 438, 67. Mid ðý ðe hine mon bere oþþe on wǽne ferige, Lchdm. iii. 30, 29. Stígan on wægn, Exon. Th. 404, 17; Rä. 23, 9. Hí gegear-wodon wægen (*carrum*) and on ásetton ða fǽmnan, Bd. 3, 9; S. 534, 9. Wæs gold on wǽn hladen, Beo. Th. 6260; B. 3134. Twégra wǽna gangweg *via,* Wrt. Voc. i. 37, 38. Tuégra uuegna gang (v. wægn-gang), Cod. Dip. B. i. 344, 12. On wǽnum *in curribus,* Ps. Spl. 19, 8. Ðæt hig nymon wǽnas (*plaustra*), Gen. 45, 19, 27. ¶ with special reference to what is carried, in the phrase *wægnes, wægna gang,* the going to fetch wood, v. Kemble's Saxons in England, ii. pp. 70, 71 :—.ii. wǽna gang mid cyninges wénum tō Bleán ðem wiada (cf. .iiii. carris trans-ductionem in silba regis sex ebdomadas a die Pentecosten, hubi alteri homines silbam cedant, 122, 8), Chart. Th. 119, 16. An ic twéga wǽna gang on clætinc tō wudurédenne, Cod. Dip. Kmbl. vi. 36, 15. [Tuége waine gong wudes, iv. 282, 15. Tō wayne gong tō wude, 282, 28.] ¶ re-ferring to the constellation Charles' *wain.* v. carles wǽn :—Wǽnes ðísl (waegne[s] þixl, 100, 72) *archtoes,* Wrt. Voc. ii. 7, 23. Tunglu ðe we hátaþ wǽnes ðísla, Bt. 39, 3; Fox 214, 19: Met. 28, 10. [*O. L. Ger.* reidi-wagan *currus: O. Frs.* wain, wein: *O. H. Ger.* wagan *plaustrum, carra, carrum, vehiculum: Icel.* vagn.] v. fyrd-, hors-, hræd-, rǽd-, ryne-, scrid-, wíg-wægn (-wǽn).

-wægnan. v. be-wægnan.

wægnere, es; *m. A driver of a carriage, a waggoner, charioteer :*—Scridwísa *vel* wǽnere *auriga,* Wrt. Voc. i. 39, 38. Wénere, ii. 4, 57.

wǽgnere, es; *m. A deceiver :*—Sponera, wǽgnera *lenonum,* Wrt. Voc. ii. 52, 42. v. wǽgnian.

waegne-þixl. v. wægn.

wægn-faru, e; *f. A chariot-journey :*—Fiscalis reda (= rheda) ge-bellícum wæg[n]fearu, Wrt. Voc. ii. 108, 64. *Fiscalis rǽde* gafellícum wǽnfare, 35, 56.

wægn-gehrado *a waggon-plank :*—Wǽngehrado *tabula plaustri,* Wrt. Voc. i. 284, 53.

wægn-geréfa, an; *m. A wain-reeve, one who has charge of carriages :*—Wǽngeréfa *carpentarius,* Wrt. Voc. i. 284, 44: ii. 16, 66.

wægn-gewǽde, es; *n. A waggon-cloth, covering for a waggon :*—Man sceal habban wǽngewǽdu, Anglia ix. 264, 4.

wǽgnian. v. ge-wǽgnian.

wægn-scilling, es; *m. A toll of a shilling on each waggon standing to be loaded at a salt-pan :*—Se wægnscilling and se seámpendjng gonge tō ðæs cynges handa swá hě ealning dyde æt Saltwíc (cf. sine aliquo tri-buto dominatoris gentis praedictae, id est statione siue inoneratione plaus-

trorum, 125, 30–32), Cod. Dip. Kmbl. v. 143, 70. v. Kemble's Saxons in England, ii. pp. 70, 71, 329.

wægn-þoll, es; *m. A cart-pin :*—Wǽnðoll *aries,* Wrt. Voc. ii. 3, 72. v. þoll.

wægn-treów, es; *n. A perquisite of a log of wood from each load to the labourer loading and leading the waggon* (? cf. wægn-scilling) :—On sumere þeóde gebyreþ . . . æt wudulǽde wǽntreówes, æt cornlǽde hreáccopp, L. R. S. 21; Th. i. 440, 27.

wægn-weg, es; *m. A cart-road, carriage-road :*—On ðone wǽnweg, Cod. Dip. Kmbl. vi. 8, 37. On ðone brádan wǽnweg, iii. 37, 26.

wægn-wyrhta, an; *m. A wain-wright, cart-wright, carriage-maker :*—Wǽnwyrhta *carpentarius,* Wrt. Voc. i. 19, 9: 66, 50: ii. 128, 68.

wǽg-pundern *a steel-yard, weighing-machine :*—Ǽlc burhgemet and ælc wǽgpundern beó be his (*the bishop's*) dihte swíðe rihte, L. I. P. 7; Th. ii. 312, 20. Hě sceal habban wǽipundern, Anglia ix. 263, 9. Cf. pundern *perpendiculum,* Hpt. Gl. 476, 77, *and* pundar.

wǽg-scealu, e; *f. The scale of a balance :*—Wǽgscala *lances,* Wrt. Voc. ii. 53, 7.

wǽg-stæp, es; *n. A shore, bank :*—Cwom .lx. monna tō wǽgstæþe rídan, Exon. Th. 404, 3; Rä. 23, 2.

wǽg-streám, es; *m. The sea :*—Ðæt feórþe cyn wód on wǽgstreám (*the Red Sea*), Cd. Th. 197, 22; Exod. 311.

wǽg-sweord, es; *n. A sword with wavy ornamentation* (v. Woor-saae's Primeval Antiquities, p. 40) :—Wrætlíc wǽgsweord, Beo. Th. 2982; B. 1489.

wǽg-þel, es; *n. A wave-plank, a ship :*—Hě álǽdde of wǽgþele (*the ark*) wrāðra láfe, Cd. Th. 90, 16; Gen. 1496. Nóe tealde ðæt se hrefn hine sécan wolde on wǽgþele, 87, 9; Gen. 1446. On wǽgþele *on board,* Andr. Kmbl. 3418; An. 1713. Under earce bord eaforan lǽdan, weras on wǽgþel, Cd. Th. 82, 6; Gen. 1358.

wǽg-þreá *the chastisement by the waters* (the deluge), Cd. Th. 90, 5; Gen. 1490.

wǽg-þreát, es; *m. A wave-host, the waters of the deluge :*—Ic wille mid wǽgþreáte ǽhta and ágend eall ácwellan, Cd. Th. 81, 29; Gen. 1352.

wæl, es; *n.*   I. in a collective sense, *the slain, the dead, a number of slain,* (a) generally of death in battle :—Wæl feól on eorðan, Byrht. Th. 135, 31; By. 126: 140, 45; By. 303. Ðæs wæles wæs geteald six hund manna mid ðām fȳrenum flānum ofsceotene *of those who died they counted six hundred shot with the fiery arrows,* Homl. Th. i. 506, 6. Ðá hě his bróðor slege ofáxode, ðá férde hě tō ðam wæle his líc sécende, ii. 358, 6. Ðá gelæhton his gebróðra his líc of ðam wæle, Homl. Skt. ii. 25, 673. Ðá sóhte hě on ðam wæle his líc, Bd. 4, 22; S. 591, 17. Hě on wæle lǽge, Byrht. Th. 139, 65; By. 279: 140, 39; By. 300. Hit næs ná gesǽd hwæt Pirruses folces gefeallen wǽre, for ðon hit næs þeáw ðæt mon ǽnig wæl on ða healfe rímde ðe wieldre wæs (*mos est, ex ea parte quae vicerit occisorum non commemorare numerum*), Ors. 4, 1; Swt. 156, 21. Ǽr hě ðæt wæl bereáfian mehte, 3, 9; Swt. 128, 9: Beo. Th. 2429; B. 1212: 6047; B. 3027. On wæl feallan *to die in battle,* Cd. Th. 123, 2; Gen. 2038. On wæll fyllan *to kill in battle,* Bd. 1, 12; S. 481, 24.   ¶ as object of verbs of slaying :—Ðǽr wæs micel wæl geslægen on gehwæþre hond *many were killed on both sides,* Chr. 871; Erl. 74, 11: 833; Erl. 64, 20. Ne weard wæl māre folces gefylled, 937; Erl. 115, 14. Ðǽr was ungemetlíc wæl geslægen Norþanhymbra, sume binnan, sume būtan, 867; Erl. 72, 15: Ors. 2, 5; Swt. 80, 26. Hí mid mycel wæl on geslógan *magnam eorum multi-tudinem sternens,* Bd. 1, 12; S. 481, 30. Ne weard wæl māre ge-slógon on hæþnum herige ðe wě secgan hiérdon óþ ðisne andweardan dæg, Chr. 851; Erl. 68, 4. Hě menigfeald wæl felde and slóh, Guthl. 2; Gdwin. 14, 7. (b) in other connections :—Ðá geát mon ðær ātter ūt on ðone sǽ, and raþe ðæs ðǽr com upp micel wæl deádra fisca, Ors. 6, 3; Swt. 258, 17.   II. *a single corpse, a slain person :*—Hě mě habban wile dreóre fáhne, gif mec deáð nimeþ, byreþ blódig wæl, Beo. Th. 900; B. 448. Ðonne walu feóllon, 2089; B. 1042. Crungon walo, Exon. Th. 477, 17; Ruin. 26.   III. in an abstract sense, (a) of destruc-tion in war, *slaughter, carnage :*—Wæl on gefeohte *strages,* Ælfc. Gr. 9, 27; Zup. 53, 5. Mycel wæl (wælfill, MS. A.) geweard on Brytene æt Wódnesbeorge, Chr. 592; Erl. 19, 34. Hě hí on gelícnysse ðæs trāiscan wæles (*caedis*) wundade, Bd. 3, 1; S. 523, 30. Mid grimme wæle and herige *saeva caede,* 4, 15; S. 583, 26. Of wæle *strage, occisione,* Hpt. Gl. 427, 60. (b) in other connections, *destruction :*—Com mycel wæl and monncwyld godcundlíce gesended *supervenit clades divinitus missa,* Bd. 4, 3; S. 567, 10. Hě hí fram ðam mánfullan wæle (*clade;* destruc-tion by famine) generede, 4, 14; S. 582, 17. Wæle *strage; occisione* (destruction of the soul by sin. v. Ald. 7), Hpt. Gl. 415, 22. [Þat wæl (heap, 2nd MS.) wes þe more, Laym. 4111. He lette al þæt wel weorpen an ane dich, 6427. Ic heo wulle biwinnen oðer an wæle liggen, 9497. *O. Sax.* wal (*in* wal-dād) : *O. H. Ger.* wal *strages, clades : Icel.* valr *the slain.*] v. ecg-, ungemet-wæl.

wǽl, es; *m. n. A weel* (e. g. Mode *weel* (*wheel*), Lanc.), *a deep pool, gulf, deep water of a stream* or *of the sea :*—Wǽl *gurges,* deópnys

abyssus, Wrt. Voc. i. 54, 34 : 80, 65. Sume weriaþ on gewitlocan wîs-dómes stréam, ðæt hê on unnyt ût ne tóflóweþ, ac se wæl wunaþ on weres breóstum dióp and stille, Past. 65 ; Swt. 469, 4. *Hic gurges* ðis (ðis *with* e *over* i, MS. F.: ðes, MSS. D. O.) wæl, ðæt is, deóp wæter, Ælfc. Gr. 9, 26 ; Zup. 52, 9. Wæles stæð *alvei* (the Nile) *marginem*, Hpt. Gl. 492, 70. Scymriendes wæles *cerulei gurgitis*, Germ. 401, 10. Wê æthrynon mid ûrum ârum ða ŷðan ðas deópan wælis, wê gesáwon eác ða muntas ymbe ðære sǽ strande, Anglia viii. 299, 38. þweálu clǽnes wæles (*gurgitis*), Hymn. Surt. 52, 13. On wæle fûlum þweán, sume wrôhte getácnaþ, Lchdm. iii. 206, 10. Fugel uppe sceal lácan on lyfte, leax sceal on wæle mid sceóte scrîðan, Menol. Fox 538 ; Gn. C. 39. Of wæle getogen *gurgite ductus*, Hymn. Surt. 70, 27 : 25, 6. Aðuah in ðær uéle (*natatoria*), Jn. Skt. Lind. 9, 7. In ðæt uoel î in ðæt fiscpôl *in piscinam*, 5, 4. On wælum ádrenctum *profundis pelagi flustris suffocato* (Ald. 12), Hpt. Gl. 426, 22. Weálu (*rubicundi oceani*) *gur-gites*, 409, 64. Ðû gedréfest deópe wælas *conturbas profundum maris*, Ps. Th. 64, 7. [With weel of þi liking *torrente voluntatis tuae*, Ps. 35, 9. Þai sink in þat wele (*v. l.* pitt), þar neuer man sank þat was o sele, C. M. 2903. Wel (*rimes with* sel), Misc. 149, 89. v. Jamieson's Dict. s. v. wele. *O. L. Ger.* wâl *abyssus*.]

wǽlan; *p.* de *To vex, torment, afflict* :—Ðæt hŷ his lîchoman leng ne môstan wîtum wǽlan, Exon. Th. 127, 34 ; Gû. 396. Dogter mîn is yfle from deófle wæled *filia mea male a daemonio vexatur*, Mt. Kmbl. Rush. 15, 22. Hé is yfle wǽlid *male torquetur*, 8, 6. [*Cf. Icel.* veill *diseased, ailing ;* veilindi *disease.*] v. ã-, be-, ge-wǽlan.

wæl-bed[d], es ; *n. The bed of the slain* :—Ic hine heardan clammum on wælbedde wríþan þôhte *I had thought to bind him on the couch of the slain* (i. e. *to kill him*), Beo. Th. 1932 ; B. 964. Hwæt befealdest ðû folmum ðînum on wælbedd brôðor ðînne? Cd. Th. 62, 8 ; Gen. 1011. v. wæl-rest.

wæl-ben[n], e ; *f. A wound inflicted by the sea*, v. wæl :—Gârsecg wedde . . . egesan stôdon, weóllon wælbenna (wæl- ?) (*the reference is to the death of the Egyptians in the Red Sea*), Cd. Th. 208, 30 ; Exod. 491.

wæl-bend, e ; *f. A deadly, mortal band* :—Wælbende handgewriþene *deathband hand-twisted* (i. e. *death at a person's hands*), Beo. Th. 3876 ; B. 1936. v. wæl-clamm.

wæl-bleát, *adj. Causing mortal weakness, deadly, mortal* :—Benne, wunde wælbleáte, Beo. Th. 5443 ; B. 2725.

wæl-ceald; *adj. Deadly cold* :—Hê him helle gescôp, wælcealde wíc (*cf.* Ðǽr (*in hell*) cymð forst fyrnum cald, Cd. Th. 20, 28 ; Gen. 316), wintre beðeahte, Salm. Kmbl. 937 ; Sal. 468.

wæl-ceásiga, an ; *m. A chooser of the slain, a raven* :—Wonn wæl-ceásega, Cd. Th. 188, 6 ; Exod. 164. v. wæl-cyrige.

wæl-clam[m], es ; *m. A fatal bond* :—Forgif mê mennen ðe ðû áhreddest wera wælclommum (*captivity in which they might have been slain ?*), Cd. Th. 128, 17 ; Gen. 2128. v. wæl-bend.

wæl-cræft, es ; *m. A deadly power, power which causes death* :—Ðonne mîn hláford wile láfe þicgan ðara ðe hê of life hêt wælcræf[tum] áwrecan (*of those whom he has ordered to be slain*), Exon. Th. 498, 11 ; Rä. 87, 11.

wæl-cwealm, es ; *m. A death-pang, pain of violent death* :—Rêcas stîgaþ ofer hrófum, hlin bið on eorþan, wælcwealm wera, Exon. Th. 381, 8 ; Rä. 2, 8.

wæl-cyrge, -cyrige, -cyrie, an ; *f. A chooser of the slain.* According to the mythology, as seen in its Northern form, the Val-kyrjur were the goddesses who chose the slain that were to be conducted by them to Odin's hall—Val-halla : 'þær ríða jafnan at kjósa val.' Something of the old idea is still shewn in the following glosses, in which the word renders a Fury, a Gorgon, or the goddess of war :—Uualcyrge *Tisifone*, Wrt. Voc. ii. 122, 34 : *Eurynis*, 107, 43. Walcrigge *Herinis*, 110, 34. Wælcyrge, 43, 2 : *Bellona*, 94, 15 : 12, 12. Wælcyrige *Allecto*, 5, 72. Wælcyrie *Tisiphona*, i. 60, 21. Ða deór habbaþ wælkyrian eágan *hae bestie oculos habent Gorgoneos*, Nar. 34, 6. But elsewhere it is used apparently with the sense of *witch* or *sorceress* :—Wyccan and wælcyrian and unlyb-wyrhtan, Wulfst. 298, 18. Wiccan and wælcerian, 165, 34. Wiccean and wælcyrian, Chart. Erl. 231, 10. [Clerkes out of Caldye . . . wychez & walkiries . . . deuinores of demorlaykes . . . sorsers & exorsismus, Allit. Pms. 85, 1577. *Icel.* val-kyrja.]

wæl-cyrging, es ; *m. One that belongs to the race of the wælcyrgan* :—*Gorgoneus*, ðæt is wælkyrging (-cyrginc, *v. l.*), Nar. 35, 6.

wæl-deáþ, es ; *m. A violent death* :—Hié wældeáð (*death at Grendel's hands*) fornam, Beo. Th. 1395 ; B. 695.

wæl-dreór, es ; *m. The blood of the slain* :—Wæter wældreóre fág, Beo. Th. 3267 ; B. 1631. Eorðe wældreóre (*the blood of Abel*) swealh of handum ðínum (*Cain's*), Cd. Th. 62, 19 ; Gen. 1016. Ic fylde mid folmum ordbanan Abeles, eorðan sealde wældreór weres, 67, 9 ; Gen. 1098.

wæl-fǽhþ, e ; *f. Deadly feud, hostility that leads to slaying* :—Hê wælfǽhða dǽl, sæcca gesette, Beo. Th. 4061 ; B. 2028.

wæl-fæðm, es ; *m. A deadly embrace* :—Brim wælfæðmum sweóp,

---

fǽge crungon (*of the overwhelming of the Egyptians in the Red Sea*), Cd. Th. 208, 9 ; Exod. 480.

wæl-ráh; *adj. Deadly hostile* (?) :—Wælfágne winter (*winter when the earth seems dead*). Beo. Th. 2260 ; B. 1128.

wæl-feall, es ; *m.* (?) *The fall of the slain, destruction* :—Tô wælfealle and tô deáðcwalum Deniga leódum, Beo. Th. 3427 ; B. 1711. [*Icel.* val-fall ; *n.* stages.] Cf. wæl-fill.

wæl-fel; *adj. Cruel to the slain* (?) or *very cruel.* Cf. wæl-hreów :—Hræfen uppe gôl, wan and wælfel, Elen. Kmbl. 105 ; El. 53.

wæl-feld, es ; *m. The field of the slain, the battle-field* :—Hî on wælfelda plegodan, Chr. 937 ; Erl. 114, 17

wæl-fill, es ; *m. Slaughter, carnage* :—Wælfill *cedes*, Wrt. Voc. ii. 15, 67. Wælfyl *statis* (*stragis*, v. Ald. 173, 3), 93, 52. Hêr micel wælfill wæs æt Wôddesbeorge (Wôdnes-, MS. E.), Chr. 592 ; Erl. 18, 30. Blôd-gyte, wællfyll weres, morð mid mundum, Cd. Th. 92, 11 ; Gen. 1527. Heó underbæc beseah wið ðæs wælfylles (*the destruction of Sodom and Gomorrah*), 154, 29 ; Gen. 2563.

wæl-fús; *adj. Ready to be slain*; referring to Beowulf before the fight in which he was mortally wounded :—Him wæs geómor sefa, wæfre and wælfús, wyrd ungemete neáh, se sceolde sécean sáwle hord, sundur gedǽlan líf wið líce, Beo. Th. 4831 ; B. 2420.

wæl-fyll, e : -fyllu(-o) ; *indecl. f. Abundance of slain* :—Grendel on reste genam þrítig þegna, ðanon eft gewát tô hám faran mid ðære wæl-fylle, Beo. Th. 250 ; B. 125.

wæl-fŷr, es ; *n.* I. *a fire that slays, deadly fire* :—Beorges weard (*the fire-drake*) wearp wælfýre, wíde sprungon hilde leóman, Beo. Th. 5157; B. 2582. II. *a fire that burns the slain, a funeral pile* :—Hêt Hildeburh hire selfre suna on bǽl dôn . . . wand tô wolcnum wælfýra mǽst, Beo. Th. 2243 ; B. 1119.

wæl-gǽst (-gǽst ?), es ; *m. A deadly guest* (*spirit ?*), *a murderous guest* :—Wælgǽst (*Grendel*), Beo. Th. 3994 ; B. 1995 : (*Grendel's mother*), 2666 ; B. 1331.

wæl-gár, es ; *m. A deadly spear* :—Wælgár slîteþ, Exon. Th. 354, 46; Reim. 61. Ðær wæs heard plega, wælgára wrixl, wígcyrm micel, Cd. Th. 120, 5 ; Gen. 1990.

wæl-gífre; *adj.* I. *eager to slay,* (a) *of persons* :—Ðá com hæleða þreát (*those who wished to kill St. Andrew*) wadan wælgífre, Andr. Kmbl. 2543 ; An. 1273. Deáð, wiga wælgífre, Exon. Th. 231, 8 ; Ph. 486 : 162, 7 ; Gû. 972. (b) *of things* :—Wǽpen wælgífru, Exon. Th. 292, 16 ; Wand. 100. II. *eager to prey on the dead* :—Se grǽga mǽw wælgífre wand, Andr. Kmbl. 743 ; An. 372. Se wanna hrefn, wælgífre fugel, Judth. Thw. 24, 25 ; Jud. 207. Wulfum tô willan, and eác wælgífrum fuglum tô frôfre, 25, 37 ; Jud. 296. v. wæl-grǽdig.

wæl-gim[m], es ; *m. The word seems to be an epithet for the sheath of a sword, which is called in the riddle the sword's byrne* :—Byrne is mîn (*a sword's*) bleófág, swylce beorht seomað (-d, MS.) wîr ymb ðone wælgim, ðe mê waldend geaf, Exon. Th. 400, 20 ; Rä. 21, 4.

wæl-grǽdig; *adj. Greedy for the slain* (an epithet of cannibals) :—Hæfdon hié áwriten wælgrǽdige wera endestæf, hwænne hié tô môse meteþearfendum weorðan sceoldon, Andr. Kmbl. 269 ; An. 135. v. wæl-gífre.

wæl-grim[m]; *adj. Cruel, destructive* —Wælgrim, *unhêre funestus, crudelis, perniciosus*, Wrt. Voc. ii. 151, 63 : *violentus*, Germ. 399, 467. (1) *of living things, bloodthirsty, cruel* :—Hwæt standest ðû (*the devil*) wælgrim (the MS. breaks off here) . . . ? *quid adstas cruenda bestia ?* Blickl. Homl. 227, 26. Wælgrim wiga, Exon. Th. 396, 21 ; Rä. 16, 8. Heó wæs ǽryst hǽðen and wælgrim, Shrn. 139, 5. Ðone Iacóbum se wælgrimma hyrde (*Herod*) ácwealde mid sweorde, 108, 23. Hí wæl-grimme wyrmas slîtaþ, Wulfst. 139, 10 : Dôm. L. 210. (2) *of other than living things, cruel, dire, destructive* :—Hunger se hearda, wælgrim werum, Cd. Th. 109, 1 ; Gen. 1816. Nið wæs rêðe, wællgrim werum, 83, 23 ; Gen. 1384. Hê geseah wíde fleógan wælgrimme rêc (*the smoke from the burning cities of the plain*), 155, 26 ; Gen. 2578. Wælgrimme wyrd (*the fall of man*), 61, 12 ; Gen. 996. Ðê sind heardlícu, wundrum wælgrim (wel-, MS.) wîtu geteohhad, Exon. Th. 258, 12 ; Jul. 264. Gefylstan of ðám wælgrimnum tintregum, L. E. I. proem. ; Th. ii. 396, 4. Þolian wælgrim wîtu, Andr. Kmbl. 2829 ; An. 1417. Wæs ðis gefeoht wælgrimre and strengre eallum ðám ǽrgedônum *strages cunctis crudeliores prioribus*, Bd. 1, 12 ; S. 481, 24. Cf. wæl-hreów.

wæl-grimlíce; *adv. With the utmost bitterness* :—Hí wælgrimlíce gefuhton. Ðær wæs se mǽsta blódgyte on ǽgðere healfe, Ors. 4, 2 ; Swt. 160, 31.

wæl-gryre, es ; *m. The terror that comes from danger of falling in battle* :—On fyrd hyra (*the Israelites*) fǽrspell (*the tidings of the approach of the Egyptian army*) becwom ; egsan stôdan, wælgryre weroda, Cd. Th. 186, 11 ; Exod. 137.

wæl-here, (ig)es ; *m. A slaughtering host* :—Fôron tôsomne wráðe wælherigas, Cd. Th. 119, 21 ; Gen. 1983.

wæl-hlem[m], es ; *m. A deadly onslaught* :—Hyne Wulf wǽpne gerǽhte, ðæt him for swenge swát ǽdrum sprong . . . ; næs hê forht

swáðéh, ac forgeald hraðe wællhlem ðone, Beo. Th. 5931 ; B. 2969. Cf. hilde-hlem.

wæl-hlenca or -hlence, an; *m.* or *f. A slaughter-link, a link of a coat of mail :*—Wriðene wælhlencan, Elen. Kmbl. 47 ; El. 24. Gúðweard gumena grímhelm gespeón, . . . [h]wælhlencan sceóc, Cd. Th. 188, 31 ; Exod. 176.

wæl-hreów, -hreáw, -reów, -ræw ; *adj. Cruel, barbarous, blood-thirsty :*—Wælhreów *crudelis,* Ælfc. Gr. 9, 28 ; Zup. 54, 12 : *atrox,* 9, 66 ; Zup. 72, 1 : *trux,* 9, 67 ; Zup. 72, 9. Wælhreówe *crudeli,* Wrt. Voc. ii. 23, 22. Ða wælhreówan *funestam,* 38, 20. (1) of living beings :—Wælhreów werod, Cd. Th. 219, 11 ; Dan. 53. Hé (*Nero*) wælhriów wunode, Met. 9, 38. Hé wæs wælhreáw cwellere cristenra manna, Homl. Th. ii. 308, 4. Welhrióu *crudelis,* Kent. Gl. 367. Irtacus wælreów cyning, Apstls. Kmbl. 137 ; Ap. 69. Wælreów wiga *a warrior who would not spare his foe,* Beo. Th. 1262 ; B. 629. Hé wunaþ wælræw deófol, Homl. Th. i. 192, 21. Se wælhreówa Antecrist, 6, 16. Se wælhreówa cyning, Þeódríc, Bt. 1 ; Fox 2, 24. Wælhreówes (*Nero's*) gewéd, Met. 9, 5. Ne læt ðú on ðæs wælhreówan hond (*crudeli*) ðín geár, Past. 36 ; Swt. 249, 11 : Homl. Th. i. 80, 31. Ne mæg ic mínne feónd lufian, ðone ðe ic wælhreówne tógeánes mé geseó, 54, 31. Ðone wælhreówan feónd ðisse menniscan gecynd[e], Blickl. Homl. 31, 31. Ðé wælreówe wítum belecgaþ, Andr. Kmbl. 2423 ; An. 1213 : Exon. Th. 380, 10 ; Rä. 1, 6. Ða wælhreówan wyþersacan Annas and Caiphas, Nicod. 7 ; Thw. 3, 32. Earn beheóld wælhreówra wíg, Elen. Kmbl. 223 ; El. 112. Wælreówra (-e, MS.) *carnificum,* Hpt. Gl. 483, 60. Ða áne ætwundon ðínum wælhreáwum handum, Homl. Th. ii. 308, 25. Hwæt is wælhreówre betwux næddercynne ðonne draca ? i. 486, 31. Ðú wælhreówasta wímman, Homl. Skt. i. 7, 182. (2) of things :—Ðæt wíf gelýfde his wælhreówum geðeahte, Homl. Th. ii. 30, 15. Mid wealhreówre † deóflicre mihte *tyrannica potestate,* Hpt. Gl. 434, 3. Mid wealreówre grimnysse *crudescente atrocitate,* 515, 23. On þysum wælhreówan cwearterne, Nicod. 26 ; Thw. 15, 1. Forgripen mid wælhreówe (*crudeli*) deáþe, Bd. 5, 19 ; S. 638, 24. Tó þrowienne wælhreówne deáð, Homl. Skt. i. 4, 117. Mid wælhreówum dǽdum, 11, 354. Geþeówode þurh wælhreówe unlaga, Wulfst. 158, 14. [þa welreowen (*those who seized Christ*), O. E. Homl. i. 229, 25.] v. wæl-grim.

wælhreówlíce ; *adv.* I. *cruelly :*—Se wælhreówlíce (*crudeli caede*) wæs ofslægen, Bd. 3, 14 ; S. 539, 14. Æt ðæm cirre wurdon Ahténiense swá wælhreówlíce forslagen *quam pugnam atrociorem fuisse ipse rerum exitus docuit,* Ors. 3, 7 ; Swt. 118, 22. Hí woldon habban ðone hálgan Eásterdæg geblódegodne wælhreówlíce (wel-, *v. l.*) mid ðæs Hǽlendes blóde, Homl. Ass. 68, 62. Swá ðæt hé wælhreáwlíce wurde áhangen, Homl. Th. ii. 252, 22. Hé ðæt suíðe wælhreówlíce (*crudeliter*) gecýðde on Urias slæge, Past. 3 ; Swt. 35, 23. Ðæt hé ne weorðe wælhreó[w]líce (-reówlíce, Cott. MSS.) *crudeliter*) gefangen mid ðæm grinum uncysta, 43 ; Swt. 313, 12. Wælhreówlíce swingan, Homl. Th. i. 424, 12. Hí áxodon, hwí hí swá wælhreówlíce dydon, ðæt hí freónda ne róhton, Homl. Skt. i. 5, 44. II. *horribly, atrociously :*—Ðæt cild wolde wyrian wælhreáwlíce Drihten, Homl. Th. ii. 326, 10.

wælhreówness, e ; *f. Cruelty :*—Wælhreównys *crudelitas,* Ælfc. Gr. 9, 25 ; Zup. 50, 12 : Bd. 1, 14 ; S. 482, 23 (wæll-, Bd. M. 48, 28). Ðara cyninga wælhreównes wæs tó ðam heard, Bt. 29, 2 ; Fox 104, 33. Weard Iulianus for his wælhreównysse ofslægæn, Homl. Skt. i. 7, 419. Wé sceolon déman mildheortlíce bútan wælhreównysse, Homl. Ass. 9, 222. Sceal his steór beón mid lufe gemetegod, ná mid wælhreáwnysse oferdón, Homl. Th. ii. 532, 13. Wé witon hwelce wælhriównessa Neron weorhte, Bt. 16, 4 ; Fox 58, 1.

wæl-hwelp, es ; *m. A dog that slays, a dog for hunting :*—Ic (*a badger*) mé siþþan (*after getting to my hole*) ne þearf wælhwelpes wíg wiht onsittan, Exon. Th. 397, 21 ; Rä. 16, 23.

Wælisc, wæll-. v. Wilisc, wæl.

wæl-líc (?) ; *adj. Deep* (of water) :—On deópum † in welicum (= wæl-lícum. v. wæl) grunde sæwe *in fundo maris,* Hpt. Gl. 452, 23.

wælm. v. wilm.

wæl-mist, es ; *m. A mist that covers the bodies of the slain :*—Hreám wæs on ýðum, wæter wǽpna ful, wælmist ástáh (*the passage refers to the destruction of the Egyptians in the Red Sea*), Cd. Th. 206, 12 ; Exod. 450. Sum sceal on galgan rídan . . . hé, blác on beáme, bídeþ wyrde bewegen wælmiste, Exon. Th. 329, 30 ; Vy. 42.

wæl-net[t], es ; *n. The net of destruction* (?), Cd. Th. 190, 20 ; Exod. 202.

wæl-níþ, es ; *m. Deadly hate, mortal enmity :*—Ðæt ys sió fǽhðo, and se feóndscipe, wælníð wera, Beo. Th. 5992 ; B. 3000. Æfter wælníðe, 170 ; B. 85. Áwehte ðone wælníð Nabochodonossor, Cd. Th. 218, 28 ; Dan. 46. Weallaþ wælníðas, Beo. Th. 4136 ; B. 2065.

wæl-not, es ; *m. A fatal mark, a mark that brings death, a rune that brings death.* v. Kemble in Archæologia, vol. 28, p. 336. See for baleful influence of runes, Egils Saga, c. 75 ; Grettis Saga, c. 81 ; see also Corpus Poeticum Boreale, vol. i. pp. 40, 41, for the virtues of runes :—Hwílum hié (*fiends*) gefeteraþ fæges monnes handa, gehefegaþ ðonne

hé æt hilde sceall wið láð werud lífes tiligan ; áwrîtaþ hié on his wǽpne wælnota heáp, bealwe bócstafas, Salm. Kmbl. 324 ; Sal. 161.

wæl-píl, es ; *m. A deadly dart, death-pang :*—Wæs his mondryhtne endedógor, . . . áwrecen wælpílum wló ne meahte oroð up geteón, Exon. Th. 171, 15 ; Gú. 1127.

wæl-rǽs, es ; *m. A deadly attack, an attack in which men are slain :*—Wæs sió swátswaðu Sweóna and Geáta, wælrǽs wera, wíde gesýne, Beo. Th. 5886 ; B. 2947. Æfter wælrǽse wunde gedýgan, 5055 ; B. 2531. Æfter ðam wælrǽse (*the fight in which Grendel was mortally wounded*), 1652 ; B. 824. Mé ðone wælrǽs wine Scyldinga leánode, 4208 ; B. 2101.

wæl-rǽw. v. wæl-hreów.

wæl-ráp, es ; *m. A rope that binds the deep, a rope with which frost binds the water :*—Ðonne forstes Fæder onlǽteþ, onwindeþ wælrápas, Beo. Th. 3224 ; B. 1610. v. wæl.

wæl-reáf, es ; *n.* I. *what is taken from the slain, spoil taken in war, spoil, prey :*—Waelreáf (wael-, uuel-reáb) *manubium,* Txts. 77, 1277. Wælreáf, Wrt. Voc. ii. 54, 44 : *manubia* (the passage is : Vesperi dirimens manubias (v. Gen. 49, 27), Ald. 26), 78, 48. Hé under segne sinc ealgode, wælreáf werede, Beo. Th. 2414 ; B. 1205. Ic sceal langne hám ána gesécan, lǽt mé on láste líc eorðan dǽl wælreáf wunigean weormum tó hróðre, Apstls. Kmbl. 189 ; Ap. 95. Hé (*the phoenix*) gebringeþ ǽdes láfe (*what is left after it is burnt*) eft ætsomne and ðæt wælreáf (*exuvias suas*) wyrtum biteldeþ, Exon. Th. 216, 24 ; Ph. 273. II. as a technical term, *robbing the slain :*—Walreáf is níðinges dǽde, L. Ath. iv. 7 ; Th. i. 228, 3. Cf. Qui aliquem quocunque modo perimit, videat ne weilref faciat. Weilref dicimus, si quis mortuum refabit armis aut vestibus, aut prorsus aliquibus, aut tumulatum aut tumulandum, L. H. I. 83, 2 ; Th. i. 591, 12, and see two following sections. [*O. H. Ger.* wala-raupa (*de vestitu mortuorum, quod* walaraupa *dicimus*) : *Icel.* val-rauf *spoils ;* val-rof *the plundering the slain on the battle-field.*] Cf. here-reáf.

wæl-réc, es ; *m. Deadly reek :*—' Mé is leófre ðæt mínne líchaman gléd fæþmie ' . . . Wód ða þurh ðone wælréc, Beo. Th. 5315 ; B. 2661.

wæl-regn, es ; *m. A deadly rain* (the rain that caused the Flood) :—Ic on andwlitan sígan lǽte wælregn ufan wídre eorðan ; fǽhðe ic wille on weras stǽlan, and mid wǽgþreáte eall ácwellan, Cd. Th. 81, 24 ; Gen. 1350.

wæl-reów. v. wæl-hreów.

wæl-rest, -ræst, e ; *f. The rest or bed of the slain :*—Wælræste wunian to be dead, Beo. Th. 5796 ; B. 2902 : Exon. Th. 184, 10 ; Gú. 1342. Wælreste ceósan *to die,* Cd. Th. 99, 8 ; Gen. 1643 : Byrht. Th. 135, 5 ; By. 113. Sceal fǽge flǽschoma foldærne biþeaht wunian wælræste (*inhabit the grave*), Exon. Th. 164, 3 ; Gú. 1006. Sió ród foldan getýned wunode wælreste (*lay buried*), Elen. Kmbl. 1444 ; El. 724.

wæl-rún, e ; *f. The secret of approaching slaughter :*—Fyrdleóð ágól wulf on walde, wælrúne ne máð (*proclaimed the coming carnage*), Elen. Kmbl. 56 ; El. 28.

wæl-sceaft, es ; *m. A deadly shaft,* Beo. Th. 801 ; B. 398.

wæl-scel *slaughter, the slain :*—Cirdon cynerófe wíggend on wiþertrod wælscel oninnan, reócende hrǽw, Judth. Thw. 26, 6 ; Jud. 313. v. scelle.

wæl-seax, es ; *n. A war-knife, a sword* or *dagger used in fight :*—Hé wælseaxe gebrǽd, ðæt hé on byrnan wæg, Beo. Th. 5400 ; B. 2703.

wæl-sliht, -sleaht, es ; *m. Slaughter in battle, slaughter, carnage :*—Hér wæs micel wælsliht (-sleht, MS. E.) on Lundenne, Chr. 839 ; Erl. 66, 16. Ðér wearþ micel wælsliht on gehwæþere hond, 871 ; Erl. 74, 32. Wǽpna wælslihtes, Cd. Th. 198, 25 ; Exod. 328. Gemyndig wælsleahta, Exon. Th. 286, 27 ; Wand. 7 : 291, 32 ; Wand. 91. Wæs on healle wælslihta gehlyn, Fins. Th. 57 ; Fin. 28. [Grickes hit (*Troy*) biuunnan mid heora wælslahte (bitere slahtes, 2nd MS.), Laym. 1369.]

wæl-slítende ; *adj. Corpse-rending, that rends the dead :*—Ðæt líc ðǽr (*in the grave*) tó fúlnesse weorðeþ and ðám wælslítendum wyrmum weorðeþ tó ǽte, Wulfst. 187, 14. On helle mid deóflum and mid dracum and mid wælslítendum wyrmum, 241, 12.

wæl-spere, es ; *n. A battle-spear, spear with which slaughter is to be wrought :*—Oft hé gár forlét, wælspere windan on ða wícingas, Byrht. Th. 141, 14 ; By. 322. Syx smiðas sǽtan wælspera worhtan, Lchdm. iii. 52, 31. [Forwunded mid walspere brade, Laym. 28577.]

wæl-steng, es ; *m. A spear :*—Feówer scoldon on ðǽm wælstenge weorcum geferian Grendles heáfod, Beo. Th. 1638.

wæl-stów, e ; *f. The place of the slain,* (1) *a battle-field :*—God ána wát hwá ðǽre wælstówe wealdan móte *God only knows who shall be master of the field,* Byrht. Th. 134, 36 ; By. 95 : Beo. Th. 4108 ; B. 2051 : 5960 ; B. 2984 : Chr. 937 ; Erl. 66, 9 : 871 ; Erl. 76, 7. Æþelwulf cyning gefeaht wiþ .xxxv. sciphlæsta, and ða Deniscan áhton wælstówe geweald, 840 ; Erl. 66, 19. Hié ðǽr nán licgende feoh ne métten, swá hié ǽr bewuna wǽron ðonne hié wælstówe geweald áhton, Ors. 3, 7 ; Swt. 116, 33. On here cringan, on wælstówe wundum sweltan, Byrht. Th. 140, 24 ; By. 293 : Chr. 937 ; Erl. 114, 9. (2) *any*

place where there is slaughter :—Him Loth gewāt of byrig (*Sodom, about to be destroyed*) gangan, wælstōwe fyrr, Cd. Th. 156, 23 ; Gen. 2593. [Cf. *O. H. Ger.* wal-stat : *Dan.* val-plads *battle-field*, beholde valpladsen *to remain master of the field.*]

wæl-strǣl ; *m. f. A fatal shaft :*—Bād se ðe sceolde endedógor āwrecen wælstrǣlum (*the pangs of mortal disease*), Exon. Th. 179, 11 ; Gū. 1260.

wæl-streám, es ; *m. A destructive stream :*—Ðonne wælstreámas (*the waters of the Deluge*) werodum swelgaþ, sceaðum scyldfullum, Cd. Th. 78, 30 ; Gen. 1301.

wæl-sweng, es ; *m. A murderous stroke :*—Æfter wælswenge (*the stroke which killed Abel*), Cd. Th. 60, 25 ; Gen. 987.

wælt *apparently some part of the thigh, a sinew* (?) :—Gif wælt wund weordeþ, .iii, scillingas gebéte, L. Ethb. 68 ; Th. i. 18, 19. (The preceding section deals with wounds to the thigh. As regards the form of the word, it might be compared with *O. H. Ger.* walza *decipula, pedica.*) **wæltan.** v. wiltan.

wæl-wang, es ; *m. A plain of slaughter :*—Ðǣr wæs secg manig on ðam wælwange (*the place at which were assembled those who maltreated St. Andrew*) wíges oflysted, Andr. Kmbl. 2453 ; An. 1228.

wæl-weg (= hwæl-weg or wæl-weg) *the sea :*—Hweteþ on wælweg ofer holma gelagu, Exon. Th. 309, 26 ; Seef. 63.

wæl-wulf, es ; *m.* **I.** as an epithet of a warrior, *a war-wolf, one who is as fierce to slay as is a wolf :*—Wódon wælwulfas, wícinga werod, Byrht. Th. 134, 38 ; By. 96. **II.** as an epithet of a cannibal, *a fierce cannibal, one who preys on the dead like the wolf :*—Wælwulfas bánhringas ábrecan Jóhton, tólýsan líc and sáwle, and ðonne tódǣlan werum tó wiste fǣges flǣschoman, Andr. Kmbl. 297 ; An. 149.

wǣm[m], wǣman. v. wem[m], wéman.

wǣmbede ; *adj. Having a great belly* ; ventriculosus, Wrt. Voc. i. 45, 37.

wǣn, wǣn, wǣn[n], wǣnan, wǣnge, wǣnian, Wǣnte, wǣnys (Hpt. 438, 70), wǣpan. v. wǣpen, wǣgn, wen[n], wénan, wenge, wenian, Wintan-ceaster, wácness, wépan.

wǣpen, wǣpn, es ; *n.* **I.** *a weapon :*—Steng oððe wǣpen *clava*, Wrt. Voc. ii. 20, 63. Mé sceal wǣpen niman, ord and íren, Byrht. Th. 139, 11 ; By. 252. Ðis (*the bridle into which the nails from the cross were put*) bið unoferswíðed wǣpen, Elen. Kmbl. 2375 ; El. 1189. Ælces wǣpnes ord *mucro*, Wrt. Voc. i. 35, 35. Swurdes ord oððe óðres wǣpnes, 84, 22. Wǣpnes ecge, Cd. Th. 109, 30 ; Gen. 1830. Gehealdan heardne méce, wǣpnes wealdan, Byrht. Th. 136, 48 ; By. 168. Gif hé folcgemót mid wǣpnes brýde árǣre, L. Alf. pol. 38 ; Th. i. 86, 16. Be ðám monnum ðe heora wǣpna tó monslyhte lǣnaþ. Gif hwá his wǣpnes óðrum onlǣne ðæt hé mon mid ofsleá, 19 ; Th. i. 74, 1-4. Wǣpnes spor *a wound*, Exon. Th. 280, 2 ; Jul. 623. Áwrítaþ hié on his wǣpne wælnota heáp, Salm. Kmbl. 323 ; Sal. 161. Ic ðý wǣpne gebrǣd, Beo. Th. 3333 ; B. 1664. Hé ðæs beran ceaflas tótær búton ǣlcum wǣmne, Ælfc. T. Grn. 7, 16. Ðæt man wǣpn ábregde ðǣr mæn drincen, L. H. E. 13 ; Th. i. 32, 11. Deáh hwá his ágen spere sette tó óðres mannes húses dura . . . oþþon gif man óðer wǣpn lecge . . . and hwilc man ðæt wǣpn gelæcce, L. C. S. 76 ; Th. i. 418, 6. Hé wǣpen hæfenade be hiltum, Beo. Th. 3151 ; B. 1573. Nolde íc sweord beran, wǣpen tó wyrme, 5031 ; B. 2519 : 5367 ; B. 2687. Gif sweordhwíta óðres monnes wǣpn tó feormunge onfó, oððe smið monnes andweorc, L. Alf. pol. 19 ; Th. i. 74, 9. Sum mæg stýled sweord, wǣpen gewyrcan, Exon. Th. 42, 29 ; Cri. 680. Hé wǣpen up áhóf, bord tó begeorge, Byrht. Th. 135, 39 ; By. 130. Wǣpnu *arma*, Ælfc. Gr. 36 ; Zup. 215, 15. Wǣpna *arma*, wǣpna hús *armamentarium*, Wrt. Voc. i. 35, 1, 2. Eorlas fornóman wǣpen wælgífru, Exon. Th. 292, 16 ; Wand. 100. Wǣpen *arma*, Ps. Surt. 56, 5. Se hálga héht his heordwerod wǣpna onfón, Cd. Th. 123, 5 ; Gen. 2040. Hé ne mihte wǣpna gewealdan, Beo. Th. 3022 ; B. 1509. Byrht. Th. 139, 50 ; By. 272. Wǣpna wyrpum, Exon. Th. 35, 28 ; Cri. 565. Wǣpna wundum, 119, 15 ; Gū. 255. Wǣpna wælslihtes, Cd. Th. 198, 25 ; Exod. 328. Seó wǣpna láf *those whom the sword spared*, 121, 5 ; Gen. 2005 : 220, 20 ; Dan. 74. Se helm hæfelan werede . . . hine worhte wǣpna smið, Beo. Th. 2908 ; B. 1452. Ðá fór hé mid eallum his folce and mid eallum his wǣpnum *omnis equitatus Pharaonis, currus ejus et equites*, Ex. 14, 23. Gif man mannan wǣpnum bebyreþ ðǣr ceás weorð, L. Ethb. 18 ; Th. i. 6, 19. Ðæt folc com mid wǣpnum (woepnum, Lind.) *venit cum armis*, Jn. Skt. 18, 3 : Andr. Kmbl. 2140 ; An. 1071. Gegearwod wǣpnum, Elen. Kmbl. 95 ; El. 48. Wǣpnum geweorðad, Beo. Th. 505 ; B. 250 : 667 ; B. 331. Ælc þing ðe orðode, hé ácwealde mid wǣpnum *omne, quod spirare poterat, interfecit*, Jos. 10, 40. Wǣpnum áswebban, Apstls. Kmbl. 138 ; Ap. 69. Leohtum wǣpnum (*lenibus armis*) gegyrwan, Nar. 10, 27. Scearpum wǣpnum, Exon. Th. 385, 30 ; Rä. 4, 52. Mid gǣstlicum wǣpnum, 112, 24 ; Gū. 148. Gescyldend wið sceaðan wǣpnum, Andr. Kmbl. 2584 ; An. 1298 : Exon. Th. 48, 22 ; Cri. 775. Hí wurpon hyra wǣpen of dúne, Judth. Thw. 25, 33 ; Jud. 291. Wǣpen and gewǣdu, Beo. Th. 589 ; B. 292. Wǣpen healdan, méce, gár and gód swurd, Byrht. Th. 138, 45 ; By. 235. Wépen and sceldas *arma et scuta*, Ps.

Surt. 45, 10. Ealle his wǣpnu (woepeno, Lind.: wépeno, Rush.) hé him áfyrð, Lk. Skt. 11, 22. Hé áwearp his wǣmna, Ælfc. T. Grn. 18, 31. Hié him ealle hiera wǣpeno ágeáfen *arma traderent*, Ors. 4, 13 ; Swt. 210, 21. Hié wǣpna náman *arma sumunt*, 1, 10 ; Swt. 44, 32. Nimaþ eówre wǣpn *ponat vir gladium super femur suum*, Ex. 32, 27. Gegríp (gefóh, Ps. Th.) wǣpen (wépen, Ps. Surt.) and scyld *apprehende arma et scutum*, Ps. Spl. 34, 2. Uoepeno, Rtl. 168, 1. Ðeáh ðe hí wǣpen ne beran *quamvis arma non ferant*, Bd. 2, 2 ; S. 504, 3. Hé ða gǣstlícan wǣpnu ne mæg áberan, Basil admn. 2 ; Norm. 36, 27. **II.** *membrum virile :*—Teors *veretrum*, teors, ðæt wǣpen *vel* lim *calamus*, Wrt. Voc. i. 283, 56. Wǣpen, gecynd (*printed* wepen-gecynd ; *but see* gecynd, **II**) *veretrum*, 44, 58. [Whiles þow art jonge, and þi wepne kene, wreke þe with wyuynge, Piers P. 9, 180.] v. wǣpen-líc, -mann, wǣpned. [*Goth.* wépna ; *pl. arma : O. Sax.* wápan : *O. Frs.* wépin : *O. H. Ger.* wáfan *gladius, framea, telum, falx, scutum : Icel.* vápn.] v. beadu-, camp-, heoru-, here-, hilde-, sige-, weoruld-, wíg-wǣpen.

wǣpen-berend, es ; *m. An armed man :*—Se stronga woepenberend (wépend-, Rush.) gehealdaþ ceafertún his *fortis armatus custodit atrium suum*, Lk. Skt. Lind. 11, 21 ; p. 7, 5. [*O. Sax.* wápan-berand.]

wǣpen-bora, an ; *m. One who bears arms, a warrior :*—Wǣpnbora *armiger*, Ælfc. Gr. 8 ; Zup. 27, 17 : Wrt. Voc. i. 84, 14. Wǣpenbora, 35, 9 : *bellator*, ii. 125, 35. Wǣpenboran *pugiles, gladium portantes, gladiatores*, Hpt. Gl. 424, 15.

wǣpen-getǣc, -tak, es ; *n. A wapentake*, a term used in northern England where in the south *hundred* was used : 'Quod alii vocant hundredum, supradicti comitatus (*counties northward from Northamptonshire*) vocant wapentagium,' L. Ed. C. 30 ; Th. i. 455. The word, which seems to be of Danish origin (cf. Icel. *vápna-tak*, though this is used in a different sense), is thus explained in the document above cited : Cum quis accipiebat prefecturam wapentagii, die statuto in loco ubi consueverant congregari, omnes majores natu contra eum conveniebant, et, descendente eo de equo suo, omnes assurgebant ei. Ipse vero erecta lancea sua, ab omnibus, secundum morem, foedus accipiebat : omnes enim quotquot venissent cum lanceis suis ipsius hastam tangebant, et ita se confirmabant per contactum armorum, pace palam concessa. Anglice vero arma vocantur wapen, et taccare confirmare, quasi armorum confirmacio, vel ut magis expresse, secundum linguam Anglicam, dicamus wapentac, i. e. armorum tactus : wapen enim arma sonat, tac tactus est. Quamobrem potest cognosci quod hac de causa totus ille conventus dicitur wapentac, eo quod per tactum armorum suorum ad invicem confoederates sunt. On this explanation see Stubbs' Const. Hist. i. 99 sq :—Wé willaþ ðæt man namige on ælcon wǣpengetǣce .ii. trýwe þegnas, L. N. P. L. 57 ; Th. ii. 298, 31. Ælc ðara ceápa ðe hé bicgge oððe sylle áðer oþþe [on] burge oþþe on wǣpengetǣce, L. Edg. 5, 6 ; Th. i. 274, 14. On wǣpentake, L. Eth. iii. 1 ; Th. i. 292, 8 : iii. 3 ; Th. i. 294, 3, 8.

wǣpen-geþræc [?], es ; *n. A weapon :*—Ofsend uoepengiðræcc (uoepen, giðræcc ?) *effunde frameam*, Rtl. 168, 5. Cf. Geþrece *apparatu*, Wrt. Voc. ii. 1, 24 : 76, 53 ; Hpt. Gl. 424, 77. Geþræce, 512, 9.

wǣpen-gewrixl, -gewrixle, es ; *n. A passage of arms, an exchange of blows, a conflict, fight :*—Gif hit geweorðe, ðæt wǣpngewrixl weorðe gemǣne þegene and þrǣle, Wulfst. 162, 7. Ðæt heó beaduweorca beteran wurdun on campstede, gárrǣittinge, gumena gemótes, wǣpengewrixles, Chr. 937 ; Erl. 114, 17. [Cf. Icel. vápna-skipti, -viðskipti.]

wǣpen-hete, es ; *m. Armed hate, hate that resorts to arms :*—Ædele sceoldon ðurh wǣpenhete weorc þrowian *the noble ones were to be slain by their foes*, Apstls. Kmbl. 159 ; Ap. 80.

wǣpen-hús, es ; *n. An armoury :*—Wǣpenhús *armamentarium*, Wrt. Voc. ii. 6, 17. [*O. H. Ger.* wáfan-hús.]

wǣpen-leás ; *adj. Without arms, unarmed :*—Ðam wǣpenleásan menn ne mihton ða wælhreówan mid wǣpnum wiðstandam, Homl. Skt. ii. 29, 175. Fram wǣpenleásre fémnan *e virgine inermi*, Wrt. Voc. ii. 144, 38. Gehwilce wǣpenleáse *inermes* (*sine armis*) *quosque*, Hpt. Gl. 423, 48. [*Icel.* vápn-lauss.]

wǣpen-líc ; *adj. Male, masculine :*—Ðæt wǣpenlíce lim *calamus*, Wrt. Voc. ii. 16, 58. Ða wǣpenlícan limo *preputia*, 68, 60 : 69, 16.

wǣpen-mann (wǣp-), es ; *m. A male, a man :*—Wǣpnmann *mas*, Anglia xiii. 366, 23. Éghuelc hé 1 woepenmon (wépenmon, Rush.: wæpned, W. S.) *omne masculinum*, Lk. Skt. 2, 23. Wer oððe wǣpman *vir*, Wrt. Voc. i. 73, 11. Ðes wǣpman *hic mas*, Ælfc. Gr. 9, 25 ; Zup. 50, 15. Ne scríde nán wíf hig mid wǣpmannes reáfe (*veste virili*), ne wǣpman (*vir*) mid wífmannes reáfe, Deut. 22, 5. Woepenmon 1 hee *masculum*, Mk. Skt. Lind. 10, 6. Hé worhte wǣpman (woepenmonn *masculum*, Lind.), Mt. Kmbl. 19, 4. Synna wið wǣpman oððe wífman, L. de Cf. 6 ; Th. ii. 262, 23. Riht is ðæt ǣnige wǣpnmen on mynecena beóderne ne etan ne drincan, Wulfst. 269, 9. Wǣpmen (wǣpned-, *v. l.*) ge wífmen, Bd. 3, 5 ; S. 527, 7. Wǣpmen, Homl. Ass. 27, 73. xx m wífmanna and wǣpmanna (wǣpned-, *v. l.*), Ors. 3, 7 ; Bos. 61, 30 : Homl. Th. i. 442, 1 : Ælfc. Gr. 6 ; Zup. 24, 5. Mægdhád is ǣgðer ge on wǣpmannum ge on wífmannum, Homl. Th. i. 148, 14. [*O. E. Homl.* wap-man *vir* : *Laym.* wap-, wep-mon : *Orm.*

wepp-mann : *Kath.* wep-man : *O. and N.* wep-mon : *Gen. and Ex.* wap-man.] v. wǽpen, II, wǽpned, wǽpned-mann.

wǽpen-strǽl, es ; *m. An arrow to be used as a weapon :*—Synd mē manna bearn mihtigum tóðum wǽpenstrǽlas *filii hominum dentes eorum arma et sagittae*, Ps. Th. 56, 5.

wǽpen-þracu ; *gen.* -þrǽce ; *f. Force of arms :*—Hine monige on winnaþ mid wǽpenþrǽce, Cd. Th. 138, 12 ; Gen. 2290. Hē hēht wígend weccan and wǽpenþrǽce, Elen. Kmbl. 212 ; El. 106. [Cf. *O. Sax.* wāpan-threki.]

wǽpen-þrǽge *arms* (?) :—Sum mæg wǽpenþrǽge (-þrǽce (?), cf. (?) wǽpen-geþræc), wíge tó nytte, módcræftig smið, monige gefremman, ðonne hē gewyrceþ tó wera hilde helm oððe hupseax, oððe heaþubyrnan, scírne mēce, oððe scyldes rond fæste gefēgan wið flyge gáres, Exon. Th. 296, 34 ; Crä. 61.

wǽpen-wífestre, an ; *f. A hermaphrodite ;* hermafroditus, Wrt. Voc. i. 45, 28.

wǽpen-wiga, an ; *m. An armed warrior :* — Ic wæs wǽpenwiga (wǽpen wigan ? *the subject of the riddle is a horn*), nū mec þeceþ geong hagostealdmon golde and sylfore, Exon. Th. 395, 1 ; Rä. 15, 1.

wǽp-mann, wǽpn. v. wǽpen-mann, wǽpen.

wǽpned ; *adj. Male ;* used substantively, *a male, a man :*—Ælc wǽpned gecyndlim ontýnende *omne masculinum adaperiens uuluam*, Lk. Skt. 2, 23. Micel gedál is on wǽpnedes and wífes líchoman, Lchdm. ii. 84, 16. Se ðe mid wǽpnedum men hǽme *qui cum viro coiverit*, L. Ecg. C. 16 ; Th. ii. 144, 7. Wǽpned and wíf geworhte hiæ God *masculum et feminam fecit eos*, Mt. Kmbl. Rush. 19, 4. Wíf and wǽpned, Cd. Th. 12, 33 ; Gen. 195 : 166, 9 ; Gen. 2745. Wífes meoluc ðe wǽpned fēde, Lchdm. ii. 338, 8. v. wǽpen, II, *and following compounds.*

wǽpned-bearn, es ; *n. A male child, a boy :*—For wǽpnedbearne ... for wífcilde *pro masculo ... pro femina*, Bd. 1, 27 ; S. 493, 14.

wǽpned-cild, es ; *n. A male child, a boy :*—Tó ðan ðæt wíf cenne wǽpnedcild, Lchdm. i. 344, 22 : 346, 3. Ða þínena heóldon ða wǽpned-cild (*mares*), Ex. 1, 17 : *pueros*, 1, 18.

wǽpned-cyn[n], es ; *n. The male kind or sex :*—Wǽpnedcyn *masculinum*, Wrt. Voc. ii. 56, 4. Ælc þing wǽpnedcynnes *omne generis masculini*, Ex. 34, 19 : Cd. Th. 139, 19 ; Gen. 2312 : 142, 35 ; Gen. 2372 : 189, 21 ; Exod. 188. Wið ðon ðe mon oððe nýten wyrm gedrince ; gyf hit sý wǽpnedcynnes . . ., Lchdm. iii. 10, 11. Hwylce wihta beóð óðre tíd wífcynnes, and óðre tíd wǽpnedcynnes, Salm. Kmbl. p. 202, 13 : Exon. Th. 419, 22 ; Rä. 39, 1. Ðæt hí mā of ðam wíf-cynne him cyning curan ðonne of ðam wǽpnedcynne *ut magis de feminea regum prosapia quam de masculina regem sibi eligerent*, Bd. 1, 1 ; S. 473, 22.

wǽpned-hád, es ; *m. The male sex :*—Swá hwæt swá wǽpnedhádes beó ácenned *quidquid masculini sexus natum fuerit*, Ex. 1, 22 : Num. 1, 2. Ærfeweard wǽpnedhádes, Chart. Th. 483, 17.

wǽpned-hand, a ; *f. The male side, male line :*—Hý fóð tó mínum ðe ic syllan mót swá wífhanda swá wǽpnedhanda, swaðer ic wylle, Chart. Th. 491, 32.

wǽpned-healf, e ; *f. The male side :*—Ðonne is mē leófast, ðæt hit gange on ðæt [bearn] strýned on ða wǽpnedhealfe, ða hwíle ðe ǽnig ðæs wyrðe sý, Chart. Th. 491, 16.

wǽpned-mann, es ; *m.*   **I.** *a male, a man :*—Þriwa on gēre ǽlc wǽpnedman (*omne masculinum tuum*) ætýwð beforan Drihtne, Ex. 23, 17 : Num. 34, 23. Wǽpnedman (-men ?) *mares*, Wrt. Voc. ii. 58, 50. Se cyning wæs gód wǽpnedman *rex erat vir bonus*, Bd. 3, 7 ; S. 529, 39. Ðū (*Eve*) scealt wǽpnedmen wesan on gewealde, Cd. Th. 56, 29 ; Gen. 919. Wæs se gryre læssa efne swá micle swá bið wiggryre wífes be wǽpnedmen, Beo. Th. 2573 ; B. 1284. God hī geworhte wǽpnedman and wímman (wǽpman and wýfman, MS. A.: wǽpned and wímman, MS. B.: wēpnedmenn and wífmenn, Rush.) *masculum et feminam fecit eos Deus*, Mk. Skt. 10, 6. Heó eode tó ðære wǽpnedmanna stówe (*ad locum virorum*), Bd. 3, 11 ; S. 536, 19. xx m wífmonna and wǽpnedmonna *viginti millia puerorum ac foeminarum*, Ors. 3, 7 ; Swt. 116, 31. Ðara manna eallra, mid wífmannum and wǽpnedmannum, Blickl. Homl. 79, 19. Hiora wíf ofslógan eall ða wǽpnedmen ðe him on neáweste wǽron, Ors. 1, 10 ; Swt. 48, 1, 6, 8.   **II.** *of plants, a male :*—Gif man scyle mugcwyrt tó lǽcedóme habban, ðonne nime man ða reádan wǽpnedmen and ða grēnan wífmen, Lchdm. iii. 72, 20. v. wǽpen-mann.

wǽpnian ; *p.* ode *To provide with weapons, to arm :*—Ic wǽpnige ðē *armo te*, Ælfc. Gr. 19 ; Zup. 122, 16 : 36 ; Zup. 215, 16. Ic wǽpnige sumne man *armo*, 43 ; Zup. 257, 12. Uoepnedum *armata*, Rtl. 99, 20. [Wepne þine cnihtes, Laym. 17945. He hæhte wepnien (wepni, 2nd MS.) his uolc, 20347. Heo wepnede hire mid bileaue, Kath. 188. Itt þatt wæpnedd iss wiþþ trowwþe on Criste, Orm. 677. *O. Frs.* wēpened : *O. H. Ger.* wáfenen *armare : Icel.* vápna.] v. be-, ge-wǽpnian.

wǽpnung, e ; *f. Armour, arms :*—[Gástlī]cere weápnunge *spiritalis armaturae*, Hpt. Gl. 423, 65. Ymbscrýdaþ eów mid Godes wǽpnunge *induite vos armaturam Dei* (Eph. 6, 11), Homl. Th. ii. 218, 2. Næs Petrus gewunod tó nánre wǽpnunge, 248, 3. Golias geảru tó ánwíge

---

mid ormēttre wǽpnunge, Homl. Skt. i. 18, 21. Iudas com mid ðām cwealmbǽrum mid ormǽtere wǽpnunge (*with an immense amount of weapons*), Homl. Ass. 74, 44 : Homl. Th. ii. 302, 4.

wæps, wǽsp, es ; *m. A wasp :*—Waefs *fespa*, Txts. 63, 859. Waefs *vel* hurnitu (uaeps, Erf. Gl.) *crabro*, 55, 603. Wæps *vespa*, Wrt. Voc. i. 23, 66 : *fe*[s]*þa*, ii. 35, 27. Wæsp, 148, 17 : *vespis*, i. 281, 37. Weaps *vespa*, 77, 49. Uuaefsas (waeffsas, Ep. Gl.) *vespas* (uuaeps *vespa*, Erf. Gl.), Txts. 105, 2098. [*O. H. Ger.* wafsa, wefsa.]

wǽr ; *adj.*   **I.** *ware, aware, having knowledge* of something which is to be guarded against :—Ðá wurdon ða landleóde his (*a band of Danes*) wære and him wiþ gefuhton, Chr. 917 ; Erl. 102, 17. Hē eode nihtes, ðæt hē his lífe geburge, ac ða hǽðenan wurdon wǽre his fare, Homl. Skt. i. 22, 230.   **II.** *ware, prepared for, on guard* against something that might be hurtful, (a) *absolute :*—Beó gē wære *uos estote parati*, Lk. Skt. 12, 40. Ús is mycel þearf, ðæt wē geornlíce wacian and wære beón, Btwk. 220, 27. Se Hǽlend ús warnode, for ðam ðe hē wyle, ðæt wē ware beón, Homl. Ass. 55, 113. Man sceal wacigean and warnian symle, ðæt man geara weorðe . . . Leófan men, utan beón ðe wærran, Wulfst. 90, 10. (b) *with gen.* :—Ús is micel þearf, ðæt wē wære beón ðæs eges-lícan tíman, ðe nū tówærd is, Wulfst. 191, 25. (c) *with preposition :*—Wes ðū giedda wís, wǽr wið willan, Exon. Th. 302, 26 ; Fä. 42. Sóna wyrð deófol inne ; is micel þearf ðæt manna gehwylc wið swylc wǽr sý, Wulfst. 280, 11. Ðæt wē geornlíce wacian and á wære beón wið deófles costnunga, Btwk. 220, 35. Woruldmenn wǽron ware wið heora fýnd, Homl. Skt. i. 13, 150. Wosas gē wære fram monnum *cavete ab hominibus*, Mt. Kmbl. Lind. 10, 17.   **III.** *ware, careful* to avoid something, *on guard* against doing something, (a) *with gen.* :—Wēnde ic ðæt ðū ðý wærra weorþan sceolde swylces gemótes, Exon. Th. 267, 34 ; Jul. 425. (b) *with preposition :*—Beó wǽr æt ðam, ðæt ðū nǽfre mínne sunu þyder ne lǽde *cave, ne quando reducas filium meum illuc*, Gen. 24, 6. (c) *with a clause :*—Mín bearn, beó ðē wǽr ðæt ðū ne drince of ðam wíne, Homl. Th. ii. 170, 17. Wǽrne ðē beón, ðæt ðū náht unrihtes ne dō getácnaþ, Lchdm. iii. 214, 25.   **IV.** *ware, observant* of, *attentive* to a warning :—Ðæt hí wære ðæs cwydes, Wulfst. 7, 6 : L. I. P. 19 ; Th. ii. 330, 2.   **V.** *wary, cautious, sagacious, prudent, cunning :*—Wǽr *cautus*, i. *sagax, prudens, acutus*, Wrt. Voc. ii. 130, 5. Wǽr geápnis *argumentum*, 125, 1. Hē bið scarp and biter and swíðe wǽr on his wordum, Lchdm. iii. 162, 13. Wǽr (*printed* þǽr) weorðe worda and dǽda, Exon. Th. 96, 32 ; Cri. 1583. Deófol gedēð, ðæt unsǽlig man wísdómes ne gýmeþ, and gyt gedēð, ðæt hē talaþ hine sylfne wǽrne and wísne, Wulfst. 52, 29. Beó gē swá ware suá suá nædran *estote prudentes sicut serpentes*, Past. 35 ; Swt. 237, 20. Hig sint wǽre and cunnon þénunga, and hig cennaþ ǽr ðam ðe wyt cumon tó him *ipsae obstetricandi habent scientiam, et priusquam veniamus ad eas pariunt*, Ex. i. 19. Se wísdóm gedēþ his lufiendas wíse and wǽre, Bt. 27, 2 ; Fox 98, 1. Werra bið *astutior fiet*, Kent. Gl. 509. Gielpaþ hié suelce hí sién micle wǽrran and wísran ðonne hié *quasi praestantius ceteris prudentes se esse gloriantur*, Past. 35 ; Swt. 243, 25. Ðæt se bið on geþance wǽrast and wísast, se ðe óðerne can raðost ásmeágan, Wulfst. 55, 21. Se þincð nū wǽrrest and geápest ðe óðerne mæig beswícan, Shrn. 17, 23. [*Goth.* wars wisan *to be ware* : *O. Sax.* war wesan wiðar : *O. H. Ger.* gi-war *providus, solers, gnarus, intentus, adtentus, vigilans : Icel.* varr.] v. ge-, un-wǽr.

wǽr *the sea :*—Wē ðissa leóda land gesóhton wǽre bewrecene, Andr. Kmbl. 537 ; An. 269. Hū ðū wǽgflotan, wǽre bestēmdan, sǽhengeste, sund wísige, 974 ; An. 487. [*Icel.* wer; *n.* (poet.) *the sea.*]

wǽr, e ; *f. A covenant, compact, agreement, pledge :*—Wǽr is ǽt-somne Godes and monna, gǽsthálig treów, Exon. Th. 36, 29 ; Cri. 583. [Gewemme]dere wǽre *violati foederis* (*pacti*), Hpt. Gl. 496, 3 : Cd. Th. 186, 18 ; Exod. 140. Wǽre gemyndig, 143, 1 ; Gen. 2372. Wǽre (cf. *Icel. use in pl.*) *foedus*, i. *pactum, conjunctio*, Wrt. Voc. ii. 148, 43. Clam oððe wed oððe wǽra *clasma*, 21, 2. Wǽra *foedera*, i. *pacta amicitiae, certa amicitia*, 148, 38. Ðære sibbe wǽre (*cujus foedera pacis*) betwyh ða ylcan cyningas and heora ríce áwunedon, Bd. 4, 21 ; S. 590, 25. Be-weddedum wǽrum *pactis sponsalibus*, Hpt. Gl. 439, 19. Se cyng mid his folce hiene gesóhte. Ac Agathocles gedyde untreówlíce wið hiene, ðæt hē hiene on his wǽrum (MS. L. *has* warum) beswác and ofslóg *rex pactus est cum Agathocle communionem belli. Sed postquam in unum exercitus junxerunt per Agathoclem insidiis circumventus occisus est*, Ors. 4, 5 ; Swt. 170, 10. Wǽre genóman *foedus fecerunt*, Wrt. Voc. ii. 39, 25. Ðæt ic ða wǽre forlǽte ðe ic tó swá myclum cyninge genom *ut pactum, quod cum tanto rege inii, ipse primus irritum faciam*, Bd. 2, 12 ; S. 513, 24. Wēre trume fæstnie *pactum firmum feriat*, Txts. 172, 8. Ic ðē wǽre míne selle, Cd. Th. 132, 33 ; Gen. 2202 : 171, 22 ; Gen. 2832. Ic ðē bidde, ðæt ðū treówa selle, wǽra ðína, 170, 24 ; Gen. 2818. Gewríþ sibbe wǽre ł wedd *asstringe pacis federa*, Hymn. Surt. 29, 3. Pehta cynn hafaþ sibbe and wǽre mid Angelðeóde *Pictorum natio foedus pacis cum gente habet Anglorum*, Bd. 5, 23 ; S. 646, 34. Haldende wēre *servantes pactum*, Ps. Surt. 118, 158. Utan wē ða drihtenlícan wǽra gehealdan, Wulfst. 253, 3. Wǽre healdan, Cd. Th. 216, 22 ; Dan. 10. Wið Waldend wǽre healdan, fæste treówe, 204, 19 ; Exod. 421 : Andr.

Kmbl. 426; An. 213: Elen. Kmbl. 1643; El. 823: Exon. Th. 339, 28; Gn. Ex. 101. Hé ða wǽre and ða winetreówe lǽstan wolde, 475, 19; Bo. 50: 172, 17; Gü. 1145: Cd. Th. 93, 8; Gen. 1542: 139, 10; Gen. 2307: 142, 23; Gen. 2366. Ðæt ǽnig mon wordum ne worcum wǽre ne brǽce, Beo. Th. 2205; B. 1100. Heó his (*Joseph's*) mægwinum morðor fremedon, wǽre frǽton, Cd. Th. 187, 7; Exod. 147. Hé lyt wǽre gewonade, Exon. Th. 148, 19; Gü. 747. Wé sceolon ús geearnian ða siblecan wǽra Godes and manna, Blickl. Homl. 111, 3. [O. H. Ger. wára *foedus*; Icel. várar; *pl.*] v. freoðo-, friðo-wǽr.

wǽr (?); *adj. True*:—Ic gelýfe ðæt hit from Gode cóme, bróht from his bysene, ðæs mé ðes boda sægde wǽrum wordum, Cd. Th. 42, 31; Gen. 681. [The word, found here only, if at all, occurs in that part of the Genesis, which seems to show Old Saxon influence, and the phrase *wǽrum wordum* may be the equivalent of that found often in the Héliand, e. g. Gumon, thea ús gôdes so filu gehêtun fon heðankuninge wárun wordun, 569. But perhaps *wærum* (v. wær, **V**; and see last passage under *wær-líc*) might be read. Cf. Heó geleáfan nom ðæt hé ða bysene from Gode brungen hæfde ðe hé hire swá wǽrlíce (= O. Sax. wárlíko; or? wǽrlíce *cunningly*) wordum sægde, iéwde hire tácen, and treówa gehêt, Cd. Th. 41, 5; Gen. 652.] [O. Sax. wár: O. Frs. wêr, weer: O. H. Ger. wár, wári *verus, verax: Lat.* vērus.]

wǽrc, wræc, es; *m. Wark* (in Northern dialects), *ache, pain*:—Mé sár gehrán, wærc in gewód, Exon. Th. 163, 29; Gü. 1001. Seó reádnes and bryne ðæs swyles and wærces *rubor tumoris ardorque*, Bd. 4, 19; S. 589, 31. Wið magan wærce ... Wið wambe wærce, Lchdm. ii. 318, 4, 15: 356, 19, 22. From wærc deáðes *a dolore mortis*, Jn. Skt. p. 2, 3. Wærco ł ádla *dolorum*, Mt. Kmbl. 24, 8. Wærceo, Mk. Skt. Lind. 13, 8. *The word occurs mostly in compounds*, v. bán- (Wrt. Voc. ii. 128, 83), blǽder- (Lchdm. ii. 320, 3), breóst- (Lchdm. ii. 4, 23), ceol- (Lchdm. ii. 312, 2), cneó-, eág-, eár-, felle-, fylle-, fót-, heáfod-, heals- (Lchdm. ii. 312, 5), heort-, lenden-, lifer-, lið-, milte-, rysel- (Lchdm. ii. 318, 15), sculdor-, síd-, stic-, sweor-, tóþ-, þeóh-, þeór-wærc (-wræc). [On eðelich stiche, oðer on eðelic eche (oðer warch, MS. T.), A. R. 282, 12. For evel and werke in bledder, Rel. Ant. i. 51, 34: Icel. verkr: Dan. værk.]

wǽrc (?):—Cuneus wecg ... *cunicellus* lytel wærc (wǽcg?), Wrt. Voc. ii. 137, 28–31.

wǽrcan; *p.* wærhte. **I.** (used impersonally) *to pain*:—Gif hine innan wærce, Lchdm. ii. 272, 11. Gif ða þeóh wærce, 312, 7. Ðonne monnes wambe wærce oððe rysle, 318, 20. **II.** *to suffer pain* (?), *be troubled*:—Ic werhte eom *exercitatus sum* (if werhte *can be taken as the past tense of the verb*, eom *is superfluous*), Ps. Spl. 76, 3. [v. Jamieson's Dictionary, wark, werk *to ache: Dan.* værke, det værker i mit Hoved *my head aches.*]

wǽrc-sár, es; *n. Pain*:—Fruma wercsáre *initium dolorum*, Mk. Skt. Rush. 13, 8.

-wǽre, -wǽred, wǽrelíce. v. on-wǽre, ge-wǽred (Wrt. Voc. ii. 148, 37), wearglíce.

wǽr-fæst; *adj. Faithful*, (1) as an epithet of the Deity:—Waldend gemunde wǽrfæst (*faithful to his covenant*) Abraham árlíce, Cd. Th. 156, 8; Gen. 2585. Ús Hǽlend God wǽrfæst onwráh *Jesus, faithful to the covenant, has revealed God to us*, Exon. Th. 24, 13; Cri. 384. Wǽrfæst Metod, Cd. Th. 79, 33; Gen. 1320: 175, 23; Gen. 2900. (2) of men:—Se eádega Loth, wǽrfæst, Waldende leóf, Cd. Th. 156, 29; Gen. 2596. Hálig, wǽrfæst (*Juliana*), Exon. Th. 256, 27; Jul. 238. Wǽrfæst (*St. Andrew*), Andr. Kmbl. 261; An. 1312: (*Abraham*), Cd. Th. 1091, 7; Gen. 1819. Fæder Abrahames, wǽrfæst hæle, 104, 24; Gen. 1740. Ne lét ðú (*Abraham*) ðé ðín mód ásealcan, wǽrfæst willan mínes (*faithful in observing my will*), 130, 31; Gen. 2168. Wǽrfæstne rinc (*Abel*), 62, 9; Gen. 1011. Wǽrfæstne hǽleð (*St. Andrew*), Andr. Kmbl. 2548; An. 1275. Ða (*the three children*) wǽron wǽrfæste, wiston Drihten écne, Cd. Th. 227, 29; Dan. 194. Wǽrfæstra wera (*Abraham and Lot*), 113, 34; Gen. 1897. (3) of things:—Ðǽr sceal lufu uncer wǽrfæst wunian, Exon. Th. 173, 19; Gü. 1163.

wǽrg, wǽrgan, wǽr-geápnis (Wrt. Voc. ii. 125, 1), wǽr-genga, wǽrg-olness, wǽrgþu, wǽriht. v. wearg, wirgan, wǽr, **V**, wer-genga, weargolness, wirgþu, wearriht.

Wǽring-wíc *Warwick*:—On ðison geáre wæs Wǽrincwíc getimbrod, Chr. 915; Th. i. 189, col. 2. Æt Wǽringwícon (-um), 913; Th. i. 186, col. 2, 187, col. 1.

Wǽringwíc-scír, Wǽring-scír, e; *f. Warwickshire*:—Tó Wǽrincwícscíre (Wǽringscíre, p. 277, cols. 1, 2), Chr. 1016; Th. i. 276, cols. 1, 2.

-wǽrlǽcan. v. ge-wǽrlǽcan.

wǽrlan; *p.* de *To wend, turn*:—Ðona foerde ł mið ðý wǽrlde *praeteriens*, Jn. Skt. Lind. 9, 1. v. bi-, ge-, ymb-wǽrlan.

wǽr-leás; *adj. Faithless, false*:—Wǽrleás mon ... and ungetreów, Exon. Th. 343, 24; Gn. Ex. 162. Se feónd, wrǽcca wǽrleás, 263, 17; Jul. 351: 267, 26; Jul. 421. Wǽrleás werod (*the fallen angels*), Cd. Th. 5, 5; Gen. 67. Wǽrleásra weorud (*the wicked at the day of judgement*), Exon. Th. 98, 27; Cri. 1614: (*the cannibal Mermedonians*), Andr. Kmbl. 2139; An. 1071.

wǽr-líc; *adj. Cautious, prudent, wise, circumspect*:—Wǽrlíc *cauta, sollicita*, Wrt. Voc. ii. 129, 70. Wǽrlíc bið ðæt man ǽghwilce geáre sóna æfter Eástron fyrdscipa gearwige, L. Eth. vi. 33; Th. i. 324, 3. Wǽrlíc mé þinceþ ðæt gé wæccende wið hettendra hildewóman wearde healden, Exon. Th. 282, 12; Jul. 662. Wísdómes beþearf, worda wǽrlícra, and witan snyttro, se ðǽre ǽðelan sceal andwyrde gifan, Elen. Kmbl. 1083; El. 544. [Icel. var-ligr.] v. ful-, un-wǽrlíc.

wǽrlíce; *adv.* **I.** where there is danger of receiving hurt, *warily, cautiously, circumspectly*, (1) *in a way that guards against surprise*:—Faraþ eów wǽrlíce, ðe lǽs ðe eów gemêton ða ðe eów æfter rídon, Jos. 2, 16. Nimaþ and lǽdaþ hine wǽrlíce (*caute*), Mk. Skt. 14, 44. Ðæt man Malchum suíðe wǽrlíce heólde, ðæt hé ne ætburste, Homl. Skt. i. 23, 644. Áhyld hit wǽrlíce, ðonne gesihst ðú hwæt ðæroninnan sticaþ, Homl. Th. ii. 170, 18. Wé mótan swýðe wǽrlíce ús healdan, gyf wé ús sculan wið deófol gescyldan, Wulfst. 38, 3. Wé sculon wið ðam fǽrscyte symle wǽrlíce wearde healdan, Exon. Th. 48, 5; Cri. 767. Hié sindon suá micle wǽrlícor tó oferbúganne suá mon ongiet ðæt hié on máran ungewitte beóð *qui tanto caute declinandi sunt, quanto insane rapiuntur*, Past. 40; Swt. 295, 21. Hú hý ðam deófle wǽrlícast magan wiðstandan, Wulfst. 80, 3. (2) *in a way that guards against an ill result, safely*:—Námon hí tó rǽde, ðæt him wǽrlícor wǽre, ðæt hí sumne dǽl heora londes wurðes æthæfdon *they came to the conclusion, that it would be safer for them to keep back some part of the price of their land*, Homl. Th. i. 316, 23. Wǽrlícor bið se man gehérod æfter lífe ðonne on lífe *there is less danger of mistake in praising a man after his death than while he is alive*, ii. 560, 14. **II.** where there is danger of doing wrong, *carefully, heedfully, prudently*:—Hwílum bið gód wǽrlíce tó míðanne his hiéremonna scylda *aliquando subjectorum vitia prudenter dissimulanda sunt*, Past. 21; Swt. 151, 8. Behalde hé hine geornlíce ðæt hé wǽrlíce sprece *sub quanto cautelae studio loquatur, attendat*, 15; Swt. 93, 18. Ðætte sié wǽrlíce gehealden sió ánmódnes ðæs godcundan geleáfan *ut unitatem fidei cauta observatione teneatis*, Swt. 95, 14. Wǽrlíce ic mé heóld *caute me tenui*, Coll. Monast. Th. 34, 9. Mǽst þearf is ðæt ǽghwelc mon his áld and his wed wǽrlíce healde, L. Alf. pol. 1; Th. i. 60, 3: Wulfst. 167, 4. Cristendóm wǽrlíce healdan, 78, 8. Is suíðe micel ðearf ðæt hé suá micle wǽrlícor hine healde wið scylde *necesse est, ut tanto se cautius a culpa custodiant*, Past. 28; Swt. 191, 10. [Wearliche to biwiten us seoluen wið þe unwiht of helle, O. E. Homl. i. 245, 17. Þa cheorles warliche heom hudden, Laym. 12300. Temien hire fleschs wislíche and warliche, A. R. 138, 8. Ha heold hire hird wislíche and warliche *familiam pervigili cura gubernabat*, Kath. 82. O. Sax. waralíko: Icel. varliga: O. H. Ger. gi-waralícho *vigilanter, diligenter, solerter*.] v. un-wǽrlíce, *and next word*.

wǽrlíce *truly; or* wǽrlíce *cunningly*. v. wǽr true.

wǽrlícness, e; *f. Caution, care, carefulness*:—Ús is micel wǽrlícnys getácnad and æteówed on ðǽre onfangennysse úres Drihtnes líchaman, Homl. Ass. 163, 263.

wǽr-loga, an; *m. One who is false to his covenant, a faithless, perfidious person*:—Ðonne mánsceaða fore Meotude on ðam dóme standeþ, bið se wǽrloga fýres áfylled, Exon. Th. 95, 25; Cri. 1562. Hám Eormanríces, wráþes wǽrlogan, 319, 8; Víd. 9. Ðone wǽrlogan, láðne leódhatan (*Holofernes*), Judth. Thw. 22, 22; Jud. 71. Hér syndan wedlogan and wǽrlogan *in this land are men false to their pledges and to their covenants*, Wulfst. 165, 37. Wǽrlogan (*the cannibal Mermedonians*), Andr. Kmbl. 141; An. 71: 215; An. 108. Wǽrlogona *the people of Sodom*) sint firena hefige, Cd. Th. 145, 22; Gen. 2409. On wǽrlogum wrecan torn Godes, 152, 33; Gen. 2530. Mid ðyssum wǽrlogan, 151, 4; Gen. 2503. On wǽrlogan (*the people before the flood*) wíte settan, 76, 32; Gen. 1266. Hé sceal wedlogan and wǽrlogan hatian and hýnan, Wulfst. 266, 29. ¶ applied to spirits:—Se atola gást, wráð wǽrloga, Andr. Kmbl. 2595; An. 1299. Hié hýrdon tó georne wráðum wǽrlogan, 1225; An. 613. Wíc æt ðam wǽrlogan *a dwelling with the devil*, Exon. Th. 362, 15; Wal. 37: 269, 24; Jul. 455. Hwílum cyrdon mánsceaþan on mennisc híw, hwílum brugdon áwyrgde wǽrlogan on wyrmes bleó, 156, 31; Gü. 883: 120, 9; Gü. 269: 139, 18; Gü. 595. Hé sceóp ðam wǽrlogan (*the apostate angels*) wræclícne hám, Cd. Th. 3, 16; Gen. 36. [This Dragon of Dissait (*the devil*) ... þis warloghe ... with wilis ynoghe mannes saule to dissaiue, Destr. Tr. 4436–45. A warlow (*a monster*), Alex. (Skt.) 1706. Snakis and oþire warlaȝes wild, þat in þe wod duelled, 3795. To þe way of wickidnes be warlaȝes (*devils*) gidid, 4425. He warded þis wrech man (*Jonah*) in warlowes gutteȝ, Allit. Pms. 99, 258. Þaa warlaus (*v. ll.* deuils, fendes), C. M. 23250. The foulle warlawes of helle, Halliw. Dict.]

wǽr-lot, es; *n. Craft, cunning*:—Wǽrlotes *astus*, Wrt. Voc. ii. 9, 33.

wǽrming, wǽrna. v. wirming, wrænna.

wǽrness, e; *f. Prudence, circumspection, caution*:—Mid wǽrnyssa (*cautela*) in gangende, ðæt óþre gebiddende hé ná gelette, Anglia xiii. 378, 188. Hæfde hé miccle lufan and ealle wǽrnesse tó ǽlcum men (*he was very considerate to everybody*), ... and ðeáh ðe hé on lǽwedum háde beón sceolde, hweðre hé tó ðon wǽrnesse hæfde on eallum ðingum (*he was so circumspect in all things*), ðæt hé munuclíce swíþor lifde ðonne

lǽwedes mannes, Blickl. Homl. 213, 6 11. [*Wick.* warnesse *prudentia.*]
v. un-wærness; wær-scipe.

wærness *cursing,* wǽrnian, wǽrnung, wǽrriht. v. weargness, warenian, wirnung, wearriht.

wǽr-sagol; adj. *Cautious in speech, careful of what one says :*—Se ðe wǽre leássagol, weorðe se sóðsagol; se ðe wǽre bæcslitol, weorðe se wǽrsagol; se ðe wǽre stuntwyrde, weorðe se wíswyrde, Wulfst. 72, 17.

wǽr-scipe, es; m. *Prudence, caution, circumspection, wisdom,* in a bad sense, *cunning, astuteness :*—Wǽrscipe *cautela,* i. *astutia,* Wrt. Voc. ii. 129, 77. Ðæt hié geícen ða gód hira ánfealdnesse mid wǽrscipe, and suá tilige ðǽre orsorgnesse mid ðǽre ánfealdnesse ðætte hé ðone ymbeðonc ðæs wǽrscipes ne forlǽte . . . Ðǽre culfran biliwitnesse sceal gemetgian ðære nǽdran wǽrscipe, ðȳ læs hine se wǽrscipe gelǽde on tó micle hátheortnesse *ut simplicitatis bono prudentiam adjungant, quatenus sic securitatem de simplicitate possideant, ut circumspectionem prudentiae non amittant . . . Debet serpentis astutiam columbae simplicitas temperare, quatenus nec seducti per prudentiam calleant,* Past. 35; Swt. 237, 15–24. Wísdóm is se héhsta cræft, and hæþ on him feówer óþre cræftas, ðara is án wǽrscipe, Bt. 27, 2; Fox 96, 34: 34, 6; Fox 140, 35: Shrn. 175, 27. Á geríst bisceopum wísdóm and wǽrscype, L. I. P. 9; Th. ii. 314, 28. Se swicola hæfð éce wíte, for ðan ðe h.s wǽrscype ne dohte, Homl. Skt. i. 19, 177. Þúhte wísast se ðe wæs swicolost . . . ac wá heom ðæs wǽrscipes, Wulfst. 268, 19. Hý lǽtaþ ðæt tó wǽrscype, ðæt hý óðre magan swicollíce pǽcan, 55, 2, 15. Mid micelum wǽrscype lufian *cum magna cautela diligere,* Anglia xiii. 374, 125. For wísdóme and wǽrscipe *consilio,* Past. 20; Swt. 149, 16. Búton wǽrscipe *unadvisedly,* Homl. Skt. i. 11, 361. Mid máran fultume and mid máran wǽrscipe *circumspectiore cura ac magis instructo adparatu,* Ors. 3, 8; Swt. 120, 25. Hé hæfde Ýrlande mid his werscipe gewunnon, and wiðútan ælcon wǽpnon, Chr. 1086; Erl. 222, 18. Ongiet mínne wísdóm and mínne wǽrscipe (*prudentiam*), Past. 38; Swt. 273, 'g. Ðes sunderhálga hæfde opene eágan tó ælmesdǽdum, ac hé næfde nǽnne wǽrscype ðæt hé ða sóðan eádmódnysse on his weldǽdum geheólde (*he had not the wisdom to observe true humility in his benefactions*), Homl. Th. ii. 432, 1. [Belin wes swíðe wis, and warscipe him folweden, Laym. 5603. Dumbe bestes habbeð þeos warschipe, þet hwon heo beað assailed, heo þrungeð alle togederes, A. R. 252, 6. Warschipe aзaines unþeawes, H. M. 41, 7. Warsipe and wisedom wið deuel, Misc. 14, 426.] v. un-wǽrscipe.

wǽrst-líc, wǽrtere. v. wrǽst-líc, weardere.

wǽrþu(-o); indecl. f. *Sagacity, cunning, cleverness :*—Gif him lífes weard of móde ábrít ðæt micle dysig ðæt hit oferwrigen mid wunode lange, þonne ic wát ðæt hí ne wundriaþ mæniges þinges ðe monnum nú wǽrþo and wunder þynceþ (*many a thing that now seems very clever and wonderful*) *cedat inscitiae nubilus error, cessent profecto mira videri,* Met. 28, 82. v. wǽr, **V**.

wǽr-word, es; n. *A word of caution, forewarning :*—Wǽrwordum *antefatis* (as if from *ante-fatus* = spoken before, cf. *antefata* forewyrde, 100, 28; but the Latin is *ante fatis*. Cf. Hpt. Gl. 529, 40 *fatis* gewyr[dum]), Wrt. Voc. ii. 88, 34: 5, 42.

wǽr-wyrde; adj. *Cautious of speech, prudent in speech, careful of one's words :*—Wǽrwyrde sceal wísfæst hæle breóstum hycgan, nales breahtme hlúd, Exon. Th. 303, 22; Fä. 57. Cf. hrǽd-wyrde.

wæsc *washing :*—Reáfa wæsc *uestimentorum ablutio,* Anglia xiii. 441, 1085. v. ge-wæsc.

wæscan, wacsan, waxan, wacxan, waxsan; p. wósc, wócs, wóx, weóx; pp. wæscen, wacsen, waxen *To wash :*—Heó wæsceþ his hrægl, Exon. Th. 339, 24; Gn. Ex. 99. Ðæt man cláðas waxe, Wulfst. 296, 7. Wicþénas on ðone Sætresdæg ǽgðer ge fata þweán, ge wætercláðas wacsan (waxsan, waxan, *v. ll.*), R. Ben. 59, 7. Wacxon hig hira reáf, Ex. 19, 10. Waxan hig ðæt innewerde, Lev. 1, 9, 13. Ðá hig hira reáf wóxon (*lavissent*), Ex. 19, 14. Ðæt hi heora hrægel weócsan and clǽnsodon, Bd. 1, 27; S. 496, 5. Hé wolde his reówan and hwítlas on sǽ wacsan (wæscan, MS. T.), 4, 31; S. 610, 11. Línene cláðas waxan, Lchdm. iii. 206, 29. Hí sculan waxan sceáp, Chart. Th. 145, 13. [*O. E. Homl.* waschen, weschen; *p.* wosch, wesch: *Laym.* wascen: *Orm* wasshenn; *p.* wessh: *A.R.* waschen; *p.* weosch: *O.L.Ger.* wascan; *p.* wósc: *O.H.Ger.* wascan; *p.* wuosc: *Icel.* vaska; *p.* vaskaði.] v. á-, ge-wæscan (-wacsan); un-wæscen, unáwæscen.

wæsc-ærn, -ern, es; n. *A wash-house :*—Wæscern *lautorium,* Wrt. Voc. i. 58, 22.

wæsce, an; f. *A washing-place.* v. sceáp-wæsce.

wæscing (?) *washing* in weascing-weg *a road leading to a sheepwashing place* (?) :—Tó weascingwege nioðeweardun, Cod. Dip. Kmbl. v. 78, 17: 138, 4.

wæser ? :—Wæser *bubimus* (? *bulimus*; cf. bulimus *vermis similis lacertae in stomacho ñominis habitans,* Corp. Gl. Hessels, 26, 209), Wrt. Voc. ii. 126, 62.

wæsma ?, wæsp. v. here-wæsmum, wæps.

wæstling, es; m. *A coverlet :*—Wæstling *lodix,* Wrt. Voc. i. 59, 34: *stragula,* 25, 46. Wæstlinge, 81, 58. Bedreáf: genihtsumiaþ hwítel

and weslinc (*lena*) and heáfudrægel, R. Ben. Interl. 93, 3. Wæstlinga *stragularum,* Hpt. Gl. 430, 66. [Cf. *Goth.* wasti *clothing.*]

wæstm (-em, -im, -um), es; *m. n.*: e; *f. Growth, increase :*—Wæstm *crementum,* i. *augmentum,* Wrt. Voc. ii. 136, 65. **I.** *growth, produce,* (1) *fruit of the earth or of a vegetable* (lit. or fig.), *plant, fruit :*—Wæstm *fructus,* Wrt. Voc. i. 80, 1. Ofet, wæstm *fruges, frumenta,* ii. 151, 31. Rǽdrípe wæstm *praecoquus fructus,* i. 39, 22. Oftost on treówcynne beóð ða treówa getealde *feminini generis,* and se wæstm *neutri generis,* Ælfc. Gr. 6, 9; Zup. 20, 15. Beó ðínes landes wæstm (*fructus*) gebletsod, Deut. 28, 4, 18. Se ðæs wæstmes.(*the fruit of the tree of knowledge*) onbát, Cd. Th. 30, 21; Gen. 470. Ðæs wæstmes yrþ *illius frugis seges,* Bd. 4, 28; S. 605, 38. Bútan wæstme *sine fructu,* Mk. Skt. 4, 19. Weastme, Mt. Kmbl. 13, 22. Ða beámas wǽron gewered mid wæstme, Cd. Th. 30, 5; Gen. 462. Treów wæstm (westm, v. 12) wircende *lignum faciens fructum,* Gen. 1, 11. Seó eorðe wæstm bereþ *terra fructificat,* Mk. Skt. 4, 28. Hé geseah geblówen treów wæstm berende, Blickl. Homl. 245, 8. Sume sealdon weastm (wæstm, MSS. A. B., Lind.: wæstm, Rush.) *alia dabant fructum,* Mt. Kmbl. 13, 8. Ælc treów ðe gódne wæstm (woestim, Rush.) ne bringð *omnis arbor, quae non facit fructum bonum,* 3, 10. Dóð medemne weastm (wæstm, MS. A., Lind.: wyrþe westem, Rush.), 3, 8. Wæstim gódne, Lk. Skt. Rush. 3, 9. Beámas ða ðe mæst and wæstm mannum bringaþ *ligna fructifera,* Ps. Th. 148, 9. Eorðe salde westem his *terra dedit fructum suum,* Ps. Surt. 66, 7. Ðæt fíctreów, on ðæm hé nánne wæstm ne funde; ðæt getácnaþ ða synfullan ðe nabbaþ nánne wæstm gódra weorca, Blickl. Homl. 71, 35. Wæstm *frumentationem,* Blickl. Gl. Ða wæstmas beóð þurh ágne gecynd eft ácende, Exon. Th. 215, 19; Ph. 255. Fægre land ðonne ðeós folde seó, ðǽr wæstmas scínaþ beorhte, Cd. Th. 277, 34; Sat. 214. Bearwas wurdon tó axan, eorðan wæstma, 154, 10; Gen. 2553. Cumaþ (-eþ?) eádilíc wæstm on wangas, weordlíc on hwǽtum *convalles abundabunt frumento,* Ps. Th. 64, 14. Of ðam twige ludon láðwende, réðe wæstme, Cd. Th. 60, 31; Gen. 990. [Dec] wæstem (wæstme ?) weordian *let earth's fruits honour thee* (cf. benedicite universa germinantia in terra Domino, Hym. T. P. 76), Exon. Th. 190, 28; Az. 80. Weastma (wæstma, MSS. A. B., Lind., Rush.) tíd *tempus fructuum,* Mt. Kmbl. 21, 34. Wæstma, Ex. 23, 15; Met. 20, 101. Hig ǽton of ðæs landes wæstmum (*de frugibus terrae*), Jos. 5, 11. Welig on wæstmum and on treówum opima frugibus atque arboribus, Bd. 1, 1; S. 473, 13: Cd. Th. 81, 3; Gen. 1339. Eówres landes wæstmas (*fruges*), Deut. 28, 42: 1, 25. Westmas, 32, 13: Bt. 33, 4; Fox 130, 7. Wæstmas (wæstmo, Lind.) *fructus,* Lk. Skt. 12, 17. Him eorðe syleþ æþele wæstme, Ps. Th. 66, 6: 67, 15, 16. Dú Adame sealdest wæstme, ða inc wǽron forbodene, Cd. Th. 55, 13; Gen. 894. (2) *fruit* of the body, *offspring, progeny :*—Beó ðínes innoðes wæstm (*fructus*) gebletsod and ðínra nýtena wæstm, Deut. 28, 4, 18. Innoðes wæstm (wæstem, Rush.), Lk. Skt. 1, 42. Se wæstm ðínes innoðes is gebletsad, Blickl. Homl. 5, 21. Ic eom búton westme, ne furðum án spearca mínes cynrenes nis mé forlǽtan, Homl. Skt. ii. 30, 205. Hé weordlíche wæstm gesette, ðe of his innaðe ágenum cwóme, ofer ðín heáhsetl *de fructu ventris tui ponam super sedem meam,* Ps. Th. 131, 12. Ic his cynn gedó brád bearna túdre wæstmum spédig, Cd.Th. 169,19; Gen. 2802. Módor ne bið wæstmum geeácnod þurh weres frige, Elen. Kmbl. 681; El. 341. Wæstmas fédan, Cd. Th. 59, 8; Gen. 960. (3) including the two preceding meanings :—Sceáwode Scyppend úre his weorca wlite and his wæstma ǽlcad níwra gesceafta, Cd. Th. 13, 24; Gen. 207. (4) *fruit* of action, *result :*—For hwan gǽst ðú búton wæstme ðínes gewinnes? Blickl. Homl. 249, 5. Míura gewinna wæstm gefullan, 191, 23. Of wæstmum weorca ðínra *de fructu operum tuorum,* Ps. Th. 103, 12. (5) *fruit, that which may be enjoyed :*—Hine Metod mundbyrde heóld, wilna wæstnum, and worulddugeðum, lufum and lissum, Cd, Th. 117, 3; Gen. 1948. Ic lisse selle, wilna wæstme, ðám ðe ðé wurðiaþ, 105, 24; Gen. 1758. (6) *produce of money, usury.* v. wæstm-sceatt :—Of wæstme *ex usuris,* Ps. Spl. 71, 14. **II.** *growth, growing,* (1) of the growth of plants :—Seó sunne tempraþ ða eordlícan wæstmas ge on wæstme ge on rípunge, Lchdm. iii. 250, 18. (2) *growing* as opposed to diminishing, *increase :*—Seó sǽ and se móna beóð geféran on wæstme and on wanunge, Homl. Th. i. 102, 27: Anglia viii. 327, 26. (3) *growth, thriving :*—Mannum becymð rén ofer eorðan eów tó wæstme (*that you may thrive*), Homl. Skt. i. 18, 64. **III.** *growth, condition reached by growing, stature, form;* the plural is sometimes used when a single person is referred to :—On ealdlícum geárum bið ðæs mannes wæstm gebíged, Homl. Th. i. 614, 13. Úre fulfremeda wæstm is swá swá middæg, ii. 76, 17. Se man áua gǽd uprihte . . . hé sceal smeágan embe ðæt éce líf . . . swiðor ðonne embe ða eordlícan þing, swá swá his wæstm him gebícnaþ, Homl. Th. i. 61. Ðé weorð wæstm ðȳ wlitegra, Cd. Th. 33, 14; Gen. 520. Swá wynlíc wæs his wæstm, ðæt him com from Drihtne, 17, 5; Gen. 255. Cniht, stranglíc on wæstme, Ælfc. T. Grn. 16, 41. Ða beóð on wæstme fíftȳne fóta lange and on brǽde tȳn fótmǽla *homines longi pedum .xv. lati pedum .x.,* Nar. 37, 10 note. Hí (*the Innocents*) wǽron gehwǽde ácwealde, ac hí árísaþ mid fullum wæstme, Homl. Th. i. 84, 22. On geðungenum wæstme, ii. 76, 26. Ðæt seax

áfealleþ, ðe ǽr wæs fæger on híwe and on fulre wæstme, Wulfst. 148, 5. Sió hæfde wæstum wuudorlícran, Exon. Th. 413, 13; Rä. 32, 5. Ðe is ungelíc wlite and wæstmas, siððan ðú mínum wordum getrúwodest, Cd. Th. 38, 27; Gen. 613. Wé gesáwon of ðam entcynne Enachis bearna micelra wæstma (procerae staturae), Num. 13, 34. Wundriaþ weras wlite and wæstma, Exon. Th. 221, 9; Ph. 332. He wæs lytel on wæstmum statura pusillus erat, Lk. Skt. 19, 3. Óðer wæs idese onlícnes, óþer on weres wæstmum, Beo. Th. 2708; B. 1352: Exon. Th. 214, 11; Ph. 237. Sum bið wlitig on wæstmum, 295, 18; Crä. 35. Se ðe he oft ǽr mid wlite and mid wæstmum fægerne geseah, Blickl. Homl. 113, 17. [Fǽla untime on corne and on ealle westme, Chr. 1124; Erl. 252, 33. Westmes þord uuele wederas scal forwurðan, O. E. Homl. i. 13, 28. Wastmes and wederes-sele, Laym. 32108. Brohhte ʒho þe wasstme forþ off wambe, Orm. 1937. He was þogen on wiutre and on wastme, O. E. Homl. ii. 127, 16. Marherete schan of wlite ant of wastum, Marh. 2, 34. Hire wliti westum vultus ipsius claritas, Kath. 310. On westme fæir, Laym. 15698. O. Sax. wastum fruit, growth, stature, form. Cf. Goth. wahstus: Icel. vöxtr: O. H. Ger. wahsmo fructus, statura.] v. bere-, eorð-, fold-, frum-, hwǽte-, lim-, ó-, on-, treów-, un-, up-wæstm.

wæstm-bǽre; adj. Fruitful, fertile, productive:—Wæstmbǽre ferax, Ælfc. Gr. 9, 60; Zup. 69, 5: frugalis, Wrt. Voc. ii. 34, 31. Wæstmbǽru fecunda, 38, 22. (1) referring to inanimate things:—Ðæt wæstmbǽre land campi uberes, Ors. 1, 3; Swt. 32, 2. Sceáwiaþ ðæt land, hwæðer hit wæstmbǽre sí considerate terram, qualis sit, bona an mala, humus pinguis an sterilis, Num. 13, 19. Land ðe ys wæstmbǽre ǽgðer ge on hunie ge on meoluce terram fluentem lacte et melle, Ex. 33, 3. Eletreów westembǽre oliva fructifera, Ps. Surt. 51, 10. Eorðan westembǽre terram fructiferam, Sáwan wæstmbǽre land serere ingenuum agrum, Bt. 23; Fox 78, 21: Met. 12, 1. Treó westembǽru ligna fructifera, Ps. Surt. 148, 9. Wæstmbǽre tyrf feraces glebas, Wrt. Voc. ii. 147, 51. Hwæt bið wæstmbǽrre ðonne meox? Homl. Th. ii. 408, 34. (2) referring to living creatures:—On hire is wæstmbǽre mægðhád, Homl. Th. i. 438, 25. (3) figurative:—Se bið cwealmbǽre, se ðe on yfelnysse ǽfre grówende and wæstmbǽre bið, Homl. Th. ii. 406, 20. Uton beón wæstmbǽre on gódum weorcum, 408, 26. v. un-wæstmbǽre.

wæstmbǽrian. v. ge-wæstmbǽrian fecundare, Wrt. Voc. ii. 148, 48.

wæstmbǽrness, e; f. Fruitfulness, fertility, productivity:—Wæst[m]bērnys fertilitas, Wrt. Voc. i. 76, 80. Wæstmbǽrnes fertilitas, i. habundantia, ii. 147. 77. Wæstmbǽrne[s] ubertas, 151, 33. Wæstembiornis fertilitas, Txts. 180, 19. (1) referring to inanimate things:—Wæstmbǽrnys on eorþan, Homl. Skt. ii. 28, 162. Hí hēton secgan ðysses landes wæstmbǽrnysse (insulae fertilitatem), Bd. 1, 15; S. 483, 15: Homl. Th. i. 286, 19. Wæstmbǽrnesse, Ors. 1, 5; Swt. 34, 9. (2) referring to living creatures:—Nis on nánum óðrum men mægðhád, gif ðǽr bið wæstmbǽrnys, ne wæstmbǽrnys, gif ðǽr bið ansund mægðhád, Homl. Th. i. 438, 27. He him geheóld wæstmbǽrnysse tuddres (fecunditatem sobolis), Bd. 1, 27; S. 493, 8. v. un-wæstmbǽrness.

wæstmbǽru (-o); indecl. f. Fertility:—Ðás eorþan ealle hiere wæstmbǽro he gelytlade terra haec sterilitate suorum fructuum castigatur, Ors. 2, 1; Swt. 58, 20.

wæstm-berende; adj. Fruit-bearing, fertile, fruitful, productive, (1) referring to inanimate things:—Se dǽl se ðæt flód ne grétte ys gyt wæstmberende on ǽlces cynnes blǽdum, Ors. 1, 3; Swt. 32, 13. Seó wæstmberendeste (fertilissima) eorþe, Nar. 5, 20. (2) referring to living creatures:—Mid ðý ne is ǽnig syn wæstmbǽrendes (-beorendes, M. 74, 24) líchoman cum non sit culpa aliqua foecunditas carnis, Bd. 1, 27; S. 493, 2. (3) figurative:—He wæs gefultumiende ðæt heora lár wǽre wæstmberende ipse praedicationem ut fructificaret adjuvans, Bd. 2, 1; S. 501, 38. Ðone æþelan Albanum seó wæstmberende (fecunda) Bryton forþbereþ, 1, 7; S. 476, 34. Woestimberende fructiferum, Rtl. 34, 14. Ðá wǽron ða wæstmberendan breóst ðæs eádigan weres mid ðam láreówdóme ðæs heán magistres Godes gefyllede, Guthl. 2; Gdwin. 18, 8.

wæstm-berendlíc. v. un-wæstmberendlíc.

wæstmberendness, e; f. Fertility, fecundity:—Mid ðý nis ǽnig synn wæstmberendnesse líchoman cum non sit culpa aliqua foecunditas carnis, Bd. 1, 27; M. 74, 24 note. v. un-wæstmberendness.

wæstm-fæst, -fæstness. v. un-wæstm-fæst, -fæstness.

wæstmian; p. ode To bring forth fruit (lit. or fig.), fructify:—Eorðo wæstmiaþ (wæstmas, Rush.) terra fructificat, Mk. Skt. Lind. 4, 28. Ic wæstmede fructificavi, Rtl. 3, 20. Manig yfel we geáxiaþ wæstmian, Blickl. Homl. 109, 2.

wæstm-leás; adj. Without fruit (lit. or fig.):—Ðæt word westemleás geweorðæd verbum sine fructu efficitur, Mt. Kmbl. Rush. 13, 22. Ðí læs ðe se Hláford ús wæstmleáse gemête, Homl. Th. ii. 408, 27. [Itt liþ uss wasstmeleas off alle gode dedess, Orm. 13858.]

wæstm-líc; adj. Fruitful:—Wæstimlíc fructuosus, Rtl. 18, 25.

wæstm-sceatt, es; m. Usury, interest:—Wæstmsceat usura, Wrt. Voc. i. 20, 71. Westemsceat, Ps. Surt. 54, 12. Wæstmscettes fenoris, Germ. 389, 45. Se ðe his feoh tó unrihtum wæstmsceatte (tó westemscette ad usuram, Ps. Surt.) ne syleþ, Ps. Th. 14, 6. Of westemsceattum ex usuris, Ps. Surt. 71, 14.

wǽt; adj. I. wet, moist, damp, consisting of moisture:—Ðæt wæter is wǽt and ceald, Bt. 33, 4; Fox 128, 35: Met. 20, 77. Hyra blód byð wǽt and wearm, Anglia viii. 299, 29. Ðú ðam wættere wǽtum and cealdum foldan tó flóre gesettest, Met. 20, 90. Mid wǽttere rude roseo (purpurei cruoris) rubore (Ald. 61), Hpt. Gl. 507, 63. Gecyrred on wǽtne deáw, Homl. Skt. ii. 30, 441. II. wet, moist, having moisture:—Sié lyft is ǽgðer ge ceald ge wǽt ge wearm, Bt. 33, 4; Fox 128, 35: Anglia viii. 299, 28. Se wǽta wong roscida tellus, Exon. Th. 417, 7; Rä. 36, 1. In wǽtan sihtran; of ðam wǽtan síce; . . . in ðæt wǽte sícc, Cod. Dip. Kmbl. iii. 386, 10-16. Loca humentia, ðæt beóð wǽte stówa, Wulfst. 249, 17. On smeþum landum and on wǽtum, Lchdm. i. 90, 4. On wǽtum (v. ll. wǽtum) stówum, 222, 18. Wǽtum udis, Hpt. Gl. 482, 42: Wrt. Voc. ii. 82, 1. Nǽfre he wǽt and wǽtan hrægel and ða cealdan ásettan wolde nunquam ipsa vestimenta uda atque algida deponere curabat, Bd. 5, 12; S. 631, 24. II a. referring to the humours or juices of bodies:—Ðonne sió wamb swíðe wǽtre gecyndo biþ, ne þrowaþ seó þurst ne hefignesse metta, and gefihð wǽtum mettum, Lchdm. ii. 220, 19-21. Be (wambe) cealdre and wǽtre gecyndo . . . and ðæt hǽmedþing ne sceþeþ hátum líchoman ne wǽtum, 162, 17-20: 222, 1, 2. Eal ða wǽtan þing and ða smerewigan sint tó forbeódanne, 210, 27: 246, 3. III. of weather, wet, rainy:—Lengtentíma ys wǽt, Anglia viii. 299, 27. Of untídlícan gewideran, ðæt is, of wǽtum sumerum and of drýgum wintrum, Ors. 3, 3; Swt. 102, 5. [O. Frs. wēt: Icel. vátr.]

wǽt, es; n. I. wet, moisture:—Se cyle geþrowode wið ða hǽto, and ðæt wǽt wiþ ðam drýgum, Bt. 33, 4; Fox 128, 33: Met. 20, 74. II. liquor, drink:—He ǽna gereorde, and be dǽle ǽt and wǽt gewanod sý reficiat solus, sublata ei portione sua de vino, R. Ben. 69, 14. He ne mæg ǽtes oððe wǽtes brúcan, Homl. Th. i. 66, 9. He fæste, swá ðæt he ne onbyrigde ǽtes ne wǽtes on eallum ðam fyrste, 166, 11: ii. 490, 11: Wulfst. 103, 1. Nán ðing tó ðigenne ne on ǽte ne on wǽte nec quicquam cibi aut potus presumere, R. Ben. 69, 19: 76, 18: Homl. Th. i. 360, 13: ii. 590, 21. Búton ǽte and búton wǽte, H. R. 11, 27. [Þis halwende wet (the blood of Jesus), O. E. Homl. i. 187, 31. Gifernesse deð þet man to muchel nimeð on ete oðer on wete, 103, 7. Lokenn himm fra luffsumm æte and wæte, Orm. 7852.] v. next word.

wǽta, an; m.: wǽte, an; f. I. wet, moisture:—Wǽta humor, Wrt. Voc. i. 76, 78. Hwílum flíht se wǽta ðæt drýge, Bt. 39, 13; Fox 234, 11: Prov. Kmbl. 71. Seó lyft sýcð ǽlcne wǽtan up tó hyre, . . . se wǽta gǽð up swylce mid miste, and gyf hit sealt byð . . . hit byð . . . tó ferscum wǽtan áwend, Lchdm. iii. 278, 7-12. Ðá forscranc ðæt sǽd, for ðan ðe hit nǽfde nǽnne wǽtan. Swá dóð sume menn . . . se wǽta ne fæstnode heora wyrtruman, Homl. Th. ii. 90, 30-35. Wǽte humor vel mador, Wrt. Voc. i. 53, 44. Snáw cymð of ðam þynnum wǽtan, ðe up átogen mid ðære lyfte, Lchdm. iii. 278, 23. Hit wǽtan nǽfde non habebat [h]umorem, Lk. Skt. 8, 6. Hwílum ðæt dríge drífð ðone wǽtan, Met. 29, 48. Hí feallan lǽtaþ seáw of bósme, wǽtan of wombe, Exon. Th. 385, 21; Rä. 4, 48. Wǽtan he (snow) oferhrægeþ, gebryceþ burga geatu, Salm. Kmbl. 612; Sal. 305. II. a liquid:—Wynsum wǽta (water) út flówende, Blickl. Homl. 209, 2. Æfter sóðum gecynde ðæt wæter is brosniendlíc wǽta, Homl. Th. ii. 270, 5. Wolde ðæt folc ðæt fýr ádwæscan, gif hit ǽnig wǽta wanian mihte, 140, 17. Hit wæs mid wǽtan (blood) bestémed, Rood Kmbl. 44; Kr. 22. II a. a liquid that may be drunk or used in cookery, medicine, etc., liquor, drink:—Wǽta liquor, Wrt. Voc. i. 27, 49 (in a list 'de generibus potionum'). Mete cibus, ðrenc potus, wǽta liquor, 82, 47. Úre wǽta wæs olfenda miolc, Shrn. 38, 18. Dó on hunig and on wín . . . dó ðæt se wǽta mǽge oferyrnan ða wyrta, Lchdm. ii. 306, 27. Gesamna tú ámbru hrýþra micgean . . . wylle óþ ðæt se wǽta sié twǽde on bewylled, 332, 17. Ægru sint tó forgánne, for ðon ðe hira wǽte bið fæt and máran hǽto wyrcð, 210, 23. Geðicge ðæs wǽtan (hot water and wine) þreó full fulle, i. 76, 25. Þeáh hý him wǽtan bǽdan, dryncges gedreahte, Exon. Th. 92, 14; Cri. 1508. Wæs glæsen fæt ðæt ðæs wynsuman wǽtan onféng. Ðær wæs gewuna ðǽm folce, ðæt hié tó ðǽm fǽte ástigon and ðǽre heofonlícan wǽtan onbyrigdon, Blickl. Homl. 209, 4-9. Wǽtan (byrele? cf. wín-byrele caupo, 21, 13; or brytta? cf. wín-bryttum cauponibus) caupo, Wrt. Voc. ii. 22, 81. Wǽtan heó ne swelgeþ, ne wiht iteþ, Exon. Th. 439, 27; Rä. 59, 10. Tó leohtum drence (a number of plants then follow), tó wǽtan (for liquor) healf háligwæter, healf eala, Lchdm. ii. 274, 4. Gif mon sié mid wǽtan forbærned, 324, 14. Gif lytel fearh áfealle on wǽtan (liquorem), and cucu sig upp átogen, sprenge man ðone wǽtan mid háligwætere, and þicge man ðone wǽtan; gif hit deád sig, and man ne mǽge ðone wǽtan gesyllan, geóte hine man út, L. Ecg. C. 39; Th. ii. 164, 3-7. Nǽnne wǽtan he ne cúþon wið hunige mengan, Bt. 15; Fox 48, 10. Ne he cealdne wǽtan ne þicge, Lchdm. i. 190, 2: 238, 9. Drince wucan æfter ðon beónbroð and mænige (nǽnige?) óþre wǽtan; óþre wucan . . . , and náne óþre wǽtan . . . ; þriddan wucan . . . náne óþerne wǽtan, iii. 216, 11-15. Ða wyrte wið ðone wǽtan gemencge, drince ðonne, iii. 18, 20. Ne dranc he wínes drenc, ne nán ðæra wǽtena ðe druncennysse styriaþ, Homl. Th. ii. 298, 18. III. moisture in an animal body, humour:—Ðonan cymeþ sió mettrymnes ðǽm healedum,

ðe se wǽta ðæra innoða (*humor viscerum*) ástígð tô ðæm lime, Past. 11; Swt. 73, 9. Ðonne bið se deáðbǽra wǽta (*humor mortiferus*) on ðæm menn ofslægen mid ðæm biteran drence, 41; Swt. 303, 16. Gif ðú wille ðæt yfel swyle and æterno wǽte út berste, Lchdm. ii. 16, 14. Gif sió wamb biþ windes full, ðonne cymð ðæt of wlacre wǽtan; sió cealde wǽte wyrcþ sâr an, 224, 24. Wið ealle gegaderunga ðæs yfelan wǽtan of ðam líchoman, i. 236, 18. Gifernes ârist of ðæs hores wǽtan ðe of ðam magan cymð, ii. 196, 3. Of yfelum wǽtan slítendum ðone magan, . . . gif se seóca man áspíwð ðone yfelan bítendan wǽtan áweg, 60, 20-23. Of yfelre wǽtan slítendre, 4, 30. Wiþ yflum wǽtan and swile . . . hit eal ðæt worms and ðone yfelan wǽtan ádrífþ, 72, 12-15. Hyt ealne ðone wǽtan (*dropsical humour*) út átýhþ, i. 204, 3. **III a.** *water, urine* :—Genim eoferes blǽdran mid ðam micgan, áhefe upp, and ábíd ôþ ðæt se wǽta of áflôwen sý, Lchdm. i. 360, 6. **IV.** *moisture of plants, juice, sap* :—Nim ǽnne sticcan . . . forbærn ðone ôderne ende, ðonne gǽð se wǽta (*v. l.* wǽte) út æt ðam ôðrum ende, Lchdm. iii. 274, 5. Sæp ł wǽte *succus*, Hpt. Gl. 450, 13. Hé bær ða wǽtan ðære uncystan in ðam telgan ðone hé getýhþ ǽr of ðam wyrtruman *portat in ramo humorem vitii, quem traxit ex radice*, Bd. 1, 27; S. 495, 26. [He þoleð hwile druie, and hwile wete, O. E. Homl. ii. 123, 6. Hwo þet bere a deorewurðe licur, oðer a deorewurðe wete in a feble uetles, A. R. 164, 14. Ifulled mid attere, weten alre bitterest, Laym. 19769. *Icel.* væta *wet, rain*.] v. hærfest-wǽta.

wǽtan; *p.* te *To wet, moisten* :—Ic ðweá *lauo, lauas* : ic wǽte *lauo, lauis*, Ælfc. Gr. 37; Zup. 220, 6. Ic mín beád wǽte (wêtu, Ps. Surt.) mid teárum *lacrymis stratum meum rigabo*, Ps. Th. 6, 5. Wǽteþ *ingurgitat*, Wrt. Voc. ii. 90, 59 : 47, 19. Ne is ðæt wín tô þicgenne ðætte hǽteþ and wǽteþ ðone innoþ, Lchdm. ii. 246, 5. Mec (*an animal's skin*) brýd wǽteþ in wǽtre, Exon. Th. 393, 34; Rä. 13, 10. Heó genam ðæs gehálgodan sealtes, and wǽtte, Guthl. 22; Gdwin, 98, 2. Wǽt ðæt gewrit on ðam drence, Lchdm. ii. 350, 15. Wǽt wulle mid biccean hlonde, i. 362, 17. Wǽt ðæt liþ mid ecede, ii. 134, 9. Wǽt mid ðínum scytefingre, Techm. ii. 126, 2. Hí ða lifre wǽten, Lchdm. i. 346, 23. Hé wylle mid ðam seáwe his eágan hreppan and wǽtan, 128, 13. Wǽtan *rigare, humectare*, Hpt. Gl. 421, 54. Wǽtende *humectans*, Wrt. Voc. ii. 43, 28 : Lchdm. ii. 156, 20. Wǽtendum *rorantibus, tingentibus*, Hpt. Gl. 439, 55. [*Icel.* væta *to wet*.] v. ge-wǽtan; wǽtian.

wǽte. v. wǽta.

wǽter, es; *n.* (*the word seems to be feminine in* on ðisse wætere, Blickl. Homl. 247, 25; *see also* Ps. Th. 17, 11 : *and a weak genitive plural* wæterena *is found in* Ps. Th. 31, 7.) **I.** *water* :—Wæter *aqua*, hlúttor wæter *limpha*, Wrt. Voc. i. 54, 17, 18. Wæter *limphale*, ii. 52, 19. Ðæt wæter is brosniendlíc wǽta, Homl. Th. ii. 270, 5. Blôd flêwð ofer eorðan swá swá wæter, Blickl. Homl. 237, 6. Byrneþ wæter swá weax, Exon. Th. 61, 23; Cri. 989. Blôd and wæter ætsomne út bicwôman, 68, 33; Cri. 1113. Ealle gewítaþ swá swá wolcn, and swá swá wæteres streám, Blickl. Homl. 59, 20. Úre líchoma wæs gesceapen of feówer gesceaftum, of eorþan and of fýre and of wætere and of lyfte, 35, 13. Hí forweorðan wæteres gelícost, ðonne hit yrnende eorðe forswelgeþ, Ps. Th. 57, 6. Þegn winedryhten his wætere gelafede, Beo. Th. 5438; B. 2722. Wætre, 5700; B. 2854. Ðætte hé gewǽte his ýtemestan finger on wættre, Past. 43; Swt. 309, 7. Wættre gelícost, Andr. Kmbl. 1906; An. 955. **I a.** *water for drinking* :—Ðæt wæter ásceortode ðe wæs on ðam buturuce, Gen. 21, 15. Ánne drinc cealdes wæteres (wætres, Lind.: wættres, Rush.), Mt. Kmbl. 10, 42. Wæteres (wætres, Lind., Rush.), Mk, Skt. 9, 41 : Andr. Kmbl. 44; An. 22. Hé gehálgode wín of wætere, 1173; An. 587. Wætre, Ps. Th. 123, 3. Hwæt drincst ðú? Ealu, gif ic hæbbe, oþþe wæter, gif ic næbbe ealu, Coll. Monast. Th. 35, 11. **I b.** *water in the sky, rain* :—Ðá hangode swíðe þýstru wæter on ðam wolcnum, and on ðære lyfte, Ps. Th. 17, 11. Ne wæter fealleþ lyfte gebysgad *nec cadit ex alto turbidus humor aquae*, Exon. Th. 201, 25; Ph. 61. Hit wǽron míne wæter, ða ðe on heofenum wǽron, Wulfst. 260, 4. **II.** *where a considerable volume of water is referred to, water of a river, sea, etc.* :—Ic sleá ðises flôdes wæter and hyt byð geworden tô blôde, Ex. 7, 17. Hé funde wynleásne wudu; wæter under stôd, Beo. Th. 2837; B. 1416 : Blickl. Homl. 211, 1. Faraþ ealle ðære eorðan sceátas emne swá wíde swá wæter bebúgeþ, Andr. Kmbl. 666; An. 333. Síd wæter *ocean*, Cd. Th. 7, 2; Gen. 100. Sealt wæter, 13, 6; Gen. 198. Ádô mé of deópe deorces wæteres ðe læs mé besencen sealte flôdas, Ps. Th. 68, 14. Ofer wæteres hrycg *across the sea*, Beo. Th. 947; B. 471. On wæteres ǽht, 1037; B. 516. Hé stilde wæteres wælmum, Andr. Kmbl. 903; An. 452. Wætres swêg, Blickl. Homl. 65, 19. Wætres (*the Deluge*) brôgan, Cd. Th. 84, 10; Gen. 1395: Exon. Th. 200, 16; Ph. 41. Ic hine of wætere genam, Ex. 2, 10. Wætre, Lind.: wættre, Rush.), Mt. Kmbl. 3, 16. Gestreón bewrigen wætere oððe eorðan, Met. 8, 59. Wið wǽge, wætre windendum, Exon. Th. 61, 9; Cri. 982. Ðú ðam wættere foldan tô flôre gesettest, Met. 20, 90. Geót ðæt blôd on yrnende wæter, Lchdm. ii. 76, 15. Se ðe gǽð on deóp wæter, Salm. Kmbl. 448; Sal. 224. Deúp wæter *ocean*, Beo. Th. 3812; B. 1904. Ofer wíd wæter, 4937; B. 2473. Swá wê on laguflôde ofer cald wæter líðan, Exon. Th. 53, 17; Cri. 852 :

Andr. Kmbl. 401; An. 201. **II a.** *water as in* Derwent*water, a body of water, a stream, lake, sea* :—Heó wolde hig þweán æt ðam wætere (*in flumine*) and hyre mêdenu eodon be ðæs wæteres ôfre (*per crepidinem alvei*), Ex. 2, 5. Hé becom tô Iordanes ôfrum ðæs wæteres *he came to the shores of the river Jordan*, Homl. Skt. ii. 23 b, 664, 678 : (*the Danube*), Elen. Kmbl. 119; El. 60. On wætere *in amne*, Coll. Monast. Th. 23, 35. Hé geseah ofer ðæm wætere bǽrne stân, Blickl. Homl. 209, 31. Ðás ðe on ðís wætere (*a flood*) syndon eft hié libbaþ . . . Ða ðe on ðisse wætere syndon, 247, 21, 25. Eástreámas feówer wǽron ádǽlede ealle of ánum wætre, Cd. Th. 14, 17; Gen. 220. Hyra (*the Egyptians'*) wæter wurdon tô blôde, Ors. 1, 7; Swt. 36, 25. Ða þreó wæter, Cd. Th. 133, 16; Gen. 2211. Swá swá ealle wæteru cumaþ of ðære sǽ, and eft ealle cumaþ tô ðære sǽ, Bt. 24, 1; Fox 80, 23. Wætera *laticum*, Wrt. Voc. ii. 52, 17. Hé tô Iordane becom ealra wætera ðam hálgestan, Homl. Skt. ii. 23 b, 63. Sǽs and wætra heá holmas, Exon. Th. 193, 16; Az. 122. Fiscwyllum wæterum *fluviis multum piscosis*, Bd. 1, 1; S. 473, 15. Hí witon on hwelcum wæterum hí sculun sêcan fiscas, Bt. 32, 3; Fox 118, 19. Ðæt folc fôr betwux ðám twám wæterum (*the two parts of the Red Sea*), Wulfst. 293, 16. Seó eorðe wæs wætrum weaht, lagostreámum leoht, Cd. Th. 115, 19; Gen. 1922. Mid bricgum ofer deópe wæteru, L. Edg. C. 14; Th. ii. 282, 10. Lǽt forð ðíne willas and tôdǽl ðín wætru æfter herestrǽtum, Past. 48; S. 373, 13, 15. Áþene ðíne hand ofer ealle Egipta wætro and flôdas, ge ofer burnan ge ofer meras and ofer ealle wæterpyttas, Ex. 7, 19. **II b.** *in plural, waters, implying abundance or great extent, waters of a great river, of a sea, etc.* :—Ða fixas ðe synd on ðam flôde ácwelaþ, and ða wæteru forrotiaþ, Ex. 7, 18. Ðǽr wǽron manega wætro (uætro, Lind.: wæter, Rush.) *there was much water there*, Jn. Skt. 3, 23. Ðé wæter sceáwedon and ðé gesáwon sealde ýpa . . . wæs swég micel sealtera wætera, Ps. Th. 76, 13. Swá ǽr wæter fleówan, flôdas áfýsde, Exon. Th. 61, 16; Cri. 985 : Andr. Kmbl. 3105; An. 1555. Ðæt lêg miclade, and him nǽnig mon mid wætra onweorpnesse wiþstondan meahte, Bd. 2, 7; S. 509, 20. Ofer wætera geðring, ofer hwæles êðel, Chr. 975; Erl. 126, 21 : Exon. Th. 351, 13; Sch. 351. Ýða gelaac, wíd gang wætera, Ps. Th. 118, 136. Ðæt flôd ðæra myclena wæterena, 31, 7. Wætrum bisencte, Exon. Th. 271, 9; Jul. 479 : Cd. Th. 88, 4; Gen. 1460. Ða scíran wæter *liquidas lymphas*, Wrt. Voc. ii. 50, 11. Hát mê cuman tô ðé ofer ðás wæteru (wætra, Lind.: ðæt wæter, Rush.), Mt. Kmbl. 14, 28. Hú heó mihte Iordanes wæteru ofersaran, Homl. Skt. ii. 23 b, 680. Wǽtru, 684. Hé gegaderode eall sǽ wætru. (*aquas maris*), Ps. Th. 32, 6. **II c.** *in reference to the surface of water* :—Ðæt hié nǽren .x. fôta heá bufan wætere *decem pedum altitudine a mari aberant*, Ors. 5, 13; Swt. 246, 11. Under wætere, Beo. Th. 3316; B. 1656. [*O. Sax.* watar : *O. Frs.* weter : *O. H. Ger.* wazzar. Cf. *Goth.* watô : *Icel.* vatn.] v. font- (fant-), hálig-, hreód-, neáh-, weorold-, wille-wæter; wæter-ordâl.

wæter-ádl, e; *f. Dropsy* :—Se ðe him seó wæterádl, Lchdm. i. 354, 8. Wið wæterádle . . . seó wæterádl út áflôweþ, 364, 19-20, 11. v. wæter-seócness.

wæter-ǽdre, an; -ǽder, e; *f.* (*in the first passage given the word is made neuter*). *A vein of water, a spring* :—Gewemmed weterêdre *uena corrupta* (Prov. 25, 26), Kent. Gl. 973. Hé hêt ða heardnysse holian onmiddan ðære flôre, and ðæt wæterǽddre ðá wynsum ásprang, werod on swæcce, Homl. Th. ii. 144, 4. Án lamb bícnode mid his swýðran fêt, swilce hit ða wæterǽddran geswutelian wolde. Clemens cwæð : 'Geopeniaþ ðás eorðan' . . . Æt ðam forman gedelfe swêgde út ormǽte wyllspring, i. 562, 10. Ealle wyllspringas and eán þurh hig (*the earth*) yrnaþ. Swá swá ðám ádran licgeaþ on ðæs mannes líchaman, swá licgaþ ðás wæterǽddran geond ðás eorðan, Lchdm. iii. 254, 23. On stemne wæterǽdrena (-ǽdrana, Ps. Lamb. *cataractorum*) ðínra, Ps. Spl. 41, 9 : Blickl. Gl. Wæterǽdra, Ps. Th. 41, 8. Wæterǽddrum *cataractis*, Hpt. Gl. 418, 63. Seó gýtsung hyre gold betweoh ða wæterǽdran rǽt *avaritia aurum inter arenas legit*, Gl. Prud. 55.

wæterælf-ádl, e; *f. Some form of illness* :—Gif mon biþ on wæterælfádle, ðonne beóþ him ða handnæglas wonne and ða eágan teárige, and wile lôcian niþer, Lchdm. ii. 350, 21 : 304, 8.

wæter-ælfen[n], e; *f. A water-elf, water-nymph* :—Wæterælfenne *nymfae*, Wrt. Voc. ii. 62, 31.

wæter-berend, es; *m. A water-bearer* :—Wæterberendra *lixarum* (*mercenariorum qui aquam portant*), Hpt. Gl. 427, 14. v. next word.

wæter-berere, es; *m. A water-bearer* :—Mid wæterbererum *cum lixarum* (*coetibus*, Ald. 13; *the passage is the same as that glossed in the preceding word*), Wrt. Voc. ii. 76, 74 : 18, 2. Wæterberere (-a?) *lixarum*, 52, 73.

wæter-bóg (-bôh), es; *m. A bough with moisture in it* :—Wæterbôh *surculus*, Wrt. Voc. i. 39, 16.

wæter-bolla, an; *m. Dropsy* :—Of ðære ádle cymð ful oft wæterbolla, Lchdm. ii. 202, 5 : 206, 11. Wiþ wæterbollan, 108, 4 : 10, 17 : 204, 13.

wæter-brôga, an; *m. Terror caused by water, the terror of the deep* :—Engel ðín con sealte sǽstreámas, warodfaruða gewinn and wæterbrôgan, Andr. Kmbl. 394; An. 197 : 912; An. 456. Cf. wæter-egesa.

wæter-búc, es; *m. A pitcher* :—Án man mid wæterbúce *homo am-*

þoram aquae portans, Lk. Skt. 22, 10. Gedeon hêt heora ǽlcne geniman ǽnne ǽmtigne sester oðða ǽnne wæterbúc *Gedeon dedit in manibus eorum lagenas vacuas*, Jud. 7, 16.

wæter-bucca, an; *m. An aquatic insect, a water-spider :*—Wæterbuc[c]a *vel* [wæter]gât *tippula*, Wrt. Voc. i. 24, 14.

wæter-burne, an; *f. A stream of water :*—Ic âna sæt innan bearwe . . . ðǽr ða wæterburnan swêgdon and urnon, Dôm. L. 3.

wæter-byden, e; *f. A water-cask ; dolium*, Wrt. Voc. ii. 82, 76.

wæter-clâþ, es; *m. A towel :*—Ðære kycenan wicþênas wæterclâðas wacsan, ðe hý heora handa and fêt mid wîpedan *linthea, cum quibus sibi fratres manus aut pedes tergunt, lavet*, R. Ben. 59, 7 : R. Ben. Interl. 66, 1.

wæter-crôg, es; *m. A pitcher :*—Watercrôg *lagenam*, Wrt. Voc. ii. 74, 28.

wæter-crûce, an; *f. A water-pot :*—Waetercrûce *urciolum*, Wrt. Voc. ii. 124, 19.

wæter-del[l], es; *n. m.* (?) *A dell in which there is water :*—Norð tô wæterdellæ, Cod. Dip. Kmbl. iii. 126, 14.

wæter-denu, e; *f. A valley with water in it :*—Andlang weterdene west tô ðære deópan dene, Cod. Dip. Kmbl. v. 365, 33.

wæter-furh; *f. A trench :*—On ða wæterfurh innan smalan brôc, Cod. Dip. Kmbl. v. 105, 17.

wæter-egesa, an; *m. Terror caused by water :*—Wæteregesa sceal liðra wyrðan *the terrors of the deep shall lose their force*, Andr. Kmbl. 870; An. 435. Wæteregsa, 750; An. 375. Grendles môdor wæteregesan wunian sceolde, cealde streámas *Grendel's mother must live among the dreadful waters, the cold streams*, Beo. Th. 2524; B. 1260. Cf. wæter-brôga.

wæter-fæsten[n], es; *n. A place protected by water :*—Hê gewîcode ðǽr ðǽr hê niéhst rýmet hæfde for wudufæstenne ond for wæterfæstenne *he encamped as near to the Danes as the wood and water, which protected their position, would allow him to find sufficient room*, Chr. 894; Erl. 90, 10.

wæter-fæt, es; *n. A vessel for water, a water-pot :*—Wæterfæt *ydria*, Ælfc. Gr. 9, 56; Zup. 68, 4 : *ydria vel soriscula*, Wrt. Voc. i. 25, 12. Ðæt wíf forlêt hyre wæterfæt (*hydriam*), Jn. Skt. 4, 28. Ðǽr wǽron âset six stǽnene wæterfatu (*hydriae*), 2, 6 : Homl. Th. ii. 56, 5, 21. Ða six wæterfatu getâcnodon six ylda ðyssere worulde, 58, 1. Ðâ hira wæterfatu fulle wǽron *impletis canalibus*, Ex. 2, 16. [*O. H. Ger.* wazzarfaz *hydria*.]

wæter-flasce, -flaxe, an; *f. A water-flask, a pitcher :*—Sum man berende sume wæterflaxan *homo lagenam aquae baiulans*, Mk. Skt. 14, 13.

wæter-flôd, es; *m. n. A flood, deluge ; in plural, floods, waters.* Cf. wæter, II b :—Swilce ôðer wæterflôd swâ fleów heora blôd, Homl. Skt. i. 23, 74. On ðæs Ambictiones tîde wurdon mycele wæterflôd (*inluvies aquarum*) geond ealle world, Ors. 1, 6 ; Swt. 36, 7. Hine storm ne mæg âwecgan, ne wæterflôdas brecan brondstæfne, Andr. Kmbl. 1006 ; An. 503. Hî mê ymbsealdan swâ wæterflôdas (*sicut aqua*), Ps. Th. 87, 17. On wæterflôdum *in aquoso*, 62, 2.

wæter-full; *adj. Dropsical :*—Wæterfull *hydropicus* (v. Lk. 14, 2), Wrt. Voc. ii. 73, 57 : 43, 21.

wæter-fyrhtness, e; *f. Fear of water, hydrophobia :*—Wæterfirhtnys *ydrofobam vel limphatici*, Wrt. Voc. i. 19, 25.

wæter-gât. v. wæter-bucca.

wæter-geblǽd *a blister with water in it* (?) ; *or a blister made by boiling water* (?), Lchdm. iii. 36, 21.

wæter-gelâd, es; *m. A water-way, an aqueduct :*—Wætergelâda *aquae ductuum*, Wrt. Voc. ii. 1, 16.

wæter-gelǽt, es; *n. A water-course, an aqueduct :*—Wætergelǽt *colimbus*, Wrt. Voc. ii. 134, 69. v. wæter-þeóte.

wæter-gewæsc, es; *n. Land formed by the washing up of earth :*—*Circumlutus locus* mid wæter ymbtyrnd stede, *alluvium* wætergewæsc, Wrt. Voc. i. 59, 15, 16.

wæter-grund, es; *m. The bottom of the sea, the depth of the sea :*—On wætergrundum *in profundo*, Ps. Th. 106, 23.

wæter-gyte, es; *m. A pouring of water, a water-course :*—Endlyfta is *aquarius*, ðæt is wætergyte (-scyte, MS. R.), oðða se ðe wæter gýt, Lchdm. iii. 246, 4.

wæter-hæfern, es; *m. A water-crab :*—Genim wæterhæfern gebærnedne, Lchdm. ii. 44, 19.

wæter-hâlgung, e; *f. Blessing or hallowing of water ; aquae benedictio :*—Waeterhâlguncge, Rtl. 117, 1.

wæter-ham[m], es; *m. Land surrounded by a ditch* (?) :—Andlang burnan on wæterweg; of ðan wæterwege on waterhammes ; of ðau hamman on grênan beorh, Cod. Dip. Kmbl. v. 374, 31. Cf. flôdhammas, i. 289, 18.

wæter-helm. v. wegan, III (1).

wæterian; *p.* ode *To water, supply with water,* (1) *to water animals, give drink to living creatures :*—Hê wæterode hig *adquavit eos*, Ps. Spl. 77, 18. Hê wæterode hire heorde *adquavit gregem*, Gen. 29, 10. Hî heora orf wæterodon *refectis gregibus*, 29, 3. Orf wæterian, Ex. 2, 16. Oxan wæterian, Coll. Monast. Th. 20, 1. Ðâ hêt ic wætrigan ûre hors

and ûre niéteno, Nar. 12, 12. Tô wætranne, Lk. Skt. Lind. Rush. 13, 15 : p. 8, 15. (2) *to water plants :*—Se man ðe plantaþ wyrta, hê hî wæteraþ, Homl. Th. i. 304, 26. Sumu treówu hê watrode, Past. 40 ; Swt. 293, 4. (3) *to water land, to irrigate :*—Hê land wæteraþ *arua rigat*, Scint. 118, 14. Ða feówer eán ealne ðisne embhwyrft wæteriaþ, Homl. Skt. i. 15, 177. Ân wyll âsprang ð ðære eorðan wætriende (*irrigans*) ealre ðære eorðan brâdnysse . . . Ðæt flôd . . . tô wætrienne (*ad irrigandum*) neorxena wang, Gen. 2, 6, 10. [Cf. *Icel.* vatna *to water*.] v. ge-wæterian.

wæterig; *adj. Watery :*—Wæterig æcer *alluvius ager*, Wrt. Voc. i. 37, 52. Gif se ûtgang sié windig and wætrig and blôdig, Lchdm. ii. 236, 7. Seó wamb ðe bið wæterigre gecyndo, 220, 26. On wæterigum *in aquoso*, Blickl. Gl. : Ps. Spl. 62, 3. Mid ðam wæterian bleó, Scrd. 21, 27. Rixe weaxst on wæterigum stôwum, Homl. Th. ii. 402, 10 : Lchdm. i. 98, 26. v. un-wæterig.

wæter-leás; *adj. Without water, dry :*—Hig dydon hine on ðone wæterleásan pytt *miserunt eum in cisternam, quae non habebat aquam*, Gen. 37, 24. Hê gâð ðerh stôwa (-e, Rush.) wæterleása (-e, Rush.) *perambulat per loca inaquosa*, Lk. Skt. Lind. 11, 24. [*O. H. Ger.* wazzerlôs *sine aqua*.]

wæter-leást, e; *f. Want of water :*—Ðæt folc weard geangsumod on môde for ðære wæterleáste, Homl. Ass. 108, 177.

wæter-lic; *adj. Aquatic :*—Wæterlíce *aquatiles*, Germ. 394, 243. [*O. H. Ger.* wazzar-lih *aquaticus*.]

wæter-mêle, -mǽle, es; *m. A water-cup :*—Wætermêle *pelvis*, Ælfc. Gr. 9, 78 ; Zup. 75, 15. Wætermǽle *pulvis*, Wrt. Voc. i. 85, 68.

wæter-nædre, an; *f. A water-snake :*—Wæternædre *anguis*, Wrt. Voc. ii. 8, 21 : i. 285, 3 : *salamandra*, 289, 29. Wæternedrum [*h*]*ydris*, ii. 97, 2. [A watyrnedyre *hic idrus*, Wrt. Voc. i. 223, 2. A wateradder *agguis*, 255, 4. Wateraddur *vipera*, 177, 37 (all 15th cent.). *O. H. Ger.* wazzar-natra *natrix, ydrus*.]

wæter-ordâl, es; *n. The ordeal by boiling water :*—Hæbbe se teónd cyre, swâ wæterordâl swâ ýsenordâl, L. Ath. iv. 6 ; Th. i. 224, 15. Cf. Ælc tiónd âge geweald swâ hwæðer hê wille swâ wæter swâ îsen, L. Eth. iii. 6 ; Th. i. 296, 4. *See* ordâl.

wæter-pund. v. pund, III.

wæter-pyt[t], es; *m. A water-pit, well :*—Of ðam wege on ðone wæterpytt ; of ðam pytte on dene, Cod. Dip. Kmbl. vi. 186, 19. On ðone wæterpyt ; of ðam wæterpyt, iii. 359, 15. Heó geseah sumne wæterpytt *videns puteum aquae*, Gen. 21, 19. Ðone wæterpytt *puteum illum* (cf. wyllspring, v. 7), 16, 14. Gif hwâ âdelfe wæterpyt (*cisternam*, Ex. 21, 33), oþþe betýnede ontýne, L. Alf. 22 ; Th. i. 50, 6. Ofer ealle wæterpyttas *super omnes lacus aquarum*, Ex. 7, 19. Hig dulfon wæterpyttas *they dug for water*, 7, 24.

wæter-rîþe, an; *f. A stream of water :*—Wæterîþan *laticem*, Hpt. Gl. 418, 25.

wæter-sceát, es; *m. A napkin ; mappa*, Wrt. Voc. i. 27, 1. v. wæter-scîte.

wæter-scipe, es; *m. A body of water, a piece of water, water :*—Gif hit beón mæg, swâ sceal mynster beón gestaþelod, ðæt ealle neádbehêfe þing ðær binnan wunien, wæter, mylen, wyrtûn (*aqua, molendinum, ortus*), R. Ben. 127, 5. On ðære neáwiste næs nän wæterscipe, Jud. 15, 8. Ðis is se wæterscipe, ðe ûs God tô frôfre gehêt . . . ðæs wæterscipes welsprynge is on hefonríce, Past. 65 ; Swt. 467, 28. Wæterscipes hûs *colimbus*, i. *aquaeductus*, Wrt. Voc. i. 57, 56. Ðâ cwômon ðǽr scorpiones swâ hié ǽr gewunelíce wǽron ðæs wæterscipes *scorpiones consuetam petentes aquationem*, Nar. 13, 11. Ðæt monnum wǽre ðý êþre to ðam wæterscipe tô ganganne *ut facilior aquatoribus esset accessus ad flumen*, 12, 20. Wæs swiþe wynsum wǽta ût fiówende . . . Wæs ongeán ðyssum wæterscipe glæsen fæt, Blickl. Homl. 209, 4. Wæs ðam gebrôðrum micel frêcednys tô âstigenne tô wæterscipe, and cômon tô ðam hâlgan were biddende ðæt hê ða mynstra gehendor ðam wæterscipe timbrian sceolde, Homl. Th. ii. 160, 29-31. Hê heora wæterscipe mid weardmannum besette *constituit centenarios per singulos fontes*, Anglia x. 94, 172. Ðone weterscype ðe hie into Nîwan mynstre be ðes cinges leáfan geteáh, Chart. Th. 232, 3. Hwalas . . . ða ðe lagostreámas, wæterscipe wecgaþ, Cd. Th. 240, 19 ; Dan. 389. Ûre Drihten gesceóp ealle wæterscypas and ða wîdgillan sǽ, Hexam. 4 ; Norm. 6, 24.

wæter-scîte, an; *f. A towel :*—Hê weard bewǽfed mid ânre wæterscýtan (*linteo*, Jn. 13, 4), Homl. Th. ii. 242, 25. v. wæter-sceát.

wæter-scyte, es; *m. A rush of water.* v. wæter-gyte.

wæter-seáþ, es; *m. A water-pit, well, reservoir :*—Ðâ wæs ðǽr on ôþre sîdan ðæs hlâwes gedolfen swylce mycel wæterseáð wǽre, Guthl. 4 ; Gdwin. 26, 8. Wæterseáðes *cisternae*, Hpt. Gl. 418, 27. [Myrige wæterseáðes ðǽr âbûten standeþ, Shrn. 13, 17.]

wæter-seóc; *adj. Dropsical :*—Ðâ wæs sum wæterseóc man *homo quidam hydropicus erat*, Lk. Skt. 14, 2 : Homl. Skt. i. 5, 145. Wæterseóc *lymphaticus*, Hpt. Gl. 514, 30. Ydropicus byð se wæterseóca, Ælfc. Gr. 9, 56 ; Zup. 68, 3. Wæterseóces mannes þurst gecêlan, Lchdm. i. 146, 13. Hit fremaþ ðam wæterseócan, 204, 2. Wæterseóce *hydropicorum*, Hpt. Gl. 478, 3. Heó gehnǽceþ ða anginnu ðam wæterseócum,

Lchdm. i. 272, 15. Hé ða wæterseócan gedrígeþ, 284, 2. [O. H. Ger. wazzar-siuh *hydropicus.*]

wæter-seócness, e; f. *Dropsy:*—Ðeós wæterseócnyss *hic ydrops,* Ælfc. Gr. 9, 56; Zup. 68, 2: Homl. Th. i. 86, 9. Wið wæterseócnysse, Lchdm. i. 122, 19: 144, 21: 202, 19: 234, 5: 272, 13: 276, 13: 322, 5. [Cf. O. H. Ger. wazzar-suht *hydrops.*] v. wæter-ádl, -bolla.

wæter-slæd, es; n. *A valley with water in it:*—On wæterslædes díc, Cod. Dip. Kmbl. v. 297, 11. On ðæt wæterslæd, iii. 394, 17. v. slæd.

wæter-spring, es; m. *A springing up of water:*—Upcyme, wæter-sprync wylla, Cd. Th. 240, 13; Dan. 386.

wæter-steal[1], es; m. *Standing water, a pool:*—Ðær synd unmǽte móras, hwílon sweart wætersteal, hwílon fúle eáríþas yrnende (*sometimes black stagnant water, sometimes foul streams running,* Guthl. 3; Gdwin. 20, 5.

wæter-stefn, e; f. *The voice* or *sound of water:*—Fram wæter-stefnum wídra manigra *a vocibus aquarum multarum,* Ps. Th. 92, 4.

wæter-streám, es; m. *A stream of water:*—Hé wæterstreámas wende tó blóde *convertit in sanguinem flumina eorum,* Ps. Th. 77, 44. [Waterr-stræm, Orm. 18092.]

wæter-þeóte, an; f. *A water-channel, conduit:*—Wæterþeóte *aqua-gium* (aquagium *aquaeductus, canalis,* Migne), Wrt. Voc. i. 22, 23: *canalis vel colimbus vel aquaeductus,* 61, 22. Ðære heofenan wæter-þeótan wǽron geopenode *cataractae coeli apertae sunt,* Gen. 7, 11: 8, 2: Homl. Th. i. 22, 4. On stefne wæterþeótena ðíura *in voce cataractarum tuarum,* Ps. Lamb. 41, 8. [Weterþeotan of þer mycele niwelnisse, O. E. Homl. i. 225, 23. O. H. Ger. wazzar-dioza *cataracta.*]

wæter-þisa (?), an; m. *A water-rusher, what rushes through the water,* applied to a ship and to the whale:—Hé wǽghengest wræc, wæterþisa (-þiswa, MS., *but the* w *is marked for erasure)* tór snel, Exon. Th. 182, 1; Gú. 1303. Hé (*the whale*) hafaþ ó|re gecynd, wæterþisa wlonc, 363, 7; Wal. 50. [Cf. Icel. þeysa *to rush, storm;* þeysir *a rusher, stormer.*] Cf. mere-þyssa.

wæter-þrúh *a water-pipe, conduit:*—Uueterþrúh, uua[e]terthrúch, uaeterthrouch *caractis,* Txts. 47, 367. Wæte[r]þrúh, Wrt. Voc. ii. 129, 1. Þeótan, wæterþrúh *cataractae,* 13, 15. Waeterðrúm *canalibus,* 102, 68.

wæter-þrýþe; *pl. f. Water-hosts, great waters:*—Ða ðe wyrceaþ weorc mænig on wæterðrýþum *qui faciunt operationem in aquis multis,* Ps. Th. 106, 22.

wæter-tyge, es; m. *An aqueduct:*—Wætertige *aquaeductus, canalis,* Hpt. Gl. 418, 50.

wæterung, e; f. *Watering, providing with water,* (1) *providing water* for people:—Sume ða hǽðenan on heora ðeówte leofodon tó wudunge and tó wæterunge (*as hewers of wood and drawers of water*), Homl. Th. ii. 222, 29. (2) *watering* of plants:—Syððan ða wyrta grówende beóð, hé geswýcð ðære wæterunge, i. 304, 27.

wæter-wǽdlness, e; f. *Poverty of water, lack of water:*—For ðyses wéstenes wæterwǽdlnysse, Homl. Skt. ii. 23 b, 538.

wæter-weg, es; m. *A water-way, a channel connecting two pieces of water* (?):—Wæterweg *tramites,* Wrt. Voc. i. 37, 43. Andlang burnan on wæterweg; of ðan wæterwege on wæterhammas, Cod. Dip. Kmbl. v. 374, 30. [Water-wey *meatus,* Prompt. Parv. 518.]

wæter-will, es; m. *A spring of water:*—Ðæt man weorðige wæter-wyllas oþþe stánas, L. C. S. 5; Th. i. 378, 20.

wæter-write, es; m. (or ? -write, an; f.) *A vessel measuring time by the running of water:*—Wæterwrite *clepsydra,* Wrt. Voc. ii. 22, 12.

wæter-wyrt, e; f. *Water-fennel:*—Wæterwyrt *callitriche,* Wrt. Voc. i. 67, 18: *gallitricum,* ii. 42, 38: *gallitricium,* Wülck. Gl. 298, 25 (omitted by Wright). Wæterwyrt. Genim ðás wyrte ðe man *calli-tricum* (*gallitricum,* MS. V.) and óðrum naman wæterwyrt nemneþ, Lchdm. i. 152, 4–6.

wæter-ýþ, e; f. *A wave of water, a wave:*—Beorh wunode on wonge wæterýðum neáh, Beo. Th. 4477; B. 2242.

wǽd, -wǽda, wǽde. v. wǽd, here-wǽða, wáþ.

wǽdan; p. de *To hunt:*—Ic wiht (*a rake*) geseah . . . seó ðæt feoh fédeþ, hafaþ fela tóþa . . . wǽþeþ geond weallas, wyrte séceþ aa, Exon. Th. 416, 27; Rä. 35, 5. Winde gelícost, ðonne hé hlúd ástígeþ, wǽðeþ be wolcnum, Elen. Kmbl. 2545; El. 1274. Brim wíde wǽðde, wæl-fǽðmum sweóp, Cd. Th. 208, 8; Exod. 480. Hwæþer gé willen wǽþan mid hundum on sealtne sǽ (cf. hwæþer gé eówer hundas út on sǽ lǽdon, ðonne gé huntian willaþ, Bt. 32, 3; Fox 118, 14), Met. 19, 15. [O. H. Ger. weidôn *venari, errare, pascere;* Icel. veiða *to hunt.*] v. wáþ.

wǽde-burne (?), an; f. *A fishing-stream* (?):—Of ðæm geate on wǽdeburnan; andlang wǽdeburnan, Cod. Dip. Kmbl. iii. 79, 27. [Cf. Icel. veiði-vatn *a fishing-lake:* O. H. Ger. weida *piscatio.*] v. preceding word.

wǽtian; p. ode *To become wet:*—Ðániaþ and wǽtigaþ *madescunt,* Wrt. Voc. ii. 57, 39. v. wǽtan.

wǽting(-ung), e; f. *Wetting, moistening:*—Ðara breósta biþ deáwig wǽtung (v. wǽtian), swá swá sié geswát, Lchdm. ii. 258, 17. Mid wǽtingum (v. wǽtian) and mettum gelácnian, 222, 8.

wǽtla, an; m. *A bandage:*—Ðonne ðú hit sníþe, ðonne hafa ðé línenne wǽtlan gearone ðæt ðú ðæt dolh sóna mid forwríðe; and ðonne ðú hit eft má lǽtan wille, teóh ðone wǽtlan of, Lchdm. ii. 208, 20–23. Cf. watel.

Wǽtlinga-ceaster, e; f. *St. Alban's:*—Wæs hé ðrowigende se eádiga Albanus ðý teóþan dæge Kalendarum Iuliarum neáh ðære ceastre ðe Rómáne héton Verolamium, seó nú fram Angelðeóde Werlamaceaster oþþe Wæclingaceaster (uaetlingacæstir, -cester, uetlinguacaester, Lat. versions, Txts. 133, 13–14) is nemned, Bd. 1, 7; S. 479, 5. Neáh ðære ceastre ðe Bryttwalas nemdon Uerolamium and Ængla þeód nemnaþ nú Wætlingaceaster, Shrn. 94, 3. Uerulamium, quod nos uulgariter dicimus Wætlingaceaster, Cod. Dip. Kmbl. iii. 248, 31. In loco qui solito æt Uueatlingaceastre nuncupatur uocabulo, 297, 7.

Wǽtlinga-strǽt, e; f. *Watling Street,* the Roman road running from Dover, through Canterbury, Rochester, London, St. Alban's, Dunstable, Fenny Stratford, Towcester, Weedon, Wroxeter to Chester. [From Douere in to Chestre tilleþ Watlingestrete, R. Glouc. 8, 1. According to Trevisa it went ' besides Wrokecestre, and then forth to Stratton, and so forth by the myddell of Wales unto Cardykan, and endeth atte Irisshe see.' Polychron. bk. i. c. 45. Florence of Worcester, in his Chronicle under the year 1013, gives a mythical explanation of the word, that it was the road which the sons of King Weatla made across England] :—Ðis sint ða landgemǽra ðara landa tó Baddanbyrig (*Badby*) and tó Doddanforda (*Dodford*) and tó Eferdúne (*Everdon*) (*all three places are in North-amptonshire, a little to the west of Watling Street*) . . . Súð on gerihte andlang Wætlinga strǽt on ðone weg tó Weóduninga gemǽre (*Weedon,* Cod. Dip. Kmbl. ii. 250, 7: iii. 421, 29. Ðis sint ða landgemǽro intó Stówe (*Stowe in Bucks*). Ǽrest of ðam hálgan wylles forda súð andlang Wætlinga strǽte, 443, 4. Hii sunt termini hujus terrae [*land* at Teobban-wyrðe (*Tebworth, Beds*).] Ðær se díc sceót in Wæclinga strǽte; andlanges Wæxlinga strǽte . . . æfter díce in Wæxlingga strǽte, v. 187, 21–31. Ðis syndon ða landgemǽra tó Hámstede. Of Sandgatan . . . west tó Wætlinga strǽte, vi. 106, 1. On Weaclinga strǽt (*the place is the same as in the first passage given*), 213, 22. Ðonne on gerihte tó Bedanforda, ðonne up on Úsan óð Wætlinga strǽt, L. A. G. 1; Th. i. 152, 10. Hé com ofer Wæclinga strǽte, Chr. 1013; Erl. 148, 6. [In one charter the word occurs in boundaries of land ' æt Eástún,' which Kemble places in Hampshire, the gift of the land being made at Glaston-bury. If this identification is correct the word seems to have been used of more than one road:—Of ðære strǽte in Ebban mór . . . in ðone díc on Uppinghǽma gemǽra (*Upham? Hants*); andlang díces on Wætlinga strǽte, Cod. Dip. Kmbl. iii. 124, 18. [In later English the word was applied to the Milky Way:—The Galaxye, which men clepeth the Milky Wey . . . and somme callen hit Watlinge Strete, Chauc. H. of Fame, ii. 431. Wattelynge strete *lactea, galaxias vel galaxia,* Cath. Angl. 410, and see note.]

wǽtness, e; f. *Wetness, moisture:*—Óðer ne hæbde wǽtnise *aliud non habebat umorem,* Lk. Skt. Lind. 8, 6.

wǽtri[g]an. v. wæterian.

wǽwærd-líc; adj. *Good* (?):—Semis ys swýðe wæwærdlíc tó ongy-tanne, swá hit gerǽd ys on ðære bóc ðe ys Exodus genemned : ' Habuit arca testamenti duos semis cubitos longitudinis.' Héræfter wé wyllaþ geopenian uplendiscum preóstum ðæra geréna æfter Lýdenwara gesceáde, Anglia viii. 335, 30. v. next word.

wǽwærdlíce; adv. *Well, successfully* (?):—Of ðissum syx tídum wihst *se quadrans* swýðe wæwerdlíce, and forð stæpð wel orglíce swylce hwylc cyng of his giftbúre stæppe geglenged, Anglia viii. 298, 34. Nú þincð ðe wærra and micele ðe snotera, se ðe can mid leásungan wæwerd-líce (-werdlíce [e *from* æ], -wyrdlíce, v. ll.) werian, and mid unsóðe sóð oferswíðan, Wulfst. 169, 1.

wæx. v. weax.

wafian; p. ode *To look with wonder, be amazed,* (1) absolute :—Ic wafige *stupeo,* Ælfc. Gr. 26, 2; Zup. 154, 13. Wafede *obstupuit,* Hpt. Gl. 510, 23. Hæled wafedon, Cd. Th. 182, 20; Exod. 78. Ðá wundon hé wundriende and wafiende *cum quasi adtonitus maneret,* Bd. 4, 3; S. 568, 4. Ðæt ðú gange wafiende for hira þinge and ege *sis stupens ad terrorem eorum,* Deut. 28, 34. Ðæt folc wafigende him sáh onbútan, Homl. Skt. i. 23, 650. Wafiende wæferséne *theatrali* (*visibili*) *spectaculo,* Hpt. Gl. 411, 77. Hí swíðe wundredon and wafiende cwǽdon, Lchdm. iii. 436, 7. (2) with gen. *to wonder at, be amazed at:*—Hwá ne wafaþ ðæs, ðonne se fulla móna wyrþ ofertogen mid þióstrum? . . . Ðises hí wundriaþ, Bt. 39, 3; Fox 214, 29. Heora dysige men wafiaþ, 14, 2; Fox 44, 3. Eówre fýnd wafiaþ eówre *stupebunt super ea inimici vestri,* Lev. 26, 32. Ealle men wafedon his ánes, Homl. Skt. i. 23, 616. Ða ðe Símónes wundordǽda wafodan, Blickl. Homl. 173, 22. Hwá ne mæge wafian ælces steorran? Met. 28, 44. Hæfde hé mé gebunden mid ðære wynsumnesse his sanges, ðæt ic his wæs swíþe wafiende *cum me stupentem carminis mulcedo defixerat,* Bt. 22, 1; Fox 76, 7. (2a) case uncertain :—Hwæt is ðeós wundrung ðe gé wafiaþ, Exon. Th. 6, 25; Cri. 89. (3) with prep. v. wafung, II :—Duguð wafade on ðære fǽmnan wlite, Exon. Th. 252, 13; Jul. 162. (4) with a clause :—Þeóda wlítaþ, wun-drum wafiaþ, hú seó wilgedryht wildne weorþiaþ, Exon. Th. 222, 1; Ph.

342. Wafiaþ weras, ðæt . . ., 493, 24; Rä. 81, 86. Hwá is ðæt ne wafige ðæt . . ., Met. 28, 18. Hwá is ðæt ne wafige (cf. hwá ne wundraþ ðæs, ðæt . . ., Bt. 39, 3; Fox 214, 25) hú . . ., 28, 31.

wafian; *p.* ode *To wave* :—Wafa mid ðínum handum, Lchdm. ii. 318, 17. Þeáh ðe man wafige wundorlíce mid handa, ne bið hit þeáh bletsung búta hé wyrce tácn ðære hálgan róde, Homl. Skt. ii. 27, 151.

wafor-líc; *adj.* *Spectacular, theatrical* :—Hí heora waforlícan plegan forléton and heora baða belucon, Ap. Th. 6, 12. v. wæfer-líc, wæfer-sín, wafian, *and following words.*

wafung, e; *f.* **I**, *glossing* spectaculum. v. *two following words* :—Wafung *spectaculum*,. Wrt. Voc. i. 55, 44. On openre wafunge (the passage is: Martyres in Circi *spectaculo* cuparum gremiis includuntur, Ald. 48), Hpt. Gl. 488, 71. Wafunge *spectaculum* (mirum mundo spectaculum exhibuit, Ald. 62), 509, 33. **II**. *amazement, wonder, astonishment* :—On ðære gesihðe hine gestód wundorlíc wafung . . . eall hé wæs ful wundrunge and wafunge, Homl. Skt. i. 23, 501–509. Him an geófor swiðlíc wafung on swá wuldorfæstan wuldre, ii. 23b, 691. Ðá arn ðæt folc tó for wafunge, i. 12, 206. Hit hí mid swá mycelre fyrhto and wafunge (*tanto stupore*) geslóh, Bd. 4, 7; S. 575, 7. Hí sceáwodon ðæt heáfod mid swiðlícre wafunge, Homl. Ass. 112, 331 : Jud. 16, 25. God hæfþ geéced mínne ege and míne wafunga *stuporem meum Deus exaggerat*, Bt. 39, 2; Fox 214, 1. v. webbung.

wafung-stede, es; *m.* *A place for spectacles* (v. wafung, **I**), *a theatre, an amphitheatre* :—Wafungstede *theatrum*, Wrt. Voc. i. 36, 45. Syneweald wafungstede *amphitheatrum*, 37, 1.

wafung-stów, e; *f.* *A place for spectacles, a theatre, an amphitheatre* :—On plegstówe oððe on wafungstówe, Lchdm. iii. 206, 16. v. wæferstów, *and preceding word.*

wág (-h), wǽg, es; *m.* *A wall*, mostly of a building :—Wáh *paries*, Wrt. Voc. i. 81, 8: 290, 7: Ælfc. Gr. 9, 26; Zup. 52, 12. Ælces húses wáh biþ fæst ǽgþer ge on ðære flóre ge on ðæm hrófe, Bt. 36, 7; Fox 184, 12. Him ne wiðstent nán ðing, náðer ne stǽnen weall ne brýden wáh (*a wattled wall*; cf. wága *cratium*, Wrt. Voc. ii. 136, 55, *and next passage*; *and* v. bréden), Homl. Th. i. 288, 4. Graticium wág *flecta* (cf. *flecta* hyrdel, 149, 43), Wrt. Voc. ii. 110, 15. Wág, Exon. Th. 476, 18; Ruin. 9. Ælc wág (*paries*) bið gebiéged twiefeald on ðæm heale, Past. 35; Swt. 245, 13. 'Ðurhðyrela ðone wág (wáh, Cott. MSS.). Ðá ic ðá ðone wáh ðurhðyreludne hæfde . . . Ealle ða hearga wǽron átiéfrede on ðæm wǽge' . . . Hwæt is sió ðyrelung ðæs wáges? 21; Swt. 153, 17–25. On áne studu ðæs wáges (*the wall of the hall*), Bd. 3, 10; S. 534, 29 : (*the wall of a church*), Blickl. Homl. 207, 16. Seó wrǽpstolu ðam wáge (*the wall of the chamber*) tó wrǽþe geseted wæs, Bd. 1, 17; S. 544, 24, 32. Hé wende-hine tó wáge (*the wall of the chamber*), Homl. Th. i. 414, 19. On ðínre healle wáge, ii. 436, 10: Cd. Th. 261, 8; Dan. 723: Andr. Kmbl. 1428; An. 714: Beo. Th. 3328; B. 1662. Wǽge, Exon. Th. 394, 17; Rä. 14, 4. Hé slóg mid his heáfde on ðone wág, ðonne hé on his setl sæt, Ors. 5, 15; Swt. 250, 12. Wáh, Ps. Th. 61, 3. Ða wágas (*the walls of a church*) nǽron rihte, Blickl. Homl. 207, 18 : (*the walls of a palace*), Nar. 4, 24. Ne mó ne lyst mid glase geworhta wága, Bt. 5, 1; Fox 10, 17. Ne beó wé tó weallum oððe tó wágum geworhte on ðære gástlícan gebytlunge, Homl. Th. ii. 582, 14. Web æfter wágum, Beo. Th. 1994; B. 995. Ðæt cyricgrid stande ǽghwǽr binnan wágum, L. I. P. 25; Th. ii. 338, 35. On wágum ðæra húsa ðe wið dúna standaþ, Lchdm. i. 124, 16. Wið wágas, 116, 21. Hí heora heáfdu slógan on ða wágas, Blickl. Homl. 151, 5 : Homl. Th. i. 106, 14. [Wahes, O. E. Homl. i. 247, 17. Þare halle wah, Laym. 25887. Waȝes (*walls of temples*) wowes (2nd MS.), 10182. Wah (wach) oðer wal, A. R. 104, 5. Wiðinnen þe uour woawes, 172, 21. Fra wah to waȝhe, Orm. 1015. Tweȝȝenn waȝhess, 6825. Wowes, O. and N, 1528. Woȝ, Ayenb. 72. Woughe, Wyck. Ps. 61, 4. Wowes, Piers P. 3, 61. O. Frs. wách : *Goth.* waddjus : *Icel.* veggr.] v. cyric-, grund-, súþ-wág (-wǽg).

wǽg *a balance.* v. wǽg.

wág-hrægel, es; *n.* *A wall-covering, a curtain, veil* (of the temple) :—Wághrægl (-hræl, Rush.) temples *velum templi*, Mk. Skt. Lind. 15, 38. Wághræl (-hrægl, Rush.), Lk. Skt. Lind. 23, 45. Wághruhel, Mt. Kmbl. Lind. 27, 51. Bitwih wághræle (wǽghrægle, Rush.), Lk. Skt. Lind. 11, 51. v. wág-rift.

wagian; *p.* ode *To move* (intrans.). **I**. *to wag, wave, shake, move backwards and forwards* :—Hé mihte hearpian ðæt se wudu wagode, Bt. 35, 6; Fox 166, 32. Ða wudubeámas wagedon and swégdon, Dóm. L. 7. Wagedan búta, Exon. Th. 436, 25; Rä. 55, 6. Hreád ðæt wagende, Mt. Kmbl. Rush. 12, 20. **II**. *of that which threatens to fall, to shake, totter* :—Hornsalu wagiaþ, weallas beofiaþ, Exon. Th. 383, 10; Rä. 4, 8. Wagaþ, áslád and gefióll *labat*, Wrt. Voc. ii. 50, 62. Weagat, 112, 43. Wagiende *nutabunda*,.77, 75 : 60, 57. Ðý wagigendan *nutabunto*, 83, 71. **III**. *to shake, be loose.* v. wagung :—His téð ne wagedon *nec dentes illius moti sunt*, Deut. 34, 9. Wið tóþa sáre and gyf hý wagegen (wagigan, wagion, *v. ll.*), Lchdm. i. 126, 15. [Ðe se is eure wagiende, O. E. Homl. ii. 175, 19. Deor gunnen waȝeȝen (pleoye, 2nd MS.), Laym. 26941. *O. H. Ger.* wagón *to be moved.*] v. wecgan, wegan.

wág-rift, es; *n.* *A wall-covering, a curtain, veil* (of the temple) :—

Wǽgryft *curtina*, Wrt. Voc. ii. 105, 68 : 15, 57. Wágrift ðes temples *velum templi*, Ps. Surt. ii. p. 203, 17. Wáhrift, Mk. Skt. 15, 38. Wáhrýft (wǽg-, Rush.), Mt. Kmbl. 27, 51 : Lk. Skt. 23, 45 : Homl. Th. ii. 258, 3. Wáhreft *velum*, Wrt. Voc. i. 74, 2. On ðæs temples wáhrift *contra velum sanctuarii*, Lev. 4, 6. Godweb tó wefanne of seolce wáhrift tó ðam temple, Homl. Ass. 132, 548. Ðǽr synt eác wáhriftu, sum ðe hyre wyrðe bið, Chart. Th. 538, 29. Wágryfta *curtinarum, velarum*, Wrt. Voc. ii. 77, 11 : 18, 6. Wáhrefta, Hpt. Gl. 430, 66. Hé hæfð ðiderynn gedón . . . .ii. wáhræft, Chart. Th. 429, 29. [An waȝherifft wass spredd fra wah to waȝhe, Orm. 1014.] v. heall-wáhrift.

wág-þiling, e; *f.* *Wall-planking, wainscoting* :—Wáhþyling *tabulatorium*, Wrt. Voc. i. 38, 15. [Cf. *Icel.* vegg-þili *wainscoting.*]

wág-þyrel (?) *a door-way* :—Swá swá wáge t wágþeorles áhyldum *tamquam parieti inclinato*, Ps. Lamb. 61, 4.

wagung, e; *f.* *Shaking, looseness.* v. wagian, **III** :—Wið tóþa sáre and wagunge, genim ðás ylcan wyrte, syle etan fæstendum, heó ða téþ getrymeþ, Lchdm. i. 210, 11 : 334, 6.

wáh *a wall.* v. wág.

wáh; *adj.* *Fine* :—Genim wáh mela hæsles oþþe alres, ásift ðonne ful clǽne tela micle hand fulle, Lchdm. ii. 270, 22. [Cf. (?) *O. H. Ger.* wáhi :—Uuáhes prótes *laboratae cereris.*]

wál (?) *some part of a helmet* [cf. *M. H. Ger.* wæl, wæle *contrivance for fastening the crest of a helmet*] :—Ymb ðæs helmes hróf heáfodbeorge wírum bewunden wál an útan (walan utan, MS.). heóld *about the helm's top a 'wál' wire-girt guarded on the outside the head's defence* (i. e. the helmet), Beo. Th. 2067; B. 1031.

wala, (an); *m.* *A root* (?) :—Ad (æt ?) walan *to the root of a matter, to certainty* ; ad liquidum, Wrt. Voc. ii. 2, 46. v. weall-, wyrt-wala.

wala, walas, walca, walch, walc-spinl, wald-, walde, wald-mora, wale. v. wela, wealh, wealca, wealh, wealc-spinl, weald-, willan, wealh-more, weale.

waled; *adj.* *Coloured* (?) :—Waledra *histriatarum* (histriatus *historiis sculptus vel depictus*, Migne), Wrt. Voc. ii. 43, 14. v. (?) walu.

walh. v. wealh.

wá-líc; *adj.* *Woeful, miserable* :—Is ðes wálíc hám (*hell*) wítes áfylled, Cd. Th. 271, 3; Sat. 100. [*O. H. Ger.* wé-líh *miser, dirus, atrox.*] v. weá-líc.

Waller-wente; *pl.* *The Celtic inhabitants of Cumbria* :—Nime hé his mága .xii. and .xii. Wallerwente, L. N. P. L. 51; Th. ii. 298, 8. v. Wente.

walu, e; *f.* *The mark left by a blow, a wale* :—Walu *vibex*, wala *vibices*, Hpt. Gl. 487, 59. Wale *vibice, livore*, 516, 16. Wala *vibices*, 510, 41. Stiðra wala swipa *asperae invectionis mastigias*, 527, 26. [Wale or strype *vibex*, Prompt. Parv. 514. A wale *vibix*, Wülck. Gl. 619, 16.]

walu, e; *f.* *A ridge, bank* (?) :—In stán wale ; andlang ðære wale on ðone portweg, Cod. Dip, Kmbl. iv. 98, 28. Of ðam beorge súþ on ða ealdan wale . . . súþ be wale on ðære díce hyrnan, 31, 2–4. [Wale of a schyppe *ratis*, Prompt. Parv. 514.] v. díc-, stán-walu.

walwian, wam. v. wealwian, wamm.

wamb, e; *f.* **I**. *of living things*, (a) *a belly, stomach* :—Wamb *venter*, Wrt. Voc. i. 71, 21. Seó inre wamb *alvus*, 44, 38. Seó útre wamb *venter*, 45, 21. Gif sió wamb wund bið, Lchdm. ii. 162, 13. Is seó womb (*of the Phenix*) neoþan wundrum fæger, Exon. Th. 219, 14; Ph. 307. Be wambe coþum, Lchdm. ii. 220, 1. Be wambe missenlícre gecyndo, 14. Wiþ wambe wærce, 318, 15. Wiþ wambe heardnesse, 358, 3. Be windigre wambe, 162, 23. Ic wiht (*a sow*) geseah féran, hæfde feówere fét under wombe, Exon. Th. 418, 11; Rä. 37, 3. Eall ðæt on ðone múð gǽð, gǽð on ða wambe (*womb*, Lind. : wombe, Rush. *ventrem*), Mt. Kmbl. 15, 17 : Lchdm. ii. 186, 23. Wambe gefyllan *ventrem implere*, Lk. Skt. 15, 16 : Exon. Th. 494, 22; Rä. 83, 5. Hé hæfð áne wambe and þúsend manna bigleofan, Homl. Th. i. 66, 1. Be cilda wambum and oferfyll, and gif him mete tela ne mylte, Lchdm. ii. 240, 12. (b) *where there is reference to the bringing forth of young, a womb* :—Westem wombe (wambe, Ps. Spl. C.) *fructus ventris*, Ps. Surt. 126, 3. Ðú átuge mé of wombe (*ventre*) . . . Of wambe (wambe, Ps. Spl. C. *ventre*) módur mínre, 21, 10–11. Ða wombe (wombo, Lind. *ventres*) ða ðe ne ácendun, Lk. Skt. Rush. 23, 29. **II**. *of inanimate things* :—Ic wiht (*bellows*) geseah, womb wæs on hindan, Exon. Th. 419, 6; Rä. 38, 1. Hí (*clouds*) feallan lǽtaþ seáw oi bósme, wǽtan of wombe, 385, 21; Rä. 4, 48. Ic seah wiht (*a cask*), wombe hæfde micle, 495, 2; Rä. 84, 1. **III**. in the following passage giving the boundaries of some land, Kemble takes the word to mean a *hollow* :—Ondlong ðære hegerǽwe ; ðæt on Ondoncilles wombe, Cod. Dip. Kmbl. iii. 52, 14. [*Goth.* wamba γαστήρ, κοιλία, *venter, uterus* : *O. L. Ger.* wamba *venter, uterus* : *O. Frs.* wamme : *O. H. Ger.* wamba *venter, ventriculus, uter, vulva* : *Icel.* vömb *belly.*]

-wamb; *adj.* v. þyrel-wamb.

wamb-ádl, e; *f.* *Disease of the stomach* :—Hér sint tácn be wambe coþum and ádlum, and hú mon ða yfelan wambe ðære wambe lácnian scyle. Ðonne wambádl tóweard sié, ðonne beóþ ða tácn . . ., Lchdm. ii. 216, 19.

wamb-hord, es; *m.* *A womb-hoard*, used of the weapons contained in a fortified place :—Mé (*the fortified place*) of hrife fleógaþ hylde pílas; hwílum ic sweartum swelgan onginne brúnum beadowǽpnum; is mín innað til, wombhord wlitig, Exon. Th. 399, 12; Rä. 18, 10.

wamb-seóc; *adj. Diseased in the stomach:*—Ða wambseócan men þrowiaþ on ðam bæcþearme and on ðam niþerran hrife, Lchdm. ii. 232, 12: 164, 10.

wamm, es; *m. n.*   I. in a physical sense, (a) *a spot, mark, blot, stain:*—Wam *livor*, Wrt. Voc. ii. 50, 17. Wommum *nevis*, 61, 39.   (b) *filth, impurity, corruption:*—Wyrms oððe wom *lues*, Ælfc. Gr. 9, 27; Zup. 53, 7. Cwealmbǽrne wom *letiferam luem* (gipsae crudelitas, quae letiferam civibus luem inferebat, Ald. 69), Hpt. 518, 41. Wom *illuviem, immunditiam (carceris*, Ald. 48), 488, 31. Gold ðæt in wylme bið womma (woman, Kmbl. *but MS. has* woma) gehwylces geclǽnsod, Elen. Kmbl. 2618; El. 1310.   II. fig. (a) *a blot, disgrace, damage, hurt:*—Wom *dispendium*, Wrt. Voc. ii. 106, 40: 28, 11. *Dispendium*, i. *damnum, impedimentum, defectio, periculum, detrimentum* æfwerdla, wonung, wom, wana, *vel* hênþa, 140, 68. Wæs him ful strang wom and wítu (cf. *O. Sax.* al gethlôian wíties endi wammes, Hél. 1536), Cd. Th. 278, 24; Sat. 227. Wam *maculam* (qui arguit impium, sibi maculam generat, Prov. 9, 7), Kent. Gl. 292. Hellbendum fæst, wommum gewítnad (*grievously punished*), Beo. Th. 6138; B. 3073. (b) *moral stain, impurity, uncleanness, defilement:*—Idese mid widle and mid womme besmítan, Judth. Thw. 22, 12; Jud. 59. Fram wæmme leahtra *a labe criminum*, Hymn. Surt. 63, 5. Womme labe (qui genitus mundum miseranda labe resolvit, Ald. 182), Wrt. Voc. ii. 94, 43: 52, 63. Wom *nevum* (moribus castis vivunt, ut spurcum vitarent pectore nevum, Ald. 168), ii. 92, 82. Synrust þweán and ðæt wom ǽrran wunde hǽlan, Exon. Th. 81, 11; Cri. 1322: 94, 23; Cri. 1544. Óþ ðæt hafaþ ǽldes leóma woruldwidles wom forbærned, 62, 25; Cri. 1007. (c) *evil, sin, shameful word* or *deed:*—Nǽfre wommes tácn in ðam eardgearde eáwed weorþeþ, ac ðé firina gehwylc feor ábúgeþ, Exon. Th. 4, 18; Cri. 54. Eorl óðerne mid teónwordum tǽleþ behindan, spreceþ fǽgere beforan . . . Byð ðæs wommes gewita weoruda Dryhten, Fragm. Kmbl. 12; Leás. 7. Genere mé fram ðam were ðe wom fremme *a viro iniquo eripe me*, Ps. Th. 139, 1. Wom dydon yldran ûsse, ðín bebodu brǽcon, Exon. Th. 186, 10; Az. 17: Cd. Th. 234, 25; Dan. 297: Exon. Th. 68, 4; Cri. 1098. Of ðám welerum ðe wom cweðen *a labiis iniquis*, Ps. Th. 119, 2. Heó mé wom spreceþ, firenaþ mec wordum, Exon. Th. 402, 22; Rä. 21, 33. Nǽnig bihelan mæg on ðam heardan dæge wom unbéted, ðǽr hit ða weorud geseóð, So, 25; Cri. 1312. Wer womma leás, Cd. Th. 233, 29; Dan. 283: Menol. Fox 415; Men. 209: Exon. Th. 89, 4; Cri. 1452. Clǽne, womma leáse, 12, 19; Cri. 188: 450, 27; Dóm. 94. Womma clǽne, 103, 26; Cri. 1694. Ne ic culpan in ðé ǽfre onfunde womma geworhtra, and ðú ða word spricest, swá ðú sié synna gehwylcre gefylled, 12, 1; Cri. 179. Hié wǽron womma ðríste, inwitfulle, Cd. Th. 77, 9; Gen. 1272. Dú tó fela synna gefremedes; wé ðé nú willaþ womma gehwylces leán forgieldan, Exon. Th. 137, 15; Gú. 559. Áþweah mé of sennum, sáule fram wammum, Ps. C. 38. Ic eom dǽdum fáh, gewundod mid wommum, Cd. Th. 274, 20; Sat. 157. Riht ágyldan ealles ðæs ðe hé on worlde tó wommum gefremede, Blickl. Homl. 113, 4. Wídgongel wíf mon wommum bilihð, hæleð hý hospe mǽnaþ, Exon. Th. 337, 16; Gn. Ex. 65. Mánsceaða, wommum áwyrged, 95, 24; Cri. 1562: Cd. Th. 211, 26; Exod. 532. Unriht dón, wommas wyrcean, 217, 17; Dan. 24. Se ðe warnaþ him wommas worda and dǽda, Exon. Th. 304, 32; Fä. 79. [*Goth.* wammé; *gen. pl. macularum: O. Sax.* wamm *evil, wrong: O. Frs.* wamm *a blemish: O. H. Ger.* wamm *damnum: Icel.* vamm; *n. a blemish*.] v. mán-, white-wamm.

wamm; *adj.*   I. *foul:*—Ic under eorþan sceáwige wom wræcscrafu (? wrað-, MS.) wráþra gésta, Exon. Th. 424, 18; Rä. 41, 41.   II. *evil, wicked:*—Ná ðú be gewyrhtum, Wealdend, ûrum, wommum wyrhtum woldest ûs dón *non secundum peccata nostra fecit nobis*, Ps. Th. 102, 10. [*O. Sax.* wamm (dǽd): cf. *Goth.* ga-wamms *communis*; un-wamms *immaculatus, sine macula*.]

wamm-cwide, es; *m. Evil speaking, reviling, slander, blasphemy:*—Him (*the devils*) wæs wráð geworden for womcwidum, Cd. Th. 282, 6; Sat. 282. Ne wíte ic him ða womcwidas, þeáh hé his wyrðe ne sié tó álǽtanne ðæs fela hé mé láðes spræc, 39, 7; Gen. 621.

wamm-dǽd, e; *f. An evil deed, a misdeed, trespass, crime:*—Swá swá wé forlǽtaþ leahtras on eorðan ðám ðe wið ûs oft ágyltaþ, and him womdǽda wítan ne þencaþ '*as we forgive them that trespass against us*,' Hy. 6, 25. Him (*David*) sáwla Neriend secgan hét ymb his womdǽda Waldendes dóm, Ps. C. 19: Exon. Th. 270, 18; Jul. 467. [*O. Sax.* wam-dád: Ef gí ne willeat weron wamdádi álátan, Hél. 1624.]

wamm-freht, es; *n. Divination:*—Ða ðæt womfreht réniaþ *ariolorum*, Wrt. Voc. ii. 82, 8. Womferht, 5, 16. Cf. frihtere, frihtrung.

wamm-full; *adj. Evil, guilty, criminal, flagitious:*—Ǽr se unsýfra (*Holofernes*) womfull onwóce, Judth. Thw. 22, 24; Jud. 77. Synfulra here . . . womfulra scolu, Exon. Th. 94, 5; Cri. 1535. Womfulle, scyldwyrcende (*the fallen angels*), Elen. Kmbl. 1519; El. 761.

wamm-lust, es; *m. A foul pleasure, an allurement, seduction:*—Womlustas *lenocinia*, Anglia xiii. 28, 19.

wamm-sceaþa, an; *m. An evil-doer, a sinner, criminal:*—Áwyrged womsceaða (*the devil*), Exon. Th. 255, 8; Jul. 211. Womsceaþan (*the wicked, at the day of judgement*), 75, 23; Cri. 1226: 96, 7; Cri. 1570.

Áwyrgede womsceaðan, leáse leódhatan, árleasra sceolu, Elen. Kmbl. 2595; El. 1299. [*O. Sax.* wam-skaðo.]

wamm-scyldig; *adj. Sinful, criminal:*—Ne mæg ðǽr (*paradise*) inwitfull ǽnig geféran, womscyldig mon, Cd. Th. 58, 20; Gen. 949.

wamm-wlite, es; *m. A wound on the face:*—Swá hwylc man swá óðrum womwlite on gewyrce, forgylde him ðone womwlite, and his weorc wyrce óð ðæt seó wund hál sig *quicunque homo alio vulnus in faciem inflixerit, emendet ei vulnus, et opus ejus operetur, donec vulnus sanetur*, L. Ecg. C. 22; Th. ii. 148, 18. v. wlite-wamm.

wamm-wyrcende *working iniquity:*—Ðæt weorþeþ þeódum tó þreá, ðám ðe þonc Gode, womwyrcende, ne cúþun ðæs ðe hé on ðone hálgan beám ahongen wæs, Exon. Th. 67, 23; Cri. 1093.

wan *wan.* v. wann.

wan, es; [*n.* (?) cf. *Icel.* vant (*neut. of* vanr) *with gen.*] *Want, lack:*—Ne byð mé nánes gódes wan *nihil mihi deerit*, Ps. Th. 22, 1. Hí habbaþ ǽghwæs genóh, nis him wihte won, Exon. Th. 352, 9; Sch. 95. On ðám ðingum ðe hí won hæfdon *in eis quae minus habuerat*, Bd. 5, 22; S. 644, 15. v. wana; *m.,* and next word.

wan; *adj.*   I. *wanting, absent:*—Ða getreówde hé in godcundne fultom, ðǽr se mennisca wan wæs *confidens in divinum, ubi humanum deerat, auxilium*, Bd. 2, 7; S. 509, 23. Him won (wona, MS. Ca.) ne wæs seó monung ðære godcundan árfæstnesse *non defuit admonitio divinae pietatis*, 4, 25; S. 599, 23. Ne wiht mé wonu bið *nihil mihi deerit*, Ps. Surt. 22, 1: 33, 10. Ǽr ðon ðe Drihten on heofenas ástige, þonon hé nǽfre won wæs þurh his godcundnesse miht, Blickl. Homl. 131, 17.   II. *lacking, not possessed of:*—Wé tíres wone á bútan ende sculon ermþu dreógan, Exon. Th. 17, 15; Cri. 270.   III. with numerals (v. læs), *less.* Cf. wana; *adj.* IIIa:—Ðæt ríce hé hæfde ánes won ðe twêntig wintra, Bd. 4, 1; M. 252, 9. Ánes won þe syxtig wintra, 3, 24; M. 238, 2. Ánes won þe twêntig wintra, 5, 1; M. 386, 25. Gewurþad mid ðám æþelestum ceastrum ánes won ðe ðrittigum, 1, 1; S. 473, 26 note. [*Goth.* wans *wanting* (Tit. 1, 6): *O. Sax.* wan: *O. Frs.* won: *O. H. Ger.* wan wesan *deesse: Icel.* vanr.] v. wana; *adj.*

wana, an; *m.*   I. *want, lack, absence:*—Mé ys feós wana *deest mihi pecunia*, Ælfc. Gr. 32; Zup. 202, 12. Hláfes wæs wana *panis deerat*, Gen. 47, 13. Ðonne wana (wona, Hatt. MS.) bið ðæs ðe hié habban woldon *hae cum desunt*, Past. 18; Swt. 126, 22. Hit nán mon ne mæg eall habban, ðæt him ne sié sumes þinges wana, Bt. 34, 9; Fox 146, 19. Ðú mænst gif ðé ǽnies willan wana biþ, 11, 1; Fox 30, 22: 26, 1; Fox 90, 22: 29, 1; Fox 102, 18. Ðonne is sum gód full ǽlces willan and nis nánes gódes wana, 34, 1; Fox 134, 27: Homl. Th. i. 272, 13: ii. 400, 11: Ps. Th. 33, 9: Shrn. 202, 11. Gif hwǽm ðara twêgra hwæderes wana biþ, Bt. 36, 3; Fox 176, 7. Ðam bið gomenes wana ðe ða earfeða dreógeþ, Exon. Th. 183, 17; Gú. 1328. Mé is wana æt ðam scýrgesceatte ðus micelys ðe míne foregengan hæfdon, Cod. Dip. Kmbl. iii. 327, 4. Swá ic feós bidde swá ic wanan hæbbe ðæs ðe mé N. behét (*I have not got what N. promised me*), L. O. 10; Th. i. 182, 7.   I a. in connection with numerals. v. wana; *adj.* III a:—Hire daga rím gefylled wæs, ðæt is ánes geáres wana sixtigra wintra (*there wanted one year of sixty*; undesexaginta annorum), Bd. 3, 24; S. 557, 6 note.   II. *want of necessaries, lack, want, defect:*—*Dispendium*, i. *damnum, impedimentum, defectio, periculum, detrimentum* æfwerdla, wonung, wom, wana, *vel* hênþa, Wrt. Voc. ii. 140, 69. Wanan *inopiam* (cum panis copia plebis inopiam refocillantes, Ald. 53), Hpt. Gl. 497, 26. [Ðet ich þurh to muche wone ne falle i fulðe of sunne . . . ðet ich mote underuon boðe wone and weole þe ine cwemnesse, O. E. Homl. i. 213, 28–32. And tah þu wone hefdest oðer drehdest ani derf, H. M. 29, 8. Uor wone of witnesse, A. R. 68, 8.] v. for-wana; wan.

wana; *adj. generally indeclinable.*   I. *wanting, lacking, absent,* (a) *with substantive verb*, wana wesan or *be wanting:*—Ic eom wana of ðam getele *desum*, Ælfc. Gr. 32; Zup. 202, 11. Án þing ðé is wana (wona, Lind., Rush.) *unum tibi deest*, Lk. Skt. 18, 22: Mk. Skt. 10, 21. Wæs eów ǽnig þing wana? *numquid aliquid defuit vobis*? Lk. Skt. 22, 35. Hwæt ys mé gyt wana (gwona, Lind.: wuon, Rush.)? *quid mihi deest*? Mt. Kmbl. 19, 20. Ðæt ic wite hwæt wana (wone, Ps. Surt.) sý mé, Ps. Spl. 38, 6: Bt. 33, 3; Fox 126, 20. Ðam biþ anweald wana (anwaldes wana, Cott. MS.), 36, 3; Fox 176, 13. Mé wana is ǽgþer ge spadu ge mattuc, Homl. Skt. ii. 23, b, 765. Synn wana ná byð *peccatum non deerit*, Scint. 78, 4: Kent. Gl. 335. Wana sié *absit*, Wrt. Voc. ii. 3, 57. Mé synd wana penegas *desunt mihi nummi*, Ælfc. Gr. 32; Zup. 202, 13. Ne heora martyrháda wana wǽron heofonlícu wundru *nec martyrio eorum coelestia defuere miracula*, Bd. 5, 10; S. 625, 4. (b) *in connection with numerals, wanting for the completion of a number:*—Ðæs hærfest cymþ ymb óðer swylc bútan ánre wanan *after one less than the same number of days comes autumn*, Menol. Fox 280; Men. 141. .X. geár bútan .xv. wucan wanan (*fifteen weeks were wanting to complete the ten years*), Chr. 1068; Erl. 206, 17.   II. *wanting, destitute* of, *without* something:—Se ne ongyteþ ða þeóstra his ágenra synna, wite hé ðæt hé bið wana ðæs ǽcan leóhtes, Blickl. Homl. 17, 36.   III. *wanting, not complete, deficient:*—Gif nán wuht full nǽre, ðonne nǽre nán wuht wana; and gif nán wuht wana nǽre, ðonne nǽre

nân wuht full; for ðý biþ ǽnig full þing, ðe sum biþ wana, and for ðý biþ ǽnig þing wana, ðe sum biþ full, Bt. 34, 1; Fox 134, 20-23. Genóg sweotol hit is ðæt ðæt fulle gôd wæs ǽr ðam ðe ðæt wana *omnia perfecta minus integris priora esse claruerunt*, 34, 2; Fox 136, 12. **III a.** with numerals, *wanting, save* (cf. *Goth.* fidwôr tiguns ainamma wanans, 2 Cor. 11, 24). v. wana; *m.* **I a,** wan; *adj.* **III.** As appears especially in the first of the following passages, the word and the numerals which precede and follow it as much form a compound as do the words which give the number they express in modern English :—Hê wæs âne-wana-xxx-wintre (xxix wintra eald, *co*. 3), Chr. 972; Th. i. 225, col. 1. Ânes wana fíftig, Andr. Kmbl. 2079; An. 1040. Ânes wona sixtig wintra *undesexaginta annorum*, Bd. 3, 24; S. 557, 6. Gewurþad mid ðam æðelestum ceastrum anes wana ðrittigum, 1, 1; S. 473, 26. Ðæt ríce hê hæfde ânes wona .xx. wintra (ân læs ðe twêntig, MS. B.), 4, 1; S. 563, 15. Hê Norþanhymbra ðeóde ânes wana .xx. wintra fore wæs *genti Nordanhymbrorum decem et novem annis praefuit*, 5, 1; S. 614, 21. [Ful lutel þer wæs wone, þat Corineus nas ouercome, Laym. 1905. Him ne schal beo wone nouht (no þing, *v. l.*) of his wille, Misc. 104, 57. Hem was ðat water wane, Gen. and Ex. 3353. Wane or wantynge *absens, deessens*, Prompt. Parv. 515. ¶ *with numerals* :—On wane of an hundred *ninety-nine*, Gen. and Ex. 1028. Twa wone of twenti *duo de viginti*, Kath. 67.] v. wan; *adj.*

wana-beám. v. wanan-beám.

wan-ǽht, e; *f. Scant possession* :—Náh ic fela goldes . . . ic mê sylf ne mæg fore mínum wonǽhtum willan âdreógan, Exon. Th. 458, 19; Hy. 4, 103. Cf. wan-spéd.

wanan-beám, es; *m. A spindle-tree* (v. English Plant Names, E.E.T.S. Pub., and cf. *O. H. Ger.* spinnel-boum *fusarius*) :—Wananbeám (uuanan-, uuonan-) *fusarius*, Txts. 65, 935 : Wrt. Voc. ii. 39, 5. Wanabeám *fussarius*, 36, 58 : *fursarius*, i. 286, 3.

wancol; *adj. Unstable, uncertain, fickle, fluctuating* :—Hió hit gecýþ self mid hire hwurfulnesse ðæt hió biþ swíþe wancol *se instabilem mutatione demonstrat*, Bt. 20; Fox 70, 35. Nû ðú hæfst ongyten ða wanclan (wonclan, *v. l.*) treówa ðæs blindan lustes *deprehendisti caeci numinis ambiguos vultus*, 7, 2; Fox 18, 3. [Ðis wunder (*the mermaid*) wuneð in wankel stede, Misc. 18, 566. This worlde is wondur wankille, Halliw. Dict. *O. Sax.* wankol (hugi) : *O. H. Ger.* wanchal *lubricus, infidelis*. Cf. *O. L. Ger.* wankil-heidî *fluctuatio*.]

wand[, e, f.?] *a mole* :—Wond (wand, uuond) *talpa*, Txts. 101, 1973. v. wande-weorpe.

-wand. v. ge-wand.

wande-weorpe, an; *f. A mole* (cf. later English mold-werp, *still used in some dialects* : *O. H. Ger.* mu-werfo *talpa*, Grff. i. 1040 : *M. H. Ger.* molt-werf : *Ger.* maul-wurf : *Icel.* mold-varpa) :—Wondeuueorpe (uuandaeuui[o]rpae, uuondæuuerpe) *talpa*, Txts. 101, 1975. Wandewurpe *talpa* vel *palpo*, Wrt. Voc. i. 22, 60 : *talpa*, 78, 19. v. wand.

wandian; *p.* ode. **I.** *to turn aside* from something (*gen.*) :—Ne beforan manegon sóðes ne wanda *nec in judicio plurimorum acquiesces sententiae, ut a vero devies*, Ex. 23, 2. **II.** *to turn aside* from a task, purpose, duty, etc., *to hesitate, shrink, flinch*, (a) absolute :—Ic wandige (âwandige, *v. l.*) *uereor*, Ælfc. Gr. 27; Zup. 162, 2. Hê wandode ðá git (*dissimulante illo*); ac hig gelǽhton hys hand and his wífes hand and gelǽddon hig út of ðære byrig, Gen. 19, 16. Wandode se wísa (*Daniel*), hwæðre hê worde cwæð tô ðam æðelinge, Cd. Th. 250, 24; Dan. 550. Hê ne wandode nâ æt ðam wígplegan, Byrht. Th. 139, 42. Ne mæg nâ wandian se ðe wrecan þenceþ freán, 139, 22; By. 258. Oft mon bið suíðe wandigende æt ǽlcum weorce and suíðe lætrǽde *agendi tarditas*, Past. 20; Swt. 149, 14. (b) where the grounds for turning aside are given, *to care for, be influenced* by :—Ðú ne wandast for nânon menn *non est tibi cura de aliquo*, Mt. Kmbl. 22, 16. Ðú for nânon men ne wandast *non accipis personam*, Lk. Skt. 20, 21. Ne wandaþ hê for rícum ne for heánum *qui personam non accipit*, Deut. 10, 17. For hira feónda yrre ic wandode *propter iram inimicorum distuli*, 32, 27. Ne hit for ðæm bryne wandode ðæs hátan léges *nec ignium tardatus ardoribus*, Nar. 15, 20. Ne wanda ðú for rícum ne for heánum ne for nânum scette *non accipies personam nec munera*, Deut. 16, 19. Nô wandige hê for ðan yflan willan *non consideret malam voluntatem*, R. Ben. 92, 11. (c) where that which is turned aside from is given, (a) by a clause :—Sume synna beóþ swíþe unsýferlíce, ðæt se man wandaþ ðæt hê hí ǽfre âsecgge, Blickl. Homl. 43, 17. Ðonne ðú behât behǽtst, ne wanda ðú ðæt ðú hit ne gelǽste *cum votum voveris, non tardabis reddere*, Deut. 23, 21. Ne wanda ðú, ðæt ðú dínum frýnd ne helpe, 15, 10. (β) by the dat. infin. :—Hí ne wandiaþ tô licgenne on stuntnysse, Homl. Th. ii. 554, 2. Hê ne wandode nâ him metes tô tylienne, Chr. 1052; Erl. 183, 20. (d) with the constructions of (b) and (c. a) :—Ðæt hyra nân ne wandode ne for mínan lufan ne for mínum ege, ðæt hý ðæt folcriht ârehton, Chart. Th. 486, 23. Ne wandige nâ se mæssepreóst nô for ríces mannes ege, ne for feó, ne for nânes mannes lufon, ðæt hê him symle riht dême, Blickl. Homl. 43, 9. (e) with the constructions of (b) and (c. β) :—Ða bydelas ðe for ege oððe lufe oððe ǽnigre worldscame eargiaþ and wandiaþ Godes riht tô sprecanne, Wulfst. 191, 6. **III.**

to turn aside from punishing, injuring, etc., *to refrain* from, *spare* a person or thing (*dat.*). (a) absolute :—Ðæt man nǽnne ne slôge . . . búton hê fleón wille oþþe hine werian; ðæt man ne wandode ðonne, L. Ath. v. 12, 3; Th. i. 242, 10. Suelce hê hine wandigende ofersuíðe *quasi parcendo superare*, Past. 40; Swt. 297, 15 : 295, 12. Næs wandigendre ðonne hit gedafenlíc sié *non plus quam expediat, parcens*, 17; Swt. 127, 4. (b) with dat. :—Ne wandode ic nâ mínum sceattum ða hwíle ðe eów unfrið on handa stôd *I did not spare my treasures while you had hostilities on hand*, Chart. Erl. 229, 27. Ða ðe heora Drihtne wiðsacan noldon, ðám man nân þingc ne wandode, ac hí tô ealre yrmðe getucode, Homl. Skt. i. 23, 71. Ne wanda ðú nân ðing ne âra ðú nânum ríce *non parcet oculus tuus ulli regno*, Anglia x. 88, 47. Se wilnaþ suíður ðæt mon lufge sôðfæsdnesse ðonne hine selfne, se ðe wilnaþ ðæt mon nânre ryhtwísnesse fore him ne wandige *ille se ipso amplius veritatem desiderat amari, qui sibi a nullo vult contra veritatem parci*, Past. 19; Swt. 145, 17. (c) with a clause :—Sanctus Paulus geliéfde, ðæt hê swâ micele unscyldigra wǽre his niéhstena blôdes swâ hê læs wandade ðæt hê hira unðeáwas ofslôge *Paulus eo se a proximorum sanguine mundum credidit, quo feriendis eorum vitiis non pepercit*, Past. 49; Swt. 379, 11. [Love wol love—for no wight wol hit wonde, Ch. L. G. W. 1187. Wolde I wonde for no sinne, Gow. i. 332, 7. For us ne schalt þou wonde, Jos. 399. To love nul i noht wonde, Spec. 29. Sche wold for no man wond, that sche no wol to him fond, Am. and Amil. 550. He wonded no woþe of wekked knaueȝ, þat he ne passed þe port, Allit. Pms. 63, 855. For to speke alle vilanie nel nu no kniht wonde for shame, P. S. 335, 262. Lust whi ihc wonde bringe þe Horn to honde, Horn 337. Jhon her son sche wolde nought wonde, Rich. 228.] v. â-, for-wandian; un-wandiende.

-wandigendlíce. v. un-forwandigendlíce.

wandlung, e; *f. Changing, mutation* :—Hié beheúldon on ðé heora âgen gecynd, and on heora wandlunga hié gecýþdon heora fæstrǽdnesse *servavit circa te propriam in ipsa sui mutabilitate constantiam*, Bt. 7, 2; Fox 16, 31. [*O. H. Ger.* wandelunga *mutatio*, cf. *O. L. Ger.* wandlôn *to change*.]

-wandodlíc, -líce. v. un-forwandodlíc, -líce.

wandrian; *p.* ode *To wander, rove, roam* :—Wandriendu *ludivaga*, Wrt. Voc. ii. 54, 26. **I.** in a physical sense :—Se steorra (*Saturn*) wandraþ ofer ôþrum steorran, Bt. 36, 2; Fox 174, 13 : Met. 24, 23. Wandraþ *vagatur*, Hpt. Gl. 412, 56. Hí maciaþ eall be luste, woriaþ and wandriaþ, and ealne dæg fleardiaþ, L. I. P. 14; Th. ii. 322, 24. Hræfen wandrode, Fins. Th. 69; Fin. 34. Wandrigende pucan *uagantes demonas*, Germ. 388, 37. **II.** figurative. (a) *to leave one's proper work* :—Ðonne gǽð Dine út sceáwian ða elðíódigan wíf, ðonne hwelces monnes môd forlǽt his ǽgne tilunga, and sorgaþ ymb ôðerra monna wísan, ðe him náuht tô ne limpð, and gǽð swâ wandriende from his hâde and of his endebyrdnesse. Sihhem geniédde ðæt mǽden ðá hê hié gemétte swâ wandrian *Dina, ut mulieres videat extraneae regionis, egreditur, quando unaquaeque mens sua studia negligens, actiones alienas curans extra habitum atque extra ordinem proprium vagatur. Quam Sichem opprimit; quia inventam in curis exterioribus diabolus corrumpit*, Past. 53; Swt. 415, 19-23. (b) *to proceed without plan, follow an uncertain course* :—Swâ ða sélestan men swíþor ðæs eorþlican ðing forseóþ, swâ hí læs réccaþ hú sió wyrd wandrige, Bt. 39, 7; Fox 222, 25. Ðíós wandriende wyrd ðe wê wyrd hátaþ, 39, 6; Fox 220, 5. [*M. H. Ger.* wandern.]

wandung, wan-fáh, -feax, -fóta, -fýr. v. for-wandung, wann-fáh, -feax, -fóta, -fýr.

wang, es; *m.* **I.** the word, which is almost confined to poetry, may be rendered by words denoting the surface of the ground taken in their most general sense, *field, plain, land, country, place* :—Wonge (wongc?) *arvum*, Wrt. Voc. ii. 10, 51. Mec se wǽta wong wundrum freórig of his innaþe cende *roscida me genuit gelido de viscere tellus* (Ald.), Exon. Th. 417, 7; Rä. 36, 1. Se wong seomaþ eádig and onsund. Is ðæt æþele lond blôstmum geblôwen, beorgas ðǽr ne muntas steápe ne standaþ . . . ne dene ne dalu *illic planicies tractus diffundit apertos, nec tumulus crescit, nec cava vallis hiat*, Exon. Th. 199, 2; Ph. 19. Wlitig is se wong . . . ǽnlíc is ðæt íglond, 198, 8; Ph. 7. Wynsum wong, wealdas grêne, 198, 20; Ph. 13. Se hálga wong *Paradise*, 227, 5; Ph. 418. Brúcan wonges, . . . neótan londes frætwa, 268, 1; Ph. 149. Hwæþere him ðæs wonges wyn (cf. londes wyn, 130, 15; Gú. 438) swêðrade *whether the land grew less delightful to him*, 123, 15; Gú. 123. Ic ða stôwe ne can ne ðæs wanges (*the place where the cross was buried*) wiht ne ða wísan cann, Elen. Kmbl. 1364; El. 684. On ðam wange, ðǽr hê sorge gefremede *on the scene of his wrong-doings*, Beo. Th. 4010; B. 2003. Hí gesêgon wyrm on wonge licgean *he saw the serpent lying on the ground*, 6070; 3039. On wonge, wæterýðum neáh, 4476; B. 2242: Cd. Th. 113, 4; Gen. 1882 : Exon. Th. 485, 21; Rä. 72, 1. Næs ðǽr hláfes wist werum on ðam wonge (*the island of Mermedonia*), Andr. Kmbl. 43; An. 22. Hê sceal ðý wonge (*the island in the fens where St. Guthlac's hermitage was*) wealdan, Exon. Th. 144, 6; Gú. 674. Hý ðone grênan wong ofgiefan sceoldan, 130, 34; Gú. 448. Hê wang sceáwode fore burggeatum *he reconnoitred the place*, Andr. Kmbl. 1678; An. 841 :

Beo. Th. 2831; B. 1413: 4809; B. 2409: 6139; B. 3073. Hí on wang stigon *they landed*, 456; B. 225. Ofer wong faran *to go across country*, Exon. Th. 481, 10; Rä. 65, 1. Hryre wong gecrong *the ruin sank to earth*, 477, 30; Ruin. 32. Đone wlitigan wong *Paradise*, 228, 16; Ph. 439. Wangas blóstmum blówaþ *fields bloom with flowers*, Menol. Fox 178; Men. 90. Wangas gréne, 410; Men. 206. Đás foldan bearm, gréne wongas, Exon. Th. 482, 21; Rä. 67, 5: Cd. Th. 100, 1; Gen. 1657. Wangas, eorðe ælgréno, Met. 20, 77: Exon. Th. 51, 5; Cri. 811: 451, 32; Dóm. 112. Him wíc curon, ðær him wlitebeorhte wongas geþúhton, Cd. Th. 108, 11; Gen. 1804: Beo. Th. 4915; B. 2462. Sum con wonga bígong, wegas wídgielle *one knows the world, ways wide-spreading*, Exon. Th. 42, 30; Cri. 680. Dæg se georstenlíca God besceáwede on wangum *dies hesterna Deum conspexit in arvis*, Hymn. Surt. 47, 10. On sumeres tíd stincaþ on stówum, wynnum æfter wongum wyrta geblówene, Exon. Th. 178, 24; Gú. 1249. Cumaþ wæstm on wangas weordlíc on hwætum *convalles abundabunt frumento*, Ps. Th. 64. 14. Ic foldan slíte, gréne wongas, Exon. Th. 393, 18; Rä. 13, 2. Wíde geond wongas, 491, 8; Rä. 80, 11. **II.** *the earth, the surface of the earth :*—Ic (*creation*) eorþan eom æghwær brædre, and wídgelra ðonne ðes wong gréna (cf. *O. Sax.* gróni wang *the earth*), Exon. Th. 426, 34; Rä. 41, 83. Cýþan werum on wonge, 474, 2; Rä. 32, 14: 439, 11; Rä. 59, 2. Seó heá miht on ðysne wang ástág, Blickl. Homl. 105, 14. Đú eorðan wang ealne gesettest, Hy. 10, 3. Se Ælmihtiga eorþan worhte wlitebeorhtne wang, Beo. Th. 186; B. 93. Gangan ofer foldan wang, Menol. Fox 225; Men. 114. **III.** fig. of any surface :—Ic (*a cup for cupping*) eom stíð and steáp wong, staþol wæs in þá wyrta wlitetorhta, Exon. Th. 484, 4; Rä. 70, 2. [Casteles and tunes, wodes and wonges, Havel. 397. Wonge of londe *territorium*, Prompt. Parv. 532. *Goth.* waggs *paradisus* (2 Cor. 12, 4): *O. Sax.* wang *field, plain, country : O. H. Ger.* holz-wang *campus nemoreus : Icel.* vangr (poet.) *field.*] v. beadu-, deáð-, fold-, freoðo-, græs-, grund-, medu-, metud-, sæ-, sæl-, sige-, stán-, staþol-, stede-, wæl-, wil-wang, neorxna wang, *and* wang-turf.

wang, es; *m.* : wange, wænge, wenge, an; *n. A cheek, side of the face :*—Đæt wange wið ða ceócan ufan *mandibula*, Wrt. Voc. ii. 58, 3. Đæs wonges locfeax *cesaries*, 22, 57. Smire ðæt hále wonge mid, Lchdm. ii. 338, 9. Bind on ðæt wænge, 20, 10. Smyre ðæt wenge, 20, 18. Gif hwá ðé sleá on ðín swýðre wenge (gewenge, *v. l.*, wonge I céke, Rush.) *si quis te percusserit in dextera maxilla tua*, Mt. Kmbl. 5, 39. Benedictus slóh ðone munuc under ðæt wencge mid anre handa, Homl. Th. ii. 180, 10. T him ða wongan briceþ, Salm. Kmbl. 192; Sal. 95. Ic ða wangas mid teárum ofergeát, Homl. Skt. ii. 23 b, 556. [Wete weoron his wongen, Laym. 30268. I wette my wenges, Jos. 647. *O. Sax. O. L. Ger. O. H. Ger.* wanga; *wk. n. maxilla : Icel.* vangi; *wk. m.*] v. þun-wang, -wange, -wenge, ge-wenge.

wang-beard, es; *m. A whisker :*—Teóh him ða loccas, and wringe ða eáran, and ðone wangbeard twiccige, Lchdm. ii. 196, 13.

wange. v. wang *a cheek.*

wangere, es; *m. A pillow, bolster :*—Wangere *cervical* (v. Mk. 4, 38), Wrt. Voc. ii. 73, 29: 17, 53: i. 25, 45: *capitale*, ii. 128, 44. Bolster *vel* wongere *cervical*, i. *capitale*, 130, 26. Fram dæle ðæs heáfdes mihte wongere (*cervical*) betwih geseted beón, Bd. 4, 11; S. 580, 16. [His helm wæs his wonger, Chauc. Sir Th. 2102. A man waggarja *super cervical*, Mk. 4, 38: *O. H. Ger.* wangâri; *m. plumatium.*]

wang-stede, es; *m.* **I.** *a place in open country, a place :*—Forlæt of ðam wangstede (cf. stópon tó ðære stówe, on ða dúne up, 1428; El. 716) réc ástígan, Elen. Kmbl. 1584; El. 794: 2205; El. 1104. Stenc út cymeþ of ðam wongstede (cf. hé séceþ dýgle stówe under dún-scrafum, 357, 31; Pa. 37), Exon. Th. 358, 13; Pa. 45. On ðam wong-stede (*the place of the last judgement*) wérig bídan, 50, 18; Cri. 802. Hwæðer hé cwicne gemétte in ðam wongstede (cf. wong. 4809; B. 2409) Wedra þeóden, Beo. Th. 5565; B. 2786. Se ðás wongstedas gróf æfter golde (cf. se ðe ða eorþan ongan delfan æfter golde, Bt. 15; Fox 48, 23), Met. 8, 56. **II.** *a town on a plain* (wang) ? :—Hé eode in burh hraðe, ... stóp on stræte ... swá him nænig gumena ongitan ne mihte; hæfde sigora weard on ðam wangstede (cf. hé wang sceáwode fore burg-geatum, 1678; An. 841. *But perhaps* wangstede = wang, *and the passage means that St. Andrew was unseen as he passed across the space* (wang) *between the sea and the town.* Cf. stede-wang) wære betolden leófne leódfruman ... Hæfde ða se æðeling in geþrungen carcerne néh, Andr. Kmbl. 1975; An. 990.

wang-tóþ, es; *m. A wang-tooth* (in northern dialects. v. e.g. Lancashire Gloss. in E. E. D. S. Pub.), *molar tooth :*—Gif mon óðrum tóð of ásleá, gif hit sié se wongtóð geselle .iiii. scitt. tó bóte, L. Alf. pol. 49; Th. i. 94, 11. Wangtéð *molares vel gemini*, Wrt. Voc. i. 43, 32. Wongtoeð (-téþ, Ps. Spl. C.) *molas*, Ps. Surt. 57, 7: [Wangeteth *les messeleres*, Wrt. Voc. i. 146, 22. Out of a wangtooth sprang a welle (v. Wick. Jud. 15, 19, where the word is used), Chauc. M. T. 3234. Wangetoothe *molaris*, Prompt. Parv. 515. Wangtoth *geminus*, Cath. Angl. 407. Wayngetothe *geminus, maxillaris*, 406 (see note). Wongtothe *uteelaris*, Wrt. Voc. i. 207.]

wang-turf; *gen.* -tyrf; *f. Turf, grass-land :*—Đæt ic móte ðis gealdor

tóðum ontýnan ... wlitigan ðás wancgturf (cf. *the beginning of the article :* Hér ys seó bót hú ðú meaht ðíne æceras bétan gif hí nellaþ wel wexan, 398, 1), Lchdm. i. 400, 7.

wan-hæfelness. v. wan-hafolness.

wan-hæfenness, e; *f. Want, need :*—Wanhæfænysse and meteláeste *famis inedia*, Hpt. Gl. 480, 33.

wan-héle; *adj. Having bad health :*—Ealle ða ðe wonnhæle wæron, healtte and blinde, dumbe and deáfe, Nar. 48, 31. [*O. H. Ger.* wan-heili *semianimis, debilis, mancus.*] v. wan-hál.

wan-hélp, e; *f. Defective health, weakness, sickness :*—þurh wanhælde *per inbecillitatem*, Scint. 54, 19. [Cf. *O. H. Ger.* wana-heilí *debilitas.*] v. wan-hálness.

wan-hafa, an; *m. A poor person :*—Wanhafa and þearfa ic eom *inops et pauper sum ego*, Ps. Spl. 85, 1.

wan-hafness, e; *f. Poverty, want :*—Nis wanhafnes (*inopia*) ondrædendum hine, Ps. Spl. 33, 9.

wan-hafol; *adj. Needy, destitute :*—Him embe stódon wépende wydewan and wanhafele þearfan, Homl. Skt. i. 10, 65. Widewena bigleofa and wanhafolra manna, ii. 25, 765. Gehelp wanhafolum mannum mid ðínum ágenum spédum, i. 21, 363.

wan-hafolness, e; *f. Need, want, destitution :*—Nis wanhafolnes (*inopia*) ondrædendum hine, Ps. Lamb. 33, 10. Úre wanhæfelnesse *inopiae nostrae*, 43, 24.

wan-hál; *adj. Imperfect as regards health* or *soundness of body, weak, sick, maimed, infirm, unsound :*—Wanhál *inbecillis*, Wrt. Voc. i. 51, 23. Betere ðé ys ðæt ðú gá wanhál (*debilis*) oððe healt tó lífe, Mt. Kmbl. 18, 8: Mk. Skt. 9, 43. Hí God mærsodon swá oft swá ænig wanhál mann wurde gehæled, Homl. Skt. i. 21, 229. Đæt wanhál wæs and álewed, ðæt gé áwurpan *quod debile erat proicebatis*, R. Ben. 51, 15. Đý læs ðe án wanhál scép ealle ða eówde besmíte, Homl. Th. i. 124, 32. Swá hwylc man swá on gecynde ðderne wanhálne (*debilem*) dó, L. Ecg. C. 22; Th. ii. 148, 17. Đa ðe limseóce wæron, wérige, wanhále, Andr. Kmbl. 1159; An. 580. Wonhále, Exon. Th. 92, 13; Cri. 1508. Næs ðær wínes drenc búton wanhálum mannum, Homl. Th. ii. 506, 22: Homl. Skt. ii. 26, 202. Hé wolde gehelpan þearfum and wanhálum, 26, 276: Elen. Kmbl. 2057; El. 1030. Clypa þearfan and wanhále and healte and blinde *uoca pauperes, debiles, clodos, caecos*, Lk. Skt. 14, 13, 21. [*Icel.* wan-heill *unsound, disabled, ill.*] v. wan-hæle.

wan-hálian; *p.* ode *To weaken, impair the health* or *soundness of something [:*—þurh ðisne drync beóð ægðær ge ða sáwle ofslagene ge ða líchaman gewanhálode, Homl. Ass. 146, 51. [*O. H. Ger.* wana-heilen *debilitare;* ka-wanaheilit *debilitatus.*]]

wan-hálness, e; *f. Weakness, sickness, unsoundness, infirmity :*—Đæm abbode is á tó behealdenne heora (*fratrum infirmorum*) wanhálnes (*imbecillitas*), R. Ben. 75, 11. Wanhálnysse (*debilitate*) ealles líchaman, Scint. 38, 7. Dysig æfter untrumnysse his ongyt, and æfter wanhálnysse (*inbecillitatem*) gecyndes his wát, 97, 15. Bróþor se untruma gif hé gefrét hys weaxan wanhálnysse (*inbecillitatem*), Anglia xiii. 442, 1102. Cf. wan-hælþ.

wan-hlyte; *adj. Not having a share* in something, *destitute* of :—Wanhlytne *expertem*, Wrt. Voc. ii. 33, 8. [Cf. *Icel.* van-hluta; *adj. unfairly dealt with;* van-hlutr *an unfair share.*] v. or-hlyte.

wan-hoga, an; *m. One who is wanting in understanding, a foolish, imprudent person :*—Hí lifiaþ him in máne, heáhgestreón healdaþ georne, ... and wénaþ wanhogan ðæt hý wile God gehýran, Salm. Kmbl. 639; Sal. 319. Ic ðíne weogas wanhogan lærde, ðæt hié árleáse eft gecerdan tó hiora sáula hiorde, Ps. C. 105. v. un-hoga, *and following words.*

wan-hygd, -hygdu(-o) [cf. ofer-hygd] *want of mind, folly, rashness, recklessness, imprudence :*—For wlence and for wonhygdum hí ceastre worhton, and tó heofnum up hlædræ rærdon, Cd. Th. 100, 33; Gen. 1673. Grendel for his wonhýdum wæpna ne récceþ; ic ðæt ðonne forhicge ðæt ic sweord bere, Beo. Th. 872; B. 434. [Cf. *Icel.* van-hyggja *want of forethought.*]

wan-hygdig, -hýdig; *adj. Foolish, imprudent, thoughtless, careless, reckless :*—Wonhýdig wer *vir insipiens*, Ps. Th. 91, 5: Exon. Th. 95, 14; Cri. 1557: 343, 25; Gn. Ex. 162. Ne sceal wína nó tó hátheort, ne tó hrædwyrde, ne tó wác wiga, ne tó wanhýdig, 290, 19. Ne mid swiðran his nele brýsan wanhýdig gemód Wealdend engla, ne ðone wlacan smocan wáces flæsces wætere gedwæscan, Dóm. L. 50. Wonhýdige (*the apostate angels*), Elen. Kmbl. 1522; El. 763. [Cf. *Icel.* van-hugað *ill-considered.*]

wanian; *p.* ode. **I.** *trans.* (1) *To make less, lessen, diminish, curtail :*—Hí sculon ælce dæg eácan ðæt mon ælce dæg wanaþ, Bt. 26, 2; Fox 94, 1. Symble hé bið gyfende, and hé ne wanaþ nán þing his, Homl. Skt. i. 1, 46: L. Edg. S. 1; Th. i. 272, 10. Hwæt tó bóte mihte æt ðæm færcwealme ðe his leódscipe swýðe drehte and wanode, Th. i. 270, 10. Hé leóde míne wanode and wyrde, Beo. Th. 2678; B. 1337. (*The last two passages might be taken under* (3).) Wirceaþ ealle ða þing ðe Drihten eów bebeád, and ne íce gé nán þing ne ne waniaþ (*nec addas quidquam nec minuas*), Deut. 12, 32. Ne sý ðæs nagutimbres gemet ofer eorþan, gif hí ne wanige se ðás woruld· teóde, Exon. Th. 335, 15;

Gn. Ex. 34. Ne íce gé nán þing . . . ne gé wanion *non addetis . . . nec auferetis*, Deut. 4, 2. Godes dómas náwþer ne ná wanian ne ne écan, Blickl. Homl. 81, 4. (2) *to bring within narrower limits, to abate, check, reduce*. v. (4):—Wona ðæt ondspyrnisse *minue offendiculum*, Rtl. 11, 13. Wé sceolon ða fúlan gálnysse symle wanian, Homl. Th. i. 96, 22. Dæghwomlíce wé sceolon úre synna wanian ; for ðan ðe hí beóð gegaderode tó micelre hýpan, gif wé hí weaxan lǽtaþ, ii. 466, 6. Ðá wolde ðæt folc ðæt fýr ádwæscan, gif hit ǽnig wǽta wanian mihte, 140, 17. (3) *to weaken, impair, injure*. v. wanung, I. (3):—Windas bláwaþ brecende, weccaþ and woniaþ woruld mid storme, Exon. Th. 59, 13 ; Cri. 952. Hé bebeád ðæt mon nǽnne mon ne slóge, and eác ðæt man nánuht ne wanade ne ne yfelade ðæs ðe on ðǽm ciricum wǽre *dato praecepto, ut si qui in sancta loca confugissent, hos inviolatos securosque esse sinerent*, Ors. 6, 38 ; Swt. 296, 32. (3 a) *to weaken, reduce by medical treatment*. Cf. wanung, I. (3 a):—Lǽcas lǽrdon ðæt nán man on ðam mónþe ne drenc ne drunce, ne áhwǽr his líchoman wanige, bútan his nýdþearf wǽre, Lchdm. ii. 146, 12. Manega nellaþ heora ding wanian on Mónandæg (cf. þrý dagas (*the last Monday in April, the first Mondays in August and January*) syndon on ðám for nánre neóde ne mannes ne neátes blód sý tó wanienne . . . Se ðe on ðysum dagum his blód gewanige, sý hit man, sý hit nýten, ðæs ðe wé secgan gehýrdan, ðæt on ðam forman dæge oþþe ðam feórþan dæge his líf geændaþ, Lchdm. iii. 76, 11–22), Homl. Th. i. 100, 25. (4) *to cause to cease or fail, to bring to nought, destroy, frustrate*:—Ic wífe ábelge, wonie hyre willan, Exon. Th. 402, 21 ; Rä. 21, 33. Mon scel ðone unþeáw of mynstre wanian and mid ealle áwyrtwalian *hoc vitium radicitus amputandum est de monasterio*, R. Ben. 56, 16. (5) *to put in an inferior position*:—Ðú wanodest (*minuisti*) hine lytle læs fram ænglum, Ps. Spl. 8, 6. II. intrans. (1) *To wane, become less, decrease, diminish*:—Ne wexþ his welena (wela ná ?), ne eác nǽfre ne wanaþ, Bt. 42 ; Fox 256, 29. His wered wanode ǽfre ðe leng ðe swíðor, Chr. 1052 ; Erl. 181, 4. Ða wæteru wanedon *aquae decrescebant*, Gen. 8, 5. Þeáh ús úre spéda wanodon, Shrn. 167, 13. Ðæt sweord ongan wanian . . . hit eal gemealt, Beo. Th. 3218 ; B. 1607. Ða wæteru begunnon tó wanigenne *aquae coeperunt minui*, Gen. 8, 3. (1 a) *of the moon's phases*:—Ðonne se móna wanaþ, Blickl. Homl. 17, 24. Dæghwamlíce ðæs mónan leóht býð weaxende and waniende, Lchdm. iii. 242, 7. Ðás wyrte ðú scealt niman on wanigendum mónan, i. 320, 3. (2) *to wane, become inferior, decline, decay*:—Ðes middangeard wanaþ and weaxeþ, Fragm. Kmbl. 60 ; Leás. 32. Hit gebyraþ ðæt hé weaxe and ðæt ic wanige *illum oportet crescere, me autem minui*, Jn. Skt. 3, 30. Wanige his weordscipe, L. Ath. v. 9 ; Th. i. 306, 23. Gesihð hé ða dómas wonian and wendan of woruldryhte, ða hé gesette, Exon. Th. 105, 24 ; Gú. 28. Nán þing ne, biþ swelce hit wæs siððan hit wanian onginþ, Bt. 34, 9 ; Fox 148, 9. Ðæs ealdigendan mannes mægen bið wanigende, Homl. Th. ii. 76, 21. [O. Frs. wania : O. H. Ger. wanôn : Icel. vana *to diminish ; to spoil, destroy*.] v. á-, gewanian ; wan ; *adj.*, wana ; *adj*.

wánian ; *p.* ode *To lament, deplore*, (1) absolute :—Ðæt synfulle mancynn wépaþ and wániaþ, Wulfst. 183, 2. Ðonne gräniaþ and wániaþ ða ðe hér blissedon and fægnedon, 245, 3 : Anglia viii. 336, 41. Beornas grétaþ, wépaþ wánende, Exon. Th. 61, 31 ; Cri. 993. Ða wánigendran welras (wániendan, Wulfst. 139, 8) *os lugens*, Dóm. L. 208. (2) with reflexive dative :—Hé wánode him sylfum : ' Wá is mé earmum . . . ,' Homl. Skt. i. 11, 223. (3) with acc. :—Sár wánigean, Beo. Th. 1579 ; B. 787. Wánian, Exon. Th. 166, 22 ; Gú. 1046. Ongan hé sár cwánian, wyrd wánian, wordum mælde . . . , 274, 24 ; Jul. 538. (4) with reflex dat. and (a) acc. :—Hé him wæs wániende ǽgðer ge his ágene heardsǽlþa ge ealles ðæs folces *ipse nunc suam, nunc publicam infelicitatem deflet*, Ors. 4, 5 ; Swt. 166, 20. (b) a clause :—Hé him wæs swiþe wániende ðæt hé tó him cucan ne com, Ors. 5, 12 ; Swt. 244, 4. [Heo weop for hire weisið, wanede hire siðes, Laym. 25847. Weape and wony (weinen, 1st MS.), 25827. Wepenn and wanenn for hiss sinne, Orm. 5653. Hit cumeþ weopinde and weowinde iwiteþ . . . þeo moder greoneþ and þ bearn woaneþ, Fragm. Phlps. 5, 32–41. Heo woneþ and groneþ day and nyht, Misc. 152, 187. Scholde euch mon woni and grede, O. and N. 975. O. H. Ger. weinôn *flere, lacrymare, ejulare, vagire*: Icel. veina *to wail*. Cf. Goth. wainags *unhappy*.]

wanigend, es ; *m. One who diminishes, weakens, impairs, injures, spoils*, etc. v. wanian :—Gyf him þince ðæt hé on reádum heorse ríde, ðæt býð his góda wanigend (wanung, MS. T.) *if he dreams that he is riding on a bay horse, that means there will be a spoiler of his goods*, Lchdm. iii. 172, 29.

wani[g]end-líc ; *adj. Diminutive* (as a grammatical term), *expressing diminution* :—Sume naman synd *diminutiva*, ðæt synd waniendlíce, ða geswuteliaþ wanunge, Ælfc. Gr. 5 ; Zup. 16, 17. *Diminutiva* syndon wanigendlíce. *Clam* is ðígellíce and of ðam is wanigendlíc *clanculum* hwónlícor ðígellíce, ÆLfc. Gr. 38 ; Zup. 231, 1–3.

waniht. v. wanniht.

wann ; *adj. Dark, dusty, sable, lurid, livid* :—Wann *bruntus*,Wrt.Voc. i. 46, 40. Wonn, ii. 12, 58. Won, 127, 28. Ðı sweartan *lurida*, wan

and flæc *luridus*, 53, 16. Ða wannan *libida* (but the Latin is *livida* (*vibex*), Ald. 77–8), 88, 3 : 50, 33. Ðære wannan *cerula*, 24, 58. Ða wonnan aetrinan *livida toxica*, 112, 63 : 50, 80. Ða wonnan *lividas*, 53, 1. (1) *blue-black, livid* :—Ðonne se dǽl ðæs líchoman sié gewended blæc oþþe won oþþe swilces hwæt, Lchdm. ii. 82, 12. Gif ðæt blód swíðe reád sié oþþe won, 254, 10. Swearte ł wan[ne] wale *caerulea* (*nigra, tetra, tunsa*) *vibice* (*livore*), Hpt. Gl. 516, 14. Gif ða ómihtan wannan þing oþþe ða reádan sýn útan cumen, Lchdm. ii. 82, 21. (2) *of the colour of living creatures, swarthy, dusty, dark-hued* :—Se wonna þegn, sweart and saloneþ, Exon. Th. 433, 8 ; Rä. 50, 4. Bið se wǽrloga (*the wicked at the judgement day*) won and wliteleás, hafaþ werges bleó, 95, 30 ; Cri. 1565. Deóful ætýwde wann and wliteleás, hæfde weriges híw, Andr. Kmbl. 2339 ; An. 1171. Hræfen gól wan and wælfel, Elen. Kmbl. 105 ; El. 53. Se wonna hrefn, Beo. Th. 6041 ; B. 3024. Wanna, Judth. Thw. 24, 25 ; Jud. 206 : Cd. Th. 119, 22 ; Gen. 1983. Bearg won, Exon. Th. 428, 12 ; Rä. 41, 107. (3) *of the colour of material, dark, dingy* :—Sý mín bæc wonn, Exon. Th. 496, 13 ; Rä. 85, 14. Wonnum hyrstum gefrætwed, 436, 1 ; Rä. 54, 7. Mec mon biþeahte mid þearfan wǽdum, and mec on þeóstre álegde biwundenne mid wonnum cláþum, 87, 12 ; Cri. 1424. (4) as a (poetical) epithet of shade, cloud, night, etc. :—Gif him (*the stars*) wan fore wolcen hangaþ (cf. ðonne sweartan wolcnu him beforan gáþ, Bt. 6 ; Fox 14, 22) ne mægen hí leóman aneódan *nubibus atris condita nullum fundere possunt sidera lumen*, Met. 5, 4. Sceadu wann under wolcnum, Rood Kmbl. 109 ; Kr. 55. Seó deorce niht won gewíteþ, Exon. Th. 204, 17 ; Ph. 99 : 292, 23 ; Wand. 292. Ðá se æþela glǽm setlgong sóhte, swearc norðrodor won under wolcnum, 178, 34 ; Gú. 1254. In ðisse wonnan niht, 163, 30 ; Gú. 1001. On wanre niht scríðan, Beo. Th. 1409 ; B. 702. Hé geseah deorc gesweorc semian sweart, wonn and wéste, Cd. Th. 7, 22 ; Gen. 110. Ða wonnan niht móna onlíhteþ (cf. se móna líht on niht, Bt. 21 ; Fox 74, 25), Met. 11, 61. Færeþ sunne in ðæt wonne genip under wætra geþring, Exon. Th. 351, 12 ; Sch. 79. Wolcnu wann, Cd. Th. 14, 5 ; Gen. 214. Sceadu sweðerodon wonn under wolcnum, Andr. Kmbl. 1673 ; An. 839. Wan, Beo. Th. 1306 ; B. 651. Won, Exon. Th. 384, 33 ; Rä. 4, 37. Wonnum nihtum, 496, 3 ; Rä. 85, 8. (5) as a (poetical) epithet of water (cf. Myn is the drenchyng in the see so wan, Chauc. Kn. T. 1598) :—Ýðgeblond ástígeþ won tó wolcnum *the troubled waves mount dark to heaven*, Beo. Th. 2752 ; B. 1374. Wonn, Exon. Th. 383, 34 ; Rä. 4, 20. Hé þeahte bearn middangeardes wonnan wǽge *he covered earth's children with the dark wave*, Cd. Th. 83, 13 ; Gen. 1379. Gewát se wilda fugel ofer wonne wǽg, 88, 8 ; Gen. 1462. Hé wolde ðæt wanne wæter tó wíne áwendan, Homl. Th. ii. 58, 16. Sweart wæter, wonne wælstreámas, Cd. Th. 78, 30 ; Gen. 1301 : 86, 13 ; Gen. 1430. Gársecg þeahte sweart synnihte wonne wǽgas *black everlasting night covered ocean, the dark waves*, 8, 4 ; Gen. 119. (6) as a.(poetical) epithet of fire. v. wann-fýr :—Nú sceal gléd fretan, wyrdan wonna lég, wigena strengel, Beo. Th. 6221 ; B. 3115. Se wonna lég, Cd. Th. 309, 24 ; Sat. 715. v. brún-wann.

wann-fáh ; *adj. Dark-hued* :—Wonfáh wale, Exon. Th. 435, 11 ; Rä. 53, 6.

wann-feax ; *adj. Dark-haired, with raven-black tresses* :—Wonfeax wale, Exon. Th. 393, 30 ; Rä. 13, 8.

wann-fóta, an ; *m. A bird with dark feet* (?) :—Stángella *vel* wanfóta *pelicanus* (cf. *porfyrionis, pellicanus*, Corp. Gl. ed. Hessels 94, 498), Wrt. Voc. i. 63, 20.

wann-fýr, es ; *n. Lurid fire* :—Wonfýres wælm, se swearta líg *lurid fire's glow, the dark flame*, Exon. Th. 60, 7 ; Cri. 966.

wann-hǽwe ; *adj. Dark-blue, blue-black* :—Ða wonhǽwan *cerula*, Wrt. Voc. ii. 20, 66.

wannian. v. á-wannian.

wanniht ; *adj. Livid* :—Ða wan[n]ihtan *lividas*, Wrt. Voc. ii. 50, 32. v. wann.

wan-sǽlig ; *adj. Unblest, miserable, evil* :—Grendel, wonsǽlig wer, Beo. Th. 210 ; B. 105. Wineleás, wonsǽlig genimeþ him wulfas tó geféran, Exon. Th. 342, 24 ; Gn. Ex. 147. In ðisse wonsǽlgan worulde lífe, 158, 33 ; Gú. 919. Weras wansǽlige mé (*Christ*) slógon and swungon, Andr. Kmbl. 1925 ; An. 965. Wonsǽlige, Elen. Kmbl. 953 ; El. 478. Fróde sace sémaþ, sibbe gelǽraþ, ða ǽr wonsǽlge áwegen habbaþ, Exon. Th. 334, 24 ; Gn. Ex. 21. Werum wansǽligum (*the Jews*), Elen. Kmbl. 1952 ; El. 978.

wan-sceaft, e ; -sceafte(-a ; *m.?*), an ; *f.* I. *misfortune, misery, unhappiness* :—Hí sorge ne cúðon, wonsceaft wera, wiht unhǽlo, Beo. Th. 240 ; B. 120. Ic ne wrecan meahte on wigan feore wonnsceaft míne, ac ic ealle þolige, Exon. Th. 499, 16 ; Rä. 88, 16. Lǽð biþ ǽghwǽr fore his wonsceaftum wineleás hǽle, 329, 10 ; Vy. 32. II. *some form of disease* :—Hú mon sceal ða wǽtan and wonsceafta (ða wonsceaftan *in the section*, 246, 6, *where no other malady than ða wǽtan is referred to except* ða áheardodan swilas) útan lácnian, Lchdm. ii. 166, 22. [Cf. O. Sax. than wôpiat thâr wanskefti thie hér ér an wunnion sind, Hél. 1352.]

wan-scrýd[d] ; *adj. Imperfectly clothed, ill-clad* :—Hé wæs swíðe

geswǽs eallum swincendum, and on mislicum yrmðum mannum geheólp, wǽdligum and wanscrýddum, Homl. Th. ii. 500, 17.

wan-seóc; *adj. Epileptic, having the falling sickness, frenzied, lunatic :—*Wanseóce *comitiales, lunaticos,* Hpt. Gl. 519, 43. v. brǽc-, fylle-, gebrǽc-, mónaþ-seóc; brǽc-coþu.

wansian; *p. ode [the word seems to occur only late, and perhaps is due to Scandinavian, cf. Icel. vansi want: wanian is the usual word] To diminish :—*Swá hwá swá úre gife ðuþer ððre gódene manne gyfe wansiaþ, wansie him seó heofenlíce iateward on heofonríce, Chr. 656; Erl. 32, 17. *The compound* á-wansian *also occurs :—*If áni man ðis ilk forward breke and áwansige, Cod. Dip. Kmbl. iv. 243, 6. [Marrchess nahhtess wannsenn and Marrchess daȝhess waxenn, Orm. 1901. Worldes catel wacset and wansit as te mone, P. R. L. P. 234, 7. Wansoñ, wansyn *evaneo, decresco,* Prompt. Parv. 515.]

wan-spéd, e; *f. Poverty, indigence :—*þurh wanspéde *per inopiam,* Scint. 226, 6. On ðǽm gefeohte wæs ǽrest anfunden Sciþþia wanspéda *ea res primo fidem inopiae Scythicae dedit,* Ors. 3, 7; Swt. 116, 34. Cf. wan-ǽht.

wan-spédig; *adj. Poor, indigent :—*Sum ǽhta onlíhð; sum bið wonspédig, Exon. Th. 295, 11; Crä. 31. Ðín wanspédiga mǽg *attenuatus frater tuus,* Lev. 25, 25. Ðás lǽssan lác, ðe wǽron wannspédigra manna lác, Homl. Th. i. 140, 6. Uton dón þearfum and wannspédigum sume híðde úre góda, ii. 100, 35. Se gýtsere berýpð ða wannspédigan, i. 66, 11.

wanspédigness, e; *f. Indigence, poverty :—*Of neóde oþþe wanspédignysse *ex necessitate uel indigentia,* Scint. 198, 5.

wanung, e; *f.*   I. *a making less,* (1) *diminution.* Cf. wanian, I. (1) :—Sume naman synd *diminutiva,* ða geswuteliaþ wanunge, Ælfc. Gr. 5; Zup. 16, 18. Ða word habbaþ hwílon *sincopam,* ðæt ys, wanunge: *amauisti vel amasti,* hér ys se *ui* áwege, 25; Zup. 146, 17. (2) *abatement, reduction, checking.* v. wanian, I. (2) :—Hwæt getácnaþ ðæs fylmenes ofcyrf on ðam gesceape búton gälnysse wanunge? Homl. Th. i. 94, 33. (3) *a weakening, an impairing, hurt, injury.* v. wanian, I. (3) :—Wonung *detrimentum,* Wrt. Voc. ii. 106, 29. *Dispendium,* i. *damnum, impedimentum, defectio, periculum, detrimentum* æfwerdla, wonung, wom, wana, *vel* hénþa, 140, 68. Gyf him þince ðæt hé hæbbe rúh líc, ðæt byð his góda wanung, Lchdm. iii. 170, 24. Góda wanigend (wanung, MS. T.), 172, 29. Wanunge *dispendio,* Wrt. Voc. ii. 28, 37. Ðæt nádær ne þǽ ne ús God ne þurfa oncunnan for ðæræ waniungæ on úrum dæge *quatinus nec tibi nec nobis Deus debeat imputare hanc imminutionem diebus nostris actam,* Chart. Th. 163, 26. Nalæs bútan mycelre wonunge his weoredes *non sine magno exercitu sui damno,* Bd. 2, 2; S. 504, 7. Is nýð ðæt sume mid wonunge heora woruldǽhta synd gerihte *necesse est ut quidam damnis corrigantur,* 1, 27; S. 490, 10. Hé mycle wonunge and ǽwyrdlan wæs wyrcende ðære mǽrwan cyrican weaxnesse *magno tenellis ecclesiae crementis detrimento fuit,* 2, 5; S. 506, 37. Mid ðám hefigestum wonungum his ríces fram his feóndum geswenced *gravissimis regni sui damnis ab hostibus adflictus,* 3, 7; S. 530, 18. (3 a) *a weakening, reducing the strength* of something. Cf. wanian, I. (3 a) :—Flǽsces wonunge *carnis maceratione,* Rtl. 14, 33.   II. *a growing less,* (1) *a decrease* in number, size, etc. v. wanian, II. (1) :—Dæghwamlíce geleáffulle men nimaþ ðæt sand, and ne biþ nǽnig wonung on ðæm sande, Shrn. 81, 6. Symle bið háligra manna getel geeácnod þurh árleásra manna wanunge, Homl. Th. i. 536, 25. (1 a) *waning* of the moon. v. wanian, II. (1 a) :—Ǽfre seó sǽ and se móna beóð geféran on wæstme and on wanunge, Lchdm. iii. 268, 13: Homl. Th. i. 102, 28. (2) *decline, decay.* v. wanian, II. (2) :—Ðonne se móna wanaþ, ðonne tácnaþ hé ðisse worlde wanunge, Blickl. Homl. 17, 24.   III. *a lack, want, defect :—*Wanunge *defectu,* Wrt. Voc. ii. 28, 43.

wánung, e; *f. Wailing, lamentation :—*Wánung *threnum,* Wrt. Voc. i. 28, 20. Ðǽr (*in hell*) is wánung and gránung and á singal sorh, Wulfst. 26, 8. Hǽðenra gránung and reáfera wánung, 186, 13. Wóp and wánung and heófung and endeleás cwylming, Homl. Th. i. 592, 16. Geómrung and wánung, Homl. Skt. i. 23, 104. Se láce cyrfð oððe bærnð, and se untruwa hrýmð, þeáhhwæðere ne miltsaþ hé ðæs óðres wánunge, Homl. Th. i. 472, 16. *Uae* getácnaþ hwílon wánunge, Ælfc. Gr. 48; Zup. 278, 12. Gesaeh ðæt wánung (*tumultum*) and woepende and máriende, Mk. Skt. Lind. 5, 38. Se álárde his hláford licgan heáfodleásne and hé ða mid wánunge wende út ongeán *videns cadaver absque capite Holofernis exclamavit voce magna cum fletu,* Anglia x. 101, 365. Mid hreówlícere wánunge, Homl. Th. i. 466, 33. [Heui is his greoning and seorhful is his woaning, Fragm. Phlps. 5, 25. Wanung and wow, O. E. Homl. i. 173, 231. After al þis cumeð of þat bearn iboren þus wanunge and wepnunge, H. M. 37, 9. Þer wes muchel waning, heortne graning, Laym. 17796. Wop and wonynge and bymenynge, Mirc. 74, 55. Þu telst . . . al mi (*the owl's*) reorde is woning, O. and N. 311.]

wan-wegende; *adj. (ptcpl.) Waning :—*On wanwegendum mónan, Lchdm. i. 100, 20. Wanwægendum, 98, 17.

wápe(-a ? m.), an; *f. A cloth, rubber (? cf. wípian) :—*Gif ðú sceát habban wille oððe wápan, ðonne sete ðú ðíne twá handa ofer ðínum bearme and tóbrǽd hí swilce sceát ástrecce, Techm. ii. 122, 23. [Cf. (?) *Icel. veipa a woman's hood.*]

wapol (-ul, -el) *foam :—*Wapul *famfaluca* (cf. faam, leásung *famfaluca,* 17), Wrt. Voc. ii. 108, 20 : 35, 4 (cf. leásung oððe fám *famfaluca,* 24, 75). v. next word.

wapolian; *p. ode To foam, bubble up, pour forth (intrans. and trans. ?), abound, swarm :—*Wapolaþ *ebullit* (os fatuorum ebullit stultitiam, Prov. 15, 2), Kent. Gl. 505. Wapolode *vaporat,* Germ. 398, 220. Up ábrǽcan, wapeladan *ebulliebant, emergebant* (cadavera horrida vermium examina ebulliebant, Ald. 48), Hpt. Gl. 488, 11. Wapeledan ł up ábrǽcan *bullirent, exundaverunt* (cum Ethnae montis incendia favillis scintillantibus bullirent, Ald. 55), 499, 46. Ingá forrotednys on bánum mínum and under mé heó wapelige *ingrediatur putredo in ossibus meis et subter me scateat,* Cant. Habac. 16. v. preceding word.

wár.   I. *sea-weed, waur* (v. E. D. S. Pub. Plant Names, in which other forms are given, *ware, woare, woore, ore :* see also Jamieson's Dict. *ware*) :—Waar, uaar, uár *alga,* Txts. 39, 120. Wár, Wrt. Voc. ii. 6, 46 : i. 285, 12.   II. *sand, strand.* Cf. sondhyllas *alga,* Txts. 39, 125 :—Streámas weorpaþ on stealc hleoþa stáne and sande, wáre (*or under* I ?) and wǽge, Exon. Th. 382, 8; Rä. 3, 8. Wára *sablonum,* strand *sablo* (mentis fundamina nequaquam arenosis sablonum glareis ultro citroque nutabundis subdiderat, Ald. 57), Hpt. Gl. 502, 76 : (*printed* wasa) 465, 8. Wárum *sablonibus,* 449, 30. v. sǽ-wár.

wara, an; *m. An inhabitant. The word is used mostly in the plural, and as the second part of compounds; but the singular in composition is found in* ceaster-weara *civis,* Bd. 3, 22; S. 552, 32 (cf. ceaster-gewara *civis,* Ælfc. Gr. 5; Zup. 11, 16), *and the independent word in the following instances :—*Heofenlícra warena *supernorum civium* (*habitatorum*), Hpt. Gl. 498, 23. Hié here samnodon ceastre (*printed* ceaster) warena, Andr. Kmbl. 2251; An. 1127. Warum *civibus,* Hpt. Gl. 518, 40. *In composition both* -waran *and* -ware *occur* (cf. Seaxe *and* Seaxan), *and also* -waras, v. Sigel-waras. *The forms are united with common nouns,* v. burh-, ceaster-, eorþ-, hell-, heofon-waran, -ware; *or with proper names, native or foreign,* e. g. Lunden-, Róm-waran, -ware, Bæx-warena land (cf. Bex-leá, 13), Cod. Dip. B. i. 295, 5, Cant-ware, Wiht-ware, Sodom-ware, Syr-ware: *see also* Up-ware. Cf. *the Icelandic* Róm-verjar, *and Latin forms like* Angri-varii. v. -waru.

-ware. v. preceding word.

warenian, warnian, wearnian; *p.* ode.   I. *intrans.* (1) *To take heed, beware, be on guard :—*Warniaþ and waciaþ *uidete, vigilate,* Mk. Skt. 13, 33. Hé wolde warnian on ǽr *he would take precautions,* Gen. 6, 6. Man sceal wacigean and warnian symle, Wulfst. 90, 2. (2) *to take heed of, guard against, abstain* from (cf. *Icel.* varna við *to abstain from*) :—Warniaþ fram beorman Fariséorum *cavete a fermento Pharisaeorum,* Mt. Kmbl. 16, 6, 11, 12. Warniaþ fram bócerum *cavete a scribis,* Mk. Skt. 12, 38. Warniaþ (warnigeaþ, *v. l.*) wið Farisêa *attendite a fermento Pharisaeorum,* Lk. Skt. 12, 1. Ðæt man wið leahtras warnie (warnige, *v. l.*), Wulfst. 68, 14. (3) *to take heed* that something is not done, does not happen (expressed in a clause) :—Warna ðæt ic ðé leng ne geseó *cave ne ultra videas faciem meum,* Ex. 10, 28. Warna ðæt ðæt leóht ðe ðé on is ne sýn þýstru *vide ne lumen quod in te est tenebrae sint,* Lk. Skt. 11, 35 : Homl. Th. i. 120, 16. Warniaþ (*videte*) ðæt gé hyt nánum men ne secgeon, Mt. Kmbl. 9, 30 : 18, 10. Warnigeaþ ðæt gé ne beón gedréfede, 24, 6. Se man mót geornlíce warnian, ðæt hé eft ðám yfelum dǽdum ne geedlǽce, Homl. Th. ii. 602, 23. Hé mé warnian hét, ðæt ic on ðone deáðes beám bedroren ne wurde, Cd. Th. 33, 29; Gen. 527. Is mycelum tó warnienne ðæt man . . . menn blód ne lǽte, Lchdm. iii. 152, 33. (4) *to take heed* that something does happen :—Wel is eác tó warnianne ðæt man wite, ðæt hý þurh mǽgsibbe tó gelænge ne beón, L. Edm. B. 9; Th. i. 256, 9.   II. *trans.* (1) *To put on guard, to warn :—*Bútan ic eów warnige, ic sceal ágyldan gesceád mínre gýmeleáste, Homl. Skt. i. 17, 72. Ðæt wyrreste þingc ðú didest, ðæt ðú mé warnodest, Ap. Th. 8, 15. Se Hǽlend ús warnode ðus, for ðan ðe hé wyle, ðæt wé wære beón, Homl. Ass. 55, 112. Wé ágan þearfe, ðæt wé wið swylcne ege wære beón and eác ða warnian, ðe swylc nyton swylc tówerd is, Wulfst. 101, 11. Men ða leófestan, wé willaþ eów warnian, and ús sylfe álýsan, Homl. Ass. 144, 18. Ðá sende Ælfríc and hét warnian ðone here, Chr. 992; Erl. 130, 31. (1 a) where no object is expressed :—Swefnu beóð onwrigene tó warnienne, Lchdm. iii. 196, 24. (1 b) *to warn* against something, *give notice* of something :—Benedictus warnode ða gebróðra wið ðæs deófles tócyme, Homl. Th. ii. 166, 17. Ðæt hý Godes folc warnian wið ðone egesan, ðe mannum is tówerd, Wulfst. 79, 14. (1 c) where the matter to which the warning refers is given in a clause :—Ic eów warnode, ðæt gé wíglunge mid ealle forlǽtan, Homl. Skt. i. 17, 68. Wé ágan þearfe, ðæt wé godcunde heorda warnian, hú hý Antecriste wærlícast magan wiðstandan, Wulfst. 80, 2. (2) used reflexively, *to be on one's guard, to look to one's self, take heed to one's self, take warning :—*Ðurh gítsunge forlýst oft se árleása his líf, ðonne hé gewilniað ðara æhta, and ne warnaþ hine sylfne, Basil admn. 9 ; Norm. 54, 2 : Cd. Th. 40, 6; Gen. 635. Gif ðú ðín ágen myrre, ne wít ðú hit ná Gode, ac warna ðé silfne, Prov. Kmbl. 51. Warniaþ eów sylfe *uidete uosmetipsos,* Mk. Skt. 13, 9, 23. Ðé is micel þearf ðæt ðú ðé warnige, for ðam ðé ðú eart fordémed, Ap. Th. 8, 1. Utan warnian ús

georne, Wulfst. 101, 21. Ðû noldest ðê warnian þurh ðînes fæder ðreále *thou wouldst not take warning by thy father's punishment*, Homl. Th. ii. 436, 7. (2 a) *to guard, be on one's guard* against something :—Gif hê hine ne warenaþ wiþ ða unþeáwas, Bt. 29, 3; Fox 106, 27. Wærnaþ (warenaþ, Cott. MS.) hê hine wiþ ðæt weder, 41, 3; Fox 250, 16. Hié oft gesyngiaþ giet wyrs on ðæm ðæt hî hî wareniaþ wið ða lytlan scylda ðonne hî dôn on myclum scyldum; for ðæm ðe hî lícettaþ hié unscyldge, ðonne hî hî wæreniaþ wið ða lytlan, Past. 57; Swt. 439, 18–20. Ic mê [wið] his hete berh and wearnode (warnode, *v. l.*: warenode, Bd. M. 128, 9) *hostium vitabam insidias*, Bd. 2, 12; S. 513, 28. Warniaþ eów wið oferfylle, Homl. Th. ii. 22, 16. Is ðæm tô cŷðanne, ðæt hî hié warenigen ǽgðer ge wið ða ungemetlícan blisse ge wið ða ungemetlícan unrôtnesse. . . . Is micel niédþearf ðæt mon hiene wið ðæt irre and wið ða ungemetlícan sǽlða warenige (warnige, Cott. MSS.), Past. 27; Swt. 189, 1–6. Ic bidde ðæt ǽlc mann hine sylfne georne wið ðisne curs warnige, Chart. Th. 445. 8: Wulfst. 101, 16. Utan warnian ûs wið his unlâra, 80, 4. (2 b) where what is to be guarded against is expressed in a clause :—Warnode hê hine ðŷ læs hî on hwylc hûs tô him in eodan *caverat ne in aliquam domum ad se introirent*, Bd. 1, 25; S. 486, 39. Hê hêt hine warnian (*or* I. 3), gif hê wolde libban, ðæt hê nǽre on ðam mynstre nǽfre eft gesewen, Homl. Skt. i. 6, 211. (3) *to keep* something from a person, *to ward off* (cf. *Icel.* varna einum eins *to deny a person something*) :—Snyttra brúceþ ðe fore sâwle lufan warniaþ him wommas worda and dǽda *he uses wisdom, that for love of his soul wards off from himself* (avoids) *sins of word and deed*, Exon. Th. 304, 32; Fä. 79 : 305, 9; Fä. 85. Ic mê warnade hyre onsŷne *I avoided seeing her, denied myself her presence*, 173. 6; Gû. 1156. Óþ ðæt hê geseah his gehŷrend ðone Eástordæg onfôn, ðone hî symle ǽrðan wearnedon (warenedon, Bd. M. 474, 20) *donec illum in Pascha diem, suos auditores, quem semper antea vitabant, suscipere videret*, Bd. 5, 22; S. 644, 44. Eall hê wearnige (weornige, MS.) swâ fŷr (syer, MS.) wudu wearnie (weornie, MS.) *let him avoid it all, as wood avoids fire*, Lchdm. i. 384, 13. [*O. H. Ger.* warnôn *munire, prospicere, admonere, instruere, attendere*; *Icel.* varna (*see* I. 2, II. 3 *above*); cf. varan *a warning*; *shunning.*] v. be-, ge-warenian (-warnian, -wearnian), un-warnod; wirnan; warian.

warenung, warnung. wearnung, e; *f.* **I.** *a taking heed, caution.* v. warnian, **I** :—Hwæðer wære strengra wyrd ðe warnung, Salm. Kmbl. 855; Sal. 427. **II.** *a putting on guard, a warning, admonition.* v. warnian, **II** :—Hit ys Godes sprǽc and his warnung and seó tíd cymð hrædlíce, Gen. 41, 32. Wísdômes bigspell and warnung wið disig, Ælfc. T. Grn. 7, 38. Hér is rihtlíc warnung and sôðlíc myngung ðeóde tô ðearfe, gŷme se ðe wille, Wulfst. 167, 26. Ðæt mæg wítes tô wearninga, ðam ðe hafaþ wîsne geþôht, Exon. Th. 57, 21; Cri. 922. [*O. H. Ger.* warnunga *munimentum, defensio, monimentum.*]

warian; *p.* ode **I.** *intrans.* (*or uncertain*) *To beware* :—Warat *cavet*, Kent. Gl. 364. Wara *cave*, Germ. 393, 136. Warige (warnige, *v. l.*) hê ðæt hit nâ forealdige, L. Edg. C. 38; Th. ii. 252, 6. **II.** *trans. To make ware*, (1) *to warn* :—Mid ðæm wordum hê ûs warode and lǽrde *quibus verbis pastoribus praecavetur*, Past. 18; Swt. 137, 21. Môtan ða hyrdas beón swîðe wacole, ðe wið ðone þeódscaðan folc sculon warian, Wulfst. 191, 13. (2) used reflexively, (a) *to be on one's guard, guard against evil* :—Forlǽtaþ ðone ǽnne beám, wariaþ inc wið ðone wæstm, Cd. Th. 15, 20; Gen. 236. Hê gelǽre ðæt hŷ hî wið ðæt warien, ðæt hŷ hǽr ne cumen, Shrn. 203, 3. (b) *to be careful* to do what is necessary, *take a precaution* :—Warige hîne se ðe his âgen befôd, ðæt hê tô ǽlcan teáme hæbbe getrŷwne borh, L. Eth. ii. 9; Th. i. 290, 6. **III.** *to guard, hold* :—Mîn hord waraþ feónd, Exon. Th. 499, 27; Rä. 88, 22: 414, 17; Rä. 32, 21. Hê hǽðen gold waraþ, Beo. Th. 4543; B. 2277. **III a.** *to hold* a place, *occupy, inhabit* :—Hié dŷgel lond warigeaþ, Beo. Th. 2720; B. 1358. Hê wêsten warode, 2534; B. 1265. Goldsele Grendel warode, 2511; B. 1253. **III b.** *to take possession of* (cf. gisehan thana hêlagon gêst ênigan man warôn, Hêl. 1003 :—Waraþ hine wræclâst, nales wunden gold, Exon. Th. 288, 17; Wand. 32. **IV.** *to ward off.* v. warenian, II. 3 :—Ðæt wit unc wíte warian sceolden, Cd. Th. 49, 33; Gen. 801. [They bad him he scholde warye (*be on his guard*), Alis. 4083. Heo mot warien hwon me punt hire, A. R. 418, 1. Iosep cuðe him biforen waren, Gen. and Ex. 2154. Ware the what thou do, Gow. ii. 388, 27. Ware þe fram wanhope, Piers P. 5, 452. *O. Sax.* warôn: *O. Frs.* waria: *O. H. Ger.* bi-warôn : *Icel.* vara *to warn* ; varask *to beware of, be on one's guard against, shun.*] v. be-, ge-warian; werian, warenian.

warian; *p.* ode *To remain, continue* :—Ne him gâst waraþ gômum on nûðe *neque est spiritus in ore ipsorum*, Ps. Th. 134, 19. Waraþ hê windes full, Salm. Kmbl. 49; Sal. 25. [*O. Sax.* warôn *to last, continue.*] v. werian *to remain.*

wârig; *adj. Stained with sea-weed, dirty* :—Biþ his ceól cumen and hyre ceorl tô hâm, and heó hine in laðaþ, wæsceþ his wârig hrægl, Exon. Th. 339, 24; Gn. Ex. 90. [Hu maht þu iseon þine sceadewe in worie watere, O. E. Homl. i. 29, 4. Schir heorte . . . wori heorte, A. R. 386, 7.] v. next word.

wâriht; *adj. Full of sea-weed* :—Wârihtum ârena tîum *algosis remorum tractibus* (Ald. 3), Hpt. Gl. 406, 68 : Wrt. Voc. ii. 75, 14 : 4, 63. **warnian**, warnung. v. warenian, warenung.

waroþ (-uþ, -aþ, -eþ), wearoþ, weroþ, warþ, es; *m. A shore, strand* :—Ic geseah men standende be ðam waruðe (werode, *v. l.*), Homl. Skt. ii. 23 b, 370. Bî waraðe (nêh warðe *secus littus*, Lind.) sittende, Mt. Kmbl. Rush. 13, 48. Seó mænigeo stôd on ðam waroðe (waraþe, Rush.: wearðe, Lind. *litore*), Mt. Kmbl. 13, 2 : Shrn. 150, 20. Ðû gemêtst scip on ðæm waroðe, Blickl. Homl. 231, 30 : Andr. Kmbl. 525; An. 263. On ðæs sǽs waroþe, Bd. 1, 12; S. 481, 11. Feówer swulung ond ân lǽs on waruðe gebyreð inn tô Raculfe, Cod. Dip. Kmbl. iii. 429, 16. On waruðe, Andr. Kmbl. 479; An. 240. Hê geseah scip on ðæm warþe, Blickl. Homl. 233, 1. On ðæm warðe (worðe, Rush.) *in litore*, Jn. Skt. Lind. 21, 4. Gewât him tô waroðe rídan þegn Hrôðgâres, Beo. Th. 473; B. 234. Ða lîchoman cômon tô ðam waroðe, Shrn. 54, 23. Oð ðone mǽran wearoð (*of Sicily*), Met. 1, 14. Nǽnig cêpa ne seah ellendne wearoð (·od, MS.) *nec nova littora viderat hospes*, 8, 30. Weroþ, Bt. 15; Fox 48, 13. Ðǽr (*at the Red Sea*) wǽron ða wareðas drîge, Ps. Th. 105, 9. Ofer waroða geweorp, Andr. Kmbl. 611; An. 306. Wereþum, Lchdm. i. 390, 11. Sǽwong tredan, wîde waroðas, Beo. Th. 3934; B. 1963. [þe whal wendeȝ and a warþe fyndeȝ, Allit. Pms. 102, 339. At vche warþe oþer water, Gaw. 715. *O. H. Ger.* warid, werid *insula.*] v. sǽ-waroþ.

wâroþ, es; *n. Sea-weed* :—Ic eom wyrslîcre ðonne ðes wudu fûla oððe ðis wâroð, ðe hêr âworpen ligeþ in corþan, Exon. Th. 424, 34; Rä. 41, 49. v. wâr.

waroþ-faroþ, es; *m. A shore-wave, a breaker* :—Waroðfaruða gewinn, Andr. Kmbl. 393; An. 197.

waroþ-gewinn, es; *n. The strife of waves near the shore, the surge* :—Wê on sǽbâte ofer waruðgewinn wada cunnedon faroðrîdende, Andr. Kmbl. 877; An. 439.

waru, e (*but acc.* waru, Ps. Th. 118, 17) ; *f. Watchful care*, (1) *observance, keeping* of a command, etc. :—Ic on lífdagum lustum healde ðînra worda waru *vivam et custodiam sermones tuos*, Ps. Th. 118, 17. (2) where need for caution is implied, *heed, care* :—Ða wiðerwinnan wurdon oferswîðde þurh ðæs engles gewinne and ware, Homl. Th. ii. 338, 2. Antiochus giémde hwæt hê hæfde monna gerîmes, and ne nom nâne ware hûlíce hié wǽron, Ors. 5, 4; Swt. 224, 22. (3) *care* for the safety of others :—Se hŷra ne bið nâðor ne mid ware ne mid lufe âstyred, Homl. Th. i. 240, 28. Paulus ne êhte geleáffulra manna ðurh andan, ac ðurh ware ðære ealdan ǽ, 390, 6. (4) *safe-keeping, custody, keeping from injury, guard* :—Stôd se grêna wong in Godes wære, Exon. Th. 146, 32; Gû. 718: 143, 17; Gû. 662: Andr. Kmbl. 1648; An. 825. Ðê God hæfde wære bewunden *God kept thee on every side*, 1069; An. 535. Wǽre betolden, 1976; An. 990. Him Scyld gewât on Freán wære, Beo. Th. 54; B. 27. In Godes wære, Menol. Fox 79; Men. 39. Hê h's gâst âgeaf on Godes wære, 432; Men. 217. Hêr Eádward kingc sende sâwle tô Criste on Godes wæra, Chr. 1065; Erl. 196, 23. (5) *defence, protection* against attack, *guard* :—Geísneum beláðiendlícre ware [scilde] wiðþyddende leásere wrôhte arwan *ferrato apologeticae defensionis clypeo retundens strophosae accusationis catapultas*, Hpt. Gl. 505, 61. Tô ware *ad tutelam* (*defensionem*) (leo ad tutelam virginis Dei nutu dirigitur, Ald. 45), 484, 49. Nân man ne dorste for ðæra deóra ware ðam hâlgum geneálécan, Homl. Skt. ii. 24, 56, 60. Scealt ðû for ware ûra goda wîta ðrowian *for the protection of our gods thou shalt suffer punishments*, Homl. Th. i. 594, 4. Cyninge gebyraþ ðæt hê sŷ on ware and on wearde Cristes gespeliga, L. I. P. 2; Th. ii. 304, 23. Hié ealle ongeán hiene wǽron feohtende and ðone weg lêtan bûtan ware (*they left the road unguarded*), ðæt seó fierd þǽr þurhfôr *in se omnes pugnando convertit, donec exercitus angustias transiret*, Ors. 4, 6; Swt. 172, 22. Hié wǽron ða burg hergende and sleánde bûton ǽlcre ware (*without any defence being offered*), 2, 8; Swt. 92, 16. Ware ł gescildnysse *defensionem*, Hpt. Gl. 471, 61. Ðû mê behête fulle wære (ware, *v. l.*) wið æftersprǽce *thou didst promise complete protection against claim*, L. O. 7; Th. i. 180, 23. Hŷ ðæs ware cunnon, healdaþ hine twâ hund wearda, Salm. Kmbl. 518; Sal. 258. His ware *munitiones ejus*, Blickl. Gl. [To hæbbe som gret cite or castel me to ware (*for my defence*), R. Glouc. 115, 9. Goth. warei *astutia* : *O. Sax.* wara *heed* (wara niman) : *safe-keeping* (wara Godes sôkean): *O. Frs.* ware: *O. H. Ger.* wara (wara neman, tuon) *heed, care.*] v. niht-, ût-waru.

waru, e (*but the declension seems partly u-stem*) ; *f. Ware, merchandise* :—Mangere *mercator*, waru *merx*, Wrt. Voc. i. 73, 73. Hî wurpon heora waru oforbord *they cast forth the wares that were in the ship into the sea* (Jonah 1, 5), Homl. Th. i. 246, 2. Ða gelamp hit æt sumum sǽle, swâ swâ gyt for oft déð, ðæt Englisce cŷpmenn brôhton heora ware tô Rômâna byrig, and Gregorius eode ðe ðære strǽt tô ðam Engliscum mannum heora ðing sceáwigende. Ða geseah hê betwux ðam warum cŷpecnihtas gesette, ii. 120, 14–18. [Chæpmen bunden heore ware, Laym. 11356. þe wreche peoddare more noise he makeð to ȝeien h's sope, þen a riche mercer al his deorewurðe ware, A. R. 66, 19. Ðe chapmen into Egipte ledden ðat ware, Gen. and Ex. 1990. *O. Frs.* were: *Icel.* vara; *f.*]

4 F

-waru, a form occurring only in compounds with a collective force, *the inhabitants* of a place. It is used with common nouns, v. burh-, ceaster-, eorþ-, hell-, heofon-, land-waru; and with proper names, native or foreign, e.g. Lunden-waru, Chr. 1016; Erl. 159, 22: Hierosolim-waru *Hierosolyma,* Mt. Kmbl. 3, 5; Sychem-ware *Sicinorum,* Wrt. Voc. ii. 73, 66. v. wara.

waru *wearing ?,* waru, Cod. Dip. Kmbl. iii. 429, 16, warum, Ors. 4, 5; Swt. 170, 10, wāsa. v. scrúd-waru, waroþ, wǽr *a covenant,* wudu-wāsa.

Wascan; *pl. m.* The Gascons, Ors. 1, 1; Swt. 22, 32, 34. [*O. H. Ger.* Wascun *Uacca.*]

wascan. v. wæscan.

wāse, an; *f. Ooze, mud, slime :—*Wāse *caenum,* Wrt. Voc. ii. 103, 2: 13, 35. *Cenum,* i. *luti vorago,* vel *lutum sub aquis fetidum,* i. wāse *vel* fæn, 130, 75. Wāsan *ceni* (squallentis ceni contagia, Ald. 49), 82, 63: 18, 39. ¶ the word occurs in several charters dealing with land in the north of Berkshire, and seems to refer to a marsh or stagnant piece of water :—On Wāse; of Wāsan (the Ock, the Thames, and Fyfield are mentioned in this charter), Cod. Dip. Kmbl. iii. 466, 17. On Wāsan; andlang Wāsan (with mention of the Ock and Fyfield), v. 386, 33. Ongeán ða díc ðe scýt tó Wāsan; siððan andlang Wāsan (with mention of the Thames and Appleton), 275, 15. Of ðære mēðe ūt tó Wāsan; of Wāsan ūt tó Eá (with mention of Buckland), 392, 32. Eást tó Wāsan (with mention of Sandford), vi. 9, 7. On Wāse; of Wǽse (with mention of the Thames and Cumnor), 84, 24. [William ... stombled at a nayle, into the waise he tombled, R. Brun. 70, 16. A wase, wayse *alga,* Cath. Angl. 409, and see note. Alle we byeþ children of one moder, þet is of erþe: and of wose (*or* v. wōs ?), Ayenb. 87, 22. As weodes wexen in wose (*v. l.* muk) and in donge, Piers P. C. 13, 229. Wose, slype of the erthe *gluten, bitumen,* Prompt. Parv. 532, and see note. *O. Frs.* wāse *mud, slime : Icel.* veisa *a pool of stagnant water.*] v. wāse-scite.

wāsend, es; *m. The weasand, gullet :—*Wāsend *rumen,* Wrt. Voc. i. 43, 43: 64, 61: 282, 81: *ingluvies,* Hpt. Gl. 490, 11. Wāsende *ingluvie,* 464, 15. Lǽcedōmas wið gealhswile and þrotan and wāsende, Lchdm. ii. 44, 8: 46, 7. In ðane wāsend *ingluviem,* Wrt. Voc. ii. 45, 30. [Weysande *isophagus,* Wülck. Gl. 590, 40. Waysande, 635, 19. Wesande, 676, 24. Wesawnt, 748, 19. *O. Frs.* wāsende (-ande) : *O. H. Ger.* weisont (-unt) *arteriae.*]

wāse-scite (cf. (?) scītan), an ; *f.* or *-scyte* (-scite ?), es ; *m. The cuttle-fish ;* or *the liquid ejected by the cuttle-fish :—*Cudele *vel* wāsescite *sepia,* Wrt. Voc. i. 56, 6. v. scyte, wæter-scyte, *and other compounds of* scyte.

watel, es; *m. A wattle, interwoven twigs :—*Watul *teges,* Ælfc. Gr. 9, 26; Zup. 52, 13. Hē mycelne aad gesomnode on beámum and on ræftrum and on wāgum and on watelum and on ðacum *advexit plurimam congeriem trabium, tignorum, parietum, virgeorum, et tecti fenei,* Bd. 3, 16; S. 542, 23. Ða ástigon hig uppan ðæne hróf þurh ða watelas (*for tegulas*) and hine mid ðám bedde ásendon, Lk. Skt. 5, 19. [v. wattle (*subst. and vb.*) in Baker's Northants Gloss.: wattle *to tile,* Halliwell's Dict.: watteled, Piers P. 19. 323.]

wāþ, e; *f.* I *wandering, roving :—*Deóra gesīð of wāðe cwom, Nabochodonossor, Cd. Th. 257, 26; Dan. 663. Fēðan sǽton, reste gefēgon, wērige æfter wǽðe, Andr. Kmbl. 1185; An. 593. Ic (*a storm*) beámas fylle ... wrecan on wāþe wīde sended *I fell trees ... sent driving a-wandering far* (cf. Aldhelm's Ego rura perago), Exon. Th. 381, 14; Rä. 2, 11. Hý sīð tugon, wīde wāðe, lyftlácende, 110, 29; Gú. 116. Hē sīðfæt sægde sínum leódum, wīde wāðe, ðe hē mid wilddeórum āteáh, Cd. Th. 256, 33; Dan. 650. Hý of wāþum wērge cwōman, restan ryneþrāgum, Exon. Th. 115, 1; Gú. 183. Wāþum strong, fugel feþrum wlonc, 204, 18; Ph. 99: 208, 26; Ph. 161. II. *hunting :—*Deāð, egeslíc hunta ābít on wāðe, nyle hē ǽnig swæð ǽfre forlǽtan *death, dread hunter, persists in his hunting, never will he abandon any track,* Met. 27, 13. [Myght we not fynde ffor to wyn as for waithe, Destr. Tr. 2350. Here is wayth fayrest þat I seз þis seuen зere, Gaw. 1381. *O. H. Ger.* weida *venatio, piscatio : Icel.* veiðr *hunting, fishing;* fara á veiðar *to go a-hunting.*] v. gamen-wāþ; wǽðan.

waþem(-um), es ; *m. A wave, billow :—*Ic þonan wōd ofer waþema gebind *I crossed the band of billows,* Exon. Th. 288, 1; Wand. 24. Waðema streám, sincalda sǽ, Cd. Th. 207, 24; Exod. 471. v. next word.

waþema(-uma), an ; *m. Moving water, wave, flood :—*Ða cwom wōpes hring ūt faran, weóll waðuman streám, and hē worde cwæð, Andr. Kmbl. 2561; An. 1282. Tungol beóþ āhýded, gewiten under waþeman westdǽlas on, Exon. Th. 204, 13; Ph. 97.

wāþol (v. wāþ); *adj. Wandering :—*Scýneþ ðes mōra wāþol under wolcnum (cf. *wandering* as an epithet of the moon in Shakspere), Fins. Th. 14; Fin. 8. [Grein takes waþol=*full moon.* v. Grmm. D. M. 674-5.]

wāwa, an ; *m. Woe, misery :—*On ðære wǽron āwritene heófunga and leóð and wāwa (*scriptae erant in eo lamentationes et carmen et uae*) ... se wāwa getācnaþ ðone ēcan wāwan, ðe ða habbaþ on hellewīte, ðe nū God forseóþ, Ælfc. Gr. 48; Zup. 279, 1-8. Ðonne sceal eów weaxan tō hearme wǽdl and wāwa, Wulfst. 133 3. Ceósan gōdes and yfeles,

welan and wāwan, Cd. Th. 30, 12; Gen. 466. On ǽlcum wāwan hí wǽron gebyldige, Homl. Skt. ii. 28, 130. *Uae* getācnaþ wāwan, Ælfc. Gr. 48; Zup. 278, 17. Sume hí wyrcaþ heora wōgerum sumne wāwan, ðæt hí hí tō wīfe habbon, Homl. Skt. i. 17, 158. Ðæt gē swā earme eów sylfe fordōþ on wīton and on wāwon, 23, 186. Hí gesāwon ða mænigfealdon wāwan ðe Cristes ða gecorenan þoledon, 23, 124. [To þolien wawe mid douelen, O. E. Homl. i. 73, 11. For ðon muchele wawen þet hi iðoleden, 87, 12. Of þan wowe alse of þe wele, ii. 197, 8. Mochel wowe (seorwen, 1st MS.), Laym. 6268. Þolemod aзean alle wowes, A. R. 198, 26. To þolenn alle wawenn, Orm. 13349. Al þat heo singeþ hit is for wowe, O. and N. 414. *O. H. Ger.* wēwo; *m.;* wēwa ; *f. dolor, poena, malum.*] v. weá.

wāwan; *p.* weów ; *pp.* wāwen *To blow, be moved by the wind :—*Hnescre ic eom micle halsrefeþre, seó hēr on winde wǽweþ on lyfte, Exon. Th. 426, 30; Rä. 41, 81. [Mine lokes ... me wes lef to showen, þe wind hem wolde towowen, Anglia iii. 279, 89. *Goth.* waian *to blow* (of the wind) : *O. H. Ger.* wājan (wāen) *ventilare, spirare.*] v. bi-wāwan.

waxan *to wash,* wax-georn. v. wæscan, weax-georn.

wē; *pron.* We. I. used of more than one person, (1) dual :—Ic and ðæt cild gāð unc tō gebiddenne and wē syððan cumaþ eft tō eów, Gen. 22, 5. Wē willaþ ðæt ðū ūs dō swā hwæt swā wē biddaþ (cf. wyt magon, v. 39), Mk. Skt. 10, 35. (2) plural :—Hwī fæste wē (woe, Lind.)? Mt. Kmbl. 9, 14. Wē þonne synt ðe fylgeaþ *it is we that follow,* Blickl. Homl. 81, 33. Wē men sculon, Exon. Th. 46, 33; Cri. 746. Wē selfe cūþen, 147, 7; Gú. 723. Wē ealle wǽron ðē fylgende, and ðū eart ūre ealra fultum ða ðe on ðē gelýfaþ, Blickl. Homl. 229, 20. Uton wē ealle wynsumian on Drihten, wē ðe his ǽriste mǽrsiaþ, 91, 8. Getīþa ūs ðæt ðe wē ðē ætforan āgyltan ... *anue nobis ut quæ* (qui has been glossed) *te coram deliquimus* ..., Hymn. Surt. 124, 30: Exon. Th. 2, 27; Cri. 25. (2 a) used by a king in reference to himself and his counsellors :—Wē (*Ine and the witan*) bebeódaþ, L. In. 1; Th. i. 102, 14. Wē (*Alfred*) lǽraþ, L. Alf. pol. 1; Th. i. 60, 2. Wē (*Aðelstan*) cwǽdon, L. Ath. i. 2; Th. i. 200, 5. Wē (*Cnut*) willaþ, L. C. E. 6; Th. i. 364, 5. II. used of one person, (1) by a writer or speaker :—Nū hæbbe wē scortlíce gesǽd (cf. scortlíce ic hæbbe nū gesǽd, 10, 3). Ors. 1, 14, 26: 22, 1 : 24, 23. Swā wē ǽr cwǽdon (cf. swā ic ǽr cwæþ, 8, 14), 24, 32. Wē mihton ðǽs rǽdinge menigfealdlícor trahtnian, Homl. Th. i. 556, 13. Hwæt wille wē eów swīðor secgan be ðisum symbeldæge, ii. 444, 13 : Blickl. Homl. 115, 28. (2) by a prince :—Beówulf maþelode : 'Wē ðæt ellenweorc fremedon,' Beo. Th. 1920; B. 958: 3308; B. 1652. [*Goth.* weis: *O. Sax. O. Frs.* wī : *O. H. Ger.* wir : *Icel.* vēr.] v. ūs, wit.

weá, an ; *m.* I. *woe, misery, evil, affliction, trouble :—*Weá his *sufficit diei malitia sua,* Mt. Kmbl. Rush. 6, 34. Weá wæs ārǽred, tregena tuddor, Cd. Th. 60, 26; Gen. 987. Mec ðín weá æt heortan gehreáw, Exon. Th. 91, 18; Cri. 1493. Weá biþ wundrum clibbor, Menol. Fox 485; Gn. C. 13. Weán on wēnum *in expectation of evil,* Cd. Th. 63, 4; Gen. 1027: 191, 11; Exod. 213: Exon. Th. 378, 32; Deór. 25: Cd. Th. 146, 6; Gen. 2418. Ne ic ðē weán ūðe *nor did I wish you ill,* 163, 3; Gen. 2692. Nysses ðū weán ǽnigne dǽl *you knew nothing of misery,* Exon. Th. 85, 3; Cri. 1385. Ne lǽd ðū ūs tō wīte in weán sorge, Hy. 6, 27. Hē þearfende of wǽdle weán ālýsde *adjuvabit pauperem de inopia,* Ps. Th. 106, 40. Gif ðē ǽnig mid weán grēteþ *if any man afflict thee,* Homl. Th. 105, 18; Gen. 1755. Hē heóld his ǽhta him tō weán, Blickl. Homl. 53, 9. Biþ hē on ēcne weán bedrifen, 95, 5. Ðæt ða yfelan bióþ micle gesǽligran ðe on ðisse worulde habbaþ micelne weán and manigfeald wíte for hyra yfelum, ðonne ða sién ðe nāne wræce nabbaþ *feliciores esse improbos supplicia luentes, quam si eos nulla justitiae poena coerceat,* Bt. 38, 3; Fox 200, 3. Hí mē weán [íhton, cf. 77, 31] mínra wunda sār *super dolorem vulnerum meorum addiderunt,* Ps. Th. 68, 27. Weán, sār and sorge, Cd. Th. 5, 20; Gen. 74: 267, 22; Sat. 42. Ic fleáh weán wana wilna gehwilces, 137, 11; Gen. 2272: 109, 7; Gen. 1819. For hwon wāst ðū weán, gesyhst sorge, 54, 12; Gen. 876. Gedígan weán and wrǽcsīð, Beo. Th. 4573; B. 2292. Gesamna ūs of wīdwegum, ðǽr wē weán dreógaþ, Ps. Th. 105, 36 : Cd. Th. 276, 7; Sat. 185. Hē for wlenco weán āhsode, Beo. Th. 2417 ; B. 1206: 851; B. 423. Wyrd wōp wecceþ, heó weán hladeþ, Salm. Kmbl. 874; Sal. 436. Eal sār and sace, hungor and þurst, wōp and hreám, and weána mā ðonne ǽniges mannes gemet sý ðæt hié āríman mæge, Blickl. Homl. 61, 36. Fela ic weána gebād, heardra hilda, Fins. Th. 51; Fin. 25. Wēn ne brúceþ, ðe can weána lyt, sāres and sorge, Runic pm. Kmbl. 340, 30; Rún. 8. Weána dǽl *a deal of trouble,* Exon. Th. 379, 17; Deór. 34: Beo. Th. 2304; B. 1150. Ic ðē wið weána gehwam wreó, Cd. Th. 131, 2; Gen. 2170: Beo. Th. 2796; B. 1396. Ic ǽnigra mē weána ne wēnde bōte gebīdan, 1870; B. 933. Hié ealle worlde weán oforhogodan, Blickl. Homl. 119, 15. Weallende weán, Exon. Th. 139, 2; Gú. 587. II. *evil, wickedness, malice.* v. weádǽd :—Nǽfre on his weorþige weá āspronge, mearce mā scyte mán inwides *non defecit de plateis ejus usura et dolus,* Ps. Th. 54, 10. Weá bið in mōde, siofa synnum fāh, gefylled mid fācne, Fragm. Kmbl. 27; Leás. 15. Ðæt gelamp for weán and for yfelnesse ðara eardiendra (*a malitia inhabitantium*), Bd. 4, 25; S. 599, 22. Hý magon weána tō fela geseón on

him selfum, synne genóge, Exon. Th. 77, 30; Cri. 1264. [Hu stont ham þ beoð þere ase alle wo and weane is, A. R. 80, 11.] v. wáwa, weó.

weacen. v. wacen.

weá-cwánian; p. ode To lament, wail :—Deófla weácwánedon mán and morður, Cd. Th. 284, 12; Sat. 320. [Cf. Goth. wai-fairhwjan ejulare : Ger. weh-klagen.]

weá-dǽd, e; f. A deed of woe, an ill-deed :—Hé (Stephen) bæd þrymcyning ðæt hé him ða weádǽd tó wræce ne sette (cf. Domine, ne statuas illis hoc peccatum, Acts 7, 60), Elen. Kmbl. 987; El. 495. Árísaþ weádǽda, Fins. Th. 15; Fin. 8. [Cf. Goth. wai-dêdja a malefactor.]

weá-gesíþ, es; m. A companion in misery or in wickedness :—Tó ðam symle sittan eodon ealle his (Holofernes') weágesíþas, Judth. Thw. 21, 13; Jud. 16. Hé ðone deófol on helle mid his weágesíðum ofþrihte, Wulfst. 145, 4. Ða deorcan and ða dimman stówe helle tintrego, ðe deófol an wunaþ mid his weágesíþum and mid ðám áwergdum sáulum, 225, 33.

weal a wall ; weala. v. weall; wela, wealh.

weá-láf, e; f. A remnant spared by calamity, those who remain after evil times, the survivors of calamity :—Land hý áwéstaþ and burga forbærnaþ and ǽhta forspillaþ and eard hý ámirraþ. And ðonne land wurðeþ for sinnum forworden and ðæs folces duguð swíðost fordwíneþ, ðonne féhð seó weáláf sorhful and sárigmód synna bemǽnan erit terra uestra deserta et ciuitates uestre destructe. Et, cum deserta fuerit terra propter peccata populi, et ipsi, qui remanserint tabescentes pronuntiabunt peccata sua, Wulfst. 133, 13: Met. 1, 22. Dæt hé ða weáláfe árum heólde, Beo. Th. 2200; B. 1098: 2172; 1084.

Wealas, wealand, -wealc. v. wealh, wealh-land, ge-wealc.

wealca, an; m. **I.** a roller, a wave, billow (cf. fretum, i. feruor maris a walke, Wülck. Gl. 584, 36). v. ge-wealc :—Streám út áweóll, fleów ofer foldan, fámige walcan eorðan þehton, miclade mereflód, Andr. Kmbl. 3047; An. 1526. **II.** a garment that may be rolled round a person, a muffler, wrap, veil. v. wealcian :—Ðá dyde heó of hire wydewan reáf and nam hire walcan (theristrum), Gen. 38, 14.

wealcan; p. weólc; pp. wealcen To roll, toss. **I.** of the movement of water; v. wealca, I, ge-wealc. (1) trans. :—Se fisc getácnaþ geleáfan, for ðan þe his gecynd is, swá hine swíðor ða ýða wealcaþ, swá hé strengra bið, Homl. Th. i. 250, 17. (2) intrans. :—Wealcynde eá fluctus, Wrt. Voc. i. 54, 28. Hé gehýrde ðæt gebrec ðara storma and ðæs weallendes (v. l. wealcendan) sǽs audito fragore procellarum ac ferventis oceani, Bd. 5, 1; S. 614, 4. Wealcendre sǽ flódas ferventis oceani flustra, Hpt. Gl. 464, 59. **I a.** fig. :—Hé hine sylfne betweox ðises andweardan middaneardes (wǽlum ? v. wǽl) weólc and welode inter fluctuantis saeculi gurgites jactaretur, Guthl. 2; Gdwin. 14, 14. **II.** of other movement, (a) literal :—Hægl hwyrft of heofones lyfte, wealcaþ hit windes scúras, Runic pm. Kmbl. 341, 6; Rún. 9. (b) metaph. (1) of action :—Godwine eorl and ealle ða yldestan menn on West-Seaxon lágon ongeán swá hí lengost mihton, ac hí ne mihton nán þing ongeán wealcan (another MS. has hí náht ná gespǽddan) Earl Godwin and the chief men of Wessex resisted as long as ever they could, but they could put no obstacle in the way, Chr. 1036; Erl. 165, 3. (2) of thought, (a) trans. To turn over in the mind, to revolve, consider :—Ða getýdde munuccild ðæt heom betweónan oft wealcaþ, Anglia viii. 314, 35. Hé hine beþóhte and ða hellícan pínunge on his mód weólc, Homl. Th. i. 448, 17. Dæt éce líf on his móde hé wealce vitam aeternam animo suo revolvat, R. Ben. Interl. 29, 2: Hymn. Surt. 121, 9. Wé witon ðæt iunge clericas ðás þing ne cunnon, þeáh ða scolieras ðisra þinga gýmon and gelómlíce heom betwux wealcun, Anglia viii. 335, 44. Hí nellaþ on heora móde wealcan ðæs Hǽlendes beboda, Homl. Skt. ii. 25, 53. For ðæra gelǽredra manna þingum, ðe ðás þing ne behófiaþ betweox heom tó wealcynne, Anglia viii. 300, 4. (β) with a preposition :—Wealce hé on his mód weólc, ðæt éce líf vitam aeternam animo suo revolvat, R. Ben. 24, 3. (γ) intrans. :—Ða ingeðoncas ðe wealcaþ in ðæs monnes móde quando cogitationes volvuntur in mente, Past. 21; Swt. 155, 22. (3) to turn over, deal with :—Þeáh ðe hí Moyses ǽ on heora múðe wealcon, and nellaþ understandan bútan ðæt steaflíce andgit, Homl. Skt. ii. 25, 72. [Hí walked (toss) weri up and dun se water deþ mid winde, O. E. Homl. i. 175, 240. He walkeþ and wendeþ and woneþ . . . on his bedde, Fragm. Phlps. 5, 33. Þa scipen ʒeond þa sæ weolken, Laym. 12040. Þat folc was walkende (going) toward Ierusalem, O. E. Homl. ii. 51, 13. He (Christ) weolc bimong men, Kath. 914. Welk, Pr. C. 4390. Ihc habbe walke wide, Horn. 953. An hundred winter welken (rolled by), Gen. and Ex. 568. O. H. Ger. ge-walchen concretus.] v. and-, ge-, on-wealcan; wealcian, wealcol.

wealc-basu. v. wealh-basu.

wealcere, es; m. A walker (v. E. D. S. Pub. Lancashire Gloss. s.v. walk-mill), a fuller :—Wealceres fullones (-is ?), Wrt. Voc. ii. 38, 3. [Fullere or walkere of cloth, Wick. Mk. 9, 3. A walker hic fullo, Wrt. Voc. i. 212, col. 2 (cf. walkyng lanugo, 238, col. 1. To walke clothe fullare, Cath. Angl. 406, where see note. Cloth ytoukid (v. l. ywalked), Piers P. 15, 447). O. H. Ger. walchare coagitator, compressor : Ger. walker a fuller ; walken to full.]

wealcian; p. ode To roll up, muffle up :—Hefeldþrǽdum liða weal-

cedon liciis articulos obvolverent, Hpt. Gl. 489, 56. [Þe sipes in see walkede, Laym. 12040, 2nd MS. Generally the word = to walk, go :— Hu me schal liggen, slepen, walkien, A. R. 4, 8. Ðe desert he walkeden ðurg, Gen. and Ex. 3882. Ihesu walkide in to Galilee, Wick. Jn. 7, 1. I haue walked ful wide, Piers P. 5, 537. Icel. valka (wk.) to roll.] v. wealcan.

wealcol; adj. That turns or rolls easily :—Wealcol mobilis, Germ. 399, 441.

wealc-spinel, e; f. A curling-iron, crisping-pin :—Walcspinl calamistrum, Wrt. Voc. ii. 127, 75. Cf. þráwing-spinel, and see wealcan.

weald, es; m. High land covered with wood (v. weald-genga), wood, forest. [The word is left in the phrase the weald of Kent and Sussex, the earlier woodland character of which district is shewn by its local names (v. Taylor's Names and Places, pp. 244–5); and in wold, e. g. the wolds of Lincolnshire, Cotswold, though from the changed condition of the country this word no longer implies the presence of wood: in Bailey's Dictionary wold is defined 'a down or champian ground, hilly and void of wood.' See, too, the examples from Mid. English given below] :—Se weald Pireni Pyrenaei saltus, Ors. 1, 1; Swt. 24, 10. Gif hí (birds) ðæs wuda benugen . . . þincþ him wynsumre ðæt him se weald oncweþe, and hí gehíran óþerra fugela stemne si nemorum gratas viderit umbras . . . silvas tantum moesta requirit, silvas dulci voce susurrat, Bt. 25; Fox 88, 20: Met. 13, 92. Wudes ne feldes, sandes ne strandes, wealtes ne wæteres, Lchdm. iii. 288, 1. Wealdes treów (the cross), Rood Kmbl. 34; Kr. 17. Án wind of Calabria wealde de Calabris sal'ibus aura, Ors. 3, 3; Swt. 102, 8. Se Limene múþa is on eásteweardre Cent, æt ðæs miclan wuda eástende ðe wé Andred hátaþ . . . seó eá lið út of ðæm wealda. On ða eá hí tugon up hiora scipu oþ ðone weald iiii míla fram ðæm múþan útanweardum, Chr. 893; Erl. 88, 26–32. On wealda, Cod. Dip. Kmbl. ii. 216, 4. In Limenwero wealdo and in burhwaro uualdo, Cod. Dip. B. i. 344, 10, 11. Wulf on wealde, 937; Erl. 115, 14. Wulf on walde, Elen. Kmbl. 55; El. 28: Judth. Thw. 24, 25; Jud. 206. 'Uton gán on ðysne weald, innan on ðisses holtes hleó.' Hwurfon hié . . . on ðone grénan weald, Cd. Th. 52, 6–10; Gen. 839–41. Dæt is wynsum wong, wealdas gréne, rúme under roderum, Exon. Th. 198, 21; Ph. 13. Gewát him se æþeling wadan ofer wealdas, Cd. Th. 174, 30; Gen. 2886. ¶ using the name of the whole for a part :—Hié heora líchoman leáfum beþeahton, weredon mid ðý wealde, 52, 19; Gen. 846. [He is bicumen hunte and flihð ouer bradne wæld (feld, 2nd MS.), Laym. 21339. Þe wald þe is ihaten Hedfeld, 31216. Fluʒen ouer þe woldes (feldes, 2nd MS.), 20138. Liðen heo bi straten and bi walden, 12832. Wilde deor þ on þeos wilde waldes (forests) wunied, Marh. 10, 4. Elpes togaddre gon o wolde, Misc. 19, 666: O. and N. 1724. On ðe munt quor men Aaron in biriele dede . . . ðor hé lið doluen on ðat wold, Gen. and Ex. 3892. Þe holy gost hyne ledde up into þe wolde for to beon yuonded of sathanas, Misc. 38, 27. Yᵉ walde alpina, Cath. Angl. 406. O. Frs. O. Sax. wald wood : O. H. Ger. walt, wald silva, saltus, nemus, eremus : Icel. völlr a field, plain.] v. út-, wudu-weald.

weald power :—Se wæs on his wealde (gewealde, MS. L.), Ors. 4, 11; Bos. 97, 23. [He haueð his soule weald, O. E. Homl. ii. 79, 14. A neuere nane walde ne mihte swa mochel folc halde, Laym. 5253. Unnderr þe deofless walde, Orm. 38. Hine þet alle þing haueð on wealde, Anglia i. 31, 186. To don swilc dede adde he no wold, Gen. and Ex. 2000. O. Frs. wald : Icel. vald.] v. án-, and-, ge-, on- (an-) weald ; wealdes, and next word.

weald; adj. Powerful, mighty :—Mid ðære wealdestan [lufe] ferventissimo amore, R. Ben. 117, 5. [v. án-, eal- (al-) wealda ; adj. O. Sax. ala-, alo-waldo : O. H. Ger. al-walto.] v. on-weald, wealda ; m.; wilde.

weald is found as the second part of many proper names. Cf. Icel. -valdr, e. g. Ás-valdr = English Ós-wald. v. for a list of such names, Txts. pp. 491–3.

weald; adv. conj. **I.** in independent clauses, with þeáh, perhaps, may be :—Nyte ge ða micclan deópnysse Godes gerýnu ; weald þeáh him beó álýfed gyt behreówsung, Homl. Th. ii. 340, 9. Ðis godspel ðincð dysegum mannum sellíc, ac wé secgaþ swá ðeáh ; weald ðeáh hit sumum nien lícige, 466, 10. Wén ys ðæt hé sig on gáste up áhafen, and onuppan nuntum geset ; ac uton ða muntas eondfaran ; weald þeáh wé hyne gemétan magon, Nicod. 19; Thw. 9, 25, 31. **II.** in dependent clauses, with indefinite pronouns or adverbs (cf. gif), in case :—Bid nú wíslícor ðæt gehwá ðis wite and cunne his geleáfan, weald hwá ða mycclan yrmðe gebídan sceole in case any one have to experience that great misery, Homl. Th. i. 6, 19. Bisceopum gebyreþ ðæt mid heom wunian welgeþungene witan . . . ðæt heora gewitan beón on ǽghwylcne tíman, weald hwæt heom tíde in case anything befall them, L. I. P. 10; Th. ii. 316, 25. Hí námon tó rǽde, ðæt him wærlícor wǽre, ðæt hí sumne dǽl heora landes wurðes æthæfdon, weald [hwæt?] him getímode, Homl. Th. i. 316, 24. Man sceal wacigean and warnian symle, ðæt man geara weorðe tó ðam dóme, weald hwænne hé us tó cyme ; wé witan mid gewisse, ðæt hit ðærtó neálécð people ought to watch and be ever on guard so that they may get ready for the judgement, in case any time it come to us ; we know with certainty that we are getting near to it, Wulfst. 90, 3.

wealda, an; *m. A ruler.* v. ân-, an-, Bret-, bryten-, eal-wealda. [*O. Sax.* ala-waldo : *O. H. Ger.* -walto : *Icel.* valdi.] ¶ as a proper name (?) :— Innan Wealdan hrícg on Eádríces gemǣre, Cod. Dip. B. ii. 259, 9. [*O. H. Ger.* Walto, Waldo : *Icel.* -valdi *in cpd. names.*] v. weald ; *adj.*

wealdan ; *p.* weóld, *pl.* weóldon ; *pp.* wealden *To have power* over :— Wealdeþ *imperitat,* Wrt. Voc. ii. 44, 43. Ælc mon biþ wealdend ðæs ðe hé welt ; nǽfþ hé nâne anweald ðæs ðe l.ê ne welt *quod quisque potest, in eo validus : quod non potest, in hoc imbecillis esse censendus est,* Bt. 36, 3 ; Fox 176, 17. I. *to control the movements of* that which is moved, *to regulate, wield* a weapon, (a) with gen. :—Sió eax welt ealles ðæs wǣnes, Bt. 39, 8 ; Fox 224, 6. Ða hwíle ðe hí wǣpna wealdan môston, Byrht. Th. 134, 13 ; By. 83 : 139, 50 ; By. 272. Wǣpnes wealdan, 136, 48 ; By. 168. Gif hé his wordcwida wealdan me.hte, Exon. Th. 171, 26 ; Gû. 1132. (b) with dat. or inst. :—Swâ hé selfa bæd, þenden wordum weóld wine Scyldinga, Beo. Th. 59 ; B. 30. Se ðe wǣtrum weóld þeahte bearn middangeardes wonnan wǣge, Cd. Th. 83, 9 ; Gen. 1377. Þenden hié ðâm wǣpnum wealdan môston, Beo. Th. 4083 ; B. 2038. II. *to control* that which moves itself, *to have control of* a person, an emotion, &c., *to govern,* (a) with gen. :—Be cnihtum, on hwylcere yldo hí môton hyra sylfra wealdan (*se ipsos gubernare*), L. Ecg. C. 27, tit. ; Th. ii. 130, 12. (b) with acc. :—Sume wealdaþ ealle uncysta and leahtras on him sylfum, Homl. Th. i. 344, 34. III. of the control exercised by one in authority, *to rule, govern, have dominion over, bear sway, wield* power, (a) with gen. :—Þenden ic wealde wídan ríces, Beo. Th. 3722 ; B. 1859. Dryhten, ðû ðe ealle gesceafta gesceópe, and heora weltst *qui mundum gubernas,* Bt. 33, 4 ; Fox 128, 6, 24. Wealdest, Met. 20, 7, 50. Waldest, Hy. 3, 5. Ðû heora wylst *reges eos,* Ps. Th. 2, 9. Wealdeþ (*dominabitur*) God manna cynnes, 58, 13. Wealdeþ, Met. 29, 77. Se ðe waldeþ ealra ôðra eorðan cyninga, 24, 35. Hé welt (wilt, *v. l.*) ealles, Bt. 35, 3 ; Fox 158, 23. Welt, 25 ; Fox 88, 3. Wylt, 5, 3 ; Fox 14, 3. Wealt, 35, 4 ; Fox 160, 14. Wealt (welt, *v. l.*), 39, 2 ; Fox 214, 13. Wealt (wylt, *v. l.*), 35, 3 ; Fox 158, 19. Ðâm ðe ðyses middangeardes waldaþ *hujus mundi potestatibus,* Past. 15 ; Swt. 89, 22. Ealdormenn wealdaþ hyra þeóda *principes gentium dominantur eorum,* Mt. Kmbl. 20, 25 : Lk. Skt. 22, 25. Hé him ealles ðæs anwaldes weóld Mæcedonia ríces, Ors. 3, 11 ; Swt. 148, 24 : Cd. Th. 258, 19 ; Dan. 678. Wíold, Met. 9, 38. Hí heora weóldan *dominati sunt eorum,* Ps. Th. 105, 30. Þeáh hé ðæs ealles wealde, Bt. 29, 3 ; Fox 106, 25 : Met. 16, 16. Geléfst ðû ðæt seó wyrd wealde ðisse worulde, Bt. 5, 3 ; Fox 14, 3. Abbod, ðe ðæs wyrðe sý, ðæt hé mynsteres wealde *abba, qui preesse dignus est monasterio,* R. Ben. 10, 9. Walde, Elen. Kmbl. 1598 ; El. 801. Hé wæs tô ðam swýðe upâhafen, swylce hé weólde ðæs cynges and ealles Englalandes, Chr. 1052 ; Erl. 181, 25 : Homl. Th. i. 488, 14 : Bt. 35, 2 ; Fox 156, 25–27. His fæder ne wolde him lǽtan waldan his eorldómes, Chr. 1079 ; Erl. 216, 21. God ne beþearf nânes ôþres fultumes his gesceafta mid tô wealdanne, Bt. 35, 3 ; Fox 158, 15. (b) with dat. or inst. :—Ðû waldes (wyldst, Ps. Spl.) mæhte sǽs *tu dominaris potestati maris,* Ps. Surt. 88, 10. Hé eorðrícum eallum wealdeþ *regnum ipsius omnibus dominabitur,* Ps. Th. 102, 18 : 75, 9. Waldeþ, Met. 25, 15. Hû hé welt eallum his gesceaftum, Bt. 21, tit. ; Fox xiv, 3. Ic weóld folce Deniga, Beo. Th. 935 ; B. 465. Hé eallum sûðmægþum weóld *cunctis australibus provinciis imperavit,* Bd. 2, 5 ; S. 506, 11. Hé weóld Walum and Scottum, Chr. 1065 ; Erl. 196, 28 : Exon. Th. 319, 26 ; Víd. 18 : Beo. Th. 4747 ; B. 2379. Hié burgum weóldon, Cd. Th. 216, 19 ; Dan. 9. Wíoldon, Met. 1, 48. (c) with acc. :—Ðû wealdan miht eall eorðan mægen, wind and wolcnu ; wealdest ealle on riht, Hy. 9, 5–7. Hé welt ealle gesceaftu, Bt. 39, 13 ; Fox 234, 22. (d) with a preposition :—Se ofer deóflum wealdeþ, Cd. Th. 263, 21 ; Dan. 765. Se ofer mægna gehwylc waldeþ, Exon. Th. 255, 32 ; Jul. 223. (e) absolute :—Wylt *presidet,* Wrt. Voc. ii. 67, 45. Wealdendum *imperantibus* (*Valeriano et Gallieno,* Ald. 67), Hpt. Gl. 515, 45. III a. fig. where the subject is an abstract noun, (a) with gen. :—Ðý lǽs mín ǽnig unriht wealde *non dominetur mei omnis injustitia,* Ps. Th. 118, 133. Sió gesceádwísnes sceal ðære wilnunge waldan, Met. 20, 198. (b) with acc. :—Unsôðfæstnys ealle wealde, Ps. Th. 54, 9. (c) with a preposition :—His mægen wealdeþ ofer eall manna cyn, Ps. Th. 65, 6. IV. *to have power over things, to possess, be in possession of, have at command, be master of,* (a) with gen. :—Hé sǽs wealdeþ *ipsius est mare,* Ps. Th. 94, 5. Hí wealdaþ eorðan *possederunt terram,* Ps. Spl. C. 43, 4. Ðonne wealdaþ hý heom sylfum weorðscypes *then shall they command for themselves respect,* L. I. P. 23 ; Th. ii. 336, 23. Manigra folca gestreónes hie wieóldon *labores populorum possederunt,* Past. 50 ; Swt. 391, 4. Hí weóldon wælstôwe *they were masters of the field,* Beo. Th. 4108 ; B. 2051. Wælstôwe wealdan, 5961 ; B. 2984 : Byrht. Th. 134, 37 ; By. 95 : Ps. Th. 90, 11. For worulde weorðscypes wealdan *to command the respect of the world,* L. I. P. 16 ; Th. ii. 324, 4. (b) with dat. or inst. :—Hé sceal ðý wonge wealdan ; ne magon gê him ða wíc forstondan, Exon. Th. 144, 6 ; Gû. 674. Ðara ðe life weóldon *of those who lived,* 118, 14 ; Gû. 239. Beáhhordum leng wyrm wealdan ne môste, Beo. Th. 5647 ; B. 2827 : Vald. 2, 31. (c) with acc. :—Heofonas ðú wealdest *tui sunt coeli,* Ps. Th. 88, 10. Habban hí and wealdan Hornemeres hunred on hyre âgenre andwealde *habeant et possideant hundredum de Hornemere*

in sua propria potestate, Cod. Dip. Kmbl. iv. 200, 7. V. *to have power to decide* or *choose* what shall take place, *to determine, ordain, have the deciding* or *control of* matters, (a) with gen. :—Se ðe lífa gehwæs lengu wealdeþ *he that determines the length of every life,* Exon. Th. 133, 2 ; Gû. 483. Wealde se cyning þreóra ǣnes (*the king shall have power to ordain one of three courses*) ; oþ þe hine man cwelle, oþþe ofer sǣ selle, oþþe hine his wergelde âlése, L. Wih. 26 ; Th. i. 42, 16. Se ðe útlages weorc gewyrce, wealde se cyningc ðæs frides, L. C. S. 13 ; Th. i. 382, 18. Sume secgaþ ðæt sió wyrd wealde ǽgþer ge gesǽlþa ge ungesǽlþa ǽlces monnes, Bt. 39, 8 ; Fox 224, 13. Ðæt hí ne geþafian, gyf his waldan magan, ðæt ðǣr ǽnig unriht up âspringe, L. I. P. 7 ; Th. ii. 312, 36. Gif hí ðæs wealdan mihton, Wulfst. 185, 3. (b) with dat. or inst. :—Seó weóld hyra (*two buckets*) síþe, Exon. Th. 435, 12 ; Rä. 53, 6. Segl síðe weóld, Cd. Th. 184, 10 ; Exod. 105. Ðǣr hé ðý fyrste wealdan môste, Beo. Th. 5141 ; B. 2574. (c) with a clause :—Petre ðæne ealdorscipe hé betǣhte, and hêt, ðæt hé weólde be manna gewyrhtum, hwâ ðǣr in môste and hwâ nâ ne môste, Wulfst. 176, 16. Wé ðé magon sélre gelǣran, ǣr ðú gúðe fremme, weald hû ðé sǽle (*decide thou how it shall happen to thee*) æt ðam gegnslege, Andr. Kmbl. 2710 ; An. 1537. (d) absolute :—Ðæt ne geþafodon ða ðe micel weóldon on ðisan lande (hit him ne geþafode Godwine eorl, ne éc ôþre men ðe mycel mihton wealdan, col. 1) *those who very much had the control of affairs in this land would not allow that,* Chr. 1036 ; Th. i. 292, col. 2. Gif lâd forberste, bisceop ðonne wealde and stiðlíce dême, L. C. S. 54 ; Th. i. 406, 10. Gif man wealdan mǣge (*if it can be managed*), ne dýde man nǣfre on Sunnandæges freólse ǽnigne forwyrhtne, L. E. G. 9 ; Th. i. 172, 13 : L. C. S. 45 ; Th. i. 402, 10 : Anglia ix. 260, 11. Binnan cirictûne ǽnig hund ne cume, ðæs ðe man wealdan mǣge, L. Edg. C. 26 ; Th. ii. 250, 8. Hé wille, gif hé wealdan môt, leóde etan, Beo. Th. 889 ; B. 442. Ne beóð wé leng somed, gif ic wealdan môt, Cd. Th. 168, 22 ; Gen. 2786. VI. *to have power* that brings something to pass, *to cause, be the cause, author, source* of something, (1) of persons, (a) with gen. :—Ðæs ðú wealdest *this is thy doing,* Elen. Kmbl. 1517 ; El. 761. Hé nínre geðylde wealdeþ *ab ipso est patientia mea,* Ps. Th. 61, 5. Gif hwelc folc bið mid hungre geswenced, and hwâ his hwǣte gehýt and ðhielt, hú ne wilt hé hiera deáðes ? *si populus fames attereret et occulta frumenta ipsi servarent, auctores procul dubio mortis existerent,* Past. 49 ; Swt. 377, 9. Syndon cyrcan wâce gegriðode . . . wâ ðam ðe ðæs wealt, L. I. P. 25 ; Th. ii. 340, 14. Ðæs ic seolfa weóld, Cd. Th. 281, 21 ; Sat. 275. Gif ðú hwæt ón druncen misdô, ne wît ðú hit ðam ealoðe, for ðam ðu his weólde ðé silf, Prov. Kmbl. 39. Ðæt hé sigora gehwæs âna weólde (wolde, MS.), Exon. Th. 276, 7 ; Jul. 562. Ic wile wealdan eów blisse and micelre lisse, Wulfst. 132, 23. (b) with dat. or acc. :—Ðæt his môd wite, ðæt migtigra wíte wealdeþ, ðonne hé him wið mæge, Cd. Th. 248, 33 ; Dan. 523. (2) of things, with gen. :—Ús unwidera for oft weóldon unwæstma, Wulfst. 129, 4. (3) of motives :—Mid ðý se willa mâ wealdeþ on ðam weorce ðære gemengdnysse, Bd. 1, 27 ; S. 495, 38. VII. *to have power to do, be able :*—Búton hí hit gebêton, ðæs ðe hí wealdan magon (*as far as lies in their power*), Wulfst. 301, 20. Þeáh fýr wið ealla sié gemenged weoruldgesceafta, þeáh waldan ne môt ðæt hit ǽnige fordô (cf. ðeáh ne mæg nâne ðara gesceafta ofercuman, Bt. 33, 4 ; Fox 130, 17), Met. 20, 129. [To walden (*weld,* 2nd MS.) kinериche, Laym. 2966. Wealden *possidere,* O. E. Homl. ii. 79, 11 : H. M. 39, 20. Welden, O. E. Homl. i. 163, 55. Goth. waldan garda οἰκοδεσποτεῖν : O. Sax. O. L. Ger. waldan *dominari* : O. Frs. walda : O. H. Ger. waltan *dominari, regnare, protegere* : Icel. valda *to wield, rule ; to cause.*] v. ge-, ofer-wealdan ; wealdende, ge-wealden ; wealdian.

weald-bǣre, es ; *n. A place where trees grow affording mast for swine :*—Ad hoc terram pertinent in diuersis locis porcorum pastus, id est uuealdbaera, Cod. Dip. Kmbl. i. 184, 1. v. den-bǣre.

wealdend, es ; *m.* I. *one who exercises power over* persons or things, *a controller, master :*—Ælc mon biþ wealdend ðæs ðe hé welt, næfþ hé nâune anweald ðæs ðe ne welt *quod quisque potest, in eo validus : quod non potest, in hoc imbecillis esse censendus est,* Bt. 36, 3 ; Fox 176, 17. Hí hine heom for god hæfdon, and hý sǽdon ðæt hé wǣre ealles gewinnes waldend (cf. hans (*Odin's*) menn trúðu því, at hann ætti heimilan sigr í hverri orrostu, Ynglinga Saga, c. 2), Ors. 1, 6 ; Swt. 36, 21. Wé witon hé úre wæs wealdend *we knew he was master of us,* Blickl. Homl. 243, 18. Se ðe ðæs weddes waldend sý, L. Edm. B. 6 ; Th. i. 254, 22. Ðú wéndest ðæt steórleáse men wǣron gesǽlige and wealdendas ðisse worulde *nequam homines potenteis felicesque arbitraris,* Bt. 8, 3 ; Fox 14, 1. Hé wolde ðætte ealle men wǣran ealra ôþra gesceafta wealdandas *ille genus humanum terrenis omnibus praestare voluit,* 14, 2 ; Fox 44, 33. II. *one who exercises dominion, a ruler, governor, sovereign :*—Ðes and ðæs wealdend *hic et haec praesul,* Ælfc. Gr. 9, 10 ; Zup. 39, 12. Cum mid ús for ðon ðe ðú eart úre wealdend, Blickl. Homl. 239, 9. Eádgâr, Engla waldend, Chr. 973 ; Erl. 124, 9. Eádweard, hæ'eða wealdend, 1065 ; Erl. 196, 27. Englalandes wealdend, Cod. Dip. Kmbl. iv. 232, 3. Ne sint wé nâne waldendas eówres geleáfan *non dominamur fidei vestrae,* Past. 17 ; Swt. 115, 24. Ne sint wé nâne waldendas ðisses folces *non dominantes in clero,* Swt. 119, 24. Ðeóda kyningas beóð ðæs folces

waldendas *principes gentium dominantur eorum*, Swt. 120, 3. Hié wéron seolfe wuldres waldend, Cd. Th. 266, 18; Sat. 24. Wealdendras *imperatores*, Scint. 215, 9. Ealdormen and þeóde wealdendras, Cod. Dip. Kmbl. iii. 350, 25. **II a.** applied to the Deity:—Ân sceppend is and se is wealdend heofones and corþan and ealra gesceafta, Bt. 21; Fox 72, 29: 35, 3; Fox 158, 25: 39, 12; Fox 232, 11. Wealdend Drihten *Dominus*, Ps. Th. 65, 16. Úre fæder, ealles wealdend, cyning on wuldre, Hy. 7, 1. God ðe is wealdand and wyrhta ealra gesceafta, L. Eth. vi. 42; Th. i. 326, 13. Ân is éce cyning, wealdend and wyrhta ealra gesceafta, L. I. P. 1; Th. ii. 304, 2. Se is waldend windes and goldes, Blickl. Homl. 133, 30. Wit Waldendes word forbrǽcon, Cd. Th. 49, 26; Gen. 798. Ðæt hé Wealdende, écean Dryhtne, gebulge, Beo. Th. 4648; B. 2329. **III.** *a possessor, master, lord:*—'Gewit ðú (*Hagar*) ðinne waldend sécan; wuna ðǽm ðé ágon.' Heó gewát engles lárum hire hláfordum, Cd. Th. 138, 17; Gen. 2293. Se wela ne mæg his wealdend gedón nó ðý weorþron, Bt. 27, 2; Fox 98, 13: 16, 3; Fox 56, 3, 17. Se wela and se anweald náuht ágnes gódes nabbaþ, ne náuht þurhwuniendes heora wealdendum sellan ná magon, 27, 4; Fox 100, 22. [*Creatorem celi et terre* scuppende and weldende of heouene and of orðe, O. E. Homl. i. 75, 26. Wealdende, ii. 17, 32. Godd, domes waldend, Laym. 28205. Wealdende (weldende, 2nd MS.), 25568. **Goth.** garda-waldands οἰκοδεσπότης: O. Sax. waldand (*used of the Deity*): O. H. Ger. Waltant (*proper name*): Icel. valdandi.] v. eal[l]- (al-), ofer-, þrym-wealdend, *and next word.*

wealdende; *adj.* (*ptcpl.*) *Ruling, powerful:*—Mihtig God, ... waldende God, Exon. Th. 62, 34; Cri. 1011: 71, 27; Cri. 1162. Se wealdenda Drihten, Homl. Th. i. 328, 11. Se anweald ne mæg gedón his wealdend wealdendne, Bt. 16, 3; Fox 56, 3, 17. Hwæþer ðú nú wéne ðæt ðæs cyninges geférræden and se wela and se anweald ðe hé gifþ his deórlingum mæge ænigne mon gedón welige oððe wealdendne? *an vero regna regumque familiaritas efficere potentem valent?* 29, 1; Fox 102, 4. Waldendne, 29, tit.; Fox xvi, 2. Nis under mé ǽnig óþer wiht waldendre, ic eom ufor ealra gesceafta, Exon. Th. 427, 6; Rä. 41, 87. v. eal[l]-, ge-, þrym-wealdend[e]; wealdan.

wealdend-god, es; *m. The Lord God:*—Ic cleopige tó Heáhgode and tó Wealdendgode ðe mé wel dyde *clamabo ad Deum altissimum, et ad Dominum qui bene fecit mihi,* Ps. Th. 56, 2. Se is wealdendgode wellíc-endlíc *beneplacitum est Deo,* 67, 16. [*O. Sax.* waldand-god.]

wealdes; *adv. Of one's own accord, purposely, voluntarily:*—Gif him wealdes (gewea ldes, Hatt. MS.) gebyrige oððe ungewealdes, Past. 28; Swt. 198, 22. [Þu forschuppeste selfwilles and waldes in to hare cunde, H. M. 27, 2. Heo suneged deadliche ðe bruche, ȝif heo hit brekeð willes and woldes, A. R. 6, 26.] v. ge-wealdes.

weald-genga, an; *m. A weald-goer* (v. weald), *bandit, brigand:*—Hé wolde beón yldest on ðam yfelan flocce, and geworhte his geféran tó wealdgengum calle on widgillum dúnum ... 'Hé is geworden tó wealdgengan and ðæra sceaðena ealdor, ðe hé him sylf gegaderode, and wunaþ on áure dúne mid manegum sceaðum.'. . . Ðá ætstód se wealdgenga ... and áwearp his wǽmna, Ælfc. T. Grn. 17, 30-18, 31. [Cf. wald-scaðe (wode-scaþe, 2nd MS.), Laym. 25859; the same creature is referred to in these previous lines: Isihst þu þe munt and þene wude muchele, þer wuneð þe scaðe inne, þa scendeð þas leode? 25689-92.]

wealdian; *p.* ode *To rule, command:*—Ic wealdige *vel* ofer bebeóde *imperito,* Wrt. Voc. i. 54, 52. [*O. Sax.* gi-waldón.] v. wealdan.

weald-leðer, es; *n. A rein:*—Hí ne móton swíþor styrian ðonne hé him ðæt gerúm his wealdleðeres tó forlǽt, Bt. 21; Fox 74, 8. Se gemetgaþ ðone brídel and ðæt wealdleþer ealles ymbhweorftes heofenes and eorþan *orbis habenas temperat,* 174, 19. Ðá gelǽhton ða weardmen his wealdleðer fæste, Ælfc. T. Grn. 18, 15. Heó wæs on gyldenum scryd, and æt ðam wǽron gyldene hors, and on ðam wǽron ða wealdleðer swá up getíged, swá swá hig urnon tó heofenum up, Shrn. 156, 12. v. ge-weald-leðer.

weald-more. v. wealh-more.

wealdness, e; *f. Rule, dominion:*—Waldnis ðín *dominatio tua,* Ps. Surt. 144, 13.

weald-stapa, an; *m. A grasshopper, locust:*—Waldstapan *locustas,* Mk. Skt. Rush. 1, 6.

weald-swaþu, e; *f. A forest-track:*—Lástas wǽron æfter waldswaþum wíde gesýne *the steps were to be seen far along the forest-tracks,* Beo. Th. 2810; B. 1403.

weale, an; *f. A female slave, servant:*—Wonfeax wale, ... nennen, Exon. Th. 393, 30; Rä. 13, 8. Wonfáh wale weóld hyra (*two buckets*) síþe, 435, 11; Rä. 53, 6. v. wealh.

weale-wyrt. v. wealh-wyrt.

wealg; *adj. Nauseous* (? Halliwell gives *wallow* = flat, insipid; *wallowish* = nauseous):—Se wearma weld on gódum cræftum, ðý læs hé sié wealg for wlæcnesse, and for ðæm weorðe út áspiwen (*ne evomatur tepidus*), Past. 58; Swt. 447, 18. [Þi muð is bitter and walh al þat tu cheowest, and hwit mete se þi mahe hokerliche undorfeð, þat is wið unlust, warpeð hit eft ut, H. M. 35, 30. Walhwe swete *supra in* bytter swete, Prompt. Parv. 515. *Icel.* válgr, volgr *warm, lukewarm.*]

-wealg (-wealh). v. on-wealh.

wealh *an implement that rolls things over* (?), *a harrow:*—Wealh occa, Wrt. Voc. ii. 79, 25. Walh, 62, 63. [Cf. *Goth.* us-walugjan περιφερεῖν: O. H. Ger. bi-walagón *volutare.*]

wealh; *gen.* weales; *m.* **I.** *a foreigner,* properly *a Celt* (cf. the name *Volcae,* a Celtic tribe mentioned by Caesar):—Walch *barbarus,* Wrt. Voc. ii. 12, 75. Ic (*an axle-tree*) síþade wíddor, mearcpaþas wala (walas, MS.) træd, móras pǽde, Exon. Th. 485, 7; Rä. 71, 10. [*Icel.* Valir; *pl. the Celtic* people in France.] ¶ wealh *is found in many proper names.* v. Txts. 489. See also the compounds in wealh-. **I a.** *a Celt of Britain;* the word occurs mostly in pl., Wealas; *gen.* Weala, Walena, *the British, the Welsh,* or *Wales:*—Wealh gafolgelda .cxx. scitt. ... Weales hýd twelfum, L. In. 23; Th. i. 118, 3. Wealh, gif hé hafaþ fíf hýda, hé bið syxhynde (cf. for relative importance of the Celt and the Englishman, L. R. 2; Th. i. 190, 15-18), 24; Th. i. 118, 10. Gif þeów Wealh Engliscne monnan ofslihð, 74; Th. i. 148, 14. Hér Hengest and Æsc gefuhton wiþ Walas (cf. Brettas, l. 17) ... and ða Walas flugon ða Englan swá fýr, Chr. 473; Erl. 12, 26. Hér Æðelfrið ofslóh unrím Walena (-ana, *v. l.*), and swá weard gefyld Augustinus wítegunge, ðe hé cwæð: 'Gif Wealas nellaþ sibbe wið ús, hí sculan æt Seaxana handa farwurþan.' Ðár man slóh .cc. preósta, ða cómon ðyder ðæt hí scoldon gebiddan for Walena here, 607; Erl. 20, 29. Hí ofslógon .ii. þúsendo Wala (Walana, *v. l.*), 614; Erl. 20, 37. Wala (Weala, *v. l.*) cyning, 710; Erl. 44. 4. Hér wæs Wala (Weala, *v. l.*) gefeoht and Defna æt Gafulforda, 823; Erl. 62, 14. Wiþ ðæs landes gewrixle ðe on Wealum is æt Pendyfig *pro commutatione alterius terre que sita est in Cornubio, ubi ruricole illius pagi barbarico nomine appellant Pendyfig,* Chart. Erl. 192, 5. Hí ofslógon monige Wealas (Walas, *v. l.*), Chr. 477; Erl. 12, 31. ¶ the word is found as part of place-names, v. Cod. Dip. Kmbl. vi. Index. v. Bret- (Bryt[t]-), Corn-, Norþ-, West-Wealas (-Walas). **I b.** *a Roman:*—Weala *sunderriht jus Quiritum* (cf. Rómwara sundorriht, Wrt. Voc. ii. 49, 11, reht Rómwala, Rtl. 189, 13, which translate the same phrase), Wrt. Voc. i. 20, 64. [*O. H. Ger.* walah *Romanus.*] **II.** *a slave, servant.* Cf. the derivation of *slave* from the name of a people:—Mín weal sprecð *meum mancipium loquitur,* mines weales sunu, mínum weale ic timbrige hús, mínne weal ic beládige, eá lá ðú mín weal, sáw wel, fram mínum weale ic underféng fela gód, mine wealas (*mancipia*) eriaþ, mínra þeówra manna (*mancipiorum*) æceras, Ælfc. Gr. 15; Zup. 101, 13-21. Ðes wísa weal (*mancipium*), 6, 4; Zup. 19, 8: 6, 3; Zup. 18, 16. Ðæs weales (*v. ll.* weles, wieles; ðrǽles, Lind.: esnes, Rush.) hláford *dominus servi illius,* Mt. Kmbl. 24, 50: Shrn. 154, 22. Ðrittegum geárum ne gestilde nǽfre stefen cearciendes wǽnes ne ceoriendes wales *for thirty years the sound of creaking wain and chiding thrall never ceased,* Lchdm. iii. 430, 34. Ne hý ne wé ne underfón óðres wealh ne óðres þeóf, L. Eth. ii. 6; Th. i. 288, 4. Wealas *servi,* Gen. 21, 25. Ðis folc ðe úre wealas syndon, Ex. 14, 5. Wé ðe nǽron wurðe beón his wealas gecígde, Homl. Th. ii. 316, 23. Weala wín *crudum vinum,* . . . hláforda wín *honorarium vinum,* Wrt. Voc. i. 27, 55, 57. Genam Abimelech wealas and wylna (*servos et ancillas*), Gen. 20, 14. Ic (*a skin which furnishes thongs*) fæste binde swearte wealas (*slaves or strangers, captives;* Aldhelm's riddle has: Nexibus horrendis homines constringere possum), hwi!um séllan men, Exon. Th. 393, 22; Rä. 13, 4. [Ælc þrel and ælc wælh wurðe iuroeid, Laym. 14852.] v. hors-, hund-, scip-wealh; weale, wilh. **II a.** *a shameless person.* v. wealian, wealh-word:—Walana *protervorum,* Hpt. Gl. 527, 22.

weal-hát. v. weall-hát.

wealh-basu (-o) *foreign scarlet, vermilion:*—Wealhbaso *vermiculo,* Wrt. Voc. ii. 77, 21. Wealhbasu, Anglia xiii. 29, 56. [*The passage glossed in both is* Ald. 15. *In glossing the same passage* wealcbasewere (weolc-(?) v. weoloc-basu; *but cf.* wealc-stód *for* wealh-stód, 463, 42) *occurs,* Hpt. Gl. 431, 32.]

Wealh-cyn[n], es; *n. The Celtic race:*—Ða land ðe ic on Wealcynne (*the Celts of the south-west*) hæbbe bútan Triconscíre, Chart. Th. 488, 26. Hig gegaderadan mycle fyrde mid Walkynne (*the Celts of Wales*), Chr. 1055; Erl. 188, 33. Griffin wæs kyning ofer eall Wealcyn, 1063; Erl. 195, 12. v. Norþ-Wealhcynn.

Wealh-fǽreld, e; *n. A 'Welsh' expedition,* a term applied to forces defending the Welsh Marches (?):—Liberabo monasterium (*Blockley, Worcestershire*) a pastu et refectione illorum hominum quos Saxonice nominamus Walhfæreld and heora fæsting, Cod. Dip. Kmbl. ii. 60, 29. v. next word.

Wealh-geféra, -geréfa, an; *m. A count of the Welsh Marches* (?), *the commander of the* Wealh-fǽreld (?):—Ðý ilcan gére forðférde Wulfríc cynges horsðegn; se wæs eác Wealhgeféra (*other MSS. have* -geréfa. Kemble, taking the latter reading, says: 'I am disposed to believe that he was a royal reeve to whose care Alfred's Welsh serfs were committed, and who exercised a superintendence over them in some one or all of the royal domains,' Saxons in England, ii. 179. See the first passage under Wealh-cyn), Chr. 897; Erl. 96, 17, and note.

wealh-hafoc, es; *m. A foreign hawk, a gerfalcon;* herodius (**v. erodius**

gerfawcune, Wrt. Voc. i. 188, col. 2 : jarfawkon, 220, col. 2) :—Walh-habuc *falc(o)*, Txts. 61, 826. Walchhabuc, uualhhaebuc, uualh[h]ebuc, ualchefuc *herodius*, 67, 1016. Gôshafuc *accipiter*, wealhhafuc *herodius*, spearhafuc *alietum*, Wrt. Voc. i. 280, 18–20 : ii. 42, 67. Wealhhafoces hûs *herodii domus*, Ps. Spl. 103, 19. Ða fugelas *nocticoraces* hâtton wæron in wealhhafoces gelîcnesse (*vulturibus similes*), Nar. 16, 13. Wealhhafeca *falconum*, Wrt. Voc. ii. 87, 68 : 37, 23. [*O. H. Ger.* waluc-l apuh *herodius.*]

wealh-hnutu; *gen.* -hnyte; *f. A foreign nut, walnut* :—Hnutbeam oððe walhhnutu *nux*, Wrt. Voc. ii. 60, 23. [On a walnot withoute is a bitter barke, Piers P. 11, 251. Walnote *avelana*, Prompt. Parv. 574. A walnotte *auellanum*, a walnott-tree *auellanus*, Cath. Angl. 407 (see note). Walnot *auelena*, Wülck. Gl. 647, 25. Walnottre *auelana*, 646, 15. A walnutte and the nutte *avelana*, 715, 26. A walnote *moracia*, 596, 38. Cf. A walshenote shale, Chauc. H. F. 1281. *Icel.* val-hnot.]

wealh-land, es; *n.*   I. *a foreign land* :—Æghwær eorðan ðær wit earda leás mid wealandum wunian (winnan, MS.) sceoldon (cf. mê ellþeódigne, l. 20), Cd. Th. 163, 30; Gen. 2706.   II. *Normandy* (cf. *Icel.* î Vallandi er sîðan var kallat Norðmandi) :—Com Eádweard hider tô lande of Weallande (fram begeondan sæ, *v. l.*), Chr. 1040; Erl. 167, 27. [*O. H. Ger.* Walho-lant *Gallia*.]

wealh-more(-u), -mora, an; *f. m. A foreign root, carrot, parsnip* :—Walhmore, uualhmorae *pastinaca*, Txts. 85, 1502. Wealmore, Wrt. Voc. ii. 67, 62 : i. 286, 27 : Lchdm. i. 120, 8. Wealmora, Wrt. Voc. i. 79, 58 : *daucus*, 31, 43. Waldmora *cariota*, 31, 46. v. wilisc.

wealh-sâda(?), an; *m. A noose for binding a captive* or *slave* (? cf. Exon. Th. 393, 22; Rä. 13, 4, *given under* wealh, II) :—Forhýddan oferhygde mê inwitgyrene, wrâðan wealsâdan *absconderunt superbi laqueos mihi*, Ps. Th. 139, 5.

wealh-stod, es; *m. An interpreter* :—Wealhstod *interpres*, Wrt. Voc. i. 86, 60: Ælfc. Gr. 9, 26; Zup. 51, 14.   I. *one who serves as a medium between speakers of different languages* :—Se cyning gerehte his witan on heora âgenum gereorde ðæs bisceopes bodunge, and wæs his wealhstod, for ðan ðe hê wel cûþe Scyttysc, Homl. Skt. ii. 26, 67. Walhstod, Bd. 3, 3; S. 526, 2. Hê (*Jerome*) is se fyrmesta wealhstod betwux Hebrêiscum and Grêcum and Lêdenwarum, Homl. Th. i. 436, 16. Se hâlga biscop hine hâdode tô messepreóste, and his wealhstod tô diácone, Homl. Skt. i. 3, 525. Nôman hî him wealhstodas (*interpretes*) of Franclande, Bd. 1, 25; S. 486, 23: Homl. Th. ii. 128, 19.   II. *an interpreter of written language, a translator* :—Ælfrêd kuning wæs wealhstod ðisse bêc, Bt. proem.; Fox viii. 1. Ðæra hundseofontigra wealhstoda gesetnyssa, Anglia viii. 336, 4. Wealcstoda *interpretum* (*praestantissimus, Hieronymus*, Ald. 33), Hpt. Gl. 463, 42. Hié hié (*books*) wendon ðurh wîse wealhstodas on hiora âgen geðióde, Past. pref.; Swt. 7, 4.   III. *an interpreter* of a subject, *an expounder* :—Wealhstod *interpres* (*divinae legis*, Ald. 64), Wrt. Voc. ii. 85, 79 : 47, 2. Lîfes wealhstod, Cd. Th. 211, 7; Exod. 522.   IV. *a mediator* :—Se wealhstod Godes and monna, ðæt is Crist *Dei hominumque mediator*, Past. 3; Swt. 33, 11.   V. the word occurs as a proper name :—Ðâm folcum ðe eardiaþ be westan Sæferne is Wealhstod biscop *eis populis qui ultra amnem Sabrinam ad occidentem habitant,Valchstod* (Uual-, *v. l.*) *episcopus*, Bd. 5, 23; S. 646, 21.

Wealh-peód, es; *f. The Welsh people* :—Ðis is seó gerædnes ðe Angelcynnes witan and Wealhþeóde rædboran gesetton, L. O. D. proem.; Th. i. 352, 1.

wealh-word, es; *n. A wanton word* :—Ic eom ondetta ðæt ic onfêng on mînne mûð wealworda, Anglia xi. 98, 37. v. wealh, II a, wealian.

wealh-wyrt, e; *f. Wall-wort, dwarf elder*; the word glosses *ebulum* and *intula* :—Walhwyrt, uualhuyrt, ualuyrt *ebulum*, *elleus*, Txts. 59, 714. Wealwyrt *ebulum*, Wrt. Voc. ii. 28, 75. Walwyrt, i. 30, 58. Wealwyrt ł ellenwyrt *ebule* ł *eobulum*, Lchdm. iii. 302, col. 1. Wælwyrt *vel* ellenwyrt. Genim ðâs wyrte ðe man *ebulum* and ôðrum naman ellenwyrte nemneþ, and eác sume men wealwyrt hâtaþ, i. 202, 3–6. Uualhwyrt *intula*, Txts. 69, 1075. Wealewyrt, Wrt. Voc. ii. 48, 71. Walwyrt, Wülck. Gl. 299, 8 (*this* gloss is omitted by Wright) Lchdm. iii. 303, col. 1. Wealwyrt, ii. 64, 27 : 70, 2. Wælwyrt, iii. 30, 13. Wealwyrte wyrttruman, ii. 108, 7. Wealwyrte moran, 264, 20. Wælwyrte, i. 354, 13. Genim wealwyrt, 66, 14. Nime wealwyrt nioþowearde, 118, 2. Wælwyrt, 38, 17. [Walwurt *ebulum*, Wülck. Gl. 555, 10. Walwort *ebulus*, 579, 33. Walwortte *ebolus*, 712, 24. Wallewurte *ebula*.]

wealian; *p.* ode *To be impudent, bold, wanton*. v. wealh, II a :—Hê wealode mid wordum, and sæde ðæt hê wolde his wîfes brûcan on ðam unâlýfedum tîman, Homl. Skt. i. 12, 48.

weá-lîc; *adj. Miserable* :—Sumum ðæt gegongeþ, ðæt se endestæf weálîc weorþeþ; sceal hine wulf etan, Exon. Th. 328, 4; Vy. 12. v. wâ-lîc.

wealig. v. welig.

weall, es; *m.*   I. *a wall* that is made, *wall* of a building, of a town, *side* of a cave :—Weal *murus*, Wrt. Voc. i. 36, 35 : Exon. Th. 281, 23 ; Jul. 650. Ofer wealles hrôf *super muros*, Ps. Th. 54, 9. Wealles rihtungþrêd *perpendiculum*, Wrt. Voc. i. 39, 64. Seó heánnes ðæs walles (*parietis*), Bd. 2, 14; S. 517, 31. Heora gewinnan tugan hî âdûn of ðam

---

wealle (*de muris*) . . . Hig ðâ forlætan ðone wall (*relicto muro*), I, 12 ; S. 481, 22. Andweorc tô wealle *cimentum*, Wrt. Voc. i. 85, 27. Tô wealle *ad moenia*, Kent. Gl. 287. Hê æfter recede wlât, hwearf be wealle, Beo. Th. 3150; B. 1573. Ofer mînre burge weall (*murum*), Ps. Th. 17. 28 : Cd. Th. 101, 3 ; Gen. 1676 : Judth. Thw. 23, 38 ; Jud. 161. Wið ðone weall *murotenus*, Wrt. Voc. ii. 57, 63. Wið ðæs recedes weal, Beo. Th. 658 ; B. 326. Wall îserne, Cd. Th. 231, 15 ; Dan. 247. Tô hwý tôwurpe ðû weal (*maceriam*) his, Ps. Spl. 79, 13. Ðâ gewrohte hê weall mid turfum (*vallum*, v. Bd. 1, 5) and brêd weall ðær onufan, Chr. 189; Erl. 9, 25. Weallas *moenia*, Wrt. Voc. ii. 54, 62 : *muri*, Jos. 6, 20. Ðæt wæter stôd an twâ healfa ðære stræte swilce twêgen hêge weallas *erat aqua quasi murus*, Ex. 14, 22. Under wealla hleó, Cd. Th. 259, 13 ; Dan. 691. Binnan ðære ylcan cyricean weallum (*muris*), Bd. 5, 20 ; S. 641, 43. On ceastre weallum beworhte *in civitatem munitam*, Ps. Th. 59, 8 : Cd. Th. 145, 21 ; Gen. 2409. Ofer ðære burge wallas (*muros*), Bd. 3, 16 ; S. 543, 2. Ðû hî betweónum wætera weallas læddest, Ps. Th. 105, 9. Ealle his weallas *omnes macerias ejus*, 88, 33. Uallas *menia*, Rtl. 124, 3.   II. *a natural wall, a steep hill, a cliff*. v. weall-clif (cf. *O. Sax.* :—Hwô sie ina fan ênumu kliðe wurpin, oðar enna berges wal, Hêl. 2676. Fan themu walle niðar werpan, 2684. Sie an hôhan wal stigun, stên endi berg, 3117) :—Munt is hine ymbûtan, geáp gylden weal, Salm. Kmbl. 511 ; Sal. 256. Cwom wundorlîcu wiht (*the sun*) ofer wealles hrôf (*over the mountain top*), Exon. Th. 412, 1 ; Rä. 30, 7. Draca beorges getrûwode, wîges and wealles (*the cliff in which the firedrake's cave was*), Beo. Th. 4635 ; B. 2323. Norð-Denum stôd egesa, ânra gehwylcum ðara ðe of wealle wôp gehýrdon (*to each that heard the cry coming from the hill on which the hall stood* (?)), 1574; B. 785. Nô wyrm on wealle leng bîdan wolde (*the serpent would not longer wait in the hill, in its cave*, 4604; B. 2307. Geseah hê mâððumsigla fela, gold glitinian grunde getenge, wundur on wealle, 5511; B. 2759. Se ðe inne gehýdde wræte under wealle, 6112; B. 3060: 6197; B. 3103. Æt wealle, 5045; B. 2526. Geseah he wealle stondan stânbogan, streám ût þonan brecan of beorge, 5077; B. 2542: 5425; B. 2716. Of wealle (*the sea-clif*) geseah weard, se ðe holmclifu healdan scolde, 463; B. 229. Winnel wæg wið wealle, Exon. Th. 383, 33; Rä. 4, 20. Æniges monnes wîg forbûgan oððe on weal fleón (*flee to the hill*) lîce beorgan, Vald. 1, 15. Weallas him wiþre healdaþ, Exon. Th. 336, 24; Gn. Ex. 54. Ic sænæssas geseón mihte, windige weallas (*wind-beaten cliffs*), Beo. Th. 1148; B. 572: Cd. Th. 214, 19; Exod. 571. Ic wiht (*a rake*) geseah, seó wæþeþ geond weallas (*among the hills* (?)), wyrte sêceþ, Exon.Th. 416, 27 ; Rä. 35, 5. [*O. Sax. O. Frs.* wal *a wall*. *From Latin* vallum.] v. bord-, breóst-, burh-, ceaster-, eorþ-, fore-, grund-, holm-, port-, sæ-, scîd-, scild-, stæd-, stân-, streám-weall.

weall, e; *f. Fervour* :—Wealle, wylm *fervorem, ardorem* (devotionis fervorem, Ald. 34), Hpt. Gl. 465, 37. v. weall-hât.

weall, es; *n.* (?) *Boiled* or *mulled wine* :—Defrutum, i. *vinum* medo geswêt *vel* weall (cf. gesoden wîn *defrutum vinum*, i. 27, 62. Coerin *defrutum*, cyren oððe âwylled wîn *dulcisaþa*, ii. 25, 10, 69. Âsodenes wînes *careni*, Hpt. Gl. 408, 42), Wrt. Voc. ii. 138, 24. Niwes ł gesodenes wealles *defruti* ł *medoni*, Hpt. Gl. 414, 1. Wealle *defruto, vino*, 520, 38.

weallan; *p.* weóll, *pl.* weóllon; *pp.* weallen.   I. *of water, &c.* issuing from a source, *to well, bubble forth, spring out, flow* :—Ic wealle *bullio*, Ælfc. Gr. 30, 5; Zup. 192, 3. Of ðæm neáhmunte wealleþ hlûter wæter, ðonne drincaþ ða menn ðæt *cadente rivo puram ex vicino monte potant aquam*, Nar. 31, 7. Of ðæm beorgum wildt seó eá Eufrates *fluvius Euphrates de radice montis effusus*, Ors. 1, 1 ; Swt. 14, 10, 29. Ðær hió (*the Nile*) ærest up wieldt *prope fontem*, Swt. 12, 24. [Ðæt treów ðæt man on heorþe leges, for ðare mycele hæten ðe ðæt treów barned beoþ, þære wylþ ût of ðan ende water, Lchdm. iii. 128, 6.] Rêcels of ðæra treówa telgan weól, Nar. 26, 22. Swât ýðum weóll *the blood welled out in streams*, Beo. Th. 5380; B. 2693 : Andr. Kmbl. 2552 ; An. 1277 : 2482 ; An. 1242. Weól, Exon. Th. 182, 23 ; Gû. 1314. Wiþ ðon ðe men blôd upp wealle þurh his mûð, Lchdm. i. 74, 14. Hê lêt teáras geótan, weallan wægdropan, Exon. Th. 165, 17 ; Gû. 1030 : Andr. Kmbl. 3005 ; An. 1505. Mon geseah weallan blôd of eorþan *sanguis e terra visus est manare*, Ors. 4, 3 ; Swt. 162, 6. Geseah ic balzamum of ðæm treówum ût weallan *video opobalsamum arborum ramis manans*, Nar. 27, 23.   II. *of the source, to well* with, *flow* with, (1) with a noun :—Ân wielle weól blôde *flumen sanguine effluxit*, Ors. 4, 7 ; Swt. 184, 21. Flôr âttre weól, Cd. Th. 284, 8 ; Sat. 318. Flôd blôde weól, Beo. Th. 2848 ; B. 1422. Weóll, 4282 ; B. 2138. Wið ðon ðe mon blôde wealle þurh his mûð, Lchdm. iii. 44, 22. Wæs on blôde brim weallende, Beo. Th. 1699; B. 847. (2) absolute :—Benna weallaþ *wounds bleed*, Andr. Kmbl. 2810 ; An. 1407. Hit ongan rînan . . . and seó eorðe weóll ongeán ðam heofonlîcan flôde *it began to rain . . . and the earth sent forth its waters to meet the waters of heaven*, Wulfst. 206, 21. Weóllon wælbenna, Cd. Th. 208, 30 ; Exod. 491.   III. *implying abundance*, (1) *to swarm, exist in large numbers* :—Him weóllon maðan geond ealne ðone lîchaman, Homl. Th. i. 472, 30. (2) *of production in large numbers or great quantity, to swarm* with, *flow* with :—Land ðe weóll meolce

and hunie *terra quae lacte et melle manabat*, Num. 16, 13. His gesceapu maðan weóllon, Homl. Th. i. 86, 10: Homl. Skt. i. 4, 212. Weallende *scaturiens* (*vermibus*, Ald. 70), Hpt. Gl. 519, 34: *scatens* (*vermibus*, Ald. 202), Wrt. Voc. ii. 96, 7. **IV.** of violent movement, *to boil, rage, heave :*—Geofon ýþum weól wintres wylme, Beo. Th. 1035; B. 515. Holm storme weól, 2267; B. 1131. Hreðer ædme weóll *his breast heaved*, 5180; B. 2593. Ða ýþa weóllan and wéddan ðæs sǽs *furentibus undis pelagi*, Bd. 3, 15; S. 541, 39, 42. Brim weallende, Andr. Kmbl. 3147; An. 1576. Ðæt gebrec ðæs weallendes (*ferventis*) sǽs, Bd. 5, 1; S. 614, 4. Wado weallende, Beo. Th. 1096; B. 546. **V.** of movement in liquids caused by heat, *to boil* (intrans.), *to be hot :*—Dó ofer fýr, áwyl; ðonne hit wealle, sing iii Pater noster, Lchdm. ii. 358, 11. Scenc fulne weallendes wæteres, 130, 1. Seóð on weallendon wætere, i. 204, 23. Mid weallendum ele, Homl. Th. i. 58, 27: Ælfc. T. Grn. 16, 16. Weallende wǽte *fervida flumina*, Hpt. Gl. 499, 51. **V a.** used of a vessel in which a liquid boils :—Seó ǽrene gripu ofer glǽda gripe gífrust wealleþ (-aþ, MS. B.), Salm. Kmbl. 98; Sal. 48. Bæð háte weól, Exon. Th. 277, 16; Jul. 581. **VI.** of other than liquids, *to be hot, burn, blaze, rage :*—Wið ðone weallendan bryne ðe weallaþ (-eþ?) on helle, L. C. E. 6; Th. i. 364, 13. Him on breóstum weóll áttor, Beó. Th. 5422; B. 2714. An ðæra dǽla is weallende (*the torrid zone*), Lchdm. iii. 260, 21. Se weallenda lég *furens flamma*, Bd. 2, 7; S. 509, 22. Hé hæfþ weallendene lég, Blickl. Homl. 61, 35. Weallende fýr, Cd. Th. 153, 22; Gen. 2542. Weallendum lígum *flammis ferventibus*, Bd. 5, 12; S. 627, 37. Weallende axan, Lchdm. i. 178, 6. Þurh ða weallendan sond *per ferventes sole arenas*, Nar. 6, 9. **VII.** figuratively, of persons, passions, emotions, *to be fervent, to burn, rage, to be strongly moved :*—Ic wealle *ferueo*, Ælfc. Gr. 26, 5; Zup. 156, 9. Weáll *fervet*, Kent. Gl. 665. Hé weld on gódum cræftum *in virtutibus inardescit*, Past. 58; Swt. 447, 18. Hé metta mid cystignesse wcald *aescarum largitate feruescit*, Scint. 56, 2. Hyge hearde wealleþ, Salm. Kmbl. 126; Sal. 62. Wyrd bið wended hearde, wealleþ (*is zealous*) swíðe geneahhe, 872; Sal. 435. Feóndscipe wealleþ *hatred burns hot*, Exon. Th. 354, 60; Reim. 68. Weallaþ wælnīðas, Beo. Th. 4136; Beo. 2065. Brand-háta nīð weóll on gewitte, Andr. Kmbl. 1537; An. 770. Hreðer innan weóll, beorn breóstsefa *their hearts burnt within them*, Exon. Th. 34, 9; Cri. 539: Beo. Th. 4233; B. 2113. Breóst innan weóll þeóstrum geþoncum, 4652; B. 2331. Weóll him on innan hyge ymb his heortan, Cd. Th. 23, 4; Gen. 353. Se nyle wearmian óð hé wealle (*ut ferveat*), Past. 58; Swt. 447, 8. Suá sculon ða hierdas weallan ymb ða geornfulnesse ðære inneran ðearfe his hiéremonna *sic pastores erga interiora studia subditorum suorum fervent*, 18; Swt. 137, 11. Hire oninnan ongan weallan wyrmes geþeaht, Cd. Th. 37, 15; Gen. 590. Weallende *furibundus*, Wrt. Voc. ii. 36, 37: *fervidus*, 147, 84: Lchdm. iii. 188, 25. Se mǽra wæs háten weallende wulf (cf. (?) Wóden), Salm. Kmbl. 423; Sal. 212. Lég, weallende wiga, Exon. Th. 61, 15; Cri. 985. Hé wæs weallende on geleáfan (*fide fervens*), Bd. 3, 2; S. 524, 17. Weallende spelboda, Blickl. Homl. 165, 33. Manegum wæs hát æt heortan hyge weallende, Andr. Kmbl. 3415; An. 1711. Ðeós gitsunc weallende byrnð, Met. 8, 45. Mid weallendre lufe, Wulfst. 286, 11. Sorge weallende, Beo. Th. 4919; B. 2464. Weallende weán, Exon. Th. 139, 2; Gú. 587. Hé geseah calle witon on þeáwum scínende and on gáste weallende, Homl. Skt. ii. 23 b, 86. **VIII.** trans. (= willan?) *To roll, turn :*—Hine on lyfte lífgetwinnan sweopum seolfrenum swíðe weallaþ, óð ðæt him bán blícaþ, blédaþ ædran, Salm. Kmbl. 288; Salm. 143. [*O. Sax.* wallan *to well; to boil, burn* (fig.): *O. Frs.* walla: *O. H. Ger.* wallan *scatere, bullire, fervescere : Icel.* vella *to boil; to swarm.*] v. á-, be-, ge-weallan; heoru-weallende, for-weallen.

weall-clif, es; *n. A steep cliff :*—Hí scufon wyrm ofer weallclif, léton wǽg niman, Beo. Th. 6255; B. 3132. v. weall, **II.**

weall-díc (?), e; *f. A walled ditch* (?) :—Andlang ðære wealdíc, Cod. Dip. Kmbl. v. 346, 21, 22. Cf. Usque la diche walle; et sic per fossatum, iii. 408, 10.

weall-dor, es; *n. A door in a wall :*—Ðú eart ðæt wealldor; þurh ðé Freá on ðás eorþan út síðade, Exon. Th. 21, 1; Cri. 328.

weall-fæsten[n], es; *n.* **I.** *a walled stronghold, a fortress :*—Ða gesceádaþ ðæt land westan and eástan óð ðæt weallfæsten, Cod. Dip. Kmbl. ii. 86, 27. Hé ongan ceastre timbran, ðæt wæs weallfæstenna ǽrest, Cd. Th. 64, 31; Gen. 1058. **II.** *a wall for defence, a bulwark :*—Forhwan ðú tówurpe weallfæsten his? *quid deposuisti maceriam ejus?* Ps. Th. 79, 12. Wicon weallfæsten, wǽgas burston, Cd. Th. 208, 14; Exod. 483. Wyrceþ wæter wealfæsten (*erat aqua quasi murus a dextra eorum et laeva*, Ex. 14, 22), 195, 27; Exod. 283.

weall-geat, es; *n. A gate in a wall :*—Hié gegán hæfdon tó ðam weallgeate *they had reached the city's gate*, Judth. Thw. 23, 26; Jud. 141. Tó weallgeatum, Andr. Kmbl. 2407; An. 1205.

weall-gebrec, es; *n. A breaking down of a wall :*—Hié noldon ðæs weallgebreces geswícan *donec perfractis muris*, Ors. 3, 9; Swt. 134, 30.

weall-geweorc, es; *n. Wall-work*, (1) *wall-building :*—Gang 16 ðínum weallgeweorce (*a monastery was being built*), Homl. Skt. i. 6, 173. Sí hit ælces þinges freoh bútan ferdfare and walgeworc (cf. burh-bót) and

brycgeworc, Cod. Dip. Kmbl. iii. 5, 13. Hé gesette hí tó his weallgeweorcum, ðæt hí worhton his burga (*in aedificationibus urbium suarum*), Anglia x. 91, 96. (2) *the destruction of walls :*—Aries byð ram betwux sceápum and ram tó wealgeweorce, Ælfc. Gr. 5; Zup. 12, 5. v. weall-weorc.

weall-hát; *adj. Boiling hot, red-hot :*—Ácéle ðú wealhát ísen ðonne hit furþum sié of fýre átogen on wíne, Lchdm. ii. 256, 15. [He bed bringen forð brune wallinde bres, and healden hit se walhat up on hire heaued, Jul. 31, 4. Wiþþ wallhat herrtess lufe, Orm. 14196.]

weallian *to wall.* v. ge-weallod.

weallian; *p. ode.* **I.** *to wander, roam :*—Weallaþ swá niéten feldgangende, feoh bútan gewitte, se þurh ðone cantic ne can Crist geherian, Salm. Kmbl. 44; Sal. 22. **II.** *to go as a pilgrim :*—Of earde weallige hé wíde and dǽdbóte dó ǽfre ða hwíle ðe hé libbe *a patria longe peregrinetur, et poenitentiam usque agat, quamdiu vivet*, L. M. I. 44; Th. ii. 276, 31. Deóplíc dǽdbót bið ðæt lǽwede man his wǽpna álecge and weallige bærfót wíde, L. Pen. 10; Th. ii. 280, 18. Oferbecumendum wealligendum þearfum se abbud mid gebróþrum gearwian hýrsumnysse *supervenientibus peregrinis pauperibus abbas cum fratribus exhibeant obsequium*, Anglia xiii. 439, 1060. [*O. H. Ger.* wallón *errare, ambulare, meare, pervagari: Ger.* wallen *to travel;* wall-fahrt *pilgrimage : Icel.* vallari *a tramp, vagrant.*]

weall-lím, es; *m. Mortar :*—Hig hæfdon tygelan for stán and tyrwan for wealliim *habuerunt lateres pro saxis et bitumen pro caemento*, Gen. 11, 3.

weall-stán, es; *m. A stone for building :*—Ðú eart se weallstán ðe ða wyrhtan wiðwurpon tó weorce (*lapidem, quem reprobaverunt aedificantes*, Mt. 21, 42), Exon. Th. 1, 2; Cri. 2. Wrætlíc is ðes wealstáu *marvellous is this masonry*, 476, 1; Ruin. 1. Ceastra, wrætlíc weallstána geweorc *cities, wondrous works of stones*, Menol. Fox 465; Gn. C. 3.

weall-steall, es; *m. A place where there are buildings :*—Ðisne wealsteal *this spot where the walls stand* (cf. weallas stondaþ, 291, 3; Wand. 76), Exon. Th. 291, 26; Wand. 88.

weall-steáp; *adj.* **I.** *high as regards its walls* or *buildings, with lofty walls :*—Hié on weallsteápe burg (cf. seó steápe burh on Sennar, 102, 15; Gen. 1700) wlítan meahton, Cd. Th. 145, 7; Gen. 2402. **II.** *with lofty cliffs, lofty.* v. weall, **II** :—Hié oferfóran weallsteápan hleoðu, Cd. Th. 108, 8; Gen. 1803.

weall-stellung, -stilling, -stylling, e; *f. The putting a wall in order, repairing of a wall.* v. burh-bót :—Tó ánes æceres brǽde on wealstillinge (cf. weall-geweorc) and tó ðære wǽre gebirigeaþ xvi. hída; gif ǽlc híd byþ be ánum men gemannod, ðonne mæg man gesettan ǽlce gyrde mid feówer mannum. Ðonne gebyreþ tó twéntigan gyrdan on wealstillinge hundeahtig hída, and tó ðam furlange gebyrgeaþ óþer healf hund hída and x hída . . . Tó fíf furlangum gebyreþ ymbeganges eahta hunda hída on wealstyllinge . . . Tó eahta furlangum ymbeganges wealstyllinge hund eahtig hída and .xii. hund hída *for one acre's breadth* (22 yds.) *in the matter of repairing a wall and for the keeping of it* 16 *hides are requisite; if each hide is assessed at one man, then four men can be appointed to each pole.* 80 *hides are requisite for the putting in order of twenty poles of wall and for the furlong* 160 *hides . . . For a circuit of five furlongs* 800 *hides are necessary . . . For a circuit of eight furlongs* 1280 *hides*, Hickes' Diss. p. 109.

weall-þrǽd, es; *m. A plumb-line :*—Waldrǽd *perpendicula*, Wrt. Voc. ii. 91, 68. v. rihtung-þrǽd.

weallung, e; *f.* **I.** *agitation :*—Se drænc is gód wið heáfodece and wið brægenes hwyrfnesse and weallunge *the potion is good against headache and against giddiness and cerebral excitement*, Lchdm. iii. 70, 20. **II.** *fervour :*—Wyrðelícre wallunge lufes *digno fervore fidei*, Rtl. 64, 26.

weall-wala, an; *m. A wall-foundation* (?) :—Hygeróf gebond weallwalan wírum wundrum tógædere, Exon. Th. 477, 9; Ruin. 21.

weall-weg (?), es; *m. A walled road* (?) :—On ðane ealdan walweg, Cod. Dip. Kmbl. v. 78, 17: 138, 4.

weall-weorc, es; *n. Wall-work, building :*—Ða gebróðra eodon tó ðam weallweorce, Homl. Th. ii. 166, 14, 25. v. weall-geweorc, *and next word*.

weall-wyrhta, an; *m. A wall-wright, a mason, builder :*—Wealwyrhta *cimentarius*, Wrt. Voc. i. 19, 15: 85, 27. Fram wealwyrhtan (-wyrhtum, Wrt. Voc. ii. 79, 6 = a cementario, Ald. 31) *a cimentario*, Anglia xiii. 32, 106. Weallwyrhtan *cimentarii*, Wrt. Voc. ii. 15, 83.

weal-more(-u, -a), wealowigan *to fade*, wealowigan *to roll*, wealsáda, -wealt [*Icel.* valtr], -wealtian, -weálu. v. wealh-more, wealwian *to fade*, wealwian *to roll*, wealh-sáda, seonu-, un-wealt, seonuwealtian, wǽel.

wealwian; *p.* ode *To fade, wither* (Halliwell gives *wallow* = to fade away, as a Somerset word) :—Hæfd se Ælmihtiga ðæt gewrixle geset, ðe nú wunian sceal, wyrta grówan, leáf grénian, ðæt on hærfest eft hrést and wealowaþ (cf. fealwaþ, Bt. 21; Fox 74, 23), Met. 11, 58. Ðǽr ðér hit gefrét ðæt hit hraþost weaxan mæg and latost wealowigan (wealowian, Cott. MS.) *ubi quantum earum natura queat, cito exarescere atque interire non possint*, Bt. 34, 10; Fox 148, 22. [Welewen *marcescere*,

Wick. Is. 19, 6. Man welewith as flouris of hay, P. R. L. P. 173, 56. Al welwed and wasted þo worþelych leues, Allit. Pms. 106, 475. See also *welewed* in Halliwell's Dict.] v. un-forwealwod.

wealwian; *p.* ode *To wallow, roll* (intrans.) :—Ðonne tyht hié ðæt ierre ðæt hié wealwiaþ on ða wédenheortnesse *impellente ira in mentis vesaniam devolvuntur*, Past. 40; Swt. 289, 6. Hé wealwode on ðæm gedrófum wætere *in lutosa aqua semetipsum volvit*, 54; Swt. 421, 8. His hors feól wealwigende geond ða eorðan . . . mid ðam ðe hit swá wealwode, Homl. Skt. ii. 26, 207. Ða felga hangiaþ on ðám spácan, þeáh hí eallunga wealowigen on ðære eorþan, Bt. 39, 7; Fox 222, 14. Ðæt hors ongan walwian and on gehwæþere sídan gelómlíce hit oferweorpan (*in diversum latus vicissim sese volvere*), Bd. 3, 9; S. 533, 40. Micel stán wealwiende of ðam heáhan munte, Bt. 6; Fox 14, 28. [Hie seched to þe fule floddri and þaron walewed, O. E. Homl. ii. 37, 27: H. M. 13, 34. They walweden as pigges in a poke, Chauc. Reeves T. 358. Þe grete wawes walweth (walketh, *v. l.*), Piers P. 8, 41.] v. be-wealwian; wilwian.

weal-word, -wyrt. v. wealh-word, -wyrt.

weá-mét[t], e: -méttu(-o); *indecl. f. Anger, wrath, passion, irascibility* :—Se feórða heáfodleahter is weámét, Homl. Th. ii. 218, 21. Se feórða leahtor is weámét, ðæt se man náge his módes geweald, ac búton ælcere foresceáwunge his yrsunge gefremaþ, 220, 12. Wé sceolon oferwinnan weámétte mid wíslícum geðylde, 222, 21. Ne gerísaþ heom hræde weámétta, L. I. P. 10; Th. ii. 318, 32. [Cf. Heo weore god ʒif heo neore to wamed. Anan se he wes wrað wið eni he hine wolde slæn, Laym. 6368.]

weá-mód; *adj. Angry, wrathful, choleric, passionate* :—Se ðe wære weámód, weorðe se geþyldmód, Wulfst. 70, 7. Ne réce ðú ná weámódes wífes worda *you are not to care for an angry woman's words*, Prov. Kmbl. 48. Ða weámódan and ða grambæran *iracundi*, Past. 40; Swt. 289, 4: Wulfst. 40, 17. Weámódum *turbulentis*, Germ. 395, 13. [Ne beo þu wemod ne ouermod, O. E. Homl. i. 5, 26. Pellican is a leane fowel, so weamod and so wreðful þet hit slead ofte uor grome his owune briddes, A. R. 118, 8.]

weámódness, e; *f. Anger, passionateness, irascibility* :—Se feórða leahtor is *ira*, ðæt is on Englisc weámódnyss, Homl. Skt. i. 16, 286: Wulfst. 68, 15. Ðonne hié beræsaþ on suelce weámódnesse hié sindon tó oferbúganne *qui in eodem furoris impetu declinandi sunt*, Past. 40; Swt. 295, 20. Forlýst se yrsigenda wer his ágene sáwle þurh weámódnysse, Homl. Skt. ii. 28, 149: Anglia xi. 113, 32, 38. Ðære sáwle miht is ðæt heó sylf beó geðyldi and ælce weámódnysse fram hire áwyrpe, Basil admn. 3; Norm. 38, 27. [*Ira*, þet is on Englisc wemodnesse, O. E. Homl. i. 103, 19.]

wear. v. wearr.

weard, es; *m.* **I.** *a guard, warder, watchman, sentinel* :—Ðara wearda sum geseah ðæt of heofonum com án læs feówertig wuldorbeága . . . ðá gecerde se weard tó Criste, Shrn. 62, 5–8. Weard Scyldinga, se ðe holmclifu healdan scolde, Beo. Th. 464; B. 229: Ps. 126, 2. Se weard (*the angel at the gate of Eden*), Cd. Th. 58, 21; Gen. 949. Ða weardas *custodes*, Mt. Kmbl. 28, 4, 11. Ða weardas heóldon ðæs cwearternes duru, Homl. Th. ii. 382, 4. Snelle gemundon weardas wígleoð, Cd. Th. 191, 27; Exod. 221. Hine twégen ymb weardas wacedon, Exon. Th. 109, 6; Gú. 86. Ða byrgene besettan mid wacelum weardum (*custodibus*), Homl. Th. ii. 262, 8: Mt. Kmbl. 27, 66: Blickl. Homl. 177, 29. Salomones reste wæs mid weardum ymbseted, ðæt wæs mid syxtigum werum, 11, 16. Hé sette him weardas ofer, Jos. 10, 18: Homl. Skt. i. 11, 210. **I a.** *fig.* :—Him oninnan oferhygda dǽl weaxeþ, ðonne se weard swefeþ, sáwele hyrde, Beo. Th. 3487; B. 1741. Geác, sumeres weard, Exon. Th. 309, 8; Seef. 54. Bánhúses weard *the mind*, Cd. Th. 211, 9; Exod. 523. **II.** *a guardian, protector, lord* :—Ðære cneórisse wæs Cainan aldordéma, weard and wísa, Cd. Th. 70, 22; Gen. 1157. Ðú (*Nebuchadnezzar*) hæleðum eart ána eallum eorðbúendum weard and wísa, 251, 19; Dan. 566. Engla weard (*Lucifer*), 2, 20; Gen. 22. Cyning, beáhhorda weard, Beo. Th. 1847; B. 921. Ríces weard, 2784; B. 1390. Folces weard, 5019; B. 2513. ¶ the term is often used of the Deity :—Weard *servatorem* (*animae tuae*, Prov. 24, 12), Kent. Gl. 932. Rodera weard, Cd. Th. 1, 2; Gen. 1. Lífes weard, 9, 20; Gen. 144. Sigores weard, Exon. Th. 15, 29; Cri. 243. Wuldres weard, 33, 17; Cri. 527. Heofonríces weard, Andr. Kmbl. 104; An. 52. [*Goth.* daura-wards: *O. Sax.* ward *a guard, a guardian*: *O. H. Ger.* wart *custos*: *Icel.* vörðr.] v. bát-, botl-, brego-, brycg-, burh-, carcern-, cweartern-, dæg-, drihten-, duru-, edisc-, eorþ-, éðel-, fore-, forþ-, freoðu-, gold-, gúþ-, hæg-, heáfod-, healf-, hearg-, heofon-, hof-, hord-, hýð-, irfe-, land-, lást-, leác-, leáctún-, lid-, mearc-, mere-, mylen-, niht-, regn-, regol-, scip-, sele-, stig-, stóc-, wudu-, wyrt-weard; *also such proper names as* Æþel-weard, Eád-weard.

weard, e; *f.* **I.** *ward, guard, watch* :—Gefangen on hergiunge oþþe æt wearde *utrum exploratem an in praelio captus*, Ors. 4, 11; Swt. 206, 5. Healdaþ wearde dæges and nihtes *die ac nocte manebitis observantes custodias*, Lev. 8, 35. Weras wæccende wearde heóldon, Judth. Thw. 23, 26; Jud. 142: Beo. Th. 616; B. 305. Wið wráð werod

wearde healdan, 644; B. 319: Exon. Th. 48, 6; Cri. 767: 282, 16; Jul. 664. Weardum *excubitis*, Wrt. Voc. ii. 30, 12. Lux et tenebre ðe ðás werþeóda weardum healdaþ, Exon. Th. 192, 5; Az. 101. Wǽrda *excubias*, Hpt. Gl. 476, 29. **I a.** *a watch, a body of men keeping watch* :—Hí besetton his birgene mid wearde, Jud. Thw. p. 161, 12. **II.** *guardianship, protection, keeping* :—Heora feorh generede mihtig Metodes weard, Cd. Th. 230, 18; Dan. 235. Cristenum cyninge gebyraþ ðæt hé sý on fæder stæle cristenre þeóde, and on ware and on wearde Cristes gespeliga, L. I. P. 2; Th. ii. 304, 23. [*O. H. Ger.* warta *speculatio, cura, custodia, excubiae*: *Icel.* vörðr; *m. ward, watch, protection*.] v. ǽg-, fird-, flód-, fore-, heáfod-, hors-, leód-, sǽ-weard; or-wearde.

weard; *adv. Ward* in to-*ward*; the form occurs in combination with *tó* (v. tó-weard; *prep.* **II.** 3) and *wiþ* (v. wiþ, **IX**) :—Hié wǽron wið ðæs fýres weard, Ors. 4, 10; Swt. 200, 16. Hé wið Róme weard farende wæs, 5, 11; Swt. 236, 9, 15, 21. Ðá ongan seó leó fægnian wið ðæs ealdan weard, Homl. Skt. ii. 23 b, 778. Heó teáh hyne wiþ hyre weard, Judth. Thw. 23, 1; Jud. 99. v. eást-, for-, forþ-, hám-, hider-, hindan-, norþ-, súþ-, þider-, west-weard.

-weard *the second component of many adjectives denoting position or direction.* v. æf-, æftan-, æfte-, æfter-, and-, eáste-, for-, fore-, forþ-, fram-, from-, heonon-, hider-, hinde-, hinder-, innan-, inne-, midde-, neoþan-, neoþe-, niþer-, norþ-, norþan-, norþe-, on-, ongeán-, súþe-, þanan-, tó-, ufan-, ufe-, up-, útan-, úte-, westan-, weste-, wiþer-weard. [*O. Sax.* -ward: *O. H. Ger.* -wart. Cf. *Goth.* -wairþs: *Icel.* -verðr.]

wearda (?), wearde (?), an; *m.* or *f. A watchman* or *a watch* :—Óð weardan hylle; fram weardan hylle (*the beacon-hill?* Cf. *Icel.* varða *a beacon*; varð-berg *a look-out place*: *O. H. Ger.* wart-perg), Cod. Dip. Kmbl. v. 191, 34. Cf. On weardæs beorh, 291, 23: 112, 32. Weardan *excubiae*, Ælfc. Gr. 13; Zup. 84, 16. [*Goth.* wardja *a guard*: *O. H. Ger.* warto.] v. next word.

weard-dún, e; *f. A beacon-hill* (? cf. weardan hyll. v. wearda) :—On wearddúne, ðær ðæt Cristes mǽl stód, Cod. Dip. Kmbl. iii. 465, 31.

weardere, es; *m. One who holds a country, an inhabitant* :—Columba com tó Pyhtum; ðæt synd wærteras þe norðum mórum Columba came to the Picts; they are the people who hold the country to the north of the hills (cf. Bd. 3, 4: Venit Columba Brittaniam praedicaturus verbum Dei provinciis Septentrionalium Pictorum, hoc est, eis quae arduis atque horrentibus montium jugis ab Australibus eorum sunt regionibus sequestratae), Chr. 565; Erl. 16, 37. [*O. H. Ger.* wartari *custos*.] v. weardian, **IV**.

weardes; *adv. Wards* in to-*wards* :—Ðá smearcode heó wið his weardes, Homl. Skt. ii. 23 b, 590. Swá eode heó wið his weardes, 684. Ðá arn se ealda wið hyre weardes, 599. v. eást-, from-, hám-, niþer-, norþ-, ongeán-, súþ-, þider-, tó-, up-, út-weardes.

weardian; *p.* ode. **I.** *to guard, keep, defend* :—Æðele getrym eorðan weardaþ *erit firmamentum in terra*, Ps. Th. 71, 16. Heofon weardiaþ ufan wætra ðrýðe *tegis in aquis superiora coeli*, 103, 3. Hý (*Seraphim*) mid hyra fiþrum Freán ælmihtiges onsýne weard (weardiað? v. Isaiah 6, 2), Exon. Th. 25, 5; Cri. 396. [Se heáhengel geong weardode (*l.* geondweardode *presented*) ðære eádigan Marian sáwle beforan, Drihtne, Blickl. Homl. 157, 9.] **I a.** *with gen.* (cf. *O. Sax.* wardón *with gen. to have charge of something*) :—Ða Englisce men ðe wærdedon ðære sǽ *the Englishmen that had charge of the sea*, Chr. 1087; Erl. 225, 26. **II.** *to act as guardian to, to rule* :—Him on láste Seth weardode, éþelstól heóld, Cd. Th. 68, 36; Gen. 1128. Nabochodonossor weardode wíde ríce, heóld hæleða gestreón, 257, 29; Dan. 665. Ríce geréfa rondburgum weóld, eard weardade, Exon. Th. 243, 33; Jul. 20. **III.** *to keep, have charge of* :—Búton hit under ðæs wífes cǽglocan gebróht wǽre, sý heó clǽne; ac ðæra cǽgean heó sceal weardian, L. C. S. 77; Th. i. 418, 21. **IV.** *to hold a country, to occupy a place, inhabit.* v. weardere :—Ðone wudu weardaþ fugel *hoc nemus avis incolit*, Exon. Th. 203, 16; Ph. 85: 208, 25; Ph. 161: 209, 10; Ph. 168. Hwílum hygegeómor healle weardaþ (*keeps the house*), Salm. Kmbl. 762; Sal. 380. Ðonne færð se deófol intó his móder innoðe, and ðær hé hine healt, and weardaþ inne, Wulfst. 193, 10. Hé heánne beám wunaþ and weardaþ, Exon. Th. 209, 17; Ph. 172. In ðam hálge wíc weardiaþ, 228, 34; Ph. 448. Him férend on fæste wuniaþ, wíc weardiaþ, 361, 27; Wal. 26. Hí dreám weardiaþ, 100, 15; Cri. 1642. Frýnd sind on eorþan, leger weardiaþ, 443, 23; Kl. 34. Ealle ða ðe on feldum eard weardiaþ *omnia quae in campis sunt*, Ps. Th. 95, 12. Ðǽr sylfǽtan eard weardigaþ, éðel healdaþ, Andr. Kmbl. 351; An. 176. Fífelcynnes eard wer weardode, Beo. Th. 211; B. 105. Reced weardode unrím eorla, 2479; B. 1237. Heó gefylled wæs wísdómes gife; hálig gást hreðer weardode, Elen. Kmbl. 2288; El. 1145: Exon. Th. 169, 30; Gú. 1102. Wé sele weardodon, Beo. Th. 4157; B. 2075. Sume stede weardedon ymb Danúbie, Elen. Kmbl. 270; El. 135. Þenden wé on eorðan eard weardigen, Exon. Th. 48, 16; Cri. 772. Ðær hig ǽnne sculan eard weardian *habitare in unum*, Ps. Th. 132, 1: Exon. Th. 356, 13; Pa. 11. Eard weardigan, án lond búgan, 473, 19; Bo. 17: Andr. Kmbl. 1198; An. 599. Wíc weardian, Exon. Th. 248, 7; Jul. 92. Staþol weardian, 496, 19; Rä. 85, 17. **IV a.** *in the phrases* lást, swaðe weardian *to keep a track*, (1) *to follow* :—Hýrde ic ðæt ðám

frætwum feówer mearas lást weardode *I heard that four steeds followed in the train of these equipments*, Beo. Th. 4335; B. 2164. (2) *to remain behind* :—Hé onweg losade, hwæþre him sió swíðre swaðe weardade hand on Hiorte *he escaped, yet his right hand remained behind in Heorot*, Beo. Th. 4203; B. 2098. Cyning úre gewát ... ðǽr hý tó ségun, ða ðe leófes ðá gén lást weardedun (*those who still remained where he had been*). Exon. Th. 31, 16; Cri. 496. Se ðe his mondryhten lífe bilidene lást weardian wiste *he who knew that his dead lord remained behind*, 182, 19; Gú. 1312. Hé his folme forlét lást weardian, Beo. Th. 1947; B. 971. Sáula sculon eft tó ðé, sceal se líchama lást weardigan eft on eorþan, Met. 20, 241. [Sicnesse warded toȝein þeo sunnen þet weren touwardes, A. R. 182, 14. Wel heo wardith heom bothe, Alis. 909. Þilke tyme þat Samuel þe prophete wardede (*ruled*) þat folc of Israel, R. Glouc. 27, 16. *O. Sax.* wardôn *to guard, to have charge of* : *O. Frs.* wardia : *Icel.* varða *to guard, defend.* Cf. *O. H. Ger.* wartên.] v. á-, be-weardian; ge-wardod.

weard-mann, es; *m. A guard, watchman, keeper* :—Nyte wé hweþer se weardmann wǽre ǽfre gefullod, Homl. Skt. i. 11, 293. Ealle ða weardmenn wǽron geswefode búton heora ánum, 11, 200 : 4, 419. Ða weardmenn ðe bewiston Cristes líc, Homl. Ass. 79, 175. Hé geseah ðæra sceaþena fær and tó ðam weardmannum becom. Ðá gelæhton ða weardmen his wealdleðer, ðæt hé mid fleáme ne burste, Ælfc. T. Grn. 18, 15. Wylsce menn geslógan mycelne dǽl Englisces folces ðæra weardmanna, Chr. 1053; Erl. 188, 10. Nytendum ðam weardmannum ic árîse *clam custodibus surgo*, Ælfc. Gr. 47; Zup. 272, 1 : Homl. Skt. i. 4, 217 : Homl. Ass. 78, 152 : Anglia x. 99, 311. Hé heora wæterscipe mid weardmannum besette *constituit centenarios per singulos fontes*, 94, 172.

weard-seld, es ; *n. A guard-house* :—Weardseld *excubias*, Wrt. Voc. ii. 108, 1.

weard-setl, es ; *n. A place where guard is kept : those who keep watch, a guard* :—On weardsetl ; of weardsetle, Cod. Dip. Kmbl. v. 48, 11. Andlang herpaðes tó weardsetle, 284, 23. On weardsetl, Cod. Dip. B. iii. 682, 24. Seofon weardsetl wacodon ofer ðone cásere. ... Ðá férde his gást and mid wǽpne ðone Godes feónd ofstang, his weardsetlum on lócigendum, Homl. Th. i. 452, 13–31. Æt ðǽm weardsetlum *ad excubias*, Wrt. Voc. ii. 3, 16. Weardsetl *excubias*, 81, 20 : 30, 11 : 71, 11. Hí ofereodon ða twá weardsetl *transeuntes primam et secundam custodiam* (Acts 12, 10), Homl. Th. ii. 382, 11.

weard-steall, es ; *m. A watch-tower* :—Weardsteal *specula* vel *conspicilium*, Wrt. Voc. i. 55, 42 : *spectacula*, 39, 35.

weard-wîte, es ; *n. A fine for neglecting to keep guard*, Chart. Th. 411, 31.

wearf. v. hwearf.

wearg(-h), es ; *m.* **I.** of human beings, *a villain, felon, scoundrel, criminal* :—Wearg *furcifer*, Wrt. Voc. ii. 37, 66. Wearh, 152, 2. Wearh sceal hangian, fǽgere ongildan ðæt hé ǽr fácen dyde manna cynne, Menol. Fox 572 ; Gn. C. 55. Hí hêton mé (*the cross*) heora wergas hebban, Rood Kmbl. 62 ; Kr. 31. **II.** of other creatures, *a monster, malignant being, evil spirit* :—Under ðæm stáne wæs niccra eardung and wearga, Blickl. Homl. 209, 34. Wé sceolun þrowian weán 7 (and ; *prep.?* or = on) wergum, nalles wul[d]res blǽd habban in heofnum *we must suffer woe with accursed ones, not have glorious honour in heaven*, Cd. Th. 267, 22 ; Sat. 42. [Þe wari of þeos wordes warð wrað, Marh. 4, 12. Ic am unwurð as weri (*v. l.* wari) þet is anhonged, A. R. 352, 21. Ich wulle hine anhon haxst alre warien, Laym. 28215. *Goth.* launawargs *an unthankful person* : *O. H. Ger.* ubiles, palowes warc *tyrannus* : der warch *diabolus* : *Icel.* vargr *a wolf* ; *an outlaw*. Graff quotes the latinized form wargus = *expulsus, latrunculus*. See Grmm. R. A. p. 733.] v. heoru-wearh, *and next word.*

wearg, werg, werig, wyrig; *adj. Evil, vile, malignant, accursed*, (1) of human beings :—Sum sceal on galgan rídan ... bið him werig noma, Exon. Th. 329, 31 ; Vy. 42. Ðú (*the body*) werga (weriga, Soul Kmbl. 43), 368, 15 ; Seel. 22. Ðú woldest brúcan ungemetlícre wrǽnnesse. Ac ðé willaþ ðonne forseón Godes þeówas, for ðam ðe dín werige flǽsc hafaþ dín anweald ... Hú mæg mon earmlícor gebǽron, ðonne mon hine underþeóde his weregan flǽsce *voluptariam vitam degas. Sed quis non spernat vilissimae fragilissimaeque rei, corporis, servum?* Bt. 32, 1 ; Fox 114, 20–24 : Met. 26, 14. Bearn Godes brýda on Caines cynne sécan, wergum folce, Cd. Th. 75, 34 ; Gen. 1250. Gé dyslíce dǽd gefremedon, werge wrǽcmǽcgas, Elen. Kmbl. 773 ; El. 387. Werige, Andr. Kmbl. 1229; An. 615. Fealleþ ðé on ða wynstran wergra þúsend, Ps. Th. 90, 7. Ðú mé áweredest wyrigra gemôtes *protexisti me a conventu malignantium*, 63, 2. Werigra, Cd. Th. 232, 30 ; Dan. 268. Werigum wrôhtsmiðum, Andr. Kmbl. 171 ; An. 86. Hé gelǽdde wǽrge weorod *adducto maligno exercitu*, Bd. 4, 12 ; S. 580, 40. (2) of evil spirits :—Ðú (*the serpent*) scealt werg dínum breóstum bearm tredan bråd[r]e eorðan, Cd. Th. 56, 3 ; Gen. 906. Se werga gǽst, Exon. Th. 129, 16 ; Gú. 422. Se werga, 268, 8 ; Jul. 429. Sió werge sceolu (*the fallen angels*), Elen. Kmbl. 1523 ; El. 763. Se weriga gǽst *serpens*, Bd. 1, 27 ; S. 497, 14 : *malignus spiritus*, 497, 19, 26. Se weria feónd *hostis malignus*, 3, 19 ; S. 549, 4. Hafaþ werges bleó, Exon. Th. 95, 31 ; Cri.

Weriges, Andr. Kmbl. 2340 ; An. 1171. Lást wergan gǽstes (*Grendel*), Beo. Th. 266 ; B. 133. Wergan gǽstes *the devil's*, 3499; B. 1747. Ðǽm wergan gǽste wiþstondan, Blickl. Homl. 135, 11. Werigan, Cd. Th. 309, 17 ; Sat. 711. Wið ðone wergan gǽst, Exon. Th. 373, 30; Seel. 117. Weregan, Cd. Th. 306, 24 ; Sat. 669. Hí sculon werge wiðra wrǽce þrowian, Exon. Th. 455, 29 ; Hy. 4, 57. Werige, Cd. Th. 6, 18 ; Gen. 90: 304, 15 ; Sat. 630. Wergan gǽstas, Exon. Th. 23, 4 ; Cri. 363. Ða werigan gǽstas *spiritus maligni*, Bd. 3, 11 ; S. 536, 36, 40 : Cd. Th. 310, 23 ; Sat. 731. Manna cynn and eác werigra gǽsta, Blickl. Homl. 83, 12. (3) of things :—Ðone werigan sele *that accursed hall* (Hell), Cd. Th. 285, 4 ; Sat. 332. [*O. Sax.* warag (*applied to Judas*).] v. preceding word.

wearg-berende ; *adj. Villainous, rascally* :—Ða weargberendan *furcifera*, Wrt. Voc. ii. 38, 1.

wearg-brǽde (wearge- [wearg-ge- (?)], wearh-), an ; *f. Some form of disease* ; the word translates *impetigo, ulcus, carcinoma* :—Wearhbrǽde *impetigo*, Wrt. Voc. i. 43, 62. Weargebrǽde, ii. 45, 39 : *nevum*, 62, 29. Werhbrǽde, i. 61, 16. Gif hwylcum weargbrǽde (wearh-, MS. B. ; *the Latin has* ulcus) weaxe on ðam nosum oððe on ðam hleóre, Lchdm. i. 86, 1. Wið ðæt wearhbrǽde (*the Latin has* carcinomata) hwam on nosa wexe, 116, 11. Gif nægl sié of handa and wiþ wearhbrǽdan (*probably* πτερύγιον, Cockayne), nim hwǽtecorn, meng wið hunig, lege on þone finger, ii. 80, 20, 24.

wearg-cwedol, -cwidol; *adj. Given to evil speaking* or *cursing* :—Ðeáh ðe wyrigcwidole (wærgcweodole, Bd. M. 356, 26) Godes ríce gesittan ne magon, hwæþere is gelýfed ðæt ða ðe be gewyrhtum wyrgede wǽron for heora árleásnysse, ðæt hí hraðe ðurh Drihtnes wrǽc heora scylde wíte ðrowedon *quamvis maledici regnum Dei possidere non possint, creditum est tamen quod hi qui merito impietatis suae maledicebantur, ocius Domino vindice poenas sui reatus luerent*, Bd. 4, 26 ; S. 602, 11. Ðæt hí nó áfyrhte ðæt gewin ðæs síþfætes ne wyrigcwydolra (wyrgcweodulra, Bd. M. 56, 14) manna tungan ne brégde *nec labor vos itineris nec maledicorum hominum linguae deterreant*, 1, 23 ; S. 486, 1.

wearg-cwedolian ; *p.* ode *To curse, speak evil* :—Wergcweoðelade mec *maledixit me*, Ps. Surt. ii. p. 183, 27. Gif feónd mín wergcweoðelade mé *si inimicus meus maledixisset mihi*, Ps. Surt. 54, 13.

wearg-cwedolness, e ; *f. Cursing* :—Lufade wergcweodulnisse *dilexit maledictionem*, Ps. Surt. 108, 18.

wearg-cweþan ; *p.* -cwæþ, *pl.* -cwǽdon *To curse* :—Wergcweoðaþ *maledicent*, Ps. Surt. 108, 28. Wergcweódon *maledicebant*, 61, 5. Wercweoðende *maledicentes*, 36, 22.

wearg-líc (werig-); *adj. Vile, mean, wretched* :—Sint ðæt werilíce welan ðisses middangeardes, ðonne hí nán mon fullíce habban ne mæg, ne hié nánne mon geweligian ne magon, búton hié óþerne gedôn tó wǽdlan *O! igitur angustas, inopesque divitias, quae nec habere totas pluribus licet, et ad quemlibet sine ceterorum paupertate non veniunt*, Bt. 13 ; Fox 38, 36. v. next word.

wearglíce ; *adv. Vilely, meanly, wretchedly* :—Gif ðú ðé wilt dôn manegra beteran and weorþran, ðonne scealt ðú ðé lǽtan ánes wyrsan. Hú ne is ðæt sum dǽl ermþa, ðæt mon swá wærelíce (werelíce, *v. l.*) scyle culpian tó ðam ðe him gifan scyle *qui praeire ceteros honore cupis, poscendi humilitate vilesces*, Bt. 32, 1 ; Fox 114, 15. v. preceding word.

weargness (werg-, werig-, wirig-, wyrig-), e ; *f. Evil* :—Wel mæg ðæm dæg werignise his *sufficit diei malitia sua*, Mt. Kmbl. Lind. 6, 34. Feala wyrgnessa wráde feóndas dínum ðam hálgum hefige brohtan *quanta malignatus est inimicus in sanctis*, Ps. Th. 73, 4. v. wearg, wirgness a curse.

weargol ; *adj. Evil* :—Ðis is seó wyrt ðe wergulu (*the crab apple* ; pirus malus, Cockayne) hátte, Lchdm. iii. 34, 14.

weargolness, e ; *f. A curse* :—Ic syngede swíðe þurh áðsware and þurh wærgolnesse *ego peccavi nimis per juramentum et maledictiones*, Confess. Peccat.

wearg-rôd, e ; *f. A gallows, gibbet* :—Waergrood *furcimen*, Txts. 65, 930. Uuergród, uaergród *furca*, 62, 409. Wearhród, Wrt. Voc. ii. 36, 68 : 70, 24 : 152, 1 : *eculeus* vel *catasta*, i. 55, 52. We[rg]ród *catasta*, ii. 22, 23. Of ðam þorne on ða wærhróda ; of ðam ródun, Cod. Dip. Kmbl. v. 345, 5. v. wearge-treów.

wearg-trǽf, es ; *m. A house of the accursed* :—Of ðam weahtreafum ic áwecce wið ðé óðerne cyning *from the tents of the accursed* (hell) *I will raise up against thee another king*, Elen. Kmbl. 1850 ; El. 927.

wearg-treów, es ; *n. The accursed tree, a gallows, gibbet, cross* :—Tó ðe waritreo, Cod. Dip. Kmbl. iii. 375, 25. [Nu raise þai up þe rode ; setis up þe warhtreo, O. E. Homl. 283, 9. Doð up and waritreo, þer on heo scullen winden (hongy, 2nd MS.), Laym. 5714. Me ledde him uorte hongen o waritreo, A. R. 122, 8. Let heom don adun of þe waritreo, Misc. 51, 491. *Icel.* varg-trê *a gallows*.] v. wearg-rôd.

wearh, -weariht. v. wearg,-wearriht.

wearm ; *adj. Warm* :—Swá swá ðæt cealde ǽrest onginð wlacian, ǽr hit fúl wearm weorðe, swá eác ðæt wearme wlacaþ, ǽr hit eallunga ácealdige *sicut a frigore per teporem transitur ad calorem, ita a calore*

per teporem reditur ad frigus, Past. 58 ; Swt. 447, 5. Wedercondel wearm *the sun*, Exon. Th. 210, 17 ; Ph. 187 : 179, 25 ; Gū. 1267. Sié lyft is ǽgđer ge ceald ge wǽt ge wearm, Bt. 33, 4 ; Fox 128, 36. On sumera hit biþ wearm, 21 ; Fox 74, 23 : Exon. Th. 340, 19 ; Gn. Ex. 113. Wearm weder, 198, 30 ; Ph. 18. Đeáh đē wel lyste wearmes mustes, Bt. 5, 2 ; Fox 10, 32. For đære wearmau *pro aprico*, Wrt. Voc. ii. 91, 62 : 9, 23. Swā weax melteþ, gif hit byđ wearmum neáh fŷre gefæstnad, Ps. Th. 57, 7. Wring on wermōd wearmne, Lchdm. ii. 310, 10. Đa sceolon beón wearme *offerrent eam calidam*, Lev. 6, 21. Wearme wederdagas, Exon. Th. 191, 30 ; Az. 96. Sumor æfter cymeþ, wearm gewideru, Met. 11, 61. Wearme gewyderu, Menol. Fox 177 ; Men. 90. [*O. Sax. O. Frs.* warm : *O. H. Ger.* warm (waram) *calidus, apricus : Icel.* varmr.] v. cū-wearm.

wearme ; *adv. Warmly :*—Genim þreó snǽda, gerest æfter wearme *take three slices, go to bed afterwards and keep warm*, Lchdm. ii. 52, 23. Bewreóh đē wearme *wrap yourself up warmly*, 116, 20 : 118, 10. Be-binde þonne genōh wearme, 270, 9. Beþe đæt heáfod swā wearme *use as warm fomentations as possible for the head*, 154, 18.

wearmian ; *p. ode To get warm :*—Ic wearmige *caleo*, Ælfc. Gr. 26, 2 ; Zup. 154, 3. *Caleo* ic wearmige and of đam *calesco* ic onginne tō wearm-igenne, 35 ; Zup. 212, 2. Gif wund ācólod sý . . . lege on đa wunda, heó cwicaþ sōna and wearmaþ, Lchdm. i. 194, 26. Wyrta wearmiaþ, Exon. Th. 212, 20 ; Ph. 213. Wearrhode ł gehǽt wæs ł āhātode heorte mīn *concaluit cor meum*, Ps. Lamb. 38, 4. Hí (*the clothes which he wore while standing in the river*) on his līchaman wearmodon, Homl. Th. ii. 354, 20. Se đe nyle đæt wlæce oferwinnan and wearmian ōđ hē wealle *quisquis nequaquam tepore superato excrescit, ut ferveat*, Past. 58 ; Swt. 447, 7. Se cealda đencđ tō wearmianne, 447, 17. v. ge-wearmian ; wirman.

wearm-líc ; *adj. Warm :*—Wearmlíc wolcna scūr *the warm rain from the clouds*, Cd. Th. 238, 5 ; Dan. 350.

wearmness, e ; *f. Warmness, warmth :*—Hē wolde hine bađian on þam wlacum wætere, ac hē gewāt sōna swā hē đæt wæter hrepode, and wearđ seó wearmnys him áwend tō deáđe, Homl. Skt. i. 11, 160.

wearn, es ; *m.* (?) *A multitude, a great number* or *quantity, a great deal :*—Þunie (þu me, Th.) him gewinnes wearn ofer wealles hróf and heom on midle wese mān and inwit *circumdabit eam super muros ejus iniquitas, et labor in medio ejus*, Ps. Th. 54, 9. Þeáh đe đa ealle đe mē āfeódon wordum wyrigen and wearn sprecan *si is, qui oderat me, super me magna locutus est*, 54, 12. Hió innwit feala ŷwdan on tungan, and mē wrāđra wearn worda sprǽcon *locuti sunt adversum me lingua dolosa, et sermonibus odii circumdederunt me*, 108, 2. Ic on unriht oft lōcade and wiđercwyda wearn gehŷrde *vidi iniquitatem et contradictionem*, 54, 8. Hí his wundra wearn gesāwon on wætergrundum *ipsi viderunt mirabilia ejus in profundo*, 106, 23. Þeáh đe eów wealan tō wearnum flówen *divitiae si affluant*, 61, 11. Hē synfulle tōdrīfeþ wearnum ealle *omnes peccatores disperdet*, 144, 20. Ful oft mon wearnum (*or from* wearn ; *f.*) tīhđ eargne đæt hē elne forleóse *full often the coward is freely* (or *with difficulty*) *accused of losing his courage*, Exon. Th. 345, 13 ; Gn. Ex. 187. v. wearn-mǽlum, and cf. worn.

wearn, e ; *f.*   **I.** *a hindrance, obstacle, difficulty.* v. wearn-wíslíce :—Wearne ł remmincge *obstaculo, impedimento*, Hpt. Gl. 455, 48. Đæt mód hæfđ fulfremedne willan tō đære wrænnesse būtan ælcre steóre and wearne *ejus animus voluptate luxuriae sine ullo repugnationis obstaculo delectatur*, Past. 11 ; Swt. 73, 8. Gif hē geþyldelíce forbyrđ ǽgđer ge hosp ge edwītu and on đære wearne þurhwunaþ þeáh and eádmódlíce bit, đæt him mon inféres tīþige, sý hē underfangen *si veniens perseveraverit pulsans, et inlatas sibi injurias et difficultatem ingressus visus fuerit patienter portare et persistere petitioni sue, annuatur ei ingressus*, R. Ben. 97, 7.   **II.** *a refusal.* v. wirnan :—Hŷ bēnan synt đæt hié wiđ đē mōton wordum wrixlan, nō đū him wearne geteóh đīnra gegncwida *they are petitioners that they may exchange words with thee, give them not a refusal of thy words in reply*, Beo. Th. 738 ; B. 366. [*Icel.* vörn *a defence.*] v. un-wearnum.

wearnian, wearnung. v. warenian, warenung.

wearn-mǽlum ; *adv. In flocks, in crowds :*—Wearnmélum *gregatim*, Wrt. Voc. ii. 110, 9.

wearn-wíslíce ; *adv. With difficulty :*—Wearnwíslíce *difficile*, Wrt. Voc. ii. 106, 47 : 25, 53.

wearoþ. v. waroþ.

wearp, es ; *n.*   **I.** *the warp, thread stretched lengthwise in a loom :*—Wearp *stamen*, Wrt. Voc. ii. 121, 34 : i. 59, 32 : 66, 21 : 282, 4. Línen wearp *linostema*, 40, 8. Be cembum wearpe *de stuppe stamineo* (*de stuppae stamine*, Ald. 51 and v. Hpt. Gl. 494, 1), ii. 83, 15 : 26, 62. Of wearpe *de stamine*, Hpt. Gl. 494, 1. Wundene mē hē beóđ wefle, ne ic wearp (uarp, Txts. 151, 5) hafu, Exon. Th. 417, 16 ; Rä. 36, 5. Wyllene wearp *lanea stamina*, Hpt. Gl. 417, 28. Wearpum *stamina*, 430, 74.   **II.** *a pliant twig* that may be used in basket-making. v. wearp-fæt :—Wearp *vimen*, Wrt. Voc. ii. 123, 73. [Warp, threde for webbynge *stamen, licium*, Prompt. Parv. 517. *O. H. Ger.* warf, waraf *stamen : Icel.* varp *a casting*.]

wearp-fæt, es ; *n. A wicker-basket :*—Corbis vel *cofinus* wylige, *spor-tella* tænel, *cartallum* windel, *calathus* (cf. wearp, II, and Ovid : Calathos e vimine textos) wearpfæt, Wrt. Voc. i. 86, 2-5 : 40, 42. [A warpe-fatte *alveolus*, Cath. Angl. 409.]

wearr, es ; *m. A piece of hard skin* (particularly on the hands or feet), *callosity :*—Wear *callus*, Wrt. Voc. ii. 14, 12. War, i. 291, 8. Wær *callositas*, Hpt. Gl. 490, 33. Đa wearras and đa swylas đe beóđ on mannes handum ođđe on ōđrum limum, Lchdm. i. 356, 16. Wiþ weartum and wearrum on lime, ii. 148, 26 : Homl. Skt. i. 5, 139. Fram þysum heardum wearrum, 5, 198. Weorras vel ill *callos*, Txts. 49, 400. Uarras, 111, 13 : *callos, tensam cutem*, 114, 93. Wearras, ilas *callos*, Wrt. Voc. ii. 13, 48 : *calces*, 127, 45. Wiþ wearras and wiþ swylas, Lchdm. i. 356, 11. Wearras and weartan on weg tō dōnne, 362, 17 : ii. 150, 1. [Warre or knobbe of a tre *vertex*, Prompt. Parv. 516, and see note.]

wearr, es ; *m. A cup, bowl :*—Clæfran seáwes .ii. lytle bollan fulle mid lytle hunige gemengde, dō wear fulne gehættes wínes tō, sele drincan þrý dagas, Lchdm. ii. 214, 12.

wearrig ; *adj. Callous :*—Hē gelōme đingode for đæs folces gyltum, bígende his cneówu on gebedum symle, swā đæt him weóxon wearrige ylas, on olfendes gelícnysse, on his līđegum cneówum, Homl. Th. ii. 298, 26.

wearriht ; *adj.*   **I.** of living beings, *having hard skin, leprous :*—Wærrehte ł hreóflige *elephantinosa, leprosa* (elephantinosa corporis incommoditas, Ald. 28), Hpt. Gl. 455, 35. Hreófe ođđe wearrihtum *callosi* (corpore calloso venere leprosi, Ald. 175), Wrt. Voc. ii. 93, 72 : 19, 53. Đa wearrihtan *callosa* (calloso corpore lepram, Ald. 201), 96, 6 : 20, 2. Wearihte *callosa*, 127, 53.   **II.** of trees, *gnarled, knotted :*—On đonæ wearrihtan stocc, Cod. Dip. Kmbl. iii. 176, 4 : v. 221, 4. In đa wæriht āc ; of đæt wærriht āc, iii. 390, 16. v. wearr.

wearrihtness, e ; *f. Hardness of skin, roughness of skin* as in leprosy :—Rūh wærihtnys *callositas*, wearrihtnys, rūh wærihtnys *scabredo* (leprosi, quos dira cutis callositas elephantino tabo deturpans, Ald. 49), Hpt. Gl. 490, 33-36. Unsmēđnes ođđe wearrihtnes *callositas*, Wrt. Voc. ii. 18, 36. Wearihtnes, 127, 54.

wearte, an ; weart(?), e ; *f. A wart :*—Uearte, uuertae, uaertae *berruca*, Txts. 45, 288. Wearte, Wrt. Voc. ii. 11, 4 : 126, 2. Wearte, uueartae, uearte *papula*, Txts. 83, 1485. Wearte, Wrt. Voc. i. 288, 73 : ii. 67, 57. Wearte, uueartae, uuertae *verruca*, Txts. 105, 2088. Wearte *verruca* . . . weartena (-e, MS.) heáp *satiriasis*, Wrt. Voc. i. 20, 7, 9. Wearte vel bŷl *furunculus*, ii. 151, 75. Wearte (*pl.* ?), bŷle *fruncúlas* (*-us* ?), 151, 34. Wiđ weartan, genim đysse wyrte meolc, dō tō đære weartan, hit đa weartan gehǽleþ, Lchdm. i. 224, 6-8 : 130, 20-21. Wiþ weartum . . . dō on đa weartan, ii. 148, 26 : 322, 12. Wiþ weartan . . . lege tō đam weartan, hē hý fornimeþ, i. 256, 1-2. Wearras and weartan on weg tō dōnne . . . wriđ on đa weartan and on đa wearras, 362, 17. Wiđ scurfedum nægle, nim gecyrnadne sticcan, sete on đone nægl wiđ đa wearta (-an ?), ii. 150, 5. [*O. H. Ger.* warta ; *f. verruca, papilla* (the word has both strong and weak forms) : *Icel.* varta *a wart*.]

weás ; *adv. By chance, by accident, fortuitously :*—Weás *casu*, Txts. 181, 54. Ic his wundrode micle đŷ læs, gif ic wiste đæt hit weás gebyrede būton Godes willan and būton his gewitnesse *minus mirarer, si misceri omnia fortuitis casibus crederem*, Bt. 39, 2 ; Fox 212, 32 : 214, 6 : 39, 3 ; Fox 216, 3 : Met. 28, 72. Witan hwæt wyrd sié, and hwæt weás ge-byrige *de fati serie, de repentinis casibus quaeri*, Bt. 39, 4 ; Fox 216, 30. Ic wolde witan hwæþer đæt áiunh sié đæt wē oft gehióraþ đæt men cweþaþ be sumum þingum đæt hit scyle weás gebyrian. . . . Hit nis náuht đæt mon cwiþ đæt ǽnig đing weás gebyrige ; for đam ælc þing cymþ of sumum đingum, for đŷ hit ne biþ weás gebyred ; ac đǽr hit of náuhte ne cóme đonne wǽre hit weás gebyred *quaero an esse aliquid omnino, et quidnam esse casum arbitrere. . . . Nihil est, quod vel casus, vel fortuitum jure appellari queat*, 40, 5 ; Fox 240, 13-30. Men cwǽdon đonne him hwæt unwēnunga gebyrede, đæt đæt wǽre weás gebyrede *quoties aliquid cujuspiam rei gratia geritur, aliudque quibusdam de causis, quam quod intendebatur, obtingit, casus vocatur*, 40, 6 ; Fox 242, 5, 9. Gif him weás gebyreþ, đæt him wyrþ sume hwíle đara þènunga oftohen, 37, 1 ; Fox 186, 13 : Met. 25, 31. Gif him weás (wealdes, Hatt. MS.) gebyrige ođđe usingewealdes, đæt hē on đæs hwæt befoo, đe wiđ his willan sié *siquando contra eos lingua labitur*, Past. 28 ; Swt. 198, 22.

weascing. v. wæscing.

weás-gelimp, es ; *n. What happens by chance, accident, chance :*—Mid weásgelimpe *fortuitu*, Wrt. Voc. ii. 34, 35.

weá-spell, es ; *n. A tale of woe :*—Æfter weáspelle (*the news of Æschere's death*), Beo. Th. 2634 ; B. 1315.

weá-tácn, es ; *n. A sign of misery, a woeful signal :*—Nis þær on đam londe, ne wōp ne wracu, weátácen nán, yldu ne yrmđu, Exon. Th. 201, 5 ; Ph. 51. Wæs đæt weátácen geond đa burh bodad, đæt hié đæs cnihtes cwealm gesōhton, Andr. Kmbl. 2239 ; An. 1121.

weá-þearf, e ; *f. Grievous need :*—Ic mē fēran gewāt folgađ sēcan, wineleás wrǽcca, for mínre weáþearfe, Exon. Th. 442, 10 ; Kl. 10.

weax, es ; *n. Wax :*—Weax *cera*, Wrt. Voc. i. 81, 33 : *cerea*, 284, 32.

Âsoden weax *obrizum metallum*, ii, 65, 14. Swâ weax melteþ, gif hit byð wearmum neáh fýre gefæstnad *sicut cera liquefacta*, Ps. Th. 57, 7: 67, 2: Exon. Th. 61, 23; Cri. 989. Swâ swâ eles gecynd bið ðæt hé beorhtor scîneþ þonne wex on sceafte, Blickl. Homl. 129, 1. Ða fótlâstas wæron swutole, swâ hié on wexe wæron âðýde, 205, 1. God hét wæpen wera wexe gelîcost formeltan, Andr. Kmbl. 2292; An. 1147. Mon ðæt weax âgæfe tô cirican, Cod. Dip. Kmbl. i. 293, 20. Ontend .iii. candella, drýp ðæt weax, Lchdm. i. 392, 11. On gemelt weax gedôn, ii. 72, 7. Ic gefrægn weax (*dough?*) nât hwæt þindan and þunian, Exon. Th. 431, 16; Rä. 46, 1. [*O. L. Ger. O. H. Ger.* wahs: *O. Frs.* wax: *Icel.* vax.]

weax-æppel, es; *m. A wax apple, a ball of wax:*—Se Pater Noster mæg âna eala gesceafta on his ðære swîðran hand on ânes weaxæpples onlicnisse gedýn and gewringan, Salm. Kmbl. p. 150, 33.

weaxan, weacsan, weahsan, weahxan, wexan, wehsan; ic weaxe; ðû wyxt; hé weaxeþ, weaxþ, weaxt, waexit, weaxst, wexeþ, wexþ, wixt, wihst, wihxþ, wyxþ, wyxt, wyxst, wycxþ; *p.* weóx, weócs, weóhs, *pl.* weóxon, weóhson, weóxson; *pp.* weaxen *To wax, grow.* **I.** glossing the following Latin words:—Ic weaxe *glesco*, weaxeþ *glescit*, Wrt. Voc. ii. 41, 60, 57. Weaxð *gliscit*, Hymn. Surt. 132, 6. Waexit *surgit*, Txts. 99, 1955. Weacsaþ *pullulant*, Kent. Gl. 1163. Weóx *matureseceret*, Wrt. Voc. ii. 90, 40: *floruerit*, Hpt. Gl. 460, 63: *pollesceret*, 466, 59. Wehsan *crescere*, Wülck. Gl. 252, 39. Weaxende *pubescentem*, Wrt. Voc. ii. 82, 64: 66, 20. Wexende, Hpt. Gl. 491, 15: *crebrescens*, 499, 13. Mid wexendre *praepollente*, 459, 30. **II.** *to grow, be produced*, (1) of animals or plants:—Of ðam weaxeþ wyrm *hinc animal sine membris fertur oriri*, Exon. Th. 213, 29; Ph, 232. Deós wyrt wihst (cf. deós wyrt bið cenned, 96, 13, *and often*) on begânum landum, Lchdm. i. 94, 6. Rixe weaxst on wæterigum stôwum, Homl. Th. ii. 402, 9. Wexeþ, Runic pm. Kmbl. 342, 9; Rûn. 15. (2) of other things, (a) concrete:—Ðæt land ðær gold wixt *terra, ubi nascitur aurum*, Gen. 2, 11. Hwæðer gê nû sêcan gold on treówum? . . . Ealle men witon ðæt hit ðær ne weaxt, ðe mâ ðe gimmas weaxaþ on wîngeardum, Bt. 32, 3; Fox 118, 8–11. Wexð, Met. 19, 8. Him wyxþ wind on ðære heortan, Lchdm. ii. 60, 7. (b) abstract:—Of ðissum syx tîdum wihst se quadrans, Anglia viii. 298, 34. Of irsunge wyxt seófung, Prov. Kmbl. 23. Him on innan oferhygda dæl weaxeþ and wridaþ, Beo. Th. 3486; B. 1741. Of mistlicum dryncum onwæcxaþ (cf. weaxaþ, Met. 25, 40) sió wóde þrâg ðære wrænnesse. . . . Þonne weaxaþ (cf. þonan cymeþ, Met. 25, 43) ða ofermêtta and ungeþwærnes, Bt. 37, 1; Fox 186, 19. Seó gâlnyss weóhs on him, Hexam. 17; Norm. 26, 3. Him weóxon ofermêtto, Past. 17; Swt. 113, 6. Ðonne sceal eów sôna weaxan tô hearme wædl and wâwa, sacu and wracu, Wulfst. 133, 2. Hé hêht geond ðæt rædleáse hof weaxan wîtebrôgan, Cd. Th. 3, 33; Gen. 45. Ne sceolon unc betweónan teónan weaxan, 114, 11; Gen. 1902. **III.** of growth in animals or plants, *to grow, grow up:*—Hé (the phenix) on sceade weaxeþ, Exon. Th. 214, 5; Ph. 234. Þonne hit wyxð (wexeþ, Rush.), hit is ealra wyrta mæst *cum creverit, majus est omnibus holeribus*, Mt. Kmbl. 13, 32. Seó wyrt weóx, and ðone wæstm brôhte, 13, 26. Ðæt cild weóx and weard gewened, Gen. 21, 8: Cd. Th. 167, 25; Gen. 2771. Ðæt cild swîþe weócs, Jud. 13, 24. His feax weóx swâ swâ wîmmanna, Homl. Th. ii. 434, 8. Sumu hê cearf ðonne him dûhte ðæt hié tô swîðe weóxen (weóxsen, Hatt. MS.). . . Sumu hê leahte mid wætre, ðonne hié tô hwôn weóxon (weóxson, Hatt. MS.), Past. 40; Swt. 292, 5–8. Ða þornas weóxon (wôxon, Lind.: wêxon, Rush.), Mt. Kmbl. 13, 7. Swâ elebeámas weaxen, Ps. Th. 127, 4. Lætaþ ægþer weaxan (wexan, Rush.), Mt. Kmbl. 13, 30. Ðîne teóðan sceattas gongendes and weaxendes âgyf ðû Gode, L. Alf. 38; Th. i. 52, 32. **IV.** *to grow, increase, wax:*—Se môna dêð ægþer, ge wycxð ge wanaþ: healfum mônðe hê bið weaxende, healfum hê bið wanigende, Homl. Th. i. 154, 27. Ðes saltus lune wyxst wundorlîce æfter bôccræfte, Anglia viii. 308, 24. Gif ðæt ne wexð ðæt hié tiohhiaþ tô dônne, ðonne wanaþ ðæt ðæt hî ær dydon, Past. 58; Swt. 445, 8. Æghwylces lâreówes lâr wihst (wihxð, Hatt. MS.) ðurh his gedýlde, 33; Swt. 216, 1. Wesaþ and weaxaþ ealle werþeóde, lifgaþ bi ðam lissum, Exon. Th. 192, 30; Az. 113. Weóx and wridade mægburg Semes, Cd. Th. 102, 18; Gen. 1702. Seó âdl dæghwamlîce weóx, Bd. 4, 30; S. 609, 25: 5, 12; 627, 12. Weóx wæteres þrym, Andr. Kmbl. 3070; An. 1538. Æðelinge weóx word and wîsdôm, 1136; An. 568: 3351; An. 1679. Æðelinges weóx rîce, Elen. Kmbl. 24; El. 12. Windas weóxon, Andr. Kmbl. 745; An. 373. Wægas weóxon, 3088; An. 1547. Wex and beó gemænigfyld on þeóda and mægþa, Gen. 35, 11. Weahxaþ and beóþ gemenigfylde, 9, 1. Wexaþ, Cd. Th. 13, 1; Gen. 196. Weaxaþ, 92, 21; Gen. 1532. Weaxe sió bót be ðam were, L. Alf. pol. 11; Th. i. 70, 2: L. In. 76; Th. i. 150, 14. Gif ðû gesihst timbrian hûs ðîn, feoh ðîn wexan hit getácnaþ, Lchdm. iii. 214, 33. Sió gîtsung wile weahsan mid ungemete, Past. 11; Swt. 71, 16. Hê lêt weaxan heora rîmgetel, Cd. Th. 166, 28; Gen. 2754. Sceal weaxan wonna lêg, Beo. Th. 6221; B. 3115. Ne tæce wê nâ ðæt hê leahtras fyrðrige and weaxan (wehsan, *v. l.*) læte, ac ðæt hé hý simle wanige *non dicimus, ut permittat nutriri vitia sed ea amputet*, R. Ben. 121, 8. Gif sió âdl sié git weaxende, Lchdm. ii. 218, 1. Weaxende spêd, Cd. Th. 100, 7;

Gen. 1660. **IV a.** *to grow in honour, grow great, flourish, prosper:*—Ic gedô ðæt ðû wyxt *faciam te crescere*, Gen. 17, 6. Ðes middangeard wanaþ and weaxeþ, Fragm. Kmbl. 60; Leas. 32. Hit gebyraþ ðæt hé weaxe and ðæt ic wanige, Jn. Skt. 3, 30. Þeáh hwâ wexe mid micelre æþelcundnesse his gebyrda, and þeó on eallum welum, Bt. 19; Fox 68, 30. Se hlîsa ðæt wære sum ancra, ðæt missenlîcum mægnum for Gode weóhse, Guthl. 12; Gdwin. 58, 14. **V.** *to be productive:*—Ær ðon eówre treówu telgum blôwe, wæstmum weaxe *priusquam producant spinae vestrae rhamnos*, Ps. Th. 57, 8. Hêr ys seó bót hû ðû meaht ðîne æceras bêtan, gif hî nellaþ wel wexan, Lchdm. i. 398, 2. Hê ða weaxendan wende eorðan on sealtne mersc *terram fructiferam in salsuginem*, Ps. Th. 106, 33. **VI.** *to grow, take shape:*—Hyre weaxan ongon under gyrdelse, ðæt oft gôde men mid feó bicgaþ, Exon. Th. 436, 21; Rä. 55, 10. [*Goth.* wahsjan: *O. L. Ger. O. H. Ger.* wahsan: *O. Frs.* waxa: *Icel.* vaxa.] v. â-, be-, for-, forþ-, ge-, ofer-, under-weaxan; ful-, un-weaxen.

weax-berende *bearing a wax candle*; the word (in the form *uæx biorende*) glosses *cerarius* in the passage: Accoluthus grece, cerarius ad recitandum evangelium (cf. *Acolitus* is gecweden se ðe candele oððe tapor byreþ þonne mann godspell ræt, Ælfc. C. 14; Th. ii. 348, 4), Rtl. 195, 16.

weax-bred, es; *n.* **I.** *a table, tablet* for writing on:—Ðâ wrât hê gebedenum wexbrede (wæx-, Lind.) *postulans pugilarem scribsit*, Lk. Skt. 1, 63. Sýn gesealde from ðæm abbode ealle neádbehêfe þing, ðæt is . . . græf, . . . weaxbreda *dentur ab abbate omnia quae sunt necessaria, id est . . . gravium, . . . tabule*, R. Ben. 92, 4. God âwrât ða ealdan æ on ðâm stænenum weaxbredum. . . . Ða stænenan weaxbredu getácnodon ðæra Iudêiscra manna heardheortnysse, Homl. Th. ii. 204, 1–13. Wexbredu, 196, 32. Wexbreda *tabulas*, Ex. 31, 18. Ne bôc, ne weaxbreda, ne græf, R. Ben. 56, 20. Ðonne ðû græf habban wille, ðonne sete ðû ðîne þrî fingras tôsomne, swilce ðû græf hæbbe, and styra ðîne fingras swilce ðû wrîte. Gyf ðû gehwæde wæxbreda habban wille, ðonne strece ðû ðîne twâ handa, and sete hý neoþan tôsomne and feald tôgædere and feald tôgædere swilce ðû weaxbreda fealde. Ðonne ðû micel weaxbreda habban wille . . . , Techm. ii. 128, 6–12. **II.** *a table, list:*—Seó forme abecede ys bûtan pricon, and seó óðer ys gepricod on ða swýðran healfe, and seó þrydde on ða wynstran healfe, swâ ûs hêr æfter gelustfullaþ tô âmearkianne on þissum æfterfyligendum wexbredum, ðe se ârwurða Bêda gesette, Anglia viii. 332, 45. [God wrate þas lage in stanene waxbredene, O. E. Homl. i. 235, 27. Cf. *O. H. Ger.* wahs-tavala *tabula: Icel.* vax-spjald.]

weax-candel[1], e; *f. A wax candle:*—Waexcondel *funalia, cerei*, Wrt. Voc. ii. 109, 45. Weaxcandel, 36, 26. Wexcandel *cereus*, 130, 16: *funalia*, i. *candelabra*, 151, 56. Genim âcmela and beolonan sæd and weax, meng tôsomne, wyrc tô weaxcandelle, and bærn, Lchdm. ii. 50, 18.

weax-georn; *adj. Eager to grow (?), eating much with the desire of growing (?):*—Swîþe waxgeorn eart ðû (*the boy*) ðonne ðû ealle þingc etst ðe ðê tôforan gesette synd *valde edax es, cum omnia manducas quae tibi apponuntur*, Coll. Monast. Th. 34, 31.

weax-gescot, es; *n. A contribution of wax*, due to a church:—Swâ hwæt swâ witan tô ðearfe geræðan, hwilum weaxgescot, Wulfst. 171, 1. [*O. Frs.* wax-skot, -schot. Cf. *Icel.* vax-tollr *a tithe in wax*, payable to a church. See Grimm R. A. 315.]

weax-hlâf, es; *m. A cake of wax:*—On weaxhlâfes wîsan on âled, Lchdm. ii. 46, 2. Dô ðonne weax on ðæt ele ðætte ðæt eall weorðe tô hnescum weaxhlâfe, 234, 10: 82, 14. [*O. H. Ger.* wahs-leip *formella* (*formella cerae* circulus cereus, eadem origine qua caseus formella dicitur, quod nempe in forma struatur, Migne).]

weaxhlâf-sealf, e; *f. A salve consisting of a cake of wax:*—Wið weaxhlâfsealfe gemeng, Lchdm. ii. 246, 9. v. weax-sealf.

weaxness, e; *f. Growth, increase, waxing:*—Gyf man mête ðæt hê his hûs timbrie, ðæt byð his weaxnes (cf. 214, 33), Lchdm. iii. 170, 12. Ðonne ðæs sæes flôdes weaxnes biþ *quando rheuma oceani in cremento est*, Bd. 5, 3; S. 616, 16. Hê mycle wonunge and æwyrdlan wæs wyrcende ðære mærwan cyrican weaxnesse *magno tenellis ecclesiae crementis detrimento fuit*, 2, 5; S. 506, 38. v. ge-weaxness.

weax-sealf, e; *f. A salve made of wax:*—Wexsealf *cerotum, unguentum de cera*, Wrt. Voc. ii. 130, 41. Weaxsealf wiþ wyrme; weaxsealf: butere, pipor, hwît sealt, meng tôsomne, smire mid, Lchdm. ii. 124, 11.

weaxung, e; *f.* **I.** *waxing, growing, increase:*—Ðonne se môna beó týn nihta eald, and nâ ðænne his leóht beó ærest on weaxunge, Anglia viii. 323, 5. Nû hæfð se eádiga wer ûs geopenod ymbe ðæs saltus weaxunge, 308, 40. **II.** *increase* of prosperity:—Eormas strange habban wexinge hit getácnaþ, Lchdm. iii. 198, 32. On hûse his offrian wexingce oððe blisse hit getácnaþ, 202, 21: 210, 4.

web(b), es; *n. A web, woven stuff:*—Web *telum*, webb, uueb *textrina*, Txts. 101, 2004, 2005. Web *textrina*, *telum*, Wrt. Voc. i. 281, 72: *textrina*, 66, 9: *tela vel peplum*, 82, 5: *peblum*, 59, 30. Lang web *tela*, 59, 20. Webb byþ gefylled mid þrædum *tela consummatur filis*, Scint.

216, 2. Webbes *pepli*, Hpt. Gl. 459, 26. Goldfág scinon web æfter wágum *shot with gold shone the work of the loom along the walls*, Beo. Th. 1994; B. 995. Webbum *peplis*, Hpt. Gl. 507, 12. Webbu swá hwilc swá wyfð, and blisse gesihð, gód ærende getácnaþ, Lchdm. iii. 210, 28. [O. Sax. webbe: O. H. Ger. weppi *tela, lodix: Icel.* vefr; *m.*] v. god- (gode-) web, á-, ó-web.

webba, an; *m. A weaver*:—Webba *textor*, Wrt. Voc. i. 59, 48. Hér kýð on ðissere béc ðæt Willelm cwæð saccles Wulwærd ðane webba, Chart. Th. 648, 3. [The webbes ant the fullares (*of Flanders*), P. S. 188, 14. *Chauc.* webbe: *Piers P.* webbe *a (female) weaver.*]

webbe, an; *f. A female weaver.* v. freodu-webbe, *and see preceding word.*

web-beám, es; *m.* I. *a weaver's beam*:—Lorh *vel* webbeám *liciatorium*, Wrt. Voc. i. 59, 19: 281, 73. II. *the treadle of a loom*:—Webbeámas *insubula*, 59, 43: *insubuli*, ii. 49, 56. [A webbeme *laciatorium*, Wrt. Voc. i. 218, 3 (15th cent.). O. H. Ger. weppi-boum *liciatorium.*] Cf. web-sceaft.

webbestre, an; *f. A female weaver*:—Webbestre *textrix*, Wrt. Voc. i. 59, 49. [Webstere *texens*, Wick Job 7, 6. Webstere *textor*, Wülck. Gl. 629, 1: 652, 23. Webster, 685, 29: *textrix*, 692, 26: 795, 8. Webstar *textor, textrix*, Prompt. Parv. 519 (all 15th cent. glossaries).]

webbian; *p.* ode To weave, contrive :—Hé wróht webbade, Andr. Kmbl. 1343; An. 672. Gé inwitþancum wróht webbedon, Elen. Kmbl. 617; El. 309. Ne beó inwit tó leóf, ne wróhtas tó webgenne, ne seaio tó rénigenne, Blickl. Homl. 109, 29. [Webboð or webbe clothe of lynnyne *linifico*, webboñ clothe of wulle *lanifico*, Prompt. Parv. 519.] v. webbung.

webbung, e; *f. A spectacle*:—Uuebung *scena*, Wrt. Voc. ii. 120, 13. Gereónedes geltes wæbbunge Arsenius geypte *concinnati sceleris scenam Arsenius prodidit (ostendit)*, Hpt. Gl. 474, 65. Cf. wafian, wafung, *and* cpds. of wæfer-.

webbung, e; *f. A weaving, contriving, plot*:—Webbung (*printed* hwebbund) *conspiratio, conjuratio*, Hpt. Gl. 476, 20. [Webbynge of wullyne clothe *lanificium*, webbynge of lynnyne *linificium*, Prompt. Parv. 519.] v. webbian.

web-geréþru (-o)? The word occurs in lists of terms connected with weaving, and glosses *tala, tara*:—Webgeréþro *tala*, Wrt. Voc. i. 282, 9. Webgeréþru *tara*, 59, 45: 66, 26. v. next word.

web-geródes *glosses* tala, Wrt. Voc. ii. 122, 9. v. preceding word.

web-geweorc, es; *n. Weaving*:—Hió (*the Virgin Mary*) on hyre mægdenháde dyde fela wundra on webgeweorce, Shrn. 127, 16. Heó wolde beón fram ðære þriddan tíde óð ða nigoþan tíd ymbe hyre webbgeweorc, Homl. Ass. 127, 348.

webgian. v. webbian.

web-hóc, es; *m. Some implement used in weaving, a tenter-hook* (?):—Webhóc *apidiscus*, Wrt. Voc. i. 59, 41: 66, 24: 282, 7: ii. 7, 70.

web-líc; *adj. Of weaving*:—Weblíc gewurc *textrinum opus*, Hpt. Gl. 431, 4. Ðæt weblíce *textrinum*, Wrt. Voc. ii. 77, 17.

web-sceaft, es; *m. A weaver's beam*:—Websceaft *liciatorium*, Wrt. Voc. i. 66, 10. Cf. web-beám.

web-tawa *thread for weaving*:—Webtawa *linea*, Wrt. Voc. ii. 51, 11. Cf. next word.

web-teáh, -teág, e; *f. Thread for weaving*:—Waebtaeg *linea*, Wrt. Voc. ii. 113, 4.

webung. v. webbung.

web-wyrhta, an; *m. A fuller*:—Webwyrhta *fullo*, Wülck. Gl. 245, 33. Swylcne gerelan swylcne nǽnig fulwa, ðæt is nǽnig webwyrhta, ðæt mihte dón, Shrn. 56, 10. Ðone Iacóbum Iudǽa leorneras ofslógan mid webwyrhtan róde, 93, 12.

weccan; *p.* weahte, wehte; *pp.* weaht, weht To wake, waken. I. *to rouse* from sleep :—Geseh hé beornas swefan on slǽpe; hé sóna ongann wígend weccean, Andr. Kmbl. 1699; An. 852. I a. *to rouse* from the sleep of death :—Býman weccaþ of deáde eall monna cynn, Exon. Th. 55, 21; Cri. 887. Ic gé ðætte of slépe ic wecce hine, Jn. Skt. Rush. 11, 11. Ne húru wundur wyrceaþ deáde; oþþe hí lǽceas weccean *numquid mortuis facies mirabilia; aut medici suscitabunt?* Ps. Th. 87, 10. II. *to rouse* from unconsciousness or torpor, *to enliven, stimulate, refresh* :—Hé wehte hine wætre, Beo. Th. 5700; B. 2854. Ealdes mannes eágan beóþ unscearpsýno; þonne sceal hé ða eágan weccan mid gnídingum, Lchdm. ii. 30, 28. Seó wæs wætrum weaht and wæstmum þeaht, Cd. Th. 115, 19; Gen. 1922. III. *to rouse* from repose, *to excite, stir up* :—Se kok, ǽr ðam ðe hé cráwan wille, hefð up his fiðru, and wecð hine selfne, Past. 64; Swt. 461, 14. Drihten windas weccaþ *Dominus ventos excitat*, Bd. 4, 3; S. 569, 22. Bíþ sǽ smilte þonne hý wind ne weceþ, Exon. Th. 336, 27; Gn. Ex. 56. Ne bið ðé rest witod, ac ðec regna scúr weceþ and wreceþ, Cd. Th. 252, 11; Dan. 577. Windas weccaþ woruld mid storme, Exon. Th. 59, 13; Cri. 952. Nalles sceal hearpan swég wígend weccean, Beo. Th. 6040; B. 3024. IV. *to raise* what is depressed :—Hé of eorðan mǽg ðone unágan weccan *suscitans a terra inopem*, Ps. Th. 112, 6. V. *to give life to, to cause, give rise to, produce, raise* :—Feorheáccno cynn,

ða ðe flód wecceþ, Cd. Th. 13, 18; Gen. 204. Wyrd wóp wecccþ, Salm. Kmbl. 873; Sal. 436. Sunnan glǽm on lenctenne lífes tácen weceþ, Exon. Th. 215, 17; Ph. 255. Ðás windas and ðás regnas ða ðe eorþan wæstmas weccaþ, Blickl. Homl. 51, 21: Exon. Th. 38, 20; Cri. 609. Hí ǽled weccaþ *they kindle a fire*, 361, 18; Wal. 21. Wec ðú cléne hiortan in mé *cor mundum crea in me*, Ps. C. 50, 88. Ðæt his bróðor nime his wíf and his bróðor sǽd wecce (*resuscitet*), Mk. Skt. 12, 19. Wæcce, Mt. Kmbl. Rush. 22, 24. Unrǽd fremman, wefan and weccean, Cd. Th. 3, 5; Gen. 31: Beo. Th. 4098; B. 2046. Bǽlfýra mǽst weccan, 6279; B. 3144. Weccean, Cd. Th. 175, 26; Gen. 2901. [*Goth.* us-wakjan: O. H. Ger. wecchen: *Icel.* vekja.] v. á-, tó-weccan; wacan, waciau.

weccend, es; *m. One who rouses, incites*:—Weccend *incitator*, Germ. 393, 67.

wece-drenc, es; *m. An emetic*:—Wecedrenc . . . sele ðæt lytlum súpan . . . óþ ðæt hé spíwe, Lchdm. ii. 268, 31: 170, 8.

wecen. v. wæcen.

wecg, es; *m.* I. *a wedge*:—Waecg *cuneus*, Wrt. Voc. ii. 105, 70. Wecg, 15, 49: 137, 29. Treówes on óste nægel oððe wecg on tó fæstnigenne ys *arboris nodo clauus aut cuneus infigendus est*, Scint. 103, 10. II. *a mass of metal*:—Ælces cynnes wecg *vel* óra oððe clyna *metallum*, Wrt. Voc. i. 34, 67. Wecg *metallum, massa*, Hpt. Gl. 417, 20. Ðætte ðǽr wǽre ðæt héhste gód, ðǽr ðǽr ða gód ealle gegæderode bióþ, swelce hí síen tó ánum wecge gegoten, Bt. 34, 9; Fox 146, 20. Hí behwyrfdon heora áre on sumum gyldenum wecge, and ðone on sǽ áwurpan, Homl. Th. i. 60, 29. Berende on wecga órum, áres and ísernes, leádes and seolfres *venis metallorum, aeris, ferri, et plumbi, et argenti faecunda*, Bd. 1, 1; S. 473, 23. Seó eorðe is cennende wecga óran *terra parens metallorum*, Nar. 2, 15. On smǽtum goldórum ł (gold-?) wecgum *in obrizum auri metallum*, Hpt. Gl. 449, 14. Nis ná Godes wuninge on ðám grǽgum stánum, ne on ǽrenum wecgum, Homl. Skt. i. 7, 136. Lǽt ús ámyltan ða sylfrenan godas and eác swylce ða gyldenan, dǽlan siððan wǽdligum ða ámoltenan wæcgas, 5, 234. III. *a piece of money*:—Nim ðone ǽrestan fisc . . . ðú finst ǽnne wecg (*staterem*) on him, Mt. Kmbl. 17, 27: Homl. Th. i. 512, 4. [O. H. Ger. wecki *cuneus: Icel.* veggr.]

wecgan; *p.* de, ede To wag (trans.), *move, shake* :—Hwílum mec wonfeax wale wegeþ and þýð, Exon. Th. 393, 31; Rä. 13, 8: 403, 10; Rä. 22, 5. Hí wecgaþ heora heáfdu *moverunt caput*, Ps. Th. 21, 6. Wecggeaþ, 43, 16. Hwalas and heofonfuglas lyftlácende, ða ðe lagostreámas wecgaþ (cf. fiscas and fuglas, ealle ða ðe onhrèraþ hreó wǽgas, Exon. Th. 194, 18; Az. 141), Cd. Th. 240, 19; Dan. 389. Hwý gé ǽfre scylen unrihtfióungum eówer mód drèfan, swá swá mereflódes ýþa hrèraþ íscalde sǽ, wecggaþ for winde (cf. swá swá ýþa for winde ða sǽ hrèraþ, Bt. 39, 1; Fox 210, 25), Met. 27, 4. Hig wegdan, hrèrdan heora heáfod *moverunt capita sua*, Ps. Th. 108, 25. Hí wegedon mec of earde, Exon. Th. 485, 30; Rä. 72, 5. Ðonne ðú antiphonariam habban wille, ðonne wege ðú ðíne swíþran hand, Techm. ii. 119, 3, 5, 10, *and often*. Wege ðú medemlíce ðín reáf mid ðínre handa, 119, 19: 120, 3. Tácn ys ðæt mon wecge his hand, 119, 7. Wæcge, 121, 9. Þeáh hit wecge (cf. ástyroð, Bt. 12; Fox 36, 19) wind, Met. 7, 35. [Swa þe hæzde wude þenne wind weieð hine, Laym. 20137. *Goth.* wagjan *agitare, movere: O. H. Ger.* wegen *agitare, movere, vibrare, quatere*.] v. á-wecgan; wagian, wegan.

-wéd. [Cf. O. H. Ger. wuoti *insania: Icel.* œði.] v. ge-wéd.

wed[d], es; *n.* I. *a pledge, what is given as security* :—Wed *vel* álǽned feoh *pignus*, gylden wed *vel* feoh *arra*, wed *vel* wedlác *arrabona vel arrabo*, Wrt. Voc. i. 21, 5–7. Wed *pignus*, ii. 82, 25. Þeós gerýnu is wedd and híw; Cristes líchama is sóðfæstnyss. Ðis wed wé healdaþ gerýnelíce óð ðæt wé becumon tó ðære sóðfæstnysse, and ðonne bið ðis wedd geendod, Homl. Th. ii. 272, 6–8. Hié onféngon fulwihte and freoðuwǽre, wuldres wedde, Andr. Kmbl. 3260; An. 1633. Ic ða wǽre gelǽste ðe ic ðé sealde frófre tó wedde, Cd. Th. 139, 13; Geu. 2309: 124, 29; Gen. 2070. Ða ylcan his dohter Criste tó gehálgianne ðam biscope tó wedde gesealde, ðæt hé ðæt gehát gelǽstan wolde *in pignus promissionis implendae, eandem filiam suam Christo consecrandam episcopo adsignavit*, Bd. 2, 9; S. 511, 39: Beo. Th. 5989; B. 2998. Gif man hrægl tó wedde selle, L. Alf. 36; Th. i. 52, 25. Gif hwá þeóf clǽnsian wylle, lecge án .c. tó wedde, L. Eth. iii. 7; Th. i. 296, 7. Se Hálga Gást wæs onsended tó wedde ðæs heofonlícan éþles, Blickl. Homl. 131, 14. Nafa ðú nánes þearfan wedd (*pignus*) mid ðé nihtlangne fyrst, Deut. 24, 12. Gif ðú wed nime æt ðínum nǽhstan *si pignus a proximo tuo acceperis*, Ex. 22, 26. Genime mon .vi. sciłł. weorð wed, L. In. 49; Th. i. 132, 13. Æt cynges spǽce lecge man .vi. healfmearc wedd, æt eorles .xii. óran wedd, L. Eth. iii. 12; Th. i. 296, 25–6. Heora ǽlc sylle .vi. healfmearc wedd, 3; Th. i. 294, 7. Wed undón *to redeem a pledge*, L. O. D. 1; Th. i. 352, 9. Wed *pignora*, Wrt. Voc. ii. 94, 20. I a. *a dowry*:—Wed, gifu *vel* fædren feoh dos, Wrt. Voc. ii. 141, 80. Mid wedde *dote*, 27, 18. I b. fig. :—Worda wed gesyllan (v. the same phrase in the passages from the laws), eallra unsnyttro ǽr gesprecen-a *to be responsible for all that has been said before*, Elen.

Kmbl. 2566; El. 1284. **II.** *a pledge, solemn promise, engagement, covenant, compact:*—Wed oððe wǣra *clasma,* Wrt. Voc. ii. 21, 2. Ða stǣnenan bredu, on ðām wæs ðæt wedd ðe Drihten wið eów gecwæð *tabulis pacti, quod pepigit vobiscum Dominus,* Deut. 9, 9. Ðis ys ðæt wedd (*pactum*), ðæt gē healdan sceolon betwux mē and eów, Gen. 17, 10. Ðis bið ðæt tācen mínes weddes *hoc signum foederis,* Gen. 9, 12, 13, 15. Se ðe ðæs weddes waldend sý, L. Edm. B. 6; Th. i. 254, 21. Beó mín wedd (*pactum*) on eówrum flǣsce on ēcum wedde (*in foedus aeternum*) . . . hē áídlode mín wedd (*pactum*), Gen. 17, 13–14. Hí mid wedde and mid āþum fryþ gefæstnodon, Chr. 926; Erl. 111, 44: 1016; Erl. 159, 4. Mid worde and mid wædde, 1014; Erl. 150, 14. Trymme hē eal mid wedde ðæt ðæt hē behāte, L. Edm. B. 5; Th. i. 254, 17. On (*in*) wedde[ge]syllan *to give on covenant, to engage to do:*—Ðā cwæð ic ðæt ic him wolde fylstan on ða gerāda ðæt hē his mē ūðe, and hē mē ðæt in wedde gesealde . . . Hē mē ða bóc āgeaf swā hē mē on ðon wedde ǣr geseald hæfde *then I said that I would help him on condition that he would make a grant of the land to me, and he engaged to do that . . . He gave me the deed, as he had before covenanted in the engagement,* Cod. Dip. Kmbl. ii. 134, 9–20. Hæfdon Eoforwícyngas hyre gehāten, and sume on wedde geseald, sume mid āþum gefæstnod, ðæt hí on hire rǣdinge beón woldon, Chr. 918; Erl. 105, 29: L. Edm. B. 1; Th. i. 254, 5. Hí sǣdon, and on wedde sealdon, hwæt hý hyre syllan woldon *they stated what they would give her, and engaged to pay it,* Homl. Ass. 196, 24. God behēt ús wedd *Deus pepigit nobiscum foedus,* Deut. 5, 2. Ic sette mín wedd tō ðē *ponam foedus meum tecum,* Gen. 6, 18. Ic sette mín wedd tō eów *ego statuam pactum meum vobiscum,* 9, 9. Hig slógon heora wedd ǣgðer tō ōðrum, ðæt hig ǣfre wurdon getrýnd *percusserunt ambo foedus,* 21, 27. Geþence hē word and wedd ðe hē Gode betǣhte, L. Eth. v. 5; Th. i. 306, 5. Sealde God his wedd Abrame *pepigit Dominus foedus cum Abram,* Gen. 15, 18. Uton syllan wedd *inemus foedus,* 31. 44: Chart. Th. 485, 37. Ðæt ða witan ealle sealdan heora wedd ðam arcebisceope, L. Ath. v. 10; Th. i. 238, 34: v. 8, 6; Th. i. 236, 35. Be āðum and be weddum. Ðæt ǣghwelc mon his āð and his wed wǣrlíce healde, L. Alf. pol. 1; Th. i. 60, 1–3: L. C. E. 19; Th. i. 372, 1: Wulfst. 113, 1. Hí wið ðone cyning hí getreówsoden, and binnan litlan fæce hit eall ālugon, ge wed ge āðas, Chr. 947; Th. 118, 14: L. In. 13; Th. i. 110, 12. Gif hwā his āð and his wæd brece, ðe eal þeód geseald hæið, L. Ed. 8; Th. i. 164, 2. Ðæt man āðas oððe wedd tōbrece, Chart. Erl. 231, 6. Gif gē cōð mín wedd for nāht *si ad irritum perducatis pactum meum,* Lev. 26, 15: Deut. 31, 16. Ǣlc geréfa nāme ðæt wedd on his āgenre scíre, L. Ath. v. 10; Th. i. 240, 1: v. 11; Th. i. 240, 15. Ða āðas and ða wedd and ða borgas synt ealle oferhafene and ābrocene, L. Ath. iv. proem.; Th. i. 220, 14. Hí ðæt mid hiera weddum (cf. cum se exsecrationibus devovissent, sacramentisque obstrinxissent) gefæstnod hæfdon, Ors. 1, 14; Swt. 56, 23: L. Ath. v. proem.; Th. i. 228, 7: v. 8, 5; Th. i. 236, 30. [Ic wille settan mí wed betwuxe me and eow, O. E. Homl. i. 225, 28. Mi lond ich wulle sette to wedde, Laym. 25172. Him þet leid his wed ine Giwerie, A. R. 394, 3. To legge a wedde, Piers P. 5, 244. His nekke liþ to wedde, Chauc. Kn. T. 360. Wedde or thynge leyyd yn plegge *vadium, pignus,* Prompt. Parv. 519. *Goth.* wadi *pignus : O. Frs.* wed : *O. L. Ger.* weddi *pignus : O. H. Ger.* wetti *pignus, pactum, stipulatio : Icel.* veð.] v. an-, under-wed[d].

wēdan; *p.* de *To be mad* or *furious, to rage, rave :*—Ic wēde *saeuio* and *insanio,* Ælfc. Gr. 30, 5; Zup. 192, 3. *Furo* ic wēde macaþ *insaniui* of *insanio* ic wēde, 33; Zup. 203, 9. Ic wēde *grasso,* Engl. Stud. xi. 66, 44. Wētt *saeuit,* Wülck. Gl. 255, 16. Wēt *furit, irascitur,* 245, 19. Wēdende *funeste,* Wrt. Voc. ii. 151, 65. **I.** *to be mad, out of one's senses :*—Cwæþ se cyning: 'Ne wille ðú swā sprēcan; gescoh ðæt ðú teala wite.' Cwæþ hē: 'Ne wēde ic (*non insanio*), Bd. 5, 13; S. 632, 32. Deófol is on him, and hē wēt (*insanit*), Jn. Skt. 10, 20. Se man wēt ðe wyle habban ǣnig þincg ǣr anginne, Homl. Skt. i. 1, 17. Ðā wēdon hí ðæt hē tela ne wiste, ac ðæt hé wēde *vulgus aestimabat eum insanire,* Bd. 2, 13; S. 517, 11. Woedendi *limphaticus,* Wrt. Voc. ii. 112, 75: *lymphatico,* 113, 36. Wēdende, 53, 66. Ðone wēdendan *insanum,* 48, 1. Hwā mæg ðam wēdendan gýtsere (*dives qui sese credit egentem*) genōh forgifan? Bt. 7, 4; Fox 22, 33. **II.** *to act with violence, be furious, rage,* (a) *of persons :*—Ðonne se deófol ðús wētt, Wulfst. 198, 5. Hē wēt swíðe and wynd on ða Cristenan, Homl. Skt. i. 16, 225. Heó geseah hú Decius wēdde and hrýmde dæges and nihtes ǣr ðon hé deád wǣre, Shrn. 149, 6. Hē wēdde on gewitte swā wilde deór, Exon. Th. 278, 13; Jul. 597. Hí wēddon þearle and tōtǣron hí sylfe mid heora āgenum tōðum, Homl. Skt. i. 6, 194. Hē (*Antichrist*) onginð deóflíce tō wēdenne, Wulfst. 200, 1. Woedende *debachatus,* Wrt. Voc. ii. 86, 21: 26, 74. Seó wēdende meniu ofslógon ðone Victor, Homl. Skt. ii. 28, 113. For wēdendre heortan ðæs leódhatan Brytta cyninges *propter vesanam Brittonici regis tyrannidem,* Bd. 3, 1; S. 524, 1. Uuoedende *bachantes,* Wrt. Voc. ii. 101, 52. Hí (*the Jews*) on Criste hosplíce word wēdende sprǣcon, Homl. Th. ii. 232, 31. Wrōhtsmiðas (*evil spirits*) wēdende swā wilde deór, Exon. Th. 156, 23; Gú. 8, 9. (b) *of animals :*—Leún wēdan (gesihð). gestric ge(tācnaþ), Lchdm.

iii. 206, 32. Wēdende hund, Bt. 37, 1; Fox 186, 8. Wulfas woedende *lupi rapaces,* Mt. Kmbl. Rush. 7, 15. (c) *of things, abstract* or *concrete :*—Gýtsung openlíce wēt *auaritia palam saeuit,* Scint. 99, 17. Wēdde stíðnes *exarsit acerbitas,* Hpt. Gl. 517, 15. Gārsecg wēdde, Cd. Th. 208, 27; Exod. 489. Ða ýða weóllan and wēddan ðæs sǣs *furentibus undis pelagi,* Bd. 3, 15; S. 541, 39. Þeáh ðeós woruld wēde and windige ēhtnysse āstyrige ongeán Cristes geladunge, Homl. Th. ii. 388, 9. Ðonne wind wēdende fǣreþ, Elen. Kmbl. 2546; El. 1274. Mid wēdendum and egislícum gehlýde *bacchanti et furibundo strepitu,* Hpt. Gl. 495, 75. Wēdende reóhnysse *tumentem insaniam,* 465, 20. Wēdende ýða *frementes* (*furentes*) *fluctus,* 464, 74. Hyt ða wēdendan bitas gehǣlcþ, Lchdm. i. 370, 14. [Biginned þe deoflen to weden, A. R. 264, 9. As mon þ bigon to weden and to wurðen ut of his ahne witte *indignatus cum furore nimio,* Kath. 1257. Fra þatt grediȝnesse þatt doþ þe mann to wedenn rihht to winnenn erþlic ahhte, Orm. 14140. Þe kyng ferde for wraþþe as he wolde wede, R. Glouc. 53, 10. *O. Sax.* wōdian : *O. H. Ger.* wuoten *furere, grassari, insanire, bacchari, fremere : Icel.* œðask *to become furious.*] v. ā-, ge-wēdan; wōd.

wed-brōðer; *m. One who is pledged to act as a brother to another, a confederate :*—Ðā luuede Wulfere hit swíðe for his brōðer luuen Peada, and for his wedbrōðeres luuen Oswí, Chr. 656; Erl. 30, 1. Cōman bēgen ða cyningas tōgædre and wurdon feólagan and wedbrōðra, and ðæt gefæstnadan ǣgðer mid wedde and eác mid āðan, 1016; Th. i. 284, 1, col. 1. [Send after mine sune Octa, and æfter Ebissa his wedbrōðer, Laym. 14469. *Icel.* veð-brōðir. Cf. eið-brōðir.]

wed-bryce, es; *m. Breach of a pledge* or *engagement :*—Gif hē ðæs weddie, ðe hym riht sý tō gelǣstanne, and ðæt ālcóge . . . bête ðone wedbryce swā him his scrift scrife, L. Alf. pol. 1; Th. i. 60, 6–21. Eác syndan wíde þurh āðbrycas and ðurh wedbrycas and ðurh mistlíce leásunga forloren and forlogen mā ðonne sc de, Wulfst. 164, 7. Wedbricas, 130, 6. [Cf. With wedbrek *cum adulteris,* Ps. 49, 18.]

wedd. v. wed[d].

weddian; *p.* ode *To engage, covenant, undertake :* — Weddodon *pepigere,* Germ. 396, 137. **I.** *to engage to do something,* (a) with gen. of that for which the engagement or pledge is given :—Be ðon ðe ordāles weddigaþ. Gif hwā ordāles weddige *if any one engage to undergo an ordeal,* L. Ath. i. 23; Th. i. 210, 25. Gif hē ðæs weddie, ðe hym riht sý tō gelǣstanne, L. Alf. pol. i; Th. i. 60, 6. Is tō witanne hwam ðæt fōsterleán gebyrige, weddige se brýdgum eft ðæs *let the bridegroom engage to furnish this,* L. Edm. B. 2; Th. i. 254, 9. Ðæt se slaga mōte sylf wæres weddian, L. Edm. S. 7; Th. i. 250, 17. (b) with gerundial infin. :—Hig him weddedon feoh tō syllenne *pacti sunt pecuniam illi dare,* Lk. Skt. 22, 5. **II.** in reference to either taking or giving in marriage, *to wed, betroth, espouse :*—Gif hý ælces þinges sammǣle beón, ðonne fōn māgas tō and weddian heora māgan tō wífe and tō rihtlífe ðam ðe hire girnde, L. Edm. B. 6; Th. i. 254, 20. Gif man mǣdan oþþe wíf weddian wille, 1; Th. i. 254, 2. [Þat mæiden he weddede, Laym. 4432. Wifmann to weddenn, Orm. 10407. Weddedd wiþþ an weppmann, 1942. He moste weddy wyf, R. Glouc. 331, 13. I wedde myne eres, Piers P. 4, 146. *Goth.* ga-wadjōn *despondere : O. Frs.* weddia *to promise, pledge : Icel.* veðja *to wager.*] v. be-, for-, ge-weddian.

weddung, e; *f. Betrothal, espousal :*—Ðā cwæþ Pilatus tō ðam folce, ða ðe sǣdon ðæt he of forligere wǣre ācenned : 'Ðeós sprǣc nys nā sōþ ðæt gē sprecaþ, for ðon seó weddung wæs beweddod, eal swā eówre āgene ðeóda secgaþ,' Nicod. 7; Thw. 3, 31. [Or men win-man to louerd giue for wedding or for morgengiwe, Gen. and Ex. 1428.] v. be-weddung.

wēde; *adj. Furious, in a rage, mad, fierce.* v. wēdan, **II** :—Nælle ðú mē woede (cf. gram, W. S. version) wosa *noli mihi molestus esse,* Lk. Skt. Lind. 11, 7. Woedo (gram, W. S.) wæs mē ðió widiua *molesta est mihi haec vidua,* 18, 5. Wið wēdes (wēde, MS. B. v. wēde-hund) hundes slite, Lchdm. i. 362, 23. Cf. wōd.

wēde-berge, an; *f. A plant that is used against madness, hellebore :*—Woedeberge, woedibergæ *elleborus,* Txts. 59, 736. Woidibergæ *helleborus,* 67, 1017. Wēdeberge, Wrt. Voc. ii. 29, 21 : 32, 20. Ðeós wyrt ðe man *elleborum album* . . . and eác sume men wēdeberge hātaþ, Lchdm. i. 258, 23.

wēde-hund, es; *m. A mad dog :*—Gif wēdehund man tōslíte, Lchdm. i. 86, 13. Wið wēdehundes (cf. wōdes [*printed* woden] huudes, 4, 8) slite, 78, 17: 92, 12: 138, 13: 198, 8: 370, 12, 15: ii. 144, 9. Hē reþigmōd rǣst on gehwilcne wēdehunde (*printed* reðe hunde, *but cf.* wēdende hund, Bt. 37, 1; Fox 186, 8) wuhta gelícost, Met. 25, 18. v. wēde.

wēden-heort, es; *n. Madness, frenzy, fury :*—Lǣcedōmas wið feóndseócum men . . . and wið bræcseócum men, and wiþ wēdenheorte, Lchdm. ii. 14, 7 : 138, 14. Drenc wiþ wēdenheorte, 356, 4 : 304, 15. Ðæt hrýðer him þúhte on wēdenheorte *the beast seemed to him mad,* Blickl. Homl. 199, 11.

wēden-heort; *adj. Mad, frenzied, furious :*—Wēdenheortra synna *furiarum,* Wrt. Voc. ii. 36, 30. v. next word.

wēdenheortness, e; *f. Madness, frenzy, fury :*—He gelōmlíce mid wēdenheortnesse mōdes ðrycced wæs *crebra mentis vesania premebatur,*

Bd. 2, 5; S. 507, 3. Wiþ wēdenheortnesse Macedones *contra vesaniam Macedonii*, 4, 17; S. 585, 45. For wēdenheortnesse đæs leódhatan *propter vesanam tyrannidem*, 3, 1; S. 524, 1. Hí ongunnon đæt hí his wēdenheortnysse gestildon *motus ejus insanos comprimere conati*, 3, 11; S. 536, 22. Hié wealwiaþ on đa wēdenheortnesse *in mentis vesaniam devolvuntur*, Past. 40; Swt. 289, 6. Wēdenheortnessum *furiis*, Wrt. Voc. ii. 37, 50. In woedenheortnisse leáse *in insanias falsas*, Ps. Surt. 39, 5.

weder, es; *n.* I. *weather, condition of the atmosphere* :—Uueder *temperies*, Wrt. Voc. ii. 122, 27. Gif hit sié gód weder, Lchdm. ii. 182, 10. Hyt byđ smylte weder *serenum erit*, Mt. Kmbl. 16, 2: Bt. 23; Fox 78, 26. Đonne wind ligeþ, weder biđ fæger, Exon. Th. 210, 8; Ph. 182. Hreóh weder *tempestas*, Mt. Kmbl. 16, 3. Rēn, swylce hagal and snáw, weder unhióre, Met. 29, 65. Hit wæs ceald weder, Ors. 6, 32; Swt. 286, 31: Met. 26, 28. Forstas and snáwas, winterbiter weder, Cd. Th. 239, 32; Dan. 379. Wearm weder, Exon. Th. 198, 30; Ph. 18. Rēnig weder, 380, 18; Rä. 1, 10. Wederes blæst, hādor heofonleóma, Andr. Kmbl. 1674; An. 839. Líþes weđres, Met. 12, 13. Wedere gelícost . . . on sumeres tíd, Cd. Th. 237, 34; Dan. 347. Đa sæ đe wæs smylte wedere glæshlútru, Bt. 6; Fox 14, 24. þeáh hine (*a sick man*) mon on sunnan læde, ne mæg hē be đȳ wedre wesan (*he can't stand the weather*), þeáh hit sȳ wearm on sumera, Exon. Th. 340, 18; Gn. Ex. 113. Hē ús giefeþ weder líþe, Exon. Th. 38, 12; Cri. 605. Winter bringeþ weder ungemetcald, swifte windas, Met. 11, 59. On sumera đonne đa hátostan weder synd, Lchdm. ii. 252, 10. Weder coledon heardum hægelscúrum, Andr. Kmbl. 2514; An. 1258. Wuldortorhtan weder, Beo. Th. 2276; B. 1136. Wedera cealdost, 1097; B. 546. Wedera cyst, Cd. Th. 238, 6; Dan. 350. Niht biđ wedera þeóstrost, Salm. Kmbl. 621; Sal. 310. Đeóf sceal gangan in đȳstrum wederum, Menol. Fox 544; Gn. C. 42. Hwȳ hí ne scínen scírum wederum, Met. 28, 45. Holmegum wederum, Cd. Th. 185, 6; Exod. 118. I a. *good weather.* v. weder-dæg :—Hine ne went nāđor ne weder ne unweder of đam đe him gecynde ys, Lchdm. iii. 268, 3. Winter sceal geweorpan, weder eft cuman, sumor swegle hāt, 338, 12; Gn. Ex. 77. Wedres on luste, 361, 28; Wal. 26. Rēn cymđ, đonne eówre wæstmas wederes beþorftan, Wulfst. 297, 11. II. *wind, storm, breeze, air* :—Weder *aura*, Wrt. Voc. i. 76, 43: 52, 59. Smylte wedere *aure tenuis*, ii. 4, 56: 6, 20. Blóstme fægerust raþe tó leohtum forscrincþ wedere *flos pulcherrimus cito ad leuem marcescit auram*, Scint. 70, 3. Wedre gesomnad, Exon. Th. 412, 19; Rä. 31, 2. In wedr *in auram*, Blickl. Gl. Weder, Ps. Surt. 106, 29. [Wurdon ormǣtlíca wædera mid þunre, Chr. 1117; Erl. 246, 15.] Wintregum wederum *cum saevis aquilonibus stridens campus inhorruit*, Bt. 5, 2; Fox 10, 31. Styrmendum wederum, 7, 3; Fox 22, 5. II a. *in reference to sailing, weather* (*as in* wea*ther-*bow, *-*bound), *wind.* v. weder-fæst :—Đá gestód hine beáh weder and storm sæ, wearþ đá fordrifan on ān íglond *vela Neritii ducis eurus appulit insulae*, Bt. 38, 1; Fox 194, 10. Đá him weder com, and Godwine and đa đe mid him wæron wendan tó Brycge, Chr. 1052; Erl. 181, 19. Wearđ đæt wæder swíđe strang, đæt đa eorlas ne mihton gewitan hwet Godwine eorl gefaren hæfđe, Erl. 183, 3. Hē đæs wederes ābād, 1094; Erl. 229, 36: 1097; Erl. 234, 20. Hē wearđ þurh weder gelet, Erl. 233, 34. Gód scipstýra ongit micelne wind on hreóre sæ ǣr ǣr hit geweorþe . . . warenaþ hē hine wiþ đæt weder, Bt. 41, 3; Fox 250, 17. [*O. Sax.* wedar *weather, storm: O. Frs.* weder: *O. H. Ger.* wetar: *Icel.* veđr.] v. ge-, ofer-, un-weder, un-geweder.

Wederas; *pl. The Geats*, a tribe of southern Scandinavia :—Wedera leóde, Beo. Th. 455; B. 225. Wedera leód (*Beowulf*), 687; B. 341. Wedra đeóden, 5305; B. 2656. v. Weder-Geátas.

weder-blác; *adj. Weather-pale, pale from exposure to weather* (?). Cf. flód-blác :—Wederblác *palus*, healfhár *semicanus*, fulhár *canus* (these glosses are omitted after Wrt. Voc. i. 45, 34), Anglia viii. 451.

weder-burh; *f. A town exposed to storms, a weather-beaten city* :—Him Dryhten bebeád, đæt hē đa wederburg wunian sceolde, Andr. Kmbl. 3390; An. 1699.

weder-candel; *f. The candle of the open air, the sun* :—Wedercandel swearc, Andr. Kmbl. 744; An. 372. Wedercondel wearm weorodum lýhteþ, Exon. Th. 210, 17; Ph. 187. Cf. heofon-, sweg -candel.

weder-dæg, es; *m. A day of fine weather, a fine day.* v. weder, I a :—Beorht sumor, wearme wederdagas, Exon. Th. 191, 30; Az. 96. [Cf. *Icel.* einn góđan veđrdag *one fine day, once on a time.*]

weder-fæst; *adj. Weather-bound* :—Đá gewendon hí west tó Peueneseá and lægen đær wederfeste, Chr. 1046; Erl. 174, 6. [*Icel.* veđrfastr.]

Weder-Geátas; *pl. The Geats* :—Weder-Geáta leód (*Beowulf*), Beo. Th. 2989; B. 1492: 3229; B. 1612. Hē Weder-Geátum weóld, 4747; B. 2379. v. Wederas.

wederian; *p.* ode *To be* (*good* or *bad*) *weather* :—Cweđaþ sume men, đæt se móna hine wende be đan đe hit wuderian (wedrian, widrian) sceal on đam mónđe; ac hine ne went nāđor ne weder ne unweder of đam đe him gecynde ys, Lchdm. iii. 268, 2. [*Icel.* viđra *to be such and such weather.*] v. ge-wederian, wederung.

weder-líce. v. unweder-líce.

Weder-mearc, e; *f. The district occupied by the Wederas* :—Oþ đæt eft byreþ ofer lagustreámas leófne mannan wudu wundenheals tó Weder-mearce, Beo. Th. 602; B. 298.

weder-tácen, es; *n. A sign of fine weather.* v. weder, I a :—Eástan cwom dægrēdwōma, wedertácen wearm, Exon. Th. 179, 25; Gū. 1267. [Cf. *Ger.* wetter-zeichen *prognostic of a storm.*]

wederung, e; *f. Weather* :—Đæs ilcan geáres wæs swíđe hefelíc geár . . . swá mycel ungelimp on wæderunge swá man náht ǣþelíce geþencean ne mæg; swá stór þunring and lægt wes, swá đæt hit ācwealde manige men, Chr. 1085; Erl. 219, 21. [Gif ȝe mine bibode healded, þenne sende ic eou rihte widerunge, O. E. Homl. i. 13, 17. We shul preyen . . . for alle trewe shipmen, þᵗ godd ȝeue hem wederyng . . .; for þe fruyte of þe londe and þe wederyng, E. G. 23, 18, 20. Wederynge of þꝛ eyre *temperies*, Prompt. Parv. 519.] v. wederian.

weder-wolcen, es; *n. A fine weather cloud.* v. weder, I a, weder-dæg, -tácen :—Hæfde wederwolcen (*the pillar of cloud*) eorđan and uprodor efne gedǣied, Cd. Th. 182, 13; Exod. 75. [Cf. *Ger.* wetter-wolke *a tempestuous cloud.*]

wed-fæstan; *p.* te *To pledge* [:—Geuuetfaestae *subarrata*, Wrt. Voc. ii. 121, 52.] [Cf. *Icel.* veđ-festa *a pledge.*]

wēding, e; *f. Madness, frenzy* :—Wēding *frenesis*, Wrt. Voc. ii. 39, 10. [*O. H. Ger.* wuotunga *furor.*]

wed-lác, es; *n.* I. *a pledge, security* :—Wed *vel* wedlác *arrabona vel arrabo*, Wrt. Voc. i. 20, 7. Wedlác *arrabo*, 50, 31. II. in reference to marriage, v. weddian, II, *wedlock, espousals* :—Wedlác wiđsacende *pacta sponsalia refutans*, Hpt. Gl. 498, 44. [The latter is the usual sense in Middle English :—Under wedlac iboren, Laym. 395. Bute one ine wedlake, A. R. 206, 14. Wass soþ weddlac haldenn, Orm. 2499. In lele wedlayk born, Pr. C. 8261. Heo þat her wedlac brekeþ, Misc. 150, 105. þei wrouȝt wedlokes aȝein goddis wille, Piers P. 9, 152. Wedlok *matrimonium*, Prompt. Parv. 520. Wedloke *maritagium*, Wulck. Gl. 595, 5.]

wed-loga, an; *m. One who is false to a pledge* or *engagement* :—On đison gēre swác Harđacnut Eádulf eorl under his gride, and hē wæs đá wedloga, Chr. 1041; Erl. 166, 33. Ic đē eom andetta mínra synna . . . ic eom wedloga, Anglia xii. 501, 19. Đæt gē ne beón wedlogan ne wordlogan, Wulfst. 40, 10: 165, 36. Cristen cyning sceal wedlogan and wǣrlogan hatian and hȳnan, 266, 29. [þu (*the body*) were wedlowe and monsware, Fragm. Phlps. 7, 27.]

wedrian, weel, Wrt. Voc. ii. 95, 79, -wef. v. wederian, þel, ge-, ó-wef.

wefan; *p.* wæf, *pl.* wǣfon; *pp.* wefen. I. *to weave* a web :—Ic wefe *texo*, Wrt. Voc. i. 59, 47. Đú wyfst and wǣda tylast, Homl. Th. i. 488, 25. Đín wyln welð *tui ancilla texit*, Ælfc. Gr. 15; Zup. 104, 13. Webbu swá hwylc swá wyfð, Lchdm. iii. 210, 28. Hí smalo hrægel wefaþ and wyrceaþ *texendis subtilioribus indumentis operam dant*, Bd. 4, 25; S. 601, 16. Đa of đæs treówes leáfum and of his flȳse spunnon and swá eác tó godewebbe wǣfon and worhtan *gens foliis arborum ex siluestri uellere uestes detexunt*, Nar. 6, 19. Đá onfēng Maria hwít godweb tó wefanne . . . Đá sprǣcon hí : 'þú eart úre gingast, đe miht wefan đæt hwíte godeweb,' Homl. Ass. 132, 550. Wefen wæs *ordiretur* (*colobium de stuppae stamine*, Ald. 51), Wrt. Voc. ii. 83, 18. From đæm weofendan *a texente*, Ps. Surt. ii. p. 184, 34. Fram wefendum wífe, Cant. Ez. 12. II. in a more general sense, lit. or fig. *to weave, construct, put together, arrange, plan, contrive* :—Swá đæt wuldor wifeþ, Exon. Th. 493, 8; Rä. 81, 27. Đus ic fród wordcræft wæf and wundrum læs, Elen. Kmbl. 2473; El. 1238. Ic wef *intexui* (*funibus lectulum meum*, Prov. 7, 16), Kent. Gl. 199. Wefan *contexere* (*coronam*), Hpt. Gl. 439, 68. Wefan *texuisse* (*oraculorum seriem*), 442, 39. Đæs engles mōd đe đone unrǣd ongan ǣrest fremman, wefan and weccean, Cd. Th. 3, 5; Gen. 31. Đonne seo þrág cymeþ wefen wyrdstafum, Exon. Th. 183, 10; Gū. 1325. [*O. H. Ger.* weban : *Icel.* vefa. Cf. *Goth.* bi-waibjan *to wind about.*] v. ā-, be-, ge-wefan; þyn-wefen.

wefl, e; wefle (-a; *m.*?), an; *f.* I. *weft, woof, thread which crosses the warp* :—Weft *vel* ówef, uuefl *cladica, caldica*, Txts. 51, 482. *Cladica* wefl ođđe ówef ođđe *claudica*, Wrt. Voc. ii. 14, 43. Wefl *vel* óweb *cladica*, 131, 59. Wefl *cladica*, 31 : i. 66, 13: 281, 76. Uuefl *panuculum*, ii. 116, 29 : titica (cf. *O. H. Ger.* below), 122, 33. Wefian *penniculae* (the passage is : Nisi panniculae diversis colorum varietatibus fucatae inter densa filorum stamina ultro citroque decurrant, Ald. 15), Hpt. Gl. 430, 69. Wefla *panucla* (this is a gloss to the same passage as the preceding), Wrt. Voc. ii. 77, 13. Wundene mē (*a coat of mail*) ne beóđ wefle (ueflæ, Txts. 151, 5), ne ic wearp hafu *the threads of the woof are not twisted for me, nor have I a warp*, Exon. Th. 417, 15; Rä. 36, 5. Wǣfla *pannicularum* (colobium cum sine pompulenta pannicularum varietate ordiretur, Ald. 51), Hpt. Gl. 494, 9. Weflum *panniculis* (*pannulis*, Wrt. Voc. ii. 65, 61, in a gloss to the same passage : Lanea filorum stamina ex glomere et panniculis revoluta, Ald. 8), 417, 20. II. *an implement for weaving* (-l suffix in words denoting implements, cf. scofl), *a shuttle* (?) :—Hē sceal habban fela towtóla . . . pihten, wefle,

wefle (*or under* **I** ?), wulcamb, Anglia ix. 263, 13. [*O. H. Ger.* wefal (-el, -il) *datica, subtemen, stamen.*] v. next word.

wefta, an: weft, es; *m. Weft, woof:*—Wefta *vel* weft *deponile,* Wrt. Voc. i. 59, 38. Wefta, 66, 14: 281, 77. Wefta *deponile,* uueftan *depoline,* Txts. 55, 642. Wefta *depo[nile],* weftan *deponile,* Wrt. Voc. ii. 138, 85, 86. Wefta *depoline,* 25, 19: *clatica,* 131, 68. [Weft *subtegmen,* Wick. Ex. 39, 3. A wefte *trama,* Wulck. Gl. 696, 21. *Icel.* veftr, vifta.] v. preceding word.

wefung, e; *f. Weaving:*—Weofung *textura,* Wrt. Voc. ii. 77, 12.

weg (wig, Kent. Gl. 207: 475: 772; *pl.* weogas, 21), es; *m. A way.* **I.** of the direction in which motion (lit. or fig.) takes place:— Ða tungelwitegan ðurh ôðerne weg tô heora earde gecyrdon. Ûre eard is neorxnawang, tô ðam wê ne magon gecyrran ðæs weges ðe wê cômon, Homl. Th. i. 118, 20–23. Þonne rîdeþ ǽlc hys weges, Ors. 1, 1; Swt. 21, 4. Hî wendon him sûðweard ôðres weges, Chr. 1016; Erl. 154, 15. Wæges, 1006; Erl. 140, 22. Hê mê eft lǽdde ðý sylfan wegge ðe wê ǽr tô côman, Bd. 5, 12; S. 629, 41. Hig gewendon him ofer langne weg, ðæt hig ðæt land embfêrdon, Num. 21, 4: Cd. Th. 35, 13; Gen. 554: 43, 13; Gen. 690. Hié ofer feorne weg ceólum lâcaþ, Andr. Kmbl. 504; An. 252: 2348; An. 1175. Fôre gefremman on feorne weg, 382; An. 191. Nân man ne mihte faran þurh ðone weg (woeg, Lind.: wæge, Rush.), Mt. Kmbl. 8, 28. Sceáweras, ðæt cŷðon ûs, on hwilcne weg wê faran sceolon (*per quod iter debeamus ascendere*), Deut. 1, 22. Ðû weg nimest geond deóp wæter, Cd. Th. 80, 16; Gen. 1329. Wǽrun wegas ðine on wîdne sǽ *in mari viae tuae,* Ps. Th. 76, 16. Onbûgan of ðæs gewealde, ðe mê wegas tǽcneþ, Exon. Th. 383, 26; Râ. 4, 16. Tôfôran on feówer wegas æðelinga bearn, Cd. Th. 102, 9; Gen. 1697. **I a.** with the idea of access or passage:—Ðâ gesette God æt ðam infære engla hyrdrǽdene and fŷren swurd tô gehealdenne ðone weg tô ðam lîfes treówe, Gen. 3, 24. Ic mê weg ryhtne gerŷme, Exon. Th. 479, 24; Râ. 63, 3. Hê sceolde gearcian and dæftan his weig, Homl. Th. i. 362, 8. Wegas syndon drŷge, haswe herestrǽta, Cd. Th. 195, 28; Exod. 283. **II.** *a road* (lit. or fig.) *made for passengers, a path commonly used:*—Weg *via,* Wrt. Voc. i. 53, 56. On eástan ealles folces weg, and an sûðan se weg se ðe lið tô ðam ilcan lande, Cod. Dip. B. i. 586, 15. Swâ swâ se weg lið, wê faraþ *via regia gradiemur,* Num. 21, 22. Ðæt geat is swŷðe wîd, and se weg is swîðe rûm, ðe tô forspilldenesse gelǽt, Mt. Kmbl. 7, 13, 14. On ðam wege, ðe lið tô Euphrate *in via, quae ducit Euphratam,* Gen. 35, 19. Se assa eode of ðam wege. Hwæt ða Balaam beót ðone assan, wolde ðæt hê eode innan ðone weg *asina avertit se de itinere et ibat per agrum; quam cum verberaret Balaam et vellet ad semitam reducere,* Num. 22, 23. Sum sacerd fêrde on ðam ylcan wege (woege, Lind.), Lk. Skt. 10, 31. Gif feorrancumen man oþþe frǽmde bûton wege gange, L. Wih. 28; Th. i. 42, 23. Gif ðû wyrfst on wege rihtum up tô ðam earde, Met. 24, 44. Gif ðû cymst on ðone weg and tô ðære stôwe, Bt. 36, 2; Fox 174, 21. Hê leóde lǽrde on lîfes weg, Andr. Kmbl. 340; An. 170: 3357; An. 1682. Sume feóllon wið weg (æt strǽt ł woeg, Lind.: bi wæge, Rush.), Mt. Kmbl. 13, 4. Wegas, ðæt ġeweorc, strǽte stânfâge, Andr. Kmbl. 2470; An. 1236. Nǽron Metode ðâ gyt wîdlond ne wegas nytte, Cd. Th. 10, 13; Gen. 156. Betŷndan wega gelǽtan *competa clausa,* wega gelǽtum *competis, terminis,* Wrt. Voc. ii. 132, 52: 19, 55. Ðæt wîf, ðet æt ðæra wega gelǽte sæt *mulier, quae sedebat in bivio,* Gen. 38, 21. Tô wega (ðære wegara ł ðæra wegana, Lind.: weogas, Rush.) gelǽtum *ad exitus viarum,* Mt. Kmbl. 22, 9. Wega gemittung *compitum,* Wrt. Voc. i. 55, 8. On wega gemôtum *in competis,* ii. 46, 12. Ágildgifu gefreóde Ælfgiðe on feówer wegas (v. Earle's note, p. 468, on manumission at four cross-roads), Chart. Erl. 255, 20: 254, 29. Ungerydu beóð on smêðe wegas (woegum, Lind.), Lk. Skt. 3, 5. Gôdige hê folces fǽr mid bricgum ofer deópe wæteru, and ofer fûle wegas, L. Edg. C. 14; Th. ii. 282, 10. Ðurh ðrióstrie weogas *per vias tenebrosas,* Kent. Gl. 21. **II a.** of what resembles a path, as in Milky Way. v. Îringes weg. **III.** *space to be traversed, a journey:*—Eáðfêre weg *iter* vel *itus,* lang and stearc weg *itiner,* Wrt. Voc. i. 37, 35, 36. Gif se weg swâ lang beó, ðæt ðû ðîne þing bringan ne mage, Deut. 14, 24. Hig hæfdon sumne dǽl weges gefaren *processerant paululum,* Gen. 44, 4: *aliquantulum itineris confecissent,* Bd. 1, 23; S. 485, 30. Mê wæs Rachel deád be wege *mortua est Rachel in itinere,* Gen. 48, 7. Hê tô ðam cyng gewǽnde. Ðâ com Sparhafoc be weg[e] tô him, Chr. 1048; Erl. 177, 19. Fela þûsenda be wæge forfôran, 1096; Erl. 233, 21. Heó forðfêrde be Rôme wege (*in itinere Rome*), 888; Erl. 87, note 10. Mid ðý ðe ðæt mîn werod gestilled wæs, ða fêrdon wê forð ðý wege ðe wê ǽr ongunnon *quae res quam anime quietiorem fecisset exercitum, ceptum iterum institui,* Nar. 8, 18: 17, 5. Gif mon fram longum wege geteorod siê, Lchdm. ii. 150, 19: 16, 16. Áris and et, ðû hæfst swŷþe langne weg, Homl. Skt. i. 18, 168. On eallum ðâm wegum ðe gê fôron, Deut. 1, 31. **IV.** in reference to conduct, action, practice, *manner, mode, method, plan:*—Geriht mînne weg (se weg is mîn weorc), Ps. Th. 5, 8. Ealle his wegas sint dômas, Deut. 32, 4. Gehwelci wega (uuaega, uuegi) *quocumque modo,* Txts. 91, 1700. Hê his wegas dyde cûðe *notas fecit vias suas,* Ps. Th. 102, 7. Unrihte wegas, 118, 104. **V.** *way,* in al-wiy, -ways:—Under

his tungan byð ealne weg ôþera manna sâr, Ps. Th. 9, 28. Ðæt edleán ðe ðû ealne weg gehête, Bt. 3, 4; Fox 6, 19. Ealne weg (symle, Met. 8, 18) hî ǽton ǽne on dæg, Bt. 15; Fox 48, 8. Ic wât ðû wêne ðæt hî on heora âgenre cýþþe ealne weg mægen *inter eos, apud quos ortae sunt, num perpetuo perdurant?* 27, 4; Fox 100, 11: 29, 1; Fox 102, 10. Ic simles wæs on wega gehwam willan ðînes georn on môde, Andr. Kmbl. 129; An. 65. Wel mon sceal wine healdan on wega gehwylcum, Exon. Th. 342, 19; Gn. Ex. 145. **VI.** in the plural, in some compounds, the word has the sense of *parts, regions.* Cf. *Icel.* -vegir. v. eást-, norþ-, sûþ-, sîd-, wîd-wegas. [*Goth.* wigs: *O. Sax. O. H. Ger.* weg: *O. Frs.* wei: *Icel.* vegr.] v. â-, ærne-, bæþ-, beám-, burh-, dîc-, eást-, eorþ-, fær-, feor-, flôd-, flot-, fold-, forþ-, gang-, here-, hîg-, holm-, hors-, horu-, hrycg-, hwæt-, hwyrft-, lîf-, mǽr-, mid-, mold-, norþ-, on-, or-, riht-, sîd-, sîdling-, sîþ-, stân-, stapol-, stîþ-, sûþ-, tûn-, twi-; þeód-, up-, wægn-, wæl-, wæter-, weall-, west-, wîd-, wil-, will-weg; ealneg.

weg (wei, wî) **lâ**; *interjection:*—Weg lâ, weig lâ *euge, euge,* Ps. Th. 69, 4. Weg lâ weg ł wâ lâ wâ ł eálâ, eálâ *euge, euge,* Ps. Lamb. 39, 16. Wî lâ wei (wei lâ wei, Cott. MS.), Bt. 35, 6; Fox 170, 12. [Cf. *Ital.* via.]

wêg *a wave.* v. wǽg.

wegan; p.ʹwæg, *pl.* wǽgon; *pp.* wegen.   **A.** *trans.* **I.** *to move, bear, carry, bring, transport:*—Ic wege oððe ic ferige *ueho,* Ælfc. Gr. 28, 5; Zup. 176, 4. (Scip) wist in wigeþ, Exon. Th. 415, 14; Râ. 33, 11. Ðone (*a dog*) on teón wigeþ feónd his feónde, 433, 28; Râ. 51, 3. Hâm wegaþ *advehunt,* Wrt. Voc. ii. 1, 5. Hê ða frætwe wæg ofer ŷða ful, Beo. Th. 2419; B. 1207. Hê com tô ðam forwundodum, and wæh hine hâm tô his inne, Homl. Ass. 47, 559. Mec wǽgun feðre on lifte, feredon mid liste, Exon. Th. 409, 19; Râ. 28, 3. Micel mænigeo elpenda ða ðe gold wǽgon and lǽddon *elephanti qui aurum uehebant,* Nar. 9, 6. Mîn weorod goldes micel gemet mid him wǽgon and lǽddon, 7, 1. Wâgon, Judth. Thw. 26, 14. Gesâwon hié weallas standan. . . . þurh ða heora beadosearo wǽgon, Cd. Th. 214, 21; Exod. 572. Wêgon, Byrht. Th. 134, 43; By. 98. Gûðspell wegan *to carry news of the war,* Cd. Th. 126, 18; Gen. 2097. Wegen on wægne, Exon. Th. 403, 15; Râ. 22, 8. **I a.** *fig.* where the object is abstract, *to bring, cause:*— Geáp stæf wigeþ biterne brôgan, Salm. Kmbl. 250; Sal. 124. **II.** *to bear, support:*—Eahta sweras syndon ðe rihtlîcne cynedôm trumlîce up wegaþ, L. I. P. 3; Th. ii. 306, 20. **III.** *to bear, carry,* (1) *to have* as part of one's equipment, *bear arms, wear:*—Sigegyrd ic mê wege, Lchdm. i. 388, 15. Ic (*a sword*) sinc wege, Exon. Th. 401, 4; Râ. 21, 6. Se ðe gold wigeþ *he that wears golden ornaments,* 484, 12; Râ. 70, 6. Mec (*a lance*) . . . on fyrd wegeþ, 486, 21; Râ. 72, 18. Hê heregeatowe wegeþ, Salm. Kmbl. 106; Sal. 52. Mec (*a horn*) folcwigan wicge wegaþ, Exon. Th. 395, 27; Râ. 15, 14. On ðǽm hrægle, ðe hê on his breóstum wæg, Past. 13; Swt. 77, 15. Wæs feówer geár, ðæt hê woroldwǽpno wæg, Blickl. Homl. 213, 4. Hæfde hê and wæg mid hine twigecgede handseax *habebat sicam bicipitem,* Bd. 2, 9; S. 511, 15: Beo. Th. 5402; B. 2704. Hê lîgegesan wæg, 5554; B. 2780. Rincas randas wǽgon, Cd. Th. 123, 22; Gen. 2049. Gyf him þince ðæt hê wǽpen wege, ðæt byð orsorh, Lchdm. iii. 174, 13: Beo. Th. 4497; B. 2252. Ne wæs âlŷfed, ðæt hê môste wǽpen wegan (*arma ferre*), Bd. 2, 13; S. 517, 7. On fyrd wegan fealwe linde, Cd. Th. 123, 13; Gen. 2044. Îs sceal brycgian wæter helm wegan (*water must wear a helm of ice*), Exon. Th. 338, 5; Gen. Ex. 74. Wegan mâððum *to wear a jewel,* Beo. Th. 6023; B. 3015. Ic nolde wegan ðîn wynsume geoc, Anglia xi. 112, 22. (1 a) *fig.* where the object is abstract:—Sume him ðæs hâdes hlîsan willaþ wegan on wordum and ða weorc ne dôð *some are ready to bear the reputation of being of the elect, as far as words go, and do not do the works,* Exon. Th. 105, 32; Gû. 32. (2) *to have* as part of or within one's self:—Fela geofona, ða ða gǽstberend wegaþ in gewitte, Exon. Th. 293, 18; Crä. 3. Ðone lîchoman ðe heó (*the soul*) ǽr louge wæg, 367, 21; Seel. 11. Ðæt lâmfæt ðæt hié (*the soul*) ǽr lange wæg, 375, 5; Seel. 133. Tîr unbrǽcne wǽgon on gewitte wuldres þegnas, Apstls. Kmbl. 173; Ap. 87. Ðû scealt wegan swâtig hleór, Cd. Th. 57, 27; Gen. 934. (3) *to be under the influence of* pain, joy, etc., *have* such and such feelings, *bear* a grudge:—Ic ðæs tâcen wege sweotol on me selfum, Cd. Th. 54, 31; Gen. 885. Hê lust wigeþ, Beo. Th. 1203; B. 599. Hê on breóstum wæg byrnende lufan, Chr. 975; Erl. 126, 14. Grendel hetenîðas wæg, Beo. Th. 307; B. 152. Môdþrŷðo wæg cwên, 3867; B. 1931: Cd. Th. 135, 6; Gen. 2238. Ic wæg môdceare micle, Beo. Th. 3559; B. 1777. Wedera helm heortan sorge wæg, 4919; B. 2464: Exon. Th. 162, 28; Gû. 982: 182, 13; Gû. 1309: Elen. Kmbl. 122; El. 61: 1307; El. 655. Lifge Ismael and ðê þanc wege, heardrædne hyge, Cd. Th. 141, 20; Gen. 2347. Ða ðe â wegen egsan Dryhtnes *qui timent Dominum,* Ps. Th. 113, 20. **IV.** *to bear, submit to* consequences:—Ne bið ǽngum gôdum gnorn ætŷwed, ne nǽngum yflum wel; ac ǽghwæþer ânfealde gewyrht andweard wigeþ, Exon. Th. 96, 23; Cri. 1578. Gylde hê ðæs cinges oferhŷrnesse, and wege ða ungerisenu, L. Alth. iv. 1; Th. i. 222, 6. Gif hwâ ǽnigne man ofsleá, ðæt hê wege sylf ða fǽhðe, L. Edm. S. 1; Th. i. 248, 2, 9. **V.** *to weigh,* (1) *to put something in a balance:*—Ic wege *trutino,* Ælfc. Gr. 36; Zup. 215, 18. Ǽlc ðæra ðinga, ðe man wihð (wehð, v.l.) on

wǽgan, 13; Zup. 84, 2. Man sett ða synne and ða sáwle on ða wǽge, and hý man wegeþ, swá man dêð gold wið penegas, Wulfst. 240, 2. Weh on wǽge, Lchdm. i. 374, 15. (1 a) fig.:—Teóðige on Godes ést eal ðæt hé áge, and wege hine sylfne swá hine oftost tó onhagige, L. Pen. 15; Th. ii. 282, 23. Wegende tódǽles ł gescádes ápinsunge *discretionis lance librantis (ponderantis)*, Hpt. Gl. 447, 71. (2) *to be equal to* a certain weight:—Ælc án hagelstán wegeþ fíf pund, Wulfst. 228, 7. Se sester sceal wegan twá pund, Lchdm. iii. 92, 14.   B. *intrans.* To *move* :—Ymb hine wǽgon wígend unforhte, Cd. Th. 189, 5; Exod. 180. Frǽtwed wǽgun (-m, MS.) wic[g] ofer wongum, Exon. Th. 353, 2; Reim. 6. [Heo weȝe (beore, 2nd MS.) on heore honde feouwer sweord, Laym. 24471. To teche an beore to weȝe boþe scheld and spere, O. and N. 1022. Chepinge þe me shule meten oðer weien, O. E. Homl. ii. 213, 34. To weien swuðer his sunne þen he þurfte. Weien hit to lutel is ase vuel, A. R. 336, 22. *Goth.* ga-wigan *to shake* : *O. L. Ger.* wegan *to weigh* : *O. Frs.* wega, weia *to move, weigh* : *O. H. Ger.* wegan *movere, vibrare, nutare, librare, trutinare, ponderare, pensare* : *Icel.* vega *to move, carry, weigh.*] v. á-, æt-, be-, for-, ge-, tó-wegan; sweord-, wan-wegende; un-wegan.

wégan *to delude*, wégan *to bend.* v. wǽgan, ge-wǽgan.

weg-bráde, -brǽde, an; *f. Way-bread* (v. E. D. S. Pub. Plant Names) :—Wegbráde, uuegbrádae, uegbrádae *arnaglossa*, Txts. 43, 213. Uuegbráde *plantago*, uuaegbrádae *plantago vel septenerbia*, 87, 1601. Wegbráde, Wrt. Voc. ii. 68, 71. Wegbráde *arnaglosse*, i. 67, 10. Wegbrǽde, 286, 22: ii. 8, 37, 48: Lchdm. i. 80, 8 (cf. title, 4, 14 wegbrǽd (-bráde, -brǽde, *v. ll.*). Wegbráde *plantago*, Wrt. Voc. i. 68, 40. Wegbrǽde, 79, 32: *cinoglossa vel plantago vel lapatium*, 30, 50. Ðú wegbráde, wyrta móder, Lchdm. iii. 32, 5. Wegbrǽdan seáw, i. 80, 12. Wegbrǽdan sǽd, 82, 6. Of ðære rúwan wegbrǽdan, ii. 106, 13. Genim ða rúwan wegbrǽdan nioþowearde, 292, 10. Ða smêþan wegbrǽdan, 350, 7. [*O. H. Ger.* wege-breita *centinodia, plantago.*]

wége. v. wǽge.

weg-farende; *adj. (ptcpl.) Wayfaring* :—Sum wegfarende (-férende, *v. l.*) man férde wið ðone feld ; ða weard his hors gesicclod, Homl. Skt. ii. 26, 204. Seó nædre ligeþ on ðam wege, and wyle ða wegfarendan mid hire tóðum slítan, Wulfst. 192, 23. [*Icel.* veg-farandi.] v. following words.

weg-férend, es; *m. A wayfarer, a traveller* :—Se nacoda wegférend *vacuus viator*, Bt. 14, 3; Fox 46, 29. Stunt wegférend *stultus viator*, Scint. 187, 6. Wíferend *viator*, Kent. Gl. 137. v. next word.

weg-férende; *adj. (ptcpl.) Wayfaring*; used subst. *a wayfarer, traveller.*   **I.** *travelling, on a journey* :—Gif ðú wǽre wegférende, and ðú becóme on þeófsceole, Bt. 14, 3; Fox 46, 25. Se wegférenda man, se ðe nimð ðone smêðan weg, ðe hine mislǽt, Homl. Th. i. 164, 7. Ánes wegférendes mannan nýten gehǽled wæs *jumentum cujusdam viantis curatum est*, Bd. 3, 9; S. 533, 3. Wé sind hér swilce wegférende menn, Homl. Th. i. 248, 15. Se ríca and se ðearfa sind wegférende on ðisse worulde, 254, 28.   **I a.** used substantively :—Swá swá wegférende þyrstende *sicut uiator siciens*, Scint. 225, 10. Wíferend, Kent. Gl. 137. Wegférende ðæt sǽd fortrǽdon, Homl. Th. ii. 90, 45. Se ðe ǽnig ðissa dó . . ., búton wegférende; ða móton for neóde mete ferian, L. N. P. L. 56; Th. ii. 298, 25. Nyhtlíc leóht wegférendum (*viantibus*), Hymn. Surt. 6, 14.   **II.** *going a way, passing by* :—Hí genýddon sumne wegférendne *angariauerunt praetereuntem quempiam*, Mk. Skt. 15, 21. Ða wegférendan (*praetereuntes*) hyne bysmeredon, Mt. Kmbl. 27, 39. [Sein Iulianes in, þet weiuerinde men ȝeorne secheð, A. R. 350, 16. Þe pilgrimes, and oþre wayuerinde men, Ayenb. 39, 3.] v. preceding words.

weg-fór, e; *f. A wayfaring, going away* :—On wegfóre *in provectione* (= profectione ?), Wrt. Voc. ii. 46, 29.

weg-gedál, es; *n. A place where a road divides* :—Weggedál *difortium*, Txts. 57, 672: *compitum*, Wrt. Voc. i. 53, 60.

weg-gelǽte, an; *f.: -*gelǽte, es; *n.* (v. ge-lǽte) *A place where roads meet* :—Weggelǽte *compitalia*, Hpt. Gl. 515, 27. Æt ðære wegegelǽton, Cod. Dip. Kmbl. v. 297, 29. Wegelǽton *trivium*, Wrt. Voc. i. 53, 58. Weggelǽta *compita*, 37, 45.

weg-gesíþa, an; *m. A companion* or *attendant on the road* :—Wæggesíðan *satellites*, Hpt. Gl. 426, 68.

wégi. v. wǽge.

weg-leás; *adj.*   **I.** *without a road, impassable* :—Ungefére *vel* wegleás pǽd *invium*, Wrt. Voc. i. 53, 61. Weglǽsa beara *aviaria*, *secreta nemora*, 39, 11.   **II.** fig. *out of the way, erroneous, unreasonable* :—Welise (= wílése ? cf. wig = weg, *and* wí-férend = wegférend, *both in the same glossary*) *devium*, Kent. Gl. 432. Gedwelde mid wegleásum *errore devio*, Hymn. Surt. 24, 13. [Cf. *Icel.* vega-lauss *out of the way, lost in the woods.*]

weg-leást, e; *f. Want of road* :—Dwelian hé dyde hig on wegleáste and ná on wege *errare fecit eos in invio et non in via*, Ps. Spl. 106, 40. v. next word.

weg-lísu (?); *f. Want of road* :—Welise (= wílésu ?) *devium*, Kent. Gl. 432. [Cf. *Icel.* vega-leysi *want of roads.*] v. preceding words.

weg-nest, es; *n. Food for a journey* :—Weard uncer wegnyst áfulod,

---

Shrn. 42, 4. Him siþþan sý wegnestes getíðad, and swá mid wegneste hám cyrren, R. Ben. 103, 21. Ða genámon wit twégen buccan, and wit hig ácwealdon, and gehióldan hiora flǽsc unc tó wægnyste, Shrn. 41, 30: 36, 31. ¶ the word is used of the sacrament administered tc the dying :—Gif se man on his ýtemestan dæge gyrneþ Cristes líchaman tó underfónne, ne wyrne him man ná, . . . ðæt bið his wegnyst (*viaticum*), and ǽlces ðæra manna ðe tó Godes ríce becymð, L. Ecg. P. i. 10; Th. ii. 176, 20. Heó onféng wǽgnyste ðære hálgan gemǽnsumnysse, Bd. 4, 23; S. 595, 27. Hé bútan hǽlo wegnyste of worulde gewát, 5, 14; S. 634, 33. Hé wæs hine trynmende mid ðý heofonlícum wægneste, 4, 24; S. 599, 2. [*O. H. Ger.* wega-nest (-nist) *cibaria, viaticum* : *Icel.* veg-nêst.]

weg-reáf, es; *n. Booty taken on the high road, robbery done on a road* :—Gif wegreáf sí gedón, .vi. scillingum gebéte. Gif man ðone man ofslǽhð, .xx. scillingum gebéte, L. Ethb. 19, 20; Th. i. 8, 1-2. Ðeúwæs wegreáf sé .iii. scillingas, 89; Th. i. 24, 16. Cf. wæl-reáf.

weg-twislung; *f. The forking of a road* :—Wegtwislung (*spelt -twiflung*) *diverticulum*, Wrt. Voc. i. 55, 6.

wegures, Wrt. Voc. i. 35, 47. v. wíg-gár.

wei lá wei. v. weg lá.

wel, well.   **I.** *adv. Well*, (1) with verbs, (a) marking the success or excellence of the action of the verb :—Ðæt hié heora fulwihthádas wel gehealdan, Blickl. Homl. 109, 26. Wel hearpan stirgan, Exon. Th. 42, 6; Cri. 668. Swíþe wel ðú mín hǽfst geholpen, Bt. 41, 4; Fox 250, 18. (a 1) *well, prosperously* :—Se man wæs wel dónde on eallum þingum *erat vir in cunctis prospere agens*, Gen. 39, 2. (b) marking the rightness, fitness, etc. of an action :—His nama wæs gereht ' Godes strengo.' Wel ðæt wæs gecweden, for ðon ðe se hǽfde mægen ofer ealle gesceafta, Blickl. Homl. 9, 14. Wel ðú sprecst *bona res est, quam vis facere*, Deut. 1, 14. Wel ðú cwǽde *bene dixisti*, Lk. Skt. 20, 39. Hé lim wel (woel, Lind.) andswarode, Mk. Skt. 12, 28. Hí nálæs wel dydan *non observaverunt pactum*, Ps. Th. 77, 57: 118, 126. Welan áh in wuldre se nú wel þenceþ, Exon. Th. 452, 12; Dóm. 119. Suíðe wel Dryhten ðreáde Iudéas, Past. 21; Swt. 151, 19. (c) marking kindness or goodness :—Gyf ge wel dóð ðam ðe eów wel dóð, Lk. Skt. 6, 33. Tó Gode ðe mé wel dyde *ad Dominum qui benefecit mihi*, Ps. Th. 56, 2. Gié magon him woel dóe (wel dóa, Rush.) *potestis illis bene facere*, Mk. Skt. Lind. 14, 7. Wese ðín mildheortnis wel ofer ús, Ps. Ben. 32, 18. (d) marking degree, *well, much, thoroughly, freely* :—Gecuua wel, Lchdm. ii. 322, 26. Lǽt gestandan wel *let it stand a good while*, 326, 19. Syle him ðæs ylcan wyrte wel drincan on wætere, i. 148, 19. Se cyng him eác wel feoh sealde, Chr. 894; Erl. 91, 32. Dó wel scaltes on, Lchdm. ii. 322, 17. Ðé ðissa woruldsǽlða tó wel ne lyste, Bt. 7, 3; Fox 22, 24. Ungemetes wel randwigan restan lyste, Beo. Th. 3589; B. 1792. Ðæt hié welena tó wel ne trúwodon, Blickl. Homl. 185, 14. Eal swá wel behófaþ ðæt heáfod ðæra óðera lima, swá swá ða lima behófiaþ ðæs heáfdes, Homl. Th. i. 274, 7. (e) marking favourable condition, absence of hindrance :—Hé his wel geweald áhte on ðæm scrǽfe, Past. 3; Swt. 37, 5. Eálá ðæt hé wolde, ðæt hé wel meahte ðæt unriht him eáðe forbiódan, Met. 9, 53. Hié wel meahton libban on ðam lande, gif hié wolden láre Godes fremman, Cd. Th. 49, 3; Gen. 786. (f) marking fitness of circumstance, *well, properly* :—Hý mihton wel habban wíf on ðam dagum, L. Ælfc. C. 7; Th. ii. 346, 7. (f 1) with verbs that denote fitness :—Wel ðæt gerás, ðæt heó wære eádmód . . . Wel ðæt eác gedafenaþ, ðæt hé tó eorþan ástige, Blickl. Homl. 13, 16-19. Hine man byrigde, swá him wel gebyrede, ful wurðlíce, Chr. 1036; Erl. 165, 34. (g) marking happy, pleasant, agreeable condition :—Líf ádreógan wel *to pass life pleasantly*, Coll. Monast. Th. 28, 31. Ðæt mé wel sig for ðé *ut bene mihi sit propter te*, Gen. 12, 13: Num. 11, 18: Exon. Th. 66, 32; Cri. 1080. Ne bið ðær ǽngum gódum gnorn æ̃týwed, ne nǽngum yfium wel, 96, 20; Cri. 1577. Ðám bið wel, ðe ðara blissa brúcan móton, Andr. Kmbl. 1770; An. 887. Is ðæt lá well *euge, euge*, Ps. Th. 39, 18. Wel lá wel is úrum módum *euge, euge animae nostrae*, 34, 33. Ðé wel weorðeþ on wynburgum *bene tibi erit*, 127, 2. (g 1) exclamatory, without a verb expressed :—Wel hym ðæs geweorkes, Hy. 2, 11. Wel ðám, ðe ðonne ne áwácaþ, Wulfst. 89, 19: 124, 8. Wel ðære heorde, ðe gefolgaþ ðam hyrde, L. C. S. 85; Th. i. 424, 12. (2) with adjectives, *well, very, quite, thoroughly* :—Strange cyningas and wel cristene, Bd. 4, 2; S. 565, 31: Wulfst. 29, 6: 39, 15: 127, 2. Glǽsfæt wel micel, Lchdm. ii. 252, 8. On wíne wel scearpum, 180, 16: Ps. Th. 67, 15, 16: 104, 37. Dagas wel manige, Blickl. Homl. 217, 15: 225, 10. Wyrta swiþe wel clǽne, Lchdm. ii. 336, 5. (3) with numerals :—Hé ðær þurhwunode wel twá geár *he stopped there quite two years*, Homl. Skt. i. 15, 37. Ic gesett hæbbe wel feówertig lárspella, Ælfc. T. Grn. 13, 45. (4) with adverbs, *very, quite* :—Wæs be eástan ðære ceastre wel néh *erat prope ipsam civitatem ad orientem ecclesia*, Bd. 1, 26; S. 487, 42. Wé wel neáh stódan dám bearwum, Nar. 28, 31: Guthl. 12; Gdwin. 58, 19. Wel wíde *passim, ubique*, Hpt. Gl. 512, 18.   **II.** *interjection, well, ah* :—Wel lá *heu*, Germ. 388, 11. Hé cwæð mid wópe; wel lá, Basilius, gif ðú sylf noldest, ǽre ðú git forðfaran, Homl. Skt. i. 3, 627. Wel lá, mín Drihten, hwæt ic hér nú ł reówlíce hæbbe

gefaren, 23, 575. **Wel lá** (cf. eálá, Bt. 4; Fox 8, 10), ðú éca sceppend ǽra monna cynne O! jam respice terras, Met. 4, 29. Wel lá, monna bearn, 21, 1. Wel lá, men, wel, Bt. 34, 8; Fox 144, 23. Wel gá heia, Wrt. Voc. ii. 110, 30. Weol gá, weol gá euge, euge; Ps. Surt. 69, 4. [Goth. waila: O. Sax. O. Frs. wel: O. H. Ger. wela, wola: Icel. vel.] v. for-wel, and compounds with wel as first component.

wél a pool. v. wǽl.

wela, weola, weala, an; m. I. wealth, riches:—Wela, hord, feoh gazofilacium, Wrt. Voc. ii. 74, 24. Wuldur and wela gloria et divitiae, Ps. Th. 111, 3. Geðenc nú hwæt ðínes ágnes seó ealra ðissa woruldǽhta and welena ... hwæt hæfst ðú ... æt ðám welum? Sege mé nú hwæþer se ðín wela (divitiae) ðínes þances swá deóre seó ... ða welan beóþ leóftǽlran ðonne ðonne hié mon sele, ðonne hié beón ðonne hí mon healt ... Gif nú eall ðises middaneardes wela cóme tó ánum men, hú ne wǽron ealle óþre men wǽdlan? Genóh sweotol ðæt is, ðætte gód hlísa biþ betera ðonne ǽnig wela, Bt. 13; Fox 38, 1–24. Ælc sóþ wela opes, 7, 3; Fox 20, 16. Ðæt unmǽte gestreón goldes and seolfres, oþþe eal se wela, Blickl. Homl. 99, 29. Eal eorþan wela, 51, 30. Wala divitiae, Rtl. 81, 18. Welan patrimonii, welan, spédignesse opulentia, Hpt. Gl. 491, 7–9. Ne biddan wé úrne Drihten ðyses lǽnan welan, ne ðyssa eorþlícra geofa, Blickl. Homl. 21, 11. Of ðisse worulde welan (wǽlom, Lind.) de mamona, Lk. Skt. 16, 9. Úre ieldran begeáton welan, and ús lǽfdon, Past. pref.; Swt. 5, 15. Se man áhte mycelne welan, Blickl. Homl. 197, 30. Ǽhte síne, beágas and botlgestreón, welan, wunden gold, Cd. Th. 116, 4; Gen. 1931: Exon. Th. 331, 1; Vy. 61: Andr. Kmbl. 603; An. 302. Welan bryttian, Cd. Th. 131, 19; Gen. 2178. Weolan, Chr. 1065; Erl. 197, 26: Ps. Th. 16, 9. Gif ðæt ðíne águe welan wǽron, Bt. 7, 3; Fox 20, 18: Blickl. Homl. 53, 21: 99, 24: 113, 25. Wealan (weolan, Surt.) divitiae, Ps. Th. 61, 11. Ægðer ge ðíura welona ge ðínes weorþscipes opum dignitatumque, Bt. 7, 3; Fox 20, 4. Ðæra wlenca ł walana (weolan, Rush.) divitiarum, Mt. Kmbl. Lind. 13, 22. Walana ł weala (willana, Rush.), Mk. Skt. Lind. 4, 19. Wiþsacaþ ðám leásum welum ... and ðám unálýfdum gestreónum, Blickl. Homl. 53, 23. Hé weorþode his deórlingas mid miclum welum, Bt. 28; Fox 100, 29: Andr. Kmbl. 1509; An. 756. Weolum divitiis, Nar. 4, 7: Bd. 4, 11; S. 579, 8. Welum (walum, Lind.), Lk. Skt. 8, 14. Ða welan dǽlan earmum monnum, Blickl. Homl. 49, 32. I a. abundance, wealth:—Hærfest cymþ, wæstmum hladen, wela byð geypwd, Menol. Fox 282; Men. 142. Welan neótan, londes frætwa, Exon. Th. 208, 2; Ph. 149. Mid wuldres welan cum gloria, Ps. Th. 72, 19. Mid welan bewunden, Cd. Th. 27, 19; Gen. 420: 42, 2; Gen. 668. Beóð ðínes wífes welan gelíce swá on wíngearde weaxen berigean uxor tua sicut vitis abundans, Ps. Th. 127, 3. Búwa eorðan and féd ðé on hyre welum (weolum, Surt.) inhabita terram, et pasceris in divitiis ejus, 36, 3. II. weal, prosperity, happy estate:—Bið him se wela onwended, and wyrð him wíte gegearwod, Cd. Th. 28, 5; Gen. 431. Wæs him beorht wela, þenden ðæt folc mid him hiera fæder wǽre healdan woldon, 216, 20; Dan. 9: 96, 32; Gen. 1603. Dó hiá onduéardlíc geféaiga uale fac eos praesenti gaudere prosperitate, Rtl. 70, 1. Onceósan gódes and yfeles, welan and wáwan, Cd. Th. 30, 12; Gen. 466. Hí móton him ðone welan ágan ðe wé on heofonríce habban sceoldon, ríce mid rihte, 27, 24; Gen. 422. Hé þeóda gehwam heofonríce forgeaf, wíðbrádne welum, 40, 22; Gen. 643. God sealde welan swá wíte, swá hé wolde sylf, 256, 23; Dan. 645: Exon. Th. 85, 9; Cri. 1385. [O. E. Homl. Laym. O. and N. wele, weole: A. R. weole: Gen. and Ex. wale: Pr. C. Chauc. Piers P. Gow. wele: O. Sax. welo: O. H. Ger. wela, wola, wolo riches, prosperity.] v. æht-, ǽr-, ǽt-, ár-, blǽd-, bold-, botl-, burg-, eád-, eorþ-, fóddur-, fold-, grund-, hord-, land-, líf-, máðum-, náwiht-, weoruld-wela.

Wéland, es; m. A character in old Teutonic legends celebrated for his skill as a smith. Allusion to him is found in Middle English poetry: 'My sword ... thorrow Velond wroght yt wase,' Torrent of Portugal, ed. Halliwell, l. 428 (v. preface, pp. vii sqq.), and a trace of the legend is preserved in the name Wayland Smith's Cave, in Berkshire (v. infra). Perhaps, too, the same may be said of the river-name Welland (but see Weolud), which occurs in Latin charters as aqua de Uueeland, Cod. Dip. Kmbl. i. 78, 10, aqua de Uueland, 304, 6: ii. pp. 90, 281, 416:—Wéland him wrǽces cunnade, earfoþa dreág, Exon. Th. 377, 9; Deór. 1. Wélandes geworc ne geswíceþ monna ǽnigum, Wald. 2; Vald. 1, 2. Wélandes bearn, 74; Vald. 2, 9. Beaduscrúda betst, Wélandes geweorc, Beo. Th. 914; B. 455. Hwǽr sint nú ðæs foremǽran and ðæs wísan goldsmiðes bán Wélondes ubi nunc fidelis ossa Fabricii (cf. faber) jacent? Bt. 19; Fox 70, 1. Wélandes, Met. 10, 33, 35, 42. ¶ in local names of England:—Ðis sint ðæs landes gemǽre æt Cumtúne (Compton Beauchamp, Berkshire) ... hit cymð on ðæt wíde geat be eástan Wélandes smiððan, Cod. Dip. Kmbl. v. 332, 23. Andlang strǽte on Wélandes stocc (boundaries of land at Princes Risborough, Bucks), Cod. Dip. B. ii. 259, 13. [O. H. Ger. Wielant, Wiolant: Icel. Völundr.] v. Kemble's Saxons in England, i. 420 sqq; Stephens' King Waldere's Lay, pp. 35 sqq.; Grmm. D. M. 350.

wel-besceáwod; adj. Considerate, prudent:—Welbesceáwod consideratus, cordatus, Wrt. Voc. ii. 133, 71. Sý hé á foregleáw and welbesceáwod sit providus et consideratus, R. Ben. 121, 15.

wel-boren; adj. Well-born, noble:—Welboren nobilis, Mk. Skt. Lind. Rush. 15, 43. Monn sum welboren homo quidam nobilis, Lk. Skt. Lind. Rush. 19, 12. Ic nam wíse menn and welborene (nobiles), Deut. 1, 15.

wel-dǽd, e; f. I. a good deed:—Wé sceolon on úrum weldǽdum blissian mid sóðre eádmódnysse, and úrum Drihtne ðancian his gife, ðæt hé ús geúðe, ðæt wé móston his willan gewyrcan þurh sume weldǽde. Ne mæg nán man náht tó góde gedón búton Godes gife, Homl. Th. ii. 432, 6–10. Dó well on eallum ðínum lífe, and wé siððan æfter ðínum weldǽdum ðé eft genimaþ tó ús, 346, 17: i. 414, 30: Homl. Skt. i. 1, 148. Wlitige gewyrtad mid hyra weldǽda, Exon. Th. 234, 21; Ph. 543. Sprec ofter ymb ðóðres monnes weldǽda ðonne ymb ðíne ágene, Prov. Kmbl. 10. II. a benefit, favour, kindness:—Weldǽd benefitium, Cod. Dip. B. i. 155, 19. Hé ús gelǽde tó his Fæder, ðe hine sealde for úrum synnum tó deáðe. Sý him wuldor and lof ðǽre weldǽde, Homl. Th. ii. 282, 27. Weldǽdum beneficiis, Scint. 16, 5. Uton brúcan godcundum weldǽdum, 133, 6: Anglia xiii. 370, 74: Homl. Th. i. 562, 7. Hé wið monna bearn wyrceþ weldǽdum (acts beneficently), Exon. Th. 191, 12; Az. 87. Wé ðínum weldǽdum wurdan áhæfene in beneplacito tuo exaltabitur cornu nostrum, Ps. Th. 88, 14. Nele God ús wítnian for his weldǽdum, oððe his milde mód mannum áfyrran. 76, 7. Weldǽda wítes merita (beneficia) martyrii, Hpt. Gl. 489, 50. Ús God mǽre weldǽda getíðaþ nobis Deus magna beneficia prestet, Scint. 16, 8: Homl. Th. ii. 298, 12: 418, 23. Wé ne magon ásecgean his weldǽda on ús, Basil admn. 4; Norm. 42, 3. Hí ofergeáton weldǽda (-déda, Surt.) his obliti sunt benefactorum ejus, Ps. Spl. 77, 14. III. an office, service:—Be reáfiácum fremedum ælmyssan dón nys weldǽd miltsunge de rapinis alienis elemosinam facere non est officium miserationis, Scint. 159, 16. His éðhylde weldǽde suo contentus officio, 133, 3. Cumlíþnysse and manscipes weldǽdum underþeódde hospitalitatis atque humanitatis offitiis deditos, Cod. Dip. B. i. 154, 38. [Weldede good deeds, O. E. Homl. i. 133, 1. Heo cunnen us unðonc for ure weldede (the good we do them), Laym. 3306. Heom (the gods) wurðen for heore weldǽde (benefits), 8052. Leueþ to writen in wyndowes of 30wre weldedes, Piers P. 3, 70. Goth. waila-déds beneficium: O. H. Ger. wola-tát beneficium, meritum: Ger. wohl-that.]

wel-dón to satisfy, please:—Hé walde ðæm folce weldón (satisfacere), Mk. Skt. Lind. 15, 15.

wel-dónd, -dóend, es; m. A benefactor:—For weldóndum pro benefactoribus, Anglia xiii. 370, 72: 394, 411. Weldóndan, 384, 275. Fore weldóendum mínum, Rtl. 125, 9.

wel-dónde; adj. (ptcpl.) Doing well, acting rightly:—Hú se reccere sceal bión ðæm weldóndum monnum for eádmódnesse geféra ut sit rector bene agentibus per humilitatem socius, Past. 17; Swt. 107, 5.

wel-dónness, e; f. Kindness, benignity:—Weldónnis benignitas, Rtl. 13, 33.

weled. v. wilwian.

weler (-ur, -or), weolor (-ur, -er), es; m.: e; f. A lip, (1) masculine or uncertain:—Weler labium, Wrt. Voc. i. 70, 48. Wæler labrum, 64, 53. Welor labium, 282, 69: ii. 51, 67. Neoðera welor album, 7, 79. Weolure labio, Lchdm. i. lxx, 4. Weleras labia, Ps. Spl. 11, 2, 4: 65, 12: Ps. Th. 62, 5: 65, 12: Kent. Gl. 1002. Weleras (weloras, Cott. MSS.), Past. 15; Swt. 91, 17. Weleras (welras, v. l.), R. Ben. 2, 22. Weoloras, Ps. Th. 30, 20. Welera labiorum, Ps. Spl. 20, 2. Welerum labiis, 62, 6: 119, 2: Mt. Kmbl. 15, 8: Mk. Skt. 7, 6: Homl. Th. ii. 450, 26: labellis, Wrt. Voc. ii. 51, 68. Wælerum labiis, Rtl. 174, 17. Walerum, 179, 11. Welrum buccis, buccellis, Wrt. Voc. ii. 126, 66: labellis, Hpt. Gl. 507, 46. Weolorum labiis, Ps. Th. 11, 2: 20, 2. Wiþ sárum weolorum, gesmire mid hunige ða weoloras, Lchdm. ii. 54, 20. Weleras labia, Ps. Spl. 11, 3: Homl. Th. i. 568, 33: Exon. Th. 363, 15; Wal. 54. Weoloras, Ps. Th. 11, 3. (2) in Ps. Surt., and occasionally elsewhere, the word is feminine:—Wégende welere lying lips; labium mentiens (cf. [wele]ra labium, 418), Kent. Gl. 596. Welure labia, Ps. Surt. 11, 3. Weolure, 62, 6: 65, 14: 70, 23. Weolere, 30, 19: 62, 4. Weolre, 11, 5: 118, 171. Weolera labiorum, 20, 3: 58, 13. Weolerum labiis, 58, 8: 118, 13: 119, 2: 139, 3. Weolure labia, 11, 4. Ic ne wirne míne welora labia mea non prohibebo, Past. 49; Swt. 380, 10. Gif mannes núð sár sié, genim betonican ... lege on ða weolore, Lchdm. ii. 48, 29. [Goth. wairilô.]

wel-frem[m]ende (-fremende, Rush.) geceiged bíðon benefici vocantur, Lk. Skt. Lind. 22, 25.

wel-fremming, e; f. A well-doing, benefit, kindness:—Uelfremming beneficium, Rtl. 187, 39.

wel-fremnes, e; f. A benefit:—Uelfremnisum beneficiis, Rtl. 58, 31. Uelfremnisse benefica, 39, 19. Uoelfremnisse, 73, 3: 77, 41.

wel-gecwéme glosses beneplacitus, Ps. Spl. 118, 108: 146, 12.

wel-gecwémedlíc glosses beneplacitus, Ps. Spl. 149, 4.

wel-gecwémness, e; f. Well-pleasingnes, good pleasure: — In

4 G

welgecuoemnise (*beneplacito*) áucendes bearnes ðínes, Rtl. 174, 33: 173, 25.

wel-gedón *well done:*—Gif hwæt welgedónes bið *si qua bene gesta sunt*, Past. 17; Swt. 111, 3. Suíðe suíðe wé gesyngiaþ, gif wé óðerra monna welgedóna dǽda ne lufigaþ *valde peccamus, si aliena bene gesta non diligimus*, 34; Swt. 231, I. *The word also glosses* beneficium:—Welgidoeno *beneficia*, Rtl. 23, 7.

wel-gehwǽr; *adv. Everywhere:*—Hí welgehwǽr hergedon and bærndon, Chr. 1001; Erl. 136, 2. v. wel-hwǽr.

wel-gelǽred; *adj. Well-instructed:*—Larwas ł welgilǽrde Godes *docibiles Dei*, Jn. Skt. Rush. 6, 45.

wel-gelícod *glosses* beneplacitum :—In welgelícodum heara *in beneplacitis eorum*, Ps. Surt. 140, 5.

wel-gelícwirþe *glosses* beneplacitus, V. Ps. 118, 108.

wel-gelícwirþniss *glosses* beneplacitum, V. Ps. 140, 7.

wel-geþungen; *adj. Of great excellence:*—Welgeþungene witan, L. I. P. 10; Th. ii. 316, 23. v. wel-þungen.

welgian. v. weligian.

wel-hǽwen; *adj. Beautifully blue:*—Ðæt bleóh ðæs welhǽwnan iacintes bið betera ðonne ðæs blácan carbuncules *coerulei coloris hyacinthus praefertur pallenti carbunculo*, Past. 52; Swt. 411, 28.

wel-hwá; *pron. Every one, every thing:*—Mé ðás woruldsǽlða welhwæs blindne (*altogether blind*) on ðis dimme hol forlǽddon, Met. 2, 10. Hé þenceþ ðæt his wíse welhwam þince eal unforcúþ, Exon. Th. 315, 13; Mód. 30. Weódmónað on tún welhwæt bringeþ, Menol. Fox 274; Men. 138.

wel-hwǽr; *adv. Everywhere, generally, commonly :*—Welhwǽr *passim*, Wrt. Voc. ii. 67, 22 : *vulgo*, 79, 36. Unriht gewuna welhwǽr is árisen, Bd. 1, 27; S. 493, 33. Swá gelǽrede biscepas, swá swá welhwǽr (well-, Cott. MSS.) siendon, Past. pref.; Swt. 9, 4. Wæs wíde and welhwǽr Waldendes lof áfylled, Chr. 975; Erl. 126, 11. Wíod ða ðe willaþ welhwǽr derian clǽnum hwǽte, Met. 12, 4. Mæniges þinges ðe monnum wunder welhwǽr þynceþ, 28, 82. v. ge-welhwǽr.

wel-hwilc; *pron. Every :*—Hit (*reason*) nǽnig hafaþ neát . . . hæfð ða wilnunga welhwilc néten, Met. 20, 191. Hine gearwe geman witena welhwylc, Beo. Th. 537; B. 266. Welhwylc gecwæð ðæt hé fram Sigemunde secgan hýrde, 1753; B. 874. Se ðe eów welhwylcra wilna dohte, 2692; B. 1344. v. ge-welhwilc.

welig (-eg); *adj. Wealthy, rich, opulent,* (1) *of persons, in respect to material or non-material riches :*—Welig *dives*, Wrt. Voc. i. 74, 18 : *pecuniosus*, 54, 53. Sum welig man wæs *homo quidam erat dives*, Lk. 16, 1, 19. Sum weli (welig, MS. A.: wælig, Lind.) mann, Mt. Kmbl. 27, 57. Hé wæs swíðe welig (weolig, Rush.), Lk. Skt. 18, 23. Sum welig mon *vir quidam, privatis opibus reipublicae vires superans*, Ors. 4, 5; Swt. 166, 24. Hé wæs swíðe welig þearfum, and him sylfum swíðe hafenleás, Homl. Th. ii. 148, 33. Swíðe welig cn golde and on seolfre and on orfe and on geteldum, Gen. 13, 5. Forseó ðysse worulde wlenco, gif ðú wille beón welig on dínum móde, Prov. Kmbl. 50. Ðes and ðeós welega *hic et haec dives*, Ælfc. Gr. 6, 2 ; Zup. 18, 12. Earfoðlíce se welega (-iga, Rush.) gǽð on Godes ríce, Mt. Kmbl. 19, 23 : Ps. Th. 71, 12 : Blickl. Homl. 51, 2. Se welega man, 197, 28. Weliga, Exon. Th. 245, 1 ; Jul. 38. On ðæs rícan neáweste and ðæs welegan, Blickl. Homl. 53, 5. Hwæt bið ðæm welegan (welgan, Bt. 26, 3 ; Fox 94, 12) woruldgítsere ðe bet, Met. 14, 1. Ðæm welgan, Mt. Kmbl. Rush. 19, 24. Welige *dites, divites*, Wrt. Voc. ii. 27, 46. Manega welige (wealigo, Lind. : weolge, Rush.) torfudon fela, Mk. Skt. 12, 41. Weolie, Ps. Surt. 33, 11. Ða welegan, Past. 26 ; Swt. 181, 3. Gongan tó byrgenne weligra manna, Blickl. Homl. 99, 13. Wǽ iúh weligum, Lk. Skt. Lind. 6, 24. Geceósan welige yldran, Blickl. Homl. 23, 25. Ge ða welegan ge ða þearfan, 107, 12. Ne clypa ðú ðíne welegan (weligo, Lind. : wealigo, Rush.) néhhebúras, Lk. Skt. 14, 12. Ða welegan (weligo, Lind. : weolige, Rush.), 21, 1. Swá mycele swá se mann biþ weligra on ðisse worlde, swá him se uplíca Déma tó sécþ, Blickl. Homl. 95, 32. Weolegrum *ditiori*, Kent. Gl. 834. Weliogran (= wiolegran) *ditiores*, 377. Welegost, Bt. 26, 1 ; Fox 92, 7. (2) *of places where wealth is accumulated :*—On ðære welegan byrig (*Rome*), Met. 1, 37. Wícstede welig ne, Beo. Th. 5207; B. 2607. Hé wolde oferwinnan sume welige burh, Homl. Skt. ii. 25, 532. Néron ðá welige hámas, Bt. 15; Fox 48, 4: Met. 8, 8. Setl wuldorspédum welig, Cd. Th. 6, 11 ; Gen. 87. Babylonia ðe ðá welegre wæs ðonne ǽnigu óþeru burg *Babyloniam, urbem tunc cunctis opulentiorem*, Ors. 2, 4; Swt. 72, 26. Sidonem, seó wæs welegast (*opulentissima*) on ðǽm dagum, 3, 5; Swt. 104, 30. (3) *of places or things which produce abundantly, of seasons in which there is abundance :*—Ðæt wiolie *opimum*, Wrt. Voc. ii. 64, 64. Eorðan ðú gefyllest éceum wæstmum, ðæt heó welig weorþeþ *multiplicasti locupletare terram*, Ps. Th. 64, 9. Hit is welig, ðis eálond, on wæstmum and on treówum *opima frugibus atque arboribus insula*, Bd. 1, 1 ; S. 473, 12. Hwæðer hit nú ðínes gewealdes sié ðæt se hærfest sié swá welig on wæstmum *an tua in aestivos fructus intumescit ubertas?* Bt. 14, 1 ; Fox 40, 28. Wæstmbǽre geár and welige *ubertatis anni*, Gen. 41, 26. Swíðe wæstmbǽre geár and swíðe welige *anni fertilitatis*,

41, 29. (3 a) *fig.* :—Mid ðam gelǽredan biscope hé wunode on weligre láre tó langum fyrste *with that learned bishop he continued for a long time, engaged in learning which was rich in results*, Homl. Th. ii. 502, 21. [*Laym.* weoli: *C. M.* weli: *O. L. Ger.* welag *ditis*: *O. H. Ger.* welac *ditis*.] v. folc-, mód-welig.

welig; *m. A willow :*—Welig *salix*, Wrt. Voc. i. 285, 62. Weliges leáf, Lchdm. ii. 156, 1. Welies, 154, 22. Ǽrest on ðone welig; of ðam welige, Cod. Dip. Kmbl. iii. 223, 23. Tó ðam greátan welige, 438, 3. On ðone ealdan myl[en] ðǽr ða welegas standaþ, ii. 250, 10. On welgum *in salicibus*, Blickl. Gl. [*Chauc.* wilwe: *Prompt. Parv.* wylowe, wilwe. Welogh *salix*, Wrt. Voc. i. 228, col. 2 (15th cent.).] v. wiliht.

weligian; *p.* ode. I. *to make rich, enrich :*—Ic weligie beo, ic welegode *beavi*, Ælfc. Gr. 24; Zup. 137, 1. II. *to become rich or abundant, to abound :*—Tír welgade, Exon. Th. 353, 58; Reim. 34. v. ge-welgian.

welig-stedende; *ptcpl. Making rich :* — Uoeligstydende (*printed* uoeglig-) *locupletans*, Rtl. 98, 18. Cf. stede.

Welisc, well, wellcumian, welle, wellere. v. Wilisc, will, wilcumian, wille, wellyrge.

wel-libbende; *adj.* (*ptcpl.*) *Of good life, living aright :*—Ðæt mynster hé gelógode mid wellybbendum mannum, Homl. Th. ii. 506, 16. Ongeán ða gódan and ða wellibbendan *bene viventibus*, Past. 17 ; Swt. 107, 14.

wel-lícung, e; *f. Well-pleasing :*—Wellícunga *beneplaciti*, Ps. Spl. T. 68, 16.

wellung. v. willung.

wellyrge, wellere *are glosses of* sinus :—Wellyrgae (uuellyrgae *sinus, simus*, Ep. Erf.) *smus* (for *sinus*), Txts. 97, 1876. Wellere *sinus*, Wrt. Voc. i. 289, 34. [*The form* wellyrgae *looks as if taken from a Latin form* velluria (?).]

welm, welode. v. wilm, wilwian.

wel-rúmlíce; *adv. Kindly, benignantly;* benigne, Rtl. 41, 11 : 46, 14 : 109, 4.

wel-rúmmód; *adj. Kind, benignant :*—Uelrúmmódo *benigni*, Rtl. 12, 39.

wel-stincende; *adj.* (*ptcpl.*) *Fragrant, sweet-smelling :*—Wyrta swíðe welstincenda *olera bene olentia*, Past. 57 ; Swt. 439, 33.

wel-swégende; *adj.* (*ptcpl.*) *Melodious, sonorous :*—Heriaþ hine on cimbalum welswégendum *laudate eum in cymbalis bene sonantibus*, Ps. Spl. 150, 5.

weltan. v. wiltan.

wel-þungen; *adj.* (*ptcpl.*) *Well-thriven, able, good, proficient, excellent :*—Hygd wæs swíðe geong, wís, welþungen, Beo. Th. 3858 ; B. 1927 : Menol. Fox 309 ; Men. 156. v. wel-geþungen.

weluc. v. weoloc.

welwan (?) *to seize :*—Wyleþ (*printed* wylcþ ; *but see* Lchdm. iii. 373, col. 1 *under* wylan, *where also Cockayne notes that the Latin is* captat, *not* raptat) *captat* (*printed* raptat), Germ. 389, 42. [*Goth.* wilwan; *p.* walw *to seize*.]

wel-weorþ; *adj. Of high esteem, of great account :*—Hé swá wuldorfulle and Gode swá welweorþe (wel weorþe? v. weorþ, III a) leóde geneósian wolde, Lchdm. iii. 432, 31.

wel-willedness, e; *f. Benevolence, kindness :*—Mǽre ys welwyllednyss ðænne ðæt ys geseald . . . nys sóðlíce mildheortnyss ðǽr nys welwillednyss *maior est beniuolentia quam quod datur . . . non est enim misericordia ubi non est beniuolentia*, Scint. 160, 4–6.

wel-willende; *adj.* (*ptcpl.*) I. *of good will, benevolent, benignant, kind :*—Welwillende *beniuolus*, Ælfc. Gr. 14; Zup. 87, 17. Ic ðé hálsie, ðú árfæsta, welwilende and welwyrcende Dryhten, Shrn. 169, 19. Swá him gewissode se welwillenda God, Jud. 6, 14 : Homl. Ass. 55, 122. Se wellwillenda bisceop Æðelwold (cf. Adelwoldus benevolus et venerabilis presul, Homl. Th. i. 1, 3), Chr. 984 ; Erl. 130, 1. Se welwillenda man wyle eáðe forberan gif hine man áhwǽr týnd, Basil admn. 4; Norm. 44, 17. Hé hit þearfum dǽlde mid welwillendum móde, Homl. Skt. ii. 26, 59. Tó ðam welwillendan Hǽlende, Homl. Th. ii. 230, 11: Homl. Ass. 80, 186 : 101, 329. Wynsum ús byð ðæt wé welwyllende beón, 10, 267. Gebyreþ ðætte sume, ða ðe welwillende beóð, on monegum weorcum unfæste beóð ongietene *contigit, ut quidam cum cordis innocentia in nonnullis suis actibus infirmi videantur*, Past. 34; Swt. 235, 17. Ða welwillendan *benevoli*, Swt. 229, 10. II. *of right will, right-minded :*—Ðá Dauid ðysne sealm sancg, ðá gealp hé and fægnode Godes fultumes wið his feóndum; and swá deð ǽlc welwillende man, ðe ðisne sealm singð, Ps. Th. 4, arg. [Þe dol, þet God 3efþ to his welwilynge . . . þet is to alle guode herten, Ayenb. 112, 11. Welewyllynge or of god wylle, welwyllyd *benevolus*, Prompt. Parv. 521.]

welwillendlíce; *adv. Benevolently, kindly :*—Wellwillendlíce dó, Drihten *benigne fac, Domine*, Ps. Lamb. 50, 20. Wópas welwillendlíce underfóh *fletus benigne suscipe*, Hymn. Surt. 29, 17. Wolde se heofenlíca lǽce ðæt geswell heora heortan welwyllendlíce gelácnian, Homl. Th. i. 338, 23 : Homl. Skt. i. 3, 64 : Wulfst. 295, 2.

welwillendness, e; *f. Benevolence, benignity, kindness:*—God wolde for his welwillendnysse ûs earmingas âlŷsan, Hexam. 18; Norm. 26, 27. Se cyngc blissode on his dohtor welwillendnysse, Ap. Th. 16, 11. On ðinre welwyllendnysse, Homl. Th. ii. 598, 17. Ofer welwillendnysse *super benignitatem*, Ps. Lamb. 51, 5: Homl. Skt. ii. 31, 44: Anglia xi. 114, 94. Wellwillendnysse, 84, 13: Basil admn. 9; Norm. 54, 16. Wellwyllendnysse, 5; Norm. 44, 22.

welwilness, e; *f. Good will, kindness, goodness:*—Welwilnes, Shrn. 175, 28. Đû ûs gescyldest mid ðam scylde ðinre welwilnesse *ut scuto bonae voluntatis tuae coronasti nos*, Ps. Th. 5, 13. Hym ic mê befeste and hys welwylnesse ic mê bebeóde, Shrn. 189, 34.

wel-wyrcende *well-doing:*—Ic ðê hâlsie, ðû ârfæsta, welwilende and welwyrcende, Shrn. 169, 19. Ælcum welwyrcendum God myd beó midwyrhta, 179, 29. Se freódóm ðæs deófollîcan onwaldes wæs seald eallum welwyrcendum, Blickl. Homl. 137, 14.

wéman; *p. de To allure, attract, persuade, entice*, (1) in a good sense:—Đa gesetednessa ðe tô hâlgum mægenum wêmaþ, Lchdm. iii. 440, 24. Hine mon georne wême ðæt hê wununge healde *suadeatur ut stet*, R. Ben. 109, 22. Đæt wê tô ælcan rihte ûs sylfe wenian and wêman, Wulfst. 266, 6. Hwær ic findan meahte ðone ðe mec frêfran wolde, wêman (wenian? *q. v.*) mid wynnum, Exon. Th. 288, 10; Wand. 29. (2) in a bad sense:—Đa teolunga ðe hine fram Gode wêmaþ, Homl. Th. ii. 288, 24. Hî (*devils*) dugude beswîcaþ and on teosu tyhtaþ tilra dæda, wêmaþ on willan, ðæt hŷ sêcen frôfre tô feóndum, Exon. Th. 362, 11; Wal. 35. v. ge-wêman.

wêmere, es; *m. One who allures* or *entices, a pander:*—Wêmere *vel* tihtere *leno*, Wrt. Voc. i. 50, 55.

wem-lîc. v. un-wemlîc.

wemm (?) *a spot:*—Wið wemme (cf. 34, 9 *which has* wenne) on eágum, Lchdm. ii. 2, 8. [*A. R. Chauc. Piers P. Wick.* wem.]

wemman; *p. de.* **I.** *to spot, mar, spoil, disfigure*, (a) lit.:—Unwlitig swile and atelîc his eágan bregh wyrde and wemde *tumor deformis palpebram oculi foedaverat*, Bd. 4, 32; S. 611, 18. (b) fig.:—Ic hâliges lâre wordum wemde (*I have not given a good account of the saint*), Andr. Kmbl. 2958; An. 1482. Wordum wemman *to reproach, blame* (cf. *Goth.* ana-wammjan *vituperare*):—Stefn æfter cwom, wordum wemde, Andr. Kmbl. 1479; An. 741. Đec (*the body*) ðîn sâwl sceal oft gesêcan, wemman mid wordum (cf. nemnan ðê mid wordum, Squl Kmbl. 127), Exon. Th. 370, 24; Seel. 64. **II.** *to defile, pollute, profane:*—Gyf rihtwîsnys mîn hî wemmaþ *si justitias meas profanaverint*, Ps. Spl. 88, 31. Gif hê ôðres ceorles wîf wemme (*maculaverit*), L. Ecg. C. 14; Th. ii. 142, 12. [Ho of hire meidenhad nawiht ne wemde, O. E. Homl. i. 83, 8. Ȝho ne shollde nohht ben wemmedd, Orm. 2326. He wolde þys tendre þyng wemmy foule, R. Glouc. 206, 1. Wemmed *maculatus*, Wick. Deut. 12, 15. *Goth.* ana-wammjan *to blame: O. H. Ger.* bi-, gi-wemmen.] v. ge-wemman; un-wemmed.

-wemme, -wemmedlîc, -wemmedlîce, -wemmedness. v. un-wemme, ge-wemmedlîc, ge-wemmedlîce, ge-wemmedness.

wemmend, es; *m. A fornicator, adulterer:*—Wemmend *scortator, adulter, fornicator*, Hpt. Gl. 484, 61. v. ge-wemmend.

-wemmendlîc. v. ge-wemmendlîc.

wemming, e; *f. Pollution, defilement:*—Wemmincge (wêmincge? v. wêman) *lenocinii, seductionis*, Hpt. Gl. 507, 20. [Wiðute wemmunge, H. M. 13, 24.] v. ge-, un-wemming.

wemness, e; *f. Pollution*, Shrn. 183, 21. v. ge-, un-wemness.

wên, e; *f.* **I.** *supposition, opinion, thought, idea:*—Hî fleóð swâ hrædlîce swâ is wên ðætte hî fleógon *longe fugiunt quasi putes eos volare*, Nar. 37, 15. Đû (*Joseph*) fæder cweden woruldcund bi wêne (cf. Jesus erat, ... ut putabatur, filius Joseph, Lk. 3, 23), Exon. Th. 13, 33; Cri. 212. Woeno *opiniones*, Mt. Kmbl. Lind. 24, 6: Mk. Skt. Lind. 13, 7. **II.** *hope, expectation:*—Hié cwædon ðæt heó rîce âgan woldan ... Him seó wên geleáh, Cd. Th. 4, 5; Gen. 49: 87, 10; Gen. 1446: Andr. Kmbl. 2150; An. 1076: Beo. Th. 4636; B. 2323. Đæs ic wên hæbbe *as I hope*, 772; B. 383. Wêna mê dîne (*the unsatisfied hopes of seeing thee*) seóce gedydon, ðîne seldcymas, Exon. Th. 380, 25; Râ. 1, 13. Sibbe oflyste, wynnum and wênum, 464, 4; Hö. 82. Wênum *hopefully, expectantly*, 380, 17; Râ. 1, 9. **II a.** *with expectation of what is hoped for or expected:*—On ðam is godcundnesse wên ðe manna ingehygd wât *divinity may be expected in him who knows men's hearts*, Blickl. Homl. 179, 25: Exon. Th. 302, 21; Fä. 39. Wistfylle wên, Beo. Th. 1472; B. 734. Is leódum wên orleghwîle, 5813; B. 2910: Exon. Th. 384, 16; Râ. 4, 28. Mê ðæs wên næfre forbirsteþ, ðe ic gefeán hæbbe, 236, 1; Ph. 567. Him wæs bêga wên, Beo. Th. 3751; B. 1873. Weán on wênum *in expectation of misery*, Cd. Th. 63, 4; Gen. 1027: 191, 11; Exod. 213: 163, 18; Gen. 2700: Andr. Kmbl. 2176; An. 1089. Đîn on wênum, Exon. Th. 474, 12; Bo. 28. Bêga on wênum, endedôgores and efscymes, Beo. Th. 5783; B. 2895. **III.** *likelihood, probability, chance:*—Nû is wên micel ðæt heó mec eft wille gehŷnan *there is now a great probability that she will again humiliate me*, Exon. Th. 280, 21; Jul. 632. Is mê on wêne geþuht ðæt ðê untrymnes oysgade *it seems to me in all likelihood that*

sickness has troubled you, 163, 6; Gû. 989. Wên ic talige, gif ðæt gegangeþ, ðæt se gâr nimeþ ealdor ðînne *I reckon there is likelihood, if that comes to pass, that the spear will carry off thy prince*, Beo. Th. 3695; B. 1845. **III a.** *in phrases such as* wên is (ðæt) = *perhaps, perchance, may be, probably:*—Wênunge, wên is *forsitan*, i. *forsan, fortasse*, Wrt. Voc. ii. 150, 24. Gyf gê mê cûþon, wên is ðæt gê cûþon mînne fæder *si me scieretis, forsitan et patrem meum scieretis*, Jn. Skt. 8, 19: Ps. Th. 123, 2, 3. Gif ðû wistes, ðû uoen is (woen is mâra, Rush.) gif ðû gegiuuedes *si scires, tu forsitan petisses*, Jn. Skt. Lind. 4, 10. Cum mid ûs, ðŷ læs wên is hî ûs eft genimon *come with us, lest haply they take us again*, Blickl. Homl. 239, 9. Đŷ læs wên sié ðæt hine God gefreólsige, 243, 19: 247, 2. Wên is ðæt ic gefyrenode *perhaps I have sinned*, 235, 32: 239, 29: Homl. Th. i. 92, 30. Ne biþ his lof nâ ðŷ læsse, ac is wên ðæt hit sié ðŷ mâre *his praise will not be the less, but may be the greater*, Bt. 40, 3; Fox 238, 11. Him bið forboden ðæt hê offrige, forðæm hit is wên ðæt se ne mæge ôðerra monna scylda of âðueán, Past. 11; Swt. 73, 17. Hit is þêh wên ðæt feala manna þence hwylcum edleáne hê onfô æt Drihtne, Blickl. Homl. 41, 14. Hwæðer hyt wên sig ðæt ðû sig se ylca Hælend ðe Satan ûre ealdor ymbe spæc? (*perhaps thou art that Jesus of whom Satan spoke*, Gospel of Nicodemus 17, 12), Nicod. 28; Thw. 16, 35. Mâra woen is *quanto magis*, Mt. Kmbl. Lind. 7, 11: 12, 12: Lk. Skt. Rush. 11, 13 (Mâra woen, Lind.). Mâra woen *alio quin*, Mk. Skt. Lind. Rush. 2, 22. Nys hit wêne sôþ ðæt wê gelŷfan sceolon ðâm cempon ..., ac ys bet wên ðæt (*more likely*) his cnyhtas cômon and heom feoh geáfon (*perhaps his disciples gave them money*, Gospel of Nicodemus 10, 29), Nicod. 19; Thw. 9, 13. Hû mæg ic hit gefaran? ac nâ wên is ðæt ðû onsende ðînne engel *how can I do the journey? but more likely thou mayst send thine angel*, Blickl. Homl. 231, 23. Nimðe wên wære *ni forsan*, Wrt. Voc. ii. 93, 3. Cômon hî tô Eald-Seaxna mægþe gif wên wære ðæt hî ðær ænige ðurh heora lâre Criste begitan mihte (*si forte aliquos ibidem praedicando Christo adquirere possent*), Bd. 5, 10; S. 624, 13. [Of þine kume nis na wene (*expectation*), Laym. 28141. Hit bið a muchele wæne it is *very doubtful*, 13503. Wen is þatt (*probably*) he wass forrdredd, Orm. 7152. Efter monnes wene *as men suppose*, A. R. 390, note e. *Goth.* wêns *spes: O. Sax. O. L. Ger.* wân *hope: O. Frs.* wên *opinion: O. H. Ger.* wân *opinio, existimatio, aestimatio, suspicio, spes: Icel.* vân *hope, expectation*.] v. next word.

wêna, an; *m.* **I.** *supposition, opinion, thought, idea, imagination:*—Se leása wêna and sió rædelse ðara dysigena monna *hominum fallax opinio*, Bt. 27, 3; Fox 98, 32. Swâ sume wênaþ, ðæt sió sunne dô, ac se wêna nis wuhte ðe sôþra, Met. 28, 35. Gewyrd nis nân ðing bûton leás wêna. ... Gê habbaþ nû gehŷred þe ðan leásan wênan, ðe ŷdele men gewyrd hâtaþ, Homl. Th. i. 114, 13-34. Sume men wênaþ, ðæt ...: ac gif heora wêna sôþ wære, ðonne ..., 124, 18. Se ðe wæs Crist geteald mid ungewissum wênan, 358, 3. Be wênan (*as a matter of opinion*) hî healdaþ God ælmihtigne, R. Ben. 135, 24. For dysiges folces wênan *falsis vulgi opinionibus*, Bt. 30, 1; Fox 108, 4. Hê ongeat ðæt hié wæron onstyrede mid ðæm wênan ðæt hî ðæs endes suâ neáh wêndon *commotos eos vicini finis suspicione cognoverat*, Past. 32; Swt. 213, 23. Đæt hié ne læten hiera geðeaht and hiera wênan suâ feor beforan ealra ôðerra monna wênan *nequaquam cunctorum consilia suae deliberationi postponerent*, 42; Swt. 306, 1-2. Gif ðæt ondgit ongiett ðæt hit self dysig sié, ðonne gegrîpð hit ðurh ðone wênan ðæt andgit ðære incundan byrhto, 11; Swt. 69, 21. Hit is betere, ðæt ælc mon âdrŷge of ôðerra monna môde ðone wênan be him ælces yfeles *cum prava aestimatio ab intuentium mente non tergitur*, 59; Swt. 451, 23. Đâ befrân hê, hû woruldmenn be him cwyddedon ... hê wolde âdwæscan ðone leásan wênan dweligendra manna, Homl. Th. i. 366, 8. Wênena *suspicionum*, Hpt. Gl. 471, 26. **II.** *hope, expectation:*—Ne weorðe ðê næfre tô ðæs wâ, ðæt ðû ne wêne betran andergilde; for ðam ðe se wêna ðe næfre tât forweorðan, Prov. Kmbl. 41. Ætes on wênan, Cd. Th. 188, 9; Exod. 165: 119, 2; Gen. 1985: Elen. Kmbl. 1165; El. 584: Exon. Th. 378, 32; Deór. 25. v. preceding word.

wênan; *p. de.* **I.** *to ween, suppose, think, imagine, opine, believe*, (1) absolute:—Ic wêne *autumo*, Wrt. Voc. ii. 4, 68. Wênð *opinatur*, 62, 53. Hê wênð *estimat*, Kent. Gl. 870. Hwilum ic gewîte, swâ ne wênaþ men (cf. Aldhelm's riddle: Cernere me nulli possunt), Exon. Th. 381, 24; Râ. 3, 1. Wênde *metitur*, Wrt. Voc. ii. 58, 31. Wêndau *autumant*, 95, 69. Ne meahton hié, swâ hié wêndon ær, Elen. Kmbl. 954; El. 478. Wênde *arbitraretur*, Wrt. Voc. ii. 3, 36. (2) with accusative:—Hwæt wênst ðû? hwæt is ðes? *quis putas est iste?* Mk. Skt. 4, 41. Hwæt wêne gê? *quid putatis?* Jn. Skt. 11, 56. Đæs ðe hê wênde *according to his belief*, Chart. Th. 140, 7. Ûs gedafenaþ ðæt wê hit wênon swiðor ðonne wê unrædlîce hit gesêþan ðæt ðe is uncûð bûton ælcere fræcednysse *it befits us to hold this as an opinion, where absence of certain knowledge is without any peril, rather than to assert it unadvisedly*, Homl. Th. i. 440, 31. Nis ðæt nô lîchomlîce tô wênanne, ac gâstlîce *that is not to be estimated corporeally, but spiritually*, Bt. 42; Fox 258, 13. (2 a) with acc. pron. and appositional clause:—Ic ðæt wênde and witod tealde, ðæt ic ðê meahte âhwerfan, Exon. Th. 263, 29;

Jul. 357. (3) with genitive:—Ne wēne ic his nô, ac wât geara, Bt. 38, 6; Fox 208, 13. Gif hē wyrsa ne bið, ne wēne ic his nā beteran, Met. 25, 29. Hié ðæt fǽge þēgon, þeáh ðæs se ríca ne wēnde, Judth. Thw. 21, 16; Jud. 20. Onstyrede mid ðæm wēnan ðæt hí ðæs endes suā neáh wēndon *commotos vicini finis suspicione*, Past. 32; Swt. 213, 24. Hí wēndon his beteran ðonne hē wǽre, Bt. 30, tit.; Fox xvi, 5. Hwæðer ðū wēne ðæt ǽnig mon sié swā andgetfull, ðæt hē mæge ongitan ǽlcne mon on ryht hwelc hē sié, ðæt hē náuþer ne sié ne betera ne wyrsa ðonne hē his wēne? *num ea mentis integritate homines degunt, ut quos probos improbosve censuerint, eos quoque, uti existimant, esse necesse sit?* 39, 9; Fox 226, 3. (3 a) with gen. and *tô*:—Ðonne scencð hē ða scylde ǽlcum ðara ðe him ǽnges yfles tô wēnð. For ðæm hit gebyreþ oft, ðonne hwā ne rēcð hū micles yfeles him mon tô wēne . . . *cunctis mala credentibus culpa propinatur. Unde plerumque contigit, ut, qui negligenter de se mala opinari permittunt* . . . , Past. 59; Swt. 451, 24–27. Him is ðeáh leófre ðæt hē leóge, ðonne him mon ǽnigra ungerisna tô wēne *eligit bona de se vel falsa jactari, ne mala possit vel minima perpeti*, 33; Swt. 217, 16. Ðæs ilcan is tô wēnanne tô eallum ðām gesǽldum ðe seó wyrd brengð *de cunctis fortunae muneribus illud etiam considerandum puto*, Bt. 16, 3; Fox 54, 24. (3 b) with gen. and appositional clause:—Wē ðæs wēnaþ, ðæt ús God mæge bringan tô beód gegearwad *numquid poterit Deus parare mensam?* Ps. Th. 77, 20. Wēnaþ ðæs sume, ðæt ic on seáð mid fyrenwyrhtum feallan sceolde *aestimatus sum cum descendentibus in lacum*, 87, 4. Ic ðæs wēnde, ðæt ic ongitan mihte *existimabam ut cognoscerem hoc*, 72, 13. Wēnde ðæs formoni man, ðæt wǽre hit úre hláford, Byrht. Th. 138, 52. Ne wēne ðæs ǽnig, ðæt ic lygewordum leóð somnige, Exon. Th. 234, 26; Ph. 546. Ne þurfan wē nā ðæs wēnan, ðæt hē ús nolde ðæra leána gemānian, Wulfst. 261, 18. (4) with a clause, (a) introduced by *ðæt*:—Ic wēne, ðæt nān mon ne sié *neminem esse hominum arbitror*, Ors. 2, 1; Swt. 58, 13. Hwam wēne (woeno, Lind.) ic ðæt hit beó gelíc? *cui simile esse existimabo?* Lk. Skt. 13, 18, 20. Wēn ic, ðæt . . ., Beo. Th. 681; B. 338: 888; B. 442. Hig wēnaþ (woenas, Lind.: woenaþ, Rush. *putant*), ðæt hí sín gehýrede, Mt. Kmbl. 6, 7. Ðonne wēnaþ hí swā ungewitfulle, ðæt hí habban ða sôþan gesǽlþa, Bt. 32, 3; Fox 118, 30: Met. 19, 34: Exon. Th. 360, 25; Wal. 11: Cd. Th. 109, 22; Gen. 1826. Wēndes ðū, ðæt ðū āhtest alra onwald, 268, 22; Sat. 59. Ðā wēnde hē (*suspicatus est*), ðæt hit wǽre sum myltystre, Gen. 38, 15: Blickl. Homl. 175, 6: Chr. 911; Erl. 100, 21: Cd. Th. 44, 20; Gen. 712. Nalles hē wēnde, ðæt hié hit wiston, 249, 14; Dan. 530. Wēndun gē and woldun, ðæt gē Scyppende sceoldan gelíce wesan, Exon. Th. 141, 30; Gú. 635. Hí wēndon, ðæt hig sceoldon māre onfón *arbitrati sunt quod plus essent accepturi*, Mt. Kmbl. 20, 10. Wēndon (woendon Lind.: woendun, Rush.) *putaverunt*, Mk. Skt. 6, 49: Jn. Skt. 11, 13: Lk. Skt. 3, 23. Wēndon, ðæt hē on heora gefére wǽre *existimantes illum esse in comitatu*, 2, 44. Wēndan, Exon. Th. 460, 8; Hö. 14. Ne wēne gē, ðæt . . . *nolite arbitrari quia* . . ., Mt. Kmbl. 10, 34. Ðeáh gē nū wēnen and wilnian, ðæt gē lange libban scylan *si putatis longius vitam trahi*, Bt. 19; Fox 70, 14: Met. 10, 63. Nelle gē wēnan (woenæ, Lind.), ðæt . . . *nolite putare quoniam* . . ., Mt. Kmbl. 5, 17. Ne þurfon gē wēnan, ðæt . . ., Blickl. Homl. 41, 12: Met. 29, 39: Exon. Th. 142, 16; Gú. 645. Nis tô wēnanne ðætte wolde God hiora gāsta mid him gýman non *est creditus cum Deo spiritus ejus*, Ps. Th. 77, 10: Bt. 16, 3; Fox 56, 28. (b) not introduced by *ðæt*:—Ic wēne (*arbitror*), ne mihte ðes middaneard ealle ða béc befón, Jn. Skt. 21, 25. Ic wēne (woeno, Lind., *aestimo*), se ðe hē māre forgef, Lk. Skt. 7, 43. Ic wēne, wit sýn oferswíþede, Blickl. Homl. 181, 29. Wēne wē, sý ðis se? 85, 16. Wēnst ðū hwæt is ðes? *quis putas hic est?* Lk. Skt. 8, 25. (5) with acc. and infin.:—Wēn ealle uferan beón ðē *aestima omnes superiores esse tibi*, Scint. 22, 2. (6) with a preposition:—Ðā ongan ic ofer ðæt georne wēnan *I began to make conjectures on the circumstance*, Homl. Skt. ii. 23 b, 420. II. *to hope, expect, look for*, (1) absolute:—On ðam dæge ðe hē nā ne wēnd (woenas, Lind.) *in die, qua non sperat*, Mt. Kmbl. 54, 50. Ðonne hý lǽst wǽnaþ (wēnaþ, Cott. MS.), Bt. 7, 1; Fox 16, 13. Ðe lǽs ðe wē forweorðan, ðonne wē lǽst wēnan, Wulfst. 76, 1. (1 a) with preps. marking the direction of the expectation or hope:—Geþyld hafa, swā ic ðē wēne tô, Beo. Th. 2797; B. 1396. Swā wē wēnaþ on ðē *sicut speravimus in te*, Ps. Ben. 32, 18. (2) with acc. of what is hoped for or expected and dat. of person for whom:—Ic wēne mē, and eác ondrǽde, dóm ðý réþran, Exon. Th. 49, 22; Cri. 789. Ic mē bættran hām ǽfre ne wēne, Cd. Th. 268, 5; Sat. 50. Hē wile syllan unábeden ðæt, ðæt wē ús ne wēndon, Homl. Th. ii. 372, 16. (3) with gen. of what is expected, (a) alone:—Ic ðǽr heaðufýres hātes wēne, Beo. Th. 5038; B. 2522. Ðín líf geendaþ, ðonne ðú his ne wēnest, Wulfst. 260, 24. Hwæs wēneþ se, ðe nyle gemunan? Exon. Th. 74, 1; Cri. 1200. Ðǽr wē úres feores ne wēnaþ *where we despair of our life*, Blickl. Homl. 51, 28. Ðā fór hē (*Saul*) forð bí ðæm scræfe ðæt hē (*David*) oninnan wæs, and hē his ðǽr nó ne wēnde, Past. 28; Swt. 197, 14. Hē ðæs mǽldæges ne wēnde, Cd. Th. 141, 4; Gen. 2340. Far ðǽr ðú freónda wēne, Exon. Th. 119, 29; Gú. 262. Geworpene on hlǽw, ðǽr hiora gemynde

men ne wēnan *projecti in monumentis, quorum non meministi amplius*, Ps. Th. 87, 5. Hwonon hié ðæs wēnan sculon, Past. 11; Swt. 67, 2. Nū swýðe raðe his (*Antichrist*) man mæg wēnan, Wulfst. 19, 5. Lífes ne wēnan, Exon. Th. 98, 22; Cri. 1611. Ne wē ðære wyrde wēnan þurfon, 6, 9; Cri. 81: Blickl. Homl. 63, 2: Cd. Th. 62, 31; Gen. 1023. Ne hí edcerres ǽfre môton wēnan, 293, 8; Sat. 451. Hwǽr hē ðara nægla swíðost on ðam wangstede wēnan þorfte, Elen. Kmbl. 2206; El. 1104. Ðéh ðe hē wēnende wǽre anwealdes, Ors. 4, 10; Swt. 194, 22. (b) with appositional clause:—Ðæs ne wēndon witan, ðæt hit manna ǽnig tôbrecan meahte, Beo. Th. 1560; B. 778. Ne þearf ðæs nān mon wēnan, ðæt hine ôþer mon mæge a̓lésan, Blickl. Homl. 101, 13: 109, 30: Cd. Th. 272, 5; Sat. 115. Frôfre ne wēnaþ, ðæt gē wrǽcsíða wyrpe gebíden, Exon. Th. 132, 28; Gú. 479. Ne þearf hæleþa nān wēnan ðæs weorces, ðæt he wísdôm mæge wið ofermétta gemengan, Met. 7, 7: 13, 24: 26, 114. (c) with dat. of object for which something is expected:—Ne wēndest ðū ðē ðínes feores *thou wouldst despair of thy life*, Bt. 14, 3; Fox 46, 26. Him mon ðæs lífes ne wēnde *proximus morti fuit*, Ors. 3, 9; Swt. 124, 32: Bd. 3, 27; S. 558, 39: 5, 3: S. 616, 9. Hē wēnde him þräge hnägre, Elen. Kmbl. 1333; El. 668. Hié sendon æfter fultume, ðǽr hié him ǽniges wēndon, Ors. 4, 1; Swt. 154, 23: 4, 5; Swt. 166, 13: 6, 13; Swt. 268, 13. Wēnaþ eów ǽlcere blisse, Homl. Th. i. 554, 30. Ðǽr ðú ðē hleahtres wēne, Guthl. prol.; Gdwin. 4, 8. Ǽr hē hym ðæs feferes wēne, Lchdm. i. 84, 7. Ne mæg ic mē nānes ôðres wēnan, Homl. Skt. i. 23, 576. (d) with preposition marking direction of expectation, *to look to a person for something*:—Wēne ic tô ðē wyrsan geþingea, Beo. Th. 1054; B. 525. Ne ic tô Sweóðeóde sibbe oððe treówe wihte wēne, 5838; B. 2923. Hē sæcce ne wēneþ tô Gār-Denum, 1205; B. 600. Ne wēndon hig nānes fleámes tô unc, Shrn. 40, 29. Nǽnig wihta wēnan þorfte beorhtre bôte tô banan folmum, Beo. Th. 317; B. 157. (e) where (c) and (d) are combined:—Wēne ic wē wrade tô ðē *ego in te sperabo*, Ps. Th. 55, 3. Ða dysegan nānwuht nyllaþ onginnan ðæs ðe hí him áwþer mægen tô wēnan oððe lofes oððe leána, Bt. 36, 5; Fox 180, 11. (f) where (a) or (d) is accompanied by a clause [v. (4)]:—Hig ðæs æðelinges eft ne wēndon, ðæt hē sigehréðig sēcean côme nǽrne þcóden, Beo. Th. 3197; B. 1596. Ne þorftan ða þegnas tô ðam frumgáre feohgestealde wēnan, ðæt hý beágas þēgon, Exon. Th. 283, 26; Jul. 686. (4) with a clause:—Ic wēne mē hwænne mē Dryhtnes rôd gefetige, Rood Kmbl. 268; Kr. 135. Wíscton and ne wēndon, ðæt hié heora winedrihten gesáwon, Beo. Th. 3212; B. 1604. (5) with infinitive:—Ic ǽnigra mē weána ne wēnde bôte gebídan, 1870; B. 933. [*Goth.* wēnjan *sperare*: O. Sax. wānian *to suppose, hope* (with gen., infin., and clause): O. Frs. wēna: O. H. Ger. wān[n]en *opinari, putare, censere, arbitrari, suspicari, aestimare, credere, sperare* (with gen., clause, infin., acc. and infin., preposition): Icel. væna *to suppose, hope* (v. á-, ge-wēnan; un-wēned.]

wen-býl or -býle *some kind of boil*:—Wiþ wenbýle, Lchdm. ii. 128, 16. Lǽcedômas tô wenbýlum, 12, 19: 128, 6.

wencel, wincel, es; *n. A child*:—Gif his hláford him wíf sylle and hig suna hæbbon and dohtra, ðæt wíf and hire winclo (*liberi*) beóð ðæs hláfordes. Gif se wiel cwið: 'Mē ys mín hláford leóf and mín wíf and míne winclo,' Ex. 21, 4, 5. Se eorðlíca kempa bið ǽfre gearo, swā hwyder swā hē faran sceal tô gefeohte mid ðam kininge, and hē for his wífe ne for his wenclum ne dearr hine sylfne belādian, Basil adm. 2; Norm. 34, 20. Weodewuun (and) wenceluun hē wel onfehð *pupillum et viduam suscipiet*, Ps. Th. 145, 8. [Juw iss borenn an wennchell þatt iss Iesu Crist, Orm. 3356. Men and wummen and children (*v.l.* wen and wif and wenchel), A. R. 334, 25. Quelæn þa wifmen, quelen þa wanclen, Laym. 31834. *The later form is* wenche, e.g. Wicklif, Mt. 9, 24.]

wencge. v. wang.

wend *a course, an alternative, a case*:—Ðonne gerecce hē, gif hē mæge, oþer twēga, oððe ðara spella sum leás oððe ungelíc ðære sprǽce ðe wit æfter spyriaþ; oððe þridde wend (*a third course or alternative*) ongite and geléfe ðæt wit on riht spirien, Bt. 38, 2; Fox 198, 26. Gif hit gebirie ðæt Alhmund swā ða freóndrēddene healdan nolde, oððe hine mon oferricte ðæt hē ne môste londes wyrðe beón, oððe þridda wend, gif hē ǽr his ende gesǽlde, Chart. Th. 141, 13. [Cf. A pryve went *a secret passage*, Chauc. T. and C. ii. 738. O. Frs. wend *a case*.] v. ed-wend.

wendan; *p.* de To turn. I. *trans.* (1) To cause to move, alter the direction or position of something (lit. or fig.):—God on gesyhðe wæs . . . mín on ða swíðran, ðanon ic ne wende onsión míne, Elen. Kmbl. 696; El. 348. Swā hwá swā his môd went tô yflum, Bt. 35, 6; Fox 170, 20. Ic āwyrgde fram mē wende and cyrde, Ps. Th. 100, 4. Ðam ðe slihþ on ðín gewenge, wend ôðer āgén *qui te percutit in maxillam, praebe et alteram*, Lk. Skt. 6, 29. Wendaþ mín heáfod ofdúne, Blickl. Homl. 191, 2. Byð his horn wended on wuldur *cornu ejus exaltabitur in gloria*, Ps. Th. 111, 8. Wyrd bið wended hearde *the course of fate is hard to turn*, Salm. Kmbl. 871; Sal. 435. (2) *to turn round* or *over*. Cf. wending, I:—Ðæt wērige môd wendað ða gyltas swíðe mid sorgum *caeca scelerum mergit vertigine mentem*, Dôm. L. 244. Se ðe wende wriþan, Exon. Th. 440, 19; Rä. 60, 5. Tô eáhsealfe . . . wende man ǽlce dæge (*let the paste be turned every day*), Lchdm. iii. 16, 24. Wend-

ende *convolvens*, Wrt. Voc. ii. 21, 27. Hé (*a cup*) in healle wæs wylted and wended wloncra folmum, Exon. Th. 441, 16; Rä. 60, 19. (3) *to turn* from one condition to another, *to change, alter, convert*:— Hé wendeþ stán on wídne mere *convertit solidam petram in stagnum aquae*, Ps. Th. 113, 8. God ús éce biþ, ne wendaþ hine wyrda, Exon. Th. 333, 24; Gn. Ex. 9. Hé ða weaxendan wende eorðan on sealtne mersc, Ps. Th. 106, 33. Hé heora wæter wende tó blóde *convertit aquas eorum in sanguinem*, 104, 25. Hí wendan unriht tó rihte, L. I. P. 11; Th. ii. 318, 23. Wend ðás stánas tó hláfum, Homl. Th. i. 168, 22. Ða yldu wendan tó life, Exon. Th. 211, 2; Ph. 191. Ða gewitnesse wendan *to pervert the testimony*, 147, 21; Gú. 730. Ðær hé hit wendan (-en, MS.) meahte *if he could have changed it*, 276, 23; Jul. 570: Elen. Kmbl. 1955; El. 979. God giet settende is and wendende ælce onwaldas and ælc ríce tó his willan, Ors. 2, 1; Swt. 64, 2. Hí beóð wended *mutabuntur*, Ps. Th. 101, 23. Wese heora beód wended on grine *fiat mensa eorum in laqueum*, 68, 23. (3 a) *to turn* from one language to another, *to translate, interpret*. v. wendere:—Ælfréd kuning wæs wealhstod ðisse béc and hié of béclédene on Englisc wende, Bt. proem.; Fox viii, 2. Ic ðé secge worda gerýnu, ða ðú wendan (or *alter?*) ne miht, Cd. Th. 262, 21; Dan. 747. **II.** reflexive, (1) *to move one's self, take one's way, go, proceed, wend* (lit. or fig.):—Ic wende mec on wæteres hricg, Salm. Kmbl. 37; Sal. 19. Wendeþ hé hine under wolcnum, wígsteall séceþ, 207; Sal. 103. Ða innoþas hí wendaþ mid heora hefignesse, and on ða sídan feallaþ ðe hé on licgeaþ, Lchdm. ii. 258, 11. Hé wende hine lythwón fram him and weóp, and wende eft tó him *avertit se parumper et flevit; et reversus est ad eos*, Gen. 42, 24. Se cyning hine west wende, Chr. 894; Erl. 92, 5. Hé wende hine ðanon, Cd. Th. 31, 31; Gen. 493: 34, 33; Gen. 547. Hé wende hine of worulde *he departed this life*, Elen. Kmbl. 877; El. 440. Wend ðé from wynne, Cd. Th. 56, 28; Gen. 919. (2) *to turn, direct the attention*:—Ic wolde ðæt wit unc wendon tó ðises folces spræce, Bt. 40, 1; Fox 236, 11. **III.** intrans. (1) *To wend, go, proceed* (lit. and fig.):—Se ðe bið on æcere, ne went hé on bæc *qui fuerint in agro, non redeant retro*, Lk. Skt. 17, 31. Went nú fulneáh eall moncyn on tweónunga, Bt. 4; Fox 8, 17: Met. 13, 55. Him eal worold wendeþ on willan *all the world goes well with him*, Beo. Th. 3482; B. 1739. For hwí hit swá went swá hit nú oft déþ *why things go as now they often do*, Bt. 39, 2; Fox 212, 26. Ða wende hé on scype ágén *ascendens nauem reversus est*, Lk. Skt. 8, 37. Se here eft hámweard wende, Chr. 895; Erl. 93, 25. Hé grundsceát sóhte, wende tó worulde, Exon. Th. 41, 3; Cri. 650. Ða bóceras ðe wendon (*descenderant*) fram Hierusalem, Mk. Skt. 3, 22. Hig wendon tó Hierusalem *regressi sunt in Hierusalem*, Lk. Skt. 24, 33. Hí wendon ðá tó horsum ... Hí wendon him fram, and heora wǽpna áwurpon, Homl. Skt. ii. 25, 425, 435. His feónda wǽmna wendon on hí sylfe, Jud. Thw. 162, 9. Ðær hwendon forð wlance þegenas, Byrht. Th. 137, 52; By. 205. Úre yldran swultan and ús from wendan, Blickl. Homl. 195, 27. Ðæt ic hám sídie, wende fram wíge, Byrht. Th. 139, 10; By. 252. Ǽr hé hionan wende *ere he depart*, Met. 18, 11. Hwí sió wyrd swá wó wendan sceolde, Met. 4, 40. Wendan of (*to depart from*) woruldryhte, Exon. Th. 105, 24; Gú. 28. Ðæt his sciperes woldon wændon fram him, Chr. 1046; Erl. 174, 13. (1 a) with reflexive dative:—Cnut wende him út, Chr. 1016; Erl. 154, 5. Hí wendon him tó ðære burge weard, 1048; Erl. 177, 40. (2) *to turn round*:—Swylce ex wendende *quasi axis versatilis*, Scint. 97, 4. (3) *to turn* from one condition to another, *to change, alter*:—Hí on wiðerméde wendan and cyrdan *conversi sunt in arcum perversum*, Ps. Th. 77, 57: Exon. Th. 73, 7; Cri. 1186. Hé gehálgode wín of wætere, and wendan hét on ða beteran gecynd, Andr. Kmbl. 1174; An. 587. Ðæt wile wendan on wæterbollan, Lchdm. ii. 248, 7. (4) *to change, shift, vary, be variable*:—God ne went nó swá swá wé dóþ, Bt. 42; Fox 258, 20. Wendeþ, Exon. Th. 379, 13; Deór. 379. Geseah ic ðæt beácen wendan wǽdum and bleóm; hwílum hit wæs mid wǽtan bestémed, hwílum mid since gegyrwed, Rood Kmbl. 43; Kr. 22. [*Goth.* wandjan: *O. Sax.* wendian: *O. Frs.* wenda: *O. H. Ger.* wenten: *Icel.* venda.] v. á-, be-, ed-, ge-, mis-, on-, óþ-, tó-, under-, ymb-wendan; un-áwendende, un-áwend(-wended); windan.

wendan (? *or* wennan? Cf. winnan); *p.* de To labour:—Ðá wende (*other MSS.* have wann, wonn) hé swýþe, ðæt hé ða ðe mid hine cóman geheólde *laboravit multum, ut eos, qui secum venerant, contineret*, Bd. 2, 9; S. 511, 5. [Cf. *Icel.* vanda *to take pains in a work*.]

-wende. v. hál-, hát-, hwíl-, lád-, leóf-, luf-wende.

-wendedlíc, -wendedlícness, -wend(ed)ness. v. á-, on-wendedlíc, á-wendedlícness, á-, and-, on-wendedness, ge-unwendness.

Wend(e)las (-e?), a; *pl. The people of Vendil* (the northern part of Jutland, *Icel.* Vendill)?, *the Vandals?*:—Wulfgár maþelode, ðæt wæs Wendla leód, Beo. Th. 702; B. 348. Mid Wenlum ic wæs and mid Wærnum, Exon. Th. 322, 6; Víd. 59. v. Grmm. Gesch. D. S. 332 sqq.: P. B. xii. 7.

Wendel-sǽ (*generally masc.*) *the Mediterranean.* In Alfred's Orosius the word is used to translate several Latin terms denoting the Mediterranean or parts of it:—Andlang Wendelsǽs (*mare Nostrum, quod Magnum*

generaliter dicimus), Ors. 1, 1; Swt. 8, 12. Wendelsǽ *mare Nostrum*, 12, 14: 26, 28: 8, 23. Óþ ðone Wendelsǽ, 10, 36. Se Wendelsǽ *mare Magnum*, 24, 26. On ðæm Wendelsǽ *per totum Magnum pelagus*, 28, 24. Seó ús fyrre Ispania, hyre is be westan gársecg, and be norðan Wendelsǽ *Hispania ulterior habet a septentrione Oceanum, ab occasu Oceanum*, 24, 8. Se Wendelsǽ ðe man hǽt Atriaticum, 22, 14: 28, 9. Andlang ðæs Wendelsǽs is Dalmatia on norðhealfe ðæs sǽs *Dalmatia habet a meridie Adriaticum sinum*, 22, 12. Hió hǽfð be norðan ðone Wendelsǽ, ðe man hǽt Adriaticum *habet a septentrione mare Siculum vel potius Adriaticum*, 26, 7. Se Wendelsǽ *mare Tyrrhenum*, 8, 25: 28, 15: 24, 3. Italia land belíð Wendelsǽ ymb eall útan búton westannorðan *Italia habet ab Africo Tyrrhenum mare, a borea Adriaticum sinum*, 22, 18. Be súðan Narbonense is se Wendelsǽ (*mare Gallicum*), 22, 29, 20. Wendelsǽ ðe man hǽt Libia Æthiopicum *mare Libycum*, 26, 1. Begeondan Wendelsǽ *citra Pontum*, Wrt. Voc. ii. 24, 52. Féng Carl tó allum ðam westríce behienan Wendelsǽ and begeondan ðisse sǽ, Chr. 885; Erl. 84, 11. On án íglond út on ðære Wendelsǽ, Bt. 38, 1; Fox 194, 11. Æt Wendelsǽ on stæde (*the Italian shore*), Elen. Kmbl. 462; El. 231. On Wendelsǽ ðǽr Apollines dohtor wunode, Met. 26, 31: Salm. Kmbl. 406; Sal. 203. [*O. H. Ger.* Wentil-séo *oceanus*. Cf. wendel-meri *oceanus*.]

-wenden. v. ed-wenden.

wendend, es; *m. That which turns round*:—Wendend *vertigo* (teres vertigo coeli, Ald. 13), Wrt. Voc. ii. 76, 32. Cf. hweorfa.

-wendendlíc, -wendendlíce. v. á-wendendlíc, á-wendendlíce.

wendere, es; *m. A translator, interpreter.* v. wendan, I. 3 a:—Wenderum *translatoribus, interpretes*, Hpt. Gl. 525, 32. [*O. H. Ger.* misse-wendari.]

wending, e; *f. Turning.* **I.** *a turning round, revolution.* Cf. wendan, I. 2:—On ánre wendinge, ða hwíle ðe hé (*the firmament*) æne betyrnð, gǽð forð feówor and twéntig tída, Hexam. 5; Norm. 8, 30. **II.** *a turning up* or *over*:—Gif ðær sié ðæs hrifes wendung *if the stomach be upset* (?), Lchdm. ii. 228, 24. **III.** *changing, mutation*:—Ne wyrð ðisses nǽfre nán wending *non movebor de generatione in generationem*, Ps. Th. 9, 26. Wendincg, 29 6. Earfoðe ys fǽrlíc wendincg *difficilis est subita permutatio*, Scint. 63, 20. Hit gedéð hit self him selfum suíðe ungelíc for ðære gelómlícan wendinge *mutabilitate se varium exhibet*, Past. 42; Swt. 306, 17. Orsorg líf lǽdaþ woruldmen wíse búton wendinge (cf. unonwendendlíce, Bt. 12; Fox 36, 24), Met. 7, 41. [Dyaþ is a wendinge, and þet ech wot, Ayenb. 70, 34. At the wendyng *at the turn* (versura), Pall. 44, 12.] v. á-wending.

wéne, adj. **I.** *hopeful.* v. or-, un-wéne. **II.** *fair, beautiful.* v. wén-líc:—Wénre (? wende, MS.) *formosior*, Hpt. Gl. 417, 23. [*Icel.* væn *hopeful; fair, beautiful*.]

wenge. v. wang.

wenian; *p.* ede To accustom. **I.** *to accustom, train, prepare, fit*, (1) with prep. *tó* marking the end of the training:—Lǽrde hé ða leóde on geleáfan weg, wenede tó wuldre weorod unmǽte, tó ðam hálgan hám, Andr. Kmbl. 3360; An. 1684. Hine his goldwine wenede tó wiste, Exon. Th. 288, 24; Wand. 36. Hié lǽrdon hira tungan and wenedon tó leásunga *docuerunt linguam suam loqui mendacium*, Past. 35; Swt. 239, 19. Ðæt hé cristen man his bearn tó cristendóme geornlíce wænige, L. Edg. C. 17; Th. ii. 248, 9. Wenian tó gefeohte, Homl. Skt. ii. 25, 571. Tó ælcan rihte ús sylfe wenian and wéman, Wulfst. 266, 5. Godes folc wenian tó ðam ðe heom þearf sý, 154, 13. (1 a) with prep. *tó*, and *mid* marking the means used:—Ðæt éce líf geearnian ðe hý ús tó weniaþ mid láre and mid bysene gódra weorca *to merit that life eternal, to which they are training us by teaching and by the example of good works*, L. Edg. S. 1; Th. i. 272, 22. Man mæg ylpas wenian tó wíge mid cræfte, Hexam. 9; Norm. 16, 10. Utan ús sylfe mid gódan geþance wenian tó rihte, Wulfst. 76, 2. (2) with prep. *in*, marking end attained by training:—Leorna láre, wene ðec in wísdóm *train yourself so that you may be wise*, Exon. Th. 303, 32; Fä. 62. (3) with instrumental:—Dó á ðætte duge ... wene ðec ðý betran (cf. *Icel.* venjask *with dat. to be accustomed to do a thing*) *always choose the better part*, Exon. Th. 300, 17; Fä. 7. **II.** *to draw, attract,* (1) *to draw to*:—Ðæt æt feohgytum Folcwaldan sunu dógra gehwylce Dene weorþode, Hengestes heáp hringum wenede (*he should attach them to himself by presents*), efne swá swíðe swá hé Fresena cyn byldan wolde, Beo. Th. 2187; B. 1091. Ðone ðe mec fréfran wolde, wenian (wéman? *q.v.*: *but cf.* Sulík folk laðóian, wennian mid willeon, Hél. 2818) mid wynnum, Exon. Th. 288, 10; Wand. 29. (2) *to draw from*:—Wene and teóh ðæt blód fram ðære ádeádedan stówe, Lchdm. ii. 84, 3. Hú mon ðæt deáde blód áweg wenian scyle, 8, 15. (2 a) *to wean*:—Swá módor déþ hyre bearn, ðonne hió hit fram hire breósta gesoce weneþ, R. Ben. 22, 21. [*O. Sax.* wenian, wennian: *O. H. Ger.* wennen *assuefacere: Icel.* venja *to accustom* to (*dat.* or *við*).] v. á-, æt-, be-, ge-, mis-wenian; for-, ofer-wened.

wéning, e; *f.* **I.** *supposition, doubtful-thought, doubt*:—Se Godes man ne sceolde be ðan morgendæge þencean, ðý læs ðæt wǽre, ðæt hé þurh ðæt ænig ðara góda forylde, ðe hé ðonne ðý dæge gedón mihte, and

(þurh) ða wēninge hweðer hē eft ðæs mergendæges gebîdan môste *the man of God ought not to think of the morrow, lest it should come to pass, that through it he should put off any of the good that he might do then on the day, and through the doubt whether he may live to see the morrow,* Blickl. Homl. 213, 24. **II.** *hope, expectation :*—Bæd heó swîþe lange ðone cyningc, ðæt hē hî forlǣte on mynstre Criste þeówian, ðæt heó ða wēnunge æt nýhstan ðurhteáh (*so that at last her hope was realized*), Bd. 4, 19 ; S. 587, 39. **III.** *chance :*—In woenunga *forte,* Mt. Kmbl. Lind. 13, 29. [Aboue onderstandingge and wenynge (*imagination*), Ayenb. 113, 6. It is a wrongful wenynge (*opinion*), Chauc. Boeth. 172, 28. *O. H. Ger.* ana-wânunga *existimatio ;* bi-wânunga *deliberatio.*] v. wenunga.

weninga. v. wēnunga.

wēn-lîc ; *adj.*   **I.** *fair, handsome, comely :*—Stranglîc on wæstme and wēnlîc on nebbe, Ælfc. T. Grn. 16, 41. Heó wæs swîðe wlitig and wēnlîces hîwes *erat eleganti aspectu nimis,* Homl. Ass. 108, 205.   **II.** the word glosses *conveniens* in the following passages :—Ne wæs woenlîc (þæslîc (*q v.*), W. S.) gecýðnisse hiora *non erat conveniens testimonium illorum,* Mk. Skt. Lind. 14, 59. Woenlîca (weonlîce, Rush.) gecýðnise *conuenientia testimonia,* 14, 56. [Swo warð iturnd þat folc of ateliche to wenliche *ita facta est Niniue speciosa que prius turpis existebat,* O. E. Homl. ii. 83, 9. Hwu lie mai hire seluen wenlukest makien, 29, 12. þe mon þe on his ȝouhþe ȝeorne leorneþ wit and wisdom, he may beon on elde wenliche lorþeu, Misc. 108, 105. *O. Sax.* wân-lîk *fair : Icel.* væn-ligr *hopeful, promising, fine.*] v. un-wēnlîc.

wēnlîce ; *adv. Fairly, in comely fashion :*—September and December mid heora seofon gefērum gladiaþ wēnlîce swýðe, Anglia viii. 302, 4. [*O. Sax.* wân-lîko *beautifully : Icel.* vænliga.]

wenn, es ; *m. A wen :*—Eágan wenn *impetigo,* Wrt. Voc. ii. 45, 39 : i. 43, 62. Wið wenne (τύλος) on eágon, Lchdm. ii. 34, 9. Wǣnne, 34, 3. Wiþ sceótendum wenne, 324, 25. Gif men synd wænnas gewunod on ðæt heáfod foran oððe on ða eágan, iii. 46, 21. Sealf wið wennas, 12, 22. Wið wennas æt mannes heortan, 40, 4. v. þeór-wenn.

-wēnness. v. or-wēnness.

wen-sealf, e ; *f. A salve for wens :*—Wensealf, Lchdm. ii. 128, 13, 19. Ðâs wyrta sceolon tô wensealfe, i. 382, 15 : ii. 128, 6 : 12, 19.

wen-spring (-spryng), es ; *m. A mole :*—Wensprynga *nevorum,* Wrt. Voc. ii. 59, 50.

Wente ; *pl.*   **I.** *the people of Gwent* (the district comprising Monmouth and Glamorgan) :—Ealle ða cyngas ðe on ðyssum îglande wǣron hē (*Athelstane*) gewylde ; ǣrest Huwal West-Wala cyning, and Constantin Scotta cyning and Uwen Wenta cyning, Chr. 926 ; Erl. 111, 43.   **II.** *the same as* Waller-wente q. v. :—Nemne man him ealswâ micel Wente swâ cyninges þegne, L. N. P. L. 52 ; Th. ii. 298, 11 : 53 ; Th. ii. 298, 14. v. Went-sǣte.

wēnþ (?) *beauty.* v. wēn-lîc :—Wēnðe *cum formosior,* Hpt. Gl. 417, 23. v. wēne.

Went-sǣte ; *pl. The inhabitants of Gwent :*—Be Wentsǣtum and Dûnsǣtum. Hwîlon Wentsǣte hýrdon intô Dûnsǣtan, ac hit gebyreþ rihtor intô West-Sexan, þyder hý scylan gafol and gîslas syllan, L. O. D. 9 ; Th. i. 356, 17–20. v. Wente.

wēnunga (-inga) ; *adv. Perhaps, haply, by chance :*—Wēnunge (-a) *forsan, forsitan, fortassis, fortasse,* Ælfc. Gr. 38 ; Zup. 229, 1 : Wrt. Voc. ii. 150, 23. Wēnunga *forsitan,* Ps. Spl. 80, 13. Wēnunga hine hig forwandiaþ, ðonne hig hine geseóþ *forsitan cum hunc uiderint uerebuntur,* Lk. Skt. 20, 13. Ne hit nǣfre næs tô geopenigenne bûton wēnunga hwilc munuc ût fôre *unless it happened that a monk had to go out,* Homl. Skt. ii. 23 b, 104. Ðe læs wēnunga *ne forte,* Lk. Skt. 14, 8. Wenðe mē Drihten gefultumede, wēnincga mîn sâwl sôhte helle *nisi quia Dominus adjuvasset me, paulominus habitaverat in inferno anima mea,* Ps. Th. 93, 16. Woenunga *forte,* Mk. Skt. Lind. 11, 13 : Lk. Skt. Lind. Rush. 9, 13. Woenunge, Mk. Skt. Lind. 14, 2 : *forsitan,* Jn. Skt. Lind. 5, 46. Woeninga, Ps. Surt. 123, 4 : 138, 11. v. un-wēnunga ; wēning.

wen-wyrt, e ; *f. The name of some plant supposed to be good for wens* [*two kinds are mentioned,* seó clufihte wenwyrt, Lchdm. ii. 128, 17 : 336, 3 : 128, 7 : 266, 26 ; *and* seó cneóehte wenwyrt, ii. 140, 8] :—Wyrc sealfe of wenwyrte, Lchdm. ii. 52, 4. Gesmire mid wenwyrte, 62, 27. Wensealf; ontre, reáde netlan, twâ wenwyrta, 128, 14.

weó *the upper part of the throat :*—Tunge *lingua,* weó *faus,* mûðes hrôf *palatum,* Wrt. Voc. i. 64, 57. Cf. (?) weohlan.

weó, ôn (?) ; *f. Woe, misery :*—Daroþas wæron weó (weá ?) ðære wihte, Exon. Th. 438, 9 ; Rä. 57, 5. [Cf. *O. H. Ger.* wêwa ; *f. dolor, pena, supplicium.*] v. weá, wáwa.

weó-bed, -bud. v. wîg-bed.

weóce, an ; *f. The wick of a lamp* or *candle :*—Weóce *licinius,* Wrt. Voc. ii. 54, 19. Leóhtfæt *lucernarium,* candelsnytels *emunctorium,* weóce *papirus,* i. 26, 56. Weócan (*papyrum*) settan *to put a wick to a lamp,* Lchdm. iii. 348, col. 1. Ðonne ðû blæcernes behôfige . . . wǣt mid ðînum scytefingre on midden, swylce ðû weócan settan wylle, Techm. ii. 126, 3. Riscene weócan *fila scirpea,* Germ. 391, 15.

Weócan *accendilia,* Wrt. Voc. i. 66, 46 : *cicindilia,* 284, 26. Wiócum *cicindilibus, stuppulis,* Hpt. Gl. 470, 77. Weócum, Wrt. Voc. ii. 80, 43 : 131, 13. [Wex on þe candele sene, þe wueke wiðinnen unsene *in candela cera exterius, luminulum interius,* O. E. Homl. ii. 47, 32. As wex and a weke were twyned togideres. . . . And as wex and weyke . . ., Piers P. 17, 204, 205. Weyke of a candel *lichinius,* weyke of a lampe *ticendulum* (l. *cicendulum.* v. Cath. Angl. 412); Prompt. Parv. 520. The weke of a candele *lichinus,* Wülck. Gl. 592, 30 : 721, 43. *M. Du.* wieke : *M. H. Ger.* wieche *licinia.* Cf. *O. H. Ger.* wioh *lucubrum.*] v. candel-, claþ-weóce.

weoc-steall. v. wîg-steall.

weód, es ; *n. f.* (?) *A useless* or *injurious plant, a weed :*—Æceres weód, ðæt ðe bið on ofen âsend *faenum agri, quod in clibanum mittitur,* Mt. Kmbl. 6, 30. Hwonan hæfd hit ðæt weód (*zizania*)? Mt. Kmbl. Rush. 13, 27. Is âwriten ðæt hē sēwe ðæt weód on ða gôdan æceras, Past. 47 ; Swt. 357, 17. Ðâ æteáwde ða weód, Mt. Kmbl. Rush. 13, 26, 25, 29, 30. Môtan ealle weóda nû wyrtum âspringan, Lchdm. iii. 36, 26. Swâ hwâ swâ wille sâwan westmbǣre land, âtió ǣrest of ealle ða weód ðe hē gesió, ðæt ðâm æcerum derigen, Bt. 23 ; Fox 78, 23 : Met. 12, 4, 28. [Forgrouwen mid brimbles, and mid þornes, and mid iuele wiedes, O. E. Homl. ii. 129, 25. Wo þat mygte weoden abbe and þe roten gnawe, R. Glouc. 404, 11. Weed or wyyld herb *herba silvestris* vel *herba nociva,* Prompt. Parv. 519. *O. Sax.* wiod.] v. un-weód.

weód, e ; *f. ? :*—Wið cneówærce genim weóde wîsan, Lchdm. iii. 16, 16.

weodewe. v. widuwe.

weód-hôc, es ; *m. A weed-hook, a hoe :*—Uueódhôc (uueád-, Ep. Erf.) *sarculum,* Txts. 95, 1764. Weódhôc (*printed* weodhoclu *sarcum*), Wrt. Voc. i. 289, 2 : Anglia ix. 263, 5. [þe wyedhoc of þe gardine, þet uordeþ al þet kueade gers, Ayenb. 121, 27. Weodhook, Wick. Is. 7, 25. A wedehoke *sarculum,* Wülck. Gl. 609, 22. Wedhoc, 724, 30 (both 15th cent.).]

weódian ; *p.* ode *To weed, clear the ground of weeds :*—Me mæig on sumera . . . weódian, Anglia ix. 261, 12. [Wede corne or herbys *runco, sarculo,* Prompt. Parv. 519. To wede *sarrio,* Wülck. Gl. 609, 24. To wedy *vello,* 618, 31.] v. â-weódian ; weódung.

Weód-mônaþ, es ; *m. August :*—Agustus mônaþ on ûre geþeóde wē nemnaþ Weódmônaþ, for ðon ðe hî on ðam mônþe mǣst geweaxaþ, Shrn. 110, 33 : 124, 14 : Menol. Fox 273 ; Men. 138.

weodu-binde. v. wudu-binde.

weódung, e ; *f. Weeding :*—Weódung *runcatio,* Wrt. Voc. i. 15, 12.

weoduwe, weofung, weogas. v. widuwe, wefung, weg.

Weogorna-, Weogora-ceaster, e ; *f. Worcester.* The first part of the name is found in the following forms:—Weogorna, Cod. Dip. Kmbl. ii. 131, 14: 100, 8 : i. 35, 21. Weogerna, 114, 15 : 152, 7 : ii. 150, 4. Weogurna, i. 315, 27. Wiogorna, 176, 5. Wiogoerna, 279, 11. Wiogerna, iii. 166, 7 : 186, 4. Wiogerne, 261, 5. Wiogurna, 50, 18 : ii. 384, 17. Wiogurnae, iii. 49, 29. Wiogurne, 36, 6. Wegorne, i. 171, 13 : 259, 32. Wegerna, 38, 17 : 171, 33. Wegrinan, 109, 21. Wegrin, 201, 4. Wigorna, 108, 5 : ii. 111, 36. Wigornae, i. 185, 33. Wigerna, 150, 32 : iii. 91, 33 : iv. 235, 28 : Chr. 992 ; Erl. 130, 38. Wigurna, Cod. Dip. Kmbl. ii. 385, 14 : iii. 52, 3. Wigeran, ii. 108, 37 : iv. 234, 27. Uigran, i. 80, 14. Wigrinnan, 154, 15. Wygerna, iii. 260, 33. Wygerne, 262, 6 : 263, 7. Wygoran, vi. 215, 7. Weogerie, ii. 405, 26. Wiogora, Past. pref. ; Swt. 3, tit. Wiogre, Cod. Dip. Kmbl. ii. 405, 5. Wigera, iv. 137, 21 : 262, 21 : Chr. 992 ; Erl. 131, 37. Wihgera, Cod. Dip. Kmbl. iv. 263, 14. Wigra, iii. 95, 28 : vi. 126, 25. Wigra, Wygra, Chr. 1047 ; Erl. 171, 30, 31. Wigre, Cod. Dip. Kmbl. i. 168, 15 : 186, 9. Wihgra, iv. 72, 22. Wigar, Chr. 959 ; Th. i. 219, col. 3. Cf. also Wiricestria, Cod. Dip. Kmbl. iv. 161, 25, and the Latin adjective forms, which shew the same variety, e.g. Weogernensis, Cod. Dip. Kmbl. i. 99, 29 : Wiornocensis, iii. 366, 29 : Wigorcestrensis, i. 167, 18 : Wigorcensis, v. 142, 16.

Weogornaceastre-scîr, e ; *f. Worcestershire :*—On Wigeraceastre-scîre, Cod. Dip. Kmbl. iv. 138, 1. Wigraceasterscîre (Wihracestrescîre, *v.l.*), Chr. 1039 ; Erl. 167, 10. Wigercestresîre, Cod. Dip. Kmbl. iv. 192, 1. Wigeceastrescîre, 263, 4. Wireceastrescîre, 56, 8. Wircestre-scîre, 193, 4.

weohlan ; *pl. The jaws :*—Tuxlas t geahlas (weohlan, MS. T.) leóna tôbrycð Drihten *molas leonum confringet Dominus,* Ps. Spl. 57, 6. v. weó.

weohlere, weoh-steall, weola, weolc. v. wîglere, wîgsteall, wela, weoloc.

weolc (? weolcen) ; *adj. Scarlet, purple :*—Twigedeágade deáge t weolcere (weolcenre ?) t wealcbasewere *bis tincto cocco,* Hpt. Gl. 431, 31. v. next word.

weolcen-reád ; *adj. Scarlet, purple :*—Se wolcnreáda wǣfels *the scarlet robe,* Homl. Th. ii. 254, 4. Hî scrýddon hyne mid weolcenreáde scyccelse, Mt. Kmbl. 27, 28. Wolcnreádum, Homl. Th. ii. 252, 25. Gif eówere synna wǣron wolcnreáde *si fuerint peccata vestra ut coccinum,* 322, 10. Wolcnereádum deáhum *conchiliis,* Hpt. Gl. 524, 57. Ðeós wyrt hæfð wolcnreáde blôstman, Lchdm. i. 244, 5. v. wcoloc-reád.

weoler. v. **weler.**

weolma, an; *m. Desire* (?), *what of its kind is most to be desired* (?), *what is best.* Cf. cyst :—Siþþan hē Marian, mægða weolman (*best of maidens*), mǣrre meówlan, mundheáls geceás, Exon. Th. 28, 12; Cri. 445. Cf. wil-.

weoloc, es; *m. A kind of shell-fish, a whelk, cockle; also the dye obtained from such fish* :—Wioloc coccum, Txts. 55, 594. Uulluc, uuluc *involucus,* 71, 1115. Weoluc, Wrt. Voc. ii. 45, 56 : cochlea, i. 65, 72. Weoloc, 281, 50: ii. 16, 29: conquilium, i. 291, 27. Wurma, weoloc *murice,* ii. 56, 62. Weluc *murice* vel *conchyleum,* i. 56, 8. Weoloces scyll *conquilium,* 34, 11. Fiscdeáh, weolces *conchilii,* Hpt. Gl. 524, 19. Lytle snæglas *vel* weolocas *cocleas,* Wrt. Voc. ii. 135, 45. Hēr beóþ swýþe genihtsume weolocas, of ðām biþ geweorht se weolocreáda tælhg *sunt et cochleae satis superque abundantes, quibus tinctura coccinei coloris conficitur,* Bd. 1, 1; S. 473, 19. Uuiolocas, uuylocas *cocleas,* Txts. 53, 542. Wilocas, Wrt. Voc. ii. 14, 81.

weoloc-basu; *adj. Purple* :—Uuylocbaso *purpuram,* Txts. 113, 66. v. **wealh-basu.**

weoloc-reád; *adj. Of the red colour that is got from the weoloc, scarlet, purple* :—Wiolocreád, wilocreád *coccum bis tinctum,* Txts. 51, 496. Weolocreád, Wrt. Voc. ii. 135, 43 : cocco, 77, 20. Weolcreád *coccum,* 14, 57; *coccum rubicundum bis tinctum,* i. 34, 10. Weol[c]rǣd *coccinea,* Hpt. Gl. 526, 33. Weolocas, of ðām biþ geweorht se weolocreáda tælhg *cochleae, quibus tinctura coccinei coloris conficitur,* Bd. 1, 1; S. 473, 19. Wolcreádum *coccineo,* Hpt. Gl. 523, 77: Anglia xiii. 29, 53. Weolocreáde *coccineas,* Wrt. Voc. ii. 89, 30. Wolcreáde, Hpt. Gl. 524, 55: Lchdm. i. 244, 5, note. v. **weolcen-reád.**

weoloc-scill, e; *f. A shell-fish, a whelk, cockle* :—Wilocscel (uuiluc-, uuyluc-) *conquilium,* Txts. 51, 499. Wiolucscel (*but* Ep. Erf. *have* iiugsegg) *papilivus,* 83, 1487. Hēr beóþ oft numene missenlícra cynna weolcscylle and muscule *exceptis variorum generibus conchyliorum, in quibus sunt et musculae,* Bd. 1, 1; S. 473, 17.

weoloc-telg, es; *m. The scarlet dye got from the* weoloc :—Wiolctælges *conquilini,* Wrt. Voc. ii. 20, 41.

Weolud *the river Welland* :—Him cirde tô þurferþ eorl and ða holdas and eal se here ðe tô Hámtûne hiérde norþ óþ Weolud, Chr. 921; Erl. 107, 29. v. **Wéland.**

weóningas (?); *pl. m. Bindings for the legs* :—Weóningas (meón-ingas? v. meó) *fascellas* (fascella = fasciola = fasciae crurales, Migne), Wrt. Voc. ii. 146, 53.

Weonod-land, es; *n. The country of the Wends* :—Weonodland him wæs on steórbord, Ors. 1, 1; Swt. 19, 34. Weonodland, Swt. 20, 4, 6. Of Weonodlande, 7. Of Winodlande, 11. [*Icel.* Vind-land.] v. **Winedas.**

weor *bad.* v. **weorr.**

weorc, es; *n. Work;* opus. **I.** *work, operative action, operation* :— Godes willa is weorc *God's will is operative,* Hexam. 6; Norm. 10, 24. Ðæt Godes weorc (uoerc, Lind. : werc, Rush.) wǣre geswutelod on him, Jn. Skt. 9, 3. Gesweotula þurh searocræft ðín sylfes weorc, and sôna forlæt weall wið wealle, Exon. Th. 1, 17; Cri. 9. **II.** *working, doing, performance* :—Be rihtes weorce betweox Wealum and Englum *concerning the doing of justice between Welsh and English,* L. O. D. 1; Th. i. 352, 14. v. **V a, V b. III.** *in a collective sense, work, doings, actions,* (1) *what a person does* :—Se ðe ôþrum forwyrneþ wlitigan wilsíþes, gif his weorc ne deág, Exon. Th. 2, 19; Cri. 21. Weorc ánra gehwæs beorhte blíceþ in ðam blíþan hām, 238, 3; Ph. 598. Ðæt hē ne forleóse his weorces wlite, 97, 9; Cri. 1588. Hē getrymede heora geleáfan mid ðon heofonlícon weorce, Blickl. Homl. 17, 8. Ðis is wæstm wíses and goodes ðe his sôðfæst weorc symble læste, Ps. Th. 57, 10. (2) *what happens* :—Ðæs dæges weorc byð egesfull eallum gesceaftum, Wulfst. 182, 7. **IV.** *work, labour, occupation, employment, any form of long-sustained* or *habitual activity* :—Weorc *opus,* cræftca *opifex,* Wrt. Voc. i. 73, 37. Towlíc weorc *weaving;* textrinum opus, 26, 13: 82, 11. Hí môtan bletsian eal Cristen folc, and him godcunde lác fore-bringan . . . ðis weorc biþ deófum se mǣsta teóna, Blickl. Homl. 47, 6. Hē nǣfre Godes weorces ne áblon, ah hē ealle niht þurhwacode on hálgum gebedum, 227, 6. God geswác hys weorces (*the work of creation*), Gen. 2, 3. Weorces (*the building of the tower of Babel*) wîsan, Cd. Th. 101, 28; Gen. 1689. Ût færð man tô weorce his, Ps. Spl. 103, 24. Hî sôhton weras tô weorce (*building*), Cd. Th. 100, 30; Gen. 1672: Exon. Th. 1, 4; Cri. 3. Ðû leóda feala forlǣrdest, nû leng ne miht gewealdan ðý weorce, Andr. Kmbl. 2729; An. 1367. Yrþlingc, hû begǣst ðû weorc ðín? Coll. Monast. Th. 19, 11. Sum mæg wrætlíce weorc áhycg-an heáhtimbra gehwæs, Exon. Th. 296, 1; Crä. 44. Weorc gebannan, Beo. Th. 149; B. 74. **IV a.** *a particular act of labour* :—Wirc six dagas ealle ðín weorc, Ex. 20, 9. Gif hý ût an æcere wurc (*v. l.* weorc) hæbben *si opera in agris habuerint,* R. Ben. **IV b.** *work-manship* :—Wæs ðæt hûs hwemdragen, nalas æfter gewunan mennisces weorces, ðæt ða wâgas wǣron rihte, Blickl. Homl. 207, 18. **V.** *a work, deed, any action* :—Dǽd ɫ wærc *opus,* Jn. Skt. p. 1, 6. Hwæt dô wē ðæt wē wyrceon Godes weorc (uuerco, Lind. : werc, Rush.)? Ðā

andswarode se Hǣlend : Ðæt is Godes weorc (uerc, Lind. : werc, Rush.), ðæt gē gelýfan on ðone ðe hē sende, 6, 29. Wēnan ðæs weorces, ðæt hē wísdom mæge wið ofermētta gemengan, Met. 7, 7. Hý weorces (*taking the forbidden fruit*) onguldon, Exon. Th. 153, 22; Gû. 829. Wērig ðæs weorces, 436, 20; Rä. 55, 10. Tô hwon syndon gē ðyses weorces swá hefige? gôd weorc heó wæs wyrcende on mē, Blickl. Homl. 69, 15. Nis eów ðæs weorces þearf, ðæt gē ða ciriecan hálgian, 205, 36. Wrǣclícne hâm weorce tô leáne, Cd. Th. 3, 18; Gen. 37. Ða ðe ðý worce gefǣgon, 232, 31; Dan. 268. Mon mæg ðý ilcan weorce (*ipso facto*) cweþan ðæt nētenu send gesǣlige, gif man cwiþ, ðæt ða men sēn gesǣlige, ða heora líchoman lustum fyligeþ *to say that those men are happy, who follow their body's lusts, is at the same time to say that beasts are happy,* Bt. 31, tit. ; Fox xvi, 9. Ân weorc (uoerc, Lind. : werc, Rush.) ic worhte, Jn. Skt. 7, 21 : Blickl. Homl. 71, 30. He Godes eorre þurh his selfes weorc áfunde, Ps. C. 25. Gif hē ðonne git mâre weorc geworht hæbbe *if then he have committed a greater crime,* L. C. S. 30; Th. i. 394, 12. Hwylce ðæs gôdan mannes weorc and his dǣda wǣron, Blickl. Homl. 55, 13. Weorcu *opera,* Scint. 20, 19. Wæstm gôdra weorca, Blickl. Homl. 71, 36 : Exon. Th. 66, 31; Cri. 1080. Eargra weorca, 80, 8; Cri. 1304. Dǣdum georn, wís in weorcum, 185, 7; Az. 4 : 159, 4; Gû. 921. Weorcum fáh, Elen. Kmbl. 2484; El. 1246. Mid ælmessan and mid mildheortum weorcum, Blickl. Homl. 37, 19: 73, 16. Leánigean æfter his weorcum and dǣdum, 123, 34. Ne dô gē nâ æfter heora worcum (*v.l.* weorcum: wærcum, Rush.) . . . Ealle heora worc (*v.l.* weorc: werca, Lind. : wærc, Rush.) hig dôð, ðæt menn hî geseón, Mt. Kmbl. 23, 3–5. Weorc (uoerca, Lind. : werc, Rush.), Jn. Skt. 9, 4. Uoerco, Lind. 10, 32. God gesihþ ealle ûre wyrc (weorc, Cott. MS.), Bt. 41, 4; Fox 252, 1. **V a.** *where action is contrasted with speech* or *thought* :—Gif hwâ hǣðendôm weorðige wordes oððe weorces, L. E. G. 2; Th. i. 168, 2. Ic dô swâ ic ne sceolde, hwíle mid weorce, hwíle mid worde, Hy. 3, 44. Ðonne on ûrum môde bið ácenned sum ðing gôdes, and wē ðæt tô weorce áwendaþ, Homl. Th. i. 138, 23. Ðæm synfullan nâuht ne helpaþ his gôdan geðôhtas, for ðæm ðe hē hæfð gearone willan tô ðæm weorce, Past. 54; Swt. 423, 27 : 11; Swt. 73, 4. Bið sió costung ǣresð on ðæm môde, ðonne fēreþ ûtweardes tô ðære hýde, ôððæt hió ût âsciét on weorc, Swt. 71, 8. Sínra weorca wlite and worda gemynd, Exon. Th. 64, 15; Cri. 1038. Gescád witan worda and worca, Beo. Th. 583; B. 289. Wordum ne worcum, 2204; B. 1100. Word-um and weorcum, Cd. Th. 278, 17; Sat. 223. Wercum, 267, 34; Sat. 48. Mid wordum oððe mid weorcum cýðan, Past. 21; Swt. 157, 21. Se ðe ðás ǣ mid sprǣcon and mid wordum gefylð and nele mid worcum, Deut. 27, 26. Swilce hē mid weorcum hî gesprǣce, Homl. Th. ii. 290, 2. Sume him ðæs hádes hlísan willaþ wegan on wordum, and ða weorc ne dôð, Exon. Th. 105, 33; Gû. 105. **V b.** *of action that gives effect to anything* :—Hwæðer hig gefyllaþ mid weorce ðone hreám, oððe hit swâ nys, Gen. 18, 21. Hwæðer mín word beó mid weorce ge-filled, Num. 11, 23. Hwî hē nolde gehýrsumian his hǣsum mid weorce, Homl. Skt. i. 21, 61. Hē wolde his gebeót mid weorcum gefremman, 25, 621. Ðæt ðû mid weorcum gefille ealle ða ǣ, Jos. 1, 7. Se ðe mægna gehwæs weorcum (*actually, indeed*) wealdeþ, Exon. Th. 121, 3; Gû. 283. Ðíu gewitnes is weorcum geleáfsum, Ps. Th. 92, 6. **VI.** *a work, what is wrought* :—Weorc *machina,* Wrt. Voc. ii. 57, 53. Ðâ wæs geforðad ðín fægere weorc, Hy. 9, 24. Nânwuht nis fæste stondendes weorces â wuniende, Bt. 9; Fox 26, 21 : Met. 6, 17. Bisiuuidi uuerci (uerci, werci) *opere plumario,* Txts. 80, 699. Weorce *fabrica,* Wrt. Voc. ii. 38, 35. Is ðam weorce þearf, ðæt se cræftga cume, and gebête, Exon. Th. 1, 21; Cri. 11. Com God wera weorc sceáwigan, beorna burhfæsten and ðæt beácen somod, Cd. Th. 101, 9; Gen. 1679. Se wealdend ðe ðæt weorc (*the universe*) staðolade, Andr. Kmbl. 1598; An. 800 : Exon. Th. 43, 19; Cri. 691. Mē glíwedon wrætlíc weorc smiþa, 408, 18; Rä. 27, 14. Mycel wǣrun ðíne weorc, Ps. Th. 103, 23. Ðâ sceáwode Scyppend ûre his weorca wlite, Cd. Th. 13, 23; Gen. 207 : 239, 2; Dan. 364: Met. 20, 21. **VI a.** *a strong building, fortress* :—Babylonia ðe ǣr wæs ealra weorca fæstast and wunderlecast and mǣrast, Ors. 2, 4; Swt. 74, 24. Bewrigene mid weorcum, Cd. Th. 218, 24; Dan. 44. **VI b.** *work, what is done, effect produced* :—Ða flǣsclícan willan cumaþ oft þurh deófles sceónessa ǣr tô manna heortan, ǣr Drihtnes weorc ðær wunian môte, Blickl. Homl. 19, 8. **VII.** *pain, travail, grief.* v. **weorcsum** :—Ðæt ðam weligan wæs weorc tô þolianne, Exon. Th. 276, 21; Jul. 569. Ðæt wæs weorc Gode, Cd. Th. 217, 18; Dan. 24. Ne hié sorge wiht, weorces ne wiston, 49, 2; Gen. 786 : Andr. Kmbl. 2556; An. 1279. Wæs hē tô ðæs árfæst, ðæt him wæs on weorce, ðæt he leng from Cristes onsýne wǣre, Blickl. Homl. 225, 28. Hē ðæs weorc gehleát, frēcne wîte, Cd. Th. 166, 10; Gen. 2745. Hē ðæs gewinnes weorc þrowade, leóðbealo longsum, Beo. Th. 3447; B. 1721 : Apstls. Kmbl. 160; Ap. 80: Rood Kmbl. 155; Kr. 79. Ic weorc þrowade, earfoða dǣl, Exon. Th. 485, 12; Rä. 71, 12. Worc, Cd. Th. 19, 24; Gen. 296. ¶ *the instrumental or da:ive is used in the phrase* weorce wesan *with the dative of the person* = *to be painful to a person* (cf. torne; *adv.*) :—Mē næs se hrædlíca ende mínes lífes swâ miclum weorce, swâ mē wæs ðæt ic lǣs mǣrðo gefremed hæfde, ðonne

mín willa wǽre, Nar. 32, 27. Him wæs on móde myccle weorce (cf. on weorce, 225, 28 *supra*) and mycel tweó, hwæt hié be ðære dorstan dón, Blickl. Homl. 205, 9. Him wæs ðæt swíþe myccle weorce, ðæt hé swá ungefulwad forðféran sceolde, 217, 22. Ðá wæs him ðæt swíþe sár and myccle weorce, 219, 14. Mé ða fraceðu sind on módsefan mǽste weorce, Exon. Th. 247, 2; Jul. 72. Ne mé weorce sind wítebrógan, 250, 30; Jul. 135. Wæs Abrahame weorce on móde, ðæt hé on wræc drife his selfes sunu, Cd. Th. 168, 31; Gen. 2791. Denum eallum wæs weorce on móde tó geþolianne, Beo. Th. 2841; B. 1418. [O. Sax. werk *work, pain:* O. Frs. werk: O. H. Ger. werah *opus, operatio, fabricatio, materia, opera: Icel.* verk.] v. æcer-, and-, beadu-, bóc-, cræft-, dǽd-, dæg-, ellen-, firen-, frum-, fyrn-, ge-, gúð-, hand-, heáh-, heaðo-, here-, in-, irre-, láð-, mægen-, mǽr-, mán-, mis-, níþ-, niht-, ofer-, orleg-, sigor-, stán-, þeów-, þreá-, þrýþ-, unriht-, untíd-, unwit-, weall-, weorold-, wic-, wundor-weorc.

-weorc; *adj.,* weorcan. v. mán-weorc, wyrcan.

weorc-dǽd, e; *f. A working, operation:*—Uoercdédo deáðberendo *operationes mortiferas,* Rtl. 125, 35.

weorc-dæg, es; *m. A work-day, any day, not a 'freólstíd,' of the week bar Sunday:*—Weorcdæg *feria,* Wrt. Voc. ii. 148, 4. Sealmas tó weorcdæge (*ad feriam*) gebyrigende, Anglia xiii. 402, 532. Ðam syxtan weorcdæge *sexta feria,* 404, 563. Worcdæge, 389, 348. Búton drihtenlícum and freólsum háligra weorcdagas þeáwe gewunelícum beón haldene *exceptis dominicis et festinitatibus sanctorum feriales more solito teneantur,* 396, 451. Freólsdæg *festivitas,* weorcdagas *fasti,* Wrt. Voc. i. 37, 14. Hú dægrédsangas on weorcdagum (*privatis diebus*) tó healdenne sýn (v. the whole chapter, and cf. the title of the previous one: Hú dægrédsangas on freólstídum tó healdenne sýn), R. Ben. 37, 4, 5. [ʒif hit is werkedei ..., ʒif hit is halidei ..., A. R. 20, 7. ʒure wuke gifeþþ ʒuw sexe werrkedaʒhess, but iff þatt aniʒ messedaʒʒ ..., Orm. 11315. Werkday *feria,* Prompt. Parv. 522. *Icel.* verk-dagr *a work-day.*]

weorce, weorcean. v. weorc, VII ¶, wyrcan.

weorc-full glosses gestuosus:—Wíf weorcfull *mulierem gestuosam,* Scint. 169, 1. [Workuol *active,* Ayenb. 199, 9.]

weorc-geréfa, an; *m. An overseer of work:*—Ða weorcgeréfan *praefecti operum,* Ex. 5, 10, 13. Sidrac, Misac, and Abdenago, ðe Nabochodonosor gesette him tó weorcgeréfan, Homl. Th. ii. 68, 5.

weorc-hús, es; *n. A workshop:*—Weorchús *officina,* Wrt. Voc. i. 58, 23: *ergasterium vel operatorium,* 59, 6. Werchús *ergasterium,* 34, 54. [Werkehowse *artificina, opificium,* Prompt. Parv. 522. A shoppe or a werkehous *operarium,* Wülck. Gl. 599, 11.]

weorc-líc; *adj. Working, busy.* [O. 'L. Ger.* werk-lík *operosus: Icel.* verk-ligr *working.*] v. un-weorclíc.

weorc-mann, es; *m. A workman, labourer:*—Wercmonn *operarius,* Mt. Kmbl. Lind. 10, 10. Woercmonn (werc-, Rush.), Lk. Skt. Lind. 10, 7. Wercmenn *operarii,* Mt. Kmbl. Lind. 10, 2. Woercmenn, 20, 1: Lk. Skt. Lind. 10, 2. Ǽlc riht cynestól stent on þrým stapelum ... *laboratores* syndon weorcmenn, Wulfst. 267, 14. Cyning sceal hæbban gebedmen, and fyrdmen, and weorcmen, Bt. 17; Fox 58, 33. [O. H. Ger. werah-man *operarius: Icel.* verk-maðr.]

weorc-rǽden[n], e; *f. Work, labour:*—Of Dyddanhamme gebyreþ micel weorcrǽden (*the work is then defined*), Cod. Dip. Kmbl. iii. 450, 31.

weorc-sige, es; *m. Success in work:*—Sigegyrd ic mé wege, wordsige and worcsige, Lchdm. i. 388, 15.

weorc-stán, es; *m.* **I.** *stone for building:*—Ne bið ðes stýpol getimbrod mid ǽnigum weorcstáne, Basil admn. 2; Norm. 38, 14. Hí man mid weorcstáne on ǽghwilce healfe ealle cuce ðǽrinne forwyrce, Homl. Skt. i. 23, 322. **II.** *a stone for building, a large stone:*—Weorcstán *saxum,* Wrt. Voc. i. 85, 20. Hét se cásere áhón ánne weorcstán on hyre swuran, Homl. Skt. i. 2, 389. Ðá geseah hé hwǽr ða weorcstánas (cf. 322 *supra*) lágon ofer eall, 23, 490. On ðam fenlande synd feáwa weorcstána, 20, 77. Hé hét ðæs scræfes ingang mid weorcstánum forwyrcan, 23, 316. Mid ormǽtum weorcstánum, Homl. Th. ii. 424, 27. Hé sprǽc ná tó ðám weorcstánum (*the stones of Jerusalem*) oððe tó ðære getimbrunge, i. 402, 10: Homl. Skt. ii. 27, 106. Hé hét wilian tó ðam scræfe micele weorcstánas (*saxa ingentia*), Jos. 10, 18, 27.

weorc-sum; *adj. Grievous, noxious:*—Deáðes beámes weorcsumne wæstm, Cd. Th. 37, 23; Gen. 594. v. weorc, VII.

weorc-þeów, es; *m.* : e; *f. A slave who works, a bondman, a bondwoman, a slave, a thrall:*—Ðá wearð unbliðe Abrahames cwén hire worcþeówe, Cd. Th. 136, 18; Gen. 2260. Nabochodonosor him dyde Israéla bearn, wǽpna láfe, tó weorcþeówum (*si quis evaserat gladium, ductus in Babylonem servivit regi,* 2 Chron. 36, 20), 220, 21; Dan. 74. Ðá Abimæleh Abrahame his wíf ágeaf, sealde him gangende feoh and weorcþeós (= -þeówas) (cf. (?) *Northumbrian forms under* þeów: *MS. has* feos. The passage in Genesis is: Tulit Abimelech oves et boves et servos et ancillas et dedit Abraham, reddiditque illi Saram uxorem suam, 20, 14), 164, 25; Gen. 2720. [Cf. *Icel.* verk-þræll.]

weorc-wísung, e; *f. The direction of work:*—Bisceopes dæg-

weorc ... weorcwísung be ðam ðe hit neód sý, L. I. P. 8; Th. ii. 314, 22.

weord, weored. v. wyrd, weorod.

weorf, es; *n. A young ass:*—Weorf *asellus,* assa *asinus,* Wrt. Voc. ii. 10, 45. Be ǽlces nýtenes weorðe gif hí losiaþ. Hors mon sceal gyldan mid .xxx. sciłł. ... wilde weo-f mid .xii. sciłł., oxan mid .xxx. þ., L. O. D. 7; Th. i. 356, 4. Ungewylde weorf, nýten t hors *indomitos subjugales,* Hpt. Gl. 458, 1. v. next word.

weorf-tord, es; *n. Dung of beasts:*—Hé mæg of woruftorde ðone þearfendan áreccan *de stercore erigens pauperem,* Ps. Th. 112, 6. v. preceding word.

weorh, Lchdm. iii. 42, 3 *read* dweorh, cf. i. 364, 13.

weorld, weorm, weorn a *multitude.* v. weorold, wyrm, worn.

weorn (wearn?) *an admonition* (?):—Hét ðá of ðam lige lifgende bearn Nabocodonossor neár æt gangan; ne forhogodon ðæt ða hálgan, siþþan hí woruldcyninges weorn gehýrdon, Exon. Th. 197, 5; Az. 185. Cf. warenian, warenung.

weornian; *p.* ode *To wither, fade, pine away:*—Ic eom hége gelíc ðam ðe hraðe weornaþ, ðonne hit byð ámówen, Ps. Th. 101, 4, 9. Ða blóstman blówaþ ðonne óþre wyrta scrincaþ and weorniaþ, Lchdm. i. 204, 13. Ic weorode *tabescebam,* Ps. Spl. 118, 158. Seó wlitige fægernes heora geógoðhádes weornode and wanode, Homl. Skt. i. 23, 127. Weornodon, Cd. Th. 294, 9; Sat. 468. Wurniende *marcescens,* Hpt. Gl. 430, 62. Seó sáwul, gif heó næfð ða hálgan láre, heó bið weornigende and mægenleás, Homl. Th. i. 168, 33. v. for-weornian; wisnian.

weorod (-ud, -ed, -ad), werod (-ud, -ed), worud (-ad), word, es; *n.* **I.** *a host, troop, band, multitude, crowd:*—Weorod *agmen,* Wrt. Voc. ii. 99, 58. Werod, 6, 42. Werud *cetus, i. congregatio, conventus, multitudo,* 130, 79. Ðæt æfterfylgende weorod *the multitudes* (turbae, Mt. 21, 9) *which followed,* Blickl. Homl. 81, 14. Ðá cwom ðǽr micel mænigeo elpenda of ðæm wudo ungemetlíc weorod ðara dióra *uenire e siluis elephantorum immensos greges,* Nar. 21, 19. Engla þreát, weorud wlitescýne, Exon. Th. 31, 9; Cri. 493: 101, 5; Cri. 1654. Leóde, weorud willhréðig, Elen. Kmbl. 2231; El. 1117. Ðǽr gewyrð ðurh Godes mihte raðe tóscaden ðæt weord (-od, v. l.) on twá, Wulfst. 26, 2. Eall werod (-ed, v. l.) ðæs folces *omnis multitudo populi,* Lk. Skt. i, 10. Ðá com ðæt wered (*turba*), 22, 47. Mycel wered (later MS. werd) his leorningcnihta, 6, 17. Ðæs welegan mannes ungeendod word and unárímed mengeo on hrýðrum, Blickl. Homl. 199, 1. Ðá com hæleða þreát weorodes brehtme, Andr. Kmbl. 2544; An. 1273. Se Hǽlend geann his twelf þegnas sundor of ðam weorode, Blickl. Homl. 15, 7. Mid ðý unárímedan weorode háligra martyra, 25, 35. Weorude, Exon. Th. 57, 2; Cri. 912. Mid engla weorede *cum agmine angelorum,* Bd. 4. 3; S. 570, 1. On weorede *in coetu,* Kent. Gl. 785. Ðǽr hit ða weorud geseóð, Exon. Th. 80, 26; Cri. 1312. Stódon twá heofonlíce werod ætforan ðære cytan dura, Homl. Th. ii. 548, 10. Weredu *examina,* Germ. 396, 180. Lytle worado *pauci,* Lk. Skt. Lind. 13, 23. Weoroda heáp, Andr. Kmbl. 1739; An. 872: Exon. Th. 66, 11; Cri. 1070. Hé ofer weoruda gehwylc scíneþ, 82, 7; Cri. 1335. Wereda, Cd. Th. 42, 8; Gen. 671. Ðǽm englícum weorodum, Blickl. Homl. 131, 19. Fore weorodum *before the multitudes,* Andr. Kmbl. 1471; An. 737: Apstls. Kmbl. 109; Ap. 55. Weorudum, 121; Ap. 61. Werodum, Cd. Th. 78, 31; Gen. 1301. Mycelum weredum (*turbis*) him embe standendum, Lk. Skt. 12, 1. **II.** *a people:*—Ðæs weorudes (*the Mermedonians*) ða wyrrestan, Andr. Kmbl. 3182; An. 1594. Werodes aldor, Cd. Th. 74, 33; Gen. 1231. Werodes rǽswa, Babilone weard, 246, 31; Dan. 487. Weredes weard, 250, 25; Dan. 552. Ðám werude (*the Jews*), 216, 28; Dan. 13: 217, 23; Dan. 27. Hé sægde him wereda gesceafte, 225, 27; Dan. 160. Faraþ geond ealne yrmenne grund, weoredum bodiaþ, 30, 22; Cri. 482. **III.** *where numbers are associated for a special purpose or arranged in regular order.* (1) *in military matters, a host, army, troop, band.* v. weorod-líst:—Werod oððe here *exercitus,* Ælfc. Gr. 11; Zup. 79, 4. Ðá wearþ snellra werod gegearewod tó campe, Judth. Thw. 24, 21; Jud. 199: Cd. Th. 184, 1; Exod. 100. Ðæt werod gefór, 218, 25; Dan. 44. Werud, 190, 24; Exod. 204. Wered *cuneus,* Wrt. Voc. ii. 15, 49. his wered wanode ǽfre, Chr. 1052; Erl. 181, 4. Fram ðám monnum ðæs feóndlícan weoredes *a viris hostilis exercitus,* Bd. 4, 22; S. 591, 3. Mycelnes heofonlíces werydes (-edes, v. l.) *multitudo coelestis militiae,* Lk. Skt. 2, 13. Man ofslóh Theódbald mid eallan his weorude, Chr. 603: Erl. 21, 15. Litle weorode, 937; Erl. 112, 34. Mid ealle his weorude *cum suo exercitu,* Bd. 3, 1; S. 523, 27. Weorede, 1, 9; S. 479, 40. Werode, Chr. 1004; Erl. 139, 31. Hé (*king Alfred*) lytle werede unieþelíce æfter wudum fór, 878; Erl. 78, 33. Wærede, 823; Erl. 63, 18. Síde worude (worulde, MS.), Cd. Th. 118, 11; Gen. 1963. Hié sceoldan ðæt hǽþene weorod geflýman, Blickl. Homl. 221, 30. Hé gesamnode weorod (werod, v. l.), Chr. 380; Erl. 11, 5. Weored, 449; Erl. 13, 10. Heora feónda werod (wærod, v. l.), 999; Erl. 134, 10. Werod (-ed) *cohortem,* Mk. Skt. 15, 16. Wered *manum* (the reference is to the Gothic host), Hpt. Gl. 513, 10. Ðegna uorud *cohortem,* Jn. Skt. Lind. 18, 3. Weredu *castra,* Ps. Spl. 26, 5.

Wælgryre weroda, Cd. Th. 186, 11 ; Exod. 137. Ðú cásere ... hyt
byþ gód ðé and ðínum weorudum (werudum, v. l.), Lchdm. i. 330, 11.
Hí ofslógon .iiii. werad (iiii wera, feówer werod, v. ll.), Chr. 456 ; Erl.
13, 28. ¶ in epithets applied to the Deity, the Lord of hosts :—
Weoruda Dryhten, Andr. Kmbl. 345; An. 173: 869; An. 435. Weorada,
Ps. C. 17 ; Hy. 8, 1. Drihten weoroda, Cd. Th. 301, 14 ; Sat. 581 :
Exon. Th. 27, 10 ; Cri. 428. Weorada ealdor, 15, 1 ; Cri. 229.
Weoroda God, 332, 31 ; Vy. 93. (The passage is printed weorod anes
God ... monna cræftas ; Mr. Bradley suggests that nes is merely an
alternative inflexion for the na of monna, and written above it. v.
Academy, 1893, p. 83.) Weoruda God, 293, 19 ; Crä. 3 : 126, 5 ; Gú.
366 : 273, 13 ; Jul. 515. Weruda, Ps. Th. 76, 11. Weoruda helm,
byrnwíggendra, Elen. Kmbl. 446 ; El. 223. Weoruda waldend, Exon.
Th. 96, 6 ; Cri. 1570 : 137, 28 ; Gú. 566 : Andr. Kmbl. 775 ; An. 388.
Sigora waldend, weoruda wilgiefa, Exon. Th. 229, 34 ; Ph. 465 : Andr.
Kmbl. 123 ; An. 62 : 2565 ; An. 1284. Weoroda wuldorcyning, Exon.
Th. 10, 32 ; Cri. 161. Weroda, Cd. Th. 213, 4 ; Exod. 547. Wereda,
1, 3 ; Gen. 2. Weoruda wuldorgeofa, Elen. Kmbl. 1358 ; El. 681.
Wereda, Hy. 10, 48. (2) where a large number is arranged in regular
companies :—Hé gesceóp týn engla werod, ðæt siud englas ... seraphim.
Hér sindon nigon engla werod ... Ðæt teóðe werod ábreáð, Homl. Th. i.
10, 12–18. (3) a body of servants, retainers, followers, associates :—
Ðis is hold weorod, Beo. Th. 586 ; B. 290. Gif se getihtloda man máran
werude beó ðonne twelfa sum, ðonne beó ðæt ordál forad, L. Ath. i. 23 ;
Th. i. 212, 8. Ðá geáscode hé ðone cyning lytle werode (wyrede, v. l.)
æt Merantúne, Chr. 755 ; Erl. 48, 29. Reste hé ðær méte weorode, Rood
Kmbl. 138 ; Kr. 124. Ðá gesamnodan hié (Peter and Paul) heora weorod
wiþ Simone, Blickl. Homl. 173, 9. Ðá gesamnode hé mycel weorod his
manna, 199, 12. Hwyder gewiton ða mycclan weorod ðe him (the rich)
ymb férdon and stódan? 99, 25. Oft wæron teónan weredum (the
servants of Abraham and those of Lot), Cd. Th. 114, 1 ; Gen. 1897.
(4) a company, assembly :—Wealhþeów fore ðæm werede (the company
in the hall) spræc, Beo. Th. 2435 ; B. 1215. Werede sinagoge, Kent.
Gl. 101. (5) a crew of a ship, ship's company. v. scip-weorod :—Sum
streámráde con, weorudes wísa ofer wídne holm, Exon. Th. 296, 22 ;
Crä. 55. [He ȝescop tyen engle werod oðer hapes, O. E. Homl. i.
219, 9. Niene englene ordres (weoredes, v. l.). A. R. 30, 19. Heouene
riche wordes, Marh. 22, 25. Bruttene weored (ferde, 2nd MS.), Laym.
19922. Engel wird agen him cam, als it were wopnede here, Gen. and
Ex. 1786.] v. burh-, eorl-, eorþ-, fird-, flet-, hell-, heofon-, heorþ-, leód-,
lind-, man-, scip-, þegen-, þegnung-, wuldor-, wyn-weorod.

weorod, werod (-ed) ; adj. Sweet :—Werod (word, v. l. late) dulcis,
Ælfc. Gr. 9, 28 ; Zup. 54, 5. Wæter ... werod on swæcce, Homl.
Th. ii. 144, 4. Hwæðer hit bið ðe wered ðe biter ðe wé ðicgaþ, 372,
29 : Ex. 15, 25. Weredre mulsae, Hpt. Gl. 413, 40. Þurh weredre
pro dulci, 462, 66. Weredre vel wynsumre dulcione, i. blanda,
weredum beóbreáde vel swæsum dulci favo, Wülck. Gl. 225, 17, 20.
Werede ðigene nectareum edulium, Hpt. Gl. 413, 38 : mulsum, 417, 56.
Werede mulsa, 408, 32 : dulcia, Kent. Gl. 179. Ða leáf beóð werede
on swæcce, Lchdm. i. 302, 21. Heó is weredre (rather sweet) and
on byringce, 108, 2 : 276, 10. Ælcum men þincð huniges biobreád ðý
weorodra, gif hé hwéne ǽr biteres onbirigþ, Bt. 23 ; Fox 78, 25.
Weorodran ofer hunig dulciora super mel, Ps. Lamb. 18, 11. v. þurh-,
un-werod, and next word.

weorod, wered, es ; n. A sweet drink :—Hé scencte scír wered, Beo.
Th. 996 ; B. 496. v. preceding word, and weorod-ness.

weorodian ; p. ode To grow sweet :—Hé is swíðe biter on múþe, and
hé ðé tirþ on ða ðrotan, ðonne ðú his ǽrest fandast ; ac hé werodaþ
(-edaþ, v. l.) syðþan hé innaþ, and biþ swíþe líþe on ðam innoþe (interius recepta dulcescant), Bt. 22, 1 ; Fox 76, 30.

weorodlǽcan. v. ge-weorodlǽcan.

weorodlíce ; adv. Sweetly :—Uton singan werodlíce canamus dulciter,
Hy. Surt. 7, 38. Werudlíce dulcione, Anglia xiii. 427, 887.

weorod-líst, e ; f. Want of troops. v. weorod, III. 1 :—Rómwara
cyning ríces ne wénde for werodlíste, hæfde wigena tó lyt, Elen. Kmbl.
125 ; El. 63.

weorod-ness, e ; f. Sweetness :—Ðeós werodnys (weorodnes, v. l.)
hoc nectar, Ælfc. Gr. 9, 16 ; Zup. 42, 7. Weorodnyss dulcedo, Hy. Surt.
98, 17. Weredness, Wrt. Voc. ii. 142, 10. Werodnes, Ps. Lamb. 30, 20.
Werednesse dulcedinem, Anglia xiii. 369, 48. Him ne lícaþ on his
gecorenum náne lustfullunga oððe werodnyssa ðyssere worulde, Homl.
Th. ii. 212, 3. [Salt ȝiueð mete wordnesse (smech v. l.), A. R. 138, 12.]

weorold (-uld), weorld, worold (-uld, -eld), world, e ; f. (but se woruld,
Prov. Kmbl. 40 : worldes, Lk. Skt. I, 70 : ðissum worulde, Met. 10, 70)
A world :—Ealra worulda scippend, Hy. 3, 23. I. the material
world :—Ðeáh ðú ealle gesceafta áne naman genemde, ealle ðú nemdest
tógædere and héte woruld, and þeáh ðone ánne noman ðú tódǽldest on
feówer gesceafta ; án ðæra is eorþe, óþer wæter, þridde lyft, feówerþe fýr,
Bt. 33, 4 ; Fox 128, 28 : Met. 20, 57. Weoruld, 20, 62, 171. Hire
þúhte eall ðeós woruld wlitigre, Cd. Th. 38, 9 ; Gen. 604. Þenden
standeþ woruld under wolcnum, 56, 22 ; Gen. 916 : Exon. Th. 203, 25 ;

Ph. 89. Ðeós world eall gewíteþ and eác ðe hire on wurdon átýdrede,
Elen. Kmbl. 2552 ; El. 1277. Weorulde sceátum, Met. 20, 251 : 24,
34 : 30, 14. Worulde, Cd. Th. 13, 9 ; Gen. 199. Ofer worulde hróf,
241, 20 ; Dan. 407. Worolde dǽlas, Beo. Th. 3469 ; B. 1732. Eall
ðætte gróweþ, wæstmas on weorolde, Met. 29, 71. Hé grundsceát sóhte,
wende tó worulde he came to the earth, Exon. Th. 41, 3 ; Cri. 650 : Cd.
Th. 30, 20 ; Gen. 420 : 32, 29 ; Gen. 510. Næron geond weorulde
welige hámas, Met. 8, 8. Ðú weorulde geworhtest, 20, 24. Weoruld,
28, 26 : 31, 14. Geond ðás wídan weoruld, 8, 41. Worulde, 11, 45.
Woruld, 13, 65 : Cd. Th. 36, 2 ; Gen. 565. Wuldres wyrhta woruld
staþelode, Exon. Th. 206, 22 ; Ph.130. Ðú woruld gesceópe, Met. 20, 4.
Swearc norðrodor, woruld miste oferteáh, Exon. Th. 178, 35 ; Gú. 1254.
Ofer ealle woruld, Hy. 9, 34. Wurdon mycele wæterflód geond ealle
world, Ors. 1, 6 ; Swt. 36, 7. I a. earth as opposed to heaven :—Ic
wæs on worulde wǽdla, ðæt ðú wurde welig on heofonum, Exon. Th. 91,
22 ; Cri. 1496. II. a state of existence, (1) the present state, (a)
with reference to time. v. VI :—Ǽr woruld wǽre ante secula, Ps. Th.
73, 12. World, 89, 2. Worulde (woruldes; Lind.: weorulde, Rush.)
endung consummatio saeculi, Mt. Kmbl. 13, 39, 40. Woreuldes, Lind.
24, 3. From fruman worulde, Exon. Th. 73, 20 ; Cri. 1192. Ðone
forman dæg ðyssere worulde (seculi), Lchdm. iii. 238, 16. Se æftera
worolde dæg, Shrn. 63, 4. Of worldes frymðe (from weorlde, Rush.)
a saeculo, Lk. Skt. 1, 70. Ǽr worulde (worlde, Cott. MSS.) ante secula,
Past. 3 ; Swt. 33, 13. Ætforan wurulde, Ps. Spl. 54, 21. God behét
gefyrn worulde Abrahame, Homl. Th. ii. 12, 23. Se cásere ðe ðú embe
áxast, hé wæs gefyrn worulde, and swíðe fela geára synd nú ágáne syððan
hé gewát of ðysan lífe, Homl. Skt. i. 23, 727. On worulde ǽr, Elen.
Kmbl. 1118 ; El. 561. (b) as the state of existence of all men :—Hié
ne dooð him nán gód ðisse worolde eis necessaria praesentis vitae non
tribuunt, Past. 18 ; Swt. 137, 5. Ðisse worolde (worlde, Hatt. MSS.)
praesentis saeculi, 1 ; Swt. 27, 2. Ælc wlíte tó ende onetteþ ðisse weorlde
lífes, Blickl. Homl. 57, 29. Worulde, Beo. Th. 4675 ; B. 2343 : Exon.
Th. 158, 5 ; Gú. 904. Télnisse weorlde aerunnas saeculi, Mk. Skt. Rush.
4, 19. Worulde, Cd. Th. 270, 22 ; Sat. 94 : Exon. Th. 122, 19 ; Gú. 308.
Moncyn winþ on ðám ýðum ðisse worulde, Bt. 4 ; Fox 8, 22 : 33, 4 ; Fox
132, 28 : Met. 4, 56. Worulde gedál death, Beo. Th. 6128 ; B. 3068.
Worulde brúcan to live, 2129 ; B. 1062. Gád worolde wilna, 1904 ; B. 950.
Worlde geweorces, 5415 ; B. 2711. Hé unæþele á forð þanan wyrð on
weorulde, Met. 17, 29. Worulde, Cd. Th. 35, 7 ; Gen. 551 : 160, 25 ; Gen.
2655. Hé on weorolda (worulda, v. l.) hér wunodæ þrágæ, Chr. 1065 ;
Erl. 197, 23. Hér on worulde, Cd. Th. 30, 29 ; Gen. 474. Ðín módor
gewíteþ of weorulde þurh scondlícne deáð and heó lígeþ unbebyrged mater
tua miserando exitu sepultura carebit, Nar. 31, 29. Worulde, Elen.
Kmbl. 877 ; El. 440. Seó burh Iericho mid hire seofon weallum getácnode ðás áteorigendlícan world, ðe tyrnð on seofon dagum, and hí symle
geedlǽcaþ, óð ðæt seó geendung eallum mannum becume, Homl. Th. ii.
214, 29. Hí ðær hyra gecynda on weorold bringaþ ibi prolem reddunt,
Nar. 35, 27. Woruld, Cd. Th. 137, 35 ; Gen. 2284. On woruld cenned,
12, 20 ; Gen. 188 : 57, 5 ; Gen. 923. In worold wacan, Beo. Th. 119 ;
B. 60. World oflǽtan, 2371 ; B. 1183. Ðás woruld þurh gást gedál
ofgyfan, Cd. Th. 68, 32 ; Gen. 1126. Hé woruld ofgeaf, 71, 2 ; Gen.
1164. ¶ where the present state is contrasted with the future,
where the temporal is contrasted with the eternal :—Ðysse worulde
(woreldes, Lind. : weorulde, Rush.) bearn ... Ða ðe synt ðære worulde
(weorlde, Rush. worulde) wyrðe, Lk. Skt. 20, 34, 35. Se ðe ða écan
ágan wille gesǽlða, hé sceal swíðe flión ðisse worulde wlite, Met. 7, 31.
Ne byð hyt hym forgyfen, ne on ðisse worulde (worold, Lind. : weorlde,
Rush. saeculo), ne on ðære tóweardan, Mt. Kmbl. 12, 32. Forgife ðé
Dryhten willan on worulde, and in wuldre blǽd, Andr. Kmbl. 711 ; An.
356 : 1895 ; An. 950. Se éca deáþ æfter ðisse worulde, Met. 10, 70.
Ðæt God ðé on worlde (in mundo) ðíne synna forgyfe, and æfter worlde
(post mundum) éce reste, L. Ecg. P. iv. 66 ; Th. ii. 226, 18. Ðás dagas
tácniaþ ðás ondweardan weorld, and ða Eásterlícan dagas tácniaþ ða écean
eádignesse, Blickl. Homl. 35, 31. Ðám ðe him willaþ ðás woruld úttor
lǽtan ðonne ðæt éce líf, Exon. Th. 109, 27 ; Gú. 96. On ðás þeóstran
weorulde .. æfter hingonge hreósan in helle, 86, 18 ; Cri. 1410. (c) of
temporal things as distinguished from spiritual :—Ðisse worulde (woruldes,
Lind., saeculi) bearn synd gleáwran ðises leóhtes bearnum, Lk. Skt. 16, 8.
Nó ic eów sweord ongeán óðberan þence, worulde wǽpen, Exon. Th. 120,
21 ; Gú. 275. Hé ðás woruld forhogde, 146, 22 ; Gú. 713. ¶ in the
phrases æfter, for worolde according to the standard of the world, in
respect to temporal matters :—Wæs sum cempena ealdorman æfter worulde
swíðe æþelboren, Homl. Skt. ii. 30, 3. Mon monþwære and for worulde
gód vir summae mansuetudinis et civilitatis, Bd. 1, 8 ; S. 479, 29. For
weorulde wís, Met. 1, 51. For Gode oððe for worulde gyltig, Lchdm.
iii. 442, 35. Ðæt folc wolde hine áhebban tó cyninge, ðæt hé wǽre
heora heáfod for worulde, Homl. Th. i. 162, 5. Ðá forlét hé eal ða
ðing ðe hé for worulde hæfde, Bd. 3, 19 ; S. 549, 33 : Exon. Th. 276,
22 ; Jul. 570. Gif hé récþ ǽniges weorþscipes hér for worulde, Bt. 40, 3 ;
Fox 238, 15 : Homl. Skt. i. 12, 102. Ðær ðú gemunan woldest hwylcra
burgwara ðú wǽre for worulde, oþþe eft gástlíce hwilces geférscipes ðú

wǽre on ðínum móde, Bt. 5, 1; Fox 10, 4: Homl. Skt. i. 21, 87. Hé ne mæg geðyldgian ðæt hé for ðisse worlde (worulde, Hatt. MS.) sié forsewen *despici in mundo hoc non patitur*, Past. 33; Swt. 216, 7: Exon. Th. 457, 5; Hy. 4, 79. (2) *the next world, the future state:*—Fæder ðære tóweardan worulde, Homl. Th. ii. 16, 8. v. (1 b ¶). **III.** *men, people:*—Woruld is onhrēred, Exon. Th. 104, 16; Gū. 8. Ic ðæt for worulde geþolade, lytel þúhte ic leóda bearnum, 87, 13; Cri. 1424. Hí biddaþ God áre ealre þeóde, ðonne ðú him tíðast, swá ðu eádmód eart ealre worlde, Hy. 7, 57. Hé woruld álýseþ, eall eorðbúend, Exon. Th. 45, 14; Cri. 718: Elen. Kmbl. 607; El. 304. **IV.** *earthly things, temporal possessions:*—Ne won hé æfter worulde, ac hé in wuldre áhóf módes wynne, Exon. Th. 126, 12; Gū. 370: 109, 34; Gū. 100. Lamech woruld bryttade, Cd. Th. 74, 22; Gen. 1226. Hié woruld bryttedon, sinc ætsomne, 103, 27; Gen. 1724. **V.** *men and things upon earth:*—Wuldorcyning worlde and heofona, Cd. Th. 242, 31; Dan. 427. Cyningas ðe weoruld heóldan, Ps. Th. 135, 19. Him God sealde gumena ríce, world tó gewealde, Cd. Th. 254, 7; Dan. 608. Wéndes ðú ðæt ðú woruld áhtest, 268, 23; Sat. 59. **VI.** *an age:*—Weorld *seculum*, Wrt. Voc. i. 76, 50. Woruld, 52, 67. Hí gesáwon ðæt beorhte leóht æfter ðære langan worolde (*the time between Adam's death and Christ's descent into hell*), Shrn. 68, 15. Fram worulde *of old* (?); a saeculo, Gen. 6, 4. Worulde *secla*, Wülck. Gl. 255, 21. Wé sind ðá ðe woruida geendunga on becómon *in quos fines saeculorum devenerunt* (1 Cor. 10, 11), Homl. Th. ii. 372, 10. God ǽr ealle woruida, 280, 13. ¶ In expressions equivalent to *for ever:*—Óð on weorulde *usque in saeculum*, Ps. Spl. 17, 52. Stændan tó worulde, Bt. 21; Fox 74, 3. Tó worulde *in seculum seculi*, Ps. Th. 51, 7. Á weoruld *in secula*, 43, 10. On woruida woruld *in seculum seculi*, 78, 14. On ealra weoruida woruld, 110, 5. **VI a.** used to give emphasis, as in ' what in the *world*.' Cf. what-*ever:*—Nǽnig wæs weorð on weorulde, Met. 8, 37. Ne gehýrde wé nǽfre on worulde *a saeculo non est auditum*, Jn. Skt. 9, 32. Nis mé on worulde mód ǽniges þegnscipes, Cd. Th. 51, 32; Gen. 835: 32, 16; Gen. 504: Ps. Th. 71, 12. Eall ðæt heó on weorulde hæfde *omnia quaecumque habuerat*, Bd. 4, 23; S. 593, 10. Hwá is on weorulde, ðæt ne wundrige? Met. 28, 40, 18. On hwam mæg ǽfre ǽnig man on worolde swíðor God wurðian ðonne on circan? L. Eth. vii. 25; Th. i. 334, 25. **VII.** *a person's lifetime:*—Gif gé mægen on eallre eówerre worulde geearnian, ðæt gé habban gódne hlísan æfter eówrum dagum, Bt. 18, 3; Fox 66, 3. Gé winnaþ eówre woruld *ye labour all your life*, 18, 1; Fox 62, 18. Hé swincþ ealle his woruld æfter ðam welan, 33, 2; Fox 124, 1. Ða eldran gnorniaþ ealle heora woruld, 11, 1; Fox 32, 10. Hí winnaþ heora woruld æfter ðæm, 24, 2; Fox 82, 4. Hí búton wærscipe heora woruld ádreógaþ, Homl. Skt. i. 11, 361. **VIII.** *a person's world, conditions of life:*—Hwæðer Boetie eall his woruld lícode ðá hé gesǽlgost wæs, Bt. 26, tit.; Fox xiv, 18: 26, 1; Fox 90, 23. Hyra woruld wæs gehwyrfed, Cd. Th. 21, 3; Gen. 318. Fremdre worulde, Met. 3, 11. **IX.** *the course of human affairs:*—Him eal worold wendeþ on willan, Beo. Th. 3481; B. 1738. Nafa ðú tó yfel ellen, ðeáh ðé sum unwilla on become; oft brincð se woruld ðone willan ðe bið eft, Prov. Kmbl. 40. Onwendeþ wyrda gesceaft weoruld under heofonum, Exon. Th. 292, 31; Wand. 107. [*O. Sax.* werold *world; men; lifetime: O. Frs.* warld, wrald: *O. H. Ger.* weralt *mundus, orbis, terra, seculum, aevum: Icel.* veröld.] v. ǽr-, gewin-, wræc-, wundor-weorold, *and following compounds.*

weorold-ǽht, e; *f. Worldly property, worldly possession or good:*—Is nýd ðæt sume mid wonunge heora woruldǽhta synd gerihte *necesse est ut quidam damnis corrigantur*, Bd. 1, 27; S. 490, 10. Ðone teóðan dǽl his woruldǽhta gesyllan, Wulfst. 283, 26: Bt. 13; Fox 38, 2. Ðæt hí þolian woruldǽhta (world-, *v.l.*), L. Edm. E. 1; Th. i. 244, 13. Hé mót his fæstan álýsan mid his woruldǽhton (*mundanis suis possessionibus*), L. Ecg. P. iv. 60; Th. ii. 220, 27: 63; Th. ii. 224, 13. Micclode God his woruldǽhta, Homl. Ass. 119, 59. [Weorelldahhtess spedd, Orm. 12079.]

weorold-afol (-el), es; *n. Worldly power:*—Ǽnigne man ðe hé (*the priest*) tó bóte gebígan ne mæge oþþe ne durre for worldafole, L. Edg. C. 6; Th. ii. 246, 2. Entas and strece woruldmen ðe mihtige wurdan on woruldafelum, Wulfst. 106, 1.

weorold-ár, e; *f.* **I.** *worldly honour:*—Ðurh ða wilnunga ðære woroldáre (world-, Hatt. MS.) *per concupiscentiam culminis*, Past. 3; Swt. 33, 9. Ða ðe woroldáre wilniaþ, 50; Swt. 387, 1. Hé wilnaþ micle woroldáre habban, 1; Swt. 27, 5. Gif hé woroldáre hæbbe, 9; Swt. 55, 16. Woruldáre, Bt. 7, 3; Fox 20, 11. Woroldáre, Beo. Th. 34; B. 17. Gewonie him God his weorldáre ond eác swá his sáwle áre, Chart. Th. 483, 31. **II.** *worldly property, property not belonging to the church:*—Ðæt mon ælles ðises freólses áre ǽfre for ám híde werian scolde; for ðam ðe Godes ár ǽfre freogre beón sceal ðonne ǽnig woruldár, Cod. Dip. Kmbl. v. 113, 35. [*O. H. Ger.* weralt-ēra *populares honores.*]

weorold-bearn, e; *n. A child of earth, a man*, Exon. Th. 493, 9; Rä. 81, 27.

weorold-bisegu; *f. Worldly, secular business:*—Ða þrig dagas ðe man fæste, forlǽte man ælce worldbysga, L. P. M. 3; Th. ii. 286, 30.

Riht is ðæt munecas hý symle ásyndrian fram woruldbysegan, L. I. P. 14; Th. ii. 322, 5.

weorold-bisegung, e; *f.* **I.** *worldly occupation:*—Nys nánum mæssepreóste álýfed, ne diácone, ðæt hí ymbe náne worldbysgunge ábysgode (*mundano negotio ullo occupati*) beón, L. Ecg. P. iii. 8; Th. ii. 198, 21. **II.** *care of this world, anxiety of this life:*—Ða strongan stormas weoruldbisgunga, Met. 3, 4.

weorold-bismer, es; *n. m. Worldly reproach:*—For woroldbismere ánum *per contumaciam*, Past. 10; Swt. 61, 10.

weorold-bliss, e; *f. Worldly bliss, earthly joy:*—Hé his líchoman wynna forwyrnde and woruldblissa, Exon. Th. 111, 32; Gū. 135.

weorold-bót, e; *f. 'Bót' prescribed by the secular power* in contrast with ' godcund bót,' that prescribed by the church :—Ða woruldbóte hig gesetton . . . swá hwár swá man nolde godcunde bóte gebúgan mid rihte tó bisceopa dihte, L. E. G. proem.; Th. i. 166, 16.

weorold-broc, es; *n. Worldly affliction, trouble of this life:*—Ðæt sár ðære suingellan ðissa woruldbroca (world-, Hatt. MSS.), Past. 36; Swt. 259, 2.

weorold-broc, es; *n. Use for secular purposes:*—Ðes pápa gesette ðæt mæssepreóstas and diáconas ne sceoldon brúcan gehálgodra mæssehrægla tó nǽnegum woroldbroce, ne nó búton on cyrcean áure, Shrn. 112, 20.

weorold-búende; *pl. The dwellers in this world, men :*—Ne furþum wundne wer weoruldbúende gesáwon under sunnan, Met. 8, 35. God is wísdóm and ǽ woruldbúendra, 29, 83: Judth. Thw. 22, 7; Jud. 82. Ðætte rinca gehwylc óþrum gulde weorc be geweorhtum weoruldbúendum, Met. 27, 27.

weorold-camp, es; *m. Worldly warfare :*—Godes þeówas nágon mid wigge ne mid worldcampe tó faren[n]e, ac mid gástlícan wǽpnan campian wið deófol, L. Ælfc. P. 51; Th. ii. 388, 4.

weorold-candel[1], e; *f. This world's candle, the sun :*—Woruldcandel scán, sigel súðan fús, Beo. Th. 3935; B. 1965.

weorold-cearu, e; *f. Worldly care, care about things of this world :*—Woruldcara and welan and flǽsclíce lustas forsmoriaþ ðæs módes ðrotan, Homl. Th. ii. 92, 10. Beóð wǽre ðæt eówere heortan ne beón gehefegode mid woruldcarum, 22, 19. Twá mynecena wǽron . . . ðam gewínode sum eáwfæst wer on woruldcarum, 174, 7. Aidan ealle woruldcara áwearp fram his heortan, nánes þinges wilnigende bútan Godes willan, Homl. Skt. i. p. 6, 55: L. I. P. 13; Th. ii. 320, 35.

weorold-cempa, an; *m. A warrior of this world, an earthly (not a spiritual) soldier :*—Se woruldkempa weraþ woruldlíce wǽpna ongeán his gelícan, ac ðú habban scealt ða gástlícan wǽpna ongeán ðone gástlícan feónd, Basil admn. 2; Norm. 34, 31. Woruldcempa, 36, 17. Se woruldcempa sceall winnan wið úre fýnd, and se Godes þeówa sceall symle for ús biddan . . . Nú ne sceolon ða woruldcempan tó ðam woruldlícum gefeohte ða Godes þeówan neádian fram ðam gástlícan gewinne, Homl. Skt. ii. 25, 820–8.

weorold-cræft, es; *m. A secular craft or art :*—Ne sí nán man swá dysig, ðæt hé ðæs gelícnysse tó ǽnigum hálgum þinge áwende, for ðan ðe ðis (*grammar*) is woruldcræft (weorld-, *v. l.*), Ælfc. Gr. 41; Zup. 246, 2. Ðé gebletsige woruldcræfta wlite and weorca gehwilc, Cd. Th. 239, 1; Dan. 364. Warniaþ ðæt gé beón wísran on eówrum gástlícan cræfte . . . ðonne ða worldmen sindon on heora worldcræftum, L. Ælfc. P. 46; Th. ii. 384, 15. Ðæt him God onsende wíse geþóhtas and woruldcræftas, Exon. Th. 294, 29; Crä. 22. [Cf. *O. H. Ger.* weralt-kraft *ciliarchus, tribunus.*]

weorold-cund; *adj.* **I.** *earthly, temporal :*—Fæder woruldcund *an earthly father*, Exon. Th. 13, 33; Cri. 212. On ðás tíd wé sceolan habban godcunde blisse and eác worldcunde, Blickl. Homl. 83, 20. Mid hú heardum brocum ús swingaþ úre worldcunde fædras, Past. 36; Swt. 253, 25. Ðonne hié eallinga ágiémeleásiaþ ðone ymbhogan woruldcundra ðinga *cum curare corporalia funditus negligunt*, 18; Swt. 137, 2. Hlǽfdige wuldorweorudes and worl[d]cundra háda under heofonum and helwara, Exon. Th. 18, 18; Cri. 285. Ðætte gé fore uueorolde sién geblitsade mid ðém weoroldcundum gódum, and hiora sáula mid ðém godcundum gódum, Cod. Dip. Kmbl. i. 293, 35. **II.** *secular, profane* as opposed to *sacred:*—Gelǽred ge on godcundum gewritum ge on weoruldcundum *literis sacris simul et saecularibus instructi*, Bd. 4, 2; S. 565, 24. **III.** *secular* as opposed to *ecclesiastical:*—Ðis is seó weoruldcunde (weorld-, *v. l.*) gerǽdnes, L. Edg. ii. 1; Th. i. 266, 2. Woruldcunde (world-, *v.l.*), L. C. S. proem.; Th. i. 376, 4. Hwelce wutan wǽron geond Angelkynn ǽgðer ge godcundra háda ge woruldcundra, Past. pref.; Swt. 2, 3. Woroldcundra, Chart. Th. 132, 2.

weoroldcundlíce; *adv. In a worldly manner :*—Hé brýcð ðære godcundan áre worldcundlíce (*seculariter*), Past. 9; Swt. 57, 7. Ðeáh hié woroldcundlíce drohtigen *cum terrena agunt*, 18; Swt. 135, 17.

weorold-cyning, es; *m.* **I.** *an earthly king :*—Án woruldcyningc hæfð fela þegna, Homl. Skt. i. p. 6, 59. Of ðam leódfruman árísað ríces hyrdas, woruldcyningas, Cd. Th. 140, 29; Gen. 2335. Woroldcyninga ðǽm sélestan, Beo. Th. 3373; B. 1684. Woruldcyninga, 6343; B. 3181. **II.** *a king of all the earth, a supreme monarch :*—Woruld-

cyninges (cf. him Gód sealde gumena rîce, world tô gewealde, Cd. Th. 254, 7 ; Dan. 608), Exon. Th. 197, 4 ; Az. 185. [Weoreldking (worlich king, 2nd MS.), Laym. 6328. *O. Sax.* werold-kuning *an earthly king, a powerful king :* O. H. Ger. weralt-kuning *an earthly king.*]

weorold-dǽd, e ; *f. A worldly deed, a deed which is concerned only with affairs of this world :*—Hé hyne sylfne ǽgđer ge wiđ woroldsprǽce ge wiđ worolddǽda warnige, L. E. I. 21 ; Th. ii. 414, 38. [O. H. Ger. weralt-tât seculi actus.]

weorold-deád; *adj. Dead as far as this life is concerned, dead as regards the body :*—Hí mé on deorce stôwe settan, samed anlíce swá đú worulddeáde wrige mid foldan *collocavit me in obscuris sicut mortuos seculi,* Ps. Th. 142, 4.

weorold-déma, an ; *m. A secular judge :*—Be eorlum. Eorlas and heretogan and đás worulddéman ágan nýdþearfe đæt hí riht lufian, L. I. P. 11 ; Th. ii. 318, 20. Bisceop sceall saca sehtan mid đám worulddéman đe riht lufian, 7 ; Th. ii. 312, 15, 36.

weorold-dóm, es ; *m. A secular judgment, judgment by a secular court :*—Sum wer wæs betogen đæt hé wǽre on stale, and hine man gelǽhte and æfter worulddóme dydon him út đa eágan, Homl. Skt. i. 21, 267.

weorold-dreám, es ; *m. Joy of this life :*—Hé worulddreáma breác, Cd. Th. 74, 10 ; Gen. 1220 : 180, 9 ; Exod. 42. Þenden ic wunige on worulddreámum *quamdiu ero,* Ps. Th. 103, 31 : Exon. Th. 184, 1 ; Gú. 1337.

weorold-dryhten, es ; *m. The Lord of the world, the Deity :*—Gif đú wilnige weorulddrihtnes heáne anwald ongitan *si vis celsi jura tonantis cernere,* Met. 29, 1.

weorold-duguþ, e ; *f. Worldly good :*—Wilna brytta and worulddugeđa bróđrum sínum, Cd. Th. 97, 30 ; Gen. 1620. Wilna wǽstmum and worulddugeđum, lufum and lissum, 117, 4 ; Gen. 1948.

weorold-earfeþe, es ; *n. Labour or trouble of this life :*—Strong wind woruldearfoþa, Met. 7, 26, 35, 49.

weorold-ege, es ; *m. Worldly fear, fear of the world :*—Hý sculan Godes ege habban on gemynde and ne eargian for woruldege ealles tô swýđe, L. I. P. 6 ; Th. ii. 310, 20.

weorold-ende, es ; *m. The end of the world :*—Đæt hé léte hyne licgean đǽr hé longe wæs, wícum wunian óđ woruldende, Beo. Th. 6159; B. 3083. [O. H. Ger. weralt-enti.]

weorold-fægerness, e ; *f. Earthly fairness :*—Seó hine lǽrde đæt hé nǽfre Godes geleáfan forléte, and đæt nǽnig woruldfægernes ǽfre his geđóht oncerde, Shrn. 59, 31.

weorold-feoh; *gen. -feós ; n. Worldly wealth, this world's goods :*—Nis woruldfeoh đe ic mé ágan wille sceat ne scilling (*I will not take from a thread even to a shoe-latchet,* Gen. 14, 23), Cd. Th. 129, 12 ; Gen. 2142.

weorold-folgoþ, es ; *m. A worldly service, service with an earthly lord :*—Sceolde Sanctus Martinus néde beón on đære geférædenne ciniges đegna . . . Næs ná đæt hé his willan on đæm woruldfolgađe wǽre . . . Đá hé wæs týnwintre, and hine hys yldran tô woruldfolgađe tyhton, đá fleáh hé tô Godes ciricean, Blickl. Homl. 211, 22–29. Đá forlét hé đone woroldfolgađ, and đá gewát tô Sancte Hilarie đæm bisceope, 217, 1.

weorold-frǽt[e]wung, e ; *f. Worldly ornament, earthly decoration :*—Ne mid golde, ne mid seolfre, ne mid nǽnigre worldfrætwunga, Blickl. Homl. 125, 36.

weorold-freónd, es ; *m. An earthly friend :*—Weoruldfrýnd míne, Met. 2, 16. Wé witan đæt ús forlǽtaþ and níde sculon ealle úre worldfrýnd, Wulfst. 127, 31. Ealle úre weoruldfreónd, 122, 7.

weorold-friþ, es ; *n. Peace that is maintained by the temporal power.* Cf. cyric-friþ :—Đæt woroldfriđ stande betweox Æđelréde cynge and eallum his leódscipe, and eallum đam here đe se cyng đæt feoh sealde, L. Eth. ii. 1 ; Th. i. 284, 9.

weorold-fruma, an ; *m. One of the world's great men :*—Đá gemunde hé đa strangan dǽda đara unmanna (iumanna ?) and đæra woruldfrumena *valida priscorum heroum facta reminiscens,* Guthl. 2 ; Gdwin. 12, 28.

weorold-gálness, e ; *f. Desire for worldly pleasures :*—Đara bócera đe nellaþ godspel sæcgan Godes folce for hiora gémeleáste and for woruldgálnesse, Wulfst. 219, 14.

weorold-gebyrd[u] ; *f. Birth (natural not spiritual) :*—Hé wæs on his móde æþelra đonne on woruldgebyrdum *erat animo quam carne nobilior,* Bd. 3, 19 ; S. 547, 26. Wæs heó æþele in weoruldgebyrdum, đæt heó wæs đæs cyninges nefan dohtor *nobilis natu erat, hoc est, filia nepotis regis,* 4, 23 ; S. 593, 2. v. ge-byrd.

weorold-gedál, es ; *n. Parting from the world, death :*—Tô woruldgedále, Elen. Kmbl. 1159 ; El. 581.

weorold-gefeoht, es ; *n. An earthly fight :*—Sigefæste on woroldgefeohtum, Shrn. 61, 29.

weorold-geflit, es ; *n. A secular dispute :*—Gif him þince đæt hé æt woruldgeflitum sî, đæt tácnaþ him ádl tówerd, Lchdm. iii. 174, 19.

weorold-gerǽdness, e ; *f. A secular ordinance :*—(Eádgáres cyninges gerǽdnes, MS. D.), L. Edg. ii. 1 ; Th. i. 266, 1.

weorold-geriht, es ; *n. A secular or civil right :*—Woruldgerihta ic wille đæt standan on ǽlcum leódscipe swá góde swá hý mon on betste áredian mæge . . . And ic wille đæt woruldgerihta mid Denum standan be

swá gódum lagum swá hý betst geceósan mægen, L. Edg. S. 2 ; Th. i. 272, 23–31.

weorold-gerisene, es ; *n. Worldly propriety :*—Æfter Godes rihte and æfter woroldgerysnum *as religion and the world require,* L. O. 1 ; Th. i. 178, 5 : L. Edm. B. 1 ; Th. i. 254, 4. Woruldgerysenum, L. I. P. 24 ; Th. ii. 336, 38.

weorold-gesǽlig; *adj. Blessed with this world's goods, prosperous :*—Wís ealdorman, woruldgesǽlig, Byrht. Th. 138, 13 ; By. 219. [Cf. O. H. Ger. weralt-sâlig *abundans in seculo.*]

weorold-gesǽlþa; *pl. f. This world's goods, earthly blessings :*—Eálá! hwæþer gé men ongiton hwelc se wela sié, and se anweald, and đa woruldgesǽlþa, Bt. 16, 2 ; Fox 50, 36 : 16, 3 ; Fox 54, 16. Đa getreówan treónd, ic secge seó đæt deórweorđeste đyng eallra đissa woruldgesǽlþa, 24, 3 ; Fox 82, 29. Tô upáhafen for woruldgesǽlþum, Met. 5, 34. Đeáh hý sýn on þyson woroldgesǽlþon đa unspédgestan, Ors. 1, 2 ; Swt. 30, 4. Ælc đara đe đás woruldgesǽlþa hæfþ, Bt. 11, 2 ; Fox 34, 23. v. weorold-sǽlþa.

weorold-gesceaft, e ; *f.* I. *the created world :*—Óđ đæt đeós woruldgesceaft þurh word gewearđ wuldorcyninges, Cd. Th. 7, 23 ; Gen. 110. II. *created things, creatures :*—God wolde đæt him eorđe and uproder and síd wæter geseted wurde woruldgesceafte on wráđra gield, Cd. Th. 7, 4 ; Gen. 101. III. *a creature of this world, an earthly creature :*—Đa unstillan woruldgesceafta, Met. 11, 19, 101. Hé waldeþ weoruldgesceafta, 29, 78. Woruldgesceafta, 11, 84. Fægerust woruldgesceafta (*the sun*), Menol. Fox 227 ; Men. 115. Weroda Waldend, woruldgesceafta, Cd. Th. 237, 4 ; Dan. 332 : 53, 19 ; Gen. 863. Đæt fýr is yfemest ofer eallum đissum woruldgesceaftum, Bt. 33, 4 ; Fox 128, 39. Wiđ ealle weoruldgesceafta, Met. 20, 129.

weorold-gestreón, es ; *n. Worldly gain, this world's wealth :*—Wéndest đú, gif đú mé sealdest ówiht đínes, đæt đé đonne wǽre đín woruldgestreón eall gelytlad ? Wulfst. 260, 19. Đás woruldgestreón, Exon. Th. 106, 15 ; Gú. 41. Sum hér ofer eorþan ǽhta onlíhđ, woruldgestreóna, 295, 10 ; Crä. 31. Ofergrǽdige woruldgestreóna (*cupidi,* 2 Tim. 3, 2), Wulfst. 81, 14. Hé breác mondreáma hér, woruldgestreóna, Cd. Th. 71, 27 ; Gen. 1177. Swíđan woruldgestreónum, 164, 19 ; Gen. 2717. Eádge eorđwelan . . . and heora woruldgestreón, 112, 32 ; Gen. 1879: Exon. Th. 215, 18 ; Ph. 255. Feor lá sí đæt Godes cyrice . . . weoruldgestreón séce (*lucra quaerere*), Bd. 1, 27 ; S. 490, 26.

weorold-geswinc, es ; *n. Worldly labour or toil :*—Sió friđstów æfter đissum weoruldgeswincum, Met. 21, 18. Đynçđ him gesuinc đæt hé biđ bútan woruldgesuincium (worldgeswincum, Hatt. MS.) *laborem deputant, si in terrenis negotiis non laborant,* Past. 18 ; Swt. 129, 1.

weorold-geþóht, es ; *m. A worldly thought :*—Cristes þegnas đeossa worda nán ongeotan ne mehton, ac hié wǽron him bedíglede, for đon đe hié wǽron đágyt mid worldgeþóhtum bewrigene, Blickl. Homl. 15, 14.

weorold-geþyngþ[u] ; *f. Worldly dignity :*—Ælc heáh ár hér on worulde biđ mid frécnessum embeseald ; efne swá đa woruldgeþincþa (-geþingþa, *v. l.*) beóđ máran, swá đa frécnessa beóđ swíđran, Wulfst. 262, 3.

weorold-gewinn, es ; *n. Earthly war :*—Hit biđ swýđe derigendlíc, đæt Godes þeówan Drihtnes þeówdóm forlǽtan, and tô woruldgewinne (weorold-, worold-, *v. ll.*) búgan, đe him náht tô ne gebyraþ, Homl. Skt. ii. 25, 832.

weorold-gewritu; *pl. n. Profane literature :*—On weoruldgewritum gelǽred *saeculari literatura instructus,* Bd. 4, 1 ; S. 564, 11. Đá lǽrde se hyne godcunde gewritu; đá forlét hé đa woruldgewrytu, Shrn. 152, 20.

weorold-gewuna, an ; *m. The custom of the world :*—Hé ásmeáde đæt godcunde be woruldgewunan *he considered the religious question from a secular standpoint,* L. Edg. S. 1 ; Th. i. 270, 15.

weorold-gifu, e ; *f. A gift of temporal things :*—Sende se eádiga pápa Gregorius Æđelbyrhte cyninge woroldgife monige, Bd. 1, 32 ; S. 498, 20. Woruldgiua, Chr. 995 ; Th. i. 244, 17.

weorold-gilp, es ; *m. Worldly glory :*—Đǽm upáhæfenum is tô cýđanne hwelc náwuht đes woruldgielp (worldgilp, Cott. MSS.) is *elatis intimandum est, quam sit nulla temporalis gloria,* Past. 41 ; Swt. 299, 6. For đære gewilnunga woruldgielpes and giétsunga *appetendis lucris temporalibus honoribusque,* 21 ; Swt. 157, 2. Wé đurh đa ne wilniaþ woruldgielpes *per eam humanas laudes assequi minime ambimus,* 48 ; Swt. 375, 11. *Largitas* . . . đæt is đæt man wíslíce his ǽhta áspende, ná for woruldgylpe, Homl. Skt. i. 16, 327, 330. [For weorldʒelpe, worldʒelpe, O. E. Homl. i. 105, 14, 13.]

weorold-gimenn (?). v. weorold-sorh (*last passage*).

weorold-gítsere, es ; *m. One who is covetous of this world's goods :*—Hwæt biđ đæm welegan woruldgítsere (cf. gítsere, Bt. 26, 3 ; Fox 94, 13) on his móde đe bet, þeáh hé micel áge goldes and gimma and gooda gehwæs, Met. 14, 1.

weorold-gítsung, e ; *f. Greed for this world's goods, covetousness :*—Ne mæg fira nán wísdóm timbran, đǽr đǽr woruldgítsung (cf. gítsung, Bt. 12 ; Fox 36, 12) beorg oferbrǽdeþ, Met. 7, 12. Hí cumaþ of woruldgítsunga, Bt. 7, 1 ; Fox 16, 15.

weorold-gleng, es or e; *m.* or *f.* *Worldly pomp* :—Se blinda ne bæd goldes, ne seolfres, ne worldglenga, Blickl. Homl. 21, 6. Se snotera wer ne gewilnaþ ðara woruldg'enga, ne ðæs lîchaman wlîte, ac gewilnaþ ðære sâwle, Basil admn. 8; Norm. 52, 14. Heora yldran on worolde ne wurdan welige ne wlance þurh woroldglænge, L. Eth. vii. 4; Th. i. 334, 4. Ðâ forlêt hê ealle ðâs woruldglenga, Guthl. 2; Gdwin. 16, 18.

weorold-gôd, es; *n.* *A temporal good, worldly good* :—Eówre woruldgôd *vestra bona*, Bt. 14, 2; Fox 46, 1. Ða getreówan freónd ne sint tô woruldgôdum tô· tellanne, ac tô godcundum, 24, 3; Fox 82, 29. Eall ða weoruldgôd ðe him fram cyningum and fram weligum mannum ðisse weorulde gegyfne wæron *euncta quae sibi a regibus vel divitibus saeculi donabantur*, Bd. 3, 5; S. 526, 24.

weorold-hâd, es; *m.* *A secular, lay condition* :—In wéoruldhâde drohtiende *in saeculari habitu conversata*, Bd. 4, 23; S. 592, 42. In weoruldhâde geseted, 4, 24; S. 597, 3. Weoruldhâd forlætan, 598, 2: 4, 23; S. 593, 7.

weorold-hlâford, es; *m.* *An earthly master, a temporal lord* :—Se ðe gyfð ge ðæs worldhlâfordes freóndscype ge his âgenne, Shrn. 177, 6. Se esne ðe ærendaþ his woroldhlâforde wîfes, Past. 19; Swt. 143,. 2. Beó manna gehwylc hold and getrýwe his worldhlâforde, Wulfst. 74, 9. Hî ic wille wyrðian swâ swâ man worldhlâford sceal, Shrn. 196, 32. Woruldhlâfordas môston ðære fiohbôte onfôn, L. Alf. 49; Th. i. 58, 7. Beóð gê underðeódde eówrum woroldhlâfordum *obedite dominis carnalibus*, Past. 29; Swt. 201, 21. Wê læraþ þæt Godes þeówas beón geornlîce Gode þeówigende . . . and ðæt hî beón â heora ealdre holde and gehýrsume . . . and ðæt hî beón heora worldhlâfordum eác holde and getrýwe, L. Edg. C. 1; Th. ii. 244, 5.

weorold-hlîsa, an; *m.* *Worldly fame, earthly renown* :—Habbon hî ðone woruldhlîsan ðe hî sôhton, nâ ða êcan mêde ðe hî ne rôhton, Homl. Th. ii. 566, 6.

weorold-hyht, es; *m.* *Earthly joy* :—Ðû lætest wæter wynlîco tô woruldhyhte of clife clænum, Exon. Th. 194, 10; Az. 136.

weorold-irmþ[u]; *f.* *Misery of this life* :—Wê nû gehýraþ hwær ûs hearmstafas onwôcan, and woruldyrmðo, Cd. Th. 58, 3; Gen. 940. Hî hêton eft Iôhannes gebringan æt his mynstre, fram ðâm woruldyrmþum ðe hê hwîle on wæs, Ors. 6, 10; Bos. 120, 36.

weorold-læce, es; *m.* *A physician for the body* :—Nis se woruldlæce wælhreów, ðeáh ðe hê ðone gewundodan mid bærnette gelâcnige, Homl. Th. i. 472, 13.

weorold-lagu, e; *f.* : -laga, an; *m.* *Law relating to secular matters, civil law* as distinguished from ecclesiastical :—Woruldcunde bôte sêce man be woruldlage, L. C. S. 38; Th. i. 398, 22. Hlâfordes searwu æfter woruldlagu is bôtleás þing, Wulfst. 274, 24. Wîse woroldwitan ðe gesettan tô godcundan rihtlagan worldlaga, L. Eth. vii. 24; Th. i. 334, 22. Leófan menn, lagiaþ gode woruldlagan, Wulfst. 274, 7.

weorold-leân, es; *n.* *Worldly reward* :—Ða ðe Godes þances hwylcne cuman underfôn, ne wilnigen hig ðær nânra woruldleána, L. E. I. 25; Th. ii. 422, 13.

weorold-lîc; *adj.*   I. *worldly, earthly, temporal, mundane* :—Nâuht woruldlîces fæstes and unhwearfiendes beón ne mæg, Bt. 8; Fox 26, 11 note. Ne seó eorþe ænigre worldlîcre frætwednesse onfôn wolde, seoþþan hire ða hâlgan fêt ûres Drihtnes on stôdan, Blickl. Homl. 127, 3. On woruldlîcum wuldre scînende, Homl. Th. i. 62, 27. Tô forsewennysse woruldlîcra æhta, 60, 25: Exon. Th. 126, 20; Gû. 374. Hê sceolde woroldlîcum wæpnum onfôn, Blickl. Homl. 213, 2. Ðæt hwâ woruldlîce spêda forhogige, Homl. Th. i. 60, 32. Worldlîce tintrega, Blickl. Homl. 119, 19. Ealle worldlîcu þing, 109, 3. Gewilnian ða woruldlîcan þingc, Boutr. Scrd. 22, 44.   II. *natural, physical* :—Nis ðeós woruldlîce niht nân þing bûton ðære eorþan sceadu, Lchdm. iii. 240, 18. For ðam ungewunan woruldlîces gesceádes, Boutr. Scrd. 18, 28. Woruldlîce ûðwitan *natural philosophers*, 18, 25 : Lchdm. iii. 240, 20.   III. in contrast with religious or ecclesiastical, *worldly, secular, civil* :—From woruldlîcum luste hearte his giscilde *a seculari desiderio cor ejus defendat*, Rtl. 96, 11. Neádian preóstas tô woruldlîcum gecampe, Homl. Skt. ii. 25, 834, 827. Woroldlîcra weorca on ðam hâlgan dæge geswîce man georne, L. Eth. vi. 22; Th. i. 320, 12. Woruldlîcra, L. C. E. 15; Th. i. 368, 18. Se ðe Gode sceal þeówigan ne sceal hê hyne nâ âbysgian worldlîcra bysgunga *qui Deo vult servire, non debet occupari mundanis negotiis*, L. Ecg. P. i. 7; Th. ii. 174, 27. Bôt æt woroldlîcan þingan, L. Eth. v. 20; Th. i. 308, 31. [*O. H. Ger.* weralt-lîh *mundanus, secularis, carnalis, civilis.*]

weoroldlîce; *adv.*   I. *secularly, civilly* :—Ne sind ealle cyricean nâ gelîcre mæðe weoruldlîce wurðscipes wyrðe, þeáh hig godcundlîce hâlgunge habban gelîce, L. C. E. 3; Th. i. 360, 16. Worldlîce, L. Eth. ix. 5; Th. i. 340, 26.   II. *after the manner of this world* :—Weoroldlîce and wîslîce gê dyde ðætte mannum bedîgled wæs on eorðan ðæt gê ðæt on heofenas tô Gode sôhtan *ye acted with worldly wisdom in seeking in heaven of God what was hidden from men on earth*, Blickl. Homl. 199, 36. [*O. H. Ger.* weraltlîcho *carnaliter.*]

weorold-lîf, es; *n.*   I. *life in this world, life on earth* :—Ðæt ðû mê forgyfe ðæt mînes worldlîfes bletsung anstande *ut tu mihi condones*

ut mundanae meae vitae benedictio permaneat, L. Ecg. P. iv. 67; Th. ii. 228, 3. Ða ðe unrihtes on weoruldlîfe worhtan, Ps. Th. 91, 6. Nis him onwendednes on woruldlîfe *non est illis commutatio*, 54, 20: 114, 7: 118, 92; Cd. Th. 222, 12; Dan. 103: Exon. Th. 172, 11; Gû. 1142: 294, 15; Crä. 15: Wulfst. 258, 15. Hê self lifde on gneáðum woroldlîfe *he* (*bishop Lupus*) *lived a very frugal life on earth*, Shrn. 110, 5. Ðæt hió ne wunian on worldlîfe *ita ut non sint*, Ps. Th. 103, 33: 61, 12: Exon. Th. 427, 7; Rä. 41, 87.   II. *the period of the world's duration, the while the world lasts* :—Ealle on weoruldlîfe weorþaþ gedrêfde *conturbentur in seculum seculi*, Ps. Th. 82, 13. Nele God wið ende æfre tô worulde his milde môd mannum âfyrran on woruldlîfe wera cneórissum *numquid Deus in finem misericordiam suam abscindet a seculo et generatione?* 76, 7. Ðû eart âna God ðe æghwylc miht wundor gewyrcean on woruldlîfe, 76, 11.   III. *worldly life, secular life* :—Hê mynsterlîf ðam weoruldlîfe forbær *monasticam saeculari vitam praetulit*, Bd. 5, 19; S. 637, 8. Hê ôþer lîf mâ lufode ðonne ðæt woruldlîf, S. 638, 7. [Þiss weorelldlif iss wel þurrh nihht bitacnedd, Orm. 2978.]

weoruld-lufu, e, an; *f.* *Love of the world, love of worldly things* :—Wê nellaþ bûgan fram ðyssere andweardan woruldlufe, Homl. Th. i. 580, 3. Se cwyrnstân, ðe tyrnð singallîce, and nænne færeld ne ðurhtîhð, getâcnaþ woruldlufe, ðe on gedwyldum hwyrftlaþ, and nænne stæpe on Godes wege ne gefæstnaþ, 514, 21. Se man ðe ânrædlîce wile his synna geswîcan, dæle on Godes êst eal ðæt hê âge, and forlæte eard and êðel and ealle ðâs worldlufu, L. Pen. 17; Th. ii. 284, 19.

weorold-lust, es; *m.* *Worldly pleasure, pleasure that comes from things of this world* :—Hû ne is ðe genôg openlîce gecýwad ðara leásena gesælþa anlîcnes; ðæt is ðonne æhta and weorðscipe and anweald and woruldlust. Be ðam woruldluste Epicurus sæde . . . ðæt se lust wære ðæt hêhste gôd *habes igitur ante oculos propositam fere formam felicitatis humanae, opes, honores, potentiam, voluptates. Quae considerans Epicurus sibi summum bonum voluptatem esse constituit*, Bt. 24, 3; Fox 84, 19–23: 24, 4; Fox 86, 29. For ðam ðe hê mæg ðurh ðæt tô anwealde cuman oððe tô sumum woruldluste *vel potentiae caussa, vel delectationis*, 24, 3; Fox 82, 34. [*O. Sax.* werold-lust: *O. H. Ger.* weralt-lust *terrena concupiscentia*.]

weorold-mæg, es; *m.* *A kinsman according to the flesh* :—Mê æfter sculon mîne woruldmâgas welan bryttian, Cd. Th. 131, 18; Gen. 2178.

weorold-mann, es; *m.*   I. in a general sense, *a man upon earth, a man* :—Orsorg lîf lædaþ woruldmen wîse (cf. se wîsa mon, Bt. 12; Fox 36, 24), Met. 7, 41. Ân ðara tungla woruldmen hâtaþ (cf. wê hâtaþ, Bt. 39, 3; Fox 214, 19) wænes þîsla, 28, 10. Weoruldmen (cf. folc, Bt. 39, 3; Fox 216, 2) wênaþ, 28, 72. Hû yfele mê dôþ manege woruldmenn . . ic eom getogen tô fremdum þeáwum ðurh ða ungefyldan gîtsunge woruldmonna (*inexpleta hominum cupiditas*), Bt. 7, 3; Fox 20, 19–26. Hwâ is weoruldmonna ðæt ne wafige (cf. hwâ ne wundraþ, Bt. 39, 3; Fox 214, 25), Met. 28, 31. Woruldmonna seó unclæne gecynd, Exon. Th. 63, 8; Cri. 1016. Ic wât ðætte wile woruldmen tweógan geond foldan sceát bûton feá âne (cf. went nû ful neáh eall moncyn on tweónunga, Bt. 4; Fox 8, 18), Met. 4, 52.   II. *a man employed, or interested, in worldly affairs; a man of the world* :—Se Hælend befrân hû woruldmenn be him cwyddedon . . . Drihten ðâ befrân : 'Hwæt secge gê ðæt ic sý? swylce hê swâ cwæde: "Nû woruldmenn ðus dwollîce mê oncnâwaþ, ge ðe godas sind, hû oncnâwe gê mê,"' Homl. Th. i. 366, 5–14. Hê hine wið eallum ðæm wæpnum geheóld, ða ðe woruldmen fremmaþ on menniscum ðingum, Blickl. Homl. 213, 6. Ðonne hê from woruldmonnum (world-, Cott. MSS.) bið ongiten suelce hê sié ældiédig on ðiosum middangearde, Past. 19; Swt. 141, 18. Ða hæþenan fêngon tô wurðienne mistlîce entas and strece woruldmen, ðe mihtige wurdan on woruldafelum, Wulfst. 105, 34.   II a. *a man engaged in secular, as opposed to ecclesiastical, affairs, a layman* :—Nalæs ðæt ân ðæt ðâs ðing dyden weoruldmen (*saeculares viri*), ac eác swylce ðæt Drihtnes eówde, Bd. 1, 14; S. 482, 25. Ða lâfe ðæs gereordes, ðæt sind ða deópnyssa ðære lâre ðe woroldmen understandan ne magon, ða sceolon ða lâreówas gegaderian, Homl. Th. i. 190, 6. Munuclîf wæron gehealdene, and ða woruldmenn wæron wære wið heora fýnd, Homl. Skt. i. 13, 150: 20, 120. Woruldmanna gebeórscypas *secularium conuiuia*, Anglia xiii. 375, 133. [For nane weorldmonne *for no man on earth*, Laym. 28131. Þe wisdom of þeos wise worldmen *sapientia sapientium*, Kath. 486. *O. H. Ger.* weralt-mann *a man.*]

weorold-mêd, e; *f.* *Worldly recompense* :—Ne sceal nân man woruldmêde wilnian æt ðam cuman, for ðam ðe him is gehâten êce gefeá fore on Godes rîce, L. E. I. 25; Th. ii. 422, 15.

weorold-nîd, -neód, e; *f.* *Secular need, need in worldly matters, temporal necessity* :—Se cyngc beóðeþ eallum his gerêfan, ðæt gê ðâm abbodan æt eallum worldneódum beorgan swâ ge mæst magon, L. Eth. ix. 32; Th. i. 346, 30. [*O. H. Ger.* weralt-nôt *tribulatio.*]

weorold-nytt, e; *f.* *Use in this world, temporal advantage* :—Âweccan ðâs wæstmas ûs tô woruldnytte, Lchdm. i. 400, 6: Cd. Th. 59, 7; Gen. 960: 62, 18; Gen. 1016.

weorold-prýt, -prýd, e; *f.* *Worldly pride* :—Næs heó, swâ nû æðelborene men synt, mid ofermêttum âfylled, ne mid woruldprýdum, Lchdm. iii. 428, 32.

weorold-rædenn, e; f. *The rule* or *way of the world* :—Hé ne forwyrnde woroldrædenne, Beo. Th. 2289; B. 1142.

weorold-rica, an; *m. A man of great worldly power* or *wealth* :—Gif him ænig heáfodman hwilces þinges forwyrnde . . . him sóna getídode his Scyppendes árfæstnys ðæs ðe se woruldríca him forwyrnde on ǽr, Homl. Th. ii. 514, 17. Ne cyning ne woruldríca, Lchdm. iii. 442, 36. Unrihtwíse déman and geréfan and ealle ða wóhgeornan woruldrícan mid heora golde and seolfre and godwebbum and eallum ungestreónum, Wulfst. 183, 8. v. next word.

weorold-ríce; *adj. Having worldly power* or *wealth* :—Sum dýre bið woruldrícum men, Exon. Th. 295, 26; Crä. 39. Nænigum woruldrícum men ne eininge sylfum, Blickl. Homl. 223, 27. Worldrícum men, ðe áhte on ðysse worlde mycelne welan and swíðe módelíco gestreón and manigfealde, 113, 5. Worldrícra manna deáþ, 107, 29.

weorold-ríce, es; *n.* I. *the kingdom of this world, this world* :—Ne·þearf ic ænigre áre wénan on woruldríce, Cd. Th. 62, 32; Gen. 1024: 67, 33; Gen. 1110: 99, 4; Gen. 1641. Eordcyninga se wísesta on woruldríce, 202, 25; Exod. 393: 201, 1; Exod. 365. Bibeád ic eów ðæt gé bróþor míne in woruldríce wel ǽretten, Exon. Th. 91, 32; Cri. 1501: 275, 12; Jul. 549: 290, 14; Wand. 65: 442, 16; Kl. 13. Hú wolde ðæt geweorðan on woruldríce? Elen. Kmbl. 910; El. 456. In worldríce, 2095; El. 1049. Hé hét ðæt on worldríce wunian éce *fundavit eam in secula*, Ps. Th. 77, 68. Ne beó nænig man hér on worldríce on his gepóhte tó módig, Blickl. Homl. 109, 27. For hwam winneþ ðis wæter geond woruldríce? Salm. Kmbl. 785; Sal. 392. II. *a kingdom of this world, an earthly kingdom, earthly power* :—Náuht woruldríces fæstes beón ne mæg, Bt. 8; Fox 26, 11. Ic ongite ðætte ælces gódes genóg nis on ðisum woruldwelan, ne æltæwe auweald nis on nánum woruldríce *video nec opibus sufficientiam, nec regnis potentiam posse contingere*, 33, 1; Fox 120, 3. Hé hine (*Nebuchadnezzar*) ásceád of ðam woroldríce (world-, Cott. MSS.), Past. 4; Swt. 39, 21. Woruldríce, Cd. Th. 253, 2; Dan. 589. Ðú woruldrícum wealdest eallum, Ps. Th. 144, 13. On worldrícum, 77, 2. Geond woruldrícu, 113, 9. [Wha wolde wenen a þissere weorldriche, Laym. 15179. Þe laþe gast himm bæd all weorelldrichess ahhte, Orm. 11800. O. *Sax.* weorold-ríki *the world; earthly power* : O. H. Ger. weralt-ríchi *orbis terrarum*.]

weorold-riht, es; *n.* I. *right in worldly matters, civil* or *secular law* :—Wylle wé ærest, ðæt Godes riht ford gá and woruldriht syððan, Wulfst. 274, 20. Beó on ðære scíre bisceop and se ealdorman, and ðær ægðer tǽcan ge Godes riht ge woruldriht, L. Edg. ii. 5; Th. i. 268, 5. II. *the law that should govern the world* :—Dryhten sceáwaþ hwǽr ða eardien ðe his ǽ healden ; gesihð hé ða dómas wonian and wendan of woruldryhte, ða hé gesette, Exon. Th. 105, 25; Gú. 28.

weorold-sacu, e; f. *A dispute about worldly matters* :—Ælce wígwæpna and æghwylce woruldsaca lǽte man stille, Wulfst. 170, 9. [O. *Sax.* werold-saka *a worldly matter* : O. H. Ger. weralt-sahha *mortalis res*.]

weorold-sǽlþa; *pl. f. This world's goods, earthly blessings* :—Eálá hwæþer gé nételícan men ongiton hwelc se wela sié and se anweald and ða woruldsǽlþa? Bt. 16, 2; Fox 50, 36 note. Nis ðé náuht swíþor ðonne ðæt ðú forloren hæfst ða woruldsǽlþa ðe ðú ǽr hæfdest (*fortunae prioris affectu tabescis*). Ic ongite ðæt ða woruldsǽlþa óleccaþ ðæm módum ðe hí willaþ beswícan, 7, 1; Fox 16, 8–12: 8; Fox 26, 5, 8. Mé áblendan ðás ungetreówan woruldsǽlþa *dum levibus malefida bonis fortuna faveret, paene caput tristis merserat hora meum*, 2; Fox 4, 9: Met. 2, 10. Se ymbhoga ðyssa woruldsǽlþa, 7, 54. Woruldsǽlþa, Bt. 12; Fox 36, 29. Swá his mód ǽr swíðor tó ðam woruldsǽlþum gewunod wæs, Fox 4, 1. Ic wolde ðæt wit máre sprǽcan ymbe ða woruldsǽlða *vellem pauca tecum fortuna ipsius verbis agitare*, 7, 3; Fox 20, 1. [O. H. Ger. weralt-sálida *fortuna, terrena felicitas*.] v. weorold-gesælþa.

weorold-sceaft, e; f. *A creature of this world, an earthly creature* :—Wuldres Waldend and woruldsceafta, Exon. Th. 188, 20; Az. 48. Woruldsceafta wuldor, 190, 16; Az. 74. v. weorold-gesceaft.

weorold-sceamu, e; f. *Worldly shame, disgrace among men* :—Wála ðære woruldscame, ðe nú habbaþ Engle. . . . Oft twégen sǽmen oððe þrý drífaþ ða dráfe cristenra manna fram sǽ tó sǽ . . . ús eallum tó woruldscame, Wulfst. 163, 3–7. Ða ðe for ege oððe lufe oððe ænigre worldscame eargiaþ and wandiaþ Godes riht tó sprecanne, 191, 5. For woruldsceame, L. I. P. 12; Th. ii. 320, 22. Gif wíf he óðrum were forlicge, and hit open weorðe, geweorðe heó tó woruldsceame hire sylfre, L. C. S. 54; Th. i. 406, 7. Tó woroldscame, Wulfst. 168, 14. [Æfter muchel weorldscome (worliche same, 2nd MS.) wurðscipe, Laym. 8323.]

weorold-scipe, es; *m. A worldly affair, an affair of this life* :—Ne scyle nán Godes ðeów hine selfne tó ungemetlíce bindan on woruldscipum (world-, Cott. MSS.), ðý læs hé mislícige ðæm ðe hé ǽr hine selfne gesealde *nemo militans Deo implicat se negotiis secularibus, ut ei placeat, cui se probavit*, Past. 18; Swt. 131, 2. [Himm þatt ledenn shall þiss lif, himm birrþ all weorelldshipe flen, Orm. 6322.]

weorold-snotor; *adj. Wise in earthly matters* :—Ægelwíg se woruldsnotra abbod on Eofeshamme, Chr. 1078; Erl. 215, 29. Woroldsnottre men (*naturalists*) secgaþ, ðæt ða ficsas sýn on sǽ hundteóntiges cynna and ðreó and fíftiges, Shrn. 65, 31. Weoroldsnottrum *gymnosophistis*, Wrt.

Voc. ii. 81, 52. Ne weorþeþ on woruldе ǽnig worldsnotera (woruld-, *v. l.*) þonne hé wyrðeþ *there shall be none in the world that has more worldly cunning than he* (*Antichrist*) *has*, Wulfst. 54, 21.

weorold-sorh; *gen.* -sorge; f. *Worldly care, care of this life* :—Hwonon wurde ðú mid ðissum woruldsorgum ðus swíþe geswenced ? . . . Gewítaþ nú, áwirgede woruldsorga, of mínes þegenes móde, Bt. 3, 1; Fox 4, 20–23. Ðæt gemearr ðære woruldsorga *curarum secularium impedimentum*, Past. 51; Swt. 401, 21. Bæd heó ðæt heó móste weoruldsorge and gýmenne forlǽtan *postulans ut saeculi curas relinquere permitteretur*, Bd. 4, 19; S. 587, 38.

weorold-spéd, e; f. I. *worldly wealth*; generally in plural, *this world's goods* :—Syllan ðone teóþan dæl úre worldspéda, Blickl. Homl. 35, 20. Mid hire æhtum and worldspédon *possessionibus suis et mundanis opibus*, L. Ecg. P. ii. 16; Th. ii. 188, 3. Weoroldspédum, Bd. 1, 27; S. 489, 27. Ða ðe habbaþ weoruldspéde *habentes subsidia*, S. 490, 8. Hé him weoruldspéde and æhte (*locus facultasque*) forgeaf, 3, 24; S. 556, 42. Dé Drihten geaf welan and wiste and woruldspéde, Andr. Kmbl. 636; An. 318. Ðonne hié wilniaþ ðæt hié hira woruldspéda (world-, Cott. MSS.) ícen ðonne weorðaþ hié bedǽlede ðæs écean éðles úres Fæder *dum hic multiplicari appetunt, illic ab aeterno patrimonio exheredes fiunt*, Past. 44; Swt. 333, 5. On ðara mánfulra forþforlǽtenesse on ðás woruldspéda, Bt. 5, 1; Fox 10, 23. Nolde hé him geceósan welige yldran, ac ða ðe hæfdon lytle worldspéda, Blickl. Homl. 23, 26: 37, 36. II. *worldly success* :—Syndon ðíne willan on woruldspédum rihte, Cd. Th. 234, 11; Dan. 290: Exon. Th. 185, 20; Az. 10.

weorold-spédig; *adj. Rich in this world's goods, wealthy* :—Se ðe wilnaþ ðæt wolde on ðam angienne his lifes woroldspédig (woruld-, Cott. MSS.) weorðan *qui in principio hereditari festinant*, Past. 44; Swt. 333, 2.

weorold-sprǽc, e; f. *Worldly speech, conversation on worldly matters* :—Ne forlǽte preóst his godcundnysse, ne ne fó tó woruldsprǽcum, L. Ælfc. C. 30; Th. ii. 354, 2. Gé lufiaþ woruldsprǽca, 34; Th. ii. 356, 20. Hyne sylfne ægðer ge wið woruldsprǽce ge wið woruld- dǽda warnige hé and healde, L. E. I. 21; Th. ii. 414, 38.

weorold-steór, e; f. *A secular penalty* :—Gif for godbótan feohbót áriseþ . . . ðæt gebyreþ . . . næfre tó woroldlícan ídelan glengan, ac for woroldsteóran tó godcundan neódan, L. Eth. vi. 51; Th. i. 328, 9.

weorold-strengu; f. *Physical strength* :—Mec feónda sum feore besnyþede, woruldstrenga binom, Exon. Th. 407, 30; Rä. 27, 2.

weorold-strúdere, -strútere, es; *m. A spoiler of this world's goods* :—Ne mót mid rihte nán preóst beón gítsiende mangere, ne worldstrútere on geréfscipe, L. Ælfc. P. 49; Th. ii. 386, 7. Tó helle sculan gítseras, rýperas and reáferas and woruldstrúderas, Wulfst. 26, 17: 165, 36. Cristen cyning sceal rýperas and reáferas and ðás woruldstrúderas hatian and hýnan, L. I. P. 2; Th. ii. 304, 19.

weorold-stund, e; f. *Time spent in this world* :—Mé ne woldon folc oncnáwan, ðeáh ic fela for him æfter woruldstundum (*in the hours I spent on earth*) wundra gefremede, Elen. Kmbl. 725; El. 363. [O. *Sax.* werold-stunda.]

weorold-þearf, e; f. *What is needed for the life of this world* :—Swá swá hé gehet him andlyfne and heora wecrulddearfe forgifan, eác swylce lýfnesse sealde ðæt hí móstan Cristes geleáfan bodian *eis, ut promiserat, cum administratione victus temporalis, licentiam quoque praedicandi non abstulit*, Bd. 1, 25; S. 487, 19.

weorold-þearfa, an; *m. One who is needy in the matter of this world's goods* :—Ic eom wǽdla and worldþearfa *ego egenus et pauper sum*, Ps. Th. 69, 6.

weorold-þearfende; *adj. Deficient in this world's goods, needy* :—Earme men, woruldþearfende, Exon. Th. 83, 4; Cri. 1351.

weorold-þeáwas; *pl. m. Conduct in the affairs of this world* :—Se wæs on woruldþeáwum se rihtwísesta *in the conduct of his life he was most righteous*, Bt. 1; Fox 2, 13.

weorold-þegen, es; *m. A secular thane* :—Mæssepreóstes áð and woruldþegenes is on Engla lage geteald efendýre, L. O. 12; Th. i. 182, 14: L. Wg. 5; Th. i. 186, 10.

[**weorold-peówdóm**, es; *m. Secular service* :—Hí hit freódon wið ealle weoruldþeóðom, Chr. 963; Erl. 121, 31.]

weorold-þing, es; *n. A worldly thing, matter, affair* :—Ne sý nán sacerdhádes man ðe durre geþrístlæcan, ðæt ænig ðara fata, ðe tó godcundum bígonge gehálgod bið, tó ǽnigum woruldþinge bið (*put it to any secular use*), L. E. I. 18; Th. ii. 412, 30. Mid ungerisenlícum gewilnungum ðissa worolddinga (world-, Cott. MSS.) *ambitione inhonesta*, Past. 21; Swt. 157, 9. Sió úterre ábisguug ðissa worolddinga ðæs monnes mód gedréfð *cor externis occupationum tumultibus impulsum*, 22; Swt. 169, 13. Woruldðinga, pref. 5, 3. Hé wæs hwón giernende ðissa worolddinga and micelra onwalda *vir tranquillissimus*, Ors. 6, 30; Swt. 280, 29. Hwæðer ðæt nú sié tó talianne wáclíc and unnyt, ðætte nytwyrþost is ealra worulddinga, ðæt is anweald? *num imbecillum, ac sine virtutibus aestimandum est, quod omnibus rebus constat esse praestantius ?* Bt. 24, 4; Fox 86, 17. Ðonne hé fægnaþ ðæt hé sié ábisgod mid worolddingum *dum se urgeri mundanis tumultibus gaudent*, Past. 18; Swt. 129, 3. Freom in weorolddingum *in saeculi rebus strenuus*, Bd. 4,

2; S. 566, 18. On woruldþingan, L. I. P. 14; Th. ii. 322, 17. Of wurðfulre mægðe æfter woruldþingum *of a family honourable from a worldly point of view*, Homl. Skt. ii. 31, 14. Wē forlēton ealle woruld-ðing *nos dimisimus omnia* (Mk. 10, 28), Homl. Th. i. 392, 32, 28. Ðā ðā his geōgoð æfter gecynde woruldðing lufian sceolde, ii. 118, 23. [Ʒif we forleosað þas lenan worldþing, O. E. Homl. i. 105, 30. He hadde michel of wereldþinge, ii. 127, 16. To geornenn aftterr weorelldþing, Orm. 2966.]

weorold-wǣpen, es; *n. A weapon used in this world's warfare:*— Ðā wæs feówer geár ǣr his fulwihte, ðæt hē woroldwǣpno wæg (*he bore this world's arms*), Blickl. Homl. 213, 4.

weorold-wæter, es; *n. An ocean:*—Saga mē, hū fela is woruldwætra? Ic ðē secge, twā sindon sealte sǣ, and twā fersce, Salm. Kmbl. p. 186, 24.

weorold-wela, an; *m. Worldly wealth, worldly good:*—Se woruldwela (*pompa*) his frǣtewunga āweorpende fleáh, Gl. Prud. 52 a. Sume mægon habban ælles woruldwelan genōg *huic census exuberat*, Bt. 11, 1; Fox 30, 30. Hē wilnaþ hwæthweg ðises woruldwelan, 26, 2; Fox 94, 3. Hī geleáfan ceósaþ ofer woruldwelan, Exon. Th. 230, 30; Ph. 480. Ne wearð ǣnig eordlic cyning mǣrra ðonne Salomon wearð þuruh ǣghwylcne woroldwelan, Wulfst. 277, 23. Ða woruldwelan synt gesceapene tō biswice ðām monnum ðe beóþ neátenum gelíce, Bt. 14, 1; Fox 42, 2. Swylcra fela weoruldwelena (cf. ealne ðisne andweardan welan, Bt. 32, 3; Fox 118, 20), Met. 19, 26. Waa ieów welegum, ðe iówer lufu eall and tōhopa is on eówrum woruldwelum, Past. 26; Swt. 181, 24. Dios-sum woruldwelum, 45; Swt. 239, 6. Ðǣs land beóð neáh ðǣm burgum ðe beóð eallum woruldwelum gefylled *hic est ciuitas uicina diues, omnibus bonis plena*, Nar. 34, 33. [Ʒif þu best aihteles . . . ac gef þu hauest woreldwele . . . , O. E. Homl. ii. 29, 28. *O. Sax.* werold-welo: *O. H. Ger.* werald-wolun; *pl. mammona*.]

weorold-weorc, es; *n.*   **I.** *worldly work, secular occupation:*— Ðǣm tídum þonne gē ða rǣdinge hāligra bōca forlǣten and ða gebeda, þonne sculon gē on sum nytlíc weoroldweorc fōn, L. E. I. 3; Th. ii. 404, 10. Nǣnig mon ne geþrístlǣce on ðone hālgan dæg on nān weoruld-weorc befōn, 24; Th. ii. 420, 22.   **II.** *in a special sense, mechanics:*— *Mechanica*, ðæt ys weoruldweorces cræft, Shrn. 152, 16.

weorold-weorþscipe, es; *m. Worldly honour, civil dignity:*—Hæbbe hē (*the priest*) Godes miltse, and tō woroldweorðscipe ðæt hē sȳ þegen-weres and þegenrihtes wyrðe (*his civil status is that of a thane*), L. Eth. v. 9; Th. i. 306, 20. Tō woruldwurðscipe sī hē þegenlage wyrðe, L. C. E. 6; Th. i. 364, 16: Wulfst. 270, 32.

weorold-widl, es; *n. Worldly pollution, defilement contracted in this life:*—Ðæt fȳr georne āsēceþ eorðan sceátas, ōþþæt eall hafaþ ældes leóma woruldwidles wom wælme forbærned, Exon. Th. 62, 25; Cri. 1007.

weorold-wig, es; *n. The warfare of this world:*—Ne gebyraþ him (*the priest*) nāðor ne tō wífe ne tō woruldwíge, L. Edg. C. 60; Th. ii. 256, 35. Worldwíge, L. Eth. ix. 30; Th. i. 346, 23.

weorold-willa, an; *m. A worldly good:*—Monige habbaþ ælces woroldwillan genōg, Bt. 11, 1; Fox 30, 30 note.

weorold-wilnung, e; *f. Worldly desire:*—Ðæt líf ðæra gesinhíwena, ðeáh hit ful wundorlíc ne sié on mægenum weoruldwilnungum tō wid-standanne, hit mæg ðeáh bión orsorglíc ǣlcra wíta, Past. 51; Swt. 399, 21. Fram weoruldwilnungum hine sceal gehwā fremdian *a seculi actibus se facere alienum*, R. Ben. 17, 4.

weorold-wís; *adj.*   **I.** *worldly wise, having knowledge of the ways of the world:*—On ōðre wísan mon sceal manian ða woroldwísan (cf. ða ðe ðisse worulde lotwrenceas cunnon and ða lufigeaþ, 30; Swt. 203, 5), on ōðre ða dysegan *aliter hujus mundi sapientes admonendi sunt, aliter hebetes*, Past. 23; Swt. 175, 16. Ðonne hē gesyhð ða welegan and ða weoruldwísan sweltan *cum viderit sapientem morientem*, Ps. Th. 48, 8.   **II.** *having secular knowledge, learned:*—Ðone hys yldran be-fæston on hys cnyhthāde sumum woruldwȳsan men, ðæt hē æt ðam leornode ða seofon cræftas, Shrn. 152, 11. Hēton woroldwíse menn wordsāwere ðone æðelan lāreów Paulus *ab hujus mundi sapientibus prae-dicator egregius seminiverbius est vocatus*, Past. 15; Swt. 97, 4. [Þe king sende æfter witien, æfter worldwise monne, ða wisdom cuðen, Laym. 15496. *O. H. Ger.* weralt-wís *mundi sapiens, gymnosophista, maleficus*.]

weorold-wísdóm, es; *m. Secular knowledge, science, learning:*—Ða dohtor befæste se fæder tō lāre, ðæt heó on woruldwȳsdóme wǣre getogen æfter Grēciste ūðwȳtegunge and Lǣdenre getingnysse, Homl. Skt. i. 2, 20. His fæder and his frȳnd hine befæstan tō lāre tō woruldwísdóme, 3, 5. Ða ðe woldon woruldwísdóm gecneordlíce leornian, Homl. Th. i. 60, 27. Ungetogene menn geceás Drihten him tō leorningcnihtum, and hí swā geteah, ðæt heora lār oferstāh ealne woruldwisdóm, 576, 30. Ða seofon cræftas on ðām beóþ gemēted ealle weoruldwȳsdómas, Shrn. 152, 12. [*O. H. Ger.* weralt-wístuom *sapientia*.]

weorold-wíse, an; *f. What is usual in the world, a fashion of the world:*—Hē bæd ðæt Godes yrre ofer hí ne cóme, ne him wǣre hwæs (hwæt?) gneáðes ne ōþerra worldwísena. Ða com stefn of heofonum and seó cwæð : . . . 'Gif hwilc man on micelre neádþearfnesse bið ðín ge-myndig . . . ic gefremme ðæs mannes nédþearfnesse' *he prayed that God's anger should not come upon tkem, nor that aught of penury or of other*

ills that are fashions of this world might be theirs. Then came a voice from heaven, and it said: . . . If any man in great need shall be mindful of thee . . . I will perform that man's need, Shrn. 77, 1–9.

weorold-wita, an; *m. A secular* or *lay councillor:*—Gif feohbót āríseþ, swā swā wise woroldwitan tō steóre gesettan, L. Eth. vi. 51; Th. i. 328, 5. Wíse eác wǣron woroldwitan ðe ǣrest gesettan tō godcundan rihtlagan worldlaga, vii. 24; Th. i. 334, 21. Worldwitan, ix. 348, 13.

weorold-wíte, es; *n.*   **I.** *a punishment suffered in this world, a punishment on earth:*—Forgield mē ðín líf, ðæs ðe ic ðe mín þurh woruld-wíte weorð gesealde, Exon. Th. 90, 22; Cri. 1478.   **II.** *a secular* (in contrast with an ecclesiastical) *punishment, secular penalty, money-fine:*—Sunnandaga cȳpinga forbeóde man georne be fullan worldwíte, L. Eth. ix. 17; Th. i. 344, 8. Gif hǣðen cild binnon .ix. nihton þurh gímelíste forfaren sí, bētan for Gode būton worldwíte; and gif hit ofer nigan niht gewurðe, bētan for Gode and gilde .xii. ōr, L. N. P. L. 10; Th. ii. 292, 7.

weorold-wlencu (-o); *indecl.:* -wlenc, e; *f. Worldly pride, worldly pomp:*—Bisceopum gebyreþ, ðæt hí woruldwlence ne hēdan tō swȳðe, L. I. P. 10; Th. ii. 316, 30. Hí lǣccaþ of manna begeátum lōc hwæt hí gefón magan . . . Syððan hȳ hit habbaþ, hí glencgaþ heora wíf mid ðam ðe hí weofoda sceoldan, and maciaþ eall heom sylfum tō woruld-wlence, 19; Th. ii. 328, 9. Ða mon sceal swā micle mā hātan ðonne biddan suā man ongiet ðæt hit for ðissum woruldwlencum (world-wlencium, Cott. MSS.) bióð suiður upāhafene and on ofermēttum āðundene *talibus rectum tanto rectius jubetur, quanto in rebus transitoriis altitudine cogitationis intumescunt*, Past. 26; Swt. 181, 21.

weorold-wrenc, es; *m. A worldly wile, a trick of this world:*—Ða ðe woruldmonnum ðynceaþ dysige, ða geciésð Dryhten, for ðæm ðæt hē ða lytegan, ðe mid ðissum woroldwrencium bióð upāhæfene, gescende *quae stulta sunt mundi, elegit Deus, ut confundat sapientes*, Past. 30; Swt. 203, 24.

weorold-wuniend, es; *m. or* -wuniende; *adj. A dweller in this world;* or *dwelling in this world:*—Būton moncynne, ðara micles tō feola woruldwuniendra winð wið gecynde, Met. 13, 17.

weorpan (wurpan, wyrpan); *p.* wearp, *pl.* wurpon; *pp.* worpen.   **I.** *to cast, throw, fling.* (1) with acc. of what is thrown:—Heó wearp twēgen feorðlingas *misit duo minuta*, Mk. 12, 42. Hē wearp wundenmǣl, ðæt hit on eorðan læg, Beo. Th. 3066; B. 1531. Hí wurpon tān betweox him, Homl. Th. i. 246, 3. Swā swā mid unmǣtnesse micles stormes worpene beón *quasi tempestatis inpetu jactari*, Bd. 5, 12; S. 627, 40. (1 a) where further the direction or end of throwing is marked, (a) by the dative:— Weorpaþ hit hundum, Ex. 22, 31. Nis nā gōd ðæt man nime bearna hláf and hundum weorpe (worpe, *v. l.*), Mk. Skt. 7, 27. Ðā hēt hē hine wurpan deórum, Homl. Skt. ii. 29, 245. (β) by prepositions or adverbs:— Ic wyrpe max míne on eá, and angil ic wyrpe . . . Ic wyrpe ða unclǣnan ūt, Coll. Monast. Th. 23, 9–17. Hira tū sǣ on lond wearp, Chr. 897; Erl. 96, 9: 1009; Erl. 142, 6. Se deófol wearp ǣnne stān tō ðære bellan, Homl. Th. ii. 156, 9. Hí wurpon heora waru oforbord, i. 246, 2. Hig tōdǣldon hys reáf, and wurpon hlot ðǣr ofer, Mt. Kmbl. 27, 35. Hí wurpon hine on ðone bāt, Chr. 1046; Erl. 174, 17. Ofen esnas wurpon wudu oninnan, Cd. Th. 231, 10; Dan. 245. Hí wurpon hyra wǣpen ofdūne, Judth. Thw. 25, 33; Jud. 291. 'Wurp (*projice*) hig on eordan.' And hē wearp, Ex. 4, 3. Wurp hym mete tōforan, Lchdm. i. 246, 3. Weorp hit ūt, Mt. Skt. 9, 47. Weorp ðone beám of ēgo ðín, Mt. Kmbl. Lind. 7, 5. Weorp (wurp, *v. l.*) ðinne angel ūt, Mt. Kmbl. 17, 27. Wurpaþ hit ūt on ðæt wæter, Ex. 1, 22. Ðæt hē wurpe his cynehelm and gecneówige æt ðæs fisceres gemynde, Homl. Th. i. 578, 6. Hwylc eówer sī synleás weorpe (wurpe, *v. l.*) stān on hí, Jn. Skt. 8, 7. Be ðære coþe þe se mon his ūtgang þurh ðone mūð him fram weorpe, Lchdm. ii. 236, 13. Swylce mon wurpe (worpe, MS. A.: worpað, Lind.: worpes, Rush., *jaceat*) gōd sǣd on his land, Mk. Skt. 4, 26. Ic hēt hit weorpan on fȳr, Ex. 32, 24. Hēt twelf weras nyman twelf stānas . . . and habban forð mid eów tō eówere wícstōwe and wurpan hig ðær *praecepe eis, ut tollant . . . duodecim lapides, quos ponetis in loco castrorum*, Jos. 4, 3. Worp-ende ða scillingas in temple *projectis argenteis in templo*, Mt. Kmbl. Lind. 27, 5. Heora líchoman on ða eá worpene wǣron, Bd. 5, 10; S. 625, 6. (2) with dat. of what is thrown. Cf. *Icel.* verpa *with dat.*:—Hē teoselum weorpeþ, Exon. Th. 345, 9; Gn. Ex. 185. Beorges weard wearp wælfȳre, Beo. Th. 5157; B. 2582: Exon. Th. 478, 11; Ruin. 39. (2 a) where the direction or end of throwing is marked:—Streámas weorpaþ on stealchleoþa stāne and sonde, Exon. Th. 382, 5; Rä. 3, 6.   **I a.** *to throw* (as in *to throw* open):—Mycel wynd wearp upp ða ðuru, Homl. Skt. i. 3, 347.   **II.** where a (forcible) change of a person's place or condition is made (lit. or fig.), *to cast into prison, cast off, out, throw into a form, drive out:*—Ic ne weorpe (wyrpe, wurpe, *v. ll.*) ūt ðone ðe tō mē cymð, Jn. Skt. 6, 37. Gif ðū worpes ūsig *si eicis nos*, Mt. Kmbl. Lind. 8, 31. Ðū wurpe þeóde *ejecisti gentes*, Ps. Th. 79, 8. Hē wearp lósep on cweartern, Gen. 39, 20: Cd. Th. 20, 7; Gen. 304. Hē wearp hine on ðæt morðer innan, 22, 18; Gen. 342. Hē wearp hine of ðan heán stóle, 19, 33; Gen. 300. Hē wearp hine

on wyrmes líc, 31, 26; Gen. 491. Hé út weorpe earme þearfan *ejiciantur*, Ps. Th. 108, 10. Men sǽdon ðæt hió sceolde mid hire drýcræft weorpan men an wildedeóra líc, Bt. 38, 1; Fox 194, 31. Hié worpene beóð in helle grund, Elen. Knibl. 2606; El. 1304. **III.** *to move* a thing from one position to another, in the phrase *weorpan tó handa* to hand over :—Weorpe hé ðone ceáp tó handa, L. In. 56; Th. i. 138, 12: L. Alf. pol. 21; Th. i. 74, 19: 24; Th. i. 78, 9. Sceal se ðe hine áh weorpan hine tó handa hláforde and mǽgum, L. In. 74; Th. i. 148, 15. **IV.** in metaphorical senses :—Drihten ádrífd fram eów ǽlc yfel and wyrpd ongén eówere fýnd *auferet Dominus a te omnem languorem, et infirmitates pessimas non inferet tibi, sed cunctis hostibus tuis*, Deut. 7, 15. Ðonne hió wyrpd (wirpd, Cott. MSS.) on ðæt gedóht hwæthugu tó bigietenne *dum adipiscenda quaeque cogitationi objicit*, Past. 11; Swt. 71, 22. Ne andswarast ðú nán ðing ágén ðæt ðás ðe on weorpað (wurpað, *v. l.*) *non respondis quicquam ad ea quae tibi objiciuntur ab his?* Mk. Skt. 14, 60. Him man wearp on, ðæt hé wæs ðes cynges swica *he was charged with being a traitor to the king*, Chr. 1055; Erl. 189, 3. Ðý læs ǽfre cweðan óðre þeódæ : 'Hwær com eówer God?' and ús ðæt on eágum worpen þær manna wese mǽst ætgædere *nequando dicant in gentibus : 'Ubi est Deus eorum?' et innotescant in nationibus coram oculis nostris*, Ps. Th. 78, 10. **V.** *to reach an object by throwing*, *to throw and hit*, *to strike* with something, (1) with gen. of what is thrown :—Hé hine ongon wæteres weorpan *he threw water upon him*, Beo. Th. 5575; B. 2791. (2) with a preposition :—Gif men cídað and hira óðer hys néxtan mid stáne wirpð oððe mid fýste slicð *si rixati fuerint viri et percusserit alter proximum suum lapide vel pugno*, Ex. 21, 18. Seó clǽnnys wyrpð ða gálnysse mid stáne *pudicitia libidinem cum saxo percutit*, Gl. Prud. 12 b. Seó sýfernes mid stáne wearp ða gálnesse on ðone mid sobrietas lapidem iacit et percutit os luxuriae, 48 a. [O. E. Homl. werpen : Laym. weorpen, werpen, worpen ; 2nd MS. werpe, wearpe : Orm. werrpenn : A. R. weorpen, worpen : Gen. and Ex. werpen : O. and N. werpe, worpe : Goth. wairpan : O. Sax. werpan : O. Frs. werpa : O. H. Ger. werfan : Icel. verpa.] v. á-, be-, for-, ge-, of-, ofer-, on-, tó-, wið-, ymb-weorpan ; worpian.

weorpe. v. wande-weorpe, seale-weorpan (?), Cod. Dip. Kmbl. iii. 78, 15.

weorpere, es ; *m. A thrower* (cf. *to throw* as a wrestling term) :—Ic (*mead*) eom weorpere, efne tó eorþan ealdne ceorl (cf. Aldhelm's riddle : Pedum gressus titubantes sterno ruina), Exon. Th. 409, 27 ; Rä. 28, 7.

weorpness. v. on-weorpness.

weorr ; *adj. Bad, grievous* :—Ðæt wæs ðam weorode weor tó geþoligenne (cf. sár tó geþolienne, 3375 ; An. 1691), Andr. Kmbl. 3317 ; An. 1661. v. wirsa.

weorras, weorþ *a place.* v. wearr, worþ.

weorþ, weorþe, worþ, wurþ, wyrþ, es ; *n.* **I.** *worth, value,* (1) of things :—Underwed ðæt sý ðæs orfes óðer healf weord *a security that is half as much again as the value of the cattle,* L. O. D. 1 ; Th. i. 352, 9. Be ðæs ceápes weorðe (wyrðe, *v. l.*), L. In. 49 ; Th. i. 132, 16. Be ´éwes weorðe (wyrðe, *v. l.*), 55 ; Th. i. 138, 6. Be his wlites weorðe . . . swá man ðæt weord up áræran mihte, L. Ath. v. 6, 2 ; Th. i. 234, 6–10. Gilde ðæs pyttes hláford ðæra nýtena wurð, Ex. 21, 34. (2) of persons, *worth, worthiness* :—Ðæt be ðære cennendra gefyrhtum ðæs bearnes weorþe ongyten wǽre, Blickl. Homl. 163, 27. **II.** *price* of anything sold, *amount paid* for purchase or redemption :—Hig cwǽdon : 'Hyt is blódes weord' (*v. l.* wurð, word, Lind. : weord, Rush., *praetium sanguinis*), Mt. Kmbl. 27, 6, 7, 9. Noldon hig nánes wurðes onfón, ac forgeáfon him ða birgene, Gen. 23, 6. Hí sumne dǽl heora landes wurðes æthæfdon, Homl. Th. i. 316, 24. Hire innoþ ðú gefyldest mid ealles middangeardes weorþe (cf. Homl. Skt. ii. 27, 120 infra, and next passage), Blickl. Homl. 89, 19. He áhongen wæs fore moncynnes mánforwyrhtum, ðæt he lífes ceápode mid ðý weorðe, Exon. Th. 68, 3 ; Cri. 1098. Hé monige mid weorþe álýsde *he redeemed many by purchase,* Bd. 3, 5 ; S. 527, 15. Gebycge hé ða lond æt hire mid halfe weorðe *let him buy the lands of her at half price,* Cod. Dip. Kmbl. ii. 120, 28. Giboht worðe miclum, Rtl. 27, 1. Ðú becýptest folc ðín búton weorðe, Ps. Spl. 43, 14 : Ps. Surt. 43, 13. Geseald tó myclum weorðe (wurðe, wyrðe, *v. ll.*), Mt. Kmbl. 26, 9. 'Ic sille eów hundteóntig þúsenda mittan hwǽtes tó ðam wurðe ðe ic hit bebohte.' . . . Ðæt wyrð ðe hé mid ðam hwǽte genam hé ágeaf ágeán tó ðare ceastre bóte, Th. Ap. 10, 1–9. Fæder gesealde bearn wið weorðe (wurðe, *v. l.*), Wulfst. 161, 7. Mon áceorfe ða tungan of, ðæt hié mon ná undeórran weorðe móste lésan ðonne hié mon be ðam were geeahtige, L. Alf. pol. 32 ; Th. i. 82, 2. Syle ðú hig wið wurðe and bring ðæt wurð tó ðære stówe, and bige mid ðam ylcan feó swá hwæt swá ðe lícige, Deut. 14, 25–26 : 24, 7. Ðæt hé ðæt weord ágife tó álýsnesse his sáwle *pretium redemtionis animae suae,* Ps. Th. 48, 7 : Bd. 4, 22 ; S. 592, 14. Álésan wé úre sáule ða hwíle ðe wé ðæt weorþ on úrum gewealde habban, Blickl. Homl. 101, 10. Tó berenne ealles middaneardes wurþ (cf. Blickl. Homl. 89, 19 *supra*), Homl. Skt. ii. 27, 120. Forgelde hé ðæt lond, and ðæt wiorth gedaele, Cod. Dip. Kmbl. i. 234, 33. Wurð, Ex. 21, 35 : Homl. Th. i. 62, 3 : 316, 11. Him man his weord ágefe *let the price of the chattel be*

returned to him, L. H. E. 16 ; Th. i. 34, 11. Nán man nán þing ne bycge ofer feówer peninga weord (*that costs more than fourpence*), L. C. S. 24 ; Th. i. 390, 3. Þéh ðe hé hié sume wið feó gesealde, hé ðæt weord nolde ágan ðæt him mon wið sealde, Ors. 4, 10 ; Swt. 198, 17. Ðæs hwǽtes wurd ðe hé ðé sealde, Gen. 44, 2. Weord, Exon. Th. 90, 23 ; Cri. 1478. **III.** *amount* to be paid in compensation :—Mid weorðe forgelde man, L. Ethb. 32 ; Th. i. 12, 1. Gif esne óðerne ofsleá, ealne weorðe forgelde, 86 ; Th. i. 24, 11. Gif esnes eáge and fót of weorðeþ áslagen, ealne weorðe hine forgelde, 87 ; Th. i. 24, 14. **IV.** *worth,* as in penny-*worth, amount of a certain value* :—Nabbaþ hí genóh on twégera hundred penega weorðe (wurþe, *v.l.*) hláfes *ducentorum denariorum panes non sufficiunt eis,* Jn. Skt. 6, 7. Sceóte man æt ǽghwilcre híde pænig oððe pæniges weord, Wulfst. 181, 5 : L. O. 11 ; Th. i. 182, 10. Ðæt hyra ǽgðer hæbbe .lx. penenga wyrd . . . ðæt sý .xxx. penega wyrd, Cod. Dip. Kmbl. vi. 133, 23, 24. [Goth. wairþa galaubamma usbauhtai *pretio empti* : O. Sax. O. L. Ger. O. Frs. werd ; *n.* : O. H. Ger. werd ; *n. pretium, aestimatio* : Icel. verð ; *n.*] v. mann-, or-, pening-weorþ ; wirþa.

weorþ, worþ, wurþ, wirþ, wyrþ, wirþe, wierþe, wyrþe, weorþe ; *adj.* **I.** *worth, of value,* (1) referring to saleable things :—Éwe bið mid hire giunge sceápe scitt. weord, L. In. 55 ; Th. i. 138, 7. Oxan horn bið .x. pæninga weord, 58 ; Th. i. 138, 21. Hú mycel feós hit wǽre wurd, Chr. 1085 ; Erl. 218, 33. Næs án híd landes, ðæt hé nyste hwæs heó wurd wæs, 1086 ; Erl. 222, 11. Ðæt yrfe ðæt wǽre .xxx. pænig wyrd, L. Ath. v. 2 ; Th. i. 230, 19. Genime man .vi. scitt. weord (wurd, *v.l.*) wed, L. In. 49 ; Th. i. 132, 13. Ágife man án ram weorðe .iiii. peningas, L. Ath. i. proem. ; Th. i. 198, 7. (2) in other cases where money is to be paid :—Gif mon óðrum wongtód of ásleá, geselle .iiii. scitt. tó bóte. Monnes tux bið .xv. scitt. weord, L. Alf. pol. 49 ; Th. i. 94, 13. Ðæt man finde of ðam yrfe æt Ceorlatúne healfes pundes wyrðne sáulsceat, and healfes pundes sáulscet fram Cynnuc, Cod. Dip. Kmbl. vi. 131, 11–14. (3) in cases where a scale expressed in money can be fixed :—Pundes weorðne áð, L. C. S. 30 ; Th. i. 394, 2. Wurðne, L. Eth. i. 1 ; Th. i. 280, 17. **II.** *possessed of honours, honourable* or *noble* as regards position, *great* :—Swá weord man wíne druncen *quasi potens crapulatus a vino,* Ps. Th. 77, 65. Wyrðro ðec *honoratior te,* Lk. Skt. Lind. Rush. 14, 8. Ða gíslas ðe on ðam here weorþuste wǽron, Chr. 876 ; Erl. 79, 10. Ðara monna ðe in ðam here weorþuste wǽron, 878 ; Erl. 80, 21. **III.** *honoured, highly thought of, held in esteem, valued, dear* :—Nǽnig wæs weord, gif mon his willan ongeat yfelne (cf. yfelwillende men nǽnne weorþscipe næfdon, Bt. 15 ; Fox 48, 17), Met. 8, 37. Ic nǽfre ne geseah nǽnne wísne mon ðe má wolde bión wrecca and earm and ælþiódig and forsewen, ðonne welig and weorþ and ríce and foremǽre on his ágenum earde, Bt. 39, 2 ; Fox 212, 17 : Lchdm. iii. 156, 24. Ðín word wunað weorþ on heofenum, Ps. Th. 118, 89. His noma wæs á seoþþan weord and mǽre geworden, Blickl. Homl. 219, 4. Deófolgild ðe mid ðæm hǽdnum mannum swíðe weord and mǽre wæs, 221, 7. Weorðiaþ his naman forðon hé wyrðe is (*quoniam suavis est*), Ps. Th. 134, 3. Unwís folc ne wát ðínne wyrðne naman, 73, 17. Ic ðíne gewitnesse wyrðe lufade, 118, 119. Hé ðæm bátwearde swurde gesealde, ðæt hé syðþan wæs máþme ðý weorþra (*he was the more thought of* (or v. **IV**?) *for having such a treasure*), Beo. Th. 3809 ; B. 1903. **III a.** with dat. of person to whom a thing seems honourable, *precious* to, *dear* to, *prized* by, *held honourable* by, *honoured* by :—Hé eallum ðisse worulde ealdormonnum wæs leóf and weord *omnibus principibus saeculi honorabilis,* Bd. 3, 15 ; S. 541, 23 : Blickl. Homl. 213, 12. Móyses se ðe wæs Gode swá weord, ðæt hé oft wið hine selfne sprǽc, Past. 18 ; Swt. 131, 11 : Lchdm. iii. 162, 1. Weord Denum, Beo. Th. 3633 ; B. 1814. Twá ðing mæg se weorþscipe and se anweald gedón, gif hé becymþ tó ðam dysgan ; hé mæg hine gedón weorþne óþrum dysgum. Ac þonēcan ðe hé ðone anweald forlǽt, oððe se anweald hine, ðonne ne biþ hé ðam dysgan weorþ *dignitates honorabilem cui provenerint reddunt,* Bt. 27, 1 ; Fox 94, 18–22. Ic (*mead*) eom weord werum, Exon. Th. 409, 14 ; Rä. 28, 1. Nis hé ná Gode wyrd, Wulfst. 52, 5. Synd mé wíc ðíne weorðe and leófe *quam amabilia sunt tabernacula tua,* Ps. Th. 83, 1. Gé wyrðe wǽron wuldorcyninge, Dryhtne dýre, Elen. Kmbl. 581 ; El. 291. Ne beó gé mé heononforð swá wurðe ne swá leófe swá gé ǽr wǽron, ac fram mé gé beóð áscyrede, Homl. Skt. i. 23, 181. Nǽron hý ðý weorþran witena ǽnegum, Met. 15, 12. Wurðran, Cd. Th. 27, 23 ; Gen. 422. Ðæt hé sié his geférum weorþost *reverendi civibus suis,* Bt. 24, 2 ; Fox 82, 6. Ðú, seó ðýreste and seó weorþeste wuldorcyninge, Exon. Th. 257, 16 ; Jul. 248. Ys mé ðín gewitnes weorðast and rihtast, Ps. Th. 118, 144. Mid ðæm cræfte ðín ðá scondlícost wæs, þéh hé him eft se weordesta wurde, Ors. 2, 8 ; Swt. 90, 29. **IV.** *worthy, honourable, noble, excellent* :—Wæs hé mid clǽnsunge forhæfednesse weorþ and mǽre *erat abstinentiae castigatione insignis,* Bd. 4, 28 ; S. 606, 39. On weorcum ælmesdǽda weorþ and mǽre, 4, 29 ; S. 608, 16. Áhsiaþ hwá sí wyrðe (*dignus*), Mt. Kmbl. 10, 11. Míne gewitnesse weorðe and getreówe *testamentum meum fidele,* Ps. Th. 88, 25. Habban ða mid wynne weorðe blisse ða ðe sécean Drihten *exultent et laetentur qui quaerunt te,* 69, 5. Ða ðe geláðode

wǽron ne synt wyrðe (*digni*), Mt. Kmbl. 22, 8. Hwelc gesceádwís mon mihte cwéþan ðæt hé á þý weorþra wǽre, þeáh hé hine weorþode *quis illos putet beatos, quos miseri tribuunt honores?* Bt. 28; Fox 100, 31. Eard wæs ðý weorþra ðe wit on stódan, hyrstum ðý hýrra, Exon. Th. 495, 20; Rä. 85, 6. Se anweald and se wela ne mæg his wealdend gedón nó ðý weorþron, Bt. 27, 2; Fox 98, 13. V. *worthy of* something, *deserving* of, (1) with gen.:—Sceal bám gelíc, mon tó gemæccan, máþþum óþres weorð (*one gift deserves another in return*), Exon. Th. 343, 11; Gn. Ex. 155. Mín uurihtwísnysse is máre ðonne ic forgifenysse wyrðe sý' *major est iniquitas mea, quam ut veniam merear*, Gen. 4, 13: Cd. Th. 81, 19; Gen. 1347. Se wyrhta ys wyrðe hys metes (*dignus cibo suo*), Mt. Kmbl. 10, 10: Homl. Skt. i. 23, 52. Heó nis nánes lofes wyrþe, Bt. 70, 24: 24, 4; Fox 86, 10: Lchdm. iii. 162, 5. Hwæs bið ðæt unwæstmbǽre treów wyrðe búton scearpre æxe? Homl. Th. ii. 408, 16. For his cræftum hé bið anwealdes weorþe, gif hé his weorþe biþ, Bt. 16, 2; Fox 50, 25. Ne onmun ðú mé nánre áre wyrþne, Blickl. Homl. 183, 1. Ðæs cynedómes Crist God weorðne munde, Ps. C. 155. Ða ðe ic ðǽr tó geladode nǽron his wyrðe, Homl. Th. i. 526, 11. Ða láreówas beóþ dómes wyrþe, Blickl. Homl. 47, 23: Met. 10, 56. Hwæþerne woldest ðú déman wítes wyrþran? Bt. 38, 6; Fox 208, 15. (2) with infin. forms:—Wé ðe nǽron wurðe beón his wealas gecígde, Homl. Th. ii. 316, 23. Ða ðing ðe weorðe sindon in gemyndum tó habbanne, Nar. 4, 9. (3) with a clause:—Wyrþe ðú eart, ðæt ðú onfó wuldor, Blickl. Homl. 75, 1. Ðæt his lár nǽre wyrþe, ðæt hí mon gehýrde, 41, 3. Ðeós woruld nǽre wyrðe, ðæt man tó hire lufe hæfde tó swíðe, Wulfst. 273, 13. Ic neom wyrðe, ðæt ic beó ðín sunu nemned *non sum dignus uocari filius tuus*, Lk. Skt. 15, 19. Se bið wurðe, ðæt hine man árwurðian, Homl. Th. ii. 560, 10. Ðæt gé weorðe (wurðe, *v.l.*: wyrðo, Lind.: wyrðe, Rush.) sýn, ðæt gé ðás tóweordan þing forfleón *ut digni habeamini fugere ista omnia quae futura sunt*, Lk. Skt. 21, 36. (4) with gen. and clause:—God is ðæs wyrðe, ðæt hine werþeóde and eal engla cynn hergen, Exon. Th. 281, 8; Jul. 643. (5) with gen. and dat. infin.:—Þeáh hé his wyrðe ne sié tó álǽtanne, Cd. Th. 39, 8; Gen. 621. (6) with other constructions:—Hine man byrigde ful wurðlíce, swá hé wyrðe wæs, Chr. 1036; Erl. 165, 36. Hé nát hwæðer hé wurðe is intó ðam écan ríce, Homl. Th. i. 532, 25. VI. *fit, meet, becoming, proper*:—Wé sculon simle secgan Gode ðoncas for eów, bróður, suá suá hit wel wierðe (wyrðe, Cott. MSS.) is (*ita ut dignum est*), Past. 32; Swt. 213, 10. Wyrcaþ wæstim wyrðne tó hreównisse, Lk. Skt. Rush. 3, 8. VII. *worthy of, fit* for or to, *properly qualified* for, (1) with gen.:—Ðæt Martinus wǽre wyrðe ðæs hádes, Homl. Th. ii. 506, 8. Ne fleáh hé ðý ríce ðý his ǽnig mon bet wirðe (wyrðe, Hatt. MS.) wǽre, Past. 3; Swt. 32, 17. (2) with dat. or inst.:—Templ Gode weorþe, Blickl. Homl. 163, 14. Nys hé mé wyrðe *non est me dignus*, Mt. Kmbl. 10, 37. Ðæt hé wǽre his biscopháde wel wyrþe, Bd. 5, 19; S. 639, 31. Ic mé sylfne nǽfre ðý háde wyrþe (wyrþne, *v.l.*) démde, 4, 2; S. 566, 7. (3) with dat. infin.:—Hálig treów ðe wyrþe (wurðe, *v.l.*) wǽre tó berenne ealles middaneardes wurþ, Homl. Skt. ii. 27, 119. Ne wat ic wyrðe tó unbindanne ðuongas sceóea his *non sum dignus soluere corrigiam calciamentorum eius*, Lk. Skt. Lind. Rush. 3, 16. (4) with a clause:—Ne eom ic wyrðe, ðæt ðú in gange under míne þecene, Mt. Kmbl. 8, 8. Ne eom ic wyrðe, ðæt ic his sceóna þwanga uncnytte, Mk. Skt. 1, 7. Se man ðæt can rihtne geleáfan, þonne biþ hé wyrðe, ðæt hé fulluht underfó, Wulfst. 33, 6: 155, 12. (5) with gen. and clause:—Hé bit ðæ re tíde hwonne hé ðæs wierðe (wyrðe, Cott. MSS.) sié, ðæt hé hine besuícan móte *aptum deceptionis tempus inquirit*, Past. 33; Swt. 227, 12. His weorc sceolon beón ðæs weorðe (wierðe, Cott. MSS.), ðæt him óðre menn onhyrien *si imitabilem ceteris in cunctis, quae agit, insinuat*, 10; Swt. 61, 18. Swá hwá swá ðæs wyrþe biþ, ðæt hé on heora ðeówdóme beón mót, Bt. 5, 1; Fox 10, 13. Hwá is ðæs wyrðe, ðæt ástíge on Godes munt *quis ascendet in montem Domini?* Ps. Th. 23, 3. Ne eom ic ðæs wyrþe, ðæt ic swá on róde gefæstnod beó, Blickl. Homl. 191, 7. Ða ðe ðæs wyrðe beóþ, ðæt híe heofoncining on heora heortum beran, 79, 32. (5 a) with impersonal construction:—Wæs ðæt ðæs wyrðe, ðæt seó stów swá fæger wǽre *it was fitting that the place should be so fair*, Bd. 1, 7; S. 478, 23. Ðæt is ðæs wyrðe, ðæt ðæt werþeóde secgen Dryhtne þonc duguða gehwylcre, Exon. Th. 38, 1; Cri. 600. For ðon is ðæs wyrðe, ðæt ðú ðæs weres frige ne forlǽte, 248, 29; Jul. 103. VIII. mostly in a legal sense, (1) *having a right* to, *entitled* to, *properly qualified* for, *possessed* of, (a) with gen.:—Gif ceorl geþeáh, ðæt hé hæfde fíf hída ..., ðonne wæs hé þegenrihtes weorðe (wyrðe, *v.l.*), L. R. 2; Th. i. 190, 18: 5; Th. i. 192, 8: 6; Th. i. 192, 11. Se wæs syþþan mǽðe and munde swá micelre wurðe, swá ðam háde gebirede mid rihte, 7; Th. i. 192, 14. Sié hé feores wyrðe and folcryhtne bóte, L. Alf. 13; Th. i. 46, 24: L. Ath. iv. 4; Th. i. 224, 3. Ne beó hé áðes wyrðe *he shall not have the right to make oath*, L. C. S. 36; Th. i. 398, 7. Ða hwíle ðe God wille ðæt ðeara ǽnig sié ðe londes weorðe sié and land gehaldan cunne, Cod. Dip. Kmbl. i. 310, 10: 311, 17. Ich cweðe eóu ðæt ich wille ðæt Gyse biscop beó ðisses biscopríches uurðe *significamus uobis nos uelle quod episcopus Giso episcopatum possideat*, iv. 198, 6. Ic bidde míne hláford ðæt ic móte beón nínes cwydes wyrðe *I pray my lord that I may*

have the right to dispose of my property by will, iii. 293, 29. Ðæt heó móte beón hyre cwydes wyrðæ, 359, 34. Gif hwá him ryhtes bidde . . . and ábiddan ne mæge, and him wedd mon sellan nelle, gebéte .xxx. scitt. and binnan .vii. nihton gedó hine ryhtes wierðne (wyrðe, *v.l.*) (*let justice be done him*), L. In. 8; Th. i. 108, 2. Forlǽt mé mínes wyrðe (weorðe, *v.l.*) wesan ðæs ðe ic mé sylf begiten hæbbe *leave me in undisturbed possession of mine own, that I myself have got*, Wulfst. 254, 21. Ne hyne micles wyrðne Drihten gedón wolde, Beo. Th. 4377; B. 2185. Ðæt hí rihtes wyrðe léte ðone leódscipe, Met. 1, 67. Ðæt hí móstan heora ealdrihta wyrðe beón, Bt. 1; Fox 2, 9: Met. 1, 37. Wé synt álýsde lífes wyrðe *nos liberati sumus*, Ps. Th. 123, 7. Gedó úsic ðæs wyrðe *make us partakers* (*of glory*), Exon. Th. 3, 2; Cri. 30. (b) with gen. and clause:—Nime se hláford twégen þegenas and swerian, . . búton hé ðone geréfan hæbbe ðe ðæs wyrðe sý ðe ðæt dón mæge (*a reeve properly qualified for doing it*), L. Eth. i. 1; Th. i. 280, 14. (c) with acc. (?):—Behét man him ðæt hé móste wurðe beón ǽlc ðæra þinga ðe hé ǽr áhte, Chr. 1046; Erl. 173, 1. Hí gerndon tó him ðæt hí móston beón wurðe ǽlc ðæra þinga ðe heom mid unrihte of genumen wæs, 1052; Erl. 185, 8. (2) *deserving* of punishment, etc., *subject* to, *liable* to (with gen.):—Ðæs ilcan dómes sié hé wyrðe *simili sententiae subjacebit* (Ex. 21, 31), L. Alf. 21; Th. i. 50, 3. Ðæt hý siþþan áðwyrðe nǽron ac ordáles wyrðe *that afterwards they might not make oath but had to submit to the ordeal*, L. Ed. 3; Th. i. 160, 21. Sý hé ðæs þeówweorces wyrðe, 9; Th. i. 164, 12. Wé cwǽdon hwæs se wyrðe wǽre ðe óðrum ryhtes wyrnde, 2; Th. i. 160, 10. Beó se leása gewita ðæs ilcan wyrðe ðe hé wolde ðæt se óðer wǽre *reddent ei, sicut fratri suo facere cogitavit*, Deut. 19, 19. Gif hý swá ne dón, ðonne sýn hý ðæs wyrðe ðe on ðam canone cwæð, L. Edm. E. 1; Th. i. 244, 12. [*Goth.* wairþs: *O. Sax. O. Frs. O. L. Ger.* werth: *O. H. Ger.* werd: *Icel.* verðr.] v. ár-, áþ-, bót-, deór-, fyrd-, mót-, róde-, tǽl-, þanc-, un-, un-leahtor-, wel-weorþ(e), -wirþe.

weorþan (wurþan, wyrþan); p. wearþ, pl. wurdon; pp. worden. I. absolute, (1) *to come to be, to be made, to arise, come, be*:—Gif bánes blice weorðeþ, L. Ethb. 34; Th. i. 12, 4. Gif bánes bite weorð, 35; Th. i. 12, 5. Ende nǽfre ðínes wræces weorþeþ, Andr. Kmbl. 2765; An. 1385. Hwá wæs ǽfre, oþþe is nú, oððe hwá wyrþ get æfter ús? Bt. 11, 1; Fox 30, 24. Hlynn weard on ceastrum, Cd. Th. 153, 30; Gen. 2546. Hwí ne wundraþ hí hwí ðæt ís weorþe, Bt. 39, 3; Fox 214, 25. Ðe læs tó mycel styrung wurde on ðam folce *ne forte tumultus fieret in populo*, Mt. 26, 5. Héht lífes weard on mereflóde middum weorðan hyhtlíc heofontimber, Cd. Th. 9, 22; Gen. 145. (2) *to come to pass, to be done, to happen, to take place, befall, come, be*:—Ðæt weorþeþ for ðyses folces synnum, ðæt ealle ðás getimbro beóþ tóworpene, Blickl. Homl. 77, 35. Daga egelícast weorþeþ in worulde, Exon. Th. 63, 21; Cri. 1023. Huu wordes ðis *quomodo fiet istud?* Lk. Skt. Lind. Rush. 1, 34: 23, 31. Ðǽr wearþ micel gefeoht, Chr. 800; Erl. 60, 7: 868; Erl. 72, 28. On ðam gemótan, þeáh rǽdíce wurdan on namcúðan stówan, Lk. Skt. iiv. 37; Th. i. 348, 17. Hwæðer ǽfre wurde þus gerád þing *si facta est aliquando hujuscemodi res*, Deut. 4, 32. Eálá ðæt hit wurde, ðæt . . ., Met. 8, 39. Sceal se dæg weorþan, Exon. Th. 447, 5; Dóm. 34. Þurh hwæt his worulde gedál weorðan sceolde, Beo. Th. 6129; B. 3068. Ðætte ríces gehwæs sceolde gelimpan, eorðan dreámas ende wurðan, Cd. Th. 223, 6; Dan. 115. Sceal feorhgedál æfter wyrðan, Andr. Kmbl. 364; Ass. 182: 430; An. 215. (2 a) when the object affected by what happens is given :—Ne wyrð him nán orne, Lchdm. iii. 16, 4. Ic wát ealne ðysse worulde wurðeþ ende *omni consummationi vidi finem*, Ps. Th. 118, 96. Dómas ðe wǽdlum weorðaþ, 139, 12. Tácnu wurðaþ on eów *erunt in te signa*, Deut. 28, 46. Hwæt weard eów? Andr. Kmbl. 2685; An. 1345. Ðæt ðé sceates ðearf ne wurde, Cd. Th. 32, 16; Gen. 504. Unc sceal weorðan swá unc wyrd geteóð, Beo. Th. 5045; B. 2526. II. *to become, be made, be*, (1) with predicative substantive:—Ða hwíle ðe hé ðǽr stód, hé wearþ fǽringa geong cniht, and sóna eft eald man, Blickl. Homl. 175, 2. On ðam dæge wurdun Heródes and Pilatus gefrýnd; sóðlíce hig wǽron ǽr gefýnd, Lk. Skt. 23, 12. Wá heom ðæs sídes ðe hí men wurdon, Wulfst. 27, 4. Weorðan his bearn steópcild, and his wíf wyrðe wydewe *fiant filii ejus orphani, et uxor ejus vidua*, Ps. Th. 108, 9. Ðæt wé ðæs mordres meldan ne weorðen, Elen. Kmbl. 856; El. 428. (2) with predicative adjective, *to get, grow*:—Gif ðú lárna ðínra éste wyrðest, Andr. Kmbl. 965; An. 483. Gif eáre þirel weorðeþ, L. Ethb. 41; Th. i. 14, 6. Gif hé healt weorð, 65; Th. i. 18, 14. Ðé weorð on ðínum breóstum rúm, Cd. Th. 33, 13; Gen. 519. Gif ða cearwylmas cólran wurðaþ, Beo. Th. 570; B. 282. Ða deáde ne weorðaþ (*v.l.* wurðaþ) *qui non gustabunt mortem*, Lk. Skt. 9, 27. Ðá weard hé druncen *inebriatus est*, Gen. 9, 21. Ðæt wíf weard wráð ðam geongan cnapan *mulier molesta erat adolescenti*, 39, 10. Weard hé swíðe yrre *iratus est valde*, 39, 19. Hwelc siððan weard herewulfa síð, Cd. Th. 121, 23; Gen. 2014. Ða fixas wurdon deáde *pisces mortui sunt*, Ex. 7, 21. Mierce wurdon cristne, Chr. 655; Erl. 28, 1. Mé milde weorð *miserere mei*, Ps. Th. 56, 1: 66, 1. Monigfaldge worðe *habundaverit*, Mt. Kmbl. Lind. 5,.20. Nænges þinges máre þearf ðonne his unriht yppe wurde, Blickl. Homl. 175, 10. Eálá ðæt úre tída nú ne

mihtau weorðan swilce, Bt. 15; Fox 48, 18. Sǽne weorðan, Andr. Kmbl. 408; An. 204. Wyrðan, 874; An. 437. Wurðan, Cd. Th. 27, 8; Gen. 414. Wæs ðǽre ǽghwilc worden mǽgburh fremde, 102, 3; Gen. 1694; 135, 2; Gen. 2236. Weard hē acol worden, 223, 24; Dan. 124. Eal cristen folc is þurh geleáfan geleáful worden, Wulfst. 279, 30. Ða dysegan sint wordene blinde, Met. 19, 29. (3) with prepositional phrase:—Heó weard mid cilde, Homl. Th. i. 24, 26. Ðæt ic tó ðínum willan weorþan móte *that I may be to thy liking*, Ps. C. 104. (4) with adverb, (a) where the subject is given:—Heó wyrð glædlíce on hyre heortan, Anglia viii. 324, 16. Óþ ðæt dín fót weorðe fæste on blóde *ut intinguatur pes tuus in sanguine*, Ps. Th. 67, 22. (b) with impersonal construction:—Weard mē on hige leóhte, Cd. Th. 42, 20; Gen. 676. Ðá wearþ hyre rúme on móde, Judth. Thw. 22, 39; Jud. 97. Gif men fǽrlíce wyrde unsófte, Rtl. 114, 24. **III.** with prepositions (see also **IV**), (a) weorþan *of to come from, be caused by, be produced from or by*:—Wiþ geswelle ðam ðe wyrð of fylle oððe of slege, Lchdm. ii. 72, 22. Hwý ðæt ís mæge weorþan of wætere, Met. 28, 60. (b) weorþan on, (1) *to get into a state of being, feeling, to become* the adjective connected with the noun, *get*:—Gif hē wyrþ on ungeþylde *if he gets impatient*; cum dederit impatientiae manus, Bt. 11, 1; Fox 32, 33. Weorþeþ (-aþ, MS.) oft on wón se sído *in hoc hominum judicia depugnant*, 39, 9; Fox 226, 4. Ðá wearþ Holofernus on gytesálum *he grew merry, as the wine flowed*, Judth. Thw. 21, 17; Jud. 21. Wurdan gesweoru on seledreáme *exultaverunt colles*, Ps. Th. 113, 6. Hié weorðen on ungeðylde, Past. 45; Swt. 341, 3. (2) *to get into a state of action, to come to be doing something, to fall to* an action, *to take to*:—Hē wierd (wird, Hatt. MS.) swíðe hræde on fielle *citius corruit*, Past. 39; Swt. 286, 17. Wênst ðú ðæt ðú ðæt hwerfende hweól, ðonne hit on ryne wyrþ (*when it gets a-running*), mæge oncyrran *tu volventis rotae impetum retinere conaris*? Bt. 7, 2; Fox 18, 36. Hē on fylle weard, Beo. Th. 3093; B. 1544. Hē weard on fleáme, Andr. Kmbl. 2771; An. 1388. Hí weorðan on slǽpe, Homl. Skt. i. 18, 161. Hí on slǽpe wurdon, 23, 249. Hig wurdon on fleáme *terga verterunt*, Jos. 7, 4. Hié weorðen on murcunga *they fall a-grumbling*; ad murmurationem proruunt, Past. 45; Swt. 341, 3. (3) *to come to be something, become, turn into*:—Mē weord on God þeccend and on trume stówe *esto mihi in Deum protectorem, et in locum munitum*, Ps. Th. 70, 2. Ðæt heó on sealtstánes wurde anlícnesse, Cd. Th. 154, 32; Gen. 2564. Hē mē ys worden on hǽlu *factus est mihi in salutem*, Ps. Th. 117, 14. (c) weorþan tó, (1) of change in material condition, *to become, turn to*:—Ðú eart dust, and tó duste wyrst *pulvis es, et in pulverem reverteris*, Gen. 3, 19. Weorðeþ tó duste, Ps. Th. 89, 6. Tó wætere weorðeþ, 147, 7: Met. 28, 63. Se wyrm wyrd tó eorþan, Lchdm. ii. 44, 16. Weorp díne girde beforan Pharaone, and heó wird tó nǽddran (*vertetur in colubrum*), Ex. 7, 9. Seó eá ðǽr wyrþ tó miclum sǽ, Ors. I, 1; Swt. 12, 28. Weorðaþ hig tó acxan *fatiscunt in cinerem*, I, 3; Swt. 32, 15. Bearwas wurdon tó axan, Cd. Th. 154. 8; Gen. 2552. Sume wurdon tó wulfan, Bt. 38, 1; Fox 194, 36: Met. 26, 79. On eorþan gangan and tó eorþan weorþan, Blickl. Homl. 123, 10. Seó eá ne mæg weorþan tó ǽwelme, ac se ǽwelm mæg weorþan tó eá, Bt. 34, 1; Fox 134, 15. (2) of the state or condition to which things come, of the event of matters, *to become, have as issue, come to*:—Ǽlc þing wyrþ tó náuhte, Bt. 34, 1; Fox 134, 13. Hí weorþaþ him selfe tó náuhte, 21; Fox 74, 36. Tó hwan weard hondræs hæleþa *what was the event of the combat*, Beo. Th. 4149; B. 2071. Ðonne hié ne giémaþ tó hwon ðerra monna wíse weorðe *when they do not care to what a state other men get*, Past. 5; Swt. 41, 24. Hē ðóhte ðæt hē hine ofslóge, wurde siððan tó ðam þe hit meahte (*be the event what it might*), 34; Swt. 235, 10. Lyt ðú geþóhtes tó hwon ðínre sáwle síd siþþan wurde, Exon. Th. 368, 12; Seel. 20. Hí bidon tó hwon his ðing weorþan sceolde *quem res exitum haberet exspectantes*, Bd. 3, 11; S. 536, 32. Tó hwon sculon wit weorðan *what is to become of us*? Gen. 815. Eall mín mægen is tó náuhte worden, Ps. Th. 21, 11. (3) where a character or function is taken by anything, *to become, turn, turn to*:—Mē tó aldorbanan weorðe wrádra sum *some fell one will become the destroyer of my life*, Cd. Th. 63, 18; Gen. 1034. Hē wierd tó dæs onlícnesse ðe áwriten is *usque ad ejus similitudinem ducitur, de quo scriptum est*, Past. 17; Swt. 111, 21. Ne wyrd nán tó láfe *none shall become a remnant, i. e. none shall be left*; non remanebit ex eis ungula, Ex. 10, 26. Gif þegen geþeáh ðæt hē weard tó eorle, L. R. 5; Th. i. 192, 7. Se tó deófle weard, Cd. Th. 20, 9; Gen. 305. Ic tó meldan weard *I turned informer*, Exon. Th. 279, 30; Jul. 621. Weard hē Headoláfe tó handbonan, Beo. Th. 924; B. 460. Hwonne liffreá weorðe ússum móde tó mundboran, Exon. Th. 2, 32; Cri. 28. Ðeáh þrǽla hwylc of cristendóme tó wícinge weorðe, Wulfst. 162, 6. Ðý lǽs sió upáhæfenes him weorðe tó wege micelre scylde ne cær *ne elatio via fiat ad foveam gravioris culpae*, Past. 57; Swt. 439, 11. (4) where a result is brought about, *to become, prove a source of*:—Seó ofering ðē wurþ tó sáre, Bt. 14, 1; Fox 42, 16. Hit him wyrþ tó teónan, Blickl. Homl. 51, 9. Fú wurde mē tó hǽlu *factus es mihi in salutem*, Ps. Th. 117, 27. Hió weard mongum tó frófre, Exon. Th. 421, 17; Rä. 40, 18. Tó blisse, Blickl. Homl. 123, 2. Tó aldorceare, Beo. Th. 1817; B 905. Hē manegum weard maruum tó hróðre, wer-

þeódum tó wrǽce, Elen. Kmbl. 30; El. 15. Ða byrig, ðe ǽr gafol guldon, wurdon Ciruse tó monegum gefeohtum *civitates, quae tributariae erant, a Cyro defecerunt; quae res Cyro multorum bellorum causa et origo exstitit*, Ors. I, 12; Swt. 54, 14. Ðe lǽs úre deáþ úrum feóndum tó gefeán weorþe, Blickl. Homl. 101, 33. Tó hleó and tó hróþer hæleþa cynne weorðan, Exon. Th. 73, 31; Cri. 1198. Tó frófre weorðan, Beo. Th. 3419; B. 1707. (5) *to become, be an object of*:—Ic eom worden mannum tó leahtrunge and tó forsewennesse *ego sum opprobrium hominum*, Ps. Th. 21, 5. **IV.** implying movement, change of position, (1) literal, *to come, get*, (a) with prepositions:—Ðonne hē (*the moon*) betwux ús and hire (*the sun*) wyrþ, Bt. 4; Fox 8, 2. Of ðǽre sǽ cymþ ðæt wæter innon ða eorþan; cymþ ðonne up æt ðam ǽwelme, wyrþ ðonne tó bróce, ðonne tó eá, ðonne andlang eá, óþ hit wyrþ eft tó sǽ, 34, 6; Fox 140, 17–20. Se regn ðæt deófol on ufan wyrðeþ, Salm. Kmb'. p. 148, 5. Swá swá wē of ðisse weorulde weorðaþ, Shrn. 202, 4. Gif hí on ðam wuda weorþaþ *if they get in the wood*, Bt. 25; Fox 88, 16. Gif hí on treówum weorþaþ, Met. 13, 36. Hē weard him on ánon scipe *he got him* (reflex.) *on board a ship*, Chr. 1052; Erl. 187, 13. Sebastianus geseah hú ða Godes cempan ongunnon hnexian, and weard him tómiddes (*he came amongst them*), Homl. Skt. i. 5, 52. Gif nægl of honda weorðe *if a nail come off the hand*, Lchdm. iii. 58, 7. Ðú mihtest ðē féran betwyx ðám tunglum, and ðonne weorþan on ðam rodore, Bt. 36, 2; Fox 174, 11. On ðǽm rodere ufan weorþan, Met. 24, 18. (b) with adverbs:—Gif eáge of weord *if an eye comes out*, L. Ethb. 43; Th. i. 14, 8. Gif fót of weorðe *if a foot comes off*, 69; Th. i. 20, 1: 70; Th. i. 20, 2: 72; Th. i. 20, 5. Hē weard him áwege *he went away, got off*, Homl. Skt. ii. 25, 228. Hié sume inne wurdon *some of them got inside*, Chr. 867; Erl. 72, 14. Móste ic áne tíd úte weorðan, Cd. Th. 23, 34; Gen. 369. (2) figurative:—Adames cynn onfehð flǽsce, weorþeþ foldræste æt ende *Adam's race shall receive flesh, shall come to the end of its rest in earth*, Exon. Th. 63, 34; Cri. 1029. Búton monnum and sumum englum, ða weorþaþ hwílum of hiora gecynde *except men and some angels, who sometimes depart from their nature*, Bt. 25; Fox 88, 8. His ǽhta weorþaþ on ðæs onwealde ðe hē wyrrest úþe, Blickl. Homl. 195, 3. Ic nó ne wearþ of ðam sóþan geleáfan *nec unquam fuerit dies, qui me ab hac sententia depellat*, 5, 3; Fox 12, 6. Hwí ðæt ís for ðǽre sunnan scíman tó his ágnum gecynde weorþe, 39, 3; Fox 216, 1. Ðæt gē of feónda fæðme weorden *that ye get out of the foes' grasp*, Cd. Th. 196, 20; Exod. 294. Ðæt ne loc of heáfde tó forlore wurde *that not a hair from the head should come to destruction*, Andr. Kmbl. 2846; An. 1425. **V.** as an auxiliary with participles, (1) present:—Gif him hwilc yfel gelimpd, ic wurðe syððan geómriende, Gen. 42, 38. (2) past, (a) of transitive verbs, forming a passive voice:—Eów weorþeþ forgifen hwæt gē sprecaþ, Blickl. Homl. 171, 19. Ne weorþeþ sió mǽgburg gemicledu eaforan minum, Exon. Th. 401, 31; Rä. 21, 20. Hē him ábolgen wurðeþ, Cd. Th. 28, 4; Gen. 430. Hú wurþ hē elles gelǽred *how else shall he get taught?* Bd. pref.; S. 471, 18. Hí weorþaþ bereáfode ǽlcre áre, Bt. 29, 2; Fox 104, 16. Ðá weard Faraones heorte gehefegod *ingravatum est cor Pharaonis*, Ex. 8, 32. Ðá him gerýmed weard, ðæt hié wælstówe wealdan móston, Beo. Th. 5959; B. 2983. Swá his maudrihten gemǽted weard, Cd. Th. 225, 21; Dan. 157. Ðý lǽs hié eft weorðen (wyrðen, Hatt. MS.) gedémde, Past. 28; Swt. 190, 15. Seó burh sceolde ábrocen weorþan, Blickl. Homl. 77, 29. Ne mihte him bedyrned wyrðan ðæt his engyl ongan ofermód wesan, Cd. Th. 17, 18; Gen. 261. (b) of intransitive verbs:—Ðē sunu weorðeþ cumen, Cd. Th. 132, 19; Gen. 2195. Ða geongan leoþu geloden weorþaþ, Exon. Th. 327, 20; Vy. 6. Hē sóna weard hál geworden, Blickl. Homl. 223, 26: Cd. Th. 223, 23; Dan. 124. Denum weard willa gelumpen, Beo. Th. 1851; B. 823; 2473; B. 1234. Ðá weard áfeallen ðæs folces ealdor, Byrht. Th. 137, 46; By. 202. Ðá weard se líchama tóslopen, Homl. Th. i. 86, 24: Jos. 5, 1. Ðæt hí forwordene weorðen *ut intereant*, Ps. Th. 91, 6. [*Goth.* wairþan: *O. Sax.* werdan: *O. Frs.* wertha: *O. H. Ger.* werdan: *Icel.* verða.] v. for-, ge-, mis-weorþan.

weorþ-apulder. v. worþ-apulder.

weorþe; *subst.* or *adj.*: weorþe; *adv.*, weorþe-líce. v. weorþ; *subst.* or *adj.*, un-weorþe, weorþ-líce.

weorþere, es; *m.* A *worshipper*:—Godes uordare *Dei cultor*, Jn. Skt. Lind. 9, 31. Sóðo uordares *ueri adoratores*, 4, 23.

weorþ-full; *adj.* **I.** *having worth, worthy, honourable, glorious, excellent*:—Beó preóst, swá his háde gebyraþ, wís and weorðfull, L. Edg. C. 58; Th. ii. 256, 17. Búton gē ondrédon Drihtnes wurðfullan naman *nisi timueris nomen ejus gloriosum*, Deut. 28, 58. Wurþfulle gegedriende *honesta colligentes*, Anglia xiii. 368, 46. Wurðfulleste *praestantissimus, dignissimus, sublimissimus*, Hpt. Gl. 463, 44. Hē manna wæs wígend weorðfullost, Beo. Th. 6189; B. 3099. **II.** *having honour* with others, *held in honour, honoured, esteemed, prized, dear*:—Se bið on eallum þingum wurþfull (cf. weorþ mannum, 162, 1), Lchd:n. iii. 158, 3. Ða hálgan weras, ðe gáde weorc beeodon, hí wurðfulle wǽron on disse worulde, Ælfc. T. Grn. 1, 9. Ðe lǽs sum weorðfulra (wurð-, *v.l.*) sig yn gelaðod fram hym *ne honoratior te sit innitatus ab eo*, Lk. Skt. 14, 8. **II a.** with dat. of person to whom another seems honourable:—

Daniel wunude on Chaldéa wurðfull ðám ciningum, Ælfc. T. Grn. 9, 43. His welwillende môd, and Gode swíðe wurðful, Homl. Skt. ii. 30, 20. III. *having honours, worshipful, noble, illustrious, magnificent:*—Án woruldcyningc . . . ne mæg beón wurðful cyningc, búton hé hæbbe ða geþincðe ðe him gebyriaþ, Homl. Skt. i. pref., 60. Se cyng Willelm wæs swíðe wís man and swíðe ríce, and wurðfulre and strengere ðonne ænig his foregengra wære . . . Hé wæs swýðe wurðful; þriwa hé bær his cynehelm ælce geáre, Chr. 1086; Erl. 221, 14–27. IV. *worthy, suitable, fitting:*—Beón wurðful wunung ðæs Hálgan Gástes, Homl. Th. ii. 600, 17. Munecas hé gestaþolode tó weorþfulre þenunge Hælendes Cristes, Lchdm. iii. 440, 13. [Helyas wass an wurrþfull prophete, Orm. 5195. His wundri werkes and wurðful, Kath. 1017. Ʒet he is wurþful and aht man, O. and N. 1481. Of prede þe dyeul begyleþ þe riche and þe wyse and þe hardi and þe worþuolle, Ayenb. 16, 33.]

weorþful-líc; *adj. Noble, magnificent:*—Hwæt rúmedlíces oððe micellíces oððe weorþfullíces hæfþ se eówer gilp *quid habet amplum magnificumque gloria?* Bt. 18, 1; Fox 62, 21.

weorþfullíce; *adv.* I. *of moral worth, worthily, honourably, excellently:*—Ic wilnode weorþfullíce tó libbanne ða· hwíle ðe ic lifede, Bt. 17; Fox 60, 15. II. *nobly, in a way that is highly esteemed:*—Swá swá men wurðlícor lybbaþ ðonne treówu, swá hý eác weorðfulícor árísaþ on dómes dæge, Shrn. 168, 26. III. *in a way that shews respect, with honour:*—Ðá onféng Dioclitianus Galerius weorðfullíce a Diocletiano plurimo honore susceptus est Galerius, Ors. 6, 30; Bos. 126, 19. IV. *in a fitting manner, worthily, properly:*—Ðæt hé gebéte Gode *digne satisfaciat Deo*, R. Ben. Interl. 42, 6.

weorþfulness, e; *f. Nobleness, magnificence:*—Gesceáwode se án engel ðe ðær ænlícost wæs, hú fæger hé silf wæs, and hú scínende on wuldre, and him wel gelícode his wurðfulniss, Ælfc. T. Grn. 2, 34. For swá miceles freólses wurþfulnesse *ob tante festivitatis honorificentiam*, Anglia xiii. 401, 522. Bróhton Rómáne ðone triumphan angeán Pompeius mid micelre weorþfulnesse (wyrð-, *v.l.*), Ors. 5, 10; Swt. 234, 29.

weorþ-georn; *adj. Desirous of honour, noble-minded, excellent:*—Se wísa and se weorðgeorna and se fæstræda folces hyrde . . . Caton, Met. 10, 48. Hý weorðgeornra sælða tóslítaþ, Salm. Kmbl. 696; Sal. 347. Lá wísan menn, gáþ on ðone weg ðe eów læraþ ða foremæran bisna ðara gódena gumena and ðæra weorþgeornena wera ðe eár wæron (*ite nunc fortes, ubi celsa magni ducit exempli via*). Eálá gé eargan and ídelgeornan . . . hwý gé nellan ácsien æfter ðám wísum monnum and æfter ðám weorþgeornum . . . ðe ær eów wæron . . . hí wunnon æfter wyrþscipe on ðisse worulde, and tiledon gódes hlísan, Bt. 40, 4; Fox 238, 28–240, 5. Ða menn ðe on hiora dagum foremæroste and weorþgeornoste wæron *clarissimos suis temporibus viros*, 18, 3; Fox 64, 36.

weorþian, wurþian, wyrþian; *p.* ode. I. *to set a value upon,* (1) *of money value:*—Be ðam ðe se man hit weorðige ðe hit áge *according to the value the owner may set upon it*, L. Ath. v. 6; Th. i. 232, 26. (1 a) *to fix interest on a loan* (?), *to lend at interest* (?) :—Wiorþigende *foenerator*, Ps. Spl. T. 108, 10. (2) *in other cases, to value, esteem, hold in honour, venerate:*—Wæs ðær gild ðe ða hæþenan men swíðe weorðodan (*held in the highest honour*), Blickl. Homl. 221, 20. Uton rihtne cristendóm geornlíce weorðian, and ælcne hæðendóm mid ealle oferhogian, L. Eth. ix. 44; Th. i. 350, 11. Wénst ðú ðæt se anweald and ðæt geniht seó tó forseónne, oððe eft swíþor tó weorþianne ðonne óþre gód (*rerum omnium veneratione dignissimum*). Ðá cwæþ ic: Ne mæg nænne mon ðæs tweógan, ðætte anweald and genihtis tó weorþianne, Bt. 33, 1; Fox 120, 22–25. Ðæs engles mægen and his wundor ðær ðonne weorðod bið and oftost æteówed, Blickl. Homl. 209, 21. II. *to honour, shew honour to, treat with reverence or respect:*—Ðú weorðast ðíne suna má ðonne mé *honorasti filios tuos magis quam me*, Past. 17; Swt. 123, 7. Ðis folc mé mid welerum weorðaþ (wurðaþ, *v.l.*: worðas, Lind.) *populus hic labiis me honorat*, Mt. Kmbl. 15, 8. Weorðas (worðias, Lind.), Mk. Skt. Rush. 7, 6. Gé ne weorðiaþ (wurðiaþ, *v.l.*: worðiges *honorificavit*, Lind.), fæder and módor, Mt. Kmbl. 15, 6. Ic lisse selle ðam ðe [ðé] wurðiaþ, Cd. Th. 105, 25; Gen. 1758. Hí hine weorþodan swá cinige geríseþ, Blickl. Homl. 69, 31. Wurðodon, Chr. 975; Th. i. 227, 13. Weorða (wurða, *v.l.*: worðig, Lind.) ðínne fæder *honora patrem tuum*, Mt. Kmbl. 15, 4. Worða, Mk. Skt. Rush. 7, 10. Cyning wyrþiaþ *regem honorificate*, Scint. 64, 10. Ðæt hí Godes þeówas werian and weorðian, L. Eth. vi. 45; Th. i. 326, 23. Hé gesiehð ða weorþigan (weorðian, Cott. MSS.) ðe ær wel ongunnon, ðá ðá hé ídel wæs *eorum palmas respiciant, in quorum nunc laboribus otiosi perdurant*, Past. 34; Swt. 229, 21. II a. *in reference to subjects divine or sacred,* (1) *of honour shewn to a god, to worship, adore:*—Næfre ðú gelærest ðæt ic deófolgieldum gaful onháte, ac ic weorðige wuldres ealdor, Exon. Th. 251, 30; Jul. 153. Gif ðú worðas (worðias, Lind.) bifora mec *si adoraueris coram me*, Lk. Skt. Rush. 4, 7. 'Gif ðú feallest tó mé, and mé weorþast.' Eálá sóþlíce se áfealleþ, se ðe deófol weorþeþ . . . Ðæt mánfulle wuht wolde ðæt hé (*Christ*) hine weorþode . . . hine

(*Christ*) ealle hálige weorþiaþ . . . Swá wé sceolan hine mid wordum weorþian, Blickl. Homl. 31, 1–11. Hig mé weorðiaþ (wurðiaþ, *v.l.*) *colunt me*, Mt. Kmbl. 15, 9. Worðiaþ (worðas, Rush.), Mk. Skt. Lind. 7, 7. Ða ðe weorðiaþ wuldres aldor *adorabunt coram te, Domine*, Ps. Th. 85, 8: Ps. Surt. 71, 11: Exon. Th. 150, 1; Gú. 772. Menn ús wurðiaþ for godas, Homl. Th. i. 462, 28. Ða þing ðe hig wurðiaþ *ea quae colunt Aegyptii*, Ex. 8, 26. Gást is God, and ða ða worðigas (*adorant*) hine, in gáste gidæfnaþ tó worðanne (*uorðia adorare*, Lind.), Jn. Skt. Rush. 4, 24. Wyrðade *oraret*, Wrt. Voc. ii. 64, 56. Gé wurðodon ðæt cealf for god, Deut. 9, 16. Ðám godum ðe hira fæderas ne wurðodon (*coluerunt*), 32, 17. Weorþedon, Ors. 4, 4; Swt. 162, 26. Wurðedon, Cd. Th. 227, 5; Dan. 182. Hiora cyningas hí weorþodon for godas, Bt. 38, 1; Fox 194, 16: Met. 26, 45: Wulfst. 98, 24. Hý wurðedon him for godas ða sunnan and ðone mónan, 105, 13. Ða tungelwítgan cuðmon tó ðon ðæt hié Crist weorþedon (wurðoden, *v.l.*), Chr. 2; Erl. 4, 29. Nánes cynnes andlícnyssa ne wurða (*non adorabis et non coles*), Deut. 5, 9. Weorþa ðínne Drihten God, Blickl. Homl. 27, 20. Weorþian wé Drihtnes godcundnesse, Blickl. Homl. 33, 36. Weorðian Waldend, Exon. Th. 25, 1; Cri. 394. Wíg weorðian, Apstls. Kmbl. 95; Ap. 48. Wurðigean, Cd. Th. 228, 24; Dan. 208. Hú hine man wurðian scyle *ritum colendi*, Ex. 18, 20. Ic ðone Déman wille weorþian wordum and dædum, Exon. Th. 139, 10; Gú. 591. Gif ðú fallas tó worðenne ł tó worðianne mec *si cadens adoraveris me*, Mt. Kmbl. Lind. 4, 9. (2) *of reverence shewn to sacred things, to worship, adore:*—Ic ðín tempel weorðige *adorabo ad templum sanctum tuum*, Ps. Th. 137, 2. Heó on cneów sette, lác (*the cross*) weorðade, Elen. Kmbl. 2272; El. 1137. Ðæt ic móte ðone sigebeám weorðian, Rood Kmbl. 255; Kr. 129: Blickl. Homl. 97, 13. (3) *of reverence shewn to holy persons or religious seasons, to celebrate, commemorate,* (a) *of persons:*—On ðisum dæge wé wurðiaþ on úrum lofsangum and on freólse ðone mæran apostol Iacóbum, Homl. Th. ii. 412, 18. Se (*St. Michael*) ðe is tó weorþienne and tó wuldrienne, Blickl. Homl. 197, 6. (β) *of seasons:*—Be ðære árwyrðnesse ðisse hálgan tíde, ðe wé nú weorþiaþ, Blickl. Homl. 115, 30. Weorðiaþ, Menol. Fox 349; Men. 176. Ðæt hié weorðeden ðone mæran dæg, Elen. Kmbl. 2442; El. 1222. Eal folc wurþodon symbelnysse, Homl. Skt. ii. 30, 152. Weorþian wé nú tódæg ðone tócyme ðæs Hálgan Gástes, Blickl. Homl. 131, 11: 171, 3. Be ðisse hálgan tíde weorþunga ðe wé mærsian sceolan and weorðian . . . ús is ðes dæg swíþe tó mærsienne and tó weorþienne, 161, 5–8. Ða dagas ðe gé sceolun Drihtne hálgian and wurðian *feriae Domini, quas vocabitis sanctas*, Lev. 23, 2. Ðære abbudissan gemynddæg on myclum wuldre weorþad is *cujus natalis solet in magna gloria celebrari*, Bd. 3, 8; S. 532, 40. (4) *used intransitively, to celebrate* (*a service*):—Se bisceop ðær gesette ciricean þegnas, ða ðær dæghwamlíce mid gelimplícre endebyrdnesse weorðode, Blickl. Homl. 207, 33. III. *to honour in words, speak in honour of, magnify, praise, celebrate, glorify:*—Ic Drihten wordum weorðige *in Domino laudabo sermonem*, Ps. Th. 55, 9. Hé wæs Drihtne fylgende, and hine herede and weorþode, Blickl. Homl. 15, 28. Hé Dryhten herede, weorðade wordum, Andr. Kmbl. 2537; An. 1270. Wyrðode, 109; An. 55. Wyrðude, 1076; An. 538. Se eádga (*Abraham*) Drihtnes noman weorðade, Cd. Th. 113, 13; Gen. 1886. Hæled hálgum stefnum cyning weorðodon, Andr. Kmbl. 2112; An. 1057. Wordum weorðedon, 1611; An. 807. Wurðedon, Cd. Th. 232, 15; Dan. 260. Weorðiaþ his naman *psallite nomini ejus*, Ps. Th. 134, 3. Wé naman ðínne weorðien *honorificabo nomen tuum*, 85, 11. Úre Hælend wæs weorþod and hered from Iudéa folce, Blickl. Homl. 67, 4. Hé wæs of cilda múþe gecnáwen and weorþad, 71, 33. IV. *to honour, pay respect to, heed, attend to* (cf. Icel. virða *to give heed*):—Hé hét mé his word weorðian and wel healdan, læstan his láre, Cd. Th. 34, 13; Gen. 537: 21, 24; Gen. 329. Wurðian, 23, 3; Gen. 353. Heó his dæd and word noldon weorðian, 20, 16; Gen. 310. IV a. *to pay court to a person:*—Weorðiaþ *colunt* (multi colunt personam potentis, Prov. 19, 6), Kent. Gl. 671. IV b. *to bestow labour upon, take pains with:*—Ðam gelícost ðe wile gyldenu fatu and sylfrenu forsewen, and treówenu mon weorþige *si vilia vasa colerentur, pretiosa sordescerent*, Bt. 36, 1; Fox 172, 20. IV c. *to care about:*—Hé mistlíce fugela sangas ne wurþode swá oft swá cnihtlícu yldo begæð *he did not care about the various songs of birds, as often is the usage of such a boyish age*; non variarum volucrum diversos crocitus, ut adsolet illa aetas, imitabatur, Guthl. 2; Gdwin. 12, 18. V. *to honour, bestow honour upon, grace:*—Swá hý his weorc weorþaþ, Exon. Th. 43, 19; Cri. 691. Gif se abbod his geearnunge swylce ongyte, hé hine mót be suman dæle furþor weorðian (wyrðian, *v.l.*), and him innor tæcan stede and setl, R. Ben. III, 4. V a. *to honour with something,* (1) *where the subject is inferior to the object:*—Godes þeówum ðe ða cyrican mid godcundum dreámum weorðiaþ, Blickl. Homl. 41, 27. Weorþiaþ gé eówerne Drihten God mid gedafenlícum þingum *honora Deum de tua substantia* (Prov. 3, 9), 41, 9. Heó hét mé fremdne god welum weorþian, Exon. Th. 247, 9; Jul. 76. (2) *where the subject is not inferior to the object, to grace, favour, honour by bestowing* something:—God geofum unhnéawum, cræftum weorðaþ eorþan tuddor, Exon. Th. 43, 12; Cri. 687. Hé

weorþode his deórlingas mid miclum welum, Bt. 28; Fox 100, 29. Drihten his folc wurðode mid ðara Egiptisean gestreóne *Dominus dedit gratiam populo coram Aegyptiis, ut commodarent eis,* Ex. 12, 36. Hé hine miclum and his geféran mid feó weorðude, Chr. 878; Erl. 80, 25. Æt feohgyftum hé Dene weorþode, Beo. Th. 2185; B. 1090. Ic ðine leóde weorðode weorcum, 4198; B. 2096. Is gesýne ðæt ðu ðyssum hysse hold gewurde, and hine geofum wyrðodest, Andr. Kmbl. 1102; An. 551. Hé hí welum weorðode, 1509; An. 756. Ðam werode ðe hé wurðode wlite and wuldre, Cd. Th. 3, 14; Gen. 35. Hé hí wolde swíþe weorþian mid éce ríce, Bt. 41, 3; Fox 248, 11. **VI.** *to make worthy, to ennoble :*—Weorða ðé selfne gódum dædum, Wald. 1, 40; Vald. 1, 22. [God wurþian, O. E. Homl. i. 11, 26. Sunnedei wurþien, 45, 36. Wurðien (weorþi, 2nd MS.), Laym. 9510. To lofenn Godd and wurrþenn, Orm. 208. He wurðede ðe ton . . . ðe was wurði wurðed to ben, Gen. and Ex. 1010. *Goth.* wairþôn *to fix the value of: O. Sax.* gi-werðôu : *O. H. Ger.* werðôn *appreciare, venerari : Icel.* virða *to fix the value of.*] v. á-, ár-, ge-, mis-, un-weorþian.

weorþig. v. worþig.

weorþing P :—Andlang streámes in wiððan weorðing (weording, Cod. Dip. Kmbl. iii. 391, 19), Cod. Dip. B. ii. 41, 2.

weorþ-leás ; *adj. Worthless, of no value :*—Wurðleás *depretiatus,* Wrt. Voc. i. 28, 59.

weorþ-líc ; *adj.* **I.** *of value, valuable :*—Ælc seldsýnde fisc ðe weorðlíc byð, Cod. Dip. Kmbl. iii. 450, 27. Weorðlíc reáf gedǽlan *dividere spolia,* Ps. Th. 67, 12. **II.** *worthy, noble, distinguished, excellent, splendid :*—Gif ðu ǽnigne mon cúþest ðara ðe hæfde ǽlces þinges anweald, and ǽlcne weorþscipe . . . geþenc hú weorþlíc and hú foremǽrlíc ðé wolde se mon þincan, Bt. 33, 1; Fox 120, 34. Bið him weorðlíc setl *sedes ejus sicut sol,* Ps. Th. 88, 31. Weorðlíc wlite wuldres ðínes *magnificentia,* 95, 6. Wæs his ríce brǽd, wíd and weorðlíc, Exon. Th. 243, 11; Jul. 9. Treów in ðé weorðlícu wunade, 6, 12; Cri. 83. Ðín heáhsetl is heáh and mǽre, fæger and wurðlíc, Hy. 7, 40. Wé ðé þanciaþ ðínes weorðlícan wuldordreámes, 8, 10. Hí mid weorðlícan weorode and wynsaman dreáme hine feredan, Chr. 1023; Erl. 163, 26. Drihten hine mid weorðlíce wlite gegyrede *Dominus praecinxit se virtute,* Ps. Th. 92, 1 : 103, 2. For ðam wyrðlícan *propter dignitosam (innocentiae palmam,* Ald. 72), Hpt. Gl. 521, 64. Weorðlíce sige *vere laudandum victoriam,* Ors. 3, 10; Swt. 140, 3. Ðǽm folce ðe on clǽnum selda weorðlícne sige gefeohtaþ *his, qui per fortitudinem in campo victores sunt,* Past. 33; Swt. 227, 25. Weorðlícne wæstm, Ps. Th. 131, 12. Hí worhton wurðlíce cyrcan, Homl. Skt. i. 19, 143. He wurðlíc lác ge-offrode; ðæt wæs án gylden calic on fíf marcon swíðe wundorlíces geworces, Chr. 1058; Erl. 193, 21. Cumaþ wæstm on wangas weorðlíc on hwǽtum *convalles abundabunt frumento,* Ps. Th. 64, 14. Hí ðam wurðlícum godum náne lác ne offredon, Homl. Skt. i. 23, 297. Ða weorðlícan godas, 23, 302. Ðu selest weorðlíca ginfæsta gifa, Met. 20, 226. Weorþlíce, Bt. 33, 4; Fox 132, 19. Gebeorh Godes bringeþ tó genihte wæstme weorðlíce and wel þince *montem Dei, montem uberem: mons coagulatus, mons pinguis,* Ps. Th. 67, 15. Wundor ðín weorðlíc *mirabilia tua,* 70, 16. His weorðlícu weorc *opera Dei,* 77, 9. Ealra þinga weorþlícost and mǽrlícost *omni celebritate clarissimum,* Bt. 33, 1; Fox 120, 31. **III.** *worthy, meet, fit, becoming :*—Heom bið weorðlíc, ðæt hí á habbon árwurðe wísan on eallum heora þeáwum, L. I. P. 10; Th. ii. 318, 33. Wyrðelícum tóhigunge *digno effectu,* Rtl. 35, 37. Wyrðelícum gimērsiga oeste *digna celebrare devotione,* 81, 31. Ðiód weorðlíc dǽdbóte wǽstmas *facite fructus dignos poenitentiae,* Lk. Skt. 3, 8. [Ðu ert wel don man and þarto wurðlich, O. E. Homl. ii. 29, 16. Wurðliche wepnen, Laym. 28923. Hwite wurðliche men *viros dealbatos, quorum vultus inspicere pre claritate non poteram,* Kath. 1576. *O. H. Ger.* werd-líh *celeber, munificus : Icel.* virði-ligr *noble, splendid.*] v. ár-, or-, un-weorþlíc (-wirþ-, -wurþ-).

weorþlíce, weorþelíce ; *adv. Worthily, honourably :*—Ðe weorðelícor *dignius,* Wrt. Voc. ii. 27, 8. **I.** *nobly, excellently, splendidly, magnificently, gloriously :*—Weorþlíce getýd on Grécisc gereorde *Graecae linguae peritissimus,* Bd. 4, 1; S. 563, 33. Hí brǽde weóxan weorðlíce wíde greówan *multiplicati sunt nimis,* Ps. Th. 106, 37. Ðu ymb ðinne esne dydest wel weorðlíce *bonitatem fecisti cum servo tuo,* 118, 65. Swíðe mycel cyrice . . . geworht swá fægre and swá weorþlíce swá hit men on eorþan fægrost and weorþlícost geþencean meahton, Blickl. Homl. 125, 22; Rood Kmbl. 33; Kr. 17. Swá weorðlíce, wíde tósáweþ Dryhten his duguþe, Exon. Th. 299, 30; Crä. 110 : 121, 27; Gú. 295. Fægere, weorðlíce, Menol. Fox 317; Men. 160. Eleutherius onféng biscopdóm and ðone wurþlíce (cf. wuldorfæstlíce, 8, 14) xv winter geheóld, Chr. 167; Erl. 9, 20. Hé his sincgyfan wurðlíce geaf, Byrht. Th. 139, 64; By. 279. Ne gefrægn ic nǽfre wurðlícor sixtig sigebeorna sel gebǽran, Fins. Th. 74; Fins. 37 : Cd. Th. 126, 12; Gen. 2094. Men wurðlícor lybbaþ þonne treówwu *the life of men is more excellent than that of trees,* Shrn. 168, 24. Swá hit weorðlícost foresnotre men findan mihton, Beo. Th. 6304; B. 3162. **II.** *in a way that shews honour to a person, honourably, with honour :*—Ða onféng Dioclitianus Galerius weorðlíce *(plurimo honore),* Ors. 6, 30; Swt. 280, 16. Hí swíðe weorðlíce hine

of heora gryðe sendon, Chr. 1075; Erl. 212, 33. Hí mid mycclan þrymme and blisse and lofsange ðone hálgan arcebiscop feredon, and swá wurðlíce intó Cristes cyrcan bróhton, 1023; Erl. 163, 30. Hine mau byrigde ful wurðlíce, 1036; Erl. 165, 35. **III.** *in a fitting manner, worthily :*—Wé willaþ offrian wurðlíce ûrum Drihtne, Ex. 10, 9. [Ðo þu iseie þine sune . . . so wurðliche stien to his blisse, A. R. 40, 7. Wel and wurrþlike gemmde, Orm. 1033. *O. Sax.* werð-líko: *O. H. Ger.* werð-lîhho : *Icel.* virði-liga.] v. ár-, un-weorþlíce.

weorþ-mynd (-mynt) ,es; *m.* -e; *f.* : -myndu(-o); *indecl. f. Honour :*—Favor, i. *fama, honor, laus, laetitia, testimonium laudis* wyrþmynd, Wrt. Voc. ii. 147, 13. **I.** *honour, respect* shewn to an object, *celebration* of an event :—Sý ûrum Drihtne lof and wuldor and weorþmynd, Blickl. Homl. 65, 25. Wurðmynt, Homl. Th. i. 76, 23. Ðam ánum is éce weorðmynd, Exon. Th. 240, 10; Ph. 636. On weorðmynde ðara twelfa apostola, Lchdm. ii. 138, 22. Ðære dǽde tó weorðmynte *in honour of the deed,* Ors. 6, 25; Swt. 276, 15. Freólsiaþ ðone seofoðan dæg Gode tó wurðmynte, Ex. 35, 2. Gode tó lofe and ðam hálgan arcebiscope tó wurðmynte, Chr. 1023; Erl. 163, 35. Gé weorðmyndu Dryhtne gieldaþ, Exon. Th. 130, 7; Gú. 434. Eodan hié him tógeánes mid blówendum palmtwigum heora siges tó wyorþmyndum, Blickl. Homl. 67, 11. Seó mǽre burh ðe ic geworhte tó wurðmyndum *Babylon magna quam ego aedificavi in gloria decoris mei* (Dan. 4, 27), Cd. Th. 254, 12; Dan. 610. Hwæt wit tó willan and tó wordmyndum árna gefremedon, Beo. Th. 2377; B. 1186. **II.** *honour* bestowed on an object, *favour, grace :*—Seó mennisce gecynd mæg ðæm Scyppende lof and wuldor secgean ðara ára and ðara weorþmenda ðe Drihten mancynne forgeaf . . . Hú mihte mannum mára weorðmynd geweorþan, ðonne him on ðyssum dæge gewearþ? Blickl. Homl. 123, 3–15. Wurðment *privilegium,* Hpt. Gl. 527, 68. Ic hæfde gemynt ðé tó árwurðienne on ǽhtum and on feó, ac God ðé benǽmde ðæs wurðmintes *decreveram magnifice honorare te, sed Dominus privavit te honore disposito,* Num. 24, 11. For synderlícum wurðmente *propter privilegium (singularem honorem),* Hpt. Gl. 411, 31. Frumgife í wurðmente *praerogativam,* 457, 29. Hit nán wundor nys ðæt sé hálga cyningc untrumnysse gehǽle, nû hé on heofonum leofaþ . . . hæfð hé ðone wurðmynt *(the privilege of healing sickness)* for his gódnesse, Homl. Skt. ii. 26, 277. Syndrige wyrðmenta *privilegia,* Hpt. Gl. 517, 2. Ic wát hwá mé wyrðmyndum *(graciously)* on wudubáte ferede ofer flódas, Andr. Kmbl. 1809; An. 907. **III.** *honour, decoration, ornament :*—Uueorðmynd *infula,* Wrt. Voc. ii. 110, 66. Weorþmynd *infulas,* 43, 60. Gif ðu wénst ðætte wundorlíc gerela hwelc weorþmynd sié *(pulcrum variis fulgere vestibus putas ?),* ðonne telle ic ða weorðmynd ðǽm wyrhtan ðe hié worhte, Bt. 14, 1; Fox 42, 18. Yr byð æðelinga gehwæs wyn and wyrðmynd, Runic pm. Kmbl. 344, 31; Rún. 27. Hé geseáh sigeeádig bil, wigena weorðmynd, Beo. Th. 3122; B. 1559. Wel bið ðam eorle ðe him oninnan hafaþ rúme heortan, ðæt him bið for worulde weorðmynda mǽst, Exon. Th. 467, 18; Alm. 3. **IV.** *honour, glory, fame :*—Byð ðé weorðmynd (wurðmynt, *v.l.*) beforan midsittendum *erit tibi gloria coram simul discumbentibus,* Lk. Skt. 14, 10. Ðæt hié witen ðæt mín þrym and mín weorðmynd máran wǽron ðonne ealra óþra kyninga, Nar. 33, 4. On his winestran handa wǽre wela and wyrðmynt *(gloria)* . . . Hé mæt ðone welan and ðone wyrðmynd tó ðære winestran handa, Past. 50; Swt. 389, 17–19. Wæs Hróðgáre heresped gyfen, wíges weorðmynd *(glorious success in war),* Beo. Th. 130; B. 65. Ðý lǽs hié ormóde wǽron, and ðý sǽnran mínes willan and weorðmynde *(the slower to do my will and promote my glory),* Nar. 32, 24. Ic *(Eve)* wæs mid weorþmende on neorxna wange *I lived glorious in Paradise,* Blickl. Homl. 89, 8. Hé heóld ðone arcestól mid mycclan weorðmynte, Chr. 1068; Erl. 206, 16. Sió eáðmódnes iernð beforan ðæm gilpe, and hió cymð ǽr ǽr ða weorðmyndu (wyrðmyndu, Hatt. MS.) *gloriam praecedit humilitas,* Past. 41; Swt. 298, 16. Dryhtne ðe hyre weorþmynde geaf, mǽrþe on moldan ríce, Judth. Thw. 26, 25; Jud. 343. Wé hæfdon wlite and weorðmynd, Cd. Th. 274, 10; Sat. 152. Him God sealde weorðmynda dǽl, Beo. Th. 3509; B. 1752. Hé wæs for weorulde wís, weorðmynþa georn, Met. 1, 51. Ðæt ðu gefeó in ðæm fromscipe mínes lífes, and eác blissige in ðǽm weorðmyndum, Nar. 32, 32. Hé weorþmyntum þáh *he throve gloriously,* Beo. Th. 16; B. 8. **V.** *honour, dignity, honourable position* or *office :*—Ne gedafenaþ ná munuce ðæt hé ǽniges worldlíces wyrðmyntes gyrne *non convenit monacho mundanum quemquam honorem desiderare,* L. Ecg. P. iii. 10; Th. ii. 198, 30. Ða weorðmynde cynehádes hé fleáh *rex fieri noluit,* Past. 3; Swt. 33, 20. Tyddre weorþmyntas *fragiles honores,* Wrt. Voc. ii. 150, 38. Tó weorðmyndum *ad fasces,* 99, 35 : 4, 47. Wyrþmyndum *titulis,* 95, 47. Ne bidde wé ná leáse welan ne gewítenlíce wurðmyntas, Homl. Th. i. 158, 26. **VI.** *dignity, nobleness :*—Seó wlitige, weorðmynda full, heáh and hálig heofuncund þrýnes, Exon. Th. 24, 2; Cri. 378. Ára mé for hire wuldres weorþmyndum, Blickl. Homl. 89, 22. Wolde reordigean ríces hyrde hálgan stefne, werodes wísa wurðmyndum *(nobly, with dignity)* sprǽc, Cd. Th. 194, 10; Exod. 258. [Habban þene eche wurðment mid Gode, O. E. Homl. i. 107, 21. Ilǽsten scal is wordmunt (me wole of him telle, 2nd MS.), Laym. 18851. Si Drihhtin wurrþmiunt and loff and wu'llderr, Orm. 3379. Ȝef þu hit ȝulde to his wurðmunt þe scheop

þe, Kath. 216. Cf. He cweð þet he wolde hit wurðmiuten and arwurðen, Chr. 656 ; Erl. 30, 3.]

weorþness, e ; *f.* **I.** *worthiness, honourable character :*—For his geearnunge wurþnys[se] (wyrðnesse, Bd. M. 194, 34) hē wæs fram eallum monnum lufad *ob meritorum dignitatem ab omnibus diligebatur*, Bd. 3, 14 ; S. 540, 10. Tó lifes wyrþnysse *ad vite honestatem*, Anglia xiii. 368, 48. **II.** *dignity, nobility, honourable* or *honoured condition :*—Werðnes *dignitas*, Kent. Gl. 582. Æþele æfter ðysse worulde wurþnysse *ad saeculi hujus dignitatem nobilis*, Bd. 4, 9 ; S. 577, 2. Ðū ðe menisc gicynd bufa frumes frumcendnisse eft boetest wyrðnise *qui humanam naturam supra prime originis reparas dignitatem*, Rtl. 35, 13. **III.** *dignity, honourable office :*—Hæfde se cyning efenhlētan ðære cynelícan wurþnysse (*regiae dignitatis*), Bd. 3, 14 ; S. 539, 30. **IV.** *dignity, state, imposing show :*—Hē fērde tó Róme mid micelre weorþnesse, Chr. 855 ; Erl. 68, 28. **V.** *honour shewn to an object :*—On wurþnysse ðínre *in honore tuo*, Ps. Spl. 44, 10. Ne is wítge būta worðnis (*sine honore*) būta on oeðel his, Mk. Skt. Lind. 6, 4. v. un-weorþness.

weorþscipe, es ; *m.* **I.** *worship, honour shewn to an object :*—Gif hwa biþ mid hwelcum welum geweorþod, hū ne belimpþ se weorþscipe tó ðam ðe hine geweorðaþ ; ðæt is tó herianne hwǣne rihtlícor si *quod ex appositis luceat, ipsa quidem, quae sunt apposita, laudantur*, Bt. 14, 3 ; Fox 46, 12. Ða dysiende wēnaþ ðætte ðæt ðing sié ælces weorþscipes betst wyrþe ðætte hī medemæste ongiton magon *labuntur hi, qui quod sit optimum, id reverentiae cultu dignissimum putant*, 24, 4 ; Fox 86, 10. Nys nān wítega būtan weorðscype (wurð-, *v. l.*) (*sine honore*), būton on his earde, Mt. Kmbl. 13, 57 : Mk. Skt. 6, 4. Hī wunnon æfter weorðscipe (wyrþ-, *v. l.*) on ðisse worulde, and tiledon gódes hlīsan, Bt. 40, 4 ; Fox 240, 5. Ealne ðæne bysmor wē gyldaþ mid weorðscype ðam ðe ūs scendaþ, Wulfst. 163, 10. Mid wurðscipe underfón, Chr. 785 ; Erl. 57, 19 : Nicod. 20 ; Thw. 10, 26. Him cōmon lāc tó wurðscipe, Ælfc. T. Grn. 7, 32. Yfelwillende men nǣnne weorþscipe næfdon, Bt. 15 ; Fox 48, 17. Uordscip, Lind. : wordscipe, Rush., *honorem*, Jn. Skt. 4, 44. **II.** *honour, honourable* or *honoured condition, dignity, honours :*—Se weorþscipe and se anweald, gif hē becymþ tó ðam dysigan, hē mæg hine gedón weorþne *dignitates honorabilem, cui provenerint, reddunt*, Bt. 27, 1 ; Fox 94, 18. Benumen ǣgþer ge ðínra welona ge ðínes weorþscipes, 7, 3 ; Fox 20, 5. Welan and weorþscipes hī willniaþ *opes, honores ambiant*, 32, 3 ; Fox 118, 29 : Met. 19, 44. Hwæt mæg ic ðē mǣre secgan be ðam weorþscipe and be ðam anwealde ðisse worulde . . . Gē ne ongitaþ ðone heofoncundan anweald and ðone weorþscipe, se is eówer ǣgen . . . Hwæt se eówer wela and se eówer anweald ðe gē nū weorþscipe hātaþ, gif hē beoymþ tó ðam eallra wyrrestan men *quid de dignitatibus potentiaque disseram, quas vos, v rae dignitatis ac potestatis inscii, coelo exaequatis? quae si in improbissimum quemque ceciderint?* Bt. 16, 1 ; Fox 48, 27–34. Mann ðā ðā hē on wurðscype (*in honore*) wæs, Ps. Spl. 41, 21. Hē (*Joseph*) heóld his fæder on fullum wurðscipe ðǣr mid eallum his bróðrum, Ælfc. T. Grn. 5, 7. **III.** *honour, glory :*—Mine fýnd mínne weorðscipe tó duste gewyrcen *inimicus gloriam meam in pulverem deducat*, Ps. Th. 7, 5. **IV.** *honour, state, magnificence :*—Hē fērde tó Róme mid myccium wurðscipe, Chr. 855 ; Erl. 69, 18. **V.** *dignity* of behaviour :—Móderlícere stæððinysse ł wurðscipe *materna gravitate ł dignitate*, Hpt. Gl. 469, 38. **VI.** *worthiness, excellence, nobleness :*—Weorþscipe *vel* gefungenes *dignitas*, i. *honestas, excellentia, fastigium*, Wrt. Voc. ii. 140, 25. Sittende hē tǣhte ; ðæt belimpð tó wurðscipe lāreówdómes, Homl. Th. i. 548, 25. Hié ālýsde for his weorþscipe Eádmund cyning, Chr. 942 ; Erl. 116, 18. **VII.** *an honour, a dignity, an honourable office* or *position :*—Ealdordómas *vel* ða hēhstan wurðscipas *fasces*, biscoplíc wurðscipe *flamininus honor*, Wrt. Voc. i. 59, 53, 54. Swelce wræccan woldon underfón ðone weorðscipe and eác ða byrdenne *infirmus quisque, ut honoris* (plebium ducatus) *onus percipiat, anhelat*, Past. 7 ; Swt. 51, 23. Se ðe wel þēnaþ, hē gódne wyrðscipe him sylfum gestryñd *qui bene ministraverit, gradum bonum sibi adquirit*, R. Ben. 54, 18. **VII a.** *pl. Dignities, persons in office* (?) :—Wyrþscipas *comitia* (cf. weorþung-dæg). Wrt. Voc. i. 21, 65. **VIII.** *an honour, ornament, decoration :*—Wurðscipe *infula*, Hpt. Gl. 458, 24. Gifu gumena byð gleng and heofens, wraðu and weorðscype, and wræcna gehwam ār and ætwist, Runic pm. Kmbl. 340, 25 ; Rūn. 7. Mid twām wurðscipum geglængde se ælmihtiga Scyppend ðæs mannes sāwle ; ðæt is mid ecnysse and eádignysse, Homl. Skt. i. 1, 150. **VIII a.** *honour, cause of an object being honoured* or *honourable :*—Hit geweard ðæt ðam wīsan men com tó lofe and tó wyrðscype ðæt se unrihtwīsa cyning him teohhode tó wīte *ita cruciatus, quos putabat tyrannus materiam crudelitatis, vir sapiens fecit esse virtutis*, Bt. 16, 2 ; Fox 52, 26. **IX.** *what is honoured* or *prized, an excellent thing, a good :*—On swelcum and on óþrum swelcum lǣnum and hreósendum weorþscipum (*riches, fame, power*, etc., *have been enumerated* ; cf. ðam lǣnum gódum, l. 1), Bt. 24, 3 ; Fox 82, 21. v. un-, weorold-weorþscype.

weorþung, e ; *f.* **I.** *honouring, shewing of honour* to an object, *honour, reverence :*—Ðǣm is simle wuldor and weorðung, Blickl. Homl. 169, 28. Ne is wítga būta worðunge (*sine honore*), būta on oeðel his, Mk. Skt. Rush. Lind. 6, 4. For ðínre weorþunge *in honore tuo*, Ps. Th.

44, 10. Gif hē on ríce becymð, for ðære weorðunge ðæs folces hē bið on ofermēttu āwended and gewunaþ tó ðæm gielpe si *ad regiminis culmen eruperit, in elationem protinus usu gloriae permutatur*, Past. 3 ; Swt. 35, 12. Leóhtfæt bið ā byrnende for ðara swaþa weorþunga, Blickl. Homl. 127, 31. Wē habbaþ on Godes naman weorðunge bisceop gebletsode, Wulfst. 176, 2. Hē bið on gódre weorþunge *he will be highly respected*, Lchdm. iii. 158, 10. **I a.** in religious matters, (1) *worship* of a god, *divine worship, religious service :*—Tídsangas *canonica*, weorþung *canor*, Wrt. Voc. ii. 128, 27. Dægrēdsanges weorþung is þus tó healdenne *matutinorum solempnitas ita agatur*, R. Ben. 37, 5. Ne dear man forhealdan lytel ne mycel ðæs ðe gelagod is tó gedwolgoda weorðunge, Wulfst. 157, 14. Drihtne tó wurðunga, Lev. 2, 2. Idola wurðinge, L. N. P. L. 48 ; Th. ii. 298, 1. (2) *honouring* of a person, thing, or season, *celebration, commemoration, festival :*—Mycel is þeós weorþung ðæs hālgan Sancte Iōhannes gebyrde, Blickl. Homl. 167, 13. On ðæm dæge ðe seó tíd bið and his (*S. Michael*) weorðung, 209, 17. Be ðisse hālgan tíde weorþunga ðe wē tó dæg mǣrsian sceolan and weorþian, 161, 4. Be ðyses dæges (*Pentecost*) weorþunga, 133, 12. Æt eallra hāligra weorðunge *at the feast of All Saints*, L. Alf. pol. 43 ; Th. i. 92, 8. Hē ða weorþunge Eástrena on riht ne heóld ne nyste *de observatione Paschae minus perfecte sapiebat*, Bd. 3, 17 ; S. 545, 2. Weorðunga, Blickl. Homl. 137, 8. Hī tó Hierusalem faran woldon for ðære hālgan róde wurðunga, ðe man æfter nāht manegum dagum wurðian sceolde, Homl. Skt. ii. 23 b, 350. **II.** *nobleness, glory, excellence :*—Ðæt wuldres bearn on ðysne middangeard āstāg, and seó heofoncunde weorþung ðone fǣmnlícan bōsm Sancta Marian gefylde, Blickl. Homl. 165, 27. Mycel is se hāligdom and seó weorþung Sancte Iōhannes, ðæs mycelnesse se Hǣlend sylfa tācn sægde . . . Hē on his mægenes weorþunga oferswīþ ealra óþerra martira wuldor, 167, 16–25. Him wile God miltsian for heora mægena weorþunga, and for eorþlícra manna gebedum, 47, 8. Næs riht on ðære stōwe ǣnigne tó ācwellanne for ðære stōwe weorþunga, Nar. 30, 3. Apostola ðínra wurðunge folc ðín giwynsumia *apostolorum tuorum Petri et Pauli honore plebs tua exultet*, Rtl. 59, 33. Ðæt hē Sanctus Iōhannes líðes weorþunga gesecgan mæge, Blickl. Homl. 163, 36. **III.** *ornament, decoration :*—Crist com tó wlitignesse and tó weorþunge his brýde, Blickl. Homl. 11, 31. Godwebba cyst, ðæt ðam hālgan hūse sceolde tó weorþunga weorud sceáwian, Exon. Th. 70, 11 ; Cri. 1137. [Þat folc sungen heore leofsong ure Helende to wurðinge, O. E. Homl. i. 7, 10. Godes laȝe bit eo mon wurðie his feder mid muchelere wurþunge, 109, 27. Ðe, God, to wurðinge, Gen. and Ex. 33. O. H. Ger. werdunga *solemnitas, celebritas, dignitas : Icel.* virðing *worship, reputation, honour.*] v. breóst-, dæg-, hāls-, hām-, hord-, hring-, mann-, neód-, sinc-, stān-, sundor-, tíd-, treów-, un-, wíg-, will-weorþung.

weorþung-dæg, es ; *m.* **I.** *a day for the bestowing of honours* or *offices :*—Ārdagas *vel* weorðungdagas (weordung-, Wrt.) *comitiorum dies, honorum dies*, Wrt. Voc. ii. 132, 29. [**II.** *a day for worship* or *celebration :*—Setteres dei wes heore Sunedei, and bet heo heolden heore wurðingdei þene we doð, O. E. Homl. i. 9, 9.]

weorþung-stōw, e ; *f. A place for worship :*—On ðære hālgan wurðungstōwe *de tabernaculo testimonii*, Lev. 1, 1.

weorud, weoruld, weosan, weosend, weosnian, weosule, weota, weotan, weoðo-bān, weoðo-bend, weotian, weotuma. v. werod, weorold, wesan, wesend, wisnian, wesle, wita, witan, wiþo-bān, wiþo-bend, witian, wituma.

weoxian ; *p.* ode *To* wipe, *make clean :*—Ðacian, ðecgan and fald weoxian, Anglia ix. 261, 18. Hūs gódian, rihtan and weoxian, 262, 19. [Cf. O. H. Ger. wisken *tergere*.]

wēpan ; *p.* weóp, wēp (wǣpde, Lind.), *pl.* weópon, wēpon ; *pp.* wōpen *To weep, wail, mourn, lament :*—Ic wēpe *fleo*, ðū wēpst (wǣpst, *v. l.*) *fles*, ic weóp *fleui*, gewōpen *fletum*, Ælfc. Gr. 26, 1 ; Zup. 152,18. **I.** *intrans.* (1) of persons :—Maria stód and weóp (hrēmende ł uoeþende *plorans*, Lind.) ; and ða heó weóp (gewǣp *fleret*, Lind.), heó ābeáh nyðer . . . Ða englas cwǣdon tó hyre : 'Wíf, hwí wēpst (uoepæs, Lind. : woepes, Rush., *ploras*) ðū ?' Jn. Skt. 20, 11–13. Hē geseah mycel gehlýd wēpende (*flentes*) . . . Hē cwæþ : 'Hwí wēpaþ (*ploratis*) gē ?' Mk. Skt. 5, 38, 39. Beornas grētaþ, wēpaþ wānende wergum stefnum, Exon. Th. 61, 31 ; Cri. 993. Hē weóp (*ploravit*) bityrlíce, Mt. 26, 75 : Andr. Kmbl. 2799 ; An. 1402. Hē weóp (gewǣp *fleuit*, Lind.) ofer hig, Lk. Skt. 19, 41. Wē heófdun and gē ne weópun (wǣpde gié *plorastis*, Lind.), 7, 32. Ne ceara ðū ne ne wēp, Blickl. Homl. 143, 4. Wēpan *ploremus*, Ps. Th. 94, 6. Wēpan wē and geþencan hū Drihten cwæð : 'Eádige beóþ ða ðe nū wēpaþ (*lugent*, Mt. 5, 5),' Blickl. Homl. 25, 19. Gif ðū wistest hwæt ðē tóweard is, ðonne weópe ðū nid mē, Homl. Th. i. 404, 27. Ðā ongan hē wēpan (woepa *flere*, Lind., Rush.), Mk. Skt. 14, 72. Mid wēpendre bēne *lacrymosis precibus*, Bd. 1, 12 ; S. 480, 26. Mid wǣpendre stefne *flebili voce*, 480, 37. Wēpendre, Blickl. Homl. 87, 26, 8. Drihten hýrde míne wēpendan stefne (*vocem fletus mei*), Ps. Th. 6, 7. Ða ðe wǣpende (*flentes*) sǣton, Bd. 5, 12 ; S. 627, 14. Heófendum and wēpendum (wōpendum *flentibus*, Lind.), Mk. Skt. 16, 10. Hē gemētte swīþe manige wēpende, and wǣron cweþende : 'Wā ūs lā . . .' And ðā him swā wēpendum, ðā com ðara sacerda ealdorman, Blickl. Homl.

153, 25-33. Hí ofslógon weras and wífmen and ða wépendan cild *interfecerunt omnia a viro usque ad mulierem, ab infante usque ad senem*, Jos. 6, 21. (2) of other than human beings :—Weóp eal gesceaft, Rood Kmbl. 110; Kr. 55. **I a.** where tears are shed :—On mínum bedde ic síce and wépe *lavabo lectum meum*, Ps. Th. 6, 5. Hé sægde ðæt ða hálgan triów swíðe wépen and mid micle sáre instyred wǽron (*uberibus lacrimis commoueri*), Nar. 28, 11. Hé ongan wépan hlúttrum teárum. Ðá fræng hine his mæssepreóst for hwon hé weópe *coepit ad lacrymarum profusionem effici. Quem dum presbyter suus, quare lacrymaretur interrogasset*, Bd. 3, 14; S. 541, 3-5. Hé wæs wépende mid teárum, Blickl. Homl. 151, 20: Andr. Kmbl. 117; An. 59. **II.** *trans.* (a) with accusative, *to mourn, lament, bewail, deplore*, (1) of persons :—Hé weóp his sunu *lugens filium suum*, Gen. 37, 34. Hit wæs þeáw, ðæt man sceolde wépan ǽlcne deádne mann; and ðæt folc hyne weóp (*flevit eum Aegyptus*) hundseofontig daga, 50. 3. (Hí) weópan wyrde (*prolis*) *luxerunt fata (parentes*, Ald. 176), Wrt. Voc. ii. 94, 5 : 51, 34. Ne wép ðone wræcsíð, Andr. Kmbl. 2861; An. 1433. Wræcsíð wépan, Exon. Th. 166, 23; Gú. 1047 : 443, 30; Kl. 38. Ðæt ðætte óðre menn unálésedes dót hé sceal wépan suá suá his ágne scylde *illicita perpetrata ab aliis ut propria deplorat*, Past. 10; Swt. 61, 15. Ðá hé hine ealle wépende geseah *when he saw all mourning him*, Blickl. Homl. 225, 22. Wópene *lamentatae*, Blickl. Gl. (2) of other than human beings :—Ne wæl wépeþ wulf se grǽga, Exon. Th. 343, 2; Gn. Ex. 151. (b) with gen. *to mourn for, be grieved at* :—Hwá is swá heardheort ðæt ne mæg wépan swylces ungelimpes? Chr. 1085; Erl. 219, 40. [O. E. Homl. wiep; *p.*: A. R. weop : Laym. weop, wep : Will. wep, wepte : Chauc. weep, wepte; *pp.* wopen : Piers P. wept : Goth. wôpjan; *p.* wôpida *to cry*: O. Sax. wôpian; *p.* wôp, wêp *to mourn*: O. L. Ger. wôpan; *p.* wiep : O. Frs. wêpa : O. H. Ger. wuofan; *p.* wiof *flere, plorare, plangere, lacrimari, deflere*; wuofen; *p.* uuofta *plorare, flere, lugere* : Icel. œpa; *p.* œpta *to cry, scream*.] v. be-, ge-wépan.

wépend-líc; *adj. Lamentable, mournful* :—Reówlíc and wépendlíc tíd wæs ðæs geáres, ðe swá manig ungelimp wæs forðbringende, Chr. 1086; Erl. 220, 22. Wépendlíce *flebiles (and* wépendlíc *flebilis.* v. Wülck. Gl. 240, 16), Wrt. Voc. ii. 149, 41. [O. H. Ger. wuofant-lîh *luctuosus*.] v. be-wépendlíc.

wépendlíce; *adv. Lamentably, mournfully, grievously* :—Wépendlíce tó bewépenne synd *flebiliter deplorandi sunt*, Scint. 77, 3.

wer, es; *m.* **I.** *a man, a male person* :—Wer oððe wǽpman *vir*, Wrt. Voc. i. 73, 11. Wer wintrum geong (*Isaac*), Cd. Th. 174, 34; Gen. 2888. Wíffæst wer *a married man*, L. C. S. 55; Th. i. 406, 14. Se Godes wer Sanctus Martinus, Blickl. Homl. 213, 36. Se eádiga wer, 215, 31. Se weor (wer, Rush.) *uir*, Lk. Skt. Lind. 8, 38. Woer (wer, W. S., Rush.), 9, 38. Of ðæs weres (*viri*) handa ic ofgange ðæs mannes (*hominis*) líf, Gen. 9, 5. On weres háde, Elen. Kmbl. 144; El. 72: Apstls. Kmbl. 53; Ap. 27. Ðæs weres tíd scí Symforiani, Shrn. 119, 17. Gelíc ðam wísan were (*viro*), Mt. Kmbl. 7, 24. Ic nǽnigne wer (*uirum*, Lk. 1, 34) ne ongeat, Blickl. Homl. 7, 21. Wundne wer (cf. gewundodne mon, Bt. 15; Fox 48, 16), Met. 8, 35. Gé Galiléiscan weras *uiri Galilei*, Blickl. Homl. 123, 20. Niniuetisce weras (wǽras *viri*, Lind.), Mt. Kmbl. 12, 41. Týn hreófe weras (wǽras, Lind.: wearas, Rush. *uiri*), Lk. Skt. 17, 12. Fíftig rihtwísra wera *quinquaginta justos*, Gen. 18, 26. Wælrǽs weora, Beo. Th. 5886; B. 2947. Fíf ðúsendo wǽro ł wǽrana (weorona, Rush. *uirorum*), Mk. Skt. Lind. 6, 44. Ymbseted mid syxtigum werum ðǽm. strengestum ðe on Israhélum wǽron, Blickl. Homl. 11, 17. Hálige weoras *viros sanctos*, Bd. 5, 10; S. 623, 41. **I a.** in conjunction with words denoting a woman :—Óðer wæs idese onlícnes, óþer on weres wæstmum, Beo. Th. 2708; B. 1352. Ðeós bið gecíged fǽmne, for ðam ðe heó ys of were genumen *haec vocabitur virago, quoniam de viro sumpta est*, Gen. 2, 23. Gif wíf be óðrum were forlicge, L. C. S. 54; Th. i. 406, 6. Gif oxa ofhníte wer oþþe wíf (*virum, aut mulierem*), L. Alf. 21; Th. i. 48, 27: Exon. Th. 225, 24; Ph. 394. Weras mid wífum, Cd. Th. 104, 20; Gen. 1738. Weras, wíf samod, Andr. Kmbl. 3330; An. 1668. Weras and wíf, Exon. Th. 448, 26; Dóm. 60. Weras and idesa, 176, 7; Gú. 1205. Eall wífa cynn and wera, Blickl. Homl. 5, 24 : Beo. Th. 1990; B. 993. Twá hund and eahta and feówertig wera, and nigon and feówertig wífa, Blickl. Homl. 239, 14. Bletsunge gemǽne werum and wífum, Exon. Th. 7, 14; Cri. 101. Ge weras ge wíf, Blickl. Homl. 107, 11. ¶ in the plural the word seems sometimes to include women as well as men :—Hé wolde for wera synnum eall áðéðan, Cd. Th. 77, 23; Gen. 1279. Folcdryht wera, sáwla gehwylce, Exon. Th. 66, 5; Cri. 1067. Wera endestæf (cf. Blickl. Homl. 239, 14 *supra*), Andr. Kmbl. 270; An. 135. Heofones gim, wyncondel wera, Exon. Th. 174, 31; Gú. 1186. In wera lífe, 26, 13; Cri. 416. Wera cneorissum, 347, 4; Sch. 7. Ðú ne wilnast weora ǽniges deáð, Ps. C. 54. Feówertig daga níð wæs wællgrim werum, Cd. Th. 83, 23; Gen. 1384: 109, 1; Gen. 1816. Lencten on tún geliden hearde werum tó wícum, Menol. Fox 58; Men. 29. Næs ðǽr hláfes wist werum, Andr. Kmbl. 43; An. 22. Fǽhðe ic wille on weras stǽlan, eall ácwellan ða beútan beóð earce bordum, Cd. Th. 81, 28; Gen. 1352. **II.** *a man, a male that has reached man's* estate :—Ðá áworden ic am uoer ic giídlade ða ðe uoeron lytles *quando factus sum vir, evacuavi quae erant parvuli*, Rtl. 6, 19. Fíf þúsenda wera (wearana, Lind.: weora, Rush., *virorum*) bútan wífum and cildum, Mt. Kmbl. 14, 21. Ic mægen wera (*virorum*) eom, and litlincgas nellaþ forbígean mé, Coll. Monast. Th. 29, 1. Weras and wífmen and ða wépendan cild, Jos. 6, 21. **III.** *a being in the form of a man* :—Grendel, wonsǽlig wer (cf. 2708; B. 1352 *supra*), Beo. Th. 210; B. 105. Twégen weras (wǽras, Lind.: wearas, Rush., *uiri*) Móysés and Hélias, Lk. Skt. 9, 30. Abraham geseah þrí weras standende him gehende, Gen. 18, 2. **IV.** *a married* or *a betrothed man, a man* (as in *man* and *wife*), *a husband.* v. wer-leás :—Swá micel swá ðæs wífes wer (*maritus mulieris*) girnþ, Ex. 21, 22. Hereríc hire wer (*vir ejus*), Bd. 4, 23; S. 594, 44. Be ðon ðe ryhtgesamhíwan bearn hæbben, and ðonne se wer gewíte, L. In. 38; Th. i. 126, 2. Wer and wíf beóð in ánum líchoman, Bd. 1, 27; S. 491, 13 : Exon. Th. 327, 11; Vy. 2 : Blickl. Homl. 185, 26. Ðæt hé hý healdan wille swá wær his wíf sceal, L. Edm. B. 1; Th. i. 254, 7. Iósep hyre wer (*vir*, Mt. Kmbl. 1, 19. Weard seó módor gegremod æfter hire weres forðsíðe fram hire cilde, Homl. Th. ii. 30, 4. Geong wuduwe mót eft ceorlian æfter hire weres forðsíðe, L. Ælfc. P. 43; Th. ii. 382, 32. Heó leofode mid hyre were seofan gér of hyre fǽmnháde, Lk. Skt. 2, 36 : Cd. Th. 134, 1; Gen. 2218. Gif mon hǽme mid monnes wífe, gebéte ðam were, L. Alf. pol. 10; Th. i. 68, 9: Exon. Th. 153, 6; Gú. 821. Gif wuduwe binnan geáres fæce wer geceóse, L. C. S. 74; Th. i. 416, 8. Wær, L. Edm. B. 4; Th. i. 254, 16. Iósep, Marian wer (wær, Rush., *virum*), Mt. Kmbl. 1, 16. Hié noldan heora wera rǽstgemánan sécean, Blickl. Homl. 173, 16. Heora wíf him sǽdon, ðæt hié him woldon óðerra wera ceósan (*sobolem se a finitimis quaesituras*), Ors. 1, 10; Swt. 44, 22. Wíf ic lǽrde ðæt hié heora weras lufedan, Blickl. Homl. 185, 23. **V.** *a male*, (1) of human beings :—Wer and wíf hé gesceóp hí *masculum et feminam creavit eos*, Gen. 5, 2. (2) of plants :—Ys ðeós wyrt twégea cynna, ðæt is wer (wær, *v. l.*) and wíf, Lchdm. i. 204, 9. Ðeós wyrt is twéga cynna, óðer ys wíf, óðer wer, 252, 20. **VI.** in grammar, *masculine gender :* —*Participia* belimpaþ tó þrým cynnum, tó were and tó wífe and tó nádrum cynne, Ælfc. Gr. 39; Zup. 243, 19. [Orm. O. and N. Gen. and Ex. were : Laym. were (*dat.*) : Goth. wair : O. Sax. O. Frs. O. H. Ger. wer : Icel. verr : Lat. vir.] v. dryht-, folc-, húsel-, leód-, riht-wer.

wer *and* were, es; *m.* [*The word seems to be interchangeable with* wer-gild (q. v.), e. g. :—Gif hé geþeó ðæt hé hæbbe hiwisc landes . . . þonne bið his wergild .cxx. scill.; and gif hé ne geþeó búton tó healfre híde, þonne sí his wer (were, *v. l.*) .lxxx. scill., L. Wg. 7; Th. i. 186, 14. Wergildes (*v. l.* weres) . . . Se wer, 1; Th. i. 186, 3, 4. Bið cynges ánfeald wergild .vi. þegna wer (wergyld, *v. l.*), L. M. L.; Th. i. 190, 4.] *The price set upon a man according to his degree* :—Be fullan were, sý swá boren swá hé sý, L. Edm. S. 1; Th. i. 248, 4. Twelfhyndes mannes wer is twelf hund scyllinga. Twyhyndes mannes wer is twá hund scill. . . . Eal man sceal æt cyrliscum were be ðære mǽðe dón ðe him tó gebyreþ, swá wé be twelfhyndum tealdan, L. E. G. 13; Th. i. 191, 13-14, 16, 3. Gif wylisc mon hæbbe híde londes, his wer bið .cxx. scill.; gif hé hæbbe healfe, .lxxx. scill.; gif hé nænig hæbbe, .lx. scillinga, L. In 32; Th. i. 122, 9 (cf. Wealh, gif hé hafaþ fíf hýda, hé bið syxhynde, 24; Th. i. 118, 10. Wealh gafolgelda, .cxx. scill.; his sunu, .c.; ðeówne, .lx.; somhwelcne, fíftegum, 22; Th. i. 118, 3). **I.** when a person was wrongfully (for other cases v. ǽ-gilde) slain, the *wer* of the slain man could be claimed from the slayer (cf. wergild, I), who was bound to furnish security for the payment, and the date for the first instalment of such payment was fixed. According to a law of Cnut the slain man must have been in a hundred and in a tithing to make the claim for the *wer* valid :—Gif man ofslægen weorðe, gylde hine man swá hé geboren sý. And riht is ðæt se slaga, siþþan hé weres beweddod hæbbe, finde ðærtó wǽrborh . . . be ðam ðe ðærto gebyrige; ðæt is æt twelfhyndum were gebyriaþ twelf men tó werborge, .viii. fæderenmǽgðe, and .iiii. médrenmǽgðe. Ðonne ðæt gedón sý, ðonne rære man cyninges munde. (*Then at intervals of twenty-one days* healsfang, manbót, fyhtwíte *respectively were to be paid.*) Ðæs (*the payment of* fyhtwíte) on .xxi. nihtan ðæs weres ðæt frumgyld, and swá forð ðæt forgolden sý on ðam fyrste ðe witan geriǽden, L. E. G. 13; Th. i. 174, 15-29. Be fǽhðe. Ǽrest æfter folcrihte slaga sceal his forspecan on hand syllan, and se forspeca mágum, ðæt se slaga wille bétan wið mǽgðe. Ðonne syþþan gebyreþ ðæt man sylle ðæs slagan forspecan on hand, ðæt se slaga móte mid griðe nýr and sylf wæres weddian. (*The proceedings are then as in the preceding extract, with the exception that* fyhtwíte *is not mentioned; so that the first payment of* wer *is made twenty-one days earlier*), L. Edm. S. 7; Th. i. 250, 12-21. Wé wyllaþ ðæt ǽlc freó man beó on hundrede and on teóðunge gebrôht ðe láde wyrðe beón wylle oþþe weres wyrðe, gif hine hwá áfylle ofer .xii. wintre, L. C. S. 20; Th. i. 386, 21. Be swá ofslægenes monnes were. Gif mon ðæs ofslægenan weres bidde, L. In. 21; Th. i. 116, 3-4. Gif mon twyhyndne mon unsynnige mid hlóðe ofsleá, gielde se ðæs sleges andetta sié wer . . . Gif hit sié syxhynde . . . se slaga wer . . . Gif hé sié twelfhynde . . . se slaga wer . . . Gif hlóð ðis gedó . . . ealle forgielden ðone wer gemǽmum

hondum, L. Alf. pol. 29-31; Th. i. 80, 6-17: 36; Th. i. 84, 13, 14. Gif mon beforan cyninges ealdormen on gemóte gefeohte, bête wer and wíte swá hit ryht sié, and beforan ðam .cxx. scitt. ðam ealdormen tó wíte, 38; Th. i. 86, 14. **I a.** of those who were concerned in the receiving of the *wer* the following passages speak; see also wer-gild, **I a:** —Se wer (*a king's*) gebiraþ mágum, L. Wg. 1; Th. i. 186, 4 : L. M. L.; Th. i. 190, 8. Gif mon elþeódigne ofsleá, se cyning áh twǽdne dǽl weres, þriddan dǽl sunu oþþe mǽgas. Gif hé mǽgleás sié, healf kyninge, healf se gesíð, L. In. 23; Th. i. 116, 15. Se ðe dearnenga bearn gestriéneþ and gehileþ, náh se his deáðes wer, ac his hláford and se cyning, 27; Th. i. 120, 3. Se forspeca sceal mágum on hand syllan, ðæt se slaga wille bétan wið mǽgðe, L. Edm. S. 7; Th. i. 250, 15. **I b.** those concerned in the payment of the *wer* are referred to in the following :—Gif fædrenmǽga mǽgleás mon gefeohte and mon ofsleá, and ðonne gif hé médrenmǽgas hæbbe, gielden ða ðæs weres þriddan dǽl, þriddan dǽl ða gegyldan, for þriddan dǽl hé fleó. Gif hé médrenmǽgas nǽge, gieldan ða gegildan healfne, for healfne hé fleó, L. Alf. pol. 27; Th. i. 78, 22. **I c.** of the form in which payment might be made see the following; see also wer-gild, **I b :**—En la were purra il rendre cheual pur .xx. sot., e tor pur .x. sot., e uer pur .v. sot., Wil. I, 9; Th. i. 470, 16. **II.** in cases other than death the whole or part of the injured person's *wer* could be claimed :—Gif se hund má (*more than three*) misdǽda gewyrce, and hé (*the owner*) hine hæbbe, bête be fullan were, L. Alf. pol. 23; Th. i. 78, 7. Gif man æt unlagum man bewǽpnige . . . and gif hine man gebinde, forgilde be healfan were, L. C. S. 61; Th. i. 408, 20. **III.** in case of certain crimes the *wer* of the criminal was exacted as a penalty; see also wer-gild, **II :**—Æt nánum bótwyrðum gylte ne forwyrce man máre ðonne his wer, L. Edg. ii. 2; Th. i. 266, 13. Gif mon sié wertyhtlan betogen . . . bíde mon . . . óþ ðæt se wer gegolden sié, L. In. 71; Th. i. 148, 4. Gif hwá æt þeófe médsceatt nime, and óðres ryht áfylle, beó hé his weres scyldig, L. Ath. i. 17; Th. i. 208, 16. Gif hwá flýman feormige, sý hé his weres scyldig, bútan hé hine ládian durre be ðæs flýman were, ðæt hé hine flýman nyste, 20; Th. i. 210, 12. Gielde hé hine (*the fugitive*) his ágenum were, L. In. 30; Th. i. 122, 1. Gif hwá ǽnigra godcundra gerihta forwyrne . . . and gif hé wigie and man gewundie, beó his weres scyldig, L. E. G. 6; Th. i. 170, 9. Gif hwá cristendóm wyrde oþþe hǽðendóm weorðige, wordes oþþe weorces, gylde swá wer swá wíte swá lahslitte, L. E. G. 2; Th. i. 168, 2 : L. Eth. v. 31; Th. i. 312, 10. Ða bæd Byrhferð ealdorman Æðelstán his wer for ðam têmbyrste, Chart. Th. 207, 3. **III a.** to whom, and by whom, the *wer* was paid is seen in the following :—Gif hé fúl wurðe bête ðam hláforde his were . . . Gif hé út hleáþe, . . . gilde se borh ðam hláforde his were (*if the lord had a share in the escape, the* wer *went to the king:* Fó se cyning tó ðam were) . . . Gif hé (*a lord's man*) út óðhleáþe, gylde se hláford ðæs mannes were ðam cyninge . . . Gif him (*the lord*) seó lád byrste, gilde ðam cynge his were, L. Eth. i. 1; Th. i. 280, 21-282, 14 : L. C. S. 30; Th. i. 394, 7-23. Beó hé his weres scyldig wið ðone cyning, and gif hé hit eft wyrde, gylde tuwa his were, L. C. S. 84; Th. i. 422, 10. Ðæt hé (manslaga binnon ciricwǽgum) his ágenne wer gesylle ðam cyninge and Criste, L. Eth. ix. 2; Th. i. 340, 11. Ic ágife ðinne wer ðam cynge, Chart. Th. 207, 11, 33 : 208, 28. **III b.** the payment of the *wer* is in some cases an alternative; see also wer-gild, **II a :**—Gif þeóf sié gefongen, swelte hé deáðe, oþþe his líf be his were man áliése, L. In. 12; Th. i. 110, 8. Sý hé (*a false accuser*) his tungan scyldig, búton hé hine mid his were forgilde, L. C. S. 16; Th. i. 384, 26 : L. Alf. pol. 32; Th. i. 82, 2. **IV.** the *wer* served as a standard by which other matters might be regulated; see also wer-gild, **III :**—Cyninges geneát, gif his wer bið twelf hund scitt., hé mót swerian for syxtig hída, L. In. 19; Th. i. 114, 10. Bútan hé hine ládian durre be ðæs flýman were, L. Ath. i. 20; Th. i. 210, 13. Be his ágnum were geládige hé hine, L. In. 30; Th. i. 120, 18. Hé hine be his were geswicne, 15; Th. i. 112, 3. Æt twyhyndum were mon sceal sellan tó monbóte .xxx. scitt.; æt .vi. hyndum, .lxxx. scitt.; æt twelfhyndum, .cxx. scitt., 70; Th. i. 146, 13. Gielden ealle án wíte, swá tó ðam were belimpe, L. Alf. pol. 31; Th. i. 80, 18. Gif hé (*a thief*) ða hand lésan wille, . . . gelde swá tó his were belimpe, 6; Th. i. 66, 6. Weaxe sió bót be ðam were, 11; Th. i. 70, 2 : L. In. 76; Th. i. 150, 15. v. þegen-wer, wer-gild.

wer (were?), es; m. n. (?) *A guard* (? cf. werian, waru), *a troop, band :*—Were *manipulo* (coelestis militiae manipulo, Ald. 50), Wrt. Voc. ii. 83, 2 : 56, 75. Iu ic wæs cempena láreów, and mid mycclum were ymbseald, nú ic eom ána forlǽten, Homl. Skt. ii. 30, 195.

wer, es; m. **I.** *a weir, a dam :*—Salomon sǽde ðætte suíðe deóp pól wǽre gewered on ðæs wísan monnes móde, and suíðe lytel unnyttes út fleówe. Ac se se ðe ðone wer bricð, and ðæt wæter út forlǽt, se bið fruma ðæs geflites *dicitur :* 'Aqua profunda verba ex ore viri,' Prov. 18, 4. Qui ergo dimittit aquam, caput est jurgiorum, Past. 38; Swt. 279, 16. **II.** often the *wer* is connected with fishing, and the word seems sometimes to be used of the water that is kept in by the dam :—*Captura* (captura *locus piscosus, ubi capiuntur pisces,* Migne), *detentio, captio hæft vel* wer, Wrt. Voc. ii. 128, 31. Ðis is ðæs hagan bóc on Winceastre and

ðæs healfan weres æt Brægentforda and ðæs æcersplottes ðe ðærtó lið (cf. dimidium cuiusdam piscarii uadum ad capturam piscium æt Bræge decurrentem, ad Uetus monasteriam pertinentem, cum unius iugeris sibi adiacentis portione, 134, 31-34), Cod. Dip. Kmbl. vi. 136, 11. Hé wundrude and ealle ða ðe mid him wǽron on ðam were (*in captura*) ðara fixa, Lk. Skt. 5, 9. Terram cum omnibus ad se pertinentibus hoc est, in siluis, in campis, in captura etiam piscium quae terrae illi adjacet, ubi sunt scilicet duo quod nostratim dicitur waeres, Cod. Dip. Kmbl. i. 64, 10. Æt ælcum were, ðe binnan ðam .xxx. hídan is, gebyreþ æfre se óðer fisc ðam landhláforde, iii. 450, 25. Andlang Úse tó Kekan were; of Kekan were andlang Úse tó Caluwan were, 170, 31. Mid were and mid mylene, 243, 10. Be eá tó Brihtwoldes were; of ðam were tó ðære díc, 424, 19. On Eádmundes wer; of Eádmundes were, vi. 31, 14, 34. [Ic gife þa landes and þas wateres and meres and fennes and weres, Chr. 656; Erl. 31, 5. Ic gife þa twa dæl of Witlesmere mid watres and mid wæres and feonnes, 963; Erl. 122, 15. He set in weres (ðam, *v. l.*) of watres wildernes *posuit desertum in stagnum aquae,* Ps. 106, 35. *M. H. Ger.* wer : *Ger.* wehr *a weir, dam.* Cf. *Icel.* vörr; *f. a fenced-in landing-place;* ver; *n. a fishing-place.*] v. cyt-, fisc-, ford- (Cod. Dip. Kmbl. iii. 437, 11), hæc-, mylen-wer.

wer-bǽre, es; n. *A weir where fish are caught :*—Se mylenstede and ðæt land ðæt ðe ðærtó hýrð . . . and ða werbǽra and seó mæd be norðan eá, and ða hammas, Cod. Dip. Kmbl. v. 383, 17. Tó Cranemere, and ðære gebyraþ tó six wærbǽre, iii. 344, 2.

wer-beám, es; m. *A strong man, warrior :*—Ða slóh mid hálige hand heofonríces weard werbeámas (*the Egyptians in the Red Sea*), wlance ðeóde, Cod. Th. 208, 20; Exod. 486. Cf. the epithets derived from words denoting trees which are applied to men in Icelandic poetry. v. Corpus Poeticum Boreale, ii. 476.

wer-bold, es; n. *Weir-building :*—Se gebúr sceal his riht dón . . . tó werbolde .xl. mǽra oððe án fóðer gyrda, Cod. Dip. Kmbl. iii. 450, 37.

wer-borh; gen. -borges; m. *A security for the payment of* wer. v. first two passages under wer, **I.**

werc glosses nanus, Wrt. Voc. ii. 60, 45 : 71, 36. [*Elsewhere* nanus *is rendered by* dweorh, *for which* werc *is perhaps wrongly written. Or (?)* werc *might be for* wearh. v. wearg.]

werc, wercan, wer-cweþan. v. weorc, wyrcan, wearg-cweþan.

wer-cyn[n], es; n. *Mankind :*—World wendeþ . . . wercyn (wen-, MS.) gewíteþ, Exon. Th. 354, 45; Reim. 61. Cf. wer-þeód.

werdan. v. wirdan.

werde glosses opes, Kent. Gl. 864. (*For* prêde ? cf. opes superbe ofermóde prêde, 249.)

were, wered *a troop,* wered *sweet,* were-mód. v. wer, weorod *a troop,* weorod *sweet,* wer-mód.

were-wulf, es; m. *A wer-wolf, a fiend :*—Ðæt se wódfreca werewulf tó swýðe ne slíte, ne tó fela ábíte of godcundre heorde, L. C. E. 26; Th. i. 374, 30 : L. I. P. 6; Th. ii. 310, 30 : Wulfst. 191, 16.

wer-fǽhþ, e; f. *Slaying, in pursuing the feud, under circumstances that call for the payment of* wer [cf. L. Alf. pol. 42 : Be fǽhðum . . . Gif hé (*a man's foe*) wille on hond gán and his wǽpenu sellan, and hwá ofer ðæt on him feohte, gielde swá wer swá wunde, swá hé gewyrce, Th. i. 90, 19] :—Be werfǽhðe tyhtlan. Se ðe bið werfǽhðe betogen, and hé onsacan wille ðæs sleges mid áðe, L. In. 54; Th. i. 136, 9-11. Ælc mon mót onsacan werfǽhðe gif hé mæg oþþe dear, 46; Th. i. 132, 1.

werg, wergan *to defend,* wergan *to curse,* wergend *a protector,* wergend *malignans.* v. wearg, werian, wirgan, weriend, wirgend.

wer-genga, an; m. *A stranger who seeks protection in the land to which he has come :*—Deóra gesíð, wildra wærgenga, Nabochodonossor the beasts' comrade, the stranger that sought shelter among wild beasts, Nebuchadnezzar, Cd. Th. 257, 25; Dan. 663. Gif eów Dryhten Crist lýfan wylle, ðæt gé his wergengan (*Guthlac, who had Christ's protection in the wilderness.* Cf. Ic mé frið wille æt Gode gegyrnan . . . mec Dryhtnes hond mundaþ . . . hér sceal mín wesan eorðlíc eþel, 117, 23-30; Gú. 228-232. Nú ic ðis. lond gestág . . . mé friðe healdeþ . . . se ðe mægna gehwæs wealdeþ, 120, 28-121, 3; Gú. 278-283) in ðone ládan lég lǽdan móste, Exon. Th. 137, 29; Gú. 536 : 144, 28; Gú. 685. [*The Latinized* wargangus *occurs in the Lombard laws :* Omnes wargangi, qui de exteris finibus in regni nostri finibus advenerint. *And* wargengus *among the Franks :* Si quis wargengum occiderit. v. Grff. iv. 103 : Grmm. R. A. 396. Cf. *Icel.* verð-gangr (ver-) *going about asking for food* (verðr).] v. waru, werian.

wergian *to curse,* wérigian *to grow weary.* v. wirgan, wérigian.

wer-, were-gild, es; n. [*The word seems interchangeable with* wer (q.v.), *which in the later laws is the more frequent form.*] *The price set upon a man according to his degree :*—Twelfhyndes mannes wergyld bið six ceorla wergyld, L. O. 13; Th. i. 182, 21. Ceorles wergild (were-gild, l. 20) is .cc. and .lxvi. þrimsa, ðæt bið .ii. hund scitt. be Myrcna lage, L. Wg. 6; Th. i. 186, 11. Norðleóda cynges gild .xxx. þúsend þrymsa, fíftêne þúsend þrymsa bið ðæs wergildes (wæres, l. 16), 1; Th. i. 186, 2. (The wergilds for other ranks are given in the sections of this article.)

Ceorles wergild is on Myrcna lage .cc. sciłł. Đegnes wergild is syx swá mycel, ðæt bið .xii. hund sciłł. Đonne bið cynges ánfeald wergild .vi. þegna wer be Myrcna lage, ðæt is .xxx. þúsend sceatta, and ðæt bið ealles .cxx. punda. Swá mycel is ðæs wergildes on folces folcrihte be Myrcna lage, L. M. L.; Th. 1: 190, 2–7. Cyninges horswealh, se ðe him mæge geæærendian, ðæs wergield bið .cc. sciłł., L. In. 33; Th. i. 122, 14. **I.** when a person was wrongfully slain the *wergild* of the slain man could be claimed from the slayer. Cf. wer. **I** :—Gif man leud ofsleá an þeófðe, licge bútan wyrgelde, L. Wih. 25; Th. i. 42, 13. Se .vii. nihta móna is gód on tó fixianne, and æðeles monnes wergild an tó manianne, Lchdm. iii. 178, 14. **I a.** for those who were concerned in the receiving of the *wergild* see wer, **I a**, and the following :—Gif man his mæn freólse gefe, ... freólsgefa áge his erfe ænde wergeld, L. Wih. 8; Th. i. 38, 16. (See also the cases quoted under **IV.**) **I b.** as to the form which the payment might take see wer, **I c**, and the following :—Mót hé gesellan monnan and byrnan and sword on ðæt wergild, L. In. 54; Th. i. 138, 1. (Cf. for similar payment : Mid .lx. sciłł. gebéte ... and ðæt sié on cwicæhtum, and mon næenigne mon on ðæt ne selle, L. Alf. pol. 18; Th. i. 72, 12.) Tó ðam ðæt hió hyre bróðra wergild gecure on swylcum þingum swylce hyre and hire nýhstan freóndum sélost lícode. And hió ðá swá dyde ðæt hió ðæt wergeld geceás on ðam íglande ðe Teneð is nemned, ðæt is hundeahtatig hída landes ðe hió ðæt æt ðæm cyninge onfeóng, Lchdm. iii. 426, 16–21. **II.** in case of certain crimes the criminal's *wergild* was exacted as a penalty, v. wer, **III** :—Gif frí man wið fríes mannes wíf geligeþ, his wergelde ábicge, L. Wih. 31; Th. i. 10, 6. Forgielde hé hine selfa be his wergilde, L. Alf. pol. 7; Th. i. 66, 12. **II a.** the payment of the *wergild* is in some cases au alternative, v. wer, **III b** :—Sí þreóra án for his feore ... wergild, éce þeówet, hengenwítnung, L. Eth. vii. 16; Th. i. 332, 18. Þolige hé lifes oþþe wæregildes (were-, *v.l.*), L. C. S. 62; Th. i. 408, 23. Wealde se cyning þreóra ænes; oþþe hine man cwelle, oþþe ofer sæ selle, oþþe hine his wergelde áliése, L. Wih. 26; Th. i. 42, 17. Hé hine be his wergilde áliése, oþþe be his were geswicne, L. In. 15; Th. i. 112, 2. Hé bið feorhscyldig, nimþe him se cyning álýfan wille ðæt man wergylde álýsan móte, L. Eth. vii. 15; Th. i. 332, 15. **II b.** of the uses to which *wergild* paid as a fine in religious matters (cf. L. E. G. 2; Th. i. 168, 1–3) could be applied see the following :—Gif for godbótan feohbót áríseþ, ðæt gebyreþ rihtlíce ... tó godcundan neódan (*these are enumerated in the section*); hwílum be wíte, hwílum be wergilde (*at times the feohbót is in the form of* wergild), L. Eth. vi. 51; Th. i. 328, 4–10. **III.** the *wergild* served as a standard by which other matters might be regulated, v. wer, **IV** :—Se ðe on ðære fóre wære ðær mon monnan ofslóge, getriéwe hine ðæs sleges, and ða fóre gebéte be ðæs ofslegenan wergilde. Gif his wergield sié .cc. sciłł., gebéte mid .l. sciłł., and ða ilcan riht dó man be ðám deórborenum, L. In. 34; Th. i. 124, 1. Twelfhyndes mannes áð forstent .vi. ceorla áð, for ðam ... his wergyld bið six ceorla wergyld, L. O. 13; Th. i. 182, 21. Gif hé hine selfne triówan wille, dó ðæt be cyninges wergelde, L. Alf. pol. 4; Th. i. 64, 2. Gif hé ládian wille, dó ðæt be ðæs cynges wergilde, oþþe mid þryfealdan ordále, L. Eth. v. 30; Th. i. 312, 7. Gylde ðam cyninge be his wereigilde (wer-, *v.l.*), L. C. S. 67; Th. i. 410, 17. In the following case the *wergild* seems to have suggested the amount of a bequest to the church :—Hió (*the testator's wife*) gebrenge æt Sancte Petre mín twá wergild, gif ðet Godes wille seó ðæt heó ðæt færeld áge, Chart. Th. 481, 10. **IV.** instances of the payment of *wergild* are the following. The two young princes Æþelred and Æþelbriht were slain by Thunor, and to their sister eighty hides of land was given as *wergild*, Lchdm. iii. 424–6. In the war between Ecgfriþ and Æþelred the former's brother was slain. Theodore brought about peace between them 'ðæt næniges mannes feorh tó lore wearþ, ne máre blódgyte wæs for ðam ofslægenan cyninges bréðer, ac hé mid feó wiþ hine gebingode, ðæt heora sib wæs,' Bd. 4, 21; S. 590, 24. In 687 Mul, Ceadwalla's brother, was burnt in Kent: in 694 'Cantware gepingodon wiþ Íne, and him gesaldon xxx m̄., for ðon ðe híe ǽr Mul forbærndon, Chr. 694; Erl. 42, 15. [O. Frs. wer-geld, -ield : O. H. Ger. wer-, weri-gelt *fiscus*, *pretium*. Cf. Icel. mann-gjöld ; *pl.*] Cf. leód, leód-gild ; *and see* Kemble's Saxons in England, vol. i. c. x, Grmm. R. A. 650.

wergild-þeóf, es ; *m. A thief whose* wergild *was paid as a punishment for his crime* [cf. Gif þeóf sié gefongen, swelte hé deáðe, oþþe his líf be his were man áliése, L. In. 12; Th. i. 110, 8] :—Be wergeldþeófes forefonge. Gif mon wergildþeóf gefehð, and hé losige ðý dæge ðam monnum ðe hine gefóð, þeáh hine mon gefó ymb niht, náh him mon máre æt ðonne ful wíte, L. In. 72; Th. i. 148, 5–8. At omni tributo publicalium rerum et ab expeditionalibus causis a cunctis operibus uel regis uel principis sit pars in perpetuam libera, itã ut nec pontem nec arcem facere debeant, nec de furtis aliquam poenam soluere, nec etiam fures illos quos Saxonice uuergeldtheouos alicui foras reddant; sed si capiantur, in illorum dominio sunt habendi, Cod. Dip. Kmbl. i. 172, 7 : 14. ¶ the word is also used to denote the right to receive the wergilds paid in cases of theft; cf. the preceding passage :—Huic libertati concedo additamentum, in qua, ut ab omnibus apertius et plenius intelligatur, nomina consuetudi-

num Anglice praecepi ponere: scilicet, mundbryche, ... flýmena fyrmðe, wergeldþeóf, úðleáp (cf. wer, III a), ... fyrdwíte ..., aliasque omnes leges et consuetudines quae ad me pertinent, Chart. Th. 411, 26–34.

wergness, wergulu, wergum, Cd. Th. 267, 22 ; Sat. 42, wergþu, werguṇg. v. weargness, wirgness, weargol, wearh ; *m.* (?), wirgþu, wirgung.

wer-hád, es ; *m. The male sex* :—Werhád oðde wífhád *sexus*, Ælfc. Gr. 11 ; Zup. 78, 16 : Wrt. Voc. i. 50, 7 : 70, 19. Werhádes man *mas vel masculus*, 70, 17. Ælc werhádes man *omne masculinum* ... se werhádes man *masculus*, Gen. 17, 12, 24. Ealle werhádes men *omnes viri*, 17, 27. Werhádes and wífhádes hé gesceóp hig *masculum et feminam creavit eos*, 1, 27. Werhádes men ongunnon ðone dreám, and wífhádes men him sungon ongeán, Homl. Th. ii. 548, 11. Ðæt hí heora clǽnnesse healdan be heora háde, swá werhádes swá wífhádes, swá hwæðer swá hit sý, L. Edm. E. 1 ; Th. i. 244, 11.

werh-bræde, werhte, weria. v. wearg-bræde, wærcan, wearg.

werian, wergan ; *p.* ede. **I.** *to hinder, check, restrain* :—Stán sépte sacerdas sweotolum tácnum, witig werede, and worde cwæð, Andr. Kmbl. 1485; An. 744. Egesan stódon, weredon wælnet (*deadly toils hampered* (?)), Cd. Th. 190, 20 ; Exod. 202. Ic wylle ðæt ælc man hæbbe symle ða men gearowe on his lande, ðe lǽden ða men ðe heora ágen sécan willen, and hý for nánum médsceattum ne werian, L. Ed. 7 ; Th. i. 162, 25. **I a.** *to dam* water. **v.** wering :—Sume weriaþ on gewitlocan wísdómes stream, welerum gehæftaþ, ðæt hé on unnyt út ne tóflóweþ, Past. 65 ; Swt. 469, 2. **II.** *to keep off, drive away* :— Wereth *abiget*, Wrt. Voc. ii. 98, 18. **II a.** *to keep off* something from a person (*dat.*), *to keep* a person (*dat.*) *from* something (*acc.*). **v.** warian, **IV** :—Ic mínum fótum fǽcne síðas werede *ab omni mala via prohibui pedes meos*, Ps. Th. 118, 101. Ægðer óðrum trymede heofonríces hyht, helle wítu wordum werede (cf. gihét im heðanríki endi helleógethwing werida mid wordun, Hél. 2082), Andr. Kmbl. 2107; An. 1055. **III.** *to defend, resist attack upon* :—God geseah his (*St. Paul's*) geðanc, ðæt hé éhte geleáffulra manna ðurh ware ðære ealdan ǽ, and hine gespræc :—'Saule ... ic eom seó sóðfæstnys ðe ðú werast,' Homl. Th. i. 390, 8. Hé unheánlíce hine werede, Chr. 755; Erl. 48, 33. His ríce hé heardlíce werode ða hwíle ðe his tíma wæs, 1016; Erl. 155, 6. Hú his seó mycle hand on gewindæge werede and ferede *qua die manus ejus liberavit eos de manu tribulantis*, Ps. Th. 77, 42. Hé under segne sinc ealgode, wælreáf werede, Beo. Th. 2414; B. 1205. Wé on orlege hafelan weredon, 2658; B. 1327. Hí céne hí weredon, Byrht. Th. 140, 5 ; By. 283. Ðá hé (*Peter*) his Drihten werian wolde, L. Ælfc. P. 51 ; Th. ii. 386, 22. Gif hine hé hine werian wille, L. Ath. i. 1 ; Th. i. 198, 20 : v. 12, 1 ; Th. i. 240, 29 : 3 ; Th. i. 242, 10. Utan líf and land ealle werian, L. Eth. v. 35 ; Th. i. 312, 22 : Chr. 1010; Erl. 144, 8. Burh werian, Blickl. Homl. 79, 16. Wígsteal wergan, Exon. Th. 315, 31; Mód. 39. Ealle ða ðe hié wergan noldon, Chr. 921; Erl. 107, 4. **III a.** *to defend* against, (1) with dat. :—Đonne hand wereþ feorhhord feóndum, Wald. 99; Vald. 2, 21. Hí woldon burh wráðum werian, Cd. Th. 119, 7 ; Gen. 1976. Wergan eþelstól Ætlan leódum, Exon. Th. 325, 34 ; Vid. 121. (2) with prep. *wið* :—Ða hí fæstlíce wið ða fýnd weredon, Byrht. Th. 134, 11 ; By. 82. Wit unc wið hronfixas werian þóhton, Beo. Th. 1086 ; B. 541. Breóstnet wera wið feónd folmium weredgan, Cd. Th. 192, 26 ; Exod. 237. **III b.** *to defend* at law :—Se ðe on gemóte mid wiðertihtlan hine sylfne oþþe his man werige, L. C. S. 27 ; Th. i. 392, 6. Se Englisca hine werige mid orneste oþþe mid írene ... Gif se Englisca nele hine werian mid orneste oþþe mid gewitnesse, hé ládige hine mid írene, L. W. ii. 2 ; Th. i. 489, 13–19. Werige hine se Frǽncisca mid unforedan áþe, 3 ; Th. i. 489, 24. Se ðe can mid leásungan wæwerdlíce werian, and mid unsóðe sóð oferswíðan, Wulfst. 169, 1. **III c.** in the phrase *werian land* the word refers to the performance of services that might be demanded from the holders of land :—Werige (*the Latin version has* adquietet) se cotsetla his hláfordes inland, gif him man beóde, æt sǽwearde and æt cyniges deórhege and æt swilcan ðingan swilc his mæd sý, L. R. S. 3 ; Th. i. 432, 27. v. Kemble's Saxons in England, i. 323. ¶ the phrase commonly occurs where an assessment is made for a smaller numoer of hides than those actually held, and is retained in Domesday Book in the Latin *defendere pro* (a certain number of hides) :—Hé geúðe ðæt man ðæt land on eallum þingon for áne híde werode, swá swá his yldran hit ǽr gesetton and gefreódon, wære ðær máre landes, wære ðær læsse ... Ealles ðæs landes is án hund hída : ac ða gódan cynegas ... ælc æfter óðran, ðæt ylce land swá gefreódon Gode tó lofe and his þeówan tó bryce intó fósterlande, ðæt hit man ǽfre on ende for áne híde werian sceolde, Cod. Dip. Kmbl. iii. 112, 5–24. Nú wille ic ðæt hit man on eallum þingon for áne híde werige ... sý ðær máre landes, sý ðær lesse (*there were* 578 hides), 203, 16. Hé werige for twá hída, iv. 262, 15. Ic wylle ðæt Æðelnóð arcebisceop werige his landáre mid, ealswá hé dyde ǽr Ælgelríc wære gerefa, vi. 187, 19. Ðæt mon ælles ðises freólses áre ǽfre for áne híde werian scolde ; for ðam ðe Godes ár ǽfre freogre beón sceal ðonne ænig woruldár, v. 113, 33. **IV.** *to protect, guard from wrong* or *injury*, (1) of persons :—God, se ðæs fyrd wereþ, Cd. Th. 195, 10; Exod. 274. Gif man ofsleá óþerne for neóde ðær hé his hláfordes ceáp werige *si quis alium occiderit ex necessitate,*

ubi rem domini sui tuebatur, L. Ecg. C. 24; Th. ii. 150, 5. Ðæt hé
(*a king*) Godes cyrcan weorþige and werige, L. I. P. 2; Th. ii. 304, 26.
Ðæt hí Godes þeówas symle werian and weorðian, L. Eth. vi. 45; Th.
i. 326, 23. Hý sculan cyrican wyrðian and werian, L. I. P. 11; Th. ii.
318, 25: 25; Th. ii. 338, 30. Manig strec man wyle, gif hé mæg and
mót, werian his man swá hwæðer him þincð ðæt hé hine eáð áwerian
mæge, L. C. S. 20; Th. i. 388, 2. (1 a) with dat. :—Ðú mé weredest
wráþum feóndum, ðe mé woldon yrre on ácýðan, Ps. Th. 137, 7. (2)
of things :—Beaduscrúda betst, ðæt míne breóst wereþ, Beo. Th. 911;
B. 453. Se hwíta helm hafelan werede, 2901; B. 1448. **V.** *to hold,*
occupy. v. warian, **III a** :—Ða ðe onhǽle eardas weredon, Exon. Th.
123, 14; Gú. 322. [Ich wolle ðat Gyso bisschop werie (*possideat*) now
hiss lond also his forgenge aforen hym er dude, Cod. Dip. Kmbl. iv. 195,
14.] [Ic eou wulle werien wið elcne herm, O. E. Homl. i. 13, 20.
I compe hine werien, Laym. 8288. Weorien heom mid wepnen, 21289.
Þu mihht werenn þe fra þeȝȝm, Orm. 1406. Scheld to werien ham mide,
A. R. 52, 5. Were þe agean me, 400, 7. Foyne if him lust on foote
himself to were, Chauc. Kn. T. 1692. *Goth.* warjan *prohibere : O. Sax.*
werian : *O. Frs.* wera : *O. H. Ger.* werien *prohibere, cohibere, inhibere,*
resistere, defendere, vetare, abnuere, abigere : Icel. verja *to defend.*]
v. á-, be-, ge-werian; un-wered; warian.

werian ; *p.* ede, ode. **I.** *to clothe* with a garment :—Líc ðæt hé ǽr
werede mid wǽdum, Exon. Th. 374, 14; Seel. 126. Hié heora líchoman
leáfum beþeahton, weredon mid ðý wealde, Cd. Th. 52, 19; Gen. 846.
Hwæt sindon gé searohæbbendra byrnum werede, Beo. Th. 481; B. 238:
5052; B. 2529. Hí lisgaþ á leóhte werede, Exon. Th. 237, 26; Ph.
596. **II.** *to wear* a garment, *wear* or *bear* a weapon, etc. :—Ðæt
hálie reáf, ðæt Aaron wereþ *vestem sanctam, qua utetur Aaron*, Ex. 29,
29. Se woruldkempa weraþ woruldlíce wǽpna, Basil adm. 2; Norm.
34, 31. Ðe má ðe se wer weraþ wimmanna gyrlan, L. Ælfc. C. 35;
Th. ii. 358, 10. Hit næs þeáw mid him ðæt ænig óþer purpuran werede
búton cyningum, Ors. 4, 4; Swt. 164, 35: 6, 31; Swt. 284, 23. Heó
wyllen weorode, Homl. Skt. i. 20, 44. Ðæt reáf, ðæt se Hǽlend werede,
Homl. Ass. 189, 249. Seó cwén werode cynehelm on heáfode, 93, 38.
Ða purpuran álecgan, ða hié weredon, Ors. 6, 30; Swt. 280, 21. Ðam
folce wæs gewunelíc, ðæt hí weredon býman on ælcum gefeohte, Jud. 7,
16. Deóplíc dǽdbót bið ðæt lǽwede man . . . wyllen werige, L. Pen. 10;
Th. ii. 280, 20. Werige gehwá swá his háde tó gebyrige, ðæt se preóst
ne werige munuccscrúd, ne lǽwedra manna, L. Ælfc. C. 35; Th. ii. 358,
7-9. Ne preóst wǽpna ne werige, 30; Th. ii. 354, 3. Ne preóst
wǽpnu werian mid rihte . . . Nú secgaþ sume preóstas ðæt hí for neóde
wǽpn móton werian, L. Ælfc. P. 50, 51; Th. ii. 386, 13-21. Gyldenne
hring werian, Ors. 4, 9; Swt. 190, 15. Gyrlan werian, Homl. Ass. 115,
427. Wǽpen wegan (werian, *v. l.*) *arma ferre*, Bd. 2, 13; S. 517, 7.
Reáf tó werigenne *vestimentum ad induendum*, Gen. 28, 20. Hrægl tó
werianne, L. Alf. 36; Th. i. 52, 25. **II a.** in reference to the hair,
to wear a beard, etc. :—Leófgár . . . Haroldes eorles mæssepreóst werede
his kenepas on his preóstháde oð ðæt hé wæs biscop. Se forlét . . . his
gástlícan wǽpna, and fénig tó his spere and tó his sweorde æfter his biscup-
háde, Chr. 1056; Erl. 190, 24. [The verb is weak in Chaucer and
Wicklif. *Goth.* wasjan *to clothe : O. H. Ger.* werien *vestire : Icel.* verja
to clothe.] v. ge-werian; for-, scír-, swegel-wered (-od).

werian ; *p.* ode *To remain, continue, live* :—Ic cýðe eów, ðæt ic wylle
ðæt Giso bisceop weryge on his lande æt Chyw alswó hys foregenga
ætforen him ǽr dyde *sciatis me uelle quod Giso episcopus possideat terram*
suam apud Chyw sicut fecerunt praedecessores sui, Cod. Dip. Kmbl. iv.
196, 24. (Cf. werian *to defend*, **V.**) [*O. L. Ger.* werôn *esse, subsistere :*
O. H. Ger. werên *manere, remanere, subsistere, durare : Ger.* währen.]
v. warian *to remain* ; wesan.

weriend, werigend, es; *m. A defender, protector* :—Ic eom ðín wergend
ego protector tuus sum, Gen. 15, 1. Utan lufian úre cyrican, for ðam
heó bið úre friðiend and werigend, Wulfst. 239, 7. Hig woldon sumne
weriend habban, ðe hí geheólde wið ðæt hǽðene folc, Ælfc. T. Grn. 6, 43.
v. be-werigend.

werig. v. wearg.

wérig ; *adj.* **I.** physical, *weary, tired, exhausted, fatigued* :—Ðá
hé wæs wérig (uoerig, Lind. : woerig, Rush.) gegán *fatigatus ex itinere*,
Jn. Skt. 4, 6 : Bd. 3, 9; S. 534, 10. Sesirra arn óð ðæt hé wérig becom
tó ánum wífmen æt néhstan, Jud. 4, 17 : Cd. Th. 88, 9; Gen. 1462.
Wérig sceal se wiþ winde rôweþ, Exon. Th. 345, 12; Gn. Ex. 187 :
307, 26; Seef. 29. Ne forlǽt ðú ðæs blódes tó fela on ǽnne síþ, ðý les
se seóca man tó wérig (*exhausted*) weorðe oððe swylte, Lchdm. ii. 208,
19. Wǽgdeóra gehwylc wérig swelteþ, Exon. Th. 61, 22; Cri. 988.
Móyses willa ne áteorode, ac se wériga líchama, Homl. Skt. i. 13, 40.
Móises handa wǽron wérige (*graves*), Ex. 17, 12. Féðan sǽton, reste
gefégon wérige æfter wǽðe, Andr. Kmbl. 1185; An. 593. Wérge, Exon.
Th. 115, 2; Gú. 183. Limseóce, wérige, wanhále, Andr. Kmbl. 1159;
An. 580. Wérge, Exon. Th. 92, 13; Cri. 1508. Ða wéregan neát ðe
man drífeþ and þirsceþ, Elen. Kmbl. 714; El. 357. **I a.** where the
source of weariness is given, (1) with gen., *weary of* or from doing something :—Wérig ðæs weorces, Exon. Th. 436, 20; Rä. 55, 10. Síþes

wérig, Beo. Th. 1162; B. 579. Síðes wérgum, feorrancundum, 3593;
B. 1794. (2) with dat. inst., *exhau-ted* by suffering :—Íserne wu.d,
beadoweorca sǽd, ecgum wérig, Exon. Th. 388, 5; Rä. 6, 3. Wundum
wérig, Andr. Kmbl. 2557; An. 1280. Wítum wérig, Cd. Th. 274, 30;
Sat. 162 : 291, 9; Sat. 428. Wítum wérige, 285, 25; Sat. 343.
Wígend cruncon wundum wérige, Byrht. Th. 140, 44; By. 303. Wundum
wérge, Beo. Th. 5866; B. 2937. **II.** *weary* at heart, *sad, grieved* :—
Ne mæg wérig mód wyrde wiðstondan, ne se hreó hyge helpe gefremman
a soul that is sad may not stand against fate, nor the mind that mourns
minister help, Exon. Th. 287, 16; Wand. 15. On wérigum sefan, 74,
18; Cri. 1208. Sendau wérigne sefan, 289, 33; Wand. 57. Hé háfaþ
wilde mód, wérige heortan, Salm. Kmbl. 756; Sal. 377. Woldan wérigu
wíf wôpe bimǽnan æþelinges deáð, Exon. Th. 459, 23; Hö. 4. Wérigra
wraþu, 183, 34; Gú. 1337. Eálá ðú ðe eart sió héhste frófer eallra
wérigra môda O! *summum lassorum solamen animorum*, Bt. 22, 1; Fox
76, 9. **III.** *that expresses sadness, weary, grievous* :—Hé wépende
wéregum teárum his sigedryhten sárgan reorde grétte, Andr. Kmbl. 118;
An. 59. Beornas wépaþ wérgum stefnum, heáne, hygegeómre, Exon.
Th. 61, 32; Cri. 993. **IV.** *weary, impatient of the continuance* of
anything painful :—Sunu mín, ne ágiémeleása ðú Godes suingan, ne ðú
ne beó wérig for his ðreáunge (*neither be weary of his correction* ; neque
fatigeris, cum ab eo argueris, Prov. 3, 11), Past. 36; Swt. 253, 3.
[*O. Sax.* síð-wôrig *weary with travel : O. H. Ger.* wôrag *crapulatus.*]
v. ádl-, deáþ-, drinc-, ferhþ-, fýl-, gúþ-, heaðu-, hrá-, lid-, lim-, medu-,
mere-, rád-, sǽ-, slǽp-, symbel-, un-wérig.

werig(e)an *to curse*, werigend. v. wirgan, weriend.

wérig-ferhþ ; *adj. Weary-hearted, disconsolate, depressed* :—Ongan
geómormód tô Gode cleopian . . . weóp wérigferð, Andr. Kmbl. 2799;
An. 1402. Hí hreówigmóde wurpon hyra wǽpen of dúne, gewitan him
wérigferhþe on fleám sceacan, Jud. Thw. 25, 24; Jud. 291. Wérigferðe
. . . reónigmóde, Exon. Th. 361, 14; Wal. 19.

wérigian ; *p.* ode *To grow weary, get exhausted* :—Ðonne ðæt deófol
swíðe wérgaþ, hit séceþ scyldiges mannes nýten, oððe unclǽne treów,
Salm. Kmbl. p. 148, 8. Hingrian, ðyrstan, hátian, célan, wérigean
(wǽrigean, Bd. M. 78, 22), eall ðæt is of untrumnysse ðæs gecyndes
esurire, sitire, aestuare, algere, lassescere, ex infirmitate naturae est, Bd.
1, 27; S. 494, 15. Ðá ongan his hors semninga wérian (wérgian, Bd.
M. 178, 19) and gestandan *equus subito lassescere et consistere coepit*, 3, 9;
S. 533, 31. Hweriende *aegrotantibus, infirmantibus*, Hpt. Gl. 478, 37.

wérig-líc, -líce. v. wearg-líc, -líce.

wérig-mód ; *adj. Weary in spirit* :—Ic wérigmód wann and cleopode
laboravi clamans, Ps. Th. 68, 3 : Andr. Kmbl. 2732; An. 1368 : Beo.
Th. 1692; B. 844 : 3090; B. 1543. Mín freónd siteþ under sánhlíðe,
. . . wine wérigmód . . dreógeþ se mín wine micle módceare, Exon. Th.
444, 18; Kl. 49. Gewíteþ wérigmód, wintrum gebysgad, 227, 24; Ph.
428. Gewítaþ áwyrgde, wérigmóde, 117, 19; Gú. 226.

wériness, e; *f. Weariness, lassitude* :—Móyses wérignyss (v. Ex. 17,
12), Homl. Skt. i. 13, 44. Gehwǽr is on úrum lífe áteorung and wérig-
nys, Homl. Th. i. 490, 7. Ðæt hors ðý gewunelícan þeáwe horsa æfter
wérinysse (*post lassitudinem*) ongan walwian, Bd. 3, 9; S. 533, 39.
Hwæt elles is tó secanne wiþ wérignysse nymþe reste, 1, 27; S. 494, 17.

wering, e; *f. A dam* :—Ðæt wæter, ðonne hit bið gepynd, hit fundaþ
wið ðæs ðe hit ǽr from com . . . Ac gif sió pynding wierð onpennad,
oððe sió wering wirð tóbrocen, ðonne tôflêwð hit eall, Past. 38; Swt.
277, 22. v. werian, **I a** ; be-werung.

wer-lád, e ; *f.* A 'lád' (q.v.) *in which the number of those who sup-*
ported the accused by their oaths is determined by the 'wer' *of the accused.*
[See passages under wer, **IV**, wer-gild, **III**, and L. H. I. 64, 4; Th. i.
566, 18 : Si quis de homicidio accusetur, et idem se purgare velit, secundum natale suum perneget, quod est werelada.] :—Búton hé geládige
hine mid werláde, L. C. S. 39; Th. i. 400, 1. ¶ The equivalent Latin
forms werelada negare or pernegare occur several times in L. H. I.; see
12, 3; Th. i. 523, 7 : 66, 1; Th. i. 569, 4 : 74, 1; Th. i. 578, 22 : 92,
14; Th. i. 604, 14. Other instances of the Latinized form werelada
are :—Wereláda fiat, 85, 4; Th. i. 592, 17 : 88, 9; Th. i. 595, 35.
Triplicem wereladam habere, 64, 1; Th. i. 566, 3.

wer-leás ; *adj. Without a husband.* v. wer, **IV** :—Sitte ǽlc wydewe
.xii. mónað werleás ; ceóse syþþan ðæt heó sylf wille, L. Eth. v. 21; Th.
i. 310, 3 : vi. 26; Th. i. 322, 3 : L. C. S. 74; Th. i. 416, 6 : Wulfst.
271, 20.

wer-líc ; *adj.* **I.** marking sex, *male.* Cf. wer-hád :—Wer *uir*,
werlíc *virilis*, Ælfc. Gr. 5; Zup. 17, 17. Of werlícum folman *sine viri*
vola, Hpt. Gl. 442, 72. Hié ǽghwelcum cnihtcilde ymbsnidon ðæt werlíce lim, Shrn. 47, 20. Ða werlícan *virilia*, Wrt. Voc. i. 283, 54. **I a.**
marking gender, *masculine* :—Æfter gecynde syndon twá cyn on namum,
masculinum and *femininum*, ðæt is wer-cyn and wíf-cyn. Werlíc cyn byð *hic*
uir ðes wer. Gemǽne cyn, ðæt is ǽgðer ge werlíc ge wíflíc . . . *Neutrum*
is náðor cynn, ne werlíces ne wíflíces, Ælfc. Gr. 6, 1-3; Zup. 18,
5-15. **II.** marking age, *that has reached man's estate.* v. wer,
II :—Ðá hé wæs in werlícre giúguðe *in his early manhood*, Shrn. 119,
20. **III.** marking married condition, *of a husband, marital* :—

Werlícere wrǽnnysse *maritalis lasciviae*, Hpt. Gl. 434, 61. Tó werlícum gemánan *ad maritale consortium*, 502, 23 : 442, 74. Werlícre beclyppincge *maritali complexu*, 442, 75.

werlíce ; *adv.* I. *after the manner of a male :*—Se ðe ðis werlíce déð *qui hoc virili modo fecerat*, L. Ecg. P. iv. 68, 6 ; Th. ii. 228, 18. II. *like a man, manfully :*—Wer *uir*, werlíce *uiriliter*, Ælfc. Gr. 232, 16. Werlíce dó ðú *uiriliter age*, Ps. Spl. 26, 20 : Ps. Surt. 26, 14. Ðǽr wǽron getealde æt ðam gereorde fíf ðúsend wera ; for ðon ðe ða menn, ðe tó ðam gástlícan gereorde belimpaþ, sceolon beón werlíce geworhte, swá swá se apostol cwæð : ' Beóð wacole, and standaþ on geleáfan, and onginnaþ werlíce (*quit you like men ;* viriliter agite, 1 Cor. 16, 13).' Ðeáh gif wífmann bið werlíce geworht, and strang tó Godes willan, heó bið ðonne getealde tó ðam werum ðe æt Godes mýsan sittaþ, Homl. Th. i. 188, 28–34 : 360, 13 : 542, 25. [*Goth.* wairaleikô taujaiþ ἀνδρίζεσθε, 1 Cor. 16, 13.] v. eal-werlíce.

wér-loga. v. wǽr-loga.

wer-mǽgþ, e ; *f.* A *tribe* or *family of men :*—Of Cames cneórisse wóc wermægða fela, Cd. Th. 98, 30 ; Gen. 1638 : 101, 29 ; Gen. 1689 Cf. wer-þeód.

wer-met, es ; *n.* A *man's measure, stature of a man :*—Tó wermete *ad staturam*, Wrt. Voc. ii. 72, 23 : 8, 70. (In both cases *stauram* is printed ; but the former is a gloss on Mt. 6, 27. v. Wülck. Gl. 479, 23.)

wermód, es ; *m. Wormwood :*—Wermód (uuermód, uermódae) *absinthium*, Txts. 37, 35 : Wrt. Voc. ii. 4, 11 : i. 79, 29. Weremód, 67, 23. Ic eom wráþre ðonne wermód sý, Exon. Th. 425, 23 ; Rä. 41, 60. Wermód. Ðeós wyrt ðe man *absinthium* and óþrum naman wermód nemneþ, Lchdm. i. 216, 17. Se fúla wermód, ii. 312, 18. Dríges wermódes blóstman, 250, 3. Gif hit sié sumor, dó wermódes sǽdes dust tó . . . gif hit sié winter, ne þearft þú ðone wermód tó dón, 180, 27. Grénne wermód oððe dríge, 206, 24 : 296, 13. Wring on wermód wearmne, 310, 10. Nim wermód nioþoweardne, 326, 10. Wærmód, i. 206, 10. Wyrmód, iii. 50, 17, 20. Súþerne wermód (*artemisia abrotanon*), ii. 34, 27 : 178, 26. Ðone súþernan wermód, ðæt is prutene, and óþerne wermód, 236, 20. Twégra cynna wermód, i. 374, 6. Wyrmód, iii. 4, 9. Wermód drincan sace hefige getácnaþ *to drink wormwood in a dream betokens grievous strife*, 198, 24. [Wermod *absinthium*, Wülck. Gl. 554, 11 (13th cent.): 560, 12 (15th cent.). Wormode, 645, 35 (15th cent.). Wormwod, 711, 24 (15th cent.). *Wick.* wermod : *Pall.* wermode : *O. H. Ger.* wermuota (weri-) *absinthium : O. L. Ger.* wermuode.]

werna. v. wrænna.

wer-nægel, es ; *m. A* warnel or *wornil.* [Bailey's Dictionary gives ' *warnel worms*, worms on the backs of cattle within the skin' ; and in Johnson's Dictionary, ed. Latham, is quoted the following : ' In the backs of cows in the summer are maggots generated, which in Essex we call *wornils*, being first only a small knot in the skin.' Halliwell explains *wornil* as ' the larva of the gadfly growing under the skin of the back of cattle.'] :—Án æþelboren wíf weard micclum geswenct mid langsumre untrumnysse, and hire ne mihte nán lǽcecræft fremian. Ðá lǽrde hí sum man ðæt heó náme ǽnne wernægel of sumes oxan hricge, and becnytte tó ánum hringe mid hire snóde, and mid ðam hí tó nacedum líce begyrde, Homl. Th. ii. 28, 17.

wernan, werod *a band,* werod *sweet,* weród *catasta,* werold, werp, werrest, wersa, wer-scipe *prudence,* werta. v. wirnan, weorod *a band,* weorod *sweet,* wearg-ród, weorold, wirp, wirrest, wirsa, wær-scipe, wyrhta.

wer-scipe, es ; *m. Married state, estate of matrimony :*—Gebodene werscipe *oblatam matrimonii sortem*, Hpt. Gl. 490, 60.

wer-stede, es ; *m. A* weir-stead*, place where there is a weir :*—Of ðam wege on ða eá, and se werstede be súðan hreódbricge, Cod. Dip. Kmbl. iv. 105, 11.

wertacen ? :—Sagaþ Scs. Iôhannis sóðum wordum wíslíce and wǽrlíce swá se wertacen (*a later rendering of the passage has* swá se wyrhte cann, 476, 66, *as if the word* = werhta cann), Engl. Stud. viii. 478, 75.

wer-þeód, e ; *f.* I. *a people, nation ; pl. nations, men :*—Wé ðé freóndlíce on ðisse werþeóde wíc getǽhton, Cd. Th. 162, 26 ; Gen. 2687 : Elen. Kmbl. 1283 ; El. 643. On ðære werþeóde, Andr. Kmbl. 273 ; An. 137. Ðú ðás werþeóde gesóhtest, Cd. Th. 149, 21 ; Gen. 2478 : 171, 2 ; Gen. 2822. In ðære folcsceare geond ða werþeóde, Elen. Kmbl. 1934 ; El. 969. Ongunnon wercan werþeóda (cf. leáse men, Bt. 38, 1 ; Fox 194, 30) spell, Met. 26, 73. Werþióda, 29, 28. Werðeóde, Cd. Th. 211, 1 ; Exod. 519. Ðæt is ðæs wyrðe, ðætte werþeóde secgen Dryhtne þonc, Exon. Th. 38, 2 ; Cri. 600 : 281, 9 ; Jul. 643. Waldend werþeóda, 45, 4 ; Cri. 714 : Cd. Th. 202, 4 ; Exod. 383. Hé manegum weard geond middangeard mannum tó hróðre, werþeódum tó wræce, Elen. Kmbl. 33 ; El. 17. Werþeódum Filistina, Salm. Kmbl. 424 ; Sal. 212. Se ðe waldeþ giond werþióda ealra óþra eorþan cyninga, Met. 24, 35. Wutun hí tówyrpan geond werþeóda *disperdamus eos ex gente*, Ps. Th. 82, 4 : 105, 19 : 59, 1 : Cd. Th. 61, 2 ; Gen. 991. Geond wǽrðeóda, Menol. Fox 252 ; Men. 127. Geond ealle werðeóda, Ps. Th. 90, 16. Geond ðás werþeóde *in omnibus gentibus*, 66, 2. Ofer werþeóda, 104, 6. Ge néh ge feor is ðín nama hálig ofer werþeóda, Andr. Kmbl.

1086 ; An. 543. Wíde geweorðod ofer werþeóda, Apstls. Kmbl. 30 ; Ap. 15 : Beo. Th. 1802 : B. 899 : Exon. Th. 243, 12 ; Jul. 9 : Lchdm. iii. 36, 24. Werþióde, Met. 9, 21. Ofer ealle werþeóde *inter gentes*, Ps. Th. 104, 1. II. *men, the world,* cf. weorold, VI a :—Hú mihte ðæt gewyrdan in werþeóde (*how in the world did it happen ?*), ðæt ðú ne gehýrde Hǽlendes miht ? Andr. Kmbl. 1146 ; An. 573. ¶ Werðeóde glosses *nixu*, Wrt. Voc. ii. 114, 73. [*Icel.* ver-þjóð *mankind, men.*]

wer-tihtle, an ; *f.* An *accusation where the crime of which a person is accused involves the payment of the* wer ; *the crime itself :*—Be wertyhtlan. Gif mon sié wertyhtlan betogen . . . bíde mon mid ðære wíterǽdenne óþ ðæt se wer gegolden sié, L. In. 71 ; Th. i. 148, 1–4.

werud, weruld, werung. v. weorod, weorold, wering.

wésa, an ; *m. A* soaker*, one that drinks intemperately :*—Wésan oþþe eteras *commessatores* (Prov. 28, 7), Kent. Gl. 1044. v. wésan ; ealowósa.

wesan ; *p.* wæs, *pl.* wǽron To be :*—Wesan and beón *fore*, Wrt. Voc. ii. 34, 61. I. as an independent verb, (1) denoting existence *to be, exist :*—Wesendum, beóndum *existentibus*, Wrt. Voc. ii. 32, 63. (a) of animate objects, *to exist, live :*—Wesaþ and weaxaþ ealle weróde, lifgaþ bi ðam lissum ðe ús Dryhten sette, Exon. Th. 192, 30 ; Az. 113. On frymðe wæs word, Jn. Skt. 1, 1. God ðe ǽr worulde wæs, Ps. Th. 54, 19. Ða hwíle ðe hé wæs *while he lived*, Chart. Th. 167, 9. Manige hálge wítgan wǽran ǽr Sancte Iôhanne, Blickl. Homl. 161, 12. Ðæt hé his móste brúcan, ða hwíle ðe hé wǽre, Chart. Th. 140, 30. Swaðer uncer leng wǽre (cf. swaðer uncer leng lifede, 38), 485, 29. Swilce hé áwár wǽre, ǽr ðan ðe hé geboren wǽre, ac . . . him betere wǽre, ðæt hé nǽfre nǽre, ðonne hé yfele wǽre, Homl. Th. ii. 244, 19. Ne mæg ic hér leng wesan, Beo. Th. 5595 ; B. 2801. Hé bið á wesende, Blickl. Homl. 19, 26. (b) of inanimate objects :—Him is eall andweard, ge ðætte ǽr wæs, ge ðætte nú is, ge ðætte æfter ús bið, Bt. 42 ; Fox 256, 28. Ǽr woruld wǽre, Ps. Th. 73, 12. Seó þ̄ǽg gewát, swá heó nó wǽre, Exon. Th. 292, 9 ; Wand. 96. Hé him tó frófre lét forð wesan hyrstedne hróf, Cd. Th. 58, 33 ; Gen. 955. (2) where an object exists, and so may be found ; where in modern English *there* precedes the verb :—Wæs ðara manna . . . endleofan síþum hund teóntig þúsenda, Blickl. Homl. 79, 17. Wǽron monge, ða ðe Meotude gehýrdun, Exon. Th. 228, 24 ; Ph. 443. Ðá wǽron monige ðe his mǽg wridon, Beo. Th. 5956 ; B. 2982. Him þúhte ðæt ðanon wǽre tó helle duru hund þúsenda míla, Cd. Th. 310, 7 ; Sat. 722. (3) denoting presence, stay of longer or shorter duration, *to be, stand, have place, dwell :*—On ðære gesihðe wesaþ ealle geleáffulle, Blickl. Homl. 13, 28. Ic wæs (*I have been*) sixtýne síðum on sǽbáte, Andr. Kmbl. 977 ; An. 489. Ic ongiten hæbbe ðæt ðú on faroðstrǽte feor ne wǽre, 1796 ; An. 900. Wǽre ðú mid ðínum fæder ? Blickl. Homl. 151, 26. Wóp wæs wíde, Cd. Th. 180, 8 ; Exod. 42. Ðæt hé lǽte hyne licgean, ðǽr hé longe wæs, Beo. Th. 6157 ; B. 3082. Ðæt word wæs mid Gode, Jn. Skt. 1, 1. Heó wæs mid twám werum *she lived with two husbands*, Blickl. Homl. Skt. i. 20, 3. Ðonne wæs hé mid his ágnum cynne, Bt. 5, 1 ; Fox 10, 10. Wé mid englum uppe wǽron, Cd. Th. 289, 2 ; Sat. 391. Ða ðe dǽr ǽr inne wǽron, Bd. 4, 24 ; S. 598, 35. Ða ðe him on neáwes-t wǽron, Ors. 1, 10 ; Swt. 46, 2. Ðær manna wese mǽst ætgædere, Ps. Th. 78, 10. Wese ús beorhtnes ofer, 89, 19. Wesan lí wið Drihtne, 108, 19. Wǽre ðǽr hé wǽre, Bt. 5, 1 ; Fox 10, 9, 10 : Elen. Kmbl. 317 ; El. 159. Gelimplíc wæs ðæt ða ætgædere wǽron on ǽcre stówe, Blickl. Homl. 133, 24. Ðæt hié ongieton mín mægen on ðé wesan, 241, 15. Ðara cynna monige hé wiste on Germanie wesan, Bd. 5, 9 ; S. 622, 14. Ne mæg hé be ðý wedre wesan *he cannot stop in the open air*, Exon. Th. 340, 18 ; Gn. Ex. 113. Gód is ús hér tó wossanne, Mt. Kmbl. Lind. 17, 4 : Mk. Skt. Lind. 9, 5. Wosanne (wosane, Rush.), Lk. Skt. Lind. 9, 33 : Mk. Skt. Rush. 9, 5. (4) where motion takes place :—Ðá wǽron wit twégen on ánum olfende þurh ðæt rúme wésten, and wit unc simble ondrédon hwonne wit sceoldon feallan of ðam olfende, Shrn. 38, 14. Hí wǽron heom tó Lundene weard, Chr. 1052 ; Erl. 185, 4. (5) denoting condition, (a) nature of persons, *to be, live :*—Ne wosas gé swǽ lēgeras, Mt. Kmbl. Lind. 6, 5. Him betere wǽre ðæt hé nǽfre nǽre, ðonne hé yfele wǽre, Homl. Th. ii. 244, 21. Ðonne gé fæston, nellon gé wesan (wosa, Lind.) swylce leáse líceteras, Mt. Kmbl. 6, 16. (v. III c.) (b) condition or state of things :—Se hálga heáp wæs sprecende mid eallum gereordum ; and eác, ðæt wunderlícor wæs, ðá ðá heora án bodade mid ánre sprǽce, ǽlcum wæs geþúht, ðe ða bodunge gehýrde, swilce hé sprǽce mid his gereorde, Homl. Th. i. 318, 26. Wese swá, Ps. Th. 71, 20 : 88, 45. Lǽtaþ ðis ðus wesan, Blickl. Homl. 69, 17 : 75, 31. (6) *to be, to be done, come to pass, happen :*—On ðǽm dagum wæs ðæt Liber Pater oferwan Indéa ðeóde, Ors. 1, 6 ; Swt. 36, 17. On ðære tíde wæs sió ofermycelo hǽto, 1, 7 ; Swt. 40, 3. On ðǽm geáre ðe ðiss wæs, 2, 1 ; Swt. 60, 17 : Chr. 1048 ; Erl. 180, 19. Git ðæt wæs, ðæt hé tó cyninges simbla gelaþod wǽre, Bd. 3, 5 ; S. 527, 2 : Blickl. Homl. 11, 23 ; Wulfst. 9, 11 : 12, 14. Hwæt wille gé nú hwæt ic hire doo ? . . . Wese hit nú be eówrum dómum, Blickl. Homl. 157, 7. Ðý læs ðæt wǽre, ðæt hé ǽnig ðara góda forylde, 213, 23. Tó wosanne onginnaþ *fieri incipient*, Lk. Skt. Lind. 21, 7. (7) *to be, have result, turn*

out (v. wâ, I):—Se hálga gebæd for ðæt seóce cyld, and him wæs sóna bet (*it was better with him at once*, i. e. *he was better*), Homl. Skt. i. 3, 311. Nâmon tô ræde, ðæt him wærlícor wære, ðæt hí sumne dæl heora landes wurðes æthæfdon *they resolved that keeping back part of the price of the land would turn out more safely for them*, Homl. Th. i. 316, 24. Hê ðôhte hine him tô yrfewearde gedôn. Ac ðæt hwæþere swâ wesan ne mihte, Bd. 5, 19; S. 638, 23. (8) with dat. of person, (a) *to belong to, for a person to have something*:—Him wæs beorht wela, Cd. Th. 96, 32; Gen. 1603: 216, 20; Dan. 9. Ðam wæs Crist nama, Andr. Kmbl. 2646; An. 1324. Ne him wese ænig fultum, Ps. Th. 108, 12. Swâ him dagas deorce and feáwe, 108, 8. Ðæt ðâm gengum gâd ne wære wiste ne wæde, Cd. Th. 222, 10; Dan. 102. (b) *to affect, be the matter with*:—Ðâ frægn hê hine hwæt him wære, Bd. 4, 25; S. 600, 32. **II.** with a predicative noun or pronoun, *to be*:—God wæs ðæt word, Jn. Skt. 1, 1. Ðæt wæs gôd cyning, Beo. Th. 22; B. 11. Wæs hira Matheus sum, Andr. Kmbl. 22; An. 11. Ðæt mon mæg gesión ðæt hí gió men wæron, Bt. 37, 3; Fox 192, 3. Wes ûs freónd, Cd. Th. 165, 1; Gen. 2725. Ic mæg wesan god, 18, 35; Gen. 283. Se ðe wæs leorningcniht on lâde ongann wesan lâreów on martyrdôme, Homl. Th. i. 50, 6. Hwæt wile ðis wesan? Blickl. Homl. 239, 29. Sæde hê ðæt hê hine cniht wesende gesáwe *quod fanum se in pueritia vidisse testabatur*, Bd. 2, 15; S. 518, 36: Exon. Th. 320, 34; Víd. 39. On ðæm cniht wesendum ðâ ðis hælo wundur geworden wæs *in quo tunc puero factum erat hoc miraculum sanitatis*, Bd. 3, 12; S. 537, 17. Umbor wesendum, Beo. Th. 2378; B. 1187. Ic hine cúðe cniht wesende, 750; B. 372. **III.** with a predicative adjective or participle:—Hê edgeong weseþ, Exon. Th. 224, 10; Ph. 373. Ðû ðe wære reód, and ic mê wæs blâc; ðû wære glæd, and ic mê wæs unrôt, L. E. I. proem.; Th. ii. 398, 14. Se beág wæs of ðornum geworht, Exon. Th. 88, 27; Cri. 1446. Þeód wæs oflysted, Andr. Kmbl. 2226; An. 1115. Cyning wæs âfyrhted, Elen. Kmbl. 112; El. 56. Ðâ wæs gesýne ðæt sige forgeaf cyning ælmihtig, 287; El. 144. Wes ðu behýdig and geniyndig, Blickl. Homl. 67, 32. Hâl wæs ðú *aue*, Mt. Knbl. 27, 30. Hâl westû, Blickl. Homl. 143, 17. Westû gearo, Bd. 5, 19; S. 640, 44. Hâle wese gê (wosaþ gié, Lind.) *auete*, Mt. Kmbl. 28, 9. Wesaþ hâle *valete*, Wrt. Voc. ii. 88, 61. Wesaþ þancfulle, Blickl. Homl. 169, 16. Wîsfæsto wossaþ gié *perfecti estote*, Rtl. 13, 19. Wese hê hrægle gelíc, Ps. Th. 108, 19. Hit næs geséne hweðer hê seóc wære (*had been*), Homl. Skt. i. 6, 259. Ðæt Adam leng âna wære, Cd. Th. 11, 5; Gen. 170. Ofermôd wesan, 17, 20; Gen. 262. Uossa oestig *esse devota*, Rtl. 15, 21. Giscroepo uossa *aptas fieri*, 117, 14. ¶ used impersonally:—Ðâ wæs on ofne windig and wynsum, Cd. Th. 237, 31; Dan. 346. Settan mê ðær mê unswæsost wæs *posuerunt me in abominationem sibi*, Ps. Th. 87, 8. Ðær him leófost wæs, Byrht. Th. 132, 29; By. 23. Swâ him gemêdost wæs, Andr. Kmbl. 1188; An. 594. (In the last three passages the superlatives might be taken as adverbs. Cf. I. 7.) **III a.** with a predicative genitive:—Ðâ sôna wæs Eþelwald ðæs wordes, ðæt hê nô ðæs rihtes wiðsacan wolde, Chart. Th. 140, 10. Wæs seó eorla geðriht ânes môdes, Cd. Th. 197, 10; Exod. 304. His þegnas wæron flæsclíces môdes, Blickl. Homl. 17, 5. **III b.** with prepositional phrases, (1) prep. and noun:—Ic wæs mid weorþmende on neorxna wange, and ic ðæt ne ongeat, Blickl. Homl. 89, 8. Ðæs cyning on hreón môde, Beo. Th. 2617; B. 1307. Sôna wæs hê on sunde, 3240; B. 1618. Ðû on sælum wes, 2345; B. 1170. Wesan him on wynne, Cd. Th. 23, 29; Gen. 367. ¶ used impersonally:—Ðâ wæs ofer midde niht, ðæt hê frægn *cum jam mediae noctis tempus esset transcensum, interrogavit*, Bd. 4, 24; S. 598, 35. (2) with gerundial infinitive:—Ne wæs ðæt tô wundrianne, Bd. 3, 12; S. 537, 17. Hwæt him be ðam tô dônne wære, Homl. Th. i. 502, 24: 506, 24. **III c.** with a clause:—Wæs ðæt hê wolde wyrcan æghwylc ðara weorca ðe ðâm ôðrum brôðrum wæs heard and hefig, Shrn. 145, 18 (cf. I. 5a). **IV.** with participles, (1) with present participles:—Swâ ic him secgende wæs, Andr. Kmbl. 1898; An. 951. On æfenne ðære nihte ðe hê of worulde gangende wæs *nocte qua de saeculo erat exiturus*, Bd. 4, 24; S. 598, 30. Wæs se engel sprecende, Blickl. Homl. 5, 2. Hê wæs Drihtne fylgende, 15, 28: Beo. Th. 321; B. 159. Hê in byrgenne bídende wæs, Elen. Kmbl. 966; El. 484. Se hálga wer hergende wæs Metodes miltse, Cd. Th. 237, 8; Dan. 334. Hí ðær stondende wæron, Blickl. Homl. 11, 23. Hí on ðæt folc winnende wæron, Ors. 1, 10; Swt. 46, 6: 44, 19. Wóeron (wêrun, Rush.) sprecende *erant loquentes*, Mk. Skt. Lind. 9, 4. Hwæðer iscende sæflôd ðâ gyt wære, Cd. Th. 86, 29; Gen. 1438. Wríðende sceal mægðe ðînre monrím wesan, 105, 33; Gen. 1763. (2) with past participles, (a) of transitive verbs forming the passive:—Ðonne wesaþ ðíne handa sôna geedneówode, and beóþ swâ hié ær wæron, Blickl. Homl. 153, 11. Wær ðû gewurðod, Cd. Th. 127, 7; Gen. 2107. Hwær âhangen wæs rodera Waldend, Elen. Kmbl. 409; El. 205. Deós geofu on heora heortan âlegd wes, Blickl. Homl. 137, 4. Ealle þing wæron geworhte (*facta sunt*) ðurh hyne, and nân þing næs geworht bûtan him, Jn. Skt. 1, 3. Ðâ ðe ðurh geleáfan gehælede wæron *qui credendo salvati sunt*, Bd. 4, 16; S. 584, 20. Wesaþ gê fram Gode gebletsade *benedicti vos a Domino*, Ps. Th. 113, 23. Ðæt ic wese gelæded *quis deducet me?* 107,

9. Wese heora beód wended on grine *fiat mensa eorum in laqueum*, 68, 23. Wesan ealle gedrêfde *turbabuntur*, 67, 5. Ne wesen hí mid sôðfæstum âwritene *cum justis non scribantur*, 68, 29. Ðæt wæron âlýsede leófe ðíne *ut liberentur dilecti tui*, 59, 4. Se magorinc sceal wesan Ismahêl hâten, Cd. Th. 138, 3; Gen. 2286. Forgifen weosan, Bd. 4, 22; M. 330, 16: 4, 23; M. 340. 15. (b) of intransitive verbs:—Ðû wære geworden . . . cild âcenned, Exon. Th. 14, 8; Cri. 216. Ðâ wæs ðæs folces fela on ân fæsten ôþflogen (*confugerant*), Ors. 4, 11; Swt. 206, 12. Ðâ wæs forð cumen geóc æfter gyrne, Andr. Kmbl. 3167; An. 1586. Ðâ wæs first âgân; 293; An. 147: Elen. Kmbl. 1; El. 1. Ðâ wæs geworden ðæt . . ., Blickl. Homl. 15, 15. Giwêdo his giwordne wêrun scínende, Mk. Skt. Rush. 9, 3. Gif ic ðæs sægde, ðæt mín sylfes fôt âsliden wære *si dicebam: ‘Motus est pes meus*,’ Ps. Th. 93, 17. [Goth. wisan: O. Sax. wesan: O. Frs. wesa: O. H. Ger. wesan: Icel. vera.] v. fore-, ge-wesan, nesan; efen-wesende.

wêsan; *p.* de. **I.** *to steep, soak*; *inficere, conficere*:—Genim grêne rudan, cnuca smale and wês mid doran hunige, Lchdm. iii. 4, 24. Heoretes sceafeþan of felle âscafen mid pumice and wêse mid ecede, 44, 11: ii. 100, 15: 246, 13. v. ge-wêsan; wêse, wêsing. **II.** *to ooze, suppurate*:—Ðonne ærest onginne se healsgûnd wêsan (wesan?), Lchdm. ii. 44, 11. [Wese, N. P. 65. *See Halliwell* wese, *and Jamieson* weese, weeze *to ooze, distil gently*.] v. wôs.

wêse; *adj. Soaked, moist with soaking*:—Sý crocca âsett on eorþan, and ðâs wyrta sýn gedôn innan ðam croccan; onuppan ðâm sý gedôn wæta, ðæt hí þearle wel wêse beón, Lchdm. iii. 292, 6. v. wôs, *and preceding word*.

wesend; *m. A bison, buffalo, wild ox*:—Weosend, uusend, wesand *bubalis*, Txts. 47, 337. Wesend, Wrt. Voc. ii. 11, 40: *bubalus*, 126, 60: *urus*, i. 22, 45. [O. H. Ger. wisunt (-ant, -ent, -int) *bubalus: Icel.* vísundr.] v. next word.

wesend-horn; *m. A buffalo-horn*:—Ælfwolde hyre twêgen wesend-hornas, Chart. Th. 536, 1. v. preceding word.

-wesenness. [Cf. *O. L. Ger.* ge-wesannussi *substantia*.] v. tô-wesness.

wêsing, e; *f. Soaking, steeping*:—Wêsing, gemangcennys ł mencinge *confectio*, Hpt. Gl. 450, 28. Wêsing ł gemang *confectio*, 449, 61. v. wêsan.

wesle, an; *f. A weasel*:—Uueosule, uuesulae *mustela*, Txts. 79, 1345. Wesle, Wrt. Voc. i. 23: 78, 18: ii. 56, 53: 71, 25: Ælfc. Gr. 6, 5; Zup. 19, 14. Gif on hwylcne mycelne wætan mûs oððe wesle (*mustela*) on befealle, and ðær deâd sig, sprenge mid hâligwætere and þycge, L. Ecg. C. 39; Th. ii. 164, 11: 40; Th. ii. 166, 6, 9. [O. H. Ger. wisala (-ula, -ela, -ila) *mustela*.]

weslinc, -wesness. v. wæstling, ge-, tô-wesness.

[**west**]; *spve.* west[e]mest; *adj. Westerly, situated in the west*:—Rômâna onweald, se is mæst and westmest, Ors. 6, 1; Swt. 252, 19. On ðæm síþmestan onwalde and on ðæm westemestan, Swt. 254, 2. Ðis sindon ðæs landes gemæra ðe gebyriaþ into ðære westmestan hîde, Cod. Dip. Kmbl. iii. 262, 18. On ðone westmestan mylengear . . . eft on ðæm westemestan mylengeare, Cod. Dip. B. ii. 305, 23-30. ¶ westan *in combination with prepositions, governing dative or adverbial*:—Bewestan Hai *ab oriente habens Hai*, Gen. 12, 8. Ðâm folcum ðe eardiaþ be-westan Sæferne *eis populis qui ultra amnem Sabrinam ad occidentem habitant*, Bd. 5, 23; S. 646, 21. Be-westan Sealwuda, Chr. 894; Erl. 92, 19: 709; Erl. 42, 28: Ors. 1, 1; Swt. 22, 7, 12, 26. Ðonne heóld man fyrde be-westan (cf. wonyng fer by weste, Chauc. Prol. 388), Chr. 1010; Erl. 144, 5. On-westan ðære cyrican *ad occidentem ecclesiae partem*, Bd. 3, 17; S. 543, 29. Is on-westan medmycel duru, Blickl. Homl. 127, 8. [*Icel.* vestari; *cpve.*; vestastr; *spve. more, most westerly*.]

west; *adv. West, westward, to the west, in a westerly direction*, (1) marking the direction of movement:—Hêr fôr se here west ðe eást gelende, Chr. 886; Erl. 84, 24: 918; Erl. 102, 23: Cd. Th. 219, 12; Dan. 53. West fêran, 220, 25; Dan. 76: Exon. Th. 412, 7; Rä. 30, 10. Hê west gewîteþ, 208, 27; Ph. 162. Wôdon wælwulfas west ofer Pantan, Byrht. Th. 134, 41; By. 97. Ðâ wende hê hine west wið Exanceastres, Chr. 894; Erl. 91, 10. Se sciphere sigelede west ymbûtan, 877; Erl. 78, 17. Ðonne heofones gim west onhylde, Exon. Th. 174, 32; Gû. 1186. (2) marking relative position:—Seó burh is west ðonon from ðære stôwe on ânre míle *the town is a mile to the west of the place*, Blickl. Homl. 129, 3. Ðonne se æfensteorra biþ west gesewen, Bt. 39, 13; Fox 232, 34: Met. 29, 28. Hê wið ðone here ðær wæst âbisgod wæs, Chr. 894; Erl. 92, 9. Sûð, eást and west, Met. 9, 42: 14, 7. Ðæt hê west and norð trymede getimbro, Cd. Th. 18, 18; Gen. 275. Ðæt is ðrittiges míla lang east and west, Bd. 1, 3; S. 475, 19. Wes[t]mest ân íglond ligð ût on gârsecg, Met. 16, 11. [Cf. *O. Sax.* westor: *O. Frs.* wester: *Icel.* vestr *westwards*. v. norþ-, sûþ-west.

westan; *adv. From the west*, (1) marking the direction of movement:—Ðæm fultume ðe him westan com, Chr. 894; Erl. 91, 15. Monige from eástan and westan (weosta, Lind.) cunaþ, Mt. Kmbl. Rush. 8, 11. Cymeþ westa (woesta, Lind.), Lk. Skt. Rush. 13, 29. Fêrde se æðeling wæston, Chr. 1052; Erl. 152, 6. Westan brôhton, Elen. Kmbl. 2030; El. 1016. Somnaþ sûþan and norþan, eástan and westan, Exon. Th. 220,

24; Ph. 325. Se þridda heáfodwind hátte *zephirus;* se blǽwð westan, Lchdm. iii. 274, 20: Cd. Th. 50, 10; Gen. 806. Ðonne blǽwð súþan and westan wind, Met. 6, 8. Swinsiaþ súþan and norþan, eástan and westan, Exon. Th. 55, 19; Cri. 886. Gesáwon wě westan ðone leóman sunnan, and se leóma gehrán ðǽm treówum ufonweardum, Nar. 28, 23. (2) marking the direction of measurement :—Is seó stów ǽghwanon mid sǽ ymbseald bútan westan *est locus ille undique mari circumdatus praeter ab occidente,* Bd. 4, 13; S, 583, 10. Se cyng hæfde funden ðæt him mon sæt wið on súþhealfe Sæfernmúþan westan from Wealum eást óþ Afene múþan, Chr. 918; Erl. 104, 4. [*O. Sax.* westan : *O. Frs.* westa : *Icel.* vestan.] v. norþan-, súþan-westan ; westane.

wéstan ; *p.* te *To lay waste, devastate, desolate :*—Hine wilde deór wéstaþ and frettaþ *singularis ferus depastus est eám,* Ps. Th. 79, 13. Hí his wícstede wéstan *locum ejus desolaverunt,* 78, 7. Hié wǽron ðæt lond herigende and wéstende, Ors. 1, 10; Swt. 44, 20. [Heo westen þat lond, Laym. 1754. *O. Sax.* á-wôstian : *O. H. Ger.* wuosten *vastare.*] v. á-, ge-, on-wéstan,

westane ; *adv. From the west, in the west :*—Ða beorgas onginnaþ westane fram ðǽm Wendelsǽ in Narbonense ðǽre ðeóde, and endiaþ eást in Dalmatia ðǽm lande æt ðǽm sǽ *Alpes a Gallico mari exsurgentes, primum Narbonensium fines, deinde Galliam Rhetiamque secludunt, donec in sinu Liburnico defigantur,* Ors. 1, 1 ; Swt. 22, 19. Dioclitianus and Maximianus bebudon ëhtnesse cristenra monna, Dioclitianus eástane, Maximianus westane *(in occidente),* 6, 30; Swt. 280, 18. [*O. Sax.* westana : *O. H. Ger.* westana *ab occidente.*] v. westan.

westan-norþan. I. *adv. From the north-west.* Cf. westan (2) :— Hit (*Italy*) belið Wendelsǽ ymb eall útan bútan westannorðan, Ors. 1, 1 ; Swt. 22, 18. II. in phrases (or compounds) marking position, *to the north-west :* — Be-westannorðan ðǽre byrig, Ors. 1, 1 ; Swt. 22, 5.

westan-súþan *in* be-westansúþan *to the south-west :*—Be-westansúðan Corinton, Ors. 1, 1 ; Swt. 22, 10, 24, 27.

westansúþan-wind, es ; *m. A south-west wind :*—Westansúðanwind *austrum,* Ps. Spl. C. 77, 30.

westan-weard ; *adj. Westward :*—Mín þrym is from eástewearde middangearde óþ ðæt westanweardne *majestas mea peruenit ab occidente usque in orientem,* Nar. 25, 25.

westan-wind, es ; *m. A west wind :*—Hé bád westanwindes and hwôn norþan, and siglde ða eást, Ors. 1, 1 ; Swt. 17, 15.

West-Centingas ; *pl. m. The people* or *the district of West Kent :*— Hí forneáh ealle West-Kentingas (Weast-Centingas, *v. l.*) fordydon, Chr. 999; Erl. 134, 28.

west-dǽl, es ; *m.* I. *a western part, the extreme western point :*— Westdǽles *Hesperiae,* Hpt. Gl. 466, 67. Manega cumaþ fram eástdǽle middangeardes, and fram westdǽle tó heofenan ríce . . . Þurh ða twégen dǽlas, eástdǽl and westdǽl, sind getácnode ða feówer hwemmas ealles middangeardes, Homl. Th. i. 130, 17–21. Ðín ofspring byð fram eástdǽle óð westdǽle, Gen. 28, 14. Se heofon tóbyrst from ðǽm eástdǽle óþ ðone westdǽl, Blickl. Homl. 93, 23 : Mt. Knibl. 24, 27. Hé gesealde him westdǽl middaneardes, Bd. 1, 6 ; S. 476, 18. Ne se steorra gestígan wile westdǽl wolcna, Met. 29, 13. Tungol beóþ gewiten under waþeman westdǽlas on, Exon. Th. 204, 14 ; Ph. 97. II. *the west :*—Be-heald . . . tó westdǽle *vide . . . ad occidentem,* Gen. 13, 14: Deut. 3, 27. God sende wind fram westdǽle, Exod. 10, 19. Se steorra ne cymþ nǽfre on ðam westdǽle, Bt. 39, 13; Fox 232, 30. Breoton is geseted betwyh norþdǽle and westdǽle *Brittania inter septentrionem et occidentem locata est,* Bd. 1, 1 ; S. 473, 9. II a. with special reference to the sun's setting :—On westdǽle geendaþ se dæg, Homl. Th. i. 130, 27. Se ðe ástáh ofer westdǽl (*super occc ′′m*), Ps. Spl. 67, 4. [Wesstdale off all þiss werelld iss Dysiss, Orm. 1646. Cf. *O. H. Ger.* wester-teil.]

West-Dene ; *pl. m. The West-Danes :*—Tó West-Denum, Beo. Th. 771 ; B. 383 : 3161 ; B. 1578.

wéste ; *adj.* I. of open country, *waste, uncultivated and uninhabited, desert :*—Ðara Terfinna land wæs eal wéste, búton ðǽr huntan gewícodon, oþþe fisceras, Ors. 1, 1 ; Swt. 17, 29 : 1, 10; Swt. 48, 25. Ðeós stów ys wéste *desertus est locus,* Mt. Kmbl. 14, 15. Is sǽd ðæt ðæt land wéste (*desertus*) wunige, Bd. 1, 15; S. 483, 27. Eall (*all of the earth*) ðæt on eallum ðeódum wéstes ligeþ, Bt. 18, 1; Fox 62, 15. On wéstere (wéstre, *v.l.*) stówe, Lk. Skt. 9, 12. On wéstum lande *in terra deserta,* Deut. 32, 10. Hé férde on wéste stówe, Mk. Skt. 1, 35 : 6, 31, 32 : Lk. Skt. 4, 42 : 9, 10: Exon. Th. 209, 12 ; Ph. 169. Hé sealde him wéste land, Ps. Th. 77, 55. Hé ne mihte on ða ceastre gán, ac beón úte on wéstum stówum, Mk. Skt. 1, 45. Of ðissum wídum, wéstum mórum *a desertis montibus,* Ps. Th. 77, 6. II. *waste, empty, unused :*—Seó grundleáse swelgend hæfþ swiþe manegu wéste holu on tó gadrianne, Bt. 7, 4; Fox 22, 32. III. *waste, useless, unproductive :*—Hé geseah deorc gesweorc semian sweart under roderum, wonn and wéste, Cd. Th. 7, 22; Gen. 104. IV. of habitations, *waste, deserted, desolate :*—Byð eówer hús eów wéste (*deserta*) forlǽten, Mt. Kmbl. 23, 38. Wese wíc heora wéste (woestu, Ps. Surt.) and ídel, Ps. Th. 68, 26. Wéste (wôstu, Ps. Surt.), 108, 7. Hié gedydon on ánre

wéstre ceastre, Chr. 894; Erl. 93, 5. Hé gesyhð wínsele wéstne, Beo. Th. 4903; B. 2456. On wéste wíc, Cd. Th. 128, 25; Gen. 2132. Babylonia, seó ðe mǽst wæs and ǽrest ealra burga, seó is nú lǽst and wéstast, Ors. 2, 4; Swt. 74, 23. V. *waste, spoiled :*—Ðonne ealle ðisse worulde wela wéste stondeþ, Exon. Th. 290, 33 ; Wand. 74. VI. *deprived, devoid* (with gen.) :—Bið on eorðan wéste (wésðe, *v. l.*) wísdómes, se þurh ðone cantic ne can Crist geherian, Salm. Kmbl. 43; Sal. 22. [*O. Sax.* wôsti : *O. Frs.* wôste : *O. H. Ger.* wuosti *solus, desertus, solitarius, vastus.*]

westemest, v. west ; *adj.*

wésten, wésten[n], wéstern (*in northern dialect*), es, e ; *m. f. n. A desert, wilderness :*—Wésten *desertum vel heremus,* Wrt. Voc. i. 53, 62. Wǽsten, 80, 35. Wíd is ðes wésten, Exon. Th. 120, 5 ; Gú. 267. Andlang ðæs wéstenes, Jos. 8, 16. Wéstennes (on wéstenne, *v. l.*) weard, Salm. Kmbl. 167; Sal. 83. Woesternes *exterminii,* Rtl. 86, 18. Hig cômon tó ðam wéstene (*in solitudine*), Gen. 21, 14. On wéstenne, Cd. Th. 137, 17; Gen. 2275. Tó Sinai wéstene *in solitudinem Sinai,* Ex. 19, 1. On wéstenne, Cd. Th. 178, 7; Exod. 8: 185, 15; Exod. 123. Tó ðam wéstene Sin *in desertum Sin,* Num. 20, 1. On wéstene (woestenne, Ps. Surt.) *in solitudine,* Ps. Th. 54, 7. On ðisum wéstene (woestenne, Ps. Surt.) wídum and sídum *in deserto,* 77, 20. On wéstenne, 77, 40. On ðam wéstene (woestenne, Rush.: woestern, Lind.), Mt. Kmbl. 3, 1. Wéstene (wéstinne, Rush.), 3, 3. On ðisum wéstene (woestenne, Rush.: woestern, Lind.) *in solitudine,* Mk. Skt. 8, 4. On ðis wéstene (wǽstenne, Rush.: woestern, Lind.) *in deserto,* Mt. Kmbl. 15, 33. Tó wéstenne, Blickl. Homl. 165, 3 : 169, 4. Se hrefen fédde Héliam, ðam eode hé tó ðam wésterne (-nne?), and him þenode, Salm. Kmbl. p. 202, 9. On woesterne, Rtl. 56, 27. Ofer wéstenne (*chaos*), Cd. Th. 8, 16; Gen. 125. On ðæt wésten *in desertum,* Ex. 4, 27 : *in solitudinem,* 5, 3. On án wésten, 15, 22. On wésten (woestenne, Rush.: woestern, Lind.) *in desertum,* Mt. Kmbl. 4, 1: Blickl. Homl. 35, 6. Hé wæs geond ðæt wésten sundorgenga, 199, 5. Wildeóra wésten, Cd. Th. 255, 10; Dan. 622. Þurh wésten *per devia,* Wrt. Voc. ii. 94, 76. On ðæt wídgille wésten, Homl. Skt. ii. 23 b, 729. Ofer ða wéstenne (-u, *v.i.*), Ors. 1, 1; Swt. 16, 35. Mid mistlícum wéstenum, Bt. 18, 2 ; Fox 62, 36. On wéstennum, Exon. Th. 107, 2 ; Gú. 52. Þurh wéstenas, Ps. Th. 77, 52. Geond wéstena, 67, 8. Geond wéstenu, 10, 1. On ða wéstenu middangeardes *in desertas orbis terrarum solitudines,* Nar. 6, 5. Gynd wéstnu *per auia,* Germ. 391, 40. [A westene *in the wilderness,* O. E. Homl. i. 245, 5. *O. Sax.* wôstun (*dat.* wôstunni) ; wôstunnia (-innia) ; *f.*: *O. L. Ger.* wôstinna ; *wk. f.*: *O. Frs.* wôstene, wéstene : *O. H. Ger.* wuostinna (-unna) ; *f.*] v. wudu-wésten.

wésten ; *adj. Desert :*—Seó stów wæs swá wésten and swá dígle, ðæt nǽs ná ðæt án ðæt heó wæs ungewunelíc, ac eác swilce uncúð ðam landleódum him sylfum, Homl. Skt. ii. 23 b, 105. Hé férde him ðanon tó ánum wéstenum earde, Homl. Ass. 66, 24 : 71, 166.

wéstend, es ; *m. A waster, destroyer, devastator :*—Wéstend, tólýsend *desolator, vastator,* Wrt. Voc. ii. 139, 34. Wéstend, ýtend *exterminator, vastator,* 145, 64. v. á-wéstend.

west-ende, es ; *m. The west end, western extremity* of anything :— Hire on westende is Scotland, Ors. 1, 1; Swt. 8, 27. Ðæt hire ǽwielme sié on westende Affrica, Swt. 12, 21. Hine man byrigde æt ðam westende, ðam stýple ful gehende, Chr. 1036; Erl. 165, 37. Æt ðam westænde, Cod. Dip. B. iii. 659, 30. v. riht-westende.

wésten-gryre, es ; *m. The terror of the wilderness, terror inspired by the wilderness,* Cd. Th. 185, 4; Exod. 117.

wésten-setla, an ; *m. A dweller in a wilderness, a hermit, an anchorite :*—Wéstensetla *eremita,* Wrt. Voc. i. 42, 28 : 72, 2. Wéstensetla (*printed* -seda) *eremita, anachoreta,* Hpt. Gl. 465, 24. Sum wéstensetla on ðǽm eálande ðe Liparus is nemned, Homl. Ass. 195, 1. Wé willaþ wrítan be sumum wéstænsetlan (*solitarius quidam*), Homl. Ass. 195, 1. Óþer cyn is muneca, ðæt is wéstensetlan, ðe feor fram mannum gewítaþ, and wéste stówa and ánwununge gelufiaþ . . . Swilce wéstensetlan . . . on wéstenes wununge gelustfulliaþ, R. Ben. 134, 11–16. Óþer cyn is ancrena, ðæt is wéstensetlena, 9, 5. [*O. H. Ger.* wuostan-sedalo *solitarius.*]

wésten-staþol, es ; *m. A waste place, a deserted place :*—Wurdon hyra wígsteal wéstenstaþolas, Exon. Th. 477, 22 ; Ruin. 28.

westerne ; *adj. Western :*—Ðá ástáh westerne wind and bleów *flante favonio,* Bd. 5, 19; S. 635, 20 note. Com Æþelmǽr ealdorman þider and ða weasternan (westenan, *v.l.*) þegnas, Chr. 1013; Erl. 148, 16. [*O. Sax. O. H. Ger.* westrôni: *Icel.* vestrænn.] v. súþ-, súþan-westerne.

weste-weard ; *adj. Westward, west, western* part of the noun to which the word refers :—Se westsúþende Európe landgemirce is in Ispania westeweardum et ðǽm gársecge *Europae in Hispania occidentalis oceanus terminus est,* Ors. 1, 1 ; Swt. 8, 24. Ðá ðá hé wæs on eásteweardum ðissum middangearde, ða from him ondrēdan ðe wǽron on westeweardum . . . Him ða swiþe hiene ondrēdan ðe on westeweardum ðisses middangeardes wǽron, 3, 9 ; Swt. 136, 6–23. On ðone westmestan mylengear westeweardne, Cod. Dip. B. ii. 305, 23. Eall ðes middangeard from eásteweardum óð westeweardne, Bt. 16, 4 ; Fox 58, 11 : 29. 3;

Fox 106, 22. From eásteweardan ðisses middangeardes óð westeweardne, 18, 2; Fox 62, 1. Gehergade Ecgbryht cyning on West-Walas from eásteweardum óþ westewearde, Chr. 813; Erl. 62, 2.

west-healf, e; f. The western side :—On westhealfe ab occasu, Ors. 1, 1; Swt. 12, 13: ad occidentem, Num. 3, 23. On westhealfe ðære cyrican ad occidentalem ecclesiae partem, Bd. 3, 17; S. 543, 34 : Ors. 1, 1; Swt. 8, 17: Chr. 1016; Erl. 155, 10. [O. H. Ger. west-halba. Cf. Icel. vestr-hálfa.]

wêstig; adj. Waste, desert, desolate :—Of Angle se á syððan stód wêstig (desertus, Bd. 1, 15), Chr. 449; Erl. 13, 16. Wêstig is stów desertus est locus, Mk. Skt. Rush. 6, 35. Wêstig (woestig, Rush.), Mt. Kmbl. Lind. 23, 38. Woestihg (woestig, Rush.), 14, 15. On woestigum stówe, Lk. Skt. Lind. 4, 42. In wêstige stówe, Mk. Skt. Rush. 1, 35. Woestig 6, 32.

west-lang; adj. Lying in a westerly direction :—On ðone westlangan hlinc ; of ðes westlangan hlinces ende, Cod. Dip. Kmbl. iii. 135, 25. Ða westlangan díc, v. 334, 22. v. next word.

west-lang; adv. With the length measured in a westerly direction :—Se wudu is eástlang and westlang hundtwelftiges míla lang the length of the wood measuring east and west is one hundred and twenty miles, Chr. 893; Erl. 88, 28. Se þridda sceáta is án hund and syfan and hundsyfantig míla westlang, Ors. 1, 1 ; Swt. 28, 9. v. preceding word.

westmest. v. west ; adj.

West-môringas; pl. m. The people of Westmoreland :—Westmôringa land, Chr. 966; Erl. 125, 2.

West-mynster, es; n. Westminster :—Hér forðférde Harold cyning, and hê wæs bebyrged æt Westmynstre, Chr. 1039; Erl. 167, 13. Willelm com tó Westmynstre, and Ealdréd arcebiscop hine tó cynge gehálgode, 1066; Erl. 203, 8. Hér man wrægde ðone biscop Ægelríc and sende hine tó Westmynstre, 1069; Erl. 207, 7. Icc habbe gifen Sainte Petre intó Westminstre, Cod. Dip. Kmbl. iv. 190, 12, 26. Ða gebróðere on Westminstre, 192, 5. The word occurs often in charters of Edward the Confessor. The Latin form Westmonasterium is found in a doubtful charter of the reign : Locum qui dicitur Westmonasterium quod a tempore sancti Augustini institutum, multaque ueterum regum munificentia honoratum, propter uetustatem et frequentes bellorum tumultus pene uidebatur destructum, 176, 1. The place is mentioned in a (doubtful) charter of Offa of the year 785 : In ioco terribili, quod dicitur æt Uuestmunstur, i. 180, 3.

wêstness, e; f. Desolation :—Woestenisse hire desolatio ejus, Lk. Skt. Lind. 21, 20. v. ā-wêstness.

west-norþ; adv. North-west :—Þonan westnorð is ðæt lond ðe mon Ongle hæt, Ors. 1, 1; Swt. 16, 6.

westnorþ-lang; adv. or adj. [cf. west-lang] With the length lying north-west (and south-east) :—Þonne is Italia land westnorðlang and eástsúðlang Italiae situs a circio in eurum tenditur, Ors. 1, 1 ; Swt. 22, 17.

westnorþ-wind, es ; m. A north-west wind :—Westnorðwind eircius, Wrt. Voc. ii. 104, 4: 24, 26. [Cf. O. H. Ger. westernort-wint chorus.]

west-ríce, es; n. A western kingdom or empire :—Ðá ðæt eástríce in Asiria gefeóll, ðá eác ðæt westríce in Róma árás, Ors. 2, 1; Swt. 62, 8. Ðȳ ilcan geáre fêng Carl tó ðam westríce, and tó allum ðam westríce behienan Wendelsǽ and begeondan ðisse sǽ, swá hit his þridda fæder hæfde, Chr. 885 ; Erl. 84, 10. [Cf. O. H. Ger. westar-ríchi occidens.]

west-rihte; adv. Due west :—Seó stów is tȳn mílum westrihte fram Cetrihtworþige locus est a vico Cataractone decem millibus passuum contra solstitialem occasum secretus, Bd. 3, 14; S. 539, 41. Seó is fram Cantwarabyrig on feówer and .xx. mílum westrihte (ad occidentem), 2, 3; S. 504, 26. Scȳt se sǽearm of ðam sǽ westrihte, Ors. 1, 1; Swt. 22, 4. Westryhte, Swt. 14, 9.

west-rodor, es; m. The western heavens :—Fram upgange sunnan óð ðæt heó wende on westrodur a solis ortu usque ad occasum, Ps. Th. 112, 3. Heó gewîteþ on westrodur, 106, 3. Westrodor, Exon. Th. 350, 24; Sch. 68.

west-sǽ; f. m. A west sea, sea on the west coast of a country :—Hê (a Norwegian) búde on ðæm lande norþweardum wiþ ða westsǽ, Ors. 1, 1; Swt. 17, 3. Hí (the Saxons in Britain) hergodon fram eástsǽ óð westsǽ (ab orientali mari usque ad occidentale), Bd. 1, 15; S. 483, 40. Fram eástsǽ óþ wæstsǽ a mari ad mare, 1, 12; S. 481, 8.

west-sceáta, an; m. A western angle or promontory :—Sicilia is ðryscȳte . . . ðone westsceátan man hæt Libéum Sicilia tria habet promontoria . . . tertium, quod adpellatur Lilybaeum, dirigitur in occasum, Ors. 1, 1; Swt. 28, 5.

West-Seaxe, -Seaxan (Wes-); pl. m. The West-Saxons; Wessex :—Hér cuômon West-Seaxe in Bretene, Chr. 514; Erl. 14, 20. Of Eald-Seaxon cômon Eást-Sexa and Súð-Sexa and West-Sexan (-Sexa, v.l.), 449; Erl. 12, 11. West-Seaxan, Bd. 1, 15; S. 483, 24. Weast-Seaxan, 5, 18; S. 635, 15. West-Seaxna biscop, S. 635, 22. West-Seaxna ríce, lond, Chr. Erl. 2, 9, 10. West-Seaxna (-Seaxena, v.l.) cyning, L. Alf. 49; Th. i. 58, 28. Wes-Seaxna, Chr. Erl. 2, 18, 23 : 4, 20. Wes-Seaxena kyning, L. In. proem.; Th. i. 102, 2. Wæst-Sæxna, Chr.

836; Erl. 65, 23. West-Sexena landes is hund þúsend hída, Cod. Dip. B. i. 415, 1. On Wes-Seaxum (Weast-, v.l.), Chr. 560; Erl. 16, 24. Hér Birinus biscop bodude West-Seaxum (Weast-, v.l.) fulwuht, 634 ; Erl. 24, 9. Hér cuom se here tó Reádingum on West-Seaxe, 871 ; Erl. 74, 5.

westsúþ-ende, es; m. The south-west extremity :—Se westsúþende Eurôpe, Ors. 1, 1; Swt. 8, 23.

westsúþ-wind, es; m. A south-west wind :—Westsúðwind affricus, Wrt. Voc. ii. 99, 51 : 6, 40 : favonius, 35, 6 : faonius, 108, 22. Westsúþwind, 39, 7. [Cf. O. H. Ger. westersunder-wint africus.]

West-Wealas; pl. m. The Celts of Cornwall; Cornwall :—Huwal West-Wala cyning, Chr. 926; Erl. 111, 42. Ðȳ geáre gehergade Ecgbryht cyning on West-Walas, 813; Erl. 62, 1. Hér cuom micel sciphere on West-Walas (Wæst-Wealas, v.l.), 835; Erl. 64, 24.

west-weard; adv. Westward, in a westerly direction :—Sume (adverbs) synd localia . . . westweard occidentem uersum, Ælfc. Gr. 38; Zup. 225, 10. Fôr se here of ðæm eástríce westweard, Chr. 893; Erl. 88, 22 : 1052; Erl. 183, 15. Ðá hê ðá hámweard tó ðære ié com, ðe hê ǽr westweard (when marching westward) hêt ða ofermǽtan brycge ofer gewyrcan, Ors. 2, 5; Swt. 84, 3. Ðás seofon tunglan gáð ǽfre eástwerd ongeán ða heofenan ; ac seó heo:en[e] is strenge and ábriet hí ealle underbæc westweard mid hire ryne ; and is for ðí mannum geþúht swilce séo sunne and ða foresǽdan tunglan gangon westweard. Sôð ðæt is westweard hí gáð unþances, Boutr. Scrd. 18, 39–42. Ða seofon steorran . . . gangende eástan westweard, Lchdm. iii. 270, 26. Affrica onginð eástan westwerd (starting from the east and coming westward) fram Egyptum æt ðære eé ðe man Nilus hæt, Ors. 1, 1; Swt. 24, 32.

west-weardes; adv. Westwards :—Hê man geseah westweardes on ðæt wésten efstan, Homl. Skt. ii. 23 b, 174.

west-wegas; pl. m. The west :—Eástan ne cymeþ gumena ǽnig, ne of westwegum neque ab oriente, neque ab occidente, Ps. Th. 74, 6. [Cf. Icel. vestr-vegir the West (the British Isles).]

West-Wille (-as ?); pl. m. The people of some district in England :—West-Willa landes is syx hund hȳda, Cod. Dip. B. i. 414, 29.

west-wind, es; m. A west wind :—Ðá bleów westwind flante favonio, Bd. 5, 19; S. 639, 20. [Cf. O. H. Ger. wester-wint favonius.]

West-Wixan; pl. m. The people of some district in England :—West-Wixna landes is syx hund hȳda, Cod. Dip. B. i. 414, 20.

wêþan; p. de To make calm, gentle, mild :—Blíþe weorðaþ ða ðe brimu wêþaþ laetati sunt quod (fluctus) siluerunt, Ps. Th. 106, 28. v. next word.

wêþe; adj. Sweet, gentle, mild, pleasant :—Ðone swêg ðæs swêtan (wêþan, MSS. O. T.) sauges sonum cantilenae dulcis, Bd. 5, 12 ; S. 630, 23. Ðone scȳnan wlite, wêðne mid willum, Exon. Th. 57, 9 ; Cri. 916. Wegas wêþe pleasant paths, 102, 15 ; Cri. 1673. [Goth. wôþeis sweet (savour) : O. Sax. wôdi.] v. wêþnes.

wôðel. v. wǽdl.

weþer; es; m. A wether, a ram :—Weþer vervex vel manto, Wrt. Voc. i. 23, 56. Weðer aries, ii. 10, 42. Ða habbaþ swá micle hornas swá weðeras habentes cornua similia arietibus, Nar. 34, 19. Tú eald hríðeru odðe .x. weðeras, L. In. 70 ; Th. i. 146, 18 : Chart. Th. 40, 7. Weðras, 468, 25. Is nú irfæs ðæs ðæs stranga winter lǽfed hæfð nigon eald hríðru . . . and fíftig wæþæra, 163, 4. Weðera vervecum, Hpt. Gl. 524, 17. His bigleofa wæs ǽlce dæg . . . hundteóntig weðera (centum arietes, 1 Kings 4, 23), Homl. Th. ii. 576, 33. [Goth. wiþrus (Guþs) agnus (Dei) : O. L. Ger. wither aries : O. H. Ger. widar aries, vervex, multo : Icel. veðr.]

wêþness, e; f. Sweetness, gentleness, mildness :—Biluitnisse and uoednisse mansuetudo et lenitas, Rtl. 100, 13. Ða miclan geniht ðínre wêðnesse (suavitatis tuae), Ps. Th. 144, 6. v. ge-wêþness.

wex, wexen. v. weax, wixen.

wî = weg. v. weg lá, weg-férend, weg-leás.

wibba, an; m. A worm or beetle :—Se glisigenda wibba the glow-worm; cicindela, Wrt. Voc. i. 23, 77. v. scearn-wibba; wifel.

wî-bed, wibil, wic cariscus. v. wíg-bed, wifel, wice.

wíc. The word is generally neuter, but as it is often used in the plural where a singular night express the meaning, the similarity of neuter plural and feminine singular accusatives seems to have caused the word to be taken sometimes as feminine, e. g. tó ânre wíc, Homl. Th. i. 402, 22. A weak form also seems to be used, Chart. Th. 446, 29. **I.** a dwelling-place, abode, habitation, residence, lodging, quarters :—Hê tó him wilniende wæs ðætte heó him funden swylce londáre swylce hê mid árum on beón mehte, and his wíc ðaer on byrig beón mihte on his lífe, Chart. Erl. 69, 23. In locum qui dicitur cynges uuíc (cf. in villa regali qui dicitur Werburging-wíc, i. 275, 3), Cod. Dip. Kmbl. iii. 373, 8. Syndon sume dígol wíc (mansio quaedam secretior) mid wealle and mid bearuwe ymbsealde . . . habbaþ ða wíc gebedhús, Bd. 5, 2 ; S. 614, 31. Synd mê wíc díne (tabernacula tua) leófe, Ps. Th. 83, 1. Beóð him wíc gestaþelad in wuldres byrig, Exon. Th. 230, 19; Ph. 474. Sindon bitre burgtûnas, wíc wynna leás, 443, 18 ! Kl. 32. Sceldes fordas boec and ðeara wíca on byrg, Txts. 443, 10. Londbôc mínra wíca, 458, 8. Hê gewát hám faran, wíca neósan, Beo. Th. 251; B. 125 : 2255; B. 1125. Hê wæs

on ðám foresprecenan wícum (*in praefata mansione*) wuniende, Bd. 4, 3; S. 567, 15, 33. Hí hine nǽnige ðinga of his wícum and of his stówe tó him gelaþian mihton *nequaquam suo monasterio posset erui*, 4, 28; S. 606, 9. Ðæt nán biscop ne nán mæssepreóst næbbe on his wícan ne on his húse wunigende ǽnigne wífman, L. Ælfc. P. 31; Th. ii. 376, 21. Of Lambhyrste tó huntan wícan (*huntsman's lodge*), Cod. Dip. Kmbl. iii. 219, 9. On ðám wícum his fæder Abrahames feorh gesealde, Cd. Th. 104, 21; Gen. 1738: 94, 17; Gen. 1563. Hé dráf of wícum idese of earde, 169, 23; Gen. 2804: Ps. Th. 77, 55: Menol. Fox 48; Men. 24. On ðám wícum (*in Heaven*), Exon. Th. 238, 28; Ph. 611. Wunian in wícum, 316, 9; Mód. 46: Cd. Th. 113, 20; Gen. 1890. Rǽsbora wícum wunode, 108, 26; Gen. 1812: Beo. Th. 6158; B. 3083. Ða ðe on carcerne hleóleásan wíc wunedon, Andr. Kmbl. 261; An. 131: 2621; An. 1312. Ic wíc búge, Exon. Th. 396, 22; Rä. 16, 8: 120, 10; Gú. 269. Wíc eardian, Beo. Th. 5172; B. 2589. Hé bróhte wíf tó háme, ðǽr hé wíc áhte, Cd. Th. 103, 21; Gen. 1721. Ðonne ic ðás ilcan wíc ge-éce, 144, 23; Gen. 2394. Hé him wíc geceás fædergeardum feor, 64, 17; Gen. 1051: 164, 29; Gen. 2722: Ph. 448. Fērend fæste wuniaþ, wíc weardiaþ, Exon. Th. 361, 27; Wal. 26: 228, 34; Ph. 448. Hé him helle gesceóp wælcealde wíc, Salm. Kmbl. 937; Sal. 468. Ic him selle on mínum húse and binnan mínum wealle wíc (*locum*), Past. 52; Swt. 407, 35. Hé him synderlíce wíc getimbrede *ipse sibi monasterium construxit*, Bd. 3, 19; S. 547, 30. Heó hire ðǽr wíc ásette ðæt heó Gode in lifede *ibi sibi mansionem instituit*, 4, 23; S. 593, 26. II. *a place where a thing remains*:—Heó (*Lot's wife*) sceal on ðám wícum wyrde bídan, Cd. Th. 155, 9; Gen. 2570. III. *a collection of houses, a (small) town, a village, a street.* v. wíc-gerēfa:—Wíc *vel* lytel port *castellum*, Wrt. Voc. i. 34, 34: 84, 42: *vicus*, 36, 27. Seó gelaþung fērde of ðǽre byrig tó ánre wíc, Homl. Th. i. 402, 22. Hí cómon tó ánre wíc *processerunt vicum unum* (Acts 12, 10), ii. 382, 13. Tǽme hé tó wíc tó cyngæs sele ... gekyþe hé ... ðæt hé ðæt feoh in wíc gebohte, L. H. E. 16; Th. i. 34, 6–10. Andlanges ðære eá tó ðære wíc; fram ðære wíc tó ðære cortan, Cod. Dip. Kmbl. vi. 217, 6: 148, 24. Hé lǽdde hine bútan ða wíc (*extra vicum*), Mk. Skt. 8, 23. 'Gáþ on ða wíc (*castellum*, Mt. 21, 2) ðe beforan inc stondeþ' ... Hwæt Drihten ða cynelícan burh forhogodlíce naman nemde; for ðon oft wíc beóþ on monegum stówum medmyccle gesette, Blickl. Homl. 77, 22–24. On wícum *in vicis*, Mt. Kmbl. 6, 2. Gá on ða strǽta and on wíc ðisse ceastre *exi in plateas et uicos ciuitatis*, Lk. Skt. 14, 21. Far geond ða strǽta and wíc, Homl. Th. ii. 374, 26. Hé begeat ... Penhyll and Grimanleáh and .ii. hína wícan, Chart. Th. 446, 29. IV. *a temporary abode, a camp, place* where one stops, *station*:—Ðá wæs feórðe wíc, randwigena ræst, be ðan Reádan Sǽ, Cd. Th. 186, 4; Exod. 133: 183, 6; Exod. 87. Ic hét ða fyrd ðǽr wícian ... wǽron ða wíc (*castra*) on leugo .l. furlanga long, Nar. 21, 10. Wæs in wícum wóp, Cd. Th. 190, 16; Exod. 200: 124, 12; Gen. 2061. Hé fōr of ðám wícum, Chr. 878; Erl. 80, 12. Restaþ incit hēr on ðissum wícum (cf. exspectate hic cum asino, Gen. 22, 5), Cd. Th. 174, 20; Gen. 2881. Onmiddan ða wíc *in medio castrorum*, Ps. Th. 77, 28. Tó ðon ðæt hié on ða úre wíc feohtan *ad expugnanda castra*, Nar. 21, 21. ¶ the word occurs in local names, some of which are still found shewing -*wich* or -*wick*:—In Lunden-wíc, L. H. E. 16; Th. i. 34, 3. Tó ðam porte ðe is nemned Cwento-wíc *ad portum cui nomen est Quentavic*, Bd. 4, 1; S. 564, 45. In loco qui vocatur Hremping-wiic, et alia nomine Hafingseota, Cod. Dip. Kmbl. i. 211, 11. Hēr wæs Wǽrinc-wíc getimbred, Chr. 915; Erl. 103, 19. Æt Wæring-wícon, -wícum, 913; Th. i. pp. 186, 187. Hēr wæs Gypeswíc gehergod, 991; Erl. 130, 19. Æt Gipeswíc, 1010; Erl. 143, 17. Cf. too : On gerihte tó hreúdwícan on ða ealdan strǽt; and-lang strǽt tó norðwícan; of norðwícan eft andlang strǽte tó Billesham, Cod. Dip. Kmbl. iii. 449, 14–17. In loco qui dicitur Childesuuicuuon (cf. Cildesuicoque, 75, 13), i. 66, 6. Iuxta marisco qui dicitur biscopes-uuíc, 104, 2 : v. 46, 13. [Of æuerelche huse þat husbonde wunede and his biweddede wíf weore on þere ilke wike, Laym. 31960. Fra wíc to wíc i tune, Orm. 8512. Þar was wonand wiðin a wike tua men, C. M. 7917. Canntyrbery, that noble wyke, Rel. Ant. ii. 93, 1. Ich can loki manne wike, O. and N. 604. O. Sax. wík : O. Frs. wík ; *f.*: O. H. Ger. wích ; *m. vicus.* From Latin.] v. deáþ-, eard-, fird-, here-, hrá-, sceáp-, sealt-, stóc-, wíþig-wíc.

wícan ; *p.* wác, *pl.* wicon ; *pp.* wicen To yield, give way :—Wicon weallfæsten, wǽgas burston, multon meretorras, Cod. Th. 208, 14 ; Exod. 483. [O. Sax. wíkan : O. Frs. wíka : O. H. Ger. wíchan *cedere* : Icel. víkja.] v. ge-, on-wícian.

wíc-bora. v. wíg-bora.

wicca, an ; *m. A wizard, soothsayer, sorcerer, magician* :—Wicca *ariolus*, Wrt. Voc. i. 57, 40: 60, 30. Drēas and wiccan *arioli et conjectoris* (in similitudinem arioli et conjectoris, Prov. 23, 7), Kent. Gl. 869 Drýmen and feóndlíce wiccan and óðre wígeleras, Homl. Th. ii. 330, 28 : Wulfst. 27, 1. Be wiccum, wíglerum, etc. Gif wiccan oþþe wigleras ..., L. E. G. 11; Th. i. 172, 20 : L. Eth. vi. 7; Th. i. 316, 20 : L. C. S. 4 ; Th. i. 378, 7. Wiccum *a pyrhonibus*, Hpt. Gl. 504, 66. Hí áxoden æt wyccum and æt wísum drýum, Homl. Skt. i. 2, 108. Ða fǽmnan ðe

gewuniaþ onfón wiccan, L. Alf. 30 ; Th. i. 52, 10. Ne áxa náne wicca[n] rǽdes *nec sit qui pythones consulat nec divinos*, Deut. 18, 11. [Symou þe wicche *Simon Magus*, Jul. 40, 9. Ðe wicches *the magicians*, Gen. and Ex. 3028. Uor ane wychche þet hette Symoun, Ayenb. 41, 28. Somme saide he was a wicche, Piers P. 18, 69. Wytche, wyche *magus, sortilegus*, Prompt. Parv. 526. Wyche *hic sortilegus*, Wülck. Gl. 652, 12 (15th cent.).] v. next word, to which perhaps some of the passages given above might belong.

wicce, an ; *f. A witch, sorceress* :—Wycce *phytonyssa*, Wrt. Voc. i. 74, 42. Nú cwyð sum wíglere, ðæt wiccan oft secgaþ swá swá hit ágǽð ... Nú secge wē ... ðæt se deófol ... geswuteláþ ðære wiccan hwæt heó secge mannum ... Ne sceal se cristena befrínan ða fúlan wiccan be his gesundfulnysse, þeáh ðe heó secgan cunne sum ðincg þurh deófol, Homl. Skt. i. 17, 108–126. Ánimaþ ða rēðan wiccan, seó ðe ðus áwent þurh wiccecræft manna mód, 7, 209. Wiccan *pythonissam*, Hpt. Gl. 451, 70. Wiccean and wælcyrian, Chart. Erl. 231, 10. Wiccan, Wulfst. 165, 34. Wiccena *parcarum*, Anglia xiii. 31, 104. v. Grmm. D. M. p. 985.

wicce-cræft, es ; *m. Witchcraft, sorcery, magic art* :—Wiccecræft *necromantia*, Hpt. Gl. 501, 66. Ða heáfodleahtras sind ... hǽðengyld, drýcræft, wiccecræft, Homl. Th. ii. 592, 7. Se cristena man ðe his hǽlde sēcan wyle æt unálýfedum tilungum, oððe æt wyrigedum galdrum, oþþe æt ǽnigum wiccecræfte, ðonne biþ hē ðám hǽðenum mannum gelíc, i. 474, 22 : Homl. Ass. 28, 99. Be wiccecræfte (*veneficio*) ðær man corn bærnað, L. Ecg. C. 32, tit. ; Th. ii. 130, 20. Be wífes wiccecræfte *de veneficio mulieris*, 33, tit. ; Th. ii. 130, 22. Se man ðe begá wiccecræft *vir in quo pythonicus vel divinationis fuerit spiritus*, Lev. 20, 27 : Wulfst. 71, 2. Hǽdenscipe bið ðæt man ... wiccecræft (wiccan cræft, *v. l.*) lufige, L. C. S. 5 ; Th. i. 378, 21 : L. N. P. L. 48 ; Th. ii. 298, 1. Wiccecræft álecgan, O. E. Homl. i. 302, 36. Seó wicce ðe áwent þurh wiccecræft manna mód, Homl. Skt. i. 7, 210. Eówer nán ne áxie þurh ǽnigne wiccecræft be ǽnigum ðinge, 17, 26. Ne gýman gē galdra ne idelra hwata ne wígelunga ne wiccecræfta, Wulfst. 40, 14. Be wiccecræftum. Wē cwǽdon be ðǽm wiccecræftum and be liblácum ... gif man ðǽr ácweald wǽre, and hē his ætsacan ne mihte, ðæt hē beó his feores scyldig, L. Ath. i. 6 ; Th. i. 202, 9–12. Wiccecræftas *prestigias*, Wrt. Voc. ii. 66, 25.

wicce-dóm, es ; *m. Witchcraft, sorcery, magic* :—Nǽfre nán man ne geþrístlǽce ǽnigne deófles bigencg tó dónne, ne on wíglunge, ne on wiccedóme, ne on ǽnegum ídelum anginne, Homl. Ass. 143, 123.

wiccian ; *p.* ode To practise witchcraft :—Gif hwá wiccige ymbe ǽniges mannes lufe, and him on ǽte sylle, oððe on drince, oððe on ǽniges cynnes gealdorcræftum, ðæt hyra lufu for ðon ðe máre beón scyle ... Gif hit bið cleric ... *si quis veneficiis utatur, alicujus amoris gratia, et ei in cibo dederit, vel in potu, vel per alicujus generis incantationes, ut eorum amor inde augeatur ... Si clericus sit* (cf. Com a modi clarc, to mi douter his love beed ... he ne miʒtte his wille have ... Thenne bigon the clerc to wiche, An. Lit. 11, 3–8), L. Ecg. P. iv. 18 ; Th. ii. 208, 31 : L. M. I. P. 39 ; Th. ii. 274, 31. [þe steven wicchand (wiecand, *v. l.*) *vocem incantantium*, Ps. 57, 6. Wytchon (wychyn, wycchyn) wythe sorcerye *ariolor, fascino*; wytchyn or charmyn *incanto*, Prompt. Parv. 527.] v. Grmm. D. M. p. 985.

wíc-cræft. v. wícg-cræft.

wiccung, e ; *f. Witching, witchcraft* :—Gif hwylc wíf wiccunga begá *si mulier aliqua veneficia exerceat*, L. Ecg. C. 29 ; Th. ii. 154, 26. [Oðer unriht inoh, wicching and swikedom, O. E. Homl. ii. 213, 15.]

wiccung-dóm, es ; *m. Witchcraft, sorcery, magic* :—Hē hēt tósomne sínra leóda ða wiccungdóm wídost bǽron (*praecepit rex, ut convocarentur arioli, et magi, et malefici, et Chaldaei*, Dan. 2, 2), Cd. Th. 223, 17 ; Dan. 121.

wic-dæg (wicu-, wuce-), es ; *m.* I. *a day of the week* :—Ðam æftran dæge (*the day after Sunday*), on ōþrum witodlíce wucedæge *die sequenti, secunda uidelicet feria*, Anglia xiii. 387, 319. Ðæt hí ðý feórþan wicdæge and ðý syxtan (*quarta et sexta Sabbati*) fæston, Bd. 3, 5 ; S. 527, 9. Ðý drihtenlícan dæge and ðý fíftan wicdæge *die dominica et quinta sabbati*, 4, 25 ; S. 599, 30 : 600, 17. II. *a week-day, a day on which business may be done* :—Wicdaga *nundinarum*, Wrt. Voc. ii. 59, 63. [O. H. Ger. wehha-tag : Icel. viku-dagr.]

wice (and wíc?), es ; *m. A wich-elm* :—Cuicbeám, uuice *cariscus*, Wrt. Voc. ii. 102, 65. Wice, 13, 21 : i. 285, 45 (at 42 *virecta* is glossed by *wice*, but perhaps *cwice* should be read, cf. *virecta* quice, ii. 123, 62). Wíc *vel* cwicbeám *cariscus*, ii. 129, 7. Tó ðam wíc ... of ðam wíce tó ðære hapuldre ... of ðam alre tó ðám twám wycan standaþ on gerēwe eal swá ðæt gemēre gǽð ; swá up tó ðam wíce stynt beneoðan bælles wæge ; of ðam wíce ... is ðe hæge tó ealdan wycan tó ðam wealle, Cod. Dip. Kmbl. iii. 424, 5–30. Genim ... wíce, ác, bircean ... and ælces treówes dǽl, ðe man begitan mæg, Lchdm. ii. 86, 7. ¶ perhaps the word is found in the place name occurring in the following :—Uno in eo loco cui uocabulum est æt Griman laeg ... Tertio æt Wícan, Cod. Dip. Kmbl. ii. 407, 22 (cf. Ðis syndon ðara halfe híde londgemǽru æt Wícan, iii. 464, 2). Ad villam quae uocatur Uuican, i. 153, 27 (cf. Ðis

synd ða langemǽra intó Wican, iii. 382, 4'. [Wyche *ulmus*, Prompt. Parv. 526.]

wíce, an ; *f. An office, a duty, function* :—Ic dó ðæt gē (hyrdas) geswícaþ ðære wícan (*cessare faciam eos* (*pastores*) *ut ultra non pascant gregem*, Ezech. 34, 10), Homl. Th. i. 242, 13. Bydele gebyraþ ðæt hē for his wýcan sý weorces frigra ðonne óðer man, L. R. S. 18 ; Th. i. 440, 6. Ðá hēt se cásere lǽtan león and beran tó ðam cynegum . . . and betǽhte ða wícan ðam wælhreówan Ualeriane, Homl. Skt. ii. 24, 31. Ne gedyrstlǽce nán lǽwede man ðæt hē wissunge oððe ealdordóm healde ofer Godes ðeówum. Hū dear ǽnig lǽwede man him tó geteón Cristes wícan ? Homl. Th. ii. 592, 28. Þonne hig bysega nabbon on heora wícum *quando vacant*, R. Ben. 84, 19. [Stiwardas and burþenas and byrlas and of mystlicean wícan, Chr. 1120 ; Erl. 248, 10. Dou wiken *to do good offices*, O. E. Homl. i. 137, 11. Inne here muðes wike (*officio*), ii. 91, 19. Hie here wiken hem binimeð ðe hie ar noteden, 183, 1. Ure archebiscop mid wurðscipe mucle haldeð his wike, Laym. 29752. He me (*the prefect*) walde warpen ut of mine wike, Jul. 24, 6. No beggeris blod brynge on hygh wyke, Bote he wolde him seolf byswyke, Alis. 4608. Ich can do wel gode wike, For ich can loki manne wike, O. and N. 603.] v. wícnian.

wíc-eard, es ; *m. A dwelling-place* :—Hē on wēstenne wíceard geceás, Exon. Th. 158, 12 ; Gú. 907.

wicel? :—Wicelre (micelre? *the next article is* : Gif ðú lytel drencefæt habban wylle) blede tācen is ðæt ðú árǽre up ðíne swýþran hand and tósprǽd ðíne fingras, Techm. ii. 125, 9.

wíce-weorc. v. wíc-weorc.

wíc-freoþu ; *f. Peace among dwellings* :—Geríseþ gárníþ werum wíg tówiþre wícfreoþu healdan *the strife of the spear beseems men to meet war and keep peace among their dwellings*, Exon. Th. 341, 21 ; Gn. Ex. 129.

wicg, es ; *n.* (a poetical word) *A steed* :—Bið se hwæteádig (ðe) ðæt wicg byrð, Elen. Kmbl. 2390 ; El. 1196. Wycg, Exon. Th. 395, 10 ; Rä. 15, 5. Wicgce ł meare *cornipede, equo,* Hpt. Gl. 406, 21. Wicge wegan, Exon. Th. 395, 27 ; Rä. 15, 14. Wicge rídan, Beo. Th. 474 ; B. 234. Hē on meare rád, on wlancan ðam wicge, Byrht. Th. 138, 54 ; By. 240. Exon. Th. 489, 14 ; Rä. 78, 7. On wicge sittan, Beo. Th. 578 ; B. 286 : Runic pm. Kmbl. 345, 1 ; Rún. 27. Gúðbeorna sum wicg gewende, Beo. Th. 635 ; B. 315. Ongunnon stígan on wægn weras and hyra wicg somod, Exon. Th. 404, 18 ; Rä. 23, 9 : 405, 11 ; Rä. 23, 21. Onweald wicga and wǽpna, Beo. Th. 2094 ; B. 1045. Wicgum rídan, Exon. Th. 404, 4 ; Rä. 23, 2. Beornas cómon wiggum gengan, on mearum mōdige, Andr. Kmbl. 2192 ; An. 1097. Þrió wicg, Beo. Th. 4355 ; B. 2174. [He (*Jesus*) sende after þe alre unwurþeste wig one to riden, and þat is asse, O. E. Homl. ii. 89, 15. *O. Sax.* wigg : *Icel.* vigg (poet.).]

wicga, an ; *m. Some kind of insect* :—Wicga *blatta* (*elsewhere* blatta *is glossed by* nihtbuttorfléoge, *and* eárwicga), *lucifuga,* lytel wicga *bruuinus,* Wrt. Voc. ii. 127, 11, 32. Genim hwætenes meluwes smedman and wicggan innelfe, gníd tósomme, Lchdm. ii. 134, 4. v. eár-wicga.

wicg-cræft, es ; *m. Steed-craft, skill in connection with horses* :—Sum bið meares gleáw, wiccrǽfta wís, Exon. Th. 297, 18 ; Crä. 70.

wíc-geréfa, an ; *m. The reeve of a* wíc. v. wíc, III. From the Latin words which are translated by *wícgeréfa,* it seems that the official so denominated was concerned in collecting taxes, and from a passage in the laws that it was one of his duties to act as witness at sales. As a *wíc-geréfa* of Winchester is mentioned in the Chronicle, wíc cannot be confined to small towns :—Wícgeréfa *publicanus,* Wrt. Voc. i. 18, 47. Se (*St. Matthew*) wæs *theloniarius,* ðæt is gafoles moniend and wícgeréfa, Shrn. 131, 24. Beornulf wícgeréfa (*so three MSS., the fourth has* wíc-geféra ; *Florence of Worcester has* praepositus Wintoniensium) on Wintanceastre, Chr. 897 ; Th. i. 174, 175, 30. Gif Cantwara ǽnig in Lundenwíc feoh gebycge, hæbbe him twēgen oþþe þreó unfácne ceorlas tó gewitnesse, oþþe cyninges wícgeréfan . . . gekýþe hē mid his gewytena ánum, oþþe mid cyninges wícgeréfan, ðæt hē ðæt feoh in wíc gebohte, L. H. E. 16 ; Th. i. 34, 3–10. Uuícgeroebum *teloniarius,* Wrt. Voc. ii. 122, 28. See Kemble's Saxons in England, ii. p. 175.

wíc-herpaþ, es ; *m. A public road to a* wíc (q. v.) :—Be ðam yrðlande óð hit cymð tó ðam wícherpaðe, ðonne andlang ðæs wícherpaðes tó ðam stǽnenan stapole, Cod. Dip. Kmbl. iii. 418, 27. Cf. wíc-weg.

wícian ; *p.* ode. **I.** *to lodge, take up one's quarters.* v. wíc, **I** :—Eallum ūs leófre ys wíkian (*hospitari*) mid ðam yrþlinge þonne mid ðē ; for ðam se yrþling sylþ ūs hláf and drenc, Coll. Monast. Th. 31, 1. An his manna wolde wícian æt ānes būndan hūse, Chr. 1048 ; Erl. 177, 36. **II.** *to camp, encamp.* v. wíc, **IV.** (1) *to stop in the course of an expedition or march* :—Hē ástyrede his fyrdwíc forð tó Iordanen and wícode þreó niht wið ða eá *movit castra, veneruntque ad Jordanem, et morati sunt ibi tres dies,* Jos. 3, 1 : Elen. Kmbl. 130 ; El. 65. Hig fóron fram Sochoþ and wícodon æt Etham (*castrametati sunt in Etham*), Ex. 13, 20 : 15, 27 : Jos. 4, 19. Wícedon, Elen. Kmbl. 167 ; El. 38. Ðū cans eal ðis wēsten and wásð hwǽr wē wícian magon *tu nosti, in quibus locis per desertum castra ponere debeamus,* Past. 41 ; Swt.

304, 16. Ðá hēt ic mīne fyrd restan and wícian *ego jussi castra poni,* Nar. 8, 26. Ðá com Eustachius mid his here tó ðam tūne . . . Wæs seó wunung þǽr swýþe wynsum on tó wícenne, and his geteld wǽron gehende hire wununge geslagene, Homl. Skt. ii. 30, 315. (1 a) *of an object that moves* :—Nihtweard (*the pillar of fire*) nýde sceolde wícian ofer weredum, Cd. Th. 185, 3 ; Exod. 117. (2) *to occupy a position for a time* :—Ðá wícode se cyng on neáweste ðare byrig ða hwíle ðe hié hiera corn gerypon, Chr. 896 ; Erl. 94, 5. Hē wícode ðǽr ða hwíle ðe man ða burg worhte, 913 ; Erl. 102, 6. Tó ðǽm monnum ðe on eásthealfe ðære ē wícodon, 894 ; Erl. 92, 30. Seó eorþe tóbærst ðǽr ðǽr hí wícodon mid wífum and mid cyldum on heora geteldum, Homl. Skt. i. 13, 226. **III.** *in case of travel by water, to land* :—Þyder hē cwæð ðæt man mihte geseglian on ānum mōðe, gyf man on niht wícode . . . and ealle ða hwíle hē sceal seglian be lande, Ors. 1, 1 ; Swt. 19, 13. Ðá hí ofersegledon, hí cōmon tó Genesar and ðǽr wícedon *cum transfretassent, peruenerunt in terram Gennesareth, et applicuerunt,* Mk. Skt. 6, 53. [Wikien ȝe scullen here (wonieþ nou here, 2nd MS.), Laym. 18102.] v. ge-, ymb-wícian.

wícing, es ; *m. A pirate, sea-robber* :—Wícing (wigcing, *v. l.*) oððe scegðman *pirata,* Ælfc. Gr. 7 ; Zup. 24, 9 : *pirata vel piraticus vel cilix,* Wrt. Voc. i. 18, 59. Wícing oððe flotman *pirata,* 73, 74 : *archipirata,* Hpt. Gl. 501, 35. Yldest wícing, Wrt. Voc. i. 18, 60. Philippus scipa gegaderode and wícingas wurdon, and sóna ān .c. and eahtatig ceápscipa gefēngon *Philippus, ut pecuniam praedando repararet, piraticam adgressus est. Captas centum et septuaginta naves mercibus confertas distraxit,* Ors. 3, 7 ; Swt. 116, 3. Metellus fōr on Belearis ðæt lond, and oferwan ða wícingas ðe on ðæt land hergedon *Metellus Baleares insulas bello pervagatus edomuit, et piraticam infestationem compressit,* 5, 5 ; Swt. 226, 23. ¶ in passages dealing with English affairs the word refers to the Northmen :—Ðeáh þrǽla hwylc hláforde æthleápe and of cristendóme tó wícinge weorðe (*become a pirate, go over to the Danes*), Wulfst. 162, 6. Hē stang wlancne wícing, Byrht. Th. 135, 56 ; By. 139. Ða flotan, wícinga fela, 133, 60 ; By. 73 : 134, 40 ; By. 97. Ðý geáre gegaderode ðn hlóþ wícenga (-inga, *v.l.*), Chr. 879 ; Erl. 80, 28. Ðá mētton hié .xvi. scipu wícenga (-inga, *v l.*), 885 ; Erl. 82, 28. Gegaderode micel here hine of Eást-Englum, ǽgðer ge ðæs landheres ge ðara wícinga ðe hié him tó fultume āspanen hæfdon, 921 ; Erl. 107, 15. Weard wícingum wiþerleán āgifen, Byrht. Th. 135, 10 ; By. 116. Ðæt mynster æt Westbyrig weard þurh yfele men and wícingas eall āwēst (cf. bereáfode þurh Densce men, 446, 6), Chart. Th. 447, 8. [*Icel.* víkingr. Cf. *O. Frs.* witsing, wising.] v. sǽ-, ūt-wícing.

wícing-sceaþa, an ; *m. A pirate* :—Uuícingsceaðan *piraticum,* Txts. 84, 736. Wícingsceaþan, sǽsceaþan, æscmen *piratici,* Wrt. Voc. 68, 12. v. next word.

wícing-sceaþe (?), an ; *f. Piracy* :—Wícincsceaðan (*the Erfurt Glossary has* uuícingsceadae) *piraticam,* Txts. 87, 1579.

wícnere, es ; *m. An officer, a minister, steward, manager* :—Wícnere *dispensator,* Hpt. Gl. 453, 47. Be ðam men ðe ðone wífman fram his hláforde āspaneþ, ðe his wícnere (*villicus*) bið, L. Ecg. P. ii. 14, tit. ; Th. ii. 180, 25. Hē clipode him tó his yldestan geréfan (*servum seniorem domus suae*), ðe ealle his þing bewiste . . . Ðá cwæð se wícnere (in v. 9 geréfa is again used, in v. 10 wícnere), Gen. 24, 5. Ðá cwǽdon hig tó ðam wícnere (v. geréfan, v. 16 ; *in each case the Latin is* dispensatorem), 43, 19. Setton him ðá ǽnne wícnere getreówne . . . æt ðam wæs gelang eall heora fóda ; se heom on ealre hwíle metes tilian sceolde, Homl. Skt. i. 23, 217. Nys nānum mæssepreóste ālýfed, ne diácone, ðæt hí geréfan (*praefecti*) beón, ne wícneras (*procuratores*), L. Ecg. P. iii. 8 ; Th. ii. 198, 21. Ic nelle ðæt ǽnig mann āht ðǽr on teó būton hē (*the archbishop*) and his wícneras (cf. the similar document of Henry II : Mine agene wicneres (*ministri*) . . . hi and heara wicneras (*ministri*) ðe hi hit betechan willað, 347, 1–4), Chart. Erl. 233, 7. Se cyngc beóðeþ his geréfan, ðæt gē ðām abbodan beorgan, and fiistan heora wícneran, L. Eth. ix. 32 ; Th. i. 346, 32. Án woruldcyningc hæfð fela þegna and mislíce wícneras, Homl. Skt. i. pref., 60. [He king wæs and his wikenares chæs, Laym. 18175. He sende word bi his beste wukeneren (one of his cnihtes, 2nd MS.), 6704.] v. next word.

wícnung, e ; *f. Discharging of an office, service, stewardship* :—Be gehādodra manna wícnungum *de ordinatorum hominum procurationibus,* L. Ecg. P. iii. 8, tit. ; Th. ii. 194, 32. v. wícnere.

wíc-sceáwere, es ; *m. A harbinger* :—Ðæs Cristes wícsceáwere (*John the Baptist*), Blickl. Homl. 163, 12.

wíc-steall, es ; *m. A camp* :—Leóde ongēton, ðæt ðǽr cwom weroda Drihten wícsteal metan, Cd. Th. 183, 16 ; Exod. 92.

wíc-stede, es ; *m. A dwelling-place, habitation* :—Þúhte him eall tó rūm, wongas and wícstede, Beo. Th. 4915 ; B. 2462. Hē gemunde ða āre, wícstede welige, 5207 ; B. 2607. Hí his wícstede wēstan *locum ejus*

desolaverunt, Ps. Th. 78, 7. Ic éþelstól hæleþa hrére, hornsalu wagiaþ, wera wícstede, weallas beofiaþ, Exon. Th. 383, 11; Rä. 4, 9.

wíc-stów, e; *f.* **I.** *a dwelling-place :*—Ðis ða wyrta sind, ða se wilda fugel somnaþ tó his wícstówe, ðær hé nest gewyrceþ, Exon. Th. 230, 6; Ph. 468. Ðá hé geseah ða wícstówa ðara ryhtwísena Israhéla *justorum tabernacula respiciens*, Past. 54; Swt. 423, 13. **II.** *a camp, an encampment;* both singular and plural forms are used to translate *castra :*—Hé nemde ðære stówe naman Manaim, ðæt is wícstów (*castra*), Gen. 32, 2. Ðá hét ic ða fyrd wícian; wæs seó wícstów on lengo xxes furlonga long, Nar. 4, 15. Hé of ðære wícstówe áfór, Ors. 2, 4; Swt. 76, 13. Bútan ðære wícstówe *extra castra*, Lev. 4, 21: 8, 17: Num. 11, 32: 12, 15; Ex. 33, 11. Bútan hira wícstówe, 33, 7. Bútan wícstówe, Lev. 10, 4. Ceósaþ eów wícstówe *castra ponetis*, Ex. 14, 2. On ðæm wícstówum *in castris Persarum*, Ors. 3, 9; Swt. 126, 5. Ær hé ða wícstówa bereáfian mehte, Swt. 128, 9. Siþþan hé wícstówa náme, 2, 4; Swt. 76, 10: Num. 11, 31.

wíc-þegen, es; *m. A brother in a monastery who performs the duties of an office for a week :*—Wicþegn *betica*, Wrt. Voc. ii. 125, 45. Be wicþénum (*de septimanariis coquine*). Gebróðru gemǽnelíce heom betwyh þénien, and nǽnig sý beládod fram ðære kycenan þénunge . . . Ðære kycenan wicþénas on ðone Sætresdæg ægðer ge fata þweán ge wæterclæðas wacsan . . . þweán on ðan sylfan dæge ealra gebróðra fét ægðer ge ðære wucan wicþénas ge ðære tóweardan . . . Ða wicþénas (cf. ða wucan þegnas *septimanarii*, R. Ben. Interl. 66, 6) ánre tíde ær gemǽnum gereorde gán tó hláfe . . . Æfterfylige ðære tóweardan wucan wicþén, R. Ben. pp. 58–60. Se diácon wucþén *diaconus hebdomadarus*, Anglia xiii. 415, 721. Fram mæssepreóste wucþéne *a sacerdote ebdomadario*, 395, 435. Gebróðru wucþénas *fratres epdomadarii*, 391, 375. Þa wucþénas *epdomadarii ministri*, 415, 714.

wíc-þegnung, e; *f. Service which lasts for a week :*—Se ðe ða ærran wicþénunga geendod hæbbe, þonne hé út of ðære wicþénunge fære, cweþe ðis fers . . . and swá mid bledsunge of ðære wicþénunge fare. Æfterfylige ðære tóweardan wucan wicþén, and þus cweþe . . . and swá mid bletsunge his wicþénunge beginne, R. Ben. 59, 21–60, 8.

wíc-tún, es; *m. A court :*—Hine weorðiaþ on wíctúnum mid lofsangum *intrate atria ejus in hymnis*, Ps. Th. 99, 3. Ingangaþ on his wíctúnas (*atria*), 95, 8. [Þar beoþ þeos gode wiketunes, O. and N. 730.]

wicu, wucu, an; *f. A week :*—Wucu *ebdomada*, Ælfc. Gr. 5; Zup. 14, 17: Wrt. Voc. i. 76, 56: *ebdomada vel septimana*, 53, 19. On ðan seofoðan dæge God geendode his weorc and seó wucu wæs ðá ágán, Lchdm. iii. 234, 16: Anglia viii. 310, 23. Seó wucu on Grécisc hætte *ebdomada* and on Lýden *septimana;* seofon daga ryne ys seó wucu, and feówer wucan wyrcaþ ánne mónð, 319, 3. Án wucu ðæs fæstenes *una quadrigesimae septimana*, Bd. 5, 3; S. 615, 3. Ðeós wucu is geteald tó ánum dæge, Homl. Th. ii. 292, 21. Ymb fyrst wucan bútan ánre niht, Menol. Fox 172; Men. 87. Hé ǽlcere wucan dæg mid nihte ætgædere áfæste *in omni septimana diem cum nocte jejunus transiret*, Bd. 3, 27; S. 559, 12. On ðære seofoðan wiecan (wucan, *v. l.*) ofer Eástron, Chr. 878; Erl. 80, 8. Tuwa on ucan (wucan, *v. l. :* wico, Lind.: wica, Rush.) *bis in sabbato*, Lk. Skt. 18, 12. Ða fullan wican (wucan, *v. l.*) ǽr Séta Marian mæssan, L. Alf. pol. 43; Th. i. 92, 7. Ymb wucan *after a week*, Cd. Th. 88, 14; Gen. 1465: 167, 21; Gen. 2769. On ðam geáre synd getealde twá and fíftig wucena, Lchdm. iii. 246, 12. Hié fela wucena sǽton on twá healfe ðære é, Chr. 894; Erl. 92, 25. vi. wicum (wucan, *v. l.*) ǽr hé forþférde, 887; Erl. 84, 35. Wucum, 901; Erl. 98, 6: Bd. 5, 4; S. 617, 7. Ðæs ymb .iii. wiecan (wucan, *v. l.*), Chr. 878; Erl. 80, 19. Wucan, 941; Erl. 116, 5: Menol. Fox 30; Men. 15. [*Goth.* wikó: *O. L. Ger.* wika: *O. Frs.* wike: *O. H. Ger.* wehha, wohha: *Icel.* vika.] v. Eáster-, fæsten-, gang-, lencten-, palm-, ymbren-wicu (-wuce).

wicu-bót, e; *f. A week's penance :*—Mót tó bóte stíðlíc dǽdbót, and hit man mót sécan be ðæs mannes mihtum, sumon geárbóte . . . sumon wucubóte, sumon má wucena, L. Pen. 3; Th. ii. 278, 13.

wíc-weg, es; *m. The road to a wíc* (q. v.) :—Tó ðæm midlestan wícwege; ondlong ðæs weges eft tó ceastergeate, Cod. Dip. Kmbl. iii. 260, 11. Cf. wíc-herpaþ.

wíc-weorc, es; *n. Weekly work, work done for the lord by the tenant so many days a week :*—On sumen lande is ðæt hé (*the gebúr*) sceal wyrcan tó wicweorc .ii. dagas swilc weorc swilc him man tǽcð ofer geáres fyrst ǽlcre wucan, and on hærfest .iii. dagas tó wicweorce, and of Candelmæsse óð Eástran .iii., L. R. S. 4; Th. i. 434, 5–8. *Consuetudines in Dyddanhamme* . . . Se gebúr sceal his riht dón; hé sceal erian healfne æcer tó wiceweorce . . ., Cod. Dip. Kmbl. iii. 450, 35. Cf. Hér synd gewriten ða gerihta ðe ða ceorlas sculan dón tó Hysseburnan . . . Hí sculan ǽlce wucan wircen ðæt hí man háte bútan þrím, án tó middanwintra, óðera tó Eástran, þridde tó gangdagan, v. 147, 26. v. Seebohm's English Village Community, s. v. week-work.

wíd; *adj.* **I.** in reference to the dimensions of an object, *wide, of (a certain) width :*—Se arc wæs fíftig fæðma wíd, Boutr. Scrd. 21, 4. Fær gewyrc fíftiges wíd, ðrittiges heáh, þreó hund lang elngemeta, Cd.

Th. 79, 7; Gen. 1307. Wite ðú hú wíd and síd helheoðo dreórig, and mid hondum ámet, 308, 29; Sat. 699. Is ðár on ðære myclan ciricean geworht emb ða lástas útan, hwǽne wíddre ðonne byden, fæt up óþ mannes breóst heáh, Blickl. Homl. 127, 6. **II.** where there is a considerable distance between the extremities or sides of an object, *wide, of great width, broad :*—Wíd strǽt *platea*, Wrt. Voc. i. 36, 33. Ðæt geat is swýðe wíd and se weg is swíðe rúm *lata porta et spatiosa via*, Mt. Kmbl. 7, 13. Se mereweard (*the whale*) múð ontýneþ, wíde weleras . . . hí ðær in faraþ, óþ ðæt se wída ceafl gefylled bið, Exon. Th. 363, 13–27; Wal. 53–60. Hí deópne seáð dulfon wídne, Ps. Th. 56, 8. Óþ ða wýde strǽte, súð andlang strǽte, Cod. Dip. Kmbl. ii. 265, 32. **III.** of great surface, *wide, vast, spacious, broad, ample :*—Ðes wída grund, Cd. Th. 7, 11; Gen. 104. Ýða gelaac, wíd gang wætera, Ps. Th. 118, 136. Wíd is ðes wésten, wræcsetla fela, Exon. Th. 120, 5; Gú. 267. Wæs his ríce brád, wíd and weorðlíc, 243, 11; Jul. 9. Þenden ic wealde wídan ríces, Beo. Th. 3723; B. 1859. On andwlitan wídre eorðan, Cd. Th. 81, 25; Gen. 1350. In ðære wídan byrig, 258, 10; Dan. 673. On egeslícere stówe and on wídum wéstene *in loco horroris et vastae solitudinis*, Deut. 32, 10. Ofer wídne holm, Exon. Th. 296, 23; Crä. 55. Ofer wíd wæter, Beo. Th. 4937; B. 2473. Geond ðás wídan weoruld, Met. 8, 41. Ic hæbbe wíde wombe, Exon. Th. 399, 20; Rä. 19, 3. Hí gesetton Sennar wídne and sídne, Cd. Th. 99, 33; Gen. 1655. Setl wíde sídon, 6, 12; Gen. 87. Of ðissum wéstum wídum mórum, Ps. Th. 74, 6. Hæfde wederwolcen wídum fæðmum eorðan and uprodor gedǽled, Cd. Th. 182, 14; Exod. 75. **III a.** of that which is spread over a wide surface. Cf. wíd-folc :—Wé ne magon rím witan; ðæs wíde sind fugla and deóra wornas wídscoope, Exon. Th. 355, 42; Pa. 4. **IV.** *wide, having no limit near, open*, cf. wíd-sǽ :—Sume hí wǽron on wíddre sǽ besencte, Homl. Th. i. 542, 29. **V.** fig. *not confined within narrow limits, of far-reaching power :*—Ne behwylfan mæg heofon and eorðe his wuldres word wíddra and síddra ðonne befæðman mæge eorðan ymbhwyrft and uprodor, Cd. Th. 204, 31; Exod. 427. **VI.** of travel, *that traverses many lands, distant, far and wide :*—Sceal ic wreclástas settan, síðas wíde, Cd. Th. 276, 16; Sat. 189. Wíde síðas, 55, 36; Gen. 905: Beo. Th. 1759; B. 877. **VII.** of the duration of time, *long, lasting long*, in phrases equivalent to *ever, always.* v. wíde-feorh, -ferhþ :—Gé sceolon ádreógan wíte tó wídan ealdre, Exon. Th. 92, 27; Cri. 1515: Cd. Th. 62, 16; Gen. 1015. Tó wídan ealdre, éce mid englum, Andr. Kmbl. 3439; An. 1723. Á tó wídan feore sý úrum Drihtne lof, Blickl. Homl. 65, 24: 103, 29. Ða ðe gewordun wíde feore from fruman worulde, Exon. Th. 272, 33; Jul. 508. Wídan feore *as long as life lasts*, 301, 23; Fä. 23. Ne seah ic wídan feorh *never in all my life have I seen*, Beo. Th. 4033; B. 2014. Ðú sceaft wídan feorh ěcan ðíne yrmðu, Andr. Kmbl. 2766; An. 1385. [*O. Sax. O. Frs.* wíd: *O. H. Ger.* wít *amplus, latus, vastus, spatiosus, capax:* *Icel.* víðr.]

wídan; *adv. From (far and) wide, from a distance :*—Hé his witan wídan gesomnod hæfde . . . Ealle ða ðegnas ðe ðær wídan gegaderode wǽron, Cod. Dip. Kmbl. iii. 315, 9, 36. Óðer sinoð wæs eft óðer healf hund biscopa wídan gesamnod . . . Se feorða sinoð wæs six hund biscopa and .xxx. sacerda swýðe wídan gegaderode, L. Ælfc. P. 26, 28; Th. ii. 374, 7, 22. Ðæt wæs háligdóm se mǽsta of gehwilcum stówum wýdan and sýdan gegaderod, Cod. Dip. B. ii. 389, 23.

wíd-brád; *adj. Wide-spread, far-spreading, ample :*—Hé þeóda gehwam hefonríce forgeaf, wídbrádne welan (cf. hwó man himilríki gehalón skoldi, wídbrédan welon, Hél. 1841), Cd. Th. 40, 22; Gen. 643. [Cf. *O. H. Ger.* wít-preiten *spargere*.]

wíd-cúþ; *adj. Widely known, well known*, (1) of persons, *noted :*—Wídcúþes wíg, Beo. Th. 2088; B. 1042. Húnferð, wídcúðne man, 2983; B. 1489. Sume beóþ swíðe æþele and wídcúþe on heora gebyrdum *hunc nobilitas notum facit*, Bt. 11, 1; Fox 30, 32. Twégen becómon tó ús, wídcúðe þurh heora yrmðe, Homl. Th. ii. 30, 30. (2) of things :—Mid ðý ðe se cyninge gehírde ðæt Apollonius ðone rǽdels swá rihte árædde, ðá ondréd hé ðæt hit tó wídcúð wǽre, Ap. Th. 5, 2. Ðæt gesýne weard, wídcúþ werum, ðæt wrecend ðá gyt lifde, Beo. Th. 2516; B. 1256. Wídcúðne weán, 3986; B. 1991.

wíde, an (wídu); *indecl.* ? cf. brǽdu, lengu, *and O. H. Ger.* wítî); *f. Width :*—Heora wíde (*longitudo*) is .cc. míla, Nar. 36, 28.

wíde; *adv.* **I.** where there is measurement, *widely, far :*—Bearwas wurdon tó axan efne swá wíde swá ða wítelác gerǽhton, Cd. Th. 154, 11; Gen. 2554. Swá wíde swá wæter bebúgeþ, Andr. Kmbl. 665; An. 333: 2469; An. 1236. **II.** with the idea of a great space between extremities, *widely, to a great width :*—Múð ic ontýnde mínne wíde, Ps. Th. 118, 131. Hý tódǽlden unc ðæt wit gewídost (*very far apart*) in woruldríce lifdon, Exon. Th. 442, 15; Kl. 13. **III.** where there is the idea of diffusion, distribution, *widely, in different places, on all sides :*—Wíde *passim*, Wrt. Voc. ii. 85, 75. Wel wíde *passim, ubique*, Hpt. Gl. 512, 18. Fela óðra deófles manna wíde wǽran, Wulfst. 100, 20. Manncwealmas bcóð wíde geond land *erunt pestilentiae per loca*, Mt. Kmbl. 24, 7. Fáh ic eom wíde, Exon. Th. 401, 24; Rä. 21, 16. Ða moldan men wíde geond eorþan lǽdaþ tó reliquium, Blickl.

Homl. 127, 15 : Beo. Th. 538 ; B. 266 : 6190 ; B. 3099. Tóférde se here wíde swá hé ǽr gegaderod wæs, Chr. 1012 ; Erl. 147, 8. Ða cóman tógædere þreóhund biscopa and eahtatýne biscopas wíde gesamnode, L. Ælfc. P. 23 ; Th. ii. 372, 28. Ic ðysne sang fand, samnode wíde, Apstls. Kmbl. 4 ; Ap. 2. Ic eom wíde funden, brungen of bearwum and of burghleoþum, of denum and of dúnum, Exon. Th. 409, 15 ; Rä. 28, 1. Ic geondférde fela londa . . . folgade wíde (*I have served in many a land*), 321, 29 ; Víd. 53. Ehtatýne sýþum hundteóntig þúsenda hí tósendon, and wið feó sealdon wíde intó leódscipas, Blickl. Homl. 79, 23. Hí tóweorp wíde *disperge eos*, Ps. Th. 53, 5 : Exon. Th. 16, 24 ; Cri. 258. Wíde tósáweþ Dryhten his duguþe, 299, 31 ; Crä. 110. Hí bráde weóxan, wíde greówan *multiplicati sunt nimis*, Ps. Th. 106, 37. Leád wíde sprong, Exon. Th. 277, 24 ; Jul. 585. Wæs on Myrceon wíde and welhwǽr Waldendes lof áfylled, Chr. 975 ; Erl. 126, 11. Hé geseah dríge stówe wíde æteówde, Cd. Th. 10, 31 ; Gen. 165. Ðú meaht swá wíde ofer woruld ealle geseón, 36, 1 ; Gen. 565. Ðǽr is wóp wíde gehéred (*heard on all sides*), 285, 6 ; Sat. 333 ; Andr. Kmbl. 3107 ; An. 1556. Ðæt wæs wíde cúþ, hú hé his dagas geendode, Chr. 946 ; Erl. 117, 24 : Cd. Th. 170, 17 ; Gen. 2814. Ða eá geond folc monig weras Eufraten wíde nemnaþ, 15, 17 ; Gen. 234 : Met. 8, 15. Ða wíde springaþ *crebrescunt*, Hpt. Gl. 517, 4. Gif ðeós sprǽc tó wýde spryngþ Nicod. 17 ; Thw. 8, 17. Woruldcyningas wíde mǽre, Cd. Th. 140, 30 ; Gen. 2335. His lof secgaþ wíde under wolcnum wera cneórisse, 117, 7 ; Gen. 1950. Is se apostolhád wíde geweorðod ofer werþeóda, Apstls. Kmbl. 29 ; Ap. 15. Wíde geond eorðan, Menol. Fox 350 ; Men. 176: Dreám geríst wel wíde gehwǽr, 118 ; Men. 59. Se ðe his wordes geweald wíde hæfde, Beo. Th. 159 ; B. 79. Hé wíde (*in all his ways, in all things*) bær herewósan hige, Cd. Th. 255, 23 ; Dan. 628. Swá hit beorna má uncre wordcwidas wíddor ne mǽnden, Exon. Th. 472, 17 ; Rä. 61, 17.   **IV.** where a great distance is traversed, *widely, far, to a distance :*—Fíor † wíde *longiuscule*, Hpt. Gl. 517, 3 : Wrt. Voc. ii. 50, 31. Wíde *longius*, 50, 39. Hig férdon swá wíde landes swá hig faran mihton, Cod. Dip. B. ii. 389, 20. Him féran gewát geond ða folcsceare Abraham wíde, Cd. Th. 106, 36 ; Gen. 1782. Bana wíde scráð, 180, 3 ; Exod. 39. Wíde ásent *relegatus*, Wrt. Voc. i. 51, 42. Ic lástas sceal wíde lecgan, Cd. Th. 63, 5 ; Gen. 1027. Lástas wǽron wíde (*for a great distance*) gesýne ofer myrcan mór, Beo. Th. 2811 ; B. 1403. Seó culufre wíde fleáh, Cd. Th. 88, 15 ; Gen. 1465. Wíde rád ofer holmes hrincg hof séleste (*the ark*), 84, 3 ; Gen. 1392. Mec wíde wolcna strengu ofer folc byreþ, Exon. Th. 390, 3 ; Rä. 8, 5. Hrá wíde sprong, Beo. Th. 3181 ; B. 1588. Ic sceal hweorfan ðý wídor, wadan wræclástas, Cd. Th. 272, 16 ; Sat. 120. Ic wíddor meahte síþas ásettan, Exon. Th. 391, 25 ; Rä. 10, 10 : 485, 6 ; Rä. 71, 9. Ða ðe wræclástas wídost lecgaþ, 309, 15 ; Seef. 57.   **IV a.** of degree, *far :*—Þeáh gé eów eác gewyrce wídor sæce, Exon. Th. 120, 14 ; Gú. 271. Hé hét tósomne sínra leóda ða wiccungdóm wídost bǽron, Cd. Th. 223, 18 ; Dan. 121.   ¶ where the word occurs with words of similar meaning :—Feor and wíde (*longe lateque*) gemǽrsode, Bd. 3, 10 ; S. 535, 2 : 4, 27 ; S. 604, 2 : 5, 12 ; S. 628, 3. Hé férde feorr and wíde geond middangeard, Shrn. 90, 23. Síde and wíde *longe lateque*, Wrt. Voc. ii. 53, 59 : Cd. Th. 8, 3 ; Gen. 118 : Exon. Th. 230, 5 ; Ph. 467. Ðá gesamnodon weras wíde and síde, Andr. Kmbl. 3273 ; An. 1639 : Ps. 56, 6, 13 : Exon. Th. 25, 2 ; Cri. 394 : 155, 3 ; Gú. 854. Wíde oððe síde, Hy. 1, 7. [*O. Sax.* wído : *O. H. Ger.* wíto *spaciose, late, passim* : *Icel.* víða.]

wíde-feorh *long life, an age ;* the word occurs only in the accusative with adverbial force, *for a long time, for ever.* v. wíd, VII :—Wé sceolon leánum hleótan, swá wé wídefeorh (*through all time*) weorcum hlódun, Exon. Th. 49, 11 ; Cri. 784. Á forð heonan wídeferh *for ever*, 36, 28 ; Cri. 583. Swá áwa sceal wesan wídeferh, 142, 12 ; Gú. 643 : 350, 1 ; Sch. 57 : 255, 32 ; Jul. 223. Ic him wille wídeferh wesan underþýded, 138, 12 ; Gú. 375 : 420, 23 ; Rä. 40, 8 : 421, 20 ; Rä. 40, 21. Wídeferg, 270, 19 ; Jul. 467. Ðonne hé gást ofgiefþ, syþþan hine gærsbedd sceal wunian wídefyrh (*so the MS. ; -fyrhþ* (?) *as Thorpe reads*), Ps. Th. 102, 15. v. next two words.

wídefeorh-líc ; *adj. Perpetual, eternal :*—Wídefeorlíc *vel éce aevum vel aetas perpetua*, Wrt. Voc. i. 21, 60.

wíde-ferhþ, -ferþ, *long life, an age ;* the word occurs only in the accusative, alone or with *eall*, with adverbial force, *for a long time, for ever, for all time :*—Heora noma leofaþ wídeferhþ in écnesse *nomen eorum vivet in generationes et generationes*, Bd. 5, 8 ; S. 621, 29. Mihtig God manna cynnes weóld wídeferhð, Beo. Th. 1408 ; B. 702. Hié ne wéndon ðæt hié wídeferhð landgeweorc beweredon, 1879 ; B. 937. Ðú scealt wídeferhð ðínum breóstum bearn tredan eorðan (*super pectum tuum gradieris cunctis diebus vitae tuae*, Gen. 3, 14), Cd. Th. 56, 2 ; Gen. 906. Ðæs ðe hié wídeferð wyrnan þôhton, 180, 26 ; Exod. 51. Ðú wunast wídeferð mid waldend Fæder, Exon. Th. 10, 36 ; Cri. 163. Hafast ðú geféred, ðæt ðé feor and neáh ealne wídeferhð (*through all time*) weras ehtigaþ, Beo. Th. 2448 ; B. 1222. Wese swá, wese swá þurh eall wídeferhð (*through all ages*), Ps. Th. 105, 37. v. two preceding words.

-widere, widerian. v. ge-, mis-, un-, unge-widere, wederian.

Wíderiggas ; *pl. m. The name of some people in England :*—Wíderigga (Witherigga, 416, 11) landes is syx hund hýda, Cod. Dip. B. i. 414, 28.

wíd-fǽðme ; *adj. Broad-bosomed :*—Wídfæðme wǽg, Andr. Kmbl. 1065 ; An. 533. Wídfæðme scip, 480 ; An. 240. [*Icel.* víð-faðmr ; víð-feðmir *a name of one of the heavens.*] Cf. síd-fæðme.

wíd-farende ; *adj. (ptcpl.) Wide-faring, wandering :*—Ðone wídfarendan lǽd on ðín hús *vagos induc in domum tuam*, Past. 43 ; Swt. 315, 14. v. wíd-férende.

wíd-férende ; *adj. (ptcpl.) Wide-journeying, far-travelling :*—On ðam (*the ocean*) wuniaþ, wídférende síðe on sunde, seldlícra fela, Exon. Th. 193, 32 ; Az. 130. Ne magon ðǽr gewunian wídférende, ne ðǽr elþeóðige eardes brúcaþ, Andr. Kmbl. 558 ; An. 279. v. wíd-farende.

wíd-floga, an ; *m. A wide-flier, one that takes wide flights :*—Se wídfloga (*the fire-drake*), Beo. Th. 5652 ; B. 2830. Oferhogode fengel ðæt hé ðone wídflogan weorode gesóhte, 4681 ; B. 2346. [Cf. *Icel.* víð-fleygr.]

wíd-folc, es ; *n. A wide-spread folk :*—Of ðam wídfolc, cneórím micel, cenned wǽron, Cd. Th. 98, 31 ; Gen. 1638. Cf. síd-, unrím-folc.

wíd-gal ; *adj. Wandering, roving :*—Se mé wídgalum wísaþ hwílum sylfum tó ríce, Exon. Th. 401, 1 ; Rä. 21, 5. v. wíd-gil[1], *and next word.*

wídgalness, e ; *f.*   I. *vastness, extensiveness :*—Be ðære wídgalnisse his síðfata and his fóra ðe hé (*Alexander*) geond middaneard férde, Nar. 1, 6.   II. *discursiveness, wandering :*—Wídgalnys módes *vagatio mentis*, Greg. Dial. 2, 3. v. wídgilness.

wíd-gangol ; *adj. Rambling, roving, wandering :*—Wídgongel wíf word gespringeþ, oft hý mon wommum bilihd, hæleð hý hospe mǽnaþ, Exon. Th. 337, 15 ; Gn. Ex. 65. Ðonne wé sittaþ innan ceastre, ðonne wé ús betýnaþ binnan ðǽm locum úres módes, ðý lǽs wé for dolsprǽce tó wídgangule weorðen *in civitate considemus si intra mentium nostrarum nos claustra constringimus, ne loquendo exterius evagemur*, Past. 49 ; Swt. 385, 7.

wíd-gil(1), -giel, -gel, *and* **-gille ;** *adj. Wide-spreading, spacious, vast, broad :*—Wídgil *passiva, vasta*, Hpt. Gl. 527, 52. Þeáh ðeós eorðe unwísum wídgel (cf. rúm, Bt. 19 ; Fox 68, 23) þince, Met. 10, 10. Ðæt is swíðe rúm weg and wídgille *lata et spatiosa via est*, Past. 18 ; Swt. 133, 20. Ðæt fenn mid menigfealdan bígnyssum wídgille and lang þurhwunaþ on norðsǽ, Guthl. 3 ; Gdwin. 20, 8. Sió wídgille *passivus*, Wrt. Voc. ii. 65, 55. Wídgilles fæces *spatiosae intercapedinis*, Hpt. Gl. 434, 46. Wídgilles embhwerftes *vasti orbis*, Hymn. Surt. 104. 7. Ðæs wídgillan wéstenes ða ungearwan stówe, Guthl. 3 ; Gdwin. 20, 10. On stówe wídgylre *in loco spatioso*, Ps. Spl. 30, 10. Tó gódum lande and wídgillum *in terram bonam et spatiosam*, Ex. 3, 8. Hwider arn ðæt wæter of ðam wídgillan flód . . . ? Wén is ðæt ðæt wæter gewende tó ðære wídgillan niwelnysse, Boutr. Scrd. 21, 13-14. Tó ánre wídgyllan byrig, Homl. Skt. i. 3, 82. On ðam wídgillan lande, Num. 21, 25 : Homl. Th. ii. 222, 29. Geond ðone wídgillan munt, Blickl. Homl. 199, 12 : Homl. Skt. ii. 26, 207. Ða wídgillan sǽ, Hexam. 4 ; Norm. 6, 24. Ofer ðæt wídgille wésten, Ælfc. T. Grn. 5, 40 : Jos. 11, 16. Behealde hé hú wídgille ðæs heofenes hwealfa bíþ (hú wídgil sint heofenes hwealfe, Met. 10, 6) *late patentes aetheris cernat plagas*, Bt. 19 ; Fox 68, 22. Wídgille *passivos*, Hpt. Gl. 405, 64. Sum con wonga bigong, wegas wídgielle, Exon. Th. 42, 31 ; Cri. 681. Ic eom brǽdre and wídgielra ðonne ðes wong gréna, 425, 4 ; Rä. 41, 51. Wídgelra, 426, 33 ; Rä. 41, 83. v. wíd-gal.

wídgilness, e ; *f. Vastness, spaciousness, vast expanse :*—Hí him menigfeald þing sǽdon be ðære wídgilnysse ðæs wéstenes, Guthl. 3 ; Gdwin. 20, 16. Seó eorðe stód mid manegum wudum on hire wídgilnysse, Hexam. 6 ; Norm. 12, 5. Ða díglan wídgilnysse *abstrusam vastitatem*, Hpt. Gl. 471, 70. Behealdaþ ða wídgilnesse and ða fæstnesse and ða hrædlérnesse ðisses heofenes *respicite coeli spatium, firmitudinem, celeritatem*, Bt. 32, 2 ; Fox 116, 5. Wé beóð ful swyíte tó farenne geond ealle wídgylnyssa (*vast expanses*) Godes ríces, Homl. Th. ii. 296, 34. v. wídgalness.

wíd-herian, -hergan ; *p.* ede *To celebrate, spread abroad the praise* of a person :—Ðeáh hí for micel gód ne dôn, hí wilniaþ ðæt hí micel ðyncen, and hí mon wídherge *quamvis implere maxima praetermittant, ea tamen minima observant, quae humano judicio longe lateque redoleant*, Past. 57 ; Swt. 439, 34. Cf. wíd-mǽrsian.

widl *filth, pollution :*—Ælc wídðil *omnis pollutio*, Rtl. 98, 24. Idese mid widle and mid womme besmítan, Judth. Thw. 22, 12 ; Jud. 59. Widl and fúl *inluviem*, Wrt. Voc. ii. 44, 53. Geseah síðe sǽlwongas synnum gehladene, widlum gewemde, Cd. Th. 78, 16 ; Gen. 1294. v. wcorold-widl.

wíd-land, es ; *n.*   I. *broad land, the face of the earth.* Cf. wíd-sǽ :—Nǽron Metode wídlond (*or under* II) ne wegas nytte, ac stód bewrigen folde mid flóde, Cd. Th. 10, 13 ; Gen. 156. Ic on middangeard nǽfre egorhere eft gelǽde, wæter ofer wídland, 92, 33 ; Gen. 1538 : 85,

9; Gen. 1412: Andr. Kmbl. 395; An. 198. Hê ûs giefeþ welan ofer wîdlond, Exon. Th. 38, 11; Cri. 605. **II.** *a broad, spacious land :*— Geaf ic welan ofer wîdlonda gehwylc, Exon. Th. 85, 2; Cri. 1385. [Cf. *Icel.* vîð-leudr *having broad lands.*] Cf. sîd-land.

wîd-lâst, es; *m. A track that stretches far, a wanderer's track :*— Wulfes ic mînes wîdlâstum (*far wanderings*) wênum dogode, Exon. Th. 380, 16; Rä. 1, 9. Gê (*the apostles*) sindon earme ofer ealle menn, wadað wîdlâstas (*wide are your wanderings*), weorn geséraþ earfoðsîða, Andr. Kmbl. 1353; An. 677.

wîd-lâst; *adj. Making a track that stretches far, wide-wandering :*— Ðû (*Cain*) fiêma scealt wîdlâst wrecan (*vagus et profugus eris super terram,* Gen. 4, 12), Cd. Th. 62, 28; Gen. 1021. (Wer) wîdlâst ferede rôfne hafoc, Exon. Th. 400, 8; Rä. 20, 6.

wîdlian; *p.* ode *To defile, pollute, violate, profane :*— Ne ðæt ingaas in mûð widlas (*coinquinat*) ðone monno, Mt. Kmbl. Lind. 15, 11. Measapreóstas sunnadæg widlas (*violant*), 12, 5. Ðäs yflo wîdlað (widlas, Rush., *communicant*) ðone monno, Mk. Skt. Lind. 7, 23. Hî (*the apostate angels*) heofon widledan (wid lædan, MS.), Exon. Th. 317, 4; Mód. 60. Se ðe âwiht þicge ðæs ðe wesle widlige (wið licge, MSS.) *qui comederit âliquid de eo quod mustela inquinaverit,* L. Ecg. C. 40; Th. ii. 166, 7. Se ðe mid ænige unclæne þinge sý besmiten . . . bête hê be ðæs widlodes mæðe (*juxta pollutionis gradum*), L. Ecg. P. addit. 10; Th. ii. 234, 2. v. â-, ge-widlian; un-widlod.

wîd-mǽran. v. ge-wîdmǽran, *and next word.*

wîd-mǽre; *adj. Far-famed, famous, celebrated;* in a bad sense, *notorious.* (1) of persons :—Sume teohhiaþ ðæt ðæt betst sý, ðæt mon seó foremǽre and wîdmǽre *quibus optimum quiddam claritas videtur,* Bt. 24, 2; Fox 82, 10. Wîdmǽre wer . . . hê moncynnes mǽste hæfde mægen and strengo, Cd. Th. 98, 14; Gen. 1630. Wîdmǽre cynn, 158, 16; Gen. 2618. (2) of things :—Ân wundorlîc tâcn gelamp, swâ wîdmǽre ðæt feáwa wǽron on ðære neáwiste ðe ðæt ne gesáwe, oððe ne gehŷrde, Homl. Th. ii. 28, 35. Hû Caudenes Furculus sió stôw wearþ swîþe wîdmǽre for Rômâna bismere, Ors. 3, 8, tit.; Swt. 3, 10. Wîdmǽre gewin (*the war of the apostate angels*), Exon. Th. 317, 1; Mód. 59. Wîdmǽre blǽst (*the fire that shall consume the world*), 60, 27; Cri. 976. Swâ gê sweotolran and wîdmǽrran gedôð eówre tælweorðlîcnesse *tanto foedior vestra reprehensibilitas appareat,* Past. 8; Swt. 53, 15. Hafaþ se cantic wîdmǽrost word, Salm. Kmbl. 101; Sal. 50. [*O. H. Ger.* wît-mâri *insignis.*]

wîd-mǽrsian; *p.* ode *To spread abroad the knowledge* or *fame* of an object, *to proclaim, publish, celebrate :*—Ðâ spræc man ofer eall and wîdmǽrsude, ðæt Iôsepes brôðru cômon tô Pharaone *auditum est et celebri sermone vulgatum in aula regis : Venerunt fratres Joseph,* Gen. 45, 16. Hê ongan bodian and wîdmǽrsian ðâ sprǽce *ille coepit praedicare et diffamare sermonem,* Mk. Skt. 1, 45. Heó nolde wîdmǽrsian Cristes dîgelnesse, Homl. Th. i. 42, 18. Wîdmǽrsiende *crebrescens,* Hpt. Gl. 512, 21. v. ge-wîdmǽrsian.

wîd-mǽrsung, e; *f. Proclamation, publication :*—Openung mûþes his wîdmǽrsung (*infamatio*) ys *he openeth his mouth like a crier* (Ecclus. 20, 15), Scint. 96, 11.

wîdness, e; *f. Width :*—Heora wîde (wîdnes, *v.l.,* v. Anglia i. 335) is .cc. mîla *longitudo eorum .cc. stadia sunt,* Nar. 36, 28. Ðæs temples længc wæs syxtig fæðma, and seó wîdnes wæs twêntig fæþma, and his heáhnys wæs þrîtyg fæþma, Anglia xi. 9, 27. Ðæt tempel wæs . . . on wîdnysse twêntig fæðma . . . Ðæt eástportic wæs on lenge twêntig fæðma be ðæs temples wîdnysse, and wæs tŷn fæðma wîd, Homl. Th. ii. 578, 10–13.

wîd-nett, es; *n. A drag-net :*—Wîdnyt (wîd nyt?) *funda,* Wrt. Voc. i. 22, 21.

wido-bâne, widrian. v. wiþo-bân, wederian.

wîd-rynig; *adj. Wide-streaming :*—Hâteþ heofona cyning ðæt ðû forð onsende wæter wîdrynig, geofon geótende, Andr. Kmbl. 3012; An. 1509.

wîd-sǽ; *f. m. Open sea, ocean :*—Deós wîdsǽ *pelagus,* Ælfc. Gr. 8; Zup. 28, 21 : 13; Zup. 84, 1 : Wrt. Voc. i. 70, 14. Him wæs â wîdsǽ on ðæt bæcbord, Ors. 1, 1; Swt. 17, 27 : 19, 26. Fǽmendre wîdsǽ *spumantis pelagi,* Hpt. Gl. 409, 69. Wîdsǽs *cataclismi,* Wrt. Voc. ii. 23, 75. On wîdsǽwes grund, Shrn. 54, 21. Mid his fôtum gangan on wîdsǽ, 111, 28. Wurpan on wîdsǽ, 57, 4. Gif massere geþeáh, ðæt hê fêrde þrige ofer wîdsǽ, L. R. 6; Th. i. 192, 9. Hê lêt him ealne weg ðæt wêste lond on ðæt steórbord, and ða wîdsǽ on ðæt bæcbord, Ors. 1, 1; Swt. 17, 10.

wîd-scofen; *adj.* (*ptcpl.*) *Pushed far, extreme :*—Weá wîdscofen, Beo. Th. 1876; B. 936.

wîd-scop, -sceop; *adj. Widely distributed (?) :*—Fugla and deóra wornas wîdscope swâ wæter bibûgeþ, Exon. Th. 356, 3; Pa. 8.

wîd-scriþol (-el, -ul); *adj. Wide-wandering, roving, rambling :*—Hlûd and wîdscriðel *garrula et vaga,* Kent. Gl. 188. Ðæt feórðe muneca cyn is wîdscriþul (wîdscriþel *gyrovagum,* R. Ben. Interl. 10, 16) genæm-ned, R. Ben. 9, 21. Hit is yfel, ðæt sume (munecas) synd tô wîdscriþole, L. I. P. 14; Th. ii. 322, 13. Fîfte cyn muneca is wîdscriþelra hleápera,

ðe under muneces gegyrlan æghwyder scrîþaþ; ða þurh nânes mannes sande ne faraþ, faraþ þeáh geond missenlîce þeóda, néfre staþolfeste, næfre wuniende, nâhwâr sittende, R. Ben. 135, 20. Wîþscriþole renas tunglena *vagos recursos siderum,* Hymn. Surt. 22, 29.

wîd-sîþ, es; *m. A far journey, long travel :*—Môdor ne rǽdaþ, ðonne heó magan cenneþ, hû him weorðe geond woruld wîdsîð sceapen, Salm. Kmbl. 744; Sal. 371. Wêrig winneþ, wîdsîð onginneþ, Exon. Th. 354, 26; Reim. 51. ¶ the word occurs also as a name for one who has travelled much :—Wîdsîð maðolade, se ðe mǽst mǽrþa ofer eorþan, folca geondférde, Exon. Th. 318, 19; Wîd. 1.

widu. v. wudu.

widuwa; *m. A widower :*—Ðæt bið rihtlîc lîf ðæt cniht þurh-wunige on his cnihthâde, ôð ðæt hê on rihtre mǽdenǽwe gewîfige; and habbe ða syððan, ða hwîle ðe seó libbe : gif hire ðonne forðsîð gebyrige, ðonne is rihtost ðæt hê þananforð wydewa þurhwunige, L. I. P. 22; Th. ii. 332, 32. [Zaynte Paul zayþ to wodewon (*non nuptis et viduis*) : Huo þet guod is, he him hyealde ine þe stat of wodewehod; and zef hit him naʒt ne lykeþ, he him wyui, Ayenb. 225, 14. *O. H. Ger.* witwo *celebs.*] v. next word.

widuwe, widewe, weoduwe, weodewe, wuduwe, wudewe, wydewe, widwe, an; *f. A widow.* v. wîf, III a :—Wudewe (wuduwe, *v.l.* : widuwe, Rush.: widiua, Lind.) *vidua,* Lk. Skt. 18, 3. Widewe, Wrt. Voc. i. 73, 15. Weodewe, Gen. 38, 11. Wydewe (wudewe, Ps. Spl. : weoduwa, Ps. Lamb.: widwe, Ps. Surt.), Ps. Th. 108, 9. Widwe, Lk. Skt. Rush. 2, 37 : 18, 5. Anna seó hâlige wuduwa, Lchdm. iii. 428, 19. Paula wæs gehâlgod wydewe, Homl. Th. i. 436, 9 : Shrn. 112, 31. Sî ælc wydewe (wuduwe, *v.l.*) on Godes griðe and on ðæs cynges; and sitte ælc .xii. mônað werleás; ceóse syþþan ðæt heó sylf wille, L. Eth. v. 21; Th. i. 310, 1. Be wudewan . . . Sitte ælc wuduwe werleás twelf mônað . . . Ne hâdige man ǽfre wudewan tô hrædlîce. And gelǽste ælc wuduwe ða heregeatu binnan twelf mônðum, L. C. S. 74; Th. i. 416, 3–17. Geong wuduwe môt eft ceorlian æfter hire weres forðsîðe, L. Æthb. 75; Th. ii. 382, 32. Mund ðære betstan widuwan eorlcundre, L. Æthb. 75; Th. i. 20, 10. Dînes wuduwan hâdes *viduitatis tuae,* Past. 31; Swt. 207, 12. Wudewan gierela *viduitatis theristrum* (Ald. 76), Wrt. Voc. ii. 87, 46. Wîf gif hire forman were forðsîð gebyrige, be leáfe heó nime ôðerne, gif heó ðæt ceósan wyle; and gif heó ðone oferbyt, wunige heó â syððan on wudewan hâde, L. Ecg. P. ii. 20; Th. ii. 190, 6. Iudith þurhwunode on hire wudewan hâde, Homl. Ass. 114, 399. Ne môston nâ wîfian on nânre wuduwan, L. Ælfc. P. 39; Th. ii. 380, 16. Bûton earmre wudewan, L. Ath. v. 2; Th. i. 230, 19. Gif man widuwan unâgne genimeþ, L. Æthb. 76; Th. i. 20, 13. Gif hwâ wydewan nŷdnǽme, gebête ðæt deópe, L. Eth. vi. 39; Th. i. 324, 25. Wæs gesett ðæt se ðe widewan nâme, oððe âworpen wîf, ðæt hê ne wurde nǽfre syððan tô nânum hâde genumen, L. Ælfc. C. 8; Th. ii. 346, 13. Heora widwan (wudwan, Ps. Spl.), Ps. Th. 77, 64. Fǽmnan and wuduwan, Cd. Th. 121, 14; Gen. 2010. Wydywyna (wudewena, *v.l.* : widuena, Lind.: widwa, Rush.) hûs, Lk. Skt. 20, 47. Weodewena (widwena, Ps. Surt.), Ps. Spl. 67, 5. Widewum, Deut. 27, 19. Weodewum, Ps. Th. 145, 8. Wydewum, 67, 5 : Blickl. Homl. 45, 1. Ða wuduwan (wydewan, wyd-wan, *v.ll.*), L. Alf. 34; Th. i. 52, 16. Earme wydewan, Cd. Th. 128, 27; Gen. 2133. [*Goth.* widuwô : *O. Sax.* widowa : *O. Frs.* widwe : *O. H. Ger.* witawa (-ewa, -uwa, -wa).]

wîd-wegas; *pl. m. Distant regions, regions lying far and wide :*—Ûs gesamna ðâ wîdwegum *congrega nos de nationibus,* Ps. Th. 105, 36. Hê synfulle tôdrîfeþ geond wîdwegas *omnes peccatores disperdet,* 144, 20. Faraþ geond ealne yrmenne grund, geond wîdwegas, bodiaþ geleáfan (*euntes in mundum universum praedicate evangelium,* Mk. 16, 15), Exon. Th. 30, 21; Cri. 482. Férdon folctogan feorran and neán geond wîdwegas, Beo. Th. 1684; An. 840. Blǽd is ârǽred geond wîdwegas, ofer þeóda gehwylce, 3412; B. 1704. Cf. sîd-wegas.

wiel, wielm, wiergan, wiers, wieta, wietan. v. wilh, wilm, wirgan, wirs, wita, witan.

wîf, es; *n.* **I.** *a woman, a female person :*—Wîf *mulier,* wîf ðe wer hæfð *uxor,* Wrt. Voc. i. 73, 12, 14. Wîf ðe hæfð ceorl *uxor,* Ælfc. Gr. 9, 21; Zup. 47, 8. Ald uuif *anus,* Wrt. Voc. ii. 100, 38 : i. 73, 17 : *anula vel vetula,* 50, 48. Ðæt wîf (*mulier*) wæs gehǽled, Mt. Kmbl. 9, 22. Gif hwylc wîf (*mulier*) hire wîfman (*ancillam suam*) swingð, L. Ecg. P. ii. 4; Th. ii. 182, 32. Cwên Hrôðgâres, freólîc wîf, Beo. Th. 1234; B. 615. Wîdgongel wîf word gespringeþ, Exon. Th. 337, 15; Gn. Ex. 65. Wæs sum wîf, seó (ðæt wîf ðió *mulier quae,* Lind.) hæfde untrumnesse gâst, Lk. Skt. 13, 11. Wæs sôna gearu wîf, swâ hire weoruda helm beboden hæfde, Elen. Kmbl. 445; El. 223. Sǽde ðæt wîf hire wordum selfa, Cd. Th. 160, 10; Gen. 2648. Wîfes sceós *baxeae,* Wrt. Voc. i. 26, 20. Ðæt hî nâgan mid rihte þæt hǽmedþing wîfes gemânan, L. Eth. v. 9; Th. i. 306, 19. For ðære synne ðæs ǽrestan wîfes, Blickl. Homl. 5, 5. Freá wîf âweahte, and ða wrade sealde leófum rince, Cd. Th. 11, 12; Gen. 174. Ðæt æðele wîf (*Eve*), 294, 19; Sat. 473. Ðǽr wǽron manega wîf (wîfo, Lind., *mulieres*), Mt. Kmbl. 27, 55 : Lk. Skt. 8, 2 : 24, 22. Betwyx wîfa bearnum *inter natos mulierum,* Mt. Kmbl. 11, 11. Betuh eall wîfa cynn, Blickl. Homl. 5, 21. Rîccra

4 I

(-æ, MS.) wífa (-e, MS.) wǽfels *regillum vel peplum vel palla*, Wrt. Voc. i. 40, 32. Seó ǽrest wífa (*feminarum*) is sǽd in Norþanhymbra mǽgþe ðæt heó munuchâde onfênge, Bd. 4, 23; S. 593, 22. **II.** *a being in the form of a woman :*—Wíf unhýre (*Grendel's mother*), Beo. Th. 4247; B. 2120. Ðǽr ða mihtigan wíf hyra mægen berǽddon, and hý gyllende gâras sǽndan, Lchdm. iii. 52, 21. **III.** *a married woman, a wife :* —His wíf *sua uxor*, Ælfc. Gr. 15; Zup. 104, 2. Câseres wíf *imperatrix vel Augusta*, 42, 10. Abram and Nachor wífudun; Abrames wíf hâtte Sarai, and Nachores wíf Melcha, Gen. 11, 29 : 16, 1 : Cd. Th. 167, 30; Gen. 2773. Gûð sceal in eorle geweaxan, and wíf geþeón leóf (lof, MS.) mid hyre leódum, leóhtmôd wesan, rûne healdan, rûmheort beón, Exon. Th. 338, 28; Gn. Ex. 85. Se man geþeót hine tô his wífe (*uxori*), Gen. 2, 24 : Mt. Kmbl. 19, 5. Se cyning mid his wífe and twâm sunum, Homl. Th. i. 468, 1. Æt his mêder ðe wǽre tô ǽwum wífe forgifen his fæder, L. Alf. pol. 42; Th. i. 90, 29. Ðe wíf hæfð *uxoratus*, Wrt. Voc. i. 50, 44. Ceorl ðe wíf hæfð *maritus*, 73, 13. Ðanon ic mê âfêde, and mín wíf and mínne sunu, Coll. Monast. Th. 27, 23. Ðâ ðâ hê mann wolde beón, hê ne geceás nâ him wíf tô mêder, ac geceás clǽne mǽden, Homl. Th. ii. 6, 34. Sume tiliaþ mid micelre geornfulnesse wífa, for ðam ðæt hí þurh ðæt mǽge mǽst bearna begitan, Bt. 24, 3; Fox 82, 26. Wôhhǽmed mid ôþerra ceorla wífum, Blickl. Homl. 61, 15. His wífum twǽm sǽgde Lameh, Cd. Th. 66, 26; Gen. 1090. Hí him wíf curon, 76, 1; Gen. 1250. Hié hæfdon wíf and cyfesa, Blickl. Homl. 99, 20. ¶ the following passages will illustrate some points connected with the position of women in relation to marriage :—Be ðon ðe mon wíf bycgge, L. In. 31; Th. i. 122, 3. Wê lǽraþ ðæt ǽnig cristen mann . . . ne gewífie . . . on ðæs wífes nêdmâgan ðe hê sylf ǽr hæfde . . . hê nâ mâ wífa ðonne ân hæbbe, and ðæt beó his beweddode wíf, L. C. E. 7; Th. i. 364, 21–28. Wer môt his wífe on fulwihte onfôn, and ðæt wíf ðam were, L. Ecg. C. 18, tit.; Th. ii. 128, 31. Gif ceorl bûton wífes wísdôm deôflum gelde . . . Gif bûtwû deôflum geldaþ, sión hió healsfange scyldigo, L. Wih. 12; Th. i. 40, 4. Gif hwâ stalie swâ his wíf nyte and his bearn, geselle .lx. sciłł. tô wíte. Gif hê stalie on gewitnesse ealles his hírêdes, gongen hié ealle on þeówot, L. In. 7; Th. i. 106, 15. Gif ceorl ceáp forsteld . . . ðonne bið se his dǽl synnig, bûtan ðam wífe, forðon heó sceal hire ealdore hiéran, 57; Th. i. 137, 17. Ðæt ða (*criminals*) ealle beón gearwe mid him silfum and mid wífe and mid ǽrfe tô farenne þider ic wille, L. Ath. iv. proem.; Th. i. 220, 6. Gif be cwicum ceorle wíf hig be ôðrum were forligce, and hit open weorðe . . . heó þolige nase and eárena . . . , L. C. S. 54; Th. i. 406, 6. Mon môt feohtan orwíge, gif hê gemêteþ ôðerne æt his ǽwum wífe, L. Alf. pol. 42; Th. i. 90, 26. Gif frí man wið fríes mannes wíf geligeþ . . . ôðer wíf (hê) his âgenum scætte begete and ðǽm ôðrum gebrenge, L. Ethb. 31; Th. i. 10, 7. Gif ceorl âcwyle be libbendum wífe and bearne, riht is ðæt ðæt bearn mêdder folgie, L. H. E. 6; Th. i. 30, 3 : L. In. 38; Th. i. 126, 3. Gif hwâ cwydeleás of ðyssum lífe gewíte . . . beó be ðæs hlâfordes dihte seó ǽht gescyft swýðe rihte wífe and cildan and nêhmâgon, L. C. S. 71; Th. i. 414, 1. Ðǽr se bônda sæt uncwyd and unbecrafod, sitte ðæt wíf and ða cild on ðam ylcan unbesacen, 73; Th. i. 44, 23. **III a.** *a woman who has been married and lost her husband* (by death or divorce) :—Lâf *vel* forlǽten wíf *derelicta*, Wrt. Voc. i. 50, 46. Wífian on nânre wuduwan, ne on forlǽtenum wífe, L. Ælf. P. 39; Th. ii. 380, 16. Ælc man ðe his wíf forlǽt . . . se ðe ðæt forlǽtene wíf nimð, se unrihthǽmð, Lk. Skt. 16, 18. Gif man mǽdan oþþe wíf (cf. *the old Latin version :* virginem vel viduam) weddian wille, L. Edm. B. 1; Th. i. 254, 2. Ne nýde man nâðer ne wíf ne mǽden tô ðam ðe hyre sylfre mislícige (cf. *passages from the Laws under* widuwe, *and* L. H. I. 1, 3; Si, mortuo marito, uxor ejus remanserit, . . . eam non dabo marito, nisi secundum velle suum, Th. i. 499, 15), L. C. S. 75; Th. i. 416, 20. **IV.** *a female.* v. wer, **V :**—Ælc mon, ge wíf ge wǽpned, Ors. 3, 6; Swt. 108, 27. Ða forman twâ, fæder and môder, wíf and wǽpned, Cd. Th. 12, 33; Gen. 195. **IV a.** *as a grammatical term, feminine.* v. wer, **V a.** [O. Sax. O. Frs. wíf : O. H. Ger. wîp : Icel. víf (*poet.*).] v. aglǽc-, gesíþ-, hǽmed-, mere-, riht-, sige-, síþ-, unriht-wíf, *and next word.*

wífa (?), an; *m. A woman :*—Gif ríce wíf and earm âcennaþ tôgædere, gangon hí ǽweig, nâst ðû hwæðer bið ðæs rícan wífan (-es?) cild, hwæðer ðæs earman, Homl. Th. i. 256, 14.

wíf-cild, es; *n. A female child :*—For wǽpnedbearne sceolde cennende wíf hí âhabban fram Godes hûse ingange ðreó and ðrittig daga, and for wífcilde (*femina*) syx and syxtig daga, Bd. 1, 27; S. 493, 16.

wíf-cyn[n], es; *n.* **I.** *woman-kind, women :*—Ðæt hí of ðam wífcynne him cyning curan *ut de feminea regum prosapia regem sibi eligerent*, Bd. 1, 1; S. 474, 22. Ðû eart gebletsod betuh ealle wífcyn (*in mulieribus*, Lk. 1, 28), Blickl. Homl. 143, 18. [Wiðuten wifkin and childre *besides women and children*, Gen. and Ex. 656.] **II.** *female sex :*—Ôþer ðara is wǽpnedcynnes, sunnan trió, ôþer wífkynnes, ðæt mônan trió *quarum lignum virile est solis, alterum est femineum lune*, Nar. 25, 18. Hwylce wihta beóð ôðre tíd wífcynnes, ôðre tíd wǽpnedcynnes? Salm. Kmbl. p. 202, 12 : Lchdm. iii. 10, 12.

wíf-cýþ[þ], e; *f. A visit to a woman, familiarity with a woman :*—

Ðâ geáscode hê ðone cyning on wífcyþþe (-cyððan, *v.l.*), Chr. 755; Erl. 48, 29.

wifel, es; *m. A weevil, a beetle :*—Wibl *panpila*, Txts. 85, 1498. Wifel *papila*, Wrt. Voc. ii. 67, 59. Wibil, uuibil *cantarus*, Txts. 49, 398. Wifel, Wrt. Voc. ii. 13, 47. Wifel *cantarus (animal)*, 128, 11 : *scarebius*, i. 281, 43. Is ðæs gores sunu gonge hrædra, ðone wê wifel nemnaþ, Exon. Th. 426, 13; Rä. 41, 73. Æfter ðam wifele, Lchdm. ii. 320, 2. Weorp ofer bæc ðone wifel (tordwifel, l. 15) on wege; beheald ðone ne lôcige æfter, 318, 19. ¶ the word seems to occur in several local names. v. Cod. Dip. Kmbl. vi. 352. [Wevyl, wyvyl *or* malte boode (bowðe) *gurgulio*, Prompt. Parv. 523 and 531. O. L. Ger. gold-uuivil *cicendela :* O. H. Ger. wibil *scarabaeus, cantarus :* Ger. wiebel : *Icel.* tord-yfill.] v. scearn-, tord-wifel.

wifel, wifer *an arrow, dart, javelin :*—Gafeluca ł wibere *jaculo, sagitta*, gâre ł wifele *spiculo*, Hpt. Gl. 432, 45, 53. Gâra *jaculorum*, gaflucas *catapultas, sagittas*, wifera *sagittarum*, gâras *spicula*, 405, 52–55. [Wyfle, wepene *bipennis*, Prompt. Parv. 526, and see note.]

wi-fêrend, -wífestre. v. weg-fêrend, wǽpen-wífestre.

wíf-fæst; *adj. Married :*—Gif wíffæst wer (*uxoratus*) hine forligce be his âgenre wylne, L. C. S. 55; Th. i. 406, 14. Cf. wíf-leás.

wíf-feax, es; *n. A woman's hair :*—Wíffex *cesaries*, Wrt. Voc. i. 282, 43 : ii. 16, 46.

wíf-gál; *adj. Incontinent, licentious :*—Swâ lǽren hí ða wífgálan gesinscipe, swâ hí ða forhæbbendan ne gebrengen on unryhthǽmde *sic incontinentibus laudetur conjugium, ut tamen jam continentes non revocentur ad luxum*, Past. 60; Swt. 453, 30.

wíf-gehrine, es; *m. Contact with woman :*—Gif ðíne geféran beóð clǽne from wífgehrine (*femineo contactu*), Nar. 27, 8.

wíf-gemǽdla, an; *m. A woman's fury :*—Wiþ wífgemǽdlan; geberge on neaht rædices moran, ðý dæge ne mæg ðe se gemǽdla sceþþan, Lchdm. ii. 342, 10. v. ge-mǽdan.

wíf-gemâna, an; *m. Mulieris consortium :*—Wífgemânan tô âweccanne . . . ðæt âwecceþ wífgemânan lust, Lchdm. i. 336, 15–17.

wíf-geornness, e; *f. Incontinence :*—Uifgiornis *adulteria*, Mt. Kmbl. Lind. 15, 19.

wíf-gifta; *pl. f. Nuptials, marriage :*—Wæs se weliga ðæra (-e, MS.) wífgifta georn on môde, ðæt him mon fǽmnan gegyrede brýd tô bolde, Exon. Th. 245, 2; Jul. 38.

wíf-hâd, es; *m.* **I.** *womanhood :*—Wê sprecaþ be ðære heofonlícan cwêne æfter wífhâde *we speak of the heavenly queen as woman*, Homl. Th. i. 546, 14. **II.** *female sex :*—Wífhâd *femininum sexus*, Wrt. Voc. ii. 148, 19. Wífhâdes man *femina*, i. 70, 18 : Homl. Th. ii. 10, 12 : 94, 30. Se ðe handlaþ wífhâdes mannes líc, Basil admn. 7; Norm. 50, 11. God âna wât hû his gecynde biþ, wífhâdes oððe weres, Exon. Th. 223, 9; Ph. 357. Se ðreát ðæra Godes ðeówa in wífhâde *ancillarum Dei caterva*, Bd. 4, 7; S. 574, 34. [O. H. Ger. wíp-heit *sexus.*] *See other instances under* wer-hâd.

wíf-hand, a; *f. The female side, female line :*—Mín yldra fæder hæfde gecweden his land on ða sperehealfe, næs on ða spinlhealfe; ðonne gif ic gesealde ǽnigre wífhanda ðæt hê gestrýnde, ðonne forgyldan míne mâgas . . . for ðon ic cwêðe ðæt hí hit gyldan, for ðon hý fôð tô mínum ðe ic syllan môt swâ wífhanda swâ wǽpnedhanda swâðer ic wylle, Cod. Dip. Kmbl. ii. 116, 16–24. v. next word.

wíf-healf, e; *f. The female side, female line :*—On ða gerâd ðæt hí gecuron heora kynecinn aa on ða wífhealfa, Chr. Erl. 3, 16. (Cf. wíf-cynn, I.) v. preceding word.

wíf-hearpe (?), an; *f. A woman's harp :*—On glígbeáme (owifhearpan = on wífhearpan? MS. C.) *in tympano*, Ps. Spl. 150, 4.

wífian; *p.* ode *To take a wife, to marry,* (1) *without an object :*—Nân wer ne wífaþ, ne wíf ne ceorlaþ, Homl. Th. i. 238, 1. Is geset swíðe micel dǽdbôt swylcum mannum tô dônne, ðe eft wífiaþ; and eác is sêlcum preóste forboden, ðæt hí beón ne môton on ða wísan ðe hí ǽr wǽron æt ðam brýdlâcum, ðǽr man ôðre síðe wífaþ. Be ðam man mæg witan, ðæt hit riht nis, ðæt wer wífige oððe wíf ceorlige oftur ðonne ǽne, Wulfst. 304, 28–305, 3. Ne wífiaþ hig, ne hig ne ceorliaþ *neque nubent, neque nubentur*, Mt. Kmbl. 22, 30 : Ne wífiaþ hí, ne ne gyftigeaþ, Mk. Skt. 12, 25. Ðysse worulde bearn wífiaþ and beóð tô giftum gesealde, Lk. Skt. 20, 34. Hí ne wífiaþ, ne hí beóð hâmbrôhte, Hpt. Gl. 436, 40. Ðæt se cniht heólde hine sylfne clǽne ôð ðæt hê wífode, Homl. Ass. 20, 149. Abraham and Nachor wífudun (*duxerunt uxores*), Gen. 11, 29. Wífodon, Lk. Skt. 17, 27. Wífian *nubere*, Hpt. Gl. 485, 72 : Homl. Skt. i. 4, 6. Mê is gesǽd ðæt eówer ancor sægd, ðæt hit sý âlýfed ðæt mæssepreóstas wel môton wífian, Homl. Ass. 13, 6. Ne fremaþ nânum menn tô wífienne (wífigæ, Lind.) *non expedit nubere*, Mt. Kmbl. 19, 10. Wífigende and gyfta syllende *nubentes et nubtum tradentes*, 24, 38. (2) *with an object governed by on :*—Be ðam men ðe wífaþ on twâm geswystrenum *de homine qui duas sorores in matrimonium ducit*, L. Ecg. P. ii. 11, tit.; Th. ii. 180, 18. Be ðam men ðe on his mâgan wífaþ *de homine qui inter cognatas suas uxorem ducit*, 18, tit.; Th. ii. 180, 30. Se ðe wífaþ on ðam forlǽtenum wífe, Homl. Th. ii. 322, 34. Tô his âðumum ðe woldon wífian on his dohtron (*qui accepturi erant filias ejus*),

Gen. 19, 14. Hé ne móste bútan áene wífigan, ne hé ne móste on wydewum wífigan, L. Ælfc. C. 7 ; Th. ii. 346, 5. Wífian, L. Ælfc. P. 39 ;
Th. ii. 380, 16. Is nýd dæt cristene menn on dære driddan cneórisse
odde on dære feórþan him betwih wífian sceole *necesse est ut tertia vel
quarta generatione fidelium licenter sibi jungi debeat,* Bd. 1, 27 ; S. 491, 8.
[Iudas wiuede o Thamar, A. R. 308, 13. To late here sones wyue, R.
Glouc. 35, 9. To wyui *nubere,* Ayenb. 225, 17. Wyvyñ or weddyñ a
wyfe *uxoro,* Prompt. Parv. 531.] v. ge-wífian.

wíf-lác, es ; *n. Intercourse with women :*—Gif hwá openlíce Lengctenbryce gewyrce . . . þurh wíflác (*concubitum,* Lat. vers. Cf. qui in Quadri-
gesima ante Pascha nupserit, .i. annum peniteat, L. Ecg. E. 108; Th. ii.
113, 3. Eác is gesynscipum micel þearf, dæt hí hig on dás hálgan tíd (*Lent*)
cláenlíce healdan, bútan álces hámedes besmytennysse, L. E. I. 43; Th.
ii. 440, 2), L. C. S. 48; Th. i. 402, 30. Ealle synodas forbudon áefre
álc wíflác (v. wífung) weófodþénum, L. I. P. 23; Th. ii. 336, 12 :
Wulfst. 270, 21.

wíf-leás ; *adj. Without a wife, unmarried :*—Gif hwylces weres forme
wíf bid deád, dæt hé be leáfe óder wíf niman móte, and gif hé da oferbýt,
wunige hé á syddan wífleás (*coelebs*), L. Ecg. P. ii. 20 ; Th. ii. 190, 3.
[Wyyfles or not weddyd *agamus,* Prompt. Parv. 526.] See also next
word.

wíf-leást, e ; *f. Lack of women :*—Menn hæfdon on frymde heora
mágan tó wífe, and swá wel mósten for dære wífleáste, Homl. Skt. i. 10,
216.

wíf-líc ; *adj.* I. *womanly, of a woman, female, feminine :*—
Wíflíc *muliebris,* Ælfc. Gr. 5 ; Zup. 17, 17. Wíflíces *femineis,* Wrt.
Voc. ii. 148, 20. Wíflícum lícome of woeres dú saldest líchome fruma
femineo corpore de viri dares carne principium, Rtl. 109, 15. Bútan
wíflícre bysnunge *without an example among women,* Homl. Th. i. 198, 5.
Mid wíflíce níde *with all a woman's hate,* Ors. 1, 2 ; Swt. 39, 18. Dæt
hé ne fordon wíflíce háde árede *ut ne sexui quidem muliebri parceret,*
Bd. 2, 20 ; S. 521, 24. Áwyrp mé hyder dínne scyccels, dæt ic mæge
da wíflícan týddernysse oferwreón, Homl. Skt. ii. 23 b, 211. I a.
as a grammatical term, *feminine* (gender) :—Æfter gecynde syndon twá
cyn on namum, *masculinum* and *femininum,* dæt is werlíc and wíflíc ;
wíflíc cyn byd *haec femina* dis wíf . . . *Neutrum* is náder cynn, ne
werlíces ne wíflíces, Ælfc. Gr. 6 ; Zup. 18, 5–15. II. *wifely,
matronly :*—Wíflícre *matronalis,* Hpt. Gl. 505, 36. Wíflícere, 520, 2.
Da wíflícan, Wrt. Voc. ii. 58, 22. [O. H. Ger. wíp-líh *muliebris,
femineus.*]

wíflíce ; *adv. Like a woman :*—Wíflíce *muliebriter,* Ælfc. Gr. 38;
Zup. 232, 17 : Hpt. 504, 30. Dú wunodest áefter dínum were wíflíce on
cláennysse *after your husband's death you continued in womanly purity,*
Homl. Ass. 114, 392.

wíf-lufu, an ; *f. Love for a woman :*—Se hálga wer dære wíflufan (*the
love of Herod for Herodias*) wordum stýrde, unryhtre áe, Exon. Th. 260,
12 ; Jul. 296. Ingelde weallaþ wælnídas, and him wíflufan cólran weordaþ,
Beo. Th. 4137 ; B. 2065. Cf. wíf-myne.

wíf-mann (wím-, wim-?), es ; *m.* (*but* seó wífman *occurs*). I.
a woman :—Wé láeraþ dæt áenig wífman neáh weófode ne cume da hwíle
de man mæssige, L. Edg. C. 45 ; Th. ii. 254, 3. Dara manna sum wæs
bescoren preóst, sum wæs láewede, sum wæs wífmon (*femina*), Bd. 5, 12 ;
S. 628, 35. Minutia hátte án wífmon, de on heora wísan sceolde nunne
beón. Seó hæfde geháten . . . dæt heó wolde hiere líf on fáemnháde álibban
Minucia, virgo vestalis, Ors. 3, 6 ; Swt. 108, 15. Seó wífman (seó wímman, vv. 18, 22) *Jahel,* Jud. 4, 21. Wífmannes loccas *crines,* Wrt. Voc.
i. 42, 49. Wífmannes innod *matrix, uterus,* 44, 39. Ne scríde nán
wáepman mid wífmannes reáfe (*veste feminea*), Deut. 22, 5. Be wífmannes
beweddunge, L. Edm. B. 1 ; Th. i. 254. Be ungewintredes wífmannes
nédháemde. Gif mon ungewintráedne wífmon tó niédháemde geþreátige,
L. Alf. pol. 26 ; Th. i. 78, 16. Nú cwede gé dæt gé ne magon beón
bútan wímmannes þénungum, L. Ælfc. C. 6 ; Th. ii. 344, 19. God
geworhte dæt ribb tó ánum wífmen (*in mulierem*), Gen. 2, 22. Dæt
bisceop . . . næbbe on his húse náenne wífman, búton hit sý his módor . . .
L. Ælfc. C. 5 ; Th. ii. 344, 13. Gif hwá wille wid wífman (*cum
muliere*) unrihtlíce háeman, L. Edg. C. 33 ; Th. ii. 274, 10. Þeówne
wímman *ancillam,* L. Ecg. C. 25 ; Th. ii. 150, 18. God hí geworhte
wáepnedman and wímman (wýfman, *v. l.,* hiuu ꞇ wífmon, Lind. : wífmenn,
Rush.) *masculum et feminam fecit eos Deus,* Mk. Skt. 10, 6. Wépmen
ge wífmen *viri ac feminae,* Bd. 3, 5 ; S. 527, 7. Wífmenn, Exon. Th.
460, 12 ; Hö. 16. Hæleþa gemót, wítgena weorod, wífmonna þreát,
fela fáemnena, folces unrím, 462, 7 ; Hö. 48. Wáepmanna sang and
wífmanna sang, Homl. Th. i. 442, 1. Wæs micel ege from dáem wífmonnum (*the Amazons*), Ors. 1, 10 ; Swt. 46, 27. I a. *a serving-
woman :*—Gif hwylc wíf (*mulier*) hire wífman (*ancillam suam*)
swingd, and heó þurh da swingle wyrd deád . . . fæste seó hláefdige
(*domina*) .vii. geár, L. Ecg. P. ii. 4 ; Th. ii. 182, 32 : ii. 4, tit. ; Th.
ii. 180, 6. Heó freóde Hægelfláede hire wímman, Chart. Erl. 253, 16.
God gewítnode ealle his wímmen (*uxorem ancillasque suas*), Gen. 20,
18. II. *applied to plants, female :*—Gif man scyle mugcwyrt tó
láecedóme habban, donne nime man . . . da grénan wífmen, Lchdm. iii.

72, 21. [*Laym.* wifmon, wimmon : *Orm.* wifmann, wimmann : *A. R.*
wummon : *Ayenb.* wyfman.]

wíf-myne, es ; *m. Love for a woman :*—Drihten weard Faraone yrre
for wífmyne (*love for Sarah*), Cd. Th. 111, 25 ; Gen. 1861. Cf.
wíf-lufu.

-wifre. v. gange-wifre.

wíf-scrúd, es ; *n. Clothing for a woman, woman's dress, female
attire :*—Ic geann mínre yldran dehter . . . ánes wífscrúdes ealles. And
mínre gyngran dehter ic geann ealles dæs wífscrúdes de tó láfe bid, Chart.
Th. 530, 14–25.

wift, e ; *f. Some implement used in weaving :*—Hé sceal habban
fela towtóla . . . pihten, timplean, wifte, wefle, wulcamb, Anglia ix.
263, 12.

wíf-þegen, es ; *m. A pander ;* leno, Wrt. Voc. i. 66, 31 : 284, 14 :
ii. 51, 63.

wíf-þing ; *pl. n. Matters connected with women, marriage, intercourse :*—Tó wífþingum foxes tægles se ýtemæsta dæl on earm áhangen ;
dú gelýfest dæt dis sý tó wífþingum on bysmær (*irritamentum ad coitum*)
gedón, Lchdm. i. 340, 22 ; 368, 16. Wífþing, gifta, háemed *hymeneos,*
Wrt. Voc. ii. 43, 13. Be dam men de gelómlíce wífþing begéd *de homine
qui crebras nuptias conciliat,* L. Ecg. P. ii. 20, tit. ; Th. ii. 180, 32.
[He weddede þat máeiden, and nom heo to his bedden ; þer wes wífding
riche, Laym. 31128.] Cf. brýd-þing.

wífung ; *f.* I. *taking a wife, marriage :*—Be gehádodra
manna wífunge (*matrimonio*), L. Ecg. P. iii. 1, tit. ; Th. ii. 194, 25 :
Gen. 24, 9. Ús sceamaþ tó secgenne ealle da sceandlícan wíglunga de
gé dwáesmenn drífaþ on wífunge, Homl. Skt. i. 17, 102. Se dridda
cwæd : 'Ic hæbbe gewífod . . .' þurh da wífunge sind getácnode dæs
líchaman lustas, Homl. Th. ii. 374, 19. Ádas and wífunga sindan tócwedene heáhfreólsdagum, L. Eth. vi. 25 ; Th. i. 320, 24. Dás sinodas
forbudon álce wífunga áefre weófodþénum, L. Ælfc. P. 30; Th. ii. 374,
35. II. in plural, *wives ; matrimonia :*—Eów preóstum þingd, dæt
eów nán sin ne sý dæt gé mid wífungum swá libban swá láewede men,
L. Ælfc. P. 32 ; Th. ii. 376, 28. v. frum-, unriht-wífung.

wig *a way,* wíg *an idol.* v. weg, wíh.

wíg, es ; *n.* I. *fight, battle, war, conflict :*—Wíg odde gefeoht
mavors, Wrt. Voc. ii. 55, 37. Donne wíg cume, Beo. Th. 46 ; B. 23 :
5737 ; B. 2872. Wíg ealle fornam, 2165 ; B. 1080 : Exon. Th. 291,
11 ; Wand. 80 : Elen. Kmbl. 262 ; El. 131. Wæs dæs wyrmes wíg wíde
gesýne, nearofáges níd neán and feorran, hú se gúdsceada Geáta leóde
hatode and hýnde, Beo. Th. 4621 ; B. 2316. Ful oft dér wíg ne álæg
there was constantly war, Exon. Th. 325, 30 ; Víd. 119. Wíges on
wénum *expectant of battle,* Cd. Th. 188, 30 ; Exod. 176. Wíges bídan,
Beo. Th. 2541 ; B. 1268. Se wyrm getrúwode wíges and wealles *the
dragon trusted to battle* (*or under* II ?) *and bulwark,* 4635 ; B. 2323.
Him wæs hild boden, wíges wóma, Elen. Kmbl. 37 ; El. 19 : Andr. Kmbl.
2709 ; An. 1357 : Exon. Th. 277, 5 ; Jul. 576. Sumum wíges spéd hé
giefeþ æt gúþe, 42, 16 ; Cri. 673. Wæs Hródgáre herespéd gyfen, wíges
weordmynd, Beo. Th. 130 ; B. 65. Hé hafaþ wíges leán, bláed bútan
blinne, Elen. Kmbl. 1647 ; El. 825. Sum bid wíges heard, beadocræftig
beorn, Exon. Th. 295, 27 ; Crä. 39 : (*Ulysses*) Met. 26, 13 : (*Sigemund*)
Beo. Th. 1776 ; B. 886 : (*St. Andrew*) Andr. Kmbl. 1677 ; An. 841.
Wíges oflysted, 2454 ; An. 1228. Wíges hrémige, Chr. 937 ; Erl. 115, 8.
Wíges sáed, Erl. 112, 20. Him wíge forstód fæder frumsceafta, weard
him seó feohte tó grim, Exon. Th. 317, 14 ; Mód. 65. Heald mé here-
wáepnum wid unholdum, and wíge belúc feóndum *effunde frameam, et
conclude adversus eos,* Ps. Ben. 34, 3. Wigge, Beo. Th. 3545 ; B. 1770.
Wígge under wætere, 3316 ; B. 1656. Æt wíge cringan, 2679 ; B. 1337.
Æt wíge sigecempa, Ps. C. 9. Æt wíges spéd, sigor æt sæcce, æt
gefeohte frid, Elen. Kmbl. 2362 ; El. 1182. Hé mid wíge ácwealde
done cyning and dæt folc *percusserunt urbem et omnes habitatores ejus,*
Jos. 10, 30. Hí mid wíge ácwealdon eall dæt hí dér fundon *percussit in
ore gladii universas animas, quae in ea fuerant,* 10, 37. Gif hwá mid
wíge godcundra gerihta forwyrne . . . Gif hé man gewundige . . . Gif hé
man áfylle . . . Gif hé gewyrce dæt man hine áfylle, L. C. S. 49 ; Th. i.
404, 6–12. Hé gewann mid wíge done eard *cepit omnem terram,* Jos.
11, 23 : Homl. Th. ii. 216, 1. Seó burhwaru heóldan mid fullan wíge
ongeán, Chr. 1013 ; Erl. 148, 12. Hú him speów áegder ge mid wíge
ge mid wísdóme, Past. pref. ; Swt. 3, 8. Giefe on wíge, Exon. Th.
299, 25 ; Crä. 107. Hé on wígge (*in bello*) áfeallen wæs, Chart. Th.
201, 27. Céne tó wíge, Jud. p. 162, 30. Dæt folc wurdon gewexene
tó wíge ful strange, Homl. Th. ii. 212, 18. Man beónn ealle Cantware tó wigge, Chart. Th. 201, 21. Æghwylc óþerne bylde tó wíge,
Byrht. Th. 138, 44 ; By. 235. Tó wígge faran, Chart. Th. 201, 22.
Hié giredon hié tó wíge, Ors. 3, 5 ; Swt. 106, 17. Wigge, Elen. Kmbl.
95 ; El. 48. Hé sende twelf þúsenda gewáepnodra manna tó dam wíge
(*ad pugnam*), Num. 31, 6. Hí beód gewáepnode on da wísan de man
hors gewáepnaþ, donne man tó wíge þencd, Wulfst. 200, 11. Hié heora
land oferhergodan, and him dæs náenige bóte dydon, búton ofermódlíce
wíg and þreátunge, Blickl. Homl. 201, 24. Abraham sealde wíg tó wedde,
nalles wunden gold, Cd. Th. 124, 29 ; Gen. 2070. Oft ic (*a shield*)

wíg seó, frécne feohtan, Exon. Th. 388, 6; Rä. 6, 3.   Wælhwelpes wíg, 397, 21; Rä. 16, 23.   Gesécean wíg, Beo. Th. 1374; B. 685.   Wíg gefeohtan, 2170; B. 1083.   An wíg gearwe, 2499; B. 1247.   II. *fighting force* (abstract or concrete), *valour; troops* :—Wæs his módsefa manegum gecýðed, wíg and wísdóm, Beo. Th. 705; B. 350.   Næfre on óre læg wídcúþes wíg, ðonne walu feóllon, 2088; B. 1082: Exon. Th. 338, 27; Gn. Ex. 85.   On Móyses hand weard wíg gifen, wigena mænieo, Cd. Th. 216, 11; Dan. 5.   Hé mid ðam óðrum flocce tó ðære birig férde beótlíce mid wíge *ascendit cum senioribus in fronte exercitus, vallatus auxilio pugnatorum*, Jos. 8, 10.   Ðanon hé gewende mid wíge tó Lebna and oferwann ða burh *transivit cum omni Israel in Lebna et pugnabat contra eam*, 10, 29.   Offór hiene (*Philip*) óðere Sciþþie mid lytelre firde . . . Philippus him dyde heora wíg unweorð (*made light of their force*), Ors. 3, 7; Swt. 118, 2.   Ne hé him ðæs wyrmes wíg for wiht dyde, easoð and ellen, Beo. Th. 4685; B. 2348.   [He scheldede his scalken al se heo to wiȝe solden, Laym. 4728.   Com mid muchle wiȝe (*a great force*) Irtac, 25365.   To werchen wi *to fight*, Gen. and Ex. 3220.   O. Sax. wíg: O. Frs. wích: O. H. Ger. wíc (ch, g) *bellum, proelium, pugna, militia*: Icel. víg; *n.* Cf. Goth. waihjô *pugna*.] v. án-, and-(Exon. Th. 112, 22; Gú. 147), camp-, féðe-, þræc-, weorold-wíg; or-wíge.   The word is found in proper names. v. Txts. p. 631.

wíg(?); *adj.* v. wíg-heafola.

wiga, an; *m.   I. one who fights, a (fighting) man, a warrior* :—Wiga *heros*, Ælfc. Gr. 9, 31; Zup. 57, 11.   Wiga oððe wígstrang *bellipotens*, Wrt. Voc. ii. 12, 45.   lung wiga *tyro*, i. 18, 16.   Wiga wintrum geong, Byrht. Th. 137, 62; By. 210.   Wælreów wiga (*Beowulf*), Beo. Th. 1262; B. 629.   Wiga ellenróf, Wald. 79; Vald. 2, 11.   Wác wiga, Exon. Th. 290, 18; Wand. 67.   Wigan wígheardne, Byrht. Th. 133, 64; By. 75: Cd. Th. 189, 22; Exod. 188.   Wigan unforhte, módige twégen, Byrht. Th. 134, 5; By. 79.   Wigan on gewinne, 140, 42; By. 302: Cd. Th. 197, 23; Exod. 311: 219, 22; Dan. 58.   Ðær wigan sittaþ on beórsele blíðe ætsomne, Runic pm. Kmbl. 342, 4; Rún. 14.   Wigena æscberendra, Byrht. Th. 123, 6; Gen. 2040.   Wigena mænieo, 216, 12; Dan. 5.   Wigena strengest (*Beowulf*), Beo. Th. 3091; B. 1543.   Hí sendon máran sciphere strengran wihgena *mittitur classis prolixior armatorum*, Bd. 1, 15; S. 483, 16.   Wigum and wæpnum, Beo. Th. 4779; B. 2395.   ¶ in phrases denoting a chief or leader :—Wigena hláford (*Byrhtnoth*), Byrht. Th. 135, 49; By. 135.   Wigena baldor (*Holofernes*), Judth. Thw. 22, 5; Jud. 49.   Dauid cyning, wigena baldor, Elen. Kmbl. 688; El. 344.   Wigena hleó . . wigena weard (*Constantine*), Elen. 300–306; El. 150–153.   Wigena strengel (*Beowulf*), Beo. Th. 6222; B. 3115.   *Similarly the Deity is called* wigena wyn, Exon. Th. 281, 4; Jul. 641.   I a. *used of that which destroys* :—Wiga wælgifre (*death*), Exon. Th. 162, 7; Gú. 972: 231, 8; Ph. 486.   Wiga unlæt láces, 164, 4; Gú. 1006.   Fýr swearta lég, weallende wiga, 61, 15; Cri. 985.   Wiga (*a dog? fire?*) is on eorþan wundrum ácenned, 433, 23; Rä. 51, 1.   II. *a noble, strenuous man* :—Se ðe mid wætere oferwearp wuldres cynebearn, wiga weorþlíce, Menol. Fox 317; Men. 160.   Wigan unsláwne (*St. Andrew*), Andr. Kmbl. 3419; An. 1713.   Wigena tíd (*the day of St. Simon and St. Jude*), Menol. Fox 370; Men. 186.   [Gaw. Allit. Pms. wyȝe; *pl.* wyȝes: Alex. (Skt.) wee; *pl.* wees, wies: Piers P. wy, wye.   Cf. O. H. Ger. Wigo (*proper name*).]   v. æsc-, beorn-, byrn, cumbol-, folc-, gár-, gúð-, lind-, ord-, ræde-, rand-, ríd-, scild-, wǽpen-, þeód-wiga.

wígan [*p.* wág, *pl.* wigon; *pp.* wigen] *to fight, do battle* :—Nú sceal hond and heard sweord ymb hord wígan, Beo. Th. 5012; B. 2509.   Móises getealde ðæs folces meniu wígendra manna *numeravit Moyses omnem summam filiorum Israel a viginti annis et supra*, Num. 26, 1.   Six hund þúsenda wígendra manna, Homl. Th. ii. 194, 14: Homl. Skt. ii. 25, 367: Homl. Ass. 103, 54.   [Goth. weihan (weigan? v. Lk. 14, 31); *p.* waih *to fight*: O. H. Ger. wíhanteru *bellantium*.   Cf. Icel. vega; *p.* vá *to fight*.]   v. ofer-wígan, wígend, wigian.

wí-gár. v. wíg-gár.

wíg-bǽre; *adj. Warlike, martial, eager for fighting* :—Wígbǽre *bellicosus, pugnandi cupidus*, Wrt. Voc. ii. 125, 36.

wíg-bealu, wes; *n. War-bale, harm caused by war* or *the calamity of war* :—Wígbealu weccean *to kindle the wasting flame of war*, Beo. Th. 4098; B. 2046.

wíg-bed, wí-bed, wió-bed, -bud, wié-bed, weó-bed, -bud, weófod (-ed, -ud), wéfod, es, *also* -beddes; *n.* (*generally, but se weóbud, Past. 33; Swt. 217, 21, and pl. wíbedas, Bd. 5, 20; S. 641, 42) An altar* [*from* wíg (wíh) *and* beód; *some forms, e. g.* wígbeddes, weóbedd, *suggest that the word was thought to be derived from* bed] :—Weófod *altar* vel *ara*, Wrt. Voc. i. 26, 51.   Hé scolde ðone Godes altar habban uppan áholodne, ðæt hé meahte on healdan ða lác ðe mon bróhte tó ðam weóbude; for ðæm, gif se weóbud ufan hol nære, and ðær wind tó cóme, ðonne tóstencte hé ða lác.   Hwæt elles getácnaþ ðæt weóbud búton ryhtwísra monna sáula? . . . Wæs eall sió offrung uppe on ðæt wiébed (wióbud, Cott. MSS.) bróht, Past. 33; Swt. 217, 19–25.   Ðæt weóbud, 219, 3.   Wígbed, Bd. 2, 3; S. 504, 39.   Ðæt weófud (-od, MS. A.: wígbed, Lind.: wíbed, Rush.), Mt. Kmbl. 23, 19.   Wígbedes hornas *cornu altaris*, Ps.

Th. 117, 25: Ps. Lamb. 117, 27.   Tó wígbedes ðénunge, Bd. 2, 20; S. 522, 9: 5, 10; S. 624, 34.   Wígbedes (weófodes, col. 1), 3, 17; S. 544, 3, col. 2.   Weófodes (wígbeddes, Lind.: wí-bedes, Rush.), Lk. Skt. 1, 11.   Weófodes þen, Homl. Ass. 22, 206.   Weóuedes (weófedes), R. Ben. 55, 2.   On wígbede tó hálsienne *ariolandi*, Wrt. Voc. ii. 9, 15.   An dǽl ðam wíbede (wígbede, *v. l.*), L. E. B. 12; Th. ii. 242, 18: Bd. 3, 23; S. 555, 14.   Tó wíbede, Ps. Surt. 42, 4.   Tó weófode (wígbed, Lind.: weófud í wíbede, Rush.), Mt. Kmbl. 5, 23.   On wígbed ðín, Ps. C. 138.   Tó wígbede (beforan ðæt weófud í wíbed, Rush.) *ad altare*, Mt. Kmbl. Lind. 5, 24.   Ic ymbgaa wíbed ðín, Ps. Surt. 25, 6: Cd. Th. 107, 18; Gen. 1791: 108, 14; Gen. 1806: 113, 5; Gen. 1882.   Weóbedd, 172, 8; Gen. 2841.   Uppan ðæt weófod, Ex. 24, 6: 29, 20.   Lege under weófod, Lchdm. ii. 138, 28: 142, 8.   Wígbedu (wíbed, Surt.: weófod, Spl.: wiébed, Spl. T.) ðín *altaria tua*, Ps. Th. 83, 4.   Tó wígbedum, Bd. 1, 27; S. 488, 38.   Wíbedum (*v. l.* weófodum), 1, 15; S. 484, 1.   Tó Godes weófedan, L. Eth. vii. 26; Th. i. 334, 30.   Tó hálgum wéfodum, Coll. Monast. Th. 36, 5.   Ðæt tempel and ða weófedu (wígbedo, Bd. M. 136, 18) . . . ða wígbed and ða heargas *templa et altaria . . . aras et fana*, Bd. 2, 13; S. 516, 33–39.   Ða wígbed (*v.l.* weófedu), S. 517, 18.   Hé wíbedas sette, 5, 20; S. 641, 42.   Wíbedu *arulas*, Germ. 394, 259.   Paulus sceáwode ða weófoda, óþ ðæt hé funde án weófod ðe ðis gewrit on stód: *Deo ignoto*, ðæt is on Englisc, 'Uncúðum gode is ðis weófod hálig,' Homl. Skt. ii. 29, 21.   Hig ðær gedydon twá weófedu, Blickl. Homl. 205, 15.   [Laym. weofed (wefð, 2nd MS.), weofd; *dat.* wæfde (wefde, 2nd MS.): A. R. Kath. weoued: Ps. R. Glouc. weved: Ayenb. wieved.]

wígbed-bót, e; *f. A fine paid to the bishop for the injury done to the church by doing wrong to one in holy orders* :—Gif man preóst gewundige, gebéte man ða wyrdlan, and tó weófodbóte for his háde sylle .xii. ór.; æt diácone .vi. ór. tó weófodbóte, L. N. P. L. 23; Th. ii. 294, 4–6.   Gif man preóst ofsleá, forgilde man hine be fullan were, and biscope feówer and .xx. ór. tó weófodbóte; æt diácone .xii. ór. tó weófodbóte, 24; Th. ii. 294, 7–9.   Gif hwá gehádodne man bende oððe beáte oþþe swýðe gebysmrige, béte wið hine swá hit riht sý, and bisceope weófodbóte be hádes mǽde, L. C. S. 42; Th. i. 400, 23.   In the laws of Henry I it is called *emendacio altaris*, 11, 8; Th. i. 521, 7: 66, 3; Th. i. 569, 13.

wígbed-heorþ, es; *m. The altar-hearth, the part of the altar where the offering is burnt* :—Hé genom on ðam wíbedheorðe ðæs dustes dǽl, Lchdm. iii. 364, col. 1.

wígbed-hrægel, es; *n. An altar-covering* :—Hé sende ða ðing eall ða ðe tó cyrican ðénunge nýdþearflíce wǽron, húselfatu and wígbidhrægl (-bed-, Bd. M. 90, 2) (*vestimenta altarium*), Bd. 1, 29; S. 498, 9.

wígbed-sceát, es; *m. An altar-cloth* :—Bewindan ða mágas ðæs cildes hand on ðæs altares weófodsceáte (*in palla altaris*), R. Ben. 103, 14.   Ðis syndon ða cyrican mádmas on Scírburnan.   Ðær synd . . . ii. mæssereáf and iii mæssehakelan and ii weóvedsceátas and ii overbrǽdels, Cod. Dip. B. iii. 660, 33.   Hit gedafenlíc is ðæt his (*the priest's*) reáf ne beó horig, and his weófodsceátas beón wel behworfene, L. Ælfc. C. 22; Th. ii. 350, 21.   Hé hæfst ðiderynn gedón . . . .v. wællene weófodsceátas and .vii. oferbrædelsas, Chart. Th. 429, 25.   Gif hwá wyle wyrcan weófodsceatas oððe óðre reáf of his. ealdum cláðum, gesylle ða ealdan, and geceápige níwe, Homl. Ass. 35, 284.   v. next word.

wígbed-sceáta, an; *m. An altar-cloth* :—On weófodsceátan *in palla altaris*, R. Ben. Inter. 99, 10.

wígbed-steall, es; *n. The part of the church where the altar stands* :—Wé lǽraþ ðæt mæssepreósta ǽnig ne cume binnan weófodstealle búton his oferslipe, ne húru æt nán weófode ðæt hé ðær þénige búton ðære wǽde, L. Edg. C. 46; Th. ii. 254, 9 note.   v. wíg-steall.

wígbed-þegen, es; *m. A minister of the altar, an ecclesiastic who performs service at the altar* :—Gif weófodþén, ðæt is, biscop oððe mæssepreóst oððe diácon, gewífode . . . hí forbudon ǽlc wíflác weófodþénum, L. I. P. 23; Th. ii. 336, 3–13: Wulfst. 270, 21.   Gif weófodþén his ágen líf rihtlíce fadige, ðonne sí hé fulles þegnweres wurðe, L. Eth. ix. 28; Th. i. 346, 17.   Be gehádedum mannum.   Gif weófodþegen manslaga wyrðe, L. C. S. 41; Th. i. 400, 13.   Gif man freóndléasne weófodþén mid tihtlan belecge, L. Eth. ix. 22; Th. i. 344, 22: L. C. E. 5; Th. i. 362, 18: L. C. S. 39; Th. i. 398, 25.   Weófodþéna mǽde medemige man for Godes ege, L. Eth. ix. 18; Th. i. 344, 9.

wígbed-þegnung, e; *f. Service at the altar* :—Wé forbeódaþ ðæt ǽnig preóst óðre[s] cirican náðer ne gebicgæ ne geþicgæ, búton hine hwá mid heáfodgylte forwyrce, ðæt hé weófodþénunge wyrðe ne sí, L. N. P. L. 2; Th. ii. 290, 8.

wígbed-wíglere, es; *m. One who divines from the sacrifices, a diviner, soothsayer* :—Wígbedwíglere *ariolus* (as if from *ara*), Wrt. Voc. i. 17, 11.

wíg-bil[1], es; *n. A battle-blade, a sword* :—Ðæt sweord ongan æfter heaþoswáte hildegicelum, wígbil wanian, Beo. Th. 3218; B. 1607.

wíg-blác; *adj. Splendid with warlike equipment* :—Werud wæs wígblác (cf. beran beorht searo, 191, 23; Exod. 219.   Wígbord scinon, 207, 14; Exod. 466), Cd. Th. 190, 24; Exod. 204.

wíg-bora, an; *m. A belligerent* :—Wígbora *belliger*, Ælfc. Gr. 8; Zup. 27, 16.

wíg-bora, an ; *m. An image-bearer :*—Wícbora (wióbora, Anglia xiii. 35, 214) *signifer*, Hpt. Gl. 495, 71. v. wíh.

wíg-bord, es ; *n. A shield :*—Hé héht him gewyrcean eallírenne wígbord; wisse hé gearwe, ðæt him holtwudu helpan ne meahte, lind wið líge, Beo. Th. 4667 ; B. 2339. Wígbord scinon, Cd. Th. 207, 14 ; Exod. 466.

wíg-cirm, es ; *m. The din of battle :*—Ðær wæs wígcyrm micel, hlúd hilde swég, Cd. Th. 120, 6 ; Gen. 1990.

wíg-cræft, es ; *m.* **I.** *war-craft, military skill :*—Pirrus wæs gemǽrsad ofer ealle óþere cyningas, ǽgðer ge mid his miclan fultume, ge mid his rǽdþeahtunge, ge mid his wígcræfte *Pyrrhus in se, ob magnitudinem virium consiliorumque, summam belli nomenque traduxit*, Ors. 4, 1 ; Swt. 154, 27. Hý him grimme forguldon ðone wígcræft ðe hý æt him geleornodon *vincere, dum vincitur, edocuit*, 1, 2 ; Swt. 30, 7. Hé hæfde Higeláces hilde gefrunen, wlonces wígcræft (or **II** ?), Beo. Th. 5898 ; B. 2953. **I a.** *a warlike art, a warlike engine :*—Hý wurdon geráde wígcræfta, Ors. 1, 2 ; Swt. 30, 6. Mid scotum, ge mid stána torfungum, ge mid eallum heora wígcræftum *vis magna telorum*, 3, 9 ; Swt. 134, 16. Wígcræftum *machinis*, Wrt. Voc. ii. 58, 33. **II.** *warlike force, military power* (abstract or concrete) :—On Thessali hé ðæt gewinn swíþost dyde for ðære gewilnunge ðe he wolde hí him on fultum geteón for heora wígcræfte, for ðon hié cúþon on horsum ealra folca feohtan betst *Thessaliam ambitione habendorum equitum Thessalorum, quorum robur ut exercitui suo admiscerit, invasit*, Ors. 3, 7 ; Swt. 112, 3. Hé (*Christ*) mihte, gif hé wolde, wígcræft habban sóna genóhne (cf. Mt. 26, 53), L. Ælfc. P. 51 ; Th. ii. 386, 34. Ðá beþôhtan hié ealle heora wígcræftas Exantipuse *Xanthippum, cum auxiliis accitum, ducem bello praefecerunt*, Ors. 4, 6 ; Swt. 174, 30.

wíg-cræftig ; *adj. Strong in war :*—Hé ðone gúðwine (*a sword*) gódne tealde, wígcræftigne, Beo. Th. 3626 ; B. 1811.

wígend, wíggend, es ; *m.* **I.** *a fighting man, a warrior, soldier :*—Wígend weorðullost (*Beowulf*), Beo. Th. 6189 ; B. 3099. Ðæm wígende (*Constantine*), Elen. Kmbl. 1964 ; El. 984. Ðone wíggend (*Holofernes*), Judth. Thw. 25, 13 ; Jud. 258. Wígend cruncon wundum wérige, Byrht. Th. 140, 43 ; By. 302 : Beo. Th. 6279 ; B. 3144. Wígend unforhte, Cd. Th. 189, 6 ; Exod. 180. Wígend, cêne under cumblum, Andr. Kmbl. 2408 ; An. 1205. Wíggend, Judth. Thw. 22, 20 ; Jud. 69 : 23, 26 ; Jud. 141. Wígendra scolu (*Ulysses and his men*), Met. 26, 31. Wíggendra, Andr. Kmbl. 2191 ; An. 1097. Hé ðæt word ácwæþ tô ðám wíggendum, Judth. Thw. 25, 29 ; Jud. 283. Wígend weccean, Beo. Th. 6040 ; B. 3024 : Elen. Kmbl. 211 ; El. 106. **II.** *a noble, strenuous man :*—Se wígend, Nergendes þegen, Mathias, Menol. Fox 49 ; Men. 24. Ða wígend, cempan coste (*St. Andrew and St. Matthew*), Andr. Kmbl. 2108 ; An. 1055. Wuldres wynn, wígendra þrym, 1774 ; An. 889. Wígend (*St. Andrew's disciples*), 1699 ; An. 852. Gelǽdde ða wígend (*those in the ark*) weroda Drihten, Cd. Th. 85, 7 ; Gen. 1411. ¶ *in the phrase* wígendra hleó = *a lord, chief :*—Wígendra hleó, freáwine folca (*Hrothgar*), Beo. Th. 863 ; B. 429 : (*Sigemund*), 1803 ; B. 899 : (*the Deity*), Andr. Kmbl. 1011 ; An. 506 : (*St. Andrew*), 1792 ; An. 898. Ðú eart weoroda God, wígendra hleó, helm alwihta, Exon. Th. 25, 31 ; Cri. 409. Wíggendra hleó, Eádmund cyning, Chr. 942 ; Erl. 116, 18. [*O. Sax. O. Frs.* wígand : *O. H. Ger.* wigant *bellator, pugnator, mars, armatus.*] v. burg-, byrn-, gár-, lind-, rand-, sweord-wígend (-wígende).

wígende ; *adj.* (*ptcpl.*) *Fighting, able to fight.* v. wígan.

Wígere-ceaster. v. Weogorna-ceaster.

Wígestas (-e ?) ; *pl. m. The name of some people in England :*—Wígesta landes is nygan hund hýda, Cod. Dip. B. i. 414, 20.

wíg-freca, an ; *m. A warrior :*—Wyrsan wígfrecan, Beo. Th. 2428 ; B. 1212 : 4985 ; B. 2496.

wíg-fruma, an ; *m. A leader in war, a chieftain :*—Wígfruma (*Hrothgar*), Beo. Th. 1332 ; B. 664. Æfter wígfruman *after the chieftain's death*, 4514 ; B. 2261.

wíg-gár, es ; *m. A lance :*—Wígár *lancea*, wegures (wígáres ?) gewrið *amentum*, Wrt. Voc. i. 35, 46–47. Cf. wíg-spere.

wíg-gebed, es ; *n. Prayer to an idol* (?) :—Wíggebed (wigg-bed ?) *ara*, Wrt. Voc. ii. 9, 43. v. wíg-bed.

wíg-getawa (-e) ; *pl. f. War-equipments :*—On wíggetawum, Beo. Th. 741 ; B. 368.

wíg-gild (wíh-), es ; *n. An idol :*—Hié onhnigon tô ðam herige, hǽdne þeóde wurðedon wíhgyld, Cd. Th. 227, 5 ; Dan. 182. Cf. deófol-gild.

wíg-gryre, es ; *m. Terror caused by war :*—Wíggryre wífes *the terror inspired when a woman makes war*, Beo. Th. 2572 ; B. 1284.

wíg-haga, an ; *m. A phalanx :*—Hé mid bordum hét wyrcan ðone wíhagan, and ðæt werod healdan fæste wið feóndum, Byrht. Th. 134, 50 ; By. 102.

wíg-heafola (?) :—[Hé] wôd þurh ðone wælréc wíg[hea]folan bær freán on fultum, Beo. Th. 5316 ; B. 2661. *Hea* is the reading of Thorkelin's transcripts, but now the MS. shews only quite uncertain traces of *h*, and *ea* is entirely gone (Zupitza). Wíg-heafola is taken to mean *a helmet* by some editors : Grein suggests wígneafolan = *umbonem*

bellicum i. e. *clypeum.* Could the reading be wígne afolan ? Cf. *Icel.* vígr *in fighting state, serviceable for fighting*, and afli *strength* ; so that the passage would mean he had *or* brought strength that might serve to help his lord in battle.

wíg-heáp, es ; *m. A war-troop, a band of warriors :*—Is mín fletwerod, wígheáp gewanod, Beo. Th. 958 ; B. 477.

wíg-heard ; *adj. Stout in fight, hardy :*—Wigan wígheardne, Byrht. Th. 133, 64 ; By. 75. [*Icel.* víg-harðr (*poet.*).]

wíg-hete, es ; *m. Hate that leads to war :*—Sunu deáþ fornam, wíghete Wedera *death took off her son, the Weders' hate that found its vent in war*, Beo. Th. 4246 ; B. 2121.

wíg-hryre, es ; *m. Fall in fight :*—Se ðe æt sæcce gebád wíghryre wráðra *he that in strife had lived to see the fall in fight of fierce foes*, Beo. Th. 3242 ; B. 1619.

wíg-hús, es ; *n.* (in Wrt. Voc. i. 36, 41 it is masc.) *A war-house, a tower, fortification :*—Ðis wíghús *haec arx*, Ælfc. Gr. 9, 75 ; Zup. 73, 14 : 3 ; Zup. 7, 9. Se híhsta wíghús *arx*, Wrt. Voc. i. 36, 41. Wíghús *propugnaculum*, Hpt. Gl. 499, 61. On ælcum ylpe wæs án wíghús getimbrod, and on ælcum wíghúse wǽron þrittig manna, Homl. Skt. ii. 25, 561. Wíghúses *turris*, Wrt. Voc. ii. 84, 28. Wíghús *propugnacula*, i. 36, 40. Wíghúsum *turribus*, ii. 91, 25 : Ps. Th. 47, 11 : Past. 33 ; Swt. 229, 5. Se weall is mid stǽnenum wíghúsum (*habitaculis defensorum*) beworht, Ors. 2, 4 ; Swt. 74, 21. Menn wyrcaþ wíghús him (*elephants*) on uppan, and of ðám feohtaþ, Hex. 9 ; Norm. 16, 11. [*O. H. Ger.* wíc-hús *turris, propugnaculum.*]

wíg-hyrst, e ; *f. The trappings of war :*—Beorn monig goldbeorht wíghyrstum scán, Exon. Th. 478, 3 ; Ruin. 35.

wigian ; *p.* ode *To fight :*—Gif hé wigie and man gewundie, L. E. G. 6 ; Th. i. 170, 8. [Cf. *Goth.* waihjó *strife.*] v. wígan.

wígle (wigle ?), es ; *n. Divination, heathen practice :*—Wíglum *ceremonias* (the passage is : Ad tortas simulacrorum ceremonias, Ald. 41), Anglia xiii. 33, 162. [þurh Merlines wiȝel (craft, 2nd MS.), Laym. 19250. He (*a devil*) makeð þe unbilefulle man to leven swilche wigeles, swo ich ar embe spac, Rel. Ant. i. 131, 27. His (*the devil's*) wiȝeles and his wrenches, A. R. 300, 5. Wieles, 92, 21 : Fragm. Phlps. 8, 54. Wiheles, Marh. 13, 9.] v. steor-wigle ; wíglere.

wíg-leóþ, es ; *n. A war-song, the trumpet's summons :*—Gemundon weardas wígleóþ . . . býman gehýrdon flotan, Cd. Th. 191, 27 ; Exod. 221.

wíglere (wiglere ?), weohlere, es ; *m. A diviner, soothsayer, augur, sorcerer :*—Wíglere *augur*, Wrt. Voc. i. 74, 37. Ðes and deós wiglere *hic et haec augur*, Ælfc. Gr. 9, 22 ; Zup. 49, 2. Nú cwyð sum wíglere, ðæt wiccan oft secgaþ swá swá hit ágáeð mid sóðum ðincge, Homl. Skt. i. 17, 108. On gelícnysse wígleres and rǽdendes (*arioli et coniectoris*), Scint. 75, 12. Wýgleras *auspices*, Germ. 398, 79. Be wiccum, wíglerum, etc. Gif wiccan oððe wígleras, oþþe morðwyrhtan . . ., L. E. G. 11 ; Th. i. 172, 20 : L. C. S. 4 ; Th. i. 378, 7. Wiccan oþþe wígleras, scíncræftigan . . ., L. Eth. vi. 7 ; Th. i. 316, 20. Wiccan and wígleras (wígeleras, *v. l.*), Wulfst. 27, 1. Drýmen, and wiccan and óðre wígeleras beóð tô helle bescofene for heora scíncræftum, Homl. Th. ii. 330, 28. Wígulera *magorum, hariolorum*, Hpt. Gl. 502, 51. Tunglera † wi[g]lera *Chaldaeorum* . . . wíhlera (? printed wineena) *hariolorum*, 483, 5–10. Ðonne man tô wiccan and tô wígleran tilunge séce æt ǽnigre neóde, Wulfst. 171, 11. Hé wiccan fordyde, and wígleras áflígde, and drýcræft tówearp, Homl. Skt. i. 18, 464. [Wielare *augur*, Wrt. Voc. i. 89, 20. Þe wielare (*the devil*) makeð a swote smel cumen, ase þauh hit were of heouene, A. R. 106, 2. *M. Du.* wijcheler.] v. fugel-, gebyrd-, wígbed-wíglere (-weohlere), *and next word.*

wíglian ; *p.* ode *To practise divination or sorcery :*—Wígliaþ stunte men menigfealde wígelunga on ðisum dæge æfter hǽðenum gewunan, swylce hí magon heora líf gelengan, oþþe heora gesundfulnysse, Homl. Th. i. 100, 19. Ne sceal nán cristen mann nán þincg þe ðam mónan wíglian, Lchdm. iii. 266, 17. [*M. Du.* wijchelen. v. Grmm. D. M. 985.] v. wíglung ; wigol.

wíg-líc ; *adj. Warlike, martial :*—Ðæt wæs wíglíc werod, Cd. Th. 192, 17 ; Exod. 233. Wíglíc *bellica*, Wrt. Voc. ii. 125, 42. Wíglíce tól *instrumenta bellica*, Hpt. Gl. 424, 28. Wíglíce *bellicosas*, 425, 7. Wǽpna wíglíce *arma bellica*, Hymn. Surt. 135, 23. [*O. H. Ger.* wíc-líh *bellicus, bellicosus : Icel.* víg-ligr.]

wíglíce ; *adv. In a warlike manner, by fighting :*—Bellatores syndon wígmen, ðe eard sculon werian wíglíce mid wǽpnum, L. I. P. 4 ; Th. ii. 306, 37 : Wulfst. 267, 11. v. án-wíglíce.

wíglung, e ; *f. Divination, soothsaying, sorcery, augury :*—Wilung *divinatio*, Kent. Gl. 554. Wé gehýrdon seggon, ðæt nán mann ne leofode gif hé gewundod wǽre on ealra hálgena mæssedæg. Nis ðis nán wíglung, ac wíse menn hit áfunden þurh ðone hálgan wísdóm, Lchdm. iii. 154, 5. Gif treówa beóð on fullum mónan geheáwene, hí beóð heardran, and langférran tô getimbrunge . . . Nis ðis nán wíglung, ac is gecyndelíc ðincg, Homl. Th. i. 102, 25. Hleótan man mót bútan wiccecræfte . . . gif hí hwæt dǽlan willaþ ; ðis nis nán wíglung, ac bið wissung for oft, Homl. Skt. i. 17, 87. Wígelunge *divinatione*, Hpt. Gl. 467, 69. Deófles bígencg, ne on wíglunge ne on wiccedóme, Homl. Ass. 143, 122. Ðæt

hé tǽld tó unálýfedlícere wíglunge, gif hwá ða wyrta on him becnitte, búton hé hí tó ðam dolge gelecge, Homl. Th. i. 476, 4. Se ðe gelýfð wíglungum oððe be fuglum, oððe be fnorum, oððe be horsum, oððe be hundum, ne bið hé ná cristen . . . Se ðe hwider faran wille . . . clypige hé tó his Dryhtne . . . and sídige orsorh þurh Godes gescyldnysse bútan ðæra sceoccena wíglunga. Ús sceamaþ tó secgenne ealle ða sceandlícan wíglunga ðe gé dwæsmenn drífaþ, oþþe on wífunge, oððe on wadunge, oððe on brýwláce, oððe gif man hwæs bitt, ðonne hí hwæt onginnaþ, oþþe him hwæt bið ácenned, Homl. Skt. i. 17, 88–104. Ne gýman gé galdra ne ídelra hwata, ne wígelunga ne wiccecræfta, Wulfst. 40, 14. Gé cépaþ dagas and mónðas mid ýdelum wíglungum (Gal. 4, 10), Homl. Th. i. 102, 19, 11, 15 : Homl. Ass. 28, 99. Hé sum þing hæfde ðe his hæle hremde þurh reðe wíglunga (wígelunga, v. l.), Homl. Skt. i. 5, 259. [King scal wicchecreft aleggan and wiȝelunge ne geman, O. E. Homl. i. 115, 22. Monies godes monnes child heo (incubii demones) biccharreð þurh wigeling, Laym. 15791.] v. ge-, líc-, steor-wíglung ; wíglian.

wíg-mann, es; m. A man of war, a fighting man, soldier :—Bellatores syndon wígmen ðe eard sculon werian wíglíce mid wǽpnum, L. I. P. 4; Th. ii. 306, 36 : Wulfst. 267, 15. [O. H. Ger. wíc-mann pugnator, pugil, bellator : Icel. víg-maðr.]

wignoþ (?), es; m. Warfare :—Wignoþes (? printed -roþes), duguðe militiae, Wrt. Voc. ii. 55, 18. v. wigian.

wigol; adj. Adapted to augury :—Wigole fugelas oscines aves, Wrt. Voc. i. 30, 8. [Cf. (?) O. H. Ger. wihil, wigil alciones.] Cf. wíglian.

wíg-plega, an; m. The game of war, battle :—Hé ne wandode ná æt ðam wígplegan, Byrht. Th. 139, 43; By. 268 : 141, 2; By. 316. Hé sumum dæleþ gúþe blǽd, gewealdenne wígplegan, Exon. Th. 331, 16; Vy. 69.

wíg-rád (?), e; f. A war-road, road along which an army passes :—Gewát him Abraham on ða wígróde (-ráde? -trode? v. wíg-trod) wiðertrod seón láðra monna Abraham betook himself to the way where they had gone and saw the track of their retreat, Cd. Th. 125, 24; Gen. 2084.

wíg-rǽden[n], e; f. Warfare, Wald. 39; Vald. 1, 22.

wíg-sigor, es; m. Victory in battle :—Hé hæfde wígsigor, Cd. Th. 121, 1; Gen. 2003. Hálig God geweóld wígsigor (Óðinn átti heimilan sigr í hverri orrostu, Ynglinga Saga, c. 2), Beo. Th. 3112; B.1554.

wíg-síþ, es; m. A warlike expedition :—Nǽfre mon lytle werede ðon wurdlícor wígsíð áteah, Cd. Th. 126, 13; Gen. 2094.

wíg-smiþ, es; A war-smith, war-maker, warrior, a man (poet.) :—Engle and Seaxe, wlance wígsmiðas, Wealas ofercóman, Chr. 937; Erl. 115, 21 : Exon. Th. 314, 14; Mód. 14. Ic wígsmiðum sægde, ðæt Sarra mín sweostor wǽre, Cd. Th. 163, 24; Gen. 2703.

wíg-smiþ, es; m. An idol-smith, a maker of idols :—Deófulgild . . . ða hér menn worhtan, wígsmiðas mid folmum simulacra . . . opera manuum hominum, Ps. Th. 113, 12.

wíg-spéd, e; f. Success in war, victory :—Hé mé tír forgeaf, wígspéd wið wráðum, Elen. Kmbl. 329; El. 165. Him Dryhten forgeaf wígspéda gewiofu, Beo. Th. 1398; B. 697.

, **wíg-spere**, es; n. A war-spear :—Wígspere falarica vel fala, Wrt. Voc. i. 35, 48.

wíg-steall, es; n. A defensive position, a bulwark, bastion, defence :—Wígsteal propugnaculum, Hpt. Gl. 487, 17 : 530, 3. Hé lǽteþ inwitflán brecan ðone burgweal, ðe him bebeád Meotud ðæt hé ðæt wígsteal wergan scealde, Exon. Th. 315, 30; Mód. 39. Hé wígsteall séceþ, heolstre behelmed, Salm. Kmbl. 208; Sal. 103. Wurdon hyra wígsteal wéstenstaþolas, brosnade burgsteal, Exon. Th. 477, 21; Ruin. 28. Wígstealla propugnacula, Hpt. Gl. 426, 73.

wíg-steall, es; n. The part of a church where the altar stands :—Weocsteall absida, Engl. Stud. xi. 64, 6. Wé lǽraþ ðæt mæssepreósta oþþe mynsterpreósta ǽnig ne cume binnan weohsteall (weófodsteall, v. l.) búton his oferslipe, ne hūru æt ðam weófode, ðæt hé ðǽr þénige búton ðære wǽde, L. Edg. C. 46; Th. ii. 254, 9.

wíg-strǽt, e; f. A high-road, public road :—An ðara wístrǽte, Cod. Dip. Kmbl. ii. 89, 4. [Cf. O. H. Ger. heri-stráza via publica.] Cf. here-paþ.

wíg-strang; adj. Powerful in war :—Wígstrang bellipotens, Wrt. Voc. ii. 12, 45.

wig-telgode for twig-telgode, Ps. Spl. C. 108, 28. v. twi-telged.

wíg-þracu, gen. -þræce; f. Violence of war, warfare :—Hwǽr ðæt hálige treó beheled wurde æfter wígþræce (the violent death of the crucifixion), Elen. Kmbl. 859; El. 430. Wé ða wíggþræce (the Trojan war) on gewritu setton, 1312; El. 658.

wíg-preát, es; m. A military troop :—Ðæs hiofenlícan werodes wígþreátas coelestis exercitus militiae, Lchdm. i. lxviii, 8.

wíg-þríst; adj. Bold in battle, daring :—Ðú mé saga hú ðú wurde þus wígþríst, ðæt ðú mec þus fæste fetrum gebunde, Exon. Th. 268, 14; Jul. 432.

wíg-trod [?], es; n.: -trodu (? v. wíg-rád), e; f. A war-track, the road along which an army has passed :—Wítrod (= wígtrod) gefeól heáh of heofonum handweorc Godes on to the track where the host of Israel had passed fell from the heavens the lofty walls raised by God's hand

(cf. se ágend up árærde reáde streámas . . . syndon ða foreweallas gestépte óð wolcna hróf, 196, 28; Exod. 298), Cd. Th. 208, 31; Exod. 491.

wíg-wægn, es; m. A war-chariot :—Se kyningc Pharon hæfde syx hund wígwægna (curruum), Ors. 1, 7; Swt. 38, 24, 35.

wíg-wǽpen, es; n. A weapon of war :—Ǽlce wígwǽpna and ǽghwylce woruldsaca lǽte man stille, Wulfst. 170, 8.

wíg-weorþung, e; f. Honour to idols :—Búton ðú forlǽte ða leásinga, weohweorðinga, and wuldres God ongyte gleáwlíce, Exon. Th. 253, 14; Jul. 180. Hwílum hié gehéton æt heargtrafum wígweorþunga, Beo. Th. 353; B. 176.

wíh (wih?), weoh; gen. wíges (weós?); m. An idol :—Hié gecwǽdon ðæt hié ðæs wíges (the golden image) ne róhton, ne hié tó ðam gebede mihte gebǽdon hǽðen heriges wísa, Cd. Th. 228, 12; Dan. 201. Hié ne willaþ ðysne wíg wurðigean, 228, 24; Dan. 208. Hé (St. Bartholomew) ne wolde wíg weorðian (cf. the account in Shrn. 120, 17–32), Apstls. Kmbl. 95; Ap. 48. Hé hǽþengield ofer word Godes, weoh gesôhte, Exon. Th. 244, 6; Jul. 23. Wóden worhte weós, 341, 28; Gn. Ex. 133. [Cf. O. Sax. wíh a temple : Icel. vé: Goth. weihs holy : O. H. Ger. wíh holy.] v. wíg-bed, -bora (signifer), -gild, -smiþ, -steall, -weorþung.

wihgena, Wihg(e)ra-ceaster, wíh-gyld. v. wiga, Weogorna-ceaster, wíg-gild.

wiht, e; f.: es; n.   I. a wight, creature, being, created thing :—Nis nán wuht (cf. nán gesceaft, 22) ðe mæge oððe wille swá heágum Gode wiþcweþan :—ne wéne ic ðæt ǽnig wuht (cf. gesceaft, 24) sié ðe wiþwinne non est aliquid, quod summo huic bono vel velit, vel possit obsistere. Non . . . arbitror, Bt. 35, 4; Fox 160, 29. Manig wyht is mistlíce férende geond eorþan quam variis terras animalia permeant figuris, 41, 6; Fox 254, 23. Ælc uht, ðæs ðe hió (an asp) ábítt, scel his líf on slǽpe geendian, Ors. 5, 13; Swt. 246, 27. Ic (a leather bottle) eom wunderlícu wiht, Exon. Th. 399, 16; Rä. 19, 1 (the word occurs often in the riddles). Úr . . . is módig wuht, Runic pm. Kmbl. 339, 12; Rún. 2. Nánre wuhte líchoma ne beoð téderra ðonne ðæs monnes, Bt. 16, 2; Fox 52, 8. Se hrycg fǽrð æfter ælcre wuhte, Past. 1; Swt. 29, 14. Wiþerweardnes wuhte gehwelcre, Met. 11, 78. Ðǽre wihte, Exon. Th. 438, 9; Rä. 57, 5. Ne mæg ic náne cwica wuht (animalia) ongitan, ðara ðe wite hwæt hit wille, oððe hwæt hit nylle, ðe ungenéd lyste forweorþan, for ðam ǽlc wuht (animal) wolde bión hál and libban, ðara ðe mé cwica ðincð; búte ic nát be swylcum gesceaftum swylce náne sáwle nabbaþ, Bt. 34, 10; Fox 148, 13–17. Sôð is ǽghwylc ðara ðe ymb ðás wiht wordum bécneþ; ne hafaþ heó ǽnig lim, leofaþ se þeáh, Exon. Th. 421, 30; Rä. 40, 26. Hí geségon syllicran wiht, wyrm on wonge, Beo. Th. 6669; B. 3038. Ic ða wihte geseah . . . heó wæs wundrum gegierwed, Exon. Th. 483, 5; Rä. 68, 1. Hwylce wihta beóð óðre tíd wífcynnes, and óðre tíd wǽpnedcynnes? Salm. Kmbl. p. 202, 12. Ic geseah ða anlícnessa ealra creópendra wuhta (reptilium) . . . Ða creópendan wuhta getácniaþ . . . , Past. 21; Swt. 155, 14. Swilca wuhta (fleógan, gnættas, loppe) him deriaþ, Bt. 16, 2; Fox 52, 14. Manega wuhta (animalia), Met. 31, 2. Ðé sculon moldwyrmas ceówan, slítan swearte wihta (wihte, Exon. Th. 371, 10), Soul Kmbl. 146; Seel. 72. Ðíne wihte animalia tua, Ps. Th. 67, 11. Ða wihte twá, Exon. Th. 429, 38; Rä. 43, 16. Flǽsc lytelra wuhta, smælra fugla, Lchdm. ii. 180, 13. Wihta Wealdend, Cd. Th. 272, 25; Sat. 125. Ne meahte ðǽr drincan wihta ǽnig, Ps. Th. 77, 44. Ealra wihta gehwam omne animal, 144, 17. Wuhta gehwylc, Met. 11, 52. Earmost ealra wihta, ðara ðe cenned wǽre, Exon. Th. 421, 7; Rä. 40, 14. Wihta gehwylce, deóra and fugla, 61, 10; Cri. 982. Cynna gehwylc cucra wuhta, ðara ðe lyft and flód fédaþ, feoh and fuglas, Cd. Th. 78, 23; Gen. 1297. Dréam cwicra wihta, Exon. Th. 411, 5; Rä. 29, 8. Ðeós lyfte byreþ lytle wihte, 438, 26; Rä. 58, 1.   I a. of evil beings :—Yfel wiht phantasma, Mt. Kmbl. Lind. 14, 26: Mk. Skt. Lind. Rush. 6, 49. Wiht unhǽlo (Grendel), Beo. Th. 241; B. 120. Werge wihta (devils), Exon. Th. 455, 29; Hy. 4, 57. Unfǽle men, wudewásan, unfǽle wihtu satiri vel fauni, Wrt. Voc. i. 17, 20. Ðás fúlan wuhta (wizards) ðú sceoldest áwurpan of ðínum ríce, Homl. Th. ii. 488, 12.   II. a whit, thing; ǽnig wiht aught, anything, (a) without a negative :—Ðǽr hí ǽnige wuht ágnes gódes an heora anwealde hæfden, Bt. 27, 3; Fox 100, 4. Ic eom swíðe gefíonde ðæt gé ǽfre woldon ǽnige wuht (ǽnig wuht (ǽnig-wuht?), Hatt. MS.) eów selfum wítan, ǽr ic hit eów wíte, Past. 31; Swt. 206, 19.   (b) with a negative, aught. See also III.   (1) alone :—Ne bið him wiht tó sorge, Exon. Th. 238, 29; Ph. 611. Ne wendaþ hine wyrda, ne hine wiht (or acc.?) dreceþ, ádl ne yldo, 334, 1; Gn. Ex. 9. Nis ðæt onginn wiht, 119, 2; Gú. 248. Nó hé him ðæs wyrmes wíg for wiht dyde, Beo. Th. 4685; B. 2348.   (2) with a genitive :—Ne bið wiht forholen monna gehygda, Exon. Th. 65, 14; Cri. 1054. Ne him wiht gescód ðæs ðe hý him tó teónan þurhtogen hæfdon, 127, 35; Gú. 396. Ne ðǽr hleonaþ unsméðes wiht, 199, 15; Ph. 26. Ne magon wé geleánian him mid láðes wihte, Cd. Th. 25, 15; Gen. 394. Ne for feóndscipe, ne for wihte ðæs ic ðé weán úðe I did it not from enmity, or from aught of ill will, 163, 2; Gen. 2692. Hé nele láþes wiht geæfnan, Exon. Th. 357, 22; Pa. 32: Cd. Th. 16, 13; Gen. 242. Ic ðínra worda ne mæg wuht

oncnáwan, 34, 8; Gen. 534. Wiht, Elen. Kmbl. 1364; El. 684.
Wonhýdig wer ðæs wiht ne cann *vir insipiens non cognoscet*, Ps. Th. 91,
5. Hí náne wuht ongitan ne cunnon ðara gǽstlecena beboda, Past. 1;
Swt. 25, 23. **III.** cases (with or without preps.) with adverbial
force, (a) without a negative:—Gif wé hit mægen wihte (*anyhow*) áþencan,
Cd. Th. 26, 2; Gen. 400. Gif hit eówer ǽnig mæge gewendan mid
wihte, ðæt hié word Godes forlǽten, 27, 35; Gen. 428. Ne wé wénaþ,
ðæt hé wihte mæge ðis folc áfédan, Ps. Th. 77, 22. (b) with a nega-
tive:—Nis mé wihtæ þearf (*there is no need at all*) hearran tó habbanne,
Cd. Th. 18, 25; Gen. 278. Hié ðæs wíges wihte ne róhton, 228, 13;
Dan. 201. Ic ðé bæd ðæt ðú ðone wælgæst wihte ne (*in no wise*) grétte,
Beo. Th. 3995; B. 1995: Andr. Kmbl. 3320; An. 1663. Næs word-
latu wihte (*at all*) ðon máre, 3043; An. 1524. Wuhte, Met. 14, 10:
16, 14. Næs him wihte ðe sél *it was not a whit the better for him*, Beo.
Th. 5368; B. 2687. Nát ic hit be wihte (*at all*; cf. be dǽle *in part*),
Exon. Th. 468, 7; Phar. 4. Ic mid wihte (cf. mid ealle) ne mæg of
ðissum lioðobendum *I am utterly unable to escape from these bonds*, Cd.
Th. 24, 22; Gen. 381. Wit ðus baru ne magon wesan tó wuhte (*at
any rate*), 52, 5; Gen. 839. Ic ne forhtige wiht (*or under* II (b)) *non
movebor amplius*, Ps. Th. 61, 2: 113, 13. Him wiht ne speów *they did
not at all succeed*, Judth. Thw. 25, 9; Jud. 274. Him wiht ne sceód
grim gléda nið, Cd. Th. 245, 17; Dan. 464. Nó hé wiht fram mé
fleótan meahte hraþor on holme, Beo. Th. 1087; B. 541. Ne beóð
winter ðín wiht ðe sǽmran *anni tui non deficient*, Ps. Th. 101, 24. Hwæt
wilt ðú cweþan, gif hwá wuht nylle wiþwinnan, ac mid fullan willan
forlǽt ǽlc gód and fulgǽþ ðam yfele, Bt. 36, 6; Fox 182, 6. Hié noldon
beón ábisgode náne wuht on eorðlícum ðingum *rebus exterioribus nullatenus
occupentur*, Past. 18; Swt. 137, 1. [*Goth.* waihts; *f. res*; ni waiht *nihil*:
O. Sax. wiht; *m. a thing, whit*; wihtī, *pl. evil spirits*: *O. H. Ger.* wiht;
n. substantia, animal, res: *Icel.* vættr; *f. a being*; especially *a supernatural being.*] v. ǽ-, ǽnig- (?), hel-, ná-, nán-, sǽ-wiht; ǽl-, eall-wihta.
wiht (e; *f.?*) *weight*:—Wiht *pondus*, Kent. Gl. 344. Wihte *pondere*,
Wülck. Gl. 237, 27. Genim ægþres gelíce micel be wihte (gewihte, *v.l.*),
Lchdm. i. 146, 20. Má hundred punda seolfres; ðæt hé nam be wihte,
and mid mycelan unrihte, Chr. 1086; Th. i. 355, 31. Genim of ǽlcere
ðisre wyrte .xx. penega wiht, Lchdm. i. 374, 21. [For his æfne wiht of
golde, Laym. 30835. Wiþþ fife wehhte of sillferr, Orm. 7812. *Ayenb.*
wiȝte: *Chauc.* wighte, weihte, weiȝte: *Piers P.* weȝt, weght: *Icel.* vætt;
f.] v. ge-wiht.

Wiht, Wiht-land, Wiht (Wihte) eáland *the Isle of Wight*:—Seó mǽið
ðe nú eardaþ on Wiht, Chr. 449; Th. i. 20, col. 1: Cod. Dip. Kmbl.
iii. 431, 16, 24: v. 82, 19: vi. 196, 8. Cómon sex scipu tó Wiht, Chr.
897; Th. i. 176, 7. Intó Wiht (Wihtlande, *v.ll.*), 1006; Th. i. 257,
col. 2. Tó Wiht (Wihtlande, *v.l.*), 1022; Th. i. 286, col. 1. On
Wihtlande, 998; Th. i. 246, 24. Intó Wihtlande, 1001; Th. i. 250, 13.
Hé on Wiht gehergade, 661; Th. i. 54, 24. Hié Wieht (Wiht, *v.l.*)
forhergedon, 681; Th. i. 62, col. 1. Hér Cerdic and Cynríc genámon
Wihte eáland (Wihtland, Wiht ðæt eáland, *v.ll.*), 530; Th. i. 26, 33.
Hié sealdon hiera nefum Wiht eáland (Wihte eáland, Wiht ðæt égland,
Wihtland, *v.ll.*), 534; Th. i. 28, col. 1. Ymbe Wiht ðæt ígland (Wiht-
land, *v.l.*) *Vectae insulae*, Bd. pref.; S. 472, 14. Seó ðeód ðe Wiht ðæt
eálond (Wihtland, *v.l.*) oneardaþ *gens quae Vectam tenet insulam*, 1, 15;
S. 483, 22. [*From Latin* Vecta *or* Vectis.]

Wiht- *in proper names.* v. Txts. 512.

Wihtgáras; *pl. m. The name of some people in England*:—Wihtgáras
(Wightgóra, 416, 7) landes is syx hund hýda, Cod. Dip. B. i. 414, 22.

Wihtgáres (-as) burh, Wihtgára burh *Carisbrooke*:—On Wihtgáras
(-gára, *v.l.*) byrg, Chr. 530; Th. i. 26, col. 1. Wihtgára (-gáras, -gáres,
v.ll.) byrg, 544; Th. i. 28, col. 1.

Wiht-land. v. Wiht.

Wihtmǽres wyrt *spoonwort* (?):—Witmǽres wyrt nioþoweard,
Lchdm. ii. 32, 10. [Uihtméres wyrt † heauen hindele *brittannica*, iii.
300, col. 2.]

wiht-mearc, e; *f. A weight-mark, a plumb-line*:—Of þunder, of
wihtmearce *perpendiculo*, Hpt. Gl. 476, 75. v. pundar.

Wiht-sǽtan, -sǽte; *pl. m. The inhabitants of the Isle of Wight*:—
Of Geáta fruman syndon Wihtsǽtan (*Victuarii*), ðæt is seó ðeód ðe Wiht
ðæt eálond oneardaþ, Bd. 1, 15; S. 483, 22. v. next word.

Wiht-ware; *pl. m. The people of the Isle of Wight*:—Cantware and
Wihtware (-wara, *v.l.*), Chr. 449; Th. i. 20, col. 1. Hé bróhte Wihtwarum (-an, *v.l.*) fulwiht ǽrest, 661; Th. i. 54, col. 1. v. preceding word.

wiisc. v. wýsc.

wil. v. wil[l].

wíl *a wile, a device.* [He wolde þurh his micele wiles ðeor beon, Chr.
1128; Erl. 257, 14. To lokenn himm fra þeȝȝre laþe wiless, Orm.
10317. Þe wrenchful feont wið his wiles, Kath. 891. Þe world ledes
▲ man with wrenkes and wyles, Pr. C. 1360. Wyle or sleythe *cautela,
astucia*, Prompt. Parv. 528.] v. flíge-wil.

wíl-bec *a stream of misery* (?):—Wuniendo wær wílbec biscær, Exon.
Th. 353, 42; Reim. 26. [Cf. *Icel.* víl *misery, wretchedness*; víl-stígr
a path of misery.]

wil-boda, an; *m. A welcome messenger*:—Mec meahtig Meotudes
þegn (*an angel*) gesóhte, and mé sára gehwylc gehǽlde, wuldres wilboda,
Exon. Th. 176, 34; Gú. 1220. Cf. wil-spell.

wil-cuma, an; *m. One whose coming is pleasant, a welcome person* (or
thing):—Mé is ðín cyme on myclum ðonce, and ðú eart leóf wilcuma
gratus mihi est multum adventus tui, et bene venisti, Bd. 4, 9; S. 577,
22. Leóf wilcuma Frysan wífe, Exon. Th. 339, 17; Gn. Ex. 95. Hé
wilcumaþ (*Christ come to hell*) grétte: ' Ðé ðæs þonc sié, ðæt ðú ús sécan
woldest,' 462, 26; Hö. 58. Ðegnas cwómon, geségon wilcuman heofones
Waldend, 35, 7; Cri. 554. Gé sind wilcuman, Cd. Th. 303, 22; Sat.
617: Beo. Th. 794; B. 394. Hié synt wilcuman Deniga leódum, 782;
B. 388: 3792; B. 1894. Ic hæleþum bodige wilcumena fela (*many
welcome things*) wóþe mínre, Exon. Th. 391, 4; Rä. 9, 11. [Wulcume
(welcome, 2nd MS.) ært þu, swíðe leof þu ært me, Laym. 8528. His
lauerd alse wilcume wæ he weóren his sune, 4901. Cum aȝean, wilkume
schaltu beon me, A. R. 394, 17. Ich am hire wel welcome, O. and N.
1600. Ðu and ðin trume ben to me welcome, Gen. and Ex. 1830.] v.
next word.

wil-cume (-a); *interj. Welcome*:—Wilcume *evax*, Wrt. Voc. i. 61,
29. Wilcymo *euge*, Mt. Kmbl. Lind. 25, 23. ['A!' seið warschipe,
'Welcume liues luue!' O. E. Homl. i. 259, 11. *O. H. Ger.* Heilo *aut*
willicomo *osianna*.] v. next word.

wilcumian; *p.* ode *To welcome, bid welcome, greet, salute*:—Gyf gé
ðæt án dóð, ðæt gé eówre gebróðra wylcumiaþ (welcumieð (*later version*);
hǽlo beádas † wilcyma, Lind.) *si salutaveritis fratres vestros tantum*, Mt.
Kmbl. 5, 47. Ðæt folc . . . wellcumiaþ Fénix, Engl. Stud. viii. 478, 45.
Basilius sende him tógeánes, and hine wylcumode, Homl. Skt. i. 3, 507.
Hine wylcumede se cásere, and cwæð him tó mid blysse, 7, 339. Wil-
cumiga (wilcymogie (wilcymo gié? v. preceding word), Lind.) † groeta
salutari, Mk. Skt. Rush. 12, 38. [He wilcumede hine to londe, Laym.
10957. To wulcumen Mærlin, 17098. Þe lilie wolcumeþ (wel-, *v.l.*)
me, O. and N. 440. Faiger welcumede he Eliezer, Gen. and Ex. 1396.]
v. ge-wilcumian, *and preceding word.*

-wild. v. ge-wild.

wil-dǽd, e; *f. An acceptable deed, favour, benefit*:—Móna se ændlefta,
wyldǽda (wel-? v. wel-dǽd) biddan nytlíc is, Lchdm. iii. 188, 24.

wil-dæg, es; *m. a welcome day*:—On ðam wildǽge, Exon. Th. 29,
7; Cri. 459. [Muchel wes þa murðe þe þat folc makode, and heo Godd
thonkeden þat heo heora wildaȝes wælden weoren, Laym. 1798.]

wildan; *p.* de. **I.** *to tame, subdue*:—Wylde *domuit*, i. *vicit,
mitigavit*, Wrt. Voc. ii. 141, 74. **II.** *to make submissive, have
dominion over, rule, control*:—Hit is swytol, ðæt man tó hwón wylde
(wilde, gewilde, *v.ll.*) and woruldlíce stýrde ðám ðe oftost for Gode syngodon and scendan ðás þeóde, Wulfst. 168, 2. Wille ic ðæt . . . ic and
míne þegnas wyldan úre preóstas tó ðan ðe úre sáula hyrdas ús tǽcaþ,
ðæt syndon úre bisceopas, L. Edg. S. 1; Th. i. 272, 17. Se ðe ðone
mǽran noman abbodes underféhð, hé sceal mid twyfealdre láre ða wyldan
and týn, ðe him underþeódde synt *qui suscipit nomen abbatis duplici debet
doctrina suis preesse discipulis*, R. Ben. 11, 12. Gyf mín hí ne beóþ
wyldde *si mei non fuerint dominati*, Ps. Spl. 18, 14. **III.** *to take
into one's power, to seize*:—Ne dýde man on Sunnandæges freólse ǽnigne
forwyrhtne man . . . ac wylde (wylde man hine, *v.l.*; *the old Latin version has* capiatur) and healde, ðæt se freólsdæg ágán sý, L. C. S. 45; Th.
i. 402, 12: L. E. G. 9; Th. i. 172, 14. v. ge-wildan (-wyldan), wilding.

wild-cyrfet *bryony*; brionia, Wrt. Voc. i. 32, 17. v. wilde.

wild-deór, wildeór, es; *n. A wild animal, wild beast*:—Wilddeór
fera, Wrt. Voc. i. 22, 39. Ðis wilddeór (wildeór, *v.l.*) well fremaþ,
Lchdm. i. 330, 7. Wildeór *fera*, Wrt. Voc. i. 77, 76. Ne mæg hit
wæter ne wildeór beswícan, Salm. Kmbl. 571; Sal. 285. Wildiór *leena*,
Kent. Gl. 989. Wildeór *bestiae*, Bd. 3, 23; S. 554, 24: Coll. Monast.
Th. 22, 23. Swá hwæt swá wilddeór ábiton, Gen. 31, 39: 37, 20.
Wildeór, Blickl. Homl. 95, 16: Ex. 22, 13. Wildeór *bestiae agri*, 23,
11. Ealra wuda wildeór *omnes ferae sylvarum*, Ps. Th. 49, 11: 103, 19.
Wilddeóra *ferarum*, Wrt. Voc. ii. 38, 32. Wilddeóra holl and denn
lustra, i. 59, 10: Soul Kmbl. 164; Seel. 82. Wildeóra þeáw, Cd. Th.
252, 2; Dan. 352: 255, 10; Dan. 622. Uildeár (-deára? -dera?)
bestiarum, Rtl. 117, 4. Anweald ofer wilddeórum, Hexam. 11; Norm.
18, 16. Hé mid wilddeórum (*cum bestiis*) wæs, Mk. Skt. 1, 13: Cd.
Th. 256, 34; Dan. 650. Wildeórum, Exon. Th. 146, 21; Gú. 713.
Wildiórum gelícran ðonne monnum, Bt. 38, 5; Fox 208, 1. Ic áfyrre
yfel wilddeór (*malas bestias*), Lev. 26, 6. Ealle yfele wilddeór, Lchdm.
i. 202, 13. Wildeór, Lev. 26, 22: *feras*, Ps. Th. 67, 27. Nétena oððe
wildeór, Bt. 38, 2; Fox 196, 18. Hwylce wildeór (*feras*) swýþost geféhst ðú? Ic gefó heortas, and báras, and rǽgan, and hwílon haran, Coll.
Monast. Th. 21, 29. Wyrmas and wildeór, Beo. Th. 2864; B. 1430.
v. wild-deór, *and following words.*

wild-deóren; *adj. Of wild beasts*:—Mid wilddeórenum tóþum *cum
feralibus dentibus*, Scint. 99, 7.

wilddeór-líc; *adj. Wild beast-like, brutish, brutal, bestial*:—Se
wísdóm is eorðlíc and wildeórlíc (-diór-, Hatt. MS.) *est ista sapientia terrena, animalis*, Past. 46; Swt. 346, 25. Seó wildeórlíce árleásnes Bretta

cyninges *feralis impietas regis Brittonum*, Bd. 3, 9; S. 533, 7. Đa wildeórlícan *ferinam*, Wrt. Voc. ii. 34, 10. Hié be sumum dǽle wildiórlíce (*bestiales*) bióđ, Past. 17; Swt. 108, 23. v. wilder-líc.

wilddeórlíce; *adv. After the manner of wild beasts, brutishly :*—Đǽr ǽr wildeór oneardodan, oþþe men gewunedon willdeórlíce (*bestialiter*) lifian, Bd. 3, 23; S. 554, 25.

wilde; *adj. Wild :*—Wildæ *agrestis*, Wrt. Voc. ii. 99, 53 : i. 17, 41. Wilde *indomitus*, ii. 111, 78. Untamed, wilde *edomitus*, 142, 40. Wudulíce ođđe wilde *agrestes*, 4, 61. *As in this gloss the word seems used in* wylde (*or cf.* weald?) elfen *hamadryades* (cf. feldelfenne *amadriades*, ii. 8, 14), i. 60, 17. **I.** in reference to animals, *wild, not domestic, not tamed, not broken in :*—Rêþra þonne ǽnig wilde deór, Blickl. Homl. 95, 31: Homl. Th. i. 486, 28: Bt. 39, 1; Fox 212, 3. Wilde oxa *bubalus*, Wrt. Voc. i. 22, 46. Wilde bâr *aper*, tam bâr *verres*, 22, 70. Assa *asinus*, wilde assa *onager*, 23, 27. Se getemeda assa . . . Se wilda fola, Homl. Th. i. 208, 20–22. Wilde goos *cente*, Wrt. Voc. ii. 103, 68 : *gente*, 109, 63. Wilde gós *cante*, 14, 21. Wæs sum wilde hrem, Homl. Th. i. 162, 21. Se wilda fugel (*the Phenix*), Exon. Th. 211, 21; Ph. 201. Hafuc sceal on glofe wilde gewunian, wulf sceal on bearowe, Menol. Fox 495; Gn. C. 18. Sió wilde beó, Met. 18, 5. Seó leó ge-monđ đǽs wildan gewunan hire eldrana, Bt. 25; Fox 88, 12. Sum sceal wildne fugel âtemian, Exon. Th. 332, 14; Vy. 85: 222, 3; Ph. 343. Hâlig feoh and wilde deór, Cd. Th. 13, 13; Gen. 202. Eoferas and wilde đeór *aper et singularis ferus*, Ps. Th. 79, 13. Wildu diór, Met. 27, 20: Cd. Th. 91, 22; Gen. 1516. Wildu deór and neáta gehwylc, 240, 20; Dan. 389. Cômon wilde beran and wulfas, Homl. Th. i. 244, 18. Wildra deóra đæt grimmeste, Exon. Th. 371, 28; Seel. 82. Wildera deóra têđ *dentes bestiarum*, Deut. 32, 24. Hyre dǽl đera wildera (*not broken in*) horsa, Chart. Th. 538, 33. Wildra, 548, 10. Wildu hors *equos in-domitos*, Past. 41; Swt. 303, 9. Fiówer wildo hors, Shrn. 71, 34. Đa stælhrânas beóđ swýđe đýre mid Finnum, for đæm hý fóđ đa wildan hrânas mid, Ors. 1, 1; Swt. 18, 12. **I a.** *not under restraint; un-controlled :*—Đâ wæs culufre eft sended wilde, Cd. Th. 88, 14; Gen. 1465. **II.** in reference to plants, *wild, not cultivated :*—Wilde cyrfet *colochintida*, hwît wilde wîngeard *brionia*, wilde (v. Wülck. Gl. 133, 12) wîngerd *labrusca*, Wrt. Voc. i. 30, 12–15. Wilde popig *saliunca*, . . . wilde næp *nap silvatica*, 31, 8, 27. Wilde næp *diptamnus vel bibulcos*, . . . wilde lactuce *sarrabum*, 32, 5, 24. *Oleastrum* đæt is wilde elebeám, Lchdm. ii. 90, 20. Wildre magþan wyrttruman, 206, 15. Wildre mealwan seáw, 214, 14. Unwæstm î wilde fôter *zizania*, Mt. Kmbl. Lind. 13, 27. **III.** of places, *wild, uncultivated, uninhabited :*—Licgaþ wilde môras emnlange đǽm bŷnum lande, Ors. 1, 1; Swt. 18, 27. Đone eard (*East Anglia*) iii mônþas hî hergodon and bærndon, ge furđon on đa wildan fennas hî fêrdon, Chr. 1010; Erl. 143, 27. Com se biscop tô đare mynstre (*Peterborough*) . . . ne fand đǽr nân þing bûton ealde weallas and wilde wuda, 963; Erl. 121, 28. **IV.** of fire, *wild, that spreads over a country* (like a prairie fire) [cf. Icel. villi-eldr] :—Hér wæs se dría sumor, and wilde fŷr com on manega scíra and forbærnde fela tûna, and eác manega burga forburnon, Chr. 1078; Erl. 215, 36. On đissum geáre atŷwde đæt wilde fŷr, đe nân mann ǽror nân swylc ne gemunde, and gehwǽr hit derode on manegum stôwum, 1032; Erl. 164, 1. Hér wæs swíđe mycel mancwealm and orfcwealm, and eác đæt wilde fŷr on Deórbŷscíre micel yfel dyde, and gehwǽr elles, 1049; Erl. 173, 19. **IV a.** figurative of a disease :—Wylde fŷr *erisipilas*, Wrt. Voc. i. 20, 3. [v. *wildfire* in Halliwell's Dictionary, and cf. Germ. das wilde feuer *St. Anthony's fire, erysipelas.*] **V.** in a moral sense, *wild, turbulent, ungoverned :*—Hê geong fareþ, hafaþ wilde môd, Salm. Kmbl. 755; Sal. 377. [*Goth.* wilþeis : *O. Frs.* wilde : *O. L. Ger. O. H. Ger.* wildi : *Icel.* villr.]

wilde ; *adj. Having power, powerful, strong :*—Hit næs þeáw đæt mon ǽnig wæl on đa healfe rîmde đe đonne wieldre wæs *mos est, ex ea parte quae vicerit occisorum non commemorare numerum*, Ors. 4, 1; Swt. 156, 22. Beó â seó mildheortnys wyldre đonne se rihta dôm *semper superexaltet misericordiam judicium*, R. Ben. 118, 27. Đæt đæt gesceád beó wylldre đonne seó yfele gewilnung, Basil admn. 3; Norm. 40, 3. Útancumene men beóđ wildran đonne gê and oðre genyđriaþ *advena ascendet super te eritque sublimior; tu autem descendes et eris inferior*, Deut. 28, 43. Hié wyldran wǽron đonne hié, and hié mid ealle of đǽm earde âdrifon *urbem suo generi vendicant, patrimonia dominorum sibi usurpant, extorres dominos procul abigunt*, Ors. 4, 3; Swt. 162, 18: Blickl. Homl. 151, 3. [Freo of heorte, of wisdom wilde, Misc. 96, 94. þet þe mon lete his iwit weldre þene his wređđe, O. E. Homl. i. 105, 19.] v. weald.

-wilde. v. ge-wilde (-wylde), wildan.

wilde-cyn[n], es; *n. A wild species :*—Wildecynnes hors *equifer* (cf. *hic equiferus* a wyld hors, 187, col. 1), Wrt. Voc. i. 23, 4.

wilde-deór, es; *n. A wild beast :*—Weorpan hî an wildedeóra líc, Bt. 38, 1; Fox 194, 31. Hê wæs miđ wildedeórum *erat cum bestiis*, Mk. Skt. Lind. Rush. 1, 13. [*Icel.* villi-dŷr.]

wilder (-or? cf. wildor-líc. v. next word) (*and* wild? cf. þa men tuhten to þan deoren, and duden of þan wilden al heora iwilla, Laym.

1129, At þe fyrst quethe of þe quest quaked þe wylde, Gaw. 1150. Went we to wod the wilde for to cacche, Destr. Tr. 2347. *O. H. Ger.* wild; *dat. pl.* wildiran; *and the declensions of* lamb, cild), es; *n. A wild beast :*—þurh đæs wildres (*the panther's*) mûđ, Exon. Th. 358, 10; Pa. 43. Đæt flǽsc, đæt wildro âbiton *carnem, quae a bestiis fuerit prae-gustata*, Ex. 22, 31. Weorpan on wildra líc, Met. 26, 76: Exon. Th. 356, 10; Pa. 9: Cd. Th. 257, 25; Dan. 663. Spêdig man on wildrum, Ors. 1, 1; Swt. 18, 9.

wilder-líc (?); *adj. Wildbeast-like, brutish :*—Hié be sumum dǽle wildorlíce (wildiórlíce, Cott. MSS.) beóđ *ex qua parte bestiales sunt*, Past. 17; Swt. 109, 23.

wild-gós, e; *f. A wild goose :*—Wildgoos *gente*, Wrt. Voc. ii. 109, 60.

-wildian. v. â-wildian.

wilding, e; *f. Dominion :*—On ǽlcere stôwe wylddingce his *in omni loco dominationis ejus*, Ps. Lamb. 102, 22. Wylding, Ps. Spl. M. 102, 22.

wildness (?), e; *f. Wildness, licentiousness :*—Gâlre wild[nesse?] *petulantis lasciviae*, Hpt. Gl. 515, 10.

wildor, wildro; *wilege, wile-wíse.* v. wilder; wilige, wilig-wíse.

wil-fægen; *adj. Having one's desire, satisfied, glad :*—Wilfægen *voti compos*, Wrt. Voc. ii. 82, 59 : *compos*, Ælfc. Gr. 9, 31; Zup. 58, 1. Wilfangen (*l.* -fægen) *voti compos*, Engl. Stud. xi. 67, 96. Ongan hê wilfægen æfter đam wuldres treó eorđan delfan, đæt hê funde behelede, Elen. Kmbl. 1652; El. 828. On eallum đǽm mid đý hê willfægen wæs gefremed, hê eft hwearf tô godcundre lâre *in quibus omnibus cum sui voti compos esset effectus, ad praedicandum rediit*, Bd. 5, 11; S. 625, 40. Mid eádigum wilfægene *cum beatis compotes*, Hymn. Surt. 36, 30. Crist ûs êcum gefeánum dô beón wilfægene *Christus nos sempiternis gaudiis faciat esse compotes*, 123, 11. [Cf. *M. H. Ger.* wille-vagunge *satisfactio poenitentiae.*] v. wil-hrêmig, -hrêþig, -tygþe.

wilfullíce; *adv. Willingly, voluntarily, with a good will :*—Wil-ful[l]íce *sponte*, Hpt. Gl. 435, 66. [Alle þet for þi luue pouerte wilfulliche þolien, O. E. Homl. i. 279, 8. þe ournemen of boзsannnesse ys þet me bouзe wiluolliche, Ayenb. 140, 19. Wylfully *voluntarie, spontanee*, Prompt. Parv. 528.]

wil-gæst, es; *m. A desirable, welcome guest :*—Godes âgen bearn, wilgest on wícum, Exon. Th. 313, 28; Môd. 7. Cf. wil-cuma.

wil-gebróþor; *pl. m. Brethren pleasant in their lives :*—Freólícu twâ frumbearn, Cain and Abel . . . willgebróđor, Cd. Th. 59, 30; Gen. 971. Cf. wil-gesweostor.

wil-gedryht, e; *f. A glad band :*—Seó wilgedryht wildne weorþiaþ *turba prosequitur munere laeta pio*, Exon. Th. 222, 2; Ph. 342. Wes đû, Andreas, hâl mid đâs willgedryht, Andr. Kmbl. 1828; An. 916.

wil-gehlêþa, an; *m. A pleasant comrade :*—Hwîlum ic (*a horn*) tô hilde bonne wilgehlêþan, Exon. Th. 395, 9; Rä. 15, 5.

wil-gesíþ, es; *m. A pleasant companion :*—Wilgesîþas, Beo. Th. 45; B. 23. Willgesíđđas, Cd. Th. 120, 31; Gen. 2003.

wil-gesteald, es; *n. A desirable possession :*—Đý lǽs đû eft cweđe đæt ic wurde willgestealdum (-gesteallum, MS. ; *but cf. the pairs of words* (*as here*) æht-gesteald, æht-gestreón ; feoh-gesteald, feoh-gestreón) eádig on eorđan ǽrgestreónum *ne dicas: Ego ditavi Abram* (Gen. 14, 23), Cd. Th. 129, 20; Gen. 2146.

wil-gesweostor; *pl. f. Gracious sisters :*—Idesa, willgesweostor (*Lot's daughters*), Cd. Th. 157, 16; Gen. 2607. Cf. wil-gebróþor.

wil-geþofta, an; *m. A pleasant associate :*—Đæt inwitspell Abraham sægde freóndum sínum, bæd him fultumes willgeþoftan, Cd. Th. 122, 14; Gen. 2026.

wil-gifa, -giefa, -geofa, an; *m. A giver of what is desirable, a giver of good*, (1) as epithet of an earthly prince :—Wilgeofa Wedra leóda, dryhten Geáta (*Beowulf*), Beo. Th. 5792; B. 2900. Đæs wilgifan (*Constantine's*) word, Elen. Kmbl. 441; El. 221. (2) as an epithet of the Deity, *the giver of all good :*—Sigora Waldend, weoruda wilgiefa, Exon. Th. 229, 34; Ph. 465. Bearn Godes, weoroda willgifa, Elen. Kmbl. 1626; El. 815. Dryhten God, weoruda willgeofa, Andr. Kmbl. 2565; An. 1284. Gumena brego, weoruda wilgeofan, 123; An. 62. God, hyra wilgifan, Exon. Th. 34, 4; Cri. 537. Willgifan, Elen. Kmbl. 2221; El. 1112.

wilh (wiel) ; *gen.* wiles; *m. A slave, servant :*—Gif se wiel (*servus*) cwiđ : 'Mê is mîn hlâford leóf,' Ex. 21, 5. Ne wilna đû đínes nêhstan wyeles, 20, 17. Đæs weles (wieles, weales, *v. ll.*) hlâford *dominus servi illius*, Mt. Kmbl. 24, 50. Se đe his wiel (*servum*) slicđ mid girde, ođđe his wylne, Ex. 21, 20, 32. v. wealh.

wil-hrêmig; *adj. Having one's desire, satisfied, exultant :*—Wil-hrêmig (*printed* -hranig, *but see* Wülck. Gl. 376, 26) *compos*, Wrt. Voc. ii. 20, 69. v. wil-fægen, *and next word.*

wil-hrêþig; *adj. Satisfied, exultant :*—Weorud willhrêđig sægdon wuldor Gode, Elen. Kmbl. 2231; El. 1117. v. wil-fægen, *and preceding word.*

wilian *to roll*, wilie. v. wilwan, wilige.

wilige (*and* -a ; *m.?*), an; *f. A basket :*—Wilige *cophinus*, Wrt. Voc. i. 25, 3. Wilige *vel* leáp, 55, 37. Wylige ođđe meoxbearwe *corbis vel cofinus*, 86, 2. Wylige (wilige, *v. l.*) ođđe windel *corbis*, Ælfc. Gr. 9,

28; Zup. 55, 13. Wiligan *corbes*, wiliga *corbis*, Hpt. Gl. 497, 41. On wylegan *in cophino*, Ps. Spl. 80, 6 : Blickl. Gl. Hí hine on ánre wilian (*in sporta*, Acts 9, 25) áléton ofer ðone weall, Homl. Th. i. 388, 9. Hú fela wyligena (-egena, *v. l.*) *quot cophinos*, Mk. Skt. 8, 19, 20. Wylegena, Mt. Kmbl. 16, 9, 10. Wiligum *corbibus*, Hpt. Gl. 468, 27. Seofon wilian fulle *septem sportas plenas*, Mt. Kmbl. 15, 37 : Mk. Skt. 8, 8 : Homl. Th. i. 182, 22. Wylian, ii. 396, 6 : Jn. Skt. 6, 13.

wilig-wíse, an ; *f. Basket-wise* :—Seó cyrice is sinhwyrfel on wilewísan geworht, Blickl. Homl. 125, 21.

wiliht; *adj. Having willows* :—On wylihte mǽdwan (*the meadow with willows in it*) ; of wylihte mǽdwan, Cod. Dip. Kmbl. iii. 235, 16.

wilincel (-uncel), es ; *n. A (young) slave* :—Wiluncel *mancipium*, Germ. 401, 30. v. wealh, *and next word* : cf. þeówincel.

wilisc ; *adj.* **I.** *foreign, not English* :—Wylisc moru *carrot* (cf. wealh-moru) . . . Englisc moru *parsnip*, Lchdm. ii. 312, 16–21. Wǽlisc *opratanum* (= *abrotanum*, cf. súperne), Wrt. Voc. ii. 65, 46. Se wǽlisca (heafoc) (cf. wealh-hafoc), Exon. Th. 332, 24 ; Vy. 90. Ðá hæfdon ða welisce menn gewroht ǽnne castel . . . Ðá wǽron ða wǽlisce men (*quidam de Normannis*; cf. Icel. Valskr *Norman*) ǽtforan mid ðam cynge, Chr. 1048 ; Erl. 178, 15, 24. **I a.** *referring to the Celts of England, Welsh* :—Ðe Wilisces monnes londhǽfene. Gif Wylisc mon hæbbe híde londes, L. In. 32 ; Th. i. 122, 9. Englisc . . . Wilisc, 46 ; Th. i. 130, 16 : L. Wg. 7 ; Th. i. 186, 13. Náh náðer tó farenne ne Wylisc man on Ænglisc land, ne Ænglisc on Wylisc ðe má, L. O. D. 6 ; Th. i. 354, 23. Tremerin se Wylisca (Wylsca, *v. l.*) biscop (*bishop of St. David's*), Chr. 1055 ; Erl. 191, 11. Cómon upp on Wylisce Axa .xxxvi. scypa and ðǽr ábútan hearmas dydon mid Gryfines fultume ðæs Wǽliscan cynges, 1050 ; Th. i. 310, 19. Welscan (Wyliscean, l. 36), 1052 ; Erl. 186, 17. Ðæt ylce ðe man ðam Wyliscean þeófe dyde, L. Ath. v. 6, 3 ; Th. i. 234, 13. Ðone Wyliscan cining, Chr. 1056 ; Erl. 191, 22. Wíteþeówne monnan Wyliscne, L. In. 54 ; Th. i. 138, 3. .xii. lahmen scylon riht tǽcean Wealan and Ænglan, .vi. Englisce and .vi. Wylisce, L. O. D. 3 ; Th. i. 354, 10. Wylisce menn geslógan mycelne dǽl Énglisces folces, Chr. 1053 ; Erl. 188, 9. Ðá Wylisce menn hí gegaderodon, and wið ða Frencisce ðe on Walon wǽron gewinn up áhôton, 1094 ; Erl. 230, 32. Hengest and Æsc gefuhton wiþ Walas, and .xii. Wilisce (Wilsce, *v. l.*) aldormenn ofslôgon, 465 ; Erl. 12, 22. ¶ the word is used of some kind of ale :—.xii. ámbra Wilisces ealaþ, .xxx. hlûttres, L. In. 70 ; Th. i. 146, 17. Twá tunnan fulle hlûtres aloð and cumb fulne liðes aloð and cumb fulne Welisces aloð, Cod. Dip. Kmbl. i. 203, 9. .xxx. ômbra gódes Uuelesces aloð ðæt limpeð tó .xv. mittum, 293, 13. Wǽlsces, ii. 46, 27. Geworht of Wiliscum ealað, Lchdm. ii. 78, 23. Drence on Welscum ealað, 136, 1. Dô ealle ðás wyrta on Wylisc ealo, 120, 6. **II.** *servile* :—Hé on ðreó tówearp ða cneór[d]nesse, ðæt wæs wǽlisc (*the race of Ham*; cf. Onwôcon of Chame .xxx. theóda mycelra, and eác ðæt cynn wæs geseald ðám óðrum cynnum twám on heaftneád and on þeówdôm, 2, 51), and on cyrlisc cynn, and on gesýðcund cynn, Anglia xi. 3, 62. [*O. H. Ger.* Walahisc *romanus, latinus* : *Icel.* Valskr *foreign, esp. French.*] v. wealh.

wil[1], es ; *n.* **I.** *will, pleasure* :—Se cyng geseah ðæt hé nán þincg his willes ðǽr gefordian ne mihte *the king saw that he could carry out nothing of his purpose*, Chr. 1097 ; Erl. 234, 6. Hé nolde his willes (*of his own accord*) heora geférrǽdene forlǽtan, Homl. Th. ii. 334, 25 : Ap. Th. 4, 5. Wylles, Nicod. 11 ; Thw. 6, 7. Gif hwá hine sylfne besmíte his ágenes willes (*sua sponte*), L. M. I. P. 36 ; Th. ii. 274, 20 : Homl. Ass. 62, 255. Gif þeówa and þeówen hyra bégra wylles hig gesomnigon *si servus et ancilla mutua voluntate se conjunxerint*, L. Ecg. C. 25 ; Th. ii. 150, 15. Be ðínum ágenum wille ðú férdest tó ðínes fæder híwrǽdene *ad tuos ire cupiebas et desiderio erat tibi domus patris tui*, Gen. 31, 30. Hí móston ðæs cynges wille folgian, Chr. 1086 ; Erl. 222, 33. Ne fornime incer nóðer óðer ofer wil bútan geþafunge *nolite fraudare invicem, nisi ex consensu*, Past. 51 ; Swt. 399, 34. Hé genam ðæt wíf ofer ðæs cynges wil, Chr. 1015 ; Th. i. 276, 4, col. 2. **II.** *a pleasant or desirable thing* :—Willa (wilna ? v. willa, **VI a**) spédum, dugeða gehwilcre stépan, Cd. Th. 142, 18 ; Gen. 2363. [Þin aȝen wil, O. E. Homl. i. 61, 119. Liues wil and eche pleie, 193, 62. Þe onnesse of o luue and of o wil, A. R. 12, 7. Al his wil to don, Laym. 2793. Ðu wurchest mi wil, Kath. 2108. Þat wil, Shor. 16. *Icel.* vil ; *n.*] v. ge-, self-, un-wil[1], willes ; willa.

will, well, wyll, es ; *m. A well, spring, fountain* (lit. and fig.) :—Well *fons*, Wrt. Voc. i. 54, 29. Án wyll (*fons*) ásprang of ðære eorðan, Gen. 2, 6. Ðǽr wæs Iacóbes wyl (wyll, *v. l.*). Se Hǽlend sæt æt ðam wylle, Jn. Skt. 4, 6. Bið on him will (wyll, *v. l.*) forðrǽsendes wætres, 4, 14. Wyl, Bd. 1, 7 ; S. 478, 27. Hió áweoll of ánum wille (welle, Cott. MSS.) *non a diverso fonte emanavit*, Past. 7 ; Swt. 49, 11. Lǽt forð ðíne willas (wyllas, Cott. MSS.) . . . Ðæt is ðætte se láreów ǽrest sceal self drincan of ðam wille his ágenre láre *deriventur fontes tui foras . . . Rectum est, ut ipse prius bibat*, 48 ; Swt. 373, 14. Of ðam geate tó wille ; fram ðan wille, Cod. Dip. Kmbl. iii. 172, 37. Áþweah ða eágan on clǽnum wylle, Lchdm. ii. 32, 16. Hwílum gehátaþ hý ælmessan tó wylle (wille, welle, *v. ll.*), Wulfst. 12, 3. Gif hwylc man his ælmessan

geháte oððe bringe tó hwylcon wylle (*ad fontem aliquem*), . . . fæste .iii. geár on hláfe and on wætere, L. Ecg. P. ii. 22 ; Th. ii. 190, 24. Gif hwá his wæccan æt ǽnigum wylle hæbbe, iv. 19 ; Th. ii. 210, 12. Hlûterra wella wæter hí druncon *potum dabat lubricus amnis*, Bt. 15 ; Fox 48, 12. Wylla, Cd. Th. 240, 13 ; Dan. 386. Willas *fontes*, Ps. Spl. 103, 11. Wyllas, 73, 16. Ne weorðian gé wyllas, Wulfst. 40, 15. [Cnihtes þane wel dutte, Laym. 19812 (2nd MS.).] v. wæter-will ; willa, wille.

willa, wella, wylla, an ; *m. A well, spring, fountain* (lit. and fig.) :—Wæs ðér wǽlla (*fons*) . . . ðe Hǽlend sæt ofer ðæm wǽlla, Jn. Skt. Rush. 4, 6, 14. In ðæm wǽlla, 9, 7. Tó ðé ðam willan ealles wísdômes *ad te fontem omnis sapientiae*, Bd. 5, 24 ; S. 649, 3. Mid ðam willan fulluhte bæþes *fonte baptismatis*, 5, 7 ; S. 620, 33. Ðiosne pytt ł uælla *puteum*, Jn. Skt. Lind. Rush. 4, 12. [Heo ȝeoten i þan welle (wille, 2nd MS.) ; þa wes þa welle mid attre bigon, Laym. 19771.] v. wille, will.

willa, an ; *m.* **I.** *will, the faculty of willing* :—Gé hwæthwega godcundlíces on eówerre sáule habbaþ, ðæt is andgit and gemynd and gesceádwíslíca willa, Bt. 14, 2 ; Fox 46, 26. Sáwul is *voluntas*, ðæt is wylla, ðonne heó hwæt wyle, Homl. Skt. i. 1, 187. Ðæs mannes sáwl hæfð on hire . . . gemynd and andgit and willa . . . Of ðam willan cumaþ geþôhtas and word and weorc, ǽgðer ge yfele ge góde . . . þurh ðone willan heó wile swá hwæt swá hire lícaþ, Homl. Th. i. 288, 18–30. Se willa sceal beón ǽfre frig, Ælfc. Gr. 32 ; Zup. 200, 2. Mid ðínum ágenum willan and mid ðínum ágenum anwealde ðú ealle ðing geworhtest, Bt. 33, 4 ; Fox 128, 12. **II.** *in case of one who has authority, will, purpose, design, command* :—Gewurðe ðín willa *fiat voluntas tua*, Mt. Kmbl. 6, 10. Hé eall gedéð, swá his willa byð *omnia, quaecumque voluit, fecit*, Ps. Th. 113, 11. Bið ðam ôþrum ungelíce willa geworden *God's will will turn out very differently for the others*, Exon. Th. 77, 29 ; Cri. 1264. Hæfde se heorde, se ðe of heofonum cwom, feóndas áfyrde. Hwylc wæs fægerra willa geworden ? *what fairer instance of God's will taking effect has there been ?* 147, 3 ; Gú. 721. Gebéte hit God elmihtiga, ðonne his willa sý, Chr. 1085 ; Erl. 219, 24. Gecyrron wé tó Drihtnes willan, Blickl. Homl. 101, 35. Hwyder magon gyt gangan from mínum willan ? 187, 25. Him eal worold wendeþ on willan, Beo. Th. 3482 ; B. 1739. Ðás fíf andgitu gewisseþ seó sáwul tó hire wyllan, Homl. Skt. i. 1, 201. Willan *nutum*, Wrt. Voc. ii. 93, 26 : 61, 5. Ic þurh his willan hider ásend wæs *Dei voluntate huc missus sum*, Gen. 45, 8. Ðǽne þeów ðe his hláfordes willan (*voluntatem*) wiste, and ne dyde æfter his hláfordes willan, Lk. Skt. 12, 47. Hé Drihtnes willan sôhte, Blickl. Homl. 225, 30. Ic dô willan mínes Drihtnes, 243, 22 : Cd. Th. 9, 15 ; Gen. 142. Hí his willan wyrcean *qui facitis voluntatem ejus*, Ps. Th. 102, 20. Heó Alwaldan brǽc word and willan, Cd. Th. 38, 1 ; Gen. 600. Mid gebedum ealne deófles willan oferswíþan, Blickl. Homl. 61, 20. Ic tó him gebeáh and his willan geceás *I submitted to him and swore allegiance to him*, L. O. 1 ; Th. i. 178, 9. Wið ðam ðæt heó his willan gecéóse *on condition of her becoming his wife*, L. Edm. B. 3 ; Th. i. 254, 12. Syndon ðíne willan rihte, Cd. Th. 234, 10 ; Dan. 290. **II a.** *with reference to the disposition of property* :—Ic Abba cýðe and wrítan háte hú mín willa is ðæt mon ymb mín ǽrfe gedóe æfter mínum dæge. Ǽrest ymb mín lond . . . is mín willa, gif mé God bearnes unnan wille, ðæt hit fôe tó londe æfter mé, Chart. Th. 469, 27 : 470, 3. **III.** *will, determination, resolution* :—Hwilc anwilnys and geortrúwad wylla, þurh ða ðeós fægre geógað nú forwurðan sceall, Homl. Skt. i. 4, 310. **IV.** *will* in contrast with power or performance, *intention, purpose, desire to act* :—Twá ðing sindon ðe ǽlces monnes ingeþanc tó fundaþ, ðæt is willa and anweald *duo sunt, quibus omnis humanorum actuum constat effectus* ; *voluntas scilicet, ac potestas*, Bt. 36, 3 ; Fox 176, 7. Þeáh hí ðæt weorc ne mægen fulfremman, hí habbaþ ðeáh fúlne willan, and se untweofealda willa bioþ tó tellenne for fullfremod weorc . . . þeáh willaþ ða yfelan wyrcan ðæt, ðæt hí lyst, . . . ne forleósaþ hí eác ðone willan, ac habbaþ his wíte . . . Se yfla willa hiora welt, 36, 7 ; Fox 184, 23–29. Se yfela willa biþ tóstenced, gif mon ðæt weorc þurhtión ne mæg, 38, 2 ; Fox 196, 31. Ic nǽfre ne teolode sittan on ánum willan mid ðám árleásum *cum impiis non sedebo*, Ps. Th. 25, 5. Arn hé inn mid sceandlícum willan, Homl. Skt. i. 7, 170. Se man se ðe wylle ôþerne ofsleán, and ne mæg his wyllan þurhteón, L. Ecg. P. ii. 1 ; Th. ii. 182, 14 : Past. 11 ; Swt. 71, 14. **V.** *will, desire, wish* :—Ic læs mǽrðo gefremed hæfde þonne mín willa wǽre, Nar. 32, 29. Wé witon ðæt ðæt is ðínes módes willa, ðæt ðú môte ðás world forlǽtan, Blickl. Homl. 225, 19. Ic beó gearo sôna willan ðínes *I will consent to your wish*, Exon. Th. 245, 26 ; Jul. 50. Hé cwæþ ðæt ðæt ídel wǽre ðæt hí wilnedon, ac æt nýhstan mid ánmóde willan monigra hé wæs oferswíþed, Bd. 5, 6 ; S. 619, 3. Tó willan (*ad votum*) ðæs weres heó eardigendlíc wæs geworden, 4, 28 ; S. 605, 20. His heorte ongann wendan tó hire willan, Cd. Th. 44, 30 ; Gen. 717. Hié ðæs ðone willan næfdon, ðæt hié heora noman hié benámon, Ors. 2, 8 ; Swt. 94, 7 : Cd. Th. 36, 9 ; Gen. 569. Hí forléton ðone willan tó ágenne, Homl. Th. i. 394, 5. Se

cyning geþafode ðam þegne his willan, Homl. Skt. i. 6, 224: Beo. Th. 1274; B. 635. Ðæt mē God gefylle feores ingeþanc, willan mīnne, Elen. Kmbl. 1359; El. 681. Willum ic wilnade *desiderio desideraui*, Lk. Skt. Lind. Rush. 22, 15. **V a.** *desire in an unfavourable sense:*—Nýtenu ... heora willa tō nānum ōþrum þingum nis āþenod būton tō gīfernesse and tō wrǣnnesse *pecudes ... quorum omnis ad explendam corporalem lacunam festinat intentio*, Bt. 31, 1; Fox 112, 7. Weres wylla, Lchdm. i. 358, 18. Sió hātheortness ðæt mōd gebringð on ðām weorce ðe hine ǣr nān willa tō ne spōn *mentem impellit furor, quo non trahit desiderium*, Past. 33; Swt. 215, 10. Fæste for ðam unrihtan wyllan *pro illa prava cupidine jejunet*, L. Ecg. P. iv. 10; Th. ii. 206, 20. Ic him on onsende in breóstsefan bitre geþoncas þurh mislīce mōdes willan, Exon. Th. 266, 31; Jul. 406. Ða flǣsclīcan willan, ða cumaþ þurh deófles sceónessa tō manna heortan, Blickl. Homl. 19, 6. **VI.** *pleasure, delight :*—Willa *voluptas*, Wülck. Gl. 253, 44. Se willa ðæs līchoman *voluptas carnis*, Bd. 1, 27; S. 493, 19, 21. Ðā cwæþ hē: 'Mē bið willa gif ðū miht' *multum delector, si potes*, 5, 3; S. 616, 30 note. Se ysela willa unrihthǣmedes *voluptas*, Bt. 31, 2; Fox 112, 24: Met. 18, 2. Wē sprǣcon ǣr be ðām fíf geselþum, ðæt is ... willa (cf. fífte beoþ seó blis, 33, 1; Fox 122, 6), Bt. 33, 2; Fox 124, 19, 22: Wulfst. 11, 7. Hire se willa gelamp, ðæt heó on ǣnigne eorl gelýfde, Beo. Th. 1257; B. 626: 1653; B. 824. Hwý ne miht ðū gehersan gif on ǣnegum ðissa eorþlīcena gōda ǣniges willan and ǣniges gōdes wana is, ðonne is sum gōd full ǽlces willan and nis nānes gōdes wana *si est quaedam boni fragilis imperfecta felicitas, esse aliquam solidam, perfectamque, non potest dubitari*, Bt. 34, 1; Fox 134, 24–27. Ðæt wíf onfēhð ðæs (*from that*) willan on ðæm hǣmede, Lchdm. i. 350, 11. Ðū tíres mōst, willan brūcan, Andr. Kmbl. 212; An. 106: Exon. Th. 151, 24; Gū. 800. Gif ðæt mōd ðæm willan ne wiðbrītt *dum in cogitatione voluptas non reprimitur*, Past. 11; Swt. 71, 8. Hē hine on ðæm willan gehielt ðæt hē mid ealre ēstfulnesse lufaþ ðæt ēce líf *sub aeterna ejus beatitudine tota devotione continetur*, 50; Swt. 389, 15. Se wer ðe his bebod healdeþ mid willan *the man that delighteth in his commandments*, Ps. Th. 111, 1. Ðeáh ðe hē lēte wæter on willan, wynnum flówan, 77, 21. Ne weóx hē him tō willan, ac tō wælfylle and tō deáðcwalum, Beo. Th. 3426; B. 1711. Tō willan and tō wordmyndum *to please and honour him*, 2376; B. 1186. Nafast ðū tō manna mægene willan *non in viribus equi voluntatem habebit*, Ps. Th. 146, 11. Þurh ungelýfedre willan *per inlicitam voluptatem*, Bd. 3, 19; S. 548, 29. Forgife ðē Dryhten willan on worulde and in wuldre blǽd, Andr. Kmbl. 711; An. 356. Heó wīde hire willan sōhte, nō hwæðere reste fand, Cd. Th. 87, 28; Gen. 1455. Ða willan and ða getǣsu ðe him on ðisse worulde becumaþ, Past. 50; Swt. 387, 15. Hwǣr cumaþ his willan and his fyrenlustas? Blickl. Homl. 113, 1. Ða ðe ðisses middangeardes wilna and welena wilniaþ, Past. 50; Swt. 387, 7. Mið walum and willum lífes *divitiis et uoluptatibus uitae*, Lk. Skt. Lind. 8, 14. Willum neótan, Exon. Th. 82, 26; Cri. 1344. Hē brūcan mōt wonges mid willum, 208, 1; Ph. 149. Willum biscyrede, 93, 3; Cri. 1520. Tō hira willum *ad suos libitos*, Wrt. Voc. ii. 3, 10. Willan *libitos, luxus*, Hpt. Gl. 480, 60. **VI a.** *a pleasant, desirable thing, a good, what gives pleasure, what is desired :*—Ic eom ǽpelinges ǽht and willa, Exon. Th. 488, 19; Rä. 77, 1. Nānes willan wana, nāþer ne weorþscipes, ne anwealdes, ne foremǣrnesse, ne blisse, Bt. 24, 4; Fox 86, 30. Gif ðē ǣnies willan wana biþ, ðeáh hit lytles hwæt sié, 11, 1; Fox 30, 22: 26, 1; Fox 90, 22. Nǣron hī bescyrede sceattes willan *non sunt fraudati a desiderio suo*, Ps. Th. 77, 29. Siððan hē ðæs welan full biþ, ðonne þincþ him ðæt hē hæbbe ealra willan, gif hē hæbbe anweald, Bt. 33, 2; Fox 124, 10. Oft brincð se woruld ðone willan ðe bið eft *time often brings the unattained desire*, Prov. Kmbl. 40. Gif man mægðman nēde genimeþ, ðam āgende .l. scillinga, and æft æt ðam āgende sīnne willan (*the object he had desired i. e. the maiden*) ætgebicge, L. Ethb. 82; Th. i. 24, 4. Losewest willana *deceptio divitiarum*, Mk. Skt. Rush. 4, 19. Wilna brūcaþ, āra on eorðan, Cd. Th. 92, 22; Gen. 1532. Wilna geniht, 113, 21; Gen. 1890. Wilna brytta, 97, 29; Gen. 1620. Wilna gehwilces weaxende spēd, 100, 6; Gen. 1660. Wana wilna gehwilces, 137, 12; Gen. 2272. Hié lǽddon eorðwelan, wíf and willan and heora woruldgestreón, 112, 31; Gen. 1879. **VII.** *will, disposition :*—On eówrum fæstendagum bið ongieten eówer willa *in diebus jejuniorum vestrorum inveniuntur voluntates vestrae*, Past. 43; Swt. 315, 3. 'Sý on eorðan sibb ðām mannum ðe synd gōdes willan.' Ne bið nān lāc Gode swā gecwēme swā se gōda willa ... Hwæt is gōd willa būton gōdnys ... Hwæt is ǣnig lāc wið ðisum willan? Homl. Th. i. 582, 33–584, 10: Hy. 8, 6. Hē (*Titus*) wæs swā gōdes willan ðæt hē sægde ðæt hē forlure ðone dæg ðe hē nōht on tō gōde ne gedyde, Ors. 6, 8; Swt. 264, 2. Mid gōdum willan fæstan, Blickl. Homl. 37, 27: 97, 27. Gode underþeódde on gōdum willan, 79, 32. On fæstendagum bið gesýne hwilcne willan gē habbaþ, L. E. I. 42; Th. ii. 438, 35. Nǣnig wæs weorð, gif mon his willan begeat yfelne, Met. 8, 37. Gelícnyssa willena *qualitates affectionum*, Scint. 28, 18. **VII a.** *good will, favourable disposition :*—Swā micel beón scyl gebiddendes embe Gode willa *tantus esse debet orantis erga Deum affectus*, Scint. 33, 8. Willa belimpð tō

blisse simle *voluntas ad laetitiam pertinet*, Past. 43; Swt. 315, 5. Se Hālga Gāst is willa and sōð lufu ðæs Fæder and ðæs Suna; sōðlīce willa and lufu getācniaþ ān ðing, Homl. Th. i. 282, 2–4: 228, 24. In ārfæstnesse willan *in devotione pietatis*, Bd. 4, 22; S. 592, 25. Gē earme men willum onfēngun, on mildum sefan, Exon. Th. 83, 5; Cri. 1351. **VIII.** *in reference to voluntary or to permitted action, will, accord, consent, pleasure :*—Gif ðam Pāpan ðæt līcode and ðæt his willa wǣre and his leáf *si Papae hoc ut fieret, placeret*, Bd. 2, 1; S. 501, 32. Gif beweddod mǣden nele tō ðam ðe heó beweddod bið, and wæs hire willa *si puella desponsata cum eo esse nolit, cui voluntate sua desponsata erat*, L. Ecg. C. 20; Th. ii. 148, 29. Selflices willan *spontaneae voluntatis*, Hpt. Gl. 436, 76. Ágnes willan hē bið gebunden, Homl. Th. i. 212, 16: 224, 23. Ða yfelan nellaþ heora willan gehýran Godes beboda, L. Ælfc. P. 4; Th. ii. 364, 20. Be willan *ultro*, Wrt. Voc. ii. 92, 74. Wæs sió fǣmne mid hyre fæder willan biweddad, Exon. Th. 244, 24; Jul. 32: Met. 24, 54: Andr. Kmbl. 2802; An. 1403. Eallra gesceafta āgnum willan (-um, Cott. MS.) God rícsaþ ofer hī, Bt. 35, 4; Fox 160, 12. Hwæðer ǣnig gesceaft seó, ðe hire willan (-um, Cott. MS.) nylle ealne weg bíon, ac wile hire āgnum willan (-um, Cott. MS.) forweorþan, 34, 9; Fox 148, 11. Mid fullan willan *volens*, 36, 6; Fox 182, 7. Nō genēded, ac mid his wyllan, Blickl. Homl. 29, 16. Mid hyra bēgra wyllan *cum consensu amborum*, L. Ecg. C. 25; Th. ii. 150, 20. His āgnum willan (willum, *v. l.*) hē com tō rōde gealgan, Past. 3; Swt. 33, 19. Ungeniédde, mid eówrum āgenum willan (willum, Cott. MSS.), 18; Swt. 137, 20. Be his āgenum willan, Homl. Th. i. 228, 30. Mid his sylfes willan *ultro*, Bd. 1, 7; S. 477, 22. Mid nænigum nēde gebǣded, ac mid his sylfes willan, Blickl. Homl. 83, 32. Hē genam ðæt wíf ofer ðes cynges willan, Chr. 1015; Erl. 152, 5. Hē ofer willan gióng *he went against his will*, Beo. Th. 4810; B. 2409: Exon. Th. 412, 6; Rä. 30, 10. Him nānwuht wið his willan ne sié, Bt. 11, 1; Fox 30, 25. Hē mid ðara wietena willum ðǣm cynedōme ne mehte tō cuman, Ors. 4, 5; Swt. 166, 26. Ic gestāg willum mínum, Exon. Th. 91, 16; Cri. 1493. Ðæs ðe ðū nǣfre þínum willum ālǣtan woldest, Bt. 11, 2; Fox 34, 13. Ðæt ǣnegu þeód ōþre hiere willum friþes bǣde, Ors. 1, 10; Swt. 48, 29. Gif hié hiera willum ūs tō noldon *si uoluntate sua nollent procedere*, Nar. 10, 23: L. Wih. 1; Th. i. 36, 16: Bt. 11, 1; Fox 32, 29: Ps. Th. 17, 43: 44, 16. Hannibal his āgnum willum hine selfne mid ātre ācwealde *Annibal veneno sese necavit*, Ors. 4, 11; Swt. 206, 30: Bt. 34, 11; Fox 150, 30. Hī hiora āgnum willum hí sylfe unþeáwum underþeódaþ, 40, 7; Fox 242, 29: 35, 4; Fox 160, 16. Hī sēcaþ sylfra willum hāmas on heolstrum, Exon. Th. 107, 4; Gū. 53. Oðer hiene his selfes willum gebeád, Past. 7; Swt. 49, 3. Mid his sylfes willum *ultro*, Bd. 1, 7; S. 477, 15. **IX.** *sake, account* (cf. *Ger.* meinetwillen) :—Hē āscade hié for hwý hié nolden geþencan ealle ða brocu and ða geswinc ðe hē for hira willan and eác for hiera niédþearfe fela wintra dreógende wæs, Ors. 5, 4; Swt. 224, 28. Hē ǣfre wan for willan ðæs Ælmihtigan, Homl. Skt. ii. 25, 683. **X.** *will, one's own way :*—Saga mē hwæt ðam men sí leófust on his life and lāðost æfter his deáðe. Ic ðē secge his willa, Salm. Kmbl. p. 204, 44. Ic hī lifian hēt æfter hiora willum *ibunt in voluntatibus suis*, Ps. Th. 80, 12. [*Goth.* wilja: *O. Sax.* willio: *O. Frs.* willa: *O. H. Ger.* willo *voluntas, voluptas, affectus, affectio, votum, placitum, intentio, nutus, propositum, arbitrium, mens, animus, ratio: Icel.* vili.] v. hyht-, un-, weorold-willa; wil[l].

willan; *prs.* ic, hē wille, wile, ðū wilt, *pl.* wē willaþ; *p.* wolde, walde: *part. prs.* willende *To will, wish :*—*Volo* ic wylle, *uis* ðū wylt, *uult* hē wyle, *uolumus* wē wyllaþ ... *utinam uellem* eálā gyf ic wolde; *utinam uelim* eálā gyf ic wylle ... *uelle* wyllan, Ælfc. Gr. 32; Zup. 199, 14–200, 6. **I.** *to will, exercise the faculty of willing :*—Ic undergyte ðæt ic wylle undergytan and gemunan, and ic wylle ðæt ic undergyte and gemune; ðǣr ðǣr ðæt gemynd bið, ðǣr bið ðæt andgyt and se wylla, Homl. Skt. i. 1, 120. Þurh ðone willan seó sāwul wile swā hwæt swā hire līcaþ, Homl. Th. i. 288, 29. Ælc mon hæfþ ðone friódōm, ðæt hē wāt hwæt hē wile, hwæt hē nele *ipsis inest volendi, nolendique libertıs*, Bt. 40, 7; Fox 242, 20. **II.** *where the will of the subject determines its own action, to will, purpose, think, mean, intend*, (a) with an infinitive :—Ic wille mid flóde folc ācwellan, Cd. Th. 78, 20; Gen. 1296. Ic reste on ðē āgan wylle, 254, 16; Dan. 612. Ic ðēc for sunu wylle freógan, Beo. Th. 1899; B. 947. Hwyder wilt ðū gangan? Ic wille gangan tō Rōme, Blickl. Homl. 191, 16. Ne wille ic leng his geongra wurþan, Cd. Th. 19, 15; Gen. 291. Hē wile eft gesettan heofona rīce, 25, 20; Gen. 396: 176, 30; Gen. 2919. Wē hine willaþ ācwellan and ūs tō mete dōn, Blickl. Homl. 231, 14. Hū geweard ðē ðæs, ðæt ðū sǣbeorgas sēcan woldes māðmum bedǣled? Andr. Kmbl. 616; An. 308. Wolde hē hiene selfne on ðæm gefeohte forspillan *mori in bello paratus*, Ors. 3, 9; Swt. 128, 6: Cd. Th. 176, 2; Gen. 2905. Ne wellaþ (willaþ, Wrt. Voc. ii. 71, 64) cweþan *ne velitis dicere*, Mt. Kmbl. Rush. 3, 9. Hwí forcwið hē ... būton hē cueðan wielle (wille, Cott. MSS.), ðæt hē ne lufige ðone Hláford, Past. 5; Swt. 43, 7. Hwæþre him Alwalda wille wyrpe gefremman, Beo. Th. 2633; B. 1314. Ðæt ðū for sunu wolde hererinc habban, 2355; B. 1175. Ðū him ðæt gehēte, ðæt ðū hyra frumcyn īcan wolde, Cd. Th. 236, 8; Dan. 318. (b) with an accusa-

tive:—God symble wyle gôd, and næfre nân yfel, Homl. Skt. i. 1, 48. From ðære tungan ðe teosu wylle *a lingua dolosa*, Ps. Th. 119, 2. Ðæt heó hî frûne hwæt hî sôhton, oþþe hwæt hî ðær woldon, Bd. 3, 8; S. 531, 39. (c) with a clause:—Wêndon gê and woldun, ðæt gê Scyppende sceoldan gelîce wesan, Exon. Th. 141, 30; Gû. 636. Wêndon and woldon, ðæt hié on elþeódigum æt geworhton, Andr. Kmbl. 2145; An. 1074. (d) absolute, (1) of purpose to go:—Nû wille ic ðam lîge neár, Cd. Th. 47, 14; Gen. 760. Ðâ hê him from wolde, Past. 3; Swt. 35, 19. Gif hé eów âxie hwæder gê willon (*quo vadis?*), Gen. 32, 17. Ðâ hî tô scipan woldon, Chr. 1009; Erl. 142, 28. Ðâ salde se here âþas ðæt hié of his rîce uuoldon, 878; Erl. 80, 17. Ðâ woldan hié on ēcnesse hæle and trume wið deófla nîþum, and wundorlîce deáþ geþrowodan, Blickl. Homl. 171, 30. Mid ðæm ðe hî hié getrymed hæfdon, and tôgædere woldon, Ors. 4, 2; Swt. 160, 28. (2) of purpose to do:—Hê cŷdde his syrewunge, hû hê ymbe wolde (*how he had intended to act*), Homl. Th. i. 82, 18. (3) of things, *to tend*:—Hwæðer ðû nû ongite hwider ðiós spræce wille? Ðâ cwæþ ic: Sege mê hwider hió wille *jamne igitur vides, quid haec omnia, quae diximus, consequatur? quidnam? inquam*, Bt. 40, 1; Fox 234, 32. **III.** where the will of the subject determines the action of another, *to will, ordain, order, command*, (a) with an accusative:—Se ealdorman gewât ðâ ðâ hit wolde God, Homl. Skt. i. 20, 13. Ðâ ðâ hê wolde ðæt ðæt hê wolde, Met. 11, 15. Hwæþer wê ænigne frýdôm habban, ðe sió godcunde foretiohhung oþþe sió wyrd ûs nêde tô ðam ðe hî willen, Bt. 40, 7; Fox 242, 16. (b) with a clause:—Ic wylle (uillo, Lind.: willo, Rush.) ðæt hê wunige ðus, Jn. Skt. 21, 22. Wyltû (wylt ðû, *v.l.*) wê secgaþ ðæt fŷr cume of heofene, Lk. Skt. 9, 54. Hê wolde ðæt ða cnihtas cræft leornedon, Cd. Th. 221, 4; Dan. 83. Hê wolde ðæt him eorðe geseted wurde, 6, 35; Gen. 99: Met. 11, 16. (c) absolute:—Hê cunnian wolde his Drihtnes wyllan, hû hê wolde be him (*what he would have him do*) . . . Cwæð se Hælend, ðæt hê sceolde underfôn mæden, Homl. Skt. i. 4, 7–13. **IV.** *to will, wish, want, desire*, (a) with infinitive:—Ic wielle heora cŷpan hêr lusîcor ðonne ic gebicge ðær, Wülck. Gl. 97, 2. Wilt ðû, gif ðû môst, wesan aldordēma? Cd. Th. 149, 26; Gen. 2480. Ðe wile beorna sum him geâgnian, 109, 26; Gen. 1828. Se ðe wyle sôð sprecan, Beo. Th. 5721; B. 2864. Ðê sælîðend secgan willaþ, ðæt wê fundiaþ Higelâc sêcan, 3641; B. 1818. Wê willaþ beón bylewite *volumus esse simplices*, Coll. Monast. Th. 33, 7. Ic ðîne bebodu wolde gegân *concupivi mandata tua*, Ps. Th. 118, 40. Swâ fela swâ hê habban wolde, Chr. 877; Erl. 78, 24. On hwilce healfe ðû wille hwyrft dôn, Cd. Th. 115, 12; Gen. 1918: 139, 20; Gen. 2312. Gesecgan mid hû micle elne æghwylc wille synrust þweán, Exon. Th. 81, 4; Cri. 1318. Se biscop ðe wile onfôn Godes mildheortnesse, Blickl. Homl. 45, 7. Gif ænig man wolde heora ôðrum fylstan, Homl. Skt. ii. 27, 56. (a 1) where an infinitive may be supplied from the context:—His nêxtan þe his mihte gehelpan, and ofer his mihte wyllan *to help his neighbour according to his power, and to wish to help him beyond his power*, Homl. Th. i. 584, 9. (b) with an accusative:—Ðæt ðæt ðû wylt, ðæt ðû lufast, Homl. Th. i. 282, 5. Hwæt wille gê? Coll. Monast. Th. 32, 23 : Blickl. Homl. 155, 35. For ealle ðe willaþ ðæt hê wile, L. Ath. iv. 3; Th. i. 222, 20. Hê cwæþ: 'Hwæt wilt ðû ðæt ic ðê do?' Næs ðæt nâ ðæt hê nyste hwæt se blinda wolde, Blickl. Homl. 19, 33. Hig dydon ymbe hyne swâ hwæt swâ hî woldon (waldon, Lind.: waldun, Rush.), Mt. Kmbl. 17, 12. Bide mê swâ hwæt swâ ðû wylle (willt ł wælle, Lind.) . . . ic ðê sylle swâ hwæt swâ ðû mê bitst, þeáh ðû wylle healf mîn rîce, Mk. Skt. 6, 22, 23. Behreówsunge mâ wyllan ðænne deáð *penitentiam malle quam mortem*, Anglia xi. 119, 66. ¶ the present participle used with force of Latin forms in *-dus*:—Gefeán ðære willendan gesynto *cupitae sospitatis gaudia*, Bd. 4, 3; S. 570, 22. (c) with a clause:—Wilt ðû ðæt ic ðê secge? Salm. Kmbl. 506; Sal. 253. Wilt ðû ðæt ic gelŷfe? Blickl. Homl. 179, 35. Æghwylc mon wile ðæt him Drihten selle ealle his þearfe, 51, 15. Hû hê wolde ðæt mon him miltsode, Past. 16; Swt. 101, 10. Hê walde ðæt hî wæren gedrêfde, 58; Swt. 443, 11. Wolde, Exon. Th. 74, 7; Cri. 1203. Wê woldun ðû gesâwe ðæt . . . , 130, 16; Gû. 439. Hî woldun, ðæt : . . , 123, 17; Gû. 324. Hî willen ðæt him Dryhten tô hyra earfeða ende gerýme, 115, 25; Gû. 195. For ðý ic wolde ðæt hit ealneg æt ðære stôwe wæren, Past. pref.; Swt. 9, 5. For ðon hê ðis dyde ðæt hê wolde ðæt hié ne wæron gedrêfede, Blickl. Homl. 17, 1. **IV a.** *to like* (where there is an expressed or implied condition):—Ic wolde ðe âcsian hwæþer wê ænigne frýdôm habban, 40, 7; Fox 242, 13. Wolde ic freóndscipe ðinne, gif ic mihte, begitan, Andr. Kmbl. 956; An. 478. Wolde ic ânes tô ðê cræftes neósan, 966; An. 483. Gif hæleþa hwone hlîsan lyste, ðonne ic hine wolde biddan, Met. 10, 3. Wolde ic, ðæt ðû funde ða, Elen. Kmbl. 2157; El. 1080. Eall þing habbaþ ænne willan, ðæt is ðæt hî woldon â bión, Bt. 34, 12; Fox 152, 29. **V.** *to will, be willing* to do something, (a) with an infinitive expressed or implied:—Gyf ðû wylt, ðû miht mê geclænsian . . . Ic wylle; beó geclænsod, Mt. Kmbl. 8, 2–3. Gif ðû þeáh mînum wilt wordum hŷran, Cd. Th. 35, 24; Gen. 559. Wylt, Beo. Th. 3709; B. 1852. Ne wylt ðû ofergeottul weorðan *noli oblivisci*, Ps. Th. 102, 2: 118, 31. Ne wile Sarran gelŷfan wordum mînum, Cd. Th. 144, 11;

Gen. 2388: 161, 7; Gen. 2661. Gif wit him geongordôm læstan willaþ, 41, 27; Gen. 663. Gif git ðæt fæsten fŷre willaþ forstandan, 152, 17; Gen. 2521. Wille gê beón beswungen on leornunge? Coll. Monast. Th. 18, 18. Gif ðû woldest myltsian, and swâ þeáh ne mihtest . . . ðæt ðû ne mæge myltsian, þeáh ðû wylle, Homl. Skt. i. 3, 184–188. Ne ðurfon wê ðæs wênan, ðæt wuldorcyning æfre wille eard âlêfan, Cd. Th. 272, 7; Sat. 116: 281, 25; Sat. 277. Ne willaþ eów andrædan, 194, 25; Exod. 266. Ic nû suna mînum syllan wolde (*should be ready to give*) gûðgewædu, ðær mê gifeðe ænig yrfeweard æfter wurde, Beo. Th. 5452; B. 2729. Ðær ðû fromlîce freónda lârum hŷran wolde, Exon. Th. 129, 22; Gû. 425. Hié gehêton ðæt hiera kyning fulwihte onfôn wolde, Chr. 878; Erl. 80, 18. Hê cwæð ðæt hê wolde ðam wîfe gemyltsian, Homl. Skt. i. 3, 179. Hê getrûwode ðæt hié his giongorscipe fylgan wolden, Cd. Th. 16, 27; Gen. 249: 46, 15; Gen. 744. Nymðe hié friðes wolde wilnian, 229, 9; Dan. 214. Ðær hŷ hit tô gôde ongietan woldan *if they had been willing to understand it aright*, Exon. Th. 68, 2; Cri. 1107. Gif ðû mînum wilt, wîf, willende wordum hŷran, Cd. Th. 35, 25; Gen. 560. ¶ along with negative forms of the verb:—Saga him swâ hê wille, swâ hê nelle, hê sceal cuman *dic illi quia, velit nolet, debet venire*, Bd. 5, 9; S. 623, 11. Wê sceolon, wylle wê, nelle wê, ârîsan, Homl. Th. i. 532, 7. Wê sceolon beón nêde geþafan, sam wê willan, sam wê nyllan, Bt. 34, 12; Fox 154, 7. Se brym hine bær, wolde hê, nolde hê, Homl. Th. ii. 388, 20. (b) with accusative, *to allow, permit, grant, consent to*:—Ne willaþ hié rûmor unc landriht heora, Cd. Th. 114, 27; Gen. 1910. Se câsere hine ðreátade ðæt hê Criste wiðsôce. Ðâ hê ðæt ne walde, Shrn. 71, 33. Ne ðæt wille God, Cd. Th. 114, 13; Gen. 1903. (c) with a clause:—Nô God wolde, ðæt sió sâwl sâr þrowade, Exon. Th. 126, 29; Gû. 378. **VI.** *to be disposed, to have such and such a will*:—Ðæt man his Scyppend lufige and ða men ðe wel willaþ (*the men that are of good will*), Homl. Skt. i. 16, 254. Ðæt hê wiðstonde ðæm ðe on wôh wiellen (cf. ðâm unrhytwillendum, 89, 22), Past. 15; Swt. 91, 1. Ðæt hê geornlîce fylste ðam ðe riht willan, and â hetelîce stŷre ðâm ðe þwyres willan, L. I. P. 2; Th. ii. 304, 17. **VII.** of habitual action:—Ða ingeðoncas ðe wealcaþ in ðæs monnes môde, ðe æfre willaþ licgean on ðæm eorðlîcum gewilnungum *quando cogitationes volvuntur in mente, quae a terrenis desideriis numquam levantur*, Past. 21; Swt. 63, 22. Hê wolde æfter ûhtsange oftost hine gebiddan and on cyrcan standan on syndrigum gebedum, Homl. Skt. ii. 26, 114. Hwæþer gê willen on wuda sêcan gold? Met. 19, 4. Ðæt se láreów sceolde beón miehtig tô tyhtanne on hâlwende lâre, and eác tô ðreánne ða ðe him wiðstondan wiellen *ut potens sit exhortari in doctrina sana, et eos, qui contradicunt, arguere*, Past. 15; Swt. 91, 16. **VIII.** *to will, profess, claim*:—Hê wæs swâ upâhafen, ðæt hê wolde beón god . . . wolde rênas wyrcan, swylce hê sylf god wære, Homl. Skt. ii. 27, 27–33. **IX.** as an auxiliary for the future, *will, shall, to be about to*:—Gif hê mê cûð ne bið, ic wille him suîðe ræde andwyrdan (*protinus respondemus*), Past. 10; Swt. 63, 4. Hwæt wille ic ðissum wiðersacan geandwyrdan? Homl. Th. i. 378, 11. Hê wæs cweþende: 'Ic mê wille nû onhwyrfan tô ðisse bære . . .' Ðâ wæs hê gongende, Blickl. Homl. 151, 14. Ic mîne sâwle wylle Gode underþeódan *nonne Deo subdita erit anima mea?* Ps. Th. 61, 1. Gif ðû ûre unriht wilt behealdan *si iniquitates observaveris*, 129, 3. Gif ðû æfre cymst tô ðære stôwe, ðonne wilt ðû cweþan (*dices*), Bt. 36, 2; Fox 174, 22: Met. 24, 48. Se ðe wyle sprecan *loquuturus*, Ælfc. Gr. 41; Zup. 247, 15, 11: 248, 6. Gif hiere ne bið sôna gestiéred, hió wile weahsan mid ungemete (*sine mensura dilatatur*), Past. 11; Swt. 71, 16. Ðæt wile þincan ungeleáflîc eallum ðæm ðe ða stôwe on ufe_um tîdum geseoð, Lchdm. iii. 438, 14. Hwæþer hit hysecild ðe mædencild beón wille, ii. 172, 18. Hê wyle naman ðinne herian *laudabunt nomen tuum*, Ps. Th. 73, 20. Hwâ wyle mê gelædan? *quis deducet me?* 59, 8. Hwæt wille wê cweþan be ðînum sunum? *quid dicam liberos?* Bt. 10; Fox 28, 30: Homl. Skt. ii. 28, 117: Homl. Th. ii. 448, 13. Hit wolde dagian *the day was about to break*, Homl. Skt. i. 21, 123. Hit æfnian wolde, 23, 245. His âdumum ðe woldon wîfian on his dohtron *generos suos, qui accepturi erant filias ejus*, Gen. 19, 14. Wât ic ðæt ðû wile gilpan, Salm. Kmbl. 409; Sal. 205. Hwæt God mælan wille *quid loquatur Deus*, Ps. Th. 84, 7. Ic wât ðætte wile woruldmen tweógan (cf. went eall moncyn on tweónunga, Bt. 4; Fox 8, 17), Met. 4, 51. Ne hê sôðfæste læteþ ðæt tô unrihte willen handum ræcean *ut non extendant justi ad iniquitatem manus suas*, Ps. Th. 124, 4. Wên is ðæt hî ûs lifigende wyllen forsweolgan *forsitan vivos deglutissent nos*, 123, 2. Swâ ic ær sægde ðæt ic dôn wolde, Blickl. Homl. 183, 29. Ðeáh ðû onsôce ðæt ðû sôð godu lufian wolde, Exon. Th. 242, 10; Jul. 195. Ðâ Darius geseah ðæt hê oferwonnen beón wolde *Darius cum vinci suos videret*, Ors. 3, 9; Swt. 128, 6: Blickl. Homl. 15, 34. Hû wolde ðæt geweorðan? Elen. Kmbl. 909; El. 456. Wên is ðæt hî ûs woldan gesûpan *forsitan absorbuissent nos*, Ps. Th. 123, 3. **IX a.** without an infinitive:—Hwænne ðû mê wylle tô *quando venies ad me*, Ps. Th. 100, 1. Ær him se fefer tô wille, Lchdm. ii. 134, 24, 22. **IX b.** as optative:—Wolde hûru se earming hine sylfne beþencan, Homl. Skt. i. 19, 161. [*Goth.* wiljan; *p.* wilda: *O. Sax.* willian, wellian; *p.* welda: *O. Frs.*

willa, wella; *p.* wilde, welde, wolde :. *O. H. Ger.* wellan, wollan; *p.* wolta : *Icel.* vilja; *p.* wolda.]

willan; *p.* de.　　　I. *to boil* (trans.):—Wyl (wel, *v.l.*) on wætere . . . wyl on ealdan wíne, Lchdm. i. 72, 7, 23. Wel on buteran, ii. 22, 25. Wæl, i. 374, 8. Wæll, 378, 3.　　II. fig. *to torment, agitate, with violent feelings* (cf. *figurative uses of* weallan *and* seóþan):—Hé wylleþ hine on-ðam wíte, wunaþ unlustum *he gives himself no peace in that pain, lives unpleasingly,* Salm. Kmbl. 537; Sal. 268. [Þe caliz þet was imelt iðe fure and stroncliche iwelled, A. R. 284, 20. A chetel of iwelled bras, Jul. 82, 54. Welled led muchen lead, H. R. 59, 501. *Icel.* vella *to boil* (trans.).] v. ā-, be-, ge-, ofer-, on-willan (-welian, -wyllan).

wille, es; *m. A well, spring, fountain*:—Se wylle *fluvius,* Bd. 1, 7; S. 478, 29 note. Hé is se libbenda wylle (-a?) *fons vivus,* Ps. Th. 41, 2. An tuddeles þorn, and an hróces wylle; . . . þonne an lawernwylle . . . On hróces wylle, þanne up on ðæne weg . . .; þanon on oðen wielle . . .; þanon on eabbincgwylle, þanne on riscbróc, Cod. Dip. Kmbl. ii. 54, 6–15. On ðone fúlan wylle; of ðam wylle, vi. 213, 16–23. v. wīþigwille; will, *and next word.*

wille, wielle, welle, wylle, an; *f. A well, spring, stream, fountain* (lit. and fig.):—Án wielle weól blóde *flumen sanguine effluxit,* Ors. 4, 7; Swt. 184, 21. Welle *fontana,* Wrt. Voc. ii. 149, 79. Ðǽr com upp wætres welle, Shrn. 93, 36. Seó wylle *fluvius,* Bd. 1, 7; S. 478, 29. Is sǽd ðæt wylle (án welle, *v.l.*) (*fons*) áweólle, seó wæter geóteþ, 5, 10; S. 625, 23. Lífes wylle (waelle, Ps. Surt.) *fons vitae,* Ps. Th. 35, 9: Basil admn. 4; Hex. 42, 16. Ealle ða námon Ændor wylle and Cisone clǽne hlimme, Ps. Th. 82, 8. Waelle lēhtes *fons luminis,* Ps. Surt. ii. p. 200, 35. Uælle *fons,* Jn. Skt. Lind. 4, 6. On saltere wellan; of saltere wellan, Cod. Dip. Kmbl. iii. 206, 31. In fúle wellan; of ðære wellan, 366, 31. Swā·culfre ðonne heó baðaþ hí on smyltum wætre on hlúttre wǽllan, Shrn. 85, 22. Ða hālwendan wellan (*fonte*) fulwihtes bæþes,.Bd. 2, 5; S. 507, 17. Wyllan, 3, 22; S. 552, 35 : 4, 13; S. 582, 13.. Hí druncon burnan wæter, calde wellan *potum dabat lubricus amnis,* Met. 8, 29. Of denum yrnaþ deópe wyllan (waellan *fontes,* Ps. Surt.), Ps. Th. 103, 10. In stówum ðǽr ðe hlúttre wyllan (*lucidi fontes*) urnon, Bd. 2, 16; S. 520, 4. Tó waellum wætra *ad fontes aquarum,* Ps. Surt. 41, 2. On·cwicu wæteres wellan *in fontes aquarum,* Ps. Th. 113, 8. Hé him forlēt feówer wellan sceótan (*the reference is to the milk from a cow's udder*), Exon. Th. 419, 26; Rä. 39, 3. [Cf. *O. H. Ger.* wella *fluctus, unda : Icel.* vella *boiling heat.*] v. ed-, sealt-wille. [The word is also found in place-names.]

-wille (cf. wille *a well*). v. cwic-, deád-, fisc-, líf-wille.

-wille (cf. willa *will*). v. án-, druncen-, on-, self-wille.

-Wille. v. Eást-, West-Wille.

wille-burne, an; *f. A bubbling burn, running stream*:—Lago yrnende, wylleburne, Cd. Th. 14, 1; Gen. 212. Drihten lēt willeburnan on woruld þringan of ǽdra gehwǽre, 83, 1; Gen. 1373.

wille-cærse, an; *f. Well-kerse* (v. Jamieson's Dict.), *water-cress*:—Wyllecyrse *foenum graecum,* Wrt. Voc. ii. 67, 76. Willecærse *britia,* 286, 28. Wyllecærse *fenegrecio,* ii. 38, 77. Seóð mid wyllecærsan (-cersan, *v.l.*), Lchdm. i. 140, 12. Nim wyilecærsan (-en, MS.), iii. 134, 2. [Were ne leuere lyue by wellecarses (ete watercrasses, *v.l.*), Piers P. C. 7, 292.]

-willedness, -willend, -willende. v. wel-willedness; riht-, unriht-willend; un-, wel-, yfel-willende, *and* willan, **IV b, V a.**

willendlíce; *adv. Diligently*:—Willendlíce *diligenter,* Wrt. Voc. ii. 140, 41. [*In the following passage the better reading is* hwílwendlíce :— Hé cwæþ ðæt hé gehyhte swā swā hé on his ðeóde willendlíce (hwílwendlíce, M. 248, 22) rícsode, ðæt hé swā on tóweardnesse ēcelíce mid Criste rícsian·móste *sperans ut sicut in sua gente regnat, ita et cum Christo in futuro conregnare,* Bd. 3, 29; S. 561, 22.] v. welwillendlíce.

-willendness. v. yfel-willendness.

willes; *adv. Willingly, voluntarily, of one's own accord*:—Be ðam men ðe willes man ofslihð *de homine qui voluntate aliquem occidit,* L. Ecg. P. ii. 1, tit.; Th. ii. 180, 1. Ne scylan hyg ǽnig unriht willes geþafian, L. I. P. 6; Th. ii. 310, 18. Hé willes deáð þrowade, R. Ben. 26, 15. Geneádod tó ánre míle gange, gang willes twá, 28, 3. Hwílum willes, hwílum geneádode, Homl. Ass. 145, 45. Gif hit geweorðeþ ðæt man un-willes oþþe ungewealdes ǽnig þing misdēd, nā bið ðæt nā gelíc ðam ðe willes and gewealdes sylfwilles misdēð, L. Eth. vi. 52; Th. i. 328, 22 : L. Ed. 7; Th. i. 162, 26: L. O. 1; Th. i. 178, 6. [Gif heo hit brekeð willes and woldes, A. R. 6, 26. Þu þat forschuppes te self willes and waldes, H. M. 27, 2.] v. self-willes, un-wil[l], wil[l].

wille-stréam, es; *m. A bubbling, running stream*:—Ðær se eádga (*the Phenix*) mót neótan wyllestreáma wuduholtum in, Exon. Th. 223, 19; Ph. 362. Se æþela fugel æt ðam æspringe wunaþ wyllestreámas, 204, 30; Ph. 105. [In ane wallestream, Laym. 2849.]

wille-wæter, es; *n. Spring-water*:—Þweah mid wyllewætre, Lchdm. ii. 308, 11. Wyrc ðæt bæþ of ðam ilcum wyrtum on cealdum wylle-wætre, 74, 27. Seóðe on yrnendum wyllewætere, i. 330, 14. [Þe ter þet mon wepð for laðe of þisse liue is inemned wellewater (*aqua fontis*), for he welleð of þe horte swa doð water of welle, O. E. Homl. i. 159, 12. Cæld wellewater (welles water, 2nd MS.), Laym. 19792.]

will-flód, es; *n. m. The waters of the deluge*:—Willflód ongan lytligan, Cd. Th. 85, 10; Gen. 1412. Cf. wille-burne.

will-gespryng, es; *n. A spring*:—Ðeós corþe is berende missenlícra fugela and sǽwihta and fiscwyllum wæterum and wyllgespryngum *avium ferax terra marique generis diversi, fluviis quoque multum piscosis, ac fontibus praeclara copiosis,* Bd. 1, 1; S. 473, 16. Of ðam wilsuman wyllgespryngum beorgeþ *e vivo gurgite libat aquam,* Exon. Th. 205, 8; Ph. 109.

willian; *p.* ode.　　I. *to will*:—Gode willigende *Deo volente,* Guthl. 20; Gdwin. 78, 20.　　II. *to desire,* (a) with a genitive :—Mæg snottor guma his· gǽste forð weges willian, Exon. Th. 104, 15; Gū. 8. Ne sceolde nán wís man willian (wilnian, *v.l.*) sēftes lífes, Bt. 40, 3; Fox 238, 13. (b) with infinitive :—Hwelc is mon se wile líf and willaþ gesián dæges góde? *quis est homo qui vult vitam et cupit videre dies bonos?* Ps. Surt. 33, 13. Gē wylladon (wilniaþ, *v.l.*) ūs ða ðing gemǽnsuman *ea nobis communicare desiderastis,* Bd. 1, 25; S. 487, 13. (c) with gerundial infin.:—Ongit hū unmihtige ða yfelan men beóþ, nū hí ne magon cuman þider ðider ða ungewittigan gesceafta williaþ (wilniaþ, *v.l.*) tó tó cumenne *vide quanta vitiosorum hominum pateat infirmitas, qui ne ad hoc quidem pervenire queunt, ad quod eos naturalis ducit, ac pene compellit intentio,* Bt. 36, 5; Fox 180, 4. (d) with a clause :—Ic willio and wille ðæt hió sión getrymed, Cod. Dip. Kmbl. ii. 121, 23. (e) absolute:—Wer se ðe in bibodum his willaþ (*cupiet*), Ps. Surt. 111, 1. [He wyllede mest of alle þynge to hym enlyance, R. Glouc. 12, 18. Naзt ne willieþ more þanne uor to þe uorlore to þe wordle, Ayenb. 142, 15. Þu willest of briddes to knowe, Piers P. 12, 221. *O. H. Ger.* wil-lôn *desiderare.*] v. ge-willian; willung.

wil-líc; *adj. From a fountain or well*:—Willícan *fontona* (*fontana flumina,* Ald. 161), Wrt. Voc. ii. 92, 10. Wyllícan, 37, 30 : 149, 79.

willíce; *adv. Willingly, voluntarily*:—Ōþre gehwylce ða wyllíce wē onfēngon *cetera queque quae uoluntarie suscepimus,* Anglia xiii. 375, 138. [*O. L. Ger.* willíco *voluntarie.*]

willnian, willnung. v. wilnian, wilnung.

will-spryng *and* -sprynge, es; *m. A well-spring, fountain, source* (lit. *and* fig.):—Welspryng *latex,* Wrt. Voc. i. 54, 30. Seó sōðe lufu is wylspring and ordfruma ealra gódnyssa, Homl. Th. i. 52, 12. Ðæs wæter-scipes welsprynge is on hefonríce, Past. Swt. 467, 31. Welsprinces *fontis,* Hpt. Gl. 418, 43. Mid dǽwigun wylsprince *roscidis fontibus,* 421, 67. Ðás synd ða feówer eán of ánum wyllspringe, Ælfc. T. Grn. 13, 3. Wyllspringas ðære micelan niwelnisse *fontes abyssi magnae,* Gen. 7, 11. Wilsprinȝas, 8, 2. Wæs ðæt wæter and ealle wyllspringas gehálgode þurh Cristes líchaman, Homl. Th. i. 28, 28. Wilspringum *fontibus,* Hpt. Gl. 509, 18. Tó wyllsprengum wætra *ad fontes aquarum,* Ps. Lamb. 41, 2. [An angel tagte hire (*Hagar*) ðor a wellespring, Gen. and Ex. 1243.]

will-sum. v. wil-sum.

willung, e; *f. Desire*:—Ðurh unrihte willunge *per ambitionem,* Bd. 4, 5; S. 573, 11. [My willing is as ye wole, Chauc. Cl. T. 319.] v. ge-willung; willian.

willung, e; *f. Boiling, heat*:—Wyl[l]inc *fervor* (*autumni*), Hpt. Gl. 419, 77. v. samod-willung.

will-weorþung, e; *f. Worship paid to springs*:—Wē lǽraþ ðæt preósta gehwilc ǽlcne hǽðendóm ādwæsce, and forbeóde wilweorðunga (cf. Hǽðenscipe biþ . . . ðæt man weorðige hǽðene godas, and sunnan oþþe mónan, fýr oþþe flód, wæterwyllas oþþe stánas, L. C. S. 5; Th. i. 378, 20. *See also* will), L. Edg. C. 16; Th. ii. 248, 3.　See Grmm. D. M. c. 20.

wilm, wielm, welm, wælm, wylm, es; *m.*　　I. *that which wells.* v. weallan. (1) of fluid, *a fount, stream, water that surges or boils, that moves in waves*:—Wæs ðære burnan wælm headofýrum hát *the burn's surging stream was hot with fierce fires,* Beo. Th. 5086; B. 2546. Fisca welm, wildeóra holt *the fishes' flood, the wild beasts' wood,* Salm. Kmbl. 165; Sal. 82. Ne foldan stán, ne wæteres wylm, ne wudutelga, 843; Sal. 421. Geofon ýþum weól, wintres wylm (*the boiling flood of winter*), Beo. Th. 1036; B. 516. Ic ðæs wælmes grundhyrde fond, 4276; B. 2135. Hé drincð of ðǽm wielme his ágnes pyttes *bibit sui fluenta putei,* Past. 48; Swt. 373, 10. Of swēttestum wylme (wylme?) *de dulcissimo fonte,* Scint. 18, 3. Gān ofer flódes wylm *to go over the tossing waves of the sea,* Andr. Kmbl. 734; An. 367. Ofer ýða wylm, 1726; An. 865. Hí stede wícedon ymb ðæs wæteres wylm (*by the surge of the sea*), Elen. Kmbl. 77; El. 39. In ðæs leádes wylm scúfan *to thrust into the boiling font of lead,* Exon. Th. 277, 20; Jul. 583. Heortan wylmas *veins, blood-vessels* (?), Beo. Th. 5008; B. 2507. (2) of fire, *surging fire, flames*:—Won fýres wælm, se swearta líg, Exon. Th. 60, 7; Cri. 966. His bān brondes wylm forþylmde, 217, 21; Ph. 283. In ðæs wylmes grund befæsted, Elen. Kmbl. 2596; El. 1299. Gold ðæt in wylme bið geclǽnsod, 2617; El. 1310 : 1527; El. 765. Wunian in wylme, Salm. Kmbl. 933; Sal. 466. God wylme gesealde Sodoman, sweartan líge, Cd. Th. 115, 26; Gen. 1925. Gedúfan in ðone deópan wælm, 266, 31; Sat. 30. Helle, grundleásne wylm, Exon. Th. 362, 34; Wal. 46 : Salm. Kmbl. 149; Sal. 74. Hátan ofnes wylm

þurhwódon, Cd. Th. 245, 16; Dan. 464. In fýrbaðe, wælmum bi-wrecene, Exou. Th. 52, 11; Cri. 832. **II.** *heat, fervent heat, fiery heat :*—Wylm *fervor,* i. *calor,* Wrt. Voc. ii. 147, 82. Gif sumeres welm (wylm, *v. l.*) tó swíðlíc bið *si aestatis fervor nimius fuerit,* R. Ben. 65, 20. Ne mihte heora wlite gewemman wylm ðæs wæfran líges, Cd. Th. 231, 2; Dan. 241. Ðæs unmætan wylmes ðære sunnan hǽto, Homl. Skt. ii. 23 b, 573. Wilme and bryne *fervore,* Wrt. Voc. ii. 33, 42. Flór is on welme, Cd. Th. 267, 17; Sat. 39. Þrowigean frécne fýres wylm, 229, 8; Dan. 214. ¶ fig.? :—Oððæt deáðes wylm hrán æt heortan *until the hot touch of death was at the heart,* Beo. Th. 4531; B. 2269. **II a.** *boiling, roasting :*—Wylm *vel* hyrsting *frixura,* Wrt. Voc. ii. 150, 84. Gif hit wæter sý, hǽte man hit ðá hit hleówe tó wylme, L. Ath. iv. 7; Th. i. 226, 14. On welme weorðan *fervere,* Past. 58; Swt. 447, 9. **II b.** *inflammation :*—Se wielm ðæs innoðes út ábierst and wierð tó sceabbe, Past. 11; Swt. 71, 9. Ðæs welmes sár ón ðære lifre, Lchdm. ii. 206, 3. Æfter ðdle welme . . . of ðara ómena welme, 82, 2, 20. Ða welmas ða ðe beóþ gehwǽr geond ðone líchoman, 204, 14. **III.** *violent movement, violence, raging, tempestuous movement* of water :—Oððe fýres feng, oððe flódes wylm, Beo. Th. 3533; B. 1764. Gestilde seó sǽ fram ðam wylme *pontus suo quievit a fervore,* Bd. 3, 15; S. 542, 3. Him ðæs endeleán þurh wæteres wylm Waldend sealde, Beo. Th. 2390; B. 1693: Exon. Th. 283, 14; Jul. 680. Hí feorh áléton þurh ǽdra wylm (*by the surging of the blood from the veins*), 271, 6; Jul. 478. Hé ýdum stilde, wæteres wælmum, Andr. Kmbl. 903; An. 452. **IV.** of mental emotion, (1) *fervour, ardour :*—Hié wénaþ ðæt hiera undeáw sié sumes ryhtwíslíces andan wielm *his suum vitium quasi virtus fervens videtur,* Past. 40; Swt. 289, 20. On ðæm welme ðære sóþan lufan Godes, Blickl. Homl. 29, 10. Wylme, Homl. Th. ii. 128, 3. Wylme *fervore, ardore,* Hpt. Gl. 469, 56. Wælme lufu ðínre *fervore dilectionis tuae,* Rtl. 95, 27. Seó hálíge cyrice sum ding ðurh wælm (*per fervorem*) receþ, Bd. 1, 27; S. 491, 30. Wylm, Hpt. Gl. 465, 37. (2) *heat, fury, rage, passion :*—Wrǽðo ðín and wælm (*furor*) ðín, Rtl. 11, 1. In uælme ðíuum *in furore tuo,* 183, 2. Mid ðam welme ðære hátheortnesse, Bt. 37, 1; Fox 186, 20: Met. 25, 46. Mid miclum wylme and yrre onstyred *nimio furore commotus,* Bd. 1, 7; S. 477, 41. Of lufe nalæs of wylme, 1, 27; S. 490, 13. Yrre ne lǽt ðe wylme besmítan, Exon. Th. 305, 8; Fä. 85. [Fouuer walmes of watere sprungen ut, O. E. Homl. i. 141, 17. In the welmes ben founde stones, Map. 355, 14.] v. ǽ-, bǽl-, breóst-, brim-, bryne-, cear-, ed-, ege-, flód-, frum-, fýr-, heáfod-, heaðo-, holm-, hyge-, sár-, sǽ-, sorg-, stream-wilm (-wælm, -wylm).

wilm-fýr, es; *n. Fierce fire, flaming fire :*—Fore Dryhtne færeþ wælmfýra mǽst, hlemmeþ háta lég, Exon. Th. 58, 7; Cri. 932.

wilm-hát; *adj. Burning hot :*—Him brego engla wylmhátne líg tó wræce sende, Cd. Th. 156, 5; Gen. 2584. [He het fecchen a ueat, and wið pich fullen, and wallen hit walmhát, Jul. 69, 20.]

wiln, e; *f. A maid-servant, a hand-maid :*—Mín wyln (wiln, *v. l.*) *mea ancilla,* mínre wylne *meae ancillae,* míne wylne *meam ancillam,* míne wylna *meae ancillae,* mínra wylna *mearum ancillarum,* Ælfc. Gr. 15; Zup. 100, 20-101, 7. Wyln *ancilla, serva, abra, dula,* Wrt. Voc. i. 50, 14 : 73, 2. Pínen, wyln *abra,* i. *ancilla,* 17, 26. Heó ýs ðín wyln (*ancilla*) under ðínre handa; þreá hig lóca hú ðú wylle, Gen. 16, 6. Seó sáwl is ðæs flǽsces hlǽfdige, and hire gedafnaþ ðæt heó simle gewylde ða wylne, ðæt is ðæt flǽsc, tó hyre hǽsum. Þwyrlíce færð ðæt ðam húse ðær seó wyln bið ðære hlǽfdian wissigend, and seó hlǽfdige bið ðære wylne underðeódd, Homl. Skt. i. 17, 8-12. Oft on ánre tíde ácend seó cwén and seó wyln . . . and ðære wylne sunu wunaþ eal his líf on ðeówte, Homl. Th. i. 110, 27 : Gen. 21, 13. Ne wilna ðú ðínes néhstan wylne, Ex. 20, 17. Gif wíffæst man hine forlicge be his ágenre wylne, L. C. S. 55; Th. i. 406, 14. Ádó ðás wylne (*ancillam*) heonon, Gen. 21, 10 : Ex. 21, 20, 32. Hé genam wealas and wylna (*servos et ancillas*), Gen. 20, 14 : Lev. 25, 44. v. wealh.

wilnian; *p.* ode. **I.** of animate objects, (1) *to desire, ask for* (the source from which marked by *tó*), (a) with gen. or uncertain :— Wilnigaþ monige men anwealdes . . . Se ealra forcúþesta wilnaþ ðæs ylcan, Bt. 18, 1; Fox 60, 27. Hwí wilnige wé ǽnigre óþre sage? *quid adhuc egemus testibus?* Mt. Kmbl. 26, 65. Ða nétenu, and eác ða óþre ge-sceafta, má wilniaþ ðæs ðe hí wilniaþ for gecynde ðonne for willan, Bt. 34, 11; Fox 152, 6. Ealle tó ðé ǽtes wilniaþ *omnia a te expectant, ut des illis escam,* Ps. Th. 103, 25. Wuhta gehwilc wilnaþ tó eorðan, sume nédþearfe, sume neódfræce, Met. 31, 14. Ealle þider willniaþ oþþe ðæs ðe hí lyst, oþþe ðæs ðe hí beþurfon, Bt. 41, 6; Fox 254, 29. Heó hiere feores tó him wilnade (*pro vita precans*), Ors. 3, 11; Swt. 150, 33. Ðæt wæter ðe hé tó Gode wilnade *aquam quam a Deo petierat,* Bd. 1, 7; S. 478, 28. Hé wilnode him tó Gode sumre frófre *he asked of God for some comfort for himself,* Ps. Th. 15, arg. Helpan nánum ðara ðe ðú him áre wilnodan, Blickl. Homl. 223, 3. Ne wilna ðú ðínes néhstan húses ne his wífes *non concupisces domum proximi tui nec desiderabis uxorem ejus,* Ex. 20, 17. Hwelc fremu is ðé, ðæt ðú wilnige ðissa and-weardena gesǽlþa ofer gemet? Bt. 14, 1; Fox 42, 8. Þonne hí tó his húse hleówes wilnian, Ps. Th. 108, 10. Tó ðæm heáhengle ðæt hié him

fultomes wilnodan, Blickl. Homl. 201, 28. Friðes wilnian, Andr. Kmbl. 2258; An. 1130. Willnian (wilnian, *v. l.*) ðæs ðe hé næfþ, Bt. 36, 3; Fox 176, 12. Him wilnian lofes *to desire praise for himself,* Past. 62; Swt. 457, 26: Exon. Th. 119, 28; Gú. 261. Tó Rómánum friþes wilnian *a Romanis pacem petere,* Ors. 4, 6; Swt. 178, 7: Cd. Th. 229, 10; Dan. 215: Exon. Th. 48, 18; Cri. 773. Wylnian, Wulfst. 277, 19. (b) with accusative :—Ðú ne wilnast weora ǽniges -deáð (cf. nolo mortem impii, Ezech. 33, 11), Ps. C. 54. Ða æþelingas wilniaþ, Exon. Th. 433, 14; Rä. 50, 7. Eall hwæt hí willniaþ hí begitaþ, Bt. 40, 7; Fox 242, 22. Ðæt sáwul mín wilnaþ (*concupivit*), ðæt ic ðín word móte healdan, Ps. Th. 118, 20. Ealle hié ðæt wilnodan, ðæt hié his word gehýran móston, Blickl. Homl. 219, 35. Hwæt (hwæs, *v. l.*) hé wilnian sceal, Bt. 40, 7; Fox 242, 18. (c) with infinitive :—Ða ðe wilniaþ fretan mín folc, Ps. Th. 13, 9. Willniaþ ealle þurh mistlíce paþas cuman tó ánum ende *diverso calle, sed ad unum finem nititur pervenire,* Bt. 24, 1; Fox 80, 8. Hé wilnode hine geseón *erat cupiens videre eum,* Lk. Skt. 23, 8. Gif ðú wilnige oncnáwan, Bt. 6; Fox 14, 31. Gif ðú willnige ongitan *si vis cernere,* 39, 13; Fox 232, 24. Wilnige, Met. 29, 1. Gebida wilnando *petere uolentes,* Lk. Skt. p. 6, 12. (d) with gerundial infinitive :—Ðæt hié wielnian (wilnian, Cott. MSS.) tó wietanne ðæt ðæt hié nyton *ut appetant scire, quae nesciunt,* Past. 30; Swt. 203, 8. (e) with genitive and gerundial infinitive :—Ǽlc mód wilnaþ sóþes gódes tó begitanne *est mentibus hominum veri boni inserta cupiditas,* Bt. 24, 2; Fox 80, 32. Hí wilniaþ welan and weorþscipes tó gewinnanne *opes, honores ambiant,* Met. 19, 43. (f) with a clause :—Gif ðú wilnast ðæt ðú mæge oncnáwan, Met. 5, 24. Wilnaþ God tó ǽlcum -men ðæt hé sié oððe wearm oððe ceald *aut calidus quisque esse, aut frigidus quaeritur,* Past. 58; Swt. 447, 15. Hé wilnode ðæt his lícræst sceolde beón æt Cridiantúne, Chr. 977; Erl. 127, 37. Heó ealle tó mé wilnodon ðæt ic hine lǽte æt mé ðæt land begeotan, Chart. Th. 167, 38. Hié wilnedon tó him ðæt hié mósten on his ríce mid friðe gesittan, Ors. 6, 34; Swt. 290, 20. Nis nán gesceaft ðara ðe ne wilnige ðæt hit þider cuman mæge, Bt. 25; Fox 88, 30. Wilnie, Met. 13, 69. Ðeáh hí wielnien (wilnien, Cott. MSS.) ðæt hié andrysne sién, Past. 17; Swt. 109, 18. Tó Sancte Michaele ðæt hié wilnodan ðæt God gecýþde ðæt mannum bemíden wæs, Blickl. Homl. 199, 32. Wilniende ðæt hí ǽlcum gewinne óðflogen hæfdon *credentes quod se a congressu totius humanae habitationis abstraherent,* Ors. 1, 4; Swt. 32, 21. Wæs hé wilniende tó Gode, ðæt hé ǽghwylcum gemildsode, Wulfst. 278, 12. (g) absolute or uncertain :—Hé hí ne gewemde, eal swá heó tó Gode wilnode, Homl. Skt. ii. 30, 221. Wilnig from mé ðætte ðú willt *pete a me quod uis,* Mk. Skt. Lind. 6, 22. Tó wilnanne *ad concupiscendam,* Mt. Kmbl. Lind. 5, 28. Án ðære sáwle gecynda is ðæt heó biþ wilnigende, Bt. 33, 4; Fox 132, 4. Wilnigendum *flagitante,* Wrt. Voc. ii. 34, 23. ¶ present participle with force of Latin form in -*dus* :—Uilnende ginyhtsumuise *desideratam abundantiam,* Rtl. 73, 32. (2) *to desire* to *go :*—Ðú wilnast ofer wídne mere, Andr. Kmbl. 565; An. 283. Swá heort wilnaþ tó wætre *sicut cervus desiderat ad fontes aquarum,* Ps. Th. 41, 1. Wuhta gehwilc wilnaþ þiderweard, ðær his mægðe bið mǽst ætgædere, Met. 20, 159. **II.** of inanimate objects, *to tend* to an end (*gen.*) :— Swíðe lytlum siceraþ ðæt wæter on ðæt hlece scip, and ðeáh hit wilnaþ ðæs ylcan ðe sió blúde ýd deð *hoc agit sentina latenter excrescens, quod patenter procella saeviens,* Past. 57; Swt. 437, 14. Sege mé hwelces endes ǽlc angin wilnige *dic mihi, quis sit rerum finis, quove totius naturae tendat intentio,* Bt. 5, 3; Fox 12, 19. Ðú cýþdest ðæt ðú nestest hwelces endes ǽlc angin wilnode, Fox 12, 35. [*Laym.* wilnien : *A. R.* wilnen : *Orm.* willnenn : *Ayenb.* wilni : *Piers P.* wilne : *Icel.* vilna.] v. ge-, yfel-wilnian.

-wilni[g]endlíc. v. ge-wilnigendlíc.

wilnung, e; *f. Desire :*—Úpwitan secgaþ ðæt sió sáwul hæbbe ðrió gecynd. An ðara gecynda is ðæt heó biþ wilnigende . . . Twá ðara gecynda habbaþ nétenu ; . . . óþer ðara is wilnung . . . Seó gesceádwísnes sceal wealdan ðære wilnunga, Bt. 33, 4; Fox 132, 3-10 : Met. 20, 186. Worldlíce wilnung *desiderium mundanum,* L. Ecg. P. i. 5; Th. ii. 174, 10. Fram gebrosnunge lícumlícre willnunge clǽne *a corruptione concupiscentiae carnalis inmune,* Bd. 3, 8; S. 532, 36. Unálýfedre will-nunge mód biþ geþeóded, 1, 27; S. 495, 9. Unrihtes willan willnunge *cupidine voluptatis,* 495, 33. Sum hláw, ðone men for feós willnunga gedulfon, Guthl. 4; Gdwin. 26, 6. Ábisgod on ðisse worulde willnunga (wilnunga, *v. l.*), Bt. 41, 3; Fox 246, 31. For ðære wilnunge (gewill-nunge, *v. l.*) ðe hé wolde hý him on fultum geteón *ambitione habendorum equitum Thessalorum, quorum robur ut exercitui suo admiscerit,* Ors. 3, 7; Bos. 59, 14. Hí nóhwæþere heora willnunge habban ne ðurhteón magan *in neutro cupitum possunt obtinere propositum,* Bd. 5, 23; S. 647, 2. Drihten gehýrd ða wilnunga his þearfena *desiderium pauperum exaudivit Dominus,* Ps. Th. 9, 37. Willnungum *petitionibus,* Mt. Kmbl. 14, 20. [Þuruh wilnunge of hereword, A. R. 148, 20. Worsipe haue þou for þine wilninge, Laym. 3160.] v. ge-, ungemet-, unriht-, weorold-wilnung.

Wil-sǽtan, -sǽte; *pl. The people of Wiltshire :*—Ðá mette hine Weoxtan aldorman mid Wilsǽtum, . . . and Wilsǽtan (-sǽte, *v. l.*) námon

sige, Chr. 800; Erl. 60, 6–9. Sumorsǽte alle and Wilsǽtan (Willsǽte, *v. l.*), 878; Erl. 80, 10.

wil-sele, es; *m. A pleasant hall :*—Weorðeþ his hús (*the nest of the Phenix*) onhǽted, willsele stýmeþ, Exon. Th. 212, 21; Ph. 213.

wil-síþ, es; *m. A desired journey, a wished for, welcome journey :*—Eádga ús siges, wlítigan wilsíþes, Exon. Th. 2, 18; Cri. 21. Ðæs sǽs smyltnys eów blíþe on eówerne willsíþ hám forlǽteþ *serenitas maris vos cupito itinere domum remittet*, Bd. 3, 15; S. 541, 36. Gewát Matheus menigo lǽdan on gehyld Godes, weorod on wilsíð (*he was leading them out of prison*), Andr. Kmbl. 2093; An. 1048. Elene ne wolde ðæs síðfǽtes sǽne weorðan, . . . ac wæs sóna gearu wíf on wilsíð, Elen. Kmbl. 445; El. 223. Sunnan wilsíð, Exon. Th. 2, 29; Cri. 26.

wil-spell, es; *n. Welcome news, glad tidings :*—Wæs him frófra mǽst æt ðam willspelle (*the news of the finding of the cross*), Elen. Kmbl. 1985; El. 994. Wilspella mǽst gesecgan, 1965; El. 984. [A steoresman ham talde wilspel, þ he Spaine isæih, Laym. 1350. *O. Sax.* wil-spel.]

wil-sum; *adj.*   **I.** *desirable, pleasant :*—Ðam bið gæst Godes ágen bearn, wilsum in worlde, Exon. Th. 318, 11; Mod. 81. Eorðan wilsume *terram desiderabilem*, Ps. Surt. 105, 24. Ðæt willsume weorc onginnaþ *desideratum opus inire*, Bd. 5, 11; S. 625, 33. Wilsum *desiderabilia*, Ps. Surt. 18, 11. Of ðám wilsuman wyllgespryngum *from the pleasant well-springs*, Exon. Th. 205, 7; Ph. 109.   **II.** *willing, voluntary, spontaneous :*—Wilsumne regn *pluviam voluntariam*, Ps. Th. Spl. 67, 10 : Blickl. Gl. Ðone wilsuman *spontaneum*, Wrt. Voc. ii. 32, 65. Him (*a child whose father is dead*) man an his fæderingmágum wilsumne (*willing, ready to undertake the guardianship ; or under* I (?), *desirable, suitable, sufficient*) berigean geselle his feoh tó healdenne, L. H. E. 6; Th. i. 30, 5. Mid selfwillum † wilsumum *ultroneis, voluntariis*, Hpt. Gl. 435, 64. Wilsum múðes mínes *voluntaria oris mei*, Ps. Surt. 118, 108.   **III.** *devout, devoted :*—Gode se willsuma wer *vir Deo devotus*, Bd. 4, 11; S. 579, 5. Gode seó willsume fǽmhe, 4, 26; S. 603, 5. Gode willsumra wífmonna láreów, 4, 6; S. 574, 16 : 4, 19; S. 588, 2. Hé sylfa wæs se wilsumesta (*devotissimus*) lǽstend, 5, 22; S. 644, 4. v. ge-, un-wilsum.

wilsum-líc; *adj.*   **I.** *desirable, pleasant :*—Wilsumlíc *desiderabilis*, Wrt. Voc. ii. 139, 18. Mon willsumlícre yldo and fægernesse *juvenis amantissimae aetatis et venustatis*, Bd. 5, 19; S. 636, 32. Hé monig ðing ge egeslíce ge willsumlíce (*desideranda*) geseah, 5, 12; S. 627, 29.   **II.** *voluntary, spontaneous :*—Hé geleornade ðæt Cristes ðeówdóm sceolde beón wilsumlíc, nalæs genédedlíc *didicerat servitium Christi voluntarium, non coactitium esse debere*, Bd. 1, 26; S. 488, 18. On wilsumlícre ðearfednesse *voluntariae paupertatis*, 4, 3; S. 569, 2. Wilsumlíce múðes mínes *voluntaria oris mei*, Ps. Spl. C. T. 118, 108.

wilsumlíce; *adv.*   **I.** *willingly, voluntarily, spontaneously :*—Hé wilsumlíce (*sponte*) hine geþeódde tó ðam cyninge, Bd. 3, 7; S. 529, 44. Se ðe ne wyle cyricean duru wilsumlíce (*sponte*) geeádmóded ingangan, se sceal nýde on helle duru unwilsumlíce geníþerad gelǽded beón, 5, 14; S. 634, 19. Wilsumlíce (*voluntarie*) ic onsecg[e] ðé, Ps. Surt. 53, 8.   **II.** *devoutly, devotedly :*—Lifde se man his líf Gode swýþe willsumlíce *ducens vitam multum Deo devotam*, Bd. 4, 25; S. 599, 29. v. un-wilsumlíce.

wilsumness, e; *f.*   **I.** *devotion, devoutness :*—Byrnende wilsumnes móðes *ardens devotio mentis*, 4, 7; S. 478, 11. Hí ánre wilsumnesse wǽron *erant unius devotionis*, 5, 10; S. 624, 14 : 5, 20; S. 642, 14. Hé smyltre willsumnesse (*tranquilla devotione*) Drihtne ðeówde, 4, 24; S. 599, 9. On willsumnesse (*devotioni*) háligra gebeda gecneord, 4, 28; S. 606, 33.   **II.** *a vow :*—Wilsumnessa *votorum*, Hpt. Gl. 404, 8.

wiltan; *p.* te To roll (trans.) :—Se ðe welt *qui volvit* (*lapidem*), Kent. Gl. 1006. Hé wylte (tówælte, Lind. : áwælte, Rush.) ánne stán tó ðære byrgenne dura *aduoluit lapidem ad ostium monumenti*, Mk. Skt. 15, 46. Hé (*a cup*) in healle wæs wylted and wended wloncra folmum, Exon. Th. 441, 16; Rä. 60, 19. [Walles he welte downe, D. Arth. 3152. *M. H. Ger.* welzen : *Ger.* wälzen : *Icel.* velta. *Goth.* waltjan *to roll* (intrans.).] v. á-, ge-wiltan (-wæltan, -wyltan); wealt, *and next word*.

-wilte in éd-wilte *that rolls* or *moves easily :*—Édwiltum *versatili, volubili, mobili*, Hpt. Gl. 433, 69. v. preceding word.

wil-þegu, e; *f. A grateful repast :*—Tólýsan líc and sáwle, and þonne tódǽlan werum tó wiste and tó wilþege fǽges flǽschoman, Andr. Kmbl. 306; An. 153.

Wil-tún, es; *m. Wilton in Wiltshire :*—Ælfréd cyning gefeaht wiþ alne ðone here lytle werede æt Wiltúne, Chr. 871; Erl. 76, 5. Hér forðferde Ælfgár cinges mǽg on Defenum, and his líc rest on Wiltúne, 962; Erl. 120, 3. Swegen lǽdde his here into Wiltúne, 1003; Erl. 139, 14. Hió becwið án pund tó Wiltúne ðám híwum, Cod. Dip. Kmbl. vi. 131, 1. ¶ the name occurs in several Latin charters :—In uilla regali qui appellatur Uuiltún, Cod. Dip. Kmbl. i. 320, 15. In uico regio æt Wiltúne. iii. 278, 32. In palacio nostro quod dicitur Wiltún, ii. 15, 13.

In monasterio quod dicitur Wiltún, 306, 30. Ad monasterium sanctae Dei genitricis Mariae quod dicitur Wiltúne, v. 214, 14. Uenerabili collegio Christicolarum in illo celebri loco qui dicitur Wiltún ad aecclesiam Sanctae Mariae, 227, 6. Ad usum sanctimonialium in Wiltúne degentium, iii. 23, 15. v. next words.

Wiltúnisc; *adj. Belonging to Wilton :*—Wiltúnisc *Wiltuniensis*, Ælfc. Gr. 5; Zup. 13, 5.

Wiltún-scír (Wiltúnes-), e; *f. Wiltshire :*—Æþeréd Wiltúnscíre biscop wearþ gecoren tó ærcebiscope tó Cantuareberi, Chr. 870; Erl. 74, 4. Æðelm Wiltúnscíre ealdorman, 898; Erl. 96, 18. Féng Ælfríc Wiltúnscíre bisceop tó ðam arcebiscopríce, 994; Erl. 134, 2. Ánes scipes Ælfríc arcebisceop geúðe ðam folce tó Cent and óðres tó Wiltúnesscíre, Cod. Dip. Kmbl. iii. 352, 18. Ða gegaderode man swíðe mycele fyrde of Wiltúnscíre, Chr. 1003; Erl. 139, 5 : 1011; Erl. 144, 29 : 1015; Erl. 152, 12. On ðam ylcan geáre forðférde Ælfstán bisceop on Wiltúnscíre, 981; Erl. 128, 18. Sum ungeréad mann wæs mid Ælfstáne bisceope on Wiltúnscíre on hírède, Homl. Skt. i. 12, 42. Brihtwold biscop féng tó ðam ríce on Wiltúnscíre, Chr. 1006; Erl. 140, 2. Hér gefór Brihtwold biscop on Wiltúnescíre, and man sette Hereman on his setle, 1046; Erl. 171, 23. Hereman biscop forðférde; se wæs biscop on Beorrucscíre and on Wiltúnscíre and on Dorsǽtan, 1078; Erl. 215, 32.

wil-tygþe, -týþe; *adj. Having one's desire, satisfied, glad :*—Wiltíðe *voti compos*, i. *laetos* † *hilares*, Hpt. Gl. 458, 62. Wiltíðe *voti compotes, hilares*, 490, 47. v. wil-fægen.

wiluncel, wílung. v. wilincel, wíglung.

wilwan, wilwian, wilian; *p.* wilwede, wilede.   **I.** *to roll* (trans.) :—Ic áwende oððe wylewige (wylwige, *v. l.*) *uoluo*, Ælfc. Gr. 28; Zup. 177, 9. Hé wylede ðone stán fram ðære byrgenne duru, Blickl. Homl. 157, 8. Hé wylode hine sylfne on ðám þornum and netelum, Homl. Th. ii. 156, 28. Hé hét wilian tó ðam scræfe nicele weorcstánas *praecepit : 'Voluite saxa ingentia ad os speluncae*,' Jos. 10, 18. Hé hí swá nacode hét wylian on ðam fýre, Homl. Skt. i. 8, 170.   **Ia.** fig. :—Sibb áfíymð saca, anda tógædre wilaþ hí *pax effugat discordias, inuidia copulat eas*, Scint. 11, 8. Hé hine sylfne betweox ðises andweardan middangeardes (wælum ?) weólc and welode *inter fluctuantis saeculi gurgites jactaretur*, Guthl. 2; Gdwin. 14, 14.   **II.** *to join, compound, compose :*—Byð wylyd ealswá middangeardes boga, Lchdm. iii. 82, 18. [Welwyñ or rollyñ al thyngys þat may not be borne *volvo*, Prompt. Parv. 521. *Goth.* walwjan *to roll* (trans.).] v. á-, be-, ge-wilwan; wealwian.

wil-wang, es; *m. A pleasant plain, pleasant land :*—Ðone wudu weardaþ fugel (*the Phenix*) . . . , eard bihealdaþ . . . , nǽfre him deáþ scepeþ on ðam willwonge, Exon. Th. 203, 24; Ph. 89.

wil-weg, es; *m. A pleasant way, a desirable way :*—Syndan wé nú eft ámearcode tó ðam gefeán neorxnawanges ; ne gelette ús ðæs síðes se fǽcna feónd, ne ús ne forwyrne ðæs wilweges, ne ús ða gata ne betýne, ðe ús opene standaþ, Wulfst. 252, 17. Ðæt hí ðé heóldan, ðæt ðú wilwega wealdan móstest *ut custodiant te in omnibus viis tuis*, Ps. Th. 90, 11.

wím-(wim-)man. v. wíf-mann.

wimpel, winpel, es; *m. An article of woman's dress, a wimple :*—Winpel *vel* orl *ricinum*, Wrt. Voc. i. 17, 1. Winpel *anabala* (cf. anaboladium *amictorium lineum feminarum, quo humeri operiuntur*, Migne), 26, 1. Wimple goldgewefenum *cyclade auro texta*, Hpt. Gl. 506, 63. Wimplum *cycladibus*, 480, 71 : 486, 41 : *mafortibus*, i. *velaminibus*, 526, 52 : Anglia xiii. 37, 293. [Sum seið þ hit limpeð to ene wummon cundeliche forte were wimpel. Nai : wimpel . . . ne nemned hali write, ah wriheles of heuet . . . Wrihen, þe Apostel seið, naut wimplin, A. R. 420, note a. Hyre body wyþ a mantel, a wympel aboute her heued, R. Glouc. 338, 4. Ful semely hire wympel ipynched was, Chauc. Prol. 151. *O. H. Ger.* wimpal *theristrum : Icel.* vimpill *a hood, veil.*]

win. v. winn.

win[n] (?), e; *f. Pasture :*—Of ðære díc tó wynne mǽduan be ðære strǽt, Cod. Dip. Kmbl. iii. 263, 29. [*Goth.* winja *pasture : O. H. Ger.* winne *pastum : Icel.* vin *a meadow.*]

wín, es; *n. Wine :*—Wín *vinum, merum*, geswét wín *mellicratum*, níwe wín *mustum*, ælces kynnes gewring bútan wíne and wætere *sicera*, ðæt seleste wín *falernum*, weala wín *crudum vinum*, geolo wín *succinacium vinum*, hláforda wín *honorarium vinum*, gewyrtod wín *compositum vinum vel conditum*, gesoden wín *defrutum vinum*, Wrt. Voc. i. 27, 36–62. Áwilled wín *dulcisapa*, gesweted wín *defrucatum*, 290, 56, 58. Ðonne wín hweteþ beornes breóstsefan, breahtme stígeþ cirm on corþre, Exon. Th. 314, 23; Mód. 18. Wǽron hí (*the Danes*) swýðe druncene, for ðam ðær wæs gebróht wín súðan, Chr. 1012; Erl. 146, 15. Wín *Bachus*, wínes *Bachi*, Wrt. Voc. ii. 12, 25, 36. Wínes *defruti*, 27, 32 : *meri*, 87, 13. Wínes god *Bachus*, 61, 6. Ðæm folce (*the Scythians*) seldsiéne and uncúðe wǽron wínes drencas . . . Hié búton gemetgunge ðæt wín drincende wǽron óð hí heora selfra lytel geweald hæfdon, Ors. 2, 4; Swt. 76, 11–19 : Homl. Th. i. 352, 6 : ii. 298, 18. Wǽre ðú (*the body*) ðé wiste wlonc and wínes sǽd, Exon. Th. 369, 11; Seel. 39. Wínes glæd, 449, 28; Dóm. 78. Wíne *temeto*, Wrt. Voc. ii. 88, 42.

Ne gemunde hé ðæt hé ǽr gespræc, wine druncen, Beo. Th. 2938; B. 1467. Wine gewǽged, Exon. Th. 315, 34; Mód. 41. Hé ofer ealne dæg dryhtguman síne drencte mid wíne, Judth. Thw. 21, 21; Jud. 29. Wer sæt æt wíne, Exon. Th. 431, 25; Rä. 47, 1. Wín nectar, Wrt. Voc. ii. 61, 31. Hé bróhte hláf and wín, Gen. 14, 18. Hwilc þinc gelǽdst ðú (the merchant) ús? Wín and ele, Coll. Monast. Th. 27, 9. Hwæt drincst ðú (the boy)? Ealu, gif ic hæbbe, oþþe wæter, gif ic næbbe ealu. Ne drincst ðú wín? Ic ne eom swá spédig ðæt ic mæge bicgean mé wín; and wín nys drenc cilda, ne dysigra, ac ealdra and wísra, 35, 9-22. Ðonne ðú wín habban wille, ðonne dó ðú mid ðínum twám fingrum swilce ðú tæppan of tunnan onteón wille, Techm. ii. 120, 9. Byrelas sealdon wín of wundorfatum, Beo. Th. 2328; B. 1162. [The word made its way into all Teutonic speeches from Latin.] v. æppel-, mæsse-wín.

wín-ærn, es; n. I. *a place where wine is stored :*—Wínærn apotheca, Wrt. Voc. ii. 6, 6. v. wín-hús. II. *a place where wine is sold and drunk, a tavern :*—Wínaern taberna, Wrt. Voc. ii. 122, 3. Wínærn, i. 290, 52. III. *a hall where wine is drunk, where there is feasting.* Cf. wín-ræced :—Grétte Hróðgár Beówulf, and him hǽl ábeád, wínærnes geweald: 'Nǽfre ic ǽnegum men ǽr álýfde ðrýþærn Dena ... Hafa nú and geheald húsa sélest, Beo. Th. 1312; B. 654.

wín-beám, es; m. *A vine-pole :*—Wínbeám partica, Wrt. Voc. i. 290, 4: *trabs uinee,* Wülck. Gl. 245, 20.

wín-beger, es; n. *A grape :*—Ðæt wínbeger uuam, Lk. Skt. Lind. 6, 44. Wínbegær uvas, Mt. Kmbl. Rush. 7, 16. Wíntrog, ðǽr monn tred ða wínbegera torcular, Lind. 21, 33.

wín-belg, es; m. *A wine-skin, wine-bottle :*—Ne menn geótaþ wín niówe in wínbelgas (utres) alde, Mt. Kmbl. Rush. 9, 17. [Icel. vín-belgr.]

wín-berige, -berie, -berge, an; f. *A grape :*—Wínberge uva, Wrt. Voc. i. 285, 72. Wínberge te hunige áwylled medus, ii. 59, 34. Hire wínberie ys gealla *uva eorum uva fellis,* Deut. 32, 32. Ne hig wínberian (uuam) on gorste ne nimaþ, Lk. Skt. 6, 44. Gesoden[e] wínberigan (-en, MS.) fecula, Wrt. Voc. i. 27, 63. Hit wæs ðá se tíma, ðæt wínberian rípodon *erat autem tempus, quando jam praecoquae uvae vesci possunt,* Num. 13, 21: Scint. 154, 2. Wínberigena bacciniorum, Hpt. Gl. 524, 21. Genim ðæs ylcan wyrte mid wínberian (-berium, -bergan, v. ll.), Lchdm. i. 282, 9. Wínberigean uvas, Gen. 40, 9. Ic nam ða wínberian and wrang on ðæt fæt, 40, 11: Lchdm. iii. 114, 4. Wínberian (-bergean, v.l.) uvas, Mt. Kmbl. 7, 16. [Ofte druie sprintles bereð winberien? A.R. 276, 12. Goth. weina-basi; n.: O. Sax. wín-beri; n.: O.H. Ger. wín-beri(-peri); n.: Icel. wín-ber; n.]

wín-bóh; gen. -bóges; m. *A branch of a vine :*—Wínbóga palmitum, Hpt. Gl. 468, 17: 496, 74: Homl. Th. ii. 74, 6. Of ðám wínbógum mid berium mid eallum *palmitem cum uva sua,* Num. 13, 24.

wín-brytta, an; m. *A wine-dealer, wine-seller, vintner, tavern-keeper :*—Tæppere, wínbrytta caupo, tabernarius, Wrt. Voc. i. 28, 10. Wínbryttum cauponibus, Wrt. Voc. ii. 79, 79: 18, 21.

wín-burh; f. I. *a town where wine is drunk, where there is feasting, where a prince feasts his followers, a chief town.* Cf. medu-burh, wín-ærn, III :—Wínburge cyning (the king of Babylon; cf. Belshazzar's feast), Cd. Th. 255, 11; Dan. 622. Wuna in ðære wínbyrig salu sinchroden, Andr. Knibl. 3340; An. 1674. Wínburh wera (Jerusalem), 219, 21; Dan. 58. Geond ða wínburg (the town of the Mermedonians), Andr. Kmbl. 3272; An. 1639. Se ðe wínburga geweald áhte, Exon. Th. 323, 11; Vid. 77. Wlonce wígsmiþas wínburgum in sittaþ æt symble, 314, 15; Mód. 14: 247, 23; Jul. 83. II. *a walled vineyard :*—For hwan ðú tówurpe weallfæsten his? wealdeþ his wínbyrig eall, ðæt on wege færð *ut quid deposuisti maceriam ejus; et vindemiant eam omnes, qui transeunt viam?* Ps. Th. 79, 12.

wín-byrele, es; m. *A vintner :*—Wínbyrele caupo, Wrt. Voc. ii. 21, 13. [Icel. vín-byrli *a cup-bearer.*]

wince, an; f. *A winch :*—Wince gigrillus (=gigrillus; cf. girgillus a reel, Wülck. Gl. 586, 30), Wrt. Voc. 42, 29.

-wince, Win-ceaster, wincel. v. hleápe-wince, Wintan-ceaster, wencel.

wincel (?) *a corner;* cf. place-names, e.g. Wincel-cumb, Homl. Skt. i. 21, 33: Cod. Dip. Kmbl. vi. 354; and modern Aldwinkle (Northants). [O. H. Ger. winkil angulus. The word is found in place-names. v. Graff. i. 721.]

wincettan; p. te *To wink :*—Ða ðe mé hatiaþ bútan scylde and wincettaþ mid heora eágum þetwuh him *qui oderunt me gratis, et annuunt oculis,* Ps. Th. 34, 19.

wincian; p. ode. I. *to wink, make a sign :*—Ic wincie annicto vel annuto, Wrt. Voc. i. 22, 27. II. *to close the eyes, blink :*—Ic wincige conniveo, Wrt. Voc. i. 34, 14: Ælfc. Gr. 26, 5; Zup. 156, 14. Se ðe ágimeleásaþ ðæt hé ðence ǽr ðæm ðe hé dó, se stæpþ forð mid ðam fótum and wincaþ mid ðæm eágum *qui negligit considerando praevidere, quod facit, gressus tendit, oculos claudit,* Past. 39; Swt. 287, 16. Lamena hé is lǽce, leóht wincendra (winciendra, v. l.), dumbra tunge, Salm. Kmbl. 156; Sal. 77. [Waryn wisdome wynked uppon Mede, Piers P. 4, 154. Or mans eghe may open or wynk, Pr. C. 4970.]

Twynkyn wythe the eye, or wynkyn conniveo, nicito, nicto, Prompt. Parv. 505. Wynkyn conniveo, 530. O. H. Ger. winchen nutare, nictare, oculo annuere.]

wín-clyster, es; n. I. *a bunch of grapes :*—Wínclyster botrus, Scint. 154, 2. II. *a row of vines :*—Wínclystra antes, Engl. Stud. xi. 64, 3.

wín-cóle, an; f. *A tub into which the juice pressed from the grapes runs :*—Wínmere sive wíncóle lacus ubi frugum liquor decurrit, Wrt. Voc. ii. 54, 13.

wind, es; m. I. *wind, air in motion :*—Seó lyft, þonne heó ástyred is, byð wind. Se wind hæfð mistlíce naman on bócum ... Feówer heáfodwindas synd. Se fyrmesta is eásterne wind ... Ðás feówer heáfodwindas habbaþ betweox him on ymbhwyrfte óðre eahta windas, ǽfre betwyx ðám heáfodwindum twégen windas ... Is án ðæra eahta winda aquilo geháten ...; ealne ðone cwyld ðe se súðerna wind auster ácænd, ealne hé tódrǽfð, Lchdm. iii. 274, 10-276, 8. Sæge mé, huona gebláwaþ wind? Ðæt is of Serafin, of ðon is ácweden Serafin windana, Rtl. 192, 33. Gif hús full ungemetlíc wind gestent, Bt. 12; Fox 36, 16. Swift wind, Met. 7, 20. Se stearca wind, 12, 14. Winneþ wind wið wǽge, 25, 58. Ðonne wind styreþ láð gewidru, Beo. Th. 2753; B. 1374. Ðonne wind ligeþ, weder bið fæger, Exon. Th. 210, 7; Ph. 182. Biþ sǽ smilte, ðonne hý wind ne weceþ, 336, 27; Gn. Ex. 56. Nó wǽgflotan wind ofer ýdum síðes getwǽfde, Beo. Th. 3819; B. 1907. Bærn eal tósomne on ða healfe ðe se wind sý, Lchdm. iii. 56, 7. Se wind strongra geswinca ... se wind ðara earfoþa, Bt. 12; Fox 36, 18, 28. Wæs mycel ýst windes geworden, Mk. Skt. 4, 37. Hwyrft hægel of heofones lyfte, wealcaþ hit windes scúra, Runic pm. Kmbl. 341, 6; Rún. 9. Holm storme weól, won wið winde, Beo. Th. 2268; B. 1132. Winde gelícost, ðonne hé hlúd ástígeþ, wǽdeþ be wolcnum, wédende færeþ, and eft semninga swíge gewyrðeþ, Elen. Kmbl. 2542; El. 1272. Winde biwáune weallas, Exon. Th. 291, 2; Wand. 76. Wérig sceal se wiþ winde róweþ, 345, 12; Gn. Ex. 187. Winde gefýsed flota, Beo. Th. 440; B. 217. Ðá sende Drihten micelne wind, Ex. 14, 21. Ðú ðe ða treówa þurh ðone stearcan wind norþan and eástan on hærfesttíd heora leáfa beréafast, and eft on lencten óþru leáf sellest þurh ðone smyltan súþanwesternan wind *quas Boreae spiritus aufert, revehat mitis Zephyrus, frondeis,* Bt. 4; Fox 8, 5-8. Þurh ðone láðran wind, Met. 4, 24. Theodosius hæfde ðone wind, ðæt his fultum mehte mǽstra ǽlcne heora flána on hiora feóndum áfæstnian, Ors. 6, 36; Swt. 294, 26. Ðǽr bleówun windas, Mt. Kmbl. 7, 25. Wedercandel swearc, windas weóxon, Andr. Kmbl. 745; An. 373. Swógaþ windas, bláwaþ brecende bearhtma mǽste, Exon. Th. 59, 10; Cri. 950. Hé fleáh ofer winda fiðeru, Ps. Th. 17, 10. Hé bebýt ge windum ge sǽ, Lk. Skt. 8, 25. II. *wind, flatulence.* v. windig, II :—Gif sió wamb biþ windes full, ðonne cymð ðæt of wlacre wǽtan, Lchdm. ii. 224, 23. Wambe wind, 168, 20. III. *wind, breath :*—Ic (a horn) winde sceal swelgan of sumes bósme, Exon. Th. 395, 28; Rä. 15, 14. [Goth. winds: O. Sax. O. Frs. wind: O. H. Ger. wint: Icel. vindr: Lat. ventus.] v. eástan-, eástansúþan- (under eástan), eástnorþ-, heáfod-, norþ-, norþan-, norþaneástan-, norþanwestan-, súþ-, súþan-, súþaneástan-, súþanwestan-, west-, westan-, westansúþan-, westnorþ-, westsúþ-wind.

wind, es; m. *winding, wrapping :*—Gif preóst ordál misfadige, gebéte ðæt. Gif preóst searwaþ be winde, gebéte ðæt *if a priest do not conduct an ordeal rightly, let him make 'bót.' If a priest uses deceit in respect to the wrapping up of the hand or arm exposed to the ordeal* (cf. in the descriptions of the proceedings at the ordeal: Inseglige man ða hand, L. Ath. iv. 7; Th. i. 226, 30. Beón þreó niht ǽr man ða hand undó, i. 23; Th. i. 212, 4), *let him make 'bót,'* L. N. P. L. 39, 40; Th. ii. 296, 9-10. [Icel. vindr a winding.]

wind? :—Uuind sclabrum, Txts. 97, 1841. Windum slabris, 181, 72. [Cf. (?) O. H. Ger. winta flabrum, ventilabrum; or (?) O. H. Ger. winta: Ger. winde a pulley, reel.] Cf. windung.

wind-ǽdre, an; f. *A windpipe :*—Góma palatum, sweora collum, hracan fauces, windǽddran arteriae, þrotu guttur, Wrt. Voc. i. 43, 35-39. [Icel. vind-æð.]

win-dæg. v. winn-dæg.

windan; p. wand, pl. wundon; pp. wunden. I. *intrans.* (1) of motion that results from a blow, swing, or other impetus, *to fly, leap, start :*—Sió æcs wint of ðam hielfe and eác ús of ðære handa ... Sió æcs wient of ðam hielfe *securis manu fugit ... Ferrum de manubrio prosilit,* Past. 21; Swt. 167, 7-9. Sum óðer hine wolde sleán mid ísene, ac ðæt wǽpen wand áweg mid ðam slege of ðæs rédan handum, Homl. Th. ii. 510, 22. Ðá slóg hé ánes monnes hors mid his sweorde, ðæt him wand ðæt heáfod of ad unum gladii ictum desecuisset, Ors. 5, 2; Swt. 216, 21. Slóh ides ðone hǽþenan hund, ðæt him ðæt heáfod wand forþ on ðá flóre, Judth. Thw. 23, 8; Jud. 110. Bærst sum sagol intó ánes beáteres eágan swá ðæt his eáge wand út mid ðam slæge, Homl. Skt. i. 4, 144. Heó weard mid swurde gewundod, ðæt hire wand se innoð út, 9, 127: Jud. 3, 22. (2) *to fly, wheel, spring.* Cf. wendan. (a) of the movement of living things :—Sume fótum foldan peðþaþ, sume fleógende windaþ (-eð, MS.) under wolcnum *sunt quibus alarum levitas vaga ...*

liquido longi spatia aetheris enatet volatu, Met. 31, 12. Hē wand him up þanon, hwearf him þurh ða helldora, Cd. Th. 29, 7; Gen. 446. Ðā wand se of his swuran *he sprang from his neck*, Homl. i. 336, 17. Hornfisc plegode, and se grǣga mǣw wælgīfre wand (*flew circling round*), Andr. Kmbl. 743; An. 372. Hremmas wundon, Byrht. Th. 134, 59; By. 106. Hē mid feðerhoman fleógan meahte, windan on wolcne, Cd. Th. 27, 15; Gen. 418. (b) of inanimate things :—Dægscealdes hleó (*the pillar of cloud*) wand ofer wolcnum, Cd. Th. 182, 23; Exod. 80. Mid ðam worde wand fȳr of heofonum *at those words fire flew from heaven*, Homl. Skt. i. 18, 249. Wand tó wolcnum wælfȳra mǣst, Beo. Th. 2242; B. 1119. Ða spearcan wundon wið ðæs hrófes *the sparks flew whirling towards the roof*, Homl. Skt. ii. 26, 229. Hē forlēt wælspere windan on ða wīcingas, Byrht. Th. 141, 14; By. 322. (c) of abstract subjects :—Sió æcs wint of ðam hielfe, and eác ūs of ðære honda ðonne ðonne sió lār wint on rēðnesse *securis manu fugit, cum sese increpatio in asperitatem pertrahit*, Past. 21; Swt. 167, 8. (3) of twisting, rolling movement, (a) of living things :—Hē wand swā swā wurm *he writhed like a serpent*, Homl. Th. i. 414, 17. Hwīlum nacode men windaþ (winnaþ, MS.) ymbe wyrmas (cf. Canto xxv of the Inferno), Cd. Th. 273, 13; Sat. 136. Hē wearp hine ðā on wyrmes līc, and wand him ymbūtan ðone deáðes beám *he twined round the tree of death*, Cd. Th. 31, 27; Gen. 491. (b) of inanimate things :—Þūfas wundon ofer gārfare *the banners fluttered above the battalions*, Cd. Th. 199, 22; Exod. 342. Streámas wundon *the waters rolled*, Beo. Th. 430; B. 212. Staþelas wið wǣge, wætre windendum, Exon. Th. 61, 9; Cri. 982. (4) fig. *to waver*. Cf. wandian :—Gearo wæs Gūðlāc; hine God fremede on ondsware and on elne strong; ne wond hē for worde (*he did not waver on account of what was said to him*), Exon. Th. 120, 1; Gū. 265. II. *trans.* (1) *to twist, roll* :—Ðæt hors on misenlíce dǣlas hit wond and ðrǣste *cum equus diversas in partes se torqueret volutando*, Bd. 3, 9; S. 533, 36. (2) *to brandish, wave* :—Hē wand wācne æsc, Byrht. Th. 132, 68; By. 43. (3) *to twist, plait, weave* :—Wundun intexunt, Wrt. Voc. ii. 110, 74. Hī wundon cynehelm of þornum *plectentes coronam de spinis*, Mt. Kmbl. 27, 29: Jn. Skt. 19, 2. Windan *plumemus*, Wrt. Voc. ii. 83, 78. Windan manigne smicerne wǣn, Shrn. 163, 15. Windende *plectentis*, Wrt. Voc. ii. 74, 32. Wundene mē ne beóð wefle, Exon. Th. 417, 15; Rä. 36, 5. Wundne loccas, 428, 7; Rä. 41, 104. Wundnum rāpum fótas gefæstnian, Ps. Th. 139, 5. (4) *to twist, give a curved form to* (mostly as an epithet of gold made into ornaments; cf. *O. Sax.* wundan gold) :—Bunden, wunden (*applied to a winecask*), Exon. Th. 410, 26; Rä. 29, 5. Him wæs wunden gold geeáwed . . . hringas, healsbeága mǣst, Beo. Th. 2391; B. 1193: 6259; B. 3134: Exon. Th. 288, 17; Wand. 32: Cd. Th. 124, 30; Gen. 2070. Beágas, welan, wunden gold, 116, 4; Gen. 1931: 258, 9; Dan. 673. Wunden gold, . . . feoh and frætwa, 128, 18; Gen. 2128. Wunden gold (*the ornament of a sheath*), Exon. Th. 437, 6; Rä. 56, 3. Ic ðē leánige eáldgestreónum, wundnum golde, Beo. Th. 2768; B. 1382. Wundnan golde, Exon. Th. 326, 16; Víd. 129. [Þat we mosten ouer sæ winden mid seile (away wende, 2nd MS.), Laym. 20818. Stanes heo letten winden, 27461. He smat an Arðures sceld, þat he wond (fleh, 2nd MS.) a þene feld, 23964. Þe sparke þet wint up, A. R. 296, 13. Gif dust windeð up, 314, 8. In to reste his sowle wond, Gen. and Ex. 4136. 3ho wand himm i winndeclut, Orm. 3320. *Goth.* bi-windan *involvere*; us-windan *plectere* : *O. Sax.* windan *to fly; to roll; to plait* : *O. H. Ger.* wintan *torquere; rotari* : *Icel.* vinda *to twist, wind; to thrust; to hurl; to turn*.] v. ā-, æt-, be- (bi-), ge-, on-, ōþ-, un-, ymb-windan; un-wunden.

wind-bland *tumult of winds* :—Windblond gelæg, Beo. Th. 6284; B. 3146.

wind-cyrice, an; *f. A round church* (? cf. seonu-wealt, I) :—Ic Eádwerd cinig begeat æt Deneulfe biscepe on Winteceastre ða windcirican, Cod. Dip. Kmbl. v. 163, 12.

winde (?); *adj. Curly* :— Winde loccas (windeloccas?) *cincinni*, Wrt. Voc. ii. 20, 43: 14, 27: 104, 6. [Cf. *Icel.* vindr *awry, twisted*.] v. windan.

-winde. [Cf. *O. H. Ger.* winta : *Ger.* winde : *Icel.* vinda *a hank*.] v. ed-, gearn-, næder-, wudu-winde.

-winde; *adj.* v. ge-winde.

windel, es; *m. A basket* :—Windil *cartellus*, Wrt. Voc. ii. 102, 42. Windel, 13, 9 : *cartellus, fiscella*, 128, 78 : *cistella vel cartellum*, i. 24, 56 : *cartallum*, 86, 4. Wylige oððe windel *corbis*, Ælfc. Gr. 9, 28; Zup. 55, 13. Ða hlāfas on ðam windle (*canistro*), Ex. 29, 32. Ic geseah swefen, ðæt ys, ðæt ic hæfde þrī windlas (*canistra*) ofer mīn heáfod, and on ðam ufemystan windle (*canistro*) wǣre manegra cynna gebæc, Gen. 40, 17. Ða nam heó ánne riscenne windel (*fiscellam scirpeam*) on scipwīsan gesceapenne, Ex. 2, 3, 5, 6. Man sceal habban wilian, windlas, systras, sǣdleáp, Anglia ix. 264, 12.

windel-stān, es; *m. A tower with a winding staircase* :—Windelstān *coclea*, gewind *circuitus ascensus* (the word occurs in a list of names of buildings), Wrt. Voc. i. 37, 3. [*O. H. Ger.* wentil-stein *cochlea, turris in qua per circuitum scanditur*.]

windel-streáw, -streów, es; *n. Windle-straw, some kind of coarse grass or reed* (v. windle-straws, E. D. S. Pub. Plant Names) :—Eár *spica*, egle *aresta*, windelstreów *calmum*, Wrt. Voc. i. 287, 22: ii. 16, 74. Genim ðæt micle greáte windelstreáw twyecge, ðæt on worþium wixð, Lchdm. ii. 44, 4. v. windel.

windel-treów, es; *n. A wild olive* :—Windeltreów *oleaster*, Wrt. Voc. i. 285, 74 : ii. 64, 6. v. windel.

wind-fana, an; *m. A cloth for winnowing with, a fan* :—Windfona *scabellum*, Wrt. Voc. ii. 119, 71 : i. 289, 22. His fone ł windfone (fonnae ł windgefonnae, Lind.) in honda his and clǣnsaþ bereflōr his *cujus uentilabrum in manu ejus et purgauit aream suam*, Lk. Skt. Rush. 3, 17. [Cf. *Ventilabrum* . . . a sayle or a wynde clothe. A wyndowe clothe *ventilabrum*, Prompt. Parv. 529, note 5. See also Cath. Angl. 419, note 3.]

wind-filled; *adj. Wind-felled, blown down by the wind* :—Wuduwearde gebyreþ ǣlc windfylled treów, L. R. S. 19; Th. i. 440, 10.

wind-gerest, e; *f. A windy resting-place* (?), *a hall open to the winds* (?) :—Hē gesyhð sorhcearig on his suna būre wīnsele wēstne, windgereste (wind gereste, MS.: windge reste, Grein) *he sees the hall deserted, the resting-place of men open to the winds* (? For the hall as a sleeping-place, cf. Monig snellic sǣrinc selereste gebeáh, 1385; B. 690), Beo. Th. 4904; B. 2456. Cf. wind-sele.

wind-hladen; *adj. Wind-laden, windy*, Lye.

wind-hreóse (?), es; *m. A storm of wind* :—Swā swā gód scipstȳra ongit micelne windhreóse ǣr ǣr hit weorþe, Bt. 41, 3; Fox 250, 14. Cf. wind-rǣs.

windig; *adj.* I. *windy* :—Ðā com windi (wyndig, *v.l.*) ȳst *descendit procella uenti*, Lk. Skt. 8, 23. Windig sumer, Lchdm. iii. 162, 30. Windig lengten, 164, 5. Wæs on ðam ofne, ðǣr se engel becwom, windig (*breezy, airy*) and wynsum, Cd. Th. 237, 33; Dan. 347. Windig wolcen, Exon. Th. 201, 24; Ph. 61. Ðes windiga sele (*Hell*), Cd. Th. 273, 14; Sat. 136. Heora wyrtruma bið swā swā windige ysla (*ashes blown by the wind*, sic radix eorum quasi favilla erit, et germen eorum ut pulvis ascendet, Is. 5, 24), Homl. Th. ii. 322, 20. Torras stódon, windige weallas, Andr. Kmbl. 1685; An. 845. Windige holmas, Exon. Th. 53, 26; Cri. 856. Ic sǣnæssas geseón mihte, windige weallas, Beo. Th. 1148; B. 572: 2721; B. 1358. Swā sīde swā sǣ bebūgeþ windge eardweallas (wind geard weallas, MS.), 2452; B. 1224. I a. fig. :—Ðeáh ðeós weoruld wēde, and windige ēhtnysse āstyrige ongeán Cristes geladunge, Homl. Th. ii. 388, 9. II. *windy, flatulent*. v. wind, II :—Gif se ūtgang sié windig and wætrig, Lchdm. ii. 236, 6. Be windigre wambe, 162, 23. Wiþ windigre āþundenesse, 166, 25 : 188, 22. Wiþ ða þing ðe windigne æþm on men wyrcen, 214, 3. [*Icel.* vindugr.]

Windles-óra, an; *m. Windsor* :—Æt Windlesóran, Chr. 1061; Erl. 194, 3. Wæs se cyng on Windlesóran, 1095; Erl. 231, 22. Ðis writ wæs gemaced æt Windlesóren, Cod. Dip. Kmbl. iv. 209, 27. Ic habbe gegefan Criste and Sancte Petre intó Westmynstre Windlesóran and Stāne, 227, 6: 178, 19.

wind-rǣs, es; *m. A storm of wind* :—Windrǣs *procella*, Mk. Skt. Lind. 4, 37.

wín-drenc, es; *m. Wine* :—Wīndrenc (-dred, l. 10, -drend, l. 12, MS.) *vinum*, R. Ben. Interl. 72, 10, 12. Ða cempan him budon drincan gebitrodne wīndrenc, Homl. Th. ii. 254, 16. v. wín-drync.

wín-druncen; *adj. Drunken with wine, drunken* :—Wīndruncen *vinolentus*, R. Ben. Interl. 20, 13. Wīndruncen gewit, Cd. Th. 262, 32; Dan. 752. Wīndruncnyes *temulenti*, Kent. Gl. 985. Wīndruncne uinolentae, ebriae, Germ. 394, 250. [Gumen weoren windrunken (dronge of wine, 2nd MS.), Laym. 8126. *O. H. Ger.* wīn-trunchan *temulentus* : *Icel.* wín-drukkinn.]

wín-drync; es; *m. Wine* :—Heortan manna must and wīndrinc myclum blissaþ *vinum laetificet cor hominis*, Ps. Th. 103, 14. Wē þeáh rǣdaþ ðæt munecum tó wīndrince (-drynce, -drence, *v.ll.*) nāht ne belimpe *licet legamus uinum monachorum non esse*, R. Ben. 64, 21. [*Icel.* vīn-drykkr.] v. wín-drenc, -gedrinc.

wind-scofl, e; *f. A fan* :—Winds(c)obl *ventilabrum*, Wrt. Voc. ii. 71, 66. [Cf. *O. L. Ger.* wind-scūfla *ventilabrum*; *O. H. Ger.* wint-scūvala *ventilabrum*.] v. windwig-scofl.

wind-sele, e; *m. A windy hall* :—Wīde geond windsele (*Hell*; cf. Ðes windiga sele, 273, 14; Sat. 136) Cd. Th. 284, 11; Sat. 320: 288, 23; Sat. 386.

wind-swingla, an; *m. A fan* :—Windswingla *pala* vel *ventilabrum*, Wrt. Voc. i. 41, 36.

windu-mǣr. v. wudu-mǣr.

windung, winnung, e; *f. Something woven or plaited, a hurdle* (cf. *plecta* hyrdle, Hpt. Gl. 497, 70) :—Windonge *plecta* (cf. gewind *plectas*, 68, 71 : plecta *quilibet nexus ex virgulis, vel papyro, vel carecto*, Migne), Wrt. Voc. ii. 83, 77.

windung, winnung, e; *f. What is winnowed, chaff, straw* :—Ða winnunga *zizania*, Mt. Kmbl. Lind. 13, 38. Wynnunga, 26. Wynnung, 25. Ða halm ł ða windungo (winnunge, Rush.) *paleas*, Lk. Skt. Lind.

3, 17. Bisin of winnuncum *parabolam de zizaniis*, Mt. Kmbl. p. 17, 4. [Cf. *O. H. Ger.* wintôn *ventilare: Goth.* winþi-skauró *ventilabrum.*] Cf. next word.

windwian; *p.* ode *To winnow*; ventilare, (1) literal :—Hé corn ðærsc and windwode, Shrn. 61, 19. (2) figurative :—Fiónd úre wé windwiaþ *inimicos nostros ventilavimus*, Ps. Surt. 43, 6. Ic windwade (*ventilabam*) in mé gást mínne, 76, 7. [God wule windwin þet er wes iþorschen, O. E. Homl. i. 85, 22. Ane wummon windwede hweate, A. R. 272, 20. Drihtin windweð his hweate, Jul. 79, 15. Winndell forr to winndwenn, Orm. 10483. Windewe, winewe, Wick. Jer. 49, 36. Wynwyn *ventilo*, Prompt. Parv. 530.] v. ā-, ge-windwian; *and* cf. *preceding word.*

windwig-ceaf, es; *n. A husk winnowed from the grain* :—Windwig-ceafum *paleis* (the passage is : Non te hordeo alam, sed paleis et fame conficiam, Ald. 34), Hpt. Gl. 464, 9.

windwig-scofl, e; *f. A winnowing-fan* :—Windiuscoful (windui-? = windwig-) *ventilabrum*, Mt. Kmbl. Rush. 3, 12. v. wind-scofl.

windwig-sife, es; *n. A winnowing-sieve* :—Windwigsyfe *ventilabrum*, Wrt. Voc. i. 34, 44.

wine, es; *m. A friend.* (1) applied to an equal :—Wine mín, Húnferð, Beo. Th. 1065; B. 530. Hé ongan winas manian, frýnd and geféran, Byrht. Th. 138, 31; By. 228. (2) applied to one who can help or protect, *a friendly lord, a (powerful) friend* :—Wine fród ... Geared lǽfde land and leódweard (cf. Geared gold brittade ..., his freómágum leóf, 72, 8; Gen. 1183), Cd. Th. 72, 29; Gen. 1194. Ne þurfon mé hæleð ætwítan, nú mín wine (*Byrhtnoth*) gecranc, ðæt ic hláfordleás hám síðie, Byrht. Th. 139, 7; By. 250. Wine Scyldinga, leóf landfruma, Beo. Th. 60; B. 30: 298; B. 148. Wine Scyldinga, ríces hyrde, 4057; B. 2026. Wine cwǽð, mín wine (*my lord*), mêce gecnáwan, 4100; B. 2047. Wine Ebréa (*Abraham*), Cd. Th. 170, 20; Gen. 2816. Æfter wines (cf. freán úserne, leófne maunan (*Beowulf*), 6206; B. 3107) dǽdum, Beo. Th. 6184; B. 3096. Hé sôhte holdne wine (*Hrothgar*), 758; B. 376. Wine Deniga, freán Scyldinga, 706; B. 350. Ne sint mé winas (*friends who will protect*) cûðe eorlas elþeódige, Andr. Kmbl. 396; An. 198. Ðǽr him folcweras fremde wǽron, wine uncúðe, Cd. Th. 110, 32; Gen. 1847. Ic fela folca gesôhte, wina uncúðra, 163, 14; Gen. 2698. Hé (*Pharaoh*) hêht him (*Abraham*) wine (*or sing.?*) ceósan, ellor æðelingas, ôðre dugeðe, 112, 8; Gen. 1867. Ne ceara incit duguða ellor sécan, winas uncúðe, ac wuniaþ hêr, 165, 19; Gen. 2734. (3) used of a husband or lover :—Mín freónd siteþ, wine wérigmód ... Dreógeþ se mín wine micle môdceare, Exon. Th. 444, 15–22; Kl. 47–51. Ðú meaht hit mé (*Eve*) wítan, wine mín, Adam, Cd. Th. 51, 10; Gen. 824. Wine *amatore* (the passage is : Ab alio amatore (*Christ*) praeventa sum, qui me annulo fidei suae subarravit, Ald. 60), Hpt. Gl. 506, 55. (4) applied to an inferior or subordinate, one to whom favour or protection may be shewn :—Hrôðgár maþelode : 'Wine mín, Beówulf ...,' Beo. Th. 919; B. 457: 3413; B. 1704. Se eádga wer (*Guthlac*) ágeaf andsware : 'Hwæt, ðú (*the disciple*) mé, mín wine, frignest,' Exon. Th. 175, 25; Gú. 1200. Ða gástas ðus ðone líchoman grétaþ : 'Wine leófesta,' Soul Kmbl. 266; Seel. 137: Andr. Kmbl. 614; An. 307: 2862; An. 1433. Dryhten wine sínne (*St. Andrew*) grétte, 2926; An. 1466. Wine leófestan (*Guthlac's disciple*), Exon. Th. 166, 1; Gú. 1036. Winiga (wunga, MS.) hleó (*Guthlac*), 184, 4; Gú. 1339. Winia bealdor (*Beowulf*), Beo. Th. 5127; B. 2567. Winigea (or under I?) leásum, 3332; B. 1664. [Brutus þe wes mi deore wine, mi drihliche lauerd (mi louerd deore, 2nd MS.), Laym. 2289. Og eurilc wurðen stedefæst his wine, Misc. 12, 374. *O. Sax.* wini: *O. H. Ger.* wini *amicus, sodalis, dilectus: Icel.* vinr.] v. freá-, freó-, gold-, gúð-, iu-, mǽg-, sundor-, un-wine.

wín-eard. v. wín-geard.

Winedas; *pl. m. The Wends* :—Wineda lond, Ors. 1, 1; Swt. 16, 9. Se port Hæþum stent betuh Winedum, and Seaxum, and Angle, Swt. 19, 23. Mid Winedum, Exon. Th. 322, 8; Víd. 60. Winedas and Burgendan, Ors. 1, 1; Swt. 16, 30. [*O. H. Ger.* Winidā *Vandali: Icel.* Vindir *the Wends.*] v. Weonod-land.

wine-dryhten, es; *m. A friendly, gracious lord.* v. wine (2) :—Ongan his magu frignan : 'Hú geweard ðé ðus, winedryhten (*Guthlac*) mín, fæder, freónda hleó?' Exon. Th. 162, 32; Gú. 984. Se ðe sceal his winedryhtnes leófes lárcwidum forþolian, 288, 27; Wand. 37. Ongon hé tô his winedryhtne mǽdlan : 'Ic ðec hálsige, hæleða leófost, 174, 9; Gú. 1175. Winedrihtne, Beo. Th. 726; B. 360. Ðæt mon his winedryhten wordum herge, 6332; B. 3176 : 5437; B. 2722 : Judth. Thw. 25, 23; Jud. 274. Hé winedryhten (*the Deity*) frægn, Andr. Kmbl. 1838; An. 921. Hié winedrihten wiht ne lôgon, glædne Hrôðgár, ac wæs ðæt gôd cyning, Beo. Th. 1728; B. 862. Hié wíscton and ne wêndon, ðæt hié heora winedrihten selfne gesáwon, 3213; B. 1604 : Byrht. Th. 139, 3; By. 248.

wine-geómor; *adj. Sad for the loss of friends* :—Ealle hié deáð fornam, and se án leóda dugeðe, se ðǽr lengest hwearf, weard winegeómor, Beo. Th. 4470; B. 2239.

wine-leás; *adj. Friendless.* v. wine :—Láð biþ ǽghwǽr fore his wonsceaftum wineleás hæle, Exon. Th. 329, 11; Vy. 32. Wineleás

wonsǽlig mon genimeþ him wulfas tô geféran, 342, 24; Gn. Ex. 147. Earm biþ se ðe sceal ána lifgan, wineleás wunian, 344, 15; Gn. Ex. 174. Wineleás guma, 289, 9; Wand. 45. Ic mê féran gewát folgaþ sécan, wineleás wrǽcca, 442, 9; Kl. 10. Ánhoga leódwynna leás, wineleás wrǽcca, 457, 27; Hy. 4, 90. Cain gewát gongan Gode of gesyhðe, wineleás wrecca, Cd. Th. 64, 16; Gen. 1051. Wreccan wineleásum bana, Beo. Th. 5219; B. 2613. Se ðec (*Nebuchadnezzar*) wineleásne on wræc sendeþ (*they shall drive thee from men*, Dan. 4, 25), Cd. Th. 251, 25; Dan. 569. [*Icel.* vin-lauss *friendless.*]

wine-mǽg, es; *m. A loving kinsman* :—Him his winemágas georne hýrdon, Beo. Th. 131; B. 65: Byrht. Th. 140, 50; By. 306: Elen. Kmbl. 2029; El. 1016. Hé winemága lyt, freónda hæfde, Cd. Th. 158, 31; Gen. 2625. Winemǽga hryre, Exon. Th. 287, 1; Wand. 7: 184, 2; Gú. 1338. Winemágum bidroren, 306, 31; Seef. 16. Flêma, winemágum láð, Cd. Th. 62, 29; Gen. 1021. [Wreke we ure winemæies, Laym. 5831.]

wine-scipe, es; *m. Friendship* :—Winescipe *collegio* (the passage is : Inseparabili angelicae sodalitatis collegio perfrui, Ald. 15), Wrt. Voc. ii. 77, 10: 18, 5. Lǽst wǽre and winescype, word ða wit sprǽcon, Exon. Th. 172, 17; Gú. 1145. [*Icel.* vin-skapr *friendship: Dan.* ven-skab. Cf. *O. H. Ger.* wine-scaft, -scaf *amor, foedus.*]

winestra; *adj. Left*; the feminine form is used substantively = *left hand.* v. swíþ, II :—Ðæt sió winestre hand ne scyle witan hwæt sió suíðre dô, Past. 44; Swt. 323, 14. 'Dryhtnes winestre hand is under mínum heáfde.' Sió winestre hand Godes, hé cwæð, wǽre under his heáfde ... Eft wæs gecueden ðætte on his winestran handa wǽre wela ... Godes fiónd, ðeáh hí on ðære winstran handa bión gedígene, hí beóð mid ðære swiðran tôbrocene, 50; Swt. 389, 10–25. Him wæs gelíce gewylde his wynstre and his swíðre ... Ða ábrǽd Aoth his swurd mid his wynstran handa, Jud. 3, 15, 21. Nyte ðín wynstre (winstra, Lind.: se winstrae hond, Rush., *sinistra*) hwæt dô ðín swýðre, Mt. Kmbl. 6, 3. Wiþ ðære winestran sídan sáre, Lchdm. ii. 64, 4. On ðam winestran earme, 254, 5. Gif ðú færst tô ðære winstran hælfe, ic healde ða swíðran healfe, Gen. 13, 9. On his wynsteran (wynstran, *v.l.*: on ða winstran healfe, Rush.: of winstrum, Lind., *a sinistris*) healfe, Mt. Kmbl. 25, 33. On ða winstran hond, Exon. Th. 75, 28; Cri. 1228. Wynstran, 83, 31; Cri. 1364: Gen. 48, 14. On ða wynstran healfe *ad sinistram*, Deut. 5, 32. Fealleþ ðé on ða wynstran þúsend, Ps. Th. 90, 7. On ðæt wynstre weorud, Exon. Th. 449, 22; Dôm. 75. [*O. Sax.* winistro: *O. Frs.* winstera: *O. H. Ger.* winstero, winistro, winstro: *Icel.* vinstri.] v. Grmm. Gesch. D. S. c. xl.

wine-þearfende; *adj. Friendless* :—Hé (*Guthlac's disciple*) ne máð fæges (*Guthlac's*) forðsíð wineþearfende, Exon. Th. 183, 2; Gú. 1321. Andreas wineþearfende mǽlde : 'Næbbe ic gold ... ðæt ic ðé mæge lust áhwettan,' Andr. Kmbl. 599; An. 300.

wine-treów, e; *f. Faith between friends (between husband and wife*; cf. wine (3)) :—Ðæt hé (*the man*) ða wǽre and ða winetreówe be him lifgendum lǽstan wolde, ðe git on ǽrdagum of gesprǽcon, Exon. Th. 475, 20; Bo. 50. [Lêste thu (*Joseph*) inka winitrewa, hald inkan friundskepi, ne lât thu sie (*Mary*) thi thiu lêðaron, Hêl. 321.]

wine-wincla (-e; *f.?*), an; *m.* (?) *A periwinkle* :—Sǽsnǽl *vel* winewinclan *chelio, testudo, vel marina gagalia*, Wrt. Voc. i. 24, 32. Winewinclan *torniculi*, 6, 21. (v. Lchdm. ii. 240, note 1, for these two passages.) Cwice winewinclan gebǽrnde tô ahsan, Lchdm. ii. 28, 25. Þicgen hié ostran and winewinclan, 254, 23. Sǽ-winewinclan gebǽrnde and gegnidene, 240, 4.

wín-fæt, es; *n. A wine-vat* :—Wínfæt *enophorum*, Wrt. Voc. i. 25, 5: *apotheca*, ii. 100, 54: 6, 6. [*O. H. Ger.* wîn-faz *vas vinale: Icel.* vín-fat.]

wín-gál; *adj. Flown with wine, wanton with wine* :—Onwôc wulfheort, se ǽr wíngál swæf, Cd. Th. 223, 8; Dan. 116. Wlonc and wíngal '*flown with insolence and wine*,' Exon. Th. 307, 25; Seef. 29 : 478, 2; Reim. 35.

wín-geard, -eard, es; *m.* I. *a vineyard, a place where vines grow, the vines growing in such a place*; vinea :—Hé út eode áhýrian wyrhtan on his wíngeard (-eard, *v.l.*) ... Hé ásende hig on hys wíngeard (-eard, *v.l.*) ... Ða sǽde se wíngeardes (ðære wíngearde, Lind.) hláford, Mt. Kmbl. 20, 1, 2, 8: 21, 40. Sum man hæfde án fictreów geplantod on his wíngearde (-georde, Rush.), Lk. Skt. 13, 6. Dô swá on ðínum wíneearde and on ðínum elebeámon *ita facies in vinea et in oliveto tuo*, Ex. 23, 11. Hé gesette him wíneard *plantavit vineam*, Gen. 9, 20. Wíngeard, Cd. Th. 94, 8; Gen. 1558. Wíngeord, Mk. Skt. Rush. 12, 1. Ðú sealdest ús landæhta and wíneardas *dedisti nobis possessiones agrorum et vinearum*, Num. 16, 14. I a. *a place where other plants than vines grow* :—Wínegeardes *palmeti* (cf. the gloss of same passage in Wrt. Voc. ii. 75, 77 : Palmberwes *palmeti*, Hpt. Gl. 496, 62. II. *a vine*; vitis, vinea :—Elebeám *oliva*, wíngeard *vinea*, wínberge *uva*, Wrt. Voc. i. 285, 71. Gescreáded wíngeard *sarpta vinea*, 54, 65. Hwít wilde wíngeard *brionia vel ampelos leuce*, wíngerd *labrusca*, blac wíngeard *brabasca vel ampelos male*, 30, 14–16. Hwít wíngeard *brionia*, 32, 17: *aminea vitis*, 39, 1. Ic eom sôð wíneard (*vitis*), Jn. Skt. 15,

1, 5. Đeós wyrt hafaþ leáf swylce wîngeard, Lchdm. i. 316, 8. Wîngeardes twiga, ii. 190, 11. Of đises wîngeardes (-eardes, *v.l.*) cynne *de
generatione vitis*, Lk. Skt. 22, 18. Swā on wîngearde weaxen berigean
sicut vitis abundans, Ps. Th. 127, 3. Of wîngearde *de vite*, Wrt. Voc.
ii. 27, 53. Se gesibsuma wer byđ đam wînearde gelic đe byrđ gôde
wæstmas, Basil admn. 6; Norm. 46, 24. Đû ût ālæddest wîngeard
(*vineam*) . . . and his wyrtruman settest, Ps. Th. 79, 8. Ic geseah wîneard (*vitem*), on đam wæron þreó clystru, Gen. 40, 9. On sumum
stôwum wîngeardas (*vineae*) grôwaþ, Bd. I, 1 ; S. 473, 14 : Ps. Th. 104,
29. Wîngearda hôcas đe hî mid bindaþ đæt him nêhst biđ *capreoli vel
cincinni vel uncinuli*, wîngearda hringa[s] *corimbi*, Wrt. Voc. i. 38, 59-
60. Wîngearda gewind *capreoli*, 39, 10. Đe mā đe gimmas weaxaþ
on wîngeardum *nec vite gemmas carpitis*, Bt. 32, 3 ; Fox 118, 11 : Met.
19, 9. [He plantede winiærd, Chr. 1137; Erl. 263, 19. Wingeardes
vineae, A. R. 294, 29. Winyard, Misc. 33, 20. Goth. *weina-gards*:
O. Sax. *wîn-gardo*: O. H. Ger. *wîn-gart, -garto*: Icel. *vîn-garđr*.] v.
following words.

wîngeard-bôh(-g), es; *m. A vine-tendril:*—Wingeardbôgas *capreoli*,
Wrt. Voc. i. 22, 15. v. wîn-geard, II, *and next word.*

wîngeard-hocgas (*for* wîngeard-bôgas, v. preceding word; *or* wîn-
geard-hôcas, cf. wîngearda hôcas *capreoli*, Wrt. Voc. i. 38, 59) *caprioli
dicti quod capiant arbores*, Wrt. Voc. ii. 129, 61.

wîngeard-hring, es; *m. A cluster of grapes:*—Wîngeardhringas (cf.
Wîngearda hringa[s] *corimbi*, i. 38, 60) *vel bergan vel croppas corimbi*, i.
viti racemi vel botriones vel circuli, Wrt. Voc. ii. 135, 74.

wîngeard-seax, es; *n. A pruning-knife:*—Wîngeardseax *falx*, Wrt.
Voc. ii. 146, 76.

wîn-gedrinc, es; *n. Wine-drinking, wine:*—Hié wlenco onwôd and
wîngedrync, Cd. Th. 155, 28; Gen. 2579. Of ungemete ælces þinges
wiste and wæda, wîngedrinces, Met. 25, 39. Wîngedrince *nectare*, Wrt.
Voc. ii. 61, 32. Hié tô đam symle sittan eodon, wlance tô wîngedrince,
Judth. Thw. 21, 12; Jud. 16. v. wîn-drync.

wîn-getred, es; *n. A place where the juice is trodden out of the
grapes:*—Wîngetred *forus, ubi uva calcatur*, Wrt. Voc. ii. 39, 66.

wîn-hâte, an; *f. A feast:*—Gefrægn ic Olofernus wînhâtan wyrcean,
and eallum wundrum þrymlíc girwan up swǣsendo; tô đâm hêt se gumena baldor ealle đa yldestan þegnas (the Latin is: *Holofernes fecit cenam
servis suis*, Judith 12, 10), Judth. Thw. 21, 6 ; Jud. 3.

wîn-horn, es; *m. A wine-horn, drinking-cup:*—Gyf đû wînhorn habban wille, đonne dô đû mid đînum swiđran scytefingre on đîne wynstran
hand swilce đû tæppan teón wille, and rær up đinne scytefinger be đînum
heófede, Techm. ii. 120, 11. [In Wrt. Voc. ii. 6, 6 *apotheca* wînfæt,
wînærn ho is written above ærn. v. Wülck. 348, 2.]

wîn-hûs, es; *n. A wine-house:*—Wînhûs *apotheca*, Wrt. Voc. i. 58,
18. Wînhûsum *apothecis*, Hpt. Gl. 468, 40. Ne môt mid rihte nân
preóst drincan æt wînhûsum ealles tô gelóme, L. Ælfc. P. 49; Th. ii.
386, 8. [Icel. *vîn-hûs*.] Cf. wîn-ærn.

wînian; *p.* ode *To gather grapes:*—Hiá wînigaþ *uindemiant*, Lk.
Skt. Lind. 6, 44.

wîning, es; *m. A band for the leg:*—Winincg *fascia*, wyncgas
(= winincgas?) *vallegias*, Wrt. Voc. i. 26, 7, 9. Đonne đû wynyngas
habban wille, đonne dô đû mid đînum twâm handum onbûtan đîne
sceancan, Techm. ii. 127, 10. [Cf. (?) O. L. Ger. *winding* : O. H. Ger.
winting fascia, fasciola, fasciale: Icel. *vindingr strip wound round the
leg instead of hose*.]

wîn-leáf, es; *n. A vine-leaf:*—Wînleáf *pampinus*, Engl. Stud. xi.
66, 73.

wîn-lîc; *adj. Of wine:*—Hê wæter âwende tô wînlícum drence, Ælfc.
T. Grn. 13, 37. Hê gemêt đæt wæter tô wînlícum swæcce âwend (cf.
l. 16), Homl. Th. ii. 58, 31 : 64, 29.

wîn-mere; v. wîn-côle.

winn, es; *n.* **I.** *labour:*—Nêđđarf woerces ł đæs wynnes *necessitas
laboris*, Lk. Skt. p. 2, 8. Đæt hî gemǣne win (*v.l.* gewin, M. 98, 18)
onfênge godcunde lâre tô lǣranne on Angelđeóde *ut communem evangelizandi gentibus laborem susciperent*, Bd. 2, 2 ; S. 502, 9. In wynn (giwinne, Rush.) hiora *in laborem eorum*, Jn. Skt. Lind. 4, 38. **II.**
strife, conflict:—Hê ongan him winn up âhebban wiđ đone hêhstan
heofones wealdend, Cd. Th. 17, 14 ; Gen. 259. [þa þe ledden here lif in
werre and in winne, O. E. Homl. i. 175, 246. Devel wecchheđ among
hem flite and win, Rel. Ant. i. 128, 32. Þar aros wale and win, Laym.
404. Đe watres win, Gen. and Ex. 598. Jeolpen for þere winne (of
þan winne, 2nd MS.) *to boast of the gain*, Laym. 12072. Þin rihhte
swinnkes winn (*gain*), Orm. 6118.] v. ge-, wiþer-winn.

winna, an; *m. An opponent:*—Đa þeóda đe hyra winnan (wiþer-
ge-winnan, *v.ll.*) wæron, Ors. 6, 35; Bos. 130, 44. v. ge-, wiþerwinna.

winnan; *p.* wann, *pl.* wunnon; *pp.* wunnen. **A. intrans. I.**
to labour, toil, work:—Swā ic þrymful þeów winne, Exon. Th. 386, 26;
Rä. 4, 67. In îdelnisse winnaþ đa timbriaþ đa *in vanum laborant qui
aedificant eam*, Ps. Surt. 126, 1. Hê mid his handum wonn and worhte
đa đing đe nýdþearflícu wæron *operi manuum studium impendebat*, Bd.

4, 3; S. 567, 30. Hê won and worhte, wîngeard sette, Cd. Th. 94, 7 ;
Gen. 1558. Đerh alle næht wê wunnon *per totam noctem laborantes*,
Lk. Skt. Lind. 5, 5. Đû sylest ûrum leomum ræste, for đon đe hié on
đînum noman wunnon, Blickl. Homl. 141, 12. Đeáh đe hê wunne on
his lâre *quamvis illo laborante in verbo*, Bd. 2, 9; S. 511, 9. Đû winnan
scealt, and on eorđan đê đine andlifne selfa gerǣcan, Cd. Th. 57, 23;
Gen. 932. Winnende *vel* swǣtende *desudans*, i. *laborans*, Wrt. Voc. ii.
139, 36. Ic geseah winnende wiht, Exon. Th. 438, 3 ; Rä. 57,
2. **I a.** *to labour, endeavour, strive after:*—Ælc winđ be his andgites mǣþe, đæt hê hine wolde ongitan gif hê mihte, Bt. 41, 4; Fox 250,
25. Â đû wunne æfter eorđlícum welum, Wulfst. 140, 24. Nô won hê
æfter worulde, ac hê in wuldre âhôf môdes wynne, Exon. Th. 126, 12 ;
Gû. 370. Đâ wann (*laboravit*) hê swýþe, đæt hê his geféran geheólde,
đæt hî ne âsprungan fram heora geleáfan, Bd. 2, 9; S. 511, 5. Hî
wunnon æfter wyrþscipe, and tiledon gôdes hlîsan, Bt. 40, 4; Fox 240,
4. Đæt hê wunne æfter worulde, Exon. Th. 109, 34 ; Gû. 100. Winnan æfter snytro, Salm. Kmbl. 778 ; Sal. 388. **I b.** *to labour, struggle,
be troubled:*—Moncyn winþ on đâm ýđum đisse worulde *homines quati-
mur fortunae salo*, Bt. 4; Fox 8, 22. For hwam winneþ đis wæter . . .
ne môt on dæg restan? Salm. Kmbl. 785; Sal. 392. Gê winnaþ and
â embe đæt sorgiaþ, đæt wê ûrne lîchoman gesyllan . . . Ûs is mycele
mâre nêdþearf, đæt wê winnon ymbe ûre sâule þearfe, Blickl. Homl. 99,
6-11. Ealle gê đe winnaþ (*laboratis*), and gebyrde sindun, Mt. Kmbl.
Rush. 11, 28. On worulde ýþum wynnaþ and swincaþ earme eorđwaran
(v. Fox 8, 22 *supra*), Met. 4, 56. Ic wêrigmôd wann and cleopode
laboravi clamans, Ps. Th. 68, 3. Đû in wræc wunne, wuldres blunne,
Andr. Kmbl. 2759 ; An. 1382. Sió his innaþ wan wætere gelíc, Ps. Th.
108, 18. Hê sceal winnan and sorgian, Blickl. Homl. 97, 25. Hê wolde
đǣm winnendum fultmian, and earme frêfran, 213, 17. Đǣm winnendum brôþrum on sǣ *laborantibus in mari fratribus*, Bd. 5, 1 ; S. 613,
7. **I c.** *to labour* under, *suffer* from:—Heó đǣre ylcan hefignesse
âdle unâblinnendlíce won *eadem molestia laborare non cessabat*, Bd. 4,
23; S. 595, 18. Horsum and ǣlcum fiþerfêtum neáte đe on wôle
winnen (cf. wôles gewinn, 330, 4), Lchdm. i. 328, 13. Longsumum
ermđum winnende *diuturnis calamitatibus laborantem*, Rtl. 41, 29. **II.**
to strive, contend, fight:—Ic wan *pugnavi*, Wrt. Voc. ii. 130, 29. Winnende *congrediens, certando*, 133, 43. Winn for sâwle đíne . . . winn
for rihtwîsnysse *agonizare pro anima tua* . . . *certa pro justitia*, Scint.
73, 14, 15. (1) of hostile action towards a person:—Gif Satanas winđ
ongên hine sylfne *si Satanas consurrexit in semetipsum*, Mk. Skt. 3, 26.
Se fæder winđ wiđ his âgenne sunu, Homl. Skt. i. 13, 296. Hû đa synna
him wiđ winnaþ, Past. 21 ; Swt. 163, 2. Gê wunnon ongeán Drihten
adversum Dominum contendisti, Deut. 9, 7. Ne wynne gê ongên đa đe
eów yfel dôđ *non resistere malo*, Mt. Kmbl. 5, 39. Heó (*Hagar*) ongan
wiđ Sarran winnan, Cd. Th. 135, 12; Gen. 2241. (1 a) of competi-
tion:—Eart đû se Beówulf, se đû wiđ Brecan wunne, ymb sund flite, Beo.
Th. 1017; B. 506. (2) of opposition to things:—Đû winsđ wiþ đam
hlâfordscipe đe đû self gecure, Bt. 7, 2; Fox 18, 29. Is micel đearf,
đonne him mon hwæđer ondrætt suiđar đonne ôđer, and wiđ đæt wienđ
(winđ, Cott. MSS.), đæt hê suâ suîđe wiđ đæt winne, suâ hê on đæt ôđer
ne befealle, đe hê him læs ondrêd *ne dum pugnat contra hoc, quod tolerat,
ei a quo se liberum aestinat, vitio succumbat*, Past. 27 ; Swt. 189, 10 :
46; Swt. 347, 12. Gif hê winđ mid gebedum ongeán, Boutr. Scrd. 20,
16. Hî winnaþ him (*vices*) tôgeánes, Homl. Skt. i. 17, 63. Monige
lâreówas winnaþ mid hira đeáwum wiđ đa gǣsđlecu bebodu, Past. 2;
Swt. 29, 21. Hê weard âhangen on rôde . . . , and hê ongeán nân đyngc
ne wan (*he made no resistance to being crucified*), L. Ælfc. P. 51 ; Th.
ii. 386, 37. Wê wiđ đam winde and wiþ đam sǣ campodon and wunnan
cum vento pelagoque certantes, Bd. 5, 1 ; S. 613, 28. Winn ongên *resist
(temptation)*, Homl. Skt. ii. 30, 137. Đæt gehwâ winne wiđ his lîchaman unrihtlustas *ut quisquis cum corporis sui pravis cupiditatibus certet*,
L. Ecg. P. iv. 63 ; Th. ii. 224, 4: Bt. 36, 6 ; Fox 182, 5. Đæt hê for
lícuman tiédernesse wiđ đa scîre ne winne *nec per imbecillitatem corpus
repugnat*, Past. 10; Swt. 61, 11. Nis nân gesceaft đe wþ hire Scippendes
willan winne, bûton dysig mon, Bt. 35, 4 ; Fox 160, 22. Hê đâm unþeáwum nyle furþum wiþ winnan, 37, 1; Fox 186, 30 : Met. 25, 67.
(3) of the action of inanimate objects:—Fâmig winneþ wǣg wiđ wealle,
Exon. Th. 383, 32; Rä. 4, 19. Ælc his gesceafta winþ wiþ ôþer . . .
ge hié betwux him winnaþ, ge eác fæste sibbe betwux him healdaþ, Bt.
21 ; Fox 74, 10-15 : Met. 11, 45 : 20, 74. Seó tunglena heofon tyrnđ
eásten westweard, and hire winnaþ ongeán đa seofon dweligendan tunglan,
Boutr. Scrd. 18, 29. Holm won wiđ winde, Beo. Th. 2268 ; B. 1132.
Oft ic (*an anchor*) sceal wiþ wǣge winnan, and wiþ winde feohtan, Exon.
Th. 398, 1 ; Rä. 17, 1. (4) *to make war* (lit. or fig.), *fight:*—Mec gesette Crist tô compe . . . Hwîlum ic frêfre đa ic ǣr winne on, Exon. Th.
389, 14; Rä. 7, 7. Ælc đǣra on gecampe winđ, Homl. Th. ii. 86,
22. Đeód winđ ongên þeóde *consurget gens in gentem*, Mt. Kmbl. 24,
7. Wê winnaþ for hǣlo ûre đeóde *pro salute gentis nostrae bella suscepimus*, Bd. 3, 2 ; S. 524, 24. Se lîchama and seó sâwl winnaþ him
betweónan, Homl. Skt. i. 17, 8. Wildu diór đa winnaþ betwuh, Met.
27, 20. Hine monige on wrâđe winnaþ mid wǣpenþræce, Cd. Th. 138,

11; Gen. 2290. Đú wealdest đises ríces đe đú æfter wunne, Guthl. 21; Gdwin. 96, 7. Đú wiđ Criste wunne and gewin tuge, Exon. Th. 267, 26; Jul. 421. Hé wann mid đam (*a sword*) on ælcum gefeohte, Homl. Skt. ii. 25, 296. Đá wan him on Amalech, i. 13, 4. Hé wonn on Sciþþie *regi Scytharum bellum intulit*, Ors. 2, 5; Swt. 78, 8. Fæht hine on and won Penda *impugnatus a Penda*, Bd. 3, 14; S. 539, 18. Đá wann him ongeán Maxentius, Homl. Th. ii. 304, 5. Hé gelómlíce uppon đone eorl wann, Chr. 1095; Erl. 231, 10. Đá won wiþ hine Cadwalla *rebellavit adversus eum Caedualla*, Bd. 2, 20; S. 521, 7. Hé feaht and won wiþ his eþle, 3, 24; S. 556, 28; Chr. 597; Erl. 20, 4. Grendel wan wiđ Hróđgár, Beo. Th. 305; B. 151. Hí wunnon him betwýnan, Homl. Th. ii. 356, 24. Wunnon hý wiđ Dryhtnes mihtum, Salm. Kmbl. 655; Sal. 327. Đa Bryttas wunnon heom wiđ đa castelmenn, Chr. 1067; Erl. 204, 5. Win him on swýđe, Homl. Skt. i. 13, 8. Seó æ đe đú under hire tæcinge winnan wylt and campian *lex sub qua militare uis*, R. Ben. 96, 23. Æfter ríce winnan, Chr. 685; Erl. 40, 16. On winnan *ingruere*, Hpt. Gl. 427, 42: Bd. 1, 12; S. 480, 23. Đonne hé on óđer folc winnan sceal, Past. 18; Swt. 129, 9. Đæt hí uppon hæđene þeódan winnan woldan, Chr. 1096; Erl. 233, 14. On gehwelc lond tó winnanne, Ors. 3, 7; Swt. 116, 8. Hé him on winnende wæs, 1, 2; Swt. 30, 5. Worhte Ælfréd cyning lytle werede geweorc æt Æþelinga eigge, and of đam geweorce wæs winnende wiþ đone here, Chr. 878; Erl. 80, 6. (4 a) of the action of inanimate objects :—Se winterlíca wind wan mid đam forste *the winter wind warred along with the frost*, Homl. Skt. i. 11, 144. (4 b) with cognate accusative :—For đæm gewinne đe hé wiþ God wan, Blickl. Homl. 63, 4. Winn gód gewinn *certa bonum certamen*, Scint. 214, 16. **III.** *to win* (v. Jamieson's Dictionary), *make one's way* :—Hwæt is đæt wundor, đæt geond đás woruld fareþ . . ., winneþ oft hider ? Salm. Kmbl. 568; Sal. 283. **B.** *trans.* **I.** *to labour at, bestow labour upon* :—Ic wann wunnunise mín *laboravi habitationem meam*, Rtl. 68, 28. Ic sende iúh gehrioppa đætte gié ne wunnon *ego misi uos metere quod uos non laborastis*, Jn. Skt. Lind. Rush. 4, 38. **II.** *to labour under, suffer, undergo* :—Ic đæt geþolade . . . læg on heardum statue . . . ic đæt earfeþe wonn, Exon. Th. 87, 21; Cri. 1428. Â ic wíte wonn mínra wræcsíþa, 441, 26; Kl. 5. Ic â þolade géara gehwylce gódes ealles, won ic módearfoþa (þonc móđ earfoþa, Th.) má đonne on óþrum, fyrhto in folce, 457, 19; Hy. 4, 86. Mid đý đá se bróþor langre tíde đyllíc ungescréþo wonn (woon, MS.) *cumque tempore non pauco frater tali incommodo laboraret*, Bd. 4, 32; S. 611, 22. Đú đæs cwealmes scealt wíte winnan and on wræc hweorfan, Cd. Th. 62, 14; Gen. 1014. Hí áwo sculon, wræc winnende, wærgđu dreógan, Exon. Th. 78, 10; Cri. 1272. **III.** *to win, get, attain* :—Đú wunne reste â óþ ende mid hálgum fæmnum, Nar. 49, 1. Hí wéndon đæt hí sceoldon winnon eall đæt land, Chr. 1070; Erl. 207, 27. [Ierusalem and Babilonie flited eure and winned bitwinen hem . . . þe king of Babilonie wan Ierusalem, O. E. Homl. ii. 51, 11–25. Iob wan wiđ þa wurse, 187, 26. Heo wunnen agean, A. R. 238, 17. Đanne sumer and winter winnen, Misc. 17, 521. He iwon (won, 2nd MS.) al þis lond, Laym. 2560. Winnenn heoffness kinedom, Orm. 801. He wan to William, Will. 2498. *Goth.* winnan παθεῖν: *O. Sax.* winnan *to strive; to suffer; to gain: O. Frs.* winna *to gain: O. H. Ger.* winnan *laborare; jurgare, decertare, dimicare: Icel.* vinna *to work; to withstand; to suffer; to win.*] v. â-, ge-, ofer-, wiþer-winnan; on-winnende.

winn-dæg, es; *m. A day of labour* or *of struggle* :—Fela sceal gebídan leófes and láþes se đe longe hér on đyssum windagum worulde brúceþ, Beo. Th. 2128; B. 1062. v. gewin-dæg.

-winnend, -winnendlíc. v. ofer-, wiþ-winnend, un-oferwinnendlíc.

winn-stów, e; *f. A wrestling-place* :—Winstówe *scammatis*, Hpt. Gl. 405, 40. On winstówe *in scammate*, 489, 59. Winstówe *palaestrarum*, 478, 50. v. gewin-stów.

winnung, winpel. v. windung, wiþ-winnung, wimpel.

win-ræced, es; *m. n. A house where there is feasting, a palace* :—Wínreced, goldsele gumena (*Hrothgar's palace*), Beo. Th. 1433; B. 714. Đæt wínreced, gestsele, 1991; B. 993. Hornsalu wunedon wéste wínræced, Andr. Kmbl. 2319; An. 1161. Cf. wín-ærn.

win-reáfetian *to take grapes* :—Plucciaþ ł wínhreáfetiaþ *vindemiant*, Ps. Lamb. 79, 13.

win-repan *to gather grapes* :—Wínreopad đæt *vendemiant eam*, Ps. Surt. 79, 13. v. repan.

win-sæd; *adj. Wine-sated, having had one's fill of wine* :—Yrrum ealowósan, wære wínsadum, Exon. Th. 330, 12; Vy. 50. Weras wínsade (cf. hé oferdrencte his duguđe ealle, 21, 22; Jud. 31; and the Latin c. 13, 2 : Erant omnes fatigati a vino), Judth. Thw. 22, 21; Jud. 71.

win-sæl, es; *n. A wine-hall, a hall where there is feasting* :—Wóriaþ đa wínsalo, Exon. Th. 291, 6; Wand. 78. v. next word.

win-sele, es; *m. A wine-hall, a hall where there is feasting* :—Nis hér (*in Hell*) wloncra wínsele, ne worulde dreám, Cd. Th. 270, 21; Sat. 94. Se wínsele (*Hrothgar's hall*), Beo. Th. 1547; B. 771. In đæm wínsele, 1394; B. 695. Beóre druncne . . . hí in wínsele sáwle forlétan, Exon. Th. 271, 25; Jul. 487: 283, 27; Jul. 686. Gesyhđ on his

suna búre wínsele wéstne, Beo. Th. 4903; B. 2456. [*O. Sax.* wínseli.]

win-sester, es; *m. A wine-can* :—Wínsester *cantarus*, Wrt. Voc. i. 24, 37.

win-stów, winstre, winsum. v. winn-stów, winestra, wynsum.

win-tæppere, es; *m. A wine-seller, tavern-keeper* :—Wíntæpperum *cauponibus*, Hpt. Gl. 468, 42.

Wintan-ceaster (Wintun-, Winta (-e, -i), Win-), e : Wænte, an; *f. Winchester.* [The name is got from the earlier *Venta* of Roman Britain. This form occurs in Latin works, e. g. : In Venta civitate, Bd. 4, 15 : Cod. Dip. Kmbl. iii. 300, 16. Monasterium in Wenta positum, vi. 29, 16. Also the adjective Wentanus (Uentanus, Bd. 5, 18), e. g. : Wentanus episcopus, v. 82, 14. Wentana ecclesia, ii. 210, 3 : v. 45, 3. Wentana civitas, ii. 140, 9 : 220, 28. Urbs Wentana, iii. 326, 10 : iv. 45, 7. Wentana sedes, v. 169, 16. And Wentana is used as the name of the place, e. g. : Wentana monasterium, iii. 8, 13. But Latinized forms of the English word are used; Wintonia is often found in the charters; the form Wincestria occurs v. 167, 7, and the adjective Wintancestrensis 90, 29.]—In ciuitate opinatissima quae Wintecaester nuncupatur, Cod. Dip. Kmbl. ii. 195, 35. Belumpon hí (*the South Saxons*) ær tó Wintanceastre biscopscíre *ad civitatis Ventanae parochiam pertinebant*, Bd. 5, 18; S. 639, 14. Daniel Wæntan biscop, Chr. 731; Erl. 47, 11. Intó Wintanceastre, Cod. Dip. Kmbl. ii. 114, 26 : iii. 111, 29. Gange ân gemet swilce man on Lundenbyrig and on Wintanceastre (Winta-, *v. l.*) healde, L. Edg. ii. 8; Th. i. 270, 2. Seó gerædnys đe Cnut cyninge gerædde on Wintanceastre (Win-, *v. l.*), L. C. E. proem.; Th. i. 358, 7. Cénwalh hét átimbran đa ciricean on Wintunceastre (Wintan-, *v. l.*), Chr. 643; Erl. 26, 9. Hér Danihel gesæt on Wintanceastre, 744; Erl. 48, 1. Hedde heóld đone biscopdóm on Wintanceastre (Wintan-, *v. l.*), 703; Erl. 42, 22. Hí West-Seaxna bisceopum underþeódde wæron, đa đe on Wintanceastre wæron, Bd. 4, 15; S. 583, 35. Tó Wintanceastre (Winte-, *v. l.*) .vi. myneteras, L. Ath. i. 14; Th. i. 206, 31 : Cod. Dip. Kmbl. iii. 326, 16. Winteceastre, ii. 176, 11 : v. 163, 11. Tó ealdan mynstære tó Wintíceastræ, ii. 127, 12. Syþþan đæt gemót wæs on Wintaceastre, L. C. S. 30; Th. i. 392, 26. Đes Swíđún wæs bisceop on Winceastre, Homl. Skt. i. 21, 14. Æđelwold biscop on Winceastre, Cod. Dip. Kmbl. vi. 207. 6.

winter, es; *m.* (*in pl. a neuter form* wintru *occurs, as well as masculine* wintras; winter: *the dat. sing.* wintra *is a trace of earlier* u-*stem declension*). **I.** *a season of the year*, winter :—Feówer tída syndon getealde on ánum geáre, đæt synd uer, aestas, autumnus, hiems . . . Hiems is winter, Lchdm. iii. 250, 12. On đone .vii. dæg đæs mónđes (*November*) bið wintres fruma; se winter hasaþ tú and hundnigontig daga, Shrn. 146, 7. Winter bringeþ weder ungemetceald, swifte windas, Met. 11, 59. Winter bið cealdost, Menol. Fox 470; Gn. C. 5. Hengest wælfágne winter wunode mid Finne . . . Holm storme weól, winter ýpe beleác ísgebinde, óþ đæt óþer com geár in geardas . . . Đa wæs winter scacen, fæger foldan bearm, Beo. Th. 2259–2278; B. 1127–1137. Đæt hit wære wintres tíd, and se winter wære grim and ceald and fyrstig and mid íse gebunden, Bd. 3, 19; S. 549, 26. Is đær nú irfæs đæs đæs stranga wintær læfæd hæfđ, Chart. Th. 163, 1. Nys hit swá stearc winter đæt ic durre lutian æt hám, Coll. Monast. Th. 19, 17. Sam hit sý sumor sam winter, Ors. 1, 1; Swt. 21, 17. Wintres *brumae*, Wrt. Voc. ii. 12, 43. On wintres tíman, đæt is fram đan anginne đæs mónđes, þe is November gehaten, þo Eástran, R. Ben. 32, 10. Siđđan (*after the first of November*) wintres dæg (*winter*; cf. Icel. á vetrardag *in the winter*) wíde gangeþ on syx nihtum, sigelbeorhtne genimđ hærfest mid herige hrímes and snáwes, Menol. Fox 401; Men. 202. Hé (*petra oleum*) is gód tó drincanne on wintres dæge, for đon đe hé hæfđ swíđe micle hæte; for đý hine mon sceal drincan on wintra, Lchdm. ii. 288, 16. Beámas gréne stondaþ wintres and sumeres, Exon. Th. 200, 7; Ph. 37. Mid đý storme đæs wintres *hiemis tempestate*, Bd. 2, 13; S. 516, 20. Geofon weól wintres wylme, Beo. Th. 1036; B. 516. Wintres wóma, Exon. Th. 292, 22; Wand. 103. Biddaþ đæt eówer fleám on wintra (wintre, Rush.) ne gewurđe, Mt. Kmbl. 24, 20. On wintra hit biþ ceald, Bt. 21; Fox 74, 24. Se oftræda rén leccaþ đa eorþan on wintra, 39, 13; Fox 234, 17. Wíciaþ Finnas on huntoþe on wintra, and on sumera on fiscaþe, Ors. 1, 1; Swt. 17, 6. Hí (*the hawks*) fédaþ hig sylfe and mé on wintra, Coll. Monast. Th. 26, 1. Beád Swegen full gild and metsunga tó his here đone winter, Chr. 1013; Erl. 149, 3: Exon. Th. 306, 19; Seef. 15. Wintras *hiemes*, Germ. 388, 26. *See also* midd, II. **I a.** *wintry weather, cold* :—Hé (*the sparrow*) sóna of wintra in winter eft cymeþ, Bd. 2, 13; S. 516, 21. Hé him helle gescóp, wælcealde wíc, winter beđeahte, Salm. Kmbl. 938; Sal. 468. Se wind (*zephirus*) tówyrpđ and đáwaþ ælcne winter, Lchdm. iii. 274, 22. **II.** *a year* :—Beóđ his winter wynnum íced *annos ejus in diem seculi adjicies*, Ps. Th. 60, 5. Úre winter *anni nostri*, 89, 10. God ána wát hwæt him weaxendum winter bringaþ, Exon. Th. 327, 26; Vy. 9. Hí wæron on Egipta lande feówer hund wintra and þritig wintra, Ex. 12, 40. Ymb þrittig wintra, Bt. 39, 3; Fox 214, 25. Hú seó ádl ær feówertigum ođđe fíftigum wintra on men ne becume *how the disease does not attack a man before*

he is forty or fifty, Lchdm. ii. 284, 10, 20. Ne mæg weorþan wîs wer ǽr hê âge wintra dǽl in woruldrîce, Exon. Th. 290, 14; Wand. 65. Hê wintra hæfde fîf and hundteóntig, Cd. Th. 69, 4; Gen. 1130: 74, 32; Gen. 1231. Wintra fela . . . geára mengeo, 103, 26; Gen. 1724. Twelf wintra tîd, Beo. Th. 296; B. 147. Ðæt swâ fyrn geweard wintra gangum, Elen. Kmbl. 1262; El. 633. Wintra gerîmes þreó and þrítig geára, Cd. Th. 296, 15; Sat. 502: Chr. 1065; Erl. 196, 26. Ic eom gomel wintrum, Ps. Th. 70, 16. Wintrum frôd, Beo. Th. 3452; B. 1724. Wintrum yldre, Cd. Th. 158, 2; Gen. 2611. Wintrum geong, 174, 34; Gen. 2888: Byrht. Th. 137, 62; By. 210. Ðæt wîf on blôdryne twelf winter (wintra twelfe, Rush.: wintrum twoelfum, Lind., *annis duodecim*) wæs, Mk. Skt. 5, 25. Hî besǽtan ða burg .x. winter (*per decem annos*), Ors. 1, 14; Swt. 56, 19. Hê geheóld rîce fîftig wintru, Beo. Th. 4424; B. 2209. Siððan hê strŷnde seofon winter suna and dohtra, Cd. Th. 69, 21; Gen. 1139. **II a.** used in singular with a collective force:—Adam wæs on þrýtiges wintres ylde, Anglia xi. 2, 27. Wine frôd wintres, Cd. Th. 72, 29; Gen. 1194. [*Goth.* wintrus; *m.* winter; *a year*: *O. Sax. O. H. Ger.* wintar; *m.*: *O. Frs.* winter; *m.*: *Icel.* vetr; *m.*] v. mid-, midde-winter; -wintre.

winter-biter; *adj. Having the bitterness of winter*:—Forstas and snâwas, winterbiter weder *frosts and snows, weather with winter's bitterness,* Cd. Th. 239, 32; Dan. 379: Exon. Th. 192, 12; Az. 105.

winter-burna, an; *m. A stream that is full in winter* (?), *a stream that has the fullness of winter* (?), *a torrent*:—Ofer ðæt burna ł uinter-burna *trans torrentem,* Jn. Skt. Lind. 18, 1. ¶ the word occurs as a local name, and is found often in the Charters, e. g.: In Winter-burnan . . . swâ on ôðerne Winterburnan, Cod. Dip. Kmbl. iii. 405, 22. See vi. 354, col. 2.

winter-ceald; *adj. Wintry-cold, cold with the cold of winter*:—Ic him gromheortum winterceald oncweþe, Exon. Th. 387, 18; Rä. 5, 7. Hê dreág wintercealde wræcce, 377, 15; Deór. 4. Wintercealdan niht, Andr. Kmbl. 2531; An. 1267. [*O. Sax.* wintar-kald snéu.]

winter-cearig; *adj. Sad from age or from the gloom of winter*:— Ic heán wôd wintercearig (*sad with the load of years* (?), cf. Gemon hê hû hine on geóguðe his goldwine wenede tô wiste, 288, 22; Wand. 35: or *depressed by gloomy winter* (?), cf. Ic earmcearig îscealdne sǽ winter wunade wræccan lâstum, 306, 27; Seef. 14) ofer waþema gebind, Exon. Th. 287, 34; Wand. 24.

winter-dæg, es; *m. A winter-day*:—Ðû ðâm winterdagum selest scorte tîda, Bt. 4; Fox 8, 4: Met. 4, 20.

winter-dûn, e; *f. A down or hill on which there is pasturage for sheep during the winter* (?):—On manegum landum tild bið redre ðonne on ôðrum, ge yrðe tîma hrædra, ge mǽda rædran, ge winterdûn (*the sheep can be sent on to the hills earlier* (?), cf. Sunt pascua ouium in meósdûne pertinentia ad Tangmere, Cod. Dip. Kmbl. iii. 373, 23) eác swâ, Anglia ix. 259, 11. [Cf. *Icel.* vetr-beit *winter-pasture;* vetr-hagi *a winter-pasture.*]

winter-feorm, e; *f. A Christmas feast*:—On sumere ðeóde gebyreþ winterfeorm, Eásterfeorm (the Old Latin version translates: In quibusdam locis datur firma Natalis Domini, et firma Paschalis), L. R. S. 21; Th. i. 440, 25. Cf. Eallum æhtemannum gebyreþ midwintres feorm and Eástorfeorm, 9; Th. i. 436, 33.

Winter-fylleþ *the month of October.* Bede, speaking of the months, says: Antiqui Anglorum populi . . . annum totum in duo tempora, hiemis et aestatis dispertiebant, sex menses . . . aestati tribuendo, sex reliquos hiemi; unde et mensem, quo hiemalia tempora incipiebant, Wintirfyllith appellabant, composito nomine ab hieme et plenilunio, quia videlicet a plenilunio ejusdem mensis hiems sortiretur initium . . . Wintirfyllith potest dici compositio novo nomine hiemi plenium. Cf. winter, I :—Se teóða mônð, October, Winterfylleð, swâ hine cîg[a]ð îgbuende, Engle and Seaxe, Menol. Fox 365; Men. 184. Ðone teóðan mônð mon nemneþ on Lêden Octember, and on ûre geðeóde Winterfylleð, Shrn. 136, 31: 143, 32.

winter-gegang, es; *m. What happens as the years pass* :—Winter-gegonge *fato* (cf. wyrde oððe gegonges *fati,* 33, 65), Wrt. Voc. ii. 37, 9.

winter-gerîm, es; *n.* **I.** *numbering by years*:—Ymb þrítig wintergerîmes *after thirty units of a numbering which takes a year as the unit,* i. e. after thirty years (cf. ymb þrittig wintrâ, Bt. 39, 3; Fox 214, 25), Met. 28, 26. **II.** *a number of years*:—Gê ða wintergerîm on gewritu setton, Elen. Kmbl. 1304; El. 654.

winter-getæl, es; *n. A number of years*:—Ða âgân wæs wintergeteles (-tæl-, *v. l.*) seofon and twêntig, Chr. 973; Erl. 124, 22. [*O. Sax.* wintar-gital.]

winter-gewǽde, es; *n. A wintry weed, wintry garment* :—Forst and snâw eorþan þeccaþ wintergewǽdum *frost and snow cover earth with winter's weeds,* Exon. Th. 215, 8; Ph. 250.

winter-geweorp, es; *n. A winter-cast, storm of snow or hail, tempest*:—Nis ðær ne wintergeweorp ne wedra gebregd *non ibi tempestas, nec vis furit horrida venti,* Exon. Th. 201, 16; Ph. 57. Snâw eorðan

band wintergeweorpum, weder côledon heardum hægelscûrum, Andr. Kmbl. 2513; An. 1258.

winter-lǽcan; *p.* lǽhte *To draw near to winter* :—Swâ seó sunne sûðor bið swâ hit swîþor winterlǽcð *the further south the sun is, the nearer are we to winter,* Lchdm. iii. 252, 2. Ða hit winterlǽhte, ða fêrde seó fyrd hâm, Chr. 1006; Th. i. 256, 15.

winter-lîc; *adj. Of winter, winter* :—Winterlîc dæg oððe niht *hiemalis dies vel nox,* Wrt. Voc. i. 53, 30: 76, 64. Se winterlîca wind *the winter wind,* Homl. Skt. ii. 11, 144. Se winterlîca cyle, Lchdm. iii. 252, 3. Winterlîces cyles *hybernalis algoris,* Anglia xiii. 397, 461. Fram heánesse ðære winterlîcan sunnan uppgange *ab alto brumalis exortus,* Bd. 4, 3; S. 567, 42. Tô ðam winterlîcan sunnstede, Lchdm. iii. 250, 24. Hî ongynnaþ heora geár æfter hǽðenum gewunan on winterlicere tîde, 246, 16. Ða winterlîcan *brumalia,* Wrt. Voc. ii. 12, 41. [*O. H. Ger.* wintar-lîh *hiemalis: Icel.* vetr-ligr.]

winter-rǽdingbôc; *f. A lectionary for the winter* :—i. winterrǽding-bôc, Chart. Th. 430, 16. [Cf. *Icel.* vetrar-bôk *a missal for the winter.*] v. rǽding-bôc.

winter-rîm, es; *n. A number of years* :—Heora winterrîm *anni eorum,* Ps. Th. 89, 5.

winter-scûr, es; *m. A winter shower* :—Ne mæg ðǽr wearm weder ne winterscûr wihte gewyrdan, Exon. Th. 198, 31; Ph. 18.

winter-selde, an; *f. A winter-house* :—Winterselde *zetas hyemales* (cf. *zeta* a chambyre, 235, col. 2), Wrt. Voc. i. 57, 48. v. selde, sumer-selde.

winter-set, es; *n. A place to stop in for the winter;* in pl. *winter-quarters* :—Se here . . . ðǽr wintersetu (-sætu, *v. l.*) nâmon, Chr. 886; Th. i. 156, cols. 2, 3. [Cf. *Icel.* vetr-seta *winter-quarters.*]

winter-setl, es; *n. A place to stop in for the winter, winter-quarters* :—Se consul wênde ðæt hê bûton sorge mehte on ðæm wintersetle ge-wunian ðe hê ðâ on wæs, Ors. 4, 8; Swt. 188, 5. Hiê ðǽr sceoldon wintersetl habban, 4, 10; Swt. 200, 11. Hiê wintersetl (-setle, *v. l.*) nâmon on Eást-Englum, Chr. 866; Th. i. 130, cols. 1, 2, 3: 868; Th. i. 132, cols. 1, 2, 3: 886; Th. i. 156, col. 1. Wintersetle, 1009; Th. i. 263, col. 1.

winter-steall, es; *m. A yearling foal* (?) :—Hors mon sceal gyldan mid .xxx. sciłł., myran mid .xx. sciłł., and wintersteal ealswâ, L. O. D. 7; Th. i. 356, 3. [*For similar use of* winter cf. *Icel.* vetr-gemlingr *a sheep a year old.*]

winter-stund, e; *f. A year's space* :—Môste ic âne tîd ûte weorðan, wesan âne winterstunde, Cd. Th. 23, 35; Gen. 370.

winter-sufel, es; *n. Provisions, other than bread, for the winter* :—Ðeówan wîfmen .i. sceáp oððe .iii. p̃. tô wintersufle (*the Old Latin version has* ad hiemale companagium), L. R. S. 9; Th. i. 436, 31. v. sufel.

winter-tîd, e; *f. Winter-time, winter* :—Hit is wintertîd nû, and ic wundrie þearle hwanon þes wyrtbrǽd þus wynsumlîce stême, Homl. Skt. i. 4, 35. Swâ gelîc swâ ðû æt swǽsendum sette mid ðînum ðegnum on wintertîde (*brumali tempore*), and sŷ fŷr onǽled, Bd. 2, 13; S. 516, 16. Ðâs wyrte ðû scealt niman on wintertîde, Lchdm. i. 148, 2. [*O. H. Ger.* wintar-zît *hyemis tempus.* Cf. *Icel.* vetrar-tîð.]

wîn-þegu, e; *f. Wine-taking* (v. þicgan, II), *drinking, feasting* :— Sum bið gewittig æt wînþege, beórhyrde geð, Exon. Th. 297, 27; Crä. 74. Hiê wlenco anwôd æt wînþege, Cd. Th. 217, 4; Dan. 17.

wîn-tiber, -tifer, es; *n. An offering of wine, a libation* :—Wîntifer *libatio,* Wrt. Voc. i. 28, 52.

-wintran. v. ofer-wintran.

-wintre. The form is combined with the cardinals to make adjectives denoting the age of the object to which the adjective is applied. v. e. g. ân-, fîf-, sixtîne-, sixtig-, hundseofontig-, hundtwêgentig-wintre. [*Icel.* -vetra.] v. winter, II.

-wintred. v. ge-wintred.

wîn-tredd (-tredde, an; *f.?* cf. wîn-wringe) *a wine-press, a place where the juice is trodden out of the grapes* :—Wîntreddum *torcularibus,* Hpt. Gl. 468, 31. [Cf. *O. H. Ger.* wîn-trota *torculare.*]

wîn-treów, es; *n. A vine* :—Wîntreów *vitis,* Wrt. Voc. i. 33, 52: 80, 27: Ælfc. Gr. 5; Zup. 14, 10: Ps. Surt. 127, 3. Wîntreó, Jn. Skt. Lind. Rush. 15, 1, 5. Hwylc treów is ealra treówa betst? Wîn-treów, Salm. Kmbl. p. 188, 10. Wæstma ðæs wîntreówes, Mt. Kmbl. Rush. 21, 34. Of ðissum cynne wîntreós (-trées, Lind.) *de hoc genimine vitis,* 26, 29. [Ic stod at a wintre, Gen. and Ex. 2059: C. M. 4465. *Goth.* weina-triu: *Icel.* vîn-trê.] v. next word.

wîntreówig; *adj. Of a vine* :—Wîntreówige *vitea,* Germ. 390, 53.

wintrig; *adj. Wintry, winter* :—Swâ dêþ se ðe wintregum wederum wile blôsman sêcan *numquam purpureum nemus lecturus violas petas, cum saevis aquilonibus stridens campus inhorruit,* Bt. 5, 2; Fox 10, 30. On ðæm wintregum tîdum wyrþ se mûþa fordrifen foran from ðæm norþernum windum *tempestivis auctus incrementis,* Ors. 1, 1; Swt. 12, 34. [*O. H. Ger.* in wintirigia zît *in winter.*]

wîn-trog, es; *m. A wine-press* :—Wîntrog, ðær monn tred ða wîn-begera *torculas,* Mt. Kmbl. Lind. 21, 33.

wín-tunne, an; *f. A wine-cask*:—Ne hē ne drince æt wíntunnum, swā swā woroldmenn dóð, L. Ælfc. C. 30; Th. ii. 354, 4. [*Icel.* vín-tunna.]

wín-twig, es; *n. A vine-twig, shoot of a vine*:—Wíntwiges *palmite*, Wrt. Voc. ii. 89, 41. Wíntwiga plantung *propaginatio*, i. 39, 5.

wín-wringe, an; *f. A wine-press*:—Frymþa wínwringan *dínre primitias torcularis tui*, Scint. 109, 3. Tó wínwringan *ad praelum* (*ad torcular*), Hpt. Gl. 468, 29: Wrt. Voc. ii. 2, 59. Hē sette wínwringan (*torcular*), Mt. Kmbl. 21, 33. Dīne wínwringan *torcularia tua*, Kent. Gl. 35.

wín-wyrcend, es; *m. A vine-dresser*:—Uínwirccendum *vinitoribus*, Mt. Kmbl. p. 19, 3.

wío-, wió-bora, wiodu, wiota. v. weo-, wíg-bora, wudu, wita.

wípian; *p.* ode *To wipe*:—Ic wípige *tergo*, Ælfc. Gr. 26, 3; Zup. 155, 11: 28, 4; Zup. 172, 8. Ic geseó Godes engel standende ætforan ðē mid handcláðe, and wípaþ ðíne swátigan limu, Homl. Th. i. 426, 30. Sum synful wíf his fēt aþwóh and mid hyre fexe wípode, Homl. Ass. 41, 436. Wæterclāðas ðe hý heora handa and fēt mid wípedan, R. Ben. 59, 8. Lege on hunig ðreó niht, nim þonne and wípa ðæt hunig of, Lchdm. iii. 4, 20.

wír *myrtle*:—Uuír, uuýr *myrtus*, Txts. 79, 1356. Wír, Wrt. Voc. i. 285, 51: ii. 55, 83. Ele on ðam ðe wære wír gesoden, Lchdm. ii. 70, 15. Genim wír, 86, 7. v. wír-treów.

wír, es; *m. Wire, metal thread*; *often used apparently in ornamental work, so, an ornament made of wire*. Cf. *Icel.* víra-virki *filigree work*:—Beorht seomað (-ad, MS.) wír ymb ðone wælgim, Exon. Th. 400, 20; Rä. 21, 4. Hæleð gierede mec (*a book*) mid golde, for ðon mē glíwedon wrætlíc weorc smiþa wíre bifongen, 408, 19; Rä. 27, 14. Wíre geweorþad, 484, 9; Rä. 70, 5. Eorðsele wæs innan full wrætta and wíra (*ornaments made of gold or silver wire*), weard unhióre goldmāðmas heóld, Beo. Th. 4817; B. 2413. Næbbe ic fæted gold, ... ne wíra gespann, landes ne locenra beága, Andr. Kmbl. 604; An. 302: Elen. Kmbl. 2267; El. 1135. Wírum gewlenced, 2525; El. 1264: Exon. Th. 402, 19; Rä. 21, 32. Ic eom fægerre frætwum golde, þeáh hit mon āwerge wírum útan, 424, 31; Rä. 41, 47. Wírum bewunden, Beo. Th. 2066; B. 1031. Hygeróf gebond weallwalan wírum wundrum tógædre, Exon. Th. 477, 7; Ruin. 21. [Gold wir, Laym. 7048. Fetislich hir fyngres were fretted with golde wyre, Piers P. 2, 11. *Icel.* vírr.]

wír-boga, an; *m. Bent wire used in ornamenting an object*:—Mec (*a horn*) þeceþ geong hagostealdmon golde and sylfore, wōum wírbogum, Exon. Th. 395, 5; Rä. 15, 3.

wircan, wircness. v. wyrcan, wyrcness.

wird, e; *f. An offence*:—Gehēndon hine ða hēhsacerdas on monigum ðingum ꞇ woerdum *accusabant eum summi sacerdotes in multis*, Mk. Skt. Lind. 15, 3. v. following words.

wirdan; *p.* de *To injure, hurt, annoy*:—Werdit *officit*, Wrt. Voc. ii. 115, 43. Wyrde *officit*, 63, 36. I. *of physical hurt*:—Ne wyrt ðæt ða seón, Lchdm. ii. 26, 14. Ne bēt hē hit, ac wyrt, 212, 20. Ða gnættas mid swiþe lytlum sticelum him deriaþ, and eác ða smalan wyrmas ðone mon werdaþ (wyrdaþ, *v.l.*), and hwílum fulneáh deáðne gedóþ, Bt. 16, 2; Fox 52, 12. Mec unsceafta innan slítaþ, wyrdaþ mec be wombe, Exon. Th. 497, 6; Rä. 85, 25. Ða menigo ðec gedringaþ and woerdaþ (*affligunt*), Lk. Skt. Lind. 8, 45. Sum mon wæs, ðam unwlitig swile his eágan brēgh wyrde and wemde *cui tumor deformis palpebram oculi foedaverat*, Bd. 4, 32; S. 611, 18. Se weolocreáda tælhg, ðone ne mæg ne sunne blēcan, ne ne rēn wyrdan *tinctura coccinei coloris, cujus rubor nullo solis ardore, nulla valet pluviarum injuria pallescere*, Bd. 1, 1; S. 473, 20. Wæron eágan míne mid wæcceum werded swýþe, Ps. Th. 76, 4. II. *to injure, do wrong to, violate a law, hinder*:—Hwæt is ðis manna, ðe mínne folgað wyrdeþ, ýceþ ealdne níð? Elen. Kmbl. 1805; El. 904. Hine teóne (teonode, MS., *with a line below* od) wyrde (wyrgde?) Chus *Chus did him wrong with abusive words*, Ps. Th. 7, arg. Grendel leóde míue wanode and wyrde, Beo. Th. 2678; B. 1337. Hwilcan geþance mæg ænig man ðæt dón, ðæt hē hine on cirican gebidde, and ær oþþon æfter, inne oþþe úte, cirican berýpe, and wyrde oþþe wanige ðæt tó circau gebyrige, L. Eth. vii. 26; Th. i. 334, 31. Gif hwā Cristendóm wyrde, oþþe hǽðendóm weorðige, L. E. G. 2; Th. i. 168, 1. Se ðe ðás laga wyrde ... gif hē hit eft wyrde ... gif hē ... hit þriddan síðe wyrde (ābrece, *v.l.*), L. C. S. 84; Th. i. 428, 4-24, 1. Gif hwā Godes lage oþþe folclage wirde, gebéte hit georne, L. N. P. L. 46; Th. ii. 296, 22. Forbeádende ꞇ woerdende gæfelo tó seallanne *prohibentem tributa dari*, Lk. Skt. Lind. 23, 2. Woerdendra *vitiorum*, Rtl. 37, 9. [Þu ne mahht nohht lufenn God and hatenn menn and werdenn, Orm. 5185. Ne birrþ þe shendenn nani mann, ne weordenn, 6249. Gif anig mann þe sheudeþþ oþerr werdeþþ, 6255. *Goth.* fra-wardjan *to corrupt, disfigure*: *O. Sax.* ā-wardian, -werdian *to spoil, destroy*: *O. H. Ger.* warten *exulcerare*: far-warten *laedere*.] v. ā-, ge-wirdan (-wyrdan).

wirde (?), es; *m. An observer*. v. circol-wirde.

-wirdelsa. v. æf-werdelsa.

wirding, e; *f. Injury, hurt*:—Woerding *lesio*, Rtl. 102. 9. v. ā-wirding.

wirdla, wirdlian. v. ǽ-, æf-werdla (-wyrdla); ge-wyrdlian.

wirdness, e; *f.* I. *injury, hurt, annoyance*:—Mið woerdnisse *affligendo*, Rtl. 16, 13. From woerdnissum *a noxiis*, 17, 15. II. *a vice*:—From sceððendum woerdnisum *a noxiis vitiis*, Rtl. 16, 25. -wíred. v. ge-wíred.

Wíre-múþa, an; *m. Wearmouth*:—On ðære stówe ðe mon hāteþ æt Wíremúðan *juxta ostium fluminis Viuri*, Bd. 4, 18; S. 586, 27: 5, 21; S. 642, 35: Shrn. 50, 30: 61, 14. Æt Wíramúðan *ad Viuraemuda*, Bd. 5, 24; S. 647, 20.

wirgan, wirigan, wirian; *p.* de, ede. I. *to curse; maledicere*:—Ic wyrge *devoto*, Wrt. Voc. i. 28, 79. Ic wyrge (wyrige, *v. l.*) *maledico*, Ælfc. Gr. 37; Zup. 222, 4. Riht ðū dēst, gif ðū ealle ðíne cild wyrigst ... wyrig hí ealle, Homl. Th. ii. 30, 10-14. Se ðe his hwǽte hýt, hiene wiergð ðæt folc (*maledicetur in populis*), Past. 49; Swt. 376, 13. Gif mē mín feónd wyrgeþ (wyrigde, Ps. Spl.) *si inimicus meus maledixisset mihi*, Ps. Th. 54, 11. Se ðe wyrigð (woerges, Lind.: wærge, Rush.) hys fæder, Mt. Kmbl. 15, 4: Homl. Th. ii. 36, 10. Hē ðe on ansýne wyrigð *he will curse thee to thy face*, 448, 33. Se man ðe wirigð Drihtnes naman *qui blasphemaverit nomen Domini*, Lev. 24, 16. Wergiaþ hig and ðū bletsast, Ps. Lamb. 108, 28. Ða ðe hine wyrgeaþ (ða wirgendan, Ps. Lamb.) *maledicentes illum*, Ps. Th. 36, 21. Bletsiaþ ða ðe eów wyrgeaþ (wiriaþ, *v.l.*: ðǽm woergendum, Lind.), Lk. Skt. 6, 28. Wyrigeaþ (wyriaþ, *v.l.*: wærgaþ, Rush.), Mt. Kmbl. 5, 11. Ðæt fíctreów ðe ðū wyrgdyst (wyrigdest, *v.l.*), Mk. Skt. 11, 21. Ðæt ðū míne fýnd wirigdest, Num. 23, 11. Wyrgde *devotaret*, Wrt. Voc. ii. 27, 29: 96, 57. Wirigde *maledixisset*, Lev. 24, 11. Wyrigde, Homl. Th. ii. 326, 15. Gē wergdon ðane ðe eów on wergðe lýsan þóhte, Elen. Kmbl. 588; El. 294. Mid heora heortan hig wergdon (wyrgedan, Ps. Th.: wyrigdon, Ps. Spl.), Ps. Lamb. 61, 4. Unārímedlíca mengeo wyrgdon ðone cásere, Blickl. Homl. 191, 10. Wyrgdan *devotabant*, Wrt. Voc. ii. 26, 48: 80, 53. Ðone hláford ðæs folces ne wyrg (werig, *v.l.*: wirig, Ex. 22, 28) ðū, L. Alf. 37; Th. i. 52, 30. Wyrig God and swelt, Homl. Th. ii. 452, 30. Ðone hláford ðæs folces ne werge ðū, L. Alf. 37; Th. i. 52, 30. Ðæt ðū hig wirige, Num. 23, 27. Se ðe werge (wyrge, wyrie, *v.ll.*), L. Alf. 15; Th. i. 48, 8. Wirige, Gen. 27, 29. Ealle ðe mē wordum wyrigen, Ps. Th. 54, 12. Hē Israhēla folc wiergean (wirgean, Hatt. MS.) wolde, Past. 1; Swt. 256, 17. Ongan hē his selfcs bearn wordum wyrgean, Cd. Th. 96, 13; Gen. 1594. Bletsian and wyrian, Homl. Th. ii. 36, 7: 326, 10. Wergendi *devotaturus*, Wrt. Voc. ii. 105, 78. Wiergende, 89, 9. Wyrgende, 27, 4. He cōme mā wítgiende ðonne wyrgende, Ps. Th. 34, arg. Bið wereged *maledicetur*, Kent. Gl. 382. Ða ðe be gewyrhtum wyrgede wǽron for heora árleásnesse *hi qui merito impietatis suae maledicebantur*, Bd. 4, 26; S. 602, 12. II. *to do evil*:—Nylle ðū onhyrgan ðæt ðū wyrge. For ðam ða ðe wyrgaþ beóþ geteorode *noli aemulari ut maligneris. Quoniam qui malignantur exterminabuntur*, Ps. Spl. 36, 8-9. In wítgum mínum nyllaþ wergan (wirigan, Ps. Spl.: wyrian ꞇ yfel wilnian, Ps. Lamb) *in prosetis meis nolite malignari*, Ps. Surt. 104, 15. [ȝif he his feder werieð, O. E. Homl. i. 109, 27. þe weregede gastes, 239, 9. An wereged gost, þ̄ is þe deuel, Rel. Ant. i. 131, 25. With hair her þai weried, Ps. 61, 5. Ge ne schulen ne warien ne swerien, A. R. 70, 20. Euch waried weoued, Kath. 201: Gen. and Ex. 544. Þai ealle wery þe tyme þat þai war wroght, Pr. C. 7422. Corozaym God weried, 4202. Curse or warie, Wickl. Rom. 12, 14. This sowdanesse, whom I thus blame and warye, Chauc. M. of L. T. 372. Waryyn or cursyñ *imprecor, maledico, execror*, Prompt. Parv. 516, and see note 5. *Goth.* ga-waɼgjan *to condemn*: O. H. Ger. far-wergen *maledicere*. Cf. *O. Sax.* gi-waragean *to punish a criminal*.] v. ā-, ge-wirgan; wirged, wirgend, wirgende.

wirged, es; *m. An accursed being, the devil*:—Cymeþ se wærgad *venit malus*, Mt. Kmbl. Rush. 13, 19. v. preceding word.

wirgedness, e; *f. Cursing*:—Hē lufode wyrgednesse *dilexit maledictionem*, Ps. Spl. 108, 16.

wirgen. v. grund-wyrgen.

wirgend, es; *m.* I. *a curser*. v. wirgan, I:—Wyrgendras, ðæra mūð bið symle mid wyrigunge āfylled, Homl. Skt. i. 17, 42. II. *an evil-doer, a malignant person*. v. wirgan, II:—Míne wergend gehýrde ðín āgen eáre *insurgentes in me malignantes audivit auris tua*, Ps. Th. 91, 10.

wirgende; *adj.* (*ptcpl.*) *Given to cursing*:—Ne ǽnig man ne gewunie, ðæt hē mid yfelum wordum tó wyriende (wyrgende, *v. l.*) weorðe, Wulfst. 70, 18.

wirgness, e; *f. Cursing, a curse*:—Wergnes *devotatio*, Wrt. Voc. i. 29, 1. Wirgnes, ii. 26, 2. Sig seó wirignys ofer mē *in me sit ista maledictio*, Gen. 27, 13. Of wirignysse mūð full is *maledictione os plenum est*, Ps. Spl. 9 second, 8. Ic sette beforan eów bletsunga and wirignissa (*maledictionem*) ... wiriginissa gif gē ne gehírað Drihtnes bebodum, Deut. 11, 26, 28. Wirinysse, 30, 19. Swā nū āwa sceal wesan ðæt gē wærnysse (wærh-?), brynewylm hæbben, nales bletsunga, Exon. Th. 142, 13; Gū. 643. Hē sceal lǽtan his wyrignesse and lufian his gebedu, Wulfst. 239, 19. His mūð býð symle full wyrignessa *cujus os maledictione plenum est*, Ps. Th. 9, 27: 13, 6. Hí ús mid heora wiþer-

wordum onbēnum and wyrinessum ēhtaþ *adversis nos inprecationibus perseguuntur*, Bd. 2, 2 ; S. 504, 4. v. wirgan.

wîr-grǽfe, an; *f. A myrtle-grove :*—Wîrgrǽfen (-an?) *mirteta*, Wrt. Voc. ii. 90, 18 : 57, 5. Cf. þorn-grǽfe.

wirgþu (-o); *indecl.:* wirgþ, e; *f.*   I. *condemnation, curse, punishment :*—Gē wergdon ðane ðe eów of wergðe lӯsan þōhte . . . eów seó wergðu for ðan scedðeþ scyldfullum, Elen. Kmbl. 588–619 ; El. 294–310. Wergðu dreógan *to be damned*, 422 ; El. 211 : 1901 ; El. 952. Werhðo dreógan, Beo. Th. 1182 ; B. 589. Hý grim helle fȳr, gearo tō wîte, seóð, on ðam hî āwo sculon wærgðu dreógan, Exon. Th. 78, 11 ; Cri. 1272. Wergðu wyrcean *to afflict, hurt*, Ps. Th. 108, 17. Ne sceolon gē on mīne wîtegan wergðe settan *in prophetis meis nolite malignari*, 104, 13. Ic hine wergðo on mīne sette *my curse shall be upon him*, Cd. Th. 105, 19 ; Gen. 1755. Is Euan scyld eal forpynded, wærgða āworpen, Exon. Th. 7, 8 ; Cri. 98.   II. *evil, wickedness :*—Ðē firina gehwylc feor ābūgeþ, wærgðo and gewinnes, Exon. Th. 4, 23 ; Cri. 57.   III. *cursing :* maledictio :—Hē hine gegyrede mid wyrgðu *induit se maledictionem*, Ps. Th. 108, 18. [*Goth.* wargiþa *condemnation.*]

wirgung, e ; *f. Cursing, a curse :*—Uae getācnaþ hwîlon wyrigunge (wyriunge, *v. l.*) . . . On wyrigunge : Uae tibi sit wā ðē sî, Ælfc. Gr. 48 ; Zup. 278, 12–16. Wyrgendras, ðæra mūð bið mid wyrigunge (wyriunge, *v. l.*) āfylled, Homl. Skt. i. 17, 43. Hē fordêd his sāwle mid ðære mānfullan wyriunge . . . Ûre tunge is gesceapen tō Godes herungum, nā tō deófollîcum wyriungum, Homl. Th. ii. 36, 3–6. Wyrgunge *maledictionem*, Ps. Lamb. 108, 18. Heó wolde ðone sunu ðe hî getirigde mid wyriungum gebindan, Homl. Th. ii. 30, 6. Tǽlincga oððe wærginga hit getācnaþ, Lchdm. iii. 214, 16. [Ne wrec þu þe mid wussinge ne mid warienge, O. E. Homl. ii. 179, 23. Wariunge, A. R. 200, 28. War-ryynge *malediccio, imprecacio*, Prompt. Parv. 516, and see note.]

wirgung-galere, es ; *m. One whose incantations are curses, a sorcerer :*—Wyrincgalere *Marsum* (the passage is : Marsum, qui virulentas matrices ad sacrae Virginis laesionem incantationum carminibus irritabat, Ald. 70), Hpt. Gl. 519, 46. v. wyrm-galere, -galdere.

wîr-hangra, an ; *m. A meadow where myrtles grow :*—Æt wîrhangran, Cod. Dip. Kmbl. v. 297, 18. Cf. sealh-hangra.

Wîr-healh; *gen.* -heales ; *pl.* -healas ; *m. Wirral, the peninsula between the Dee and the Mersey :*—Fōr se here of Wîrheale (-healan, *v. l.*) in on Norð-Wealas, Chr. 895 ; Th. i. 170, 171. Hié fōron ðæt hié gedydon on ānre wēstre ceastre on Wîrhealm; seó is Lēgaceaster gehāten, 894 ; Th. i. 170, 171.

wirian, wirigness. v. wirgan, wirgness.

wirman; *p.* de *To warm, make warm :*—Ic wyrme mē *calefacio*, Ælfc. Gr. 37 ; Zup. 218, 5. Ic mē wyrme, 222, 1. Ðæt wyrmð and heardaþ ðone magan, Lchdm. ii. 188, 18. Heó mec wǽteþ in wætre, wyrmeþ hwîlum tō fӯre, Exon. Th. 393, 35 ; Rä. 13, 10. Se cyning gestōd æt ðam fӯre and hine wyrmde *rex coepit consistens ad focum calefieri*, Bd. 3, 14 ; S. 540, 34. Hé wyrmde (wærmde, Lind.: wermde, Rush.) hine *calefaciebat se*, Mk. Skt. 14, 54 : Jn. Skt. 18, 25. Ða þeówas wyrmdon (uearmdon, Lind.) hig, for ðam hit wæs ceald, 18, 18. Cnuca mid wîne, and wyrm hit, Lchdm. i. 108, 7. Wyrm tō fӯre, 374, 10. Wirman *fovere*, Wrt. Voc. ii. 33, 34. For ðý hé cwæð þe ðam cōlan wætere, ðæt nān man ne ðorfte hine belādian, ðæt hé fæt næfde, on hwý hé hit wyrman mihte, Homl. Ass. 141, 84. Tō wyrmanne ðone cealdan magan, Lchdm. ii. 188, 22. Heó geseah Petrum wyrmende (wærmigende, Lind.: wermende, Rush., *calefacientem*), Mk. Skt. 14, 67. Mid wyrmendum þingum lācnian, swilc swā pipor is, and ōþra wermenda wyrta, Lchdm. ii. 62, 2–3. [*Goth.* warmjan: *O. Sax.* wermian: *O. H. Ger.* warmen : *Icel.* verma.] v. ge-wirman ; wearmian.

wirming, e ; *f. Warming :*—Se cyning gestōd æt ðam fӯre and hine wyrmde ; and ðā betwih ða wærminge (werminge, M. 196, 27) (*inter calefaciendum*) gemunde hē ðæt word, Bd. 3, 14 ; S. 540, 34.

wirn, e; *f. A hindrance, obstacle, difficulty :*—Gif hē gedyldelîce forbyrð ægðer ge hosp ge edwît, and on ðære wirne þeáh þurhwunaþ and eádmōdlîce bitt, ðæt him man infæres tîðige, sӯ hē underfangen *si veniens perseveraverit pulsans, et inlatas sibi injurias et difficultatem ingressus visus fuerit patienter portare et persistere petitioni sue, annuatur ei ingressus*, R. Ben. 96, 7. Færð ðæt fӯr ofer eall . . . ne nān man næfð ðæra mihta, ðæt ðær ænige wyrne dō *the fire will go everywhere . . . and no one will be able to hinder it*, Wulfst. 138, 7. v. wearn, wirnan.

wirnan; *p.* de.   I. *to refuse, refrain from granting* a prayer, claim, grant, etc., (a) with gen. of what is refused :—Se ðe ne wiernð (wirnð, Hatt. MS.) ðæs wînes his lāre ða mōd mid tō oferdrencanne ðe hiene gehiéran willaþ *vino eloquii auditorum mentem inebriare non desinit*, Past. 49 ; Swt. 380, 6. Cyning ne wyrneþ wordlofes, wîsan mǽneþ mîne for mengo, Exon. Th. 401, 13 ; Rä. 21, 11. Hî swenga ne wyrnaþ, deórra dynta, Salm. Kmbl. 244 ; Sal. 121. Hē swenges ne wyrnde, Byrht. Th. 135, 15 ; By. 118. Ætsōc Goda ðæs feós ǽgiftes, and ðæs landes wyrnde (*he refused to give up the land*), Chart. Th. 201, 30. Myrce ne wyrndon heardes hondplegan, Chr. 937 ; Erl. 112, 24. Se

hlāford ðe ryhtes wyrne, L. Ath. i. 3 ; Th. i. 200, 14. (b) with dat. of person to whom a refusal is given :—Syle ðam ðe ðē bidde, and ðam ðe æt ðē borgian wylle, ne wyrn ðū him (*volenti mutuari a te ne avertaris*) Mt. Kmbl. 5, 42. Biddaþ ðæs ðe riht sié, for ðam hē eów nyle wyrnan, Bt. 42 ; Fox 258, 24. (c) with the constructions of (a) and (b) :—Gif ðū ðam frumgāran brӯde wyrnest, Cd. Th. 161, 4 ; Gen. 2660. Eal hit him wyrþ tō teónan ðæm ðe his Gode wyrneþ, Blickl. Homl. 51, 10. Ðā wyrnde him mann ðera gîsla, Chr. 1048; Erl. 180, 13. Gif hē him ryhtes wyrnde, L. Ath. i. 3 ; Th. i. 200, 19. Hî Mōyse and hys folce ðæs ûtfæreldes wyrndon, Ors. 1, 7 ; Swt. 38, 19. Ne beó ðū swā heard-heort, ðæt ðū him ðînes gōdes wyrne *non obdurabis cor tuum, nec contrahes manum*, Deut. 15, 7. Sele him scearpne wyrtdrenc, wyrne him metes, Lchdm. ii. 46, 25. For hwan ðū woldest ðînre gesihðe mê wyrnan ? Ps. Th. 87, 14.   II. *to prevent, prohibit, keep from*, (a) absolute :—Gif hǽto oþþe meht ne wyrne, lǽt him blōd, Lchdm. ii. 254, 4. (b) with gen. of what is prohibited :—Ðū wāst ðæt ic ne wyrne mînra welera (wirne mîne welora, Cott. MSS.) *labia mea non prohibebo*, Past. 49 ; Swt. 381, 10. (c) with gen. of what is prohibited, and dat. of that to which the prohibition is given :—Se lîchoma getācnaþ ðone engel ðe him tōgênes stent, and him wiernð his unnyttan færelta, Past. 36 ; Swt. 257, 9. Āwierged bið se mann se ðe wirnð (wyrnð, Cott. MSS.) his sweorde blōdes *maledictus, qui prohibet gladium suum a sanguine*, 49 ; Swt. 379, 1. Mē ðæs hyhtplegan wyrneþ se mec on bende legde, Exon. Th. 402, 13 ; Rä. 21, 29. Hié wyrnan þōhton Mōyses māgum leófes sîðes, Cd. Th. 180, 27 ; Exod. 51. (d) with dat. of person prevented, and a clause giving that which is prevented :—Hē ûs ne wyrnþ (wernþ, *v. l.*), ðæt wē yfel dōn, Bt. 41, 4 ; Fox 252, 4. Georne is tō wyrnanne bearneácenum wîfe, ðæt hió āht sealtes ete oððe swētes, Lchdm. ii. 330, 6. (e) with acc. See II b. [ʒif he hit wul auon, ich hit wulle wernen, Laym. 30310. He ne mei uor reouðe wernen hire, A. R. 330, 11. An hwet þ tu ne maht nawt wearnen (wernin, *v. l.*) mid rihte *quod negare jure non potes*, Kath. 769. Ne mai ich mine songes werne, O. and N. 1358. He him wernde his elmesse, Ayenb. 189, 6. He taketh mete, whan men hym werneth, Piers P. 20, 12. He that wol werne a man to light a candel at his lanterne, Chauc. W. of B. T. 330. *O. Sax.* wernian: *O. Frs.* werna.] v. for-wirnan ; warenian, II. 3, and next word.

wirnung, e ; *f. Refusal, denial :*—Be ryhtes wærnunge. Se hlāford ðe ryhtes wyrne, L. Ath. i. 3 ; Th. i. 200, 13.

wirp, wierp, es ; *m. A throw, a blow with a missile :*—Ðā wearð hiere mid ānum wierpe (wyrpe, *v. l.*) ān ribb forod, ðæt hió siþþan mægen ne hæfde hié tō gescildanne, ac raðe ðæs hió wearð ofslagen *hic serpens ad unius saxi ictum cessit, ac mox facile oppressus est*, Ors. 4, 6 ; Swt. 174, 11. v. wyrp.

wirp, e ; *f. A change for the better, recovery* from sickness, *improvement* in circumstances :—Hē tilaþ ðæs gewundedan werpe ðe hē bewitan sceal *vulnerati sui, cui medicamentum adhibet, vitam servat*, Past. 62 ; Swt. 457, 16. Lege on lǽcedōmas ða ðe ût teón ða yfelan wǽtan, ðonne biþ ðær wyrpe wēn (*hope of recovery*), Lchdm. ii. 46, 27. Gē frōfre ne wēnaþ, ðæt gē wræcsîða wyrpe gebîden *ye look not for comfort, that ye may live to see redemption from exile*, Exon. Th. 132, 30 ; Gū. 480. Gē sceolon dreógan deáþ and þӯstro, ðæs wyrpe gebîdaþ (*never will that lot be bettered*), 140, 11 ; Gū. 608. Se mon ne þearf tō ðisse worulde wyrpe gehycgan *man need not look to this life to mend his lot*, 105, 5 ; Gū. 18. Is ðæt bearn cymen tō wyrpe weorcum Ebrēa (*the child is come to alleviate the afflictions of the Hebrews*, S, 9 ; Cri. 67. Se Waldend him (*the blind man*) mæg wyrpe syllan, hǽlo on heáfodgimme (of heofodgimme, MS.), 336, 5 ; Gn. Ex. 43. Se snotera bād hwæþre him Alwalda æfre wille æfter weáspelle wyrpe gefremman (*make his lot better*), Beo. Th. 2635 ; B. 1315. v. next word.

wirpan; *p.* de *To recover :*—Wyrpton hié wērige, wiste genǽgdon mōdige meteþegnas, hyra mægen bēton, Cd. Th. 185, 29 ; Exod. 130. Sōna ic wæs wyrpende and mē sēl wæs statim melius habere incipio, Bd. 5, 3 ; S. 616, 34. Ðā sōna gefelde ic mē b[e]ōtiende and wyrpende (batiende and werpende, Bd. M. 404, 1) *confestim me melius habere sentirem*, 5, 6 ; S. 620, 12. v. ā-, ge-wirpan, -wyrpan, ge-edwyrpan, and preceding word.

wirping. v. ed-wirping.

wirrest. v. wirs, wirsa.

wîr-rind, e ; *f. Myrtle-bark :*—Tō hāligre sealfe sceal wyirrind, Lchdm. iii. 24, 3. Nim wîrrinde, ii. 98, 8 : 332, 8 : iii. 14, 2.

wirs; *cpve.:* wirrest, wirst ; *spve.; adv. Worse, worst*, (1) in reference to moral ill :—Wyrs dêð se ðe lȳhð, Salm. Kmbl. 364 ; Sal. 181. Ðonne hié wēnen ðæt hié hæbben betst gedōn, ðæt wē him ðonne secgen ðæt hié hæbben wierst (wyrst, Cott. MSS.) gedōn *cum ea, quae bene egisse se credant, male acta monstramus*, Past. 32 ; Swt. 209, 17. (2) marking an inferior degree of what is desirable or proper :—Ðæt hié wiers ne dōn ðonne him man bebeóde *ne minus, quae jubentur, impleant*, Past. 28 ; Swt. 189, 18. Ðӯ lǽs hira lufu āslacige, and hē him ðe wirs lîcige, Past. 19 ; Swt. 143, 10. Se æfterra anweald git wyrs lîcode ðonne se ǽrra, Bt. 16, 2 ; Fox 50, 13. Ic mîn fulluht wyrs geheóld ðonne ic

behête, L. Edg. C. 9; Th. ii. 264, 8: L. Ath. iv. proem.; Th. i. 220, 2. Hé ðý wyrs meahte þolian ða þráge, Met. 1, 76. Se arcebiscop wênde ðæt hit sum ôðer mann âbiddan wolde, ðe hê his wyrs trûwude and ûðe, Chr. 1043; Erl. 169, 28. Oft hit gesæleþ ðæt his æhta weorþaþ on ðæs onwealde, ðe hê ær on his lîfe wyrrest ûþe, Blickl. Homl. 195, 4. (3) marking unfavourable condition, a higher degree of what is unpleasant or improper:—Ðý læs him ðý wirs (wiers, Cott. MSS.) sié, gif hié ða trumnesse ðære Godes giefe him tô unnyte gehweorfaþ, Past. 36; Swt. 247, 7. Hî wyrs gefêrdan (gefêrdon mâran hearm and yfel, v. ll.), ðonne hî æfre wêndan, Chr. 994; Th. i. 241, col. 2. Eôw wyrs gelomp, Exon. Th. 142, 1; Gú. 637. Ne wæs hyra ænigum ðý wyrs, ne sîde ðý sárra, 394, 19; Rä. 14, 5. Hê bið on ðæt wynstre weorud wyrs gesceáden ðonne hê on ða swîþran hond swîcan môte *he will be assigned to the host on the left hand by a sentence too stern to allow him to pass to the right hand*, 449, 23; Dôm. 75. Hit ðê wyrs ne mæg hreôwan ðonne hit mê dêð *you cannot repent it more bitterly than I do*, Cd. Th. 51, 12; Gen. 825. Heora weóldan ða him wyrrest ær on feóndscipe gestôdon, Ps. Th. 105, 30. Wyrst, Met. 24, 60. [*Goth.* wairs: *O. Sax.* wirs: *O. H. Ger.* wirs: *Icel.* verr; *cpve.*; verst; *spve.*]

wirsa (wirra *occurs once in the Chronicle*); *cpve.*; wirrest, wirst; *spve.* adj. *Worse, worst*, (1) in a moral sense:—For hwam lifaþ se wyrsa leng? Salm. Kmbl. 716; Sal. 357. Ne weard nân wærsa dæd gedôn ðonne ðeós wæs, Chr. 979; Erl. 129, 4. Gif wê ðæt ne dôþ, ðonne wyrce wê ûs myccle synne; and ûs is wyrse ðæt wê ûrne ceáp teóþian, gif wê willaþ syllan ðæt wyrste Gode, Blickl. Homl. 41, 7. Hî for nânum ermþum ne byóð nô ðý betran, ac ðý wyrsan, Bt. 39, 11; Fox 230, 17. Ða gæð hê and him tô genymð seofun ôðre gástas wyrsan (*nequiores*) ðonne hê . . . and wurðaþ ðæs mannes ýtemestan wyrsan (*pejora*) ðonne ða ærran, Mt. Kmbl. 12, 45: Wrt. Voc. ii. 72, 59. Ðes wyrresta cyning Neron, Homl. Th. i. 384, 3. Se wyresta sceaþa (*Judas*), Blickl. Homl. 69, 10. Ðis is manna se wyrresta, 185, 2. Se eallra wyrresta mon, Bt. 14, 3; Fox 46, 20. Dê þûhte ðæt eallra ðinga wyrrest, 38, 4; Fox 204, 9. Ðæs wyrrestan eorðcyninga, Cd. Th. 235, 13; Dan. 305. On werrestre dæde *in actione pessima*, Confess. Peccat. Swâ byð ðisse wyrrestan (wyrsesta, Lind., *pessimae*) cneórysse, Mt. Kmbl. 12, 45. Wirestan, Deut. 1, 35. Ða wyrstan (*pessimam*) ingewitnesse mê ic geseó, Bd. 5, 13; S. 632, 32. Ða wyrrestan, fâ folcsceaðan, Andr. Kmbl. 3183; An. 1594. Ðâm wyrrestum wîtes þegnum, Exon. Th. 251, 28; Jul. 152. On werstum ðingum *in rebus pessimis*, Kent. Gl. 23. (1 a) of an unfitting condition of things:—And ðæt git wyrse is, ðæt wê witon manige foremære weras forþgewitene ðe swîþe feáwa manna â ongit, Bt. 19; Fox 70, 11: Met. 10, 57. (2) of the physical condition of persons or things:—Hê tôbrycð hys stede on ðam reáfe, and se slite byð ðe wyrsa, Mt. Kmbl. 9, 16. Sió wund bið ðæs ðe wierse, Past. 17; Swt. 123, 18. Heó wæs ðe wyrse *deterius habebat*, Mk. Skt. 5, 26. Seó frecednes dæghwamlîce wæs wyrse and wyrse, Bd. 4, 32; S. 611, 24. (2 a) where injury is done to a person in respect to his well-being:—Se ðe ôðerne mid wô forsecgan wille, ðæt hê áðer oþþe feó oþþe sreme ðâ wyrse sý, L. C. S. 16; Th. i. 384, 24. Hî dydan mycelne hearm âbûtan Hâmtûne . . . swâ ðæt seó scîr and ða ôðra scîra, ðæ ðær neáh sindon, wurdon fela wintra ðe wyrsan, Chr. 1065; Erl. 197, 11. (3) of the condition of affairs, of an (unfavourable) circumstance or event:—Mê ðær wyrse gelamp, ðonne ic tô hyhte âgan môste, Cd. Th. 275, 22; Sat. 175. Hit him wyrse gelomp, 272, 26; Sat. 125. Wæs æfre heora æftra sýð wyrse ðonne se ærra, Chr. 1001; Erl. 137, 14. Swâ weard hit fram dæge tô dæge lætre and wyrre, 1066; Erl. 202, 17. Ne weard wyrse dæd (*more disastrous act*) monnum gemearcod, Cd. Th. 37, 24; Gen. 594. Hê âwende hit him tô wyrsan þinge, 17, 13; Gen. 259. Hê tæhte Absalone ôðerne ræd wyrsan tô his willan, Homl. Skt. i. 19, 206. Wêne ic tô ðê wyrsan þinga, gif ðû Grendles dearst bîdan, Beo. Th. 1055; B. 525. (4) of that which is harmful, painful, etc.:—Hî næfre wyrsan handplegan on Angelcynne ne gemitton *they never met with harder fighting in England*, Chr. 1004; Erl. 138, note 7. Ðý læs God ûs sende on wyrsan tintrego, Blickl. Homl. 243, 20. Ðæra synfullena deáþ byð se wyrsta (wyrresta, Ps. Surt.: wyrst, Ps. Spl., *pessima*), Ps. Th. 33, 21. Wilddeóra ðæt wyrreste (grimmeste, Exon. Th. 371, 29) . . . wyrmcynna ðæt grimmeste (wyrreste, Exon. Th. 371, 32), Soul Kmbl. 164-167; Seel. 82-84. Se deófol slôh Iôb mid ðære wyrstan wunde (*with the most grievous disorder*), Homl. Th. ii. 452, 26. Mid ðý werrestan âttre *with the most virulent poison*, Shrn. 84, 28. On ðone wyrrestan deáð *to the most cruel death*, Andr. Kmbl. 172; An. 86. We[r]stum gedrecenyssum *saevissimis afflictionibus*, Hpt. Gl. 409, 59. Getogen tô ðæm wyrstan tintregum, Blickl. Homl. 245, 1. Ða werrestan tintrega, 229, 28. Wyrrestan, Exon. Th. 257, 20; Jul. 250: Elen. Kmbl. 1860; El. 932. (5) marking inferiority:—Hê bið swîðe gelîc sumum ðara gumena ðe him þringaþ ymbe ûtan; gif hê wyrsa ne bið, ne wêne ic his nâ beteran, Met 25, 20. Ælc man sylþ ærest gôd wîn, and ðonne hig druncene beóð ðæt ðe wyrse (wyrrest, Rush.: wurresta, Lind., *deterius*) byð, Jn. Skt. 2, 10. Hê ðæt betere geceás, and ðam wyrsan wiðsôc, Elen. Kmbl. 2078; El. 1040. On ðone wyrsan dæl scyrede, Exon. Th. 75, 24; Cri. 1226. On ða wyrsan hand, Salm. Kmbl. 998;

Sal. 500. Onwendan heora wuldor on ðæne wyrsan hâd styrces, Ps. Th. 105, 17. Wyrsan wîgfrecan, Beo. Th. 4985; B. 2496. Buccena flæsc is wyrrest, Lchdm. ii. 196, 17. Gif wê willaþ syllan ûre ðæt wyrste Gode, Blickl. Homl. 41, 8. [*Goth.* wairsiza; *cpve.*: *O. Sax.* wirsa; *cpve.*: wirsista; *spve.*: *O. Frs.* wirra; *cpve.*: *O. H. Ger.* wirsiro; *cpve.*; wirsisto; *spve.*: *Icel.* verri; *cpve.*; verstr; *spve.*] v. weorr.

wirsian; *p.* ode *To get worse:*—Hit fareþ yfele ealles tô wîde. Swâ swýðe hit wyrsaþ, ðæt ðæs hâdes men, ðe hwýlum wæron nyttoste, ða syndon nû unnyttaste, L. I. P. 14; Th. ii. 322, 18. And aa hit wyrsode mid mannan swîðor and swîðor, Chr. 1085; Erl. 219, 23. Wyrsadon *deterioraverunt*, Wrt. Voc. ii. 139, 37. Folclaga wyrsedan ealles tô swýðe, Wulfst. 158, 6. Hê sceolde beón âscyred fram manna neáwiste, gif his hreófla wyrsigende wære, Homl. Th. i. 124, 26. [Þet his licome, ðe feble wes, ne sceolde noht wursien, O. E. Homl. i. 47, 26. Þe wunde þet euer wursed, A. R. 326, 23. Þenne wursede (wersede, 2nd MS.) ich on crafte, Laym. 18931. Werihede þet makeþ þane man worsi, Ayenb. 33, 18.]

wirs-lîc; adj. *Mean, vile:*—Ðysse worulde wela is wyrslîc and yfellîc and forwordenlîc, Wulfst. 263, 13. Ic eom wyrslîcre ðonne ðes wudu fûla, oðòe ðis warod, ðe hêr âworpen ligeþ on eorþan, Exon. Th. 424, 32; Rä. 41, 48.

wirþig; adj. *Worthy, fitting:*—Wyrþigre wrace h'é forwurdon ðâ, ðæt ðâ heora synna sceoldon hreówsian and dædbôte dôn, swiþor ðonne heora plegan begân, Ors. 6, 2; Swt. 256, 11. [Wurrþi to winnenn Cristess are, Orm. 2705. Wurði wurðed to ben, Gen. and Ex. 1012. Wurði to hauen same, Misc. 14, 447. *O. Sax.* wirðig: *O. H. Ger.* wirdig *dignus, meritus*: *Icel.* verðugr.]

wirþu; *indecl.*: wirþ, e; f. *Honour, decoration, dignity:*—Uyrðo *infula*, Wrt. Voc. ii. 111, 75. Cf. weorþ-mynd, III. [*O. H. Ger.* wirdî *dignitas, infula.*] v. or-wirþu.

wir-treów, es; n. *A myrtle-tree:*—Wîrtreów *myrtus*, Wrt. Voc. ii. 55, 83. Cnuca mid rosan wôse oðòe wýrtreówes, Lchdm. i. 232, 12. **wir-treówen**, -trîwen; adj. *Of a myrtle-tree, myrtle:*—þweah mid wearmum wýrtrýwenum (-treówenum, *v. l.*) wôse, Lchdm. i. 236, 1.

wís *a manner.* v wîse.

wís; adj. I. *wise, discreet, judicious:*—Wîs *sapiens*, Wrt. Voc. i. 76, 10: *fronimus*, 47, 34. (1) of persons:—Ne scyle nân wîs monn (*vir sapiens*) forhtigan, Bt. 40, 3; Fox 238, 8, 13, 15. Ne mæg weorþan wîs wer ær hê âge wintra dæl in woruldrîce, Exon. Th. 290, 12; Wand. 64. Ðû eart gleáw and scearp, wîs on ðînum gewitte and on ðînum worde snottor, 463, 30; Hö. 78. Cyninges ræswa, wîs and wordgleáw, Cd. Th. 242, 12; Dan. 418. Ne hýrde ic snotorlîcor guman þingian. Ðû eart wîs wordcwida, Beo. Th. 3694; B. 1845. Azarias Dryhten herede, wîs in weorcum, Exon. Th. 185, 7; Az. 4. Se wîsa mon eall his lîf læt on gefeân, ðonne hê forsihþ ðis eorþlican gôd, Bt. 12; Fox 36, 24. Se wîsa spræc sunu Healfdenes, Beo. Th. 3401; B. 1698. Ðis is wæstm wîses and goodes, ðe his sôðfæst weorc symble læste *est fructus justo*, Ps. Th. 57, 10. Gelîc ðam wîsan were (*viro sapienti*), Mt. Kmbl. 7, 24. Se cyning him ceóse sumne wîsne man and glæwne (*virum sapientem et industrium*), Gen. 41, 33. Hié sædon ðæt hié wæren wîese (wîse, Cott. MSS.), and ðâ wurdon hié dysige, Past. 11; Swt. 71, 2: 30; Swt. 203, 10. Wîn nys drenc cilda ne dysigra, ac ealdra and wîsra, Coll. Monast. Th. 35, 21: Ps. Th. 106, 42. Mæg ic wîsran findan, ðonne ðû eart? Gen. 41, 39; Andr. Kmbl. 197; An. 474. Swelce hî sién micle wærran and wîsran, Past. 35; Swt. 245, 1. Swelc eówer swelce him selfum ðynce ðæt te wîsasð sié on ðæm lotwrencum, weorðe ðæs ærest dysig, ðæt hê mæge ðonan weorðan wîs, 30; Swt. 203, 20. Mid his ealdormannum, ða ðe hé wîseste and snotereste wiste, hé gelômlîce ðeahtade, Bd. 2, 9; S. 512, 10. (2) of animals:—Sió wilde beó ðeáh wîs sié, Met. 18, 5. Wîsran *sapientiora* (v. Prov. 30, 24), Kent. Gl. 1101. (3) of things:—Worde and gewitte, wîse þance, Cd. Th. 118, 1; Gen. 1958. Wîsne wordcwide, 249, 28; Dan. 537. Ðâm ðe hafaþ wîsne geþôht, Exon. Th. 57, 22; Cri. 922: 150, 2; Gú. 772. Wîsne geleáfan, Ps. Th. 77, 36. On wîsne weg worda ðînra, 118, 32. þurh wîs gewit, Exon. Th. 73, 21; Cri. 1193. Ealle mîne wegas wîse syndan on ðînre gesihðe, Ps. Th. 118, 168. Ðonne hê ðîne wîsan word gehealde, 118, 9. I a. *in a bad sense, cunning:*—Wille gê wesan prættige? Wê nellaþ swâ wesan wîse, Coll. Monast. Th. 33, 1. Hî ân geþeaht ealle ymbsæton, and gewitnesse wið ðê wîse gesettan (*adversum te testamentum disposuerunt*), Ps. Th. 82, 5. II. *wise, learned, skilled, expert:*—Wîs *sophus* vel *sophista*, Wrt. Voc. i. 47, 40. Se wîsa *gnarus*, ii. 40, 31. Hond bið gelæred, wîs and gewealden sele âsettan, Exon. Th. 296, 4; Crä. 46. Sum bið meares gleáw, wic(g)cræfta wîs, 297, 18; Crä. 70. Wordcræftes wîs *an able speaker*, Elen. Kmbl. 1180; El. 592. Wîs sâwle rædes, Frag. Kmbl. 79; Leás. 41. Se wîs oncneów (*he, being a skilful man, knew*) ðæt hê Marmedonia mægðe hæfde gesôhte, Andr. Kmbl. 1686; An. 845: Ps. Th. 106, 16. Ðû mê gewurde wîs on hælu *factus es mihi in salutem*, 117, 20, 21, 27. Ðæs wîsan goldsmiðes bân Wêlondes, Bt. 19; Fox 70, 1. Micel is tô hycgenne wîsum wôðboran, hwæt sió wiht sié, Exon. Th. 414, 22; Rä. 32, 24. Wîse men *learned men*, Cd. Th. 201, 24; Exod. 377. Hê

feára sum gengde wîsra monna wong sceáwian, Beo. Th. 2830; B. 1413. Geceós wîse men (*viros potentes*), Ex. 18, 21. Unrihtwîse habbaþ on hospe ða ðe him sindon rihtes wîsran, Met. 4, 45. Eorðcyninga se wîsesta (*Solomon*), Cd. Th. 202, 24; Exod. 393. Tómiddes ðara wietena ðe wîsoste wæron *in medio doctorum*, Past. 49; Swt. 385, 22. Ða wîsestan, ða ðe snyttro cræft þurh fyrngewrit gefrigen hæfdon, Elen. Kmbl. 306; El. 153: 337; El. 169: 645; El. 323. Sum from æs wîsistum *quidam ex legis peritis*, Lk. Skt. Lind. Rush. 11, 45, 46. Hê sende tô Egipta wîsustan witun, Gen. 41, 8.   **III.** *known* :—Dô mê wegas ðine wîse *vias tuas notas fae mihi*, Btwk. 208, 6: Ps. Ben. 24, 3: Ps. Th. 102, 7. [*Goth.* weis: *O. Frs.* O. *Sax.* O. *H. Ger.* wîs: *Icel.* vîss.] v. and-, brægd-, fore-, gesceád-, getæl-, med-, riht-, sâm-, sundor-, un-, unriht-, weoruld-, wrenc-wîs. (*Some of these compounds may be connected with* wîse.)

wis *certain.* v. wiss.

wisa :—Ân wisa (wihta? cf. ða wrætlîcan wiht, 505; Sal. 253) is on woruldrîce, ymb ða mê fyrwet bræc L wintra, Salm. Kmbl. 491; Sal. 246.

wîsa, an; *m. A leader, director, captain* :—Wæs Cainan æfter Enose aldordêma, weard and wîsa, Cd. Th. 70, 22; Gen. 1157. Ðu eart eallum eorðbûendum weard and wîsa, 251, 19; Dan. 566. Enoch ealdordôm âhôf, folces wîsa, 73, 2; Gen. 1198. Leóda aldor, herges wîsa, freom folctoga, 178, 18; Exod. 13: 228, 16; Dan. 203. Mægenes wîsa, 26, 2; Dan. 703. Elamitarna ordes wîsa, 121, 3; Gen. 2004. Rîces wîsa, werodes wîsa, 194, 9; Exod. 258: Beo. Th. 523; B. 259: Exon. Th. 296, 22; Crä. 55. Þeóda wîsan, 196, 9; Az. 171. Weorces wîsan, Cd. Th. 101, 28; Gen. 1689. [*O. Sax.* balu-wîso (*the devil*): *O. H. Ger.* wîso *dux* : *Icel.* vîsi (*poet.*) *a guide, leader, captain.*] v. brim-, camp-, cræt-, ealdor-, fyrd-, heáfod-, here-, hilde, mægen-, scrid-wîsa.

wîsan; *p.* de *To shew* :—Ðeóden wîsðe herepað tô ðære heán byrig eorlum elðeódigum, Cd. Th. 218, 5; Dan. 35. [*O. Sax.* wîsian; *p.* wîsda *to shew* (he im te heðanrîkea thena weg wîsit, Hel. 1872): *O. H. Ger.* wîsen; *p.* wîsta: *Icel.* vîsa; *p.* vîsti.] v. gin-wîsed, wîsian.

wîs-bôc, e; *f. A book in which the state of things is described, a record* :—Eágan ðine gesâwon ðæt ic wæs unfrom on ferhþe; eall ðæt forþ heonan on dînum wîsbôcum âwriten standeþ *imperfectum meum viderunt oculi tui, et in libro tuo omnes scribentur*, Ps. Th. 138, 14.

wîse *a marsh* (?) :—Concedo terram in loco qui dicitur Fearnleág (*Farleigh, in Kent, by the Medway*) & an myclan wisce vi. æceres mæde (*and in the big marsh vi. acres of meadow* (?)), Cod. Dip. Kmbl. ii. 128, 33. Cf. Wiscleágeat, v. 179, 34. [Cf. (?) *O. H. Ger.* Wisicha (*place-name*).]

wîsan. v. wŷscan.

wîschere (?), es; *m. A wizard* :—Mannum is tô witenne ðæt manega drŷmen maciaþ menigfealde dydrunga þurh deófles cræft, swâ swâ wischeras dôð, and bedydriaþ menn, swylce hî sôðlîce swylc þincg dôn, Homl. Skt. i. 21, 466.

wîs-dôm, es; *m.*   **I.** *wisdom, discretion* :—Wîsdôm (*sapientia*) ys gerihtwîsud fram heora bearnum, Mt. Kmbl. 11, 19: Lk. Skt. 11, 49. Wæs his môdsefa manegum gecŷðed, wîg and wîsdôm, Beo. Th. 705; B. 350. Ðæt hê ða yldestan lærde ðæt heó wîsdômes word oncneówan *ut senes prudentiam doceret*, Ps. Th. 104, 18. Iosue weard gefilled mid wîsdômes gâste (*spiritu sapientiae*), Deut. 34, 9: Exon. Th. 273, 15; Jul. 516. Gleáwhydig, wîsdômes ful, Elen. Kmbl. 1875; El. 939. Hê wîsdômes beþearf, worda wærlicra, 1082; El. 543. Hié nâhton foreþances, wîsdômes gewitt, 713; El. 357: Andr. Kmbl. 1289; An. 645. Wê willaþ wesan wîse. On hwilcon wîsdôme (*sapientia*)? Wê willaþ beón bylewite, and wîse, ðæt wê bûgon fram yfele and dôn gôda, Coll. Monast. Th. 32, 27. Hê wîsdôme heóld êðel sînne, Beo. Th. 3923; B. 1959. Ic healde ðînra worda waru mid wîsdôme, Ps. Th. 118, 17. Hine God þurh his worda wîsdôm âhôf, 104, 15. Hê sette on hî sôðne wîsdôm worda and weorca, 104, 23. Ðæt se sâwle weard lîfes wîsdôm forloren hæbbe, se ðe nû ne giémeþ hwæþer his gæst sié earm þe eádig, Exon. Th. 95, 4; Cri. 1552. Ýwaþ wîsdôm weras, wlencu forleósaþ, 132, 17; Gû. 474.   **II.** *knowledge, cognizance* :—Gif ceorl bûton wîfes wîsdôme deóflum gelde, L. Wih. 12; Th. i. 40, 4.   **III.** *wisdom, knowledge, learning, philosophy* :—Swilc is se wîsdôm ðæt hine ne mæg nân mon ongitan swilcne swilce hê is . . . Ac se wîsdôm mæg ûs ongitan swilce swilce wê sind . . . for ðæm se wîsdôm is God. Hê gesihþ eall ûre wyrc, Bt. 41, 4; Fox 250, 24. Hê lærde hig, swâ ðæt hig cwædon: 'Hwanon ys ðysum ðes wîsdôm?' Mt. Kmbl. 13, 54: Mk. Skt. 6, 2: Andr. Kmbl. 1137; An. 569: Elen. Th. 169, 33; Gû. 1104. Rûmran geþeaht wîsdôm onwreáh . . . mê lâre onlâg mægencyning, Elen. Kmbl. 2483; El. 1243. Wundorlîc is geworden ðîn wîsdôm (*scientia tua*), Ps. Th. 138, 4. Ðû mê lær wîsdômes word *scientiam doce me*, 118, 66. Sefa deóp gewôd, wîsdômes gewitt, Elen. Kmbl. 2379; El. 1191. Wîsdômes gife, 1189; El. 596: Exon. Th. 178, 1; Gû. 1220. Wîsdômes *philosophiae*, Wrt. Voc. ii. 66, 28. Wæs se wer in wîsdôme (*scientia*) gewrita wel geláred, Bd. 5, 8; S. 621, 33. Hê wîsdôm hâligra gewrita

from him nom, 4, 27; S. 603, 40. *Philosophus* is se ðe lufaþ wîsdôm: of ðam is *philosophor* ic smeáge embe wîsdôm, Ælfc. Gr. 36; Zup. 215, 6–8. Ða scearpþanclan witan ðe ðone twydæledan wîsdôm tôcnâwaþ, ðæt is andweardra þinga and gâstlicra wîsdôm, Lchdm. iii. 440, 29. Tô gehŷranne Salomones wîsdôm, Mt. Kmbl. 12, 42: Andr. Kmbl. 1299; An. 650: Elen. Kmbl. 667; El. 334. Gif hê hafaþ ofer ealle men wîsdôm, Exon. Th. 299, 16; Crä. 103. Wîsdôm swelgan, 147, 31; Gû. 735. Wîsdôm cŷpan, 500, 19; Rä. 89, 9. Wîsdôm onwreón, Elen. Kmbl. 1344; El. 674. Se ðe men læreþ micelne wîsdôm *qui docet hominem scientiam*, Ps. Th. 93, 10. Swâ ûs gleáwe wîtgan þurh wîsdôm on gewritum cŷþaþ, Exon. Th. 199, 23; Ph. 30. Ða mîne þeówas sindon wîsdômas and cræftas (*sciences and arts*), Bt. 7, 3; Fox 20, 33. ¶ throughout the Boethius, in which Philosophy personified is a speaker, the word used in the translation is wîsdôm. [*O. Sax.* O. *Frs.* wîs-dôm: *O. H. Ger.* wîs-tuom: *Icel.* vîs-dômr.] v. un-, weoruld-wîsdôm.

wîse, an; *f.*   **I.** *a wise, way, manner, mode, fashion* :—Hit is ælces môdes wîse, ðæt sôna swâ hit forlæt sôþcwidas, swâ folgaþ hit leásspellunga *eam mentium constat esse naturam, ut quoties abjecerint veras falsis opinionibus induantur*, Bt. 5, 3; Fox 14, 15. Maniges mannes wîse bið, ðæt hê wile tô his nêhstan sprecan ða word ðe hê wênþ ðæt him leófoste sŷn tô gehŷrenne . . .; deófles wîse bið, ðæt hê ðone unwaran man beswîcan mæge, Blickl. Homl. 55, 19–23: Exon. Th. 362, 5; Wal. 32: 315, 12; Môd. 30: 489, 20; Rä. 78, 10: 419, 4; Rä. 37, 14. Seó wîse (*that manner of treatment*) hine hæleþ, Lchdm. i. 328, 21. Hié hine lîchomlîce gesâwon, and him æfter eorþlîcre wîsan hŷrdon, Blickl. Homl. 135, 20. Ðæt Léden and ðæt Englisc nabbaþ âne wîsan on ðære spræce fadunge. Æfre se ðe âwent of Lédene on Englisc, æfre hê sceal gefadian hit swâ ðæt ðæt Englisc hæbbe his âgene wîsan, elles hit biþ swîþe gedwolsum tô rædenne ðam ðe ðæs Lédenes wîsan ne can, Ælfc. Gen. Thw. 4, 7–11. On ða ylcan wîsan (*juxta quem ratum*) nymaþ ticcenu, Ex. 12, 5: Ps. Th. 30, arg. Ðû gesettest ælcere þeóde þeáw and wîsan, Hy. 7, 22. Ic healde mîne wîsan, Exon. Th. 390, 19; Rä. 9, 4: 401, 14; Rä. 21, 11: 483, 12; Rä. 69, 1. Ðû hafast ofer witena dôm wîsan gefongen *thou hast taken a course opposed to the judgment of understanding men*, 248, 20; Jul. 98. Biscopum gebiraþ ealdlîce wîsan, L. I. P. 10; Th. ii. 318, 29. Gif hê ne cunne his dæda andettan, âcsa hine his wîsena, L. de Cf. 3; Th. ii. 260, 21. Ic ðînra ne mæg worda ne wîsna (*words or ways*; or v. **III**) wuht oncnâwan, sîðes ne sagona, Cd. Th. 34, 7; Gen. 534. Is ðes middangeard missenlîcum wîsum gewlitegad, Exon. Th. 413, 7; Rä. 32, 2. Mîn gebed him on wîsum is wel lŷcendlîce *est oratio mea in beneplacitis eorum*, Ps. Th. 140, 8. Wîsum clæne, Exon. Th. 312, 16; Seef. 110. Se his godcundnesse mid sôþum wîsum gerŷmeþ, Blickl. Homl. 179, 24. Sendon hié Amilchor, ðæt hê Alexandres wîsan besceáwode (*ad perscrutandus Alexandri actus*), Ors. 4, 5; Swt. 168, 13. Hê sorgaþ ymb ôðerra monna wîsan *actiones alienas curans*, Past. 53; Swt. 415, 20. Nân nyste ôþres wîsan oþþe dæda, Homl. Skt. ii. 23 b, 133. Ðæt wê forlætan ða wîsan ðe wê langere tîde mid ealle Angelðeóde heóldan, Bd. 1, 25; S. 487, 10. Hê forlêt ða wæpna and ða woruldlîcan wîsan, Shrn. 61, 16. Ealle ûre wîsan rædlîce fadian, Wulfst. 143, 22: L. I. P. 10; Th. ii. 318, 12. On feala wîsan (*multis modis*) ic beswîce fugelas, Coll. Monast. Th. 25, 11. ¶ in adverbial phrases as in other-*wise* :—Mid suman gemete † wîsan *quodammodo*, Hpt. Gl. 435, 59. Ðâs cŷþnesse Drihten nam of ðisse wîsan, Blickl. Homl. 31, 16. Ne dyde hê ða wîsan (so) beforan mê, 181, 4. On ælce wîsan 163, 2. On ænige wîsan, Wulfst. 158, 1: L. C. S. 5; Th. i. 378, 22. On ænige ôðre wîsan *aliter*, Wrt. Voc. ii. 2, 56. On nâne wîsan, Bt. 16, 2; Fox 54, 5. On ðâs word ic becom, ðe læs de ôðre wîsan ænig man leóge, Blickl. Homl. 177, 33. On ôðre wîsan hit ys *aliter est*, Gen. 42, 12: Bd. 1, 27; S. 492, 3, 6. Hit feor on ôðre wîsan wæs *longe aliter erat*, 3, 14; B. 539, 42. On ôðre wîsan *secus*, Wrt. Voc. ii. 81, 74. Wê ongîtaþ mon on ôðre wîsan hine God ongit, Bt. 39, 10; Fox 226, 29. *Bifariam* on twâ wîsan, *omnifariam* on ælce wîsan, *multifarie* on manega wîsan, Ælfc. Gr. 38; Zup. 237, 14–17. *Meapte* on mîne wîsan, *tuapte* on ðîne wîsan, *nostrapte* on ûre wîsan, 16; Zup. 107, 17. On hire wîsan *suatim, suo more*, Hpt. Gl. 435, 21. On ða betstan wîsan ðâ ðe dêmest, Blickl. Homl. 189, 35. Ealde wîsan *as of old*, Beo. Th. 3735; B. 1865.   **Ia.** the word is found with strong forms :—Onwendan mîne wîse (wîsan, Th.), Exon. Th. 485, 29; Rä. 72, 5.   **II.** *state, condition* :—Ðonne hié ðenceaþ hû hié selfe scylen fullfremodeste weorðan, and ne giémaþ tô hwon ôðerra monna wîse weorðe *cum sua et non aliorum lucra cogitant*, Past. 5; Swt. 41, 24. Ðæt hié oncnâwen tô hwæm hiera âgen wîse wirð *ut ad cognitionem sui revocentur*, 37; Swt. 265, 24. Ne scyle nân wîs monn gnornian tô hwæm his wîse weorþe, oððe hwæþer him cume þe rêþu wyrd þe lîþu *vir sapiens moleste ferre non debet, quoties in fortunae certamen adducitur*, Bt. 40, 3; Fox 238, 8. Heora wîse on nænne sæl wel ne geôn, nâþer ne innan fram him selfum, ne ûtane fram ôþrum folcum *nulla unquam tempora vel foris prospera vel domi quieta duxerunt*, Ors. 4, 4; Swt. 164, 13: L. I. P. 7; Th. ii. 312, 28. Ðonne ðê ðîn wîse lîcie *cum bene tibi fuerit*, Gen. 40, 14. Hæte hym man bæþ

swâ hraþe swâ hys wíse gódige, Lchdm. iii. 122, 8. Gehýre hû his wíse gerâd sî, L. de Cf. 2; Th. ii. 260, 17. **III.** *an arrangement, instruction, a disposition, direction, condition* :—Worda mê dînra wíse onleóhteþ *declaratio sermonum tuorum illuminat ne*, Ps. Th. 118, 130. Ðæra manna naman ðe ðeosse wísan (*a will*) geweotan sindon, Chart. Th. 483, 36. Hebfað hiá ðâs wísan ðûs fundene, 465, 26: 473, 22. Ðæt is tô þafianne on ða wísan, ðæt man gíslas sylle, L. A. G. 5; Th. i. 156, 4. Ða wísan âbeád weoroda ealdor : ' Nú sié geworden gefeá,' Exon. Th. 14, 34; Cri. 229. Wísna fela, lâre longsume, wítgena wôðsong, 3, 28; Cri. 43. **IV.** *a thing* ; res, negotium :—Seó wíse wæs mîne on twâ healfa unêþe *quae res dupliciter me torsit*, Nar. 9, 23 : 10, 32 : Blickl. Homl. 33, 5. Ðær seó wíse on tweón cyme *ubi res perveniret in dubium*, Bd. 1, 1; S. 474, 21. Gelimp wísan *eventum rei*, Hpt. Gl. 457. 45. On ðysse wísan *hac in re*, Bd. 1, 27; S. 490. 9. Be ðære wísan ðe mín môd gedréfed hæfþ, Bt. 39, 4; Fox 216, 11. Ðá hê hæfde ða wísan onfangene *suscepto negotio*, Bd. 4, 24; S. 597, 36. Hwanon hê ða wísan (*rem*) cûþe, 4, 25; S. 600, 39: Exon. Th. 20, 11; Cri. 316: Elen. Kmbl. 1365; El. 684. Ne sette ic mê fore eágum yfele wísan (*rem malam*), Ps. Th. 100, 3. Secgan ymb sume wísan, Salm. Kmbl. 852; Sal. 425. Ne syndon tô lufianne ða wísan fore stôwum, ac for gôdum wísum stôwe syndon tô lufianne *non pro locis res, sed pro bonis rebus loca amanda sunt*, Bd. 1, 27; S. 489, 41. Wísena Sceppend alra *rerum Creator omnium*, Ps. Surt. ii. p. 202, 28. Swâ on ðam ende ðara wísena ætýwed is *sicut rerum exitus probavit*, B. 1, 14; S. 482, 42. Feala ðû ætýwdest folce ðínum heardra wísan (wísna ?) *ostendisti populo tuo dura*, Ps. Th. 59, 3. Hû hê his wísna trûwade on ðære dimman âdle *how he expected matters would be with him in his illness*, Exon. Th. 171, 30; Gû. 1134. Ðes biscop is swíðe mihtig on frêcnum wísum gescyldnesse tô biddanne, Shrn. 70, 9. Hê ne conn ôðre lêran ða godcundan wísan ðe hê lêran scolde *interna, quae alios docere debuerat, ignorent*, Past. 18; Swt. 129, 3. Wê oft ymb ungedafenlíce wísan smeágeaþ, Swt. 139, 22. Hê hæfde his wísan beþôht tô Seleucuse *he had entrusted his affairs to Seleucus*, Ors. 3, 11; Swt. 150, 16. **IV a.** *a cause, reason* ; res :—For ðære wísan (*pro qua re*) hê wæs heáfde becorfen, Bd. 1, 27; S. 491, 18. For ðære wísan *quare*, 4, 5; S. 583, 32 : *quamobrem*, 4, 18; S. 587, 3. For ðisse wísan *pro hac re*, 1, 27; S. 491, 27. Be ðisse wísan *hinc*, S. 496, 12. Of hwylcere wísan hit gegange *ex qua re accidat*, S. 496, 35. For hwylcre wísan côme ðú tô mê synfulre, Homl. Skt. ii. 23 b, 249. [*O. Sax.* wísa (*wk.* and *str.*) : *O. Frs.* wís : *O. H. Ger.* wîsa (*wk.* and *str.*) *modus, mos, consuetudo, usus, ratio, modulatus* : *Icel.* vísa *a stanza* ; ôðru-vís (-vísa (-u, -i)) *otherwise.*] v. cniht-, cyne-, fyrd-, hring-, hyse-, leóþ-, mann-, munuc-, mynster-, riht- (?), sceáwend-, scip-, tungolcræft-, unriht- (?), weorold-, wilig-wíse.

wíse, an; *f. A sprout, stalk* :—Streáwbergean wíse, Lchdm. ii. 36, 12 : 334, 11. Genim streáwberian wísan nioþowearde, 34, 24, 27. Nim hwíteclæfran wísan, 326, 21. Hæþbergean wísan, 344, 10. Weóde wísan, iii. 16, 16. Eallwíte wýsan *gesie*, Wrt. Voc. ii. 42, 16. [Wyse of strawbery or pesyn *fragus* (cf. a streberytre *fragus*, Wülck. Gl. 584, 29), Prompt. Parv. 531. Take the wyse of tormentile, and bray it, Halliwell's Dict. Cf. *Icel.* vísir *a sprout.*] v. streáwberige-wíse.

wíse; *adv. Wisely, with wisdom* :—Ðú worhtest wíse hælu, Ps. Th. 73, 12 (cf. 117, 20). Ða ðe wyllaþ his gewitnesse wíse smeágan *qui scrutantur testimonia ejus*, 118, 2 : 36, 79. Ic ðé wegas mîne wíse secge *vias meas enuntiavi tibi*, 118, 26. Ic wegas ðíne wíse þence tô fêrenne *cogitavi vias tuas*, 118, 59. Gemune ðínes môdes, ða miclan geniht ðínre wêðnesse wíse sæcgenum roccette, and ræd sprece *memoriam abundantiae suavitatis tuae eructabunt*, 144, 7.

wísere, es; *m. A sign-post* (?) :—Tô Afene ; on wísere ; on ða fûlan lace, Cod. Dip. Kmbl. iii. 301, 36.

wís-fæst; *adj.* I. *wise, discreet, judicious*, (1) of persons :—Hió grētte Geáta leód, Gode þancode, wísfæst wordum, Beo. Th. 1256; B. 626. Wísfæstne wer, wordes gleáwne, Andr. Kmbl. 3294; An. 1650. Is nú þearf micel, ðæt wê wísfæstra wordum hýran, 2335; An. 1169. (2) of things :—Ðæt heó his wísfæst word efnan *ut faciant mandata ejus*, Ps. Th. 102, 17. II. *wise, having knowledge* or *skill, learned* :—Esaias, wísfæst wítga, Exon. Th. 19, 25; Cri. 306. Ðis ys se dæg, ðe hine Drihten ûs wísfæst geworhte, Ps. Th. 117, 22 : Menol. Fox 122; Men. 61. Micel is tô hycganne wísfæstum menn, hwæt seó wiht sý, Exon. Th. 411, 15; Rä. 29, 13. Swâ wítgan wísfæste sægdon, 5, 3 ; Cri. 64. Sume bôceras weorþaþ wísfæste, 331, 22; Vy. 72. Ðæs ðe wísfæste weras on gewritum cýþan, 356, 19 ; Pa. 14 : Elen. Kmbl. 627 ; El. 314. Ðæt is tô geþencanne wísfæstum werum, hwæt seó wiht sý, Exon. Th. 429, 5; Rä. 42, 9. **II a.** *intelligent, rational* (?) :—Hê wile on dômes dæg on ðysne middangeard cuman, and hê eallum wísfæstum gesceaftum ecn[e] dôm gesetton (*he will pass an eternal sentence on all intelligent creatures*), Blickl. Homl. 121, 20. v. next word.

wís-fæst (v. wíse, and cf. þeáw-fæst) ; *adj. Perfect* :—Gif ðú wilt wísfæst (*perfectus*) wosa, Mt. Kmbl. Lind. Rush. 19, 21. Wísfæst êghwelc bið *perfectus omnis erit*, Lk. Skt. Lind. 6, 40. Folc wísfæst *plebem perfectum*, 1, 17. Wísfæsto (*perfecti*) wossað gié, Rtl. 13, 19. [Perhaps these passages might be put under I of preceding word.]

wísfæst-líc; *adj. Wise* :—Hê him wísfæstlíc word onsende, þurh ðæt hî hrædlíce hælde wæron *misit verbum suum et sanavit eos*, Ps. Th. 106, 19.

wís-hycgende *thinking wisely, having wise thoughts* :—Hê wíshycgende gesæt on sesse, seah on enta geweorc, Beo. Th. 5426; B. 2716.

wís-hygdig; *adj. Wise-minded* :—Him ðâ wíshýdig Abraham gewât, Cd. Th. 109, 2 ; Gen. 1816. Ongan his brýd wíshýdig wer wordum lêran, 109, 15 ; Gen. 1823 : 123, 29; Gen. 2053 : 136, 8; Gen. 2255.

wísian; *p.* ode. **I.** where movement takes place, *to shew the way, guide, direct,* (1) absolute :—Hê stôp on stræte, stíg wísode, Andr. Kmbl. 1970; An. 987. Hê lêt his francan wadan þurh ðæs hysses hals, hand wísode, Byrht. Th. 135, 61 ; By. 141. Snyredon ðær secg wísode, Beo. Th. 810; B. 402. Hê hêt him fýrenne beám beforan wísian, Ps. Th. 104, 34. (2) with dat. :—Ic eów wísige, Beo. Th. 590 ; B. 292 : 6198; B. 3103. Ic fêre swâ mê wísaþ feónd, Exon. Th. 403, 4 ; Rä. 22, 2. Hê fêrde swâ him God wísode, Gen. 35, 5 : Num. 10, 28. Ísernhergum ân wísode, Cd. Th. 199, 34; Exod. 348 : Ps. Th. 77, 16. Stíg wísode gumum ætgædere, Beo. Th. 646; B. 320. Se ðæm headorincum hider wísade, 746; B. 370. Him seleþegn forð wísade, 3595; B. 1795. Ðæt heáfod sceal wísian ðæm fôtum, ðæt hié stæppen on ryhtne weg, Past. 18 ; Swt. 131, 24. (3) with dat. of person and acc. of way :—Hwâ ðam sæflotan sund wísode, Andr. Kmbl. 762; An. 381. Hû ðû sæhengeste sund wísige, 976 ; An. 488. (4) with acc. of person :—Swâ mec wísaþ, se mec wrǽde on legde, Exon. Th. 383, 19 ; Rä. 4, 13. (5) with acc. of that to which the way is shewn, *to shew the way to, shew, point out* :—Secg wísade, lagucræftig mon, landgemyrcu, Beo. Th. 422; B. 208. Hê sceolde wong wísian (*act as guide to the place*), 4809; B. 2409. **II.** figurative, (1) absolute, *to shew the course to be followed, guide, direct, indicate* :—Ic Werferð cýðe, swâ mê Alchûn sægde, and eác mîne gewrytu wísodon, Chart. Th. 166, 6. Eorðcyningas ðe folcum fore wísien, Ps. Th. 148, 11. (2) with dat. :—Swâ ic ðé wísie, Cd. Th. 35, 32; Gen. 563. Se ðe him hâlig gæst wísaþ, Exon. Th. 124, 1 ; Gû. 333. Swâ mê wísaþ tô ríce, 401, 2 ; Rä. 21, 5. Hê wíf gefette, swâ hyne his hláford hêt and him God wísode, Gen. 24, 15 : Beo. Th. 3331; B. 1663. Him se eorl wísade (*compulit illos*, Gen. 19, 3), Cd. Th. 147, 24 ; Gen. 2444. Him se Dryhtnes dôm wísade tô ðam nýhstan nýdgedâle, Exon. Th. 129, 3; Gû. 415. Úre Drihten beád Môyse ðam heretogan, ðæt hê folce wísode (folc wissode, *v. l.*), Wulfst. 132, 11. Ðus him gewísede se mon ða gemǽru, swâ him ða ealdan bêc ryhtan and wísedon, Chart. Th. 142, 15 : 141, 18. Hwæt mæg ic dôn, bûton mê God wísige? Gen. 41, 16. Swâ him ryht wísie, L. Alf. pol. 1 ; Th. i. 60, 20 : 3; Th. i. 62, 9. Se consul sceolde him eallum wísian and beón heora yldost tô ânes geáres fyrste, Jud. Thw. p. 161, 23. (3) with acc. :—Ðæt wê ǽgðer ge ûs sylfe, ge ða ðe wê wísian sceolan, swâ gewísian môtan, swâ swâ ûre ealra þearf sý, L. I. P. 21 ; Th. ii. 332, 24. (4) with clause stating what is pointed out :—Hié lêton tân wísian hwylcne hira ǽrest ôðrum sceolde tô fôddurþege feores ongildan, Andr. Kmbl. 2200; An. 1101. (5) with dat. of person and acc. (or clause) of what is pointed out :—Hálgan heápe hlýt wísode ðær hié Dryhtnes ǽ dêman sceoldon, Apstls. Kmbl. 18 ; Ap. 9. Mê ða treahteras tala wísedon on ðam micelan bêc, Salm. Kmbl. 10 ; Sal. 5. [Heȝe Diana, wíse mi, Laym. 1200. Hwi nultu wisi heom hu engles singeþ, O. and N. 915. Thut lond wel to wise, R. Glouc. 524, 8. *Q. Sax.* wísian : *O. Frs.* wísa : *O. H. Ger.* wísen *monstrare, ducere, regere, docere* : *Icel.* vísa.] v. ge-, riht-wísian ; wíssian.

Wísle, an ; *f. The Vistula* :—Weonodland wæs ûs ealne weg on steorbord ôð Wíslemûðan. Seó Wísle is swýðe mycel eá, and hió tôlíð Witland and Weonodland ; and seó Wísle líð ût of Weonodlande, and líð in Estmere . . . Ðonne cymeþ Ilfing eástan, and Wísle sûðan, and benimð Wísle Ilfing hire naman . . . ; for ðý hit man hæt Wíslemûða, Ors. 1, 1 ; Swt. 20, 6-13.

Wísle-land, es ; *n. The land in which the Vistula rises, part of Poland* :—Be eástan Maroara londe is Wíslelond, Ors. 1, 1; Swt. 16, 17.

Wísle-mûþa, an ; *m. The mouth of the Vistula.* v. Wísle.

wís-líc; *adj. Wise, discreet, prudent, sagacious* :—Mê ðynceþ wíslíc, gif ðú geseó ða þing beteran, ðæt wê ðam onfôn, Bd. 2, 13 ; S. 516, 10. Is wíslíc ræd, ðæt manna gehwylc geornlíce smeáge, Wulfst. 4, 21. Wíslíc wærscipe, L. I. P. 10 ; Th. ii. 318, 37. Ðín mildheortnes wíslíc standeþ, deórust and gedéfust, Ps. Th. 102, 16. Mid wíslícum geðylde, Homl. Th. ii. 222, 21. Hê him wíslíce andsware sende *ille ei prudens responsum misit*, L. Ecg. P. iii. 14; Th. ii. 200, 20. Ðú ǽghwylces canst worda wíslíc andgit, Andr. Kmbl. 1018; An. 509. Wera gehwylcum wíslícu word gerísaþ, Exon. Th. 343, 34; Gen. Ex. 166. Ongan se biscop lustfullian ðæs iungan snyttro and his wíslícra worda *delectabatur antistes prudentia verborum juvenis*, Bd. 5, 19 ; S. 637, 47. Drihten wordum wíslícum herian, Ps. Th. 65, 1. Ræd forð gǽð, hafaþ wíslícu

word on fæðme, Cd. Th. 211, 14; Exod. 526. Wíslícu wundur oncnáwan, Ps. Th. 87, 11. Swá dême hé swá him wíslícost þince *judicet pro ut ipsi prudentissimum videbitur*, L. Ecg. C. 32; Th. ii. 156, 20. [*O. Sax.* wís-lík: *O. H. Ger.* wís-lîh *sagax, urbanus.*] v. un-wíslíc.

wis-líc *certain.* v. wiss-líc.

wíslíce; *adv.* I. *wisely, sagaciously, with wisdom, prudently:—* Sapienter wíslíce . . . *sapienter loquor* wíslíce ic sprece, Ælfc. Gr. 38; Zup. 223, 15: Past. 15; Swt. 93, 24: Homl. Th. i. 236, 8: Ps. Th. 46, 7. Hé him wíslíce (*sapienter*) andwyrde, Mk. Skt. 12, 34. Wíslíce spyrian, Bt. 18; Fox 60, 27. Beþencan heora dǽda wíslíce and wærlíce, L. I. P. 10; Th. ii. 318, 35: Chr. 1067; Erl. 204, 34: Blickl. Homl. 97, 2. Wíslíce gé dyde, 201, 1: Homl. Skt. i. 5, 42: Exon. Th. 348, 2; Sch. 22: Ps. Th. 77, 12. Hé wíslíce rǽdde for Gode and for worulde his þeóde, Chr. 959; Erl. 119, 26. Hit ða téð getrymeþ, gif his man wíslíce brúceþ, Lchdm. i. 334, 10. Bið nú wíslícor ðæt gehwá ðis wite, Homl. Th. i. 6, 18. II. *wisely, skilfully, cunningly:—* Se wolcnreáda wæfels wíslíce getácnode úres Drihtnes deáð mid ðære deáge híwe, Homl. Th. ii. 254, 5. Hé Adam funde, wíslíce geworht, and his wíf, Cd. Th. 29, 26; Gen. 456: Ps. Th. 138, 13. Ða wíslíce áwriten standaþ, 101, 16. Ðú unstilla gesceafta wíslíce ástyrest, Met. 20, 15. Daniel sægde him wíslíce wereda gesceafte ðætte sóna ongeat cyning, Cd. Th. 225, 26; Dan. 160. [*O. Sax.* wísliko; *O. H. Ger.* wíslícho *sapienter, mature, sophistice.*] v. un-wíslíce.

wisness. v. un-wísness.

wisnian, weosnian; *p.* ode *To wizen, dry up:—*Wisnaþ (-eþ, Lind.) *aruit*, Jn. Skt. Rush. 15, 6. Ðá wisnode hé on Cristes háligra heortum, and is nú on úrum heortum blówende, Blickl. Homl. 115, 13. Weosniendre *arida*, Wrt. Voc. ii. 3, 53. [*O. H. Ger.* wesanên *arescere, marcescere: Icel.* visna *to wither.*] v. á-, for-wisnian.

wiss; *adj. Certain:—*Ðeáh ðe hé wís (gewiss, M. 412, 5) geworden wære ðurh ða ætýwnesse ðære gesihðe *tametsi certus est factus de visione,* Bd. 5, 19; S. 623, 15. ¶ *in the phrase* tó wissum :—Tó wissum *profecto, omnino,* Hpt. Gl. 431, 15. Tó wissan *praesertim, maxime, saltim,* 416, 41. Wite gé tó wissan ðæt se deófol ne mæg mannum derian bútan Drihtnes geþafunge, Homl. Skt. i. 17, 174. [He seʒʒde him to wisse whillc ende he shollde sekenn, Orm. 8460. þ wite þu to wisse, Kath. 1532. Hi wenden to wisse of here lif misse, Horn 121. He is here fader mid wisse, O. E. Homl. ii. 25, 23. *O. Sax. O. Frs.* wiss: *Icel.* viss. Cf. *Goth.* du unwisamma *in incertum.*] v. ge-wiss.

wisse (?); *adv. Certainly:—*Sculan wé wrecan wordum forð, wisse gesingan, ðæt . . ., Menol. Fox 140; Men. 70. [As wis ase . . . ase wis . . . *as certainly* . . . *so certainly* . . ., O. E. Homl. i. 187, 36. Alse wis alse . . ., A. R. 38, 8. Ʒho wass wiss allre manne mast off lufe filledd, Orm. 2597. Þatt wass wiss to soþe þe maste þing, 2866. *O. H. Ger.* wisso *profecto.*] v. ge-wisse.

wís-sefa, an; *m. A wise-minded person:—*Him mæg wíssefa wyrda gehwylce gemetigan, gif hé bið módes gleáw, Salm. Kmbl. 877; Sal. 438.

wissian; *p.* ode. I. *to shew a way* (acc.) *to a person* (dat.):—Ðæt ðú nyme ðé ládmenn, ðæt ðé wegas wissigeon, Gen. 33, 15. II. *fig. to shew the way, guide, direct, rule,* (1) *absolutely* (see also (2), (3)):—Gif swá gesceád wissaþ *si ita ratio dictaverit,* Anglia xiii. 443, 1116. Ða ðe him betǽhte sindon tó wissianne, Wulfst. 108, 16. Wissiendum *gubernante,* Hpt. Gl. 453, 39. (2) *with dat.* (or uncertain):—*Rego* ic wissige, of ðam cymð *rex* cyning, ðe rihtlíce wissaþ his folce, Ælfc. Gr. 28, 5; Zup. 173, 6. Ða ðe heora synna bétaþ swá swá hym man wissaþ, Wulfst. 104, 14. Hé ðé wissaþ, Gen. 24, 7. Hé wítegode swá him wissode God, Num. 33, 8. *Rex* cyning is gecweden *a regendo* . . ., for ðan ðe se cyning sceal mid micelum wísdóme his leóde wissian, Ælfc. Gr. 50, 18; Zup. 293, 9. Cyning sceal wissigan mid wísdóme his folce, O. E. Homl. i. 302, 28. On ðæra (ðære, MS.) gewitnysse, ðe ðú wissian scealt on ðissere gelaðunge, Ælfc. T. Grn. 17, 39. (3) *with acc.* :—Ælces mannes weorc cýðaþ hwilc gást hine wissaþ. Godes gást wissaþ tó hálignesse; deófles gást wissaþ tó leahtrum, Homl. Th. i. 324, 27. Úre Drihten beád Móyse ðam heretogan, ðæt hé folc wissode, Wulfst. 132, 11. Weard ðæt mæden hohful, hú heó ǽfre wæras wissian sceolde, Homl. Skt. i. 2, 122. Hú mæg úre gegaderungc búton geþeahtynde beón wissod (*regi*)? Coll. Monast. Th. 30, 9. III. *to declare, make known:—*Se cræft sceolde wissian gewisslíce þe steorrum hwæt gehwilcum menn gelumpe on his lífes endebyrdnysse, Homl. Skt. i. 5, 253. [Ure Drihten cweð to Moyses þet he scolde wissien his folc, O. E. Homl. i. 13, 15. Witen þat lond and wissien þa leoden, Laym. 5280. Antenor ʒam ladde, wissede and radde, 1365. To wissenn himm, Orm. 10823. Crist, that kan wisse and rede, Havel. 104. Crist þe wisse, Horn 1457. Coudestow wissen us þe weye? Piers P. 5, 540. Wyssyñ or ledyñ *dirigo,* Prompt. Parv. 530. Cf. *O. H. Ger.* wissen.] v. ge-, mis-wissian; wísian, *and next word.*

wissigend, es; *m.* I. *a director, guider* of that which moves:—Cræt and his wissigend *currus et auriga ejus,* Homl. Skt. i. 18, 295. II. *a director, ruler* :—Wissiend *gubernator, rector* (*ecclesiae*), Hpt. Gl. 459, 54: Gesceafta Sceppend and wissigend. (*rector*) úre,

---

Hymn. Surt. 20, 25. *Rex* wé cwæþaþ cyning, ðæt is gecweden wissigend, O. E. Homl. i. 302, 27. Þwyrlíce færð æt ðam húse ðær seó wyln bið ðære hláfdian wissigend, Homl. Skt. i. 17, 11.

wiss-líc; *adj. Certain:—*Ne heora wítes bið wislíc trymnes *nec est firmamentum in plaga eorum,* Ps. Th. 72, 3. Dryhten eorle monegum áre gesceáwaþ, wislícne blǽd, sumum weána dǽl, Exon. Th. 379, 16; Deór. 34. v. un-gewislíc, *and next word.*

wisslíce; *adv. Certainly:—*Hí wisslíce witon *scient,* Ps. Th. 58, 13. Wislíce, 99, 2. [Wenndenn þeʒʒ þatt he wisslike wære Crist, Orm. 10330. He falleþþ wissliʒ for þatt gillt, 928. Alse wislíche alse hie þis dai was hoven into hevene, Rel. Ant. i. 130, 37: Kath. 185. Wislike for soth, Havel. 274. I wot wislike, Will. 2947. Also wisly God my soule blesse, Chauc. C. T. B. 2112.] v. ge-wislíc.

wissung, e; *f.* I. *shewing of the way, guidance, direction:—*Hwænne ðú eáðelícost miht tó ðam folce becuman be mínre wissunge, Homl. Ass. 110, 259. II. *fig. direction, instruction, teaching:—*Hleótan man mót mid geleáfan, gif hí hwæt dǽlan willaþ; ðis bið wissung, Homl. Skt. i. 17, 87. Hé mót lǽtan hí lybban be heora bóca wissunge and heora gástlican ealdres tǽcunge, Homl. Th. ii. 594, 2. Hí (*the apostles*) ða láre on bócum áwriton be Godes ágenre wissunge, L. Ælfc. P. 20; Th. ii. 370, 29: Homl. Ass. 20, 156. Hí heóldon Godes ǽ æfter Móyses wissunge, 101, 319. Hí ðurhwunedon swá þurh his wissunge, 30, 149. Þurh góde wissunge, Wulfst. 32, 13: Homl. Th. ii. 482, 1: Homl. Skt. i. 3, 104. Hí wurðan swýðe bliþe ðurh swilce wissunge, Chr. 995; Th. i. 244, 23. III. *rule, government, direction* of one in authority:—Wissung *regimen,* Hpt. Gl. 412, 69. Wissunge *regimine,* 453, 49. Wæs wuniende Israël on friðe feówertig wintra be Gedeones wissunge *quievit terra per quadraginta annos, quibus Gedeon praefuit,* Jud. 8, 28. Under abbodes wissunge, Homl. Ass. 39, 382. Hí leofodon be heora ágenum dihte, be nánes ealdres wissunge, 44, 502. Ne gedyrstlǽce nán lǽwede man ðæt hé wissunge oððe ealdordóm healde ofer Godes ðeówum, Homl. Th. ii. 592, 25. Dathan and Abiron forsáwon Móyses wissunge, Homl. Skt. i. 13, 224 note. [Hit wes iloked bi Godes wissunge, þet mon scule childre fulhten, O. E. Homl. i. 73, 29. Hiss wissing and his lare, Orm. 11830. Al þe world is iwald þurh his wissunge, Kath. 187.] v. ge-wissung.

wist, e; *f.* (and *m.? v.* big-, dæg-, hús-, neáh-wist.) I. *being.* v. æt-, ed-, gador-, gegador-, hús-, los-, mid-, neáh-, on-, sam-, stede-wist. II. *subsistence* :—Wist *vel* anleofa *stips,* Wrt. Voc. i. 17, 8. Wiste *stipis,* Anglia xiii. 36, 248. II a. *sustenance, food, provisions* :—Næs ðǽr hláfes wist, ne wæteres tó brúcanne; ah hié blód and fel þegon, Andr. Kmbl. 42; An. 21. Hé næfþ ða neódþearfe áne, ðæt is wist and wǽda, Bt. 33, 2; Fox 124, 17. Of ungemete wiste and wǽda, Met. 25, 39: Cd. Th. 222, 11; Dan. 103. Welan and wiste, 59, 29; Gen. 971. Hé smeáde hwǽr hí bigleofan biddan sceoldon, ðá ðá hí ða fare térdon búton wiste, Homl. Th. ii. 138, 34: Cd. Th. 185, 30; Exod. 130. Gif feohbót áríseþ, ðæt gebyreþ tó wǽde and tó wiste ðam ðe Gode þeówian, L. Eth. vi. 51; Th. i. 328, 7. Tódǽlan werum tó wiste fǽges flǽschoman, Andr. Kmbl. 305; An. 153: Menol. Fox 388; Men. 195: Soul Kmbl. 49; Seel. 25. Genóh wære ðam wǽdlan his untrumnys, þeáh ðe hé wiste hæfde, Homl. Th. i. 330, 16. Mon tó andleofne eorðan wæstmas hám gelǽdeþ, wiste wynsume, Exon. Th. 214, 26; Ph. 245: Cd. Th. 81, 4; Gen. 1340. Nafast ðú hláfes wiste, ne hlútterne drync, Andr. Kmbl. 623; An. 312: Elen. Kmbl. 1231; El. 617. Forlǽt eal ðæt ðú áge bútan wiste and wǽda, Prov. Kmbl. 80. Næbbe ic welan ne wiste, Andr. Kmbl. 603; An. 302: 635; An. 318. Hé áfédde of fixum twám and of fíf hláfum fira cynnes fíf þúsendu; wiste þegon menn, 1186; An. 593. Waldend ðé wist gife, heofonlícne hláf, 776; An. 388. Hunig, wynsume wist, Frag. Kmbl. 40; Leás. 22. Fóddurwelan, wiste, Exon. Th. 415, 14; Rä. 33, 11. Sylle him mon wist and wǽdo, 336, 12; Gn. Ex. 48. Wistum gehladen, 492, 16; Rä. 81, 16. Mid wistum þénian *to serve with food,* Homl. Skt. i. 7, 390. Ic welan and wista gife eów genóge, Wulfst. 132, 15. III. *dainty food, a feast.* v. wistfullian:—Deós wist *epulum,* Ælfc. Gr. 13; Zup. 86, 6. Wist *epulae,* keninga wist *vel* éstas *dapes,* Wrt. Voc. i. 41, 12, 13. Wiste wlonc and wínes sæd, Exon. Th. 369, 10; Seel. 39. Ðonne ðú dést wist oððe feorme *cum facis prandium aut caenam,* Lk. Skt. 14, 12. Æt hám findaþ witode him wiste and blisse, Exon. Th. 430, 14; Rä. 44, 8. Wista *dapes,* Wrt. Voc. i. 26, 63. Hwǽr beóþ ðonne his welan and his wista? Blickl. Homl. 111, 33. Wista *epularum,* Hpt. Gl. 481, 15: Exon. Th. 130, 6; Gú. 434. Gebytlu mid wistum áfyllede, Homl. Th. i. 68, 3: 74, 27. In wistum mínum *in delitiis meis,* Ps. Surt. 138, 11. Wystu *delicias, epulas,* Hpt. Gl. 480, 76. Wista *delicias,* Wrt. Voc. ii. 28, 69. Wiste *epulas,* Kent. Gl. 787. Hié hæfdon wiste and plegan, Blickl. Homl. 99, 21. IV. *eating, feasting* :—Nelle ðú grǽdig beón on ealre wiste (*epulatione*), Scint. 169, 17. Hí on druncennysse and on wiste hiora wombe þeówiaþ, L. E. I. 45; Th. ii. 442, 1. Wunaþ hé on wiste, Beo. Th. 3474; B. 1735. Hine his goldwine wenede tó wiste, Exon. Th. 288, 24; Wand. 36. Hé ǽlce dæge symblede and mid micelre wiste wǽre gefeormod, Past. 45; Swt. 337, 24. [*Goth.* wists; *f. natura: O. Sax.* wist; *m. food: O. H. Ger.*

wist; *f. substantia; alimentum, stipendium: Icel.* vist; *f. abode; food.*]
v. and-, big-, dæg-, ofer-wist (*for other compounds see* I).

wist-full; *adj. Abounding in food, productive:*—Ðis wæs swiðe gód geár and swiðe wistfull on wudan and on feldan, Chr. 1112; Erl. 243, 38.

wistfullian; *p.* ode *To feast:*—Ic wistfullige *epulor,* Ælfc. Gr. 25; Zup. 146, 1. Tíma is ðæt ðú mid ðínum gebróðrum wistfullige on mínum gebeórscipe, Homl. Th. i. 74, 15. Utan wistfullian *epulemur,* Wrt. Voc. ii. 143, 62. Se apostol tǽhte ðæt wé sceoldon wistfullian ná on yfelnysse beorman, ac on þeorfnyssum sýfernysse (*epulemur, non in fermento malitiae, sed in azymis sinceritatis,* I Cor. 5, 8), Homl. Th. ii. 278, 24. v. ge-wistfullian.

wistfullíce; *adv. Sumptuously:*—Wistfullíce *sumptuosius* (si tu te sumptuosius comas, Ald. 75), Wrt. Voc. ii. 87, 24.

wistfulli[g]end, es; *m. One that feasts:*—Swég wistfulgend[es] *sonus epulantis,* Ps. Spl. 41, 5.

wistfullness, e; *f. Luxury in eating:*—His wistfullnys him wyrðeþ tó biternysse, Basil admn. 8; Norm. 50, 25.

wistfullung, e; *f. Feasting:*—Wistfullunga *epulas,* Hpt. Gl. 452, 4. v. ge-wistfullung.

wist-fyllu; *indecl.* -fyll, e; *f. Abundance of food:*—Him álumpen wæs wistfylle wén, Beo. Th. 1472; B. 734.

wist-gifende; *adj.* (*ptcpl.*) *Yielding food, fertile:*—Ðære wistgifendan *opulenti,* Wrt. Voc. ii. 62, 47.

-wistian, -wistlǽcan. v. ge-wistian, -wistlǽcan.

wistle, an; *f. A hollow reed:*—Wistle *avena,* Wrt. Voc. i. 285, 5: ii. 8, 26: *fistula,* 90, 25: 37, 26. v. wóde-wistle.

-wistlíc. v. ofer-wistlíc.

wist-líce *for* wíslíce, Anglia xi. 108, 14: 109, 46. Cf. Wulfst. 51, 15: 52, 28.

wist-mete, es; *m. Food for sustenance:*—Ic eom áféded of ðam genihtsumestan wistmettum mínre fylle, Homl. Skt. i. 23 b, 582.

wísung, e; *f. Direction, guidance:*—Scylon hý gán tó heora scriftan and hym hys synna ealle geandettan, and ealle be his wísunge gebétan Homl. Ass. 141, 71. Dathan and Abiron mycele teónan Móyse gedydon and forsáwon his wísunge, Homl. Skt. i. 13, 224. v. ge-, weorc-wísung.

wís-wyrdan; *p.* de *To be wise in speech:*—Wýswyrdan *philosophari,* Anglia xiii. 38, 301. v. next word.

wís-wyrde; *adj. Wise in speech:*—Se ðe wǽre stuntwyrde, weorðe se wíswyrde, Wulfst. 72, 18.

wit; *pers. pron. We two,* (I) *I and thou,* (a) alone:—Ðæt hí sýn án swá wyt sýn án, Jn. Skt. 17, 22. Abram cwæð tó Lothe: ‘Wyt sind gebróðru,’ Gen. 13, 8. Wit, Cd. Th. 114, 14; Gen. 1904. Geþenc hwæt wit sprǽcon, Beo. Th. 2957; B. 1476: Exon. Th. 172, 18; Gú. 1145. Wit baru standaþ, Cd. Th. 50, 20; Gen. 811. (b) with numeral forms:—Wit bútú sprecaþ, Cd. Th. 36, 20; Gen. 574: 52, 3; Gen. 838. Ne forlǽte ic ðé, þenden wit lifiaþ bú, 136, 11; Gen. 2256. (2) *I and he* (*she*), (a) alone:—Ðá becóme wit tó ðam inneran dǽle ðæs wéstenes ðǽr uncer hlǽfdige wæs, and wit wǽron belocene in carcerne, Shrn. 38, 20: Gen. 41, 12. Rincas míne, restaþ incit hér, wit (*Isaac and I*) eft cumaþ, Cd. Th. 174, 21; Gen. 2882: 152, 31; Gen. 2529: Beo. Th. 1074; B. 535. (b) with numeral forms:—Ic wæs gehloten mid ánum wífe in ánes ceorles þeówdóme. Ðá wǽron wit twégen on ánum olfende, Shrn. 38, 14. Ic gean intó Élig, ðér mínes hláfordes líchoma rest, ðara þreó landa ðe wit bútá geheótan Gode, Chart. Th. 524, 20. Ðá bær unc mon liþ forþ, and wit bú druncan, Bd. 5, 3; S. 616, 31. (c) with the name of the person associated with the speaker:—Wit Scilling for uncrum sigedryhtne song áhófan, Exon. Th. 324, 31; Víd. 103. (d) with name and numeral:—Wit Adam twá eaples þigdon, Cd. Th. 290, 6; Sat. 411. [Gif þu me dest woh and wit beon anes lauerdes men, O. E. Homl. i. 33, 1. þe bet wit (*he and I*; we, 2nd MS.) mawen libben, Laym. 9515. Wit (we, 2nd MS.) tweie, 23653. Witt ne muȝhenn tæmenn, Orm. 202. Wit beoð ifestnet and þe cnotte is icnut bituhhen unc tweien, Kath. 1512. Ðo quat Laban: ‘Frend sule wit ben and trewðe pligt nu unc bitwen, Gen. and Ex. 1775. *Goth.* O. Sax. wit: *Icel.* vit.] v. unc, wé.

wit(t), es; *n.* I. *right mind, wits:*—Wóde hé gehǽlde, and on witte gebróhte, Homl. Skt. i. 15, 7. II. *wit, intelligence, understanding:*—Ðæs ðú scealt werhðo dreógan, þeáh ðín wit duge, Beo. Th. 1183; B. 589. III. *the mind:*—Ðeós gítsunc hafaþ gumena gehwelces mód ámerred, ðæt hé máran ne recð, ac hit on witte weallende byrnð, Met. 8, 45. [*O. Frs.* wit: *O. H. Ger.* wizzi *ingenium, ratio: Icel.* vit *consciousness, sense, understanding: Goth.* un-witi *ignorance, foolishness.*]
v. ge-wit; bil-, fyr-wit.

-wit (-wid). v. in-wit.

-wít. v. ed-wít.

wita, an; *m.* I. *one who knows, a person of understanding* or *learning, a wise man:*—Wita (-e, MS.) *sophista,* Wrt. Voc. i. 47, 41. Fród wita, snottor ár, beorn bóca gleáw, Exon. Th. 313, 16; Mód. 1. Se ðe wita (*sapiens*) is, mid feáum wordum geswytelaþ, R. Ben. 30, 15. Wita sceal geþyldig, ne sceal nó tó háþeort, ne tó hrædwyrde, Exon. Th. 290, 15; Wand. 65. Ðissere worulde hæl is ðæt heó witan hæbbe, and swá

---

má witena beóð, swá hit bet færð. Ne bið se ná wita ðe unwíslíce leofaþ, ac bið open sott, Homl. Skt. i. 13, 131. Mé com swíðe oft on gemynd, hwelce wiotan (wutan, Cott. MSS.) iú wǽron giond Angelcynn, ǽgðer ge godcundra háda ge woruldcundra, Past. pref.; Swt. 3, 3. Wín gedéð, ðæt furðon witan (*sapientes*) oft misfóð, R. Ben. 65, 4. Filistina witan, *the wise men of the Philistines,* Salm. Kmbl. 861; Sal. 430. Ða ǽlaruwas ł aldo uuto *Pharisaei,* Lk. Skt. Lind. 5, 17. Witena *peritorum,* Wrt. Voc. ii. 67, 37. Ofer witena dóm, Exon. Th. 248, 19; Jul. 98. Hit witena nán þider ne séceþ, Met. 19, 7: 20, 3: Runic pm. Kmbl. 340, 8; Rún. 4. Hé (*Nero*) wæs ǽlcum witum láþ and unweorþ, Bt. 28; Fox 100, 28. Ðæt Godes hús wíslíce fram witum (*sapientibus*) sig gefadod, R. Ben. 84, 24. I a. with special reference to taking part in deliberations:—Ðis witena gemót *haec sinodus,* Ælfc. Gr. 8; Zup. 30, 8. Bǽdon ðæt eft óþer seonaþ wǽre, and hí ðonne woldan mid má heora witena gesécean, Bd. 2, 2; S. 502, 37. Wurdon monega seonoðas háligra biscepa and eác óðerra geþungenra witena, L. Alf. 49; Th. i. 58, 5. II. *one able to give counsel, a counsellor:*—Se wæs wita and geþeahtere ðæs Pápan *consiliarius erat Papae,* Bd. 5, 19; S. 638, 14. II a. *one able to give counsel in affairs of state, one who takes part in the councils of a nation, a leading man:*—Sum in mǽðle mæg módsnottera folcrǽdenne forð gehycgan, ðær witena biþ worn ætsomne, Exon. Th. 295, 34; Crä. 43. (1) in reference to other than Teutonic people:—Se ríca Rómána wita Brutus, Met. 10, 44. Hié sendon .x. hiera ieldstena wietena (*decem principes*), Ors. 4, 7; Swt. 182, 11. Witena, 4, 10; Swt. 196, 29. Hí hæfdon ǽlce dæge heora witena gemót (-met, Thw.), and wǽron gesette synderlíce tó ðam ða senatores, ðæt synd þeódwitan, Jud. Thw. p. 161, 31. Wiþ ðám Rómániscum witum, Bt. 1; Fox 2, 15. Hé ofslóg ealle ða witan (*in Thrace*), Ors. 3, 7; Swt. 114, 20. Crêca witan, Met. 1, 66. Ða rícostan Rómána witan, 9, 25. (2) in reference to England. See also *gemót:*—Ðyssum wordum óðer ðæs cyninges wita and ealdorman (*alius optimatum regis*) geþafunge sealde, Bd. 2, 13; S. 516, 12. Gif hwá on ealdormannes húse gefeohte, oþþe on óðres geþungenes witan, L. In. 6; Th. i. 106, 6. Ðis is seó gerædnes ðe Engla cyng and ǽgðer ge gehádode ge lǽwede witan gecuran (cf. ðe Engla rǽdgifan gecuran, vi. 1; Th. i. 314, 3), L. Eth. v. tit.; Th. i. 304, 4. Miercna cyning and his weotan, Chr. 868; Erl. 72, 23. Eádweard cyng and his witan, 911; Erl. 100, 18. Se cyng ond his biscopas ond his aldormenn ond alle ða wioton ðisse ðióde ðǽr gesomnade wǽron, Chart. Th. 70, 15. Cynewulf benam Sigebryht his ríces and West-Seaxna wiotan, Chr. 755; Erl. 48, 19. Bútan ðæs cyninges leáfe and his witena, 901; Erl. 96, 28. Eádmund cyning cýþ . . . ðæt ic smeáde mid mínra witena geþeahte ge hádedra ge lǽwedra, L. Edm. S. proem.; Th. i. 246, 19. Ic Ælfréd West-Seaxna cyning eallum mínum witum ðás geeówde, and hié ðá cwǽdon, ðæt him ðæt lícode eallum tó healdenne, L. Alf. 49; Th. i. 58, 28. Ic Íne . . . mid eallum mínum ealdormonnum and ðǽm ieldstan witum mínre þeóde and eác micelre gesomnunge Godes þeówa wæs smeágende be ðære hǽlo úrra sáwla and be ðam staðole úres ríces, L. In. proem.; Th. i. 102, 6. Ædelréd wæs mid mycelum geseán Angelcynnes witon gehálgod tó cyninge, Chr. 979; Erl. 129, 30. Weotum, Chart. Th. 480, 16. (3) in reference to other Teutonic people:—Witan Scyldinga, Beo. Th. 1561; B. 778. Hé ða weálǽfe weotena dóme árum heólde, 2201; B. 1098. III. *an elder, a chief person, senior* (cf. fród *for double sense of* wise *and* old):—Beón gesette án oðþe twégen ealde witan (*unus aut duo seniores,* R. Ben. 74, 14. Ældo ł uuto ðæs folces *seniores populi,* Mt. Kmbl. Lind. 21, 23. On gemóte heora witena *in conventu seniorum,* Bd. 3, 5; S. 527, 23. Wutuna (uutuna ł ældra, Lind.) *patrum,* Lk. Skt. Rush. 1, 17. Cwæð se Hǽlend tó ðam witum (*ad seniores*), Lk. Skt. 22, 52. Hé ge fram ðam witum ge fram his efenealdum (*a senioribus et coaetaneis suis*) mid rihtre lufan lufad wæs, Bd. 5, 19; S. 637, 18. Hé geseah ealle witon on þeáwum and dǽdum scínende, Homl. Skt. ii. 23 b, 85. IV. *one who has knowledge, a witness:*—Eall mín mǽgð mé is tó witan, Homl. Skt. i. 8, 42. Leáse uuta *falsi testes,* Mt. Kmbl. Lind. 26, 60. Mid sægene unrím geleáffula witena (*testium*) ða ðe ða ðing wiston, Bd. pref.; S. 472, 25. Gif hé hit nǽbbe beforan gódum weotum (witum, *v.l.*) geceápod, L. In. 25; Th. i. 118, 14. V. a *wise man, one professing supernatural knowledge:*—Hé sende tó Egipta wísustan witan *misit ad omnes conjectores Egypti cunctosque sapientes,* Gen. 41, 8. [Witene imot, Laym. 11545. Beon weote and witnesse þerof, A. R. 204, 24. þe wite (Helyas þe prophete, 8628), Orm. 8673. *O. Frs.* wita *a witness: O. H. Ger.* wizzo *gnarus, sapiens; divinus.* Cf. *Goth.* un-wita *foolish; ignorant.*] v. æ-, burh-, folc-, fyrn-, ge-, láh-, lár-, leód-, rǽd-, rún-, scír-, síg-, un-, þþ-, weorold-wita.

witan; *prs.* ic, hé wát, ðú wást, wǽst, *pl.* wé witon; *p.* wiste; *pp.* witen. I. *to wit, know, have knowledge, be aware,* (1) absolute:—*Noui* ic can oððe ic wát, *noui* ic wiste, Ælfc. Gr. 33; Zup. 205, 8. Oft wé oferswíðon swá swá ðú sylf wistest, Homl. Skt. i. 11, 27. Ne meahte hire Iudas, ne ful gere wiste, sweotole gecýðan, Elen. Kmbl. 1717; El. 860. Ne ongeátan hí, ne geara wistan *nescierunt, neque intellexerunt,* Ps. Th. 81, 5. Giefe monigfealdran ðonne ǽnig mon wite, Exon. Th. 177, 4; Gú. 1221. Hé wæccende ðóhte ðæt hé nó witende (*nesciens*)

āræfnode, Bd. 1, 27 ; S. 497, 8. Ic oft swór mǽne áðas ge weotende ge nytende, Anglia xi. 99, 65. Ða gáð libbende and witende on helle *ad infernum viventes sentientesque descendunt*, Past. 55 ; Swt. 429, 27. Hié æt níhstan witende mid deófolcræftum sóhton hú hí hit gestillan mehte, Ors. 3, 10 ; Swt. 140, 7. Ða ðe him ne ondrǽdaþ witende (*sciendo*) syngian, Bd. 1, 27 ; S. 491, 37. Ne weotendum (*nescientibus*) oþþe nó gýmendum ðǽm hyrdum ðære stówe, 4, 3 ; S. 570, 12. (2) with acc., *to know something, have knowledge of, be aware of*:—Hwanun wát (witto, Lind., wito, Rush. *sciam*) ic ðis ? Lk. Skt. 1, 18. Ne ic ǽniges wát hæleða gehygdo, Andr. Kmbl. 398 ; An. 199. Ðú wást ða menniscan týddernysse, Blickl. Homl. 243, 30 : Ps. C. 31. Ðú wæst and const ánra gehwylces earfeðsíðas, Andr. 2566 ; An. 1284. Crist ealle wát góde dǽde, Exon. Th. 449, 7 ; Dóm. 67 : Blickl. Homl. 19, 33. Ǽlc here hæfð ðý lǽssan cræft ðonne hé cymð, gif hine mon ǽr wát (*if people know of it*), ǽr hé cume, Past. 56 ; Swt. 433, 28. Hé manna ingehygd wát and can, Blickl. Homl. 179, 26. Ne magon wé hit ná dyrnan, for ðam ðe hit Drihten wát, Hy. 7, 93 : Exon. Th. 183, 11 ; Gú. 1325. Manna geþóhtas nǽnig mon ne wát, Blickl. Homl. 181, 11 : Ps. Th. 73, 17 : Hy. 3, 32. We gewislíce witon (*know of*) unrím ðara monna ðe ða ēcau geséɫða sóhtan *multos scimus beatitudinis fructum quaesisse*, Bt. 11, 2 ; Fox 36, 2 : Elen. Kmbl. 1285 ; El. 644. Rincum ðe béc witan, Exon. Th. 429, 19 ; Rä. 43, 7. Ic wiste hira sár *sciens dolorem ejus*, Ex. 3, 8. Hé heora lotwrencas wiste, Mk. Skt. 12, 15. Hé wiste sprǽca fela, Cd. Th. 29, 5 ; An. 445. Ðeáh ðe hé hit ǽr wisðe, Past. 35 ; Swt. 243, 3. Hé ne wisse word ne angin, Cd. Th. 223, 25 ; Dan. 125. Ealle ða ðe ðone gylt mid him wiston *conscii servi*, Ors. 4, 4 ; Swt. 164, 2. Ða swíðe lytle fíorme ðara bóca wiston *very little profitable matter in those books did they know*, Past. pref. ; Swt. 5, 11. Yldo bearn ǽr ne cúðon, þeáh hié fela wiston, Cd. Th. 179, 16 ; Exod. 29. Metod hié ne cuþon, ne wiston hié Drihten God, Beo. Th. 365 ; B. 181. Gé sweltaþ deáðe, nymþe ic dóm wite swefnes, Cd. Th. 224, 29 ; Dan. 143. Nis ðæt eówer ðæt gé witan ða þráge and ða tíde *non est vestrum nosse tempora et momenta*, Blickl. Homl. 117, 24. Ða mildestan ðara ðe men witen, Exon. Th. 255, 1 ; Jul. 207. Gé ne magon witan ðæra tída tácnu, Mt. Kmbl. 16, 3 : Exon. Th. 339, 11 ; Gn. Ex. 92. Gerím witan heardra heteþonca, 261, 13 ; Jul. 261. Ús Hǽlend God onwráh, ðæt wé hine witan mótan, 24, 14 ; Cri. 384 : Beo. Th. 509 ; B. 252. Ǽghwæþres sceal scearp scyldwiga gescád witan worda and weorca, 582 ; B. 288. Wytan, Hy. 3, 17. Dó hit mon ús tó witanne, Past. 46 ; Swt. 357, 5. Béc ða ðe niédbeðearfosta sién eallum monnum tó wietonne, pref. ; Swt. 7, 7. Tó wietenne, 15 ; Swt. 92, 26. Witende (*scientes*) ǽgðer ge gód ge yfel, Gen. 3, 5. Witendum (weotendum, Ps. Surt.) ðe *scientibus te*, Ps. Spl. 35, 11. Nán þing nis behýdd ðæt ne sý witen (*quod non sciatur*), Lk. Skt. 12, 2. (3) with acc. and infin.:—Ðǽr ic seomian wát ðínne sigebróðor bendum fæstne, Andr. Kmbl. 365 ; An. 183. Wé witun ðe bilewitne wesan, Coll. Monast. Th. 18, 22. Ðara cynna monige hé wiste on Germanie wesan, Bd. 5, 9 ; S. 622, 14 : Exon. Th. 182, 20 ; Gú. 1313 : 248, 16 ; Jul. 91. Wisse, 436, 15 ; Rä. 55, 1 : 324, 28 ; Víd. 101. Ðǽr ðú wite elenan standan, Lchdm. ii. 346, 10. (4) with acc. and complementary word or phrase:—For ðære byldo ðe ic tó him wát, Blickl. Homl. 179, 21. Ic ðé on ðyssum hýnðum wát wyrmum tó wiste, Soul Kmbl. 303 ; Seel. 155. Ic mé ðæt wát tó helpe, Ps. Th. 51, 7. Ic wát heáhburh hér áne neáh, Cd. Th. 152, 8 ; Gen. 2517. Ic hine goodne wát, Ps. Th. 53, 6 : 106, 1 : Beo. Th. 3731 ; B. 1863 : Hy. 1, 3. Ne wát ic mé beworhten wulle flýsum, Exon. Th. 417, 11 ; Rä. 36, 3. Ðú wǽst ðé bǽles cwealm hátne in helle, Andr. Kmbl. 2374 ; An. 1188. Hé wát his sincgiefan bihelode, Exon. Th. 183, 13 ; Gú. 1326: 311, 15 ; Seef. 92. For ðam ǽrende ðæt hé tó ús eallum wát, 451, 34 ; Dóm. 113. Hý him in wuldre witon Waldendes giefe, 76, 23 ; Cri. 1244: 107, 20 ; Gú. 61. Ðú wýsctest ðæt ðú wistest Crist on róðe áhangene, Elen. Kmbl. 2404 ; El. 1203 : Cd. Th. 3, 26 ; Gen. 41 : Beo. Th. 1297 ; B. 646. Hé wende hine ðǽr hé wiste handgeweorc heofencyninges, Cd. Th. 31, 32 ; Gen. 494: 169, 3 ; Gen. 2793 : 259, 1 ; Dan. 685 : Exon. Th. 162, 16 ; Gú. 976. Hé aldorþegn gesáwon wiste, Beo. Th. 2623 ; B. 1309 : Cd. Th. 249, 25 ; Dan. 535. Ðæt hé wiste hine scyldigne, Chart. Th. 166, 33. Hié ðone here tóweardne wiston, Blickl. Homl. 79, 13 : Shrn. 86, 3 : Exon. Th. 459, 20 ; Hö. 2. Ne mé unrihtes on áwiht wistan, Ps. Th. 58, 3. Wiston him be súðau Sigelwara land, Cd. Th. 182, 1 ; Exod. 69. Wisson, Beo. Th. 498 ; B. 246. Gif hé hine sylfne wite ðæs clǽnne, L. C. E. 5 ; Th. i. 362, 9. (5) with a clause, (a) without connecting word:—Ic wát ðú eart Godes hálga, Mk. Skt. 1, 24 : Cd. Th. 24, 30 ; Gen. 385 : 35, 8 ; Gen. 551. Ic wát þeáh, gif ðé ǽfre gewyrð ðæt . . . þonne gesihst ðú . . . , Bt. 36, 2 ; Fox 174, 24. Wé witon hé úre wæs wealdend, Blickl. Homl. 243, 17. Wiste hé his bearn on myclum ymbhygdum wǽron, 131, 26. Wite ðé be ðissum, gif ðú eádmódne eorl geméte, ðam bið gǽst gegǽderad Godes ágen bearn, Exon. Th. 318, 4 ; Mód. 77. (b) with introductory ðæt :—Ic wát (*novi*), ðæt ðú eart wlitig, Gen. 12, 11. Ic wát (*scio*), ðæt ðú swá didest, 20, 6. Ic wát (*cognovi*), ðæt Drihten ys mǽre, Ex.

18, 11. Ic wát (uát, Lind., wátt, Rush. *scio*), ðæt seó cýðnes is sóð, Jn. Skt. 5, 32 : Cd. Th. 35, 22 ; Gen. 558. Ic wát and cað, ðæt ðú mín God wǽre *agnovi quoniam Deus meus es tu*, Ps. Th. 55, 8. Ic wát geare, ðæt . . . , Beo. Th. 5306 ; B. 2656. Ðú wást, ðæt ic eom untýmende, Gen. 16, 2 : Jn. Skt. 21, 15 : Ps. Th. 68, 6. Wé weotan, ðæt wé ðæs ðearfe nabbaþ, Bd. 2, 5 ; S. 507, 21. Se hellsceaða wiste, ðæt hié Godes yrre habban sceoldon, Cd. Th. 43, 23 ; Gen. 695. Wisse, Beo. Th. 4668 ; B. 2339. Ða men wisson (wisston, Bt. 38, 1 ; Fox 196, 8), ðæt . . . , Met. 26, 100. Westan, Judth. Thw. 24, 26 ; Jud. 207. Wé witon (uutton, Lind., wutun, Rush. *scimus*), ðæt hé is synful, Jn. Skt. 9, 24, 29, 31. Wé wuton (wutan, Rush.), Mt. Kmbl. Lind. 22, 16. Wite gé (wutas gié, Lind., gé wutan, Rush. *scitis*), ðæt . . . , 26, 2 : Mk. Skt. 10, 41. Wit ðú, ðæt ic eom drý, Blickl. Homl. 183, 17. Witaþ gé, ðæt hit swá nis *scitote, quia non est mea*, Bd. 4, 8 ; S. 576, 2. Witaþ (wutas gié, Lind.), ðæt . . . , Mt. Kmbl. 24, 43 : Lk. Skt. 10, 11 : Ps. Th. 99, 2. Wite gé, ðæt . . . *scitote, quoniam* . . . , Ps. Th. 4, 4 ; Blickl. Homl. 191, 36. Wite ðú, ðæt . . . *scito, quod* . . . , Gen. 15, 13 : Jud. 6, 14 : Blickl. Homl. 181, 11 : 183, 18 : Elen. Kmbl. 1889 ; El. 946. Wite hé, ðæt hé bið wana, Blickl. Homl. 17, 36. Suá suá hié selfe wieten, ðæt hié hit for Gode dón, Past. 28 ; Swt. 191, 2. Witen, Met. 19, 13. Se reccere sceal geornlíce wietan, ðætte . . . , Past. 20; Swt. 149, 1. Ðǽm láreówe is tó wietanne, ðæt . . . , 63 ; Swt. 459, 6. Tó witenne, Blickl. Homl. 63, 5 : 129, 26 : 209, 19. (bb) with *for ðon ðe*:—Crist ða wiste, for ðon ðe (*quia*) se hálga ða slép, Blickl. Homl. 235, 13. (c) with indirect interrogative forms:—Ic wát hwæt hé þenceþ, Blickl. Homl. 181, 10 : Cd. Th. 34, 10 ; Gen. 535. Ic wát hwá mé ferede, Andr. Kmbl. 1808 ; An. 906. Ic ne wát hwǽr ðú eart, Blickl. Homl. 241, 7 : Exon. Th. 496, 21 ; Rä. 85, 18. Ðú cans eal ðis wésten, and wásð hwær wé wícian magon, Past. 41 ; Swt. 304, 16. Ðú wást and canst hú ðú lifian scealt, Cd. Th. 56, 23 ; Gen. 916. Wást ðú hú ðeós ádle scyle ende gesettan ? Exon. Th. 163, 16 ; Gú. 994. Eówer Fæder wát hwæs eów þearf biþ, Blickl. Homl. 21, 1. Wé witon hwelce wælhríównessa Neron weorhte, Bt. 16, 4 ; Fox 56, 36. Ne wuti gé of hwelc tíd hláferd ǐwer tó cymmende sié, Mt. Kmbl. Lind. 24, 42. Gif ic wiste hú . . . , Beo. Th. 5032 ; B. 2519. Ðæt ðú wisse (wistest, *v. l.*) hwæs ðú wundredest, Bt. 41, 4 ; Fox 252, 14. Ne wiste hé hwonne him fǽmnan tó brýde wǽron, Cd. Th. 157, 5 ; Gen. 2600. Ðæt wé wissen (wiston, *v. l.*) hwæt hé wǽre, Bt. 42 ; Fox 256, 2. Ðæt ic wite gearwe on hwylcne weg ic gange, Ps. Th. 142, 9. Wite ðú hú wíd and síd helheoðo, and mid hondum ámet, Cd. Th. 308, 27 ; Sat. 699. Wé witon magon hú swíþe ús is ðes ðæg tó mǽrsienne, Blickl. Homl. 161, 7 : 47, 21. Gif ðú witan wille hwæt gedón wæs, 177, 1. (d) with *gif*:—Ðú wást gif hit is swá wé secgan hýrdon, Beo. Th. 550; B. 272. (6) with the construction of (2) and of (5). (a) (2) and (5 b):—Án þing ic wát, ðæt ic wæs blind, ðæt ic nú geseó, Jn. Skt. 9, 25. Wát ic ðæt nú ðá, ðæt hé bið alles leás ēcan dreámes, Cd. Th. 275, 34 ; Sat. 181 : Andr. Kmbl. 866 ; An. 433. Ðú wást míne geheówunga, ðæt ic eom dust, Blickl. Homl. 89, 15. Ðæt ðú wást, ðæt ic wæs deád, 183, 13. Ða ðe hit witon, ðæt hié him þeówiaþ, Bt. 21 ; Fox 72, 32. Ðæt ic gearwe wiste, ðæt . . . , Exon. Th. 196, 7 ; Az. 170: Cd. Th. 24, 31 ; Gen. 386. Ðis wutaþ gié, ðætte geneólēcaþ ríc Godes, Lk. Skt. Lind. 10, 11. Ǽr hé wát, ðæt ða synfullan sáwla sticien helle tó middes, Salm. Kmbl. 342 ; Sal. 170. Hú mæg ic hit witan, ðæt ic hit ágan sceal *unde scire possum, quod possessurus sim eam*? Gen. 15, 8. (b) with (2) and (5 c):—Ðú ðæt ána wást, hú mé módor gebær, Ps. C. 61. Gif ðú hit wást, hú ðú mǽre eart, Hy. 3, 20. Ðæt ne wát ǽnig, hú ða wísan sind wundorlíce, Exon. Th. 223, 10 ; Ph. 357. Ðæt ne wiste hé, hwæt se manna wæs, Andr. Kmbl. 521 ; An. 261. Ðæt hié ðæt wiston, hwonne hé ðisse worlde ende gesettan wolde, Blickl. Homl. 119, 9. Nis nǽnig mon ðe ðæt án wite, hú lange hé ðás gedón wille, hwæþer ðis þúsend sceole beón scyrtre ðe lengre, 119, 5. Ðæt ðú sóð wíte, hú ðæt geeode, Exon. Th. 28, 6 ; Cri. 442. Ðæt ic sóð wite, hwæðer . . . , Andr. Kmbl. 1206 ; An. 603. Ne mæge wé ðæt witan, hú ðú æðele eart, Hy. 3, 13. (c) with (2) and (5 d):—Gif hé synful is, ðæt ic nát, Jn. Skt. 9, 25. (7) with preposition *be*:—Wé witon bí monnum, se se ðe bitt ðone monn ðæt him ðingie wið oðerne ðe hé bið ierre, ðæt irsigende mód hé gegremeð, Past. 10 ; Swt. 63, 11. Hié wiston ge be heora sige, ge eác be ðara hǽþenra manna fleáme *they knew both about their victory and about the heathens' flight*, Blickl. Homl. 203, 3. Se consul heora ungemet ofslóg and sige hæfde ; be ðam mon mehte witan, ðá hé and ða consulas hié ætellan ne mehton *quot millia hominum interfecta, quot capta sint, ipse consul ostendit; qui, cum multitudinem capti populi referre vellet, numerum explicare non potuit*, Ors. 3, 10 ; Swt. 140, 30. Be ðam ǽfteran is tó witanne, ðæt hé wæs tó biscope gehálgod *de secundo intimandum, quod in episcopatum fuerit ordinatus*, Bd. 4, 23 ; S. 594, 11. **II.** *to be wise, be in one's senses*:—Ða wéndon hí ðæt hé tela ne wiste, ac ðæt hé wédde *vulgus aestimabat eum insanire*, Bd. 2, 13 ; S. 517, 10. 'Geseoh ðæt ðú teala wite.' Cwæþ hé : 'Ne wéde ic' *vide ut sanum sapias. Non, inquit, insanio*, 5, 13 ; S. 632, 32. **III.** *to be conscious of, to know* fear, pity, etc., *to feel, shew* respect, honour, etc.:—Wát ic sorga ðý má, Cd. Th. 54, 33 ; Gen. 886. For hwon

wâst ðû weán? 54, 12; Gen. 876. Hit mâre ne wât bûton gnornunge, Met. 3, 9. Se wyrsa ne wât on his mǽgwinum mâran âre, Salm.
Kmbl. 717; Sal. 358. Hió him tô litelne ege tô witan *they feel too
little awe of him*, Wulfst. 220, 27. Ðû ðæs þonc ne wisses *thou knewest
(feltest) no gratitude for it*, Exon. Th. 85, 5; Cri. 1386: 90, 15; Cri.
1474. Ðone ðe in meoduhealle mine wisse (*should feel affection*),
oþþe mec frēfran wolde, 288, 7; Wand. 27. Ic lǽrde ðæt ǽlc on
ôþrum ârwyrþnesse wiste, Blickl. Homl. 185, 13. Hē him forgeaf ðone
nîð ðe hē tô him wiste, Ors. 5, 15; Swt. 250, 15. Hē sâr ne wiste,
Cd. Th. 12, 3; Gen. 179. Hiê sorge wiht, weorces ne wiston, 49, 2;
Gen. 786. Ðæt is tô wundrianne, ðæt ða Egipti swâ lytle þoncunge
wiston Iôsepe, Ors. 1, 5; Swt. 34, 32. Ðæt hî nǽnige incan tô him
wiston *se mentem ad illum ab omni ira remotam habere*, Bd. 4, 24; S.
598, 41. Wite mâran þanc ðæs ðe ðû hæbbe, ðonne ðæs ðe ðû wêne,
Prov. Kmbl. 22. Ǽlc ðe gescâd wite *every rational person*, L. C. S.
75; Th. i. 424, 19. Ðæt hē on ðam griðe micle mǽðe wite *that he
shew great respect to that 'grið,'* L. Eth. vii. 31; Th. i. 336, 14. Gif
man on Godes griðe mǽðe witan wolde, Wulfst. 161, 3. Ne wolde hē
ǽnige âre witan on ðære Cristenan æfestnysse *nec religioni Christianae
aliquid intendebat honoris*, Bd. 2, 20; S. 521, 29. Ðû mē noldest þanc
witan mînra gôda, Wulfst. 261, 10. Nâ heáge witende *non alta sapientes*,
Scint. 19, 2. [He wald ha witen (witten, wist, *v. ll.*), C. M. 10793.
Goth. witan; *prs.* wait, *pl.* witun; *p.* wissa: *O. Sax.* witan; *prs.* wêt,
pl. witun; *p.* wissa: *O. Frs.* wita; *prs.* wêt, *pl.* witen, witath: *O. H.
Ger.* wizzan; *prs.* weiz, *pl.* wizzun; *p.* wissa, wista, westa; *pp.* wizzan: *Icel.* vita; *prs.* veit, *pl.* vitu; *p.* vissa; *pp.* vitinn.] v. be-, gewitan, nytan, un-witende.

witan; *p.* wât, *pl.* witon; *pp.* witen. **I. to see to, take heed to,
guard, keep,** (1) absolute :—God wîteþ on ðam hēhstan heofna rîce ufan
Alwalda, Cd. Th. 32, 31; Gen. 511. [He (*God*) witeð and wialdeð
alle þing, Anglia i. 11, 40. Ihesu, wel þu witest hem, Jul. 51, 15. We
is him þat waked and witeð wel him seoluen, 74, 6. Swuch wardein
(*God*), þet wit and wereð us ever, A. R. 312, 8. Þe vif wittes, þet
witeð þe heorte alse wakemen, 14, 6. Wite mine Bruttes a to þines
lifes, Laym. 28604. Crist . . . wite his soule, Havel. 405. To witen
ant to welden, Marh. 2, 23. To wyten us wyþ þan unwihte, Misc. 72,
4.] (2) with acc. :—Ðæt bið gôd swefen, wite ðû ðæt georne on ðînre
heortan, Lchdm. iii. 154, 19. (3) with a clause :—Wite ðû georne, ðæt
ðû dô ealle ða tâcn *vide, ut omnia ostenta facias*, Ex. 4, 21. Wîte ðæt
ðîn geþanc ne losige, Lchdm. iii. 154, 20. Wŷte ðæt ðû swâ dô, Nicod.
26; Thw. 14, 23. Wîte se ôðer, ðæt hē hit bête, L. C. S. 76; Th. i.
418, 13. Wē willað âwerian ûs; wîte gē hwæt gē dôn siddan, L. Ælfc.
P. 1; Th. ii. 364, 13. Wē beóð unscildige, gif wē hit secgaþ eów; wîte
gē hwæðer gē silfe eówrum sâwlum beorgan willan, 43; Th. ii. 382, 27.
[Wite ʒe þet ʒe ʒemen þenne halie sunnedei, O. E. Homl. i, 11, 29. Cf.
Goth. þu witeis σὺ ὄψῃ, Mt. 27, 4.] **II. to lay** to a person's *charge,
lay the blame of* something on a person or thing, *impute.* (1) absolute :—
Wîte *imputet*, Germ. 400, 560. (2) with dat. of person :—Ðæt hē him
ne wîte, Bt. proem.; Fox viii. 12. (3) with dat. of person and acc. of
charge :—Mînum âgnum scyldum ic hit wîte, Ps. Th. 21, 2. Ne wîte
ic him ða womcwidas, Cd. Th. 39, 7; Gen. 621. Hwæt wîtst ðû ûs
what do you lay to our charge? Bt. 7, 5; Fox 22, 36: Homl. Th. ii.
164, 28. Mē Freá wîteþ sume ðara synna, Exon. Th. 456, 32; Hy. 4,
75: Salm. Kmbl. 885; Sal. 442. Hwæt wîte ðû mē? Soul Kmbl. 43;
Seel. 22. Ic nyste hwæt hî mē witon, Ps. Th. 34, 15. Hié wîtan
Claudiuse ðone hunger, and hē weard him grom (*imperator convitiis in-
festatus*), Ors. 6, 4; Swt. 260, 22. Ne wît ðû heom ðâs synna *ne
statuas illis hoc peccatum*, H. R. 9, 29. Gif ðû hwæt on druncen misdô,
ne wît ðû hit ðam ealoðe, Prov. Kmbl. 39: 18: 54. Gif hē hwylc
hleahterlîc word onfinde, ðæt hē ðæt ûs ne wîte, Guthl. prol.; Gdwin. 2,
13: Ps. Th. 65, 16. Hwæþer Rômane hit wîten nû ǽnegum men tô
secganne hwæt hiera folces forwurde? Ors. 5, 2; Swt. 220, 9. Hwæt
sió syn wǽre, ðe him seó cwên wite, Elen. Kmbl. 832; El. 416. Ic eom
swîðe gefiónde ðætte gē woldon ǽnige wuht eów selfum wîtan (wiétan,
Hatt. MS.), ǽr ic hit eów wite. Hit is gôd ðæt gē hit nû wiétun (wîton,
Hatt. MS.), Past. 31; Swt. 206, 19. Æfter ðæm ðe him swâ oftrǽdlîce
mislamp, hié angunnan hit wîtan heora lâtteówum and heora cempum
heora earseþa, Ors. 4, 4; Swt. 164, 25 : Cd. Th. 51, 9; Gen. 824: Hy.
6, 25; Beo. Th. 5475; B. 2741. (4) with prep. governing person, and
charge expressed in a clause :—Gif ðû mē on wîte, ðæt ic unrihtlîce
ðone biscopdôm onfenge, Anglia x. 141, 22. [Gif þu witest eni þing
þine sunne bute þi sulueð, A. R. 304, 10. Schal he hit wite me? O. and
N. 1248. If that I mysspeke wyte it the ale of Southwerk, Chauc.
Mill. Prol. 32. Wytyn *imputo*, Prompt. Parv. 531. *O. Sax.* wîtan:
O. H. Ger. wîzan *imputare, statuere.* Cf. *Goth.* fra-, in-weitan.] **III.
to go, depart** :—Nylle ic ǽfre bionan ût wîtan, ac ic symle hēr sôfte wille
standan, Met. 24, 52. [Witeð ge awariede gastes into þat eche fir *ite
maledicti in ignem eternum*, O. E. Homl. ii, 5, 36. He heðen wit, 123,
4. Þe wolf to witeþ, Laym. 21311. Herode wass witenn ut off life,
Orm. 8222. Ne wite þou noʒt fra me *ne discesseris a me*, Ps. 21, 12.]
v. æt-, ed-, ge-, ôþ-wîtan.

wîte, es (*a weak gen. pl.* wîtena *occurs*); *n.* **I. *punishment, pain
that is inflicted as punishment, torment* :—Wîte *poena vel supplicium*,
Wrt. Voc. i. 86, 35. Tintregung *vel* wîte *tormentum*, Wülck. Gl. 178,
20. Heó (*Eve*) hæfde hire sylfre geworht ðæt mǽste wîte and eallum
hire cynne, ge ðæt wîte wæs tô ðæs strang, ðæt ǽghwylc man sceolde
sâre on ðâs world cuman, and hēr on sorhgum beón, and mid sâre
of gewîtan, Blickl. Homl. 5, 27: Cd. Th. 28, 6; Gen. 431. Hié (*Lot's
wife*) strang begeat wîte, 155, 5; Gen. 2568. Rêðe wîte (*the deluge*),
79, 30; Gen. 1319. Wæs ðæt wîte (*the destruction of Jerusalem*) swâ
strang, swâ Godes geþeld ǽr mycel wæs, Blickl. Homl. 79, 27. Hwæþer
ðû ongite ðæt ǽlc yfelwillende mon sié wîtes wyrþe? Bt. 38, 6; Fox
208, 9, 13 : 39, 2; Fox 212, 25. Wŷtes, 39, 9; Fox 226, 5. Sweartne
lîg weorutn tô wîte, Cd. Th. 153, 21; Gen. 2542: Hy. 6, 27. Hié
âhôfon hine of ðam hefian wîte (*crucifixion*), Rood Kmbl. 121; Kr. 61.
Licgeþ lonnum fæst . . ., wylleþ hine on ðam wîte, Salm. Kmbl. 537;
Sal. 268. Hē wîte wealdeþ *he is the disposer of punishment*, Cd. Th.
248, 33; Dan. 523. Wîte *poenam, vindictam*, Hpt. Gl. 496, 7: Blickl.
Homl. 77, 28. Ðæt ðû inc meaht wîte bewarigan, Cd. Th. 35, 31;
Gen. 563. Ðû ðæs cwealmes scealt wîte winnan, 62, 14; Gen. 1014.
Ic wîte þolade, Exon. Th. 89, 5; Cri. 1452: 240, 25; Ph. 644: Elen.
Kmbl. 1038; El. 520. Freá wolde on wǽrlogan wîte settan, Cd. Th. 76,
33; Gen. 1265. Geseah hē engles hand wrîtan Sennara wîte, 261, 17;
Dan. 727. Ðê sind wîtu weotud be gewyrhtum, Andr. Kmbl. 2730; An.
1367: Exon. Th. 258, 13; Jul. 264. Ne ondrǽde ic mē dômas ðîne,
ne ðînra wîta bealo, 255, 9; Jul. 211. Hē weorna feala wîta geþolode,
Andr. Kmbl. 2979; An. 1492. Manigra wîta (wiéta, Hatt. MS.) hié
beóð wyrðe, Past. 28; Swt. 190, 7. Wîtena *tormentorum, poenarum*,
Hpt. Gl. 485, 10. Ne bist ðû orhlŷte ðæra wîtena, Homl. Th. ii. 310,
27. Wîtum *cruciatibus, poenis*, Hpt. Gl. 487, 12. Ðâ heó wæs tô ðâm
wîtum (*ad poenam*) gelǽdd, Gen. 38, 25. Tô manegum wîtum geworht
put to many tortures, Bt. 16, 2; Fox 52, 20. Wîtum belecgan, Andr.
Kmbl. 2424; An. 1213. Mid wîtum swingan, Exon. Th. 279, 22; Jul.
617. Forniman mid wîtum, Blickl. Homl. 189, 31. Wîta *tormenta, supplicia*, Hpt. Gl. 499, 34. Wîtu, Andr. Kmbl. 2829; An. 1417. ¶ referring to the punishment of hell :—Ðæt ungeendode wîte, Blickl. Homl.
25, 24: 51, 31: Andr. Kmbl. 1778; An. 891: Exon. Th. 446, 8;
Dôm. 19. Is ðæs wǽlica hâm, wîtes âfylled, Cd. Th. 271, 4; Sat. 100.
Wîtes fŷr, Exon. Th. 39, 21; 625. Grim helle fŷr tô wîte, 78, 7; Cri.
1270. Synna tô wîte, 77, 2; Cri. 1250. In wîte bîdan, Cd. Th. 268,
1; Sat. 48. Gelǽded ðe tô wîte þe tô wuldre, Blickl. Homl. 97, 22.
Se gâst nimeþ ǽt Gode swâ wîte swâ wuldor, swâ him ǽr ðæt eorðfæt
geworhte, Soul Kmbl. 13; Seel. 7: Blickl. Homl. 23, 6. In êce wîte
gefeallan, 57, 21. Se ðæt wîte ǽr tô wrece gesette, Cd. Th. 295, 28;
Sat. 494. Hē ðæs ôþres sâule of wîtum generede, and of tintregum
âlêsde, Blickl. Homl. 113, 33. Hē bið mid wîtum þreád æfter his deáþe,
49, 25. On êcum wîtum wunian, 83, 18. Ic sceal weán and wîtu and
wrace dreógan, Cd. Th. 276, 7; Sat. 185. **I a.** *a means or implement
of punishment* :—Wundor on wîte (*the fiery furnace*) âgangen, Cd. Th.
233, 3; Dan. 270. Wîta cyn *catastarum*, Wt. Voc. ii. 85, 58: 18, 64:
20, 34. **I b.** *a fine.* v. wîte-rǽden :—Sié ðæt wîte .lx. sciłł., ôð
ðæt ângylde ârîse tô .xxx. sciłł. . . . siþþan sié ðæt wîte .cxx. sciłł., L.
Alf. pol. 9; Th. i. 68, 3–5. Gilde se borh ðam hlâforde his were ðe his
wîtes wyrðe sî, L. Eth. i. 1; Th. i. 282, 4. Se hlâford gesette .xxx.
sciłł. tô wîte, L. In. 3; Th. i. 104, 4: 6; Th. i. 106, 7: 7; Th. i. 106,
16: 10; Th. i. 118, 16. Be wîte, Th. i. 118, 15:
L. E. G. 3; Th. i. 168, 6. Hē âge healf ðæt wîte, L. Wih. 11; Th. i.
40, 3. Gylde swâ wer swâ wîte, L. E. G. 2; Th. i. 168, 2. Gif hwâ
æfter ðam wîte crafige, L. C. S. 70; Th. i. 412, 24. Beó se cyng ǽlces
ðæra wîta wyrðe ðe ða men gewyrcen ðe bôcland hæbben, L. Eth. i. 1;
Th. i. 282, 16. See Kemble's Saxons in England, ii. 53. **II. in a**
general sense, *torment, plague, disease, evil, pain* :—Wîte *malum*, Wrt.
Voc. ii. 58, 63. Ðæt wîte geswâc *plaga cessavit*, Num. 16, 48, 46.
Ðis ylce wîte (*plaga*) Hibernia gelîce wǽle slôh and cwylmde, Bd. 3,
27; S. 558, 19: 4, 7; S. 574, 35. Wæs ðæt wîte (*famine*) tô strang,
Cd. Th. 109, 8; Gen. 1819. Drihten slôh ðæt folc mid swîðe micclum
wîte (*plaga magna nimis*), Num. 11, 33. Of ðam wîte gehǽled *sanata
a plaga*, Mk. Skt. 5, 29. Ne ondrǽd ðû ðê deáð tô swîðe for nânum
wîte, Prov. Kmbl. 49. Ne bið him hyra yrmðu ân tô wîte, ac ðara
ôþerra eád tô sorgum, Exon. Th. 79, 20; Cri. 1293. Waldend him ðæt
wîte (*blindness*) teóde, 336, 4; Gn. Ex. 43. God sealde gumena gehwelcum welan swâ wîte, Cd. Th. 256, 23; Dan. 645. Wîte âwinnan
to be tormented, Exon. Th. 130, 18; Gû. 440: 441, 26; Kl. 5. Wîte
lecgan on *to torment, plague*, 144, 29; Gû. 685. Syndon hyra wîta
scytelum cilda onlîcost *sagittae parvulorum factae sunt plagae eorum*,
Ps. Th. 63, 7. Wraðu wannhâlum wîta gehwylces, sæce and sorge,
Elen. Kmbl. 2058; El. 1030. Hē monge gehǽlde hefigra wîta, Exon.
Th. 155, 9; Gû. 857. Wanhâle wîtum gebundene, Andr. Kmbl. 1160;
An. 580. Hē gehǽlde manega of wîtum (*plagis*), Lk. Skt. 7, 21. Sleá
ðê Drihten mid ðâm Egiptiscan wîton *ulcere Aegypti*, Deut. 28, 27.
Ic sende eall mîn wîto (*plagas*) ofer ðê, Ex. 9, 14. Wē geâxiaþ unge-
cyndelîco wîtu, Blickl. Homl. 107, 26. Nis nó ðæt ân ðæt hē him ûre

wîtu (*the pains that we inflict*) ne ondrǽde, 85, 15 : Cd. Th. 289, 3 ; Sat. 392. [His wîte abideð on þere oðre weorlde, O. E. Homl. i. 103, 32. Mid ærmlíce witen (in ȝoure bendhuse, 2nd MS.), Laym. 1046. Uppe wite of feowerti punden, 5118. *O. Sax.* wíti *punishment, torment* : *O. Frs.* wîte : *O. H. Ger.* wîzi *poena, supplicium, tormentum, passio, damnatio, judicium, crux* : *Icel.* víti *a punishment, fine.*] v. blôd-, dol-, fiht- (fyht-), fyrd- (ferd-), helle-, leger-, weard-, weorold-, wræc-wîte.

wîte-ærn, es ; *n. A house of punishment, a prison* :—Wîtern *carcer*, Wrt. Voc. ii. 128, 62.

wîte-bend *a toŕturing lond, a prison-bond* :—Ðé wǽrlogan wîte-bendum swencan môton, Andr. Kmbl. 216 ; An. 108. Wé ellþeódigne on carcerne clommum belegdon, wîtebendum, 3120 ; An. 1563.

wîte-brôga, an ; *m. Penal horror, a horrid punishment* or *torment* :—Ne mé weorce sind wîtebrôgan, ðe ðú tô mé beótast, Exon. Th. 250, 31 ; Jul. 135. Eal ðæt man ús foreseget embe helle wîtebrôgan (cf. Wende him God fro heuene riche into helle witerbrogen (hellewites brogen ?), Chart. Th. 581, 3), Wulfst. 151, 24. Hé ðec sendeþ in ða sweartestan and ða wyrrestan wîtebrôgan, Elen. Kmbl. 1861 ; El. 932 : Cd. Th. 3, 33 ; Gen. 45.

wited-lîc. v. witod-lîc.

wîte-dôm, es ; *m.* **I.** *knowledge derived from a superhuman source, prophecy, foreknowledge* :—Wîtedôm *profetia,* Kent. Gl. 1064. Se Godes wer ðurh wîtedômes gást (*per prophetiæ spiritum*) ðone storm tôwardne foreseah, Bd. 3, 15 ; S. 542, 4. Ðæt heó ðurh wîtedômes gást ða ádle forecwǽde, 4, 19 ; S. 588, 15 : 4, 28 ; S. 606, 20. Ðæt wundor, ðæt þurh wîtedômes cræft [hé] wiste and him cýdde, Guthl. 17 ; Gdwin. 70, 2. Wîtedôme *vaticinatione,* Hpt. Gl. 520, 17. Hí þurh wîtedôm eal ánemdon, Exon. Th. 104, 24 ; Gú. 12. **II.** *a statement of what is known through superhuman agency, a prophecy* :—Wæs gefylled se wîtedôm (*præsagium*) Agustinus, Bd. 2, 2 ; S. 504, 8 : 3, 14 ; S. 541, 9 : Blickl. Homl. 71, 3 : Exon. Th. 14, 1 ; Cri. 212. Wæs se wîtedôm beforan sungen, Elen. Kmbl. 2304 ; El. 1153. Æfter ðam wîtedôme *secundum vaticinium* (*prophetiam*), Hpt. Gl. 493, 48. Æfter Esaias wîtedôme, Bd. 3, 23 ; S. 554, 22. Gehýraþ wîtedôm Iôbes gieddinga, Exon. Th. 234, 31 ; Ph. 548. Wîtedômas *oracula,* Hpt. Gl. 409, 50. Wîtedôma † godcundra spréca *oraculorum,* 442, 36. Mid wîtedômas *presago,* Wrt. Voc. ii. 66, 32. v. *next word,* and cf. wîteg-dôm.

wîtedôm-lîc ; *adj. Of superhuman knowledge, prophetical* :—Wîtedômlîc wundor *a miracle which displayed a knowledge communicated by God* (cf. him God ealle ða díglan þingc cúð gedyde, l. 12), Guthl. 11 ; Gdwin. 54, 1. Gûðlác wîtedômlîce gáste (*in prophetic spirit*) weóx, and hé ða tôweardan mannum cýdde swá cúðlíce swá ða andweardan, 13 ; Gdwin. 60, 19. Wîtedômlîce mûðe hé sang, 4 ; Gdwin. 28, 19.

wîte-fæst ; *adj. In slavery as a punishment for crime.* v. wîte-þeów :—Hé wyle ðæt man freóge æfter his dæge ǽlcne wîtefæstne man ðe on his tíman forgylt wǽre *si quis, secundum patriæ Angliæ morem, in aliquam incurrisset servitutem tempore suæ potestatis, libertate sibi penitus contributa, relaxatus ejus jussu est,* Chart. Th. 551, 14. Ic gean ðæt man gefreóge ǽlcne wîtefæstne man ðe ic on sprǽce áhte, 557, 21.

wîtega, an ; *m.* **I.** *a wise man, one who has knowledge* :—Hé is wîtgan (cf. *the epithets applied to Simon,* eald ǽwita, 907 ; El. 455, guma gehðum frôd, 1059 ; El. 531, *and the whole passage in which these forms occur*) sunu, Elen. Kmbl. 1181 ; El. 592. Swá ús gefreogun gleáwe wîtgan þurh wísdôm on gewritum cýþaþ, Exon. Th. 199, 23 ; Ph. 30. **II.** *one who has knowledge from a superhuman source,* (1) *a prophet* :—Wîtega *propheta vel vates,* Wrt. Voc. i. 41, 69 : *propheta,* 71, 68. (I a) *in the biblical sense* :—Swá se wîtega sang, Menol. Fox 119 ; Men. 59. Wîtga, Exon. Th. 41, 4 ; Cri. 650 : 316, 18 ; Môd. 50. Se wîtiga (*the Psalmist ;* v. Ps. 28, 3) cwæð, Fragm. Kmbl. 13 ; Leás. 8. Oð ðæt wîtga cwom, Daniel tô dôme, Cd. Th. 225, 5 ; Dan. 149. Wîtga (*Isaiah*), Exon. Th. 19, 26 ; Cri. 306. Iônas tácn ðæs wîtegan (*prophetæ*), Mt. Kmbl. 12, 39. On ðæs wîtegan bêc Isaiam, Mk. Skt. 1, 2. Sunu Dauides wîtgan (*Nathan the prophet*) lárum getimbrede tempel, Cd. Th. 202, 19 ; Exod. 390. Ðæt fram Drihtne gecweden wæs þurh ðone wîtegan (wîtgo, Lind., witgu. Rush.), Mt. Kmbl. 1, 22 : 2, 15. Twelf wîtegan syndon ðe twelf bêc áwriton . . . Wǽron eác ôðre wîtegan ðe ne writon náne bêc, Ælfc. T. Grn. 10, 8, 28. 'Eówre wîtgan (*prophetæ*) eów wîtgodon dysig' . . . Ða gôdan láreówas beóð oft genemnede on hálgum gewritum wiétgan (wîtgan, Cott. MSS.), for ðæm hié gereccaþ ðis andwearde líf fleónde and ðæt tôwearde gesweotoligeaþ, Past. 15 ; Swt. 91, 3-7. Wîtigan, Cd. Th. 293, 26 ; Sat. 460 : Blickl. Homl. 105, 9. Ðæt in fyrndagum wîtegan sǽdon, 293, 32 ; Sat. 464. Ða wîtigan þrý (*Abraham, Isaac, and Jacob*), Andr. Kmbl. 1602 ; An. 802. Hú on worulde ǽr wîtgan sungon, gásthálige guman, be Godes bearne, Elen. Kmbl. 1119 ; El. 561 : Exon. Th. 5, 3 ; Cri. 64. Æ and wîtegena bebod (wîtgas † wîtgo, Lind., wîtgu, Rush.) *lex et prophete,* Mt. Kmbl. 7, 12. Wîtgena word, Exon. Th. 29, 27 ; Cri. 469. Heáhfædera nán, ne wîtgena, 273, 12 ; Jul. 515. Siteþ Waldend mid wîtegum, Cd. Th. 301, 25 ; Sat. 587. Né wé sweotul tácen ús geseóð ǽnig, ne wé wîtegan habbaþ, ðæt ús andgytes má secgen, Ps. Th. 73, 9. (2) *a wise man, diviner, soothsayer* :—Wîtgan, Caldêa cyn, Cd. Th. 218, 19 ; Dan. 41.

Andswarode cyning wîtgum sínum (*the wise men of Babylon, the magicians, and the astrologers, and the sorcerers, and the Chaldeans,* Dan. 2, 2, 12), 224, 13 ; Dan. 135. Uuîtgan *divinos,* Wrt. Voc. ii. 106, 57 : 25, 42 : *divinos, ariolos,* 141, 55. **III.** *applied to things, a presage* :—Ætýwdon twêgen steorran . . . Hí wîtegan (*præsagæ*) wǽron grimmes wæles, Bd. 5, 23 ; S. 645, 26. Wîtegum *præsagminibus,* Hpt. Gl. 448, 64. [Dauid þe halie witege, O. E. Homl. i. 43, 16. Se witiȝe, 233, 12. Ðe lorðew þe tehte Salemon and alle wise witege here wisdom, ii. 83, 36. Dauid þe witeȝe, H. M. 5, 2. Teilesin heo heolten for witie, Laym. 9094. Merlin þe witeȝe, 17415. Tweolue of þine witiȝn, of þine wisuste monnen, 4368. He þeos word seide þurh an of his witeȝen *propheta clamabat dicens,* Kath. 483. *O. H. Ger.* wîzago *propheta ; pitho, divinus, ariolus.*] v. deófol-, tungol-wîtega.

wîteg-dôm, es ; *m.* **I.** *prophecy* :—Ðæt sié gefylled wîtigdôm (*prophetia*) Essaies, Mt. Kmbl. Rush. 13, 14. Ðæt uîtgadôm and allra canône cuido ða ðe ymb Cristes ðroung ácueden uæs † wêron, Jn. Skt. Lind. 19, 30 margin. **II.** *divination* :—Ne meahte seó manigeo þurh wîtigdôm wihte áþencean, ne áhicgan, Cd. Th. 224, 34 ; Dan. 146. [*O. H. Ger.* wîzag-tuom *prophetia, divinatio.*] Cf. wîte-dôm, wîtegung.

wîtege, an ; *f. A prophetess* :—Anna ðió witga Anna *prophetissa,* Lk. Skt. Lind. Rush. 2, 36. [*O. H. Ger.* wîzaga *prophetissa.*] v. wîtegestre.

wîte-geard (?), es ; *m. A place of punishment* :—Wîtehûses † wyerteardes (wîtegeardes ?) *amphitheatri,* Hpt. Gl. 484, 47.

wîtegend-lîc ; *adj. Prophetic* :—Ic (*Elisha*) bidde ðé (*Elijah*), ðæt ic beó áfylled mid ðam wîtegendlícum gáste ðe on ðé nú wunaþ, Homl. Skt. i. 18, 282. Ðæt cild on his môdor innoðe . . . mid wîtigendlícre fægnunge getácnode ðone tôcyme úres Álýsendes, Homl. Th. i. 352, 28. Hí wiston ða tôwerdan ðing, and mid wîtigendlícre gyddunge bododon, 540, 25.

wîtegestre, an ; *f. A prophetess* :—Anna wæs wîtegystre (*prophetissa*), Lk. Skt. 2, 36. Týn mǽdena wǽron on hǽðenum folcum, ðe man hêt Sibillas, ðæt synd wîtegestran, and hí wîtegodon ealle be Criste, Ælfc. T. Grn. 10, 31.

wîtegian, wîtgian ; *p.* ode *To prophesy,* (1) *absolute* :—Ða hig wîtegodon (*prophetarent*), ða arn án cnapa and cwæð : 'Eldad and Meldad wîtegiaþ (*prophetant*),' Num. 11, 27. Wîtigaþ, Cd. Th. 246, 16 ; Dan. 480. Wîtgas, Mt. Kmbl. p. 7, 10. Zacharias wæs mid hálegum gáste áfylled and hé wîtegode (*prophetauit*), Lk. Skt. 1, 67 : Num. 23, 8. Hú ne wîtegode wé on ðínum naman ? Mt. Kmbl. 7, 22. Ealle wîtegan wîtegudun (wîtgadun, Rush.) oð Iôhannes, 11, 13. Mid wîtegiende mûðe, Guthl. 5 ; Gdwin. 36, 19. (2) *with an object,* (a) *an accusative* :—Ðæt hé him wîtgode wyrda geþingu, Cd. Th. 250, 13 ; Dan. 546. 'Eówre wîtgan eów wîtgodan dysig' . . . Hié scolden leásunga wîtgian, Past. 15 ; Swt. 91, 3-8. Hié eal, ðæt tôweard wæs, beforan wîtgodan, Blickl. Homl. 161, 15. Se swêg wæs þurh wîtgan wîtgod, 133, 31. (b) *with a clause* :—Hé wîtgode, ðæt se Hǽlend sceolde sweltan for ðære þeóde, Jn. Skt. 11, 51. Hé wîtgode suá suá hit geweorðan sceolde, Past. 1 ; Swt. 29, 11. Wîtigan wîtigodan, ðæt se wolde cuman, Blickl. Homl. 105, 9. Wîtga (*prophetiza*) ús, huá is se ðe ðec ofslôg, Mt. Kmbl. Lind. 26, 68. (c) *with constructions of* (a) *and* (b) :—Heora fæderas ðæt wîtgodan, ðæt him God wolde sendan his sunu, Blickl. Homl. 177, 10. (3) *with a preposition* :—Anna wîtegode be him . . . swá hálig wíf wæs ðæs wyrðe, ðæt heó môste wîtigian embe Crist, Homl. Th. i. 146, 27-29. Hé wîtgode be ðære ácennednesse Cristes, Ps. Th. 8, arg. : Blickl. Homl. 133, 28. Wîtgade, 83, 24. Wîtegan ðe wîtegodon ymbe Crist, Ælfc. T. Grn. 2, 18. [Þis witeȝede Dauid . . . þis he witeȝede bi Drihtene, O. E. Homl. i. 7, 13-15. Minna þen sculen witeȝan, 91, 5. *O. Frs.* wîtgia : *O. H. Ger.* wîzagôn *prophetare, vaticinari, auguriari, divinare.*] v. fore-, ge-wîtegian.

wîtegung, e ; *f.* **I.** *prophecy* :—Án ðæra gecýdde Cristes tôcyme mid sealmsange, and ôðer mid wîtegunge. Sind sealmsang and wîtegung, swylce hí syflinge wǽron . . . tô ðam fíf ǽlicum bôcum, Homl. Th. i. 188, 19. Ða wæs gefylled Hieremias wîtegung, ðe ðus wîtegode, 80, 18. Esaias wîtegung (wîtigung *prophetia,* Lind.), Mt. Kmbl. 13, 14. Wîtgiung, p. 16, 15. In stefne wîtgeonges *in uoce prophetiæ,* Mk. Skt. p. 1, 8. Swá swá Isaias se wîtega hit on bêc sette on his wîtegunge, Ælfc. T. Grn. 2, 22. **II.** *divination* :—Þurh eorþan wîtegunga *geomantia,* Wrt. Voc. ii. 42, 23. Þurh deáþes wîtgung *nicromantia,* 62, 30. [He ȝifð summe wîtegunge, O. E. Homl. i. 97, 19. All þatt witeȝhunnge þatt hallȝhe witess writenn, Orm. 15149. *O. H. Ger.* wîzagunga *divinatio, vaticinium, auspicium.*] v. fore-wîteguug.

wîtegung-bôc ; *f. A book containing prophecies, a prophetical book* :—Hit is áwriten be mê on wîtegungbôcum, Homl. Skt. ii. 24, 115. Ic geliornod hæbbe on eówer wîtegungbôcum, ðæt gê wǽron fram frymðe gecorene fram Criste selfum, H. R. 7, 11, 30.

wîte-hrægel, es ; *n. A garment worn as a punishment, sackclo'h* :—Ic míne gewǽda on wîtehrægl cyrde *posui vestimentum meum cilicium,* Ps. Th. 68, 11.

wîte-hûs, es ; *n. A house of punishment* or *torment,* (1) *a prison* :—Wîtehûsa *ergastulorum,* Hpt. Gl. 516, 8. (2) *an amphitheatre* in which the Christians were martyred :—Wîtehûses *amphitheatri,* Hpt. Gl. 484, 47. On wîtehûse *in amphitheatrum* (the passage is : In amphitheatrum

sanctos ferreis collariis connexos cruentus carnifex imperat duci, Ald. 49), 489, 69. (3) *hell :*—Hê hêht ðæt wîtehûs wræcna (*the fallen angels*) bîdan, Cd. 3, 21; Gen. 39 : 304, 11; Sat. 628. On wrâþra wíc . . . , on wîtehûs, Exon. Th. 94, 7; Cri. 1536.

wîte-lâc, es; *n. Punishment, torment, pain :*—Wurdon tô axan eorðan wæstma, efne swâ wîde swâ ða wîtelâc (*the burning and terror at the destruction of Sodom and Gomorrah*) geræhton, Cd. Th. 154, 12; Gen. 2554. Weras bâsnedon wîteloccas (wîtelâces, Grn.) weán under weallum, 146, 5; Gen. 2417.

wîte-leás; *adj. Not having to pay a fine :*—Gelæste ælc wuduwe ða heregeata binnan twelf môndum, bûton hire ær tô onhagige, wîteleás, L. C. S. 74; Th. i. 416, 18.

wîtend-lîc *prophetic.* v. wîtiend-lîc.

wîtendlîce; *adv. Surely, certainly :*—Witendlîce hê getrymde ymbhwyrft eorðan *etenim firmavit orbem terrae,* Ps. Spl. 92, 2 : 40, 10 : 88, 6. Cf. witodlîce.

wîter, wîtter; *adj. Knowing, wise :*—Hê wîslíce hine beþôhte, swâ hê full wîtter wæs, Chr. 1067; Erl. 204, 35. [Heo wes wîter, heo wes wis, Laym. 9600. þeo weoren þa alre witereste þe wuneden on Bruttene, 15204. Full witerr takenn *a manifest token,* Orm. 4013. Wurð ðe child witter and war, Gen. and Ex. 1308. Wex he witter and wyse, Alex. Skt. 629. *Icel.* vitr *wise.*]

wîte-ræden[n], e; *f.* I. *punishment :*—Ðes cyning bebeád ðæt feówertiglîce fæsten healden beón ær Eástrum be wîteræðenne *jejunium quadraginta dierum observari praecepit . . . in transgressores dignas et competentes punitiones proposuit,* Bd. 3, 8; S. 531, 11. II. *fine.* v. wîte, I b :—Ut sit tuta . . . regalibus tributis majoribus et minoribus, sive taxationibus quod nos dicimus wîteréden, Cod. Dip. B. ii. 84, 7. Ego Tûnburht episcopus aliquam partem terrae donabo liberam ab omnibus terrenis difficultatibus omnium gravitudinum . . . a taxationibus quod dicimus wîteræðenne, Cod. Dip. Kmbl. v. 121, 25. Bîde mon mid ðære wîteræðenne oþ ðæt se wër gegolden sié, L. In. 71; Th. i. 148, 4. Nâh hê ðær nâne wîteræðenne *he cannot exact any fines,* 50; Th. i. 134, 4.

wîte-scræf, es; *n. A den of torment, hell :*—Gewît ðû âwyrgda in ðæt wîtescræf, Cd. Th. 308, 12; Sat. 691.

wîte-steng, es; *m. A pole used for punishment* or *torture :*—Wîtestengces, rôde *eculei,* wîtestenges *eculei, gabuli,* Hpt. Gl. 478, 70–73 : Anglia xiii. 34, 169. v. þrîpel.

wîte-stôw, e; *f. A place of punishment* or *torment, hell :*—Upp cômon sume ðara ðýstra gâsta of ðære neowolnesse and of ðære wîtestôwe (*de abysso illa flammivoma*), Bd. 5, 12; S. 628, 41. Nis hêr (*in hell*) nû nænig wôp, swâ hit ær gewunelíc wæs on ðisse wîtestôwe, Blickl. Homl. 85, 29.

wîte-swinge, an; *f. A stroke given as a punishment, chastisement :*—Ongæt gumena aldor hwæt him Waldend wræc wîteswingum, Cd. Th. 112, 2; Gen. 1864.

wîte-þeów, es; *m. One who had been condemned to slavery for crime, or from inability to pay the fines incurred for violation of the law.* For cases which involved loss of freedom, v. þeów. (1) literal :—Gif hwelc man biþ wîteþeów (or *adj.*? v. next word) nîwan geþeówad, L. In. 48; Th. i. 132, 7. (2) figurative, *one in hell :*—Bring ûs hælo líf wêrigum wîteþeówwum, Exon. Th. 10, 12; Cri. 151.

wîte-þeów; *adj. In slavery as a consequence of crime :*—Be wîteðeówes monnes slege. Gif witeþeów Englisc mon hine forstalie, hô hine mon, L. In. 24; Th. i. 118, 6. Gif ðær hwylc wîteðeów man sý ðe hió geðeówede, hió gelýfst tô hýre bearnon ðæt hî hine willon lýhtan for hyre sâulle, Cod. Dip. Kmbl. vi. 132, 8. Wîteðeówne monnan Wyliscne mon sceal bedrîfan be twelf hîdum tô swingum, L. In. 54; Th. i. 138, 3. Ic wullan ðæt man gefreógen ælcne wîteðeówne man on ælcum ðæra landæ ðæ ic mînon freóndon hæbbæ, Cod. Dip. Kmbl. iii. 128, 10. Ðæt man freóge on ælcum tûnæ ælcne wîtæþæównæ mann ðæ undær hiræ geðeówuð wæs, 360, 6. Ðis is Ælfsiges biscopes cwide. Ðæt is ærest, ðæt ic wille ðæt man gefreóge ælcne wîteþeówne mannan ðe on ðam biscopríce sié for hine and for his cynehlâford, Cod. Dip. B. ii. 329, 17 : L. Ath. i. proem.; Th. i. 198, 9. Wêron ðær þreó wîteþeówe men and þreó þeówberde, Cod. Dip. Kmbl. v. 152, 8. Be wîteðeówwum mannum, L. In. 48; Th. i. 132, 6. Cf. wîte-fæst, and see Kemble's Saxons in England, i. 200, Grmm. R. A. 328.

witga-dóm, wîtgian. v. wîteg-dóm, wîtegian.

wiþ; *prep.* (*adv. conj.*). I. with gen. (1) determining the direction of motion or action, (a) marking an object towards which motion is directed, *towards, to, in the direction of :*—Wende hê hine west wið Exanceastres, Chr. 894; Erl. 91, 10. Râd ût wið Lygtûnes, 917; Erl. 102, 16. Hê âfaren wæs wiþ þara scipa, Ors. 6, 36; Swt. 292, 30. On ðone ealdan weg wið huîtan stanes, Cod. Dip. Kmbl. ii. 29, 5. Fleógan wið ðæs holtes, Byrht. Th. 131, 14; By. 8. Wið ðæs fæstengeates folc onette, Judth. Thw. 23, 38; Jud. 162 : 25, 7; Jud. 248. Hê irneþ wið his eardes, Met. 5, 15. Heó stigþ wiþ hire uprynæs, Bt. 25; Fox 88, 27. Hê him bebeád, swâ hié feohtan angunnen, ðæt hié wið his flugen, Ors. 3, 7; Swt. 116, 28. Hê wið ðæs beornes stôp, Byrht. Th. 135, 41; By. 131. Líget fleáh wið ðæs hæðenan folces, Homl. Th. i. 504, 29. Ðæt wolcn leát wið his and hine genam fram heora gesihðum,

296, 2. Ðâ se hâlga wer ne com, ðâ cômon hí eft wið his (*they made their way to him*), ii. 172, 22. Sum fæmne âsende wið his, 506, 6. Hî âsendan twêgen weras wið his (tô him, *v. l.*), Homl. Skt. i. 10, 61. Ne gemêt hê hine, ne rihtne weg wiþ his ne âredaþ, Bt. 33, 3; Fox 128, 2. (b) marking an object towards which an action is directed, *towards, to, at :*—Hê hnâh tô eorðan, âleát wið ðæs engles (*he bowed to the angel*), Num. 22, 31 : Homl. Th. i. 120, 2. Hí luton wið heora, 38, 21. Grîp wið ðæs grundes *clutch at the bottom,* Cd. Th. 308, 31; Sat. 701. Se lêg læhte wið ðæs lâþan, 309, 25; Sat. 716. Beseah hê hine underbæc wiþ ðæs wîfes, Bt. 35, 6; Fox 170, 14 : Cd. Th. 154, 29; Gen. 2563. (c) marking the object of an operation, purpose, aim, feeling, *with, towards, to, at, against :*—Gif gebyrige ðæt heora hwilc wið ðære bige habban wille (*wants to come to us to buy*), oþþe wê wið heora, L. A. G. 5; Th. i. 156, 3. Hê beseah wið mín *respexit me,* Ps. Th. 39, 1. Hê wrigaþ wiþ his gecyndes, Bt. 25; Fox 88, 24, 28 : Met. 13, 67. Wiþ ðæs, ic wât, ðû wilt higian, Bt. 11, 2; Fox 34, 7. Mê wære liófre ðæt ic onette wiþ ðæs, ðæt ic ðê môste gelæstan ðæt ic ðê ær gehêt *festino debitum promissionis absolvere,* 40, 5; Fox 240, 16. Hwî murcnast ðû wið mín? 7, 3; Fox 20, 3. Deófles anda bið âstyred wið ðín, Homl. Skt. ii. 30, 115. (2) marking position, *over against, opposite to :*—Sætt se Hælend wið (*contra*) ðæs dores, Mk. Skt. Lind. 12, 41. (3) marking an object against which there is protection, *against, from :*—Hê hié wið ðæs hêhstan brôgan gefriðode, Judth. Thw. 21, 3; Jud. 4. Wið hungres hleó, Elen. Kmbl. 1228; El. 616. Wið yfela gefreó ûs feónda gehwylces, Hy. 6, 31. II. with dat. (1) marking local relations, (a) proximity, *by, near, against, beside :*—Æt Alre, and ðæt is wiþ Æþelingga eige, Chr. 878; Erl. 80, 22. Hire lîchama resteþ wið Rômebirig on ðam wege ðe man nemneþ Latina, Shrn. 31, 28. Sæweall uplang gestôd wið Israhêlum, Cd. Th. 197, 8; Exod. 303. (b) extension, *unto :*—Wið wolcnum *usque ad nubes,* Ps. Th. 56, 12. (c) contact, *at, against :*—Heald wiþ wætan (or acc. ?), Lchdm. ii. 150, 7. Him on hreþre langað beorn wið blôde (*burnt against the blood, heated his blood?*), Beo. Th. 3764; B. 1880. (d) collision or impact, *with, against, on :*—Scearp cymeþ sceó wiþ ôþrum, ecg wið ecge, Exon. Th. 385, 8; Rä. 4, 41. Ic hnîtan sceal hearde wið heardum, 497, 23; Rä. 87, 5. Streámas wundon sund wið sande, Beo. Th. 431; B. 213. Hê wið âttorsceaðan oreðe geræsde (*rushed and met the breath*), 5670; B. 2839 : Cd. Th. 126, 14; Gen. 2095. Hire wið halse heard grâpode, Beo. Th. 3136; B. 1566. Mid grâpe fôn wið feónde *to lay hands on the foe,* 882; B. 439. Ne sceal mon nô mid openlíce edwîte him wið sleán *non aperta exprobratione sunt feriendi,* Past. 40; Swt. 295, 11. (e) confronting, *over against, opposite :*—Ongan ic steppan forð âna wið englum *I stepped forth and alone confronted the angels,* Cd. Th. 280, 1; Sat. 249. Be norðan is se sæ, ðe ægþer is ge nearo ge hreóh wið Italia ðam lande (*opposite Italy*), Ors. 1, 1; Swt. 28, 12. (f) obstruction, *against, in the way of :*—Bordrand onswâf wið ðam gryregieste, Beo. Th. 5113; B. 2560. (2) marking association, combination, *with.* v. III. 2 :—Gesweotula ðín sylfes weorc, and forlæt weall wið wealle (*let wall join with wall*), Exon. Th. 1, 20; Cri. 11. Hê teofanade æghwylc wiþ ôþrum, 349, 10; Sch. 44. Sand is geblonden, grund wið greóte, Andr. Kmbl. 849; An. 425. Mengan lyge wið sôðe, leóht wið þýstrum, Elen. Kmbl. 613; El. 307. Hî wið mânfullum mengdan þeóde *commisti sunt inter gentes,* Ps. Th. 105, 26. Swâ gæð þeóstru wið leúhte *sicut tenebrae ejus, ita et lumen ejus,* 138, 11. Ðâ beád heó hire wer ðæt hé wið hire wylne týman sceolde, Boutr. Scrd. 22, 23. (3) marking separation, *with* (as in part *with*), *from.* v. III. 3 ; and see wiþ-faran, -ferian, -lædan :—Tôsceádene mid Trêntan streáme wiþ Norþ-Myrcum *discreti fluvio Treanta ab Aquilonalibus Mercis,* Bd. 3, 24; S. 557, 37. Hê gesundrode leóht wið þeóstrum, sceade wið scîman, Cd. Th. 8, 21; Gen. 127 : 10, 27; Gen. 163. Hwonne se dæg cume ðe hê sceole wið ðæm lîchomon hine gedælon, Blickl. Homl. 97, 20. Gedælan líf wið líce, Beo. Th. 4837; B. 2423 : Apstls. Kmbl. 73; Ap. 37. Nô hé hine wið monna miltse gedælde, Exon. Th. 122, 7; Gû. 302 : 146, 18; Gû. 711. Swâ nô man scyle his gâstes lufan wið Gode dælan, Cd. Th. 217, 12; Dan. 21. Ðam ðe his gâst wile meltan wið morðre, mergan of sorge, âsceádan of scyldum, Salm. Kmbl. 111; Sal. 55. (4) marking exchange or return, (a) buying (lit. or fig.), marking the object for which a price is paid, *for, in return for, as payment for :*—Abraham sealde feówer hund scillinga seolfres wið ðæm æcere and wið ðam scræfe, Gen. 23, 16 : Chart. Th. 232, 13. Twa and twêntig þûsend punda goldes and seolfres mon gesealde ðam here of Ængla-lande wið friðe, L. Eth. ii. 7; Th. i. 288, 12. Cantware him feoh gehêton wiþ ðam friþe, Chr. 865; Erl. 70, 33. Sendan beágas wið gebeorge, Byrht. Th. 132, 44; By. 31. Ðâ beád hê ealle his æhta wiþ his feore, Bt. 29, 2; Fox 104, 21. Ðæt mihte beón geboden wið clænum legere, Chart. Th. 208, 30. Hê sealde ælcon ænne penig wið hys dæges worce, Mt. Kmbl. 20, 2. Hê bæd ðæt hé him ðæs siþfætes lâtteów wære, and him mycel feoh wið ðon gebeád, Bd. 4, 5; S. 571, 35. (b) selling (lit. or fig.), marking the payment which is received, *for, in consideration of :*—Hwî ne sealde heó ðâs sealfe wiþ þrîm hundred penegon? *quare hoc ungentum non uenit trecentis denariis?* Jn. Skt. 12, 5. Hí him ðæt land sealdon wiþ .iii. pundon, Chart. Erl. 235, 27. Hê

gesealde wiþ feó heofones Hláford, Blickl. Homl. 69, 13 : Chr. 1036 ; Erl. 164, 34. Ðæt nán preóst ne dó his hálgan þénunge wiþ sceattum, L. Ælfc. C. 27 ; Th. ii. 352, 18 : Cd. Th. 262, 14 ; Dan. 744. Wið ðam golde grið fæstnian, Byrht. Th. 132, 52 ; By. 35. Gé ne réccaþ þeáh hweþer gé áuht tó gode dón wiþ ænegum óþrum þingum búton wið ðam lytlan lofe ðæs folces and wiþ ðam scortan hlísan, Bt. 18, 4 ; Fox 66, 21. Ðý læs men wénan ðæt ðú náne treówe næbbe búton wið hlísan (unless you can get reputation for it), Prov. Kmbl. 76. (c) exchanging (lit. or fig.), for, in exchange for :—Ðes landes boec ðet Eðelbearht cyning sealde his ðegne wið óðrum sué miclum lande, Cod. Dip. Kmbl. ii. 66, 17. Se ðe ealle his æhta·behwyrfde wið ánum gyldenum wecge, Homl. Th. i. 394, 12. (d) redemption, for :—Beád Darius healf his ríce wið þam wífmonnum, Ors. 3, 9 ; Swt. 126, 7. (e) reward or requital, for, in reward of, in return for :—Ic sylle Wulfsíge wið his holdum mægene and eádmódre hérnesse ánes hídes lond, Cod. Dip. B. ii. 268, 8. Forþ gewát ðurh martyrdóm Laurentius ; hæfþ nú líf wiþ ðan mid Wuldorfæder weorca tó leáne, Menol. Fox 290 ; Men. 146. Hí mé yfel settan á wið goode posuerunt adversum me mala pro bonis, Ps. Th. 108, 4. (f) reply, in answer to :—Suwade Crist wið ðæs wífes clypunge, Homl. Th. ii. 182, 7. (g) compensation, for, as compensation for. v. III. 4 :—Sylle líf wið (pro) lífe, tóð wið teð, hand wið handa, fót wið fét . . . lǽl wið lǽle, Ex. 21, 23–25. Gif hwá forstele óðres sceáp . . . selle feówer sceáp wið ánum. Gif hé næbbe hwæt hé selle, sié hé self beboht wið ðam fió, L. Alf. 24 ; Th. i. 50, 15. (h) where the condition, in consideration of which something takes place, is given, in consideration of, in return for, on condition of :—Hit Scipia nolde him áliéfan wið nánum óþrum þinge búton hié him ealle hiera wǽpeno ágeáfen Scipio would not grant it them on any other condition than that of giving up all their weapons to him, Ors. 4, 13 ; Swt. 210, 20. Ælces mannes þeówetlingas ða dríȝ dagas weorces beón gefreóde wið cyricsócne, and wið ðam ðe hý ðæt fæsten te lustlícor gefæsten, Wulfst. 171, 20 : 181, 19. ¶ wiþ ðam ðe or ðæt, introducing a clause that contains the condition or considera- tion :—Sende hé ǽrendracan tó him and mycel feoh wiþ ðon ðe hé hine ofslóge misit nuncios, qui Redualdo pecuniam multam pro nece ejus offer- rent, Bd. 2, 12 ; S. 513, 9. Se cásere him beád gold and seolfor wið ðon ðe hý forléton Cristes geleáfan, Shrn. 134, 5. Hé cwæð ðæt hé heom hold hláford beón wolde, . . . wið ðam ðe hí ealle tó him gecyrdon, Chr. 1014 ; Erl. 150, 12 : L. O. 1 ; Th. i. 178, 7. Nolde hé syllan ealle his æhta, wið ðan ðe hé libban móste? Homl. Skt. i. 12, 118. Hé forlǽt manigne woruldlust, wiþ ðam ðe hé ðone welan begite, Bt. 33, 2 ; Fox 124, 2. Hwelc wíte sceal ús tó hefig ðyncan, wið ðæm ðe wé mægen geearnian ðone hefenlícan éðel? Past. 36 ; Swt. 255, 3. Hú micle suiðor sculon wé beón gehiérsume, wið ðæm ðæt wé móten libban on écnesse, 255, 9. Hé wolde ungerím feós syllan, wið ðam, gif hé hit gebicgan mihte, ðæt hé hér lybban móste, Homl. Skt. i. 12, 102. (5) marking balance, counterpoise, against (as in to set one thing against another), as a set-off. v. III. 5 :—Swelce hié setten ða synne wið ðære ælmessan, Past. 45 ; Swt. 341, 20. (6) marking comparison, by the side of, compared with. v. III. 6 :—Hwæt is ǽnig lác wið ðisum willan ? Homl. Th. i. 584, 10. Nǽre ðeós blis ðe gelícre ðære écean myrhðe, ðonne bið ðam menn ðe sit on cwearterne, wið ðam menn ðe címéð feond land, Homl. Skt. i. 12, 109. (7) marking contrast, in contrast with :—Wiþ ðon e contrario, Bd. 5, 13 ; S. 633, 34 : 5, 14 ; S. 634, 42. (8) marking address, with, to. v. III. 7 :—Drihten wið Abrahame spræc, Cd. Th. 139, 2 ; Gen. 2303. Reordode ríces hyrde wið ðære fǽmnan fæder, Exon. Th. 246, 25 ; Jul. 67. Hyre se wrǽcmæcga wið þingade, 258, 5 ; Jul. 260. Him Andreas wið mælde, Andr. Kmbl. 598 ; An. 299. Wé habbaþ word gearu wið ðam ǽglǽcan, 2717 ; An. 1361. Ne heó wiþ monnum sprǽce hafaþ, Exon. Th. 421, 3 ; Rä. 40, 10. (9) marking dealing, with. v. III. 8 :—Hǽþen here genámon friþ wiþ Cantwarum, Chr. 865 ; Erl. 70, 32 : Ors. 3, 5 ; Swt. 106, 22. (10) marking hostility, with, against, to. v. III. 14 :—Hé feaht and won wiþ his éþle (contra patriam), Bd. 3, 24 ; S. 556, 28 : Exon. Th. 398, 1–2 ; Rä. 17, 1. Hí gefuhton wiþ hǽþnum herige, Chr. 853 ; Erl. 68, 17. Hié gefuhton wiþ Walum (Walas, MS. E.). 495 ; Erl. 14, 11. Holm won wið winde, Beo. Th. 2268 ; B. 1132 : Cd. Th. 5, 26 ; Gen. 77. Hilde gefremman wiþ ealdfeóndum, Exon. Th. 35, 32 ; Cri. 567. Wǽpen áhebban wið hetendum, Elen. Kmbl. 35 ; El. 18. Wið fírenum in gefeoht gearo, Exon. Th. 298, 24 ; Crä. 30. Hé honda árǽrde wið ðam herge, Cd. Th. 4, 9 ; Gen. 51. Hé wið ðam wyrme gewegan sceolde, Beo. Th. 4791 ; B. 2400. Se wið mongum stód, Exon. Th. 121, 26 ; Gú. 294. Swincan wið synnum, 150, 21 ; Gú. 782. For ðære synne ðe hé wið Sarran gefremede, Cd. Th. 166, 4 ; Gen. 2742 : Elen. Kmbl. 831 ; El. 416. Hié wið Godes bearne níð áhófon, 1671 ; El. 837. Wé wið Gode oft ábylgeaþ, Hy. 6, 21. Hié him ondrǽden wið (for, Cott. MSS.) hiera wordum and dǽdum hiera geférena tǽlinge, Past. 38 ; Swt. 273, 7. Hospcwide fremman wið Godes bearne, Elen. Kmbl. 1048 ; El. 525 : Andr. Kmbl. 1120 ; An. 560. Wæs yrre fæder wið dehter, Exon. Th. 251, 7 ; Jul. 141. Gód sceal wyð yfele, Menol. Fox 561 ; Gn. C. 50. Hé him ðǽr wiþ gefeaht, Chr. 871 ; Erl. 74, 8. Ðæt hí him wiþ ne winnan, Bt. 41, 5 ; Fox 254, 1. Ðæt inigtigra wíte

wealdeþ, ðonne hé him wið mǽge, Cd. Th. 249, 1 ; Dan. 523. (11) marking friendly relation, with, for, v. III. 15 :—Hé forget ðone fieóndscipe wið Israhéle, Past. 54 ; Swt. 423, 17. Hié wið Rómánum (-e, v.l.) sibbe heóldon civitatem amicam populi Romani, Ors. 4, 8 ; Swt. 186, 3. Hwæt is mannes sunu ðæt hit gemet wǽre, ðæt ðú him aht wið hæfdest (that thou shouldst have consideration for him ; quoniam reputas eum), Ps. Th. 143, 4. (12) marking protection, defence, salvation, against, from, for. v. III. 16 :—Ic ðé wið weána gehwam wreó and scylde, Cd. Th. 131, 2 ; Gen. 2170 : Exon. Th. 47, 27 ; Cri. 761. Ðú eart gescyldend wið sceaðan wǽpnum, Andr. Kmbl. 2584 ; An. 1293. Wið ælfsylcum éþelstólas healdan, Beo. Th. 4731 ; B. 2371 : 6000 ; B. 3004. Ealle ða wócre ðe hé wið wætre belcác, Cd. Th. 85, 4 ; Gen. 1049. Gefæstnod wið flóde, 80, 3 ; Gen. 1323. Hǽle and trume wið deófla níþum, Blickl. Homl. 171, 30. Wið fǽrscyte wearde healdan, Exon. Th. 48, 4 ; Cri. 766. Him holtwudu helpan ne meahte wið líge, Beo. Th. 4671 ; B. 2341 : 358 ; B. 178 : Elen. Kmbl. 369 ; El. 185. Sang se mæssepreóst orationem ða ðe wiþ ðære ádle áwritene wǽron, and ða ðing dyde ðe hé sélust wiþ ðon cúþe dicebat presbyter exorcismos, et quaeque poterat pro sedando miseri furore agebat, Bd. 3. 11 ; S. 536, 23–24. Wið eágena sáre, Lchdm. i. 2, 7 (and often). Godes módor hí áhredde wið heora feóndum, Chr. 994 ; Erl. 133, 16. Mé wið blód-hreówes weres bealuwe gehǽle de viris sanguinum salva me, Ps. Th. 58, 2. Wið níþum genergan, Exon. Th. 116, 24 ; Gú. 212 : Cd. Th. 233, 22 ; Dan. 279. (13) marking contrary motion or action, against, contrary to, in opposition to. v. III. 17 :—Wiþ winde rówan, Exon. Th. 345, 12 ; Gn. Ex. 187. Se wið ðínum willan wyrceþ, Met. 4, 28 : Bt. 14, 2 ; Fox 44, 9 : Blickl. Homl. 25, 15. Ic sceolde wiþ gesceape mínum on bonan willan búgan, Exon. Th. 486, 2 ; Rä. 72, 6. (14) marking the instrument, with. v. III. 19 :—Hiora in ánum weóll sefa wið sorgum, Beo. Th. 5193 ; B. 2600. (15) in reference to time, at :—Weard gesewen wið sunnan setlunge geond ealne ðone eard yrnende mere up on ðám wolcnum, Homl. Th. ii. 302, 2.   III. with accusative, (1) marking local relations, (a) where one object is near to or in contact with another, against, beside, by, at :—Wið ðone weg iuxta uiam, Ælfc. Gr. 47 ; Zup. 269, 15 : secus uiam, 271, 2. Wið ðone weall muro tenus, Wrt. Voc. ii. 57, 63. (α) of the position occupied by one body in rela- tion to another at rest :—Ðá hé wæs wið ða stówe (secus locum), Lk. Skt. 10, 32. Hé stód wið ðone mere (secus stagnum), 5, 1, 2. Hé gestód wið steápne rond, Beo. Th. 5126 ; B. 2566. Æteówde án engel wið hine (by him) Homl. Skt. i. 5, 88. Se geatweard sceal cytan habban wið ðæt geat (juxta portam), R. Ben. 126, 19. Tó ðám hátum baðum wið ðæt botl Salustii, Homl. Th. i. 428, 10. Sittan lǽte ic hine wið mé sylfne, Cd. Th. 28, 19 ; Gen. 438. Hé mé wið his sylfes sunu setl ge- tǽhte, Beo. Th. 4030 ; B. 2013. Wið ðæt dómsetl ic sitte pone tribunal sedeo, Ælfc. Gr. 47 ; Zup. 269, 16. Wið (secus) ðone ford hé sit, 271, 2. Sittende wið (juxta) ðone pitt, Gen. 29, 2 : Ex. 2, 15. Heó sæt wið (secus) ðæs Hǽlendes fét, Lk. Skt. 10, 39. Seó bóc lið wiþ (juxta) ðé, Ælfc. Gr. 38 ; Zup. 225, 2. Eal ðæt hé (Norway) man áþer oððe ettan oððe erian mæg, ðæt lið wið ða sǽ, Ors. 1, 1 ; Swt. 18, 26. Hé búde on ðǽm lande norþweardum wiþ ða westsǽ, 17, 3. Hé is ðǽr byrged wið Cnut cyng, Chr. 1046 ; Erl. 175, 4. Hí wacodon wið ða byrgene, Homl. Skt. i. 21, 120. Hig gewícodon wið ðone munt, Num. 20, 22. Heó wið wágas weaxan wylle, Lchdm. i. 116, 21. Ðeós wyrt bið cænned wið wegas, 224, 14. Hié wið eorðan fǽdm þúsend wintra eardodon they had remained on the ground a thousand years, Beo. Th. 6091 ; B. 3049. (β) of the position which is reached after movement :—Sume feóllon wið (secus) weg, Mt. Kmbl. 13, 4 : Lk. Skt. 8, 5. Wið ðone weg circa uiam, Mk. Skt. 4, 4. Hí setton scyldas wið ðæs recedes weal, Beo. Th. 658 ; B. 326. Hé heora fela gesette wið ðone sǽ pluri- mos ad mare habitare praecepit, Ors. 3, 5 ; Swt. 104, 26. Nim sticcan, sete on ðone nægl wið ða wearta . . . Heald wiþ wǽtan, Lchdm. ii. 150, 4–7. (γ) giving the direction of movement by reference to a body at rest :—Se Hǽlend eode wið (juxta) ða sǽ, Mt. Kmbl. 4, 18. Férde sum man wið hine quidam iter transiens uenit secus eum, Lk. Skt. 10, 33. Sum man férde wið ðone feld (cf. sum mon rád be ðære stówe (juxta locum), Bd. 3, 9 ; S. 533, 30, the incident being the same in both passages), Homl. Skt. ii. 26, 204. Heó wið ða eorðan (along the ground) hyre telgran tóbrǽdeþ, Lchdm. i. 324, 3. (δ) giving the direction of movement by reference to a moving body :—Wið ðone segn foran manna þengel rád, Cd. Th. 188, 23 ; Exod. 172. (b) marking position in connection with the parts of an object, by, against, at :—Gif monnes sconca bið of áslagen wið ðæt cneóu, L. Alf. pol. 72 ; Th. i. 98, 19. Forborn bord wið rond the shield burned against the rim, Beo. Th. 5339 ; B. 2673. (c) marking extension, unto :—Wið heofenas usque ad coelos, Ps. Th. 56, 12. (2) marking association, combination, with. v. II. 2 :—Drihten lét rínan hagol wið fýr gemenged and hig férdon ætgædere pluit Dominus grandinem, et grando et ignis mista pariter ferebantur, Ex. 9, 24 : Lchdm. ii. 30, 2 : Met. 7, 8 : Bt. 12 ; Fox 36, 9. Se yfela willa næfþ nænne geférscipe wiþ ða geselþa, 36, 7 ; Fox 184, 32. Ðis leóht wé habbaþ wið nýtenu gemǽne, Blickl. Homl. 21, 13. Hú ðone cumbolwigan wiþ ða hálgan mægþ hæfde geworden, Judth. Thw. 25, 14 ;

Jud. 260. Hē wolde dǣlan rîce wið God ælmihtigne, Wulfst. 306, 27: Homl. Th. i. 172, 1. Ðæt hié healfne geweald wið Eotena bearn âgan môston, Beo. Th. 2180; B. 1088. Hē gemôt wið hî habban wolde, Homl. Skt. i. 23, 21: Exon. Th. 334, 20; Gn. Ex. 19. Se hrefn wið wulf (*sic MS.*) wæl reáfode, Beo. Th. 6046; B. 3027. (3) marking separation, *from.* v. II. 3 :—Ne mæg mîn lîchoma wið ðás lǣnan gesceaft deáð gedǣlan (*my body cannot separate death from this frail condition in which it is created*, i.e. death is a condition inseparable from the frailty of the body), ac hē gedreósan sceal, Exon. Th. 124, 24; Gû. 342. (4) marking compensation, *for.* v. II. 4 g :—Sylle eáge wið eáge, Ex. 21, 24. (5) marking balance, counterpoise, *with, against* (as in to weigh one thing *with* or *against* another). v. II. 5 :—Genim ácmistel, gegnîd tô meluwe, âweh ðonne wið ǣnne pening, Lchdm. ii. 88, 6. Man sett ða synne and ða sáwle on ða wǣge, and hý man wegeþ, swâ man dêð gold wið penegas, Wulfst. 240, 2. Hiora birhtu ne bið âuht tô gesettanne wið ðǣre sunnan leóht, Met. 6, 7. (6) marking comparison, *in comparison with.* v. II. 6 :—Heora dýre gold ne bið nähte wurð wið ða foresǣdan mâðmas, Homl. Skt. i. 21, 55. (7) marking address, conversation, *with, to.* v. II. 8 :—Hē sprǣc heardlîcor wið hig ðonne wið fremde men *quasi ad alienos durius loquebatur*, Gen. 42, 7: 45, 15. Hû stîðe se landhlâford sprǣc wið hig *locutus est nobis dominus terrae dure*, 42, 30. Sprǣcan twégen weras wið hyne *duo uiri loquebantur cum illo*, Lk. Skt. 9, 30. Ongan Waldend wið Abraham sprecan, sǣgde him unlytel spell, Cd. Th. 145, 13; Gen. 2405. Hē wordum wið hý Waldend sprǣc, 155, 22; Gen. 2576. Heó ne mæg wordum wrixlan wið ðone wergan gâst, Exon. Th. 373, 30; Seel. 117. Wið ðone rǣdde Chromatius, Homl. Skt. i. 5, 323. (8) marking dealing, arrangement, where terms are come to with a person, *with.* v. II. 9 :—Wiþ ðone here se cyning friþ nam, Chr. 876; Erl. 78, 9. Swegen griðode wið ðone cyng, 1046; Erl. 172, 6. Ða foreword ðe Ælfwerd and se hîred worhtan wið Æðelmǣr, Chart. Erl. 235, 26. Hē sibbe ne wolde wið manna hwone mægenes Deniga feó þingian, Beo. Th. 315; B. 155. Tô þingienne þiódum sînum wið ðane Sceppend, Ps. C. 8: Exon. Th. 39, 4; Cri. 617: 254, 15; Jul. 197. þingeras wið ðone ælmihtigan þrym, Wulfst. 240, 10. Bûton hē gebête wið God, 271, 27: Homl. Skt. i. 12, 160. Ðâ rǣdde se cyng wið his witan (*the king settled with the 'witan'*), ðæt man sceolde mid scipfyrde faran, Chr. 999; Erl. 135, 29. (9) marking action affecting a person, (to deal) *with*, (act) *towards* :—Hē wið monna bearn wyrceþ weldǣdum, Exon. Th. 191, 11; Az. 86. Hwî dêst ðû wið mē swâ? Gen. 12, 18. For ðǣre ûtdrǣfe ðe hē gedyde wið hî, Homl. Skt. i. 21, 85. Begaa hē ða ryhtwîsnesse ðæs lâreówes wið ða gyltendan, Past. 17; Swt. 123, 23. Ic lufan symle lǣste wið eówic, Exon. Th. 30, 10; Cri. 477. Beó ðu hâlig wið ða hâlgan, and hwyrf ðe wið ða forhwyrfdan, Ps. Th. 17, 25. Men mihton tôcnâwan his mihte wið God, Homl. Skt. i. 19, 114. (10) marking action having reference to a person :—Hē wolde lîcettan wið Dauid *he would dissemble with David*, Homl. Skt. i. 12, 250. Âlýse ic mē sylfne wið God, 17, 75. Ðæt rîce and ðone anwald hē nâ ne angeat wið Cornelius (*in the case of Cornelius*), Past. 17; Swt. 115, 18. Þēh ðe he hit wið ða senatus hǣle *though he concealed it from the senate*, Ors. 4, 10; Swt. 196, 16. Hē bediglode his fær wið ðone wîtegan, Homl. Th. i. 400, 22. Nis mîn bân wið ðē deópe behýded *non est occultatum os meum abs te*, Ps. Th. 138, 13. (11) marking action directed to a person :—His gerêfa weard wið hine forwrēged *his steward was accused to him*, Lk. Skt. 16, 1. (12) marking position or attitude in regard to a person, *with, in respect to* :—Ne bið heó nâ wið God un-scyldig *non erit insons coram Deo*, L. Ecg. P. ii. 17; Th. ii. 188, 12. Scyldig wið God, Homl. Skt. ii. 27, 171: Cd. Th. 250, 20; Dan. 549. Beó hē ûtlah wið God and wið ðone cyningc scyldig ealles ðæs, ðe he âge, Wulfst. 271, 24: 296, 10. Hió hit hæbben unbesacen wið ǣlce hand, Cod. Dip. Kmbl. ii. 150, 23. Land unbecwedene and unforbodene wið ǣlcne man, Chart. Th. 209, 1. Ðæm ðeówan is tô cýðonne ðæt hē wiete ðæt hē nis freoh wið his hlâford, Past. 29; Swt. 200, 19. (13) *with* a person by whom something is held :—Býð ðē meord wið God, Andr. Kmbl. 550; An. 275. Hē wið ælda mæg eádes hleótan, Exon. Th. 305, 16; Fä. 89. Wið Drihten dýrne *dear in God's eyes*, Cd. Th. 32, 22; Gen. 507. (14) marking hostility, *with, against, to.* v. II. 10 :—Hié wiþ ðone here gefuhton, Chr. 871; Erl. 74, 10: Byrht. Th. 139, 61; By. 277. Wið his Waldend winnan, Cd. Th. 19, 28; Gen. 298. Simle hē feaht and won oþþe wiþ Angelcynn oþþe uuiþ Walas, Chr. 597; Erl. 20, 4. Hē wolde gecompian wiþ ðone âwerigdan gâst ... hē wolde deófol gelaþian tô campe wiþ hine, Blickl. Homl. 29, 17, 20. Þeáh wē fǣhþo wið ðec gefremed hæbben, Exon. Th. 23, 14; Cri. 368: Andr. Kmbl. 2773; An. 1389. Gylt, ðe wið Metod men gefremeden, Cd. Th. 61, 18; Gen. 999. Næbbe ic synne wið hié gefremed, 160, 15; Gen. 2650. Dǣdbôte dôn ðæs mycclan yfeles and mânes ðe hié wið heora Drihten gedydon, Blickl. Homl. 79, 6. Swâ hwæt swâ ðes middangeard wiþ hine æbyligða geworhte, 9, 12 : Elen. Kmbl. 1024; El. 513. Hē spræc heálig word wið Drihten sînne, Cd. Th. 19, 22; Gen. 295. Hē rêsade ðæt hē hæfde ǣrendo sum wiþ Francena rîce (*contra regnum*), Bd. 4, 1; S. 565, 1. Ic eom fâh wið God, Cd. Th. 270, 28; Sat. 97: Beo. Th. 1627; B. 811. For heora heardheortnesse wið ðone Hǣlend,

Homl. Skt. ii. 25, 529. Hē wæs strengesð wið scylda, Past. 17; Swt. 115, 17. (15) marking friendly relation, *with, to.* v. II. 11 :—Ic sibbe wið hine healdan wille, Exon. Th. 145, 2; Gû. 688. Treów ðū wið rodora weard healdest, Cd. Th. 127, 31; Gen. 2119. Ðæt ðū wið Waldend wǣre heólde, 204, 18; Exod. 421: Andr. Kmbl. 425; An. 213. Ðæt frið wið hý gefreoþad wǣre, Exon. Th. 127, 6; Gû. 382. Uton beón rihtwîse on ûrum môde wiþ ôþre men, Blickl. Homl. 95, 28. Beó wið Geátas glæd, Beo. Th. 2350; B. 1173. (16) marking protection, defence, salvation, *against, from, for.* v. II. 12 :—Ðæt hē ûs gescylde wiþ ða cræftas deófles, Blickl. Homl. 19, 16: Cd. Th. 245, 6; Dan. 458. Unc wið hronfixas werian, Beo. Th. 1085; B. 540. Wið wrâð werod wearde healdan, 643; B. 319. Hit ǣr hit nolde behealdan wið unnyt word, Past. 38; Swt. 279, 4. Geheald ðîne heortan wið unþeáwas, Wulfst. 247, 3. Ðæt man wið fûlne gálscipe warnie, 308, 2: Cd. Th. 15, 20; Gen. 236. Ðæt manna gehwylc wið swylc wær sý, Wulfst. 280, 11. Hē hine wiþ ða mânfullan âhreddan, Homl. Skt. ii. 29, 233. Wið swýðlîcne blôdryne of nosum, Lchdm. i. 2, 11, and often. Wið ðæt mannes innoð tô fæst sý, 2, 16, and often. (17) marking contrary motion or action, *against, contrary to.* v. II. 13 :—Wiþ Godes gife *contra gratiam Dei*, Bd. 1, 10; S. 480, 2. Wið mînes môdes willan *contra animi voluntatem*, Nar. 30, 26. Ǣr gē sceonde wið gesceapu fremmen, Cd. Th. 149, 4; Gen. 2469. (18) marking objection, *against* :—Ða geweddodan fǣmnan hire yldran hî ne môton syllan ôðrum men, bûton heó eallunga ðone (*the man to whom she is betrothed*) wið cweðe, ðæt heó hine nelle (*unless she bring the objection against him, that she does not wish to have him*), L. Ecg. C. 20; Th. ii. 146, 22. (19) marking the instrument, *by, through.* v. II. 14. (a) personal :—Hē sende ân tyccen wið his hirde *misit hoedum per pastorem suum*, Gen. 38, 20. Hē ðæt wið yfele englas sende *immissiones per angelos malos*, Ps. Th. 77, 49. (b) in the phrases sittan wið earm, &c., to rest on the arm :—Âîâs ânra gehwylc and wið earm gesæt, hleonade wið handa, Cd. Th. 291, 18; Sat. 432. Hē wið earm gesæt, Beo. Th. 1503; B. 749. (20) in reference to time, *till.* v. also VI :—Wið ða hwîle *donec*, Mt. Kmbl. Lind. 5, 18. IV. with dat. and acc. in the same passage :—Gesæt ðâ wið sylfne se ðe sæcce genæs, mǣg wið mǣge (v. III. 1 a a, II. 1 a, e), Beo. Th. 3958; B. 1977. Nǣfre Ismaël wið Isace, wið mîn âgen bearn yrfe dǣleþ (v. II. 2, III. 2), Cd. Th. 168, 24; Gen. 2787. Ðæt hî wurdon ðe geheortran wið ðam âwyrgedan strangan and ðone ealdan wiðerwinnan (v. II. 10, III. 14), Homl. Skt. i. 23, 241. Breóstnet wið ord and wið ecge ingang forstôd (v. III. 16, II. 12), Beo. Th. 3102; B. 1549. Ðîn mildheortnes is mycel wið heofenas (*usque ad coelos*), is ðîn sôðfæstnes wið wolcnum (*usque ad nubes*), Ps. Th. 56, 12. V. with the instrumental, cf. II. 2 :—Gemeng wiþ ðý leáce, Lchdm. ii. 34, 5. VI. not unfrequently the form of the word governed by *wiþ* does not shew the case: as instances of this are given the following passages in which the word is used with force of *till, to* :—Wið ende *usque in finem*, Ps. Th. 67, 16: *in finem*, 73, 10, 11. Wið oryldu *usque in senectam et senium*, 70, 16. Wið sefo sîða *usque septies*, Mt. Kmbl. Lind. 18, 21. Wið nû *usque nunc*, 11, 12: *usque modo*, Jn. Skt. Lind. Rush. 16, 24. Uið tô ðises (ðisse, Rush.) ꝥ uið nû ꝥ uið ðâgeána *usque athuc*, 2, 10. Wið tô ðæm dæge *usque ad eum diem*, Mt. Kmbl. Lind. 24, 38: 11, 13. VII. used adverbially; see also compounds with *wiþ* :—Meng hwîtcwudu wiþ (v. II. 2, III. 2), Lchdm. ii. 54, 3. Ðæt ǣnig wiþerweard ðing beón gemenged wiþ ôðrum wiþerweardum, oððe ǣnige geférrǣdenne wið habban (v. II. 2, III. 2), Bt. 16, 3; Fox 54, 13. Nâuþer ne ðone anweald, ne eác ðæt ðæt hē wiþ sealde (v. II. 4 a), 33, 2; Fox 124, 15. Hē cwæð jâ wið (v. II. 4 f), Chr. 1067; Erl. 204, 23. Gif hwâ forstele ôðres oxan ... sette twēgen wið (v. II. 4 g), L. Alf. 24; Th. i. 50, 15. Him cômon ongeán .vi. cyningas, and ealle wið trýwsodon, ðæt hî woldon efenwyrhton beón (v. III. 8), Chr. 972; Erl. 125, 12. Heald ðē elne wið (v. II. 10, III. 14), Exon. Th. 303, 9; Fä. 50. Wilna brûceþ, and nô wið spriceþ (v. II. 13, III. 17), 411, 10; Rä. 29, 10. VIII. as conjunction. v. III. 20, *until* :—Wið gē ðona geonga *donec exeatis*, Mt. Kmbl. Lind. 10, 11, 23: 24, 39. IX. combined (1) with *weard* (q.v.) (a) with gen. :—Hundas rǣsdon wið Petres weard, Homl. Th. i. 376, 34. Âstrehte hē hine sylfne tô eorðan wið his weard, ii. 168, 24. Ðâ ðâ hî wið his werd wǣron, Homl. Skt. i. 3, 102: ii. 23 b, 136. Hî wið ðæs heres weard wǣron, Chr. 1003; Erl. 139, 5. (b) with acc. :—Hē beheóld wið heofonas weard, Homl. Th. i. 46, 29: 382, 9: 464, 29. (2) with *weardes.* v. weardes. [O. Sax. wið: O. Frs. with: Icel. við.] v. þǣr-wiþ.

wiþ-æftan; prep. adv. *Behind.* I. prep. (1) with dat. :—Heó hym tô geneálæhte wyðæftan hym, Homl. Ass. 182, 48. Hî cômon tô Wiht, and nâmon ðǣr ðæt him ǣr wiðæstan wæs (*what had been left behind them*), Chr. 1052; Erl. 183, 25. (2) with acc. or doubtful :—Heó com wiðæstan ða menigu *uenit in turba retro*, Mk. Skt. 5, 27. Sette syrwa wiðæstan ða burh *pone insidias urbi post eam*, Jos. 8, 2. Heó stôd wiðæstan his fêt *stans retro secus pedes eius*, Lk. Skt. 7, 38. Ðū âwurpe mîne word wiðæstan ðē, R. Ben. 12, 3. II. adv. :—Ân wîf geneálæhte wiðæstan *mulier accessit retro*, Mt. Kmbl. 9, 20. Fîf scipu

belifan wiðæftan, Chr. 1047; Erl. 175, 12. Wiðeftan *posse* (=*post se* filios derelinquet*, Prov. 20, 7), Kent. Gl. 735.

wiþ-bláwan; *p.* -bleów *To strain at* :—Ðæt hí wiðbleówen ðære fleógan and forswulgun ðone olfend *liquantes culicem, camelum autem glutientes*, Past. 57; Swt. 439, 24.

wiþ-bregdan, -brédan; *p.* -brægd, -bréd, *pl.* -brugdon, -brúdon *To withhold, restrain, check, hold back* :—Gif ðæt mód ðæm willan ne wiðbritt *dum in cogitatione voluptas non reprimitur*, Past. 11; Swt. 71, 8. Godes feónd wiðbritt ðæm untruman móde ðære sibbe (*dilectionem proximorum vulneratis cordibus subtrahens*) ðe hé self forlét, 47; Swt. 361, 2. Ðá ðá hé wolde árwierðra monna mód from ðisses middangeardes geférrædenne áteón, suíðe suíðe hé him wiðbræd, ðá hé cuæð *Paulus religiosorum mentes a mundi consortio contestando, ac potius conveniendo suspendit, dicens*, 18; Swt. 131, 1. Hé hét heora ælcum fiftig scyllinga tó sceatte syllan, ðæt hí heora handa fram ðam blódes gyte ne wiðbrúdon, Homl. Th. i. 88, 5. Hit is micel ðearf, ðæt mon hire suíðe hrædlíce wiðbregde *festinare necesse est, ut repugnatione vincantur*, Past. 13; Swt. 79, 21. [Bute þu wiðbride þe, H. M. 9, 9.]

wiþ-ceósan; *p.* -ceás, *pl.* -curon; *pp.* -coren *To reject* :—Hé wiðceóseþ (-círt) *reprobat*, Blickl. Gl. Stán ðone wiðcurun timbrende *lapidem quem reprobaverunt aedificantes*, Ps. Surt. 117, 22. Wiðcurun, Mt. Kmbl. Rush. 21, 42. v. next word.

wiþ-coren; *adj.* (*ptcpl.*) *Reprobate* :—Ðá ongeat hé ðæt se wæs Gode wiðcoren, se ðe on ðæt bæþ eode, Shrn. 62, 8. Ðæt yfel wræc cóme ofer ða wiþcorenan *ut veniret contra improbos malum*, Bd. 1, 14; S. 482, 41. v. wiþer-coren.

wiþ-cwedenness, e; *f. Gainsaying, contradiction, opposition* :—Hí woldon hine besyrewian æt his lífe, and habban syþðan his ríce bútan ælcre wiðcweðenesse, Chr. 1002; Erl. 137, 36 note. Tó wetre wiðcwedenisse *ad aquas contradictionis*, Ps. Surt. 105, 32. Of wiðcweðenisse, 17, 44. Wiðcwedennysse, Ps. Spl. C. 30, 26: 79, 7.

wiþ-cweþan; *p.* -cwæþ, *pl.* -cwædon; *pp.* -cweden. **I.** *to reply.* v. wiþ, VII :—Ðá wiþcwæþ him se engel *contradicens angelus*, Bd. 3, 19; S. 549, 6. Com Swegen tó Eádwerde cinge, and gyrnde tó him landes. Ac Harold his bróðor wiðcwæð, and Beorn eorl, ðæt hig noldon him ágyfan nán þingc ðæs ðe se cing heom gegyfen hæfde, Chr. 1049; Erl. 172, 31. Cwæð sum wyln, ðæt hé mid ðam Hælende wære, and hé wiðcwæð, ðæt hé hine ne cúðe, Homl. Th. ii. 248, 31. **II.** *to gainsay, contradict, maintain an opposite opinion* :—Ða hálgan apostolas heredon ða clænnysse . . . Se ðe him wiðcwyð, hé ne byð ná wita, ac gedwola, Homl. Ass. 22, 198. Hé sægð ðæt hit sý álýfed, ðæt massepreóstas móton wífian, and míne gewritu wiðcweðþ ðysum, 13, 7. Ða mágas setton ðam cilde naman Zacharias, ac seó módor him wiðcwæð mid wordum, and se dumba fæder mid gewrite, Homl. Th. i. 354, 25. Drihtne ðrowende him cuoeðende wiðcuoeð *Domino passurum se dicenti contradicit*, Mk. Skt. p. 4, 2. Hí eallum his wordum wiðcwædon, Bd. 2, 2; S. 503, 17. Ne mæg ic ná wiþcweþan ne andsacigan ðæt ðe ðú mé ær sædest, Bt. 10; Fox 26, 24. Ic ne mæg nó wiðcweþan, ne furþum ongeán ðæt geþencan, 34, 1; Fox 134, 29. Ic sylle eów múð and wísdóm ðam ne magon eówer wiðerwinnan wiðstandan and wiðcweðan (-cuoeða, Lind., -cweoða, Rush. *contradicere*), Lk. Skt. 21, 15. Hé begann tó wiðcweðenne ðam geleáfan ðe se apostol tæhte, Homl. Th. ii. 412, 28. **III.** *to contradict, oppose, resist* :—Se man, ðe wiðcwið ðínum wordum *qui contradixerit ori tuo et non obedierit cunctis sermonibus tuis*, Jos. 1, 18. Éghwoelc se ðe hine cyning wyrcið wiðcuoeðæs (wiðcuoeðes, Rush. *contradicit*) ðæm cáser, Jn. Skt. Lind. 19, 12. ‘Ne stala ðú.’ Ðis bebod wiðcweð ælcum reáfláce, Homl. Th. ii. 208, 24, 27 : 210, 1. Tó ðcum forwyrde ðam ðe him (*Antichrist*) onbúgaþ, and tó ðcere myrhðe ðam ðe him wiðcweðaþ, i. 4, 35. Wiðcwæð *reluctaretur*, Hpt. Gl. 509, 16. Cristes naman, ðam hí ær wiþcwædon *nomen Christi, cui contradixerant*, Bd. 3, 30; S. 562, 16. Hé wæs ofer eall gemett stearc ðam mannum ðe wiðcwædon his willan, Chr. 1086; Erl. 221, 18. Nis nán wuht ðe mæge oððe wille swá heágum góde wiþcweþan *non est aliquid, quod summo huic bono vel velit, vel possit obsistere*, Bt. 35, 4; Fox 160, 30. On tácen ðam ðe wiðcweden byð *in signum cui contradicetur*, Lk. Skt. 2, 34. **IV.** *to refuse, reject, not to allow* :—Hé wiðcwyð ðeódegendra folce and hé wiðcwyþ geþeaht ealdrum *reprobat cogitationes populorum et reprobat consilia principum*, Ps. Spl. 32, 10. Ða com Sparhafoc tó ðam arcebiscope, tó ðam ðet hé hine hádian sceolde. Ða wiðcweð se arcebiscop, and cwæð ðet se pápa hit him forboden hæfde, Chr. 1048; Erl. 177, 21. Wiðcwæð *renunciaverit*, Hpt. Gl. 512, 72. Seó burhwaru wolde ðone hálgan geniman, and Pictauienscisce þearle wiðcwædon *the citizens wanted to take the saint, and the Poitevins absolutely refused to allow it*, Homl. Th. ii. 518, 20. Ða begann se cyngc gyrnan his sweostor him tó wífe, ac hé and his menn ealle lange wiðcwædon, and eác heó sylf wiðsóc, 1067; Erl. 204, 17. Ða þrý cnihtas wiðcwædon his hæþenscipe, Homl. Ass. 70, 131. Sume sind gedwedene *vitia*, ðæt synd leahtras, on manegum wísum miswritene oððe miscwedene; ðam eallum wé sceolon wiðcweðan, gyf wé cunnon ðæt gesceád, Ælfc. Gr. 50, 23; Zup. 294, 15. **IV a.** with dat. of person to whom a refusal is given :—Hé wolde ðæt hé ána wære heora cyning, ac ealle ða leódscipas ánmódlíce

him wiðcwædon, Homl. Ass. 103, 34. **IV b.** with dat. of person and gen. of thing refused :—Him ða burgleóde ðæs wiðcwædon, Ors. 3, 7; Swt. 116, 8. Gif inc hwá ðæs wiþcweþe, Blickl. Homl. 71, 1. [Ealle munechades men hit wiðcwæðen . . . Ealle þa biscopas him underfengen, him wiðcwæðen muneces and eorles, Chr. 1123; Erl. 250, 17, 24. He wiðquað (*respondit*, Lk. 3, 16), and sede, O. E. Homl. ii. 137, 30.] v. wiþer-cweþan.

wiþ-drífan; *p.* -dráf *To repel* :—Næfre wiðdríseþ Drihten úre his ágen folc *non repellet Dominus plebem suam*, Ps. Th. 93, 13 : 94, 4.

wiþ-eástan; *prep. adv. To the east*, (1) prep. :—Wyðeástan Constantinopolim Créca byrig iš se sæ Proponditis, Ors. 1, 1; Swt. 22, 2. (2) *adv.* :—Seó eá wiðeástan út on ða sæ flóweþ, Swt. 8, 20.

wiþer; *prep. adv.* (1) prep. with acc. *Against* :—Míne ágen word wiðer (*adversum*) mé wæran georne, Ps. Th. 55, 5. Uiððir ða *adversus eos*, Rtl. 168, 5. (2) *adv. Against, in opposition* :—Wiþer *infensus*, Germ. 394, 366 cf. he wæh Wiðer king þe wiðer wes an compe, Laym. 9287. [*Goth.* wiþra : *O. Sax.* wiðar : *O. Frs.* withir : *O. H. Ger.* widar : *Icel.* viðr.] v. wiþere.

wiþer (?), es; *n. Opposition, resistance* :—Hé hæfde Higeláces hilde gefrúnen, wlonces wígcræft; wiðres ne trúwode, ðæt hé sæmannum onsacan mihte, Beo. Th. 5899; B. 2953. [Ða ich wer i wide sæ, wiðer com toʒenes, þet weder wes swa wilde, Laym. 4678. Cf. His wiðerfulle hine, þo ben deules on helle, O. E. Homl. ii. 51, 21. Wiðerfulle cheorles, Laym. 21520.] v. wiþere, *and preceding word.*

wiþer-breca, an; *m. An adversary* :—On eallum dædum Godes wiþerbreca, Blickl. Homl. 175, 8. Nis ðe wiðerbreca nymðe Metod ána, Cd. Th. 251, 20; Dan. 566. Se (*Ishmael*) bið wiðerbreca wera cneórissum, 138, 7; Gen. 2288. Wæs ðú geðafsum wiðerbracæ (*adversario*) ðínum; ðý læs gesellæ ðec ðe widerbracæ tó dóme, Mt. Kmbl. Lind. 5, 25. Gif ðæ wiðerbraca (*Satanas*) ðone wiðerbraco drífes, 12, 26. Wiðerbrecan *obpositum*, Wrt. Voc. ii. 65, 24. Hé his wiþerbreocum sorge gesægde, Exon. Th. 120, 2; Gú. 265. Ðú forbriccest wiþerbrecan *conteruisti adversarios*, Cant. Moys. Ex. 15, 7 : Cd. Th. 4, 35 ; Gen. 64. [Cf. *O. H. Ger.* widar-brechinta *repugnantem*.] v. next word.

wiþer-broca, an; *m. An adversary* :—Wiðyrbroca *adversarius*, Ps. Spl. C. 73, 11. Forhtiaþ wiðerbrocan (*adversarii*) his, Ps. Surt. ii. p. 186, 36 : 187, 25. Wiðerbrocum *adversariis*, 194, 37. Ðú slóge alle wiðerbrocan mé *tu percussisti omnes adversantes mihi*, Ps. Surt. 3, 8. v. preceding and following words.

wiþer-brocian; *p.* ode *To oppose, be adverse to, be against* :—Se feónd bismerad wiðerbrocaþ noman ðínne *inimicus inritat adversarius nomen tuum*, Ps. Surt. 73, 10. Ða ðe wiðerbrociaþ mé *qui adversantur mihi*, Ps. Spl. C. 34, 22. v. preceding word.

wiþer-bróga, an; *m. Terror caused to an adversary* (?) :—Nú sind duguþum bidæled deófla cempan; ne meahtan wiþerbrógan, wíge spówan (*they could not succeed in being terrible to their adversaries, could not succeed in war*), siþþan wuldres cyning hilde gefremede wiþ his ealdfeóndum, Exon. Th. 35, 26 ; Cri. 564.

wiþer-cirr, es; *m. A going against, resistance* :—Ic gehýned eom, fáh and freóndleás; ic findan ne can wiðercyrr wið ðan of ðám wearhtreafum *I am humiliated, proscribed and friendless; against this I can devise no resistance from hell*, Elen. Kmbl. 1849; El. 926. [Cf. *O. H. Ger.* widar-ker; *m. conversio:* widar-kera; *f. controversio.*] Cf. ed-, ofer-cirr.

wiþer-cora, an; *m.* **I.** *an adversary, opponent, rebel* :—Wiþercora *contrarius*, Wrt. Voc. ii. 140, 75. Ne sý hé sacerd geteald, ac Godes wiþercora (wiþersaca, *v. l. rebellio*), R. Ben. 113, 13. Freónd hé wæs ðurh geleáfan, and wiþercora þurh weorc, Homl. Th. i. 530, 5. Gesamnodon gehwylce ðwyrlíce wiðercoran, and wrehton ðone cyning tó his bréðer, 468, 5. Wiþercorum *rebellibus*, Wülck. Gl. 256, 31. **II.** *a reprobate person* :—Wiðercora *reprobus*, R. Ben. Interl. 13, 8. Mid micelre geornfulnysse gewilniaþ ða wiðercoran (*the wicked in hell*) ðæt hí móton of ðære susle ðe hí on cwylmiaþ, Homl. Th. i. 332, 19.

wiþer-coren; *adj.* (*ptcpl.*) **I.** *reprobate, wicked* :—Elles wiðercoren hé is, líchamlícere wrace hé sig underþeód *sin autem improbus est, vindicte corporali subdatur*, R. Ben. Interl. 56, 2. For ðissum lænan lífe ðæt unlæne, for ðyssum ungecorenum (wiðercorenum, *v.l.*) ðæt gecorene, Wulfst. 264, 19. Ðæt yfel wræc cóme ofer ða wiþercorenan (*improbos*), Bd. 1, 14; S. 482, 41 note. **II.** *rejected* from heaven, *reprobate* as opposed to elect :—Ðæt ða gecorenan ðý geleáffulran wæron; and ða wiðercorenan náne beládunge nabbaþ, Homl. Th. i. 406, 35 : ii. 568, 33. Ðær beóð feówer werod æt ðam dóme, twá gecorenra manna, and twá wiðercorenra, i. 396, 17 : 332, 23, 29 : 536, 32. Sam ðe gecorenra tó reste, sam ðe wiþercoreura tó deáþe *siue electorum ad requiem, siue reproborum ad mortem*, Scint. 226, 14. [Cf. *O. H. Ger.* widar-kiusan *reprobare*.] v. wiþ-coren.

wiþer-corenness, e; *f. Reprobateness* :—Swá fela manna gebúgaþ tó geleáfan on ðissere andwerdan geladunge, ðæt hí sume eft út berstaþ ðurh wiðercorennysse and leahtrum heora ðwyran lífes, Homl. Th. ii. 290, 19.

wiþer-crist, es; *m. An antichrist* :—Wiðer ł leáso cristo *pseudochristi*, Mk. Skt. Lind. 13, 22.

wiþer-cwedolness, e; *f. Contradiction;* contradictio: — Wiþer-

cwedolnesse *contradictionis*, Blickl. Gl. Wiðercwydelnysse, Ps. Spl. 80, 7 : Ps. Lamb. 105, 32. Wiþercwedulnisse, Blickl. Gl. : Ps. Spl. 79, 7. Widercwidelnyssum, Ps. Lamb. 17, 44.

wiþer-cwedung, e ; *f. Gainsaying* :—Word wyþercwedunga *verba praecipitationis*, Ps. Spl. 51, 4.

wiþer-cweþan; *p.* -cwæþ, *pl.* -cwǽdon *To resist*, cf. wiþ-cweþan, III, wiþer-cwide :—Gemágnesse wiðsacende wiðercweðan (-en, MS.) *importunitatem refutando frustrari* (*contradicere*), Hpt. Gl. 491, 32. Widercwiðendum *resistentibus*, Ps. Lamb. 16, 8. [*O. H. Ger.* widarquedan *contradicere*.]

wiþer-cweþness, e ; *f. Contradiction* :—Hine mon ne cnysð mid nánre rednesse ne nánre widercuednisse (-cwed-, Cott. MSS.) *quem nulla asperitas contradictionis pulsat*, Past. 19 ; Swt. 143, 20.

wiþer-cwida, an ; *m.* I. *a contradicter* :—Ungeleáful widercwyda *incredulus negator, infidelis contradictor*, Hpt. Gl. 451, 11. II. *a rebel* :—Widercwyda *rebellio*, Wrt. Voc. i. 18, 19. [*O. H. Ger.* widar-queto *a contradicter*.]

wiþer-cwide, es ; *m. Resistance, opposition, contest* :—Ðæt twelf hída land bútan ælcum widercwide seó ágefen tô Wigornacestre, Chart. Th. 131, 25 : Chart. Erl. 162, 1. Gif hwá openne widercwyde ongeán lahriht gewyrce, L. Eth. v. 31 ; Th. i. 312, 8. Ðær hí widercwyde wæteres hæfdon *ad aquas contradictionis*, Ps. Th. 105, 25. Ic on unriht lôcade and widercwyda wearn gehýrde *vidi iniquitatem et contradictionem*, 54, 8. [Cf. *O. H. Ger.* widar-queta *contradictio*.]

wiþer-dúne *glosses* angusta *in* :—Hú naru t widerdúne geate *quam angusta porta*, Mt. Kmbl. Rush. 7, 14.

wiþere ; *adv. prep. Against* :—Weallas him (*the waves*) wiþre healdaþ *the cliffs hold out against the waves*, Exon. Th. 336, 24 ; Gn. Ex. 54. Cf. ʒif þe king wolde wið heom widerheolden, Laym. 9175. [*O. H. Ger.* widari, wid[i]ri.] v. tô-wiþere, wiþer.

wiþer-feohtend, es ; *m. An adversary* :—Gáð fromlíce, ðæt gê widerfeohtend wíges gehnægan, Andr. Kmbl. 2367 ; An. 1185. Ðæt gê wearde healden, ðý læs eów wiþerfeohtend weges forwyrnen tô wuldres byrig, Exon. Th. 282, 17 ; Jul. 664. v. wiþ-feohtend.

wiþer-flita, an ; *m. An adversary, opponent* :—Magan hiora sprǽce gemetgian ða ðe ðæs cristendômes wiþerflitan sint, Ors. 2, 1 ; Swt. 64, 14. Widerflitan, 2, 5 ; Swt. 84, 26 : 3, 3 ; Swt. 102, 15. Cf. wiþ-flítan.

Wiþer-gild, es ; *m. A man's name* :—Weóldon wælstôwe, syððan Wiðergyld læg (cf. syððan Heardred læg, 4766 ; B. 2388), æfter hæleþa hryre, hwate Scyldingas, Beo. Th. 4109 ; B. 2051. Sôhte ic Wiþergield and Freoþeríc, Exon. Th. 326, 5 ; Víd. 124. [For a form similar to this, but used as a common noun, in other languages, v. Grmm. R. A. 652.]

wiþer-habban ; *p.* -hæfde *To resist* :—Hwæt mæg mê widerhabban ? *quid mihi restat?* Ps. Th. 72, 20. [*O. H. Ger.* widar-habēn *reniti, retundere, resultare*.] v. wiþ-habban.

wiþer-hlinian ; *p.* ode *To lean against* :—Widerhlingende, uuidirhliniendae, uuidirlinienti *innitentes*, Txts. 71, 1098. Wiþerhlyniende, Wrt. Voc. ii. 48, 78.

wiþer-hycgende ; *adj. Having hostile thoughts* or *purpose* against another, *of evil intent* :—Emulus, i. *contrarius* gewinna, wiþerwinna, æfstig, wiþerhycgende, Wrt. Voc. ii. 143, 48. Ongan meldigan helle hinca ðone hálgan wer, widerhycgende, Andr. Kmbl. 2345 ; An. 1174. Ðú (*the devil*) scealt, widerhycgende (*the adversary of God and man*), wergðu dreógan, Elen. Kmbl. 1900 ; El. 952. Ðê leán sceolan wiþerhycgende (*opponent of the gods*), witebrôgan æfter weorþan, Exon. Th. 254, 12 ; Jul. 196. Wendun gê (*the devils*) and woldun, wiþerhycgende (*rebellious*), ðæt gê Scyppende sceoldan gelíce wesan, 141, 31 ; Gú. 635. Wendon and woldon, widerhycgende (*having evil designs upon the strangers*), ðæt hié on elþeódigum æt geworhton, Andr. Kmbl. 2146 ; An. 1074. v. wiþ-hogian, -hycgan.

wiþer-hygdig, -hýdig ; *adj. Hostilely disposed, adverse* :—Hé áhôf wôðe widerhýdig *he raised his voice with mind adverse*, Andr. Kmbl. 1349 ; An. 675.

wiþerian, wiþrian ; *p.* ode. I. *to be against, be hostile* :—Ic widerige *adversor*, Ælfc. Gr. 25 ; Zup. 145, 18. Ða ðe widriaþ mê *qui adversantur mihi*, Ps. Lamb. 34, 19. Ða widrigendan (widriende, Ps. Spl.) mê *adversantes mihi*, 3, 8. II. *to strive with, against* (*wiþ*, *ongeán*), *struggle, dispute* :—Ic widerige *controuersor*, Ælfc. Gr. 37 ; Zup. 219, 9. Beó ðú gebeogul ðínum widerwinnan, ðe læs ðe ðín widerwinna, gif ðú wiðerast wið hine, ðê betǽce ðam dêman, Homl. Ass. 4, 95. For ði synd ða gesibsuman Godes bearn, for ðan ðe nán ðing on him ne wiðeraþ ongeán God, Homl. Th. i. 552, 22. Mislára, ða úrum ongeán wiþeriaþ andgytum *suggestiones, quae nostris obstrepunt sensibus*, Scint. 33, 20. Ðás twá burh widriaþ betwux him, Homl. Th. ii. 66, 28. Hé ne widerode ongeán, ne ne feaht, 40, 17. Hí wiðerodon ongeán Cristes láre, 224, 30. Gif preóst ongeán biscopes gerǽdnesse widerige, L. N. P. L. 45 ; Th. ii. 296, 18. Se ðe sôðlíce God lufaþ, nele hé widerian ongeán his bebodum, Homl. Th. ii. 522, 18. III. *to resist, oppose* :—Wiðstôd t widerode *refragabatur, resistebat*, Hpt. Gl. 426, 41. Hig wǽron gemǽste and widerodun (*recalcitravit*), Deut. 32, 15. Eal folc hine tô ðære geðincðe geceás, þeáh ðe hé mid eallum mægne

widerigende wǽre, Homl. Th. ii. 122, 23. IV. *to make hostile, to provoke* :—Se ðe gecyrredne búton liðnysse lǽrð, widerian (*exasperare*) má dænne þreágean cann, Scint. 61, 12. V. *to become provoked* :—Gebíg fram unwitan, and ðú ná wiþerast (*exacerbaberis*) on stuntnysse his, Scint. 188, 11. [He seð þo þe wideried togenes him, O. E. Homl. ii. 123, 36. So hit unmedluker is, heo wunnen (widereð, *v. l.*) agean þe uestluker, A. R. 238, 17. Wrestlin ant widerin wið ham seoluen, Marh. 14, 13. Shep ... þær mann cwelleþþ itt, ne wiþþreþþ itt nohht swiþe, Orm. 1181. Fleges ... wideren in ðæt web, Misc. 15, 475. *O. H. Ger.* widarôn *abnuere, renuere, reniti, obviare, reluctari*.]

wiþer-lécan ; *p.* -lǽhte *to deprive* :—Wyþerlécaþ *privabit*, Ps. Spl. T. 83, 13.

wiþer-leán, es ; *n. Recompense, retribution* :—Weard wícingum wiþerleán ágifen ; gehýrde ic ðæt Eádweard ánne slôge, Byrht. Th. 135, 11 ; By. 116. Deápes háliges wiþerleáne (*as recompense*) líf eádig geáhniaþ *mortis sacre compendia vitam beatam possident*, Hymn. Surt. 130, 9. Wunde widerleán *retribution for sin*, Soul Kmbl. 187 ; Seel. 94. [*O. L. Ger.* withir-lôn *retributio* : *O. H. Ger.* widar-lôn *recompensio, recompensatio*.]

wiþerling, es ; *m. An adversary* :—Ðú forbriccest wiþerlingas (*adversarios*), Cant. Moys. Ex. 15, 7. [Iesu cristes wiþerling (wiþering, ed. Lumby), K. Horn 154 (ed. Ritson).]

wiþer-mál, es ; *m. A case against* (*in reply to*, or (?) *by way of accusation*), *defence, prosecution* (?) :—Man útlagode Swægn eorl, his ôðerne sunu. Ða ne onhagode him tô cumenne tô widermále ongeán ðone cyng, and ágeán ðone here ðe him mid wæs *his* (*Godwin's*) *other son, Swegen, was outlawed. Then it did not suit him to come to meet the king and the army that was with him in order to defend himself* (or? *in order that the case against him might be brought*; cf. Geornde se eorl griðes ðæt hé môste hine betellan æt ælc ðæra þinga ðe him man on léde, Erl. 180, 12), Chr. 1052 ; Erl. 181, 7.

wiþer-méde ; *adj.* I. *contrary-minded, contrary, adverse, hostile, opposed* :—Se widerméda (*the devil*), Andr. Kmbl. 2391 ; An. 1197. Gif huoelc uidirmoedo (*contraria*) sindon in húse esnes ðínes, Rtl. 123, 12. II. *opposed to good, perverse, depraved* :—Ic (*Eve*) wæs widerméde and unwísum nêtenum gelíc geworden, Blickl. Homl. 89, 9. [*O. H. Ger.* widar-muoti *injuriosus*.] v. wiþer-médu, wiþer-môd, *and next word*.

wiþer-médness, e ; *f.* I. *adversity* :—From ælcum widermoednise (*adversitate*) giscild ðú, Rtl. 89, 24. Nængum widirmoednisum (*adversitatibus*) áðryht, 106, 15. From allum uidirmoednesum (*adversis*) áscildad, 75, 7. II. *perversity, depravity* :—Widirmoednise *pravitate*, Rtl. 34, 9.

wiþer-médu(-o) ; *indecl.* : -méd, e ; *f.* I. *hostility, disfavour* :—His hyldo is unc betere tô gewinnanne ðonne his widermédo, Cd. Th. 41, 22 ; Gen. 660. II. *adversity, injury* :—Allum widirmoedum (*adversitatibus*) in líchome, Rtl. 52, 22. III. *perversity, depravity* :—Hí on widerméde wendan and cyrdan *conversi sunt in arcum perversum*, Ps. Th. 77, 57. [*O. H. Ger.* widar-muotî ; *f.*; -muoti ; *n. injuria, sinistrum, detrimentum, malum*.]

wiþer-metan ; *p.* -mæt, *pl.* -mǽton ; *pp.* -meten *To compare* :—Hine widermet *equat*, Wrt. Voc. ii. 90, 77 : *equiparat*, i. *coequat*, i. *imitatur, assimilat*, 143, 70. Widermeten is *confertur*, 90, 46 : *adsimilatum est*, Mt. Kmbl. Rush. 18, 23. [*O. H. Ger.* widar-mezan *comparare, rependere, compensare*.] Cf. wiþ-metan.

wiþer-môd ; *adj. Having the mind set against* something, *adverse, hostile, contrary* :—Ðæt wê hié widermôde ne gedôn ús mid ðære tǽlinge *that we may not set them against us with the blame*, Past. 32 ; Swt. 212, 1. [*O. Sax.* widar-môd.] v. wiþer-méde.

wiþer-môdness, e ; *f. Adversity, contrary fortune* :—Hine ne gedrêfe nán wuht widerweardes, ne hine ne gedrysce nán widermôdnes tô ormôdnesse *non hunc adversa perturbent, non aspera ad desperationem premant*, Past. 14 ; Swt. 83, 19. Cf. wiþer-médu.

wiþer-rǽde ; *adj. Adverse, contrary* :—Aduersus is nama þwyr oððe widerrǽde, Ælfc. Gr. 38 ; Zup. 240, 1. Wiþerrǽde *contrarius*, 47 ; Zup. 275, 6. I. *where there is ill-will, at variance, hostile* :—Ðæra Persisca cyning wæs ðam Câsere wiþerrǽde, Jud. Thw. 162, 24. Ongeán ðam wíslícan rǽde, ðe of Godes ágenre gyfe cymð, se widerrǽda deófol (*the devilish adversary*) sǽwð receleásnesse, Wulfst. 53, 7. Wurdon ða widerrǽde se cyng and se eorl, Chr. 1104 ; Erl. 239, 24. Woldon ða wiþerrǽdan hæþenan mid micelre fyrde faran on hergoþ on ðæs Câseres anwealde, Jud. Thw. 162, 36. Þeówum Godes ealle ðyses middaneardes wiþerrǽde *servis Dei cuncta hujus mundi contraria sunt*, Scint. 62, 4. II. *where there is opposition to duty, rebellious, contumacious* :—Gif hé gyt widerrǽde bið, hé líchamlíce wrace mid swingelle þolige *sin improbus est, vindicte corporali subdatur*, R. Ben. 48, 11. Ðæt Israhéla folc weard on ðam wéstene widerrǽde ongeán God, Homl. Th. ii. 238, 10. Ne beó gê wiþerrǽde wið eówerne Drihten *nolite esse rebelles contra Dominum*, Num. 14, 9. Se câsere wolde gewyldan mid wíge ða leóda ðe wiþerrǽde wǽron, and his ríce forsáwon, Homl. Skt. ii. 28, 4. III. *out of harmony, repugnant, offensive, disagreeable* :—Widerrǽde ðú eart mê *scandalum es mihi*, Mt. Kmbl. 16, 23. Nis nán ǽ widerrǽde þus

geworhtum mannum, Homl. Skt. i. 17, 60. Ðeós wyrt bið ðam góman stíð and wiðerræde for mete geþiged, Lchdm. i. 300, 10. Wulfes tǽsl hafaþ leáf wiþerræde (*unpleasant, rough?*) and þyrnyhte, 282, 15. **IV.** *adverse, not fitted to further the good of anything, unfavourable, disadvantageous:*—Mín wíf is for manegum wintrum untrum, ðam wæs ǽlc lǽcecræft wiðerræde (*no medicine suited her*), Homl. Th. i. 22, 44. Ðeós wyrt byþ cenned on wiþerrǽdum stówum (*in places not favourable to growth*) wið wegas and hegas, Lchdm. i. 228, 17. On feldum and on wiðerrǽdum stówum, 304, 3. Rihtwís þoligende wiþerræde *justus tolerando aduersa*, Scint. 12, 7. **V.** *contrary, of an opposite nature:*—Stán is gesett ongeán ðone hláf, for ðam ðe heardmódnys is wiðerræde sóðre lufe, Homl. Th. i. 252, 19. Twá wiðerræde ðing geðeódde Drihten on ðisum cwyde, ðæt sind ymhídignyssa and lustas, ii. 92, 13. Hæfst se yfela gást seofonfealde ungifa, and ða syndan wiðerræde mid ealle ðyssum gódum Godes gyfum, Wulfst. 52, 10. v. wiþer-rǽdness; wiþ-ræde.

wiþer-rǽdlíc; *adj. Adversative:*—Sume (*conjunctions*) sind *adversativae,* ðæt sind wiþerrǽdlíce, Ælfc. Gr. 44; Zup. 264, 1.

wiþer-rǽdness, e; *f. Contrariety, opposition:*—Wiðerrǽdnys *contrarietas, contrauersio,* Ælfc. Gr. 47; Zup. 275, 7. **I.** *hostility; ill-will.* v. wiþer-ræde, I:—Wið hunda rédnysse and wiðerrǽdnysse; se ðe hafaþ hundes heortan mid him, ne beóð ongeán hine hundas céne, Lchdm. i. 372, 3. **II.** *unfavourableness, disadvantage.* v. wiþer-ræde, IV:—Ðæs fyrhýses hlýwing[e] winterlíces cyles and ungetemprunge wiþerrǽdnes sí gelýht *caumene refugio hybernalis algoris et intemperei adversitas leuigetur,* Anglia xiii. 397, 462. **III.** *oppositeness of nature.* v. wiþer-ræde, **V:**—On wiþerǽdnysse went *in contrarium uertit,* Scint. 55, 3.

wiþer-ræhtes; *adv. Opposite:*—Hí geségan wyrm on wonge wiðerræhtes licgean, Beo. Th. 6071; B. 3039.

wiþer-riht, es; *n. Recompense, compensation:*—Wiðerriht *vel* edleán *hostimentum,* Wrt. Voc. i. 22, 24.

wiþer-saca, an; *m.* **I.** *an adversary, opponent, enemy:*—Anticristus is on Lǽden *contrarius Cristo,* ðæt is on Englisc Godes wiðersaca, Wulfst. 78, 13: Homl. Th. i. 376, 16. Ǽlc ðæra ðe hyne tó cynge déð ys ðæs cáseres wiðersaca (*contradicit Caesari*), Jn. Skt. 19, 12. Hér sýn on earde Godes wiðersacan, apostatan ábroðene, Wulfst. 164, 10. Wiðersa[cena] *contrariorum, inimicorum,* Hpt. Gl. 471, 74. **I a.** *a rebel:*—Ne beó hé ná sacerd geteald, ac Godes wiðersaca *non sacerdos sed rebellio judicetur,* R. Ben. 112, 13. **I b.** *an adversary at law, a prosecutor* (?):—Ðá andsweredon Pílate ða twégen wælhreówan wyþersacan, Annas and Caiphas, and cwǽdon: 'Lá, leóf déma, eall ðeós mænio secgaþ ðæt hé wæs of forligre ácenned,' Nicod. 7; Thw. 3, 32. **II.** *one who renounces or denies, an apostate:*—Wiðersaca *apostata,* Hpt. Gl. 493, 26. Wiðersaca (*pervicax fidei*) *refragator vel negator,* 502, 65: Homl. Skt. i. 3, 413. Gif munuc oþþe mæssepreóst wiðersaca wurðe mid ealle, hé sí ámánsumod ǽfre, búton hé ðe rǽdlícor gebúge tó his þearfe, L. Eth. ix. 41; Th. i. 348, 31. Iúdas se wiþersaca, Mt. Kmbl. 26, 14: Mk. Skt. 14, 10, 43. Under Juliane ðam árleásan wiðersacan (*Julian the apostate*), Homl. Skt. ii. 31, 19. Wé beódaþ ðæt wiðersacan and útlagan Godes and manna of earde gewítan, L. C. S. 4; Th. i. 378, 11. Hý synt genemnede sarabagite oððe renuite, ðæt ys sylfedéman and wiðersacan, R. Ben. 136, 11. Wiðersacena *apostatorum,* Hpt. Gl. 510, 54: *apocryphorum, falsorum scriptorum,* 452, 58. [Þat heðene cun is Goddes wiðersake, Laym. 12620. O. Sax. wiðar-sako: O. L. Ger. wither-sacco *adversarius:* O. H. Ger. wider-sacho *adversarius.*]

wiþer-sacian; *p.* ode. **I.** *to blaspheme:*—Ðam ðe wiðersacaþ ongén hálige gást, ne bið ðam forgyfen *ei, qui in spiritum sanctum blasphemauerit, non remittetur,* Lk. Skt. 12, 10. Swá hwylc man swá wyþersacaþ (*blasphemes,* v. Gospel of Nicodemus c. 4, v. 7) ðam Cásere, hé byþ deáþes scyldig, Nicod. 10; Thw. 5, 23. Wiþersacendra *blasphemantium,* Scint. 209, 5. **II.** *to be apostate:*—Wiðersaca[n]dan *apostataverant,* Hpt. Gl. 510, 49. Wiðersacedan *apostatarent,* 513, 24. Wiðersacian *apostatare,* 493, 25: *apostare,* 477, 68. [Cf. O. H. Ger. widar-sachan *recusare.*]

wiþer-sacung, e; *f.* **I.** *blasphemy:*—Wiþersacung *blasphemia,* Scint. 102, 16. 'Wylt ðú hys wyðersacunge gehýran?' Ðá cwæþ Pílatus: 'Gif seó sprǽc wyþersacung ys ðe hé spycþ, nymaþ hyne and lædaþ hyne tó eówre gesomnunge,' Nicod. 10; Thw. 5, 31. **II.** *apostasy:*—Wiðersacunge *apostasiae,* Hpt. Gl. 477, 69: 515, 69.

wiþer-sæc, es; *n.* **I.** *striving, opposition, contradiction:*—Æt ðæs wiðersæces wæterum *ad aquas contradictionis,* Deut. 32, 51: Ps. Spl. 105, 31. Fram wiðersace tungana *a contradictione linguarum,* 30, 16. Genera mé of wiðersacum (*contradictionibus*) folces, 17, 45. **II.** *denial:*—Hé (*Peter*) gemunde his micclan gebeótes, and mid biterum wópe his wiðersæc behreówsode, Homl. Th. ii. 248, 35. Heó worda gehwæs wiðersæc fremedon, ðæt heó frignan ongan; cwædon ðæt heó on aldre áwiht swylces ne ǽr ne síð ǽfre hýrdon, Elen. Kmbl. 1135; El. 569. **III.** *apostasy, recusancy:*—Ðæt heora (*the Northumbrians*) geleáfa wurde áwend eft tó Gode fram ðam wiþersæce ðe hí tó gewende wǽron, Homl. Skt. ii. 26, 63. Weard geopenad his earman wífe his

mánfullan behát ðam deófle . . . . Heó cýdde Basilie hyre cnihtes wiþersæc, i. 3, 408. [O. Sax. widar-sak, -saka *contradiction.*]

wiþer-sæc; *adj. Adverse, unfavourable:*—Hlinunge and hligiunge wiþersæc, Lchdm. ii. 258, 20.

wiþer-sínes (-sýnes); *adv. Withershins* (v. *widder-sinnis* in Jamieson's Dictionary), *backwards:*—Steorran yrnaþ wiþersýnes *the course of the stars shall be reversed,* Blickl. Homl. 93, 19.

wiþer-stæger; *adj. Hard to mount, steep, abrupt:*—Wiðerstægre *prerupti,* Wrt. Voc. ii. 68, 59: 69, 15. v. stæger.

wiþer-standan; *p.* -stód *To withstand, resist:*—Fram ðam wyderstandendum swýþran ðínre *a resistentibus dexterae tuae,* Ps. Spl. 16, 9. [O. H. Ger. widar-standan *resistere.*] Cf. wiþ-standan.

wiþer-steall, es; *m. Resistance, opposition:*—Wiðerstal *obvix,* Wrt. Voc. ii. 115, 22: 63, 22. Fǽrd ðæt fýr ofer eall, ne bið ðær nán widersteall (cf. foresteall, Dóm. L. 146, *where the Latin is:* Ignis ubique suis ruptis regnabit habenis), ne nán man næfð ðæra nihta, ðæt ðǽr ǽnige wyrne dó, Wulfst. 138, 6. Næs Petrus gewunod tó nánre wǽpnunge, ac ðǽr wǽron twá swurd gebróhte tó ðam widersteallе, gif hit Crist swá wolde, Homl. Th. ii. 248, 4. Mé hwílum biþ forwyrned þurh wiþersteall willan mínes, Exon. Th. 268, 32; Jul. 441. Cf. wiþ-steall.

wiþer-sýnes. v. wiþer-sínes.

wiþer-talu, e; *f. Reply, defence:*—Hé ðærrihte ádumbode, for ðan ðe æt Godes dóme ne bið nán beládung ne wiþertalu, Homl. Th. i. 530, 6.

wiþer-tihtle, an; *f. A counter-charge, cross-action:*—Gif ænig yfelra manna wǽre ðe wolde óðres yrfe tó borge settan for widertihtlan, ðæt hé gecýðe mid áðe, ðæt hé hit for nánum fácne ne dyde, L. Ed. 1; Th. i. 160, 5, and see note. Se ðe on gemóte mid widertihtlan hine sylfne oþþe his man werige, hæbbe ðæt eall forspecen, and geandwyrde ðam óðrum swá hundrede riht þynce, L. C. S. 27; Th. i. 392, 5. Cf. Si quis in placito per justiciam posito sui vel suorum causam injustis conterminacionibus (*v. l.* concriminationibus) vel contraposicionibus difforciet, hanc perdat, et de cetero rectum faciat, sicut hundreto vel judicibus videbitur ydoneum, L. H. I. 34, 5; Th. i. 537, 6-10.

wiþer-tíme (-týme); *adj. Troublesome, grievous:*—Apozeus ys ðam foresprecenan híwe genóh wyðertýme, Anglia viii. 331, 14. Ðá ðá mé widertýme ł hefigtýme hí wǽrun *cum mihi molesti essent,* Ps. Lamb. 34, 13.

wiþer-trod, es; *n. Return, retreat:*—Cirdon cyneröfe wíggend on wiþertrod *they turned to march back,* Judth. Thw. 26, 6; Jud. 313. Widertrod seón láðra monna *to see the retreat of the foe,* Cd. Th. 125, 25; Gen. 2084.

wiþer-weard (-word, -wurd), *and* -wierde; *adj.* **I.** *of direction, contrary:*—Him wæs widerweard (-word, Lind., Rush.) wind *erat ventus contrarius eis,* Mk. Skt. 6, 48: Mt. Kmbl. Lind. 14, 24. Wiþerward wind ástígeþ . . . ástigon wiþerwarde windas, Bd. 3, 15; S. 541, 33, 39. **II.** *of hostility or conflict, adverse, hostile;* used substantively, *an adversary, enemy, opponent, a fiend:*—Ǽlc hús ðe byð widerweard ongeán hyt sylf *omnis domus divisa contra se,* Mt. Kmbl. 12, 25. Se widirwearda god diúl *Asmadeus demon,* Rtl. 146, 37. Gá ðú onbæcling, wiþerwearda (*Satanas*), Blickl. Homl. 27, 20. Se ilca wiþerwearda ðe him ǽr ða synna lærde, 61, 17. Se wiðerwearda (-worda, Rush.) *Satanas,* Mk. Skt. Lind. 8, 33. Ðe wiðerworda, 4, 15: Lk. Skt. Lind. Rush. 13, 16. Bysmraþ se widerwearda (*adversarius*) naman ðínne, Ps. Spl. 73, 11. Ðæm wiþerweardan (*the devil*) beóþ ðæs mannes synna gecwémran ðonne eal eorþlíc goldhord, Blickl. Homl. 43, 20, 24. From ðæm widerwearda (-e, Lind.) *a Satana,* Mk. Skt. Rush. 1, 13. Mid widerweardum *cum emulo,* Wrt. Voc. ii. 74, 63: 17, 61. Beó ðú gemód ðínum ðæm wiþerwearde (*adversario tuo*), ðý læs se widerwearde ðec selle doeme, Mt. Kmbl. Rush. 5, 25. Mid widerworde ðínum (widerwordne ðínne, Rush.), Lk. Skt. Lind. 12, 58: 18, 3. Widerweardne wið hine *adversum se,* Past. 32; Swt. 211, 2. On ðam geáre wurdon ða Gallie Rómánum widerwearde *eodem anno Galli novi exstitere hostes,* Ors. 4, 7; Swt. 180, 24. Ealle ða ðe mé widerwearde wǽron *omnes adversantes mihi,* Ps. Th. 3, 6. Mé widerwearde wǽron ealle, ða him sǽton on portum *adversum me exercebantur qui sedebant in porta,* Ps. Th. 68, 12: 123, 3: 139, 8: Blickl. Homl. 223, 18: Past. 21; Swt. 161, 23. Ða men ðe hié ongeáton ðæt widerwearde wǽron Godes beboda, Blickl. Homl. 135, 12. Naman ðínne bysmriaþ ða wiþerweardan (*adversarius*), Ps. Th. 73, 10. Alle wiðiwordas (widerworda, Rush.) ł fióndas iúra *omnes adversarii uestri,* Lk. Skt. Lind. 21, 15. Se wæs on dǽle ðara wiþerweardra *in parte erat adversariorum,* Bd. 3, 24; S. 556, 27. Mínra widerweardra, Ps. Th. 17, 4. Mid wiþerwordum (*adversis*) onbénum, Bd. 2, 2; S. 504, 3. ii land ðe wǽron bereáfodon þurh Densce men and widerwearde (*hostile;* or *evil,* v. **IV**) déman út of ðam mynstre, Chart. Th. 446, 7. Heó heora ða wiþerweardan (*adversarios*) feor ádrífan, Bd. 1, 15; S. 483, 3. **II a.** *hostile to rightful authority, rebel:*—Nis nán gesceaft ðe wiþ hire Scippendes willan winne, búton dysig mon, oþþe eft ða wiþerwierdan (-weardan, -*v. l.*) englas, Bt. 35, 4; Fox 160, 25. Wiþerwyrd *perduelles, ualde rebelles,* Germ. 393, 53. **III.** *of hindrance, contrary, opposed, that presents an obstacle, obstinate:*—Nis áhwǽr geméted on hálgum bócum ðæt ðysse frignysse wiþerword sí gesawen

nequaquam in sacris eloquiis invenitur quod huic capitulo contradicere videatur, Bd. 1, 27 ; S. 490, 32. Hé oft wolde ðæt eorþlíce ríce forlǽtan, gif him ne wiþstóde ðæt wiþerwarde mód (*obstinatus animus*) his wífes, 4, 11 ; S. 579, 10. Wiðerwurdra *contrariarum* (omnes rerum contrariarum machinas exterminans, Ald. 57), Hpt. Gl. 502, 26. IV. *opposed to what is right, arrogant, perverse, depraved, reprobate, false* ; in special senses, *heretic, apocryphal* :—Wiðerweard heorte *cor pravum*, Ps. Th. 100, 3. Manega mid mannum synd getealde gecorene and mid Gode wiþerwyrde (*reprobi*), and fela mid mannum wiþerwyrde synd and mid Gode gecorene ; nán hine getelle gecorene, ðe lǽs ðe hé mid Gode sý wiþerwyrd, Scint. 74, 13–16. Wiðerwurde *importunus, improbus*, Hpt. Gl. 425, 59. Ðæt wiðerwurde *importuna, improba*, 444, 22. Mid wiðerwurde *protervo, contrario*, 434, 12. Betera geþyldig wiþerwyrdum *melior patiens arrogante*, Scint. 8, 18. Fela ðúsenda folgeaþ Criste, þeáh ðe hí sume (*the Jews*) wunian wiðerwerde, Homl. Skt. ii. 25, 526. Wiðerwearde crist *pseudo-cristi*, Mt. Kmbl. Lind. 24, 24. Wiðerworde criste and wiðerworde wítgu, Mk. Skt. Rush. 13, 22. Alle wiðerweardra gedwola *omnes apocryphorum naenias*, Mt. Kmbl. p. 10, 9. Wiðerwordra lárwa[s] sǽda *haereticorum semina*, 8, 19. Wiðerwurdra *perfidorum, impiorum*, Hpt. Gl. 415, 45. From wiðirwordum lárwum *ab ereticis*, Rtl. 198, 19. V. *opposed to the good or pleasure of anything, unfavourable, adverse, hurtful, pernicious, disagreeable* :—Nánwuht ne byð yfel, ǽr mon wéne ðæt hit yfel seó ; and þeáh hit nú hefig seó and wiþerweard, þeáh hit biþ gesǽlþ gif hit mon geþyldlíce árǽfnþ *nihil est miserum, nisi cum putes ; contraque beata sors omnis est aequinamitate tolerantis*, Bt. 11, 1 ; Fox 32, 31. Seó wiþerwearde wyrd *adversa fortuna*, 20 ; Fox 70, 29. [Nán þing] swá wiðerweard þén is [cristenum monnum] swá swá oferfylle *nihil sic contrarium est omni christiano quomodo crapula*, R. Ben. Interl. 71, 7. Hé álýseþ mé fram worde wiðerweardum (*a verbo aspero*), Ps. Spl. 90, 3 : Blickl. Gl. Alle wiðerwǽrda hǽles mennisces wyrttruman *omnes adversas salutis humani radices*, Rtl. 125, 33. Gif huoelc sindon wiðirworda in húse esnes ðínes *si qua sunt adversa in domo famuli tui*, 123, 13. Ðonne ðé for worulde wiþerwearda mǽst þinga þreáge, Met. 5, 36. Þolemód on heardum and on wiþerweardum (*contrariis*) þingum, R. Ben. 26, 18. Lufian wé hine nǽs nó on gesundum þingum ánum, ac eác swylce on wiðerweardum þingum, Blickl. Homl. 13, 8. Wið wiþerweard hǽr ; gif ðú nimest wulfes mearh and smyrest mid hraðe ða stówe ðe ða hǽr beóð of ápullud, ne geþafaþ seó smyrung ðæt hý eft wexen, Lchdm. i. 362, 8. Wala middangeardes getéla, and nǽngo his wiðirweardo (*adversa*) onscynia, Rtl. 50, 6. Geþyld gódu gehealt, áweg nýt wyþerwerde, Scint. 13, 10 : 62, 5. Wyþerwyrde, 62, 2. VI. *of diversity, contrary, opposite* in nature, action, etc. :—Ðæt gecynd nyle nǽfre nánwuht wiþerweardes lǽtan gemengan . . . Nú ðonne nú ǽlc gesceaft onscunaþ ðæt, ðæt hire wiþerweard biþ . . . hwelce twá synd wiþerweardran betwux him ðonne gód and yfel? Bt. 16, 3 ; Fox 54, 35–56, 7. Ða wiþerweardan gesceafta ǽgþer ge betwux him winnaþ, ge eác fǽste sibbe betwux him healdaþ, swá nú fýr déþ and wæter . . . Ac á sceal ðæt wiðerwearde ðæt óðer wiðerwearde gemetgian, 21 ; Fox 74, 13–20 : Met. 11, 49, 52. Ðæt mé þincþ wiþerweard þing *in contrarium relapsa res est*, Bt. 26, 2 ; Fox 92, 24. Hé náwyht wiðerweardes (*contrarium*) ðære sóðfæstnysse ðæs geleáfan Créca ðeáwe on Angelcynnes cyricean on gelǽdde, Bd. 4, 1 ; S. 564, 20. Hí monig óþer ðing ðære cyriclícan ánnesse wiþerweard hæfden, 2, 2 ; S. 502, 12. On monegum ðingum gé wiþerwearde wǽron úrum gewunan *in multis nostrae consuetudini contraria geritis*, S. 503, 18. From wiðerwordum lárwum *a diversis auctoribus*, Mt. Kmbl. p. 7, 4. [Wið al folc he wes wiðerword, Laym. 6875. Wiþerrwarrd onnȝænes Godd, Orm. 9667. Þis king him his wel wiðerward agen ðis folc, Gen. and Ex. 2935. Goth. wiþra-wairþs *that is over against* ; *contrary* : O. Sax. widar-ward, -word *hostile* ; *displeasing* : O. H. Ger. widar-wart, -wert *contrarius, adversus, adversarius*.] v. un-wiþerweard.

wiþerweardian ; p. ode *To oppose, be adverse to* :—Ða ðe wiþerweardiaþ mé *qui adversantur mihi*, Ps. Spl. 34, 22. [O. H. Ger. widar-wartôn, -wertôn *obviare, adversari, contraire, fraudare*.] v. ge-wiðerworded ; wiþerwirdan.

wiþerweard-líc ; adj. *Unfavourable, adverse, hurtful*. v. wiþer-weard, V :—Nis cristenum monnum nán ðing swá wiðerweardlíc and hefigtýme swá swá oferfyl *nihil sic contrarium est omni christiano quomodo crapula*, R. Ben. 63, 20. Warna ðé ðæt ðú nán þing wiðerwerdlíces ne sprece ongén Jacob *cave ne quidquam aspere loquaris contra Jacob*, Gen. 31, 24. [O. H. Ger. widarwart-líh *tyrannicus*.]

wiþerweardlíce ; adv. *Detrimentally, against the interests of any one* :—Þurh ðæt ðe ðú ðysne wuldres cyning áhénge, ðú dydest wyþerwerdlíce ongeán ðé and eác ongeán mé (*thou hast acted against thine own interests and against mine*. v. Gospel of Nicodemus c. 18, v. 11), Nicod. 29 ; Thw. 17, 10.

wiþerweard-ness, e ; f. I. *hostility, contention, opposition*. v. wiþer-weard, I :—Nis ðǽr ege, ne geflit, ne yrre, ne nǽnig wiþerweardnes, Blickl. Homl. 25, 32. Hé weard grǽdig ðæs gódan deáþes bútan ǽlcre scylde and ǽlcre wiðerweardnesse wið hine *he (David) was greedy for the death of the good man (Uriah), who was without any crime against*

---

him and had shewn no hostility to him, Past. 3 ; Swt. 37, 2. Mid wiðerwurdnessa *cum aemulo*, Hpt. Gl. 405, 32. Sume sace wyðerwyrdnesse hit getácnaþ, Lchdm. iii. 198, 13. II. *perversity, frowardness, depravity, arrogance*. v. wiþer-weard, IV :—Wiðirweardnis ł wyrs *perversius*, Mt. Kmbl. p. 2, 1. Wiþerwerdnysse *arrogantie*, Anglia xiii. 371, 83. Wyþerwyrdnysse, 369, 56. Wiðirwordnisum *pravis*, Rtl. 91, 24. III. *unfavourable condition, adverse circumstance, adversity*. v. wiþer-weard, V :—Seó wiþerweardnes *adversa fortuna, adversitas*, Bt. 20 ; Fox 72, 5, 9, 12. Nán yfel ne mæg ðé geneálǽcan, ac ǽlc wiðerweardnys gewíteþ fram ðínre sáwle, Basil admn. 1 ; Norm. 34, 10. Ðú ðé ne anhebbe on ofermétto on ðínre gesundfulnesse, ne eft ðé ne geortrýwe nánes gódes on ðínre wiþerweardnesse, Bt. 6 ; Fox 16, 1. On wiþerweardnesse *in aduersitate*, Wülck. Gl. 252, 4. Wiðirwordnise, Rtl. 14, 20. Hé ðisses middangeardes orsorgnesse ne gímð, ne him náne wiðerweardnesse ne andrǽt ðisse worolde *qui prospera mundi postposuit, qui nulla adversa pertimescit*, Past. 10 ; Swt. 61, 8 : 33 ; Swt. 219, 2. Hé sǽde ge hwylce wiþerwardnesse (-wordnesse, Bd. M. 330, 10), ge eft hwylce frófre on ðám wiþerweardnessum (-wordnissum, Bd. M.) him becom, Bd. 4, 22 ; S. 592, 17. Geþyld on wiðerwerdnyssum, Scint. 12, 12. On wiþerwerdnyssum *in adversitatibus*, 62, 2. Ða getreówfullan for Godes ege ealle lífes wiðerweardnesse (*universa contraria*) forþyldigian scylun, R. Ben. 27, 7. IV. *contrariety, diversity*. v. wiþer-weard, VI :—Seó wiþerweardnes ðe wé ǽr ymbe sprǽcon, Bt. 21 ; Fox 74, 32 : Met. 11, 78.

wiþer-wierde. v. wiþer-weard.

wiþer-winn, es ; n. *Contest, conflict* :—Wiþerwinnes *exercitationis* (qui laboriosi certaminis coronam difficillimis propriae exercitationis viribus nanciscuntur, Ald. 2), Hpt. Gl. 405, 20.

wiþer-winna, an ; m. *An adversary, opponent, enemy* :—*Emulus*, i. *contrarius*, gewinna, wiþerwinna, Wrt. Voc. ii. 143, 45. Beó ðú onbúgende ðínum wiþerwinnan (*adversario tuo*) . . . ðe lǽs ðe ðín wiðerwinna ðé sylle ðam déman, Mt. Kmbl. 5, 25 : Homl. Ass. 4, 95. Ðý lǽs hé sié ongieten ðæt he sié wiðerwinna on ðære diegelnesse his gedóhtes ðæs ðe hé bið gesewen ðeów on his ðénunge *ne inveniatur ei, cui servire per officium cernitur, occulta cogitationis tyrannide resultare*, Past. 19 ; Swt. 147, 16. Ðǽr (*in heaven*) ne wunaþ nán wiþerwinna, Homl. Ass. 78, 145. Úre wiðerwinna is se deófol. . . . Is óðer wiðerwinna, ðæt is Godes word, ðæt word wind on ús, 5, 120–128 : 52, 53. Ðæt hálige Godes word is ðín freónd, and ðú wyrcst ðé sylfne ðé tó wiðerwinnan, 6, 138. Ðonne ðú gǽst on wege mid ðínum wiðerwinnan (*cum adversario tuo*) tó hwylcum ealdre, Lk. Skt. 12, 58. Wrec mé wið mínne wiðerwinnan, 18, 3. Wiþerwinnan *conluctatorem*, i. *oppugnatorem*, Scint. 151, 4. Ðam ne magon ealle eówer wiðerwinnan (*aduersarii uestri*) wiðstandan and wiðcweðan, Lk. Skt. 21, 15. Ða Godes wiðerwinnan, Homl. Ass. 178, 306. Ða þeóda ða hiora wiðerwinnan wǽron, Ors. 6, 35 ; Swt. 292, 7. Wiðerwinnena *aemulorum, contrariorum, inimicorum*, Hpt. Gl. 424, 22 : 471, 72 : 475, 70. Gescylde mé wiþ mínum wiþerwinnum, gesewenlícum and ungesewenlícum, Bt. 42 ; Fox 260, 10. Nigon x hund þúsenda of Persa ánra anwealde búton hiera wiþerwinnum, ǽgþer ge of Scippium ge of Crécum, Ors. 2, 5 ; Swt. 84, 30. [Þe wyþerwynne (*the devil*), Misc. 74, 77. Forgive us ure sinne als we don ure wiðerwinnes, Rel. Ant. i. 235, 18. O. H. Ger. widar-winno.]

wiþer-winnan *to oppose, resist* :—Wiþerwinnende *rebelles*, Germ. 389, 88. [O. H. Ger. uuidar-uuinanten *conluctantem*.] v. wiþ-winnan.

wiþer-winning, e ; f. *Contest, controversy* :—Bútan wiþerwennincge (-winninge ?) *sine controversia*, Scint. 146, 15.

wiþerwirdan ; p. de *To oppose, be adverse to* :—Ealle ða ðe wiþerwyrdaþ *omnes qui tibi aduersantur*, Scint. 165, 4. v. wiþerweardian.

wiþer-word, -wurd ; wiþe-winde. v. wiþer-weard ; wiþo-winde.

wiþ-faran ; p. -fór *To escape*. v. wiþ, II. 3 :—Siððan hié ðam [herge] wiðfóron, Cd. Th. 214, 23 ; Exod. 573. v. wiþ-ferian, wiþ-gangan, II.

wiþ-feohtan *to fight against, contend with* :—Hé gefeaht mid ða ǽ ðæs módes, ðære wiþfeaht (wiðflát, *v. l.*) seó ǽ ðe on his limum wæs *pugnabat legi mentis, cui lex, quae in membris est, repugnabat*, Bd. 1, 27 ; S. 497, 39. Wiðfeohtan *certare*, Wrt. Voc. ii. 22, 17. v. wiþ, II. 10.

wiþ-feohtend, es ; m. *An adversary, opponent, enemy, a rebel* :—Hió self fieht wið hié selfe tó fultome ðæm wiðfeohtende (*adversario*), Past. 38 ; Swt. 279. 1. Ðone mángengan and ðone wiþfeohtend *rebellem et sacrilegum*, Bd. 1, 7 ; S. 477, 18. Betweoh ða elreordan and ða wiþfeohtend Cristes geleáfan *inter rebelles fidei barbaros*, 2, 5 ; S. 507, 33. v. wiþer-feohtend.

wiþ-feolan ; p. -fealh *To apply one's self to* :—Ða hé ða ongeat ðæt hé ðære godspellícan láre georne wiþfealh, and ða ðeóde tó Cristes geleáfan gecyrréd hæfde *qui ubi prosperatum ei opus evangelii comperit*, Bd. 3, 22 ; S. 552, 43.

wiþ-ferian ; p. ede *To carry off, to rescue*. v. wiþ, II. 3 :—Ðú wiðferedes (fæðeras, MS.) Israhéla bearn of Ægyptum *redemisti filios Israel et Joseph*, Ps. Th. 76, 12. Hé of heofenum hider onsende, ðe mé

álýsde, láþum wiðferede *misit de caelo, et liberavit me*, 56, 3. Míne sáwle álýs, and wiðfere láþum feóndum *animam meam libera : propter inimicos meos eripe me*, 68, 18. Ðæt ðú symle sáwle míne álýse, láðum wiðferige *liberabit in pace animam meam ab his qui adpropiant mihi*, 54, 18. Ðú áwurpe hí ða hí wēndan, ðæt hí wǣron álýsde, láðum wiðferede *dejecisti eos dum allevarentur*, 72, 14. v. wiþ-faran, wiþ-lǽdan, wiþ-teón, III.

wiþ-flítan; *p.* -flát *To contend with :*—Oferstǽleþ oððe wiðflíteþ *confutat*, Wrt. Voc. ii. 15, 31. v. wiþ-feohtan.

wiþ-fón; *p.* -fēng *To lay hold on, seize on.* Cf. wiþ, II. 1 d :—Hé uplang ástód, and him fæste wiðfēng, Beo. Th. 1524; B. 760. Cf. wiþ-grípan.

wiþ-foran; *prep. with dat. acc. Before :*—Hé feaht him wiðforan, Jos. 8, 22. Hé ofirnþ ða sunnan hindan, and cymþ wiþforan ða sunnan up, Bt. 39, 13; Fox 234, 4. ¶ wiþ . . . foran :—Ðone mist ðe wið ða eágan foran usses módes (cf. beforan úres módes eágum, Bt. 33, 4; Fox 132, 32) hangode, Met. 20, 265. Hwý hí (*stars*) ne scínen beforan ðære sunnan, swá hí dóð wið ðone mónan foran (beforan ðam mónan, Bt. 39, 3; Fox 214, 30), 28, 47. Wið ðone segn foran, Cd. Th. 188, 23; Exon. 172.

wiþ-gán *to go against, act in opposition to, in contravention of.* Cf. wiþ, II. 13 :—Nǽfre míne lástweardas geðrístlǽcen ðæt heó hit (*a grant*) onwenden oððe ðon wiðgǣn, Chart. Th. 29, 14. v. next word.

wiþ-gangan. I. *to go against :*—Ic ne meahte mægnes cræfte gúðe wiðgongan (*I could not go and meet the foe in fight*), ac ic sceal sēcan cempan sǣmran, Exon. Th. 266, 4; Jul. 393. II. *to go off, withdraw, fail :*—Byð mē eágon wiðgangen *defecerunt oculi mei*, Ps. Th. 68, 3. v. wiþ-faran.

wiþ-gemetness, e; *f. Comparison :*—In ða wiþgemetnesse wæs lytel gesewen *in comparatione tenuissima videbatur*, Bd. 5, 12; S. 629, 36. v. wiþ-metenness.

wiþ-geondan; *prep. Beyond :*—Eal ðæt ríce wiðgeondan Iordanen *omnis regio circum Iordanen*, Mt. Kmbl. 3, 5.

wiþ-gínan; *p.* de *To reply (? cf. Icel.* gegna *to reply) ; to repel, reject (? v. gynde,* Homl. Skt. ii. 25, 636):—Ða cwæð hē eft tó him sylfum : ' Tó sóðan ne þincð mē nǣfre ðæt hit sóð sý ðæt ðys sý Efesa byrig . . .' Ða wiðgýnde hē eft his geðance, ond him þus andwyrde (*he replied to his thought, or he rejected the idea, and answered himself thus*) : ' Ac ic nát eftsóna, ne ic nǣfre git nyste ðæt ǽnig óþer byrig ús wǣre gehende búton Ephese ánre,' Homl. Skt. i. 23, 541.

wiþ-grípan; *p.* -gráp *To seize on :*—Gif ic wiste hú wið ðam áglǣcan elles meahte gripe wiðgrípan, swá ic wið Grendle dyde, Beo. Th. 5035; B. 2521. v. wiþ-fón.

wiþ-habban; *p.* -hæfde *To hold out against, to withstand, resist :*—Gif ðæs synfullan ingehýd bið gehrepod mid fyrhte ðæs upplícan dómes, ðonne wiðhæfð hē ðam unlustum, Homl. Th. i. 494, 9. Ðæt wæs wundor micel, ðæt se wínsele wiðhæfde heaðodeórum, Beo. Th. 1548; B. 772. Þurh ða gedurstignysse ðe folces men wiðhæfton (-hæfdon ?) ðære gelómlícan mynegunge ðe úre láreówes dydon, L. Edg. S. 1; Th. i. 270, 24. Se ðe him ǽr geþúhte, ðæt him nán sǽ wiþhabban ne mehte, ðæt hē hiene mid scipum and mid his fultume áfyllan ne mehte, Ors. 2, 5; Swt. 84, 13. Næs nán ðæs stronglíc . . . ðæt wiþere ðam miclan mægne wiðhabban, Cd. Th. 297, 18; Sat. 519. v. wiþer-habban.

wiþ-heardian *to make obdurate :*—Nylle gē wiðheardian (*obdurare*) heortan eówre, Ps. Spl. 94, 7.

wiþ-hindan; *prep. (adv.) Behind :*—Hé feaht him wiðforan and his geféran wiðhindan, Jos. 8, 22.

wiþ-hogian; *p.* ode *To be adverse in thought or purpose, to be disposed to resist :*—Abraham . . . nalles Nergendes hǣse wiðhogode (*had no thought of disobeying the command*), Cd. Th. 173, 20; Gen. 2864. v. next word.

wiþ-hycgan; *p.* -hogde *To be adverse in thought or purpose, to set one's self against :*—Heó ðæs beornes lufan fæste wiðhogde *her heart was fast closed against the man's love*, Exon. Th. 245, 9; Jul. 42. Gē wiðhogdun hálgum Dryhtne *your hearts were hostile to the holy Lord*, 139, 34; Gú. 603. Ðæt hē stán nime, hláfes ne gýme, ða wiste wiðsæce, beteran wiðhyccge (*the food refuse, set himself against the better*), Elen. Kmbl. 1232; El. 618. v. wiþer-hycgende, *and previous word.*

wíþig, wíþing (?), es; *m. A withy, willow :*—Ðes wíþig *salix*, Ælfc. Gr. 9, 63; Zup. 70, 10: Lchdm. ii. 86, 6: Wrt. Voc. i. 33, 53. Wíðig, 80, 28. Wíðies rinde, Lchdm. ii. 150, 2. On ðone háran wíðig . . .; of ðam wíþige, Cod. Dip. Kmbl. iii. 457, 8, 10: 313, 27: 399, 21: 400, 2. On ðone ealdan wíðig; ðonne of ðam wíðige, vi. 35, 33. On ðone wíðig, 10, 25. In ǽnne wíðing, 391, 27. v. wíþig-mere. [Cf. *O. H. Ger.* wída *salix :* Icel. víðir *a willow.*] See the following words.

wíþig-bed[d], es; *n. A bed of willows, an osier-bed :*—On ðæt wíðigbed, Cod. Dip. Kmbl. iii. 437, 21.

wíþig-bróc, es; *m. A brook by which willows grow :*—In wíðibróc, Cod. Dip. Kmbl. iii. 380, 2. On wíðigbróch, 202, 3.

wíþig-ford, es; *m. A ford by which willows grow :*—On wíðigford, of wíðigford, Cod. Dip. Kmbl. iii. 135, 14 : 252, 20, 36.

wíþig-gráf, es; *m. A willow-grove :*—Of weardsetle on wíðiggráfas; of wíðiggráfan, Cod. Dip. Kmbl. v. 328, 11 : 48, 11.

wíþig-leáh; *gen.* -leás; *m. A meadow where willows grow* (a place-name) :—Ðis synt ða landgemǣro tó Wíðileá . . . Ðis is ðæra feówer hýda landbóc æt Wíðigleá, Cod. Dip. Kmbl. iii. 457, 13–23.

wíþig-mǽd; *f. A meadow where willows grow :*—Ǽrest æt wíðigmǣde . . . ðæt eft on wíðigmǣde, Cod. Dip. Kmbl. iii. 464, 18–30.

wíþig-mere, es; *m. A mere with willows on the banks :*—On wíðimǣre, Cod. Dip. B. iii. 188, 29. In wíðingmere, ii. 41, 4.

wíþig-mór, es; *m. A moor where willows grow :*—On wíðigmór, Cod. Dip. Kmbl. iii. 412, 21.

wíþig-pól, es; *m. A pool with willows on the banks :*—On wíðepól, Cod. Dip. B. iii. 188, 30.

wíþig-pyt[t], es; *m. A pit with willows by it :*—On wíðigpytt, Cod. Dip. B. iii. 336, 21.

wíþig-rǽw, e; *f. A row of willows :*—On ða wíðigrǽwe, Cod. Dip. Kmbl. iii. 48, 5.

wíþig-rind, e; *f. Willow-bark :*—Nim wíþigrinde, Lchdm. ii. 98, 9.

wíþig-slǽd, es; *n. A slade (v. slǽd) where willows grow :*—Tó wíðigslǽde, Cod. Dip. Kmbl. iii. 457, 16.

wíþig-þýfel, es; *m. A willow-copse :*—On wíðigðýfel, Cod. Dip. B. iii. 336, 21. Andlang díche foren ongēn wíðigþéuel, Cod. Dip. Kmbl. iii. 418, 2. Anlang bróke on ánne wíðigþefele, þiers ouer ðane mersc, 426, 26. Tóemnes ðám wíðigðýfelum bewestan flódan, v. 194, 32.

wíþig-wíc, es; *n. A dwelling-place by which willows grow :*—Wíðigwíc, Cod. Dip. Kmbl. ii. 195, 18.

wíþig-will, es; *m. A spring by which willows grow :*—On ðone fúlan wylle . . . on wýðigwylle, Cod. Dip. Kmbl. vi. 213, 16–21.

wíþing. v. wíþig.

wiþ-innan; *adv. prep. Within.* (1) as adverb :—Gehrepod mid heortan sárnisse wiðinnan (*intrinsecus*), Gen. 6, 6. Fácn wiþinnan (*intus*) týddriende swá swá bergyls wiþinnan (*intus*) full stence, Coll. Monast. Th. 32, 33, 35. Ðú clǣmst wiðinnan and wiðútan (*intrinsecus et extrinsecus*) mid tyrwan, Gen. 6, 14. Hí ofslógon ǽgðer ge wiðinnan ge wiðútan má þanne .xx. manna, Chr. 1048; Erl. 178, 1. Symle wē beóð fram Gode gesewene ǽgðer ge wiðútan ge wiðinnan, Homl. Th. i. 604, 19. (2) as preposition :—Ealle ða ðe wiðinnan mē (*intra me*) synd, Ps. Spl. 102, 1: 108, 21. v. wiþ-útan.

wiþ-lǽdan; *p.* de *To lead away, carry off, take away.* v. wiþ, II. 3 :—Ðú ðe Jóseph swá sceáp gramum wiðlǣddest *qui deducis velut ovem Joseph*, Ps. Th. 79, 1. Ðú míne sáwle of swyltdeáðes láþum wiðlǣddest *eripuisti animam meam de morte*, 55, 11. Ða ðú wiðlaeddun ús *qui abduxerunt nos*, Ps. Surt. 136, 3. Cneóris mín wiðlaeded is *generatio mea ablata est*, ii. p. 184, 30. Wiðlaedde eam *ablatus sum*, 108, 23 : Ps. Spl. C. 108, 22. Cf. wiþ-ferian.

wiþ-licgan; *p.* -læg, *pl.* -lǣgon *To be obstructive, object, oppose.* Cf. wiþ-standan :—Behēt man him ðæt hē móste wurðe beón ǽlc ðæra þinga ðe hē ǽr áhte. Ða wiðlæg (wiðcwæð, MS. D.) Harold, Chr. 1046; Erl. 173, 2. Ða eorlas gerndon tó ðam cynge ðæt hí móston beón wurðe ǽlc ðæra þinga ðe heom of genumen wæs. Ða wiðlæg se cyng sume hwíle, 1052; Erl. 187, 1.

wiþ-metan; *p.* -mæt, *pl.* -mǣton; *pp.* -meten *To compare :*—Wiðmeteþ *equiperat*, Wrt. Voc. ii. 83, 70: 31, 23. Hine wiðmete *equat*, 31, 49. Wiðmeten is *confertur*, 19, 27. Wiþmeten *comparatus, assimilatus*, 132, 77. Bión wiðmetene *comparari*, Kent. Gl. 42 : 1023. (1) with dat. :—Hwylcum bigspelle wiðmete wē hit ? *cui parabolae comparabimus illud ?* Mk. Skt. 4, 30. Ðeáh ðe hē nó sí his foregengan tó wiþmetenne *tametsi praedecessori suo minime comparandus*, Bd. 5, 8; S. 621, 35 : Homl. Th. i. 486, 25, 29. Beón wiðmeten ðínre strengðe *comparari fortitudini tuae*, Deut. 3, 24 : Bd. 1, 34 ; S. 499, 21 : Ps. Spl. 48, 12 : Homl. Th. ii. 200, 33 : 456, 13. (2) with prep. :—Ða cræftas ne sint tó wiþmetanne (metanne, *v. l.*) wiþ ðære sáwle cræfta ǽnne, Bt. 32, 1 ; Fox 116, 2. v. wiþer-metan.

wiþ-métedness, e; *f. An invention ;* adinventio. v. métan :—Wiðmētednyssa heora *adinventionum ipsorum*, Ps. Spl. 27, 5. On wiðmētednysse heora, 80, 11. Wiðmētednyssa, 98, 9.

wiþ-metendlíc; *adj. Comparative :*—Wiðmetendlíce naman *comparativa nomina*, Ælfc. Gr. 9, 21 ; Zup. 45, 14. v. next word.

wiþ-metenlíc; *adj. Comparative :*—Hí synd *comparativa*, ðæt synd wiðmetenlíce, Ælfc. Gr. 9; Zup. 15, 15. v. un-wiþmetenlíc.

wiþ-metenlíce. v. un-wiþmetenlíce.

wiþ-metenness, e; *f. Comparison :*—Wiþmetenes *comparatio*, Wrt. Voc. ii. 132, 79. Ðyslíc mē is gesewen ðis andwarde líf tó wiþmetenyse ðære tíde ðe ús uncúþ is *talis mihi videtur vita praesens ad comparationem ejus quod nobis incertum est temporis*, Bd. 2, 13 ; S. 516, 14. Wiðmetennysse, Homl. Th. ii. 430, 18. On wiðmetenysse *in comparatione*, Hpt. Gl. 420, 22. On his wiðmetennysse *in comparison with it*, Homl. Th. i. 618, 20. Næs hē geteald tó ðyssere wiðmetennysse *he was not included in this comparison*, ii. 38, 3. Sume naman synd *diminutiva*, ða geswuteliaþ wanunge, ná wiðmetennysse, Ælfc. Gr. 5; Zup. 16, 18.

wiþ-meting, e ; f. *Comparison :*—Wiđmetincg *comparatio,* Scint. 194, 13. Of wiđmetincge *ex comparatione,* 103, 9.

wiþ-neoþan ; *adv. Beneath :*—Wiđneođan (-nioþan, -nyđan, *v. ll.*) *infra,* Ælfc. Gr. 38 ; Zup. 225, 5 : 240, 10. Duru đū setst be đære sīdan wiđneođan *ostium pones ex latere deorsum,* Gen. 6, 16. Beón hí beworpene mid wuda wiđneođan, Homl. Skt. i. 18, 106.

wiþo-bán, es : -bâne (? cf. Icel. -beina), an ; *n. A collar-bone :*—Gif widobâne gebroced weorđeþ, L. Ethb. 52 ; Th. i. 16, 5. Ofer ealle đa sīdan āstīhþ ōþ đæt wiþobân and ōþ đone swîþran sculdor đæt sâr, Lchdm. ii. 198, 18. Stingende sâr ōþ đa wiþobân ōđ đa eaxle, 204, 26. Hwîlum ofer ealle đa sīdan biþ đæt sâr, hwīlum becymđ on đa weoþobân, and eft ymb lytel đa gesculdru đæt sâr grēt, 258, 5. [*O. Frs.* widu-bēn : *Icel.* við-beina ; *n. a collar-bone.*]

wiþo-bend *wood-bine :*—Nim weoþobend, Lchdm. ii. 312, 12. [A bordun ibounde with a brod lyste, in a wethebondes wyse iwrithen aboute, Piers P. A. 6, 9.] Cf. wudu-bend, *and next word.*

wiþo-winde (wiþ-), an ; f. *Withy-wind, with-wind* (v. E. D. S. Pub. Plant Names), *convolvulus :*—Wiþewinde *involuco,* Wrt. Voc. ii. 49, 2. Widwinde *viticella,* i. 33, 13. Genim wiþowindan twigu, Lchdm. ii. 34, 17. Wiþowindan leáf, 52, 6. Wiþewindan, 122, 18. [In a withewyndes (weythwynde, MS. C.) wise ywounden, Piers P. 5, 525.]

wiþ-rǽdan *to act against, be an antidote :*—Đære wyrte wyrttruma on wætere geđyged wiđrǽd īceom and næddrum, Lchdm. i. 144, 15.

wiþ-rǽde ; *adj. Contrary :*—Wiđrǽde *contraria,* R. Ben. Interl. 13, 7. v. wiþer-rǽde.

wiþre. v. wiþere.

wiþ-reótan ; *pp.* -roten *To clamour against* (?) :—Gē đam rihte wiđroten hæfdon, onscunedon đone scīran Scippend, Elen. Kmbl. 738 ; El. 369.

wiþ-sacan ; *p.* -sōc, *pl.* -sōcon ; *pp.* -sacen *To deny, refuse, reject :*—Ic wiþsace *recuso,* Ælfc. Gr. 28, 6 ; Zup. 178, 13. Sume (*adverbs*) syndan *abnegativa,* đæt synd wiđsacendlíce, mid đâm wē wiđsacaþ, 38 ; Zup. 226, 4. Wē wiđsacaþ *diffitemur,* Wrt. Voc. ii. 28, 21. Ic ne wiþsōc *non abnui,* 60, 32. Wiđsōc *refragatur,* 87, 37. Wiđsōcan *refragabantur,* 78, 8. **I.** *to say no* to a request, *to refuse permission :*—Đā ongunnon đa iungan biddan đone biscop, đæt hē him ālýfde, đæt hí ærnan mōstan. Đā wiþsōc (*negavit*) se biscop, Bd. 5, 6 ; S. 619, 1. **II.** where an offer or command is expressed or implied or choice is possible, *to refuse, reject, decline,* (1) absolute :—Bæd se gesíþ hine, đæt hē eode on his hūs ; wiþsōc (*rennit*) se biscop, Bd. 5, 4 ; S. 617, 11. Begann se cyngc gyrnan his sweostor him tó wīfe . . . heó sylf wiđsōc, Chr. 1067 ; Erl. 204, 17. Wiđsōc *refragabatur* (*oblatam matrimonii sortem,* Ald. 49), Hpt. Gl. 490, 65: *exhorruit,* 504, 10. Wiđsacende *refutans* (*carnalis luxus lenocinia,* Ald. 9), 420, 69: *refutando* (*obstinatam importunitatem,* Ald. 49), 491, 29. (2) with dat. of what is refused :—Gif ic sié đīnum folce nēdþearflíc tó hæbbenne, þonne ne wiđsace ic đæm gewinne, Blickl. Homl. 225, 27. Wiđsæcest đū sylfre rædes đīnum brýdguman, Exon. Th. 248, 21 ; Jul. 99. Ic wiđsōc sáwle mīnre frōfre *negavi consolari animam meam,* Ps. Th. 76, 3. Ætfæste hē mē mīne efenþeówene, đā wiđsōc ic hire, Shrn. 39, 9. Đā bæd hē đa cempan, đæt hí onfēngon gereorde mid him ; geþafode đæt ōþer, ōđer đam wiþsōc, 129, 32. Ōđer hiene gebeád tó đæm færelte ; ōđer him wiđsōc (*pergere recusavit*), Past. 7 ; Swt. 49, 5. Hē-đæt betere geceás, and đam wyrsan wiđsōc, Elen. Kmbl. 2078 ; El. 1040. Mid đon đe hē Egypte oferwon . . . hē heora godgieldum eallum wiđsōc, and hié mid ealle tōwearp *cunctam Aegypti religionem abominatus, ceremonias ejus et templa deposuit,* Ors. 2, 5 ; Swt. 78, 5. (3) with acc. :—Đæt hē đone stân nime and đa wiste wiđsæce, Elen. Kmbl. 1231 ; El. 617. For hwan đū mīn gebed woldest wiđsacan? *quid repellis orationem meam?* Ps. Th. 87, 14. Foregehēht brengende him lytla ne wiđsaca *praecepit oblatos sibi parvulos non repelli,* Mt. Kmbl. p. 18, 10. (4) with a clause :—Hē wiþsōc đæt hē đone Godes andettere slóge (*ferire recusavit*), Bd. 1, 7 ; S. 478, 40. Se wiþsōc đæt hē geleáfan onfēnge and đam gerýne đæs heofonlīcan cyninges *et fidem ac sacramenta regni coelestis suscipere renuit,* 3, 7 ; S. 529, 27. (5) with dat. and clause in apposition :—Ne wiđsace ic đon, đæt ic on đæm campe leng sié, Blickl. Homl. 225, 32. **III.** where a claim is made or implied, *to deny, refuse to acknowledge* a person, (1) absolute :—Tó wiđsacenne *ad negandum* (*Deum*), Kent. Gl. 1080. (2) with gen. of what is denied :—On đissere nihte đū wiþsæcst mīn (*me negabis*) . . . Ne wiđsace ic đīn (*non te negabo*), Mt. Kmbl. 26, 34, 35 : Mk. Skt. 14, 30. (3) with acc. (or uncertain) :—Se đe mē wiđsæcđ, ic wiđsace hyne, Mt. Kmbl. 10, 33. Đū mē wiđsæcst, 26, 75 : Jn. Skt. 13, 38. Se đe mē wiđsæcđ beforan mannum, se byđ wiđsacen beforan Godes englum, Lk. Skt. 12, 9. **IV.** where a statement is made or implied, *to deny, reject, refuse assent.* (1) absolute :—Hē wiđsōc (*negavit*) and cwæđ : 'Nât ic hwæt đū segst,' Mt. Kmbl. 26, 70 : Jn. Skt. 18, 27. Wē wiđsōcun ǽr mid leásingum, Elen. Kmbl. 2242 ; El. 1122. Wiđsacende *post tergum ponentes, abjicientes,* Hpt. Gl. 428, 65. (2) with gen. :—Hí wiđsacaþ Cristes tōcymes, Homl. Th. i. 144, 23. (3) with dat. :—Wiđsacest đū sōđe and rihte ymb đæt līfes treów, Elen. Kmbl. 1322 ; El. 663. Gē wiđsōcon sōđe and rihte,

đæt in Bethleme bearn Wealdendes cenned wǽre, 779 ; El. 390. (4) with a clause :—Hí wiđsócon, đæt hē God wǽre . . . Sume wiđsócon, đæt hē deádlíc flǽsc underfēnge, Homl. Th. i. 116, 16-19. (4 a) where the clause is put negatively :—Đā wiđsóc Crist, đæt hē deofol on him næfde ; ac hē ne wiđsóc, đæt hē nǽre Samaritanisc, Homl. Th. ii. 230, 1-2. **V.** where a claim has been acknowledged or a relation has been established, *to renounce, reject, give up,* (1) absolute :—Heó wiđsóce *respuerit* (*mundi opes gloriamque,* Ald. 65), Hpt. Gl. 512, 69. Wiđsacan *abdicare* (*apocriphorum deliramenta,* Ald. 26), 452, 62. (2) with gen. v. (5). (3) with dat. :—Ǽlc of eów đe ne wiđsæcđ (*renuntiat*) eallum þingum đe hē āh, Lk. Skt. 14, 33. Đū wiđsóce sōþum criste *tu repulisti christum tuum,* Ps. Th. 88, 32. Hē wiđsóc (*repulit*) snytru hūse, wæs his ágen hūs, 77, 60, 67. Hē đīnum wiđsóc aldordóme, Elen. Kmbl. 1531 ; El. 767. Đǽm englum đe Gode wiþsócan, Blickl. Homl. 49, 8. Būton hí đam deófolgylde geoffrodon and Drihtne wiđsócon, Homl. Skt. i. 23, 114. Monige wiþsócan đære unsýfernysse deófolgylda *abrenunciata sorde idolatriae,* Bd. 3, 21 ; S. 551, 21. Wiþsacaþ nū đām leásum welum, Blickl. Homl. 53, 23. Đæt đū heofoncyninge wiđsóce, Exon. Th. 264, 8 ; Jul. 361. Deófulgyldum wiþsacan *abrenunciatis idolis,* Bd. 2, 9 ; S. 511, 35. (4) with acc. :—Đæt đū wiđsæcest đone cyning, đam đū hýrdest ǽr, Elen. Kmbl. 1863 ; El. 933. Læsse ys wiđsacan (*abnegare*) đæt hē hæfđ, swýþe micel ys wiđsacan đæt hit ys (*abnegare quod est*), Scint. 60, 13. (5) with gen., dat., and acc. in the same sentence :—Se fæder wiđsóc his bearne, and đæt bearn wiđsóc đone fæder, and æt nēxtan ǽlc freónd wiđsóc ōđres, Homl. Skt. i. 23, 110. **V a.** of self-renunciation :—Gyf hwā wylle fyligean mē, wiđsace (*abneget*) hyne sylfne, Mt. Kmbl. 16, 24. **VI.** *to refuse, withhold, not to give :*—Wæs Eþelwald đæs wordes, đæt hē nō đes rihtes wiđsacan wolde . . . and hit mildlíce āgeaf đan biscope, Chart. Th. 140, 12. **VII.** *to declare hostility* (?) :—Hí hiene (*Mucius*) secgan hēton, hū fela đæra manna wǽre đe wiđ đæm cyninge Tarcuime swíđost wiđsacen hæfde, Ors. 2, 3 ; Swt. 68, 24. [Wiđsaken cristindom (heþene beo, 2nd MS.), Laym. 10898. þ iherde Uortiger, and fastliche hit wiđsoc, 13000. Hit is so wide ibrouht forth, ich hit ne mei nout wiđsaken, A. R. 88, 11.]

wiþ-sacendlíc ; *adj. Negative, expressing negation :*—Sume (*adverbs*) syndan *abnegativa,* đæt synd wiđsacendlíce, Ælfc. Gr. 38 ; Zup. 226, 3.

wiþ-sacung, e ; f. *Renunciation :*—Nâht ūs framaþ wiđsacing (*abrenuntiatio*) līchaman būtan wiđsacinge geþances, Scint. 60, 14.

wiþ-scorian ; *p.* ode *To refuse :*—Se đe đeónde biđ on cræftum, and đonne tó swíđe wiđscoraþ (-sceoraþ, Hatt. MS.) đæm ealdordóme (*si omnino renititur*), healde hine đæt hē ne cnytte đæt underfongne feoh on đam swātlíne, đæt Xrist ymbe spræc, Past. 9 ; Swt. 58, 12. [Yef þou louest to bi sobre, wyþscore and wyþdraз þine willes, Ayenb. 254, 26.]

wiþ-scúfan ; *p.* -sceáf, *pl.* -scufon ; *pp.* -scofen *To push back* or *away, to repel, drive away, refute ; repellere, expellere, praecipitare :*—Wiđscyfs đū *precipitas,* Wrt. Voc. ii. 68, 67. Ūs drīfaþ đa ællreordan tó sǽ, wiþscúfeþ (*repellit*) ūs seó sǽ tó đam ællreordum, Bd. 1, 13 ; S. 481, 44. Hē oft stormas wiþsceáf (*repellere consueverat*), 2, 7 ; S. 509, 33. Gif hwylc monn his ágen wīf wiþscúfe (*expulerit*), 4, 5 ; S. 573, 17. Hwī willaþ gē wiþscúfan (*repellere*) đone đe gē ǽr onfēngon, 3, 19 ; S. 549, 4. Wiþscúfan (*refutare*) đa đe gedyrstigedon, đæt hí Eástran heóldan būtan heora rihte tíde, 5, 21 ; S. 642, 39. Fultum tó wiþscúfanne hergunge (*ad repellendas inruptiones*), 1, 14 ; S. 482, 37.

wiþ-secgan ; *p.* -sægde *To renounce :*—Eardlíco lusto wiđsæcgende *terrena desideria respuentes,* Rtl. 34, 20. [þ hit beo so open sunne, þ he hit ne mei wiđsiggen (*deny*), A. R. 86, 7. Wiđsuggen (-segge, 2nd MS.), Laym. 13237. Manig mann þiss merrke shall wiþþstanndenn and wiþþseggenn (*contradicere*), Orm. 7646. No men ne mygt wel it wyþsegge, R. Glouc. 106, 3. No þing to hele, no þing wyþzigge *to conceal nothing, to deny nothing,* Ayenb. 175, 4. Whoso wole my juggement withseie, Chauc. Prol. 805. Wytheseyne or geyneseyne *contradico,* Prompt. Parv. 530.]

wiþ-seón ; *p.* -seah, *pl.* -sáwon *To plot against* (?) :—Hié sume heora þeówas gefreódon. . . . Đā ofþūhte heora ceorlum đæt mon đa þeówas freóde, and hí nolde. Đā wiđseon hié đæm hlāfordum, and đa þeówas mid him, ōþ hié wyldran wǽron þonne hié *cum servos suos passim liberos facerent, libertini in partem potestates recepti plenitudinem per scelus usurpare meditati sunt. Itaque conspirantes in facinus libertini correptam urbem suo tantum generi vendicant,* Ors. 4, 3 ; Swt. 162, 14-18.

wiþ-setness, e ; f. *A placing opposite* or *something placed opposite :*—Uuitsetnis *objectus,* Wrt. Voc. ii. 115, 26.

wiþ-settan ; *p.* te *To oppose, resist :*—Sende hē him fultum þurh sumne dēman, đe wiđsette heora feóndum, and hí ālīsde of heora yrmđe, Ælfc. T. Grn. 6, 25. Fram ansýne ārleásra đa đe mē geswenctun ꝥ wiđsettun (*afflixerunt*), Ps. Lamb. 16, 9. [Wythesettyn obsto, obsisto, Prompt. Parv. 530.]

wiþ-sleán ; *p.* -slóh *To counteract :*—Hí woldon đæra hālgena líc besencan on flóde, ac se ælmihtiga Scyppend wiđslóh đam unrǽde, Homl. Skt. ii. 29, 324.

wiþ-sprecan ; *p.* -spræc, *pl.* -sprǽcon ; *pp.* -sprecen *To speak against,*

to revile :—From stefne edwĕtendes and wiðspreocen[des] *a voce exprobrantis et obloquentis*, Ps. Surt. 43, 17.

wiþ-spurnan; *p.* -spearn *To dash against:*—Ðȳ læs ðū wiðspurne wið stāne fōt ðīnne *ne forte offendas ad lapidem pedem tuam*, Mt. Kmbl. Lind. 4, 6.

wiþ-standan; *p.* -stōd, *pl.* stōdon; *p.* -standen.   **I.** *of opposition to force or compulsion, to withstand, resist,* (1) absolute :—Wiðstōd *reluctaretur*, Wrt. Voc. ii. 85, 45. Ðet landfolc hardlīce wiðstódon *the people offered a stout resistance,* Chr. 1046; Erl. 171, 4. Wiðstōde *disputans,* Mt. Kmbl. p. 17, 1. (2) with dat. :—Gif hwylc eów wiþstondeþ (*restiterit*), ðonne gefultumiaþ wē eów, Bd. 1, 1; S. 474, 17. Him man swīðe fæstlīce wiðstōd and heardlīce, Chr. 1001; Erl. 137, 8: Exon. Th. 156, 15; Gū. 875. Hē galdorcræftum wiðstōd stranglíce, Andr. Kmbl. 333; An. 167. Wiðstōd *refragabatur* (*decalogi sanctionibus,* Ald. 12), Hpt. Gl. 426, 40. Hē wolde ðæt gyld ábrecan. Ðā wiðstōdan him ða hǣþenan men, Blickl. Homl. 221, 21. Wǣpen wyrcean and heora feóndum wiþstondan (*resistere*), Bd. 1, 12; S. 481, 14. Ðæm sloegende wiðstonda, Mt. Kmbl. p. 14, 18. From ðæm wiðstondendum (*resistentibus*) ðere swiðra ðīnre, Ps. Surt. 16, 8.   **II.** *to stand against, succeed in opposing, be a match for, refute :*—Se nama tācnaþ ðone sige ðe Drihten wiþstōd deófle, Blickl. Homl. 67, 15. Eftforefundeno wiðstōd *reprehensores redarguit,* Mt. Kmbl. p. 16, 13. Ðæt hí ðām yrmðum ne wiðstanden *in miseriis non subsistent,* Ps. Th. 139, 10. Ne mæg eów nān þing wiðstandan (*resistere*), Jos. 1, 5 : 10, 8 : Nicod. 26; Thw. 14, 10 : Ps. Th. 75, 5. Wyrde wiðstondan, Exon. Th. 287, 17; Wand. 15 : 161, 32; Gū. 967 : 278, 18; Jul. 599. Wísdōm, ðam ne magon ealle eówer wiðerwinnan wiðstandan and wiðcweðan, Lk. Skt. 21, 15 : Blickl. Homl. 161, 17.   **III.** *to stand in the way, be a hindrance, obstruct, prevent, be a preventive,* (1) absolute :—Wið blōdryne of nosum ; ádrȳg gāte blōd and gníd tō duste, dō on ðæt næsþyrl ; hyt wiðstandeþ (*it acts as a preventive*), Lchdm. i. 352, 4. (2) with dat. :—Him nǣnig wiþstōd *nullo prohibente,* Bd. 1, 15; S. 483, 41. In swā micclum heápe ðæra ðe ðær wǣron ūt gongende, hira nǣnig ðām in gangendum ne wiðstōd, Shrn. 41, 10. Ða þióstro ðīnre heortan willaþ mīnre lāre wiðstondan, Met. 5, 22. (3) with dat. of that which is hindered and gen. of that in respect to which the hindrance occurs :—Micel stān ðone brōc tōdǣlð and him his rihtrynes wiþstent; swā dōð nū ða þeóstro ðīnre gedrēfednesse wiþstandan mīnum lārum, Bt. 6; Fox 14, 30. Hē ðē oft wiðstōd willan ðīnes, Exon. Th. 268, 5; Jul. 427.   **IV.** *to stand off* (cf. wiþ *in* wiþ-faran), *keep away, be absent :*—Fearr dióules fācon uiðstonde *procul diaboli fraus absistat,* Rtl. 98, 22. Be ðon ðe mon wíf bycgge and ðonne sió gift wiðstande. Gif mon wíf gebycgge and sió gyft forð ne cume, L. In. 31; Th. i. 122, 4 note.   **V.** *to be hostile :*—Ic wiðstande ongēn eów *ponam faciem contra vos,* Lev. 26, 17. Cf. wiþer-standan.

wiþ-steall, es ; *m.*   **I.** *a defence :*—Ic ingehygd eal geondwlíte, hū gefæstnad sȳ ferð innanweard, wiðsteall geworht *I scan the mind to see how the soul is fortified within, how its defences are built,* Exon. Th. 266, 20; Jul. 401.   **II.** *an obstruction, obstacle :*—Wiðsteallas *obstacula* (nimborum obstacula rupit, ut fluerent imbres, Ald. 143), Wrt. Voc. ii. 89, 71 : 64, 39. Cf. wiþer-steall.

wiþ-steppan *glosses* praetergredi, Ps. Lamb. 79, 13.

wiþ-stunian; *p.* ode *To dash against :*—Eallum ðū wiðstōde and wiðstunedest . . . stunaþ heó wærce, wiðstunaþ heó ǣttre, Lchdm. iii. 32, 13-24.

wiþ-styllan; *p.* de *To leap back, retreat :*—Wiðstylde *descivit, pedem retraxit,* Wrt. Voc. ii. 106, 25. v. stellan *to leap.*

wiþ-styltan; *p.* te *To hesitate, doubt :*—Gif gié hæbbe leáfo and gié ne wiðstylte *si habueritis fidem et non haesitaveris,* Mt. Kmbl. Lind. 21, 21.

wiþ-teón; *p.* -teáh, *pl.* -tugon; *pp.* -togen.   **I.** *with acc. to withdraw, draw back :*—Swā micel wæð seó sǣ heó mǣst wiðteóhð *as far as ever the sea withdraws itself* (*recedes*), Chart. Th. 318, 9.   **II.** *with dat. to draw back, restrain :*—Balaham wolde fēran ðǣr hiene mon bæd, ac hís ēstfulnesse wiðteáh (wit-, Hatt. MS.) se esol ðe hē onuppan sæt *Balaam pervenire ad propositum tendit, sed ejus votum animal, cui praesidet, praepedit,* Past. 36; Swt. 254, 22. Ōðerne hē drāf suíðe geornfullíce mid sticele, ōðrum hē wiðteáh mid brídle *illum stimulo impellere nititur, hunc freno moderatur,* Swt. 293, 1.   **III.** *to draw away,* cf. wiþ-ferian :—Wiþtugon *detrahebant,* Ps. Spl. T. 108, 3.   **IV.** *to draw to :*—Wiðtíhþ *attrahit* (other Latin versions have *abstrahit*), Ps. Lamb. second 9, 9. [Wiðteod giu of þe fleslische lustes *abstinete uos a carnalibus desideriis,* O. E. Homl. ii. 137, 18. Þat he us wissie to wiðtien of alle fleslische lustes, 79, 4.]

wiþþe, an ; *f.* *A with* (v. Jud. 16, 9 where Wicklif has *wiþþis*), *a thong, cord :*—Wiðde *loramentum* vel *tormentum,* Wrt. Voc. i. 57, 26 : *lorumentum,* ii. 53, 39. Wiðde *circus* vel *circulus,* rāp *funiculus* vel *funis,* i. 15, 18-19 : 75, 3-4. Hē hēt hí (*Agatha*) on hencgene āstreccan, and ðrāwan swā swā wiððan, Homl. Skt. i. 8, 113. Hē bebeád ðām cwellerum, ðæt hí hine mid wiððum handum and fótum on ðǣre rōde gebundon, Homl. Th. i. 594, 31 : 596, 21. [Nimeð me þene ilke mon,

and doð widðe (raketeȝe, 2nd MS.) an his sweore, Laym. 22833. Twælf swine iteied tosomne, mid wiðen swiðe grete ywriðen al togadere, 25973. Crist himm wrohhte an swepe all alls itt wære off wiþþess, Orm. 15563. Þe þief . . . þet heþ nieȝ þe wyþþe ine þe nykke, Ayenb. 135, 25. Witthe, wythth *boia,* Prompt. Parv. 531. *O. Frs.* withthe : *Icel.* viðja, *and* við ; *gen.* viðjar.] v. cyne-wiþþe.

wiþ-þyddan; *p.* de *To thrust back :*—Wiðþyddende *retundens,* Hpt. Gl. 505, 52.

wiþ-tremman; *p.* de *To step back :*—Ðonne wiðtremð hē and onhupaþ *gressum post terga revocet,* Past. 58; Swt. 441, 27. v. trem *a step.*

wiþ-ufan; *adv. prep. Above,* (1) as adverb :—Sume (*adverbs*) synd *localia . . . super* wiðufan, Ælfc. Gr. 38; Zup. 225, 5 : *supra,* 240, 9. On ðære bytminge wæs se arc rūm, and wiðufan genyrwed, Homl. Th. i. 536, 15. Hēr wiðufan on ðyssere rǣdinge, 608, 15 : ii. 228, 7. Swā swā wiðufan gecweden hit is *sicut supra dictum est,* Ath. Crd. 27 : Lchdm. iii. 438, 7. Hē bebeád wolcnum wiþufan *mandavit nubibus desuper,* Ps. Lamb. 77, 23 : Hymn. Surt. 24, 31. (2) as prep. :—Tō grǣwan stāne, ðonon wiðufan ðæs wælles heáfod, Cod. Dip. Kmbl. ii. 29, 4.

wiþ-ūtan; *adv. prep. Without.*   **I.** as adverb :—Gē clǣnsiaþ ðæt wiðūtan ys caliceas and dixas . . . Clǣnsa ǣryst ðæt wiðinnan ys calices and disces, ðæt hit sí clǣne ðæt wiðūtan ys *mundatis quod deforis est calicis et parapsidis . . . Munda prius quod intus est calicis et parapsidis, ut fiat et id, quod deforis est, mundum,* Mt. Kmbl. 23, 25-26. His líchama barn wiðūtan mid langsumere hǣtan, Homl. Th. i. 86, 4. Man scolde fandian gif man mihte betræppan ðane here āhwār wiþūtan, Chr. 992; Erl. 130, 43.   **II.** as preposition. (1) with dat. (a) *without* (the opposite of *within*), *outside of :*—Wiðūtan ðæm díce is geworht heáh weall, Ors. 2, 4; Swt. 74, 19. Ðā cwæð man mycel gemōt wiðū:an Lundene, Chr. 1052; Erl. 187, 16. Se cyng gefeaht tōgeánes his sunu wiðūtan Normandíge, 1079; Erl. 216, 7. (b) *without* (the opposite of *with*) :—Hē hæfde Ȳrlande gewunnon wiðūtan ǣlcon wǣpnon, Chr. 1086; Erl. 222, 18. (2) with acc., *without, to the outside of :*—Lēd ūt ðone hirwend wiðūtan ða wícstōwe *educ blasphemum extra castra,* Lev. 24, 14. Hig āwurpon hyne wiðūtan ðone wíngeard (*extra vineam*), Mt. Kmbl. 21, 39. v. wiþ-innan.

wiþ-weorpan; *p.* -wearp, *pl.* -wurpon; *pp.* -worpen *To reject :*—Ðū eart se weallstān ðe ða wyrhtan wiðwurpan, Exon. Th. 1, 4; Cri. 3. Gē ðære snyttro [stān (? cf. Lk. 20, 17)] unwíslíce wiðweorpon, Elen. Kmbl. 587 ; El. 294.

wiþ-winde. v. wiþo-winde.

wiþ-winnan; *p.* -wann, *pl.* -wunnon *To strive against, resist :*—Went hē mid ealle cræfte ongēn ðæs ōðres geðyld, ðe him ðonne giet wiðwinð (*eum obsistentem fortiter*), Past. 33; Swt. 227, 7. Eallum his wordum hí wiðcwǣdon and wiþwunnan *cunctis quae dicebat contradicere laborabant,* Bd. 2, 2; S. 503, 17. 'Nis nān wuht ðe mæge swā heágum gōde wiþcweþan.' Ðā cwæþ ic : 'Nē wēne ic ðæt ǣnig wuht sié ðe wiþwinne' *non est igitur aliquid, quod summo huic bono possit obsistere. Non, inquam, arbitror,* Bt. 35, 4; Fox 160, 31. Hwæt wilt ðū cweþan, gif hwā nylle wiþwinnan, 36, 6; Fox 182, 6. Ðone anwald mæg wel reccan se ðe ǣgðer ge hine habban cann ge wiðwinnan *potentiam bene regit qui et tenere illam noverit et impugnare,* Past. 17; Swt. 113, 21. Ðeáh ðe hē swȳþe wiþwinnende wǣre *quamvis multum renitens,* Bd. 4, 28; S. 606, 17. Ða biscepas sǣdon ðæt ealle godas him irre wǣren and wiðwinnende, Ors. 3, 7; Swt. 114, 4. v. wiþer-winnan, wiþ, II. 10.

wiþ-winnend, es ; *m. An opponent :*—Wiðwinnend *refragator* (*-ur,* MS.), Wrt. Voc. ii. 84, 62.

witian, witiendlíc. v. be-witian, witod, fore-witiendlíc.

wītiend-líc (wītend-) ; *adj. Prophetic :*—Wītiendlícere mihte *prophetica virtute,* Hpt. Gl. 492, 22. Wītendlícum wītedōme *prophetica vaticinatione,* 520, 16. Wītendlícere, 505, 3. Wītenlícere, 443, 58. Wītiendlícum *propheticis,* 416, 55. [Cf. *O. H. Ger.* wízôn *prophetizare, vaticinari, divinare.*] v. wīte-dōm ; wītegend-líc *and* wīteg-dōm.

witig, wittig; *adj.*   **I.** *having knowledge, wisdom, sense ; sagacious, wise :*—Stān witig werede and worde cwæð, Andr. Kmbl. 1485 ; An. 744. Swilce wittige ł gleáwe leorneras *velut sagaces* (*prudentes*) *gymnosophistas,* Hpt. Gl. 404, 76. ¶ as an epithet of the Deity (cf. witte *of* witty God, Piers P. 15, 126) :—Witig God, Cd. Th. 182, 24 ; Exod. 80 : Exon. Th. 14, 29 ; Cri. 226 : Beo. Th. 1375 ; B. 685 : 2116 ; B. 1056. Witig Drihten, 3113 ; B. 1554 : Hy. 4, 6 : Exon. Th. 379, 12 ; Deór. 32 : Cd. Th. 179, 8 ; Exod. 25 : 241, 14 ; Dan. 404. Wittig (wigtig, MS.), Beo. Th. 3687 ; B. 1841. Witig Wuldorcyning, Cd. Th. 242, 30 ; Dan. 427.   **II.** *in one's wits, in one's right mind :*—Wearð his suna wittig, Homl. Skt. i. 7, 428. [Wygar þe witeȝe (wittye, 2nd MS.) wurhte, Laym. 21134. Mine wise and mine witie (wittye, 2nd MS.) men, 15829. Witti and wise, Kath. 315. Ich am witi and wot al þat to cumen is, O. and N. 1189. Ȝe wise men and witty of the lawe, Piers P. C. 10, 51. *O. Sax.* wittig, wittig : *O. H. Ger.* wizíg, wizzíg *solers, sapiens : Icel.* vitugr.] v. for-, fore-, ge-, un-witig, -wittig.

witiga, wītig-dōm. v. wítega, wíteg-dōm.

witigness, e ; *f. Sagacity, prudence :*—Wyttinysse *industriam* (saga-

cissimam animi industriam, Ald. 3: cf. gleáunes *industria*, Wrt. Voc. ii. 46, 2), Hpt. Gl. 407, 71.

wíting-stów. v. wítung-stów.

wit-leás; *adj. Witless, senseless* :—On ðam fíftan mónþe hé (*the fœtus*) biþ cwica and weaxeþ and seó módur líð witleás, Lchdm. iii. 146, 12. [*Ne wurðe non so witleas*, A. R. 256, 25. *Giff þin macche iss wis and god and tu wittlæs and wicke*, Orm. 6197. *Nis neure mon redles ar his heorte beo witles*, O. and N. 692. *Ine foles, and yne wytlease, þet ne habbeþ nenne skele*, Ayenb. 86, 13. *Icel.* vit-lauss *witless, foolish, mad.*] v. gewit-leás, *and next word.*

wit-leást, e; *f. Senselessness, folly* :—His (*Job's*) wífes witleást (gewitleást, Homl. Th. ii. 456, 4), Job. Thw. 167, 32. [Cf. *Icel.* vit-leysi *madness.*] v. gewit-leást.

wítnere, es; *m. A punisher, tormentor* :—Wítnere *lictor*, Wrt. Voc. ii. 52, 59. Se déma betǽcð ða unrihtwísan ðam unmildheortan wítnere, and se wítnere hí gebrincð on cwearterne, Homl. Ass. 8, 205. Ðonne beóð ða hire (*the soul's*) wítneras, ða ðe hí tó ðam leahtrum forspeónon, Homl. Th. i. 410, 31. Se hláford sealde hyne ðám wítnerum (*tortoribus*), Mt. Kmbl. 18, 34. Se heretoga cwæþ: 'Gé beóð gewítnode' ... Ða swór se déma, ðæt hí þurh drýcræfte ða stánas áwendon tó heora wítnerum, Homl. Skt. i. 11, 110. [*O. L. Ger.* wítneri *tortor* : *O. H. Ger.* wízinári *ultor, tortor, lictor.*]

witness, e; *f.* I. *knowledge* :—Fore wísdóm ł witnesse *propter scientiam*, Rtl. 194, 37. II. *witness, cognisance, knowledge* :—Menigo óðro béceno worhte se Hǽlend on witnesa (*in conspectu*) ðara ðegna, Jn. Skt. 20, 30. III. *witness, testimony* :—Ásceacaþ eówer fóta dust ofer hig on witnesse (gewytnysse, *v. l.*) (*in testimonium*), Lk. Skt. 9, 5. In cýðnisse ł witnesa *in testimonium*, Mt. Kmbl. Lind. 8, 4. Leása witnesa *falsa testimonia*, 15, 19. IV. *a person who gives testimony, a witness* :—Monigo leáse witnesa (*testes*), Mt. Kmbl. Lind. 26, 60. In múð tuoe witnesa (*testium*), 18, 16. Tó witnesum *testibus*, 26, 65. v. ge-witness.

wítnian; *p.* ode *To punish, torment, plague* :—Ic wítnie *multo*, Engl. Stud. xi. 66, 58. Uuítnath *multabitur*, Wrt. Voc. ii. 114, 42. Wítnað *plectit*, 90, 12. Wítnode *multavit, punivit*, Hpt. Gl. 455, 15. Déme ðæt se bisceop and wítnige be ðam (*juxta hoc puniatur*), L. Ecg. C. 16; Th. ii. 144, 7. Wítnian *vapulare, multare, flagellare*, Hpt. Gl. 477, 27. Wítniende *multans*, Wrt. Voc. ii. 57, 31. Wítniendra þiówa *lictorum*, 52, 77. Déman wíðnigendne *judicem punientem*, Scint. 38, 3. (1) with acc. of person :—Hé wítnaþ ða scyldigan *injusti punientur*, Ps. Th. 36, 28. Ðæt ða bióþ gesǽlegran ðe mon wítnaþ ðonne ða bión ðe hí wítniaþ *infeliciores eos esse, qui faciunt, quam qui patiuntur injuriam*, Bt. 38, 6; Fox 208, 6. Hwæþerne woldest ðú déman wítes wyrþran, ðe ðone ðe ðone unscyldgan wítnode, ðe ðone ðe ðæt wíte þolode? *cui supplicium inferendum putares, eius qui fecisset, an qui pertulisset injuriam?* Fox 208, 16. Ðone blacan Heáwald hí lange cwylmdon and ðurh lima wítnadon *Nigellum Hewaldum longo suppliciorum cruciatu et horrenda membrorum omnium discerptione interemerunt*, Bd. 5, 10; S. 624, 41. Ðæt man ðás menn wítnige and cwelle, Blickl. Homl. 183, 2. Nele God ús wítnian, Ps. Th. 76, 7. Ða unrihtwísan beóð wítnade (*punientur*), Ps. Surt. 36, 28. Hí wǽron wítnade *virgis caesi*, Ors. 4, 1; Swt. 160, 14. (1 a) with the means of punishment expressed :—Ic wítnige eów seofon wíton *corripiam vos septem plagis*, Lev. 26, 28. Se uultor ne slát ða lifre Tyties, ðe hine ǽr mid ðý wítnode, Bt. 35, 6; Fox 170, 4. Wítna mid tintregum ðínne sunu, Homl. Skt. i. 4, 205. Ðæt se hí móte mid mycclum wítum wítnian, Blickl. Homl. 61, 18. Hé hí wolde wítnian mid deáþe, Bt. 41, 3; Fox 248, 12. Hé biþ wítnad manegum wítum *vapulabit multis*, Lk. Skt. 12, 47, 48. (2) with acc. of fault :—Ðæt hí heora synna wítnade and bétte, Bd. 4, 25; S. 599, 24. Sume wyllaþ wítnian stíðlíce ða læssan gyltas on heora underþeóðdum, and nella[?] wítnian mid nánre wrace ða máran synna on him sylfum, Homl. Ass. 7, 182. Ðý læs hit him sié wítnod *lest it be punished in him*, Past. 9; Swt. 59, 17. [*O. Sax.* wítnón : *O. Frs.* wítnia : *O. H. Ger.* wízinón *damnare, dijudicare, vexare, angere, plectere, torquere.*] v. ge-wítnian; un-wítnod.

wítnigend-líc; *adj.* I. *that punishes* or *torments* :—Seó ðwyre sáwul gǽð tó ðam wítnigendlícum fýre, Homl. Th. i. 408, 23. Wítniendlícum fýre, ii. 344, 12, 17 : 590, 13. II. *that deserves to be punished* :—Ne gemétst ðú on mé áht wítniendlíces, Homl. Th. ii. 518, 4. Cf. un-wítniendlíce.

wítnung, e; *f. Punishment, torment, pain* :—Mægmorðres wítnung *parricidii actio*, ... gebohte scíre wítnung *ambitus judicium*, Wrt. Voc. i. 21, 10, 12. Geligra wítnung *incerta* (*incesti?*) *judicium*, ii. 49, 29. Ðæs ic geléfe, ðætte ǽlc unriht wítnung sié ðæs yfel ðe hit déð, næs ðæs ðe hit þafaþ *apparet, illatam cuilibet injuriam non accipientis, sed inferentis esse miseriam*, Bt. 38, 6; Fox 208, 20. Ðæt hé on wítnunge stówe swungen wǽre, oþ ðæt hé swylte, Blickl. Homl. 193, 3. Ðære synne tó wítnunge mínre unhýrsumnesse *ad puniendam inobedientiae meae culpam*, Bd. 5, 6; S. 619, 22. Ðonne seó sáwul bið tó hire wítnunge geléd ... betǽht tó écere wítnunge, Homl. Th. i. 410, 24, 30. Helpan ðám forðfarenum ðe on wítnunge beóð, ii. 356, 12. Gefyl hié nú mid ðære

---

wítnunga ðe ðú him geteohhod hæfdest, Ps. Th. 16, 13. Búton wítnunge *without exacting a penalty*, L. Edg. S. 1; Th. 1. 270, 19. Swerie hé (*a criminal who has been punished*) ðæt hé ǽfre wítnunge ne wrece, L. Eth. vii. 17; Th. i. 332, 22. Áwend nú fram mé ðíne wítnunga (*plagas tuas*), Ps. Th. 38, 11. v. ge-, hengen-, un-wítnung.

wítnung-stów, e; *f. A place of punishment* :—Seó micele byrnende dene is wítnungstów, in ðære beóð manna sáwla gewítnode and geclǽnsode, Homl. Th. ii. 352, 20. Oft men wurdon of ðisum lífe gelǽdde, and eft tó lífe árǽrde, and hí fela wítnungstówa and eác hálgena wununga gesáwon, 354, 28. v. wítung-stów.

witod; *adj.* (*ptcpl.*) I. *appointed, ordained, assured, certain* :—Him is unhyldo Waldendes witod, nú hié wordcwyde his forléton, Cd. Th. 45, 21; Gen. 730. Ðé is gedál witod líces and sáwle, 57, 19; Gen. 930: 252, 9; Dan. 576: Andr. Kmbl. 1777; An. 891. Ðonne bið ús seó méd æt Drihtene witod, L. E. G. 21; Th. ii. 418, 20. Mé bið gyrn witod . . . bearnum biþ deáþ witod, Exon. Th. 396, 18, 28; Rá. 16, 6, 11 : 494, 13; Rá. 82, 7 : Fins. Th. 53; Fin. 26. Mé bið witod, ðæt ic þolian sceal bearngestreóna, Exon. Th. 402, 3; Rá. 21, 24. Ðé is súsl weotod, Cd. Th. 308, 14; Sat. 692 : Andr. Kmbl. 1902; An. 953. Here bád witodes willan, Cd. Th. 213, 12; Exod. 551. Witodre fyrde, 207, 23; Exod. 471. Sceal ic witodes bídan *I must await my certain fate*, 137, 18; Gen. 2275. Dóm wutedne *judicium certum*, Rtl. 92, 18. Wé ús nytan witod líf óð ǽfen *we are not sure of life until the evening*, Wulfst. 241, 16 : 240, 18 : 151, 17. Nú hæbbe ic ðíne hyldo mé witode geworhte, Cd. Th. 45, 15; Gen. 727. Weotude, Andr. Kmbl. 2149; An. 1076. Fleág fugla cyn, ðǽr hý feorhnere witude fundon (*where they were sure of finding food*), Exon. Th. 157, 11; Gú. 890. Witode, 430, 13; Rá. 44, 8. Béc bodiaþ weotedne willan, Salm. Kmbl. 475; Sal. 238. Ne cýþ ðú witod on wén ðín *do not feel sure of your expectation*, Prov. Kmbl. 22. Se ealda man him mæg gewislíce witod witan, ðæt him se deáð geneáláecð *the old man may surely know, that for him the approach of death is certain*, Wulfst. 147, 26. Hí eác wénan ne þurfon, ac witod witan, ðæt hig yfel leán habban scylan, 270, 26. Ic ðæt wénde and witod tealde, ðæt ic ðé meahte áhwyrfan from hálor, Exon. Th. 264, 1; Jul. 357. Him tó wǽron witode geþingþo, Cd. Th. 30, 30; Gen. 475. Ðé sind wítu weotud be gewyrhtum, Andr. Kmbl. 2731; An. 1368. Feohgestealda witedra wénan, Exon. Th. 283, 26; Jul. 686. Hé him wælbende weotode tealde, Beo. Th. 3877; B. 1936. Uuteod *certa*, Rtl. 171, 41. II. *with much the same force as* **witodlíce**, (a) *with definite sense, it is certain, certainly, assuredly* :—Witod, se ðe his broces bóte secð, búton tó Gode sylfum, hé drýhð deófles wyllan, Wulfst. 12, 11 : 85, 14. Án þing ic eów secge tó gewisse, ðæt witod sceal geweorðan godspel gecýþed geond ealle worulde ǽr worulde ende, 89, 21. Se ðe forsyhð eów, witod hé forsyhð mé, 177, 15. (b) *in a less definite sense, indeed, surely* :—Allo wuted iornaþ *omnes quidem currunt*, Rtl. 5, 35. Ða heordas wutud gisprécun betwih him, Lk. Skt. Rush. 2, 15. Witud *quidem*, Anglia xiii. 392, 383: *nam*, 368, 40: *itaque*, 379, 194. [*O. Sax.* witod :—Nadra, ðár siu iro níðskepies witodes wánit *where it thinks hostility intended*, Hél. 1880. Cf. *Goth.* witóþ *law* : *O. H. Ger.* wizod, wizzod *lex, jus.*] v. ge-, un-witod, *and next word.*

witod-líc; *adj. Certain* :—Wutudlíce sindun wítga ðætte wǽre *certi sunt prophetam esse*, Lk. Skt. Rush. 20, 6. [Cf. *O. H. Ger.* wizzod-líh *legalis.*]

witodlíce; *adv.* I. *certainly* :—Witodlíce (*amen*) ic secge eów, Mt. Kmbl. 26, 21. Wéne ic ful swíðe and witodlíce, Exon. Th. 461, 5; Hö. 30. II. *with a somewhat indefinite sense, translating many Latin words, indeed, surely, truly* :—Witodlíce (wotetlíce, Lind.) *autem*, Mt. Kmbl. 1, 21. Wiototlíce, Lind. 2, 3. Wutedlíce (wutudlíce, Rush.), Mk. Skt. Lind. 2, 10. Witodlíce *enim*, Mk. Skt. 1, 38. Wiotudlíce *ergo*, Jn. Skt. Rush. 18, 3. Witedlíce *etenim*, Ps. Spl. 15, 6. Witudlíce, Anglia xiii. 365, 3. Witodlíce *igitur*, Gen. 4, 11 : Mt. Kmbl. 12, 28: *inquam*, Kent. Gl. 945. Wutudlíce *itaque*, Mk. Skt. Rush. 18, 4. Witodlíce *nam*, Anglia xiii. 386, 302: *quippe* and *nempe*, Ælfc. Gr. 38; Zup. 227, 2: *quidem*, Mt. Kmbl. 9, 37. Witedlíce, Ps. Spl. 34, 23. Uutetlíce, Mt. Kmbl. Lind. 26, 24. Witudlíce *quoque*, Anglia xiii. 397, 459: *utique*, 366, 19. Wutudlíce, Jn. Skt. Rush. 14, 28. Witodlíce *vero*, Mt. Kmbl. 8, 24. Wiotudlíce, Mk. Skt. Rush. 1, 8. Witodlíce *videlicet*, Anglia xiii. 387, 318. Wietodlíce, Past. 35; Swt. 239, 20. [*O. H. Ger.* wizzodlícho *quidem*.]

-witol, -wittol. v. fore-witol, Chr. 1067; Erl. 204, 28, un-wittol.

witon, wuton (-an, -un), **uton** (-an, -un); *interjectional form with an infinitive, the combination being the equivalent of a subjunctive—let us* . . . :—Uton (wuton, Cott. MS.) ágifan ðæm esne his wíf, Bt. 35, 6; Fox 170, 6. Wuton wuldrian weorada Dryhten, Hy. 8, 1. Uuton nú gehýran, Blickl. Homl. 83, 30. Wutan cuman ealle, and úre mágas mid ús wutun þyder habban, Ps. Th. 73, 8. Wutun cuman ealle and hí tówyrpan *venite et disperdamus eos*, 82, 4; Beo. Th. 5290; B. 2648. Gǽ wé ł wutun (wutu, Rush.) geonga, Mk. Skt. Lind. 1, 38: 14, 42. Uton gán (uutun geonga, Lind.) *eamus*, Jn. Skt. 11, 16. Uton wircean *faciamus*, Gen. 1, 26: 2, 18: 11, 3: Cd. Th. 26, 8; Gen. 403: 278, 6; Sat. 217. Ða cwæþ hé: 'Uton geécan ðone anweald ...' Ða

cwæþ ic : 'Uton ðæs,' Bt. 33, 1 ; Fox 120, 28. Utan biddan God, Bd. 2, 2 ; S. 502, 18 : 3, 2 ; S. 524, 21 : Exon. Th. 48, 14 ; Cri. 771. Utun faran *transeamus*, Lk. Skt. 2, 15. ¶ the word was originally a tense of the verb wítan, and its verbal character is occasionally still marked by the use of the pronoun :—Wuton wē ðæt·gemunan, Blickl. Homl. 125, 2. Uutun uē geonga (uton gan, W. S., wutun gonga, Rush.) *eamus*, Jn. Skt. Lind. 14, 31. [Uten don elmessen, O. E. Homl. i. 107, 6. Uten we heom to liðe, Laym. 20635. Ute we to him fare, O. and N. 1779. *O. Sax.* wita.]

witran *to make certain* (?), *to inform* :—Witro *veror*, Wrt. Voc. ii. 123, 23. [Wise mi and witere (witte me, 2nd MS.), whuder ic mæi liðan, Laym. 1200. Wite me and were and witere and wisse þurh þi wisdom to wite me wið sunne, Jul. 33, 13. Ho has witered hire of þis, and ho has hire kenned, Jos. 466. Ho watȝ wytered bi wyȝes what watȝ þe cause, Allit. Pms. 85, 1587. Cf. *Icel.* vitra *to manifest, reveal*.] v. witer.

wi-trod. v. wíg-trod.

wit-seóc ; *adj. Lunatic, possessed* :—Hrýmde sum wód mann ðurh deófles gást ... Weard se mann geclǽnsod fram ðam fúlan gáste ... Ðá geáxode se cyning be ðam witseócum menn, Homl. Th. i. 490, 2–8. Hí deóflu fram wittseócum mannum áflígdon, ii. 490, 23. *Exorcista* is se ðe ræt ofer ða witseócan men, L. Ælfc. P. 34 ; Th. ii. 378, 7 : Homl. Skt. i. 7, 392. v. gewit-seóc.

witt, witter, wittiend-líc, wittig, witud. v. wit, witer, witiend-líc, witig, witod.

wituma, an ; *m. A dowry* :—Wituma *vel* wetma, uuituma *dos*, Txts. 57, 704. Weotoma *dote* (the line is : Ne metuas juvenis sortiri dote puellam, Ald. 170), Wrt. Voc. ii. 93, 28 : 27, 18. Lócige hē ðæt hió hæbbe ðæt weorð sié hire mægðhádes, ðæt is se weotuma (wituma, *v.l.*) *pretium pudicitiae non negabit* (Ex. 21, 10), L. Alf. 12 ; Th. i. 46, 18. Ágife hē ðæt fioh æfter ðæm weotuman (*juxta modum dotis, quam virgines accipere consueverint*, Ex. 22, 17), 29 ; Th. i. 52, 8. In Anglia xiii. 30, 82, wytuma *paranymphus* seems a mistake for witumbora. v. next word. [*O. Frs.* wetma, witma, v. Richthofen : *O. H. Ger.* widemo *dos*.]

witum-bora, an ; *m. A bridesman ; paranymphus*, Hpt. Gl. 448, 25.

witung. v. fore-witung.

witung-stów, e ; *f. A place of torment* or *punishment* :—Ðæt is eác cúþ, ðæt for ðæs dæges weorþunge, ðæt ða sáuwla onfóþ reste, ða ða beóþ on wítincgstówan, Wulfst. 219, 34. v. wítnung-stów.

wit-word, es ; *n. A statement which bears witness* to anything, *testament, covenant* :—Witword and gewitnes, ðæt ðæt stande ðæt hit nán man ne áwende, L. Eth. iii. 3 ; Th. i. 294, 1. Wē willaþ ðæt . . . witword and getrýwe gewitnes . . fæste stande, L. N. P. L. 67 ; Th. ii. 302, 5. Ofer ðæm landum ðe Ealdrēd ærcebiscop hæfð siðþan begitan on witword oððe on caupland (*by testament or purchase*), Chart. Th. 439, 4. [His witeword *testamentum ejus*, Ps. 24, 14. Alle þat felle to me . . . of my lordes witeword, witnes þerof haf I, R. Brun. 152, 9. Fulfille I salle in dede þe kynges witworde, 153, 2. Cf. *Swed.* wits-ord *witness, testimony. Icel.* vit-orð *knowledge*.]

Wixan ; *pl. The name of some people in some district in England* :—Eást-Wixna is þryú hund hýda, West-Wixna syx hund hýda, Cod. Dip. B. i. 414, 19. Cf. on wixena bróc, Cod. Dip. Kmbl. iii. 78, 1.

wixen ; *adj. Of wax* :—Hláf wexenne, Lchdm. iii. 210, 1. [*M. H. Ger.* wehsin.]

wlacian ; *p.* ode. **I.** *to be* or *get lukewarm* :—Ic wlacige *tepeo*, Ælfc. Gr. 26, 2 ; Zup. 154, 4. Swá swá ðæt cealde ǽrest onginð wlacian, ǽr hit ful wearm weorðe, swá eác ðæt wearme wlacaþ, ǽr hit eallunga ácealdige *sicut a frigore per teporem transitur ad calorem, ita a calore per teporem reditur ad frigus*, Past. 58 ; Swt. 447, 4. **II.** *to make lukewarm* :—Ic wlacige *tepefacio*, Ælfc. Gr. 37 ; Zup. 218, 6. v. á-, ge-wlacian ; wleccan.

wlacu *and* **wlæc ;** *adj. Lukewarm, tepid* :—Mid wlæcre *tepida* (*lepida*, MS.), Wrt. Voc. ii. 50, 43. (1) in a physical sense :—Gedó ðæt sió wyrt wlacu (blacu, MS.) sý, and þyge hý, Lchdm. i. 80, 13. Wlece hyt, ðæt hyt wlæc beó, and habbe on hys múþe swá wlac, iii. 106, 2–4. Gif sió wamb biþ windes full, cymð ðæt of wlacre wǽtan ; sió cealde wǽte wyrcþ sár an, ii. 224, 23. Hié beóð mid wlacum wætre on hǽlo gebróhte *aegros ad salutem tepens aqua revocavit*, Past. 37 ; Swt. 269, 25 : Homl. Skt. i. 11, 158. On wlacum ele, Homl. Th. i. 86, 23. Syle hyt him wlacu súpan, Lchdm. i. 196, 19. Genim ðæt swá wlacu, ii. 40, 5. Sete him wlacu wæter drincan swíþe hát, 62, 11. Wlaco, 40, 9 : 192, 10. Gewyrm hyt and swá wlæc drýpe on ðæt eáre, i. 178, 25 : 188, 7 : 210, 9. On wlæc wín, ii. 24, 28. (2) in a figurative sense :—Hē is wlaco (*tepidus*), and nis náuðer ne hát, ne ceald . . . Se biþ wearm, nalles wlaco . . . Swá eác se ðe wyrð wlacra treówa, and nyle ðæt wlæce oferwinnan (*nequaquam tepore superato*) . . . Se ðe tó lange wunaþ on ðǽm wlacum treówum . . . hē wlacu bið . . . Se ðe tó lange wlæc bið, Past. 58 ; Swt. 447, 1–14. Gif wēn sí ðæt hē on strengo þeódscipes tó wlæc (*tepidus*) sý, Bd. 1, 27 ; S. 492, 18. Oft ða monðwǽran weorðaþ suā besolcne and suā wlace and suā sláwe *saepe mansueti dissolutionis*

torpescunt taedio, Past. 40 ; Swt. 289, 15. [Ðe wop ðe cumeð of þe wlache heorte *lacrima tepida*, O. E. Homl. ii. 151, 9. Torpor is þe uorme, þet is wlech heorte, A. R. 202, 4. Wlech weater, Jul. 31, 11.]

wlæcce (?), an ; *f. Lukewarmness* :—Wlæccan *frigum*, Germ. 397, 448.

wlæclíce ; *adv. Lukewarmly* :—Wlæclíce *tepide, enerviter*, Hpt. Gl. 420, 39. For hwon segdes ðú Ǽcgbrihte swá gēmeleáslíce and swá wlæclíce (*tam negligenter ac tepide*) ða ðing, ðe ic bebeád him tó secganne, Bd. 5, 9 ; M. 410, 33. [*In Ps. Th.* 148, 5 wlæclíce *seems a mistake for* wræclíce.]

wlæcness, e ; *f. Lukewarmness* :—Wlæcnesse *teporis* (wlætnesse *leporis*, MS.), Wrt. Voc. ii. 50, 45. Ðý lǽs hē for wlæcnesse sié út áspiwen *ne tepidus evomatur*, Past. 58 ; Swt. 447, 16, 18.

wlæffetere, es ; *m. A stammerer, one who speaks imperfectly* :—Wlæffetera *uilium bauilorum*, Germ. 403, 910. [Cf. Ich ne ssolde by bote a wlaffere, ne zigge þing to þe uolle, Ayenb. 262, 1. A checun mot l'un balbeye (*wlaffes*), Wrt. Voc. i. 173, 8. Som useþ strange wlaffyng, chyteryng, harryyng & garryng, Trev. c. 59.]

wlǽta, wlǽtta, an ; *m.* **I.** *nausea, loathing* :—Wið spiwðan and wlǽttan, Lchdm. i. 358, 24. Wiþ wlǽttan, ðam men ðe hine ne lyst his metes ne líþes, ii. 62, 15. Wiþ unluste and wlǽttan ðe of magan cymð, 184, 5. Wlǽtan, 158, 12. Gif hwá on scipe wlǽttan þolige, i. 206, 9. Ðone wlǽttan ðæs magan, 204, 20. Ne yrne hē, ðe læs hē mid ðæs rynes edgunge hwylcne wleáttan (wlǽttan, *v.l.*) and sogeðan on his heortan ne ástyrige, R. Ben. 68, 3. **II.** *what produces nausea, an object of loathing* :—Oð hit gǽð þurh eówre næsþyrlu and sí gewend tó wlǽttan (*vertatur in nauseam*), Num. 11, 20. Bútan hláfe ǽlc mete tó wlǽttan byþ gehwyrfed, Coll. Monast. Th. 28, 35. Seó ofering ðē wurþ oþþe tó sáre oððe tó wlǽttan, Bt. 14, 1 ; Fox 42, 16. Wlǽttan *sentina* (ab omni spurcitiae sentina immunes, Ald. 10), Anglia xiii. 28, 28. Fúlne wlǽttan *foetidam nauseam* (*sentinam*) (the passage is : Cum falsae garrulitatis incestum velut foetidam melancholiae nauseam de recessibus falsi pectoris evomuisset, Ald. 40), Hpt. Gl. 475, 50. Wlǽtan *nausiam* (the gloss belongs to the passage given in the preceding), Wrt. Voc. ii. 81, 9. **III.** *defilement, disfigurement.* v. an-wlǽta, -wláta ; á-, ge-wlǽtan :—Wlǽtta *deformatio* (venusti capitis deformatio, Ald. 62), Hpt. Gl. 510, 6. [Þu miht mid wlate þe este bugge, O. and N. 1506.]

wlǽtan. v. á-, ge-wlǽtan.

wlǽtung, e ; *f.* **I.** *sickness, nausea* :—Mid micelre wlǽtunge gewíteþ ðæt sár on weg, Lchdm. i. 80, 14 note. v. morgen-wlǽtung, Lchdm. iii. 44, 19. **II.** *defilement, disfigurement.* v. wlǽta, III :—Wlēttuncg *deformatio*, Hpt. Gl. 510, 6.

wlanc ; *adj.* **I.** *proud, high-spirited, bold.* v. wlencu, I :—Wlanc Wedera leód, Beówulf, Beo. Th. 687 : B. 341. Wlonc hæleþ, 668 ; B. 331. Wæterþisa wlonc, Exon. Th. 363 7 ; Wal. 50. Ðær wlanc manig on stæde stódon, Elen. Kmbl. 461 ; El. 231. Duguþ eal gecrong wlonc, Exon. Th. 291, 10 ; Wand. 80. Hē hæfde Higeláces hilde gefrunen, wlonces wígcræft, Beo. Th. 5898 ; B. 2953. Wlance þegenas, unearge men, Byrht. Th. 137, 53 ; By. 205 : Cd. Th. 188, 19 ; Exod. 170. Wlance wígsmiðas, eorlas árhwate, Chr̄. 937 ; Erl. 115, 21. Men módum wlonce, Exon. Th. 325, 4 ; Víd. 100. Hē in healle wæs wended wloncra folmum, 441, 17 ; Rä. 60, 19. I ega wlancum, ðǽr wígan sittaþ, Runic pm. Kmbl. 342, 3 ; Rún. 14. **I a.** applied to animals :—On wlancan ðam wicge, Byrht. Th. 138, 54 ; By. 240 : Exon. Th. 489, 13 ; Rä. 78, 7. Sum sceal wildne fugel wloncne ætemian, hafoc on honda, 332, 15 ; Vy. 85. **II.** in an unfavourable sense, *proud, bold, arrogant, haughty, insolent.* v. wlencu, II :—Hē (*a dog*) leánaþ grimme ðe hine wloncne weorþan lǽteþ, Exon. Th. 434, 13 ; Rä. 51, 10. Ða wlanca[n] scamlēstan *frontosam* (*elationis*) *impudentiam*, Hpt. Gl. 526, 5. Tó manege weorðaþ tó wlance and ealles tó rance and tó gylpgeorne *erunt homines elati, superbi* (2 Tim. 3, 1), Wulfst. 81, 15 : L. I. P. 14 ; Th. ii. 322, 12. Ne wlance (*elati*) synd eágan míne, Ps. Spl. 130, 1. Wlancra (wancla, MS.) manna *protervorum*, Hpt. Gl. 526, 70. Oð ðæt wlance (*the Egyptians*) forsceáf mihtig engel, Cd. Th. 190, 25 ; Exod. 204. **III.** *proud, elate, exultant* :—Se ðe áh lífes wyn, wlonc and wíngál, Exon. Th. 307, 25 ; Seef. 29 : 478, 2 ; Ruin. 35. Hē mid gáre stang wlancne wícing, ðe him ða wunde forgeaf, Byrht. Th. 135, 56. **IV.** *splendid, great, high, august, magnificent, rich.* v. wlencu, III :—Welig † wlonc *diues*, Lk. Skt. Lind. 12, 21 : 16, 22. Wlonc *dives* . . ðe wlonca *divitem*, Mt. Kmbl. Lind. 19, 23, 24. Summ monn wlong *quidam homo dives*, 27, 57. Ðú, weliga, ðínne Drihten ne lufadest . . . Hwæt, wēndest ðú, wlanca, gif ðú mē sealdest ówiht ðínes, ðæt ðē ðonne wǽre ðín woruldgestreón gelytlad? Wulfst. 260, 18. Wereda Wuldorgifa, wlanc and ēce God *great and eternal*, Hy. 10, 48. Se wlonca dæg *the great and terrible day of the Lord*, Exon. Th. 448, 7 ; Dóm. 50. Monnes wloncas (wlonches, Rush.) lond *hominis diuitis ager*, Lk. Skt. Lind. 12, 16. Of beád ðæs wlonces *de mensa diuitis*, 16, 21. Se Hǽlend cwæð tó ðam wlancan : 'For hwí wǽre ðú swá fæsthafol mínra góda, ðe ic ðē sealde?' Wulfst. 258, 12. Ðam wlancan *to the great king* (Nebuchadnezzar), Cd. Th. 221, 30 ; Dan. 96. Ða ðe heora yldran on worolde ne wurdan welige ne wlance þurh woroldglænge *those whose*

forefathers were not wealthy or great through worldly splendour, L. Eth. vii. 21; Th. i. 334, 3: Wald. 116; Vald. 2, 30. Ealle gelíce on woruld cumaþ, wlance and heáne (*high and low*), Met. 17, 6. Wlance *the grandees of Egypt*, Cd. Th. 109, 20; Gen. 1825. Wloncra wín-sele, 270, 21; Sat. 94. Hé feorgbona weorþeþ wloncum and heánum, Exon. Th. 362, 27; Wal. 43. Ic lǽrde wlance men and heáhgeþungene ðæt hié ne ástigan on ofermēdu, ne welena tó wel ne trúwodon, Blickl. Homl. 185, 13. **IV a.** where the circumstance, in which the splendour, etc., consists, is given:—Fugel feþrum wlonc *the bird splendid of plumage*, Exon. Th. 204, 19; Ph. 100. Draca on hlǽwe frætwum wlanc, Menol. Fox 513; Gn. C. 27. Wǽre ðú wiste wlonc and wínes sǽd *thou wast sumptuous in food, sated with wine*, Exon. Th. 369, 10; Seel. 39. Ǽse wlanc (*abundantly provided*), fylle gefrægnod, Beo. Th. 2668; B. 1332. Mādmǽhta wlonc *rich in treasures*, 5659; B. 2833. Weras duguðum wlance Drihtne guldon gōd mid gnyrne, Cd. Th. 146, 8; Gen. 2419. [He wes prud and wlonc, O. E. Homl. i. 35, 16. Neuer upen eorþe to wlonk þu ny uurþe, Misc. 112, 184. Godelike on horse, wlanc on werge, and unwurþ on wike, 121, 315. þat child (*Christ*) þat is so milde and wlong, 197, 11. Ȝe beoð toswollen wið wind of wlonke wordes, Kath. 842. My wodbynde so wlonk þat wered my heued, Allit. Pms. 106, 486. Al my weole wlonke, P. S. 156, 17. Sumeres tide is al to wlonc, O. and N. 489. Asked Crist, quethir thai yed to se sain Ion in wlanke wede, Met. Homl. 42, 2. þe wlonkest wedes, Gaw. 2025. [*O. Sax.* wlank.] v. fela-, gold-, hyge-, mod-, symbel-wlanc.

wlanc, es; *n.* Pride:—Wlanc *typhus*, Wrt. Voc. ii. 97, 9. [For wlaunke (*rimes with* ranke), P. S. 341, 5.] v. wlencu.

wlancian; *p.* ode *To grow proud, great*:—Wlancaþ *insolescat* . . . wlancende *indruticans*, Wrt. Voc. ii. 44, 5–8. Wlancude *adolesceret*, wlancige *adolesco*, Hpt. Gl. 508, 12–14. Wlancode, Wrt. Voc. ii. 3, 42. v. á-wlancian.

wlanclíce; *adv.* Proudly, arrogantly:—Uulanclíce *adrogantissime*, Txts. 42, 112. Wlanclíce, Wrt. Voc. ii. 7, 53.

-wlát, -wláta, wlátend, wlát-ful. v. on-wlát, an-wlǽta, ymb-wlátend, neb-wlátful.

wlátian; *p.* ode; *impers. To cause a person* (acc.) *loathing*:—Mé wlátaþ *nauseo*, Ælfc. Gr. 26, 6; Zup. 158, 7. Ús wlátaþ for ðisum mete *anima nostra nauseat super cibo isto*, Num. 21, 5. Ðonne hié mete þicgeaþ and drincaþ, ðonne wlátaþ hié, Lchdm. ii. 220, 5. Gif man sý innan unhál, oþþe hyne wlátige, i. 76, 9. Búton ðú git tó full sý ðæs ðe ðé lǽfed is, ðæt ðé for ðý wlátige, Bt. 11, 1; Fox 30, 20. [Gif heo hit stunken, ham wolde wlatien þer agean, A. R. 86, 19. Overfulle makeþ wlatie, O. and N. 354. Menslaers Laverd wlate sal (*abhominabitur*), Ps. 5, 7. Me wlateȝ withinne, Allit. Pms. 47, 305. Him wlatis, H. S. 3541. Surfet us wlattis, Alex. (Skt.) 4277. It wold haue wlated any wee, 5634.] v. wlǽtan.

wlátian; *p.* ode *To gaze, look*:—Hraðe wæs æt holme hýðweard, se ðe ǽr lange tíd feor wlátode, Beo. Th. 3837; B. 1916. Ðæt is gefylled, ðæt se fróda mid eágum on wláte, Exon. Th. 20, 34; Cri. 327. [*Goth.* wlaitōn *circumspicere*.] v. be-, ymb-wlátian; wlítan.

wlátung, e; *f.* Nausea, loathing:—Uulatung (-ing, -unc) *nausatio, vomitus*, Txts. 78, 667. Mid micelre wlátunge gewíteþ ðæt sár, Lchdm. i. 80, 14. Wiþ wlátunge, ii. 62, 18. Wlátunge *nausiam*, Wrt. Voc. ii. 59, 67. [Habbeð wlatunge of þe muðe þet speowed ut atter, A. R. 80, 25. Lest heo suppose þow make þat fare for wlatynge, Mirc. 894.]

wlátung. v. ymb-wlátung.

wleccan; *pp.* wleced, wlecced, wleht *To make lukewarm*:—Wlece listum on wearmum glēdum, Lchdm. ii. 26, 8: 30, 13. Wlece hyt eall tógadere, ðæt hyt wlæc beó, iii. 106, 2. Ælc wæter bið ðý unwerodre tó drincanne, æfter ðæm ðe hit wearm bið, gif hit eft ácólaþ, ðonne hit ǽr wǽre, ǽr hit mon ó ongunne wleccan, Past. 58; Swt. 447, 21. v. ge-wleccan; wlacian.

wlencan *to make* wlanc (*q. v.*) [Ech man is strong ðe awelt is lichame, and wlencð his soule, O. E. Homl. ii. 189, 27. Leaf þi lease wit þ tu wlenchest te in *depone false sapientie supercilium*, Kath. 1010.] v. for-, ge-wlencan, ofer-wlenced.

wlencu (-o); *indecl.*: wlenc, e; *f.* **I.** *pride, high spirit.* v. wlanc, **I**:—Wénic ðæt gé for wlenco, nalles for wræcsíðum, ac for higeþrymmum Hróðgár sóhton, Beo. Th. 681; B. 338. Þrym sceal mid wlenco, þríste mid cēnum, Exon. Th. 337, 7; Gn. Ex. 61. **II.** in an unfavourable sense, *pride, arrogance, haughtiness, insolence.* v. wlanc, **II**:—Him wlenco gesceód, oferhýd egle, Cd. Th. 258, 20; Dan. 678. Hié wlenco onwód, ðæt hié firendǽda tó frece wurdon, 155, 27; Gen. 2579: 217, 3; Dan. 17. Uulencu *fastu*, Wrt. Voc. ii. 108, 32. Wlenceo, 35, 12. Git for wlence wada cunnedon, and for dolgilpe on deóp wæter aldrum néþdon, Beo. Th. 1020; B. 508: Exon. Th. 114, 27; Gú. 179. Ðý lǽs hé for wlence, wuldorgeofona ful, of gemete hweorfe, and forhycge heánspēdigran, 294, 32; Crä. 24: Cd. Th. 100, 32; Gen. 1673. For wlenco, Beo. Th. 2416; B. 1206. Wlence *insolentiam*, Wrt. Voc. ii. 44, 6. Þeódum ýwaþ wísdōm weras, wlencu forleósaþ, Exon. Th. 132, 18; Gú. 474. **II a.** used of an animal:—Se fear ðæs hyrdes dráfe forhogode and him on ðæt wēsten gewunode . . . .

Ðá ðæt se hláford geáhsod, ðæt ðæt hrýþer swá on wlencu geond ðæt wēsten fērde, Blickl. Homl. 199, 10. **III.** *distinction* of various kinds, *splendour, pomp, dignity, magnificence, wealth, greatness.* v. wlanc, **IV**:—Ða tída ða áne burg welge gedydan . . . þurh ðære ánre burge wlenco (*wealth*) wurdon ealle ōþra tó wǽdlan gedōne, Ors. 5, 1; Swt. 214, 10. Forseó ðysse worulde wlenco, gif ðú wille beón welig on ðínum mōde, Prov. Kmbl. 50. Æghwylce wlence and ídele rence forhogian *to despise all pomp and vanity*, L. I. P. 14; Th. ii. 322, 9. Ðæt mennisce mōd bið oft upáhafen, ðeáh hit mid náne onwalde ne sié underlēd; ac hú micle má wēnst ðú ðæt hit wolde, gif ða wlencea (wlenca, Hatt. MS.) and se anwald ðǽr wǽre tó gemenged, Past. 17; Swt. 114, 1. Hié wǽron welige on ðyssum middangearde, and heora wlenca wǽron swíþe monigfealde on landum and on wíngeardum, and heora hordernu wǽron mid monigfealdum wlencum gefylde, Blickl. Homl. 99, 14–17: 101, 7. Hwǽr beóþ ðonne his welan and his wista? hwǽr beóþ ðonne his wlencea and his anmēdlan? 111, 34. Hé is wyrma wlence *it is the pride of serpents*, Salm. Kmbl. 165; Sal. 82. Ðæra wlenca ł walana *divitiarum*, Mt. Kmbl. Lind. 13, 22. Hé breác longe ǽr wlencea under wolcnum (cf. his mōd ǽr tó ðám woruldsǽlþum gewunod wæs, Bt. 1; Fox 4, 1), Met. 1, 76. Ic cwæð on mínum wlencnm and on mínre orsorhnesse *ego dixi in abundantia mea*, Ps. Th. 29, 6: Past. 65; Swt. 465, 15. Ne ðyrfe hé bión tó upáhæfen for nánum wlencum ne for nánre orsorgnesse *non hunc prospera elevent*, 14; Swt. 83, 16. Ðone naman ic sceolde habban, ðæt ic wǽre wela and weorþ-scipe; ac hié hine habbaþ on mē genumen, and hine habbaþ gesealdne heora wlencum and getehhod tó heora leásum welum, Bt. 7, 3; Fox 20, 30: Blickl. Homl. 53, 9. Þeáh hwá wexe mid micelre æþelcundnesse his gebyrda, and þeó on eallum welum and on eallum wlencum *magna titulis fulgeat claris domus*, Bt. 19; Fox 68, 32: Met. 10, 28. Ðú forlǽtan scealt ídle ofersǽlþa . . . ne ðú ðē ǽfre ne lǽt wlenca gewǽcan, Met. 5, 31. v. gold-, ofer-, weorold-wlencu.

wlisp, wlips; *adj. Speaking inarticulately, lisping, stuttering, stammering*:—Balbus, qui vult loqui et non potest wlips, Wrt. Voc. ii. 125, 11. Wlisp *balbus*, 101, 50: 10, 71: i. 288, 8: *balbutus*, ii. 101, 56: 10, 75. Wlips *blessus*, i. 75, 38. Stamerum and wlipsum *balbis et blaesis*, Hpt. Gl. 478, 15: *blessis*, Wrt. Voc. ii. 81, 42.

wlita, an; *m.* **I.** *face, countenance*:—Hleór *vel* wlita *frons*, wlitan *frontes*, Wrt. Voc. ii. 151, 4–5. **II.** *beauty*:—Wlitan (*or from* wlitu? v. wlite) *decore* (in pulcherrimo pubertatis decore, Ald. 71), Hpt. Gl. 520, 22. [Heo wes a wliten alre vairest, Laym. 2934.] v. and-wlita; wlite.

wlítan; *p.* wlát, *pl.* wliton *To look, gaze*, (1) absolute:—Þeóda wlítaþ, Exon. Th. 221, 28; Ph. 341. (2) with prep. (adv.):—Ðú on magan wlítest, Cd. Th. 144, 26; Gen. 2395. Wuhta gehwylc on weoruld wlíteþ, Met. 31, 14. Hé wlít ofer ealle ða ðe ealre eorðan ymbhwyrft búiaþ *respexit super omnes qui habitant orbem*, Ps. Th. 32, 12. Ðissum ídesum ðe wē on wlítaþ, Cd. Th. 150, 32; Gen. 2500. On ða synwyrcend wlítaþ, Exon. Th. 68, 18; Cri. 1105. Wlát wítga geond þeódland, óþ ðæt hé gestarode, ðǽr gestaþelad wæs æþelíc ingong, 19, 25; Cri. 306. Hió wlát ofer ealle, Elen. Kmbl. 770; El. 385. Hé tó heofenum wlát, Byrht. Th. 136, 56; By. 172. Hé æfter recede wlát, Beo. Th. 3149; B. 1572. Ða ðe on holm wliton, 3189; B. 1592. Wlítan on Wílǽf, 5696; B. 2852: Cd. Th. 145, 8; Gen. 2402. Heó swá wíde wlítan meahte ofer heofonríce, 38, 18; Gen. 608. Wlítan in wuldre *to see heaven*, 290, 2; Sat. 409. Fleóhnet, ðæt hé mihte wlítan ðurh on æghwylcne, and on hyne nǽnig monna cynnes, Judth. Thw. 22, 5; Jud. 49. (2 a) amplified by the addition of *eágum*:—Hé ofer ealle þeóde eágum wlíteþ *oculi ejus super gentes respiciunt*, Ps. Th. 65, 6. Hý geseoð hyra cyning, eágum on wlítaþ, Exon. Th. 352, 7; Sch. 94. On ðone eágum wlát cining, Cd. Th. 7, 15; Gen. 106. Wlít (hǽwa, Bt. 4; Fox 8, 20) on moncyn mildum eágum, Met. 4, 54. Hý wēnaþ ðæt hý on eálond sum eágum wlíten, Exon. Th. 360, 28; Wal. 12. Eágum wlítan on, Cd. Th. 107, 25; Gen. 1794: 109, 19; Gen. 1825. [*Icel.* líta *to look*.] v. be-, geond-, þurh-wlítan; wlátian.

wlite, es; *m.* : wlitu, e (*and* ? an; *v.* wlita, **II**); *f.* **I.** *aspect, countenance, looks, appearance, shape, form*:—Wlite his *vultus ejus*, Ps. Spl. 10, 8. Cristes onsýn, æþelcyninges wlite, Exon. Th. 56, 27; Cri. 907: Beo. Th. 506; B. 250. Se wlite ðæs wundorlícan líchoman *species corporis gloriosi*, Bd. 4, 9; S. 576, 35. Ðeáh ðe him se wlite cwēme *though the looks (of the sword) please him*, Salm. Kmbl. 332; Sal. 165. Sceal on leóht cuman sínra weorca wlite, Exon. Th. 64, 15; Cri. 1038. Wæs gelícnes horses and monnes, hundes and fugles, and eác wífes wlite, 418, 28; Rä. 37, 12. Ðeós wlitu *haec species*, Ælfc. Gr. 12; Zup. 82, 11. Gilde he his (*a horse's*) wlites wyrðe, L. Ath. v. 6, 1; Th. i. 232, 25. Be his (*a slave's*) wlites weorðe, 2; Th. i. 234, 6. Be his wlite, L. In. 26; Th. i. 118, 20. Mid wlite and mid wæstmum fæger *fair in face and form*, Blickl. Homl. 113, 16. Ðeáh ðe ðú wǽre eallra monna fægrost on wlite, Bt. 32, 1; Fox 114, 27. Hí ealle tó ðæs mannes wlite gesceapene synd *they are all made in his likeness*, Boutr. Scrd. 19, 22. Wlit ł onsión *personam*, Mt. Kmbl. Lind. 22, 16. Wundriaþ weras wlite and wæstma, Exon. Th. 221, 9; Ph. 332.

Nǽnig mæg wlite and wísan wordum gecýþan, 491, 30; Rä. 81, 7. **II.** *good looks, beautiful appearance, beauty, glory, ornament :*—Hwæþer nú gimma wlíte eówre eágan tó him getió, heora tó wundrianne? seó duguð ðæs wlites ðe on ðám gimmum bið, biþ heora, næs eówre *an gemmarum fulgor oculos trahit? si quid est in hoc splendore praecipui, gemmarum est lux illa, non hominum*, Bt. 13; Fox 40, 1: Cd. Th. 239, 1; Dan. 364: Exon. Th. 82, 32; Cri. 1347. Fealwe blóstman wudubeáma wlite, 202, 25; Ph. 75. Ðínes wuldres wlite *gloria tua*, Ps. Th. 56, 13. Weorðlíc wlite wuldres ðínes *magnificentia*, 95, 6. Se wlite his andwlitan *species decoris ejus*, 49, 2: Cd. Th. 278, 18; Sat. 223. Ælc wlite and ælc fægernes ðisse weorlde lífes. . . . Se wlite and seó fægernes ðære sáule, Blickl. Homl. 57, 28–31: 59, 6. Priscianus se ðe ys ealre Lédenspræce wlite gehâten, Ælfc. Gr. 15; Zup. 94, 3. Wlites wealdend, Ps. Th. 67, 12. Se fulla móna wyrð wlites bereáfad, Met. 28, 42. Wlittes *decoris*, Rtl. 92, 10. Wlite *decore*, 97, 16: *stemmate*, Wrt. Voc. ii. 88, 46. Ðe læs ðe hé for hire (*Sarah's.* v. Gen. 12, 11) wlite wurde ofslagen, Boutr. Scrd. 22, 1. Sunnan beorhtra, æþeltungla wlite, Exon. Th. 181, 4; Gú. 1288. Him mid sídedon twǽgen scínende englas mid wundorlícre wlite swá hé sylf wæs geglenged, Homl. Skt. ii. 25, 775. Heó nalles on goldes wlite and on seolfres ne scíneþ, Blickl. Homl. 197, 9: Elen. Kmbl. 2636; El. 1319: Exon. Th. 238, 24; Ph. 506. Of wlite wendaþ wæstma gecyndu, 104, 29; Gú. 15. Sió micle Babilon ðe ic self átimbrede mé selfum tó wlite and wuldre (*in gloria decoris mei*), Past. 4; Swt. 39, 18: Exon. Th. 70, 18; Cri. 1140. Wlítan *decore* (*see* wlita), Hpt. Gl. 520, 22. Drihten hine mid weordlíce wlite gegyrede *Dominus decorem induit*, Ps. Th. 92, 1: Cd. Th. 3, 15; Gen. 36. Myceine wlite. (*decorem*) ðú ásetst ofer hine, Ps. Spl. 20, 5. Gehéald ðínne wlite and ðíne fægernesse *speciem tuam et pulcritudinem tuam intende*, Ps. Th. 44, 5: Hpt. Gl. 523, 60. Ðes middangeard wæs tó ðon fæger, ðæt hé teáh men tó him þurh his wlite and þurh his fægernesse, Blickl. Homl. 115, 11: Met. 7, 31: Exon. Th. 86, 10; Cri. 1406: Cd. Th. 132, 10; Gen. 2191: 13, 23; Gen. 207. Ðære rosan wlite, Bt. 9; Fox 26, 20: Met. 6, 13. Sprǽcon ymb ðæs wífes wlite monige, Cd. Th. 110, 34; Gen. 1848. Heora wlite gewemman *to mar their beauty*, 231, 1; Dan. 240. Gif hé hafaþ ofer ealle men wlite and wísdóm, Exon. Th. 299, 16; Crä. 103. Gǽstes wlite, 53, 11; Cri. 849: 96, 29; Cri. 1581. His weorces wlite, 97, 9; Cri. 1588. Bringaþ Drihtne wlite and áre, wuldor ðridde *afferte Domino gloriam et honorem*, Ps. Th. 95, 7. Gegyrede mid eallum mistlícum hrægla wlitum *circumcincta varietate*, 44, 15. Hé hié gegierep nyd ðám winsumestan wlitum, and eft geungewlitegaþ, Shrn. 195. 11. [Haueden men ispeken of hire mucla fæira wlita (of hire mochele fairsipe, 2nd MS.), Laym. 3139. Kerueð of hire neose, and heore wlite ga tó lose, 22844. Gif itt seþ þe wlíte off ennglekinde, Orm. 666. Þi wlite *speciem tuam*, Ps. 44, 5. Ne sal þu þi wif bi hire wlite chesen, Misc. 119, 249. Min hew fælewidþ, and min wlite is wan, 135, 580. O schene nebschaft . . . areow þi wlite, Kath. 1452. Þe lilie mid hire faire wlite, O. and N. 439. Al his wlite wurð teres wet, Gen. and Ex. 2288. *Goth.* wlits *face, form : O. Frs.* wlite: *O. Sax.* wliti *form, beauty : Icel.* litr *hue, countenance*.] v. and-, mǽg-, on-, wamm-wlite, neb-wlitu.

wlite-andett, es; *n.* (?) *A confession of splendour :*—Ðú ðé weordlíce wliteandette góde gegyredest *confessionem et decorem induisti*, Ps. Th. 103, 2.

wlite-beorht; *adj. Of splendid beauty, beautiful,* (1) of persons :—Wlitebeorht ides (*Sarah*), Cd. Th. 103, 34; Gen. 1728. Hié (*Adam and Eve*) wlitebeorht wǽron on woruld cenned, 12, 19; Gen. 188. (2) of things :—Dǽg, wlitebeorhte gesceaft, 8, 28; Gen. 131. Of ánum wætre wlitebeorhtum, 14, 17; Gen. 220. Eorþan, wlitebeorhtne wang, Beo. Th. 186; B. 93. Hí him wíc curon, ðær him wlitebeorhte wongas geþúhton, Cd. Th. 108, 10; Gen. 1804. Eorþan cyningas monegum and mislícum wǽdum wlitebeorhtum scínaþ, golde gegerede and gimcynnum *reges purpura claros nitente*, Met. 25, 4. Wlitebeorhte wæstmas, Cd. Th. 94, 11; Gen. 1560.

wlite-full; *adj. Beautiful, handsome, comely :*—Ofermódig gif hé wlitefull (*decorus*) sí geþúht on gesihþe, swá þeáh on weorcum wác ys, Scint. 21, 8.

wlite-leás; *adj. Without beauty, uncomely, hideous :*—Deóful ætýwde wann and wliteleás, Andr. Kmbl. 2339; An. 1171.

wlitelíce; *adv. Beautifully, in comely fashion :*—Hí weófod wlitelíce geworhtan and gegyredon, Blickl. Homl. 205, 6.

wlite-sceáwung, e; *f. The word is used to translate Sion :*—Bið gesegen God in wlitesceáwunge (-scéwunge, Bd. S. 547, 39) *uidebitur Dominus in Sion*, Bd. 3, 19; M. 212, 11.

wlite-scíne; *adj. Of brilliant beauty, splendid, beauteous :*—Engel ælbeorht, wlitescýne wer, Cd. Th. 237, 15; Dan. 338: Elen. Kmbl. 143; El. 72. Weorud wlitescýne, Exon. Th. 31, 9; Cri. 493: 35, 6; Cri. 554. Seó wlitescýne wuldres condel (*Juliana*), 269, 22; Jul. 454. Wlitescíene wíf (*Eve*), Cd. Th. 33, 28; Gen. 527. On mǽrum dæge hé on wlitescénan dæge *insigni die*, Ps. Lamb. 80, 4. Weoruda wlitescýnast, Exon. Th. 101, 27; Cri. 1665. [*O. Sax.* wliti-skóni.]

wlite-seón, -sín, e; *f. A sight to gaze on, a spectacle :*—Wæs be

---

feaxe on flet boren Grendles heáfod, and ðære idese mid, wliteseón wrǽtlíc, Beo. Th. 3304; B. 1650. Cf. wæfer-sín, wundor-seón.

wlite-torht; *adj. Brilliant, splendid :*—Wlitetorht scíneþ sunna, Met. 28, 60. Wyrta wlitetorhtra, Exon. Th. 484, 5; Rä. 70, 3.

wlite-wamm, es; *m. A disfigurement of the face, personal disfigurement :*—Æt ðam læsestan wlitewamme .iii. scillingas, and æt ðam máran vi. scill., L. Ethb. 56; Th. i. 16, 15. Wlitewomma *nevorum* (nullis naevorum maculis deformatos, Ald. 10), Wrt. Voc. ii. 76, 27: 60, 56. [*O. Frs.* wlite-wam; cf. *also* wlite-wimelsa.]

wlitig; *adj. Beautiful, comely, fair :*—Wlitig *speciosus* vel *decorus*, Wrt. Voc. i. 72, 17: *formosa*, ii. 33, 57. *Elegans*, i. *speciosus, gratus, pulcher* wynsum, wlitig *praecipuus, magnus*, 142, 81. Wlitigre *formosior*, 34. 59. **I.** of beauty that appeals to the senses, (1) appearance in persons or things, (a) of earthly beauty :—Ðæt wíf wæs swíðe wlitig (*pulchra*), Gen. 12, 14. Sum bið wlitig on wæstmum, Exon. Th. 295, 18; Crä. 35. He (*the Phénix*) is wlitig and wynsum, wuldre gemearcad *regali plena decore*, 220, 10; Ph. 318. Onlícnes wlitig. Andr. Kmbl. 1463; An. 732. Ðæt treów wæs wlitig on eágum (*pulchrum oculis*), Gen. 3, 6: Cd. Th. 30, 16; Gen. 467: 247, 18; Dan. 499. Wlitig is se wong, Exon. Th. 198, 8; Ph. 7. Ðeós wlitige gesceaft, heofon and eorþe, Andr. Kmbl. 2873; An. 1439. Ðis leóhte beorht cymeþ eástan wlitig and wynsum, Exon. Th. 350, 13; Sch. 63. Smicere on gearwum cymeþ wlitig scríðan Maius, Menol. Fox 152; Men. 77. Hærfest, wlitig, wæstmum hladen, 281; Men. 142. Ðære wlitegan byrig weallas, Judth. Thw. 23, 24; Jud. 137. In ðam wlitegan træfe, 25, 11; Jud. 255. Hús wlitig and wynsum, Exon. Th. 211, 25; Ph. 203. Wlitig sweord, Beo. Th. 3329; B. 1662. Manna dohtra wǽron wlitige (*pulchrae*), Gen. 6, 2. Gelíce hwítum byrgenum, ða þinceaþ mannum útan wlitige (wlittig, Lind. *speciosa*), Mt. Kmbl. 23, 27. Ne seleþ ðé wæstmas eorþe wlitige, Cd. Th. 62, 18; Gen. 1016. Ðás wlitegan tungl, Met. 28, 6. Ðé weorð wæstm ðý wlitegra, Cd. Th. 33, 14; Gen. 520. Þúhte ðeós woruld wlitigre, 38, 9; Gen. 604. Wífa wlitegost, 39, 17; Gen. 627. Mid ðam wlitegostum nebbe, Homl. Th. i. 430, 14. Ðeáh hé hine gescyrpte mid eallum ðám wlitegestum wǽdum *quamvis se Tyrio ostro comeret*, Bt. 28; Fox 100, 26. (b) of celestial beauty, *beauteous, glorious :*—Hé (*Christ*) bið ðám gódum glædmód on gesihþe, wlitig, Exon. Th. 57, 1; Cri. 912: 232, 33; Ph. 516. Seó wlitige þrýnes, 24, 1; Cri. 378. Wlitig weoroda heáp and wuldres þreát, Andr. Kmbl. 1739; An. 872. Wlitig wuldres boda, Elen. Kmbl. 153; El. 77. Sió wlitige stów (*heaven*), Met. 20, 279. Wlitig, wuldorfæst, Exon. Th. 151, 2; Gú. 789: Cd. Th. 277, 33; Sat. 214. Him is engel mid, ne mæg him bryne sceþþan wlitigne wuldorhoman, Exon. Th. 196, 24; Az. 179. (2) of sound :—Hyre stefn oncwæð wlitig of wolcnum, Exon. Th. 259, 16; Jul. 283. Swég eallum songcræftum swétra and wlitigra, 206, 26; Ph. 132. Wóða wlitegaste, Elen. Kmbl. 1494; El. 749. (3) of scent :—Ðæt wæs swéte stenc, wlitig and wynsum, Exon. Th. 359, 19; Pa. 65. **II.** of beauty that appeals to the mind :—Wynsum and wlitig herung *jocunda decoraque laudatio*, Ps. Spl. 146, 1. Ðeáh ðe ne beó wlitig lof on ðæs synfullan múðe, hwæðere ne geswíce hé ðære herunge, Homl. Th. i. 448, 5. Þúhte fæger and wlitig heora líf, Blickl. Homl. 107, 30. Is ðín nama mǽre, wlitig and wuldorfæst, Cd. Th. 234, 3; Dan. 286. Wlitigan wilsíþes, Exon. Th. 2, 18; Cri. 21. Gǽst weorcum wlitigne, 180, 11; Gú. 1278. Ðæt gé eówer ðæt wlitige líf magon generian, Homl. Skt. i. 23, 189. Ðonne hé ús seld micle getyngnesse and wlitige sprǽce ymb sóðfæsðnesse tó cýðanne *cum nobis luce veritatis plena eloquia subministrat*, Past. 48; Swt. 369, 14. Hine wlitegum wordum herigeaþ, Ps. Th. 146, 1. Wlitige and unclǽne, tile and yfle, Cd. Th. 303, 8; Sat. 609. Wlitegran *formosiore* (venustate formosiore fretus virginitate, Ald. 71), Hpt. Gl. 520, 24. [He awundrede him of hire wliti westum, Kath. 310. *O. Sax.* wlitig.] v. sunn-, un-, un-ge-wlitig.

wlitige; *adv. Beautifully, fairly, splendidly :*—Hálge gǽstas stígaþ tó wuldre, wlitige gewyrtad mid hyra weldǽdum, Exon. Th. 234, 20; Ph. 543. His blǽd scíneþ wlitige in wuldre, Andr. Kmbl. 3438; An. 1723.

wlitig-fæst; *adj. Beauteous, glorious :*—Swá se æþela fugel wlitigfæst wunaþ wyllestreámas, Exon. Th. 204, 29; Ph. 105.

wlitigian; *p.* ode. I. *to make beautiful :*—Ða hé geðwǽraþ and wlitegaþ, hwilum eft unwlitegaþ and on óþrum híwe gebrengþ, Bt. 39, 8; Fox 224, 9. Simle ðæt unwlitige wlitigaþ ðæt wlitige *ever the beautiful beautifies the unbeautiful*, Shrn. 165, 35. Hit worulde wlitigaþ, Exon. Th. 493, 17; Rä. 81, 32. Fyl nú ða frumsprǽce, wlitega ðíne wordcwidas (*give glorious effect to thy words*), and ðín wuldor ús gecýð, 188, 9; Az. 43. Wlitiga ðínne wordcwyde and ðín wuldor on ús gecýð, Cd. Th. 236, 26; Dan. 327. Ðæt ic móte áweccan ðás wæstmas ús tó woruldnytte, wlitigigan ðás wancgturf, Lchdm. i. 400, 7. Wlitigende *decorans*, Hymn. Surt. 140, 14. **II.** *to grow beautiful :*—Byrig fægriaþ, wongas wlitgiaþ (wlitigaþ, MS.), Exon. Th. 308, 33; Seef. 49. v. ge-, un-wlitigian.

wlitigness, e; *f. Beauty, comeliness, adornment :*—Seó wlitignes heora ræsta and setla, Blickl. Homl. 99, 32. Crist com tó wlitignesse and tó weorþunge his brýde, 11, 31. v. un-wlitignes.

wlitigung, wlitu. v. un-wlitigung, wlite.

wlố; _adv._ (?) _Readily, easily_ :—Hế ấwrécen wælpílum wlố ne meahte oroð up geteón (cf. sốna ne meahte oroð up geteón, 163, 20 ; Gú. 997), ellenspræce hleóþor ãhebban, Exon. Th. 171, 16 ; Gú. 1127. v. next word.

wlôh (; _gen._ wlêh ; _f._ ?) _A hem, fringe_ :—Næs him gewemmed wlite, ne wlôh of hrægle ãlýsed, ne loc of heáfde, Andr. Kmbl. 2941 ; An. 1473. Seó hãlge stôd ungewemde wlite, næs hyre wlôh ne hrægl, ne feax ne fel, fýre gemæled, Exon. Th. 277, 34 ; Jul. 590. Wlôh wêdes his _fimbriam vestimenti ejus,_ Mt. Kmbl. Lind. 9, 20 : 14, 36. Wglôana (wlogana ?) mið ðý gehrân _fimbriae tactu,_ p. 17, 10. Hiá miclas wloeh _magnificant fimbrias,_ 23, 5. [Clothes wel neiȝ forwerd, & the wlon offe, Pl. Cr. 736.] v. an-, ge-wlô.

wlott (?) _a blemish_ :—Wlotta, smyttena _naevorum, notárum,_ Hpt. Gl. 421, 55. v. (?) wlæta, III.

wố; _adv._ _Wrongly, perversely, unequally_ :—Hwî sió wyrd swã wố wendan sceolde, Met. 4, 40. v. wôh.

Wocen- (Wrocen- ? v. Wreocen-sǽte) **sǽte,** -sǽtan ; _pl. The name of the occupants of some district in England_ :—Wocensǽtna land is syfan þúsend hîda, Cod. Dip. B. i. 414, 16.

wocig (?), e ; _f. A snare, noose_ :—Wocie _tendiculum, decipulam, laqueum,_ Hpt. Gl. 429, 18. Wociga _catenarum,_ 489, 72.

wốðor, e ; _f. Increase, fruit, offspring_ :—Sceal fæsl wesan cwiclifigendra cynna gehwilces on ðâm wudufæsten, wôcor gelæded eorðan tûdres, Cd. Th. 79, 17 ; Gen. 1312. Fêd feora wôcre, 81, 9 ; Gen. 1342. Ðã gemunde God sunu Lameches, and ealle ða wôcre ðe hê wið wætre beleác, 85, 3 ; Gen. 1409. Hiwan læd ðû, and ealle ða wôcre ðe ic nerede, 90, 4 ; Gen. 1490. [_Goth._ wôkrs τόκος : _O. Frs._ wôker _interest_ : _O. H. Ger._ wuochar _augmentum, incrementum, fructus, fecunditas, germen_ : _Icel._ ôkr _interest._]

wocorlîce. v. wacorlîce.

wôd ; _adj. Mad_ :—Wôd _rabidus vel insanus,_ Wrt. Voc. i. 45, 70 : 75, 56. (1) in reference to persons :—Ðû eart wôd _daemonium habes,_ Jn. Skt. 8, 48, 49, 52 : Homl. Th. ii. 232, 17. Hwâ is swã wôd, ðæt hê dyrre cweðan, ðæt God ne sê æce, Shrn. 176, 32. Ne synt nã ðis wôdes mannes word, Jn. Skt. 10, 21. Wôdan gewittes, Cd. Th. 255, 22 ; Dan. 628. Tô biddenne hire wôdan dehter gesundfulnysse . . . seó dohtor on wôdum dreáme læg dweligende, Homl. Th. ii. 110, 15-19 : 50, 27. Fela wôde menn heora gewit underfêngon, Homl. Skt. i. 27, 130. Hê wôdum mannum gewitt forgeaf, Homl. Th. i. 480, 14. Hê ða deôflu ãflîgde of ðâm wôdum wyrhtum, Homl. Skt. i. 6, 205. (1 a) _raving, blasphemous._ v. wodlîce, II, wôdness, II, _and cf._ woffian :—Mûð wôdne sôðfæstnysse andsware genyþerian _os blasphemum ueritatis responsione dampnare,_ Scint. 9, 11. (2) of animals :—Wið wôdes hundes slite, Lchdm. i. 4, 8. His hors feól wealwigende geond ða eorðan wôdum gelícost, Homl. Skt. ii. 26, 206. (3) of things, _mad, raging, furious_ :—Heom on becom swîðe hreóh weder, and seó wôde sæ and se stranga wind hî on ðæt land ãwearp, Chr. 1075 ; Erl. 212, 23. Wôd _effera_ (_fluctuum ferocitas,_ Ald. 42), Hpt. Gl. 478, 60. Sió wôde þrâg ðære wrænnesse, Bt. 37, 1 ; Fox 186, 18 : Met. 25, 41. [_Laym. Orm. A. R. Ayenb._ wod : _Chauc._ wood : _Prompt. Parv._ wood, ooth : _Goth._ wôds : _O. H. Ger._ wuot : _Icel._ ôðr.] v. ellen-, tung-wôd ; wêde.

wôd _madness_ :—Wôd (wôdnesse ?) _rabiem, insaniem,_ Hpt. Gl. 476, 32. v. ellen-wôd.

wôda, an ; _m. A madman, an insane person, one possessed_ :—Wôda _epilepticus,_ Wrt. Voc. ii. 107, 30 : _demoniaticus, insanus, amens,_ Wülck. Gl. 218, 41. Wôdan _limphaticum,_ Wrt. Voc. ii. 53, 56. Hê eode ût tô ðâm earmum wôdum, Homl. Skt. i. 6, 203. Wôdan _inergumenos,_ Wrt. Voc. ii. 110, 57 ; 45, 8. [_O. H. Ger._ wuoto.]

wôda, an ; _m. Danger_ (?) :—Ðâ gyrnde hê ðæt hê môste macian ænne hwerf wið ðon (_Kemble reads_ ðone, Cod. Dip. iv. 58, 1) wôdan tô werianne, Chart. Th. 341, 8. [_Cf._ (?) _Icel._ vâði (vôði) _a danger, a dangerous object._]

wôddor (= wôþ-dor ?), es ; _n. The gate of speech_ (?), _the mouth_ (?) :—T hine teswaþ, and hine on ða tungan sticaþ, wræsteþ him ðæt wôddor, and him ða wongan briceþ, Salm. Kmbl. 191 ; Sal. 95.

Wôden, es ; _m. Woden, one of the Teutonic deities._ Among the Roman gods Mercury seems to have been thought most nearly to correspond, and _Wôden_ is rendered by _Mercurius,_ e. g. :—Wôden _Mercurium,_ Wrt. Voc. ii. 114, 4. Cf. Saga mê hwã ǽrost bôcstafas sette. Ic ðê secge, Mercurius se gýgand, Salm. Kmbl. p. 192, 7 : 200, 24. The name is of rare occurrence in the literature :—Wôden worhte weós, wuldor alwalda rûme roderas, Exon. Th. 341, 28 ; Gn. Ex. 133. Wyrm com snícan, tôslât hê man ; ðã genam Wôden viiii. wuldortânas, slôh ðã ða næddran, ðæt heó on viiii tôfleáh, Lchdm. iii. 34, 23. ¶ Woden is found in most of the genealogies of the old English royal families :—Ðæs (Wihta) fæder wæs Wôden nemned, of ðæs strýnde monigra mǽgþa cyningcynn fruman lǽdde, Bd. 1, 15 ; S. 483, 30. Fram ðan Wôdne ãwôc eall ûre cynecynn, and Sûðan-Hymbra eác, Chr. 449 ; Erl. 13, 20 : 547 ; Erl. 16, 13 : 560 ; Erl. 16, 32 : 855 ; Erl. 70, 9. See Grimm's Teutonic Mythology, Stallybrass's translation, vol. i. p. 163, vol. iv. pp.

1709 sqq. ¶ the word is found in place-names, e. g. Wôdnes beorg, Wôdnes den, Wôdnes dîc, Cod. Dip. Kmbl. vi. 355. See also Wôdnes-dæg. [We (_the Saxons_) habbeð godes gode . . . þe þridde hæhte Woden . . . Woden hehde þa hæhste laȝe, Laym. 13897-13921. _O. L. Ger._ Wôdan : _O. H. Ger._ Wuotan : _Icel._ Ôðinn.] v. Ôðen.

wôden-dreám, es ; _m. Madness, fury_ :—Rêþnes, wôdendreám (cf. wêden-heort ; _or_ (?) wôden dreám ; cf. on wôdum dreáme, v. wôd, (1)) _furor animi,_ Wrt. Voc. ii. 151, 69.

Wôdening, es ; _m. A son of Woden_ :—Bældæg Wôdening, Chr. pref. ; Erl. 2, 7 : 547 ; Erl. 16, 13 : 552 ; Erl. 16, 21 : 560 ; Erl. 16, 31 : 855 ; Erl. 70, 9. Wôdning, 449 ; Erl. 13, 20.

wôde-wistle, an ; _f. Hemlock_ :—Wôdewistle, uuôdaewistlae, uuôdewislae _cicuta,_ Txts. 51, 463. Wôdewistle (_printed_ -þistle) _elleborum vel veratrum,_ Wrt. Voc. i. 31, 56. Wôdewistle (_printed_ -þisele, but see Wülck. Gl. 297, 8) _cicuta,_ 67, 37.

wôd-frec; _adj. Furiously greedy, raging, ravening_ :—Ðæt se wôdfreca werewulf (_the devil_) tô swýðe ne slîte, ne tô fela ne ãbîte of godcundre heorde, L. C. E. 26 ; Th. i. 374, 30. Wôdfræca, Wulfst. 191, 16. [Cf. _O. H. Ger._ wuot-grimm _tyrannus_ : _Icel._ ôð-fúss, -gjarn _madly eager._]

wôd-hen[n], e ; _f. A quail_ :—Wôdhae[n] _coturno,_ Txts. 53, 583. Wôdhen, Wrt. Voc. ii. 15, 30.

-wôdian. v. ellen-wôdian ; wêdan.

wôd-lic; _adj. Mad, furious, frantic_ :—Benedictus manode ðone rêðan êhtere ðæt hê ðære wôdlican rêðnysse geswice, Homl. Th. ii. 182, 1. Heó ne rôhte his worda for ðæra wôdlícan ontendnysse, Homl. Skt. i. 3, 397. Se sceocca fordwân mid swîðlícum reáme, swã ðæt ða munecas wurdon ãwrehte ðurh his wôdlícan stemne, 6, 318. [_Icel._ ôðligr _vehement._]

wôdlíce; _adv._ I. _madly, furiously, franticly_ :—Ðâm unþeáwfæstum ðe wôdlíce drincaþ, and heora gewitt âmyrraþ, Homl. Ass. 6, 145 : Homl. Skt. i. 13, 76 : L. Ælfc. C. 35 ; Th. ii. 356, 43. Wôdlíce ãstyrode wið ðone hãlgan, Homl. Ass. 79, 162. Wôdlíce geyrsod, Homl. Skt. ii. 25, 616. Ðû þus wôdlíce wilnast ceorles, i. 3, 396. Hê môt wôdlíce derian, Wulfst. 85, 5. Ðam wulfe gelíc ðe wôdlíce ãbîteþ ða sceáp, Basil admn. 6 ; Norm. 46, 23. II. _blasphemously._ v. wôd (1 a) :—Ðæt ôðer ðæra hospworda hê wiðsôc, ðæt hê deófol hæfde ; ac hî wǽron witodlíce mid deófle âfylled, ðâ ðâ hí swã wôdlíce tô ðam Hælende sprǽcon, Homl. Th. ii. 230, 11. [He mochul þa wodeloker wilnede þeos mæidenes, Laym. 3201. He schal scheten woodlich or fersliche, Halliw. Dict. _Icel._ ôðliga _rashly._]

Wôdnes-dæg, es ; _m. Wednesday_ :—Wôdnesdæges nama wæs of Mercurio, Anglia viii. 321, 16. On Wôdnesdæg, Mt. Kmbl. Rubric 3, 1, 13 _and often_ : Homl. Skt. i. 12, 1 : R. Ben. 65, 16 : Wulfst. 180, 25. On ðone Wôdnesdæg ofer Pentecosten, Mt. Kmbl. Rubric 5, 17. .iiii. Wôdnesdagas on .iiii. Ymbrenwican, L. Alf. pol. 43 ; Th. i. 92, 8. [Woden we ȝefue wendesdei, Laym. 13925 (2nd MS.). _A. R._ Wodnesdei : _Kath._ Wednes-dai : _Piers P._ Wodnes-, Wednes-dai : _O. Frs._ Wernsdei : _M. Du._ Woens-dach : _Icel._ Ôðins-dagr.]

Wôdnes-niht, e ; _f. The night between Tuesday and Wednesday._ v. Sunnan-niht :—Gebyreþ ðæt hig hyra clænnysse healdon ǽfre Sunnannihte and Wôdnesnihte, L. Ecg. P. ii. 21 ; Th. ii. 190, 19. Sunnannihtum ne mæssenihtum ne Wôdnesnihtum, Wulfst. 305, 23.

wôdness, e ; _f._ I. _madness, fury, frenzy, rage_ :—Wôdnys _rabies,_ Wrt. Voc. i. 45, 71 : 75, 58. Ðâ geãxode se cyning be ðam witseócum menn, hû se apostol hine ðære wôdnysse ãhredde, Homl. Th. i. 458, 9. Wurdon ãflîgde deófla fram mannum, ðâ ðe on wôdnysse ǽr wǽron gedrehte, Homl. Skt. ii. 26, 199. Hê of his gewitte wearð, and hine se feónd swýþe swencte mid ðære wôdnysse, Guthl. 12 ; Gdwin. 56, 15. Ðæt wîf wearð mid mâran wôdnysse (_with greater fury_) ãstyrod, Homl. Th. ii. 30, 15 : Homl. Ass. 72, 170. His sãwul is ðurh deófol gedreht ; him is neód ðæt hê his ãgene wôdnysse tôcnâwe, Homl. Th. ii. 110, 29. On wôdnessum ł gewytlýstum leásum _in insanias falsas,_ Ps. Lamb. 39, 5. Wôdnyssa and rêðnyssa _furias atque ferocia,_ Hymn. Surt. 132, 18. II. _blasphemy._ v. wôd (1 a) :—Ðâ sǽt hê tǽlende ðone Hælend . . . His wôdnys wearð gewrecen ðurh God, Homl. Ass. 60, 212. [Wodnesse _insania, furia, furor,_ Prompt. Parv. 531 : _Chauc._ woodnesse : _O. H. Ger._ wuotnessa _dementia._]

wôd-scipe, es ; _m. Madness, fury_ :—Wôdscipe _furia, insania, amentia,_ Wrt. Voc. ii. 151, 72. [_Ira furor brevis est_ wreððe is a wodschipe, A. R. 120, 14.]

wôd-þrâg, e ; _f. A mad fit or time, madness, fury_ :—Weaxeþ ðære wrænnesse wôdþrâg (wôd þrâg ? v. þrâg, II) micel, Met. 25, 41. Oft ða wôdþrâga ðæs ungewitfullan monnes se lǽce gestilð and gehǽlð mid ðæm ðæt hê him ôlecð æfter his ãgnum willan . . . Ðonne Saule se wiðerwearda gǽst on becom, ðonne gefêng Dauid his hearpan, and gestillde his wôdþrâga. . . . Dauid mid his sange gemetgode ða wôdþráge Saules _furor insanorum saepe ad salutem medico blandiente reducitur . . . Cum Saulem spiritus adversus invaderet, apprehensa Dauid cithara ejus vesaniam sedabat . . . David canente ejus vesania temperatur,_ Past. 26 ; Swt. 183, 21-185, 5.

woede, woel. v. wêde, wǽl.

woepe ? :—*Catasta, genus supplicii, vel* woepe (þrēpel ? v. þrīpel), *eculeo simile,* Wrt. Voc. ii. 129, 44.

woerd, woerdan, woestig. v. wird, wirdan, wēstig.

woffian; *p.* ode *To rave, blaspheme* :—Ðǣr wæs sum dysig mann plegol ungemetlíce, and tô ðám mannum cwæð, swylce for plegan, ðæt hê Swýðun wǣre ... Hê woffode ða swá lange mid wordum dyslíce, óð ðæt hê feóll geswôgen, Homl. Skt. i. 21, 298. Woffode *debacchatur,* Hpt. Gl. 506, 76. Woffie *insolescat, superbiat,* 461, 59. Woffigende *blasphemantem,* Scint. 9, 9. v. ā-woffian, *and cf.* wôd (1 a).

woffung, e; *f. Raving, blasphemy* :—Woffung *insania,* Greg. Dial. i. 9. Ðäs word wǣron geþúhte beforan him swá woffung (*deliramentum*), Lk. Skt. 24, 11. Hwæt is ðes ðe sprycþ woffunga (*blasphemia*), 5, 21.

wōg. v. wôh.

wōgere, es; *m. A wooer, suitor* :—Wôgere *procus,* Wrt. Voc. i. 50, 36 : 73, 7 : Hpt. Gl. 501, 58. Wôgeres (*printed* fogeres) *proci,* 498, 42. Wôgere *proco,* 503, 70. Wôghere (*printed* foghere), 506, 45. Wôgere (*printed* fogere), 498, 72. Basilla hæfde ēnne hǣðene wôgere ... Heó ðone hǣðenan wôgere habban nolde, Homl. Skt. i. 2, 349, 353. Sume wíf wyrcaþ heora wôgerum drencas, 17, 157. [He, ase noble woware, com uor to preouen his luue, A. R. 390, 21. Þise woweres þat wedde none wydwes, Piers P. 11, 71. Woware *procus,* Prompt. Parv. 532.]

wōgian; *p.* ode *To woo, marry* :—Nâht framaþ flǣsc habban mǣden gif on geþance ǣnig wôgaþ *nihil prodest carnem habere uirginem si mente quis nupserit,* Scint. 70, 7. Bearn worulde ðissere wôgiaþ (*nubunt*) ... hí ne wôgiaþ (*nubunt*), ne hí ne lǣdaþ wíf, 68, 14, 17. [Hwi ne con ic þe wouian wiþ swete luue, O. E. Homl. i. 187, 19. Ase a mon þet woweþ (wohes, *v. l.*), A. R. 388, 13. Crist wowude ure soule, 390, 20. Uorte wowen hire, 388, 17. *Chauc.* woweth : *Piers P.* wowede : *Destr. Tr.* woghit; *pp.*] v. ā-wôgian.

wōgung, e; *f. Wooing* :—Sum heretoga âwôgode ðæs cāseres dohtor; weard se cāsere tô ðǣre wôgunge ástyrod, Homl. Skt. i. 7, 301. [Mid wouhinge, A. R. 204, 25. Wowunge efter Godes grome, 116, 12. Wowynge *procacio,* Prompt. Parv. 533.]

wōh; *adj.* I. *not straight, bent, crooked, twisted, oblique* :—Wiþ lyftâdle, gif se mûð sié wôh, Lchdm. ii. 338, 5. Gif mûð oððe eáge wôh weorðeþ, L. Ethb. 44; Th. i. 14, 9. Hlâford mín (*the plough's*) wôh fǣreþ, weard æt steorte, Exon. Th. 403, 7; Rä. 22, 4: 483, 14; Rä. 69, 2. Sió micle nosu and sió woo (*tortus*), Past. 11; Swt. 67, 5. Mid wôgum bígelse *obliqua curvatura,* Hpt. Gl. 458, 72. Mid ânum wôgan íserne, Lchdm. i. 318, 18. Wiþ wôuum mûþe, genim ompran ... sele on ðone wôn dǣl, ii. 54, 22. Wôn *obunca* (*arpagine*), Anglia xiii. 37, 296. Tô ðam wôn stocce, Cod. Dip. Kmbl. ii. 73, 22. Tô ðære wôhgan apeldran, iii. 389, 32. Tô wôhan (wôgan, *v. l.*) ǣc, Cod. Dip. B. i. 417, 16. Ða oncierde ðæt scip on wônne síþfæt þurh deófles beswicennesse, Shrn. 60, 8. Gif hê hæfde wô (*tortum*) nosu, Past. 11; Swt. 65, 4. Woo, 67, 7. On ðæt wô treów (*printed* wottreów), Cod. Dip. Kmbl. iii. 130, 31. Mistlíce wôge wegas *divortia, diverticula,* Wrt. Voc. i. 37, 44. Wôum wírbogum, Exon. Th. 395, 5 ; Rä. 15, 3. Tô ðǣm wôn *ad tortas,* Wrt. Voc. ii. 2, 68. Wôge hylcas *obliquos* (*curvos, flexos*) *anfractus,* Hpt. Gl. 486, 71 : *anfractus,* 448, 20. Hí hæfdon wôh nebb, and wôge sceancan, Guthl. 5; Gdwin. 34, 22–27. II. *not right, perverse, froward, wrong, unfair* :—Hit is riht ðæt mon yfelige ða yfelan, and hit is wôh (wôg, *v. l.*) ðæt hí mon lǣte unwîtnode, Bt. 38, 3; Fox 202, 6. Forlǣten ða ðístro ðæs wôn weorces (*actionis pravae*), Past. 55; Swt. 429, 13. Mid wôre twiefealdnesse *duplicitatis perversitate,* 35; Swt. 245, 15. On wôre heortan *pravo corde,* 47; Swt. 357, 21. Mid wôre lâre *perversa praedicatione,* 48; Swt. 367, 15. Be wôhre gewitnesse. Gif man áfinde ðæt heora ǣnig on wôhre (wôre, *v. l.*) gewitnesse wǣre, L. Ath. i. 10; Th. i. 204, 22. Gescynded on heora wôn willan *abominabiles in voluntatibus suis,* Ps. Th. 13, 1. Mid wô mûðe *ore perverso,* Past. 47; Swt. 357, 20. Ðæt hyne gehwâ wið wô gewitnysse gehealde, L. E. G. 27; Th. ii. 422, 35. Wið ǣlc wôh gestreón (*but see* wôhgestreón) beorge man, ac strýne mid rihte, Wulfst. 70, 2 : L. I. P. 7; Th. ii. 312, 29. Hú micle unrôtnesse se hæfþ ðe ðone wôn willan hæfþ on ðisse worulde, Bt. 31, 1; Fox 110, 31. Ða wôn (woon, *v. l.*) wyrd on ðara unrihtwísra anwealda heánesse, 5, 1; Fox 10, 20. Ðurh ðæt woo (wô, Cott. MSS.) weorc hê forliést ðone wlite ôðerra gôdra weorca, Past. 11; Swt. 71, 25. Ðæra gerēsena unriht and wô dômas (v. wôhdôm) and prættas, Anglia viii. 336, 40. False gewihta and wôge gemeta, L. Eth. v. 24; Th. i. 310, 13 : Wulfst. 70, 3. Wôh wyrda gesceapu *the unequal decrees of fate,* Exon. Th. 421, 26; Rä. 40, 24. Hine gebindaþ ða wôn wilnunga, Bt. 16, 3; Fox 56, 18. Hê wiste him sprǣca fela wôra worda, Cd. Th. 29, 6; Gen. 446. Mid ðǣm gewunan ðara wôna weorca, Past. 11; Swt. 69, 7. Wôm wundorbebodum wergan gǣstes, Beo. Th. 3498; B. 1747. Mid wôm wilnungum, Past. 11; Swt. 69, 9. From hiera woom (wôn, Cott. MSS.) wegum, 37; Swt. 267, 5. Wôn, 11; Swt. 73, 13 : 37; Swt. 267, 12, 16 : L. E. G. proem.; Th. ii. 400, 20. Woeum *pravis,* Rtl. 52, 24. Mîne wôn wîsan, Exon. Th. 393, 10; Rä. 12, 8. Hí him wôh godu worhtan, Ps. Th. 77, 58. [He mid woȝe dome benimeð him his beliue, O. E. Homl. i. 179, 16. Þat is woh and na wiht riht, Laym. 4333. Of woh inwit, A. R. 2, 12. Ure woȝhe dededs, Orm. 1375. Þe fox can paþes rihte and woȝe (wowe, *v. l.*), O. and N. 815. *Goth.* un-wâhs.]

wōh; *gen.* wôges, wôs; *dat.* wôge, wô; *n. Wrong, perversity, injustice, error* :—Englas nânes wôges (wôs, *v. l.*) ne willniaþ, Bt. 40, 7 ; Fox 242, 23. Gif wê wilnigen ðæt hié ðæs wôs geswícen *hos cum conamur instruere, ne perversa sentiant,* Past. 48; Swt. 367, 23. Wôes ł wôhfulnise *nequitia,* Mt. Kmbl. Lind. 22, 18. Ic him wolde fylstan tô ryhte and nǣfre tô nânan wô, Cod. Dip. Kmbl. ii. 134, 10. From ǣlcum wôe *ab omni pravitate,* Rtl. 34, 9 : 37, 23. Hié nyllaþ wietan mid hwelcum woo (wô, Cott. MSS.) hié hit gestriéndon, Past. 45; Swt. 343, 23. Mid wôge (wô, *v. l.*) forsecgan, L. Edg. ii. 4; Th. i. 266, 22. Mid wô fordêman, Cod. Dip. Kmbl. ii. 114, 3. Sceal gehwâ gerihtlǣcan ðæt ðæt hê ǣr tô wôge gebígde, Homl. Th. ii. 2, 25. Tô wôge gebringan *to render incorrectly,* Ælfc. Gr. pref.; Zup. 3, 23 : Ælfc. T. Grn. 24, 32. Paulus hine âwende of wôge tô rihte, Homl. Skt. ii. 29, 8. Hí wǣron on woo besmitene *propter injustitias suas humiliati sunt,* Ps. Th. 106, 16. Micel yfel dêð se unwritere, gyf hê nele his wôh gerihtan, Ælfc. Gr. pref.; Zup. 3, 25. Ða sǣde him hîora ân, ðæt hê wôh bude, Ors. 6, 10; Swt. 264, 28. Gif hwâ ǣnigum preóste ǣnig wôh beóde, L. N. P. L. 1 ; Th. ii. 290, 2. Ðæt hê ðurh hine nân wôh ne bodige *ut ab eis prava nullo modo proferantur,* Past. 15 ; Swt. 95, 16. Hí wôh meldiaþ *pronuntiabunt iniquitatem,* Ps. Th. 93, 4. Se haftere ðe ôðrum wôh dême, L. Edg. ii. 3 ; Th. i. 266, 15. Hê ðæt mǣste wôh dyde wið ða Godes þeówas, Ors. 6, 34 ; Swt. 290, 18. Ne dô wê eác wôh, Past. 45; Swt. 337, 21. Se ðe niman wôh tô dônne, 19; Swt. 145, 12 : Bt. 41, 3; Fox 246, 19 : Ps. Th. 61, 9. Wôh fremian, 54, 20. Wê ðæt wôh ne worhton, ðæt wê ðîne ǣ forlēton *inique non egimus in testamento tuo,* 43, 19. Gif hwâ wôh wyrce, L. Edg. ii. 6 ; Th. i. 268, 9. Ic nylle ðæt gê mê hwæt mid wôh (cf. unrihte, l. 17) begytaþ, L. Ath. i. proem.; Th. i. 196, 31. Mid wôh fordôn, iv. 1 ; Th. i. 220, 23. On wôh spanan, Salm. Kmbl. 1002 ; Sal. 502. Sôna swâ sacerda hwylc hwone on wôh gesyhð, hê sceal tilian ðæt hé hyne on rihtum gebrynge, L. E. I. 28 ; Th. ii. 424, 26. Weorþeþ (-aþ, MS.) swîþe oft on wôn (*in error*) se sido, Bt. 39, 9 ; Fox 226, 4. On wôn gebringan *destruere,* Past. 2 ; Swt. 31, 24 : 28 ; Swt. 191, 8. ¶ on wôh *wrongfully, wrongly* :—On wôh cierran *deviare,* Wrt. Voc. ii. 25, 40 : 139, 57. Ða ðe on wôh dêmaþ, and rihte dômas onwendaþ, Blickl. Homl. 61, 26. On wôh dôn *perverse agere,* Past. 2 ; Swt. 31, 12. Ðeáh ûs þince ðæt hit on wôh fare *tametsi confusa omnia perturbataque videantur,* Bt. 39, 8 ; Fox 224, 21. On wôh yrsian, Ps. Th. 4, 5. On wôh lǣran, 25, arg. On wôh libban, Blickl. Homl. 45, 11, 19. On wôh niman, 61, 22. Ða gôdan ðæt gôd on riht sêcaþ, and ða yfelan on wôh, Bt. 36, 3 ; Fox 178, 6. [Gif þu me dest woh, O. E. Homl. i. 33, 1. Mid woȝe, Laym. 24811. Þu hauest wouh, A. R. 54, 1. Hatenn woh and sinne, Orm. 5555. Meanen him of wohe, Kath. 1236. Þay laften ryȝt and wroȝten woghe, Allit. Pms. 19, 621. Mid wowe ne myd ryhte, Misc. 49, 412. *O. Sax.* wâþ *evil.*] v. ā-, folc-wôh.

wōh-bogen; *adj. Bent, crooked* :—Wyrm wôhbogen *the crooked* (cf. Job 26, 13) *serpent,* Beo. Th. 5646; B. 2827.

wōh-ceápung, e; *f. The fine to be paid for trading contrary to the regulations of a market* :—Ge wôhceápung, ge ǣlc ðæra wônessa ðe tô ǣnigre bôte gebyrie, ðæt hit âge healf ðære cyrcean hlâford, swâ swâ hit mon tô ceápstôwe gesette, Cod. Dip. Kmbl. v. 143, 22.

wōh-dǣd, e; *f. A wicked deed, crime* :—Manua wôhdǣda sind swîþe gemonigfealdode, Blickl. Homl. 107, 24. Gif mon ne mihte hî tô rihte gecyrron, ðæt hí heora wôhdǣda geswícan woldan, ðonne sceal ǣghwylc man bêtan his wôhdǣda be his gyltes andefne, 45, 26–29. Ne byð ðær nân stefen gehýred, bûton wôp and wânung for wôhdǣdum, Wulfst. 139, 4. v. wôh, II.

wōh-dôm, es; *m. An unjust judgement* :—Ðurh leóde unlaga and ðurh wôhdômas, Wulfst. 166, 24. v. wôh, II.

wōh-fôtede; *adj. Crook-footed, splay-footed, club-footed; peduncus,* Wrt. Voc. i. 45, 45.

wōh-fremmende; *adj.* (*ptcpl.*) : or wôh-fremmend, es; *m. Wrong-doing;* or *a wrong-doer* :—Nalles sorgode hwæþer mihtig Drihten âmetan wolde wrece be gewyrhtum wôhfremmendum, Met. 9, 36.

wōh-full; *adj. Wicked, evil* :—Wôhgfull *nequam,* Mt. Kmbl. Lind. 20, 15. Suno sindon yfelwyrcende ł wôhfulra (*nequam*), 13, 38. Mid unrehtuísum ł wôhfullum *iniquis,* Mk. Skt. Lind. 15, 28. From gâstum wôhfullum ł yflum ł unrehtwísum (*malignis*), Lk. Skt. Lind. 8, 2. Wôhfulro *nequiores,* Mt. Kmbl. Lind. 12, 45.

wōh-fulness, e; *f. Wickedness, iniquity* :—Wôhgfulnis *nequitia,* Rtl. 120, 33. Wôghfulnisse his *nequitias ejus,* 113, 40. Wôhfulnise, Mt. Kmbl. Lind. 22, 18. Wôghfulnisse *nequitias,* Rtl. 122, 16.

wōh-georn; *adj. Loving iniquity* :—Ða wôhgeornan woruldrícan mid heora ungestreónum forweordaþ, Wulfst. 183, 8.

wōh-gestreón, es; *n. Wrongful gain, ill-gotten gain* :—Þurh rícra reáflâc and þurh gítsunge wôhgestreóna, Wulfst. 166, 24. Ðæt mancyn, ðe nû is on synnlustum and in ðâm wôhgestreónum goldes and seolfres beswicen, 182, 13. v. wôh, II.

wōh-god. v. wôh, II.

wôh-hǽmed, es; n. Adultery, fornication :—Se yfla willa wôhhǽmetes (cf. unrihthǽmedes, Bt. 31, 2 ; Fox 112, 24), Met. 18, 2. Sió hreófl getácnaþ þæt wôhhǽmed per scabiem luxuria designatur, Past. 11 ; Swt. 71, 5. Ða ðe wôhhǽmed begangaþ mid ôþerra ceorla wífum, Blickl. Homl. 61, 14.

wôh-hǽmende; adj. (ptcpl.); or wôh-hǽmend, es; m. Adulterous, fornicating; or an adulterer, a fornicator :—Ða wôhhǽmendan fornicatores, Past. 51 ; Swt. 401, 27. Ðú dydest ðé tô ðám wôhhǽmendum cum adulteris portionem tuam ponebas, Ps. Th. 40, 19.

wôh-hǽmere, es; m. An adulterer, a fornicator :—Ðæm wôhhǽmerum dêmeþ Dryhten fornicatores et adulteros judicabit Deus, Past. 51 ; Swt. 401, 30.

wôh-handede; adj. Crook-handed, having a maimed hand; mancus, Wrt. Voc. i. 45, 44.

wôh-líc; adj. Wrong, perverse, evil :—Hit ys swíðe wôlíc, ðæt ða geworhtan gesceafta ðam ne beón gehírsume, ðe hí gesceóp and geworhte, Ælfc. T. Grn. 2, 1. Mid wôlícum obliqua (invidia), Hpt. Gl. 527, 1. On wôlícum dǽdum, Blickl. Homl. 107, 28. Wôlíce inritos, Germ. 402, 76.

wôhlíce; adv. Wrongly, unjustly, perversely, wickedly :—Gif hié on ǽnigum dǽle wôlíce libban heora líf, Blickl. Homl. 109, 19. Ða ðe ǽwbryce ne wyrceaþ wôlíce (wôhlíce, v. l.) and sceamlíce, Homl. Ass. 19, 140: 29, 127. Mêdsceattas áwendaþ wôlíce ða rihtan dômas, Ælfc. T. Grn. 20, 32 : Basil admn. 9; Norm. 52, 20. Swá lagu lahlíce gewítnode ða ðe wôlíce singodon, L. Ælfc. P. 8 ; Th. ii. 366, 23. Hé cwæd ðæt hé wurde wôlíce swá getúcod, Homl. Skt. i. 21, 276. Nú dô wê swýðe wôlíce, gif wê ne wurðiaþ God, 13, 180: 17, 233 : Wulfst. 105, 9 : Homl. Ass. 29, 264: 102, 6.

wôhness, e; f. I. crookedness (lit. or fig.), a crooked place :—Ic gerihte sume wôhnysse dirigo, Ælfc. Gr. 28, 5 ; Zup. 173, 9. Ealle wôhnyssa beóð gerihte erunt prava in directa (Is. 40, 4), Homl. Th. i. 360, 33. II. wrongdoing, iniquity, perversity, depravity, wickedness :—Heora wôhnys on ðam regole his rihtwísnysse ætspearn, Homl. Th. ii. 158, 10. Ic wæs on wônysse geeácnod in iniquitatibus conceptus sum, Bd. 1, 27; S. 495, 24. Fram langre wônesse and ungesǽlignysse álýsde a longa iniquitate atque infelicitate liberatam, 2, 15 ; S. 519, 10. Wônessa iniquitates, Bd. 5, 13 ; S. 633, 38: Blickl. Homl. 107, 24. Heora wôhnyssa forgyfennys, Homl. Ass. 136, 668. Ælc ðæra wônessa (crimes) ðe tô ǽnigre bôte gebyrie, Cod. Dip. Kmbl. v. 143, 23. On wônyssum in iniquitatibus, Bd. 1, 27; S. 495, 25. Gif hié on ǽnigum dǽle wôlíce libban heora líf, sýn hié from heora wônessum onwende, and fram heora unrihtum oncyrran, Blickl. Homl. 109, 20.

wôhsum; adj. Wicked, evil :—Wôgsum nequam, Rtl. 27, 17.

woide-berge. v. wêde-berge.

wôl, es; m. : e; f. Pest, pestilence, plague, murrain :—Ádle and wôle luem, Wrt. Voc. ii. 53, 3. (1) in a physical sense in reference to men or animals :—Wôl (pestis) wæs æfter fyligende, Bd. 1, 13; S. 482, 6. Mycel wôl and grim acerba pestis, 1, 14; S. 482, 29. Ðætte nó mid him getió mê wôl (mortalitas) ðyses geáres, Lchdm. i. lxviii, 3 : 330, 1. Ǽr ðæm ðe seó wôl geendod wǽre cessatum a mortibus non est, Ors. 2, 4; Swt. 70, 12. Seó monigfealdeste wôl pestilentia gravis, praecipue mulieres pecudesque corripiens, 4, 1 ; Swt. 158, 17. On ða tíd ðær miclan wôles and moncwylde ðe Breotona eálond mid mycle wôle forhergode tempore mortalitatis quae Brittaniam lata strage vastavit, Bd. 3, 13; S. 538, 15: 3, 23; S. 555, 9. His hýd is brýce eallum fiþerfêtum nýtenum wið wôles gewinne on tô ðônne, Lchdm. i. 330, 4. Ǽlcum fiþerfêtum neáte ðe on wôle winne, 328, 13. For ðæm wôle (pestis) ðe on ðæt lond becom, Ors. 1, 5; Swt. 34, 15. (2) figurative :—Hwelc is wyrsa wôl oððe ǽngum men mâre daru ðonne wôl hæbbe on his geférrǽdenne feónd on freóndes anlícnesse? quae pestis efficacior ad nocendum, quam familiaris inimicus? Bt. 29, 2 ; Fox 106, 13. Wôl lues, Bd. 1, 14; S. 482, 23. Ðæs Pelagianiscan wôles (pestis), 1, 21 ; S. 485, 5. On wôles setle in cathedra pestilentiae, Past. 56 ; Swt. 435, 21. Wôle, Anglia xiii. 33, 146. Fram ðysses gemetes wôle (labe; the heresy of Eutyches) clǽne, Bd. 4, 17; S. 585, 12. Wênst ðú ðe ic nyte ðone wôl ðínre gedrêfednesse (perturbationum morbum), Bt. 5, 3; Fox 12, 17. Ic ðone wôl (witchcraft) eów forbeóde, Homl. Skt. i. 17, 72. [O. Sax. wôl: O. H. Ger. wôl clades, strages.]

wôlbǽrness, e; f. Pestiferousness, destructivity :—Ic wolde ðæt ða ongeáten, ðe ða tída úres cristendômes leahtriaþ, hwelc mildsung siþþan wæs, siþþan se cristendôm wæs, and hú monigfeald wôlbǽrnes ðære worulde ǽr ðæm.wæs (with how many kinds of plagues the world was afflicted before Christianity), Ors. 2, 1 ; Swt. 62, 34.

wôl-berende; adj. Pestiferous, pestilential, pernicious, (1) physical :—Æteówde wôlberende lyft, Nar. 15, 31. Se wôlberenda (pestifer) stenc ðære lyfte monige ðúsendo monna and neáta fordilgade, Bd. 1, 13; S. 482, 8. Ne sceþþeþ ðê wôlberendes áwiht, Lchdm. i. 326, 19. Wæs ðæra wyrma oroð swíðe deáðberende and ǽterne (quorum halitus erat pestifer) and for hiora ðæm wôlberendan oroðe monige men swulton, Nar. 14, 17: 16, 2. Hé onsent ofer hig wôlberende windas, Ps. Th. 10, 7. (2) figurative :—On heora wôlberendum setle in cathedra

pestilentiae, Ps. Th. 1, 1: Past. 56 ; Swt. 435, 22. On ðæm wôlberendan setle, 435, 19. Forspend hê hit mid ðære wôlberendan ðliccunge mentem securitatis pestiferae blanditiis seducit, 53; Swt. 415, 12.

wôlberend-líc; adj. Pestilential :—Geweard swíðe wôlberendlíc geár on ðissum lande, Chr. 1086 ; Erl. 219, 29.

wôl-bryne, es ; m. Deadly violence :—Weard micel wundor on heofonum gesewen, swelce eal se hefon birnende wæs. Ðæt tâcen weard on Rômánum swíþe gesweotolad mid ðæm miclan wôlbryne monncwealmes, ðe him rađe ðæs æfter com Romae gravis pestilentia per universam civitatem violenter incanduit, ut merito praecedente prodigio coelum ardere visum sit, quando caput gentium tanto morborum igne flagravit, Ors. 2, 6 ; Swt. 86, 24.

wolcen, wolc (wolc), es ; n.: also wolcne, an; f. A cloud :—Wolcn nubes, Wrt. Voc. i. 76, 46. Ealle ða gewítaþ swá swá wolcn, Blickl. Homl. 59, 20. Nalas ðæt wolcn ðý forþ com ðe úre Drihten ðæs wolcnes fultomes þearfe hæfde, oþþe ðæt wolcn hiene up âhôfe, ac hê ðæt wolcn him beforan nam, and hê on ðæm wolcne from heora gesihþe gewât, 121, 11–17. Regn wolcen bringcgeþ, Ps. Th. 67, 10. Wolcen the pillar of cloud, 77, 16. Beorht wolcn (wolcen, Lind.: wolken, Rush.) nubes lucida, Mt. Kmbl. 17, 5. Blôdig wolcen, Blickl. Homl. 91, 32. Wan wolcen, Met. 5, 4. Windig wolcen, Exon. Th. 201, 24; Ph. 61. Se ðe him ǽlc wolcn ondrǽdt. . . . Hwæt getácnaþ ðæt wolc (wolcn, Cott. MSS.) ? . . . Se wind drífeþ ðæt wolcn, Past. 39 ; Swt. 285, 18–21. 'Send mê ðínne engel on fýrenum wolcne.' . . . Fýren wolc âstâh of heofonum, Blickl. Homl. 245, 30. Ðonne ða wolcnan sceótaþ betweón ðære sunnan and ðê. . . . Þeáh nân wolcne sí betweón ðê and hyre, Shrn. 201, 27. Wǽt wolcnes tiér, Met. 20, 81. Sealdon wolcnes stefne vocem dederunt nubes (the translator has read nubis?), Ps. Th. 76, 14. Ic cume tô ðê on sweartum wolcne (in caligine nubis), Ex. 19, 9 : Cd. Th. 27, 15; Gen. 418. Wolcan nubem, Ps. Surt. 103, 3. Ðonne sweartan wolcnu him beforan gâþ, Bt. 6 ; Fox 14, 22. Ðás ðe fleógaþ swâ swâ wolcnu, Homl. Th. i. 584, 28. Wolcnu scríþaþ, Menol. Fox 486 ; Gn. C. 13. Nalles wolcnu ofer rûmne grund regnas bǽron, Cd. Th. 14, 2; Gen. 212. Bletsiaþ weolcnu Drihtne, Hymn. T. P. 73. Wolgceno, Rtl. 81, 24. Wolcna nimborum, Wrt. Voc. ii. 59, 57. Wolcna strengu, Exon. Th. 390, 4; Rä. 8, 5. For ðæra wolcna ðicnysse, Lchdm. iii. 232, 16. Wolcna scûr, Cd. Th. 238, 5 ; Dan. 350. Ðonne ic oferteó heofenan mid wolcnum (nubibus), ðonne æteówd mín boga on ðâm wolcnum, Gen. 9, 14. On heofones wolcnum, Mt. Kmbl. 26, 64 : Cd. Th. 303, 5 ; Sat. 608. Wind wǽdeþ be wolcnum, Elen. Kmbl. 2545; El. 1274. Wolcnum beþehte, Andr. Kmbl. 2094; An. 1048: Rood Kmbl. 105 ; Kr. 53. Môna waþol under wolcnum, Fins. Th. 14; Fin. 8. Se ðe him ða wolc (wolcn, Cott. MSS.) ondrêde, Past. 39; Swt. 285, 24. Hê fram ðysse eorðan ende lǽdeþ wolcen wrǽclicu educens nubes ab extremo terrae, Ps. Th. 134, 7 : 77, 25 : Cd. Th. 265, 11 ; Sat. 6. Seó lyft âbyrð ealle wolcnu (-a, v. l.), Lchdm. iii. 274, 9, 24: Bt. 36, 2 ; Fox 174, 9. ¶ in pl. (1) the clouds, the heavens, the sky :—Ðá árâs se wind and ða wolcnu sweartodon, and com ormǽte scûr of ðære lyfte (coeli contenebrati sunt, et nubes, et ventus, et facta est pluvia grandis, 1 Kings 18, 45), Homl. Skt. i. 18, 151. Hwâ is unlǽredra ðe ne wundrige wolcna fǽreldes, rodres swifto (cf. ðæs roderes fǽreldes and his swiftnesse, Bt. 39, 3 ; Fox 214, 15), Met. 28, 2 : Cd. Th. 255, 15 ; Dan. 624. Oð wolcna hróf to the skies, 196, 28 ; Exod. 298. Ofer wolcna hróf above the clouds, Elen. Kmbl. 178 ; El. 89. Wið wolcnum usque ad nubes, Th. 56, 12. Tô wolcnum, Beo. Th. 2242; B. 1119. Hyre stefn oncwæð of wolcnum, Exon. Th. 259, 16; Jul. 289. Hwæðer sincende sǽflôd wǽre under wolcnum, Cd. Th. 86, 29 ; Gen. 1438: Beo. Th. 3266; B. 1631 : Met. 7, 26 : Exon. Th. 199, 17 ; Ph. 27. Scip wíde râd wolcnum under, Cd. Th. 84, 4 ; Gen. 1392. Oþ wolcen (wolcnu, Ps. Surt.) usque ad nubes, Ps. Th. 107, 4. Oþ ða wolcnu (wolcen, Ps. Surt.), 35, 5. (2) the clouds of night :—Oþ ðe nípende niht scríðan cwôme, wan under wolcnum, Beo. Th. 1306; B. 651: 1432 ; B. 714: Salm. Kmbl. 207; Sal. 103: Andr. Kmbl. 1673; An. 839 : Exon. Th. 178, 34; Gû. 1254: Rood Kmbl. 109; Kr. 55. (3) in the phrases under wolcnum, under wolcna hrófe under heaven, on earth :—Ðenden hê on ðysse worulde wunode under wolcna hrófe, Judth. Thw. 22, 19; Jud. 67. Â þenden standeþ woruld under wolcnum, Cd. Th. 56, 22; Gen. 916: 64, 30; Gen. 1058: 117, 7 ; Gen. 1950: Exon. Th. 14, 28 ; Cri. 226. Hê weóx under wolcnum, Beo. Th. 15; B. 8. Ic Hring-Dena weóld under wolcnum, 3544; B. 1770 : Met. 1, 76. Landes frætwe gewítaþ under wolcnum, Elen. Kmbl. 254† ; El. 1272. [Ða scipen foren wide mid wolcnen and mid wedere, Laym. 102. Com winden mid ðam weolcnen a drake, 25592. In the later English, however, the word seems used mostly in the sense of sky, welkin :—Fir weax up to þam wolcne, and se wolcne undide on fower healfe and faht þær togeanes, Chr. 1122; Erl. 249, 22. Þa wolcne gon to dunien, þa eorðe gon to biuien, Laym. 27452: 4575. Þere weolcne (wolkne, 2nd MS.) he wes swiðe neh, 2883. Bonen þurled þe weolcne oratio penetrat nubes, A. R. 246, 24: Marh. 7, 3. We sitteþ under weolcne (welkne, v. l.) bi nihte, O. and N. 1682. On the welkne shoon the sterres, Chauc. Cl. T. 1124. Al þe wyde worlde bothe

welkne (wolkne, þe welkene, welken, *v. ll.*) and þe wynde, water and erþe, Piers P. 17, 160. *O. Sax.* wolkan; *n. a cloud*: *O. Frs.* wolken: *O. H. Ger.* wolchan; *n. nubes.*] v. heofon-, weder-wolcen.

wolcen-faru, e; *f. The cloud-host, the moving clouds*:—Đec forstas and snāwas, winterbiter weder and wolcenfaru (cf. wolcna genipu, Exon. Th. 192, 13; Az. 105) lofige on lyfte, Cd. Th. 239, 33; Dan. 379. Ic (*a storm*) wolcnfare wrēge, Exon. Th. 386, 33; Rä. 4, 71.

wolcen-gehnāst, es; *n. The collision of clouds*; Exon. Th. 386, 12; Rä. 4, 60.

wolcen-reád. v. weolcen-reád.

wolcen-wyrcende; *adj. Cloud-producing*:—Wolcenwyrcende *nubigenu* (-a?), Wrt. Voc. ii. 62, 13. [*The glosser seems to have mistaken* (?) *the word, which would be more nearly rendered by* wolcen-geworht. Cf. *O. H. Ger.* wolc-poran *nubigena.*]

wolc-reád. v. weoloc-reád.

wōl-dæg, es; *m. A day of pestilence, a day of death*:—Cwōmon wōldagas; swylt eall fornom secgrōf wera, Exon. Th. 477, 18; Ruin. 26.

wōl-gewinn, es; *n. A conflict where there is a great mortality*:—Gif hié gemunan willaþ hiora ieldrena unclǣnnessa, and heora wōlgewinna, and hiora monigfealdan unsibbe *recolant majorum suorum tempora, bellis inquietissima, sceleribus exsecrabilia, dissensionibus foeda*, Ors. 2, 1; Swt. 64, 15.

wō-lic. y. wōh-lic.

wollen-teár; *adj. Having hot tears, with hot tears*:—Weorod eall ārās, eodon unbliðe, wollenteáre, wundur sceáwian, Beo. Th. 6056; B. 3032. [Cf. (?) *Icel.* ollinn; *pp.* of wella.]

wōlness, e; *f. Pest, pestilence, plague*:—Wōlnes, fefor, ādl *pestis, febris, langor*, Lchdm. i. lxxiii, 1.

wom. v. wamm.

wōm, es; *m. Sound, noise*:—Wunian ðone werigan sele, ðær is wōm and wōp wíde gehēred, and gristbítunge, and gnornunge mecga, Cd. Th. 285, 5; Sat. 333. [*Icel.* ōmr *sound.*] v. next word.

wōma, an; *m. Sound, noise* (cf. hilde-wōma *and* hilde-swēg):—Se wōma (*the noise of battle*) cwom, Cd. Th. 190, 21; Exon. 202. Siððan tō reste gehwearf ríce þeóden, com on sefan hwurfan swefnes wōma, 222, 26; Dan. 110: Elen. Kmbl. 142; El. 71. Hríð hreósende, wintres wōma, Exon. Th. 292, 22; Wand. 103. Hē secgan ongan swefnes wōman, Cd. Th. 249, 33; Dan. 539. Hebban herebȳman hlúdan stefnum, wuldres wōman, 183, 31; Exod. 100. Ǣr ðū gúðe fremme, wíges wōman, Andr. Kmbl. 2709; An. 1357. Wíges wōmum, Exon. Th. 277, 5; Jul. 576. [Cf. *Icel.* Ómi, *one of the names of Odin*; a personification of the wind; ōma *to resound*; ōman *sound, voice.* Grimm says: Scheint mir der grund weshalb *wóma* mit *hild, wíg, dæg, dægréd, swefen* verbunden wird, anzuzeigen, dass das alterthum sich hierunter lauter persönliche wesen dachte, die rauschend nahten, And. u. El. xxx.] v. dæg-, dægréd-, heofon-, hilde-wōma, *and preceding word.*

wome, Chart. Th. 483, 30, *read* wonie.

won-, wōness. v. wan-, wōhness.

wōp, es; *m.* I. *a whoop, cry.* v. here-wōp. II. *mostly a cry of grief, wailing, lamentation, weeping*:—Hlúde swēgde ðæra muneca wōp on Martines deáðe, Homl. Th. ii. 518, 16. Wōp (*fletus*) and tōþa gristbítung, Mt. Kmbl. 8, 12: 13, 42. Wōm and wōp, Cd. Th. 285, 2; Sat. 333. Nis nǣnig wōp ne nǣnig heáf gehȳred, Blickl. Homl. 85, 28: Exon. Th. 164, 32; Gú. 1020. Hreám and wōp, Blickl. Homl. 115, 15. Đara cirm and wōp tō mē āstāh, 249, 7. Ne sorg ne wōp, 103, 36: Exon. Th. 201, 4; Ph. 51. Hlúd wōp, 62, 9; Cri. 999. Wæs wōp up āhafen, atol æfenleóð, Cd. Th. 190, 17; Exod. 200: Beo. Th. 257; B. 128. Wōp, hlúd heriges cyrm, Andr. Kmbl. 2311; An. 1157. Đa gesíðas, wōp and hleahtor, Salm. Kmbl. 695; Sal. 347. *Coragium*, i. *virginale funus vel* wōp, Wülck. Gl. 213, 33. Eall ðæt folc hyne weóp hundseofontig daga. Đá wæs wōpes dagas āgāne wǣron (*expleto planctus tempore*), Gen. 50, 4. On wōpe and on unrōtnesse hē leofaþ, Blickl. Homl. 59, 36. Mid swiðlíce heáfe and wōpe *luctu*, Ors. 4, 5; Swt. 166, 12. Wōpe cwíðan, Cd. Th. 61, 13; Gen. 996. Wōpe besingan, Exon. Th. 139, 3; Gú. 517. Wōpe bimǣnan, 459, 24; Hö. 4. Wōpe bewunden, Beo. Th. 6283; B. 3146. Wōpe gewǣged, wreccea giōmor *flebilis*, Met. 2, 3. Đara ðe wōp gehȳrdon galan Godes andsacan, sār wānigean, Beo. Th. 1575; B. 785. Wōp dreógan, Exon. Th. 140, 10; Gú. 608. Wōp þrowian, heáf under heofonum, Salm. Kmbl. 934; Sal. 466. Đurh fæsten and ðurh wōpas (*fletus*) and ðurh gebedo, Bd. 4, 25; S. 599, 25. II a. *where shedding of tears is referred to*:—Ús wōpe forcymenum, bitrum bryneteárum, Exon. Th. 10, 13; Cri. 151. Mid myclum wōpe (cf. wēpende wēregum teárum, Andr. Kmbl. 117; An. 59), Blickl. Homl. 229, 19. Ne rēce ðú nā weámōdes wífes worda, for ðam heó wile oft mid wōpe geswigian (*be silent and burst into tears*), Prov. Kmbl. 48. Se wæs ðurh micelne wōp āblend, Homl. Th. i. 420, 31. *See* wōpes hring *under* hring. [Hæleð ðe iherde ðesne weop, Laym. 11991. Muchel wes þa wop (wepinge, *2nd MS.*), 5970. Cullfren sang lic lic wiþþ wop, Orm. 7931. His moderes wop (ream. *v.l.*), and þe oðres Maries, ꝥ melten al of teares, A. R. 110, 15: Kath. 2332: R. Glouc. 34, 15. Þer is wop and grindinge of teþ, Ayenb.

265, 5. *O. Sax.* wōp: *O. H. Ger.* wuof *fletus, luctus, ploratus, planctus, gemitus.*] v. feld-, here-wōp.

wōp-dropa, an; *m. A tear*:—Hwæt is ðæt wundor ðæt geond ðás woruld styrnenga gǣð, āweccaþ wōpdropan? Salm. Kmbl. 567; Sal. 283.

wōpen. v. wēpan.

wōperian; *p.* ode *To wail, lament*:—Đá cleopode seó ungesǣlige wōperiende him tō: 'Eálā, help mín, wildeór mē habbaþ forneán tōslyten,' Homl. Ass. 196, 32.

wōpig; *adj. Mournful, doleful*, (1) of persons expressing grief:—Đæt ic wōpig sceal teárum mǣnan, Exon. Th. 285, 9; Jul. 711. (2) of things which are the expression of grief:—Hē hine on ða eorþan āstrehte, mid wōpegum teárum hlúde clypigende, Homl. Skt. ii. 23 b, 601.

wōp-leóþ, es; *n. A mournful lay, a tragedy*:—Wōpleóð *tragoediam*, Hpt. Gl. 488, 57.

wōp-líc; *adj. Mournful, doleful, lamentable*:—Wōplíc *flebilis*, Ælfc. Gr. 9, 28; Zup. 55, 4. (1) of persons expressing grief:—Wōplíc (*printed* -lie) *lacrimabundus*, Hpt. Gl. 472, 66. In faran ðæt tungla wōplícan heofones eáhþerl ðú eart geworden *intrent ut astra flebiles, coeli fenestra facta es*, Hymn. Surt. 76, 5. (2) of that which is an expression of grief:—Hē sprǣc mid wōplícre stemne, Homl. Th. i. 402, 9: Homl. Ass. 196, 29: 198, 121. Mid wōplícre ceorunge, Homl. Skt. i. 2, 355. Mid wōplícum murcnunge *flebilibus questibus*, Hpt. Gl. 518, 25. Mid wōplícum siccitungum *lacrimosis singultibus*, 504, 62. (3) of that which occasions grief:—Se dæg is heora sōðe ācennednys; nā wōplíc, swā swā seó ǣrre, ac blissigendlíc tō ðam ēcum lífe, Homl. Th. i. 354, 10. [*O. H. Ger.* wuof-líh *lugubris.*]

wōplíce; *adv. Mournfully, with lamentations*:—Wē healdaþ heora gemynd, nāteshwōn wōplíce, swā swā man bewēpð deáðne, Homl. Ass. 77, 124.

word, es; *n.* I. *a word, a single part of speech*; in pl. *words* forming connected speech:—Bútan ðam stafum ne mæg nān word beón āwriten, Ælfc. Gr. 2; Zup. 5, 12. *Barbarismus*, ðæt is ānes wordes gewemmednyss ... *Solocismus*, ðæt is miscweden word on endebyrdnysse ðære rǣdinge ... *Barbarismus* bið on ānum worde, and *solocismus* bið sum leás word on ðam ferse, 50; Zup. 294, 4–10. Ðæs wordes andgit is swā mon cweþe þingere oþþe frērend, Blickl. Homl. 135, 33. Seó ceaster ealde worde is nemned Wiltaburh, Bd. 5, 11; S. 626, 26. Hí ígbúend ōðre worde Baðan nemnaþ, Chr. 973; Erl. 124, 12. Ðær wæs hæleþa hleahtor, word wǣron wynsume, 1228; B. 612. Wera gehwylcum wíslícu word gerísaþ, Exon. Th. 343, 34; Gn. Ex. 166. Ealle ða ídlan word ðe hē út forlēt, Blickl. Homl. 59, 19. Ealle 'ða word sint sōþe ðe Paulus sægþ, 187, 2. Ðæt sindon ða word, swā ús gewritu secgaþ, Exon. Th. 241, 12; Ph. 655. Gif ðás word sind sōþ, 247, 24; Jul. 83: Beo. Th. 1282; B. 639. Nó ðæs fela Daniel gesprǣc sōðra worda, Cd. Th. 253, 13; Dan. 595. Engel wrāt in wāge worda gerȳnu, baswe bócstafas, 261, 9; Dan. 723. Gif hē his wordcwida wealdan meahte, ðæt he him onwrige worda gongum, hú ..., Exon. Th. 171, 29; Gú. 1134. Meaht ðú worda gewe aldan, 163, 5; Gú. 989. Worda tō hræd, 330, 13; Vy. 50. Worda gleáw, 415, 20; Rä. 33, 14. Hē wile tō his nēhstan sprecan ða word ðe hē wēnþ ðæt him leófoste sȳn tō gehȳrenne, and ðonne þencþ hú hē hine beswícan mæge þurh ða swētnesse ðara worda, Blickl. Homl. 55, 22. Hí þeossa worda nān ongeotan ne mehton, 15, 13: Exon. 246, 6; Jul. 57. Swā hē bæd, þenden hē wordum weóld, Beo. Th. 59; B. 30. Wordum wísfæst, Exon. Th. 418, 4; Rä. 36, 14. Ne wile Sarran gelȳfan wordum mínum, Cd. Th. 144, 13; Gen. 2389. Æfter ðissum wordum, Blickl. Homl. 135, 34: Andr. Kmbl. 175; An. 88. ¶ *wordum* is often used pleonastically with verbs of saying or writing. Cf. *worde* under II. 1:—God cwæð him wordum tō *Dominus ait*, Jud. 6, 14: Cd. Th. 148, 16; Gen. 2457. Ðæt wíf wordum sægde, 44, 11; Gen. 707. Wordum sprǣcon monige, 110, 33; Gen. 1847. Wordum herian, 1, 4; Gen. 1594. Ðone wē wifel wordum nemnaþ, Exon. Th. 426, 14; Rä. 41, 73. Se ongan godspell wordum wrítan, Andr. Kmbl. 25; An. 13. Ic ne mæg word sprecan, Exon. Th. 399, 16; Rä. 19, 1. Hē lǣteþ word út faran, 315, 35; Mōd. 41: Beo. Th. 5096; B. 2551. Hē word æfter cwæþ; 'Mǣl is mē tō feran,' 636; B. 315: 688; B. 341: Cd. Th. 204, 11; Exod. 417. Hē word āhóf, Andr. Kmbl. 832; An. 416: 2993; An. 1499: Elen. Kmbl. 1445; El. 724. Ic ðás word sprece, Exon. Th. 457, 12; Hy. 4, 82: Blickl. Homl. 191, 29: 205, 23. Ða word ðæs heofonlícan gerȳnes, 17, 7. Éces lífes word (wordo, Lind.) *uerba uitae aeterne*, Jn. Skt. 6, 68. Wordu, Scint. 94, 8. Hié þrý cwǣdon þurh gemǣne word, Cd. Th. 238, 30; Dan. 362: 149, 14; Gen. 2474. I a. *a verb*:—*Verbum* is word, and word getācnaþ weorc oððe þrowunge oððe geþafunge ... *Adverbum* is wordes gefēra, Ælfc. Gr. 9; Zup. 9, 2–8. On ðisum eahta dǣlum synd ða mǣstan and ða mihtigostan *nomen et verbum*, ðæt is nama and word. Mid ðam naman wē nemnaþ ealle ðing and mid ðam worde wē sprecaþ be eallum ðingum, Zup. 11, 8–11. I b. *a written word*:—Moððe word fræt, Exon. Th. 432, 4; Rä. 48, 1. II. *a word, a group of words forming a phrase, clause, sentence or sentences*, (1) *a saying, sentence, anything said, words*:—Hē ðæs geanwyrde wes, þeáh him ðæt word ofscute his

unnþances (*the words escaped him involuntarily*), Chr. 1055; Erl. 189, 6. Ðæt word belimpð synderlíce tó Gode ánum, 'Ic eom,' Homl. Th. ii. 236, 11. Him andswarode God swá ðæt ne wiste, se ðæs wordes (*the answer*) bád, Andr. Kmbl. 522; An. 261. On ðam worde: 'Uton wyrcan,' . . . on ðam worde: 'Tó úre anlícnysse,' Boutr. Scrd. 19, 13. For ðam worde hé wæs geunrét *he was sad at that saying*, Mk. Skt. 10, 22. 'Ic hit eom.' Hí mid ðam worde wendon underbæc . . . Eft áxode se Hǽlend . . . Hí eft andwyrdon mid ðam ǽrran worde . . . Ðá andwyrde hé mid ðam ylcan worde, Homl. Th. ii. 246, 15–20: Cd. Th. 31, 35; Gen. 495: 165, 4; Gen. 2726. ¶ *worde* is often used pleonastically with verbs of saying, cf. *wordum* under I :—Ðá hé worde cwæð, ðæt . . ., Cd. Th. 3, 6; Gen. 31. Hé worde cwæð: 'Témaþ and wexaþ,' 12, 34; Gen. 195: Andr. Kmbl. 1432; An. 716. Swá ðú worde becwist, 386; An. 193: Exon. Th. 123, 32; Gú. 331. Ðá worde frægn wuldres Aldor Cain hwǽr Abel wǽre, Cd. Th. 61, 24; Gen. 1002. Hé ðæt word gecwæð, ðæt hit aa hæfde ofer Godes ǽst ðe hit hæfde bútan ðære cyrcan hláforde, Chart. Th. ii. 41, 1. Hé ðæt word ácwæð, ðæt ðæt micle mord menn ne þorfton þolian, Cd. Th. 40, 14; Gen. 639. Hé ðæt word ácwæð:' Ic ðé mæg secgan . . .,' Exon. Th. 20, 12; Cri. 316: Andr. Kmbl. 2722; An. 1363 (and often). Ðis is sceortlíce gesǽd; uton secgan word gyt, Homl. Th. ii. 330, 23. Wǽron ðás word gewídmǽrsode *these sayings were noised abroad*, Lk. Skt. 1, 65. Ðá áhsode hé hine manegum wordum *interrogabat illum multis sermonibus*, 23, 9. On ðam twám formum wordum *in the two first sentences* (of the Lord's prayer), Homl. Th. i. 262, 22. Hé rihte ǽ getácnode on týn wordum [*or* (6)], Andr. Kmbl. 3023; An. 1514. (2) *a saying, maxim* :—Hí cweþaþ ðæt tó worde, ðæt se bið un gepance wǽrast and wísast, se ðe óðerne can raðost ásmeágan; cweþaþ eác tó worde ða ðe syndan stunte, ðæt mycel forhæfednes lytel behealde, ac ðæt mete wǽre mannum gescapen, tó ðam ánum, ðæt men his scoldan brúcan, Wulfst. 55, 20–25. (3) *a tale, story* :—Ðá hæfdon monige unwíse menn him tó worde and tó leásungspelle, ðæt sió hǽte . . . wǽre for Fétontis forscapunge *quidam . . . suas inanes ratiunculas conquirentes, ridiculam Phaetontis fabulam texuerunt*, Ors. 1, 7; Swt. 40, 8. (4) *a report, tidings* :—Ðam cynge com word (*word came to the king*), ðæt unnfrið-scipa lǽgen and hergodon, Chr. 1046; Erl. 173, 5. Sóna swá ðæt word becom tó Neróne, Blickl. Homl. 173, 35. Ðá sprang ðæt word, ðæt hé on ðam holte dwelode, óð ðæt hine wulfas tótǽron, Homl. Th. i. 384, 9. (5) *fame, name,* (*good*) *word,* (*good*) *report* :—Gód word and gód hlísa ǽlces monnes biþ betera ðonne ǽnig wela, Bt. 13; Fox 38, 23. Ðá ásprang his word wíde geond land, hú se mǽra man manna fét áðwóh, Homl. Skt. i. 7, 388. Úre word sprang wíde geond ðás eorðan, 13, 151 : Shrn. 17, 9. Æðelinge (*Christ*) weóx word and wísdóm (cf. Lk. 2, 52), Andr. Kmbl. 1137; An. 569 : 3352; An. 1680. Hé þóhte ðæt hé him myceles wordes wircean sceolde (wolde geearnian him hereword, *v.l.*), Chr. 1009; Erl. 142, 2. Uton ús selfum betst word and longsumast æt úrum ende gewyrcan *Spartanos admonet, de gloria plurimum, de vita nihil sperandum*, Ors. 2, 5; Swt. 82, 2. Widgongel wíf word gespringeþ (*gets a* (*bad*) *name*), hæled hý hospe mǽnaþ, Exon. Th. 337, 15; Gn. Ex. 65. (6) *a command, an order, ordinance* :—Word hleódrode: 'Ne wép ðone wræcsíð,' Andr. Kmbl. 2860; An. 1432 : Cd. Th. 173, 14; Gen. 2861. Eoppa þe Wilferþes worde bróhte Wihtwarum fulwiht, Chr. 661; Erl. 34, 17. Ðú lífes word lǽstan noldes, ac mín bibod brǽce þe ðínes bonan worde, Exon. Th. 85, 21; Cri. 1394. Ðæt hié ðæt on wendon, ðæt hé mid his worde bebeád, Cd. Th. 26, 11; Gen. 405. Ðá sende se cyng Leófsig, and hé ðæs cynges worde gríð gesætte, Chr. 1002; Erl. 137, 25 : L. Ath. v. 10; Th. i. 238, 36 : Exon. Th. 99, 19; Cri. 1627. Ic ne mæg áwendan Godes word . . . God cwæð: 'Dó ðæt ic ðé bebeóde,' Num. 22, 18–20. Hý brǽcon cyninges word, beorht bóca bibod, Exon. Th. 99, 26; Cri. 1630: Cd. Th. 38, 1; Gen. 600 : 49, 27; Gen. 798. Word gehyrwan, Elen. Kmbl. 442; El. 221. Ofer Drihtnes word, Cd. Th. 37, 21; Gen. 593 : Rood Kmbl. 70; Kr. 35. Ðæt ðú Drihtnes word healde, and ðæs cininges bebod begange, Elen. Kmbl. 2334; El. 1168. Þurh his word *at his command*, Cd. Th. 10, 17; Gen. 158 : 7, 24; Gen. 111 : 82, 15; Gen. 1362. Ǽr áwǽged sié worda ǽnig, Andr. Kmbl. 2877; An. 1441. Hé com be Honorius wordum ðes pápan, Chr. 634; Erl. 25, 28. Hwæt Drihten him bebeád, Ex. 34, 28 : Deut. 10, 4. (7) *a message, an announcement* :—Hé word ábeád : 'Eów hét secgan sigedrihten mín, ðæt hé eówer æþelu can, Beo. Th. 786; B. 390. (8) *word, solemn statement* :—Biscopes word and cyninges sié unlægne búton áðe, L. Wih. 16; Th. i. 40, 12. Ðú ðæt gehéte þurh ðín hálig word, Andr. Kmbl. 2836; An. 1420. (9) *promise, oath.* v. word-fæst, -loga :—Man freóndscipe gefæstnode mid worde and mid wædde, Chr. 1014; Erl. 150, 14. Ðe þence hé word and wedd ðe hé Gode betǽhte, L. Eth. v. 5; Th. i. 306, 5 : vi. 3; Th. i. 314, 24. Hwǽr syndon ðíne word, on ðám ðú ús gestrangodest, and ðú cwǽde: 'Gif gé mé gehýraþ, ne án loc of eówrum heáfde forwyrð,' Blickl. Homl. 243, 31. (10) *an* (*expressed*) *intention or opinion* :—Ðara ǽlces ðe ðæs wordes wǽre ðæt hí from Rómebyrg þóhte *of every one that talked of leaving Rome*, Ors. 4, 9; Swt. 190, 25. Hié wǽron ðæs wordes, ðæt him leófre wæs se cristendóm tó begánne ðonne

his scíra tó habbanne *omnes officium quam fidem deserere maluerunt*, 6, 31; Swt. 286, 6. Ðá wæs ǽlc ðæs wordes, ðæt him leófre wǽre ðæt hé land foreode, ðonne hé ðæne hád underfénge, Chart. Th. 167, 32. Wæs Eþelwald ðæs wordes, ðæt hé nó ðes rihtes wiðsacan nolde *Ethelwald declared his intention of not opposing the right*, 140, 11. **III.** *speech, language, words* :—Word spearcum fleáh, Cd. Th. 274, 31; Sat. 162. Scóp him Heort naman se ðe his wordes geweald hæfde (*who had power to name things as he pleased*), Beo. Th. 158; B. 79. Wordes ord *the first word*, 5576; B. 2791. Rǽdsnotteran, wordes wísran, Andr. Kmbl. 947; An. 474. Wordes gleáwne, 3295; An. 1650. Weras wordes cræftige, Elen. Kmbl. 628; El. 314: 837; El. 419. Of eallum ðæm worde ðe gǽþ of Godes múþe, Blickl. Homl. 27, 9. Mid ðon worde ðæs godcundan gewrites hé hine oferswíþde, 33, 20. On worde mid nǽnigre mihte gewelgode, Blickl. Homl. 179, 15. Wís on ðínum gewitte and on ðínum worde snottor, Exon. Th. 463, 31; Hö. 78. Men ðú sealdest word and gewitt, Hy. 9, 56. **III a.** *language, style* :—Ǽrest Eroico metro, and æfter fæce geræde worde (*plano sermone*) ic áwrát, Bd. 5, 24; S. 648, 27. **III b.** where speech is contrasted with act or thought :—Lufige man Godes riht wordes and dǽde, L. Eth. v. 26; Th. i. 310, 20: L.C.E. 19; Th. i. 372, 4. Wordes oððe weorces, L.E.G. 2; Th. i. 168, 2: L. Eth. vi. 30; Th. i. 322, 23. Móde and dǽdum, worde and gewitte, Cd. Th. 117, 23; Gen. 1958. Hé men of deáðe worde áwehte, Andr. Kmbl. 1167; An. 584: Elen. Kmbl. 1888; El. 946. Mihtig mid worde eal tó dónne, Blickl. Homl. 235, 36. Scyndan mid worde oþþe weorce, L. Eth. vii. 27; Th. i. 334, 36. Þurh geþóht and þurh word and þurh weorc, Blickl. Homl. 35, 14. **IV.** *word* (in *word* of God) :—Se ðe sǽwð, word hé sǽwð . . . Hí ðæt word gehýraþ, Mk. Skt. 4, 14–20. Wé wǽron gesamnode ðǽr wé gehérdan Godes word, Blickl. Homl. 141, 27. Gif heó ne bið mid Godes worde féded, 57, 11. **IV a.** translating *verbum* in Jn. 1, 1. v. word-cennend. [*Goth.* waurd: *O. Sax.* *O. Frs.* word: *O. H. Ger.* wort: *Icel.* orð.] v. beót-, cyne-, galdor-, gilp-, gleó-, gnorn-, gram-, heoru-, here-, hosp-, husc-, lást-, leóþ-, lyge-, lygen-, mǽðel-, mán-, óleht-, orgel-, sceand-, sóþ-, sorh-, teón-, torn-, þanc-, þrýþ-, wær-, wealh-, wit-, wuldor-word.

-word; *adj.* in róf-word (?), Exon. Th. 353, 21; Reim. 16. [*Icel.* -orðr.] Cf. -wyrde.

word-beót, es; *n. A promise* :—Ðá com féran Freá tó Sarran swá hé self gecwæð, hæfde wordbeót leófum gelǽsted, Cd. Th. 167, 6; Gen. 2761. Wáciaþ wordbeót, Exon. Th. 469, 22; Hy. 11, 6. v. word-gebeót, *and next word.*

word-beótung, e; *f. Promising, a promise* :—Ðec biddan hét se ðisne beám ágróf, ðæt ðú gemunde on gewitlocan wordbeótunga, Exon. Th. 473, 14; Bo. 14. v. preceding word.

word-cennend, es; *m. The begetter of the Word* (Jn. 1, 1) :—Ó milda Wordcennend *pie Verbigena*, Germ. 389, 2.

word-cræft, es; *m. The art of speaking or writing* :—Wordcræftes wís *a good speaker*, Elen. Kmbl. 1180; El. 592. Ic wordcræft wæf *I composed poetry*, 2473; El. 1238.

word-cwide, es; *m.* **I.** *a saying, words* :—Fyl nú frumsprǽce, wlitiga ðínne wordcwyde (*what thou hast said*), ðíne wordcwidas, Exon. Th. 188, 9; Az. 43), Cd. Th. 236, 26; Dan. 327. Ne lengde leóda aldor wítegana wordcwyde, 256, 27; Dan. 647. Ðý ðíne wordcwidas weorðan gefelde (*ut justificeris in sermonibus tuis*), ðæt ðú ne wilnast weora ǽniges deáð, ac ðú synfulle símle lǽrdes ðæt . . ., Ps. C. 53. Ðé ða wordcwydas Drihten on sefan sende . . . Ðú eart on móde fród, wís wordcwida, Beo. Th. 3686–3694; B. 1841–1845. Gif hé his wordcwida wealdan meahte *if he could talk*, Exon. Th. 171, 25; Gú. 1132. Wís on wordcwidum, 294, 11; Crä. 31: Andr. Kmbl. 1104; An. 552. Wuldorcyninges word hleódrode. . . . Æfter wordcwidum wuldorcyninges *after the words*, 2892; An. 1449: Beo. Th. 5499; B. 2753. Uncre wordcwidas *what we said to one another*, Exon. Th. 472, 16; Rä. 61, 17. Cleopaþ se alda, wriceþ wordcwedas, Cd. Th. 267, 8; Sat. 35. **II.** *speech, language* :—On ðam (*Daniel*) Drihtenweard wisse sídne geþanc and wísne wordcwide, Cd. Th. 249, 28; Dan. 537. Sum mæg searolíce wordcwide wrítan *one is a clear writer*, Exon. Th. 42, 15; Cri. 673. [*O. Sax.* word-quidi *a saying, a speech.*]

word-fæst; *adj. Adhering to what one says, keeping one's word* :—Se hláford sceal beón egesfull ðam dysegum, ðæt hé heora dysig álecge; and hé sceal beón wordfæst and witan hwæt hé clypige (he sceal beon weordfeste and wise lare lusten, 111, 32), O. E. Homl. i. 301, 13. Cf. word-loga.

word-full; *adj. Wordy, verbose, talkative* :—Mann wordfull (*verbosus*) ásyndraþ ealdras, Scint. 134, 12. Wordful *verbosa* (*garrulorum loquacitas*), Hpt. Gl. 528, 49. Wordfulle *uerbosi*, Scint. 78, 1.

word-gebeót, es; *n. A promise* :—Hé his wordgebeót gemunde *memor fuit testamenti sui*, Ps. Th. 105, 34. v. word-beót.

word-gecwide, es; *n. An expressed agreement, a formal contract* :—Eal ic him gelǽste ðæt ic him scolde swá forð swá uncre wordgecwydu fyrmest wǽron, L. O. 11; Th. i. 182, 11. Gif hit heó gehaldeþ mid ðare clǽnnisse ðe uncer wordgecwǽdu seondan, Chart. Th. 481, 8.

word-gemearc, es; *n. A limit fixed in words, a term* :—Sceal sóð

forð gán wyrd æfter ðissum wordgemearcum (*according to these terms*), Cd. Th. 142, 2; Gen. 2355. [Cf. Wrítan wordgimerkiun hwat sie that barn hêtan skoldin, Hêl. 233.]

word-gerȳne, es; *n. A mystery expressed in words, a deep saying* :—Him tácna fela tíres brytta onwráh wordgerȳnum, Exon. Th. 29, 16; Cri. 463. Sum biþ listhendig tó áwrítanne wordgerȳnu, 299, 3; Crä. 96. ‘Ic ðæt ongiten hæbbe þurg wítgena wordgerȳno on Godes bócum, Elen. Kmbl. 578; El. 289: 646; El. 323.

word-gid[d], es; *n. A lay* :—Cyning mǽnan, wordgyd wrecan and worn sprecan, Beo. Th. 6325; B. 3173.

word-gleáw; *adj. Prudent in speech* :—Cwæð se ðe wæs cyninges rǽswa, wís and wordgleáw, Cd. Th. 242, 12; Dan. 418.

word-hleóþor, es; *m. The sound of speaking, voice* :—Wordhleóðor ástág háliges láre *the voice of the holy one's teaching rose up*, Andr. Kmbl. 1416; An. 708. Wearð gehýred heofoncyninges stefn, wordhleóðres swég mǽres þeódnes, 186; An. 93.

word-hord, es; *n. A word-hoard, store of words* :—Him Andreas þurh andsware wordhord onleác, Andr. Kmbl. 632; An. 316. Weges weard wordhord onleác, beald reordade, 1202; An. 601: Beo. Th. 524; B. 259: Met. 6, 1: Exon. Th. 318, 20; Víd. 1. Mé fród wita sægde sundorwundra fela, wordhord ouwreáh, 313, 20; Mód. 3.

wordian; *p. ode To speak* :—Wurdiaþ (*but changed to* wurdliaþ) *rhetoricamur, loquimur*, Hpt. Gl. 527, 58. Wordiende *concionandi, loquentes*, 461, 36. [Þe king wordede þus, Laym. 18052. Þei wordeden wyseli a gret while togideres, Piers P. 4, 46. *Icel.* orða. Cf. *Goth.* waurdjan.]

wordig; *adj. Wordy, verbose* :—Wordig gehlýd *verbosa garrulitas*, Hpt. Gl. 439, 58. [*Icel.* orðigr.]

word-láo, es; *n. A speech* ; loquela :—Nǽron wordlácu ne sprǽcu ðara ðe ne wǽron gehérde stefna heora *non sunt loquelae neque sermones quorum non audiantur uoces eorum*, Ps. Lamb. 18, 4.

word-lapu, e; *f. Speech, discourse* :—Sumum hé wordlaþe wíse sendeþ on his módes gemynd, Exon. Th. 41, 31; Cri. 664. Mín hyge blissaþ þurh ðíne wordlæðe, Andr. Kmbl. 1270; An. 635.

word-latu, e; *f. Delay in speaking* :—‘Ðú scealt hrǽðe cýðan, gif ðú his ondgitan ǽnige hæbbe.’ Næs ðá wordlatu (*there was no delay in the answer*), Andr. Kmbl. 3042; An. 1524. (Cf. búton late *sine mora*, R. Ben. 55, 12, *omitted under* latu.)

word-leán, es; *n. A reward for words* (*a song*) :—Oft ic wóðboran wordleána sum ágyfe æfter giedde, Exon. Th. 489, 18; Rä. 78, 9.

wordlian, wurdlian; *p. ode To talk, discourse* :—Wurdliaþ (*changed from* wurdiaþ) *rhetoricamur, loquimur*, . . . snytrian ł wurdlian *philosophari*, Hpt. Gl. 527, 58–63. Epactas ðe wíse preóstas oft ymbe gerádlíce wurdliaþ, Anglia viii. 300, 45. Hyt geríst ðæt wé ymbe ða epactas wurdlion, 305, 19: 308, 16. Se sceop in gebringþ óðre hádas, ðe wið hine wurdlion swylce hig him andswarion, 330, 43. Uton nú on Englisc ymbe ðys be dǽle wurdlian, 303, 14. [Gewurdlud *vel* gesprecen, 320, 16.] Wordlian *sermocinari*, Hpt. Gl. 461, 38. [*O. H. Ger.* wortalón; wortalónti *verbosus* ; wortalári *verbosus*.] v. wordlung; wordrian.

word-loc, es; *n. A conclusion expressed in words* :—Wordlocum *dialectica* (the passage is: Ut tomus dialectica dogmata rerum disceret, Ald. 170), Wrt. Voc. ii. 93, 23: 27, 16.

word-loca, an; *m. The storehouse of words* :—Ongan hé reordigan, wordlocan onspeónn, Andr. Kmbl. 940; An. 470.

word-lof, es; *n. Praise in words, praise* :—Cyning mec weorþaþ, ne wyrneþ wordlofes, wísan mǽneþ míne for mengo, Exon. Th. 401, 13; Rä. 21, 11. [*Icel.* orð-lof *praise*.]

word-loga, an; *m. One who is false to his word* :—Ðæt gé ne beón wedlogan ne wordlogan, Wulfst. 40, 10. v. word, II. 9.

wordlung, e; *f.* I. in a good sense, *discourse, conversation* :—His wordlunc *sermocinatio ejus* (*cum simplicibus*, Prov. 3, 32), Kent. Gl. 61. II. in a bad sense, *idle talk, babbling, chattering* :—Ðæt sidefulle wíf wordlunge ne lufaþ (cf. idele weord ne luuað, 111, 21), O. E. Homl. i. 301, 2. Ásolcennys ácenð ídelnysse, gemágnysse and wordlunge, Homl. Th. ii. 220, 26. [Cf. *O. H. Ger.* wortal *verbosus*.] v. wordlian.

word-mittung, e; *f. Collation* :—Wordmittung *vel* wordsomnung *collatio*, Wrt. Voc. i. 54, 51.

wordrian; *p. ode To speak, discourse* :—Wordriendra, bænnendra, maðeliendra *concionatorum, locutorum, rhetorum*, Hpt. Gl. 460, 70. Cf. wordlian, wordian.

word-riht, es; *n.* I. *a law expressed in the form of a command* (v. word, II. 6), *an ordinance* ; or *a law expressed in spoken words, a spoken law* :—Móyses dómas, wrǽclíco wordriht, Cd. Th. 177, 31; Exod. 3. II. *a statement of what is right ; or* (?) *a duty which one has given his word to perform* (v. word, II. 9) :—Wíglâf maðelode, wordrihta fela sægde gesíðum (*told them much of what they ought to do; or* (?) *told them much of what they had promised to do.* Cf. wé gehêton, 5261; B. 2634), Beo. Th. 5256; B. 2631.

word-samnere, es; *m. A collector of words* :—Wordsomnere *cocologus* (= καταλογεύς?), Wrt. Voc. ii. 135, 42.

word-samnung. v. word-mittung.

word-sáwere, es; *m. A word-sower* :—Hêton menn wordsáwere ðone æðelan láreów *praedicator egregius seminiverbius est vocatus*, Past. 15; Swt. 97, 4.

word-sige, es; *m. Success in speaking* :—Sigegyrd ic mé wege, wordsige and worcsige, Lchdm. i. 388, 15.

word-snotor; *adj. Expert in speech, eloquent, learned* :—Óslác, gamolfeax hæleð, wís and wordsnotor, Chr. 975; Erl. 126, 21. Lýfing se wordsnotera biscop, 1047; Erl. 171„28. Wordsnoteran (*Homerum*), Hpt. Gl. 463, 53. Wordsnotere *oratores, rhetores, grammatici*, 481, 72. Wordsnoterum *sapientium*, 503, 67. Ne weorþeþ on worulde ǽnig wordsnotera ne on wordum getingra, ðonne hé (*Antichrist*) wyrðeþ, Wulfst. 54, 21.

word-snotorung, e; *f. A sophism* :—Wordsnoterung *sophisma*, Hpt. Gl. 459, 61.

word-wís; *adj. Wise in speech, learned* :—Ðæs wordwísan *sophiste*, Wrt. Voc. ii. 78, 39. [*O. Sax.* word-wís: *Icel.* orð-víss.]

word-wynsum; *adj. Pleasant in speech, affable* ; affabilis, Wrt. Voc. i. 61, 37.

wór-hana, an; *m. A pheasant* :—Wórhona, uuórhana, -hona *fasianus*, Txts. 61, 830. Wórhana, Wrt. Voc. ii. 34, 71 : *fusianus*, i. 280, 29 : *fursianus* (*fursianus* is glossed by mórhana, Hpt. Zeit. 33, 240, 27), 62, 24.

wór-hen[n], e; *f.* The word glosses *cracinus*, Wrt. Voc. ii. 22, 75 : 136, 59.

wórian; *p. ode To wander about* :—Ic wórige *uagor*, Ælfc. Gr. 25; Zup. 145, 13. (1) literal, *to wander about, ramble, be a vagabond* :—Ic wórige and beó áflýmed geond ealle eorðan *ero vagus et profugus in terra*, Gen. 4, 14. Hí lufiaþ ídele blisse, wóriaþ and wundriaþ, and ealne dæg fleardiaþ, L. I. P. 14; Th. ii. 322, 24. Is ðæs (*the whale's*) híw gelíc hreófum stáne, swylce wórie bi wǽdes ófre, Exon. Th. 360, 21; Wal. 9. Seó rípung ðæs geatweardes gestæþþignesse sý swr̄ : ðæt hine ne wórian ne scríðan ne lyste (*eum non sinat uagari*), R. Ben. 126, 17. Ðú færsð wórigende (*vagus*), Gen. 4, 12 : Boutr. Scrd. 20, 43 : 19, 2. Ne fêrde heó wórigende geond land, ac wæs wunigende binnan Godes temple, Homl. Th. i. 148, 3 : ii. 160, 21. Wórigende geond wudas and feldas, 188, 14. Eówre bearn beóð wórigende on ðisum wéstene *filii vestri erunt vagi in deserto*, Num. 14, 33 : Homl. Th. ii. 30, 27. Ǽfre unstaþolfæste and wóriende, R. Ben. 9, 23. (1 a) of the movements of the planets :—Hí (*the planets*) synd wórigende gecwedene, for ðan ðe ǽlc gǽð on his ágenum ryne, Boutr. Scrd. 18, 29. (2) figurative in various senses :—Wóraþ *fluctuat, estuat*, i. *vacillat, dubitat, anxiat*, Wrt. Voc. ii. 149, 60. Wóriaþ ða winsalo *the halls totter* (*are ruinous*), Exon. Th. 291, 6; Wand. 78. Gangas rihte dóþ, ðæt ná healtigende wórige (*erret*), Scint. 186, 4. Bútan sóþre lufe, ná gán (*ambulare*) magan menn ac wórian (*errare*), 3, 8. Wer unsnoter and wórigende (*errans*) þencþ stunte, 138, 18. Wóriende *vagi* (*sunt gressus ejus*, Prov. 5, 6) i. *vagabunda* (*rumorum praeconia*, Ald. 64), Hpt. Gl. 512, 51. His eágan ne fêrdon wórigende geond mistlíce lustas, Homl. Th. i. 168, 13. Wórigende sefan (*vagos sensus*) hé þreáge, Hymn. Surt. 114, 15.

wóriend, es; *m. A vagabond* ; vagabundus, Hpt. Gl. 484, 64.

world. v. weorold.

worms, worsm, wurms, wursm, es; *n. Corrupt matter* :—Worms *pus*, Wrt. Voc. ii. 68, 52. Uuorsm, Txts. 86, 777. Wurms *virus*, Hpt. Gl. 520, 41. Ðæt worms (worsm, Cott. MSS.) ðara wunda, Past. 36; Swt. 259, 15. Ðæt worsm *putredo*, 38; Swt. 273, 22. Bið seó micge lyswen swilce worms, Lchdm. ii. 198, 27. Se swile and ðæt worms, 208, 11. Wið ða gerynnincge ðæs worsmes (wormses, *v. l.*), i. 292, 8. On ða ádle ðe mon wormse spíweþ, ii. 200, 21 : 208, 5. Hreófeligum wormse *elephantino tabo*, Hpt. Gl. 490, 38. Áflêwð ðæt sár of ðære wunde mid ðý wormse *mala livor vulneris abstergit*, Past. 36; Swt. 259, 2. Heó ðæt worsm (worms, *v. l.*) út átýhþ, Lchdm. i. 100, 13. Ðæt worms, ii. 72, 14. Ðæt wursm, 202, 25. Eall ðæt folc wæs on blǽdran and ða wǽron swíðe hreówlíce berstende and ða worms út siónde *vesicas effervescentes, ulceraque manantia*, Ors. 1, 7; Swt. 38, 7. [Mine wunden gedered neowe wrusum (wursum, *v. l.*), A. R. 274, 3. Wrusum *sanies*, 322, 11. Worsum, C. M. 11835. Wirrsenn, Orm. 4782.] v. wyrms.

worms-gemang (?), es; *n. A mixture in which there is corrupt matter* :—Wið ðæt man blód and worsmgemang (worsm gemang ?) hrǽce, Lchdm. i. 250, 7. Cf. blód-gemang.

worn, weorn, es; *m. A swarm, band, flock, crowd, multitude, many, a great number, a great quantity*, (1) of animate objects :—Seó wilgedryht wildne weorþaþ, worn æfter óþrum (*flock* (*of birds*) *following flock*) *turba prosequitur*, Exon. Th. 222, 4; Ph. 343. Folc onette, weras wíf somod, wornum and heápum, ðreátum and þrymmum, þrungon and urnon, Judth. 23, 39; Jud. 164. Mǽgen wêrge monna cynnes wornum hweorfaþ on wídne lêg, Exon. Th. 59, 25; Cri. 958. (1 a) with gen. pl. :—Wæs ðér worn (*grex*) swína michil, Mk. Skt. Lind. Rush. 5, 11, 13. Ðǽr witena biþ worn ætsomne, 295, 35; Crä. 43 : Salm. Kmbl. 802; Sal. 400. Wearð Seme suna and dohtra worn áféded, Cd. Th. 99, 5; Gen. 1641. Hé worn gestrýnde suna and dohtra, 74, 11; Gen. 1220. Oft se snáw gecostaþ

wildeóra worn, Salm. Kmbl. 611 ; Sal. 305. Fugla and deóra wornas, Exon. Th. 356, 3 ; Pa. 6. (1 b) with gen. sing. of a collective noun :— Lǽð æfter láðum, leódmægnes worn, þúsendmælum, Cd. Th. 190, 7 ; Exod. 195. (2) of inanimate objects, abstract or concrete. (a) alone, *much, many things* :—Hē worn gemunde, Beo. Th. 4235 ; B. 2114. Hē worn eall gespræc, 6180 ; B. 3094. Se gomola sægde eaforan worn, Exon. Th. 304, 7 ; Fä. 66. Ongan worn sprecan, 319, 9 ; Víd. 9. (b) with adj. :—Dū worn fela spræce *you have said many, many things*, Beo. Th. 1064 ; B. 530. (c) with gen. pl. :—Árleásta fela, misdǽda worn, Met. 9, 7. (Wintra) worn, twā hund oððe mā, Elen. Kmbl. 1263 ; El. 633. Ymb wintra worn, Cd. Th. 79, 32 ; Gen. 1320 : 236, 22 ; Dan. 325 : Beo. Th. 533 ; B. 264. Missera worn, Cd. Th. 71, 10 ; Gen. 1168. Ymb worn daga, 86, 30 ; Gen. 1438 : 142, 10 ; Gen. 2359 : Menol. Fox 336 ; Men. 169. Ic spræc worda worn, Andr. Kmbl. 1807 ; An. 906. Se ðe ealdgesegena worn gemunde, Beo. Th. 1744 ; B. 873. Hē wundra worn cýðde, Andr. Kmbl. 1623 ; An. 813. Worn sárcwida, Exon. Th. 11, 11 ; Cri. 169 : 291, 32 ; Wand. 91 : 315, 19 ; Mōd. 33. Weorn earfoðsíða, Andr. Kmbl. 1354 ; An. 677. Fæstena worn, Cd. Th. 181, 5 ; Exod. 56. Se ðe worna fela gūða gedígde *he that from numbers and numbers of battles escaped*, Beo. Th. 5078 ; B. 2543. Hē weorna feala wíta geþolode, Andr. Kmbl. 2978 ; An. 1492. (d) with adj. and gen. pl. :—Unc sceal worn fela máþma gemǽnra *we two shall have many, many treasures in common*, Beo. Th. 3571 ; B. 1783. (e) with gen. sing. :—Gewát dægrímes worn *a great number of days passed*, Cd. Th. 60, 1 ; Gen. 975. Hē wunode dægrímes worn, 156, 31 ; Gen. 2597 : 80, 20 ; Gen. 1331 : Met. 26, 33. Hē ðæs wítes worn gefēlde *he felt the multitudinous pain*, Cd. Th. 269, 23 ; Sat. 77. Hē worna fela sorge gefremede, yrmðe, Beo. Th. 4011 ; B. 2003. v. wearn.

worn-gehát (? word-gehát ; cf. word-beót, -gebeót), es ; *n. A promise of a numerous progeny* :—Ðē beóþ worngehát (cf. patrem multarum gentium constitui te, Gen. 17, 5) mín gelǽsted, Cd. Th. 144, 24 ; Gen. 2394.

-worpenness. v. á-, on-, tó-worpenness.

worpian ; *p.* ode.

v. weorpan, I. *to throw* with something at an object. I. 2 a :—Ðonne hié forwandigaþ ðæt hié mid ðǽm kycglum hiera worda ongeán hiera ierre worpigen (worpien, Cott. MSS.) *cum contra irascentem dissimulat verborum jacula reddere*, Past. 40 ; Swt. 297, 2. **II.** *to throw and strike* with something. v. weorpan, V :—Worpaþ hine deófol of blācere liðran īrenum aplum, Salm. Kmbl. 50 ; Sal. 25. Stephanus wæs stanum worpod, Elen. Kmbl. 982 ; El. 492 : 1646 ; El. 825. [*O. H. Ger.* worfōn *projicere*.]

worsm. v. worms.

worþ, weorþ, wurþ, wierþ, wyrþ, e ; *f.* : es ; *m.* : wyrþe, wirþe (v. wyrþe-land, *and first extract under* I), es ; *m.* I. *a close* (?), *an enclosed place* (?) :—Ūt on rigewyrðe (*the rye-close?*) westeweardne, Cod. Dip. Kmbl. iii. 437, 35. Uppan rigeweorðe on ða ealdan díc ; of ðære díc ūt on rigewurðe heal, v. 377, 21. On lindwyrðe, iii. 375, 6. **II.** *an enclosed homestead, a habitation with surrounding land* :—Be hagan on weorðe hege ; forð be ðan hege on weorðapeldre, Cod. Dip. Kmbl. v. 381, 30. Tó ealdan wyrðe . . . wið westan ealdan wyrðe, 195, 3-5. Ondlang híweges tó Ecgguines wyrðe, iii. 437, 32. Tó Cumbran weorðe ; of Cumbran weorðe tó ðære mǽran æc, 78, 35. ¶ perhaps in the last two passages *weorp* may be regarded as the second part of a compound name : such expressions as 'in loco ubi soliculae illius regionis Ægeleswurð nomen imposuerunt' are not uncommon in the Charters, and such names seem to have remained. In the index of places given in Cod. Dip. Kmbl. vi. 251 sqq. about 70 combinations with *weorp* occur, and for many of these modern representatives terminating in *-worth* are found. Already places whose names contain the form (cf. those with *tún*), when they are mentioned in the Charters, may have extended beyond their original limits and have become properties, whose area was considerable (e. g. Hē gean ðæra hundtwýntiga hída æt Wyrðæ, Cod. Dip. Kmbl. iv. 167, 15. Brinkewurða terra est .v. hidarum, iv. 167, 1. Æt Æscmæres-weorðæ (-wyrðe, l. 14), .x. hída, v. 218, 22), whose boundaries consequently had to be defined (e. g. Ðis syndon ða landgemǽro tó wyrðe, vi. 8, 25. Tó Ceorles-wyrðe, iii. 458, 3. Tó Ægeles-uurðe, 428, 18. Tó Æscmæres-wierðe, v. 173, 36. Tó Peádanwyrðe, 383, 8), and upon which a number of persons resided (e. g. .xxx. mansas illic ubi Anglica appellatione dicitur æt Wurðe (Weorðe, 329, 32), v. 395, 13. Quarta terra .iii. manentium, et uocatur Gislheresuuyrth, i. 44, 11. Monasterium quod situm est in loco qui dicitur æt Baedricesworth, ii. 258, 25 : iii. 272, 10 : 305, 11. In Blacewyrðe .v. mansas). Various Latin words are used in speaking of such places ; Wealawyrð is a *uillula*, iii. 347, 11 : v. 346, 33 : Æbbewyrð is a *uiculus*, iv. 164, 8-10 : Æscmeresweorð is a *uilla*, v. 216, 10 : Gislheresuuyrth is a *terra*, i. 44, 11, so also Brinkeuurða, iv. 167, 1, and Deceuurthe, ii. 367, 22-23 : Ceolwurð is spoken of as *aliquantulum terrae*, ii. 135, 16, 22 : and Oswald grants *aliquam telluris partem* æt Bynnyncgwyrðe, iii. 177, 23. Corresponding to these last terms are the English forms with *land* : Ic gean ðara twēgra landa Cæorlesweorþæ and Cochanfelde, iii. 274, 4. Ic gean ðara twēgra landa æt Cohhanfeldæa and æt Cæorlesweorþæ, 272, 8.

Ðæt land æt Ægeleswyrðe, 125, 10. Some passages are added which may further illustrate the different forms and the variation in gender :— In loco quae dicitur Meranworð, Txts. 437, 10. Ab occidente Hodoworða, Cod. Dip. Kmbl. ii. 49, 18. Oslanwyrð and eall ðæt ðærtó gebyreþ, 267, 36. Andlang Ædeleswyrðe, 195, 3. Tó Lulleswyrðe hyrnan, iii. 343, 31. Tó Uffawyrða gemǽre, 428, 22. Tó ðan norðran Denceswurðe . . . ða þreó hída on ðan norðran Denceswurðe, v. 310, 34-36. Denceswyrðe, 400, 12. On Cwicelmeswyrðe eástwearde, iii. 344, 7 : v. 121, 6. Oð Bulonweorðe ; of Bulanweorðe, iii. 343, 37. On Hananwurðe, 403, 11. On tūnlesweorþ eastweardne, 425, 22, 28. On Wulfrēdeswyrð ; of Wulfrēdeswyrðe, iv. 103, 13. **III.** *a place enclosed by buildings, a court or hall of a house, a place or street of a town* :—Hē sæt úta in worðe *sedebat foris in atrio*, Mt. Kmbl. Lind. 26, 69 : Mk. Skt. Rush. 14, 66. Oð tó on worðe *usque in atrium*, 14, 54 : Jn. Skt. Lind. Rush. 18, 15. Bifora ðone (þ, Lind.) word *ante atrium*, Mk. Skt. Rush. 14, 68. On word (*atrium*) ðæs dómernes, 15, 16. In hwommum worþana (huommum ðara plæcena l worðum, Lind.) *in angulis platearum*, Mt. Kmbl. Rush. 6, 5. On worðum *in plateis eorum*, Ps. Th. 143, 18 : Mt. Kmbl. Rush. Lind. 12, 19. Cf. In plægiword l on plæcum *in plateis*, Rtl. 36, 7. [*O. Sax.* wurð :—Thār that korn gikrund haƀad ende imu thiu wurð bihagōd, Hēl. 2478. *M. L. Ger.* word, wurd *an enclosed homestead*, v. Leo, A. S. Names of Places, p. 60 : Jellinhaus, Die Westfälische Ortsnamen, p. 134.] v. worþig, wyrþe-land.

worþ-apulder, e ; *f. An apple-tree growing by a homestead* (?) :—Be hagan on weorðe hege ; forð be ðan hege on weorðapeldre, Cod. Dip. Kmbl. v. 381, 31.

worþ-cærse, an ; *f. The name of some plant* :—Wordcærsa *grissa garina*, Wrt. Voc. ii. 42, 31. v. worþig-cærse.

worþig, weorþig, wurþig, wyrþ'g [*Ps. Surt.* has forms as from wordign ; one such form is found in Ps. Spl. C., and a dative worðine *occurs in* Bd. S. 539, 42], es ; *m.* **I.** this word, which remains in proper names in the form *-worthy*, has much the same meaning as *worþ* (q. v.), and seems sometimes to exchange with it (cf. In Beniguurthia, Cod. Dip. Kmbl. i. 70, 27, with : In loco qui dicitur Benninguuyrð, ii. 152, 19). In its simplest application it seems to mean *an enclosed homestead* :—Be Ceorles weorðige (worðige, *v. l.*). Ceorles weorðig (weorði, wurðig, *v. ll.*) sceal beón wintres and sumeres betýned. Gif hē bið untýned, and recð his neáhgebúres ceápe in on his ágen geat, nāh hē æt ðam ceápe nánwuht, L. In. 40 ; Th. i. 126, 12-16. But it is found also in connection with land of considerable extent (e. g. Trium cassatorum in loco qui dicitur Worði (cf. tó Wordie, 34), Cod. Dip. Kmbl. v. 109, 7. Ðis synd ða landgemǽra tó Worðige, 110, 32. .v. cassatos in loco qui appellatur æt Worðige (Worðie, 120, 5), 118, 31), and where there are habitations of considerable importance (e. g. Ego Offa rex sedens in regali palatio in Tamouurðige, i. 172, 19. Tamouuordie, 171, 6. In loco celeberrimo quae a vulgo vocatur Tomeworðig, 238, 11). Various Latin words are used in reference to places in whose names the word occurs :— In uico celeberrimo qui vocatur Tomouuorðig, i. 256, 24. In uilla omnibus notissima quae Worðig nuncupatur, v. 199, 10. Rura . . . Tantun . . . , Uuorðig, . . . Stoke, iii. 155, 27. .viii. mansas agelluli, ibidem ubi uulgares prisco more uocitant æt Worðige, v. 240, 9. Worðig, vi. 244, 13, is agellus in the body of the Charter, iv. 150, 26. In Bd. 3, 14 a vico Cataractone is in English fram Cetrihtworðige (-worðine, *v. l.*), S. 539, 42. Other instances of the use of the word in reference to localities are the following :—Unam mansam loco qui celebri æt Monowyrðige appellatur. . . . Ðis synd ðære ánre hýde landgemǽru tó Monawurðige. Ǽrust on Monawurðiges forde, vi. 57, 9-15. Ofer ðæt hǽð wið Cyblesweorðiges, Cod. Dip. Kmbl. iii. 392, 5. Sūð tó Elle-wurðie, vi. 194, 11. Of ðam ealdan lace on Burhgeardesweorðig, iii. 412, 12. Instances of the independent use of the word are the following :— Wurðig (worþig, weorþi, *v. ll.*) *fundus*, Ælfc. Gr. 8 ; Zup. 28, 12. Worþig *predium*, Wrt. Voc. i. 84, 59. Hió an ðæs worðiges, Cod. Dip. Kmbl. vi. 133, 35. Of ðære rōde on Heaðeburhe weorðyg ; of ðæm worðige ondlong hrycges, iii. 77, 10. Sancte Andreas cirican and ðone worðig ðe ðærtó gaunnan wes, v. 163, 20. At Sunemannes wyrðige ; ond of ðam wyrðige . . . on Sunemannes weorðig, vi. 62, 16-31. Wē wrítaþ him ða circan and ðone circstall and ðone worðig tó ðære burnan and ðone croft be sūðan ðære burnan, iii. 53, 1. Ðæt se gídsere his weorðig (worðig, Hatt. MS.) and his land mid unryhte rýme *cum multiplicare large habitationis spatia cupiunt*, Past. 44 ; Swt. 328, 21. Hygelāce wæs gecýðed ðæt ðǽr on worðig (*into the precincts of the palace*) wígendra hleó cwom tó hofe gongan, Beo. Th. 3948 ; B. 1972. Æt Hunigburnan twēgen weorðias and .xi. æceras earðlandes, vi. 219, 1. Ðæt greáte windelstreáw ðæt on worþium wixð (*that grows in yards about houses?*), Lchdm. ii. 44, 5. On worþigum, 92, 26 : iii. 56, 1. Twelf æceras mǽde ðe licgaþ on sūðhealf weges into ðām þreom worðigan (cf. agellorum, v. 150, 26), 244, 13. Ða worðias æt Æscwícan (v. preceding passage), iv. 171, 7. **II.** *a place surrounded by buildings, a place or street of a town* ; platea :—Hē sǽde ðam cyninge, ðæt æghwanone cōman micel menigo ðearfena, ðæt se weorþig full sǽte *indicavit regi quia multitudo pauperum undecumque adveniens maxima*

per plateas sederet, Bd. 3, 6; S. 528, 18. Næfre on his weorþige (*or under* I?) weá áspringe *non defecit de plateis ejus usura*, Ps. Th. 54, 10. Fenn worðigna *lutum platearum*, Ps. Spl. C. 17, 44: Ps. Surt. 17, 43. Of wurðigum *de plateis*, Ps. Spl. C. 54. 11. Worðignum, Ps. Surt. 54, 12: 143, 14. Hweorfaþ ymb Sion . . . and dǽlaþ hire weorðias *circumdate Sion . . . et distribuite gradus ejus*, Ps. Th. 47, 11.

worþig-cærse, an; *f. The name of some plant:*—Uorthigcearse *grissa garina*, Lchdm. iii. 303, col. 1. v. worþ-cærse.

worþig-netele, an; *f. A nettle that grows by a homestead* (?):— Sió micle worþignetle, Lchdm. ii. 116, 2.

woruld. v. weorold.

wŏrung, e; *f. Wandering about, rambling:*—Hé hêt ðæt hé wunode bûtan wŏrunge on mynstre, Homl. Skt. i. 6, 99. Ásolcennys ácenð îdelnysse . . . , wŏrunge and fyrwitnysse, Homl. Th. ii. 220, 26.

wŏs, es; *n. Moisture, juice:*—Ofetes wŏs *ydromellum*, Wrt. Voc. i. 27, 43. Genim ðysse wyrte wŏs, Lchdm. i. 200, 15. Genim rosan wŏs, 214, 1. Genim leáf, wyl on wætere and wring ðæt wŏs (*press 'he moisture out of the leaves*), 72, 8. Genim ðæs wyrte, cnuca hý swá grêne, wring ðæt wŏs, 126, 7: 208, 12: iii. 102, 14. Wring ðæt wŏs on eced, i. 200, 15. Genim cetel, dô þriddan dǽl ðara rinda and ða wyrta, wyl on wætre swîþe; dô ðonne of ða rinda and dô nîwe on innan ðæt ilce wŏs, ii. 86, 16. [He thrast hom as men dos crapbys, thrastyng owt the wos, Halliwell's Dict.] v. pere-wŏs; wêsan, wôsig.

-wŏsa. v. ealo-, here-wôsa; wêsa.

wŏsig; *adj. Juicy, succulent:*—Deós wyrt is wel wôsig, Lchdm. i. 270, 21. Genim ðæs wyrte swá wôsige gecnucude, 278, 23. Ða beóð fulle of gehwǽdum leáfum wel wôsigum, 258, 3.

wŏp, e; *f.*   I. *a sound, cry, noise:*—Weard breahtm hæfen, wôð up ástâg, cearfulra cirm, cleopedon monige, Exon. Th. 118, 4; Gû. 234: 125, 31; Gû. 362. Hý mislîce, mongum reordum, wôðe hôfun, hlûdne herecirm, 156, 8; Gû. 871.   II. *of articulate or melodious sound, voice, song, speech:*—Woorð, uuóþ *lepor*, Txts. 73, 1196. Wôþ *facundia*, i. *eloquentia*, Wrt. Voc. ii. 35, 3. Mid ðære getyngan wôð *lepida*, 50, 44. Ic hǽleþum bodige wilcumena fela wôþe mînre, Exon. Th. 391, 5; Rä. 9, 11. Hé áhôf wôðe: 'Hwæt! ge sind earme,' Andr. Kmbl. 1349; An. 675. Hî singaþ heofoncyninges lof, wôða wlitegaste, and ðás word cweðaþ, Elen. Kmbl. 1494; El. 749. Swêghleóþor cymeþ, wôþa wynsumast, þurh ðæs wildres mûð, Exon. Th. 358, 9; Pä. 43. [Cf. (?) *Goth.* weit-wôdei *witness: Icel.* ôðr; *m. mind; song.*] v. heáfod-wôþ.

wŏþ-bora, an; *m. A (good) speaker, orator, poet, prophet, philosopher:*—Sum biþ wôðbora, giedda giffæst, Exon. Th. 295, 19; Crä. 35. Sægde sum wôðbora, Esaias 19, 18; Cri. 302. Ic wôðboran wordleána sum ágyfe æfter giedde, 489, 17; Rä. 78, 9. Micel is tô hycgenne wîsum wôðboran, hwæt sió wiht sié, 414, 22; Rä. 32, 24. Wilt ðú wîsne wôðboran wordum grêtan, biddan ðê gesecge gesceafta cræftas, 346, 21; Sch. 2. Cræftgleáwe men, wîse wôþboran, Chr. 975; Erl. 126, 27. Wôðborum *rhetoribus*, Wrt. Voc. ii. 81, 53.

wôþ-cræft, es; *m. The art of poetry or song:*—Wôðcræfte, beorhtan reorde, Exon. Th. 206, 15; Ph. 127. Ne wêne ænig ðæt ic lygewordum leóð somnige, wrîte wôðcræfte, 234, 30; Ph. 548. Ic wille wôðcræfte wordum cŷþan bi ðam hwale, 360, 7; Wal. 2.

wôþ-dor (P). v. wôd-dor.

wôþ-gifu, e; *f. The gift of song:*—Hyre (*a musical instrument*) is on fôre fæger hleóþor, wynlîcu wôðgiefu . . . seó wiht mæg wordum lâcan þurh fôt neoþan, Exon. Th. 414, 10; Rä. 32, 8.

wôþ-sang, es; *m. Song:*—Wîtgena wôðsong, Exon. Th. 4, 1; Cri. 46.

woxo (= oxan) *bovem*, Lk. Skt. Lind. 13, 15.

wracian; *p.* ode *To be in exile:*—Wracode *exulat*, Wrt. Voc. ii. 81, 13: 31, 14. Hé ge mid Scottum ge mid Pehtum wracode *apud Scottos sive Pictos exulabat*, Bd. 3, 1; S. 523, 17. Hé on Gallia wracode (wrecca wæs, *v.l.*), 3, 18; S. 545, 38. Wracade, 4, 23; S. 594, 44. His menn ða ðe mid wracedon *suos homines qui exules vagabantur*, 4, 13; S. 583, 9. Wraciende *exulans*, Wrt. Voc. ii. 31, 15.

wracnian, wræcnian; *p.* ode *To be or travel in a foreign country, be a pilgrim or stranger:*—Ic wræcnige *peregrinor*, Ælfc. Gr. 25; Zup. 145, 19 note. Ic wracnode mid Labane *apud Laban peregrinatus sum*, Gen. 32, 4. Ephron, ðær wracnode Abraham *Hebron, in qua peregrinatus est Abraham*, 35, 27: 37, 1. Wræcnede *exulat, peregrinatus est*, Hpt. Gl. 476, 3. Chanaan land, ðe hig on wracnodon and útancymene wǽron *Chanaan, terram peregrinationis eorum, in qua fuerunt advenae*, Ex. 6, 4. Gif mæssepreóst manslaga wurðe, ðonne þolige hé ǽgðres, ge hádes ge eardes, and wrǽcnige swá wîde swá pâpa him scrîfe, L. Eth. ix. 26; Th. i. 346, 6. Þolige hé ðeles, and wræcnige, L. C. S. 41; Th. i. 400, 15. Þolige se, ðe hit on gelang sŷ, ǽlcere eardwununge, and wrǽcnige of earde, oððon on earde swîðe deópe gebête, swá biscop him tǽce, Wulfst. 120, 13: 300, 25. v. for-wracned, R. Ben. 82, 2.

wracu, e; *f.*   I. *pain, suffering, misery:*—Is fela yfela and mistlîcra gelimpa wîde mid mannum; and eal hit is for synnum; and gyt weorþeþ mâre, ðæs ðe bêc secgaþ, wracu and gedreccednes, ðonne ǽfre ǽr wǽre on worulde, Wulfst. 91, 7. Nis mê wracu ne gewin, ðæt ic God sêce, Exon.

Th. 167, 2; Gû. 1054. Nis ðær lâð genîðla, ne wôp ne wracu, weátácen nân, yldu ne yrmðu . . . ne sâr wracu *non huc exangues morbi, non aegra senectus . . . luctus acerbus abest*, 201, 2-11; Ph. 50-54. Him com swá hrædlîc sâr and wracu swá ðam cennedan wîfe cymð fǽrlîc sâr *ibi dolores sicut parturientis*, Ps. Th. 47, 6. His þegnas for hiora eardes lufan and for ðære wrace (cf. for ðæm yrmþum eardes lyste, Met. 26, 71) tihodon hine tô forlǽtanne, Bt. 38, 1; Fox 194, 29. Him ðæt tô longsumere wrace côme, ðær hié ðe raðor gesêmed ne wurden *actum de Romano nomine intestina pernicie foret, nisi reconciliatio subrepsisset*, Ors. 2, 4; Swt. 70, 5. Hé weard werþeódum tô wræce, Elen. Kmbl. 33; El. 17. Hé hæfde him tô gesîþþe sorge and longað, wintercealde wræce, weán oft onfond, Exon. Th. 377, 15; Deór. 4. Wræce bisgodon fǽge þeóda *miseries troubled the doomed peoples*, Cd. Th. 76, 29; Gen. 1264.   II. *suffering that comes as punishment, retributive punishment, vengeance, retribution:*—Seofonfeald wracu (*ultio*) bið gesealde for Cain, Gen. 4, 24: Cd. Th. 63, 35; Gen. 1042. Hwylc wracu him forhogiende æfter fyligde *quae illos spernentes ultio sècuta sit*, Bd. 2, 2; S. 502, 4. Swá micele mâre byþ êhtnysse grama, swá micele rihtwîsre gewyrþ and hefigre of êhtnysse wracu *quanto major fuerit persecutionis injuria, tanto justior fiet et grauior de persecutione vindicta*, Scint. 212, 5. Swingella wracu *verberum vindicta*, R. Ben. 52, 7. Ðám eardum beon ôðer wracu siððan, Ælfc. T. Grn. 8, 14. Ǽr ðam ðe seó wracu (*the destruction of Jerusalem*) côme, Homl. Th. i. 402, 24: 408, 12. Synna wracu, Exon. Th. 98, 14; Cri. 1607: Cd. Th. 309, 18; Sat. 711. Ðis synt wrace dagas *dies ultionis hi sunt*, Lk. Skt. 21, 22. On dæge wræce *in die ultionis*, Scint. 178, 11. On dæge wræce (*vindicte*), 179, 6. Áhebban hine ofer ða scyldgan mid andan and mid wræce *se peccantibus zelo ultionis anteferre*, Past. 17; Swt. 115, 5. Hé gecýðde his nîð and his onwald mid ðære wræce *zelus ultionis jus aperuit potestatis*, 115, 22. Swá wê for monnum orsorglîcor ungewîtnode syngiaþ bûton ǽlcre wrace *quanto apud homines inulte peccamus*, 117, 24. Bûtan ǽlcre ôðerre wrace *inulte*, Past. 44; Swt. 327, 17. Ðære ceastre tôworpennysse, ðe gelamp for ðære wrace heora mândǽda, Homl. Th. i. 402, 8. Ða gesceafte ðe synd þwyrlîce geðúhte, hî sind tô wrace gesceapene yfeldǽdum, 102, 3. On gelîcre wrace (*vindicta*) dædbête hê, R. Ben. 50, 14. Hê lîchamlîce wrace mid swingelle þolige *vindicte corporali subdatur*, 48, 11. Ða yfelan bióþ micle gesǽligran ðe on ðisse worulde habbaþ micelne weán and manigfeald wîte for hyra yfelum, ðonne ða sáin ðe náne wræce nabbaþ, ne nân wîte on ðisse worulde for hiora yfle *feliciores esse improbos supplicia luentes, quam si eos nulla justitiae poena coerceat*, Bt. 38, 3; Fox 200, 4. Ðær sceal ǽghwylc man onfôn ðam rihtan dôme his ágenra gewyrhta, . . . swá wrace, swá êce wîte, swá êce lîf, Wulfst. 136, 8. Hê ðolaþ þeóstra ðurh wrace, Homl. Th. ii. 556, 21. Wræce, Exon. Th. 37, 15; Cri. 593: 455, 30; Hy. 4. 57. Wrace, Andr. Kmbl. 1230; An. 616.   II a. *where the punishment or vengeance is attributed to the Deity:*—Seó wracu (*ultio*) is mîn and ic hit ágilde, Deut. 32, 35. Sôðcyninges seofonfeald wracu, Cd. Th. 67, 14; Gen. 1100. Waldendes wracu, hungor, Chr. 975; Erl. 126, 28. Open wracu ys on his yrsunga *ira in indignatione ejus*, Ps. Th. 29, 4. Him becom seó godcundlîce wracu, Homl. Th. i. 86, 1. Him com on Godes wracu (*irato Deo*) an gefeohtum tôeácan ôþrum yflum, Ors. 4, 4; Swt. 164, 22. Wraco, 1, 3; Swt. 32, 9. Cymð se Dryhtnes dômes dæg and wrace (*vindictae*), Past. 35; Swt. 245, 18. Hê ðæt eal for Godes wræce fordyde, Blickl. Homl. 79, 26. Hwæt him se Waldend tô wrace sette, Exon. Th. 98, 4; Cri. 1602. Tô wræce, Cd. Th. 156, 6; Gen. 2584. Hê bæd þrymcyning, ðæt hê him ða weáðǽd tô wræce ne sette, Elen. Kmbl. 988; El. 495. Ðæt gê witon mîne wrace (*ultionem*), Num. 14, 34. Wrace (wrece, Ps. Surt.) *vindictam*, Ps. Spl. 57, 10: Ps. Th. 78, 13: Cd. Th. 235, 21; Dan. 309. Drihten sende on hié mâran wræce ðonne ǽfre ǽr ǽnigu ôþru gelumpe, Blickl. Homl. 79, 9. Hwæðer Drihten ámetan wolde wrece be gewyrhtum, Met. 9, 36. Wracena (wraca, Ps. Spl.: wreca, Ps. Surt.) God *Deus ultionum*, Ps. Th. 93, 1.   II b. *where the punishment takes the form of exile:*—Hié ádrǽfdon ðone consul on elþeóde . . . Hit wæs swîþe ofþyncende ðam ôþrum consulum . . . þeh ðe hié mid ðære wrace (*in the matter of his banishment*) ðæm ádrǽfdan on nânum stale beón ne mehton, Ors. 5, 9; Swt. 232, 22. Hê wîtgode be ðære wræce . . . , ðæt wæs ðá hit tô Babilonia gelǽdde wǽron, Ps. Th. 30, arg. Heó on wrace seomodon swearte sîðe, Cd. Th. 5, 14; Gen. 71. Ic sceal wrace dreógan . . . sceal nû wreclástas settan, sîðas wîde, 276, 8; Sat. 185.   III. *persecution, hostility, active enmity:*—Of ðære wræce mînra feónda álýs me, Ps. Th. 16, 12. Ic wræce fêre geond foldan, folcsalo bærne, ræceð reáfige, Exon. Th. 381, 1; Rä. 2, 4. Gif hé monna dreám of ðam orlege eft ne wolde gesêcan, . . . lǽtan wræce stille, 114, 10; Gû. 170. His sunu hätte Mars, se macode ǽfre gewinn, and saca and wraca hê styrede gelôme, Wulfst. 106, 26.   IV. *where hurt is inflicted in return for hurt suffered, vengeance, revenge:*—Wracu sceal heardum men, Exon. Th. 343, 7; Gn. Ex. 153. Onginþ him leógan se tôhopa ðære wræce, Bt. 37, 1; Fox 186, 23: Met. 25, 51. Hé gesette ða men on ǽnne truman ðe mon hiora mǽgas ǽr slôg, and wiste ðæt hié woldon geornfulran beón ðære wrace (*or under* III? *see the Latin* certaminis) þonne ôþere men, and hié swá wǽron *illi quorum cognati occubuerant, certaminis extitere*

principium, Ors. 2, 5; Swt. 80, 21. Gif hwā wrace dó, ǽr hé him ryhtes bidde, L. In. 9; Th. i. 108, 4. Se ðe þeóf wrecan wille . . . Gif hé man ofsleá on ða wrace, L. Ath. v. 1, 5; Th. i. 230, 12. Ne mæg wé ðæs wrace gefremman, Cd. Th. 25, 14; Gen. 393. Ðæs hé wræce leornode *he learnt how to take revenge for the wrong*, Beo. Th. 4660; B. 2336. **IV a. with gen. of person for whose sake vengeance is taken :—** Gif hwylc man for his mǽges wræce (*in ultione propinqui*) man ofsleá . . . Se ðe man ofsleá on his mốdor wrace (*in ultione matris suae*), L. Ecg. P. iv. 68, 18, 19; Th. ii. 230, 18–21. Hét se cyning ofsleán mycel wæll on ðære byrig on ðæs abbodes wrece, ðe hí ǽr ofslốgon, Chr. 952; Erl. 118, 29. Ne déð God his gecorenra wrace (wraco, Rush. : ðæt wræcce, Lind. *uindictam*)? . . . Ic eów secge ðæt hé raþe hyra wrace déð, Lk. Skt. 18, 7, 8. [Þa ilke wrake þe ic dude þe, þu scoldest dôn me, O. E. Homl. i. 9, 18. Scal eou gewaxen muchele wrake and sake, 13, 26. His swerd, þat is his wrake, ii. 61, 23. Wrake wes on londe, Laym. 4040. Min is te wrake *mihi vindictam*, A. R. 186, 1. We schule sechen efter wrake on alle þeo þat te biwiteð, Jul. 51, 10. Schal beo niþ and wrake, O. and N. 1194. Wele after wrake (wowe, *v. l.*), Misc. 111, 142. Wrake or weniawnce *vindicta, ulcio*, Prompt. Parv. 533. *Goth.* wraka *persecution : O. Sax.* wraka, wreka.] v. gnyrn-, gring-, gyrn-, níd-, níþ-, sár-, syn-, torn-, þeóf-. þeów-wracu; wræc.

wracu, Bt. 20; Fox 72, 6 *for* wacru? v. wacor :—Sió wiþerweardnes biþ simle untælu and wracu (wæru, *v.l.*) āscirred (-scirped?) mid ðære styringe hire āgenre frécennesse. The Latin is : Adversam fortunam videas sobriam, succintamque, et ipsius adversitatis exercitatione prudentem.

wræc, es ; *n.* **I. wrack, misery, suffering :**—Ðæt cûþ is ðæt ðæt mid Drihtnes mihte gestihtad wæs, ðæt yfell wræc cốme ofer ða wiþcorenan *quod Domini nutu dispositum esse constat, ut veniret contra improbos malum*, Bd. 1, 14; S. 482, 41. Ðæt wæs wræc micel wine Scyldinga, mốdes brecða, Beo. Th. 342; B. 170. Wéland wræces cunnade, earfoþa dreág, Exon. Th. 377, 10; Deór. 1. Oft sceal eorl monig wræc ādreógan, Beo. Th. 6148; B. 3078. **II. suffering that comes as punishment, retributive punishment, vengeance :**—Ǽlc wræc and nâ wræc *omnis uindicta et non uindicta*, Scint. 223, 5. Ðæt unâsecgenlîce wræc and ðæt ungeendode wîte, ðæt ðon unlǽdon ðǽr geteohhod biþ, Blickl. Homl. 25, 24. Mid ðý wîte ðæs forespecenan wræces slægene wǽron *praefatae ultionis sunt poena multati*, Bd. 4, 25; S. 601, 30. Ende nǽfre ðînes wræces weorþeþ, Andr. Kmbl. 2765; An. 1385. Ne wæs ungelíc wræce (*ultioni*) ðam ðe Chaldéas bærndon Hierusaleme weallas, Bd. 1, 15; S. 483, 41. God ðæt wîte tô wrece gesette, Cd. Th. 295, 29; Sat. 494. Hí wǽron þurh heora handa deáþes wræc ðrowiende *per horum manus ultionem essent mortis passuri*, Bd. 2, 2; S. 503, 31. Hí ðurh Drihtnes wræc heora scylde wîte ðrowedon *Domino vindice poenas sui reatus luerent*, 4, 26; S. 602, 13. Hé ðæt wîte and ðæt éce wræc âsette on ðone aldor deófla and mancyn freólsode, Blickl. Homl. 83, 23 ; Andr. Kmbl. 2759; An. 1382 : Exon. Th. 78, 10; Cri. 1272 : 92, 28; Cri. 1515. Wræccum *plagis*, Lk. Skt. Rush. Lind. 12, 48. **III.** where the punishment or misery is exile or banishment :—Hié (*Adam and Eve*) wǽron on helle fíf þûsend wintra and twâ hund wintra ǽr ðon God wolde heó ðæs wræces unbindan, Anglia xi. 2, 24. Se Hǽlend wolde Adam gefreólsian of ðam langan wræce, Blickl. Homl. 29, 21. Wilfriþ æfter langum wræce (*post longum exilium*) wæs eft onfangen on bisceophâd, Bd. 5, 3; S. 615, 37. Dauid sang ðysne sealm gebiddende tô Drihtne for his hâmcyme of ðam wræce and of ðam earfoðan, ða hé ðâ on wæs, Ps. Th. 30, arg. Ðone kyning ðe hine (*David*) on suâ heardum wræce gebrôhte, and of his earde ādrǽfde, Past. 3; Swt. 37, 4. Hé wæs ðǽr iii. geár on wrece (wræce, wreccesíð, *v.ll.*) ; hæfde hine Penda ādrifenne, Chr. 658; Erl. 34, 4. Hé (*Abraham*) on wræce lifde *he lived a wanderer*, Cd. Th. 202, 5 ; Exod. 383. Ðû scealt ôðerne éðel sécean, and on wræc hweorfan, 57, 15; Gen. 928 : 62, 15; Gen. 1014. Ðæt hé on wræc drife his selfes sunu (*Ishmael*), 168, 32; Gen. 2791. Metod ðec (*Nebuchadnezzar*) āceorfeþ of cyningdôme, and ðec wineleásne on wræc sendeþ, 251, 26; Dan. 569.

wræc *what is driven :*—Wraec *actuarius*, Txts. 37, 62.

wræc *pain*, Lchdm. i. 338, 9. v. wærc.

wræc, e; *f. Vengeance*. This form seems to be implied by later English forms, e.g. þatt was mikell wræche, þatt all follc for till helle, Orm. 19; don wreche (*rimes with* speche, leache, teche), Misc. 143, 56; tak wreche (*rimes with* preche), Alis. 2858 : but there appears to be no instance in Old English of a nominative *wræc* which is certainly feminine ; where the gender of a nominative *wræc* is marked it is neuter. Some of the oblique cases given under *wracu* and *wræc* might belong to the word and perhaps the following passage :—Hió cwǽdon : ' Sió his blôd and his blôdes wræc ofer ûs and ofer ûre bearn, H. R. 7, 23. [Gif þu heuedest wreche inumen, O. E. Homl. i. 197, 107. Unwreste þu best, gef þu wreche ne secst, ii. 29, 25 : Laym. 29581. Min is þe wreche (wrake, *v.l.*) *mihi vindictam*, A. R. 186, 1. Cain on werlde wreche and wrake, Gen. and Ex. 552. He heþ ynome tô lite wreche, Ayenb. 45, 28. *O. H. Ger.* rāhha *vindicta, ultio*. Cf. *Goth.* wrêkei *persecution*.]

-wræc (-wræce?). v. god-, mân-, sceþ-wræc.

wrǽcca, wrǽccan, wrǽcce-ness. v. wrecca, wreccan, wrec-ness.

wræc-fǽc (?), es ; *n. A time of misery*, Exon. Th. 354, 51 ; Reim. 64. Cf. wræc-hwíl.

wræc-full; *adj. Wretched, miserable :*—Ðæt ân líf is wræcful, ðæt ôðer is eádig ; ân hwílwendlíc, ôðer éce, Homl. Th. ii. 440, 4. Bisceáwiaþ hû wræcfull ðis andwyrde líf is, i. 488, 21. Fram ðisum wræcfullum lífe, 84, 1 : ii. 370, 19. Ðis wræcfulle líf, ðe wé on sind, wé lufiaþ, 540, 12 : 490, 15.

wræc-hwíl, e ; *f. A period of misery or exile, the present life :*—Ðǽr ða eádgan beóð æfter wræchwíle weorcum bifongen, Exon. Th. 233, 19; Ph. 527.

wræc-lást, es ; *m. An exile-track :*—Waraþ hine wræclást, Exon. Th. 288, 16; Wand. 32. Hé wunode wræclástum *he lived in exile*, Chr. 1065; Erl. 196, 36. Hé wræclástas træd *he wandered an outlaw*, Beo. Th. 2709; B. 1352. Ic sceal hweorfan ðý wídor, wadan wræclástas, wuldre benémed, duguþum bedéled, Cd. Th. 272, 17; Sat. 121. Hé longe sceolde hréran mid hondum hrímcalde sǽ, wadan wræclástas (*to wander an exile*), Exon. Th. 286, 23; Wand. 5. Hwæt ða sume dreógaþ ðe ða wræclástas wídost lecgaþ *what some of those suffer whose exiled steps go furthest*, 309, 14; Seef. 57. Ic sceal wreclástas settan, síðas wíde, Cd. Th. 276, 14; Sat. 188. Wreclástas wunian *to live in exile*, 280, 21; Sat. 259.

wræc-lástian; *p.* ode *To be in exile or banishment :*—Wræclástaþ *exulat*, i. *captivatur, peregrinatur, alienatur, fugatur, expellitur*, Wrt. Voc. ii. 146, 25. Wræclástode *exulat*, 96, 3.

wræc-líc; *adj.* **I. strange, wonderful :**—Fæste is ðîn templ éce and wræclíc âwa tô feore *sanctum est templum tuum, mirabile in aequitate*, Ps. Th. 64, 5 ; Exon. Th. 26, 12; Cri. 416. Geseah cyning wundor on wîte āgangen ; him ðæt wræclíc þûhte, hyssas hâle hwurfon in ðam hâtan ófne, Cd. Th. 233, 4 ; Dan. 270. Wundor ðîn wræclíc *mirabilia tua*, Ps. Th. 88, 4 : 74, 2. Ðîne weorc wǽron wræclíce swýþe *mira opera tua*, 138, 12. Wræclíce syndon wǽgea gangas *mirabiles elationes maris*, 92, 5. Weorca wræclícra *mirabilium*, 76, 9 : 105, 11. Ǽðele cyningas, weras wræclíce *reges mirabiles*, 135, 19. Wundur wræclícu *magnalia*, 70, 18 : *mirabilia*, 118, 18. Wolcen wræclícu, 134, 7. Môyses dômas, wræclíco wordriht, Cd. Th. 177, 31 ; Exod. 3. **II. wretched, miserable :**—Wé on ðâs wræclican woruld ācende wurdon, Wulfst. 1, 12. Hý wurdon of ðære myrhðe āworpene, ðe hý ǽr on wǽron, and on ðis wræclíce líf bescofene and on earfoðan and on geswince wunedon, 9, 13. Hé sceóp ðam wérlogan wræclícne hâm, Cd. Th. 3, 17 ; Gen. 37.

wræclíce; *adv.* **I. abroad, to foreign parts :**—Hé férde wræclíce (*peregre*) on feorlen ríce, Lk. Skt. 15, 13. **II. strangely, wonderfully :**—Wræclíce *mirabiliter*, Ps. Th. 75, 4. Gemunaþ hû hé mænig wundor worhte wræclíce, 104, 5 : 105, 7 : 148, 5.

wræc-mæcg, es ; *m. A wretch :*—Wræcmæcgas, ða ðe ne bimurnaþ monnes feore, Exon. Th. 109, 35; Gú. 100. Gé dyslíce dǽd gefremedon, werge wræcmæcgas, Elen. Kmbl. 773 ; El. 387. Hyne wræcmæcgas ofer sǽ sôhton, Beo. Th. 4748; B. 2379. ¶ *used of evil spirits :*—Wræcmæcgas, . . . Godes andsacan, Exon. Th. 116, 5; Gú. 202 : 118, 3; Gú. 234 : 135, 26; Gú. 530. Wræcmæcgas, wergan gǽstas, 23, 3; Cri. 363.

wræc-mæcga, an ; *m. A wretch :*—Se wræcmæcga *the devil*, Exon. Th. 258, 4 ; Jul. 260.

wræc-mann, es ; *m. A fugitive :*—Wræcmon gebâd lâðne lâstweard *the fugitive (the Israelites) awaited the hated pursuer (the Egyptians)*, Cd. Th. 186, 12 ; Exod. 137.

wræcnian. v. wracnian.

wræc-setl, es ; *n. An exile-abode :*—Wíd is ðes wésten, wræcsetla fela, eardas onhǽle earmra gǽsta, Exon. Th. 120, 6; Gú. 267.

wræc-síþ, es ; *m.* **I. travel in a foreign land, peregrination, pilgrimage :**—Hí noldon geþafian ðæt swâ getogen mann (*Gregory*) ða burh forléte, and swâ fyrlen wræcsíð genâme, Homl. Th. ii. 122, 15. Gif hwâ weófodþén āfylle, sý hé útlah, búton hé þurh wræcsíð (wrec-, *v.l.*) gebéte, L. C. S. 39; Th. i. 398, 26. Â ic wîte wonn mínra wræcsíþa, Exon. Th. 441, 27; Kl. 5. Ic wépan mæg míne wræcsíþas, earfoþa fela, 443, 31 ; Kl. 38. **I a. fig. of absence from heaven :**—Gé (*evil spirits*) frôfre ne wénaþ, ðæt gé wræcsíða wyrpe gebíden, Exon. Th. 132, 29; Gú. 480. **II. exile, banishment :**—Wræcsíð *exilium*, Wrt. Voc. i. 21, 21 : 51, 36. Wræcsíþ, ii. 32, 18. Wræcsíðe *liminio* (*postliminium*, Ald. 37), 80, 30. Wræcsíþe, 52, 33 : *exiliata*, 144, 80. Dauid sang be his gehwyrftnesse of his wræcsíðe, Ps. Th. 22, arg. Hé is swíþe sárig for ðínum earfoþum and for ðínum wræcsíþe, Bt. 10; Fox 28, 19. Of wræcsíðe lǽdan, Exon. Th. 143, 12 ; Gú. 660. Hé bebeád ðæt mon Iôhannes gebrôhte on Bothmose on wræcsíþe from ôþrum cristenum monnum *Ioannes in Patmum relegatur fuit*, Ors. 6, 9; Swt. 264, 11. Ðâ āsende hé hine on wræcsíð tô ânum ígeoðe, Homl. Th. i. 58, 31 : 560, 20. Seó wæs gelǽded from Rôme on wræcsíð on ða ceastre seó is nemned Piceno, Shrn. 102, 35. Wén ic ðæt gé for wlenco, nalles for wræcsíðum, Hrôðgâr sôhton, Beo. Th. 682; B. 338. Heora lâtteówum and heora cempum hié bebudan, ðæt hié on wræcsíþas

fóran, and on ellþéode *ducem suum et milites exsulare jusserunt*, Ors. 4, 4; Swt. 164, 26.   **II a.** fig. of living out of heaven:—Nis ðeos woruld nã ûre êðel, ac is ûre wræcsíð, Homl. Th. i. 162, 17. Ðam bið wræcsíð witod, ðe sceal heán hwearfian, ðonne heonon gangaþ, Andr. Kmbl. 1777; An. 891. Gê in wræcsíðe longe lifdon, swegle benumene, Exon. Th. 139, 19; Gû. 595. Wræcsíð wêpan in ðam deáðsele (*hell*), 166, 23; Gû. 1047: 466, 24; Hö. 126. Wê synd on ðisse worlde ælþeódige . . .; for gylte wê wæron on ðysne wræcsíþ sende, Blickl. Homl. 23, 5.   **III.** misery, wretchedness:—Uton gangan ðæt wê bysmrigen bendum fæstne, óðwiton him his wræcsíð, Andr. Kmbl. 2715; An. 1360. 'Ic nû þrý dagas þolian sceolde wælgrim wítu. . . .' 'Ne wêp ðone wræcsíð,' 2861; An. 1433. Mæg unfǽge eáðe gedígan weán and wræcsíð, Beo. Th. 4573; B. 2292. [*O. Sax.* wrak-síð *pilgrimage ; exile.*]

wræc-síþian; *p.* ode *To be* or *travel in a foreign country, to be in exile:*—Ic wræcsíðige *peregrinor*, Ælfc. Gr. 25; Zup. 145, 19. Ðæt hine mann ásende ofer sǽ on wræcsíð tô sumum wêstene, on ðam ðe cristene menn for geleáfan fordêmde wræcsíðedon, Homl. Th. i. 560, 22. Tô wræcsíðienne *peregrinandi, vagandi*, Hpt. Gl. 412, 59: *ad incolatum peregre*, 413, 12.

wræc-stów, e; *f.*   **I.** *a place of exile:*—Seó stôw ðe ðû nû on hæft eart, and ðû cwist ðæt ðín wræcstôw sý, heó is ðam monnum êþel ðe ðæron geborene wæran *hic ipse locus, quem tu exsilium vocas, incolentibus patria est*, Bt. 11, 1; Fox 32, 27.   **II.** *a place of misery* or *punishment:*—Siððan wræcstôwe (*or* I?) werige gástas under hearmlocan heáne gefôran, Cd. Th. 6, 17; Gen. 90. Wræcstôwa *ergastula*, Lchdm. i. lxii, 4.

wræc-weorold, e; *f.* *A world of misery* or *exile:*—Adam wæs gesceapen on neorxnawonge, and for his sylfes synnum ðanan ádrǽfed on ðás wræcworuld, and on eall ða earfeðu, ðe wê siððan drugon, Wulfst. 1, 2.

wræc-wíte, es; *n.* *Punishment:*—Seó ǽreste môdor ðyses menniscan cynnes wræcwíte middangearde bróhte, ðã heó Godes bebodu ábræc, and on ðis wræcwíte áworpen wæs, Blickl. Homl. 5, 24-26.

wrǽd, wrǽd, es; *m.*   **I.** *a bandage, band, fillet:*—Wrǽda *fasciarum*, wrǽd *fascia*, Wrt. Voc. ii. 34, 21-22: 39, 69. Wrǽd sceal wunden, Exon. Th. 343, 6; Gn. Ex. 153. Sió wund wile tóberan, gif hió ne bið gewriðen mid wrǽde (wrǽðe, Cott. MSS.), Past. 17; Swt. 123, 16: Lchdm. ii. 306, 18. Se mec wrǽde on furþum legde, bende and clomme, Exon. Th. 383, 20; Rä. 4, 13. Genim nioþowearde wrætte, dô on reádne wrǽd, binde ðæt heáfod mid, Lchdm. ii. 304, 26. Wrǽdas *redimicula*, Hpt. Gl. 527, 7. Wrǽda *fasciarum, vinculorum*, 488, 48. Sume heora fnada and wrǽdas gemiccliaþ, R. Ben. 135, 27.   **II.** *what is bound together, a bundle:*—Wrǽda *fascis*, Hpt. Gl. 529, 4.   **III.** *a band, company, flock.* Cf. wrǽd-mǽlum:—Wrǽd *grex*, Mt. Kmbl. Rush. 8, 32. Wrǽda *manipulorum* (*innumeris manipulorum milibus equitatu et peditatu*, Ald. 76), 525, 24. v. beado-wrǽd (Lchdm. ii. 350, 29); wríþan.

-wrǽde, wrǽdel. v. un-samwrǽde, under-wrǽdel.

wrǽd-mǽlum; *adv.* *In bands :*—Heápmǽlum oððe wrǽdmǽlum *gregatim*, Wrt. Voc. ii. 40, 18. Cf. wrǽd, III.

-wrǽdness, wrǽnan. v. sam-wrǽdness, á-wrǽnan.

wrǽne; *adj.* *Lascivious, libidinous, salacious, wanton:*—Uuraeni urêni *petulans vel spurcus*, Txts. 90, 835. Wraene *petulans*, 87, 1569. Wrǽne *petulcus, luxuriosus*, Hpt. Gl. 484, 55: *libidinosus*, 514, 4. Hê (*Sardanapalus*) wæs swíþe furþumlíc mon, and hnesclíc, and swíþe wrǽne, swã ðæt hê swíðor lufade wífa gebǽro þonne wǽpnedmonna, Ors. 1, 12; Swt. 52, 1. Gif mon sié tô wrǽne, Lchdm. ii. 144, 19: Prov. Kmbl. 54. Wrǽnre *lascivae*, Hpt. Gl. 505, 37. Wrǽnre *petulantis*, 515, 9. Ða wrǽnan *lascivam*, 463, 71. Tarcuinius wæs ǽgðer ge eargast, ge wrǽnast, ge ofermôdgast, Ors. 2, 2; Swt. 66, 28. [Cf. *O. H. Ger.* reino *emissarius, admissarius:* reinisc *admissarius: Icel.* reini *a stallion: Dan.* vrinsk *rank: Swed.* wrensk *lascivious.*] v. un-wrǽne.

wrǽnna. v. wrenna.

wrǽn-ness, e; *f.* *Wantonness, licentiousness, lasciviousness, lust:*—Wrǽnnes *lascivia, ferventia*, Hpt. Gl. 432, 32. Wrǽnnyss *petulantia*, Hymn. Surt. 126, 28. Wrǽnnes *luxuria*, Past. 43; Swt. 309, 1. Wrǽnnes, seó bið ǽlcum men gecynde *gignendi opus, quod natura semper appetit*, Bt. 34, 11; Fox 152, 12. Ðû woldest brúcan ungemetlícre wrǽnnesse *voluptariam vitam degas*, 32, 1; Fox 114, 21. Sió wôde þrág ðære wrǽnnesse *iibido*, 37, 1; Fox 186, 18: Met. 25, 41. Werlícere wrǽnnysse *maritalis lasciviae* (*luxuriae* ł *petulantiae*), Hpt. Gl. 434, 61. Se anga ðære wrǽnnesse *aculei libidinis*, Past. 43; Swt. 309, 16. Ðæt môd hæfð fulfremedne willan tô ðære wrǽnnesse *ejus animus voluptate luxuriae delectatur*, 11; Fox 7, 13. Heó mid ungemetlícre wrǽnnesse (*libidine ardens*) mænigfeald geligre fremmende wæs, Ors. 1, 2; Swt. 30, 28: Ps. Th. 7, 13: L. E. I. 32; Th. ii. 428, 33. Nýtena willa tô nánum ôþrum þingum nis áðenod bûtan tô gífernesse and tô wrǽnnesse *pecudum omnis ad explendam corporalem lacunam festinat intentio*, Bt. 31; Fox 112, 8. v. sin-wrǽnness.

---

wrênsa, an; *m.* *Lasciviousness:*—Wrênsan *lasciviae, luxuriae*, Hpt. Gl. 461, 51. Cf. gǽlsa.

wrênscipe, es; *m.* *Wantonness:*—Wrênscipe *petulantia*, Hpt. Gl. 527, 74.

wrǽsen. v. hilde-wrǽsen.

wrǽsnan; *p.* de *To twist, change the character of* something:—Ic (*a woodpecker*) eom wunderlícu wiht, wrǽsne míne stefne, hwílum beorce swã hund, hwílum blǽte swã gát, hwílum grǽde swã gôs (cf. Ic (*a woodpigeon*) þurh mûþ sprece mongum reordum *based on the Latin:* Vox mea diversis variatur pulcra figuris, 390, 13; Rä. 9, 1), Exon. Th. 406, 15; Rä. 25, 1.

wrǽst, wrǽste, wrǽst; *adj.*   **I.** *delicate, elegant, splendid:*—Wrǽst *delicatus*, Txts. 55, 630. Wrǽstum (*urastum*) *delicatis*, 55, 645. Hê hine wǽdum wrǽstum geteóde, Ps. Th. 108, 18. Óð wígbedes wrǽste hornas, 117, 25. Ne ðê on ðínum selegescotum swíðe lícaþ, þeáh ðe weras wyrcean wrǽst um eorðan, 146, 11. Rose wynlíc weaxeþ; ic eom wrǽstre þonne heó, Exon. Th. 423, 23; Rä. 41, 26.   **II.** *noble, excellent:*—Ðû ût álǽddest wrǽstne wíngeard. . . . Ðû him his wyrtruman wrǽstne settest, Ps. Th. 79, 8-9. Nolde ic ðíne gewitnesse wrǽste forlǽtan, 118, 157. Hê on his welan spêde wrǽste getrúwode, 51, 6. Hí ne wiston wrǽstran rǽd *they knew not a more excellent way*, Cd. Th. 227, 6; Dan. 182. v. un-wrǽst.

wrǽstan; *p.* te.   **I.** *to wrest, twist:*—T hine on ða tungan sticaþ, wrǽsteþ him ðæt wôddor, and him ða wongan briceþ, Salm. Kmbl. 191; Sal. 95.   **II.** *to move the strings of the harp in playing*, Cf. wreste of an harpe *plectrum*, Prompt. Parv. 533 :—Sum sceal mid hearpan æt his hláfordes fôtum sittan, and á snellíce snere wrǽstan, lǽtan scralletan, Exon. Th. 332, 9; Vy. 82. [Iulius þat sweord wraste (wreste, 2nd MS.), Laym. 7532. Wresten *to struggle*, wrestle, A. R. 374, 7. Wrestoñ *plecto*, wrestyñ and wrythyñ aȝen *replecto*, Prompt. Parv. 533 : *Icel.* reista *to twist.*] v. á-, ge-wrǽstan.

wrǽstan (?); *p.* te *To be* or *make elegant.* v. wrǽst :—Wrǽstende *indruticans* (but the passage is : Ista (*mulier nupta*) stolidis ornamentorum pompis infruticans, Ald. 17), Wrt. Voc. ii. 77, 44: 44, 8: 110, 58.

wrǽste; *adv.* *Delicately, elegantly:*—Ne hafu ic in heáfde hwíte loccas wrǽste gewundne, Exon. Th. 427, 30; Rä. 41, 99.

wrǽstlere, es; *m.* *A wrestler:*—Wrǽstlere *luctator* (*-ur*, MS.), Wrt. Voc. ii. 50, 37. [Iacob speleð wrastlare, A. R. 374, 4. Wrestelare *luctator*, Prompt. Parv. 533.] Cf. wraxlere.

wrǽstlian; *p.* ode *To wrestle.* [To wreastlene, Laym. 1858. Summe heo wræstleden, 24699. To wrastlen aȝein þes deofles swenges, A. R. 80, 7. Wrestlin and wiðerin wið ham seoluen, Marh. 14, 13. Hwerto wultu wreastlin (wrestlen, v.l.) wið þe worldes wealdent *quid contra Deum eluctaris?* Kath. 2035. Ðor wrestlede an engel wið, Gen. and Ex. 1803. *M. Du.* wrastelen.] v. next word, *and* cf. wraxlian.

wrǽst-líc; *adj.* *Pertaining to wrestling:*—Ðǽm wǽrstlícum *palestricis*, Wrt. Voc. ii. 69, 3 : 74, 54.

wrǽst-líc (wrást-); *adj.* *Delicate, elegant :*—Ðære wrǽstlícan *delicate*, Wrt. Voc. ii. 77, 29 : 26, 44. Wrǽstlícum *delicatis* (*ornamentis vestium delicatis decorari*, Ald. 73), 87, 17.

wrǽstlíce. v. un-wrǽstlíce.

wrǽstliend, es; *m.* *A wrestler:*—Wrǽstliendra *luctatorum*, Wrt. Voc. ii. 50, 36.

wrǽstlung, e; *f.* *Wrestling:*—Wrǽstlunge *palaestram*, Hpt. Gl. 515, 56. [Wes muchel folc æt þere wrastlinge, Laym. 1871. Bitternesse in wrastlunge aȝean uondunges. . . . Þeos wrastlunge is ful bitter to monie, A. R. 374, 2-5. Ȝif tweie men goþ to wrastlinge, O. and N. 795. Wrestelynge *colluctacio*, Prompt. Parv. 533.] Cf. wraxlung.

wrǽt[t], e; *f.* *A work of art, a jewel, an ornament :*—Se (*the cave*) wæs innan full wrǽtta and wíra, weard unhióre goldmáðmas heóld, Beo. Th. 4817; B. 2413. Wundenmǽl wrǽttum gebunden, 3067; B. 1531. Is ðes middangeard wísum gewlitegad, wrǽttum gefrætwad, Exon. Th. 413, 8; Rä. 32, 2 : 414, 27; Rä. 33, 2. Hê ðone grundwong ongitan meahte, wrǽte (wiræce, MS.) geondwlítan, Beo. Th. 5535; B. 2771. Ðam ðe inne gehýdde wrǽte (wræce, MS.) under wealle, 6112; B. 3060.

wrǽt[t], es; *m.* : e; *f.* *Crosswort:*—Wrǽttes cíð, Lchdm. iii. 12, 28 : 24, 4. Mid wrǽte, ii. 306, 18. Genim nioþowearde wrǽtte, 304, 26. Cf. *Warantia* wret (12th cent.?), i. 376, note. *Vermiculum* warance, wrotte (13th cent.), Wrt. Voc. i. 140, 2.

wrǽþ *a band*, wrǽþ *anger.* v. wrǽd, wrǽþu.

wrǽþan; *p.* de *To be angry, get angry:*—Se ðe uraeðes bróðere his *qui irascetur fratri suo*, Mt. Kmbl. Lind. 5, 22. Wraeðde hláford *iratus dominus*, 18, 34. Se cynig wrǽðde *rex iratus est*, Rtl. 107, 29. Urǽdde *fremuit*, 197, 31. [He wile wreðe wið þe, O. E. Homl. i. 33, 8. He bigon to wreðen (cf. he wreððede him, 10, 4), Jul. 11, 6. African wreaðede and swor, 13, 7. Cf. *O. Sax.* wrêðian (*with reflex. acc.*): *Icel.* reiðask *to get angry.*] v. ge-wrǽþan; wríþian.

wrǽþian. v. wreþian.

wrǽþ-studu, -stuþu, e; *f.* *A support, prop, buttress, stay :*—Seó

wræþstudu *destina*, Bd. 3, 17; S. 544, 17, 24. Wræþstuþum *fulcris*, Wülck. Gl. 245, 28. Wreðstuþum, Exon. Th. 422, 6; Rä. 41, 2.

wrǽþþu (-o); *indecl.* : wræþþ, e; *f.* I. *wrath, anger*:—Wrǽððo *ira*, Lk. Skt. Lind. Rush. 21, 23 : Jn. Skt. Lind. Rush. 3, 36 : *indignatio*, Rtl. 12, 35. Uráððo *iracundiae*, 8, 37. Mið wrǽððo *cum ira*, Mk. Skt. Lind. 3, 5. Hæbbe hé Godes curs and wrǽððe ealra hálgena, Chart. Erl. 253, 14. II. *injury* :—Ðú in wrǽððo giscildnise *tu in injuria defensio*, Rtl. 105, 9. v. next word.

wrǽþu (-o); *indecl.* : wræþ, e; *f. Wrath, anger* :—Wrǽðo ðín *ira tua*, Rtl. 11, 1. Hí wǽran intinga ðare wrǽðe ðe wæs betwyx him and ðan cinge, Chr. 1051; Erl, 182, 28. Hæfþ eal folc micele wræþe æt Gode þurh his ǽnne gilt, þe hé nolde healdan ða þincg, Wulfst. 174, 27. From tóweard wuráðo *a futura ira*, Mt. Kmbl. Lind. 3, 7. [Icel. reiði.]

wrǽt-líc; *adj.* I. *wondrous, curious* :—Grendles heáfod and ðære idese mid, wliteseón wrǽtlic, Beo. Th. 3304; B. 1650. Stefn cwom þurh heardne [stán] . . . , wrǽtlíc þúhte stánes ongin, Andr. Kmbl. 1480; An. 741. Ic eom wrǽtlíc wiht, on gewin sceapen, Exon. Th. 405, 14; Rä. 24, 2 : 483, 11; Rä. 69, 1. Wiht wrǽtlícu, 415, 23; Rä. 34, 2. Mé ðæt þúhte wrǽtlícu wyrd, 432, 6; Rä. 48, 2. Wrǽtlíc mé þinceþ, hú seó wiht mæge wordum lácan þurh fót neoþan, 414, 11; Rä. 32, 18. Ðæt is wrǽtlíc þing tó gesecganne, 421, 27; Rä. 40, 24. Wrǽtlícne wyrm, Beo. Th. 1786; B. 891. Wrǽtlíce gecynd wildra, Exon. Th. 356, 9; Pa. 9. Ða wrǽtlícan wiht, Salm. Kmbl. 505; Sal. 253. Se mé on flíteþ wordum wrǽtlícum, Andr. Kmbl. 2401; An. 1202. Ic seah wrǽtlíce wuhte feówer, Exon. Th. 434, 15; Rä. 52, 1 : 429, 8; Rä. 43, 1. Hé hafaþ óþre gecynd wrǽtlícran, 363, 8; Pa. 50. II. *of wondrous excellence, beautiful, noble, excellent, elegant* :—Ceastra . . . orðanc enta geweorc, . . . wrǽtlíc weallstána geweorc, Menol. Fox 465; Gn. C. 3 : Exon. Th. 476, 1; Ruin. 1. Wrǽtlíc is seó womb neoþan, wundrum fæger, scír and scýne, 219, 14; Ph. 307 : 356, 29; Pa. 19. Heofoncyninges stefn wrǽtlíc, Andr. Kmbl. 185; An. 93. Syndon ða foreweallas fægre gestépte, wrǽtlícu wǽgfaru, Cd. Th. 196, 27; Exod. 298. Ðæs wrǽtlícan hringes, Exon. Th. 441, 12; Rä. 60, 17. Healsbeáh, . . . wrǽtlícne wundormáþdum, Beo. Th. 4352; B. 2173. Wrǽtlíc wígsweord, 2982; B. 1489 : 4668; B. 2339 : Exon. Th. 437, 5; Rä. 56, 3. Wundrum wrǽtlíce wyllan, 202, 1; Ph. 63. Wrǽtlíc weorc smiþa, 408, 18; Rä. 27, 14. Wordum wrǽtlícum, . . . beorhtan reorde, 32, 7; Cri. 509 : Andr. Kmbl. 1259; An. 630. Wrǽtlícra, ǽnlícra and fægerra, Exon. Th. 357, 12; Pa. 27. Heó wæs on sangum wrǽtlícre, ðonne heora ǽnig ǽr wǽre, Homl. Ass. 127, 365.

wrǽtlíce; *adv.* I. *wondrously, curiously* :—Hé (*the phenix*) eft cymeþ, áweaht wrǽtlíce, wundrum tó lífe, Exon. Th. 223, 29; Ph. 367 : 224, 19; Ph. 378. Seó wiht wæs wrǽtlíce, wundrum gegierwed, 418, 8; Rä. 37, 2 : 422, 14; Rä. 41, 6 : 427, 2; Rä. 41, 85 : 428, 2; Rä. 41, 102. II. *wondrously, excellently, beautifully, elegantly, nobly* :—Ðær wrǽtlíce symle telgan gehladene gréne stondaþ, Exon. Th. 202, 26; Ph. 75. Is him ðæt heáfod hindan gréne, wrǽtlíce wrixled, wurman geblonden, 218, 13; Ph. 294. Swá wrǽtlíce weoroda God monna cræftas sceóp and scyrede, 332, 30; Vy. 93. Mé on gescyldrum scínan mótan ful wrǽtlíce wundne loccas, 428, 6; Rä. 41, 104. Ða ðe wrǽtlícost wyrcan cúðon stángefógum, Elen. Kmbl. 2037; El. 1020.

wrǽtte. v. wrǽt[t].

wrǽxliende. v. wraxlian.

wrang; es; *n. Wrong* :—Unrihtdéman, ðe wendaþ wrang tó rihte and riht tó wrange, Wulfst. 203, 26 : 298, 20. [Ealle sǽidon þet se king heold his broðer mid wrange on heftnunge, and his sunu mid unrihte aflemde, Chr. 1134; Erl. 252, 30. Cf. *Icel.* rangr : *adj. Wrong.*]

wrang, wranga *the hold of a ship* :—Wranga (*printed* pranga) *cavernamen* (in a list of nautical words), Wrt. Voc. i. 56, 50. Wrong, ii. 129, 65. [Wrangis *the ribs* or *floor-timbers of a ship*, Jamieson's Dict. : *Icel.* röng *a rib in a ship.*]

wrásan. v. next word.

wrásen, e; *f. A band, tie* :—Wrásan (=? wrásen ; for suffix cf. (?), bodan *fundus*, 98, 10), óst *nodus* (cf. *nodos* bende, 95, 27, *nodorum* rápa, 61, 68), Wrt. Voc. ii. 114, 79. v. fetor-, freá-, inwit-wrásen ; wríþan.

wrást, wrást-líc. v. wrǽst, wrǽst-líc.

wráþ, es; *n.* I. *cruelty* :—Wráð *crudelitas*, Hpt. Gl. 518, 35. II. *what is grievous, the painful* :—Ðæt nán wiht ne sý, . . . ne ðæs heardes ne ðæs hnesces, ne ðæs wráðes ne ðæs wynsumes, . . . ðæt hig þonne mihte fram úres Drihtnes lufan ásceádan, Wulfst. 184, 20.

wráþ; *adj.* I. *wroth, angry, incensed* :—Gram ł wráð *furibundus*, Hpt. Gl. 510, 37. Weard se cyng swíþe gram (wráð, *v. l.*) wið ða burhware, Chr. 1048; Erl. 178, 6. Crist him wurðe wráð, ðe hí geþýwie, Chart. Erl. 253, 17. Bið úre Drihten ðám synfullum swíðe wráð æteówed, and ðám sóðfæstum hé byð blíðe gesewen, Wulfst. 184, 2. Ðín yrre fram ús oncyrre, ðæt ðú ús ne weorðe wráð on móde, Ps. Th.

84, 4 : Cd. Th. 26, 12; Gen. 405 : 46, 17; Gen. 745. Unblíðe, wráð on móde, 136, 19; Gen. 2260. Weard yrre God, and ðam werode wráð, 3, 13; Gen. 35. Ðe cynig wuráð wæs *rex iratus est*, Mt. Kmbl. Lind. 22, 7. Wráð, Lk. Skt. Lind. Rush. 14, 21. Wráð wæs *indignatus est*, Lind. 15, 28. II. *fierce, cruel, grievous, hostile, bitter, fell, evil, malignant*, (1) of living creatures, often used substantively :—Eormanríces, wráþes wǽrlogan, Exon. Th. 319, 8; Víd. 9. Wið wráð werod wearde healdan *to keep watch and ward against foes*, Beo. Th. 643; B. 319. Wráðe wælherigas, Cd. Th. 119, 21; Gen. 1983. Ða þe wydewum sýn wráðe æt dóme, Ps. Th. 67, 5. Mé tó aldorbanan weorðeþ wráðra sum *some fell one shall be my life-destroyer*, Cd. Th. 63, 18; Gen. 1034 : 109, 29; Gen. 1830. Wráðra gryre *the horror of fierce foes*, 178, 32; Exod. 20 : Beo. Th. 3242; B. 1619 : Andr. Kmbl. 2547; An. 1275 : 2635; An. 1319. Burh wráðum werian, Cd. Th. 119, 7; Gen. 1976. Torn gewrecan on wráðum, 123, 1; Gen. 2038 : Elen. Kmbl. 329; El. 165 : Ps. Th. 104, 34. Wráþþum forstolen áhreddan, flýman feóndsceaþan, Exon. Th. 396, 2; Rä. 15, 17. Andsware findan wráþum tówiðere *to find an answer against bitter adversaries*, 12, 13; Cri. 185. Ðú mé weredest wráþum feóndum, Ps. Th. 137, 7. Wráþum wyrmum, Exon. Th. 94, 30; Cri. 1548. (1 a) of evil spirits :—Se atola gást, wráð wǽrloga, Andr. Kmbl. 2595; An. 1299 : Cd. Th. 43, 6; Gen. 686. Þurh ðæs wráþan geþanc, þurh ðas deófles searo, 39, 25; Gen. 631. Hié hýrdon wráðum wǽrlogan, Andr. Kmbl. 1225; An. 613. Waca wið wráþum (*Grendel*), Beo. Th. 1324; B. 660 : 1421; B. 708. Hé wráðne gegrípeþ feónd be ðam fótum, Salm. Kmbl. 226; Sal. 112. Wreceþ heó wráðan, Lchdm. iii. 32, 25. Wráðe wræcmæcgas, Exon. Th. 135, 26; Gú. 530. On wráþra wíc (*hell*), 94, 4; Cri. 1535. Wráðra, Cd. Th. 7, 5; Gen. 101. Wráþra gǽsta, Exon. Th. 424, 19; Rä. 41, 41. Wíte mid wráþum, 37, 18; Cri. 595. Hé gráp on wráðe, Cd. Th. 4, 30; Gen. 61. (2) of things :—Hú sárlíc and hú sorhful and hú geswincful and hú teónful ðis líf is, hú tealt and hú wráð (*grievous or evil*), Wulfst. 273, 7. Is him on welerum wráð sweord and scearp, Ps. Th. 58, 7. Se yfla unrihta wráþa (*evil*) willa wóhhǽmedes, Met. 18, 2. Wráðan (*fierce*) yrres, Ps. Th. 77, 50. On ðam wráðan dæge *diem tentationis*, 94, 9. Wráþe hægle, 77, 47. Wráð yrre ðín, 78, 5. þurh wráð (*evil*) gewitt, Elen. Kmbl. 915; El. 459. Hearmstafas wráðe (*bitter*) and woruldyrmðo, Cd. Th. 58, 2; Gen. 940. Wráþe wyrde, Exon. Th. 468, 14; Phar. 8. Ic eom wráþra láf, fýres and feóle, 484, 6; Rä. 70, 3. Gemyndig wráþra wælsleahta, 286, 27; Wand. 7. Wíta wráðra, 253, 9; Jul. 177 : 261, 7; Jul. 311. Feala ic gebiden hæbbe wráðra wyrda, Rood Kmbl. 101; Kr. 51. Wráðum teárum, Ps. Th. 55, 11. Folmum ðínum wráðum, Cd. Th. 62, 8; Gen. 1011. Hí mid wráðum wordum trymmaþ *firmaverunt sibi verbum malum*, Ps. Th. 63, 4 : Met. 26, 76. Wráþe firene, Exon. Th. 80, 28; Cri. 1323 : 272, 30; Jul. 507. Ic mínum fótum fǽcne síþas, ða wráþan wegas werede *ab omni via mala prohibui pedes meos*, Ps. Th. 118, 101. Ic eom wráþre (*bitterer*) þonne wermód sý, Exon. Th. 425, 22; Rä. 41, 60. [He answare ȝaf, eorlene wraðest (wroþliche swiþe, 2nd MS.), Laym. 18583. *Also in the sense* bad, evil :—To wraðere (wroþere, 2nd MS.) hele, 29556 : A. R. 102, 8 : Marh. 10, 11 : Misc. 148, 27. þu were ibore o wraðe time (*in an evil hour*), Jul. 57, 3. Wraðe werkes wurchen aȝein Godes wille, Kath. 171. *O. Sax.* wréð : *Icel.* reiðr : *O. H. Ger.* reid *crispus.*] v. and-wráð, *and next word.*

wráþe; *adv.* I. *angrily, with* or *in anger, with indignation* :—Eów se Waldend wráðe (*in his wrath*) bisencte, Exon. Th. 142, 3; Gú. 638. Ondsworade ðæs folches aldor wráðe (wráðlíce, Lind.) *respondens archesynagogus indignans*, Lk. Skt. Rush. 13, 14. II. *fiercely, cruelly, greviously, bitterly* :—Woroldlaga syndan innan ðysan earde wráðe forhwyrfde (*grievously perverted*), Wulfst. 268, 5. Him grimme on woruldsælþa wind, wráðe bláweþ . . . hine se ymbhoga ðyssa woruldsælþa wráðe drecce, Met. 7, 51–54 : 29, 89, 91. Hí wráðe tóweorp *destrue eos*, Ps. Th. 58, 11 : 61, 4 : 72, 14, 15. Ða wiðerwearde mé wráðe hycgeaþ *cogitaverunt adversum me*, 139, 8 : Cd. Th. 284, 4; Sat. 316. Hine monige on wráðe winnaþ, 138, 11; Gen. 2290. Wé synd wráðe geswæncte, Homl. Skt. i. 4, 156 : Exon. Th. 443, 19; Kl. 32. Wráþe geworhtra wíta, 252, 32; Jul. 172. Ðú ðé sylfne swíþe wráðe bepæcst *you deceive yourself most grievously*, Homl. Skt. i. 12, 99. Ðæs wráðe ongeald, hearde mid híwum, hægstealdra wyn, Cd. Th. 111, 26; Gen. 1861. III. *evilly, perversely, wickedly* :—Hé ða gehát swíðe yfele gelǽste, and swíðe wráðe geendode mid manegum máne, Bt. 1; Fox 2, 10. Gé on heortan hogedon inwit, worhton wráðe *in corde iniquitates operamini*, Ps. Th. 57, 2. Ys hyra múðes scyld mánworda feala, ða hí mid welerum wráðe ásprǽcan *delicta oris eorum sermo labiorum ipsorum*, 58, 12 : Elen. Kmbl. 587; El. 294. IV. *with an intensive force to qualify an unfavourable idea* :—Syndon gewordene heora willan wráðe besmitene (*horribly defiled*), Ps. Th. 52, 1. Ðæt bið forwisnad wráðe sóna (*terribly soon*), 128, 4. [On two wise, wel and wroðe (*ill*), O. E. Homl. ii. 193, 28. In helle smyche acoryen hit ful wraþe (*very grievously*), Misc. 75, 96. þunne ischrud and ifed wroþe thinly clad and badly fed, O. and N. 1529. Ich habbe more þan þi sostren boþe yloued þe one, and þou ȝeldest now my loue wroþe,

R. Glouc. 31, 10. Þou hest euele and wroþe yloked hire festes, Ayenb. 20, 23.]

wráþian; p. ode *To be angry :*—Ða tēnu ongunnun wráðiga (wuræðia, Lind. *indignari*), Mk. Skt. Rush. 10, 41. [He wrathed in his wyt, Allit. Pms. 94, 74. Þe king bigon to wradden (wredden, *v. l.*) *stomachatur tirannus,* Kath. 745. Wraþen *irasci,* Wick. Prov. 18, 14. *The verb is used also in the sense of* to anger, e. g. Þa sæ þe wind wraðede, Laym. 4577.] v. ge-wráþian; wræþan.

wráþ-líc; adj. *Cruel, dire, bitter.* v. wráþ, II :—Hát wæs him útan wráðlíc wíte, Cd. Th. 23, 7; Gen. 355. Hí sculon onfón in fýrbaðe wráþlíc andleán, Exon. Th. 52, 12; Cri. 832.

wráþlíce; adv. *Cruelly, direly, bitterly :*—Sió fǽhð geweard gewrecen wráðlíce, Beo. Th. 6116; B. 3062.

wráþ-mód; adj. *Angry-hearted, incensed :*—Unc is God wráðmód, Cd. Th. 50, 27; Gen. 815 : 34, 33; Gen. 547. [O. Sax. wréð-mód.]

wráþ-scræf, es; n. *An evil cave, a den :*—Wom wráðscrafu wráþra gésta *the foul dens of evil spirits,* Exon. Th. 424, 18; Rä. 41, 41.

wraþu, e; f. *A prop, stay, support :*—Wraþe *fulcimentum,* i. *adminiculum,* Wülck. Gl. 245, 27. (1) literal :—Se biscop hine onhylde tó ánre ðæra studa ðe útan tó ðære cyrican geseted wæs ðære cyrican tó wraþe (*pro munimine*), Bd. 3, 17; S. 543, 40. Ðam wáge tó wraþe *in munimentum parietis,* S. 544, 24. Tó wealles wraðe, Ps. Th. 117, 21. Wraðe *cimento,* Wrt. Voc. ii. 23, 51. Wraðum *columnis,* 21, 70. (2) figurative, *support, assistance :*—Wérigra wraþu (St. Guthlac), Exon. Th. 183, 34; Gú. 1337. Wísdómes wraðu, and witena frófur, Runic pm. Kmbl. 340, 7; Rún. 4. Gyfu gumena byð . . . wraðu and weordscype, 340, 25; Rún. 7. Ðǽr bið á gearu wraðu wannhálum wíta gehwylces, Elen. Kmbl. 2057; El. 1030. Wéne ic mê wraðe tó ðe *ego in te sperabo,* Ps. Th. 55, 3. Wyrta ðú geworhtest tó wraðe manna *producens herbam servituti hominum,* 103, 13. Se mê wraþe healdeþ, Exon. Th. 117, 7; Gú. 220. Ðǽr hí wraðe métaþ, 215, 3; Ph. 247. Ðæt hý wraþe sécen, frófre tó feóndum, 362, 12; Wal. 35. Ðǽr ðú wraðe findest, Elen. Kmbl. 168; El. 84. Him Freá ælmihtig fultum tióde, wíf áweahte, and ða wraðe sealde leófum rince, Cd. Th. 11, 13; Gen. 174. Mæg secgan se ðe wyle sóð sprecan, ðæt se mondryhten, . . . ðonne hê gesealde helm and byrnan healsittendum, . . . gúðgewǽdu, wraðe (*armour that ought to have proved of assistance to him; or* wraðe; *adv.?*) forwurpe, Beo. Th. 5736; B. 2872. v. líf-wraþu; wræþ-studu, wreþian.

wraxlere, es; m. *A wrestler :*—Wraxlerum *agonothetis,* Hpt. Gl. 489, 46. Cf. wræstlere.

wraxlian; p. ode *To wrestle :*—Ic wraxlige *luctor,* Ælfc. Gr. 19; Zup. 122, 12. Ða wraxlode (*luctabatur*) án engel wið hine, Gen. 32, 24. Ic wæs on unmǽtum costnungun winnende and wraxligende, Hml. Skt. ii. 23 b, 578. Wraxliendum *luctando,* Germ. 389, 91. Pleglícum ł wræxliendum *palaestricis,* Hpt. Gl. 405, 6. [To wraxli, Laym. 1858 (2nd MS.). Somme wraxlede, 24699. Thaugh Coūetyce wolde with the poure wraxle, Piers P. C. 17, 88. O. Frs. wraxlia.] Cf. wræstlian.

wraxlung, e; f. *Wrestling;* luctatio, Wrt. Voc. i. 39, 34. [Was mochel folk at þare wraxlinge, Laym. 1871 (2nd MS.).] Cf. wræstlung.

wrec[c]; adj. *Wretched, miserable :*—Ða wreccan munecas lágon onbúton ðam weófode, Chr. 1083; Erl. 217, 21. Ðá ða wreccæ men lǽgen fordrifene full neáh tó deaðe, 1086; Erl. 219, 36. [Uppon his wreccea folc, Chr. 1104; Erl. 239, 37. Crist ræde for þa wrecce muneces of Burch and for þ wrecce stede, 1131; Erl. 260, 15. Ich æm a wrecche mon, Laym. 3474. A wrecche, sunful mon, A. R. 56, 18. Godd wurrþenn wass wrecche child off wrecche kinn, Orm. 3878. Þu wrecche wiht, O. and N. 556.] v. wrecca.

wreca. v. ǽrend-wreca.

wrecan; p. wræc, pl. wrǽcon; pp. wrecen *To drive, press :*—Wraec *aegit,* Wrt. Voc. ii. 99, 43. Uurac *torquet,* 122, 50. Wrǽc *egit,* i. *ducit, compulit,* Wülck. Gl. 227, 6. Wrecende, drífende *agens,* Wrt. Voc. ii. 1, 3. I. *to drive, force to move :*—Hwílum mec mín freá sendeþ under sǽlwonge, and on bíd wriceþ, Exon. Th. 382, 29; Rä. 4, 3. Ýða wrǽcon árleásra feorh of flǽschoman, Cd. Th. 83, 25; Gen. 1385. Hwá mec on síð wrǽce, Exon. Th. 380, 39; Rä. 2, 2. Weard ecgum sweorda blondenfexa on bíd wrecen, Beo. Th. 5917; B. 2962. Mê þurh hrycg wrecen hongaþ under án orþoncpíl, Exon. Th. 403, 21; Rä. 22, 11. Sume wurdon wrecen of lande, Chr. 1076; Erl. 214, 38. I a. *to drive out, expel :*—Wreceþ heó wráðan, weorpeþ út áttor, Lchdm. iii. 32, 25. Ferh ellen wræc, Beo. Th. 5406; B. 2706. I b. *to drive out words, to express* in words, *utter, recite :*—Ic ðis giedd wrece . . ., mínre sylfre síð, Exon. Th. 441, 18; Kl. 1. Cleopaþ se alda, wriceþ wordcwedas, Cd. Th. 267, 8; Sat. 35. Hí sittaþ æt symble, sóðgied wrecaþ, Exon. Th. 314, 17; Mód. 15. Hê gyd wefer wræc, Beo. Th. 4315; B. 2154. Hwæt mid gieddungum guman oft wrǽcan, Exon. Th. 347, 14; Sch. 12. Ðonne hê gyd wrece, sárigne sang, Beo. Th. 4884; B. 2446. Monna gehwone ðe ðis gied wrǽce, Exon. Th. 285, 25;

Jul. 719. Sculan wê martira gemynd áreccan, wrecan wordum forð, wisse gesingan, Menol. Fox 139; Men. 70: Beo. Th. 1750; B. 873 : 6325; B. 3173 : Exon. Th. 306, 2; Seef. 1. Wæs gid oft wrecen, Beo. Th. 2135; B. 1065: Andr. Kmbl. 3094; An. 1550. I c. *to drive in, impress, inlay* [:—Goldwreken spere, Chart. Th. 556, 24. *Icel.* gull-rekinn *inlaid with gold*]. I d. *to drive, practise, carry out* or *on :*—Sóð líf ys on ðam ðæt man wrece his willan *vita in voluntate ejus,* Ps. Th. 29, 4. I e. *to drive* (intrans.), *press on :*—Stópon stíðhýdige, stundum wrǽcon, þrungon þræchearde, Elen. Kmbl. 242; El. 121. Ðú scealt wídlást wrecan, Cd. Th. 62, 28; Gen. 1021. Wrecan on wáðe wíde sended *sent driving* (?) *wide on the chase,* Exon. Th. 381, 14; Rä. 2, 11. II. *to wreak* anger, etc. :—Hí tredaþ ðec, and hyra torn wrecaþ, Exon. Th. 119, 24; Gú. 259. Ne wrec ðú ðíne yrre *ut non irascaris,* Ps. Th. 84, 5. Wrecaþ Godes yrre on ðam mannum, Ex. 32, 27. Ðæt wê magon wrecan Godes yrre on ðám mannum, 32, 26 : Cd. Th. 152, 34; Gen. 2530. Ðás folc sleán, and his torn wrecan, 151, 13; Gen. 2508. III. *to punish,* (a) absolute :—Ælc wyrd is nyt ðara ðe áuþer déþ, oðde lǽrþ, oþþe wricþ *fortuna, quae aut exercet, aut corrigit, prodest,* Bt. 40, 2; Fox 236, 16. (b) *to punish* a person (*dat.*) :—Parcas, ða hí secgaþ, ðæt on nánum men nyton náne áre, ac ǽlcum menn wrecan be his gewyrhtum, Bt. 35, 6; Fox 168, 26. Ðæt mon wrǽce and wítnode hwone for his yfle, 41, 3; Fox 248, 7. Mid ðý hí wrecan þenceaþ wráðum cynnum *ad faciendam vindictam in nationibus,* Ps. Th. 149, 7. (c) *to punish* a fault :—Se ðe ungemetlíce wricð ða scylda *immaniter feriendo quod delinquitur,* Past. 20; Swt. 149, 23. Suá wê hér hiera synna wrecaþ suíðor *quanto hic eorum culpas sine vindicta disciplina nostra non deserit,* 17; Swt. 119, 1. Ic wundrige . . . for hwý God yfel sóna ne wrece (*mala impunita praetereant*), Bt. 36, 1; Fox 172, 7. Gif him mon hýran nelle, þonne mót se mæssepreóst hit wrecan, swá hit hér beboden is, Blickl. Homl. 49, 3. (d) *to punish* a person (*dat.*) for a fault (*acc.* or *clause*) :—Ongæt gumena aldor hwæt him Waldend wræc wíteswingum, Cd. Th. 112, 1; Gen. 1864. Títus com mid herige and him wræc ðæt hié heora cyning on ðán áhéngon, Blickl. Homl. 79, 11. Ðæt sceal wrecan swefyl and sweart líg sáre and grimme hǽðnum rince, Cd. Th. 145, 32; Gen. 2414. IV. where hurt is inflicted on account of injury, *to avenge,* (a) absolute :—Se wrecenda brynæ *vindex ardor,* Dóm. L. 154. Be ðam wrecendan ǽr hê him rihtes bidde, L. In. 9; Th. i. 108, 3. Wrecendum lígum *flammis ultricibus,* Bd. 5, 14; S. 634, 27. (b) *to avenge* a person :—Ic wrece (wræco, Lind.) hig *vindicabo illam,* Lk. Skt. 18, 5. Nú hine man wricð *en sanguis ejus exquiritur,* Gen. 42, 22. Hê wræc ðone aldormon, Chr. 755; Erl. 48, 24. Sélre bið ǽghwæm ðæt hê his freónd wrece, ðonne hê fela murne, Beo. Th. 2774; B. 1385. Ða ðe þeóf wrecon, L. Ath. i. 6; Th. i. 202, 19. Ðæt beorna gehwylc Byrhtnóð wræce, Byrht. Th. 139, 21; By. 257. Se ðe þeóf wrecan wille, and nánne man ne gewundige, L. Ath. i. 6; Th. i. 202, 20: 20; Th. i. 210, 10: v. 1, 5; Th. i. 230, 10. Gif man ðone twelfhyndan man wrecan sceolde, hê bið fulwrecan on syx ceorlum, L. O. 13; Th. i. 182, 20. Hine nolden his eordlícan mágas wrecan, Chr. 979; Erl. 129, 12. Heora weras wrecan *ultionem caesorum conjugum consequi,* Ors. 1, 10; Swt. 44, 32: Beo. Th. 2683; B. 1339: 3097; B. 1546. (b 1) *to avenge* a person on some one :—Hê his sincgyfan on ðám sǽmannum wrec, Byrht. Th. 139, 64; By. 279. Wrec (wræc, Lind.) mê wið mínne wiðerwinnan *uindica me de aduersario meo,* Lk. Skt. 18, 3. Ðæt mê bearn wrǽce on bonan feore, Exon. Th. 401, 27; Rä. 21, 18. Se ðe wrecan þenceþ freán on folce, Byrht. Th. 139, 23; By. 258. (c) *to avenge* a wrong :—Se wer hit wreceþ, gif his wíf hié forhealdeþ, Blickl. Homl. 185, 26. Ic wræc Wedera níð, Beo. Th. 850; B. 423 : 3343; B. 1669. Heó ða fǽhðe wræc, ðe ðú Grendel cwealdest, 2671; B. 1333. Wrecaþ ealdne níð, Exon. Th. 280, 3; Jul. 623. Swerie hê ðæt hê wítnunge ne wrece, L. Eth. vii. 17; Th. i. 332, 22. Ða dǽde wrece swíðe deópe se ðe cyning sý on þeóde, L. E. G. 12; Th. i. 174, 10 : L. Eth. ix. 34; Th. i. 348, 10. Hê sceal Cristes ábilgðe wrecan swíþe georne, 2; Th. i. 340, 15 : L. C. S. 40; Th. i. 400, 10. Wê úrne teónan wrecan, and ðone þeóf lecgean, L. Ath. v. 8, 3; Th. i. 236, 18. Se ðe úre ealra teónan wrǽce, v. 7; Th. i. 234, 21. Nán man ðæt ne wrǽce, ne bóte ne bidde, L. Eth. ii. 6; Th. i. 288, 3. (c 1) *to avenge* a wrong on some one :—Ða wræc hê his æfþancas on his feóndum, Guthl. 2; Gdwin. 14, 4. Ðæt hig wrecon mihton heora teónrǽdenne mid tintergum on him *ut reddamus ei, quae in nos operatus est,* Jud. 15, 11. Ic sceal on ðære grimmestan godscyld wrecan, Exon. Th. 254, 29; Jul. 204. Nó ic wrecan meahte on wigan feore wonnsceaft míne, 499, 14; Rä. 88, 15. (d) *to take vengeance* (on) :—Ic wreocu in him *vindicabor in eis,* Ps. Surt. 117, 12. Ic wræc on him *ultus sum in eos,* Ps. Spl. 117, 11. Wreocende ic eam híe *ultus sum eos,* Ps. Surt. 117, 10, 11. ¶ of the action of the Deity :—Dryhten cwæð : 'Ic wrice on eów (*visitabo super vos*) æfter eówrum geðeahte.' Ðý ne wricð Dryhten nó gelíce ða gesiredan synne and ða fǽrlíce ðurhtogenan, Past. 56; Swt. 435, 12. Ic wrece fædera unrihtwísnysse on bearnum *ego sum Deus visitans iniquitatem in filios,* Ex. 20, 5 : Deut. 5, 9. Ic hit wrece on eów *ego ultor existam,* 18, 19. Ðú wrices *vin[di]cas,* Ps. Surt. 50, 6. Hê

wricð his þeówas *sanguinem servorum suorum ulciscetur*, Deut. 32, 43. Drihten wreceþ þearfendra *faciet Dominus vindictam pauperum*, Ps. Th. 139, 12. God hit suíðe hrædlíce wræc *vox illius irae vindictam aperte pertulit*, Past. 4; Swt. 39, 20. Hygeteónan wræc Metod on monnum, Cd. Th. 83, 16; Gen. 1380. Wrec ágen blód esna ðínra *vindica sanguinem servorum tuorum*, Ps. Th. 78, 11. Hú ne wræce hit God? *nonne Deus requiret ista?* 43, 22. Ðú miht wrecan æghwylcne mann *Deus ultionum*, 93, 1. Wrecende (*ulciscens*) on eallum widmétednyssa heora, Ps. Spl. 98, 9. Tó wreoganne hine hí gecýgdon *ad aemulationem eum provocaverunt*, 77, 64. Wreocende *vindicans*, Ps. Surt. 98, 8. Dryhten wreocende wes *Dominus zelatus est*, ii. p. 193, 27. [(1) Ðe bones ut of ðe erðe wroken, Gen. and Ex. 3191. Þou waʒ wroken fro uch a woþe, Allit. Pms. 12, 375. He his ssel wreke out of his uelaʒrede, Ayenb. 189, 33. Huerout he wrek þo þe zyalde and boʒte ine þe temple, 215, 7. Þe deuel fram hir for to wreke, Greg. 216. (2) Heo hine wreken wolden, wreken hine of his unwines, Laym. 1627. Heo wreken heore cun, 13749. Godd wollde himm wrekenn o þe preost, Orm. 914. For te wreken þe, A. R. 286, 13. On him for to ben wreken, Gen. and Ex. 2028. Leste þu wreoke mine sunnen on me, O. E. Homl. i. 209, 30. Þat micte hire bale wreken, Havel. 327. **Goth.** wrikan *to persecute*: **O. Sax.** wrecan *to punish, avenge*: **O. Frs.** wreka: **O. H. Ger.** rechau *ulcisci, vindicare, retribuere, punire*: **Icel.** reka *to drive; to take vengeance*.] v. á-, be-, for-, ge-, ofer-, on-, tó-, þeód-, þurh-wrecan; un-wrecen, scyld-wreccende.

wrecca, wræcca, an; *m.* **I.** *one driven from his own country, a wanderer in foreign lands, an exile, a stranger, pilgrim* :—Wræcca *exul*, Wrt. Voc. ii. 33, 27: Bd. 2, 14; S. 517, 38. Wæs hé wrævca on Gallia lande *cum exularet in Gallia*, 2, 15; S. 519, 1. Ðá wæs mid him án wræccea of Læcedamania *Demaratus Lacedaemonius apud Xerxem exsulabat*, Ors. 2, 5; Swt. 78, 33. Com se foresprecena wrævca . . . , hine se kyning hider and þider wíde áflýmde, Guthl. 19.; Gdwin. 76; 12. Wundorlíc wrævca (*Nebuchadnezzar*), Cd. Th. 256, 1; Dan. 634. Ic mé féran gewát folgað sécan, wineleás wrævca, Exon. Th. 442, 9; Kl. 10: 457, 27; Hy. 4, 90. Aldbryht wræccea (wrecca, *v. l.*) gewát on Súþ-Seaxe, Chr. 722; Erl. 44, 28. Ðá hé wrecca wæs *dum exularet*, Bd. 3, 18; S. 545, 39. Wrecca (wreccea, *v. l.*), Bt. 5, 3; Fox 12, 33. Fundode wrecca, gist of geardum, Beo. Th. 2279; B. 1138. Wineleás wrecca (*Cain*), Cd. Th. 64, 16; Gen. 1051. Ðú ðás werðeóðe wræccan láste feorran gesóhtest, 149, 22; Gen. 2478: 171, 3; Gen. 2822: Exon. Th. 306, 30; Seef. 15. Wreccan, 420, 24; Rä. 40, 8. Hé ða scíre gesealde ánum wræccean of Ahténa (*Atheniensem virum, qui apud Cyprum exsulabat*), Ors. 3, 1; Swt. 96, 24. Wræccan *extorrem*, Wrt. Voc. ii. 32, 64. Wreccan *advenam*, Ps. Spl. 93, 6: Blickl. Gl. Wræccan *extorres*, Wrt. Voc. ii. 107, 83: *expulsi*, 30, 9. Wreccean *extranei*, 146, 6. Gifu byð wræcna gehwám ár and ætwist, Runic pm. Kmbl. 340, 26; Rún. 7. Wreccena mærost, Beo. Th. 1800; B. 898. Wreccena feormunge, L. Alf. pol. 4; Th. i. 62, 16. Hé bebeád ðæt mon ealle ða wræccan an cýþþe forléte *jussit omnes exsules patriae restitui*, Ors. 3, 11; Swt. 144, 14. Drihten gehealdeþ wreccan (*advenas*), Ps. Spl. 145, 8: Wulfst. 295, 1. **I a.** applied to a hermit :—Mantat ancer, Godes wrævca, Cod. Dip. Kmbl. vi. 192, 3. **I b.** figurative :—Wræccan (*those in Hades, exiles from Heaven*), Exon. Th. 461, 28; Hö. 42. Hé héht ðæt wítehús wrævna (*the angels driven from Heaven*) bídan, Cd. Th. 3, 22; Gen. 39. Ðæt ðú helpe gefremme wérgum wreccan, . . . and ðíu hondgeweorc móte cuman tó ðam upcundan ríce, 17, 2; Cri. 264. Ðonne gesihst ðú ða unrihtwísan cyningas bión swíþe earme wreccan *cernes tyrannos exsules*, Bt. 36, 2, Fox 174, 28. **II.** *a wretch, an evil person* :—Se feónd, wrævca wærleás, Exon. Th. 263, 17; Jul. 351. Mé ceigendæ ðæt ic sié Godes wracco *me clamans esse sacrilegum*, Mt. Kmbl. p. 1, 9. **III.** *a wretched person, a miserable, feeble creature* :—God selfa tyhte Móyses on ðone folgoð, swáðeáh hé him ondréd; and nú fandiaþ swelce wræccan and teóð tó, woldon underfón ðone weordscipe and eác ða byrðenne *Moyses suadente Domino trepidat, et infirmus quisque, ut honoris onus percipiat, anhelat*, Past. 7; Swt. 51, 22. **IV.** *a wretched, unhappy, miserable, poor person* :—Dohtor se Babilónisca wrævca *filia Babilonis misera*, Ps. Lamb. 136, 8. Ða lióð ðe ic wrecca geó lustbærlíce song, ic sceal nú heofiende singan, Bt. 2; Fox 4, 6: Met. 2, 3. Ne mæg mon ænne wræccan his cræftes beniman, 10, 38. Heó áhredde ða húþe, and tó hám bedræf wreccan (*the hapless wight?*) ofer willan, Exon. Th. 412, 6; Rä. 30, 10. Wræccena reáflác is on heora hámum (*rapina·pauperis in domo vestra*, Is. 3, 14), Wulfst. 45, 18. [Heo scullen wræcchen (*expelled?*) to heoren scipen liðen, sæilien ouer sæ, Laym. 20887. Wrecce *thou wretch*, Kath. 2049. Ðat folc unseli, sinne wod, ðo sori wrecches of yuel blod, Gen. and Ex. 1074. Drihten alesde þene wrechan *liberauit pauperem*, O. E. Homl. i. 129, 14. Wiþþ usell wrecche dælenn, Orm. 10140. Þer wes moni wrehche iworden riche, Laym. 5932. Þes wrecche ayhte nabbeþ, Misc. 75, 103. Ich nam non aswunde wrecche, O. and N. 534. **O. Sax.** wrekkio (*used of the three kings from the East*): **O. H. Ger.** reccho *exul, extorris, profugus, incola*.]

wreccan; *p.* wreahte, wrehte; *pp.* wreaht, wreht. **I.** *to raise,*

---

lift up :—Wreceþ tó ræde Drihten ðara manna bearn ðe ær man gebræc *Dominus erigit elisos*, Ps. Th. 145, 7. **II.** *to take up, undertake* :—Ðæm hé hæfde beboden ðæt hé scolde þearfena and earmra monna ærendo wreccan *cui suscipiendorum inopum erat cura delegata*, Bd. 3, 6; M. 166, 4. **III.** *to rouse* :—Ðú ðe ært fæder ðæs suna ðe ús áwehte, and gyt wrehð of ðam slæpe úre synna, Shrn. 166, 9. Ðec regna scúr weceþ and wreceþ, swá wildu deór, Cd. Th. 252, 11; Dan. 577. Wé feóllan on slæpe, ac hé læg þurhwacol, and wræhte ús sicðan, Homl. Skt. i. 11, 241. Hié wrehton cumbolwigan, Judth. Thw. 25, 5; Jud. 243: 24, 37; Jud. 228. Ne sceal hé nó ðæt án dón ðæt hé ána wacie, ac hé sceal eác his friénd wreccan *non solum ut ipse vigilet, sed etiam ut amicum suscitet, admonetur*, Past. 28; Swt. 193, 21. Héht hé mid ærdæge wígend wreccan, Elen. Kmbl. 211; El. 106. [He of his eyre briddes wrahte (wraʒte, wrauhte, *v. ll.*), O. and N. 106.] v. á-, ærend- (Bd. 2, 9; S. 511, 20) wreccan; wrehtend.

wreccan (?); *p.* wreahte; *pp.* wreaht, wræht, wreht *To twist*, (1) *to strain* [:—Gif hors bið gewræht, Lchdm. iii. 62, 12.] (2) fig. *to torment* :—Ic hálsigo ðec ne mec ne wrecce (*for* wrece?; wuræcce, Lind.) *adiuro te ne me torqueas*, Mk. Skt. Rush. 5, 7.

wrecel (?). v. spor-wrecel.

wrecend, es; *m. An avenger* :—Hwæt hwá óðrum tó wó gedó, God his bið wrecend, L. E. I. 35; Th. ii. 432, 27: Chr. 979; Erl. 129, 17. Ðæt gesýne wearð, ðætte wrecend ðágyt lifde æfter láþum, Beo. Th. 2517; B. 1256. Hí habbaþ eác wrecend (*ultorem*), Scint. 39, 13.

wrecness, e; *f. Vengeance* :—Dagas wrecnisse (wræcnisse, Lind.) *dies ultionis*, Lk. Skt. Rush. 21, 22.

-wrecness. v. god-wrecness.

wrec-scipe, es; *m. Exile, living in a foreign land* :—Mín wrecscype *incolatus meus*, Blickl. Gl.: Ps. Spl. T. 119, 5.

wrégan (wrécan); *p.* wrégde, wréhte; *pp.* wréged, wréht *To bewray, accuse, denounce*, (1) absolute :—Ne ðú ne wréi *nec accuses*, Kent. Gl. 1083. Wíte ł wréce *imputet*, Germ. 400, 560. Wroegde *defert, meldadun vel* wroegdun (roactum, Erf.) *defferuntur*, Txts. 57, 663, 652. Wrégde, wrégdan, Wrt. Voc. ii. 25, 35, 26. Wrégian *insimulare*, 81, 7. Ðæs wrégendan *mussantes*, 58, 34. (2) *to accuse* a person :—Ðysne man ic wrége *hunc hominem accuso*, Ælfc. Gr. 7; Zup. 22, 22. Ic ðé wrége beforan Crystes þrymsetl, Shrn. 154, 9. Híe yfel gewitnes ne wrégde, Blickl. Homl. 163, 1. Man wrægde ðone biscop, Chr. 1069; Erl. 207, 6. Mid ðý ðe hyne wrégdon ðæra sacerda ealdras *cum accusaretur a principibus sacerdotum*, Mt. Kmbl. 27, 12: Jn. Skt. 8, 10. Ðæt hí wréhton hyne *ut accusarent eum*, Mt. Kmbl. 12, 10: Homl. Th. i. 570, 21. Ðé wrég te accusa, Scint. 165, 1. Gif ænig mann óðerne wrége and him hwilcne gilt on secge *si steterit testis mendax contra hominem accusans eum praevaricationis*, Deut. 19, 16. Hit is betre ðæt mon wrége ðone scyldigan, Bt. 38, 7; Fox 210, 5. Hí águnnon hyne wrégan (*accusare*), Lk. Skt. 23, 2. Heó begann hí tó wrægenne, Homl. Skt. i. 2, 184. Mid micelre wróhte hine wrégende *bringing a heavy accusation against him*, Homl. Th. ii. 250, 10. (2 a) *to accuse* a person *to* (*tó, wiþ*) somebody :—Ne wéne gé ðæt ic eów wrége tó fæder (*apud patrem*). Se is ðe eów wrégð Móyses, Jn. Skt. 5, 45. Hié mon wrégde tó ðæm cásere, Blickl. Homl. 173, 10. Wréhte, Homl. Skt. ii. 25, 597. Wrégdon ða óðre cræftigan hý tó ðam cásere, Shrn. 146, 20. Hí wréhton ðone cyning tó his bréðer, Homl. Th. i. 468, 6. Ðe læs hé wrége ðé tó Drihtene *ne clamet contra te ad Dominum*, Deut. 15, 9. Ðeáh ðín wíf ðé hwane tó wrége, ne gelýf ðú ná tó hraðe, Prov. Kmbl. 4. Ongan hé hí wrégean tó ðam cyninge, Lchdm. iii. 424, 21. Ðíne æhta mid stylre stemne wyllaþ ðé wrégan tó ðínum Drihtne, Homl. Th. ii. 410, 21. Hé began ðæt cynn tó wrégenne wið ðone cyning, Homl. Ass. 96, 148. (2 b) *to accuse* a person *of* something :—Of ðam ðe gé hine wrégaþ *ex his in quibus eum accusatis*, Lk. Skt. 23, 14. Lóca hú mycelum hí ðé wrégeaþ *uide in quantis te accusant*, Mk. Skt. 15, 4. Ða wrégdon hine ða heáhsacerdas on manegum þingum, 15, 3. (3) *to denounce* something to a person :—Ða onféng ðære þeóde kyning fulwihte; ðá fóron ða hæþnan bisceopas and ðæt wrégdon tó ðæs kyninges bréþer, Shrn. 120, 34. Ðæt folc wréhton his módignysse tó ðam cásere, Homl. Th. i. 478, 17. [Mon schal wreien him suluen ine schrifte, A. R. 304, 1. Gif þu wreiest þe seoluen to þine scrifte, O. E. Homl. i. 27, 36. He ne wollde unnshaþiʒ wimmann wreʒhenn, Orm. 2889. Þaʒr syns sal wreghe þam, Pr. C. 5462. Naʒt him to defendi, ne nenne oþrenne wraye, Ayenb. 175, 5. Fund mann nan þing uppon hemm to wreʒenn, Orm. 416. **Goth.** wrôhjan *to accuse*: **O. Sax.** wrôgian: **O. Frs.** wrógia, wreia: **O. H. Ger.** ruogen: **Icel.** rægja.] v. for-, ge-wrégan.

wrégan; *p.* de *To rouse, excite* :—Hwílum ic streámas styrge, hwilum wolcnfare wrége, Exon. Th. 386, 33; Rä. 4, 71. Hwílum ic (*a storm*) sceal ýþa wrégan, [streámas] styrgan, 383, 28; Rä. 4, 17. v. ge-wrégan.

wrégend, es; *m. An accuser, a denouncer* :—Wrégend *accussator*, Scint. 39, 14: Wrt. Voc. ii. 8, 71: 72, 57. His wrégend and gesacan *accusatores ejus*, Bd. 5, 19; S. 640, 13. Hyra wrégendras, Scint. 29, 4. Wrégendum *delatoribus*, Wrt. Voc. ii. 28, 13.

wrégend-líc; adj. Accusative :—Accussativus ys wrégendlíc, Ælfc. Gr. 7; Zup. 22, 20.

wrégere, es; m. An accuser :—Wrégere accusator, Wrt. Voc. i. 83, 63 : Homl. Th. ii. 236, 22 : 340, 22. [Wreiere ne beo þu, O. E. Homl. i. 57, 49. Wreieres and wrobberes, Havel. 39.]

wrégestre, an; f. A female accuser :—Seó leáse wrǽgistre, Homl. Skt. i. 2, 208.

wréging, e; f. Accusation :—Wréginc accusatio, Wrt. Voc. i. 83, 65. [Wreiunge, A. R. 200, 22. Wreynge ant gret blame that byth, Rel. Ant. i. 267, 3. O. Frs. wróginge.]

wrehtend, es; m. One who excites :—Wrehtend, tyhtend incentor, Wrt. Voc. ii. 44, 62. Cf. weccan.

wrehtend, es; m. An accuser :—Wrēhten his selfes accusator sui, Kent. Gl. 650. Cf. wrégan.

wrenc, es; m.   I. a trick, artifice, wile, stratagem :—On swā hwylcum wrence (arte) worda ǽnig swerige, Scint. 136, 18. Wrencum modis, Wrt. Voc. ii. 57, 61. Đæt leáse lot, đe beoþ mid đām wrencum bewrigen fraus, mendaci compta colore, Bt. 4; Fox 8, 17 : Met. 4, 47. Tó fela manna weard mid þyllícan wrencan þurh deófol forlǽred, Wulfst. 54, 12. Tógeánes đæs deófles wrencum, 198, 12. Đā sceolde Ælfríc lǽdan đa fyrde, ac hē teáh ford đā his ealdan wrenceas, . . . gebrǽd hē hine seócne, Chr. 1003; Erl. 139, 7.   I a. a stratagem in war :—Hē hié mid dǽm ilcan wrence beswác, đe hē æt heora ǽrran mētingue dyde, Ors. 4, 9; Swt. 188, 32 : 6, 36; Swt. 294, 21. Siþþan Rōmāne gesāwan đæt him mon swelcne wrenc tó dyde, đā flugon hié, 4, 1; Swt. 156, 8.   II. a modulation of the voice :—Biþ đæs hleóđres swēg eallum songcræftum swētra, and wynsumra wrenca gehwylcum, Exon. Th. 2c6, 28; Ph. 133. Ic þurh mūþ sprece mongum reordum, wrencum singe (vox mea diversis variatur pulcra figuris), . . . ic bûgendre stefne styrme, 390, 15; Rä. 9, 2. [Þurh his micele wrences beiæt he þone ærcebiscoprice, Chr. 1127; Erl. 156, I. Gif þær wære an unwreste wrenc þ he mihte get beswicen anes Crist, 1131; Erl. 260, 4. Paris mid pret wrence biwon Elene, Laym. 81. Þis sacrament unwrihð his (the devil's) wrenches, A. R. 270, 10. Swikele men and ful of vuele wrenche, Misc. 66, 247. With wrenkes and wyles, Pr. C. 1360. He (a wrestler) can his wrenches wel forhele, O. and N, 798. Þis is þe soþe wei, withouten eny wrenche, R. Glouc. 55, 2. His wyly wrenches thou ne mayst nat flee, Chauc. Ch. Y. T. 1081. Wrenche or sleythe of falsheed dolositas, fraudulencia, cautela, Prompt. Parv. 533, and see note. Ger. rank a trick.] v. lot-, nearu-, searu-, siru-, smeá-, un-, weorold-wrenc.

wrencan; p. te.   I. to turn, twist (intrans.) :—Is đæs horderes tācen, đæt mon wrǽnce mid his hande, swilce hē wille loc unlūcan, Techm. ii. 118, 12.   II. to practise wiles, use tricks. v. wrenc :—Biþ ōþer swice, . . . wrenceþ hē and blenceþ, worn geþenceþ hinderhóca, Exon. Th. 315, 18; Mōd. 33. [Þu ne mihtes nohwider wrenche fra þa duntes, O. E. Homl. i. 281, 30. Ich chulle wrenchen hire þideweard ase heo mest dredeð, A. R. 222, 16. Þu ne maht wenden me ne wrenchen ut of þe weie, Marh. 4, 27. Some gase wrynchand to and fra, Pr. C. 1538. Germ. renken.] v. be-wrencan.

wrenc-wís; adj. Unjust, unrighteous :—Wer wrencwis vir iniquus, Rtl. 10, 30. Cf. riht-wís.

wrenna, wrænna, werna, wærna, an; m. : wrenne, an; f. A wren :—Wrenna vel hicemáse parrax, Wrt. Voc. i. 29, 56. Wrenne (wrænna, v.l.), 77, 46. Wrenna bitorius vel pintorus, 29, 27 : bitorius, bitriscus, ii. 126, 37. Wrænna biturius, 12, 62 : bitorius, i. 62, 41. Werna birbicariolus, ii. 101, 76. Wærna bitorius, 11, 12 : litorius, 51, 59 : i. 281, 12. [Hwat dostu godes among manne na mo þene deþ a wrecche wranne (wrenne, v.l.), O. and N. 564. Wrenne regulus, Wrt. Voc. i. 221, 7.]

Wreocen-sǽte, -sǽtan (Wrocen-); pl. The occupants of the district about the Wrekin :—Gesta est hujus libertatis donatum anno incarnationis DCCC.Lvᵒ, in loco qui uocatur Oswaldes dūn, quando fuerunt pagani in Uureocensētun, Cod. Dip. Kmbl. ii. 59, 35. In prouincia Wrocensētna, vi. 60, 2. Cf. Wocen-sǽte.

wreogan, Ps. Spl. 77, 64. v. wrecan, IV ¶.

wreón (from wríhan); p. wráh, wreáh, pl. wrigon, wrugon; pp. wrigen, wrogen To cover.   I. to put a covering on something, (1) literal :—Se đe wrígð wæterum đa uferan his qui tegit aquis superiora ejus, Ps. Spl. 103, 3. Ōþer eáre hī him underbrēdaþ and mid ōðran hī wreóð (se cooperiunt), Nar. 37, 12. Hē wreáh and þeahte mānsǽhðu bearn wonnan wǽge, Cd. Th. 83, 10; Gen. 1377. Reste hē hine sófte, and wreó hine wearme : . . . læt drincan . . . , and wreóh hine wearme, Lchdm. ii. 292, 6–14. Swā đū woruldedeáde wrige mid foldan as you would cover the dead with earth, Ps. Th. 140, 4. Ongunnun summe gehýdæ ꝉ wríga (uelare) onsíone his, Mk. Skt. Lind. 14, 65. (1 a) to cover with clothes, to clothe :—Ic wreó mē wǽda leásne, leáfum þecce, Cd. Th. 53, 26; Gen. 867. Ic wæs nacod, and gē clǽddon ꝉ wrigon (wriogan, Rush.) meh eram nudus, et operuistis me, Mt. Kmbl. Lind. 25, 36. Hē ne mihte hine handum self mid hrægle wryón, Cd. Th. 95, 2; Gen. 1572. (1 b) to cover a book, to bind a book :—Mec (a book) wráh hæled hleóbordum, Exon. Th. 408, 13; Rä. 27, 11. (2) with the idea

of concealment, to conceal, hide :—For hwon wāst đū weán and wríhst sceome, Cd. Th. 54, 13; Gen. 876. Đa word đe gē wrigon under womma scealum, Elen. Kmbl. 1162; El. 582. (3) with the idea of protection :—Ic đē wið weána gehwam wreó and scylde folmum mínum, Cd. Th. 131, 3; Gen. 2170.   II. to serve as a covering to something, be spread over, (1) literal :—Flód ealle wreáh heá beorgas, Cd. Th. 83, 28; Gen. 1386. Niht lagustreámas wreáh, 147, 34; Gen. 2449. Mec (an oyster) ȳþa wrugon, Exon. Th. 488, 5; Rä. 76, 2. Sió filmen biþ þeccende and wreónde đa wambe, Lchdm. ii. 240, 17. (1 a) of clothing :—Woede tó wriánne vestem ad operiendam, Rtl. 103, 42. Of hwon wē bidon wrigen (gewrigene, Rush.) quo operiemur, Mt. Kmbl. Lind. 6, 31. (2) with the idea of concealment :—Ȳþa mec (a storm) wrugon, Exon. Th. 382, 23; Rä. 3, 15. (3) with the idea of protection :—Ic læbbe mē on hrycge đæt ǽr hādas wreáh foldbûendra, Exon. Th. 381, 17; Rä. 2, 11. Unc holt wrugon, wudubeáma helm, wonnum nihtum, scildon wið scúrum, 496, 1; Rä. 85, 73. [Þu mihtes wrihe þine banes, O. E. Homl. i. 279, 2. Wummon schal wrihen hire heauet. Wrihen, he seið, naut wimplin. Wrihen ha schal hire scheome . . . Gef ei þing wrihed þi neb, A. R. 420, note. To wrien and te helien, . . . he heled hit and wrihð, 84, 14–17. Þe uikelares wreoð and helieð, 88, 19. Þis scheld þet wreih his Godhed, 390, 26. Ane cheste wreon mid golde, Laym. 27859. Þa Irisce wriȝen al þa feldes, 17349. Wreoð wel þene king, 17762. Wrugen, p. pl., P. L. S. viii. 81. Uor to wry his confusioun, Ayenb. 258, 18. Þe sseld him wriȝþ, 167, 10. Hi wreþ þe uelþes of zenne, 61, 4. O. H. Ger. int-ríhan revelare.] v. ā-, be-, ge-, in-, ofer-, on-, un-wreón.

wreótaþ, wreoþen-hilt. v. reótan, wriþen-hilt.

wreþian; p. ede; pp. ed To prop, stay, support, sustain :—Wreþeþ fulcit, Wrt. Voc. ii. 38, 28. Wreðeð (-ed ?) nisa, 61, 59. Heora ǽlc wind wiþ ōþer, and þeáh wræþeþ ōþer, đæt hié ne mōton tóslûpan, Bt. 21; Fox 74, 11. Hiora ǽghwilc wið ōþer wind, and þeáh winnende wreþiaþ fæste ǽghwilc ōþer, Met. 11, 34. Se đe rodor āhóf and gefæstnode folmum sínum, worhte and wreðede, Andr. Kmbl. 1045; An. 523. Cypressus styde hié ûtan wreþedon nitebant[ur] testudinibus cupressinis, Nar. 5, 9. Wreþian fulcire, Wrt. Voc. ii. 34, 68. Gif đær sié ȝyred mid tó đreágeanne, sié đǽr eác stæf mid tó wreðianne si est districtio virgae, quae feriat, sit et consolatio baculi, quae sustentet, Past. 17; Swt. 127, 2. His đa untruman limo mid his cricce wreðiende imbecilles artus baculo sustentans, Bd. 4, 31; S. 610, 28. Biþ seó mōdor wundrum wreþed, Exon. Th. 492, 15; Rä. 81, 16. [Euerichon wreoðeð him bi oðer, A. R. 252, 13. Alle þeos writeres writes þ ȝe wreoðieð ow on, Kath. 857. O. Sax. wreðian to prop, stay, support.] v. ā-, ge-, under-wreþian; wraþu.

wreþung. v. under-wreþung.

wríd (cf. ' A ride of hazle or such like wood, is a whole plump of spriggs or frith growing out of the same root,' E. D. S. Pub. Old Farming Words, no. III. Here is an heelful thing, a wonder wride (rimes with abyde), Pall. 51, 207), es; m. A shoot, stalk, plant, bush :—Uurýd culmus, Txts. 52, 252. Genim æscþrote ǽnne wríd, Lchdm. i. 216, 11. Genim đysse wyrte wríd, 224, 1. Bedelf ǽnne wríd cileþenigan moran, iii. 38, 9. v. hæsel-wríd; ge-wríd, and next word.

wrídan, wríþan; p. de To put forth shoots, be productive :—Weaxaþ and wrídaþ, . . . fyllaþ eówre fromcynne foldan sceátas, teámum and tûdre, Cd. Th. 92, 21; Gen. 1532. Wríðende sceal mǽgđe đinre monrím wesan, 105, 33; Gen. 1762. Geunne đē éce Drihten æcera wexendra and wrídendra, Lchdm. i. 402, 4. v. preceding and following words.

wrídian, wríþian; p. ode To put forth shoots, be productive, grow, flourish :—Þúfaþ and wrídaþ frutescit, Wrt. Voc. ii. 38, 13. (1) literal, of vegetable growth :—Se æþela feld wrídaþ, wynnum geblówen, Exon. Th. 199, 17; Ph. 27. On đære eá ófre stód hreód and þintreów and abies đæt treówcyn ungemetlícre grýto and micelnysse đý clife weóx and wrídode (wríðode, Cockayne; but see Anglia i. 509) cujus ripas pedum sexagenum harundo uestiebat pinorum abietumque robora uincens grossitudine, Nar. 8, 22. (2) figurative, of growth in things abstract or concrete :—Hē wrídaþ on wynnum, đæt hē bið wæstmum gelíc ealdum earne, and æfter đon feþrum gefrætwad, swylc hē æt frymđe wæs, beorht geblówen reformatur qualis fuit ante figura, et Phoenix ruptis pullulat exuviis, Exon. Th. 214, 10; Ph. 237. Mín hyge blissaþ, wynnum wrídaþ my mind rejoices, blossoms with joyous thoughts, Andr. Kmbl. 1269; An. 635. Him oninnan oferhygda dǽl weaxeþ and wrídaþ, Beo. Th. 3486; B. 1741. Mān wrídode geond beorna breóst, Andr. Kmbl. 1534; An. 768. Weóx đa and wrídode mægburg Semes, Cd. Th. 102, 19; Gen. 1702. Ne sceal unc betweónan bearno weaxan, wrôht wrídian, 114, 12; Gen. 1963. v. ā-, ge-wrídian, and two preceding words.

wríga to cover. v. wreón.

wrigedness, wrigenness. v. un-wrigedness, un-wrigenness.

wrigels, es; m. n.   I. a covering :—In wrigelse fiðra đínra in velamento alarum tuarum, Ps. Surt. 60, 5 : 62, 8. Wæs him wrigils fuit illis in velamento, Rtl. 92, 26. God āfyrde hym đone unrihtan wrigels of heora heortan, Wulfst. 252, 4.   II. a garment, veil :—Hí mon mid

ðæm hálgan wrigelse bewríhþ, Blickl. Homl. 61, 16. Hálgum wriilcse *sacro velamine*, Rtl. 106, 4. Be ðý wyrgelse ofer Cristes nesðyrlum, Anglia xi. 173, 9. Ðú hí onwendest swá man wrigels (*opertorium*) dêð, Ps. Th. Surt. 101, 23. Hé his wrigels geopenode, Homl. Ass. 196, 56. [Adam & Eue makeden wrieles of leaues, A. R. 322, 19. Wriheles, 420, note. Wriels *velamen*, Wick. Job 24, 8.] v. ofer-, unriht-wrigels.

wrigian; *p.* ode *To turn, wend, hie, go, move* :—Þeáh ðú teó hwelcne bôh of dúne tô ðære eorþan, swelce ðú bégan mæge, swá ðú hine álætst, swá sprincþ hê up, and wrigaþ (cf. went on gecynde, Met. 13, 55) wiþ his gecyndes . . . Swá dêþ ælc gesceaft, wrigaþ wiþ his gecyndes, and gefagen biþ gif hit æfre tô cuman mæge *validis quondam viribus acta pronum flectit virga cacumen; hanc si curvans dextra remisit, recto spectat vertice coelum . . . Repetunt proprios quaeque recursus, reditaque suo singula gaudent*, Bt. 25; Fox 88, 22–29: xiv, 14. Ælc gesceaft wrigaþ and higaþ wið his gecyndes, Met. 13, 65. Wuhta gehwilc wrigaþ tôheald wið ðæs gecyndes . . . þinga gehwilc þiderweard fundaþ, 13, 10. Weard æt steorte (*the ploughtail*) wrigaþ on wonge *the ploughman pushes his way over the field*, Exon. Th. 403, 9; Rä. 22, 5. [That feyre founden me mete ant cloht, hue wrieth awey as hue were wroht, Spec. 48, 22. With hir heed sche wriede fast awey, Chauc. Mill. T. 97. Hwenne so wil to wene wriedh, R. S. 3, 7. Þy face from hyre þou wry, Mirc. 888.]

wrinclod. v. ge-wrinclod.

wringan; *p.* wrang, *pl.* wrungon; *pp.* wrungen *To wring*, (1) *to twist* :—Teóh him ða loccas, and wringe ða eáran, and ðone wangbeard twiccige, Lchdm. ii. 196, 13. (2) *to squeeze out* moisture from something :—Ic nam ða wínberian and wrang on ðæt fæt *tuli uvas et expressi in calicem*, Gen. 40, 11. Genim ðære ylcan wyrte leáf, ðonne heó grénost beó, wyl on wætere, and wring ðæt wôs, Lchdm. i. 72, 7. Wring ðæt seáw, ii. 110, 26: 240, 8. Ne miht ðú wín wringan on midne winter, Bt. 5, 2; Fox 10, 31. Tô wringen[n]e *ad exprimendos*, Hpt. Gl. 468, 32. [O. H. Ger. ringan *rixari, luctari*.] v. á-, ge-wringan.

wring-hwæg *the whey pressed out of cheese* :—Ðæt heó of wringhwæge buteran macige tô hláfordes beóde, L. R. S. 16; Th. i. 438, 31.

wrislan. v. wrixlan.

wrist, e; *f.* *A wrist* :—Gif hit ânfeald tyhtle sý, dúfe seó hand æfter ðam stáne óð ða wriste; and gif hit þryfeald sý, óð ðæne elbogan, L. Ath. iv. 7; Th. i. 226, 17. [Wryst or wyrste of an hande *fragus*, Prompt Parv. 534. The wryste or a knokyl *fragus*, Wülck. Gl. 584, 27. A wyrste, 678, 40. O. Frs. hand-wirst: M. H. Ger. rist, riste *wrist, instep: Icel.* rist *instep: Dan.* vrist.] v. cneów-wyrste, hand-wyrst (-wrist).

writ, es; *n.* I. *a writing* :—Ðæs ne ûs leorneras, wordum secgaþ, and writu cýþaþ, Exon. Th. 227, 19; Ph. 425. II. *writ* (as in holy *writ*), *scripture* :—Cwoeð ðió writ *dicit scriptura*, Rtl. 79, 11. Ne writ ðiús (*scripturam hanc*) leornada gié, Mk. Skt. Lind. 12, 10. Æfter ðon ðe hálige writu sprecaþ, Bd. 2, 20; S. 522, 28. Wrioto wîtgana *scripturae prophetarum*, Mt. Kmbl. Lind. 26, 56. Wriotto, 26, 54. Writto, Mk. Skt. Lind. 14, 49. Writta *scripturarum*, Mt. Kmbl. p. 1, 1. Wriottana, Jn. Skt. p. 2, 4. Ða wrioto *scripturas*, Lk. Skt. Lind. 24, 45. [Þe king nom þat writ on hond, Laym. 484. Ase holi writ seið, A. R. 98, 7: Misc. 36, 3. Þatt broþerr þatt tiss Ennglissh writt wrat, Orm. dedic. 331. Þis writ shal henge bi him, Havel. 2486: Gen. and Ex. 1974. *Icel.* rit.] v. ge-, ofer-writ; hreód-writ.

wrítan; *p.* wrát, *pl.* writon; *pp.* writen *To write.* I. *to cut* a figure on something :—Wrít ðysne circul mid ðínes cnífes orde on ânum stáne, Lchdm. i. 395, 3. I a. where the figures are letters :—Genim hæslenne sticcan, wrít ðinne naman on, . . . gefylle mid ðý blôde ðone naman, Lchdm. ii. 104, 7. Ræd sceal mon secgan, rúne wrítan, Exon. Th. 342, 7; Gn. Ex. 139. Hróðgár hylt sceawode, on ðæm wæs ôr writen (*or* I?) fyrngewinnes, Beo. Th. 3381; B. 1688. ¶ of the writing on the tables of stone :—On ðe wrát wuldres God gerýno, Andr. Kmbl. 3018; An. 1512. II. *to draw* a figure. v. wrítere, I :—Nim sume tigelan, and wrít on hiere ða burg Hierusalem *sume tibi laterem, et describes in eo civitatem Jerusalem*, Past. 21; Swt. 161, 3, 10. Wrít ðam horse on ðam heáfde foran Cristes mæl, and on leoþa gehwylcum ðe ðú ætfeolan mæge, Lchdm. ii. 290, 23. Wrít him Cristes mæl on ælcum lime, 346, 6. Wrít .iii. crucem mid oleum infirmorum, . . . nim ðæt gewrit, wrít crucem mid ofer ðam drence, 350, 9-11. III. *to form* letters, *to write* :—Mycel yfel dêð se ðe leás wrít, Homl. Th. ii. 2, 23. Hê wrát mid his fingre on ðære eorþan, Jn. Skt. 8, 6, 8. Engel wrát in wáge baswe bôcstafas, Cd. Th. 261, 8; Dan. 723. Geseah hê engles hand in sele wrítan Sennara wîte. Ðæt gyddedon hæleð, hwæt seó hand write, Cd. Th. 261, 15–21; Dan. 727–9. Weard gesewen swilce ânes mannes hand wrítende on ðære healle wáge, Homl. Th. ii. 434, 33. IV. *to write* a book, narrative, etc., *to compose, be the author of* :—Ne wêne ænig ðæt ic lygewordum leóð somnige, wríte wôþcræfte, Exon. Th. 234, 30; Ph. 548. Wríteþ *digerit*, Wrt. Voc. ii. 27, 50. Wrítat *caraxabimus* (*mentionem*), 85, 33. Wrítaþ, 18, 61: Hpt. 507, 76. Ðæt cyriclíce stær ûres eálondes and ðeóde ic wrát on fíf béc, Bd. 5, 24; S. 648, 31. Se ðe ðás bóc wrát, Lchdm. ii. 114, 5. Se ongan godspell ærest wordum

wrítan, Andr. Kmbl. 25; An. 13. Sum mæg searolíce wordcwide wrítan, Exon. Th. 42, 15; Cri. 673. IV a. with preps. *to write* about a subject :—Ða wríteras and ða ðe hí ymbe writon, Bt. 18, 3; Fox 66, 1. Be ðam þingum wrítende ðe ic gehýrde, Homl. Skt. ii. 23 b, 16. IV b. *to write* to a person, *write* with the intention of sending what is written :—Mé geþúhte wrítan ðé, ðú se sêlusta Theophilus, Lk. Skt. 1, 3. IV c. *to write, state* in a book :—Wríteþ Eutropius ðæt Constantinus wære on Breotene âcenned, Bd. 1, 8; S. 479, 31. Ptolomeas wrát ealles ðises middangeardes gemet on ânre béc, Bt. 18, 1; Fox 62, 6. IV d. where many persons assent to a written statement, *to write, get a thing written* :—Hí on heora sinoþe ðus writon be him, Bd. 5, 19; S. 639, 39. Hér sindan ða naman ðere monna ðe ðis (*the charter*) wreotan and festnedan, Cod. Dip. Kmbl. ii. 47, 10. V. *to convey by charter* :—Wé him wrítaþ ða mædue æt Pirigforda, Cod. Dip. Kmbl. iii. 32, 23. Ðam him wrítaþ ðone hagan his dæg, and twám ôðrum æfter him . . . Eác wé wrítaþ him ða circan and ðone circstall and ðone wordig, 52, 5–37. Wé wrítaþ him ðone croft, 258, 27. [O. Sax. wrítan *to cut, wound; to write:* O. H. Ger. rîzan *scindere, scribere: Icel.* ríta *to cut, scratch; to write.* Cf. Goth. writs κεραία.] v. â-, be-, for-, ge-, mis-, under-, ymb-wrítan; un-writen, wrítere, wrítian.

writ-bred, es; *n.* *A writing-tablet* :—Writbred (printed -brec; but see gyrdel-bred, i. 288, 75, *and* weax-bred) *pugillarem* (Lk. 1, 63), Wrt. Voc. ii. 74, 36. Ðá álýfde se ðám cnihtum ðæt hí hyne ofslôgen mid heora writbredum and hine ofsticodon mid hira writýrenum, Shrn. 117, 29.

-write, -writenness. v. wæter-write, tô-writenness.

wrítere, es; *m.* I. *a draughtsman, painter.* v. wrítan, II :—Lôca hú wlitigne monnan ic hæbbe âtæfred, swá unwlitig wrítere swá swá ic eom *pulchrum depinxi hominem pictor foedus*, Past. 65; Swt. 467, 19. II. *a writer, scribe, copyist* :—Wrítere *scriptor*, gewrit *scriptura*, Wrt. Voc. i. 75, 7: *antigrafus*, 61, 5. Wrýtere *librarius, scriba*, 37, 12. Se wrítere (*scriptor*), gif hê ne dilegaþ ðæt hê ær âwrát, ðeáh hê næfre má náuht ne write, ðæt bið ðeáh undilegod, ðæt hê ær wrát, Past. 54; Swt. 423, 32. Mîn tunge ys gelícost ðæs wríteres feþere ðe hraðost wrít, Ps. Th. 44, 2. Oft gehwá gesihð fægre stafas âwritene, þonne heraþ hê ðone wrítere and ða stafas, and nát hwæt hí mænaþ, Homl. Th. i. 186, 3. Wrítera strican *notariorum characteres*, Hpt. Gl. 473, 12. Wríterum *antiquariis* (antiquariis describentibus, Ald. 79), Wrt. Voc. ii. 88, 16: 5, 40. Siððan mîn on Englisc Ælfréd kyning âwende worda gehwelc, aud mé his wríterum sende súð and norð; hêht him swelcra má brengan bi ðære bisene, Past. pref.; Swt. 9, 14. Ðe læs ðe wé þurh gýmeleásum wríterum geleahtrode beón, Homl. Th. ii. 2, 22. Þurh gýmeleúse wríteras, i. 8, 12: Ælfc. T. Grn. 24, 32. III. *a writer, author* :—Se gyt ôþ tô dæg, cwæþ se wrítere, lifigende is, Bd. 5, 18; S. 636, 11. Tô geáre, ðá Brihtferð wrítere ðis âwrát, Anglia viii. 327, 11. Ða wríteras and ða ðe hí ymbe writon, Bt. 18, 3; Fox 66, 1. Gebyrede þurh ða heardsælþa ðara wrítera ðæt hí for heora slæwþe forleton unwriten ðara monna þeáwas and hiora dǽda, ðe foremæroste wæron, Fox 64, 33. IV. *a scribe* in the Biblical sense :—Esdras se wrítere âwrát âne bóc, Ælfc. T. Grn. 10, 37. Rihtwîsnyss ðæra wrítera (*scribarum*), Mt. Kmbl. 5, 20. Folces wríteras *scribas populi*, 2, 4. v. eald-, ge-, not-, stæf-, stær-, tíd-, un-, wyrd-wrítere.

wríþa, an; *m.* I. *a band, collar* :—Þeówan yfelwillendum wriþa and fôtcopsas *seruo maliuolo tortura et compedes*, Scint. 190, 6. Mid wriþan treówenum gewriþen grundweall getimbrunge nâ byþ tôslopen *loramento ligneo conligatum fundamentum aedificii non dissoluitur*, 200, 8. Smeaþancollíce wriþan þ cnottan cræftelícum *sertaque mystica*, Germ. 389, 28. Hí becnytton ânne wriþan eall onbútan his swuran, Homl. Skt. i. 23, 608. II. *a ring* :—Ic gesleá ænne wriðan on his nosu *ponam circulum in naribus tuis* (2 Kings 19, 28), Homl. Th. i. 568, 33. Ic geseah in healle hring gyldenne (*a cup*) men sceáwian, . . . friþospéde bæd God gæste sínum se ðe wende wriþan, Exon. Th. 440, 19; Rä. 60, 5. v. beáh-, heals- (*where misprinted* -wiþa) wriþa; wríþan.

wríþan; *p.* wráþ, *pl.* wriþon; *pp.* wriþen. I. *to twist, give a curved form to* :—Ic wríðe *torqueo*, Ælfc. Gr. 26, 3; Zup. 155, 14. Wríðene (cf. wriþa, II) wælhlencan, Elen. Kmbl. 47; El. 24. II. *to bind up, wrap round, bandage* :—Hê wráð (*alligavit*) his wunda, Lk. Skt. 10, 34. Ða ðe forbrocene wæron, ða gé ne wriþon *quae fractae erant, eas non ligavistis*, L. Ecg. P. iii. 16; Th. ii. 202, 26. Ðá wæron monige ðe his mæg wriðon, Beo. Th. 5957; B. 2982. Ðá bebeád hê him ðæt hê ða tôlýsdan geþeódnesse mínre heáfudwunde gesette and wriþe *dissolutam mihi emicranii juncturam componere atque alligare jussit*, Bd. 5, 6; S. 620, 14. Hé weard wriþen ofer wunda, Exon. Th. 435, 27; Rä. 54, 7. III. *to bind* one thing to another :—Nim ða sylfan wyrt, lege on ðone naflan, and wríð ðærtô swýðe fæste, Lchdm. i. 82, 25. IV. *to bind, fetter* :—Oft wíf hine (*a dog*) wríð, Exon. Th. 434, 3; Rä. 51, 5. Ic hine heardan clammum wriþan þôhte, Beo. Th. 1933; B. 964. [Of one wrase of þornes he wryþen hym one crune, Misc. 48, 383. Wrythyñ *idem quod* wrestyñ *torqueo*, wrythyñ *or* wrethyn *tortus, torsus*, Prompt. Parv. 534. Me dide cnotted strenges abuton here hæued and uurythen to ð it gæde to þe hærnes, Chr. 1137; Erl. 262, 6.

Ʒe mote uaste heom wriðen mid strongen sæilrapen, Laym. 17394. *O. H. Ger.* gi-rîdan *contorquere : Icel.* rîða *to twist, knit, wind.*] v. ā-, be-, ge-, ofer- (Lchdm. ii. 130, 10), tô-, un-wríþan.

wríþan *to flourish.* v. wrídan.

wríþels *a bandage :*—Seaxclâð oððe wræd, wriðels *fascia*, Wrt. Voc. ii. 39, 69.

wríþen-hilt; *adj. Having a hilt bound round* ['In some specimens of swords the handles are wound round with gold wire,' Worsaae's *Primeval Antiquities*, p. 29] :—Ðæt sweord, írena cyst, wreoþenhilt and wyrmfāh, Beo. Th. 3400 ; B. 1698.

wríþian, wríþung. v. wrídian, on-wríþung.

writ-breód (?), es; *n. A reed for writing :*—Hreódwrit (writhreód ? cf. writ-îren) *calamus scribae*, Ps. Spl. C. 44, 2.

wrítian; *p.* ode. **I.** *to draw a figure.* v. wrítan, **II** :—Ðonne wercaþ hió of weaxe, wrítiaþ Fênix, mêtaþ Fênix *they make waxen images of the Phenix, and drawings and paintings*, Engl. Stud. viii. 478, 49. **II.** *to write, compose.* v. wrítan, **IV a** :—Wrítigaþ and singaþ onbûtan him ælc on his wîsan, Engl. Stud. viii. 478, 42. [(*later*) Icel. ríta ; *p.* rítaði.]

wríting, e; *f. Writing :*—Wrítinge fyðer *calamus scribae*, Ps. Spl. 44, 2. v. on-wríting.

wríting-feþer, e; *f. A pen :*—Wrítingfeþere *calamus*, Ps. Spl. T. 44, 2.

writ-íren, es; *n. A style, an iron implement for writing :*—Hí hyne ofsticodon mid hira wrítýrenum, Shrn. 117, 30.

writ-seax, es; *n. A pen :*—Mid pinn ł writtseax *calamo*, Mt. Kmbl. p. 2, 17.

wrixend-líc; *adj. Mutual*, Greg. Dial. 2, 7.

wrixendlíce; *adv. In turn, one after the other :*—Hí wrixendlíce (*vicissim*) hine bǽdon, Bd. 4, 24 ; S. 598, 42. Ða wrixendlíce (*vicissim*) on twā halfe gesewene wǽron, swā swā mid unmǽtnesse miceles stormes worpene beón, 5, 12 ; S. 627, 39. Wrixendlíce *singillatim*, Ps. Surt. 32, 15.

wrixl, e; *f.* **I.** *change, alteration, vicissitude :*—Ðæt is wrǽclíc wrixl in wera lífe, ðætte moncynnes Scyppend onfêng æt fǽmnan flǽsc, and sió weres friga wiht ne cûþe, Exon. Th. 26, 12 ; Cri. 416. God, ðû ðe gimetgaþ giscæfta wrixla (*rerum vices*), Rtl. 164, 12. **II.** *where there is alternation, alternation, exchange :*—Mid ðý hí ðysse ungesǽligan wrixle (*hac infelici vicissitudine*, i. e. the passing from heat to cold and vice versa) ðrǽste wǽron, Bd. 5, 12 ; S. 628, 2. **III.** *where there is reciprocal action, interchange :*—Ðǽr wæs heard plega, wælgāra wrixl, Cd. Th. 120, 5 ; Gen. 1990. **IV.** *where one thing takes the place of another, place, stead :*—Ðonne sculon hié gadrian óðer ierfe on ðæs wrixle (wriexle, Hatt. MS.) ðe hê ǽr sealdon, Past. 45 ; Swt. 340, 18. Hæfdon hí mid him gehālgode tabulan on wîgbedes wrixle *habentes secum tabulam altaris vice dedicatam*, Bd. 5, 10 ; S. 624, 34. **V.** *a loan :*—Borge oððe wrixle *mutuo*, Wrt. Voc. ii. 56, 5. Wrixle *mutuum* (Lk. 6, 34), 74, 43. **VI.** *what is given in return, return, requital :*—Hê forgeald wyrsan wrixle wælhlem ðone ; ... hê him on heáfde helm gescær, Beo. Th. 5930 ; B. 2969. v. ge-wrixl.

wrixlan, wrixlian; *p.* ede. **I.** *to change, vary, alter :*—Is him ðæt heáfod hindan grēne, wrǽtlíce wrixleþ wurman geblonden (*the head shews shifting colours*), Exon. Th. 218, 13 ; Ph. 294. **I a.** *with dat. of that in which change is made :*—Ic þurh mûþ sprece mongum reordum, wrencum singe, wrixle geneahhe heáfodwôþe (*I change my voice*; cf. the Latin riddle : Vox mea diversis variatur pulcra figuris), Exon. Th. 390, 16 ; Rä. 9, 2. Se fugel swinsaþ and singeþ swegle tógeánes ... wrixleþ wôðcræfte beorhtan reorde *incipit illa saeri modulamina fundere cantus, et mira lucem voce ciere novam*, 206, 15 ; Ph. 127. Bleóm wrixleþ *changes colour*, Elen. Kmbl. 1515 ; El. 759. **II.** *to change, alternate :*—Ðás feówer tíman (*the seasons*) wrixliaþ wyð mancynne, Anglia viii. 312, 34. Wrixliende *alterna*, Wrt. Voc. ii. 9, 56. Ða wrixliende on twā halfe gesewene wǽron worpene beón *vicissim hinc inde videbantur jactari*, Bd. 5, 12 ; S. 627, 39 note. **III.** *of reciprocal, mutual action, to exchange, deal :*—Hê cwæð, ðæt him tô micel ǽwisce wǽre ðæt hê swa emnlíce wrixleden (*that they should deal on equal terms*; the terms being that each side should return the captives, and then peace be maintained by each side), Ors. 4, 6 ; Swt. 178, 16. Wrixlindum *reciprocis*, unrixlende *reciprocatu*, Wrt. Voc. ii. 119, 5, 13. **III a.** *with dat. of what is exchanged, fig. of conversation, intercourse.* v. **IV a** :—Wîgsmiþas sittaþ æt symble, wordum wrixlaþ, Exon. Th. 314, 18 ; Mód. 16. Ðǽr hæleðas wrixlaþ sprǽce, Runic pm. Kmbl. 343, 8 ; Rûn. 19. Hý bênan synt, ðæt hié wið ðē môton wordum wrixlan *they beg that they may have interchange of words with thee*, Beo. Th. 737 ; B. 366 : Exon. Th. 373, 29 ; Seel. 117. Wrixlian, Soul Kmbl. 226. Gleáwe men sceolon gieddum wrixlan, Exon. Th. 333, 14 ; Gn. Ex. 4. **IV.** *to lend :*—Wrixlan *mutuare*, Wrt. Voc. ii. 56, 10. Wrislan, 72, 18. **IV a.** *with dat. of what is lent, fig. of words, to speak :*—Secg eft ongan sîð Beówulfes snyttrum styrian, ... wordum wrixlan, Beo. Th. 1752 ; B. 874. Lyt ic wēnde ðæt ic ǽfre sceolde mûðleás sprecan, wordum wrixlan, Exon. Th. 472, 2 ; Rä. 61, 10. [Say me, ...

what wrixlit þi wit & þi wille chaunget, Destr. Tr. 2061. Þai hade laisure þere likyng to say, and wrixle þere wit & þere wille shewe, 3120.] v. be- (Ps. Th. 43, 14), ge-wrixlan(·ian).

wrixlung, e; *f.* **I.** *change, alternation.* v. wrixlan, **II.** [Bi his cloðes wrixlunge, nu red, nu hwit, him on hokerunge, O. E. Homl. i. 207, 3.] **II.** *a loan.* v. wrixlan, **IV** :—Wrixlung *mutuum*, Wrt. Voc. i. 21, 3 : 58, 60. v. ge-wrixlung.

wrocen, e; *f.?* :—Of ðam byrcelse on wrocene ; andlang wrocene in Uppinghǽma gemǽre, Cod. Dip. Kmbl. iii. 124, 13. Be eástan wrocena stybbe ; ðæt swā tô wrocena stybbe, v. 297, 26.

wróht, e; *f.* : es ; *m.* **I.** *accusation :*—Wróht *accusatio* (ex-, MS.), Wrt. Voc. ii. 146, 15. Wróhte *insimulatio*, 44, 74. *Hic susurro* ðes rûnere oððe wróht, Ælfc. Gr. 36 ; Zup. 217, 3. Leásere wróhte *strophosae accusationis*, Hpt. Gl. 505, 55. Wróhte *insimulatione, accusatione*, 517, 55. Uuróctae, uuróchtae, Txts. 70, 524. Mid micelre wróhte hine wrēgende, Homl. Th. ii. 250, 10. Hwylce wróhte (*accusationem*) bringe gē ongeán ðysne man? Jn. Skt. 18, 29. Mid leásum wróhtum beswicene *falsis criminationibus seducti*, Scint. 136, 11. Ða werian gāstas wróhta onsægdon *sequuntur accusationes malignorum spirituum*, Bd. 3, 19 ; S. 548, 35. **II.** *a false accusation, slander, calumny.* v. wróht-bora, **II** :—Heó (*the Egyptians*) his (*Joseph's*) mǽgwinum morðor fremedon, wróht berénedon (cf. Ex. 1, 9–11), manum treówum woldon hié ðæt feorhleán fācne gyldan, Cd. Th. 187, 6 ; Exod. 147. Huscworde ongan þurh inwitþanc ealdorsacerd herme hyspan, wróht webbode ; hē on gewitte oncneów, ðæt wē sôðfæstes swaðe folgodon, Andr. Kmbl. 1343 ; An. 672. Gē inwitþancum wróht webbedon, Elen. Kmbl. 617 ; El. 309. Ne beó nǽnig man bregda tô full, ne inwit tô leóf, ne wróhtas tô webgenne, ne searo tô rēnigenne, Blickl. Homl. 109, 29. **III.** *what is an occasion for accusation, fault, crime, offence :*—Wǽg heora wróht biþ him *via illorum scandalum ipsis*, Ps. Spl. T. 48, 13. Wróhtes wyrhtan (*the devil*), fyrnsynna fruman, Exon. Th. 263, 7 ; Jul. 346. Hé gewrēgde his bróðru tô hira fæder ðære mǽstan wróhte *accusavit fratres suos apud patrem crimine*, Gen. 37, 2. Gangende on eallum his bebodum bûtan wróhte (*sine quaerela*), Lk. Skt. 1, 6. Ðone wróht *abominationem*, Mk. Skt. Lind. Rush. 13, 14. Ðú wróhte onstealdest, Cd. Th. 56, 12 ; Gen. 911 : 57, 22 ; Gen. 932. Hwæt sceal ic mā rîman yfel endeleás ? Ic eall gebæ wrāþe wróhtas geond werþeóde, ða ðe gewurdon from fruman worulde, Exon. Th. 272, 30 ; Jul. 507. **IV.** *a quarrel, strife :*—Weard micel ungeþwǽrnes, ... swā nān mon nyste hwonon sió wróht com, Ors. 6, 4 ; Swt. 260, 21. Wǽron ðá gesôme ða ðe swegl bûan, wróht wæs āsprungen, Cd. Th. 6, 4 ; Gen. 83 : 114, 12 ; Gen. 1903. Wæs wróht gemǽne, herenið hearda, Beo. Th. 4938 ; B. 2473 : 5819 ; B. 2913 : 4564 ; B. 2287 : Exon. Th. 125, 30 ; Gū. 362. Tô ðǽm sǽde ðære wróhte *ad seminanda jurgia*, Past. 47 ; Swt. 358, 3. Bið ðæt deófol on wróhte onlícnisse ; ... bið se Pater Noster on sibbe onlícnisse, Salm. Kmbl. p. 146, 20. Ðú worhtest wróhte betwuh ðē and ðínre môdor suna óðrum *adversus filium matris tuae ponebas scandalum*, Ps. Th. 49, 21. Sume ic geteáh tô geflite, ... ic him byrlade wróht of wēge, Exon. Th. 271, 24 ; Jul. 487. Hí wróht āhôfan, heardne heresîþ, 317, 2 ; Mód. 59. Hé in wuldre wróhte onstalde, Cd. Th. 287, 19 ; Sat. 369. Mars macode ǽfre gewinn and wróhte, Wulfst. 106, 25. Ða ðe wróhte sāwaþ *seminantes jurgia*, Past. 47 ; Swt. 357, 14, 22. **V.** *cause of complaint, injury, hurt :*—Næs hyra wlite gewemmed, ne nǽnig wróht on hrægle, Cd. Th. 243, 17 ; Dan. 437. Ne bið him on ðām wícum wiht tô sorge, wróht ne weþel ne gewindagas, Exon. Th. 238, 30 ; Ph. 612. Rǽhton wîde geond werþeóda wróhtes telgan, hrinon hearmtānas hearde drihta bearnum, Cd. Th. 61, 3 ; Gen. 991. Ðú woldest lǽðlíce þurh ðæt wíf on mē wróhte ālecgean, ormǽte yfel, 162, 21 ; Gen. 2684. [*O. Sax.* wróht *strife*. Cf. *Goth.* wrôhs *accusation : Icel.* rôg *slander* ; *poet.* strife.]

wróht-berend, es ; *m. An accuser :*—Wróhtberend *excussor, accusator*, Wrt. Voc. ii. 146, 14. Bearn wróhtberendra (wôrhtberendra, Ps. Lamb.) *filii excussorum*, Ps. Spl. M. 126, 5.

wróht-bora, an ; *m.* **I.** *an accuser, informer.* v. wróht, **I** :—Wróhtbora *delator*, Wrt. Voc. i. 49, 19. Ðonne wróhtbora (*the devil*) in folc Godes forð onsendeþ biterne stræl, Exon. Th. 47, 31 ; Cri. 763. **II.** *one who brings false accusations, a malicious person.* v. wróht, **II** :—Wróhtbora *factiosa* (cf. ða fǽcnan *factiosam*, 77, 46), *falsa*, Wrt. Voc. ii. 146, 68. [Cf. *Icel.* rôg-beri *a slanderer.*]

wróht-dropa, an ; *m. A drop which brings strife* (v. wróht, **IV**) *or crime* (v. wróht, **III**) :—Weard fǽðþo fyra cynne, siþþan swealg eorðe Abeles blóde, ... of ðam wróhtdropan wíde gesprungen, micel mān (mon, MS.) ældum, monegum þeódum bealoblonden níþ, Exon. Th. 345, 26 ; Gn. Ex. 196.

wróht-georn ; *adj. Quarrelsome, contentious, eager for strife.* v. wróht, **IV** :—On óðre wísan sint tô monianne ða wróhtgeornan, on óðre ða gesibsuman. ... Hé be ðǽm wróhtgeornan secgean wolde *quomodo admonendi qui jurgia serunt, et pacifici. ... Quem seminantem jurgia dicere voluit*, Past. 47 ; Swt. 357, 12, 23. [Cf. *Icel.* rôg-girni *a disposition to slander.*]

wróht-getíme, es ; *n. A series of crimes* (? Cf. teám, getýme) :—

Hǽfdon hié wrôhtgetéme wið God gesomnod *they had heaped up crimes against God*, Cd. Th. 3, 34 ; Gen. 45.

wrôht-hangra (?) :—On wrôhthangran ; of wrôhthangran, Cod. Dip. Kmbl. vi. 120, 12.

wrohtian (?) :—Hé sǽ bedráf, ðǽr ðe heó wrohtaþ (frohtaþ = forhtaþ? *the Latin has*: Mare formidat, 210, 90) dæges and nihtes, Homl. Ass. 173, 105.

wrôht-lác, es ; *n.* (?) *Calumny, slander* :—Unrihtlíce mǽst ǽlc ôþerne ǽftan heáweþ mid scandlícan onscytan and mid wrôhtlácan, Wulfst. 160, 5 note.

wrôht-sáwere, es ; *m. A sower of strife* :—Gehíren ða wrôhtsáweras (*jurgiorum seminatores*) hwæt áwriten is on ðæm godspelle : 'Eádige beóð ða gesibsuman,' Past. 47 ; Swt. 359, 9, 18.

wrôht-scipe, es ; *m. Crime* :—Hí sôhton weras tô weorce (*the building of Babel*) and tô wrôhtscipe, Cd. Th. 100, 31 ; Gen. 1672.

wrôht-smiþ, es ; *m. A worker of crime, a criminal* :—Wrôhtsmiðas (*evil spirits*), Exon. Th. 156, 19 ; Gú. 877. Ðæt ðú mé ne gescyrige mid scyldhetum, werigum wrôhtsmiðum (*the cannibal Mermedonians*), on ðone wyrrestan deáð ofer eorðan, Andr. Kmbl. 171 ; An. 86.

wrôht-spitol ; *adj. Slanderous* :—Wrôhtspitel *susurio*, Txts. 99, 1943.

wrôht-stafas ; *pl. m. Accusations* :—Ic eom fáh and freóndleás, gén ic findan ne can þurh wrôhtstafas wiðercyr wið ðam *I am proscribed and friendless ; still I can by accusations* (cf. vv. 1813-1830, where the devil complains of unfair treatment (nis ðæt fæger sið)) *devise no resistance to my fate* ; i. e. *complaints are useless*, Elen. Kmbl. 1848 ; El. 926.

wrong. v. wrang.

wrôt *a snout, trunk* :—Wrôt *bruncus* (in a list 'de suibus'), Wrt. Voc. i. 286, 54 : ii. 11, 47 : 102, 23 : 127, 27. Ylpes bile *vel* wrôt *promuscida*, i. 22, 45. [Mi drivil druith, and mi wrot wet, Rel. Ant. ii. 210, 29.]

wrôtan ; *p.* wreót *To turn up with the snout, root up* :—Wrôtu *subigo*, Wrt. Voc. ii. 121, 64. Ic wrôte *subigo*, Ælfc. Gr. 28, 6 ; Zup. 176, 12. Hine ûtan of wuda eoferas wrôtaþ *exterminavit eam* (*vineam*) *aper de sylva*, Ps. Th. 79, 13. Swín on bôcwuda wrôtende, Exon. Th. 428, 12 ; Rä. 41, 107. [Swin þe uulied and wroted and sneuied, O. E. Homl. ii. 37, 25. Þat wilde swin þat wroted ȝeond þan grouen, Laym. 469. Schullen wormes wroten (*verrunt*) on the skin, Rel. Ant. ii. 216, 18. As a sowe wroteth in everich ordure, Chauc. Pers. T. A were . . . ȝowre walles with to wrote, Min. 19, 32. Wrotyñ as swyne *verror*, Prompt. Parv. 534. O. H. Ger. ruozan (*used of the action of the plough*) : *Icel.* rôta *to root up*, as swine.]

wude-. v. wudu-.

wudere, wudi[g]ere, es ; *m. A bearer of wood* :—Wuderas *calones* (the passage is : Ejusdem militiae calones et clientes cum lixarum coetibus ad inferiorem gradum pertinentes, Ald. 13. In another gloss on the same passage is the note : Calones sunt qui ligna militibus portant, Hpt. Gl. 427, 4), Wrt. Voc. ii. 76, 72 : 17, 73. Wudu *silva*, wudieras *calones*, i. 33, 55. Wudigeras, 39, 54. v. wudian.

wuderian. v. wederian.

wudian ; *p.* ode *To cut wood* :—Gelamp on Môyses dagum ðæt Môyses fôr þurh ánne wudu mid his werode. Ða gesáwan hié ǽnne ceorl, hwǽr hé stôd and wudede him, Wulfst. 220, 11, 15. Hé him bebeád ðæt hí bǽron wæter tô ðæs folces neóde and wudedon him simble *decrevit eos esse in ministerio cuncti populi, caedentes ligna et aquas comportantes*, Jos. 9, 27. Me mæcg on sumera wudian, Anglia ix. 261, 11. v. wudere, wudung.

wudi[g]ere, wudi[g]ung. v. wudere, wudung.

wudig ; *adj. Woody, full of woods* or *trees* :—Waldend scôp wudige môras, Exon. Th. 193, 12 ; Az. 120.

wudiht ; *adj. Full of woods* or *trees* :—Wudihtes *silvosi*, Germ. 402, 72.

wudu (-o), widu, wiodu ; *gen.* wuda, wudes ; *dat.* wuda, wudu (-o), wyda ; *acc.* wudu, wuda ; *pl.* wuda, wudas ; *m.* I. *wood,* (1) *the substance of growing trees* :—Hû ne miht ðú gesión ðæt ǽlc wyrt and ǽlc wudu (-a, *v. l.*) (*cum herbas atque arbores intuearis*), wile weaxan on ðæm lande sêlost, ðe him betst gerist. . . . Sumra wyrta oððe sumes wuda eard biþ on dúnum. . . . Nim swá wudu (-a, *v. l.*) swá wyrt, of ðære stôwe ðe his eard biþ on tô weaxanne, and sette on uncynde stôwe him, ðonne ne gegréwþ hit ðǽr náuht, for ðam ǽlces landes gecynd is, ðæt hit him gelíce wyrta and gelícne wudu týdrige, Bt. 34, 10 ; Fox 148, 19-29. Ðæt treów wæs on wynne, wudu weaxende, Exon. Th. 435, 19 ; Rä. 54, 3. (1 a) *a tree* :—Wudu môt him weaxan, tánum lǽdan, Exon. Th. 458, 21 ; Hy. 4, 104. Þeáh ðú hwilcne bôh býge wið eorðan, hé bið upweardes, swá ðú ánforlǽtest widu on willan, Met. 13, 55. Hé ðæs foldan ásíów sǽda monegum wuda and wyrta (cf. treówa and wyrta, Bt. 33, 4 ; Fox 132, 27), 20, 251. Smicere on gearwum, wudum and wyrtum cyneþ wlitig scríþan on tún Maius, Menol. Fox 151 ; Men. 77. (2) (*hewn*) *wood, the material obtained from trees* :—Drîge wudu *ligna*, Wrt. Voc. i. 80, 31. Ic eom wyrslícre ðonne ðes wudu fûla, Exon. Th. 424,

33 ; Rä. 41, 48. Hér ys wudu (*ligna*) and fýr, Gen. 22, 7. Be wuda onfenge bûtan leáfe, L. In. 44 ; Th. i. 130, 1. Wuda and wætres nyttaþ, ðonne him biþ wîc álýfed, Exon. Th. 340, 12 ; Gen. Ex. 110. .C. fôðra uuido, Cod. Dip. B. i. 344, 11. Hé hí bewǽg mid wuda ûtan and forbernde mid fýre, Bt. 39, 4 ; Fox 216, 25. Ic on wuda stonde, Exon. Th. 496, 14 ; Rä. 85, 14. Hét ic, of ðæm wudo ðe ðǽr gefylled wæs, ðæt mon fýr onǽlde, Nar. 12, 28. Hé hét Isaac beran ðone wudu (*ligna*), Gen. 22, 6 : Cd. Th. 174, 31 ; Gen. 2886 : 231, 10 ; Dan. 245. Wé heáwaþ ðone wudu *ligna succidimus*, Past. 21 ; Swt. 167, 6. Se ðe ðone wuda (wudu, Cott. MSS.) hiéwð *qui ligna percutit*, 167, 15. (2 a) *wood which forms something, something made of wood* :—Wudu (*a ship*) wundenheals, Beo. Th. 601 ; B. 298 : Exon. Th. 384, 8 ; Rä. 4, 24. Secgan hû se wudu (*a sheathe*) hätte, 437, 32 ; Rä. 56, 16 : (*a loom*), 438, 10 ; Rä. 57, 5. Liþendum wuda (*a ship*), 392, 9 ; Rä. 11, 5. Wudu bundenne, Beo. Th. 438 ; B. 216 : 3842 ; B. 1919. Lǽtaþ hildebord hér onbîdan, wudu (*spears*), wælsceaftas, 801 ; B. 398. Wido (*part of a loom*), Exon. Th. 438, 4 ; Rä. 57, 2. ¶ used of the cross. Cf. beám, treów :—Ongan sprecan wudu sêlesta : 'Ic wæs áheáwen holtes on ende,' Rood Kmbl. 54 ; Kr. 27. II. *wood, forest* :—Wudu *silva*, Wrt. Voc. i. 33, 55. (1) in a generic or collective sense, *wood, the wood, woods* :—Hé mihte hearpian ðæt se wudu (-a, *v. l.*) wagode and ða stanas hí styredon, Bt. 35, 6 ; Fox 166, 32. Wudu (cf. se weald, Bt. 25 ; Fox 88, 20) eallum oncwyð, Met. 13, 50. Ne reccaþ hí ðara metta, gif hí ðæs wuda benugon, Bt. 25 ; Fox 88, 19. Wildeór wuda *bestiae sylvae*, Ps. Spl. 103, 21. On feldum wudes *in campis sylvae*, 131, 6. Wǽstmas wudes (cf. treówa, Bt. 15 ; Fox 48, 9) and wyrta, Met. 8, 20. Eofor of wuda *aper de sylva*, Ps. Spl. Th. 79, 14 : Ps. Th. 67, 27. Hí ðearfendum life on wuda (*in silvis*) and on heán clifum wunedon, Bd. 1, 15 ; S. 484, 8. Gif hí on ðam wuda weorþaþ, Bt. 25 ; Fox 88, 16. Hé teáh tô wuda, 35, 6 ; Fox 168, 7 : Met. 19, 5, 18 : Coll. Monast. Th. 26, 3. Stôw mid wuda (*silvis*) ymbseald, Bd. 4, 13 ; S. 582, 22. Ða myneteras ðe inne wuda wyrcaþ, L. Eth. iii. 16 ; Th. i. 298, 13. On wudu *in sylva*, Ps. Th. 73, 5 : *saltu*, Wrt. Voc. ii. 94, 35. Greát beám on wyda (wuda, *v. l.*), Bt. 38, 2 ; Fox 198, 9. Nôht elles bûton ða wêstan feldas and wudu and dúna, Nar. 20, 10. Fæsten Crêca, wudu Egipta, Salm. Kmbl. 387 ; Sal. 193 : Exon. Th. 381, 9 ; Rä. 2, 8 : Ps. Th. 82, 10. Fýr ðe bærnð wuda (*sylvam*), Ps. Spl. 82, 13. (2) *a wood* :—Hé hét lǽdan hí on ðone wudu ; se wæs genemned *silua nigra*, se swearta wudu, Shrn. 89, 10 : Exon. Th. 200, 8 ; Ph. 37 : Beo. Th. 2732 ; B. 1364. Be wuda bærnette. Gif mon ôðres wudu bærneþ oþþe heáweþ, L. In. 12 ; Th. i. 70, 3. Of ðæs wuda midle, Exon. Th. 202, 6 ; Ph. 65. Anlanges wudes, Cod. Dip. Kmbl. iii. 172, 33. Ðis is ðæs wudes gemǽre . . . tô ðæs wudes efese, 389, 22, 27. Wudæs, vi. 33, 31. Bútan ðem wuda, ii. 66, 23 : Cod. Dip. B. ii. 202, 9. Wiada, Cod. Dip. Kmbl. ii. 64, 29. Wuda, iii. 390, 4. Ðonne mon beám on wuda forbærne. . . . Gif mon áfelle on wuda wel monega treówa, L. In. 43 ; Th. i. 128, 17, 20. Hé rǽsde intô ðam wudu, Homl. Skt. ii. 30, 31. Of ðam wudu, Cod. Dip. Kmbl. iii. 390, 1. Wæs hé eall mid wudu (*silva*) beweaxen, Nar. 12, 8. Of ðæm wudu, 21, 19. Hét ic ceorfan ða bearwas and ðone wudu fyllan *jubeo cedi nemus*, 12, 19. On þicne wudu, Bt. 35, 5 ; Fox 164, 13. Gif feorcund man bûtan wege geond wuda gonge, L. In. 20 ; Th. i. 116, 1 : Byrht. Th. 137, 29 ; By. 193 : Beo. Th. 2836 ; B. 1416. Ðurh ðane wioda, Cod. Dip. B. ii. 202, 10. On ðone wuda ; ofer ðone wuda, Cod. Dip. Kmbl. v. 317, 29. Hé hearpode ðæt ða wudas bifodon, and ða stôdon *silvas currere mobiles, amnes stare coëgerat*. Bt. 35, 6 ; Fox 168, 8. Ealra wuda wildeór *omnes ferae silvarum*, Ps. Th. Spl. Surt. 49, 11. Ealle treówa wuda *omnia ligna sylvarum*, Ps. Spl. Surt. 95, 12. On wudum *in sylvis*, Coll. Monast. Th. 22, 23. Betwyx ðam twám wudum, Cod. Dip. Kmbl. vi. 218, 25. Geond wudas and feldas, Homl. Th. ii. 188, 14. Wuda *silvas*, Ps. Surt. 82, 15. ¶ in several instances of compounds with *wudu* it may be rendered by *wild*; e. g. wudu-bucca, -cerfille, -hunig, -rose. [O. H. Ger. witu *lignum: Icel.* viðr *wood*; *a tree*; *a wood.*] v. ác-, bǽl-, bôc-, bord-, brêmber-, brim-, camp-, flôd-, furh-, gamen-, gár-, heal-, holm-, holt-, mægen-, sǽ-, sund-, þræc-wudu. The word occurs in many local names, v. Cod. Dip. Kmbl. vi. Index.

wudu-ælfen[n], -elfen[n], e ; *f. A wood-elf, wood-nymph* :—Wuduelfen *dryades*, Wrt. Voc. i. 60, 15. Wuduælfenne *oreades*, ii. 65, 44.

wudu-æppel, es ; *n. A wild apple, crab* :— Gesodene wuduæpla, Lchdm. ii. 190, 14.

wudu-bǽre ; *adj. Woody*; *silvestris* :—Wudebǽre gerda *vimina silvestria*, Hpt. Gl. 449, 7.

wudu-bærnet[t], es ; *n. Burning trees in a wood* : — Be wudubærnette. Ðonne mon beám on wuda forbærne, L. In. 43 ; Th. i. 128, 16.

wudu-bát, es ; *m. A wooden ship* :—Hwá mé on wudubâte ferede ofer flôdas, Andr. Kmbl. 1810 ; An. 907.

wudu-beám, es ; *m. A forest tree* :—Wudubeám wlitig, wyrtum fæst, Cd. Th. 247, 18 ; Dan. 499. Wyrtruman ðæs wudubeámes, 248, 21 ; Dan. 516 : Exon. Th. 328, 27 ; Vy. 24. Ða wudubeámas wagedon and swêgdon, Dôm. L. 7. Wudubeáma wlite, Exon. Th. 202, 25 ; Ph.

75. Wudubeáma helm, 496, 2; Rä. 85, 8. Wudubeámum, 277, 6; Jul. 576.

wudu-bearu (-o), wes; *m. A grove of trees, a wood :*—Wudubearwes weard (*the Phenix*), Exon. Th. 208, 7; Ph. 152. On wudubearwe (cf. on holtwuda, l. 16), 209, 11; Ph. 169. Ðæt treów, ðe wexeþ on ðam wudubearwe, Wulfst. 262, 6. Wǽrun wudubearwas on wyndagum *exultabunt omnia ligna sylvarum,* Ps. Th. 95, 12: Exon. Th. 191, 5; Az. 83. On feldum, and on mǽdum, and on wudubearwum, and on sealtum merscum, Cod. Dip. Kmbl. iii. 350, 7. Ða wudubearwas *nemora,* Nar. 22, 15. Drihten sende ceferas, ða ádilegedan ealle wudebeorwas, Wulfst. 221, 17.

wudu-bend *wood-bine :*—Wudubend. Genim ðysse wyrte wyrttruman ðe man *capparis* and óþrum naman wudubend hátaþ, Lchdm. i. 302, 11. Wuduhunig ðæt wæxeþ on wudebendum, Mk. Skt. Rush. 1, 6. v. wudubind, *and* cf. wiþo-bend.

wudu-bil[l], es; *n. A wood-bill :*—Wudubil (uuidu-) *falcis,* Txts. 63, 834 : *falcastrum,* 836: Wrt. Voc. ii. 35, 1, 5. Uudubil *falcastrum, ferramentum curvum a similitudine falcis vocatum,* 146, 82.

wudu-bind, es; *m. :* -binde, an; *f. :* -bindele, an; *f. Woodbine :*—Uuidubindae *volvola, herba similis hederae, quae vitibus et frugibus circumdari solet,* Txts. 104, 1059 : *viburna,* 106, 1082. Wudubind *hedera nigra,* Wrt. Voc. ii. 43, 51. Wudebinde, i. 32, 22. Weodubinde *viticella,* 69, 10. Uudubinde, uududbindlae, uuidubindlae *involuco,* Txts. 71, 1116. Wudubindes leáf, Lchdm. ii. 34, 26 : 306, 24 : 326, 11. Wudubindan leáf, iii. 14, 2 : 30, 8. Wuduhunig ðæt wæxes on wudubinde, Mk. Skt. Lind. i. 6. [*Mater silva* chevefoil, wudebinde, Wrt. Voc. i. 140, 19. *Caprifolium* wodebinde, Wülck. Gl. 570, 31. Woodebynde *caprifolium, viticella,* Prompt. Parv. 531.] v. wudu-bend, *and* cf. wiþo-winde.

wudu-binde, an; *f. A bundle of wood :*—Uuidubinde *lignarium* (lignarium *bois à brûler,* Migne), Txts. 35, 18.

wudu-bléd, e; *f. A forest fruit :*—Steám swéttra swæcca gehwylcum, wyrta blóstmum and wudublédum, Exon. Th. 358, 18; Pa. 47. Hé somnaþ wyrta wynsume and wudubléda *colligit succos et odores divite silva,* 211, 8; Ph. 194.

wudu-bora (?), an; *m. One who carries wood for fuel :*—Wudubior (-bora ?) *calo militum,* Hpt. Gl. 427, 7. v. wudere.

wudu-bucca, an; *m. A wild goat :*—Firginbucca, ðæt ys wudubucca, Lchdm. i. 348, 2. Wudubuccan gealla, 348, 6. v. wudu-gát.

wudu-cerfille, an; *f. Wild chervil :*—Wuducerfille *brassica,* Wrt. Voc. i. 67, 4. Wuducerefille *brassica sylvatica,* 68, 74. Wuducerfille *pastinace,* 19. Wuducarfille *speragus,* 46. Wuduceruille. Genim ðysse wyrte wyrttruman ðe man *sparagi agrestis,* and óðrum naman wuducerfillu nemneþ, Lchdm. i. 188, 19–22. Nim cerfillan and wuducerfillan, ii. 152, 15 : 268, 14. v. wudu-fille.

wudu-cocc, es; *m. A woodcock :*—Wudecocc *aceta,* Wrt. Voc. i. 29, 52. Wuducoc *acega,* 280, 3. Wudecocc, Hpt. 33, 240, 1: 28. [In later English the word translates several Latin words: *castrimargus,* Wülck. Gl. 571, 17: 625, 2 : 701, 38 : 762, 2 : *gallus sylvestris,* 625, 3 : *fornix,* 639, 36 : *orna,* 639, 37 : *castrimargus, gallus sylvestris,* Prompt. Parv. 531.]

wudu-croft, es; *m. A croft with trees on it (?) :*—On wudecrofte; of ðam crofte, Cod. Dip. Kmbl. iii. 376, 7.

wudu-culfre, an; *f. A wood-pigeon :*—Wuduculfre *palumba,* Wrt. Voc. i. 77, 21 : *palumbes,* 62, 27. Wudeculfre *palumbus,* 29, 26.

wudu-cunelle, an; *f. Wild thyme :*—Wuducunille, Lchdm. ii. 96, 22. Wuducunellan, 320, 14.

wudu-cyn[n], es; *n. A kind of wood :*—Wuducynn ł wyrtcynn *nardi pistici,* Jn. Skt. Lind. 12, 3.

wudu-docce, an; *f. Sorrel :*—Wududocce. Genim ðás wyrte ðe man *lapatium,* and óðrum naman wududocce nemneþ, Lchdm. i. 132, 215.

wudu-fæsten[n], es; *n.*   I. *a place rendered secure by woods, a wood as a place of security :*—Ðær gewexen is wudufæstern micel *there has grown a great wood which affords shelter,* Cod. Dip. B. ii. 376, 4. Hé gewícode ðær ðǽr hé niéhst rýmet hæfde for wudufæstenne *he pitched his camp in the nearest spot allowed by the woods,* Chr. 894; Erl. 90, 9. Ða flugon ða Bryt-Walas tó ðam wudufærstenum (cp. silvis sese obdidere, Bd. 1, 2), pref.; Erl. 5, 12.   II. *a place of security built of wood :*—Sceal fæsl wesan cwiclifigendra cynna gehwylces on ðæt wudufæsten (*Noah's ark*) geléded, Cd. Th. 79, 16; Gen. 1312.

wudu-feld, a, es; *m. A field of the wood :*—On wudufeldum *in campis silvae,* Ps. Th. 131, 6.

wudu-feoh; *gen.* -feós; *n. A wood-tax, tax on forests :*—Wudefeoh *lucar* (vectigal quod ex lucis contrahitur, Du Cange), Lchdm. i. lxiii, 2. Cf. land-feoh.

wudu-fille, an; *f. Wild chervil :* — Wudufille, Lchdm. iii. 24, 7. Nim wudufillan, ii. 312, 14: 340, 2. Wudafillan, 4, 27. Ða reádan wudufillan (*sparagia agrestis*), 50, 1. v. wudu-cerfille.

wudu-fín, e; *f. A heap of wood :*—Wudufin *strues,* Ælfc. Gr. 9, 27; Zup. 53, 5. Wudefíne, Wrt. Voc. i. 39, 53. Wudufíne *strue, congerie,* Hpt. Gl. 464, 30. [*O. H. Ger.* witu-uîna *strues.*]

wudu-fugel, es; *m. A bird of the woods :*—Wudufuglas, ðeáh hí beón wel átemede, gif hí on ðam wuda weorþaþ, hí forseóð heora láreówas, and wuniaþ on heora gecynde, Bt. 25; Fox 88, 15: Met. 13, 35.

wudu-gát, e; *f. A wild goat :*—Wudugáte geallan, Lchdm. i. 348, 13, 18. v. wudu-bucca.

wudu-gehæg, es; *n. An enclosed wood (?) :*—Of ðam hwítan stoccæ þurh ðæt wudugehæg, Cod. Dip. Kmbl. iii. 176, 1.

wudu-hana, an; *m. A woodcock :*—Uuduhona *pantigatum,* Wrt. Voc. ii. 116, 56.

wudu-heáwere, es; *m. A hewer of wood, woodcutter :*—Búton wudu-heáwerum *exceptis lignorum caesoribus,* Deut. 29, 11.

wudu-herpaþ, es; *m. A public road through a wood :*—On ðone wuduherpaþ, Cod. Dip. Kmbl. iii. 213, 2.

wudu-holt, es; *m. A grove :*—Ðǽr is se fægere wuduholt ðe is on bócum gehaten *Radians saltus,* Engl. Stud. viii. 477, 12. Sunbearo, wuduholt wynlíc *solis nemus, et consitus arbore multa lucus,* Exon. Th. 200, 1; Ph. 34. Wuduholtum, 223, 20; Ph. 362.

wudu-hunig, es; *n. Wild honey :*—Wuduhunig *mel silvestre,* Mt. Kmbl. 3, 4. Wudehunig, Homl. Th. i. 352, 7. Be wyrtum and be wuduhunige, Blickl. Homl. 167, 36. Wuduhunig hé æt, Mk. Skt. 1, 6.

wudu-lád, e; *f. Carting wood :*—Æt wuduláde wǽntreów, L. R. S. 20; Th. i. 440, 27.

wudu-lǽs, we; *f. Forest pasture :*—Seó útlǽs and seó wudulǽs mid óðrum mannum gemǽne, Cod. Dip. Kmbl. vi. 214, 22. Wæs tiolo micel sprǽc ymb wudulǽswe tó Súðtúne; waldon ða swángeréfan ða lǽswe forður gedrífan ond ðone wudu .geþiogan ðon hit aldgeryhto wéron, i. 278, 32.

wudu-land, es; *n. Wood-land, forest-land, forest :*—Ægðer ge etelond ge eyrðlond ge eác wudoland, Cod. Dip. Kmbl. ii. 95, 14. Ðæt wudæland, ðæ mín fæder geúþæ, iii. 273, 27. Him wǽre forneh eall ðæt wudulond on gereáfad . . . ðæt Æðelbald cyning gesealde tó mǽstlonde and tó wudulonde, v. 140, 17. Feldlondes and wudulandes, iii. 262, 19: vi. 219, 5. Hí hine geond ealle eorðan sóhton, ge on dúnlandum ge on wudulandum, Ap. Th. 7, 14. [Þa wilde bær i þon wodelonde, Laym. 1699.]

wudu-leáctric, es; *m. Wild lettuce :*—Wuduléctric. Ðeós wyrt ðe man *lactucam sylvaticam,* and óðrum naman wuduléctric nemneþ, Lchdm. i. 128, 6–8. Wuduléhtric, iii. 2, 21.

wudu-líc; *adj.*   I. *of a wood :*—Wudulíc *siluester,* Ælfc. Gr. 9, 18; Zup. 44, 16. Wudelícra treówa *arborum silvestrium,* Hpt. Gl. 419, 42.   II. *wild :*—Wudulíce oððe wilde *agrestes,* Wrt. Voc. ii. 4, 60.

wudu-mǽd. v. mǽd.

wudu-mǽr *echo :*—Wudumer (uuydu-) *echo,* Txts. 59, 715. Wudumær, Wrt. Voc. ii. 29, 17: 70, 7 (windu-, MS.). v. Grmm. D. M. pp. 452, 1412 (Stallybrass' Trans.).

wudu-merce, es; *m. Wood-marche :*—Wudumerce *apis sylvatica,* Wrt. Voc. i. 67, 29. Wudemerce *apiaster,* 31, 9. Genim wudumerce, Lchdm. ii. 22, 16: 66, 18: 326, 9. [Wudemerch *saniculum,* Wrt. Voc. i. 139, 6.]

wudung, e; *f.*   I. *cutting wood :*—Ða hæðenan on heora ðeówte leofodon tó wudunge and tó wæterunge, Homl. Th. ii. 222, 29.   II. as a technical term referring to the right of cutting timber in a wood :—Ðis is seó wudung ðe ðærtó gebyreþ, ælce geáre fíftig fóðra and án hund of ðæs cinges ácholte, and húsbót, Cod. Dip. Kmbl. vi. 243, 11. Seó wudung on gemǽnan gráse tó Ðorndúne, iii. 463, 9. Uuidigung, uuidiung, Cod. Dip. B. i. 344, 13, 17. v. wudian.

wudu-rǽden[n], e; *f. Woodcutting, right of cutting timber in a wood :*—Ánan esne gebyreþ tó metsunge .xii. pund gódes cornes, and wudurǽden be landside (*the amount of wood that he may cut is to be determined by local custom*), L. R. S. 8; Th. i. 436, 27. Twá hund swína mæsten and wudurǽden loca hwæs man beþurfe, Cod. Dip. Kmbl. iv. 20, 5. An íc twǽga wǽna gang tó wudurédenne, vi. 36, 16. Heó hæbbe ða wudurǽddenne in ðæm wuda ðe ða ceorlas brúcaþ, and éc ic hire léte tó ðæt ceorla gráf, ii. 100, 14.

wudu-réc, es; *m. The smoke from burning wood :*—Wuduréc ástáh sweart, Beo. Th. 6280; B. 3144.

wudu-rima, an; *m. The edge of a wood :*—West be wuduriman, Cod. Dip. Kmbl. iii. 34, 15. [To mine lauerde i þon woderime, þer he under rise lið, Laym. 739.]

wudu-rofe, -rife (cf. Jamieson's Dict. *wood-rip*), an; *f. Woodruff :*—Wuderofe *astula regia,* Wrt. Voc. i. 30, 31. Wudurofe. Genim ðysse wyrte seáw ðe man *astula regia,* and óðrum naman wudurofe nemneþ, Lchdm. i. 132, 6–9. Wuduhrofe. Genim ðysse wyrte wyrttruman ðe Grécas *malochin agria,* and Rómáne *astula regia* nemnaþ, and eác Ængle wudurofe hátaþ, 156, 8–11. Wel wudurofan, ii. 54, 2: 108, 19: 324, 13. Wudurifan, 64, 5.

wudu-rose, an; *f. Wild rose :*—Genim wudurosan, Lchdm. ii. 90, 16.

wudu-sníte, an; *f. The name of some bird :*—Uudusníte *cardicus,* Wrt. Voc. ii. 103, 46. Wudusníte, 14, 9. v. sníte.

wudu-telga, an; *m.* *A branch of a tree* :—Ne foldan stān, ne wudu-telga, Salm. Kmbl. 844; Sal. 421.

wudu-þistel, es; *m.* *Wood-thistle* :—Wuduþistel. Ðeós wyrt ðe man *carduum sylvaticum,* and óðrum naman wuduðistel nemneþ, Lchdm. i. 224, 11 : iii. 28, 21. Wuduþistles ðone grēnan mearh ðe biþ on ðam heáfde, ii. 358, 1.

wudu-treów, es; *n.* *A tree of the woods, a forest tree* :—Nān man ne mōt his ælmessan behātan tō wylle ne tō wydetreówe, Wulfst. 303, 18. Wrǣtlíc wudutreów, Exon. Th. 437, 5; Rä. 56, 3. Ðæt man weorðige ǣniges cynnes wudutreówa, L. C. S. 5; Th. i. 378, 20. Wudutreówu, Wulfst. 40, 15.

wudu-wāsa, an; *m.* *A satyr, a faun* :—Satiri, vel fauni, vel celini, vel fauni ficarii unfǣle men, wudewāsan, unfǣle wihtu, Wrt. Voc. i. 17, 20. *Satyri* vel *fauni* unfǣle men, *ficarii* vel *invii* wudewāsan, 60, 23-24. Wudewāsan *faunos,* Germ. 394, 242. [Sumwhyle wyth wodwos he werreȝ, þat woned in þe knarreȝ, Gaw. 721. A vestoure wroȝt full of wodwose, and oþer wild bestis, Alex. (Skt.) 1540. Wodewese, woodwose *silvanus, satirus,* Prompt. Parv. 531, and see note. A wodewose *silvanus,* Wülck. Gl. 612, 2. Wright, in a note to the second of the passages cited above from the Vocabularies, quotes from Withal's *Dictionarie* (ed. 1608) 'a woodwose *satyrus.*']

wudu-weald, es; *m.* *High ground covered with wood, a wooded height* :—On wuduwaldum *in saltibus,* Wrt. Voc. ii. 47, 71.

wudu-weard, es; *m.* *A wood-keeper, forester* :—Be wuduwearde. Wuduwearde gebyreþ ǣlc windfylled treów, L. R. S. 19; Th. i. 440, 9. [The wodeward *le verder,* Wrt. Voc. i. 164. The wodeward waiteth us wo that loketh under rys, P. S. 149, 17. Wodewarde or walkare in a wode for kepynge *lucarius,* Prompt. Parv. 531.]

wudu-weaxe, an : -weax, es; *n. Wood waxen, wood wex* (v. E. D. S. Pub. Plant Names); genista tinctoria :—Wuduweaxe, Lchdm. ii. 66, 11. Weoduweaxe, iii. 30, 13. Wuduweaxan gódne dǣl, ii. 324, 21 : iii. 28, 28. Nime wuduweaxan nioþoweard, wealwyrt nioþowearde, ii. 118, 2. Genim gearwan and weoduweaxan and hrǣfnes fót, iii. 30, 4. Nim gearwan and wuduweax (cf. weax, iii. 24, 4) and hrefnes fót, ii. 324, 25.

wudu-wēsten *wild woodland* :—Ðā flugon ða Bryt-Walas tō ðam wuduwēstenum, Chr. pref.; Erl. 5, 12 note. v. wudu-fæsten[n].

wudu-winde, an; *f. Woodbine* :—Uuduuuinde, wi[d]windae *volvola,* Txts. 107, 2158. Uuduuuinde, uuiduuuindae *viburna,* 2129. Uudu-winde, uuiduuuindae *edera,* 59, 717. Wuduwinde, Wrt. Voc. ii. 29, 2 : *vivorna,* i. 286, 1. v. wudu-binde, wiþo-winde.

wudu-wyrt, e; *f. A wild plant* :—Ða swētan stencas ðara wudu-wyrta, Blickl. Homl. 59, 3.

Wuffingas ; *pl. m. The patronymic of the royal house of East-Anglia* :—Wuffa fram ðam Eást-Engla cyningas Wuffingas wǣron nemde, Bd. 2, 15; S. 518, 38.

wuhhung, wuhung, e; *f. Fury* :—Him (*Nero*) ðǣre wuhhunge ge-steóran *vertere insani rabiem Neronis,* Bt. 16, 4; Fox 58, 14. Wuhunga *furias,* Wrt. Voc. ii. 38, 74.

wuht, wul, wulder. v. wiht, wull, wuldor.

wuldor (-ur, -er), es; *n. Glory.* (1) in reference to earthly subjects :—Woruldsceafta wuldor, Exon. Th. 190, 16; Az. 74. Hǣlo mīne and wuldor (*gloria*) mīn, Ps. Spl. 61, 7. Him wuldor (*gloria*) and wela wunaþ æt hūse, Ps. Th. 111, 3. Wuldres gim (*the sun*), Andr. Kmbl. 2538; An. 1270. Tō ðínes folces wuldre (wulder, Lind.: wuldur, Rush.), Lk. Skt. 2, 32. Hē fērde ūt on huntaþ mid eallum his werode and his wuldre, Homl. Skt. ii. 30, 25. Ne beseoh tō ðínum ǣrran wuldre, 30, 121. Eodon of ðam fȳre feorh unwemme, wuldre ge-wlitegad, Exon. Th. 197, 8; Az. 187. Hē (*the Phenix*) is wlitig and wynsum, wuldre gemearcad *regali plena decore,* 220, 11; Ph. 318. Ge-wīteþ mid ðȳ wuldre mǣre tungol (*the sun*), 350, 23; Sch. 68. Tempel wuldre gewlitegod, Andr. Kmbl. 1337; An. 669. Ðā sceáwode ic mīne gesǣlinesse and mín wuldor, Nar. 7, 22. (1 a) in a bad sense, *vain-glory.* v. wuldor-full, II :—On wlence ic fērde þurh ðæt īdele wuldor, Anglia xi. 113, 50 : Exon. Th. 107, 12; Gū. 57. Wēndes ðū ðurh wuldor, ðæt ðū woruld āhtest, alra onwald, Cd. Th. 268, 22; Sat. 59. (1 b) applied to persons or things :—Wīfa wuldor (*the Virgin Mary*), Menol. Fox 295; Men. 149. Receda wuldor, Salomones templ, Cd. Th. 219, 23; Dan. 59. (2) of celestial or spiritual glory :—Godes wuldor *gloria Domini,* Lev. 9, 23. Gode sȳ wuldor, Lk. Skt. 2, 14 : Ps. Spl. 103, 32. Him wīdeferh wuldor stondeþ, Exon. Th. 350, 2; Sch. 57. Stefn of heofonum, wuldres hleóðor, Cd. Th. 204, 10; Exod. 417. Se wyrhta þurh his wuldres gāst sette, 265, 28; Sat. 14. In wuldres wlite, 279, 5; Sat. 233 : 285, 26; Sat. 343. Wuldres rǣst *the rest of heaven,* Exon. Th. 103, 19; Cri. 1690. Wuldres neótan *to enjoy heaven,* 365, 15; Wal. 89. Wuldres eard āgan, 74, 8; Cri. 1203. Wuldres wynlond, 317, 13; Mōd. 65. Wuldres bearnum (*angels*), Cd. Th. 1, 22; Gen. 11. Wuldres þegn, engel Drihtnes, 137, 1; Gen. 2266 : 95, 6; Gen. 1574. Wuldres þegnas (*St. Matthew and St. Andrew*), Andr. Kmbl. 2052; An. 1028. Seó fǣmne, wuldres wynmǣg (*Guthlac's sister,* seó Cristes þeówe, Guthl. 20; Gdwin. 92, 2), Exon. Th. 182, 32;

Gū. 1319: (*St. Juliana*), 278, 20; Jul. 600: 269, 23; Jul. 474. Se eorðan dǣl . . . se wuldres dǣl *the body . . . the soul,* 184, 11; Gū. 1342. Wuldres treó *the cross,* Elen. Kmbl. 177; El. 89 : Rood Kmbl. 28; Kr. 14. Wuldores stæf, Salm. Kmbl. 225; Sal. 112. Mannes sunu cumende mid mycelum wuldre, Mk. Skt. 13, 26. For Godes wuldre (uldre, Lind.), Jn. Skt. 11, 4. In wuldre *in heaven,* Andr. Kmbl. 712; An. 356 : Elen. Kmbl. 1491; El. 747. Tō wuldre, Exon. Th. 3, 3; Cri. 30 : Andr. Kmbl. 3360; An. 1684. Ðæt ēce wuldor geearnian, Homl. Th. ii. 284, 31. Wē gesāwon his wuldor (uuldor, Lind.), Jn. Skt. 1, 14. Godes wuldor (uulder, Lind.), 11, 40 : Ps. Spl. 18, 1. Ealles ðæs Iudith sægde wuldor Dryhtne, Judth. Thw. 26, 24; Jud. 343. Wulder, R. Ben. 4, 4. Sāule sōdfæstra wuldrum hrēmge, Exon. Th. 4, 17; Cri. 54. ¶ in phrases denoting the Deity :—Wuldres aldor, Cd. Th. 40, 15; Gen. 639: 91, 12; Gen. 1511. Wuldres weard, 58, 4; Gen. 941. God, wuldres hyrde, Beo. Th. 1867; B. 931. Wuldres āgend, Andr. Kmbl. 420; An. 210. Wuldres bearn (*Christ*), Cd. Th. 301, 26; Sat. 589. Wlitig wuldres gim, Exon. Th. 232, 33; Ph. 516. (2 a) applied to the Deity :—Drihten, wulder mīn, Ps. Spl. 3, 3. Metod, cyninga wuldor, Judth. Thw. 23, 34 : Jud. 155 : Andr. Kmbl. 342; An. 171. Ðæt ðe wealdend God ācenned weard, cyninga wuldor, Elen. Kmbl. 10; El. 5. Dryhten, hæleða wuldor, Andr. Kmbl. 2925; An. 1465. Ðȳ þriddan dæge beorna wuldor of deáðe ārās, Dryhten ealra hæleða cynnes, Elen. Kmbl. 372; El. 186. [Þin wombe was þin God and þin wulder . . . echeliche wunien in alre wuldre, Fragm. Phlps. 7, 20, 58. Si Drihhtin loff and wullderr, Orm. 3379. Cf. *Goth.* wulþus *glory.*] v. heofon-, sigor-, swegel-wuldor.

wuldorbeágian ; *p.* ode *To crown* :—Tō wuldorbeágienne mid Criste, Homl. Th. i. 84, 32. v. ge-wuldorbeágian.

wuldor-beáh ; *gen.* -ges; *m. A crown* :—Wuldorbeáh *corona,* Ps. Spl. 64, 12. Wuldorbeáh, Wrt. Voc. i. 43, 5 : Hpt. Gl. 438, 24. [For] wuldurbeága *pro corona,* 458, 22. [Tō] wuldurbēge *ad coronam,* 460, 5. Wulderbeáge *tropheo,* 508, 64. Wuldorbeág *coronam,* Ps. Lamb. 20, 4. Hē (*Stephen*) hæfð ðone ēcan wuldorbeáh, Homl. Th. i. 50, 13. Ān læs feówertig wuldorbeága, Shrn. 62, 7. Hī wuldorbeágum beóð gewelgode scínendum *laureis ditantur fulgidis,* Hymn. Surt. 133, 1.

wuldor-blǣd, es; *m. Glorious success* :—Eów ys wuldorblǣd torhtlíc tōweard, and tír gifeþe, Judth. Thw. 23, 35; Jud. 156.

wuldor-cyning, es; *m. The king of glory, the Deity* :—Wuldor-cyning þeóda gehwylce hāteþ ārīsan, Exon. Th. 63, 22; Cri. 1023 : Cd. Th. 272, 6; Sat. 115. Se wuldorcyning, 10, 32; Gen. 165. Ælmihtig God, wuldorcyning, 242, 30; Dan. 427 : Salm. Kmbl. 640; Sal. 319. Wuldorcyning, fæder frymða gehwæs, Exon. Th. 211, 12; Ph. 196. Wuldorcyning (*Christ*), 227, 9; Ph. 420. Ðæt wæs þonne ðæt se wuldorcyning on middangeard cwom forþ of ðam innoþe ðære ā clǣnan fǣmnan, Blickl. Homl. 9, 32. Se hālga Dryhten, ðū . . . mín wuldor-cyning, 452, 16; Hy. 4, 2. Ðū, weroda wuldorcyning, Met. 20, 162. Hié yrfes brūcaþ wuldorcyninges, Elen. Kmbl. 2639; El. 1321 : Andr. Kmbl. 835; An. 418 : Exon. Th. 153, 5; Gū. 821. Heó Gode þancode, wuldorcyninge, Elen. Kmbl. 1922; El. 963. Wuldurcyninge, ēcum Dryhtne, Beo. Th. 5582; B. 2795. Ðæt wē rodera weard, wereda wuldorcyning herigen, Cd. Th. i. 3; Gen. 2 : 213, 4; Exod. 547. Fāh wið wuldorcyning, Exon. Th. 364, 7; Wal. 67.

wuldor-dreám, es; *m. Joy in the glory of heaven, celestial joy* :—Wē ðe þanciaþ, þióda Waldend, ðínes weorðlícan wuldordreámes, Hy. 8, 10. In ðínne wuldordreám, Exon. Th. 455, 2; Hy. 4, 43.

wuldor-fæder ; *m. The father of glory, the heavenly Father* :—Weorc wuldorfæder (wuldurfadur, Txts. 149, 3) *facta Patris gloriae,* Bd. 4, 24; S. 597, 21. Líf mid wuldorfæder, Menol. Fox 291; Men. 147. Mid ðínne wuldorfæder, Exon. Th. 14, 11; Cri. 217.

wuldor-fæst ; *adj. Glorious* :—Wuldurfest *gloriosus,* Ps. Surt. ii. p. 188, 1. Wuldorfæst, Cd. Th. 234, 3; Dan. 286. Wuldorfæst cyning (*Solomon*), 202, 18; Exod. 390. Ðes wuldorfæsta kyning *rex gloriae,* Ps. Th. 23, 8, 10 : Nicod. 29; Thw. 16, 38. Ðæt wuldorfæste líf ðætte englas on Drihtnes onsȳne wuniaþ, Blickl. Homl. 103, 32. Ða stōwe ðínes wuldorfæstan temples *locum tabernaculi gloriae tuae,* Ps. Th. 25, 8. For ðære swētnesse ðære wuldorfæstan gesihðe, Homl. Skt. ii. 23 b, 179. Wulderfæstan, 236, 8. Heora (*the stars'*) wuldorfæstne wlite, Cd. Th. 132, 10; Gen. 2191. His ðone wuldorfæstan gāst, Blickl. Homl. 85, 4. His ða wuldorfæstan onsȳne, 103, 29. Ða wuldorfæstan Godes weorc, Homl. Skt. ii. 23 b, 11. Wuldorfæstan wíc (*heaven*), Cd. Th. 2, 30; Gen. 27.

wuldorfæste ; *adv. Gloriously* :—Hió Gode þancode ðæs geleáfan ðe hió swā leóhte oncneów, wuldorfæste, in ðæs weres breóstum, Elen. Kmbl. 1930; El. 967.

wuldorfæstlíce ; *adv. Gloriously* :—Hēr Eleutherius on Rōme onfēng biscopdōm, and ðone wuldorfæstlíce .xii. winter geheóld, Chr. 167; Erl. 8, 14.

wuldorfæstlícness, e; *f. Gloriousness, glory* :—Sȳ ðū gebletsod, Drihten God, ðe mē æteówdest ðā wuldorfæstlícnysse ðe ðū ondrǣdendum gyfest, Homl. Skt. ii. 23 b, 603.

wuldor-full ; *adj.* I. *glorious* :—Gif ðū eádmōdnysse healtst

wuldorful (*gloriosus*) ðū byst, Scint. 22, 4. Wuldorfull mid ēcum wurð-mynte, Homl. Ass. 77, 125. Ðes Dauid wæs wuldorful cyning, Homl. Skt. i. 18, 32. Wulderfull ðrowung, Homl. Th. i. 360, 20. Se wuldor-fulla (*gloriosus*) Eádgār, Anglia xiii. 365, 3. Se wuldorfulla cyning (*Christ*), Nicod. 28 ; Thw. 16, 6. For his wuldorfullan sige oretlofes *propter ejus gloriosissimi victoriam triumphi*, Anglia xiii. 400, 497. Seó wunung on ðam wuldorfullum dreáme, Homl. Ass. 43, 481. Tō ðære wuldorfullan byrig Hierusalem, H. R. 7, 4. Wuldorfulle on mægðhāde, Homl. Ass. 44, 499. Wuldorfulle lofu *glorificum melos*, Hymn. Surt. 57, 24. Babilonia ðe hwīlon wæs wuldorfullost burh ealra burha, Wulfst. 194, 10. II. *vainglorious*. v. wuldor (I a) :—Betere ys þearfa and behōfigende him þænne wer wulderfull (*gloriosus*) and genihtsumigende hlāfe, Scint. 178, 15. Wuldorfull, 180, 6.

wuldorfullian ; *p.* ode To *glorify* :—Hī wuldorfulliaþ (*glorificabunt*) naman ðīnne, Ps. Spl. 85, 8.

wuldorfullíce ; *adv. Gloriously* :—On eallum ðām ðe wuldorfullíce (*gloriose*) fram him gewurdon, Lk. Skt. 13, 17. Hī wurdon wuldor-fullíce gemartyrode, Homl. Th. i. 80, 29. Heó tō dæg wuldorfullíce of ðam līchaman gewāt, 440, 12.

wuldor-gāst, es ; *m. A spirit of glory, glorious spirit, an angel* :—Of roderum wuldorgāst Godes wordum mælde (*angelus Domini de coelo clamavit*, Gen. 22, 11), Cd. Th. 176, 15 ; Gen. 2912.

wuldor-geflogena, an ; *m. A fugitive from glory, an evil spirit* :—Magon ðās .viiii. wyrta wið nygon wuldorgeflogenum, Lchdm. iii. 36, 15.

wuldor-gesteald ; *pl. n.* I. *glorious possessions* :—Gold and godweb, Iōsepes gestreón, wera wuldorgesteald, Cd. Th. 215, 24 ; Exod. 588. Ða gerēno and se reáda telg and ða wuldorgesteald (*the binding of a book*?), Exon. Th. 408, 22 ; Rä. 27, 16. II. *glorious mansions* :—Fæder and Sunu and frōfre Gāst on þrinnesse wealdeþ wuldorgestealda, Andr. Kmbl. 3369 ; An. 1688. God bescyrede his wiðerbrecan wuldor-gestealdum, Cd. Th. 4, 36 ; Gen. 64.

wuldor-gifu, e ; *f. A glorious gift, a gift of heaven* :—Wuldorgife, Hy. 9, 44. Ðȳ læs hē for wlence, wuldorgeofona ful, mon mōde swīð, of gemete hweorfe, Exon. Th. 294, 33 ; Crä. 24. Ðē beorht Fæder geweorðaþ wuldorgifum, cræfte and mihte, Andr. Kmbl. 1875 ; An. 940. Gāstes mihtum, wuldorgifum, Elen. Kmbl. 2141 ; An. 1072.

wuldor-gimm, es ; *m. A glorious gem, the sun* :—Wynsum wuldor-gimm, Exon. Th. 492, 23 ; Rä. 81, 20.

wuldor-hama, an ; *m. A glorious garb* :—Engel ælbeorht, wlitescȳne wer on his wuldorhaman, Cd. Th. 237, 16 ; Dan. 338 : Exon. Th. 189, 2 ; Az. 53. Him is engel mid, hafaþ beorhtne blæd, ne mæg him bryne sceþþan, wlitigne wuldorhaman, 196, 24 ; Az. 179.

wuldor-helm, es ; *m. A crown* :—Mōyses onfēng scīnendum wuldor-helme, Blickl. Homl. 49, 11.

wuldor-leán, es ; *n. A glorious reward, the reward of heaven* :—Bið hyra meaht and gefeá swīðe gesǽliglíc sāwlum tō gielde, wuldorleán weorca, Exon. Th. 66, 31 ; Cri. 1080. In ðam ēcan gefeán niman weorca wuldorleán, 184, 20 ; Gū. 1347.

wuldor-líc ; *adj.* I. *glorious* :—Wuldurlíc *gloriosus*, Rtl. 181, 27. Wuldorlíc, Exon. Th. 62, 33 ; Cri. 1011. Hū wuldorlíc (*admirabile*) ðīn nama ys, Ps. Th. 8, 9 : Ps. Spl. 8, 1. Hē wītgode be ðære wuldor-lícan ácennednesse Cristes, Bt. 5, arg. Be his ðære wuldorlícan ǽriste, Blickl. Homl. 117, 3. Þurh wuldorlícne martyrdōm, Shrn. 30, 32. Hē wæs hæbbende wuldorlícne beág on his heáfde, 106, 10. Wuldorlícne wlite, Salm. Kmbl. 115 ; Sal. 57. Wǽrun wuldurlíce wið ðe ácweðene *gloriosa dicta sunt de te*, Ps. Th. 86, 2. II. the word glosses *orthodoxus*, Wrt. Voc. i. 288, 54 : ii. 64, 17.

wuldorlíce ; *adv. Gloriously* :—Hē ðæt setl ðære apostolícan cyrican wulderlíce (*gloriosissime*) heóld and rehte, Bd. 2, 1 ; S. 500, 10. Wuldor-líce, Blickl. Homl. 211, 31. Hié on manegum godcundum mægenum swíþe wuldorlíce áscinon, 161, 19.

wuldor-māga, an ; *m. A man who will attain the glory of heaven, an heir of heaven* :—Se wuldormāga (*St. Guthlac*), Exon. Th. 167, 28 ; Gū. 1067. v. next word.

wuldor-magu, a ; *m. A son of glory, an heir of heaven* :—Se wuldor-mago, eádig, Exon. Th. 179, 25 ; Gū. 1267.

wuldor-micel ; *adj. Gloriously great, magnificent* :—Gewitnesse beóð wuldormicele heofonwaru and eorðwaru, helwaru þridde, Hy. 7, 94.

wuldor-nyttung, e ; *f. Glorious use* :—Wuldornyttingum (woruld-? cf. weorold-nytt), Exon. Th. 492, 22 ; Rä. 81, 19.

wuldor-spéd, e ; *f. Glorious abundance* :—Setl wuldorspédum welig (*heaven*), Cd. Th. 6, 11 ; Gen. 87.

wuldor-spédig ; *adj. Glorious* :—Gingran sīne, wuldorspédige weras, Andr. Kmbl. 855 ; An. 428.

wuldor-tān, es ; *m. A glory-twig, a plant with medicinal virtues* :—Ðā genam Wōden .viiii. wuldortānas, slōh ðā ða næddran, ðæt heó on viiii. tōfleáh, Lchdm. iii. 34, 24.

wuldor-þrymm, es ; *m. Glorious majesty* :—Wealdend and wyrhta wuldorþrymmes, ēce God, Andr. Kmbl. 650 ; An. 325 : 1404 ; An. 702. Godes wuldorþrymmas mannum cȳþan, Blickl. Homl. 111, 17.

wuldor-torht ; *adj. Gloriously bright, splendid* :—Wuldortorht heofon-weardes gāst, Cd. Th. 8, 5 ; Gen. 119. Hādor sægl wuldortorht gewāt, Andr. Kmbl. 2912 ; An. 1459 : Cd. Th. 174, 7 ; Gen. 2874. Beácen wuldortorht, 167, 21 ; Gen. 2769. Wuldortorhtan weder, Beo. Th. 2276 ; B. 1136.

wuldor-weorod, es ; *n. The host of heaven* :—Ðæt ðū sié hlǽfdige wuldorweorudes, and worl[d]cundra hāda under heofonum, and helwara, Exon. Th. 18, 17 ; Cri. 285.

wuldor-word, es ; *n. A glorious word* :—Ðū, ealra cyninga þrym, clypast ofer ealle ; bið ðīn wuldorword wīde gehȳred, Hy. 7, 46.

wuldrian (and wuldran ?) ; *p.* ode. I. *to glorify*, (1) *to ascribe glory to* :—Ic wuldrige (*glorificabo*) naman ðīnne, Ps. Spl. 85, 11. Gif ic wuldrige (wuldria, Lind.: wuldrigo, Rush. *glorifico*) mē sylfne, Jn. Skt. 8, 54. Hī lofiaþ leóflícne, and wuldriaþ ordfruman ealra gesceafta, Exon. Th. 25, 16 ; Cri. 401. Hē God wuldrode *Deum magnificans*, Lk. Skt. 5, 25 : Homl. Skt. i. 3, 662. Wē sculon wuldrian and herian ūrne Dryhten, Homl. Th. i. 44, 2 : Hy. 8, 1. Se is tō weorþienne and tō wuldrienne, Blickl. Homl. 197, 6. God wuldriende (wuldrigendo, Lind.: wuldrende, Rush.) and heriende, Lk. Skt. 2, 20. (2) *to make glorious, bestow glory on* :—Wuldra (uuldra, Lind.) ðū mec *clarifica me*, Jn. Skt. Rush. 17, 5. Hē wolde ðone cyning mid ðyssum hwīlendlícum ārum wuldrian *temporalibus honoribus regem glorificare satagens*, Bd. 1, 32 ; S. 498, 22. Wē sié wuldræd *gloriemur*, Rtl. 41, 41. II. *to glory* in respect to something :—Ic wuldrige *glorior*, Ælfc. Gr. 25 ; Zup. 145, 11. Hwet wuldras (*gloriaris*) ðū in hete, Ps. Surt. 51, 3. Wuldraþ *gloriatur*, Hpt. Gl. 501, 55. Hȳ wuldriaþ (*gloriabuntur*) on ðē, Ps. Spl. 5, 14. Ða ðe ðære mycelnesse hiora spēda gylpaþ and wuldraþ (-iaþ ?), Ps. Th. 48, 6. Ða anlícnyssa ðe ðū on wuldrodest, Homl. Skt. i. 4, 382. Se brȳdguma ðe Agnes on wuldrode, 7, 77. Ða Iudēiscan wuldrodon on heora ǽlícum offrungum, Homl. Th. ii. 470, 24. Se ðe wuldrige, wuldrige on God ælmihtigne, and nō on hine sylfne, R. Ben. 4, 8. Gedafenaþ ðæt hī wuldrion on gedrēfednessum, Homl. Th. i. 554, 24. Þeáh ðe ic wylle wuldrian (*gloriari*), ne beó ic nā unsnoter, Ælfc. Gr. 44 ; Zup. 262, 7. Heó ongan wuldrian on God, Blickl. Homl. 157, 18. III. *to receive glory, be glorified* :—God wuldraþ (*glorificatur*) in geðæhte hāligra, Ps. Surt. 88, 8. Hē wuldraþ mid Gode on ðam heofenlícum setle, Homl. Th. ii. 552, 25. Basilius ðe ðā wuldrode mid Gode, Homl. Skt. i. 3, 661. v. ge-wuldrian.

wuldrig ; *adj. Glorious* :—Foreðingunge wuldrigo *intercessio gloriosa*, Rtl. 49, 34. Beodum wuldrigum *precibus gloriosis*, 72, 18.

wuldrung, e ; *f.* I. *glorifying* :—Wuldrung *glorificatio*, Rtl. 57, 6. II. *glorying* :—Wuldor and wuldrung *gloria et gloriatio*, Scint. 65, 4 : 191, 14.

wulf, es ; *m.* I. *a wolf* :—Wulf *lupus*, Wrt. Voc. ii. 113, 32 : i. 77, 77 : *licos*, 22, 61 : *lupa*, ii. 51, 29. Hwonne of heortan hunger oððe wulf sāwle and sorge somed ābregde, Cd. Th. 137, 20 ; Gen. 2276. Wulf sceal on bearowe, Menol. Fox 496 ; Gn. C. 18. Sceal hine wulf etan, hār hǽðstapa, Exon. Th. 328, 5 ; Vy. 12. Se hāra wulf, 291, 15 ; Wand. 82. Wulfes gehlēþan, 499, 30 ; Rä. 88, 23. Reáfiende wulfas, Mt. Kmbl. 7, 15. Ic (*the shepherd*) stande ofer mīne sceáp mid hundum ðe læs wulfas forswelgen hig, Coll. Monast. Th. 20, 15. Wulfa geþot *ululatus*, Wrt. Voc. i. 287, 24. Sume wurdon tō wulfan ; ða ðuton. ðonne hí sprǽcan sceoldon, Bt. 38, 1 ; Fox 194, 36 : Met. 26, 79. Swā sceáp gemang wulfas, Mt. Kmbl. 10, 16. Wineleás, wonsǽlig mon genimeþ wulfas tō gefēran, Exon. Th. 342, 25 ; Gn. Ex. 147. ¶ in battle-scenes the wolf is a frequent figure :—Ne wæl wēpeþ wulf se grǽga, morþorcwealm mæcga, ac hit ā māre wille, Exon. Th. 343, 3 ; Gn. Ex. 151. Ðæs se hlanca gefeah wulf in walde, Judth. Thw. 24, 25 ; Jud. 206. Ðæt grǽge deór, wulf on wealde, Chr. 937 ; Erl. 115, 14. Fyrd-leóð āgōl wulf on walde, Elen. Kmbl. 55 ; El. 28. Wulf sang āhōf, holtes gehlēða, 224 ; El. 112. Wulfas sungon atol ǽfenleóð ǽtes on wēnan, Cd. Th. 188, 7 ; Exod. 164. Se mǽsta dǽl ðæs heriges læg, hilde gesǽged, wulfum tō willan, Judth. Thw. 25, 36 ; Jud. 296. ¶ an early admiration for the wolf seems shewn by the frequency of *wulf* in proper names ; see e.g. Txts. 554 sqq. ; and its presence in early England is marked by the numerous place-names ; see e.g. Cod. Dip. Kmbl. vi. Index. I a. *in the phrase* wulfes heáfod (v. wulfheáfod-treów), *used in reference to outlaws* :—Si postea repertus fuerit et teneri possit, vivus regi reddatur, vel caput ipsius, si se defenderit ; lupinum enim caput gerat a die utlagacionis sue, quod ab Anglis *uulueshened* nominatur. Et hec sententia communis est de omnibus utlagis, L. Ed. C. 6 ; Th. i. 445, 4. [Gamelyn wolueshened was cryed and maad, Gam. 700. Cf. wearg, *and see* Grmm. R. A. 734.] II. applied to a cruel person :—Se biscop cwæþ tō ðæm hǽþnan kāsere : 'Ne gang ðū nā on Godes hūs ; ðū hafast besmitene handa; and ðū eart deófles wulf,' Shrn. 58, 9. Se āwyrgda wulf (*the devil*), Exon. Th. 16, 21 ; Cri. 256. [*Goth.* wulfs : *O. Sax.* wulf : *O. H. Ger.* wolf : *Icel.* ulfr.] v. heoru-, here-, hilde-, wæl-wulf ; wylf, wylfen[n].

wulfes-camb, es ; *m. Wild teazle* :—Wulfes-camb *cameleon*, Wrt. Voc. i. 31, 3 : *camellia*, 67, 9 : *camellea*, ii. 102, 50 : 13, 12. Se brāda wulfes-camb *camemelon alba*, i. 67, 26. Wulfes-camb. Genim ðysse wyrte seáw ðe man *chameaelae*, and ōðrum naman wulfes-camb nemneþ,

Lchdm. i. 122, 12. Heó hafaþ leáf swā wulfes-camb, 278, 14. Wið eágena dymnesse nim wulfes-camb neoðeweardne, iii. 4, 19.

wulfes-tǽsl. v. tǽsl.

wulfheáfod-treów, es ; *n. A cross* (?) :—Ealle naman habbaþ ánne, wulfheáfedtreó, Exon. Th. 437, 23 ; Rä. 56, 12. Cf. wulf, I a *and* weargtreów (*where add O. Sax.* warag-treó *a cross*).

wulf-heort ; *adj. Wolf-hearted, cruel* :—Onwóc wulfheort, se ǽr wīngāl swæf, Babilone weard, Cd. Th. 223, 7 ; Dan. 116. Wulfheort cyning, 224, 12 ; Dan. 135 : 231, 14 ; Dan. 247.

wulf-hlip, es ; *n. A hill where the wolf has its den* :—Hié dȳgel lond warigeaþ, wulfhleoþu, Beo. Th. 2720 ; B. 1358.

wulf-hol, es ; *n. A wolf's den* :—Uulfholu *lupinare*, Wrt. Voc. ii. 113, 34. Wulfholu, 51, 13.

wulf-seáþ, es ; *m. A wolf-pit* :—Be eástan ðæm wulfseáðe, Cod. Dip. Kmbl. iii. 264, 5.

wull, e : wulle, an ; *f. Wool* :—Uul *lana*, Wrt. Voc. ii. 112, 44. Wul, i. 66, 29 : 82, 7. Wull, ii. 51, 61. Unāwaxen wul *lana sucida*, 54, 6. Unāwæscen wul, i. 61, 8. Rammes wul (wull, *v. l.*), Lchdm. i. 356, 11. Ða loccas hire heáfdes wǽron swā hwīte swā wull, Homl. Skt. ii. 23 b, 177. Gā seó wǣge wulle tō .cxx. þ., L. Edg. ii. 8 ; Th. i. 270, 3. Wulle flȳs *lanam*, Ps. Th. 147, 5. Ne wāt ic mec beworhtne wulle flȳsum (uullan fliúsum, Txts. 150, 3), Exon. Th. 417, 12 ; Rä. 36, 3. Hī beóð gegyrede gódre wulle, Ps. Th. 64, 14. Mid liuene clāðe oððe mid eówocigre wulle, Lchdm. ii. 182, 5. Mid hnesce wulle oferwrīðe ealle ða scearpan, 130, 10. Nim wǣte wulle, i. 312, 12 : 362, 17 : Ps. Surt. 147, 16. Wullan (? *the MS. has* wulla *with a stroke after the* a) *lanam*, Kent. Gl. 1135. Wulla *lanas*, Hpt. Gl. 524, 14. [*Goth.* wulla : *O. H. Ger.* wolla : *Icel.* ull.] v. wyll.

wull-camb, es ; *m. A wool-comb* :—Hē sceal fela towtōla habban, ... wulcamb, Anglia ix. 263, 13. [*O. H. Ger.* wolla-champ *tradula* : *Icel.* ull-kambr.]

wull-fleós, -flȳs, es ; *n. A fleece of wool* :—Wulflȳs *cana vellus*, Wrt. Voc. ii. 128, 17.

wull-hnoppa, an ; *m. Wool-nap, the wool on a fleece* :—Wullhnoppa (*printed* -knoppa ; *but cf.* hnoppian *vellere*, Wrt. Voc. ii. 72, 56 : noppe *detuberare*, a noppe of clothe *tuberus*, Cath. Angl. 256) *lanugo*, Wrt. Voc. ii. 51, 66.

wullian ; *p.* ode *To wipe with wool* :—Wið scurfum ; rammes smeoru ; and meng ðǽrtō sōt and sealt and sand, and hyt wulla on weg, Lchdm. i. 356, 24.

wull-mod (-mōd ?) *a distaff* :—Wulmod *colus*, Wrt. Voc. i. 281, 80 : ii. 16, 32 : *colum*, 25, 9 : 134, 59. Uuilmod (wulf-) *colus*, Txts. 54, 306. [Cf. *O. H. Ger.* wolla-meit *colus*.] v. Anglia xix. 496.

wull-tewestre, an ; *f. A female wool-carder* :—Mǽden milde, wultewestre, Lchdm. iii. 188, 20. Mǽden grǽdig, wulltewestre, 196, 2.

wulluc. v. weoloc.

wull-wǽga ; *pl. f. Scales for weighing wool* :—Momentana lytle wǽga, campana wulwǣga, Wrt. Voc. i. 38, 43.

wund (*printed* pund, Wrt. Voc. i. 289, 61) *talpa*, Wülck. Gl. 279, 11, *read* wand.

wund, e ; *f. A wound* ; *vulnus*, Wrt. Voc. i. 85, 49. **I.** in a physical sense, (1) *a wound, an injury caused by a blow* :—Sió wund, ðe him se eorðdraca geworhte, Beo. Th. 5416 ; B. 2711. Blōdig wund, Andr. Kmbl. 2945 ; An. 1475 : Exon. Th. 143, 33 ; Gū. 670. Hér sindon dolhsealfa tō eallum wundum ... Sceád on ða wunde ... Wiþ ealdre tōbrocenre wunde ... lācna swilce wunda. Tō wunde clǽnsunge ... smire ða wunde mid, ðonne fullaþ hió ... Wiþ innanwunde, Lchdm. ii. 90, 23–92, 21. Wæs se cyning gehǣled fram ðære wunde ðe him ǽr gedōn wæs (*a vulnere sibi pridem inflicto*), Bd. 2, 9 ; S. 512, 1. Sylle wunde wið wunde *reddat vulnus pro vulnere*, Ex. 21, 25. Wīcing ðe him ða wunde forgeaf, Byrht. Th. 135, 57 ; By. 139. Hē ofer benne spræc, wunde wælbleáte, Beo. Th. 5443 ; B. 2725. Wundum āwyrded, Beo. Th. 2230 ; B. 1113. Wundum wērge, 5866 ; B. 2937. Se wīðfloga wundum stille hreás, 5653 ; B. 2830. Wundum sweltan, Byrht. Th. 140, 25 ; By. 293. Wǣpna wundum, Exon. Th. 119, 15 ; Gū. 255. Ða ealdan wunde and ða openan dolg on hyra Dryhtne, 68, 23 ; Cri. 1108. Swǣtge wunde, 89, 19 ; Cri. 1459. Hē wrāð his wunda (wundo *uulnera*, Lind.), Lk. Skt. 10, 34. (2) *a sore caused by disease.* v. wundig, wundiht :—Gif wambe bið oninnan wund, Lchdm. ii. 220, 3. Hē (*the itch*) wundaþ and sió wund sāraþ, Past. 11 ; Swt. 71, 20. Se deofol slōh Iób mid ðære wyrstan wunde ... Iób sæt eal on ánre wunde, Homl. Th. ii. 452, 25–28. Óðer wæs wæterseóc, óðer eall on wundum, ac hī wurdon gehǣlede fram heora untrumnysse, Homl. Skt. i. 5, 145. Heó wæs swýðe unhāl, and on eallum limum egeslíce wunda hæfde ... 'Ðū scealt underfón ðinra wunda hǽle.' ... On hire līce næs gesýne āht ðæra sárra wunda, 7, 265–278. **II.** in a figurative sense :—Feónda fǣrsearo, ðæt bið frēcne wund, Exon. Th. 48, 12 ; Cri. 770. Ðæt wom ǣrran wunde hǣlan, 81, 12 ; Cri. 1322. Wunde *cicatrice*, Hpt. Gl. 504, 35. Ic ofslōh wer on mīne wunde (*in vulnus meum*), Gen. 4, 23. Gāstes wunde *sins*, Ps. C. 51. Beóð wunde (wunda, Soul Kmbl. 177) onwrigene, ða ðe firenfulle men geworhton, Exon. Th. 372, 9 ; Seel. 89. Mīnra

wunda sār *dolorem vulnerum meorum*, Ps. Th. 68, 27. Synna wundum, Exon. Th. 263, 25 ; Jul. 355. Geseón on ussum sāwlum synna wunde, 80, 30 ; Cri. 1314. [*O. Sax.* wunda : *O. Frs.* wunde : *O. H. Ger.* wunta *vulnus, ulcus, plaga* : *Icel.* und.] v. cancor-, feorh-, in-, innan-, innoþ-, līc-, sweord-wund, *and next word.*

wund ; *adj. Wounded.* **I.** in a physical sense, (1) *of a wound inflicted.* v. wund, I. 1 :—Ða hwīle ðe hē wund wæs *dum convalescit a vulnere*, Ors. 3, 7 ; Swt. 118, 9. Gif wælt wund weorðeþ, L. Ethb. 68 ; Th. i. 18, 19. Wund wearð Wulfmǣr, wælreste geceás, Byrht. Th. 135, 4 ; By. 113. Se wyrm swefeþ sāre wund, Beo. Th. 5485 ; B. 2746 : Apstls. Kmbl. 121 ; Ap. 61. Gewāt him wund hǣleð gangan, Fins. Th. 86 ; Fin. 43. Geddung ðæs wundes *parabolam vulnerati*, Lk. Skt. p. 6, 19. Wundum dryhtne, heaðosiócum, Beo. Th. 5500 ; B. 2753. (1 a) *where the place of the wound is given* :—Gif mon bið on eaxle wund (gewunded, *v.l.*), L. Alf. pol. 53 ; Th. i. 94. 22. On breóstum wund, Byrht. Th. 136, 1 ; By. 144. Wund on ōþran earme *brachio saucius*, Ors. 4, 1 ; Swt. 158, 2. Wund þurh ōþer cneów *transfixo femore*, 4, 6 ; Swt. 180. 6. (1 b) *where the instrument with which the wound is inflicted is given* :—Īserne wund, Exon. Th. 388, 2 ; Rä. 6, 1. Mīn heáfod is searopīla wund, 497, 17 ; Rä. 87, 2. Mēcum wunde, Beo. Th. 1135 ; B. 565 : 2154 ; B. 1075. (2) *of disease.* v. wund, I. 2 :—Dolhsealf wið lungenādle ... mid ðȳ sceal mon lācnian ðone man ðe biþ lungenne wund, Lchdm. ii. 92, 21. Be wambe coþum, and gif hió innan wund biþ hū ðæt mon ongitan mæge and gelācnian, 220, 1. **II.** figurative. v. wund, II :—Hwider hweorfaþ wē hlāfordleáse, synnum wunde, gif wē swīcaþ ðē ? Andr. Kmbl. 813 ; An. 407. [*Goth.* wunds : *O. Sax.* wund : *O. H. Ger.* wunt (*in cpds.*) : *Ger.* wund.] v. dolg-, hrif-, þurh-wund.

wundel, e : wundle, an ; *f. A wound* (lit. or fig.), *sore* :—Gif hwylc lǣwede man ōðerne wundige, gebēte wið hine ða wunde (wundlan, wundlāc, *v.ll.*), L. Ecg. P. iv. 22 ; Th. ii. 210, 25. Wið nīwe wundela (wunda, *v.l.*), Lchdm. i. 8, 14 : 10, 9 : 92, 21 : 100, 1 : 108, 19 : 206, 6, 17. His sáule wundela (*vulnera*) gehǣlan, R. Ben. 72, 7. v. wyndle.

wunden-feax ; *adj. With plaited mane* :—Wicg wundenfeax, Beo. Th. 2804 ; B. 1400.

wunden-heals ; *adj. With twisted prow* :—Wudu wundenheals, Beo. Th. 601 ; B. 298.

wunden-locc ; *adj. With braided locks* :—Wīf wundenlocc, Exon. Th. 407, 26 ; Rä. 26, 11. Slōh wundenlocc (*Judith*) ðone feóndsceaþan, Judth. Thw. 23, 3 ; Jud. 103. Seó cneóris, wlanc, wundenlocc, 26, 13 ; Jud. 326.

wunden-mǣl ; *adj. Having curved markings*, applied to a sword :—Wearp wundenmǣl (wundel-, MS.), ðæt hit on eorðan læg, stīð and stȳlecg, Beo. Th. 3066 ; B. 1531. Cf. hring-mǣl.

wundenness. v. ofer-wundenness.

wunden-stefna, an ; *m. A ship with curved prow* :—Wundenstefna gewaden hæfde, ðæt ða līðende land gesāwon, Beo. Th. 445 ; B. 220.

wunder. v. wundor.

wundian ; *p.* ode *To wound* :—Se ðe mann wundaþ and wyle hine ofsleán *qui percusserit hominem volens occidere*, Ex. 21, 12. Hē (*the itch*) wundaþ and sió wund sāraþ, Past. 11 ; Swt. 71, 20. Beón hwīlum wundiaþ, Fragm. Kmbl. 41 ; Leás. 22. Wǣpenstrǣlas mē wundedon, Ps. Th. 56, 5. Indisce mȳs ūre feþerfōt niétenu wundedon and monige for hiora wundum swultan, Nar. 16, 8. Gif hwylc lǣwede man ōðerne wundige, gebēte wið hine ða wunde, L. Ecg. P. iv. 22 ; Th. ii. 210, 24. Ða cwōman tō ðon ðæt hié woldon ūs wundigan, Nar. 22, 17. Se cempa ongon Waldend wundian, Exon. Th. 260, 2 ; Jul. 291. Swelce hē nacodne hine selfne eówige tō wundigeanne his feóndum, Past. 38 ; Swt. 277, 17. Ungehēredre leoma tōslitnysse wundade (*lacerati*), Bd. 1, 7 ; S. 479, 14. [*Goth.* ga-wundōn : *O. H. Ger.* wuntōn : *Icel.* undaðr *wounded.*] v. for-, ge-wundian ; un-wundod.

wundig ; *adj. Ulcerous, full of sores* :—Wundie *ulcerosos*, Germ. 396, 267.

wundiht ; *adj. Ulcerous* :—Wundihtum *ulcerosis*, Germ. 396, 153.

wund-lāc, es ; *n. A wound* ; *see first passage under* wundel.

wund-líc ; *adj. That inflicts wounds* :—Wundlíce *uulnificum*, Germ. 402, 51.

wundor, es ; *n.* **I.** *a wonder*, (1) *a circumstance or act that excites astonishment* :—Ðæt is wundor tō cweþanne *mirum dictu*, Bd. 3, 6 ; S. 528, 10 : Beo. Th. 3453 ; B. 1724. Wundor mē ðincð eówer ðingrǣden, Homl. Th. ii. 484, 14. Mē þincþ wundor, Blickl. Homl. 179, 13 : 175, 13. Ðā wæs wundor micel, ðæt se wīnsele wiðhæfde, Beo. Th. 1546 ; B. 771 : Cd. Th. 37, 26 ; Gen. 595. Ðæt folc wundraþ ðæs ðe hit seldost gesihþ, ðeáh hit læsse wundor sié, Bt. 39, 3 ; Fox 216, 3. Ne þincþ ūs ðæt nān wundor, Blickl. Homl. 33, 7. Ne þincþ mē ðæt wundur wuhte ðe læsse, Met. 20, 117. Ac nis nā wunder *sed quid mirum* ? Hpt. Gl. 473, 44. Nis ðæt nān wundor, Bt. 31, 2 ; Fox 110, 9 : Met. 17, 7. Næs ðæt nānþing wundor, ðæt Drihten wæs ðam folce gram *nec miranda indignatio in populum*, Deut. 1, 37. Nis ǽnig wundor, hū ..., Exon. Th. 63, 7 ; Cri. 1016. Wundor wearð on wege ; wæter weard tō bāne, 483, 9 ; Rä. 68, 3. Ðæt is wundres dǣl, ðam ðe swylc ne conn, hū ...,

472, 3; Rä. 61, 10. Ic ðæt wundor gefrægn, ðæt se wyrm forswealg wera gied sumes, 432, 7; Rä. 48, 2. Ne ic on mægene miclum gange, ne wundor ofer mê wuniaþ ǽnig *neque ambulavi in magnis, neque in mirabilibus super me*, Ps. Th. 130, 2. Ðǽr bið wundra má, ðonne hit ǽnig mǽge áþencan, Exon. Th. 61, 24; Cri. 989. Ðæt wæs wundra sum, ðæt ðæt sweord gemealt íse gelícost, Beo. Th. 3219; B. 1607. (2) *a circumstance that excites astonishment as being out of the usual course of nature, a prodigy, portent* :—Gewurdon on Rôme ða yfelan wundor *obscoena et dira prodigia vel visa Romae vel nunciata sunt*, Ors. 4, 2; Swt. 160, 17. Wundra ł forebeácna *prodigia*, Hpt. Gl. 488, 34. (3) of the works of Divine power, *a wonder, miracle* :—For fyrwetgeornnesse ðæs wundres (*the raising of Lazarus*), Blickl. Homl. 69, 22. Á mæg God wyrcan wundor æfter wundre, Beo. Th. 1866; B. 931. Eal ðæt folc ðe ðis wundor (*the giving sight to the blind man*) geseah, Blickl. Homl. 15, 29. Ic bebeóde wundor geweorðan, Andr. Kmbl. 1459; An. 730: Cd. Th. 245, 31; Dan. 471: Elen. Kmbl. 2241; El. 1122. Gemunaþ hû hê mænig wundor geworhte *mementote mirabilia ejus, quae fecit*, Ps. Th. 104, 5. Mǽre synd his wundur ofer manna bearn *mirabilia ejus filiis hominum*, 106, 30. Swá fela wundra, swá wê gehýrdon gedône on Cafarnaum, Lk. Skt. 4, 23. Ðis worhte fruma ðara wundra (uundra, Lind.) ðe Hǽlend *hoc fecit initium signorum Jesus*, Jn. Skt. Rush. 2, 11: Blickl. Homl. 105, 25: Andr. Kmbl. 1138; An. 569. Ðæt (*the turning of Lot's wife into a pillar of salt*) is wundra sum, ðara ðe geworhte wuldres Aldor, Cd. Th. 155, 14; Gen. 2572. On eallum mínum wundrum *in cunctis mirabilibus meis*, Ex. 3, 20. For ðínum wundrum *a signis tuis*, Ps. Th. 64, 8. Mænigu wundur hê geworhte, 77, 5. Wundor, Cd. Th. 246, 4; Dan. 474: Blickl. Homl. 17, 10. Wundru, Past. 16; Swt. 103, 13: Ps. Th. 87, 12: Ex. 12, 12. Wundro, Blickl. Homl. 81, 10. Uundra, Jn. Skt. Lind. 11, 47. (3 a) *of supernatural power working through a human being, a miracle* :—Eft gelamp ôþer wundor ðissum onlíc, Blickl. Homl. 219, 6. His (*Oswald's*) wuudor wǽron miclo, Shrn. 114, 5: Elen. Kmbl. 1650; El. 827. Synd ðás wundru (*virtutes*) gefremede on him, Mt. Kmbl. 14, 2. Wæs ðis ðara wundra ǽrest ðe ðes eádiga wer geworhte, Blickl. Homl. 219, 2. Þurh ða wundor ðe heó geseah æt ðam bisceope, Shrn. 115, 6. Hêhbiscopes micla wundra *pontificis magnalia*, Rtl. 77, 19. (4) *a wonderful object, wondrous thing* :—Wên is ðæt hwilc wundor ineode on ðæt carcern and ða hyrdas ácwælde, Blickl. Homl. 239, 30. Heofonbeácen . . . ôðer wundor syllíc . . byrnende beám, Cd. Th. 184, 17; Exod. 108. Hwæt is ðæt wundor, ðæt geond ðás woruld fareþ ? . . . Yldo, Salm. Kmbl. 563; Sal. 281. Fêrdon folctogan wunder sceáwian, láþes lástas, Beo. Th. 1685; B. 840. Wundur, 6057; B. 3032: 6197; B. 3103. Hine wundra fela swencte on sunde, 3023; B. 1509. II. *wonderful, miraculous power* :—Ðæs engles mægen and his wundor weorðod bið, Blickl. Homl. 209, 20. Heofenas andettaþ hû wundor ðín standeþ, Ps. Th. 88, 4. Mycel ys his wundur ofer manna bearn, 106, 20. Eal ðis wæs geworden tô ðon ðæt wê sceoldan úres Drihtnes wundor oncnáwan, Blickl. Homl. 71, 23. III. *wonder, admiration* :—Þeóda wlítaþ, wundrum wafiaþ, hû seó wilgedryht wildne weorþiaþ, Exon. Th. 222, 1; Ph. 342. Ðysne wíg ðe ðú ðê tô wundrum (*as the object of thy adoration ?* cf. Ðam gyldnan gylde ðe hê him tô gode geteóde, l. 19) teódest, Cd. Th. 228, 25; Den. 208. ¶ cases, with or without prepositions, used adverbially or adjectively :—Ðæt of ðê ácenned bið, ðæt bið on wundra (*shall be a source of wonder*) eallum folcum, Homl. Ass. 121, 138. Gê mec tô wundre (*so as to excite wonder, wonderfully*), wǽgan môtun, Exon. Th. 124, 21; Gû. 341: Homl. Skt. i. 23, 652. Wundrum monigo *very many*, Mk. Skt. Lind. Rush. 7, 8. Wundrum lytel *wonderfully little*, Bt. 11, 1; Fox 32, 21. Swýþe wundrum well, Lchdm. i. 80, 21. Wundrum fæger, Exon. Th. 214, 1; Ph. 232: 202, 1; Ph. 63. Wundrum gegierwed, 483, 8; Rä. 68, 2. Eallum wundrum ðrymlíc girwan up swǽsendo, Judth. Thw. 21, 7; Jud. 8. Hû woruld wǽre wundrum geteód, Cd. Th. 222, 28; Dan. 111. [O. Sax. wundar : O. H. Ger. wuntar *mirum, prodigium, portentum, mirabile, miraculum, magnale, stupor* : Icel. undr.] v. eall-, fǽr-, hand-, lyft-, mægen-, niþ-, searo-, suudor-, swegel-, þeód-wundor.

wundor *hostimen*, Wrt. Voc. ii. 43, 20: 70, 32, *for (?)* pundor.

wundor-ágræfen ; *adj. (ptcpl.) Wondrously graven* :—Hê wundor-ágræfene anlícnesse engla sínra geseh, Andr. Kmbl. 1424; An. 712.

wundor-beácen, es ; *n. A wondrous sign* :—Swá hí on wege wyrcean sceoldon wundorbeácen, Ps. Th. 73, 5.

wundor-bebod, es ; *n. A monstrous command* :—Hê him bebeorgan ne con wôm wundorbebodum wergan gǽstes, Beo. Th. 3498; B. 1747.

wundor-bleó *a wondrous colour* :—Ðæs temples segl wundorbleóm geworht, Exon. Th. 70, 17; Cri. 1140.

wundor-clam[m], es ; *m. A wondrous clasp* :—Wæs gebunden deóran since duru ormǽte, wundurclommum bewriþen, Exon. Th. 19, 33; Cri. 310.

wundor-cræft, es ; *m.*   I. *wondrous skill, great cunning* :—Hê lǽmen fæt biwyrcan hêt wundorcræfte, Exon. Th. 277, 4; Jul. 575.   II. *miraculous power* :—Hê cyninges brôðor áwehte wundor-

craæfte þurh Drihtnes miht, ðæt hê of deáðe árás, Apstls. Kmbl. 110; Ap. 55. Godspell wrítan wundorcræfte, Andr. Kmbl. 26; An. 13: 1290; An. 645: Exon. Th. 427, 3; Rä. 41, 85. [Heo dude uundercraeftes, þe scucke hire fulste . . . to hire weoren iwoned þa uundercreftie men, Laym. 1147.]

wundor-dǽd, e ; *f. A deed of magic* :—Ealle ða men ða ðe Símônes wundordǽda wafodan, Blickl. Homl. 173, 22. [Þet folc com to se þys wonderdede (*the ordeal of Queen Emma*), R. Glouc. 337, 6. *M. H. Ger.* wunder-tát : *Ger.* wunder-that.]

wundor-deáþ, es ; *m. A wondrous death* :—Wedra þeóden (*Beowulf, killed by the fire-drake*) wundordeáðe swealt, Beo. Th. 6067; B. 3037.

wundor-fæt, es ; *m. A wondrous vessel* :—Byrelas sealdon wín of wunderfatum, Beo. Th. 2328; B. 1162.

wundor-full ; *adj. Wonderful, glorious* :—Mid wundurfulre (wundum-, MS.) wæfersêne *stupendo spectaculo*, Hpt. Gl. 470, 75. Wynsum is seó wunung on ðam wuldorfullum (wunderfullum, *v. l.*) dreáme, Homl. Ass. 43, 481. Wundorfulla (*gloriosa*) gecweden synd be ðê, Ps. Spl. 86, 2. [Wonderfol to telle, Laym. 280, 2nd MS. Þis ilke best zuo wonderuol and dreduol, Ayenb. 15, 4.]

wundor-gehwyrft *a wonderful turn* :—Of wundorgehwerfte *vice mirifica*, Germ. 390, 161.

wundor-geweorc, es ; *n. A wonderful work, a miracle* :—Þurh ðæt wundorgeweorc ðe hê Lazarum áwehte of deáþe, Blickl. Homl. 67, 6. Gelômlícu wundurgeweorc (*sanitatum miracula*) gewordene wǽron, Bd. 3, 9; S. 533, 3. Áwritene gemang ðara apostola wundorgewurcum, H. R. 13, 12. Hê hié tô heofona ríce laþode þurh his wundorgeweorc, Blickl. Homl. 7, 10. v. wundor-weorc.

wundor-gifu, e ; *f. A wondrous gift, wondrous capacity* :—Sumum wundorgiefe þurh goldsmiþe gearwad weorþeþ, ful oft hê gehyrsteþ wel brytencyninges beorn, Exon. Th. 331, 23 ; Vy. 72.

wundor-líc ; *adj. Wonderful, exciting admiration or surprise* :—Is wundorlíc (*mirabilis*) Drihten, Ps. Th. 92, 5: Met. 20, 3. Wunderlíc, Bt. 33, 4; Fox 128, 4. Mín (*an angel's*) nama is mycel and wundorlíc, Blickl. Homl. 137, 29. Wundorlíc (*mirabilis*) is geworden ðín wísdôm, Ps. Th. 138, 4: 118, 129. Hit is wundorlíc, ðæt ic secgan wille, Bt. 20; Fox 70, 27. Ðys is fram Drihtne geworden, and hit ys wundorlíc (wundurlíc, Lind.: wunderlíc, Rush. *mirabile*) on úrum eágum, Mt. Kmbl. 21, 42: Ps. Th. 117, 21. Ðæt is wundorlíc, ðæt gê nyton hwanon hê is, Jn. Skt. 9, 30: Met. 20, 86. Cymeþ wundorlíc Cristes onsýn, Exon. Th. 56, 25; Cri. 906. Wundorlíc wrǽcca (*Nebuchadnezzar*), Cd. Th. 256, 1; Dan. 634. Wundorlíc wǽgbora, Beo. Th. 2884; B. 1440. Ic eom wunderlícu wiht, Exon. Th. 399, 16; Rä. 19, 1: 400, 14; Rä. 21, 1. Ðæt wæs wunderlícu gemetgung *miro modo*, Past. 17; Swt. 113, 16. Wunderlíc gestreón *mirandum negotium*, Hpt. Gl. 469, 3. Oft hwǽm gebyreþ ðæt hé hwæt mǽrlíces and wundorlíces gedêð, Past. 4; Swt. 39, 6. Hê (*Samson*) weard swíðe ofþyrst for ðam wundorlícan slege, Jud. 15, 18. Wundorlícre hrædnysse hê bið álýsed, Lchdm. i. 288, 16. On wundorlícre mycelnesse, Blickl. Homl. 181, 20. Gesáwon hié wundorlíce wyrd, ðone man lifgendne ðone ðe hié ǽr deádne forlêton, 217, 36. Wundorlíc tácn, 205, 31. Ic ðê sǽde swíðe lang spell and wundorlíc, Bt. 35, 5; Fox 166, 2. Hû ða wísan sind wundorlíce, Exon. Th. 223, 14; Ph. 359. Hû his ða goodan weorc syndon wundorlíce *quam terribilia sunt opera ejus*, Ps. Th. 65, 2. Gif ðú wênst ðætte wundorlíce gerela[n ?] hwelc weorþmynd sié, Bt. 14, 1; Fox 42, 18. Hwonon him ða wundorlícan gereordo côman, Blickl. Homl. 153, 8. Eorðe brengð wæstma fela wundorlícra, Met. 20, 101. Sió hæfde wæstum wundorlícran, Exon. Th. 413, 14; Rä. 32, 5. Ðá cwôman ðǽr nædran wundorlícre ðonne ða ôþre wǽron and egeslícran . . . wǽron hié wunderlícre micelnisse, Nar. 14, 1–3. Seó burg ðe ǽr wæs ealra weorca fæstast and wunderlecast and mǽrast, Ors. 2, 4; Swt. 74, 24. Is ðæt eác ealles wundorlícost, ðæt . . ., Blickl. Homl. 127, 14. [*Laym.* wunder-lic : *Orm.* wunnderr-like : *Wick.* wondir-li : *O. Sax.* wundar-lík : *O. H. Ger.* wuntar-líh : *Icel.* undr-ligr.]

wundorlíce ; *adv. Wonderfully*, (1) *with adjectives* :—Ðǽr weard gegaderod wundorlíce micel folc, Homl. Skt. i. 23, 616. Hê hine gesette in wundorlíce micle cyrcean, Shrn. 121, 3. (2) *with verbs* :—Wundurlíce *mirabiliter*, Ps. Surt. 75, 5. Wundorlíce *mire*, Hymn. Surt. 70, 5. Drihten hine swá wundorlíce of eallum his earfoþum gefriþode, Ps. Th. 32, arg.: Ex. 11, 7: Past. 54; Swt. 423, 4: Bt. 33, 4; Fox 130, 35: Met. 20, 162: 13, 5. Wunderlíce, Bt. 33, 4; Fox 128, 5. Hié wundorlíce deáþ geþrowodan for Godes naman, Blickl. Homl. 171, 31. Wundurlíce heó hǽleþ, Lchdm. i. 194, 22. Wundorlíce, 220, 20. Hí wurdon wundorlíce áfirhte *timuerunt valde*, Gen. 20, 8. Ic ne fêrde on mǽrðum ne wundorlíce mid getote be mê ne bodude *neque ambulavi in magnis, neque in mirabilibus super me*, R. Ben. 22, 17. Hê hine gescerpte wlitegum wǽdum wundorlíce, Met. 15, 3. In ðis tô uundranne ł uundorlíce is *in hoc mirabile est*, Jn. Skt. Lind. 9, 30. And eác ðæt wunderlícor wæs, ðá ða heora án bodade mid ánre sprǽce, ǽlcum wæs geðúht swilce hê sprǽce mid his gereorde, Homl. Th. i. 318, 26. Se fugel wrixleþ wôðcræfte wundorlícor ðonne ǽfre byre monnes hýrde, Exon. Th. 206, 16; Ph. 127. [*O. H. Ger.* wuntarlíhho *mirabiliter*.]

wundor-máþþum, es; _m. A wondrous treasure_:—Ðone healsbeáh, wrǽtlícne wundormáþðum, Beo. Th. 4352; B. 2173.

wundor-seón, e; _f. A wondrous spectacle_:—Wundorsióna fela, Beo. Th. 1995; B. 995. [_O. H. Ger._ wuntar-siuni _spectaculum: Icel._ undr-sjónir; _pl. f. a spectacle._]

wundor-smiþ, es; _m. A smith who makes wonderful things_ or _who works by wondrous art_:—Gylden hilt, . . . enta ǽrgeworc, . . . wundor-smiþa geworc, Beo. Th. 3366; B. 1681.

wundor-tácen, es; _n. A wondrous sign_:—Wundortácna and fore-beácna _signorum et prodigiorum_, Ps. Th. 104, 23. [_O. Sax._ wundar-tékan: _O. H. Ger._ wuntar-zeichen _miraculum_.]

wundor-weorc, es; _n. A wondrous work, a miracle_:—Hé (Christ) óðerra unrím cýðde wundorworca, Andr. Kmbl. 1409; An. 705. Manige wítgan ǽr Sancte Iôhanne on swíþe manegum godcundum mǽgenum ealra wundorweorcum swíþe wuldorlíce áscinon, Blickl. Homl. 161, 19. Ðæt cwyce secgeaþ his wundorweorc ofer ealle werþeóde _annuntiate inter gentes opera tua_, Ps. Th. 104, 1. [War stod þat wonderworc (seolkuð werc, 1st MS.), Laym. 17376: _Ger._ wunder-werk.] v. wundor-geweorc.

wundor-weorold, e; _f. The wondrous world_:—Geond ðás wundor-woruld, Exon. Th. 421, 12; Rä. 40, 17.

wundor-wyrd, e; _f. A wondrous case_:—Be ðám nǽglum frignan ongan cwén, Cyriacus bæd, ðæt hire gástes mihtum ymb wundorwyrd willan gefylde, Elen. Kmbl. 2139; El. 1071.

wundrian; _p._ ode.   **I.** _to wonder at, to regard with surprise_ or _admiration_. (1) absolute:—Ealle gé wundriaþ (wundrigeaþ, _v. l._) _omnes miramini_, Jn. Skt. 7, 21. Se Hǽlend wundrode (wundriende wæs, Rush.) _Jesus miratus est_, Mt. Kmbl. 8, 10. Hig wundrodun (wund-radan, Rush.), 19, 25: 21, 20. Hí wundrodon mycelre wundrunge _obstupuerunt stupore maximo_, Mk. Skt. 5, 42. Ðá ongan ic wundrigan, Bt. 40, 1; Fox 236, 9. Is se godcunda anweald tó wyndrianne, 32, 2; Fox 116, 16. Tó wundranne (uundranne, Lind.) is _mirabile est_, Jn. Skt. Rush. 9, 30. Nis ðæt tó wundrigenne, þeáh ðe hé wǽre costod, Blickl. Homl. 33, 12. Ða leóda beheóldon swíðe wundrigende, Homl. Skt. ii. 26, 186. (2) with gen. :—Ðæt ungestæððige folc wundraþ ðæs ðe hit seldost gesihþ, ðeáh hit lǽsse wundor sié, Bt. 39, 3; Fox 216, 2: Met. 28, 49. Hwæt stondaþ gé hér and ðyses wundriaþ? Blickl. Homl. 123, 22. His wundriaþ ða ðe him underðiédde bióþ, Past. 4; Swt. 39, 7: Met. 28, 66. Ðá wundrade ic swíðe swíðe ðara gódena wiotona, Past. pref.; Swt. 5, 19. Ðá wundrode ðæt folc his láre, Mt. Kmbl. 7, 28. Hé wundrade Godes wundra, Ps. Th. 8, arg. Hwá is on weorulde, ðe ne wundrige fulles mônan? Met. 28, 40. Hwæþer gimma wlite eówre eágan tó him getió heora tó wundrianne? Bt. 13; Fox 40, 2. Hé férde wundrigende ðæs ðær geworden wæs, Lk. Skt. 24, 12. (3) with acc. :—Wundrað weras wlite and wǽstma, Exon. Th. 221, 7; Ph. 331. Ic ða wynsumnesse and fægernesse ðæs londes wundrade, Nar. 26, 26: 28, 1. Dý lǽs ðæt wundredan weras and idesa, Exon. Th. 176, 6; Gú. 1205. (4) with a clause :—Hwá ne wundraþ ðætte sume tunglu habbaþ scyrtran hwyrft ðonne sume habban? Bt. 39, 3; Fox 214, 17. Hí ne wundriaþ ðætte . . ., Met. 28, 50. Hí ne wundriaþ, hú hit on wolcnum þunraþ, þrágmælum eft ánforlǽteþ, 28, 54. Ðá wundrade se ðeng for hwon hé ðæs bǽde, Bd. 4, 24; S. 598, 31. Gif hwá wundrie, hú hit gewurðan mihte, Jud. 15, 19. (5) with gen. and clause :—Hwá ne wundraþ ðæs, ðæt sume steorran gewítaþ under ða sǽ? Bt. 39, 3; Fox 214, 26. Hwá ungelǽredra ne wundraþ ðæs roderes færeldes, hú hé ǽlce dæg úton ymb-hwyrfð ealne ðisne middaneard? 214, 15. Hwá wundraþ ðæs, oððe óðres eft, hwý ðæt ís mǽge weorðan of wætere, Met. 28, 58. Hwý ne wundriaþ hí ðæs, ðæt hit hwílum þunraþ, hwílum ná ne onginþ, Bt. 39, 3; Fox 214, 33. Hwá is ðe ne wundrige wolcna færeldes, roderes swifto, hú hý ǽlce dæge úton ymbhwerfaþ eallne middangeard? Met. 28, 2. Ǽlc wile ðæs wundrian for hwý hí swá dón, Bt. 39, 9; Fox 226, 14. (6) with prepositions :—Hé wundrode æfter ðære gesihþe, Blickl. Homl. 153, 35. Wundradun ða mengu se láre his _ammirabantur turbae super doctrinam ejus_, Mt. Kmbl. Rush. 7, 28. Hí wundrodon on his láre _admirabantur in doctrina ejus_, Mk. Skt. 6, 2. Ðá wǽron ða apostolas swíþe wundrigende fram him, and wǽron cweþende tó him hwonon him ða wundorlícan gereordo cóman, Blickl. Homl. 153, 7. (7) with preposition and clause :—Wundrade heó ymb ðæs weres snyttro, hú hé swá geleáfful on swá lytlum tíde ǽre wurde, Elen. Kmbl. 1914; El. 959. Hwá is moncynnes, ðæt ne wundrie ymb ðás tungl, hú hý sume habbaþ scyrtran ymbhwearft? Met. 28, 6.   **II.** _to make wonderful, magnify_ (?) :—Hé wundraðe (_mirificavit_) ealle willan míne, Ps. Spl. 15, 2. [_O. Sax._ wundrôn: _O. H. Ger._ wuntarôn: _Icel._ undra.] v. á-, ge-, of-wundrian.

wundrigend-líc; _adj. Expressing admiration_ or _astonishment_:—O is tóclypigendlíc _adverbium_ . . . hé is eác wundrigendlíc: O _qualis facies_, Ælfc. Gr. 38; Zup. 241, 16.

wundrum. v. wundor.

wundrung, e; _f._   **I.** _wondering, wonder, admiration, astonishment_:—Hwæt is ðeós wundrung ðe gé wafiaþ? Exon. Th. 6, 24; Cri. 89. Eall hé wæs ful wundrunge and wafunge; and eác ða byrig hé

geseah eall on ôþre wísan gewend, on ôþre heó ǽr wæs, Homl. Skt. i. 23, 509. Heó mid wundrunge wearð befangen, 2, 251. Hé þearle siððan Maurum wurðode, and on wundrunge hæfde (_held him in admiration_), 6, 185. Ðǽr heó lið ôð ðis on mycelre árwurðnysse mannum tô wundrunge (_to the admiration of men_), 20, 101. Hé on ðære micclan his môdes wundrunge ðǽr gestôd dreórig _in the great bewilderment of his mind he stood there downcast_, 23, 627. Hí wundrodon mycelre wund-runge _obstupuerunt stupore maximo_, Mk. Skt. 5, 42. Pape geswutelaþ wundrunge, Ælfc. Gr. 5; Zup. 11, 3. _Interjectio_ getácnaþ hwílon ðæs môdes blisse, hwílon sárnysse, hwílon wundrunge, 48; Zup. 278, 6.   **II.** _a wonderful sight, a spectacle_:—Wundrunge _spectaculi_, Hpt. Gl. 508, 27.

wund-swaþu, e; _f. The trace of a wound, a scar_:—Wundsweðe míne _cicatrices meae_, Ps. Surt. 37, 6.

wund-wácu (?) _a wound-weakness, a wound, sore_:—Swá benne ne burnon ne burston, ne fundian ne fcologan ne hoppetan, ne wundwáco sían (_sores may not run_), ne dolh dióþian, Lchdm. ii. 352, 2. Cf. wác; _n._

wune-líc (wun-, wunu-); _adj. Wonted, usual, accustomed_:—Wunlícre árfeastnisse _solita pietate_, Rtl. 35, 21. Wunulíco rúmmôdnise _solita clementia_, 180, 10. v. ge-wunelíc.

wune-ness (wunu-), e; _f._   **I.** _a dwelling, habitation_:—Hé him wunonesse stówe (_locum mansionis_) sealde, Bd. 5, 11; S. 626, 13. In wununise _in habitaculo_, Rtl. 58, 5: _habitatione_, 68, 20. Ða geworhte hé him nearo wíc and wunenesse (_mansionem angustam_), Bd. 4, 28; S. 605, 23. Ðá sealde se cyning him wununesse and stówe (_mansionem_) in Cautwarabyrig, 1, 25; S. 487, 18. Hé him sylfum wununesse and wíc geceás _ipse locum mansionis elegit_, 4, 26; S. 602, 38. Wit oferférdon ðás wununesse (_has mansiones_) ðara eáðigra gásta. . . . Wit becôman tô ðám blíþan wunenyssum (_ad mansiones laetas_), 5, 12; S. 629, 31, 43.   **II.** _continuance, perseverance_:—Wununise (_perseverantiam_) éces hehstaldnisse, Rtl. 105, 36. v. in-wuneness.

wung, Wrt. Voc. ii. 129, 26. v. pung.

wunian; _p._ ode _To dwell, remain_:—Wunat _inmoratur_, Wrt. Voc. ii. 111, 76. Wunaþ _constat_, Kent. Gl. 1176. Wunian _consistere_, 190.   **I.** _of living creatures, to dwell, abide, stay, remain, live_, (1) of dwelling in a place or with a person, (a) with preps. or adverbs :—Ðú geond holt wunast _thou shalt live in the woods_, Cd. Th. 252, 6; Dan. 574. Ðú in heánnissum wunast mid Waldend Fæder, Exon. Th. 10, 36; Cri. 163. Se þeów ne wunaþ (_manet_) on húse on écnesse; se sunu wunaþ on écnesse, Jn. Skt. 8, 35. Pellicane gelíc, se on wéstene wunaþ, Ps. Th. 101, 5. Monna gehwylc cwic þendan hér wunaþ, Exon. Th. 37, 8; Cri. 590. Ða hwíle ðe wé on ðysse worlde wuniaþ, Blickl. Homl. 103, 24. Him (_the whale_) ða férend on fæste wuniaþ, wíc weardiaþ, Exon. Th. 361, 26; Wal. 25. Mislíce wildeór wuniaþ (_morantur_) on wudum, Coll. Monast. Th. 22, 23. On heán muntum heortas wuniaþ, Ps. Th. 103, 17. Ic on wéstene wunode lange _mansi in solitudine_, 54, 7. Wunude, 83, 1. Hé wunode ðǽr on mynstre, Homl. Skt. i. 6, 99. Hé on ðæm lande feala wintra wunode, Blickl. Homl. 13, 13. Hé wunode be Iordane, Cd. Th. 116, 5; Gen. 1931. Hé ðǽr wunode mid him, Blickl. Homl. 239, 18: 249, 16: Exon. Th. 162, 8; Gú. 972: Beo. Th. 2261; B. 1128. Wé cômun tô ðam ðæt wé wunedon on ðínum lande, Gen. 47, 4. Ðá hig wunedon on Galiléa _conversantibus eis in Galilaea_, Mt. Kmbl. 17, 22. Hié ealle onyppan wunedon, bídende ðæs Hálgan Gástes, Blickl. Homl. 133, 26. Wunedon on ðám wícum Abraham and Loth, Cd. Th. 113, 20; Gen. 1890. Wuna mid úsic and ðé wíc geceós, 164, 29; Gen. 2722. Medmicel fæc nú gyt wuna mid ús, Blickl. Homl. 247, 33. Wuna in ðære wínbyrig, Andr. Kmbl. 3340; An. 1674. Wuniaþ (wunas, Lind.: wynigaþ, Rush. _manete_) ðǽr, Mt. Kmbl. 10, 11. Wunigaþ on ðam ylcan húse, Lk. Skt. 10, 7. Eal ðæt manegu ðe him mid wunige, Andr. Kmbl. 1890; An. 947. Hí on his neáweste wunian, Ps. Th. 148, 14. Beón, gif hí man acwellaþ, cwelle hig man raþe, . . . ðæt hig ofer niht ðǽron ne wunigon (_restent_), L. Ecg. C. 39; Th. ii. 164, 3. Ic wylle tó-dæg on ðínum húse wunian (tô wunianne, Lind.: tô wuniganne, Rush. _manere_), Lk. Skt. 19, 5. Wunian on éðle, Cd. Th. 294, 27; Sat. 477. Mid wuldorcyninge wunian, 283, 30; Sat. 312. In worulde wunian, Exon. Th. 51, 21; Cri. 819. Wunian in wícum, 316, 9; Mód. 46. Wunigan in wuldre mid weoroda God, 22, 5; Cri. 347: Blickl. Homl. 25, 35. Hé leng mid him líchomlíce wunian nolde, 135, 22. Hé on his môdor bôsme wunigende wæs, 165, 18. Wǽron ealle ða apostolas wunigende on ánre stówe, 133, 15. (b) with dat. (inst.) :—Hé wícum wunode, Cd. Th. 108, 26; Gen. 1812. Abraham wunode éðeleardum Cananéa, 116, 32; Gen. 1945. Wuna ðǽm ðe ágon _abide with those own thee_, 138, 18; Gen. 2293. Wícum wunian, Beo. Th. 6158; B. 3083. (c) with acc. _to inhabit a place, live in_ or _on_:—Hé heánne beám wunaþ, Exon. Th. 209, 17; Ph. 172. Ic ísceáldne sǽ winter wunade, 306, 29; Seef. 15. Ða ðe hleóleásan wíc hwíle wunedon, Andr. Kmbl. 262; An. 131. Wunian wíc unsýfre, 2621; An. 1310. Wederburg wunian, 3391; An. 1699. Seó ðe wunian sceolde cealde streámas, Beo. Th. 2525; B. 1260: Cd. Th. 280, 22; Sat. 259: 282, 36; Sat. 297. Ic (_the soul_) ðé (_the body_) wunian sceolde, Soul Kmbl. 86; Seel. 43. ¶ in

figurative expressions :—Se fugel wunaþ wyllestreámas (*bathes*), Exon. Th. 204, 29 ; Ph. 105. Wunian wælreste *to lie dead*, 184, 10 ; Gú. 1342 : Beo. Th. 5796 ; B. 2902 : *to be buried*, Elen. Kmbl. 1444 ; El. 724. Reste wunian *to sleep*, Cd. Th. 223, 22 ; Dan. 123 : Rood Kmbl. 6 ; Kr. 3. (2) *to live, be* in certain conditions or circumstances, (a) with prep. or adv. :—Þenden ic wunige on worulddreámum *quamdiu ero*, Ps. Th. 103, 31. Seó sáwel ðe wunaþ on heofena ríces gefeán, Blickl. Homl. 57, 31. Wunaþ hé on wiste, ne hine wiht dréfeþ, Beo. Th. 3474 ; B. 1735. Ða menigo ðe wuniaþ on nearonédum, Andr. Kmbl. 202 ; An. 101. On fýrbæðe ðú wunodest, Elen. Kmbl. 1897 ; El. 950. Hé in yrmðum wunade, Andr. Kmbl. 326 ; An. 163. Wé wunodon on wynnum, Cd. Th. 279, 12 ; Sat. 237. Hí wunedon ætsomne, Met. 20, 243. Wunian on écean wuldre, Blickl. Homl. 105, 1. In wynnum wunade, Cd. Th. 299, 26 ; Sat. 556: Exon. Th. 140, 2 ; Gú. 604. Wunian in wylme, Salm. Kmbl. 933 ; Sal. 466. . Adam wæs wunigende on ðisum lífe mid geswince, Homl. Th. i. 20, 6. (b) with dat. (inst.) :—Hé wunaþ unlustum, Salm. Kmbl. 538 ; Sal. 268. Heó helltregum wunodon, Cd. Th. 5, 19 ; Gn. 74. Eádig weorþan, wunian wyndagum, Exon. Th. 330, 34 ; Vy. 61. (c) with noun or adj. :—Borhhond [hé] wunade *fidejussu exstitit*, Kent. Gl. 743. Wunude *extitit* (*praestantior*), Hpt. Gl. 511, 60. Ðæt ðú langlíf ofer eorðan wunie, Homl. Th. ii. 36, 2. Ána lífgan, wineleás wunian, Exon. Th. 344, 15 ; Gn. Ex. 174. (3) *to abide, be present* with a person to comfort or help :—Ic ðé mid wunige, Andr. Kmbl. 198 ; An. 99 : Exon. Th. 30, 12 ; Cri. 478. God wunaþ on him *est in ipsis Dominus*, Num. 16, 3. II. of things abstract or concrete, *to be, rest, reside, remain*, (1) in respect to locality, *occupy a position* :—Wunaþ *morabitur*, Kent. Gl. 481 : *commorabitur*, 540. (a) with prep. or adv. :—Se hálga stenc wunaþ geond wynlond, Exon. Th. 203, 10 ; Ph. 82. Ðær se wísdóm wunaþ on gemyndum, Met. 7, 39. Lyft on middum wunaþ, 20, 79. Mid ðam wítegendlícum gáste ðe on ðé wunaþ, Homl. Skt. i. 18, 282 : Cd. Th. 56, 7 ; Gen. 908. Wæter ðe wuniaþ gyt under fæstenne folca hróses, 10, 6 ; Gen. 152. Eorðe and wæter wuniaþ on fýre, Met. 20, 148. Beorh wunode on wonge, Beo. Th. 4476 ; B. 2242. Se monlíca stille wunode, ðær hié begeat wíte, Cd. Th. 155, 3 ; Gen. 2567. Egesa on breóstum wunode, 173, 24 ; Gen. 2866. Treów on ðé wunade, Exon. Th. 6, 12 ; Cri. 83 : 126, 4 ; Gú. 366. Hwæþer him yfel ðe gód under wunige, 82, 4 ; Cri. 1333. Þeh mín líchama on niðerdælum eorðan wunige, Ps. Th. 138, 13. Saga mé hwær seó ród wunige, Elen. Kmbl. 1244 ; El. 624. Tó manna heortan, ǽr Drihtnes weorc ðǽr wunian móte, Blickl. Homl. 19, 8 : 111, 5. Ǽnigne dǽl secgas geségon on sele wunian, Beo. Th. 6248 ; B. 3128. Geweoton hí mearcland tredan, forlǽton moldern wunigean, open eorðscræfu, Andr. Kmbl. 1605 ; An. 803. Lazarus, ðe Crist áwehte ðý feorþan dæge ðæs ðe hé on byrgenne wæs fúl wunigende, Blickl. Homl. 75, 5. (b) with acc. :—Ðæt treów sceolde wésten wunian, Cd. Th. 251, 5 ; Dan. 559. Hine gærsbedd sceal wunian, Ps. Th. 102, 15. (2) of state or circumstance :—Inc sceal sealt wæter wunian on gewealde, Cd. Th. 13, 7 ; Gen. 199. Ðá ðá ðis ígland wæs wunigende on sibbe, Homl. Skt. i. 13, 148. III. *to consist* of or in, *subsist, exist*. v. wunung, III :—On wordum Godes ríce ne wunaþ, ac on ánwylnysse ðæs hálgan geleáfan, Guth. prol. ; Gdwin. 2, 15. Ðeós lyft ys án ðæra feówer gesceafta, ðe ǽlc líchamlíc ðing on wunaþ. Feówer gesceafta synd, ðe ealle eorðlíce líchaman on wuniaþ, Lchdm. iii. 272, 11–13. Ða hálgan þrynnysse on ánre godcundnysse ǽfre wunigende, Homl. Skt. i. 15, 216 : 16, 1. Nis ná se Hálga Gást wunigende on his gecynde swá swá hé gesewen wæs, for ðan ðe hé is ungesewenlíc, Homl. Th. i. 322, 17. Him (*man*) is gemǽne mid stánum ðæt hé beó wunigende ; him is gemǽne mid treówum ðæt hé lybbe, 302, 20. Hí nǽron ǽfre wunigende, ac God hí gesceóp, 276, 15. God is þurh hine sylfne wunigende, ii. 236, 18. IV. where there is permanence, continuity, *to remain, last, continue, endure* :—Ðú wunast *tu permanebis*, Ps. Th. 101, 23. Ðú on écnesse wunast *tu in aeternum permanes*, 101, 10 : 92, 3. Ðǽr nóht elles ne wunaþ, Blickl. Homl. 101, 5. Ðínne naman ðe wunaþ on ealra worlda world, 143, 31 : Ps. Th. 111, 8. Seó ðe ǽfre wæs and eác nú wunaþ, Homl. Skt. i. 15, 217. Þenden ðær wunaþ húsa sélest, Beo. Th. 574 ; B. 284. Swá hwylc swá on elne óþ his ende wunaþ, se bið hál, Blickl. Homl. 171, 26 : Homl. Th. ii. 502, 21. Ðære wylne sunu wunaþ eal his líf on ðeówte, i. 110, 29. Wuniaþ ða wácran, Exon. Th. 311, 4 ; Seef. 87. Gif hé wunode ofer middæg *if he continued to live past noon*, Homl. Skt. i. 3, 595. Lucia on ðære ylcan stówe wunode ðe heó ofslagen wæs *Lucia remained lying in the same place that she was struck down*, 9, 146. Se snáw leng ne wunede ðonne áne tíde, Nar. 23, 21. Hí wunedun (wéren wungiende *mansissent*, Lind.) óð ðysne dæg, Mt. Kmbl. 11, 23. Ðæt hió ne wunian on worldlífe *ita ut non sint*, Ps. Th. 103, 33. Herenes Drihtnes hér sceal wunian on worulda woruld *laudatio ejus manet in seculum seculi*, 110, 8 : 118, 44, 91. Hæfð hé ðæt gewrixle geset ðe nú wunian sceal, Met. 11, 56. On sáre his líchoma sceal hér wunian, Blickl. Homl. 61, 1 : Exon. Th. 7, 19 ; Cri. 103. Eallum rihtgelýfdum mannum wunigendum for his noman, Blickl. Homl. 171, 14. IV a. with a complementary word or phrase :—Heó wæs fǽmne ǽr hire beorþre, and heó wunaþ fǽmne æfter hire beorþre, Blickl. Homl. 155, 33. Ðú un-

stilla gesceafta ástyrest and ðé self wunast swíðe stille, Met. 20, 16. God ána unáwendendlíc wunaþ, Bt. 35, 5 ; Fox 166, 9. Gescylded á wunaþ ungewyrded, þenden woruld stondeþ, Exon. Th. 210, 5 ; Ph. 181. Hió dumb wunaþ, 414, 7 ; Rä. 32, 16. Heó wæs mid twám werum and swá ðeáh wunode mǽden, Homl. Skt. i. 20, 3. Se hearda hyge hálig wunode, Exon. Th. 135, 1 ; Gú. 517. Hí ðágyt hǽdene wunodon, Homl. Th. ii. 502, 23. Is sǽd of ðære tíde ðe hí þanon gewiton óþ tódæge ðæt ðæt land wéste wunige (*manere desertus perhibetur*), Bd. 1, 15 ; S. 483, 27. Ðis ungefremed wunie, L. Ath. i. proem. ; Th. i. 198, 13. Hé hét wunian wyrtruman eorðan fæstne, Cd. Th. 248, 20 ; Dan. 516. Sceal lufu uncer wǽrfæst wunian, Exon. Th. 173, 19 ; Gú. 1163. Abrames wíf wæs ðágit wuniende bútan cildum *she remained still childless*, Gen. 16, 1. V. *to be wont* :—Ic gewunige *soleo*, wunigende *solens*, Ælfc. Gr. 41 ; Zup. 247, 5. [O. E. Homl. wunian : A. R. Kath. wunien : Laym. wunien, wonien : Orm. wunenn : Gen. and Ex. wunen : Ayenb. wonie : Chauc. Piers P. Wick. wone : O. Sax. wonôn, wunôn : O. Frs. wona, wuna : O. H. Ger. wonên *habitare, morari, conversari, manere, solere*.] v. á-, ge-, on-, þurh-wunian ; án-, dryht-, weorold-wuniende.

wunigend, es ; m. *An inhabitant* :—Gyf wé gefyllaþ wunigendes þenunge *si compleamus habitatoris officium*, R. Ben. Interl. 5, 11.

wunigend-líc, wunn, wununess. v. un-wunigendlíc, wyn[n], wuneness.

wunung, e ; f. I. *dwelling, living* :—Gif hé hine sylfne tó mynstres wununge gefæstnian wyle *if he will settle to living in a monastery* ; the Latin is : Si voluerit stabilitatem suam firmare, R. Ben. 108, 13. II. *a dwelling, habitation, place to live in* :—Feala muneca wunung *coenobium*, Wrt. Voc. i. 59, 5. Wunung *mansio*, 86, 46. Sý wunung (*habitatio*) heora onwést, Ps. Spl. 68, 30. Com Eustachius mid his here tó ðam túne ðe heó ðá on wæs. Wæs seó wunung ðǽr swýþe wynsum on tó wícenne, and his geteld wǽron gehende hire wununge geslagene, Homl. Skt. ii. 30, 315. Wé wendaþ ús eástweard, þonne wé ús gebiddaþ ... ; ná swylce on eástdǽle synderlíce sý his wunung ... on rihtwísum mannum is Godes wunung ... Swá eác se fordóna man bið deófles templ, and deófles wunung, Homl. Th. i. 262, 5–18. Ne biþ ðǽr Cristes eardung ne his wunung on ðære heortan, Blickl. Homl. 13, 24. On ðære fíftan flёringe wæs ðæra manna wunung gelógod, Boutr. Scrd. 21, 10 : Homl. Th. i. 536, 16. Wununge *contubernio, habitaculo*, Hpt. Gl. 468, 63. Tó móderlícum wununge ł bósme *ad maternum gremium*, 504, 12. Ðæt hí sceoldon habban ða fægeran wununge ðe se feónd forleás, Ælfc. T. Grn. 3, 6. Hé him ðǽr wununge getimbrode, Shrn. 13, 16. Spyrian hwǽr ða mánfullan wununge habban, L. Eth. ix. 40 ; Th. i. 348, 26. Ða habbaþ hundfealde méde and ða mærestan wununge, Homl. Ass. 21, 187. Wununga *sedes*, Hpt. Gl. 412, 33. Se Hælend sǽde, ðæt on his Fæder húse syndon manega wununga, Homl. Ass. 42, 454. On muntum and on feldlícum wunungum, Jos. 10, 40. Ðú wircst wununge (*mansiunculas*) binnan ðam arce, Gen. 6, 14. Gé begeáton eów ðeósterfulle wununga on heora heortum Criste sylfum, Homl. Th. i. 68, 5. Geleáffulle menn gearwiaþ clǽne wununga on heora heortum Criste sylfum, Blickl. Homl. 73, 12. III. *being, existence, living*. v. wunian, III :—Wunung *essentia*, i. *aeternitas, natura*, Wrt. Voc. ii. 144, 20. Þeáh se líchama geendige, ðe sceal eft þurh Godes mihte árísan tó écere wununge, Homl. Th. i. 20, 6. Se is lybbende God ðe hæfð líf and wununge ðurh hine sylfne, 366, 33. Gesceafta nabbaþ náne wununge þurh hí sylfe, ac ðurh God, se ðe ána is þurh hine sylfne wunigende, ii. 236, 17. Yfel nis náu þing þurh hit sylf, and náne wununga næfð búton on sumum gesceafta, Boutr. Scrd. 20, 44. [Wreððe hafð wununge on þes dusian bosme *ira requiescit in sinu stulti*, O. E. Homl. i. 105, 24. Hore wunynge naued no ʒet *habitatio eorum non habet januam*, A. R. 74, 12. Þe wununge of euch wunne *quietis eterne mansio*, Kath. 2423. God woning (hæh bold, 1st MS.), Laym. 7094. His (*the Reeve's*) wonyng was ful fair upon an hethe, Chauc. Prol. C. T. 606. O. L. Ger. wununga *habitatio* : O. H. Ger. wununga *mansio*.]

wurdian, wurdlian, wurm. v. wordian, wordlian, wyrm.

wurma, wyrma, an ; m. : wurme, an ; f. *A shell-fish from which a purple dye was obtained, a purple dye* ; also *woad, a plant from which a dye is got* :—Wurma *murex*, wurma, weoloc *murice*, Wrt. Voc. ii. 56, 64, 62. Wurma, reád godwebb *ostrum*, 64, 10. Wyrma *ostrum*, i. 286, 34. Wurman *murice*, ii. 114, 46. Ungemæccre wurman *dispari murice*, 141, 19. Mid unilícere wurman, Hpt. Gl. 431, 42. Ungelícum wurman, Anglia xiii. 29, 58. Wolcreádum wurman *bistincto cocco sive vermiculo*, 29, 56. Twyhíwedum wurman, Hpt. Gl. 431, 31. Mid reádre wurman *croceo luto* (v. Ald. 75), 524, 40 : Wrt. Voc. ii. 52, 48 (*printed* wurmaman). Wurman geblonden, Exon. Th. 218, 14 ; Ph. 294. Wyrman *murice*, Wrt. Voc. ii. 77, 23 : 89, 29. Wyrman (*printed* wyrmaman, cf. 52, 48) *luto*, 87, 32. Wyrman (*purpureo*) *ostro*, Hpt. Gl. 522, 5. Genim myrran and hwít récels and safinam and saluiam and wurman, Lchdm. ii. 294, 24. Wurmum *muricibus*, Wrt. Voc. ii. 55, 20 : Hpt. Gl. 524, 27. Wurman, 431, 47. [O. Frs. worma : O. H. Ger. wurmo *vermiculus*.] v. cor-, corn-, feld-, stán-wurma ; wurm-reád.

wurmille, an ; f. *Marjoram* :—Wurmille, uurmillae *origanum*, Txts. 83, 1452. Wurmilla, Wrt. Voc. ii. 65, 27. Wurmille, 64, 11. Wyr-

melle (printed war-), i. 32, 11. Wyrmella, 286, 35. [Cf. O. H. Ger. wurmeli vermiculus.]

wurm-reád; adj. Scarlet :—Wurmreádne þræd coccinum, Gen. 38, 28. Wurmreádne basing pallium coccineum, Jos. 7, 21. Cf. wyrm-basu.

wurms. v. worms.

wurmsig; adj. Purulent :—Wurmsi purulentus, Wrt. Voc. ii. 67, 29. v. wyrmsig.

wurmsihtig; adj. Purulent :—Wurmsihtig purulentus, Wrt. Voc. i. 22, 1.

wurpan. v. weorpan.

wurpol (-ul); adj. That throws down :—Wurpul ster[n]ax, Wrt. Voc. ii. 121, 42.

wurpþ, wursm, wurst, wurþ, wurt-mete. v. wirpan, worms, wirsa, weorþ, wyrt-mete.

wúsc-bearn, es; n. A beloved or an adopted child :—Uúscbearn (wuso, Rush.) filioli, Jn. Skt. Lind. 13, 33. [Cf. Icel. óska-barn a chosen, adopted child; ósk-mögr a beloved son: M. H. Ger. wunsch-kint; see Grmm. D. M. p. 139 (Stallybrass' trans.), and s. v. wunsch. Cf. too the proper name Wúsc-freá, Bd. 2, 14; S. 518, 1.] v. wýsc.

Wúse, wuso, wutan (-on). v. Úse, wúsc-bearn, witon.

wyde-treów. v. wudu-treów.

wyla (? hyla v. hylu) :—Wyla ł hola cabearum, Hpt. Gl. 489, 71.

wylcþ, Germ. 389, 42. v. welwan.

wylf, e; f. A she-wolf :—Wylf lupa, Txts. 75, 1260. Fǽddæ hiǽ wylif in Rómæcæstri, 127, 2. [Cf. O. H. Ger. wulpa lupa: Icel. ylgr a she-wolf.] v. brim-wylf, and next word.

wylfen[n], e; f. A she-wolf, (1) literal :—Gif heó drinceþ wylfene meolc, Lchdm. i. 362, 13. Wylfene beluae, bestiae maris, Wrt. Voc. ii. 125, 43. (2) figurative :—Wylfen Bellona, i. furia, dea belli, mater Martis, Wrt. Voc. ii. 125, 41. Reþre wylfenne dire parce, 140, 53. [Wummone wroð is wuluene, and mon wroð is wulf, A. R. 120, 9. Leoun or uulf, uuluine or bere, Havel. 573. M. H. Ger. wulfinne: Ger. wölfinn.]

wylfen; adj. Wolfish, fierce :—Wé geáscodan Eormanríces wylfenne geþóht; ðæt wæs grim cyning, Exon. Th. 378, 24; Deór. 22.

wylinc, Hpt. Gl. 419, 77. v. willung.

wyll, e; f. Wool :—Ða wylle and ða horna hý dóð heom tó nytnysse lanam et cornua in usum suum convertunt, L. Ecg. C. 40; Th. ii. 166, 31. v. wull.

wyllen; adj. Woollen, of wool :—Wyllen laneum, línen wearp vel wyllen áb linostema, Wrt. Voc. i. 40, 7, 8. Hé náðor ne wyllenes hrægles ne línenes brúcan nolde, Guthl. 4; Gdwin. 26, 11. Bind mid wyllenan þreáde, Lchdm. ii. 310, 22. Hí mid willenum reáfe heora líchoman gegearwiaþ, R. Ben. 139, 14. Ða wyllenan (? the word is printed wylnenan and put as a gloss to vetulae; the passage is: Cygnaeam vetulae senectutis caniciem, Ald. 25) hárnysse (in the margin is ða grǽgan hárnysse) cygneam canitiem, Hpt. Gl. 450, 62. Wyllene wearp lanea stamina, 417, 27. Ne hé wyllenra hrægla breác, ac línenra ealra, Shrn. 93, 7: 94, 28. Heó nǽfre línenum hrægium brúcan wolde, ac wyllenum, Bd. 4, 19; S. 588, 6. ¶ used substantively, woollen stuff :—Heó wyllen weorode, Homl. Skt. i. 20, 44: L. Edg. C. 10; Th. ii. 280, 19. Nime man wyllen tó líce, Wulfst. 170, 10. [O. H. Ger. wullín laneus.]

wylnenan. v. preceding word.

wyn[n], e; f. **I. delight, pleasure** :—Wyn luxus, Wrt. Voc. ii. 71, 11. Wynn luxoria, wynne luxus, 49, 67, 65. Genihtsumere wynne opulenti luxus, Hpt. Gl. 413, 71. Wyn eal gedreás, Exon. Th. 288, 25; Wand. 36. On Gode standeþ wuldor mín and wyn mycel, Ps. Th. 61, 7. Mín wynn álæg, 119, 5. Nis hearpan wyn, Beo. Th. 4517; B. 2262. Hwæþere him ðæs wonges wyn (his delight in the country) swéðrade, Exon. Th. 123, 16; Gú. 323. Sý æt him sylfum gelong eal his worulde wyn, 444, 12; Kl. 46. Ðǽr wæs wuldres wynn . . . næs ðǽr ǽnigum gewinn, Andr. Kmbl. 1773; An. 889. Ágan mé ðæs dreámes gewald, wuldres and wynne, Cd. Th. 275, 21; Sat. 175: Exon. Th. 230, 31; Ph. 480. On wynne in laetitia, Ps. Th. 104, 38. Wend ðé from wynne, Cd. Th. 56, 28; Gen. 919. In lifgendra londes wynne, Exon. Th. 27, 28; Cri. 437: 151, 5; Gú. 790. Wenne, Ps. C. 157. Weorod wæs on wynne, Beo. Th. 4032; B. 2014: Exon. Th. 462, 21; Hö. 55. Ðæt treów wæs on wynne, wudu weaxende, 435, 18; Rä. 542. Beóð on wenne ða bán ðe on hǽndum wǽron exultabunt ossa humiliata, Ps. C. 80. Habban ða mid wynne weorðe blisse exultent et laetentur, Ps. 69, 5. Wé sealmas him singan mid wynne in psalmis jubilemus ei, 94, 2. Ic mé on ðé gehálgode hús tó wynne, Exon. Th. 90, 31; Cri. 1482: 76, 26; Cri. 1245. Wé sceolan þrowian weán, nalles habban héhselda wyn, Cd. Th. 267, 25; Sat. 43: Exon. Th. 142, 31; Gú. 652. Se ðe áh lífes wyn gebiden in burgum, 307, 22; Seef. 27. Wynna gewítaþ, Runic pm. Kmbl. 345, 18; Rún. 29. Líðsa and wynna hám, Cd. Th. 58, 13; Gen. 945. Ealra ðæra wynna ðe ic on worulde gebád, Byrht. Th. 136, 58; By. 174. Hé his líchoman wynna forwyrnde and woruld-blissa, Exon. Th. 111, 31; Gú. 135: 122, 20; Gú. 308. Ídelra eágena wynna, 112, 2; Gú. 137. Wíc wynna leás, 443, 18; Kl. 32.

Hine yldo benam mægenes wynnum, Beo. Th. 3778; B. 1887. Hæleþ beóþ on wynnum the men are joyous, Exon. Th. 361, 20; Wal. 22: 464, 19; Hö. 89. Hé sunbeorht gesetu séceþ on wynnum, 217, 11; Ph. 278. Wé ðǽr wunodon on wynnum, Cd. Th. 279, 12; Sat. 237: 296, 26; Sat. 508. Þurh leáslíce líces wynne, earges flǽschoman ídelne lust, Exon. Th. 79, 28; Cri. 1297: 364, 12; Wal. 69. Ídle lustas, lǽne lífes wynne, 352, 19; Sch. 100. Ðás eorþan wynne, ðás lǽnan dreámas, 102, 4; Cri. 1667. God seleþ him on éþle eorþan wynne tó healdanne hleóburh wera, Beo. Th. 3465; B. 1730: 5447; B. 2727. Worolde wynne, 2164; B. 1080. Hé ðæt betere geceás, wuldres wynne, Elen. Kmbl. 2077; El. 1040. ¶ wynnum delightfully, pleasantly :—Is se wong wynnum geblissad mid ðám fægrestum stencum, Exon. Th. 198, 9; Ph. 7: 199, 18; Ph. 27. Ðín gemynd on ealra worulda woruld wynnum standeþ, Ps. Th. 134, 13. **Ia.** with prep. tó, marking object in which delight is taken :—Ne biþ him tó hearpan hyge, ne tó wífe wyn, Exon. Th. 308, 25; Seef. 45. Wæs mé wyn tó ðon, 380, 22; Rä. 1, 2. Næs him tó máðme wynn, Andr. Kmbl. 2228; An. 1115: 2326; An. 1164. Ða forweorþaþ ðe hira wynne tó ðé habban noldan qui elongant se a te, peribunt, Ps. Th. 72, 22. Hé genom him tó wildeórum wynne, Exon. Th. 146, 21; Gú. 713. **II. a delight, that which causes pleasure** :—Eh byð æðelinga wyn, Runic pm. 343, 4; Rún. 19: 344, 31; Rún. 27. Fugles wyn (a quill), Exon. Th. 408, 5; Rä. 27, 7. Him leófedan londes wynne, bold on beorhge, 110, 20; Gú. 110. Gæst inne swǽf óþ ðæt hrefn blaca heofenes wynne bodode, Beo. Th. 3607; B. 1801. **II a.** as an object of delight, (1) of human beings :—Hægstealdra wyn (Pharaoh), Cd. Th. 111, 28; Gen. 1862. Winemǽga wyn (Guthlac), Exon. Th. 184, 2; Gú. 1338. Eorla wyn, 174, 17; Gú. 1179. Wynn, 168, 22; Gú. 1081. Ǽðelinga wynn (St. Andrew), Andr. Kmbl. 2447; An. 1225. Wunn, 3423; An. 1715. (2) of the Deity :—Lífes wynn, . . . tíreádig cyning, Hy. 3, 1. Mægna God, . . . æþelinga wyn, Exon. Th. 286, 12; Jul. 730: 466, 15; Hö. 121. Neoman ús tó wynne weoroda Drihten, Cd. Th. 277, 2; Sat. 198. Wigena wyn, . . . heofonengla God, Exon. Th. 281, 4; Jul. 641. **III.** the best of a class, the pride of its kind. Cf. cyst :—Án engla þreát, heápa wyn (best of troops), Exon. Th. 460, 16; Hö. 18. Hleóþra wyn most excellent of melodies, 198, 18; Ph. 12. Gimma gladost, æþeltungla wyn, 218, 5; Ph. 290. Laguflóda wynn, 202, 16; Ph. 70. Eálá wífa wynn, fǽmne freólicast ah, pride of womankind, maiden most noble, 5, 18; Cri. 71. Ðú eart se æðela, ðe on ǽrdagum ealra fǽmnena wyn (the Virgin Mary) ákende, Hy. 3, 26. **IV.** the name of the w-rune $\overset{uu}{\text{Þ}}$ uyn, Archæologia, vol. 28, plate 15, fig. 7. In the following passages the symbol is put instead of the word wyn :—Þ is geswiðrad, gomen æfter geárum, Elen. Kmbl. 2526; El. 1264. Biþ se[ó] Þ scæcen eorþan frætwa, Exon. Th. 50, 23; Cri. 805. Þ sceal gedreósan, Anglia xiii. 9, 5. Wenne (Hickes prints Þ Þne) brúceþ ðe can weána lyt, and him sylfa hæfð blǽd and blisse, Runic pm. Kmbl. 340, 29; Rún. 8. On wuldres Þ (Kemble writes wealdend in place of the rune in the MS.; but cf. wuldres wynn, Andr. Kmbl. 1773; An. 889), Elen. Kmbl. 2177; El. 1090. In Ps. Vos. 99, 1 jubilate is rendered by Þ sumiaþ. See also mod-wên (l. mód-wyn). v. Cynewulf's Christ, ed. Gollancz, pp. 173 sqq., Anglia xiii. 1 sqq., Zacher, Das Gothische Alphabet, p. 9. [Laym. wunne, winne, wonne: A. R. wunne: Havel. winne: O. Sax. wunnea: O. H. Ger. wunna, wuuní delectatio, voluptas, jubilatio, jocunditas. Cf. Goth. un-wunands moestus: Icel. unaðr delight; yndi charm, delight.] v. éðel-, hord-, hyht-, leód-, líf-, lyft-, mód-, symbel-wyn[n].

wynan ? :—Eóh bið útan unsméðe treów . . . wynan (wyn, wynn ?) on éðle, Runic pm. Kmbl. 341, 31; Rún. 13.

wyn-beám, es; m. A tree that causes delight, an epithet of the cross :—Wuldres wynbeám, Elen. Kmbl. 1684; El. 844.

wyn-burh; f. A town where life is pleasant, a delightful town :—Þú eádig leofast, and ðé wel weorðeþ on wynburgum, Ps. Th. 127, 2.

wyn-candel(l), e; f. A lamp that gives delight, an epithet of the sun :—Wyncondel wera west onhylde, Exon. Th. 174, 31; Gú. 1186.

wyncgas. v. wining.

wyn-dæg, es; m. A day of gladness, a joyous time :—Wǽrun wudu-bearwas on wyndagum exultabunt omnia ligna sylvarum, Ps. Th. 95, 12. Ne móstun gé á wunian in wyndagum, ac scofene wurdon in éce fýr, Exon. Th. 140, 3; Gú. 604. Eádig weorþan, wunian wyndagum, and welan þicgan, 330, 34; Vy. 61.

wynde-cræft, es; m. An art of weaving :—Uuyndecreft ars plumaria, Txts. 43, 217. Uyrmas mec ni áuefun uyndicraeftum (uyrdi-, MS.), 151, 9.

wyndle, an; f. A wound :—Gif man preóst gewundige, gebéte man ða wyndlan, L. N. P. L. 23; Th. ii. 294, 4. v. wundel.

wyn-dreám, es; m. A joyful sound, jubilation :—Wyndreámes jubilationis, Ps. Lamb. Spl. Blickl. Gl. 150, 5. On wyndreáme in jubilo, Ps. Spl. 46, 5: in jubilatione, Blickl. Gl. Wyndreám jubilationem, Ps. Spl. Lamb. 88, 15.

wyndrian. v. wundrian.

wyn-ele, es; *m. Pleasant oil:*—Wynele se ðe bānes byrst bēteþ and hǽleþ, Ps. Th. 108, 18.

wyn-fæst; *adj. Joyous:*—Ðætte Sione dūn sigefest weorðe, and weallas Sion wynfeste getremed, Ps. C. 133.

wyn-gesíþ, es; *m. A pleasant companion, a companion in whom one delights:*—Næs mē wyngesíð wiðerweard heorte *non adhaesit mihi cor pravum*, Ps. Th. 100, 3.

wyn-gráf, es; *m. n. A pleasant grove:*—Mid wynngráfe weaxaþ geswiru *exultatione colles accingentur*, Ps. Th. 64, 13.

wynian. v. wunian.

wyn-land, es; *n. A land of delight, a happy, pleasant land:*—Se hālga stenc wunaþ geond wynlond, Exon. Th. 203, 10; Ph. 82. Wuldres wynlond (*heaven*), 317, 13; Mōd. 65.

wyn-leás; *adj. Joyless, dreary:*—Wynleásne wudu, Beo. Th. 2836; B. 1416. Wynleás wíc, 1641; B. 821. Ōðerne ēðel, wynleásran wíc, Cd. Th. 57, 14; Gen. 928.

wyn-líc; *adj. Delightful, pleasing, agreeable, charming:*—Hæfde hē hine swā hwītne geworhtne, swā wynlíc wæs his wæstm, Cd. Th. 17, 5; Gen. 255. Onstæl wynlíc, fæger and gefeálíc, Exon. Th. 151, 17; Gū. 796. Sunbearo, wuduholt wynlíc, 200, 1; Ph. 34: 423, 22; Rä. 41, 26. Ōðer wæs swā wynlíc, wlitig and scēne, ðæt wæs lífes beám, Cd. Th. 30, 15; Gen. 467. Fæger hleóðor, wynlícu wōðgiefu, Exon. Th. 414, 10; Rä. 32, 18. Ic ðē swā scíénne gesceapen hæfde, wynlícne geworht, 85, 8; Cri. 1388. Wynlíce wætera þrýðe, Ps. Th. 77, 18. Wæter wynlíco, Exon. Th. 194, 9; Az. 136. Hē gemon tō oft wynlícran wíc, 444, 24; Kl. 52. [Was imaked an wunlic fur, Laym. 8090. *O. H. Ger.* wunni-líh *amoenus, jucundus.*]

wynlíce; *adv. Pleasantly, delightfully:*—Ðæt ic wynlíce on psalterio ðē singan mōte, Ps. Th. 107, 2: 149, 4: Exon. Th. 82, 30; Cri. 1346.

wyn-lust, es; *m. Sensual pleasure:*—Ic wilnode mid him tō farenne, ðæt ic ðe mā emnwyrhtena on ðære þrówunge mínes wynlustes hæfde, Homl. Skt. ii. 23 b, 359. Hér synt disse weorolde wynlustas, ac ðǽr synt ða ēcan tintregu, L. E. I. proem.; Th. ii. 394, 8. Gif hwam hwæt yfeles gedōn bið, ðæt hē ne mæge hys wynlusta brūcan, Lchdm. i. 330, 13.

wyn-mǽg, e; *f. A beloved kinswoman:*—Seó fǽmne, wuldres wynmǽg (*the kinswoman in whom he had delighted*), Exon. Th. 182, 32; Gū. 1319.

wynnung. v. windung.

wyn-psalterium *a joyous psaltery:*—Āris, wynpsalterium *exurge, psalterium*, Ps. Th. 56, 10.

wyn-rōd, e; *f. A joy-giving cross:*—Wynrōd (*the cross*), sōðfæstra segn, Salm. Kmbl. 470; Sal. 235.

wyn-sang, es; *m. A joyous song, jubilant song:*—Ðǽr is wynsang, Wulfst. 265, 31.

wynstra. v. winestra.

wyn-sum; *adj.*   I. *winsome, agreeable, pleasant:*—Wynsum *suavis,* Ælfc. Gr. 9, 28; Zup. 54, 5. Wynsum, wlitig *elegans,* i. *speciosus, gratus, pulcher, praecipuus, magnus,* Wrt. Voc. ii. 142, 80. Ða wynsuman *amoena,* 1, 6. (1) *pleasant* to the senses or to the mind:—Treów tō brūcenne wynsum *lignum ad vescendum suave,* Gen. 2, 9. Wæs swíþe wynsum wǽta ūt flōwende. . . . Seó wǽte wæs wynsumu on ðǽre onbyrignesse, Blickl. Homl. 209, 2–9. Hē ys Drihtne wynsum onsægednys *oblatio est Domino odor suavissimus,* Ex. 29, 18. Wynsum stenc, Exon. Th. 363, 16; Wal. 54. Swēte stenc, wlitig and wynsum, 359, 19; Pa. 65. Wlitig and wynsum, wuldre gemearcad *regali plena decore,* 220, 10; Ph. 318: 350, 13; Sch. 63: Cd. Th. 277, 33; Sat. 214. Ðes middangeard, fæger and wynsum, Blickl. Homl. 115, 13. Wæs on ðam ofne windig and wynsum, Cd. Th. 237, 33; Dan. 347. Wynsum gefeá, Exon. Th. 77, 8; Cri. 1253. Hū wynsum (*iocundum*) is ðæt mon eardige on ðara gebrōðra ānnesse, Blickl. Homl. 139, 29. Þincð him wynsum ðæt se weald oncwyð, Met. 13, 46. Mē swēte and wynsum wæs ðæt ic oþþe leornode oþþe lǽrde *aut discere aut docere dulce habui,* Bd. 5, 24; S. 647, 27. Mín geoc is wynsum *jugum meum suave est,* Mt. Kmbl. 11, 30. Wynsum gamen *sales,* Wrt. Voc. i. 21, 54. Wynsum glíw *facetiae,* 61, 19. Wensum *lepida* (*sermonum series*), Hpt. Gl. 512, 55. Wynsumere † fǽgere *venustae,* 456, 41. Hwæt þincþ ðē on ðam welan and on ðam anwealde wynsumes *quid est, quod in se pulcritudinis habeant?* Bt. 27, 4; Fox 100, 20. Mid wynsume wíne, Ps. Th. 59, 3. Tō wynsumum stence in *suavem odorem,* Lev. 1, 9. Hunig, wynsume wist, Fragm. Kmbl. 40; Leás. 22. Wynsumne rēc, Elen. Kmbl. 1585; El. 794. Wynsumne wlite, Cd. Th. 111, 13; Gen. 1855. Scip, wudu wynsuman, Beo. Th. 3842; B. 1919. Wynsume *cantabiles,* Wrt. Voc. ii. 128, 9. Wæter wynsumu *dulces aquae,* Exon. Th. 202, 5; Ph. 65. Ðeós wyrt byþ cenned on wynsumon stōwum (παραδείσοις), Lchdm. i. 280, 13: 290, 6. Wyrta wynsume, Exon. Th. 211, 7; Ph. 194. Hí his weorc wynsum wíde sægean *annuntient opera ejus in exultatione,* Th. 106, 21. Wensumre *suavior* (*panis absconditus*), Kent. Gl. 310. Wynsumra steám, Exon. Th. 358, 14; Pa. 45. Swēg swētra and wlitigra and wynsumra, 206, 27; Ph. 133. Eal innanweard wæs ǽnlícra and wynsumra, ðonne hit mæge stefn āreccan, se stenc and se swēg, 181,

18; Gū. 1295. 'Is ðis winsum spell ðæt ðū nū segst.' Ðā cwæþ hē: 'Nis nān wuht winsumre ðonne ðæt þing ðæt ðis spell ymbe is,' Bt. 34, 5; Fox 140, 11. Biþ micle ðe winsumre sió sōþe gesǽld tō habbenne æfter ðām eormþum ðisses lífes, 23; Fox 78, 30. Wynsumre, Met. 12, 20. Þincþ him wynsumre ðæt him se weald oncweþe, Bt. 25; Fox 88, 20. Wōþa wynsumast, Exon. Th. 358, 9; Pa. 43. His englas, ealra folca mǽst, wereda wynsumast, Cd. Th. 42, 8; Gen. 671. (2) in reference to the conduct of living creatures:—Swǽs *vel* wynsum *eucharis,* Wrt. Voc. i. 61, 17: ii. 32, 52. Wynsum (*suavis*) is Dryhten, Ps. Surt. 33, 9: Ps. Th. 85, 4. Eálá ðū wynsuma man, Wulfst. 246, 2. Sum sceal wildne fugel ātemian, ōþþæt seó heoroswealwe wynsum weorþeþ, Exon. Th. 332, 18; Vy. 87. León, wynsume wiht, wel ātemede, Met. 13, 19. Eálá gē gōde cildra and wynsume (*venusti*) leorneras, Coll. Monast. Th. 35, 33. Hē wæs se swētesta lāreów and se wynsumesta *doctor suavissimus,* Bd. 5, 22; S. 644, 3.   II. *joyous.* v. wynsumian:—Beóð gefylde mid gefeán mūðas ūre, beóð ūre tungan teala wynsume *repletum est gaudio os nostrum, et lingua nostra exultatione,* Ps. Th. 125, 2. [*O. Sax.* wun-sam: *O. H. Ger.* wunni-sam *jucundus, amoenus, amabilis.*] v. un-, word-wynsum; wynsumness, *and next word.*

wynsum, es; *n. The pleasant:*—Ðæt nān wiht ne sy ðæs wynsumes, Wulfst. 184, 20.

wynsumian; *p.* ode *To rejoice, exult, be joyful:*—Ic fægnie and wynsumige and blissige *exultabo et laetabor,* Ps. Th. 30, 7. Wynsumaþ woesten *exultet desertum,* Rtl. 1, 17: Blickl. Homl. 7, 3: Wulfst. 254, 5. Ða eádigan ceasterwaran gefeóð and wynsumiaþ on lisse and on blisse and on ēcum gefeán, 265, 12: Shrn. 118, 4. Heora heortan and líchoman wynsumedon (*exultaverunt*) on God, Bd. 4, 13; S. 582, 37. Nā wynsuma ðū (*non iocunderis*) on bearnum ārleásum, Scint. 176, 6. Wynsumiaþ Gode *jubilate Deo,* Ps. Surt. 65, 1. Gefeáþ and wynnsumiaþ *gaudete et exultate,* Mt. Kmbl. Lind. 5, 12. Wynsumiaþ, Ps. Th. 31, 13: Blickl. Homl. 191, 35. Gedō ðæt mín gást wynsumige on ðínre hǽlo, 159, 2. Wynsumian *jocundari,* Bd. 5, 12; S. 630, 16: Blickl. Homl. 91, 8. Wæs heó swíþe wynsumiende, 137, 33. Wynsumigende, 143, 25. Mid micclum wynsumigendum gefeán, Homl. Skt. ii. 23 b, 678. Wynsumiende *letantem,* Rtl. 97, 16. [*O. H. Ger.* wunnisamōn *exultare.*] v. ge-wynsumian.

wynsum-líc; *adj. Pleasant, agreeable:*—Hē bið ðām gōdum glædmōd on gesihþe, wlitig, wynsumlíc weorude ðam hālgan, Exon. Th. 57, 1; Cri. 912. Wynsumlíc *votivum, acceptum, desiderativum,* Hpt. Gl. 446, 49. Þūhte fæger and wlitig heora líf and wynsumlíc, Blickl. Homl. 107, 30. Eall ðæt him hér on worlde wynsumlíc wæs, 111, 26: 115, 11. v. ge-wynsumlíc.

wynsumlíce; *adv.*   I. *pleasantly, agreeably.* v. wynsum, I. 1:—Wynsumlíce stēman, Homl. Skt. i. 4, 36: ii. 27, 113. Sume tiliaþ wífa, for ðam ðæt hí þurh ðæt mæge mǽst bearna begitan, and eác wynsumlíce libban *uxor, ac liberi, qui jucunditatis gratia petuntur,* Bt. 24, 3; Fox 82, 27. Engla werod wynsumlíce sungon, Homl. Skt. ii. 29, 297. Ðe eáþelícor and ðe wynsumlícor ða myclan byrþenne āberan, Blickl. Homl. 135, 7.   II. *pleasantly, graciously.* v. wynsum, I. 2:—Wē gelýfaþ ðæt Drihten sylf hire tōgeánes cōme, and wynsumlíce mid gefeán tō him on his þrymsetle hí gesette, Homl. Th. i. 442, 15.   III. *gladly, joyously.* v. wynsum, II:—Wynsumlíce (*voluntarie*) ic ofrige ðē, Ps. Spl. 53, 6. Āwend ðíne nosu fram unālýfedum stencum, ðæt ðū mæge wynsumlíce cweðan: 'Sýn wē æðele stencas beforan Godes gesihðe,' Wulfst. 246, 13.

wynsumness, e; *f.*   I. *pleasantness, agreeableness, delight.* v. wynsum, I. 1:—Wynsumnisse orcerd *paradisum voluptatis,* Gen. 2, 8. Of stōwe ðære winsumnisse *de loco voluptatis,* 2, 10. Ðære wynsumnysse brǽd *odorem suavitatis,* 8, 21. Woruldlícere wensumnesse *mundanae suavitatis, secularis dulcedinis,* Hpt. Gl. 413, 67: Confess. Peccat. Ēces wynsumnisse *aeterne jocunditatis,* Rtl. 103, 24. Hæfde hē mē gebunden mid ðære wynnsumnesse his sanges *me carminis dulcedo defixerat,* Bt. 22, 1: Fox 76, 6. Hē on wynsumnesse lifde, Blickl. Homl. 113, 7. Se middangeard wæs blōwende on swýþe manigfealdre wynsumnesse . . . and teáh men tō him þurh his wlite and þurh his fægernesse and wynsumnesse, 115, 7–12. Ða wynsumnesse and fægernesse ðæs londes wundrade, Nar. 26, 25. On ðære stōwe wynsumnesse in *amoenitatem loci,* Bd. 5, 12; S. 629, 39. Geseón ealles ðysses middangeardes wynsumnessa, ge on golde, ge on deórwyrþum hræglum, Blickl. Homl. 31, 3.   I a. *pleasantness* which affects the eye, *fairness, beauty:*—Wynsum[nysse] *venustate,* Hpt. Gl. 526, 22.   II. *pleasantness* of behaviour. v. wynsum, I. 2:—God ūs lǽrð sibbe and wynsumnesse, and deófol ūs lǽrð unsibbe and wrōhte, Homl. Ass. 168, 111.   III. *joyousness, exultation.* v. wynsum, II:—Wynsumnis mín *exultatio mea,* Ps. Surt. 31, 7. Weolure wynsumnisse *labia exultationis,* 62, 6. In wynsumnisse *in jubilatione,* 32, 3. Mid wynsumnesse *exultatione,* Blickl. Gl.: Rtl. 50, 19.   IV. *devotion.* v. wynsumian:—Mid wynsumnysse heortan (wilsume heortan, Bd. M. 228, 6) *devoto corde,* Bd. 3, 22; S. 553, 22. On micelre wynsumnesse (wilsumnisse, Bd. M. 376, 11) gebeda *orationis devotione,* 4, 30; S. 609, 5. v. un-wynsumness.

wyn-weorod, es; *n. A joyous band:*—Wynwerede *choro,* Blickl. Gl.

wyn-wyrt, e; f. *A pleasant plant* :—Đǽr wynwyrta weóxon and bleówon, Dóm. L. 5.

wyrcan, weorcan ; p. worhte; pp. worht. I. *to work, labour,* (1) absolute :—Mín fæder wyrcđ (*operatur*) óþ đis, and ic wyrce (wyrco, Lind., Rush. *operor*), Jn. Skt. 5, 17. Efne swā hē wyrceþ *secundum opera ejus*, Ps. Th. 61, 12. Hē won and worhte, wīngeard sette, Cd. Th. 94, 7; Gen. 1558. Gā and wyrce (wyrc, Rush. : wuirc, Lind. *operare*) on mínum wīngerde, Mt. Kmbl. 21, 28. Gāđ and wircaþ, Ex. 5, 18. Wyrceaþ eów syx dagas, L. Alf. 3; Th. i. 44, 10. Gif þeów mon wyrce on Sunnandæge, L. In. 3; Th. i. 104, 2, 4, 6: L. E. G. 7; Th. i. 170, 17. Se đe hors nabbe, wyrce đam hláforde đe him fore rīde, L. Ath. v. 5; Th. i. 232, 20. Hwý sceal ǽnig monn bión ídel, đæt hē ne weorce (wyrce, *v. l.*)? Bt. 41, 3; Fox 248, 24. Sió hond sceal wyrcean for đa wambe, Past. 34; Swt. 233, 9. Hē đǽr wircean sceolde, Gen. 2, 15. Niht cymþ đonne nān man wyrcan (*operari*) ne mæg, Jn. Skt. 9, 4. (1 a) where the instrument or material of work is given :— Hē wiđ monna bearn wyrceđ weldǽdum, Exon. Th. 191, 12; Az. 87. Đa đe wyrcan cūđon stāngefōgum, Elen. Kmbl. 2038; El. 1020. (2) with acc., (a) of that on which the work is done. v. wín-wyrcan :— Se đe werđ qui *operatur* (*terram suam*, Prov. 12, 11), Kent. Gl. 404. Đæt hē đa eorđan worhte *ut operaretur terram*, Gen. 3, 23. Se đe wille wyrcan wæstmbǽre lond *qui serere ingenuum volet agrum*, Met. 12, 1. Hē began tō wircenne đæt land *coepit exercere terram*, Gen. 9, 20. (b) of the work :—Hē áxode hwæt hig wyrcean cūđon (*quid habetis operis?*). Hig answaredon: 'Wē synd scēphyrdas,' Gen. 47, 3. II. *to make,* (1) with acc., (a) *to make, form, construct,* (α) where the agent is a person :—Ic tōwurpe mīne bernu and ic wyrce (*faciam*) māran, Lk. Skt. 12, 18. Wirc đē ǽnne arc . . . and đū wircst wununge binnan đam arce. . . . Đū wircst hine đus. . . . Đū wircst đǽron ēhþirl, Gen. 6, 14– 16. Đū wercest sumurlange dagas, đǽm winterdagum sceorta tīda getiohhast, Met. 4, 18. Mid đīs andweardan welan mon wyrcþ oftor feónd đonne freónd, Bt. 24, 3; Fox 84, 3. Ic worhte (*feci*) earce of sethimtreówum, Deut. 10, 3. On đære bēc đe ic weorhte, Bd. 3, 17; S. 545, 4. Đū đa scīran gesceaft sceópe and worhtest, Hy. 10, 2. Nān neóđđearf đē ne lǽrde tō wyrcanne đæt đæt đū worhtest, Bt. 33, 4; Fox 128, 12. Worhtes, Met. 20, 22 : Exon. Th. 15, 23 ; Cri. 240. Se đe on fruman worhte (worohte, Lind.), hē worhte wǽpman and wīfman, Mt. Kmbl. 19, 4: Cd. Th. 12, 11 ; Gen. 183. Se đe đas bōc worhte, Blickl. Homl. 169, 25. Đæt folc worhte mycele gesomnunga, Nicod. 20; Thw. 10, 1. Đa sundorhālgan worhton geþeaht, Mt. Kmbl. 12, 14. Hig wrohton gemōt, 27, 7. Hig wrohton (worhton, *v. l.* : uorhtun, Lind.) him beórscipe, Jn. Skt. 12, 2. Æfter đǽm formālan đe hī worhton, L. Eth. ii. 1 ; Th. i. 284, 12. Ne wirc đū đē āgrāfene godas, Ex. 20, 4. Đonne wyrce wē manega bēc, Homl. Th. ii. 28, 12. Uton wircean man tō ūre gelícnisse, Gen. 1, 26. Wyrcan, Hexam. 11 ; Norm. 18, 8, 19. Scip wyrcan, Cd. Th. 78, 33 ; Gen. 1302. Wǽpen wyrcean, Bd. 1, 12 ; S. 481, 14. Burg wyrcean, Ors. 5, 5 ; Swt. 226, 18. Wyrcan đone wīhagan, Byrht. Th. 134, 50 ; By. 102. Wyrcan spell, Bt. 38, 1 ; Fox 194, 30. Wercan, Met. 26, 73. Sealfe weorcean, Lchdm. iii. 6, 31. (β) where the agent is not a person, *to be the source,* or *cause of, to produce* :—Seó eá wyrcđ đæt fen, Ors. 1, 1 ; Swt. 8, 18. Seó eá đǽr wyrcđ micelne sǣ, Swt. 12, 23. Hit wyrcđ feóndscipe, Past. 11 ; Swt. 71, 24. Sum feóll on gōde eorđan, and worhte hundfealde wæstm, Lk. Skt. 8, 8. Grōwende gærs and sǣd wircende. . . . Treów westm wircende, Gen. 1, 11, 12. (b) *to make, constitute* :—Ic wolde witon hwæþer đū wēndest đæt hwylc đā đara fíf gōda worhte đa sōþan gesǽlþe and siđđan đa feówer good wǽron hire gōd, swā swā nū sāwl and líchoma wyrcaþ ǽnne mon, Bt. 34, 6 ; Fox 140, 23–28. Feówer wucan wyrcaþ ǽnne mōnđ, Anglia viii. 319, 4. (c) as a verb of incomplete predication, (a) with adj. :—Ic tō wīdan feore wyrce đīn heáhsetl hrōr and weorđlíc swā heofones dagas *ponam in seculum seculi semen ejus, et thronum ejus sicut dies coeli*, Ps. Th. 88, 26. Hwilcne wyrcst đū đe sylfne (đone đec seolfne wyrcas (wyrces, Rush.) *quem te ipsum facis?* Jn. Skt. Lind. 8, 53)? Homl. Th. ii. 234, 1. (β) with prepositional phrase :—Hē lǽdeþ wolcen, wind and līget, and đa tō regne wyrceþ (*fulgura in pluviam fecit*), Ps. Th. 134, 7. Nywolnessa hē him tō gewǽde woruhte, 103, 7. Worhte man hit him tō wīte, Cd. Th. 21, 2 ; Gen. 318. (2) with gen. :—Se đeóden ongan gedinges wyrcan, Cd. Th. 245, 25 ; Dan. 468. III. *to work, do, perform,* (1) absolute :—Swā đū worhtest tō mē, Exon. Th. 370, 25 ; Seel. 64. (2) with acc. :—On hwylcum anwealde ic đas þing wyrce, Mt. Kmbl. 21, 24. On hwylcre mihte wyrcsđ (wyrcst, *v. l.* : wircest, Rush.) đū đas þing? 21, 23. Swā hwæt swā se gesǽnelíca líchama dēþ oþþe wyrceþ, eal đæt dēþ seó ungesýnelíce sāwl þurh đone líchoman, Blickl. Homl. 21, 24. Werđ *operabitur* (*stultitiam*, Prov. 14, 17), Kent. Gl. 486. Eallum đe unriht wyrceaþ *omnibus, qui operantur iniquitatem*, Ps. Th. 58, 5. Tō mannum đe mildheortnesse wyrceaþ, Blickl. Homl. 169, 2L. Đære scame đe đū worhtes, Past. 31 ; Swt. 207, 11. Đa hand đe hē đæt fūl mid worhte, L. Ath. i. 14 ; Th. i. 206, 21, 24. Đa mǽran weorc đe hē worhte, Deut. 11, 7. Đa dǽda đa đe hē worhte, Blickl. Homl. 33, 6. Ne worhte (wrohte, Rush.) Iōhannes nān tācn, Jn. Skt. 10, 41. Hī blōdgyte worhtan, Exon. Th. 44, 26 ; Cri.

708. Ealle đe unriht worhtan *omnes peccatores*, Ps. Th. 100, 8. Wirc six dagas ealle đīne weorc *sex diebus facies omnia opera tua*, Ex. 20, 9. Lǽr mē hū ic đīnne willan wyrce and fremme, Ps. Th. 142, 10. Đæt đū furþur mē fraceþu ne wyrce, Exon. Th. 274, 31 ; Jul. 541. Gif esne þeów weorc wyrce, L. Wih. 9 ; Th. i. 38, 18. Đæt mon ōđrum riht wyrce, L. O. D. 2 ; Th. i. 352, 17. On đa gerād, wyrce đæt hē wyrce, đæt đæt land sī unforworht, Cod. Dip. Kmbl. ii. 383, 32. Weorce, 384, 21. Monig gōd weorc wyrcan, Past. 9 ; Swt. 55, 20. Wyricean, Blickl. Homl. 75, 13. Gōd wyrcan, Ps. Th. 52, 4. Yfel wyrcean, Blickl. Homl. 181, 34. Wundor wyrcean, Beo. Th. 1865 ; B. 930 : Ps. Th. 85, 9. Lof sceolde hē Drihtnes wyrcean, Cd. Th. 17, 8 ; Gen. 256. Đa heápas frugnon, hwæt hié wyrcean mihton đæt hié Godes erre beflugon, Blickl. Homl. 169, 11. Godes willan wercan, 67, 34. Mē geꞁyraþ tō wyrceanne đæs weorc đe mē sende, Jn. Skt. 9, 4. (2 a) *to perform* a rite, *keep* a season :—Mīn tīma ys gehende đæt ic mid đē wyrce mīne Eástro, Mt. Kmbl. 26, 18. (3) with gen. :—Ic mē đæs wyrce, đæt ic gange on hūs Godes, Ps. Th. 83, 11. Ealle đe unrihtes wyrceaþ *omnes qui operantur iniquitatem*, Ps. Th. 52, 5 : 58, 2 : 73, 19. Hē him đæs worhte tō, Cd. Th. 143, 11 ; Gen. 2377. Đa đe unrihtes worhtan, Ps. Th. 91, 6, 8. IV. *to work, effect* a purpose, *attain* an object, (1) with acc. or gen. :—Heó wēnde đæt heó hyldo heofoncyninges worhte mid đām wordum *she thought to win the favour of heaven's king with those words*, Cd. Th. 44, 22 ; Gen. 713. Đæt hī lifgen on geleáfan, and ā lufan Dryhtnes wyrcan in đisse worulde, Exon. Th. 448, 6 ; Dóm. 50. Hié sculon lufe wyrcean . . . ond habban his hyldo forđ, Cd. Th. 39, 12 ; Gen. 624. (2) with gen. :—Ā đīn dōm sý gōd and genge ; đū đæs wyrcest (*thou wilt bring that to pass*), Exon. Th. 192, 22 ; Az. 110. Wē đæs lifgende worhton in worulde, 186, 9 ; Az. 17. Wyrce se đe mōte dōmes ǽr deáđe *let him that may do deeds deserving of glory ere he die*, Beo. Th. 2779 ; B. 1387. Til sceal on ēđle dōmes wyrcean, Menol. Fox 501 ; Gn. C. 21. Hē þōhte đæt hē him myceles wordes wirceau sceolde (wolde geearnian him hereword, MS. F.), Chr. 1009 ; Erl. 142, 2. Se hæfde moncynnes leóhteste hond lofes tō wyrcenne (*to call forth praise*), Exon. Th. 323, 2 ; Víd. 72. (3) with a clause :—Is đæt wundorlíc, đæt đū mid geþeahte đīnum wyrcest, đæt đū đǽm gesceaftum mearce gesettest and hī gemengdest eác, Met. 20, 87. [Goth. waurkjan ; p. waurhta : O. H. Ger. wurchen, wirchen ; p. worhta : Icel. yrkja ; p. orti : O. Sax. wirkian ; p. warhta : O. Frs. werka ; p. wrochte.] v. ā-, be- (bi-), for-, fore-, ful-, ge-, in- (Exon. Th. 337, 21 ; Gn. Ex. 68), ofer-, ōþ-, sām-, un-, ymb-wyrcan (-weorcan) ; firen-, scyld-, syn-, unriht-, wam-, wel-, wolcen-, yfel-wyrcende ; wyrcend.

wyrce. v. ge-wyrce.

wyrcend, es ; m. I. *a worker, labourer.* v. efen-, fore-, wīn-wyrcend, *and* wyrcan, I. II. *a maker.* v. wyrcan, II :—Ic gelýfe on ǽnne God, wyrcend heofenan and eorđan, Homl. Th. ii. 596, 25. Heó wǽron đām wyrcendum gelíce *similes illis fiant qui faciunt ea*, Ps. Th. 113, 17. III. *a doer.* v. wyrcan, III :—Þurh đa unrōtnesse đe is deáđes wyrcend, Anglia xi. 113, 43. Ealle ic feóđe fācnes wyrcend *facientes praevaricationes odivi*, Ps. Th. 100, 3. v. leás-wyrcend, Homl. Th. i. 102, 1.

wyrcness, e ; f. I. *work, labour, operation.* v. wyrcan, I :—Dōnde wircnisse (*operationes*) in wǽtrum miclum, Ps. Surt. 106, 23. II. *working, doing, operation.* v. wyrcan, III :—Đurh swā hwylces bēne swā hē gehǽled sī, đysses geleáfa and wyrcnes (*operatio*) sī gelýfed Gode andfenge, Bd. 2, 2 ; S. 502, 23. His geearnunge oft đurh godcunde wyrcnesse (*operationem*) mid miclum mægenum scīnaþ, 3, 19 ; S. 550, 16. Đa đe lǽrdon ǽnne willan and āne wyrcnesse beón on Drihtne Hǽlende, 5, 19 ; S. 639, 34. II a. *working, performance* of something :—Wyrcnes heofonlícra mægena *operatio virtutum*, Bd. 1, 7 ; S. 479, 9. Mid wundra wyrcnesse, 2, 3 ; S. 505, 1 : 3, 13 ; S. 538, 39.

wyrcung, e ; f. *Working, doing* ; operatio, Rtl. 15, 42 : 31, 1 : 170, 3.

wyrd, e; f. *What happens, fate, fortune, chance.* I. the word is used to gloss the following Latin words :—*Casibus* wyrdum, Wrt. Voc. ii. 85, 1 : 18, 29 : 81, 45. *Eventus* wyrd, 75, 61 : 30, 71. *Fati* wyrde ođđe gegonges, 33, 65. *Fata* wyrde, 94, 6. *Fatis* wyrdum, 37, 54. *Fors* wyrd, 109, 5 : 83, 43 : 37, 14. *Fortuna* wyrd, 108, 78 : 33, 78. *Fortunae* wyrde, 33, 77 : 79, 61. *Sortem* wyrd, 120, 76. *Fatu* (*statu?* v. Ald. 30) wyrde, 78, 77. II. *fate, the otherwise than humanly appointed order of things* :—Đæt đætte wē hātaþ Godes foreþonc and his foresceáwung, . . . siđđan hit fullfremed biđ, đonne hātaþ wē hit wyrd. . . . Hī sint twā đing, foreþonc and wyrd. . . . Đæt đæt wē wyrd hātaþ, đæt biþ Godes weorc đe hē ǽlce dæg wyrcþ, ǽgđer ge đæs đe wē geseóþ, ge đæs đe ūs ungesewenlic biþ. . . . Sió wyrd dǽlþ eallum gesceaftum and-wlitan and stōwa and tīda and gemetgunga. Ac sió wyrd cymþ of đam foreþonce Godes, Bt. 39, 5 ; Fox 218, 21–220, 1. Điós wandriende wyrd, đe wē wyrd hātaþ, færþ æfter his foreþonce. . . . Siþþan wē hit hātaþ wyrd, syđđan hit geworht biþ ; ǽr hit wæs Godes foreþonc. Đa wyrd hē wyrcþ oþþe þurh đa gódan englas, oþþe . . ., 39, 6 ; Fox 220, 5–23. Đæt wē hātaþ wyrd, đonne se gesceádwisa God hwæt wyrcþ ođđe

geþafaþ ðæs ðe wé ne wénaþ *fit illud fatalis ordinis insigne miraculum, cum ab sciente geritur, quod stupeant ignorantes*, 39, 10; Fox 226, 24. Ðé sceal on woruld bringan Sarra sunu, sóð forð gán wyrd æfter ðiosum wordgemearcum, Cd. Th. 142, 1; Gen. 2355. Gǽd á wyrd swá hió sceal, Beo. Th. 915; B. 455. Ne wæs wyrd, ðæt hé má móste manna cynnes ðicgean, 1473; B. 734. Wǽron sume gedwolmen ðe cwǽdon, ðæt ælc man beó ácenned be steorrena gesetnyssum, and þurh heora ymbryna him wyrd gelimpe, Homl. Th. i. 110, 8. Sceal heó (*Lot's wife*) wyrde bídan, Drihtnes dómes, Cd. Th. 155, 10; Gen. 2570: Exon. Th. 329, 29; Vy. 41. Hí wyrd ne cúþon, Beo. Th. 2471; B. 1233. **III.** in a personal sense, *one of the Fates* (the *weird* sisters):—Wyrde *Parcae*, Wrt. Voc. ii. 116, 9: 67, 55. **III a.** as a personification, *fate, fortune*:—Wyrd biþ swíþre, Meotud mihtigra, ðonne ænges monnes gehygd, Exon. Th. 312, 27; Seef. 115. Wyrd byð swíþost, Menol. Fox 469; Gn. C. 5: Salm. Kmbl. 855; Sal. 427: 886; Sal. 442. Wyrd bið ful árǽd, Exon. Th. 286, 24; Wand. 5: Salm. Kmbl. 871; Sal. 435. Sume úþwitan secgaþ ðæt sió wyrd wealde ægðer ge gesǽlþa ge ungesǽlþa ælces monnes, Bt. 39, 8; Fox 224, 13. Weord (wyrd, *v. l.*), 5, 1; Fox 8, 30. Swá him wyrd ne gescráf, Beo. Th. 5142; B. 2574: Elen. Kmbl. 2092; El. 1047: Met. 1, 29. Behindan beleác wyrd mid wǽge, Cd. Th. 206, 25; Exod. 457. Eorlas fornóman wǽpen wælgífru, wyrd seó mǽre, Exon. Th. 292, 17; Wand. 100: Beo. Th. 2415; B. 1205. Hié wyrd forswéop, 959; B. 477: 5621; B. 2814. Wyrd ðone gomelan grétan sceolde, 4832; B. 2420. Hwý ðú ǽfre woldest ðæt seó wyrd swá hwyrfan sceolde? Heó þreáþ ða unscildigan, Bt. 4; Fox 8, 12: Met. 4, 34: Andr. Kmbl. 1226; An. 613: 3121; An. 1563. Wyrd oft nereþ unfǽgne eorl, Beo. Th. 1149; B. 572: Exon. Th. 165, 18; Gú. 1030. Tó eallum ðám gesǽldum ðe seó wyrd brengð, Bt. 16, 3; Fox 54, 25: 14, 1; Fox 40, 31. Ne wén ðú nó ðæt ic tó ánwillíce winne wiþ ða wyrd (*fortunam*) . . . hit oft gebyraþ ðæt seó leáse wyrd náuþer ne mæg ðam men dón ne fultum, ne nænne dem, 20; Fox 70, 22. Wyrde wiðstondan, Exon. Th. 287, 17; Wand. 15. **IV. an event**, (1) with the special idea of that which happens by the determination of Providence or fate:—Ne wile Sarran gelýfan wordum mínum; sceal seó wyrd swá ðeáh forð steallian, Cd. Th. 144, 14; Gen. 2389. Wyrd wæs geworden, swefen geséðed, swá ǽr Daniel cwæð, 257, 5; Dan. 653. God éce biþ; ne wendaþ hine wyrda, ne hine wiht dreceþ ádl ne yldo, Exon. Th. 333, 24; Gn. Ex. 9: Salm. Kmbl. 666; Sal. 332. Wyrda Waldend, Cd. Th. 205, 7; Exod. 432: Andr. Kmbl. 2113; An. 1058: Elen. Kmbl. 159; El. 80: Exon. Th. 455, 1; Hy. 4, 43. Wyrda gerýnu, Cd. Th. 225, 5; Dan. 149. Wyrda geþingu, 250, 14; Dan. 546. Wyrda gesceaft, 224, 6; Dan. 132. Onwrigen is wyrda bigang, Elen. Kmbl. 2245; El. 1124. Gif ic ðé ðone [. . . age, *the MS. is here imperfect*] gesecge ðises feores, ýþelíce ðú ða wyrde oncyrrest and his hond beféhst *si mortis tuae tibi insidiatorem prodidero, sublato eo facile instantia fata mutabis, mihique tres irascentur sorores, Clotos, Lachesis, Atropos*, Nar. 31, 24. (2) in a general sense, *an event, occurrence, circumstance, incident, fact*:—Nænigne tweógean ne þearf, ðæt seó wyrd on ðás ondweardan tíd geweorþan sceal, ðæt se Scyppend gesittan wile on his dómsetle, Blickl. Homl. 83, 10. Ða gelamp wundorlíc wyrd, ðæt se lég ongan sleán ongeán ðone wind, 221, 11. Ðæt is mǽro wyrd, Cd. Th. 84, 18; Gen. 1399: Menol. Fox 107; Men. 53. Egeslíc wyrd, Rood Kmbl. 148; Kr. 74: Exon. Th. 432, 6; Rä. 48, 2. Seó wyrd gewearð (*it happened*) ðæt ðæt wíf geseah Ismael plegan, Cd. Th. 168, 3; Gen. 2777. Is seó wyrd mid eów open *the event is patent among you*, Andr. Kmbl. 1516; An. 759: Apstls. Kmbl. 84; Ap. 42. Ne wé ðære wyrde wénan þurfon tóweard in tíde, Exon. Th. 6, 8; Cri. 81. Wénan ðære wyrde, ðæt heó híre taman healde, Met. 13, 24: 26, 114: Ps. Th. 119, 5. Hé wyrde bídeþ, hwonne God wille ðisse worlde ende gewyrican, Blickl. Homl. 109, 32. On ðæm dæge gewíteþ heofon and eorþe. . . . Swá eác for ðære ilcan wyrde gewíteþ sunne and móna, 91, 22. Ða gesáwon hié wundorlíce wyrd—ðone man lífgendne, ðone ðe hié ǽr deáðne forléton, 217, 36: Cd. Th. 61, 12; Gen. 996: 245, 30; Dan. 471. Hé ða wyrd ne máð, fǽges forðsíð, Exon. Th. 182, 33; Gú. 1319. Hé wyrd ne ful cúþe freóndrǽdenne hú heó from hogde *he did not fully know the circumstance, how her heart was turned from loving him*, 244, 26; Jul. 33. Dígle wyrd *an obscure circumstance*, Elen. Kmbl. 1077; El. 541: 1163; El. 583. Ymb ða mǽran wyrd, 2126 El. 1064. Geopenigean uncúðe wyrd, hwǽr hé ðara nægla wénan þorfte, 2202; El. 1102. Hé ðé mæg onwreón wyrda gerýno *he can disclose to thee the secrets of events* (can tell thee of events which are a secret to most men), 1174; El. 589: 1623; El. 813. Hé ne leág fela wyrda ne worda, Beo. Th. 6052; B. 3030. **V.** *what happens to a person, fate, fortune, lot, condition*:—Ic wille secgan ðæt ælc wyrd (*omnis fortuna*) bió gód, sam hió monnum gód þince, sam hió him yfel þince. . . . Ælc wyrd, sam hió sié wynsum, sam hió sié unwynsum, for ðý cymþ tó ðæm gódum ðæt hió . . . hine þreátige tó ðon ðæt hé bet dó, . . . oððe him leánige ðæt hé teala dyde, Bt. 40, 1; Fox 224, 33—226, 5. Ða graman gydena, ðe folcisce men hátaþ Parcas, ða hí secgaþ ðæt wealdan ælces monnes wyrde, 35, 6; Fox 168, 27. For hwý ætwíte gé eówerre wyrde ðæt hió nán geweald náh, 39, 1; Fox 210, 26. Him ne wæs nǽnig earfoþe ðæt

líchomlíce gedál on ðære neówan wyrde (*in their new condition*), Blickl. Homl. 135, 31. Under wyrd *sub condicione*, Jn. Skt. p. 5, 10. Ne meaht ðú nó mid sóþe getǽlan ðíne wyrd and ðíne gesǽlþa, swá swá ðú wénst *quod tu falsae opinionis supplicium luis, id rebus jure imputare non possis*, Fox 28, 1. Wyrd wánian, Exon. Th. 274, 24; Jul. 538. Unc sceal weorðan swá unc wyrd geteóð Metod manna gehwæs *to us shall it befall, as the Lord of every man decrees to us our fate*, Beo. Th. 5046; B. 2526. Nýd bið wyrda heardost, Salm. Kmbl. 622; Sal. 310. Him mæg wíssefa wyrda gehwylce gemetigan, 877; Sal. 438. Gnornsorga mǽst, wyrda láðost, Elen. Kmbl. 1953; El. 977: Rood Kmbl. 101; Kr. 51. **V a.** *fate, death.* See also **III a** :—Wille forgieldan gǽsta Dryhten willum æfter ðære wyrde, ðam ðe his synna nú sáre geþenceþ, Exon. Th. 450, 3; Dom. 82. **VI.** *chance, accident*:—Ðæt wille ic gecýþan, ðæt ðu rícu of nánes monnes mihtum swá gecræftgade ne wurdon, ne for nánre wyrde, búton from Godes gestihtunge *ut omnia haec profundissimis Dei judiciis disposita, non autem humanis viribus, aut incertis casibus accidisse perdoceam*, Ors. 2, 1; Swt. 69, 23. Sprecan wiþ ða ðe secgaþ ðæt ða anwaldas sién of wyrda mǽgenum geworddene, Swt. 62, 10. [Worþe hit wele, oþer wo, as þe wyrde lykeȝ hit hafe, Gaw. 2134. Þe same þat sett is be wirde, Alex. (Skt.) 443. Wyrdis (wyrde systres) *Parce*, Cath. Angl. 420, and see note. To dreȝe his wyrdes, Allit. Pms. 74, 1224. Heo biupeð hire wurdes, H. M. 33, 24. Is þi werid (werd, *v. l.*) to þe wissid, Alex. (Skt.) 689. Out of wo into wele ȝoure wyrdes shul chaunge, Piers P. C. 13, 209. Þe soroful werdes of me olde man, Chauc. Boet. 4, 10. *O. Sax.* wurd *fate, death* : *O. H. Ger.* wurt *fatum, fortuna, eventus* : *Icel.* urðr (*poet.*) *fate ; one of the Norns.* v. Grmm. D. M. pp. 376 sqq.] v. deáþ-, eft-, fǽr-, for-, ge-, tó-, un-, wundor-wyrd.

-wyrd *speech*, wyrdan *to injure.* v. ge-wyrd, wirdan.

-wyrdan *to speak.* [*Goth.* -waurdjan : *O. Sax.* -wordian : *O. Frs.* -wardia : *O. H. Ger.* -wurten.] v. and-, torn-, wís-wyrdan.

-wyrde ; *n. Speech.* [*Goth.* -waurdi : *O. Sax.* -wurdi, -wordi : *O. H. Ger.* -wurti.] v. and-, bí-, ge-wyrde.

-wyrde ; *adj.* [*Goth.* -waurds : *O. H. Ger.* -wurti.] v. beald-, biter-, fæger-, fela-, hócor-, hræd-, snotor-, stunt-, swǽs-, swét-, wær-, wís-wyrde.

wyrd-gesceap, es ; *n. A decree of fate* or *of fortune* :—Wyrdgesceapum *fortuiter*, Wrt. Voc. ii. 34, 5. [*O. Sax.* wurdi-giskapu ; *pl.*]

wyrdig ; *adj. Wordy* :—Werdi *verbosus*, Kent. Gl. 576. v. gearo-, twi-wyrdig.

-wyrding. v. and-wyrding.

wyrdness, e ; *f. Condition, state* :—Se godcunda foreþonc heaþeraþ ealle gesceafta ðæt hí ne móton tóslúpan of heora endebyrdnesse (wyrdnesse, *v. l.*), Bt. 39, 5 ; Fox 218, 32.

wyrd-stæf, es ; *m. A decree of fate* :—Ðonne seó þrág cymeþ wefen wyrdstafum *when comes that season fixed by fate's decrees*, Exon. Th. 183, 101 ; Gú. 1325.

wyrd-wrítere, es ; *m. One who writes an account of events, a historian, historiographer* :—Wurdwrítere *historiographus*, Hpt. Gl. 453, 1 : 468, 65. Andromachus se wyrdwrítere, Anglia viii. 307, 9. Se wyrdwrítere Iósephus áwrát on ðære cyrclícan gereccednesse, ðæt Heródes lytle hwíle æfter Ióhannes deáðe ríces weolde, Homl. Th. i. 488, 12. Wyrdwríteras secgaþ, 80, 5 : 454, 11 : Homl. Skt. i. 3, 21 : ii. 25, 676. Wyrdwrítera *historiographorum*, Hpt. Gl. 410, 54.

wyrgan ; *p.* de *To worry* (as an animal does), *strangle, throttle* :—Wyrgeþ *vel* smoraþ *st[r]angulat*, Wrt. Voc. ii. 121, 32. [Wolwes þat wald worow men (the whilke wol a man strangly and destrye, *v. l.*), Pr. C. 1229. Ilc wirwed lay, als it were dogges þat weren henged, Havel. 1921. Werewed, 1915. A wolf wolde lambes wery, R. R. 6267. Wolues that wyryeþ (wyrhyeþ, *v. l.*) men, Piers P. C. 10, 226. Wyrwyñ, worowen *strangulo, suffoco*, Prompt. Parv. 530. *O. Frs.* wergia : *O. H. Ger.* wurgen *strangulare, suffocare.* Cf. *O. Sax.* wurgil *a halter* : *O. L. Ger.* wurgarín *strangulatrix*.] v. á-wyrgan.

wyrgan *to curse*, wyrgels, wyrgedness, wyrgen, wyrgness, wyrgþu. v. wirgan, wrigels, wirgedness, wirgen, wirgness, wirgþu.

wyrht, e ; *f. Doing, work* :—Ná ðú be gewyrhtum úrum, wommum wyrhtum, woldest ús dón *non secundum peccata nostra fecit nobis*, Ps. Th. 102, 10. [Betere þenne we habbeð wrihte, O. E. Homl. i. 69, 251. Bi mine wrihte, ii. 217, 19. Æfftterr hise wrihhte, Orm. 8240. *O. H. Ger.* wuruht *meritum*.] v. for-, ge-, leóþ-, stán-wyrht.

wyrhta, an ; *m.* **I.** *a wright, workman, artificer, labourer, one who works at some trade* :—Wyrhta *operarius*, Wrt. Voc. i. 73, 25 : *opifex*, 47, 10. Yldest wyrhta *architectus*, 19, 14 : 47, 11. Se wyrhta (*operarius*) ys wyrðe hys metes, Mt. Kmbl. 10, 10. Wyrihte *faber*, Mk. Skt. Lind. 6, 3. Wrihtes *fabri*, p. 3, 8. Micel gedál is on ðam mægene ðæs dæghwamlícan wyrhtan and ðæs ídlan, Lchdm. ii. 84, 18. Hé wæs ðæs wyrhtan sunu (*the carpenter's son*), Homl. Th. i. 488, 12. Smiðes ł wyrchta (*fabri*) sunu, Mt. Kmbl. Lind. 13, 55. Hond bið gelǽred, wís and gewealden, swá bið wyrhtan ryht, sele ásettan, Exon. Th. 296, 5 ; Crä. 46. Gif ðú wénst ðætte wundorlíce gerela hwelc weorþmynd sié, ðonne telle ic ða weorþmynd ðæm wyrhtan ðe hié worhte, næs ná

ðe (*ingenium mirabor artificis*), Bt. 14, 1; Fox 42, 19. Mon sceal simle tó beregafole ágifan æt ánum wyrhtan (*the labourer who is the tenant of land*. Cf. Hér synd gewriten ða gerihta ðæ ða ceorlas sculan dón tó Hysseburnan. Æt ælcan híwisce ... þreó pund gauolbæres, Chart. Th. 145, 1.) six pund wǽga, L. In. 59; Th. i. 140, 5. Eálá góde wyrhtan (*operarii*) ... ðis geþeaht ic sylle eallum wyrhtum, ðæt ánra gehwylc cræft his geornlíce begange; for ðam se ðe cræft his forlǽt, hé byþ forlǽten fram ðam cræfte, Coll. Monast. Th. 31, 21-35. Ðone stán, ðe hine wyrhtan áwurpan *lapidem quem reprobaverunt aedificantes*, Ps. Th. 117, 21: Exon. Th. 1, 3; Cri. 2. Micel ríp ys, and feáwa wyrhtena (*operarii pauci*), Mt. Kmbl. 9, 37. Áhýrian wyrhtan on his wíngeard, 20, 1. **II.** *a maker, producer, author, creator, fabricator* :—Wróhtes wyrhtan, fyrnsynna fruman (*the devil*), Exon. Th. 263, 7; Jul. 346. Wyrhtan *fabricatores* (*falsitatum*), Hpt. Gl. 505, 64. On wyrhte gileáfes *in auctorem fidei*, Rtl. 27, 29. **II a.** *used of the Deity, the Creator, Maker* :—Se wyrhta, Cd. Th. 8, 17; Gen. 125: 265, 27; Sat. 14. Werhta *operator*, Kent. Gl. 808. Drihten, ælmihtiga God, wyrhta and wealdend ealra gesceafta, Bt. 42; Fox 260, 1: L. Eth. vi. 42; Th. i. 326, 13: L. I. P. 1; Th. ii. 304, 2: Cd. Th. 301, 21; Sat. 585. Wuldres wyrhta, Exon. Th. 206, 21; Ph. 130. Wealdend and wyrhta wuldorþrymmes, Andr. Kmbl. 649; An. 325: 1403; An. 702. Wyrhta and Sceppend weorulde þisse, Met. 29, 82. **III.** *a doer, worker* :—Cwealmes wyrhta *a murderer*, Cd. Th. 61, 29; Gen. 1004. Ealle ða ðe unrihtes wǽran wyrhtan *omnes discedentes a justificationibus tuis*, Ps. Th. 118, 118. Mánes wyrhtan *peccatores*, 100, 8. [*O. Sax.* wurhtio: *O. H. Ger.* wurhto.] v. ceaster-, efen-, esne-, firen-, for-, ge-, gegader-, heáfod-, hróf-, ísen-, leþer-, lyge-, mán-, meter-, mid-, scip-, sealm-, smeá-, stán-, teld-, tigel-, treów-, unlyb-, unriht-, wægn-, weall-, web-wyrhta.

wyrhte (?), an; *f.* *A female worker*, in cýs-wyrhte :—Be cýswyrhte. Cýswyrhtan gebyreþ hundred cýse, and ðæt heó buteran macige, L. R. S. 16; Th. i. 438, 30. [*O. H. Ger.* wurhta.]

wyrian (wyrigan), **wyrig, wyriginess.** v. wirgan, wearg, wirgness.

wyrm, wurm, weorm, es; *m.* **I.** *a reptile, serpent* :—Mé nædre beswác, táh wyrm þurh fægir word, Cd. Th. 55, 24; Gen. 899. Se wyrm (*the fire-drake*) onwóc, Beo. Th. 4563; B. 2287. Ðæs wyrmes wíg, 4621; B. 2316. Hé wearp hine on wyrmes líc, Cd. Th. 31, 26; Gen. 491. Ne wirce gé eów náne andlícnissa wurmes ne fisces (*reptilium sive piscium*), Deut. 4, 18. Hé wyrm ácwealde, hordes hyrde ... Ðæt swurd þurhwód wyrm ... draca morðre swealt, Beo. Th. 1777-1789; B. 886-892. Wyrmas *reptilia*, Blickl. Gl. Froxan ... swá fela ðæt man ne mihte nánne mete gegyrwan, ðæt ðara wyrma nǽre emfela ðæm mete *ranae per omnia reptantes*, Ors. 1, 7; Swt. 36, 28. Wyrma þreát, dracan and nædran, Cd. Th. 285, 12; Sat. 336. Wyrma slite, Exon. Th. 77, 4; Cri. 1251. Wyrmum bewunden in helle bryne, Judth. Thw. 23, 10; Jud. 115. Ic sende wildera deóra tèð on hig mid wurmum and næddrum *dentes bestiarum immittam in eos atque serpentium*, Deut. 32, 24. Wurmum tó ǽte, Wulfst. 145, 10. Aspidas, ǽtrene wyrmas, Ps. Th. 139, 3. Nicras, wyrmas and wildeór, Beo. Th. 2864; B. 1430. **I a.** fig. :—Brandháta níð weóll on gewitte, weorm blǽdum fág, Andr. Kmbl. 1538; An. 770. **II.** *a creeping insect, a worm* :—Wyrm *vermis*, Wrt. Voc. i. 78, 24. Wyrm ðe boraþ treów *termes vel teredo*, 24, 8. Hundes wyrm *ricinus*, 24, 33. Se wyrm (*a book-worm*) forswealg wera gied sumes, Exon. Th. 432, 8; Rä. 48, 3. Ðes lytla wyrm ðe on flóde gǽð fótum drýge, 426, 20; Rä. 41, 76. Of ðam weaxeþ wyrm *hinc animal sine membris fertur oriri, sed fertur vermi lacteus esse color*, 213, 29; Ph. 232. Hyra wyrm (*vermis*) ne swylt, Mk. Skt. 9, 44: Cd. Th. 212, 9; Exod. 536: Exon. Th. 373, 31; Seel. 118. Weorðan wyrme tó hróþor, 267, 17; Jul. 416. Wiþ ðam smalan wyrme, Lchdm. ii. 122, 18. Dó on ðæt eáre; þeáh ðær beón wyrmas on ácennede, hí þurh ðis sceolon beón ácwealde, i. 200, 22. Rib reáfiaþ rèþe wyrmas, Exon. Th. 373, 22; Seel. 113. Wyrmas, ða ðe geolo godwebb geatwum frætwaþ, 417, 23; Rä. 36, 9. Wyrma gif *food for worms* (the body), 368, 16; Seel. 22. Weormum tó hróðre, Apstls. Kmbl. 190; Ap. 95. Wið weormum, Lchdm. iii. 4, 5. Wið wyrmas on innoðe, i. 272, 10. **II a.** fig. :—Ic eam wyrm (*vermis*) and nales mon, Ps. Surt. 21, 7. wyrm (weorm, *v.l.*), R. Ben. 29, 13. [*Goth.* waurms *a serpent*: *O. Sax.* wurm *a serpent; a worm*: *O. H. Ger.* wurm *serpens, coluber, anguis, hydra; vermis, vermiculus, batis*: *Icel.* ormr *a serpent*.] v. cáwel-, deáw-, fág-, fíc-, flǽsc-, hand-, leáf-, mold-, must- (? Wrt. Voc. i. 23, 74), regn-, reng-, seoluc-, síd-, slá-, smeá-, tóþ-, treów-, twín-, þeór-wyrm.

wyrma. v. wurma.

wyrmǽte, an; *f.* *Wormeatenness* :—Ða treówa ðe beóð áheáwene on fullum mónan beóð heardran wið wyrmǽtan ðonne ða ðe beóð on níwum mónan áheáwene, Lchdm. iii. 268, 10. v. next word.

wyrmǽte; adj. *Worm-eaten* :—Wiþ wyrmǽtum líce, Lchdm. ii. 12, 15: 126, 4. [Frut ne is naȝt guod huanne hit is uorroted and wermethe, Ayenb. 229, 25. Cf. *O. H. Ger.* wurmázih *cariosus*.]

wyrmaman, wyrman. v. wurma, wirman.

wyrm-basu; adj. *Scarlet* :—Wyrmbaso *coccus*, Txts. 113, 67. v. wurma.

wyrm-cyn[n], es; *n.* **I.** *the genus reptile, reptiles, serpents* :—Hí gesáwon æfter wætere wyrmcynnes fela, sellíce sǽdracan, sund cunnian, Beo. Th. 2855; B. 1425. Betwux dracum and aspidum and eallum wyrmcynne, Homl. Th. i. 488, 1. Betwux eallum deórcynne and wyrmcynne, 102, 6. On ðam fíftan dæge hé gesceóp eall wyrmcynn, and eall fisccynn, Lchdm. iii. 234, 11. **II.** *a species of reptile or serpent* :—Scorpio, ðæt is án wyrmcynn, Lk. Skt. 11, 12. Wyrmcynn, Nar. 13, 10. Nis nán wyrmcynn ne wildeóra cynn on yfelnysse gelíc yfelum wífe, Homl. Th. i. 488, 10. On wéstennum wildeóra and wyrmcynna missenlícra, Ors. 3, 9; Swt. 136, 25: Exon. Th. 371, 31; Seel. 84. [*O. H. Ger.* wurm-chunni.]

wyrmelle. v. wurmelle.

wyrm-fáh; adj. *Having serpentine ornamentation* :—Ðæt sweord wreoþenhilt and wyrmfáh, Beo. Th. 3400; B. 1697. v. Worsaae's Primeval Antiquities, p. 49.

wyrm-galdere, es; *m.* *A serpent-charmer, sorcerer* :—Hé hèt sumne wyrmgaldere micle næddran hire in tó gelǽdan, ðæt seó hí ábítan sceolde. Ðá stód seó fǽmne forð on hire gebede, and seó næddre stód be hire; ðonne seó fǽmne onleát, ðonne onleát seó næddre. Ðá gelýfde se wyrmgaldere tó Gode þurh ðæt wundor, Shrn. 103, 5, 9. v. wyrm-galere.

wyrm-galdor, es; *n.* *A charm against worms* (?) :—Ðæt wyrmgealdor (cf. ðis ylce galdor mæg mon singan wið smeógan wyrme, 10, 17), Lchdm. iii. 24, 25.

wyrm-galere, es; *m.* *A serpent-charmer, sorcerer* :—Wyrmgalere *marsum* (the word occurs in reference to the incident given under *wyrm-galdere*), Wrt. Voc. ii. 96, 11: 55, 11. Wyrmgalera *marsorum* (*Chaldaeorum et hierophantorum phantasmata, simulque ariolorum et marsorum machinas*, Ald. 45), 82, 9: 56, 74: Hpt. Gl. 483, 14. v. wirgung-galere, wyrm-hælsere.

wyrm-geard, es; *m.* *An enclosure full of snakes* :—Wyrmgeardas, atol deór monig, ... blace næddran, Salm. Kmbl. 940; Sal. 469. [*Icel.* orm-garðr. Cf. in the story of Gunnar's fate in Atla kviða: Nú es sá ormgarðr yðr um folginn, v. 68: í garð þann es skriðinn vas innan ormom, 121. See, too, the stories of the deaths of Ragnar Lodbrog and Roderick, the last Gothic king of Spain.] v. wyrm-sele.

wyrm-geblǽd, es; *n.* *A blister raised by a snake-bite* (?), Lchdm. iii. 36, 21.

wyrm-hǽlsere, es; *m.* *A serpent-charmer, sorcerer* :—Wyrmhǽlseras *marsi* (printed *maris*), Wrt. Voc. ii. 55, 15. v. wyrm-galere.

wyrm-híw, es; *n.* *The form of a reptile or serpent* :—Hé sceolde hí áwendan of ðam wyrmhíwe (cf. *serpentia terrae*, Acts 10, 12), Homl. Skt. i. 10, 104.

wyrm-hord, es; *n.* *A treasure held by a serpent*, Beo. Th. 4447; B. 2222.

wyrm-líc, es; *n.* *The body of a serpent or of a worm*, (1) of carving on a wall. Cf. wyrm-fáh :—Weal wundrum heáh, wyrmlícum fáh, Exon. Th. 292, 13; Wand. 98. (2) fig. cf. wyrm, II a :—Ic eom ofersfongen mid synnum tó wyrmlíce, Anglia xii. 501, 22.

wyrm-melu (-o), wes; *n.* *Dust of dried worms powdered* (cf. 'Dry fair large earthworms before the fire, or in an oven, which when thorough dry, beat into powder,' Salmon's English Physician, quoted by Cockayne. See also: Eft angeltwæccan, gegníd swíþe, Lchdm. iii. 44, 4) :—Wyrc sealfe ... of wyrmmeluwe, Lchdm. ii. 78, 15. Nim wyrmmelu, 150, 10. Wyrmmelo, 238, 30. [In *O. H. Ger.* wurmmelo = *caries*.]

wyrms, es; *n. m.* *Corrupt matter* :—Ðis wyrms *hoc uirus*, Ælfc. Gr. 8; Zup. 29, 1. Wyrms *lues*, 9, 27; Zup. 53, 7: *colera*, Wrt. Voc. ii. 134, 54. Wið eárena sáre ... gif ðær wyrms inne bið, hyt ðæt út áwyrpð, Lchdm. i. 354, 16. Wyrms (worms, *v.l.*), 358, 16. Sáh út wyrms (of ðam geswelle), Homl. Skt. i. 20, 64. Hé áscræp ðone wyrms of his líce, Homl. Th. ii. 452, 28. ¶ figurative :—His wuldor is wyrms and meox, Homl. Skt. i. 25, 261. v. worms; ge-wyrms; adj.

wyrmsan, wyrsman; *p.* de *To produce corrupt matter* :—Ðonne se lǽce on untíman lácnaþ wunde, hió wyrmseþ and rotaþ, Past. 21; Swt. 153, 3. Sió wund wolde hálian, æfter ðæm ðe heó wyrmsde (wyrsmde, Cott. MSS.), 36; Swt. 259, 1. Gif hit wille wyrsman, Lchdm. ii. 102, 4. v. ge-wyrmsed.

wyrm-sele, es; *m.* *A serpent-hall* [cf. the hall, thick swarming now, With ... scorpion, and asp ... Cerastes horned, hydrus, and elops drear, And dipsas, Par. Lost 10, 522 sqq.], *a place where there are serpents* (hell) :—Ne þearf hé hopian ðæt hé þonan móte, of ðam wyrmsele, Judth. Thw. 23, 13; Jud. 119. v. wyrm-geard.

wyrms-hrécung; *f.* *The expectoration of corrupt matter* :—Wyr[m]shrǽcing *vel* wyr[m]sútspíung *phthisis*, Wrt. Voc. i. 19, 39.

wyrmsig; adj. *Purulent* :—Ðæm wyrmsigum *purulentis*, Wrt. Voc. ii. 78, 56. v. wurmsig.

wyrm-slite, es; *m.* *A snake-bite* :—In weán and on wyrmslitum betweónan deádum and deóflum, in bryne and on biternesse, Wulfst. 188, 1.

wyrms-útspíung. v. wyrms-hrécung.

wyrm-wyrt, e; *f.* *Worm-grass* (v. E. D. S. Pub. Plant Names); *sedum album*, Lchdm. ii. 94, 18: 104, 3: 128, 3: 308, 16. [*O. H. Ger.* wurm-wurz *aganoe*.]

wyrn, wyrnan. v. wirn, wirnan.

wyrp, es ; *m. A throw, cast, the distance which a thing may be thrown* :—Swā mycel swā is ānes stānes wyrp (weorp ł wyrp, Lind.) *quantum jactus est lapidis,* Lk. Skt. 22, 41. [Þurh on eie wurp to one wummon, A. R. 56, 14. Iesus from heom iwende þe uurp of o ston, Misc. 41, 155. *O. H. Ger.* wurf *jactus, ictus.*] v. æ-wyrp.

wyrp *recovery,* wyrpan *to recover,* wyrpan *to throw,* -wyrpan, -wyrpe. v. wirp, wirpan, weorpan, (be-, ge-)sceat-wyrpan, ge-, lang-wyrpe.

wyrpel, es ; *m. A vervel,* a *ring put on a falcon's leg.* Thorpe in his note on the following passage quotes from Roquefort the explanation of the French *vervelle: Large anneau qu'on passoit au pied d'un faucon pour le retenir* :—Sum scea. wildne fugel ātemian, heafoc on honda . . . dēþ hē wyrplas on, fēdeþ swā on feterum fiþrum dealne (cf. the description of a falcon's equipment given in a M. H. Ger. poem, Haupt Zsch. 7, 341, quoted by Leo: Lancvezzel, würfel und hoselín, daz waren diu kleit sîn), Exon. Th. 332, 19 ; Vy. 87.

-wyrplíc, -wyrpness, wyrra, wyrrest, wyrs, wyrs-hrǽcing, wyrsm. v. scort-wyrplíc, for-, tô-wyrpness, wirsa, wirs, wyrms-hrǽcung, wyrms.

wyrst. v. wrist.

wyrt, e ; *f.* I. *a wort* (e. g. St. John's *wort*), *plant, herb* :—Gærs *vel* wyrt *herba,* Wrt. Voc. i. 30, 35 : 78, 71. Þeós wyrt, ðe man betonicam nemneþ, Lchdm. i. 70, 1 : 90, 2, *and often.* Seó wyrt (*herba*) weóx, Mt. Kmbl. 13, 26. Gemolsnad wyrt, Ps. Th. 89, 6. Wyrta wynsume, Exon. Th. 233, 23 ; Ph. 529. Sumra wyrta eard biþ on dūnum, sumra on merscum, sumra on mórum, Bt. 34, 10 ; Fox 148, 22. Mid missenlícum blóstmum wyrta āfægrod *variis herbarum floribus depictus,* Bd. 1, 7 ; S. 478, 22 : Exon. Th. 358, 17 ; Pa. 47. Gif mon sié wyrtum forboren, Lchdm. ii. 114, 8, 12. Hē getimbreþ tānum and wyrtum nest, 227, 29 ; Ph. 430. God geworhte eall gærs and wyrta (*omnem herbam*), Gen. 2, 5. Ðū ytst ðære eorðan wyrta, 3, 18 : Ps. Th. 103, 13. Werta, Kent. Gl. 687. I a. *a garden herb, herb for food* :—Gē teóþiaþ mintan and ælce wyrte (alle wyrte, Rush. *omne holus*), Lk. Skt. 11, 42. Wyrta *olera,* Wrt. Voc. i. 82, 31 : *fordalium* (cf. wyrt-mete), ii. 150, 20. Hit is ealra wyrta mǽst *majus est omnibus holeribus,* Mt. Kmbl. 13, 32. Tô wertum *ad olera,* Kent. Gl. 524. Gif gē mē (*the cook*) ūt ādrífaþ fram eówrum gefērscype, gē etaþ wyrta (*olera*) eówre grēne, Coll. Monast. Th. 29, 11 : 34, 27. II. *a root* :—Wudubeám wæs wyrtum fæst, Cd. Th. 247, 19 ; Dan. 499 : Beo. Th. 2732 ; B. 1364 : Exon. Th. 209, 18 ; Ph. 172 : 417, 2 ; Rä. 35, 7. [*Goth.* waurts *a plant, a root: O. Sax.* wurt *a plant, root : O. H. Ger.* wurz *herba, olus : Icel.* urt *a herb.*] v. wudu-wyrt. The word occurs in the names of many plants, see the lists of plant-names given in Wrt. Voc. i. pp. 30~, 66~, 78~, 286~, and in Lchdm. iii. 311 sqq.

wyrt, e ; *f. Wort* (in brewing) :—Wyrt *sandix* (the word occurs in a list of terms 'de mensa,' and among a number denoting various kinds of drink. Cf. sandix, genus frugi, Corp. Gl. Hessels, 105, 103), Wrt. Voc. i. 290, 64 : 289, 9 : ii. 87, 33. Bewylle on hwǽtene wyrte, Lchdm. ii. 268, 12. [Wurte *idromellum,* Wrt. Voc. i. 257, col. 2. *Ger.* würze : *Swed.* wört.] v. leáh-mealt-, mǽsc-, mealt-wyrt.

wyrt-bed[d], es ; *n. A garden-bed* :—Ðeós wyrt biđ cenned on be-gānum stówum and on wyrtbeddum and on mǽdum, Lchdm. i. 96, 22 : 184, 6. [*O. H. Ger.* wurz-betti *areola.*]

wyrt-brǽþ, es ; *m. A perfume from plants, an odour, aroma* :—Mid brǽðe āfylled swylce ðǽr lǽgon lilie and rose. Ða cwæð Basilissa: 'Ic wundrie hwanon ðes wyrtbrǽð ðus wynsumlíce stēme,' Homl. Skt. i. 4, 36. Ne mihte nán wyrtbrǽð swā wynsumlíce stēman, ii. 27, 113. Āgeótende wyrtbrǽð (*aroma*) of rinde, Hymn. Surt. 79, 13. Orþiende wyrtbrǽða swētnyssa *spirans odorum balsama,* 98, 19. Seó cwēn com tô Salomone mid lácum on golde, and on deórwurðum gymstānum and wyrtbrǽðum . . . Seó geleáfulle geladung offraþ Criste wyrtbrǽðas þurh gebeda, Homl. Th. ii. 586, 6-11.

wyrt-cyn[n], es ; *n. A species of plant* or *vegetable* :—Ǽghwylc wyrt-cyn *omne genus holitorum,* i. *holerum,* Wrt. Voc. i. 55, 29. Wyrtcynn (wyrta cynn, Rush.) *nardus pisticus,* Jn. Skt. Lind. 12, 3 : *aloes,* 19, 39 : *unguentum,* Ps. Th. 132, 2.

wyrt-cynren, es ; *n. The genus plant, plants, herbs* :—Wyrtcynren *herbam,* Ps. Lamb. 146, 8.

wyrt-drenc, es ; *m. A herb-drink, potion made from herbs* :—Wyrt-drenc *antidotum,* Wrt. Voc. ii. 6, 70 : 100, 31. Wyrtdrenc wiđ ātre *sityriaca* (= theriaca), 77, 4. Biter wyrtdrenc *picra,* wyrtdrenc *catartica,* i. *purgatoria,* i. 20, 19, 21. Mid onðounge wyrtdrences þurh horn oððe pīpan sió wamb biþ tô clǽnsianne, Lchdm. ii. 260, 11. Dô ealle ða wyrta tô wyrtdrence, 22, 17. Æfter ðon sceal man wyrtdrenc sellan, 22, 2. Wyrtdrencas *antidota,* Wrt. Voc. ii. 2, 4. Lǽcedómas wiþ ðære healfdeádan ādle, and onlegena and wyrtdrencas, Lchdm. ii. 172, 8.

wyrt-eceddrenc, es ; *m. An acid potion made with herbs* :—Be ðam sūþernan wyrteceddrence, Lchdm. ii. 172, 11.

wyrtel (?) *a plant.* v. biscop-wyrtil. [*O. H. Ger.* wurzala *radix.*]

wyrt-fæt, es ; *n. A scent-bottle* :—Wyrtfata *olfactoriola* (cf. *olfactoriola* ðe hiera eleoealfa on wǽran, Wrt. Voc. ii. 64, 35), Hpt. Gl. 517, 27.

wyrt-forbor, es ; *n. Restraint from an action by the operation of*

herbs :—Wiþ wyrtforbore (cf. Gif mon sié wyrtum forboren, 114, 8) **and** yfium gealdorcræftum, Lchdm. ii. 306, 12. Cf. next word.

wyrt-gælstre, an ; *f. A woman who uses herbs for charms* :—Mǽden yfeldǽda and wyrtgælstre (*malefica et herbaria*), Lchdm. iii. 186, 11. Cf. previous word.

wyrt-geard, es ; *m. A kitchen-garden* :—Wyrtgeardas *promptuaria,* Ps. Spl. C. 143, 16. [*Wick.* wort-ȝerd *hortus olerum.*]

wyrt-gemang, es ; *n. A spice* :—Wyrtgemangc *myrra,* Ps. Lamb. 44, 9. Maria nam ān pund deórwyrðre sealfe mid ðam wyrtgemange ðe hig nardus hātaþ *Maria accepit libram ungenti nardi pistici preciosi,* Jn. Skt. 12, 3. Wyrtgemang and alewan *mixturam murrae et aloes,* 19, 39. Myrre and gutta and cassia . . ̇ Ða wyrtgemang getâcniaþ mistlícu mægen Cristes, Ps. Th. 44, 10. Wyrta oððe wyrtgemangu *herbae vel pigmenta,* Scint. 36, 11. Wyrtgemanga strengðe *pigmentorum uim,* 120, 13. Mid wyrtgemangum *cum aromatibus,* Jn. Skt. 19, 40 : Anglia xiii. 427, 885. Hig bǽron mid him ða wyrtgemang (*aromata*), Lk. Skt. 24, 1. Hig gearwodon wyrtgemang (wyrta gemong, Lind. *aromata*), 23, 56 : Mk. Skt. 16, 1. v. next word.

wyrt-gemengness, e ; *f. A spice* :—Hig bebyrigdon Andreas líchaman myd wyrtgemengnyssum and myd swētum stencum, Shrn. 153, 17 : Wulfst. 263, 5. v. preceding word.

wyrþe. v. weorþ.

wyrþe-land, es ; *n. Land that has lain fallow, land ploughed for the first time, a cultivated field* :—Wyrðelandum *novalibus* (tellus millenos animarum manipulos in fructiferis ecclesiae novalibus protulit, Ald. 32), Wrt. Voc. ii. 79, 26 : 77, 50 : 59, 56. v. worþ, *and next word* (?).

wyrþen *a field* (?) :—Wyrþenna *leti* (the passage in which the gloss occurs is : Graculus, qui segetum glumas, et laeti cespites occas depopulare studet, Ald. 142. Perhaps *wyrþenna* should be taken as a gloss to *occas,* v. wyrþing), Wrt. Voc. ii. 89, 57 : 52, 20. v. preceding word (?).

wyrþian, wyrþig. v. weorþian, wirþig.

wyrþing *a cultivated field* (?) :—Wealh (fealh ?) oþþe wyrðing *occa* (the passage is : Anthonius coelestis aratri stivarius . . . a quo primitus per Aegyptum fertilis coenobiorum seges et foecunda conversationis occa granigeris germinavit spicis, Ald. 32), Wrt. Voc. ii. 79, 25. v. wyrþen.

wyrþo. v. wirþu.

wyrtian ; *p.* ode *To season, spice* :—Ic wyrtige *condo,* Wrt. Voc. ii. 21, 39. v. ge-wyrtian.

wyrtig ; *adj. Full of herbs* :—On ānum wyrtigan hamme, Homl. Skt. ii. 30, 312.

wyrt-mete, es ; *m. Vegetable food, food consisting of herbs* :—Wyrt-mete *clerius cibus,* Wrt. Voc. i. 290, 40 : ii. 17, 23. Gesoden wyrtmete *fordalium,* 38, 56 : 150, 2. Wurtmete mid meluwe *polentum,* i. 27, 25.

wyrt-stenc, es ; *m. A perfume from a plant* :—Hūs gefylled wæs wyrtstence (*odore*) ðære smirnisse, Jn. Skt. Rush. 12, 3.

wyrt-truma (wyrtruma, an : -trum, es ; *m.* : -trume, an ; *f.* (v. Be ðare wyrtruman, Cod. Dip. Kmbl. iv. 93, 7). I. *the root* of a plant :—Wyrttruma *radix,* Wrt. Voc. i. 33, 11 : 80, 8 : 285, 79 : Cd. Th. 252, 20 ; Dan. 581. Is seó æx āsett tô ðæs treówes wyrtruman, Lk. Skt. 3, 9. Be ðam wyrttruman, Lchdm. i. 172, 10. Wyrttruman *radicem,* Ps. Spl. 51, 5. Hig næfdon wyrttruman (wyrttrum, *v. l.*), Mt. Kmbl. 13, 6 : Mt. Skt. 4, 6 : Lk. Skt. 8, 13. Hyt næfð ðone wyrttruman (wyrttrum, *v. l.*), Mt. Kmbl. 13, 21. Wyrttruman ðæs wudu-beámes eorðan fæstne, Cd. Th. 248, 20 ; Dan. 516 : Exon. Th. 328, 28 ; Vy. 24. Treów wyrttrumum underwreðyd, Runic pm. Kmbl. 341, 30 ; Rūn. 13. Wyrttruman *radices,* Ps. Spl. 79, 10. Oþ ða wirttruman *usque ad radices,* Num. 22, 4. I a. *the root* of a tooth :—[Ða grindigtēþ ðe ālc mid feówer wyrttrume gefæstned byð, and ðanne hý hero wurtruma forleátaþ, ðanne sweartigeþ hý, and fealleþ, Lchdm. iii. 104, 15.] I b. *figurative* :—Ne næfð ænig bôh grēnnysse gôdes weorces, se ðe nā wunaþ on wyrttruman (*radice*) sóðre lufe, Scint. 3, 19. Ða ðe heora heortan wyrttruman on ðisum andwerdum lífe plantiaþ, Homl. Th. i. 132, 7. II. *the root, source, origin* :—Hē cuæð ðæt ælces yfeles wyrttruma (wyrtruma, Cott. MSS.) wǽre ðæt mon wilnode hwelcre gîtsunge, Past. 11 ; Swt. 73, 3. Seó grǽdignys is wyrtruma ælces yfeles, and seó sóðe lufu is wyrttruma ǽlces gôdes, Homl. Th. ii. 410, 3. Ðætte of wyrttruman besmitenes geþôhtes ācenned bið, Bd. 1, 27 ; M. 80, 13. II a. *a stock* :—Hwæt limpeþ ðæs tô ðē of hwylcum wyrttruman ic ācenned sí *quid ad te pertinet qua sim stirpe genitus?* 1, 7 ; S. 477, 28. III. *this word and the word of like meaning, wyrtwala* (q. v.), seem to be used in reference to local relations in the sense of *foot, lower side,* the opposite of *heáfod* or *heáfdu,* e. g. Of ðes pôles hēuede on gerigte tô ðane ellene ; of ðane ellene on gerigte â be wertuualen on ðe herestrāte, Cod. Dip. Kmbl. v. 17, 10. Tô ðan heáfdan . . . tô uurtwalan, vi. 2, 4-6. Andlang fyrh on ða heáfda ; andlang heáfda on ðæne grēnan pæð . . . andlang fyrh on ða wyrtwale ; swā be ðære wyrtwale, iv. 19, 17-28. Cf. *too:* Be ðām heáfdon, iii. 378, 22. Â be heáfdan, 438, 29. Tô ðam heáfde ; big ðam heáfde tô ðere fureh, 384, 16, with similar uses of *wyrttruma* and *wyrtwala* :—Of ðam seáðe swā wyrttruna sceát oð Ramleáhweg, Cod. Dip. Kmbl. iii. 455, 22. On ðone feld ; ðæt andlang wyrttruman on Hildes hlǽw, 170, 27. On dinningc-

gráfes wyrttruman; of dynningcgráfes wyrttruman eall swá se díc sceót, 208, 5 : 34, 14. On wiðigleás wyrttruman; ðonne ealling be wyrttruman ðð ácleá, v. 230, 1. On wiðigleás wyrttruman; on eatan beares wyrtruman; ðð leás eástende; norð be wyrttruman, 334, 25–27. On loxanwuda wyrttruman; of wyrttruman on þiccan stánas, 345, 5. Oð ða dúnæ ufewearde on ða ðeðenan byrigelsas; swá áðún be wyrttruman æft tó gemíðum, 346, 20. Innan leá; ðanne be wurtruman anlanges wudes, iii. 172, 33. Oð ða lége; ðonne be wyrttruman, 406, 28, 33 : v. 358, 18. Forð be wyrttruman, iii. 422, 1 : vi. 33, 37. Bæ ðam wyrttruman, v. 191, 32. Wyrttrumman, iii. 135, 8. Tó wuda; swá be ðan eald wyrtruman, 279, 31. Be wyrttrume, v. 100, 20. Wirtrume, iii. 440, 33. Ofer ðane sceagan; ðonne forð á be wyrttruman, 460, 2. Of ðan hamme á be wurtruman, vi. 137, 22. Â be ðare wyrttruman, iv. 93, 7. On wyrttruman, iii. 390, 26. On feld on wyrttruman oð gráfes suðende, v. 334, 34.

wyrttrumian; p. ode *To take root :—*Sēd wyrttrumiaþ (wyrttrymaþ, Rush.) *semen germinet,* Mk. Skt. Lind. 4, 27. v. ge-wyrttrumian.

wyrt-tún, es; m. *A garden :—*Wyrttún *botanicum* vel *viridarium, cucumerarium,* Wrt. Voc. i. 30, 17, 18. Wyrtún *viridiarium,* 84, 54: *hortus,* ii. 42, 51 : Jn. Skt. 18, 1 : 19, 41. On wyrttúne *in cucumerario,* Wrt. Voc. ii. 48, 24. On wyrtúne (wyrttúne, *v. l.*) *in horto,* R. Ben. 71, 18. Syle mē ðinne wíneard mē tó wyrtúne, Homl. Skt. i. 18, 173. Wyrttún ne sáw ðú, Lchdm. iii. 184, 19 : Lk. Skt. 13, 19. Wyrtúna *hortorum,* Wrt. Voc. ii. 83, 27. Hé nemde ða undiórestan wyrta ðe on wyrttúnum weaxe, and ðeáh swíðe welstincenda, Past. 57 ; Swt. 439, 32 : Lchdm. i. 94, 7.

wyrtung, e; f. *Seasoning with herbs :—*On scírum wíne . . . : ge on wíne ge on wyrtunge, Lchdm. i. 342, 26. v. wyrtian.

wyrt-wala, an; m.: -wala, e; f.   **I.** *the root of a plant :—*Swá fela bóga treówes of ánum wyrtwalan (*radice*) spryttaþ, Scint. 3, 17. Genim wegbrǽdan wyrtwalan, Lchdm. ii. 82, 19 : 90, 6, 23 : 94, 19, 23. Wyrtwalan *radices,* Ps. Surt. 79, 10. Andlang pæþes on ða wyrtwalan; of ðam wyrtwalan on heortsole, Cod. Dip. Kmbl. iv. 19, 25. Andlang strǽt wiðútan ða wyrtwalan, 20, 2. Andlang ríðe on ða wurtwalan, vi. 1, 26.   **I a.** fig. :—God út álúceþ wirtwelæ ðinne of lande lyfigendra, Ps. Spl. T. 51, 5.   **II.** *a root, source :—*Wyrd, ealra firena fruma, fǽhðo módor, weána wyrtwela, wópes heáfod, Salm. Kmbl. 889; Sal. 444.   **III.** *the foot* of a hill, *lower side of a wood, field,* etc. v. wýrt-truma, III :—Swá ðe wyrtwala scádet tó turlan homme (cf. *first passage under* wyrttruma, III), Cod. Dip. Kmbl. v. 267, 33. Forð andlang wyrtwale on ða róde, 356, 4. Forð be gráfes wurtwale, iii. 405, 29. On ða eorðburg; ðæt forð be wurtwalan tó mearcwege, vi. 43, 18. Uuirtwalan, wyrtwalan, 93, 33, 34 : v. 86, 18. Of ðære leáge be wyrtwalan, iii. 464, 21 : v. 148, 14 : 298, 16. Be wirtwalan on ða efsan; and ðan on ðone wiðig; and swá be wirtwalan on ðone mēreþorne, 226, 16, 17. Ðurh henna leáh, ðð hit cymeþ tó ðære efese; ðonne á norþ be wyrtwalan, ii. 172, 23 : iii. 380, 25 : 437, 33 : v. 330, 33 : 336, 27. Wurtwalan, vi. 41, 20. Weortwalan, v. 389, 15, 16. Â be wyrtwale . . . on hel ufewearden æfter wyrtwalan, iii. 48, 11–16. On heáfdbeorh; ðonne on wyrtwalan on ðæs hagan ende . . . ; andlang herpaðes tó ðære efise, ðonon eft on wyrtwalen, v. 300, 8–13.

wyrtwalian; p. ode.   **I.** *to plant :—*Ongelǽdde wyrtwǽlæs (*plantabis*) hié on dúne yrfeweærdnesse ðíne, Cant. Moys. 21 (= Ex. 15, 17). Wirtwǽledæst *plantasti,* Ps. Spl. T. 43, 3. Wyrtwalodes, 79, 10.   **II.** *to root up, eradicate :—*Ic wyrtwalige (áwyrtwalige, *v. l.*) *uello, uellico,* Ælfc. Gr. 36; Zup. 214, 16. Wyrtwalod, Shrn. 184, 3. v. á-, under-wyrtwalian.

wyrt-weard, es; m. *A gardener :—*Heó wēnde ðæt hit se wyrtweard (*hortulanus*) wǽre, Jn. Skt. 20, 15.

wýsc, es; m. *Choice :—*Ðá him wiisc [wūsc ?] seald wæs *optione data,* Bd. 5, 19; S. 638, 40. [Wusche *exoptatio,* Prompt. Parv. 535. *O. H. Ger.* wunsc; m. *optio: Icel.* ōsk; f.] v. wúsc-bearn.

wýscan; p. te *To wish.* (1) with gen. *to wish for, desire :—*Hē helle wísceþ, ðæs engestan éðelríces, Salm. Kmbl. 212; Sal. 105. Hý ðæs betran lífes wýscaþ and wēnaþ, Exon. Th. 106, 26; Gū. 47. Wíscaþ, 115, 24; Gū. 194. Hié his tócymes wýscton, Blickl. Homl. 103, 12. (1 a) *to wish* something to or for a person :—Ða apostolas hǽlo eów wýscaþ, L. Alf. 49; Th. i. 56, 13. Ne cuæð hé ðæt for ðý ðe hé ǽnegum men ðæs wýscte oððe wilnode *non optantis animo,* Past. 1; Swt. 29, 11. Ne wyrige nán man óðerne, ne yfeles ne wísce, Homl. Th. ii. 34, 27. (2) with acc. :—Ic sceal his róde sigor swíðor wíscan ðonne ondrǽdan, Homl. Th. i. 594, 20. (3) in a precatory or imprecatory sense, = utinam, (a) with clause :—Ic wýsce ðæt heorte healde lufe *utinam cor teneat amorem,* Scint. 25, 1. Ðæt ic eác swylce wísce forþ sié on leornunge úra stafa *quod utinam exhinc etiam nostrarum lectione litterarum fiat,* Bd. 5, 14; S. 635, 7. Ic wísce ðæt Ysmahel libbe ætforan ðē *utinam Ismael vivat coram te,* Gen. 17, 18. Ic wísce ðæt hig wiston *utinam saperent,* Deut. 32, 29. Gif ic ðe ne geþence, ic wísce ðæt ic eft forlidennesse gefare, Ap. Th. 12, 10. Wē wísceaþ ðæt wē wǽron ǽr deád *utinam mortui essemus,* Num. 14, 3. Hié wýscaþ

ðæt hié nǽfre nǽron ácennede, Blickl. Homl. 93, 27. Ic oft wíscte and wolde ðæt hyra læs wǽre swá gewinfulra *que utinam minus fuissent laboriosa,* Nar. 2, 28. Ðú wýsctest ðæt ðú wistest Crist on róde ahangenne, Blickl. Homl. 85, 33. Hé oft wýscte ðæt ealle Rōmāne hæfden ǽnne sweoran *exclamasse fertur: 'Utinam populus Romanus unam cervicem haberet,'* Ors. 6, 3; Swt. 256, 26 : Exon. Th. 378, 33; Deór. 25. Wíscte, Ps. Th. 14, arg. Hí wíscton ðæt hí móston swá wunian ðð ende, Homl. Skt. i. 5, 401. (b) where the words of the wish are given :—Alexander ðá wíscte: 'Eálá gif ðú wǽre hund,' Homl. Th. ii. 308, 13. (c) with gen. and appositional clause :—Ic ðæs wísce, ðæt wegas míne on ðínum willan weorþan gereahte *utinam dirigantur viae meae,* Ps. Th. 118, 5. [*O. H. Ger.* wunscen *optare: Icel.* œskja.] v. ge-wýscan.

-wýscedness, -wýscendlíc, -wýscendlíce, -wýscing. v. ge-wýscedness, ge-wýscendlíc, ge-wýscendlíce, ge-wýscing.

Y

For the Runic Y see ýr.

ýce, an; f.: ýce, es; m. *A (poisonous) frog :—*Ýce *botrax* vel *botraca,* Wrt. Voc. i. 24, 19: *botrax,* 45, 26 : ii. 13, 2 : 126, 57 : *rana,* 71, 15. **Ýcean roboete** (the passage is :—Regulorum et aspidum venena, ad quae quadrupedis *robetae* et spalangii pestifera confectio humanae naturae nocitura habebatur, Ald. 25), 78, 44. Ðæt ilce biþ nyttol íces slite oþþe hundes, Lchdm. ii. 86, 2 (see note). Ðære wyrte wyrttruma on wætere geðyged wiðrǽd íceom and næddrum, i. 144, 15. Ýcan i froggan *ranas,* Ps. Lamb. 104, 30. ¶ Yce *parruca,* Wrt. Voc. ii. 67, 69, *seems to be for* hyce; v. hicae *paruca,* 116, 50. [*M. H. Ger.* üchen *ranas.*] v. fen-ýce.

yfel, es; n. *Evil, ill :—*Gód *bonum,* yfel *malum,* Wrt. Voc. i. 74, 49.   **I.** in a moral sense :—Ða ðe him biþ unwítnode eall hiora yfel on ðisse worulde, Bt. 38, 3; Fox 200, 26. Hwæt yfeles dyde þes ? Mt. Kmbl. 27, 23. Dǽdbóte dón ðæs mycelan yfeles and mānes, ðe hié wið heora Drihten gedydon, Blickl. Homl. 79, 6. Ðone besmítan ðe ðú nánwiht yfeles on nystest, 85, 36. Yfeles ordfruma, Cd. Th. 288, 1; Sat. 374. Hē ðæs yfeles geswíce, Hy. 2, 8 : Met. 9, 52. Forðhealde tó yfele *in malum prona,* Gen. 8, 21 : Hy. 7, 113. For rihtwísnysse hē sceal habban andan tó hira yfele, Past. 12 ; S. 75, 14. Beó nú on yfele, noldæs ǽr teala, Cd. Th. 310, 26 ; Sat. 733. Ic syngade and mycel yfel beforan ðē ic gedyde, Blickl. Homl. 87, 30. Leahtra gehwylcne, yfel unclǽne, Exon. Th. 80, 21 ; Cri. 1310 : Past. 21 ; Swt. 157, 23. Ealle ðæt yfel and ðæt unnet, ðæt hē ǽr on his móde hæfde, Bt. 35, 1 ; Fox 154, 26. Gýtsung, mān, . . . stuntscipe, ealle ðás yfelu of ðam innoðe cumaþ, and ðone man besmítaþ, Mk. Skt. 7, 23. Ðæt hē feala yfla sægde, Blickl. Homl. 173, 20. Yfla gehwylc, grimme gieltas, Exon. Th. 229, 25 ; Ph. 460. Tó eácan ðþrum unárímedum yflum hē ðone pāpan hēt ofsleán, Bt. 1 ; Fox 2, 11. Hié nǽnige bóte dón noldan, ah hié on heora yfelum þurhwunedon, Blickl. Homl. 79, 8 : Ps. Th. 105, 25. Ic ðē þreáge and ðē cýðe eal ðás yflu, 49, 23.   **II.** what is hurtful, grievous :—Hú mycel yfel ðē gelamp for ðínre gítsunge, Blickl. Homl. 31, 13. Nú is æghwanon yfel and slege, 115, 16 : 181, 32. Is mín yfel twyfeald *I am doubly injured,* 175, 13. Is sáwl mín sáres and yfeles gefylled *repleta est malis anima mea,* Ps. Th. 87, 3 : 106, 38. Ægðer hyra óðrum yfles hogode, Byrht. Th. 135, 45 ; By. 133 : Exon. Th. 54, 28; Cri. 875. Bydelas ðæs ēcan yfeles, ðe yfelum mannum becymð, Homl. Th. ii. 538, 23. Ða fuglas ús nǽnige láðe ne yfle ne wǽron *aves non nobis perniciem ferentes,* Nar. 16, 18. Hē wile hit him mid grimnesse and mid yfele forgyldan, Blickl. Homl. 55, 25. Hē nǽnigum yfel wiþ yfele geald, 223, 33 : Elen. Kmbl. 983; El. 493. Hí ðær mycel yfel gedydon, Chr. 897; Erl. 95, 18 : 993; Erl. 133, 3. Ælc yfel man him gedeð, 1036; Erl. 165, 22. Ðām ðe mē syrwedan yfel *qui quaerunt mala mihi,* Ps. Th. 70, 12. Ealle ðe mē yfel hogedon *qui cogitant mihi mala,* 69, 3. Heó þolian ne wolde yfel, Cd. Th. 136, 26 ; Gen. 2264 : Past. 36; Swt. 261, 4 : Exon. Th. 77, 9 ; Cri. 1254. Fela yfelu sceolon foreyrnon ǽr seó geendung ðissere worulde cume, Homl. Th. ii. 538, 22. Ðú mē yfela feala oft oncnyssedest, Ps. Th. 70, 19. Ðē gehealde Drihten wyð yfela gehwam *Dominus custodit te ab omni malo,* 120, 6. Yfla gehwylc, Exon. Th. 356, 27 ; Pa. 18. Ic earfeþa dreág, yfel ormǽtu, 280, 10; Jul. 627. Ic gegaderie ofer hig yflu (*mala*), Cant. M. ad fil. 23.   **II a.** of disease :—Wiþ ðam wǽtan yfle ðæs miltes, Lchdm. ii. 246, 9. Hit mæg wið æghwilcum uncúþum yfele, iii. 288, 17. Wið lungenādle and wið gehwylce yfelu ðe on ðam innoðe dereþ, i. 280, 18.   **II b.** of abusive speech :—Hié wyrgdon ðone cāsere and him yfel cwǽdon, Blickl. Homl. 191, 10. ¶ *the word often occurs in contrast with* gód :—Swá ðæs gódan gódnes biþ his ágen gód, swá biþ eác ðæs yfelan yfel his ágen yfel, Bt. 37, 3 ; Fox 190, 15. Hwæþer him yfel þe gód under wunige, Exon. Th. 82, 3; Cri. 1333. Ealles ðæs ðe wē geweorhtan gódes oððe yfles, 447, 21 ; Dōm. 43 : Blickl. Homl. 51, 26. Treów ingehýdes gódes and yfeles, Gen. 2, 9. Gódes and yfeles, welan and wāwan, Cd. Th. 30, 10; Gen. 465. Gódes and ýfles ic cunnade, Exon. Th. 321, 25; **Vid.**

51. Hig woldon gildan gôd mid yfele, Gen. 44, 4. Swâ gôd swâ yfel swâ hê ǽr dyde, Blickl. Homl. 101, 30. Se Hâlga Gâst hié ǽghwylc gôd lǽrde and him ǽghwylc yfel bewerede, 131, 30. Ðurh ða gesceádwîsnesse wê tôcnâwaþ good and yfel, and geceósaþ ðæt gôd and âweorpaþ ðæt yfel, Past. 11; Swt. 65, 22. Geþenc ðæt ðú gôd onfênge and gelîce Lazarus onfêng yfel, Lk. Skt. 16, 25. [Goth. ubil: O. Sax. ubil: O. Frs. evel: O. H. Ger. ubil.]

yfel; adj. Evil, ill, bad :—Yfel malus, Wrt. Voc. i. 74, 47. I. in a moral sense :—Yfel mann of yfelum goldhorde bringð yfel forð, Mt. Kmbl. 12, 35. Hié nǽnigo firen ne gewundode, ne yfel gewitnes (witness of wrong-doing) ne wrêgde, Blickl. Homl. 161, 33. Ðæt ðǽr mæge yfelu uncyst on eardian, 37, 10. Se yfela dêma onfêhþ medmycclum feó, and onwendeþ ðone rihtan dôm, 61, 30. Se yfela þeów, Mt. Kmbl. 24, 48. Se yfela willa, Bt. 36, 7; Fox 184, 31. Se yfela unrihtwîsa cyningc, Met. 15, 1. Fram yfelum menn ab homine malo, Ps. Th. 139, 1. Yflum, Exon. Th. 96, 20; Cri. 1577. From ðære inwitfullan yflan tungan a lingua dolosa, Ps. Th. 119, 3. Ne sette ic mê fore eágum yfele wîsan (rem malam), 100, 3. Synnigra cirm, yfele sprǽce, Cd. Th. 145, 20; Gen. 2408. Âfyr fram ðê ða yfelan sǽlþa and ða unnettan, and ðone yflan ege disse worulde, Bt. 6; Fox 14, 33. Yfele men magon yfel dôn, 36, 7; Fox 184, 4: Blickl. Homl. 45, 23. Yfele geþancas, Mk. Skt. 7, 21. Hyra weorc wǽron yfele, Jn. Skt. 3, 19. Se anweald ðara yflena cymþ of unþeáwum, Bt. 36, 7; Fox 182, 26. Ic tô yflum cwæð dixi iniquis, Ps. Th. 74, 4 : Exon. Th. 57, 15; Cri. 919. Yfflum, Blickl. Homl. 33, 22. Yfelum wordum, 39, 3. Hê wile gesceáwian wlitige and unclǽne, tile and yfle, Cd. Th. 303, 10; Sat. 610. Hê âraþ ða gôdan, and hê wîtnaþ ða yfelan, Bt. 41, 2; Fox 246, 20. II. of things, bad, not good of its kind :—Sió yfele gillestre and ðæt yfele blôd, Lchdm. ii. 148, 6. Gif eáran willen âdeáfian, oþþe yfel hlyst sié . . . Gif mon yfelne hlyst hæbbe (cf. wiþ yfelre hlyste, 2, 13), 40, 22–26. Swâ gôd treów ðe yfelne wæstm deð, ne nis yfel treów gôdne wæstm dônde, Lk. Skt. 6, 43. Heó is on onsýne ûtan yfeles heówes, Blickl. Homl. 197, 11. Ðá gecuron hig ða gôdan (fiscas) on hyra fatu, ða yflan hig âwurpon ût, Mt. Kmbl. 13, 48. III. of what is grievous, hurtful, etc., (1) of animate objects :—Yfel wiht phantasma, Mk. Skt. Lind. 6, 49. Hê sealde yfelan wyrme hiora wyrta, Ps. Th. 77, 46. Hî ǽtan yfle tostan, 77, 45. Hê gehǽlde manega of yfelum (yflum, Lind.) gâstum, Lk. Skt. 7, 21. (2) of things :—Yfel gesihð malus oculus, Mk. Skt. 7, 22. Yfel wyrd bad fortune, Bt. 40, 2; Fox 236, 22. Him ðæs æfter becwom yfel endeleán, Cd. Th. 227, 15; Dan. 187. Ðorn byð þearle scearp, anfengys yfel (bad to take hold of), Runic pm. Kmbl. 340, 1; Rûn. 3. Ic hit mid yfelre bysene inc forgylde, Blickl. Homl. 189, 25. Yfele habban sorge, Exon. Th. 376, 32; Seel. 163. Earmne gehýnan yflum yrmþum, 280, 24; Jul. 634. Ic wîte þolade, yfel earfeþu, 89, 6; Cri. 1453. [Goth. ubils : O. Sax. ubil : O. Frs. evel : O. H. Ger. ubilo.] v. wirsa.

yfel-âdl glosses cacexia, Wrt. Voc. i. 19, 43.

yfel-cund; adj. Of evil nature, malignant :—Se yfelcunda malignus, Ps. Lamb. 14, 4. Yfelcundra malignantium, 21, 17.

yfel-cweþan (yfle-) glosses maledicere :—Se ðe yflecuoeðas ł woerges (maledixerit) ðæm feder, Mt. Kmbl. 15, 4. Fîcbeám ðæm ðú yflecuoede (maledixisti), Mk. Skt. 11, 21. Yfelcweþende hine maledicentes ei, Ps. Spl. 36, 23.

yfel-dǽd, e; f. I. an evil deed, misdeed, sin. v. yfel, I :—Dôn sôðe bôte ûre yfeldǽda, Blickl. Homl. 99, 1 : Exon. Th. 285, 12; Jul. 713. Ðú scealt andettan yfeldǽda mâ, 269, 27; Jul. 456. Ða gesceafta ðe sind þwyrlîce geðúhte, hî sind tô wrace gesceapene yfeldǽdum, Homl. Th. i. 102, 4. Cweðaþ stunte men ðæt hî ne gewyrde lybban sceolon, swylce God hî neádige tô yfeldǽdum, 110, 31. II. an injurious deed, injury, mischief. v. yfel, III :—Gesete sâwle mîne fram yfeldǽdum heora restitue animam meam a malignitate eorum, Ps. Spl. 34, 20. [O. H. Ger. ubil-tât.] Cf. yfel-weorc.

yfel-dǽde; adj. : yfel-dǽda, an; m. Of evil deeds; a person of evil deeds :—Gif hê nǽre yfeldǽda (malefactor), ne sealde wê hine ðê, Jn. Skt. 18, 30. Ða fêng his sunu tô his rîce swýðe yfeldǽda, Homl. Skt. i. 18, 228. ¶ with special reference to magical practices :—Gif hwylc yfeldǽde man þurh ǽnigne æffpancan ôþerne begaleþ, Lchdm. i. 190, 9. Unlybwyrhta veneficus, yfeldǽda maleficus, drý magus, Wrt. Voc. i. 74, 40. Swâ swâ yfeldǽda ut magus (maleficus), Hpt. Gl. 487, 61. Mǽden wyrst swelt, for ðî yfeldǽda (malefica) and wyrtgælstre, Lchdm. iii. 186, 11. Ðæra manna naman ðe wǽron entas and yfeldǽda, Homl. Th. i. 22, 31. v. unriht-dǽde, -dǽda. Cf. yfel-weorc.

yfel-dônd, -dôend, es; m. An evil-doer, malefactor :—Yfeldôend malefactor, Jn. Skt. Lind. 18, 30. Cf. wel-dônd.

yfel-dônde; adj. (ptcpl.) Evil-doing :—Drihtenes ondwlita bið ofer ða yfeldôndan men tô ðon ðæt hê hig forspille, L. E. I. 28; Th. ii. 424, 22.

yfel-dysig glosses stultomalus, Wrt. Voc. i. 47, 44.

yfele; adv. Evilly, badly, ill :—Yfele male, Ælfc. Gr. 38; Zup. 235, 1. I. in a moral sense :—Yfele gê dydon pessimam rem fecistis, Gen. 44. 5. Hit is gecweden, ðæt him betere wǽre ðæt hê nǽfre wǽre, ðonne hê yfele wǽre, Homl. Th. ii. 244, 21. II. badly, imperfectly,

improperly :—Seó bóc wæs yfele of Grêcisce on Lêden gehwyrfed (badly translated), Bd. 5, 24; S. 648, 23. Hê ða gehât swîðe yfele gelǽste, Bt. 1; Fox 2, 9. Gif ic yfele sprǽce si male locutus sum, Jn. Skt. 18, 23. Gif se esne his hlâforde hýreþ yfle, Exon. Th. 430, 18; Rä. 44, 10. III. where there is hurt or suffering :—Mîn dohtor ys yfle (yfele, v.l.) mid deófle gedreht (grievously afflicted), Mt. Kmbl. 15, 22. Ðú eart, Babilone, bitere ætfæsted, ænge and yfele, hire earm dohter filia Babylonis misera, Ps. Th. 136, 8. Fremde þeóde ðîn hûs yfele gewemdan, Ps. Th. 78, 1. Wel tô dônne hweþer ðe yfele; sâwla gehǽlan hweþer ðe forspillan? Mk. Skt. 3, 4. Se abbot dyde heom yfele, Chr. 1087; Erl. 217, 7. Wæs Godes yrre þurh ða dǽde yfele genîwod, Wulfst. 10, 1. Ic him yfle ne môt, Exon. Th. 491, 5; Rä. 80, 9. III a. of bodily suffering :—Gif men sié fǽrlîce yfele if it suddenly goes badly with a man, Lchdm. ii. 294, 15. Ðes lǽcedôm sceal tô ðam menn ðe byð yfele on ðam breóstum, iii. 120, 1. IV. marking ill-success :—Yfele dêð him sylfum (he does badly for himself) ðe mid swîcdôme his tilaþ, and hê bið sceaðena gefêra ðe man sceandlîce wîtnaþ, Homl. Skt. i. 19, 172 : Cd. Th. 49, 13; Gen. 791. Ðý lǽs wên sié ðæt wê yfele forweorþon lest perhaps we perish miserably, Blickl. Homl. 247, 2 : Ps. Th. 79, 15 : 106, 26. Hê ða yfele and earmlîce geendode, Homl. Skt. ii. 25, 546. On ðære fare heom yfele gelamp, Chr. 1075; Erl. 212, 22. Sceolde unc Adame yfele gewurðan ymb ðæt heofonrîce, Cd. Th. 25, 2; Gen. 387. V. of injurious speaking :—Ic wyrige oððe yfele secge maledico, Ælfc. Gr. 37; Zup. 222, 4. Nis nân ðe on mînum naman mægen wyrce, and mæge raðe be mê yfele specan (male loqui de me), Mk. Skt. 9, 39. Oft mê feala cwǽdon feóndas yfele dixerunt inimici mei mala mihi, Ps. Th. 70, 9. [O. Sax. ubilo : O. H. Ger. ubilo.]

yfel-full; adj. Wicked, evil :—On dǽdum yfelfullum in factis malitiosis, Anglia xi. 116, 13.

yfelgeornness, e; f. Evil, wickedness :—Yfelgiornisse nequitiae, Rtl. 98, 24. Ofer yfelgiornise super malitia, 5, 12 : 12, 25.

yfel-hǽbbende; adj. (ptcpl.) Sick, ill :—Ealle yfelhæbbende missenlîcum âdlum, Mt. Kmbl. 4, 24.

yfelian; p. ode. I. to do evil to, to maltreat, afflict, injure, wrong :—Ða þingeras þingiaþ ðǽm ðe læssan þearfe âhton, þingiaþ ðǽm ðe man yflaþ, and ne þingiaþ ðám ðe ðæt yfel dôþ; ðæm wǽre mâre þearf, ðe ða ôþre unscyldige yfelaþ (yflaþ, v.l.), ðæt him mon þingode tô ðǽm rîcum pro his, qui grave quid, acerbumque perpessi sunt, miserationem judicum excitare conantur oratores, cum magis admittentibus justior miseratio debeatur, Bt. 38, 7; Fox 208, 25–29. E hine yflaþ, Salm. Kmbl. 193; Sal. 96. Íne gelîcre geswencednysse ða mǽgþe yfelade Ini simili provinciam illam adflictione mancipavit, Bd. 4, 15; S. 583, 31. Hê bebeád ðæt mon nǽnne mon ne slôge, and eác ðæt man nânuht ne wanode ne ne yfelade ðæs ðe on ðám ciricum wǽre dato praecepto, ut si qui in sancta loca confugissent, hos inviolatos securosque esse sinerent, Ors. 6, 38; Swt. 296, 32. Se ilca Dauid forbær ðæt hê ðone kyning ne yfelode, ðe hine of his earde âdrǽfde David ferire deprehensum persecutorem noluit, Past. 3; Swt. 37, 3. Ic wolde helpan ðæs ðe unscyldig wǽre, and hênan ðone ðe hine yfelode (yflode, v.l.), Bt. 38, 6; Fox 208, 18. Hî yfeledon and slôgan Cristene men affligi interficique Christianos praeceperunt, Bd. 1, 6; S. 476, 21. Yfeladan, Ps. Th. 82, 3. Hit is riht ðæt mon yfelige ða yfelan, and hit is wôh, ðæt hî mon lǽte unwîtnode, Bt. 38, 3; Fox 202, 5. Gif hwâ cyrican gesêce, and hine man ðǽr yflige, L. Edm. S. 2; Th. i. 248, 17. Hî ðara nânne yflian noldan ðe tô ðæm Godes hûse ôðflugon, Ors. 2, 8; Swt. 94, 8 : Nar. 25, 27. Ða ðe willaþ Godes cyricean yfelian and strûdan, Blickl. Homl. 75, 24. II. to get bad, (1) of persons :—Hié beóð swîðe ungesǽlige, ðonne hié yfeliaþ (yfliaþ, v.l.) for ðǽm ðe ôðre menn godigaþ quantae infelicitatis sint, qui melioratione proximi deteriores fiunt, Past. 34; Swt. 231, 18. (2) of things or circumstances :—Aa æfter ðam hit yfelode swîðe things got very bad, Chr. 975; Erl. 127, 33. Â syððan hit yflade swîðe, wurðe gôd se ende, ðonne God wylle, 1066; Erl. 202, 41. Gif blôddolg yflige . . . oððe gif ðú ne mæge blôddolg âwrîþan, Lchdm. ii. 16, 4 : 148, 8. Nýde hit sceal on worulde for folces synnan yfelian swýðe, Wulfst. 81, 8: 156, 7. [Wæstmes ne synd swâ gôde swâ heó iu wǽron, ac yfelað swýðe eall eorðe wæstme, Shrn. 17, 21. Ne scal us na mon uuelien, O. E. Homl. i. 15, 13.] v. ge-yflian.

yfel-lǽrende inciting to evil :—Yfelonbecweþende oþþe yfellǽrende malesuada, Germ. 390, 113.

yfel-lîc; adj. Bad, foul, rotten. v. yfel, II :—Ðysse worulde wela is gebrosnadlîc and yfellîc and forwordenlîc, Wulfst. 263, 13. Twêgen león âdulfan his byrgenne on ðæs wêstenes sande; ðǽr resteþ Paules lîchoma mid yfelîce duste bewrigen, ac on dômes dæge hê ârîseþ on wuldor, Shrn. 50, 18. Seó byrgen is bewrigen mid dimmum stânum and yfellîcum, 66, 25.

yfelness, e; f. Evil, wickedness, badness :—Yfelnys malignitas, Ælfc. Gr. 9, 25; Zup. 10. I. in a moral sense :—Micel yfelnys (malitia) manna wæs ofer eorðan, Gen. 6, 5. Hê (Antichrist) neádaþ þurh yfelnysse ðæt men sceolon bûgan fram heora Scyppendes geleáfan tô his leásungum, se ðe is ord ǽlcre leásunge and yfelnysse . . . on ðam tîman bið micel yfelnyss and þwyrnys betwux mancynne, Homl. Th. i. 4, 27–

33. Sume burgon heora feore and ámeldodon heora cristenan mágas . . . Ðeós yfelnys bið eác on Antecristes tócyme, ii. 542, 24. Bydelas ðæs écan yfeles, ðe yfelum mannum becymð for heora ánwillan yfelnysse, 538, 24. Yfelnysse (*malitiam*) ná hé hatude, Ps. Spl. 35, 4 : 51, 3. Ðurh yfelnysse (*nequitiam*) uurihtes willan, Bd. 1, 27 ; S. 495, 13. Hé áwearp yfelnysse and ða unrihtan biggengas ðæra leásra goda, Homl. Skt. i. 18, 461 : Chr. 1086 ; Erl. 223, 2. God gesihð úre yfelnyssa and úre gyltas forðyldgað, Homl. Th. ii. 84, 2. **II.** *malignity, cruelty.* v. yfel, **III:**—Hé slóh and tó sceame túcode ða Norðhymbran leóde, óþ ðæt Óswold his yfelnysse ádwæscte, Homl. Skt. ii. 26, 13. **III.** *misfortune, ill fortune:*—Oxan grasiende gesihð sige ceápas getácnaþ ; oxan slápende gesihð yfelnysse ceápes getácnaþ, Lchdm. iii. 200, 10. [Hé forbere monna hufelnesse þurh his liðnesse, O. E. Homl. i. 95, 14.]

yfel-onbecweþende. v. yfel-lǽrende.

yfel-sacian ; *p.* ode To *calumniate :*—Ðe lǽs hé mé yfelsacode wið God, Blickl. Homl. 189, 24.

yfel-sacung, e ; *f.* Calumny, *vituperation :*—On yfylsacunge heora *in malitia eorum,* Ps. Spl. C. 93, 23. Módignys ácenð andan and yfelsacunge, ceorunge and gelómlíce tála, Homl. Th. ii. 222, 7. Þurh yfelsacunge *per blasphemiam,* Confess. Peccat. Hé him rehte hwylce searwa and yfelsacunga se drý árefnde, Blickl. Homl. 173, 8.

yfelsian ; *p.* ode To *blaspheme :*—Yfelsaþ, tǽleþ *blasuemiat* (v. Mk. 2, 7), Wrt. Voc. ii. 73, 21. In the Northern Gospels the same Latin verb is translated by the following forms :—Ebalsas (hefalsaþ, Rush.) *blasphemat,* Mt. Kmbl. Lind. 9, 3. Ebolsas (heofolsaþ, Rush.), Mk. Skt. Lind. 2, 7. Ebolsas (eofolsas, Rush.), 3, 29. Ebolsaþ (eofolsigaþ, Rush.), Lk. Skt. Lind. 12, 10. Eofolsade (efalsade, Rush.) *blasphemavit,* Mt. Kmbl. Lind. 26, 65. Ebolsadon (eofulsadon, Rush.), Mk. Skt. Lind. 3, 28. Ebolsande (wéron) *blasphemabant,* 15, 29. Eofolsende, Jn. Skt. Rush. 10, 36. v. ge-ebolsian, *and next word : cf. also, two preceding words,* and eoful-sæc.

yfel-sprǽce ; *adj.* Of evil speech, *evil-speaking :*—Ða yfelsprǽcan tungan *linguam maliloquam,* Ps. Th. 11, 3.

yfel-sprecende ; *adj. (ptcpl.)* Evil-speaking :—Tungan yfelspreccende *linguam maliloquam,* Ps. Surt. 11, 4.

yfelsung, eofulsung, e ; *f.* Blasphemy :—Dionysius cwæð, ðæt ðæt yfelsang (-ung ?) wǽre on God Dionysius *dixit blasphemiam id esse in Deum,* L. Ecg. C. 41 ; Th. ii. 166, 12. Ic ondette eofulsunge, Anglia xi. 98, 33. In the Northern Gospels and Durham Ritual *blasphemia* is glossed by the following forms :—Ebolsung *blasphemia,* Rtl. 12, 37. Ebolsung í efalsongas (efulsung, Rush.), Mt. Kmbl. Lind. 12, 31. Ebolsung (hefalsunge, Rush.) *blasphemiae,* 15, 19. Ebolsungas, Mk. Skt. Lind. 3, 28. Efolsong (eofulsongas, Rush.) *blasphemia,* 7, 22. From ðæm ebolsong (eofolsonge, Rush.), Jn. Skt. Lind. 10, 33. Efolsungas (efalsunge, Rush.) *blasphemiam,* Mt. Kmbl. Lind. 26, 65. Ðæt ebolsung (ða eofulsunge, Rush.), Mk. Skt. Lind. 14, 64. Ebolsongas *blasphem(i)as,* Jn. Skt. Lind. 10, 36.

yfel-tyhtend, es ; *m.:* -tyhtende ; *adj. (ptcpl.)* One who incites to evil ; *inciting to evil :*—Deófol is yfeltihtend and leáswyrcend, Homl. Th. i. 102, 1. Ungeleáffulle and yfeltihtende sind mid ðé, 528, 3.

yfel-weorc ; es ; *m. Work of magic :*—Yfeluoerc *maleficium,* Rtl. 103, 1. Cf. yfel-dǽd, yfel-dǽde, ¶.

yfel-willende ; *adj. (ptcpl.)* Ill-disposed, *wicked :*—Hwæþer ðú ongite ðæt ǽlc yfelwillende mon and ǽlc yfelwyrcende ðe ðone wítes wyrþe ? . . . Hú ne is se ðonne yfelwillende and yfelwyrcende ðe ðone unscyldigan wítnaþ ? *omnem improbum num supplicio dignum negas? . . . Infelices esse, qui sint improbi, liquet,* Bt. 38, 6 ; Fox 208, 8–11. Mid ðé ne wunaþ se yfelwillenda *non habitabit juxta te malignus,* Ps. Th. 5, 4 : 9, 18. Ðæt yfelwillende mód *malitiosa mens,* Past. 35 ; Swt. 243, 7. Se ðe nele wunian on yfelwyllende sáwle, ne eác on ðam líchaman ðe lið under synnum, Homl. Th. ii. 326, 1. Yfelwillende men nǽnne weorþscipe nǽsdon, Bt. 15 ; Fox 48, 17. Se Drihten tóstencð ða geþeaht yfelwillendra kynna *Dominus dissipat consilia gentium,* Ps. Th. 32, 9. Fram gegaderunge yfelwillendra (*malignantium*), Ps. Lamb. 63, 3. On yfelwillendum *malignantibus,* 91, 12 : Ps. Spl. 36, 1. Hit náuht unriht wǽre ðæt mon ða yfelwillendan men (*vitiosos*) héte nétenu, Bt. 38, 2 ; Fox 198, 17.

yfelwillendness, e ; *f. Evil, wickedness :*—Hwæt wuldrast ðú on yfelwyllendnysse (*malitia*)? Ps. Spl. 51, 1.

yfel-wilnian ; *p.* ode *glosses* malignari :—Hú fela yfelwilnode (*malignatus est*) fýnd on hálgum, Ps. Lamb. 73, 3. Nelle gé wyrian í yfelwilnian *nolite malignari,* 104, 15.

yfel-wyrcende ; *adj. (ptcpl.)* **I.** of persons, *evil-doing, wicked :*—Hwæþer ðú ongite ðæt ǽlc yfelwillende and ǽlc yfelwyrcende sié wítes wyrþe ? . . . Hú ne is se ðonne yfelwillende and yfelwyrcende ðe ðone unscyldigan wítnaþ ? Bt. 38, 6 ; Fox 208, 8–11. Gif ne wére ðes yfelwyrcende (*malefactor*), ne ðé wé gisaldun hine, Jn. Skt. Rush. 18, 30. Yfelwyrcende *nequam,* Mt. Kmbl. Lind. 6, 23 : 13, 38. Mið yfelwyrcendum and synfullum *cum publicanis et peccatoribus,* 9, 11. Hé hataþ ða yfelwyrcendan and ða unrihtwísan, Homl. Skt. i. 1, 48. **II.** of

things, *noxious, hurtful, mischievous :*—Ðerh wyrto yfelwyrcendo *per herbas malificas,* Rtl. 103, 1.

yfemest, yfmest ; *adv. Upmost, highest, in the highest position* or *degree :*—Hió cymþ swá up swá hire yfemest gecynde bið *it* (*the sun*) *mounts up to the highest point at which it is natural for it to be,* Bt. 25 ; Fox 88, 28. Ðǽr hire yfemest bið eard gecynde, Met. 13, 63. Ðæt fýr is yfemest ofer eallum ðissum woruldgesceaftum, Bt. 33, 4 ; Fox 128, 38 : Met. 20, 84. Yfmest, 24, 20. Saturnus yfemest wandraþ ofer eallum óðrum steorrum, 24, 23. Uton habban úre mód up swá swá wé yfemest mǽgen wið ðæs heán hrófes ðæs héhstan andgites, Bt. 41, 5 ; Fox 254, 15. Ǽresð alra glengea and ymesð scolde scínan gold on his hrægle *in sacerdotis habitu ante omnia aurum fulget,* Past. 14 ; Swt. 85, 2. v. ufor.

yfera ; *cpve.:* yfemest ; *spve. adj. Upper, higher ;* of time, *later, after : upmost, highest :*—Yferan hýse *triclinio,* Wrt. Voc. ii. 72, 66. Sioððan yferran dógre, Cod. Dip. Kmbl. i. 310, 29. Cyng ah ðone uferan (yferan, *v.l.*), and bisceop ðone nyðeran, L. E. G. 4 ; Th. i. 168, 16. Þurh his upstige tó ðam yfemystan þrymsetle, Homl. Th. i. 308, 9. Of ðǽm yfemestum (ymestum, Hatt. MS.) tó ðǽm nieðemestan, Past. 18 ; Swt. 134, 24. v. ufera.

yfes-drype, es ; *m. Eaves-drip :*—Ðǽr ne gebyreþ an ðam lande an folcæs folcryht tó léfænnæ rúmæs bútan twígen fýt tó yfæsdrype, Chart. Erl. 141, 16, where see note.

yl-ful (=ild-ful) *morosus,* Hpt. Gl. 529, 9.

ylp (elp), es ; *m. An elephant :*—Ylp *elefans,* Wrt. Voc. i. 78, 11 : 22, 41. Ylp is ormǽte nýten, mǽre þonne sum hús, Homl. Skt. ii. 25, 566. Ylpes bile *promuscida,* Wrt. Voc. i. 22, 42. Ylpes bán *ebur,* Coll. Monast. Th. 27, 9 : Lchdm. iii. 204, 2, 3. Hé sende þrittig ylpas tó wíge gewenode . . . and on ǽlcum ylpe wæs án wíghús getimbrod, Homl. Skt. ii. 25, 561. Hé (*the unicorn*) fiht wið ðone myclan ylp, and hine oft gewundaþ on ðære wambe óþ deáþ, Wrt. Voc. i. 78, 1. Gif hé ylp gesihð láðne oððe gramne, sume wróhte hit getácnaþ, Lchdm. iii. 204, 1. Ða ylpas beóð swá mycele swylce óðre muntas, Hexam. 9 ; Norm. 16, 9. Hé geworhte ða ormǽtan ylpas, Norm. 14, 34. [Elpes arn in Inde riche, Misc. 19, 604. White so alpes bon, L. N. F. 248, 282.] v. elpend.

ylpen-bán (elpend-), es ; *n. Ivory :*—Ðis ylpenbán *hoc ebur,* Ælfc. Gr. 9, 22 ; Zup. 49, 9. Mid ylpenbáne and mid báres tuxe, Lchdm. i. 244, 8. Genim ylpenbán, 368, 19. v. elpend-bán (*where these passages should be put*).

ylpen-bænen, -bánen (elpend-) ; *adj. Ivory :*—Mid ylpenbánenon (-bǽn-, *v.l.*) stæfe, Lchdm. i. 244, 24. Ylpenbánene *eburna,* Germ. 403, 19. v. elpen(d)-bǽnen (*where these passages should be put*).

ylpend, Wrt. Voc. ii. 142, 81. v. elpend.

yltst = **ildest,** Mt. Kmbl. 23, 11 : Ex. 17, 5.

yltwist (?) *fowling :*—Yltwist *aucupium,* Wrt. Voc. ii. 7, 50.

ymb, ymbe, umbe, embe, emban ; *prep. About, by :*—Ymb *erga,* Wrt. Voc. ii. 32, 62. **I.** with acc., (1) local, *about, round :*—Ymbe ða dúne *circum montem,* Ælfc. Gr. 47 ; Zup. 269, 8. (a) marking an object which forms a centre for others :—Ymb ðone écan æðele stondaþ hæleð ymb héhseld, Cd. Th. 267, 32 ; Sat. 47 : Beo. Th. 804 ; B. 399 : Elen. Kmbl. 519 ; El. 260 : Judth. Thw. 25, 19 ; Jud. 268. Ymb ðæt *circumquaque* (turmas circumquaque cum simulacro debacchantes, Ald. 52), Wrt. Voc. ii. 83, 33 : 18, 47. Ymb hine wǽgon wígend unforhte, Cd. Th. 189, 5 ; Exod. 180. Hí ymb þeódenstól þringaþ, Exon. Th. 25, 7 ; Cri. 397. Hié ymb ða gatu feohtende wǽron, Chr. 755 ; Erl. 50, 26. Hý fuhton stíðlíce ymbe ða hálgan sáwle, Wulfst. 236, 23. ¶ *in combination with* útan :—Fuglas þringaþ útan ymbe æþelne, Exon. Th. 209, 1 ; Ph. 164 : Andr. Kmbl. 1741 ; An. 873. Ymb ðæt líc útan stondan, Blickl. Homl. 217, 21. (b) marking an object near to which are others :—Geségon hý englas twégen ymb ðæt frumbearn blícan, Exon. Th. 32, 3 ; Cri. 507. Hine twégen ymb weardas wacedon, 109, 5 ; Gú. 85. Ealle ða ðe ymbe mé standaþ, Blickl. Homl. 141, 1. Hine ymb monig sǽrinc selereste gebeáh, Beo. Th. 1383 ; B. 689. Mycel menegeo ymbe Tírum (*circa* Tyrum), Mk. Skt. 3, 8. Cynewulf and Offa gefuhton ymb Benesingtún and Offa nam ðone tuun, Chr. 777 ; Erl. 54, 1. Ymbe Brúnanburh, 937 ; Erl. 112, 5 : Hy. 10, 23. (b) *about a person, in attendance upon :*—Hé sundernytte beheóld ymb aldor Dena, Beo. Th. 1340 ; B. 668. (c) marking an object which is surrounded or enclosed :—Hié worhton fæsten ymb hié selfe, Chr. 885 ; Erl. 82, 21. Wall ymb ðǽste, Cd. Th. 231, 16 ; Dan. 248. Ymbe ða herehúþe hlemmeþ tógædre grimme góman, Exon. Th. 363, 29 ; Wal. 61. Hí ymb his heáfod gebígdon beág þyrnenne, 69, 25 ; Cri. 1126 : 400, 20 ; Rä. 21, 4. Wæs fleóhnet ymbe ðæs folctogan bed áhongen, Judth. Thw. 22, 4 ; Jud. 47. ¶ *in combination with* útan :—Is ðǽr cyrice ymb ða stówe útan getimbred, Blickl. Homl. 125, 20 : 127, 32. Ymb ðinne beód útan *in circuitu mensae tuae,* Ps. Th. 127, 4. (d) marking an object along whose border others are placed :—Unc módige ymb mearce sittaþ, Cd. Th. 114, 21 ; Gen. 1907. Hié wícedon ymb ðæs wæteres wylm, Elen. Kmbl. 77 ; El. 39 : 271 ; El. 136 : Exon. Th. 188, 2 ; Az. 39. Ymb ða gifhealle *round the walls of the hall,* Beo. Th. 1680 ; B. 838. Ðæt ríce súð licgeþ ymbe Gealboe and ymb Geador, Salm. Kmbl. 383–4 ; Sal. 191. Is hyge ymb heortan gerúme,

Cd. Th. 47, 11; Gen. 759: 23, 5; Gen. 354: Exon. Th. 306, 21; Seef. 11. (e) marking an object throughout or along which there is position or movement:—Ic .lǽrde sibbe ymb ða burh Hierusalem and manige þeóda, Blickl. Homl. 185, 11. Hé ymb ðæs wæteres stæð werod samnode, Elen. Kmbl. 119; El. 60: 453; El. 227. Æfter dúnscræfum, ymb stânhleoðo, Andr. Kmbl. 2467; An. 1235: 3152; An. 1578. Ymb ða weallas scínað engla gástas, Cd. Th. 305, 25; Sat. 652. Ymbe hárne stán tigelfágan trafu stódan, Andr. Kmbl. 1682; An. 843. Ymbe Sanere feld, Salm. Kmbl. 417; Sal. 209. Ymb healfa gehwone, Exon. Th. 4, 31; Cri. 61. Sár eft gewód ymb ðæs beornes breóst, Andr. Kmbl. 2495; An. 1249. (f) marking an object round which anything moves:—Faraþ ymbe ða burh *circuite urbem*, Jos. 6, 3, 12: Beo. Th. 6319; B. 3170. Hý him ymb hond flugon, Exon. Th. 146, 14; Gú. 709. Hí ymb ða eaxe hwearfaþ, Bt. 39, 3; Fox 214, 23: Met. 28, 22. ¶ *in combination with* útan:—Hé ymb ðás útan hweorfeþ, Exon. Th. 422, 13; Rä. 41, 5. (2) temporal, (a) *at*:—Ymbe (embe, *v.l.*) underntíde ... ymbe ða sixtan and nigoðan tíde *circa horam tertium* ... *circa sextam et nonam horam*, Mt. Kmbl. 20, 3, 5. Ymbe underntíd, ðá ðá se bróðor wæs gewunod tó mæssigenne, Homl. Th. ii. 358, 20. Ofer Eástron ymbe gangdagas oððe ǽr, Chr. 891; Erl. 88, 16. On ðýs geáre ymb Martines mæssan, 913; Erl. 100, 33. Ymb ðone tiéman wǽren micel snáwgebland, Ors. 4, 8; Swt. 186, 33. Þeáh hine mon gefó ymb niht, L. In. 72; Th. i. 148, 8. (b) *after*:—Ic sende rén nú ymb seofon niht ofer eorðan *adhuc et post dies septem ego pluam super terram*, Gen. 7, 4. On ðisse tíde nú ymbe twelf mónað *tempore isto in anno altero*, 17, 21. Næs hit lengra fyrst ac ymb áne niht, Beo. Th. 270; B. 135. Ymb twá niht, Cd. Th. 181, 18; Exod. 63. Ymb healfa gé mónað wucan bútan ánre niht, Menol. Fox 172; Men. 87. Ymb fíftig nihta æfter ðære gecýþdan ǽriste, Blickl. Homl. 133, 13. Hé forþférde ymb .xx. wintra his ríces, bútan án ne wæs ðǽgyt gefylled *defunctus est anno regni sui vicesimo necdum impleto*, Bd. 5, 18; S. 635, 19. Ymb hwíle, Blickl. Homl. 217, 30. Ymb long, L. In. 21; Th. i. 116, 7: Bt. 39, 2; Fox 214, 8. Ymb tela micel fæc, Chr. 942; Erl. 116, 21. Ymbe án lytel, Jn. Skt. 16, 16. Ymb lytel fæc, Elen. Kmbl. 543; El. 272. Ymbe geára rina, Chr. Pref.; Erl. 3, 17. Ymb wintra hwearft, Exon. Th. 188, 5; Az. 41. (b 1) *where the point from which the time is measured is given by* ðæs, (a) *preceding*:—Ðæs ymb án geár, Ors. 3, 10; Swt. 138, 28: 3, 11; Swt. 152, 19: Chr. 871; Erl. 74, 6, 8, 14, 25: Exon. Th. 29, 21; Cri. 466. Ðæs ymb ðreó swylc bútan ánre wanan, Menol. Fox 279; Men. 141: 359; Men. 181. Ðæs ymbe lytel, Chr. 1038; Erl. 167, 6. Ðæs ymb litel fæc, Guthl. 18; Gdwin. 76, 6. (β) *following*:—Ymb feówer hunde wintra and ymb feówertig ðæs ðe Tróia áwésted wæs *anno post eversionem Trojae cccxiv*, Ors. 2, 2; Swt. 64, 20. Ymb .xxxi. wintra ðæs ðe hé ríce hæfde, Chr. 755; Erl. 48, 25. Ymbe .xli. wintra bútan ánre niht ðæs ðe Ælfréd cyning forþférde, 941; Erl. 116, 2: 606; Erl. 20, 25: 855; Erl. 68, 22: Cd. Th. 167, 21; Gen. 2769. Ymb swýðe lang ðæs ðe hine God álýsde, Ps. Th. 17, arg. (b 2) *of recurring periods*:—Saturnus ne cymþ ðǽr ǽr ymb þrittig wintra ðǽr he ǽr wæs, Bt. 39, 3; Fox 214, 25. Ælce geáre ymbe twelf mónaþ, Ors. 1, 10; Swt. 46, 9. Ǽfre ymbe ðæt feórðe geár, Lchdm. iii. 246, 13. Simble ymb þrítig nihtgerímes, Andr. Kmbl. 313; An. 157. Symble ymbe seofon niht, Soul Kmbl. 19; Seel. 10. Emb stemn *uicissim*, Germ. 388, 77: Scint. 140, 17. (c) *of past time*:—Ymb þreó niht com þegen Hǽlendes *the Saviour's servant came three days ago*, Cd. Th. 291, 5; Sat. 426. Ðæs ymb áne niht, 300, 26; Sat. 571. (3) in figurative senses, *about*, (a) marking approximation:—Ymb ðæt *plus minus*, Wrt. Voc. ii. 117, 50: 68, 24. (b) marking the object of speaking, enquiry, telling, etc.:—Hé ymb Godes word and Cristes geleáfan (Godes word ymbe Cristes geleáfan, M. 422, 9) bodude and lǽrde, Bd. 5, 11; S. 626, 29. Ymb ðín líf sprecan, Cd. Th. 32, 25; Gen. 508: 110, 34; Gen. 1848: Beo. Th. 3194; B. 1595. Hé sǽgde ymb Godes ríce, Blickl. Homl. 117, 13. Wítgan sægdon ymb ðæt æþele bearn, ðæt . . . , Exon. Th. 73, 26; Cri. 1195. Hǽlde se cyng swíðe deópe spǽce wið hys witan ymbe ðis land, hú hit wǽre gesett, Chr. 1085; Erl. 218, 23. Wé cweþaþ lof ymb hié, Blickl. Homl. 149, 32. Wé beót áhófon ymbe heard gewinn, Byrht. Th. 138, 3; By. 214. Dislíc cýðan ymb dígle wyrd, Elen. Kmbl. 1077; El. 541. Ærendgewrit ymb Cristes þrowunga, Blickl. Homl. 177, 3. Fitte ymb fisca cynn, Exon. Th. 360, 6; Wal. 1. Hé gieddade ymb his ǽriste, 236, 10; Ph. 572. Ðá ðá hí umbe óþer þing gesprecon hæfdon umbe ðæt hí sprecan woldon, Chr. 1070; Erl. 208, 12. Se esne ðe ic hér ymb sprice, Exon. Th. 430, 32; Rä. 44, 17: Met. 10, 45. Ðe ic ðé recce ymb, 17, 20. Ymb ðæt áscian, Bt. 39, 4; Fox 216, 29. Gif ðú gehýre ymb ðæt hálige treó fróde frignan, Elen. Kmbl. 881; El. 442: 1065; El. 534: Beo. Th. 712; B. 353. (c) marking the object of thought, feeling, etc.:—Giorne ymb láre, Past. pref.; Swt. 3, 10. Giémen ymb ða gehiérsuman, 12; Swt. 74, 14: Exon. Th. 267, 13; Jul. 414. Hé ná ymb his líf cearaþ, Beo. Th. 3077; B. 1536. Ymb sáwle forht, Exon. Th. 456, 10; Hy. 4, 64. Ymb ðæs geongan feorh onbryrded, Andr. Kmbl. 2236; An. 1119. Ymb ða mé fyrwet bræc, Salm. Kmbl. 493; Sal. 247. Heó wundrade ymb ðæs weres snyttro, Elen. Kmbl. 1914; El. 959. Ðæt seó forlǽtene cyrice ne hycgge ymb ða ðe on hire neáwiste lifgeaþ, Blickl. Homl. 43, 1: Exon. Th. 473,

3: Bo. 9: Menol. Fox 571; Gn. C. 55. Ic þence ymbe míne synna *cogitabo pro peccato meo*, Ps. Th. 37, 18: Cd. Th. 26, 18; Gen. 408. Ymb wundorwyrd willan gefylde, Elen. Kmbl. 2139; El. 1071. Ic nát ymbe hwæt ðú tweóst, Bt. 5, 3; Fox 12, 12. (d) marking the object with which an action is concerned:—Is máre nédþearf ðæt wé winnon ymbe úre sáule þearfe, Blickl. Homl. 99, 10. Ymb land sacan, Menol. Fox 568; Gn. C. 53: Beo. Th. 5012; B. 2509: 1019; B. 507. Ðæt hé hié ymb ðæt ríce gesémde, Ors. 3, 7; Swt. 114, 17: Salm. Kmbl. 505; Sal. 253. Hí sendon ǽrendracan ymbe frið, Ors. 3, 11; Swt. 142, 2. Hig dydon ymbe hyne (*in eo*) swá hwæt swá hig woldon, Mt. Kmbl. 17, 12. Ðú ymb ðinne esne dydest wel weordlíce *bonitatem fecisti cum servo tuo*, Ps. Th. 118, 65. Hwæt ymb hine gedón wǽre *quid erga se actum esset*, Bd. 4, 31; S. 610, 39. Hú hine mon ymbe gedón wolde *quid erga eum agere rex promisisset*, 2, 12; S. 513, 20. Ðæt hé móste dón embe ða æþelingas swá hé wolde, Lchdm. iii. 424, 27. Ymb his womdǽda Waldendes dóm, Ps. C. 19. Hé wæs ymbe Godes þeówdóm ábisgod, Blickl. Homl. 211, 31. Beón, wesan ymb *to be about* a business, *be occupied with* a matter or a person:—Gif gé ymb woruldcunde dómas beón scylen *secularia judicia si habueritis*, Past. 18; Swt. 131, 6. Hwonne hé móste beón ymbe ðæs líchaman oferfylle, Wulfst. 236, 11. Wit sculon git deóplícor ymbe ðæt beón, Bt. 5, 3; Fox 12, 12. Gif hwylc ðis dón nylle and læs ymbe beó ðonne wé gecweden habbaþ, L. Ath. i. 26; Th. i. 212, 28: iv. 1; Th. i. 222, 2. Hé byð á ymbe ðæt án, hú hé on manna sáulum mǽst gesceaðian mæge, L. C. E. 26; Th. i. 374, 25. Hé sǽde ðæt Aldberht and Alhhún wǽron ǽr ymb ðæt ylce, Chart. Th. 140, 14. Ðás feówero ymb woeson (= woeron?) ðás bóc *these four were engaged on this book*, Mk. Skt. p. i. 3. Hé cwæð ðæt hé ne mihte embe munuclíf smeágan, ac wolde beón embe his þincg, Homl. Skt. i. 6, 120. Emban ða steóran beon, L. Ath. v. 11; Th. i. 240, 17. Settaþ ða tó démerum ðæt hié striénen and stihtien ymb ða eordlícan ðing *ut ipsi dispensationibus terrenis inserviant*, Past. 18; Swt. 131, 8. Ðá gesomnodon wé ús ymb ðæt, L. Alf. 49; Th. i. 56, 19. Gif hwá wiccige ymbe ǽniges mannes lufe (*alicujus amoris gratia*), L. Ecg. P. iv. 18; Th. ii. 208, 31. Hié sieredon ymbe ðone cyning, Bt. 16, 2; Fox 52, 21: Cd. Th. 38, 15; Gen. 607. Mé seredon ymb secgas monige, 296, 6; Sat. 498: Ps. Th. 54, 18. (e) marking the object in relation to which circumstances are stated:—Ic ðé cýðe hú hit wæs ymb ðæt lond æt Funtial, Chart. Th. 169, 16. Sceolde unc yfele gewurðan ymb ðæt heofonríce, Cd. Th. 25, 3; Gen. 388. Hú ymb ðæt sceolde, Exon. Th. 378, 7; Deór. 12. Sý ymb ríce swá hit mæge, 301, 29; Fä. 26. Hú ða wísan sind wundorlíce ymb ðæs fugles gebyrd, 223, 16; Ph. 360. Ðæt wundor ymb ðone beorhtan beám, Elen. Kmbl. 2507; El. 1255: 1324; El. 664. II. with dative, (1) local:—Ða weorod ðe him ymb férdon and stódon, Blickl. Homl. 99, 25: Beo. Th. 5188; B. 2597. Ða hire midore ymbe þrǽgaþ, Met. 28, 23. Geseó ic him his englas ymbe hweorfan, Cd. Th. 42, 5; Gen. 669. Him ymb flugon engla þreátas, 300, 21; Sat. 568. (2) temporal:—Embe geára ymbrynum, Homl. Th. i. 104, 21. Ǽfre ymbe geáres ymbrynum, Lchdm. iii. 238, 25. (3) figurative:—Hé férde embe sumere neóde, Homl. Th. ii. 508, 15. III. without a case:—Ǽghwider ymb swá swá Édwines ríce wǽre *quaquaversum imperium Aeduini pervenerat*, Bd. 2, 16; S. 519, 38. Ðonon eode gehwyder ymb (*circumquaque*), 3, 17; S. 543, 26. Hié wǽron ymb eal útan mid eágum besett, Past. 38; Swt. 195, 19. Hring útan ymb bearh, ðæt heó ðone fyrdhom þurhfón ne mihte, Beo. Th. 3011; B. 1503. Hygeþ ymbe se ðe wile, Met. 19, 1. Þencð ymb se ðe wile, 20, 27. Dǽd ymbe moncynnes fruma, swá him gemet þinceþ, 29, 41. Ðá cýdde man, ðet hí man eáðe befaran mihte, gif man ymbe beón wolde, Chr. 1009; Erl. 141, 34. Hé cýdde hú hé ymbe wolde, gif hé hine gemétte *he shewed what he would have been about, if he had found him*, Homl. Th. i. 82, 18. [A. R. Kath. Marh. umbe: O. E. Homl. Laym. umbe, embe: Orm. ummbe: Piers P. um: O. Sax. umbi: O. Frs. umbe: O. H. Ger. umpi: Icel. umb, um.]

ymb-ærnan; *p.* de *To go round*:—Ðá gelamp ðætte Peahte ðeód com of Scyþþia lande and ymbærndon eall Breotone gemǽro, ðæt hí cómon on Scotland upp *contigit gentem Pictorum de Scythia, circumagente flatu ventorum, extra fines omnes Brittaniae Hiberniam pervenisse*, Bd. 1, 1; S. 474, 10. v. ymb-irnan.

ymb-bǽtan; *p.* de *To put restraint upon, curb*:—Se mid his brídle ymbebǽted hæfð ymbhwyrft ealne eorþan and heofenes *Dominus orbis habenas temperat*, Met. 24, 37.

ymb-begang. v. ymb-bígness.

ymb-beran; *p.* -bær, *pl.* -bǽron; *pp.* -boren *To surround*:—Se wæs ǽghwonan ymbboren brondum, Exon. Th. 277, 15; Jul. 581. Ymb-beara *glosses* circumferre, Mk. Skt. Lind. Rush. 6, 55.

ymb-bígness, e; *f.* A *bending round, a bend of* a river:—Ðæt mynster is of ðam mǽstan dǽle mid ymbbígness (ymbbegange [ymbegang?], *v.l.*: ymbebígnesse, M. 424, 10) Tweode streámes betýned *monasterium Tuidi fluminis circumflexu maxima ex parte clauditur*, Bd. 5, 12; S. 627, 25.

ymb-bindan; *p.* -band, *pl.* -bundon; *pp.* -bunden *To bind about*:—

Sié ymbunden ꝉ ymbsald coern tó suiro his *circumdaretur mola collo ejus*, Mk. Skt. Lind. 9, 42.

ymb-cæfed; *adj. (ptcpl.) Having embroidered gàrments :*—Ymbcæfed mid missenlícnesse *circumamicta varietatibus*, Ps. Spl. T. 44, 15.

ymb-ceorfan; *p.* -cearf, *pl.* -curfon; *pp.* -corfen *To circumcise :*—Gé ymbceorfas (-cearfas, Lind.) ꝺone monno *circumciditis hominem*, Jn. Skt. Rush. 7, 22. Tó ymbceorfanne (-cearfanne, Lind.) ꝺone cnæht *circumcidere puerum*, Lk. Skt. Rush. 1, 59. Ꝺætte ymbcorfen wére ꝺe cnæht, 2, 21.

ymb-ceorfness, e; *f. Circumcision :*—Ymbcer[f]nisse *circumcisionem*, Jn. Skt. Rush. 7, 23.

ymb-cirr, es; *m. A turning about,* (1) *going from one place to another, removal :*—In ymbcerr Babilonis *in transmigratione Babylonis*, Mt. Kmbl. Lind. 1, 11, 12, 17. (2) *turning over, moving, stirring.* v. ymb-cirran (4) :—Wætres ymbcerr (-cer, Rush.) ꝉ styrenise *aquae motum*, Jn. Skt. Lind. 5, 3. (3) the word also glosses *versutia*, Rtl. 120, 32.

ymb-cirran; *p.* de *To turn about,* (1) *to revolve round :*—Hí ꝺære eaxe útan ymbhwerfaþ (-eþ, MS.) ꝺone norꝺende, neáh ymbcerraþ (-eþ, MS.), Met. 28, 14. Saturnus hæfꝺ ymb þrítig wintergerímes weoruld ymbcyrred, 28, 26. (2) *to turn one's self round :*—Ymbcerred wæs on bægcgling *conuersa est retrorsum*, Jn. Skt. Lind. Rush. 20, 14. (3) *to turn away, avert :*—Onsión mín ne ymbcerdig (*averti*) from gispittendum on mec, Rtl. 19, 15. (4) *to turn over, move, stir, overturn :*—Hé ꝺa discas ymbcerde *mensas subvertit*, Jn. Skt. Lind. Rush. 2, 15. Engel ymbcerde (*mouebat*) ꝺæt wæter, 5, 4. Mid fynger hiora nallas ꝺa ymbcerræ (*styrgan*, Rush.) *digito suo nolunt ea movere*, Mt. Kmbl. Lind. 23, 4. (5) *to change :*—Ymbcerred *mutata*, Mt. Kmbl. p. 1, 2.

ymb-clyppan; *p.* te *To embrace, clasp,* (1) *of persons :*—Ic ymbclyppe ꝺé *complector te*, Ælfc. Gr. 19; Zup. 122, 4: *amplector*, 36; Zup. 214, 5. Ic ymbclyp[p]e *obunco*, Wrt. Voc. i. 22, 31. Ymbclypte *obuncabat* (Timotheum ulnarum gremiis procax obuncabat, Ald. 40), ii. 81, 12: 64, 28. (2) *of things :*—Æghwilc oþer útan ymbclyppeþ, Met. 11, 35. Swá swá lyft and lagu land ymbclyppaþ, 9, 40. Swá ymbclyppaþ cealda brymmas, Chr. 1065; Erl. 197, 31. Fingras þrý útan eáþe eaþe mægon mec ymbclyppan, Exon. Th. 425, 9; Rä. 41, 53. Rápas synfulra ymbclyppende syndon (*circumplexi sunt*) mé, Ps. Lamb. Surt. 118, 61. [Stringes of sinful umclipped me, Ps. 118, 61. þe cercle þat umbeclypped his croun, Gaw. 616.]

ymb-clypping, e; *f. An embrace :*—Emclippingca *amplexus*, Hpt. Gl. 511, 36.

ymb-cyme, es; *m. A convention, an assembly :*—Ꝺær wæs gesamnad eádigra geþeahtendlíc ymcyme, L. Wih. pref.; Th. i. 36, 7. Cf. ymb-þreodian.

ymb-cyrf, es; *m.* I. *circumcision :*—Miꝺ ymbcyrf *circumcisione*, Mt. Kmbl. p. 12, 11: 16, 13. II. *a cutting off :*—Of ymbcyrf liomana *de abscisione membrorum*, Mk. Skt. p. 4, 9.

ymb-dón; *p.* -dyde *To put round, encompass :*—Ic embedó *circumdo*, Ælfc. Gr. 24; Zup. 139, 13.

ymbe (imbe), es; *m.* (?) *A swarm of bees :*—Wiꝺ ymbe . . . forweorp ofer greót þonne hí swirman, and cweꝺ : ' Sitte gé sigewíf . . .,' Lchdm. i. 384, 18. ¶ Imbæs dæl *occurs* Cod. Dip. Kmbl. iii. 176, 20. [*O. H. Ger.* impi bíanó *examen apium : M. H. Ger.* imbe; *m.: Ger.* imme; *f.*] v. ymb-haga.

ymbe *about,* ymbe-. v. ymb, ymb-.

ymbeaht, es; *m. The word glosses* collatio :—Ymbeahtas *collationes* (the passage is : Haec x collationes patrum a Cassiano digestae propalabant, Ald. 13. In Hpt. Gl. 428, 7 *collationes* is glossed by *race* and explained by *narrationes*), Wrt. Voc. ii. 76, 80 : 18, 3. Elsewhere the form is identical with ambiht :—Ambechtae, oembecht *collatio*, Txts. 46, 187. Ambect, ambaect *rationatio*, 92, 866. Ambiht *office* is neuter, ymbeaht is masculine : it seems (?) as if the form had been connected with eahtian *to consider*, and the word were regarded as a compound, ymb-eaht. See Engl. Stud. xi. 492.

ymb-eardiendra glosses circumhabitantium, Ps. Surt. 30, 14.

ymb-fær, es; *n. A going about, circuit :*—Túna embefær *uillarum circuitus*, Anglia xiii. 375, 131. Mid emfare *circilo* (= *circulo*?), Hpt. Gl. 422, 14.

ymb-færeld, es; *n. m. A going round, circuit :*—Fram þénuncge embefærelde his *ab officio circuitus sui*, Anglia xiii. 434, 980. Hig férdon seofon síꝺon embe ꝺa buruh. And on ꝺam seofoꝺan ymbfærelde (*circuitu*) . . . burston ꝺa weallas, Jos. 6, 16.

ymb-fæstness glosses circumstantia, Rtl. 174, 17.

ymb-fæstnung, e; *f. A monument, tomb :*—Ymbfæstnung ꝉ byrgenn *monumentum*, Jn. Skt. Lind. 19, 41.

ymb-fæþmian; *p.* ode *To embrace, clasp :*—Ne magon hý ꝺa lífes lísan on middan ymbfæꝺmian, Salm. Kmbl. p. 152, 32.

ymb-faran; *p.* -fór *To surround :*—Hé hét ꝺæt fæste lond útan ymbfaran, ꝺæt him mon sceolde an má healsa on feohtan þonne on án, Ors. 2, 5; Swt. 80, 26.

ymb-feng, es; *m. A cover, an envelope :*—Emfencge (*librorum*) *tegmine, operimento*, Hpt. Gl. 417, 47.

ymb-féran. v. emb-féran.

ymb-fón; *p.* -féng. I. *to grasp, clasp :*—Hé fótum ymbféhꝺ fýres láfe, Exon. Th. 217, 6; Ph. 276. Heó ymbféng Drihtnes fét, Blickl. Homl. 157, 17. Ymbféng *obuncat* (moecham, quam manus tollentis obuncat, Ald. 164), Wrt. Voc. ii. 92, 39. Ymbeféng, Beo. Th. 5376; B. 2691. II. *to encompass, surround, comprehend :*—Ealle stówa hé gefylleþ and ymbféhþ, Blickl. Homl. 23, 20. Seó séleste gesælþ ꝺa óþra gesælþa ealle on innan him gegaderaþ, hí útan ymbféhꝺ, Bt. 24, 1; Fox 80, 21. Ꝺú meaht ymbfón eal folca gesetu, Exon. Th. 466, 2; Hö. 115. Ymbfónde *gyrens*, Wrt. Voc. ii. 41, 66. Hit is on ælce healfe ymbfangen mid gársecge, Ors. 1, 1; Swt. 24, 17. Ꝺínre gedréfednesse ꝺe ꝺú mid ymbfangen eart, Bt. 5, 3; Fox 12, 18. Sunu Meotodes habbaþ ealle ymbhangen mid sange, Cd. Th. 273, 30; Sat. 144. Ꝺeáh hé wære mid írne ymbfangen, Cd. Th. 297, 16; Sat. 518. II a. *to comprehend, conceive :*—Embféhþ *concipit, i. intelligit*, Wrt. Voc. ii. 136, 21. III. *to put something round an object, surround, envelope :*—Genim foxes gecynd, ymbfóh (ym-, *v. l.*) ꝺæt heáfod útan, Lchdm. i. 340, 19. Healfnacode on hiora líchaman búton ꝺæt hig wæron mid riftum ymbfangene (*but see* ymb-hón), Shrn. 38, 7.

ymb-frætwan; *p.* ode *To surround with ornament :*—Ꝺeáh ꝺe men him háton gewyrcan heora byrgene of marmanstáne, and útan emfrætewian mid reádum golde, Wulfst. 148, 21. Ymbfrætewode *circumornatae*, Ps. Lamb. 143, 12.

ymb-gán; *p.* -eode; *pp.* -gán. I. *to go round* (1) a circular course :—Ær sunne twelf mónꝺa hringc útan ymbgán hæbbe, Guthl. 21; Gdwin. 96, 6. (2) an object :—Hí útan ymbgáꝺ ceaster *circuibunt civitatem*, Ps. Spl. C. 58, 16. II. *to go about, in the neighbourhood of.* v. ymb, I. 1 b :—Ic ymbgaa weófod ꝺínre *circumdabo altare tuum*, Ps. Spl. C. 25, 6. Ic ymbgá and ic offrige onsægednessa *circumivi et immolavi hostiam*, Ps. Spl. 26, 11. III. *to go about, through.* v. ymb, I. 1 e :—Swá hundas ymbgáꝺ hwommas ceastre, Ps. Th. 58, 6, 14. Ymbeode ides Helminga duguꝺe and geógoþe, dæl æghwylcne, Beo. Th. 1244; B. 620. Ymbeade Hælend alle Galiléa *circumibat Jesus totam Galilaeam*, Mt. Kmbl. Lind. 4, 23. Ꝺa ongan heó ymbgán ꝺa hús ꝺæs mynstres *coepit circuire in monasterio casulas*, Bd. 3, 8; S. 531, 32. [I umyhode, Ps. 26, 6. Umga, 58, 7. þe laddes unbiyeden him, Havel. 1842. *O. H. Ger.* umbi-gán.]

ymb-gang, es; *m.* I. *a going round :*—Seó burh (*Jericho*) næs mid nánum wíge gewunnen, ac mid ꝺam ymgange, Homl. Th. ii. 216, 2. Is ꝺære sunnan ymgang (ymbe-, ymb-, *v. ll.*) hremming, ꝺæt se dæg ne byꝺ on ælcum earde gelíce lang, Lchdm. iii. 258, 11. Ælc mann, swá swá hé stód on ꝺam ymbgange, Jos. 6, 20. Emgange *abitum* (= ambitu, Ald. 73), Hpt. Gl. 522, 78. II. *a going about :*—Embgong *deambulacrum, circuitus*, Wrt. Voc. ii. 139, 82. III. *of extent traversed or measured, circuit, circumference :*—His ymbgong (*ambitus*) is hundseofontig míla and seofeꝺa dæl ánre míle, . . . and busan ꝺæm máran wealle ofer ealne ꝺone ymbgong hé is mid stænenum wíghúsum beworht, Ors. 2, 4; Swt. 74, 15–21. Six hund fóta and feówertig seó cyrce wæs ymbganges, Homl. Th. ii. 496, 35. Ofer ymbgang *supra pinnam* (cf. pinnaculum, circuitus templi, 71, 69), Wrt. Voc. ii. 74, 41. Læssan ymbgang hæfd se mann ꝺe gæꝺ abútan án hús, ꝺonne se ꝺe ealle ꝺa burh begæꝺ, Lchdm. iii. 248, 11. IV. *of position, on ymbgange about, around :*—Ealle ꝺe on ymbegonge hys synd *omnes qui in circuitu ejus sunt*, Ps. Spl. T. 88, 8. On ymbgeonge, Rtl. 178, 31. V. *a winding course, bend :*—Ymbgongum *anfractibus*, Wrt. Voc. ii. 9, 53. V a. figurative :—Ymbgeong *decursum*, Mt. Kmbl. p. 12, 14. VI. *a going about a business.* v. ymb, I. 3 d :—Hiora in spréc ꝺone ymbgeong cýꝺaþ *eorum in foro ambitum notat*, Mk. Skt. p. 5, 5. [In umgang *in circuitu*, Ps. 11, 9 : *circum*, 30, 14. þat was of umgang (abowte, *v. l.*) thre iorne, C. M. 9192. *O. H. Ger.* umbi-gang *circuitus, ambitus, deambulacrum, circulus, conversio : Icel.* um-gangr *circuit ; management.*] v. embe-gang.

ymb-gangan; *p.* -géng. I. *to go round :*—Hí ymbgangaþ ceaster *circuibunt civitatem*, Ps. Spl. T. 58, 16. II. *to go about, in the neighbourhood of :*—Ic ymbgonge weófod *circumdabo altare*, Ps. Spl. T. 25, 6 : Ælfc. Gr. 24; Zup. 139, 13. Hine ymbegangaþ gástas twégen, Salm. Kmbl. 973; Sal. 487. III. *to go about, over, through :*—Gé ymbgangaþ sæ and eorꝺu *circuitis mare et aridam*, Mt. Kmbl. Rush. 23, 14. [Other have hem umbiꝺonge (*circumdederunt*), Pall. 119, 437. *O. H. Ger.* umbi-gangan.] v. ymb-gán.

ymb-gearwian; *p.* ode *To clothe, dress :*—Ymbgearuad *coopertum*, Mk. Skt. Lind. 16, 5.

ymb-gedelf, es; *n. A digging round* or *about :*—Ꝺæs treówes ymbgedelf is seó eádmódnys ꝺæs behreówsiendan mannes, Homl. Th. ii. 408, 31.

ymb-gefrætwude glosses circumornatae, Ps. Spl. C. 143, 15.

ymb-geóting, e; *f. A pouring round* or *about, purification :*—þweál, ymgeóting (þrinted yn-) *lustramentum*, Hpt. Gl. 483, 20.

ymb-gerénode glosses circumornatae, Ps. Spl. 143, 15: Blickl. Gl.

ymb-gesett; *adj.* *(ptcpl.) Placed round about, neighbouring*:—Hē ꝺæt ymbgesette folc (*vulgus circumpositum*) feor and wīde . . . gȳmde tō gehwyrfanne . . . on his fōtum gongende com tō ꝺām ymbgesettum tūnum (*ad circumpositas villas*), Bd. 4, 27; S. 604, 2–13.

ymb-gyrdan; *p.* de.   **I.** *of clothing, to gird about,* (1) *to put a girdle round*:—Ic embgyrde *cingo* and *accingo* and *succingo*, Ælfc. Gr. 28, 5; Zup. 173, 17. Hē ymbgyrde hine *praecinxit se*, Jn. Skt. Lind. Rush. 13, 4. Ymbgyrdaþ eówre lendena, Anglia viii. 322, 19. Ymbgyrde wē ūre lendena, 323, 27. Ymbgyrded *amictus*, Mk. Skt. Lind. Rush. 14, 51. Ymbgyrd *circumamicta*, Ps. Spl. T. 44, 15. His lendena wæron ymbgirde, L. Ælfc. P. 17; Th. ii. 370, 12. 'Beón eówre lendena ymbgyrde.' On ꝺām ymbgyrdum lendenum is se mægꝺhād tō understandenne, Homl. Th. ii. 564, 25. (2) *to serve as a girdle*:—Ymbgyrde hine gyrdilse sōꝺfæstnisęs *circumcinxit eum zona justitiae*, Rtl. 79, 5.   **II.** *to surround, encompass, enclose*:—Ymbgyrdeþ *ambit*, Wrt. Voc. ii. 9, 61. Mid ꝺyssum gemærum hī synd ūtan ymbgyrde, Cod. Dip. Kmbl. iii. 396, 3. v. embe-gyrdan.

ymb-habban; *p.* -hæfde.   **I.** *to surround, encompass*:—Ymbhæfdan *cingebant*, Wrt. Voc. ii. 15, 73. Mid ꝺȳ unmætan weorode ymbhæfd *optimo vallatus exercitu*, Bd. 3, 18; S. 546, 31 : 2, 9; S. 511, 25 note. Emhæfd *circumseptus* (*densis agminibus*, Ald. 3), Anglia xiii. 27, 5. Ispania land is eall mid fleóte ūtan ymbhæfd, ge eác binnan ymbhæfd ofer ꝺa land ægþer ge of ꝺam gārsecge ge of ꝺam Wendelsæ *Hispania circumfusione oceani Tyrrhenique pelagi pene insula efficitur*, Ors. 1, 1; Swt. 24, 1–3.   **II.** *to include, contain*:—Befēhꝺ t emhæfd *circumgirat, circuit, complectitur*, Hpt. Gl. 422, 70. Embhæfþ *continet*, i. *habet, tenet*, Wrt. Voc. ii. 135, 16. Seó sēleste gesælþ ꝺe ꝺa ōþra gesælþa ealle oninnan him gegaderaþ and hī ūtan ymbhæfþ, Bt. 24, 1; Fox 80, 21.

ymb-haga, an; *m. An enclosure where bees are kept*:—Wrīt ꝺysne circul on ānum mealanstāne (mealm- ?), and sleah ænne stacan onmiddan ꝺam ymbhagan, and lege ꝺone stān onuppan ꝺam stacan (the words on the stone are : Contra apes ut salui sint. There are other charms connected with bees on pp. 384, 397), Lchdm. i. 395, 5. v. ymbe *a swarm of bees.*

ymb-hagian. v. ymb-hegian.

ymb-hammen; *adj. Surrounded, covered*:—Ymbhamne (*printed* -humne) *ambitiuntur* (=ambiuntur; the passage is : Manicae sericis clavate calliculae rubricatis pellibus ambiuntur, Ald. 77), Wrt. Voc. ii. 87, 59. Ymbhwyrfte, ymbhammene, 2, 14. Cf. seolfor-hammen.

ymb-hangen. v. ymb-hōn.

ymb-healdan *To encompass*, Cd. Th. 265, 14; Sat. 7.

ymb-heápian; *p.* ode *To crowd about, surround in crowds*:—Ymbheápiendum *glomerantibus*, Wrt. Voc. ii. 40, 52. Ymbheápod *glomeratus* (the passage is:—Lucifer parasitorum sodalibus vallatus et apostatarum satellitibus glomeratus, Ald. 10), 76, 31. v. ymb-hīpan.

ymb-hēdig. v. ymb-hygdig.

ymb-hegian; *pp.* od *To hedge about, surround*:—Ic ymbhegige (embhagige, *v. l.*) *saepio*, Ælfc. Gr. 30, 2; Zup. 190, 15. Emhegod mid weorodum mædena *septus choreis virginum*, Hymn. Surt. 140, 12.

ymb-hīpan; *pp.* ed *To crowd about, surround in crowds, assail*:—Ymbhīpan (*printed* -hiwan) *constipari*, Wrt. Voc. ii. 23, 21. Ꝺā wæs hē sōna æghwanon mid wæpnum ymbhȳped *cum mox ubique gladiis impeteretur*, Bd. 2, 9; S. 511, 25. Mid wæpnum and mid feóndum eall ūtan ymbhēped *cum armis et hostibus circumseptus*, 3, 12; S. 537, 28 note. Ymbhēped *faltum*, Wrt. Voc. ii. 146, 75. v. ymb-heápian.

ymb-hlennan; *pp.* ed *To crowd about, surround*:—Emhlennende *constipantes*, Hpt. Gl. 409, 3. Emhlemmende (-hlennende ?) *circumvallantes, stipantes*, 408, 62. Emhlenned *circumseptus*, 406, 47 : *vallatus, circumseptus, circumdatus*, 422, 41. Emhledned *stipatus, circumdatus, vallatus*, 406, 44. Emhlæned *circumseptus*, Anglia xiii. 27, 5.

ymb-hoga, an; *m. Care, solicitude, anxiety*:—Se ymbhoga (cf. gēmen, Bt. 12; Fox 36, 28) ꝺyssa woruldsælþa, Met. 7, 53. Se rēn ungemetlīces ymbhogan, Bt. 12; Fox 36, 19: Met. 7, 28. For ꝺære ungemetgunge ꝺæs ymbehogan ꝺara ūterra ꝺinga, Past. 18; Swt. 141, 8. On tō monigfaldum ymbehogan ꝺisse worulde *curis hujus mundi*, 43; Swt. 317, 11. Æghwylc dæg hæfꝺ genōh on hys āgenum ymbhogan *sufficit diei malitia sua*, Mt. Kmbl. 6, 34. Ꝺæt hē forlæte ælcne ymbhogan, ꝺe him unnet sié, Met. 22, 10. Hē ꝺone ymbhogan ne forlæt ꝺæs flæsclīcan beddgemānan *nec stratum carnalium sollicitudine deserit*, Past. 16; Swt. 99, 24. Ꝺonne hié āgiémeleásiaþ ꝺone ymbhogan woruldcundra ꝺinga *cum curare corporalia negligunt*, 18; Swt. 137, 2. Gif ꝺū hwilcne cræft cunne, begā ꝺone georne; swā swā sorge and ymbhogan geȳcaþ (-eꝺ, MS.) monnes mōd, swā geȳcꝺ se cræft his āre, Prov. Kmbl. 59. Wind woruldearfoþa, oꝺꝺe ymbhogena ungemet rēn, Met. 7, 36. Ymbhogona, 16, 6. Byꝺ ælc man gedrēsed on īdlum sorgum and on ymbhogum *universa vanitas omnis homo vivens*, Ps. Th. 38, 13. Ælc deáþlīc swencþ hine selfne mid manigfealdum ymbhogum *omnis mortalium cura, quam multiplicium studiorum labor exercet*, Bt. 24, 1; Fox 80, 7 : 24, 4; Fox 84, 32. Āꝺō hē of his mōde ungerisenlīce ymbhogan, 30, 3; Fox 106, 20. Unnytte ymbhogan, 35, 1; Fox 154, 22.

ymb-hogian; *p.* ode *To be solicitous, exercise the mind*:—Ic ymbhogige on wundrum ꝺīnum *exercebor in mirabilibus tuis*, Ps. Lamb. 118, 27. Ymbhochige, 48. Ðeówa ꝺīn ymbhogode on rihtwīsnessum ꝺīnum *seruus tuus exercebatur in justificationibus tuis*, 23.

ymb-hōn; *pp.* -hangen *To hang round with clothing, ornament, etc., to drape, clothe, deck*:—Þeáh wē ūs gescirpen mid ꝺȳ reádestan godwebbe and gefrætewian mid ꝺȳ beorhtestan golde and mid ꝺām deórwyrþestan gimmum ūton ymbehōn, Wulfst. 262, 23. Ymbhangen mid fægernysse *circumamicta varietatibus*, Ps. Spl. 44, 15. Healfnacode on hiora līchaman, būton ꝺæt hig wæron mid riftum ymbhangene, Homl. Ass. 202, 220. Seó fone is mid .xii. godwebbum ūtan ymbhangen, Salm. Kmbl. p. 152, 17.

ymb-hringan; *p.* de.   **I.** *to ring round, surround, encompass*:—Embhrincþ *cingit*, Wrt. Voc. ii. 135, 53. Mē ymbhringde manig yfel *circumdederunt me mala*, Ps. Th. 39, 13. Mē ymbhringdon sār and sorga and grānung, 17, 4, 5. Mē ymbhringdon swīꝺe mænige calfru, 21, 10, 14. Mīne fȳnd mē ymbhringdon ūtan on ælce healfe, 16, 9. Emhrinced *circumseptus*, Hpt. Gl. 406, 47. Embþrungen *vel* (emb)hringed *constipata, circumdata*, Wrt. Voc. ii. 133, 62. Hē wæs ymbhringed mid his feóndum *vallatus exercitu*, Bd. 3, 18; S. 546, 30. Ðonne hē biꝺ ūtane ymbhringed mid ungemetlīcre heringe *dum foris immenso favore circumdatur*, Past. 17; Swt. 111, 8. Ða ꝺe tō Gode hopiaþ beóꝺ ymbhringde mid swȳþe manegre mildheortnesse *sperantes in Domino misericordia circumdabit*, Ps. Th. 31, 12.   **II.** *to turn round in a ring, wind round*:—Ymbhringde *glomeravit* (the passage is : In spira morsum glomeravit inertem, Ald. 202), Wrt. Voc. ii. 96, 15 : 41, 48.

ymb-hringend, es; *m. A surrounder, an attendant, one of a suite*:—Ymbhringendum (ymbdringendum (=þringendum), Erfurt. 61) *stipatoribus*, Txts. 96, 929.

ymb-humne. v. ymb-hammen.

ymb-hūung, e; *f. Circumcision*:—Yymbhūungun *circumcisionem*, Jn. Skt. Lind. 7, 22.

ymb-hweorfan; *p.* -hwearf.   **I.** *to go round, revolve round*:—Se roder ælce dæg ūtou ymbhwyrfꝺ ealne ꝺisne middaneard, Bt. 39, 3; Fox 214, 16. Ymbhwyrfeþ, Met. 20, 137. Ymbhwerfeþ, 28, 4. Hī ꝺære eaxe ūtan ymbhwerfaþ (-eþ, MS.) ꝺone norꝺende, 28, 13.   **II.** *to go about, in the neighbourhood of.* v. ymb, I. 1 b:—Ic ymbehwyrfe weófod ꝺīn *circumdabo altare tuum*, Ps. Lamb. 25, 6.   **III.** *to go about, over, through.* v. ymb, I. 1 e:—Ic ymbhweorfe ꝺīn ꝺæt hālige tempel, Ps. Th. 26, 7. Ymbhwurfaþ woegas *circuite vias*, Rtl. 36, 5. Gē ymbhurfon sæ and drȳgi *circuitis mare et aridam*, Mt. Kmbl. Lind. 23, 15.   **IV.** fig. *to go about a business, be occupied with, attend to, cultivate.* v. ymb, I. 3 d; ymb-hwyrft, VII:—Hē underfēng ꝺa hālgan gesomnunga tō plantianne and tō ymbhweorfanne, suā se ceorl dēꝺ his ortgeard, Past. 40; Swt. 293, 3.   **V.** causative, *to turn round*:—Ðū ꝺe on hrædum færelde ꝺone heofon ymbhweorfest *qui rapido coelum turbine versas*, Bt. 4; Fox 6, 31. Ymbhwearfest, Met. 4, 4. Ic eom ealne ꝺone heofon ymbhweorfende *rotam volubili orbe versamus*, Bt. 7, 3; Fox 20, 25.

ymb-hweorfness, e; *f. Change, alteration*:—Tīdo ymbhuoerfnise *temporum vicissitudine*, Rtl. 37, 35.

ymb-hwirfan; *pp.* -hwirft (?). v. ymb-hammen.

ymb-hwyrft (-hwearft, -hweorft, -hwerft), es; *m.*   **I.** *a ring, circle*:—Lytel ymbhweorft *rotella vel orbiculus*, Wrt. Voc. i. 17, 44. Emhwerfte (-hferte, MS.) *gyro*, Kent. Gl. 271.   **II.** *a circular course, an orbit*:—Se mōna hæfꝺ his ryne hraꝺor āurnen on ꝺam læssan ymbhwyrfte, ꝺonne seó sunne hæbbe on ꝺam māran, Lchdm. iii. 248, 14. Hī (*certain stars*) habbaþ sceortne ymbhwyrft, Bt. 39, 3; Fox 214, 19: Met. 28, 20. Ymbhwearft, 28, 12. Ymbehwearft, 28, 8.   **III.** *circuit, surrounding space, a* (in) ymbhwyrfte *around, round about*:—On ymbhwyrfte *in giro*, Wrt. Voc. ii. 47, 63. Ealle ꝺe on ymbhwyrfte āhwær syndan *omnes qui in circuitu ejus sunt*, Ps. Th. 75, 8 : 88, 6. Fȳr onælꝺ on ymbhwyrfte (*in circuitu*) fȳnd his, Ps. Spl. 96, 3. On ymbhwyrfte stōdan hær, Bd. 5, 2; S. 614, 45. Stefn in ymbhwyrfte (*in gyro*) ymbsealde ꝺæt hūs, 4, 3; S. 567, 44. God him forgeaf sibbe on eallum ymbhwirfte *data est a Deo pax in omnes per circuitum nationes*, Jos. 21, 42. On eallum ꝺam ymbhwyrfte, 10, 21. On his ymbhwyrfte biꝺ swiꝺlīc storm, Homl. Th. i. 618, 11. Ðā eode Israhēla folc on ymbhwyrfte ꝺære byrig, ii. 212, 27. On ymbhwyrfte ondrædendum hine *in circuitu timentium eum*, Ps. Spl. 33, 7. Haldeþ heora ymbhwyrft Drihten *Dominus in circuitu populi sui*, Ps. Th. 124, 2.   **IV.** *surrounded space, extent*:—Eall swā brād seó sunne is, swā eall eorꝺan ymbhwyrft, Lchdm. iii. 236, 7. Gif ꝺu witan wilt ymbe ealre ꝺisse eorꝺan ymbhwyrft from eásteweardan ꝺisses middangeardes ōꝺ westeweardne, and fram sūþeweardum ōꝺ norþeweardne (*omnem terrae ambitum*), Bt. 18, 1; Fox 60, 31. Seó līne ꝺe wile xxxiii sīꝺa ealne eorꝺan ymbehwyrft ūtan ymblicgan, Salm. Kmbl. 152, 6.   **IV a.** *the earth, world, globe*; orbis terrarum:—Ymbhwerft *orbis vel firmamentum*, Wrt. Voc. i. 17, 43. Ꝺæt eall ymbhwyrft (-hyrft, Lind.) wære tōmearcod *ut describeretur universus orbis*, Lk. Skt. 2, 1: Homl. Th. i. 30, 2. Eorꝺe and eall hire gefyllednys, and eal ymbhwyrft and ꝺa ꝺe on ꝺam wuniaþ,

ealle hit syndon Godes æhta *Domini est terra et plenitudo ejus, orbis terrarum et universi qui habitant in eo*, Homl. Th. i. 172,9. Ymbhwyrft eorðena, Ps. Spl. Surt. 23, 1. Ymbhwyrft eorðan *orbis terrae*, Ps. Th. 89, 2. Eorðan ymbhwyrft and uprodor, Cd. Th. 205, 1; Exod. 429. Ic eom micle yldra ðonne ymbhwyrft ðes, oþþe ðes middangeard, Exon. Th. 424, 21; Rä. 41, 42. Ðæt wealdleþer ealles ymbhweorftes heofenes and eorþan, Bt. 36, 2; Fox 174, 19. Yrnð seó sunne bufon ðysum ymbhwyrfte, Lchdm. iii. 250, 14. Eallum ymbehwyrfte (ymbhuirfte, Lind.) *universo orbi*, Lk. Skt. 21, 26. Úre ieldran ealne ðisne ymbhwyrft ðises middangeardes swá swá Oceanus ymbligeþ on þreó tódældon *majores nostri orbem totius terrae, oceani limbo circumseptum, triquadrum statuere*, Ors. 1, 1; Swt. 8, 1. Hé gesette ofer hig ymbhwyrft (*orbem*, 1 Sam. 2, 8), Cant. An. 8: Ps. Lamb. 32, 8. Geond alnæ ymbhwyrft *in universo orbe*, Mt. Kmbl. Rush. 24, 14: Homl. Th. i. 76, 27. Se cásere, se ðe ealne ymbhwyrft on his anwealde hæfde, L. Ælfc. C. 2; Th. ii. 342, 22. Ástag ðæt heofonlíce goldhord on ðysne ymbhwyrft, Blickl. Homl. 11, 29. Hé ymbhwyrft eorðan folca sóðe and rihte dêmeþ *judicabit orbem terrae in aequitate, et populos in veritate sua*, Ps. Th. 95, 13. Ymbhwyrft ealne eorðan and heofenes, Met. 24, 38. Ealne ymbhwyrft and uprador, Elen. Kmbl. 1458; El. 731. Eorðan ymbehwyrft *orbem terrarum*, Ps. Th. 88, 10. **IV b.** *a district, region, world* (= part of the world occupied by a particular people):—Hí fêrdon geond eallum Rômāniscum ymbhwyrfte *they went through all the Roman world*, Homl. Th. ii. 30, 28. Gang óð ðæt ðú ðone ymbhwyrft alne canne, Cd. Th. 308, 33; Sat. 702. ¶ On ymbhwyrfte *among*:—Se ðe is on ealra ymbhwyrfte tô weorþienne *he that is to be honoured among all people*, Blickl. Homl. 197, 5. **V.** *a bend, turn*:—Nim his lifre, tôdæl, and bedealf æt ðam ymbhwyrftum ðinra landgemǽra, Lchdm. i. 328, 22. **VI.** *turn, regular course*:—His suna fêrdon, and ðênode ǽlc óðrum mid his gódum on ymhwyrfte æt his húse, Homl. Th. ii. 446, 17. **VII.** *attention, cultivation.* v. ymb-hweorfan, IV:—Gif se wîngeard næfð ðone ymbhwyrft, and ne bið onriht gescreádod, ne bið hé wæstmbǽre, ac for hraðe áwîldaþ, Homl. Th. ii. 74, 14.

ymb-hygd; *f.n.:* -hygdu; *f.* (v. ofer-hygd) *Care, anxiety*:—Wiste úre se heofonlíca Fæder his ða leófan bearn on myclum ymbhygdum wǽron æfter him; ðá wolde hé se Hǽlend hié áfrêfran, Blickl. Homl. 131, 28.

ymb-hygdig; *adj.* **I.** *feeling anxiety, careful, anxious, solicitous, attentive*:—Ymbhêdig *sollicitus*, Wrt. Voc. i. 51, 24. Emhídig ł carful *zelotypus*, Hpt. Gl. 415, 1. Emhídi, 414, 77. Emhêdig ł hohful, 459, 71. Hé mid ymbhýdie (behygdige, Bd. M. 264, 31) môde smeáde *sollerti animo scrutaretur*, Bd. 4, 3. S. 568, 4. Ne beó gê ymbehýdige eówre sáwle hwæt gê etan *nolite solliciti esse animae vestrae quid manducetis*, Lk. Skt. 12, 22. Be óðrum þingum ymbehýdige *de ceteris solliciti*, 12, 26. Ymbhýdige be reáfe, Mt. Kmbl. 6, 28. Ða sýn emhýdige and cariende embe heora ealdorscipas on eallum þingum *sollicitudinem gerant super decanias suas in omnibus*, R. Ben. 46, 10. Ymbhêdigra *sollicitorum*, Kent. Gl. 352. Hié forgytaþ ðæt hié hwǽne ǽr ymbhygdigum eárum and ingeþancum gehýrdon reccean, Blickl. Homl. 55, 27. **II.** *causing anxiety, anxious:*—Gif him þince ðæt hé geseó man mid wǽpnan gewundodne, ymbhídig sorg ðæt byð, Lchdm. iii. 174, 12.

ymb-hygdiglíc; *adj. Careful, anxious, solicitous, sedulous:*—Mid emhêdilícere geornfulnysse *sollicita (curiosa, sedula) intentione*, Hpt. Gl. 410, 9.

ymb-hygdiglíce; *adv. Carefully, sedulously* [:—Mid ðan ðe hé his salmes and his gebeden and rǽdingan embhýdiglíce smeáde, Shrn. 14, 14.]

ymb-hygdigness, e; *f. Care, anxiety, solicitude:*—Þurhwacol emhídignys *pervigil sollicitudo*, Hpt. Gl. 426, 57. Geornfulnys ł emhêdinys *diligentia, cura*, 437, 58. Se abbod mid ealre emhýdignesse (*sollicitudine*) carige embe ða gyltendan gebróðru, R. Ben. 50, 18: 54, 19: 137, 21. Ǽlc ðæra wæs hám tô his ágenum farende myd mycelre ymbhýdignysse and mid mycelum ege, Nicod. 33; Thw. 19, 26. Wê sceolon ða ymhídignysse fram ús áwurpan, Homl. Th. ii. 462, 12. Twá wiðerrǽde ðing gedeódde Drihten on ðisum cwyde, ymhídignyssa and lustas. Ymhídignyssa ofðriccaþ ðæt môd, and unlustas tólýsaþ, 92, 14. Gehyspendlíce on ymbhigdinyssum sínum (*studiis suis*), Ps. Lamb. 13, 2.

ymb-irnan; *p.* -arn. **I.** *to go round:*—Hí ymbyrnaþ ceaster *circuibunt civitatem*, Ps. Spl. 58, 7, 16. **II.** *to go about:*—Seofona gástas ymbiornas (*discurrentes*), Mt. Kmbl. p. 10, 3. v. ymb-ærnan.

ymb-lédan; *p.* de *To lead about:*—Hé ymblǽdde hine *circumduxit eum* (Deut. 32, 10), Cant. M. ad fil. 10.

ymb-lér(i)gian (?) *to surround, encompass:*—Sýn emblærg[ede] ambiuntur (cf. ymb-hammen, *which is a gloss to the same passage*), Anglia xv. 207, 289. v. lǽrig.

ymb-licgan; *p.* -læg. **I.** *to lie round, surround, encompass:*—Ealne ðisne ymbhwyrft ðises middangeardes, swá swá Oceanus útan ymbligeþ *orbem totius terrae, Oceani limbo circumseptum*, Ors. 1, 1; Swt. 8, 2. Seó líne ðe wile xxxiii síða ealne eorðan ymbhwyrft útan ymblicgan, Salm. Kmbl. 152, 6. **II.** *to lie about, along.* v. ymb, I. 1 d:—Se cyng ðæt land on ða sǽhealfe mid scipum ymbelæg, Chr. 1072; Erl. 211, 2. [To umbelyȝe Lotheȝ hous, Allit. Pms. 63, 836.]

ymb-lîþan *to circumnavigate :*—Ymblîþendre Breotone útan *circumnavigata Brittania*, Bd. 5, 9; S. 622, 17.

ymb-lôcian *to look round:*—Ymblôcade *circumspiciens*, Mk. Skt. Lind. Rush. 3, 34. [Þat leris man him umbiloke, C. M. 8468. Nedefull it es . . . þat he warely umbyluke hym þat he pryde hym noghte þareof, Rol. H. i. 319, 18.]

ymb-loflan *to praise:*—Heriaþ Drihten ealle þeóda, ymblofiaþ (*laudate*) hine ealle folctruman, Ps. Lamb. 116, 1.

ymb-lyt ? :—Hé gesette sunnan and mônan, stánas and eorðan, streám út on sǽ, wæter and wolcen ðurh his wundra miht, deópne ymblyt (ybmlyt, MS.) dene (clene, MS.) ymbhaldeþ Meotod on mihtum, Cd. Th. 265, 13; Sat. 7.

ymbren, es; *pl.* ymbrenu (*the reading* ymbren 7 fæstena, L. Eth. vi. 23; Th. i. 320, 20, *should rather be* ymbrenfæstena, *as in* Wulfst. 272, 16); *n. Ember* (in *Ember*-day), *Embring* (e. g. Keep *embrings* well and fasting days. . . . For Friday, Saturn and Wednesday, Tusser) ; the name of the four periods of fasting and prayer appointed by the Church to be observed in the four seasons of the year respectively. Each was a period of three days, a Wednesday and the following Friday and Saturday (cf. ða twelf ymbrendagas, Wulfst. 244, 20. *For the dates see the passage given under* ymbren-dæg, L. Ecg. P. addit. 21; Th. ii. 234, 33):—Ðis godspel sceal on Wôdnesdæg tô ðam ymbrene ǽr myddawyntran (cf. Ðys gebyraþ on Frigedæg tô ðam ylcan fæstene, v. 39), Lk. Skt. 1, 26 rubc. Ðis sceal on Wôdnesdæg on ðære Pentecostenes wucan tô ðam ymbrene, 9, 12 rubc. On Frigedæg on ðære Pentecostenes wucan tô ðam ymbrene, 8, 40 rubc. On Sæternesdæg on ðære Pentecostenes wucan tô ðam ymbrene, Mt. Kmbl. 20, 29 rubc. Ðis sceal tô ðam ymbrene innan hærefeste on Wôdnesdæg, Mk. Skt. 9, 17 rubc. Tô ðam ymbrene innan hærefaste on Frigedæg, Lk. Skt. 7, 36 rubc. Tô ðam ymbrene innan hærefeste on Sæternesdæg, 13, 6 rubc. Fæstaþ ða feówer ymbrenu on twelf môndum, ðe eów rihtlíce ásette synd, Wulfst. 136, 17. ¶ *the form occurs also with* riht *prefixed:*— Áðas and wîfunga ǽfre sindan tôcwedene heáhfreólsdagum and rihtymbrenum, L. Eth. vi. 25; Th. i. 320, 25: Wulfst. 117, 15 note. [Perhaps both the Latin (*jejunia*) *quatuor temporum* and the English *ymb-ryne* (q. v.) may have a share in the formation of *ymbren;* cf. Germ. *quatember* and Swed. *tamper-dagar*.] v. following words.

ymbren-dæg, es; *m. An Ember-day:*—Wê forbeóðaþ ordál and áðas freólsdagum and ymbrendagum, L. C. E. 17; Th. i. 370, 3: Wulfst. 117, 15. Ða ðe heora lencten wel gefæsten and ða twelf ymbrendagas, 244, 20. ¶ *with* riht *prefixed:*—Ða rihtymbrendagas (*legitimi quatuor temporum dies*), ðe man mid rihte healdan sceal; ðæt is, on kł. Martii, on ðære forman wucan ; and kł. Iunii, on ðære æfteran wucan ; and on kł. Septembr. on ðære þriddan wucan ; and on kł. December, on ða nêhstan wucan ǽr Cristes mæssan, L. Ecg. P. addit. 21; Th. ii. 234, 33. Áðas sindon tôcweden freólsdagum and rihtymbrendagum, L. Eth. v. 18; Th. i. 308, 25. Gyf hwylc wydewe hý forlicge, bête .i. geár, and rihtymbrendagas tô eácan ðæs geáres (*et insuper quattuor temporum legitimis anni diebus*), L. Ecg. C. 39; Th. ii. 164, 30. [Iðe Umbridawes, Wodnesdawes and Fridawes, A. R. 70, 6. Embyrday, embyr *angarium vel quatuor temporum*, Prompt. Parv. 139. *Icel.* Imbru-dagar (*taken from English*).]

ymbren-fæsten, es; *n. The fast of the Ember-days:*—Ðæt man ǽlc beboden fæsten healde, sî hit ymbrenfæsten, sî hit lengctenfæsten, L. C. E. 16; Th. i. 368, 22. Ðæt ymbrenfæsten byð on ðissum mônþe (*December*), Anglia viii. 311, 39. On ðam lenctenfæstene and on ǽlcum ymbrenfæstene, Homl. Th. ii. 608, 17. Feówer ymbrenfæstenu beóð on twelf môndum, eallswá feówer tîman beóð, Anglia viii. 312, 14. Ymbrenfæstena healde man rihte, swá swá Scs. Gregorius Angelcynne sylf hit gedihte, Wulfst. 272, 16: L. Eth. vi. 23; Th. i. 320, 20.

ymbren-wicu, an; *f. A week in which Ember-days fall :*—.iiii. Wôdnesdagas on .iiii. ymbrenwican, L. Alf. pol. 43; Th. i. 92, 9. [Iðe ymbri wikis Wodnesdawes and Fridawes, A. R. 70, 6 note. *Icel.* Imbru-vika.]

ymb-ryne, es; *m.* **I.** *course of a moving body:*—Wǽron sume gedwolmen ðe cwǽðon, ðæt man beó ácenned be steorrena gesetnyssum, and þurh heora ymbryna him wyrd gelimpe, Homl. Th. i. 110, 8. **II.** *course* of time, *revolution, period:*—Ðes geárlíca ymryne ús gebrincþ efue nú ða clǽnan tíd Lenctenlíces fæstennes, Homl. Th. ii. 98, 24. Gyf hê (*the 29th of February*) byð forlǽten unteald, ðærinhte áwent eall ðæs geáres ymbryn[e] (-rene, *v. l.*) þwyres, Lchdm. iii. 264, 12. Emrynes *lustrationis, circuli, curriculo annorum*, Hpt. Gl. 455, 6. Áurnenum (wucan) emrene *emenso hebdomadis curriculo*, 428, 72. Se dæg bið ofer Eástrum on ymbryne ðæs geáres, Homl. Skt. ii. 19, 58. Iond ðære wucan emrene *per septimane circulum*, R. Ben. Interl. 52, 4. Yrnende geond gǽres ymbrene *currens per anni circulum*, Hymn. Surt. 39, 29. Dægena embrynum *dierum circulis*, 27, 1. Embrenum *lustris*, Hpt. Gl. 415, 67. Ymrynum, 493, 62. Ǽfre ymbe geáres ymbrynum, Lchdm. iii. 238, 25. Be ðæs geáres ymbrenum *de temporibus*, 232, 5.

ymb-sǽlan *to bind round, tie round:*—Sié unbunden (ymbunden, Lind.) ł unsǽled (= ymbsǽled) *circumdaretur*, Mk. Skt. Rush. 9, 42.

ymb-sætnung, e; f.   I. *a siege :*—Emsǽtnungum *obsidione,* Hpt. Gl. 525, 40.   II. *a sedition :*—Mið ðȳ gié gehēreþ gefehto and ymbsǽtnungo (-e, Rush.) ymb burgum (v. ymb, **I.** 1 e, *and* sǽtnung: *or under* I P) *cum audieritis proelia et seditiones,* Lk. Skt. Lind. 21, 9.

ymb-sceáwian *to look round, to behold :*—Ymbsceáwade (-sceówade, Rush.) tô geseánne hiá *circumspiciebat uidere eam,* Mk. Skt. Lind. 5, 32. Ymbsceáwde (-sceówadun, Rush.) hiá *circumspiciens eos,* 3, 5. Ymbsceáude (-sceówade, Rush.) hine *intuitus eum,* Jn. Skt. Lind. 1, 42.

ymb-sceáwiendlíce; *adv. Circumspectly :*—Mid ðȳ hé swā gemetfæstlíce and swā ymbsceáwiendlíce hine sylfne on eallum ðingum beheóld *cum ita se modeste et circumspecte in omnibus gereret,* Bd. 5, 19; S. 937, 5.

ymb-sceáwung, e; f. *Beholding :*—Embeþonc *vel* (embe)sceáwung *circumspectio,* Wrt. Voc. ii. 131, 27. Wer se ðe giðenceþ ymbsceáwung (*circumspectionem*) Godes, Rtl. 46, 5 : 84, 27.

ymb-scínan; p. scán *To shine round, surround with brightness :*—Ðæs Hēhstan mægen ðe ymbscíneþ, Blickl. Homl. 7, 36. Seó sunne ymbscínð ðone blindan, and se blinda ne gesihð ðære sunnan leóman, Homl. Th. ii. 446, 32. Berhtnise Godes ymbsceán hiá (him ymbesceán, W. S.) *claritas Dei circumfulsit eos,* Lk. Skt. Lind. Rush. 2, 9. Hié leóht ymbscán, Andr. Kmbl. 2034; An. 1019. [þe schyre sunne hit umbeschon, Allit. Pms. 105, 455.]

ymb-scríþan; p. -scráþ *To go round, revolve round :*—Rodor ymbscríþeþ dôgora gehwilce ðisne middangeard, Met. 20, 208.

ymb-scrýdan; p. de *To clothe :*—Ymbscrýdaþ eów mid Godes wæpnunge *induite vos armaturam Dei* (Eph. 6, 11), Homl. Th. ii. 218, 2. Mid hwam gē sȳn ymbscrýdde *quid induamini,* Mt. Kmbl. 6, 25.

ymb-sellan; p. -sealde *To surround;* circumdare:—Ic ymbsylle *circumdabo,* Ps. Spl. 25, 6. Ðū ymbseles *circumdas,* Rtl. 76, 1. Hé mid eallum ðyssum ða burh ymbsealde (*circumdedit*), Bd. 3, 16; S. 542, 24 : Ps. Th. 114, 3. Fȳren wolcen ymbsealde ealle ða ceastre, Blickl. Homl. 245, 31. Se sang in ymbhwyrfte ymbsealde ðæt hūs, Bd. 4, 3; S. 567, 45. Mē ymbsealdon (*circumdederunt*) þeóde, Ps. Th. 117, 10, 12. Ymbsaldun (ym-, Lind.), Mt. Kmbl. Rush. 27, 28. Hī mē ūtan ymbsealdan, Ps. Th. 87, 17. Ūton ymbsele *circumda,* Kent. Gl. 157. Ymbselle *circumdet,* Rtl. 34, 7. Ymbsyllendum mē *circumdantibus me,* Ps. Spl. 31, 9. Seó fǽmne wæs ymbseald mid ðon campweorode, Blickl. Homl. 11, 24. Sondbeorgum ymbseald, Exon. Th. 360, 23; Wal. 10. Ða ymbsealde sint mid sixum eác fiðrum gefrætwad, Elen. Kmbl. 1480; El. 742.

ymb-seón *to behold, look :*—Ic hine wolde biddan, ðæt hē sweotole ymbsáwe sūð, eást and west (cf. behealde hē on feówer healfe his, Bt. 19; Fox 68, 21), hū wīdgil sint heofones hwealfe, Met. 10, 5.   [For þi oure soile or þou seke umse þe betyme, Alex. (Skt.) 3728.]

ymb-seón *beholding.* v. ymb-sín.

ymb-set, es; n. *Siege, blockade :*—Ðæt gēr ymbsetes ðære Beadonescan dūne *annum obsessionis Badonici montis,* Bd. 1, 16; S. 484, 22. Hé ne mihte nā mid gefeohte ne mid ymbsete (*obsidione*) ða burh ābrecan ne gegān, 3, 16; S. 542, 19. [O. H. Ger. umbi-sez *obsidio.*]

ymb-seten[n], e; f. *A row of vines :*—Oemsetinne wiingeardes *amtes* (=*antes* ?), Wrt. Voc. ii. 100, 17. v. ymb-settan, **II** ; seten, **II.**

ymb-setenness, e; f. *Besieging, siege :*—Ðæt hý sceoldon ðam Gode þancian ðe hý gefriðode fram ðære ymbsetennesse, and fram ðære hergunge ðara twēga kyningca, Ps. Th. 45, arg.

ymb-sétnung. v. ymb-sǽtnung.

ymb-settan; p. te.   I. *to set round, put round, surround :*—Hē ymbseteþ ūtan líc and feþre on healfe gehware hālgum stencum, Exon. Th. 212, 3; Ph. 204. Beád hē ūt scypfyrde and landfyrde, and ðæt land eall ūtan embsette, Chr. 1072; Erl. 210, 31. Giarn ān and gifylde copp mid æcede ymbsette and tô rôde ða drinca salde him *currens unus et implens spongiam aceto circumponensque calamo potum dabat ei,* Mk. Skt. Rush. Lind. 15, 36. Ymbsetton (ymsettun, Rush.) ł ymbuundun *circumponentes,* Jn. Skt. Lind. 19, 29. Salomones reste wæs mid weardum ymbseted, Blickl. Homl. 11, 16. Ymbseted mid ðǽm wāgum his misdǽda, L. E. I. 32; Th. ii. 430, 14. Ymbsett mid fǽgnesse *circumdata varietate,* Ps. Lamb. 44, 10. Mid hwilcum feóndum heó ymbset bið, Homl. Th. i. 410, 9. Emset *glomeratus, circumseptus,* Hpt. Gl. 422, 47. Ða heargas ðara deófolgylda mid heora hegum ðe hí ymbsette wǽron *fana idolorum cum septis quibus erant circumdata,* Bd. 2, 13; S. 516, 39.   II. *to plant* with something. v. ymb-seten:—Ic embsette *consero,* Wrt. Voc. ii. 133, 56. Eá mid treówum ymbset *amnis,* i. 54, 16. [How Iuus Iesu oft umsette (bisette, *v.l.*), C. M. 195. Alle umset with enmys, Pr. C. 1250. *O. H. Ger.* umbi-sezzen.]

ymb-sín (-seón), e; f. *Beholding, regard :*—Clǽnum gisceáwiga wē ymbseáne *puro cernamus intuitu,* Rtl. 35, 37.

ymb-sirwan; p. -sirwde, -sirede.   I. *to deliberate about* an evil deed :*—Swā micel tôsceád is betwuh ðære beðōhtan synne, ðe mon longe ymbsireþ, and ðære ðe mon fǽrlíce ðurhtíehð, swā ðætte se se ðe ða synne gesireþ, ǽgðer ge gesyngaþ ge eác hwīlum on ormódnesse gewīt . . . For ðæm sint tô manianne ða ðe lange ymbsieriaþ ðæt hí ongieten hū micel

---

wīte hí sculun habban beforan ðǽm ōðrum *hoc ergo praecipitatione lapsi per consilium pereuntes differunt, quod, cum hi a statu justitiae peccando concidunt, plerumque simul et in laqueum desperationis cadunt . . . Admonendi ergo sunt, ut hinc colligant, qui in culpa etiam se per consilium ligant,* Past. 56; Swt. 435, 4–31.   II. *to lie in wait for :*—Se ðe hine ne ymbsyrede (-syrwde, ymbesierede, *v.ll.*) *qui non est insidiatus* (Ex. 21, 13), L. Alf. 13; Th. i. 46, 24.

ymb-sittan; p. -sæt, pl. -sǽton; pp. -seten.   I. *to sit* or *be round,* (1) *to sit at* table, meat, etc. :—Ðæt hié mē þēgon, symbel ymbsǽton, Beo. Th. 1132; B. 564. Hȳ twēgen sceolon tæfle ymbsittan, Exon. Th. 345, 2; Gn. Ex. 182. Ða ymbsittendan *circumsedentes,* Bd. 4, 9; S. 577, 31 : *convivae,* 5, 5; S. 618, 16 : Ap. Th. 15, 6 : 17, 4.   (1 a) *to sit at* council, *be engaged about :*—Hī ān gepeaht ealle ymbsǽtan *cogitaverunt consensum in unum,* Ps. Th. 82, 5.   (2) *to be around, be neighbouring.* v. ymb-sittend:—Ðām ðe ūs ymbsittaþ *his qui in circuitu nostro sunt,* Ps. Th. 43, 15.   Hī þreátiaþ gehwider ymbsittenda ōþra þeóda, Met. 25, 14.   II. *to beset :*—Ic ymbsitte *obsideo,* Ælfc. Gr. 26, 5; Zup. 157, 3.   Fearras fǽtte ymbsǽton mē *tauri pingues obsederunt me,* Ps. Lamb. 21, 13.   Ða com micel werod werigra gāsta and ðis hūs ūtan ymbsǽtan (*domum hanc et exterius obsedit*), Bd. 5, 13; S. 633, 2.   II a. as a term of war, *to besiege, invest :*—Ðíne fȳnd ðē ymbsittaþ mid ymbtrymminge *circumdabunt te inimici tui uallo* (Lk. 19, 43), Homl. Th. i. 408, 35. Hé ymbsǽt ða burh (*circumdedit Eglon*), Jos. 10, 34. Eádmund ymbsǽt Anlāf cyning and Wulfstán arcebiscop on Legraceastre, Chr. 943; Erl. 117, 16. Ælle and Cissa ymbsǽton Andredescester, 491; Erl. 14, 5 : 885; Erl. 82, 20. Hié ymbsǽton ān geweorc, 894; Erl. 91, 7. Ymbesǽtan, 1011; Erl. 145, 8. Ymbsittaþ ða burg suīðe gebyrdelíce *ordinabis adversus eam obsidionem,* Past. 21; Swt. 161, 19. Ða hié hæfdon Cirinen ða burh ymbseten, Ors. 2, 2; Swt. 66, 18. Hē besierede ðæt folc ðe hié ymbseten hæfden, 4, 5; Swt. 170, 2 : Ps. Th. 12 arg.   [O. H. Ger. ymbi-sizzan *obsidere.*]   v. emb-sittan.

ymb-sittend, es; m. *One living on the borders of another's country, a neighbour :*—Gif ic ðæt gefricge, ðæt ðec ymbsittend (*those that sit on thy borders*) egesan þȳwaþ, Beo. Th. 3658; B. 1827. Him ǽghwylc ðara ymbsittendra hȳran sceolde, 18; B. 9: Elen. Kmbl. 65; El. 33. Ymbesittendra, Beo. Th. 5461; B. 2734. Wē synd gewordene eallum edwītstæf ymbsittendum *facti sumus in opprobrium vicinis nostris,* Ps. Th. 78, 4: 88, 34.

ymb-smeá(g)ung. v. embe-smeágung.

ymb-snidenness, e; f. *Circumcision :*—Wēn is ðæt eówer sum nyte hwæt sȳ ymbsnidennys, Homl. Th. i. 92, 30. Se intinga ðære æftran ymbsnidennysse, Jos. 5, 6. Beóð ēstfulle heortan mid dæghwonlícere ymbsnidenysse áfeormode, Homl. Th. i. 98, 14. Mōyses eów sealde ymbsnydenisse, Jn. Skt. 7, 22. Ða ealdan ymbsnidenysse, Shrn. 47, 18.

ymb-snídan; p. -snāþ, pl. -snidon *To circumcise :*—On restedæge gē ymbsnídaþ mann, Jn. Skt. 7, 22. Abraham ymbsnād his sunu, Gen. 17, 23. Ðæt stǽnene sex, ðe ðæt cild ymbsnāð, Homl. Th. i. 98, 10. Hié ǽghwelcum cnihtcilde ymbsnidon ðæt werlíce lim, Shrn. 47, 20. Hé hine lǽt ymbsnídan mid scearpum flinte, Wulfst. 195, 9. Ymsníþan (ymbsnýðan, *v.l.*), Lk. Skt. 1, 59. Ðæt ðæt cild emsnyden (ymb-, *v.l.*) wǽre, 2, 21. Ymbsniden, Homl. Th. i. 90, 14, 18, 30. Heora fæderas wǽron ymbsnidene, Jos. 5, 4. [Embsníþeu mid ane ulintsexe, O. E. Homl. i. 81, 27.]

ymb-spannan *to span round :*—Swyle tô ðon swíþe āswollen ðæt hine mon nā mid twām handum ymbspannan (*circumplecti*) mihte, Bd. 5, 3; S. 616, 7.

ymb-sprǽc, e; f. *Speech about a subject, talk :*—Be ðysum is oft mycel ymbsprǽc (ymbe-, emb-, *v.ll.*) *there is often much discussion about this,* Lchdm. iii. 266, 9. Ne beó gē áfyrhte ðurh geswince ðæs langsuman fǽreldes, oððe þurh yfelra manna ymbesprǽce *nec labor vos itineris, nec maledicorum hominum linguae deterreant* (Bd. 1, 23), Homl. Th. ii. 128, 2.

ymb-sprǽce; adj. *Talked about :*—Geond ðās eorþan ǽghwǽr sindon hiora gelícan hwōn ymbsprǽce, Met. 10, 59.

ymb-sprecan *to speak about something :*—Alle yfle ymbsprēcon *omnes murmurabant,* Lk. Skt. 19, 7.

ymb-standan; p. -stód, pp. -standen.   I. *to stand about* or *around :*—Ðis folc ðæt hēr ymbstandeþ, Blickl. Homl. 143, 7 : Jn. Skt. Rush. 11, 42. Eall seó gesomnung brōþra and sweostra on twā healfe singende ymbstōdon (*circumstaret*), Bd. 4, 19; S. 589, 30.   ¶ pres. part. used substantively :—Hē sceal grētan his ymbstandendan, and hig sceolon andswarian, L. E. I. 7; Th. ii. 406, 23.   II. *to surround :*—Mē ymbstōdan strange manige *circumdantes circumdederunt me,* Ps. Th. 117, 11. Hý habbaþ mē ūtan ymbstanden *circumdederunt me,* 16, 10. Hī bióþ ūton ymbstandene (*printed* -standende, *but see* ūtan ymbestandne, Met. 25, 7) mid miclon gefērscipe hiora þegna *septos tristibus armis,* Bt. 37. 1; Fox 186, 4. [He saw how þe laddes wode Hauelok his louerd umbistode, Havel. 1875. *O. H. Ger.* umbi-standan *circumstare, circumdare.*] v. next word.

ymb-standend, es; m. *A by-stander :*—Hí ānra gehwilcum ymbstandendra forsǽton heáfodsiéna, Cd. Th. 150, 8; Gen. 2488. Ðá cwæð

se cásere tó đám embstandendum, Homl. Skt. i. 23, 275. Sum mon of đæm ymbstondendum *quidam de circumstantibus*, Mk. Skt. Rush. Lind. 14, 47.

ymb-standenness glosses *circumstantia*, Ps. Lamb. Surt. 140, 3: Rtl. 179, 9: 182, 16: Ps. Surt. 30, 22.

ymb-strícan; *p.* -strác *To rub round* so as to smooth :—Gif đæs dolges ófras synd tó heá ymbstríc mid háte ísene *if the edges of the wound are too high, pass a hot iron round*, Lchdm. ii. 96, 5.

ymb-styrian *to stir about, upset* :—Ymbstyreþ đæt hús *evertit domum*, Lk. Skt. Lind. 15, 8.

ymb-swǽpe, an; *f. A roundabout way, digression.* Cf. ymb-swápan, I :—Ymbsuaepe *ambages*, Wrt. Voc. ii. 100, 13. [Cf. O. H. Ger. umbi-suaifan: Ger. um-schweif.]

ymb-swápan; *p.* -sweóp; *pp.* -swápen. I. *to sweep about* (of the motion of waves) :—Đa ýþa weóllan and ymbsweópan and ǽghwonene đæt scyp fyldon *verrentibus undique et implere incipientibus navem fluctibus*, Bd. 3, 15; S. 541, 42. II. *to wrap round* :—Ymbswápen *circumamicta*, Ps. Surt. Spl. C. 44, 10, 15. Emswápen *circumamicta, circumdata*, Hpt. Gl. 430, 45. [*M. H. Ger.* umbe-swief; *p.*]

ymb-þanc, es; *m. n.* : -þanca, an; *m. Thought about a matter, consideration, attention* :—Embeþonc *circumspectio*, Wrt. Voc. ii. 131, 26. Mid micelum embeþance *magna animaduersione*, Anglia xiii. 373, 106. Hié eallneg rǽswaþ and ondrǽdaþ đæt hí mon tǽlan wille and beóđ eallneg mid đæm ymbeđoncan (-đonce, Cott. MSS.) ábisgode and ofdrǽdde *dum deprehendi metuunt, semper pavidis suspicionibus agitantur*, Past. 35; Swt. 239, 7. Đæt hí ongieten mid wærlíce ymbeþonce *ut cauta circumspectione considerent*, 58; Swt. 44, 5. Đætte hé đone ymbeþonc đæs wærscipes ne forlǽte *ut circumspectionem prudentiae non amittant*, 35; Swt. 237, 17. Hwæt sceolan ús, oþþe hwæt dóþ ús đara worda ymbþonc? Tó morgenne wé beóþ gesémde *of what use are considerations of the words, or what will they do for us? To-morrow we shall be at one on the matter*, Blickl. Homl. 183, 12. [Clene wasshen of þe embeþonke of fleshliche lustes *a mollitie fluxae cogitationis purgata*, O. E. Homl. ii. 87, 2.]

ymb-þeahtian; *p.* ode *To deliberate, consider* :—Đa đe longe ǽr ymbđeahtigeaþ, and hit đonne on lásđ đurhtióđ *qui consulto peccant*, Past. 56; Swt. 429, 31. Đa đe ǽr đenceaþ tó syngianne and ymbđeahtiaþ ǽr hí hit đurhtión *qui in culpa ex consilio ligantur*, 433, 32. Hí beóđ đæs đe lator đe hí oftor ymbđeahtiaþ *tardius peccatum solvitur, quod et per consilium solidatur*, 435, 2. Đæt leóht him đa stówe wæs ontýnende, đe (đǽr, Bd. S. 575, 12) heó ǽr ymbþeahtedon, Bd. 4, 7; M. 284, 20.

ymb-þencan; *p.* -þóhte *To consider* :—On óđre wísan sint tó manienne đa đe mid fǽrlíce luste bióđ oferswíđde, on óđre đa đe lange ymbþenceaþ and đeahtiaþ and swá weorđaþ beswicene *aliter admonendi sunt, qui repentina concupiscentia superantur, atque aliter qui in culpa ex consilio ligantur*, Past. 56; Swt. 429, 34. Ic ymbđóhte *decernam*, Mt. Kmbl. p. 1, 3. Ne beó gé ymbeþencende hú óđđe hwæt gé specon *nolite solliciti esse qualiter aut quid respondeatis*, Lk. Skt. 12, 11. [Patt te birrþ ummbeþennkenn hu þu mihht cwemenn þín Drihhtin, Orm. 1240. He umthoght him what was best, Met. Homl. 79, 26.]

ymb-þreodian *to deliberate* :—Embđrydiendra *circumvenientium*, Wrt. Voc. ii. 131, 30. v. þreodian *and next word, and cf.* ymb-cyme.

ymb-þreodung, e; *f. Deliberation* :—Ymbđriodung (-dritung, Erfurt Gl.) *deliberatio*, Txts. 55, 644. Ymbþriodung, Wrt. Voc. ii. 25, 20. Ymbþrydung, 138, 45.

ymb-þringan; *p.* -þrang; *pp.* -þrungen *To throng round, crowd round, surround* :—Hine F and M útan ymbđringaþ, Salm. Kmbl. 256; Sal. 127. Hí ymbđrungon mé *circumdederunt me*, Ps. Lamb. 16, 11. Ymbeþrungon, 21, 17. Ymþrungon, 16, 9. Ic mé ná ondrǽde þúsendu folces, þeáh hí mé útan ymbþringen *non timebo millia populi circumdantis me*, Ps. Th. 3, 5. Embþrungen *constipata, circumdata*, Wrt. Voc. ii. 133, 62. [O. H. Ger. umbi-dringan *stipasse*.]

ymb-þringend. v. ymb-hringend.

ymb-trymian, -trymman; *p.* -trymede, -trymde. I. *to surround* :—Engla werod embtrymmaþ đone mǽran kyning mihte and đrymme, Wulfst. 137, 15. Ymbsyllende ymbtrymedon mé *circumdantes circumdederunt me*, Ps. Spl. 117, 11. Ymbtrymdon, 17, 5, 6: Ps. Lamb. 16, 9: 21, 13. Mid micelum fǽmnena heápe ymbtrimed, Ap. Th. 23, 16. II. *to fortify, protect, support* :—Ic ymbtrymme *munio*, Ælfc. Gr. 30; Zup. 192, 1. Hig ymbetrymedon đa byrgene *munierunt sepulchrum*, Mt. Kmbl. 27, 66. Hwæt getácniaþ đa truman ceastra bútan hwurfulu mód, getrymedu and ymbtrymedu mid lytelícre lǽdunge? *quid per civitates munitas exprimitur, nisi suspectae mentes et fallaci semper defensione circumdatae?* Past. 35; Swt. 245, 8. Hiericho seó buruh wæs mid weallum ymbtrymmed and fæste belocen *Jericho clausa erat atque munita*, Jos. 6, 1. Ic eom embtrymed *fulcior, sustentor*, Wülck. Gl. 245, 26.

ymb-trymming, e; *f. A fortification* :—Ymbtrymming ođđe fæstnys *munimen*, Ælfc. Gr. 9, 12; Zup. 41, 3. Đíne fýnd đé ymbsittaþ mid ymbtrymminge *circumdabunt te inimici tui uallo* (Lk. 19, 43), Homl. Th. i. 408, 35.

---

ymb-týnan; *p.* de *To enclose, surround* :—Đeáh man đone gársecg mid ísene útan ymbtýnde, Wulfst. 146, 27.

ymb-tyrnan (1) *to turn round* :—Feówer and twéntig tída beóđ ágáne, ǽr đan đe heó beó ǽne ymbtyrnd, Lchdm. iii. 254, 15. (2) *to surround* :—Mid wæter ymbtyrnd stede *circumlutus locus*, Wrt. Voc. i. 59, 15.

ymb-útan *about, around, without.* I. *prep.* (1) local, (a) with dat. :—Đam nis nán wuht bufan, ne nán wuht benyþan, ne ymbútan, Bt. 36, 5; Fox 180, 19. Hú wídgil sint wolcnum ymbútan heofones hwealfe, Met. 10, 6. (b) with accus. :—Geseah se Hǽlend mycle menigeo ymbútan hyne (*circum se*), Mt. Kmbl. 8, 18. Hé wand him ymbútan đone deáđes beám, Cd. Th. 31, 27; Gen. 492. Ymbútan đone weall is se mǽsta díc, Ors. 2, 4; Swt. 74, 17. Suǽ suǽ se here sceolde bión getrymed onbútan Hierusalem, suǽ sculon beón getrymed đa word đæs sacerdes ymbútan đæt mód his hiéremonna *quasi obsidio circa civitatem Jerusalem voce praedicatoris ordinatur*, Past. 21; Swt. 163, 1. Hwí séce gé ymbútan eów đa geselþa đe gé oninnan eów habbaþ geset? *quid extra petitis intra vos positam felicitatem?* Bt. 11, 2; Fox 34, 4. Onginne hé sécan oninnan him selfum đæt hé ǽr ymbúton hine sóhte, 35, 1; Fox 154, 22: Met. 22, 7. Munt is hine ymbútan, gylden weal, Salm. Kmbl. 510; Sal. 255: Ps. Th. 124, 2. Licgaþ mé ymbútan grindlas, Cd. Th. 24, 24; Gen. 382. Hine ymbútan hálge herefeđan blícaþ, Exon. Th. 62, 35; Cri. 1012. Standan ymbeútan đa eardungstówe *stare circa tabernaculum*, Num. 11, 24: Ex. 29, 20: Lev. 3, 2. (2) *about, concerning* :—Hí ne gesáwon sundbúende, ne ymbútan hí ne herdon, Met. 8, 14. II. *adv.* (1) *alone* :—Fýr bið ymbútan on ǽghwylcum, Cd. Th. 280, 34; Sat. 264. Him on healfa gehwone heofonengla þreát ymbútan faraþ, Exon. Th. 58, 1; Cri. 929. Hit bið sinbyrnende and ymbútan hit óđra stówa forbærnd, Met. 8, 53. Glídeþ ǽg ymbútan, 20, 171. (2) *with other adverbs* :—Se sciphere sigelede west ymbútan, Chr. 877; Erl. 78, 17. Sum hund scipa fóron súđ ymbútan, and sum feówertig scipa norþ ymbútan, 894; Erl. 91, 6. Tó farenne útan ymbútan, Ors. 6, 36; Swt. 292, 29. Đá ongon hé sprecan swíþe feorran ymbúton *velut ab alio orsa principio disseruit*, Bt. 39, 5; Fox 218, 11. For đam folce đe hér ymbútan stent, Jn. Skt. 11, 42. Ealla đa neáhstówa đǽr ymbútan, Bt. 15; Fox 48, 22: Cd. Th. 154, 3; Gen. 2550. Hú sunnu đǽr scíneþ ymbútan, 286, 15; Sat. 352. Ymbeútan, Mk. Skt. 14, 47. v. ymb.

ymb-wǽfan; *p.* de *To wrap round, to clothe* :—Ymbwǽfd mid fǽgnyssum *circumamicta uarietatibus*, Ps. Lamb. 44, 15. [Þe brawden bryne umbeweuid þat wy3, Gaw. 581.]

ymb-wǽrlan; *p.* de *To turn round* :—Ymbwærlde tó đæm wífe *conversus ad mulierem*, Lk. Skt. Lind. 7, 44, 9. Ymbwærlde (-wælde, Lind.), Rush. 9, 55.

ymb-weaxan *to surround* :—Seó burh wæs ungemettan fæste mid cludum ymbweaxen *saxum mirae asperitatis et altitudinis*, Ors. 3, 9; Swt. 132, 10. [Ger. um-wachsen.]

ymb-wendan *to turn round, convert, avert, move, change* :—Ymbuoendest *conversas* (but the Latin is p. part. ac. pl. f.), Rtl. 114, 34: Ymbuoende on bæggcgling *conuersa est retrorsum*, Jn. Skt. Lind. 20, 14. Ymbwoend *averte*, Rtl. 8, 37: 15, 25. Ymbwoendendum *vellentibus*, 19, 15. Sié ymbuoended *inmutatur*, 96, 13. Sié umbuoendedo *moveantur*, 167, 1.

ymb-wendedlíc. v. un-ymbwendedlíc.

ymb-wending glosses *vegetatio*, Rtl. 17, 1: *conversatio*, 63, 8.

ymb-weorpan; *p.* -wearp *To throw round, surround* :—Þurh lyftgelác léges blǽstas weallas ymbwurpon, Andr. Kmbl. 3104; An. 1555. [Cf. Ger. um-werfen.]

ymb-wícian *to encamp about* a place :—Hêht ymbwícigean Æthanes byrig mearclandum on (*castrametati sunt in Etham in extremis finibus solitudinis*, Ex. 13, 20), Cd. Th. 181, 22; Exod. 65.

ymb-windan; *p.* -wand. I. *to wind* (intrans.) *round, encompass* :—Rápas synfulra ymbwundon mé *funes peccatorum complexi sunt me*, Ps. Spl. T. 118, 61. II. *to wind* (trans.) *about, wind round* :—Ymbuundun *circumponentes*, Jn. Skt. Lind. 19, 29. [O. H. Ger. umbi-wintan *amicire*.]

ymb-wlátend, es; *m. A spectator, observer* :—Beó đú emwlátent đín *esto catascopus tui*, Lchdm. i. lx, 11. Emwlátend(o)um *spectatoribus*, Hpt. Gl. 488, 64. v. tíd-ymbwlátend.

ymb-wlátian *to contemplate, observe* :—Ic ymbwlátige *contemplor*, Ælfc. Gr. 25; Zup. 145, 12.

ymb-wlátung, e; *f. Contemplation, look, regard* :—Ymbwlátung *aspectus*, Ælfc. Gr. 28, 5; Zup. 175, 5. Emwlátunge *contemplationis*, Hpt. Gl. 412, 20. Spiritus gást belimpđ tó đære sáwle ymbwlátunge, Homl. Skt. i. 1, 182. Emwlátunge *spectaculum*, Hpt. Gl. 435, 49. Widgille emwlátunge *passivos oculorum obtutus*, 405, 64. v. embwlátung.

ymb-wrítan *to cut round, circumscribe* :—Hine eác ymbwrít mid sweorde on .iiii. healfa, Lchdm. ii. 346, 26.

ymb-wyrcan; *p.* -worhte. I. *to surround with works* :—Hé mid eallum đyssum đa burh on mycelre heánnesse ymbworhte (*v. l.*

ymbsealde. v. ymb-sellan), Bd. 3, 16; S. 542, 24 note. Byrig ðære ðe mid náne wealle ne bið ymbworht *urbs absque murorum ambitu*, Past. 38; Swt. 277, 21. II. *to weave :*—Ymbworhton bége *plectentes coronam*, Mt. Kmbl. Lind. 27, 29.

ymel (emel, *q. v.*), e; *f. A canker-worm :*—Ymel *gurgulio* (= curculio), Ælfc. Gr. 9, 3; Zup. 35, 7.

ymele, an; *f. A scroll, leaf* of paper :—Ymele *sceda vel scedula*, Wrt. Voc. i. 75, 15. Ymle *scedula*, 46, 69. Ðæt ðú ðás áne synne, ðe on ðyssere ymlan stent, þurh ðíne gebedu ádilige, Homl. Skt. i. 3, 642.

ymen, hymen, es; *m. A hymn* (*hymnus*) eallum háligum his, Ps. Spl. 148, 14; Ps. Surt. 64, 2. Æfter ðysum is ymen tó singenne, R. Ben. 33, 12. Mid ferse and mid ymene (imene, *v. l.*), 41, 5. Ymen *hymnum*, Ps. Surt. 39, 4: 64, 14: 118, 171. Hé wæs ymen singende, Blickl. Homl. 147, 3: 151, 9. Ymmon, Rtl. 184, 25. Hymen, Ps. Surt. 136, 3: ii. p. 203, 35. Ymenas and capitula rǽdinga sýn ánum gemete gehealdene, R. Ben. 43, 2. On ymnum *in hymnis*, Ps. Spl. 99, 4. Ymenum, Ps. Surt. 99, 4. Míne weleras ðé wordum belcettaþ ymnas *eructabunt labia mea hymnum*, Ps. Th. 118, 171. [From Latin.] v. hymen.

ymen-bóc; *f. A book of hymns :*—Ymenbéc *missenlíce metre librum hymnorum diverso metro*, Bd. 5, 24; S. 648, 36.

ymener (ymnere?), es; *m. A book of hymns* ; hymnare, hymnarium :—Thǽr synd twá Cristes béc, and i mæssebóc, and i. ymener, and i. salter, Cod. Dip. B. iii. 660, 32. Hymneres tácen is ðæt mon wæcge brádlinga his hand and rǽre up his litlan finger, Techm. ii. 121, 9. ii. salteras, and se saltere swá man singð on Róme, and .ii. ymneras, Chart. Th. 430, 13.

ymen-sang, es; *m. A hymn*, Greg. Dial. 2, 3, 4.

ymesene (-séne?) ; *adj. Sightless, blind :*—Sum ymesene man mid wópe his fét gesóhte, biddende his hǽle. Laurentius mearcode ródetácen on ðæs blindan eágan, and hé ðǽrrihte beorhtlíce geseah, Homl. Th. i. 418, 22.

ymest. v. yfera, ysemest ; *adv.*

yna, Techm. ii. 126, 14 (see under tún, I), where it is printed with a space before *y*, as if a letter were wanting in the MS. Cockayne, Lchdm. iii. 334, col. 2, takes the word as the gen. pl. of *yne* = onion.

ynce, es; *m. An inch :*—Wund ynces (inces, *v. l.*) lang, L. Alf. pol. 45; Th. i. 92, 18, 19. Gif ofer ynce scilling; æt twám yncum, twégen; ofer þrý, .iii. scill., L. Ethb. 67; Th. i. 18, 17. Hé (Adam) wæs vi and cx ynca lang, Salm. Kmbl. p. 180, 20. [Wunde feouwer unchene long, Laym. 23970. *From Latin* uncia.]

yndan *in* ða belocenan yndan wega *conpeta clausa*, Wrt. Voc. ii. 94, 11. *For* betýndan? cf. betýndan wega gelǽtan *competa clausa*, 132, 52.

yndse. v. yntse.

ynne-leác (yne-), es; *n. Onion :*—Ynnelaec, hynnilaec, ynnilěc *ascalonium*, Txts. 43, 229. Ynnilaec *cepa*, 49, 448. Ynneleác *scalonia*, Wrt. Voc. i. 66, 57: *unio*, 68, 62. Yneleác *ungio*, 286, 9. [*Latin* unio.] v. enne-leác.

yntse, yndse, an; *and* ynts (?), e; *f. An ounce :*—Genim huniges ánre yndsan gewǽge, Lchdm. i. 76, 11. Ánre yndsan (ynsan, *v. l.*) gewihte, 248, 8. Dó alwan áne yntsan tó, ii. 60, 5 : 190, 9. Áne ynsan, iii. 74, 19. Ælc wífmon hæfde áne yntsan goldes *uxores singulas auri uncias*, Ors. 4, 10; Swt. 196, 21. Fíftig yntsena seolfres *quinquaginta siclos argenti*, Deut. 22, 29. Þreóra yntsena gewihte . . . six yntsena . . . þreóra yntsena (yntsa, *v. l.*), Lchdm. i. 150, 16–18. [*Latin* uncia.]

yplen. v. ypplen.

yppan; *p.* te. I. *to bring up* or *forth :*—Ypte *depromsit* (decies senas de cespite ruris fruges depromsit, Ald. 139), Wrt. Voc. ii. 89, 18: 27, 5. II. *to disclose, reveal, declare, manifest :*—Hé ýweþ him and yppeþ earmra manna misgemynda, Salm. Kmbl. 985; Sal. 494. Hé ða unrótnesse his heortan mid his andwlitan tácnunge ypte and cýdde *tristitiam cordis vultu indice prodebat*, Bd. 4, 25; S. 600, 30. Ðæt hé þeódscipes gehyld mid his sylfes dǽde ýwde (ypte, *v. l.*) and cýdde (*propria actione praemonstraret*), 4, 27; S. 604, 40. Ypte and cídde *ederet*, Wrt. Voc. ii. 32, 5. Ðæt wé hit for ðý yppen ðæt mon God herige *ea ostendendo sunt, ut laudem coelestis Patris augeamus*, Past. 59; Swt. 451, 4. Ic ne dear yppan (*pandere*) ðé dígla úre, Coll. Monast. Th. 34, 13. Ypped *oriundus*, Wrt. Voc. ii. 62, 65. On his ágenum dagum ypped weorðeþ sóðfæstnes *orietur in diebus ejus justitia*, Ps. Th. 71, 7. Ypped eard in mægne ðínum *exortus es in virtute tua*, Ps. Surt. ii. p. 188, 9. Ðæt ypped wæs *prolatum*, Hpt. Gl. 510, 75. III. *to come forth*, (1) *to proceed :*—Of andwlitan ðínum dóm mín yppe *de vultu tuo judicium meum prodeat*, Ps. Spl. 16, 3. (2) *to be disclosed :*—Sóna ðæt ypeþ, swá hwæt swá ðé geswefnaþ, Lchdm. iii. 154, 23. v. forþ-, ge-yppan, uppan, *and next word.*

yppe; *adj. Brought to light, disclosed, manifest :*—Gif hé hit ðonne dierneþ and weorðeþ ymb long yppe, L. In. 21; Th. i. 116, 7: 35; Th. i. 124, 8. Ðonne mon beám on wuda forbærne and weorðe yppe on ðone ðe hit dyde, 43; Th. i. 128, 10. Nǽnges þinges mǽre þearf nǽre ðonne his unriht yppe wurde, Blickl. Homl. 175, 10. Ðonne him þince ðæt hé spíwe, ðæt byð swá hwæt swá hé ána wiste, ðæt hit

weorðæþ yppe (geypped, *v. l.*), Lchdm. iii. 170, 27. Mid Sigelwarum sóð yppe weard, dryhtlíc dóm Godes, Apstls. Kmbl. 128; Ap. 64. Gif ðis yppe bið, Elen. Kmbl. 870; El. 435.

yppe, an; *f. A raised place,* (1) *a look-out place :*—Yppe *vel* weardsteal *spectacula*, Wrt. Voc. i. 39, 35. (2) *a stage, platform :*—Glígmanna yppe *orcestra vel pulpitus*, Wrt. Voc. i. 39, 36. (3) *a dais, the raised floor in a hall.* Cf. Icel. pallr *for this sense* :—Eode æþeling (*Beowulf*) tó yppan, ðǽr se óþer wæs, Hróðgár grétte, Beo. Th. 3634; B. 1815. (4) *the upper part of a house, an upper chamber :*—Yppe (Ep. Gl. uppae) *in aestivo caenaculo, ubi per aestatem frigus captant*, Txts. 70, 553. Hié ealle on yppan wunedon (cf. *in coenaculo ascenderunt ubi manebant*, Acts 1, 13), Blickl. Homl. 133, 26.

ypping, e; *f.* I. *manifestation :*—Ypping *manifestatio* (*epiphania*), Rtl. 195, 24. II. *what mounts up* (?), applied to the water of the Red Sea which had risen up on either side of the track followed by the Israelites. Cf. Holmweall ástáh, merestreám módig, Cd. Th. 207, 16; Exod. 467; *and* multon meretorras, 208, 16; Exod. 484 :—Synfullra sweót sáwlum lunnon, siððan hié onbugon (on bogum, MS.) brún[e] yppinge (cf. *for the epithet* brúne ýða, Andr. Kmbl. 1038; An. 519), módewǽga mǽst *the host of sinners lost their lives, after the brown waters that had towered aloft broke over them*, Cd. Th. 209, 13; Exod. 498. Cf. ypplen.

ypping-íren, es; *n. The name of some tool, a crowbar* (?) :—Hé sceal fela andlómena habban . . . mattuc, ippingíren, scear, culter and eác gádíren, Anglia ix. 263, 3.

ypplen, yplen, es; *n. A top, summit :*—Ypplene *fastigio*, Hpt. Gl. 473, 47. Ðá ágeolewedan yplenu *crocata cacumina*, Wrt. Voc. ii. 137, 13.

ýr *the name of the rune for* y, a bow (?) :—Ýr byð æðelinga wyn and fyrdgeatewa sum, Runic pm. Kmbl. 344, 29; Rún. 27. The letter occurs Exon. Th. 50, 14; Cri. 800: 284, 28; Jul. 704: Elen. Kmbl. 2518; El. 1260. [*Icelandic has* ýr; *gen.* ýs *a yew, also a bow, as the name of the Runic* y.]

yr-. For words beginning with *yr-* see *ir-*.

yrf-cwealm, es; *m. Murrain :*—Hér com ǽrest se myccla ·yrfcwalm on Angelcynn, Chr. 986; Erl. 131, 6. v. orf-cwealm.

yrfe (cf. orf; or (?) irfe, *q. v.*), es; *n. Cattle :*—For án eówre yrfe sceal beón hér *oves tantum vestrae et armenta remaneant*, Ex. 10, 24. Gnættas wǽron gewordene on mannum and on yrfe (*in jumentis*), 8, 17. Eft hwyrfende wæs tó ðæm yrfe and tó ðæm ceápe and tó heora gesetum, Blickl. Homl. 199, 6. Ægðer ge on mannum ge on gehwelces cynnes yrfe, Chr. 910; Erl. 100, 14. Menn and yrfe (orf, *v. l.*) hí slógon, 1010; Erl. 143, 28. Ðæt land ǽrest mín láford mǽ tó lǽt, ðá wæs hit ierfelæás (*omni peccunia caruit*). . . . And ic sælf ðæt ierfæ (*peccuniam*) tó gestrínde. . . . Ðonnæ is ðǽr nú irfæs (*pecuniae*) ðæs ðæs stranga wintær lǽfæd hæfð nigon ealh hríðru, and feówer and hundændlæftig ealdra swína, Chart. Th. 162, 26–163, 4. v. irfe.

yrfe-leás; *adj. Without cattle, unstocked :*—Wæs ðæt land ierfelæás *omni peccunia caruit*, Chart. Th. 162, 28.

yrrest. v. wirrest.

yrse-binn [= ? yrsen- = ísern- : cf. Wülck. Gl. 142, 2 irsenhelm *cassis, where* Wrt. Voc. i. 35, 4 *has* iren], e; *f. An iron box :*—Yrse-binne (cf. hunigbinna, 264, 15), Anglia ix. 265, 1.

ysel, e; *an; f. A spark, cinder, an ash, ember :*—Ysle *favilla*, Wrt. Voc. i. 37, 19: 66, 44: 284, 17: ii. 36, 53. On yslan *in favillam*, Hpt. Gl. 495, 31. Hé geseah hú ða ysla up flugon mid ðam smíce *vidit ascendentem favillam*, Gen. 19, 28. Gé syndon dust and acsan and ysela, Guthl. 5; Gdwin. 38, 23. Heora wyrtruma bið swá swá windige ysla *radix eorum quasi favilla erit* (Is. 5, 24), Homl. Th. ii. 322, 20. Ða yslan *cineres*, Exon. Th. 213, 13; Ph. 224. In onlícnesse uppástígendra yselena (ysla, *v. l.*) *instar favillarum ascendentium*, Bd. 5, 12; S. 628, 23. Geong of ðam yselum (*de favilla*) eft áríseþ, Nar. 39, 7. Ic eom yslum and axum geanlícod *assimilatus sum favillae et cineri* (Job 30, 19), Homl. Th. ii. 456, 13. Bearwas wurdon tó axan and tó yslan, Cd. Th. 154, 9; Gen. 2553. Gebrineþ bán and yslan, ádes láfe, eft ætsomne, Exon. Th. 216, 21; Ph. 271: 236, 18; Ph. 576. [On asshen and on iselen *in fauilla et cinere*, O. E. Homl. ii. 65, 18. I am bot erþe ful euel and usle so blake, Allit. Pms. 60, 747. Isyl of fyre *fauilla*, Prompt. Parv. 266 and see note. M. H. Ger. usele; *and see* Grff. i. 487 : Icel. usli *a conflagration ; a field of burning embers.*]

yslende *sending forth sparks :*—Yslendra *favillantium*, Wrt. Voc. ii. 147, 20.

ysope, hysope, an; *f.:* ysopo, *indeclinable, or* ysopon *in oblique cases. Hyssop :*—Ðás wyrte sculon tó lungensealfe, bánwyrt, . . . isopo, saluie, Lchdm. iii. 16, 8. Ysopan sceaft *fasciculum hyssopi*, Ex. 12, 22. Fram ðam heágan cederbeáme tó ðære lytlan ysopan, Homl. Th. ii. 578, 5. Hysopan gelícne, Lchdm. i. 160, 12. Bespreng mé mid ðínum háligdóme swá swá mid ysopon, Ps. Th. 50, 8. Mid ysopo, Jn. Skt. 19, 29. Of butran and of weaxe and of ysopo, Lchdm. ii. 244, 20. Genim ysopan, i. 254, 20. Wyll ysopon in buteran, iii. 22, 23 : Ps. C. 73. Genim ysopo, Lchdm. i. 374, 18 : 378, 21. [From Latin.]

ýst, e; *f.:* ýste, es; *m.* (?) I. *a storm, tempest, whirlwind :*—

Mycel ŷst windes *procella magna uenti*, Mk. Skt. 4, 37. Windi ŷst, Lk. Skt. 8, 23. Mētte hié micel ŷst on sǽ, Chr. 877; Erl. 78, 18. Án mycel ŷst *atrocissimus turbo*, Ors. 3, 5; Swt. 104, 22. Hé sǽde ðæt ðǽr tó côme ðæs strongestan windes ŷste, and ðæt se swá stronglíce hrure on ða circan ðæt ealle ða men ðe ðǽr wǽron lágon áþænede on ðære eorðan, ðþ ðæt seó ondrysnlíce forð geleóreþ, Sbrn. 81, 19–27. Ðæs norþanwindes ŷst, Bt. 9; Fox 26, 21. Norðerne ŷst, Met. 6, 14. Swá swá hradu ŷst windes scip tóbrycð, Ps. Th. 47, 6. Gást ŷstes *spiritu procellae*, Ps. Spl. C. 106, 25. Mid ðý storme and mid ðære ŷste onwend *tempestate convulsa*, Past. 26; Swt. 181, 11. On ŷste mǽstre *tempestate maxima*, Scint. 15, 18. Stormes ŷste *tempestatis turbine*, Hpt. Gl. 421, 22. Mid swiftre ŷste *precipiti turbine*, Germ. 392, 73. Ŷst *procellam*, Ps. Spl. C. 106, 29. Hé ŷste mæg oncytran, ðæt hí (= heó?) windes hweoðu weorðeþ smylte *statuit procellam in auram*, Ps. Th. 106, 28. Ís and ŷste ealra gǽstas ðe his word willaþ wyrcean *glacies, spiritus procellarum, quae faciunt verbum ejus*, Ps. Th. 148, 8. Ðonne sǽ gemengaþ micla ŷsta, Met. 5, 9. Æfter ðám ŷþum úra geswinca ŷsta gehwilcre, 21, 15. Ŷsta *procellarum*, Blickl. Gl.: Ps. Spl. C. T. 10, 7. Swá ðæt twig, ðæt bið ácorfen of ðam treówe and áworpen on micclum ŷstum, Homl. Skt. ii. 30, 192, 207. Ðara geþóhta ŷstum *cogitationum procellis*, Past. 9; Swt. 59, 5. Æfter eallum ðám ŷstum and ðám ŷþum úrra geswinca, Bt. 34, 8; Fox 144, 28. II. *rough water, surge*:—Ŷst *aestus, recessus et accessus maris*, Wrt. Voc. i. 57, 10. [O. Sax. ûst *a storm of wind*: O. H. Ger. unst *procella, nimbus, tempestas, turbo*.]

ŷstan; *p.* te *To be stormy*:—Ŷstendre *ferventis (oceani)*, Hpt. Gl. 464, 57.

ŷstig; *adj.* I. *stormy, tempestuous*:—Windig sumer and ŷstig, Lchdm. iii. 162, 31. II. *of storm*:—Ŷstige gǽstas *spiritus procellae*, Ps. Th. 106, 24.

ŷtan; *p.* te *To put out*, (1) *to put out* a person from a place, *expel, banish*:—Hér man ŷtte út Ælfgár eorl, ac hé com sóna inn ongeán þurh Gryffines fultum, Chr. 1058; Erl. 192, 35. (2) *to put out* a thing from one's possession, *alienate, give away*:—Hé ná mynstres ǽhta ne ŷte, ne ná myrre *neque prodigus sit, aut stirpator substantie monasterii*, R. Ben. 55, 4. v. á-ŷtan; útian, *and next word*.

ŷtend, es; *m.* A *waster, destroyer*:—Wéstend, ŷtend *exterminator, vastator*, Wrt. Voc. ii. 147, 64. v. ŷtan.

ŷtera, cpve.: ŷtemest; spve. *adj. Outer: outmost, extreme.* I. *local*:—Of helle ŷteran *ex inferno inferiori*, Ps. Spl. T. 85, 12. On ðan ŷttren *in citeriorem*, Hpt. Gl. 492, 69. On ða ŷtran *in posteriora*, Ps. Spl. 77, 72. Ŷtemeste *extremus*, Wrt. Voc. ii. 146, 39. Ðæt ŷtemeste land, ðæt man hǽt Thila, Ors. 1, 1; Swt. 24, 20. Ðæt hé gewǽte his ŷtemystan finger, Past. 43; Swt. 309, 6. On ða ŷtemesta[n] sǽ *in extremis maris*, Ps. Spl. 138, 8. Ða ŷtemestan endas ðare seglgyrde *cornua*, Wrt. Voc. i. 63, 46. Ða ŷtmestan eorðbúende, Met. 10, 25. From feówerum foldan sceátum ðám ŷtemestum, Exon. Th. 55, 7; Cri. 880. Æt ðám ŷtmestan eorþan gemǽrum *usque ad ultimum terrae*, Blickl. Homl. 119, 25; 133, 35. Bind his ŷtmestan limo mid byndellum, Lchdm. ii. 196, 12. II. *marking order or degree, later, lower; last, lowest*:—Gif munuc hine sylfne ŷttran (*inferiorem*) and unweorðran talaþ þonne ǽnigne óþerne, R. Ben. 29, 11. Stande hé ealra ŷtemest (*ultinus*), 68, 17: Scint. 21, 19. Ealra ŷtemest *nouissima omnium*, Lk. Skt. 20, 32. Ágynn fram ðam ŷtemestan (*novissimo*) óð ðone fyrmestan, Mt. Kmbl. 20, 8, 14. On ðam ŷtemestan dæge, Jn. Skt. 6, 54. On ðam ŷtemestan dæge his lifes, Bd. 3, 17; S. 543, 18: 4, 8; S. 575, 30: Exon. Th. 172, 7; Gú. 1140. Æt ðám ŷtemestan ende, 128, 34; Gú. 414. On ŷtemestum síðe *in extremis*, Mk. Skt. 5, 23. Tó ðam ŷtemestan geléæded, Guthl. 20; Gdwin. 80, 5. Ðæm ǽtemestan háðre *suprema sorte*, Hpt. Gl. 453, 34. Ǽr ðú ágylde ðone ŷtemestan (*novissimum*) feorðlingc, Mt. Kmbl. 5, 26: Lk. Skt. 12, 59. Ŷtemystan *infimam, minimam*, Germ. 403, 31. Swá beóð ða fyrmestan ŷtemeste (*novissimi*), and ða ŷtemestan fyrmeste, Mt. Kmbl. 20, 16. Wurðaþ ðæs mannes ŷtemestan wyrsan ðonne ða ǽrran, 12, 45. On ŷtemestum *in extremis*, Scint. 46, 15. Hé ða ŷtemestan word (*ultima verba*) on his herenesse betŷnde, Bd. 4, 24; S. 599, 12. III. *external*:—Ðonne twám pundum is getácnod ǽgðer ge ðæt ŷttre andgit ge ðæt inre.... Sume lǽwede tǽcaþ riht ðæs ðe hí magon tócnáwan be ðam ŷttrum andgitum, þeáh ðe hí ne cunnon ða incundan deópnysse Godes láre ásmeágan, Homl. Th. ii. 550, 14–22. Ðan incundum *internis*, ða ŷttran *exteriora*, Wülck. Gl. 248, 7; Scint. 226, 16. Ðú miht blissigan ðæt ðære ðeóde sáwla þurh ða ŷttran wundra beóð getogene tó ðære incundan gife *gaudeas quia Anglorum animae per exteriora miracula ad interiorem gratiam pertrahuntur* (Bd. 1, 31), Homl. Th. ii. 132, 2. v. útera.

ŷteren; *adj. Made of otter's skin*:—Berenne kyrtel oððe yterenne, Ors. 1, 1; Swt. 18, 21.

ŷþ, e; *f.* I. *a wave of the sea* (lit. or fig.):—Flód oððe ŷð *fluctus*, Ælfc. Gr. 11; Zup. 79, 2. Éð *unde*, Wrt. Voc. i. 54, 23. Brim eft oncwæð, ŷð óðerre, Andr. Kmbl. 885; An. 443. Stunede sió brúne ŷð wið óðre, Met. 26, 30. Wédende ŷða *frementes fluctus*, Hpt. Gl. 464, 74. Ŷða

flustra, 478, 57. Swá swá ŷþa for winde ða sǽ hrêraþ, Bt. 39, 1; Fox 210, 25: Met. 27, 3: Cd. Th. 83, 25; Gen. 1385: Ps. Th. 77, 53: Exon. Th. 488, 5; Rä. 76, 2. Ða sylfan ŷþa wǽron áhafene ofer ðæt scip, Blickl. Homl. 235, 6. Ða ŷða swygiaþ *siluerunt fluctus*, Ps. Th. 106, 28. Ða ŷða ðara costunga, Past. 16; Swt. 103, 21. Ŷþe, Exon. Th. 188, 3; Az. 40: Cd. Th. 196, 8; Exod. 288. Ða wonhǽwan oððe swearthǽwenan oððe ŷða *cerula*, Wrt. Voc. ii. 20, 66. Wonne ŷþa, Cd. Th. 86, 13; Gen. 1430. Sealte ŷþa, Ps. Th. 76, 13. Ŷþa hlúde, 64, 7. Ŷþa ofermǽta, Exon. Th. 53, 23; Cri. 855. Ŷþa geþwǽre, 382, 22; Rä. 3, 15. Hreó wǽron ŷþa, Beo. Th. 1101; B. 548. Seó sǽ mót brúcan smyltra ŷþa, Bt. 7, 3; Fox 20, 23. Eástreám ŷða, Cd. Th. 240, 11; Dan. 385. Ŷða *ful the sea*, Beo. Th. 2421; B. 1208. Ŷða yrfeweard, Salm. Kmbl. 163; Sal. 81. Ŷða swengas, Elen. Kmbl. 478; El. 239. Ŷða ðrym, Beo. Th. 3841; B. 1918. Sǽs swêges and ŷða (ŷðana *fluctuum*, Lind. Rush.), Lk. Skt. 21, 25. Ðæt scyp wearð ofergoten mid ŷðum (*fluctibus*), Mt. Kmbl. 8, 24. Of ðám ŷðum tótorfod, 14, 24. Hé gesette ŷðum heora onrihtne ryne, Cd. Th. 10, 34; Gen. 166. Flota wæs on ŷðum, Beo. Th. 426; B. 210. Ofer ŷðum, 3819; B. 1907. Hé ŷðum stilde, Andr. Kmbl. 902; An. 451. Sealtum ŷðum, Cd. Th. 207, 26; Exod. 472. On ðám ŷðum ðisse worulde, Bt. 4; Fox 8, 22: Met. 4, 56. Æfter eallum ðám ŷstum and ŷþum úrra geswinca, Bt. 34, 8; Fox 144, 28: Met. 21, 54. Ealle ŷða ðíne *omnes fluctus tuos*, Ps. Spl. 87, 7: Mk. Skt. 4, 37. Ŷþa wrêgan, Exon. Th. 383, 28; Rä. 4, 17. Gán ofer sǽs ŷþa, Blickl. Homl. 177, 18. Fêran ofer sǽs ŷþe, Shrn. 104, 34: Exon. Th. 72, 5; Cri. 1168: Beo. Th. 91; B. 46. Winter ŷþe beleác, 2269; B. 1132. ¶ *gen. pl.* with words denoting the movement of the waves forming phrases = *the billowy sea*:—Ŷða gelaac, Ps. Th. 118, 136: Exon. Th. 442, 3; Kl. 7. Ŷða geswing, Beo. Th. 1700; B. 848: Andr. Kmbl. 703; An. 353. Ŷða geþræc, 1645; An. 824: Exon. Th. 381, 26; Rä. 3, 2: 404, 13; Rä. 23, 7. Ŷða geþring, Andr. Kmbl. 736; An. 368. Ŷða gewealc, 517; An. 259: Cd. Th. 206, 21; Exod. 455: Beo. Th. 932; B. 464; Chr. 975; Erl. 126, 19. Ŷða gewin, Beo. Th. 2942; B. 1469. Ŷða ongin, Andr. Kmbl. 931; An. 466. Ŷða wylm, 1726; An. 865. I a. *in a collective sense, the wave, water, sea*:—Ŷð, ǽdwella *flustra*, i. *unda*, Wrt. Voc. ii. 149, 67. Ŷð up færeþ, Cd. Th. 195, 25; Exod. 282. Ŷð (cf. gewinn ŷþa and landes, Bt. 39, 3; Fox 214, 35) wið lande ealneg winneþ, Met. 28, 57. Mec ŷð sió brúne beleólc, Exon. Th. 471, 25; Rä. 61, 6. I b. *applied to a quantity of any liquid, flood as in floods of tears*:—Flód ŷðum weóll, Andr. Kmbl. 3091; An. 1548. Blód ŷðum weóll, 2482; An. 1242. Swát ŷðum weóll, 2552; An. 1277: Beo. Th. 5380; B. 2693. Teagor ŷðum weóll, Exon. Th. 182, 23; Gú. 1314. II. *any liquid, water*:—Suǽ hwæd in húsum gileáffulra ðás ŷð eft ástrægde *quicquid in domibus fidelium haec unda respererit*, Rtl. 121, 36. [*Innan þan sea weren* .vii. *bittere uþe*, O. E. Homl. i. 43, 3. þe wind þa sǽ wraðede, uðen þer urnen, Laym. 4578. þet uðen (unðes, *v. l.*) ne stormes þet scip ne ouerworpen, A. R. 142, 11. Hit reled upon þe roʒe yþes, Allit. Pms. 96, 147. O. Sax. ûðia: O. H. Ger. unda: Icel. unnr, uðr.] v. ár-, flód-, geofon-, líg-, sǽ-, sealt-, wæter-ŷþ; ŷþe.

ŷþ-. v. íþ-.

ŷþan *to fluctuate.* v. ŷþian.

ŷþ-bord, es; *n.* A *ship's side*:—Ðonne sǽrófe snelle mægne árum bregdaþ ŷðborde neáh (*sitting near the side of the ship*), Exon. Th. 296, 27; Crä. 57. Ŷða eów scipweardas ofer ŷðbord (*speaking across the ship's side*; cf. over-board) unnan willaþ, Andr. Kmbl. 595; An. 298. Cf. Bord oft onfêng ŷða swengas, Elen. Kmbl. 476; El. 238.

ŷþe, an; *f.* A *wave*:—Wé æthrynon mid úrum árum ða ŷðan ðæs deópan wælis ... ða ŷðan getácniaþ ðisne deópne cræft, Anglia viii. 299, 38–41. v. ŷþ.

ŷþ-faru, e; *f.* The *wave-course, the waves, sea*:—Swá ealne middangeard mereflód þeahte, ða se aþela wong onsund wið ŷðfare gehealden stód hreóra wǽga eádig unwemme *cum diluvium mersisset fluctibus orbem, Deucalionea exsuperavit aquas*, Exon. Th. 200, 22; Ph. 44. Sume on ŷðfare wurdon wætrum bisencte, on mereflóde, 271, 7; Jul. 478: Andr. Kmbl. 1799; An. 902.

ŷþ-gebland, es; *n.* The *tossing waves*:—Ŷðgebland up ástígeþ won tó wolcnum, ðonne wind styreþ láð gewidru, Beo. Th. 2750; B. 1373: 3190; B. 1593. Wǽron ŷðgebland eal gefǽlsod, eácne eardas, 3244; B. 1620.

ŷþ-gewinn, es; *n.* The *wave-strife, the billows*:—Sumne hé feores getwǽfde ŷðgewinnes, Beo. Th. 2872; B. 1434. Holmwylme neáh, ŷðgewinne, 4815; B. 2412.

ŷþgian, ŷþgung. v. ŷþian, ŷþung.

ŷþ-hengest, es; *m.* A *wave-steed, a ship*:—Hé fêrde ðǽr hé wiste his ŷðhengestas, Chr. 1003; Erl. 139, 16. [Cf. Icel. unnar hestr *a ship* (poet.).]

ŷþ-hof, es; *n.* A *wave-house, a vessel*:—Ceólas lêton æt sǽfearoðe, ald ŷðhofu, oncrum fæste, Elen. Kmbl. 503; El. 252. Ongan ófostlíce ðæt hof (ŷðhof is suggested by Grein, which would restore the missing alliteration) wyrcan, Cd. Th. 79, 25; Gen. 1316.

ŷþian, ŷþgian; *p.* ode. I. *to overflow* (intrans.) (1) literal:—

Đá ýđode đæt flód ofer eorþan *aquae diluvii inundaverunt super terram,* Gen. 7, 10. Đæs flódes wæteru ýđedon ofer eorþan, 7, 6, 18. Burnon ýþgodon (ýđgadun, Surt.) *torrentes inundaverunt,* Ps. Spl. 77, 23. Éđiende *redundans* (*torrens*), Kent. Gl. 632. Đæt ýđigende flód, đe đa synfullan ádylegode, Homl. Th. ii. 60, 4. Swilc storm ýđigende feóll *such a storm fell in torrents,* 184, 5. (2) figurative, *to be filled :*—Đæs cyninges ríce ge foreweard ge forþgang swá monigum and swá myclum styrenessum wiþerweardra đinga ýþiaþ *cujus regni et principia et processus tot ac tantis redundavere rerum adversantium motibus,* Bd. 5, 23 ; S. 646, 4. Ic ýđgode mid synnum, swá sæ mid ýđum, Shrn. 140, 18. II. *to move in waves, to toss, roll,* (1) of the sea. v. ýþung :—Đæs ýþiendan sæs *fluctivagi ponti,* Wrt. Voc. ii. 149, 61. Hé đa ýđigendan sæ mid ánre hǽse gestilde, Homl. Th. ii. 378, 20. (2) of movement like that of the sea, *to wave :*—Sume sind gehátene *tropi* . . . swá swá is gecweden *fluctuare segetes,* đæt æceras ýđiaþ, for đan đe æceras faraþ on sumera, swá swá sæ ýđigende, Ælfc. Gr. 50 ; Zup. 295, 10. (3) figurative, *to fluctuate.* v. ýþig :—His mód biđ swíđe ýđegende (iéđegende, Hatt. MS.) and swíđe ábisgod mid eorđlícra monna wordum *valde inter humana verba cor defluit,* Past. 22 ; Swt. 168, 11. Swá biđ đis eorđlíce líf oft ýđigende swá swá sæ, 52 ; Swt. 409, 35. Seó sæ getácnode đás andwerdan woruld, đe is swíþe ýđigende for mislícum styrungum and eostnungum, Homl. Th. ii. 384, 23. Of ýđigendre sæ đyssere worulde, 290, 33. Ne syleþ hé sóđfæstum đæt him ýþende mód innan hređre *non dabit fluctuationem justo,* Ps. Th. 54, 22. [*O. H. Ger.* undeón *fluctuare, aestuare.*] v. on-ýþian.

ýþig ; *adj. Fluctuating, stormy :*—Đyssere ýđegan worulde, Homl. Skt. i. 16, 72.

ýþ-láđ, e ; *f. A way across the waves :*—Gode þancedon đæs đe him ýþláde eáđe wurdon, Beo. Th. 461 ; B. 228.

ýþ-láf, e ; *f. The shore left bare by the waves :*—Hié (the sea-beasts) mécum wunde be ýđláfe uppe lǽgon, Beo. Th. 1136 ; B. 566. Ofer ýđláfe on sæ lǽdan, Andr. Kmbl. 998 ; An. 499. Dǽlan on ýđláfe ealde máđmas, Cd. Th. 215, 18 ; Exod. 585.

ýþ-lid, es ; *n. A ship :*—Of ýđlide, Andr. Kmbl. 555 ; An. 278. Ofer ýđlid (-liđ, MS.), 889 ; An. 445.

ýþ-lida, an ; *m. A wave-traverser, a ship :*—Hé hét him ýđlidan gódne gegyrwan, Beo. Th. 399 ; B. 198.

ýþ-mearh; *gen.*-meares; *m. A wave-steed, a ship :*—Sundhengestas, ealde ýđmearas, Exon. Th. 54, 5 ; Cri. 864. Se micla hwæl bisenceþ sǽlíþende, eorlas and ýđmearas, 363, 5 ; Wal. 49.

ýþ-mere, es ; *m. The billowy main :*—Hwonne up cyme æþelast tungla ofer ýđmere éstan líxan, Exon. Th. 204, 7 ; Ph. 94.

ýþung, ýþgung, e ; *f. Movement as of waves* (v. ýþian, II. 1), *fluctuation* (v. ýþian, II. 3) :—Seó burh Naim is gereht ýđung ođđe styrung, Homl. Th. i. 492, 1. Ýđgunge, ýđgunga *fluctuationem,* Ps. Spl. C. T. 54, 25 : Ps. Surt. 54, 23 : Blickl. Gl.

ýþ-wórigende ; *adj.* (*ptcpl.*) *Wave-wandering :*—Đa ýþwórigendan húþa *fluctivagam praedam,* Wrt. Voc. ii. 149, 71.

ýting, e ; *f. A being out, away from home, a journey :*—Đa đe on ýtinge faraþ áhwyder *hi qui in via diriguntur* . . . Đa đe on ýtinge faraþ *exeuntes in viam,* R. Ben. 90, 8-12. Crist wolde on ýtinge beón ácenned, Homl. Th. i. 34, 13.

ýtmest, ýwan. ýtera, íwan.